Aberdeenshire Library and Information Service
www.aberdeenshire.gov.uk/libraries
Renewals Hotline 01224 661511

ABERDEENSHIRE
LIBRARY &
INFORMATION SERVICES

WITHDRAWN
FROM LIBRARY

Chambers
dictionary of science and
technology

D1582851

2589668

Chambers Dictionary of Science and Technology

General Editor
John Lackie

Chambers

CHAMBERS
An imprint of Chambers Harrap Publishers Ltd
7 Hopetoun Crescent
Edinburgh
EH7 4AY

This edition first published by Chambers Harrap Publishers Ltd 2007

© Chambers Harrap Publishers Ltd 2007

Previous edition published in 1999. Previously published in 1995 as *Larousse Dictionary of Science and Technology*. First published as *Chambers's Technical Dictionary* in 1940 by W&R Chambers Ltd (revised 1958, 1971, 1974 and 1984).

All rights reserved. No part of this publication may be reproduced, stored in a retrieval system, or transmitted by any means, electronic, mechanical, photocopying or otherwise, without the prior permission of the publisher.

A CIP catalogue record for this book is available from the British Library.

ISBN-13: 978 0550 100719

ABERDEENSHIRE LIBRARY AND INFORMATION SERVICES	
2589668	
HJ	585193
R503	£35.00
AD	REFB

Text design by Ken Wilson
Typeset by Macmillan India Ltd
Printed in Italy by LegoPrint S.p.A

Preface

This edition of the *Chambers Dictionary of Science and Technology* is based substantially upon the edition produced in 1999 which, for the first time, incorporated panel entries giving additional detail on a range of topics. The dictionary has its origin in 1940, when it was called *Chambers's Technical Dictionary*, although it always had a substantial basic science content. It became the *Dictionary of Science and Technology* in 1971, changed in 1984 to the *Chambers Science and Technology Dictionary*, then transformed again in 1995 into the *Larousse Dictionary of Science and Technology* (with illustrations for the first time) before reverting to the present title in 1999.

The first question I asked when I had the privilege of taking over as General Editor from Professor Peter Walker*, the architect of major changes in recent editions, was about the role of a dictionary in these days of search engines and the Internet. It is still a valid question, but I think the answer is probably that a dictionary should provide crisp, authoritative definitions that are useful to a diverse audience of non-specialists, especially those who are not connected to the Web or who do not have the time or inclination to dredge for a definition among sites that may or may not be reliable. This puts the onus on the Editor and the specialist contributors to ensure the accuracy of the content, which I hope we have managed. That said, it would be remarkable if a few mistakes had not crept through; I hope they are rare and trivial, and apologize in advance if we have perpetuated or propagated errors. Given the complexity of modern science and technology and the diversity of the technical language that is employed, it is difficult for a small number of people, no matter how catholic or eclectic in their interests, to be expert in everything. The most challenging situation is when terminology has evolved and subtly changed: only the aficionado may realize that the old definition is no longer accurate and that current usage is different – and the users are always right if the language is evolving, as it will in our constantly developing technological world. But if you are reading the older literature, the older meaning may well be correct, so we have deleted very few entries, even when there is a strong suspicion that the term (or even the technology) is obsolete.

What is new

The obvious areas where neologisms are proliferating are molecular bioscience and information technology, and these have received the most attention. Biology is emerging from a period of schismatic diversity, and the pervasive influence of molecular biology has forced a common lexicon on most bioscientists. It therefore seemed sensible to merge botany, zoology, biology and immunology (each of which had separate subject fields in the last edition); where, after all, were genetics, cell biology, virology? A new Bioscience subject field was therefore formed – it is inclusive rather than exclusive, and has gained several hundred new headwords. Indeed, much of the bioscience terminology is equally familiar to the modern medical practitioner – but we have retained Medicine, splitting off Pharmacology as a 'home' for the many drugs that we suspect are often looked up. Environmental Sciences has become the category for most ecology and meteorology terms, as well as terms relating to waste management and pollution. Psychology has emerged as a new category, incorporating some of the entries formerly labelled Behaviour (others have gone into Bioscience), but with many new additions – it is, after all, one of the most popular subjects at both school and university level. Agriculture, one of the oldest of technologies, has gained a place, and some of the details in Forestry entries have been pruned to accommodate the agricultural ones. Modern telecommunications are so dependent upon digital information processing that the former Telecommunications and Computer

---

* Sadly, although I was able to discuss the preparation of this edition with Peter, he died in early 2006. The present format of the dictionary owes much to his efforts over several earlier editions.

Science fields have been merged into Information and Communications Technology, with its inevitable three-letter acronym, ICT. This area too has gained a substantial number of new headwords.

It should be noted that these subject fields are only a rough indication of how an entry should be classified. They are important in differentiating between very different meanings (bioscientists and electrical engineers use 'cell' in rather different ways, for example), but relatively unimportant in fields that communicate extensively and have boundaries that are fuzzy (most bioscientists use pharmacological tools, often to research medically important diseases, sometimes even diseases that are partly the province of psychology).

There are a few new panels, notably on world mortality statistics and on the human brain, and some changes have been made to the appendices. Otherwise, the major change is the growth of the dictionary by the addition of some 2,000 new headwords.

Arrangement

Entries are presented in alphabetical order, apart from those that are numbers, which are presented in numerical order in a section of their own at the beginning of the dictionary. Single-letter entries appear at the beginning of each letter section. Greek letters are placed under the anglicized equivalent (omega under 'O', for example). Numbers preceding chemical names are generally ignored, so that 1,1,1-trichloroethane is under 'T', below 3:4:4'-trichloro-carbanilide and above trichloroethanoic acid. Abbreviations are placed in their alphabetical position, but the definition is usually given under the full version. Where some ambiguity might arise, a cross-reference is given. Spellings are generally British English, but where confusion could arise the US version is given in the earlier occurrences (thus, under 'heme-', the reader is directed to 'haem-'). An exception arises with the names of some drugs which have British Approved Names (BANs) generally used as the headword, but these BANs have been harmonized with the 'recommended international non-proprietary names' (rINNs) and are sometimes therefore Americanized (thus 'sulfasalazine', even though 'sulphur' is the general spelling used elsewhere). Even so, some things may well not be where the reader expects, and an alternative location should be tried.

An evolving dictionary

In deciding upon new headwords for this edition we have relied partly upon new words identified by Chambers Wordtrack, the programme used to identify new usages for other Chambers titles, notably *The Chambers Dictionary*. In addition, specialist contributors, including myself, have added terms that we felt were missing.

A more vexed question is when to omit headwords. Inevitably some terms become obsolete and are occupying space that would be better devoted to new words; but if the reader is seeking a definition of a term used in an older text, where will they find it, except perhaps in an older edition of this dictionary? In this edition we have been conservative and deleted few entries; pressure on space, and a wish that the dictionary remains a single manageable volume, may force the issue during future revisions.

Italic and bold

Italic is used for:
(1) alternative forms of, or alternative names for, the headword. Synonyms are generally given at the end of the definition;
(2) terms derived from the headword, often after 'adj' or 'pl';
(3) variables in mathematical formulae;

(4) generic and specific names in binomial classification of organisms;

(5) after 'abbrev for', 'symbol for', etc, where the meaning is defined within the entry or is obvious and there is no cross-referenced entry.

Bold is used for:

(1) cross-references, either after 'see', 'cf', 'abbrev for', 'symbol for', etc, or in the body of an entry. Cross-referencing indicates that there is a headword elsewhere which gives additional information (although the absence of a cross-reference does not necessarily mean that there is not such an entry);

(2) vector notation in formulae, etc.

Trade names

Trade and proprietary names are identified by an initial capital, the prefix TN or suffix ®, or by a statement.

Illustrations

With a few exceptions, these were drawn by Professor Peter Walker.

Acknowledgements

This edition of the dictionary relies substantially upon the content of the earlier editions mentioned above, and the efforts of earlier contributors must be acknowledged.

For this edition, Ephraim Borowski scrutinized the Mathematics entries, Alan Cooper the Chemistry ones; Brian Grout wrote the new Agriculture entries and Heather Parton checked the Food Science entries. Richard Smithson advised on and Julie MacLoughlin checked the new Psychology section, and added various entries. Peter Tiplady checked and added to the Medicine section, and wrote a new panel; Nicholas Graham merged the former Telecommunications and Computing sections and added extensively to the new ICT section, as well as writing a new panel. Adrian Webster added a few new Astronomy entries and Mike Weighell checked the Chemical Engineering entries. The Bioscience, Pharmacology and Environmental Science areas were handled by me, with help from Katie Brooks of Chambers, especially on the Immunology entries. I am also grateful to Neville Hankins, who copy-edited with a keen eye. But in the end, if there are errors, they are mine; I hope they are few and I apologize in advance.

John Lackie

Contributors

General Editor
John Lackie

Contributors
Ephraim Borowski
Katie Brooks
Alan Cooper
Nicholas Graham
Brian Grout
Julie MacLoughlin
Heather Parton
Richard Smithson
Peter Tiplady
Adrian Webster
Mike Weighell

Editorial Assistance
Katie Brooks
Lianne Vella

Copyeditor
Neville Hankins

Publishing Manager
Camilla Rockwood

Production and Prepress Controller
Karen Stuart

Contents

Panels

Subject Categories

(Abbreviations are shown in brackets)

Acoustics (*Acous*)
Aeronautics (*Aero*)
Agriculture (*Agri*)
Architecture (*Arch*)
Astronomy (*Astron*)
Automobiles (*Autos*)
Bioscience (*BioSci*)
Building (*Build*)
Chemical Engineering (*ChemEng*)
Chemistry (*Chem*)
Civil Engineering (*CivEng*)
Crystallography (*Crystal*)
Electrical Engineering (*ElecEng*)
Electronics (*Electronics*)
Engineering (*Eng*)
Environmental Sciences (*EnvSci*)
Food Science (*FoodSci*)
Forestry (*For*)
General (*Genrl*)
Geology (*Geol*)
Glass (*Glass*)
Image Technology (*ImageTech*)

Information and Communications
  Technology (*ICT*)
Mathematical Sciences (*MathSci*)
Medicine (*Med*)
Mineral Extraction (*MinExt*)
Mineralogy (*Min*)
Nuclear Engineering (*NucEng*)
Paper (*Paper*)
Pharmacology (*Pharmacol*)
Physics (*Phys*)
Plastics (*Plastics*)
Powder Technology (*PowderTech*)
Printing (*Print*)
Psychology (*Psych*)
Radar (*Radar*)
Radiology (*Radiol*)
Ships (*Ships*)
Space (*Space*)
Surveying (*Surv*)
Textiles (*Textiles*)
Veterinary Science (*Vet*)

Abbreviations

(Showing main abbreviations used in the dictionary; others will be found as headwords.)

abbrev	abbreviation	lb	pounds
ac	alternating current	Lt	(Latin)
AD	Anno Domini	m	metre(s)
adj	adjective	mg	milligram(s)
am	before noon	min	minute(s)
approx	approximately	MKS(A)	metre–kilogram(me)–second
at no	atomic number		(–ampere)
BC	Before Christ	ml	millilitre(s)
bp	boiling point	mm	millimetre(s)
BP	before present (with 1950 as	mp	melting point
	'present')	Mt	Mount
C	Celsius (centigrade)/Central	mya	million years ago
c.	circa	n	noun
cc	cubic centimetre(s)	N	north(ern)
cf	compare	NB	note well (*nota bene*)
CGS	centimetre–gram(me)–second	no.	number
cm	centimetre(s)	nos	numbers
colloq	colloquial(ly)	oz	ounce(s)
conc	concentrated	p (pp)	page (pages)
cu	cubic	%	per cent
dc	direct current	pl	plural
dim	diminutive	pm	after noon
E	east(ern)	pt	pint(s)
eg	for example	ram	relative atomic mass
esp	especially	rel. d.	relative density
etc	and so on	rms	root-mean-square
F	Fahrenheit	s	second(s)
fig.	figure	S	South(ern)
fl oz	fluid ounce(s)	SI	Système International
Fr	French		(d'Unités)
ft	foot/feet	sing	singular
g	acceleration due to gravity	sp (spp)	species (sing/pl)
g	gram(s)	spec	specific
gall	gallon(s)	sq	square
gen	genitive	stp	standard temperature and
Ger	German		pressure
Gk	Greek	TN	trade (proprietary) name
h	hours	TV	television
ie	that is	UK	United Kingdom
in	inch(es)	US	United States (of America)
K	Kelvin	v	verb
kb (kbp)	kilobase (kilobase pair)	W	west(ern)
kg	kilogram(s)	yd	yard(s)
km	kilometre(s)	yr	year(s)
l	litre(s)		

1, 2, 3...

$\frac{3}{8}$ rule (*MathSci*) See THREE-EIGHTHS RULE.

2- (*Genrl*) See TWO-

2,4,5,-T (*BioSci*) Abbrev for *2,4,5-trichlorophenoxyethanoic acid*, a selective herbicide that is widely used but with increasing public concern.

2E (*Min*) Apparent optic axial angle as measured in air.

2V (*Min*) The optic axial angle when measured in the mineral.

4:2:2 (*ImageTech*) A standard prescribing the ratio of frequencies used in digital COMPONENT VIDEO systems for sampling the LUMINANCE SIGNAL and two COLOUR DIFFERENCE SIGNALS — 13·5 MHz for luminance and 6·75 MHz for each of the colour difference signals. Extended 4:2:2 prescribes 18 MHz and 9 MHz respectively, providing greater resolution for 625-line wide-screen and 1250-line high definition. There is also a 4:1:1 standard. See FSC.

4GL (*ICT*) Abbrev for FOURTH-GENERATION LANGUAGE.

4-methyl-aminophenol (*Chem*) See entry under 'M'. $NH_2C_6H_4OH$; a solid crystalline compound, soluble in acids and alkalis and readily oxidized in air, used as a photographic developer and widely marketed under the TN Rodinal.

5-HT (*Med, BioSci*) Abbrev for *5-hydroxytryptamine*. See SEROTONIN.

7S antibody (*BioSci*) Immunoglobulin with sedimentation coefficient of about 7S. Term often used as synonym for IgG. See 19S ANTIBODY.

8 mm time code (*ImageTech*) See TIME CODE.

10 base 2 (*ICT*) A transmission standard for LOCAL AREA NETWORKS under the IEEE 802.3 standard. This system uses thin coaxial cable and is capable of baseband transmission at 10 Mbps and a maximum cable length of 185 m between REPEATERS. Also *thin Ethernet, thin-net*.

10 base 5 (*ICT*) A transmission standard for LOCAL AREA NETWORKS under the IEEE 802.3 standard. This system uses a thick coaxial cable and is capable of baseband transmission at 10 Mbps and a maximum cable length of 500 m between REPEATERS. Also *standard Ethernet, thick Ethernet*.

10 base T (*ICT*) A transmission standard for LOCAL AREA NETWORKS under the IEEE 802.3 standard. This system uses unshielded twisted pair cable and is capable of baseband transmission at 10 Mbps and a maximum cable length of 100 m between REPEATERS.

12mo (*Print*) See DUODECIMO. Also *duodecimo*.

16 mm (*ImageTech*) A narrow gauge of motion picture film, widely used for *non-theatrical* production and presentation.

16mo (*Print*) See SEXTODECIMO.

16:9 switching (*ImageTech*) The facility for a TV or VCR to sense and switch automatically to WIDE-SCREEN display or recording.

18-electron rule (*Chem*) The structures of most stable organometallic compounds of the TRANSITION ELEMENTS, eg *ferrocene*, can be rationalized by showing that 18 valence electrons can be associated with each metal atom. See ELECTRON OCTET.

18mo (*Print*) See OCTODECIMO.

19S antibody (*BioSci*) Immunoglobulin with sedimentation coefficient of about 19S. Term often used as synonym for IgM. See 7S ANTIBODY.

21 cm line (*Astron, Phys*) A line in the radio spectrum of neutral hydrogen at 21·105 cm. It is caused by the spontaneous reversal of direction of spin of the electron in the magnetic field of the hydrogen nucleus, but it may be detected only in the vast hydrogen clouds of the Galaxy.

24mo (*Print*) The 24th part of a sheet, or a sheet folded four times to make 24 leaves of 48 pages.

24-hour rhythm (*Psych*) See CIRCADIAN RHYTHM.

30 Doradus (*Astron*) See TARANTULA NEBULA.

32 bit (*ICT*) A processor or computer bus, handling data consisting of 32 binary digits.

32mo (*Print*) The 32nd part of a sheet, or a sheet folded five times to make 32 leaves or 64 pages. More than four folds are not in fact practicable.

35 mm (*ImageTech*) The standard gauge of motion picture film used for professional production and presentation. Also a popular photographic film format.

61 Cygni (*Astron*) A faint visual binary star system in the constellation Cygnus. One of the components is itself a binary system. One of the nearest stellar objects to the Sun, it was the first to have its trigonometric parallax measured. Distance 3·42 pc.

64mo (*Print*) The 64th part of a sheet or a sheet folded six times to make 64 leaves or 128 pages. More than four folds are not practicable and the sheet would be cut into four parts, each being folded four times.

70 mm (*ImageTech*) The widest gauge of motion picture film.

80 Ursae Majoris (*Astron*) See ALCOR.

100 Hz scanning (*ImageTech*) A means of reducing TV flicker by storing every FIELD and showing it twice at double rate.

802.11 (*ICT*) A family of specifications developed by the IEEE for wireless LAN technology. The 802.11 standard specifies an over-the-air interface between a wireless client and a base station or between two wireless clients. See WI-FI.

A (*BioSci*) Symbol for ADENINE.

A (*Eng*) See AE.

A (*Phys*) Symbol for AMPERE.

Å (*Phys*) Symbol for ANGSTROM.

A (*Chem*) Symbol for: (1) RELATIVE ATOMIC MASS (atomic weight); (2) HELMHOLTZ FREE ENERGY.

A (*Genrl*) Symbol for area.

A (*Phys*) Symbol for: (1) ABSOLUTE TEMPERATURE; (2) magnetic VECTOR POTENTIAL.

[A] (*Phys*) A strong absorption band in the deep red of the solar spectrum (wavelength 762·128 nm) caused by oxygen in the Earth's atmosphere. The first of the FRAUNHOFER LINES.

a (*Genrl*) Symbol for ATTO-.

a (*Phys*) Symbol for: (1) ACCELERATION; (2) AMPLITUDE; (3) linear ABSORPTION COEFFICIENT.

a- (*Genrl*) Prefix signifying *on*. Also shortened form of AB-, AD-, AN-, *ap-*.

a- (*Chem*) Abbrev for: (1) asymmetrically substituted; (2) ANA-.

α (*Genrl*) See under ALPHA.

α (*Chem*) Symbol for: (1) substitution on the carbon atom of a chain next to the functional group; (2) substitution on a carbon atom next to one common to two condensed aromatic nuclei; (3) substitution on the carbon atom next to the hetero-atom in a hetero-cyclic compound; (4) a stereoisomer of a sugar.

α (*Phys*) Symbol for: (1) ABSORPTION COEFFICIENT; (2) ACCELERATION; (3) ANGULAR ACCELERATION; (4) ATTENUATION COEFFICIENT; (5) FINE-STRUCTURE CONSTANT; (6) helium nucleus.

[α]$_D^t$ (*Chem*) Symbol for the specific optical rotation of a substance at $t°$C, measured for the D lines of the sodium spectrum.

aa (*Geol*) A term of Hawaiian origin for lava flows with a rough, jagged surface.

Aalenian (*Geol*) The oldest stage of the Middle Jurassic. See MESOZOIC.

AAMI (*Med*) Abbrev for *age-associated memory impairment*.

A-amplifier (*Acous*) An amplifier associated with, or immediately following, a high-quality microphone, as in broadcasting studios. NB Not the same as CLASS-A AMPLIFIER.

A & AEE (*Aero*) Abbrev for *Aeroplane & Armament Experimental Establishment*, at Boscombe Down, UK.

A and R display (*Radar*) See R-DISPLAY.

Ab (*BioSci*) Abbrev for ANTIBODY.

ab (*Build*) Abbrev for *as before* in eg bills of quantities.

ab- (*ElecEng*) Prefix to name of unit indicating derivation in the CGS system, eg ABAMPERE.

ABA (*BioSci*) Abbrev for ABSCISIC ACID.

abactinal (*BioSci*) Pertaining to that part of the surface of an echinoderm that lacks tube feet.

abacus (*Arch*) The uppermost part of a column capital or pilaster, on which the architrave rests.

abacus (*MathSci*) A bead frame. Used as an arithmetic calculating aid.

abambulacral (*BioSci*) Pertaining to any of the five grooves along which the tube feet of echinoderms are located.

abampere (*Phys*) Unit of electric current in the CGS electromagnetic system of units. One abampere equals 10 A.

abamurus (*Arch*) A supporting wall or buttress, built to add strength to another wall.

abandonment (*MinExt*) Voluntary surrender of legal rights or title to a mining claim.

abandonware (*ICT*) Computing slang for software that is no longer distributed by its original publisher.

abapical (*BioSci*) Pertaining to, or situated at, the lower pole: remote from the apex.

abatjour (*Arch*) An opening to admit light and generally to deflect it downwards; a skylight.

abattoir (*Agri*) A facility where animals are slaughtered and prepared for sale as meat.

abaxial (*BioSci*) Remote from the axis. Used to describe eg a leaf or petal that faced away from the axis during early development (and usually, therefore, the under-surface of an expanded leaf). Cf ADAXIAL.

abaxial (*ImageTech, Phys*) Said of rays of light which do not coincide with the optical axis of a lens system.

Abbe refractometer (*Chem*) An instrument for measuring directly the refractive index of liquids, minerals and gemstones.

ABC (*ElecEng*) Abbrev for AERIAL BUNCHED CONDUCTORS.

ABC (*ImageTech*) Abbrev for AUTOMATIC BEAM CONTROL.

ABC proteins (*BioSci*) Membrane proteins involved in active transport or regulation of ion channel function and having an *ATP binding cassette*. A well-known example is the cystic fibrosis transmembrane conductance regulator (CFTR).

abdominal air sac (*BioSci*) Posterior part of the lung in birds.

abdominal cavity (*Med*) See PERITONEAL CAVITY.

abdominal gills (*BioSci*) In the aquatic larvae of many insects, paired segmental leaf-like or filamentous expansions of the abdominal cuticle for respiration.

abdominal limbs (*BioSci*) Segmented abdominal appendages in most Crustacea which are used for swimming, setting up currents of water for feeding and/or respiration, or for carrying eggs and young. In Diplopoda, segmented ambulatory appendages on the abdomen.

abdominal regions (*Med*) Nine regions into which the human abdomen is divided by two horizontal and two vertical imaginary planes, ie right and left hypochondriac, right and left lumbar, right and left iliac, epigastric, umbilical and hypogastric.

abdominoplasty (*Med*) A plastic surgery procedure for the stomach. Also *tummy tuck*.

abducens (*BioSci*) In vertebrates, the sixth cranial nerve, purely motor in function, supplying the rectus externus muscle of the eye.

abduction (*Med*) The action of pulling a limb or part away from the median axis.

abductor (*BioSci*) Any muscle that draws a limb or part away from the median axis by contraction, eg the abductor pollicis, which moves the thumb outwards.

Abegg's rule (*Chem*) Empirical rule that the solubility of salts of alkali metals with strong acids decreases from lithium to caesium, ie with increase of relative atomic mass, and those with weak acids follow the opposite order. Sodium chloride is an exception to this rule, being less soluble than potassium chloride.

Abegg's rule of eight (*Chem*) Rule that the sum of the maximum positive and negative valencies of an element is eight, eg sulphur in SF_6 and H_2S.

Abel flash-point apparatus (*MinExt*) A petroleum-testing apparatus for determining the flash point.

Abelian group (*MathSci*) A GROUP in which the group operation is COMMUTATIVE. It is important in the study of RINGS and VECTOR SPACES.

abelite (*Chem*) Explosive, composed mainly of ammonium nitrate and trinitrotoluene.

aberrant (*BioSci*) Having characteristics not strictly in accordance with the type.

aberration (*Astron*) An apparent change of position of a heavenly body, due to the speed of light having a finite ratio to the relative velocity of the source and the observer.

aberration (*Phys*) In an image-forming system, eg an optical or electronic lens, failure to produce a true image, eg a point object as a point image. Geometrical aberrations include spherical aberration, coma, astigmatism, curvature of the field and distortion. See CHROMATIC ABERRATION.

abhesive (*Eng*) A substance which prevents two materials sticking together, eg TEFLON on frying pans.

abiogenesis (*BioSci*) The development of living organisms from non-living matter; either the spontaneous generation of yeasts, bacteria, etc, believed before Pasteur, or the gradual process postulated for the early Precambrian in modern theories of the origin of life. Also *spontaneous generation*.

abiotic (*BioSci*) Pertaining to non-living things.

ABL (*EnvSci*) Abbrev for ATMOSPHERIC BOUNDARY LAYER (panel).

ablation (*Geol*) (1) Any one of the processes by which snow and ice are lost from a glacier, mainly by melting and evaporation (sublimation). (2) Removal of surface layers of a meteorite and tektites during flight.

ablation (*Med*) Removal of body tissue by surgery.

ablative polymer (*Eng*) A material which degrades controllably in an aggressive environment, esp on re-entry spacecraft. Extreme temperatures are reached on a heat shield, so it is protected with an ablation shield made of eg silicone polymer. The same principle is used in intumescent paints for fire resistance.

Abney law (*Phys*) Rule stating that if a spectral colour is desaturated by the addition of white light, and if its wavelength is less than 570 nm, its hue then moves towards the red end of the spectrum, while if the wavelength is more than 570 nm its hue moves towards the blue.

Abney level (*Surv*) Hand-held instrument in which angles of steep sights are measured while simultaneously viewing a spirit-level bubble.

Abney mounting (*Phys*) A form of mounting for a concave diffraction grating, in which the eyepiece (or photographic plate holder) is fixed at the centre of curvature of the grating and the slit can move around the circumference of the Rowland circle, to bring different orders of the spectrum into view.

abnormal glow discharge (*Electronics*) A discharge carrying current in excess of that which is required to cover the cathode completely with visible radiation.

abnormal reflection (*ICT*) Reflection from the ionosphere of a radio wave whose frequency is greater than the CRITICAL FREQUENCY.

ABO blood group substances (*BioSci*) Large glycopeptides with oligosaccharide side chains bearing ABO antigenic determinants identical to those of the erythrocytes of the same individual, present in mucous secretions of persons who possess the secretor gene.

ABO blood group system (*BioSci*) The most important of the antigens of human red blood cells for blood transfusion serology. Humans belong to one of four groups: A, B, AB and O. The red cells of each group carry respectively the A antigen, the B antigen, both A and B antigens, or neither. Natural antibodies are present against the blood group antigen that is absent from the red cells. Thus persons of group A have anti-B, of group B have anti-A, of group O have anti-A and anti-B, and group

AB have neither. Before blood transfusion the blood must be cross-matched to ensure that red cells of one group are not given to a person possessing antibodies against them.

abomasitis (*Vet*) Inflammation of the ABOMASUM.

abomasum (*BioSci*) In ruminant mammals, the fourth or true stomach. Also *reed*.

A-bomb (*Phys*) See ATOMIC BOMB.

aboral (*BioSci*) Opposite to, leading away from, or distant from, the mouth.

abort (*ICT*) The unplanned failure of a program or a piece of software.

abort (*Space*) To terminate a vehicle's flight either by failure or by deliberate action to prevent dangerous consequences; if manned, a predetermined sequence of events is followed to ensure the safety of the crew.

abortifacient (*Med*) A term describing anything which causes artificial abortion; a drug which does this.

abortion (*Med*) (1) Expulsion of the fetus from the uterus during the first three months of pregnancy. Abortion may be spontaneous or induced. (2) Termination of the development of an organ.

abortive infection (*BioSci*) Viral infection of a cell in which the virus fails to replicate fully, or produces defective progeny; the effect can still be cytopathogenic.

abortive transformation (*BioSci*) Temporary TRANSFORMATION of a cell by a virus that does not become integrated into the host DNA.

abradant (*Eng*) A substance, usually in powdered form, used for grinding. See ABRASIVE.

abrade (*Eng*) To scratch or tear away two surfaces in contact by relative motion.

Abram's law (*CivEng*) Rule that the ratio of water to cement for chemical action to impart strength to concrete is 0·35:1.

abranchiate (*BioSci*) Lacking gills.

abrasion (*Eng*) Surface damage due to an abrasive or to rubbing contact.

abrasion (*Geol*) Mechanical wearing away of rocks by rubbing during movement.

abrasion (*Med*) A rubbed-away area of the surface covering of the body, ie of skin or of mucous membrane.

abrasion hardness (*MinExt*) Resistance to abrasive wear, under specified conditions, of metal or mineral.

abrasive (*Eng*) A hard substance, usually in powdered form, used for the removal of material by scratching and grinding, eg silicon carbide powder (*carborundum*).

abrasive blast cleaning (*Build*) A method for preparing steel for painting whereby abrasive particles, eg copper slag, are projected under air pressure through a nozzle. Very effective in removing rust and mill scale, leaving an *anchor pattern* (a pattern of minute projections) on the substrate affording good paint adhesion.

abrasive papers (*Build*) Special papers coated in grit used for flatting down. Supplied in a range of grits from very fine to coarse as either dry or waterproof abrasive papers.

abrasive wear (*Eng*) Mechanism of wear due to the presence in one or both surfaces of hard particles (eg carbide in steels), or to hard particles trapped between them.

abreaction (*Psych*) In psychoanalytic theory, an intense emotional outburst to a previously repressed experience; the therapeutic effect is known as *catharsis*.

A/B roll editing (*ImageTech*) Video editing using two SOURCE PLAYERS (A and B) enabling DUBBING from both. Necessary if scenes are to be superimposed. See EDIT CONTROLLER, EDIT RECORDER, EDIT SUITE.

A/B roll printing (*ImageTech*) A method of film printing with alternate scenes assembled in two rolls, each having black spacing equivalent in length to the omitted scene; double printing from the two allows the inclusion of fade and DISSOLVE effects and avoids visible splice marks between scenes in 16 mm printing.

ABS (*Plastics*) Abbrev for *acrylonitrile-butadiene-styrene*. See COPOLYMER.

ABS brake (*Autos*) See ANTILOCK BRAKE.

abscess (*Med*) A localized collection of pus in infected tissue, usually confined within a capsule.

abscisic acid (*BioSci*) A sesquiterpenoid plant growth substance ($C_{15}H_{20}O_4$) with a variety of reported effects, eg inhibiting growth, causing stomatal closure and promoting senescence, abscission and dormancy. Also *dormin*. Abbrev *ABA*.

abscissa (*MathSci*) For rectilinear axes of co-ordinates, the distance of a point from one axis measured in a direction parallel to another axis, usually horizontal. The sign convention is that measurements to the right from the axis of ordinates are positive, measurements to the left negative. Pl *abscissae*. Cf CARTESIAN CO-ORDINATES.

abscission (*BioSci*) The organized shedding of parts of a plant by means of an ABSCISSION LAYER.

abscission layer (*BioSci*) In the ABSCISSION ZONE, a layer of cells the disjunction or breakdown of which causes abscission. Also *separation layer*.

abscission zone (*BioSci*) Zone at the base of leaf, petal, fruit, etc, that contains the ABSCISSION LAYER and PROTECTIVE LAYER.

absolute (*MathSci*) (1) See MODULUS. (2) A conic (a quadric in three dimensions) formed by the assemblage of the points at infinity on a line (in general two points). Its form determines the metrical properties of the geometrical system being operated. Thus in Euclidean geometry, the absolute is the degenerate conic comprising the line at infinity taken twice, while in non-Euclidean geometry, the absolute is either a real conic (hyperbolic geometry) or an imaginary conic (elliptic geometry).

absolute address (*ICT*) Code designation of a specific memory location as determined by the HARDWARE. Cf RELATIVE ADDRESS.

absolute age (*Geol*) The geological age of a fossil, mineral, rock or event, generally given in years. Preferred synonym *radiometric age, isotopic age*. See panel on RADIOMETRIC DATING.

absolute alcohol (*Chem*) Water-free ethanol; rel.d. 0·793 (15·5°C); bp 78·4°C; obtained from rectified spirit by adding benzene and refractionating. Very hygroscopic.

absolute ampere (*Phys*) The standard MKS unit of electric current; replaced the international ampere in 1948. See AMPERE.

absolute block system (*CivEng*) See BLOCK SYSTEM.

absolute ceiling (*Aero*) The height at which the rate of climb of an aircraft, in standard atmosphere, would be zero; the maximum height attainable under standard conditions.

absolute coefficient (*MathSci*) A coefficient with an absolute value, ie a multiplier which is numerical rather than symbolic.

absolute configuration (*Chem*) The arrangement of groups about an asymmetric atom, esp a tetrahedrally bonded atom with four different substituents. See CHIRALITY, CAHN–INGOLD–PRELOG SYSTEM.

absolute convergence (*MathSci*) The property of an infinite series when the sum of absolute values converges; ie $\sum a_r$ is absolutely convergent if the series $\sum |a_r|$ is convergent.

absolute electrometer (*Phys*) A high-grade attracted-disk electrometer in which an absolute measurement of potential can be made by 'weighing' the attraction between two charged disks against gravity.

absolute filter (*Phys*) A filter which removes most particulate matter from gases.

absolute humidity (*EnvSci*) See VAPOUR CONCENTRATION.

absolute instrument (*Phys*) An instrument which measures a quantity directly in absolute units, without the necessity for previous calibration.

absolute lethal concentration (*BioSci*) The lowest concentration of a substance in an environmental medium which kills all the test organisms or species under defined conditions. Abbrev *LC100*.

absolute magnitude (*Astron*) See MAGNITUDE.

absolute permeability (*ElecEng*) See PERMEABILITY.

absolute potential (*Chem*) The theoretical true potential difference between an electrode and a solution of its ions, measured against a hypothetical reference electrode, having an absolute potential of zero, with reference to the same solution.

absolute pressure (*Phys*) Pressure measured with respect to zero pressure, in units of force per unit area.

absolute reaction rate (*Chem*) The reaction rate determined from statistical thermodynamics; uses the assumption of the theory of absolute reaction rates that the rate of a chemical reaction is governed by the rate of crossing an energy barrier or of forming an ACTIVATED COMPLEX. See ARRHENIUS THEORY OF DISSOCIATION.

absolute-rest precipitation tank (*Build*) A tank used for batch treatment of sewage, as opposed to one taking a continuous flow. After two- or three-hour settlement, the top water is drawn off from above and the precipitated sludge from below.

absolute temperature (*Phys*) Temperature measured with respect to ABSOLUTE ZERO, ie the zero of the KELVIN THERMODYNAMIC SCALE OF TEMPERATURE, a scale which cannot take negative values. See KELVIN, RANKINE SCALE.

absolute threshold (*Psych*) The minimal intensity of a physical stimulus required to produce a response.

absolute units (*Phys*) Units derived directly from the fundamental units of a system and not based on arbitrary numerical definitions. The differences between absolute and international units were small; both are now superseded by the definitions of SI UNITS.

absolute value (*MathSci*) See MODULUS.

absolute wavemeter (*ICT*) A wavemeter in which the frequency of the injected radio-frequency signal is by calculation of physical properties (circuit elements or dimensions) of a resonant circuit line or cavity.

absolute weight (*Phys*) The weight (or mass) of a body in a vacuum.

absolute zero (*Phys*) The least possible temperature for all substances. At this temperature the molecules of any substance possess no heat energy. A figure of $-273\cdot15°C$ is generally accepted as the value of absolute zero. See KELVIN.

absorbance (*Chem*) The logarithm of the ratio of the intensity of light incident on a sample to that transmitted by it. It is usually directly proportional to the concentration of the absorbing substance in a solution. See BEER'S LAW, TRANSMITTANCE.

absorbance (*Paper, Textiles*) The capacity of materials such as textile fibres and paper to absorb liquids. See REGAIN.

absorbed dose (*Radiol*) Quantity of energy imparted by ionizing radiation to a unit mass of biological tissue. Unit is the gray (Gy). See panel on RADIATION.

absorbency test (*Paper, Textiles*) Any test method for measuring the capacity of materials such as textile fibres and paper to absorb liquids or fluids. Results are usually expressed as the gain in weight of the test piece, the capillary rise in a test strip in given time, or the time required to reach a predetermined capillary rise.

absorber (*Phys*) Any material which converts energy of radiation or particles into another form, generally heat. Energy transmitted is not absorbed. Scattered energy is often classed with absorbed energy. See ABSORPTION COEFFICIENT.

absorber rod (*NucEng*) See CONTROL ROD.

absorbing material (*Phys*) Any medium used for absorbing energy from radiation of any type.

absorbing well (*CivEng*) A shaft sunk through an impermeable stratum to allow water to drain through to a permeable one.

absorptance (*Phys*) A measure of the ability of a body to absorb radiation; the ratio of the radiant flux absorbed by the body to that incident on the body. Formerly *absorptivity, absorptive power*.

absorptiometer (*Chem*) Apparatus for determining the solubilities of gases in liquids or the absorption of light.

absorption (*BioSci*) (1) The uptake of a drug (see ADME), or other compounds, eg nutrients from the intestinal tract, into the body. (2) The use of reagents to remove unwanted substances, eg antibodies or antigens, from a mixture.

absorption band (*Phys*) A dark gap in the continuous spectrum of white light transmitted by a substance which exhibits selective absorption. See fig. at ABSORPTION SPECTRUM.

absorption capacitor (*ElecEng*) A capacitor connected across a spark gap to damp the discharge.

absorption coefficient (*Chem*) The volume of gas, measured at STP, dissolved by unit volume of a liquid under normal pressure (ie one atmosphere).

absorption coefficient (*Phys*) (1) At a discontinuity (*surface absorption coefficient*): (a) the fraction of the energy which is absorbed; or (b) the reduction of amplitude, for a beam of radiation or other wave system incident on a discontinuity in the medium through which it is propagated, or in the path along which it is transmitted. (2) In a medium (*linear absorption coefficient*), the natural logarithm of the ratio of incident and emergent energy or amplitude for a beam of radiation passing through unit thickness of a medium. (The *mass absorption coefficient* is defined in the same way but for a thickness of the medium corresponding to unit mass per unit area.) NB *True absorption coefficients* exclude scattering losses, *total absorption coefficients* include them. See ATOMIC ABSORPTION COEFFICIENT.

absorption discontinuity (*Phys*) See ABSORPTION EDGE.

absorption dynamometer (*Eng*) A dynamometer which absorbs and dissipates the power which it measures; eg the ordinary rope brake and the Froude hydraulic brake. Cf TRANSMISSION DYNAMOMETER.

absorption edge (*Phys*) The wavelength at which there is an abrupt discontinuity in the intensity of an absorption spectrum for electromagnetic waves, giving the appearance of a sharp edge in its photograph. This transition is due to one particular energy-dissipating process becoming possible or impossible at the limiting wavelength. In X-ray spectra of the chemical elements the K-absorption edge for each element occurs at a wavelength slightly less than that for the K-emission spectrum. Also *absorption discontinuity*.

absorption hygrometer (*EnvSci*) An instrument by which the quantity of water vapour in air may be measured. A known volume of air is drawn through tubes containing a drying agent such as phosphorus pentoxide; the increase in weight of the tubes gives the weight of water vapour in the known volume of air.

absorption inductor (*ElecEng*) See INTERPHASE TRANSFORMER.

absorption lines (*Phys*) Dark lines in a continuous spectrum caused by absorption by a gaseous element. The positions (ie the wavelengths) of the dark absorption lines are identical to those of the bright lines given by the same element in emission.

absorption nebula (*Astron*) See DARK NEBULA.

absorption plant (*MinExt*) Plant where oils are removed from natural gas by absorption in suitable oil.

absorption refrigerator (*Eng*) A plant in which ammonia is continuously evaporated from an aqueous solution under pressure, condensed, allowed to evaporate (so absorbing heat) and then reabsorbed.

absorption spectrum (*Phys*) The system of absorption bands or lines seen when a selectively absorbing substance is placed between a source of white light and a spectroscope. See KIRCHHOFF'S LAW. Fig. ▷

absorption spectrum Lines on a continuum.

absorption tubes (*Chem*) Tubes filled with solid absorbent for the absorption of moisture (eg silica gel) and gases (eg charcoal).

absorption wavemeter (*ElecEng*) A wavemeter which depends on resonance absorption in a tuned circuit, constructed with very stable inductance and capacitance.

absorptive power (*Phys*) See ABSORPTANCE.

absorptivity (*Phys*) See ABSORPTANCE.

abstraction (*ICT*) A principle of object-oriented design, in which common data or functionality are removed from individual objects into a self-contained layer, making them consistently available from a single source.

abstraction (*Psych*) The mental process of arriving at an abstract idea or concept from specific examples.

AB toxin (*BioSci*) A multi-subunit toxin with two major components: an active (A) portion and a portion that is involved in binding (B) to the target cell. In well-known examples (cholera, diphtheria, and pertussis toxins), the A subunit has ADP-ribosylating activity.

abundance (*BioSci*) See FREQUENCY, RELATIVE ABUNDANCE.

abundance (*Phys*) For a naturally occurring element, the proportion or percentage of one isotope to the total. Also *abundance ratio*.

abundant number (*MathSci*) A natural number for which the sum of the proper factors is greater than the number itself; eg 18 is abundant since $1 + 2 + 3 + 6 + 9 > 18$. Cf DEFICIENT NUMBER, PERFECT NUMBER.

abura (*For*) Tropical W African hardwood (Mitragyna) with a fine, even texture.

abutment (*CivEng*) A structure provided to withstand thrust, eg end supports of an arch or bridge. See KNAPSACK ABUTMENT.

abutment load (*MinExt*) In stoping or other deep-level excavation, weight transferred to the adjacent solid rock by unsupported roof.

abutting joint (*Build*) A timber joint whose plane is at right angles to the fibres, the fibres of both joining pieces being in the same straight line.

abyssal (*EnvSci*) A term describing the ocean floor environment between c.4000 and 6000 m. Also *bathysmal*. Cf BATHYAL, LITTORAL.

abyssal deposits (*Geol*) Pelagic marine sediments, accumulating in depths of more than 2000 m including, with increasing depth, calcareous oozes, siliceous oozes and red clay (500 m).

abyssal plain (*Geol*) A flat region of the deep ocean floor with a slope of less than 1:1000.

abyssopelagic (*EnvSci*) Relating to the open waters of the ABYSSAL zone.

abzyme (*BioSci*) A catalytic antibody, one that has enzymic activity.

Ac (*Chem*) Symbol for ACTINIUM.

Ac (*Eng*) The transformation temperature on heating of the phase changes of iron or steel, subscripts indicating the designated change, eg Ac_1 is the eutectoid (723°C) and Ac_3 the ferrite/austenite phase boundary.

Ac (*EnvSci*) Abbrev for ALTOCUMULUS.

AC-3 (*ImageTech*) TN for the digital audio coding used in 35 mm motion picture film to provide six-channel SURROUND SOUND. It uses data blocks recorded optically between the PERFORATIONS, leaving room for a conventional soundtrack. It is also suitable for multi-channel TV audio, as well as video SOFTWARE and HOME CINEMA.

ac (*ElecEng*) Symbol for ALTERNATING CURRENT.

ac- (*Chem*) Abbrev indicating substitution in the alicyclic ring.

acacia (*For*) See MYRTLE.

acacia gum (*Chem*) See GUM ARABIC.

acalculia (*Psych*) Inability to make simple mathematical calculations. Also *dyscalculia*.

acamprosate (*Pharmacol*) A drug thought to be beneficial in maintaining abstinence in alcohol-dependent patients.

Acanthamoeba (*Med*) A protozoon which can survive in tapwater and may cause KERATITIS if contact lenses are washed in tapwater.

acanthite (*Min*) An ore of silver, Ag_2S, crystallizing in the monoclinic system. Cf ARGENTITE.

acantho- (*Genrl*) Prefix from Gk *akantha*, spine, thorn.

Acanthocephala (*BioSci*) A phylum of elongate worms with a rounded body and a protrusible proboscis, furnished with recurved hooks. There is no mouth or alimentary canal. The young stages are parasitic in various Arthropoda, the adults in fish and aquatic birds and mammals. Thorny-headed worms.

acanthoma (*Med*) A tumour of epidermal cells.

acanthosis nigricans (*Med*) A rare disease characterized by pigmentation and warty growths on the skin, often associated with cancer of the stomach or uterus.

acariasis (*Vet*) Contagious skin disease caused by mites (acari).

Acarina (*BioSci*) An order of small Arachnida, with globular, undivided body. The immature stages (hexapod larvae) have six legs. The order is a large worldwide group, occupying all types of habitat, and of great economic importance. Many are ectoparasitic. Mites and ticks.

acarophily (*BioSci*) A symbiotic association between plants and mites. Also *acarophytism*.

acatamathesia (*Psych*) The inability to comprehend data (objects, language, etc) presented to the senses.

acaulescent (*BioSci*) Having a short stem.

acauline (*BioSci*) Stemless or nearly so. Also *acaulose*.

ac balancer (*ElecEng*) An arrangement of transformers or reactors used to equalize the voltages between the wires of a multiple-wire system. Also *static balancer*.

ac bias (*Electronics*) A high-frequency signal applied to a magnetic tape recording head along with the signal to be recorded. This stabilizes magnetic saturation and improves frequency response, at the same time reducing noise and distortion. The bias signal frequency has to be many times the highest recording frequency.

AC-boundary layer (*Acous, Aero*) See STOKES LAYER.

accelerated ageing test (*ElecEng*) A stability test for cables using twice normal working voltage. It is claimed this gives quick results that correlate with service records.

accelerated fatigue test (*Eng*) Test which applies a cyclic loading schedule, which can be of varying frequency and/ or amplitude, to a machine or component simulating its loading in service, but at a higher rate, to determine its safe fatigue life before it is reached in service. See panel on FATIGUE.

accelerated freeze-drying (*FoodSci*) FREEZE-DRYING in which the final sublimation under vacuum is accelerated by heat. Abbrev *AFD*.

accelerate-stop distance (*Aero*) The total distance, under specified conditions, in which an aircraft can be brought to rest after accelerating to CRITICAL SPEED for an engine failure at take-off.

accelerating chain (*Electronics*) The section of an electron beam tube or system, eg cathode-ray tube or electron microscope, in which electrons are accelerated by voltages on accelerating electrodes. Also used in particle accelerators.

accelerating electrode (*Electronics*) An electrode in a thermionic valve or cathode-ray tube maintained at a high positive potential with respect to the electron source. It accelerates electrons in their flight to the anode but does not collect a high proportion of them.

accelerating machine (*Electronics*) See ACCELERATOR.

accelerating potential (*Electronics*) The potential applied to an electrode to accelerate electrons from a cathode.

acceleration (*Phys*) The rate of change of velocity, expressed in metres (or feet) per second squared. It is a vector quantity and has both magnitude and direction.

acceleration due to gravity (*Phys*) Acceleration with which a body would fall freely under the action of gravity in a vacuum. This varies according to the distance from the Earth's centre, but the internationally adopted value is $9\cdot806\,65\ m\,s^{-2}$ or $32\cdot1740\ ft\,s^{-2}$. Abbrev *g*. See HELMERT'S FORMULA.

acceleration error (*Aero*) The error in an airborne magnetic compass due to manoeuvring; caused by the vertical component of the Earth's magnetic field when the centre of gravity of the magnetic element is displaced from normal.

acceleration stress (*Space*) The influence of acceleration (or deceleration) on certain physiological parameters of the human body. Humans can withstand transverse accelerations better than longitudinal ones, which have a profound effect on the cardiovascular system. The degree of tolerance also depends on the magnitude and duration of the acceleration.

acceleration tolerance (*Space*) The maximum acceleration force that an astronaut can withstand before 'blacking out' or otherwise losing control.

accelerator (*Aero*) A device, similar to a CATAPULT, but generally mounted below deck level, for assisting the acceleration of aircraft flying off aircraft carriers. Land versions have been tried experimentally.

accelerator (*Autos*) A pedal connected to the carburettor throttle valve of a motor vehicle, or to the fuel injection control.

accelerator (*BioSci*) Any muscle or nerve that increases rate of action.

accelerator (*Build*) A hardener or catalyst mixed with synthetic resins in TWO PACK MATERIALS to speed up the hardening rate.

accelerator (*Chem*) (1) A substance which increases the speed of a chemical reaction. See CATALYSIS. (2) A substance which increases the efficient action of an enzyme. (3) Any substance effecting acceleration of the vulcanization process of rubber. The principal types are aldehyde derivatives of Schiff's bases: butyraldehyde-butylidene-aniline, di-orthotolyl-guanidine, diphenyl-guanidine, benzthiazyl disulphide, tetramethyl-thiuran disulphide and zinc dimethyl-dithiocarbamate.

accelerator (*CivEng*) Any substance mixed with cement concrete for the purpose of hastening hardening.

accelerator (*Electronics*) Machine used to accelerate charged particles to very high energies. See BETATRON, CYCLOTRON, LINEAR ACCELERATOR, SYNCHROCYCLOTRON, SYNCHROTRON.

accelerator (*ICT*) A special circuit board that is placed within a computer to speed up some aspect of its operation.

accelerator (*ImageTech*) A chemical used to increase the rate of development; eg sodium carbonate or borax.

accelerator pump (*Autos*) A small cylinder and piston fitted to some types of carburettor, and connected to the throttle so as to provide a momentarily enriched mixture when the engine is accelerated.

accelerometer (*Acous, Electronics*) A transducer used to provide a signal proportional to the rate of acceleration of a vibrating or other body, usually employing the piezoelectric principle. See PICK-UP HEAD.

accelerometer (*Aero*) An instrument, carried in aircraft, guided missiles and spacecraft, for measuring acceleration in a specific direction. Main types are *indicating*, *maximum-reading*, *recording* (graphical) and *counting* (digital, totalling all accelerations above a set value). See IMPACT ACCELEROMETER, VERTICAL-GUST RECORDER.

acceptable daily intake (*FoodSci*) Estimated daily intake in $mg\,kg^{-1}$ of a substance (eg additive, residue) regarded as having no obvious harmful effect to humans during their lifespan. Abbrev *ADI*.

acceptance angle (*Electronics*) The solid angle within which all incident light reaches the photocathode of a phototube.

acceptance test (*ICT*) The phase of the testing cycle in which an entire system is tested by a group of users, generally customers, in order to establish its acceptability for purchase or implementation.

acceptor (*Chem*) (1) The reactant in an induced reaction whose rate of reaction with a third substance is increased by the presence of the inductor. (2) The atom which accepts electrons in a co-ordinate bond.

acceptor (*Electronics*) Impurity atoms introduced in small quantities into a crystalline semiconductor and having a lower valency than the semiconductor, from which they attract electrons. In this way HOLES are produced, which effectively become positive charge carriers; the phenomenon is known as *p-type conductivity*. See DONOR, IMPURITY.

acceptor level (*Electronics*) See ENERGY LEVELS.

access charge (*ICT*) A charge for access to a computer or telecommunications network.

access eye (*Build*) A screwed plug provided in soil, waste and drain pipes at bends and junctions, to clear a stoppage.

accessorius (*BioSci*) (1) A muscle that supplements the action of another muscle. Also *accessory*. (2) In vertebrates, the eleventh cranial nerve or spinal accessory.

accessory bud (*BioSci*) A bud additional to a normal axillary bud.

accessory cell (*BioSci*) (1) Cell other than a lymphocyte that takes part in an immune reaction, eg in antigen presentation and/or by modulating the function of the lymphocyte. Usually a macrophage or dendritic cell. (2) In plants, a subsidiary cell in a stomatal complex.

accessory chromosome (*BioSci*) See SEX DETERMINATION.

accessory gearbox (*Aero*) A gearbox, driven remotely from an aero-engine, on which aircraft accessories, eg hydraulic pump and electrical generator, are mounted.

accessory glands (*BioSci*) Glands of varied structure and function in connection with genitalia, esp of Arthropoda.

accessory hearts (*BioSci*) See ACCESSORY PULSATORY ORGANS.

accessory minerals (*Geol*) Minerals which occur in small, often minute, amounts in igneous rocks; their presence or absence makes no difference to classification and nomenclature.

accessory pigments (*BioSci*) Pigments found in chloroplasts and blue-green algae that transfer their absorbed energy to chlorophyll a during photosynthesis. They include chlorophylls b, c and d, the carotenoids and the phycobilins.

accessory plates (*Min*) Quartz wedge, gypsum plate and mica plate. Used with petrological microscope to help determine the optical character of a mineral as an aid in its examination.

accessory pulsatory organs (*BioSci*) In some insects and molluscs, sac-like contractile organs, pulsating independent hearts; variously situated on the course of the circulatory system. Also *accessory hearts*.

accessory shoe (*ImageTech*) Mounting forming part of the camera body onto which separate units such as flash guns and rangefinders may be fitted.

access time (*ICT*) The time interval between the instant at which data are called from memory and the instant at which the data can be used. It can vary from microseconds with FAST STORE to minutes with MAGNETIC TAPE.

access to store (*ICT*) Entry or extraction of data from a memory location. The method and speed of access depends on the type of memory. See BACKING STORE, FAST STORE, RANDOM ACCESS MEMORY, SERIAL ACCESS MEMORY.

Accipitriformes (*BioSci*) An order comprising the diurnal birds of prey and raptors. It Includes eagles, harriers, vultures and buzzards. See also FALCONIFORMES, which are sometimes considered a separate order.

ac circuit (*ElecEng*) A circuit which passes alternating current as opposed to direct current, eg it may have a capacitor in series, which blocks direct current.

acclimatization (*BioSci*) Adaptation to environmental stress; reversible physiological adjustment in an organism when moved to a new environment, usually taking days or weeks. Also *acclimation*. See HARDENING.

accommodation (*Med*) The natural alteration of the effective focal length of the eye in order to see objects distinctly at varying distances. The range of vision for a human eye is from about 250 mm to infinity. Power of accommodation usually diminishes with advancing age.

accommodation (*Psych*) The creation of new cognitive schemas when objects, experiences or other information do not fit with existing schemas.

accommodation rig (*MinExt*) Offshore rig with sleeping, supply and recreational facilities.

ac commutator motor (*ElecEng*) An ac motor which embodies a COMMUTATOR as an essential part of its construction. See AC SERIES MOTOR, COMPENSATED INDUCTION MOTOR, REPULSION MOTOR, SCHRAGE MOTOR.

accordion (*Print*) US for CONCERTINA FOLD.

accoucheur (*Med*) A physician who practises midwifery; an obstetrician.

accretion (*Astron*) The process in which a celestial body, particularly an evolved star in a binary system, is enlarged by the accumulation of extraneous matter falling in under gravity.

accretion (*BioSci*) External addition of new matter: growth by such addition.

accretion (*Geol*) The process of enlargement of a continent by the tectonic coalescence of exotic crustal fragments.

accretion disc (*Astron*) The disc of material at the edge of a black hole, which has been attracted from a neighbouring star and which emits X-rays as its inner edge disappears into the gravitational field of the hole.

accumulated temperature (*EnvSci*) The integrated product of the excess of air temperature above a threshold value and the period in days during which such excess is maintained.

accumulation point (*MathSci*) One of a set of points, such that every neighbourhood of it includes at least one point of the set. Also *limit point*.

accumulator (*ICT*) Special storage register associated with the ARITHMETIC LOGIC UNIT, used for holding the results of a computation or data transfer.

accumulator (*ElecEng*) Voltaic cell which can be charged and discharged. On charge, when an electric current is passed through it into the positive and out of the negative terminals (according to the conventional direction of flow of current), electrical energy is converted into chemical energy. The process is reversed on discharge, the chemical energy, less losses in both potential and current, being converted into useful electrical energy. Accumulators therefore form a useful portable supply of electric power, but have the disadvantages of being heavy and of being at best 70% efficient. More often known as *battery*; also *reversible cell*, *secondary cell*, *storage battery*.

accumulator (*Eng*) Bottle or other reinforced reservoir for storing pressurized gas or fluid during moulding. Its use helps conserve energy during injection moulding cycle.

accumulator box (*ElecEng*) A vessel usually made of plastic which contains the plates and electrolyte of an accumulator.

accumulator grid (*ElecEng*) The lead grid which forms one of the plates of a lead–acid accumulator having pasted plates.

accumulator traction (*ElecEng*) See BATTERY TRACTION.

accumulator vehicle (*ElecEng*) See BATTERY TRACTION.

ACE inhibitor (*Pharmacol*) Abbrev for an ANGIOTENSIN-CONVERTING ENZYME INHIBITOR.

acellular (*BioSci*) Not partitioned into cells. Sometimes used for unicellular but also for multinucleate or coenocytic. See COENOCYTE.

acenaphthenequinone (*Chem*) Chemical, crystallizing in yellow needles, sparingly soluble in water. Forms the basis of scarlet and red vat dyes of the 'Ciba' type.

acentric (*BioSci*) Having no centromere; applied to chromosomes and chromosome segments.

acentrous (*BioSci*) Having a persistent notochord with no vertebral centra, as in the Cyclostomata.

acephalous (*BioSci*) Showing no appreciable degree of cephalization: lacking a head region, as in Pelecypoda.

acervulus (*BioSci*) A dense cushion-like mass of conidiophores and conidia formed by some fungi. Adj *acervulate*.

acet-, aceto- (*Genrl*) Prefixes from Lt *acetum*, vinegar.

acetabulum (*BioSci*) (1) In Platyhelminthes, Hirudinea and Cephalopoda, a circular muscular sucker. (2) In insects, a thoracic aperture for insertion of a leg. (3) In vertebrates, a facet or socket of the pelvic girdle with which the pelvic fin or head of the femur is articulated. (4) In ruminant mammals, one of the cotyledons of the placenta.

acetal (*Chem*) (1) *1,1-diethoxy ethane*. $CH_3CH(OC_2H_5)_2$. Bp 104°C. A colourless flammable liquid used as a solvent. (2) A term applied to any compound of the type $RCH(OR')_2$, where R and R' are organic radicals and R may be hydrogen.

acetaldehyde (*Chem*) *Ethanal*. CH_3CHO. A colourless, ethereal, pungent liquid, bp 21°C, mp −121°C, rel.d. 0·8, oxidation product of ethanol. An intermediate for production of ethanoic acid; also an important raw material for the synthesis of organic compounds.

acetaldehyde

acetal resin (*Chem*) See POLYOXYMETHYLENE.

acetals (*Chem*) The dehydration products of aldehydes, with an excess of alcohols present. Also *1,1-dialkoxyalkanes*.

acetamide (*Chem*) *Ethanamide*. CH_3CONH_2. A primary amide of ethanoic acid, crystalline, soluble in water and alcohol, mp 82°C, bp 222°C.

acetate fibres (*Textiles*) Artificial fibres; continuous filaments and staple fibres manufactured from cellulose acetate produced from cotton linters or wood pulp. Between 74 and 92% of the hydroxyl groups in the original cellulose are acetylated. For triacetate fibres the cellulose is more highly acetylated.

acetate film (*ImageTech*) Film with its photographic emulsion coated on a BASE of cellulose triacetate, of low flammability. Also *non-flam film*, *safety film*.

acetates (*Chem*) *Ethanoates*. Salts of acetic (ethanoic) acid; eg sodium acetate. Also esters of acetic acid.

acetic acid (*Chem*) *Ethanoic acid*. CH_3COOH. Synthesized from acetylene (ethyne), also obtained by the destructive distillation of wood and by the oxidation of ethanol; mp 16·6°C, bp 118°C, rel.d. (20°C) 1·0497. Main ingredient of vinegar and used as a food preservative. See PRESERVATION INDEX.

acetic anhydride (*Chem*) *Ethanoic anhydride*. $(CH_3CO)_2O$. The anhydride of acetic acid. Colourless liquid, bp 137°C. Used industrially for preparation of cellulose acetate and acetylsalicylic acid. Valuable laboratory acetylating agent.

acetic fermentation (*Chem*) The fermentation of dilute ethanol solutions by oxidation in presence of bacteria, esp *Bacterium aceti*. Acetic acid is formed.

acetin (*Chem*) *Monoacetin, glyceryl monoacetate*. $CH_3COOC_3H_5(OH)_2$. Bp 130°C, rel.d. 1·22, a colourless hygroscopic liquid; used as an intermediate for explosives, a solvent for basic dyestuffs and a tanning agent.

aceto- (*Genrl*) See ACET-.

acetogens (*BioSci*) Autotrophic bacteria that live in certain basalts. They use hydrogen gas for energy and derive carbon from inorganic carbon dioxide. They excrete simple organic compounds that other bacteria consume.

acetone (*Chem*) CH_3COCH_3. Bp 56°C, of ethereal odour, a very important solvent and basis for organic synthesis. Acetone is the simplest saturated ketone and is a useful solvent for acetylene. It is found in the blood and urine of patients with uncontrolled diabetes mellitus or in starvation. Also *propanone*.

acetone cyanhydrin (*Chem*) *2-hydroxy 2-methyl propanonitrile*. $(CH_3)_2C(OH)CN$. Addition product of acetone and hydrogen cyanide.

acetone resin (*Chem*) A synthetic resin formed by the reaction of acetone with another compound, such as phenol or formaldehyde.

acetonitrile (*Chem*) CH_3CN. A polar organic solvent.

acetonuria (*Med*) See KETONURIA.

acetophenone (*Chem*) *Phenyl methyl ketone*. $C_6H_5COCH_3$. Mp 20°C, bp 202°C, rel.d. 1·03, large colourless crystals or liquid, soluble in most organic solvents, insoluble in water; used for organic synthesis and in perfumery.

acetoxyl group (*Chem*) The group CH_3COO—.

acetylation (*Chem*) *Ethanoylation*. Reaction which has the effect of introducing an acetyl radical (CH_3CO) into an organic molecule.

acetylators (*BioSci*) Individuals who can acetylate or metabolize substances in the liver such as many common drugs. A single gene mutation determines whether an individual is a fast or slow acetylator.

acetylcelluloses (*Chem*) See CELLULOSE ACETATES.

acetyl chloride (*Chem*) *Ethanoyl chloride*. CH_3COCl. Mp −112°C, bp 51°C, rel.d. 1·105. Colourless liquid, of pungent odour, used for synthesis, in particular for introducing the acetyl group into other compounds.

acetylcholine (*BioSci*) An important neurotransmitter, particularly at the neuromuscular junction but also involved in nerve–nerve transmission (at chemical synapses) in the brain.

acetylcholine esterase (*BioSci*) An enzyme, found in the synaptic clefts of cholinergic synapses, that breaks down the neurotransmitter acetylcholine into acetate and choline, thus limiting the size and duration of the postsynaptic potential. Many nerve gases and insecticides are potent acetylcholine esterase inhibitors, and thus prolong the timecourse of postsynaptic potentials.

acetylcholine esterase inhibitors (*Pharmacol*) A class of drugs that inhibit the breakdown of acetylcholine and thus increase availability as a neurotransmitter in the central nervous system. This appears to be beneficial in treatment of Alzheimer's disease. Examples are *donezepil, rivastigmine*. Up to one-half of patients treated with these drugs show a slower rate of decline in cognition.

acetyl CoA (*BioSci*) The acetylated form of coenzyme A; important in intermediary metabolism, particularly in the TCA CYCLE.

acetylene (*Chem*) *Ethyne*. $HC{\equiv}CH$. A colourless, poisonous gas, owing its disagreeable odour to impurities; soluble in ethanol, in acetone (25 times its volume at stp) and in water. Bp −84°C, rel.d. 0·91. Prepared by the action of water on calcium carbide and catalytically from naphtha. Used for welding, torches, illuminating (historically), acetic acid synthesis and for manufacturing derivatives.

acetyl group (*Chem*) *Ethanoyl group*. CH_3CO—. The radical of acetic acid.

acetylide (*Chem*) *Ethynide*. Carbide formed by bubbling acetylene through a solution of a metallic salt, eg cuprous acetylide, Cu_2C_2. Violently explosive compounds.

acetylsalicylic acid (*Chem*) $C_6H_4(OCOCH_3)$ COOH. Mp approximately 128°C. Used in medical and veterinary practice as an analgesic, antipyretic and antirheumatic. The acid or its salts are the active components of aspirin.

aceval (*Aero*) Abbrev for *air combat evaluation*.

ACF diagram (*Geol*) A triangular diagram used to represent the chemical composition of metamorphic rocks. The three corners of the diagram are Al_2O_3, CaO and $FeO + MgO$.

ac generator (*ElecEng*) An electromagnetic generator for producing alternating emf and delivering ac to an outside circuit. See ALTERNATOR, INDUCTION GENERATOR.

achaenocarp (*BioSci*) See ACHENE.

achalasia (*Med*) Failure to relax.

achalasia of the cardia (*Med*) Failure to relax on the part of the sphincter round the opening of the oesophagus into the stomach.

achene (*BioSci*) A dry indehiscent, one-seeded fruit, formed from a single carpel and with the seed distinct from the fruit wall. Also *achaenocarp, akene*.

Achernar (*Astron*) A conspicuous blue-white star in the constellation Eridanus, the ninth brightest star in the sky. Distance 35 pc. Also *Alpha Eridani*.

Achilles tendon (*BioSci*) In mammals, the united tendon of the soleus and gastrocnemius muscles.

achlorhydria (*Med*) Absence of hydrochloric acid from gastric juice.

acholuric jaundice (*Med*) See SPHEROCYTOSIS.

achondrite (*Geol*) A type of stony meteorite which compares closely with some basic igneous rocks such as eucrite.

achondroplasia (*Med*) Dwarfism characterized by shortness of arms and legs, with a normal body and head. Adj *achondroplastic*.

achroglobin (*BioSci*) A colourless respiratory pigment occurring in some Mollusca and some Urochorda.

achroite (*Min*) See TOURMALINE.

achromatic lens (*Phys*) A lens designed to minimize CHROMATIC ABERRATION. The simplest form consists of two component lenses, one convergent, the other divergent, made of glasses having different dispersive powers, the ratio of their focal lengths being equal to the ratio of the dispersive powers.

achromatic prism (*Phys*) An optical prism with a minimum of dispersion but a maximum of deviation.

achromatic sensation (*Phys*) A visual perception of grey. Represented by the equal energy point on a chromaticity diagram.

achromatic stimulus (*Phys*) Stimulus which produces an ACHROMATIC SENSATION.

achromatopsia (*BioSci*) Absence of colour vision, the norm in many species but rare in humans; due to an absence of cones in the retina.

achylia gastrica (*Med*) Complete absence of pepsin and hydrochloric acid from the gastric secretion.

acicle (*BioSci*) A stiff bristle, or slender prickle; sometimes with a glandular tip.

aciclovir (*Pharmacol*) An antiviral drug used to treat herpes virus infections such as chickenpox, shingles, cold sores and herpes. It inhibits DNA polymerase, but does not eliminate the virus. It is effective only if given at the very outset of the infection. Formerly *acyclovir*.

acicular (*BioSci, Min*) Needle-shaped; used to describe the needle-like habit of crystals.

acid (*Chem*) Normally, a substance which: (1) dissolves in water with the formation of hydrogen ions; (2) dissolves metals with the liberation of hydrogen gas; or (3) reacts with a BASE to form a SALT. More generally, a substance which tends to lose a proton (BRÖNSTED–LOWRY THEORY) or to accept an electron pair (see LEWIS ACIDS AND BASES).

acid (*Geol*) See ACID ROCK.

acid amides (*Chem*) A group of compounds derived from an acid by the introduction of the amino group in place of the hydroxyl radical of the carboxyl group.

acid anhydride (*Chem*) Compound generating a hydroxylic acid on addition of (or derived from the acid by removal of) one or more molecules of water, eg sulphur (VI) oxide: $SO_3 + H_2O \rightarrow H_2SO_4$.

acid azides (*Chem*) The acyl derivatives of hydrazoic acid, obtainable from ACID HYDRAZIDES by treatment with nitrous acid. They are very unstable.

acid brittleness (*Eng*) The brittleness developed in steel in a pickling bath, through evolution of hydrogen. Cf HYDROGEN EMBRITTLEMENT.

acid chlorides (*Chem*) Compounds derived from acids by the replacement of the hydroxyl group by chlorine.

acid cure (*MinExt*) In extraction of uranium from its ores, lowering of gangue carbonates by puddling with sulphuric acid before leach treatment.

acid deposition (*EnvSci*) Acid compounds emitted into the atmosphere which then return to the surface either in the form in which they were discharged or as new compounds formed by reaction in the atmosphere. Includes DRY DEPOSITION, usually of sulphur and nitrogen oxides near the source, and WET DEPOSITION, which follows when acids are washed from the atmosphere by precipitation. This includes ACID RAIN and OCCULT DEPOSITION.

acid drift (*MinExt*) The process by which ores, pulps and products become acidic through pick-up of atmospheric oxygen through standing.

acid dyes (*ImageTech*) Dyes which have their colour associated with the negative ion or radical.

acid egg (*ChemEng*) A pump for sulphuric acid, of simple and durable construction, with few moving parts. The acid is run into a pressure vessel, usually egg-shaped, from which it can be forcibly expelled by compressed air.

acid esters (*Chem*) Compounds derived from acids in which part of the replaceable hydrogen has been exchanged for an alkyl radical.

acid fixer (*ImageTech*) Fixing solution (hypo) with the addition of an acid (sodium bisulphite or potassium metabisulphite) to prevent staining.

acid growth hypothesis (*BioSci*) The hypothesis that auxin-stimulated plant cell elongation results from increased proton extrusion (see PROTON-TRANSLOCATING ATPASE) with an increased wall extensibility as a result of its lower pH.

acid hydrazides (*Chem*) Hydrazine derivatives into which an acyl group has been introduced.

acid hydrolases (*BioSci*) Hydrolytic enzymes that operate best in acid conditions. Usually refers to the phosphatases, glycosidases, nucleases and lipases that are found in the lysosome and that are responsible for digesting phagocytosed material.

acidimetry (*Chem*) The determination of acids by titration with a standard solution of alkali. See TITRATION, VOLUMETRIC ANALYSIS.

acidity (*Chem*) (1) The extent to which a solution is acid, normally expressed as its pH value. Cf ALKALINITY. (2) The concentration of any species in a solution which is titratable by a strong base.

acidity regulator (*FoodSci*) A substance used to increase, reduce or stabilize the acidity of a product or an intermediate product. To retain pH within specified range, a buffer salt (usually the sodium or calcium salt of a fatty acid) may be added.

acidizing (*MinExt*) Improving the flow of oil from a limestone formation by pumping acid into it.

acid mine water (*MinExt*) Water containing sulphuric acid as a result of the breakdown of the sulphide minerals in rocks. Acid mine water causes corrosion of mining equipment, and may contaminate water supplies into which it drains.

acidophiles (*BioSci*) Bacteria able to grow at very low pH (<1). Their enzymes function under these extreme conditions and can be used in biotechnological processes involving an environment that would inhibit normal enzymes. Cf THERMOPHILES.

acidophilic (*BioSci*) (1) Cells or tissues that are easily stained with acid dyes. (2) Micro-organisms or plants that flourish in an acidic environment.

acidosis (*Med*) A condition in which the hydrogen ion concentration of blood and body tissues is increased from the normal range of 36–43 nmol l^{-1}. *Respiratory acidosis* is caused by retention of carbon dioxide by the lungs; *metabolic acidosis* by retention of non-volatile acids (renal failure, diabetic ketosis) or loss of base (severe diarrhoea).

acid process (*Eng*) A steel-making process in which the furnace is lined with a siliceous refractory, and for which iron low in phosphorus is required, as this element is not removed. See BASIC PROCESS.

acid process (*Paper*) Any pulp digestion process utilizing an acid reagent, eg a bisulphite liquor with some free sulphur dioxide.

acid radical (*Chem*) A molecule of an acid minus replaceable hydrogen.

acid rain (*EnvSci*) A form of WET DEPOSITION in which acid molecules or particles in the atmosphere are returned to the surface having been washed out by rain or snow (*acid snow*) as it falls. The unnatural acidity (pH 3–5·5) is caused mainly by the oxides of sulphur and nitrogen from the burning of coal and oil. See panel on ATMOSPHERIC POLLUTION.

acid refractory (*Eng*) See SILICA.

acid resist foils (*Eng*) Blocking foils for use in etching metal. The foil is stamped onto paper and the excess foil blocked onto the metal rule or other object which is then exposed to an acidic etching fluid such as ferric chloride.

acid rock (*Geol*) An igneous rock with >63% quartz.

acid salts (*Chem*) Salts formed by replacement of part of the replaceable hydrogen of the acid.

acid slag (*Eng*) Furnace slag in which silica and alumina exceed lime and magnesia.

acid smut (*EnvSci*) See ACID SOOT.

acid soil complex (*BioSci*) Combination of aluminium and/or manganese toxicity with calcium deficiency that affects a relatively CALCICOLE (or non-calcifuge) plant growing on an acid soil; preventable horticulturally by liming. Cf LIME-INDUCED CHLOROSIS.

acid solution (*Chem*) An aqueous solution containing more hydroxonium ions than hydroxyl ions, ie with PH value <7; one which turns blue litmus red.

acid soot (*EnvSci*) A pollutant, consisting of particles of carbon bound together by water containing sulphuric acid, formed as a by-product of the incomplete combustion of carbon-based fuel. Also *acid smut*.

acid steel (*Eng*) Steel made by an acid process.

acid stop (*ImageTech*) Weak acid processing solution used immediately after the DEVELOPER to halt its chemical activity and neutralize it before fixing. Also *stop-bath*.

acid value (*Chem*) The measure of the free acid content of eg vegetable oils and resins, indicated by the number of milligrams of potassium hydroxide (KOH) required to neutralize 1 g of the substance.

aciniform (*BioSci*) Berry-shaped; eg in spiders, the aciniform glands producing silk and leading to the median and posterior spinnerets.

acinostele (*BioSci*) A PROTOSTELE in which the xylem is star-shaped in cross-section with phloem between the arms as in the roots of most seed plants and the stems of *Lycopodium*.

Ackermann steering (*Autos*) An arrangement whereby a line extended from the track-arms, when the wheels are set straight ahead, should meet on the chassis centre-line at $^2/_3$ of the wheelbase from the front, allowing the inner stub axle to move through a greater angle than the outer.

Acker process (*Chem*) A process for production of sodium hydroxide. Molten sodium chloride is electrolysed, using a molten lead cathode, and the resulting lead–sodium alloy is decomposed by water, yielding pure lead and pure sodium hydroxide. Obsolete.

acknowledgement signal (*ICT*) A signal transmitted along a circuit from B to A when triggered by a signal from A to B.

A-class insulation (*ElecEng*) Insulating material which will withstand temperatures up to 105°C.

ACM (*ICT*) Abbrev for *Association for Computing Machinery*. US professional association.

ac magnet (*ElecEng*) Electromagnet excited by alternating current having normally a laminated magnetic circuit. See SHADED POLE.

acme screw-thread (*Eng*) A thread having a profile angle of 29° and a flat crest and root, used eg for lathe lead screw for easy engagement by a split nut.

acmite (*Min*) A variety of aegirine; also used for the $NaFe^{+3} Si_2O_6$ end-member.

ac motor (*ElecEng*) An electric motor which operates from a single or polyphase alternating current supply. See CAPACITOR MOTOR, INDUCTION MOTOR, SYNCHRONOUS MOTOR.

acne (*Med*) Inflammation of a sebaceous gland. Pimples in adolescents are commonly due to infection with the acne bacillus.

acnode (*MathSci*) See DOUBLE POINT.

acoelomate (*BioSci*) A term for an animal without a coelom, an acoelomate animal.

acoelomate triploblastica (*BioSci*) Animals with three embryonic cell layers but no coelom. They consist of the Platyhelminthes, Nematoda and some minor phyla, ie all the helminth phyla.

acoelomatous (*BioSci*) See ACOELOMATE.

acoelous (*BioSci*) Lacking a gut cavity.

acontia (*BioSci*) In Anthozoa, free threads, loaded with nematocysts, arising from the mesenteries or the mesenteric filaments, and capable of being discharged via the mouth or via special pores.

acotyledonous (*BioSci*) The embryo of a vascular plant, having no cotyledons.

acoustic absorption (*Acous*) Transfer of energy into thermal energy when sound is incident at an interface.

acoustic absorption factor (*Acous*) The ratio of the acoustic energy absorbed by a surface to that which is incident on the surface. For an open window this can be 1·00, for painted plaster 0·02. The value varies with the frequency of the incident sounds, eg for 2 cm glass fibre it is 0·04 at 125 Hz, 0·80 at 4000 Hz. Also *acoustic absorption coefficient*.

acoustical mass (*Acous*) The quantity M, where ωM is the part of the acoustical reactance which corresponds to the inductance of an electrical reactance: ω is the pulsatance, given by $2\pi f$ where f is the frequency in hertz. Also *acoustical inertia*.

acoustical stiffness (*Acous*) For an enclosure of volume V, the quantity given by $S = \rho c^2/V$, where c is velocity of propagation of sound and ρ is density. It is assumed that the dimensions of the enclosure are small compared with the sound wavelength and that the walls around the volume do not deflect.

acoustic amplifier (*Acous*) An amplifier of mechanical vibrations.

acoustic branch (*Phys*) A branch of the dispersion curve (frequency ω against wavenumber q) for crystal lattice vibrations for which ω is proportional to q for small q. For a crystal containing n atoms per unit cell, the dispersion curve has $3n$ branches of which three are acoustic branches. The branches are characterized by different patterns of movement of the atoms. See OPTIC BRANCH.

acoustic centre (*Acous*) The effective 'source' point of the spherically divergent wave system observed at distant points in the radiation field of an acoustic transducer.

acoustic compliance (*Acous*) The reciprocal of the ACOUSTICAL STIFFNESS.

acoustic construction (*Build*) Building construction which aims at the control of transmission of sound, or of mechanical vibration giving rise to sound, particularly unwanted noises. The parts of the structure are separated by air spaces or acoustic absorbing material and can be decoupled by the interposing of springs. Also *discontinued construction*.

acoustic coupler (*ICT*) A device that enables a digital signal to be transmitted over the telephone network using an ordinary telephone handset.

acoustic delay line (*ICT*) A device, magnetostrictive or piezoelectric, eg a quartz bar or plate of suitable geometry, which reflects an injected sound pulse many times within the body.

acoustic distortion (*Acous*) Distortion in sound-reproducing systems.

acoustic emission (*Eng*) Non-destructive-testing method of investigating deformation and failure processes in materials by the signals generated when the elastic waves released by them are detected at the materials' surfaces.

acoustic feedback (*ICT*) Instability or oscillation in a second reproduction system caused by the microphone or pick-up receiving vibrations from the loudspeaker.

acoustic filter (*Acous*) Filter which uses tubes and resonating boxes in shunt and series as reactance elements, providing frequency cut-offs in acoustic wave transmission, as in an electric wave filter.

acoustic grating (*Acous*) A diffraction grating for production of directive sound. Spacings are much larger than in optical gratings owing to the longer wavelength of sound waves. Both transmission and reflection gratings are used.

acoustic impedance (*Acous*) The complex ratio of sound pressure on surface to sound flux through surface, having imaginary (reactance) and real (resistance) components, respectively. Unit is the ACOUSTIC OHM.

acoustic interferometer (*Acous*) Instrument in which measurements are made by study of interference pattern set up by two sound or ultrasonic waves generated at the same source.

acoustic lens (*Acous*) A system of slats or discs to spread or converge sound waves.

acoustic microscope (*Acous*) Microscope based on acoustic waves (longitudinal compressions and rarefactions of density) at microwave frequencies: the interaction of an acoustic wave with a material is sensitive to its elastic properties. Images can be created by modulating a display with the intensity received by a detector/specimen system scanned synchronously (*ultrasonic imaging*). Coupling between electrical signals and acoustic vibrations exploits the PIEZOELECTRIC EFFECT.

acoustic model (*Acous*) A scale model of a room (eg concert hall) or structure which is used to measure qualities important for architectural acoustics and noise control, eg sound distribution. The scale is typically between 1:10 and 1:20. In order to adjust the wavelength, the frequency has to be increased by a factor 10–20.

acoustic ohm (*Acous*) Unit of acoustic resistance, reactance and impedance, equal to 10^5 Pa s m^{-3}.

acousticolateral system (*BioSci*) In vertebrates, afferent nerve fibres related to the neuromast organs and to the ear; receptors in aquatic forms of relatively slow vibrations.

acoustic perspective (*Acous*) The quality of depth and localization inherent in a pair of ears, which is destroyed in a single channel for sound reproduction. It is transferable with two microphones and two telephone ear receivers with matched channels, and more adequately realized with three microphones and three radiating receivers with three matched channels.

acoustic plaster (*Build*) Rough or flocculent plaster which has good acoustic absorbing properties and which can be used for covering walls. Added to the mix is fine aluminium, which evolves gas on contact with water and so aerates the mass. These tiny holes lower the acoustic impedance and so reduce the reflection of incident sound waves.

acoustic pressure (*Acous*) See SOUND PRESSURE.

acoustic radiator (*Acous*) A device to generate and radiate sound. The most common radiators are: (1) vibrating elastic systems (membrane, string, vocal cord) which cause a fluctuating pressure in the surrounding medium; (2) electrically driven membranes and plates (loudspeaker, sonar transducer); (3) vortices in turbulent fluid flow.

acoustic ratio (*Acous*) The ratio between the directly radiated sound intensity from a source, at the ear of a listener (or a microphone), and the intensity of the reverberant sound in the enclosure. The ratio depends on the distance from the source, the polar distribution of the radiated sound power, and the period of reverberation of the enclosure.

acoustic reactance (*Acous*) See ACOUSTIC IMPEDANCE.

acoustic resistance (*Acous*) See ACOUSTIC IMPEDANCE.

acoustic resonance (*Acous*) Enhancement of response to an acoustic pressure of a frequency equal or close to the EIGENFREQUENCY of the responding system. When a system is at resonance, the imaginary part of its impedance is zero. Prominent in Helmholtz resonators, organ and other pipes, and vibrating strings.

acoustics (*Genrl*) (1) The science of sound waves including production and propagation properties. (2) The characteristics of a room which determine the quality of sound transmission inside.

acoustic saturation (*Acous*) The aural effectiveness of a source of sound amid other sounds: it is low for a violin, but high for a triangle. The relative saturation of instruments indicates the number required in an auditorium of given acoustic properties.

acoustic scattering (*Acous*) Irregular and multi-directional reflection and diffraction of sound waves produced by multiple reflecting surfaces the dimensions of which are small compared with the wavelength; or by certain discontinuities in the medium through which the wave is propagated.

acoustic spectrometer (*Acous*) An instrument designed to analyse a complex sound signal into its wavelength components and measure their frequencies and relative intensities. See REAL-TIME ANALYSER.

acoustic spectrum (*Phys*) Graph showing frequency distribution of sound energy emitted by source.

acoustic streaming (*Acous*) Generation of constant flows by a strong sound wave. Acoustic streaming is a non-linear effect. It is responsible for the motion of the light particles (lycopodium spores) in a KUNDT'S TUBE. See QUARTZ WIND.

acoustic survey (*MinExt*) Determination of the porosity of a rock by measuring the time required for a sonic impulse to travel through a given distance.

acoustic suspension (*Acous*) Sealed-cabinet system of loudspeakers in which the main restoring force of the diaphragm is provided by the acoustic stiffness of the enclosed air.

acoustic telescope (*Acous*) An array of microphones. The signals of the microphones are added with certain phase delays so as to generate desired directivities. See DIRECTIONAL MICROPHONE.

acoustic tile (*Build*) A tile made of a soft, sound-absorbing substance.

acousto-optic modulator (*ICT*) A device in which acoustic waves in an optical medium form a grating used to diffract an optical signal and thus effectively turn it on or off.

acquired behaviour (*Psych*) Behaviour introduced in attempts to adapt to environmental change; most definitions assume long-term changes in the central nervous system and exclude short-term behaviour changes due to maturation, fatigue, sensory adaptation or habituation.

acquired character (*BioSci*) A modification of an organ during the lifetime of an individual due to use or disuse, but not inherited, contrary to the hypothesis of LAMARCK-ISM.

acquired immunity (*BioSci*) Immunity resulting from exposure to foreign substances or microbes. Cf INNATE IMMUNITY.

acquired immunodeficiency syndrome (*Med*) A severe immunodeficiency caused by a sexually tranmitted retrovirus (HIV) that infects T-helper cells, macrophages and brain cells. The failure of cell-mediated immunity removes a key part of the anti-viral system of the body; AIDS is not itself lethal but patients die of infections that would normally be resisted (particularly tuberculosis but also a variety of opportunistic infections). The ability of the virus to mutate has made vaccine development impossible so far, but a number of drugs are now available, often in combination therapies, that hold the disease in check. AIDS and AIDS-related illnesses are of epidemic proportions in many parts of the developing world. See AZT, PROTEASE INHIBITORS. Abbrev *AIDS*.

acquired variation (*BioSci*) Any departure from normal structure or behaviour, in response to environmental conditions, which becomes evident as an individual develops.

ACR (*Aero*) Abbrev for APPROACH CONTROL RADAR.

ACR (*Agri*) See AUTOMATIC CLUSTER REMOVAL.

Acrasiomycetes (*BioSci*) Cellular slime moulds. A class of slime moulds (Myxomycota) feeding phagotrophically as MYXAMOEBAE that aggregate to form a migratory pseudoplasmodium which eventually develops into a fruiting body, liberating most of the cells as spores, eg *Dictyostelium*. Also *Dictyosteliomycetes*.

acre (*Surv*) A unit of area, equal to 10 square chains (1 chain = 66 ft) or 4840 sq yd or 0·4047 hectares. The following are now obsolete: Cheshire acre, 10 240 sq yd; Cunningham acre, 6250 sq yd; Irish acre, 7840 sq yd; Scottish acre, 6150·4 sq yd.

acridine (*Chem*) $C_{13}H_9N$. A basic constituent of the crude anthracene fraction of coaltar. It crystallizes in colourless needles and has a very irritating action upon the epidermis. Chemically it may be considered an analogous compound to anthracene, in which one of the CH groups of the middle ring is replaced by N. Certain amino-acridines have valuable bactericidal powers. Used as an intermediate in the preparation of dyestuffs.

acriflavine (*Med*) A deep orange, crystalline substance possessing antiseptic (bacteriostatic and bactericidal) properties; used in wound dressings.

Acrilan (*Chem*) TN for a synthetic polyacrylonitrile fibre obtained by copolymerizing acrylonitrile (85%) with vinyl acetate (15%).

acritarch (*Geol*) A unicellular microfossil of unknown biological affinity, abundant in Precambrian and Palaeozoic strata. Used in establishing geological correlations. See appendix on Geological time.

acro- (*Genrl*) Prefix from Gk *akros*, topmost, farthest, terminal.

Acrobat (*ICT*) TN for an application that converts text, line drawings and half-tones into a stream of alphanumeric text while retaining the format of the original. Such a portable document format (PDF) file is an extension of the Adobe POSTSCRIPT language and can be read by any type of computer.

acrocarp (*BioSci*) A moss in which the main axis is terminated by the development of reproductive organs. Any subsequent growth must be SYMPODIAL GROWTH. Most are erect in habit. Cf PLEUROCARP.

acrocentric (*BioSci*) A rod-shaped chromosome with the CENTROMERE at the end.

acrocyanosis (*Med*) A vascular disorder (usually of young women) in which there is persistent blueness of the extremities.

acrodont (*BioSci*) Said of teeth that are fixed by their bases to the summit of the ridge of the jaw.

acrolein (*Chem*) *Acrylaldehyde, propenal*. $CH_2 \!=\! CHCHO$. A colourless liquid, bp 52·5°C, of pungent odour, obtained by dehydrating glycerine in the presence of a catalyst.

acromegaly (*Med*) A disease in which the hands and feet enlarge with thickening of nose, jaw, ears and brows, due to overproduction of growth hormone after the EPIPHYSES of the long bones have fused. Usually due to an adenoma (tumour) of the pituitary gland.

acromion (*BioSci*) In higher vertebrates, a ventral process of the spine of the scapula. Adj *acromial*.

acron (*BioSci*) In insects, the embryonic, presegmental region of the head.

acronical (*Astron*) US for ACRONYCHAL.

acronychal rising (*Astron*) The rising of a star at nightfall.

acronychal setting (*Astron*) The setting of a star at nightfall.

acroparaethesia (*Med*) Numbness and tingling of the fingers, tending to persist, in middle-aged women. US *acroparesthesia*.

acropetal (*BioSci*) Transport or differentiation towards the apex, away from the base.

acropodium (*BioSci*) The part of the pentadactyl limb of land vertebrates that comprises the digits and includes the phalanges.

acrosome (*BioSci*) Structure forming the tip of a mature spermatozoon. Adj *acrosomal*.

acroterium (*Arch*) A base or mounting, on the apex and/or extremities of a pediment, for the support of an ornamental figure or statuary. Pl *acroteria*.

acrotrophic (*BioSci*) In insects, said of ovarioles in which nutritive cells occur at the apex.

Acrux (*Astron*) A bright white supergiant star in the constellation Crux. A visual binary consisting of two spectroscopic binary components. Distance 80 pc. Also *Alpha Crucis*.

acrylaldehyde (*Chem*) See ACROLEIN.

acrylamide gel (*BioSci*) The clear gel formed by casting the acrylamide monomer in the form of sheets or cylinders by polymerization *in situ*; used for the electrophoretic separation of proteins and RNA. See POLYACRYLAMIDE GEL ELECTROPHORESIS, SODIUM DODECYL SULPHATE.

acrylate rubbers (*Chem*) Crosslinkable rubbers based on COPOLYMERS of ethyl acrylate and about 5% 2-chloroethyl vinyl ether as a site for VULCANIZATION. Oil- and heat-resistant speciality rubbers. Also used to describe elastomers based on acrylate–methacrylate copolymers used in rubber toughening of polyvinyl chloride for example.

acrylic acid (*Chem*) *Prop-2-enoic acid*. $CH_2 \!=\! CHCOOH$. Mp 7°C, bp 141°C, of similar odour to acetic acid; a very reactive monomer, forming polyacrylic acid.

acrylic adhesives (*Chem*) Derivatives of acrylic acid, some of which are copolymerized, which gives a variety of adhesive types, eg water-soluble, pressure-sensitive, anaerobic. See CYANOACRYLATE.

acrylic ester (*Chem*) An ester of acrylic acid or of a structural derivative of acrylic acid, eg methacrylic acid or its chemical derivatives. Monomers for acrylic polymers.

acrylic fibres (*Textiles*) Continuous filaments or, more usually, staple fibres made from linear polymers which are synthesized from several monomers containing at least 85% by weight of acrylonitrile.

acrylic resin paints (*Build*) Water-thinned paints, including emulsions, wood primers, undercoats and microporous coatings. These paints have the advantage of good adhesion, excellent durability, speed of drying and good colour retention.

acrylic resins (*Chem*) Resins formed by the polymerization of the monomer and derivatives, generally esters or amides, of acrylic acid or α-methylacrylic acid. They include polymethyl methacrylate and acrylic rubbers.

acrylonitrile (*Chem*) *Vinyl cyanide (1-cyanoethene)*. Used as raw material for synthetic acrylic fibres, eg ABS, Acrilan, Courtelle, Orlon, nitrile rubber and SAN.

ACS (*Aero*) Abbrev for: (1) ACTIVE CONTROL SYSTEM; (2) ATTITUDE CONTROL system; (3) *air-conditioning system*.

ac series motor (*ElecEng*) A series motor which operates from an ac supply with laminated field construction and usually a compensating winding.

ACT (*Aero*) Abbrev for *active control technology*. See ACTIVE CONTROL SYSTEM.

ACTH (*Med*) A protein hormone of the anterior pituitary gland controlling many secretory processes of the adrenal cortex. Used medically to stimulate cortisol production as an anti-inflammatory measure. *Adrenocorticotropic hormone, corticotropin*.

actin (*BioSci*) A globular protein (G actin) that can polymerize into long fibres (F actin). Originally discovered in muscle in association with myosin, it is now known to be widely distributed at sites of cellular movement. See fig. at MUSCLE.

actinal (*BioSci*) In Anthozoa, pertaining to the crown, including the mouth and tentacles; star-shaped. See AMBULACRA.

actin binding proteins (*BioSci*) A diverse group of intracellular proteins that bind to actin and that may stabilize F-actin filaments, nucleate filament formation, cross-link filaments, lead to bundle formation, etc.

actinic keratosis (*Med*) Thickened area of skin as a result of excessive exposure to sunlight, particularly common in those with very fair skin.

actinic radiation (*Radiol*) Ultraviolet waves, which have enhanced biological effect by inducing chemical change; basis of the science of photochemistry.

actinic rays (*ImageTech*) Electromagnetic waves of wavelength that can cause a latent image, potentially developable, in a photographic emulsion. They include an extension at each end of the visible spectrum and X-rays.

actinides (*Chem*) Radioactive elements after actinium (atomic number 89). All have similar chemical properties. Cf LANTHANIDES.

actinium (*Chem*) A radioactive element in the third group of the periodic system. Symbol Ac, at.no. 89, ram 227; half-life 21·7 years. Produced from natural radioactive decay of uranium-235 or by neutron bombardment of radium-226. Gives its name to the actinium (4*n*+3) series of radioelements (*actinides*).

actino- (*Genrl*) Prefix from Gk *aktis*, ray.

actinobacillosis (*Vet*) A chronic granulomatous disease of cattle caused by the bacterium *Actinobacillus lignieresii* and characterized by infection of the tongue ('wooden tongue') and occasionally of the stomach, lungs and lymph glands. In sheep the bacterium causes abscesses involving the head and neck or internal organs.

actinobiology (*Radiol*) The study of the effects of radiation upon living organisms.

actinodermatitis (*Med*) Inflammation of the skin arising from the action of radiation, usually applied to overexposure to ultraviolet light (sunburn).

actinodromous (*BioSci*) A term used to describe leaf venation having three or more primary veins originating at the base of the lamina and running out towards the margin (formerly *palmate* or *digitate* venation).

actinoid (*BioSci*) Star-shaped.

actinolite (*Min*) A monoclinic calcium magnesium iron member of the amphibole group, green in colour and usually showing an elongated or needle-like habit; occurs in metamorphic and altered basic igneous rocks.

actinomorphic (*BioSci*) A term describing organisms that are radially symmetric; divisible into two similar parts by any one of several longitudinal planes passing though the centre. Includes starfish, sea urchins and plants in which the stamens are helically arranged rather than whorled.

Actinomycetales (*BioSci*) An order of bacteria producing a fine mycelium and sometimes arthrospores and conidia. Some members are pathogenic in animals and plants. Some produce antibiotics, eg STREPTOMYCIN.

actinomycosis (*Med, Vet*) A chronic granulomatous infection, most commonly affecting the head and neck, caused by infection by various species of filamentous bacteria of the genus *Actinomyces*. Actinomycosis occurs in cattle, commonly affecting the lower jaw (lumpy jaw); in pigs the mammary gland is the common site of infection. The infection is not transmitted to humans.

Actinopterygii (*BioSci*) A subclass of GNATHOSTOMATA in which the basal elements of the paired fins do not project outside the body wall, the fin webs being supported by rays alone. GANOID scales are diagnostic of the group and the skeleton is fully ossified in most members. The most widespread modern group of fishes, including eg cod, herring.

actinotherapy (*Med*) Treatment of disease by means of exposure to ultraviolet, infrared and visible radiation.

Actinozoa (*BioSci*) See ANTHOZOA.

action (*ImageTech*) (1) The performance of a scene to be recorded on camera. (2) The film record of this performance as picture only, separate from the sound record.

action (*Phys*) The time integral of kinetic energy (*E*) of a conservative dynamic system undergoing a change, given by

$$2 \int_{t_1}^{t_2} E \, dt$$

action potential (*BioSci*) The potential produced in a nerve by a stimulus. It is a voltage pulse arising from sodium ions entering the axon and changing its potential from -70 mV to $+40$ mV. With a continuing stimulus the pulses are repeated at up to several hundred times a second, leading in motor nerves to continuous muscular response (tetanus).

action research (*Psych*) Research carried within the workplace by practitioners, eg teachers, with the aim of improving practice; usually involves iterations of observation, analysis, planning, implementation and further research.

action spectrum (*BioSci*) The relationship, usually plotted as a graph, between the rate of a light-dependent physiological process (eg PHOTOSYNTHESIS or PHOTOMORPHOGENESIS) and the wavelength of light. Comparison with the spectral absorption of known pigments may suggest which is involved.

activated carbon (*Chem*) Carbon obtained from vegetable matter by carbonization in the absence of air, preferably in a vacuum. Activated carbon has the property of adsorbing large quantities of gases. Important for gas masks, adsorption of solvent vapours, clarifying of liquids and in medicine.

activated cathode (*Electronics*) Emitter in thermionic devices comprising a filament of basic tungsten metal, alloyed with thorium, which is brought to the surface by a process of activation, such as heating without electric field.

activated charcoal (*Chem*) Charcoal treated eg with acid to increase its adsorptive power.

activated complex (*Chem*) A high-energy complex formed when activated molecules of the reactants collide in the Eyring theory of ABSOLUTE REACTION RATES; it can lose energy either by decomposing to reform the reactants or by rearranging to form the products.

activated sintering (*Eng*) Sintering of a COMPACT in the presence of a gaseous reactant. Also *reaction sintering*.

activated sludge (*Build*) Sludge through which compressed air has been blown, or which has been aerated by mechanical agitation as part of the sewage treatment process.

activating agent (*MinExt*) See ACTIVATOR.

activation (*BioSci*) (1) A step in the fertilization process triggered by the incorporation of the spermatozoon into the egg cytoplasm, by which the secondary oöcyte is stimulated to complete its division and becomes haploid. (2) A process by which lymphocytes or macrophages differentiate from a resting state and acquire new capacities such as the ability to secrete lymphokines, or in the case of macrophages, increased ability to kill and digest microbes.

activation (*Chem*) (1) An increase in the energy of an atom or molecule, rendering it more reactive. (2) The heating process by which the capacity of carbon to adsorb vapours is increased.

activation (*Eng*) Alteration of the surface of a metal to a chemically active state. Cf PASSIVATION.

activation (*NucEng*) Induction of radioactivity in otherwise non-radioactive atoms, eg in a cyclotron or reactor.

activation cross-section (*Phys*) The effective cross-sectional area of a target nucleus undergoing bombardment by eg neutrons for radioactivation analysis. Measured in BARNS. See CROSS-SECTION.

activation energy (*Chem*) (1) The excess energy over that of the ground state which an atomic system must acquire to permit a particular process, such as emission or reaction, to occur. (2) The energy barrier for a thermally activated physical or chemical process. See ARRHENIUS'S RATE EQUATION.

activator (*BioSci*) (1) Of an enzyme, a small molecule which binds to it and increases its activity. (2) Of DNA transcription, a protein which, by binding to a specific sequence, increases the production of a gene product. (3) Any agent, eg a hormone, whose presence stimulates biochemical or physiological changes.

activator (*Chem*) Chemical that promotes VULCANIZATION or acceleration of CROSS-LINKING.

activator (*MinExt*) Surface-active chemical used in a flotation process to increase the attraction to a specific mineral in an aqueous pulp of collector ions from the ambient liquid and increase its aerophilic quality. Also *activating agent*.

activator (*Phys*) An impurity, or displaced atom, which augments luminescence in a material, ie a sensitizer such as copper in zinc sulphide.

active array (*Radar*) Antenna array in which the individual elements are separately excited by integrated circuit or transistor amplifiers.

active centres (*Chem*) Centres of higher catalytic activity formed by peak or loosely bound atoms on the surface of an adsorbent.

active chromatin (*BioSci*) See TRANSCRIPTIONALLY ACTIVE CHROMATIN.

active component (*Phys, ElecEng*) The component of the vector representing an alternating quantity which is in phase with some reference vector; eg the active component of the current, commonly called the active current. See ACTIVE CURRENT, ACTIVE VOLTAGE, ACTIVE VOLT-AMPERES.

active control (*Acous*) Modern technique of noise or vibration control employing one or more sources that generate signals with the aim of making the resulting total signal smaller. Used eg for the control of low-frequency airborne noise and vibration of machinery. See ANTI-SOUND.

active control system (*Aero*) An advanced automatic flight control system designed to provide several special features, eg activation of flight control surfaces to minimize gust loads and bending stresses in the wing by detection and response to normal accelerations, provision of stability to a naturally unstable aircraft and implementation of pilot manoeuvre demands. All these characteristics improve aircraft behaviour and performance, but the active control system demands extensive integration between aerodynamics, structure and electronic system design to achieve these advantages with reliability and safety. Abbrev ACS.

active current (*Phys*) The component of a vector representing the ac in a circuit which is in phase with the voltage of the circuit. The product of this and the voltage gives power.

active device (*Electronics*) A component capable of controlling voltages or currents, to produce gain or switching action in a circuit; valves, diodes and transistors, and integrated circuits are all classed as active devices or components.

active electrode (*ElecEng*) The electrode of an electrical precipitator which is kept at a high potential. Also *discharge electrode*.

active filter (*Electronics*) A filter which combines amplification with conventional passive filter components (capacitance, inductance, resistance) to enhance fixed or tunable passband or rejection characteristics.

active galaxy (*Astron*) A galaxy which emits unusually large amounts of radiation from a compact central source, such as a SEYFERT GALAXY, N GALAXY, quasar or BL LAC OBJECT. See panels on GALAXY and QUASAR.

active homing (*Radar*) A guidance system where the missile contains the transmitter for illuminating the target and the receiver for the reflected energy.

active hydrogen (*Chem*) Atomic form of hydrogen obtained when molecular hydrogen is dissociated by heating, or by electrical discharge at low pressure.

active lattice (*NucEng*) The regular pattern of arrangement of fissionable and non-fissionable materials in the core of a lattice reactor.

active lines (*ImageTech*) Lines which are effective in establishing a picture.

active margin (*Geol*) A continental margin characterized by earthquakes, igneous activity and mountain building as a result of convergent- or transform-plate movements. See PASSIVE MARGIN.

active mass (*Chem*) Molecular concentration generally expressed as moles dm^{-3}; in the case of gases, active masses are measured by partial pressures.

active materials (*Phys*) (1) General term for essential materials required for the functioning of a device, eg iron or copper in a relay or machine, electrode materials in a primary or secondary cell, emitting surface material in a valve, or photocell, phosphorescent and fluorescent material forming a phosphor in a cathode-ray tube, or that on the signal plate of a TV camera. (2) A term applied to all types of radioactive isotopes.

active power (*ElecEng*) (1) See ACTIVE VOLT-AMPERES. (2) The time average over one cycle of the instantaneous input powers at the points of entry of a polyphase circuit.

active resistance (*BioSci*) Host resistance to a pathogen built up as result of the previous presence of the pathogen or its metabolites.

active satellite (*Space*) Satellite equipped for sending out probing signals and receiving returned information. A *passive satellite* only receives information on the state of the target.

active space (*BioSci*) The area surrounding an animal within which it can communicate.

active Sun (*Astron*) The Sun during periods of intense sunspot activity.

active transducer (*ICT*) Any transducer in which the applied power controls or modulates locally supplied power, which becomes the transmitted signal, as in a modulator, a radiotransmitter or a carbon microphone.

active transport (*BioSci*) See panel on ACTIVE TRANSPORT.

active voltage (*ElecEng*) The component of a vector representing the voltage which is in phase with the current in a circuit.

active volt-amperes (*ElecEng*) The product of the active voltage and the amperes in a circuit, or of the active current (amperes) and the voltage of the circuit; equal to the power in watts. Also *active power*.

ActiveX control (*ICT*) TN for a small application (applet) developed by Microsoft for downloading from a server to a client to change and enhance the appearance of Internet information. Similar to JAVABEANS. See panel on INTERNET.

activity (*Chem*) (1) See OPTICAL ACTIVITY. (2) The ideal or thermodynamic concentration of a substance, the substitution of which for the true concentration permits the application of the law of mass action.

activity (*ElecEng*) The magnitude of the oscillations of a piezoelectric crystal relative to the exciting voltage.

activity (*Radiol*) The rate at which transformations occur in a RADIONUCLIDE. Unit is the BECQUEREL. See panel on RADIATION.

activity coefficient (*Chem*) The ratio of the ACTIVITY to the true concentration of a substance.

activity constant (*Chem*) The EQUILIBRIUM CONSTANT written in terms of activities instead of molar concentrations.

actomyosin (*BioSci*) Generally, a motor system that is based on actin and myosin. Myosin makes transient contact with the actin filaments and undergoes a conformational change before releasing contact. The hydrolysis of ATP is coupled to movement, through the requirement for ATP to restore the configuration of myosin prior to repeating the cycle. More specifically, a viscous solution formed when actin and myosin solutions are mixed at high salt concentrations.

ac transformer (*ElecEng*) An electromagnetic device which alters the voltage and current of an ac supply in inverse ratio to one another. It has no moving parts and is very efficient.

Aculeata (*BioSci*) Stinging hymenoptera, eg bees, ants and some wasps.

Active transport

The movement of a molecule across a cell membrane against its concentration or electrochemical gradient. Such a movement requires the input of energy and active transport can only proceed in a metabolically active cell, ceasing when the energy supply in the form of ATP (ADENOSINE TRIPHOSPHATE) is restricted. That ATP is used is shown by the association between hydrolysis of ATP and solute flux when the sodium/potassium pump of plasma membranes is active. This establishes the ion asymmetry across the membrane by pumping sodium out of the cell and potassium in.

Solute movement which is directly linked to ATP hydrolysis is known as *primary transport* in contrast to *secondary transport*. In the latter the solute also moves against its concentration gradient, but its movement depends upon a gradient of a second solute itself established by a mechanism requiring the hydrolysis of ATP. Thus the entry of many sugars and amino acids into animal cells depends upon the initial establishment of a sodium ion gradient in the opposite direction from that which the sugar is to travel. The sugars or amino acids are taken along with the sodium ions as the latter move passively down their own concentration gradient.

Active transport systems consist of catalytic proteins, with many of the properties of enzymes which are strategically located in the membrane so as to direct the solute flux. In primary transport these proteins can be recognized by an ATPASE activity which is solute-dependent. Thus the sodium/potassium pump requires both sodium and potassium and the analogous calcium pump calcium ions. It is known that ATP hydrolysis is associated with the phosphorylation of pump proteins but the precise mechanism is not understood. Cf FACILITATED DIFFUSION.

---

aculeate (*BioSci*) Bearing prickles, or covered with needle-like outgrowths.

acuminate (*BioSci*) Having a long point bounded by hollow curves; usually descriptive of a leaf-apex. Dim *acuminulate*.

acupuncture (*Med*) The practice of puncturing the skin with needles to produce analgesia, anaesthesia or for wider therapeutic purposes; originated in China. The mechanism of action is not clear but it may stimulate the body to produce its own analgesic substances called ENDORPHINS.

acutance (*ImageTech*) Objective formulation of the sharpness of a photographic image, expressed as:

$$\bar{G}_x/(D_B - D_A)$$

where

$$\bar{G}_x^2 = \frac{\sum (\Delta D/\Delta x)^2}{N}$$

N is the number of increments between A and B, $D_B - D_A$ is the average gradient of density curve, and $\Delta D/\Delta x$ is the maximum gradient curve.

acute (*BioSci*) Bearing a sharp and rather abrupt point: said usually of a leaf tip.

acute (*Med*) Said of a disease which rapidly develops to a crisis. Cf CHRONIC.

acute angle (*MathSci*) An angle of less than 90°. Cf OBTUSE ANGLE, REFLEX ANGLE.

acute-phase proteins (*BioSci*) Proteins that appear in the blood in increased amounts shortly after the onset of infections or tissue damage. They are made in the liver and include C-REACTIVE PROTEIN, fibrinogen, proteolytic enzyme inhibitors, transferrin. The stimulus is interleukin-1 (IL-1) released by macrophages.

acute reference dose (*Med*) The largest amount of a poison which can be taken without normally causing harm. Actual ingested doses can be expressed as multiples of the reference dose.

ACV (*Aero*) Abbrev for *air cushion vehicle* (hovercraft).

acyclic compound (*Chem*) An organic chemical compound having molecules in which the carbon atoms are arranged in open chains as opposed to closed rings.

acylation (*Chem*) Introduction of an acyl group into a compound, by treatment with a carboxylic acid, its anhydride or its chloride.

acyl CoA (*BioSci*) Coenzyme A conjugated by a thioester bond to an acyl group, eg ACETYL COA, succinyl CoA. These compounds are intermediates in the transfer of acyl groups, eg the formation of citric acid by the interaction of acetyl AcA with oxaloacetic acid.

acyl group (*Chem*) The carboxylic radical RCO (R being aliphatic), eg CH_3CO.

acylic (*BioSci*) Having the parts of the flower arranged in spirals, not in whorls.

ad- (*Genrl*) Prefix from Lt *ad*, to, at.

A–D (*ImageTech*) Analogue-to-digital, referring to the conversion of signals.

ADA (*ICT*) Programming language designed for complex ON-LINE, REAL-TIME SYSTEM monitoring (eg in military applications). Named in honour of Ada Lovelace.

ADA (*Med*) Abbrev for ADENOSINE DEAMINASE.

adamantine (*Min*) See LUSTRE.

adamantine compound (*Chem*) Compound with the same tetrahedral covalently bonded crystal structure as that of DIAMOND, eg zinc sulphide (sphalerite).

adambulacral (*BioSci*) In Echinodermata, adjacent to the AMBULACRAL GROOVES.

adamellite (*Geol*) A type of granite with approximately equal amounts of alkali-feldspar and plagioclase.

Adam's apple (*BioSci*) In primates, a ridge on the anterior or ventral surface of the neck, caused by the protuberance of the thyroid cartilage of the larynx.

Adams' catalyst (*Chem*) A hydrogenation catalyst based on platinum oxide.

Adams sewage lift (*Build*) An apparatus employed to force sewage from a low-level sewer into a nearby high-level sewer by using the sewage in the latter from a point that will give the air pressure necessary to secure the lift of sewage.

adaptation (*BioSci*) (1) Any morphological, physiological or behavioural characteristic that fits an organism to the conditions under which it lives; the genetic or developmental processes by which such characteristics arise. (2) Adjustment to environmental demands through the long-term process of natural selection acting on the genotype (*evolutionary adaptation*). (3) A short-term change in the response of a sensory system as a consequence of repeated or protracted stimulation (*sensory adaptation*).

adaptation (*Psych*) In child psychology, a term used by J Piaget to describe the developmental process underlying the child's growing awareness and interactions with the physical and social world. The processes of ASSIMILATION, ACCOMMODATION and EQUILIBRATION are fundamental to this concept of psychological adaptation.

adaptation of the eye (*BioSci*) The sensitivity adjustment effected after considerable exposure to light (*light adapted*), or darkness (*dark adapted*).

adapter (*ElecEng*) A device used to connect two different types or sizes of electrical terminals.

adapter (*ImageTech*) (1) An arrangement for using types of photographic material in a camera different from that for which it was designed; eg filmpack in a plate camera, or a smaller plate than normal. (2) A device for the interchange of lenses between different types of camera.

adapter card (*ICT*) A printed circuit board that gives the computer added capability such as more memory or a colour display and plugged into an EXPANSION SLOT. Also *expansion board, expansion card*.

adapter hypothesis (*BioSci*) The prediction that some molecule would be needed to adapt the four-base genetic code to the 20 amino acid product. TRANSFER RNA fulfils the prediction.

adapters (*Psych*) Characteristic gestures or actions that are developed to allow coping with particular situations. 'Idiosyncratic gestures'.

adaptive algorithm (*MathSci*) An ALGORITHM used in EVOLUTIONARY COMPUTATION.

adaptive array (*Radar*) A radar antenna (either a PHASED ARRAY or an ACTIVE ARRAY) whose gain, directivity and sidelobes can be adjusted automatically to optimize the radar's performance under specific operating conditions.

adaptive differential pulse code modulation (*ICT*) A form of DIFFERENTIAL PULSE CODE MODULATION in which the basic step size is varied continually to suit the rate of change of the signal. A further refinement is to transmit only differences from a continually adjusted prediction of the signal. These measures greatly reduce the required bandwidth.

adaptive radiation (*BioSci*) Evolutionary diversification of species from a common ancestral stock, filling available ecological niches. Also *divergent adaptation*.

adaxial (*BioSci*) The surface of eg a leaf or petal that faced towards the axis during early development (and usually, therefore, the upper surface of an expanded leaf). Cf ABAXIAL.

ADC (*Chem*) Abbrev for AZODICARBONAMIDE.

ADC (*ICT*) Abbrev for ANALOGUE-TO-DIGITAL CONVERTER.

ADCC (*BioSci*) Abbrev for ANTIBODY-DEPENDENT CELL CYTOTOXICITY.

Adcock antenna (*ICT*) Directional antenna consisting of pairs of vertical wires, spaced by one-half wavelength or less, and fed in phase opposition; a figure-of-eight radiation pattern results, and arrays of Adcock antennas can be used for direction finding.

A/D converter (*ICT*) See ANALOGUE-TO-DIGITAL CONVERTER.

ADD (*Aero*) Abbrev for *airstream direction detector*; used for stall protection.

ADD (*Med*) Abbrev ATTENTION DEFICIT (HYPERACTIVITY) DISORDER.

ADD (*Psych*) Abbrev for ATTENTION DEFICIT DISORDER.

add-drop multiplexer (*ICT*) Equipment used to add data originating from a particular source or group of sources to a SYNCHRONOUS DIGITAL HIERARCHY data stream, or conversely to extract data destined for a particular source or group of sources.

addend (*MathSci*) See ADDITION.

addendum (*Eng*) (1) The radial distance between the major and pitch cylinders of an external thread. (2) The radial distance between the minor and pitch cylinders of an internal thread. (3) The height from the pitch circle to the tip of the tooth on a gear wheel. See fig. at GEAR WHEEL.

adder (*BioSci*) The only venomous snake in the UK, *Vipera berus*. Also *viper*.

adder (*ICT*) A device that adds digital signals. It can also be applied to an amplifier in analogue computing. See FULL ADDER, HALF ADDER.

addict (*Med*) Someone physically dependent on a drug and who will experience withdrawal effects if the drug is discontinued.

addiction (*Psych, Med*) Psychological or physiological over-dependence on a drug. See SUBSTANCE DEPENDENCE.

Addison's disease (*Med*) A disease in which there is progressive destruction of the supra-renal cortex; characterized by extreme weakness, wasting, low blood pressure and pigmentation of the skin. Not to be confused with Addison's anaemia or PERNICIOUS ANAEMIA.

addition (*MathSci*) The process of finding the sum of two quantities, which are called the *addend* and the *augend*. Denoted by the plus sign (+). Numbers are added in accordance with the usual rules of arithmetic, but addition of other mathematical entities has to be defined specifically for the entities concerned, eg by the PARALLELOGRAM RULE for the addition of two vectors.

addition agent (*ElecEng*) A substance added to the electrolyte in an electrodeposition process in order to improve the character of the deposit formed. The agent does not take part in the main electrochemical reaction.

addition formulae (*MathSci*) Formulae to express the sine, cosine and tangent of the sum of two angles A and B as functions of the individual angles:

$$\sin(A + B) = \sin A \sin B + \cos A \cos B$$

$$\cos(A + B) = \cos A \cos B - \sin A \sin B$$

$$\tan(A + B) = \frac{\tan A \cos B + \tan B}{1 - \tan A \tan B}$$

addition polymerization (*Chem*) See CHAIN POLYMERIZATION and panel on POLYMER SYNTHESIS.

addition reaction (*Chem*) The union of two molecules to form a larger molecule with no by-products.

additive (*FoodSci*) Any substance added to a food raw material or recipe to modify flavour, colour, texture, keeping properties or nutritional content. Only permitted additives can be used in foods. See E NUMBER.

additive constant (*Surv*) A term used in the computation of distance by tacheometric methods. It is that length (usually constant and small) which must be added to the product of staff intercept and multiplying constant to give the true distance of the object. See ANALLATIC LENS.

additive function (*MathSci*) A function $f(x)$ such that $f(x + y) = f(x) + f(y)$. If $f(x + y) < f(x) + f(y)$, the function is said to be *subadditive*. If $f(x + y) > f(x) + f(y)$, the function is said to be *superadditive*.

additive genetic variance (*BioSci*) The part of the genetic variance of a QUANTITATIVE CHARACTER that is transmitted and so causes resemblance between relatives.

additive printer (*ImageTech*) Photographic or motion picture printer or enlarger in which the intensity and colour of the exposing light is controlled by the separate variation of its red, green and blue components.

additive process (*ImageTech*) Colour reproduction in which the picture is presented by the combination (addition) of red, green and blue light representing these three components in the original subject; it is effectively obsolete for general photography and cinematography but is the basis for colour TV display.

additive property (*Chem*) A property whose value for a given molecule is equal to the sum of the values for the constituent atoms and linkages.

address (*ICT*) Code identifying a memory location. See ABSOLUTE ADDRESS, RELATIVE ADDRESS, SYMBOLIC ADDRESS.

addressable cursor (*ICT*) A cursor for which a program can specify the position usually by giving the co-ordinates.

address bus (*ICT*) A pathway within the processor linking the memory address register to the memory, enabling the processor to instruct the memory at which location a read or write operation will take place.

address calculation (*ICT*) The process of determining an ABSOLUTE ADDRESS from the contents of the address field in a MACHINE-CODE INSTRUCTION. See ADDRESS MODIFICATION, INDEXED ADDRESS.

address field (*ICT*) Part of a MACHINE-CODE INSTRUCTION that contains addresses. Also *operand, operand field*. See ADDRESS CALCULATION.

addressin (*BioSci*) Any of a group of cell surface proteins, especially on the surface of endothelial cells lining the blood vascular system, that provide signals for the homing of leucocytes to particular tissues.

address modification (*ICT*) The process of changing the address in a machine instruction, so that each time the instruction is executed, it can refer to a different storage location. See INDEXED ADDRESS.

address register (*ICT*) A register that holds the address part of the instruction being executed.

address space (*ICT*) The number of locations that can be addressed directly, as determined by the design of the INSTRUCTION SET.

adduct (*Chem*) The addition product of a reaction between molecules.

adductor (*BioSci*) A muscle that draws a limb or part inwards, or towards another part; eg *adductor mandibulae* in Amphibia is a muscle that assists in closing the jaws.

adelphous (*BioSci*) Said of the male part of a flower or androecium in which the stamens are partly or wholly united by their filaments.

aden-, adeno- (*Genrl*) Prefixes from Gk *aden*, gland.

adendritic (*BioSci*) Without dendrites.

adenine (*Chem*) *6-aminopurine*. One of the five bases in nucleic acids; it pairs with thymine in DNA and uracil in RNA. Symbol A. See panel on DNA AND THE GENETIC CODE.

adenitis (*Med*) Inflammation of a gland.

adeno- (*BioSci*) See ADEN-.

adenocarcinoma (*Med*) A malignant tumour of a glandular epithelium, or a carcinoma showing gland-like organization of cells.

adenohypophysis (*Med*) The glandular lobe of the pituitary gland, derived from buccal ectoderm (Rathke's pouch).

adenoid (*Med*) Generally, gland-like. More specifically, lymphoid tissue in the nasopharynx which may become enlarged as a result of repeated upper respiratory tract infection.

adenoid (*BioSci*) Gland-like.

adenoma (*Med*) A benign tumour with a gland-like structure or developed from glandular epithelium.

adenomatous polyposis coli (*Med, BioSci*) An inherited disorder in which many adenomatous polyps develop in the colon and may progress to malignancy. The defect lies in the APC gene, a tumour suppressor. See POLYPOSIS.

adenomyoma (*Med*) See ENDOMETRIOMA.

adenopathy (*Med*) Disease or disorder of glandular tissue. The term is usually used in reference to lymphatic gland enlargement.

adenosine (*BioSci*) *9-β-D-ribofuranosyladenine*. The nucleoside formed by linking adenine to ribose.

adenosine deaminase (*Med*) Any enzyme which hydrolyses adenosine to inosine. Abbrev ADA.

adenosine deaminase deficiency (*Med*) A genetically inherited condition in which the lack of this enzyme causes disorders in the immune system. Usually fatal at a young age.

adenosine triphosphate (*BioSci*) The triphosphate of the nucleotide adenosine (adenine + ribose) that serves as the common energy currency in all cells. Enzymic transfer of the terminal phosphate or pyrophosphate to a wide variety of substrates provides a means of transferring chemical free energy from anabolic to catabolic processes. Abbrev *ATP*. See panel on ACTIVE TRANSPORT.

adenovirus (*BioSci*) Any of a large group of DNA-containing viruses that cause gastrointestinal and respiratory infections in humans. Some adenoviruses induce tumours in rodents and are thus ONCOGENIC VIRUSES.

adenyl cyclase (*Chem*) An enzyme which catalyses the formation of cyclic adenylic acid from ATP.

ADF (*Aero*) Abbrev for AUTOMATIC DIRECTION FINDING.

Adhara (*Astron*) A very bright blue-white giant star in the constellation Canis Major, which is a visual binary. Distance 200 pc. Also *Epsilon Canis Majoris*.

ADHD (*Med*) Abbrev ATTENTION DEFICIT (HYPERACTIVITY) DISORDER.

adherend (*Eng*) A material which is bonded by an adhesive.

adherens junction (*BioSci*) A specialized cell–cell junction into which are inserted actin microfilaments (*zonula adherens*) or intermediate filaments (*macula adherens* or *spot desmosomes*). See DESMOSOMES.

adhesins (*BioSci*) General term for molecules involved in adhesion, but in microbiology refers only to bacterial surface components.

adhesion (*Build*) (1) The securing of a bond between plaster and backing, by physical means as opposed to mechanical keys. (2) The inherent ability of a surface coating of paint to adhere to the underlying surface. Lack of adhesion is the cause of such defects as flaking, blistering and cissing. Good surface preparation and selection of suitable primers are important factors affecting adhesion.

adhesion (*ElecEng*) Mutual forces between two magnetic bodies linked by magnetic flux, or between two charged non-conducting bodies which keeps them in contact.

adhesion (*Eng*) (1) The bonding of materials with adhesives (glues, cements, binders, etc), in which the intermolecular forces between ADHESIVE and ADHEREND provide the bonds. (2) The intimate sticking together of metallic surfaces under compressive stresses by bonds which form as a function of stress, time and temperature. The speed of formation is related to DISLOCATIONS, and may occur virtually instantaneously under high shear stresses. See COLD WELDING.

adhesion (*Med*) Abnormal union of two parts which have been inflamed; a band of fibrous tissue which joins such parts.

adhesion (*Phys*) Intermolecular forces which hold matter together, particularly closely contiguous surfaces of neighbouring media, eg liquid in contact with a solid. US *bond strength*.

adhesion molecules (*BioSci*) Generally, proteins involved in cell–cell or cell–substratum adhesion. See ADHESINS, CELL ADHESION MOLECULES, INTEGRINS, SELECTINS.

adhesion plaque (*BioSci*) A discrete region of the plasma membrane of a cell in close contact with a non-cellular substratum. The plaque, or *focal adhesion*, appears dark under the transmission electron microscope and contains MICROFILAMENTS.

adhesive (*Eng*) Agent for joining materials by adhesion, usually polymeric material. May be based on thermoplastic resin (eg polystyrene cement) or thermoset (eg epoxy resin). Viscosity is important for gap filling (high, as in epoxies) or surface penetration (low, as in cyanoacrylates). Also *binder, cement, glue*.

adhesive binding (*Print*) Unsewn binding in which the back of the sections are trimmed and roughened before adhesive is applied to bind the leaves and the cover.

adhesive-bonded non-woven fabric (*Textiles*) A fabric made from a WEB of fibres stuck together with an adhesive.

adhesive bonding (*Eng*) Joining of parts with polymeric adhesive. Widely used for assembly of complex composite products (eg helicopter rotor blade). Great potential for routine assembly of car bodies, replacing SPOT WELDING.

adhesive cells (*BioSci*) Glandular cells producing a viscous adhesive secretion for attachment use, as on the pedal disk of Hydra, the tentacles of Ctenophora and in the epidermis of Turbellaria.

adhesive wear (*Eng*) Mechanism of wear due to the welding together and subsequent shearing off of the contact areas between two surfaces sliding over one another.

ADI (*FoodSci*) Abbrev for ACCEPTABLE DAILY INTAKE.

adiabatic (*Phys*) Without loss or gain of heat.

adiabatic change (*Phys*) A change in the volume and pressure of the contents of an enclosure without exchange of heat between the enclosure and its surroundings.

adiabatic curve (*Phys*) The curve obtained by plotting pressure against volume in the adiabatic equation.

adiabatic demagnetization (*Phys*) A method of obtaining very low temperatures. A paramagnetic salt is cooled to 1 K by liquid helium. The salt is magnetized under isothermal conditions and then magnetized under adiabatic conditions. As a result, the temperature falls. Temperatures below 10^{-2} K can be obtained in this way.

adiabatic efficiency (*Eng*) (1) Of a steam engine or turbine, the ratio of the work done per unit mass of steam to the available energy represented by adiabatic heat drop. (2) Of a compressor, the ratio of that work required to compress a gas adiabatically to the work actually done by the compressor piston or impeller.

adiabatic equation (*Phys*) The equation $PV_\gamma = \text{constant}$, expressing the law of variation of pressure (P) with the volume (V) of a gas during an adiabatic change, γ being the ratio of the specific heat of the gas at constant pressure to that at constant volume. The value of γ is approximately 1·4 for air at stp.

adiabatic expansion (*Phys*) An ADIABATIC CHANGE in which a substance expands.

adiabatic heating (*Eng*) Self-heating effect which occurs in extruder or injection moulding barrel from action of rotating screw on polymer melt. Attributed to dissipation of mechanical shear forces as heat. Important in injection moulding of rubbers. Also *shear heating*. See DAMPING.

adiabatic lapse rate (*EnvSci*) The rate of decrease of temperature which occurs when a parcel of air rises adiabatically through the atmosphere.

adiabatic process (*Phys*) A process which occurs without interchange of heat with surroundings.

adiactinic (*Phys*) Said of a substance which does not transmit photochemically active radiation, eg safelights for dark-room lamps.

ad infinitum (*MathSci*) Continuing in a similar manner indefinitely (Latin).

adinole (*Min*) An argillaceous rock that has undergone albitization during contact metamorphism.

adipamide (*Chem*) *1,4-Butanedicarboxamide.* $NH_2CO(CH_2)_4CONH_2$. Used in synthetic fibre manufacture.

adipic acid (*Chem*) *Butanedicarboxylic acid.* $HOOC(CH_2)_4$ COOH. Colourless needles; mp 149°C, bp 265°C; formed by the oxidation of *cyclo*-hexanone, or by the treatment of oleic acid with nitric acid. Used in the manufacture of nylon 6,6 and polyester resin.

adipo- (*Genrl*) Prefix from Lt *adeps*, fat.

adipocere (*Med*) White or yellowish waxy substance formed by the post-mortem conversion of body fats to higher fatty acids. Also *mortuary fat*.

adipocyte (*BioSci*) A large cell used for fat storage.

adipose tissue (*BioSci*) A form of connective tissue consisting of vesicular cells (*adipocytes*) filled with fat and collected into lobules.

adiposis dolorosa (*Med*) A condition characterized by the development of painful masses of fat under the skin and by extreme weakness.

Adiprene (*Chem*) TN for polyurethane elastomers which combine high abrasion resistance with hardness and resilience.

A-display (*Radar*) Co-ordinate display on a cathode-ray tube in which a level time base represents distance and vertical deflections of beam indicate echoes.

adit (*CivEng*) A horizontal passage or tunnel into a mine. See fig. at MINING.

adjacent channel (*ICT*) A channel whose frequency is immediately above or below that of the required signal.

adjective dye (*Chem*) A dye which has no direct affinity for the particular textile fibre but can be affixed to it by a MORDANT.

adjoint (*MathSci*) Of a square matrix or determinant: the TRANSPOSE of the matrix obtained by replacing each element by its cofactor. Sometimes an untransposed adjoint is used. Also *adjugate*.

adjustable-pitch propeller (*Aero*) See PROPELLER.

adjustable-port proportioning valve (*Eng*) Air and fuel valves for oil or gas burners, motor operated in unison by automatic temperature-control equipment.

adjuvant (*BioSci*, *Med*) In general, a remedy that assists others; particularly a substance that increases the immunogenicity of antigens when administered with them, in some cases by providing a depot from which there is slow release, in other cases by activation of macrophages. See FREUND'S ADJUVANT.

adlacrimal (*BioSci*) The lacrimal bone of reptiles, so called to indicate that it is not homologous with the lacrimal bone of mammals.

ADME (*Pharmacol*) Generally used abbreviation for information about a drug relating to its absorption, distribution, metabolism and excretion; this is important for clinical development and for registration with regulatory authorities.

Admiralty brass (*Eng*) See TOBIN BRONZE.

admission (*Eng*) The point in the working cycles of a steam or internal-combustion engine at which the inlet valve allows entry of the working fluid into the cylinder.

admittance (*Phys*) A property which permits the flow of current under the action of a potential difference. The reciprocal of IMPEDANCE.

admixture (*Chem*) Property-modifying additive to eg Portland cement.

A-DNA (*BioSci*) Right-handed double-helix form of DNA with approximately eleven residues per turn.

adnate (*BioSci*) Joined to another organ of a different kind, as when stamens are fused to the petals. Cf CONNATE.

adnexa (*Med*) Appendages; usually refers to ovaries and Fallopian tubes.

adobe (*Build*) A name for any kind of mud which when mixed with straw can be sun-dried into bricks. Also *adobe clay*.

adobe (*Min*) A calcareous clay found in semi-arid plains and basins of the SW US and Mexico.

Adobe type manager (*ICT*) A proprietary method of making available well-formed type for display and printing in several graphical environments. Abbrev ATM.

adoral (*BioSci*) Adjacent to the mouth.

ADP (*BioSci*) Adenosine diphosphate. Unless otherwise specified it is the nucleotide $5'$ ADP, adenosine bearing a diphosphate (pyrophosphate) group in ribose-O-phosphate ester linkage at position $5'$ of the ribose moiety.

ADP (*ICT*) Abbrev for *automatic data processing*. See DATA PROCESSING.

ADP-ribosylation (*BioSci*) A post-translational modification of protein structure involving the transfer to the protein of an ADP-ribosyl moiety from NAD. It is believed to play a part in normal cellular regulation as well as in the mode of action of several bacterial toxins.

Adrastea (*Astron*) A tiny natural satellite of Jupiter, discovered in 1979 by the Voyager 2 mission. Distance from the planet 129 000 km; diameter 24 km.

adrectal (*BioSci*) Adjacent to the rectum.

adren-, adreno- (*Genrl*) Prefixes from Lt *ad*, to or at, and *renes*, kidney.

adrenal (*BioSci*) Adjacent to the kidney; pertaining to the adrenal gland.

adrenal cortex (*BioSci*) Outer portion of the adrenal gland that produces glucocorticoids.

adrenal gland (*BioSci*) See SUPRARENAL BODY.

adrenaline (*BioSci*) A hormone secreted, together with NORADRENALINE, by the adrenal medulla, and by adrenergic neurons in response to stress. It elicits the classic 'fight or flight'response; increased heart function, elevated blood sugar levels, paling of the skin, and raising of hairs on the neck. US *epinephrine*. See CATECHOLAMINES.

adrenal medulla (*BioSci*) The inner region of the adrenal gland that produces catecholamines.

adrenergic (*Med*) Pertaining to or causing stimulation of the sympathetic nervous system; applied to sympathetic nerves which act by releasing an adrenaline-like substance from their nerve endings.

adrenergic receptors (*BioSci*) Receptors for noradrenaline and adrenaline. All are G-protein coupled receptors linked variously to either adenylate cyclase or phosphoinositide second-messenger pathways. Three subgroups are usually recognized, α_1, α_2 and β.

adreno- (*Genrl*) See ADREN-.

adrenocorticotropic hormone (*BioSci*) See ACTH.

ADS (*Aero*) Abbrev for AIR DATA SYSTEM.

ADSL (*ICT*) Abbrev for ASYMMETRIC DIGITAL SUBSCRIBER LINE.

adsorbate (*Chem*) A substance, usually in gaseous or liquid solution, which is to be removed by ADSORPTION.

adsorbent (*Chem*) The substance, either solid or liquid, on whose surface ADSORPTION of another substance takes place.

adsorption (*Chem*) Increase in the concentration of a substance at an interface because binding lowers the free energy of the surface.

adsorption catalysis (*Chem*) The catalytic influence following adsorption of the reactants, exercised upon many reactions, often upon a specific adsorbent, attributed to the free residual valencies and hence higher reactivity of molecules at an interface (eg the ammonia oxidation process).

adsorption chromatography (*Chem*) See CHROMATOGRAPHY.

adsorption coefficient (*BioSci*) A constant, under defined conditions, that relates the binding of a molecule to a matrix as a function of the weight of matrix, eg in a column.

adsorption isotherm (*Chem*) The relation between the amount of a substance adsorbed and its pressure or concentration, at constant temperature.

adsorption potential (*Chem*) Change of potential in an ion in passing from a gas or solution phase onto the surface of an adsorbent.

adsorption surface area (*Chem*) The surface area of a powder particle determined by the mass of a specified substance that it can adsorb.

adularescence (*Min*) A milky or bluish sheen shown by moonstone.

adularia (*Min*) A transparent or milky-white variety of potassium feldspar, distinguished by its morphology.

adulteration (*FoodSci*) The illegal addition of any substance to a food product by accident or design, usually to defraud the consumer. See AUTHENTICITY.

adult respiratory distress syndrome (*BioSci*) See SEPTIC SHOCK. Abbrev ARDS.

advance (*CivEng*) The length of railway track beyond a signal which is covered by that signal.

advanced gas-cooled reactor (*NucEng*) Carbon-dioxide-cooled, graphite-moderated reactor using slightly enriched uranium oxide fuel clad in stainless steel, in use in the UK. Abbrev AGR.

advanced intelligent network (*ICT*) A form of INTELLIGENT NETWORK, developed in the USA from 1987 onwards, in which signalling, software and accounting procedures are designed to allow service providers to compete freely for network users' business.

advanced mobile phone system (*ICT*) The US forerunner of the UK TOTAL ACCESS COMMUNICATIONS SYSTEM. Abbrev AMPS. Developed by Bell in 1978, AMPS, like TACS, is an analogue cellular system using FREQUENCY MODULATION.

advance metal The p (*Eng*) Copper-base alloy with 45% nickel.

advance workings (*MinExt*) In flattish seams, mining in which the whole face is carried forward, no support pillars being left.

advantage ratio (*NucEng*) Ratio between the radiation dosage received at any point in a nuclear reactor and that of a reference position.

advection (*EnvSci*) The process of transport of a substance in air or water solely by mass motion.

advection fog (*EnvSci*) Fog produced by the ADVECTION of warm moist air across cold ground.

advection layer (*Astron*) The region immediately adjacent to the EVENT HORIZON where matter is being continuously pulled into a black hole. See panel on BLACK HOLE.

adventitia (*BioSci*) In general, accidental or inessential structures. More specifically, the outermost covering of an organ or the superficial layers of the wall of a blood vessel. Adj *adventitious*.

adventitious (*BioSci*) A term describing a plant part developed out of the usual order or in an unusual position. An *adventitious bud* is any bud except an AXILLARY bud; it gives rise to an *adventitious branch*. An *adventitious root* develops from some part of a plant other than a pre-existing root.

adventive (*BioSci*) A plant not permanently established in a given habitat or area.

advertisement (*Psych*) A conspicuous display that can involve coloration, posture or sound, and that serves to convey some information about the sender, eg age, sex, status or motivation.

adware (*ICT*) Software that is bundled with free files and programs, loaded onto a computer and that can provide information about a computer user's preferences so that advertising can be targeted at the user.

adze (*Build*) A cutting tool with an arched blade at right angles to the handle, used like a double-handed axe for shaping timber. Mainly used in boat construction.

adze-eye hammer (*Build*) Type of claw hammer with pronounced curve on the claw.

AE (*ImageTech*) Abbrev for AUTOMATIC EXPOSURE.

Ae (*Eng*) The TRANSFORMATION TEMPERATURE at equilibrium of the phase changes in iron and steel, subscripts indicating the designated change. Also *A*.

aecidiospore (*BioSci*) See AECIDIUM. Also *aeciospore*.

aecidium (*BioSci*) A cup-shaped, spore-forming structure (sorus) characteristic of some rust fungi (Uredinales) in which dikaryotic aecidiospores (*aeciospores*) are formed. Also *aecium*.

aedeagus (*BioSci*) In male insects, the intromittent organ.

aedicule (*Arch*) A shrine, set into a wall and framed by two columns, ENTABLATURE and PEDIMENT; the framing of a window or door in this manner.

aegirine (*Min*) Green sodium–iron member of the pyroxene group of minerals, essentially $NaFe^{3+}$ Si_2O_6. Characteristic of the alkaline igneous rocks. *Acmite* is a brown variety which is also used for the pure NaFe end-member.

aegirine–augite (*Min*) Minerals intermediate between AEGIRINE and AUGITE.

aegophony (*Med*) The bleating quality of voice heard through the stethoscope when fluid and air are present in the pleural cavity. Also *egophony*.

aenigmatite (*Min*) A complex silicate of sodium, iron and titanium; occurs as reddish-black triclinic crystals in alkaline igneous rocks. Also *cossyrite*.

aeolian deposits (*Geol*) Sediments deposited by wind and consisting of sand or dust (LOESS).

aeolian tone (*Acous*) A musical note set up by vortex action on a stretched string when it is placed in a stream of air. See STROUHAL NUMBER.

aeolotropic (*Phys*) Having physical properties which vary with direction or position. See ANISOTROPIC.

aeon (*Geol*) (1) A large part of geological time consisting of a number of ERAS, eg the Phanerozoic eon which includes the PALAEOZOIC, MESOZOIC and CENOZOIC eras. (2) A period of one thousand million (10^9) years. Also *eon*. See panel on GEOLOGICAL COLUMN.

aequorin (*BioSci*) Protein derived from the coelenterate, *Aequorea victoria*, that emits light in the presence of calcium.

aerated concrete (*Build*) Concrete made by adding constituents to the mix which, by chemical reaction, liberate gases which are trapped in the concrete and so reduce its density and increase its heat insulation value.

aerating root (*BioSci*) See PNEUMATOPHORE.

aerating tissue (*BioSci*) See AERENCHYMA.

aeration (*FoodSci*) Incorporation of air or gas into a liquid or solid raw material or food product either by mixing or agitation or by chemical means, eg fermentation of yeast or direct injection of CO_2 in carbonation. Often undesirable after mixing and agitation because entrapped air can oxidize the product which then needs vacuum de-aeration.

aeration test burner (*Eng*) Burner for measuring the combustion characteristics of commercial gases. Abbrev *ATB*.

aerenchyma (*BioSci*) Tissue with particularly well-developed, air-filled, intercellular spaces. Characteristic of the cortex of roots and stems of hydrophytes, where it probably facilitates gas exchange between the roots and the leaves. Cf PNEUMATOPHORE.

aerial (*ICT*) Original UK term for antenna; technical publications now usually use *antenna*. Reference is still made to aerial in domestic use, eg TV aerial, car radio aerial.

aerial bunched conductors (*ElecEng*) A method of power transmission where the three conductors are twisted into a thicker insulated cable. More expensive but better at surviving blizzard conditions than normal separate conductors. Abbrev *ABC*.

aerial fog (*ImageTech*) Fog caused by exposure of portions of the film to air in a processing machine.

aerial radiometric surveying (*Radiol*) Use of low-flying aircraft to measure gamma-ray intensity due to natural radioactive emissions or radioactive contamination over large areas. Scintillators are used with photomultipliers whose signals are fed to multichannel analysers to distinguish the energies of the gamma rays received from a wide area; typically 90% of the gamma rays can be recorded from an area with linear dimensions about five times the aircraft's height above the ground. Also *airborne radiometric surveying*.

aerial root (*BioSci*) Adventitious root rising above ground, esp the long roots hanging from some tropical EPIPHYTES, and the short roots acting as attaching organs for many climbers (eg ivy) and epiphytes. See PROP ROOT.

aerial ropeway (*CivEng*) An apparatus for the overhead transport of materials in carriers running along an overhead cable or cables supported on towers.

aerial surveying (*Surv*) A process of surveying by photographs taken from the air, the photographs being of two types: (1) those giving a vertical or plan view; and (2) those giving an oblique or bird's-eye view. See OBLIQUE AERIAL PHOTOGRAPH, VERTICAL AERIAL PHOTOGRAPH.

AERO (*Aero*) Abbrev for *Air Education and Recreation Organisation* (UK).

aero- (*Genrl*) Prefix from Gk *aer*, signifying air.

aero-acoustics (*Acous*) Branch of acoustics that treats sound generation and transmission by fluid flow.

aerobe (*BioSci*) An organism that can live and grow only in the presence of free oxygen: an organism which uses aerobic respiration.

aerobic (*BioSci*) (1) Characterized by, or occurring in, the presence of free oxygen. (2) Requiring oxygen for respiration.

aerobic plate count (*FoodSci*) The number of yeasts, moulds or bacterial colonies counted on a culture plate when a suitable growth medium has been inoculated with a sample, exposed to the air or to a surface and incubated at a specified temperature and time period (eg 37°C for 48 h).

aerodynamic balance (*Aero*) (1) A balance, usually but not necessarily in a wind tunnel, designed for measuring aerodynamic forces or moments. (2) Means for balancing air loads on flying control surfaces, so that the pilot need not exert excessive force, particularly as speed increases. The principle is to use aerodynamic forces, either directly on a portion of the control surface ahead of the hinge line, or indirectly through a small auxiliary surface with a powerful moment arm, to counterbalance the main

airloads. An example of the first is the HORN BALANCE, and of the second the BALANCE TAB.

aerodynamic braking (*Space*) Use of a planet's atmosphere to reduce the speed of space vehicles.

aerodynamic centre (*Aero*) The point about which the pitching moment coefficient is constant for a range of aerofoil incidence.

aerodynamic coefficient (*Aero*) A non-dimensional measure of aerodynamic force, pressure or moment that expresses the characteristics of a particular shape at a given incidence to the airflow. Typically the lift coefficient is given by $C_L = L/\frac{1}{2}\,\rho V^2 S$, where L is the lift, ρ is the air density, V is the air speed and S is a typical area of the body (eg wing area). Similarly for drag coefficient.

aerodynamic damping (*Aero*) The suppression of oscillations by the inherent stability of an aircraft or of its control surfaces.

aerodynamic diameter (*Genrl*) The diameter of a spherical particle, with relative density equal to unity, that has the same settling velocity in air as the particle in question.

aerodynamic heating (*Space*) The heating of a vehicle passing through the atmosphere, caused by friction and compression of air (or other gas).

aerodynamics (*Aero*) The part of the mechanics of fluids that deals with the dynamics of gases. See panel on AERODYNAMICS.

aerodynamic sound (*Acous*) See FLOW NOISE.

aerodyne (*Aero*) Any form of aircraft deriving lift in flight principally from aerodynamic forces. Includes AIRCRAFT, GLIDER, KITE, HELICOPTER. Commonly *heavier-than-air aircraft*.

aero-elastic divergence (*Aero*) Aero-elastic instability which occurs when aerodynamic forces, or moments, increase more quickly than the elastic restoring forces or couples in the structure. Generally applied to wing weakness where the incidence at the tips increases under load, so tending to twist the wings off.

aero-elasticity (*Aero*) The interaction of aerodynamic forces and the elastic reactions of the structure of an aircraft. Phenomena are most prevalent when manoeuvring at very high speed.

aero-embolism (*Med*) Release of nitrogen bubbles in the blood stream resulting from too rapid a reduction in ambient air pressure; the *bends*, encountered by undersea divers. See CAISSON DISEASE.

aero-engine (*Aero*) The power unit of an aircraft. Originally a lightweight reciprocating internal-combustion engine, usually Otto cycle, as a general rule either air-cooled radial, in-line, vee or liquid-cooled vee; gas turbines gradually superseded reciprocating engines from 1945 for large civil and military aircraft but reciprocating engines are still widely used in small aircraft. See DUCTED FAN, GAS TURBINE, RAMJET, TURBOJET, TURBOPROP, TURBORAMJET, TURBOROCKET, VARIABLE CYCLE ENGINE.

aerofall mill (*MinExt*) Dry grinding mill with large diameter and short cylindrical length in which ore is mainly or completely ground by large pieces of rock; the mill is swept by air currents which remove finished particles.

aerofoil (*Aero*) A body shaped so as to produce an aerodynamic reaction (lift) normal to its direction of motion, for a small resistance (drag) in that plane; eg a wing, plane, aileron, tailplane, rudder or elevator.

aerofoil section (*Aero*) The cross-sectional shape or profile of an aerofoil.

aerogel (*Eng*) A silicon-based solid with a porous structure with 99% of its volume as open space, used as an insulator between sheets of eg glass.

aero-isoclinic wing (*Aero*) A swept-back wing which has its torsional and flexural stiffness so adjusted that the angle of attack remains constant as the wing bends under flight loads, instead of decreasing with deflection towards the tip, which is the normal geometric effect.

aerolites (*Geol*) A general name for stony meteorites as distinct from iron meteorites.

Aerodynamics

The science of the interactions between moving air and solid surfaces as in aeronautics, or when spacecraft enter or leave the atmosphere. Movement can also be caused by heating, as in meteorology. Air molecules respond to external disturbances in many ways and there are therefore many branches of the subject.

The interaction largely depends on the physical properties of air which, at sea level, has a density (mass per unit volume) of 1.225 kg m^{-3} at 15°C, decreasing with altitude. Air pressure (the force per unit area) is 101.3 kPa at sea level. The speed of sound, equal to Mach 1, is 340.3 m s^{-1} at normal temperature and pressure and, generally, is $20.1 \sqrt{T}$ m s^{-1} where T is the local absolute temperature. Heat transfer results from viscous energy and involves specific heat, C_p, at constant pressure and C_v at constant volume.

Regions of airflow are defined by the properties of the boundary layer. This is the thin layer of air or other fluid adjacent to the surface of a body moving through it, in which viscous forces exert a major influence on the motion of the fluid. Air molecules near a surface encounter forces which cause them to stick to the surface and so reduce their airspeed to zero. This creates a shearing stress responsible for skin friction drag and, depending on conditions, characteristic airflow patterns are formed as shown in the diagram.

The laminar boundary layer occurs where the airspeed increases linearly with the distance from the surface and there are no large eddies and skin friction is low. Such a layer may occur towards the front of an aerofoil if the surface is smooth and the pressure gradient favourable. It is difficult to maintain and eventually breaks down at the transition point with the formation of eddies of many sizes. Downstream there is the turbulent boundary layer in which the skin friction increases to between three and ten times that of the laminar layer. At extremely high altitude as encountered by re-entry of spacecraft the reduced air viscosity cannot bring the moving air to rest on the surface, causing a residual surface velocity and a low skin friction.

Shock waves occur at surfaces of discontinuity across which there is either an abrupt increase in pressure, as at the rear of a local supersonic region of flow around an aircraft moving at transonic speeds, or an abrupt change in direction like the bow shock wave ahead of a body moving at supersonic speeds or in any compression corner at supersonic or hypersonic speeds.

A number of types of fluid flow are defined: *ideal* flow is that of a mathematically perfect fluid without viscosity and hence without boundary layer or heating; *viscous* flow is real flow with viscosity present; *Newtonian* flow is that in a rarefied gas in which the molecular mean free path is long compared with that of a body moving in it; *free-molecule* flow is like Newtonian but the molecules impact and rebound from the body without influencing each other.

A number of speed regimes are defined for relative motion between a body and the fluid: *subsonic* with speeds less than that of sound; *sonic* at the speed of sound; *transonic* around the speed of sound, usually taken to be between Mach 0.7 and 1.3; *supersonic* faster than the speed of sound (> Mach 1); *hypersonic* much faster than the speed of sound (> Mach 5). At very high hypersonic speeds MAGNETOHYDRODY-NAMICS becomes important because the heat dissociates the molecules and the resulting flow of free electrons behave as electric currents which can then create and be influenced by magnetic fields. *Hypervelocity* refers, in physics, to velocities approaching the speed of light and, in aerospace, to those exceeding the Earth satellite speed.

The flow of air over an aircraft

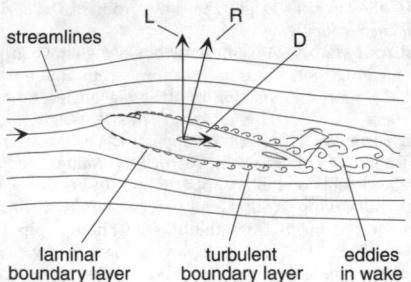

aerodynamics Flow of air over an aircraft.

L is the lift force, **D** is the drag force and **R** is the resultant of the aerodynamic pressure forces. In the diagram these forces are resolved along 'wind axes'. In guided flight 'body axes' are used with axial and normal forces resolved along and normal to the body.

See AERODYNAMIC BRAKING, AERODYNAMIC COEFFICIENT, AERODYNAMIC DAMPING, AERODYNAMIC FORCE, AERODYNAMIC HEATING, AERO-ELASTICITY, AEROTHERMOCHEMISTRY, AEROTHERMODYNAMICS, AEROTHERMO-ELASTICITY, BOUNDARY LAYER, COMPUTATIONAL FLUID DYNAMICS, WIND TUNNEL.

aerological diagram (*EnvSci*) A thermodynamic diagram used for plotting the results of upper-air soundings usually containing, as reference lines, isobars, isotherms, DRY ADIABATICS, SATURATED ADIABATICS and lines of constant saturation humidity mixing ratio.

aerology (*EnvSci*) The study of the FREE ATMOSPHERE.

aeronautical engineering (*Aero*) The branch of engineering concerned with the design, production and maintenance of aircraft structures, systems and power units.

aeronautical fixed services (*Aero*) A telecommunication service between fixed stations for the transmission of aeronautical information, particularly navigational safety and flight planning messages. Abbrev *AFS*.

aeronautics (*Aero*) All activities concerned with aerial locomotion.

aerophagy (*Med*) The swallowing of air, with consequent inflation of the stomach.

aerophone (*Acous*) Group of musical instruments in which the air in a tube-shaped resonator is excited to vibrate.

aeroplane (*Aero*) See AIRCRAFT.

aerosol (*Chem*) (1) A colloidal system, such as a mist or a fog, in which the dispersion medium is a gas. See COLLOID. (2) Pressurized container with built-in spray mechanism used eg for packaging insecticides, deodorants and paints.

aerospaceplane (*Aero*) Aircraft-like vehicle which can take off from and land on runways, manoeuvre in the atmosphere, operate in space and re-enter the atmosphere.

aerostat (*Aero*) Any form of aircraft deriving support in the air principally from its buoyancy, eg a balloon or airship.

aerothermochemistry (*Aero*) The chemical reactions which occur with airflow heating, eg a candle flame in air or the combustion of kerosine in a jet engine.

aerothermodynamics (*Space*) The branch of thermodynamics relating to the heating effects associated with the dynamics of a gas; in particular the physical effects produced in the air flowing over a vehicle during launch and re-entry.

aerothermo-elasticity (*Aero*) Aero-elasticity complicated by heating effects.

aestival (*BioSci*) Occurring in summer or characteristic of summer.

aestivation (*BioSci*) (1) Arrangement of unexpanded sepals and petals in the flower bud. Also *prefloration*. (2) The spatial organization of unexpanded leaves, sepals or petals. Types include imbricate, valvate, quincuncial, contorted and intricate (or intertwined). See QUINCUNCIAL AESTIVATION. Cf VERNATION. (3) Prolonged summer torpor, as in some insects and snails. Cf HIBERNATION.

aether (*Phys*) See ETHER.

aetiology (*Med*) The medical study of the causation of disease. US *etiology*.

AF (*ICT*) See AUDIO-FREQUENCY.

AF (*ImageTech*) Abbrev for AUTOMATIC FOCUSING.

afara (*For*) W African hardwood (*Terminalia*), pale yellow-brown to straw coloured, with black streaks in the heartwood. It has a close, straight-grained texture. Also *limba*.

AFC (*ICT*) Abbrev for AUTOMATIC FREQUENCY CONTROL.

AFCS (*Aero*) Abbrev for AUTOMATIC FLIGHT CONTROL SYSTEM.

AFD (*FoodSci*) Abbrev for ACCELERATED FREEZE-DRYING.

afebrile (*Med*) Without fever.

affect (*Psych*) General term for emotion, emotionality, mood, feeling, etc, as opposed to cognition and volition.

affect displays (*Psych*) Non-verbal cues that reveal emotional state.

affective behaviour (*Psych*) A wide range of behaviour in which the emotional aspects of social interactions are salient and often fundamental, eg mother–infant interactions.

affective disorders (*Psych*) A group of disorders whose primary characteristic is a disturbance of mood; feelings of elation or sadness become intense and unrealistic.

afferent (*BioSci*) Carrying towards, as blood vessels carrying nervous impulses to the central nervous system. Cf EFFERENT.

afferent arc (*BioSci*) The sensory or receptive part of a reflex arc, including the adjustor neuron(s).

affine (*MathSci*) Relating to a transformation that is equivalent to a linear transformation followed by a translation and can be used of curves that differ only in the scales of one or both co-ordinates. Affine geometry is that of vectors, and does not involve notions of length or angle.

affine transformation (*MathSci*) Geometrically, a transformation which preserves parallel lines. Algebraically, an invertible linear transformation. The two are equivalent.

affinity (*BioSci*) Measure of the strength of interaction or binding between antibody and antigen or between a receptor and its ligand.

affinity (*Chem*) The extent to which a compound or a FUNCTIONAL GROUP is reactive with a given reagent. See ELECTRON AFFINITY.

affinity (*Textiles*) The quantitative expression of *substantivity*, which is the attraction between fibres and dye or other substance. It results in fabric attracting colour when immersed in a solution of dye.

affinity chromatography (*BioSci*) A type of adsorption CHROMATOGRAPHY in which ligands, coupled to the solid, stationary phase, have a specific affinity for the substance to be isolated, eg a LECTIN for its specific carbohydrate, an antibody for its antigen.

affinity labelling (*BioSci*) Labelling of the active site of an enzyme or the binding site of a receptor by means of a reactive substance that forms a covalent linkage once having bound. Binding may be triggered by a change in conditions, eg in photoaffinity labelling by illuminating with light of an appropriate wavelength.

affinity purification (*BioSci*) A method of concentrating molecules in solution based upon their affinity for a specific ligand bound to a solid phase, from which they can subsequently be eluted by changing conditions or competing with a soluble ligand. In antibody purification the ligand is a specific antigen.

afforestation (*BioSci*) The establishment of forest in an area where trees did not grow formerly. This may occur naturally or be due to planting to provide commercial timber or wood crop (eg conifer plantations).

afgalaine (*Textiles*) An all-wool, plain-weave, dress cloth.

aflagellar (*BioSci*) Lacking flagella.

aflatoxins (*BioSci*) Group of secondary metabolites produced by *Aspergillus flavus* and *A. parasiticus* that commonly grow on stored food, esp peanuts, rice and cotton seed. Some are highly toxic to cattle and are suspected of causing liver cancer in Africa.

AFM (*ImageTech*) Abbrev for AUDIO-FREQUENCY MODULATION.

A-frame (*Eng*) See SHEAR-LEGS.

African blackwood (*For*) E African hardwood (*Dalbergia*) that is very dense, exceptionally hard, almost black, straight-grained and durable.

African glanders (*Vet*) See EPIZOOTIC LYMPHANGITIS.

African horse-sickness (*Vet*) A viral disease of horses and some other hoofed animals, occurring in Africa and the Middle East, and transmitted by midges (*Culicoides*). Occurs in an acute, pulmonary form called locally *dunkop* (meaning 'thin head') and due to pulmonary oedema, or in a subacute cardiac form called *dikkop* (meaning 'thick head') which is characterized by oedematous swelling of head, neck and sometimes body.

African mahogany (*For*) W African hardwood (*Khaya*) used for cabinet-making, veneering, plywood manufacture, high-class joinery and furniture.

African swine fever (*Vet*) An acute, contagious, viral disease of pigs in Africa; characterized by fever, diarrhoea and multiple haemorrhages. Warthogs are symptomless carriers of the virus.

African walnut (*For*) A tropical W African hardwood from the genus *Lovoa*; it is not a true walnut, but finds similar uses. It has a bronze orange-brown heartwood, with an interlocked, sometimes spiral grain and lustrous texture.

African whitewood (*For*) See OBECHE.

afrormosia (*For*) A very durable hardwood from W Africa, used as a substitute for TEAK.

AFS (*Aero*) Abbrev for AERONAUTICAL FIXED SERVICES.

aft cg limit (*Aero*) See CG LIMITS.

afterbirth (*Med, BioSci*) The placenta and membranes expelled from the uterus after delivery of the fetus. See DECIDUA.

afterbody (*Aero*) Rear portion of a flying-boat hull, aft of the main step.

afterburner (*Aero*) See REHEAT.

afterburning (*Autos*) In an internal-combustion engine, persistence of the combustion process beyond the period proper to the working cycle, ie into the expansion period.

afterburst (*MinExt*) Delayed further collapse of underground workings after a rockburst.

aftercooler (*MinExt*) Chamber in which heat generated during compression of air is removed, allowing cool air to be piped underground.

afterdamp (*MinExt*) The non-flammable heavy gas, carbon dioxide, left after an explosion in a coal mine. The chief gaseous product produced by the combustion of coal gas. See BLACK DAMP, CHOKE DAMP, FIRE DAMP, WHITE DAMP.

afterglow (*Electronics*) The glow of a gaseous medium immediately after the cessation of electric current or downstream of an electric discharge. See PERSISTENCE.

afterheat (*NucEng*) The heat which comes from fission products in a reactor after it has been shut down.

after-image (*Phys*) Formation of image on retina of eye after removal of visual stimulus, in colour complementary to this stimulus. See COMPLEMENTARY AFTER-IMAGE.

aftermath (*Agri*) A second crop of grass grazed or harvested from a field in a single season.

after-pains (*Med*) Pains occurring after childbirth due to contraction of the uterus.

afterpeak (*Ships*) Space abaft the aftermost bulkhead. Lower part frequently used as fresh-water tank; upper part may be used as storeroom.

after-ripening (*BioSci*) The chemical and/or physical changes that must occur inside the dry seeds of some plants after shedding or harvesting if germination is to take place when the seeds are moistened.

aftershock (*Geol*) A minor earthquake that follows a major one. Many aftershocks may occur, decreasing in magnitude.

after tack (*Build*) A paint defect in which the coating retains a slight degree of tackiness.

afwillite (*Min*) Hydrated calcium silicate occurring in natural rocks and set cements.

Ag (*Chem*) Symbol for SILVER (*argentum*).

agalactia (*Med*) Failure of the breast to secrete milk. Also *agalacia*.

agalmatolite (*Min*) See PAGODITE.

agamic (*BioSci*) See AGAMOGENESIS.

agammaglobulinaemia (*Med*) Inherited immunodeficiency disease characterized by the partial or complete absence of one or more immunoglobulin subclasses from the serum. The consequences can be mild or severe depending upon which immunoglobulins are affected.

agamogenesis (*BioSci*) Asexual reproduction. Adj *agamic*.

agamospermy (*BioSci*) Reproduction by seed formed without sexual fusion. See APOMIXIS.

agar (*FoodSci*) A stabilizer and thickening agent extracted from seaweed, dried and bleached, then processed to powder form. When hydrated it is virtually colourless, tasteless and odourless, and when solutions are cooled to below 40°C they form a transparent gel. Also used as a solidifying agent in microbiological culture media. Also *agar-agar*.

agarics (*BioSci*) The Agaricales, an order of the Hymenomycetes containing the mushrooms and toadstools; c.3000 species. The spores are borne on the surface of gills or in the lining of pores.

agarose (*BioSci*) A galactan polymer purified from agar that forms a rigid gel with high free water content. It is primarily used as a matrix for electrophoretic separation of macromolecules.

agarose gel electrophoresis (*BioSci*) A method used particularly for fractionating DNA fragments produced by restriction endonuclease digestion. Fragments migrate through the gel matrix under the influence of an electric field.

agate (*Min*) A cryptocrystalline variety of silica, characterized by parallel, and often curved, bands of colour. See panel on SILICON, SILICA, SILICATES.

agba (*For*) Tropical W African hardwood (*Gossweilerodendron*). The heartwood is pale pinkish-straw to tan coloured, has a straight to wavy grain and a fine texture.

AGC (*ICT*) Abbrev for AUTOMATIC GAIN CONTROL.

age distribution (*BioSci*) The relative frequency, in an animal or plant population, of individuals of different ages. Generally expressed as a polygon or age pyramid, the number or percentage of individuals in successive age classes being shown by the relative width of horizontal bars.

age equation (*NucEng*) See AGE THEORY.

age hardening (*Eng*) The production of structural change spontaneously after some time; normally it is useful in improving mechanical properties in some respect, particularly hardness. See PRECIPITATION HARDENING.

ageing (*Eng*) (1) Final stage of PRECIPITATION HARDENING, producing an increase in strength and hardness in metal alloys, due to precipitation of second phase particles from supersaturated solid solution over a period of days at room temperature, or several hours at an elevated temperature ('artificial' ageing). (2) Slow deterioration in polymer products due to oxygen or ozone cracking, increase in crystallinity, relaxation of internal stress, etc. (3) Deterioration of the properties of ferroelectric materials.

ageing (*NucEng*) Loss of strength in the cladding or the pressure vessel in a nuclear reactor due to irradiation. Artificial ageing would be the simulation of such processes by increasing the rate of irradiation to obtain information more rapidly.

ageing (*Phys*) Change in the properties of a substance with time. A change in the magnetic properties of iron, eg increase of hysteresis loss of sheet-steel laminations; also the process whereby the subpermanent magnetism can be removed in the manufacture of permanent magnets.

ageing (*Textiles*) The exposure of freshly printed fabrics to steam to produce fully developed colours.

agenesis (*Med*) Imperfect development (or failure to develop) of any part of the body. Also *agenesia*.

agent (*ICT*) A self-sufficient computer program that functions without user intervention and that can be sent through the INTERNET to operate in another computer.

ageotropic (*BioSci*) Not reacting to gravity. See TROPISM.

age theory (*NucEng*) In nuclear reactor theory, the slowing down of neutrons by elastic collisions. The *age equation* relates the spatial distribution of neutrons to their energy. The equation is given by

$$\nabla^2 q - \frac{\partial q}{\partial \tau} = 0$$

where q is the slowing-down density and τ is the FERMI AGE. It was first formulated by Fermi who assumed that the slowing-down process is continuous and so is least applicable to media containing light elements.

agg (*BioSci*) Abbrev for AGGREGATE SPECIES.

agglomerate (*Geol*) An indurated rock built of angular rock fragments embedded in an ashy matrix, and resulting from explosive volcanic activity. Occurs typically in volcanic vents.

agglomerate (*PowderTech*) Assemblage of particles rigidly joined together, as by partial fusion (SINTERING) or by growing together.

agglomerating value (*MinExt*) Index of the binding (SINTERING) qualities of coal which has been subjected to a prescribed heat treatment.

agglomeration (*FoodSci*) Increase of the particle size of a dry product or raw material to improve reconstitution properties by wetting an airborne stream of particles with steam and then re-drying the agglomerated particles. Used mainly for quality milk and coffee powders. Also *instantizing*.

agglutination (*BioSci*) The clumping together of red blood cells, bacteria, large molecules, etc, to form a visible precipitate, as when blood cells from two incompatible groups are mixed, or an antigen reacts with its specific antibody.

agglutination (*Chem*) The coalescing of small suspended particles to form larger masses which are usually precipitated. Cf COAGULATION.

agglutinin (*BioSci*) A molecule that causes AGGLUTINATION of cells or bacteria, eg antibodies of the human ABO BLOOD GROUP SYSTEM and lectins from plants and invertebrate animals.

aggregate (*CivEng, Eng*) (1) Assemblage of powder particles which are loosely coherent. (2) Mixture of sand and gravel or crushed rock used in making concrete. Graded aggregate has a graded size distribution so that the particles fit better together, requiring less cement in the mix. See COARSE AGGREGATE, FINE AGGREGATE.

aggregate (*Geol, Min*) A mass consisting of rock or mineral fragments.

aggregate fruit (*BioSci*) (1) The fruit-like structure formed by a single flower with several free carpels. (2) A MULTIPLE FRUIT.

aggregate ray (*BioSci*) A group of small rays closely spaced so as to appear to be one large ray.

aggregate species (*BioSci*) Abbrev *agg.* A group of two or more closely similar species denoted, for convenience, by a single shared name, eg the blackberry, *Rubus fruticosus* agg.

aggregation (*BioSci*) A type of animal and plant dispersion in which individuals are closer to each other than they would be if they were randomly dispersed. Also *contagious distribution*.

aggregator (*ICT*) A piece of software that allows a computer user to view data from various web pages in a single window.

aggressive behaviour (*Psych*) An interpersonal style where only the immediate needs of the individual are considered rather than the needs of others. Contrasts with passive or assertive behaviour. In animal behaviour, any action that intimidates or damages conspecifics, eg scent marking. See AGONISTIC BEHAVIOUR.

aggressive mimicry (*BioSci*) Resemblance to a harmless species in order to facilitate attack.

agile development (*ICT*) A systems development paradigm characterized by extreme flexibility and rapid responsiveness to changing requirements. See also EXTREME PROGRAMMING.

agitation (*ImageTech*) Vigorous movement of film and solutions during processing to ensure that fresh chemicals are brought in contact with the emulsion.

agitator (*MinExt*) Tank, usually cylindrical, which has a mixing device such as a propeller or airlift pump near the bottom. Finely ground mineral slurries (the aqueous component perhaps being a leaching solution) are exposed to appropriate chemicals for purpose of extraction of gold, uranium or other valuable constituents. Types include PACHUCA TANK or *Brown agitator*.

A-glass (*Glass*) Designation for a glass fibre of composition (percentage by weight) SiO_2 72, Na_2O 14, CaO 10, MgO 2·5, Al_2O_3 0·6, which is similar to that of the soda–lime–silica glass used for windows and bottles. Its resistance to water, mineral acids and alkalis is much less than that of C-GLASS and E-GLASS fibres.

aglomerular (*BioSci*) Devoid of glomeruli; said eg of the kidney in certain fishes.

aglossate (*BioSci*) Lacking a tongue. Also *aglossal*; n *aglossia*, congenital absence of the tongue.

agmatite (*Geol*) A MIGMATITE with a brecciated appearance.

Agnatha (*BioSci*) A superclass of eel-shaped chordates without jaws or pelvic fins. Lampreys and hagfishes.

agnathous (*BioSci*) Having a mouth without jaws, as in the lampreys. Also *agnathostomatous*.

Agnesi (*MathSci*) See WITCH OF AGNESI.

agnosia (*Psych*) Inability to recognize familiar things or people, esp after brain damage.

AGO (*Genrl*) Abbrev for AUTOMOTIVE GAS OIL.

agonic line (*Geol*) The line joining all places with no magnetic declination, ie those where true north and magnetic north coincide on a compass.

agonist (*BioSci*) A pharmacological term for a compound that acts on a receptor to elicit a response.

agonistic behaviour (*Psych*) A broad class of behaviour patterns, including all types of attack, threat, appeasement and flight, between members of the same species in response to a conflict between aggression and fear. Behaviour often alternates between attack and escape, eg across a territory boundary.

agoraphobia (*Psych*) An anxiety disorder (phobia) characterized by an intense fear of open spaces.

AGR (*NucEng*) Abbrev for ADVANCED GAS-COOLED REACTOR.

agranulocytosis (*Med*) A pathological state in which there is a marked decrease in the number of granulocytes in the blood.

agraphia (*Med*) Loss of power to express thought in writing, as a result of a lesion in the brain.

agrestal (*BioSci*) Growing in cultivated ground, but not itself cultivated, eg a weed.

agriculture (*Gnrl*) Intensive cultivation of domesticated crops and/or livestock.

agriscience (*Genrl*) Science as applied to agriculture.

agroforestry (*EnvSci*) Form of land use in which herbaceous crops and tree crops co-exist in an integrated scheme of farming. See TAUNGYA.

agrology (*Genrl*) The scientific study of soils and their influence upon crops.

agronomy (*Genrl*) Science as applied to agriculture; sometimes restricted to the science underlying plant production and land management.

agrostology (*Genrl*) The study of grasses.

AGS (*Aero*) Abbrev for AIRCRAFT GENERAL STANDARD.

ahm (*ElecEng*) Abbrev for AMPERE-HOUR METER.

AI (*ICT*) Abbrev for ARTIFICIAL INTELLIGENCE.

AIAA (*Aero*) Abbrev for *American Institute of Aeronautics and Astronautics*.

AIDS (*Med*) Abbrev for ACQUIRED IMMUNODEFICIENCY SYNDROME.

aiguille (*Build*) A stone-boring tool.

aileron droop (*Aero*) The rigging of ailerons so that under static conditions their trailing edges are below the wing trailing-edge line, pressure and suction causing them to rise in flight to the aerodynamically correct position.

ailerons (*Aero*) Surfaces at the trailing edge of the wing, controlled by the pilot, which move differentially to give a rolling motion to the aircraft about its longitudinal axis.

air absorption (*Acous*) Absorption of sound waves propagating in air, caused by molecular relaxation processes and viscosity.

air bells (*ImageTech*) Minute bubbles which have adhered to the emulsion during processing, leaving small circular spots where it has been protected from chemical action.

air bladder (*BioSci*) In fish, an air-containing sac developed as a diverticulum of the gut, with which it may retain connection by the pneumatic duct in later life; usually it has a hydrostatic function, but in some cases it may be respiratory or auditory, or assist in sound production.

air-blast circuit breaker (*ElecEng*) A form of circuit breaker or switch in which an arc is deliberately drawn between two contacts. The arc is cooled by a blast of high-pressure air which removes ions, thereby extinguishing the arc and breaking the circuit. Also *air-blast switch*.

airborne flux (*Agri*) The quantity of agrochemical crossing a given area, horizontally or vertically, in a given time.

airborne missile control system (*Space*) A method of controlling missile attacks from a secure command position. Abbrev *AMCS*.

airborne radiometric surveying (*Surv*) See AERIAL RADIOMETRIC SURVEYING.

Airborne Warning and Control System (*Genrl*) An electronic system designed to carry out airborne surveillance by means of radar carried on a specially modified aircraft, and also to manage defensive responses. Abbrev *AWACS*.

air brake (*Aero*) An extendable device, most commonly a hinged flap on wing or fuselage, controlled by the pilot, to increase the drag of an aircraft. Originally a means of slowing bombers to enable them to dive more steeply, it is an essential flight control on jet aircraft and sailplanes.

air brake (*Eng*) (1) A mechanical brake operated by air pressure acting on a piston. (2) An absorption dynamometer in which the power is dissipated through the rotation of a fan or propeller.

air break (*ElecEng*) A term describing a switch or circuit breaker with contacts in air.

air brick (*Build*) A perforated cast-iron, concrete or earthenware brick built into a wall admitting air under the floors or into rooms.

air brush (*Build*) A small spray gun used in decorative artwork and signwork.

air capacitor (*ElecEng*) A capacitor in which the dielectric is nearly all air, for tuning electrical circuits with minimum dielectric loss.

air cell (*Eng*) A small auxiliary combustion chamber used in certain types of compression-ignition engines, for promoting turbulence and improving combustion.

air chamber (*BioSci*) An air-filled cavity, eg towards the upper surface of the thallus of some liverworts, opening externally by a pore and containing photosynthetic cells; also in some hydrophytes.

air classifier (*MinExt*) Appliance in which vertical, horizontal or cyclonic currents of air sort falling ground particles into equal-settling fractions or separate relatively coarse falling material from finer dust which is carried out. Also *air elutriator*.

air cleaner (*Autos*) A filter placed at the intake of an internal-combustion engine to remove dust from the air entering the cylinders.

air compressor (*Eng*) A machine which draws in air at atmospheric pressure, compresses it and delivers it at higher pressure.

air conduction (*Acous*) The passing of noise energy along an air path, as contrasted with structure-borne conduction of vibrational energy.

air-cooled engine (*Autos*) An internal-combustion engine in which the cylinders, finned to increase surface area, are cooled by an airstream. See COWLING.

air-cooled machine (*ElecEng*) A machine, transformer or other piece of apparatus, in which the heat caused by the losses is removed solely by natural or fan-assisted airflow. Also *air-cooled transformer*.

air cooler (*Phys*) The cold 'accumulator' used in the LINDÉ PROCESS of air liquefaction for the preliminary cooling of the air.

air cooling (*Eng*) The cooling of hot bodies by a stream of cold air, instead of eg water cooling.

aircraft (*Aero*) Any mechanically driven heavier-than-air flying machine with wings of fixed or variable sweep angle. Subdivisions: landplane, seaplane (float seaplane and flying boat), amphibian.

aircraft design (*Aero*) The specification of an aircraft, following compromises between requirements of performance, economy and safety. It includes external aerodynamic shape, and the spatial arrangement of flying surfaces, engines, control surfaces and internal systems. *Gross weight*, *take-off weight*, *landing weight* and *load* are all legal values in

specification and contract, and *verification* is proof of the design by ground and flight tests. New concepts include an inherently unstable aircraft controlled continuously by on-board computers.

aircraft engine (*Aero*) See AERO-ENGINE.

aircraft flutter (*ICT*) A term used for the rapid fluctuations in very-high-frequency reception, affecting sound and vision; due to a secondary transmission path, or rapidly shifting phase, set up by reflection from an aircraft.

Aircraft General Standard (*Aero*) A term referring to small parts or items such as bolts, nuts, rivets, fork joints, etc, which are common to all types of aircraft. Abbrev *AGS*.

aircraft noise (*Aero*) Noise from propeller, engine, exhaust, and that generated aerodynamically over the surfaces; characterized by unstable low frequencies. See JET NOISE.

air data system (*Aero*) A centralized unit into which are fed the essential physical measurements for flight, eg airspeed, MACH NUMBER, Pitot and static pressure, barometric altitude, stagnation air temperature. From this central source, data are transmitted to the cockpit dials, to flight and navigational instruments and to computers. Abbrev *ADS*.

air door (*MinExt*) In a mine ventilating system, a door which admits air or varies its direction.

air dose (*Radiol*) The radiation dose in roentgens delivered at a point in free air.

airdox (*MinExt*) US system for breaking coal in fiery mine by use of injected high-pressure air.

air drag (*Space*) Resistance to the motion of a body passing through the Earth's atmosphere, most serious in the lower regions, producing changes in the geometry of the orbit, even causing the body to re-enter. More generally the term *atmospheric drag* is used in reference to other planets.

air drilling (*MinExt*) Drilling method which uses air instead of mud as the cooling and debris-removal medium. Faster and easier than mud drilling, it cannot prevent water ingress and emergency mud equipment will then be necessary. Also *gas drilling*.

air dry (*MinExt, Paper*) Said of minerals, pulp and paper in which moisture content is in equilibrium with that of atmosphere. The basis of sale for wood pulp; pulp with a conventionally accepted theoretical moisture content is usually 10% of total mass.

air drying (*Build*) The process by which a dry film is formed from oil or paints under normal atmospheric conditions only. This may occur by solvent evaporation and/or by the oxidation and polymerization on exposure to oxygen. See STOVING.

air ducts (*Eng*) Pipes or channels through which air is distributed throughout buildings or machinery for heating and ventilation.

aired up (*MinExt*) Said of an oil plunger pump which no longer sucks because gas or air has filled the suction chamber.

air ejector (*Eng*) A type of air pump used for maintaining a partial vacuum in a vessel through the agency of a high-velocity steam jet which entrains the air and exhausts it against atmospheric pressure.

air elutriator (*MinExt*) See AIR CLASSIFIER.

air engine (*Eng*) (1) An engine in which air is used as the working substance. Rapid heating from an external source expands the air in the cylinder with consequent motion being imparted to a piston. After transfer to a compression cylinder, for rapid cooling, the air is returned to the working cylinder for the next cycle. Also *hot-air engine*. See STIRLING ENGINE. (2) A small reciprocating engine driven by compressed air.

air-entraining agent (*Eng*) Resin added to either cement or concrete in order to trap small air bubbles.

air equivalent (*Phys*) The thickness of an air column at 15°C and 1 atmosphere pressure which has the same absorption of a beam of radiation as a given thickness of a particular substance.

air escape (*Build*) A device for releasing excess air from a water pipe. A valve is opened by a float when sufficient air has accumulated and closed in time to prevent loss of water.

air exhauster (*Eng*) (1) A suction fan. (2) A vacuum pump.

air filter (*Autos*) Attachment to the air intake of a carburettor for cleaning air drawn into the engine.

air-float table (*MinExt*) Shaking table in which concentration of heavy fraction in sand-sized feed is promoted by air blown up through the deck, which is porous. Used in desert work. Also *air table*.

airflow meter (*Aero*) An instrument, mainly experimental, for measuring the airflow in ducts.

air flue (*Build*) A flue which is built into a chimney stack so as to withdraw vitiated air from a room.

airframe (*Aero*) The complete aircraft structure without power plant, systems, equipment, furnishings and other readily removable items.

air frost (*EnvSci*) Situation when the temperature in the Stephenson screen reaches zero; the ground may, however, stay above freezing because it retains residual warmth. Conversely, the ground can fall below freezing when the air does not, a ground frost. See WIND FROST.

air–fuel ratio (*Eng*) The proportion of air to fuel in the working charge of an internal-combustion engine, or in other combustible mixtures, expressed by weight for liquid fuels and by volume for gaseous fuels.

air-gap (*ElecEng*) Gap with points or knobs, adjusted to break down at a specified voltage and hence limit voltages to this value.

air-gap (*Phys*) Section of air, usually short, in a magnetic circuit, esp in a motor or generator, a relay, or a choke. The main flux passes through the gap, with leakage outside depending on dimensions and permeability.

air-gap torsion meter (*Eng*) A device for measuring the twist in a shaft by causing the relative rotation of two sections to alter the air-gap between a pair of electro-magnets, the resulting change in the current flowing being indicated by an ammeter.

air gate (*Eng*) Passage from interior of a mould to allow the escape of air and other gases as the metal or plastic enters. See RISER.

airglow (*Astron*) The faint permanent glow of the night sky, due to the emission of light from atoms and molecules of sodium, oxygen and nitrogen, activated by sunlight during the day.

air-hardening steel (*Eng*) Steel with sufficient carbon and other alloying elements to allow sections over 500 mm (20 in) to harden fully when cooled in air or other gas from above its transformation temperature. Also *self-hardening steel*. See ISOTHERMAL TRANSFORMATION DIAGRAM.

air heater (*Eng*) (1) Direct-fired heater, in which the products of combustion are combined with the air. (2) Indirect-fired heater, in which the combustion products are excluded from the airflow. Both can be operated in a recirculation system, by which a proportion of the heated air is returned to and passed through the heating chamber. See AIR PREHEATER.

air hoist (*MinExt*) Air winch or other mechanical hoist actuated by compressed air.

airing (*Eng*) Removal of sulphur from molten copper in a WIREBAR furnace, together with slag-forming impurities.

air insulation (*ElecEng*) Insulation for part of an electrical circuit provided by atmospheric air; eg a high-voltage transmission line, which is suspended between transmission towers (pylons), is insulated for the section between the towers by atmospheric air.

air intake (*Aero*) Any opening introducing air into an aircraft; the opening for the main engine air is usually implied if unqualified.

air intake (*Eng*) Vent in a carburettor through which air is sucked to mix with the petrol vapour from the jet.

air-intake guide vanes (*Aero*) Radial, toroidal or volute vanes which guide the air into the compressor of a gas turbine, or the supercharger of a reciprocating engine.

air interface (*ICT*) The physical specification and operating protocols governing the radio links to and from a mobile telephone.

air jet spinning (*Textiles*) A method of converting staple fibres into yarn: they are spun together by jets of air which strike the fibres tangentially, making them rotate.

air jet texturing (*Textiles*) See TEXTURED YARN.

air jig (*MinExt*) Use of pulses of air to stratify crushed ore into heavy and light layers. Used in waterless countries.

airlance (*MinExt*) Length of piping used to work compressed air into settled sand or to free choked sections of process plant, restoring aqueous flow.

air layering (*BioSci*) Horticultural method of vegetative propagation (esp of shrubby house plants) in which an aerial shoot is induced to form roots, by wounding and packing the wound with eg *Sphagnum* moss, while still attached to the plant. It is severed and potted up after rooting.

air laying (*Textiles*) A method for forming a WEB by collecting fibres from an air stream on a mesh ready for manufacturing a NON-WOVEN FABRIC.

air leg (*MinExt*) Telescopic cylindrical prop expanded by compressed air, used to support a rock drill.

airless spraying (*Build*) A system of spray application in which paint at extremely high fluid pressure is forced though a precision orifice in the spray gun when it 'atomizes' in a cloud of fine particles.

airlift (*NucEng*) A jet of air or neutral gas used to move solid or liquid material during processing to avoid necessity for pumps, particularly in 'maintainance-free' radioactive environments.

airlift pump (*MinExt*) An air-operated displacement pump for elevating or circulating pulp in cyanide plants.

airline (*Phys*) Straight line drawn on the magnetization curve of a motor, or other electrical apparatus, expressing the magnetizing force necessary to maintain the magnetic flux across an air-gap in the magnetic circuit.

air liquefier (*Phys*) A type of gas refrigerating machine based on the 'Stirling' or hot-air engine cycle.

air lock (*CivEng*) A device by which access is obtained to the working chamber (filled with compressed air to prevent entry of water) at the base of a hollow CAISSON. The worker at surface enters and is shut in an air-tight chamber filled with air at atmospheric pressure. Pressure within this air-lock is gradually raised to that used in the working chamber, so that the worker can pass out through another door and communicate with the working chamber.

air lock (*Eng*) An air pocket or bubble in a pipeline which obstructs the flow of liquid. See VAPOUR LOCK.

air log (*Aero*) An instrument for registering the distance travelled by an aircraft relative to the air, not to the ground.

air manometer (*Phys*) A pressure gauge in which the changes in volume of a small quantity of air enclosed by mercury in a glass tube indicate changes in the pressure to which it is subjected.

airmanship (*Aero*) Skill in piloting an aircraft.

air mass (*EnvSci*) A part of the atmosphere where the horizontal temperature gradient at all levels within it is very small, perhaps of the order of 1°C per 100 km. See FRONTAL ZONE.

air mass flow (*Aero*) In a gas turbine power plant, the quantity of air which is ingested by the compressor, normally expressed in pounds or kilograms per second.

air meter (*Eng*) An apparatus used to measure the rate of flow of air or gas.

air-mileage unit (*Aero*) An automatic instrument which derives the air distance flown and feeds it into other automatic navigational instruments.

air miles per gallon (*Aero*) The number of miles flown through the air for each gallon of fuel burnt by the propulsion units.

air monitor (*Radiol*) Radiation (eg gamma-ray) measuring instrument used for monitoring contamination or dose rate in air.

air plant (*BioSci*) See EPIPHYTE.

air pocket (*Aero*) Colloq term for a localized region of rising or descending air current. Causes an abrupt vertical acceleration as an aircraft passes through it, severity increasing with speed and also with low wing loading. Also *bump*. See VERTICAL GUST.

airpore (*BioSci*) See AIR CHAMBER, LENTICEL, STOMA.

airport markers (*Aero*) Particoloured boards defining areas on an airfield, eg *boundary markers* which indicate the limits of the landing area, *taxi-channel markers* for taxi tracks, *obstruction markers* for ground hazards, and *runway visual markers*, situated at equal distances, by which visibility is gauged in bad weather.

airport meteorological minima (*Aero*) The minimum cloud base (vertical) and horizontal visibility (expressed as RUNWAY VISUAL RANGE, RVR) in which landing or take-off is permitted at a particular aerodrome. ICAO standards: category 1, 200 ft (60 m) height, 2600 ft (800 m) RVR; category 2, 100 ft (30 m) height, 1300 ft (400 m) RVR; category 3, zero height, (a) 700 ft (210 m) RVR, (b) 150 ft (45 m) RVR, (c) zero RVR.

air position (*Aero*) The geographical position which an aircraft would reach in a given time if flying in still air.

air-position indicator (*Aero*) An automatic instrument which continually indicates air position, incorporating alterations of course and speed.

air preheater (*Eng*) A system of tubes or passages, heated by flue gas, through which combustion air is passed for preheating before admission to the combustion chamber, thus appreciably raising flame temperatures and returning to the combustion chamber some heat otherwise lost. See RECUPERATIVE AIR HEATER, REGENERATIVE AIR HEATER.

air pump (*Eng*) (1) A reciprocating or centrifugal pump used to remove air, and sometimes the condensate, from the condenser of a steam plant. See AIR EJECTOR. (2) Any device used for transferring air from one place to another. A compressor increases the pressure, a VACUUM PUMP reduces the pressure and a blower causes a rapidly moving air blast.

air quality (*EnvSci*) The extent to which air is free from contaminants, conventionally taken to be the respiratory irritants nitrogen dioxide and sulphur dioxide.

air receiver (*Build, Eng*) A pressure vessel in which compressed air is stored, facilitating moisture removal and the equalizing of pressure fluctuations, before air is conveyed to eg a spray gun, drill or hammer.

Air Registration Board (*Aero*) The airworthiness authority of the UK until its functions were taken over in 1972 by the CIVIL AVIATION AUTHORITY. Abbrev *ARB*.

air route (*Aero*) In organized flying, a defined route between two aerodromes; usually provided with eg direction-finding facilities, lighting and emergency-landing grounds. See AIRWAY.

air sac (*BioSci*) (1) In insects, a thin-walled distensible dilation of the tracheae, occurring esp in rapid fliers, which increases the oxygen capacity of the respiratory system and assists flying. (2) In birds, an expansion of a blind end of certain bronchial tubes, which projects into the general body cavity, assisting in respiration and lightening the body.

airscrew (*Aero*) Any type of screw designed to rotate in air; defined in 1951. Term now obsolete and replaced by PROPELLER, a device for propelling aircraft, and FAN, a rotating bladed device for moving air in ducts or eg wind tunnels. See ROTOR.

air seal (*Eng*) Curtain of air maintained in front of kiln or furnace door to aid retention of heat or in front of a workstation to reduce dust entry.

air shaft (*CivEng*) An air passage, usually vertical or nearly vertical, which provides for the ventilation of a tunnel or mine.

airship (*Aero*) Any power-driven AEROSTAT. In a *non-rigid airship*, the envelope is so designed that the internal pressure maintains its correct form without the aid of a built-in structure; small, and used for naval patrol work. A *rigid airship* has a rigid structure to maintain the designed shape of the hull, and to carry the loads; usually a number of ballonets or gas bags inside the frame; large, used for military purposes in World War I, and having limited commercial use until 1938. A *semi-rigid airship* has a partial structure, usually a keel only, to distribute the load, and maintain the designed shape of the envelope or ballonets; intermediate size.

air shooting (*MinExt*) (1) Charging of shot-hole so as to leave pockets of air, thus reducing the shatter effect of a blast. (2) In seismic prospecting, producing an explosion in air, above the rock formation under examination, to propagate a seismic wave.

air shower (*Phys*) See CASCADE SHOWER.

air sinuses (*BioSci*) In mammals, cavities connected with the nasal chambers and extending into the bones of the skull, esp the maxillae and frontals.

air space (*Aero*) The part of the atmosphere which lies above a nation and which is therefore under the jurisdiction of that nation.

air space (*BioSci*) Air-filled INTERCELLULAR SPACES, esp large ones.

air-spaced coil (*ElecEng*) Inductance coil in which the adjacent turns are spaced (instead of being wound close together) to reduce self-capacitance and dielectric loss.

airspeed (*Aero*) Speed measured relative to the air in which the aircraft or missile is moving, as distinct from groundspeed. See EQUIVALENT AIRSPEED, INDICATED AIRSPEED, TRUE AIRSPEED.

air standard cycle (*Autos*) A standard cycle of reference by which the performance of different internal-combustion engines may be compared, and their relative efficiencies calculated.

air standard efficiency (*Autos*) The thermal efficiency of an internal-combustion engine working on the appropriate air standard cycle.

airstrip (*Aero*) Unidirectional landing area, usually of grass or of a makeshift nature.

air superiority fighter (*Aero*) Combat aircraft intended to remove hostile aircraft from a volume of air space and so establish control of the air.

air surveying (*Surv*) See AERIAL SURVEYING.

air-swept mill (*MinExt*) In dry grinding of rock in a BALL MILL, use of a modulated current of air to remove sufficiently pulverized material from the charge in the mill.

air table (*MinExt*) See AIR-FLOAT TABLE.

air-traffic control (*Aero*) The organized control, by visual and radio means, of the traffic on air routes, and into and out of aerodromes. Abbrev *ATC*. ATC is divided into general *area control*, including defined airways; *control zones*, of specified area and altitude round busy aerodromes; *approach control* for regulating aircraft landing and departing; and *aerodrome control* for directing aircraft movements on the ground and giving permission for take-off. Air-traffic control operates under two systems: VISUAL FLIGHT RULES and, more severely, INSTRUMENT FLIGHT RULES. Since World War II great advances in radar technology have enabled air-traffic controllers to be given very complete 'pictures' of the position of aircraft, not only in flight, but also when manoeuvring on the ground.

air-traffic control centre (*Aero*) An organization providing: (1) air-traffic control in a control area; and (2) FLIGHT INFORMATION in a region.

air-traffic controller (*Aero*) Someone who is licensed to give instructions to aircraft in a control zone.

Air Transport Association (*Aero*) A US organization noted particularly for its specification which sets a standard to which manufacturers of aircraft and associated equipment are required to produce technical manuals for the aircraft operator's use. The specification is accepted by the INTERNATIONAL AIR TRANSPORT ASSOCIATION as the basis for international standardization. Abbrev *ATA*.

air trap (*Build*) A trap which, by a water seal, prevents foul air from rising from eg sinks, wash basins, drains and sewers. Also *drain trap*, *stench trap*, *U-bend*.

air valve (*Build*) A valve in a spray gun which controls the flow of air by the operation of the trigger.

air volume spraying (*Build*) A method of spray application which involves higher volume and lower pressure of air than high-pressure air spraying.

air wall (*NucEng*) Wall of an ionization chamber designed to give same ionization intensity inside the chamber as in open space. This means the wall is made of elements with atomic numbers similar to those for air constituents.

airway (*Aero*) A specified three-dimensional corridor (the lower as well as the upper boundary being defined) between CONTROL ZONES which may only be entered by aircraft in radio contact with AIR-TRAFFIC CONTROL.

airway (*MinExt*) Underground passage used mainly for ventilation.

airworthy (*Aero*) (1) Fit for flight aircraft, aero-engine, instrument or equipment. (2) Complying with the regulations laid down for ensuring the fitness of an aircraft for flight. (3) Possessing a CERTIFICATE OF AIRWORTHI-NESS.

Airy disk (*ImageTech*) Circular image of a point source of light formed by a lens. After Sir George Airy.

Airy points (*Phys*) The optimum points for supporting a beam horizontally to minimize the bending deflection. The distance apart of the points is equal to $l/\sqrt{(n^2-1)}$ where l is the length of the beam and n the number of supports.

Airy's differential equation (*MathSci*) One of the form

$$\frac{d^2 y}{dx^2} = xy$$

Its general solution can be expressed in terms of BESSEL FUNCTIONS of order $\pm\frac{1}{3}$:

$$y(x) = A\sqrt{x}J_{\frac{1}{3}}\left(\tfrac{2}{3}ix^{\frac{3}{2}}\right) + B\sqrt{x}J_{-\frac{1}{3}}\left(\tfrac{2}{3}ix^{\frac{3}{2}}\right)$$

Airy's integral (*Phys*) The factor 1·22, by which the dimensions of the diffraction pattern produced by a slit must be multiplied to obtain the dimensions of the pattern due to a circular aperture.

Airy spirals (*Phys*) The spiral interference patterns produced when quartz, cut perpendicularly to the axis, is examined in convergent light circularly polarized.

aisle (*Arch*) (1) A side division of the NAVE or other part of a church or similar building, generally separated off by pillars. (2) Loosely, any division of a church, or a small building attached, including a passage between rows of seats.

akaryote (*BioSci*) A cell lacking a nucleus.

akathisia (*Psych*) A psychological condition characterized by agitation and a frequent desire to alter the posture. Also *acathisia*.

akene (*BioSci*) See ACHENE.

Akermanite (*Min*) The calcium–magnesium end-member, $Ca_2MgSi_2O_7$, of the melilite group of minerals.

akinesia (*Med*) Absence of, or diminished spontaneous movement characteristic of diseases such as PARKINSON-ISM.

akinete (*BioSci*) A non-motile thick-walled resting spore containing food reserves, formed without division by the direct modification of a vegetative cell in some Cyano-bacteria.

Akulon (*Plastics*) TN for Dutch nylon-6 polymer used for mouldings and fibres.

Al (*Chem*) Symbol for ALUMINIUM.

Ala (*Chem*) Symbol for ALANINE.

ala (*BioSci*) Any flat, wing-like process or projection, esp of bone. Adj *alar, alary*.

alabaster (*Min*) A massive form of gypsum, often pleasingly blotched and stained. $CaSO_4{\cdot}2H_2O$. Because of its softness it is easily carved and polished, and is widely used for ornamental purposes. *Oriental alabaster* (also *Algerian onyx, onyx marble*) is a beautifully banded form of stalagmitic CALCITE.

alalia (*Med*) See APHONIA.

alanine (*Chem*) 2-aminopropanoic acid. $CH_3C(NH_2)COOH$. The *L*- or *S*-isomer is a common constituent of proteins. Symbol Ala, short form A.

ALARA (*NucEng*) Abbrev for *as low as reasonably achievable*; said eg of radiation levels or decontamination.

alarm flag (*ElecEng*) See FLAG INDICATOR.

alarm reaction (*Psych*) Non-verbal cues that reveal emotional state. The first stage of the GENERAL ADAPTATION SYNDROME.

alary muscles (*BioSci*) In insects, pairs of striated muscles arising from the terga and spread out fanwise over the surface of the dorsal diaphragm. Also *aliform muscles*.

alaskite (*Min*) Leucocratic variety of alkali feldspar granite.

ala spuria (*BioSci*) See BASTARD WING.

alastrim (*Med*) A mild form of smallpox differing from it in certain features, mainly non-fatal. Also *Variola minor*.

alate (*BioSci*) (1) Winged; applied to stems when de-current leaves are present. (2) Having a broad lip (esp of shells).

Albada viewfinder (*ImageTech*) Viewfinder with a lightly silvered plano-concave objective which reflects frame marks placed on the eyepiece and at the focus of the mirror. Also *bright-line viewfinder*.

albedo (*Phys*) (1) A measure of the reflecting power of a non-luminous body, such as the surface of a planet, expressed as the ratio of energy reflected in all directions to total incident energy. (2) Ratio of the neutron flow density out of a medium free from sources, to the neutron flow density into it, ie reflection factor of a surface for neutrons.

Albers–Schönberg disease (*Med*) See OSTEOPETROSIS.

albert (*Paper*) A former standard size of notepaper, 192×102 mm (6×4 in).

albertite (*Min*) A pitch-black solid bitumen of the asphaltite group.

Albian (*Geol*) A stage of the Cretaceous system, comprising the rocks between the Aptian stage below and the Cenomanian stage above. See MESOZOIC.

albinism (*BioSci*) The state of being an ALBINO.

albino (*BioSci*) An organism deficient in pigment in eg hair, skin and eyes or, in the case of plants, chlorophyll. Adj *albinotic*.

albite (*Min*) The end-member of the plagioclase group of minerals. Ideally a silicate of sodium and aluminium, but commonly contains small quantities of potassium and calcium in addition, and crystallizes in the triclinic system.

albitization (*Geol*) In igneous rocks, the process by which a soda–lime feldspar (plagioclase) is replaced by albite (sodium–feldspar).

albumen (*BioSci*) White of egg containing a number of soluble proteins, mainly ovalbumin. Adj *albuminous*. See ALBUMIN.

albumen process (*Print*) A process in which dichromated albumen is used as a light-sensitive coating when preparing SURFACE PLATES for lithography and line blocks for relief printing.

albumin (*BioSci*) A general term for proteins soluble in water as distinct from saline. Specific albumins are designated by their sources, eg ovalbumin from egg white (ALBUMEN), serum albumin from blood serum. Used in food products as aerating and stabilizing agents; also as a fining to precipitate out tannins from wine.

albuminous (*BioSci*) Endospermic.

albuminous cell (*BioSci*) Specialized parenchyma cell in gymnosperm phloem, associated with a sieve cell but not originating from the same precursor cell. Cf COMPANION CELL.

albuminuria (*Med*) Albumin in the urine.

alburnum (*BioSci*) See SAPWOOD.

Aichlor process (*Chem*) A process used in refining lubricants, by removal of impurities with aluminium chloride.

Alclad (*Eng*) Composite sheets consisting of an alloy of the DURAL type (to give strength) coated with pure aluminium (to give corrosion resistance).

alcohol (*Chem*) A general term for compounds formed from hydroxyl groups attached to carbon atoms in place of hydrogen atoms; in particular, ETHANOL. The general formula is ROH, wherein R signifies an aliphatic radical. Hydroxyl groups attached to aromatic rings give PHENOLS. See ABSOLUTE ALCOHOL, METHANOL.

alcohol fuel (*Autos*) Volatile liquid fuel consisting wholly or partly of alcohol, able to withstand high-compression ratios without detonation.

alcoholic fermentation (*BioSci*) A form of anaerobic RESPIRATION in which a sugar is converted to alcohol and carbon dioxide. Alcoholic fermentation by yeasts is important in baking, brewing and wine making.

alcoholism (*Psych*) Disease produced by addiction to alcohol, manifesting itself in a variety of psychotic disorders, eg hallucinations, delirium tremens.

alcoholometry (*Chem*) The quantitative determination of alcohol in aqueous solutions.

Alcomax (*Eng*) UK equivalent of ALNICO permanent magnet alloy.

Alcor (*Astron*) A white star in the constellation Ursa Major which forms part of the Plough and is the optical double star of MIZAR. Distance 125 pc. Also *80 Ursae Majoris*.

Aldebaran (*Astron*) A bright red giant irregular variable star in the constellation Taurus, which is both a visual and an optical binary. Distance 21 pc. Also *Alpha Tauri*.

aldehyde acids (*Chem*) Products of the partial oxidation of dihydric alcohols, containing both an aldehyde group and a carboxyl group.

aldehyde resins (*Plastics*) Highly polymerized resinous condensation products of aldehydes obtained by treatment of aldehydes with strong caustic soda.

aldehydes (*Chem*) *Alkanals*. RCHO. A group of compounds containing the CO— radical attached to both a hydrogen atom and a hydrocarbon radical.

alder (*For*) A tree (*Alnus*) producing a straight-grained fine-textured hardwood noted for its durability under water. It is used for cabinet making, plywood, shoe heels, clogs, bobbins, wooden cogs and small turned items. Density 530 kg m^{-3}.

aldimines (*Chem*) Condensation products of phenols with hydrocyanic acid, formed in the presence of gaseous hydrogen chloride.

aldohexoses (*Chem*) The most important group of MONOSACCHARIDES, including GLUCOSE and GALACTOSE. All have a formula which can be expressed as $OHCH_2(CHOH)_4CHO$.

aldol (*Chem*) *2-Hydroxybutanal*. $H_3CCH(OH)CH_2CHO$. A condensation product of acetaldehyde.

aldolase (*BioSci*) The enzyme that catalyses the cleavage of fructose-1-6-diphosphate into glyceraldehyde-3-phosphate and dihydroxyacetone.

aldol condensation (*Chem*) The condensation of two aldehyde molecules in such a manner that the oxygen of the first molecule reacts with the hydrogen of the second, forming a hydroxyl group, with the simultaneous formation of a new link between the two carbon atoms. Water is eliminated.

aldoses (*Chem*) A group of MONOSACCHARIDES with an aldehydic constitution, eg glucose. See ALDOHEXOSES.

aldosterone (*Chem, Med*) A potent mineralocorticoid secreted by the zona glomerulosa of the adrenal cortex which promotes the retention of sodium ions and water. See panel on STEROID HORMONES.

aldoximes (*Chem*) *Hydroxyimino alkanes*. A group of compounds in which the oxygen of the aldehyde group is substituted by the radical $=NOH$, derived from hydroxylamine H_2NOH and an aldehyde by dehydration. The general formula is $RCH=NOH$.

aldrin (*Chem*) A chloro-derivative of naphthalene used as a contact insecticide, incorporated in plastics to make cables resistant to termites; persistent toxicity; formerly used in agriculture, esp against wireworm.

alecithal (*BioSci*) Of ova, having little or no yolk.

alemtuzumab (*Pharmacol*) A humanized monoclonal antibody, used in the treatment of B-cell chronic lymphocytic leukaemia that is refractory to cytopathic drugs such as fludarabine.

alendronic acid (*Pharmacol*) A bisphosphonate drug used for the treatment of osteoporosis. The drug is adsorbed onto hydroxyapatite crystals in bone, reducing the increased bone turnover which is a feature of the disease.

aleph-0 (*MathSci*) Cardinal number of the set of natural numbers; the first infinite cardinal. Written ℵ. Also *aleph-null*.

alert box (*ICT*) A box appearing on the screen of a computer to warn the user of a problem.

aleurone (*BioSci*) Reserve protein occurring in seed granules, usually in the outermost layer (the *aleurone layer*) of the endosperm, eg in cereals and other grasses.

Aleutian disease (*Vet*) A chronic, fatal disease of mink, with characteristic changes in the liver, kidneys and other organs; it occurs esp in mink homozygous for the aleutian gene controlling fur colour.

alexandrite (*Min*) A variety of chrysoberyl, the colour varying, with the conditions of lighting, between emerald green and red.

alexia (*Med*) Word blindness; loss of the ability to interpret written language due to a lesion in the brain.

alexin (*BioSci*) Obsolete term for COMPLEMENT.

Alford antenna (*ICT*) Antenna comprising a vertical cylindrical tube with longitudinal slots, often used to transmit very high or ultrahigh frequency.

algae (*BioSci*) A non-taxonomic term used to group several phyla of the lower plants, including the Rhodophyta (red algae), Chlorophyta (green algae), Phaeophyta (brown algae) and Chrysophyta (diatoms). Many algae are unicellular or consist of simple undifferentiated colonies, but red and brown algae are complex multicellular organisms, familiar to most people as seaweeds. Blue-green algae are a totally separate prokaryotic group, more correctly known as CYANOBACTERIA (Cyanophyta).

algae poisoning (*Vet*) A form of poisoning affecting farm livestock in the USA due to the ingestion of toxins in decomposing algae; characterized by nervous symptoms and death.

algal bloom (*BioSci*) A sudden growth of algae in the sea or a lake that occurs when the algae multiply faster than they are being eaten. Can also result from the enrichment of waters in plant nutrients.

algal corrosion (*Aero*) Impairment of structure and systems by algae and other micro-organisms.

algal layer (*BioSci*) A layer of algal cells lying inside the thallus of a HETEROMEROUS lichen. Also *algal zone, gonidial layer, gonimic layer*.

algebra (*MathSci*) Originally the abstract investigation of the properties of numbers by means of symbols (x, y, etc); eg the solution of equations, the summation of series, permutations and combinations and matrices. More recently extended to include the general study of sets with operations; eg mappings, groups, rings, integral domains, fields and vector spaces. See MATHEMATICS.

algebraic (*MathSci*) Of a function, number, operation, etc: able to be expressed in terms of polynomials with rational coefficients, ie able to be defined by a finite number of algebraic operations, including root extraction. Thus $\sqrt{2}$ is an algebraic number, but π is not.

Algerian onyx (*Min*) See ALABASTER.

algesis (*Med*) The sense of pain.

alginate (*FoodSci*) Principal carbohydrate component with their esters and metallic salts of the brown seaweeds *Ascophyllum* (British Isles), *Laminaria* (Europe, Japan, N America) and *Macrocystis* (US). Only the alkali metal alginates are soluble in water. Used in the food industry as binding, thickening, stabilizing or gelling agents.

alginic acid (*Chem*) *Norgine*. $C_3H_8O_6$. Occurs both in the free state and the calcium salt in the larger brown algae (*Phaeophyceae*). The sodium salt gives a very viscous solution in water even at a concentration of only 2%, and is used in the dyeing, textile, plastics and explosives industries, in making waterproofing and insulating materials, foodstuffs, adhesives, cosmetics, and in medicine, for its sodium absorption ability by a cation exchange reaction.

Algol (*Astron*) A star in the constellation Perseus which is the prototype of the eclipsing binary, where one component passes in front of the other at each revolution, causing an eclipse and a systematic fluctuation of magnitude. Also *Beta Persei*.

ALGOL-68 (*ICT*) Abbrev for *algorithmic language 1968*. Very powerful language with structured programming features. Like the earlier ALGOL-60, designed to aid the programming of algorithms.

algology (*BioSci*) The study of algae.

algorithm (*ICT*) A set of rules that specify a sequence of actions to be taken to solve a problem. Each rule is precisely and unambiguously defined so that in principle it can be carried out by machine. See FORMAL LANGUAGE THEORY.

algorithm (*MathSci*) A procedure that generates an infinite sequence of terms by applying the same formula repeatedly.

aliasing (*Acous*) Error in making real-time spectra of short signals or of directivity in sound fields. Caused by insufficient number of data points.

aliasing (*ImageTech*) Image imperfections resulting from limited detail in a RASTER display, eg resulting in diagonal lines appearing stepped.

A-licence (*Aero*) Basic private pilot's licence.

alicyclic (*Chem*) Abbrev for *aliphatic–cyclic*; used to describe a ring compound not containing aromatic groups.

alidade (*Surv*) An accessory instrument used in plane-table surveying, consisting of a rule fitted with sights at both ends, which gives the direction of objects from the plane-table station. Also *sight rule*.

alien (*BioSci*) Species of plant or animal believed on good evidence to have been introduced by humans and now more or less naturalized.

alienation (*Psych*) Generally a feeling of strangeness or separation from others but in existentialist psychology also has the connotation of dissociation from one's inner self.

alien tones (*Acous*) Frequencies, harmonic and sum-and-difference products, introduced on sound reproduction because of non-linearity in some part of the transmission path.

aliform muscles (*BioSci*) See ALARY MUSCLES.

alignment (*CivEng*) (1) The setting in line (usually straight) of eg successive lengths of a railway which is to be constructed. (2) The plan of a road or earthwork.

alignment (*Eng*) The setting in a true line of a number of points, eg the centres of the bearings supporting an engine crankshaft.

alignment (*ICT*) Adjustment of preset tuned circuits to give optimum performance.

alignment (*Phys*) A process of orientation of eg electric or magnetic dipoles when acted on by an external field. During magnetization, the alignment of domains is changed by the magnetizing field.

alignment chart (*MathSci*) See NOMOGRAM.

alimentary (*Med*) Pertaining to the nutritive functions or organs.

alimentary canal (*Med*) The passage from the mouth to the anus which receives, digests and assimilates nutrients from foodstuffs; the digestive tract; the gut. Also *alimentary tract*.

alimentary system (*Med*) All the organs connected with digestion, absorption and nutrition, comprising the digestive tract and associated glands and masticatory mechanisms.

alimentary tract (*Med*) See ALIMENTARY CANAL.

aliphatic compounds (*Chem*) Methane derivatives of fatty compounds; open-chain or ring carbon compounds not having aromatic properties.

aliquot (*Phys*) A small sample of material assayed to determine the properties of the whole, eg in process control, the representative fraction whose quantitative analysis gives information on the assay grade. Term often applied to radioactive material. See ALIQUOT PART.

aliquot part (*MathSci*) A number or quantity which exactly divides a given number or quantity.

aliquot part (*MinExt*) In sampling for process control, a representative fraction whose quantitative analysis gives information on the assay grade.

aliquot scaling (*Acous*) In a piano, the provision of extra wires above the normal wires. These are not struck, but are tuned very slightly above the octave of the struck strings below, so that by sympathetic vibration the musical quality of the note is enhanced. Also *aliquot tuning*.

Alismatidae (*BioSci*) A subclass or superorder of monocotyledons, consisting of aquatic and semi-aquatic herbs. Typically, their perianth segments are free, sometimes differentiated into sepals and petals; they are mostly apocarpous. The pollen is trinucleate, and the mature seeds mostly lack endosperm. There are c.500 spp in 16 families.

alisphenoid (*BioSci*) A wing-like cartilage bone of the vertebrate skull, forming part of the lateral wall of the cranial cavity, just in front of the foramen lacerum; one of a pair of dorsal bars of cartilage in the developing vertebrate skull, lying in front of the basal plate, parallel to the trabeculae; one of the sphenolateral cartilages.

alite (*Chem*) Tricalcium silicate, $CaOCa_2SiO_4$. The major constituent (typically 55% by weight) of Portland cement, it hydrates at a medium rate during the setting reaction. See panel on CEMENT AND CONCRETE.

alizarin (*Chem*) 1,2-dihydroxyanthraquinone. $C_{14}H_6O_2(OH)_2$. One of the most important natural and synthetic dyes; red prisms, or needles, mp 289°C, soluble in alcohol and ether, very slightly soluble in water, soluble in caustic soda; insoluble stains are formed with the oxides of aluminium, tin, chromium and iron. Alizarin can be nitrated and forms the basis of a series of other dyestuffs.

alkali (*Chem*) A hydroxide which dissolves in water to form an alkaline or basic solution which has pH>7 and contains hydroxyl ions (OH^-).

alkali basalt (*Min*) A variety of basalt with normative nepheline.

alkalic (*Geol*) Denoting rocks that contain either feldspathoids or alkali amphiboles or pyroxenes, or normative feldspathoids or acmite.

alkali disease (*Vet*) (1) *Western duck sickness*. A form of botulism causing death of wild duck and other waterfowl in North America, due to the ingestion of vegetation contaminated with toxin produced by *Clostridium botulinum*, type C. (2) A chronic disease of domestic animals, characterized by emaciation, stiffness and anaemia, due to an excess of selenium in the diet. An acute form of the disease is called *blind staggers*.

alkali feldspar (*Min*) A member of the feldspar group of minerals comprising orthoclase, microcline, sanidine, perthite, anorthoclase and the plagioclase end-member albite (anorthite 0–5%).

alkali granite (*Geol*) Granite containing alkali amphibole or pyroxene. It should not be used as a synonym for alkali feldspar granite.

alkali metals (*Chem*) The elements lithium, sodium, potassium, rubidium, caesium and francium, all metals in the first group of the periodic table. In most compounds they occur as univalent ions.

alkalimetry (*Chem*) The determination of alkalinity by titration with a standard solution of acid as in volumetric analysis. See TITRATION, VOLUMETRIC ANALYSIS.

alkaline earth metals (*Chem*) The elements calcium, strontium, barium and radium, all divalent metals in the second group of the periodic system.

alkaline phosphatase (*BioSci*) Enzyme catalysing cleavage, optimally at alkaline pH, of inorganic phosphate from a wide variety of phosphate esters. Commonly conjugated with antibodies for use in indirect immunoassay; the phosphatase catalyses the deposition of dye at the site of the bound antibody. Abbrev *AP*.

alkalinity (*BioSci*) Of eg a lake, $[HCO_3^-]+2[CO_3^{2-}]+[OH^-]-[H^+]$, where square brackets indicate molar concentrations.

alkalinity (*Chem*) The extent to which a solution is alkaline. See PH. Cf ACIDITY.

alkaloids (*Chem*) Natural organic bases found in plants; characterized by their specific physiological action and toxicity; used by many plants as a defence against herbivores, particularly insects. Alkaloids may be related to various organic bases, the most important ones being pyridine, quinoline, *iso*quinoline, pyrrole and other more complicated derivatives. Most alkaloids are crystalline solids, others are volatile liquids, and some are gums. They contain nitrogen as part of a ring, and have the general properties of amines.

alkalosis (*Med*) A decrease in hydrogen ion concentration in blood and tissue. May be respiratory owing to excessive loss of CO_2 in the lungs (hyperventilation) or metabolic owing to loss of non-volatile acids from the body, ie vomiting.

alkane (*Chem*) General name of hydrocarbons of the methane series, of general formula C_nH_{2n+2}.

methane ethane

alkanes

alkaptonuria (*Med*) The congenital absence of homogentisic acid oxidase, an enzyme that breaks down tyrosine and phenylalanine. In homozygotes the accumulation of homogentisic acid causes brown pigmentation of skin and eyes and damage to joints; the urine blackens on standing.

Alkathene (*Chem*) TN for POLYETHYLENE.

alkene (*Chem*) *Olefin.* General name for unsaturated hydrocarbons of the ethene series, of general formula C_nH_{2n}.

alkyd resins (*Chem*) Polyester thermosets derived from glycerol and phthalic anhydride (glyptal resins). Also includes diallyl esters and various polyesters used as resin binders in alkyd moulding materials. Widely used for coatings. See panel on THERMOSETS.

alkyl (*Chem*) A general term for monovalent aliphatic hydrocarbon radicals.

alkylating drug (*Pharmacol*) A cytotoxic drug which acts by damaging DNA and interfering with cell replication. Cyclophosphamide, chlorambucil, busulphan and mustine are common examples.

alkylene (*Chem*) A general term for divalent hydrocarbon radicals.

alkyne (*Chem*) An aliphatic hydrocarbon with a triple bond. The simplest is ethyne or acetylene, HC≡CH.

allanite (*Min*) A cerium-bearing epidote occurring as an accessory mineral in igneous and other rocks.

allantoic (*BioSci*) See ALLANTOIS.

allantoin (*BioSci*) A urinary excretion product resulting from purine metabolism or from the oxidation of uric acid. It is found in allantoic and amniotic fluids, in urine of certain mammals, and is also excreted by certain insects and gastropods. Used in the treatment of wounds, skin irritation and in various oral hygiene preparations, etc.

allantois (*BioSci*) In the embryos of higher vertebrates, a sac-like diverticulum of the posterior part of the alimentary canal, having respiratory, nutritive or excretory functions. It develops to form one of the embryonic membranes. Adj *allantoic.*

Allan valve (*Eng*) Once popular slide-valve design with an internal passage designed to reduce valve travel and wear.

all-burnt (*Aero*) The moment at which the fuel of a missile or spacecraft is completely consumed.

alleghanyite (*Min*) A hydrated manganese silicate, crystallizing in the monoclinic system.

allele (*BioSci*) Abbrev for *allelomorph.* Any one of the alternative forms of a specified gene. Different alleles usually have different effects on the phenotype. Any gene may have several different alleles, called *multiple alleles.* Genes are *allelic* if they occupy the same LOCUS.

all-electric signalling (*ElecEng*) A railway system in which the signals and points are controlled and operated electrically. See ELECTROPNEUMATIC SIGNALLING.

allelic exclusion (*BioSci*) The process whereby one or more loci on one of the chromosome sets in a diploid cell is inactivated (or destroyed) and not expressed in that cell or its clonal derivatives. For instance, in mammals one of the X chromosome pairs of females is inactivated early in development (see LYON HYPOTHESIS) so that individual cells express only one allelic form of the product of that locus. Since the choice of chromosome to be inactivated is random, different cells express one or other of the X chromosome products resulting in mosaicism.

allelomorph (*BioSci*) See ALLELE.

allelopathy (*BioSci*) The condition when one strain is harmful to another of the same species, eg in plants, by the production and release of a chemical inhibitor, often a terpenoid or phenolic.

allemontite (*Min*) A mineral consisting of a solid solution of antimony and arsenic.

Allen cone (*MinExt*) Conical tank used for continuous sedimentation of liquids at constant level, the solids being removed from the base of the cone and the clear liquid drawn off from the top.

Allende meteorite (*Astron*) A meteorite which fell near the village of Pueblito de Allende, Mexico, in February 1969. It scattered 5 tonnes of material, rich in carbon, and with a composition believed to typify the primitive solar system.

allene (*Chem*) *Propadiene, dimethylene methane.* $CH_2=C=CH_2$. Obtained by the electrolysis of itaconic acid.

Allen equation (*MinExt*) An equation applied to sedimentation of finely ground particles intermediate between streamline and turbulent in settling mode.

$$\rho = Kr^np\mu^{2-n_vn}$$

where ρ is fluid resistance, K is a constant for a given shape and velocity of fall, r is the radius of an equivalent sphere, n is a coefficient of velocity, v, p is the density of fluid and μ is the kinematic viscosity (given by b/p where b is the absolute viscosity).

allenes (*Chem*) Generic term for a series of non-conjugated and di-olefinic hydrocarbons, of which ALLENE is the first, and which have the general formula $C_{2n}H_{2n-2}$. They consist mostly of colourless liquids with strong garlic odour.

Allen's law (*BioSci*) An evolutionary generalization stating that feet, ears and tails of mammals tend to be shorter in colder climates when closely allied forms are compared.

Allen's loop test (*ElecEng*) A modification of the VARLEY LOOP TEST for localizing a fault in an electric cable; it is particularly suitable for high-resistance faults in short lengths of cable.

allergen (*BioSci*) Antigenic substances that provoke an allergic response (see ALLERGY). Commonly used to describe those which cause immediate type HYPERSENSITIVITY REACTIONS such as pollens or insect venoms.

allergic (*BioSci, Med*) Reacting in an abnormally sensitive manner to a substance. See ALLERGY.

allergic rhinitis (*Med*) An allergic response to inhaled substances that causes swelling of mucous membranes of the upper respiratory tract. A common seasonal form is hay fever, the allergen being pollen from grasses or trees.

allergy (*BioSci, Med*) (1) Showing altered responsiveness to an antigen as the result of previous contact with that antigen. Responsiveness is usually increased, but can be decreased. (2) The reaction of the body to a substance to which it has become sensitive, characterized by oedema, inflammation and destruction of tissue.

alliaceous (*BioSci*) Looking or smelling like an onion.

alligator (*MinExt*) See JAW BREAKER.

alligatoring (*Build*) See CROCODILING.

all-insulated switch (*ElecEng*) See SHOCKPROOF SWITCH.

allithium (*Aero*) Aluminium–lithium alloys.

all-moving tail (*Aero*) A one-piece TAILPLANE, also controlled by the pilot as is the ELEVATOR. Also *flying tail*, *stabilator*. See T-TAIL.

allo- (*BioSci*) Prefix used to indicate eg a gene product or tissue from a different individual of the same species.

allo- (*Chem*) Prefix used to show that a compound is a stereoisomer of a more common compound. See STEREO-ISOMERISM.

allo- (*Genrl*) Prefix from Gk *allos*, other.

alloantibody (*BioSci*) An antibody raised in one member of a species that recognizes genetic determinants in other individuals of the same species.

alloantigen (*BioSci*) An antigen, often of the *histocompatibility complex*, that differs between individuals of the same species. Such antigenic differences are the consequence of expression of different alleles and as a result there will be an immune response to allografts, leading to rejection without the use of immunosuppressive drugs.

allobar (*Phys*) A mixture of isotopes of an element differing in proportion from that naturally occurring.

allochromatic (*Electronics*) Having photoelectric properties which arise from micro-impurities, or from previous specific irradiation.

allochromy (*Phys*) Fluorescent reradiation of light of different wavelength from that incident on a surface. See STOKES' LAW.

allochthonous (*Geol*) A term used to describe a block of rock that is exotic to its environment, eg a block of limestone that has slid down a submarine slope into a muddy environment or a tectonically moved block of rock.

allogamy (*BioSci*) Fertilization involving pollen and ovules from: (1) different flowers (whether on the same plant or not), including GEITONOGAMY and XENOGAMY; or (2) genetically distinct individuals of the same species (ie from another GENET). See CROSS FERTILIZATION, CROSS POLLINATION. Cf AUTOGAMY.

allogeneic (*BioSci*) Said of individuals of the same species, or cell lines, that are not genetically identical. Cf ISOGENEIC, XENOGENEIC.

allograft (*BioSci*) See HOMOGRAFT.

allomeric (*BioSci*) Having the same crystalline form but a different chemical composition.

allometry (*BioSci*) The relationship between the growth rates of different parts of an organism.

allomone (*BioSci*) A chemical signal produced by one species of animal that influences the behaviour of members of another to the advantage of the signaller. Cf PHEROMONE.

allopatric (*BioSci*) A term describing two species or populations not growing in the same geographical area; unable to interbreed by reason of distance or geographical barrier. Cf SYMPATRIC.

allopatric speciation (*BioSci*) The accumulation of genetic differences in an isolated population leading to the evolution of a new species.

allophanate (*Chem*) Type of group occurring in polyurethanes, formed by reaction of excess isocyanate with a chain urethane group, so giving cross-links in the resultant material. Cf BIURET.

allophane (*Min*) Hydrous aluminium silicate, apparently amorphous.

allopolyploid (*BioSci*) A polyploid of hybrid origin containing sets of chromosomes from two or more different species, often self-fertile but not interbreeding with the parental species. Allopolyploidy is important in speciation and in the evolution of some crop plants, eg wheat, brassicas. See AMPHIDIPLOID. Cf AUTOPOLYPLOID.

allopurinol (*Med*) A xanthine oxidase inhibitor that blocks formation of uric acid from xanthine; used for the treatment and prevention of gout.

all-or-nothing response (*BioSci*) Response to stimuli that is either with full intensity or not at all; occurs in many irritable protoplasmic systems: eg in lower animals, nematocysts; in higher animals, nerve fibres, cardiac and voluntary muscle fibres.

allose (*Chem*) An aldohexose, an optical stereoisomer of glucose.

allosomes (*BioSci*) Chromosomes that can be distinguished from autosomes by their morphology and behaviour. Also *accessory chromosomes*, *heterochromosomes*, *sex chromosomes*.

allosteric protein (*BioSci*) A protein that alters its three-dimensional conformation as a result of the binding of a smaller molecule, often leading to altered activity, eg of an enzyme.

allosteric site (*BioSci*) A site on a protein that, when it binds the appropriate ligand, induces a conformational change. In the case of an enzyme the allosteric site is distinct from the active site and the induced conformational change affects the enzymic activity.

alloter (*ICT*) A uniselector used to improve the efficiency of distribution of line finders, by automatically pre-selecting and pre-connecting the first available line finder in the group to which it has access.

allotetraploid (*BioSci*) See AMPHIDIPLOID.

allotriomorphic (*Geol*) A textural term used for igneous rocks describing crystals which show a form related to surrounding previously crystallized minerals rather than to their own rational faces. Cf IDIOMORPHIC. Also *anhedral*.

allotrope (*Chem*) See ALLOTROPY.

allotropic (*Psych*) Descriptive term for people who are particularly concerned with the well-being of others.

allotropous flower (*BioSci*) A flower in which the nectar is accessible to all kinds of insect visitors.

allotropy (*Chem*) The existence of an element in two or more solid, liquid or gaseous forms, in one phase of matter, called *allotropes*.

allotype (*BioSci*) (1) In taxonomy, an additional type specimen of the opposite sex to the original type specimen or HOLOTYPE. (2) An animal or plant fossil selected as a species or subspecies, illustrating morphological details not shown in the holotype. (3) In immunology, used to describe identifiable differences between immunoglobulin molecules that are inherited as alleles of a single genetic locus. They are due to single amino acid substitutions in light or heavy chains, and are useful in population studies.

allowable deficiencies (*Aero*) Aircraft systems or certain items of their equipment, tabulated in the flight or operating manual, which even if unserviceable will not prevent an aircraft from being flown or create a hazard in flight.

allowances (*Aero*) Fuel reserves, usually specified as time factors under certain conditions, as distance plus descent, or as a percentage (by weight or volume) of the cruising fuel for a given stage.

allowed band (*Phys*) Range of energy levels permitted to electrons in a molecule or crystal. These may or may not be occupied.

allowed transition (*Phys*) Electronic transition between energy levels which is not prohibited by any quantum selection rule.

alloy (*Eng*) Mixture of atomic species exhibiting metallic properties and usually prepared by adding other metals or non-metals to solvent metal in the liquid state, but may also be formed from sintered powders or by intimate mixing by mechanical means. They may be compounds, eutectic mixtures or solid solutions. If the component metals crystallize in the same form and their atomic radii are within 15% then, provided there is no significant electrochemical difference, complete intersolubility in the solid state may result. However, normally solid solubility is restricted and a number of different solid phases may exist in the system (see PHASE DIAGRAM). Alloys are stronger than their constituent metals and may exhibit a variety of hardening mechanisms. Solid solutions exhibit lower electrical conductivity and reduced ductility compared with the pure components. The term alloy is also used to describe molecular mixtures and solutions in plastics and ceramics.

alloy cast-iron (*Eng*) CAST-IRON containing alloying elements in addition to carbon and the normal low levels of manganese and silicon, usually some combination of nickel, chromium, copper and molybdenum. These elements may be added to increase the strength of ordinary irons, to facilitate heat treatment, or to obtain martensitic, austenitic or ferritic irons.

alloying (*Eng*) The process of making an alloy.

alloy junction (*Electronics*) A junction formed by alloying one or more impurity metals with a semiconductor. Small buttons of impurity metal are placed at desired locations on a semiconductor wafer; heating to melting point and rapidly cooling again produces regions of P-TYPE CONDUCTION or N-TYPE CONDUCTION, according to choice of impurity. Also *fused junction*.

alloy reaction limit (*Eng*) Concentration in alloy of a specific component, below which corrosion occurs in a given environment.

alloy steel (*Eng*) A steel to which elements not present in carbon steel have been added, or in which the content of manganese or silicon is increased above that in carbon steel. See HIGH-SPEED STEEL, NICKEL STEEL, STAINLESS STEEL and panel on STEELS.

allozymes (*BioSci*) Different forms of an enzyme specified by allelic genes.

all-pass network (*ICT*) A network that introduces a specified phase-shift response without appreciable attenuation for any frequency.

Allström relay (*ElecEng*) See RELAY.

alluvial mining (*MinExt*) Exploitation of alluvial or placer deposits. Minerals thus extracted include tin, gold, gemstones, rare earths and platinum. Term embraces beach deposits, eluvials, riverine and offshore workings.

alluvial values (*MinExt*) Values shown by panning or assay to be recoverable from an alluvial deposit.

alluvium (*Geol*) Sand, silt and mud deposited by a river or floods; geologically recent in age.

allyl alcohol (*Chem*) 2-propen-1-ol. $H_2C=CHCH_2OH$. An unsaturated primary alcohol, present in wood spirit, made from glycerine and oxalic acid. Mp $-129°C$, bp $96°C$, rel.d. 0·85, of very pungent odour; an intermediate for organic synthesis.

allyl chloride (*Chem*) $ClCH_2CH—CH_2$. Important intermediate in the manufacture of synthetic glycerol. Formed by the high-temperature (400–600°C) chlorination of propene.

allylene (*Chem*) Propyne. $CH_3C≡CH$.

allyl group (*Chem*) The unsaturated monovalent aliphatic group $H_2C=CHCH_2—$.

allyl resin (*Chem*) A resin formed by the polymerization of chemical compounds of the ALLYL GROUP, eg CR 39.

allyl sulphide (*Chem*) $(CH_2=CHCH_2)_2S$. Oil of garlic; bp 139°C; colourless liquid, found in garlic and largely responsible for its odour. It possesses antiseptic qualities.

Almagest (*Astron*) The Arabic form (meaning 'The Greatest') of the title of Claudius Ptolemy's great astronomical treatise, *The Mathematical Syntaxis*, written in Greek around AD 140.

almandine (*Min*) Iron–aluminium garnet, occurring in mica-schists and other metamorphic rocks. Commonly forms well-developed crystals, often with 12 or 24 faces.

almandine spinel (*Min*) See RUBY SPINEL.

almond oil (*Chem*) Oil used for fruit essences, in perfumery and soap making. Two grades are known: *bitter almond oil* and *sweet almond oil*.

almucantar (*Astron*) A small circle of the celestial sphere parallel to the horizontal plane. The term is also applied to an instrument for measuring altitudes and azimuths.

Alnico (*Eng*) US TN for a high-energy permanent magnet material, an alloy of aluminium, nickel, cobalt, iron and copper.

alnöite (*Min*) A dark lamprophyre with phenocrysts of mica, olivine and augite in a groundmass of melilite, augite and other minerals.

alopecia (*Med*) Baldness, or hair loss.

alopecia areata (*Med*) A condition in which the hair falls out in patches, leaving smooth, shiny, bald areas.

Aloxite (*Eng*) TN designating a proprietary fused alumina and associated abrasive products.

alpaca (*Textiles*) The fine, strong hair of the alpaca (*Lama pacos*) of S America; the fabric made from such hair. This animal belongs to the camel family and is a close relative of the llama (*L. glama*) and the vicunìa (*L. vicugna*).

alpha-actinin (*BioSci*) A protein frequently associated with actin, serving as a terminus for actin filaments. It was first found in the Z bands of striated muscle.

alpha (α) activation (*Chem*) The influence of organic radicals and groups in directing the course of chemical reactions, eg the carbonyl function in ketones which leads to 2-halogenation predominating.

alpha-2-macroglobulin (*BioSci*) A large molecule with antiprotease activity found in blood plasma, which inhibits a wide range of proteases and thus also has experimental uses.

Alpha Aquilae (*Astron*) See ALTAIR.

Alpha Aurigae (*Astron*) See CAPELLA.

alpha–beta brass (*Eng*) Copper–zinc alloy containing 38–46% (usually 40%) zinc. It consists of a mixture of the α-constituent (see ALPHA BRASS) and the β-constituent (see BETA BRASS).

alpha blockers (*Pharmacol*) Alpha-adrenoceptor blockers, a class of vasodilatory drugs that block the effect of noradrenaline, which is a vasoconstrictor, on peripheral blood vessels. They are used in the management of hypertension. Examples are *doxazosin*, *phentolamine*, *prazosin*, *tamsulosin*.

Alpha Bootis (*Astron*) See ARCTURUS.

alpha brass (*Eng*) A copper–zinc alloy containing up to 38% zinc. Consists constitutionally of a solid solution of zinc in copper. Commercial alpha brasses of several compositions are made. All are used mainly for cold-working. See COPPER ALLOYS.

alpha bronze (*Eng*) A copper–tin alloy consisting of the alpha solid solution of tin in copper. Commercial forms contain 4 or 5% of tin. This alloy, which differs from gun metal and phosphor bronze in that it can be worked, is used for eg coinage, springs and turbine blades. See COPPER ALLOYS.

Alpha Canis Minoris (*Astron*) See PROCYON.

Alpha Carinae (*Astron*) See CANOPUS.

Alpha Centauri (*Astron*) The brightest star in the constellation Centaurus, actually three stars, the faintest of which, PROXIMA CENTAURI, is the nearest star to the Sun. Also *Rigil Kent*.

alpha chain (*BioSci*) A heavy chain of IGA. Alpha-chain disease is a rare disease in which the intestine is infiltrated by lymphoma which makes alpha chains but no light chains, owing to a deletion involving the site required to link the two.

alpha chamber (*Phys*) Ionization chamber for measurements of alpha radiation intensity. Also *alpha counter tube*.

alpha counter (*Phys*) Tube for counting alpha particles, with pulse selector to reject those arising from beta and gamma rays.

alpha counter tube (*Phys*) See ALPHA CHAMBER.

Alpha Crucis (*Astron*) See ACRUX.

alpha cut-off (*Electronics*) Frequency at which the current amplification of a transistor has fallen by more than 3 dB (0·7) of its low-frequency value.

Alpha Cygni (*Astron*) See DENEB.

alpha decay (*Phys*) Radioactive disintegration resulting in emission of alpha particle. Also *alpha disintegration*.

alpha decay energy (*Phys*) The sum of the kinetic energies of the alpha particle emitted and the recoil of the product atom in a radioactive decay. Also *disintegration energy*.

alpha disintegration (*Phys*) See ALPHA DECAY.

alpha diversity (*BioSci*) See DIVERSITY.

alpha emitter (*Phys*) Natural or artificial radioactive isotope which disintegrates through emission of alpha-rays.

Alpha Eridani (*Astron*) See ACHERNAR.

alphafetoprotein (*Med*) A plasma protein made by the fetus, but not by adults unless they have primary liver cancer or some other tumours in which fetal genes are expressed. Alphafetoprotein escapes into the maternal blood during pregnancy, and an abnormally high concentration at certain stages has been found to indicate that normal closure of the spinal canal of the fetus is incomplete (spina bifida).

Alpha Geminorum (*Astron*) See CASTOR.

alpha helix (*BioSci*) Important element of protein structure formed when a polypeptide chain turns regularly about itself to form a rigid cylinder stabilized by hydrogen bonding.

alpha iron (*Eng*) One of the polymorphic forms of iron, stable below 1179 K. Has a body-centred cubic lattice, and is magnetic up to 1041 K.

Alpha Leonis (*Astron*) See REGULUS.

Alpha Lyrae (*Astron*) See VEGA.

alphanumeric (*ICT*) From the set of characters consisting of the alphabet and the numerals 0–9. Most other typewriter characters are usually excluded and are reserved for programming, as CONTROL CHARACTERS or as PROMPTS. Also *alphameric*.

alphanumeric printer (*Print*) A printer which uses characters of the alphabet and the numbers 0–9.

Alpha Orionis (*Astron*) See BETELGEUSE.

alpha particle (*Phys*) Nucleus of helium atom of mass number 4, consisting of two neutrons and two protons and so doubly positively charged. Emitted from natural or radioactive isotopes. Often written α-particle. See panel on RADIATION.

Alpha Piscis Austrini (*Astron*) See FOMALHAUT.

alpha pulp (*Paper*) Wood pulp processed so that only a very small percentage of hemicellulose remains. Also *dissolving pulp*.

alpha radiation (*Phys*) ALPHA PARTICLES emitted from radioactive isotopes.

alpha-ray (*Phys*) Stream of ALPHA PARTICLES.

alpha-ray spectrometer (*NucEng*) Instrument for measuring the energy distribution of α-particles emitted by a radioactive source.

alpha rhythm (*Med*) The regular electro-encephalographic pattern of c.10 Hz obtained from a waking, inactive individual. The pattern is disrupted by changes such as falling asleep or concentrating. Also *alpha wave*.

Alpha Scorpii (*Astron*) See ANTARES.

Alpha Tauri (*Astron*) See ALDEBARAN.

alpha test (*ICT*) Testing of software that is carried out by a software company before it is released to a chosen group of potential users for BETA TEST.

Alpha Virginis (*Astron*) See SPICA.

alpha wave (*Med*) See ALPHA RHYTHM.

alpine (*BioSci*) Vegetation or plants of high mountains.

Alpine orogeny (*Geol*) The fold movements during the Tertiary period which led to the development of the Alps and associated mountain chains.

alstonite (*Min*) A double carbonate of calcium and barium.

Altair (*Astron*) A bright white optical double star in the constellation Aquila. Distance 5·1 pc. Also *Alpha Aquilae*.

altar tomb (*Arch*) A raised tomb or monument usually standing detached or in a position against a wall, and sometimes supporting an effigy. In appearance it resembles a solid altar, but it is never used as one.

altazimuth (*Astron*) A type of telescope in which the principal axis can be moved independently in altitude (swinging on a horizontal axis) and azimuth (swinging on a vertical axis). Used in very large optical telescopes and radio telescopes. See panel on ASTRONOMICAL TELESCOPE.

altazimuth (*Surv*) An instrument similar to the THEODO-LITE but generally larger and capable of more precise work.

alter ego (*Psych*) One's second self; a trusted, intimate friend, a confidant.

alternant (*MathSci*) A determinant whose elements are functions of $x_1, \ldots, x_n$ such that interchanging the variables x_i and x_k interchanges the ith row and the kth row and thus changes the sign of the determinant.

alternate (*BioSci*) Leaves, branches, etc, placed singly on the parent axis, ie not in pairs (opposite), not whorled. See ALTERNATE HOST.

alternate airfield (*Aero*) An airfield designated in a flight plan at which a pilot will land if prevented from alighting at the intended destination.

alternate angles (*MathSci*) The angles lying on either side of a transversal which cuts two straight lines, each angle having one of the lines as an arm, the other arm being in both cases the transversal. If the angles are between the two lines, they are *interior alternate angles*. If neither of the angles is interior, they are said to be *exterior alternate angles*.

alternate host (*BioSci*) One of the two (rarely more) hosts of a parasite that has the different stages of its life cycle in unrelated hosts. Cf ALTERNATIVE HOST.

alternate husbandry (*Agri*) Alternating arable and pasture system to produce crops and livestock, commonly on a cycle of 2–5 years.

alternating cleavage (*BioSci*) See SPIRAL CLEAVAGE.

alternating current (*ElecEng*) Abbrev *ac*. Electric current whose flow alternates in direction; the time of flow in one direction is a half-period, and the length of all half-periods is the same. The normal waveform of ac is sinusoidal, which allows simple vector or algebraic treatment. Provided by ALTERNATORS or electronic OSCILLATORS.

alternating function (*MathSci*) A function of two or more variables such that interchanging any two changes the sign but not the absolute value of the function, eg $f(x, y, z) = -f(y, x, z)$. Also *antisymmetric function*.

alternating-gradient focusing (*Electronics*) The net focusing effect achieved using a series of alternate converging and diverging lenses because, under suitable conditions, the rays will strike the diverging lenses nearer to the axis. Using magnetic or electrostatic lenses, the idea has been used for the design of electron synchrotrons and ion linear accelerators.

alternating-gradient synchrotron (*Phys*) A SYNCHRO-TRON modified by having magnetic field gradients around the orbit alternating towards and away from the centre of the orbit. This produces a focusing effect which reduces beam divergence caused by the mutual repulsion of the particles in the beam. Proton energies of up to 500 GeV and electron energies of about 10 GeV can be achieved.

alternating group (*MathSci*) The PERMUTATION group of degree n, normally denoted by A_n, which contains all even permutations of the set of n elements. It is a SIMPLE GROUP for $n \geq 5$.

alternating light (*Ships*) A navigation mark identified during darkness by a light showing alternating colours. See FLASHING LIGHT, OCCULTING LIGHT.

alternating series (*MathSci*) A series whose terms are alternately positive and negative. If the terms of an alternating series decrease monotonically then a necessary and sufficient condition for convergence is that their absolute values tend to zero.

alternating stress (*Phys*) The stress induced in a material by a force which acts alternately in opposite directions.

alternation of generations (*BioSci*) The regular alternation of two (rarely, three) types of individual in the life history of an animal or plant; typically in plants a diploid SPOROPHYTE and a haploid GAMETOPHYTE or in animals a sexually and an asexually produced form. They may be morphologically similar (*isomorphic*) or different (*hetero-morphic*). Also *metagenesis*.

alternative energy (*EnvSci*) Heat, work or electric power that is provided by unconventional means. The term is usually taken to include passive solar heating (solar panels), photovoltaic (solar) cells, geothermal heat and the harnessing of wind, tidal flow and wave motion.

alternative fuel (*EnvSci*) A fuel other than petrol for motor vehicles, eg alcohol.

alternative host (*BioSci*) One of two or more possible hosts for a given stage in the life cycle of a parasite, particularly when it is not the commonest or the most important economically. Cf ALTERNATE HOST.

alternative hypothesis (*MathSci*) The hypothesis, that there is a difference between two or more sets of data, for which a statistical test seeks support; if the test does not achieve the required SIGNIFICANCE LEVEL the default NULL HYPOTHESIS cannot be rejected.

alternative medicine (*Med*) Systems of medicine such as ACUPUNCTURE, CHIROPRACTICE, herbal medicine, HOMEOPATHY and OSTEOPATHY, which are able to alleviate symptoms for reasons which are poorly understood. The methods have not usually been subjected to test by a randomized CLINICAL TRIAL and are often not fully accepted by orthodox medical science. Also *complementary medicine*.

alternative pathway (*BioSci*) See COMPLEMENT.

alternative routing (*ICT*) The manual or automatic diversion, to a prearranged secondary route, of traffic which originates at an instant when the primary route is not available.

alternative splicing (*BioSci*) Cell or tissue-specific variation in the combination of exons from a single gene that are translated into protein. The mRNA transcript is processed to remove introns and some exons before it leaves the nucleus. A single gene can therefore produce different splice-variant proteins.

alternator (*ElecEng*) A type of ac generator, driven at a constant speed corresponding to the particular frequency of the electrical supply required from it. Also *synchronous generator*.

altimeter (*Aero, Phys*) An aneroid barometer used for measuring altitude by the decrease in atmospheric pressure with height. The dial of the instrument is graduated to read the altitude directly in feet or metres, the zero being set to ground or aerodrome level. See ENCODING ALTIMETER, RADIO ALTIMETER, RECORDING ALTIMETER.

altitude (*Aero, Surv*) The height in feet or metres above sea level. For precision in determining the performance of an aircraft, this must be corrected for the deviation of the meteorological conditions from that of the INTERNATIONAL STANDARD ATMOSPHERE. See CABIN ALTITUDE, PRESSURE ALTITUDE.

altitude (*Astron*) The angular distance of a heavenly body measured on that great circle which passes, perpendicular to the plane of the horizon, through the body and through the zenith. It is measured positively from the horizon to the zenith, from 0° to 90°.

altitude (*MathSci*) (1) The line through a vertex of a geometrical figure or solid perpendicular to its base. (2) The length of this line.

altitude level (*Surv*) Sensitive spirit level which ensures that theodolite is truly horizontal with respect to the telescope when vertical angles are measured.

altitude switch (*Aero*) A switching device generally comprising electrical contacts, actuated by an aneroid capsule which in turn is deflected by change in atmospheric pressure. The contacts are adjusted to make or break a warning circuit at the pressure corresponding to a predetermined altitude.

altitude valve (*Aero*) A manually or automatically operated valve fitted to the carburettor of an aero-engine for correcting the mixture strength as air density falls with altitude.

ALT key (*ICT*) A special key on a computer keyboard that may be used to select alternative functions of particular

keys; eg pressing ALT and P together may initiate a printout of the contents of the screen. Also *Alt key*.

altocumulus (*EnvSci*) White and/or grey patch, sheet or layer of cloud, generally with shading, composed of laminae, rounded masses, rolls, etc, which are partly fibrous or diffuse and which may or may not be merged; most of the irregularly arranged small elements usually have an apparent width of between 1° and 5°. Occurs at a height of 3000–7500 m. Abbrev *Ac*.

altostratus (*EnvSci*) Greyish or bluish cloud sheet or layer of striated, fibrous or uniform appearance, totally or partly covering the sky, and having parts thin enough to reveal the Sun at least vaguely, as through ground glass. Altostratus does not show halo phenomena and occurs at 3000–7500 m. Abbrev *As*.

altrices (*BioSci*) Birds whose young are hatched in a very immature condition, generally blind, naked or with down feathers only, unable to leave the nest, fed by the parents, eg the perching birds, Passeriformes.

altrose (*Chem*) An aldohexose, an optical stereoisomer of glucose.

altruism (*Psych, BioSci*) Behaviour that is unselfish, and may even be detrimental to the individual, but which benefits others. Observed in some non-human species.

ALU (*ICT*) Abbrev for ARITHMETIC LOGIC UNIT.

alula (*BioSci*) See BASTARD WING.

alum (*Min*) Hydrated aluminium potassium sulphate and related minerals.

Alumel (*Eng*) TN for an alloy of nickel with up to 5% aluminium, manganese and silicon, used with CHROMEL in thermocouples.

alumina (*Min*) Aluminium oxide, Al_2O_3. It is used as an abrasive, and as a structural ceramic. See CORUNDUM.

aluminate (*Chem*) Salt of aluminic acid, H_3AlO_3, a tautomeric form of aluminium hydroxide, which acts as a weak acid. Ortho-aluminates have the general formula M_3AlO_3 or $M_3Al(OH)_6$; meta-aluminates have the formula $MAlO_2$ or $MAl(OH)_4$, where M is a monovalent metal. Sodium aluminate, Na_3AlO_3, is used as a coagulant in water purification and softening. Tricalcium aluminate, $2CaOCa(Al_2O_2)_2$, is a constituent ($\approx 12\%$ by weight) of Portland cement; it hydrates at a rapid rate during the setting reaction.

alumina trihydrate (*Plastics*) $Al_2O_3 \cdot 3H_2O$. Used as a fire-retarding additive in plastics.

aluminium (*Chem*) Silver-white metallic element, forming a protective film of oxide. Symbol Al, at no 13, ram 26·9815, mp 659·7°C, bp 1800°C, rel.d. 2·58. The third commonest element in the Earth's crust after oxygen and silicon and the commonest metallic element (8·8% by mass); it does not occur as native Al, but always in silicate or other minerals. Bauxite is the principal ore of aluminium, which is prepared on a vast scale by the electrolysis of fused cryolite, particularly where electric power is cheap. In sea water it is only present to the extent of 0·01 ppm. It has numerous uses, and is the basis of light alloys for use in eg structural work; alloyed with silicon for transformer laminations, and iron and cobalt in many types of permanent magnet. Polished aluminium reflects well beyond the visible spectrum in both directions, and does not corrode in sea water. Foil aluminium is much used for capacitors. The metal can be used as a window in X-ray tubes and as sheathing for reactor fuel rods. ^{28}Al and ^{29}Al are strong gamma-ray emitters of very short half-life. As an electrode in gas-discharge tubes, Al does not sputter like other metals. US *aluminum*.

aluminium alloys (*Eng*) Alloys in which aluminium is the basis (ie predominant) metal, eg aluminium–copper and aluminium–silicon alloys. There are two principal classifications, namely casting alloys and wrought alloys, both of which are further subdivided into the categories heat-treatable and non-heat-treatable. When intricate engineering castings are needed as finished shapes with little or no further deformation, the alloy composition is selected to give good casting characteristics (eg fluidity, accurate mould reproduction, internal soundness). For this reason, 4–13% silicon is almost universally added; the greater the percentage, the better the fluidity, but higher values tend to have an embrittling effect. High-performance aluminium alloys, eg for aircraft applications, are based on wrought materials which are precipitation hardened after working to the desired form. Also *light alloys*.

aluminium anode cell (*ElecEng*) A cell with an aluminium anode immersed in an electrolyte which does not attack aluminium. The cathode may also be of aluminium or some other metal, eg lead. Such cells can be used as rectifiers or as high-capacitance capacitors. See ELECTRO-LYTIC CAPACITOR.

aluminium antimonide (*Electronics*) A semiconducting material used for transistors up to a temperature of 500°C.

aluminium–brass (*Eng*) Brass to which aluminium has been added to increase its resistance to corrosion. Used for condenser tubes. Contains 1–6% Al, 24–42% Zn, 55–71% Cu. See COPPER ALLOYS.

aluminium bronze (*Eng*) Copper–aluminium alloys which contain 4–11% aluminium, and may also contain up to 5% each of iron and nickel. These alloys have high tensile strength, are capable of being cast or cold-worked, and are resistant to corrosion. See COPPER ALLOYS.

aluminium chloride (*Chem*) Strong Lewis acid used eg in cationic polymerization.

aluminium foil (*Build*) Foil used in conjunction with plaster board for insulation purposes in walls or roofs.

aluminium leaf (*Build*) Thin foil similar to but thicker than gold leaf, used for decorative work. Normally sold in books of 25 leaves with silver leaf sizes from 82 mm to 152 mm^2.

aluminium paints (*Build*) Paints comprising paste or powder in a suitable medium, their main uses being: (1) as decorative finishing materials where their high opacity and reflectivity are important; and (2) as primary paints where their qualities of mechanical strength, moisture resistance and sealing properties are invaluable.

aluminium–steel cable (*ElecEng*) See STEEL-CORED ALU-MINIUM.

aluminium sulphate (*Chem*) An additive to drinking water that coagulates particles that cause discoloration.

aluminon (*Chem*) Ammonium aurine-tricarboxylate. Reagent for the colorimetric detection and estimation of aluminium, with which it forms a characteristic red colour.

alumino-silicates (*Chem, Min*) Compounds of alumina, silica and bases, with water of hydration in some cases. They include clays, mica, zeolites, and constituents of glass, porcelain and cement. See panel on CEMENT AND CONCRETE.

aluminothermic process (*Chem*) The reduction of metallic oxides by the use of finely divided aluminium powder. An intimate mixture of the oxide to be reduced and aluminium powder is placed in a refractory crucible; a mixture of aluminium powder and sodium peroxide is placed over this and the mass fired by means of a fuse or magnesium ribbon. The aluminium is almost instanta-neously oxidized, at the same time reducing the metallic oxide to metal. This process, also known as the *thermite process*, is used esp for the oxides of metals which are reduced with difficulty (eg titanium, molybdenum). On ignition, the mass may reach a temperature of 3500°C. Magnesium incendiary bombs have thermite as the igniting agent. Also *Thermit process*.

aluminous cement (*CivEng*) See HIGH-ALUMINA CEMENT.

aluminum (*Eng*) US for ALUMINIUM.

alums (*Chem*) A large number of isomorphous compounds whose general formula is $R'R''(SO_4)_2 \cdot 12H_2O$ or $R'_2 \ SO_4 \ R''(SO_4)_3 \cdot 24H_2O$, where R' represents an atom of a univalent metal or radical (potassium, sodium, ammo-nium, rubidium, caesium, silver, thallium) and R'' represents an atom of a tervalent metal (aluminium, iron, chromium, manganese, thallium). See PSEUDOALUMS.

alunite (*Min*) Hydrated sulphate of aluminium and potassium, resulting from the alteration of acid igneous rocks by solfataric action; used in the manufacture of alum. Also *alumstone*.

alunogen (*Min*) Hydrated aluminium sulphate, occurring as a white incrustation or efflorescence formed in two different ways: either by volcanic action, or by the decomposition of pyrite in carbonaceous or alum shales.

alveolar (*BioSci*) Having pits over the surface and resembling honeycomb. Also *alveolate*.

alveolitis (*Med*) Inflammation of the pulmonary alveoli, usually by external source, eg inhaled mouldy hay.

alveolus (*BioSci*) (1) A small pit or depression on the surface of an organ. (2) The cavity of a gland. (3) A small cavity of the lungs. (4) In higher vertebrates, the tooth socket in the jaw bone. (5) In Echinodermata, part of Aristotle's lantern, one of five pairs of grooved ossicles which grasp the teeth. (6) In Gastropoda, the glandular end-portion of the tubules of the digestive gland, secreting enzymes.

alvikite (*Min*) A medium- to fine-grained carbonatite.

Alzheimer's disease (*Med*) A degenerative brain disease, manifesting itself in premature ageing, with speech disorder. See ACETYLCHOLINE ESTERASE INHIBITORS.

Am (*Chem*) Symbol for AMERICIUM.

AM (*Phys*) Abbrev for AMPLITUDE MODULATION.

amacrine cell (*BioSci*) In the vertebrate eye, cell concerned with the processing, but not the reception, of images.

amacronics (*ICT*) Artificial replication of the image processing performed by the amacrine cells of the retina, achieved by etching lenses or mirrors onto computer chips. See DIGITAL MICRO-MIRROR DEVICE.

amagat (*Phys*) The unit of density of a gas at 0°C and 1 atmosphere pressure; usually 1 amagat = 1 mol per 22·4 dm^3.

Amagat's law of combining volumes (*Chem*) A law stating that the volume of a mixture of gases is equal to the sum of the volumes of the different gases, as existing each by itself at the same temperature and pressure.

amalgam (*Chem, MinExt*) The alloy of a metal with mercury; in the treatment of gold ores, the pasty amalgam of gold and mercury obtained from the plates in a mill and containing about $\frac{1}{3}$ gold by weight. Specially formulated solid amalgams are used for dental fillings. See ALLOY.

amalgamating table (*MinExt*) Flat sheet of metal to which mercury has adhered to form a thin soft film, used to catch metallic gold as mineral sands are washed gently over it. Also *amalgamated plate*.

amalgamation pan (*MinExt*) Circular cast-iron pan in which finely crushed gold-bearing ore or concentrate is ground with mercury, the valuable metal thus being amalgamated before separate retrieval.

amalgam barrel (*MinExt*) Small ball mill used to regrind gold-bearing concentrates, and then give them prolonged rubbing contact with mercury.

amalgam retort (*Eng*) Iron vessel in which the mercury is distilled off from gold or silver amalgam obtained in amalgamation. See AMALGAMATION PAN.

Amalthea (*Astron*) The fifth natural satellite of Jupiter, discovered in 1892. Distance from the planet 181 000 km; diameter 270 km.

amaurosis (*Med*) Blindness due to a lesion of the retina, optic nerve or optic tracts.

amazonstone (*Min*) A green variety of microcline, some-times cut and polished as a gemstone, and falsely called *Amazon jade*. Also *amazonite*.

amber (*Min*) A natural thermoplastic fossil resin known and highly valued since prehistoric times for its lustre, transparency and yellow colour. Produced by certain species of tree in Cenozoic times and occurring in eg Estonia. Rich in succinic acid. See SUCCINITE.

ambergris (*BioSci*) A greyish-white fatty substance with a strong but agreeable odour, obtained from the intestines of diseased sperm whales; sometimes found floating on the surface of the sea. It is used in perfumery as a fixative; on suitable treatment it yields ambreic acid.

amber mica (*Min*) See PHLOGOPITE.
amber mutation (*BioSci*) A base change in a CODING SEQUENCE of DNA that gives the STOP CODON UAG, resulting in a shortened gene product.
amberoid (*Min*) See AMBROID.
ambi- (*Genrl*) Lt form of Gk AMPHI-.
ambient air standard (*EnvSci*) A prescription for air quality based on the maximum concentrations of specified pollutants (commonly carbon monoxide, sulphur dioxide, ozone, nitrogen oxide and nitrogen dioxide) considered tolerable in a particular location.
ambient illumination (*ImageTech*) Background uncontrollable light level at a location.
ambient noise (*Acous*) (1) Random uncontrollable and irreducible noise at a location, or in a valve or circuit. (2) The noise existing in a room or any other environment, eg the ocean.
ambient temperature (*ElecEng*) Temperature of the surrounding air.
ambiophony (*Acous*) Technique of sound reproduction which creates an illusion to the listener of being in a very large room.
ambipolar (*Electronics*) Said of any condition or property which applies equally to positive and negative charge carriers (eg positive or negative ions, holes, electrons) in a plasma or semiconductor.
ambisexual (*BioSci*) Pertaining to both sexes; activated by both male and female hormones. Also *ambosexual*.
amblygonite (*Min*) Fluorophosphate of aluminium and lithium, a rare white or greenish mineral, crystallizing in the triclinic system and found in pegmatites.
amblyopia (*Med*) Dimness of vision, from the action of noxious agents on the optic nerve or retina.
ambosexual (*BioSci*) See AMBISEXUAL.
amboyna (*For*) A hardwood (*Pterocarpus*) from the E Indies, related to the paduak. The heartwood varies from pale yellow to brick red, with a wavy grain, fine texture and characteristic smell. Also *narra*.
ambroid (*Min*) A semisynthetic AMBER made by compressing and heating small, scrap pieces of amber and used for pipe mouthpieces, beads, etc. Also *amberoid, pressed amber*.
ambrosia (*BioSci*) (1) Certain fungi which are cultivated for food by some beetles (see AMBROSIA BEETLE). (2) The pollen of flowers collected by social bees and used in the feeding of the larvae.
ambrosia beetle (*BioSci*) Beetles of the family Scotylidae that cultivate the fungus *Monilia candida* in galleries in wood to feed their larvae and themselves.
ambulacra (*BioSci*) In Echinodermata, the radial bands of locomotor tube feet. Adj *ambulacral*.
ambulacral grooves (*BioSci*) Radially arranged grooves containing the tube feet in Asteroidea.
ambulatory (*Arch*) Covered promenade area, particularly the curved aisle behind the altar in an apse.
ambulatory (*BioSci*) Having the power of walking; used for walking.
AMCS (*Space*) Abbrev for AIRBORNE MISSILE CONTROL SYSTEM.
AMD (*Med*) Abbrev for AGE-RELATED MACULAR DEGENERATION.
ameiosis (*BioSci*) Non-pairing of the chromosomes in synapsis (meiosis).
amelia (*Med*) A congenital abnormality in which one or more limbs are completely absent. See PHOCOMELIA.
amelification (*BioSci*) The formation of enamel.
ameloblast (*BioSci*) A columnar cell forming part of a layer immediately covering the surface of the dentine, and secreting the enamel prisms in the teeth of higher vertebrates.
amenorrhoea (*Med*) Absence or suppression of menstruation. US *amenorrhea*.
ament (*Med*) One suffering from AMENTIA; a mentally deficient person.
amentia (*Med*) Mental deficiency; failure of the mind to develop normally, whether due to inborn defect, or to injury or disease.

Amentiferae (*BioSci*) See HAMAMELIDAE.
amentum (*BioSci*) A CATKIN. Adj *amentiform*.
American bond (*Build*) A form of brick bond in which every fifth or sixth course consists of headers, the other courses being stretchers. Very much used because it can be quickly laid.
American caisson (*CivEng*) See STRANDED CAISSON.
American filter (*MinExt*) See DISK FILTER.
American mahogany (*For*) Hardwood (*Swietenia*) from C and S America. The heartwood has a medium, uniform texture with colour varying from light, reddish brown to deep, rich red. Also *baywood, Brazilian mahogany, Cuban mahogany, Honduras mahogany*.
American Petroleum Institute (*MinExt*) An association of US companies which represents the interests of the oil industry and sets standards for products such as crude oils. See API SCALE.
American pitch pine (*For*) See PITCH PINE.
American plane (*For*) See SYCAMORE.
American red gum (*For*) A hardwood tree (*Liquidambar styraciflua*), of silky surface and irregular grain. Its pinkish-brown to deep red brown heartwood (known as *gum, sweet gum, bilsted* or *satin walnut*) is marketed separately from its creamy white sapwood (sap gum or hazel pine).
American Society for Testing and Materials (*Eng*) Society for developing and publishing agreed standards. Abbrev *ASTM*.
American Standard Wire Gauge (*Eng*) See BROWN AND SHARPE WIRE GAUGE.
American water turbine (*Eng*) See MIXED-FLOW WATER TURBINE.
American whitewood (*For*) A hardwood tree (*Liriodendron tulipifera*) with very wide, whitish sapwood, yellow- to olive-brown heartwood, straight grain and fine texture. Also *canary whitewood, tulip tree*.
America Online (*ICT*) A commercial information service. Abbrev *AOL*.
americium (*Chem*) Transuranic element, symbol Am, at no 95, half-life 433 years. Of great value as a long-life α-particle emitter, free of criticality hazards and γ-radiation, eg in laboratory neutron sources.
Ames test (*BioSci*) Test procedure for potential carcinogenicity of a substance. The substance being tested is mixed with animal tissue (often liver homogenate), which will generate active metabolites of the substance as in the body, and the mixture tested for mutagenicity in bacteria.
amesite (*Min*) A variety of septechlorite rich in magnesium and aluminium.
ametabolic (*BioSci*) Having no obvious metamorphosis.
amethyst (*Min*) A purple form of quartz, used as a semiprecious gemstone.
amianthus (*Min*) A fine silky ASBESTOS.
amicable numbers (*MathSci*) See FRIENDLY NUMBERS.
amicrons (*Chem*) Particles, of the order of nanometres, invisible in the ultramicroscope; they act as nuclei for larger submicron particles.
amide group (*Chem*) The —$CONH_2$— group in amides, when replacing the hydroxyl in a carboxyl group.
amides (*Chem*) Alkanamides. A group of compounds, eg polyamides, in which the hydroxyl of the carboxyl group of acids has been replaced by the amine group —NH_2. The group —CO—NH— in polymers.
amidines (*Chem*) Compounds derived from amides $RCONH_2$, $RCONHR'$ and $RCONR'_2$ in which the oxygen has been replaced by the divalent imido residue NH or NR. Amidines are crystalline bases forming stable salts but are themselves readily hydrolysed.
amido group (*Chem*) The $CONH_2$ group in amides, when replacing the hydroxyl in a carboxyl group.
amines (*Chem*) Aminoalkanes. Organic derivatives of ammonia NH_3 in which one or more hydrogen atoms are replaced by organic radicals. See AMINO GROUP.

aminoacetic acid (*Chem*) *Aminoethanoic acid, glycocoll, glycine.* NH_2CH_2COOH. Mp 230°C, colourless crystals, soluble in water, slightly in alcohol but not in ether. The simplest of the AMINO ACIDS.

amino acid (*Chem*) The basic chain unit of proteins and polypeptides. There are 20 natural amino acids and all except glycine (Gly) are chiral and comprise a tetravalent asymmetric carbon atom to which are attached four different groups (R—, H—, NH_2— and —CO_2H). The family of amino acids is created by variation of R, the simplest being glycine (R=H) and alanine (R=CH_3). The configuration of the asymmetric carbon atom is predominantly L in natural amino acids. D-amino acids are occasionally found, eg in peptide antibiotics. Some amino acids can be synthesized by animals, varying somewhat between species. Those that cannot are called *essential amino acids*. For humans, these are arginine, histidine, isoleucine, leucine, lysine, methionine, phenylalanine, threonine, tryptophan and valine. See appendix on Amino acids.

amino acid transmitters (*BioSci*) Amino acids released as neurotransmitter substances from nerve terminals and acting on postsynaptic receptors. Several, eg glutamate, are important in the mammalian central nervous system.

aminoacyl tRNA (*BioSci*) A complex of an amino acid with its transfer RNA (tRNA), formed by the action of aminoacyl tRNA synthetase.

aminoaldehydic resins (*Plastics*) See UREA RESINS.

aminoglycoside antibiotics (*Pharmacol*) Bactericidal antibiotics that inhibit bacterial protein synthesis. They are active only against aerobic gram-negative bacilli and staphylococci. Examples are *gentamicin, kanamycin, streptomycin, tobramycin*.

amino group (*Chem*) Essential component of the AMINO ACIDS.

aminoplastic resin (*Plastics*) One derived from the reaction of urea, thiourea, melamine, or allied compounds (eg cyanamide polymers and diaminotriazines) with aldehydes, particularly formaldehyde (methanal). See panel on THERMOSETS.

amino-sugar (*BioSci*) A monosaccharide in which an OH group is replaced with an amino group, often acetylated. Common examples are D-galactosamine, D-glucosamine, neuraminic acid, muramic acid. Amino-sugars are important constituents of bacterial cell walls, some antibiotics, blood group substances, milk oligosaccharides and chitin.

amiodarone (*Pharmacol*) An anti-arrhythmic drug that slows the impulses of the heart and is used for treatment of tachycardia and atrial fibrillation.

amitosis (*BioSci*) Direct division of the nucleus by constriction, without the formation of a spindle and chromosomes; direct nuclear division, occurring in the meganuclei of the *Ciliophora*. Also *amitotic division*. Cf MEIOSIS, MITOSIS.

amitriptyline (*Pharmacol*) A tricyclic antidepressant drug used in the treatment of moderate to severe endogenous depression, which also tends to sedate. It can also be used to treat post-herpetic neuralgia, and enuresis.

ammeter (*ElecEng*) An indicating instrument for measuring the current in an electric circuit.

ammines (*Chem*) Complex inorganic compounds which result from the addition of one or more ammonia molecules to a molecule of a salt or similar compound.

ammonia (*Chem*) NH_3. A colourless, pungent gas, bp −33·5°C, extremely soluble in water and very soluble in alcohol. Formed by bacterial decomposition of protein, purines and urea. Obtained mainly by the HABER PROCESS. Forms salts with most acids, and nitrides with metals. The liquefied gas is used as a refrigerant.

ammonia clock (*Phys*) An accurate clock controlled by the periodic inversion of the ammonia molecule with a frequency of $2·3786 \times 10^{10}$ Hz. See ATOMIC CLOCK.

ammonia oxidation process (*Chem*) An important process for producing nitric acid by the catalytic oxidation of ammonia gas with air on the surface of platinum–rhodium gauze at 900°C.

ammonia soda process (*Chem*) See SOLVAY'S AMMONIA SODA PROCESS.

ammonification (*BioSci*) The release of ammonia from amino acids (and ultimately protein) in decaying organic matter by soil bacteria. A step in the mineralization of nitrogen. Commonly followed, except in waterlogged and/or acid soils, by NITRIFICATION. Also *ammonization*.

ammonite (*Geol*) An extinct cephalopod belonging to the order Ammonitida. The shell is generally coiled in a plane spiral, and suture lines show complicated patterns. Fossils ranged from the Triassic to the top of the Cretaceous. See AMMONOIDS.

ammonium (*Chem*) The ion NH_4^+, which behaves in many respects like an alkali metal ion.

ammonium alum (*Chem*) See ALUMS.

ammonium chloride (*Chem*) NH_4Cl. A white salt formed by the reaction of ammonia with hydrochloric acid. See SAL-AMMONIAC.

ammonium dihydroxide phosphate (*Chem*) A piezo-electric crystal used in microphones and other transducers; it can withstand a temperature higher than can Rochelle salt.

ammonium hydroxide (*Chem*) *Ammonia hydrate.* NH_4OH. A solution of ammonia in water.

ammonization (*BioSci*) See AMMONIFICATION.

ammonoids (*Geol*) Completely extinct marine cephalopods that belong to the subclass Ammonoidea ranging from the Devonian to the Cretaceous; include goniatites and ammonites. The ammonites usually have shells coiled in one plane consisting of successive body chambers increasing in size towards the aperture of the shell. Chambers are separated by thin walls or septae which form complicated suture lines, important in identifying ammonites, allowing them to be used as zonal fossils for Mesozoic rocks. See NAUTILOIDEA.

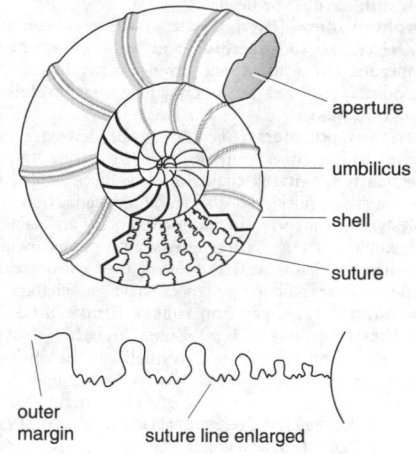

ammonoid The suture line is used for identification.

ammonolysis (*Chem*) Solvolysis in liquid ammonia solution.

amnesia (*Psych*) Loss of memory, common in dissociation states of hysteria. There is usually only partial or temporary loss of some types of information. See ANTEROGRADE and RETROGRADE AMNESIA.

amniocentesis (*BioSci*) A method for diagnosing fetal abnormalities in which fetal cells removed from amniotic fluid at about the 16th week of gestation are cultured and used in diagnostic assays, including probing with DNA sequences to detect disease-associated alleles.

amnion (*BioSci*) (1) In insects, the inner cell envelope covering and arising from the edge of the germ band.

(2) In higher vertebrates, one of the embryonic membranes, the inner fold of blastoderm covering the embryo, formed of ectoderm internally and somatic mesoderm externally.

Amniota (*BioSci*) Those higher vertebrates that possess an amnion during development, ie reptiles, birds and mammals. Adj *amniote*.

amniotic cavity (*BioSci*) In AMNIOTA, the space between the embryo and the amnion.

amniotic fluid (*BioSci*) Liquid filling the amniotic cavity.

amniotic folds (*BioSci*) Protrusions round the periphery of the blastoderm that give rise to the AMNION and the CHORION.

amoebiasis (*Med*) Disease caused by a Rhizopod parasite (*Entamoeba histolytica*), producing dysentery.

Amoebida (*BioSci*) Protozoa of the order Sarcodina, which extrude lobose pseudopodia and generally lack a skeleton, or have only a simple shell; their ectoplasm is never vacuolated.

amoebocyte (*BioSci*) (1) A metazoan cell having some of the characteristics of an amoeboid cell, esp as regards form and locomotion. (2) In Porifera, a wandering cell of varied function. (3) In Echinodermata, a wandering coelomic cell of excretory function. (4) A leucocyte.

amoeboid (*BioSci*) Of a cell, having no fixed form, creeping and putting out pseudopodia.

amoeboid movement (*BioSci*) Locomotion of an individual cell by means of pseudopodia.

Amor asteroids (*Astron*) The group of asteroids whose orbits fall within the perihelion distance of Mars but outside the orbit of the Earth.

amorphous (*Crystal*) A term describing a material without the periodic, ordered structure of crystalline solids.

amorphous carbon (*Chem*) See DIAMOND-LIKE CARBON.

amorphous head (*ImageTech*) A tape head composed of laminated amorphous alloy layers which improves the SIGNAL-TO-NOISE RATIO and reduces friction by comparison with ferrite-type heads.

amorphous metal (*Phys*) A material with good conductivity, electrical and thermal, and with other metallic properties but with atomic arrangements that are not periodically ordered as in crystalline metal solids. See METALLIC GLASS.

amorphous polymers (*Chem*) Materials having polymer chains which either cannot crystallize owing to chain irregularity (eg atactic chains) or have been cooled from the melt so quickly as to inhibit crystallization. Thus polystyrene is amorphous because its chains are atactic but polyethylene terephthalate film is amorphous owing to cooling at a high rate. Term must be used with care, since some non-crystalline polymers have a characteristic microstructure. See panel on RUBBER TOUGHENING.

amorphous regions (*Chem*) Zones in partly crystalline polymers which have not crystallized, eg owing to entanglements. If the GLASS TRANSITION temperature is below ambient, such regions are elastomeric and help toughen the material. See CRYSTALLIZATION OF POLYMERS and panel on ELASTOMERS.

amorphous selenium (*Chem*) Allotropic form of elemental selenium composed of linear chains of atoms linked together. Brittle, glassy solid at ambient temperature but becomes elastomeric above a T_g of 70°C. Like plastic sulphur and polymeric tellurium, a member of group VI of the periodic table. Used commercially in XEROGRAPHY, owing to its photoconductive properties.

amorphous semiconductor (*Phys*) Semiconductor prepared in the amorphous state. It tends to have a much lower electrical conductivity than its crystalline counterparts, and is typically made from hydrogenated AMORPHOUS SILICON or CHALCOGENIDE GLASS. See panel on SEMICONDUCTOR FABRICATION.

amorphous silicon (*Chem*) Non-crystalline silicon film made by eg CHEMICAL VAPOUR DEPOSITION, usually in fact containing considerable amounts of chemically

bonded hydrogen (hence sometimes αSi:H). Used as a semiconductor in thin-film transistors and photocells.

amorphous sulphur (*Chem*) Compound formed when sulphur vapour is cooled quickly. If the product so formed is treated with carbon disulphide and filtered, the amorphous sulphur is left on the filter as a white substance. It consists of long chains of sulphur atoms and reverts to rhombic sulphur at room temperature.

amosite (*Min*) A monoclinic amphibole form of ASBESTOS, the name embodying the initials of the company exploiting this material in the Transvaal, namely the 'Asbestos Mines of South Africa'.

amoxicillin (*Pharmacol*) A broad-spectrum beta-lactam antibiotic. Proprietary names include Amoxil® and Augmentin®. Formerly *amoxycillin*.

amp (*Phys*) Deprecated abbrev for AMPERE; A is preferred.

amperage (*Phys*) Current in amperes, esp the rated current of an electrical apparatus, eg fuse or motor.

ampere (*Phys*) SI unit of electric current. Defined as that current which, if maintained in two parallel conductors of infinite length, of negligible cross-section, and placed 1 m apart in vacuum, would produce between the conductors a force equal to 2×10^{-7} N m^{-1}. One of the SI fundamental units. Symbol A. See SI UNITS.

ampere-hour (*Phys*) Unit of charge, equal to 3600 coulombs, or 1 ampere flowing for 1 hour.

ampere-hour capacity (*Phys*) Capacity of an accumulator battery measured in ampere-hours, usually specified at a certain definite rate of discharge. Also applicable to primary cells.

ampere-hour efficiency (*Phys*) In an accumulator, the ratio of the ampere-hour output during discharge to the ampere-hour input during charge.

ampere-hour meter (*ElecEng*) A meter designed to record the product of current and time (ampere-hours) for a given circuit or passing at a given point. If the voltage is constant, the meter can be calibrated as an energy (kilowatt-hour) meter. Abbrev *ahm*.

Ampère's law (*Phys*) The relation between the magnetizing field H around a conductor, length l, carrying a current i, given by

$$\oint H \cdot dl = i$$

Ampère's rule (*Phys*) A rule giving the direction of the magnetic field associated with a current. If the conductor is grasped with the right hand, the thumb pointing in the direction of the current, the fingers will curl around the conductor in the direction of the field. Also *right-hand rule*.

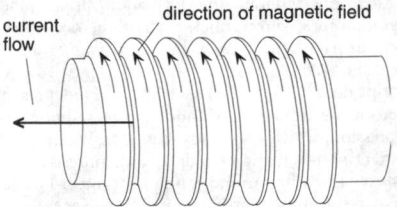

current flow   direction of magnetic field

Ampère's rule

Ampère's theory of magnetization (*Phys*) A theory based on the assumption that the magnetic property of a magnet is due to currents circulating in the molecules of the magnet.

ampere-turn (*Phys*) SI unit of magnetomotive force, which drives flux through magnetic circuits, arising from a current of 1 ampere flowing round one turn of a conductor. Symbol At.

ampere-turn amplification (*Phys*) The ratio of the load ampere-turns to the control ampere-turns in a MAGNETIC AMPLIFIER. Also *ampere-turn gain*.

ampere-turns per metre (*Phys*) SI unit of magnetizing force, magnetic field intensity.

amphetamines (*Pharmacol*) A group of drugs that act as stimulants of the central nervous system but are of limited use in clinical medicine.

amphi- (*Chem*) Prefix meaning containing a condensed double aromatic nucleus substituted in the 2,6-positions.

amphi- (*Genrl*) Prefix from Gk meaning both, on both sides (or ends) or around.

amphiaster (*BioSci*) During cell division by meiosis or mitosis, the two asters and the spindle connecting them.

Amphibia (*BioSci*) A class of semi-aquatic chordates with larvae possessing gills and anamniotic eggs. Frogs, toads and salamanders.

amphibian (*Aero*) Aircraft capable of taking off and alighting on land or water, eg seaplane or flying boat with retractable landing gear, or landplane with HYDROSKIS.

amphibious (*BioSci*) Adapted for both terrestrial and aquatic life.

amphiblastic (*BioSci*) Of ova, showing complete but unequal segmentation.

amphiboles (*Min*) An important group of dark-coloured rock-forming silicates, including hornblende, the commonest. See panel on SILICON, SILICA, SILICATES.

amphibolic (*BioSci*) (1) Description of a metabolic pathway that functions not only in catabolism, but also to provide precursors for anabolic pathways. (2) Capable of being turned backwards or forwards, eg the fourth toe of owls.

amphibolite (*Geol*) A crystalline, coarse-grained rock, containing amphibole as an essential constituent, together with feldspar and frequently garnet; eg hornblende-schist, formed by regional metamorphism of basic igneous rocks, but not foliated.

amphicoelous (*BioSci*) Having both ends concave, said eg of the centra of the vertebrae of Selachii and a few reptiles.

amphicondylous (*BioSci*) Having two occipital condyles. Also *amphicondylar*.

amphicribral bundle (*BioSci*) A vascular bundle in which a central strand of xylem is surrounded by phloem.

amphidentate (*Chem*) Of a ligand, capable of co-ordinating through one of two different atoms; eg S=C=Na.

amphidiploid (*BioSci*) An ALLOPOLYPLOID containing the diploid set of chromosomes from each of two species.

amphimixis (*BioSci*) True sexual reproduction, with the fusion of two gametes to form a zygote. Cf APOMIXIS.

Amphineura (*BioSci*) A class of bilaterally symmetrical Mollusca in which the foot, if present, is broad and flat, the mantle is undivided, and the shell is absent or composed of eight valves, eg coat-of-mail shells, chitons.

amphipathic (*Chem, Phys*) A term describing an asymmetrical molecular group, one end being hydrophilic and the other hydrophobic (wetting and non-wetting).

amphiphloic (*BioSci*) A STELE, having phloem on both sides of the xylem, eg SOLENOSTELE.

amphiphiles (*Chem*) Molecules with both polar groups, soluble in water, and non-polar, usually hydrocarbon, groups, soluble in oils. The basis for the action of detergents. They form monomolecular films on water.

amphiplatyan (*BioSci*) Having both ends flat, as in certain types of vertebral centrum.

amphipneustic (*BioSci*) Possessing both gills and lungs; in dipterous larvae, having only the prothoracic and posterior abdominal spiracles functional.

Amphipoda (*BioSci*) An order of MALACOSTRACA in which the carapace is absent, the eyes are sessile, and the uropods styliform; the body is laterally compressed. Amphipoda show great variety of habitat, being found on the shore, in the surface waters of the sea, in fresh water and in the soil of tropical forests. Some are parasitic. Includes whale lice, sandhoppers, skeleton shrimps.

amphipodous (*BioSci*) Having both walking and swimming appendages.

amphiprotic (*Chem*) Having both protophilic (ie basic) and protogenic (ie acidic) properties.

amphiprotic solvent (*Chem*) Solvent capable of showing either protophilic (ie acid-generating) or protogenic (base-generating) properties to different solutes, eg water with HCl and NH_3, respectively.

amphirhinal (*BioSci*) Having two external nostrils.

amphistomatal (*BioSci*) Said of a leaf having stomata on both surfaces. Also *amphistomatic*. Cf EPISTOMATAL.

amphistomous (*BioSci*) Having a sucker at each end of the body, as in leeches.

amphitheatre (*Arch*) An oval or circular building in which the spectators' seats surround the arena or open space in which the spectacle is presented, the seats rising away from the arena.

amphithecium (*BioSci*) Outer layer(s) of a developing sporophyte of a bryophyte giving rise to capsule wall. Cf ENDOTHECIUM.

amphitrichous (*BioSci*) Having a flagellum at each end of the cell.

amphitroph (*BioSci*) An organism that can grow either photosynthetically or chemotrophically.

amphitropous (*BioSci*) An ovule bent like a 'V' and attached to its stalk, near the middle of its concave side.

amphivasal bundle (*BioSci*) A vascular bundle in which a central strand of phloem is surrounded by xylem.

ampholines (*BioSci*) Mixtures of aliphatic amino acids with a range of iso-electric points that are used to establish the pH gradients used in ISO-ELECTRIC FOCUSING.

amphoric (*Acous*) Like the sound made by blowing across a narrow-necked vase.

amphoteric (*Chem*) Having both acidic and basic properties, eg aluminium oxide, zinc oxide, which form salts with acids and with alkalis.

ampicillin (*Pharmacol*) A semi-synthetic penicillin with a broad range of activity against those bacteria causing bronchitis, pneumonia, gonorrhoea, certain forms of meningitis, enteritis, biliary and urinary tract infections.

amplexicaul (*BioSci*) Said of a sessile leaf with its base clasping the stem horizontally.

amplexiform (*BioSci*) A type of wing-coupling formed in some Lepidoptera, whereby the wings are coupled simply by overlapping basally.

amplidyne (*ElecEng*) A rotating magnetic amplifier, widely used as a power amplification device, in which a small increase in power input to the field coils produces a large boost in power output.

amplexus (*BioSci*) The clasping of the female of certain amphibians by the male as part of the mating process; the period of this.

amplification (*BioSci*) The process by which multiple copies of genes or DNA sequences are formed.

amplified spontaneous emission (*ICT*) Unwanted noise in an ERBIUM-DOPED FIBRE AMPLIFIER arising from amplification of spontaneous as opposed to stimulated emission. Its optical power in bandwidth B centred on frequency v is $(G-1)\mu h v B$ for each polarization state of the fibre, where G is the power gain and μ the inversion factor of the amplifier (unity when ideal).

amplifier (*ElecEng, ICT*) A circuit or assembly that uses a valve, transistor or solid-state device, magnetic contrivance or any active device to increase the strength of a signal without appreciably altering its characteristics. An amplifier transfers power from an external source to the signal, unlike a transformer; therefore, an amplifier with equal input and output impedances might exhibit current or voltage gain, or it may have unity current or voltage gain with an increase or decrease in impedance from input to output. Analogous devices employing pneumatic and hydraulic systems are sometimes used.

amplitude (*MathSci*) The argument of a complex number. The term would more naturally be taken to mean the modulus, and is best avoided.

amplitude (*Phys*) The maximum value of a periodically varying quantity during a cycle; eg the maximum displacement of a vibrating particle from its mid-position, the maximum value of an alternating current (see PEAK VALUE), or the maximum displacement of a sine wave. See DOUBLE AMPLITUDE.

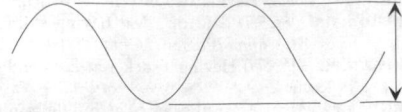

amplitude Sine wave.

amplitude discriminator (*ICT*) See PULSE-HEIGHT DISCRIMINATOR.

amplitude distortion (*ICT*) Distortion of waveform arising from the non-linear static or dynamic response of a part of a communication system, the output amplitude of the signal at any instant not having a constant proportionality with the corresponding input signal.

amplitude limiter (*ImageTech*) One which separates synchronizing signals in a TV signal from the video (picture) signal. *Also limiter.*

amplitude modulation (*ICT*) A simple method of impressing a signal on a carrier wave, in which the amplitude of the carrier is made to vary in proportion to the instantaneous value of the signal. Variation of the carrier amplitude from zero to twice its unmodulated value is termed 100% modulation; beyond this point the signal recovered from the carrier by a simple detector will be distorted. Abbrev *AM*.

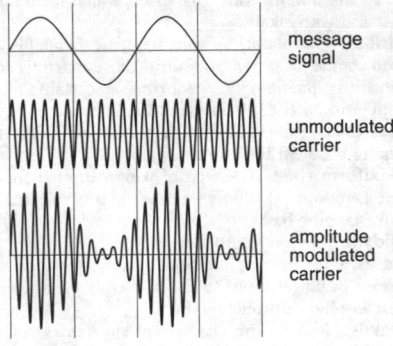

message signal

unmodulated carrier

amplitude modulated carrier

amplitude modulation

amplitude peak (*ICT*) Maximum positive or negative excursion from zero of any periodic disturbance.

amplitude shift keying (*ICT*) A form of AMPLITUDE MODULATION in which the amplitude of the carrier assumes only certain discrete values, allowing transmission of digitally coded information.

ampoule (*Med*) A small, sealed glass capsule for holding measured quantities of vaccines, drugs, serums, etc, ready for use.

ampoule tubing (*Glass*) Tubing of special composition suited to the manufacture of ampoules. It must work well in the blowpipe flame, and must resist the action of the materials stored in the ampoule.

AMPS (*ICT*) Abbrev for ADVANCED MOBILE PHONE SYSTEM/SERVICE.

ampulla (*BioSci*) (1) Generally, any small membranous vesicle or the dilated end of a canal or duct. Specifically: (2) In vertebrates, the dilation housing the sensory epithelium at one end of a semicircular canal of the ear. (3) In mammals, part of a dilated tubule in the mammary gland. (4) In fish, the terminal vesicle of a neuromast organ. (5) In Echinodermata, the internal expansion of the axial sinus below the madreporite. (6) In Ctenophora, one of a pair of small sacs forming part of the aboral sense organ. Adj *ampullary*.

AMSL (*Genrl*) Abbrev for *above mean sea level*.

amu (*Chem*) Abbrev for ATOMIC MASS UNIT.

amydricaine hydrochloride (*Pharmacol*) A compound used as a local anaesthetic, particularly in ophthalmic practice.

amyelinate (*BioSci*) A term describing nerve fibres which lack a myelin sheath.

amygdala (*BioSci*) (1) Generally, an almond-shaped part of the body, such as a lobe of the cerebellum or one of the palatal tonsils. (2) In the brain, the part of the limbic system that attaches emotional significance to information and mediates both defensive and aggressive behaviour.

amygdale (*Geol*) An almond-shaped infilling (by secondary minerals such as agate, zeolites, calcite, etc) of elongated steam cavities in igneous rocks. Also *amygdule*.

amygdalin (*Chem*) $C_{20}H_{27}O_{11}N$. Colourless prisms, mp $200°C$, a glucoside found in bitter almonds, and in peach and cherry kernels.

amygdule (*Geol*) See AMYGDALE.

amyl-, amylo- (*Genrl*) Prefixes from Gk *amylon*, starch.

amyl acetate (*Chem*) Pentyl ethanoate. $CH_3COOC_5H_{11}$. A colourless liquid, of ethereal pear-like odour, bp $138°C$, used for fruit essences and as an important solvent for nitrocellulose. *Iso*-amyl acetate is also known as *pear oil*.

amyl alcohol (*Chem*) Pentanol. $C_5H_{11}OH$. The fraction of fusel oil that distils about $131°C$. There are eight isomers possible and known: four primary, three secondary and one tertiary amyl alcohol. The important isomers are *iso*-amyl alcohol, *l*-amyl alcohol and tertiary amyl alcohol (amylene hydrate).

amylase (*BioSci*) An enzyme that hydrolyses the internal 1,4-glycosidic bonds of starch to produce reducing sugars. The amylase found in human saliva is known as *ptyalin*.

amyl group (*Chem*) The pentyl radical C_5H_{11}.

amyl nitrite (*Chem*) $C_5H_{11}ONO$. The nitrous acid ester of *iso*-amyl alcohol, a yellowish liquid, bp $98°C$, of pleasant odour. Intermediate for the preparation of nitroso- and diazo-compounds. Used in medicine as a vasodilator, esp for acute angina.

amylo- (*BioSci*) See AMYL-.

amylobarbitone (*Pharmacol*) A barbiturate, the sodium salt of which was formerly used as an intermediate-acting hypnotic and sedative.

amyloid (*BioSci, Chem*) (1) A starch-like cellulose compound produced by treatment of cellulose with concentrated sulphuric acid for a short period. (2) An insoluble fibrillar glycoprotein deposited extracellularly in AMYLOIDOSIS. The glycoprotein may derive from immunoglobulin or from serum amyloid, an acute phase protein. Should be distinguished from $β$-amyloid deposited in the brain, which is derived from AMYLOID PRECURSOR PROTEIN.

amyloidosis (*Med*) The deposition of amyloid in the organs and tissues of the body, often as a result of chronic infection.

amyloid precursor protein (*BioSci*) The protein from which amyloid fibrils are generated by proteolytic cleavage. Senile plaques in the brains of Alzheimer's disease sufferers consist of a core of amyloid fibrils surrounded by dystrophic neurites. Amyloid deposits are also found in brains of older Down's syndrome patients. Abbrev *APP*.

amylolytic (*BioSci*) Starch-digesting.

amylopectin (*BioSci*) A polymer, $α(1→4)$ linked with $α(1→6)$ branches, of glucose. A constituent of starch.

amyloplast (*BioSci*) A plant plastid involved in the synthesis and storage of starch. Such plastids are found in many cell types, but particularly in storage tissues. They characteristically contain starch grains in the plastid stroma.

amylose (*BioSci, Chem*) The sol constituent of starch paste; linear polymer of glucose units having $α(1→4)$ glucosidic bonds. Cf AMYLOPECTIN.

amylum (*BioSci*) Starch.

amyotrophic lateral sclerosis (*Med*) A nervous disease in which atrophy of the muscle follows degenerative changes in the motor fibres of the spinal cord and brain. Also *motor neuron disease*.

amyotrophy (*Med*) Wasting or atrophy of muscle.

an- (*Genrl*) Prefix from Gk *an*, not. Also *ap-*.

ana- (*Chem*) Containing a condensed double aromatic nucleus substituted in the 1·5-positions.

ana- (*Genrl*) Prefix from Gk *ana*, up, anew.

anabaric crystallization (*Chem*) A term for high-pressure (>3 kbar) crystallization of polymers, esp polyethylene, when extended chains form.

anabatic wind (*EnvSci*) A local wind blowing up a slope heated by sunshine, and caused by the difference in density between the warm air in contact with the ground and the cooler air at corresponding heights in the FREE ATMOSPHERE.

anabiosis (*BioSci*) A temporary state of reduced metabolism in which metabolic activity is absent or undetectable. See CRYPTOBIOSIS.

anabolic (*BioSci*) Metabolic events that lead to the synthesis of body constituents.

anabolic steroids (*Pharmacol*) Synthetic forms of male sex hormones (androgens) that promote tissue growth, especially of muscle. Have been misused by body-builders and athletes. *Stanozolol* is an example. All have some masculinizing action.

anabolism (*BioSci*) The chemical changes proceeding in living organisms with the formation of complex substances from simpler ones. Adj *anabolic*.

anabolite (*BioSci*) A substance participating in anabolism.

anaclitic (*Psych*) Characterized by strong emotional dependence on others.

anadromous (*BioSci*) Having the habit of migrating from more dense to less dense water to breed, generally from oceanic to coastal waters, or from salt water to fresh water; eg the salmon.

anaemia (*Med, Vet*) Diminution of the amount of total circulating haemoglobin in the blood. *Primary anaemia* is the failure to produce red blood cells or haemoglobin. *Secondary anaemia* results from blood loss. US *anemia*. See FELINE INFECTIOUS ANAEMIA, INFECTIOUS ANAEMIA OF HORSES.

anaerobe (*BioSci*) An organism that can grow in the absence or near absence of oxygen. Facultative anaerobes can utilize free oxygen; obligate anaerobes are poisoned by it. Adj *anaerobic*.

anaerobic (*BioSci*) Living in the absence of oxygen. *Anaerobic respiration* is the liberation of energy which does not require the presence of oxygen.

anaerobic adhesives (*Chem*) Adhesives based on MONOMERS which will only polymerize (ie *cure*) in the absence of oxygen. They are usually tetrafunctional esters derived from acrylic acid, such as diacrylates and dimethacrylates.

anaerobic respiration (*BioSci*) See RESPIRATION.

anaerobiosis (*BioSci*) Existence in the absence of oxygen. Adj *anaerobic*.

anaesthesia (*Med*) Correctly, loss of feeling but often applied to the technique of pain relief for surgical procedures. US *anesthesia*.

anaesthetic (*Med*) Insensible to touch (loosely, also to pain and temperature); a drug which produces insensibility to touch, pain and temperature, with or without loss of consciousness. US *anesthetic*.

anaesthetist (*Med*) One skilled in the administration of anaesthetic drugs. US *anesthesiologist*.

anafront (*EnvSci*) A situation at a front, warm or cold, where the warm air is rising relative to the FRONTAL ZONE.

anaglyph (*ImageTech*) Pair of stereoscopic images reproduced in two colours, generally red and blue-green, for viewing with corresponding colour filters, one for each eye, to give a three-dimensional sensation.

anal (*BioSci*) See ANUS.

anal cerci (*BioSci*) In insects, sensory appendages of one of the posterior abdominal somites, generally the eleventh, retained throughout life.

anal character (*Psych*) In psychoanalytic theory, an adult personality derived from unresolved conflicts (*fixation*) during the anal stage of psychosexual development. It is characterized by a reaction formation against impulsiveness, stemming from the anal stage of childhood, a person with personality traits (anal compulsive or retentive character) who transfers their unresolved anal (or control) issues into characteristics such as compulsiveness, stinginess, cleanliness, organization and obstinacy. An anal expulsive personality exhibits the opposite traits. See ANAL PHASE.

analcime (*Min*) Hydrated aluminium silicate, a member of the zeolite group, occurring in some igneous and sedimentary rocks, but particularly in cavities in lavas. Also *analcite*.

analcimite (*Min*) An alkaline volcanic rock composed essentially of analcime, pyroxene and other minerals, with little or no olivine.

analcite (*Min*) See ANALCIME.

analeptic (*Med*) Having restorative or strengthening properties.

analgesia (*Med*) Loss of sensibility to pain.

analgesic (*Pharmacol*) A drug which relieves pain.

anal-gland disease (*Vet*) Inflammation of the perianal glands, and inflammation or impaction of the canine anal sacs.

anallatic lens (*Surv*) Special lens which, when correctly placed between the object glass and the eyepiece lens of a tacheometric telescope, optically reduces the additive constant for the tacheometer to zero.

anallatic telescope (*Surv*) Telescope which, when used in tacheometry, has a zero additive constant.

anallatism (*Surv*) See CENTRE OF ANALLATISM.

analog (*Genrl*) US for ANALOGUE.

analogous organs (*BioSci*) Organs that are similar in appearance and/or function but which are neither equivalent morphologically nor of common evolutionary origin, eg foliage leaves and cladodes; the wings of birds and insects. Cf HOMOLOGOUS ORGANS.

analogue (*Chem*) A compound which may be considered to be derived from another by the substitution of saturated aliphatic groups for hydrogen, eg ethanol is an analogue of methanol. US *analog*.

analogue (*EnvSci*) A previous weather map similar to the current map. The developments following the analogue aid forecasting. US *analog*.

analogue (*ICT*) Any form of transmission of information where the transmitted signal's information-bearing characteristic (usually amplitude or frequency) is varied in direct proportion to the intensity of the sound, or brightness of pictures, etc, which it is desired to communicate. US *analog*. Cf DIGITAL.

analogue clock (*Genrl*) The traditional clock with rotating hands. Cf DIGITAL CLOCK.

analogue computer (*ICT*) A computer that uses continuous physical variables such as voltage or pressure to represent and manipulate the measurements it handles. Now usually a special purpose computer.

analogue filter (*ICT*) A filter suitable for use with analogue signals, ie those which are continuous with time. Cf DIGITAL FILTER.

analogue-to-digital converter (*ICT, ElecEng*) An electronic device that converts an analogue signal into a digital one. An example would be a smoothly varying voltage representing temperature that is converted to an 8 bit BINARY CODE. Abbrev ADC. Also *A/D converter, digitizer*.

analogue watch (*Genrl*) The traditional watch with rotating hands. Cf DIGITAL WATCH.

analogy (*BioSci*) Likeness in function but not in evolutionary origin, eg tendrils, that may be modified leaves, branches, inflorescences; the wings of birds and insects. Adj *analogous*. Cf HOMOLOGY.

analogy (*Electronics*) Correspondence of pattern or form between mechanical and electrical quantities, or vice versa; eg a network of resistance, capacitance and inductance can be made to represent a complex mechanical system, or a stretched rubber membrane for the potential distributions between electrodes in electronic tubes.

anal phase (*Psych*) In psychoanalytic theory, Freud's second stage of psychosexual development, during which the focus of pleasure is on activities related to retaining and expelling the feces; occurs in the second year of life and is often thought of as representing a child's ability to control his or her own world.

anal retentive personality (*Psych*) See ANAL CHARACTER.

anal stage (*Psych*) See ANAL PHASE.

anal suture (*BioSci*) In the posterior wings of some insects, a line of folding, separating the anal area of the wing from the main area.

analyser (*Chem*) The second polarizer (originally a Nicol prism) in a polarimeter or petrological microscope; when rotated 90° relative to the polarizer it will not allow polarized light to pass. It receives the light which has passed through the object under study which, in a petrological microscope, is a thin section of rock or mineral.

analysis (*MathSci*) The theory of limits, including sequences, series, differentiation and integration. The practical methods of calculus are applied analysis.

analysis meter (*ICT*) A registering meter used to determine the loading of groups of circuits with calls, particularly for determining the correctness or otherwise of grading.

analysis of covariance (*MathSci*) ANALYSIS OF VARIANCE, after adjustment for the effect of other, possibly related, variables (covariates).

analysis of variance (*MathSci*) The partition of the total variation in a set of observations into components corresponding to differences between and within subclassifications of the data.

analyst (*Chem*) A person who carries out any process of analysis.

analytical engine (*ICT*) The name for the first general purpose digital computer designed about 1835 by Charles Babbage but only partly built.

analytical geometry (*MathSci*) The application of the methods of algebra to geometry by utilizing the concept of co-ordinates.

analytical reagent (*Chem*) An indication of a definite standard of purity. Abbrev *AR*.

analytic continuation (*MathSci*) A process for extending the range of definition of a function of a complex variable. Starting with a first Taylor series about a point A representing the function, a second Taylor series about a point B within the first circle of convergence can be obtained and, providing the line AB does not intersect the first circle of convergence at a singularity, the circle of convergence of the second Taylor series will extend beyond the first circle. This process can be repeated indefinitely and it can be shown that, if a particular region can be reached by two different routes, the two Taylor series obtained will be identical, providing no singularities lie between the two routes. Some functions have their own *natural boundaries* beyond which analytic continuation is impossible. Such a function is

$$g(z) = 1 + \sum_{r=1}^{\infty} z^{2r}$$

which has an infinite number of singularities on every arc of its circle of convergence $|z| = 1$.

analytic function (*MathSci*) A function of a complex variable which is single-valued and differentiable at all points in a region. Sometimes a function is said to be analytic in a region if it is single-termed and differentiable at all but a finite number of points (*singularities*) of the region. In this case the term *regular* is frequently contrasted with *analytic* to mean that there are no singularities. When the singularities are all poles the term *meromorphic* is sometimes used. A function that is analytic throughout the whole complex plane, except possibly at a finite number of singularities, is called an *entire* or an *integral* function. Also *holomorphic function*, *monogenic function*.

anamnesis (*Med*) The recollection of past things; the patient's recollections of symptoms and past illnesses.

anamnestic (*BioSci*) An outmoded term for a secondary immune response or immunological memory.

Anamniota (*BioSci*) Vertebrates without an amnion during development, ie Amphibia and fish.

anamniotic (*BioSci*) Lacking an amnion during development. Also *anamniote*.

anamorph (*BioSci*) The asexual or imperfect stage of a fungus, esp Deuteromycotina.

anamorphic (*ImageTech*) A wide-screen image that has been laterally compressed or squeezed, either optically by an ANAMORPHIC LENS or electronically.

anamorphic lens (*ImageTech*) Lens with cylindrical elements giving different magnification in horizontal and vertical directions. In WIDE-SCREEN cinematography the image is compressed laterally in the camera and expanded to compensate in projection. The equivalent term *anamorphotic* is rare.

Ananke (*Astron*) The 12th natural satellite of Jupiter, discovered in 1951. Distance from the planet 21 200 000 km; diameter 30 km.

anaphase (*BioSci*) The stage in mitotic or meiotic nuclear division when the chromosomes or half-chromosomes move away from the equatorial plate to the poles of the spindle; more rarely, all stages of mitosis leading up to the formation of the chromosomes. See fig. at MITOSIS.

anaphoresis (*Chem*) The migration of suspended particles towards the anode under the influence of an electric field.

anaphylactic shock (*BioSci*) See ANAPHYLAXIS.

anaphylatoxin (*BioSci*) Peptides released from COMPLEMENT components C3 and C5 during complement activation which act on mast cells to release histamine etc, as in anaphylaxis.

anaphylaxis (*BioSci*) An acute immediate *hypersensitivity* reaction following administration of an antigen to a subject resulting from combination of the antigen with IgE on mast cells or BASOPHILS which causes these cells to release histamine and other vasoactive agents. An acute fall in blood pressure may be so severe as to be fatal. Other symptoms include bronchospasm, laryngeal oedema and urticaria. Also *anaphylactic shock*.

anaplasia (*Med*) Loss of the differentiation of a cell associated with proliferative activity; a characteristic of a malignant tumour.

anaplasmosis (*Vet*) A disease of cattle caused by infection by protozoa of the genus *Anaplasma*, and characterized by fever, anaemia and jaundice. The protozoa are found in the red blood corpuscles and are transmitted by ticks, biting flies and mosquitoes. Infections of sheep and pigs also occur. Also *gall-sickness*.

anaplerotic (*BioSci*) Denoting reactions that replenish deficiencies of metabolic intermediates, eg the formation of oxaloacetate by the carboxylation of pyruvate.

anapophysis (*BioSci*) In higher vertebrates, a small process just below the postzygapophysis that strengthens the articulation of the lumbar vertebrae.

anapsid (*BioSci*) Having the skull completely roofed over, ie having no dorsal foramina other than the nares, the orbits and the parietal foramen.

anarthrous (*BioSci*) Without distinct joints.

anasarca (*Med*) Excessive accumulation of fluid (dropsy) in the skin and subcutaneous tissues.

Anaspida (*BioSci*) A subclass of the reptiles containing the oldest known forms, characterized by the temporal region of the skull without apertures. Turtles.

anastigmat lens (*Phys*) A photographic objective designed to be free from astigmatism on at least one extra-axial zone of the image plane.

anastomosis (*BioSci*) In general, a cross-connection. Also applied to the formation of an interconnecting meshwork of blood vessels or nerves, or an artificial communication, made surgically, between any two parts of the alimentary canal. Pl *anastomoses*.

anastrozole (*Pharmacol*) A drug that inhibits the production of estrogen, used in the treatment of advanced breast cancer.

anatase (*Min*) One of the three naturally occurring forms of crystalline titanium dioxide, of tabular or bipyramidal habit. See OCTAHEDRITE.

anatomy (*BioSci*) (1) The study of the form and structure of animals and plants; it includes the study of minute structures, and thus includes HISTOLOGY. (2) Dissection of an organized body in order to display its physical structure.

anatropous (*BioSci*) An inverted ovule, so that the MICROPYLE is next to the stalk.

anaxial (*BioSci*) Asymmetrical.

ANCA (*BioSci*) Abbrev for ANTINEUTROPHIL CYTOPLASMIC ANTIBODY.

anchorage dependence (*BioSci*) The necessity for attachment (and spreading) in order that an animal cell will grow and divide in culture. Loss of anchorage dependence seems to be associated with greater independence from external growth control and correlates well with tumorigenicity *in vivo*.

anchor bolt (*Build*) A bolt used to secure frameworks, stanchion bases, etc, to piers or foundations, and having usually a large plate washer built into the latter as anchorage.

anchor clamp (*ElecEng*) A fitting attached to the overhead contact wire of a tramway or railway to support the wire, and also to take the longitudinal tension and prevent movement of the wire in a direction parallel to the track.

anchor gate (*CivEng*) A heavy gate, such as a canal lock gate, which is supported at its upper bearing by an anchorage in the masonry such as an ANCHOR BOLT.

anchor ring (*MathSci, NucEng*) See TORUS.

anchor string (*MinExt*) Length of CASING run into the top of wells and often cemented in to prevent a BLOWOUT outside the casing. It provides fixings for the well-head equipment. Also *surface casing, surface pipe, top casing*.

anconeal (*BioSci*) Pertaining to, or situated near, the elbow.

anconeus (*BioSci*) An extensor muscle of the arm attached in the region of the elbow.

AND (*ICT*) A logical operator such that (*p* AND *q*) written *p·q* takes the value TRUE if *p* is TRUE and *q* is TRUE otherwise (*p* AND *q*) takes the value FALSE. See LOGICAL OPERATION.

andalusite (*Min*) One of several crystalline forms of aluminium silicate; a characteristic product of the contact metamorphism of argillaceous rocks. Orthorhombic. See CHIASTOLITE.

AND element (*ICT*) See AND GATE.

andesine (*Min*) A member of the plagioclase group of minerals, with a small excess of sodium over calcium: typical of the intermediate igneous rocks.

andesite (*Geol*) A fine-grained igneous rock (usually a lava), of intermediate composition, having plagioclase as the dominant feldspar.

AND gate (*ICT*) A gate producing an output signal only if all inputs are energized simultaneously, ie output signal is 1 when all input signals are 1. Also *AND element*. See LOGICAL OPERATION.

andiroba (*For*) Hardwood (*Carapa*) of C and S America, light to red-brown, and straightish-grained.

andradite (*Min*) Common calcium–iron garnet. Mainly dark brown to yellow or green. See DEMANTOID, MELANITE, TOPAZOLITE.

andro- (*Genrl*) Prefix from Gk *aner*, gen *andros*, man, male.

androconia (*BioSci*) In certain male Lepidoptera, scent scales serving to disseminate the PHEROMONES which serve the purpose of sexual attraction.

androcyte (*BioSci*) Cells in an ANTHERIDIUM that will metamorphose to form antherozoids.

androdioecious (*BioSci*) A species, having some individuals with male flowers only and others hermaphrodite flowers only. Cf DIOECIOUS.

androecium (*BioSci*) The male part of a flower, consisting of one or more stamens. Cf GYNOECIUM.

androgen (*BioSci*) General term for a group of male sex hormones that eg stimulate the growth of male secondary sex characteristics. Cf ESTROGEN.

androgenesis (*BioSci*) (1) Development from a male cell. (2) Development of an egg after entry of male germ cell without the participation of the egg nucleus.

androgenic (*Med*) Having the effects of a male sex hormone.

androgynophore (*BioSci*) See ANDROPHORE.

androgynous (*BioSci*) Bearing staminate and pistillate flowers on distinct parts of the same inflorescence; having the male and female organs on or in the same branch of the thallus.

Andromeda (*Astron*) A constellation in the northern sky, one of 48 listed by Ptolemy (AD 140), named for the daughter of Cepheus and Cassiopeia; it contains the ANDROMEDA GALAXY.

Andromeda Galaxy (*Astron*) The largest of the nearby galaxies, distance approx 700 kpc. Spiral, like the Milky Way, and around 38 kpc in diameter, it is the most remote object easily visible to the naked eye. Also *Andromeda Nebula, M31*.

andromonoecious (*BioSci*) A species in which all of the plants bear both male and hermaphrodite flowers. Cf MONOECIOUS.

androphore (*BioSci*) An elongation of the receptacle of the flower between the corolla and the stamens. Also *androgynophore*.

androsporangium (*BioSci*) Sporangium in which ANDROSPORES are produced.

androspore (*BioSci*) In heterosporous plants, same as MICROSPORE. Cf GYNOSPORE.

anecdysis (*BioSci*) The intermoult period in Arthropoda.

anechoic room (*Acous*) A room in which internal sound reflections are reduced to an ineffective value by extremely high sound absorption, eg by using glass-fibre wedges. Also *dead room*.

anelasticity (*Phys*) (1) Any recoverable deformation which deviates from linear elastic behaviour. (2) Any structural inhomogeneity or discontinuity which would dampen or attenuate an elastic wave propagating in a body.

anelectric (*Phys*) A term once used for a body which does not become electrified by friction.

anemia (*Med, Vet*) US for ANAEMIA.

anemo- (*Genrl*) Prefix from Gk *anemos*, wind.

anemochorous (*BioSci*) Seeds or other PROPAGULES dispersed by wind.

anemograph (*EnvSci*) See ANEMOMETER.

anemometer (*Eng*) An instrument for measuring the rate of flow of a gas, either by mechanical or electrical methods.

anemometer (*EnvSci*) An instrument for measuring the speed of the wind. A common type consists of four hemispherical cups carried at the ends of four radial arms pivoted so as to be capable of rotation in a horizontal plane, the speed of rotation being indicated on a dial calibrated to read wind speed directly. An *anemograph* records the speed and sometimes the direction.

anemophily (*BioSci*) Pollination by means of wind. Dispersal of spores by wind. Adj *anemophilous*.

anemotaxis (*BioSci*) Orientation to an odour source based upon wind direction.

anencephaly (*Med*) A neural tube defect in which skull and cerebral hemispheres fail to develop. Adj *anencephalic*.

anergy (*BioSci*) Generally, absence of energy. In immunology, the inability to give the expected allergic responses, esp delayed-type hypersensitivity. Occurs when the lymphocytes or monocytes needed are absent or suppressed. Adj. *anergic*.

aneroid barometer (*EnvSci, Surv*) A barometer having a vacuum chamber or syphon bellows of thin corrugated

Angiosperms (flowering plants)

The group, often classified as a class, Angiospermae, or as a division, Anthophyta or Magnoliophyta, which contains those seed plants in which the ovules are enclosed within carpels (in contrast to the unenclosed ovules of the GYMNOSPERMS), the pollen germinating on a stigma and pollen tubes growing to the ovule(s) in the ovary; there is characteristically double fertilization. The carpels and stamens are usually borne in flowers, which are often more or less showy and/or fragrant and thus attractive to insect, bird or bat pollinators or, alternatively, more or less inconspicuous and wind-pollinated. The xylem usually has vessels and the phloem has sieve tubes and companion cells. There are about 220 000 species in two classes, MONOCO-TYLEDONS and DICOTYLEDONS.

Angiosperms dominate most terrestrial habitats, the most obvious exceptions being the boreal coniferous forests, many fresh-water habitats and a few intertidal marine habitats, eg EEL-GRASS.

The earliest undisputed angiosperm fossils (pollen and leaves) date from the Lower Cretaceous; angiosperms may have originated, and certainly underwent rapid diversification, in this period. Many of the Mesozoic groups of Gymnospermae have been suggested as the immediate ancestors of the angiosperms but the latter's evolutionary origins are not clear. The angiosperm's success in replacing the dominant gymnosperms of the Early Mesozoic has been ascribed to the protection offered to the ovules by the ovary, to the potential of stigmatic isolating mechanisms and of insect pollination to increase the rate of speciation, to faster growth and maturation and to the more efficient transport afforded by the more advanced phloem and particularly by the xylem vessels.

With the major exception of the conifers, important for softwood timber, most plants of economic importance are angiosperms and almost all agriculture and horticulture is based on them. They also provide food like the grains and pulses (eg Gramineae, Leguminosae), leaf and root vegetables (Cruciferae, Umbelliferae, Solanaceae), fruit (Rosaceae, *Citrus*), oil (Palamae, Cruciferae) and sugar (Gramineae, Chenopodiaceae), beverages, herbs (Labiatae, Umbelliferae) and spices, fodder (Gramineae), constructional materials such as timber (dicotyledenous trees), bamboo (Gramineae) and rattans (Palmae), fibres (Malvaceae, flax), many important drug plants (*Digitalis*, *Vinca*, *Atropa*) and much of the world's fuel.

Although perhaps 2000–5000 species are important in these ways, fewer than 200 are of major importance in world trade and fewer than 20 provide the bulk of the world's food. Many more species are cultivated as ornamental and amenity plants.

---

metal, one end diaphragm of which is fixed, the other being connected by a train of levers to a scale pointer which records the movements of the diaphragm under changing atmospheric pressure.

anesthesia (*Med*) US for ANAESTHESIA.

anesthesiologist (*Med*) US for ANAESTHETIST.

anethole (*Chem*) p-Propenyl anisole. CHMe = CHC$_6$H$_4$OMe. An ether forming the chief constituent of oil of aniseed, an essential oil.

aneuploid (*BioSci*) A cell or individual with missing or extra chromosomes or parts of chromosomes, then called a *segmental aneuploid*.

aneurysm (*Med*) Pathological dilatation, fusiform or saccular, of an artery.

angel beam (*Arch*) A horizontal member of a medieval roof truss, usually decorated with angels carved on the member.

angels (*Radar*) Radar echoes from an invisible and sometimes undefined origin. High-flying birds, insect swarms and certain atmospheric conditions can be responsible.

angina pectoris (*Med*) A condition characterized by the sudden onset of pain or crushing sensation in the chest which may radiate to the throat and arms. Frequently provoked by exercise and due to the narrowing of the coronary arteries.

angio- (*Genrl*) Prefix from Gk *angion*, denoting a case or vessel.

angioblast (*BioSci*) An embryonic mesodermal cell from which the vessels and early blood cells are derived.

angiocardiography (*Radiol*) The radiological examination of the heart and great vessels after injection of a CONTRAST MEDIUM.

angiogenesis (*BioSci*) The process of vascularization of a tissue involving the development of new capillary blood vessels.

angiography (*Radiol*) The study of the cardiovascular system by means of radio-opaque media.

angiology (*BioSci*) The study or scientific account of the anatomy of blood and lymph vascular systems.

angioma (*Med*) See HAEMANGIOMA.

angioneurotic oedema (*Med*) An immunologically mediated disease which produces dramatic and sometimes life-threatening swelling of the eyelids, lips, mucous membranes of the mouth and respiratory tract.

angioplasty (*Radiol*) A procedure where a balloon CATHETER is inserted into a blood vessel and the balloon inflated to widen a narrowed segment.

angiosperms (*BioSci*) The group of flowering plants that contains those seed plants in which the ovules are enclosed within carpels. See panel on ANGIOSPERMS (FLOWERING PLANTS).

angiotensin (*Med*) A group of peptides (angiotensin I, II and III) formed by the action of a series of enzymes on the precursor molecule. Angiotensin II causes a sharp rise in blood pressure.

angiotensin II receptor antagonists (*Pharmacol*) A group of drugs that work by blocking binding of angiotensin II to its receptor and thus have effects similar to those of angiotensin-converting enzyme inhibitors; used to treat hypertension. Examples are *candesartan*, *irbesartan*, *losartan*, *valsartan*.

angiotensin-converting enzyme inhibitors (*Pharmacol*) Drugs that inhibit the enzyme that converts the inactive form of angiotensin (I) to the active form (angiotensin II) and are used in the treatment of hypertension and heart failure. Captopril and Enalapril are common examples. Abbrev *ACE inhibitors*. See ANGIOTENSIN II RECEPTOR ANTAGONISTS.

angle (*Eng*) See ANGLE IRON.

angle (*MathSci*) The inclination of one line to another intersecting line; a measure of the rotation of one line around the common point required for it to coincide with the other. Angles are measured in degrees or radians, and a complete revolution, when the two arms of the angle coincide, is an angle of 360° or π radians.

angle bar (*Eng*) See ANGLE IRON.

angle bars (*Print*) On rotary presses, bars at an angle to transfer one or more webs of paper over each other, or the web to the other side of the press, or at right angles to its previous direction. Also *turner bars*.

angle bead (*Build*) A small rounded moulding placed at an angle formed by plastered surfaces to protect from damage.

angle bearing (*Eng*) A shaft-bearing in which the joint between base and cap is not perpendicular to the direction of the load, but is set at an angle.

angle block (*Build*) A small wooden block used in woodwork to make joints, esp right-angle joints, more rigid.

angle brace (*Build*) (1) Any bar fixed across the inside of an angle in a framework to render the latter more rigid. Also *angle tie, dragon tie*. (2) A special tool for drilling in corners where there is not room to use the cranked handle of the ordinary brace.

angle bracket (*Build*) A bracket projecting from the corner of a building beneath the eaves, and not at right angles to the face of the wall.

angle bracket (*Eng*) A bracket consisting of two sides set at right angles, often stiffened by a gusset. Also *gallows bracket*.

angle bracket (*Genrl*) A bracket (parenthesis symbol) with a pointed shape, '<' or '>'.

angle cleat (*Build*) A small bracket formed of angle iron, used to support or locate a member in a structural framework.

angle closer (*Build*) Loose term for CLOSER cut at an angle.

angle cutter (*Paper*) A machine in which the cross-cut knife is not at a right angle to the edge of the reel, for cutting sheets of paper from the reel. The parallelogram-shaped sheets were originally intended for conversion into banker envelopes.

angled deck (*Aero*) The flight deck of an aircraft carrier prolonged diagonally from one side of the ship, so that aircraft may fly off and land on without interference to or from aircraft parked at the bows. US *canted deck*.

angledozer (*CivEng*) A BULLDOZER with a blade able to be set skew to its tracks to cast aside material.

angle drilling (*MinExt*) A technique for drilling at an angle to an existing bore, achieved by special DOWNHOLE equipment, in order to straighten a bore, gather oil from a wide area to a production platform or to reach otherwise inaccessible formations, eg under a city. Also *deviated drilling*. See SLANT RIG.

angle elevation (*Surv*) The vertical angle measured above the horizontal, from the surveyor's instrument to the point observed.

angle float (*Build*) A plasterer's trowel, specially shaped to fit into the angle between adjacent walls of a room.

angle gauge (*Build*) A tool which is used to set off and test angles in a carpenter's, bricklayer's and mason's work.

angle grinder (*Eng*) Hand-held electrical tool with a rapidly rotating abrasive cutting or grinding disk used on metal or masonry.

angle iron (*Eng*) Mild-steel bar rolled to an L-shaped cross-section, used in structural work. Legs may be equal or unequal and leg lengths up to 800 mm are available. Also *angle, angle bar, angle steel, L-iron*.

angle modulation (*ICT*) Any system in which the transmitted signal varies the phase angle of an otherwise steady carrier frequency, ie phase and frequency modulation.

angle of acceptance (*Build*) The horizontal angle within which light rays should reach a window to ensure adequate penetration.

angle of advance (*Eng*) (1) The angle in excess of 90° by which the eccentric throw of a steam-engine valve gear is in advance of the crank. (2) The angle between the position of ignition and outer dead centre in a spark-ignition engine; optimizes combustion of the fuel.

angle of approach light (*Aero*) A light indicating an approach path in a vertical plane to a definite position in the landing area.

angle of arrival (*ICT*) Angle of elevation of a downcoming wave.

angle of attack (*Aero*) The angle between the CHORD LINE of an aerofoil and the relative airflow, normally the immediate flight path of the aircraft. Also, erroneously, *angle of incidence*.

angle-of-attack indicator (*Aero*) An instrument which senses the true angle of incidence to the relative airflow, and presents it to the pilot on a graduated dial or by means of an indicating light.

angle of bank (*Aero*) See ANGLE OF ROLL.

angle of bite (*Eng*) Maximum angle obtainable between the roll radius where it first contacts the metal and the line joining the centres of the two opposing rolls, when rolling metal. Also *angle of nip*.

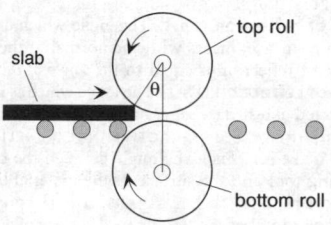

angle of bite   The angle θ above.

angle of contact (*Eng*) The angle subtended at the centre of a pulley by that part of the rim in contact with the driving belt.

angle of contact (*Phys*) The angle made by the surface separating two fluids (one of them generally air) with the wall of the containing vessel, or with any other solid surface cutting the fluid surface. For liquid–air surfaces, the angle of contact is measured in the liquid.

angle of cut-off (*Phys*) The largest angle below the horizontal at which a reflector allows the light source to be visible when viewed from a point outside the reflector.

angle of deflection (*Electronics*) The angle of the electron beam in a cathode-ray tube relative to the axis.

angle of departure (*ICT*) Angle of elevation of maximum emission of electromagnetic energy from an antenna.

angle of depression (*Surv*) The vertical angle measured below the horizontal, from the surveyor's instrument to the point observed. Also *plunge angle*.

angle of deviation (*Phys*) The angle which the incident ray makes with the emergent ray when light passes through a prism or any other optical device.

angle of dip (*Geol*) See DIP.

angle of flow (*ElecEng*) Angle, or fraction of alternating cycle, during which current flows, eg in a thyristor. Also *conduction angle*.

angle of friction (*Eng*) The angle between the normal to the contact surfaces of two bodies, and the direction of the resultant reaction between them, when a force is just tending to cause relative sliding.

angle of heel (*Ships*) The angle through which a floating vessel or pontoon tilts owing to eccentric placing of loads etc; the angle of inclination of a ship due to 'rolling' or to a 'list'. It is the angle formed between the transverse centre line of the ship when on 'even keel' and when inclined.

angle of incidence (*Aero*) Angular setting of any aerofoil to a reference axis. See ANGLE OF ATTACK and fig. at PROPELLER.

angle of incidence (*Phys*) The angle which a ray makes with the normal to a surface on which it is incident.

angle of lag (*ElecEng*) In ac circuit theory the phase angle by which the current lags behind, or leads ahead of, the voltage. Also *angle of lead*. See PHASE ANGLE.

angle of minimum deviation (*Phys*) The minimum value of the angle of deviation for a ray of light passing through a prism. By measuring this angle (θ) and also the angle of the prism (α), the refractive index of the prism may be calculated by means of the expression

$$n = \frac{\sin\frac{1}{2}(\alpha + \theta)}{\sin\frac{1}{2}\alpha}$$

angle of nip (*MinExt*) The maximum included angle between two approaching faces in a crushing appliance, such as a set of rolls, at which a piece of rock can be seized and entrained.

angle of obliquity (*Eng*) The deviation of the direction of the force between two gear teeth in contact, from that of their common tangent.

angle of pressure (*Eng*) The angle between a gear tooth profile and a radial line at its pitch point. See fig. at GEAR WHEEL.

angle of reflection (*Phys*) The angle which a ray, reflected from a surface, makes with the normal to the surface. The angle of reflection is equal to the *angle of incidence*.

angle of refraction (*Phys*) The angle which is made by a ray refracted at a surface separating two media with the normal to the surface. See REFRACTIVE INDEX, SNELL'S LAW.

angle of relief (*Eng*) The angle between the back face of a cutting tool and the surface of the material being cut.

angle of repose (*CivEng, PowderTech*) The greatest angle to the horizontal which the inclined surface of a heap of loose material (eg a powder, earth or gravel, or an embankment) can assume and remain stationary.

angle of roll (*Aero*) The angle through which an aircraft must be turned about its longitudinal axis to bring the lateral axis horizontal. Also horizontal *angle of bank*.

angle of slide (*MinExt*) Slope at which heaped rock commences to break away.

angle of stall (*Aero*) The angle of attack which corresponds with the maximum lift coefficient.

angle of twist (*Eng*) The angle through which one section of a shaft is twisted relative to another section when a torque is applied.

angle of view (*ImageTech*) The angle subtended at the centre of the lens by the limits of the image recorded; in still photography this is taken as the diagonal of the negative area but in motion picture and TV work it is the width of the frame.

angle plate (*Eng*) Cast-iron plate with the faces machined truly square and having slots on each face for clamping bolts. Used to hold work when marking off on a SURFACE PLATE or when machining on a lathe face plate or machine tool table.

angle ply laminate (*Eng*) Laminated material of wood or fibre-reinforced composites in which the angles between the orientation directions of the laminae are not 90°; commonly used angles are 30°, 45° and 60°.

angle rafter (*Build*) The rafter at the hip of a roof to receive the JACK RAFTERS. Also *angle ridge*. See HIP RAFTER.

angle shaft (*Build*) An angle bead which is enriched eg with a capital base.

anglesite (*Min*) Orthorhombic sulphate of lead, a common lead ore; named after the original locality, Anglesey.

angle staff (*Build*) A strip of wood placed at an angle formed by plastered surfaces to protect from damage. A rounded staff is called an *angle bead*.

angle steel (*Eng*) See ANGLE IRON.

angle stone (*Build*) See QUOIN.

angle support (*ElecEng*) A transmission line tower or pole placed at a point where the line changes its direction. Such a tower or pole differs from a normal tower or pole in that it has to withstand a force tending to overturn it (due to the resultant pull of the conductors).

angle tie (*Build*) See ANGLE BRACE.

Anglian (*Geol*) A glacial stage in the late Pleistocene. See QUATERNARY.

angora (*Textiles*) The hair of the angora rabbit or the soft yarn and fabric made from it.

Angst (*Psych*) German word for anxiety, anguish or psychological distress; regarded by existentialists as a fundamental aspect of confronting the reality of life, the universe, etc.

angstrom (*Phys*) Unit of wavelength for electromagnetic radiation covering visible light and X-rays. Equal to 10^{-10} m. The unit is also used for interatomic spacings. Symbol Å. Superseded by nanometre (= 10^{-9} m) but still used widely in crystallography. Named after the Swedish physicist A J Ångström (1814–74).

anguilliform (*BioSci*) Eel-like in shape.

Anguilliformes (*BioSci*) An order of OSTEICHTHYES comprising fish, elongate in form, in which the pelvic fins and girdle are absent or reduced. Eels.

angular acceleration (*Phys*) The rate of change of angular velocity; usually expressed in rad s^{-2}.

angular contact bearing (*Eng*) A ball bearing for radial and thrust loads in which a high shoulder on one side of the outer race takes the thrust.

angular diameter (*Astron*) The observed diameter of any celestial object expressed as the angle subtended by its diameter as perceived by the observer.

angular displacement (*Phys*) The angle turned through by a body about a given axis, or the angle turned through by a line joining a moving point to a given fixed point.

angular distance (*MathSci*) (between two points). The angle between the two lines from a given reference to the points in question.

angular distance of stars (*Astron*) The observed angular separation of two stars as perceived by the observer.

angular distribution (*Phys*) The distribution relative to the incident beam of scattered particles or the products of nuclear reactions.

angular divergence (*BioSci*) The angle subtended at the mid-line of an apical meristem of a shoot by the mid-point of two successive leaf primordia. This varies between species but, where the phyllotaxis is spiral, it is commonly the Fibonacci angle 137·5°.

angular frequency (*Phys*) The frequency of a steady recurring phenomenon, in rad s^{-1}, ie frequency in hertz multiplied by 2π. Symbol ω or p. Also *pulsatance*, *radian frequency*.

angular magnification (*Phys*) The ratio of the angle subtended at the eye by an image formed by an optical instrument to the angle subtended by the object at the unaided eye.

angular momentum (*Phys*) The moment of the linear momentum of a particle about an axis. Any rotating body has an angular momentum about its centre of mass, its *spin angular momentum*. The angular momentum of the centre of mass of a body relative to an external axis is its *orbital angular momentum*. In atomic physics, the orbital angular momentum of an electron is *quantized* and can only have values which are exact multiples of DIRAC'S CONSTANT. In particle physics, the angular momentum of particles which appear to have spin energy is quantized to values that are multiples of half of Dirac's constant. See MOMENTUM.

angular thread (*Eng*) See VEE THREAD.

angular velocity (*Phys*) The rate of change of angular displacement, usually expressed in rad s^{-1}.

Angus–Smith process (*Build*) An anti-corrosion process applied to sanitary ironwork; this is heated to about 316°C immediately after casting, and then plunged into a solution of four parts coaltar or pitch, three parts prepared oil and one part paranaphthalene heated to about 149°C. See BOWER–BARFF PROCESS.

anharmonic (*Electronics*) Said of any oscillation system in which the restoring force is non-linear with displacement, so that the motion is not simple harmonic.

anharmonic ratio (*MathSci*) See CROSS-RATIO.

anhedonia (*Psych*) The inability to feel pleasure; the loss of interest in formerly pleasurable pursuits.

anhedral (*Aero*) See DIHEDRAL ANGLE.

anhedral (*Geol*) See ALLOTRIOMORPHIC.

anhidrosis (*Med, Vet*) Absence of sweating. Although rare in humans, it affects horses in humid tropical countries, and is characterized by an inability to sweat after exercise. It is believed to be due to prolonged overstimulation of the sweat glands by adrenaline.

anhydrides (*Chem*) Substances, including organic compounds and inorganic oxides, which either combine with water to form acids, or may be obtained from the latter by the elimination of water.

anhydrite (*Build, Min*) Naturally occurring anhydrous calcium sulphate which readily forms gypsum and from which anhydrite plaster is made by grinding to powder with a suitable accelerator.

anhydrite process (*ChemEng*) A process for the manufacture of sulphuric acid from anhydrite $CaSO_4$. The mineral is roasted with a reducing agent and certain other minerals in large kilns, so that SO_2 gas in relatively low concentration is recovered and after cleaning is passed to a specially designed CONTACT PROCESS. The solid residue is, under normal conditions, readily converted into cement and this forms an economic factor in the process. In the UK the process has a special significance as it provides a large potential of sulphuric acid from an indigenous source of sulphur.

anhydrous (*Chem*) Containing no water. With crystalline oxides, salts, etc, it emphasizes that they contain no combined water, eg WATER OF CRYSTALLIZATION.

anhydrous lime (*Build*) See LIME.

anilides (*Chem*) N-phenyl amides. A group of compounds in which the hydrogen of the amino group in aniline is substituted by organic acid radicals. The most important compound of this class is *acetanilide*.

aniline (*Chem*) *Phenylamine, aminobenzene.* $C_6H_5NH_2$. A colourless oily liquid, mp $-8°C$, bp $189°C$, rel.d. $1·024$, slightly soluble in water; manufactured by reduction of nitrobenzene with iron shavings and hydrochloric acid at $100°C$. Basis for the manufacture of dyestuffs, pharmaceuticals, plastics (with methanal) and many other products.

aniline black (*Chem*) An azine dye, produced by the oxidation of aniline on the fabric.

aniline dyes (*Chem*) A general term for all synthetic dyes having aniline as their base.

aniline foils (*Print*) Blocking foils which contain dyestuff; used chiefly for leather.

aniline formaldehyde (*Chem*) Synthetic resin formed by the polycondensation of aniline with formaldehyde (methanal).

aniline oil (*Chem*) A coaltar fraction consisting chiefly of crude aniline.

anilinium chloride (*Chem*) *Phenylammonium chloride.* $C_6H_5NH_2HCl$. Mp $198°C$, bp $245°C$, rel.d. $1·22$, white crystals, soluble in most organic solvents and water.

anima (*Psych*) A term used in Jungian psychology to denote the unconscious feminine component in men. Cf ANIMUS.

animal charcoal (*Chem*) The carbon residue obtained from carbonization of organic matter such as blood, flesh, etc.

animal electricity (*BioSci*) A term for the ability possessed by certain animals of giving powerful electric shocks (eg electric eel).

animal field (*BioSci*) In developing BLASTULAE, a region distinguished by the character of the contained yolk granules, and representing the first rudiment of the GERM BAND.

animal pole (*BioSci*) In the developing ovum, the apex of the upper hemisphere, which contains little or no yolk; in the blastula, the corresponding region, wherein the micromeres lie.

animal-sized (*Paper*) Said of paper which has been sized by passing the sheet or web through a bath containing a solution essentially of gelatine and then drying. See SIZE.

animation (*ImageTech*) Apparent movement produced by recording step by step a series of still drawings, three-dimensional objects or computer-generated images.

animism (*Psych*) Attributing feelings and intentions to non-living things. In Piagetian theory children's thinking is characterized by animism in the years 2 to 6.

animus (*Psych*) A term used in Jungian psychology to denote the unconscious masculine component in women. Cf ANIMA.

anion (*Phys*) Negative ion, ie atom or molecule which has gained one or more electrons in an electrolyte, and is therefore attracted to an anode, the positive electrode. Anions include all non-metallic ions, acid radicals and the hydroxyl ion. In a primary cell, the deposition of anions on an electrode makes it the negative pole. Anions also exist in gaseous discharge. Cf CATION. See panel on BONDING.

anionic detergents (*BioSci*) Detergents in which the hydrophilic function is fulfilled by an anionic grouping. Important synthetic species are aliphatic sulphate esters, eg sodium dodecyl sulphate (SDS or SLS).

anionic polymerization (*Chem*) Polymerization using anionic catalyst such as butyl lithium. See CHAIN POLYMERIZATION.

anisaldehyde (*Chem*) 4-methoxybenzaldehyde. Colourless liquid; bp $248°C$, occurring in aniseed, and used in perfumery.

anisidines (*Chem*) *Amino-anisoles, methoxyanilines.* $CH_3OC_6H_4NH_2$. Bases similar to aniline. Intermediates for dyestuffs.

aniso- (*Genrl*) Prefix from Gk *an*, not; *isos*, equal.

anisocercal (*BioSci*) Having the lobes of the tail-fin unequal.

anisodactylous (*BioSci*) Of birds, having three toes turned forward and one turned backward when perching, as in the Passeriformes.

anisodesmic structure (*Crystal*) A structure giving a crystal marked difference between its bond strengths in the intersecting axial planes.

anisogamete (*BioSci*) A gamete differing from the other conjugant in form or size. Adj *anisogamous*.

anisogamy (*BioSci*) Sexual fusion of gametes that differ in size but not necessarily in form. See ISOGAMY, OÖGAMY.

anisokont (*BioSci*) Having two flagella unequal in length but otherwise more or less similar. Cf HETEROKONT, ISOKONT.

anisole (*Chem*) *Phenyl methyl ether.* $C_6H_5OCH_3$. A colourless liquid, bp $155°C$.

anisomeric (*Chem*) Not isomeric.

anisopleural (*BioSci*) Bilaterally asymmetrical.

anisotonic (*Chem*) Not isotonic.

anisotropic (*BioSci*) (1) Of ova, having a definite polarity, in relation to the primary axis passing from the animal pole to the vegetal pole. (2) Optical properties of oriented arrays of molecules, classic example being the A (anisotropic) band of the sarcomere of striated muscle. N *anisotropy*.

anisotropic (*Min, Phys*) A term describing any material whose physical properties depend upon direction relative to some defined axes (eg crystalline axes, fibre orientation, draw direction) in the material. These properties normally include elasticity, thermal and electrical conductivity, permittivity, permeability, refractive index, strength, etc. Also said of such processes as ETCHING when certain directions are preferred.

anisotropic conductivity (*Phys*) The property of a body which has a different conductivity for different directions of current flow (electrical or thermal).

anisotropic dielectric (*Phys*) Dielectric in which electric effects depend on the direction of the applied field, as in many crystals.

anisotropic etching (*Electronics*) Describes an etching process which proceeds preferentially in one direction. In semiconductor processing when dry etching is accomplished with energetic ion bombardment, the lateral etch rate may be substantially less than the vertical rate so that under-cutting is avoided, allowing narrow, steep-sided features to be defined. Cf ISOTROPIC ETCHING.

anisotropic liquid (*Chem*) See LIQUID CRYSTAL and panel on LIQUID CRYSTAL DISPLAYS.

anisotropy (*Min, Phys*) A term describing a property of a substance that depends on direction as revealed by measurement. See ANISOTROPIC.

ankerite (*Min*) A carbonate of calcium, magnesium and iron.

ankylosing spondylitis (*Med*) Rheumatoid arthritis of the spine, which may progress to cause complete spinal and thoracic rigidity.

ankylosis (*Med*) Fixation of a joint by fibrous bands within it, or by pathological union of the bones forming the joint. Also *anchylosis*.

ankylostomiasis (*Med*) An infection in the small intestine by two parasitic nematode worms (*Ankylostoma duodenale* and *A. americanum*), which produces iron deficiency for those on an inadequate diet. Also *hookworm disease*.

anlage (*BioSci*) See PRIMORDIUM.

annabergite (*Min*) Hydrated nickel arsenate, apple-green monoclinic crystals, rare, usually massive. Associated with other ores of nickel. Also *nickel bloom*.

anneal (*BioSci*) To reform the duplex structure of a nucleic acid.

anneal (*Eng*) To heat in a furnace for a period followed by slow cooling in order to bring about softening or relaxation of internal stress. Commonly applied to metals or glass processing.

annealing (*Eng, Phys*) A heat treatment process intended to bring about a soft or stress-free state in worked materials. It usually involves heating to a temperature where diffusion or stress relaxation can occur, holding for a period and then cooling slowly so as to minimize thermal gradients which could reintroduce stress by differential thermal contraction. In nuclear engineering it refers, additionally, to the process of removing the dislocations and swelling which occurs in eg graphite, under neutron bombardment. See RECRYSTALLIZATION, STRESS RELIEF ANNEALING, WIGNER EFFECT.

annealing furnace (*Eng*) Batch-worked or continuous oven or furnace with controllable atmosphere in which metal, alloy or glass is annealed.

annealing point (*Glass*) One of the reference temperatures in glass production. See panel on GLASSES AND GLASS-MAKING.

Annelida (*BioSci*) A phylum of metameric Metazoa, in which the perivisceral cavity is coelomic and there is only one somite in front of the mouth. Typically there is a definite cuticle and chitinous setae arising from pits of the skin. The central nervous system consists of a pair of preoral ganglia connected by commissures to a postoral ventral ganglionated chain; if a larva occurs it is a trochophore. Fossils do not exist but are deduced from tracks and burrows. Earthworms, ragworms, leeches.

annexin (*BioSci*) Any of a family of calcium-binding proteins that bind to cell membrane phospholipids and may be involved in exocytosis, eg LIPOCORTIN.

annihilation (*Phys*) Spontaneous conversion of a particle and its antiparticle into radiation, eg positron and electron yielding two gamma-ray photons each of energy 0·511 MeV.

annihilation radiation (*Phys*) The radiation produced by the annihilation of an elementary particle with its corresponding antiparticle.

annihilator (*MathSci*) An object, y, whose product with a given object, x, is 0. Here x and y may be elements of rings, functions, etc, and need not be the same type of object so long as xy is defined. An annihilator of a set X is y which is an annihilator of every element of X. The annihilator of x is the set of all such individual annihilators.

annite (*Min*) The ferrous iron end-member of the biotite series of micas.

annual (*BioSci*) A plant that flowers and dies within a period of one year from germination. Cf BIENNIAL, EPHEMERAL, PERENNIAL.

annual equation (*Astron*) A periodic variation in the motion of the Moon, which arises from variations in the solar attraction due to the eccentricity of the Earth's orbit. Its period is one year.

annual load factor (*ElecEng*) The load factor of a generating station, supply-undertaking or consumer, taken over a whole year.

annual parallax (*Astron*) The apparent angular displacement of a star when measured at two points separated by a distance equal to the radius of the Earth's orbit round the Sun. It is significant only for the stars with distances less than around 30 pc, and is largest for the nearest star, PROXIMA CENTAURI (0·71 arcseconds). Also *heliocentric parallax*.

annual ring (*BioSci*) A GROWTH RING formed over a year.

annular bit (*Build*) A bit which cuts an annular (ring-shaped) channel and leaves intact a central cylindrical plug.

annular borer (*CivEng*) A rock-boring tool which does the work of an ANNULAR BIT, and provides a means of obtaining a core showing a section of the strata.

annular combustion chamber (*Aero*) A gas turbine combustion chamber in which the perforated flame tube forms a continuous annulus within a cylindrical outer casing.

annular eclipse (*Astron*) See ECLIPSE.

annular gear (*Eng*) A ring in the shape of an annulus with gear teeth cut on the periphery for engagement with a pinion. Usually shrunk fit onto a mating diameter, eg starter ring on automobile flywheel.

annular space (*MinExt*) The space between the CASING and the producing or drilling bore.

annular thickening (*BioSci*) The secondary wall deposited in the form of discrete transverse rings or hoops, in tracheids and vessel elements of xylem, esp PROTOXYLEM.

annular vault (*Build*) See BARREL VAULT.

annulated column (*Arch*) A column formed of slender shafts clustered together, or sometimes around a central column, and secured by stone or metal bands.

annulus (*BioSci*) (1) Generally, any ring-shaped structure. (2) A membranous frill present on the stipe of some agarics. (3) A patch or a crest of cells with thickened walls occurring in the wall of the sporangium of ferns, and bringing about dehiscence by setting up a strain as they dry. (4) A zone of cells beneath the operculum of the sporangium of a moss, which break down and assist in the liberation of the operculum. (5) The fourth digit of a pentadactyl forelimb. (6) In Arthropoda, subdivision of a joint forming jointlets. (7) In Hirudinea, a transverse ring subdividing a somite externally. Adj annular, annulate.

annulus (*MathSci*) A plane surface bounded by two concentric circles, like the surface of a washer.

annunciator (*CivEng*) Any device for indicating audibly the passage of a train past a point.

annunciator (*ElecEng*) Arrangement of indicators which display details on operational condition and functioning of complex plant. Also *indicator*.

anode (*Electronics*) (1) In a valve or tube, the electrode held at a positive potential with respect to a cathode, and through which positive current generally enters the vacuum or plasma, through collection of electrons. US *plate*. (2) The positive electrode of battery or cell. See CATHODE, ULTOR.

anode breakdown voltage (*Electronics*) The voltage required to trigger a discharge in a cold-cathode glow tube when the starter gap (if any) is not conducting. It is measured with any grids or other electrodes earthed to cathode.

anode brightening (*Eng*) See ELECTROLYTIC POLISHING.

anode characteristic (*Electronics*) Graph relating anode current and anode voltage for an electron tube.

anode dark space (*Electronics*) Dark zone near the anode in a glow-discharge tube.

anode dissipation (*Electronics*) Generally, the energy produced at the anode of a thermionic tube and wasted as heat owing to the bombardment by electrons; specifically, the maximum permissible power which may be dissipated at the anode.

anode drop (*Electronics*) The voltage between the positive column and the anode of a gas-discharge tube. It may be positive, zero or negative, depending on the gas pressure, but not the discharge current. Also *anode fall*.

anode efficiency (*Electronics*) The ratio of ac power in the load circuit to the dc power supplied to the anode of a valve amplifier or oscillator.

anode fall (*Electronics*) See ANODE DROP.

anode feed (*Electronics*) Supply of direct current to anode of a tube, generally decoupled, so that the supply circuit does not affect the condition of operation of the tube.

anode glow (*Electronics*) Luminous zone on anode side of positive column in a gas-discharge tube.

anode modulation (*Electronics*) Insertion of the modulating signal into the anode circuit of a valve, which is oscillating or is rectifying the carrier. Also *plate modulation*.

anode mud (*Eng*) See ANODE SLIME.

anode polishing (*Eng*) See ELECTROLYTIC POLISHING.

anode saturation (*Electronics*) Limitation of current through the anode of a valve, arising from current, voltage, temperature or space charge.

anode shield (*Electronics*) Electrode used in high-power gas tubes to shield the anode from damage by ion bombardment.

anode slime (*Eng*) Residual slime left when anode has been electrolytically dissolved. It may contain valuable by-product metals. Also *anode mud*.

anode strap (*ICT*) Connecting strip between alternate anode segments of a multi-cavity magnetron. Used for mode selection and control.

anode tap (*Electronics*) Tapping point on the inductance coil of a tuned-anode circuit, to which the anode is connected. The position of the tap is adjusted so that the tube operates into the optimum impedance.

anodic etching (*ElecEng*) A method of preparing metals for electrodeposition by making them the anode in a suitable electrolyte and at a suitable current density.

anodic oxidation (*Chem*) Oxidation, ie removal of electrons from a substance, by placing it in the anodic region of an electrolytic cell. The substance to be oxidized may be either a part of the electrolyte or the anode itself. See ANODIZING.

anodic protection (*Eng*) A system for passivating steel by making it the anode in a protective circuit. Cf CATHODIC PROTECTION.

anodic treatment (*Chem*) See ANODIZING.

anodized (*Eng*) A description of metal surface protected by chemical or electrolytic action. Commonly refers to aluminium where an oxide layer is produced which acts as a barrier to corrosive agents which would otherwise attack the metal. Thin films may be dyed with bright colours before sealing to produce reflective, highly decorative finishes. Small steel items in marine construction are frequently anodized, being more convenient but less effective than GALVANIZING. Thicker films (engineering finishes) render the surface resistant to abrasion. Anodic films on titanium yield colourful interference films often applied to jewellery and art decorative work.

anodizing (*Chem*) Electrolytic process which increases the thickness of the layer of oxide on the surface of metals and alloys when these form the anode of the electrolytic cell. Result is an *anodized* surface. Also *anodic treatment*.

anodontia (*BioSci*) Absence of teeth.

anoestrus (*BioSci*) In mammals, a resting stage of the oestrus cycle occurring between successive heat periods.

anomalistic month (*Astron*) The interval (amounting to 27·554 55 days) between two successive passages of the Moon in its orbit through perigee.

anomalistic year (*Astron*) The interval (equal to 365·259 64 mean solar days) between two successive passages of the Sun, in its apparent motion, through perigee.

anomaloscope (*Phys*) An instrument for detection and classification of defective colour vision. Two colours are mixed, and the result matched with a third.

anomalous dispersion (*Phys*) The type of dispersion given by a medium having a strong absorption band, the value of the refractive index being abnormally high on the longer wave side of the band, and abnormally low on the other side. In the spectrum produced by a prism made of such a substance the colours are, therefore, not in their normal order.

anomalous magnetization (*ElecEng*) Irregular distribution of magnetization, eg when consequent poles exist as well as main poles on a magnetic circuit.

anomalous scattering (*Phys*) See SCATTERING.

anomalous secondary thickening (*BioSci*) The production of new vascular tissue in plants by a secondarily formed cambium.

anomalous viscosity (*Phys*) A term used to describe liquids which show a decrease in viscosity as their rate of flow (ie velocity gradient or shear strain rate) increases. Also *non-Newtonian fluids* or *pseudo-plastic fluids*. Advantage is taken of this behaviour when injection-moulding polymer melts.

anomaly (*Astron*) The angle between the radius vector of an orbiting body and the major axis of the orbit, measured from eg perihelion or periastron in the direction of motion.

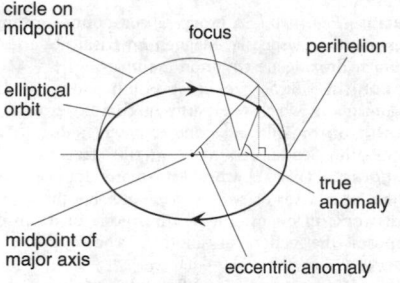

anomaly

anomaly (*Genrl*) Any departure from the strict characteristics of the type.

anomer (*Chem*) In carbohydrate chemistry, one of two isomers differing in conformation at the aldehydic carbon in a ring form.

anomeristic (*BioSci*) Of metameric animals, having an indefinite number of segments or somites.

anomie (*Psych*) Breakdown, in a society or group, of the normal social structure, often as a consequence of some catastrophic event, although also applied to highly developed societies where there is isolation and an absence of interpersonal interactions. Also *anomy*.

anonymous ftp (*ICT*) A partial implementation of a FILE TRANSFER PROTOCOL which only allows file downloading.

anorectic (*Med*) Substance that suppresses feeling of hunger.

anorexia (*Med, Psych*) Loss of appetite. See ANOREXIA NERVOSA.

anorexia nervosa (*Psych, Med*) Chronic failure to eat due to fear of gaining weight or to emotional disturbance; results in malnutrition, semi-starvation and sometimes death.

anorthic system (*Crystal*) See TRICLINIC SYSTEM.

anorthite (*Min*) The calcium end-member of the plagioclase group of feldspars; silicate of calcium and aluminium, occurring in some basic igneous and other rocks.

anorthoclase (*Min*) A triclinic sodium-rich high-temperature sodium–potassium feldspar; occurs typically in volcanic rocks, and is also known from a syenite, larvikite, from S Norway, which is widely used for facing buildings.

anorthosite (*Geol*) A coarse-grained plutonic igneous rock, consisting almost entirely of plagioclase, near labradorite in composition.

anosmatic (*BioSci*) Lacking the sense of smell. Also *anosmic*.

anosmia (*Med*) Loss, partially or completely, of the sense of smell.

anoxaemia (*BioSci*) Deficiency of oxygen in the blood; any condition of insufficient oxygen supply to the tissues; any condition which retards oxidation processes in the tissues and cells. Also *anoxemia*. See also ANOXIA.

anoxia (*BioSci*) A condition in which living cells receive little or no oxygen.

anoxybiosis (*BioSci*) Life in absence of oxygen.

Anseriformes (*BioSci*) An order of birds with webbed feet, unusual in the possession of an evaginable penis. They are all aquatic forms, living on animals found in the mud at the bottom of shallow waters and in marshes; some are powerful fliers. Geese, ducks, screamers, swans.

Anschauung (*Psych*) Direct perception through the senses; an attitude or point of view.

ANSI (*ICT*) Abbrev for *American National Standards Institute*.

ANSI keyboard (*ICT*) A keyboard configuration in which the keys are programmed to provide the set of 256 characters defined by the American National Standards Institute.

answer print (*ImageTech*) First print from the edited negative of a film shown to the producer for approval before release.

antacids (*Pharmacol*) A group of compounds given in the treatment of dyspepsia. Magnesium trisilicate and aluminium hydroxide are common examples.

antagonism (*BioSci*) A relationship between different organisms in which one partly or completely inhibits the growth of, or kills, a second, esp when due to a toxic metabolite. See ALLELOPATHY, ANTIBIOTIC.

antagonist (*BioSci*) A word with many applications, used where one factor or structure opposes another, eg a drug that works in opposition to a hormone or a muscle that opposes the action of another. The opposite of an AGONIST.

antagonizing screws (*Surv*) See CLIP SCREWS.

antapex (*Astron*) See SOLAR ANTAPEX.

Antares (*Astron*) Prominent red supergiant star in the constellation Scorpius, a visual binary system. Distance 130 pc. Also *Alpha Scorpii*.

ante- (*Genrl*) Prefix from Lt *ante*, before.

antebrachium (*BioSci*) The region between the brachium and the carpus in land vertebrates; the forearm.

antecedent (*MathSci*) (1) In logic, the term of a conditional statement on which the other depends; ie in the material implication 'if *p*, then *q*', *p* is the antecedent, and *q* is the consequent. (2) (archaic) The NUMERATOR of a ratio *a*:*b*.

antecedent drainage (*Geol*) A river system that has maintained its original course despite subsequent folding or uplift.

antechamber (*Eng*) A small auxiliary combustion chamber, used in some compression-ignition engines, in which partial combustion of the fuel is used to force the burning mixture into the cylinder, so promoting more perfect combustion.

antecubital (*BioSci*) In front of the elbow.

antefixae (*Arch*) Ornaments placed at the eaves and cornices of ancient buildings to hide the ends of the roof tiles; sometimes perforated to convey water away from the roof.

antenna (*BioSci*) (1) In Arthropoda, one of a pair of anterior appendages, normally many-jointed and of sensory function. (2) In angler fish the elongate first dorsal fin-ray, which bears terminally a skinny flap, used by the fish to attract prey. Pl *antennae*. Adjs *antennary*, *antennal*.

antenna (*ICT*) A structure for receiving or transmitting electromagnetic signals; an aerial. Pl *antennas*.

antenna changeover switch (*ICT*) Switch used for transferring an antenna from the transmitting to the receiving equipment, and vice versa, protecting the receiver.

antenna download (*ICT*) Wire running from the elevated part or conductor of an antenna down to the transmitting or receiving equipment.

antenna effect (*ICT*) (1) Errors arising when a directional antenna, used in an electronic navigation system, picks up radiation from a non-intended direction, as a result of imperfections in the radiation pattern. (2) Spurious effects in radio-direction-finding systems caused by stray capacitance between a loop antenna and earth.

antenna efficiency (*ICT*) See RADIATION EFFICIENCY.

antenna feeder (*ICT*) The transmission line or cable by which energy is fed from the transmitter to the antenna.

antenna field (*ICT*) Map showing electromagnetic field strength produced by an antenna in the form of contour lines joining points of equal field intensity; it may be in azimuth or any plane of elevation. Also *radiation pattern*.

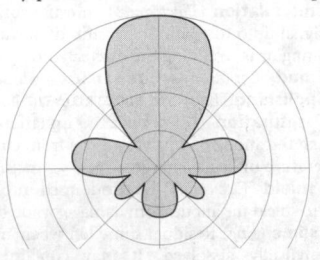

antenna field Showing main and side lobes.

antenna gain (*ICT*) The ratio of maximum energy flux from an antenna, to that which would have been received from a non-directional aerial radiating the same power. See DIRECTIONAL GAIN.

antenna impedance (*ICT*) Complex ratio of voltage to current at the point where the feeder is connected.

antennal complex (*BioSci*) Complexes of protein and light-harvesting pigment, organized into arrays (antennae) on photosynthetic membranes, that capture photon energy and transfer it to reaction centres. See PHOTOSYSTEM I, PHOTOSYSTEM II.

antennal glands (*BioSci*) The principal excretory organs of Crustacea. They open at the bases of the osmoregulatory appendages from which they take their name. Also *maxillary glands*.

antenna load (*ElecEng*) See DUMMY LOAD.

antenna noise temperature (*ICT*) The temperature of a BLACK BODY that, when placed around an antenna similar to the real one, but loss-free and perfectly matched to the receiver, produces the same noise power, within a specified frequency band, as the real antenna in its operating environment.

antenna resistance (*ICT*) Total power supplied to an antenna system divided by the square of a specified current, eg in the feeder, or at the earth connection of an open-wire antenna.

antenna-shortening capacitor (*ICT*) A capacitor connected in series with an antenna to allow operation at a frequency other than its natural resonant one. See LOADED ANTENNA.

antennule (*BioSci*) A small antenna; in some Arthropoda (eg the Crustacea) that possess two pairs of antennae, one of the first pair.

antepetalous (*BioSci*) Inserted opposite to the petals. Also *antipetalous*.

anteposition (*BioSci*) Situation opposite, and not alternate to, another plant member.

anterior (*BioSci*) (1) That end of a motile organism which goes first during locomotion. (2) In animals in which cephalization has occurred, nearer the front or cephalad end of the longitudinal axis. (3) In human anatomy, ventral. (4) The side of a flower next to the bract, or facing the bract.

antero- (*Genrl*) Prefix from *anterior*, former.

anterograde amnesia (*Psych*) Loss of memory for events after injury to the brain or mental trauma, with little effect on information acquired previously.

antesepalous (*BioSci*) Inserted opposite to the sepals. Also *antisepalous*.

anthelion (*EnvSci*) A mock sun appearing at a point in the sky opposite to and at the same altitude as the Sun. The phenomenon is caused by the refraction of sunlight by ice crystals.

anthelmintic (*Pharmacol*) A drug used against parasitic worms.

anther (*BioSci*) Fertile part of a stamen, usually containing four sporangia, and producing pollen.

anther culture (*BioSci*) The aseptic culture on suitable medium of anthers, with the possibility of the production of haploid CALLUS, embryoid and plantlets. If treated with eg *colchicine* these plantlets may give rise to diploid plants, autodiploids, that are completely homozygous.

antheridiophore (*BioSci*) A specialized branch of a thallus bearing antheridia. Also *antheridial receptacle*.

antheridium (*BioSci*) The organ that produces the male gametes in lower plants.

antherozoid (*BioSci*) A motile male gamete, spermatozoid or sperm.

anthesis (*BioSci*) The opening of a flower bud; by extension, the duration of life of any one flower, from the opening of the bud to the setting of fruit.

antho- (*Genrl*) Prefix from Gk *anthos*, flower.

Anthocerotopsida (*BioSci*) A class of the Bryophyta containing the hornworts, which differ from the liverworts (Hepaticae) in that the gametophyte is always thalloid and that the sporophyte shows a relatively long-continued growth, and spore production, from an intercalary meristem near its base. Also *Anthocerotae*.

anthocyanins (*BioSci*) A large group of water-soluble, flavonoid, glycoside pigments in cell vacuoles; responsible for the red, purple and blue colours of flowers, fruit and leaves in most flowering plants. Cf BETALAINS.

anthogenesis (*BioSci*) A form of parthenogenesis in which both males and females are produced by asexual forms, as in some aphids.

anthophilous (*BioSci*) Flower-loving; feeding on flowers.

anthophore (*BioSci*) An elongation of the floral receptacle between the calyx and corolla.

anthophyllite (*Min*) An orthorhombic amphibole, usually massive, and normally occurring in metamorphic rocks; magnesium iron silicate, of low aluminium content.

Anthophyta (*BioSci*) (1) Usually Angiospermae. See ANGIOSPERMS. (2) Rarely, Spermatophyta.

Anthozoa (*BioSci*) A class of CNIDARIA in which alternation of generations does not occur, the medusoid phase being entirely suppressed; the polyps may be solitary or colonial; the gonads are of endodermal origin. Corals, sea anemones and sea pens. Also *Actinozoa*.

anthracene (*Chem*) $C_{14}H_{10}$. Colourless, blue fluorescent crystals, mp 218°C, bp 340°C, a valuable raw material for dyestuffs obtained from the fraction of coaltar boiling above 270°C. Anthracene represents a group of polycyclic compounds with a series of three benzene rings condensed together. Carcinogenic. Useful as a scintillator in photo-electric detection of β-particles.

anthracene

anthracene oil (*Chem*) A coaltar fraction boiling above 270°C, consisting of anthracene, phenanthrene, chrysene, carbazole and other aromatic hydrocarbon oils.

anthracite (*Geol*) The highest metamorphic rank of coal. See RANK OF COAL.

anthracnose (*BioSci*) One of a number of plant diseases characterized by black, usually sunken, lesions; mostly caused by one of the fungi of the Melanconiales.

anthracosis (*Med*) 'Coal-miner's lung', produced by inhalation of coal dust.

anthraflavine (*Chem*) An anthraquinone vat dyestuff, which dyes cotton greenish yellow, obtained by heating 2-methylanthraquinone with alcoholic potassium hydroxide at 150°C.

anthranil (*Chem*) The intramolecular anhydride of anthranilic acid (2-aminobenzoic acid), intermediate in the synthesis of indigo.

anthranilic acid (*Chem*) *2-aminobenzoic acid*. $C_6H_4(COOH)NH_2$. Obtained from phthalimide by the Hofmann reaction, an oxidation product of indigo.

anthraquinone (*Chem*) *Diphenylene diketone*. $C_6H_4(CO)_2$ C_6H_4. Yellow needles or prisms, which sublime easily, mp 285°C, bp 382°C. More closely related to diketones than to quinones. Obtained by the oxidation of anthracene with sulphuric acid and chromic (VI) acid. Parent substance of an important group of dyes, including alizarin.

anthrax (*Med, Vet*) An acute infectious disease caused by the anthrax bacillus, communicable from animals to humans in whom it causes cutaneous malignant pustules and lung, intestinal and nervous system infection. A notifiable disease in animals. Also *woolsorter's disease*.

anthraxolite (*Min*) A member of the asphaltite group.

anthraxylon (*Min*) One of the constituents of coal, derived from the lignin of the plants forming the seam.

anthrop-, anthropo- (*Genrl*) Prefixes from Gk *anthropos*, denoting man or human.

anthropic principle (*Astron*) The idea that the nature of the universe is constrained due to our presence as observers.

anthropo- (*Genrl*) See ANTHROP-.

anthropogenic (*BioSci*) Anything, particularly a change, resulting from or influenced by human activities.

anthropogenic (*NucEng*) Human-made, the opposite of natural; used particularly of radiation and nuclear particles.

anthropoid (*BioSci*) Resembling humans; pertaining to, or having the characteristics of the Anthropoidea.

Anthropoidea (*BioSci*) A suborder of the primates that includes monkeys, apes and humans.

anthropomorph (*BioSci*) A conventional design of the human figure; resembling a human in form or in attributes.

anthropophyte (*BioSci*) A plant introduced incidentally in the course of cultivation.

anti- (*Genrl*) Prefix from Gk *anti*, against.

anti-aldoximes (*Chem*) The stereoisomeric form of aldoximes in which the H and the OH groups are far removed from each other. See SYNALDOXIMES.

anti-aliasing (*ImageTech*) Treatment of video picture signal elements to reduce the effects of ALIASING.

Antian (*Geol*) A temperate stage in the Pleistocene. See QUATERNARY.

anti-auxin (*BioSci*) Compound that in low concentrations will directly interfere with AUXIN action, eg tri-chlorobenzoic acid.

antibaryon (*Phys*) Antiparticle of a baryon, ie a hadron with a baryon number of −1. The term BARYON is often used generically to include both.

antibiosis (*BioSci*) A state of mutual antagonism. Cf SYMBIOSIS.

antibiotic (*BioSci*) Strictly speaking, any chemical substance produced by or derived from one organism that has the capacity in dilute solutions to destroy or inhibit the growth of other organisms. Usually, the product of bacteria or fungi, often chemically modified (semi-synthetic antibiotics) that is used to treat infectious diseases of humans or domestic animals. The first safe and effective antibiotics, the penicillins, were extracted from the fungus *Penicillium*. Antibiotics discovered subsequently can be subdivided into several different classes: the beta lactams, which include penicillins and cephalosporins; aminoglycoside antibiotics;

macrolide antibiotics; tetracyclines; quinolones; sulfonamides; and a miscellaneous group of others. Increasing antibiotic resistance means there is a constant need for new antibiotics to cope with eg MRSA.

antibiotic resistance (*BioSci*) The property of microorganisms or cells that can survive high concentrations of a normally lethal agent. Normally acquired by the selection of a rare resistant mutant in the presence of low concentrations of the agent, but can be added by GENETIC MANIPULATION (panel).

antibiotic resistance gene (*BioSci*) A gene that confers resistance to an antibiotic, often by enzymic degradation or increased excretion. Such genes are frequently found in cloning vectors like plasmids, and sometimes in natural populations of bacteria where they can spread rapidly between species.

antiblocking agent (*Chem*) Fine powder added to eg low-density polyethylene film after manufacture to prevent film sticking to itself. Examples include talc or fine silicas.

antibody (*BioSci*) Immunoglobulin with combining site able to combine specifically with antigenic determinants on an antigen. See IMMUNOGLOBULIN.

antibody-dependent cell cytotoxicity (*BioSci*) A process in which a specific antibody binds to antigen (normally pathogen-derived, can be a self-antigen in auto-immune disease) present on the surface of a cell. This activates complement and results in the lysis of the antibody targeted cell. Abbrev *ADCC*.

antibody-directed enzyme drug therapy (*Med*) The method of attacking cancer cells while avoiding damage to normal cells by attaching a drug, or its precursor, to a specific antibody.

antibonding orbital (*Phys*) Orbital electron of two atoms, which increases in energy when the atoms are brought together, and so acts against the closer bonding of a molecule.

antical (*BioSci*) The upper surface of a thallus, stem or leaf.

anti-capacitance switch (*ElecEng*) A switch designed to have very little capacitance between the terminals when in the open condition.

anti-capillary groove (*Build*) See CHECK THROAT.

anticatalyst (*Chem*) See CATALYTIC POISON.

anti-cathode (*Med, Phys*) The anode target of an X-ray tube on which the cathode rays are focused, and from which the X-rays are emitted.

anticlimb paint (*Build*) Paint specially formulated to remain permanently wet after application to deter vandals or intruders.

anticlinal (*BioSci*) Perpendicular to the nearest surface. If a cell divides anticlinally the daughter cells will be separated by an anticlinal wall. Cf PERICLINAL.

anticlinal trap (*Geol*) A petroleum reservoir in which the oil or gas migrates to the top of an anticlinal structure beneath an impervious cap rock.

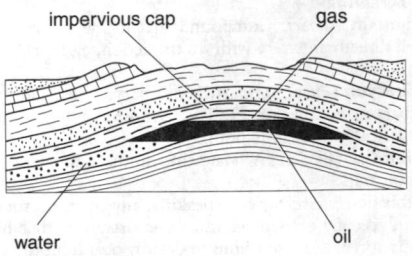

impervious cap   gas

water   oil

anticlinal trap Confining gas, oil and water.

anticline (*Geol*) A type of fold, comparable with an arch, the strata dipping outwards, away from the fold axis. See fig. at FOLDING.

anticlutter (*Radar*) A term describing a circuit or part of a radar system designed to eliminate unwanted echoes (*clutter*) and permit the display of signals which might

otherwise be obscured. Often takes the form of a gain control which automatically reduces gain immediately after the transmitted pulse and gradually restores it during the interval leading up to the anticipated return echo.

anticoagulant (*Pharmacol*) Any chemical substance which hinders normal clotting of blood, eg heparin, warfarin sodium, phenindione.

anticodon (*BioSci*) The sequence of three bases on TRNA that binds to the codon of MRNA. The complement of the coding triplet.

anticoincidence circuit (*Electronics*) One which delivers a pulse if one of two pulses is independently applied, but not when both are applied together or within the same assigned time interval.

anticoincidence counter (*Electronics*) A system of counters and circuits which record only if an ionizing particle passes through particular counters but not through the others.

anticollision beacon (*Aero*) A flashing red or blue light which is mounted above and below an aircraft to make it conspicuous when flying in CONTROL ZONES or other busy areas.

anticondensation paints (*Build*) A range of coatings specially formulated to form an isolating barrier between cold substrates and moisture laden air.

anticonvulsants (*Pharmacol*) Drugs used to prevent or reduce the severity and frequency of seizures that occur when electrical activity in the brain that controls motor systems becomes chaotic and paroxysmal. The commonest form of seizure is epilepsy but not all seizures cause convulsions, and not all convulsions are due to epileptic seizures. Commonly used drugs are *carbamazepine, phenytoin, valproate* and *diazepam*.

anticyclone (*EnvSci*) A distribution of atmospheric pressure in which the pressure increases towards the centre. Winds in such a system circulate in a clockwise direction in the northern hemisphere and in an anticlockwise direction in the southern hemisphere. Anticyclones give rise to fine, calm weather conditions, although in winter fog is likely to develop.

anticyclonic blocking (*EnvSci*) The effect caused by an area of high pressure that diverts depressions. During the winter in Northern Europe an extensive anticyclonic area over Russia and Scandinavia can develop and divert Atlantic depressions towards the Western Mediterranean region.

anticyclotron tube (*Electronics*) A type of travelling-wave tube.

antidazzle mirror (*Autos*) A mirror having a two-position setting, providing a dim partial reflection of headlamps behind for night driving.

antidepressants (*Pharmacol*) Drugs that relieve the symptoms of moderate to severe depression. The main classes are MONOAMINE OXIDASE INHIBITORS, SSRIS, TRICYCLIC ANTIDEPRESSANTS and lithium salts.

antiderivative (*MathSci*) See INTEGRAL.

antidiazo compounds (*Chem*) The stereoisomeric form of diazo compounds in which the groups attached to the nitrogen atoms are far removed from each other.

antidiuretic (*Med*) Inhibiting the formation of urine; an antidiuretic drug.

antidromic (*BioSci*) Contrary to normal direction, eg applied to nerve cells, when the impulse is conducted along the axon towards the cell body.

antiemetic drugs (*Pharmacol*) Drugs that stop vomiting and, to a lesser extent, nausea. They are used for motion sickness and for the side-effects of chemotherapy and some gastrointestinal disorders. Examples include *hyoscine, antihistamines, phenothiazines, metoclopramide* and *ondantseron*.

anti-extrusion ring (*Eng*) Nylon or acetal ring fitted to heavy-duty rubber seal to prevent extrusion through sealed gap.

antifading antenna (*ICT*) An antenna that confines radiation mainly to small angles of elevation, to minimize radiation of sky waves which are prone to fading. For medium-wave transmitters, the antenna is usually a vertical mast about 0·6 of a wavelength high. ADAPTIVE ARRAYS are also used to combat fading in higher-frequency applications.

antiferromagnetism (*Phys*) Phenomenon in some magnetically ordered materials in which there is an antiparallel alignment of spins in two interpenetrating structures so that there is no overall bulk spontaneous magnetization. Antiferromagnetics have a positive susceptibility. The antiparallel alignment is disturbed as the temperature increases until at the NÉEL TEMPERATURE the material becomes paramagnetic.

antiflood and tidal valve (*Build*) A valve consisting of a cast-iron box containing a floating ball, fitted near a drain outlet to prevent back flow.

antifouling composition (*CivEng*) A substance applied in paint form to ships' bottoms and structures subject to the action of sea water, to discourage marine growths.

antifouling paints (*Build*) Highly poisonous paints applied to the hulls of ships to minimize the accumulation of barnacles etc.

antifreeze (*Chem, Eng*) Solutes which lower the freezing point of water, usually in automobile engine cooling. Most commonly ethylene or propylene glycol, or methanol with a few per cent of corrosion inhibitor such as phosphates. A 35% solution of ethene glycol or 30% of methanol and water will not freeze at temperatures above $-5°F$ $(-20·6°C)$.

anti-friction bearing (*Eng*) A term used to describe a wide range of bearings such as ball, roller, special metallic alloy and plastic-based bearings. All designed to reduce friction between moving parts, the choice depending on the duty.

anti-friction metal (*Eng*) See WHITE METAL.

anti-g (*Space*) Resistant to the effects of high acceleration, esp of an astronaut's equipment.

antigen (*BioSci*) A substance which has determinant groups that can interact with specific receptors on lymphocytes, or on antibodies released from them. The term is often used to include substances that can stimulate an immune response, although these are more correctly termed *immunogens*.

antigenic determinant (*BioSci*) A small part of the antigen which has a structure complementary to the recognition site on a T-cell receptor or an antibody. Most antigens are large molecules with several different antigenic determinants, each of which interacts with lymphocytes carrying a different specific recognition site.

antigenic variation (*BioSci*) The development of new antigenic determinants by many viruses, bacteria and protozoa as a result of genetic mutation and selection during multiplication in their hosts. If the variation involves the antigenic component that stimulates protective immunity, the variant can cause infection in subjects who would otherwise be immune to that microbe.

antigen-presenting cells (*BioSci*) Cells that present foreign antigenic determinants or epitopes to the antigen-specific receptors found on lymphocytes. Especially those cells which present antigen in association with MHC II protein to CD4-positive T-lymphocytes. Abbrev *APCs*.

antigen processing (*BioSci*) Biochemical process by which antigen-presenting cells associate foreign antigenic determinants or epitopes with self-proteins encoded by the MAJOR HISTOCOMPATIBILITY COMPLEX.

antiglobulin (*BioSci*) General term for an antibody against an immunoglobulin. Can be used to detect the presence of immunoglobulins bound to the surface of cells or microbes. Usually class-specific and more commonly referred to as such, eg anti-IgG, anti-IgM, etc.

antigorite (*Min*) One of three minerals which are collectively known as SERPENTINE, hydrated magnesium silicate. It is abundant in the rock type, serpentinite.

anti-g suit (*Aero*) A close-fitting garment covering the legs and abdomen, which is inflated, either automatically or at will by the wearer, so that counter-pressure is applied when

blood is displaced away from the head and heart during high-speed manoeuvres. Colloq *g-suit*.

anti-g valve (*Aero*) (1) A spring-loaded mass type of air valve which automatically regulates the inflation of an ANTI-G SUIT according to the acceleration (g) loads being imposed. (2) A valve incorporated in some aircraft fuel systems to prevent engines being starved of fuel under specific g loads.

antihalation (*ImageTech*) The use of backing to reduce halation in plates or films.

antihistamine (*Pharmacol*) A substance or drug which inhibits the actions of histamine by blocking its site of action. Useful in the treatment of allergic rhinitis, especially hay fever, and can be used topically in the eye and on the skin. Taken by mouth the drug can cause drowsiness. See HISTAMINE RECEPTORS.

anti-icing (*Aero*) Protection of aircraft against icing by preventing ice formation on eg wind-shield panels, leading edges of wings, tail units and turbine engine air intakes. The most common methods are to apply continuous heating by hot air tapped from an engine, by electrical heating elements or periodically inflating rubber bags. Cf DE-ICING.

anti-idiotype (*BioSci*) Antibody that recognizes the combining site of an antibody against an antigenic determinant on an antigen. The combining site of the anti-idiotype may thus be expected to resemble the shape of the determinant on the original antigen.

anti-incrustator (*Eng*) A substance used to prevent the formation of scale on the internal surfaces of steam boilers.

anti-induction network (*ICT*) A network connected between circuits to minimize crosstalk.

anti-inflammatory drugs (*Pharmacol*) Drugs that inhibit the inflammatory response. There are two major classes: the non-steroidal anti-inflammatory drugs (NSAIDs) and the glucocorticoids.

antiknock substances (*Autos*) Substances added to petrol to lessen its tendency to detonate or 'knock' in an engine, eg tetra-ethyl lead.

antiknock value (*Autos*) The relative immunity of a volatile liquid fuel from detonation, or 'knocking', in a petrol engine, as compared with some standard fuel. See KNOCK RATING, OCTANE NUMBER.

antilepton (*Phys*) An antiparticle of a LEPTON. Positrons, positive muons, antineutrinos and tau-plus particles are antileptons.

antilock brake (*Autos*) A system which prevents the locking of road wheels under braking, giving improved control and stopping on poor road surfaces. A sensor detects over-rapid deceleration of the wheels and signals for a reduction in braking effort. Also *ABS brake* (from Ger *Anti-Blockier-System*).

antilogarithm (*MathSci*) A number whose LOGARITHM is the given number.

antilymphocytic serum (*BioSci*) Serum containing antibodies reactive with surface antigens on lymphocytes and capable of killing or otherwise suppressing their capabilities. Used as an immunosuppressive agent.

antimatter (*Phys*) See ANTIPARTICLE.

antimere (*BioSci*) A part on a bilaterally or radically symmetrical organism corresponding to a similar structure on the other side. Adj *antimeric*.

antimetabolite (*Pharmacol*) Drugs used in treatment of cancer which are incorporated into new nuclear material and prevent normal cell division. Common examples are methotrexate, cytoarabinose and fluorouracil.

antimitotic drugs (*BioSci*) Drugs that block mitosis; the term is often used of those which cause metaphase arrest such as colchicine and the vinca alkaloids. Many antitumour drugs are antimitotic, blocking proliferation rather than being cytotoxic.

antimonial lead (*ChemEng*) A lead–antimony alloy of controlled analysis with much improved mechanical strength and retaining good chemical corrosion resistance.

For sheets and pipes antimony may be as high as 12%, for machined castings up to 20–25%.

antimonial lead ore (*Min*) See BOURNONITE.

antimoniates (*Chem*) The products of antimonic acids with aqueous solutions of potassium hydroxide.

antimonite (*Min*) See STIBNITE.

antimony (*Chem*) Metallic element. Symbol Sb, at no 51, ram 121·75, mp 630°C, rel.d. 6·6. Used in alloys for cable covers, batteries, etc; also as a donor impurity in silicon. Has several radioactive isotopes which emit very penetrating γ-radiation. These are used in laboratory neutron sources. Abundance in Earth's crust 0·2 ppm; it forms stibnite and other sulphide minerals.

antimony alloys (*Eng*) Alloys containing antimony, which is an essential constituent in type metals, bearing metals (which contain 3–20%), in lead for shrapnel (10%), storage battery plates (4–12%), roofing, gutters and tank linings (6–12%).

antimony black (*Eng*) Finely powdered antimony, which gives plaster casts a metallic look.

antimony glance (*Min*) Obsolete name for STIBNITE.

antimony halides (*Chem*) Antimony (III) fluoride SbF_3 and (V) fluoride SbF_5; (III) chloride $SbCl_3$ and (V) chloride $SbCl_5$; (III) bromide and (III) iodide.

antimony hydrides (*Chem*) Two hydrides, stibine [(III) hydride] SbH_3 and the solid dihydride Sb_2H_2. See STIBINE.

antimonyl (*Chem*) The monovalent radical SbO— [Sb (III) present].

antimuon (*Phys*) Antiparticle of a MUON.

anti-muscarinic (anticholinergic) drugs (*Pharmacol*) A class of drugs that block the action of acetylcholine at the muscarinic subclass of acetylcholine receptors. Their effect is generally to relax smooth muscle of the gut (antispasmodic) or airways (bronchodilatory). An example is *tolterodine*.

antimutagen (*BioSci*) A compound that inhibits the action of a mutagen.

antineutrino (*Phys*) Antiparticle to the NEUTRINO. As for the neutrino there are three types of antineutrino, associated with the electron, muon and tau lepton.

antineutron (*Phys*) Antiparticle with spin and magnetic moment oppositely orientated to those of the neutron.

antineutrophil cytoplasmic antibody (*BioSci*) An auto-antibody specific for proteins in the cytoplasmic granules of polymorphonuclear leucocytes and monocytes. Different types are characteristic of vasculitis and glomerulonephritis and are used in the laboratory diagnosis of these conditions. Abbrev *ANCA*.

antinode (*Phys*) At certain positions in a standing-wave system of acoustic or electric waves or vibrations, the location of maxima of some wave characteristic, eg amplitude, displacement, velocity, current, pressure, voltage. At the NODES these would have minimum values.

antinuclear (*Genrl*) Opposed to the development or use of nuclear weapons.

antinuclear factor (*BioSci*) Auto-antibody reactive with nucleic acids (DNA or RNA) present in the blood of subjects with SYSTEMIC LUPUS ERYTHEMATOSUS and some other auto-immune conditions. Used diagnostically.

anti-oncogene (*BioSci*) See TUMOUR SUPPRESSOR GENE.

anti-oxidants (*Chem*, *FoodSci*) Substances which delay the oxidation of materials. Raw vegetable oils contain natural anti-oxidants which reduce the speed of drying of paints. Deliberately added anti-oxidants, generally phenol derivatives, delay the skinning of paints in the can at the cost of slightly slower drying. Similar substances added to plastics, rubbers, foods and drugs delay degradation by oxidation.

anti-ozonants (*Chem*) Materials added to rubbers, esp those with alkene bonds in main chain, to inhibit ozone cracks. Polybutadienes and natural rubber are esp liable to attack. Often based on waxes, which leach to surface.

antiparallax mirror (*Phys*) Mirror positioned on an arc adjacent to the scale of an indicating instrument, so that the parallax error in reading the indication of the pointer is avoided by aligning the eye with the pointer and its image.

antiparallel (*MathSci*) (of a pair of vectors). Having the same DIRECTION but opposite SENSE.

antiparticle (*Phys*) A particle that has the same mass as another particle but has opposite values for its other properties such as charge, baryon number or strangeness. The antiparticle to a fundamental particle is also fundamental, eg the electron and positron are particle and antiparticle. Interaction between such a pair means simultaneous annihilation, with the production of energy in the form of radiation.

antiperistaltic (*BioSci*) Said of waves of contraction passing from anus to mouth, along the alimentary canal. Cf PERISTALTIC; n *antiperistalsis*.

antiperthite (*Min*) An intergrowth of plagioclase and potassium feldspars with plagioclase as the dominant phase. See PERTHITE.

antipetalous (*BioSci*) See ANTEPETALOUS.

antiplasticization (*Chem*) Effect produced in a polymer by addition of a specific chemical; the opposite of plasticization, giving a material with higher modulus and lower elongation to break. Beyond a critical concentration, properties revert to those of conventional plasticization.

antiplectic (*BioSci*) Pattern of metachronal co-ordination of the beating of cilia, in which the waves pass in the opposite direction to that of the active stroke.

antipodal cells (*BioSci*) Wall-less cells, usually three, typically haploid, derived by mitotic division of the megaspore, lying in the embryo sac at the end remote from the micropyle.

antipodal points (*MathSci*) The pair of points at each end of a diameter of a sphere.

antipodes (*Geol*) On a sphere, eg the Earth, points on the surface at either extremity of a diameter.

antipolarizing winding (*ElecEng*) Winding on a transformer or choke which carries a direct current to neutralize the magnetizing effect of another direct current.

antiport (*BioSci*) Transport of two different ions or molecules, in opposite directions, across a cellular membrane. Energy is required if movement is against an electrochemical gradient.

antiproteases (*BioSci*) Substances that inhibit proteolytic enzymes. Also *antipeptidases*, *antiproteinases*.

antiproton (*Phys*) Short-lived particle, half-life 0·05 μs, identical to the proton, but with negative charge; annihilating with normal proton, it yields mesons. Also *negative proton*.

antipsychotic drugs (*Pharmacol*) A group of drugs sometimes referred to as major tranquillizers, used short term to calm or sedate disturbed patients, to control acute symptoms of mania and to relieve severe positive symptoms of schizophrenia. Most act by reducing levels of the neurotransmitter dopamine in the central nervous system (eg *chlorpromazine*, *haloperidol*, *flupentixol*), but the atypical antipsychotics (eg *risperidone*, *clozapine*) act by interfering with serotonin-based neurotransmission. Also *neuroleptic drugs*.

antipyretic (*Med*) Counteracting fever; a remedy for fever.

antipyrine (*Pharmacol*) Compound formerly used as analgesic and antipyretic.

antiqua (*Print*) The German name for ROMAN type.

Antiquarian (*Arch*) The general term to describe the final phase of the RENAISSANCE style when architects reverted to ancient models as a source of inspiration; Greek, Roman, Gothic and to a certain extent Egyptian architecture were studied and it became fashionable to travel and to compile portfolios of drawings of ancient ruins. It is regarded as a reaction to the flamboyant BAROQUE phase which preceded it.

antiquark (*Phys*) The antiparticle of a QUARK.

antique (*Paper*) The surface finish originally applied to machine-made papers made in imitation of handmade printings. The term is now used to describe any rough-surfaced paper which bulks well, eg book or cover paper.

antiresonance frequency (*Electronics*) Frequency at which the parallel impedance of a tuned circuit rises to a maximum.

antiretroviral (*Pharmacol*) Acting to counteract or control a *retrovirus*; an agent or drug with this property.

antiroll bar (*Autos*) Torsion bar mounted transversely in the chassis in such a way as to counteract the effect of opposite spring deflections.

antisag bar (*Build*) A vertical rod connecting the main tie of a roof truss to the ridge to support it against sagging under its own weight.

antisense (*BioSci*) The complementary strand of a coding strand of DNA or RNA. Cf SENSE STRAND.

antisepalous (*BioSci*) See ANTESEPALOUS.

antisepsis (*Med*) The inhibition of growth, or the destruction of bacteria in the field of operation by chemical agent; the principle of antiseptic treatment.

antiseptic (*Med*) Counteracting sepsis or contamination with bacteria: an agent which destroys bacteria or prevents their growth.

antiserum (*BioSci*) Serum from an individual that contains a high titre of antibodies specifically directed against a particular pathogen or foreign protein. This may occur as a result of natural exposure or have been artificially induced by immunization. Antisera are commonly used in laboratories as diagnostic and detection reagents and in some circumstances as therapeutic reagents, eg the antivenom sera used to treat snakebites.

anti-set-off spray (*Print*) Spray used to apply a layer of fine particles to the surface of each freshly printed sheet to prevent contact with the succeeding sheet so that SET-OFF does not occur.

anti-set-off tympan cover (*Print*) A top cover for the second cylinder of any perfecting press, flat-bed or rotary, consisting of a material coated with very small glass beads.

antisolar glass (*Glass*) Glass which absorbs heat from sunshine and reduces glare, but transmits most of the light.

antisound (*Acous*) Sound signal with same amplitude but opposite phase of some unwanted sound signal so that both signals cancel each other when superimposed. Used in ACTIVE CONTROL.

antispasmodic drugs (*Pharmacol*) Drugs that relax smooth muscle of the gut wall and relieve symptoms of indigestion, irritable bowel syndrome and diverticular disease. Most are antimuscarinic drugs.

anti-spin parachute (*Aero*) A small parachute, normally in a canister, which may be fixed to the tail (occasionally to the wing tips) of an aircraft or glider for release in emergency to lower the nose into a dive and so assist recovery from a spin. It is jettisoned after use. Colloq *spin chute*.

antispray film (*ElecEng*) An oil film placed on the surface of accumulator cells to prevent the formation of acid spray due to the bursting of gas bubbles during the charging process.

antistatic agent (*Textiles*) A substance applied to a textile to render it less prone to becoming charged with static electricity by friction during processing or in wear.

antistatics (*Chem*) Chemicals added to polymers to discourage build-up of static electricity on product surfaces. Include quaternary ammonium salts and polyethylene glycols. Also *antistats*.

anti-Stokes lines (*Phys*) Those in scattered or fluorescent light with frequencies greater than that in the incident radiation, because of departure of atoms or molecules from their normal states.

antisurge valve (*Aero*) A valve for bleeding off surplus compressor air to suppress the unstable airflow due to SURGE in a gas turbine engine.

antisymmetric (*Phys*) Pattern or waveform in which symmetry is complete except for one particular feature, eg sign of electric charge, direction of current, or of components in waveform. A system containing several electrons must be described quantum mechanically by an *antisymmetric eigenfunction*.

antisymmetric dyadic (*MathSci*) See CONJUGATE DYADICS.

antisymmetric function (*MathSci*) See ALTERNATING FUNCTION.

antithetic alternation of generations (*BioSci*) (1) See ANTITHETIC THEORY OF ALTERNATION. (2) Sometimes the same as heteromorphic alternation of generations.

antithetic theory of alternation (*BioSci*) The hypothesis that the sporophyte is a novel phase in the life cycle resulting from the postponement of meiosis. Cf HOMOLOGOUS THEORY OF ALTERNATION.

antithrombins (*BioSci*) Plasma glycoproteins of the α-2-globulin class that inhibit the proteolytic activity of thrombin and thus regulate the process of blood clotting.

antitoxin (*BioSci*) Antibody capable of neutralizing toxins, eg tetanus antitoxin, diphtheria antitoxin. Used in treatment when the main damaging agent is the toxin. Produced by immunizing horses or other animals, but liable to cause SERUM SICKNESS. Nowadays antibodies from pre-immunized humans are used when possible, and in future human monoclonal antibodies may be widely used.

antitrades (*EnvSci*) Winds, at a height of 900 m or more, which sometimes occur in regions where trade winds are prevalent, their direction being opposite to that of the trade winds.

anti-transmit–receive tube (*Radar*) Gas-discharge tube which isolates a pulsed radar transmitter from the antenna so that echoes can be received. Abbrev ATR tube. Cf TRANSMIT–RECEIVE TUBE.

antitranspirant (*BioSci*) Substance that reduces transpiration (and, usually, also photosynthesis), eg by causing stomatal closure or by forming a more or less impermeable surface film; of some use horticulturally when transplanting.

antitussive drug (*Pharmacol*) A cough suppressant. Most work by suppressing the cough reflex, rather than treating the cause of the cough.

antivibration mounting (*Eng*) Rubber spring designed to absorb vibrations from engines etc. Care needed in design and materials selection to match vibration frequency with main damping peak of elastomer.

antivivisectionists (*Med*) Those who oppose experiments on live animals.

Antlia (Air Pump) (*Astron*) A small southern hemisphere constellation.

Antonoff's rule (*Chem*) The interfacial tension between two liquid phases in equilibrium is equal to the difference of the surface tensions of the two phases.

antorbital (*BioSci*) In front of the orbit. In vertebrates, a small bone in the nasal region.

antrorse (*BioSci*) Directed or bent forward.

antrum (*BioSci*) A sinus, such as the maxillary sinus in vertebrates; a cavity, such as the ANTRUM OF HIGHMORE. Pl *antra*.

antrum of Highmore (*Med*) An air-containing cavity in the maxilla which communicates with the nasal cavity.

Antrycide (*Pharmacol, Vet*) Vet TN for a drug, quinapyramine, which shows low toxicity and effective trypanocidal powers in cases of *Trypanosoma congolense, T. vivax, T. evansi, T. brucei* and *T. simiae*, in cattle and various domestic animals.

anucleate (*BioSci*) Having no nucleus.

A-number (*ICT*) The telephone number from which a call originates in an INTELLIGENT NETWORK. Cf B-NUMBER, C-NUMBER.

Anura (*BioSci*) See SALIENTIA.

anural (*BioSci*) Without a tail; pertaining to Anura. See SALIENTIA. Also *anurous*.

anuria (*Med*) Complete failure to secrete urine.

anurous (*BioSci*) See ANURAL.

anus (*BioSci*) The opening of the alimentary canal by which indigestible residues are voided, generally posterior. Adj *anal*.

anvil (*Eng*) A block of iron, sometimes steel-faced, on which work is supported during forging.

anvil (*Med*) One of the three small bones (ossicles) which transmit mechanical vibrations between the outer ear drum and the inner ear. See INCUS.

anvil chisel (*Eng*) See ANVIL CUTTER.

anvil cloud (*EnvSci*) A common feature of a thundercloud, consisting of a wedge-shaped projection of cloud suggesting the point of an anvil.

anvil cutter (*Eng*) A chisel with a square shank for insertion in the hardy hole of a smith's anvil, the cutting edge being uppermost.

anxiety (*Psych*) A physiological and psychological reaction to an expected danger, whether real or imaginary.

anxiolytic (*Pharmacol*) A drug used for relieving anxiety states.

AOL (*ICT*) Abbrev for AMERICA ONLINE.

aorta (*BioSci*) (1) In Arthropoda, Mollusca and most vertebrates, the principal arterial vessel(s) by which the oxygenated blood leaves the heart and passes to the body. (2) In amphibians, the principal artery by which blood passes to the posterior part of the body, formed by the union of the systemic arteries. (3) In fish (ventral aorta), the vessel by which the blood passes from the heart to the gills, and also (dorsal aorta) the vessel by which the blood passes from the gills to the body. Adj *aortic*.

aortic arches (*BioSci*) In vertebrates, a series of pairs of vessels arising from the ventral AORTA.

aortic incompetence (*Med*) A defect in aortic valve function which allows blood to regurgitate from the aorta to the left ventricle during diastole when the valve should be tight shut. Occurs in rheumatic heart disease and in INFECTIOUS ENDOCARDITIS but may also complicate ANKYLOSING SPONDYLITIS, REITER'S SYNDROME and SYPHILIS.

aortitis (*Med*) Inflammation (usually syphilitic) of the aorta.

AP (*BioSci*) Abbrev for ALKALINE PHOSPHATASE.

AP (*Surv*) Abbrev for *Amsterdamsch Peil* (Amsterdam level), ie the datum, or mean level, used as a basis for levels in the Netherlands, Belgium and N Germany.

ap- (*Genrl*) See AN-.

α-particle (*Phys*) See ALPHA PARTICLE.

apatite (*Min*) Phosphate of calcium, also containing fluoride, chloride, hydroxyl or carbonate ions, according to the variety. It is a major constituent of sedimentary phosphate rocks, of the bones and teeth of vertebrate animals (including humana), and it is usually present as an accessory mineral in igneous rocks.

aperient (*Pharmacol*) A drug having a laxative or purgative effect.

aperiodic (*ICT*) Said of any device or circuit (eg antenna, amplifier) that does not exhibit any variation in characteristics with varying frequency of applied signals.

aperiodic (*Phys*) Said of any potentially vibrating system, electrical, mechanical or acoustic, which, because of sufficient damping, does not vibrate when impulsed. Used particularly of the pointers of indicating instruments, which, having no natural period of oscillation, do not oscillate before coming to rest in the final position, and so give their ultimate reading as fast as possible.

aperiodic antenna (*ICT*) An antenna with useful efficiency over a range of radio frequencies, terminated to minimize resonance by reflection, eg RHOMBIC ANTENNA, WAVE ANTENNA. Also *non-resonant antenna*.

aperturate (*BioSci*) Pollen grains having one or more apertures, ie areas of the wall where the outer part, or EXINE, is thinner or absent and through which the pollen tube may emerge. Cf COLPUS, PORE.

aperture (*ImageTech, Phys*) (1) The opening, usually circular, through which light enters an optical system, such as a camera lens; its area may be varied by an IRIS diaphragm to control the amount of light passing. See F-NUMBER, NUMERICAL APERTURE, STOP. (2) The rectangular opening at which motion picture film is exposed in a camera or projector.

aperture (*Radar, ICT*) The effective area over which an aerial extracts power from an incident plane wave. The aperture (A) and gain (G) are related by the equation $G = 4\pi A/\lambda^2$ where λ is the wavelength.

aperture correction (*ImageTech*) One form of enhancement of signal differences at image boundaries to increase apparent sharpness.

aperture distortion (*ImageTech*) Distortion arising from the scanning spot having finite, instead of infinitely small, dimensions.

aperture efficiency (*ICT*) The ratio of an antenna's actual directivity to the theoretical figure which would be obtained with ideal aperture illumination, ie with uniform electromagnetic field strength over its aperture.

aperture grille (*ImageTech*) The Trinitron picture tube equivalent of a shadowmask, with vertical slits instead of holes through which the electron beams pass.

aperture number (*ImageTech*) See F-NUMBER.

aperture plate (*ImageTech*) Plate carrying the opening at which film is exposed or projected.

aperture priority (*ImageTech*) A facility enabling a camera to select the shutter speed automatically once the aperture has been selected by the photographer.

aperture synthesis (*Astron*) A technique in which two or more radio telescope antennas are connected as pairs of INTERFEROMETERS. The amplitude and phase of the interference pattern is continuously recorded. The interferometer baseline is normally variable, and the rotation of the Earth changes the position angle with respect to a distant radio source. The FOURIER TRANSFORM of the amplitude and phase patterns are then used to compute a map of the radio source. By means of long baselines, achieved by linking telescopes on different continents, it is possible to achieve a RESOLVING POWER of around 0·001 arcseconds.

apetaly (*BioSci*) Absence of petals. Adj *apetalous*.

apex (*Astron*) See SOLAR APEX.

apex (*BioSci*) The end of an organ or plant part remote from its point of attachment or origin, eg root tip or shoot tip.

apex (*Genrl*) The top or pointed end of anything. Adj *apical*.

apex (*Med*) Said of the root of a tooth, of the top of the upper lobe of a lung, or of the rounded end of the left ventricle of the heart.

apex (*MinExt*) The upper edge of a vein reef or lode.

apex beat (*Med*) The point that is furthest out and furthest down the chest where the heart beat is visible or palpable.

apex law (*MinExt*) The law entitling the discoverer of an outcrop or exposure of ore to exploit it in depth beyond its lateral boundaries. Also *extra-lateral rights*.

apex stone (*Build*) Triangular stone at the summit of a gable, often decorated with a carved trefoil. Also *saddle stone*.

Apgar score (*Med*) A scoring system for assessing a baby's condition at birth of which a value of 0, 1 or 2 is given to each of five signs: colour, heart rate, muscle tone, breathing effort and response to stimulation. A score of 10 indicates the baby is in excellent condition.

aphagia (*Med*) Inability to swallow or feed.

aphakia (*Med*) Absence of the lens of the eye.

aphanitic (*Min*) A term describing the texture of an igneous rock in which the crystals are not distinguishable by the unaided eye.

aphasia (*Psych, Med*) An impairment of, or defect in, language function due to a lesion in certain association areas of the brain. It may be an inability to comprehend speech or written words (*receptive aphasia*), or a deficiency in writing or speech (*expressive aphasia*).

Aphebian (*Geol*) The lower part of the Proterozoic. See PRECAMBRIAN.

aphelion (*Astron, Space*) The farthest point from the Sun on a planet's, comet's or spacecraft's orbit round it. Pl *aphelia*.

apheliotropic (*BioSci*) Turning away from the Sun.

aphids (*BioSci*) Insects of the family Aphidae (order Hemiptera). Reproduction is either sexual or parthenogenetic, oviparous or viviparous, giving rise to a complex life cycle. Greenfly.

aphonia (*Med*) Loss of voice. Also *alalia*.

aphotic zone (*EnvSci*) The zone of the sea below about 1500 metres which is essentially dark. Cf PHOTIC ZONE.

aphototropic (*BioSci*) (1) Usually, not phototropic. (2) Less commonly and confusingly, negatively phototropic.

aphthous fever (*Vet*) See FOOT-AND-MOUTH DISEASE.

aphthous ulcer (*Med*) A small grey ulcer in the mouth.

API (*ICT*) Abbrev for APPLICATIONS PROGRAMMING INTERFACE.

Apiaceae (*BioSci*) See UMBELLIFERAE.

apical cell (*BioSci*) (1) A single cell at the apex of a filament, multicellular thallus or organ, from which all the cells of the filament or organ are descended. (2) One of a quartette of small cells found at the apex of the egg of some invertebrates, eg the limpet (*Patella*), during its cleavage.

apical dome (*BioSci*) The usually dome-shaped part of an apical meristem distal to the most recently formed leaf primordium.

apical dominance (*BioSci*) The influence of a terminal bud in inhibiting or controlling the growth of buds or lateral branches on the shoot below it, ceasing if the terminal bud is destroyed.

apical growth (*BioSci*) (1) The elongation of tubular cell or hypha by continued growth at the apex only (the normal pattern for root hairs, pollen tubes and fungal hyphae). (2) The condition in which the only transverse divisions in a filament of cells take place in the apical cell.

apical meristem (*BioSci*) A group of meristematic cells at the tip of a thallus, stem or root, which divide to produce the precursor of the cells of the thallus, or of the primary tissues of root or shoot. There may or may not be a distinct apical cell.

apical placentation (*BioSci*) In plants, the condition in which the ovule is inserted at the top of the plant ovary.

apical sense organ (*BioSci*) In Ctenophora, an elaborate sensory structure formed of small otoliths united into a morula, supported on four pillars of fused cilia and covered by a roof of fused cilia.

apicolysis (*Med*) An operation for compressing or collapsing the apex of the lung; as in the treatment of pulmonary tuberculosis.

apiculate (*BioSci*) Ending in a short, sharp point.

API scale (*Phys*) Abbrev for *American Petroleum Institute scale*. Scale of relative density, similar to Baumé scale. Degrees API = $(141\cdot5/s) - 131s$, where s is the rel.d. of the oil against water at 15°C.

apituitarism (*Med*) Absence or deficiency of pituitary gland secretion. See HYPOPITUITARISM.

Apjohn's formula (*Phys*) A formula which may be used for determining the pressure of water vapour in the air from readings of the wet and dry bulb hygrometer. The formula is: $p_t = p_w - 0\cdot00075H(t - t_w)[1 - 0\cdot008(t - t_w)]$, where p_w is the saturated vapour pressure at the temperature (t_w) of the wet bulb, H is the barometric height, and t is the temperature of the dry bulb.

APL (*ICT*) Abbrev for *a programming language*. Scientific programming language using a special character set and syntax designed to aid the programming of mathematics.

aplacental (*BioSci*) Without a placenta.

aplanatic (*Phys*) Said of an optical system which produces an image free from spherical aberration.

aplanatic refraction (*Phys*) Refraction at a surface under conditions in which there is no spherical aberration and in which the sine condition is satisfied.

aplanetic (*BioSci*) Denoting organisms or structures, especially spores, that are non-motile or lack a motile stage.

aplanogamete (*BioSci*) A non-motile gamete.

aplanospore (*BioSci*) A non-motile spore, eg autospore, hypnospore. Cf ZOOSPORE.

aplasia (*Med*) Defective structural development.

aplastic anaemia (*Med*) Anaemia due to loss of most or all of the haematopoietic bone marrow. Usually all haematopoietic cells are equally diminished in number.

aplite (*Geol*) A fine-textured, light-coloured, igneous rock composed of quartz and feldspar.

apneusis (*Med*) State of maintained inspiration.

apneustic (*BioSci*) Possessing no organs specialized for respiration as in some aquatic insect larvae without functional spiracles.

apneustic centre (*BioSci*) The part of the brain controlling the inflation of the lungs in higher vertebrates.

apnoea (*Med*) Cessation of breathing. Recognized to occur in sleep in obese and other persons. Also *sleep apnoea syndrome*.

apo- (*Genrl*) Prefix from Gk *apo*, away.

apoapsis (*Astron*) The furthest point of approach of an orbiting body (eg planet, comet, spacecraft) to the primary body. Cf PERIAPSIS.

apocarpous (*BioSci*) A gynoecium consisting of two or more free (ie not fused) carpels. Cf SYNCARPOUS.

apochromatic lens (*Phys*) A lens so designed that it is corrected for chromatic aberration for three wavelengths thus reducing the secondary spectrum.

apochromatic objective (*Phys*) Microscope objective in which spherical and chromatic aberrations have been corrected as completely as possible.

apocrine (*BioSci*) A form of secretion in which the apical portion of the cell is shed, as in the secretion of fat by cells of the mammary gland.

Apoda (*BioSci*) An order of amphibians having a cylindrical snake-like body without limbs, reduced eyes and an anterior sensory tentacle. Members are burrowing forms, living near water and feeding chiefly upon earthworms. Caecilians.

apodal (*BioSci*) Without feet or other locomotor appendages. Also *apodous*.

apodeme (*BioSci*) In Arthropoda, an ingrowth of the cuticle forming an internal skeleton and serving for the insertion of muscles; in insects, more particularly, an internal lateral chitinous process of the thorax.

apodous (*BioSci*) See APODAL.

apodous larva (*BioSci*) A type of insect larva in which the trunk appendages are completely suppressed; formed in some Coleoptera, Diptera, Hymenoptera and Lepidoptera.

apoenzyme (*BioSci*) An enzyme without its cofactor.

apogamy (*BioSci*) The development of a sporophyte directly from a cell of the gametophyte without fusion of gametes so that the resulting sporophyte has the same chromosome number as the parent gametophyte. Cf APOMIXIS, APOSPORY. Adj *apogamous*.

apogee (*Astron, Space*) The point in the orbit of the Moon or an artificial satellite which is furthest from the Earth. Also the highest altitude attained by a missile. See fig. at ORBIT.

apogee motor (*Space*) The engine fired at the apogee of an elliptical orbit to establish a circular orbit whose altitude is that of the apogee of the original orbit. Similarly a *perigee motor* for transforming a circular orbit into an eccentric one.

A-point (*Eng*) Temperature above which steel can be hardened. The equilibrium point of the TRANSFORMATION TEMPERATURE. Also *Ae point*.

Apollo asteroid (*Astron*) An ASTEROID whose orbit brings it within the orbit of the Earth.

Apollonius' circle (*MathSci*) For two fixed points A and B, the locus of the point P, which moves so that the ratio $PA:PB$ is a constant.

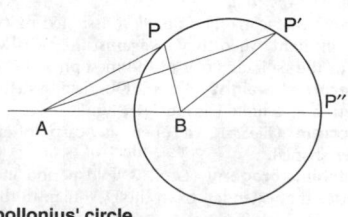

Apollonius' circle

Apollonius' theorem (*MathSci*) Theorem that if the base *BC* of a triangle *ABC* is divided by a point *P*, such that

$$\frac{BP}{PC} = \frac{r}{s}$$

then

$$s(AB)^2 + r(AC)^2 = (s+r)(AP)^2 + s(PB)^2 + r(PC)^2$$

Apollo programme (*Space*) Manned spaceflight programme of the USA (1968–72), leading to the first lunar landing, by Neil Armstrong and Buzz Aldrin, on 20 July 1969.

apolune (*Astron*) The point in lunar orbit when an orbiting object, eg a spacecraft, is furthest from the Moon (opposite to *perilune*).

apomecometer (*Surv*) Instrument, based on optical square, for measuring heights and distances.

apomixis (*BioSci*) (1) Reproduction by seeds formed without sexual fusion. Also *agamospermy*. (2) Any form of asexual reproduction, including vegetative propagation.

apomorphine (*Pharmacol*) An alkaloid of the morphine series, obtained from morphine by dehydration. It is not a narcotic, but is an expectorant and emetic.

apophyllite (*Min*) A secondary mineral occurring with zeolites in amygdales in basalts and other igneous rocks. Composition: hydrated fluorosilicate of potassium and calcium.

apophysis (*BioSci*) In vertebrates, a process from a bone, usually for muscle attachment; in insects, a ventral chitinous ingrowth of the thorax for muscle insertion.

apophysis (*Geol*) A vein-like offshoot from an igneous intrusion.

apoplast (*BioSci*) That part of the plant body which is external to the living protoplasts, ie the cell walls, the intercellular spaces and the lumina of dead cells such as xylem vessels and tracheids or in some contexts, the water-filled parts of this space. Cf SYMPLAST.

apoplexy (*Med*) Sudden loss of consciousness and paralysis as a result of haemorrhage into the brain or of thrombosis of a cerebral artery.

apoprotein (*BioSci*) The protein component of a conjugated protein, eg the *globin* of haemoglobin.

apoptosis (*BioSci*) The most common form of physiological (as opposed to pathological) cell death. Apoptosis is an active process requiring metabolic activity by the dying cell; often characterized by shrinkage of the cell, cleavage of the DNA into fragments, and by condensation and margination of chromatin. Often called *programmed cell death*, though this is not strictly accurate. Cells that die by apoptosis do not usually elicit an inflammatory response in contrast to necrotic cells.

aporogamy (*BioSci*) The entrance of the pollen tube into the ovule by a path other than through the micropyle.

aposematic coloration (*BioSci*) Warning coloration, often yellow and black as in some stinging insects.

apospory (*BioSci*) The development of a gametophyte directly from a sporophyte cell without meiosis and the formation of spores. The resulting gametophyte has the same chromosome number as the parent sporophyte. Cf APOGAMY, APOMIXIS.

apostilb (*Phys*) A unit of surface luminance used in the case of diffusing surfaces, numerically equal to 1/10 000 LAMBERT ($1/\pi$ cd m^{-2}).

apostrophe (*BioSci*) The position assumed by chloroplasts in bright light, when they lie against the radial walls of the cells of the palisade layer of the mesophyll.

apothecaries' weight (*Chem*) Obsolete system based on the grain as a unit: 1 grain = 15·4 g.

apothecium (*BioSci*) An open ascocarp, often cup- or saucer-shaped.

Appalachian orogeny (*Geol*) A fold period in eastern N America that extended from the Devonian to the Permian.

It was caused by subduction which led to the closure of the Atlantic Ocean. See SUBDUCTION ZONE.

apparent cohesion (*CivEng*) Cohesion of silts and sands due to surface tension in the enclosed films of water; these films tend to pull the silt grains together.

apparent expansion, coefficient of (*Phys*) The coefficient of expansion when the expansion of eg a dilatometer is neglected. See COEFFICIENT OF EXPANSION.

apparent horizon (*Surv*) See VISIBLE HORIZON.

apparent magnitude (*Astron*) See MAGNITUDE.

apparent particle density (*PowderTech*) The mass of a particle of powder divided by the volume of the particle, excluding open pores but including closed pores.

apparent powder density (*PowderTech*) The mass of the powder divided by the volume occupied by it under specified conditions of packing.

apparent power (*Phys*) The volt-amperes, ie the product of volts and amperes, in an ac circuit or system.

apparent resistance (*Phys*) See IMPEDANCE.

apparent solar day (*Astron*) The interval, not constant owing to the Earth's elliptical orbit, between two successive transits of the Sun over the meridian.

apparent solar time (*Astron*) Time as measured by the apparent position of the Sun in the sky, eg by a sundial.

apparent viscosity (*Phys*) A term applied to the viscosity of many non-Newtonian fluids (eg polymers). Specifically, viscosity calculated using POISEUILLE'S FORMULA.

appeasement behaviour (*Psych*) Submissive behaviour which inhibits attack by a conspecific, often by minimizing threat signals or by mimicking sexual or infantile behaviours, eg crouching or sexual invitation.

append (*ICT*) To extend a file by adding items at the end of the file. The added items may be part of or all of another file.

appendage (*BioSci*) (1) In plants, a general term for any external outgrowth that does not appear essential to growth or reproduction. (2) In animals, a projection of the trunk, such as the parapodia and tentacles of Polychaeta, sensory tentacles of Apoda, fins of fish and limbs of land vertebrates. (3) In Arthropoda, almost exclusively one of the paired, metamerically arranged, jointed structures with sensory, masticatory or locomotor function, but also used for the wings of Insecta.

appendicectomy (*Med*) The surgical removal of the APPENDIX VERMIFORMIS. Also *appendectomy*.

appendicitis (*Med*) Inflammation of the appendix vermiformis.

appendicular (*BioSci*) Pertaining to, or situated on, an appendage.

appendix (*BioSci*) An outgrowth.

appendix vermiformis (*Med*) In some mammals, the distal rudiment of the caecum of the intestine, which in humans is a narrow, blind tube of gut, 25–250 mm in length. See APPENDICITIS.

appetitive behaviour (*Psych*) The active exploratory phase that precedes the presumed goal of a behaviour sequence. Traditionally said to lead to CONSUMMATORY BEHAVIOUR.

appinite (*Min*) A group name for rocks of variable composition and texture containing conspicuous hornblende in a base of plagioclase, with or without orthoclase or quartz.

apple (*For*) European and Asian hardwood (*Malus*) whose heartwood is pinkish-buff coloured, with a straight-grained, fine texture.

Appleby–Frodingham process (*ChemEng*) A fluidized-bed process using specially prepared iron oxide for removal of sulphur from coke oven, and coal gas and vaporized oils. Removes H_2S, CS_2, mercaptans and thiophenes, converting sulphur direct to sulphur dioxide for sulphuric acid.

Applegate diagram (*Electronics*) Presentation of the BUNCHING and DEBUNCHING of an electron beam in a velocity-modulation tube, eg a klystron.

applet (*ICT*) A small application with a defined function, which can be used with a main application. Often used for transmission over the Internet. See panel on INTERNET.

Appletalk (*ICT*) A proprietary LOCAL AREA NETWORK system for connecting Apple computers together.

Appleton layer (*Phys*) See F-LAYER.

application (*ICT*) The use of a computer to carry out a specific task, eg word processing. The term is sometimes used to refer to the software involved in the computer application. Also *application program*.

application layer (*ICT*) The seventh layer of the OPEN SYSTEMS INTERCONNECTION specification that relates to application-specific services to user devices.

application program (*ICT*) See APPLICATION.

applications programmer (*ICT*) Programmer who writes for specific user applications.

applications programming interface (*ICT*) The arrangement of the controls of a particular program such as the common appearance of many window applications.

applications software (*ICT*) Specialized software designed to help carry out a real-life task such as producing spreadsheets and graphs.

applicator (*ElecEng*) Electrodes used in industrial high-frequency heating or medical diathermy; often specially shaped to fit the sample or body. See HEATING INDUCTOR.

applied geology (*Geol*) Geology studied in relation to human activity.

applied mathematics (*Genrl*) Originally the application of mathematics to physical problems, differing from physics and engineering in being concerned more with mathematical rigour and less with practical utility. More recently, also includes numerical analysis, statistics and probability, and applications of mathematics to eg biology, economics, insurance.

applied potential tomography (*Phys*) A system of medical imaging based on the measurement of the electrical impedance, at about 50 kHz frequency, between many electrodes placed around the body.

applied power (*Phys*) For an electrical transducer, the power which would be received if the load matched the source in impedance. That *applied* is not equal to the actual power received, because of the reflection arising from non-equality of impedance matching.

applied psychology (*Genrl*) The part of psychology which puts its knowledge to work in practical situations, eg in vocational guidance and assessment, education and industry.

applied stress (*CivEng*) The stress induced in a member under load.

appliqué (*Textiles*) Ornament, frequently of fabric or plastic, attached to the surface of a fabric to give a three-dimensional effect.

apposition (*BioSci*) The addition of new material to a cell wall at the surface next to the plasmalemma. Cf INTUSSUSCEPTION.

appraisal well (*MinExt*) One of a series of wells drilled near a discovery well to determine the size and nature of an oil or gas field. The results will show whether exploitation of the field is economically worthwhile. Also *step-out well*.

appressed (*BioSci*) Flattened, and pressed close to, but not united with, another organ.

appressorium (*BioSci*) A flattened outgrowth that attaches a parasite to its host, esp a modified hypha, closely applied to the host epidermis. A narrow infection hypha or penetration tube is pushed into the cell or space below the attachment.

approach–avoidance conflict (*Psych*) The conflict arising when the best positive choice will result in a negative outcome as well as the positive.

approach control radar (*Aero*) A surveillance radar which shows on a cathode-ray tube display the positions of aircraft in an aerodrome's traffic control area. Abbrev ACR.

approach lights (*Aero*) Lights indicating the desired approach to a runway, usually of sodium or high-intensity type and laid in a precise pattern of a lead-in line with cross-bars at set distances from the RUNWAY THRESHOLD.

approach speed (*Aero*) The indicated airspeed at which an aircraft approaches for landing.

approximate integration (*MathSci*) A method of approximating to the area under a curve by considering its behaviour at a small number of its ordinates. See SIMPSON'S RULE, THREE-EIGHTHS RULE, WEDDLE'S RULE.

appulse (*Astron*) Seemingly close approach of two celestial objects as perceived by an observer, particularly the close approach of a planet or asteroid to a star without the occurrence of an eclipse.

apraxia (*Psych*) Loss of the ability to perform purposeful movements but in the absence of any organic cause.

apron (*Aero*) A firm surface of concrete or 'tarmac' laid down adjacent to aerodrome buildings to facilitate the movement, loading and unloading of aircraft.

apron (*Build*) (1) The protecting slope on the downstream side of the sluices of a lock gate or dam provided to withstand the force of the falling water. (2) Blocks of concrete, masonry, etc, deposited around the toe of a sea wall to protect its base from scour caused by the returning wave. (3) The lead sheeting or other weather-resistant material raggled into a wall to divert water into a gutter or drain.

apron (*Eng*) In a lathe, that part of the saddle enclosing the gear operated by the lead screw.

apron (*ImageTech*) Flexible strip used as film support in some types of processing tank.

apron (*Paper*) A strip of rubber, metal or other material at the outlet from the flow box to seal the gap between it and the machine wire.

apron conveyor (*Eng*) A conveyor for transporting packages or bulk materials, consisting of a series of metal or wood slats (also rubber, cotton, felt, wire, etc) attached to an endless chain. Also *slat conveyor*.

apron feeder (*MinExt*) Short, endless conveyor belt, sturdily built of articulated plates, used to draw ore at regulated rate from bottom of stockpile or ore bin.

apron lining (*Build*) A lining of wrought boarding covering the APRON PIECE at a staircase landing.

apron piece (*Build*) The horizontal timber carrying the upper ends of the CARRIAGE PIECES or rough-strings of a wooden staircase. Also *pitching-piece*.

aprotic (*Chem*) A term normally restricted to solvents such as acetonitrile, which have high relative PERMITTIVITIES and hence aid the separation of electric charges but do not provide protons.

apse (*Arch*) The semicircular or polygonal recess, either arched or dome-roofed, terminating the choir or chancel of a church. Also *apsis*.

apse line (*Astron*) The diameter of an elliptical orbit which passes through both foci and joins the points of greatest and least distance of the revolving body from the centre of attraction. Also *line of apsides*.

apsis (*Arch*) See APSE.

apsis (*Astron*) In an orbit, the point of greatest or least distance from the central body.

aptamer (*BioSci*) A double-stranded DNA or single-stranded RNA molecule that binds to a specific molecular target.

apterism (*BioSci*) The condition of winglessness, either primitive or secondary, found in many insects.

apterous (*BioSci*) Without wings.

apterygial (*BioSci*) Without wings; without fins.

Apterygota (*BioSci*) A subclass of small, primitively wingless insects showing little metamorphosis. Bristletails and springtails.

Aptian (*Geol*) A stage of the Cretaceous system lying between the Barremian below and the Albian above. See MESOZOIC.

aptitude (*Psych*) A specific ability or capacity to learn. *General aptitude* refers to the capacity for acquiring knowledge in a wide range of areas, as opposed to a *specific aptitude*, such as the ability to acquire musical skills. *Aptitude testing* is an attempt to measure individual differences in potential for learning, as opposed to *achievement testing*, which measures present levels of competence in a given area.

APU (*Aero*) Abbrev for AUXILIARY POWER UNIT.

Apus (Bird of Paradise) (*Astron*) A small and inconspicuous southern hemisphere constellation, often erroneously translated as 'bee'.

apyrexia (*Med*) Absence of fever.

aq (*Chem*) Symbol for water.

aqua- (*Genrl*) Prefix from Lt *aqua*, water.

Aquadag (*Eng*) TN for a colloidal suspension of graphite in water.

aqua fortis (*Chem*) Ancient name for concentrated nitric acid.

aquamarine (*Min*) A variety of beryl, of attractive blue-green colour, used as a gemstone.

aquaporin (*BioSci*) An integral membrane protein that greatly increases water permeability. Found esp in kidney and red blood cells.

aqua regia (*Chem*) A mixture consisting of one volume of concentrated nitric acid to three volumes of concentrated hydrochloric acid.

Aquarids (*Astron*) One of two major meteor showers. The *Delta Aquarids* show maximum activity on 28 July with an hourly rate of approx 30 meteors. The *Eta Aquarids* show maximum activity on 4 May with an hourly rate of approx 20 meteors.

Aquarius (Water Bearer) (*Astron*) A large constellation in the southern sky, lying between Capricornus and Pisces.

aquatic (*BioSci*) Said of a plant living or growing in or on water. See HYDROPHYTE.

aquatint (*Print*) An intaglio printing process using a copper plate with a ground of resin particles. Half-tone effects are produced by etching and progressive stopping-out.

aqueduct (*BioSci*) A channel or passage filled with or conveying fluid: in higher vertebrates, the reduced primitive ventricle of the midbrain.

aqueduct (*CivEng*) An artificial conduit, generally elevated on columns, used to convey a water supply.

aqueductus Sylvii (*BioSci*) In vertebrates, the ventricle of the midbrain or iter.

aqueductus vestibuli (*BioSci*) In Craniata, a narrow tube arising from the auditory sac and opening on the dorsal surface of the head, as in some fishes, or ending blindly. The endolymphatic duct.

aqueous (*Chem*) Consisting largely of water; dissolved in water.

aqueous humour (*BioSci*) In vertebrates, the watery fluid filling the space between the lens and the cornea of the eye.

aqueous tissue (*BioSci*) Water-storage tissue of plants, made up of large, thin-walled, hyaline cells.

aquiculture (*BioSci*) Augmentation of aquatic animals of economic importance by direct methods: cultivation of the resources of sea and inland waters as distinct from exploitation.

aquifer (*Geol*) Rock formation containing water in recoverable quantities.

Aquila (Eagle) (*Astron*) A constellation on the celestial equator. Its brightest star is ALTAIR.

Aquitanian (*Geol*) The lowest stage of the Neogene (Miocene). See TERTIARY.

AR (*Chem*) Abbrev for ANALYTICAL REAGENT.

AR (*ImageTech*) Abbrev for ASPECT RATIO.

Ar (*Chem*) Symbol for: (1) ARGON; (2) an aryl, or aromatic, radical.

Ar (*Eng*) The TRANSFORMATION TEMPERATURE on cooling of the phase changes in iron and steel, subscripts indicating the appropriate change.

Ara (Altar) (*Astron*) A small southern hemisphere constellation.

arabesque (*Arch*) An ornamental work used in decorative design for flat surfaces; consists usually of interlocked curves which may be painted, inlaid or carved in low relief.

arabic numbers (*MathSci*) The numerals 1, 2, 3, etc, as opposed to the Roman numerals I, II, III, etc. Actually derived from India.

Arabidopsis thaliana (*BioSci*) The common wall cress. A small plant, much used as a model system for plant molecular biology, because of its small genome (7×10^7 base pairs) and short generation time (5–8 weeks).

arabinose (*Chem*) A monosaccharide belonging to the pentose group. *L-arabinose*, $C_5H_{10}O_5$, is produced by boiling gum-arabic, cherry gum or beetroot chips with dilute sulphuric acid; prisms soluble in water forming a dextrorotatory solution. Used as a culture medium for certain bacteria. See PENTOSES.

arabitol (*Chem*) $OHCH_2(CHOH)_3CH_2OH$. A pentahydric alcohol corresponding to ARABINOSE, and obtained from it by reduction with sodium tetrahydroborate.

arable farming (*Agri*) Cultivation of food crops for livestock or human use.

arachidonic acid (*BioSci*) *Eicosatetraenoic acid*. $CH_3[CH_2]_4[CH = CHCH_2]_4CH_2CH_2COOH$. An unsaturated fatty acid that is a dietary requirement. In the esterified form, a component of cell membranes and, when converted to the free acid, is the precursor of PROSTAGLANDINS and LEUKOTRIENES and is thus of importance in inflammation and hypersensitivity.

arachis oil (*Chem*) Peanut oil.

Arachnida (*BioSci*) A class of mainly terrestrial Arthropoda in which the head and thorax are continuous (PROSOMA). The head bears pedipalps and chelicerae but no antennae. There are four pairs of ambulatory legs. Includes spiders, harvest-men, mites, ticks, scorpions.

arachnidium (*BioSci*) In spiders, the spinnerets and silk glands.

arachnodactyly (*Med*) Extreme length of fingers and toes seen in MARFAN'S SYNDROME.

arachnoid (*BioSci*) A cobweb-like entanglement of hairs or fibres; pertaining to or resembling the Arachnida; one of the three membranes that envelop the brain and spinal cord of vertebrates, lying between the dura mater and the pia mater.

araeostyle (*Arch*) A colonnade in which the space between the columns is equal to or greater than four times the lower diameter of the columns. Also *araostyle*.

aragonite (*Min*) The relatively unstable, orthorhombic form of crystalline calcium carbonate, deposited from warm water, but prone to inversion into calcite; also stable at high pressures. See FLOS FERRI.

Arago point (*Phys*) The bright spot found along the axis in the shadow of a disk illuminated normally.

Arago's rotation (*ElecEng*) Experiments (conducted by Arago before the discovery of electromagnetic induction by Faraday) in which a rotating copper disk was made to cause rotation of a pivoted magnet.

Araldite (*Plastics*) TN for range of epoxy resins used for adhesives, encapsulation of electrical components, etc. See panel on THERMOSETS.

aramid fibres (*Chem, Plastics*) Fibres made by spinning liquid crystal aramid oligomers. See panel on HIGH-PERFORMANCE POLYMERS.

Araneae (*BioSci*) An order of ARACHNIDA in which the PROSOMA is joined to the apparently unsegmented abdomen or opisthosoma by a waist. Spinnerets and several kinds of spinning glands occur. The pedipalps are modified in the male for the transmission of sperm. Spiders.

araneous (*BioSci*) Cobweb-like.

araostyle (*Arch*) See ARAEOSTYLE.

ARB (*Aero*) Abbrev for AIR REGISTRATION BOARD.

arbitration bar (*Eng*) Test bar, cast with a given heat of metal, to determine whether the main casting is to specification.

arbor (*Eng*) (1) Cylindrical or conical shaft on which a cutting tool or part to be machined is mounted. (2) The axis or shaft upon which a rotatable part is mounted: the shaft upon which a gear or wheel is mounted. See MANDREL.

arboretum (*BioSci*) An area devoted to the cultivation of trees and other woody plants.

arboviruses (*BioSci*) Various single-stranded RNA viruses, not of the same taxonomic group, that are arthropod-borne and multiply in both invertebrate and vertebrate hosts; they are pathogenic when transferred to the vertebrate, causing diseases such as yellow fever and encephalitis. Three major families are recognized: Togaviridae, Bunyaviridae and Arenaviridae.

arbuscule (*BioSci*) (1) A dwarf tree or shrub of tree-like habit. (2) A much-branched HAUSTORIUM formed within the host cells by some endophytic fungi. See VESICULAR–ARBUSCULAR MYCORRHIZA.

ARC (*Aero*) Abbrev for: (1) *Aeronautical Research Council*, UK (1909–79); (2) *Ames Research Center*, USA.

arc (*ElecEng*) Ionic gaseous discharge maintained between electrodes, characterized by low voltage and high current. See MERCURY-ARC RECTIFIER.

arc (*MathSci*) A portion of a curve.

arc (*Phys*) See ARC LAMP.

arc absorber (*ElecEng*) Same as SPARK ABSORBER, but referring to a discharge likely to be destructive if not extinguished.

arcade (*Arch*) (1) A series of arches, usually in the same plane, supported on columns, eg the nave arcades in churches. When filled in with masonry, it becomes a 'blind arcade'. (2) An arched passage, esp one having shops on one or both sides.

arcade (*BioSci*) See DENTAL ARCADE.

arc-back (*Electronics*) Flow of electrons, opposite to that intended, in a mercury-arc rectifier. Caused by a heated spot on the anode acting as a cathode, leading to possible damage.

arc baffle (*Electronics*) Means of preventing liquid mercury contacting an anode in a mercury-arc rectifier. Also *splash baffle*.

arc-control device (*ElecEng*) A device fitted to the contacts of a circuit breaker to facilitate the extinction of the arc.

arccos (*MathSci*) See INVERSE TRIGONOMETRICAL FUNCTIONS.

arccosh (*MathSci*) See INVERSE HYPERBOLIC FUNCTIONS.

arc crater (*ElecEng*) Depression formed in electrodes between which an electric arc has been maintained. In arc welding, the depression which occurs in the weld metal.

arc duration (*ElecEng*) Time during which an arc exists between the contacts of an opening switch or circuit breaker. In ac circuits usually measured in cycles, varying between half a cycle and perhaps 20 cycles.

arc furnace (*ElecEng*) An electric furnace in which the heat is produced by an electric arc between carbon electrodes, or between a carbon electrode and the furnace charge.

arch (*BioSci*) A curved or arch-shaped skeletal structure supporting, covering or enclosing an organ or organs, such as the HAEMAL ARCH, NEURAL ARCH or zygomatic arch (see ZYGOMA).

arch (*CivEng*) A form of structure having a curved shape, used to support loads or to resist pressures.

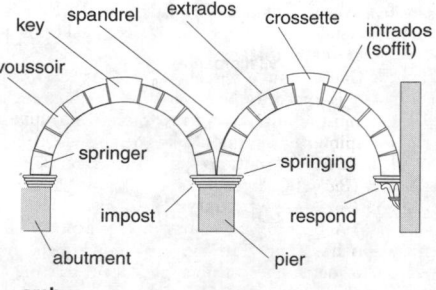

key spandrel extrados crossette intrados (soffit)
voussoir
springer springing
impost respond
abutment pier

arch

archae-, archaeo- (*Genrl*) Prefixes from Gk *archaios*, ancient. US *arche-, archeo-*.

Archaea (*BioSci*) One of two major subdivisions of the prokaryotes. There are three main Orders: extreme halophiles, methanobacteria and sulphur-dependent extreme-thermophiles. Archaea differ from Eubacteria in several important characteristics. Formerly *Archaebacteria*.

Archaean (*Geol*) The oldest rocks of the Precambrian. Usually taken to be older than c.2500 Ma. See PRECAMBRIAN. See appendix on Geological time.

archaebacteria (*BioSci*) Old name for ARCHAEA.

archaeo- (*Genrl*) See ARCHAE-.

archaeology (*Genrl*) The study of human antiquities, usually as discovered by excavation of material remains. US *archeology*.

archaeometry (*Genrl*) The use of scientific methods in archaeology. US *archeometry*.

archaeostomatous (*BioSci*) Having a persistent BLASTO-PORE, which gives rise to the mouth.

arch brick (*Build*) A brick having a wedge shape, esp one with a curved face suitable for wells and other circular work.

arch bridge (*CivEng*) A bridge that depends on the principle of the arch for its stability. See RIGID ARCH, THREE-HINGED ARCH and panel on BRIDGES AND MATERIALS.

arch dam (*CivEng*) Dam in which the abutments are solid in rock at sides of impounding area.

arche-, archeo- (*Genrl*) US for ARCHAE-, ARCHAEO-.

archecentra (*BioSci*) In vertebrates, centra formed by the enlargement of the bases of the arched elements that grow around the notochord outside its primary sheath. Cf CHORDACENTRA. Adjs *archecentrous, aricentrous, arcocentrous*.

archegonial chamber (*BioSci*) A small cavity at the micropylar end of the female gametophyte of some gymnosperms, eg cycads, into which the spermatozoids are liberated to swim to the archegonia.

Archegoniatae (*BioSci*) In some classifications, one of the main groups within the plant kingdom, including the Bryophyta and Pteridophyta. Characterized by the presence of the archegonium as the female organ, and by the regular alternation of gametophyte and sporophyte in the life cycle.

archegoniophore (*BioSci*) A specialized branch of a thallus bearing ARCHEGONIA. Also *archegonial receptacle*.

archegonium (*BioSci*) A sessile or stalked organ, bounded by a multicellular wall, and flask-shaped in general outline. It consists of a chimney-like neck containing an axial series of neck-canal cells, and a swollen venter below, containing a single egg and a ventral-canal cell. The archegonium is the female organ of Bryophyta and Pteridophyta, and, in a slightly simplified form, of most Gymnospermae.

archencephalon (*BioSci*) In vertebrates, the primitive forebrain; the cerebrum.

archenteron (*BioSci*) Cavity in the GASTRULA, enclosed by endoderm. It opens to the exterior at the BLASTOPORE.

archeo- (*Genrl*) See ARCHAE-, ARCHAEO-.

archeology (*Genrl*) US for ARCHAEOLOGY.

archeometry (*Genrl*) US for ARCHAEOMETRY.

archesporium (*BioSci*) The tissue in a sporangium that gives rise to the spore mother cells, including the region of the nucellus giving rise to the megaspore mother cells.

archetype (*BioSci*) A primitive type from which others may be derived.

archetype (*Psych*) In the psychology of Carl Jung, an emotionally laden image assumed to be present in the unconscious mind of all human beings throughout history; an aspect of the COLLECTIVE UNCONSCIOUS.

archi- (*Genrl*) Prefix from Gk *archi-*, first, chief.

Archiannelida (*BioSci*) A class of ANNELIDA, of small size and marine habit, which usually lack setae and parapodia and have part of the epidermis ciliated. The nervous system retains a close connection with the epidermis. Members of this class resemble the Polychaeta in many of their characteristics.

archiblastic (*BioSci*) A term describing an egg exhibiting total and equal segmentation.

archiblastula (*BioSci*) A regular spherical blastula, having cells of approximately equal size.

archicoel (*BioSci*) See BLASTOCOEL.

Archimedean drill (*Eng*) A drill in which to-and-fro axial movement of a nut on a helix causes an alternating rotary motion of the bit.

Archimedean screw (*Eng*) An ancient water-lifting contrivance: a screw within a pipe forming a helical tube (or a pipe wound in helix fashion around an inclined axis) which has its lower end in water so that, on rotation of the 'screw', water rises to a higher level.

Archimedes' principle (*Phys*) The principle that when a body is wholly or partly immersed in a fluid it experiences an upthrust equal to the weight of fluid it displaces; the upthrust acts vertically through the centre of gravity of the displaced fluid.

Archimedes' spiral (*MathSci*) A spiral with polar equation $r = a\theta$.

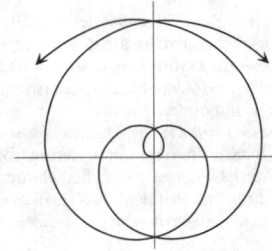

Archimedes' spiral Shows two mirror images.

archinephric (*BioSci*) In vertebrates, pertaining to the archinephros (see PRONEPHROS); in invertebrates, pertaining to the larval kidney or excretary organ.

archinephridium (*BioSci*) In invertebrates, the larval excretory organ, usually a solenocyte.

archinephros (*BioSci*) See PRONEPHROS.

archipallium (*BioSci*) In vertebrates, that part of the cerebral hemispheres not included in the olfactory lobes and corpora striata, and comprising the hippocampus and the olfactory tracts and associated olfactory matter; that part of the pallium excluding the neopallium.

archipelago (*Geol*) A sea area thickly interspersed with islands, originally applied to the part of the Mediterranean which separates Greece from Asia.

architectural acoustics (*Acous*) The study of propagation of sound waves in buildings, the results being applied to the design of studios and auditoria for optimum audition and to the noise isolation of buildings.

architecture (*ICT*) Of software, hardware or data: the desirable, consistently structured result of a rational process of systems design. See COMPUTER ARCHITECTURE.

architrave (*Arch*) The lowest part of an entablature in immediate contact with the ABACUS on the capital of a column. Also *epistyle*.

architrave (*Build*) The mouldings surrounding a door or window opening, including the lintel.

architrave block (*Build*) The block, placed at the foot of the side moulding around a door opening, into which the skirting fits.

architrave jambs (*Build*) The mouldings at the sides of a door or window opening.

archive file (*ICT*) Copy of file held for safety on a secure and stable storage medium.

archivolt (*Arch*) An ornamental moulding carried around the face of an arch.

Archosauria (*BioSci*) A subclass of diapsid reptiles that were the dominant forms in the Mesozoic. The only surviving group are the crocodiles and alligators.

arch piece (*Ships*) See STERN FRAME.

arch stone (*CivEng*) A wedge-shaped stone used as a constituent part of an arch. Also *voussoir*.

aricentrous (*BioSci*) See ARCHECENTRA.

arcing contact (*ElecEng*) An auxiliary contact fitted to a switch or circuit breaker which opens after and closes before the main contact and receives most of the damage due to arcing. Designed for easy replacement. Also *arcing tips*.

arcing-ground suppressor (*ElecEng*) See ARC SUPPRESSOR.

arcing ring (*ElecEng*) Circular or oval ring conductor, placed concentrically with a pin insulator or a string of insulators, for deflecting an arc from the insulator surface which could be damaged.

arcing shield (*ElecEng*) See GRADING SHIELD.

arcing tips (*ElecEng*) See ARCING CONTACT.

arcing voltage (*ElecEng*) Voltage below which a current cannot be maintained between two electrodes.

arc lamp (*Phys*) A form of electric lamp which makes use of an electric arc between two carbon electrodes as the source of light. It has an extremely high intrinsic brilliance, and is therefore used for searchlights and spotlights. See CARBON ARC LAMP.

arcminute (*Astron*) A unit of angular measure equal to $\frac{1}{60}°$.

arcocentrous (*BioSci*) See ARCHECENTRA.

arc of approach (*Eng*) The arc on the PITCH CIRCLE of a gearwheel over which two teeth are in contact while approaching the pitch point.

arc of contact (*Eng*) The arc on the PITCH CIRCLE of a gearwheel over which two teeth are in contact.

arc of recess (*Eng*) The arc on the pitch circle of a gearwheel over which two teeth are in contact while receding from the pitch point.

arc resistance (*ElecEng*) The ability of an insulator to withstand high-voltage sparking.

arcsecond (*Astron*) A unit of angular measure equal to $\frac{1}{3600}°$.

arcsin (*MathSci*) See INVERSE TRIGONOMETRICAL FUNCTIONS.

arcsinh (*MathSci*) See INVERSE HYPERBOLIC FUNCTIONS.

arc spectrum (*Phys*) A spectrum originating in the non-ionized atoms of an element; usually capable of being excited by the application of a comparatively low stimulus, such as the electric arc. See SPARK SPECTRUM.

arc spraying (*Eng*) A method of fusing (and thence depositing) refractory ceramic and metal powders by blowing them through an electric arc or plasma. Used for applying a variety of thin and thick film coatings. Also *plasma spraying*.

arc-stream voltage (*ElecEng*) Voltage drop along the arc stream of an electric arc, excluding the voltage drops at the anode and cathode.

arc-suppression coil (*ElecEng*) See PETERSEN COIL.

arc suppressor (*ElecEng*) A device for automatically earthing the neutral point of an insulated-neutral transmission or distribution line if an arc to ground occurs. Also *arcing-ground suppressor*.

arctan (*MathSci*) See INVERSE TRIGONOMETRICAL FUNCTIONS.

arctanh (*MathSci*) See INVERSE HYPERBOLIC FUNCTIONS.

arc therapy (*Radiol*) X-ray therapy in which the angle of rotation of therapeutic radiation is limited, to avoid sensitive organs, eg lungs or gonads.

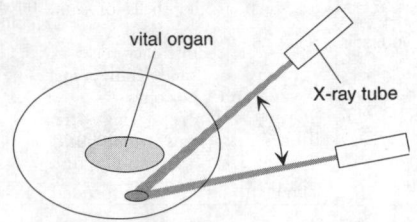

arc therapy

arc-through (*Electronics*) Overflow of electron stream into an intended non-conducting period.

Arcturus (*Astron*) A prominent red giant star in the constellation Bootes. Distance 11 pc. Also *Alpha Bootis*.

arcuate (*BioSci*) Bent like a bow.

arcus senilis (*Med*) Degeneration of the periphery of the cornea in old people. May occur earlier in HYPERCHOLES-TEROLAEMIA.

arc voltage (*ElecEng*) The total voltage across an electric arc, ie the sum of the arc stream voltage, the voltage drop at the anode and the voltage drop at the cathode. The term is frequently used in connection with arc welding, and with the arc in a switch or circuit breaker.

arc welding (*ElecEng*) A process for joining of metal parts by fusion in which the heat necessary for fusion is produced by an electric arc struck between two electrodes or between an electrode and the metal.

are (*Surv*) A metric unit of area used for land measurement. 1 are = 100 m^2 = 119·6 yd^2. See HECTARE.

area (*Build*) The sunken space around the basement of a building, providing access and natural lighting and ventilation.

area (*Genrl*) A measure of the extent of a surface. Unit m^2.

area (*Surv*) In plane surveying, the superficial content of a ground surface of definite extent, as projected onto a horizontal plane.

areal velocity (*Astron*) The rate, constant in elliptical motion, at which the radius vector sweeps out unit area.

area-moment method (*CivEng*) A method of structural analysis based on the slope and displacement of any part of the structure.

area monitoring (*Radiol*) The survey and measurement of types of ionizing radiation and dose levels in an area in which radiation hazards are present or suspected. See AERIAL RADIOMETRIC SURVEYING.

area opaca (*BioSci*) In developing avian embryos, a whitish peripheral zone of blastoderm, in contact with the yolk.

area pellucida (*BioSci*) In developing avian embryos, a central clear zone of blastoderm, not in direct contact with the yolk.

area rule (*Aero*) An aerodynamic method of reducing drag at transonic speeds by maintaining a smooth cross-sectional variation throughout the length of an aircraft. Because of the effect of the wing, this often results in a 'wasp-waist' on the fuselage or the addition of bulges to the wing or fuselage.

area vasculosa (*BioSci*) In developing vertebrate embryos, part of the extraembryonic blastoderm, in which the blood vessels develop.

Arecaceae (*BioSci*) See PALMAE.

Arecidae (*BioSci*) A subclass or superorder of monocotyledons comprising trees, shrubs, terrestrial herbs and a few free-floating aquatics. They mostly have broad, petiolate leaves that are often net-veined. The inflorescence is usually numerous small flowers generally subtended by a spathe and often aggregated into a spadix. The Arecidae contains c.6400 spp in six families including Palmae, Araceae, Pandanaceae and Lemnaceae.

arenaceous (*BioSci*) (1) Plants growing best in sandy soil. (2) Animals occurring in sand. (3) Composed of sand or similar particles, such as the shells of some kinds of Radiolaria. Also *arenicolous*.

arenaceous rocks (*Geol*) Sedimentary rocks in which the principal constituents are sand grains, including the various sorts of sands and sandstones.

Arenaviridae (*BioSci*) A family of single-stranded RNA viruses that includes Lassa virus, lymphocytic choriomeningitis virus, and the Tacaribe group of viruses.

arenicolous (*BioSci*) See ARENACEOUS.

Arenig (*Geol*) A series of rocks in the Ordovician System, taking their name from Arenig Mountain in N Wales, where they were originally described by Adam Sedgwick. See PALAEOZOIC.

arenite (*Geol*) The general term for any sedimentary rock with sand-sized grains.

areola (*BioSci*) (1) One of the spaces between the cells and fibres in certain kinds of connective tissue. (2) In the vertebrate eye, that part of the iris bordering the pupil. (3) In mammals, the dark-coloured area surrounding the nipple. Pl *areolae*.

areolar (*BioSci*) (1) Divided into small areas or patches. (2) Pitted. (3) Pertaining to an areola. Also *areolate*.

areolar tissue (*BioSci*) A type of connective tissue consisting of cells separated by extracellular matrix with bundles of white (collagen) and yellow (elastin) fibres.

areolate (*BioSci*) See AREOLAR.

areole (*BioSci*) (1) A small area delimited in some way, esp: (a) an island into which a reticulated and veined leaf is divided by the veins; and (b) an area demarcated by the network of cracks in a lichen thallus. (2) A cushion, representing a condensed lateral shoot from which spines, branches and flowers arise in cacti.

arête (*Geol*) A sharp-edged precipitous ridge of bare rock, often in mountainous country between two cirques.

arfvedsonite (*Min*) A monoclinic iron-rich alkali-amphibole.

Arg (*Chem*) Symbol for ARGININE.

Argand burner (*Eng*) A form of gas or oil burner in which air is admitted to the inside of a cylindrical wick, ensuring a large area of contact between the flame and the fuel.

Argand diagram (*MathSci*) A plane diagram in which every complex number can be represented uniquely by a single point. The diagram represents the complex number $z = x + iy$ by the point (x,y) referred to rectangular cartesian axes Oxy, the axis Ox being called the *real axis* and the axis Oy the *imaginary axis*.

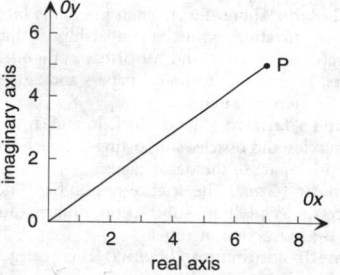

Argand diagram P(7,5) represents the complex number $z = 7 + 5i$.

argentate (*BioSci*) Of silvery appearance.

argentic [silver (II)] oxide (*Chem*) AgO, an oxide of silver.

argentiferous (*Min*) Containing silver.

argentite (*Min*) An important ore of silver, having the composition Ag$_2$S (silver sulphide); crystallizes in the cubic system. Cf ACANTHITE. Also *silver glance*.

argentous [silver (I)] oxide (*Chem*) Ag$_2$O. A lower oxide of silver.

argillaceous rocks (*Geol*) Sediments of silt or clay-particle size. Common clay minerals are kaolinite and montmorillonite.

argillicolous (*BioSci*) Living on a clayey soil.

argillite (*Geol*) A slightly metamorphosed siltstone or mudstone which lacks fissility.

arginine (*Chem*) *2-amino-5-guanidopentanoic acid.* H$_2$NC(NH)NH(CH$_2$)$_3$CH(NH$_2$)COOH. The L- or S-isomer is an essential amino acid. Symbol Arg, short form R.

argon (*Chem*) Element which forms no known compound, one of the rare gases. Symbol Ar, at no 18, ram 39·948. A colourless, odourless, monatomic gas; mp −189·2°C; bp −185·7°C; density 1·7837 g dm^{-3} at stp. Argon constitutes about 1% by volume of the atmosphere, from which it is obtained by the fractionation of liquid air. It is used in gas-filled electric lamps, radiation counters, fluorescent tubes etc. The isotope argon-40 is formed by the radioactive decay of potassium-40 and the ratio of potassium-40 to argon-40 in a rock is used for the determination of its age (abbrev *K–Ar method*).

argon laser (*Phys*) Laser using singly ionized argon. It gives strong emission at 488·0, 514·5 and 496·5 nm.

argument (*ICT*) Input parameter to a program.

argument (*MathSci*) (1) Of a function $y = f(x)$, the independent variable x with respect to which it is defined. (2) Of a complex number $x + iy$, the angle $\tan^{-1}y/x$ which is the angle in the ARGAND DIAGRAM between the line from the origin to the point representing the complex number and the positive direction of the real axis.

Argyll–Robertson pupil (*Med*) An irregular eccentric pupil of the eye which reacts to accommodation, but not light.

argyrodite (*Min*) A double sulphide of germanium and silver, the mineral in which the element germanium was first discovered.

Ariane (*Space*) European satellite launcher developed by the ESA; first launched in 1979. The Ariane 5 rocket is still (2006) being used.

arid zone (*EnvSci*) A zone of latitude 15°–30°N and S in which the rainfall is so low that only desert and semi-desert vegetation occurs, and irrigation is necessary if crops are to be grown.

Ariel (*Astron*) A natural satellite of Uranus, discovered in 1851. Distance from the planet 191 000 km; diameter 1160 km.

Aries (Ram) (*Astron*) A constellation in the northern sky, lying between Pisces and Taurus.

aril (*BioSci*) An outgrowth on a seed, formed from the stalk or from near the micropyle. It may be spongy or fleshy, or may be a tuft of hairs.

ARINC (*Aero*) Abbrev for *Aeronautical Radio Incorporated*, an US organization whose membership includes airlines, aircraft constructors and AVIONICS component manufacturers. It publishes technical papers and agreed standards, and finances research.

Aristotle's lantern (*BioSci*) In Echinoidea, the framework of muscles and ossicles supporting the teeth, and enclosing the lower part of the oesophagus.

arithmetic (*Genrl*) The science of numbers, including such processes as addition, subtraction, multiplication, division and the extraction of roots.

arithmetic continuum (*MathSci*) The aggregate of all real numbers, rational and irrational.

arithmetic–geometric (*Geol*, *Min*) An approximating principle in geology: that as the quality of an ore diminishes in an arithmetic fashion, so does the quantity of the ore increase, but geometrically. Such an over-simplification that its utility is perhaps doubtful.

arithmetic logic unit (*ICT*) Circuits within the central processing unit where arithmetic/logic operations are performed. Also *arithmetic unit*. Abbrev ALU. See fig. at MICROPROCESSOR.

arithmetic mean (*MathSci*) Of n numbers a_r, their sum divided by n, ie

$$\frac{1}{n}\sum_{r=1}^{n} a_r.$$

Also *average*.

arithmetic operation (*ICT*) An OPERATION performed using the laws of arithmetic. Cf LOGICAL OPERATION.

arithmetic operator (*ICT*) Symbol used to indicate the arithmetic operation to be performed (eg '+' in 5+7). See OPERATOR.

arithmetic progression (*MathSci*) A sequence of numbers, each term being obtained from the preceding one by the addition of a constant difference, eg a, $(a + d)$, $(a + 2d)$, ..., $[a+(n-1)d]$,... where d is the COMMON DIFFERENCE. Also *arithmetic sequence*. The sum of the sequence is the ARITHMETIC SERIES.

arithmetic register (*ICT*) Special memory location usually part of the ARITHMETIC LOGIC UNIT, used to hold operands and results temporarily during processing.

arithmetic series (*MathSci*) The sum of an ARITHMETIC PROGRESSION, ie $(a + d) + (a+2d) + ... + [a+(n-1)d] + ...$ where d is the common difference. The sum of the first n terms is

$$\sum_{r=0}^{n-1}(a + rd) = \frac{n}{2}[2a + (n - 1)d]$$

arithmetic shift (*ICT*) One where, if bits are shifted from the right of the location, they are lost and copies of the sign bit are shifted in at the opposite end. If the data represent a number, this operation preserves the positive/negative sign.

arithmetic unit (*ICT*) See ARITHMETIC LOGIC UNIT.

arkose (*Geol*) A feldspar-rich, coarse-grained sandstone, derived from the erosion of granites and gneisses.

ARM (*Aero*) Abbrev for *anti-radiation missile*.

arm (*BioSci*) (1) In Echinodermata, a prolongation of the body in the direction of a radius. (2) In Cephalopoda, one of the tentacles surrounding the mouth. (3) In bipedal mammals, one of the upper limbs.

arm (*ICT*) See BRANCH.

armature (*ElecEng*) (1) Moving part which closes a magnetic circuit and which indicates the presence of electric current as the agent of actuation, as in all relays, electric bells, sounders, telephone receivers. (2) Piece of low-reluctance ferromagnetic material (keeper) for temporarily bridging the poles of a permanent magnet, to reduce the leakage field and preserve magnetization. (3) The rotating part (*rotor*) of a dc motor or generator.

armature bars (*ElecEng*) Rectangular copper bars forming the conductors on the armature in large electric machines having only a few conductors per slot.

armature coil (*ElecEng*) An assembly of conductors ready for placing in the slots of the armature of an electric machine.

armature conductor (*ElecEng*) One of the wires or bars on the armature of an electric machine.

armature core (*ElecEng*) The assembly of laminations forming the magnetic circuit of the armature of an electric machine. The thickness of each lamination is usually of the order of 0·5 mm.

armature end connections (*ElecEng*) The portion of the armature conductors which project beyond the end of the armature core, and which are used for making the connections among the various conductors. Also *overhang*.

armature end plate (*ElecEng*) The end plate of a laminated armature core. It is of sufficient mechanical strength to enable the laminations to be clamped together tightly to prevent vibration. Also *armature head*.

armature ratio (*ElecEng*) Ratio of distance moved by the spring buffer of an electromagnetic relay to that moved by the armature.

armature reactance (*ElecEng*) A reactance associated with the armature winding of a machine, caused by armature leakage flux, ie flux which does not follow the main magnetic circuit of the machine.

armature reaction (*ElecEng*) The magnetic field in an electrical machine produced by the armature current.

armature relay (*ElecEng*) A relay operated electromagnetically, thus causing the armature to be magnetically attracted.

armature winding (*ElecEng*) The complete assembly of conductors carried on the armature and connected to the commutator or to the terminals of the machine.

Armco (*Eng*) TN for a soft iron with less than 1% impurities. Can be rolled or formed with deep corrugations as in circular culverts or traffic barriers.

armed (*BioSci*) Protected by eg prickles, thorns, spines or barbs.

armillary sphere (*Astron*) Celestial globe, first used by the Greek astronomers, in which the sky is represented by a skeleton framework of intersecting circles, the Earth being at the centre. In antiquity, of major importance for measuring star positions.

arming press (*Print*) A form of blocking press used for stamping designs on book covers.

Armorican orogeny (*Geol*) An old name for the HERCYNIAN OROGENY.

armour-clad switchgear (*ElecEng*) See METAL-CLAD SWITCHGEAR.

armour clamp (*ElecEng*) A fitting designed to grip the armouring of a cable where it enters a box. Also *armour gland, armour grip*.

armour plate (*Eng*) Traditionally, specially heavy alloy steel plate hardened on the surface; used for the protection of fighting vehicles and ships. There is also a form of armour plate based on aluminium alloy particularly suitable for fast-moving military vehicles ('Chobham armour').

Armstrong oscillator (*ICT*) The original oscillator, in which tuned circuits in the anode and grid circuits of a valve are coupled.

Arndt–Eistert reaction (*Chem*) Reaction used for converting a carboxylic acid to a higher homologue. The acid chloride is added to an excess of diazomethane to form a diazoketone. The ketone undergoes catalytic rearrangement to the higher homologue or a derivative.

aromatic compounds (*Chem*) Planar organic ring compounds such as benzene that have a cyclic cloud of delocalized electrons above and below the ring. The chemical bonding between adjacent atoms is intermediate between single and double bonds. Aromatic compounds do not undergo addition reactions as other unsaturated compounds do, but show substitution reactions in which hydrogens on the ring are replaced with other molecular species. HETEROCYCLIC COMPOUNDS such as pyridine also show aromatic properties. Lighter members of the aromatic series have characteristic 'aromatic' odours.

aromatic hydrogenation (*Chem*) Hydrogenation in the naphthalene series, of such nature that hydrogenation takes place only in the unsubstituted benzene ring.

aromatic polymers (*Chem*) Polymers possessing benzene rings either in side groups (eg polystyrene) or in the main backbone chain (eg polycarbonate, aramid fibre).

aromatic properties (*Chem*) The characteristic properties of aromatic compounds, eg reaction with concentrated nitric acid, forming nitro derivatives, reaction with concentrated sulphuric acid, forming sulphonated derivatives. The homologues of benzene differ from alkanes with regard to oxidation by readily forming benzene carboxylic acids. There are many other distinguishing characteristics between aromatic hydrocarbons and alkanes.

arousal (*Psych*) A general psychophysiological concept referring to the effect of various non-specific stimulation or motivational factors on a number of physiological variables, eg heart rate, skin resistance. It is used to describe differences in responsiveness to general stimulation, usually along a continuum from eg drowsiness to alertness.

arousal theory (*Psych*) The theory that we are motivated by an innate desire to maintain an optimal level of arousal.

ARPA (*ICT*) Abbrev for *Advanced Research Projects Agency*. Supported by US government grant money and now renamed DARPA.

ARPANET (*ICT*) A long-distance PACKET-SWITCHING US network used by research interests funded by ARPA.

array (*ICT*) (1) Set of storage locations referenced by a single identifier. Individual elements of the array are referenced by combining one or more subscripts with the identifier, eg NICK(20) is an element in the array NICK, and JOS(3,5) is an element in the two-dimensional array JOS. (2) An assembly of two or more individual radiating elements, appropriately spaced and energized to achieve desired directional properties. See BEAM ANTENNA.

array (*MathSci*) An ordered set of elements, possibly in several dimensions, such as a MATRIX or VECTOR.

array bounds (*ICT*) Limits on the number of items in an array.

array dimension (*ICT*) Number of subscripts necessary to identify an item in an array, eg CLAR(26,3) has dimension 2.

array processor (*ICT*) A processor designed to allow any machine instruction to operate on a number of data locations simultaneously.

arrectores pilorum (*BioSci*) In mammals, unstriated muscles attached to the hair follicles, which cause the hair to stand on end by their contraction.

arrested crushing (*MinExt*) Crushing so conducted that the rock falling through the machine is free to drop clear of the zone of comminution when broken smaller than the exit orifice or set.

arrested failure (*ElecEng*) The taking of a cable off voltage before failure is complete and its examination to determine the mechanism of breakdown.

arrester (*ElecEng*) See LIGHTNING ARRESTER.

arrester gear (*Aero*) (1) A device on aircraft carriers and some military aerodromes, usually consisting of a number of individual transverse cables held by hydraulic shock absorbers, which stop an aircraft when its ARRESTER HOOK catches a cable. (2) A barrier net, usually of nylon or webbing, attached to heavy drag weights, which stops fast aircraft from overrunning the end of the runway in an emergency.

arrester hook (*Aero*) A hook extended from an aircraft to engage the cable of an arrester gear, mainly on aircraft carriers.

arrest muscle (*BioSci*) See CATCH MUSCLE.

arrest points (*Eng*) Discontinuities on heating and cooling curves, due to absorption of heat during heating or evolution of heat during cooling, and indicating structural (phase) changes occurring in a metal or alloy.

Arrhenius's rate equation (*Phys*) Equation giving the rate R of a thermally activated, physical process:

$$R = R_0 \exp(E_a/kT)$$

where R_0 is a constant, E_a is the ACTIVATION ENERGY, k is BOLTZMANN'S CONSTANT and T is the absolute temperature.

Arrhenius theory of dissociation (*Chem*) The description of aqueous solutions in terms of acids, which dissociate to give hydrogen ions and bases, which dissociate to give hydroxyl ions. The product of the reaction of an acid and a base is a salt and water. The dissociation of these species gives their solutions the property of conducting electricity.

arrhenotoky (*BioSci*) Parthenogenetic production of males.

arrhythmia (*Med*) Abnormal rhythm of the heart beat.

arris (*Build*) The (generally) sharp exterior edge formed at the intersection of two surfaces not in the same plane (eg the meeting of two sides of a stone block). See EXTERNAL ANGLE.

arris edge (*Glass*) Small bevel, of width not exceeding $\frac{1}{16}$ in (1·5 mm), at an angle of approximately 45° to the surface of the glass.

arris fillet (*Build*) A small strip of wood of triangular cross-section packed beneath the lower courses of slates or tiles on a roof to throw off the water which might otherwise get under the flashing.

arris tile (*Build*) Purpose-made angular tile used to cover the intersections at hips and ridges in slated and tiled roofs. See BONNET TILE.

arris-wise (*Build*) A term used to describe the sawing of square timber diagonally.

arrow (*Surv*) Light steel wire pin, bent into ring at one end and perhaps flagged with piece of bright cloth, used to mark measured lengths in chain traversing.

arrowroot (*FoodSci*) A starch extracted from the root of arrowroot *Marantha* which grows in the West Indies. When hydrated and heated, arrowroot produces a transparent, odourless and colourless gel. Used in the food industry as a thickening or gelling agent.

arsenic (*Chem*) Symbol As, at no 33, ram 74·9216, oxidation states 3, 5. An element which occurs free and combined in many minerals. An impurity of several commercial metals. Called grey or γ-arsenic to distinguish it from the other allotropic modifications. Mp 814°C (36 atm), bp 615°C (sublimes), rel.d. 5·73 at 15°C. Used in alloys and in the manufacture of lead shot. It is important as a donor impurity in germanium semiconductor devices. The arsenic compound of commerce is the trioxide of the element (As_2O_3), also known as *white arsenic, arsenious oxide*. Obtained from the roasting of arsenical ores. It only occurs to the extent of 1·8 ppm in the Earth's crust, but it is very widely distributed being found esp in sulphide ore deposits. Present to a very minor extent as the native element but is present in a large number of minerals. It is highly poisonous, and its presence in foods and drinks is subject to severe restriction. Medical uses, once important, have much declined, but still used as a herbicide and rodenticide. In silicon, arsenic is an n-type dopant.

arsenic acid (*Chem*) H_3AsO_4. Formed by the action of hot dilute nitric acid upon arsenic, or by digesting arsenic (III) oxide with nitric acid. Also formed when arsenic (V) oxide is dissolved in water.

arsenical copper (*Eng*) Copper containing up to about 0·6% arsenic. This element slightly increases the hardness and strength and raises the recrystallization temperature.

arsenical pyrites (*Min*) See ARSENOPYRITE.

arsenic halides (*Chem*) Group of compounds including: arsenic (V) fluoride, AsF_5; arsenic (III) fluoride, AsF_3; arsenic (III) chloride, $AsCl_3$; arsenic (III) bromide, $AsBr_3$; and arsenic (III) iodide, AsI_3.

arsenides (*Chem*) Compounds formed by arsenic with most metals, eg iron arsenide ($FeAs_2$). Arsenides are decomposed by water or dilute acids with the formation of the hydride ARSINE. See GALLIUM ARSENIDE.

arsenious acid (*Chem*) Solution of arsenious oxide. See ARSENIC.

arsenites (*Chem*) Arsenates (III). Salts of arsenious acid.

arseniuretted hydrogen (*Chem*) See ARSINE.

arsenolite (*Min*) Arsenic oxide, a decomposition product of arsenical ores; occurs commonly as a white incrustation, rarely as octahedral crystals.

arsenopyrite (*Min*) Sulphide of iron and arsenic; the chief ore of arsenic. Also *arsenical pyrites, mispickel.*

arsine (*Chem*) Arsenic (III) hydride. AsH_3. Produced by the action of nascent hydrogen upon solutions of the element, or by the action of dilute sulphuric acid upon sodium or zinc arsenide. Very poisonous. Arsines are organic derivatives of AsH_3 in which one or more hydrogen atoms is replaced by an alkyl radical; other hydrogen atoms may also be replaced by halogen etc. Also *arseniuretted hydrogen.*

Art Deco (*Arch*) A style of design which took its name from the Exposition des Arts Decoratif, an international trade fair held in Paris in 1925. In the 1920s the style was characterized by curvilinear shapes and by stylized human and animal forms, but by the 1930s a commitment to industrial technology dictated shapes which tended to lend themselves to methods of mass production; hence the sleek, angular, linear forms in stainless steel, aluminium, chromed metals and colourful glazed tiles and plastics.

artefact (*BioSci*) Any apparent structure that does not represent part of the actual specimen, but is due to faulty preparation. Particularly a microscope image that has no counterpart in reality. Also *artifact.*

artefact (*Genrl*) A human-made stone, wood or metal implement. Also *artifact.*

artemisinin (*Pharmacol*) A drug extracted from the Chinese herb *Artemisia annua* (qinghaosu or sweet wormwood), used to treat uncomplicated falciparum malaria; a sesquiterpene lactone.

arterial drainage (*Build*) A system of drainage in which the flow from a number of branch drains is led into one main channel.

arterial system (*BioSci*) That part of the vascular system which carries the blood from the heart to the body.

arteriography (*Radiol*) The radiological examination of arteries following direct injection of a CONTRAST MEDIUM, eg coronary arteriography, renal arteriography, carotid arteriography.

arteriole (*BioSci*) A small artery.

arteriosclerosis (*Med*) Hardening or stiffening of the arteries due to thickening and loss of elasticity of arterial walls. Commonly but incorrectly used to imply ATHERO-SCLEROSIS.

arteritis (*Med*) Inflammation of an artery.

artery (*BioSci*) One of the vessels of the ARTERIAL SYSTEM. Adj *arterial.*

artesian (*Geol*) A term describing groundwater confined under hydrostatic pressure.

artesian well (*CivEng*) A well sunk into a permeable stratum which has impervious strata above and below it, and which outcrops at places higher than the place where the well is sunk, so that the hydrostatic pressure of the water in the permeable stratum is alone sufficient to force the water up out of the well. Named from Artois (France).

arthralgia (*Med*) Pain in a joint.

arthrectomy (*Med*) Excision of a joint.

arthritic (*BioSci*) Pertaining to the joints; situated near a joint.

arthritis (*Med*) Inflammation of a joint.

arthrodesis (*Med*) The surgical immobilization of a joint by fusion of the joint surfaces.

arthrodia (*BioSci*) A joint.

arthrodial membranes (*BioSci*) In Arthropoda, flexible membranes connecting adjacent body sclerites and adjacent limb joints, and occurring also at the articulation of the appendages.

arthrography (*Radiol*) The radiological examination of a joint cavity after direct injection of air or other CONTRAST MEDIA.

arthropathy (*Med, BioSci*) Any disease affecting a joint.

Arthrophyta (*BioSci*) A division of the plant kingdom, the horsetails and allies, here treated as the class Sphenopsida.

Arthropoda (*BioSci*) A phylum of metameric animals having jointed appendages (some of which are specialized for mastication) and a well-developed head. There is usually a hard chitinous exoskeleton; the coelom is restricted, the perivisceral cavity being haemocoelic. The phylum contains the largest number of known species and includes many fossil groups, eg trilobites, eurypterids, ostracods, ranging from the Lower Cambrian to the present day. Includes centipedes, millipedes, insects, crabs, lobsters, shrimps, spiders, scorpions, mites, ticks.

arthroscope (*Med*) A fibre optic ENDOSCOPE used in examination of joints (arthroscopy).

arthrospore (*BioSci*) Spore resulting from hyphal fragmentation.

arthrotomy (*Med*) Surgical incision into a joint.

Arthus reaction (*BioSci*) A localized inflammation due to injection of antigen into an animal that has a high level of circulating antibody against that antigen. A Type III allergic reaction first described by Arthus.

articular (*BioSci*) Pertaining to, or situated at, or near, a joint. In vertebrates, a small cartilage at the angle of the mandible, derived from the Meckelian, and articulating with the quadrate forming the lower half of the jaw hinge. Also *articulare.* Pl *articularia.*

articulated (*BioSci*) Jointed or segmented; divided into portions that may easily be separated.

articulated blade (*Aero*) A rotorcraft blade which is mounted on one or more hinges to permit flapping and movement about the DRAG AXIS.

articulation (*Arch*) The means by which an architect gives definition to the individual elements of a building.

articulation (*BioSci*) The movable or immovable connection between two or more bones.

articulation (*Eng*) The connection of two parts in such a way (usually by a pin joint) as to permit relative movement.

articulation (*ICT*) Percentage of specified speech components (usually LOGATOMS) received over a communication system. May be: (1) *word*: percentage of words correctly received; (2) *syllable*: percentage number of meaningless syllables correctly recognized; (3) *sound*: percentage number of fundamental speech-sounds (consonant, vowel, initial or final consonant) correctly recognized.

artifact (*Genrl*) See ARTEFACT.

artificial ageing (*Eng*) A method of accelerating the hardening of particularly aluminium alloys at slightly elevated temperatures. See PRECIPITATION HARDENING.

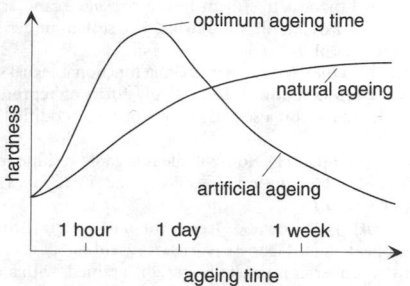

artificial ageing In aluminium alloys.

artificial antenna (*ICT*) Combination of resistances, capacitances and inductances with the same characteristics as an antenna except that it does not radiate energy. It is used in place of the normal antenna for purposes such as repair and checking of a transmitter, or for retuning of the transmitter on to a different frequency. Also *dummy antenna, phantom antenna*.

artificial classification (*BioSci*) A classification based on one or a few arbitrarily chosen characters and giving no attention to the natural relationships of the organism as was the old grouping of plants into trees, shrubs and herbs.

artificial community (*BioSci*) A plant community kept in existence by artificial means, eg a garden habitat or a cloche.

artificial daylight (*Phys*) Artificial light having approximately the same spectral distribution curve as daylight, ie having a colour temperature of about 4000 K.

artificial disintegration (*Phys*) The transmutation of non-radioactive substances brought about by the bombardment of the nuclei of their atoms by high-velocity particles, such as alpha particles, protons or neutrons.

artificial ear (*Acous*) A device for testing earphones which presents an acoustic impedance similar to the human ear and includes facilities for measuring the sound pressure produced at the ear.

artificial earth (*ICT*) See COUNTERPOISE.

artificial feel (*Aero*) In an aircraft flying control system, esp with *automatic control of flying surfaces*, in which the pilot's control actions are modified to provide forces moving the flying controls, a natural feel, opposing the pilot's actions, which is fed back from the controls. Since these forces vary mostly with dynamic air pressure as in

$$q = \frac{1}{2}\,e^{v^2}$$

artificial feel is sometimes known as *q-feel*.

artificial horizon (*Aero*) See GYRO HORIZON.

artificial horizon (*Surv*) An apparatus, eg a shallow trough filled with mercury, used in order to observe altitudes of celestial bodies with a sextant on land, ie where there is no visible horizon. The reflection of the object in the artificial horizon is viewed directly and the object itself indirectly by reflection from the index glass of the sextant.

artificial insemination (*Med*) The introduction of semen into the vagina using some form of instrument in order to facilitate conception. Abbrev *AI*.

artificial intelligence (*ICT*) (1) The concept that computers can be programmed to assume capabilities thought to be like human intelligence such as learning, reasoning, adaptation and self-correction. (2) An extensive branch of COMPUTER SCIENCE embracing COMPUTER VISION, KNOWLEDGE-BASED SYSTEMS, PATTERN RECOGNITION, ROBOTICS, scene analysis, NATURAL LANGUAGE processing, and mechanical theorem proving, with more areas being added all the time. Abbrev *AI*.

artificial kidney (*Med*) Machine which is used to replace the function of the body's own organ, when the latter is faulty or has ceased to function. The patient's blood is circulated through sterile semipermeable tubing lying in a suitable solution and is purified by dialysis.

artificial larynx (*Med*) A reed actuated by the air passing through an opening in front of throat to assist articulation of person who has undergone a tracheotomy operation.

artificial line (*ICT*) Repeated network units that have collectively some or all of the transmission properties of a line. Also *simulated line*.

artificial pneumothorax (*Med*) See PNEUMOTHORAX.

artificial radioactivity (*Phys*) Radiation from isotopes after high-energy bombardment in an accelerator by alpha particles, protons and other light nuclei, or by neutrons in a nuclear reactor. Discovered by I Joliot-Curie and F Joliot in 1933.

artificial rubber (*Plastics*) See SYNTHETIC RUBBER.

artificial satellite (*Space*) Constructed space vehicle whose velocity is sufficient to maintain it in orbit. Earth satellites are used for the purposes of observation of the surface, the atmosphere, the Sun and deep space, as communication links, in weather forecasting and for the performance of microgravity and other technological experiments.

artificial stability (*Aero*) An automatic flight control system which provides positive stability to an otherwise unstable or neutrally stable aircraft.

artificial stone (*Build*) A precast imitation of natural stone made in block moulds. The interior of the block is of concrete, the required exterior face of cement mixed with dust or chippings of the natural stone to be imitated.

artificial traffic (*ICT*) Automatically generated calls that are deliberately mixed with subscriber-originated traffic to sample the overall service provided by the switching equipment of an automatic exchange, by recording or holding faults recognized by test equipment.

artificial voice (*Acous*) Loudspeaker and baffle for simulating speech in testing of microphones.

Artinskian (*Geol*) A stratigraphical stage in the Lower Permian rocks of Russia and Eastern Europe.

artiodactyl (*BioSci*) An animal possessing an even number of digits.

Artiodactyla (*BioSci*) The one order of the mammalia containing the 'even-toed' hooved 'ungulates', ie those with a PARAXONIC FOOT. Includes pigs, peccaries, hippopotami, camels, llamas, giraffes, sheep, buffaloes, oxen, deer, gazelles and antelopes.

Art Nouveau (*Arch*) A decorative movement in European art which occurred during the late 19th and early 20th centuries. It was basically the stylized adaptation of plant forms and was used by artists and architects mainly for surface decoration, although the characteristic curvilinear forms were incorporated into the structure, notably in France and Spain.

art paper (*Paper*) Paper coated on one or both sides with one or more applications of an aqueous suspension of adhesive and mineral matter, such as china clay, to provide a surface suitable for high-class colour print reproduction.

arundinaceous (*BioSci*) Reed-like and thin.

aryl (*Chem*) A term for aromatic monovalent hydrocarbon radicals; eg C_6H_5Cl is an aryl halide.

aryl amines (*Chem*) Amino derivatives of the aromatic series, eg $C_6H_5NH_2$ (aniline).

arytaenoid (*BioSci*) (1) In general, pitcher-shaped. (2) In vertebrates, one of a pair of anterior lateral cartilages, forming part of the framework of the larynx.

As (*EnvSci*) Abbrev for ALTOSTRATUS.

ASA speed (*ImageTech*) Abbrev for *American Standards Association* photographic speed rating, expressed on an arithmetic scale. Now replaced by ISO SPEED.

asbestos (*Min*) Naturally occurring, fine (≈ 5 μm diameter) mineral fibres which are highly heat-resistant, used in brake linings, thermal insulation, fire-resistant fabrics, asbestos cement, etc, though with increasing concern for the health hazards from loose fibres and dust. Derived from chrysotile (serpentine), and the amphiboles actinolite, amosite, anthophyllite and crocidolite (blue asbestos).

asbestos cement (*Build*) An inexpensive, but brittle, fire-resisting and weatherproof, non-structural building material, made from Portland cement and asbestos. No longer made because of the dangers of asbestos use, but still widely found.

asbestosis (*Med*) Disease of the lungs due to inhalation of asbestos particles, causing severe pulmonary fibrosis; may be associated with malignant growth of the pleura (mesothelioma).

asbolane (*Min*) A form of WAD; soft earthy manganese dioxide, containing cobalt. Also *asbolite*.

ascending letters (*Print*) Letters the top portions of which rise above the X-HEIGHT, eg b, d, f, h. See fig. at TYPEFACE.

ascending node (*Space*) For Earth, the point at which a satellite crosses the equatorial plane travelling from south to north.

ascension (*Astron*) See RIGHT ASCENSION.

ascertainment (*BioSci*) In human genetics, the way by which families come to the notice of the investigator. The method of ascertainment may lead to biased data.

Aschelminthes (*BioSci*) Phylum of invertebrate animals that have in common the possession of a pseudocoelom and an unsegmented elongate body with terminal anus and a non-muscular gut.

Aschoff's nodes (*Med*) Inflammatory nodules found in rheumatic inflammation of the heart. Also *Aschoff's bodies*.

ascidium (*BioSci*) A pitcher-shaped leaf or part of a leaf.

ASCII file (*ICT*) A FILE consisting of characters represented in ASCII codes (see CHARACTER CODE) and without special formatting codes peculiar to an individual program. ASCII files are often used to transfer data between otherwise incompatible programs or incompatible computer systems. Also *text file*.

ASCII keyboard (*ICT*) Keyboard providing the full range of ASCII characters including CONTROL CHARACTERS.

ascites (*Med*) See HYDROPERITONEUM.

ascites tumour (*BioSci*) A tumour that grows in the peritoneal cavity as a suspension of cells.

ascocarp (*BioSci*) The fruiting body of the Ascomycotina consisting of a sterile wall more or less enclosing the asci. Also *ascoma*. See APOTHECIUM, PERITHECIUM.

ascolichen (*BioSci*) One of the majority of lichens in which the fungal constituent is an ascomycete.

ascoma (*BioSci*) See ASCOCARP.

ascomycete (*BioSci*) A fungus of the ASCOMYCOTINA.

Ascomycotina (*BioSci*) A subdivision or class of those Eumycota or true fungi in which the sexual spores are formed in an ASCUS, usually within ascocarps. There are no motile stages. Such fungi are usually mycelial with hyphae with simple septa; some are yeasts. Asexual reproduction is by means of conidia. Includes the Hemiascomycetes, Plectomycetes, Pyrenomycetes and Discomycetes. Also *Ascomycetes*

ascorbic acid (*BioSci*) See vitamin C in panel on VITAMINS.

ascospore (*BioSci*) Spore, typically uninucleate and haploid, formed within an ascus.

ascus (*BioSci*) Specialized, usually more or less cylindrical cell within which (usually eight) ascospores are formed

following fusion of two heterokaryotic nuclei in ascomycete reproduction.

asdic (*Acous*) Abbrev for *allied submarine detection investigation committee*. Underwater acoustic detecting system which transmits a pulse and receives a reflection from underwater objects, particularly submarines, at a distance. Also used by trawlers to detect shoals of fish. Equivalent to US *sonar*, now the preferred term.

asepalous (*BioSci*) Devoid of sepals.

asepsis (*Med*) Freedom from infection.

aseptate (*BioSci*) Not divided into segments or cells by septa.

aseptic (*FoodSci*) Free from viable organisms. Aseptic filling is the process of packing a product which has been sterilized by heat treatment into a package (can, jar, film) which is also sterile, and which is sealed under sterile conditions.

asexual (*BioSci*) Without sex; lacking functional sexual organs.

asexual reproduction (*BioSci*) Any form of reproduction not depending on a sexual process or on a modified sexual process.

ash (*Chem*, *Geol*) (1) Non-volatile inorganic residue remaining after the ignition of an organic material. (2) See VOLCANIC ASH.

ash (*For*) A hardwood tree (*Fraxinus*) with American, European and Japanese varieties, yielding a tough and elastic timber. It is generally straight-grained, with a coarse, even texture.

ash curve (*MinExt*) Graph which shows result of sink-and-float laboratory test in form of relationship between specific gravity of crushed small particles and the ash content at that gravity.

ashen light (*Astron*) (1) A faint glow sometimes seen in that part of the disk of Venus that is not directly illuminated by the Sun; thought to result from bombardment of atmospheric atoms and molecules by high-energy particles and radiation. (2) See EARTHSHINE.

ash flow tuff (*Geol*) A tuff deposited by an ash flow or gaseous cloud (*nuée ardente*).

Ashgill (*Geol*) The youngest series of the Ordovician. See PALAEOZOIC.

ashlar (*Build*) Also *broadstone*. (1) Masonry work in which the stones are accurately squared and dressed to given dimensions so as to make very good joints over the whole of the touching surfaces. (2) A thin facing of squared stones or thin slabs laid in courses, with close-fitting joints, to cover brick, concrete or rubble walling.

ash tuff (*Min*) A pyroclastic rock in which the average pyroclast size is between $\frac{1}{16}$ and 2 mm.

Asn (*Chem*) Symbol for ASPARAGINE.

Asp (*Chem*) Symbol for ASPARTIC ACID.

asparagine (*Chem*) $NH_2COCH_2CH(NH_2)COOH$. The monoamide of aspartic acid. The L-form is one of the 20 amino acids directly coded in proteins. Symbol Asn, short form N.

asparagus stone (*Min*) Apatite of a yellowish-green colour, thus resembling asparagus.

aspartame (*FoodSci*) Dipeptide of aspartic acid and phenylalanine, a permitted artificial sweetener. Not suitable for baked products as it decomposes above 150°C and it must not be used in foods for those with PHENYLKETONURIA.

aspartic acid (*Chem*) *2-aminobutanedioic acid*. $HOOCCH_2$ $CH(NH_2)COOH$. The L- or S-isomer is a constituent of proteins. Mp 271°C. An amino acid formed by the hydrolysis of ASPARAGINE. Symbol Asp, short form D.

aspect (*Aero*) See ATTITUDE.

aspect (*Astron*) The position of a planet or the Moon relative to the Sun when viewed from Earth; specific aspects include CONJUNCTION and OPPOSITION.

aspect (*BioSci*) (1) Degree of exposure to sun, wind, etc, of a plant habitat. (2) Effect of seasonal changes on the appearance of vegetation.

aspect (*CivEng*) On railways the indication given by a coloured light signal, as contrasted with that of a semaphore arm signal. A multiple-aspect signal (MAS) conveys more information.

aspect ratio (*Aero*) The ratio of span/mean chord line of an aerofoil (usually in wing); defined as S^2/A, where S is the span and A is the area. Important for INDUCED DRAG and range/speed characteristics. Normal figure between 6 and 9, lesser values than 6 being *low aspect ratios*, greater than 9 *high aspect ratios*.

aspect ratio (*Eng*) Ratio of the length of a fibre or wire to its width or diameter.

aspect ratio (*ImageTech*) The ratio of the width to the height of the reproduced picture or computer screen, eg 4×3, often expressed with the height as unity, 1·33:1. Widescreen systems have aspect ratios between 1·65:1 and 2·35:1. Abbrev AR.

aspect ratio (*NucEng*) In a Tokamak type of fusion machine, the ratio of the major to minor radii of the torus.

asperate (*BioSci*) Having a rough surface due to short, upstanding stiff hairs. Also *asperous*.

Asperger's syndrome (*Psych, Med*) A mild psychiatric disorder characterized by poor social interaction and obsessive behaviour.

aspergillosis (*Med, Vet*) A disease of the lungs caused by the fungus *Aspergillus fumigatus*. It may cause allergic reaction in the bronchioles to give asthma, or may infect old cavities to give a ball-like growth or aspergilloma. In severely immune compromised patients it may spread beyond the lung. Cause of *pneumomycosis* (brooder pneumonia) in birds; in cattle the fungus is a cause of abortion.

Aspergillus (*BioSci*) A form-genus of Deuteromycotina that includes parasites (causing eg ASPERGILLOSIS), saprophytes and food-spoilage organisms (see AFLATOXINS). Certain species are used to prepare soy sauce and for the production of industrial enzymes.

asperity (*Phys*) Slightly raised parts of a surface which form the actual points of contact between two surfaces at a microscopic level, elastically and plastically flattened to take the load (normal force).

aspermia (*Med*) Complete absence of spermatozoa.

asperous (*BioSci*) See ASPERATE.

asphalt (*Geol, Min*) One of various bituminous substances which may be: (1) of natural occurrence in oil-bearing strata from which the volatiles have evaporated; (2) a residue in petroleum distillation; (3) a mixture of asphaltic bitumen and granite chippings, sand or powdered limestone. Asphalt is used extensively for paving, road-making, damp-proof courses, in the manufacture of roofing felt and paints, and as the raw material for certain moulded plastics. See BITUMEN.

asphaltenes (*Chem*) Such constituents of asphaltic bitumens as are soluble in carbon disulphide but not in petroleum spirit. See CARBENES, MALTHENES.

asphaltite (*Min*) A group name for the organic compounds albertite, anthraxolite, grahamite, impsonite, libollite, nigrite and uintaite.

aspheric surface (*Phys*) A lens surface which departs to a greater or lesser degree from a sphere, eg one having a parabolic or elliptical section.

asphyxia (*Med*) Suffocation due to lack of inspired oxygen.

ASPIC (*Print*) Abbrev for *authors' symbolic pre-press interfacing codes*. A typographic coding system used in electronically produced text to indicate to the phototypesetter or imagesetter the desired typographic output.

aspirated psychrometer (*EnvSci*) A PSYCHROMETER which uses a forced draught of at least 12 km h^{-1} (8 miles per hour) over the wet bulb.

aspiration (*Med*) The removal of fluids or gases from the body by suction.

aspiration pneumonia (*Med*) Pneumonia due to inhalation of food, drink or gastric contents.

aspirator (*Chem*) A device for drawing a stream of air or oxygen or liquid through an apparatus by suction.

aspirin (*Pharmacol*) *Acetylsalicylic acid.* Analgesic, antipyretic and a non-steroidal anti-inflammatory drug. It inhibits both COX-1 AND COX-2 and, because it inhibits platelet clotting, is used in low doses as an antithrombotic and prophylactic for heart disease and stroke. Its major side effect is irritation of the gastric lining.

asplanchnic (*BioSci*) Having no gut.

assay (*Chem*) The quantitative analysis of a substance to determine the proportion of some valuable or potent constituent, eg the active compound in a pharmaceutical or metals in an ore. See DRY ASSAY, WET ASSAY.

assay balance (*Chem*) A balance specially made for weighing the small amounts of matter met with in assaying. See CHEMICAL BALANCE.

assayer (*Chem*) A person who carries out the process of assay. See DRY ASSAY, WET ASSAY.

assay ton (*Eng*) Used in assaying precious metals. It is equivalent to 29·160 g and 32·670 g for the short and long ton respectively. The number of milligrams of precious metal in an assay ton of ore indicates the ASSAY VALUE, since 1 mg of precious metal per assay ton equals 1 troyoz of precious metal per avoirdupois ton of ore.

assay value (*Eng*) Troy ounces of precious metal per avoirdupois ton of ore.

assemble edit (*ImageTech*) Videotape editing in which a new scene is added to follow directly on existing material.

assembler (*ICT*) Program, usually provided by the computer manufacturer, to translate a program written in ASSEMBLY LANGUAGE into MACHINE CODE. In general, each assembly language instruction is changed into one machine-code instruction. Also *assembly program*. Cf COMPILER, DISASSEMBLER.

assembly (*Eng*) Construction of product from several or many components. Methods used for attachment include welding, fastening, push-fit, snap-fit, lock-fit, adhesive bonding, ultrasonic welding etc. Many products are now designed for robotic assembly.

assembly (*ICT*) A process of converting a program written in ASSEMBLY LANGUAGE into MACHINE CODE.

assembly language (*ICT*) Low-level PROGRAMMING LANGUAGE, generally using symbolic addresses, that is translated into MACHINE CODE by an ASSEMBLER.

assembly program (*ICT*) See ASSEMBLER.

assert (*ICT*) In systems security, the act of presenting one's authority and right to gain access to a system.

assign (*ICT*) To place a value in the memory location corresponding to a given variable.

assigned frequency (*ICT*) The frequency assigned as centre frequency of a class of transmission, with tolerance, by authority.

assigning authority (*Ships*) A national body authorized to assign LOAD LINES to ships.

assimilation (*BioSci*) (1) The metabolic processes, mostly anabolic, by which the mostly inorganic substances, taken up by plants, are converted into the constituents of the plant body. Includes PHOTOSYNTHESIS. (2) Conversion of food material into protoplasm, after it has been ingested, digested and absorbed. (3) Resemblance of an animal to its surroundings, not only by coloration but also by configuration.

assimilation (*Geol*) The incorporation of extraneous material in igneous magma.

assimilation (*Psych*) Incorporating objects, experiences or information into existing schemas. In Piagetian theory, one of the two main biological forces responsible for cognitive development. See ACCOMMODATION.

assimilatory quotient (*BioSci*) See PHOTOSYNTHETIC QUOTIENT.

assisted take-off (*Aero*) Take-off in which the full power of the normal engines is supplemented by auxiliary means, which may or may not be jettisonable. Small turbojet or rocket motor units, powder, or liquid rockets may be used. See JATO, RATOG.

assize (*Build*) A cylindrical block of stone forming part of a column, or of a layer of stone in a building.

associate Bertrand curves (*MathSci*) See CONJUGATE BERTRAND CURVES.

associated emission (*Electronics*) Emission which brings about equilibrium between incident photons and secondary electrons in ionization.

associated gas (*MinExt*) A gas mixture found associated with crude oil in an underground geological formation. Volume and composition vary widely according to location.

associated liquid (*Chem, Phys*) A liquid in which molecules of the same kind form a complex structure, eg water. See HYDROGEN BOND.

association (*BioSci*) (1) A plant community usually occupying a wide area, consisting of a definite population of species, having a characteristic appearance and habitat, and stable in its duration. (2) In certain Sporozoa, adherence of individuals without fusion of nuclei: a characteristic set of animals, belonging to a particular habitat.

association (*Print*) In rotary printing, the bringing together of separate webs, after printing, to pass through the folder as a complete product.

association (*Psych*) The phenomenon in learning that information is easier to remember if paired with something familiar or with something idiosyncratic and unusual.

associationism (*Psych*) The theory that association of ideas is the basis of all mental activity.

associative (*MathSci*) Said of a binary operation * such that $(a*b)*c = a*(b*c)$ for all a, b, c in the set concerned. Thus in ordinary arithmetic + and × are associative; − and ÷ are not.

associative learning (*Psych*) Learning through the formation of associations between ideas or events based on their co-occurrence in past experience. The term originated in philosophy, but it is now most often used as a synonym for learning through both CLASSICAL CONDITIONING and OPERANT CONDITIONING procedures.

associative memory (*Psych, ICT*) A human or machine memory system in which a particular input is associated with a particular output.

associative storage (*ICT*) Storage that is identified by means of content rather than by an address. Also *content-addressable storage*.

assortative mating (*BioSci*) Non-random mating caused by eg pollinating insects, which may cause preferential inbreeding or outbreeding.

astable circuit (*Electronics*) An active circuit, having two quasi-stable states, which alternates automatically and continuously between them, eg certain MULTIVIBRATORS.

A-stage (*Plastics*) Stage at which a synthetic resin of the phenol formaldehyde type is fusible and wholly soluble in alcohols and acetone.

astatic galvanometer (*Phys*) Moving-magnet galvanometer in which adjustable magnets form an *astatic system*.

astatic system (*Phys*) Ideally an arrangement of two or more magnetic needles on a single suspension so that in a uniform magnetic field, such as the Earth's field, there is no resultant torque on the suspension.

astatine (*Chem*) Radioactive element, the heaviest halogen. Symbol At, at no 85, mass numbers 202–12, 214–19, half-lives 2×10^{-6} s to 8 h. Isotopes occur naturally as members of the actinium, uranium or neptunium series, or may be produced by the α-bombardment of bismuth.

astelic (*BioSci*) Not having a STELE.

aster (*BioSci*) A group of radiating fibrils formed of microtubules surrounding the centrosome, seen immediately prior to and during cell division, and more prominent in animal than plant nuclei.

Asteraceae (*BioSci*) See COMPOSITAE.

astereognosis (*Med*) Loss of ability to recognize, by the sense of touch, the three-dimensional nature of an object.

Asteridae (*BioSci*) A subclass or superorder of dicotyledons, comprising some trees and shrubs but mostly herbs. They are mostly gamopetalous (sympetalous), the number of stamens being equal to or less than the number of corolla lobes; they mostly have two fused carpels or an ovary made up of two or more carpels, but that appears to be just a single carpel (pseudomonomerous). Contains c.56 000 spp in 43 families including Solanaceae, Scrophulariaceae, Labiatae, Verbenaceae, Rubiaceae and Compositae.

asterism (*Astron*) A conspicuous or memorable group of stars, smaller in area than a constellation, eg the PLOUGH.

asterism (*Min*) The star effect, with four-, six- or twelve-rayed stars, seen by reflected light in gemstones, eg ruby and sapphire, cut *en cabochon*, and produced by rod-like inclusions. See GEM.

asteroid (*Astron*) One of thousands of rocky objects normally found between the orbits of Mars and Jupiter, ranging in size from 1 to 1000 km. A few, eg Eros, have passed close to the Earth. Also *minor planet, planetoid*.

asteroid (*MathSci*) See ASTROID.

Asteroidea (*BioSci*) A class of ECHINODERMATA, having a dorsoventrally flattened body of pentagonal or stellate form. The arms merge into the disk; the tube feet possess ampullae and lie in grooves on the lower surface of the arms. The anus and madreporite are aboral, and there is a well-developed skeleton. The class comprises free-living carnivorous forms. Starfish.

asthenia (*Med*) Loss of muscular strength.

asthenosphere (*Geol*) The shell of the Earth below the lithosphere. It is identifiable by low seismic wave velocities and high seismic attenuation. It is soft, probably partly molten and a zone of magma generation. It is equivalent to the *upper mantle*. See panels on EARTH and PLATE TECTONICS.

asthma (*Med*) A chronic disease characterized by difficulty in breathing, accompanied by wheezing and difficulty in expelling air from the lungs. This is due to constriction of the bronchi and their blocking by viscid mucous secretions. In *extrinsic asthma* the condition is due to a type I allergic reaction to inhaled or ingested allergens.

astigmatism (*Med*) Unequal curvature of the refracting surfaces of the eye, which prevents the focusing of light rays to a common point on the retina.

astigmatism (*Phys*) A defect in an optical system such that, instead of a point image being formed of a point object, two short line images (focal lines) are produced at slightly different distances from the system and at right angles to each other. Astigmatism is always present when light is incident obliquely on a simple lens or spherical mirror.

ASTM (*Eng*) Abbrev for AMERICAN SOCIETY FOR TESTING AND MATERIALS.

astomatous (*BioSci*) (1) Lacking stomata. (2) Without a mouth.

Aston dark space (*Electronics*) The space in the immediate vicinity of a cathode, in which the emitted electrons have velocities insufficient to ionize the gas.

Aston whole-number rule (*Phys*) Empirical observation that relative atomic masses of isotopes are approximately whole numbers. See MASS SPECTROGRAPH.

Astrafoil (*Print*) A thin, dimensionally stable transparent plastic sheet used for mounting lithographic negatives or positives.

astragal (*Build*) A small convex moulding having a semicircular cross-section sometimes plain and sometimes curved, the smaller cross-section bars in windows separating glass panes. See fig. at MOULDINGS.

astragalus (*BioSci*) In tetrapod vertebrates, one of the ankle bones, corresponding to the LUNAR in the wrist. Also *talus* (pl *tali*).

astrakhan (*Textiles*) A curled-pile woven, warp-knitted, or weft-knitted fabric designed to resemble the fleece of a stillborn or very young astrakhan lamb.

astringent (*Med*) Having the power to constrict or contract organic tissues: that which does this.

astro- (*Genrl*) Prefix from Gk *astron*, star.

astrobiology (*Astron, BioSci*) See EXOBIOLOGY.

astrobleme (*Geol*) A circular impact structure on the Earth's crust, caused by a meteorite.

astrochemistry (*Chem*) The study of the nature and evolution of molecules and radicals found in outer space.

astrocompass (*Aero*) A non-magnetic instrument that indicates true north relative to a celestial body.

astrocyte (*BioSci*) A much branched, star-shaped neuroglial cell.

astrocytoma (*BioSci*) A neuro-ectodermal tumour (*glioma*) arising from astrocytes.

astrodome (*Aero*) A transparent dome, fitted to some aircraft usually on the top of the fuselage, with calibrated optical characteristics, for astronomical observations.

astrogeology (*Geol*) The study by geological, geophysical, geochemical and related techniques of the solid bodies of the solar system, excluding the Earth but including meteorites and tektites.

astroid (*MathSci*) Four-cusped star-like curve with cartesian equation $x^{\frac{2}{3}} + y^{\frac{2}{3}} = a^{\frac{2}{3}}$. Envelope of a straight line whose ends move along the co-ordinate axes. A HYPOCYCLOID in which the radius of the rolling circle is $\frac{1}{4}$ or $\frac{3}{4}$ times the radius of the fixed circle. Also *asteroid*. Cf GLISSETTE, ROULETTE.

astroid

astrolabe (*Astron*) Ancient instrument (c.200 BC) for showing the positions of the Sun and bright stars at any time and date. If fitted with sights, also used for measuring the altitude above the horizon of celestial objects, and in this mode a 15th-century forerunner of the SEXTANT.

astrometry (*Astron*) The precise measurement of position in astronomy, deduced eg from the co-ordinates of images on photographic plates, or using VLBI.

astronaut (*Space*) A man or woman who flies in space. Also *cosmonaut* (esp a Russian astronaut), *spationaut* (esp a French astronaut).

astronautics (*Space*) The science of space flight.

astronomical clock (*Astron*) An elaborate clock showing astronomical phenomena such as the phases of the Moon and principally found in medieval cathedrals, eg Wells in Somerset, and Strasbourg. In modern observatories the term is applied to any clock displaying SIDEREAL TIME.

Astronomical Ephemeris (*Astron*) Annual handbook (*ephemeris*) published a few years in advance by Her Majesty's Nautical Almanac Office, essentially identical to the *American Ephemeris* and issued in an abridged form as *The Nautical Almanac*.

astronomical telescope (*Astron*) See panel on ASTRONOMICAL TELESCOPE.

astronomical triangle (*Astron*) Triangle on the celestial sphere formed by a celestial body S, the zenith Z and the pole P. The three angles are the hour angle at P, the azimuth at Z and the parallactic angle at S. Fig. ▷

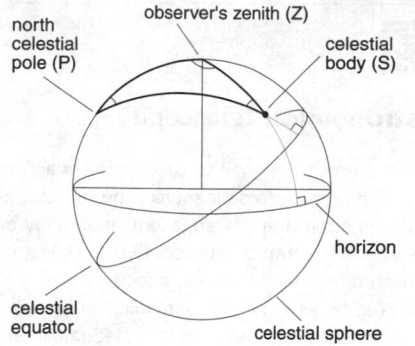

astronomical triangle

astronomical twilight (*Astron*) The interval of time during which the Sun is between 12° and 18° below the horizon, morning and evening. See CIVIL TWILIGHT, NAUTICAL TWILIGHT.

astronomical unit (*Astron*) Mean distance of the Earth from the Sun, 1.496×10^8 km or around 93 million miles. Abbreviated AU and commonly used as a unit of distance within the solar system. There are 63 240 AU in one light-year.

astronomy (*Genrl*) The science of the heavens in all its branches.

astrophotography (*Astron*) Photography of celestial bodies for astronomical study.

astrophyllite (*Min*) A complex hydrated silicate of potassium, iron, manganese, titanium and zirconium; occurs in brown laminae in alkaline igneous rocks.

astrophysics (*Astron*) Branch of astronomy which applies the laws of physics to the study of interstellar matter and the stars, eg their constitution, evolution and luminosity.

astrosclereide (*BioSci*) A sclereide with radiating branches ending in points.

asulam (*Agri*) Translocated herbicide used widely to control bracken in grasslands, upland pastures and forest plantations.

asymmeter (*ElecEng*) An instrument having three movements so arranged that any lack of symmetry when these are connected to a three-phase system can be observed by a single reading.

asymmetric (*BioSci*) Irregular in form: not divisible into halves about any longitudinal plane.

asymmetrical (*ElecEng*) Said of circuits, networks or transducers when the impedance (image impedance, or iterative impedance) differs in the two directions. Also *dissymmetrical*, *non-symmetrical*.

asymmetrical conductivity (*ElecEng*) Phenomenon whereby a substance, or a combination of substances as in a rectifier, conducts electric current differently in opposite directions.

asymmetric atom (*Chem*) An atom bonded to three or more other atoms in such a way that the arrangement cannot be superimposed on its mirror image. In particular, a carbon atom attached to four different groups. Most chiral molecules can be described in terms of specific asymmetric atoms, eg the alpha-carbon atoms in amino acids. See CHIRALITY.

asymmetric conductor (*ElecEng*) Conductor which has a different conductivity for currents flowing in different directions through it, eg a diode.

Asymmetric Digital Subscriber Line (*ICT*) A method of transmitting digital data (eg video and audio signals) over the normal copper line. It can carry up to 8 megabits per second downstream and 1.5 Mbps upstream using an additional circuit to split the digital and analogue

Astronomical telescope

An instrument designed to collect, detect and record radiation from any cosmic source. The collector may be a mirror, often dish-shaped, a lens or an array of DIPOLES (for a RADIO TELESCOPE). The collector concentrates the radiation at a focus, where it is detected by eye, in a photographic emulsion or by an electronic device. Finally this detected radiation is recorded, either photographically or digitally. Almost all telescopes therefore have three essential elements: collector, detector and recorder. The great range of telescope designs follows from the need to observe a large variety of objects (Sun, planets, stars and galaxies) across the entire ELECTROMAGNETIC SPECTRUM (see appendices), from radio waves to gamma rays.

Galileo first applied the optical refracting telescope to astronomy in 1610 and even with this crude instrument the power of the technique was immediately apparent. He resolved the MILKY WAY into stars and discovered four natural satellites of Jupiter. The first lenses suffered badly from chromatic aberration, and to avoid this fault Newton designed the reflecting telescope which used a parabolic metallic mirror to form an image (see Fig. 1).

Further development of this instrument by William Herschel and Lord Rosse led to the construction of giant reflectors in the 19th century with which the distant galaxies and nebulae could be catalogued and studied for the first time. Towards the end of that century, astronomers in California discovered the advantages of siting telescopes on mountains. The Mount Wilson Observatory in Pasadena led the way with a 2·5 metre telescope (1917), followed by the 5 metre Hale Telescope at Mount Palomar (1949). These instruments showed the immense size of the cosmos and the diversity of objects in the universe.

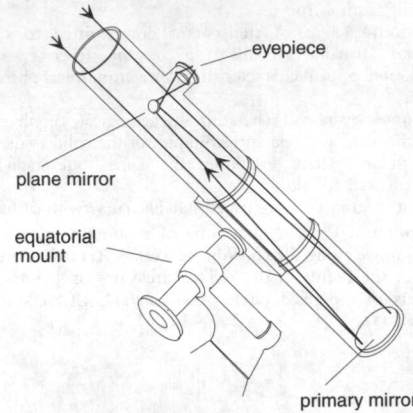

Fig. 1 **Newtonian telescope**.

A modern optical telescope functions as a multi-purpose instrument. For surveying the sky a Schmidt telescope (see Fig. 2) is generally the best; it uses a thin corrector plate to produce an undistorted field over an area of about 6° square. The *primary focus* (see Fig. 3) may be used for direct photography. The *Cassegrain focus*, behind the main mirror, is favoured for most investigations as it is simpler to mount instruments such as spectrographs or CCD detectors there. For long focal lengths, favoured for high-resolution

telephonic systems. Abbrev *ADSL*. Cf *SDSL* where upstream and downstream speeds are comparable.

asymmetric flight (*Aero*) The condition of flying with asymmetrically balanced thrust, weight, drag or lift forces, as could occur, eg with one external weapon mounted under one wing, or in a twin-engined aircraft with one engine inoperative.

asymmetric reflector (*Phys*) A reflector in which the beam of light produced is not symmetrical about a central axis.

asymmetric refractor (*Phys*) A refractor in which the light is redirected, unsymmetrically, about a central axis.

asymmetric synthesis (*Chem*) The synthesis of optically active compounds from racemic mixtures. This can be carried out in some cases by chemical methods in which one component is more reactive than the other one. In other cases asymmetric synthesis occurs in the presence of enzymes.

asymmetric system (*Crystal*) See TRICLINIC SYSTEM.

asymmetric top (*Chem*) A model of a molecule having no three- or higher-fold axis of symmetry.

asymmetry (*BioSci*) The condition of the animal body in which no plane can be found which will divide the body into two similar halves, as in snails.

asymmetry (*Genrl*) The condition of being irregular and not divisible into equal halves about any plane.

asymmetry potential (*ElecEng*) The potential difference between the inside and outside surface of a hollow electrode.

asymptote (*MathSci*) A line or curve (called a *curvilinear asymptote*) such that the distance between it and the given curve tends to zero as they tend to infinity in some direction. A *linear asymptote* can be considered as a tangent at infinity.

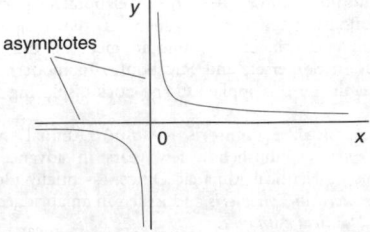

asymptote Drawn to the two curves.

asymptotic curve (*MathSci*) A curve on a surface such that at every point on it, its tangent lies in an asymptotic direction at that point.

spectroscopy, the *Coudé focus* is available. In the 1970s several 4 metre telescopes were constructed, but design changes meant that 8 metres was the preferred size in the 1990s. Lightweight mirrors and the use of computer-controlled optics have significantly reduced the costs of large ground-based instruments.

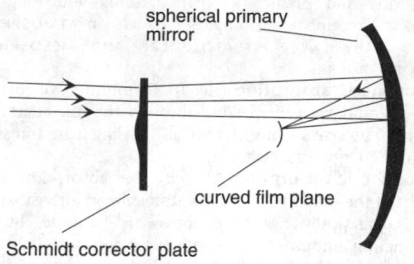

Fig. 2 **Schmidt telescope** Corrector plate with exaggerated contour.

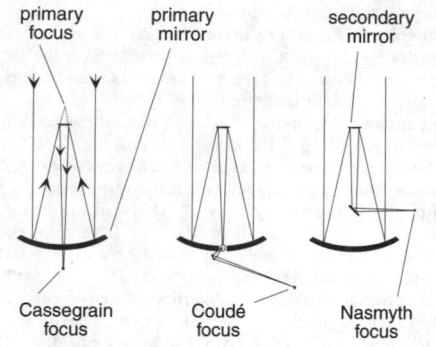

Fig. 3 **Telescope** Focal positions of astronomical reflectors.

Astronomical telescope *(Cont.)*

There are two basic ways of mounting any telescope. It must be able to swing on two axes at right angles in order to reach all parts of the sky. Until recently the EQUATORIAL MOUNT in which one axis is parallel to the axis of the Earth's rotation was universal and the telescope was driven around only this axis to follow the motion of the sky. From about 1970 computer developments have allowed astronomers to use the ALTAZIMUTH MOUNT, which has a vertical axis. It is much cheaper to build but both axes must be precisely driven when observing.

Radio telescopes

These gather radio waves from the Sun, hot gas clouds and active galaxies. They need a far larger aperture than an optical telescope in order to reach a useful resolving power. In practice this is achieved by linking single dishes into an interferometer. Baselines up to 10 000 km are available by linking observatories on different continents in very long baseline interferometry.

Progress in space technology has enabled astronomers to design X-ray and gamma-ray telescopes which must operate from a space platform as these radiations are absorbed in the atmosphere. In this part of the spectrum extremely hot interacting binary stars, neutron stars and black holes can be detected. The HUBBLE SPACE TELESCOPE operating from a space platform in the optical and ultraviolet gives unrivalled images of very faint objects that cannot be seen at ground level.

See entry on CCD ARRAY and panel on GALAXY.

---

asymptotic directions (*MathSci*) See CONJUGATE DIRECTIONS, HYPERBOLIC POINT ON A SURFACE.

asymptotic freedom (*Phys*) The property that at small distances quarks behave asymptotically as free particles. In quantum chromodynamics, STRONG INTERACTION between quarks becomes stronger with distance.

asynapsis (*BioSci*) Absence of pairing of chromosomes at MEIOSIS.

asynchronous computer (*ICT*) A computer in which operations are not all timed by a master clock. The signal to start an operation is provided by the completion of the previous operation. Also *non-synchronous computer*.

asynchronous data transmission (*ICT*) The transmission of data in which the end of the transmission of one character initiates the transmission of the next.

asynchronous motor (*ElecEng*) See NON-SYNCHRONOUS MOTOR.

asynchronous transfer mode (*ICT*) A method of data exchange where each transmitted PACKET of data has integral START BITS and STOP BITS. This means that each packet may be sent in such a way that the end of one packet initiates the transmission of another. Adopted by the CCITT for the BROADBAND INTEGRATED SERVICES DIGITAL NETWORK; the primary rate of transmission is 155·52 Mbps. Abbrev *ATM*.

asynergia (*Med*) Lack of co-ordinated movement between muscles with opposing actions, due to a lesion in the nervous system. Also *asynergy*.

asystole (*Med*) Arrest of heart contraction.

AT (*ElecEng*) Abbrev for AMPERE-TURN.

At (*Chem*) Symbol for ASTATINE.

ATA (*Aero*) Abbrev for AIR TRANSPORT ASSOCIATION.

atacamite (*Min*) A hydrated chloride of copper, widely distributed eg in S America, Australia and India in the oxidation zone of copper deposits; also occurs at St Just, Cornwall, UK.

atactic (*Chem*) Denotes the property of polymers where the substituent groups are randomly arranged along the chain. Cf STEREOREGULAR POLYMERS.

atactosol (*Chem*) A colloidal sol not containing TACTOIDS.

atactostele (*BioSci*) Stele characteristic of the stems of monocotyledons, consisting of many vascular bundles apparently scattered throughout the ground tissue.

atavism (*BioSci*) The appearance in an individual of characteristics believed to be those of its distant ancestors.

ataxia (*Med*) Inco-ordination of muscles, leading to irregular and uncontrolled movements; due to lesions in the nervous system. Also *ataxy*.

ataxia telangiectasia (*Med*) Human clinical syndrome in which spontaneous chromosome rearrangements occur at a high rate, preferentially involving non-homologous chromosomes. Also *Louis Bar syndrome*

ATB (*Eng*) Abbrev for AERATION TEST BURNER.

ATC (*Aero*) Abbrev for AIR-TRAFFIC CONTROL.

ATCRBS (*Radar*) Abbrev for *air-traffic control radar beacon system*. A direct development of the World War II IFF system. Operating at about 1 GHz, it gives air-traffic controllers three-dimensional positional information and full identification of aircraft.

atelectasis (*Med*) Failure to expand and collapse of part or all of the lung.

atenolol (*Pharmacol*) Beta receptor antagonist or blocker which acts more on the β_1-receptors. Thus its action is more towards the heart with fewer effects on airway tone.

athermal solutions (*Chem*) Solutions formed without production or absorption of heat on mixing the components.

athermal transformation (*Eng*) A solid-state reaction, eg the MARTENSITIC TRANSFORMATION of steel, in which thermal activation is not required. The transformation is driven by increasing thermodynamic instability of a metastable phase, which eventually transforms by physical shear of the crystal lattice.

atherogenic (*Med*) Causing atheroma.

atherosclerosis (*Med*) Thickening of and rigidity of the intima of the arteries, caused by a gruel-like deposition of cells. Commonest form of arterial disease in Western societies. Also *atheroma*.

athetosis (*Med*) Slow, involuntary, spontaneous, repeated writhing movements of the fingers and of the toes, due to a brain lesion.

athymic (*BioSci*) Genetically lacking a functional thymus gland. Individuals with this immune deficiency fail to mature functional T-lymphocytes and have severely compromised responses to all types of infection.

Atkins diet (*Med*) A high-protein, low-carbohydrate diet intended to cause rapid weight loss, although medical opinion is generally unfavourable.

Atkinson cycle (*Autos*) A working cycle for internal-combustion engines, in which the expansion ratio exceeds the compression ratio; more efficient than the Otto cycle, but mechanically impracticable.

Atlas (*Astron*) The 15th natural satellite of Saturn, discovered in 1980. Distance from the planet 138 000 km; diameter 40 km.

Atlas (*ICT*) See SECOND-GENERATION COMPUTER.

atlas (*BioSci*) The first cervical vertebra.

Atlas rocket (*Space*) NASA satellite launcher in use since May 1962; the fourth and most powerful version (Atlas IIAS) can lift 3500 kg into geostationary transfer orbit.

ATM (*ICT*) Abbrev for ADOBE TYPE MANAGER, ASYNCHRONOUS TRANSFER MODE, AUTOMATED TELLER MACHINE.

atm (*Phys*) Symbol for STANDARD ATMOSPHERE. See ATMOSPHERIC PRESSURE.

ATM adaptation layer (*ICT*) In an ASYNCHRONOUS TRANSFER MODE network, the equipment and procedures that interface between its standardized CELLS and the many data types to be carried. Telephony, for example, requires a constant data rate over a fixed channel, while SWITCHED MULTIMEGABIT DATA SERVICE will tolerate a variable data rate over a changeable connection path.

ATM cell (*ICT*) The basic data packet handled by an ASYNCHRONOUS TRANSFER MODE network, consisting of a 5-OCTET header followed by 48 octets of user information. The header is used to route the cell between switches, and receives new labels at each switching point. The user information is carried unchanged across the ATM network for delivery at the far terminal.

atmolysis (*Chem, Phys*) The method of separation of the components of a mixture of two gases, which depends on their different rates of diffusion through a porous partition. The basis for the enrichment of uranium-235 by the diffusion process.

atmometer (*BioSci*) Apparatus, like a POTOMETER, but designed to measure water loss from a wet, non-living surface, eg a porous pot.

atmosphere (*EnvSci*) The movement of masses of air in the atmosphere and therefore of energy, ultimately derived from the Sun, determines the weather. Its structure and the description of these movements are complex and intensively studied in attempts to describe and predict weather patterns and long-term effects like global warming. See panels on ATMOSPHERIC BOUNDARY LAYER, STRATOSPHERE AND MESOSPHERE, TROPOSPHERE.

atmospheric absorption (*Acous*) Diminution of intensity of a sound wave in passing through the air, apart from normal inverse-square relation, and arising from transfer of sound energy into heat.

atmospheric absorption (*Astron*) The absorption of the light of the stars by the Earth's atmosphere; for visible light it is practically negligible above 45° altitude, but the extinction amounts to around half a magnitude at 20°, one magnitude at 10° and two magnitudes at 4° altitude. See ATMOSPHERIC WINDOW.

atmospheric acoustics (*Acous*) The study of the propagation of sound in the atmosphere, of importance in sound ranging and aircraft noise.

atmospheric boundary layer (*EnvSci*) The region of the Earth's atmosphere that interacts directly with the Earth's surface. Abbrev *ABL*. See EKMAN SPIRAL and panel on ATMOSPHERIC BOUNDARY LAYER.

atmospheric dispersion (*EnvSci*) A natural mechanism for the removal and dilution of atmospheric pollutants whereby substances injected into the air are diluted by mixing with clean air, with which they are carried.

atmospheric electricity (*EnvSci*) That causing increasing potential with height, about 100 V m^{-1}, in calm conditions, altered considerably by thunderclouds. See LIGHTNING.

atmospheric engine (*Eng*) Earliest form of practicable steam engine, in which a partial vacuum created by steam condensation allowed atmospheric pressure to drive down the piston.

atmospheric gas-burner system (*Eng*) A natural-draught burner injector, in which the momentum of the gas passing into the injector throat inspirates part of the air required for combustion.

atmospheric line (*Eng*) A datum line drawn on an indicator diagram by allowing atmospheric pressure to act on the indicator piston or diaphragm.

atmospheric pollution (*EnvSci*) Gases and aerosols in the atmosphere that are harmful to plant and animal life. See GREENHOUSE EFFECT and panel on ATMOSPHERIC POLLUTION.

atmospheric pressure (*Phys*) The pressure exerted by the atmosphere at the surface of the Earth due to the weight of the air. Its standard value is $1 \cdot 01325 \times 10^5 \text{ N m}^{-2}$, $1 \cdot 01325$ bar or $14 \cdot 7 \text{ lbf in}^{-2}$. Variations in the atmospheric pressure are measured by means of the barometer. See BAROMETRIC PRESSURE, STANDARD ATMOSPHERE.

atmospheric radio wave (*ICT*) Any radio wave that reaches its destination after reflection from the upper ionized layers of the atmosphere.

atmospherics (*ICT*) Interfering or disturbing signals of natural origin. Also *spherics*. US *strays*. See STATIC.

atmospheric tides (*EnvSci*) The changes of atmospheric pressure arising directly from changes in temperature due to the Earth's rotation. See DIURNAL RANGE.

atmospheric waveguide duct (*ICT*) Atmospheric layer that acts as a waveguide for high-frequency (>20 MHz) radio waves under certain conditions of temperature and humidity, giving reception far outside the normal service area.

Atmospheric boundary layer

Abbrev *ABL*. The region of the atmosphere in which direct interaction with the Earth's surface predominates. The airflow in this layer is largely turbulent with eddies on scales roughly equal to their height above the surface, and with turbulent shearing stresses comparable in magnitude with BUOYANCY, CORIOLIS FORCES and PRESSURE GRADIENTS; there are significant vertical flux divergences of momentum, heat and moisture due directly to surface influence. The top of the boundary layer is often marked by a shallow stable region which normal turbulent motions cannot penetrate (although strongly convective elements may). This top is usually at a height of 1–2 km. Much nearer the ground, at a height of some tens of metres, there is frequently a sublayer called the surface-flux layer (sometimes *constant-flux layer*) where

fluxes may be assumed to be almost constant with height; in this layer the Coriolis force may be neglected.

Because of the turbulence in this layer it is extremely difficult to make accurate theoretical and mathematical treatments of the energy transfers between the ABL and the Earth's surface. One important parameter is the *surface roughness length*. This is a surface character which ought to be independent of the flow. However, roughness lengths can be different for transfers of momentum, of heat and of water vapour, and their estimation is in any case difficult; consider eg bare soil, growing crops, forests and built-up areas. Another important parameter is the non-dimensional RICHARDSON NUMBER, derived from the relative rates at which mechanical turbulence is produced. These depend on the vertical shear of the mean wind which can be enhanced or suppressed according to its thermal stability.

See panels on STRATOSPHERE AND MESOSPHERE and TROPOSPHERE.

---

atmospheric window (*Astron*) A region of the electromagnetic spectrum to which the atmosphere is essentially transparent. Such regions occur at certain optical, infrared and radio regions of the spectrum but not at any X- or gamma-ray wavelengths; these are mainly absorbed by molecules in the atmosphere.

atokous (*BioSci*) Having no offspring: sterile.

atoll (*Geol*) A coral reef usually forming a circular, elliptical or irregular chain of islets around a shallow lagoon, and surrounded by deep water of the open tropical sea.

atom (*Chem*) The smallest particle of an element which can take part in a chemical reaction. See DALTON'S ATOMIC THEORY and panel on ATOMIC STRUCTURE.

atom (*ICT*) A primitive syntactic unit, an indivisible unit of data. Used in PROLOG.

atomic absorption coefficient (*Phys*) For an element, the fractional decrease in intensity of radiation per number of atoms per unit area. Symbol μ_a. Related to the linear absorption coefficient μ by

$$\mu = \frac{1}{V} \sum_i n_i \, (\mu_a)_i$$

where the material contains n_i atoms of element i in a volume V.

atomic absorption spectroscopy (*Chem*) A method in which light from a standard source is passed through a flame into which a sample of the substance under investigation has been introduced. The outer electrons of the sample are excited and emit energy at characteristic wavelengths which, in turn, absorb those from the standard source. The resulting spectrum can identify the elements present and indicate their relative proportions.

atomic bomb (*Phys*) A bomb in which the explosive power, measured in terms of equivalent TNT, is provided by nuclear fissionable material such as uranium-235 or plutonium-239. Detonated by the rapid impaction of subcritical amounts of uranium or imploding a low density and therefore subcritical volume of plutonium. Also known as *A-bomb, atom bomb, fission bomb*. Terms using 'atom' are deprecated here because the energy released is of nuclear origin. See HYDROGEN BOMB and panel on LIFETIME STUDY OF THE NUCLEAR BOMB SURVIVORS. Fig. ▷

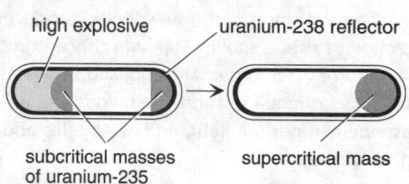

subcritical masses   supercritical mass
of uranium-235

gun method of detonation

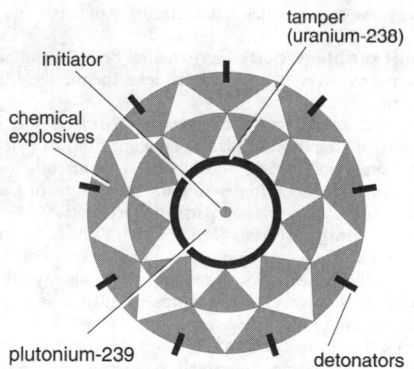

implosion type using plutonium

atomic bomb

atomic bond (*Chem*) See COVALENT BOND and panel on BONDING.

atomic clock (*Phys*) A clock whose frequency of operation is controlled by the frequency of an atomic or molecular process. The inversion of the ammonia molecule with a frequency of $2 \cdot 3786 \times 10^{10}$ Hz provides the basic oscillations of the AMMONIA CLOCK. The difference in energy between two states of a caesium atom in a magnetic field giving a frequency of 9 192 631 770 Hz is the basis of the

Atmospheric pollution

The presence in the atmosphere of gases and aerosols that are in one way or another harmful to plant and animal life, or cause damage to buildings, and either are due entirely to human activity or occur in much greater concentrations than would be natural without such activity. For instance, carbon dioxide occurs naturally, but quantities have considerably increased since the Industrial Revolution, owing to the burning of fossil fuels; similarly, ammonia and oxides of nitrogen and sulphur are produced by lightning flashes and volcanoes, but their concentrations have steadily increased owing to intensive farming practices and general industrial activity. Chlorofluorocarbons (CFCs), on the other hand, are new and have been made artificially during the last few decades for use in refrigerators and aerosol sprays; they have recently been discovered to play a part in destroying atmospheric ozone and enhancing the greenhouse effect. Photochemical smog is also new, being produced by the action of strong sunlight on high concentrations of exhaust fumes from internal-combustion engines; it is particularly common in sunny, populous and affluent areas where winds are light, eg Los Angeles and its neighbourhood.

Noxious compounds, after emission from chemical plants, power stations, etc, are diffused widely by turbulent motions in the atmospheric boundary layer, and can subsequently spread through the troposphere and even penetrate into the stratosphere. They may also reach the ground either by DRY DEPOSITION or by being dissolved in raindrops or fog droplets, WET DEPOSITION. Dry deposition occurs as a combined effect of gravitational settling, impaction of particulates on vegetation and absorption of chemically reactive gases; it is most important near the original source of pollution. The popular term 'acid rain' is a rough synonym for wet deposition. 'Natural' pre-industrial rain was itself acidic with a pH of between 5 and 6; however, since about 1970 the pH of much rain in industrial areas has fallen to about 4 and on exceptional occasions to as low as 3, about as acid as vinegar. Wet deposition is much more efficient at depositing pollution than dry deposition and can rapidly bring down high concentrations of pollutants from the upper atmosphere at great distances from the original source, as when radioactive material from the Chernobyl disaster was deposited on high ground in north Wales, the Lake District and the west of Scotland in the UK.

The effects of atmospheric pollution on animal and vegetable life are complex and difficult to unravel and there is still controversy over their precise nature.

See panels on ATMOSPHERIC BOUNDARY LAYER, STRATOSPHERE AND MESOSPHERE and TROPOSPHERE.

CAESIUM CLOCK which has an accuracy of better than one in 10^{13}.

atomic disintegration (*Phys*) Natural decay of radioactive atoms, as a result of radiation, into chemically different atomic products.

atomic displacement cross-section (*NucEng*) The probability of a neutron displacing an atom from its place in a crystalline solid. Measured in BARNS as for other cross-sections. Important in determining the lifetime of graphite moderator and structural parts of reactors.

atomic energy (*Chem*) Strictly the energy (chemical) obtained from changing the combination of atoms originally in fuels. Now often applied to energy obtained from breakdown of fissile atoms in nuclear reactors for which a more accurate term is NUCLEAR ENERGY.

atomic force microscopy (*BioSci*) A form of scanning probe microscopy, in which a microscopic probe is mechanically tracked over the surface of interest, and the force encountered at each coordinate measured with piezoelectric sensors. This provides information about the chemical nature of a surface at the atomic level.

atomic frequency (*Phys*) A natural vibration frequency in an atom used in the atomic clock.

atomic heat (*Chem*) The product of specific heat capacity and relative atomic mass in grams; approx the same for most solid elements at high temperatures.

atomic hydrogen (*Chem*) See ACTIVE HYDROGEN.

atomicity (*Chem*) The number of atoms contained in a molecule of an element.

atomic mass unit (*Chem*) Abbrev *amu*, *u*. Exactly one-twelfth the mass of a neutral atom of the most abundant isotope of carbon, ^{12}C. 1 amu = 1.660×10^{-27} kg. Before 1960, the amu was defined in terms of the mass of the ^{16}O

isotope with 1 amu = 1.6599×10^{-27} kg. Also *dalton*. See ATOMIC WEIGHT.

atomic number (*Chem*) The order of an element in the periodic (Mendeleyev) chemical classification, and identified with the number of unit positive charges in the nucleus (independent of the associated neutrons). Equal to the number of external electrons in the neutral state of the atom, and determines its chemistry. Symbol Z.

atomic orbital (*Chem*) Strictly a wavefunction defining the energy of an electron in an atom. Electron energy is quantized, the lowest main shell being K, followed by L, M, etc. Atomic orbitals are designated s, p, d and f, sub-shells which can hold 2, 6, 10 and 14 electrons respectively; s orbitals are spherical while the others are lobed and thus directional. It is the p, d and f orbitals that determine stereochemistry. See panel on ATOMIC STRUCTURE.

atomic plane (*Phys*) A solid is crystalline because its atoms are ordered in intersecting planes (*atomic planes*) corresponding to the planes of the crystal. See X-RAY CRYSTALLOGRAPHY.

atomic radii (*Chem*) Half of the internuclear distance between the nuclei of two identical non-bonded atoms at such a separation that they neither attract nor repel one another.

atomic refraction (*Chem*) The contribution made by a mole of an element to the molecular refraction of a compound.

atomic scattering (*Phys*) The scattering of radiation, usually electrons or X-rays, by the individual atoms in the medium through which it passes. The scattering is by the electronic structure of the atom in contrast to *nuclear scattering* which is by the nucleus.

Atomic structure

The electron configuration of the atoms in their normal electrically neutral state determines the chemical behaviour of the elements. Each atom consists of a heavy nucleus with a positive charge produced by a number of protons equal to its atomic number. There are an equal number of electrons outside the nucleus to balance this charge. The nucleus also contains neutrons which are electrically neutral. Protons and neutrons are collectively referred to as NUCLEONS.

The Sommerfeld model, modified by the wave mechanical concept of orbitals, describes the electron configuration of the atom. Electrons are fermions which must conform to the PAULI EXCLUSION PRINCIPLE which governs the way in which electrons can fill the available orbitals because no two electrons in the same atom can be in the same quantum state, ie have the same set of four quantum numbers. The principal quantum number (n) indicates the shell to which the orbital belongs and varies from 1 (K-shell) closest to the nucleus to 7 (Q-shell), the most remote.

For a given principal quantum number n, there are n allowed values of l, the orbital angular momentum quantum number; for each value of l, there are $(2l + 1)$ allowed values of m_l, the magnetic angular momentum quantum number; for each value of m_l, there are two values of m_s, the magnetic spin number. This makes a total of $2n^2$ orbitals ($2n^2$ electrons per shell) for a given value of n, and, as the Pauli principle allows only one electron for each set of four quantum numbers n, l, m_l, m_s, this limited number of allowed orbitals makes up the electron shell for a given n.

In general, the closer an electron is to the nucleus, the greater the coulomb attraction and so the greater the binding energy retaining the electron in the atom. Inner filled shells are therefore relatively inert and the chemical properties of the atom are determined by the electron arrangement in the outermost shell.

Nuclear binding forces tend to give greatest stability when the neutron number and the proton number are approximately equal. Owing to electrostatic repulsion between protons, the heavier nuclei are most stable when more than half their nucleons are neutrons; elements with more than 83 protons are unstable and undergo radioactive disintegration. Those with more than 92 protons are not found naturally on Earth, but can be synthesized in particle ACCELERATORS and nuclear REACTORS. These are the TRANSURANIC ELEMENTS which have short half-lives. Most elements exist with several stable isotopes and the chemical atomic weight gives the average of a normal mixture of these isotopes.

---

atomic scattering factor (*Phys*) The ratio of the amplitude of coherent scattered X-radiation from an atom to that of a single electron placed at the atomic centre. The atomic scattering factor depends on the electron-density distribution in the atom and is a function of the scattering angle.

atomic spectrum (*Phys*) Characteristic pattern of light frequencies emitted or absorbed by a given element due to electronic transitions between the discrete energy states of the atoms accompanied by the emission or absorption of *photons*. See ABSORPTION LINES.

atomic structure (*Phys*) The arrangement of the parts of an atom. See panel on ATOMIC STRUCTURE.

atomic transmutation (*Phys*) The change of one type of atom to another as a result of a nuclear reaction. The transmutation can be produced by high-energy radiation or particles and is most easily produced by neutron irradiation. The change in atomic number means the chemical nature of the atom has been changed. Also *transformation.*

atomic volume (*Chem*) The ratio for an element of the relative atomic mass to the density; this shows a remarkable periodicity with respect to atomic number.

atomic weight (*Chem*) The mass of atoms of an element in atomic mass units on the unified scale where $1\ \text{amu} = 1{\cdot}660 \times 10^{-27}$ kg. For natural elements with more than one isotope, it is the average for the mixture of isotopes. See RELATIVE ATOMIC MASS.

atomism (*Psych*) The theory that mental processes and psychological states can be analysed into simple elements.

atomized powder (*PowderTech*) A powder produced by the dispersion of molten metal or other material by spraying under conditions such that the material breaks down into powder.

atomizer (*Eng*) A nozzle through which a liquid is sprayed under pressure. Its function is to break up the substance into a fine mist which may subsequently solidify to form fine-powder particles or be deposited onto a surface and form a coating.

atony (*Med*) Loss of muscular tone.

atopy (*Med*) A constitutional or hereditary tendency to develop high levels of IgE and immediate hypersensitivity to allergens, esp those which are absorbed across the respiratory mucosa. Adj *atopic.*

atorvastatin (*Pharmacol*) A STATIN used to lower the amount of fatty substances, esp cholesterol, in the blood.

ATP (*Chem*) Abbrev for ADENOSINE TRIPHOSPHATE.

ATPase (*BioSci*) Enzyme which converts adenosine triphosphate to adenosine diphosphate. In this process the free energy change of the exergonic hydrolysis is used to drive an endergonic reaction, eg muscle myosin possesses ATPase activity and the ATP breakdown is coupled to the movement of the MYOSIN fibres relative to the ACTIN.

ATR (*Phys*) Abbrev for ATTENUATED TOTAL REFLECTION.

atrate (*BioSci*) Blackened; blackening. Also *atratous.*

atresia (*BioSci*) (1) Pathological narrowing of any channel of the body. (2) Disappearance by degeneration, eg as the follicles in the mammalian ovary. Adjs *atresic, atretic.*

atrial (*BioSci*) Pertaining to the ATRIUM.

atrial fibrillation (*Med*) Atrial arrhythmia marked by rapid randomized contractions of the atrial myocardium, causing a totally irregular, and often rapid, ventricular rate.

atriopore (*BioSci*) The opening by which the atrial cavity communicates with the exterior.

atrioventricular (*BioSci*) Pertaining to, or connecting, the atrium and ventricle of the heart; eg the atrioventricular connection, a bundle of muscle fibres that transmits the wave of contraction from the atrium to the ventricle, in higher vertebrates. Also *auriculoventricular*.

atrium (*Arch*) The entrance court in a Roman house, open to the sky in the centre, but with a roofed perimeter walkway. The concept is used in contemporary building where several storeys of offices or shops frequently enclose a void or atrium, glazed at roof level, thus allowing daylight to reach the surrounding accommodation.

atrium (*BioSci*) (1) In Platyhelminthes, a space into which open the ducts from the male and female genital organs. (2) In pulmonate Mollusca, a cavity into which the vagina and the penis open and that itself opens to the exterior. (3) In Protochordata, the cavity surrounding the respiratory part of the pharynx. (4) In vertebrates, the anterior part of the nasal tract. (5) In reptiles and birds, the cavity connecting the bronchus with the lung chambers. (6) In the vertebrate heart, one of the two smaller chambers that receives blood from either lungs (left atrium) or the body (right atrium) and opens into the ventricle (formerly *auricle*).

atrophic rhinitis (*Vet*) Infectious disease of pigs, characterized by chronic rhinitis and deformity of the snout. Various infectious agents, plus other factors, are involved. *Pasteurella multocida* and *Bordatella bronchiseptica* are commonly isolated. Incidence can be limited by vaccination.

atrophy (*BioSci*, *Med*) Degeneration or wasting, ie diminution in size, complexity or function of an organ or tissue through disuse.

atropine (*Pharmacol*) An antimuscarinic alkaloid drug, used as an antispasmodic and smooth muscle relaxant. Atropine constitutes the active principle of belladonna (obtained from *Atropa belladonna*, deadly nightshade). It has sedative effects on the central nervous system, blocks activity of the vagus nerve and causes dilatation of the pupil of the eye.

atropus (*BioSci*) See ORTHOTROPOUS.

ATR tube (*Radar*) Abbrev for ANTI-TRANSMIT–RECEIVE TUBE.

attached column (*Arch*) A column partially built into a wall, instead of standing detached.

attachment (*Psych*) The strong emotional bond a child forms with his or her primary caregiver: considered in some theories to have an evolutionary basis and that various attachment behaviours, eg smiling, crying, enhance the probability of survival.

attapulgite (*Min*) See PALYGORSKITE.

attar of roses (*Chem*) Oil distilled from fresh roses for perfumery purposes. Also *otto de rose*.

attention (*Psych*) Selective attention. That aspect of perception which implies a readiness to respond to a particular stimulus or aspects of it.

attentional (*Psych*) Relating to attention or concentration.

attention deficit (hyperactivity) disorder (*Psych*, *Med*) Increasingly recognized as a common affliction of young people who cannot concentrate on school work or intricate play. Often associated with hyperactivity and thought by some to be environmentally induced. Abbrev *ADD* or *ADHD*.

attenuated total reflection (*Phys*) Spectroscopic method of analysing thin films on reflective substrates, esp using infrared radiation.

attenuated vaccine (*Med*) Live bacterial or virus vaccine in which the microbes have been selected or otherwise treated in such a way as to diminish greatly their capacity to cause disease but still to retain their ability to evoke protective immunity, eg polio, measles and yellow fever vaccines.

attenuation (*BioSci*) (1) Reduction in the strength or intensity of a stimulus. (2) Lessening of the capacity of a pathogen to cause disease.

attenuation (*Phys*, *ICT*) General term for reduction in magnitude, amplitude or intensity of a physical quantity, arising from absorption, scattering or geometrical dispersion. The latter, arising from diminution by the inverse-square law, is not generally considered as attenuation proper.

attenuation coefficient (*Phys*) The coefficient which expresses energy losses of electromagnetic radiation due to both absorption and scattering in a medium. Relevant to narrow beam conditions. Also *total absorption coefficient*.

attenuation compensation (*ICT*) The use of networks to correct for frequency-dependent attenuation, eg in transmission lines. See PRE-EMPHASIS.

attenuation constant (*ICT*) The real part of α in the relationship $\rho = \rho \exp(-\alpha x)$, where ρ is a physical quantity, such as the amplitude of a wave propagating along a transmission path, and x is the distance along the path. The imaginary part of α is known as the PHASE CONSTANT. More simply, but less commonly defined by $\mu = \alpha\lambda$ where μ is the attenuation and λ is wavelength, ie α is the attenuation per wavelength distance of propagation. See DECIBEL, NEPER, PROPAGATION CONSTANT.

attenuation distortion (*ICT*) Distortion of a complex waveform resulting from the differing attenuation of each separate frequency component in the signal. This form of distortion is difficult to avoid, eg in transmission lines.

attenuation of X-rays (*Radiol*) Absorption and scattering of X-rays as they pass through an object.

attenuator (*ICT*) An arrangement of fixed or variable resistive elements designed to reduce the strength of any signal (audio- or radio-frequency) without reducing appreciable distortion. Attenuators also incorporate impedance matching to the transmission lines or circuits to which they are connected, regardless of the attenuation they introduce. For lower-frequency applications they may be simply variable or fixed resistances; for high frequencies they may be pieces of resistive material, introduced into transmission lines, stripline or waveguide. Fixed attenuators are sometimes referred to as *pad*.

Atterburg limits (*Min*) In fine-grained sediments and soils, the empirical moisture-content boundaries between the liquid and plastic states (the *liquid limit*) and between the plastic and semi-solid states (the *plastic limit*).

attitude (*Aero*) Of an aircraft in flight, the angle made by its axes with the relative airflow; the *aspect* is the angle made by its axes with the ground when the aircraft is on the ground.

attitude (*Psych*) An inferred disposition to feel, think and act in certain ways which is used to explain the variation between individuals in their response to similar situations; attitudes are assumed to represent the effects of past experience on behaviour through their effects on the cognitive and emotional structuring of perception.

attitude control (*Space*) The provision of a desired orientation to satisfy mission requirements; it is usually effected by a low-thrust system in conjunction with a measuring instrument, such as a star sensor, and maintained by a stabilizing device, such as a gyroscope. Attitude control can also be maintained by spinning the spacecraft about one of its axes.

attitude indicator (*Aero*) A GYRO HORIZON which indicates the true attitude of the aircraft in pitch and roll throughout 360° about these axes. See HEADING INDICATOR.

attitude scale (*Psych*) Standard procedure for measuring attitudes.

atto- (*Genrl*) Prefix denoting one million million millionth, or 10^{-18}. Symbol a.

attracted-disk electrometer (*ElecEng*) Fundamental instrument in which potential is measured by the attraction between two oppositely charged disks.

attribute (*ICT*) Information about a file stored by the OPERATING SYSTEM indicating whether the file is read-only, hidden or has been changed since a BACK-UP FILE was last made. Also *file attribute*.

attribution theories (*Psych*) Theories concerned with the way in which an individual infers the intentions or responses of others on the basis of attributed motivation or external drivers that are consistent with the behaviour.

attrition (*Geol*) The wearing down of rocks by friction, esp that between loose fragments or particles under natural processes.

attrition test (*CivEng*) A test for the determination of the wear-resisting properties of stone, particularly stone for road-making. Pieces of the stone are placed in a closed cylinder, which is then rotated for a given time, after which the loss of weight due to wear is found.

Attwood's formula (*Ships*) A formula for determining the moment of static stability at large angles of heel of a ship. Taking angle of heel θ, and the weight of the ship W, the moment

$$= W\left(\frac{vhh_1}{V} \pm BG\sin\theta\right)$$

where v is the volume of emerged wedge, hh_1 is the distance between the centres of gravity of emerged and immersed wedges, V is the volume of displacement, B is the centre of transverse buoyancy, and G is the centre of gravity.

at wt (*Genrl*) Abbrev for *atomic weight*, now RELATIVE ATOMIC MASS.

AU (*Astron*) Abbrev for ASTRONOMICAL UNIT.

Au (*Chem*) Symbol for GOLD (*aurum*).

audial (*Genrl*) Relating to hearing or sounds.

audibility (*Acous*) Ability to be heard; said of faint sounds in the presence of noise. The extreme range of audibility is 20–20 000 Hz in frequency, depending on the applied intensity; and from 2×10^{-5} N m^{-2} (rms) at 1000 Hz (the zero of the phon scale, selected as the average for good ears) to 120 dB.

audible ringing tone (*ICT*) An audible tone fed back to a caller as an indication that ringing current has been remotely extended to the called subscriber's telephone. On circuits in UK it is heard as a double beat recurring at 2 s intervals.

audio- (*Genrl*) Prefix from Lt *audire* pertaining to sound, esp broadcast sound.

audio codec (*ICT*) A CODEC for use in a MULTIMEDIA system, designed to handle a range of sound signals in addition to speech.

AU diode (*Electronics*) See BACKWARD DIODE.

audio dub (*ImageTech*) Replacing the existing audio with new.

audio-frequency (*Acous, ICT*) Frequency which, in an acoustic wave, makes it audible. In general, any wave motion including frequencies in the range of 20 Hz to 20 kHz.

audio-frequency amplifier (*ICT*) Amplifier for frequencies within the audible range.

audio-frequency choke (*ElecEng*) Inductor with appreciable reactance at audio-frequencies.

audio-frequency modulation (*ImageTech*) Audio signal that has been frequency modulated onto a carrier and then recorded with the video signal through the video heads, as with 8 mm, or separately by additional heads on the drum, as with VHS. Abbrev *AFM*. See DEPTH MULTIPLEX RECORDING.

audio-frequency shift modulation (*ICT*) A method of facsimile transmission in which tone values from black to white are represented by a graded system of audio-frequencies.

audio-frequency transformer (*ICT*) Transformer for use in a communication channel or amplifier, designed with a specified, normally uniform, response for frequencies used in sound reproduction.

audiogram (*Acous*) Standard graph or chart which indicates the hearing loss (in BELS) of an individual ear in terms of frequency. See OBJECTIVE NOISE METER, SOUND-LEVEL METER.

audiometer (*Acous*) Instrument for measurement of acuity of hearing. Specifically to measure the minimum intensities of sounds perceivable by an ear for specified frequencies. See NOISE AUDIOMETER.

auditory (*BioSci*) Pertaining to the sense of hearing or to the apparatus which subserves that sense; the eighth cranial nerve of vertebrates, supplying the ear. Also *aural*.

auditory canal (*Acous, Med*) Duct connecting the ear drum with the external ear (PINNA), by which sound waves are transmitted from outer to inner ear.

auditory ossicles (*BioSci*) Three small bones, the incus, malleus and stapes, bridging the tympanic cavity of the middle ear in mammals.

auditory perspective (*Acous*) See STEREOPHONY.

audit trail (*ICT*) Record of the file updating which takes place during a specific transaction. It enables a trace to be kept of all operations on files.

aufbau (*Chem*) Filling of successive electron shells around an atom, electron by electron. Each electron occupies the lowest energy level or atomic orbital available, so creating an electron structure for all the elements. The principle explains the structure of the periodic table.

augend (*MathSci*) See ADDITION.

augen-gneiss (*Geol*) A coarsely crystalline rock of granitic composition, containing lenticular, eye-shaped masses of feldspar or quartz embedded in a finer matrix. A product of regional metamorphism.

auger (*Build*) A tool for boring holes, esp in wood or in the earth.

Auger effect (*Phys*) Effect seen when an atom is ionized by the ejection of an inner electron; the resulting empty orbital is filled by an outer electron of higher energy and the excess energy, the difference in orbital energies, may cause the ejection of an outer electron, making the atom doubly charged. Energies of the Auger electrons emitted are characteristic of the atomic energy levels, providing a method of determining surface composition and character. Also *Auger ionization*.

Auger yield (*Phys*) For a given excited state of an atom of a given element, the probability of de-excitation by Auger process instead of by X-ray emission.

augite (*Min*) A pyroxene, a complex aluminous silicate of calcium, iron and magnesium, crystallizing in the monoclinic system, and occurring in many igneous rocks, particularly those of basic composition; it is an essential constituent of basalt, dolerite and gabbro.

augmentor (*Aero*) Means of increasing forces: (1) by afterburning in a gas turbine; (2) by induced airflow in a rocket; or (3) in a wing of STOL aircraft by ducting compressed airflow from a gas turbine into circulation-increasing slots and flaps to create high lift coefficients, thereby giving slow landing speeds.

AUI (*ICT*) Abbrev for *attachment unit interface*. Usually a 15-way 'D' connector. Described in the IEEE 802.3 specification as the INTERFACE between the ETHERNET transceiver and the NETWORK device.

Aujesky's disease (*Vet*) An encephalomyelitis affecting cattle, sheep, pigs, dogs, cats and rats, and caused by a thermostable herpes virus. Very widespread on farms but can be eradicated. Notifiable disease, controlled by vaccination or a slaughter policy. Also *infectious bulbar paralysis, mad itch, pseudorabies*.

aura (*Psych*) The subjective sensation preceding an epileptic seizure or migraine.

aural (*BioSci*) See AUDITORY.

aural masking (*Acous*) See MASKING.

auramines (*Chem*) Dyestuffs of the diphenylmethane series.

aureole (*ElecEng*) Luminous glow from the outer portion of electric arc which has a spectrum different from that of the highly ionized core.

aureole (*EnvSci*) (1) The reddish ring round the Sun or Moon, forming the inner part of a corona. (2) The bright indefinite ring round the Sun in the absence of clouds.

aureole (*Geol*) Area surrounding an igneous intrusion affected by metamorphic changes.

auric acid (*Chem*) Gold (III) oxide. Au_2O_3. An amphoteric oxide of gold. Also *auric oxide*.

auricle (*BioSci*) (1) A small ear-shaped lobe at the base of a leaf or other organ. (2) Obsolete term for an ATRIUM of the heart. (3) Ear-like extensions, projecting towards the front, of the atria of the heart that provide additional capacity. (4) The external ear of vertebrates; any lobed appendage resembling the external ear. Also *auricula*. Adjs *auricled*, *auriculate*.

auric oxide (*Chem*) See AURIC ACID.

auricularia (*BioSci*) In Holothuria and Asteroidea, a pelagic ciliated larva, having the cilia arranged in a single band, produced into a number of short processes.

auriculoventricular (*BioSci*) See ATRIOVENTRICULAR.

auriferous deposit (*Geol*) A natural repository of gold, in the general sense, including gold-bearing lodes and sediments such as sands and gravels, or their indurated equivalents, which contain gold in detrital grains or nuggets. See BANKET, LODE, PLACERS.

auriferous pyrite (*Min*) Iron sulphide in the form of pyrite, carrying gold, probably in solid solution.

Auriga (Charioteer) (*Astron*) A prominent northern hemisphere constellation in the Milky Way, containing many star clusters. Its brightest star is CAPELLA.

aurine (*Chem*) *Pararosolic acid*. $(HOC_6H_4)_2 = C = C_6H_4 = O$. Made from phenol, oxalic acid and sulphuric acid. It is similar to rosolic acid in properties. Basis of aurine dyes.

aurora (*Astron*) Luminous curtains or streamers of light seen in the night sky at high latitudes, caused when electrically charged particles from the Sun are guided by the Earth's magnetic field to the polar regions, there colliding with atoms in the upper atmosphere. In the northern hemisphere known as *aurora borealis* ('Northern Lights') and in the southern as *aurora australis* ('Southern Lights').

auroral zone (*ICT*) Zone where radio transmission is affected by aurora.

aurous (*Chem*) Containing gold (I).

aurum (*Chem*) See GOLD.

auscultation (*Med*) The act of listening to the sounds produced in the body.

ausforming (*Eng*) Working an alloy steel in the metastable AUSTENITE condition. The material is first heated to a temperature where the austenite is stable, ie above the Ac_3 temperature, and is then cooled rapidly to the region of 550°C and worked to shape before any transformation to PEARLITE or BAINITE takes place. It transforms to MARTENSITE on cooling at ambient temperature and is then tempered. Strength and toughness are enhanced compared with the same material worked conventionally in the austenite region and quenched and tempered as separate operations. See ISOTHERMAL TRANSFORMATION DIAGRAM and panel on STEELS.

austempering (*Eng*) Heating a steel to transform it to AUSTENITE, followed by cooling rapidly to a temperature above the martensitic change point, but below the critical range, so that the austenite isothermally transforms to BAINITE, which has properties resembling a quenched and tempered steel of the same composition. See ISOTHERMAL TRANSFORMATION DIAGRAM.

austenite (*Eng*) The higher-density, high-temperature, face-centred cubic, γ-form of iron and of solid solutions based on it. In pure iron it is stable between 1183 K and 1663 K.

austenite bay (*Eng*) The shape of the region around 550°C in an ISOTHERMAL TRANSFORMATION DIAGRAM which defines the zone where AUSTENITE is metastable and remains in that condition pending transformation to PEARLITE or BAINITE.

austenitic steel (*Eng*) Steel containing sufficient amounts of nickel, nickel and chromium, or manganese to retain austenite at atmospheric temperature; eg austenitic stainless steel and Hadfield's MANGANESE STEEL.

Austin Moore prosthesis (*Med*) A prosthesis which is inserted into the upper end of the femur to reconstruct the hip joint when the original head has been removed.

Australasian region (*BioSci*) One of the primary faunal regions into which the land surface of the globe is divided; includes Australia, New Guinea, Tasmania, New Zealand, and the islands south and east of Wallace's line.

Australia antigen (*BioSci*) An envelope antigen of hepatitis B virus, now known as *HBsAg*. Presence of the antigen in serum is associated with a phase of high infectivity.

Australian blackwood (*For*) A hardwood tree (*Acacia*), whose wood is golden to reddish brown, with a straight to wavy grain and a fine and even texture.

australites (*Min*) See TEKTITES.

Austrian cinnabar (*Chem*) See BASIC LEAD CHROMATE.

aut-, auto- (*Genrl*) Prefixes from Gk *autos*, self.

autacoid (*Med*) General name for an endocrine secretion; a specific organic substance formed by the cells of one organ, and passed by them into the circulating fluid, to produce effects upon other organs. Also *autocoid*. See HORMONE.

autecology (*BioSci*) The study of the ecology of any individual species. Cf SYNECOLOGY.

authentication centre (*ICT*) A node within a PERSONAL COMMUNICATIONS NETWORK containing the database files needed to check that potential users have authority to use the system.

authigenic (*Geol*) Pertaining to minerals which have crystallized in a sediment during or after its deposition.

authoring language (*ICT*) A method whereby material may be produced for COMPUTER-AIDED LEARNING without having to write a computer program. Also *authoring system*.

authoritarian personality (*Psych*) Descriptive of a person demanding subservience and obedience from others and exhibiting subordination behaviour to those in authority.

autism (*Psych*) Generally a state of self-absorption, but usually used as an abbreviation for infantile autism, a childhood psychosis originating before 30 months that is characterized by a lack of responsiveness in social relationships, language abnormality and a need for constant environmental input or sameness. Stereotypic motor habits, overactivity, and epilepsy are often associated with it. *Asperger's syndrome* is considered to be a mild form.

auto- (*Genrl*) See AUT-.

auto-adhesion (*Eng*) Bonding together of identical surfaces, as with contact adhesives.

auto-allogamy (*BioSci*) The condition of a species in which some individual plants are capable of self-pollination and others of cross-pollination.

auto-antibody (*BioSci*) An antibody that reacts specifically with an antigen present on normal constituents of the body of the individual in whom the antibody was made. B-lymphocytes able to make auto-antibodies are normally suppressed, but in auto-immune disease the regulation mechanisms break down.

auto-assemble (*ImageTech*) A system of videotape editing in which selected scenes are transferred in their required sequence according to a preselected programme of time-code information.

autocapacitance coupling (*ElecEng, ICT*) Coupling of two circuits by a capacitor included in series with a common branch.

autocast (*ICT*) To PODCAST sound files that have been automatically converted from text files.

autocatalysis (*BioSci*) A reaction that is catalysed by one of its products or an enzyme-catalysed reaction in which one of the products is an activator of the enzyme.

autocatalysis (*Chem*) The catalysis of a reaction by the product of that reaction, eg MnO_4^- and $C_2O_4^{2-}$ by Mn^{2+} ions.

autochthonous (*BioSci*) Generally, indigenous, inherited, hereditary (eg autochthonous species, autochthonous characteristics). In an aquatic community, said of food material produced within the community.

autochthonous (*Psych*) In psychology, used of ideas coming into the mind with no apparent connection to the subject's train of thought and independently of outside influences.

autocidal control (*BioSci*) A method of insect pest control by release of sterile or genetically altered individuals into the wild population.

autoclave (*BioSci*) (1) Apparatus for sterilization by steam at high pressure. (2) To sterilize, to kill all micro-organisms and heat-resistant spores by using steam at high pressure. A high temperature, eg 121°C for 15 minutes, ensures sterilization. See RETORT.

autoclave (*Chem*) A vessel, constructed of thick-walled steel (usually alloy steel or frequently nickel alloys), for carrying out chemical reactions under pressure and at elevated temperatures.

auto coarse pitch (*Aero*) The setting of the blades of a propeller to the minimum drag position if there is a loss of engine power during take-off.

autocoid (*Med*) See AUTACOID.

autocollimator (*Phys*) (1) An instrument for accurately measuring small changes in the inclination of reflecting surfaces. Principally used for engineering metrology measurements. (2) A convex mirror used to produce a parallel beam of light from a reflecting telescope. It is placed at the focus of the main mirror.

autocorrelation (*ICT*) Technique for detecting weak signals against a strong background level. Signal is subjected to controlled delay, the original delay signals then being fed to the autocorrelation unit which responds strongly only if delay is an exact multiple of signal period.

autocorrelation (*MathSci*) The correlation between successive items in a series such that their covariance is not zero and they are not independent.

autocrine (*BioSci*) A term used to describe the secretion of a substance, such as a GROWTH FACTOR, that stimulates the secretory cell itself. This may account for the difficulty of growing isolated single cells (the signal gets diluted by the medium); excess production of an autocrine growth factor can cause the loss of normal growth control.

autocue (*ImageTech*) A visual prompter which displays a script to persons in front of a TV camera. Normally mounted on the camera to give eye contact with the viewers.

autodiploid (*BioSci*) See ANTHER CULTURE.

autodyne (*ICT*) A term describing an electrical circuit in which the same elements and valves are used as both oscillator and detector. Also *endodyne, self-heterodyne*.

autodyne receiver (*ICT*) A receiver utilizing the principle of beat reception and including an autodyne oscillator.

autoecious (*BioSci*) A parasite or pest living only in a single host species. Also *autoxenous*. Cf HETEROECIOUS.

auto-erotism (*Psych*) A condition where sensual pleasure is sought and gratified in one's own person, without the aid of an external love object; eg masturbation, thumb sucking. See NARCISSISM.

autoexec.bat file (*ICT*) A BATCH FILE (set of commands) that is automatically carried out whenever the computer is booted. Commonly used with computers using the MS-DOS operating system. See BOOT.

autoflare (*Aero*) An automatic landing system which operates on the FLARE-OUT part of the landing, using an accurate radio altimeter.

autofluorescence (*BioSci*) A property of a compound or material that will fluoresce in its own right, without the addition of an exogenous fluorophore. It is a common problem in fluorescence microscopy and in assays where the read-out is fluorescence.

autofocus assist (*ImageTech*) A device which improves autofocus performance in low light by projecting a high-contrast light pattern onto the subject. See AUTOMATIC FOCUSING.

autogamy (*BioSci*) (1) Self-fertilization. (2) The fusion of sister cells, or of two sister nuclei. (3) Fertilization involving pollen and ovules from the same flower, the same plant or genetically identical individuals (same GENET or CLONE). See SELF-FERTILIZATION, SELF-POLLINATION. Cf ALLOGAMY.

autogenic (*BioSci*) Denoting changes caused by interactions between members of a plant community.

autograft (*BioSci*) A mass of tissue, or an organ, moved from one region to another within the same organism.

auto-ignition (*Autos*) The self-ignition or spontaneous combustion of a fuel when introduced into the heated air charge in the cylinder of a compression–ignition engine. See SPONTANEOUS IGNITION TEMPERATURE.

auto-immune diseases (*Med*) A group of diseases caused by antigen/antibody reactions to the host's own tissues. See panel on AUTO-IMMUNITY.

auto-immunity (*BioSci*) A condition in which T- or B-lymphocytes capable of recognizing 'self' constituents are present and activated so as to cause damage to cells by cell-mediated immunity or to release auto-antibodies, and so to cause *auto-immune diseases*. See panels on AUTO-IMMUNITY and IMMUNE RESPONSE.

auto-inductive coupling (*ElecEng*) Coupling of two circuits by an inductance included in series with a common branch.

auto-infection (*BioSci*) Reinfection of a host by its own parasites.

auto-intoxication (*Med*) Poisoning of the body by toxins produced within it.

autokinesis (*Psych*) (1) Voluntary movements induced by internal stimuli such as proprioceptive feedback. (2) A shift in cognitive mindset arising from internal, subjective factors. (3) *Autokinetic effect*: a perceptual illusion experienced as the apparent movement of a stationary spot of light in a darkened room.

autoland (*Aero*) A landing in which the descent, forward speed, FLARE-OUT, alignment with the runway and touchdown are all automatically controlled. See AUTOFLARE, AUTOTHROTTLE.

autolithography (*Print*) The drawing by an artist of a design direct on the lithographic stone or plate.

autologous (*BioSci*) Derived from an organism's own tissues or DNA. cf HETEROLOGOUS, HOMOLOGOUS.

autolysis (*BioSci*) Self-destruction of cells or tissues by their own enzymes. Important in food processing; meat and fish will decay rapidly due to proteolytic enzymes in the gut but this process is retarded if the gut is removed early and the flesh chilled or frozen. Adj *autolytic*.

automated library system (*ImageTech*) An apparatus which holds several videotape recorders and many cassettes that are selected and loaded automatically to allow the broadcasting of commercials and programmes under computer control, eg TN MARC for multiple automated record/playback cassette system. See PROGRAMME DELIVERY CONTROL.

automated teller machine (*ICT*) Machine used by banks and building societies whereby a customer may carry out transactions on an account, such as cash withdrawal, gaining secure access by means of a plastic card and a personal identification number (PIN). Abbrev *ATM*. Also *automatic teller machine*.

automatic arc lamp (*Phys*) An arc lamp in which the feeding of the carbons into the arc and the striking of the arc are done automatically, by electromagnetic or other means.

automatic arc welding (*ElecEng*) Arc welding carried out in a machine which automatically moves the arc along the joint to be welded, feeds the electrode into the arc, and controls the length of the arc.

automatic beam control (*ImageTech*) A system in a TV camera which momentarily alters the beam current in the camera tube to reduce the tailing effects on moving highlights. Abbrev *ABC*. Also *automatic beam optimizer*.

automatic blankets (*Print*) On letterpress rotary machines, a covering for the impression cylinder with a felt base and a surface of rubber on plastic.

automatic brightness control (*ImageTech*) Circuit used in some TV receivers to keep average brightness level of screen constant.

Auto-immunity

The recognition of self

An immune system evolved to recognize foreignness, characteristic of pathogens, must distinguish between 'self' and 'non-self'. Lymphocytes of each individual must tolerate that individual's own cells. This classical view incorporates the idea that tolerance of self antigens is 'learned' in the neonatal period when the immunological system is developing; antigens encountered at this stage do not elicit a response and auto-immune clones of cells are deleted (see panel on IMMUNE RESPONSE). However, auto-immune T-cells and auto-antibodies frequently occur among normal individuals, where they are subject to regulatory control. Auto-immune disease can therefore be seen as a breakdown in this control.

An interesting alternative view has been proposed by Matzinger: that whether an immune response is produced at all depends upon the context in which the antigen is encountered and that concurrent exposure to 'danger' signals, recognized by Toll-like receptors, is important. In this hypothesis, tolerance is the default condition; this is supported by the fact that non-specific exogenous signals, particularly adjuvants, are important in artificially eliciting an active immune response. The apoptotic death of cells avoids the generation of endogenous danger signals and avoids induction of an active immune response against self.

Some auto-immune diseases seem to be caused by a malfunction in apoptosis, leading to the production of endogenous danger signals and de-repressing the auto-immune response. Others may be caused by the presence of an infection in which antigens on the pathogen are similar to self antigens, leading to cross-recognition.

Tolerance is much more difficult to induce in the adult, but can be achieved by giving immunosuppressive drugs (eg ciclosporin) to stop the immune response while the antigens are present. This is the basis for methods of preventing organ rejection after surgical transplantation.

Auto-immune diseases range from rather specific illnesses affecting eg the thyroid, to rather generalized diseases such as ulcerative colitis and rheumatoid arthritis. Another abnormal immune response causes allergy, where antibodies of one particular type (IgE) come into contact with otherwise harmless substances such as grass pollen, dust mite feces and so on. This causes the release of substances such as histamine, inducing anaphylaxis or the well-known symptoms of hay fever or asthma.

---

automatic call distribution (*ICT*) An INTELLIGENT NETWORK service that takes account of factors such as time of day or caller location to route calls to the appropriate point within an organization. An example would be the use of a single national FREEPHONE number to access the branch of an organization nearest the caller during the day but a more distant emergency service at night.

automatic camera (*ImageTech*) Camera in which the focus, lens aperture and shutter speed are selected automatically; film advance by motor drive may also be included. Priority selection may be available, eg exposure based on either general or spot areas and with aperture or shutter speed limitations.

automatic circuit breaker (*ElecEng*) A circuit breaker which automatically opens the circuit as soon as certain predetermined conditions (eg an overload) occur.

automatic cluster removal (*Agri*) In a milking machine, a system by which cessation of milk flow is detected and the vacuum holding the teat cups to the udder is broken, releasing the cluster.

automatic computer (*ICT*) Obsolete term for one type of FIRST-GENERATION COMPUTER.

automatic contrast control (*ImageTech*) Form of automatic gain control used in video signal channel of a TV receiver.

automatic control (*Eng*) (1) Switching system which operates control switches in correct sequence and at correct intervals automatically. (2) Control system incorporating servomechanism or similar device, so that feedback signal from output of system is used to adjust the controls and maintain optimum operating conditions.

automatic cut-out (*ElecEng*) A term frequently applied to a small automatic circuit breaker suitable for dealing with currents of a few amperes.

automatic data processing (*ICT*) See DATA PROCESSING.

automatic direction finding (*Aero*) Airborne navigational aid tuned to radio source of known position. Using rotatable loop aerial mounted above an aircraft to detect the direction of the radio source by rotating until the signal is zero. Abbrev *ADF*.

automatic exposure (*ImageTech*) A control system using a photosensor in the camera to measure scene brightness and automatically set the lens aperture/shutter speed combination. Refinements include measuring particular areas of the scene and PROGRAM EXPOSURE MODES. A video camera uses the video signal to determine exposure. Abbrev *AE*.

automatic flight control system (*Aero*) A category of AUTOMATIC PILOT for the control of an aircraft while *en route*. It can be monitored by speed and altitude data signals, signals from an INSTRUMENT LANDING SYSTEM and VOR, has automatic approach capability, and is disengaged before landing. Abbrev *AFCS*. Cf AUTOFLARE, AUTOLAND, AUTOTHROTTLE.

automatic focusing (*ImageTech*) Control system for automatically setting the lens focus to the subject distance; in a simple form, this may be by means of a coupled rangefinder but advanced types employ completely automatic examination of the image. In an enlarger or rostrum camera, lens focus is mechanically set by the distance from the base. Abbrev *AF*.

automatic frequency control (*ICT*) Electronic or mechanical means for automatically compensating, in a receiver, frequency drifts in transmission carrier or local oscillator. Abbrev *AFC*.

automatic gain control (*ICT*) A system in amplifiers which compensates for a wide range of input signals to give a more uniform level of output and thus accommodate a wide range of conditions including fading, masking of antenna and ambient light. Abbrev *AGC*.

automatic generating plant (*ElecEng*) A small generating station, eg a petrol- or diesel-driven generator and battery, which is automatically started when the battery voltage falls below a certain value and stopped when it is fully charged. The term is also applied to the plant in small, unattended, hydroelectric generating stations.

automatic mixture control (*Aero*) A device for adjusting the fuel delivery to a reciprocating engine in proportion to air density.

automatic observer (*Aero*) Apparatus for recording, photographically or electronically, the indications of a large number of measuring instruments on experimental research aircraft.

automatic parachute (*Aero*) A parachute for personnel which is extracted from its pack by a static line attached to the aircraft.

automatic phase control (*ImageTech*) In reproducing colour TV images, the circuit which interprets the phase of the chrominance signal as a signal to be sent to a matrix.

automatic pilot (*Aero*) A device for guiding and controlling an aircraft on a given path. It may be set by the pilot or externally by radio control. Also *autopilot*. Colloq *George*.

automatic pipette (*Chem*) See PIPETTE.

automatic quiet gain control (*ICT*) Joint use of automatic gain control and muting.

automatic reel change (*Print*) On rotary machines, equipment to attach a new reel to an old web, without stopping the machine and severing the butt end of the old web. Also *autopaster, flying paster*.

automatic screw machine (*Eng*) Fully automatic single-spindle or multiple-spindle bar stock turret lathe.

automatic shutter (*ImageTech*) In a film projector, a shutter which cuts off the light when the mechanism stops, to protect the film from heat.

automatic signalling (*CivEng*) A system of railway signalling, usually with electric control, in which the signals behind a train are automatically put to 'danger' as soon as the train has passed, and held in that position until the train has attained the next section of line.

automatic stabilizer (*Aero*) A form of automatic pilot, operating about one or more axes, adjusted to counteract dynamic instability. Also *autostabilizer*. See DAMPER.

automatic starter (*ElecEng*) A starter for an electric motor which automatically performs the various starting operations (eg cutting out steps of starting resistance) in the correct sequence, after being given an initial impulse by means of a push-button or other similar device.

automatic stoker (*Eng*) See MECHANICAL STOKER.

automatic substation (*ElecEng*) A substation containing rotating machinery which, as occasion demands, is started and stopped automatically, eg by a voltage relay which operates when the voltage falls below or rises above a certain predetermined value.

automatic synchronizer (*ElecEng*) A device which performs the process of synchronization in an AC CIRCUIT automatically.

automatic tap-changing equipment (*ElecEng*) A voltage-regulating device which automatically changes the tapping on the winding of a transformer to regulate the voltage in a desired manner.

automatic teller machine (*ICT*) See AUTOMATED TELLER MACHINE.

automatic tracking (*ImageTech*) Maintenance of head–track alignment in a HELICAL SCAN VTR over a range of playback speeds. This may be achieved by control signals recorded in the video tracks (instead of a CONTROL TRACK) or by sensing variations in radio-frequency amplitude caused by AZIMUTH RECORDING, with either being used to adjust the servomechanism or the position of heads on PIEZOELECTRIC mounts. See DYNAMIC TRACK FOLLOWING.

automatic tracking (*Radar*) Servo control of radar system operated by a received signal, to keep antenna aligned on target.

automatic train stop (*ElecEng*) A catch, used in conjunction with an automatic signalling system, which engages a trip-cock on the train if the train passes a signal at danger.

automatic transmission (*Eng*) A power transmission system for road vehicles, in which the approximately optimum engine speed is maintained through mechanical or hydraulic speed-changing devices which are automatically selected and operated by reference to the road speed of the vehicle. Additional features meet special requirements for braking, reversing, parking and unusual driving conditions.

automatic trolley reverser (*ElecEng*) An arrangement of the overhead contact line of a tramway, located at terminal points, which ensures that the trolley collector is reversed when the direction of motion of the car is reversed.

automatic tuning (*ICT*) (1) A system of tuning in which any of a number of predetermined transmissions may be selected by means of push-buttons or similar devices. (2) Fine tuning of receiver circuits by electronic means, following rough tuning by hand.

automatic voltage regulator (*ElecEng*) A voltage regulator which automatically holds the voltage of a distribution circuit or an alternator constant within certain limits, or causes it to vary in a predetermined manner. See AUTOMATIC TAP-CHANGING EQUIPMENT, MOVING-COIL REGULATOR.

automatic volume compression (*ICT*) Reduction of signal voltage range from sounds that vary widely in volume, eg orchestral music. This is necessary before they can be recorded or broadcast but ideally requires corresponding expansion in the reproducing system to compensate.

automatic volume control (*ICT*) Alteration of the contrast (dynamics) of sound during reproduction by any means. By compression (compounder) a higher level of average signal is obtained for modulation of a carrier, the expansion (expander) performing the reverse function at the receiver. In high-fidelity reproduction, arbitrary expansion can be disturbing because of variation in background noise, if present. Abbrev *AVC*.

automatic volume expansion (*ICT*) Expansion of dynamic range, eg by keeping peak level constant and automatically reducing the lower levels. Used to counteract loss of dynamic range through studio or recording equipment, or during transmission.

automatic weather station (*EnvSci*) Transistorized and packaged apparatus which measures and transmits weather data for electronic computation.

automatic white balance (*ImageTech*) A self-adjusting balancing system which monitors the lighting and corrects for changes in COLOUR TEMPERATURE. Abbrev *AWB*.

automation (*Genrl*) Industrial closed-loop control system in which manual operation of controls is replaced by servo operation.

automatism (*Psych*) An automatic act done without the full co-operation of the personality, which may even be totally unaware of its existence. Commonly seen in hysterical states, such as fugues and somnambulism, but may also be a local condition as in automatic writing.

automaton (*ICT*) A device that can take a finite number of states.

automixte system (*ElecEng*) A system of operation of petrol–electric vehicles in which a battery, connected in parallel with the generator, supplies current during starting and heavy-load periods and is charged by the generator during light-load periods. Also *Pieper system*.

automorphism (*MathSci*) A one-to-one HOMOMORPHIC mapping from a set onto itself, ie an ISOMORPHISM from a set onto itself. See HOMOMORPHISM.

automotive gas oil (*Genrl*) US term for gas oil used mainly as diesel fuel; same as the UK term DERV. Abbrev *AGO*.

autonomic (*BioSci*) Independent; self-regulating; spontaneous. Also *autonomous*.

autonomic movement (*BioSci*) Movement in parts of organisms maintained by an internal stimulus, eg beating of flagella, chromosome movements, CIRCUMNUTATION, CYCLOSIS. Cf PARATONIC MOVEMENT.

autonomic nervous system (*BioSci*) In vertebrates, a system of motor nerve fibres supplying the smooth muscles and glands of the body. See PARASYMPATHETIC NERVOUS SYSTEM, SYMPATHETIC NERVOUS SYSTEM.

autonomics (*Electronics*) Study of self-regulating systems for process control, optimizing performance.

autonomous (*BioSci*) See AUTONOMIC.

autonomous vehicle (*Aero*) Generally unmanned aircraft operating without external assistance.

auto-oxidation (*FoodSci*) The most common process by which fats are oxidized. It is a self-catalysing process which goes through three phases: initiation, the removal of hydrogen from a fatty acid chain to form a free radical; propagation, the addition of oxygen to the radical to form a peroxide, which removes hydrogen from other fatty acids to form new radicals which react with peroxide to form hydroperoxide in a chain reaction; termination, in which hydroperoxides react with oxygen and other compounds to form off-flavours and odours.

autopaster (*Print*) See AUTOMATIC REEL CHANGE.

autophosphorylation (*BioSci*) Phosphorylation of a protein kinase, possibly affecting its function, by its own enzymic activity.

autopilot (*Aero*) See AUTOMATIC PILOT.

autoplasma (*BioSci*) In tissue culture, a medium prepared with plasma from the same animal from which the tissue was taken. Adj *autoplastic*. Cf HETEROPLASMA, HOMOPLASMA.

autoplastic transplantation (*BioSci*) Reinsertion of a transplant or graft from a particular individual in the same individual. Also *autotransplantation*. Cf HETEROPLASTIC, HOMOPLASTIC.

autoplate (*Print*) A machine which can deliver a curved stereoplate for rotary printing; built to suit the requirements of each particular rotary machine.

autopodium (*BioSci*) In vertebrates, the hand or foot.

autopolyploid (*BioSci*) A POLYPLOID containing three or more BASIC CHROMOSOME SETS all from the same species.

autopsy (*Med*) See NECROPSY.

autoradiograph (*ImageTech*) Photographic record, usually of a biological specimen, produced by exposure to radiation from self-contained radioactive material which has been injected or absorbed.

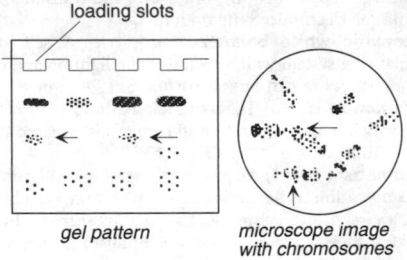

loading slots

gel pattern   microscope image with chromosomes

autoradiograph The arrows indicate the grains of photographic emulsion exposed by the decay of the label.

autoradiography (*BioSci*) Technique originally used to show the distribution of radioactive molecules in cells and tissues after injecting the organism with, or growing the cells in, a medium containing a radioactive precursor. It is now widely used to show the distribution of radiolabelled molecules separated on the basis of size, charge, etc. Photographic film or emulsion is exposed after applying it to the section, fixed cell or separating medium and the distribution of developed grains viewed directly or under the microscope. Similar procedures exist for fluorescent and other labels.

autoradiography (*Min*) Examination of radioactive minerals in thin and polished sections by allowing them to rest on photographic film to record their radioactive emissions.

auto-reclose circuit breaker (*ElecEng*) A circuit breaker which, after tripping due to a fault, automatically recloses after a time interval which may be adjusted to have any value between a fraction of a second and 1 or 2 min.

autoregressive process (*MathSci*) A process in which a random variable is regressed on past values of itself.

autorotation (*Aero*) (1) The spin; continuous rotation of a symmetrical body in a uniform airstream due entirely to aerodynamic moments. (2) Unpowered rotorcraft flight, ie a helicopter with engine stopped, in which the symmetrical aerofoil rotates at high incidence parallel with the airflow.

autosave (*ICT*) A facility that automatically saves newly recorded data at regular intervals.

autoscoper (*MinExt*) Pneumatic rock drill mounted on long cylinder with extendable ram, so as to be held firmly across opening (*stope*) when packed out by compressed air.

autoscopy (*Psych*) Hallucination of an image of one's body.

autoset level (*Surv*) A form of dumpy level for rapid operation, in which the essential features are a quick-levelling head and an optical device which neutralizes errors of levelling so that the bubbles need not be central while an observation is being made.

autoshaping (*Psych*) The classical conditioning of an OPERANT RESPONSE that is not reinforced by instrumental conditioning.

autosomal dominant (*BioSci*) A mutation in a gene located on an autosome that has a dominant effect, even though the other copy is unaltered.

autosomal recessive (*BioSci*) A mutation in a gene located on an autosome that is deleterious only in homozygotes.

autosome (*BioSci*) A chromosome that is not one of the sex-determining chromosomes. See SEX DETERMINATION.

autospasy (*BioSci*) The casting of a limb or part of the body when it is pulled by some outside agent, as when the slow-worm casts its tail. See AUTOTOMY.

autospore (*BioSci*) A non-motile spore, one of many formed within the parent algal cell and having all the characteristics of the parent in miniature before it is set free. Characteristic of some Chlorococcales.

autostabilizer (*Aero*) See AUTOMATIC STABILIZER.

autostyly (*BioSci*) In Craniata, in Dipnoi and all tetrapods, a type of jaw suspension in which the hyoid arch is broken up and the hyomandibular attached to the skull. Adj *autostylic*.

auto-suggestion (*Psych*) A mental process similar to suggestion, but originating in a belief in the subject's own mind.

autosynchronous motor (*ElecEng*) See SYNCHRONOUS INDUCTION MOTOR.

autotetraploid (*BioSci*) A polyploid containing four similar sets of chromosomes all from the same species.

autothrottle (*Aero*) A device for controlling the power of an aero-engine to keep the approach path angle and speed constant during an automatic blind landing.

autotomy (*BioSci*) Voluntary separation of a part of the body (eg limb, tail), as in certain worms, arthropods and lizards.

autotransductor (*ElecEng*) Transductor in which the same winding is used for power transfer and control.

autotransformer (*ElecEng*) Single winding on a laminated core, the coil being tapped to give desired voltages. Fig. ▷

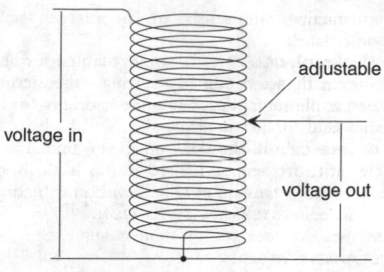

autotransformer Step-down mode shown; laminations not drawn.

autotransformer starter (*ElecEng*) A starter for squirrel-cage induction motors, in which the voltage, applied to the motor at starting, is reduced by means of an autotransformer.

autotransplantation (*BioSci*) See AUTOPLASTIC TRANSPLANTATION.

autotrophic (*BioSci*) Able to elaborate all its chemical constituents from simple, inorganic compounds (esp all its carbon compounds from CO_2). Cf HETEROTROPHIC.

autotrophic bacteria (*BioSci*) Bacteria that obtain their energy from light and inorganic compounds, and that are able to utilize carbon dioxide in assimilation.

autowinder (*ImageTech*) A device for advancing the film in a camera automatically.

autoxenous (*BioSci*) See AUTOECIOUS.

autoxidation (*Chem*) (1) The slow oxidation of certain substances on exposure to air. (2) Oxidation which is induced by the presence of a second substance, which is itself undergoing oxidation. Also *auto-oxidation*.

autoxidator (*Chem*) An alkene–oxygen compound acting as a carrier or intermediate agent during oxidation, in particular during AUTOXIDATION.

autumnal equinox (*Astron*) See EQUINOX.

autumn wood (*BioSci*) See LATE WOOD.

autunite (*Min*) Hydrated calcium uranium phosphate, yellow in colour.

auxanometer (*BioSci*) A device for recording the elongation of a plant stem, leaf, etc, traditionally by means of a lever and smoked drum.

auxiliary air intake (*Aero*) (1) An air intake for accessories, cooling, cockpit air, etc. (2) Additional intake for turbojet engines when running at full power on the ground, usually spring-loaded so that it will open only at a predetermined suction value.

auxiliary circle (*MathSci*) Of an ellipse, the circle whose diameter is the major axis of the ellipse. Of a hyperbola, the circle whose diameter is the transverse axis.

auxiliary contact (*ElecEng*) See AUXILIARY SWITCH.

auxiliary equation (*MathSci*) Of an ordinary linear differential equation $F(D)y = f(x)$, where D is the differential operator d/dx, the algebraic equation $F(m) = 0$. Also CHARACTERISTIC EQUATION.

auxiliary plant (*ElecEng*) A term used in generating-station practice to cover the condenser pumps, mechanical stokers, feed-water pumps, and other equipment used with the main boiler, turbine and generator plant.

auxiliary pole (*ElecEng*) See COMPOLE.

auxiliary power unit (*Aero*) An independent airborne engine to provide power for ancillary equipment, electrical services, starting, etc. May be a small reciprocating or turbine. Abbrev *APU*.

auxiliary rotor (*Aero*) A small rotor mounted at the tail of a helicopter, usually in a perpendicular plane, which counteracts the torque of the main rotor; used to give directional and rotary control to the aircraft.

auxiliary switch (*ElecEng*) A small switch operated mechanically from a main switch or circuit breaker; used

for operating such auxiliary devices as alarm bells, indicators, etc. Also *auxiliary contact*.

auxiliary tanks (*Aero*) See FUEL TANKS.

auxiliary winding (*ElecEng*) A special winding on a machine or transformer, additional to the main winding.

auxin (*BioSci*) A plant growth substance, indole-3-acetic acid (IAA), or any of a number of natural or artificial substances with similar effects. Auxins promote root initiation, cell elongation, xylem differentiation and may be involved in apical dominance and tropism. At high concentrations some synthetic auxins are used as herbicides. See DICHLOROPHENOXYACETIC ACID, INDOLE-3-BUTYRIC ACID, NAPHTHALENE ACETIC ACID, 2,4,5,-T.

auxochromes (*Chem*) A chromophore (or group of atoms) introduced into dyestuffs to give full effectiveness to the colouring properties. The principal auxochromes are Cl, Br, SO_3H, NO_2, NH_2, OH. Auxochromes can also permit, by being present, the formation of salts and the creation of a dyestuff. They have a selective absorption frequency for radiation, and function as colour carriers for a frequency determined mainly by the compound of which they are part.

auxocyte (*BioSci*) Any cell in which meiosis has begun; an androcyte, sporocyte, spermatocyte or oöcyte, during the period of growth.

auxometer (*Phys*) An apparatus for measuring the magnifying power of an optical system.

auxotonic (*BioSci*) Of muscle contraction, or of against increasing force.

auxotroph (*BioSci*) A variant organism requiring the addition of special nutrients or growth factors before they will grow, eg amino acid requiring bacteria. Cf PROTOTROPH.

AVA (*ICT*) Abbrev for *audiovisual aid(s)*.

availability (*ICT*) The percentage of time over a period in which a system is available for use. As distinct from DOWNTIME.

availability heuristic (*Psych*) Reliance upon information that is more readily available in memory than that which is less accessible, though the latter might possibly be more relevant.

available (*BioSci*) That part of eg water or mineral nutrient in the soil or fertilizer, which can be drawn upon by a plant. Cf UNAVAILABLE.

available light photography (*ImageTech*) Photography carried out without the use of flash or artificial lighting.

available line (*ImageTech*) Percentage of total length of scanning line on a cathode-ray-tube screen on which information can be displayed.

available potential energy (*EnvSci*) That part of the total potential energy of the atmosphere available for conversion into kinetic energy by adiabatic redistribution of its mass so that the density stratification becomes horizontal everywhere.

available power efficiency (*ElecEng*) The ratio of electrical power available at the terminals of an electro-acoustic transducer to the acoustical power output of the transducer. The latter should conform with the reciprocity principle so that the efficiency in sound reception is equal to that in transmission.

available power gain (*ICT*) The ratio of the available power output of an amplifier to the input power; equal to POWER GAIN only when the output of the device or circuit is correctly matched to the load.

available power response (*ElecEng*) For an electro-acoustic transducer, the ratio of mean square sound pressure at a distance of 1 m, in a defined direction from the 'acoustic centre' of the transducer, to the available electrical power input. The response will be expressed in dB above the reference response of $1\ \mu\text{bar}^2\ \text{W}^{-1}$ of available electrical power.

avalanche (*Phys*) Self-augmentation of ionization. See TOWNSEND AVALANCHE, ZENER EFFECT.

avalanche diode (*Electronics*) A semiconductor breakdown diode, usually silicon, in which avalanche breakdown occurs across the entire p–n junction, giving a voltage drop which is constant and independent of current. Avalanche diodes break down much more sharply than ZENER DIODES. Used in high-speed switching circuits and microwave oscillators.

avalanche effect (*Electronics*) Cumulative multiplication of carriers in a semiconductor because of avalanche breakdown. This occurs when the electric field across the barrier region is strong enough to allow production and cumulative multiplication of carriers by ionization.

avalanche photodiode (*ICT*) A photosensitive AVALANCHE DIODE used as the detector in OPTICAL FIBRE systems. Its avalanche multiplication factor improves the receiver SIGNAL-TO-NOISE RATIO by enhancing signal power without increasing thermal noise.

avalanche transistor (*Electronics*) A transistor depending on avalanche breakdown to produce hole–electron pairs. It can give very high gain in the common-emitter mode or very rapid switching.

avascular (*Med*) Not having blood vessels.

avatar (*ICT*) The graphic representation of a system user in an on-line or virtual world.

AVC (*ICT*) Abbrev for AUTOMATIC VOLUME CONTROL.

aventurine feldspar (*Min*) A variety of plagioclase, near albite-oligoclase in composition, characterized by minute disseminated particles of red iron oxide which cause fire-like flashes of colour. Also *sunstone*.

aventurine quartz (*Min*) A form of quartz spangled, sometimes densely, with minute inclusions of either mica or iron oxide. Used in jewellery. Sometimes falsely 'Indian Jade'.

average (*MathSci*) Loose term usually for ARITHMETIC MEAN.

average (*Ships*) Loss or damage of marine property, less than total: compensation payment in proportion to amount insured.

average current (*ElecEng*) That current obtained by adding together the products of currents flowing in a circuit and the times for which they flow and dividing by the total time considered, ie

$$\frac{1}{n}\sum_{j=1}^{n}|x_j - \bar{x}|$$

where the mean is as follows:

$$\bar{x} = \frac{1}{n}\sum_{j=1}^{n} x_j$$

For direct current the average value is constant: for true alternating current the average value is zero.

average curvature (*MathSci*) See CURVATURE.

average haul distance (*CivEng*) The distance between the centre of gravity of a cutting and that of the embankment formed from material excavated from the cutting.

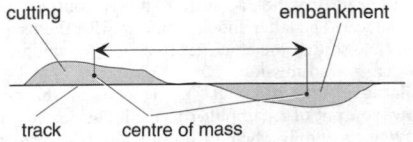

average haul distance

average power output (*ICT*) In an amplitude-modulated transmission, the radio-frequency power delivered by a transmitter, averaged over one cycle or other specified interval of the modulating signal.

aversion therapy (*Psych*) A form of behaviour therapy in which the aim is to eliminate an undesirable behaviour by pairing it with an aversive stimulus.

aversive stimulus (*Psych*) A stimulus that an animal will attempt to avoid or escape from; it can be used experimentally to punish or negatively reinforce a response.

Aves (*BioSci*) A class of CHORDATA adapted for aerial life. The forelimbs are modified as wings, the sternum and pectoral girdle are modified to serve as origins for the wing muscles, and the pelvic girdle and hind limbs to support the entire weight of the body on the ground. The body is covered with feathers and there are no teeth. Respiratory and vascular systems are modified for homeothermy. Birds. See panel on VERTEBRATE EVOLUTION.

avgas (*Aero*) Abbrev for *aviation gasoline*. See AVIATION SPIRIT.

avian big liver disease (*Vet*) See AVIAN LEUCOSIS.

avian diphtheria (*Vet*) See FOWL POX.

avian erythroblastosis (*Vet*) See AVIAN LEUCOSIS.

avian erythroid leucosis (*Vet*) See AVIAN LEUCOSIS.

avian favus (*Vet*) A fungal disease of fowls affecting the skin of the head and comb, due to *Trichophyton gallinae*. Also *white comb*.

avian influenza (*BioSci*) Also *avian flu*. See BIRD FLU.

avian gout (*Vet*) A symptom of nephritis in the domestic chicken, in which urates become deposited on the surface of internal organs (*visceral gout*) and in the joints.

avian granuloblastosis (*Vet*) See AVIAN LEUCOSIS.

avian leucosis (*Vet*) Occurs naturally only in the chicken, caused by members of the leukosis/sarcoma group of avian retroviruses. The virus is transmitted efficiently through the embryo, and infected birds become depressed prior to death. There are few typical clinical symptoms, but diffuse or nodular tumours are found in most organs. Involvement of the bursa is considered PATHOGNOMONIC and the disease is controlled by eradication. The virus is important in cancer research. Also *avian big liver disease, avian erythroblastosis, avian granuloblastosis, avian myeloblastosis, avian visceral lymphomatosis, lymphoid leukosis*.

avian monocytosis (*Vet*) An acute or subacute disease of chickens characterized by depression, loss of appetite and diarrhoea. The main pathological changes are focal necrosis of liver cells, enteritis, nephritis, and an increased number of monocytes in the blood. The cause is unknown, but a similar disease in turkeys is caused by a coronavirus. Also *blue comb, pullet disease*.

avian myeloblastosis (*Vet*) See AVIAN LEUCOSIS.

avian paratyphoid (*Vet*) A contagious disease of birds due to infection by *Bacillus aertrycke*.

avian spirochaetosis (*Vet*) Avian sleeping sickness. An acute and highly fatal septicaemia of domestic and certain wild birds, due to infection by the spirochaete *Borrelia anserina* (*B. gallinarum*); transmitted by ticks of the genera *Argas* and *Ornithodorus*.

avian typhoid (*Vet*) A contagious disease of birds due to infection by *Bacterium gallinarum*.

avian visceral lymphomatosis (*Vet*) See AVIAN LEUCOSIS.

aviation bi-phase shift keying (*ICT*) A digital MODULATION scheme in which a '1' is represented by a +90° phase transition and a '0' by a −90° transition of the CARRIER.

aviation fuels (*Aero*) Generally liquid hydrocarbons, because of high heat of combustion per unit of fuel mass (specific energy) and volume (energy density), ease of combustion, moderate volatility and viscosity, and good thermal stability and capacity. Liquid hydrogen and pentaborane (B_5H_9) have also been used experimentally. See AVIATION GASOLINE, AVIATION KEROSINE.

aviation gasoline (*Aero*) Blends of liquid hydrocarbons, almost all petroleum products boiling between 32° and 220°C, with anti-knock rating from 80 octane number to 145 performance number. Only small quantities are now used. Abbrev *AVGAS*. See AVIATION FUELS, AVIATION KEROSINE.

aviation kerosine (*Aero*) For gas turbine engines, fuel which typically boils over the range 144°–252°C. Variants include Jet A-1 (*AVTUR*), the international jet fuel; Jet B (*AVTAG*), a blend of naphtha with kerosine now being phased out except for use in cold climates; *AVCAT*, a naval jet fuel with high flash-point for safety in enclosed spaces in ships; *AVPIN*, an aviation isopropyl nitrate; and *AVGARD*, TN for an additive with anti-misting properties. See AVIATION GASOLINE.

aviation spirit (*Aero*) A motor fuel with a low initial boiling point and complying with a certain specification, for use in aircraft. Ranges from 73 to 120/130 octane rating. Abbrev *avgas*. See AVIATION KEROSINE, WIDE-CUT FUEL.

aviatrix (*Aero*) Female aviator.

avidin (*BioSci*) A protein that binds very strongly to biotin. It can be labelled by fluorescence or by attachment of enzymes, and is used to reveal antibodies to which biotin has been conjugated.

avidity (*BioSci*) A measure of the strength of binding between an antigen and an antibody. Since antigens are liable to have several combining sites the avidity is an average and less precise than affinity.

avionics (*Aero, Space*) The collective word for a spacecraft or aircraft's subsystem elements which involve electronic principles. A contraction of *aviation electronics*.

avitaminosis (*Med*) The condition of being deprived of vitamins; any deficiency disease caused by lack of vitamins.

avodiré (*For*) Tropical W African hardwood (*Turraeanthus*) whose heartwood is golden yellow with a straight to wavy grain and fine texture.

Avogadro constant (*Chem*) See AVOGADRO'S NUMBER.

Avogadro's law (*Chem*) The law that equal volumes of different gases at the same temperature and pressure contain the same number of molecules.

Avogadro's number (*Chem*) The number of atoms in 12 g of the pure isotope ^{12}C, ie the reciprocal of the ATOMIC MASS UNIT in grams. It is also by definition the number of molecules (or atoms, ions, electrons) in a MOLE of any substance and has the value $6 \cdot 02252 \times 10^{23}$ mol^{-1}. Symbol N_A. Also *Avogadro constant*.

avoidance–avoidance conflict (*Psych*) The conflict where both possible choices have an equal negative outcome. Damned if you do, and damned if you don't.

avoirdupois (*Genrl*) A system of weights in which the pound (lb) equals 16 ounces (oz).

avpin (*Aero*) Abbrev for *aviation isopropyl nitrate*.

avpol (*Aero*) Abbrev for *aviation petrol, oil and lubricant*.

Avrami equation (*Chem*) Empirical equation for describing crystallization of polymers. Relates degree of crystallinity (ξ) to time, t, as $1 - \xi = \exp(-kt)$, where k is a constant.

avtag (*Aero*) Abbrev for *aviation wide-cut turbine fuel*. See WIDE-CUT FUEL.

avtur (*Aero*) Abbrev for *aviation turbine fuel*. See AVIATION KEROSINE.

avulsion (*Med*) The tearing away of a part.

AWACS (*Aero*) Abbrev for *airborne warning and control system*.

AWB (*ImageTech*) Abbrev for AUTOMATIC WHITE BALANCE.

awl (*Build*) A small pointed or edged tool for making holes which are to receive nails or screws.

awn (*BioSci*) (1) A long bristle borne on the glumes and/or lemmas of some grasses, eg barley. (2) A similar structure on another organ.

awning deck (*Ships*) A superstructure deck, as the name implies. In its simplest form, it is the top deck of a two-deck ship, and places the ship in a certain category for scantling and freeboard.

axe (*Build*) A pointed hammer used for dressing stone.

axed arch (*Build*) An arch built from bricks cut to a wedge shape.

axed work (*Build*) Hard building stone dressed with an axe to leave a ribbed face.

axenic culture (*BioSci*) A culture of a single species in the absence of all others; pure culture.

axes (*Genrl*) Pl of AXIS.

axes (*MathSci*) (1) Of co-ordinates: the fixed reference lines used in a system of co-ordinates. (2) Of a conic: that pair of conjugate diameters which are mutually perpendicular. For an ellipse they are referred to as *major* and *minor* in accordance with their length. For a hyperbola the one

which does not cut the curve is called the *conjugate* axis, and the other the *transverse* axis.

axial (*Arch*) A term which implies that a building is symmetrical about a central axis.

axial (*BioSci*) Relating to the axis of a plant or organ; longitudinal.

axial compressor (*Eng*) A multistage, high-efficiency compressor comprising alternate rows of moving and fixed blades attached to a rotor and its casing respectively.

axial engine (*Aero*) Turbine engine with an axial-flow compressor.

axial-flow compressor (*Aero*) A compressor in which alternate rows of radially mounted rotating and fixed aerofoil blades pass the air through an annular passage of decreasing area in an axial direction.

axial-flow turbine (*Aero*) Characteristic aero-engine turbine, usually of 1–3 rotating stages, in which the gas flow is substantially axial.

axial pitch (*Eng*) The distance from any point on one thread or helix to the corresponding point on the next thread or helix measured along the axis of the screw or helix.

axial-plane cleavage (*Geol*) Cleavage parallel to the axial plane of a fold.

axial ratio (*Phys*) The ratio of major to minor axis of polarization ellipse for eg a wave propagated in waveguide, polarized light. Also *ellipticity*.

axial response (*Acous*) The response of a microphone or loudspeaker, measured with the sound-measuring device on the axis of the apparatus being tested.

axial runout (*Eng*) Variation from the plane normal to its axis of a rotating part. Its wobble, rather than its eccentricity. Cf RADIAL RUNOUT.

axial skeleton (*BioSci*) The skeleton of the head and trunk; in vertebrates, the cranium and vertebral column, as opposed to the appendicular skeleton.

axial tomography (*Med*) A technique of tomography using detectors that rotate round an axis, as of the body. See COMPUTER-AIDED TOMOGRAPHY.

axiate pattern (*BioSci*) The morphological differentiation of the parts of an organism, with reference to a given axis.

axil (*BioSci*) The upper hollow where the adaxial surface of leaf or bract attaches to a stem.

axile (*BioSci*) Coinciding with the longitudinal axis.

axilemma (*BioSci*) In medullated nerve fibres, the whole of the medullary sheath.

axile placentation (*BioSci*) In plants, placentation in which the placentas are in the angles formed where the septa along the central axis of an ovary meet the two or more locules, eg tomato.

axilla (*Med*) The arm-pit; the angle between the fore limb and the body. Adj *axillary*.

axillary (*BioSci*) Situated in or arising from an axil, esp of buds, shoots, flowers, inflorescences, etc.

axillary air sac (*BioSci*) In birds, one of the paired air sacs, lying in the axillary position. It communicates with the median interclavicular air sac.

axinite (*Min*) A complex borosilicate of calcium and aluminium, with small quantities of iron and manganese, produced by pneumatolysis and occurring as brown wedge-shaped triclinic crystals.

axiom (*MathSci*) One of the basic assumptions underlying a particular branch of mathematics. Cf POSTULATE.

axiotron (*Electronics*) Valve in which the electron stream to the anode is controlled by the magnetic field of the heating current.

axis (*Aero*) One of the three axes of an aircraft, which are the straight lines through the centre of gravity about which change of attitude occurs: *longitudinal* or *drag* axis in the plane of symmetry (roll); *normal* or *lift* axis vertically in the plane of symmetry (yaw); and the *lateral* or *pitch* axis transversely (pitch). Pl *axes*. See WIND AXES.

axis (*BioSci*) (1) A central line of symmetry of an organ or organism. (2) A stem or root. (3) A rachis. (4) In higher vertebrates, the second cervical vertebra. Pl *axes*.

axis (*ImageTech*) Of a lens, the line of symmetry of the optical system; the line along which there is no refraction. Pl *axes*.

axis (*MathSci*) A line that has a peculiar importance in relation to a particular problem or set of circumstances. Thus the *axis of symmetry* of a figure is a line which divides it symmetrically, and the *axis of rotation* is the line about which a body rotates. Pl *axes*. See also AXES.

axle (*Eng*) The cross-shaft or beam which carries the wheels of a vehicle; they may be either attached to and driven by it, or freely mounted thereon.

axle-box (*Eng*) Box-shaped housing containing the axle bearings and lubricant. Constrained laterally on guides and supports the weight of the vehicle through springs.

axle weight (*Eng*) That part of the all-up weight of a vehicle which is borne by the wheels on one particular axle.

axon (*BioSci*) The process of a typical nerve cell or neuron which transmits an impulse or action potential away from the cell body. See fig. at NEURON.

axoneme (*BioSci*) The central microtubule complex of eukaryotic cilia and flagella with the characteristic '9 + 2' arrangement of tubules when seen in cross-section. See diagram of CILIA.

axonometric projection (*Arch*) A three-dimensional representation of an architectural drawing in which all parallel lines remain parallel and to scale, unlike a perspective, the lines of which converge. Axonometric drawings fall into various categories depending on the angle of projection. See ISOMETRIC PROJECTION.

axonometry (*Crystal*) Measurement of the axes of crystals.

axoplasm (*BioSci*) The cytoplasm of a neuron, particularly that in axonal processes.

az-, azo- (*Genrl*) Prefixes from Gk *a-* (privative), *zaein* (to live) denoting nitrogen.

azeleic acid (*Chem*) *Heptan 1,7-dicarboxylic acid*. $COOH(CH_2)_7COOH$. Mp 106°C. Found in rancid fat. Prepared by the oxidation of oleic acid.

azeotropic distillation (*ChemEng*) A process for separating, by distillation, products not easily separable otherwise. The essential is the introduction of another substance, called the *entrainer*, which then forms an AZEOTROPIC MIXTURE, increasing the relative volatility of one of the compounds making the separation relatively easy.

azeotropic mixtures (*Chem*) Liquid compounds whose boiling point, and hence composition, does not change as vapour is generated and removed on boiling. The boiling point of the azeotropic mixture may be lower (eg water–ethanol) or higher (eg water–hydrochloric acid) than those of its components. Also *constant-boiling mixtures*.

azides (*Chem*) (1) See ACID AZIDES. (2) Salts of hydrazoic acid. The heavy-metal azides are explosive.

azidothymidine (*Pharmacol*) See AZT.

azimino compounds (*Chem*) Heterocyclic compounds containing three adjacent nitrogen atoms in one ring; very stable. They are prepared by the action of nitrous acid on 1,2-diamines or 1,8-diaminonaphthalenes.

azimuth (*Astron, MathSci, Surv*) The angle between the vertical plane containing a line or a curve such as the locus of a celestial body and the plane of the meridian, conventionally measured from north through east in astronomical computations, and from south through west in triangulation and precise traverse work.

azimuth (*ICT*) See BEARING.

azimuth (*ImageTech*) The angle, normally 90°, between the direction of motion of the film or tape and the slit or gap in the optical or magnetic head. See fig. at AZIMUTH RECORDING.

azimuthal power instability (*NucEng*) Abnormal neutron behaviour which results in uneven nuclear conditions in the reactor.

azimuthal projection (*Genrl*) The map projection in which a portion of the globe is projected upon a plane tangent to

it, usually at the pole, or a place which is to be the centre of the map.

azimuth angle (*Surv*) Horizontal angle of observed line with reference to true north.

azimuth marker (*Radar*) Line on radar display made to pass through target so that the bearing may be determined.

azimuth recording (*ImageTech*) Employing heads with opposed AZIMUTH angles to minimize crosstalk between adjacent tracks, each head attenuating the other recorded signal. Removes the need for GUARD BANDS. Also *slant-azimuth recording*.

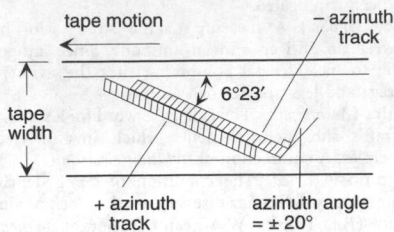

azimuth recording Two adjacent helical scans shown.

azimuth-stabilized PPI (*Radar*) Form of plan position indicator display which is stabilized by a gyrocompass, so that the top of the screen always corresponds to north.

azines (*Chem*) Organic bases containing a heterocyclic aromatic ring of four carbon and two nitrogen atoms, the nitrogen atoms being in the *para*-position with respect to one another.

azo- (*Genrl*) See AZ-.

azobenzene (*Chem*) $C_6H_5N=NC_6H_5$. Orange-red crystals, mp 68°C. Prepared by the partial reduction of nitrobenzene.

azodicarbonamide (*Eng*) Blowing agent used in structural foam moulding to create foam core. Decomposes at about 190°C to give CO, CO_2 and N_2 gases. Abbrev *ADC*.

azo dyes (*Chem, FoodSci*) Artificial colourings. Derivatives of azobenzene, obtained as the reaction products of diazonium salts with tertiary amines or phenols (hydroxy-benzenes). Usually coloured yellow, red or brown, they have acidic or basic properties. Now being superseded as food additives by natural colours.

azo group (*Chem*) The group —N=N—, generally combined with two aromatic radicals. The azo group is a chromophore, and a whole class of dyestuffs is characterized by the presence of this group.

Azoic (*Geol*) That part of the Precambrian without life.

azomethane (*Chem*) $CH_3N=NCH_3$, bp 1.5°C, a yellow liquid, obtained by the oxidation of *sym*-dimethyl hydrazine with chromic (VI) acid.

azonal soil (*EnvSci*) Immature soil. Cf INTRAZONAL SOIL.

azonium bases (*Chem*) A group of bases including AZINES and QUINOXALINES.

azoospermia (*Med*) Complete absence of spermatozoa in the semen.

azotaemia (*Med*) An excess of nitrogenous breakdown products in the blood. See URAEMIA.

azote (*Chem*) The French name for NITROGEN.

azothioprine (*Pharmacol*) A drug used to suppress the immune response as in rejection after transplantation and in a variety of connective tissue disorders.

Azotobacter (*BioSci*) A free-living genus of bacteria found in soil and water. The bacteria are able to fix free nitrogen in the presence of carbohydrates.

azoturia (*Vet*) An acute degeneration of muscles in horses of unknown cause, occurring particularly during exercise after a few days of idleness. Characterized by stiffness, lameness and paralysis, hardness of affected muscles, and urine coloured red to dark brown due to myoglobin. Also *paralytic equine myoglobinuria*.

azoxy compounds (*Chem*) Mostly yellow or red crystalline substances obtained by the action of alcoholic potassium hydroxide upon the nitro compounds, or by the oxidation of azo compounds.

AZT (*Pharmacol*) Abbrev for AZIDOTHYMIDINE, an antiviral drug derived from thymidine, used in the treatment of AIDS and conditions associated with it. It blocks the enzyme that stimulates growth and multiplication of the human immunodeficiency virus (HIV), but is not a cure for AIDS. Also *zidovudine*.

azurite (*Min*) A deep-blue hydrated basic carbonate of copper, occurring either as monoclinic crystals or as kidney-like masses built of closely packed radiating fibres.

azusa (*ICT*) US radio-tracking system for missile guidance.

azygomatous (*BioSci*) Lacking a zygomatic arch. See ZYGOMA.

azygos (*BioSci*) An unpaired structure. Adj *azygous*.

azygospore (*BioSci*) A structure resembling a zygospore in morphology, but not resulting from a previous sexual union of gametes or of GAMETANGIA. Also *parthenospore*.

B

B (*Chem*) Symbol for BORON.

B (*Phys*) Symbol for: (1) SUSCEPTANCE in an ac circuit (unit is the siemens; measured by the negative of the reactive component of the admittance); (2) MAGNETIC FLUX DENSITY in a magnetic circuit (unit is the tesla; $1\,T = 1\,Wb\,m^{-2} = 1\,V\,s\,m^{-2}$).

[B] (*Phys*) A FRAUNHOFER LINE in the red of the solar spectrum, due to absorption by the Earth's atmosphere. [B] is actually a close group of lines having a head at a wavelength of 686·7457 nm.

b- (*Chem*) Symbol for: (1) substitution on the carbon atom of a chain next but one to the functional group; (2) substitution on a carbon atom next but one to an atom common to two condensed aromatic nuclei; (3) substitution on the carbon atom next but one to the heteroatom in a heterocyclic compound; (4) a stereoisomer of a sugar.

β (*Phys*) See under BETA. Symbol for: (1) phase constant; (2) ratio of speed to speed of light.

β- (*Phys*) The intermediate refractive index in a biaxial crystal.

BA (*Eng*) See BRITISH ASSOCIATION SCREW-THREAD.

Ba (*Chem*) Symbol for BARIUM.

Babbitt's metal (*Eng*) A bearing alloy originally patented by Isaac Babbitt, composed of 50 parts of tin, 5 of antimony and 1 of copper. Addition of lead greatly extends range of service. Composition varies widely, with tin 5–90%; copper 1·5–6%; antimony 7–10%; lead 5–48·5%.

Babcock and Wilcox boiler (*Eng*) A water-tube boiler consisting in its simplest form of a horizontal drum from which is suspended a pair of headers carrying between them an inclined bank of straight tubes.

Babesia (*BioSci*) A genus of protozoal parasites which occur in the erythrocytes of mammals.

babesiosis (*Vet*) A disease caused by protozoa of the genus *Babesia*. Also *piroplasmosis*. See REDWATER.

Babinet's compensator (*Phys*) A device used, in conjunction with a Nicol prism, for the analysis of elliptically polarized light. It consists of two quartz wedges having their edges parallel and their optical axes at right angles to each other.

Babinet's principle (*Phys*) A principle that the radiation field beyond a screen which has apertures, added to that produced by a complementary screen (in which metal replaces the holes, and spaces the metal), is identical to the field which would be produced by the unobstructed beam of radiation, ie the two diffraction patterns will also be complementary.

Babinski's sign (*Med*) See EXTENSOR PLANTAR RESPONSE.

Babo's law (*Phys*) The vapour pressure of a liquid is lowered when a non-volatile substance is dissolved in it, by an amount proportional to the concentration of the solution.

baby (*ImageTech*) A small incandescent spotlight used in film and TV production.

baby beef (*Agri*) High-quality meat cattle slaughtered at between 9 and 15 months of age.

baccate (*BioSci*) Resembling a berry.

Bacillaceae (*BioSci*) A family of bacteria included in the order *Eubacteriales*. Many are able to produce highly resistant endospores. They are large Gram-positive rods and may be aerobic or anaerobic. The family includes many pathological species, eg *Bacillus anthracis* (anthrax). See CLOSTRIDIUM.

bacillaemia (*Med*) The presence of bacilli in the blood. US *bacillemia*.

Bacillariophyceae (*BioSci*) The diatoms, a class of eukaryotic algae in the division Heterokontophyta, with two orders: Centrales and Pennales. The cell wall or frustule contains silica. The majority are unicellular, with some forming colonies and chains of cells. Most are phototrophic (sometimes auxotrophic); some are heterotrophs. They are ubiquitous in both fresh water and marine environments and in soils. May be planktonic, benthic or epiphytic. Fossil deposits of frustules constitute diatomaceous earth or kieselguhr. Also *Diatomophyceae*.

bacillary necrosis (*Vet*) See NECROBACILLOSIS.

bacillary white diarrhoea (*Vet*) See PULLORUM DISEASE.

Bacille Calmette–Guerin (*BioSci*) An attenuated mycobacterium derived from *Mycobacterium tuberculosis*. The bacterium is used in tuberculosis vaccination. Extracts of the bacterium have remarkable adjuvant properties. Abbrev BCG.

bacillemia (*Med*) US for BACILLAEMIA.

bacilluria (*Med*) Presence of bacilli in the urine.

bacillus (*BioSci*) (1) A rod-shaped member of the bacteria. (2) Genus in the family *Bacillaceae*. Pl *bacilli*.

Bacillus thuringiensis (*BioSci*) A soil-living bacterium that produces an endotoxin deadly to insects. Many strains exist, each with great specificity as to its target orders of insects. The gene for the toxin has been artificially engineered into plants to make them resistant to insect pests, although the use of such genetically modified (GM) crops is controversial.

back (*Genrl*) A large vat used in various industries, such as dyeing, soap-making, brewing. Also *beck*.

back ampere-turns (*ElecEng*) That part of the armature ampere-turns which produces a direct demagnetizing effect on the main poles. Also *demagnetizing ampere-turns*.

back annealing (*Eng*) Controlling the softening of a fully work-hardened metal so as to produce the desired degree of temper by partial recrystallization. See ANNEALING, TEMPER.

back band (*Build*) The outside member of a door or window casing.

backbone (*ICT*) The major long-distance, multichannel link in a network, from which smaller links branch off.

backbone network (*ICT*) A high-capacity network that links together other networks of lower capacity. Fibre optic cables are often used to form these links.

back coupling (*ICT*) Any form of coupling that permits the transfer of energy from the output circuit of an amplifier to its input circuit. See FEEDBACK, REGENERATION.

backcross (*BioSci*) The mating of an individual to one of its parents or parental strains. In MENDELIAN GENETICS a mating of a HETEROZYGOTE to the recessive HOMOZYGOTE, producing a 1:1 ratio in the progeny.

back diode (*Electronics*) See BACKWARD DIODE.

back edging (*Build*) A method of cutting a tile or brick by chipping away the biscuit below the glazed face, the front itself being scribed.

back emf (*ElecEng*) The emf which arises in an inductance (because of rate of change of current), in an electric motor (because of flux cutting), or in a primary cell (because of polarization) or in a secondary cell (when being charged). Also *counter emf*.

back-emf cells (*ElecEng*) Cells connected into an electric circuit in such a way that their emf opposes the flow of current in the circuit.

back emission (*Electronics*) Emission of electrons from the anode.

back-fire (*Autos*) (1) Premature ignition during the starting of an internal-combustion engine, resulting in an explosion before the end of the compression stroke, and consequent reversal of the direction of rotation. (2) An explosion of live gases accumulated in the exhaust system due to incomplete combustion in the cylinder.

backfitting (*NucEng*) Making changes to nuclear (and other) plants already designed or built, eg to cater for changes in safety criteria.

back-flap hinge (*Build*) A hinge in two square leaves, screwed to the face of a door which is too thin to permit the use of a butt hinge.

back focus (*ImageTech*) The distance between the rear surface of a lens and the image of an object at infinity.

back gear (*Eng*) A speed-reducing gear fitted to the headstock of a belt-driven metal-turning lathe. It consists of a simple layshaft, which may be brought into gear with the coned pulley and mandrel when required.

background (*Phys*) Extraneous signals arising from any cause which might be confused with the required measurements; eg in electrical measurements of nuclear phenomena and of radioactivity, it would include counts emanating from amplifier noise, cosmic rays and insulator leakage. Cf SIGNAL-TO-NOISE RATIO.

background job (*ICT*) Job having a low priority within a multiprogramming system. See FOREGROUND/BACK-GROUND PROCESSING, JOB QUEUE, TIME SHARING.

background noise (*Acous*) Extraneous noise contaminating sound measurements and which cannot be separated from wanted signals. Residual output from eg microphones, pickups, lines, giving a SIGNAL-TO-NOISE RATIO. Also *ground noise*.

background radiation (*Astron*) See MICROWAVE BACK-GROUND.

background radiation (*Radiol*) Radiation coming from sources other than that being observed.

background video (*ImageTech*) A technique for overlaying video on previously recorded depth multiplex audio. Abbrev *BGV*. Also *video on sound* (abbrev *VOS*). See DEPTH MULTIPLEX RECORDING.

backhand welding (*Eng*) Welding in which the torch or electrode hand faces the direction of travel, thus post-heating the existing weld. Cf FOREHAND WELDING.

backheating (*Electronics*) Excess heating of a cathode due to bombardment by high-energy electrons returning to the cathode. In magnetrons, it may be sufficient to keep the cathode at operating temperature without external heating.

backing (*EnvSci*) The changing of a wind in an antic-lockwise direction. Cf VEERING.

backing (*ImageTech*) Light-absorbent layer on the rear surface of photographic film or plate to reduce HALATION.

backing (*Print*) The binding process by which one-half of the sections at the back of a volume are bent over to the right and the other half to the left. The projections formed are JOINTS, to which the case is hinged, by hand or machine. Some machines also simultaneously perform ROUNDING.

backing boards (*Print*) Wedge-shaped wooden boards between which an unbound book is held in the lyingpress, while the joints are being formed for attaching the case.

backings (*Build*) Wooden battens, secured to rough walls, for the fixing of wood linings etc. Also *strapping*.

backing store (*ICT*) Means of storing large amounts of data outside the main memory. Will be a combination of MAGNETIC DISK, MAGNETIC DRUM, MAGNETIC TAPE, OPTICAL DISK, containing archive files, BACK-UP FILES. Also *secondary memory*.

backing-up (*Build*) The use of inferior bricks for the inner face of a wall.

backing-up (*Print*) (1) Printing on the second side of a sheet. (2) Backing a letterpress printing plate to required height.

back inlet gulley (*Build*) A trapped gulley in which the inlets discharge under the grating and above the level of the water in the trap, so that splashing and blocking of the grating are avoided.

back iron (*Build*) The stiffening plate screwed to the cutting iron of a plane. Also *cap iron*, *cover iron*.

back joint (*Build*) The part of the back of a stone step which is dressed to fit into the rebate of the upper step.

back-kick (*Eng*) The violent reversal of an internal-combustion engine during starting, due to a BACK-FIRE.

backlash (*Eng*) The lost motion between two elements of a mechanism, ie the amount the first has to move, owing to imperfect connection, before communicating its motion to the second.

backlash (*ICT*) (1) Mechanical deficiency in a tuning control, with a difference in dial reading between clockwise and anticlockwise rotation. (2) Property of most regenerative and oscillator circuits, by which oscillation is maintained with a smaller positive feedback than is required for inception.

backlight (*Electronics*) The light source (often a cold-cathode discharge in a flat fluorescent envelope) used in some light-modulating flat-panel displays such as those based on LIQUID CRYSTALS. See panel on LIQUID CRYSTAL DISPLAY.

backlight compensation (*ImageTech*) The opening of the IRIS to expose correctly a backlit subject which would otherwise be a silhouette. Abbrev *BLC*.

back lighting (*ImageTech*) Lighting illuminating the subject from behind, opposite the camera, often to provide rim light or halo effects.

back lining (*Print*) See HOLLOWS.

back lobe (*ICT*) Lobe of polar diagram for antenna, microphone, etc, that points in the reverse direction to that required.

backlocking (*CivEng*) Holding a signal lever partially restored until completion of a predetermined sequence of operations.

backmatter (*Print*) US for END MATTER.

back-mixing (*ChemEng*) In a chemical reactor, jet engine or other apparatus through which material flows and thereby undergoes a change in some property, the turning back of some of the material so that it mixes with that which has entered the reactor after it.

back-mutation (*BioSci*) See REVERSION.

back observation (*Surv*) An observation made with an instrument on station just left. Also *back sight*.

back-off (*MinExt*) (1) To raise the drilling bit or down-hole assembly for a short distance from the bottom of the hole. (2) To unscrew drilling components.

back porch (*ImageTech*) A short period of BLACK-LEVEL signal transmitted at the end of the horizontal SYNC PULSE before the picture information.

back-porch effect (*ICT*) The prolonging of the collector current in a transistor for a brief time after the input signal (particularly if large) has decreased to zero.

back pressure (*Build*) Air pressure in drainage pipes exceeding atmospheric pressure.

back pressure (*Eng*) (1) The pressure opposing the motion of the piston of an engine on its exhaust stroke. (2) The exhaust pressure of a turbine. Increased by clogged or defective exhaust system.

back pressure (*Med*) Proximal pressure produced by an obstruction to fluid flowing through the cardiovascular or urinary systems.

back pressure (*MinExt*) The hydrostatic pressure exerted by the fluids in the bore of a well which acts against that of the oil and gas in the strata. Control of this pressure by valves and CHOKES maintains an even and productive flow of oil.

back-pressure turbine (*Eng*) A steam turbine from which the whole of the exhaust steam, at a suitable pressure, is taken for heating purposes.

back projection (*ImageTech*) (1) Projection of a picture, from film, transparency or video, onto a translucent screen to be viewed from the opposite side. (2) A form of motion picture composite photography in which the projected picture forms the background to action taking place in front of it, both being photographed together.

back rake (*Eng*) In a lathe tool, the inclination of the top surface or face to a plane parallel to the base of the tool.

back saw (*Build*) A saw stiffened by a thickened back, eg a tenon saw.

back-scatter (*Phys*) The deflection of radiation or particles by scattering through angles greater than 90° with reference to the original direction of travel. Cf FORWARD SCATTER.

backsetting (*Build*) A stone with a boasted or broached face, but with smooth border all round.

back shore (*Build*) One of the outer members of an arrangement of raking shores or props, for supporting temporarily the side of a building. The back shore supports the RAKING SHORE, or raker, which takes the thrust from the highest part of the building.

back sight (*Surv*) See BACK OBSERVATION.

backspacing (*ImageTech*) The process which maintains SYNCHRONIZATION when video recording is stopped and started, the tape being rolled back for roughly 1 second at the end of a recorded segment then switched into play to compare and synchronize the CONTROL TRACK pulses with the incoming synchronization pulses before recording begins again. See PRE-ROLL.

back-step (*Print*) See BLACK-STEP MARKS.

backstop (*ElecEng*) The structure of a relay which limits the travel of the armature away from the pole piece or core.

back stopes (*MinExt*) OVERHAND STOPES worked upslope. Cf UNDERHAND STOPES.

back titration (*Chem*) In volumetric analysis, the technique of adding a reagent in excess, to exceed the end-point, then determining the end-point by titrating the excess.

back-to-back (*Electronics*) Parallel connection of valves, with the anode of one connected to the cathode of the other, or transistors in parallel in opposite directions, to allow control of alternating current without rectification. Frequently used with THYRATRONS and IGNITRONS.

back-to-back test (*ElecEng*) The arrangement, originated by Hopkinson, for testing two substantially similar electrical machines on full load, by coupling them mechanically and loading them by regulating the electrical circuit, the total power supply accounting for total losses only; it has been extended to transformers and mechanical gearing.

back-up file (*ICT*) Copy of a file, held inside a computer system, to be used in the event of the current file being corrupted or lost.

Backus–Naur form (*ICT*) Notation for defining the syntax of a programming language. Abbrev BNF.

backward busying (*ICT*) Applying busy condition at the incoming end of a trunk or junction (usually during testing or fault clearance) to indicate at outgoing end that circuit must not be used.

backward chaining (*ICT*) An inference method whereby a system starts with a defined goal or an outcome to prove and tries to establish the facts to do so. In the context of an EXPERT SYSTEM, the system will select appropriate rules and attempt to reach its defined goal by proving the conditions of this rule are true. Cf FORWARD CHAINING.

backward diode (*Electronics*) One with characteristic of reverse shape to normal. Also *AU diode, back diode*.

backward explicit congestion notification (*ICT*) A method of dealing with overloading in a FRAME RELAY or ASYNCHRONOUS TRANSFER MODE network by setting a specific bit in the header of a FRAME or CELL sent back from the destination to the source of traffic. On receiving this frame, the source reduces its offered load.

backward hold (*ICT*) A method of interlocking the links of a switching chain by originating a locking condition in the final link and extending it successively backwards to each of the preceding links.

backward shift (*ElecEng*) Movement of the brushes of a commutating machine around the commutator, from the neutral position, and in a direction opposite to that of the rotation of the commutator, so that the brushes short-circuit zero-emf conductors when the load current, through armature reaction, results in a rotation of the neutral axis of the air-gap flux. Shifting the brushes in this way reduces sparking on the commutator. Also *backward lead*.

backward signalling (*ICT*) Signalling from the called to the calling end of a circuit.

backward wave (*Electronics*) In a travelling-wave tube, a wave with group velocity in the opposite direction to the electron stream. Cf FORWARD WAVE.

backward-wave tube (*ICT*) General term for a family of microwave TRAVELLING-WAVE TUBES in which energy on a slow-wave circuit or structure, linked closely to the electron beam, flows in the opposite direction to the electrons. They can be used as stable, low-noise amplifiers or as oscillators; as the latter, they can be easily tuned over a wide frequency range by altering the beam voltage.

backwashing (*Chem*) A method of cleaning a filter or exchange column by reversing the flow.

back-water (*Genrl*) Water dammed back in a stream or reservoir by some obstruction.

backwater (*Paper*) Water, containing fine fibres, loading and other additives, removed in the forming section of a paper or board-making machine. It is generally reused within the system or clarified in a saveall to recover suspended matter.

back-water curve (*Build*) The longitudinal profile of the water surface in a channel obstructed by a weir, when the water surface is not parallel to the invert.

back wave (*ICT*) See SPACING WAVE.

BAC library (*BioSci*) Abbrev for *bacterial artificial chromosome library*. A DNA library constructed using a vector with an origin of replication that allows it to propagate in bacteria as an extra chromosome. Relatively large DNA fragments (100–300 kb) can be incorporated.

baclofen (*Pharmacol*) A skeletal muscle relaxant used to relieve muscle spasm in trauma, multiple sclerosis and cerebral palsy.

BACS (*ICT*) Abbrev for *banks automated clearing system*. A system of electronic data interchange used by UK banks.

bacteraemia (*Med*) Presence of bacteria in the blood. US *bacteremia*.

bacteria (*BioSci*) A large group of unicellular or multicellular prokaryotic organisms, lacking chlorophyll, multiplying rapidly by simple fission, some species developing a highly resistant resting ('spore') phase. See panel on BACTERIA.

bacteria beds (*Build*) Layers of a filtering medium such as broken stone, used in the final or oxidizing stage in sewage treatment. See CONTACT BED, PERCOLATING FILTER.

bacterial artificial chromosome (*BioSci*) See BAC LIBRARY.

bacterial flagella (*BioSci*) Thin filaments, composed of *flagellin* subunits, that are rotated by a basal motor assembly and act as propellers. If the flagellum is rotating anticlockwise (as viewed from the tip) the bacterium moves in a straight path; if clockwise the bacterium 'tumbles'.

bacterial leaching (*MinExt*) See MICROBIOLOGICAL MINING.

bacterial recovery (*MinExt*) See MICROBIOLOGICAL MINING.

bactericide (*BioSci*) A substance that destroys bacteria. Examples are halogens and halogen-releasing compounds, quarternary ammonium compounds, ozone and ultraviolet radiation. Adj *bactericidal*. Cf BACTERIOSTAT.

bacteriochlorophyll (*BioSci*) Varieties of chlorophyll found in photosynthetic bacteria that differ from plant

Bacteria

A large group of unicellular or multicellular organisms that constitute the PROKARYOTA. They are distinguished from the Eukaryota in being generally small (around 1 μm) and the absence of internal subdivisions of the cell. There is no nucleus, and the DNA is circular, not associated with histones, and free within the cytoplasm. Ribosomes are 70S in size (cf eukaryotic 80S). The bacteria are now generally subdivided into two major classes, the Eubacteria (Bacteria) and the Archaebacteria (Archaea), that are considered by some to be separate kingdoms. The Archaebacteria (Archaea) are in some ways more similar to eukaryotes and are distinct in a variety of ways. Many are extremophiles, living at high temperatures (eg in hot springs or around black smokers).

Eubacteria live in a remarkably diverse range of environments, using almost all possible means of eking out a living, and are arguably the most successful organisms. There are phototrophs (particularly in the Cyanobacteria, formerly known as blue-green algae), heterotrophs, lithotrophs and all variants of these. Oxidative phosphorylation occurs across the plasma membrane. They are bounded by a cell wall situated outside and separated from the plasma membrane by a periplasmic space; the characteristics of this wall are important in classification. Gram-positive bacteria stain with Gram's method and have a simpler kind of cell wall bounded by an outer membrane containing lipopolysaccharide (endotoxin), whereas Gram-negative bacteria have a more complex wall containing teichoic acid and peptidoglycan. Further classification is often based upon nutritional requirements for growth or existence of particular antigens (serotyping). More modern approaches use DNA markers and as a result many more distinct species are being identified. In

shape they can be spherical, rod-like, spiral or filamentous. Some are motile by means of flagella and they occur in every natural habitat often in large numbers, as much as 10^8 per gram. Not only are bacteria important as pathogens but their activities are essential for eg nitrogen fixation, and by means of genetic manipulation they are now being used to produce eukaryotic proteins such as human insulin. Some species develop a highly resistant resting ('spore') phase and require high temperatures to kill them.

Bacteria can be infected with viruses called *bacteriophage* or *phage* which commonly occur in two states in their host, depending on the strains of host and phage. In the first or *lysogenic* state, the genetic material of the phage after penetrating its host either becomes incorporated into the bacterial DNA or remains as a separate *episome* in the bacterial cytoplasm. In either case the phage DNA replicates in step with its host's DNA and it does not destroy the host. In the second or *lytic* state, the infective DNA multiplies rapidly and covers itself with a protein coat, using the host's synthetic machinery, before rupturing the cell wall and dispersing into the medium. External stimuli can convert the lysogenic to the lytic state and episomes can also exist in the cytoplasm as multiple copies called *plasmids*. Lysogenic phage or plasmids can confer new properties on their hosts such as the ability to produce a toxin or resistance to further attack by related lytic phage. In the laboratory, phage can be used to transfer DNA sequences from a previous host to a new one, a process called *transduction*.

In an analogous process, *transformation*, pure DNA can be ingested and recombined with the host DNA to alter its genome. Finally, in a 'sexual' process DNA can be transferred from an a+ to a − strain in a few bacterial types. All these methods rely on selection by the research worker for the easy demonstration of gene transfer and it is common to have a marker for, say, antibiotic resistance joined to the DNA sequence being transferred to allow elimination of unaltered cells.

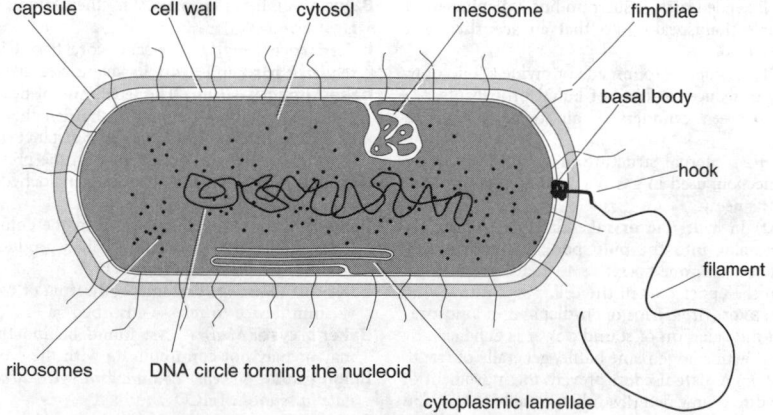

Bacteria Generalized drawing showing features found in some but not all species. The basal body, hook and filament are part of the flagellum.

chlorophyll in the substituents around the tetrapyrrole nucleus of the molecule, and in the absorption spectra.

bacteriocin (*BioSci*) A toxin that is produced by one class of bacteria and that kills another, usually related, class.

bacteriology (*Genrl*) The scientific study of bacteria.

bacteriophage (*BioSci*) A VIRUS that infects bacteria. Also *phage*.

bacteriorhodopsin (*BioSci*) A light-driven proton-pumping protein (248 residues, 26 kDa), similar to eukaryotic rhodopsin, found in 'purple patches' in the cytoplasmic membrane of the bacterium *Halobacterium halobium*. Light absorption drives the formation of a proton gradient across the membrane that is then used to drive the synthesis of ATP.

bacteriostat (*BioSci*) A substance or process that prevents the growth of micro-organisms but does not kill them.

bacteriotoxin (*BioSci*) A toxin destructive to bacteria.

bacteriotropin (*BioSci*) A substance, usually of blood serum, that makes bacteria more likely to be killed or phagocytosed (an *opsonin*).

bacteroid (*BioSci*) The enlarged, X- or Y-shaped, nitrogen-fixing form of a *Rhizobium* bacterium within a legume root nodule.

Bacteroidaceae (*BioSci*) A family of pleomorphic bacteria included in the order *Eubacteriales*. They are usually anaerobic Gram-negative rods, and inhabit the intestine and mucous membrane of higher vertebrates.

bacteruria (*Med*) The presence of bacteria in the urine.

Baculovirus (*BioSci*) Any of a group of large DNA viruses, eg nuclear polyhedrosis virus (NPV), granulosis virus (GV), that are pathogenic to insects and have been used in biological control of insect pests. One to many VIRIONS occur within a thick protein coat and are large enough (NPV) to be visible by light microscopy. By experimental insertion of appropriate DNA, the system can be genetically engineered to produce large amounts of eukaryotic proteins.

baddeleyite (*Min*) Zirconium dioxide, found in Brazil, where it is exploited as a source of zirconium.

badger (*Build*) An implement used to clear mortar from a drain after it has been laid.

badger plane (*Build*) Large wooden plane with a rebated offside edge, used for wide rebates.

badger softener (*Build*) A brush with badger hair filling, used in water colour graining.

Baeyer's tension theory (*Chem*) Theory observing that since the four valencies of the tetravalent carbon atom are symmetrically distributed in space, one may predict the strain involved in the formation of a ring compound from a chain of carbon atoms, and also estimate the stability of a ring compound from the number of carbon atoms in the ring. Also *Baeyer's strain theory*.

baffle (*Acous*) Extended surface surrounding a diaphragm of a sound source (loudspeaker) so that an acoustic short circuit is prevented.

baffle (*Aero*) (1) Any device to impede or divide a fluid flow, eg in a tank, to assist sloshing of liquid propellants. (2) Plates fitted between cylinders of air-cooled engines to assist cooling.

baffle (*Electronics*) Internal structure or electrode, with no external connection, used in gas-filled tubes to control the discharge or its decay.

baffle (*MinExt*) In hydraulic or rake classifier, a plate set across and dipping into the pulp pool; in mechanically agitated flotation cell, one so set as to reduce centrifugal movement in the upper part of the cell.

baffle loudspeaker (*Acous*) An open-diaphragm loudspeaker, in which the radiation of sound power is enhanced by surrounding it with a large plane baffle, generally of wood.

baffle plate (*Eng*) A plate used to prevent the movement of a fluid in the direction which it would normally follow, and to direct it into the desired path.

baffle plate (*ICT*) A plate inserted into waveguide to produce change in mode of transmission.

bagassosis (*Med*) Respiratory disease similar to FARMER'S LUNG occurring in persons who inhale dust from mouldy sugar cane. Due to type III hypersensitivity reaction induced by thermophilic mould spores.

bag moulding (*Eng, Plastics*) Use of a flexible membrane (the 'bag') to exert pressure, usually about 1 atmosphere, on a thermosetting composite LAMINATE or sandwich component while it is curing at ambient temperature in an open mould. Pressure can be generated either by evacuating the inside of the bag (vacuum bag moulding) or by pressurizing its outer surface (pressure bag moulding).

bag plug (*Build*) A drain plug consisting of a cylindrical canvas bag which is placed in the drain pipe and inflated.

bag pump (*Eng*) A form of bellows pump, in which the valved disk taking the place of the bucket is connected to the base of the barrel by an elastic bag, distended at intervals by rings.

bailer (*MinExt*) Length of piping closed at bottom by a check or clack valve, used to remove drilling sands and muds from borehole.

bailey (*Arch*) The open area within the fortified walls of a medieval castle and usually surmounting a mound or MOTTE.

Bailey bridge (*CivEng*) A temporary bridge made by assembling portable prefabricated panels. A 'nose' is projected over rollers across the stream, being followed by the bridge proper, with roadway. Used also over pontoons.

Bailey test for sulphur (*Chem*) A test in which, after fusion with sodium carbonate, the aqueous solution gives a blood-red colour with sodium nitroprusside.

Baily furnace (*ElecEng*) An electric-resistance furnace in which the resistance material is crushed coke placed between carbon electrodes; used for heating ingots and bars in rolling mills, for annealing, etc.

Baily's beads (*Astron*) A phenomenon in which, during the last seconds before a solar eclipse becomes total, the advancing dark limb of the Moon appears to break up into a series of bright points. First observed by F Baily in 1836, it is caused by sunlight shining through lunar valleys.

bainite (*Eng*) A microstructural product formed in steels when cooled from the austenite state at rates or transformation temperatures intermediate between those which form PEARLITE and MARTENSITE, ie between about 800 and 500 K. It is an acicular structure of supersaturated ferrite containing particles of carbide, the dispersions of the latter depending on the formation temperature. Its hardness is intermediate between that of pearlite and martensite and exhibits mechanical properties similar to those of tempered martensite in a steel of the same carbon content. See ISOTHERMAL TRANSFORMATION DIAGRAM.

baize (*Textiles*) A lightweight woollen felt used to cover tables (eg for billiards) and noticeboards.

Bajocian (*Geol*) A stage in the Middle Jurassic. See MESOZOIC.

baked core (*Eng*) A dry-sand core baked in the oven to render it hard and to fix its shape. See CORE SAND.

baked images (*Print*) The technique of heating a printing plate (mainly lithographic) to harden the printing image and thus increase the image's resistance to wear, hence lengthening the run expectancy on the press.

Bakelite (*Chem*) TN for phenol–formaldehyde resin (named after L H Baekeland).

bake-off (*ICT*) A term used esp in the computer industry for a public test of several similar products in order to establish which one performs best.

bake-out (*ElecEng*) Preliminary heating of components of a vacuum device to release adsorbed gases.

Baker's cyst (*Med*) A cyst found behind the knee, which may or may not communicate with the synovial joint.

baking soda (*Chem*) *Sodium hydrogen carbonate, bicarbonate of soda.* $Na(HCO_3)_2$.

BAL (*Pharmacol*) Abbrev for *British anti-lewisite*, dithiglycerol. Antidote for poisoning by LEWISITE and other poisons, particularly arsenic and mercury.

balance (*Acous*) Adjustment of sources of sound in studios so that the final transmission adheres to an artistic standard.

balance (*BioSci*) Equilibrium of the body; governed from the cerebellum, in response to stimuli from the eyes and extremities and esp from the SEMICIRCULAR CANALS of the ears.

balance (*Chem*) See CHEMICAL BALANCE.

balance (*ElecEng*) Said to be obtained in bridge measurements when the various impedances forming the arms of the bridge have been adjusted, so that no current flows through the detector. See CURRENT BALANCE.

balance (*Eng*) The vibrating member of a watch, chronometer or clock (with platform escapement). In conjunction with the balance spring it forms the time-controlling element.

balance bar (*Eng*) The heavy beam by which a canal-lock gate may be swung on its PINTLE, and which partially balances the outer end of the gate.

balance box (*Eng*) A box, filled with heavy material, used to counterbalance the weight of the jib and load of a crane of the cantilever type.

balance bridge (*CivEng*) See BASCULE BRIDGE.

balance crane (*Eng*) A crane with two arms, one having counterpoise arrangements to balance the load taken by the other.

balanced amplifier (*ICT*) One in which there are two identical signal-handling branches operating in phase opposition, with input and output connections balanced to earth. A PUSH–PULL AMPLIFIER is an example.

balanced-armature pickup (*Acous*) A pickup in which the reproducing needle is held by a screw in a magnetic arm, which is pivoted so that its motion diverts magnetic flux from one arm of a magnetic circuit to another, thereby inducing emf in coils on these arms.

balanced circuit (*ElecEng*) For ac and dc, a circuit which is balanced to earth potential, ie the two conductors are at equal and opposite potentials with reference to earth at every instant. See UNBALANCED CIRCUIT.

balanced current (*ElecEng*) A term used, in connection with polyphase circuits, to denote currents which are equal to all the phases. Also applied to dc three-wire systems.

balanced draught (*Eng*) A system of air supply to a boiler furnace, in which one fan forces air through the grate, while a second, situated in the uptake, exhausts the flue gases. The pressure in the furnace is thus kept atmospheric, ie is *balanced*.

balanced equation (*Chem*) The equation for a chemical reaction in which the correct relative numbers of moles of each reactant and product are shown.

balanced laminate (*Eng*) Symmetrical laminated material in which the sequence of laminae above the centre plane is the mirror image of that below it.

balanced line (*ICT*) A line in which the impedances to earth of the two conductors are, or are made to be, equal. Also *balanced system*.

balanced load (*ElecEng*) A load connected to a polyphase system, or to a single-phase or dc three-wire system, in such a way that the currents taken from each phase, or from each side of the system, are equal and at equal power factors.

balanced mixer (*ICT*) A mixer that may be made of discrete components or formed in stripline or waveguide, in which the local oscillator breakthrough in the output is minimized and certain harmonics suppressed. The contribution of local oscillator noise to the receiver's overall performance is also reduced by such a mixer.

balanced modulator (*ICT*) A modulator in which the carrier and modulating signal are combined in such a way that the output contains the two sidebands but not the carrier. Used in colour TV to modulate subcarriers, and in suppressed-carrier communication systems.

balanced network (*ICT*) A network arranged for insertion into a BALANCED CIRCUIT and therefore symmetrical

electrically about the mid-points of its input and output pairs of terminals.

balanced-pair cable (*ICT*) A cable with two conductors forming a loop circuit, the wires being electrically balanced to each other and earth (shield), eg an open-wire antenna feeder. Cf COAXIAL CABLE.

balanced pedal (*Acous*) In an organ console, the foot-operated plate, pivoted so that it stays in any position, for remote control of the shutter of the chambers in which ranks of organ pipes are situated; it also serves for bringing in all the stops in a graded series. See SWELL PEDAL.

balanced protective system (*ElecEng*) A form of protective system for electric transmission lines and now widely used domestically in which the current entering the line or apparatus is balanced against that leaving it. Any fault, such as a short circuit to earth, upsets this balance and energizes a relay which trips the faulty circuit. Also *differential protective system* or (colloq) *earth leak relay*, *earth trip*. See BEARD PROTECTIVE SYSTEM, BIASED PROTECTIVE SYSTEM.

balanced sash (*Build*) See SLIDING SASH.

balanced solution (*Chem*) A solution of two or more salts, in such proportions that the toxic effects of the individual salts are mutually eliminated, eg sea water.

balanced step (*Build*) See DANCING STEP.

balanced system (*ICT*) See BALANCED LINE.

balanced termination (*ICT*) A two-terminal load in which both terminals present the same impedance to ground.

balanced voltage (*ElecEng*) A term used, in connection with polyphase circuits, to denote voltages which are equal to all the phases. Also applied to dc three-wire systems.

balanced weave (*Textiles*) A weave in which the length of free yarn between the intersections is the same in the warp and weft directions and on both sides of the fabric.

balance equation (*EnvSci*) An equation expressing the balance between the non-divergent part of the horizontal windfield and the corresponding field of geopotential on a constant pressure surface. If ψ is the stream function, φ the geopotential and f the CORIOLIS PARAMETER, the balance equation is

$$f\,\nabla^2\psi + \nabla\psi\cdot\nabla f + 2\left[\frac{\partial^2\psi}{\partial x^2}\frac{\partial^2\psi}{\partial y^2} - \left(\frac{\partial^2\psi}{\partial x\,\partial y}\right)^2\right] = \nabla^2\varphi$$

Winds derived from the balance equation, which requires numerical solution on a computer, are closer to their actual values than those derived from the GEOSTROPHIC APPROXIMATION, esp in regions where the isobars are markedly curved.

balance gate (*Eng*) A flood gate which revolves about a vertical shaft near its centre, and which may be made either self-opening or self-closing as the current sets in or out by giving a preponderating area to one leaf of the gate.

balance pipe (*Eng*) A connecting pipe between two points at which pressure is to be equalized.

balance piston (*Eng*) See DUMMY PISTON.

balance point (*CivEng*) Any point where a MASS-HAUL CURVE cuts the datum line, showing that up to this point all excavated material has been used up in embankment.

balancer (*ElecEng*) A device used on polyphase or three-wire systems to equalize the voltages between the phases or the sides of the system, when unbalanced loads are being delivered.

balancer transformer (*ElecEng*) An autotransformer connected across the outer conductors of an ac three-wire system, the neutral wire being connected to an intermediate tapping.

balance tab (*Aero*) A TAB whose movement depends upon that of the main control surface. It helps to balance the aerodynamic loads and reduces the STICK FORCES. Cf SERVO TAB, SPRING TAB, TRIMMING TAB.

balance theories (*Psych*) Theories concerned with how an individual's attitudes and perceptions of other people, events or objects relate to each other.

balance weights (*ElecEng*) Small weights threaded on radial arms on the movement of an indicating instrument, so adjusted that the pointer gives the same indication whatever the orientation of the instrument.

balance weights (*Eng*) A weight used to counterbalance some part of a machine; eg the weights applied to a crankshaft to minimize or neutralize the inertia forces due to reciprocating and rotating masses of the engine. Commonly used on wheel rims to balance wheel and tyre assembly.

balancing (*Electronics*) See NEUTRALIZATION.

balancing (*ImageTech*) In colour reproduction, control of the levels of the three colour components to achieve a satisfactory picture without obvious colour bias, esp in the representation of neutral grey tones.

balancing (*Surv*) The process of adjusting a traverse, ie applying corrections to the different survey lines and bearings so as to eliminate the closing error.

balancing antenna (*ICT*) Auxiliary reception antenna that responds to interfering but not to the wanted signals. The interfering signals thus picked up are balanced against those picked up by the main antenna, leaving signals more free from interference.

balancing capacitance (*ICT*) See NEUTRALIZING CAPACITANCE.

balancing machine (*Eng*) A machine for testing the extent to which a revolving part is out of balance, and to determine the weight and position of the masses to be added, or removed, to obtain balance.

balancing speed (*ElecEng*) See FREE-RUNNING SPEED.

balanitis (*Med*) Inflammation of the glans penis.

balanoposthitis (*Med*) Inflammation of the glans penis and prepuce.

balanorrhagia (*Med*) Gonorrhoeal inflammation of the glans penis, with discharge of pus.

balantidiosis (*Vet*) An enteric infection, esp of pigs, by ciliate protozoa of the genus *Balantidium*.

balanus (*Med*) The terminal bulbous portion of the penis; the glans penis.

balas ruby (*Min*) A misnomer for the rose-red variety of the mineral spinel. See FALSE RUBY.

balata (*Chem*) The coagulated LATEX of the bullet tree of S America, tapped in the same way as natural rubber. It consists mainly of trans-1,4-polyisoprene together with natural resins. After removal of the resin, the material can be shaped and vulcanized. It was used for high-quality golfball covers, but has now been largely replaced by synthetic ionomer resins. See GUTTA PERCHA.

Balbach process (*Chem*) Electrolytic separation of gold and silver from a metal, by making it the anode in a bath of silver nitrate, the silver being discharged at the cathode.

Balbiani rings (*BioSci*) Very large PUFFS at specific sites on Dipteran (fruit-fly) salivary gland chromosomes. Occur when RNA, coding for secretory proteins, is being transcribed at these sites.

balbuties (*Med*) Stammering, stuttering.

balconet (*Arch*) A low ornamental railing to a door or window, projecting very little beyond the threshold or sill; mainly used in the Swiss style of architecture.

balcony (*Arch*) A stage or platform projecting from the wall of a building within or without, supported by pillars or consoles, and surrounded with a balustrade or railing.

baldacchino (*Arch*) A canopy suspended above an altar, tomb, etc, or supported in such position by columns. Also *baldachin, balaquin*.

bale breaker (*Textiles*) A machine that opens up the highly compressed cotton from a bale, producing a large number of tufts ready for subsequent blending, cleaning and further opening.

balection moulding (*Build*) See BOLECTION MOULDING.

baleen (*BioSci*) In certain whales, horny plates arising from the mucous membrane of the palate and acting as a food strainer.

baler (*Agri*) A machine driven across a mown field to lift and compress the crop into tied bales for storage. Also *pick-up baler*.

balise (*Eng*) A device mounted on a railway track that transmits information to (and sometimes receives information from) passing trains by telegraphy.

balk (*CivEng*) The material between two excavations. Also *baulk*.

Balkan frame (*Med*) A frame, with pulleys attached, for supporting the leg in the treatment of fractures.

balking (*ElecEng*) See CRAWLING.

ball-and-socket head (*ImageTech*) Camera mounting allowing universal movement in rotation and tilt before fixing by clamping; usually fitted to the top of a tripod.

ball-and-socket joint (*Eng*) A joint between two rods, permitting considerable relative angular movement in any plane. A ball formed on the end of one rod is embraced by a spherical cup on the other. Used in light control systems (eg in connecting a pair of bell-cranks which operate in planes at right angles) and in the steering mechanism of motor vehicles, in which both ball and cups are of case-hardened metals. Heavier examples allow a large base plate to be placed under a supporting column in a jack-up pontoon or modified as bridge bearings to allow some articulation. Also *ball joint*.

ball-and-socket joint (*Med*) A joint in which the hemispherical end of one bone is received into the socket of another. Also *enarthrosis*.

ballast (*CivEng*) (1) A layer of broken stone, gravel or other material deposited above the formation level of road or railway; it serves as foundation for road-metal or permanent-way respectively. (2) Sandy gravel used as a coarse aggregate in making concrete.

ballast (*Ships*) Gravel, stone or other material placed in the hold of a ship to increase its stability when floating with insufficient cargo.

ballast lamp (*ElecEng*) Normal incandescent lamp used as a ballast resistor, current limiter, alarm, or to stabilize a discharge lamp.

ballast resistance (*ElecEng*) A term used in railway signalling to denote the resistance between the two track rails across the BALLAST on which the track is laid. If allowed to fall too low, it will have the effect of shunting the signal from a train's wheels.

ballast resistor (*ElecEng*) A resistor inserted into a circuit to swamp or compensate changes, eg those arising through temperature fluctuations. One similarly used to swamp the negative resistance of an arc or gas discharge. Also *ballast tube*. See BARRETTER.

ballast tube (*ElecEng*) See BALLAST RESISTOR, BARRETTER.

ball bearing (*Eng*) A shaft bearing consisting of a number of hardened steel balls which roll in spherical grooves (ball tracks) formed in an inner race fitted to the shaft and in an outer race carried in a housing. Balls are spaced and held by a light metal or plastic cage. Also one of the balls itself. See LINEAR BALL BEARING, RECIRCULATING-BALL THREAD.

ball catch (*Build*) A door fastening in which a spring-controlled ball, projecting through a smaller hole, engages with a striking plate.

ball clay (*Geol*) A fine-textured and highly plastic clay; a reworked china clay. So named because it used to be rolled into balls. Also *potters' clay*.

ball-cock (*Build*) A self-regulating cistern-tap which, through a linkage system, is turned off and on by the rise and fall of a floating ball. Also *ball-valve*.

ball-ended magnet (*ElecEng*) A permanent magnet, consisting of a steel wire with a steel ball attached to each end; this gives a close approximation to a unit pole.

ball flower (*Arch*) An ornament, like a ball enclosed within three or four petals of a flower, set at regular intervals in a hollow moulding.

balling (*Eng*) (1) A process that occurs in the cementite constituent of steels on prolonged annealing at 650°–700°C. (2) The operation of forming balls in a puddling furnace.

ballistic circuit breaker (*ElecEng*) A very high-speed circuit breaker, in which the pressure produced by the fusing of an enclosed wire causes interruption of the circuit.

ballistic galvanometer (*ElecEng*) A galvanometer with a long swing period; the deflection measures the electric charge in a current pulse or the time integral of a voltage pulse.

ballistic method (*ElecEng*) A method of high-grade testing used in electrical engineering, a BALLISTIC GALVANOMETER being used.

ballistic missile (*Aero, Space*) See MISSILE.

ballistic pendulum (*Phys*) A heavy block suspended by strings so that its swings are restricted to one plane. If a bullet is fired into the block, the velocity of the bullet may be calculated from a measurement of the angle of swing of the pendulum.

ballistics (*Phys*) The study of the dynamics of the path taken by an object moving under the influence of a gravitational field.

ballistospore (*BioSci*) A spore that is violently projected, eg the basidiospore of the basidiomycete fungi.

ball joint (*Eng*) See BALL-AND-SOCKET JOINT.

ball lightning (*EnvSci*) A slowly moving luminous ball, which is occasionally seen at ground level during a thunderstorm. It appears to measure about 0·5 m in diameter.

ball mill (*ChemEng*) A mill consisting of a horizontal cylindrical vessel in which a given material is ground by rotation with steel or ceramic balls.

ballonet (*Aero*) An air compartment in the envelope of an *aerostat*, used to adjust changes of volume in the filler gas.

balloon (*Aero*) A general term for aircraft supported by buoyancy and not driven mechanically.

balloon (*Arch*) A spherical ball or globe crowning a pillar, pier, etc.

balloon barrage (*Aero*) An anti-aircraft device consisting of suitably disposed tethered balloons.

balloon former (*Print*) On rotary presses, an additional former mounted above the others, from which folded webs are gathered to make up the sections of multisectioned newspapers or magazines. See LENGTH FOLD COLLECTION.

ballooning of yarn (*Textiles*) The shape taken up by yarns on the spinning or doubling machines.

ballotini (*Eng*) Small, solid glass spheres or beads used as a filler for plastics and to increase reflectivity in paints and printing inks. See HOLLOW GLASS MICROSPHERES.

ballottement (*Med*) A method of diagnosing pregnancy by manual displacement of the fetus in the fluid which surrounds it in the uterus.

ball-pane hammer (*Eng*) A fitter's hammer, the head of which has a flat face at one end, and a smaller hemispherical face or pane at the other; used chiefly in riveting. Also *ball-pein, ball-peen*.

ball race (*Eng*) (1) The inner or outer steel ring forming one of the ball tracks of a ball bearing. (2) Commonly, the complete ball bearing.

ball sizing (*Eng*) Forcing a suitable ball through a hole to finish-size it, usually part of a BROACH with a series of spherical lands of increasing size arranged along it.

ballstone (*Geol*) A mass of fine unstratified limestone, occurring chiefly in the Wenlock Limestone of Shropshire, and representing colonies of corals in position of growth.

ball track (*Eng*) See BALL BEARING.

ball-valve (*Build*) See BALL-COCK.

ball-valve (*Eng*) A single non-return valve consisting of a ball resting on a cylindrical seating; used in small water and air pumps.

Balmer series (*Phys*) A group of lines in the hydrogen spectrum discovered by H H Balmer in 1885, later shown to correspond to excitation levels of the hydrogen electron. Their positions are given by the formula

$$v = R_h \left(\frac{1}{m^2} - \frac{1}{n^2} \right)$$

where *m* and *n* have various integral values, *v* is the wavenumber and R_h is the hydrogen Rydberg number ($= 1·0967\ 758 \times 10^7\ \mathrm{m}^{-1}$). A number of similar series such as the LYMAN SERIES and PASCHEN SERIES are found in other parts of the electromagnetic spectrum.

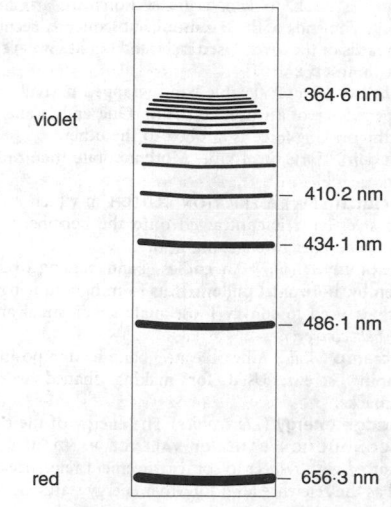

violet — 364·6 nm

— 410·2 nm

— 434·1 nm

— 486·1 nm

red — 656·3 nm

Balmer series

balneology (*Med*) The scientific study of baths and bathing, and of their application to disease.

BALPA (*Aero*) Abbrev for *British Airline Pilots Association*.

balsam of fir (*Chem*) See CANADA BALSAM.

balsam of Peru (*Chem*) An oleoresin containing esters of benzoic and cinnamic acids, obtained from a S American papilionaceous tree. Used in perfumery and chocolate manufacture.

balsam of Tolu (*Chem*) An oleoresin containing esters of benzoic and cinnamic acids, obtained from a S American evergreen tree (*Myroxylon toluiferum*). Used in perfumery, chocolate manufacture and medically as a mild expectorant.

balsa wood (*For*) The wood of *Ochroma lagopus* (W Indian corkwood); it is highly porous, and has the lowest density (average 160 kg m⁻³) of any hardwood.

BALT (*BioSci*) Abbrev for BRONCHUS ASSOCIATED LYMPHOID TISSUE.

Baltic redwood (*For*) See SCOTS PINE.

BALUN (*ICT*) Abbrev for *balance to unbalance transformer*, usually a resonating section of transmission line, for coupling balanced to unbalanced lines; eg used to transfer a signal from an UNSHIELDED TWISTED-PAIR CABLE to a coaxial cable in the context of a LOCAL AREA NETWORK. Also *bazooka*.

baluster (*Arch*) A small pillar supporting the coping of a bridge parapet, or the handrail of a staircase. Also *banister*.

balustrade (*Arch*) A coping or handrail with its supporting balusters.

bamboo (*For*) A genus of fast-growing, giant grasses (*Bambusa*) common in tropical countries, whose hollow stems become 'woody'.

BAN (*Pharmacol*) Abbrev for *British approved name* for a medicinal substance. Several BANs were recently (2003) changed to match the corresponding recommended international non-proprietary names (RINNS).

banak (*For*) See VIROLA.

banana plug (*ElecEng*) A single conductor plug which has a spring metal tip, in the shape of a banana. The corresponding socket or jack is termed a *banana jack*.

Banbury mixer (*Eng*) Type of machine used for compounding rubber with vulcanizing ingredients and carbon black.

band (*Build*) A flat horizontal member, occasionally ornamented, separating a series of mouldings or dividing a wall surface.

band (*ICT*) See FREQUENCY BAND.

band (*Print*) One of the tapes, or cords, placed across the back of a book, to which the sections are attached by sewing. The ends of the bands are subsequently secured to the boards of the cover. Used in bound books. See FLEXIBLE BAND, RAISED BAND.

band brake (*Eng*) A flexible band wrapped partially round the periphery of a wheel or drum. One end is anchored, and the braking force is applied to the other.

band chain (*Surv*) Steel tape. More accurate than ordinary chain.

band clutch (*Eng*) A FRICTION CLUTCH in which a fabric-lined steel band is contracted onto the periphery of the driving member by engaging gear.

band conveyor (*Eng*) An endless band passing over, and driven by, horizontal pulleys, thus forming a moving track which is used to convey loose material or small articles. Also *belt conveyor*.

band cramp (*Build*) A flexible steel band held in position by clamping screws. Used for making shaped work, eg chairbacks.

band edge energy (*Electronics*) The energy of the edge of the CONDUCTION BAND or VALENCE BAND in a solid, measured with respect to some convenient reference or else used as the reference level for other energy states. See BAND THEORY OF SOLIDS.

banded structure (*Geol*) A structure developed in igneous and metamorphic rocks due to the alternation of layers of different texture and/or composition.

band gap (*Electronics*) The range of energies which correspond with those values which are forbidden for delocalized states, according to the BAND THEORY OF SOLIDS. Localized states such as those associated with ionized dopants, impurity atoms or crystal imperfections exist in the gap. The generation of pairs of electrons and holes requires quanta of at least the energy of the band gap. Direct recombination likewise furnishes quanta with energies at least equal to the band gap. See OPTOELECTRONICS.

band ignitor tube (*Electronics*) A valve of mercury pool type in which the control electrode is a metal band outside the glass envelope. Also *capacitron*.

banding (*Eng*) A structural feature of wrought metallic materials revealed by etching, resulting from microstructural segregates and constitutional differences within the grain structure becoming drawn out in the direction of working.

banding (*ImageTech*) A defect in videotape recording heads causing visible horizontal bands in the picture.

banding plane (*Build*) A plane used for cutting out grooves and inlaying strings and bands in straight and circular work.

banding techniques (*BioSci*) Methods of treating chromosomes to produce patterns of BANDS characteristic of an individual chromosome, as an aid to recognition. See panel on CHROMOSOME.

band-pass filter (*ICT*) A filter that freely passes currents having frequencies within specified nominal limits, and highly attenuates currents with frequencies outside these limits.

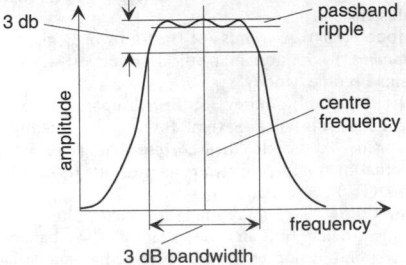

band-pass filter

band-rejection filter (*ICT*) See BAND-STOP FILTER.

bands (*BioSci*) METAPHASE chromosomes, when stained with a variety of banding techniques, exhibit a pattern of transverse bands of varying width characteristic of specific chromosomes. The bands appear to be made by distinctive interactions between specific types of DNA sequence and proteins. Factors such as base and sequence composition, eg blocks of short repeats, are involved. Polytene chromosomes (see POLYTENY) also exhibit characteristic bands, with or without staining. They are much more numerous than those of metaphase chromosomes and the more stretched state of the polytene chromosomes would be expected to show a greater number of bands. The two kinds of banding patterns may therefore be related. The combination of size, shape and banding pattern gives each chromosome a unique appearance, which has made eukaryotic gene mapping possible.

bandsaw (*Eng*) A narrow endless strip of saw-blading running over and driven by pulleys, as a belt; the strip passes a work table placed normal to the straight part of the blade. The workpiece is forced against the blade and intricate shapes can be cut. Also used for cutting animal carcases in butchery.

band spectrum (*Phys*) Molecular optical spectrum consisting of numerous very closely spaced lines which are spread through a limited band of frequencies.

band-spreading (*ICT*) (1) Use of a relatively small tuning capacitor in parallel with the main tuning capacitor of a radio receiver, so that fine tuning control can be done with the smaller; useful when the frequency band is crowded. (2) Mechanical means, like reduction gearing, to achieve the same result.

band-stop filter (*ICT*) Filter that attenuates signals having frequencies within a certain range or *band*, while freely passing those outside this range. Also *band-rejection filter*.

B and S wire gauge (*Eng*) Abbrev for BROWN AND SHARPE WIRE GAUGE.

band theory of solids (*Phys*) For atoms brought together to form a crystalline solid, their outermost electrons are influenced by a *periodic* potential function, so that their possible energies form *bands* of allowed values separated by bands of forbidden values (in contrast to the discrete energy states of an isolated atom). These electrons are not localized or associated with any particular atom in the solid. This band structure is of fundamental importance in explaining the properties of metals, semiconductors and insulators. See CONDUCTION BAND, ENERGY BAND, VALENCE BAND.

bandwidth (*ICT*) (1) The width, or spread, of the range of frequencies used for a given purpose, eg the width of individual channels allotted to speech or to TV transmissions. (2) The space occupied in the frequency domain by signals of a specified nature, eg telephone-quality speech, broadcast-quality stereophonic music, TV, radar transmission, etc. (3) The range of frequencies within which the characteristics of a device (filter, amplifier, etc) are within specified limits, often the points at which the performance has changed by 3 dB from a mean level, or the HALF-POWER POINTS.

Bang's bacillus (*BioSci*) *Brucella abortus*, the cause of contagious abortion in animals and of undulant fever in humans.

Bang's disease (*Vet*) See BOVINE CONTAGIOUS ABORTION.

banister (*Arch*) See BALUSTER.

banjo axle (*Autos*) The commonest form of rear-axle casing, in which the provision of the differential casing in the centre produces a resemblance to a banjo with two necks.

bank (*Eng*) A number of similar pieces of equipment grouped in line and connected, eg a bank of engine cylinders, coke ovens or transformers.

Banka drill (*MinExt*) A drill widely used in shallow testing of alluvial deposits. Portable, hand-operated, it consists essentially of an assembly of 5 ft pipes, 4 in in diameter, which are worked into the ground, the material traversed being recovered by a sand pump or BAILER.

banked fire (*Eng*) A boiler furnace in which the rate of combustion is purposely reduced to a very low rate for a period during which the demand for steam has ceased by eg covering the fire with slack or fine coal or banking-up. Also *banked boiler*. See DEAD BANK.

banker (*Build*) A bench upon which bricklayers and stonemasons shape their materials. Also *siege*.

banket (*Geol*) The term originally applied by the Dutch settlers to the gold-bearing conglomerates of the Witwatersrand. It is now used for any metamorphosed conglomerate, containing barren quartz pebbles cemented with siliceous matrix-bearing gold.

banking (*Aero*) Angular displacement of the wings of an aircraft about the longitudinal axis, to assist turning.

banking (*Eng*) The process of suspending operation in a smelter, by feeding fuel only into the furnace until as much metal and slag as possible have been removed, after which all air inlets are closed.

banking (*MinExt*) The operations involved in removing full trucks, tubs or wagons and replacing them by empty ones at the top of a shaft.

bank paper (*Paper*) A thin writing paper of less than 50 gm^{-2}, intended for typewriting or correspondence purposes.

bank protector (*Build*) Any device for minimizing erosion of river banks by water, eg groins, pitching.

bank switching (*ICT*) A technique used to overcome the limitation that computers can only access a finite amount of main memory. Several banks of memory are provided and the computer switches between them as required, only one being in use at once, often as SIDEWAYS ROM.

bannisterite (*Min*) A hydrated silicate of manganese, crystallizing in the monoclinic system, and occurring in manganese deposits in North Wales and New Jersey.

banquette (*Arch*) A narrow window seat in masonry, brickwork or wood.

banquette (*CivEng*) (1) A raised footway inside a bridge parapet. (2) A ledge on the face of a cutting. See BERM.

bantam (*Agri*) Any small domesticated fowl, but typically applied to chickens. Bantams may be miniatures of conventional breeds.

BAP (*BioSci*) Abbrev for 6-BENZYLAMINOPURINE.

bar (*CivEng*) A pivoted bar, parallel to a running rail, which, being depressed by the wheels of a train, is capable of holding points or giving information about a train's position.

bar (*Eng*) Material of uniform cross-section, which may be cast, rolled or extruded.

bar (*EnvSci, Phys*) Unit of pressure or stress. 1 bar = 10^5 Pa = 750·07 mm of mercury at 0°C and latitude 45°. The *millibar* (10^{-3} bar) is used for barometric purposes. The *hectobar* (1 hbar = 10^7 N m^{-2}, approx 0·6475 tonf in^{-2}) is used for some engineering purposes.

baragnosis (*Med*) Loss of the ability to judge differences between the weights of objects.

bar-and-yoke (*ElecEng*) A method of magnetic testing in which the sample is in the form of a bar, clamped into a yoke of relatively large cross-section, which forms a low-reluctance return path for the flux.

Bárány manoeuvre (*Med*) Tests for assessing the functions of the semicircular canals in the inner ear.

barathea (*Textiles*) Woven fabric used for coats and suits and made from silk, worsted or artificial fibres. Characteristic surface appearance arising from the twill or broken-rib weave used in its manufacture.

barb (*BioSci*) A hooked or doubly hooked hair or bristle-like structure; in birds, one of the lateral processes of the rachis of the feather that form the vane.

Barbados Earth (*Geol*) A siliceous accumulation of remains of Radiolaria, formed originally in deep water and later raised above sea level.

Barba's law (*Eng*) Law concerned with the plastic deformation of metal test pieces when strained to fracture in a tensile test; it states that test pieces of identical size deform in a similar manner.

barbate (*BioSci*) Bearded; bearing tufts of long hairs.

barbel (*BioSci*) In some fish, a finger-shaped tactile or chemosensitive appendage arising from one of the jaws.

barber's rash (*Med*) Infection of the beard region of the face with a bacterium or a fungus.

barbital (*Pharmacol*) *Diethyl-malonyl-urea, 5,5-diethylbarbituric acid*. White crystalline solid; mp 191°C. Once widely used as a sedative and hypnotic, usually in the form of the sodium salt. Also *barbitone, veronal*.

barbiturates (*Pharmacol*) A class of drugs that depress activity of the central nervous system. Their use has largely been superseded by BENZODIAZEPINES.

barbituric acid (*Chem*) *Malonyl urea*. CO(NHCO)$_2$CH$_2$. Crystallizes in large colourless crystals. The hydrogen atoms of the methylene group are reactive and can be replaced by halogens. Basis of important derivatives with therapeutic action.

barbule (*BioSci*) In birds, one of the processes borne on the barbs of a feather, by which the barbs are bound together.

barchan (*Geol*) An isolated crescentic sand-dune.

bar chart (*MathSci*) A graphical representation of the frequencies of the different values (sometimes known as levels) of a qualitative characteristic as lines, the length of a line being proportional to the frequency of the corresponding level.

bar code (*ICT*) Code consisting of parallel thick and thin lines. Used on a label attached to an item to give machine readable information. See UNIVERSAL PRODUCT CODE.

bar code reader (*ICT*) A scanning device that reads a BAR CODE. Also *wand*.

bar cramp (*Build*) Metal or wooden bar with a fixed and a moving jaw which can clamp two pieces of wood together for eg gluing planks edgewise. See SASH CRAMP.

bare (*Eng*) A term signifying slightly smaller than the specified dimension. Cf FULL.

bare conductor (*ElecEng*) A conductor not continuously covered with insulation, but supported intermittently by insulators, eg bus-bars and overhead lines.

bare electrodes (*ElecEng*) Electrodes used in welding that are not coated with a basic slag-forming substance.

bareface tenon (*Build*) A tenon which has a shoulder on one face only; used when jointing a rail which is thinner than the stile.

barge board (*Build*) A more or less ornamental board fixed at the gable end of a roof. It hides the ends of the horizontal timbers, and protects from the weather the underside of the BARGE COURSE.

barge couple (*Build*) The outer couples, pair of rafters, of a roof which project over the gable of a roof.

barge course (*Build*) (1) That part of the roof of a house which projects slightly over the gable end, and is made up underneath with mortar to keep out rain etc. (2) A coping course of bricks laid edge-wise and transversely on a wall.

bar generator (*ImageTech*) The source of pulse signals, giving a bar pattern for testing TV cathode-ray tubes.

bariatric (*Med*) Relating to the medical treatment of obesity.

barite (*MinExt*) See BARYTES.

barium (*Chem*) A heavy element in the second group of the periodic system, an alkaline earth metal. Symbol Ba, at no 56, ram 137·34, mp 725°C. In most of its compounds it occurs as Ba^{2+} and it is present to the extent of 390 ppm in the Earth's crust. Its mass makes it an effective absorber of high-energy particles and it is used as barytes in loaded concrete for this purpose. It is a brittle and expensive metal, difficult to machine and giving off toxic dust.

barium concrete (*Build*) Concrete containing high proportion of barium compounds, which has a high absorption for radiation and thus used as a shield.

barium enema (*Radiol*) A radiological examination of the lower gastro-intestinal tract using barium sulphate as a CONTRAST·MEDIUM.

barium feldspar (*Min*) A collective term for barium-bearing feldspars, including celsian and hyalophane.

barium ferrite (*Phys*) See FERRITE.

barium hydroxide (*Chem*) $Ba(OH)_2$. See BARIUM OXIDE.

barium meal (*Med*) A mixture of barium sulphate administered to render the alimentary canal opaque to X-rays.

barium oxide (*Chem*) BaO. When freshly obtained from the calcined carbonate it is even more reactive with water than calcium oxide and forms barium hydroxide (alkaline). Also *baryta*.

barium plaster (*Build*) A cement–sand plaster containing barium salts, used for lining hospital and experimental X-ray rooms to absorb radiation and minimize back-scattering.

barium sulphate (*Chem*) $BaSO_4$. Formed as a heavy white precipitate when sulphuric acid is added to a solution of a barium salt. Very nearly insoluble in water. Although of little pigmentary value, it is much used in paint manufacture and in the preparation of lake pigments. Used in BARIUM MEALS. See BARYTES.

barium titanate (*Chem*) $BaTiO_3$. A crystalline ceramic with outstanding dielectric, piezoelectric and ferroelectric properties. Used in capacitors and as a piezoelectric transducer. Has a higher CURIE POINT than Rochelle salt.

bark (*BioSci*) A non-technical term applied to all the tissues outside the cambium, ie the corky and other material which can be peeled from a woody stem.

bar keel (*Ships*) See KEELSON.

Barker index (*Crystal*) A method of identification of crystalline substances from measurements of interfacial angles.

barkevikite (*Min*) A member of the amphibole group, resembling basaltic hornblende but having a higher total iron content and a low ferric–ferrous iron ratio. Occurs in alkaline plutonic rocks, eg at Barkevik, Norway.

Barkhausen effect (*Phys*) The phenomenon of discontinuous changes in the magnetization of a magnetic material while the magnetizing field is smoothly varied. It is the consequence of sudden changes in the domain structure as domain walls overcome various pinning defects and to a lesser extent as domain orientations discontinuosly rotate away from preferred crystal axes. H G Barkhausen (in 1919) detected voltage pulses induced in coils surrounding a magnetic sample as it was magnetized. Analogous ultrasonic emissions are also associated with the magnetization of magnetostrictive materials. The character of Barkhausen emissions is strongly dependent on microstructure and stress.

Barkhausen–Kurz oscillator (*ICT*) Oscillator with a triode valve having its grid more positive than the anode. Electrons oscillate about the grid before reaching the anode. Output frequency depends on the transit time of electrons though the tube.

bar lathe (*Eng*) A small lathe of which the bed consists of a single bar of circular, triangular or rectangular section.

Barlow lens (*Phys*) A plano-convex lens between the objective and eyepiece of a telescope to increase the magnification by increasing the effective focal length.

bar magnet (*ElecEng*) A straight bar-shaped permanent magnet, with a POLE at each end.

bar mill (*Eng*) A rolling mill with grooved rolls, for producing round, square or other forms of bar iron of small section.

bar mining (*MinExt*) Alluvial mining of sand-banks, river bars or submerged deposits.

barn (*Phys*) Unit of effective cross-sectional area of nucleus equal to 10^{-28} m^2. So called, because it was pointed out that although 1 barn is a very small unit of area, to an elementary particle the size of an atom which could capture it is 'as big as a barn door'. See CROSS-SECTION.

Barnard's star (*Astron*) A red dwarf star in Ophiuchus, found in 1916 to have the largest proper motion yet measured, amounting to 10 arcseconds per annum.

barn door (*ImageTech*) Pair of adjustable flaps on a studio lamp for controlling the light.

Barnett effect (*Phys*) Magnetization of a ferromagnetic material by rapid rotation of the specimen. Used to measure magnetic susceptibility. See EINSTEIN–DE HAAS EFFECT.

barney (*ImageTech*) A soft cover to reduce noise from a film camera.

baroclinic atmosphere (*EnvSci*) An atmosphere which is not a BAROTROPIC ATMOSPHERE.

barograph (*EnvSci*) A recording barometer, usually of the aneroid type, in which variations of atmospheric pressure cause movement of a pen which traces a line on a clockwork-driven revolving drum.

barometer (*EnvSci, Phys*) An instrument used for the measurement of atmospheric pressure. The MERCURY BAROMETER is preferable if the highest accuracy of readings is important, but where compactness has to be considered, the ANEROID BAROMETER is often used. See ALTIMETER.

barometric corrections (*EnvSci*) Necessary corrections to the readings of a mercury barometer for index error, temperature, latitude and height.

barometric error (*Genrl*) The error in the time of swing of a pendulum due to change of air pressure. Though small, it is sometimes avoided in clocks by causing the pendulum to swing in an atmosphere of constant (low) pressure.

barometric pressure (*EnvSci*) The pressure of the atmosphere as read by a barometer. Expressed in *millibars* (see BAR), the height of a column of mercury, or (SI) in hectopascals.

barometric tendency (*EnvSci*) The rate of change of atmospheric pressure with time. The change of pressure during the previous three hours.

barophil (*BioSci*) Used of organisms that grow and metabolize as well (or better) at increased pressures than at atmospheric pressure.

barophoresis (*Chem*) Diffusion of suspended particles at a speed dependent on external forces.

Baroque (*Arch*) One of the later phases of the Renaissance style of architecture prominent in the 17th century and based on classical features but employing lighting, sculpture and painting to produce a theatrical, monumental effect. Typically, forms were manipulated to create focal points rather than being dispersed to achieve the rhythmic consistency which characterized the High Renaissance style.

baroque organ (*Acous*) A type of pipe organ in which low fundamentals are obtained by the subjective difference tones arising from pipes operating at the musical interval of the fifth.

baroreceptor (*Med*) A sensory receptor sensitive to stimulus of pressure, eg receptors located in the walls of arteries and veins responding to changes of intraluminal pressure, such as the carotid and aortic arterial, or right arterial atrial, baroreceptors.

barostat (*Aero*) A device which maintains constant atmospheric pressure in a closed volume, eg the input and output pressure of the fuel metering device of a gas turbine to compensate for atmospheric pressure variation with altitude.

barotrauma (*Med*) Injury to the ears, lungs, etc, caused by changes in atmospheric pressure.

barotropic atmosphere (*EnvSci*) An atmosphere with zero horizontal temperature gradient at all levels so that the ISOPLETHS of density and pressure coincide and the THICKNESS CHART has no pattern.

barrage (*Build*) An artificial obstruction placed in a watercourse in order to secure increased depth for irrigation, navigation or some other purpose.

barrage balloon (*Aero*) A small captive kite balloon, the cable of which is intended to destroy low-flying aircraft.

barrage-fixe (*Build*) A dam provided with sluices to control the flow of water.

Barr body (*BioSci*) The densely staining HETEROCHROMATIN of the inactive X chromosome. See X-INACTIVATION.

barré (*Textiles*) Undesirable stripes in fabrics. In weft-knitted fabrics sometimes caused by irregularities in the texturing of the yarn resulting in variation in dye affinity.

barred code (*ICT*) Any dialled code that automatic exchange apparatus is primed to reject by connecting the caller no further than number-unobtainable tone.

barred spiral galaxy (*Astron*) A galaxy with bright spiral structures emerging from each end of a straight band across the centre.

barrel (*Eng*) (1) A hollow, usually cylindrical, machine part; often revolving, sometimes with wall apertures. (2) The main cylinder in which molten polymer is prepared for extrusion or injection into moulds. See INJECTION MOULDING.

barrel (*MinExt*) US barrel of 42 US gallons (35 imperial gallons or 159·1 l), frequently employed as a unit of capacity, esp of output in the oil industry. Abbrev *bbl*.

barrel amalgamation (*MinExt*) Recovery of gold from rich concentrates by prolonged gentle tumbling with mercury in a steel barrel.

barrel bolt (*Build*) Hand-operated door fastening comprising a metal rod sliding in cylindrical guides.

barrel cam (*Eng*) A cylindrical cam with circumferential or end track.

barrel distortion (*ImageTech, Phys*) Curvilinear distortion of an optical or electronic image in which horizontal and vertical straight lines appear barrel-shaped, bowed outwards. Also *positive distortion*. See PINCUSHION DISTORTION.

barrel drain (*Build*) A cylindrical drain.

barrel etcher (*Electronics*) A device usually used to oxidize and thereby strip away hardened photoresist materials during semiconductor processing. In it a batch of wafers is exposed to a low-pressure oxygen plasma. See SEMICONDUCTOR DEVICE PROCESSING.

barrel hopper (*Eng*) A machine for unscrambling, orientating and feeding small components during a manufacturing process, in which a revolving barrel tumbles the components onto a sloping, vibrating feeding-blade.

barrel nipple (*Build*) See SHOULDER NIPPLE.

barrel plating (*ElecEng*) Electroplating of many small items by placing them in a perforated barrel revolving in a vat filled with an appropriate plating solution. The barrel is made the CATHODE in the cell and the articles tumble against each other during rotation, continually touching at different places, and so become uniformly coated with the electrodeposit.

barrel printer (*ICT*) LINE PRINTER in which the complete character set is provided at each printing position, embossed on the surface of a horizontal barrel or cylinder. See BI-DIRECTIONAL PRINTER, CHAIN PRINTER.

barrels per calendar day (*MinExt*) A measure of the output of a production unit, in which the total annual output quoted in BARRELS is divided by 365. Because it includes downtime (eg for maintenance), the value is less than BARRELS PER STREAM DAY. Abbrev *bpcd*.

barrels per stream day (*MinExt*) The output of a production unit quoted in BARRELS for 1 day of full operation (stream day). Multiplying it by 365 gives the theoretical maximum annual output. The higher the ratio between barrels per calendar day and per stream day for a particular production unit, the more efficient the unit. Abbrev *bpsd*.

barrel temperatures (*Eng*) Temperatures at which an extrusion or injection moulding barrel is kept, usually rising to a peak at the nozzle. The range is determined by the polymer type and its melt viscosity. See INJECTION MOULDING.

barrel-type crankcase (*Autos*) A petrol-engine crankcase so constructed that the crankshaft must be removed from one end; in more normal construction the crankcase is split. See SPLIT CRANKCASE.

barrel vault (*Build*) A vault of approximately semicircular cross-section, whose length exceeds its diameter. Also *annular vault, tunnel vault, wagon vault*.

barrel-vault roof (*CivEng*) A roof formed of reinforced concrete in the shape of an open cylindrical shell, generally with lateral stiffening diaphragms and edge beams. The roof itself is frequently very thin in section.

barrel winding (*ElecEng*) See DRUM WINDING.

Barremian (*Geol*) A stage in the Lower Cretaceous. See MESOZOIC.

barren (*Geol*) Without fossils.

barren solution (*MinExt*) In chemical extraction of metals from their ores, solution left after these have been removed. Cf PREGNANT SOLUTION.

barretter (*ElecEng*) Iron-wire resistor mounted in a glass bulb containing hydrogen, and having a temperature variation so arranged that the change of resistance ensures that the current in the circuit in which it is connected remains substantially constant over a wide range of voltage. Also *ballast tube*.

barrier (*ElecEng*) (1) In transformers, the solid insulating material which provides the main insulation, apart from the oil. (2) The refractory material intended to localize or direct any arc which may arise on the operation of a circuit breaker.

barrier coat (*Build*) A coating applied to a substrate to protect subsequent coating from active constituents in the substrate.

barrier layer (*Electronics, ICT*) (1) In semiconductor junctions, the DEPLETION LAYER. (2) In an optical fibre cable, an intermediate layer of glass between the low-refractive-index core and the high-refractive-index cladding.

barrier layer (*Eng*) In general a layer placed so as to inhibit interdiffusion of heat, matter, etc.

barrier layer capacitance (*Electronics*) DEPLETION LAYER capacitance.

barrier penetration (*Acous*) The passage of a sound wave, at an angle for which SNELL'S LAW predicts zero transmission, through a very thin layer.

barrier pillar (*MinExt*) A pillar of solid coal left in position to protect a main road from subsidence, or as a division, or to protect workings from flooding.

barrier reef (*Geol*) A coral reef developed parallel with the shoreline and enclosing a lagoon between itself and the land. It marks a stage between a fringing reef and an atoll.

barring gear (*ElecEng*) An arrangement for moving heavy electrical plant, using people. Rotating machines and transformers are equipped with wheels and movement is possible by inserting crowbars at suitable points and levering the equipment.

barring motor (*ElecEng*) A small motor which can be temporarily connected, by a gear or clutch, to a large machine to turn it slowly for adjustment or inspection.

Barrovian metamorphism (*Geol*) Regional metamorphism of the type first described in the Scottish Highlands by G Barrow in 1893. Zones of increasing metamorphism are characterized by the presence of a series of index minerals: chlorite, biotite, garnet, staurolite, kyanite, sillimanite. This type of metamorphism has since been recognized in many other parts of the world. Also *Barrovian zones*.

bars of foot (*Vet*) Part of the sole of the horse's foot formed by reflexion of the wall on each side of the FROG.

bar suspension (*ElecEng*) A method of mounting the motor on an electrically propelled vehicle. One side of the motor is supported on the driving axle and the other side by a spring-suspended bar lying transversely across the truck. Also *yoke suspension*.

Bartholin's duct (*BioSci*) An excretory duct of the sublingual gland.

Bartholin's glands (*BioSci*) In some female mammals, glands (corresponding to Cowper's glands in the male) lying on either side of the upper end of the vagina.

bartonellosis (*Vet*) Infection by organisms of the genus *Bartonella*, affecting humans, cattle, dogs and rodents.

Bartonian (*Geol*) A stage of the Eocene. See TERTIARY.

bar tracery (*Arch*) Window tracery characteristic of GOTHIC work, resembling more a bar of iron twisted into various forms, than stone.

bar-type current transformer (*ElecEng*) A CURRENT TRANSFORMER in which the primary consists of a single conductor that passes centrally through the iron core upon which the secondary is wound.

bar winding (*ElecEng*) An armature winding for an electric machine whose conductors are formed of copper bars.

bar-wound armature (*ElecEng*) An armature with large sectioned conductors which are insulated and fixed in position and connected, in contrast with former-wound conductors which are sufficiently thin to be inserted, after shaping in a suitable jig.

barycentre (*Astron*) The centre of mass of a system of bodies, such as a planet and its moon.

barye (*Phys*) See MICROBAR.

baryon (*Phys*) A HADRON with a baryon number of +1. Baryons are involved in strong interactions, and include neutrons, protons and hyperons.

baryon number (*Phys*) An intrinsic property of an elementary particle. The baryon number of a baryon is +1, of an antibaryon −1. The baryon number of mesons, leptons and gauge bosons is zero. Baryon number is conserved in all types of interaction between particles. QUARKS have a baryon number of $+\frac{1}{3}$ and antiquarks of $-\frac{1}{3}$.

baryta (*Chem*) See BARIUM OXIDE.

baryta paper (*Paper*) Paper coated on one side with an emulsion of barium sulphate and gelatine. Used in moving-pointer recording apparatus and for photographic printing papers.

baryta water (*Chem*) A suspension of barium hydroxide in distilled water; it is a fairly strong alkali.

barytes (*Min*) Barium sulphate, typically showing tabular orthorhombic crystals. It is a common mineral in association with lead ores, and occurs also as nodules in limestone and locally as a cement of sandstones. Used to increase the density of concrete in radiation shields and to load DRILLING MUD and thus increase the BACK PRESSURE during drilling. Also *barite*, *heavy spar*. See panel on DRILLING RIG.

barytes concrete (*NucEng*) See LOADED CONCRETE.

barytocalcite (*Min*) A double carbonate of calcium and barium, $BaCa(CO_3)_2$, crystallizing on the monoclinic system, and occurring typically in lead veins.

basal area (*BioSci*) A measure of the extent of trees in an area, being the total cross-sectional area of the trunks.

basal body (*BioSci*) (1) A cylindrical structure, found at the base of cilia composed of nine sets of triplet MICROTUBULES, that serves as a centre for the growth of microtubules in culture. In flagellate or ciliate Protozoa, zoospores or spermatozoids, it occurs as a small, deeply staining granule at the base of the locomotory organelle. Also *basal granule*, *blepharoplast*. (2) Part of a THALLUS fixed to the substrate by rhizoids. See panel on BACTERIA.

basal cell carcinoma (*Med*) A common carcinoma derived from the basal cells of the epidermis. Often a consequence of exposure to sunlight and much more common in those with fair skin, basal cell carcinomas rarely metastasize. Also *rodent ulcer*.

basal conglomerate (*Geol*) A first stage of sedimentation resting on a plane of erosion. See CONGLOMERATE.

basal ganglia (*BioSci*, *Med*) In vertebrates generally, ganglia connecting the cerebrum with other nerve centres. In humans they are localized concentrations of grey matter deep in the cerebral hemispheres and the midbrain concerned with the regulation of movement, often referred to as the extra-pyramidal system. Disease of the basal ganglia gives PARKINSON'S DISEASE and CHOREA. See panel on BRAIN STRUCTURE.

basal lamina (*BioSci*) A thin sheet of extracellular matrix underlying epithelia. It contains, in addition to collagen and other proteins, the distinctive glycoprotein laminin.

basal metabolic rate (*Med*) The minimal quantity of heat produced by an individual at complete physical and mental rest, but not asleep, 12–18 h after eating, expressed in milliwatts per square metre of body surface.

basal placentation (*BioSci*) In plants, placentation in which the ovules are attached to the bottom of the LOCULE in an ovary.

basal planes (*Crystal*) The faces representing the terminating PINACOID in all the crystal systems exclusive of the cubic system.

basal plates (*BioSci*) (1) In the developing vertebrate skull, a plate of cartilage formed by the fusion of the parachordals and the trabeculae. (2) In Crinoidea, certain plates situated at or near the top of the stalk. (3) In Echinoidea, certain plates forming part of the apical disk.

basalt (*Geol*) A fine-grained igneous rock, dark colour, composed essentially of basic plagioclase feldspar and pyroxene, with or without olivine; minor amounts of quartz and feldspathoid may be present. In the field, the term is generally restricted to lavas, but many minor intrusions of basic composition show identical characters, and therefore cannot be distinguished in the laboratory. The term probably comes from ancient Egypt. The extrusive equivalent of GABBRO. See VOLCANIC ROCKS.

basalt glass (*Geol*) See TACHYLITE.

basaltic hornblende (*Min*) A variety of hornblende with a high ferric–ferrous iron ratio and a low hydroxyl content, occurring chiefly in volcanic rocks. Also *oxyhornblende*.

basanite (*Geol*) A basaltic rock containing plagioclase, augite, olivine and a feldspathoid (nepheline, leucite or analcite). A term of great antiquity, probably Egyptian.

bascule bridge (*CivEng*) A counterpoise bridge which can be rotated in a vertical plane about axes at one or both ends. The roadway over the river rises while the counterpoise section descends into a pit or rolls in tracks along the deck of the approach spans. Also *balance bridge*.

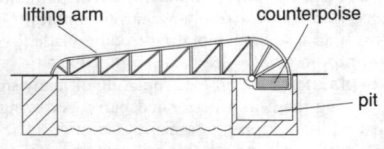

lifting arm   counterpoise   pit

bascule bridge

base (*BioSci*) The end of an organ or plant part nearest to its point of attachment or origin.

base (*Chem*) Generally, a substance which tends to donate an electron pair or co-ordinate an electron. In particular, a substance which dissolves in water with the formation of hydroxyl ions and reacts with acids to form salts.

base (*Electronics*) (1) The region between the emitter and collector of a transistor, into which minority carriers are injected. It is essentially the control electrode of the transistor. (2) The part of an electron tube which has pins, leads or terminals through which connections are made to the internal electrodes.

base (*ImageTech*) The thin flexible support on which a photographic emulsion or magnetic coating is carried.

base (*MathSci*) (1) The number whose powers are represented by the successive positions in a positional number system. When numbers are displayed without qualification, base 10 is normally understood, except in contexts such as computing in which the BINARY SYSTEM is used. Other common bases are 8 (*octal*) and 16 (*hexadecimal*); Also *radix*. (2) The number whose powers are expressed as LOGARITHMS. (3) The side of a *triangle*, *cone*, *cylinder* or other rectilinear figure or solid, on which it stands in a particular orientation.

baseband (*ICT*) The frequency band occupied by the signal in modulation.

baseboard (*Build*) See SKIRTING BOARD.

base bullion (*Eng*) Ingot base metal containing sufficient silver or gold to repay recovery, eg argentiferous lead.

base circle (*Eng*) The circle used in setting out the profiles of gearwheel teeth of involute form.

base course (*Build*) The lowest course of masonry in a building.

Basedow's disease (*Med*) Thyrotoxicosis. See GRAVE'S DISEASE.

base exchange (*Chem*) The reversible replacement of a cation by another in solution. The superficial physical structure of the solid is not affected. The ion exchange may take place in colloids, on surfaces, in crystal lattices, notably zeolites, or in interlayer crystal lattice sites. Applied in SOIL MECHANICS to chemical methods to strengthen clays by replacement of hydrogen ions by sodium ions.

base level (*Geol*) The lowest level towards which erosion progresses.

baseline (*Print*) The bottom alignment of type; below it is the *beard* which accommodates the descenders of f, g, etc.

baseline (*Surv*) A survey line the length of which is very accurately measured by precise methods; used as a basis for subsequent triangulation.

base load (*ElecEng*) The steady load, more economically produced by eg nuclear power with peaks of output carried by other more expensive methods.

basement (*Geol*) (1) A complex of igneous and metamorphic rocks covered by sediments. (2) The crust of the Earth extending downwards to the Mohorovičić discontinuity.

basement membrane (*BioSci*) The extracellular matrix lying between epithelial cells and the underlying connective tissue. It comprises two layers; the BASAL LAMINA, containing non-fibrillar collagen and secreted by the epithelium, and the reticular lamina, containing fibrillar collagen and secreted by connective tissue fibroblasts.

base metal (*Chem, Eng*) The common metals, towards the electronegative end of the ELECTROCHEMICAL SERIES, remote from the NOBLE METALS. They have a relatively negative electrode potential (on the IUPAC system).

base pair (*BioSci*) A purine and a pyrimidine, linked by hydrogen bonds, and found within double-stranded nucleic acid. The purine adenine pairs with the pyrimidine thymine (uracil in RNA), and the purine guanine pairs with the pyrimidine cytosine. Abbrev bp. See panel on DNA AND THE GENETIC CODE.

base pairing rule (*BioSci*) Rule that the bases in the opposite strands of DNA are stable when adenine pairs with thymine and guanine with cytosine.

base resistance (*Electronics*) Total resistance to base current, including spreading effect.

base station (*ICT*) A fixed radio transmission and reception station serving a single CELL of a mobile telephone network and consisting of a BASE STATION TRANSCEIVER controlled by a BASE STATION CONTROLLER.

base station controller (*ICT*) Equipment serving a single CELL of a mobile telephone network and providing the interface between its BASE STATION and a MOBILE SERVICES SWITCHING CENTRE.

base station transceiver (*ICT*) Radio transmission and reception equipment serving a single CELL of a mobile telephone network and capable of operating on several channels simultaneously.

base unit (*Genrl*) One of the seven units on which the coherent international system of units (SI UNITS) is based. All derived units are obtained from the base units by multiplication without introducing numerical factors, and approved prefixes are used in the construction of sub-multiples and multiples. There is only one base or derived unit for each physical quantity. The base units are METRE, KILOGRAM, SECOND, AMPERE, KELVIN, CANDELA and MOLE. See appendices on Units of measurement and SI derived units.

base vector (*MathSci*) In a co-ordinate system, a unit vector taken in the positive direction of a co-ordinate axis; any vector may be expressed as a linear combination of base vectors. Also *basis vector*.

basher (*ImageTech*) A small studio lamp placed close to or on the camera mounting.

Bashkirian (*Geol*) The oldest epoch of the Pennsylvanian period.

basi- (*Genrl*) Prefix from Gk *basis*, base.

BASIC (*ICT*) Abbrev for *beginners' all-purpose symbolic instruction code*. A programming language. It or its derivatives are usually available on microcomputers with an INTERPRETER.

basic chromosome set (*BioSci*) The haploid set of chromosomes as found in the gametes. Because species may have evolved by polyploidy, aneuploidy or chromosome rearrangement, it may also be possible to infer a basic chromosome set for the ancestor. See CHROMOSOME COMPLEMENT.

basicity (*Chem*) The number of hydrogen ions of an acid which can be neutralized by a base.

basic lead carbonate (*Chem*) $2PbCO_3Pb(OH)_2$ (approx composition). See WHITE LEAD.

basic lead chromate (*Chem*) $PbCrO_4Pb(OH)_2$. Used as a pigment. Produced when lead chromate is boiled with aqueous ammonia or potassium hydroxide. Also *Austrian cinnabar*.

basic lead sulphate (*Chem*) $2PbSO_4PbO$. A fine powder, obtained by roasting galena (PbS).

basic loading (*ElecEng*) The limiting mechanical load, per unit length, on an overhead line conductor.

basiconic (*BioSci*) In insects, said of certain sub-conical and immobile sensilla arising from the general surface of the cuticle.

basic process (*Eng*) A steel-making process, Bessemer, open-hearth or electric, in which the furnace is lined with a basic refractory, a slag rich in lime being formed, enabling phosphorus to be removed. See ACID PROCESS.

basicranial (*Med*) Pertaining to, or situated at, the base of the skull.

basic rocks (*Geol*) Igneous rocks with low silica content. The limits are usually placed at 45% silica, below which rocks are described as ultrabasic, and 52% silica, above which they are described as intermediate. Basic igneous rocks include basalt, the commonest type of lava, and gabbro, its plutonic equivalent.

basic six (*Aero*) The group of instruments essential for the flight handling of an aircraft and consisting of the airspeed indicator, vertical speed indicator, altimeter, heading indicator, gyro horizon, and turn and bank indicator.

basic size (*Print*) The sheet size used in the US to determine a particular paper's basis weight. Some of the more common basic sizes are: *bond* (17 in × 22 in); *coated, text, book* and *offset* (25 in × 38 in); *cover* (20 in × 26 in).

basic slag (*Eng*) Furnace slag rich in phosphorus (as calcium phosphate) which, with silicate and lime, is produced in steel-making, and ground and sold for agricultural fertilizer.

basic solvent (*Chem*) A protophilic solvent, hence one which enhances the acidic (ie proton-donating) properties of the solute.

basic steel (*Eng*) Steel which has reacted with a basic lining or additive to produce a phosphorus-rich slag and a low-phosphorus steel.

basic T (*Aero*) A layout of flight instruments standardized for aircraft instrument panels in which four of the essential instruments are arranged in the form of a T. The pitch and roll attitude display is located at the junction of the T flanked by airspeed on the left and attitude on the right. The vertical bar portion of the T is taken up by directional information.

basic weight (*Aero*) The weight of the structure (wing, body, tail unit and landing gear) of an aircraft, plus the propulsion systems and the airframe services and equipment (mechanical systems, avionics, fuel tanks and pipes). Includes residual oil and undrainable fuel but no operational equipment or payload.

basidiocarp (*BioSci*) The fruiting body of the Basidiomycotina. Also *basidioma*.

basidioma (*BioSci*) See BASIDIOCARP.

Basidiomycotina (*BioSci*) A subdivision or class of those Eumycota or true fungi in which the sexual spores are formed on a basidium. Such fungi have no motile stages. They are usually mycelial with septate hyphae. Sexual reproduction typically involves the fusion of a fruiting body in which the basidia develop. Includes the Tiliomycetes comprising the rusts (Uredinales) and smuts (Ustilaginales), the Hymenomycetes and the Gasteromycetes. Also *Basidiomycetes*.

basidiospore (*BioSci*) Spore, typically uninucleate and haploid, formed at the end of a sterigma on a basidium.

basidium (*BioSci*) Specialized, usually more or less club-shaped, cell on which (typically four) basidiospores are formed following the fusion of two heterokaryotic nuclei and meiosis in the reproduction of the Basidiomycetes.

basifixed (*BioSci*) Said of an anther which is attached by its base to the filament.

basifugal (*BioSci*) Transport, differentiation, etc, in the direction away from the base, towards the apex.

basilar (*BioSci*) Situated near, pertaining to, or growing from the base.

basilar membrane (*Acous, BioSci*) In mammals, a flat membrane forming part of the partition of the cochlea in the inner ear. It contains the auditory nerves that translate mechanical vibrations of differing frequencies into nerve impulses.

basin (*Geol*) A large depression in which sediments may be deposited. Alternatively, a gently folded structure in which beds dip inwards from the margin towards the centre.

basin-and-range (*Geol*) A structural area of fault-block mountains separated by alluvium-filled basins.

basion (*Med*) The mid-point of the anterior margin of the foramen magnum.

basipetal (*BioSci*) Transport, differentiation, etc, in the direction towards the base, away from the apex.

basiphil (*Med*) See BASOPHIL.

basiphilia (*Med*) See BASOPHILIA.

basipodium (*Med*) The wrist or ankle.

basis (*MathSci*) (1) A set of BASE VECTORS that determine a space. (2) A collection of open sets such that every open set is a union of open sets in the basis.

basis cranii (*BioSci*) In Craniata, the floor of the cranium, formed from the basal plate of the embryo.

basis vector (*MathSci*) See BASE VECTOR.

basis weight (*Paper*) US method for identifying various papers. The basis weight is the weight in pounds of a ream (500 sheets) of a particular paper in the BASIC SIZE for the grade. The metric system, the GRAMMAGE, expressed in gm^{-2} is now the preferred system.

basket coil (*ElecEng*) Coil with criss-cross layers, so designed to minimize self-capacitance.

basket-weave structure (*Eng*) See WIDMANSTÄTTEN STRUCTURE.

basophil (*Med*) See BASOPHIL LEUCOCYTE. Also *basiphil*.

basophile (*BioSci*) Having a marked affinity for basic dyes. Also *basiphile*.

basophilia (*Med*) An increase of basophil cells in the blood. Also *basiphilia*.

basophil leucocyte (*BioSci*) A white blood cell (*leucocyte*) with granules that bind basic dyes. Has properties similar to MAST CELLS and will bind IgE and release histamine and other mediators on contact with specific antigen.

bas relief (*Arch*) Sculpture or carved work in which the figures project less than their true proportions from the surface on which they are carved.

bass boost (*Acous*) Amplifier circuit adjustment which regulates the attenuation of the lowest frequencies in the

audio scale, usually to offset the progressive loss towards low frequencies.

bass compensation (*Acous*) Differential attenuation introduced into a sound-reproducing system when the loudness of the reproduction is reduced below normal, to compensate for the diminishing sensitivity of the ear towards the lowest frequencies reproduced.

bass frequency (*Acous*) A frequency close to the lower limit in an audio-frequency signal or a channel for such, eg below 250 Hz.

basswood (*For*) A N American tree (*Tilia*) that may grow to over 30 m, giving a HARDWOOD with straight-grained fine and uniform texture, creamy white to lightish brown in colour.

bast (*BioSci*) Phloem.

bastard (*Genrl*) A general term for anything abnormal in shape, size or appearance.

bastard ashlar (*Build*) (1) Stones, intended for ashlar work, which are merely rough-scabbled to the required size at the quarry. (2) The face-stones of a rubble wall selected, squared and dressed to resemble ashlar.

bastard-cut (*Build*) Describes file teeth of a medium degree of coarseness.

bastard font (*Print*) See LONG-BODIED TYPE.

bastard size (*Paper*) Paper or board not of a standard size.

bastard thread (*Eng*) A screw-thread which does not conform to any recognized standard dimensions.

bastard title (*Print*) The fly page before the full title page of a book. Often wrongly called a HALF-TITLE.

bastard tuck pointing (*Build*) Pointing in which a slight projection is given to the stopping on each joint.

bastard wing (*BioSci*) In birds, quill feathers, usually three in number, borne on the thumb or first digit of the wing. Also *ala spuria, alula*.

bast fibre (*Textiles*) Cellulose fibre obtained from the stems of various plants often by a rotting (RETTING) stage followed by BEATING (scutching). Examples are flax, hemp and jute.

bastite (*Min*) A variety of serpentine, essentially hydrated silicate of magnesium, resulting from the alteration of orthorhombic pyroxenes. Also *schillerspar*.

bastnaesite (*Min*) Fluorocarbonate of lanthanum and cerium. An ore mineral of rare earth elements.

bat (*Build*) A portion of a brick, large enough to be used in constructing a wall. See CLOSER.

batch (*Glass*) The mixture of raw materials from which glass is produced in the furnace. A proportion of cullet is either added to the mixture, or placed in the furnace previous to the charge. Also *charge*.

batch box (*CivEng*) See GAUGE BOX.

batch culture (*BioSci*) A culture initiated by the inoculation of cells into a finite volume of fresh medium and terminated at a single harvest after the cells have grown. Cf CONTINUOUS CULTURE.

batch distillation (*ChemEng*) An arrangement by which the boiler of a still is charged with a batch. The components of the batch are distilled off in order of relative volatility. The contrary process is known as CONTINUOUS DISTILLATION.

batch file (*ICT*) A file containing a set of OPERATING SYSTEM commands in the form of a program. When this file is recalled, the operating system executes the commands as if each had been typed in by the user.

batch furnace (*Eng*) A furnace in which the charge is placed and heated to the requisite temperature. The furnace may be maintained at the operating temperature, or heated and cooled with the charge. Distinguished from CONTINUOUS FURNACE.

batching sphere (*MinExt*) An inflatable hard rubber sphere, fitting a pipe, used when miscible oil fractions are being sent down the pipeline to separate the fractions.

batch mill (*Eng*) Cylindrical grinding mill into which a quantity of material for precise grinding treatment is charged and worked till finished.

batch process (*Eng*) Any process or manufacture in which operations are completely carried out on specific quantities or a limited number of articles, as contrasted to continuous or mass production. In semiconductor manufacture, one in which several wafers are treated simultaneously as distinct from stages in which wafers are processed singly.

batch processing (*ICT*) Computing jobs that are run to completion in sequence. Has the disadvantage that turn-around time is long compared with actual processing time. Cf INTERACTIVE COMPUTING, TIME SHARING.

batement light (*Arch*) A window, or one division of a window, having vertical sides, but with the sill not horizontal, as where it follows the rake of a staircase.

Batesian mimicry (*BioSci*) A defence mechanism by which an animal is protected from predators by its resemblance to another animal that is dangerous or unpalatable.

bath lubrication (*Eng*) A method of lubrication in which the part to be lubricated, such as a chain or gearwheel, dips into an oil bath.

BATNEEC (*EnvSci*) Abbrev for *best available technology not entailing excessive costs.*

batho-, bathy- (*Genrl*) Prefixes from Gk *bathys,* deep, used esp with relation to sea depths.

bathochrome (*Chem*) A radical which shifts the absorption spectrum of a compound towards the red end of the spectrum. Cf HYPSOCHROME.

bathochromic (*Chem*) Changing to a longer wavelength (red shift) in the absorption spectrum of a compound.

bathoflare (*Chem*) A radical which shifts the fluorescence of a compound towards the red end of the spectrum. Cf HYPSOFLARE.

batholith (*Geol*) A large body of intrusive igneous rock, frequently granite, with no visible floor. Also *bathylith.*

Bathonian (*Geol*) A stage in the Middle Jurassic. See MESOZOIC.

bathophilous (*BioSci*) Adapted to an aquatic life at great depths.

bathotonic (*Chem*) Tending to diminish surface tension. Cf HYPOTONIC.

BA thread (*Eng*) See BRITISH ASSOCIATION SCREW-THREAD.

bathy- (*Genrl*) See BATHO-.

bathyal (*Geol*) Refers to the ocean-floor environment between c.200 and 4000 m. The three zones of increasing depth are LITTORAL, ABYSSAL and BATHYAL. There are numerous definitions of the depth ranges of these zones.

bathybic (*BioSci*) Relating to, or existing in, the deep sea, eg plankton floating well below the surface.

bathylimnetic (*BioSci*) Living in the depths of lakes and marshes.

bathylith (*Geol*) See BATHOLITH.

bathymetric (*Geol*) Relating to the measurement of the depths of features at the bottom of the oceans, especially by echo sounding.

bathysmal (*BioSci*) See ABYSSAL.

batik dyeing (*Textiles*) Dyeing process in which the fabric is treated with wax to form a pattern that is left unaffected by a dye. The wax may then be removed and a different dye applied to give interesting colour effects. Also *batik printing.*

batiste (*Textiles*) A soft, fine plain-woven fabric often of flax or cotton.

batrachian (*BioSci*) Relating to the Salientia (ie frogs and toads).

batt (*Textiles*) See WEB.

batten (*Build*) (1) A piece of square-sawn converted timber, 2–4 in (50–100 mm) thickness and 5–8 in (125–200 mm) width, used for flooring or as a support for laths. See SLATING AND TILING BATTENS. (2) A bar fastened across a door, or anything composed of parallel boards, to secure them and to add strength and/or reduce warping.

battenboard (*Build*) See BLOCKBOARD.

battened wall (*Build*) See STRAPPING.

Batten's disease (*Med*) A severe progressive genetic disorder that causes blindness, deafness, loss of muscle control and early death.

batter (*Build*) Slope (eg of the face of a structure) upwards and backwards.

battered baby syndrome (*Med*) Describes the state of a baby or child subject to repeated assault from its parent(s) or guardian(s). The child may have multiple bruises of varying age and may have evidence of new and old fractures. Also *non-accidental injury.*

batter level (*Surv*) A form of clinometer for finding the slope of cuttings and embankments.

batter pile (*CivEng*) See RAKING PILE.

batter post (*Build*) One of the inclined side-timbers supporting the roof of a tunnel.

battery (*ElecEng*) General term for a number of objects co-operating together, eg a number of accumulator cells, dry cells, capacitors, radars, boilers, etc. See ACCUMULATOR.

battery booster (*ElecEng*) A motor–generator set used for giving an extra voltage, to enable a battery to be charged from a circuit of a voltage equal to the normal voltage of the battery.

battery coil ignition (*Autos*) High-tension supply for sparking plugs in automobiles, in which the interruption of a primary current from a battery induces a high secondary emf in another winding on the same magnetic circuit, the high tension being distributed in synchronism with the contact breaker in the primary circuit.

battery cut-out (*ElecEng*) An automatic switch for disconnecting a battery during its charge, if the voltage of the charging circuit falls below that of the battery.

battery hen (*Agri*) A female chicken housed indoors in restricted conditions and reared for intensive egg production.

battery regulating switch (*ElecEng*) A switch to regulate the number of cells connected in series in a battery.

battery spear (*ElecEng*) A special form of spike used to connect a voltmeter to the plates of the accumulator cells for battery testing under load. The voltmeter incorporates a low resistance in shunt which simulates a heavy load on the battery, thus testing its work capability. The heavy current passed for this purpose necessitates special heavy-duty battery connectors.

battery traction (*ElecEng*) An electric-traction system in which the current is obtained from batteries (accumulators) on the vehicles.

battery vehicle (*ElecEng*) See BATTERY TRACTION.

baud (*ICT*) A measure of the signalling speed in a digital communication system; the speed in bauds is the number of discrete conditions or signal events per second, eg 1 baud equals 1 bit per second in a train of binary signals. Since many digital systems transmit additional information for control and signalling, the baud rate is not necessarily the same as the DATA SIGNALLING RATE.

Baudot code (*ICT*) Code in which five equal-length bits represent one character; sometimes used for teleprinters where one start and one stop element are added to each group of 5 bits.

Baudouin reaction (*Chem*) A test for certain vegetable oils which give a characteristic red colour with alcoholic furfural and concentrated HCl, or with $SnCl_2$ and HCl.

Bauhaus (*Arch*) A school founded in 1919 by the architect Walter Gropius, in Weimar, Germany. It aimed to achieve modern design in all aspects of art, in a manner which acknowledged the advancement of industrial technology. The school is esp noted for pioneering a style of architecture which defined clearly the various functional elements of the accommodation within a building.

baulk (*CivEng*) See BALK.

baulk (*For*) A piece of timber square-sawn from the log to a size greater than 6×6 in^2 (150×150 mm^2).

Baumé hydrometer scale (*Phys*) For the continental Baumé hydrometer, the rational scale proposed by Lunge, in which $0°$ is the point to which it sinks in water and $10°$ the point to which it sinks in a 10% solution of sodium chloride, both liquids being at $12.5°C$.

Baum jig (*MinExt*) Pneumatically pulsed JIG used in coal-washing plants to lift and remove a lighter and low-ash fraction from a denser one containing shale and high-ash material (dirt), which is stratified downwards by the effect of pulsed water, and separately withdrawn.

bauxite (*Geol*) A residual rock composed almost entirely of aluminium hydroxides formed by weathering in tropical regions. The most important ore of aluminium. See LATERITE.

Baventian (*Geol*) A cold stage in the Pleistocene. See QUATERNARY.

bay (*Arch*) Any division or compartment of an arcade, roof, building, etc, or space from column to column in a building.

bay (*ICT*) (1) Unit of racks designed to accommodate numbers of standard-sized panels, eg repeaters or logical units. (2) Unit of horizontally extended antenna, eg between masts.

Bayard and Alpert gauge (*Phys*) A device for measuring very low gas pressure by collecting ions on a fine wire inside a helical grid.

Bayer pattern (*ImageTech*) A commonly used arrangement of the three colour filters over a CCD ARRAY in a digital camera.

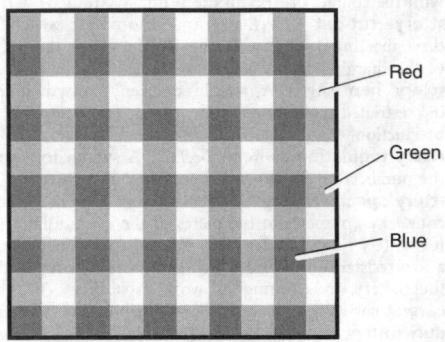

Bayer pattern

Bayer process (*Chem*) A process for the purification of bauxite, as the first stage in the production of aluminium. Bauxite is digested with a sodium hydroxide solution which dissolves the alumina and precipitates oxides of iron, silicon, titanium, etc. The solution is filtered and the aluminium precipitated as the hydroxide.

Bayesian (*MathSci*) Proceeding from a definition of probability as a measurement of belief; considering statistical inference as a process of re-evaluating such probabilities on the basis of empirical observation.

bayonet cap (*ElecEng*) A type of cap fitted to an electric lamp, consisting of a cylindrical outer wall fitted with two or three pins for engaging in L-shaped slots in a lampholder (*bayonet holder*). Within the wall are two contacts connected to the filament, which make contact with two pins in the lampholder. Abbrev *BC*. See CENTRE-CONTACT CAP, SMALL BAYONET CAP.

bayonet fitting (*Eng*) An engineering fastening similar to a BAYONET CAP.

bayonet holder (*ElecEng*) See BAYONET CAP.

bay-stall (*Build*) See CAROL.

baywood (*For*) See AMERICAN MAHOGANY.

bazooka (*ICT*) See BALUN.

B-battery (*ElecEng*) US for HIGH-TENSION BATTERY.

BBB polymers (*Chem*) Abbrev for *polybisbenzimidazoben-zophenanthroline polymers*. See POLYBENZIMIDAZOLES.

bbl (*MinExt*) Abbrev for BARREL.

BBL polymers (*Chem*) Ladder polymers with chain structure very similar to POLYBENZIMIDAZOLES, with stability to over 600°C.

BBS (*ICT*) Abbrev for BULLETIN BOARD SYSTEM.

BC (*ElecEng*) Abbrev for BAYONET CAP.

BCC (*Chem*) Abbrev for *body-centred cubic*. See BODY-CENTRED CUBIC PACKING.

BCD (*ICT*) Abbrev for BINARY-CODED DECIMAL.

B-cell (*BioSci*) See B-LYMPHOCYTE.

BCF (*Chem*) Abbrev for BROMOCHLORODIFLUORO-METHANE.

BCG (*BioSci*) See BACILLE CALMETTE GUÉRIN.

B-channel (*ICT*) See BEARER CHANNEL.

B-chromosomes (*BioSci*) Accessory non-essential chromosomes present in variable numbers in addition to the normal *A-chromosomes*. They are usually small and heterochromatic (see HETEROCHROMATIN).

B-class insulation (*ElecEng*) A class of insulating material which will withstand temperatures up to 130°C.

BCS (*ICT*) Abbrev for *British Computer Society*. Professional association.

BCS theory (*Phys*) Bardeen, Cooper and Schrieffer theory of superconductivity. See COOPER PAIR.

BCT (*Chem*) Abbrev for *body-centred tetragonal*. See BODY-CENTRED TETRAGONAL STRUCTURE.

B-display (*Radar*) Rectangular radar display with target bearing indicated by horizontal co-ordinate and target distance by the vertical co-ordinate, the targets appearing as bright spots.

BDV (*ElecEng*) Abbrev for BREAKDOWN VOLTAGE.

Be (*Chem*) Symbol for BERYLLIUM.

beaching gear (*Aero*) Floatable, detachable, temporary trolleys which enable a seaplane to be run on and off the shore or slipway.

beach marks (*Eng*) Fracture surface markings associated with fatigue crack propagation. See FRACTOGRAPHY and panel on FATIGUE.

beacon (*Aero, Ships*) (1) System of visual lights indicating fixed features, eg masts, reefs. (2) A *radio beacon*, which can be of any frequency but is usually very high frequency, and can be omnidirectional or of directional beam type. *Vertical fan marker* beacons are radio beams used to identify particular spots in control zones and on approach patterns. A *non-directional beacon* (abbrev *NDB*) is a transmitter, the bearing of which can only be determined by an aircraft equipped for direction finding. See INSTRUMENT LANDING SYSTEM.

bead (*Build*) A small convex moulding formed on wood or other material.

bead-and-quirk (*Build*) A bead formed with a narrow groove separating it from the surface which it is decorating. Also *quirk-bead*.

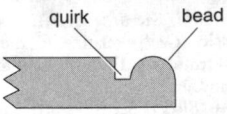

quirk   bead

bead-and-quirk

bead-coil (*Eng*) The part which keeps the tyre on the wheel rim. See panel on TYRE TECHNOLOGY.

bead-jointed (*Build*) Said of the form of jointing in which one of the butting edges has a bead.

bead-tool (*Build*) A specially shaped cutting tool used in wood-turning for forming convex mouldings.

beak (*BioSci*) See ROSTRUM.

beak (*Build*) The crooked end of a bench hold-fast.

beak iron (*Eng*) (1) The pointed, or horn-shaped, end of a blacksmith's anvil; used in forging rings, bends, etc. (2) A T-shaped stake, similarly shaped, fitting in the hardy hole of the anvil. Also *beck iron*, *bickern*, *bick iron*.

beam (*Eng*) (1) A bar or member which is loaded transversely, predominantly in bending. (2) Rolled or extruded sections of certain profiles, eg I-beam.

beam (*Phys*) A collimated, or approximately unidirectional, flow of electromagnetic radiation (radio, light, X-rays, or of particles (atoms, electrons, molecules). The angular beam width is defined by the half-intensity points.

beam (*Textiles*) A wooden or metal cylinder having large flanges at each end. Warp yarns are wound on the beam from cones or cheeses correctly arranged for inserting into the loom or warp-knitting machine. Beams are also used to furnish thread during lace-making.

beam antenna (*ICT*) Generally, any antenna that has DIRECTIVITY. Most commonly used to describe short-wave or very-high-frequency antennas, rather than microwave antennas which are almost invariably directional.

beam balance (*Chem*) A balance in which the weight of the sample contributes to the balance of moments of a beam about a central fulcrum.

beam compasses (*Eng*) An instrument for describing large arcs. It consists of a beam of wood or metal carrying two beam heads, adjustable for position along the beam, and serving as the marking points of the compasses. Also *trammels*.

beam-coupling coefficient (*ElecEng*) The ratio of the ac signal current produced to the dc beam current in beam coupling.

beam current (*Electronics*) That portion of the gun current in cathode-ray tube which passes through the aperture in the anode and impinges on the fluorescent screen.

beam-engine (*Eng*) A form of construction used in early steam-engines, now obsolete. The vertical steam cylinder acted at one end of a pivoted beam, the workload being connected to the other.

beam-filling (*Build*) Brick, masonry or concrete work used between joists carried upon a wall.

beam-forming electrode (*Electronics*) Electrode to which a potential is applied to concentrate the electron stream into one or more beams. Used in beam tetrodes and cathode-ray tubes.

beam hole (*NucEng*) Hole in shield of reactor, or that around a cyclotron, for extracting a beam of neutrons or γ-rays or to insert equipment or samples for irradiation.

beam lead (*Electronics*) An integrated circuit bonding option for high-frequency applications in which material is etched clear of part of the metallization layer to provide a short beam of metal (usually gold). The chip is then inverted and the beam is bonded direct to conducting tracks.

beam relay (*ElecEng*) An electromagnetic relay in which the contacts are mounted on a balanced beam with energizing coils acting on each end and tending to tilt it one way or the other.

beam rider (*ICT*) System in which a guided missile maintains and returns to a course of maximum signal on a radio beam. See GUIDED MISSILE.

beam splitter (*ImageTech*) Optical device for dividing a light beam into two or more paths. In particular, a prism system in a camera to produce three colour-separation images from a single objective lens.

beam system (*ICT*) A point-to-point radio system in which highly directive transmitting and receiving antennas are used.

beam tetrode (*Electronics*) Tetrode having an additional pair of plates, normally connected internally to the cathode, so designed as to concentrate the electron beam between the screen grid and anode, and thus reduce secondary emission effects.

bean (*ICT*) A reusable software component, generally one written in the Java programming language.

bearded (*BioSci*) Having an awn; bearing long hairs like a beard.

bearded needle (*Textiles*) See SPRING NEEDLE.

bearding (*ImageTech*) Picture defect in which dark image areas spread into adjacent light areas.

Beard protective system (*ElecEng*) A form of balanced protective system in which the current entering the winding of an alternator is balanced against that leaving it by passing the conductor at the two ends round the core

of a single current transformer, in opposite directions, so that there is normally no flux in the transformer core.

bearer (*ICT*) The physical medium and set of protocols used to carry useful traffic as opposed to those used merely for the control signals that set up and maintain the link.

bearer cable (*ElecEng*) See MESSENGER WIRE.

bearer channel (*ICT*) A single 64 Kbps channel within an INTEGRATED SERVICES DIGITAL NETWORK link. Also *B-channel*.

bearing (*Build*) The part of a beam or girder which actually rests on the supports.

bearing (*ICT*) Angle of direction in horizontal plane in degrees from true north, eg of an arriving radio wave as determined by a direction-finding system. Also *azimuth*.

bearing (*Surv*) The horizontal angle between any survey line and a given reference direction.

bearing current (*ElecEng*) A stray current, induced by magnetic flux linking the shaft of an electrical machine, that flows between the shaft and bearings and may injure the bearing surfaces.

bearing distance (*Build*) The unsupported length of a beam between its bearings. Also *clear span*.

bearing metals (*Eng*) Metals (alloys) used for that part of a bearing which is in contact with the JOURNAL; eg bronze or white metal, used on account of their low coefficient of friction when used with a steel shaft.

bearing pile (*CivEng*) A column which is sunk or driven into the ground to support a vertical load by transmitting it to a firm foundation lower down, or by consolidating the soil so that its bearing power is increased. Formerly of timber but now more usually reinforced concrete or steel.

bearings (*Eng*) Supports provided to locate a revolving or reciprocating shaft.

bearing surface (*Eng*) That portion of a bearing in direct contact with the JOURNAL; the surface of the journal. See BRASSES.

bearing wall (*CivEng*) The supporting or abutment wall of a bridge or arch.

beat (*ICT*) Periodic variation in the amplitude of a summation wave containing two sinusoidal components of nearly equal frequencies.

beater (*Paper*) A vat containing a heavy cylindrical roll (*beater roll*), fitted with bars, parallel to the JOURNAL, which rotates against a fixed set of bars (BEDPLATE). The paper fibres in suspension in water pass between these bars in preparation for sheet-making.

beater (*Textiles*) High-speed revolving shaft having arms equipped with blades or pins. These beat out the heavy impurities in matted raw fibres in opening and scutching processes.

beater mill (*MinExt*) In rockbreaking, a mill with swinging hammers, disks or heavy plates, which revolve fast and hit a falling stream of ore with breaking force. Also *hammer mill*. See DISINTEGRATING MILL.

beat frequency (*ICT*) Generally, the difference frequency produced by the intermodulation of two frequencies. Specifically, the intermediate frequency in a superhet receiver.

beat-frequency oscillator (*ICT*) The frequency changing stage of a superhet radio receiver. Abbrev *BFO*. See HETERODYNE OSCILLATOR.

beat-frequency wavemeter (*ICT*) See HETERODYNE WAVEMETER.

beating (*Acous*) The subjective difference tone when two sound waves of nearly equal frequencies are simultaneously applied to one ear. It appears as a regular increase and decrease of the combined intensity.

beating (*Paper, Textiles*) (1) Process for partially breaking down the cell-wall structure of cellulose fibres in water before forming paper sheet. (2) Process for removing heavy impurities from matted, raw, natural textile fibres in the opening and SCUTCHING process. (3) The spare threads available during the weaving of wool to replace missing warp threads in the mending process.

beating-up (*Textiles*) The process in weaving by which the newly inserted weft thread is pushed against the edge of the woven fabric.

Beattie–Bridgeman equation of state (*Phys*) A semi-empirical equation of state for the compressibility of gases. It is most conveniently stated in the virial form as

$$\frac{pV}{n} = RT + \frac{n\beta}{V} + \frac{n^2\gamma}{V^2} + \frac{n^3\delta}{V^3}$$

where β, γ and δ are virial coefficients.

Beaufort notation (*EnvSci*) A code of letters used for indicating the state of the weather, eg *b* stands for *blue sky*, *o* for *overcast*, *r* for *rain*.

Beaufort scale (*EnvSci*) A numerical scale of wind force, ranging from 0 for winds less than 1 knot to 12 for winds within the limits 110–118 knots.

beauty (*Phys*) See BOTTOM.

beaver board (*Build*) A soft building board made of wood-fibre material.

beaver cloth (*Textiles*) Heavy woollen overcoating simulating the lustrous nap of the skin of the beaver by milling and raising the fibres, cutting them level and laying them in the same direction.

beavertail antenna (*ICT*) An antenna producing a broad, flat, radar beam.

Bechgaard salt (*Phys*) $(TMTSF)_2X$ where X is an inorganic anion such as $(PF_6)^-$, $(AlO_4)^-$, $(ReO_4)^-$, and TMTSF is the tetramethyl selenium derivative of TTF (tetrathiofulvalene). These salts are *organic electrical conductors*.

beck (*Genrl*) See BACK.

beck iron (*Eng*) See BEAK IRON.

Becke line (*Min*) A narrow line of light seen under the microscope at the junction of two minerals (or a mineral and the mount) in contact in a microscope section. Used in refractive index determinations.

Beck hydrometer (*Phys*) Hydrometer for measuring the relative density of liquids less dense than water. Graduated in degrees Beck, where $°Beck = 200(1-rel.d.)$.

Beckmann apparatus (*Eng*) Apparatus used for measuring the freezing and boiling points of solutions (eg in the CRYOSCOPIC METHOD).

Beckmann molecular transformation (*Chem*) The transformation and rearrangement of ketoxime molecules into acid amides or anilides under the influence of reagents, such as acetyl chloride. An important reaction for determining the configuration of steroisomeric KETOXIMES.

Beckmann thermometer (*Eng*) A limited-range mercury thermometer with a large bulb. It is used to measure small changes of temperature with great precision. Its mean range can be altered by moving mercury from a reservoir in or out of the bulb.

becquerel (*Phys*) SI unit of radioactivity; 1 becquerel is the activity of a quantity of radioactive material in which one nucleus decays per second. Symbol Bq. Replaces the CURIE. $1\,Bq = 2{\cdot}7 \times 10^{-11}$ Ci. It is a very small unit and commonly used with the standard SI prefixes, a gigabecquerel (GBq or 10^9 Bq) being often needed. See panel on RADIATION.

Becquerel cell (*Electronics*) See PHOTOCHEMICAL CELL.

becquerelite (*Min*) Hydrated oxide of uranium, an alteration product of uraninite.

bed (*Build*) The upper or lower surface of a building-stone or ashlar when it is built into a wall; the horizontal surface upon which a course of bricks is laid in mortar.

bed (*Chem*) A packed, porous mass of solid reagent, adsorbent or catalyst through which a fluid is passed for the purpose of chemical reaction.

bed (*Geol*) A small rock unit; the smallest formal lithostratigraphic unit.

bedding (*Build*) Material on which an underground pipe is laid, providing support for the pipe. Can be concrete, granular material or the prepared trench bottom.

bedding (*Geol*) See STRATIFICATION.

bedding-in (*Eng*) The process of accurately fitting a bearing to its shaft by scraping the former until contact occurs uniformly over the surface.

bedding plane (*Geol*) Surface that separates layers of rock. It is caused by changes in mineralogy, grain size, colour, etc.

Bedford cord (*Textiles*) Cloth with rounded cords separated by fine sunken lines and running in the warp direction. Made from wool for riding breeches and worsted yarns for suiting materials.

bed joints (*Build*) The horizontal joints in brickwork or masonry: the radiating joints of an arch.

bed-moulding (*Build*) Any moulding used to fill up the bare space beneath a projecting cornice.

bedplate (*Eng*) A cast-iron or fabricated steel base, to which the frame of an engine or other machine is attached.

bedplate (*Paper*) A plate into which metal bars are inserted; situated beneath the roll in a beater.

bedrock (*MinExt*) Barren formation (seat earth, clay, 'farewell rock') underlying the exploitable part of a mining deposit.

beech (*For*) A tree (*Fagus*) yielding a hardwood with straight grain and uniform texture. Its colour ranges from whitish to a light reddish brown.

beef (*Geol*) Fibrous calcite occurring in veins in sedimentary rocks. Rarely, other minerals with the same structure and occurrence.

beef cattle (*Agri*) Large-bodied breeds bred and selected for maximum muscle mass of high eating quality, as well as rapid growth rate and high fertility.

Beehive (*Astron*) See PRAESEPE.

beekite (*Min*) A variety of chalcedony, commonly occurring as an incrustation on pebbles.

beer (*FoodSci*) An alcoholic drink produced by yeast fermentation of a solution of sugar derived from cereals (usually malted barley). The sugary solution (WORT) also contains vitamins and amino acids and trace elements necessary for the growth of yeast to provide ethanol and CO_2. The CO_2 produced during fermentation is mainly responsible for the effervescence in beer. See BREWING.

Beer's law (*Chem*) Law stating that the degree of absorption of light varies exponentially with the thickness of the layer of absorbing medium, its molar concentration and EXTINCTION COEFFICIENT.

Beestonian (*Geol*) A cold stage in the Pleistocene. See QUATERNARY.

beeswax (*Chem*) A white or yellowish plastic substance obtained from honeycomb of the bee, mp 63–65°C. It consists of the myricyl (melissyl) ester of palmitic acid $C_{15}H_{31}COOC_{30}H_{61}$, free cerotic acid $C_{25}H_{51}COOH$, and other homologues. Used in eg polishes, modelling and ointments.

beetle (*BioSci*) A member of the insectan order of Coleoptera.

beetle (*Build*) A heavy mallet, or wooden hammer, used for driving wedges, consolidating earth, etc. Also *mall*, *maul*.

beetle (*Textiles*) A machine consisting of a row of wooden or metal hammers, which fall on a roll of damp cloth as it revolves. The operation closes the spaces between the warp and the weft yarns, and imparts a soft glossy finish to cotton and linen.

beetle-stones (*Geol*) Coprolitic nodules akin to septaria which, when broken open, give a fancied resemblance to a fossil beetle.

beetroot red (*FoodSci*) See BETANIN.

beet sugar (*Chem*) Sucrose derived from sugar beet. Identical with sucrose derived from sugar cane.

beforsite (*Min*) Medium- to fine-grained dolomite carbonatite, mainly consisting of dolomite and occurring principally as dykes.

Beggiatoales (*BioSci*) An order of chemosynthetic sulphur-oxidizing gliding bacteria that occur mostly as filaments. Sulphur granules occur intracellularly.

Behaviourism (*Psych*) The school of psychology that considers only observable behaviour as appropriate subject matter for study, and that views with distrust explanations that refer to non-observable mental events such as consciousness and imagery.

behaviour therapy (*Psych*) The application of behavioural theory, that behaviour disorders are the result of maladaptive learning, to the treatment of, eg mental illness, by conditioning or reinforcement of normal behaviour. Behaviour therapy is based on classical conditioning principles; behaviour modification is based on operant conditioning principles.

behind the pipe (*MinExt*) Refers to a gas or oil reservoir outside the CASING string.

beidellite (*Min*) A variety of the montmorillonite group (smectites) of clay minerals.

Beilby layer (*MinExt*) Flow layer produced by polishing a metal or mineral surface, in which the true lattice structure is modified or destroyed by incipient fusion.

Beilstein test (*Chem*) Negative test for the presence of a halogen in an organic compound. The latter is heated in an oxidizing flame on a copper wire; if no halogen is present there is no green colour to the flame. Volatile nitrogen compounds also give a green colour.

bel (*MathSci*) A non-dimensional unit used to express the *common logarithm* of a number, so that the multiplication of the number by a factor can be accomplished by the addition of the logarithm of the factor to the measure in bels. If N is the measure of a in bels, $N = \log_{10} a$, the measure of $r^2 N$ is $2 \log_{10} r + \log_{10} a$ etc. Useful wherever constant coefficients are applied to variable quantities, eg amplifiers or attenuators in electrical circuits. See DECIBEL, NEPER.

belemnite (*BioSci*) An extinct cephalopod group, related to ammonites, that had an internal, chambered shell (*phragmocone*) with a terminal pencil-shaped solid region, the *rostrum* or 'guard', that may have acted as a counterweight. The fossil remains of the rostral portion of these shells are commonly found and look rather like mineral bullets, ranging from 1 to 20 cm in size. Belemnites were common from the Lower Jurassic period to the end of the Cretaceous period.

belemnoid (*BioSci*) Dart-shaped. See also BELEMNITE.

belfry (*Arch*) A tower, either detached or forming part of a building, containing suspended bells.

Belgian truss (*CivEng*) See FRENCH TRUSS.

belite (*Min*) *Dicalcium silicate.* Ca_2SiO_4. A constituent ($\approx 20\%$ by weight) of Portland cement, it hydrates at a slow rate during the setting reaction.

bell (*Acous*) Hollow metallic vessel with a flared mouth which, when struck, vibrates with a fundamental frequency determined by parameters such as its mass and dimensions.

Bellatrix (*Astron*) A bright blue-white giant star in the constellation Orion. Distance 140 pc. Also *Gamma Orionis*.

bell centre punch (*Eng*) A centre punch whose point is automatically located centrally on the end of circular work by a sliding hollow conical guide.

bell chuck (*Eng*) See CUP CHUCK.

bell-crank lever (*Eng*) A lever consisting of two arms, generally at right angles, with a common fulcrum at their junction.

bell gable (*Arch*) A gable built above the roof in a church having no belfry, and pierced to accommodate a bell.

Bellini's ducts (*BioSci*) In the kidney of vertebrates, ducts formed by the union of the primary collecting tubules and opening into the base of the ureter at the pelvis of the kidney.

bell metal (*Eng*) High-tin bronze, containing up to 30% tin and some zinc and lead. Used in casting bells. See COPPER ALLOYS.

bell-metal ore (*Min*) See STANNITE.

bell-mouthed (*Eng*) Said of a hole or bore when its diameter gradually increases towards one or both open ends, the bore profile in section being curved. Usually a manufacturing fault.

bellows (*Eng*) A flexible, corrugated tubular machine element used for pumping, for transmitting motion, as an expansion joint, etc.

bellows (*ImageTech*) The flexible connection between parts of a camera or enlarger, necessarily light-tight, to permit delicate adjustments, usually of focusing.

bell-shaped curve (*Genrl*) A normal distribution or normal curve. A bell-shaped curve is a perfect mesokurtic curve where the mean, median and mode are equal.

Bell's palsy (*Med*) Sudden paralysis of the muscles of one side of the face, due to impaired conduction in the lower part of the facial nerve. The cause is unknown, but the majority of cases recover.

bell-type furnace (*Eng*) A portable inverted furnace or heated cover operated in conjunction with a series of bases upon which the work to be heated can be loaded and then left to cool after heat treatment. Used chiefly for bright annealing of non-ferrous metals and bright hardening of steels.

belly (*Print*) The front of a type-letter which bears the nick; placed uppermost in the setting stick by the compositor.

belly tank (*Aero*) See VENTRAL TANK.

Belousov–Zhabotinsky reaction (*Chem*) An example of chemical oscillations in dissipative structures, giving rise to characteristic sustained spatial patterns in concentrations of reactants.

belt (*Build*) A projecting course of stones or bricks. Also *belt course*, *string course*.

belt (*Eng*) A strip of leather, cotton, plastic, reinforced rubber, etc, generally of rectangular cross-section, used for lifting slings and strengthening bands. In endless form used as driving, conveyor, abrasive and other belts. See VEE BELT.

Belt (*Geol*) A great thickness (perhaps 12 000 m) of younger PRECAMBRIAN rocks occurring in the Little Belt Mts, Montana, Idaho and British Columbia. Argillaceous strata predominate, accompanied by algal limestones. Correlated with the Grand Canyon Series in Colorado and the Uinta Quartzite Series in the Uinta Mts. Approx equivalent of Riphean. Also *Beltian Series*.

belt conveyor (*Eng*) See BAND CONVEYOR.

belt course (*Build*) See BELT.

belt drive (*Eng*) The transmission of power from one shaft to another by means of an endless belt running over pulleys having correspondingly shaped rims.

belt fork (*Eng*) Two parallel prongs attached at right angles to a sliding rod, used to slide a flat belt from a fast to a loose pulley and vice versa. Also *belt striker*.

Beltian Series (*Geol*) See BELT.

belting (*Eng*) A general term descriptive of materials from which driving belts are made, eg leather, cotton, balata, woven hair, plastics, etc.

belt slip (*Eng*) The slipping of a driving belt on the face of a pulley, due to insufficient frictional grip to overcome the resistance to motion.

belt striker (*Eng*) See BELT FORK.

belt transect (*BioSci*) A strip of ground marked between two parallel lines so that its vegetation may be recorded and studied. See QUADRAT, TRANSECT.

belvedere (*Arch*) A room from which to view scenery; it is built on the top of a house, cliff, etc, the sides being either open or glazed.

benazepril (*Pharmacol*) A pro-drug that is metabolized to active benazeprilat, a non-sulphydryl ACE INHIBITOR used for treating hypertension.

Bence–Jones protein (*BioSci*) A protein present in the urine of some persons with PLASMACYTOMAS; it consists of immunoglobulin light-chain dimers.

bench (*ICT*) Fixed rails with adjustable and slidable supports for a waveguide system.

benched foundation (*Build, CivEng*) A foundation which is stepped at the base to safeguard against sliding on sloping sites.

benching iron (*Surv*) A small steel plate sometimes used to provide a solid support for the staff at a change point. It is formed usually of a triangular plate, with the corners turned down so that they may be driven into the ground surface to fix the plate in position, while the staff rests upon a raised central portion.

bench mark (*Surv*) A fixed point of reference for use in levelling, the reduced level of the point with respect to some assumed datum being known. Abbrev *BM*. See ORDNANCE BENCH MARK.

benchmarking (*ICT*) The practice of measuring the performance of a system (or piece of hardware or software) by reference to a commonly agreed standard rate of processing (eg so many reads/writes per second).

bench plane (*Build*) A plane for use on flat surfaces. See JACK PLANE, SMOOTHING PLANE, TRYING PLANE.

bench screw (*Build*) The vice fixed at one end of a bench.

bench stop (*Build*) A metal or wooden stop, adjustable for height, set in the top of a bench, at one end; used to hold work while it is being planed.

bench test (*Eng*) A complete functional test of a piece of apparatus, when new or after repair, carried out in a workshop or laboratory. It is undertaken to ensure correct and satisfactory operation prior to installation in a situation where repair may be difficult.

bench work (*Build, Eng*) (1) Work executed at the bench with hand tools or small machines, as distinct from that done at the machines. (2) Small moulds made on a bench in the foundry.

bend (*Eng*) (1) To form into a curved or angular shape. (2) A curved length of tubing or conduit used to connect the ends of two adjacent straight lengths which are at an angle to one another.

bend (*ICT*) Alteration of direction of a rigid or flexible waveguide. It is *E* or *minor* when the electric vector is in plane of arc of bending and *H* or *major* when at right angles to this. Also *corner*.

Ben Day tints (*Print*) Celluloid sheets with a patterned surface, and inked impression from which is used to lay a MECHANICAL STIPPLE.

bending moment (*Eng*) At any transverse section of a beam, the algebraic sum of the moments of all the forces to either side of the section.

bending moment diagram (*Eng*) Diagram representing the variation of bending moment along a beam. It is a graph of bending moment (*y*-axis) against distance along the beam axis (*x*-axis).

bending of strata (*Geol*) See FOLDING.

bending rollers (*Print*) Rollers at the nose of rotary presses. Also *forming rollers*.

bending rolls (*Eng*) Usually three rolls with axes arranged in a triangle so that adjusting one relative to the others forms a curve on a strip or sheet of metal passed between them.

bending strength (*Eng*) The ability of a beam, or other structural member, to resist a BENDING MOMENT. Also *flexural strength*. See STRENGTH MEASURES.

bending test (*Eng*) (1) A test made on a beam to determine its deflection and strength under bending load. The most usual forms are symmetrical three-point and symmetrical four-point bending, the advantage of the latter being that a constant bending moment is imposed between the two central loading points. Also *flexural test*. (2) A forge test in which flat bars etc are bent through 180° as a test of ductility.

bending wave (*Acous*) Wave observed on thin plates and bars. The motion is perpendicular to the direction of propagation. Important for sound radiation from walls and enclosures.

bendrofluazide (*Pharmacol*) Thiazide DIURETIC used in the treatment of oedema and hypertension.

bends (*Med*) See CAISSON DISEASE.

Benedict's test (*Chem, Med*) Test for glucose (and for reducing disaccharides), involving the oxidation of the sugar by an alkaline copper sulphate solution, in the presence of sodium citrate, to give a red copper (I) oxide. Used in testing of urine in treatment of diabetes.

beneficiation (*MinExt*) See MINERAL PROCESSING.

benign tumour (*Med*) A clone of neoplastic cells that does not invade locally or metastasize, having lost growth control but not positional control. Usually surrounded by a fibrous capsule of compressed tissue.

Benioff zone (*Geol*) A plane beneath the trenches of the Pacific dipping under the continents; the site of earthquake activity.

benitoite (*Min*) A strongly dichroic mineral, varying in tint from sapphire blue to colourless, discovered in San Benito County, California. Silicate of barium and titanium.

benmoreite (*Min*) A sodic variety of trachyandesite, consisting of anorthoclase and sodic sanidine.

benomyl (*BioSci*) A pesticide and fungicide, with possible health risks, used on growing fruit and vegetables.

bent chisel (*Build*) Carving tool for recessing backgrounds, made in three types: right-angled, and right- and left-corner bent.

bent gouge (*Build*) A curved gouge for hollowing out concave work.

benthic (*BioSci*) Living at the soil–water interface at the bottom of a sea or lake.

benthon (*BioSci*) Collectively, the sedentary animal and plant life living on the sea bottom. Also *benthos*. Cf NEKTON, PLANKTON. Adj *benthic*.

bent knees (*Vet*) Flexion of the carpus of horses or dogs due to permanent contraction of the flexor tendons or to chronic arthritis.

bentonite (*Geol, MinExt*) A valuable clay, similar in its properties to fuller's earth, formed by the decomposition of volcanic glass, under water. Consists largely of MONTMOR-ILLONITE. Used as a bond for sand, asbestos, etc; to line landfill pits and in the paper, soap and pharmaceutical industries; also as a FINING agent, eg for the removal of protein haze in wines. Thixotropic properties exploited for altering the viscosity of oil DRILLING MUDS.

bent-tail carrier (*Eng*) A LATHE CARRIER having a bent shank projection into, and engaged by, a slot in the driving plate or chuck.

benzal chloride (*Chem*) *Dichloromethylbenzene*. $C_6H_5CHCl_2$. Bp 207°C. A chlorination product of toluene, intermediate for the production of benzaldehyde. Also *benzylidene chloride*.

benzaldehyde (*Chem*) *Benzene carbaldehyde, oil of bitter almonds*. C_6H_5CHO. Mp 13°C, bp 179°C, rel.d. 1·05, a colourless liquid, with aromatic odour, soluble in alcohol, ether, slightly in water. Flavouring agent.

benzaldoximes (*Chem*) $C_6H_5CH=NOH$. Formed from benzaldehyde and hydroxylamine; there are two stereo-isomeric forms. The *alpha* or *antiform*, mp 35°C, can be transformed by means of acids into the *beta* or *syn*-form, mp 125°C.

benzamide (*Chem*) *Benzene carboxamide*. $C_6H_5CONH_2$. The amide of benzoic acid, obtainable from benzoyl chloride and ammonia or ammonium carbonate; lustrous plates, mp 130°C.

benzanilide (*Chem*) *N-phenylbenzamide*. $C_6H_5CONHC_6H_5$. Colourless plates, mp 158°C; the anilide of benzoic acid or benzoyl chloride.

benzanthrone (*Chem*) Yellow, crystalline powder. An intermediate widely used in manufacture of vat dyestuffs.

benzene (*Chem*) C_6H_6. Mp 5°C, bp 80°C, rel.d. 0·879; a colourless liquid, soluble in alcohol, ether, acetone, insoluble in water. Produced from coaltar and coke-oven gas; can also be synthesized from open-chain hydrocarbons. Basis for benzene derivatives. A solvent for fats, resins, etc; very flammable. Benzene is the simplest member of the aromatic series of hydrocarbons. Carcinogenic. Its structure was established by Kekulé in 1858. See BENZOL.

benzene carboxylic acids (*Chem*) Aromatic acids originating from benzene.

benzene formula (*Chem*) The generally recognized formula for benzene, established by Kekulé, representing a closed chain of six carbon atoms, to each of which a hydrogen atom is attached, the carbon atoms being linked alternately by single and double bonds. It is a stable structure, showing resonance energy of c.160 kJ mol^{-1}.

benzene formula

benzene hexachloride (*Chem*) *1,2,3,4,5,6-hexachlorocyclo-hexane*. Abbrev *BHC*. See GAMMEXANE.

benzene hydrocarbons (*Chem*) Homologues of benzene of the general formula C_nH_{2n-6}.

benzene nucleus (*Chem*) The group of six carbon atoms which, with the hydrogen atoms, form the benzene ring. See SIDE CHAINS.

benzene ring (*Chem*) See BENZENE FORMULA, BENZENE NUCLEUS.

benzene-sulphonic acids (*Chem*) Aromatic acids formed from compounds of the benzene series by sulphonation. The acid characteristics are given by the group —SO$_2$OH. Important intermediates for dyestuffs.

benzhydrol (*Chem*) $(C_6H_5)_2$CHOH. Reduction product of benzophenone prepared by treatment with zinc and aqueous alcoholic alkali.

benzidine (*Chem*) *4-4'-diamino-biphenyl*. NH$_2$C$_6$H$_4$C$_6$H$_4$NH$_2$. Mp 127°C. White to pinkish crystals, soluble in alcohol, ether, insoluble in water. It is an important intermediate for azodyestuffs. It is a known carcinogen.

benzidine transformation (*Chem*) The transformation of benzene-hydrazo-compounds into benzidine derivatives by strong acids.

benzil (*Chem*) *Bibenzoyl, diphenyl-glyoxal*. C$_6$H$_5$COCOC$_6$H$_5$, mp 95°C, large six-sided prisms, a diketone of the diphenyl group.

benzocaine (*Chem*) *Ethyl para-aminobenzoate*. White crystalline powder, insoluble in water; used as a local anaesthetic and for internal treatment of gastritis.

benzodiazepines (*Pharmacol*) Class of drug used as an ANXIOLYTIC or HYPNOTIC. Diazepam (*Valium*) is commonly used for relieving anxiety and nitrazepam (*Mogadon*) for inducing hypnosis, although the hazard of addiction with these drugs is being increasingly recognized. Chlordiazepoxide (*Librium*) is also of this class.

benzoic acid (*Chem, FoodSci*) *Benzenecarboxylic acid*. C$_6$H$_5$COOH. Mp 121°C, bp 250°C, colourless glistening plates or needles, sublimes readily, volatile in steam. Used as a food preservative.

benzoin (*Chem*) C$_6$H$_5$CHOHCOC$_6$H$_5$. Mp 137°C, colourless prisms, a condensation product of benzaldehyde. It is both a secondary alcohol and a ketone and can react accordingly. Occurs as a natural resin obtained from a Javanese tree. Chief constituent of friar's balsam.

benzol (*Autos*) Crude benzene; has been used as a motor spirit and valued for its anti-knock properties.

benzol scrubber (*Chem*) A device for washing gases and absorbing the benzol contained therein by means of a high-boiling mineral oil.

benzonatate (*Pharmacol*) A non-narcotic oral antitussive that apparently works by anaesthetizing stretch receptors in the airways.

benzonitrile (*Chem*) *Cyanobenzene*. C$_6$H$_5$CN. Bp 191°C, the nitrile of benzoic acid. Also *phenyl cyanide*.

benzophenone (*Chem*) *Diphenyl ketone*. C$_6$H$_5$COC$_6$H$_5$. Mp 49°C, bp 307°C, colourless prisms, soluble in alcohol and ether. It is dimorphous, mp of the unstable modification 26°C.

benzoquinones (*Chem*) C$_6$H$_4$O$_2$. Two isomers. See QUINONES.

benzotropine mesylate (mesilate) (*Pharmacol*) An *anti-muscarinic* drug used in the treatment of Parkinson's disease.

benzoyl chloride (*Chem*) *Benzene carboxyl chloride*. C$_6$H$_5$COCl. A colourless liquid, of pungent odour, bp 198°C, obtained by the action of PCl$_5$ on benzoic acid, commercially prepared by chlorinating benzaldehyde.

benzoyl peroxide (*Chem*) C$_6$H$_5$COOOCOC$_6$H$_5$. Bleaching agent and catalyst for free radical reactions. Mp 108°C. Prepared by the action of sodium peroxide on benzoyl chloride.

benzpinacol (*Chem*) $(C_6H_5)_2$C(OH)C(OH)$(C_6H_5)_2$. Reduction product of benzophenone.

1,2-benzpyrene (*Chem*) A polycyclic hydrocarbon isolated from coaltar as pale-yellow crystals, mp 177°C. It has strong carcinogenic properties.

1,2-benzpyrene

benzyl (*Chem*) The aromatic group, C$_6$H$_5$CH$_2$—.

benzyl alcohol (*Chem*) *Phenylmethanol*. C$_6$H$_5$CH$_2$OH. A colourless liquid, bp 204°C, the simplest homologue of the aromatic alcohols.

benzylamine (*Chem*) *Phenylmethylamine*. C$_6$H$_5$CH$_2$NH$_2$. Colourless liquid, bp 183°C, a primary amine of the aromatic series.

benzyl chloride (*Chem*) *Chloromethyl benzene*. C$_6$H$_5$CH$_2$Cl. Colourless liquid, bp 178°C, obtained by the action of chlorine on boiling toluene. Intermediate for benzyl derivatives.

benzylidene chloride (*Chem*) See BENZAL CHLORIDE.

benzyl penicillin (*Pharmacol*) The first of the penicillins; it has the disadvantage of being inactivated by bacterial penicillinases. It remains, however, the drug of choice for streptococcal, gonococcal and meningococcal infections. Also used to treat syphilis, yaws, tetanus, anthrax, actinomycosis and diphtheria.

benzyne (*Chem*) C$_6$H$_4$. An unstable intermediate formed by the removal of two orthohydrogens from benzene.

beraunite (*Min*) Hydrated phosphate of iron, red, found in iron ore deposits.

berber (*Textiles*) A carpet square hand-woven by North Africans from hand-spun yarns from the natural coloured wool of local sheep. Commonly misused to describe machine-made carpets considered to have a similar appearance.

berberine (*Chem*) C$_{20}$H$_{19}$O$_5$NH$_2$O. Chief alkaloid present in *Hydrastis*. Has been used as an amoebicide and in the treatment of cholera. Also *jamaicin, xanthopicrite*.

bergamot oil (*Chem*) Yellow-green volatile essential oil from the rind of *Citrus bergamia* (*Rutaceae*). Used in perfumery.

Bergeron–Findeisen theory (*EnvSci*) The theory that the initiation of precipitation in a cloud consisting mainly of supercooled water droplets is due to the presence of ice crystals which grow at the expense of the droplets because the saturation vapour pressure with respect to ice is lower than that with respect to liquid water at the same temperature.

Bergmann's law (*BioSci*) Observation that in warm-blooded animals and within a species, southern forms are smaller than northern forms.

Bergstrom's method (*Aero*) A method of assessing the stresses in concrete pavements with particular reference to aerodrome runways and taxiing tracks.

beri-beri (*Med*) A disease causing peripheral nerve lesions and/or heart failure due to a deficiency of the vitamin B_1 (thiamine).

Berkefeld filter (*Build*) Domestic filter using diatomite for removing bacteria. Microporous cellulose or other material now generally used.

berkelium (*Chem*) Element, symbol Bk, at no 97, synthesized by helium ion bombardment of americium-241.

berm (*CivEng*) A horizontal ledge on the side of an embankment or cutting, to intercept earth rolling down the slopes, or to add strength to the construction. Also *bench*.

berm ditch (*CivEng*) A channel cut along a berm to drain off excess water.

Bernoulli equation (*MathSci*) A differential equation of the form

$$\frac{dy}{dx} + py = qy^n$$

where p and q are functions of x alone.

Bernoulli's law (*Phys*) The law that for a non-viscous, incompressible fluid in steady flow, the sum of the pressure, potential and kinetic energies per unit volume is constant at any point. It is a fundamental law of fluid mechanics.

Bernoulli's numbers (*MathSci*) The numbers B_1, B_2, B_3, etc, defined by the expansion

$$\frac{x}{1-e^{-x}} = 1 + \frac{1}{2}x + \frac{B_1}{2!}x^2 - \frac{B_2}{4!}x^4 + \frac{B_3}{6!}x^6 - \cdots$$

The first few values are $B_1 = \frac{1}{6}$, $B_2 = \frac{1}{30}$, $B_3 = \frac{1}{42}$, $B_4 = \frac{1}{30}$, $B_5 = \frac{5}{66}$, after which they continue to increase to infinity. Other definitions vary in initial terms, numbering and signs.

Bernoulli's polynomials (*MathSci*) The coefficients $\varphi_n(z)$ of $t^n/n!$ in the expansion

$$t\frac{e^{zt}-1}{e^t-1} = \sum_{n=1}^{\infty} \varphi_n(z)\frac{t^n}{n!}$$

Bernoulli's theorems (*MathSci*)

(1) $$\sum_{s=1}^{\infty} s^r = \frac{n^{r+1}}{r+1} + \frac{1}{2}n^r + \frac{r}{2!}B_1 n^{r-1}$$
$$- \frac{r(r-1)(r-2)}{4!}B_2 n^{r-3}$$
$$+ \frac{r(r-1)(r-2)(r-3)(r-4)}{6!}B_3 n^{r-5}$$

there being $\frac{1}{2}(r+3)$ terms if r is odd, and $\frac{1}{2}(r+4)$ if r is even.

(2) $$\sum_{s=1}^{\infty} \frac{1}{s^{2m}} = \frac{(2\pi)^{2m}}{2(2m)!}B_m, \; m \geq 1$$

where B_1, B_2, B_3, etc, are Bernoulli's numbers.

berry (*BioSci*) (1) Strictly a fleshy fruit, without a stony layer, usually containing many seeds (eg grape, tomato and blueberry) in contrast to a DRUPE. Many so-called berries are aggregate or accessory fruit. The former consist of a mass of small drupes (drupelets), each developed from a separate ovary of a single flower (eg blackberry) or ovaries of many flowers growing in a cluster (eg mulberry). Accessory fruits contain tissue derived from plant parts other than the ovary; the strawberry is actually a number of tiny achenes (miscalled seeds) supported on a central pulpy pith that is the enlarged base of the flower. (2) The eggs of lobster, crayfish and other macruran Crustacea. (3) Part of the bill in swans.

Bertrand curves (*MathSci*) See CONJUGATE BERTRAND CURVES.

bertrandite (*Min*) A hydrated beryllium silicate, occurring in pegmatites.

beryl (*Min*) A beryllium aluminium silicate, occurring in pegmatites as hexagonal colourless, green, blue, yellow or pink crystals. Important ore of beryllium, also used as a gemstone. See AQUAMARINE, EMERALD.

beryllicosis (*Med*) Chronic beryllium poisoning, the main symptom of which is serious and usually permanent lung damage. Also *berylliosis*.

beryllides (*Eng*) Compounds of other metals with beryllium.

beryllium (*Chem*) Steely uncorrodible white metallic element. From Gk and Lt *beryl*, the old mineral name. Symbol Be, at no 4, ram 9·0122, mp 1281°C, bp 2450°C, rel.d. 1·93. It is a rare element both cosmically and in the Earth, where its abundance is only 2 ppm. It occurs in a number of minerals, including the gemstone beryl, of which EMERALD and AQUAMARINE are varieties, and CHRYSOBERYL. Main use is for windows in X-ray tubes and as an alloy for hardening copper. Used as a powder for fluorescent tubes until found poisonous. The metal can be evaporated onto glass, forming a mirror for ultraviolet light. As an alloy with nickel, it has the highest coefficient for secondary electron emission, 12·3. Alpha particles projected into beryllium make it a useful source of neutrons, from which they were discovered by Chadwick in 1932. The oxide (*beryllia*) is a good reflector of neutrons and is also used in thermoluminescent dating; highly toxic. Once called *glucinum*.

beryllium bronze (*Eng*) A copper-base alloy containing 2·25% of beryllium. Develops great hardness (ie 300–400 Brinell) after quenching from 800°C followed by heating to 300°C. See PRECIPITATION HARDENING.

beryllonite (*Min*) A rare mineral, found as colourless to yellow crystals. Phosphate of beryllium and sodium.

Berzelius theory of valency (*Chem*) Theory that chemical affinity is electrical in character, and depends on the mutual attraction of positive (metallic) and negative (non-metallic) elements. A forerunner of the modern ELECTRONIC THEORY OF VALENCY.

Bessel functions (*MathSci*) (1) Of the first kind of order n:

$$J_n(x) = \sum_{r=0}^{\infty} \frac{(-1)^r}{r!\Gamma(n+r+1)} \left(\frac{x}{2}\right)^{n+2r}$$

where Γ indicates the GAMMA FUNCTION. $J_n(x)$ is a solution of BESSEL'S DIFFERENTIAL EQUATION. (2) Of the second kind of order n:

$$Y_n(x) = \frac{J_n(x)\cos n\pi - J_{-n}(x)}{\sin n\pi}.$$

There is also Hankel's type of Bessel function of the second kind, namely

$$Y_n(x) = \frac{\pi e^{n\pi i}}{\cos n\pi} Y_n(x)$$

This is a second independent solution of Bessel's differential equation. (3) Of the third kind of order n:

$$H_n^{(1)}(x) = Jn(x) + iY_n(x) \text{ and } H_n^2(x) = J_n(x) - iY_n(x)$$

These are also *Hankel functions* of the first and second kind respectively. (4) Modified Bessel functions: (a) Of the first kind of order n:

$$I_n(x) = \sum_{r=0}^{\infty} \frac{1}{r!\Gamma(n+r+1)} \left(\frac{x}{2}\right)^{n+2r}$$

(b) Of the second kind of order n:

$$K_n(x) = \frac{\pi}{2}\frac{I_{-n}(x) - I_n(x)}{\sin n\pi}$$

For all these, notations vary, however. Bessel functions have application to vibration and heat flow.

Bessel's differential equation (*MathSci*) The equation

$$x^2 \frac{d^2 y}{dx^2} + x \frac{dy}{dx} + (x^2 - n^2) y = 0$$

Satisfied by Bessel functions $J_n(x)$ and $Y_n(x)$.
Bessel's inequality (*MathSci*) The inequality which states that

$$\frac{a_0^2}{2} + \sum_{n=1}^{\infty} (a_n^2 + b_n^2) \leq \frac{1}{\pi} \int_{-\pi}^{\pi} [f(x)] \, dx$$

when $f(x)$ is approximated by FOURIER SERIES.
Bessemer converter (*Eng*) Large barrel-shaped tilting furnace, charged while fairly vertical with molten metal, and 'blown' by air introduced below through tuyères. Discharged by tilting. Now obsolete but replaced by variety of similar shaped but smaller vessels operating in slightly different ways and using oxygen in place of air.
Bessemer pig iron (*Eng*) Pig iron which has been dephosphorized in Bessemer converter lined with basic refractory material.
Bessemer process (*Eng*) A process in which impurities are removed from molten metal or matte by blowing air through molten charge in Bessemer converter. Used to remove carbon and phosphorus from steel, sulphur and iron from copper matte.
best available technology (*EnvSci*) A US term for the process giving the maximum abatement of pollution without regard to cost or proven necessity. Abbrev *BAT*. Also *best available control technology* (*BACT*).
best practical environmental option (*EnvSci*) A concept recognizing that treatment of pollutants in one medium of the environment (air, land, water) may simply transfer them into another. Thus removing sulphur dioxide from flue gases may cause the calcium sulphate produced to have polluting effects at its disposal sites. Also called the *cross-media approach*, it is a co-ordinated approach to pollution pathways, media and disposal routes. Abbrev *BPEO*.
best practical means (*EnvSci*) A term with statutory force since 1863, and the basis for control of atmospheric pollution in the UK. Defined as the best practicable means with regard to local conditions, financial implications and current technical knowledge, and includes the provision, maintenance and correct use of plant. Abbrev *BPM*.
best selected copper (*Eng*) A metal of a lower purity than high-conductivity copper. Generally contains over 99·75% of copper.
best technical means available (*EnvSci*) European Commission term requiring consideration of the economic availability of the means of pollution abatement. It approaches the UK term BEST PRACTICAL MEANS. Abbrev *BTMA*.
Beta (*ImageTech*) TN for a domestic COMPOSITE format video recorder system using $\frac{1}{2}$ in tape in a cassette. *ED* (*Extended Definition*) Beta is a HIGH-BAND Y/C version using metal tape, but only available for the NTSC system.
beta-adrenoceptor blocking drugs (*Pharmacol*) A group of drugs which block beta-adrenoreceptors of the heart, peripheral vasculature, bronchi, pancreas and liver. Used in the treatment of angina, hypertension and also migraine, thyrotoxicosis and anxiety states. Commonly subgrouped into those which are unselective (*propranolol*) and those which largely act on the β_1-receptors (*atenolol*).
beta back-scattering sedimentometer (*PowderTech*) Device in which the mass of sediment on the bottom of the sedimentation chamber is measured by the amount of radiation it scatters back from a 1 mCi ($3·7 \times 10^7$ Bq) strontium-90 radioactive source.
beta blockers (*Pharmacol*) See BETA-ADRENOCEPTOR BLOCKING DRUGS.
beta brass (*Eng*) Copper–zinc alloys, containing 46–49% of zinc, which consist (at room temperature) of the intermediate constituent (or intermetallic compound) known as β. See COPPER ALLOYS.

Betacam (*ImageTech*) TN for a sub-BROADCAST STANDARD COMPONENT format video recorder system using $\frac{1}{2}$ in tape in a cassette. A HIGH-BAND *SP* (*Superior Performance*) version uses metal tape in two sizes of cassette. *Digital Betacam* is a digital broadcast standard version, with some models capable of playing analogue Betacam recordings.
Beta Centauri (*Astron*) See HADAR.
Beta Crucis (*Astron*) See MIMOSA.
betacyanins (*BioSci*) Red pigments of the betalain type, eg the red pigment of the beetroot.
beta decay (*Phys*) Radioactive disintegration with the emission of an electron or positron accompanied by an uncharged antineutrino or neutrino. The mass number of the nucleus remains unchanged but the atomic number is increased by one or decreased by one depending on whether an electron or positron is emitted. See ELECTRON CAPTURE.
beta detector (*NucEng*) A radiation detector specially designed to measure β-radiation.
beta disintegration (*Phys*) See BETA DECAY.
beta disintegration energy (*Phys*) For electron (β^-) emission it is the sum of the energies of the particles, the neutrino and the recoil atom. For positron (β^+) emission there is in addition the energy of the rest masses of two electrons.
beta diversity (*BioSci*) See DIVERSITY.
BET adsorption theory (*Chem*) A theory postulated by Brunauer, Emmett and Teller in which multimolecular adsorption layers build up on the catalyst surface, and Langmuir's derivation for single molecular layers is extended to obtain an isothermal equation for multi-molecular adsorption.
betafite (*Min*) A hydrated niobate, tantalate and titanate of uranium.
beta function (*MathSci*) The function $B(p, q)$ defined by

$$\int_0^1 x^{p-1} (1-x)^{q-1} dx$$

Its main property is that

$$B(p,q) = \frac{\Gamma(p)\Gamma(q)}{\Gamma(p+q)}$$

where Γ is the gamma function.
Beta Geminorum (*Astron*) See POLLUX.
betaine (*Chem*) $(CH_3)_3N^+ CH_2COO^-$. Betaine crystallizes with 1 molecule of H_2O; mp of anhydrous betaine 293°C with decomposition. Occurs naturally in plants and animals.
betaines (*Chem*) A class name for the ZWITTERIONS exemplified by betaine.
beta (β) interferon (*Med*) A synthesized (*recombinant*) version of an endogenous biological compound, used to treat relapsing remittent multiple sclerosis. It is used to reduce the frequency and severity of relapses. Not all patients respond and deterioration sometimes results.
beta-iron (*Eng*) Iron in the temperature range 750–860°C, in which a change from the magnetic (alpha) state to the paramagnetic occurs at about 760°C. With carbon in solution the transition is lowered toward 720°C, and when cooling RECALESCENCE is more marked.
beta-lactam antibiotics (*Pharmacol*) A large group of bactericidal *antibiotics* that includes the penicillins. They act by inhibiting bacterial cell wall synthesis and activating enzymes that destroy the cell wall. Examples are *penicillin, ampicillin, amoxicillin*.
betalains (*BioSci*) A group of nitrogen-containing pigments functionally replacing other pigments including ANTHO-CYANINS.
Betamax (*ImageTech*) TN for a now obsolete domestic videotape recorder system using $\frac{1}{2}$ in tape in a cassette.
beta-2 microglobulin (*BioSci*) A protein that forms part of the structure of class I major histocompatibility antigens, but is present in small amounts in blood, urine and seminal plasma.

betanin (*FoodSci*) Natural colour extracted from beetroot by crushing to extract juice and then concentrating to around 68°Brix. Colour changes from red towards blue as pH is increased. Betanin is sensitive to heat processing and the presence of oxygen and its colour stability is adversely affected by high water activity. Also *beetroot red*.

Beta Orionis (*Astron*) See RIGEL.

beta-oxidation (*BioSci*) The oxidative degradation of the fatty acid chains of lipids into two carbon fragments by cleavage of the penultimate C—C bonds.

beta particle (*Phys*) An electron or positron emitted in beta decay from a radioactive isotope. Also *β*-particle.

Beta Persei (*Astron*) See ALGOL.

beta-pleated sheet (*BioSci*) Important element of protein structure resulting from hydrogen bonding between parallel polypeptide chains.

beta radiation (*Phys*) BETA PARTICLES emitted from a radioactive source.

beta-ray gauge (*Paper*) Equipment installed on a paper or converting machine to obtain an indication of grammage by measuring the absorption of beta rays passing through the web.

beta rays (*Phys*) Streams of BETA PARTICLES.

beta-ray spectrometer (*NucEng*) Spectrometer which determines the spectral distribution of energies of *β*-particles from radioactive substances or secondary electrons.

beta test (*ICT*) Testing of software carried out by potential users in association with the software development company. This is carried out before the software is generally released. Cf ALPHA TEST.

beta thickness gauge (*NucEng*) An instrument measuring thickness, based on absorption and back-scattering (reflection) by material or sample being measured of *β*-particles from a radioactive source.

betatopic (*Phys*) Said of atoms differing in atomic number by one unit. One atom can be considered as ejecting an electron (beta particle) to produce the other one.

betatron (*Phys*) Machine used to accelerate electrons to energies of up to 300 MeV in pulsed output. The electrons move in an orbit or constant radius between the poles of an electromagnet, and a rapidly alternating magnetic field provides the means of acceleration. See CYCLOTRON.

beta value (*NucEng*) In fusion, the ratio of the outward pressure exerted by the plasma to the inward pressure which the magnetic field is capable of exerting. Also *plasma beta*.

beta waves (*Med*) Higher-frequency waves (15–60 Hz) produced in human brain.

betaxanthins (*BioSci*) Yellow pigments of the BETALAIN type.

Betelgeuse (*Astron*) A prominent bright red supergiant variable star in the constellation Orion, the twelfth brightest in the sky. Distance 120 pc. Also *Alpha Orionis*.

Bethe cycle (*Astron*) See CARBON CYCLE.

Bethell's process (*Build*) A process for preserving timber, esp telegraph poles, which is first dried, then subjected to a partial vacuum within a special cylinder, and finally impregnated with creosote under pressure.

béton (*CivEng*) Fr originally for lime concrete, now for any kind of concrete.

béton armé (*CivEng*) Fr for REINFORCED CONCRETE.

BET surface area (*PowderTech*) Surface area of a powder calculated from gas adsorption data, by the method devised by Brunauer, Emmett and Teller.

Betts process (*Eng*) An electrolytic process for refining lead after DROSSING. The electrolyte is a solution of lead silica fluoride and hydrofluorsilicic acid, and both contain some gelatine. Impurities are all more NOBLE METALS and remain on the anode. Gold and silver are recovered from anode sponge.

between-lens shutter (*ImageTech*) A shutter located between the elements of a camera lens, actuated electro-magnetically or by a spring-loaded mechanism which opens a series of metal blades pivoted around the periphery of the aperture, the length of exposure being regulated by an air brake, clockwork escapement or electronic timer which controls the closing action.

between perpendiculars (*Ships*) The length between the FORWARD PERPENDICULAR and after perpendicular (after side of sternpost). Abbrev BP.

Beutler method (*ImageTech*) Use of slow film and fast developer to produce the same results as a fast film and slow fine-grain developer, given the same exposure.

BeV (*Phys*) See GEV.

Bevatron (*Phys*) A synchrotron at Berkeley, California, which gives a beam of 6·4 GeV protons.

bevel (*Build*) A light hardwood stock, slotted at one end to take the blade, which is fastened by a clamping screw passing through the stock and the slot in the blade, enabling the latter to be set at any desired angle to the former.

bevel (*Print*) The slope on the type from the *face* (see BODY) to the SHOULDER.

bevel gear (*Eng*) A system of toothed wheels connecting shafts whose axes are at an angle to one another but in the same plane.

bevelled boards (*Print*) Boards intended for case-making, with the edges at the head, foot and fore-edge cut at an angle.

bevelled-edge chisel (*Build*) Chisel with the blade bevelled all the way up to the handle. Used for light work.

bevelled halving (*Build*) A halving joint in which the meeting surfaces are not cut parallel to the plane of the timbers but at an angle, so that when they are forced together, the timbers may not be pulled apart by a force in their own plane.

Beverage antenna (*ICT*) See WAVE ANTENNA.

bezel (*Autos*) (1) A retaining rim, eg speedometer outer rim, panel light-retaining rim. (2) A small indicator light (eg for direction flashers) on instrument panel or dashboard.

bezel (*Build*) The sloped cutting edge of a chisel or other cutting tool.

bezel (*Eng*) The grooved ring holding the glass of a watch or an instrument dial.

Bézier curve (*ICT*) A curve created from its end-points and two or more control points that serve as positions for the shape of the curve. This technique was devised by Paul Bézier in 1962 for use in car body design.

BFO (*ICT*) Abbrev for BEAT-FREQUENCY OSCILLATOR.

BG (*Eng*) Abbrev for BIRMINGHAM GAUGE.

BGA (*Aero*) Abbrev for *British Gliding Association*.

b-group (*Phys*) A close group of FRAUNHOFER LINES in the green of the solar spectrum, due to magnesium.

BGV (*ImageTech*) Abbrev for BACKGROUND VIDEO.

Bh (*Chem*) Symbol for BOHRIUM.

BHA (*FoodSci*) Abbrev for BUTYLATED HYDROXYANISOLE.

BHA (*MinExt*) Abbrev for BOTTOM-HOLE ASSEMBLY.

bhang (*BioSci*) See CANNABIS.

BHC (*Chem*) Abbrev for BENZENE HEXACHLORIDE.

B(H) curve (*Eng*) Also *B–H curve*, *B/H curve*. See MAGNETIC MATERIALS.

B(H) loop (*Eng*) Also *B–H loop*, *B/H loop*. See HYSTERESIS LOOP, MAGNETIC MATERIALS.

BHN (*Eng*) Abbrev for *Brinell hardness number*, obtained in the BRINELL HARDNESS TEST. Preferred term is now H_B after the hardness number. Obtained by forcing a round steel ball into the surface of the object to be tested under a known load and subsequently measuring the diameter of the indentation so produced. See panel on HARDNESS MEASUREMENTS.

BHP (*Eng*) Abbrev for BRAKE HORSEPOWER.

bi-, bin- (*Genrl*) Prefixes from Lt *bis* denoting two, twice or double.

Bi (*Chem*) Symbol for BISMUTH.

Biacore (*BioSci*) Proprietary name for an instrument that uses SURFACE PLASMON RESONANCE to detect the rate and extent of binding of a substance to the surface of a flow chamber.

bialternant (*MathSci*) The quotient of two ALTERNANTS. A homogeneous symmetric polynomial.

bias (*Electronics*) The application of a potential difference across, or electric currents through, an electronic device to set an operating condition upon which signals are superimposed. See TRUE BIAS.

bias binding (*Textiles*) A non-fraying narrow fabric made by cutting full-width woven cloth into strips at 45° to the selvedge. The material is used for binding curved seams and garment edges.

bias current (*Electronics*) Non-signal current supplied to electrode of semiconductor device, magnetic amplifier, tape recorder, etc, to control operation at optimal working point.

biased protective system (*ElecEng*) A modification of a balanced protective system, in which the amount of out-of-balance necessary to produce relay operation is increased as the current in the circuit being protected is increased.

biased result (*Surv*) In observations, sampling, etc, the introduction of a systematic error through some malfunction of instrument or weakness in method used, so that error accumulates in a series of measurements.

biasing (*Electronics*) Polarization of a recording head in magnetic tape recording, to improve linearity of amplitude response, using dc or using ac much higher than the maximum audio-frequency to be reproduced.

biasing transformer (*ElecEng*) A special form of transformer used in one form of biased protective system.

biaxial (*Crystal, Min*) Said of a crystal having two optical axes. Minerals crystallizing in the orthorhombic, monoclinic and triclinic systems are biaxial. Cf UNIAXIAL.

biaxial orientation (*Chem*) State of polymer orientation where molecules are oriented in two orthogonal directions, esp in blow-moulded products such as polyethylene terephthalate bottles. The orientation is in the plane of the wall and helps toughen it.

bib-cock (*Eng*) A draw-off tap for water supply, consisting of a plug cock having a downward curved extension for discharge.

bibenzoyl (*Chem*) See BENZIL.

Bible paper (*Paper*) Heavily loaded, strong, thin, printing paper, generally of 20–40 gm m^{-2}.

biblio (*Print*) Details of a book's 'history', ie original publication date and dates of subsequent reprints and revised editions; usually placed on the verso of the title page.

bib-valve (*Eng*) A draw-off tap of the kind used for domestic water supply; closed by screwing down a rubber-washered disk onto a seating in the valve body.

bicarbonates (*Chem*) Hydrogen carbonates. The acid salts of carbonic acid; their aqueous solutions contain the ion $(HCO_3)^-$.

bicarpellary (*BioSci*) Said of an ovary consisting of two carpels.

biceps (*BioSci*) A muscle with two insertions. Adj *bicipital*.

bicipital (*BioSci*) See BICEPS.

bicipital groove (*BioSci*) A groove between the greater and lesser tuberosities of the humerus in mammals.

bick iron (*Eng*) Also *bickern*. See BEAK IRON.

bicollateral bundle (*BioSci*) A vascular bundle with two strands of phloem adjacent to the single strand of xylem; one placed centrifugally, the other centripetally. Cf COLLATERAL BUNDLE.

bicomponent fibre (*Textiles*) A synthetic fibre made from two fibre-forming polymers which may be arranged to lie side by side or as a sheath surrounding a core. By suitable heat treatment a crimped fibre is produced because of the different shrinkage properties of the two polymers. Advantage may also be taken of the sheath having a lower softening temperature than the core to produce a non-woven fabric.

biconcave (*ImageTech, Phys*) Said of a lens having both surfaces concave.

biconical horn (*ICT*) Two flat cones apex to apex, for radiating uniformly in horizontal directions when driven from a coaxial line. See DISCONE ANTENNA.

biconvex (*ImageTech, Phys*) Said of a lens which is convex on both surfaces.

bicuspid (*BioSci*) Having two cusps, eg the premolar teeth of some mammals. Also *bicuspidate*.

bicuspid valve (*BioSci*) The valve in the left ATRIOVENTRICULAR aperture in the mammalian heart. Also *mitral valve*.

bicyclic (*Chem*) Having two rings of atoms in a molecule.

bidentate (*Chem*) Complexing molecule or ion with two donating atoms.

bi-directional microphone (*Acous*) Microphone which is most sensitive in both directions along one axis.

bi-directional printer (*ICT*) LINE PRINTER in which the right-to-left return movement of the print head is used to print a second line, thus increasing print speed.

bi-directional waveform (*ICT*) Waveform that shows reversal of polarity; a bi-directional pulse generator produces both positive and negative pulses.

bieberite (*Min*) Hydrated cobalt sulphate, found as stalactites and encrustations in old mines containing other cobalt minerals. Also *cobalt vitriol*.

biennial (*BioSci*) A plant that flowers and dies between its first and second years from germination and which does not flower in its first year. Cf ANNUAL, PERENNIAL.

bifacial leaf (*BioSci*) A dorsiventral leaf, typically with palisade mesophyll towards the upper surface and spongy mesophyll below. The commonest sort.

bifid (*BioSci*) Divided halfway down into two lobes; forked.

bifilar micrometer (*Astron*) An instrument attached to the eyepiece end of a telescope to enable the angular separation and orientation of a visual double star to be measured.

bifilar pendulum (*Phys*) See BIFILAR SUSPENSION.

bifilar resistor (*ElecEng*) Resistor formed by winding a resistor with a hairpin-shaped length of resistance wire, thus reducing the total inductance.

bifilar suspension (*Phys*) The suspension of a body by two parallel vertical wires or threads which give a considerable controlling torque. If the threads are of length *l* and are distance *d* apart, the period of torsional vibration of a suspended body of moment of inertia *I* and mass *m* is

$$T = 4\pi \sqrt{\frac{Il}{mgd^2}}$$

bifurcate (*BioSci*) Twice-forked; forked. V *bifurcate*; n *bifurcation*.

bifurcated rivet (*Eng*) A rivet with a split shank, used for holding together sheets of light material; it is closed by opening and tapping down the two halves of the shank.

big bale (*Agri*) Cylindrical or cuboid bales of straw, hay or silage, 1 m × 1 m or larger, that require mechanical handling. They can be wrapped for silage production.

Big Bang (*Astron*) A hypothetical model of the universe which postulates that all matter and energy were once concentrated into an unimaginably dense state, from which it has been expanding from a creation event some $13–20 \times 10^9$ years ago. Main evidence favouring this model is the MICROWAVE BACKGROUND (*cosmic background radiation*) and the REDSHIFT of galaxies. See panel on COSMOLOGY.

Big Crunch (*Astron*) One possible fate of the universe, when after a long period the universe's expansion ceases and all the material in the universe collapses to a singularity. This is the predicted outcome if the density of the universe is greater than a certain critical value.

Big Dipper (*Astron*) US See the PLOUGH.

bigeminal pulse (*Med*) A pulse in which the beats occur in pairs, each pair being separated from the other by an interval; due to a disturbed action of the heart.

big-end (*Autos*) The part of the connecting rod which is attached to the crankshaft.

big-end bolts (*Eng*) See CONNECTING-ROD BOLTS.

bigeneric hybrid (*BioSci*) A hybrid resulting from a cross between individuals from two different genera, eg *Triticale*, a hybrid between wheat (*Triticum*) and rye (*Secale*).

big head disease (*Vet*) (1) Of horses: see OSTEODYSTROPHIA FIBROSA. (2) Of sheep: (a) Infection of the head and neck by *Clostridium septicum*. (b) A form of light sensitization occurring in sheep. Controlled by vaccination. Also *swelled head*. (3) Of turkeys: see INFECTIOUS SINUSITIS OF TURKEYS

biguanides (*Pharmacol*) Drugs used in the treatment of maturity onset DIABETES. They seem to act by increasing peripheral utilization of glucose and are of particular value in obese diabetics, eg *metformin*.

biharmonic equation (*MathSci*) The equation

$$\frac{\partial^4 z}{\partial x^4} + \frac{2\partial^4 z}{\partial x^2 \partial y^2} + \frac{\partial^4 z}{\partial y^4} = 0$$

The solutions of this equation are called *biharmonic functions*.

BIIR (*Chem*) Abbrev for *brominated isoprene isobutene rubber*. See BROMOBUTYL RUBBER.

bilabiate (*BioSci*) With two lips.

bilateral (*Med*) Having, or pertaining to, two sides.

bilateral cleavage (*BioSci*) The type of cleavage of the zygote formed in CHORDATA.

bilateral impedance (*ElecEng*) Any electrical or electromechanical device through which power can be transmitted in either direction.

bilateral slit (*Phys*) A slit used in a spectrometer and consisting of two metal strips whose separation can be accurately adjusted.

bilateral symmetry (*BioSci*) The condition when an organism is divisible into similar halves by one plane only. Cf RADIAL SYMMETRY.

bilateral tolerance (*Eng*) A tolerance with dimensional limits above and below the basic size.

bile (*Med*) A viscous liquid produced by the liver. Human bile has an alkaline reaction, a green or golden-yellow colour and a bitter taste. It consists of water, bile salts, mucin and pigments, cholesterol, fats and fatty acids, soaps, lecithin and inorganic compounds.

bile duct (*BioSci*) The duct formed by the junction of the hepatic duct and the cystic duct, leading into the intestine.

bile pigments (*Med*) Pigments produced by the breakdown of haemoglobin, consisting of an open chain of four substituted pyrrole nuclei joined by two methene (=CH—) bridges and one methylene (—CH₂—) bridge. The chief ones are bilirubin (reddish yellow) and its oxidation product biliverdin (green).

bile salts (*BioSci*) Sodium salts of the bile acids, a group of hydroxy steroid acids, some unsaturated, condensed with taurine or glycine. The commonest are the salts of taurocholic and glycocholic acids, breakdown products of cholesterol, secreted in the bile and important in aiding absorption of fats from the intestine by reducing surface tension and emulsifying them.

bilge (*Ships*) (1) The curved part of the shell joining the bottom to the sides. (2) The space inside this, at the sides of the cellular double bottom, into which unwanted water drains.

bilge keel (*Ships*) A projecting fin attached to the shell plating for about half the length of the ship, amidships, at the turn of the bilge, to reduce rolling.

bilharziasis (*Med*, *Vet*) Parasitic disease of humans and domestic animals caused by BLOOD FLUKES. Endemic in Africa and Far East. See SCHISTOSOMIASIS.

biliary fever (*Vet*) A form of BABESIOSIS affecting the horse and dog, caused by *Babesia equi* or *Babesia canis* respectively. It is characterized by fever, jaundice and haemoglobinuria.

bilicyanin (*Med*) An oxidation product of bilirubin, blue in colour.

bilinear transformation (*MathSci*) A one-to-one transformation between the *w*- and *z*-planes, given by

$$w = \frac{az + b}{cz + d}$$

A special form of linear transformation. Also *Möbius transformation*.

bilirubin (*Med*) A reddish pigment occurring in bile; a breakdown product of haemoglobin. See BILE PIGMENTS.

biliverdin (*Med*) A green pigment occurring in bile. It is an oxidation product of BILIRUBIN. See BILE PIGMENTS.

billet (*Build*) A piece of timber which has three sides sawn and the fourth left round.

billet (*Eng*) Semi-finished solid product which has been hot-worked by extrusion, forging and rolling. Smaller than a BLOOM.

billet mills (*Eng*) The rolling mills used in reducing steel ingots to billets. Also *billet rolls*.

Billet split lens (*Phys*) A device used to produce interference fringes. The two halves of the lens are separated so that two images of a slit source provide the coherent sources.

billiard cloth (*Textiles*) Woollen cloth manufactured from finest-quality wool, with a closely cropped dress-face finish to render it perfectly smooth and damp resistant.

billion (*Genrl*) In the USA and now generally, a thousand million, or 10^9. Previously elsewhere, a million million or 10^{12}.

billion-electron-volt (*Phys*) See GEV.

billitonites (*Geol*) See TEKTITES.

bill of quantities (*Build*, *CivEng*) A list giving the quantities of material and brief descriptions of work in an engineering or building works contract, the basis for comparing tenders.

bilocular (*BioSci*) Consisting of two loculi or chambers.

bimanous (*BioSci*) Having the distal part of the two forelimbs modified as hands, as in some primates.

bimanual (*Med*) Performed with both hands.

bimastic (*BioSci*) Having two nipples.

bimetal-fuse (*ElecEng*) A fuse element composed of two different metals, eg a copper wire coated with tin or lead.

bimetallic plates (*Print*) Lithographic plates of great durability, in which the ink-accepting and ink-rejecting parts are of two different metals, eg copper, printing image and chromium, non-printing image; called *polymetallic* when a supporting metal, eg steel, is used.

bimetallic strip (*ElecEng*, *Eng*) Bonded strip composed of two metals with differing THERMAL EXPANSION COEFFICIENTS; the strip deflects when one side of the strip expands more than the other. Used eg in thermal switches.

bimirror (*Phys*) A pair of plane mirrors slightly inclined to one another. Used for the production of two coherent images in interference experiments.

bimorph (*ElecEng*) A unit in microphones and vibration detectors in which two piezoelectric plates are cemented together in such a way that application of potential difference causes one to contract and the other to expand, so the combination bends as in a BIMETALLIC STRIP.

bin- (*Genrl*) See BI-.

binary (*Astron*) See BINARY STAR.

binary (*Chem*) Consisting of two components.

binary (*MathSci*) Of or relating to the number 2. *Binary arithmetic* is a POSITIONAL system with BASE 2; a *binary function* has two arguments. Binary arithmetic is particularly useful in computing because the figures 1 and 0 can be represented by 'on' and 'off' in electronic circuits.

binary code (*ICT*) A representation of information using a sequence of zeros and ones. See CHARACTER CODE.

binary-coded decimal (*ICT*) A method of representing a number in DENARY NOTATION whereby each digit is coded as a discrete pattern of 4 BITS. Abbrev *BCD*.

binary counter (*ICT*) Flip-flop or toggle circuit that gives one output pulse for two input pulses, thus dividing by two.

binary digit (*ICT*) See BIT.

binary file (*ICT*) A FILE that contains information in direct machine-readable form.

binary fission (*BioSci*) Division of the nucleus into two daughter nuclei, followed by similar division of the cell body.

binary frequency shift keying (*ICT*) A digital MODULATION scheme in which '1' and '0' are represented by switching the CARRIER between two different frequencies. It is 3 dB less resistant to additive white Gaussian noise interference than BINARY PHASE SHIFT KEYING.

binary notation (*ICT*) A system of representing numbers on the BINARY SCALE.

binary phase shift keying (*ICT*) A digital MODULATION scheme in which '1' and '0' are represented by reversing the phase of the CARRIER. It is 3 dB more resistant to additive white Gaussian NOISE interference than BINARY FREQUENCY SHIFT KEYING.

binary point (*ICT, MathSci*) The RADIX POINT in calculations carried out on the BINARY SCALE.

binary search (*ICT*) A strategy for locating a record in an ordered list by repeatedly comparing the key with the mid-item in the list and discarding that half of the list which cannot contain the required record.

binary star (*Astron*) A double star in which the two components revolve about their common centre of mass under the influence of their gravitational attraction. Also *binary*. See ECLIPSING BINARY, SPECTROSCOPIC BINARY, VISUAL BINARY.

binary system (*Eng*) An alloy formed by two metals; this is represented by the *binary constitutional diagram* for the system. In general, any two-component system. See PHASE DIAGRAM.

binary tree (*ICT*) A TREE where any node may have no more than two branches.

binary vapour-engine (*Eng*) A heat engine using two separate working fluids, generally mercury vapour and steam, for the high- and low-temperature portions of the cycle respectively, thus enabling a large temperature range to be used, with improved thermal efficiency.

binaural (*Acous*) Listening with two ears, the result of which is a sense of directivity of the arrival of a sound wave. Said of a stereophonic system with two channels (matched) applying sound to a pair of ears separately, eg by earphones. The effect arises from relative phase delay between wavefronts at each ear.

binder (*Build*) The medium or vehicle in a paint which retains the solids in suspension during storage and later dries to form the film.

binder (*PowderTech*) A component employed in the mix of carbon products, organic brake linings, sintered metals, tar macadam, etc, to impart cohesion to the body to be formed. The binder may have cold-setting properties, or subsequently be heat treated to give it permanent properties as part of the body or to remove it by volatilization.

binder coat (*Build*) A surface coating applied to a friable or unstable surface to provide a stable base for subsequent coatings.

bindery (*ICT*) A DATABASE file stored on a FILE SERVER that holds details of users, their access rights and other system configuration parameters in the context of a LOCAL AREA NETWORK.

binding-beam (*Build*) A timber tie serving to bind together portions of a frame.

binding energy (*Phys*) Energy required (1) to remove a particle from a system, eg electron, when it is the IONIZATION POTENTIAL; (2) to overcome forces of cohesion and disperse a solid into constituent atoms. (3) Of a nucleus, the energy which holds nuclear particles together. See FISSION, FUSION, PACKING FRACTION and panel on BINDING ENERGY OF THE NUCLEUS.

binding wire (*ElecEng*) See TIE WIRE.

Binet–Cauchy theorem (*MathSci*) Rule that if A is a matrix of order $m \times n$, and B is a matrix of order $n \times p$, then any determinant of order r of the product matrix AB is equal to a sum of terms, each of which is the product of a determinant of order r of A, and a determinant of order r of B (in that order).

Binet intelligence test (*Psych*) Early (1905) scale of intelligence that focuses on cognitive capacity. Usually refers to the much more recent modifications, originally carried out by the Stanford psychologist Lewis Terman, the Stanford–Binet Scale.

Binet's formulae for $\log \Gamma(z)$ (*MathSci*) (1) The first formula is

$$\log \Gamma(z) = \left(z - \frac{1}{2}\right) \log z - z + \frac{1}{2} \log (2\pi)$$
$$+ \int_0^\infty \left(\frac{1}{e^t - 1} - \frac{1}{t} + \frac{1}{2}\right) \frac{e^{-tz}}{t}\, dt$$

(2) The second formula is

$$\log \Gamma(z) = \left(z - \frac{1}{2}\right) \log z - z + \frac{1}{2} \log (2\pi)$$
$$+ 2 \int_0^\infty \frac{\arctan (t/z)}{e^{2\pi t} - 1}\, dt,$$
$$[\Gamma(z) > 0]$$

Bingham flow (*Phys*) See BINGHAM SOLID.

Bingham solid (*Phys*) A material which shows little tendency to flow until a critical stress is reached (eg toothpaste or modelling clay). Such materials may be NEWTONIAN, dilatant (see DILATANCY) or PSEUDO-PLASTIC. Also *Bingham flow*.

binocular camera (*ImageTech*) See STEREOCAMERA.

binoculars (*Phys*) A pair of telescopes for use with both eyes simultaneously. Essential components are an objective, an eyepiece and some system of prisms to invert and reverse the image.

binomial (*MathSci*) Of, or relating to, an expression containing the sum (or difference) of two terms, eg $2x + 3y$.

binomial array (*Radar, ICT*) A linear array in which the current amplitudes are proportional to the coefficients of a binomial expansion. Such an array has no sidelobes.

binomial coefficients (*MathSci*) See BINOMIAL THEOREM.

binomial distribution (*MathSci*) The probability distribution of the total number of outcomes of a particular kind in a predetermined number of trials, the probability of the outcome being constant at each trial, and the different trials being statistically independent.

binomial nomenclature (*BioSci*) The system (introduced by Linnaeus) of denoting an organism by two Latin words, the first the name of the genus, the second the specific epithet. The two words constitute the name of the species, eg *Homo sapiens*, *Bellis perennis*. Also *binominal nomenclature*. See SPECIES.

binomial theorem (*MathSci*) The expansion

$$(1 + x)^n = 1 + \frac{n}{1}x + \frac{n(n-1)}{1.2}x^2$$
$$+ \frac{n(n-1)(n-2)}{1.2.3}x^3$$
$$+ \frac{n(n-1)(n-2)(n-3)}{1.2.3.4}x^4 + \cdots$$

which is valid either if n is a positive integer (in which case the series terminates) or if $|x| < 1$. The coefficients of the powers of x are called *binomial coefficients*, that of x^r being written

$$\binom{n}{r}$$

The series can thus be written

$$(1 + x)^n = \sum_{r=0}^{n} \binom{n}{r} x^r$$

Binomial coefficients are used in the construction of probability models for discrete distributions. See BINOMIAL DISTRIBUTION.

binormal (*MathSci*) See MOVING TRIHEDRAL.

binovular twins (*Med*) Non-identical twins resulting from the fertilization of two separate ova.

binucleate phase (*BioSci*) See DIKARYOPHASE.

Binding energy of the nucleus

From the standpoint of nuclear energy this is the most important property of an element.

The *mass energy* of a nucleus or any other particle is the product of its mass and energy, and is defined by the mass–energy equation which Einstein deduced from the SPECIAL THEORY OF RELATIVITY:

$$E = mc^2$$

where E is the energy, m the mass and c the speed of light. In nuclear physics it is convenient to define m in terms of u, the atomic mass unit, when uc^2 becomes 931·50 MeV (mega-electron-volts).

To a high degree of accuracy the mass energy of a particular nucleus is equal to the mass energy of the atom less that of its electrons. This can be simply expressed as the mass energy of hydrogen-1 multiplied by the number of protons plus the mass energy of all the neutrons. A nucleus X will have Z protons and N neutrons and so its binding energy will be Z times the mass of a hydrogen-1 atom plus N times the mass of a neutron less the mass of the combined nucleus X which will have $Z + N$ (= A) nucleons. All of this is multiplied by c^2 as shown:

$$B = [Zm\,(^1H) + Nm\,(neutron) - m\,(^AX)]c^2$$

where B is the binding energy and m the mass of the particle within the following parentheses. The figure below shows the relation between binding energy and mass number.

This relation shows a number of important features. Although B/A stays fairly constant around the value of 8 for most elements, for those with mass numbers below about 30, B/A increases with higher values of A, but for elements with A above 100 the reverse is true. This means that for the light elements, fusing two nuclei will produce energy, while for heavy elements fission to give lighter elements gives energy. Iron-56, the most stable element, separates the two regions of the binding energy curve. It also shows why fusion reactions release more energy for a given mass of fuel than fission reactions.

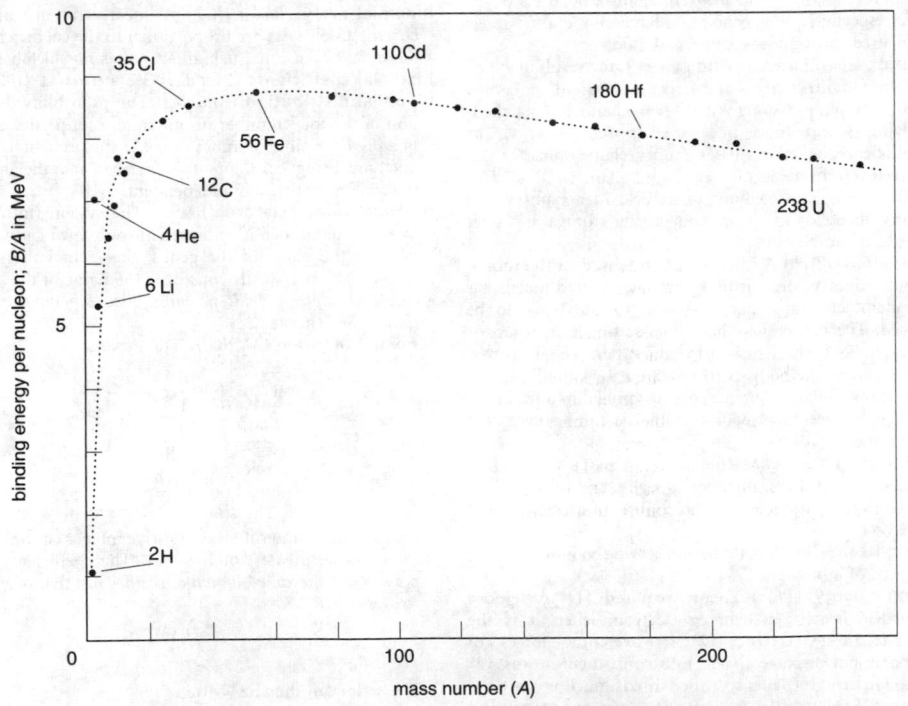

Binding energy of a nucleus A few representative nucleons shown.

bio- (*Genrl*) Prefix from Gk *bios* denoting life, living organisms and living tissue.

bioaccumulation (*BioSci*) Accumulation of substances in living organisms because the rate of intake exceeds the organism's excretory capacity. Organisms at the top of a food chain can accumulate considerable amounts of some substances, the most notorious of which was DDT. Also *bioconcentration*.

bioactivation (*BioSci*) Metabolic conversion of a xenobiotic substance to a more toxic derivative.

bio-aeration (*Build*) A system of sewage purification by oxidation; aeration of the crude sewage is effected by passing it through specially designed centrifugal pumps. See ACTIVATED SLUDGE, ACTIVATION.

bio-assay (*BioSci*) Determination of the potency (activity) of a drug or of a biological product by testing its effect on an assay system that involves live biological material, either *in vivo*, or *in vitro* (cell or organ culture, for example).

bio-availability (*Pharmacol*) The extent and rate at which the active substance in a drug is taken up by the body. Differences in tablet formulation and the rate of absorption by the gut will alter bio-availability.

biochemistry (*Genrl*) The chemistry of living things; physiological chemistry. Now considered by many to be part of a wider spectrum of 'molecular biosciences'.

bioclastic limestone (*Geol*) A carbonate made up of broken fragments of organic material, esp shells.

bioclimatology (*BioSci*) The study of the effects of climate on living organisms.

biocoenosis (*BioSci*) The association of animals and plants together, esp in relation to any given feeding area. Adj *coenotic*.

biocomposites (*Eng*) (1) COMPOSITE MATERIALS which occur in and are made by living organisms, such as bone, leather. (2) Composite materials which replace the function of living tissues or organs in mass, such as carbon fibre/epoxy resin artificial limbs.

biodegradation (*BioSci*) Breaking down of materials by bacteria, fungi and other organisms.

biodiesel (*EnvSci*) Fuel for diesel-engined vehicles manufactured from crop plants.

biodiversity (*BioSci*) The genetic, taxonomic and ecosystem variety in the living organisms of a given area, environment, ecosystem, or indeed the whole planet.

biodynamic agriculture (*Agri*) A system devised by Austrian philosopher Rudolf Steiner in 1924, viewing soil as a living entity. It uses specific and original organic practices of composting, fertilizer production and pest and weed control, employing homeopathic principles.

bio-engineering (*ChemEng*) Engineering methods of achieving biosynthesis of animal and plant products, eg for fermentation processes. Also *biological engineering*.

bio-engineering (*Med*) (1) Application of scientific study of human body to improve, aid or assist impaired limbs or organs. (2) Provision of artificial means with the use of synthetic materials, electronic devices, etc, to assist defective body functions or parts, such as hip joint implants. Cf GENETIC MANIPULATION (panel).

bioethanol (*EnvSci*) Ethanol produced from fermented plant material, used as a petrol additive or substitute.

biofeedback (*Psych*) Strictly, any feedback about bodily function, but usually refers to therapeutic procedures whereby subjects are given information about physiological functions that are not normally available to conscious experience (eg heart rate, blood pressure, etc). The object is sometimes of gaining some conscious control of these functions.

biogas (*EnvSci*) Gas, mostly methane and carbon dioxide, produced in suitable equipment by bacterial fermentation of organic matter (see BIOMASS) and used as fuel.

biogenesis (*BioSci*) The formation of living organisms from their ancestors and of organelles from their precursors.

biogeochemical cycles (*EnvSci*) The transfer of chemical elements between living organisms and the abiotic environment along approximately circular pathways. Those involving elements essential for life are called *nutrient cycles*.

biogeographic regions (*BioSci*) Regions of the world containing recognizably distinct and characteristic endemic fauna or flora.

biogeography (*BioSci*) The study of the past and present distribution of plants and animals throughout the world, and the mapping of the world into faunal and floral regions and provinces.

bioherm (*Geol*) A reef or mound made up of the remains of organisms growing *in situ*.

biological amplification (*BioSci*) See BIOLOGICAL MAGNIFICATION.

biological bench marking (*EnvSci*) Analysis of the fitness and abundance of living organisms as indicators of particular aspects of environmental quality, eg the use of lichens to indicate the concentration of sulphur dioxide in air.

biological clock (*BioSci*) An internal, physiological time-keeping system underlying eg circadian rhythms and photoperiodism.

biological constraint (*Psych*) A general term in learning theory that refers to the fact that certain behaviours are more easily learned by some organisms than by others, and conversely, that some behaviours are not easily learned by some organisms.

biological containment (*BioSci*) Alteration of the genetic constitution of an organism so as to minimize its ability to grow in a non-laboratory environment.

biological control (*BioSci*) Use of an organism to control a disease, pest or weed.

biological engineering (*ChemEng*) See BIO-ENGINEERING.

biological engineering polymers (*Eng*) Materials of biological origin (wood, bone, cotton, natural rubber, etc) which have long been used for engineering purposes, ranging from buildings to rope for ships' cables. See panel on BIOLOGICAL ENGINEERING POLYMERS.

biological form (*BioSci*) See PHYSIOLOGICAL RACE.

biological half-life (*BioSci*) Time interval required for half of a quantity of radioactive material, or other substance, absorbed by a living organism to be eliminated naturally.

biological hole (*NucEng*) A cavity within a nuclear reactor in which biological specimens are placed for irradiation experiments.

biological magnification (*EnvSci*) Process whereby the concentration of a pollutant, within living tissues, increases at each link in a food chain. Also *biological amplification*.

biological monitoring (*BioSci*) The routine recording of changes in a habitat based on estimates of the variety of species it supports and the size of their populations.

biological oxygen demand (*EnvSci*) See OXYGEN DEMAND. Abbrev BOD.

biological race (*BioSci*) A race occurring within a taxonomic species; distinguished from the rest of the species by slight or no morphological differences, but by evident differences of habitat, food preference or occupation which inhibit interbreeding. Also *physiological race*.

biological shield (*BioSci*) The heavy concrete barrier placed round a nuclear reactor or other plant to protect workers from radiation.

biological warfare (*Genrl*) The use of bacteriological (biological) agents and toxins as weapons. Cf CHEMICAL WARFARE, WAR GAS.

biology (*Genrl*) The science of living things; the life sciences collectively, including botany, ecology, anatomy, physiology and zoology. The subject heading in this dictionary, 'BioSci', also encompasses biochemistry, cell biology, genetics, immunology, microbiology, molecular biology and some experimental pathology, reflecting the modern continuum in molecular and organismal biosciences and the interchangeability of the terminology. Many medical terms are also part of 'Biosci' but are restricted to humans.

bioluminescence (*BioSci*) The production of light by living organisms, eg glow worms, some deep-sea fish, some bacteria, some fungi.

biomass (*BioSci*) (1) The total dry mass of an animal or plant population. (2) Organic matter (mostly from plants) harvested as a source of energy (by burning or BIOGAS production) and/or as a chemical feedstock.

biomaterials (*BioSci, Eng*) (1) Solid materials that occur in and are made by living organisms, such as CHITIN, FIBROIN or bone. (2) Any materials which replace the function of living tissues or organs in humans. See IMPLANT, PROSTHESIS. (3) Materials from a biological source, eg paper.

Biological engineering polymers

Materials of biological origin (wood, bone, cotton, natural rubber, etc) have long been used for engineering purposes, ranging from buildings to rope for ships' cables. Many of them still find wide application although now supplemented by an increasing range of manufactured and synthetic materials. Some are natural biocomposites, a combination of hard and stiff materials embedded in a softer, ductile matrix. Thus

bone is an intimate mixture of inorganic hydroxyapatite and the protein collagen, the mechanical properties being greater than either component alone. Others are relatively pure: cotton fibre is almost pure β-cellulose, comprising glucose units of very high molecular mass linked together by oxygen atoms.

The chains are highly crystalline and further stabilized by intra- and interchain hydrogen bonds giving a ladder polymer structure:

The chains are aligned along the fibre axis, giving a high tensile modulus when strained. In the cotton *boll*, the cellulose fibres act not as structural reinforcement (as in wood or plant stems) but as a lightweight 'wing'

Fig. 1 **The β-cellulose chain of cotton fibres**.

Fig. 2 **Hydrogen bonding between cellulose chains**.

biome (*BioSci*) The largest land community region recognized by ecologists, eg tundra, savanna, grassland, desert, temperate and tropical forest.

biomechanics (*BioSci*) The study of motion and energetics of living organisms, esp human motion; a method used in ergonomics.

biometeorology (*BioSci*) The study of the effects of atmospheric conditions on living things.

biometry (*BioSci*) Statistical methods applied to biological problems.

biomining (*MinExt*) See MICROBIOLOGICAL MINING.

bionics (*Genrl*) The various phenomena and functions which characterize biological systems with particular reference to electronic systems.

bionomic axes (*BioSci*) A mathematical representation of ecological resources (eg food, space or nest sites) for which there may be competition; used in calculations of the efficiency with which niches are exploited within an ecosystem.

bionomics (*BioSci*) A little-used term that is generally synonymous with ECOLOGY and esp with AUTECOLOGY.

bionomic strategy (*BioSci*) All those features of an organism or population that help adapt it to its environment, eg migration or hibernation.

biophysics (*BioSci*) The physics of vital processes; study of biological phenomena in terms of physical principles.

biopiracy (*BioSci*) The development and often patented use by more technically advanced countries of materials native to developing countries, eg medicinal plants, with no fair compensation to their country of origin.

Biopol (*Chem*) TN for biodegradable polyester synthesized by bacteria under industrial control. The homopolymer is *polyhydroxybutyrate* (PHB), which is highly crystalline and brittle. Copolymerization with *polyhydroxyvalerate* (PHV) reduces the degree of crystallinity and toughens the material. The material is intended to compete with synthetic thermoplastics like polyethylene etc in packaging as well as structural applications.

biopolymers (*Chem*) Naturally occurring long-chain molecules, eg polysaccharides, proteins, DNA. See panel on BIOLOGICAL ENGINEERING POLYMERS.

Biological engineering polymers *(Cont.)*

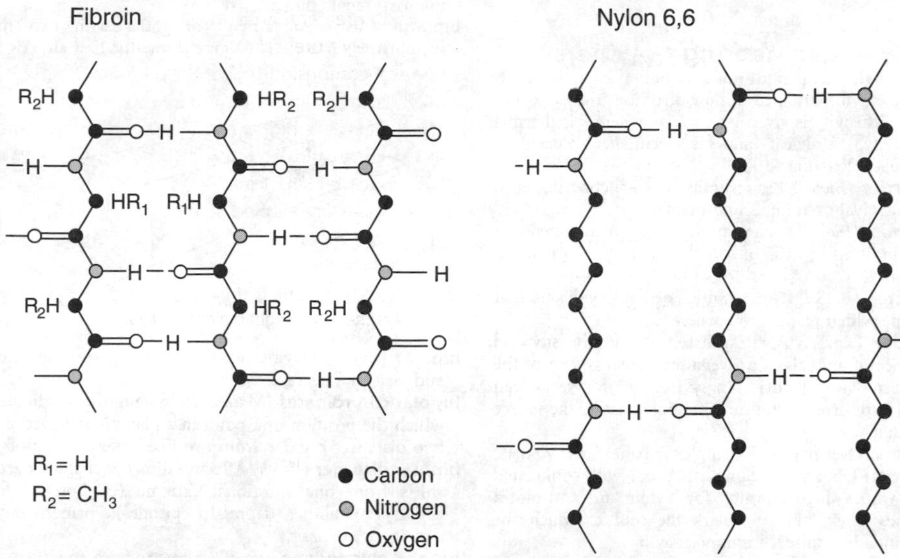

Fibroin

Nylon 6,6

$R_1 = H$
$R_2 = CH_3$

● Carbon
◉ Nitrogen
○ Oxygen

Fig. 3 **The chemical structures of silk fibroin and nylon 6,6 compared.**

for the seed pod to aid wind dispersion. The fibres are short (approx 2·5 cm) so must be spun into yarn before being woven into cloth. Both spinning and weaving increase the flexibility of the product, so modifying a very stiff fibre for practical use.

By contrast, silk is the natural product of spiders, silkworms, etc, and has the specific function of structural support for webs and as a cocoon wrapping. This strong, continuous fibre is an almost regular alternating copolymer of glycine, alanine and serine in a 3:2:1 ratio, which crystallizes into a pleated sheet structure with the main chains aligned along the fibre axes. The structure is supported laterally by hydrogen bonds very like that found in nylon 6,6.

Stiffness is not always a desirable property of a fibre: insulation is a desirable additional property to the flexibility and toughness of body hair, eg Keratin, the basic protein constituent of wool and hair, possesses a much more complex amino acid structure than silk and is less crystalline. The crystalline parts (some 40% compared with 50% in silk) are formed from α-helices loosely packed together. They can easily be unwound when stressed, so wool is much more extensible than silk. Like all natural materials, the full tertiary structure is cellular in origin, with a core surrounded by a cuticle. The cuticle consists of scales oriented in one direction, the basic origin of felting, since they act as a ratchet against the scales of neighbouring wool fibres.

bioprospect (*BioSci*) To investigate living organisms with the aim of discovering materials that can be exploited for commercial gain.

biopsy (*Med*) Diagnostic examination of tissue (eg tumour) removed from the living body.

bioremediation (*BioSci*) The use of living organisms to decontaminate soil by absorbing pollutants; cf PHYTOR-EMEDIATION.

bioscience (*Genrl*) See BIOLOGY.

BIOS (*ICT*) Abbrev for *basic input–output system*. This is an essential component of a computer OPERATING SYSTEM upon which more complex functions of the operating system such as disk operating systems are based. Stored in ROM and available on switch-on.

biosecurity (*BioSci*) The protection of living organisms from harmful effects brought about by other species, esp the transmission of disease.

biosensor (*EnvSci*) A living organism used to detect the presence of chemicals in the environment, eg to find alkaline rocks in an acidic environment.

biosphere (*EnvSci*) The part of the Earth (upwards at least to a height of 10 000 m, and downwards to the depths of the ocean, and a few hundred metres below the land surface) and the atmosphere surrounding it, which is able to support life. A term that may be extended theoretically to other planets. Cf HYDROSPHERE, LITHOSPHERE.

biostratigraphy (*Geol*) Stratigraphy based on the use of fossils, esp for correlation.

biosynthesis (*BioSci*) The synthesis of complex molecules using enzymes and biological structures like ribosomes and chromosomes either within or without the cell.

biosystematics (*BioSci*) The study of relationships with reference to the laws of classification of organisms; taxonomy.

biota (*BioSci*) Fauna and flora of a given region.

biotechnology (*BioSci*) The use of organisms or their components in industrial or commercial processes, which can be aided by the techniques of GENETIC MANIPULATION (panel) in developing eg novel plants for agriculture or industry.

Biot–Fourier equation (*Eng*) The equation representing the non-steady conduction of heat through a solid. In the one-dimensional case,

$$\frac{\partial \varphi}{\partial t} = a \frac{\partial^2 \varphi}{\partial x^2}$$

where φ is the temperature of a section at right angles to the flow, x is the distance in flow direction, t is time, and thermal diffusivity is given by $a = k/\rho s$, where k is thermal conductivity, ρ is density and s is specific heat capacity.

biotic (*BioSci*) Relating to life.

biotic barrier (*BioSci*) Biotic limitations affecting dispersal and/or survival of animals and plants.

biotic climax (*BioSci*) A community that is maintained in a stable condition because of some biotic factor, eg grazing. See CLIMAX.

biotic factor (*BioSci*) The activities of any organisms that determine which plants grow where.

biotic index (*EnvSci*) A rating based on the diversity and abundance of animals in a river used in assessment of the water's ecological quality; ranges from 10 for very clean water to 0 for gross pollution. Useful in locating the source of pollutants.

biotin (*BioSci*) See vitamin B complex in panel on VITAMINS.

biotinylation (*BioSci*) Labelling of a probe with conjugated biotin, whose high affinity for avidin or anti-biotin antibodies is exploited to mark the spot to which the probe binds by indirect immunoassay.

biotite (*Min*) Black mica widely distributed in igneous rocks (particularly in granites) as lustrous black crystals, with a perfect cleavage. Also occurs in mica-schists and related metamorphic rocks. In composition, it is a complex silicate, chiefly of aluminium, iron and magnesium, together with potassium and hydroxyl.

Biot laws (*Phys*) Laws stating that the rotation produced by optically active media is proportional to the length of path, to the concentration (for solutions) and to the inverse square of the wavelength of the light.

Biot modulus (*Eng*) The heat transfer to a wall by a flowing medium, giving the ratio of heat transfer by convection to that by conduction. Defined as $\alpha\theta/\lambda$, where α is the heat transfer coefficient, λ is the thermal conductivity of medium, and θ is the characteristic length of apparatus.

biotope (*BiosSci*) A small habitat in a large community, eg a cattle dropping on a grass prairie, whose several short seral stages comprise a microsere.

biotroph (*BioSci*) A parasite which feeds off the living cells of its host and therefore needs their continued functioning, eg rust and smut fungi. Cf NECROTROPH. See OBLIGATE PARASITE.

Biot–Savart law (*Phys*) The expression for the intensity of magnetic flux density produced at a point a distance from a current-carrying conductor.

biotype (*BioSci*) A group of individuals within a species with identical, or almost identical, genetic constitution.

bipack (*ImageTech*) Two films run through a camera in contact, emulsion to emulsion.

biparous (*BioSci*) Having given birth to two young.

bipartite graph (*MathSci*) A graph in which each vertex can be coloured white or black in such a way that vertices of the same colour are not joined by an edge. See COLOURING OF A GRAPH, GRAPH THEORY.

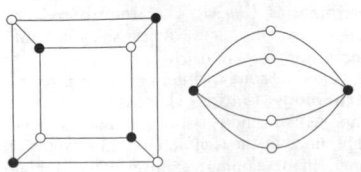

bipartite graph Both graphs are bipartite.

bipedal (*BioSci*) Using only two limbs for walking.

bi-phase (*ElecEng*) See TWO-PHASE.

bipinnate (*BioSci*) Said of a compound pinnate leaf with its main segments pinnately divided.

biplane (*Aero*) An aircraft or glider with two main supporting surfaces. The figure shows the first aircraft that ever flew.

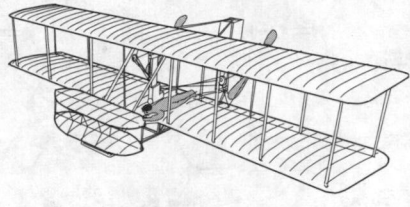

biplane The Wright biplane of 1903.

bipolar (*BioSci*) Having two poles; having an axon at each end, as some nerve cells.

bipolar co-ordinates (*MathSci*) A system of co-ordinates in which the position of a point in a plane is specified by its two distances r and r' from two fixed reference points.

bipolar disorder (*Psych*) Affective disorder characterized by shifts from one emotional extreme (euphoria, intense activity) to another (depressive episodes). Formerly *manic-depressive psychosis*.

bipolar electrode (*ElecEng*) An electrode in an electroplating bath not connected to either the anode or cathode. Also *secondary electrode*.

bipolar germination (*BioSci*) Germination of a spore by the formation of two germ tubes, one from each end.

bipolar transistor (*Electronics*) A transistor that uses both positive and negative charge carriers. Both p–n–p and n–p–n types of bipolar transistor can be manufactured, as discrete devices, or for incorporation into integrated circuits.

biprism (*ImageTech*) Two prisms of very acute angle placed side by side and used as a focusing aid on the screens of cameras. See PENTAPRISM.

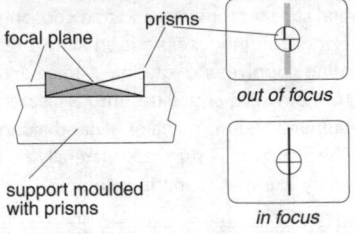

biprism Enlarged in the section with image of line on a focusing screen at right.

bipropellant (*Space*) Rocket propellant made up of two liquids, one being the fuel and the other the oxidizer, which are kept separate prior to combustion. See panel on ROCKET.

bipyramid (*Crystal*) A crystal form consisting of two pyramids on a common base, one being the mirror image of the other. Each pyramid is built of triangular faces, 3, 4, 6, 8 or 12 in number. See PYRAMID.

biquartz (*Phys*) A quartz cut perpendicular to the axis, one-half of the disk being right-handed and the other left-handed quartz. The thickness is such that each half rotates the plane of vibration of yellow light through 90° but in opposite directions. The device is used as a sensitive analyser for SACCHARIMETERS.

biradial symmetry (*BioSci*) The condition in which part of the body shows radial, part bilateral symmetry, as in some Ctenophora.

biramous (*BioSci*) Having two branches; forked, as some crustacean limbs. The two branches may have different functions. Cf UNIRAMOUS.

birch (*For*) A common European hardwood tree (*Betula*) yielding a pale or very-light-brown wood.

bird flu (*Vet*) A highly contagious strain of influenza that affects poultry and can be transmitted to humans. Also *avian flu* or *avian influenza*.

bird's beak (*Electronics*) In microelectronic fabrication, descriptive of the shape of that part of a silicon dioxide layer grown on a silicon wafer near the edge of a region which is protected from oxidation by a diffusion barrier.

bird's mouth (*Build*) A re-entrant angle cut into the end of a timber, so as to allow it to rest over the arris of a cross-timber.

birefringence (*Min*) The splitting of incident light into two rays vibrating at right angles to each other, and causing two images to appear, eg in calcite, caused by DOUBLE REFRACTION in birefringent materials whose refractive index varies with direction. A measure of birefringence is the difference, Δn, between the greatest and least value of refractive index. Arises from anisotropy in the material, eg crystalline anisotropy, molecular orientation, frozen-in or imposed strains, and flow where there are suspended anisotropic particles. The effect is used in mineralogy, in stress analysis and to determine degree of orientation in biological materials, plastics and textile fibres.

birefringence (*Textiles*) The difference between the refractive index of a fibre measured parallel to the fibre axis and perpendicular to it. This gives an expression for the orientation of the molecules in the fibre.

birefringent filter (*Phys*) Filter based on the polarization of light which enables a narrow spectral band of <0.1 nm to be isolated, ie effectively a MONOCHROMATOR; used eg for photographing solar flares.

Birmingham wire gauge (*Eng*) Systems of designating the diameters of rods and wires by numbers. Obsolescent, being replaced by preferred metric dimensions. Abbrev *BWG*. Also *Birmingham gauge (BG)*.

birnessite (*Min*) Hydrated manganese oxide with sodium and calcium, and enriched in other elements. It is one of the dominant minerals of the deep-sea MANGANESE NODULES.

Birox resistor (*Electronics*) Resistor made from a thick film of bismuth ruthenate fired with a glass; noted for stability.

birthmark (*Med*) See NAEVUS.

bischofite (*Min*) Hydrated magnesium chloride. A constituent of salt deposits, such as those of Stassfurt in Germany; decomposes on exposure to the atmosphere.

biscuit joint (*Build*) A timber joint in which oval biscuits 4 mm thick made of compressed beech are glued into matching slots made by a biscuit jointer, a small (10 cm diameter) circular saw with guide fences arranged to make plunge cuts. Unlike a dowel joint there is room for some movement along the length of the slot.

bisector (*MathSci*) A straight line which divides another line, angle or figure into two equal parts.

biseriate (*BioSci*) In two whorls, cycles, rows or series.

biserrate (*BioSci*) Of a leaf margin, having a series of saw-like teeth, which are themselves serrated.

bisexual (*BioSci*) Possessing both male and female sexual organs. See HERMAPHRODITE.

bisexuality (*Psych*) (1) Having the physical or psychological attributes of both sexes. (2) Being sexually attracted to both sexes.

Bismarck brown (*Chem*) A brown dyestuff obtained by the action of nitrous acid on *m*-phenylenediamine. It contains triaminoazobenzene, $H_2NC_6H_4N{=}NC_6H_3(NH_2)_2$, and a more complex disazo compound, $C_6H_4{=}[N{=}NC_6H_3(NH_2)]_2$.

bismite (*Min*) The monoclinic phase of bismuth trioxide. Cf SILLÉNITE.

bismuth (*Chem*) Element, symbol Bi, at no 83, ram 208·98, mp 268°C. It is very rare, with only about 0·008 ppm in the Earth's crust, but occurs as native metal and in minerals often in association with gold, sulphur, selenium or tellurium. Used as a component of fusible alloys with lead; ^{209}Bi is the heaviest stable nuclide. It is strongly diamagnetic and has a low capture cross-section for neutrons, hence its possible use as a liquid metal coolant for nuclear reactors. A high absorption for γ-rays makes it a useful filter or window for these, while transmitting neutrons.

bismuth (III) chloride (*Chem*) $BiCl_3$. Formed by the direct combination of chlorine and bismuth. Treated with an excess of water, it forms bismuth oxychloride, BiOCl, once used as a pigment called *pearl white*.

bismuth (III) hydride (*Chem*) BiH_3. Volatile, unstable compound. Also *bismuthine*.

bismuth (III) oxide (*Chem*) Bi_2O_3. Formed when bismuth is heated in air or when the hydroxide, carbonate, or nitrate is calcined.

bismuth glance (*Min*) See BISMUTHINITE.

bismuthine (*Chem*) See BISMUTH (III) HYDRIDE.

bismuthinite (*Min*) Sulphide of bismuth, rarely forming crystals. Also *bismuth glance*.

bismuth ochre (*Min*) A group name for undetermined oxides and carbonates of bismuth occurring as shapeless masses or earthy deposits.

bismuth spiral (*ElecEng*) Flat coil of bismuth wire used in magnetic flux measurements; the change of flux is measured by observing the change in resistance of the bismuth wire, which increases with increasing fields.

bismutite (*Min*) Bismuth carbonate. $(BiO)_2CO_3$. A tetragonal mineral.

bisoprolol fumarate (*Pharmacol*) A cardioselective *beta-adrenoceptor blocking drug* used for treatment of angina.

bisphenoid (*Crystal, Min*) A crystal form consisting of four faces of triangular shape, two meeting at the top and two at the base in chisel-like edges, at right angles to one another; hence the name, meaning 'double edged'.

bisphenol A (*Plastics*) *1,2-bis(4-hydroxyphenyl)-propane*. Intermediate in the manufacture of epoxy resins and polycarbonate. Also *phenol A*.

bisphosphonates (*Pharmacol*) A family of drugs primarily used to prevent and treat *osteoporosis*.

bisporangiate (*BioSci*) Said of a STROBILUS that consists of megasporophylls and microsporophylls, with megasporangia and microsporangia.

bistable (*ICT*) Said of a device that has two stable states, which can be used to represent 0 and 1. See FLIP-FLOP.

bistable circuit (*ICT*) A valve or transistor circuit that has two stable states which can be decided by input signals; much used in counters and scalers.

bistoury (*Med*) A long, narrow surgical knife for cutting abscesses etc.

bisulphites (*Chem*) *Hydrogen sulphites, hydrogen sulphates (IV)*. Acid salts of sulphurous acid. Useful as preservatives and as a source of sulphur dioxide. See SULPHUROUS ACID.

bisynchronous motor (*ElecEng*) A motor like an ordinary synchronous motor but running at twice synchronous speed.

bit (*Build*) (1) A boring tool which fits into the socket of a brace, by which it is rotated. (2) The cutting-iron of the plane.

bit (*ICT*) (1) Abbrev for *binary digit*; a digit in binary notation, ie 0 or 1. It is the smallest unit of storage or communication. (2) A unit of INFORMATION CONTENT equal to that of a message, the a priori probability of which is one-half.

bit (*MinExt*) Cutting end of length of drill steel used in boring holes in rock. See DIAMOND BIT, DRILL BIT, ROLLER BIT.

bitch (*Build*) A kind of DOG in which the ends are bent so as to point in opposite directions.

bit error ratio (*ICT*) The rate at which erroneous BITS are received over a link, expressed as a proportion of the overall bit rate. In good systems the bit error ratio can be less than 1 in 10^9.

bit gauge (*Build*) An attachment to a bit which limits drilling or boring to a given depth. Also *bit stop*.

bit-mapped display (*ICT*) A display for which an exact image is kept in MAIN MEMORY and is changed by changing the memory and then mapping it to the screen.

bit-mapped font (*ICT*) A FONT that is stored as a pattern of PIXELS. Cf VECTOR FONT.

bit-mapped graphics (*ICT*) A graphics image that is stored as a pattern of PIXELS. A disadvantage of bit-mapped graphics representation is that the image may not contain sufficient detail if it is enlarged due to the finite pixel size in the original image.

BITNET (*ICT*) Abbrev for *because it's time network*. An international co-operative network originating in the USA, mainly serving academic institutions. See EARN, NETNORTH.

bit rate (*ICT*) The number of BITS per second that may be transferred over a digital communication system. See BAUD.

bit-stock (*Build*) See BRACE.

bit stop (*Build*) See BIT GAUGE.

bitter almond oil (*Chem*) Benzaldehyde. Also occurs naturally in ALMOND OIL.

bittern (*Chem*) The residual liquor remaining from the evaporation of sea water, after the removal of the salt crystals.

Bitter pattern (*Phys*) A pattern showing boundaries of magnetic domains on the surface of a magnetic material, formed by applying a colloidal suspension of a magnetic powder. The particles accumulate where the domain boundaries intersect the surface.

bittiness (*Build*) A defect where particles of grit etc appear on a finished paint film. Can only be rectified by rubbing down when dry and recoating with clean material and tools.

bitumen (*Chem*) Tarry, non-mineralized substances of coal, lignite, etc, and their distillation residues. The heaviest residues of vacuum distillation or long-term weathering of crude oils. Some grades are hardened by oxidation at high temperature to give *blown bitumen*. Used for road surfacing as a component of ASPHALT.

bitumen (*Min*) All naturally occurring inflammable hydrocarbons, of various compositions and consistencies, from liquids to solids: petroleums, tars, asphalts, waxes, etc.

bitumen of Judea (*Min*) Natural asphalt from Middle East, with PHOTORESIST properties. Used by Niepce in 1826 to create first photograph. Also *oil of Judea*.

bitumen varnishes (*Build*) Varnishes containing asphalts, driers and aliphatic hydrocarbons.

bituminous felt (*Build*) A manufactured material incorporating asbestos, flax or other fibres, and bitumen, generally about $\frac{1}{8}$ in (3 mm) thick. It is produced in rolls, is impervious to water, and is largely used for roof coverings.

bituminous paints (*Build*) A range of surface coatings from cheap tar blacks to good-quality BLACK JAPAN with a limited colour range. Made from pitch, asphaltum or coumarone resin, they possess good anti-corrosive properties, moisture and chemical resistance together with good flexibility and durability.

bituminous plastics (*Chem*) Compression moulding materials based on natural asphalts or artificial pitches reinforced with asbestos or cotton fibre, ground wood or cork, talc, slate dust or china clay. Formerly used for products of simple shape, eg toilet cisterns, battery cases.

bituminous shale (*Geol, MinExt*) Shale rich in hydrocarbons, which may be recoverable as gas or oil by distillation.

biuret (*Chem*) Carbamoyl urea. $NH_2CONHCONH_2$. Colourless needles, crystallizes with 1 molecule of H_2O; mp of the anhydrous compound 190°C. It is formed from urea at 150–170°C with liberation of NH_3. Has a peptide bond (CONH). It is the type of group formed in polyurethanes by reaction of excess isocyanate with urea groups, so giving cross-links.

biuret reaction (*Chem*) A reaction in which the alkaline solution of biuret, or any compound with two or more peptide bonds (eg proteins), gives a reddish-violet coloration on the addition of copper (II) sulphate. Used as a colorimetric test.

bivalent (*BioSci*) One of the pairs of homologous chromosomes present during meiosis.

bivalent (*Chem*) See DIVALENT.

bivalve (*BioSci*) Said of a mollusc of the class BIVALVIA. Also *lamellibranch*.

Bivalvia (*BioSci*) A class of Mollusca with the body usually enclosed by paired shell valves joined by a hinge and closed by adductor muscles. Ctenidia or gills are used for filter feeding and there are inhalant and exhalant siphons. Such creatures are mostly sessile aquatic forms. Mussels, clams, oysters, scallops. Also *Pelecypoda*.

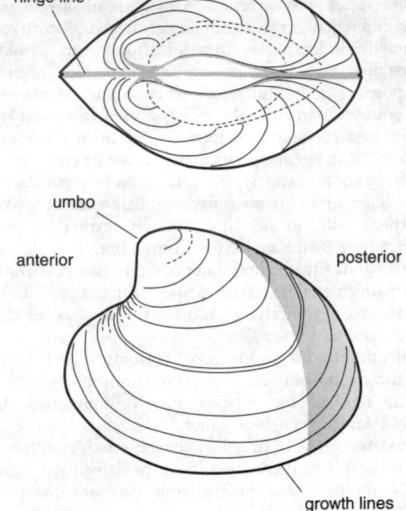

Bivalvia

bivariant (*Chem*) Having two degrees of freedom.

bivoltine (*BioSci*) Having two broods in a year. Cf UNIVOLTINE.

Bjerknes circulation theorem (*EnvSci*) For relative motion with respect to the Earth, the rate of change dC/dt of the circulation C along a closed curve always consisting of the same fluid particles is equal to the number N of isobaric–isosteric (ie pressure–density) solenoids enclosed by the curve, minus the product of twice the angular velocity of the Earth and the rate of change dA/dt of the area A defined by the projection of the curve on the equatorial plane. If V is the velocity of the fluid, dl an element of the curve, p the pressure, and ρ the density, then C and N are defined by

$$C = \oint V \cdot dl$$

and

$$N = -\oint \frac{dp}{\rho}$$

Bk (*Chem*) Symbol for BERKELIUM.

black (*Eng*) Of parts of castings and forgings not finished by machining, the dark coating of iron oxide retained by the surface.

black ash (*MinExt*) *Soda ash*. Impure sodium carbonate, containing some calcium sulphide and carbon. Important pH regulator in flotation process.

black-band iron ore (*MinExt*) A carbonaceous variety of clay ironstone, the iron being present as carbonate (siderite); occurs in the English Coal Measures.

black bean (*For*) A tree of the genus *Castanospermum*, having extremely hard wood.

Blackberry (*ICT*) A proprietary name for a mobile computing device featuring wireless network communications.

black body (*Phys*) A body which completely absorbs any heat or light radiation falling upon it. A black body maintained at a steady temperature is a full radiator at that temperature, since any black body remains in equilibrium with the radiation reaching and leaving it. Also *complete radiator*.

black-body radiation (*Phys*) Radiation that would be radiated from an ideal BLACK BODY. The energy distribution is dependent only on the temperature and is described by PLANCK'S RADIATION LAW. See STEFAN–BOLTZMANN LAW, WIEN'S LAWS and panel on QUANTUM THEORY.

black-body temperature (*Phys*) The temperature at which a BLACK BODY would emit the same radiation as is emitted by a given radiator at a given temperature. The *black-body temperature* of carbon-arc crater is about 3500°C, whereas its true temperature is about 4000°C.

black box (*Aero*) See FLIGHT RECORDER.

black box (*Genrl*) A generalized colloquial term for a self-contained unit of electronic circuitry; not necessarily black. It should produce a defined output for a defined input, without the operator needing to know its contents.

black burst (*ImageTech*) A signal without picture information but with SYNC PULSES, COLOUR BURST and BLACK LEVEL.

blackbutt (*For*) Tree of the genus *Eucalyptus*, giving a hardwood of interlocked grain and coarse but uniform texture. Its colour is light brown, with pinkish markings.

black copper (*Eng*) Impure metal, carrying some iron, lead and sulphur. Produced from copper ores by blast furnace reduction.

black crush (*ImageTech*) Tonal distortion in the picture whereby varying dark tones are all reproduced as black.

black damp (*MinExt*) Mine air which has lost part of its oxygen as result of fire, and is dangerously high in carbon dioxide. See FIRE DAMP.

black death (*Med*) Popular name for the bubonic/pneumonic plague that swept Europe in the middle of the 14th century.

black diamond (*Min*) A variety of cryptocrystalline massive diamond, but showing no crystal form. Highly prized, for its hardness as an abrasive. Occurs only in Brazil. Also *carbonado*.

black disease (*Vet*) A toxaemic disease of sheep caused by *Clostridium novyi* (*Cl. oedematiens*). Infection is associated with the presence of liver damage caused by immature liver flukes. Controlled by vaccination.

black dwarf (*Astron*) The remnant of a white dwarf which has cooled and no longer emits visible light.

blackening (*Paper*) A mottled appearance which may occur on the surface when paper of high moisture content is calendered.

black fever (*Med*) See ROCKY MOUNTAIN FEVER.

black frost (*EnvSci*) An air frost with no deposit of HOAR FROST.

black-hat hacker (*ICT*) A computer hacker who tries to break into a system to perform acts of sabotage (cf WHITE-HAT HACKER).

blackhead (*Vet*) *Infectious entero-hepatitis, histomoniasis.* An infectious disease affecting turkeys and occasionally domestic and wild fowl, caused by infection of the liver and caecae by the protozoon, *Histomonas meleagridis.*

blackheart (*Eng*) A form of malleable iron, in which the core contains rosettes of graphite which appear as a dark area on fracture surfaces.

blackheart (*For*) An abnormal black or dark-brown coloration that may occur in the heartwood of certain timbers.

black hole (*Astron*) A region of SPACE–TIME from which matter and energy cannot escape. See panel on BLACK HOLE.

blacking (*Eng*) A carbonaceous material applied as a powder or wash to the internal surface of a mould to protect the sand and improve the finish of the casting. Prepared in a blacking mill.

blacking a tape (*ImageTech*) Preparing a blank tape for the INSERT EDIT by recording a black burst signal which lays down the video tracks and CONTROL TRACK.

black iron oxide (*Build*) The only inorganic black pigment used on any scale, mainly in anti-corrosive and anti-fouling paints. It consists mainly of iron oxide Fe_3O_4.

black jack (*Min*) A popular name for the mineral sphalerite or zinc blende.

black japan (*Build*) A semi-transparent, quick-drying black varnish, based on asphaltum and drying oil.

black lava glass (*Geol*) Massive natural glass of volcanic origin, jet black and vitreous.

black lead (*Min*) A commercial form of GRAPHITE.

blackleg (*Vet*) An acute infection of cattle and sheep due to *Clostridium chauvoei*; characterized by fever and usually crepitant swelling of the infected muscles. It is controlled by vaccination. Also *blackquarter, quarter evil, quarter ill.*

black letter (*Print*) A simple style of type including Old English, Ancient Black, Tudor Black and many others.

black level (*ImageTech*) Signal level corresponding to zero illumination in the display.

black-level clamp (*ImageTech*) A circuit which holds the BLACK LEVEL of a TV picture constant regardless of variations in brightness.

black liquor (*Paper*) The reagent used for digesting a fibrous raw material at the end of the digestion process.

Blackman theory of specific heats of solids (*Phys*) A theory based on the dynamics of a crystal lattice of particles developed by Born and von Karman. It is more exact, but much more complicated than the DEBYE THEORY OF SPECIFIC HEATS OF SOLIDS which treats the crystal as a continuous isotropic medium.

black mortar (*Build*) A low-strength mortar containing a proportion of ashes. Sometimes used for pointing, where a dark colour is required.

black oil (*MinExt*) A term for heavier, darker petroleum products such as fuel oils. Defines equipment, eg tanks and tankers dedicated to these products, which would need cleaning before use for WHITE OIL.

black opal (*Min*) Any opal of dark tint, although the colour is rarely black; the fine Australian blue opal, with flame-coloured flashes, is typical.

Black Orlon (*Eng*) TN for precursor to carbon fibre.

blackout (*Med*) Temporary loss of vision, perhaps with loss of consciousness, due to sudden reduction of blood supply to the brain.

black powder (*MinExt*) Gunpowder used in quarry work. Standard contains 75% potassium nitrate, slow 59% and blasting 40%, the balance being charcoal and sulphur.

blackquarter (*Vet*) See BLACKLEG.

black red heat (*Eng*) Temperature at which hot metal is just seen to glow in subdued daylight (about 540°C).

black sand (*Eng*) A mixture of sand and powdered coal forming the floor of an iron foundry.

black smoker (*Geol*) A hydrothermal vent, at the crest of an oceanic ridge, producing a plume of large quantities of black fine-grained and very hot sulphide precipitates. There are many varieties of deposits, with iron, copper, manganese and zinc sulphides or oxides common. Cf CHIMNEY, WHITE SMOKER.

black-step marks (*Print*) Collating marks printed on the fold between the first and last pages of each section, as an alternative or addition to the SIGNATURE MARK. The recommended size is 12 pt deep by 5 pt wide and the position is stepped down for each successive section, giving an immediate visual check on the GATHERING. Sometimes *back-step.*

black-tongue (*Vet*) A disease of dogs, due to a deficiency of nicotinic acid in the diet.

blackwater (*Vet*) See REDWATER.

blackwater fever (*Med*) An acute disease prevalent in tropical regions, esp Africa, with feverishness, bilious

Black hole

A region of space–time from which matter and energy cannot escape. A black hole is formed from a star or galactic nucleus which has collapsed in on itself to the point where its gravity is so strong that nothing, not even light, can escape from it. Some *binary stars* which strongly emit X-rays may have black-hole companions. Accretion onto a black hole is also invoked as the energy source of a QUASAR (see panel).

Large stars, about ten times the mass of the Sun, explode as supernovae at the end of their lives. Enormous amounts of energy are radiated into space but the core of the star can collapse under its own gravitational field and begins to pull into itself surrounding gases and other matter including light and radiation. Some smaller stars have insufficient gravity to collapse to a singularity and instead compress their protons and electrons into a mass of neutrons only a few kilometres across to form NEUTRON STARS.

Black holes cannot be seen directly, but can be detected because well over half of all stars exist as binary stars in which each rotates around the other causing characteristic perturbations in their orbits. Since black holes and neutron stars retain their gravitational fields, these perturbations persist after the star has disappeared and an increasing number of such objects have now been identified and studied.

Surrounding each black hole is an EVENT HORIZON where the gravitational pull is just sufficient to pull radiation into the core. Outside this is an *advection layer* where interstellar gas is being pulled into the event horizon. This gas heats up and becomes ionized and emits X-rays before it disappears behind the event horizon. Many possible black holes have been identified by this characteristic X-ray emission.

Supermassive black holes
There is a second class of much larger black holes which are thought to occur near the centre of galaxies. These *supermassive black holes* are the collapsed remnants of a mass several million times that of the Sun. These supermassive objects are thought to be demonstrated by the unique spectrum of the radiation emitted by interstellar gas, which is continually pulled into the supermassive black hole, and from the movements of stars in their gravitational field. Our own Milky Way is thought to have such an object at its centre.

Black holes are the explanation which best fits all these observations and no one so far has thought of another astronomical object which explains them equally well. They may disappear gradually as the result of HAWKING RADIATION, which was predicted by S Hawking (1974) to emerge continuously from black holes.

See panels on COSMOLOGY and GALAXY.

---

vomiting, and passages of red or dark-brown urine. Associated with malignant tertian malaria infection (*Plasmodium falciparum*). Also *Haemoglobinuric (haematuric) fever.*

blackwood (*For*) See AFRICAN BLACKWOOD, AUSTRALIAN BLACKWOOD.

bladder (*BioSci*) (1) Any membranous sac containing gas or fluid; esp the urinary sac of mammals. (2) A small hollow spherical trap for catching and digesting small animals on the bladder-wort, *Utricularia*.

bladderworm (*BioSci*) See CYSTICERCUS.

blade (*BioSci*) The flattened part of a leaf (the lamina), sepal, petal or thallus.

blade (*ElecEng*) The moving part of a knife-switch which carries the current and makes contact with the fixed jaws.

blade activity factor (*Aero*) The capacity of a propeller blade for absorbing power, expressed as a non-dimensional function of the surface and by the formula

$$AF = \left(\frac{5}{R}\right)^5 \int_{0.2R}^{R} cr^3 dr$$

where R is the diameter, and c is the blade chord at any radius r.

blade angle (*Aero*) The angle between blade chord and plane of rotation at any radius. It is not constant because of the higher airspeed towards the tip, the incidence being progressively reduced to maintain optimum thrust. Change of blade angle from root to tip is called *blade twist*. See fig. at PROPELLER.

blade loading (*Aero*) The thrust of a helicopter rotor divided by the total area of the blades.

blade twist (*Aero*) See BLADE ANGLE.

Blagden's law (*Chem*) The law that the lowering of the freezing point of a given solvent is proportional to the molar concentration of the solute.

Blaine fineness tester (*PowderTech*) PERMEAMETER in which the powder bed is connected to the low-pressure side of a U-tube manometer. Air is allowed to flow through the powder bed to equalize the pressure in the two arms of the manometer. The time required for a specified movement of the manometer is used to calculate the surface area.

Blake crusher (*MinExt*) See JAW BREAKER.

blanching (*FoodSci*) (1) Heat process carried out on vegetables to inactivate enzymes causing discoloration and other spoilage and to remove gases or air from their cells. It also kills some bacteria and insect contaminants. Also carried out on shelled nuts to remove the skins.

blank (*Acous*) The lacquer-coated disk ready for placing on a recording machine for making records with a stylus.

blank (*Eng*) A piece of metal, shaped roughly to the required size, on which finishing processes are carried out.

blanket (*NucEng*) (1) Region of FERTILE material surrounding the core in a breeder reactor in which neutrons coming from the core breed more fissile fuel, eg uranium-233 from thorium. (2) The lithium surrounding a fusion reactor core within which fusion neutrons are slowed down, heat is transferred to a primary coolant and tritium is bred from lithium.

blanket (*Print*) (1) The rubber-covered clothing of the blanket cylinder on an offset press. (2) The rubber, textile, cork, plastic or composition covering of the impression cylinder on a rotary press.

blanket (*Textiles*) A thick woven, knitted or non-woven fabric giving good thermal insulation. Traditionally blankets were made from wool fabrics that were milled and raised but cotton materials of an open construction are also in common use.

blanket bog (*EnvSci*) Bog vegetation forming an extensive and continuous layer of peat over a flat and undulating landscape, deriving its mineral nutrients largely from rainfall, and found in cold wet parts of the world.

blanket clamp (*Print*) On impression or offset cylinders, the means of locating the fixed end of the blanket.

blanket cylinder (*Print*) The rubber-covered cylinder between the plate cylinder and the impression cylinder on an offset press. See fig. at LITHOGRAPHY.

blanket pins (*Print*) Pins used to attach the blanket to another sheet on the BLANKET WIND.

blanket strake (*MinExt*) Table or sluice with gentle downslope, with bottom lining of material on which heavy mineral (eg metallic gold) is caught. Originally of rough blanket, now usually corduroy with ribs across flow, or chequered rubber.

blanket-to-blanket press (*Print*) A PERFECTOR machine commonly used in web offset, whereby two blanket cylinders oppose each other and use each other to obtain printing impression. A sheet or web passing between the two blanket cylinders is thus printed on both sides.

blanket wind (*Print*) The means of attaching the loose end of a blanket round the cylinder, usually a rotating blanket bar.

blank flange (*Eng*) A disk, or solid flange, used to blank off the end of a pipe.

blank groove (*Acous*) Unmodulated groove on disk recording.

blanking (*Electronics, ImageTech*) (1) Blocking or disabling a circuit for a required interval of time. (2) Suppression of the picture information while the scanning spot of a cathode-ray tube returns after each line, ie horizontal blanking, or after each field, ie vertical blanking, taking place during the blanking interval.

blast (*Eng*) Air under pressure, blown into a furnace.

-blast, blasto- (*Genrl*) Suffix and prefix from Gk *blastos*, bud.

blast cells (*BioSci*) The proliferating and relatively undifferentiated cells in a cell lineage.

blastema (*BioSci*) Anlage (see PRIMORDIUM); a mass of undifferentiated tissue; the protoplasmic part of an egg as distinguished from the yolk.

blast furnace (*Eng*) Vertical shaft furnace into the top of which ore mineral or scrap metal, fuel and slag-forming rock (FLUX) is charged. Air, sometimes oxygen-enriched and pre-heated, is blown through from below and products are separately tapped (slag higher and metal lower). Used to smelt iron ore, copper, lead, zinc and other minerals.

iron ore, coke and limestone feed

hot gases to heat exchangers and dust precipitators

inlet control valves

steel lined with firebrick

hot air blast through tuyéres

slag

molten iron

slag tap

iron tap

blast furnace The blast is heated by the heat exchangers.

blasting (*Acous*) A marked increase in amplitude distortion due to overloading the capacity of some part of a sound-reproducing system; eg attempt to exceed 100% depth of modulation in a radio transmitter, or break of continuity in carbon granules in a carbon transmitter.

blasting (*CivEng*) The operation of disintegrating rock etc by boring a hole in it, filling with gunpowder or other explosive charge, and firing it.

blasting fuse (*CivEng*) Compound designed to burn at a regulated speed when closed in a tube, used to ignite detonator or explode blasting charge. Types include 'safety' (slow) and detonating (instantaneous).

blast joint (*MinExt*) The specially made heavy-duty joint of a producing well's tubing string, able to withstand the abrasive action of the oil entering from a high-pressure formation.

blast main (*Eng*) The main blast air-pipes supplying air to a furnace.

blasto- (*Genrl*) See -BLAST.

blastochyle (*BioSci*) Fluid in the blastocoel.

blastocoel (*BioSci*) The cavity formed within a segmenting ovum; cavity within a blastula; primary body cavity; segmentation cavity. Also *archicoel*.

blastocyst (*BioSci*) In mammalian development, a structure resulting from the cleavage of the ovum; it consists of an outer hollow sphere and an inner solid mass of cells. Also *germinal vesicle*.

blastoderm (*BioSci*) In eggs with much yolk, the disk of cells formed on top of the yolk by cleavage.

blastodermic vesicle (*BioSci*) See BLASTULA.

blastodisk (*BioSci*) In a developing ovum, the germinal area.

blastoma (*BioSci*) A neoplasm composed of immature blast cells. Pl *blastomas* or *blastomata*.

blastomere (*BioSci*) One of the cells formed during the early stages of cleavage of the ovum.

blastopore (*BioSci*) The aperture by which the cavity of the gastrula retains communication with the exterior.

blastosphere (*BioSci*) See BLASTULA.

blastospore (*BioSci*) A spore produced by budding.

blast pipe (*Eng*) Device located in the smoke-box of a steam locomotive used to improve the draft through the fire-tubes. Exhaust steam passing from the nozzle of the blast pipe reduces smoke-box pressure and induces the draft.

blastula (*BioSci*) A hollow sphere, the wall of which is composed of a single layer of cells, produced as a result of the cleavage of an ovum. Also *blastodermic vesicle, blastosphere*.

blastulation (*BioSci*) Cleavage resulting in the formation of a blastula.

blast wave (*Phys*) See SHOCK WAVE.

Blatthaller loudspeaker (*Acous*) First electrodynamically driven loudspeaker used for high-quality sound reproduction. It has a flat surface which is large compared with the wavelength of the radiated sound and it generates sound of high intensity.

Blavier's test (*ElecEng*) A method of locating a fault on an electric cable; resistance measurements are taken with the far end of the cable free, and again with it earthed.

blazar (*Astron*) A BL LAC OBJECT that varies dramatically in visible light output.

blaze (*Surv*) Temporary survey mark, such as slash on tree trunk, to guide prospector or explorer.

blazer cloth (*Textiles*) An all-wool milled and raised fabric used in the manufacture of jackets.

BLC (*ImageTech*) Abbrev for BACKLIGHT COMPENSATION.

bleach (*ImageTech*) Conversion of a developed silver image into a white or colourless compound, often a silver halide; also the chemical solution used for this process.

bleach (*Textiles*) Any chemical agent used to whiten or remove natural colouring matter from textiles and fibres, eg HYDROGEN PEROXIDE, sodium hypochlorite.

bleach–fix (*ImageTech*) Processing solution which both bleaches a silver image to silver halide and dissolves it away, as in FIXING.

bleaching (*Build*) (1) A whitening of a paint film due to exposure to light or chemical agents. (2) A preparatory treatment for wood to remove exposure stains or to balance variations in natural colour prior to staining or varnishing.

bleaching (*ImageTech*) The removal of reduced silver after development, so that the remaining silver halide, which has not been developed because of its insufficient exposure to light, can be further developed. The resulting image is a positive.

bleaching (*Paper*) That part of the pulp purification process intended to bring the raw material to the desired whiteness. Bleaching may be achieved by oxidation methods (eg using free and/or combined chlorine) or by reduction.

bleaching (*Textiles*) A series of wet processes for removing residual impurities, colour and fatty or waxy substances from fibres, yarns or fabrics. This improves the whiteness and promotes brighter colours after dyeing or printing. Hydrogen peroxide is often used for this purpose.

bleaching powder (*Chem*) Powder commercially obtained from calcium hydroxide and chlorine. The commercial value depends on the amount of available chlorine.

bleach-out process (*ImageTech*) A system of colour printing involving the decolorizing of dyes by exposing them through transparencies.

bleb (*BioSci*) A small vesicle containing clear fluid.

bleed (*Print*) (1) Illustration whose edges are cut away after printing are said to bleed off. (2) The accidental mutilation of type matter by trimming the paper too closely. (3) The running of printing ink on posters and or packages as a result of weather or chemical action.

bleeder resistor (*ElecEng*) Resistor placed across secondary of transformer to regulate its response curve, esp when the transformer is not loaded with a proper terminating resistance. One placed in a power supply or rectifier circuit to control its regulation.

bleeding (*Aero*) The tapping of air from a gas turbine compressor: (1) to prevent SURGING; or (2) to feed some other equipment, eg cabin pressurization or a de-icing system.

bleeding (*BioSci*) The exudation of xylem sap, phloem sap or latex from wounds. See ROOT PRESSURE.

bleeding (*Build*) A paint defect in which a constituent in an underlying surface discolours subsequent coatings, prevented by applying a barrier coating.

bleeding (*Eng*) (1) A method of improving the thermal efficiency of a steam plant by withdrawing a small part of the steam from the higher-pressure stages of a turbine to heat the feed-water. (2) Removing undesirable entrapped air from a hydraulic (eg braking) system.

bleeding (*ImageTech*) Diffusion of dye from an image.

bleeding (*Textiles*) In fibres, yarns or fabrics of two or more colours, the running of the darker colours, and consequent staining of the lighter colours, during finishing, washing or solvent cleaning.

bleeding edge (*ICT*) Highly innovative software or hardware that may confer great advantage to a user, but which has considerable risks, of reliability, upgrade path, life cycle, etc, associated with its use.

bleed line (*MinExt*) Pipeline used in the BLOWOUT PREVENTER stack to remove excess gas pressure.

blend (*Chem*) Physical mixture of thermoplastic polymers.

blend (*Textiles*) An intimate mixture of different qualities or kinds of natural or artificial staple fibres.

blende (*Min*) See SPHALERITE.

blender (*Build*) A signwriting brush with a short square sable or ox filling used for blending colours in shading effects.

blending (*Build*) A term used in graduation from one colour to another. This effect may be created using a BLENDER, a hair stippler or a spray gun.

blending (*Chem*) A process where mixing of polymers is achieved by adding fillers etc to polymer solution. See COWLES DISSOLVER. Term also applied to mixing of dry powders. See HENSCHEL MIXER.

blennorrhagia (*Med*) Discharge of mucus, usually from the genital organs, due to gonorrhoea. Also *blennorrhoea*, US *blennorrhea*.

blephar-, blepharo- (*Genrl*) Prefixes from Gk *blepharon*, eyelid.

blepharism (*Med*) Spasm of the eyelids.

blepharitis (*Med*) Chronic inflammation of the eyelids.

blepharo- (*Genrl*) See BLEPHAR-.

blepharoplast (*BioSci*) See BASAL BODY.

blepharoplegia (*Med*) Paralysis of the muscles of the eyelid resulting in drooping of the upper eyelid or blepharoptosis.

blepharospasm (*Med*) Spasm of the orbicular muscle of the eyelid.

BLEU (*Aero*) The *Blind Landing Experimental Unit* operated by the Royal Aircraft Establishment which developed a fully automatic blind landing system. See AUTOTHROTTLE, FLIGHT DIRECTOR.

B-licence (*Aero*) Commercial pilot's licence.

blight (*BioSci*) The common name for a number of plant diseases characterized by the rapid infection and death of the leaves or the whole plant, eg POTATO BLIGHT.

blimp (*Aero*) Colloq term for a non-rigid AIRSHIP.

blimp (*ImageTech*) A noise-reducing enclosure for a film camera.

blind apex (*MinExt*) Outcrop of mineral vein or lode where ore deposit dies out before reaching the surface (*sub-outcrop*), leaving an apparently barren deposit.

blind arcade (*Build*) See ARCADE.

blind arch (*Arch, CivEng*) A closed arch which does not penetrate the structure; used for ornamentation, to make one face of a building harmonize with another in which there are actual arched openings.

blind area (*Build*) A sunken space round the basement of a building, broken up into lengths by small cross-walls, which support the earth-retaining wall but restrict ventilation.

blind blocking (*Print*) Applying lettering or design to the case of a book without gold leaf or alternative. See BLOCKING.

blind flying (*Aero*) The flying of an aircraft by a pilot who, because of darkness or poor visibility, must rely on the indication of instruments. See GROUND-CONTROLLED APPROACH, INSTRUMENT LANDING SYSTEM.

blind-flying instruments (*Aero*) A group of instruments, often on an individual central panel, essential for BLIND FLYING. Commonly airspeed indicator, altimeter, vertical speed, turn and slip, artificial horizon and directional gyro. See BASIC SIX, BASIC T.

blind image (*Print*) A term applied in lithographic printing to a plate image which loses its affinity for ink.

blinding (*CivEng*) (1) Sprinkling small chippings of stone over a tar-dressed road surface. (2) Placing a thin layer of concrete over a foundation area before the structural concrete is placed.

blinding (*Print*) See BLIND TOOLING.

blind landing (*Aero*) The landing of an aircraft by a pilot who, because of darkness or poor visibility, must rely on the indication of instruments. See GROUND-CONTROLLED APPROACH, INSTRUMENT LANDING SYSTEM.

blind lode (*MinExt*) A lode which does not outcrop to the surface. Also *blind vein*.

blind mortise (*Build*) A mortise which does not pass right through the piece in which it is cut.

blind page (*Print*) A page which has no printed folio, but is included in the pagination; usually found in the preliminary matter.

blind rivet (*Eng*) A type of rivet which can be clinched as well as placed by access to one side only of a structure. Usually based on a tubular or semi-tubular rivet design, eg CHOBERT RIVET, EXPLOSIVE RIVET.

blindsight (*Psych*) A condition, caused by brain damage, in which a person is able to respond to visual stimuli without consciously perceiving them.

blind spot (*BioSci*) In vertebrates, an area of the retina where there are no visual cells (due to the exit of the optic nerve) and over which no external image is perceived.

blind spot (*ICT*) Point within normal range of a transmitter at which field strength is abnormally small. Usually results from interference pattern produced by surrounding objects or geographical features, eg valleys.

blind staggers (*Vet*) See ALKALI DISEASE.

blind study (*Psych, Pharmacol*) A study in which the subject has no knowledge of the anticipated results nor sometimes even the nature of the study. The subjects are said to be blind to the expected results and thus no *placebo* effect should occur. In *double-blind* trials (the gold standard for testing drugs) neither the experimenter nor the subject know whether the subject is receiving active compound or placebo and bias on the part of the experimenter is also eliminated.

blind tooling (*Print*) Applying lettering or design to the case of a book using a press with a heated relief die. Gold leaf is used for special work only; BLOCKING FOIL for EDITION BINDING. The platen may be horizontal and the press hand-operated or partly mechanized, but for long runs a press with a vertical platen and automatic feeding of case and foil(s) is used. Also *blinding*.

blind vein (*MinExt*) See BLIND LODE.

blink comparator (*Astron*) An instrument in which two photographic plates of the same region are viewed simultaneously, one with each eye, any difference being detected by a device which alternately conceals each plate in rapid succession.

blinking (*Radar*) Modification of a loran transmission, so that a fluctuation in display indicates incorrect operation.

blip (*Radar*) Spot on cathode-ray-tube screen indicating radar function.

blister (*Build*) See BLISTERING.

blister (*Eng*) A raised area on the surface of solid metal produced by the emanation of gas from within the metal while it is hot and plastic.

blister (*Med*) A thin-walled circumscribed swelling in the skin containing clear or blood-stained serum; caused by irritation.

blister (*Vet*) An irritant drug applied to the skin to cause inflammation and assist in the healing of deep-seated diseased tissues by reflex nervous action.

blister copper (*Eng*) An intermediate product in the manufacture of copper. It is produced in a converter, contains 98·5–99·5% of copper, and is subsequently refined to give commercial varieties, eg tough pitch, deoxidized copper.

blistering (*Build*) A defect in a finished paint film where areas rise away from the underlying surface, caused by lack of adhesion, trapped moisture and heat.

blister pack (*Eng*) Transparent, thin sheet of plastic thermoformed to cover product for display purposes. Also *bubble pack*.

blister steel (*Eng*) Wrought-iron bars impregnated with carbon by heating in charcoal. Before 1740 this was the only steel available. Obsolete.

blit (*ICT*) Contraction of *block transfer*. To transfer a large array of bits between different locations in a computer's memory.

blivet (*Aero*) Flexible bag for transporting fuel, often slung beneath a helicopter.

BL Lac object (*Astron*) An ACTIVE GALAXY (prototype BL Lacertae) which gives out intense radiation and has a spectrum similar to that of a quasar, but without emission and absorption features. Can display violent variability and is thought to be associated with elliptical galaxies. See panels on GALAXY and QUASAR.

bloat (*Vet*) An acute digestive disorder of ruminants in which excessive amounts of gas accumulate in the rumen, associated usually with the ingestion of lush herbage. Also *dew-blown, hoven*.

Bloch band (*Phys*) See BAND THEORY OF SOLIDS.

Bloch function (*Phys*) The solutions of the SCHRÖDINGER EQUATION for electrons in a crystalline solid which have the same periodicity as the periodic potential function under whose influence the electrons move. See BAND THEORY OF SOLIDS.

Bloch wall (*Phys*) The wall of a MAGNETIC DOMAIN.

block (*Eng*) The housing holding the pulley or pulleys over which the rope or chain passes in a lifting tackle.

block (*ICT*) Group of records treated as a complete unit during transfer to or from backing.

block (*Min*) Crystal, glass or rock fragment of disruptive volcanic origin whose generally angular shape indicates that it was solid when formed, and with an average diameter over 64 mm.

block (*MinExt*) Rectangular panel of ore defined by drives, raises and winzes, giving all-round physical access for sampling, testing and mining purposes.

block (*Print*) A term applied to any letterpress printing plate, duplicate or original, brought to type height by mounting. US *cut*.

blockboard (*Build*) Board composed of softwood strips bonded together and sandwiched between two outer layers of veneer or HARDBOARD. Also *battenboard, coreboard, laminboard*.

block brake (*Eng*) A vehicle brake in which a block of cast-iron is forced against the rim of the revolving wheel, either by hand-power, electromagnetic mechanism or fluid pressure acting on a piston. See AIR BRAKE, ELECTRO-MAGNETIC BRAKE.

block caving (*MinExt*) A method of mining in which block of ore is undercut, so that it caves in and the fragments gravitate to withdrawal points.

block clutch (*Eng*) A FRICTION CLUTCH in which friction blocks or shoes are forced inwards into the grooved rim of the driving member, or expanded into contact with the internal surface of a drum.

block coefficient (*Ships*) The ratio of the underwater volume of a ship to the volume of an enclosing rectangular block.

block copolymer (*Chem*) A polymer in which the different monomers occur in long sequences; has different properties from the equivalent RANDOM COPOLYMER. See panel on POLYMERS.

block diagram (*Eng*) An illustration in which parts of a machine or a process are represented notionally by blocks or similar symbols.

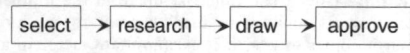

block diagram

block diagram (*ICT*) Diagram made up of squares and rectangles representing different hardware and software components with lines to show their interconnections.

blocked impedance (*Phys*) The impedance of the input of a transducer when the output load is infinite, eg when the mechanical system, as in a loudspeaker, is prevented from moving.

blocked-out ore (*MinExt*) See BLOCK.

block gauge (*Eng*) A block of hardened steel having its opposite faces accurately ground and lapped flat and parallel. The faces are separated by a specified distance, the gauge distance, used for checking the accuracy of other gauges etc.

block-in course (*Build*) A type of masonry, used for heavy engineering construction, in which the stones are carefully squared and finished to make close joints, and the faces are hammer-dressed.

blocking (*Build*) Securing together two pieces of board by gluing blocks of wood in the interior angle.

blocking (*Electronics*) Cut-off of anode current in a valve because of the application of a high negative voltage to the grid; used in GATING or BLANKING.

blocking (*Eng*) Tendency of polymer film to adhere to itself, a problem in manufacture. Inhibited with surface coating of an anti-blocking agent.

blocking (*Print*) Applying lettering or design to the case of a book using a press with a heated relief die. US *stamping*. See BLOCKING FOIL.

blocking action (*EnvSci*) See ANTICYCLONIC BLOCKING.

blocking antibody (*BioSci*) Antibody that combines preferentially with an antigen so as to prevent it from combining with IgE on mast cells, and thereby prevents type I allergic reactions. Blocking antibodies are usually IgG.

blocking capacitor (*ElecEng*) One in signal path to prevent dc continuity. Also *buffer capacitor*.

blocking course (*Build*) A course of stones laid on the top of a cornice.

blocking factor (*ICT*) Number of records in a block.

blocking foil (*Print*) A film base with a layer of gold, other metal or coloured material, used as a substitute for gold leaf. See BLOCKING.

blocking-out (*ImageTech*) The use of Indian ink or other opaque pigment for covering parts of negatives so that they print white.

blocking-out (*MinExt*) A method of estimating tonnage of mineral reserves in a volume established by drilling, with reference to the grade of the representative sample.

blocking press (*Print*) See BLOCKING.

block lava (*Geol*) Lava flows with surfaces of rough angular blocks. Similar to AA but blocks more regular in shape.

block plan (*Build*) A plan of a building site, showing the outlines of existing and proposed buildings.

block plane (*Build*) A small plane about 6 in (150 mm) long which has no cap iron and has the cutting bevel reversed; used for planing end grain, and fine work.

block prism (*ImageTech*) A cube of glass, slit along its diagonal and half-silvered, for splitting the beam in a three-colour beam-splitting camera.

block section (*CivEng*) The length of track in a railway system that is limited by stop signals.

block system (*CivEng*) The system of controlling the movements of trains by signals and by independent communication between block posts, where the instruments indicating the position of trains, condition of the block sections, and controlling levers for signals, points, etc, are situated. It is *absolute* if one train alone is permitted within a block section, and *permissive* if trains are allowed to follow into a block section already occupied by a train.

block time (*Aero*) The time elapsed from the moment an aircraft starts to leave its loading point to the moment when it comes to rest. It is an important factor in airline organization and scheduling. Also *buoy-to-buoy* (seaplanes), *chock-to-chock, flight time*.

blockwork (*Build*) Walling constructed of pressed or cast blocks with a basic constituent of cement, the other constituents being generally to improve the insulating qualities of the wall, eg clinker ash, framed blast furnace slag or an air entraining agent.

blog (*ICT*) Contraction of *weblog*. An on-line diary, originally a log of sites visited on the web by a surfer, now more usually recording the blogger's daily activities and opinions (often political), and inviting comment upon them from other surfers.

blogger (*ICT*) The keeper of an on-line diary or BLOG. Also *bloggist*.

blogosphere (*ICT*) The totality of *blogs* existing in cyberspace. By extension the marketplace of opinion and comment represented by those blogs.

blogring (*ICT*) A set of linked blogs, usually with a common theme or purpose, reached in turn via a *hyperlink* embedded in member blogs.

blogroll (*ICT*) A set of hyperlinks, usually found in the sidebar of a blog, to other weblogs the blogger believes may be of interest to the readers.

blonde (*ImageTech*) TN for a 2 kW variable-beam floodlight for studio use.

Blondel–Rey law (*Phys*) A law used to assess the apparent point brilliance B of a flashing light: $B = B_0[f/(a+f)]$, where B_0 is the point brilliance during the flash, f is the duration of the flash in seconds and a is constant of value about 0·2 s, and B is near the threshold value for white light. The law holds for flashing frequencies of less than 5 Hz.

blondin (*CivEng*) See CABLE-WAY.

blood (*Med*) A fluid circulating through the tissues of the body, performing the functions of transporting oxygen, nutrients and hormones, and carrying waste products to the organs of excretion. It plays an important role in maintaining a uniform temperature in the body in warm-blooded organisms. Its relative density in humans is about 1·054–1·060, and it has a normal pH of 7.4. Its chief constituents are: red cells, white cells, platelets, water (77·5–79%), solids including proteins, lipids, enzymes, hormones and immune bodies, blood sugar, vitamins and inorganic substances. See ABO BLOOD GROUP SYSTEM, RHESUS BLOOD GROUP SYSTEM.

blood albumin (*Med*) See SERUM ALBUMIN.

blood–brain barrier (*BioSci*) A term used to describe the fact that the blood vessels of the brain (and the retina) are much more impermeable to large molecules (like antibodies) than blood vessels elsewhere in the body. This barrier makes the brain a distinct compartment of the body, isolated from the general tissue fluid.

blood cell (*Med*) See HAEMATOBLAST.

blood-clotting factors (*Med*) An internationally agreed scale of discernible factors concerned in blood-clotting. Indicated by roman numerals, eg I, fibrinogen factor; II, prothrombin factor; III, thromboplastin factor; VIII, antihaemophilic factor A; IX, antihaemophilic factor B (Christmas factor) etc.

blood corpuscle (*Med*) A cell normally contained in suspension in the blood. See ERYTHROCYTE, LEUCOCYTE.

blood count (*Med*) The number of red or white corpuscles in the blood.

blood donor (*Med*) A volunteer who donates blood for administration to others.

blood flukes (*BioSci*) Trematodes of the genus *Schistosoma*, parasitic on humans and domestic animals via various species of water snail as intermediate host. They can attack the liver, spleen, intestines and urinary system, and occasionally the brain. See BILHARZIASIS.

blood groups (*Med*) See ABO BLOOD GROUP SYSTEM, RHESUS BLOOD GROUP SYSTEM.

blood islands (*BioSci*) In developing vertebrates, isolated syncytial accumulations of reddish mesoderm cells containing primitive erythroblasts, which give rise respectively to the walls of the blood vessels and to the red corpuscles.

blood plasma (*Med*) See PLASMA.

blood pressure (*Med*) The pressure of blood in the arteries, usually measured by sphygmomanometry. The maximum occurs in systole and the minimum in diastole. Normal young adults will have a (systolic/diastolic) blood pressure of approximately 120/80 mm Hg. See SPHYGMOMANOMETER.

blood-red heat (*Eng*) Dark-red glow from heated metal, in temperature range 550–630°C.

blood serum (*Med*) See SERUM.

bloodstone (*Min*) Cryptocrystalline silica, a variety of chalcedony, coloured deep green, with flecks of red jasper. Also *heliotrope*.

blood substitutes (*Med*) Plasma, albumin and dextran, when used to substitute volume for loss of blood. Newer substances are being investigated which may be able to transport oxygen.

blood sugar (*Med*) The level of glucose in the blood, normally between 3·2 and 5·2 mmol^{-1} in the fasting state.

blood transfusion (*Med*) See TRANSFUSION.

blood vessel (*Med*) An enclosed space, with well-defined walls, through which blood passes. See ARTERY, CAPILLARY, VEIN.

bloom (*BioSci*) (1) A covering of waxy material occurring on the surface of some leaves and fruits, resulting in a whitish cast. (2) A visible, often seasonal occurrence of very large numbers of algae in the plankton of fresh water or sea. See RED TIDE.

bloom (*Build*) (1) Efflorescence on a brick wall. (2) See BLOOMING.

bloom (*Chem*) The colour of the fluorescent light reflected from some oils when illuminated; usually different from that shown by the oil with transmitted light.

bloom (*Eng*) Semi-finished metal, rectangular in cross-section and for steel not more than twice as long as it is thick. Cf BILLET.

bloom (*Glass*) Surface film on glass: (1) the thin dielectric layers vacuum deposited on a lens to alter its reflectance properties, hence *blooming* (see COATED LENS); or (2) the film of sulphites and sulphates formed during the annealing process; or (3) the film caused by weathering. Obsolete.

bloom (*MinExt*) Efflorescence of altered metallic salt at surface of ore exposure, eg cobalt bloom (erythrite).

Bloom gelometer (*FoodSci*) A method of measuring the rigidity of a gelatine gel as the load in grams required to depress by 4 mm a 12·7 mm diameter plunger into the gel surface; 40–80 Bloom g and 250–300 Bloom g are low- and high-strength ranges respectively.

blooming (*Build*) A paint defect affecting gloss-finish materials such as varnish enamel and gloss paint, normally attributable to moisture affecting the film during the drying process.

blooming (*Electronics*) (1) Spread of spot on cathode-ray-tube phosphor due to excessive beam current. (2) Coating of dielectric surfaces to reduce reflection of electromagnetic waves.

blooming (*ImageTech*) Treatment of the glass–air surfaces of a lens with a deposit of magnesium fluoride or other substance, which reduces internal reflection and increases light transmission.

blooming mills (*Eng*) The rolling mills used in reducing steel ingots to blooms. Called *cogging mills* in UK, and not always distinguished from billet (slab) mills.

Bloom's syndrome (*Med*) A rare human autosomal recessive defect associated with genomic instability causing short stature, immunodeficiency and increased risk of all types of cancer. Caused by mutation in the gene encoding DNA helicase.

bloop (*ImageTech*) Dull thud in sound-film reproduction caused by a join in the sound track.

blooping patch (*ImageTech*) Triangular black patch applied or painted over a splice in the sound track of a film print to prevent it causing a noise on projection.

blotting (*BioSci*) A method by which biological molecules are transferred from, usually, a gel to a membrane filter, often nitrocellulose, to which they bind. In the former they can be separated physically by eg size and in the latter they can be tested for the presence of specific sequences by radioactive probes or with antibody for specific antigenic determinants (immunoblotting). In Southern blotting (named for the inventor) DNA is transferred; in Northern blotting, RNA; in Western blotting, protein; in Northwestern blotting protein is transferred but is probed with specific RNA. In dot blots the gel separation step is omitted and the test material spotted directly onto the filter.

blotting paper (*Paper*) Weak, free beaten, unsized paper intended for the absorption of aqueous inks from the surface of documents.

blow (*Eng*) In a Bessemer converter, passage of air through molten charge.

blow (*MinExt*) (1) Ejection of part of the explosive charge (unfired) from a hole. (2) Sudden rush of gas from coal seam or ore body.

blow-and-blow machines (*Glass*) Machines in which the glass is shaped in two stages, but each time by blowing, as opposed eg to pressing or sucking. Cf BLOW MOULDING.

blow back (*Autos*) The return, at low speeds, of some of the induced mixture through the carburettor of a petrol engine; due to late closing of the inlet valve during compression, or by worn or sticking valves.

blowdown stack (*MinExt*) A container into which refinery vessels can be emptied during emergencies and into which steam or water can be injected to prevent ignition of volatile components.

blower (*Eng*) (1) A rotary air compressor for supplying a relatively large volume of air at low pressure. See AIR COMPRESSOR, SUPERCHARGER. (2) A ring-shaped perforated pipe, encircling the top of the BLAST PIPE in the smoke-box, to which steam is supplied while a steam engine is standing, the jets providing sufficient draught to keep the fire going.

blower (*MinExt*) (1) A fissure or thin seam which discharges a quantity of coal gas. (2) An auxiliary ventilating appliance, eg a fan or venturi tube, for supplying air to subsidiary working places or to dead-ends.

blowfly myiasis (*Vet*) Infestation of the skin of sheep, esp in the breech region, by blowfly larvae. Also *strike*.

blowhole (*Build*) A pock mark existing on the surface of *in situ* or precast concrete when the shuttering is removed; due to air entrapped during the process of placing the concrete and not released normally due to inadequate vibration.

blowhole (*Eng*) A gas-filled cavity in a solid metal. Usually formed by the trapping of bubbles of gas evolved during solidification (see GAS EVOLUTION), but may also be caused by steam generated at the mould surface, air entrapped by the incoming metal, or gas given off by inflammable mould dressings.

blowhole (*Geol*) An aperture near a cliff-top through which air, compressed in a sea-cave by breaking waves, is forcibly expelled.

blowing agent (*Chem*) Speciality chemical (eg AZODICARBONAMIDE) added to polymers which decomposes at a specific elevated temperature during moulding to produce gas. The foam creates a lightweight core which improves the stiffness- and strength-to-weight ratio of the product. Sodium bicarbonate is a simple blowing agent for cellular rubbers and bread.

blowing a well (*MinExt*) Old method of temporary removal of pressure at the well head to allow tubing and casings to be blown free of debris, water, etc.

blowing current (*ElecEng*) The current (dc or rms) which will cause a fuse link to melt.

blowing engine (*Eng*) The combined steam or gas engine and large reciprocating air-blower for supplying air to a blast furnace.

blowing-iron (*Glass*) See BLOWPIPE.

blowing-out (*Eng*) The operation of stopping down a blast furnace.

blowing road (*MinExt*) In a colliery, the main ventilation ingress.

blowing room (*Textiles*) In a cotton-spinning mill, the room containing the bale breakers, openers and scutchers, in which revolving beaters and exhaust fans remove motes and dust from the fibres.

blow moulding (*Eng*) Two-stage route for making hollow products, eg bottles, surfboards (when filled with foam). Extruded PARISON is dropped into the split, female-only tool, and blown to shape by air pressure.

blown (*Autos*) (1) Supercharged. See BOOST. (2) Of cylinder head or manifold gaskets, having sprung a leak under pressure.

blown casting (*Eng*) Casting spoilt by porosity.

blown film (*Eng*) Polymer process where molten low-density polyethylene extrudate is blown by air pressure to make thin film for packaging etc. See EXTRUSION.

blown flap (*Aero*) A FLAP, the efficiency of which is improved by blowing air or other gas over its upper surface to maintain attached airflow even at high angles of deflection.

blown oil (*Chem*) Linseed oil treated by heating and aeration. Used in the manufacture of linoleum.

blown sand (*Geol*) Sand which has suffered transportation by wind, the grains in transit developing a perfectly spherical form (*millet-seed sand*); grain size is dependent upon the wind velocity. See SAND DUNES.

blowout (*MinExt*) The sudden eruption of gas and oil from a well, which then has to be controlled before drilling can recommence.

blowout coil (*ElecEng*) See MAGNETIC BLOWOUT.

blowout magnet (*ElecEng*) A permanent or electromagnet used to extinguish more rapidly the arc (in a switch etc) caused by breaking an electric circuit.

blowout preventer (*MinExt*) The stack of heavy-duty valves attached to the casing at a well head, designed to control the pressure in the bore when drilling or WORKING OVER. Abbrev *BOP*. Also *Christmas tree*. See figs at KELLY, ROTARY DRILL.

blow pin (*Eng*) Device through which air is blown in final stage of BLOW MOULDING. Usually ascends into base of descending PARISON.

blowpipe (*Glass*) A metal tube, some 2 m in length, with a bore of 2–4 mm and a thickened nose which is dipped into molten glass and withdrawn from the furnace. The glass is subsequently manipulated on the end of the blowpipe and blown out to shape. Also *blowing-iron*.

blow-up (*ImageTech*) A photographic enlargement of the image, esp from one gauge of film to a larger one.

blow-up (*Print*) Enlargement of copy, artwork, photographic and text.

blub (*Build*) A swelling on the surface of newly plastered work.

blubber (*BioSci*) In marine mammals, a thick fatty layer of the dermis.

blue asbestos (*Min*) Crocidolite, a fibrous variety of riebeckite. The best-known occurrences are in S Africa.

blue-backing shot (*ImageTech*) Scene in which the foreground action is shot against a uniform blue background for combination with another scene, either by travelling matte (Cine) or COLOUR SEPARATION OVERLAY (TV). See TRAVELLING MATTE SHOT.

blue billy (*Eng*) The residue left after burning off the sulphur from iron sulphide ores.

blue bricks (*Build*) Bricks of high strength and durability; highest-quality engineering bricks.

blue brittleness (*Eng*) Embrittlement of medium- and high-carbon steels during tempering in the range 205–315°C, so named because the surface of the steel becomes coated with blue-coloured oxidation film.

blue comb (*Vet*) See AVIAN MONOCYTOSIS.

blue giant (*Astron*) A very large bright-blue star in a late stage of evolution.

blue-glass lamp (*ImageTech*) An incandescent lamp with a coloured bulb giving an approximation to daylight illumination.

blue-green algae (*BioSci*) Also *blue-green bacteria*. See CYANOBACTERIA.

blue ground (*Geol, MinExt*) Peridotite, kimberlite. Decomposed agglomerate, usually found as a breccia and occurring in volcanic pipes in S Africa and Brazil; it contains a remarkable assemblage of ultramafic plutonic rock fragments (many of large size) and diamonds.

blue gum (*For*) See RED RIVER GUM.

blueing (*Eng*) The production of a blue oxide film on polished steel by heating in contact with saltpetre or wood ash, either to form a protective coating, or incidental to annealing.

blueing (*Textiles*) The process of neutralizing a yellowish tint in fabric by adding a blue colour, in order to obtain a better white appearance.

blueing salts (*Eng*) Caustic solution of sodium nitrate, used hot to produce a blue oxide film on the surface of steel.

blue john (*Min*) A massive, blue and white variety of the mineral fluorite, occurring near Castleton, Derbyshire, and worked for ornamental purposes.

blue key (*Print*) Blue images produced photographically on film from an assembly of the key negatives or positives. Used as a guide in the film assembly of separation negatives or positives required to process colour printing.

blue lead (*Chem*) A name used in the industry for metallic lead, to distinguish it from other lead products such as white lead, orange lead and red lead.

blue-line key (*Print*) Technique used when making sets of lithographic plates for colour printing. A key image of each page in position is prepared on Astrafoil using a blue dye, and the appropriate positives or negatives for each page are patched up in register with the key. When printed down to make a plate, the blue does not affect the result.

blue metal (*Eng*) Condensed metallic fume resulting from distillation of zinc from its ore concentrates. Blue tint is due to slight surface oxidation of the fine particles.

blue Moon (*Astron*) (1) The second full Moon in a calendar month in which two full moons occur. There are around 12·36 lunar months in the year, so there can be only one blue Moon every 2·7 years on average. (2) The blue Moon observed occasionally when the atmosphere contains large quantities of particles 0·8–1·8 μm in diameter, eg from volcanoes or forest fires; red light gets scattered out of the line of sight, but the blue is allowed through.

blue nose disease (*Vet*) A form of photosensitization occurring in the horse, characterized by oedema and purplish discoloration around the nostrils, nervous symptoms, and sometimes sloughing of unpigmented skin.

blue of the sky (*EnvSci*) Phenomenon produced when sunlight is 'scattered' by molecules of the gases in the atmosphere and by dust particles. Since this scattering is greater for short waves than for long waves, there is a predominance of the shorter waves of visible light (ie blue and violet) in the scattered light which we see as the blue of the sky.

blueprint (*ImageTech*) A process for reproducing plans and engineering drawings, based on the principle that ferric salts are reduced to ferrous when exposed to light. Also *cyanotype*.

blueprint paper (*Paper*) A paper coated with a solution of potassium ferricyanide and ammonium ferric citrate bound together with gelatine or gum arabic.

blue schist (*Geol*) A metamorphic rock formed under conditions of high pressure and relatively low temperature. Characteristic minerals are glaucophane and kyanite.

blueshift (*Astron*) The displacement of features in the spectra of astronomical objects towards shorter wavelength as a result of the DOPPLER EFFECT when celestial bodies have a relative motion towards the Earth.

blue stain (*For*) A form of sapstain producing a bluish discoloration; caused by the growth of fungi which, however, do not greatly affect the strength of the wood.

bluestone (*Min*) The mineral chalcanthite, hydrated copper sulphate, $CuSO_4 \cdot 5H_2O$. Also *blue vitriol*.

bluetongue (*Vet*) Malarial catarrhal fever of sheep. A febrile disease of sheep and cattle, occurring in parts of Africa and the Near East, caused by a virus and transmitted by mosquitoes; characterized by haemorrhagic inflammation of the buccal mucosa and cyanosis and swelling of the tongue.

Bluetooth (*ICT*) TN for a short-range radio technology that simplifies communication between computers, cellular phones, etc.

blue vitriol (*Min*) See BLUESTONE.

blunt-ended DNA (*BioSci*) DNA cleaved straight across the double-stranded molecule, without forming any single-stranded ends. An effect of some RESTRICTION ENZYMES.

blunt start (*Eng*) End of a threaded screw which is rounded or coned to facilitate insertion.

blurb (*Print*) A short note by the publisher recommending a book or its author. Usually printed on the dust jacket or at the beginning of the preliminary matter.

blushing (*Build*) A condition in which a cloudy film appears on a newly lacquered surface; due usually to too rapid drying or to a damp atmosphere.

Blu-tack (*Chem*) TN for filled polymer dough with tack sufficient to grip vertical surfaces.

B-lymphocyte (*BioSci*) A lymphocyte that derives from precursors in the bone marrow (or in birds the *bursa of Fabricius*, a tissue budding off from the hind gut) and that does not undergo differentiation in the thymus. B-lymphocytes when fully mature and differentiated (PLASMA CELLS) make and secrete immunoglobulins and their activities are regulated in a complex fashion by T-lymphocytes. See B-MEMORY CELLS and panel on IMMUNE RESPONSE.

BM (*Surv*) Abbrev for BENCH MARK.

BMC (*Chem*) US abbrev for *bulk moulding compound*. See DOUGH MOULDING COMPOUND.

B-memory cell (*BioSci*) A resting B-lymphocyte that is derived from an antigen-stimulated B-lymphocyte and will rapidly differentiate, with T-cell help, into an antibody-producing cell if the antigen is encountered in the future. One of the aims of prophylactic immunization is to elicit B-memory cells.

BMEP (*Eng*) Abbrev for BRAKE MEAN EFFECTIVE PRESSURE.

BMEWS (*Radar*) Abbrev for *ballistic missile early warning system*. An over-the-horizon radar system for the detection of intercontinental ballistic missiles, with linked sites in the UK, Alaska and Greenland.

BNA (*Med*) Abbrev for *Basle Nomina Anatomica*, an international anatomical terminology accepted at Basle in 1895 by the Anatomical Society to standardize terms used in describing parts of the anatomy. Since 1955, Nomina Anatomica.

BNC connector (*ICT*) A bayonet-type electrical connector used for joining screened coaxial cables particularly in LOCAL AREA NETWORKS using thin-wire Ethernet (see 10 BASE 2) and for video signals.

BNF (*ICT*) Abbrev for BACKUS–NAUR FORM.

BNFL (*NucEng*) Abbrev for *British Nuclear Fuels* plc. Organization involved in uranium enrichment, fabrication of fuel elements, reprocessing of irradiated nuclear fuel and production of plutonium. It also operates experimental reactors.

B-number (*ICT*) The telephone number dialled by a caller in an INTELLIGENT NETWORK. See A-NUMBER, C-NUMBER.

boar (*Agri*) A reproductively capable male pig, over 6 months old.

board (*For*) Timber cut to a thickness of less than 2 in (50 mm), and to any width from 4 in (100 mm) upwards.

board (*ICT*) A printed circuit board holding electronic components that carry out a discrete function within the computer system such as control of the HARD DISK. Also *card*. See MOTHERBOARD.

board (*Paper*) Stiff, thick paper, generally of 220 or 250 gm m^{-2} or more.

board (*Print*) A general term for eg millboard or strawboards, used for cases of books. Weight generally in excess of 200 GSM.

board-and-brace work (*Build*) Work consisting of boards grooved along both edges, alternating with thinner boards fitting into the grooves.

board foot (*For*) A volume measure for timber: 1 bdft = 144 in^3 = 2·36 × 10^{-3} m^3.

boarding (*Textiles*) Heat treatment of knitted garments, esp nylon stockings, on a former in order to give the desired shape and size.

board measure (*Build*) Area measure for wooden boards. See FOOT SUPER.

Board of Trade Unit (*ElecEng*) Obsolete unit of electrical energy, equal to 1 kilowatt-hour. Abbrev *BTU*.

boart (*Min*) See BORT.

boasted joint surface (*Build*) The surface of a stone which has been worked over with a BOASTING CHISEL until it is covered with a series of small parallel grooves, thus forming a key for the mortar at the joint.

boasted work (*Build*) See DROVE WORK.

boaster (*Build*) See BOASTING CHISEL.

boasting (*Build*) The operation of dressing stone with a broad chisel and mallet.

boasting chisel (*Build*) A steel chisel having a fine, broad cutting edge, 2 in (50 mm) wide; used for preparing a stone surface prior to finish-dressing with a broad tool. Also *boaster*.

boat (*Chem*) The conformation of a six-membered ring in which atoms 1, 2, 4 and 5 are essentially co-planar, while atoms 3 and 6 extend on the same side of the plane. It is less stable than the CHAIR conformation.

boat scaffold (*Build*) See CRADLE SCAFFOLD.

bobbin (*ElecEng*) A flanged structure intended for the winding of a coil. Also *spool*.

bobbin (*Textiles*) A light spool on which SLUBBINGS, ROVING or yarn is wound ready for the next process. A weft bobbin or *pirn* is loaded with yarn suitable for use as the weft of woven fabrics. A brass bobbin is formed from two brass disks riveted together to carry the threads used in the manufacture of lace, eg on Leavers machines (see LACE MACHINE). Bobbin lace is a hand-made lace produced from threads fed from small bobbins.

bobbin winding (*ElecEng*) A transformer winding in which all the turns are arranged on a bobbin instead of in the form of a disk. Generally used for the high-voltage windings of small transformers.

bobby calf (*Agri*) A calf slaughtered within 2–3 weeks of birth.

bob veal (*Agri*) Meat from calves up to 4 months old, reared under very restrictive conditions with a specialized diet.

bob-weight (*Eng*) A weight used to counterbalance some moving part of a machine. See BALANCE WEIGHTS.

BOD (*EnvSci*) Abbrev for *biological oxygen demand*. See OXYGEN DEMAND.

Bode's law (*Astron*) A numerical relationship linking the distances of planets from the Sun, discovered by J Titius (1766) and published by J Bode (1772). The basis of this relationship is the series 0, 3, 6, 12, …, 384, in which successive numbers are obtained by doubling the previous one. If 4 is added to create the new series 4, 7, 10, 16, …, 388, the resulting numbers correspond reasonably to the planetary distances on a scale with the Earth's distance equal to 10 units. It is believed to be a chance coincidence.

body (*Build*) (1) The degree of opacity possessed by a pigment. (2) The apparent viscosity of a paint or varnish. (3) The ability of a paint to give a good, uniform film over an irregular or porous surface.

body (*Print*) (1) The measurement from top to bottom of eg a type or rule. The unit is the POINT, 72 points amounting to (approx) 1 in. (2) The solid part of a piece of type below the printing surface or *face*. Also *shank*, *stem*. (3) Body of a work, the text of a volume, distinguished from the preliminary matter, such as title and contents, and the end matter, such as appendices and index.

body burden (*BioSci*) The total amount of a substance present in an organism at a given time, usually used in reference to a toxin of some sort.

body cavity (*BioSci*) The perivisceral space, or cavity, in which the viscera lie; a vague term, sometimes used incorrectly to mean COELOM.

body cell (*BioSci*) (1) A somatic cell, as opposed to a germ cell. (2) The cell that divides to give the two sperm cells in the gymnosperm pollen tube.

body-centred crystal lattice (*Crystal*) See UNIT CELL.

body-centred cubic packing (*Chem*) A crystal lattice with a cubic unit cell, the centre of which is identical in environment and orientation to its vertices. Specifically a common structure of metals, in which the unit cell contains two atoms, based on this lattice. See panels on CLOSE PACKING OF ATOMS and CRYSTAL LATTICE.

body-centred tetragonal structure (*Chem*) A distorted form of BODY-CENTRED CUBIC PACKING in a crystal formed esp in MARTENSITE in steel, the amount of distortion depending on the carbon content of the steel. Abbrev *BCT*. The BCT lattice does not have the five independent slip systems (see VON MISES CRITERION) necessary for DUCTILITY; thus it contributes to the hardness of martensite by impeding DISLOCATION movement.

body language (*Psych*) Communication of information by means of conscious or unconscious gestures, attitudes, facial expressions, etc. Also *non-verbal communication* (*NVC*).

body louse (*Med*) A surface parasite (*Pediculosis humanus corporis*) which infests humans in overcrowded insanitary conditions. It is a vector for TYPHUS and TRENCH FEVER.

body scanner (*Med*) An X-ray or ultrasound scanner for the whole body, usually with computer enhancement of the images.

body-section radiography (*Radiol*) See EMISSION TOMOGRAPHY.

body wall (*BioSci*) The wall of the perivisceral cavity, comprising the skin and muscle layers.

boehmite (*Min*) Orthorhombic form of aluminium monohydrate, AlOOH. An important constituent of some bauxites.

BOE sill (*Build*) Abbrev for BRICK-ON-EDGE SILL.

bog (*EnvSci*) Wetland vegetation forming an acid peat which is mainly composed of dead individuals of the moss *Sphagnum*. Also *mire*. See BLANKET BOG, RAISED BOG, VALLEY BOG.

boghead coal (*Min*) Coal which is non-banded and translucent, with high yield of tar and oil on distillation. Consists largely of resins, waxes, wind-borne spores and pollen cases. Originated in deeper, more open parts of the coal swamps than ordinary household coals. Essentially a spore coal. Also *parrot coal*. See TORBANITE.

bogie (*Eng*) (1) A small truck of short wheel base running on rails. Commonly used for the conveyance of eg coal, gold or other ores, concrete. (2) A four- or six-wheel undercarriage of short wheel base, which forms a pivoted support at one or both ends of a long rigid vehicle such as a locomotive or coach. Also *bogie truck*. US *truck*.

bogie landing gear (*Aero*) A main landing gear carrying a pair or pairs of wheels in tandem and pivoted at the end of the shock strut or *oleo*. This arrangement helps to spread the weight of an aircraft over a larger area and also allows the wheel size to be minimized for easier stowage after retraction.

bogie truck (*Eng*) See BOGIE.

bog iron ore (*Min*) Porous form of LIMONITE often mixed with vegetable matter. Found in marshes.

boglame (*Vet*) See OSTEOMALACIA.

bog spavin (*Vet*) Dilation of the capsule of the tibio-tarsal joint of the horse.

Bohemian garnet (*Min*) Reddish crystals of the garnet PYROPE, occurring in large numbers in the Mittelgebirge in Bohemia.

Bohemian gemstone (*Min*) Misnomer for the garnet PYROPE, the false ruby ROSE QUARTZ, and the false topaz CITRINE.

Bohr atom (*Phys*) See BOHR MODEL.

bohrium (*Chem*) An artificially manufactured radioactive chemical element, symbol Bh, at no 107, of the transactinide series. Also *unnilseptium*.

Bohr magneton (*Phys*) Unit of magnetic moment, for electron, defined for the SI system by $\mu_B = eh/4\pi\, m_e$, where e is the charge, h is PLANCK'S CONSTANT and m_e is the rest mass, so that $\mu_B = 9{\cdot}27 \times 10^{-24}$ J T^{-1}. The *nuclear Bohr magneton* is defined by

$$\mu_N = \frac{eh}{4\pi M} = \frac{\mu_B}{1836} = 5{\cdot}05 \times 10^{-27} \text{J T}^{-1}$$

M being the rest mass of the proton.

Bohr model (*Phys*) A combination of the Rutherford model of the atom with the quantum theory, based on the following four postulates. (1) An electron in an atom moves in a circular orbit about the nucleus under the influence of the electrostatic attraction between the electron and the nucleus. (2) An electron can only move in an orbit for which its orbital angular momentum is an integral multiple of $h/2\pi$, where h is PLANCK'S CONSTANT. (3) An electron moving in such an orbit does not radiate electromagnetic energy and so its total energy E remains constant. (4) Electromagnetic radiation is emitted if an electron makes a transition from an orbit of energy E_i to one of lower energy E_f, and the frequency of the emitted radiation is $v = (E_i - E_f)/h$.

Bohr radius (*Phys*) According to the Bohr model of the hydrogen atom, the electron when in its lowest energy state moves round the nucleus in a circular orbit of radius

$$a_0 = \frac{4\pi\,\varepsilon_0 \hbar^2}{m_e\, e^2} = 5{\cdot}292 \times 10^{-11} \text{m}$$

where $\hbar$ is Dirac's constant, m_e is the mass of the electron, e is the electronic charge and ε_0 the permittivity of free space. The Bohr radius is a fundamental distance in atomic phenomena.

Bohr–Sommerfeld atom (*Phys*) Atom obeying modifications of the BOHR MODEL suggested by Sommerfeld and allowing for possibility of elliptical electron orbits.

Bohr theory (*Phys*) See BOHR MODEL.

boil (*Med*) Infection (with *Staphylococcus aureus*) of a hair follicle, resulting in a painful red swelling, which eventually suppurates.

boiled oil (*Build*) Linseed oil raised to a temperature of 400–600°F (200–300°C) and admixed with driers.

boiler (*Eng*) One of a wide range of pressure vessels in which water or other fluid is heated and then discharged, eg either as hot water for heating or as high-pressure steam for power generation.

boiler capacity (*Eng*) The weight of steam, usually expressed in kilograms or pounds per hour, which a boiler can evaporate when steaming at full load output.

boiler compositions (*Eng*) Chemicals introduced into boiler feed-water to inhibit scale formation and corrosion, or to prevent priming or foaming. Examples are sodium compounds (such as soda ash), organic matter and barium compounds.

boiler covering (*Eng*) See LAGGING.

boiler cradles (*Ships*) See KEELSON.

boiler crown (*Eng*) The upper rounded plates of a boiler of the shell type.

boiler efficiency (*Eng*) The ratio of the heat supplied by a boiler in heating and evaporating the feed-water to the heat supplied to the boiler in the fuel. It may vary from 60 to 90%.

boiler feed-water (*Eng*) The water pumped into a boiler for conversion into steam, usually consisting of condensed exhaust steam and 'makeup' fresh water treated to remove air and impurities.

boiler fittings and mountings (*Eng*) See FEED CHECK VALVE, PRESSURE GAUGE, SAFETY VALVE, STOP VALVE, WATER GAUGE.

boilermaker's hammer (*Eng*) Hammer with ball or straight and cross panes; used for caulking, fullering and scaling boilers.

boiler plate (*Eng*) Mild-steel plate, generally produced by the open-hearth process; used mainly for the shells and drums of steam boilers. Latterly steel with a higher yield stress is frequently specified.

boiler pressure (*Eng*) The pressure at which steam is generated in a boiler. It may vary from little over atmospheric pressure for heating purposes, to 1500 lb in^{-2} (10 000 kN m^{-2}) and over for high-pressure turbines.

boiler scale (*Eng*) A hard coating, chiefly calcium sulphate, deposited on the surfaces of plates and tubes in contact with the water in a steam boiler. If excessive, it leads to overheating of the metal and ultimate failure.

boiler setting (*Eng*) The supporting structure on which a boiler rests; usually of brick for land boilers and of steel for marine boilers.

boiler stays (*Eng*) Screwed rods or tubes provided to support the flat surfaces of a boiler against the bursting effect of internal pressure.

boiler test (*Eng*) (1) A hydraulic-pressure test applied to check watertightness under pressure greater than the working pressure. (2) An efficiency test carried out to determine evaporative capacity and the magnitude of losses.

boiler trial (*Eng*) An efficiency test of a steam boiler, in which the weight of feed-water and of fuel burnt are measured, and various sources of loss assessed.

boiler tubes (*Eng*) Steel tubes forming part of the heating surface in a boiler. In water-tube boilers the hot gases surround the tube; in locomotive and some marine boilers (fire-tube boilers), the gases pass through the tube.

boiling (*Phys*) The very rapid conversion of a liquid into vapour by the violent evolution of bubbles. It occurs when the temperature reaches such a value that the saturated vapour pressure of the liquid equals the pressure of the atmosphere. Also *ebullition*.

boiling bed (*Chem*) A gas fluidized bed in which two separate phases are formed at high gas velocities: gas, containing a relatively small proportion of suspended solids, bubbles through a higher density fluidized phase with the result that the system closely resembles in appearance a boiling liquid.

boiling point (*Phys*) The temperature at which a liquid boils when exposed to the atmosphere. Since at the boiling point the saturated vapour pressure of a liquid equals the pressure of the atmosphere, the boiling point varies with pressure; it is usual, therefore, to state its value at the standard pressure of $101 \cdot 325 \ kN \ m^{-2}$. Abbrev *bp*.

boiling-water reactor (*NucEng*) Light-water reactor in which the water is allowed to boil into steam which drives the turbines directly. Abbrev *BWR*.

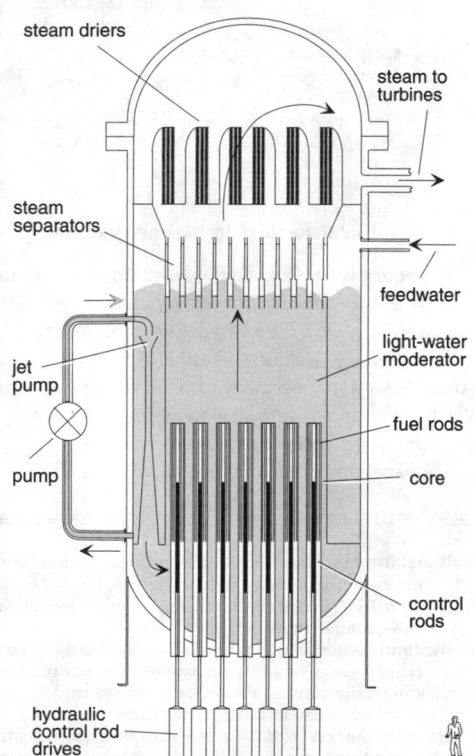

boiling-water reactor Schematic diagram.

Bok globule (*Astron*) One of many small dark nebulae (10^3–10^5 AU in diameter) in the Milky Way, thought to be regions of star formation.

bole (*BioSci*) The trunk of a tree.

bole (*Print*) A compact clay, a reddish variety of which is used in powdered form (with water and a small quantity of gilding size) as a foundation for gilt edges.

bolection moulding (*Build*) A moulding fixed round the edge of a panel and projecting beyond the surface of the framing in which the panel is held. Also *balection moulding*.

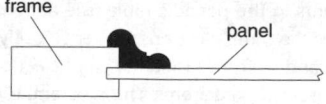

bolection moulding

bolide (*Astron*) A brilliant METEOR, generally one that explodes. Also *fireball*.

boll (*BioSci, Textiles*) A capsule, esp of cotton. It contains the cotton seeds and the attached fine fibres or LINT, and opens as ripening proceeds.

bollard (*Genrl, Ships*) (1) Generally, any short upright post used eg to prevent vehicular access. (2) On a quay or vessel, a short post round which ropes are secured for purposes of mooring.

Bollinger bodies (*Vet*) Intracytoplasmic inclusion bodies occurring in epithelial cells infected with fowl pox virus; they are made up of aggregates of smaller bodies, *Borrel bodies*, which are the actual masses of the virus collected in the cells.

bolometer (*ElecEng, ICT*) A device for measuring micro-wave or infrared energy, consisting of a temperature-dependent resistance used in a bridge circuit that gives an indication when incident energy heats the resistor. Used for power measurement, standing-wave detectors and infrared search and guidance systems.

bolster (*Build*) A form of cold chisel with a broad splayed-out blade, used in eg cutting stone slabs.

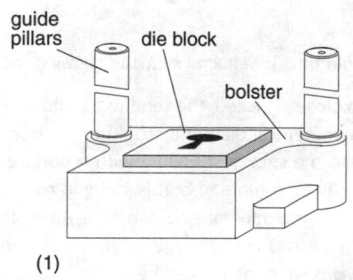

(1)

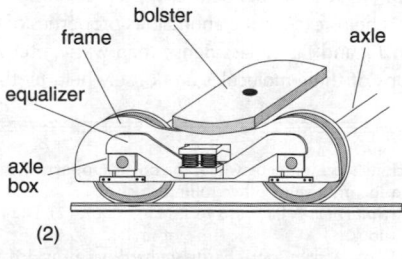

(2)

bolster (1) Part of a press tool; the top bolster with punch and the stripper are not shown. (2) Part of a railway bogie; frame and wheels opposite not shown.

bolster (*Eng*) (1) A steel block which supports the lower part of the die in a pressing or punching machine. (2) The

Bonding

Interactions between individual atoms and molecules are classified according to the way in which the outermost orbital electrons behave. Most of the elements in the periodic table (see appendices) are metals: the outer electrons are very loosely held by the nuclei and so form a mobile 'sea of electrons' between the closely packed atoms. This is essentially why metals are good electrical conductors. Pure metals are ductile and malleable, as the metallic bond is relatively weak. Alloying metals with other elements inhibits dislocation motion and so improves mechanical properties like tensile modulus and yield strength.

In energetic terms, the strongest bond is the covalent chemical bond formed by electron sharing between neighbouring atoms. It is exemplified by bonding in elemental carbon, particularly diamond and graphite. In diamond, each carbon atom forms four covalent bonds (so-called sp^3 hybrids) by sharing its four outer electrons with four neighbouring atoms in a tetrahedral configuration (Fig. 1).

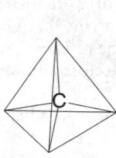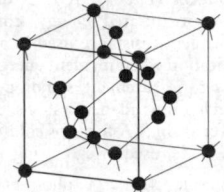

Fig. 1 **Diamond** Tetrahedral structure showing C-C bonds.

When closely packed, the tetrahedral atoms form the highly symmetrical diamond structure. Carbon–carbon bonds are the strongest known with a bond energy of about 330 kJ per mole of bonds, helping to explain the hardness of the mineral (see panel on HARDNESS MEASUREMENTS). Perhaps the weakest covalent bond is the hydrogen bond formed between hydrogen atoms which are strongly bonded to oxygen or nitrogen.

This helps explain the anomalous properties of water (high T_m and T_B, ice less dense than water, etc) where groups of three molecules are loosely held by the bond. It is also important in biological molecules such as DNA, proteins and cellulose. See panel on BIOLOGICAL ENGINEERING POLYMERS.

Ionic bonding, such as that which exists in refractories (eg MgO), ceramics (eg silicates), salts (eg sodium fluoride) and glasses, is formed by transfer of electrons between atoms of different electronegativity (see appendices) Highly electronegative non-metals like fluorine gain in electron(s) to form anions (eg F^-), with a complete set of paired electrons in their outer orbitals. Metals easily lose electrons to form cations (eg Na^+) and will supply exactly the number needed in sodium fluoride (NaF) where the numbers of cations and anions are equal. The anions are larger than the cations (see appendices) so they close pack to form a CRYSTAL LATTICE (panel), and the cations fit into the octahedral or tetrahedral interstitial sites. Unlike covalent bonds, ionic bonds which hold the solid together are non-directional and electrostatic in nature. They can be very strong, up to 280 kJ mol^{-1} in eg MgO, explaining their very high melting points which are exploited in high-temperature-resistant refractories.

Bonding types can be described more exactly by the interatomic potential energy curve (Fig. 2), a graph of interatomic distance (r) against potential energy:

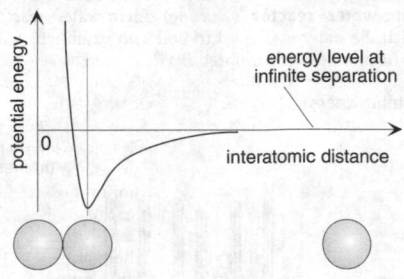

Fig. 2 **Interatomic potential energy curve**.

The equilibrium distance between atoms in solids is a result of two opposing forces: repulsion when close and attraction when separated. The depth of the potential energy well is a measure of the bond strength, so that a very weak bond like the van der Waals bond possesses a shallow well at greater equilibrium distances than a covalent bond.

See panel on CLOSE PACKING OF ATOMS.

---

rocking steel frame by which the bogie supports the weight of a locomotive or other rolling stock.

bolt (*Build*) (1) A bar used to fasten a door. (2) The tongue of a lock.

bolt (*Eng*) A cylindrical, partly screwed bar provided with a head. With a nut, a common means of fastening two parts together.

bolt (*Textiles*) A specified length of fabric as agreed between seller and buyer.

bolting (*BioSci*) Premature flowering and seed production, esp of a biennial crop plant during its first year. Also *running to seed*.

bolt-making machine (*Eng*) A machine which forges bolts by forming a head on a round bar.

bolts (*Print*) The folded edges at the head, fore-edge and tail of a sheet before cutting. US *close*.

Boltzmann equation (*Phys*) A fundamental diffusion equation based on particle conservation. The rate of losses, including leakage out of the region of interest and the rate of disappearance by reactions of all kinds, is equal to the rate of production from sources within the region and the rate of scattering into the region.

Boltzmann principle (*Phys*) See BOLTZMANN'S DISTRIBUTION.

Boltzmann's constant (*Phys*) Fundamental physical constant, given by $k = R/N_A = 1\cdot3805 \times 10^{-23}$ J K^{-1}, where R is the ideal gas constant and N_A is AVOGADRO'S NUMBER.

Boltzmann's distribution (*Phys*) Statistical distribution of large numbers of small particles when subjected to thermal agitation and acted upon by electric, magnetic or gravitational fields. In statistical equilibrium, the number of particles n per unit volume in any region is given by $n = n_0\exp(-E/kT)$, where k is BOLTZMANN'S CONSTANT, T is absolute temperature, E is the potential energy of a particle in given region, and n_0 is the number per unit volume when $E = 0$. Also *Boltzmann principle*.

Boltzmann's superposition principle (*Phys*) The principle that in a linear viscoelastic material, the accumulated viscoelastic creep strain resulting from a series of stress increments is the superposed sum of the creep responses to the individual increments.

bombardment (*Phys*) Process of directing a beam of neutrons or high-energy charged particles onto a target material in order to produce nuclear reactions.

bomb calorimeter (*Eng*) An apparatus used for determining the calorific values of solid or liquid fuels. The 'bomb' consists of a thick-walled, highly polished, steel vessel in which a weighed quantity of the fuel is electrically ignited in an atmosphere of compressed oxygen. The bomb is immersed in a known volume of water to which the combustion heat is transferred, and from the rise of temperature of which the calorific value is calculated.

bombesin (*BioSci*) A peptide neurohormone, originally isolated from the toad *Bombina*, that has also been found in the brain and gut of humans, and causes an increase in blood pressure.

bomb sampler (*PowderTech*) A device for obtaining samples of dispersed particles at predetermined depths within a suspension, consisting of a closed cylindrical vessel with an automatic valve which opens when an extension tube hits the bottom of the suspension container. The sampler fills with suspension and then closes when the vessel is lifted.

bomb, volcanic (*Geol*) See VOLCANIC BOMB.

bonanza (*MinExt*) Rich body of ore.

bond (*Build*) The system under which bricks or stones are laid in overlapping courses in a wall in such a way that the vertical joints in any one course are not immediately above the vertical joints of an adjacent course.

bond (*CivEng*) The adhesion between concrete and its reinforcing steel, due partly to the shrinkage of the concrete in setting and partly to the natural adhesion between the surface particles of steel and concrete. See MECHANICAL BOND.

bond (*Phys*) The link between atoms, considered to be electrical and arising from the distribution of electrons around the nuclei of bonded atoms. See CHEMICAL BOND.

bond angle (*Chem*) The angle between the lines connecting the nucleus of one atom to the nuclei of two other atoms bonded to it; eg in water the H—O—H angle is about 105°.

bond distance (*Chem*) See BOND LENGTH.

bonded fabrics (*Paper*) A material made by fabricating fibres into sheet form with the aid of a binder. Used for eg polishing cloths, curtains, filter cloths.

bonded-fibre fabric (*Textiles*) A non-woven fabric made from a mass of fibres held together by adhesive or by processes such as needling or stitching. See ADHESIVE-BONDED NON-WOVEN FABRIC.

bonded wire (*ElecEng*) Enamelled insulated wire coated with thin plastic; after forming a coil, it is heated by a current or in an oven or both for the plastic to set and the coil to attain a solid permanent form.

bond energy (*Chem*) The energy in joules released on the formation of a chemical bond between atoms, and absorbed on its breaking.

bonder (*Build*) See BONDSTONE.

bonding (*Aero*) (1) The electrical interconnection of metallic parts of an aircraft normally at earth potential for the safe distribution of electrical charges and currents. Protects against charges due to precipitation, static and electrostatic induction due to lightning strikes. Reduces interference and provides a low-resistance electrical return path for current in earth-return systems. (2) Joining structural parts by adhesive. May be performed at high temperature and pressure.

bonding (*ElecEng*) An electrical connection between adjacent lengths of armouring or across a joint. See CROSS-BONDING.

bonding (*Eng*) The interactions between individual atoms and molecules. See CHEMICAL BOND and panel on BONDING.

bonding (*Psych*) The forming of a close emotional attachment, esp between a mother and her newborn child.

bonding clip (*ElecEng*) A clip used in wiring systems to make connection between the earthed metal sheath of different parts of the wiring, in order to ensure continuity of the sheath.

bond length (*Chem*) The distance between bonded atoms in a molecule. Specifically used for distances between atoms in a covalent compound. Typical lengths are O—H 96 pm, C—H 107 pm, C—C 154 pm, C=C 133 pm, C≡C 120 pm. Also *bond distance*.

bond length (*CivEng*) The minimum length of reinforcing bar required to be embedded in concrete to ensure that the bond develops the full stress in the bar.

bond paper (*Paper*) A paper similar to bank paper but of 50 gm m^{-2} or more.

bondstone (*Build*) A long stone laid as a header through a wall. Also *bonder*.

bond strength (*Chem*) Strictly, bond dissociation energy or the energy needed to separate a pair of bonded atoms from their equilibrium position to infinity. Carbon–carbon bonds are the strongest, with bond energies of about 350 kJ mol^{-1} of bonds for aliphatic compounds and about 410 kJ mol^{-1} for aromatics. See panel on BONDING.

bond strength (*Phys*) US for ADHESION.

bone (*BioSci*) A structural BIOCOMPOSITE of c.70% (by weight) inorganic calcium salts embedded in collagen fibres. Most of the inorganic phase consists of hydroxyapatite (calcium phosphate) but large amounts of carbonate, citrate and fluoride amines are also present. Long bones like the femur or thigh bone are composed of a harder, compact composite outer layer surrounding a spongy interior (*cancellous bone*) which improves the stiffness- and strength-to-weight indices for the material. Like most biocomposites, bone exhibits viscoelastic properties and is therefore sensitive to eg rate of loading.

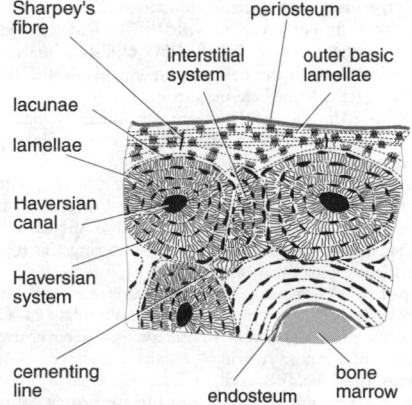

bone Cross-section; not all lamellae are drawn.

bone (*MinExt*) Coal containing ash (bone) in very fine layers along the cleavage planes. Also *bony coal*.

bone bed (*Geol*) A sediment characterized by abundant fossil bones or bone fragments, eg scales, teeth, coprolites.

bone china (*Genrl*) Form of 'soft' (ie lower firing temperature) PORCELAIN produced in the UK. Based on approximately equal amounts of bone ash (mainly calcium phosphate), China clay (KAOLIN) and Cornish stone (a potash feldspar partially converted into kaolinite, a hydrated aluminium silicate), yielding a white, translucent, low-porosity material after firing to about 1250°C.

bone conduction (*Med*) The conduction of sound waves from the bones of the skull to the inner ear, rather than through the ossicles from the outer ear.

bone-dry paper (*Paper*) Paper dried completely to contain no moisture. Also *oven-dry paper*.

bone marrow (*BioSci*) The tissue filling some bone cavities, of two types: yellow which contains fat cells, and red which contains the stem cells of all blood cell types. Particularly important in the mammalian immune system as the site of B-lymphocyte maturation before antigen encounter. T-lymphocyte stem cells also come from the bone marrow but migrate to the thymus gland for maturation.

bone marrow grafting (*Med*) Grafting or transplantation of bone marrow to patients with bone marrow failure. Used in aplastic anaemia or after therapeutic destruction of the marrow in LEUKAEMIA.

bone marrow purging (*Med*) Therapeutic treatment for some forms of either malignant or genetic disease requiring bone marrow transplantation. Bone marrow is removed from a healthy donor and is purged of any lymphocytes which might react against the patient's cells. The patient is then treated with high-intensity radiotherapy to kill off the patient's own bone marrow and the purged donor cells are transfused into the patient.

bone seeker (*Chem*) Radioelement similar to calcium, eg strontium, radium or plutonium, which can become incorporated into bone where it continues to radiate.

bone-setter (*Med*) A medically unqualified person who treats disorders of joints by manipulation. See OSTEOPATHY.

bone tolerance dose (*Radiol*) The dose of ionizing radiation which can safely be given in treatment without bone damage.

bone turquoise (*Min*) Fossil bone or tooth, coloured blue with phosphate of iron; widely used as a gemstone. It is not true turquoise and loses its colour in the course of time. Also *odontolite*.

boning-in (*Surv*) The process of locating and driving in pegs so that they are in line and have their tops also in line; carried out by sighting between a near and a far peg previously set in the gradient desired.

boning-rods (*Surv*) T-shaped rods used, in sets of three, to facilitate the process of BONING-IN; two of the rods are held on the near and far pegs to establish a line of sight between them in the desired gradient, while the third is used to fix intermediate pegs in line.

boninite (*Min*) A highly magnesian siliceous glassy lava containing pyroxene and olivine phenocrysts in a glassy base full of crystallites.

Bonne's projection (*Geol*) A derivative CONICAL PROJECTION in which the parallels are spaced at true distances along the meridians, which are plotted as curves.

bonnet (*Build*) A wire-netting cowl covering the top of a ventilating pipe or a chimney.

bonnet (*Eng*) A movable protecting cover, eg: (1) the cap of the valve box of a pump; (2) the cover plate of a valve chamber; (3) the hood of a forge; or (4) the cover over the engine of a motor vehicle.

bonnet tile (*Build*) Special rounded tile used to cover the external angles at hips and ridges on tiled roofs. See ARRIS TILE.

bony coal (*MinExt*) See BONE.

bookbinding (*Print*) Arranging in proper order the sections of a book and securing them in a cover. The bookbinder distinguishes BOUND BOOK from CASED BOOK, and both from brochure and paperback.

book cloth (*Print*) The usual covering for edition binding. There are two main kinds: thin, hard-glazed cloth containing starch and other fillings, and matt-surfaced cloth with less starch.

book gill (*BioSci*) See GILL BOOK.

book lung (*BioSci*) See LUNG BOOK.

bookmark (*ICT*) To save the URL of a favourite website, allowing easy access to it subsequently.

book plate (*Print*) A label pasted on the inside front cover of a book bearing the owner's name, crest, coat-of-arms or other peculiar device.

Boolean (*ICT*) A DATA TYPE that may take only two values, true or false. See TRUTH VALUE.

Boolean algebra (*ICT, MathSci*) An algebra, named after G Boole (1815–64), in which there are two elements, true or false, and in which the basic operations are the logical AND and OR operations, that are written '&' and '∨' respectively, and may also be symbolized as multiplication and addition respectively. Its main applications are to the design of switching networks and to mathematical logic, and in computing it has formed the basis for the development of LOGICAL OPERATIONS and the use of GATES. See TRUTH VALUE.

boom (*Acous*) Enhanced reverberation or resonance in an enclosed space at low frequencies, due to reduced acoustic absorption of the surfaces for low frequencies.

boom (*Eng, Ships*) Any long beam, esp: (1) the top and bottom members of a built-up girder; (2) the main jib of a crane; (3) the spar holding the lower part of a fore-and-aft sail; (4) a spar attached to a yard to lengthen it; (5) a barrier of logs to prevent the passage of a vessel; (6) a line of floating timbers used to form a floating harbour; (7) a pole marking a channel; (8) a device to trap spilled oil floating on water, eg at sea, in estuaries. Effective only at slow relative water speeds, ≈ 1 knot.

boom (*ImageTech*) Long movable arm to carry a microphone or lamp above the action in film or TV shooting.

boost (*Autos*) The amount by which the induction pressure of a supercharged internal-combustion engine exceeds atmospheric pressure expressed in kN m^{-2} or lbf in^{-2}.

boost control (*Aero*) A device regulating reciprocating-engine manifold pressure so that supercharged engines are not over-stressed at low altitude.

boost control override (*Aero*) In a supercharged piston aero-engine fitted with BOOST CONTROL, a device (sometimes lightly wire-locked so that its emergency use can be detected), which allows the normal maximum manifold pressure to be exceeded. Also *boost control cut-out*.

booster (*Space*) A rocket engine, or cluster of engines, forming part of a launch system, either the first stage or auxiliary stage, used to provide an initial thrust greater than the total lift-off weight. Also *booster rocket*.

booster amplifier (*Acous*) An aplifier used specially to compensate loss in mixers and volume controls, so as to obviate reduction in signal-to-noise ratio.

booster coil (*Aero*) A battery-energized induction coil which provides a starting spark for aero-engines.

booster fan (*Eng*) A fan for increasing the pressure of air or gas; used for restoring the pressure drop in transmission pipes, and for supplying air to furnaces.

booster pump (*Aero, Eng*) A pump which maintains positive pressure between the fuel tank and the engine, thus intensifying the flow. Any pump to increase the pressure of the liquid in some part of a pipe circuit.

booster response (*BioSci*) The enhanced response to readministration of an antigen, or to a microbial infection, that occurs when prior contact with the antigen has elicited B-MEMORY CELLS and HELPER T-LYMPHOCYTE cells.

booster rocket (*Aero*) See BOOSTER, TAKE-OFF ROCKET.

booster station (*MinExt*) In long-distance transport by pipeline of oils, other liquids, mineral slurries or water-carried coal, an intermediate pumping station where lost pressure energy is restored.

booster station (*ICT*) A station that rebroadcasts a received transmission directly on the same wavelength.

booster transformer (*ElecEng*) A transformer connected in series with a circuit to raise or lower the voltage of the circuit.

boost gauge (*Aero*) An instrument for measuring the manifold pressure of a supercharged aero-engine in relation to ambient atmosphere or in absolute terms. Also used for racing and other car sports.

boost transformer (*ElecEng*) See BUCK TRANSFORMER.

boot (*Glass*) See POTETTE.

boot (*ICT*) A term derived from BOOTSTRAP. A *cold boot* occurs when the electrical power to the computer is turned on and the computer goes through the complete series of power-on self-tests. A *warm boot* occurs when the computer is reset without turning off the power. In both cases the bootstrap or boot program will run and hence the process is colloquially known as booting.

boot disk (*ICT*) A floppy disk that is used to set up a computer system when it is initially switched on or restarted. The disk usually contains CONFIGURATION files and/or the OPERATING SYSTEM and is an essential means of restarting a computer in which these files have become corrupted on the hard disk.

booted (*BioSci*) Having the feet protected by horny scales.

Boötes (Herdsman) (*Astron*) A large northern hemisphere constellation. It contains ARCTURUS, the brightest star in the northern sky.

boot stage (*Agri*) The stage of grain crop development when the leaf sheath swells due to growth of the contained flower spike.

bootstrap (*Eng*) A self-sustaining system in liquid rocket engines by which the main propellants are transferred by a turbo-pump which is driven by hot gases. In turn the gas generator is fed by propellants from the pump.

bootstrap (*ICT*) A short section of a program that can be used to get the rest of the program running in a machine. After a computer has been switched off, the OPERATING SYSTEM has to be put back into the MAIN MEMORY. To reload the LOADER a bootstrap is used, a small number of instructions often loaded and executed by pushing a button. See BOOT.

bootstrap circuit (*Electronics*) A feedback circuit in which part of the output is fed back across the input, giving effectively infinite input impedance and unity gain. Often used to improve the linearity of a voltage sweep generator.

bootstrap cold-air unit (*Aero*) A unit of the compressor–turbine type in which the air charging an aircraft cabin passes through the compressor and, via an INTERCOOLER, the turbine.

bootstrapping (*Electronics*) The technique of using bootstrap feedback. See BOOTSTRAP CIRCUIT.

boot tapping (*Ships*) The demarcation line between the two main colours of a ship's paintwork, at or near the waterline.

BOP (*MinExt*) Abbrev for BLOWOUT PREVENTER.

boracic acid (*Chem*) See BORIC ACID.

boracite (*Min*) The orthorhombic, pseudocubic form of magnesium borate and chloride, found in beds of gypsum and anhydrite.

Boral sheet (*Eng*) A composite made of boron carbide crystals dispersed in aluminium and also faced with aluminium. Used as a neutron absorber.

borane (*Chem*) General name for a hydride of boron, the simplest being diborane, B_2H_6. They are high-energy compounds, yielding water and boric acid on oxidation. There are three principal series of boranes: (1) *closo*-, with formulae B_nH_{n+2} (these mainly occur as anions with fomulae $B_nH_n{}^{2-}$); (2) *nido*-, with formulae B_nH_{n+4}; and (3) *arachno*-, with formulae B_nH_{n+6}. See MULTI-CENTRED BONDING.

borates (*Chem*) See BORIC OXIDE.

borax (*Min*) Hydrated sodium borate, deposited by the evaporation of alkaline lakes.

borax bead (*Chem*) The clear glass formed by fusion of borax when heated. Fused borax dissolves some metal oxides giving glasses with a characteristic colour. The use of borax bead in chemical analysis is based on this fact.

borazole (*Chem*) $B_3H_6N_3$. A compound iso-electronic with BENZENE.

borazon (*Chem*) See BORON NITRIDE.

bord-and-pillar (*MinExt*) A method of mining coal by excavating a series of chambers, rooms or stalls, leaving pillars of coal in between to support the roof. Also *room-and-pillar*.

Bordeaux B (*Chem*) An azo-dyestuff derived from 1-naphthylamine coupled with R-ACID.

bordered pit (*BioSci*) A PIT in which the secondary wall overarches the pit membrane, markedly narrowing the cavity towards the cell lining. Characteristic of tracheids and vessel elements. Cf SIMPLE PIT.

border effect (*ImageTech*) A faint dark line on the denser side of a boundary between a lightly exposed and a heavily exposed region on a developed emulsion. Also *edge effect*, *fringe effect*. See MACKIE LINE.

border-pile (*CivEng*) A pile driven to support the sides of a coffer dam.

Bordetella pertussis (*BioSci*) A small, aerobic, Gram-negative bacillus, the causative organism of whooping cough.

bore (*Eng*) (1) The circular hole along the axis of a pipe or machine part. (2) The internal wall of a cylinder. (3) The diameter of such a hole. (4) The calibre of a gun.

bore (*Genrl*) A great tide-wave, with crested front, travelling rapidly up a river; it occurs on certain rivers having obstructed channels.

boreal (*EnvSci*) Of the north. The *boreal zone* is the geographical region where short summers and long cold winters occur. The *boreal period* in northern Europe extended from 7500 to 5500 BC and had warm summers and cold winters. *Boreal forests* are those in the boreal zone or boreal period.

borehole (*CivEng*) A sinking made in the ground by the process of BORING.

borehole (*MinExt*) The hole made by a drill for a well; its whole length.

borehole survey (*MinExt*) Check for deviation from required line of deep borehole. Methods include camera records of compass, plumbline or gyroscope.

boric acid (*Chem*) H_3BO_3. A tribasic acid. On heating it loses water and forms metaboric acid, $H_2B_4O_4$, and on further heating it forms tetraboric acid, or the so-called pyroboric acid, $H_2B_4O_7$. On heating at a still higher temperature it forms anhydrous boron (III) oxide, or boric oxide. It occurs as tabular triclinic crystals deposited in the neighbourhood of fumaroles, and known also in solution in the hot lagoons of Tuscany and elsewhere. Also *boracic acid*, *sassolite*.

boric oxide (*Chem*) *Boron (III) oxide*. B_2O_3. An intermediate oxide like aluminium oxide, having feeble acidic and basic properties. As a weak acid it forms a series of borates. See BORIC ACID.

boride (*Chem*) Any of a class of substances, some of which are extremely hard and heat resistant, made by combining boron chemically with a metal.

borine radical (*Chem*) The radical BH_3, capable of forming compounds through co-ordination of a lone pair of electrons to the electron-deficient boron, eg borine carbonyl, BH_3CO.

boring (*Eng*) The process of machining a cylindrical hole, performed in a lathe, boring machine or boring mill; for large holes, or when great accuracy is required, it is preferable to using a drill. Fig. ▷

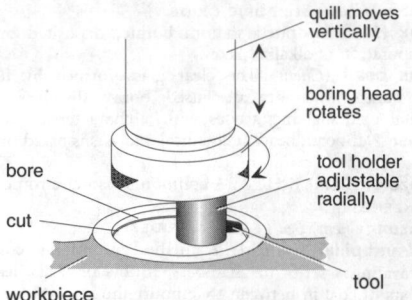

quill moves
vertically

boring head
rotates

bore

tool holder
adjustable
radially

cut

workpiece

tool

boring A boring machine can move the workpiece
in three dimensions.

boring (*Geol*) A TRACE FOSSIL, consisting of eg etchings and
grooves, caused by fossil animals or plants originally
present.

boring (*MinExt*) The drilling of deep holes for the
exploitation or exploration of oil fields. The term *drilling*
is used similarly in connection with metalliferous deposits.

boring bar (*Eng*) A bar clamped to the saddle of a lathe or
driven by the spindle of a boring machine, and carrying the
boring tool.

boring machine (*Eng*) A machine on which boring
operations are performed, comprising a head, carrying a
driving spindle, and a table to support the work.

boring mill (*Eng*) A vertical boring machine in which the
boring bar is fixed, the work being carried by the rotating
table.

boring tool (*Eng*) The cutting tool used in boring
operations. It is held in a boring bar.

Borna disease (*Vet*) An infectious encephalomyelitis of
horses, caused by a virus, and characterized by mild fever, a
variety of nervous symptoms and a high death rate. Sheep
may occasionally be affected.

borneol (*Chem*) $C_{10}H_{17}OH$. Mp 208°C, bp 212°C. A
secondary alcohol which yields camphor on oxidation.
Also *Borneo camphor*.

Born–Haber cycle (*Chem*) The thermochemical calculation
of crystal lattice energy by application of HESS'S LAW using
STANDARD HEAT OF FORMATION, ELECTRON AFFINITY and
IONIZATION POTENTIAL.

bornite (*Min*) Sulphide of copper and iron occurring in
Cornwall and many other localities. Develops a brilliant
iridescent red and blue tarnish; hence also *erubescite*,
peacock ore, *variegated copper ore*.

Born–Oppenheimer approximation (*Phys*) An approximation used in considering the electronic behaviour of
molecules. The problems of the electronic and nuclear
motion are treated separately.

bornyl chloride (*Chem*) $C_{10}H_{17}Cl$. Mp 148°C, white
crystals. Identical to pinene hydrochloride, obtained from
BORNEOL by treatment with PCl_5.

borofluorides (*Chem*) See FLUOROBORIC ACID.

borolanite (*Geol*) A basic igneous rock occurring near Loch
Borolan, Assynt, in NW Scotland; it consists essentially of
feldspar, green mica, garnet, together with conspicuous
rounded white aggregates of 'pseudo-leucite', consisting of
orthoclase feldspar and altered nepheline.

boron (*Chem*) Amorphous yellowish-brown element discovered by H Davy (1808), also J L Gay-Lussac and L J
Thénard. Symbol B, at no 5, ram 10·811, mp 2300°C, rel.d.
2·5. Can be formed into a conducting metal. Most
important in reactors, because of large cross-section
(absorption) for neutrons; thus, boron steel is used for
control rods. The isotope ^{10}B on absorbing neutrons breaks
into two charged particles ^7Li and ^4He which are easily
detected, and is therefore most useful for detecting
and measuring neutrons. There are numerous minerals
in which boron occurs, mainly as borates, or silicates,

including tourmaline. They occur in the late stages of
magmatic crystallization, in volcanic emanations and in
evaporite deposits. Boron's abundance in the Earth's crust
is 9 ppm and in sea water 4·8 ppm.

boron carbide (*Chem*) B_4C. Obtained from B_2O_3 and coke
at about 2500°C. Very hard material, and for this reason
used as an abrasive in cutting tools where extreme hardness
is required. Extremely resistant to chemical reagents at
ordinary temperatures.

boron chamber (*NucEng*) Counter tube containing boron
fluoride, or boron-covered electrodes, for the detection and
counting of low-velocity neutrons, which eject α-particles
from the isotope boron-10. Also *boron counter*.

boron nitride (*Chem*) BN, a compound iso-electronic with
elemental carbon, and having two polymorphs, one similar
to graphite and the other (*borazon*) similar to diamond.

boron trihalides (*Chem*) The compounds BF_3, BCl_3, BBr_3
and BI_3.

borosilicate glass (*Glass*) Family of glasses based on silica
and borax which have a higher resistance to thermal shock
(due to their lower THERMAL EXPANSION COEFFICIENT)
and are more chemically resistant than soda–lime–silica
glasses. Used for domestic ovenware and laboratory
glassware, typical composition (percentage by weight):
SiO_2 80·8, B_2O_3 12·0, Na_2O 4·2, Al_2O_3 2·2, K_2O 0·6, CaO
0·3, MgO 0·3. Different compositions are used for glass-to-
metal seals.

Borrel bodies (*Vet*) See BOLLINGER BODIES.

borrow pit (*CivEng*) Excavation which provides material to
serve as fill when required.

borsic (*Aero*) Boron fibre coated with silicon carbide.

bort (*Min*) A finely crystalline form of diamond in which the
crystals are arranged without definite orientation. Possessing the hardness of diamond, bort is exceedingly tough,
and is used as the cutting agent in rock drills. Also *boart*.

Bosch process (*Chem*) Production of hydrogen by the
catalytic reduction of steam with carbon monoxide at
500°C: $CO + H_2O \rightarrow H_2 + CO_2$.

Bose–Einstein condensate (*Phys*) The classical hypothesis
that atoms cooled to a few nanokelvin above absolute zero
would collapse and merge to form a 'syncytium' or pool
which behaves as a single entity. Recently this state has
been demonstrated experimentally and is the subject of
considerable research.

Bose–Einstein distribution law (*Phys*) A distribution law
of statistical mechanics which is applicable to a system of
particles with symmetric wavefunctions unchanged when
two particles are interchanged, this being the characteristic
of most neutral gas molecules. It can be stated as

$$\bar{n}_i = \frac{g_i}{\frac{1}{A}\exp\left(\frac{E_i}{kT} - 1\right)}$$

where $\bar{n}_i$ is the average number of molecules with energy E_i,
g_i is the degeneracy factor and A is a constant. See BOSON.

Bose–Einstein statistics (*Phys*) Statistical mechanics theory
obeyed by a system of particles whose wavefunction is
unchanged when two particles are interchanged.

bosh (*Eng*) The tapering portion of a BLAST FURNACE,
between the largest diameter (at the bottom of the stack)
and the smaller diameter (at the top of the hearth).

bosh (*Glass*) A water tank for cooling glass-making tools or
quenching glass. See panel on GLASSES AND GLASS-
MAKING.

boson (*Phys*) A particle which obeys BOSE–EINSTEIN
STATISTICS but not the PAULI EXCLUSION PRINCIPLE.
Bosons have a total spin angular momentum of $n\hbar$ where n
is an integer and $\hbar$ is Dirac's constant (PLANCK'S
CONSTANT divided by 2π). Photons, alpha particles and
all nuclei having an even mass number are bosons.

boss (*Eng*) A projection, usually cylindrical, on a machine
part in which a shaft or pin is to be supported; eg the
thickened part at the end of a lever, provided to give a
longer bearing to the pin.

boss (*Geol*) An igneous intrusion of cylindrical form, less than 100 km^2 in area; otherwise like a batholith.

bossage (*Build*) Roughly dressed stones, such as quoins and corbels, which are built in so as to project, and are finish-dressed in position.

bosset (*BioSci*) In deer, the rudiment of the antlers in the first year.

bossing (*Build*) The operation of shaping malleable metal, particularly sheet lead, to make it conform to surface irregularities, by tapping with special mallets, known as bossing mallets or dressers.

Bostock sedimentation balance (*PowderTech*) A SEDI-MENTATION TECHNIQUE in which a torsion balance is used directly to weigh the accumulation of particles. The high sensitivity enables low concentrations to be used, giving increased accuracy.

bostonite (*Geol*) A fine-grained intrusive igneous rock allied in composition to syenite; essentially feldspathic, and deficient in coloured silicates; found in locality of Boston, Massachusetts.

bot (*ICT*) Contraction of *robot*. A computer program designed to perform routine tasks, such as searching the Internet, with some autonomy.

bot (*Vet*) The larva of flies of the genus *Gastrophilus*; bots parasitize the membrane of the stomach of horses, rarely of other animals.

botanical pesticide (*Agri*) A pesticide based on an active compound found naturally in growing plants and expected to have less damaging environmental impact than synthetic pesticides.

botany (*Genrl*) The science of plants or the plants of an area. Also *phytology*.

botry-, botryo- (*Genrl*) Prefixes meaning like a bunch of grapes, from Gk *botrys*, bunch of grapes.

botryoid (*Min*) Of minerals, formation resembling grapes. Also *botryoidal*.

botryoidal (*BioSci*) See BOTRYOSE.

botryomycosis (*Vet*) A suppurative, granulomatous infection of horses due to *Staphylococcus aureus*. When the spermatic cord becomes infected, following castration, the term *scirrhous cord* is applied.

botryose (*BioSci*) Branched; shaped like a bunch of grapes. Also *botrytic, botryoidal*. See RACEMOSE.

bottle glass (*Glass*) Soda–lime–silica glass used for the manufacture of common bottles with a typical composition (percentage by weight) of: SiO_2 74·0, Na_2O 16·4, CaO 9·0, Al_2O_3 0·6.

bottle jack (*Eng*) A screw-jack in which the lower part is shaped like a bottle.

bottle-making machine (*Glass*) A machine to produce glass bottles; operated in various ways, the bottle is formed in two stages, ie the PARISON and the finished bottle. Wide-mouth ware may be formed by pressing the parison and then blowing; narrow-mouth by blowing and blowing or sucking and blowing. In the last method, the glass is gathered by suction into the parison mould; in the other two it is dropped by a mechanical feeding device, hence the terms suction-fed and feeder-fed machines.

bottle-nose drip (*Build*) The shaped edge formed in sheet-lead work at a step on a roof, when jointing the lead across the direction of fall.

bottle-nosed step (*Build*) A step which has the edge and ends rounded.

bottle screw (*Eng*) See SCREW SHACKLE.

bottom (*Phys*) One of the six FLAVOURS of QUARKS, with a mass of 4800 MeV and a charge of $-e/3$. The bottomness of a bottom quark is 1, of an antibottom quark -1 and of all other particles 0. Bottomness is conserved in strong and electromagnetic interactions between particles but not in weak interactions. Also *beauty*.

bottom dead-centre (*Eng*) See OUTER DEAD-CENTRE.

bottom gate (*Eng*) An INGATE leading from the RUNNER into the bottom of a mould.

bottom-hole assembly (*MinExt*) The drilling string attached to the bottom of the DRILLING PIPE. It comprises the drill bit and collars used to maintain direction and may contain several stabilizers and reamers in more difficult conditions, when it is called a *packed-hole assembly*. Abbrev BHA. See panel on DRILLING RIG.

bottom-hole pump (*MinExt*) Electric or hydraulic pump placed at the bottom of a well.

bottoming (*CivEng*) The lowest layer of foundation material for a road or other engineering works including structures.

bottoming tap (*Eng*) See PLUG TAP.

bottoms (*Eng*) (1) A term used in connection with the Orford process for nickel and copper which have separated as sulphides. When the mixed sulphides are fused with sodium sulphide, the nickel sulphide separates to the bottom. Hence 'bottoms' as distinct from 'tops'. (2) In reverberatory furnace, the heaviest molten material at bottom of pool.

bottomset beds (*Geol*) Fine-grained sediments laid down at the front of a growing delta. Cf FORESET BEDS, TOPSET BEDS.

bottom shore (*Build*) One of the members of an arrangement of RAKING SHORES to support temporarily the side of a building; it is the one nearest the wall face.

bottom structure (*Geol*) See SOLE MARK.

bottom-up programming (*ICT*) An approach taken when writing a computer program whereby segments of the program are developed independently before being brought together into the final system. Programs developed in this way often have excellent USER INTERFACES but individual segments may not integrate well with other parts of the program. Cf JACKSON STRUCTURED PRO-GRAMMING, STRUCTURED PROGRAMMING.

bottom yeast (*BioSci*) See YEAST.

botulism (*Med, Food*) Severe and often fatal poisoning due to eating food contaminated by the anaerobic, spore-forming bacterium *Clostridium botulinum*, which secretes a potent neurotoxin. The bacterium grows at temperatures as low as 3–4°C and at pH above 4.5, and the spores may survive boiling for several hours. All susceptible foods (eg canned and vacuum-packed meat, fish and high-pH products) must therefore be processed with generous safety margins. Botulism causes many cases of food poisoning in humans but all types of animals can be affected by eating contaminated food. Fish farms have outbreaks of *C. botulinum* type E. See Z-VALUE and BOTULINUM TOXIN.

botulinum toxin (*BioSci*) One of the most toxic substances known, a very potent AB-type neurotoxin that is not inactivated in the gastro-intestinal tract and causes inhibition of cholinergic neuromuscular synapses. Very small doses injected subcutaneously paralyse muscles and can be used to remove wrinkles (BoTox treatment).

Boucherot circuit (*ElecEng*) An arrangement of inductances and capacitances, whereby a constant-current supply is obtained from a constant-voltage circuit.

bouchon (*Eng*) A hollow plug, or bush, inserted in watch or clock plates to form the pivot holes.

bouclé (*Textiles*) A fabric made from fancy yarns and having a rough, textured surface, mainly used for women's garments. The yarn is made from a core thread with an outer yarn wrapped round it to give a knobbly appearance.

boudinage (*Geol*) A structure found in sedimentary series subjected to folding. It consists of strike-elongated 'sausages' of more rigid rock, enclosed between relatively plastic rocks.

bougie (*Med*) A tube or a rod for dilating narrowed passages in the body.

Bouguer anomaly (*Geol*) A gravity anomaly which has been corrected for the station height and for the gravitational effect of the slab of material between the station height and datum. Cf FREE-AIR ANOMALY.

Bouguer law of absorption (*Phys*) Law stating that the intensity p of a parallel beam of monochromatic radiation entering an absorbing medium is decreased at a constant rate by each infinitesimally thin layer db, ie

$$-\frac{dp}{p} = k\,db$$

where k is a constant that depends on the nature of the medium and on the wavelength.

boulder (*Geol*) The unit of largest size occurring in sediments and sedimentary rocks, the limit between pebble and boulder being placed at 256 mm, although some authorities recognize COBBLES between pebbles and boulders. Boulders may consist of any kind of rock, may be subangular or well-rounded, may have originated in place or have been transported by running water or ice. Accumulations of boulders are *boulder beds*.

boulder clay (*Geol*) See TILL.

boule (*Min*) The pear-shaped or cylindrical drop of synthetic mineral, commonly ruby, sapphire or spinel, produced by the VERNEUIL PROCESS.

Bouma cycle (*Geol*) A sedimentary succession of five intervals that makes up a complete turbidite deposit. Typically incomplete.

bounce (*ICT*) To reject an e-mail message by returning it to the sender without delivery to the intended recipient.

bounce mark (*Geol*) A short depression caused by an object (eg shell, pebble) bouncing over the surface of a sediment, esp a turbidite. Also *prod mark*.

bouncing-pin detonation meter (*Eng*) An apparatus for determining quantitatively the degree of detonation occurring in the cylinder of a petrol engine; used for fuel testing.

boundary film (*Eng*) A film of one constituent of an alloy surrounding the crystals of another.

boundary layer (*Aero*) The thin layer of fluid (air) adjacent to the surface in which viscous forces exert a noticeable influence on the motion of the fluid and in which the transition between still air and the body's velocity occurs. See panel on AERODYNAMICS.

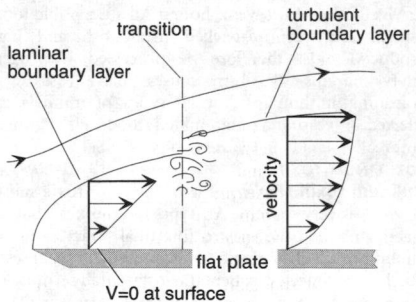

boundary layer Air velocity as a function of distance from a flat plate.

boundary layer (*BioSci*) Surface layer of gas or liquid across which molecular movement is diffusion-limited. This has a significant effect on the uptake of CO_2 by leaves and of some solutes by cells. Also *unstirred layer*.

boundary layer (*ChemEng*) The total thickness of fluid over which a solid surface exerts a differential effect when a fluid flows past it. The layer of fluid next to the solid surface is brought to rest, setting up a viscous motion in adjacent layers. The boundary layer governs the rate of transfer of heat, mass or momentum between the solid surface and the homogeneous bulk of the fluid. The velocity of the layer differs significantly from that of the main fluid stream, and is therefore of considerable importance in heat transfer problems, as in nuclear reactors.

boundary layer control (*Aero*) Modification of the airflow in the BOUNDARY LAYER to increase lift and/or decrease drag. Means include: (1) removal of the boundary layer by sucking through slots or porous surfaces; (2) use of vortex

generators to re-energize sluggish surface flow; (3) ejection of high-speed air through slits; and (4) blowing, by propulsion efflux, over wing surfaces.

boundary layer noise (*Aero*) The noise occurring at high speeds due to the oscillations in the turbulent boundary layer at many frequencies and heard in cockpit and cabin.

boundary lights (*Aero*) Lights defining the boundary of the landing area.

boundary lubrication (*Eng, Phys*) A state of partial lubrication which may exist between two surfaces in the absence of a fluid oil film, due to the existence of adsorbed monomolecular layers of lubricant on the surfaces.

boundary markers (*Aero*) See AIRPORT MARKERS.

bound book (*Print*) A book with cords or tapes which are sewn on by hand and firmly attached to the boards (by lacing-in or using split boards) before applying the covering, normally of leather. Cf CASED BOOK. See FULL-BOUND, HALF-BOUND, QUARTER-BOUND, THREE-QUARTER-BOUND.

bound charge (*ElecEng*) The induced static charge which is 'bound' by the presence of the charge of opposite polarity which induces it. Also, in a dielectric, the charge arising from polarization. Also *surface charge*. Cf FREE CHARGE.

bounded function (*MathSci*) A function which has a bounded set of values, ie a defined upper and lower limit.

bounded set (*MathSci*) (1) A set of numbers lying between two particular numbers. (2) A set of points for which the set of distances between pairs of points has an upper limit.

bounds of a function (*MathSci*) The upper and lower limits of a set of numbers. An *upper bound M* of a set S of numbers is a number such that $x \leqslant M$ for all x in S. The *supremum* of S is its least upper bound. *Lower bound* and *infimum* are defined similarly.

bound state (*Phys*) Quantum mechanical state of a system in which the energy is discrete and the wavefunction is localized, eg that of an electron in an atom, where transitions between the bound states give rise to atomic spectral lines.

bound vector (*MathSci*) A vector with a particular point of application. Also *localized vector*.

bound water (*BioSci*) Water held by matric forces. Cf MATRIC POTENTIAL.

bouquet stage (*BioSci*) See PACHYTENE.

Bourdon gauge (*Eng*) See PRESSURE GAUGE.

bourgeois (*Print*) An old type size, approximately 9 pt.

bourne (*Geol*) An intermittent or seasonal stream.

bournonite (*Min*) Lead copper antimony sulphide; commonly occurs as wheel-shaped twins, hence known as *cogwheel ore*, or *wheel ore*. Also *antimonial lead ore*.

Boussinesq approximation (*EnvSci*) An approximation to the equations of motion in which variations of density from the mean state are ignored except when they are multiplied by the acceleration of gravity.

boutonnière deformity (*Med*) A type of deformity in which a finger is bent down at the middle joint and bent back up at the end joint as the result of a buttonhole-shaped tear in a tendon.

bouyant density (*BioSci*) The density of molecules, particles or viruses as determined by flotation in a suitable liquid. A gradient of CsCl will separate DNA according to its base composition, different DNA molecules banding at discrete positions.

bouyoucous hydrometer (*PowderTech*) A special hydrometer used in particle size analysis of soil grains. It gives readings direct in concentration of soil colloids per litre of suspension in water.

bovine acetonaemia (*Vet*) Negative energy balance associated with ketonaemia and hypoglycaemia. Affects mainly lactating cows, particularly over the winter period. Ketones are detected in milk, urine and breath. Characterized by reduction in appetite and milk production, sometimes with nervous signs. Also *bovine ketosis, slow fever*.

bovine contagious abortion (*Vet*) A contagious bacterial infection with *Brucella abortus*. Infects horses, sheep, dogs and humans (undulant fever), as well as cattle. Main symptom in cattle is abortion. Eradication programmes are possible and the disease is now rare in some countries. Also *Bang's disease, bovine brucellosis*.

bovine cutaneous streptothricosis (*Vet*) A chronic, exudative dermatitis affecting cattle in tropical Africa and elsewhere, caused by the fungus *Dermatophilus congolensis*. A similar disease occurs in goats and horses.

bovine cystic haematuria (*Vet*) A disease which occurs in adult cattle grazed on bracken areas. Symptoms are of intermittent and then continuous haematuria. Various neoplasias can occur but oropharangeal papillomas are also present. Also *enzootic haematuria, urinary bladder neoplasia*.

bovine farcy (*Vet*) A disease characterized by generalized purulent and granulomatous nodular lesions. *Nocardia farcinia* is involved. Nocardial mastitis has been the predominant infection reported in cattle by this soil-borne organism, usually a chronic problem.

bovine hyperkeratosis (*Vet*) A disease of cattle, characterized by emaciation, loss of hair and thickening of the skin, due to poisoning by chlorinated naphthalene compounds. Also *X-disease*.

bovine hypomagnesaemia (*Vet*) Lactational TETANY. A reduction in blood magnesium levels. Clinical disease occurs in spring (and autumn) after turnout onto lush grass or young cereal crops. Climate associated. Symptoms include HYPERAESTHESIA, tetany and sudden death. Controlled by mineral supplementation. Also *grass disease, grass staggers, grass tetany, Hereford disease*.

bovine infectious petechial fever (*Vet*) A disease of cattle in Kenya characterized by fever, petechial haemorrhages of mucous membranes, diarrhoea and often death; believed to be caused by a rickettsia-like organism. Also *Ondiri disease*.

bovine ketosis (*Vet*) See BOVINE ACETONAEMIA.

bovine lipomatosis (*Vet*) A diffuse growth of lipomata in cattle, usually involving the abdominal mesenteries and viscera.

bovine pasteurellosis (*Vet*) See HAEMORRHAGIC SEPTI-CAEMIA.

bovine pyelonephritis (*Vet*) A specific infection of the kidneys of cattle by the bacterium *Corynebacterium renale*.

bovine spongiform encephalopathy (*Vet*) An infectious degenerative brain disease of cattle. Colloq *mad cow disease*. Abbrev *BSE*. See panel on TRANSMISSIBLE SPONGI-FORM ENCEPHALOPATHY.

bow (*ElecEng*) A sliding type of current collector, used on electric vehicles to collect the current from an overhead contact wire. It consists of a bow-shaped contact strip, mounted on a hinged framework.

bow (*Eng*) A flexible strip of whalebone or cane, the ends of which are drawn together to give tension to a thread or line which is given a single turn round a pulley of a pair of turns, drill or mandrel. It is used as a sensitive drive for these tools and was used traditionally for the making of accurate holes esp for clock pivots.

bow compasses (*Eng*) See SPRING BOWS.

Bowden gauge (*ElecEng*) Form of pressure-sensitive transducer.

Bowden–Thomson protective system (*ElecEng*) A form of protective system for feeders, in which special cables, with the cores surrounded by metallic sheaths, are employed; a fault causes current to flow in the sheath and operate a relay to trip the circuit.

Bowditch's rule (*Surv*) A rule for the adjustment of closed compass traverses, in which it may reasonably be assumed that angles and sides are equally liable to error in measurement. According to this rule, the correction in latitude (or departure) of any line is

$$\frac{\text{Length of that line}}{\text{Perimeter of traverse}} \times \frac{\text{Total error in latitude}}{\text{(or departure)}}$$

bowel oedema disease (*Vet*) A disease of pigs character-ized by nervous symptoms and oedema in many tissues;

believed to be due either to immunological hypersensitivity to *Escherichia coli* or to the production of toxins by this bacterium. Can be limited by anti-sera. Also *gut oedema*.

bowenite (*Min*) A compact, finely granular, massive form of serpentine, formerly thought to be nephrite, and used for the same purposes.

Bowen ratio (*EnvSci*) The ratio of the amount of sensible heat (enthalpy) to latent heat lost by a surface to the atmosphere, by conduction and turbulence.

Bower–Barff process (*Build*) An anti-corrosion process applied to sanitary ironwork; this, when red hot, has superheated steam passed over it in a closed space, so that a protective layer of black magnetic oxide is formed on the ironwork. See ANGUS–SMITH PROCESS.

bowk (*MinExt*) A large iron barrel used when sinking a shaft. Also *kibble*.

bowl (*Print*) The enclosed part of letters such as B and R, and the upper portion of g, distinguishing it from the loop or tail.

bowlingite (*Min*) See SAPONITE.

Bowman's capsule (*BioSci*) In the vertebrate kidney, the dilated commencement of a uriniferous tubule.

Bowman's glands (*BioSci*) In some vertebrates, serous glands of the mucous membranes of the olfactory organs.

bow nut (*Eng*) See WING NUT.

bow propeller (*Ships*) A propeller whose thrust can be directed at right angles to the ship's axis, used in docking and manoeuvring in a confined space. Angular thrust can also be provided to the stern.

bows (*Eng*) See SPRING BOWS.

bow-saw (*Build*) A thin-bladed saw which is kept taut by a bow or special frame.

bowstring bridge (*CivEng*) An arched bridge in which the horizontal thrust on the arch is taken by a horizontal tie joining the two ends of the arch.

bowstring suspension (*ElecEng*) A form of suspension for the overhead contact wire of an electric-tramway system, in which the contact wire is suspended from a short cross-wire attached to the bracket arm of the pole.

bow strip (*ElecEng*) See CONTACT STRIP.

bow wave (*Phys*) The wave disturbance emanating from the leading edge of an object moving through fluid, esp the V-shaped surface wave associated with boats moving through water.

box and whisker plot (*MathSci*) A method of displaying statistical data by means of a box representing the values between the 25th and 75th percentile, divided by a horizontal line representing the median, and two termi-nated lines representing the maximum and minimum values respectively. Also *boxplot*.

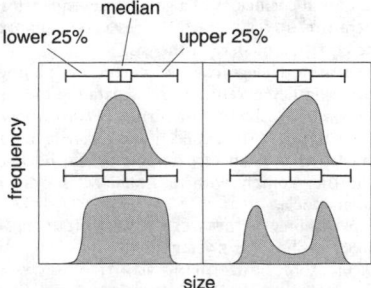

box plot Distribution with equivalent box plot and 'whiskers' above.

box annealing (*Eng*) Heating to soften work-hardened material by placing the work in a sealed box inside the furnace in order to exclude air. Also *close annealing*.

box baffle (*Acous*) Box, with or without apertures and damping, one side fitted with an open diaphragm loudspeaker unit, generally coil-driven.

box chronometer (*Ships*) The marine chronometer. The chronometer is normally supported on gimbals, inside a wooden box with a hinged lid.

box cloth (*Textiles*) A woven woollen fabric, milled and finished with a smooth surface like felt; eg BILLIARD CLOTH.

box culvert (*CivEng*) A culvert having a rectangular opening.

box drain (*Build*) A small rectangular section drain, usually built in brickwork or concrete.

boxed frame (*Build*) See CASED FRAME.

boxed mullion (*Build*) A hollow mullion in a sash window frame, arranged to accommodate the counterweights connected to the vertically moving sashes.

box-frame motor (*ElecEng*) A traction motor in which the frame is cast in one piece instead of being split.

box girder (*Build*) A cast-iron, or mild-steel, girder of hollow rectangular section. See BOX PLATE GIRDER.

box gutter (*Build*) A wooden gutter, lined with sheet lead, zinc or asphalt, and having upright sides; used along roof valleys or parapets.

box-in (*Print*) To surround text with rule, resulting in the printed matter appearing in a frame.

box nut (*Eng*) See CAP NUT.

box plate girder (*Build*) A built-up steel girder, similar to the plate girder, but having two web plates at a distance apart, so that flanges and webs enclose a rectangular space.

boxplot (*MathSci*) See BOX AND WHISKER PLOT. Also *box plot*.

box spanner (*Build*) A spanner composed of a cylindrical piece of steel shaped at one or both ends to a hexagon in order to fit over the appropriate nut. The spanner is turned by a steel rod (tommy bar) inserted through a diametrically drilled hole at the opposite end to the hexagon in use. Used for nuts inaccessible to an ordinary spanner.

box-staple (*Build*) The part on a door post into which the bolt of a lock engages.

box stones (*Geol*) Hollow concretions. Nodules of sandstone containing molluscan casts found in the Pleistocene deposits of East Anglia.

box tool (*Eng*) A single-point cutting tool, set radially or tangentially, used in automatic screw machines and in capstan and turret lathes.

box-type brush-holder (*ElecEng*) See BRUSH-BOX.

boxwood (*For*) The pale-yellow, close-grained, hard and tough wood of the box tree (*Buxus* spp).

Boyle's law (*Phys*) Statement that the volume of a given mass of gas is inversely proportional to its pressure at constant temperature. There are deviations from this law at low and high pressures and according to the nature of the gas. Also *Mariotte's law*.

Boyle temperature (*Phys*) The temperature at which the second virial coefficient of a gas changes sign. Close to this temperature, BOYLE'S LAW provides a good approximation to the equation of state of the gas.

Boy's camera (*ImageTech*) A camera for photographing lightning flashes, gyrating lenses separating the strokes.

BP (*Pharmacol*) Abbrev for *British Pharmacopoeia*. The authoritative published collection of current standards for UK medicinal substances. Compounds may be described as 'BP', ie they comply with the standards set down in the pharmacopoeia.

BP (*Ships*) Abbrev for BETWEEN PERPENDICULARS.

bp (*BioSci*) Abbrev for BASE PAIR.

bp (*Chem*) Abbrev for BOILING POINT.

BPA (*Aero*) Abbrev for *British Parachute Association*.

β-particle (*Phys*) See BETA PARTICLE.

bpcd (*MinExt*) Abbrev for BARRELS PER CALENDAR DAY.

bpi (*ICT*) Abbrev for *bits per inch*. This refers to the density of BITS stored per inch on eg MAGNETIC TAPE.

bps (*ICT*) Abbrev for *bits per second*. This is a measure of the rate at which digital data may be transmitted through a digital communication system; 1 bps is equivalent to 1 BAUD.

bpsd (*MinExt*) Abbrev for BARRELS PER STREAM DAY.

Bq (*Phys*) Symbol for the SI unit of radioactivity. See BECQUEREL.

BR (*Chem*) Abbrev for BUTADIENE RUBBER.

Br (*Chem*) Symbol for BROMINE.

Brabender mixer (*Chem*) Laboratory-scale mixer for plastics and rubbers comprising an internal chamber fitted with contra-rotating rotors, very similar in function to the industrial-scale BANBURY MIXER.

braccate (*BioSci*) Of birds, having feathered legs or feet.

brace (*Build*) A tool used to hold a bit and give it rotary motion. The bit is secured axially in a socket at one end, the other end (to which pressure is applied) being in line with it, while the middle part of the brace is cranked out so that the whole may be rotated. Also *bit-stock*.

brace (*Eng*) A member connecting two nodes of a structure out of plane with the main members, for stiffening purposes. Depending on the applied loads it may be subject to tensile or compressive forces. Also *swaybrace*.

braced girder (*Build*) A girder formed of two flanges connected by a web consisting of a number of bars dividing the girder into triangles or trapeziums and transmitting the vertical loads from one flange to another.

brace jaws (*Build*) The parts of the socket of a brace which clamp upon the shank of the brace bit to secure it while drilling.

brachi-, brachio- (*Genrl*) Prefixes from Lt *brachium*, arm.

brachial (*BioSci*) See BRACHIUM.

brachiate (*BioSci*) Branched: having widely spreading branches; bearing arms. Also *brachiferous*.

brachio- (*Genrl*) See BRACHI-.

brachiocephalic (*Med*) Pertaining to the arm and head.

Brachiopoda (*BioSci*) A phylum of solitary non-metameric Metazoa that have a well-developed coelom. The phylum comprises sessile marine forms, with a LOPHOPHORE in the form of a double vertical spiral, and usually with a bivalve shell. Brachiopods range from early geological periods up to the present time; they occur in all seas, often at great depths.

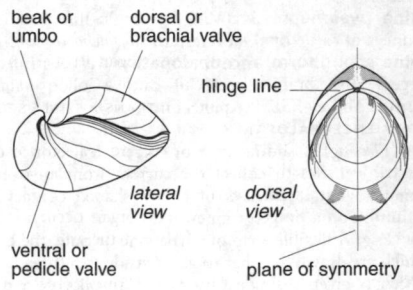

Brachiopoda

brachium (*BioSci*) The proximal region of the fore limb in land vertebrates; a tract of nerve fibres in the brain; more generally, any arm-like structure, as the rays of starfishes. Adj *brachial*.

brachy- (*Genrl*) Prefix from Gk *brachus*, short.

brachycephalic (*Med*) Short-headed; said of skulls whose breadth is at least four-fifths of the length.

brachycerous (*BioSci*) Having short antennae, as some Diptera.

brachydactyly (*Med*) Abnormal shortness of fingers or toes. Also *brachydactylia*.

brachydont (*BioSci*) Said of mammals having low-crowned grinding teeth in which the bases of the infoldings of the enamel are exposed; used also of the teeth. Also *brachydons*. Cf HYPSODONT.

brachypterism (*BioSci*) In insects, the condition of having wings reduced in length. Adj *brachypterous*.

brachysclereid (*BioSci*) A more or less isodiametric cell with a thick lignified wall, eg in the flesh of the pear fruit. Also *stone cell*. See SCLEREID.

brachyurous (*BioSci*) Said of decapodan Crustacea, in which the abdomen is reduced and bent forward underneath the laterally expanded cephalothorax by which it is completely hidden. Also *brachyural*.

bracing (*CivEng*) The staying or supporting rods or ties which are used in the construction or strengthening of a structure.

bracing wires (*Aero*) The wires used to brace the wings of biplanes and the earlier monoplanes. See DRAG WIRES, LANDING WIRES.

bracken poisoning (*Vet*) A disease occurring in cattle and horses due to the ingestion of bracken. In cattle, the main symptoms are multiple haemorrhages and high fever, associated with bone marrow damage and later tumours of the gut. In horses, nervous symptoms are shown and the disease is essentially an induced thiamin deficiency.

bracket (*Build*) A projecting support for a shelf or other part.

bracket arms (*ElecEng*) The transverse projecting arms on the poles, for supporting the overhead contact wire equipment for a tramway or railway system.

bracket baluster (*Build*) An iron baluster, bent at its foot and fixed into the side of the step, usually when the latter is made of stone or concrete.

bracketed step (*Build*) A step supported by a CUT STRING which is shaped on its lower edge to form an ornamental bracket.

bracket fungus (*BioSci*) BASIDIOMYCOTINA that have the fruiting body projecting as a rounded bracket from the side of a tree trunk or stump.

bracketing (*Build*) The shaped timber supports forming a basis for plasterwork and moulding of ceilings and parts near ceilings.

Brackett series (*Phys*) A group of spectral lines of atomic hydrogen in the infrared given by the formula

$$\nu = R_\mathrm{H}\left(\frac{1}{n_1^2} - \frac{1}{n_2^2}\right)$$

in which ν is the wavenumber, $n_1 = 4$, n_2 has various integral values, and R_H is the hydrogen Rydberg number $(1{\cdot}0967\,758 \times 10^7\ \mathrm{m}^{-1})$.

brackish (*EnvSci*) Salty, but not as salty as sea water. Brackish water occurs in estuaries, creeks and deep wells.

bract (*BioSci*) A leaf, often modified or reduced, that subtends a flower or an inflorescence.

bracteate (*BioSci*) Having bracts.

bracteole (*BioSci*) A small leaf-like organ occurring along the length of a flower stalk between the true subtending bract and the calyx.

bract scale (*BioSci*) The structure in conifers that subtends the ovule-bearing scale and may be more or less fused to it.

brad (*Build*) A nail with a small head projecting on one side, or with the head flush with the sides. See SPRIG.

bradawl (*Build*) A small chisel-edged tool, used to make holes for the insertion of nails and screws.

bradsot (*Vet*) See BRAXY.

brady- (*Genrl*) Prefix from Gk *bradys*, slow.

bradyarthria (*Med*) Abnormally slow delivery of speech.

bradycardia (*Med*) Slow heart beat.

bradykinesia (*Med*) Abnormal slowness of the movements of the body.

bradykinin (*BioSci*) An inflammatory mediator that causes dilation of blood vessels and changes in vascular permeability, partly by increasing the release of ARACHIDONIC ACID and production of PROSTAGLANDINS.

Bragg angle (*Phys*) The angle the incident and diffracted X-rays make with a crystal plane when the Bragg equation is satisfied for maximum diffracted intensity.

Bragg curve (*Phys*) A graph giving average number of ions per unit distance along beam of initially monoenergetic alpha particles (or other ionizing particles) passing through a gas.

Bragg diffraction (*Phys*) Diffraction of X-rays according to the BRAGG EQUATION.

Bragg equation (*Phys*) If X-rays of wavelength λ are incident on a crystal, diffracted beams of maximum intensity occur in only those directions in which constructive interference takes place between the X-rays scattered by successive layers of atomic planes. If d is the interplanar spacing, the Bragg equation

$$n\lambda = 2d \sin \theta$$

gives the condition for these diffracted beams; θ is the angle between the incident and diffracted beams and the planes, and n is an integer. Also applied to electron, neutron and proton diffraction.

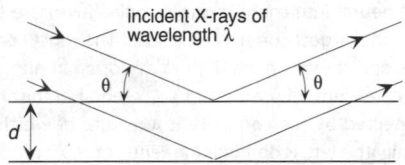

Bragg equation

Bragg rule (*Phys*) An empirical relationship according to which the mass stopping power of an element for alpha particles (also applicable to other charged particles) is proportional to (relative atomic mass)$^{-0.5}$.

Bragg's law (*Phys*) See BRAGG EQUATION.

braid (*Textiles*) A wide range of narrow fabric woven on smallware looms and used as a trimming for the dress material, upholstery or coach and car interiors. The product obtained by BRAIDING.

braided stream (*EnvSci*) A stream which consists of several channels which separate and join in numerous places. Braided streams occur where the gradient is steep and where seasonal floods are liable to occur. They generally have wide beds filled with loose detritus.

braiding (*Textiles*) The process of plaiting in which three or more threads are interlaced to give a flat or tubular fabric.

braids (*Textiles*) (1) A wide range of narrow fabrics woven on smallware looms and used as a trimming for dress material, upholstery or coach and car interiors. (2) The product obtained by BRAIDING.

brain (*BioSci*) A term used loosely to describe the principal ganglionic mass of the central nervous system; in invertebrates, the pre-oral ganglia; in vertebrates, the expanded and specialized region at the anterior end of the spinal cord, developed from the three primary cerebral vesicles of the embryo.

brain death (*Med*) A state in which electroencephalographic recording shows the absence of normal brain function. It is often used to determine whether a patient on a life-support system should continue to be maintained.

brain stem (*BioSci*) In vertebrates, regions of the brain conforming to the organization of the spinal cord, as distinct from such suprasegmental structures such as the cerebral cortex and the cerebellum.

brain stimulation (*Psych*) Technique for studying the neurophysiological basis of some behaviour patterns thought to be under central nervous system (CNS) control; involves stimulation of the CNS by electrical or chemical means, sometimes producing behaviour that appears motivated. See MOTIVATION.

brain voltage (*Med*) Electric signal waves generated in human brain. Usually classed as alpha, beta and delta waves according to frequency.

brake (*Eng*) A device for applying resistance to the motion of a body, either: (1) to retard it, as with a vehicle brake; or (2) to absorb and measure the power developed by an engine or motor.

Human brain

The human brain, the seat of complex thought and presumably consciousness, is a functionally and anatomically complex organ that defies simple description. The adult human brain weighs around 1–1·5 kg and consumes 20% of the energy of the body; it is estimated that there are about 10^{11} neurons and 10^{15} neural interconnections, significantly more than even the largest current computers. The most complex area, and the one most highly developed in humans, is the cerebrum, divided into two cerebral hemispheres connected by the corpus callosum, one of which, usually the left, is dominant in terms of motor function, and controls the right side of the body hence the predominance of right-handedness. The outer layer of the cerebral cortex (grey matter) contains neural cells and the surface area is increased by deep infoldings (sulci). Various functions such as motor association, visual information processing, language comprehension (Wernicke's area), etc can be mapped onto the cerebrum (see figure), which is divided into four lobes (frontal, parietal, temporal and occipital) each with different roles. The next largest region, posterior to the cerebrum, is the cerebellum, also divided into two hemispheres. It is the region that controls the skeletal muscles and can also be considered to act as a relay station between the cerebrum and the body. Moving backwards, the diencephalon (midbrain), which contains the thalamus and hypothalamus, links the cerebellum with the brain stem that controls basic life functions such as breathing and heart rate. The brain stem grades posteriorly into the spinal cord. Not only is the brain concerned with neural processing, it is linked through the neuroendocrine system to the hormonal system of the body; the pituitary is intimately associated with the hypothalamus and although the brain is isolated by the bloodbrain barrier, there is cross-talk between somatic and neural compartments. Simplistically the anatomy of the brain can be considered to reflect the evolution of the nervous system with the most elaborate neural activities (thought, emotion, etc) in the most anterior regions, which in humans have been grossly expanded into the cerebrum to accommodate increasing complexity. Thus the human brain has a disproportionately expanded prefrontal region, a large portion of the temporal lobe devoted to auditory stimuli and memory (in the hippocampus), and the importance of visual image processing is mirrored in the size of the portion of the occipital lobe devoted to this function.

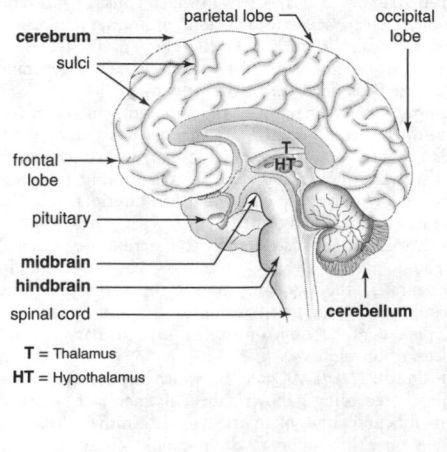

T = Thalamus
HT = Hypothalamus

Cross-sectional view

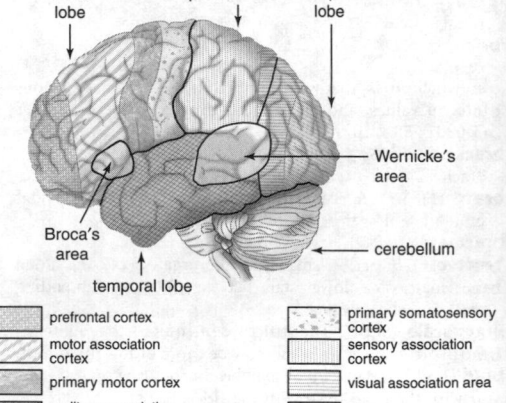

Surface view

Human brain Cross-sectional view and functional mapping.

brake (*Print*) The manual or automatic mechanism to control the tension of the reel when a rotary press is running.

brake bands (*Print*) Strips of leather, fabric or metal acting on a pulley on the reel shaft of a rotary press.

brake drum (*Eng*) A steel or cast-iron drum attached to a wheel or shaft so that its motion may be retarded by the application of an external band or internal brake shoes. See BAND BRAKE, EXPANDING BRAKE.

brake drum (*Print*) A flat or V-shaped wheel at one end of the reel spindle by which the web tension is controlled.

brake efficiency (*Autos*) The retarding force expressed as a percentage of the total vehicle weight.

brake-fade (*Autos*) Condition caused by overheating due to excessive use, resulting in decreased efficiency and sometimes in complete absence of stopping power.

brake horsepower (*Eng*) The effective or useful horsepower developed by a prime mover or electric motor, as measured by a brake or dynamometer. The preferred unit of work is now the kilowatt, but the older unit is still often used as in car engine specifications. Abbrev *BHP*. 1 BHP = 746 W or (approx) 3 BHP = 4 kW.

brake lining (*Eng*) Strips of friction fabric riveted to the shoes of internal expanding brakes in order to increase the friction between them and the drum and provide a renewable surface. Modern practice is to bond lining to shoe. See BRAKE SHOE.

brake magnet (*ElecEng*) A permanent magnet or electromagnet which produces a braking effect, either by inducing eddy currents in a moving conductor or by operating a mechanical brake by means of a solenoid.

brake mean effective pressure (*Eng*) That part of the INDICATED MEAN EFFECTIVE PRESSURE (IMEP) developed in an engine cylinder output equal to the brake horsepower of the engine; the product of IMEP and mechanical efficiency. Abbrev *BMEP*.

brake pads (*Autos*) The friction material in a disk brake, corresponding to the brake shoes in ordinary drum-type brakes.

brake parachute (*Aero*) Parachute attached to the tail of some high-performance aircraft and streamed as a brake for landing. Sometimes a ribbon canopy is used for greater strength and on large aircraft a cluster of two or three is required to give sufficient area with convenient stowage. Also *landing parachute, parabrake*.

brake shoe (*Eng*) (1) Unit which carries the renewable rubbing surface of a block brake. (2) The segmental member which is pressed against the inner surface of a brake drum.

brake thermal efficiency (*Eng*) The efficiency of an engine reckoned in terms of the brake horsepower; given by the ratio of the heat equivalent of the brake output to the heat supplied to the engine in the fuel or steam.

braking notches (*ElecEng*) Positions of the handle of a drum-type controller which apply some form of electric braking.

brammallite (*Min*) A variety of illite with sodium as the interlayer cation.

bran (*FoodSci*) The indigestible parts of the grain kernel comprising the outer and inner pericarp, the seed coat and the endosperm and separated from the grain by milling. A useful source of dietary fibre often added back to food products. Possibly protective against colorectal cancer by reducing contact with carcinogens in feces.

branch (*ICT*) Electric components comprising a minimum path between junction points of common connection in a network. Also *arm*.

branch (*ICT*) See JUMP.

branch (*Phys*) An alternative mode of radioactive decay.

branch-circuit (*ElecEng*) A circuit branched off a main circuit.

branched polymer (*Chem*) Polymer molecules possessing side chains made by a branching reaction. See CHAIN POLYMERIZATION.

branch exchange (*ICT*) See PRIVATE BRANCH EXCHANGE.

branch gap (*BioSci*) A region of parenchyma in the vascular cylinder of the stem, located above the level where the BRANCH TRACES bend out towards the branch. Cf LEAF GAP.

branchia (*BioSci*) In aquatic animals, a respiratory organ consisting of a series of lamellar or filamentous outgrowths; a gill. Adj *branchial*.

branchial arch (*BioSci*) In vertebrates, one of a series of bony or cartilaginous structures lying in the pharyngeal wall posterior to the hyoid arch; it prevents the gill slits from collapsing.

branchial basket (*BioSci*) (1) In Cyclostomata and cartilaginous fish, the skeletal framework supporting the gills. (2) In the larvae of certain dragonflies (Anisoptera), an elaborate modification of the rectum associated with respiration.

branchial chamber (*BioSci*) In Urochordata, cavity of the pharynx.

branchial clefts (*BioSci*) See GILL SLITS.

branchial heart (*BioSci*) In vertebrates, a heart such as that of the CYCLOSTOMATA, in which all the blood entering the heart is deoxygenated and passes thence directly to the respiratory organs; in Cephalopoda, special muscular dilations which pump blood through the capillaries of the CTENIDIA.

branchial rays (*BioSci*) Branches of the hyoid and branchial arches that support the gills and gill-septa.

branching (*Phys*) The existence of two or more modes or branches by which a radionuclide can undergo radioactive decay, eg copper-64 can undergo β^-, β^+ and electron-capture decay.

branching ratio (*Chem*) Where a radioactive element can disintegrate in more than one way, the ratio of the quantities of the element undergoing each type of disintegration.

Branchiopoda (*BioSci*) A subclass of Crustacea, the members of which are distinguished by the possession of numerous pairs of flattened, leaf-like, lobed swimming feet that also serve as respiratory organs. The Branchiopods are mainly fresh-water forms including the fairy shrimps, brine shrimps, tadpole shrimps, clam shrimps, water fleas.

branchiostegal (*BioSci*) Pertaining to the gill covers.

branchiostegal membrane (*BioSci*) In fish, the lower part of the opercular fold below the operculum. Also *branchiostege*.

branch jack (*ICT*) See JACK.

branch of a curve (*MathSci*) Any section or portion of a curve which is separated by a discontinuity (sometimes any singularity) from another section, eg a hyperbola has two branches.

branch of a function (*MathSci*) See RIEMANN SURFACE.

branch point (*MathSci*) See RIEMANN SURFACE.

branch switch (*ElecEng*) A term used in connection with electrical installation work to denote a switch of any type for controlling the current in a branch circuit.

branch trace (*BioSci*) The vascular bundle arising from the main stele and extending into a branch.

brandering (*Build*) The process of nailing small fillets of wood in a counter direction and on the underside of floor joists. Plasterboards or metal lath are fixed to the branders to take plaster. Also *counter lathing*.

bran disease (*Vet*) See OSTEODYSTROPHIA FIBROSA.

brand spore (*BioSci*) The thick-walled resting spore of the USTILAGINALES. It is black or brown and forms sooty masses.

branes (*Astron*) Objects in quantum superstring theory comprising membrane-like structures of one to eleven dimensions. A point-like object is a 0-brane, a string is a 1-brane, a membrane is a 2-brane and *p*-branes have spatial dimensionality *p*, which can be as large as 8 or 9. In *brane cosmology* (*M-theory*), the term is used to refer to objects similar to our four-dimensional universe, but in a higher-dimensional context.

brass (*Eng*) Primarily, name applied to an alloy of copper and zinc, but other elements such as aluminium, iron, manganese, nickel, tin and lead are frequently added. There are numerous varieties. See COPPER ALLOYS.

brasses (*Eng*) Those parts of a bearing which provide a renewable wearing surface; they consist of a sleeve or bored block of brass split diametrally, the two halves being clamped into the bearing block by a cap.

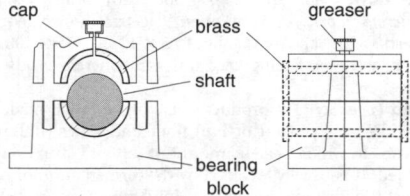

brasses Partial elevation (right) and exploded section on centre line (left).

Brassica (*BioSci*) A genus of the CRUCIFERAE that includes cabbage, broccoli, kale, rape, turnip, swede, mustard.

Brassicaceae (*BioSci*) See CRUCIFERAE.

brattice (*MinExt*) A partition for diverting air, for the purpose of ventilation, into a particular working place or section of a mine. Also *brattice cloth*.

bratticing (*Build*) Also *brattishing*. See CRESTING.

Braun-Blanquet system (*BioSci*) A method of classifying vegetation in the European school of phytosociology, first enunciated by Braun-Blanquet in 1921. With the advent of computers, its use has declined.

braunite (*Min*) A massive, or occasionally well-crystallized, ore of manganese. Composition $3Mn_2O_3MnSiO_3$.

Braun tube (*Electronics*) Original name for CATHODE-RAY TUBE, after K F Braun (1850–1918), the inventor.

Bravais lattices (*Crystal*) The 14 distinct lattices which can be formed by the array of representative points in the study of crystal structure. See panel on CRYSTAL LATTICE.

braxy (*Vet*) An acute and fatal toxaemia of sheep due to infection of the abomasum by *Clostridium septicum*. Controlled by vaccination. Also *bradsot*.

Brayton cycle (*Eng*) A constant-pressure cycle of operations used in gas turbines.

Brazilian emerald (*Min*) A misnomer for a pure-green, deeply coloured variety of tourmaline, occurring in Brazil; used as a gemstone.

Brazilian kingswood (*For*) A rare but important timber from the genus *Astronium*, used eg for high-class furniture.

Brazilian mahogany (*For*) See AMERICAN MAHOGANY, VINHATICO.

Brazilian pebble (*Min*) Brazilian quartz or rock crystal.

Brazilian peridot (*Min*) A misnomer for green crystals of tourmaline or chrysoberyl from Brazil having the typical colour of peridot (olivine).

Brazilian rosewood (*For*) Highly prized, decorative cabinet-wood from the Brazilian hardwood (*Dalbergia nigra*). Heartwood is black-streaked violet to chocolate with straight to wavy grain and coarse texture. Also *jacaranda*, *rio rosewood*.

Brazilian ruby (*Min*) A misnomer for pink topaz, or topaz which has become red after heating, or red tourmaline.

Brazilian sapphire (*Min*) TN for the beautiful clear-blue variety of tourmaline mined in Brazil; used as a gemstone, but not a true sapphire.

Brazilian topaz (*Min*) True topaz varying in colour from pure white to blue and yellow; mined chiefly in the state of Minas Geraes, Brazil.

Brazilian tulipwood (*For*) Fragrantly scented, tropical S American hardwood (*Dalbergia frutescens*) with yellowish-pink heartwood striped with shades from violet to light pink, irregular, straightish grain and fine texture. Also *jacaranda rosa*.

brazing (*Eng*) The process of joining two pieces of metal by fusing a layer of brass, SPELTER or BRAZING SOLDER between the adjoining surfaces.

brazing solders (*Eng*) Alloys used for brazing. They include copper–zinc (50–55% copper), copper–zinc–silver (16–52% copper, 4–38% zinc and 10–80% silver) and nickel–silver alloys.

BRC fabric (*Build, CivEng*) Abbrev for *British reinforced concrete fabric*. A very open, electrically welded, wire mesh with apertures about 3×12 in $(75 \times 300$ mm), used as a reinforcing medium for concrete roads, floor slabs, etc.

bread (*FoodSci*) The product of baking a dough made from cereal (usually wheat or rye) flour and water (unleavened bread) to which a small amount of salt and fat may also be added. It is usually leavened by the fermentation of bakers' yeast (*Saccharomyces cerevisiae*) and may be produced from white flour or flours containing wheatmeal, wholemeal, wheatgerm, malt and protein enrichment.

bread-crust bomb (*Geol*) A type of VOLCANIC BOMB having a compact outer crust and a spongy vesicular interior.

breadth coefficient (*ElecEng*) Also *breadth factor*. See DISTRIBUTION FACTOR.

break (*Build*) (1) Any projection from, or recess into, the surface of a wall. (2) To nail the laths so that the joints are staggered, ie not in the same vertical line. See BREAKING JOINT.

break (*ElecEng*) The shortest distance between the contacts of a switch, circuit breaker, or similar apparatus, when contacts are in the fully open position.

break (*MinExt*) (1) A jointing plane in a coal seam. (2) Optimum range of size to which ore should be ground before further processing.

breakaway (*ImageTech*) A film studio set designed to come to pieces easily.

break-before-make (*ICT*) Classification of switch and relay wipers where existing contacts are opened before new ones close.

break crop (*Agri*) A crop grown to disrupt the usual arable rotation to control weeds, reduce disease and improve subsequent yield when the rotation is restored.

breakdown (*ElecEng*) The sudden passage of current through an insulating material at BREAKDOWN VOLTAGE.

breakdown crane (*Eng*) A portable jib crane carried on a railway truck or motor lorry, for rapid transit to the scene of an accident.

breakdown diode (*Electronics*) See ZENER DIODE.

breakdown voltage (*ElecEng*) The potential difference at which a marked increase in the current through an insulator or a semiconductor occurs. Abbrev *BDV*. See DISRUPTIVE VOLTAGE.

breaker (*ElecEng*) See CIRCUIT BREAKER.

breaker (*Eng*) The tread bracing part of a tyre. See panel on TYRE TECHNOLOGY.

breaker fabric (*Textiles*) In a conveyor belt, a layer of fabric placed between the main fabric of the belt and the outer rubber or plastic surface. Such fabrics are also part of cross-ply tyres.

breaker plate (*Eng*) A device fitted in front of the extruder screw to aid mixing. See EXTRUSION.

breakeven (*NucEng*) In fusion, situation when the power produced exceeds the power input for heating and confinement.

break impulse (*ICT*) An impulse formed by interrupting a current in a circuit.

breaking (*BioSci*) The development of streaks and stripes in flowers due to virus infection in eg Rembrandt tulips.

breaking capacity (*ElecEng*) The capacity of a switch, circuit breaker, or other similar device to break an electric circuit under certain specified conditions.

breaking current (*ElecEng*) The maximum current which a switch, circuit breaker or other similar device will interrupt without being damaged.

breaking down (*MinExt*) Unscrewing the drill pipe preparatory to storage. Also *breaking out*.

breaking elongation (*Textiles*) The maximum elongation of fibre, yarn or fabric just before breaking. Also *breaking extension*.

breaking joint (*Build*) The principle of laying bricks or building stones so that vertical joints are not in line in adjacent COURSES. Also *break joint*.

breaking length (*Paper*) The length beyond which a strip of paper of uniform width would break under its own weight if suspended from one end. Usually expressed in metres.

breaking of the meres (*BioSci*) A local name for the sudden development of large masses of blue-green algae (Cyanobacteria) in small bodies of fresh water.

breaking out (*MinExt*) See BREAKING DOWN.

breaking piece (*Eng*) An easily replaceable member of a machine subject to sudden overloads; made weaker than the remainder, so that in breaking it protects the machine from extensive damage.

braking radiation (*Phys*) See BREMSSTRAHLUNG.

breaking stress (*Eng*) The stress necessary to break a material, either in tension or compression. See STRENGTH MEASURES.

break jack (*ICT*) See JACK.

break joint (*Build*) See BREAKING JOINT.

break line (*Print*) See CLUB LINE.

breakpoint (*ICT*) Temporary halt inserted in a program, in order for the programmer to inspect the contents of registers, storage locations, etc, to aid debugging.

break spinning (*Textiles*) See OPEN-END SPINNING.

breakthrough (*MinExt*) In industrial ion-exchange recovery of metal (eg uranium) from solution, the point at which traces of metal begin to arrive in the last of a series of resin-filled stripping columns.

breakwater (*CivEng*) A natural or artificial coastal barrier serving to break the force of the wave so as to provide safe harbourage behind; it differs from the bulwark in that it has the sea on both sides of it.

breakwater-glacis (*CivEng*) An inclined stone paving of piers and breakwaters, designed to take the force of impact of the waves.

breast (*Build*) The wall between a window and the floor. See CHIMNEYBREAST.

breast (*Med*) An accessory gland of the generative system, rudimentary in the male and secreting milk in the female. Extending from the third to the sixth rib in the front of the chest, it consists of fatty, fibrous and glandular tissue, the ducts of which end in the nipple.

breast (*MinExt*) (1) The working coal face in a colliery. (2) Underground working face. In flat lodes, breast stopes are those from which detached ore will not gravitate without help.

breast bone (*BioSci*) In higher vertebrates, the STERNUM.

breast lining (*Build*) Panelling between window board and skirting.

breast mouldings (*Build*) Moulding on the part of the wall between a window and the floor.

breast roll (*Paper*) Roll for carrying the wire cloth at the breast box end of the paper machine.

breastsummer (*Build*) See BRESSUMMER.

Breathalyzer (*Chem*) TN for apparatus designed to measure alcohol content of the blood by a chemical analysis of alveolar air (eg by reduction of potassium dichromate in sulphuric acid solution).

breather pipe (*Eng*) A vent pipe from the crankcase of an internal-combustion engine, to release pressure resulting from blow-by.

breathing (*BioSci*) An activity of many animals, resulting in the rapid movement of the environment (water or air) over a respiratory surface. See RESPIRATION.

breathing (*ImageTech*) Variation of sharpness of the image caused by movement of the film in and out of the correct plane of focus in a camera, printer or projector.

breathing apparatus (*MinExt*) Mine rescue equipment in which oxygen is fed to a face mask carried by the wearer, via a demand valve. See WEG RESCUE APPARATUS.

breathing root (*BioSci*) See PNEUMATOPHORE.

breccia (*Geol*) A coarse-grained clastic rock consisting largely of angular fragments of pre-existing rocks. According to its mode of origin, a breccia may be a *fault breccia*, a *crush breccia*, an *intrusion breccia* or a *flow breccia*.

Bredig's arc process (*Eng*) A process for making colloidal suspensions of metals in a liquid by striking an arc in the liquid between two electrodes of the metal.

breech block (*Eng*) A movable block used for closing and opening an aperture, originally in guns but now also in machines.

breeder (*NucEng*) A fusion reactor in which further fuel (tritium) is bred from lithium BLANKET surrounding the fusion chamber.

breeder reactor (*NucEng*) A fission reactor which produces more fissile material than is consumed in its operation. FAST REACTORS can be so designed. See panel on NUCLEAR REACTOR. Fig. ▷

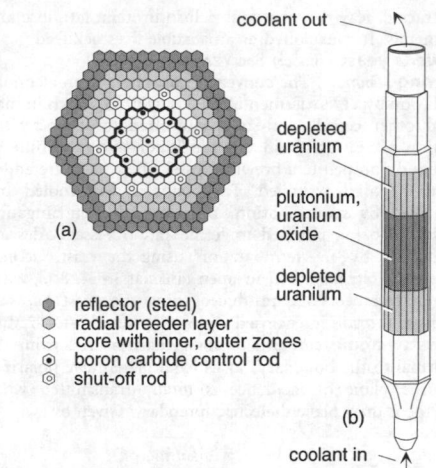

coolant out

depleted uranium

plutonium, uranium oxide

depleted uranium

(a)

● reflector (steel)
◐ radial breeder layer
○ core with inner, outer zones
⊙ boron carbide control rod
⊚ shut-off rod

(b)

coolant in

breeder reactor Cross-section of core zone (a) and fuel assembly from core zone (b), much foreshortened and cut away.

breeding ratio (*NucEng*) The number of fissionable atoms produced in fertile material per fissionable atom destroyed in a nuclear reactor. Symbol b_r. The quantity $b_r - 1$ is known as the *breeding gain*.

breeze (*Build*) A general term for furnace ashes or clinker as in coke breeze, pan breeze.

breeze block (*Build*) Lightweight building brick made from breeze, ie furnace ashes or clinker, bound with PORTLAND CEMENT.

breeze concrete (*Build*) A concrete made of three parts coke breeze, one of sand and one of Portland cement. It is cheap, and nails can be driven into it, but it has poor fire-resisting qualities.

breeze fixing brick (*Build*) A brick made from cement and breeze, built into the surface of a wall to take nails.

bregma (*Med*) The point of junction of the coronal and sagittal sutures of the skull.

breithauptite (*Min*) Nickel antimonide, occurring as bright coppery-red hexagonal crystals, widely distributed in sulphide ore deposits in small amounts.

Breit–Wigner formula (*Phys*) A theoretical expression for the dependence of the cross-section σ of a particular nuclear reaction on the energy E of the bombarding particle and the width Γ of the resonant energy E_0. σ is proportional to

$$\sigma(A, B) = (2l + 1)\frac{\lambda^2}{4\pi}\frac{\Gamma_A\Gamma_B}{(E - E_R)^2 + (\frac{1}{2}\Gamma)^2}$$

The formula has been used with considerable success for many nuclear reactions, particularly those involving neutron bombardment.

bremsstrahlung (*Phys*) Electromagnetic radiation emitted when a charged particle changes its velocity. Thus when electrons collide with a target and suffer large decelerations, the X-radiation emitted constitutes the continuous *X-ray spectrum*. From the German for *braking radiation*.

bressummer (*Build*) A beam or lintel spanning a wide opening in a wall with whose surface it is flush. Also *breastsummer*.

breunnerite (*Chem, Min*) Variety of MAGNESITE containing some iron. Used in manufacture of magnesite bricks.

brevi- (*Genrl*) Prefix from Lt *brevis*, short.

brevier (*Print*) An old type size, approximately 8 pt.

brewer's grains (*Agri*) A brewery by-product comprising barley from which the bulk of the starch has been extracted, leaving a material rich in protein, fat, fibre and minerals. It is exploited as a palatable livestock feed.

brewer's yeast (*BioSci*) See YEAST.

brewing (*FoodSci*) The conversion of starch to an alcoholic solution by YEAST fermentation. Usually, the starch in malt and other cereal adjuncts is converted to sugar by the action of amylases, and the free liquor (WORT) is filtered off and pumped to a brewing vessel where hops are added and the mixture boiled. The liquor is then cooled and clarified by sedimentation, filtration and centrifugation prior to being pumped to fermentation vessels. Ales are produced by 'top fermentation' using the yeast *Saccharomyces cerevisiae*, often in open tanks at 15·5–18°C, while lager-style beers are produced using the yeast *Saccharomyces uvarum* in enclosed fermenting vessels at 4–7°C.

Brewster angle (*Phys*) The angle θ (measured from the normal to the boundary) at which a plane wave polarized in the plane of incidence is totally transmitted when incident on a plane dielectric boundary. Given by

$$\tan \theta = \sqrt{\varepsilon_2/\varepsilon_1}$$

where ε_1, ε_2 are the permittivities of the two media. For optical wavelengths, $\tan \theta = n$, the refractive index between the media. Also *polarizing angle*. See BREWSTER WINDOW.

brewsterite (*Min*) A rare strontium–barium zeolite.

Brewster law (*Phys*) The law relating the Brewster angle (θ) to the refractive index (n) of the medium for a particular wavelength: $\tan \theta = n$. For sodium light incident on a particular glass, $n = 1·66$, θ is 51°.

Brewster's bands (*Phys*) Interference fringes which are visible when white light is viewed through two parallel and parallel-sided plates, whose thicknesses are in a simple ratio (eg 1:1, 2:1, 1:3).

Brewster window (*Phys*) A window attached in certain designs of gas laser to reduce the reflection losses which would arise from the use of external mirrors. Their operation depends on the setting of the windows at the BREWSTER ANGLE to the incident light.

Brianchon's theorem (*MathSci*) The theorem that the lines joining pairs of opposite vertices of a hexagon circumscribed about a conic are concurrent. The dual of PASCAL'S THEOREM.

brick (*Build*) A shaped and burnt block of special clay, used for building purposes.

brick-and-stud work (*Build*) See BRICKNOGGING.

brick-axe (*Build*) The two-bladed axe used by bricklayers in dressing bricks to special shapes.

brick clay (*Geol*) An impure clay, containing iron and other ingredients. In industry, the term is applied to any clay, loam or earth suitable for the manufacture of bricks or coarse pottery. See BRICK EARTH.

brick-core (*Build*) Rough brickwork filling between a timber lintel and the soffit of a relieving arch.

brick earth (*Build*) Earths used for the manufacture of ordinary bricks; they consist generally of clayey silt interstratified with the fluvioglacial gravels as in southern England, frequently exploited in brick manufacture. Also *brickearth*.

bricking (*Build*) Work on plastered or stuccoed surfaces, in imitation of brickwork.

bricklayer's hammer (*Build*) A hammer having both a hammer-head and a sharpened peen; used for dressing bricks to special shapes.

bricklayer's scaffold (*Build*) A scaffold used in the erection of brick buildings, a characteristic being that one end of the PUTLOGS is supported in holes left in the wall.

bricknogging (*Build*) The type of work used for walls or partitions which are built up of brickwork laid in spaces between timber. Also *brick-and-stud work*.

brick-on-edge coping (*Build*) A coping finish to the exposed top of a wall; formed of bricks built on the edge in cement in courses $4\frac{1}{2}$ instead of 3 in high, so that the frogs are concealed and only a few joints are exposed to the weather.

brick-on-edge sill (*Build*) An external sill to window or door, formed in the manner of the BRICK-ON-EDGE COPING. Abbrev *BOE sill*.

brick trowel (*Build*) A flat triangular-shaped tool used by bricklayers for picking up and spreading mortar.

bridge (*CivEng*) A structure spanning a river, road, etc, giving communication across it. See panel on BRIDGES AND MATERIALS.

bridge (*ElecEng*) A circuit often used for measurement of the impedance of passive components, for both ac and dc. Four arms of the bridge are arranged in a diamond-shaped configuration, three comprising accurately known impedances, and the fourth, the unknown. A voltage supply is connected to two opposite corners of the diamond and a detector between the other two. By adjusting the known component values, the bridge is balanced when the detector shows a null signal and equations are then available for the unknown in terms of the other three arms of the bridge.

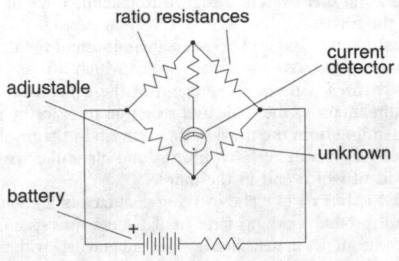

bridge Wheatstone type; by altering the ratio and adjustable resistances a wide range of unknowns can be measured.

bridge (*ICT*) An electronic device that connects together two discrete LOCAL AREA NETWORKS which use the same PROTOCOLS.

bridge bearing (*Eng*) Bearing on which the decks of bridges rest to accommodate esp horizontal thermal expansion and contraction. Usually a steel plate on steel rollers or a laminated structure of steel plates sandwiching blocks of rubber. See EXPANSION ROLLERS, LAMINATED BEARING.

bridge classification (*CivEng*) Bridge structures can be classified into five groups, each of which imposes loads on their materials in different ways. See panel on BRIDGES AND MATERIALS.

bridged-T filter (*ICT*) A filter consisting of a T-network, with a further arm bridging the two series arms; used for phase compensation.

bridge fuse (*ElecEng*) A fuse in which the fusible wire is carried in a holder, supported by spring contacts at its two ends; it is thus easily removable for renewing the fuse wire.

bridge gauge (*Eng*) A measuring device for detecting the relative movement of two parts of a machine due to wear at bearings etc.

bridge hanger (*ElecEng*) A form of hanger of small vertical dimensions, for supporting the overhead contact wire of a traction system under bridges or tunnels.

bridge-megger (*ElecEng*) A portable instrument for measuring large resistances on the Wheatstone-bridge principle. A megger contains a source of emf and the instrument dial on which the balance is indicated.

bridge network (*ICT*) See LATTICE NETWORK.

bridge oscillator (*ElecEng*) An oscillator in which positive feedback and limitation of amplitude is determined by a bridge, which contains a quartz crystal for determining the frequency of oscillation. Devised by Meachan for high stability of operation in crystal clocks etc.

bridge rectifier (*ElecEng*) Type of full-wave rectifier employing four rectifiers in the form of a bridge. The alternating supply is connected across one diagonal, and the dc output is taken from the other.

bridges and materials (*Eng*) The dependence of the length of the span of a bridge on the strength of materials. See panel on BRIDGES AND MATERIALS.

bridge transformer (*ICT*) See HYBRID COIL.

bridge transition (*ElecEng*) A method, employed in connection with the series–parallel control of traction motors, in which the change from series to parallel is effected without interrupting the main circuit, and without any change in the current flowing in each of the motors.

bridging (*Build*) The principle of reducing lateral distortion of adjacent floor joists by connecting them together with short cross-pieces or *dwangs*.

bridging (*MinExt*) Arching of jammed rock so as to obstruct flow of ore; clogging of filtering septum by tiny particles which are individually small enough to pass, but which form such arches.

bridging floor (*Build*) A floor supported by bridging joists, without girders.

bridging joist (*Build*) A timber beam immediately supporting the floorboards in a floor. Also *common joist*.

bridging ligand (*Chem*) A LIGAND that is bonded to more than one metal at a time. In Au_2Cl_6, each gold atom is co-ordinated by four chlorines, two of which bridge the two gold atoms.

Bridgman process (*Crystal*) A method of growing a large single crystal in a crucible. A seed crystal is dipped into a melt which is being slowly withdrawn from a furnace. Cf CZOCHRALSKI PROCESS, FLOAT ZONE.

bridle (*ElecEng*) A portion of an overhead contact-wire system. It extends longitudinally between supporting structures and is attached at intervals to the contact wire, in order to retain the latter in its proper lateral position.

bridle joint (*Build*) The converse of the mortise-and-tenon joint. The central part on the first member is cut away to leave two side tongues projecting, and the second member is cut away at the sides to receive these tongues.

Briggs logarithms (*MathSci*) A LOGARITHM to base 10.

bright annealing (*Eng*) The heating and slow cooling of steel or other alloys in a carefully controlled atmosphere, so that oxidation of the surface is reduced to a minimum and the metal surface retains its bright appearance. See BOX ANNEALING.

bright emitter (*Electronics*) A thermionic valve with a pure tungsten cathode, dc heated to 2600 K in order to emit electrons. Originally used on all thermionic valves; now superseded by treated cathodes which emit at much lower temperatures.

brightener (*ElecEng*) An agent added to an electroplating solution to produce bright deposits.

brightening agent (*Textiles*) A compound that on addition to a white or coloured textile material increases its brightness by converting some of the ultraviolet radiation into visible light. Also *fluorescent brightener*, *optical brightener*. Cf FLUORESCENT WHITENING AGENTS.

bright-field illumination (*BioSci*) The common method of illumination in microscopy in which the specimen appears more or less dark on a bright background. Cf DARK GROUND ILLUMINATION, INTERFERENCE MICROSCOPY, PHASE-CONTRAST MICROSCOPY.

bright-line viewfinder (*ImageTech*) See ALBADA VIEWFINDER.

brightness (*Phys*) See LUMINANCE. As a quantitative term, *brightness* is deprecated.

brightness control (*ImageTech*) Control which alters the brightness on a cathode-ray-tube screen.

bright plating (*ElecEng*) The production of a fairly bright deposit from an electroplating plant. Such surfaces require little finishing.

Bright's disease (*Med*) Old name for acute and chronic nephritis.

brilliance (*Acous*) The presence of considerable numbers of high harmonics in musical tone, or the enhancement of these in sound reproduction.

brilliant (*Print*) Old type size approximately 4 pt.

brilliant green (*Chem*) The sulphate of *tetra-ethyl-diamino-triphenyl-methanol anhydride*. A green dye used as a disinfectant.

brilliant viewfinder (*ImageTech*) Viewfinder comprising a reflector between two small lenses. An inclined mirror gives waist-height viewing; a prism gives eye-level viewing. The image is laterally inverted unless an erector lens is added.

Brillouin formula (*Phys*) A quantum mechanical analogue in paramagnetism of the Langevin equation in classical theory of magnetism.

Brillouin scattering (*Phys*) The scattering of light by the acoustic modes of vibration in a crystal, ie *photon–phonon* scattering.

Brillouin zone (*Phys*) Polyhedron in *k*-space, *k* being the position wavevector of the groups or bands of electron energy states in the band theory of solids. Often constructed by consideration of crystal lattices and their symmetries.

brindled bricks (*Build*) Bricks which, owing to their chemical composition, show a striped surface.

brine (*FoodSci*) A solution of common salt used for preserving foods, pickling and aseptic packing of vegetables. It may also contain flavouring or other preservatives.

Brinell hardness test (*Eng*) A method of measuring the hardness of a material by measuring the area of indentation produced by a hard steel ball under standard conditions of loading. Expressed as either Brinell hardness number (BHN) or, preferably, B_H following the number, which is the quotient of the load on the ball in kilogramforce divided by the area of indentation in square millimetres. See panel on HARDNESS MEASUREMENTS.

brise soleil (*Arch*) Screening in front windows to interrupt sun glare. It is used in the tropics and in Mediterranean countries and can take the form of precast concrete slats or of a permanent trellis supporting climbing plants.

brisket (*Vet*) The breast or anterior sternal region of an animal.

bristle (*Build*) The hair of the hog or boar used as a filling material in good-quality paint brushes.

bristletails (*BioSci*) A group of small wingless insects of the order Archaeognatha. They resemble silverfish (Thysanura) but, unlike silverfish, bristletails can jump distances of up to 10 cm by flexing their abdomens.

Bristol board (*Paper*) A fine-quality cardboard made by pasting several sheets together, the middle sheets usually being of an inferior grade.

Bristol diamonds (*Min*) Small lustrous crystals of quartz, ie rock crystal, occurring in the Bristol (UK) district.

Britannia metal (*Eng*) Alloy series of tin (80–90%) with antimony, copper, lead or zinc, or a mixture of these.

British Approvals Board for Telecommunications (*ICT*) A body set up under the British Telecommunications Act of 1981 to test and approve apparatus for connection to the deregulated UK telecommunications system created by that act.

British Association screw-thread (*Eng*) A system of symmetrical vee threads of 47·5° included angle with rounded roots and crests. It is designated by numbers from 0 to 25, ranging from 6·0 to 0·25 mm in diameter and from 1 to 0·07 mm pitch. Used in instrument work, but now being superseded by standard metric sizes. Even numbers are preferred sizes. Abbrev *BA thread*.

British Columbian pine (*For*) See DOUGLAS FIR.

British Standard brass thread (*Eng*) A screw-thread of Whitworth profile used for thin-walled tubing; it has 26 threads per inch for each diameter. Abbrev *BSB thread*. See BRITISH STANDARD WHITWORTH THREAD.

British Standard fine thread (*Eng*) A screw-thread of Whitworth profile, but of finer pitch for a given diameter. Abbrev *BSF thread*. Obsolescent.

British Standard pipe thread (*Eng*) A screw-thread of Whitworth profile, but designated by the bore of the pipe on which it is cut (eg $\frac{3}{8}$ in gas) and not by the full diameter, which is a decimal one, slightly smaller than

Bridges and materials

The largest bridges are spectacular structures involving huge quantities of material. The Humber Bridge (Yorkshire, UK) had in 1993 the world's longest span of 1410 m (see Fig. 1) and needed 500 000 tonnes of concrete for its towers and anchors and 30 000 tonnes of steel for its deck and cables.

Apart from the very biggest, bridges are such a common feature of our towns and landscape that they tend to be taken for granted, but there are 200 000 of them in the UK alone and more are added every year. Only about 1% have spans greater than 30 m. Their

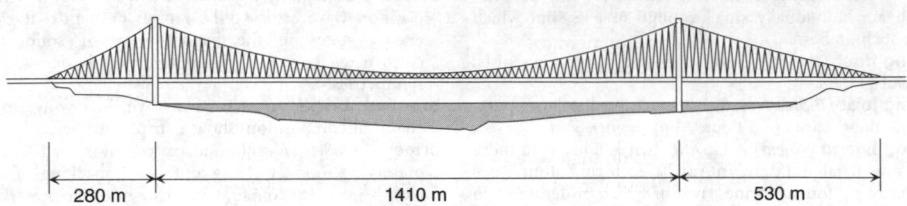

← 280 m →   1410 m   ← 530 m →

Fig. 1 **Schematic drawing of the suspension bridge over the River Humber**.

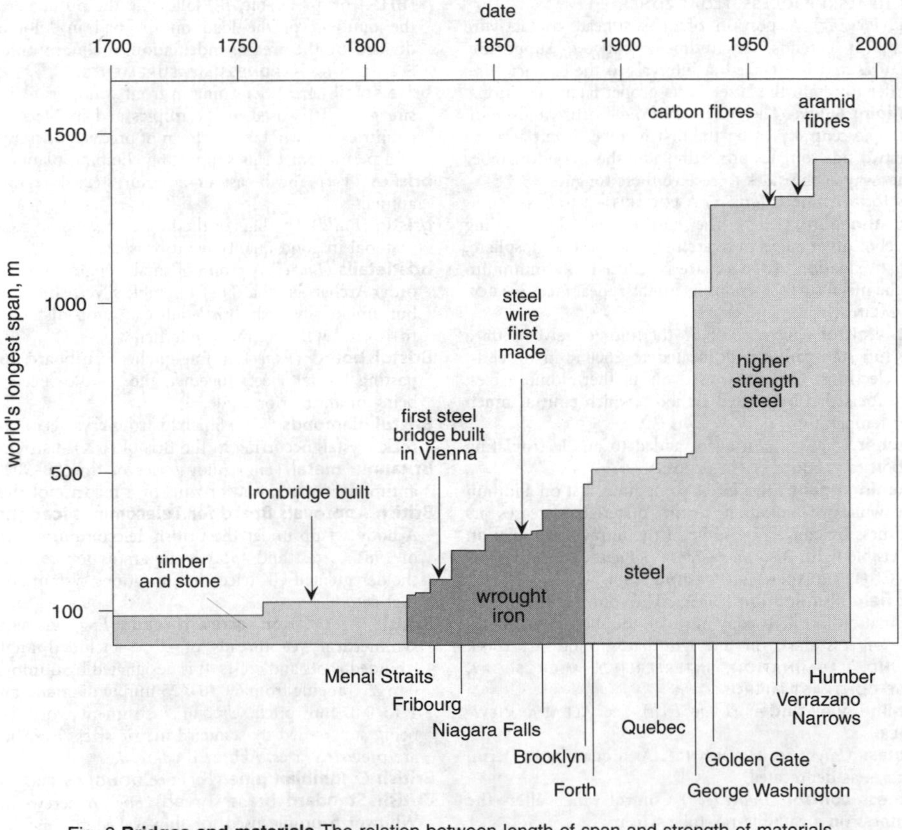

Fig. 2 **Bridges and materials** The relation between length of span and strength of materials.

that of the pipe. Abbrev *BSP thread*. Also *British Standard gas thread*. See BRITISH STANDARD WHITWORTH THREAD.

British Standards Institution (*Genrl*) UK national organization for the preparation and issue of standard specifications. Abbrev *BSI*.

British Standard specification (*Genrl*) A specification of efficiency, grade, size, etc, drawn up by the British Standards Institution, referenced so that the material required can be briefly described in a bill or schedule of quantities. The definitions are legally acceptable. Abbrev *BSS*.

Bridges and materials (Cont.)

structural forms are usually easily visible and have evolved in parallel with developments in materials and in the understanding of how to use them.

This is well illustrated by the progression of the longest spans which reflect the limits of what was feasible in their day (see Fig. 2). The main limit was and still is the highest value of tensile strength, or more precisely specific tensile strength, offered in materials that were relatively cheap, available in tonnage quantities, could be readily handled and were sufficiently durable to withstand the environment for many years (the current British Standard specifies 120 years).

Classification of bridge structures

Bridge structures can be classified into five groups (Fig. 3), each of which imposes loads on their materials in different ways. The earliest of these was the *beam*, initially just a tree trunk or slab of rock laid across a stream or gully. The loading (dead due to the weight of the structure and live due to all other forces) puts the beam into bending between the supports, inducing compressive stresses on the upper surface and tensile stresses on the lower at mid-span. The *cantilever* bridge comprises two centrally supported beams carrying a central span between them. This reverses the loading of the simple beam, with tension on the upper surface and compression on the lower.

The *arch* is the second-oldest structure, believed to have been invented in Babylonian times to provide a structure which could be built with clay bricks, which have negligible tensile strength. Thus the main load-bearing elements around the ring of the arch had to remain in compression. *Suspension* bridges (eg Fig. 1) are effectively inverted arches with the signs of the loads reversed so that their cables are in tension. Although light suspension bridges with ropes of natural fibres were long a feature of many cultures, a significant load-bearing capacity had to await the advent of high-strength materials – initially wrought-iron, then steel wires and latterly high-strength steel wire. Future developments will certainly include the synthetic, low-density, high-strength fibres such as aramid, carbon and gel-spun polyethylene. Finally there is the *cable-stayed* or *bridle-chord* bridge which is becoming increasingly popular for medium spans. These can be of the fan type or the harp type. In both, the tension in the cables induces a compression in the decking which is particularly useful during construction.

See panel on HIGH-PERFORMANCE POLYMERS.

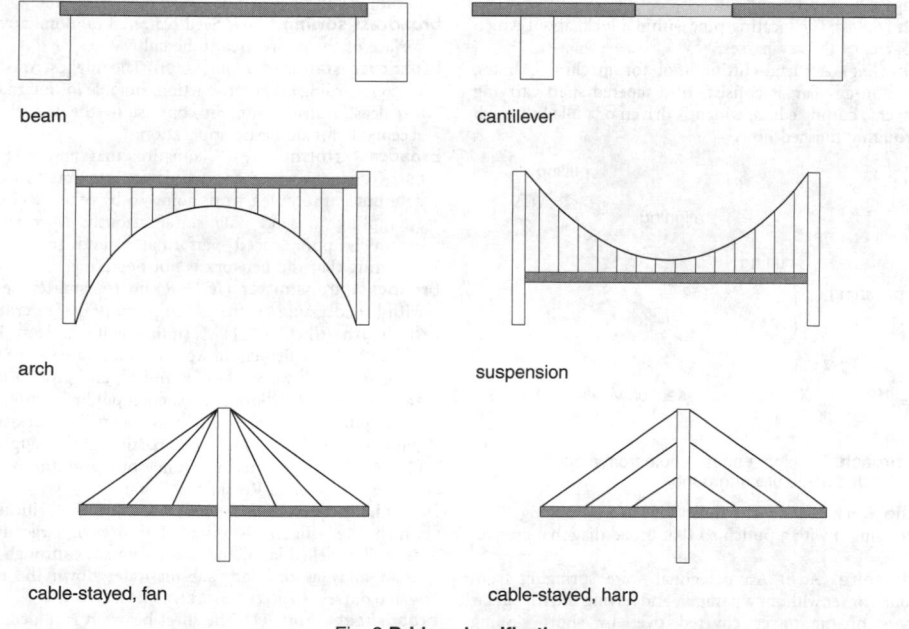

beam

cantilever

arch

suspension

cable-stayed, fan

cable-stayed, harp

Fig. 3 **Bridge classification**.

British Standard Whitworth thread (*Eng*) The pre-metric British screw-thread, still widely used in the USA, having a profile angle of 55° and a radius at root and crest of $0.1373 \times$ pitch; $\frac{1}{6}$ of the thread cut-off. The pitch is standardized with respect to the diameter of the bar on which it is cut. Abbrev *BSW thread*.

British Standard Wire Gauge (*Eng*) See STANDARD WIRE GAUGE.

British thermal unit (*Phys*) The amount of heat required to raise the temperature of 1 lb of water by 1°F (usually taken as 60–61°F). Symbol Btu. Equivalent to 252 calories, 778.2 ft lbf, 1055 J; 10^5 Btu = 1 therm.

brittle fracture (*Eng*) A fracture which occurs with no discernable plastic deformation, ie in the elastic region of the stress–strain curve. Caused by propagation of a crack as distinct from yielding. In metals it may be either intergranular or by cleavage along certain crystal planes. See STRENGTH MEASURES and panel on FATIGUE.

brittleheart (*For*) A defect in wood, esp in low-density, tropical hardwoods, in which circumferential shrinkage stresses in outer layers become large enough to exceed compressive strength of core wood; resulting yield produces shear lines in the timber.

brittle micas (*Min*) A group of minerals (the clintonite and margarite group) resembling the true micas in crystallographic characters, but having the cleavage flakes less elastic. Chemically, they are distinguished by containing calcium as an essential constituent.

brittleness (*Eng*) The tendency to brittle fracture, ie without significant plastic deformation. Loosely used as the opposite of TOUGHNESS, but more precisely means having low values of toughness or FRACTURE TOUGHNESS. See STRENGTH MEASURES.

brittle silver ore (*Min*) A popular name for STEPHANITE.

brittle temperature (*Eng*) Point at which a material changes in fracture behaviour, from ductile to brittle. For polymers, it is often a little below T_g. Sometimes denoted T_B. Sensitive to sample geometry (eg stress concentrations) and rate effects, such as occur in impact tests.

Brix (*ChemEng*) Scale of densities used in the sugar industry. Hydrometers are marked in 'degrees Brix' (°Bx), representing the density of a corresponding pure sugar solution in units equivalent to the percentage of sugar in the solution, either by volume ('volume Brix') or by mass ('mass Brix').

broach (*Arch*) The sloping timber or masonry pyramid at the projecting corner of the square tower from which springs a BROACH SPIRE.

broach (*Build*) The locating pin, within a lock, about which the barrel of the key passes.

broach (*Eng*) A metal-cutting tool for machining holes, often non-circular; it consists of a tapered shaft carrying transverse cutting edges, which is driven or pulled through the roughly finished hole.

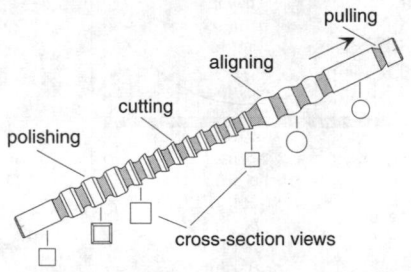

broach Forming a square hole from a circular. Shaded areas are unground.

broache work (*Build*) The finish given to a building stone by dressing it with a punch so that broad diagonal grooves are left.

broach spire (*Arch*) An octagonal spire springing from a square tower without a parapet, and having the triangular corners of the tower covered over by short sloping pyramids blending into the spire.

broad (*Build*) A wood-turning tool, often consisting of a flat disk with sharpened edges fixed at right angles to a stem; used for shaping the insides and bottoms of cylinders.

broad (*ImageTech*) A studio light source giving a wide angle of illumination.

broadband (*ICT*) (1) Said of a device (amplifier, mixer, transistor, etc) that is capable of operating with consistent efficiency over a wide range of frequencies. See WIDEBAND AMPLIFIER. (2) Used as a verb to imply the process of making a circuit or device operate over a wide range of

frequencies. (3) Description of signals, noise, interference, etc, that spreads over a wide range of frequencies.

broadband integrated services digital network (*ICT*) A form of INTEGRATED SERVICES DIGITAL NETWORK offering 30 64 Kbps bearer channels and a 16 Kbps data channel, ie a total bearer capacity of 1·92Mbps. Also *B-ISDN*, *basic rate ISDN*, *ISDN 30*. See B-CHANNEL, D-CHANNEL, NARROW-BAND INTEGRATED SERVICES DIGITAL NETWORK.

broadband network (*ICT*) A NETWORK based on a transmission medium having a wide frequency bandwidth such as coaxial cable or OPTICAL FIBRE cable. Several signals may be carried simultaneously by allocating different channels to separate frequency bands.

broad-base tower (*ElecEng*) A transmission-line tower with each leg separately anchored.

broad beam (*Radiol*) Said of a gamma- or X-ray beam when scattered radiation makes a significant contribution to the radiation intensity or dose rate at a point in the medium traversed by the beam.

broadcast channel (*ICT*) Any specified frequency band used for broadcasting; chosen with regard to freedom from mutual and other forms of interference, consistency of propagation and reception, intended range of broadcasting (ie local, international, satellite) and bandwidth of programme material (ie sound or vision).

broadcast control channel (*ICT*) An identification and organization signal emitted continuously by a BASE STATION in a mobile telephone network. Its primary function is to allow each MOBILE STATION to identify its nearest base station, but the signal also contains information used by the mobile station to decide whether it is allowed to CAMP on that base station as well as a list of the broadcast control channel frequencies of neighbouring cells.

broadcasting (*ICT*) The transmission of a programme of sound, vision or fascimile for general reception.

broadcast sowing (*Agri*) Seed scattered randomly over the surface of the entire area to be cultivated.

broadcast standard (*ImageTech*) The highest quality of video recording and reproduction, suitable for international broadcast transmission, in contrast to the lower quality acceptable for domestic application.

broadcast storm (*ICT*) A situation that may exist on a CSMA/CD network when each device attempts to transmit data but is prevented from doing so by other devices also attempting to send data simultaneously. May result in extremely poor speed performance and some devices reporting that the network is not operating.

broadcast transmitter (*ICT*) Radio transmitter designed with broadcasting as one of the primary design criteria.

broadcloth (*Textiles*) (1) A suiting cloth at least 1·35 m wide. (2) A woollen cloth, woven from fine MERINO WOOL yarns in a twill weave, heavily milled, and given a DRESS-FACE FINISH. (3) US for a lightweight poplin shirting fabric.

broad gauge (*CivEng*) A railway gauge in excess of the standard 4 ft $8\frac{1}{2}$ in (1·435 m). In particular, the gauge of 7 ft (2·134 m) laid down by Brunel but also the standard gauges of eg India, Russia and Spain.

broad irrigation (*Build*) A process of sewage purification in which the effluent is distributed over a large area of carefully levelled land, and allowed to soak through it and drain away as ordinary subsoil water down the natural watercourses. Cf INTERMITTENT FILTRATION.

broadsheet (*Print*) (1) The sheet before it is folded. (2) In rotary printing, the size of newspapers printed COLUMNS AROUND the cylinder.

broadside (*Print*) A large sheet printed on one side, such as a poster. See PAPER SIZES.

broadside antenna (*ICT*) Array in which the main direction of the reception or radiation of electromagnetic energy is normal to the line of radiating elements.

broad-spectrum (*Pharmacol*) Agent that is effective against a range of organisms or a variety of conditions. Also *widespectrum*.

broadstone (*Build*) See ASHLAR.

broad tool (*Build*) A steel chisel having a cutting edge $3\frac{1}{2}$ in (90 mm) in width, used for finish-dressing stone.

brocade (*Textiles*) Dress or furnishing fabrics produced by JACQUARD or dobby weaving. The design is developed by floating the warp and/or weft threads in irregular order on a simple ground fabric, such as satin.

Broca's aphasia (*Psych, Med*) Impairment in producing understandable speech (aphasia) associated with damage to the *Broca's area* of the brain. See panel on BRAIN.

Broca's area (*Med*) The left inferior convolution of the frontal lobe of the brain; the 'speech centre'.

brochanite (*Min*) A basic sulphate of copper occurring in green fibrous masses, or as incrustations; occurs in the oxidation zone of copper deposits.

Brockenspectre (*EnvSci*) The shadow of an observer cast by the Sun onto a bank of mist, sometimes surrounded by *glories*. The phenomenon, occasionally seen from a hilltop, gives the illusion of a gigantic form seen through the mist.

brockram (*Geol*) A sedimentary rock occurring in the Permian strata in north-west England; consists of angular blocks which probably accumulated as scree material.

broderie anglaise (*Textiles*) Machine-embroidered, lightweight woven cloth which includes holes in the pattern.

Broglie wavelength (*Phys*) See DE BROGLIE WAVELENGTH.

broiler (*Agri*) A domesticated fowl bred and reared for meat production, usually applied to chickens.

broke (*Paper*) Wet or dry paper removed during the papermaking or finishing processes and reused within the mill. Cf CULLET.

broken ends (*Textiles*) Warp threads which have broken during weaving.

broken-over (*Print*) The term used to indicate that plates or other separate sheets to be inserted in a book have been given a narrow fold on the inner edge, so that they will lie flat and turn easily when fixed.

broken picks (*Textiles*) Defects in weaving due to breaking of the weft.

broken-space saw (*Build*) A handsaw having usually six teeth to the inch with spaces between each group of teeth.

broken twills (*Textiles*) Fabrics in which the diagonal line forming the TWILL is broken, or broken and reversed in direction, at intervals.

broken wind (*Vet*) A chronic emphysema of the lungs of horses; sometimes associated with chronic obstructive pulmonary disease.

broker (*ICT*) A computer program that manages the behaviour of certain classes of operation, such as messages, requests for the instantiation of objects, or other services.

bromcresol green (*Chem*) Indicator used in determination of pH values, suitable for ranges 3·6–5·2.

bromcresol purple (*Chem*) Indicator used in determination of pH values within the range 5·2–6·8.

Bromeliaceae (*BioSci*) A family of c.2500 spp of monocotyledonous flowering plants (superorder Commelinidae). It comprises terrestrial and esp epiphytic herbs (including tank epiphytes and atmospheric plants) from tropical and subtropical America. Many are CAM PLANTS. The flowers often have showy bracts and are bird- or insect-pollinated. Includes the pineapple (the only major CAM crop plant) and some plants grown for fibre.

bromic acid (*Chem*) $HBrO_3$. With bases it forms bromates (V). A powerful oxidizing agent.

bromides (*Chem*) Salts of hydrobromic acid. Silver bromide is extensively used in photography.

bromination (*Chem*) The substitution by bromine in or addition of bromine to organic compounds.

bromine (*Chem*) A non-metallic element in the seventh group of the periodic system, one of the halogens. Symbol Br, at no 35, ram 79·909, oxidation states 1, 3, 5, 7, mp $-7·3°C$, bp 58·8°C, rel.d. 3·19. A dark-red liquid, giving off a poisonous vapour, Br_2, with an irritating smell. It occurs as a scarce element in the Earth's crust, to the extent of 2·5 ppm, with 65 ppm in sea water. Bromine forms very few minerals and it appears mainly to occur as the bromide

ion, sometimes replacing chlorine. The chief commercial source is sea water from which bromine is manufactured by treating the BITTERN with chlorine. Bromine is used extensively in synthetic organic chemistry, as an antiknock additive to motor fuel, in medicine and in halogen-quenched Geiger tubes.

bromobutyl rubber (*Chem*) A type of butyl rubber used in tyre industry for linings etc, made by treating isoprene–isobutene rubber with bromine to enhance reactivity during vulcanization.

bromochlorodifluoromethane (*Chem*) $CHBrClF_2$. Bp $-4°C$. Organic substance used as a fire-extinguishing fluid, particularly for fires in confined spaces. Low-toxicity vapour, 5·7 times as dense as air. Abbrev BCF.

bromoform (*Chem*) Tribromomethane. $CHBr_3$. Mp 5°C, bp 151°C, rel.d. 2·9; a colourless liquid, of narcotic odour. Much used in laboratory separation of minerals into floats, rel.d. less than 2·9, and sinks, greater than 2·9.

bromoil process (*ImageTech*) Printing process in which a bleached and tanned bromide print is brushed with oil pigment which adheres to the shadow portion and is repelled by the highlights.

bromoil transfer (*ImageTech*) Print made by transferring a bromoil print to another sheet of paper by passing the two in contact through a press.

bromothymol blue (*Chem*) An indicator used in acid–alkali titrations, having a pH range of 6·0–7·6, changing from yellow to blue.

bronch-, broncho- (*Genrl*) Prefixes from Gk *bronchos*, windpipe.

bronchi (*BioSci*) See BRONCHUS.

bronchia (*BioSci*) The branches of the bronchi. Adj *bronchial*.

bronchiectasis (*Med*) A chronic bronchopulmonary infection associated with an abnormal dilatation of the bronchial tree.

bronchiole (*BioSci*) One of the terminal subdivisions of the BRONCHIA.

bronchitis (*Med, Vet*) Inflammation of the bronchi. May be acute or chronic, the latter being a major cause of morbidity and mortality in the community. See also HUSK.

broncho- (*Genrl*) See BRONCH-.

bronchodilators (*Pharmacol*) Substances that cause relaxation of the smooth muscle of the wall of the bronchial airway, thus relieving breathlessness caused acutely by asthma or chronically by obstructive pulmonary disease. Administered by inhalation or intravenous injection. Examples include SYMPATHOMIMETIC DRUGS (eg *salbutamol*), ANTIMUSCARINIC DRUGS and XANTHINES (*theophylline*).

bronchogenic carcinoma (*Med*) Lung cancer arising from the epithelium of the bronchial tract.

bronchography (*Radiol*) The radiological examination of the trachea, bronchi or the bronchial tree after the introduction of a CONTRAST MEDIUM.

bronchoscopy (*Med*) Endoscopic examination of the tracheobronchial tree.

bronchus (*BioSci*) One of two branches into which the trachea divides in higher vertebrates and that lead to the lungs. Pl *bronchi*. Adj *bronchial*.

bronchus associated lymphoid tissue (*BioSci*) Subset of MUCOSAL ASSOCIATED LYMPHOID TISSUE found as lymphoid nodules in the lamina propria of the bronchus. Abbrev BALT.

Brönsted–Lowry theory (*Chem*) The theory defining as an acid every molecule or ion able to produce a proton and as a base every molecule or ion able to take up a proton. Thus acid ⇌ base+proton. The acid and the corresponding base are called *conjugates*.

Brönsted's relation (*Chem*) An expression for the catalytic activity k of acids and bases in terms of their dissociation constants:

$$k_{acid} = G_a K_a^\alpha$$

$$k_{base} = G_a K_a^\beta$$

where G_a or G_b is constant for a series of analogous catalysts of a given reaction in a given solvent and at a given temperature.

bronze (*Eng*) Primarily an alloy of copper and tin, but the name is now applied to other alloys not containing tin, eg aluminium bronze, manganese bronze and beryllium bronze. See COPPER ALLOYS.

bronzed diabetes (*Med*) See HAEMOCHROMATOSIS.

bronze powders (*Build*) Metallic powders made from alloys of copper and zinc or aluminium. Normally mixed with gold-size or bronze medium immediately prior to use in decorative work.

bronzing (*Build*) (1) Certain blue pigments, particularly of the PRUSSIAN BLUE and Monastral blue types, which exhibit a metallic lustre when ground at fairly high concentrations into paint media. (2) Application of imitation gold or other metals by mixing powders with gold-size or bronzing medium.

bronzing (*Print*) Dusting freshly printed sheets, by hand or machine, with any suitable metallic powder, bronze-coloured or otherwise.

bronzite (*Min*) A form of orthopyroxene, more iron-rich than enstatite and more magnesian than hypersthene; often has metallic sheen, due to the reflection of light from planes of minute metallic inclusions in the surface layers.

bronzitite (*Geol*) A rock composed of bronzite with smaller amounts of augite and calcic plagioclase. A common constituent of layered basic igneous intrusions, such as those of the Bushveld, S Africa and Stillwater, Montana.

brood (*BioSci*) A set of offspring produced at the same birth or from the same batch of eggs.

brookite (*Min*) One of the three naturally occurring forms of crystalline titanium dioxide.

Brouncker's series for ln 2 (*MathSci*) The series

$$\ln 2 = \frac{1}{1\cdot2} + \frac{1}{3\cdot4} + \frac{1}{5\cdot6} + \frac{1}{7\cdot8} + \dots$$

brouter (*ICT*) A NETWORK interconnection device with a BRIDGE which supports more than two LOCAL AREA NETWORK connections.

brow (*MinExt*) The top of the shaft or 'pit'; hence also *pit-brow*.

Brown agitator (*MinExt*) See AGITATOR.

brown algae (*BioSci*) See PHAEOPHYCEAE.

Brown and Sharpe wire gauge (*Eng*) A system of designating the diameter of wire by numbers; it ranges from 0000 (0·46 in) to 50 (0·001 in). Abbrev *B and S wire gauge*. Also *American Standard Wire Gauge*.

brown coal (*Min*) An intermediate between peat and true coals, with high moisture content, the calorific value in the range 4000–8300 Btu lb^{-1} (9·5–20 MJ kg^{-1}). Also *lignite*.

brown dwarf (*Astron*) A hypothetical very large planet, which is just below the critical mass (thought to be around 0·08 solar masses) needed to ignite a stellar nuclear reaction in its own interior.

brown earths (*EnvSci*) A range of brown soils, with weakly developed horizons, MULL humus and a pH of 5–7. They are formed under deciduous forests at humid, temperate latitudes. In UK, a common soil type with good potential for agriculture. Also *brown forest soils*. Cf BROWN PODZOLIC SOIL.

brown haematite (*Min*) A misnomer, the material bearing this name being LIMONITE, a hydrous iron oxide, whereas true haematite is anhydrous, Fe_2O_3. US *brown hematite*.

Brownian movement (*Phys*) Small movement of bodies such as particles in a colloid, due to statistical fluctuations in the bombardment by surrounding molecules of the dispersion medium. It may be detected by movement of a galvanometer coil. Also *Brownian motion, colloidal movement, pedesis*. See COLLOIDAL STATE.

brown nose disease (*Vet*) A form of photosensitization occurring in cattle, characterized by brown discoloration and irritation of the muzzle and teats. Also *copper nose*.

brown podzolic soil (*EnvSci*) An acid soil, usually formed from a brown forest soil in areas of high rainfall, with a pale layer from which elements and particles have been leached to a deeper zone where iron is often precipitated to form an impenetrable layer.

brown rot (*BioSci*) (1) Fungal diseases of plum and other fruit trees, infecting shoots and fruit. (2) The fungal decay of timber in which celluloses are preferentially attacked.

browser (*ICT*) A computer program that accesses the Internet and returns pages of HTML from a website to a client device, allowing the user to browse the information contained therein. Eg Firefox, Internet Explorer, Netscape, Safari, etc.

Brucellaceae (*BioSci*) A family of obligate parasites belonging to the order *Eubacteriales*. It comprises Gram-negative cocci or rods that may be aerobic or facultatively anaerobic. Many species are pathogenic, eg *Pasteurella pestis* (now known as *Yersinia pestis*, plague), *Pasteurella multocida* (fowl cholera, swine plague, haemorrhagic septicaemia), *Brucella abortus* (undulant fever in humans, contagious abortion in cattle, goats and pigs).

brucellosis (*Med, Vet*) See UNDULANT FEVER.

brucine (*Chem*) $C_{23}H_{26}O_4N_2$. A STRYCHNINE base alkaloid, mp of the anhydrous compound 178°C; it contains two methoxyl groups, and is a monoacidic tertiary base. Less physiologically active than strychnine.

brucite (*Min*) Magnesium hydroxide, occurring as fibrous masses in serpentinite and metamorphosed dolomite.

bruise (*Med*) Rupture of blood vessels in a tissue, with extravasation of blood, as a result of a blow which does not lacerate the tissue.

bruit (*Med*) A sound or murmur due to vascular blood flow, heard over heart, blood vessels and vascularized organs.

Brunt–Vaisala frequency (*EnvSci*) The frequency $N/2\pi$ of small vertical oscillations of a parcel of air about its equilibrium position in a stable atmosphere. N is given by

$$N^2 = \frac{g}{\theta}\frac{\partial\theta}{\partial z}$$

where g is the acceleration due to gravity and $\partial\theta/\partial z$ the vertical gradient of the potential temperature.

brush (*ElecEng*) A rubbing contact on a commutator, switch or relay. Also *wiper*.

brush border (*BioSci*) See MICROVILLUS.

brush-box (*ElecEng*) That portion of the brush-holder of an electrical machine in which the brush slides or in which it is clamped.

brush coating (*Paper*) A paper coating method in which the freshly applied wet coating is regulated and smoothed by means of brushes, some stationary and some oscillating, before drying.

brush contact (*ElecEng*) See LAMINATED CONTACT.

brush curve (*ElecEng*) The voltage drop between the brush arm and the segment beneath the brush at points along the brush width plotted against brush width as an indication of the correctness of compole flux density in a dc machine.

brush discharge (*ElecEng*) Discharge from a conductor when the potential difference between it and its surroundings exceeds a certain value but is not enough to cause a spark or an arc. It is usually accompanied by high hissing noise. Also *brushing discharge, corona*.

brush discharge (*MinExt*) Electrical discharge from points along a bar charged to between 18 000 and 80 000 V to create electrostatic field through which mineral particles fall and acquire polarity, in the high-intensity separation process. Also *brushing discharge*.

brushed fabric (*Textiles*) A fabric, usually woven, that has been brushed or plucked so that some of the fibres stand out from the constituent yarns. The process is also known as *raising* and is carried out by machines with rollers covered with wire hooks, emery paper or teasels.

brush gear (*ElecEng*) A general term used to denote all the equipment associated with brushes of a commutating or slip-ring machine.

brush-holder (*ElecEng*) The portion of an electrical machine or other piece of apparatus which holds a brush. See BRUSH-BOX.

brush-holder arm (*ElecEng*) The rod or arm supporting one or more brush-holders. Also *brush spindle*, *brush stud*.

brushing (*Vet*) An injury to the inside of a horse's leg caused by the shoe of the opposite foot. Also *cutting*.

brushing discharge (*ElecEng, MinExt*) See BRUSH DISCHARGE.

brush lead (*ElecEng*) See BRUSH SHIFT.

brushmarks (*Build*) A paintwork defect characterized by visible depressed lines in the direction in which the paint has been brushed on. They are due to insufficient flow or levelling of the liquid paint. Also *ropiness*.

brush plating (*Eng*) A method in which the anode carries a pad or brush containing concentrated electrolyte or gel which is worked over the surface to be plated. Similar methods are used for *brush polishing*.

brush-rocker (*ElecEng*) A support for the brushes of an electrical machine which enables them to be moved bodily round the commutator. Also *brush-rocker ring*.

brush shift (*ElecEng*) The amount by which the brushes of a commutating machine are moved from the centre of the neutral zone. Also *brush lead*. See BACKWARD SHIFT, FORWARD SHIFT.

brush spindle (*ElecEng*) See BRUSH-HOLDER ARM.

brush spring (*ElecEng*) A spring in a brush-holder which presses the brush against the commutator or slip-ring surface.

brush stud (*ElecEng*) See BRUSH-HOLDER ARM.

Brutalism (*Arch*) A term developed in UK in the 1950s to describe a style of architecture, popular at that time, characterized by expanses of unrelieved concrete and the juxtaposition of massive forms.

brute (*ImageTech*) A very large high-intensity spot light.

bry-, bryo- (*Genrl*) Prefixes from Gk *bryon*, moss.

bryology (*Genrl*) The study of mosses. Also *muscology*.

Bryophyta (*BioSci*) (1) A division of the plant kingdom containing c.25 000 spp of small, rootless, thalloid or leafy, non-vascular plants. It includes the liverworts (Hepaticopsida), the hornworts (Anthoceropsida) and the mosses (Bryopsida). There is alternation of generations in which the gametophyte is the dominant generation, the sex organs are archegonia and antheridia and the sporophyte is more or less parasitic on the gametophyte. (2) In some confusing usages, the mosses alone. See BRYOPSIDA.

bryophyte (*BioSci*) See BRYOPHYTA.

Bryopsida (*BioSci*) The mosses. A class of those Bryophyta, c.15 000 spp, that have a leafy (not thalloid) gametophyte with the leaves not strictly in two or three ranks, multicellular rhizoids and, in most, a capsule (sporophyte) with a columella and a lid (operculum). Includes *Sphagnum*. Also *Musci*. See ACROCARP, PLEUROCARP.

bryostatin (*BioSci*) A complex lactone, extracted from certain marine animals (Bryozoa), that modulates protein kinase C activity and is used in treatment of certain tumours.

Bryozoa (*BioSci*) A phylum of invertebrate animals, the 'moss animals', that are mainly marine. They usually live in colonies (zoaria) and are important as fossils from the Ordovician to Recent, but they are rarely found as complete specimens. Also *Polyzoa*. See ECTOPROCTA.

BS (*Genrl*) Standard developed and published by BRITISH STANDARDS INSTITUTION, eg BS 5750.

BSB thread (*Eng*) Abbrev for BRITISH STANDARD BRASS THREAD.

BSE (*Vet*) Abbrev for BOVINE SPONGIFORM ENCEPHALOPATHY. See panel on TRANSMISSIBLE SPONGIFORM ENCEPHALOPATHY.

BSF thread (*Eng*) Abbrev for BRITISH STANDARD FINE THREAD.

BSI (*Genrl*) Abbrev for BRITISH STANDARDS INSTITUTION.

BSP thread (*Eng*) Abbrev for BRITISH STANDARD PIPE THREAD.

BSS (*Genrl*) Abbrev for BRITISH STANDARD SPECIFICATION.

B-stage (*Plastics*) Transition stage through which a thermosetting synthetic resin of the phenol formaldehyde type passes during the curing process, characterized by softening to rubber-like consistency when heated, and insolubility in ethanol or acetone (propanone). See panel on THERMOSETS.

BSW thread (*Eng*) Abbrev for BRITISH STANDARD WHITWORTH THREAD.

BTU (*Phys*) Abbrev for *Board of Trade unit*. Equal to 1 kW h.

Btu (*Phys*) Abbrev for BRITISH THERMAL UNIT.

Bu (*Chem*) Symbol for the butyl radical, C_4H_8.

bubble (*Surv*) The bubble of air and spirit vapour within a LEVEL TUBE: loosely, the level tube itself.

bubble chamber (*Phys*) A device for making visible the paths of charged particles moving through a liquid. The liquid, often liquid hydrogen, heated above its normal boiling point, becomes superheated if the applied pressure is suddenly released. Bubbles of vapour are formed on the ions produced by the passage of charged particles through the chamber. The tracks of the particles are revealed by the bubbles, which, if suitably illuminated, can be photographed. See CLOUD CHAMBER.

bubble film (*Eng*) Duplex polymer film with regular array of bubbles thermoformed into one side, used for crush-proof packaging.

bubble-jet printer (*ICT*) An inkjet printer, in which the heated ink is projected onto the paper, usually piezo-electrically. Resolutions up to 1440 dots per inch can now be achieved by a cheap colour printer.

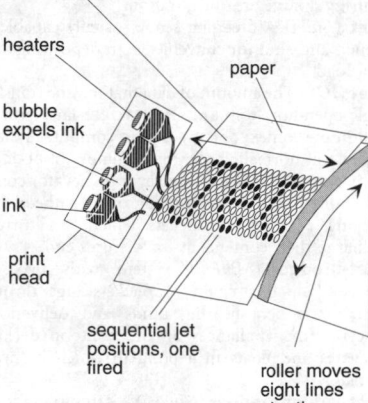

print head moves across paper heating selected heads at each position

heaters

paper

bubble expels ink

ink

print head

sequential jet positions, one fired

roller moves eight lines at a time

bubble-jet printer Schematic of thermal head, paper and roller only shown. Only four of eight or more heads shown.

bubble pack (*Eng*) See BLISTER PACK.

bubble point (*Chem*) The temperature at which the first bubble appears on heating a mixture of liquids. See DEW POINT.

bubbles, pressure in (*Phys*) See PRESSURE IN BUBBLES.

bubble stabilization (*Phys*) The tendency of polymer bubble blown after extrusion in manufacture of film to maintain a constant shape without breaking or collapsing. Depends on tension-stiffening behaviour of molten polymer. See TROUTONIAN FLUID.

bubble store (*ICT*) See MAGNETIC BUBBLE MEMORY.

bubble trier (*Surv*) See LEVEL TRIER.

bubble tube (*Surv*) See LEVEL TUBE.

bubo (*Med*) An inflamed and swollen lymphatic gland, esp in the groin.

bubonic plague (*Med*) A form of plague in which there is great swelling of a lymphatic gland, esp those in the groin. See PLAGUE.

buccal (*BioSci, Med*) Pertaining to, or situated in or on, the cheek or the mouth.

buccal cavity (*BioSci*) The cavity within the mouth opening but prior to the commencement of the pharynx.

buccal glands (*BioSci*) Glands opening into the BUCCAL CAVITY in terrestrial Craniata; the most important are the salivary glands.

buccal respiration (*BioSci*) See BUCCOPHARYNGEAL RESPIRATION.

buccinator (*Med*) A broad, thin muscle at the side of the face, between the upper and lower jaw.

bucco- (*Genrl*) Prefix from Lt *bucca*, cheek.

buccopharyngeal respiration (*BioSci*) Breathing by means of the moist vascular lining of the mouth cavity or diverticula thereof, as in some amphibians and certain fish which have become adapted to existence on land. Also *buccal respiration*.

Buchholz relay (*ElecEng*) A protective relay for use with transformers or other oil-immersed apparatus; it embodies a float which becomes displaced and operates the relay contacts if gas bubbles are generated by a fault within the equipment being protected.

buchite (*Geol*) A glassy rock which represents the result of partial fusion and recrystallization at very high temperatures of clay and shale material. It often occurs as xenoliths within igneous rocks.

Buchmann–Meyer effect (*Acous*) The special type of reflection of light from the sound track on a disk record whereby the lateral velocity of the track can be determined.

Buchner funnel (*Chem*) A stout porcelain funnel having at its base a fixed horizontal perforated plate to act as a support over which a piece of filter paper is placed, thus ensuring a large area of filtration.

bucket (*Build*) A dredging scoop, usually capable of being opened and shut for convenience in depositing and taking up a load.

bucket (*ICT*) The amount of data that may be transferred in a single operation from BACKING STORE. This term is usually used in the context of MAINFRAME computer systems and a bucket will normally be a whole number of BLOCKS.

bucket conveyor (*Eng*) A conveyor or elevator consisting of a pair of endless chains running over toothed wheels, and carrying a series of buckets which, on turning over, discharge their contents at the delivery end.

bucket-dredge (*MinExt*) System with two pontoons between which a chain of buckets digs through alluvium to mineral-bearing sands and delivers these to concentrating appliances on the pontoon decks. Dredge excavates and floats in a pond in which it traverses the deposit.

bucket-ladder dredger (*CivEng*) A BUCKET-DREDGE reaching down into the material to be dredged, and lifting it for discharge into the vessel itself or into an attendant vessel.

bucket-ladder excavator (*CivEng*) A mechanical excavator working on the same principle as a BUCKET-LADDER DREDGER but adapted for use on land. Also *dredger excavator*.

bucket valve (*Eng*) A non-return (delivery) valve fitted in the bucket or piston of some types of reciprocating pump.

bucking coil (*ElecEng*) A winding on an electromagnet to oppose the magnetic field of the main winding. Such a device is sometimes used in electromagnetic loudspeakers to smooth out voltage pulsations in the power supply. Also *hum-bucking coil*.

buckle (*Eng*) (1) To twist or bend out of shape; said usually of plates or of the deformation of a structural member under compressive load. (2) A metal strap. (3) In foundrywork, a swelling on the surface of a sand mould due to steam generated below the surface.

buckle fold (*Print*) A fold made in the paper parallel to its leading edge or to the fold previously made in it; the paper is brought to a sudden stop causing it to buckle into, and be folded by, the rollers. Cf KNIFE FOLD.

Buckley gauge (*Electronics*) A sensitive ionization gauge for measuring very low gas pressures.

buckling (*ElecEng*) A distortion of accumulator plates caused by uneven expansion, usually as a result of heavy discharges or other maltreatment.

buckling (*Eng*) Mode of deformation in which an elastic instability occurs in a plate or a structural member under compressive load, resulting in a twisting or bending out of shape. Usually leads to plastic deformation and eventual collapse. See EULER BUCKLING LIMIT.

buckling (*ImageTech*) Pile-up of film in a camera or projector mechanism as the result of a break or incorrect threading.

buckling (*NucEng*) A term in reactor diffusion theory giving a measure of curvature of the neutron density distribution. The *geometric buckling* depends only on the shape and dimensions of the assembly while the *material buckling* provides a measure of the multiplying properties of an assembly as a function of the materials and their disposition.

buckminsterfullerene (*Chem*) Allotrope of carbon, having 60 carbon atoms linked together in hexagons and pentagon rings to form a closed, near spherical and very stable structure. The first discovered fullerene. Soluble in benzene, forming deep-red solution. Named after eponymous architect of GEODESIC DOMES. Colloq *buckyballs*.

buckram (*Textiles*) A strong linen or cotton fabric stiffened by starch, gum or latex and used for linings, hats and in bookbinding.

buck saw (*Build*) A large frame saw having one bar of the frame extended to form a handle.

buck transformer (*ElecEng*) A transformer with secondary in mains circuit to regulate voltage according to a controlling circuit feeding the primary. Also *boost transformer*.

buckwheat (*Agri*) A lesser cereal, *Fagopyrum esculentum*, a source of grain (groats) for human and animal consumption.

buckwheat rash (*Vet*) See FAGOPYRISM.

buckyballs (*Chem*) See BUCKMINSTERFULLERENE.

buckytube (*Chem*) See NANOTUBE.

bud (*BioSci*) (1) An unexpanded shoot consisting of a short rudimentary stem bearing immature and primordial leaves and/or flowers. At least in extant spermatophytes, buds are expected at shoot tips (terminal or apical buds) and in leaf axils (axillary buds); other buds are accessory or adventitious. (2) An outgrowth of a parent organism that becomes detached and develops into a new individual, esp from some yeasts and other fungi. See BUDDING.

budding (*BioSci*) (1) A primitive method of asexual reproduction by growth and specialization and separation by constriction of a part of the parent. (2) Asexual reproduction esp by some yeasts. (3) Bud grafting used esp for the propagation of fruit trees and some woody ornamentals including roses, in which a bud, together with more or less of the underlying stem, from the scion variety is grafted into a suitable root stock.

buddle (*MinExt*) A shallow annular pit for concentrating finely crushed, slimed, base-metal ores.

buddle-work (*MinExt*) The treatment of finely ground tin-bearing sands by gentle sluicing, in which a heavier fraction of the fed pulp is built up (*buddled*) while the lighter fraction flows to discard. This is continued until a satisfactory concentrate is produced.

budesonide (*Pharmacol*) A synthetic *corticosteroid* with strong glucocorticoid and weak mineralocorticoid activity, used for the treatment of Crohn's disease.

bud scale (*BioSci*) A simplified leaf or stipule on the outside of a bud, forming part of a covering which protects the contents of the bud.

bud sport (*BioSci*) A shoot, branch, inflorescence or flower differing markedly from the rest of the plant with the differences persisting in vegetatively propagated offspring; due to nuclear or cytoplasmic mutation when the sport will often be chimeric, or in horticulture to a change in the structure of a pre-existing chimera.

buff (*Eng*) A revolving disk composed of layers of cloth charged with abrasive powder; used for polishing metals.

buffed crumb (*Eng*) Flakes of rubber produced by abrading treads of worn tyres for retreading; of limited use as recycled material for new tyre compounds.

buffer (*ElecEng*) (1) An electronic amplifier, often with unity gain, which is designed to decouple input from output. Normally designed to have high input impedance so that it does not load the driving stage and low output impedance such that it can provide current drive. (2) See BUFFER REAGENT.

buffer (*Eng*) A spring-loaded pad attached to the framework of railway rolling-stock to minimize the shock of collision; any resilient pad used for a resilient purpose. May be hydraulically controlled or dampened.

buffer (*FoodSci*) An ionic compound, usually the sodium, potassium or calcium salt of a fatty acid, which is added to foods to hinder significant changes in pH. Classified as acidity regulators.

buffer (*ICT*) An area of memory used temporarily to hold data being transmitted between a peripheral device and the central processor, to allow for differences in their working speeds. Buffering can also be used between two peripheral devices. See DOUBLE BUFFERING.

buffer action (*Chem*) The action of certain solutions in opposing a change of composition, esp of hydrogen ion concentration. See BUFFER SOLUTION.

buffer battery (*ElecEng*) A battery of accumulators arranged in parallel with a dc generator to equalize the load on the generator by supplying current at heavy-load periods and taking a charge during light-load periods.

buffer capacitor (*ElecEng*) See BLOCKING CAPACITOR.

buffer capacity of developer (*ImageTech*) The capacity of an alkali (eg sodium metaborate) in a developing solution to maintain a slow rate of decrease in pH value.

buffer circuit (*Acous*) The resistance–capacitor unit which determines the rate of rise or fall of the envelope of the waveform of emitted sounds which has been generated in electrostatic circuits in electronic organs.

buffer coat (*Build*) A coating applied to a surface to isolate it from a subsequent coating.

buffer reagent (*ElecEng*) A substance added to an electrolyte solution which prevents rapid changes in the concentration of a given ion. Also *buffer*.

buffer resistance (*ElecEng*) See DISCHARGE RESISTANCE.

buffer solution (*Chem*) A solution whose pH value is not appreciably changed by additions of acid or alkali. Normally it is the solution of a weak acid or base with one of its salts.

buffer spring (*Eng*) The part lending resiliency to railway buffing gear. See BUFFER.

buffer stage (*ICT*) An amplifying stage coming between the master oscillator and the modulating stage of a radio transmitter, to prevent the changing load of the modulated output from affecting the frequency of the master drive.

buffer tank (*NucEng*) A closed tank that cushions the explosive expulsion of liquid from a system connected to it by controlling the gas pressure in the tank.

buffet boundary (*Aero*) The limiting values of MACH NUMBER and altitude at which an aircraft can be flown without experiencing buffet in unaccelerated flight.

buffeting (*Aero*) An irregular oscillation of any part of an aircraft, caused and maintained by an eddying wake from some other part; commonly, tail buffeting in the downwash of the main plane, which gives warning of the approach of the STALL.

buffing (*Build*) The grinding down of a surface to remove extrusions or to expose the underlying material.

bug (*ICT*) A fault or imperfection in coding that results in a program failing or behaving in a way unintended by the author. (Allegedly originating from the discovery of an insect short-circuiting a wire connection in the databank of an early computer.) Also *glitch*.

bug-eye lens (*ImageTech*) See FISH-EYE LENS.

bug key (*ICT*) A Morse key that permits higher transmission speeds than a normal key. A lever, moved horizontally by the hand, sends dashes in one direction when held over. Dots are sent by a spring contact attached to the lever, when the lever is released from sending a dash.

buhl saw (*Build*) A kind of frame saw in which the back of the frame is so spaced from the saw itself as to allow the latter to cut well into the work.

buhr mill (*MinExt*) Mill in which material is ground by passage between a fixed and a rubbing surface. Types include old-fashioned flour mill, with a circular grindstone rotating above a fixed lower one, radially grooved to facilitate passage of grist from centre to peripheral discharge. Also, rotating cone in fixed casing, material gravitating through the intervening space. Used for softish material, eg grain, food processing, and such minerals as soft limestone. Also *burr mill*.

builders' level (*Build*) (1) A spirit-level tube set in a long straightedge; for testing and adjusting levels. (2) A simple form of dumpy or tilting level, used on building works or for running the levels of drains.

building board (*Build*) Board manufactured from various materials and supplied with various finishes; used for lining walls and ceilings. See FIBRE-BOARD.

building certificates (*Build*) Certificates made out by the architect during the progress, or after completion, of the works on a building contract, to enable the contractors to obtain payments on account or in settlement from the employer.

building line (*Build*) The line beyond which a building may not be erected on any given plot.

building paper (*Build*) Sandwich of fibre and bitumen between two sheets of heavy paper, used in damp-proofing and for insulation between the soil and road surfacing.

build-up (*Radiol*) Increased radiation intensity in an absorber over what would be expected on a simple exponential absorption model. It results from scattering in the surface layers and increases with increasing width of the radiation beam.

build-up sequence (*Eng*) The order in which successive welding runs or beads are applied in joining thick plates to achieve maximum strength with acceptable stress from heat distortion.

build-up time (*ICT*) See RISE TIME.

built-in voltage (*Electronics*) The potential difference which arises across an unbiased DEPLETION LAYER in a semiconductor device. The drift of charges in the electric field associated with it is exactly balanced by diffusion in the local charge concentration gradients. In effect the contact potential between regions of extrinsic semiconductor where there are abrupt changes in the doping.

bulb (*BioSci*) (1) In plants, an organ of storage and perennation, usually underground, consisting of a short stem bearing a number of overlapping swollen fleshy leaf bases and/or scale leaves, with or without a tunic, the whole enclosing next year's bud, eg onion. Cf CORM, RHIZOME. (2) Generally, any bulb-shaped structure. Adj *bulbar*.

bulb (*ElecEng*) The gas-tight envelope, usually glass, which encloses a vacuum or gas-filled device, eg a photomultiplier or filament lamp.

bulb bar (*Eng*) A rolled, or extruded, bar of strip form in which the section is thickened along one edge.

bulbiferous (*BioSci*) Having, on the stem, bulbs or bulbils in place of ordinary buds.

bulbil (*BioSci*) (1) Any small bulb-like structure. (2) A small bulb or tuber developing above ground from an axillary bud or in an inflorescence, and functioning in asexual reproduction. (3) A contractile dilatation of an artery.

bulbus (*BioSci*) See BULB, BULBUS ARTERIOSUS.

bulbus arteriosus (*BioSci*) In many vertebrates, a strongly muscular region following the CONUS ARTERIOSUS.

bulbus oculi (*BioSci*) The eyeball of vertebrates.

bulge cylinder (*ImageTech*) Head drum on which bulges (normally two) have been added to minimize TAPE SLAP. Also *bulge drum*.

bulimia (*Med*) An abnormal increase in the appetite, often part of the symptoms of *bulimia nervosa*, a pathological eating disorder in which binge eating is followed by depression and guilt, self-induced vomiting and purging.

bulk (*Paper*) A measure of the reciprocal of the density of paper, being the ratio of thickness to substance. A loose synonym for thickness.

bulk concrete (*CivEng*) See MASS CONCRETE.

bulk density (*PowderTech*) The value of the apparent powder density when measured under stated freely poured conditions.

bulked yarn (*Textiles*) See TEXTURED YARN.

bulk flotation (*MinExt*) Froth flotation process applied to concentrate more than one valuable mineral in one operation.

bulkhead (*Aero*) In fuselages, a major structural transverse dividing wall providing access between several internal sections, or a strengthened and sealed wall at the front and rear designed to withstand the differential pressure required for pressurization. In power plant nacelles, a structure forming a firewall.

bulkhead (*Autos*) On a public service vehicle, partition at the front between driver and passenger accommodation. The partition between engine and passenger compartments in any vehicle.

bulkhead (*CivEng*) A masonry or timber partition to retain earth, as in a tunnel or along a waterfront.

bulkhead (*Ships*) A partition within a ship's hull or superstructure. It may be transverse or longitudinal, watertight, oiltight, gastight or partially open. It may form part of the ship's subdivision for seaworthiness or otherwise.

bulkhead deck (*Ships*) The uppermost deck up to which watertight transverse bulkheads are carried.

bulkiness (*PowderTech*) A term used to describe the properties of a bed of powder. It is defined as the reciprocal of the apparent density of the powder under the stated conditions.

bulking (*Build*) An expression used to describe the increase in volume of damp sand compared with dry or saturated sand. See MOISTURE EXPANSION.

bulking factor (*Eng*) The ratio between the volume of loosely placed material and the same weight of material when compacted to a given specification.

bulk modulus (*Eng*) One of the four basic elastic constants for elastically isotropic materials, defined as the ratio of the applied, uniform triaxial stress (eg hydrostatic stress) to volumetric strain in a body. Symbol K. Related to YOUNG'S MODULUS (E), POISSON'S RATIO (v) and SHEAR MODULUS (G) by:

$$K = \frac{E}{3(1 - 2v)} = \frac{EG}{3(3G - E)}$$

bulk sample (*MinExt*) Sample composed of several portions taken from different locations within a bulk quantity of the material under test.

bulk test (*NucEng*) Test for materials having a high attenuation for use as a radiation shield.

bulla (*Med, BioSci*) (1) A blister or bleb. A circumscribed elevation above the skin, containing clear fluid; larger than a vesicle. (2) In vertebrates, with a flask-shaped tympanum, the spherical part of that bone which usually forms a protrusion from the surface of the skull.

bullate (*BioSci*) (1) Having a blistered or puckered surface. (2) Bubble-like. (3) Bearing one or more small hemispherical outgrowths.

bull beef (*Agri*) Meat from entire male cattle rather than bullocks.

bull-chain (*For*) The endless chain used in a log haul for conveying logs.

bulldog calf (*Vet*) A lethal form of achondroplasia (dwarfism) occurring mainly when Dexter cattle are mated

together; due to the inheritance of a pair of semi-dominant genes responsible for the achondroplasia trait.

bulldozer (*CivEng*) A power-operated machine, provided with a blade for spreading and levelling material. See ANGLEDOZER.

bulletin board system (*ICT*) A DATABASE system that is available to the general public who use their own computers to access the HOST computer via a telephone line. The system may offer electronic mail or message facilities. The systems are often small and operated by computer enthusiasts from home as a hobby. Abbrev BBS.

bulletwood (*For*) Timber from a tree of the genus *Mimusops*, which also yields *balata*, a soft rubber-like material used in golfballs and for impregnating belts. It is renowned for its strength and durability.

bull-headed rail (*CivEng*) A rail section once used widely in the UK, having the shape roughly of a short dumb-bell in outline, but with unequal heads, the larger being the upper part in use. See FLANGED RAIL.

bull header (*Build*) A brick with one corner rounded, laid with the short face exposed, as a quoin or for sills etc.

bullhead tee (*Build*) A tee having a branch which is longer than the run.

bull-holder (*Vet*) Forceps for grasping the nasal septum of cattle as a means of restraint.

bulliform cell (*BioSci*) An enlarged epidermal cell present, with other similar cells, in longitudinal rows in the leaves of some grasses and alleged to be motor cells causing the rolling and unrolling of leaves in response to changes of water status.

bullion (*Eng*) (1) Gold or silver in bulk, ie as produced at the refineries, not in the form of coin. (2) The gold–silver alloy produced before the metals are separated.

bullion content (*Eng*) In parcel of metal or minerals being sold, where the main value is that of the base metal which forms the bulk of the parcel, the contained gold or other precious metal of minor value included in the sale.

bullion point (*Glass*) The centre piece of a sheet of glass made by the old method of spinning a hot glass vessel in a furnace until it opened out under centrifugal action to a circular sheet. The centre piece bears the mark of attachment to the rod used to spin the sheet. The method is obsolete now, but is revived for 'antique' effects. Also *bull's eye*.

bull-nose (*Build*) A small metal rebating plane having the mouth for the cutting iron near the front.

bull-nosed step (*Build*) A step which, in plan, is half-round or quarter-round at the end.

bullock (*Agri*) A castrated male bovine destined for meat production. In more general use, also young, entire bull. Also *steer*.

bull-ring (*ElecEng*) A metal ring used in the construction of overhead contact-wire systems for electric schemes; it forms the junction of three or more straining wires.

bull's eye (*Glass*) See BULLION POINT.

bull's eye lens (*ImageTech*) A small thick lens, used for condensing light from a source.

bull stretcher (*Build*) A brick with one corner rounded, laid, with the long face exposed, eg as a QUOIN.

bull wheel (*MinExt*) The driving pulley for the camshaft of a stamp battery; one on which bull rope of drilling rig is wound.

bulwark (*CivEng*) A sea wall built to withstand the force of the waves; in some cases the reinforcement of the natural BREAKWATER.

bumblefoot (*Vet*) A cellulitis of the foot of birds due to infection by *Staphylococcus aureus*.

bumetanide (*Pharmacol*) A LOOP DIURETIC used in the treatment of OEDEMA.

bump (*Aero*) See AIR POCKET.

bumping-up (*Print*) Interlaying the half-tone areas of lithographic rotary printing plates to provide heavier impression, there being no provision for MAKING-READY on the rubber-covered impression cylinder.

Buna (*Chem*) Former name for family of polybutadienes. See panel on POLYMERS.

buncher (*Electronics*) Arrangement which velocity-modulates and thereby forms bunches of electrons in the electron beam current passing through it. Bunching would be *ideal* if the bunches contained electrons all having the same beam velocity. Also *buncher gap*, *input gap*. See CATCHER, DEBUNCHING, RHUMBATRON.

bunching (*Electronics*) The process of forming a steady electron beam into a succession of electron groups, or bunches. The result of interaction between an alternating electric field at the mouth of a cavity (see RHUMBATRON) and an electron beam passing close by. See VELOCITY MODULATION.

bunching angle (*Electronics*) Transit delay or phase angle between modulation and extraction of energy in a bunched beam of electrons.

bundle (*BioSci*) (1) Fibres collected into a band in the nervous system or in the heart. (2) See VASCULAR BUNDLE.

bundle (*NucEng*) See FUEL ASSEMBLY.

bundle cap (*BioSci*) A strand of sclerenchyma or parenchyma adjacent to the xylem and/or phloem sides of a vascular bundle.

bundle conductor (*ElecEng*) Two or more overhead line conductors, suitably spaced to avoid BRUSH DISCHARGE loss, forming a phase; replaces a single large conductor.

bundle divertor (*NucEng*) See DIVERTOR.

bundle end (*BioSci*) The much simplified termination of a small vascular bundle in the mesophyll of a leaf.

bundle of His (*Med*) In the mammalian heart, a bundle of small specialized conducting muscle fibres extending from the wall of the right atrium to the septum between the ventricles; responsible for transmitting electrical impulses from atrium to ventricle.

bundle sheath (*BioSci*) A sheath of one or more layers of parenchymatous or sclerenchymatous cells, surrounding a vascular bundle in a leaf.

bundling (*ICT*) The practice of grouping programs or software together for ease of sale, distribution, download or installation.

bungalow (*Arch*) A single-storey house.

bungarotoxin (*BioSci*) A toxic polypeptide from the venom of the snake *Bungarus* (krait) that is used experimentally as an antagonist of acetylcholine. It blocks the NICOTINIC ACETYLCHOLINE RECEPTOR at the vertebrate neuromuscular junction.

bunion (*Med*) An enlarged deformed joint of the big toe where it joins the foot, as a result of pressure of tight-fitting shoes or boots, with overlying BURSITIS.

bunker (*Eng*) A storage space for coal or oil fuel.

bunker buster (*Genrl*) A thermobaric bomb designed to penetrate thick layers of rock.

bunker capacity (*Ships*) The capacity of a space in a ship used for carrying fuel. It is calculated at a fixed rate of stowage per unit volume, according to fuel, and allowances for obstructions are made in percentage.

bunodont (*BioSci*) Said of mammalian teeth in which the cusps remain separate and rounded. Also *bunoid*. Cf LOPHODONT, SELENODONT.

Bunsen burner (*Chem*) A gas burner consisting of a tube with a small gas jet at the lower end, and an adjustable air inlet by means of which the heat of the flame can be controlled; used as a source of heat for laboratory work and formerly, in conjunction with an incandescent mantle, as the usual form of gas burner for illuminating purposes.

Bunsen flame (*Chem*) The flame produced when a mixture of a hydrocarbon gas and air is ignited in air, as in a Bunsen burner. It consists of an inner cone, in which carbon monoxide is formed, and an outer one, in which it is burnt.

Bunsen photometer (*Phys*) See GREASE-SPOT PHOTOMETER.

bunt (*Aero*) A manoeuvre in which an aircraft performs half an inverted loop, ie the pilot is on the outside where he or she experiences NEGATIVE G.

bunt (*BioSci*) A disease of cereals caused by a SMUT fungus in which the grain of infected plants is transformed into a mass of spores.

buoyancy (*Phys*) The apparent loss in weight of a body when wholly or partly immersed in a fluid; due to the upthrust exerted by the fluid. See ARCHIMEDES' PRINCIPLE, CORRECTION FOR BUOYANCY, RESERVE BUOYANCY.

buoy-to-buoy (*Aero*) See BLOCK TIME.

bupivacaine (*Pharmacol*) A powerful local anaesthetic used for regional nerve block anaesthesia, particularly epidural.

buproprion hydrochloride (*Pharmacol*) An aminoketone *antidepressant*, unrelated to tricyclics and SSRIS. It is used as an aid to the cessation of smoking.

bur (*For*) See BURR.

burden (*ElecEng*) The load on an instrument transformer. It is usually expressed as the normal rated load in volt-amperes, or as the impedance of the circuit fed by the secondary winding.

burden (*Eng*) See ON-COSTS.

burden (*MinExt*) (1) Amount of rock to be shattered in blasting between drill-hole and nearest free face. (2) See OVERBURDEN.

Burdigalian (*Geol*) A stage in the Miocene. See TERTIARY.

Burdizzo pincers (*Vet*) A castrating instrument which crushes the spermatic cord.

burdo (*BioSci*) A graft hybrid presumed to have arisen from the union of vegetative nuclei derived from the stock and scion.

burette (*Chem*) A vertical glass tube with a fine tap at the bottom and open at the top, usually holding up to 50 cm^3 of reagent solution; used in VOLUMETRIC ANALYSIS. The tube is usually graduated in tenths of a cubic centimetre, so that the amount of liquid allowed to run out through the tap can be estimated to $\frac{1}{20}$ cm^3.

Burgers vector (*Crystal*) A crystal vector which denotes the amount and direction of atomic displacement which will occur within a crystal when a DISLOCATION moves.

Burgundy mixture (*Agri*) A pesticidal solution of copper sulphate and sodium carbonate, developed in the 1800s against mildew, blight and other fungal diseases. Widely used in organic production.

burial site (*NucEng*) Place for the deposition, usually in suitable containers, of radioisotopes after use, contaminated material or radioactive products of the operation of nuclear reactors. Also *graveyard*.

buried layer (*Electronics*) A high-conductivity layer diffused into active regions of a semiconductor wafer before growth of the EPITAXIAL layer in which devices are defined. It is used to decrease the collector resistance of certain bipolar junction transistors.

Burkitt lymphoma (*Med*) A malignant tumour of B-lymphocytes, esp affecting the jaw and the gut, common in children in hot humid regions of Africa but not confined to these regions. EPSTEIN–BARR VIRUS is present and may be responsible for malignant transformation occurring in a B-cell population subject to constant antigenic stimulation. Associated with a specific chromosomal rearrangement affecting chromosome 8q24.

burl (*For*) See BURR.

burlap (*Textiles*) A coarse jute, hemp or flax fabric.

Burma lancewood (*For*) A durable wood from the genus *Homalium*.

burmite (*Min*) An amber-like mineral occurring in the upper Hukong Valley, Myanmar (Burma), differing from ordinary amber by containing no succinic acid. A variety of retinite.

burn (*Electronics*) See ION BURN.

burn (*Space*) Controlled firing of rocket engine for adjusting course and re-entry initiation.

burnable poison (*NucEng*) A neutron absorber introduced into a reactor system to reduce initial reactivity but becoming progressively less effective as burn-up proceeds. This helps to counteract the fall in reactivity as the fuel is used up. Boron-10, which is transmuted into helium by neutron capture, has been used in the form of borosilicate glass placed in empty control-rod guides.

burner firing block (*Eng*) Unit made from refractory material that fits into a furnace wall at the burner position, having a nozzle-protecting recess at back and a tunnel on the firing side. It is called *quarl* in oil-firing practice.

burner loading (*Eng*) Potential heat that can be liberated efficiently from a burner. Expressed in kilowatts or Btu h^{-1}.

burner turndown factor (*Eng*) Minimum gas rate at which a burner is capable of stable flame propagation without the flame flashing back to the air–gas mixing point or blowing off from the burner nozzle or head.

burning (*Eng*) The heating of an alloy to too high a temperature, causing local fusion or excessive penetration of oxide, and rendering the alloy weak and brittle.

burning (*MinExt*) Changing the colour of certain precious stones by exposing them to heat.

burning-in kiln (*Glass*) A kiln in which stain or enamel colour painted on glassware or sheet glass is fired to cause it to adhere more or less permanently; usually of muffle type.

burnishing (*Print*) The operation of applying a brilliant finish to gilt or coloured edges by means of a burnishing tool, which is applied under great pressure.

burn mark (*Eng*) Moulding defect found on polymer surfaces caused by adiabatic compression of gas trapped in mould cavity by advancing melt front.

burnout (*Electronics*) Sudden failure of any device, caused by excessive current, leading in turn to overheating; may also be due to failure of artificial cooling in any electronic assembly or sub-assembly.

burnout (*Psych*) Changes in thoughts, emotions and behaviour as a result of extended work-related stress. Burnout is associated with extreme dissatisfaction, pessimism, lowered job satisfaction, and a desire to quit.

burnout mask (*Print*) See PRINTOUT MASK.

burnout velocity (*Space*) The maximum velocity achieved by a rocket when all the propellant has been consumed.

burnt coal (*Min*) Sooty product of weathering of a coal outcrop.

burnt lime (*Build, Chem*) See LIME.

burnt metal (*Eng*) Metal which has become oxidized by overheating, and so is rendered useless for engineering purposes.

burn-up (*NucEng*) (1) In nuclear fuel, amount of fissile material burned up as a percentage of total fissile material originally present. (2) Of fuel element performance, the amount of heat released from a given amount of fuel, expressed as megawatt- or gigawatt-days per tonne.

burr (*BioSci*) A fruit covered with hooks to aid in dispersal by animals.

burr (*Eng*) (1) The rough edge or ridge on a material resulting from various operations like punching and cutting. (2) A rotary tool with cutting teeth like a file.

burr (*For*) A knob or knot in a tree which, when sliced, produces strong contrasts in the form and colour of the markings which are prized for their decorative effect in edge veneers. Also *bur, burl*.

burr mill (*MinExt*) See BUHR MILL.

burrs (*Build*) Lumps of brick, often mis-shapen, which have fused together in burning.

bursa (*BioSci*) Any sac-like cavity; particularly, in vertebrates, a sac of connective tissue containing a viscid, lubricating fluid, and interposed at points of friction between skin and bone and between muscle, ligament and bone.

bursa copulatrix (*BioSci*) A special genital pouch of various animals acting generally as a female copulatory organ.

bursa inguinalis (*BioSci*) The cavity of the scrotal sac in mammals.

bursa of Fabricius (*BioSci*) A sac-like structure arising as a diverticulum from the cloaca of young birds, composed of primary follicles containing B-lymphocyte precursors. The

bursa is the only source of these cells in birds and removal of the bursa at hatching (or by certain viral infections) results in a severe B-cell deficiency.

bursa omentalis (*BioSci*) In mammals, a sac formed by the EPIPLOON or great omentum.

bursattee (*Vet*) *Cutaneous habronemiasis*. A disease of the skin of horses caused by nematode larvae of the genus *Habronema*; characterized by granulomatous nodules in the skin. Also *bursati*.

bursicon (*BioSci*) In insects, a hormone produced by neurosecretory cells of the brain and released by neurohaemal organs in the thoracic and abdominal ganglia. It affects many post-ecdysal processes such as cuticular tanning.

bursiform (*BioSci*) Resembling a bag or pouch.

bursitis (*Med*) An inflammation of a bursa.

burst (*ICT*) (1) Short period of intense activity on an otherwise quiet data channel. (2) Sudden increase in strength of received radio signals caused by sudden changes in the ionosphere.

burst (*ImageTech*) See COLOUR BURST.

burst (*NucEng*) A defect, often very small, in fuel cladding or sheathing which allows fission products to escape.

burst (*Phys*) Unusually large pulse arising in an ionization chamber caused by a cosmic-ray shower.

burst binding (*Print*) Unsewn binding where the spine of the section is 'burst', or slit, at intervals, during the folding or web printing operation to allow adhesive to reach all the pages without trimming off the back. See NOTCH BINDING.

burst-can detector (*NucEng*) An instrument for the early detection of ruptures of the sheaths of fuel elements inside a reactor. Also *burst-cartridge detector, leak detector*.

burst cartridge (*NucEng*) Fuel element with a small leak, emitting fission products. Also *burst slug*.

burst-cartridge detector (*NucEng*) See BURST-CAN DETECTOR.

burstiness (*ICT*) A measure used to characterize traffic for planning purposes. It indicates the extent to which a given level of traffic occurs as short periods at a high data rate separated by longer periods at a lower rate.

bursting (*ICT*) Separating continuous stationery.

bursting disk (*ChemEng*) A protective device for process vessels in which hazardous operations are performed, consisting of a thin disk of noble or corrosion-resisting metal, carefully controlled as to thickness, and designed to burst in event of excess internal pressure, giving a large opening for rapid release of the pressure.

burst slug (*NucEng*) See BURST CARTRIDGE.

burst test (*Paper*) A physical test method to determine the limiting pressure (applied normally to the paper surface by means of a rubber diaphragm) that a test piece will withstand when fixed horizontally between two clamps, under the prescribed conditions of test. The Mullen burst tester is frequently used for this purpose.

bus (*ICT*) Physical path followed by data, particularly between a central processing unit and a peripheral device or, by extension, between the user's program and a file on backing store. Shared by signals from several components of the computer (eg all input/output devices would be connected to the IO bus). Also *highway, trunk*. Several linked buses are a *channel*.

bus (*Space*) The part of the payload of a space exploration vehicle which contains the atmospheric (re-)entry probes, or a universal platform for diverse space experiments and applications.

bus-bar (*ElecEng*) (1) Length of constant-voltage conductor in a power circuit. Normally of rigid copper construction and located in a power station or substation. (2) Supply rail maintained in a constant potential (including zero or earth) in electronic equipment.

bus-coupler switch (*ElecEng*) A switch or circuit breaker serving to connect two sets of duplicate bus-bars.

bush (*Build*) A reducing adapter or screwed piece for connecting together in the same line two pipes of different sizes.

bush (*Eng*) (1) A cylindrical sleeve, usually inserted in a machine part to form a bearing surface for a pin or shaft. (2) A hardened cylindrical insert in a drilling jig to position a drill or reamer accurately.

bushel (*Genrl*) A dry measure of 8 gallons, no longer official in UK, for grain, fruit, etc.

bush-hammering (*CivEng*) The operation of dressing the surface of stone or concrete with a special hammer having rows of projecting points on its striking face for decoration or to improve bonding to the next placement of further concrete.

bushing (*ElecEng*) An insulator which enables a live conductor to pass through an earthed wall or tank (eg the wall of a switch house or the tank of a transformer).

bushing (*Glass*) A small electric melting unit, usually made of platinum, with numerous holes (usually in multiples of 204) in the base, used for the manufacture of glass fibres.

bush sickness (*Vet*) See PINE.

bus-line (*ElecEng*) A cable, extending the whole length of an electric train, which connects all the collector shoes of like polarity. Also *power line*.

bus-line couplers (*ElecEng*) Plug-and-socket connectors to join the bus-line of one coach of an electric train to that of the next.

bustamite (*Min*) A triclinic silicate of manganese, calcium and iron.

bustle pipe (*Eng*) Main air pipe surrounding blast furnace, which delivers low-pressure compressed air to tuyères.

bus topology (*ICT*) A type of NETWORK in which all devices are connected in a line to a single cable. See fig. at NETWORK TOPOLOGY.

bus-wire coupler (*ElecEng*) A flexible connection between the coaches of an electric train for maintaining the continuity of bus-wires which run throughout the length of the train.

busy (*ICT*) A term applied to any line or equipment that is engaged. See BUSY TONE.

busy hour (*ICT*) A period of the day, often mid-morning for telephone systems, identified by the network operator as having the highest traffic level and therefore defining the required installed capacity of the network.

busy tone (*ICT*) An audible signal that is fed back to a caller to indicate that the switching equipment, line or required subscriber's instrument is already engaged.

butadiene (*Chem*) Butan 1,2:3,4 diene. $CH_2=CHCH=CH_2$. A di-alkene with conjugate linking. An isoprene homologue, important in the manufacture of synthetic rubbers etc. See panels on ELASTOMERS and POLYMERS.

butadiene rubber (*Chem*) *Polybutadiene*. Abbrev BR. See panels on ELASTOMERS and POLYMERS.

butanal (*Chem*) Butyraldehyde. $CH_3CH_2CH_2CHO$. Bp 76°C. Made by the catalytic dehydrogenation of butan-1-ol.

butane (*Chem*) C_4H_{10}. An alkane hydrocarbon, bp 1°C, rel.d. at 0°C 0·600, contained in natural petroleum, obtained from casing head gases in petroleum distillation. Used commercially in compressed form, and supplied in steel cylinders for domestic and industrial purposes, eg Calor gas.

butanoic acids (*Chem*) See BUTYRIC ACIDS.

butanol (*Chem*) See BUTYL ALCOHOLS.

butanoyl (*Chem*) The monovalent acyl radical $C_3H_7CO—$.

butenes (*Chem*) C_4H_8, alkene hydrocarbons, the next higher homologues to propylene. Three isomers are possible and known, normally gaseous, bp between 6°C and −3°C. Monomers for various polymers. Also *butylenes*.

Butex (*NucEng*) TN for diethylene glycol dibutyl ether, used for separating uranium and plutonium from fission products. Cf PUREX.

buthocrome (*Phys*) Particular groups of atoms in organic compounds which have the effect of lowering frequency of the radiation absorbed by these compounds.

butobarbitone (*Pharmacol*) A barbiturate hypnotic and sedative; 5-butyl-5-ethylbarbituric acid. TNs include *Buto-met*, *Soneryl*.

butt (*Build*) See BUTT HINGE.

butte (*Geol*) A small, flat-topped, steep-sided hill. See MESA.

butter (*FoodSci*) A fat emulsion derived from cream containing 38–42% fat. The cream is pasteurized, cooled to below 5°C to allow the fat to crystallize and then the temperature increased to between 7 and 13°C for churning which separates the buttermilk from the fat. Contains in solution sugar, albumin, salts; fats and casein are present in colloidal dispersion.

butterfat (*FoodSci*) The fat constituent of milk and milk-derived products, often regulated for particular products.

butterfly (*Eng*) A term used in metal extrusion where an open 'U' shape is first made and the sides then folded closer to make a vertically sided 'U'. This enables the die to be much stronger because the narrow section to form the inside of the 'U' can have a wider base.

butterfly (*ImageTech*) A diffuser used to soften the light from the Sun or from lamps; made from silk stretched on a frame.

butterfly curve (*Eng*) The strain versus applied field curve of ferroelectrics.

butterfly diagram (*Astron*) A graphical presentation of the occurrence of sunspots in the 11-year sunspot cycle. Also *Maunder diagram*.

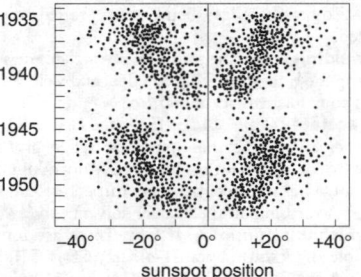

butterfly diagram

butterfly flower (*BioSci*) A flower pollinated by butterflies.

butterfly nut (*Eng*) See WING NUT.

butterfly tail (*Aero*) See VEE-TAIL.

butterfly valve (*Eng*) (1) A disk turning on a diametral axis inside a pipe; used as a throttle valve in petrol and gas engines. (2) A valve consisting of a pair of semicircular plates hinged to a common diametral spindle in a pipe; by hinging axially, the plates permit flow in one direction only.

buttering (*Build*) The operation of spreading mortar on the edges of a brick before laying it.

buttering trowel (*Build*) A flat tool similar to, but smaller than, the BRICK TROWEL; used for spreading mortar on a brick before placing it in position.

butternut (*For*) A N American HARDWOOD of the walnut (*Juglans*) family, of economic importance for its fruits as well as its wood. It is brownish in colour with normally straight grain but a somewhat coarse texture.

Butterworth filter (*ICT*) CONSTANT-K FILTER designed to give response of maximum flatness through pass band. Cf CHEBYSHEV FILTER.

butt gauge (*Build*) Gauge used in fitting the butt hinges on doors, having three markers: one for the thickness of the butt, one for the depth of the door, and one for the depth of the jamb.

butt hinge (*Build*) A hinge formed by two leaves, which are secured to the door and door frame in such a manner that when the door is shut the two leaves are folded into contact. Also *butt*.

butt joint (*Build*, *Eng*) A joint formed between the squared ends of the two jointing pieces, which come together but do not overlap. Riveted or bolted joints may be covered by a narrow strip or *strap*. Welded joints, except between thin sheets, require bevelling so that their edges form a V-shape for the filler metal.

butt-jointed (*ElecEng*) See JOINT.

buttock (*MinExt*) Coal from which an undercut has been removed, in readiness for bringing it down.

buttock planes (*Ships*) Longitudinal sectional planes drawn through a ship's form; used for laying-off in the moulding loft, and for calculation of volumes etc.

button (*ICT*) A small symbol on a screen that can be clicked to select an option.

button-headed screws (*Eng*) Screws having hemispherical heads, slotted for a screwdriver. Also *half-round screws*.

button microphone (*Acous*) Small microphone which can be fitted in the buttonhole.

buttonwood (*For*) See SYCAMORE.

buttress (*CivEng*) A supporting pier built on the exterior of a wall to enable it to resist outward thrust.

buttress root (*BioSci*) A form of PROP ROOT that thickens unevenly to produce a flat, apparently supporting, structure something like a buttress.

buttress screw-thread (*Eng*) A screw-thread designed to withstand heavy axial thrust in one direction. The back of the thread slopes at 45°, while the front or thrust face is perpendicular to the axis.

butt-welded tube (*Eng*) Tube made by drawing mild-steel strip through a bell-shaped die, so that the strip is coiled into a tube, the edges being then pressed together and welded.

butt-welding (*Eng*) The joining of two plates or surfaces by placing them together, edge to edge, and welding along the seam thus formed. See WELDING.

butyl acetate (*Chem*) *Butyl ethanoate*. $CH_3COOC_4H_9$. The commercial product has a boiling range of 124–128°C, rel.d. 0·885, is a colourless liquid of fruity odour, soluble in ethanol, ether, acetone, benzene, turpentine and slightly in water. A very important lacquer solvent.

butyl alcohols (*Chem*) C_4H_9OH. There are four isomers possible and known: butan-1-ol $CH_3(CH_2)_3OH$, bp 117°C; butan-2-ol $CH_3CH_2CH(OH)CH_3$, bp 110°C; 2-methyl propan-1-ol, bp 107°C; 2,2-dimethyl-ethanol, mp 25°C, bp 83°C.

butylated hydroxyanisole (*FoodSci*) An antioxidant used in fats, margarine, essential oils, dried soups and dried mashed potatoes. Abbrev *BHA*.

butylenes (*Chem*) See BUTENES.

butyl group (*Chem*) The aliphatic group C_4H_9—, with four isomeric forms: *primary*, *secondary*, *iso-* and *tertiary*.

butyl lithium (*Chem*) Organo-metallic compound, C_3H_7Li, used in anionic polymerization.

butyl rubber (*Chem*) A copolymer made from isobutylene and a small amount of isoprene ($\approx 1\%$) to aid VULCANIZATION. Cationically polymerized at low temperature. Outstanding properties include very low gas permeability (so used for tyre linings) and very broad DAMPING peak (so useful for antivibration products). Very low REBOUND RESILIENCE at ambient temperature.

butyraldehyde (*Chem*) See BUTANAL.

butyric acids (*Chem*) *Butanoic acids*. C_3H_7COOH. There are two isomers: *normal* and *iso*-butyric acid. Only *n*-butyric (butanoic) acid is of importance, mp −8°C, bp 162°C; it is a thick liquid of rancid odour and occurs in rancid butter.

Buxton certification (*ElecEng, MinExt*) The certification of the suitability of electrical equipment for use in an atmosphere in which fire or explosion hazards are present.

Buys Ballot's law (*EnvSci*) The rule that if an observer stands with his or her back to the wind, the lower atmospheric pressure is on the left in the northern hemisphere, and on the right in the southern hemisphere.

buzz (*Aero*) (1) Severe vibration of a control surface in transonic or supersonic flight caused by separation of the airflow due to compressibility effects. (2) To interfere with an aircraft in flight by flying very close to it.

buzz track (*ImageTech*) (1) A test film used to set the correct position of the slit in an optical sound reproducer. (2) A sound recording of low background level used to fill silent gaps in commentary or dialogue.

BVH (*ImageTech*) Abbrev for *Broadcast Video Helical scan*. TN for a C-format videotape recorder.

BVU (*ImageTech*) Abbrev for *Broadcast Video U-matic*. TN for a HIGH-BAND version of U-MATIC.

BWD (*Vet*) Abbrev for *bacillary white diarrhoea*. See PULLORUM DISEASE.

BWG (*Eng*) Abbrev for *Birmingham wire gauge*. See BIRMINGHAM GAUGE.

BWR (*NucEng*) Abbrev for BOILING WATER REACTOR.

bye-channel (*CivEng*) A waterway dug round the side of a reservoir or dam to carry off surplus water from the streams entering it.

bypass (*Build*) Any device for directing flow around a fixture, connection or pipe, instead of through it.

bypass capacitor (*ElecEng, Electronics*) A capacitor having a low reactance for frequencies of interest connected in shunt with other components so as to short-circuit them for signal frequency currents.

bypass ratio (*Aero*) The ratio of the bypassed airflow to the combustion airflow in a dual-flow turbojet having a single air intake.

bypass turbojet (*Aero*) A turbojet in which part of the compressor delivery is bypassed round the combustion zone and turbine to provide a cool, slow propulsive jet when mixed with the residual efflux from the turbine. See DUCTED FAN.

bypass valve (*ElecEng*) A switching device (SILICON-CONTROLLED RECTIFIER or, in the past, mercury-arc valve), connected across the converter switching devices of a high-voltage dc transmission system, normally not conducting but able to maintain flow of current whenever the main conducting devices have to be interrupted.

bypass valve (*Eng*) A valve by which the flow of fluid in a system may be directed past some part of the system through which it normally flows, eg an oil filter in a lubrication system.

B-Y signal (*ImageTech*) A component of colour TV chrominance signal. Combined with luminance (Y) signal, it gives primary blue component.

bysmalith (*Geol*) A form of igneous intrusion bounded by a circular fault and having a dome-shaped top; described by Iddings from Yellowstone Park. Cf PLUG.

byssinosis (*Med*) Respiratory disease among workers in the vegetable fibre industry, characterized by chest tightness on returning to work after a period of absence. Due to sensitization by substances present in the fibre dust.

byssus (*BioSci*) In Bivalvia, a tuft of strong filaments secreted by a gland in a pit (byssus pit) in the foot and used for attachment. Adj *byssogenous*, *byssal*.

byte (*ICT*) A fixed number of BITS, often corresponding to a single CHARACTER and operated on as a unit.

bytownite (*Min*) A variety of plagioclase feldspar, containing 70–90% of the anorthite molecule; occurs in basic igneous rocks.

Bz (*Chem*) Symbol for the benzoyl (benzenecarboxyl) radical C_6H_5CO.

C

C (*BioSci*) Symbol for COMPLEMENT. The component proteins for complement are designated C1–C9 and proteolytically derived fragments as, eg C5a. Activated and enzymatically active components may be shown with a bar above the number.

C (*Chem*) Symbol for: (1) CARBON; (2) CYTOSINE.

C (*ICT*) A HIGH-LEVEL LANGUAGE in which the UNIX operating system is written.

C (*Phys*) Symbol for: (1) capacitance; (2) COULOMB; (3) when used after a number of degrees, thus 45°C, a temperature on the Celsius or centigrade scale.

C (*Chem*) Symbol for: (1) CONCENTRATION; (2) (with subscript) MOLAR HEAT CAPACITY (C_p at constant pressure, C_v at constant volume).

C– (*Chem*) Containing the radical attached to a carbon atom.

C++ (*ICT*) An enhancement of the c programming language to include OBJECT-ORIENTED PROGRAMMING.

[C] (*Phys*) One of the FRAUNHOFER LINES in the red of the solar spectrum. Its wavelength is 656·3045 nm; it is due to hydrogen.

c (*Genrl*) Symbol for CENTI-.

c (*Chem*) Symbol for CONCENTRATION.

c (*Phys*) Symbol for the SPEED OF LIGHT in a vacuum.

c- (*Chem*) Abbrev for: (1) *cyclo-*, ie containing an alicyclic ring; (2) *cis-*, ie containing the two groups on the same side of the plane of the double bond or ring.

χ (*Phys*) *Chi.* Symbol for magnetic susceptibility.

χ² (*MathSci*) Symbol for chi-squared. See CHI-SQUARED TEST.

C1a inhibitor (*BioSci*) An inhibitor of activated esterase formed from complement C1. It also inhibits some other esterases activated during blood clotting.

C3a, C5a (*BioSci*) Peptide fragments that are split off from COMPLEMENT proteins C3 and C5 respectively during conversion to their enzymically active forms. The peptides are chemotactic for leucocytes and cause local increase in vascular permeability. They act as ANAPHYLATOXINS causing histamine release from local mast cells.

C3b receptors (*BioSci*) Receptors that are present on cell membranes and that can bind C3b, the activated form of complement C3, or its breakdown products, C3bi or C3d. The receptor enables immune complexes or microbes that have bound complement to become attached to the cells, thus increasing their ingestion.

C3 plant (*BioSci*) A plant in which CO_2 is fixed directly by RIBULOSE 1,5 BISPHOSPHATE CARBOXYLASE OXYGENASE into 3-phosphoglyceric acid, which is subsequently converted into sugars etc by the Calvin cycle. Most plants are C3 plants. See panel on CALVIN CYCLE.

C4 pathway evolution (*BioSci*) C4 photosynthesis has evolved apparently independently in a number of angiosperm families (eg gramineae, Euphorbiaceae, Chemopodiaceae). In some sorts of C4 plant, the amino acids, aspartate (C_4) and alanine (C_3), are shuttled between the cell types in place of malate and pyruvate, being formed from and forming oxaloacetate and pyruvate by transamination in the two cell types. See panel on C4 PHOTOSYNTHETIC PATHWAY.

C4, C4 photosynthetic pathway (*BioSci*) A form of photosynthesis in which the first products of CO_2 fixation are C_4 acids (acids with four carbon atoms) rather than the phosphoglyceric acid (three carbons) of the commoner C3 plants. See C4 PATHWAY EVOLUTION and panel on C4 PHOTOSYNTHETIC PATHWAY.

C4 plant (*BioSci*) A plant that fixes CO_2 in photosynthesis by the Hatch–Slack pathway (see panel on C4 PHOTOSYNTHETIC PATHWAY). Identified C4 plants (all are angiosperms including some crop plants such as maize and sugar cane) mostly grow in warm sunny places where they photosynthesize substantially faster than C3 plants in which photosynthesis may be limited by CO_2. Most have KRANTZ ANATOMY. See C3 PLANT, CAM PLANT.

Ca (*Chem*) Symbol for CALCIUM.

CAA (*Aero*) Abbrev for CIVIL AVIATION AUTHORITY.

CAB (*Aero*) Abbrev for *Civil Aeronautics Board*, USA.

CAB (*Plastics*) Abbrev for CELLULOSE ACETATE–BUTYRATE.

cabin (*MinExt*) A firefighter's station underground in a coal mine.

cabin altitude (*Aero*) The nominal pressure altitude maintained in the cabin of a pressurized aircraft.

cabin blower (*Aero*) An engine-driven pump, usually of displacement type, for maintaining an aircraft cockpit or cabin above atmospheric pressure. Also *cabin supercharger.*

cabin differential pressure (*Aero*) The pressure in excess of that of the surrounding atmosphere which is needed to maintain comfortable conditions at high altitude. For an aircraft flying at 9000 m this differential would be about 60 kN m^{-2}.

cabinet file (*Build*) A single-cut smooth file used by joiners and cabinet-makers.

cabinet screwdriver (*Build*) Originally a screwdriver with a round shank flattened at the end to suit a slot in the FERRULE but now with a grooved cylindrical shank inserted into the handle.

cabinet-work (*Build*) Fine joinery used in the construction of furniture and fixtures.

cabin hook (*Build*) A hooked bar and eye, serving as a fastener for doors and casements.

cabin supercharger (*Aero*) See CABIN BLOWER.

cable (*Eng*) General term for rope or chain used for engineering purposes. Specifically, a ship's anchor cable.

cable-angle indicator (*Aero*) An indicator showing the vertical angle between the longitudinal axis of a glider and its towing cable, also its yaw and roll attitude relative to the towing aircraft.

cable buoy (*Ships*) A buoy attached to an anchor and serving to mark its position.

cablecar (*CivEng*) Tram pulled by a moving underground cable, in the same manner as the CABLE RAILWAY.

cablecast (*ICT*) A broadcast on cable TV.

cable ducts (*ElecEng*) Earthenware, steel, plastic or concrete pipes containing cables.

cable form (*ElecEng*) The normal scheme of cabling between units of apparatus. The bulk of the cable is made up on a board, using nails at the appropriate corners, each wire of the specified colour identification being stretched over its individual route with adequate SKINNER. When the cable is bound with twine and waxed, it is fitted to the apparatus on the racks and the skinners connected, by soldering, to the TAG BLOCKS.

C4 photosynthetic pathway

A form of photosynthesis used by plants living in hot sunny places where there is plenty of energy available in the form of light. Here the first products of CO_2 fixation are C4 acids (acids with four carbon atoms) rather than the phosphoglyceric acid (three carbons) of the commoner C3 plants. Typically, CO_2 is fixed in the mesophyll cells of a leaf with Krantz anatomy by phosphoenolpyruvate (PEP) carboxylase into the 4-carbon oxaloacetic acid, which is then reduced to malic acid. This is then transported, probably through the plasmodesmata, to the bundle sheath cells where it is decarboxylated. The CO_2 thus released is refixed, by ribulose 1,5-bisphosphate carboxylase oxygenase in the Calvin cycle, into sugars in the normal way while the 3-carbon pyruvic acid remaining is returned to the mesophyll cells. The whole cycle acts as a pumping mechanism for CO_2, promoting CO_2 uptake and, by concentrating CO_2 in the bundle sheath cells, reducing photorespiration.

   C4 metabolism uses more energy (as ATP from light reactions) than C3, hence the need for a sunny climate. Under such conditions, C4 plants typically photosynthesize more rapidly than C3 plants and C4 crop plants like maize sorghum, sugar beet and some tropical fodder grasses can outyield C3 crops. See c4 PATHWAY EVOLUTION.

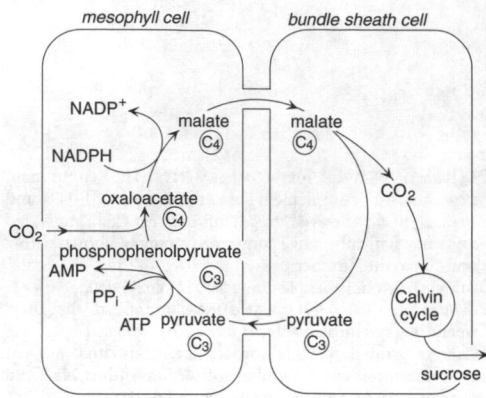

C4 photosynthetic pathway The Hatch–Slack pathway.

AMP = adenosine monophosphate; ATP = adenosine trisphosphate; $NADP^+$ oxidized-, NADPH = reduced-nicotinamide adenine dinucleotide phosphate; PP_i = inorganic pyrophosphate. The C_3, C_4 in circles indicate the numbers of carbon atoms in the compounds.

   See panel on CALVIN CYCLE.

---

cable grip (*ElecEng*) A flexible cone of wire which is put on the end of a cable. When the cone is pulled, it tightens and bites into the sheath of the cable, and can be used to pull the cable into a duct.

cable-laid rope (*Eng*) A rope formed of several strands laid together so that the twist of the rope is in the opposite direction to the twist of the strands. Cf LANG LAY.

cable-length (*Genrl*) Unit of length, originally the length of a ship's anchor cable, equal to one-tenth of a nautical mile (608 ft, 185 m).

cable railway (*CivEng*) Means of transport whereby carriages are pulled up an incline by an endless over- or underground cable.

cable release (*ImageTech*) A device for releasing a camera shutter in which the trigger is actuated by a length of stiff wire cable in a flexible tube.

cable-stayed bridge (*CivEng*) A bridge type for medium spans in which the decking is suspended by diagonal cables attached directly to the supporting tower. Can be of fan or harp design. The decking is always in compression and is self-supporting during construction. See BRIDGE and panel on BRIDGES AND MATERIALS.

cable stitch (*Textiles*) In knitting, the rope-like appearance obtained by passing groups of adjacent wales under and over one another.

cable tool drilling (*MinExt*) A method of drilling in which a heavy sharpened tool bit (*churn drill*, *percussion drill*) is reciprocated in a borehole by a steel cable attached to a WALKING BEAM.

cable tools (*MinExt*) The variously shaped drilling tools used in CABLE TOOL DRILLING.

cable TV (*ImageTech*) A TV broadcasting system in which TV signals are relayed directly to individual subscribers by means of underground or overhead cables, as opposed to being broadcast by radio waves.

cable wax (*Chem*) A solid wax formed by the ionic bombardment of the oil in a cable. It is a good insulator, and cables operate very successfully even when much wax is present. It is produced by a condensation process such as $C_6H_{14} + C_5H_{12} \rightarrow C_{10}H_{22} + CH_4$.

cable-way (*CivEng*) A construction consisting of cables slung over and between two or more towers, so that skips suspended from the cables may be moved often over long distances. It is used for transport of ore etc. Also *blondin*.

cabling (*Arch*) A round moulding used to decorate the lower parts of the flutes of columns.

cabling (*ICT*) The collection of cables required for distributing the power supplies in a telephone exchange. See TRUNKING.

cabling (*Textiles*) Twisting together two or more doubled or FOLDED YARNS. The result in most cases is a balanced cord of four, six or more yarns. Tyre cord is one example.

cache controller (*ICT*) A hardware device that controls the transfer of data to and from the CACHE MEMORY.

cache memory (*ICT*) Extremely fast part of main memory.

cachexia (*Med*) A marked wasting and emaciation of the body, with an 'earthy' complexion. May occur in some patients with cancer.

cacodyl (*Chem*) $As_2(CH_3)_4$. A colourless liquid, bp 170°C, of horribly nauseous odour. It combines directly with oxygen, sulphur, chlorine, etc. Cacodyl and cacodyl oxide form the basis for other secondary arsines. Cacodyl derivatives are important as rubber accelerators.

Cactaceae (*BioSci*) The Cactus family, c.2000 spp of dicotyledonous flowering plants (superorder Caryophillidae). Most are leafless, spiny, stem-succulent, CAM PLANTS of the semi-deserts of the Americas. They are of little economic importance other than as ornamentals, but some produce edible fruit.

CAD (*ICT*) Abbrev for COMPUTER-AIDED DESIGN.

cadastral map (*Genrl*) A map on which land ownership and boundaries are shown.

cadastral survey (*Surv*) Land survey, boundary delineation.

cadaver (*Med*) A dead body.

cadaverine (*Chem*) *1,5-diaminopentane*. $NH_2(CH_2)_5NH_2$. A colourless syrupy liquid, bp 178°C. Formed by the bacterial decomposition of diamino acids. Ring formation occurs by elimination of NH_2 with the formation of the heterocyclic base PIPERIDINE. See PUTRESCINE.

cadherins (*BioSci*) Integral membrane proteins involved in calcium-dependent cell adhesion. Three main types are recognized: *epithelial*, *neural* and *placental cadherins*.

cadmium (*Chem*) White metallic element, symbol Cd, at no 48, ram 112·40, mp 320·9°C, bp 767°C, rel.d. 8·648. It is a rare element, the Earth's crust containing only 0·16 ppm. Cadmium has considerable affinity for sulphur and, although it forms some independent minerals, it mainly occurs concealed in the crystal lattice of *sphalerite*, zinc sulphide. The smelting of zinc is thus the principal commercial source of the element. Cadmium plating is widely used as a corrosion protective for steel and its alloys. Cadmium is a powerful absorber of neutrons and is used in control elements in nuclear reactors. Films of cadmium are photosensitive in the ultraviolet between 250 and 295 nm, with a peak at 260 nm.

cadmium cell (*Electronics*) A reference voltage standard, giving 1·0186 V at 20°C. Also *Weston standard cadmium cell*.

cadmium copper (*Eng*) A variety of copper containing 0·7–1·0% of cadmium. Used for trolley, telephone and telegraph wires because it gives high strength in cold-drawn condition combined with good conductivity.

cadmium photocell (*Electronics*) A photoconductive cell using cadmium disulphide or cadmium selenide as the photosensitive semiconductor. Sensitive to longer wavelengths and infrared. It has a rapid response to changes in light intensity.

cadmium red line (*Phys*) Spectrum line formerly chosen as a reproducible standard of length, wavelength 643·8496 nm.

cadmium sulphide detector (*Radiol*) Radiation detector equivalent to a solid-state ionization chamber, but with amplifier effect (due to hole trapping).

caducibranchiate (*BioSci*) Possessing gills at one period of the life cycle only, as in some Caudata.

caducous (*BioSci*) In mammals, descriptive of the situation in which the maternal part of the placenta comes away immediately post-partum. See DECIDUATE. In plants, falling off at an early stage, as with parts of certain flowers. Cf DECIDUOUS.

CAE (*Eng*) Abbrev for COMPUTER-AIDED ENGINEERING.

caecostomy (*Med*) Formation of an artificial opening into the caecum. US *cecostomy*.

caecum (*BioSci*) Any blind diverticulum or pouch, esp one arising from the alimentary canal. US *cecum*.

Caelum (Chisel) (*Astron*) An inconspicuous southern hemisphere constellation.

caenogenesis (*BioSci*) The state where adaptations to the needs of the young develop early and disappear in the adult. Adj *caenogenetic*.

Caenorhabditis elegans (*BioSci*) A small nematode worm that is ideally suited for genetic, molecular and cellular studies of eukaryotic development.

Caerfai (*Geol*) The oldest epoch of the Cambrian period.

Caesarean section (*Med*) Delivery of a fetus through the incised abdomen and uterus. US *Cesarean section*.

caesious (*BioSci*) Bearing a bluish-grey waxy covering (bloom). Also *caesius*.

caesium (*Chem*) Metallic element, symbol Cs, at no 55, ram 132·905, mp 28·6°C, bp 713°C, rel.d. 1·88. It is a rare element with an abundance of 2·6 ppm in the Earth's crust. Since its ionic radius is large (1·65 Å), it tends to occur in the last stages of magmatic crystallization, forming the pegmatite mineral POLLUCITE. The radioactive ^{137}Cs is

used in radiotherapy and as a medical tracer, and the resonance frequency of the natural isotope [133]Cs has been used as a standard for the measurement of time. As a photosensitor it has a peak response at 800 nm in the infrared, both thermal and photoemission being high. Caesium, when alloyed with antimony, gallium, indium and thorium, is generally photosensitive. US *cesium*.

caesium cell (*ElecEng*) Cell having a cathode consisting of a thin layer of caesium deposited on minute globules of silver; particularly sensitive to infrared radiation, but generally approximating to that of the eye.

caesium clock (*ElecEng*) Frequency-determining apparatus used on caesium-ion resonance of 9 192 631 770 Hz.

caesium–oxygen cell (*ElecEng*) Cell in which the vacuum is replaced by an atmosphere of oxygen at very low pressure. It is more sensitive to red light than the caesium cell.

caesium unit (*Radiol*) A source of radioactive caesium (half-life 30 yr) mounted in a protective capsule.

caesius (*BioSci*) See CAESIOUS.

caespitose (*BioSci*) Growing from the root in tufts, eg as many grasses. US *cespitose*.

caffeine (*Pharmacol*) A weak central nervous stimulant, a methyl xanthine, found in coffee and tea.

cage (*Chem*) A regular arrangement of anions in a crystal structure within which smaller cations may be held. Four anions form a tetrahedral cage, six an octahedral one, eight a cubic one and twelve a cubeoctahedral one.

cage (*CivEng, MinExt*) The platform on which goods are hoisted up or lowered down a vertical shaft or guides; in mines the steel box used to raise and lower workers, materials or trucks (trams, tubs). May have two or three decks.

cage (*Eng*) The part of a ball or roller bearing which separates the balls or rollers and keeps them correctly spaced along the periphery of the bearing.

cage antenna (*ICT*) Antenna comprising a number of wires connected in parallel, and arranged in the form of a cage to reduce the copper losses and increase the effective capacitance.

cage rotor (*ElecEng*) A form of rotor, used for induction motors, having on it a CAGE WINDING. Also *squirrel-cage rotor*.

cage winding (*ElecEng*) A type of winding used for rotors of some types of induction motors, and for the starting or damping windings of synchronous machines. It consists of a number of bars of copper or other conducting material, passing along slots in the core and welded to rings at each end. Also *squirrel-cage winding*.

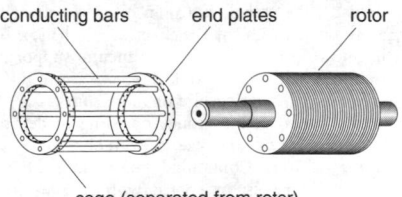

cage winding Cage consists of copper brazed to the end plates or aluminium cast *in situ*.

Cahn–Ingold–Prelog system (*Chem*) A system for describing stereoisomerism unambiguously. It is based on the identification of a *priority* among substituents, the atom of higher atomic number having higher priority. When the two atoms are of the same element, then *their* substituents are examined etc. For geometrical isomerism, the substituents of highest priority on each side of the bond in question are examined. If they are on the same side of the bond, the prefix Z- is given (from Ger *zusammen*), otherwise E- (Ger *entgegen*). Chiral centres are viewed with the substituent of lowest priority away from the viewer. The configuration is said to be R- (Lt *rectus*) if the other substituents in order of decreasing priority are arranged clockwise, otherwise S- (Lt *sinister*).

CAI (*ICT*) Abbrev for COMPUTER-ASSISTED INSTRUCTION.

Cailletet process (*Phys*) A method for the liquefaction of gases based on the free expansion of a gas from a higher to a lower pressure.

Cailletet's and Mathias's law (*Chem*) The law that the arithmetic mean of the densities of a pure unassociated liquid (d_l) and its saturated vapour (d_v) is a linear function of the temperature, T, ie $0 \cdot 5(d_l + d_v) = A + BT$, where A and B are constants.

Cainozoic (*Geol*) See CENOZOIC.

cairngorm (*Min*) Smoky-yellow or brown varieties of quartz; named from Cairngorm in the Scottish Grampians, the more attractively coloured varieties being used as semiprecious gemstones. See SMOKY QUARTZ.

caisson (*Build*) A deeply recessed sunk panel in a soffit or ceiling.

caisson disease (*Med*) Decompression sickness. Pains in the joints and paralysis, occurring in workers in compressed air who are suddenly subjected to atmospheric pressure after compression; it is due to the accumulation of bubbles of nitrogen in the nervous system. Also *the bends*, *diver's palsy*, *diver's paralysis*.

cake (*Eng*) The rectangular casting of copper or its alloys before rolling into sheet or strip.

caking coal (*MinExt*) Coal which cakes or forms coke when heated in the absence of air.

CAL (*ICT*) Abbrev for COMPUTER-ASSISTED LEARNING.

cal (*Phys*) Abbrev for CALORIE.

calamine (*Min*) (1) Former UK term for SMITHSONITE. (2) US for HEMIMORPHITE.

Calamitales (*BioSci*) An order of extinct, mainly Carboniferous, Sphenopsida. The sporophytes were mostly large trees, with substantial hollow trunks that had secondary xylem and whorled or opposite branches. They were codominant with the Lepidodendrales in the Carboniferous swamps.

calamus (*BioSci*) The proximal hollow part of the scapus of a feather; quill. Pl *calami*.

calandria (*ChemEng, NucEng*) Closed vessel penetrated by pipes so that liquids in each do not mix. In evaporating plant the tubes carry the heating fluid and in certain types of nuclear reactor, eg CANDU reactors, the sealed vessel is called a calandria.

calcaneum (*BioSci*) In some vertebrates, the fibulare, or large tarsal bone forming the heel; more generally, the heel itself; in birds, a process of the METATARSUS.

calcar (*BioSci*) (1) In insects, a tibial spine. (2) In amphibians, the prehallux. (3) In birds, a spur of the leg, or, more occasionally, of the wing. (4) In bats, a bony or cartilaginous process of the calcaneum supporting the interfemoral part of the PATAGIUM.

calcarate (*BioSci*) Bearing one or more spurs.

calcarenite (*Geol*) A limestone consisting of detrital lime particles (>50%) of sand size. See CALCILUTITE.

calcareous (*Chem*) Containing compounds of calcium, particularly minerals, or coated with calcium carbonate (lime).

calcareous clay (*Build*) See MARL.

calcareous rock (*Geol*) Sediment containing a large amount of calcium carbonate (eg limestone, chalk, shelly sandstone, calc-tufa).

calc-flinta (*Geol*) A hard fine-grained rock composed of calcium silicate minerals and produced by the contact metamorphism of an impure limestone.

calcicole (*BioSci*) Plants found on or confined to soils containing free calcium carbonate. Also *calciphile*. Cf CALCIFUGE.

calciferol (*BioSci*) See vitamin D in panel on VITAMINS.

calciferous (*BioSci*) Producing or containing calcium salts. Also *calcigerous*.

calcification (*BioSci*) The accumulation of calcium carbonate on or in tissues of either plants or animals.

calcifuge (*BioSci*) Plants not normally found on, or intolerant of, soils containing free calcium carbonate. Also *calciphobe*. Cf CALCICOLE.

calcigerous (*BioSci*) See CALCIFEROUS.

calcigerous glands (*BioSci*) In some Oligochaeta, a pair of oesophageal glands producing a limy secretion to control the acid/base balance of the body; in some amphibians, the glands of Swammerdam, calcareous concretions lying on either side of the vertebrae, close to the points of exit of the spinal nerves.

calcilutite (*Geol*) A limestone consisting of (> 50%) clay or silt particles of lime. See CALCARENITE.

calcination (*Chem*) The subjection of a material to prolonged heating at fairly high temperatures.

calcination (*MinExt*) The operation of heating ores to drive off water and carbon dioxide, frequently not distinguished from ROASTING.

calcine (*MinExt*) Ore, carbonate, mineral or concentrate which has been roasted, perhaps in an oxidizing atmosphere, to remove sulphur as SO_2 (*sweet roasting*) or carbon dioxide (*dead roasting*).

calcined powder (*PowderTech*) Powder produced or modified by heating to a high temperature.

calcinosis (*Med*) Deposits of calcium salts in various tissues of the body.

calciphile (*BioSci*) See CALCICOLE.

calciphobe (*BioSci*) See CALCIFUGE.

calcite (*Min*) The commonest crystalline form of calcium carbonate, showing trigonal symmetry and a great variety of crystal habits. It is the principal constituent of limestone and many marbles, and occurs extensively in other rocks. Also *calcspar*. See panel on TWINNED CRYSTALS.

calcitonin (*Med*) Hormone secreted by parafollicular cells of the thyroid gland in response to high blood calcium, opposing the action of a parathyroid hormone. It is of value in the treatment of hypercalcaemia and PAGET'S DISEASE OF BONE.

calcium (*Chem*) Metallic element, symbol Ca, at no 20, ram 40·08, mp 850°C, bp 1440°C, rel.d. 1·58. Occurs in nature in the form of several compounds, the carbonate predominating. Produced by electrolysis of fused calcium chloride. Used as a reducing metal and as a getter in low-noise valves.

calcium antagonists (*Pharmacol*) Compounds that block or inhibit transmembrane calcium ion movement and will block contraction of cardiac and smooth muscle. Most are potent vasodilators and some are anti-arrhythmic. Examples are *diltiazem*, *nifedipine* and *verapamil*. Also *calcium channel blockers*.

calcium carbide (*Chem*) Ethynide dicarbide. CaC_2. A compound of calcium and carbon usually prepared by fusing lime and hard coal in an electric furnace. See ACETYLENE.

calcium carbonate (*Chem*) $CaCO_3$. Very abundant in nature as chalk, limestone and calcite. Almost insoluble in water, unless the water contains dissolved carbon dioxide, when solution results in the form of calcium hydrogen carbonate, causing the temporary hardness of water.

calcium channel (*BioSci*) A 'pore' through a cell membrane that allows specifically the passage of calcium ions, eg the VOLTAGE-GATED ION CHANNEL through the membrane of the SARCOPLASMIC RETICULUM of muscle cells.

calcium channel blockers (*Pharmacol*) See CALCIUM ANTAGONISTS.

calcium chloride (*Chem, FoodSci*) $CaCl_2$. Formed by the action of hydrochloric acid on the metal and its common compounds. It absorbs moisture from the atmosphere, and is extensively used for drying gases. Tubes filled with granular calcium chloride are used in industry and laboratories for producing dry air. Also used as a source of dietary calcium, to prevent or delay the breakdown of fruit cells during processing, and to stimulate curd formation during cheese processing.

calcium fluoride (*Chem*) CaF$_2$. In the form of fluorspar it is used for the manufacture of hydrofluoric acid. It is also an important constituent of opal glass. Exhibits high thermoluminescence.

calcium phosphate precipitation (*BioSci*) A technique for introducing foreign DNA or chromosomes into cells; co-precipitation with calcium phosphate facilitates the uptake of DNA or chromosomes.

calcium plumbate (*Build*) A lead-based priming paint, not now common, with exceptional weather resistance suitable for use on ferrous metals, galvanized or zinc sheeting and composite metal/timber substrates.

calcium tungstate screen (*Electronics*) A fluorescent screen used in a cathode-ray tube; it gives a blue and ultraviolet luminescence.

calcspar (*Min*) See CALCITE.

calculus (*MathSci*) See ANALYSIS, DIFFERENTIAL CALCULUS, INTEGRAL CALCULUS.

calculus (*Med*) A concretion of mineral or organic matter in certain organs of the body, eg the kidney, the gall bladder.

calculus of variations (*MathSci*) The determination of one or more functions of one or more independent variables, in order that a definite integral of a known function of these functions and their derivatives shall be stationary.

caldera (*Geol*) A large volcanic crater produced by the collapse of underground lava reservoirs or by ring fracture, possibly as a surface expression of CAULDRON SUBSIDENCE.

Caledonian Orogeny (*Geol*) The great mountain-building episode of late Silurian to early Devonian time.

Caledonoid direction (*Geol*) The direction assumed by the Caledonian (Siluro-Devonian) mountain folds and associated structures in UK and Scandinavia. Commonly NE–SW, but subject to considerable variations.

calendar month (*Genrl*) Popularly, a month or year as defined in a calendar, particularly the GREGORIAN CALENDAR. A calendar month differs from the synodic (or lunar) month. The terms are also used for periods equivalent to a month or year, eg from 9 July to 9 August, or 9 July of one year to 9 July of the next. Also *calendar year*.

calender (*Paper, Textiles*) Rolling machine with horizontal-axis rolls of metal or fibrous composition, stacked vertically and carried in side frames. Material is fed through the gaps between the rolls, known as nips, to impart the required degree of finish or to control its thickness and compression. Used in paper, textile, plastics and rubber processing. Calenders may form part of a complete processing machine or exist separately.

calendered paper (*Paper*) Paper that has been calendered. If the calenders are part of the paper-making machine the resultant effect is known as machine-finished. A higher degree of gloss can be obtained by means of a super-calender, separate from the paper machine and containing some rolls of fibrous composition. This is known as supercalendered finish.

calf (*Agri*) A young individual of any of various mammals prior to weaning; in agriculture, typically cattle and buffalo.

calf (*Print*) Superior quality calfskin used for covering books, finished either smooth or rough.

calf diphtheria (*Vet*) Ulceration and necrosis of the mouth and pharynx of calves due to infection by the bacterium *Sphaerophorus necrophorus* (*Fusiformis necrophorus*). Also *malignant stomatitis, necrotic stomatitis*.

calf scour (*Vet*) See WHITE SCOUR.

calf tetany (*Vet*) A form of BOVINE HYPOMAGNESAEMIA, occurring in calves, caused by a deficiency of magnesium. Also *milk tetany*.

Calgon (*Chem*) TN for sodium hexametaphosphate, used in water softening because of its marked property of forming soluble complexes with calcium. Also used in textile and laundry work.

caliber (*Eng*) See CALIBRE.

calibrated airspeed (*Aero*) Indicated airspeed corrected for POSITION ERROR and instrument error only. Not to be confused with EQUIVALENT AIRSPEED or TRUE AIRSPEED. Abbrev CAS. Also *rectified airspeed*.

calibration (*Phys*) The process of determining the absolute values corresponding to the graduations on an arbitrary or inaccurate scale on an instrument.

calibre (*Eng*) (1) The internal diameter or bore of a pipe, esp the barrel of a firearm. (2) The arrangement of the various components of a watch or clock. Also, esp US, *caliber*.

caliche (*Geol*) Concretions of calcium carbonate, sodium nitrate and other minerals occurring in soil in arid regions; due to surface evaporation of subsurface waters.

calicivirus (*Med*) Any of a group of related viruses of the genus *Picornaviridae*. Caliciviruses affecting humans are a major cause of gastro-intestinal upsets; a haemorrhagic calicivirus which kills rabbits from heart and lung failure within 40 hours of infection has been used for control of rabbit populations.

calico (*Textiles*) A plain cotton cloth heavier than muslin.

Californian jade (*Min*) A compact form of green vesuvianite (idocrase) obtained from California, and used as an ornamental stone and in jewellery. Also *californite*.

Californian stamp (*MinExt*) See GRAVITY STAMP.

californite (*Min*) See CALIFORNIAN JADE.

californium (*Chem*) Artificially generated element, symbol Cf, at no 98, produced in a cyclotron. Its longest-lived isotope is ^{251}Cf with a half-life of 800 years.

calipers (*Eng*) See CALLIPERS.

calked ends (*Build*) The ends of built-in iron ties, split and splayed to provide more secure anchorage.

calking (*CivEng*) See CAULKING.

call (*ICT*) In a program, a statement that transfers CONTROL to a SUBROUTINE or CO-ROUTINE.

call-back (*ICT*) A return telephone call; an act of re-calling.

call charge splitting (*ICT*) An arrangement sometimes used in mobile-telephone networks whereby the called party is billed with part of the charge for a call.

call diversion (*ICT*) A service provided by an INTELLIGENT NETWORK in which telephone calls are automatically rerouted to a number prearranged by the called party, eg to a locum while a doctor is off duty.

caller ID (*ICT*) A facility that displays the telephone number of an incoming call.

call forwarding (*ICT*) Diversion of telephone calls in a private network. Users may arrange for all calls to be forwarded, or only those that receive no answer or for which the called number is busy.

call gapping (*ICT*) A technique used to prevent congestion in telephone systems by limiting the number of calls that can pass through the network at any time.

Callier coefficient (*ImageTech*) The ratio of SPECULAR DENSITY to DIFFUSE DENSITY in a photographic image, the difference resulting from the grain structure. Also *Callier quotient, Callier Q factor* and denoted by Q.

Callier effect (*ImageTech*) Effect which occurs because of the scattering of light by the granular structure of a photographic image; its transmission is greater for a directional (specular) beam than for a diffuse one. This results in greater image contrast from optical printing or enlarging with a condenser light source than from contact printing or use of a diffused source.

Callier quotient (*ImageTech*) Also *Callier Q factor*. See CALLIER COEFFICIENT.

calling card (*ICT*) An INTELLIGENT NETWORK service whereby a telephone caller may arrange for the call to be billed to an account other than that of the subscriber from whose terminal the call is made. Security is maintained by requiring the calling card holder to key-in a personal identity number.

calling card validation (*ICT*) The process by which database files held in an INTELLIGENT NETWORK are used to check that a CALLING CARD is still current and matches the personal identity number keyed in by its user.

calling line identification (*ICT*) A signal available from a digital telephone exchange that enables the called terminal to display or otherwise make use of the number of the calling terminal.

callipers (*Eng*) An instrument, consisting of a pair of hinged legs, used to measure external and internal dimensions. Also *calipers*.

calliper splint (*Med*) A splint fitted so that the patient may walk without any pressure on the foot, the weight of the body being taken by the hip bone.

Callisto (*Astron*) The fourth natural satellite of Jupiter, discovered in 1610 by Galileo. Distance from the planet 1 883 000 km; diameter 4800 km. Its dark surface is a mixture of ice and rocky material, and is heavily cratered.

callose (*BioSci*) A polymer of glucose, $\beta 1 \rightarrow 3$ linked, occurring eg as a plant cell-wall constituent, esp in the sieve areas of sieve elements. See also CALLUS.

callose (*Med*) See CALLOUS.

callosity (*Med*) A thickening of the skin as a result of irritation or friction. Also *callus*.

callous (*Med*) Hardened, usually thickened, and often like horn in appearance; having calluses. Also *callose*.

Callovian (*Geol*) A stage in the Middle Jurassic. See MESOZOIC.

call screening (*ICT*) Checking of outgoing telephone calls from or within a private network to establish whether they are permitted; eg out-of-hours calls from the network may be blocked unless the caller keys in an authorization code.

call sign (*ICT*) Letter and/or numeral for a transmitting station or one of its authorized channels. Used for calling or identification.

calltime (*ICT*) Time available for use in making calls on a mobile phone; the time used on a single phone call.

callus (*BioSci*) (1) Generally, wound tissue. (2) In plants, tissue consisting of large, thin-walled parenchymatous cells developing as a result of injury, as in wound healing and grafting, or in tissue culture. (3) An accumulation of CALLOSE. (4) Hard basal projection at the base of the floret or spikelet of some grasses. (5) A localized thickening of the upper layer of skin as a result of repeated friction. (6) Hard new bone that forms in an area of bone fracture. (7) Also, erroneously, callous.

calmodulin (*BioSci*) A calcium-binding protein that is virtually ubiquitous in eukaryotic cells. It plays a central role in controlling the calcium levels in cytoplasm.

calomel (*Chem*) *Mercury (I) chloride.* Hg_2Cl_2. Found naturally in whitish or greyish masses, associated with cinnabar. Used in physical chemistry as a REFERENCE ELECTRODE (ie a half-cell comprising a mercury electrode in a solution of potassium chloride saturated with calomel).

calorescence (*Phys*) The absorption of radiation of a certain wavelength by a body, and its re-emission as radiation of shorter wavelength. The effect is familiar in the emission of visible rays by a body which has been heated to redness by focusing infrared heat rays onto it.

calorie (*Phys*) The unit of quantity of heat in the CGS system. The 15°C calorie is the quantity of heat required to raise the temperature of 1 g of pure water by 1°C at 15°C; this equals 4·1855 J. By agreement, the International Table calorie (cal_{IT}) equals 4·186 J exactly, and the thermochemical calorie equals 4·184 J exactly. There are other designations, eg gram calorie, mean calorie, and large or kilocalorie (=1000 cal, used particularly in nutritional work). The calorie has now been largely replaced by the SI unit of the joule (J).

calorific value (*Eng*) The number of heat units obtained by the combustion of unit mass of a fuel. The numerical value obtained for the calorific value depends on the units used; eg lb-calories lb^{-1}, British thermal units Btu-lb or MJ kg^{-1} for solid and liquid fuels, and Btu ft^{-3} or MJ m^{-3} for gaseous fuels. In fuels containing hydrogen, which burns to water vapour, there are two heating values, gross and net (also *higher (HCF)* and *lower (LCF) calorific values*). The gross value of a fuel is the total heat developed after the products are cooled to the starting point, and the water vapour condensed. The net value is the heat produced on combustion of the fuel at any given temperature, with the flue products cooled to the initial temperature, the water vapour remaining uncondensed.

calorifier (*Eng*) Apparatus for heating water in a tank, the source of heat being a separate coil of heated pipes immersed in the water in the tank.

calorimeter (*Phys*) The vessel containing the liquid used in calorimetry. The name is also applied to the complete apparatus used in measuring thermal quantities.

calorimetry (*Phys*) The measurement of thermal constants, such as specific heat, latent heat or calorific value. Such measurements usually necessitate the determination of a quantity of heat, by observing the rise of temperature it produces in a known quantity of water or other liquid.

calorizing (*Eng*) A process of rendering the surface of steel or iron resistant to oxidation by spraying the surface with aluminium and heating to a temperature of 800–1000°C.

calotte (*Build*) A small dome in the ceiling of a room, used to increase head room.

calotype (*ImageTech*) An early wet-plate process using silver iodide, patented in 1841 by Fox Talbot.

calvarium (*Med*) The dome of the cranium, above the ears, eyes and occipital protuberance.

Calvé's disease (*Med*) Aseptic necrosis of a vertebral body, usually in children, producing a vertebral collapse.

Calvin cycle (*BioSci*) Cyclical sequence of reactions in which carbon dioxide is fixed by ribulose bisphosphate carboxylase and reduced to produce eg sugars. Occurs in all photosynthetic plants and algae and most other autotrophic organisms. Cf HATCH–SLACK PATHWAY. See PHOTOSYNTHESIS and panels on C4 PHOTOSYNTHETIC PATHWAY and CALVIN CYCLE.

calving fever (*Vet*) See MILK FEVER.

calx (*Chem*) Burnt lime or quicklime.

calx (*Med*) Calcaneum; os calcis; the heel.

calycle (*BioSci*) See CALYX.

calyon (*Build*) Flint or pebble stone used in wall construction.

Calypso (*Astron*) The 14th natural satellite of Saturn, discovered in 1980 and associated with TELESTO and TETHYS. Distance from the planet 295 000 km; diameter 30 km.

calypter (*BioSci*) One or two small lobes at the posterior margin of the base of the wing in some Diptera. Adj *calyptrate*.

calyptra (*BioSci*) The layer of cells, developed from part of the ARCHEGONIUM wall that protects the developing sporophyte in mosses and liverworts.

calyptrogen (*BioSci*) The layer of meristematic cells that gives rise to the root cap.

calyptron (*BioSci*) In calyptrate Diptera, the enlarged squama which covers the HALTERE.

calyx (*BioSci*) (1) Outer whorl of the perianth, often green and protective, composed of free or fused sepals. (2) A pouch of an oviduct, in which eggs may be stored. Also *calycle*. (3) In some Hydrozoa, the cup-like exoskeletal structure surrounding a hydroid. Also *calycle*. (4) In Crinoidea, the body as distinct from the stalk and arms. (5) In some mammals, part of the pelvis of the kidney.

calyx tube (*BioSci*) The tube formed by fused sepals.

CAM (*BioSci*) Abbrev for CELL ADHESION MOLECULE or CRASSULACEAN ACID METABOLISM.

CAM (*ICT*) Abbrev for COMPUTER-AIDED MANUFACTURE.

cam (*Eng*) Linear or rotary device, machined to a predetermined profile, whose movement imparts a linear motion to another component, the *cam follower*. The profile can be complex giving eg a variable slow forward and rapid reverse movement to a cutting tool on an automatic lathe or, with several on a shaft, opening and closing the valves of an internal-combustion engine in the desired sequence.

Calvin cycle, photosynthetic carbon reduction cycle

Abbrev *PCR cycle* (better avoided since PCR is the very common abbreviation for polymerase chain reaction). The cyclical sequence of reactions in which carbon dioxide is fixed by ribulose bisphosphate carboxylase oxygenase and reduced to produce eg sugars in all photosynthetic plants and algae and most other autotrophic organisms.

The cycle as it occurs in a mature photosynthesizing green plant leaf which is exporting sucrose is shown in the diagram. The ATP and NADPH come from the light reactions of photosynthesis.

See panel on C4 PHOTOSYNTHETIC PATHWAY.

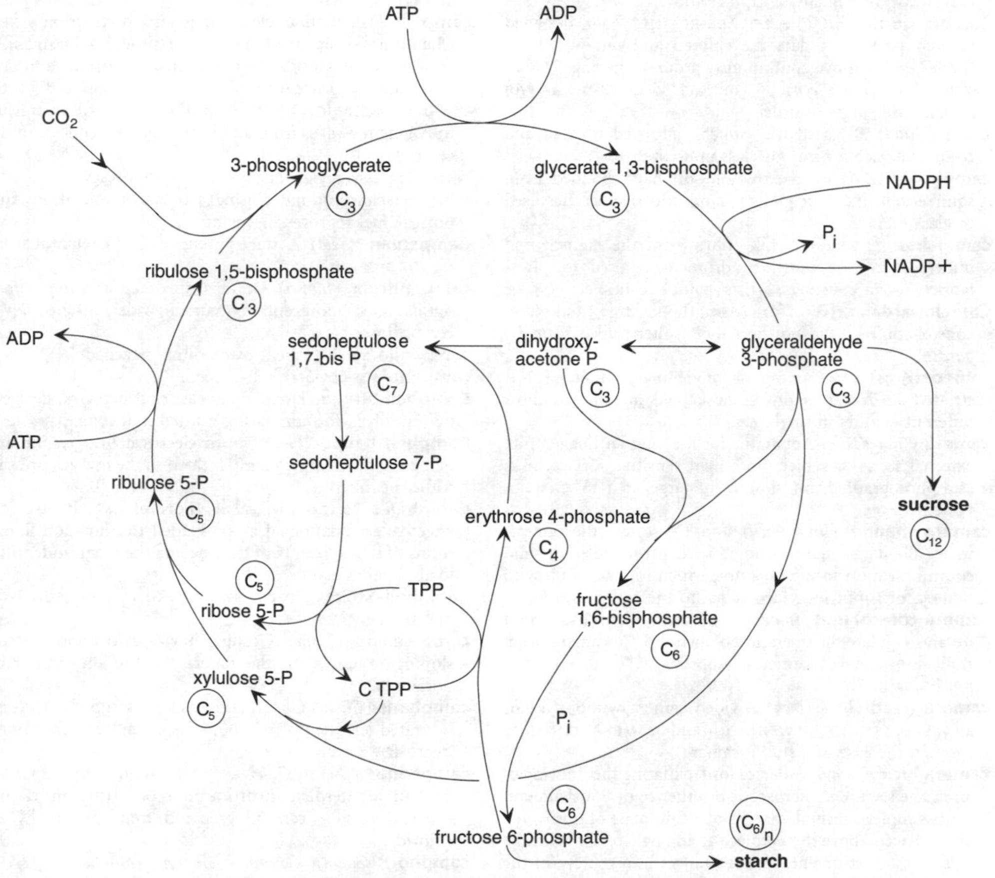

Calvin cycle ADP = adenosine diphosphate, ATP = adenosine triphosphate, NADP$^+$ = oxidized-, NADPH = reduced-nicotinamide adenine dinucleotide phosphate, P = phosphate, Pi = inorganic orthophosphate, TPP = thiamine pyrophosphate. The numbers in the circles indicate the number of carbon atoms in the compounds.

camber (*Aero*) The curvature of an aerofoil, relative to the chord line. Colloq term for the curved surface of an aerofoil.

camber (*Autos*) Inclination of the wheels, where there is an angle between the plane of the wheel and the vertical.

camber (*Build*) The recess in the side of the entrance to a basin, lock or graving dock, accommodating the SLIDING CAISSON.

camber (*CivEng*) An upward curvature to allow eg for settlement or to facilitate runoff of water.

camber (*Eng*) A convexity applied for some specific purpose, eg to girders to allow for deflection under load.

camber (*Ships*) The convexity of a deck line in a transverse section, normally 2 cm to each metre of breadth. Its purpose is to assist drainage and provide strength. Also *round of beam*.

camber arch (*Build*) An arch having a flat horizontal EXTRADOS and a cambered INTRADOS, with a rise of about 1 cm per metre of span.

camber beam (*Build*) A beam having an arched upper surface, or one sloping down towards each end, so as to form a support for roof covering on a flat roof.

camber flap (*Aero*) See PLAIN FLAP.

camber slip (*Build*) A strip of wood having a slightly cambered upper surface, upon which the brickwork of a flat arch is laid, so that after settlement the SOFFIT shall be straight.

cambial initial (*BioSci*) One of the permanently meristematic cells of a CAMBIUM.

cambium (*BioSci*) Zone of living cells in wood lying between the bark and sapwood; it is here that growth of the tree through cell division (periclinal divisions) occurs and secondary tissues are produced. See CORK CAMBIUM, VASCULAR CAMBIUM and panel on STRUCTURE OF WOOD.

Cambrian (*Geol*) The lowest division of the Palaeozoic era, covering an approx time span from 570 to 510 million years ago. Named after Cambria, the Roman name for Wales. The corresponding system of rocks.

cambric (*Textiles*) A closely woven fine linen cloth, used chiefly for handkerchiefs; the name is also applied to a plainweave fine-quality cotton cloth.

Cambridge ring (*ICT*) A LOCAL AREA NETWORK designed so that packets of data are entered and removed from frames which move continuously around the ring.

camcorder (*ImageTech*) A compact video camera with integral videotape recorder.

came (*Build*) A bar of lead suitably grooved to hold and connect adjacent panes of glass in a window.

camel hair (*Build*) An extremely soft hair obtained from squirrel tail. Used for gilder's mops and size brushes used in glass gilding.

camel hair (*Textiles*) A silky fibre from the haunch and underpart of the camel or dromedary; used for dress fabrics, warm coverings, artists' paint brushes, etc.

Camelopardalis (Giraffe) (*Astron*) A large and faint constellation in the northern hemisphere. Also *Camelopardus*.

cameo (*Arch*) (1) Carving or modelling in relief. (2) A striated shell or precious stone carved in relief to show different colours in the layers.

camera (*ImageTech*) Apparatus for forming an image of an external scene or subject on a light-sensitive surface, such as a photographic emulsion or the target in a TV CAMERA TUBE.

camera channel (*ImageTech*) In a TV studio, the camera, with all its supplies, monitor, control position and communication to the operator, which forms a unit, with others, for supplying video signal to the control room.

camera control unit (*ImageTech*) A remote unit used to set up and regulate the operation of a studio TV camera head. Built into smaller cameras. Abbrev CCU. Also *camera processing unit*.

camera head (*ImageTech*) A video camera without a lens and ELECTRONIC VIEWFINDER (and CAMERA CONTROL UNIT in the case of a studio camera).

camera lucida (*Phys*) A device for facilitating the drawing of an image seen in a microscope or other optical instrument. In its simplest form, it consists of a thin plate of unsilvered glass, placed above the eyepiece at an angle of 45° with the axis of the instrument, so as to reflect into the eye of the observer an image of the drawing surface, which is seen simultaneously with the microscope image.

camera marker (*ImageTech*) A system in a cinematograph camera for simultaneously producing a fogged area on the picture negative and a short tone on the sound record at the start of each take for subsequent synchronization.

camera obscura (*Phys*) A darkened room in which an image of surrounding objects is cast on a screen by a long-focus convex lens.

camera phone (*ICT*) A mobile phone incorporating a camera, enabling the transmission of digital images. Also *camphone*.

camera processing unit (*ImageTech*) See CAMERA CONTROL UNIT.

camera-ready copy (*Print*) Print copy suitably prepared for presentation to the camera for the production of final film for platemaking. Often referred to as CRC. US *mechanicals*.

camera tube (*ImageTech*) Tube which converts an image of an external scene into a video signal. Essential component in a video camera. US for *pick-up tube*.

camomile (*Med*) See CHAMOMILE.

camouflage (*BioSci*) The colour pattern or other physical features that enable an animal to blend with its natural environment and so avoid detection by predators.

camouflage (*Geol*) Describes the relationship between a trace element and a major element whose ionic charge and ionic radius are similar, as a result of which the trace element always occurs in the minerals of the major element but does not form separate minerals of its own. Thus gallium can be considered to be camouflaged by aluminium.

camouflage (*Radar*) Treatment of objects so that there is ineffective reflection of radar waves.

cAMP (*BioSci*) Abbrev for CYCLIC ADENOSINE MONOPHOSPHATE.

camp (*ICT*) A mobile telephone is said to camp on a CELL when it has determined that this cell provides the strongest authorized BROADCAST CONTROL CHANNEL and has initiated the procedure whereby it is registered at the VISITOR LOCATION REGISTER of this cell. It will then listen to paging messages from and initiate access requests to this cell only.

campaign (*NucEng*) The period often of several months between starting and closing a batch of operations in a nuclear fuel reprocessing plant.

Campanian (*Geol*) A stage in the Upper Cretaceous. See MESOZOIC.

campaniform (*BioSci*) Dome-shaped, eg as *campaniform sensilla*, mechanoreceptors occurring widely on the body of certain insects.

campanile (*Arch*) A bell-tower, often detached.

campanulate (*BioSci*) Bell-shaped.

Campbell bridge (*ElecEng*) An electrical network, designed by Campbell, for comparing mutual inductances.

Campbell gauge (*ElecEng*) An electrical bridge, one arm being the filament of a lamp located in the low gas pressure to be measured.

Campbell's formula (*ElecEng*) Formula which gives the effective attenuation of a coil-loaded transmission line in terms of the constants of the line and the magnitude of the loading.

Campbell–Stokes recorder (*EnvSci*) See SUNSHINE RECORDER.

camp ceiling (*Build*) A ceiling having two opposite parts sloping in line with the rafters, the middle part being horizontal.

camphane (*Chem*) $C_{10}H_{18}$. White crystals, mp 154°C. It is a saturated terpene hydrocarbon, the parent substance of the CAMPHOR group.

camphene (*Chem*) $C_{10}H_{16}$. White solid, mp 50°C. An unsaturated terpene hydrocarbon, occurring in various essential oils; it can be prepared from pinene hydrochloride.

camphor (*Chem*) Common (Japan) camphor. $C_{10}H_{16}O$. Colourless transparent prisms of characteristic odour; mp 175°C, bp 204°C, rel.d. 0·985. Can be sublimed readily and is volatile in steam. Natural camphor is obtained from the camphor tree (*Cinnamomum camphora*). Widely cultivated in Taiwan, China and Japan. Synthetic camphor (optically inactive) is derived from α-pinene. Camphor is an ingredient of various lotions and liniments; used as a solid plasticizer for cellulose nitrate in celluloid. See BORNEOL.

campimetry (*Med*) A technique for assessing the field of vision.

CAM plant (*BioSci*) A plant that photosynthesizes by means of CRASSULACEAN ACID METABOLISM.

cam profile (*Eng*) The shape of the CAM as determined by the form of the flanks and tip; in general, the cam outline.

camp sheathing (*CivEng*) An earth-retaining wall formed of timber piles placed 2–3 metres apart and connected by stout timber WALINGS; often used to support river banks.

camp sheeting (*Build*) Sheet piling used in foundation work to retain sandy earth.

camptonite (*Geol*) An igneous rock occurring in minor intrusions, and belonging to the family of lamprophyres. It consists essentially of plagioclase feldspar and brown hornblende, usually BARKEVIKITE.

CAMPUS (*Chem*) TN for database of properties of polymers supplied by a range of European manufacturers.

Campylobacter (*Med*) A genus of Gram-negative bacteria that are a common cause of food poisoning and of opportunistic infections, particularly in immuno-compromised patients.

campylotropous (*BioSci*) Descriptive of a plant ovule that is curved so that the micropyle and the stalk are approximately at right angles and the stalk appears to be attached to the side.

camshaft (*Autos, Eng*) (1) A shaft with lobed cams to operate the inlet and exhaust valves of a four-stroke engine. It is driven at half-crankshaft speed by various means. See TIMING GEAR. (2) Any shaft to which cams are keyed or formed integrally.

camshaft controller (*ElecEng*) A form of control equipment for electric motors (usually in locomotives), in which the contactors are operated mechanically by cams on a rotating shaft.

cam-type steering gear (*Autos*) Steering gear in which the steering column carries a pair of opposed volute cams, which engage with a peg or roller carried by a short arm attached to the drop-arm spindle.

can (*NucEng*) A cover for reactor fuel rods, usually metallic (aluminium, magnox, stainless steel, zircaloy). Also *cartridge, jacket*. See CLADDING.

Canada balsam (*Chem*) Balsam of fir, or Canada turpentine. A yellowish liquid, of pine-like odour, soluble in ethoxyethane, trichloromethane, benzene; obtained from *Abies balsamica*. Used for lacquers and varnishes, and as an adhesive for lenses, instruments, etc, its refractive index being approximately the same as that of most optical glasses.

Canadian asbestos (*Min*) See CHRYSOTILE.

Canadian latch (*Build*) See NORFOLK LATCH.

Canadian shield (*Geol*) The vast area of Precambrian rocks which cover 5 million square kilometres in E Canada.

Canadian spruce (*For*) Name for the wood of several trees of the *Picea* genus, the most important commercial timber in Canada. See SITKA SPRUCE, WHITEWOOD.

Canadian Standard Freeness (*Paper*) A laboratory control test of beating in which diluted pulp is allowed to drain through a sieve. A wet beaten pulp gives a low value. Abbrev *CSF*.

canal (*BioSci*) An elongated intercellular space containing air, water or a secretory product such as resin or oil.

canal (*NucEng*) Water-filled trench into which the highly active elements from a reactor core can be discharged. The water acts as a shield against radiation but allows objects to be easily inspected.

canal cell (*BioSci*) One of the short-lived cells present in the central cavity of the neck of the ARCHEGONIUM.

canaliculate (*BioSci*) Marked longitudinally by a channel or groove.

canaliculus (*BioSci*) Any small channel; in the liver, an intercellular bile channel; in bone, one of the ramified passages uniting the lacunae; in nerve cells, a fine channel penetrating the cytoplasm of the cell body. Adj *canalicular*.

canalization (*Med*) The formation of a new channel in a clot blocking the lumen of a blood vessel.

canals of Mars (*Astron*) Indistinct channel-like markings apparent on the surface of Mars when viewed through a telescope.

canard (*Aero*) See TAIL-FIRST AIRCRAFT.

canaries (*ImageTech*) Extraneous high-frequency noises reproduced from a recording channel.

canary whitewood (*For*) See AMERICAN WHITEWOOD.

cancelbot (*ICT*) A computer program that identifies and deletes unwanted articles sent to an Internet newsgroup.

cancellous (*BioSci*) Having a spongy structure, with obvious interstices. Also *cancelled*.

cancels (*Print*) Pages printed to substitute for book pages containing errors.

cancer (*Med*) Any malignant neoplasm. An uncontrolled growth of cells which exhibits invasiveness and remote growth. Adj *cancerous*.

Cancer (Crab) (*Astron*) An inconspicuous northern constellation, lying between Gemini and Leo. It contains the star cluster PRAESEPE.

cancer death rate (*Med*) The measure of deaths from cancer, usually expressed as a rate per unit of population.

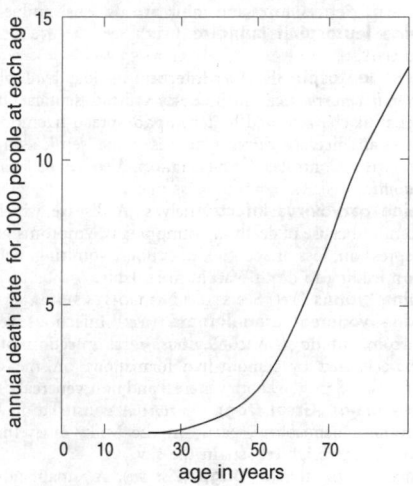

cancer death rate

can coiler (*Textiles*) A device for feeding a sliver into a revolving can in spiral form so allowing easy withdrawal for subsequent processing.

cancrinite (*Min*) A hydrated silicate of aluminium, sodium and calcium, also containing the carbonate ion. Found in alkaline plutonic igneous rocks. Cf VISHNEVITE.

cancrum oris (*Med*) A destructive ulceration of the cheek in debilitated children, usually during convalescence from an infectious disease. Also *noma*.

candela (*Phys*) Fundamental SI unit of luminous intensity. If, in a given direction, a source emits monochromatic radiation of frequency 540×10^{12} Hz, and the radiant intensity in that direction is 1/683 watts per steradian, then the luminous intensity of the source is 1 candela. Symbol cd.

candesartan (*Pharmacol*) A selective AT1-subtype *angiotensin II receptor* antagonist.

Candida albicans (*Med*) A dimorphic fungus that is an opportunistic pathogen of humans (causing *candidiasis* or *thrush*).

candidiasis (*Vet*) See MONILIASIS.

candle (*Phys*) Older unit of luminous intensity. See CANDELA.

candle-power (*Phys*) A former unit of luminous intensity, now replaced by the CANDELA.

candlewick (*Textiles*) A coarse folded yarn made from cotton. As well as being used in candles the yarn is used for the manufacture of fabrics suitable for bedspreads. See WICK.

CANDU (*NucEng*) Type of thermal nuclear power reactor developed by and widely used in Canada. It uses natural (unenriched) uranium oxide fuel canned in ZIRCALOY and HEAVY WATER as MODERATOR and coolant.

cane-sugar (*Chem*) See SUCROSE.

Canes Venatici (Hunting Dogs) (*Astron*) An inconspicuous constellation in the northern hemisphere which includes many bright galaxies.

canicola fever (*Med, Vet*) The disease in humans caused by infection by *Leptospira canicola*, the natural host of which is the dog.

canine (*BioSci*) Pertaining to, or resembling, a dog; in mammals, a pointed tooth with single cusp, adapted for tearing, and occurring between the incisors and premolars; pertaining to a canine tooth; pertaining to a ridge or groove on the surface of the maxillary.

canine distemper (*Vet*) Paramyxovirus infection with secondary bacterial complications. Symptoms include pyrexia, malaise, conjunctivitis, respiratory, enteric and nervous signs. Hyperkeratosis (hard pad) occasionally also present. Ferrets, foxes and mink are also susceptible.

canine leptospiral jaundice (*Vet*) See CANINE LEPTOS-PIROSIS.

canine leptospirosis (*Vet*) Infection of dogs by *Leptospira icterohaemorrhagiae* which causes an acute septicaemia and hepatitis characterized by fever, widespread haemorrhages and jaundice. *L. canicola* causes acute septicaemia and nephritis. Controlled by vaccination. Also *canine leptospiral jaundice, yellows*.

canine parvovirus infection (*Vet*) A disease which is a common cause of death in young pups. Symptoms include depression, loss of appetite, diarrhoea, vomiting, dehydration leading to death. Vaccines available.

canine typhus (*Vet*) See STUTTGART DISEASE.

canine venereal granulomata (*Vet*) Infectious venereal sarcoma of dogs. A contagious, viral infection of dogs characterized by tumour-like formations on the genital mucosae. Spread by both venereal and non-venereal contact.

Canis Major (Great Dog) (*Astron*) A constellation in the southern hemisphere, partly in the Milky Way. Includes SIRIUS, the brightest star in the sky.

Canis Minor (Little Dog) (*Astron*) A small northern hemisphere constellation. Its brightest star is PROCYON.

canker (*BioSci*) (1) A plant disease characterized by well-defined necrotic lesions of a main root, stem or branch in which the tissues outside the xylem disintegrate. (2) A chronic inflammation of the keratogenous membrane of the frog and sole of a horse's foot. (3) Chronic eczema of the ear of dogs. (4) An abscess or ulcer in the mouth, eyelids, ear or cloaca of birds.

cannabinoids (*Pharmacol*) A group of compounds, all derivatives of 2-(2-isopropyl-5-methylphenyl)-5-pentylresorcinol, found in cannabis. The most important members of the group are *cannabidiol, cannabinol* and various *tetrahydrocannabinols* (THCs). Cannabinoids bind to specific cellular receptors and mimic the actions of endogenous agonists.

cannabis (*Pharmacol*) The plant *Cannabis sativa* or *Cannabis indica*. The dried flowers, exuded resin and leaves are used to produce the drug hashish, marijuana or bhang. Also *Indian hemp*. It can be taken by smoking (joint or pipe) or by ingestion. Cannabis dependence is increasing, and cannabis use is often associated with the development of dependence on heroin and cocaine. Chronic use is associated with increased risk of lung cancer, reduced sperm motility, cognitive dysfunction and psychiatric illness. The active ingredients (cannabinols) have some medicinal use in the treatment of vomiting, intractable hiccups, cachexia in AIDS patients and in reducing muscular spasm and pain in multiple sclerosis, but these treatments are controversial.

cannibalism (*Astron*) (1) The merging of small galaxies into a large galaxy at the centre of a cluster of galaxies. This process is thought to create the most massive galaxies known. (2) The merging of a star into its giant companion in a close binary system.

cannibalism (*Vet*) A habit, mainly of poultry raised in captivity and characterized by varying degrees of tissue loss by picking of the vent, toes, feathers and around the head. Contributory factors thought to be overcrowding, excessive light and temperature, nutritional imbalance and variations in feeding regime. Reduced by proper husbandry.

canning (*FoodSci*) A packaging process in which foods are preserved by being hermetically sealed in steel or aluminium cans. The cans are generally lacquered internally to prevent acid corrosion and formation of sulphur compounds, which can cause discoloration, and to protect from tin or aluminium residues.

Cannizzaro reaction (*Chem*) Reaction used for the DISPROPORTIONATION of aromatic aldehydes to alcohols and acids, heating with concentrated alkali, eg benzaldehyde→benzyl alcohol plus a benzoate.

cannon bone (*BioSci*) In the more advanced ARTIODAC-TYLA, the characteristic bone formed by the fusion of the two metapodials in the limb, associated with the reduction of the number of toes to two.

cannula (*Med*) A tube, usually fitted with a TROCAR, for insertion into the body for the injection or removal of fluids or gases.

cannular combustion chamber (*Aero*) A gas turbine combustion system with individual flame tubes inside an annular casing.

canola (*Agri*) A specific edible type of oil-rich rapeseed (*Brassica napus*); the TN is registered by the Western Canadian Oilseed Crushers Association.

canonical assembly (*Phys*) Term used in statistical thermodynamics to designate a single assembly of a large number of systems which are such that the number of systems with energies lying between E and $E + dE$ is proportional to $e^{-E\theta}$, where θ is a parameter characteristic of the assembly.

canonical correlation (*Genrl*) A correlational technique used when there are two or more data categories to be correlated with two or more other categories, eg the correlation between (age and sex) and (income and life satisfaction).

Canopus (*Astron*) A prominent supergiant star in the constellation Carina, the second brightest in the sky. Distance 60 pc. Also *Alpha Carinae*.

canopy (*Aero*) (1) The transparent cover of a cockpit. (2) The fabric (nylon, silk or cotton) body of a parachute, which provides high air drag. Usually hemispherical, but may be lobed or rectangular in shape. See RIBBON PARACHUTE.

canopy (*BioSci*) The leaves, stems and branches of a plant or an area of vegetation, considered as a whole.

canopy (*Build*) A roof or balcony projecting from a wall or supported on pillars.

canopy cover (*BioSci*) The percentage of the ground occupied by the vertical projection of all the individuals of one plant species. The sum of such percentages for all plant species gives the total canopy cover.

cant (*Build*) A moulding having plane surfaces and angles instead of curves.

cant (*Surv*) The slope of rail or road curve whereby the outer radius is superelevated, to counteract centrifugal thrust of traffic.

cant bay (*Arch*) A bay window having three sides, the outer two being splayed from the wall sides.

cant board (*Build*) A board laid on each side of a valley gutter to support the sheet lead.

cant brick (*Build*) See SPLAY BRICK.

canted column (*Arch*) A column having faceted sides instead of curved flutes.

canted deck (*Aero*) US for ANGLED DECK.

canted wall (*Build*) A wall built at an angle to the surfaces of another wall.

Canterbury hammer (*Build*) A type of hammer with a thick and shallow-curved claw.

cantharadin (*Pharmacol*) A pharmaceutical product obtained from the dried elytra of the Spanish fly (spp of *Lytta* and *Mylabris*). Formerly used as an irritant (vesicant) and to treat warts. Taken orally it was believed to be an aphrodisiac.

cantilever (*CivEng*) A beam or girder fixed at one end and free at the other.

cantilever bridge (*CivEng*) A bridge formed of self-supporting projecting arms built outwards from the piers and meeting in the middle of the span, where they are connected together. See SUSPENDED SPAN and panel on BRIDGES AND MATERIALS.

cantilever deck (*CivEng*) A bridge where the deck slab is fixed above the main beams or trusses and is cantilevered beyond the outer beams or trusses.

cantilevered steps (*Build*) See HANGING STEPS.

canting (*Eng*) Tilting over from the proper position; as in the canting of a piston in its cylinder under the oblique thrust of the connecting rod.

canting strip (*Build*) A projecting sloping member of wood or masonry fitted around a building to deflect water from the wall. Also *water table*.

cantling (*Build*) The lower of two courses of burnt brick enclosing a clamp for firing bricks.

canton (*Build*) A pilaster, or quoin forming a salient corner, which projects from the wall face.

canvas (*Textiles*) A heavy closely woven fabric made from cotton, flax, hemp or jute for uses where strength and firmness are required, eg interlinings, sails, tents.

canyon (*Geol*) A deep, narrow, steep-sided valley.

canyon (*NucEng*) US for long narrow space often partly underground with heavy shielding for essential processing of wastes from reactors.

caoutchouc (*Chem*) Fr name for natural rubber, derived from Carib Indian term for coagulated latex tapped from various species of plants and trees. Also *kautschuk*.

CAP (*Agri*) See COMMON AGRICULTURAL POLICY.

CAP (*ICT*) Abbrev for *computer-aided production*. See COMPUTER-AIDED MANUFACTURE.

CAP (*FoodSci*) Abbrev for CONTROLLED ATMOSPHERE PACKING.

cap (*BioSci*) (1) A modified base added to the 5′ ends of eukaryotic messenger RNA molecules. (2) A protein added to the end of, eg a microfilament, to prevent further assembly. (3) Abbrev for *catabolite gene activator protein*.

cap (*Build*) (1) The upper member of a column. (2) A wall coping. (3) The head added to top of a pile. (4) A planted piece on top of a post for weathering or ornamentation. (5) A hand-rail supported on balusters.

cap (*EnvSci*) (1) The covering of cloud which congregates at the top of a mountain. (2) The transient top of detached clouds above an increasing cumulus. Also *pileus*.

capacitance (*Phys*) Of an isolated conductor, the ratio of the total charge on it to its potential; $C = Q/V$. See FARAD, STRAY CAPACITANCE.

capacitance bridge (*ElecEng*) An ac bridge network for the measurement of capacitance. See SCHERING BRIDGE, WIEN BRIDGE.

capacitance coefficients (*Phys*) Charges ($q_1, ..., q_n$) of a system of conductors can be expressed in terms of coefficients of electric induction (C_{ij}) by the following equations:

$$q_1 = C_{1\infty}V_1 + C_{12}(V_1 - V_2) + ... + C_{1n}(V_1 - V_n)$$
$$q_2 = C_{21}(V_2 - V_1) + C_{2\infty}V_2 + ... + C_{2n}(V_2 - V_n)$$
$$q_n = C_{n1}(V_n - V_1) + C_{n2}(V_n - V_2) + ... + C_{n\infty}V_n$$

where

$$C_{km} = -C_{mk}(m \neq k)$$

and

$$C_{m\infty} = C_{m1} + C_{m2} + ... + C_{m(n-1)} + C_{mn}$$

They are the fundamental relations for partial capacitances of a number of conductors, eg electrodes in valves, conductors in cables, variable air capacitors.

capacitance coupling (*ElecEng, Electronics*) Interstage coupling through a series capacitance or by a capacitor in a common branch of a circuit.

capacitance grading (*ElecEng*) Grading of the properties of a dielectric, so that the variation of stress from conductor to sheath is reduced. The inner dielectric has the higher permittivity. Ideally, the grading is continuous and the permittivity varies as the reciprocal of the distance from the centre. See CONDENSER BUSHING.

capacitance integrator (*ElecEng*) Resistance–capacitance circuit whose output voltage is approximately equal to the time integral of the input voltage.

capacitive load (*ElecEng*) Terminating impedance which is markedly capacitive, taking an ac leading in phase on the source emf, eg electrostatic loudspeaker.

capacitive reactance (*ElecEng*) The impedance associated with a capacitor. Has a magnitude in ohms equal to the reciprocal of the product of the capacitance (in farads) and the angular frequency of the supply (in rads s^{-1}). Also introduces a 90° phase angle such that the current through the device leads the applied voltage.

capacitor (*ElecEng*) Electric component having CAPACITANCE; formed by conductors (usually thin and extended) separated by a dielectric, which may be vacuum, paper (waxed or oiled), mica, glass, plastic foil, fused ceramic, air, etc. Maximum potential difference tolerated depends on the electrical breakdown of dielectric. Modern construction uses sheets of metal foil and insulating material wound into a compact assembly. Air capacitors, of adjustable parallel vanes, are used for tuning high-frequency oscillators. Formerly *condenser*.

capacitor bushing (*ElecEng*) See CONDENSER BUSHING.

capacitor loudspeaker (*Acous*) See ELECTROSTATIC LOUDSPEAKER.

capacitor microphone (*Acous*) See ELECTROSTATIC MICROPHONE.

capacitor modulator (*Acous*) Capacitor microphone, or similar transducer, which, by variation in capacitance, modulates an oscillation either in amplitude or frequency.

capacitor motor (*ElecEng*) A single-phase induction motor arranged to start as a two-phase motor by connecting a capacitor in series with an auxiliary starting winding. The capacitor may be automatically disconnected when the motor is up to speed (*capacitor-start motor*) or it may be left permanently in circuit for power-factor improvement (*capacitor start-and-run motor*).

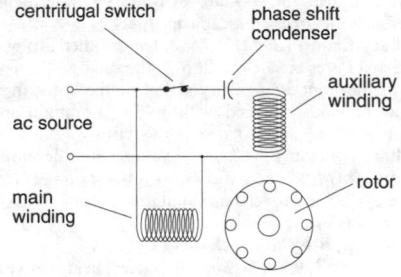

capacitor motor Switch omitted in start-and-run configuration.

capacitor–resistance law (*ElecEng*) The law relating to exponential rise or decay of charge on a capacitor in series with a resistor, and, by extension, to signal distortion on long submarine cables. Abbrev *C–R law*.

capacitor start (*ElecEng*) Starting unit for electric motor using series capacitance to advance phase of current.

capacitor terminal (*ElecEng*) See CONDENSER BUSHING.

capacitron (*ICT*) See BAND IGNITOR TUBE.

capacity (*ElecEng*) (1) The output of an electrical apparatus, eg that of a motor or generator in kilowatts. In an accumulator (secondary battery), it is measured by the ampere-hours of charge it can deliver. The capacity of a switch is the current it can break under specified circuit conditions. (2) Sometimes used to mean CAPACITANCE.

cap-and-pin-type insulator (*ElecEng*) A special form of the SUSPENSION INSULATOR.

Cape asbestos (*Min*) Blue asbestos from S Africa. See CROCIDOLITE.

capecitabine (*Pharmacol*) A drug used in the treatment of advanced cancers.

Cape diamond (*Min*) A name used in grading diamonds to designate an off-colour stone of a yellowish tint.

Capella (*Astron*) A prominent yellow giant star in the constellation Auriga, a spectroscopic triple star. Distance 13·7 pc. Also *Alpha Aurigae*.

Cape olive (*For*) See STINKWOOD.

Cape ruby (*Min*) A misnomer for the red garnet PYROPE, obtained in the diamond mines of S Africa. See FALSE RUBY.

Cape walnut (*For*) See STINKWOOD.

capillariasis (*Vet*) Inflammation of the alimentary tract of animals or birds due to infection by nematode worms of the genus *Capillaria*.

capillarity (*Phys*) A phenomenon associated with surface tension, which occurs in fine-bore tubes or channels. Examples are the elevation (or depression) of liquids in capillary tubes and the action of blotting paper and wicks. The elevation of liquid in a capillary tube above the general level is given by the formula

$$h = \frac{2T\cos\theta}{\rho gr}$$

where T is the surface tension, θ is the angle of contact of the liquid with the capillary, ρ is the liquid density, g is the acceleration due to gravity and r is the capillary radius.

capillary (*BioSci*) (1) Of very small diameter; slender, hair-like. (2) Any thin-walled vessel of small diameter, forming part of a network, that aids rapid exchange of substances between the contained fluid and the surrounding tissues, eg *bile capillaries*, *blood capillaries*, *lymph capillaries*.

capillary action (*Chem*) See CAPILLARITY.

capillary condensation (*Chem*) The hypothesis that adsorbed vapours can condense under the incidence of capillary forces to form liquid inside the pores of the adsorbate.

capillary electrometer (*Chem*) An instrument in which small electric currents are detected by movement of a mercury meniscus in a capillary tube.

capillary fitting (*Build*) A bend, tee or other fitting whose internal bores are a close fit over the tube. Solder is drawn in to the joint by capillary action or the bores may have been previously tinned with solder. Common joint in copper work. Cf COMPRESSION FITTING.

capillary pressure (*Chem*) The pressure developed by CAPILLARITY. Mathematically expressed as $p = 2T(\cos\theta)/r$, where T is the surface tension, θ the angle of contact and r the radius of the capillary.

capillary pyrite (*Min*) See MILLERITE.

capillary soil water (*EnvSci*) Water held between the particles of the soil by capillarity.

cap iron (*Build*) See BACK IRON.

capital (*Build*) The upper member of a column, pier or pilaster.

capitate (*BioSci*) Head-like.

capitate (*BioSci*) Having an enlarged tip, eg as *capitate antennae*.

capitellum (*BioSci*) An enlargement or boss at the end of a bone, for articulation with another bone; particularly, the smaller of the two articular surfaces on the distal end of the mammalian humerus, for articulation with the radius; the distal knob-like extremity of the haltere in Diptera.

capitulum (*BioSci*) (1) An inflorescence on which the sessile flowers or florets are crowded on the surface of the enlarged apex of the peduncle, the whole group being surrounded, and covered in the bud by an envelope of bracts forming an involucre; the whole inflorescence superficially appears to be one flower, as in the daisy, *Bellis*. See COMPOSITAE. (2) A terminal expansion, as that of some shaft bones, some tentacles and some hairs.

cap nut (*Eng*) A nut whose outer end is closed, so protecting the end of the screw and giving a neat appearance. Also *box nut*, *dome nut*.

capon (*Agri*) A male chicken surgically castrated before 8 weeks. This modifies behaviour, especially reducing aggression, increases growth rate and is said to improve quality of the flesh for food.

caponizing (*Vet*) Castration of a cock bird.

capped elbow (*Vet*) A swelling of the OLECRANON bursa of animals.

capped hock (*Vet*) A swelling over the point of the hock of animals.

capped lens element (*ImageTech*) A glass lens element with a moulded transparent plastic cap. Allows a more complex curvature than would be possible by grinding the glass.

capping (*Build*) A copper strip rolled over the ridge or roll and welted to the underlying sheets.

capping (*BioSci*) (1) Movement of cross-linked cell-surface material to the posterior region of a moving cell, or to the perinuclear region. (2) Intracellular accumulation of inter-mediate filament protein in the pericentriolar region following microtubule disruption by colchicine. (3) Block-ing of further addition of subunits by binding of eg a cap protein to the free end of a linear polymer such as actin.

capping (*MinExt*) (1) The fixing of a shackle or a swivel to the end of a hoisting rope. (2) Overburden lying above valuable seam or bed of mineral.

capping brick (*Build*) See COPING BRICK.

capping-plane (*Build*) A plane for giving a slight rounding to the upper surface of a wooden hand-rail.

capric acid (*Chem*) *Decanoic acid*. $CH_3(CH_2)_8COOH$.

Capricorn (*Astron*) A southern constellation, lying between Sagittarius and Aquarius. Also *Capricornus* (*Goat*).

caprification (*BioSci*) The fertilization of the flowers of fig trees by the agency of fig insects, a family of chalcids (Agaonidae), facilitated by hanging wild figs (caprifigs) in the female trees.

caprock (*Geol*, *MinExt*) An impervious rock stratum overlying a reservoir formation, thus trapping the oil or natural gas in the reservoir.

caproic acid (*Chem*) *Hexanoic acid*. $CH_3(CH_2)_4COOH$. Oily liquid, solidification point about 2°C, bp 205°C. Occurs as glycerides in milk, palm oil, etc.

caprolactam (*Chem*) Cyclic amide which polymerizes to form nylon 6.

capryl alcohol (*Chem*) *Octan-2-ol*. $CH_3(CH_2)_5CH(OH)CH_3$. It is obtained by distilling castor oil with strong alkali. Liquid, colourless, strong smell, bp 179°C. Used as a foam-reducing agent.

caprylic acid (*Chem*) *Octanoic acid*. $C_7H_{15}COOH$, bp 237°C.

CAPS (*Chem*) Abbrev for *computer-aided polymer selection* methods using large databases of polymers from different manufacturers.

capsaicin (*BioSci*) *8-methyl-N-vanillyl-6-nonenamide*, a compound in chilli peppers responsible for the hot taste. Used medically to treat phantom limb pain and other neuropathies.

caps and small caps (*Print*) Small capitals with the first letter in capitals. The first word of a chapter is often set in caps & small caps.

cap screw (*Eng*) See SOCKET HEAD SCREW.

capsid (*BioSci*) (1) The protein coat, often of icosahedral symmetry, that covers the nucleic acid core of a VIRION. (2) In zoology, an insect of the bug family Miridae.

capsomere (*BioSci*) Proteins which form regular structures on the surface of a virus.

capstan (*Eng*) A vertical drum or spindle on which rope is wound (eg for warping a ship alongside a wharf); it is rotated manually or by hydraulic or electric motor.

capstan (*ImageTech*) A roller providing the constant-speed drive in a magnetic tape recorder.

capstan-head screw (*Eng*) A screw having a cylindrical head provided with radial holes in its circumference. It is tightened by a tommy bar inserted in these holes.

capstan lathe (*Eng*) A lathe in which the tools required for successive operations are mounted radially in a tool-holder resembling a capstan; by revolving this, each tool in turn may be brought into position in exact location.

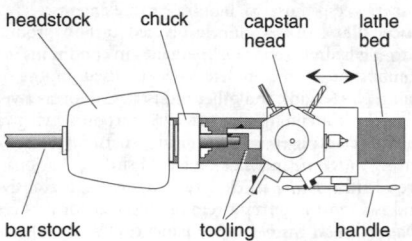

labels: headstock, chuck, capstan head, lathe bed, bar stock, tooling, handle

capstan lathe View from top; the handle advances the capstan head and slide to positions set by multiple stops and turns head after each operation.

capstan nut (*Eng*) A nut which is tightened in the same way as a CAPSTAN-HEAD SCREW.

capsular polysaccharides (*BioSci*) Polysaccharides that are present as constituents of bacterial capsules and that often allow the bacteria to avoid phagocytosis.

capsule (*BioSci*) (1) That part of the sporophyte of a bryophyte which contains the spores. (2) A fruit, dry when mature, composed of more than one carpel, that splits at maturity to release the seeds. (3) A coating of mucilaginous material outside the wall of a bacterial cell (*capsular polysaccharide*). (4) Any fibrous or membranous covering of a organ, eg the kidney. (5) Name applied to certain areas in the brain that are formed by nerve fibres. (6) A soluble case of gelatin or similar substance in which a medicine may be enclosed.

captan (*Chem*) *N-(trichloromethylthio) cyclohex-4-ene-1,2-dicarboxyimide.* A fungicide.

caption (*Print*) The descriptive wording under an illustration. Also *legend*. US *cut line*.

captive balloon (*Aero*) A balloon anchored or towed by a line. Usually the term refers to spherical balloons only. Special shapes, eg for stability, are called *kite balloons*.

captive nut (*Eng*) A nut (loosely) fastened to an adjacent machine member so as to retain it in position when the corresponding screw element is absent.

captive screw (*Eng*) A screw (loosely) fastened in position by its head or shank so as to be retained when unscrewed from the matching machine member.

captive tape (*Print*) Tapes which operate only at the slow speeds when the web is being threaded on rotary presses.

cap torque (*FoodSci*) The tightness of a screw cap measured as the product of the tangential force applied and its distance from the pivot in N cm. For containers with a vacuum or partial vacuum, the torque to break the vacuum and the torque to continue twisting the cap both have to be measured.

captopril (*Pharmacol*) An ACE INHIBITOR used to treat hypertension.

capture (*ICT*) (1) In frequency and phase modulation, the diminution to zero of a weak signal (noise) by a stronger signal. See K-CAPTURE. (2) The situation in which a piece of hardware is monopolized by a process or program, preventing other services from accessing it.

capture (*Phys*) Any process in which an atomic or nuclear system acquires an additional particle. In a nuclear radiative capture process there is an emission of electromagnetic radiation only, eg the emission of gamma rays subsequent to the capture of a neutron by a nucleus.

caput (*BioSci*) An abrupt swelling at the distal end of a structure. Pl *capita*. Adj *capitate*.

caput medusae (*Med*) Dilated subcutaneous veins around the umbilicus in cirrhosis of the liver.

car (*Aero*) In an airship, the part intended for the carrying of the load (crew, passengers, goods, engines, etc). It may be suspended below, or may be inside the hull or envelope.

caracole (*Build*) A helical staircase.

Caradoc (*Geol*) An epoch of the Ordovician period. See PALAEOZOIC.

caramel (*FoodSci*) The product of heating sugars or other carbohydrates in solution, used as a flavour and colour in food production.

caramelization (*FoodSci*) A non-enzymic browning which occurs when sugars are heated, usually a desirable feature in confectionery but sometimes an unwanted consequence of overheating.

carapace (*BioSci*) An exoskeleton shield covering part or all of the dorsal surface of an animal, eg the bony dorsal shield of a tortoise, the chitinous dorsal shield of some Crustacea.

carat (*Min*) (1) A standard mass for precious stones. The *metric carat*, standardized in 1932, equals 200 mg (3·086 grains). (2) The standard of fineness for gold. The standard for pure gold is 24 carats; 22-carat gold has two parts of alloy; 18-carat gold six parts of alloy. Also *karat*.

carbamazepine (*Pharmacol*) An anticonvulsant drug used to treat many forms of epilepsy; also used to treat trigeminal neuralgia, phantom-limb pain, and manic-depressive illness resistant to lithium.

carbamic acid (*Chem*) NH_2COOH. Not known to occur free, being known only in the form of derivatives, eg the ammonium salt, NH_2COONH_4. The esters are known as *urethanes*.

carbamide (*Chem*) Urea.

carbamyl chloride (*Chem*) NH_2COCl. Colourless needles of pungent odour; mp 50°C, bp 61°C; formed by the action of hydrochloric acid (gaseous) on cyanic acid; it serves for the synthesis of organic acids.

carbamyl phosphate (*BioSci*) The phosphate ester of carbamic acid. It is an intermediate in the biosynthesis of urea and pyrimidines.

carbanilide (*Chem*) *Diphenylurea.* $CO(NHC_6H_5)_2$.

carbanion (*Chem*) A short-lived, negatively charged intermediate formed by the removal of a proton from a C—H bond, eg in BUTYL LITHIUM.

carbazole (*Chem*) $(C_6H_4)_2NH$. Colourless plates, mp 238°C, sublimes readily; contained in coaltar and crude anthracene oil. It is the imine (intramolecular) of diphenyl and is formed from diphenylamine by passing the vapour through red-hot tubes, or by distilling *o*-aminodiphenyl over lime at about 600°C.

carbenes (*Chem*) (1) Reactive uncharged intermediates of formula CXY, where X and Y are organic radicals or halogen atoms. (2) Such constituents of asphaltic material as are soluble in carbon disulphide but not in carbon tetrachloride. See ASPHALTENES, MALTHENES.

carbenium ion (*Chem*) A (usually) short-lived intermediate in a reaction with a positive charge on a carbon atom. Also *carbocation, carbonium ion*.

carbenoxolone (*Pharmacol*) Synthetic derivative of glycyrrhizic acid which is useful in healing gastric ulcers.

carbethoxy (*Chem*) The group $—COOC_2H_5$ in organic compounds.

carbides (*Chem*) Binary compounds of metals with carbon. Carbides of group IV to VI metals (eg silicon, iron, tungsten) are exceptionally hard and refractory. In groups I and II, calcium carbide (ethynide) is the most useful. See CEMENTED CARBIDES, CEMENTITE.

carbide tools (*Eng*) Cutting and forming tools used for hard materials or at high temperatures. They are made of carbides of tungsten, tantalum and other metals held in a matrix of cobalt, nickel, etc, and are very hard with good compressive strength.

carbimazole (*Pharmacol*) A drug that inhibits thyroxine synthesis and is used in the treatment of THYROTOXICOSIS.

carbinol (*Chem*) *Methanol*. The nomenclature of alcohols is often based on their homologous relation to methanol, eg tertiary butyl alcohol, $(CH_3)_3COH$, is termed *trimethyl carbinol*.

carbocation (*Chem*) See CARBENIUM ION.

carbocyclic compounds (*Chem*) Closed-chain or ring compounds in which the closed chain consists entirely of carbon atoms. Also *isocyclic compounds*.

carbodi-imide resins (*Chem*) Thermoset foams formed by self-polymerization reaction of di-isocyanates, Cross-linking occurs by reaction between free isocyanate groups and the di-imide (—N=C=N—) group.

carbofuran (*Chem*) *2,3-dihydro-2,2-dimethylbenzofuran-7-yl methylcarbamate*; a broad spectrum pesticide and a cholinesterase inhibitor.

carbohydrates (*Chem, FoodSci*) A group of compounds represented by the general formula $C_x(H_2O)_y$. Found in plants and animals, eg sugars, starch, cellulose. The carbohydrates also comprise other compounds of a different general formula but closely related to the above substances, eg rhamnose, $C_6H_{12}O_5$. Carbohydrates are divided into DISACCHARIDES, MONOSACCHARIDES, OLIGOSACCHARIDES and POLYSACCHARIDES. The carbohydrate element in diet supplies energy, provided by the oxidation of the constituent elements.

carbolic acid (*Chem*) Phenol.

carbolic oils (*Chem*) See MIDDLE OILS.

carbomethoxy (*Chem*) The group —CH_3OCO in organic compounds.

carbon (*Chem*) Amorphous or crystalline (graphite and diamond) element, symbol C, at no 6, ram 12·011, mp above 3500°C, bp 4200°C. Its allotropic modifications are diamond, graphite and graphene. The assumption that its atom is tetravalent, the bonds being directed towards the vertices of a regular tetrahedron, is the basis for all theoretical organic chemistry (see CARBON COMPOUNDS). Widely used in brushes for electric generators and motors, and alloyed with iron for steel. Colloidal carbon or graphite is used to coat cathode-ray tubes and electrodes in valves, to inhibit photoelectrons and secondary electrons. High-purity carbon, crystallized to graphite in a coke furnace for many days, is used in many types of nuclear reactors, particularly for moderation of neutrons. See CARBON FIBRES and panel on RADIOMETRIC DATING.

carbonaceous (*Chem*) Said of material containing carbon as such or as organic (plant or animal) matter.

carbonaceous chondrite (*Min*) Friable black carbonaceous chondritic stony meteorite.

carbonaceous rocks (*Geol*) Sedimentary deposits of which the chief constituent is carbon, derived from plant residues, including peat, lignite or brown coal, and the several varieties of true coal (bituminous coal, anthracite, etc).

carbonado (*Min*) See BLACK DIAMOND.

carbon arc (*ElecEng*) An arc between carbon electrodes; usually limited to pure carbon rather than flame carbon electrodes.

carbon-arc lamp (*ElecEng*) Obsolete light source from the arc between carbon electrodes.

carbon-arc welding (*ElecEng*) Arc welding carried out by means of an arc between a carbon electrode and the material to be welded.

carbonas (*Min*) Zones of mineralization that have spread into the host rock around mineral veins (commonly tin), and have been rich enough to mine.

carbonate (*Chem*) A compound containing the acid radical of carbonic acid (CO_3 group). Bases react with carbonic acid to form carbonates, eg $CaCO_3$, calcium carbonate.

carbonate–apatite (*Min*) A variety of apatite containing appreciable CO_2.

carbonate compensation depth (*Geol*) The level in the ocean below which the rate of solution of calcium carbonate exceeds the rate of its deposition, so no

limestones etc are formed; 4000–5000 m in the Pacific, less in the Atlantic. Abbrev *CCD*.

carbonation (*FoodSci*) The addition of carbon dioxide under pressure to water in which its solubility increases at lower temperatures. Opening the container releases the pressure and causes effervescence.

carbonatite (*Geol*) An igneous rock composed largely of carbonate minerals (calcite and dolomite). Carbonatites are invariably associated with alkaline igneous rocks such as nepheline–syenite, and are rich in a number of unusual minerals, esp those of the rare earth elements.

carbon black (*Chem*) Finely divided carbon produced by burning hydrocarbons (eg methane) in conditions in which combustion is incomplete. Widely used in the rubber, paint, plastics, ink and other industries. It forms a very fine pigment containing up to 95% carbon, giving a very intense black; prepared by burning natural gas or oil and letting the flame impinge on a cool surface. Various grades are identified by particle size; HAF is 'high abrasive fine' and is a most important reinforcing filler for rubbers. Also *channel black*, *gas black*. See panel on TYRE TECHNOLOGY.

carbon brush (*ElecEng*) A small block of carbon used in electrical equipment to make contact with a moving surface.

carbon burial (*EnvSci*) The removal of carbon dioxide from the atmosphere and its consignment to long-term storage; the process by which the composition of the atmosphere has been changed from one dominated by carbon dioxide to one in which it is a minor (about 0·03% by volume) constituent. Occurs mainly via the reaction in water by which dissolved carbon dioxide is converted to bicarbonate and its subsequent reaction with calcium to form insoluble calcium carbonate, mainly as shells of marine organisms, and through the formation of fossil fuels.

carbon compounds (*Chem*) Compounds containing one or more carbon atoms in the molecule. They comprise all organic compounds and include also compounds, eg carbides, carbonates, carbon dioxide, etc, which are usually dealt with in inorganic chemistry. Carbon compounds are the basis of all living matter.

carbon contact (*ElecEng*) In a switch, an auxiliary contact designed to break contact after and to make contact before the main contact to prevent burning of the latter; it is of carbon and designed to be easily removable.

carbon cycle (*Astron*) A chain of nuclear fusion reactions, believed to take place in stars more massive than the Sun, which transmutes protons into helium nuclei, with carbon atoms effectively acting as a catalyst. The reaction rate is strongly dependent on temperature and this cycle is believed to be the main source of energy in hot massive stars. Also *Bethe cycle*, *carbon–nitrogen cycle*.

carbon cycle (*BioSci*) The biological circulation of carbon from the atmosphere into living organisms and, after their death, back again. See CARBON DIOXIDE, PHOTOSYNTHESIS. Fig. ▷

carbon dating (*Phys*) Dating method which utilizes the fact that atmospheric carbon dioxide contains a constant proportion of radioactive ^{14}C, formed by cosmic radiation. Living organisms absorb this isotope in the same proportion. After death it decays with a half-life of $5·57 \times 10^3$ years. The proportion of ^{12}C to the residual ^{14}C indicates the period elapsed since death. Also *radiocarbon dating*. See panel on RADIOMETRIC DATING.

carbon dioxide (*Chem*) CO_2. A colourless gas; density at stp $1·976$ kg m^{-3}, about 1·5 times that of air. Produced by the complete combustion of carbon, by the action of acids on carbonates (eg Kipp's apparatus), by the thermal decomposition of carbonates (eg lime burning) and during fermentation. It plays an essential part in metabolism, being exhaled by animals and absorbed by plants (see PHOTOSYNTHESIS). May be liquefied at 20°C at a pressure of $5·7$ MN m^{-2}, but at atmospheric pressure it sublimes at $-78·5$°C. Liquid and solid CO_2 are much used as refrigerants, notably for ice-cream, and in fire extinguishers.

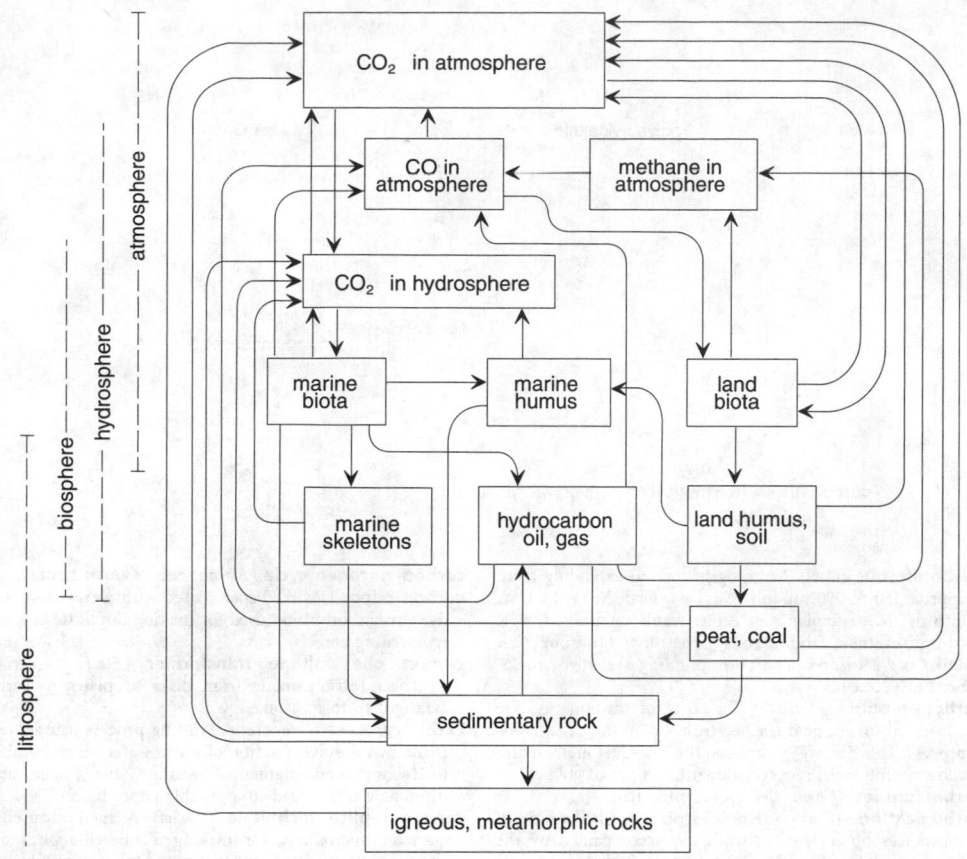

carbon cycle

CO_2 dissolves in water to form unstable carbonic acids; the pressurized solution produces the effervescent 'sparkle' in carbonated beverages. When solid it has a convenient temperature for testing electronic components. High-pressure carbon dioxide has found a considerable use as a coolant in carbon-moderated nuclear reactors.

carbon dioxide laser (*Phys*) A laser in which the active gaseous medium is a mixture of carbon dioxide and other gases. It is excited by glow discharge and operates at a wavelength of 10·6 μm. Carbon dioxide lasers are capable of pulsed output with peak power up to 100 MW or continuous output up to 60 kW.

carbon dioxide welding (*Eng*) METAL ARC WELDING using CO_2 as the shielding gas.

carbon disulphide (*Chem*) CS_2. Formed when sulphur vapour is passed over heated charcoal and combines with the carbon. Used as a solvent for sulphur and rubber. The disagreeable smell associated with commercial carbon disulphide is due to impurities.

carbon-fibre-reinforced plastic (*Chem*) A class of COMPOSITE MATERIALS comprising a polymeric MATRIX (frequently an EPOXY RESIN) reinforced with carbon fibre. Increasingly used as a structural material for its high specific stiffness and specific strength in applications varying from aerospace to sports equipment. Abbrev *CFRP*.

carbon fibres (*Chem, Textiles*) High-modulus, highly oriented fibres of about 8 μm in diameter, consisting almost exclusively of carbon atoms. Made as continuous filament by the pyrolysis in an inert atmosphere of organic fibres such as cellulose but usually of polyacrylonitrile. Used as a reinforcing material with epoxy or polyester resins to form composites which have a higher strength/weight ratio than metals. BORON, aramid and glass are alternative fibre materials. See panel on HIGH-PERFORMANCE POLYMERS. Fig. ▷

carbon film technique (*BioSci*) One of the methods used in electron microscopy to provide a supporting film for the specimen. The film is prepared by subliming carbon in a vacuum and is itself supported on a metal grid.

carbon fixation (*BioSci*) The synthesis of organic compounds from carbon dioxide, most notably in PHOTOSYNTHESIS.

carbon gland (*Eng*) A type of gland used to prevent leakage along a shaft. It consists of carbon rings cut into segments and pressed into contact with the shaft by an encircling helical spring or GARTER SPRING.

carbonic acid (*Chem*) H_2CO_3. A weak acid formed when carbon dioxide is dissolved in water.

carbonic acid derivatives (*Chem*) Carbonic acid forms both normal and acid salts. The esters, chlorides and amides form two series: (1) normal compounds, in which both hydroxyl groups are substituted; and (2) acid compounds, in which only one hydroxyl group is substituted. The acid compounds are unstable in the free state, but form stable salts.

carbonic acid gas (*Chem*) Carbon dioxide effervescing from liquids which have been saturated with carbon dioxide under pressure. The gas escapes when the pressure is withdrawn.

carbonic anhydrase (*BioSci*) An enzyme in blood corpuscles catalysing the decomposition of carbonic acid. It is essential for the effective transport of carbon dioxide from the tissues to the lungs.

carbonic anhydride (*Chem*) Carbon dioxide. See CARBONIC ACID.

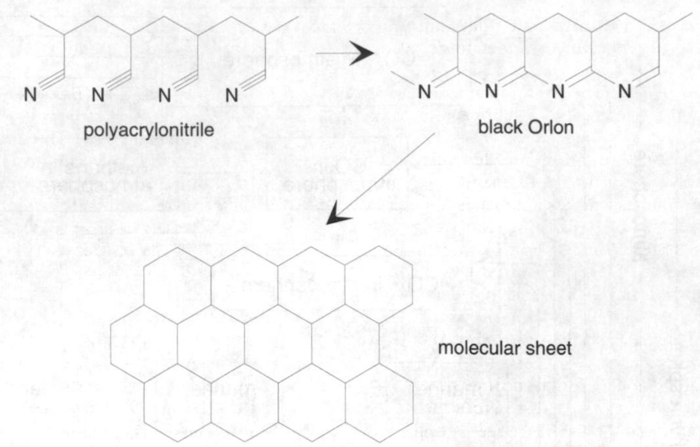

polyacrylonitrile

black Orlon

molecular sheet

carbon fibres Polymerization from polyacrylonitrile.

Carboniferous (*Geol*) A geological period extending from approx 360 to 290 million years ago. Divided in the USA into the Mississippian and Pennsylvanian periods. In UK comprises the Carboniferous Limestone, Millstone Grit and Coal Measures. The corresponding system of rocks. See PALAEOZOIC.

carbon-in-pulp (*MinExt*) The use of carbon as the adsorbent for the gold leached from ore in the CYANIDING process. This has largely replaced the zinc-dust method and allows commercial recovery down to 2 ppm of gold.

carbonium ion (*Chem*) See CARBENIUM ION.

carbonization (*Chem*) The destructive distillation of substances out of contact with air, accompanied by the formation of carbon, in addition to liquid and gaseous products.
Coal yields coke, while wood, sugar, etc, yield charcoal.

carbonization (*Eng*) See CEMENTATION.

carbonization (*Geol*) The conversion of fossil organic material to a residue of carbon. Plant material is often preserved in this way.

carbonization (*Textiles*) The steeping of wool in a dilute solution of sulphuric acid, or its treatment by hydrochloric acid gas (dry process). This converts any cellulosic impurities into carbon dust and thereby facilitates their removal.

carbonized filament (*Electronics*) Thoriated tungsten filament coated with tungsten carbide to reduce loss of thorium from the surface.

carbon microphone (*Acous*) A microphone in which a normally dc energizing current is modulated by changes in the resistance of a cavity filled by granulated carbon which is compressed by the movement of the diaphragm. The diameter of the cavity is frequently very much less than that of the diaphragm, and it is then known as a *carbon button*.

carbon monoxide (*Chem*) CO. Formed when carbon is heated in a limited supply of air, when carbon dioxide is heated in an excess of carbon, when carbon dioxide is passed over some hot metals, or by dehydration of methanoic acid. It is a product of incomplete combustion. Poisonous. Its properties as a reducing agent render it valuable in industrial processes. See CARBONYL.

carbon monoxide-haemoglobinaemia (*Med*) Haemoglobin combines with carbon monoxide instantaneously and is thus deprived of its oxygen-exchanging properties. This leads to poisoning of the body and death by asphyxia without the more normal cyanosis. Also *carboxyhaemoglobinaemia*.

carbon–nitrogen cycle (*Astron*) See CARBON CYCLE.

carbon paper (*Paper*) Paper coated with waxes containing dyes or carbon black, used for making duplicate copies in typewriting etc.

carbon pile voltage transformer (*ElecEng*) Variable electrical resistor made from disks or plates of carbon arranged to form a pile.

carbon process (*ImageTech*) Printing process using a relief made by the solvent action of warm water on an emulsion of bichromated pigmented gelatine, the image being ultimately transferred to a suitable paper base.

carbon replica technique (*BioSci*) A method used in electron microscopy for making a surface replica of a specimen. It is coated with a structureless carbon film, and the film and specimen are subsequently removed by dissolving in an appropriate solvent.

carbon resistor (*ElecEng*) Negative temperature coefficient, non-inductive resistor formed of powdered carbon with ceramic binding material. Used for low-temperature measurements because of the large increase in resistance as temperature decreases.

carbon star (*Astron*) Rare giant star showing molecular bands of carbon compounds in its spectrum.

carbon steel (*Eng*) A steel whose properties are determined principally by the amount of carbon present and contains no other deliberate alloying ingredient except those necessary to ensure deoxidation and physical quality. Also *plain carbon steel*. See STEEL and panel on STEELS.

carbon suboxide (*Chem*) *Malonic anhydride*. C_3O_2, $O{=}C{=}C{=}C{=}O$. A colourless liquid or gas, mp $-107°C$, bp $+7°C$; formed by heating malonic acid to 140–150°C. Example of a KETENE.

carbon taxes (*EnvSci*) A proposal arising from the Kyoto protocol whereby taxes would be levied on the amount of fossil fuel consumed in proportion to the carbon content and thus the potential for contributing to global warming by the release to the atmosphere of carbon dioxide.

carbon tetrachloride (*Chem*) *Tetrachloromethane*. CCl_4. A colourless liquid, bp 76°C; prepared by the exhaustive chlorination of methane or carbon disulphide. Solvent for fats and oils. Not hydrolysed by water. Because of toxicity, its use in fire extinguishers and dry-cleaning has now much declined.

carbon tissue (*ImageTech*) Paper coated with a mixture of gelatine and a pigment (sometimes carbon powder). Used in the CARBON PROCESS.

carbon tissue (*Print*) A photosensitive gelatine layer on a paper backing which is printed down with positives and the photogravure screen, and, after developing, is stripped down on the photogravure cylinder to act as an etchant resist.

carbon value (*Chem*) The figure obtained empirically as a measure of the tendency of a lubricant to form carbon when in use.

carbonyl (*Chem*) Carbon monoxide when acting as a radical, as it appears to do in many reactions. Carbon monoxide combines with certain metals to form carbonyls, eg $Co(CO)_3$, $Ni(CO)_4$, $Fe(CO)_5$, $Mo(CO)_6$. ALDEHYDES and KETONES are organic carbonyl compounds.

carbonyl chloride (*Chem*) See PHOSGENE.

carbonyl powders (*Eng*) Metal powders produced by reacting carbon monoxide with the metal to form the gaseous carbonyl. This is then decomposed by heat to yield powder of high purity.

carborundum (*Eng*) TN for SILICON CARBIDE abrasives.

carborundum wheel (*Eng*) See GRINDING WHEEL.

carboxydismutase (*BioSci*) See RIBULOSE 1,5-BISPHOSPHATE CARBOXYLASE.

carboxyhaemoglobin (*BioSci*) Haemoglobin co-ordinated with carbon monoxide – almost irreversible, hence the toxicity of CO. US *carboxyhemoglobin*.

carboxyhaemoglobinaemia (*Med*) See CARBON MONOXIDE-HAEMOGLOBINAEMIA.

carboxylase (*BioSci*) An enzyme that catalyses the incorporation of carbon dioxide into its substrate, eg ribulose bisphosphate carboxylase, PEP carboxylase.

carboxyl group (*Chem*) The acid group —COOH.

carboxylic acid (*Chem*) R—$(COOH)_n$. An organic compound having one or more carboxyl radicals. A carboxyl radical is usually designated, but shows resonance.

carboxymethylcellulose (*FoodSci*) A carbohydrate STABILIZER which enables less starch to achieve a given viscosity while reducing a tendency to SYNERESIS. Important in reduced-calorie product formulations. Abbrev *CMC*.

carboy (*Glass*) Large, narrow-necked container, usually of balloon shape, having a capacity of 20 l or more.

Carbro process (*ImageTech*) Colour print process in which relief images are forced in pigmented tissues by contact with colour-separation bromide prints, a set of three colour positives resulting for mounting in superimposition on a paper backing.

carbuncle (*Med*) Circumscribed staphylococcal infection of the subcutaneous tissues.

carbuncle (*Min*) A gem variety of garnet cut *en cabochon*. It has a deep-red colour. See ALMANDINE.

carburation (*Eng*) The mixing of air with volatile fuel to form a combustible mixture for using in internal-combustion (petrol) engines.

carburettor (*Eng*) A device for mixing air and a volatile fuel in correct proportions, in order to form a combustible mixture. It consists essentially of a jet, or jets, discharging the fuel into the airstream under the pressure difference created by the velocity of the air as it flows through a nozzle-shaped constriction. Also *carburetor, carburetter*.

carburizing (*Eng*) A method of CASE-HARDENING low-carbon steel in which the metal component is heated above its ferrite–austenite transition in a suitable carbonaceous atmosphere. Carbon diffuses into the surface and establishes a concentration gradient. The steel can subsequently be hardened by quenching either directly or after reheating to refine the grain structure. It is usually lightly tempered afterwards, producing a hard case over a tough core.

carbylamines (*Chem*) See ISOCYANIDES.

carcase (*Agri*) See CARCASS.

carcass (*Agri*) An animal body prepared for food use by removal of the skin, offal, head and feet.

carcass (*Build*) The shell of a house in construction, consisting of walls and roof only, without floors, plastering or joiner's work.

carcass (*Eng*) The body of a tyre, comprising bead coil, inner lining, side wall, breaker and plies often without tyre tread. Term also applied to used tyres from which worn

tread has been buffed ready for retreading or to green tyre prior to shaping and vulcanization. See panel on TYRE TECHNOLOGY.

carcassing timber (*Build*) Timber for the framing of a building or other structure.

carcinogenesis (*Med*) The induction of a change in a cell that will eventually cause it to become a cancer.

carcinoma (*Med*) A disorderly growth of epithelial cells which invade adjacent tissue and spread via lymphatics and blood vessels to other parts of the body. See SARCOMA.

carcinoma en cuirasse (*Med*) Extensive carcinomatous infiltration of the skin, characterized by hardness and gross thickening.

carcinomatosis (*Med*) Cancer widely spread throughout the body. Also *carcinosis*.

carcinomatous (*Med*) Of the nature of cancer.

carcinosis (*Med*) See CARCINOMATOSIS.

card (*ICT*) See BOARD, PUNCHED CARD.

card (*Surv*) The graduated dial or face of a magnetic compass to which the card and needle are firmly connected.

card (*Textiles*) Machine for carrying out the CARDING process.

Cardan mount (*Eng*) Type of gimbal mount used for compasses and gyroscopes.

card cloth (*Textiles*) Strong material (eg a fabric–rubber laminate) fitted with masses of strong, flexible wire teeth, pins or spikes. Often in the form of flats or rotating cylinders, and used for CARDING.

card deck (*ICT*) A set of PUNCHED CARDS comprising a program or a set of data. Obsolete.

carded yarns (*Textiles*) Yarns spun from SLIVERS directly after CARDING.

cardi-, cardio- (*Genrl*) Prefixes from Gk *kardia*, heart.

cardia (*Med*) The physiological sphincter surrounding the opening of the oesophagus into the stomach. Also *cardiac sphincter*.

cardiac (*Med*) (1) Pertaining to the heart. (2) Pertaining to the upper part of the stomach.

cardiac aneurysm (*Med*) A fibrous dilatation of one or other ventricle due to destruction of cardiac muscle.

cardiac arrest (*Med*) The sudden cessation of the heart's action as a pump. May be due to VENTRICULAR FIBRILLATION or the stopping of contraction (ASYSTOLE).

cardiac asthma (*Med*) A sensation of breathlessness and wheeze due to pulmonary congestion brought on by failure of the left side of the heart.

cardiac massage (*Med*) The manual procedure whereby the heart action is maintained after cardiac arrest, externally by massage on the sternum, or by massaging the heart directly with a hand around it.

cardiac muscle (*BioSci*) The contractile tissue forming the wall of the heart of vertebrates. It has conspicuous striated myofibrils but is not syncytial; cells are joined by specialized junctional complexes (*intercalated discs*).

cardiac pacemaker (*Med*) Electronic device implanted into chest wall, usually by means of a transvenously introduced catheter in the right atrium or ventricle. It senses and regulates abnormally slow heart rhythms.

cardiac sphincter (*Med*) See CARDIA.

cardiac tamponade (*Med*) The condition when fluid or blood fills the pericardial sac and interferes with the pumping action of the heart.

cardiac valve (*BioSci*) A valve at the point of junction of the fore- and mid-intestine in many insects. Also *oesophageal valve*. US *esophageal valve*.

cardinal (*BioSci*) In Bivalvia and Brachiopoda, pertaining to the hinge; more generally, primary, principal, as the *cardinal sinuses* or *veins*, being the principal channel for the return of blood in the lower vertebrates.

cardinal number (*MathSci*) A property associated with a set such that two sets X and Y have the same cardinal number if and only if the elements of X and Y can be arranged in ONE-TO-ONE CORRESPONDENCE. The cardinal number of a set differs from ORDINAL NUMBER in that the former relates to quantity and the latter to order or arrangement, although this distinction becomes of real interest only for transfinite numbers. For a finite set, the cardinal number is just the number of elements in the set. An infinite set is defined as one with the same cardinal number as some of its proper subsets.

cardinal planes (*Phys*) In a lens, planes perpendicular to the principal axis, and passing through the cardinal points of the lens.

cardinal point (*Astron*) One of the four principal points of the horizon, north, south, east and west, corresponding to AZIMUTH 0°, 180°, 90° and 270°. See QUADRANTAL POINT.

cardinal points (*Phys*) For a lens or lens system, the two *principal foci* (see FOCUS), the two NODAL POINTS and the two PRINCIPAL POINTS. For a lens used in air, the principal points coincide with the corresponding nodal points. For a lens of negligible thickness the principal points and the nodal points all coalesce at a single point at the optical centre of the lens.

carding (*Textiles*) The process of passing fibres through a machine called a *card* which disentangles them and makes them lie fairly straight, to form a light fluffy web or SLIVER.

cardio- (*Genrl*) See CARDI-.

cardioblast (*BioSci*) An embryonic mesodermal cell that will form part of the heart.

cardiocentesis (*Med*) Puncture of the heart with a needle.

cardiogram (*Med*) Trace produced by *electrocardiogram* (ECG) showing voltage waveform generated during heart beats.

cardiograph (*Med*) An instrument for recording movements of the heart. See ELECTROCARDIOGRAM.

cardioid (*MathSci*) A heart-shaped curve with polar equation $r = 2a(1 + \cos\theta)$. An epicycloid in which the rolling circle equals the fixed circle. Cf LIMAÇON.

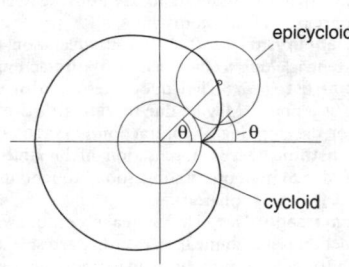

cardioid Cycloid and epicycloid here have the same radius.

cardioid directivity (*Acous*) Special shape of a directivity. It is produced by superimposing the fields of a monopole and a dipole, and has the shape of a cardioid.

cardioid reception (*ICT*) A receiving system with a directional antenna, the RADIATION PATTERN of which is heart-shaped.

cardiolipin (*BioSci*) An important phospholipid component of the inner mitochondrial membrane and thus present in relatively large amounts in energy-intensive tissues such as muscle. Cardiolipin, purified from beef heart, is the active antigen in the Wassermann reaction and other serological tests for syphilis and raised blood levels of anti-cardiolipin antibody are associated with thrombosis and heart attacks.

cardiology (*Med*) The part of medical science concerned with the function and diseases of the heart.

cardiolysis (*Med*) Operative freeing of the heart from the chest wall when it is adherent to it in chronic adhesive pericarditis.

cardiomalacia (*Med*) Pathological weakening of the heart muscle.

cardiomegaly (*Med*) Abnormal enlargement of the heart.

cardiomyopathy (*Med*) Any disease affecting the heart muscle.

cardiopulmonary bypass (*Med*) Apparatus including a pump and an oxygenator for artificial maintenance of circulation during operations on the heart.

cardiospasm (*Med*) Spasm of the cardia (or cardiac sphincter) of the stomach.

cardiotachometer (*Med*) An electronic amplifying instrument for recording and timing the heart rate.

cardiotocograph (*Med*) A record of the heart rate of a fetus.

cardiovascular (*Med*) Pertaining to the heart and blood vessels.

carditis (*Med*) Inflammation of the muscle and coverings of the heart.

cardo (*BioSci*) The hinge of a bivalve shell. Pl *cardines*.

card reader (*ICT*) Component part of an obsolete computing system that scanned PUNCHED CARDS and delivered signals corresponding to the information recorded by the holes. See CARD SWIPE.

card swipe (*ICT*) An electronic device that reads information on a credit or debit card when the card is passed through it.

caret (*Print*) A symbol (∧) used in proof correcting to indicate that something is to be inserted at that point.

Careware (*ICT*) (computing slang). Computer software that is made available in exchange for making a donation of one's services.

car-floor contact (*ElecEng*) A contact attached to the false floor of an electrically controlled lift; it is usually arranged to prevent operation of the lift by anyone outside the car while a passenger is in the lift.

caries (*Med*) Decay of bone or teeth. Adj *carious*.

carina (*BioSci*) A median dorsal plate of the exoskeleton of some CIRRIPEDIA; a ridge of bone resembling the keel of a boat, eg the STERNUM of flying birds.

Carina (Keel) (*Astron*) A southern hemisphere constellation, formerly part of the ancient constellation Argo Navis. Includes the star CANOPUS.

carinate (*BioSci*) Shaped like a keel; having a projection like a keel.

cariogenic (*FoodSci*) Substance or food which causes tooth decay.

Carme (*Astron*) The eleventh natural satellite of Jupiter, discovered in 1938. Distance from the planet 22 600 000 km; diameter 40 km.

carminative (*Pharmacol*) Relieving gastric flatulence; medicine which does this.

carnallite (*Min*) The hydrated chloride of potassium and magnesium, occurring in bedded masses with other saline deposits, as at Stassfurt, Germany. Such deposits arise from the desiccation of salt lakes. It is used as a fertilizer.

carnassials (*BioSci*) In terrestrial Carnivora, large flesh-cutting teeth derived from the first lower molar and the last upper premolars.

carnelian (*Min*) A translucent variety of chalcedony (silica) of red or reddish-brown colour. Also *cornelian*.

Carnivora (*BioSci*) An order of primarily carnivorous mammals, terrestrial or aquatic. Such creatures usually have three pairs of incisors in each jaw and large prominent canines; the last upper premolar and the first lower molar are frequently modified for flesh cutting (*carnassials*). The collar bone is reduced or absent. There are four or five claw-like digits on each limb. Cats, lions, tigers, panthers, dogs, wolves, jackals, bears, raccoons, skunks, seals, sea-lions and walruses.

carnivorous (*BioSci*) Flesh-eating.

carnivorous plant (*BioSci*) One of c.400 species belonging to several unrelated families, mostly growing on substrates poor in mineral nutrients, which trap and digest insects and other small animals. Also *insectivorous plant*.

Carnot cycle (*Eng*) An ideal heat engine cycle of maximum thermal efficiency. It consists of isothermal expansion, adiabatic expansion, isothermal compression and adiabatic compression to the initial state. Fig. ▷

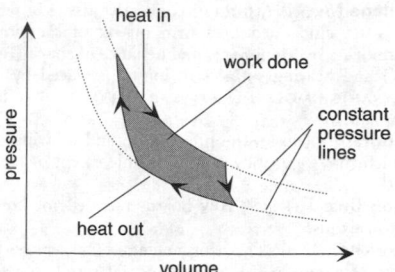

Carnot cycle

carnotite (*Min*) A hydrated vanadate of uranium and potassium, found as a yellow impregnation in sandstones. It is an important source of uranium.

Carnot's theorem (*Phys*) A theorem stating that no heat engine can be more efficient than a reversible engine working between the same temperatures. It follows that the efficiency of a reversible engine is independent of the working substance and depends only on the temperatures between which it is working.

carob (*FoodSci*) Roasted and processed powder from seeds of the Locust Bean or Carob tree which can be used as a substitute for cocoa.

carol (*Build*) A seat built into the opening of a bay window. Also *bay-stall, caroll*.

Caro's acid (*Chem*) Permonosulphuric acid.

carotenes (*BioSci*) Red to yellow carotenoids, unsaturated tetraterpene hydrocarbons ($C_{40}H_{56}$). Accessory and photo-protective pigments in chloroplasts and in chromoplasts in some fruits and in carrot roots.

carotid arteries (*BioSci*) In vertebrates, the principal arteries carrying blood forward to the head region.

carotid bodies (*Med*) Two small oval masses situated close to the carotid sinus, richly vascularized and containing chemosensory receptors composed of an epithelioid cell and associated endings of sensory fibres of the glossopharyngeal nerve. The chemoreceptors are sensitive to the tension of oxygen and carbon dioxide and to the pH of the arterial blood flowing through them, bringing about important compensatory respiratory and cardiovascular reflexes.

carotid sinus (*Med*) The dilated portion of the common carotid artery at its division into internal and external branches. The wall of the sinus is richly endowed with sensory endings of the glossopharyngeal nerve which act as baroreceptors, the nerve impulses generated in these sensory receptors producing important compensatory cardiovascular and respiratory reflexes.

carp-, carpo-, -carp, -carpous (*Genrl*) Prefixes and suffixes from Gk *karpos*, fruit.

carpal (*BioSci*) One of the bones composing the CARPUS in vertebrates. Pl *carpals, carpalia*. Also *carpale*.

carpel (*BioSci*) A female organ in a flower, bearing and enclosing one or more ovules, and forming singly or with others the gynoecium. Typically like a leaf, folded long-itudinally so the edges come together, and bearing one or more ovules on a placenta along the line of the junction. Comprises ovary, style (usually) and stigma.

carpale (*BioSci*) See CARPAL.

carpellate (*BioSci*) Of a flower, female.

carpo-, -carpous (*Genrl*) See CARP-.

carpus (*BioSci*) In land vertebrates, the basal podial region of the forelimb; the wrist.

carr (*EnvSci*) FEN vegetation with a conspicuous component of tree species.

carragheenin (*Chem, FoodSci*) Carbohydrate extracted from the edible seaweed *Chondrus crispus* (carragheen, Irish moss). Used as an emulsifying agent in foods, as a clarifying agent in drinks, and to prevent crystal growth in ice-cream. Also *carrageenan, carrageenin*.

carriage (*Build*) A timber joist giving intermediate support, between the wall string and the outer string, to the treads of wide wooden staircases. Also *carriage piece, rough-string*.

carriage (*Print*) The reciprocating assembly on a cylinder printing machine made up of a bed on which the forme lies, bearers at each side, and an ink table.

carriage-maker's plane (*Build*) Special rebate plane for wide rebates, fitted with a back iron to minimize tearing of the grain.

carriage piece (*Build*) See CARRIAGE.

carriage return (*ICT*) A CONTROL CHARACTER that has the effect of moving the current character printing position to the start of the line. See LINE FEED.

carriage spring (*Eng*) See LAMINATED SPRING.

carriage-type switchgear (*ElecEng*) See TRUCK-TYPE SWITCHGEAR.

carrier (*BioSci*) (1) In human genetics particularly, a HETEROZYGOTE for a recessive disorder. (2) An organism harbouring a parasite but showing no symptoms of disease, esp if it acts as a source of infection. (3) See CARRIER PROTEIN.

carrier (*Electronics*) A real or imaginary particle responsible for the transport of electric charge in a material. In oxide ceramics, electrons hopping between ions, diffusing oxygen ions and mobile cations can also transport charge. See CARRIERS.

carrier (*Eng*) A device for conveying the drive of a face plate of a lathe to a piece of work which is being turned between centres. It is clamped to the work and driven by a pin projecting from the face plate.

carrier (*ICT*) (1) A vehicle for communicating information, when the chosen medium itself cannot convey the information but can convey a carrier, on to which the information is impressed by MODULATION. See fig. at AMPLITUDE MODULATION. (2) In radio transmission, the output of the transmitter before it is modulated. See FREQUENCY MODULATION. (3) The frequencies chosen for sending many signals simultaneously along a single communication channel by FREQUENCY-DIVISION MULTI-PLEX.

carrier (*ImageTech*) A frame for holding a negative in an enlarger or slides in a projector.

carrier (*Phys*) Non-active material mixed with, and chemi-cally identical to, a radioactive compound. Carrier is sometimes added to carrier-free material.

carrier (*Textiles*) A compound added to a dye bath to assist the dyeing of hydrophobic fibres particularly with disperse dyes.

carrier beat (*ICT*) Audible note produced by heterodyne process between two carriers or between a carrier and a reference oscillator.

carrier-controlled approach (*Radar*) System used for landings on aircraft carriers.

carrier filter (*ICT*) Filter suitable for discriminating between currents used in carrier telephony according to their frequency, as a means of channel separation.

carrier mobility (*Phys*) The mean drift velocity of the charge carriers in a material per unit electric field.

carrier noise (*Acous*) Noise which has been introduced into the carrier of a transmitter before modulation.

carrier pre-selection (*ICT*) A mechanism for routing telephone calls through an alternative operator without entering an additional dialling code.

carrier power (*Phys*) Power radiated by a transmitter in the absence of modulation.

carrier protein (*BioSci*) (1) A protein to which a specific ligand or hapten is conjugated. (2) An unlabelled protein added into an assay system at relatively high concentrations that distributes in the same manner as labelled protein analyte present in very low concentrations. (3) Protein added to prevent non-specific interaction of reagents with surfaces, sample components and each other. (4) Any of a number of proteins found in cell membranes that facilitate transport across the membrane.

carriers (*Electronics*) In a crystal of semiconductor material thermal agitation will cause a number of electrons to dissociate from their parent atoms; in moving about the crystal they act as carriers of negative charge. Other electrons will move from neighbouring atoms to fill the space left behind, thus causing the holes where no electrons exist in the lattice to be transferred from one atom to another. As these holes move around they can be considered as carriers of positive charge. See IMPURITY.

carrier suppression (*ICT*) The process of eliminating the CARRIER in amplitude-modulated signals in order to produce *suppressed-carrier modulation*. Achieved through filtering and/or the use of balanced modulators; used in some forms of radio communication and FREQUENCY-DIVISION MULTIPLEX telephony, giving a better signal-to-noise ratio. See SUPPRESSED-CARRIER SYSTEM.

carrier system (*ICT*) See FREQUENCY-DIVISION MULTIPLEX.

carrier telegraphy (*ICT*) Use of modulated frequencies, usually in the five-unit code originated for teleprinters, and transmitted, with others, as a voice-frequency signal in telephone circuits.

carrier-to-interference ratio (*ICT*) The ratio, usually expressed in DECIBELS, of a wanted CARRIER to all other interfering carriers. In eg a cellular telephone network this ratio limits the extent of frequency reuse for a given CELL size.

carrier-to-noise ratio (*ICT*) The ratio of the received carrier signal-to-noise voltage immediately before demodulation or limiting stage.

carrier wave (*Phys*) An unmodulated radio wave produced by a transmitter on which information is carried by amplitude or frequency modulation.

carry (*ICT, MathSci*) In computing, as in arithmetic, in a POSITIONAL NUMBER SYSTEM, to transfer a multiple of the base from one position to another and compensate by adjusting the digits in both positions.

carry flag (*ICT*) A single bit that is set to one when a CARRY occurs during a binary addition.

carrying capacity (*Agri*) The maximum crop or livestock density that can be farmed without causing environmental damage likely to reduce future production.

carrying-current (*ElecEng*) See INSTANTANEOUS CARRYING-CURRENT.

carstone (*Geol*) Brown sandstone in which the grains are cemented by limonite. See IRON PAN.

cartesian co-ordinates (*MathSci*) A system of co-ordinates in which the position of a point is specified by its distances from each of a number of reference lines (the *co-ordinate axes*) measured parallel to another axis. If the axes are mutually perpendicular the system is said to be *rectangular*. Devised by Descartes and published in 1637.

cartesian diver (*Phys*) See DIVER.

cartesian hydrometer (*PowderTech*) An improved form of hydrometer used in the particle size analysis of soils, having a rubber membrane as part of the wall. Adjustment of pressure above the suspension enables the specific gravity of the hydrometer to be adjusted.

cartesian ovals (*MathSci*) Curves, named after Descartes, defined by the bipolar equation $mr \pm nr' =$ constant.

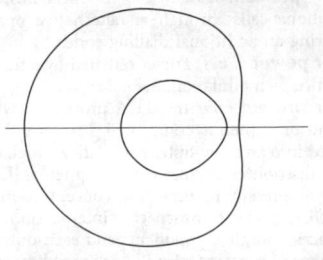

cartesian ovals

cartilage (*BioSci*) A form of connective tissue in which the cells are embedded in a firm matrix of *chondroitin*, or chondrin. Two functions: maintenance of shape (ear, nose, etc) and bearing surfaces in joints. Replaced by sintered ULTRAHIGH MOLECULAR MASS POLYETHYLENE in hip joint implant.

cartography (*Surv*) The preparation and drawing of maps which show, generally, a considerable extent of the Earth's surface.

carton board (*Paper*) Any board intended for conversion into cartons.

cartouche (*Arch*) (1) An ornamental block supporting the eave of a house. (2) An ornamental scroll to receive an inscription. Also *cartouch*.

cartridge (*ICT*) Removable module containing software, often permanently stored on an integrated circuit.

cartridge (*ImageTech*) (1) Sealed electromechanical transducer unit as in a PICK-UP DEVICE. (2) Light-tight container for loading unexposed photographic film in a camera. (3) Container for a single spool of motion picture film or magnetic tape feeding into a reproducer and taken up on a separate spool. (4) A container for a continuous loop of motion picture film or magnetic tape.

cartridge (*NucEng*) See CAN.

cartridge brass (*Eng*) Copper–zinc alloy containing approximately 30% zinc. Possesses high ductility; capable of being heavily cold-worked. Widely used for cold pressings, cartridges, tubes, etc. See COPPER ALLOYS.

cartridge-operated hammer (*Build*) Hammer using the force of a small explosive cartridge to drive nails and bolts into concrete, brickwork, etc.

cartridge paper (*Paper*) Originally a tough paper intended for winding the tubes of shotgun cartridges (ammunition cartridge). Now also a paper made for drawing purposes (drawing cartridge) or for lithographic printing (offset cartridge or matt-coated cartridge).

cartridge starter (*Aero*) A device for starting aero-engines in which a slow-burning cartridge is used to operate a piston or turbine unit which is geared to the engine shaft.

caruncle (*BioSci*) (1) An outgrowth from the neighbourhood of the micropyle of a seed. The seed is said to be *carunculate*. (2) Any fleshy outgrowth; in some Polychaeta, a fleshy dorsal sense organ; in some Acarina, a tarsal sucker; in embryo chicks, a horny knob at the tip of the beak.

caruncle (*Med*) Any small fleshy excrescence; a small growth at the external orifice of the female urethra; (pl) epithelial nodules found at the end of pregnancy on the placenta and amnion.

carvacrol (*Chem*) *1-methyl-4-isopropyl-2-hydroxy-benzene*. $C_{10}H_{14}O$. An isomer of THYMOL; mp 0°C, bp 236°C; obtained from camphor by heating with iodine; present in *Origanum hirtum*.

carvone (*Chem*) Δ-6·8-*Terpadiene-2-one*. An unsaturated ketone of the terpene series, the principal constituent of caraway seed oil; liquid of bp 228°C; readily forms CARVACROL.

cary-, caryo- (*Genrl*) See KARY-, KARYO-.

caryatid (*Arch*) A sculptured female form used as a column.

caryo- (*Genrl*) See KARY-, KARYO-.

Caryophyllaceae (*BioSci*) A family of c.2000 spp of dicotyledonous flowering plants (superorder Caryophyllidae) that are mostly herbs and mostly grow in temperate climes. Such plants typically have opposite leaves, flowers with five free petals, and a superior ovary with free-central placentation. They are of little economic importance other than as ornamentals, eg pinks and carnations (*Dianthus*).

caryophyllenes (*Chem*) A mixture of isomeric SESQUITERPENES forming the chief constituents of clove oil.

Caryophyllidae (*BioSci*) A subclass or superorder of dicotyledons, mostly herbs. Most have trinucleate pollen (binucleate is commoner in flowering plants), most have BETALAINS rather than ANTHOCYANINS (except Caryophyllaceae) and/or free-central or basal placentation. There are c.11 000 spp in 14 families including Caryophyllaceae, Chenopodiaceae and Cactaceae. (Corresponds approximately to older group Centrospermae.)

caryopsis (*BioSci*) A dry, indehiscent, one-seeded fruit, characteristic of the grasses, with the ovary wall (pericarp) and seed coat (testa) united, eg a grain of wheat.

CAS (*Aero*) Abbrev for CALIBRATED AIRSPEED.

cascade (*Build*) An outflow falling in a series of steps.

cascade (*ICT*) (1) A way of arranging open WINDOWS on a DESKTOP such that they overlap with each other with the title bar of each remaining visible. (2) A number of devices connected in such a way that each operates the next one in turn, as transistors or valves in an amplifier.

cascade (*NucEng*) The arrangement of stages in an enrichment or reprocessing plant in which the products of one stage are fed either *forwards* to the next closely similar or identical stage or *backwards* to a previous stage, eventually resulting in two more or less pure products at each end of the cascade. The classic examples are gaseous or centrifugal enrichment plants. An *ideal* cascade is the arrangement of stages in series and in parallel which gives the highest yield for a given number of units (eg centrifuges) and a given separation factor. It has the shape shown in the figure with more units in parallel in the middle of the cascade where the feedstock enters and the number tapering off towards the extremities. A *squared-off* cascade has the number of stages arranged as rectangular blocks approximately like the ideal arrangement.

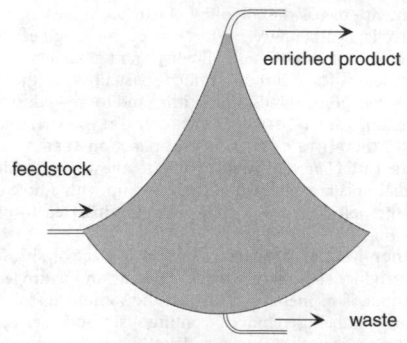

enriched product

feedstock

waste

cascade The 'ideal' arrangement of stages.

cascade generator (*Electronics*) High-voltage generator using a series of voltage-multiplying stages, esp when designed for X-ray tubes or low-energy accelerators.

cascade impactor (*PowderTech*) A device for sampling aerosols or dusty air, which automatically fractionates the particles or spray droplets by drawing the gas stream through a series of jet impactors of decreasing nozzle size.

cascade particle (*Phys*) Particle formed by a cosmic ray in a CASCADE SHOWER.

cascades (*Aero*) Fixed aerofoil blades which turn the airflow round a bend in a duct, eg in wind tunnels or engine intakes.

cascade shower (*Phys*) Manifestations of cosmic rays in which high-energy mesons, protons and electrons create high-energy photons, which produce further electrons and positrons, thus increasing the number of particles until the energy is dissipated. Also *air shower*.

cascading of insulators (*ElecEng*) Flashover of a string of suspension insulators; initiated by the voltage across one unit exceeding its safe value and flashing over, thereby imposing additional stress across the other units, and resulting in a complete flashover of the string.

cascara sagrada (*Pharmacol*) Dried bark of *Rhamnus purshiana*, a shrub of N California, formerly used (as extract) as a cathartic and laxative drug.

case (*Build*) The external facings of a building when these are of better material than the backing.

case (*Eng*) That part near the surface of a ferrous alloy which has been so altered as to allow case-hardening.

case (*Print*) A wooden tray divided into many compartments to accommodate individual letters. Originally in pairs, the UPPER CASE and LOWER CASE, then largely replaced by the double case. Now obsolete except in reference to the upper case and lower case letters in a FONT.

caseation (*Med*) The process of becoming cheese-like; eg in tissue infected with tubercle bacillus the cells break down into an amorphous cheese-like mass.

cased book (*Print*) Book for which a case has been made separately and attached to the book block as a separate operation. The usual method for EDITION BINDING.

cased frame (*Build*) The wooden box-frame containing the sash weights of a window. Also *boxed frame*.

case-hardening (*Eng*) The production of a hard surface layer in steel either: (1) by heating in a carbonaceous medium to increase the carbon content, followed by quenching and lightly tempering; or (2) by rapidly heating the surface of a medium/high-carbon steel to above the ferrite/austenite transformation temperature and then quenching and tempering, as in flame and induction hardening. See CYANIDE-HARDENING, PACK-HARDENING.

case-hardening (*For*) In seasoning, a condition in which the surface of timber becomes set in an expanded condition and remains under compression whilst the interior is in tension.

case-hardening (*MinExt*) Cement-like surface on porous rock caused by evaporation.

casein (*BioSci*) The principal albuminous constituent of milk, in which it is present as a calcium salt. Transformed into insoluble paracasein (cheese) by enzymes. Present in micellar form in skimmed milk and precipitated from it by acidification. It contains inorganic phosphate salts which aid processing into shaped products. Plasticized with water and cross-linked with formaldehyde, it was formerly used widely for decorative products, but is now mainly confined to buttons. Also used for priming artists' canvases. In the food industry caseinates are widely used for their emulsifying and water-binding properties.

Casella automatic microscope (*PowderTech*) A microscope whose stage moves automatically to enable the images of particles to be measured electronically.

casement (*Build*) Window sash hinged on one vertical edge to open out or in.

casement (*Textiles*) A plain-woven cotton (or manufactured) fabric used for curtains.

Casemix (*Med*) A database system for storing information about medical patients, based on classification according to specific predetermined characteristics such as age, severity of illness, diagnosis, etc.

caseous (*Med*) Cheese-like; having undergone CASEATION.

caseous lymphadenitis (*Vet*) A disease of sheep and goats due to infection with *Corynebacterium ovis* and characterized by nodule formation in the lymph nodes, lungs, skin and other organs. A chronic respiratory problem can ensue.

case-sensitive (-insensitive) (*ICT*) A computer program that takes (no) account of whether a text character is upper case or lower case.

cashmere (*Textiles*) Fine down-like fibres obtained from the undercoat of the Asiatic goat, or similar material obtained from goats in Australasia and Scotland. Frequently blended with wool and used for the manufacture of cardigans and sweaters.

casing (*MinExt*) (1) Piping used to line a drill hole. (2) The steel lining of a circular shaft.

casinghead (*MinExt*) The part of the well which is above the surface and to which the flow lines and control valves are attached.

casinghead gas (*MinExt*) Gas produced as a by-product from an oil well.

cask (*NucEng*) Also *casket*. See FLASK.

Casogrande hydrometer (*PowderTech*) An improved form of hydrometer used in the particle size analysis of soils. Its main feature is its smooth symmetrical outline as compared with the BOUYOUCOUS HYDROMETER.

casparian strip (*BioSci*) A band, running round the cell in which apparently the whole thickness of the primary wall is impregnated with suberin and/or lignin making it impermeable to water and solutes. Typical of the root endodermis where it occurs in all radial and transverse walls, preventing the movement of water and solutes between the cortex and the stele in the APOPLAST. Also *casparian band*.

Cassegrain antenna (*ICT*) Antenna in which radiation from a focus is collimated at one surface, eg a parabola, and reflected at another surface, eg a plane, or in reverse.

Cassegrain telescope (*Astron*) A form of reflecting telescope in which the rays after reflection at the main mirror fall on a small convex mirror placed inside the prime focus. Similar to the Gregorian telescope. See panel on ASTRONOMICAL TELESCOPE.

Cassel's yellow (*Chem*) Commercial name for lead oxychloride made by heating lead (II) oxide and ammonium chloride.

cassette (*ImageTech*) (1) A holder for magnetic tape or film which contains two reels, the tape or film moving from one to another as it passes the reading head or film gate. Reversing or rewinding the cassette allows repeated use and the holder protects tape or film outside the apparatus. (2) A holder for X-ray film, often incorporating an image-intensifying screen, preventing exposure to light but allowing exposure to X-rays.

cassia oil (*Chem*) An oil obtained from the bark of *Cinnamomum cassia*, a yellow or brown liquid of cinnamon-like odour; bp 240–260°C, rel.d. 1·045–1·063.

Cassini (*Space*) Joint NASA/ESA mission to observe Saturn and its moon Titan; launched in October 1997, transmitted pictures of Titan in October 2004.

Cassini's division (*Astron*) A gap between Saturn's A and B rings, around 4000 km wide; first observed by G D Cassini in 1675.

Cassini's ovals (*MathSci*) See OVALS OF CASSINI.

Cassiopeia (*Astron*) A large northern constellation that includes rich fields of clouds, gas, dust, and star clusters in the Milky Way; also includes radio sources CASSIOPEIA A and B.

Cassiopeia A (*Astron*) The strongest radio source in the sky, after the Sun, located in the constellation Cassiopeia. Around 3 kpc away, it is the remnant of a SUPERNOVA seen to explode around 1667.

Cassiopeia B (*Astron*) A powerful radio source in the constellation Cassiopeia which is associated with the remnant of the supernova known as TYCHO'S STAR.

cassiterite (*Min*) Oxide of tin, crystallizing in the tetragonal system; it constitutes the most important ore of this metal. It occurs in veins and impregnations associated with granitic rocks; also as 'stream-tin' in alluvial gravels. Also *tinstone*.

Cassius (*Chem*) See PURPLE OF CASSIUS.

cast (*Geol*) The impression which fills a natural MOULD. Most frequently a fossil shell, where the infilling is both a cast of the original animal and an internal mould of the shell. Term also used for impressions of sedimentary structures.

cast (*Med*) A mould of cellular or organic matter shed from tubular structures of the body, eg the bronchi or the tubules of the kidney.

caste (*BioSci*) In some social insects, one of the types of polymorphic individuals composing the community.

caster action (*Autos*) The use of inclined swivel axes or king-pins by which the steerable front wheels are given fore-and-aft stability and a self-centring tendency after angular deflection by road shocks, on the principle of the domestic caster. See TRAIL.

cast holes (*Eng*) Holes made in cast objects by the use of cores, in order to reduce the time necessary for machining, and to avoid metal wastage.

casting (*Eng*) (1) The operation of pouring molten metals into sand or metal moulds in which they solidify. (2) A metallic article cast in the shape required, as distinct from one shaped by working.

casting (*Vet*) (1) The process of throwing and securing an animal from the upright to the prone position. (2) The pellet of undigested feathers, fur or bones disgorged by a raptorial bird.

casting copper (*Eng*) Metal of lower purity than BEST SELECTED COPPER. Generally contains about 99·4% of copper.

casting ladle (*Eng*) A steel ladle, lined with refractory material, in which molten metal is carried from the furnace to the mould in which the casting is to be made.

casting resins (*Plastics*) A term applied to liquid resins which can be introduced into a moulded shape and polymerized catalytically *in situ*. Usually polymethyl methacrylate, polyester, epoxy and polyurethane resins.

casting wheel (*Eng*) Large wheel on which ingot moulds are arranged peripherally and filled from a stream of molten metal issuing from a furnace or pouring ladle.

cast-*in-situ* concrete piles (*CivEng*) A type of pile formed by driving a steel pipe into the ground and filling it with concrete, using the pipe as a mould, or by a similar method. The mould may be withdrawn as the concrete is consolidated by heavy tamping.

cast-iron (*Eng*) Any iron–carbon alloy in which the carbon content exceeds the solubility of carbon in austenite at the eutectic temperature. Widely used in engineering on account of their high fluidity and excellent casting characteristics. Carbon content usually in the range 2–4·3%. Some kinds are brittle and others difficult to machine. See DUCTILE CAST-IRON, GREY IRON, SPHERULITIC GRAPHITE CAST-IRON and panel on STEELS.

castle nut (*Eng*) A six-sided nut in the top of which six radial slots are cut. Two of these line up with a hole drilled in the bolt or screw, a split pin being inserted to prevent turning.

Castner–Kellner process (*Chem*) The electrolysis of brine to produce sodium hydroxide, chlorine and hydrogen. The cathode is of mercury, in which the sodium dissolves, the amalgam being removed continuously and treated with water to liberate mercury and sodium hydroxide.

Castner's process (*Chem*) The production of sodium cyanide from sodium. As the temperature is gradually raised, dry ammonia is blown into metallic sodium melted with charcoal, the resulting sodamide reacting quickly with the charcoal to form cyanamide. This is heated to c.850°C, and combines further with carbon to form sodium cyanide.

cast-off (*Print*) A calculation of the amount of space that a manuscript will occupy when set in a given typeface and measure.

Castor (*Astron*) A white visual triple star in the constellation Gemini. Distance 14 pc. Also *Alpha Geminorum*.

castor oil (*Chem*) Oil obtained from the seeds of *Ricinus communis*, a yellow or brown, syrupy, non-drying liquid; mp −10°C, rel.d. 0·960–0·970, saponification value 178, iodine value 85, acid value 19·21. Main constituent, ricinoleic acid. Uses: *pure*, in medicine and hydraulic fluids; *dehydrated*, as a drying oil in paints; *hydrogenated*, as a fat in cosmetics; *sulphonated*, see TURKEY-RED OIL.

castration (*Med, Vet*) Removal or surgical destruction of the testicles.

castration anxiety (*Psych*) In psychoanalytic theory, the male child's fear that his penis will be cut off as punishment for his sexual desire for the mother.

cast steel (*Eng*) Shapes that have been formed directly from liquid by casting into a mould. Formerly applied to wrought objects produced by working steel made by the crucible process to distinguish from that made by cementation of wrought-iron, but both of these methods are long obsolete. See panel on STEELS.

cast stone (*Build*) Precast artificially manufactured building components, eg blocks, lintels, sills, copes. Basically precast concrete with a facing of fine material and cement intended to look like natural stone.

cast-up (*Print*) The overall *en* content of a job after making necessary additions to the CAST-OFF for headlines, folios, etc.

cast welded rail joint (*CivEng*) A joint between the ends of two adjacent rails made in position using the THERMITE process.

casual species (*BioSci*) An introduced plant that occurs but is not established in places where it is not cultivated.

CAT (*ICT*) Digital electronic telecommunication systems, including computer networks (esp the INTERNET), cable and satellite TV, and telephone links.

cata- (*Genrl*) See KATA-.

catabolism (*BioSci*) A metabolic process of breaking down complex molecules into simpler ones and releasing energy. Adj *catabolic*.

catabolite (*BioSci*) A product of CATABOLISM.

catacaustic (*MathSci*) See CAUSTIC CURVE.

cataclasis (*Geol*) The process of rock deformation involving fracture and rotation of mineral grains without chemical reconstitution. Also *kataklasis*.

cataclysmic variable (*Astron*) A close binary system in which accretion from one star onto a white dwarf companion results in violent and irregular variability. Class includes NOVAE and SYMBIOTIC STARS.

catadioptric (*Phys*) An optical system using a combination of refracting and reflecting surfaces designed to reduce aberrations in a telescope.

catadromous (*BioSci*) See KATADROMOUS.

catalan process (*Eng*) Reduction of haematite to wrought-iron by smelting with charcoal.

catalase (*BioSci*) An enzyme that catalyses the oxidation of many substrates by hydrogen peroxide. In the absence of a suitable substrate it destroys hydrogen peroxide, converting it to water.

catalepsy (*Med*) The condition in which any posture of a limb can be maintained without movement for a period of time longer than normal; occurring in disease of the cerebellum and in hysteria, also in deep hypnotic states and in certain types of schizophrenia. In animals is used to indicate the action known as 'feigning death' which can be induced in some animals by any sudden disturbance. Adj *cataleptic*.

catalysis (*Chem*) The acceleration or retardation of a chemical reaction by a substance which itself undergoes no permanent chemical change, or which can be recovered when the chemical reaction is completed. It lowers the energy of activation.

catalyst (*Chem*) A substance which catalyses a reaction. See CATALYSIS.

catalytic antibody (*BioSci*) An antibody raised against a transition-state analogue that can catalyse the analogous chemical reaction, though not as effectively as a true enzyme.

catalytic converter (*Autos*) A device fitted into the exhaust system of petrol-engined vehicles to reduce the emissions of carbon monoxide (CO), nitrogen oxides and hydrocarbons. A *three-way catalyst* of platinum, palladium and rhodium on a ceramic lattice oxidizes CO to carbon dioxide (CO_2), hydrocarbons to CO_2 and water, and reduces nitrogen oxides to nitrogen. An *oxidation catalyst* oxidizes CO and hydrocarbons but does not affect oxides of nitrogen: these have to be reduced by engine design using lean-burn principles. See LEAN-BURN ENGINE.

catalytic cracking (*MinExt*) A process of breaking down the heavy hydrocarbons of crude petroleum, using silica or aluminium gel as a catalyst. See CRACKING.

catalytic poison (*Chem*) A substance which inhibits the activity of a catalyst. Also *anticatalyst*.

catalytic reforming (*Chem*) Petroleum-refining process to improve the OCTANE NUMBER of light hydrocarbons by reaction with hydrogen at high temperatures (500°C) and pressures over a platinum catalyst. Cf HYDROFINING.

cataphorite (*Min*) See KATOPHORITE.

cataphyll (*BioSci*) A non-foliage leaf inserted low on a shoot, eg a scale on a rhizome or a bud scale. Cf HYPSOPHYLL.

cataplexy (*Med*) Sudden attack of weakness, following some expression of emotion; the patient falls to the ground, immobile, speechless, but conscious.

catapult (*Aero*) An accelerating device for launching an aircraft in a short distance. It may be fixed or rotatable to face the wind. It is usually used on ships which have no landing deck, having been superseded on aircraft carriers by the ACCELERATOR. During World War II, fighters were carried on catapult-armed merchant ships for defence against long-range bombers. Land catapults have been tried but have been superseded by RATOG and STOL aircraft.

cataract (*Med*) Opacity of the lens of the eye as a result of degenerative changes in it.

catarrh (*Med*) Inflammation of a mucous membrane, with discharge of mucus, commonly associated with the common cold.

catastrophe theory (*MathSci*) Theory of sudden, as opposed to continuous, changes.

catastrophism (*Geol*) The theory that the Earth's geological history has been fashioned by infrequent violent events. See UNIFORMITARIANISM.

catch basin (*CivEng*) See CATCH PIT.

catch-bolt (*Build*) A door lock having a spring-loaded bolt which is normally in the locking position (ie extended), but which is automatically and momentarily retracted in the process of shutting the door.

catch crop (*Agri*) A rapidly maturing crop sown between plantings of mainstream crops to maximize land use. Such crops may be used to offset poor performance of the preceding crop.

catcher (*Electronics*) The element in a velocity-modulated ultrahigh-frequency or microwave beam tube which abstracts, or catches, the energy in a bunched electron stream as it passes through it. See BUNCHER.

catcher foil (*NucEng*) Aluminium sheet used for measuring power levels in a nuclear reactor by absorption of fission fragments.

catching diode (*Electronics*) Diode used to clamp a voltage or current at a predetermined value. When it becomes forward-biased it prevents the applied potential from increasing any further.

catch-line (*Print*) A temporary headline inserted on slip proofs etc.

catchment area (*CivEng*) The area from which water runs off to any given river valley or collecting reservoir. Also *catchment basin*.

catch mounts (*Print*) Individual mounts for bookwork printing to hold duplicate plates, and from which they can be removed and replaced without unlocking the forme.

catch muscle (*BioSci*) A set of smooth muscle fibres that form part of the adductor muscle in bivalve molluscs and are capable of keeping the valves closed by means of a sustained tonus; any set of smooth muscle fibres associated with striated muscle fibres for a similar purpose. Also *arrest muscle*.

catch net (*ElecEng*) See CRADLE.

catch pit (*CivEng*) A small pit constructed at the entrance to a length of sewer or drain pipe to catch and retain matter which would not easily pass through the pipes. Also *catch basin*. See SUMP.

catch plate (*Eng*) A disk on the spindle nose of a lathe, driving a carrier locked to the work.

catch points (*CivEng*) Trailing points placed on an up-gradient for the purpose of derailing rolling stock accidentally descending the gradient. See SPRING POINTS.

catch props (*MinExt*) In a coal mine, props put in advance of the main timbering for safety, ie watch props or safety props.

catch-up (*Print*) A term used in lithographic printing when the non-image areas of the printing plate begin to accept ink due to insufficient font solution being fed to the plate.

catch-water drain (*CivEng*) A drain to catch water on a hillside, with open joints or multiple perforations to take in water in as many places as possible.

catch word (*Print*) (1) At the foot of the page, the first word of the next, as in manuscripts and books printed prior to the 19th century. (2) At the head of the page a guide word, eg in dictionaries. (3) A guide word at the head of a galley of type.

catch work (*Genrl*) A system of water channels which may be used for flooding land.

cat cracker (*MinExt*) Refinery vessel in which hydrocarbon fractions are processed in the presence of a catalyst.

cat E (*Aero*) Category E damage to an aircraft; equivalent to a total loss or 'write-off'.

catechin (*Chem*) A water-soluble polyphenol antioxidant compound found in tea; a derivative of tannin; a major constituent of *catechu*, used in tanning and dyeing.

catechol (*Chem*) *2-hydroxyphenol, 1,2-hydroxybenzene.* $C_6H_4(OH)_2$ (1,2); colourless crystals; mp 104°C, bp 240°C. Occurs in fresh and fossil plant matter and in coaltar. Important derivatives are GUAIACOL and adrenaline (see CATECHOLAMINES).

catecholamines (*Pharmacol*) A series of compounds derived from dihydroxyphenylalanine (DOPA), dopamine, noradrenaline (norepinephrine) and adrenaline (epinephrine). They function as NEUROTRANSMITTERS, adrenaline also acting as a hormone.

category (*MathSci*) A collection of objects and arrows such that: (1) each arrow leads from an object of the collection to another (or the same one); (2) a composite arrow in the collection is defined in all cases where two arrows 'join together' (eg *f* goes from *a* to *b*, *g* goes from *b* to *c*, composite arrow goes from *a* to *c*, written *f.g* or *g.f* according to convention); (3) arrow COMPOSITION is ASSOCIATIVE; (4) for each object there is defined an identity arrow from that object to itself such that any composite of it with another arrow is equal to that other arrow. For instance, in a category of groups the objects will be groups and the arrows will be group HOMOMORPHISMS.

catenary (*MathSci*) The curve assumed by a perfectly flexible, uniform, inextensible string when suspended from its ends. When it is symmetrical about the *y*-axis its equation is $y = c \cosh(x/c)$, where *c* is the point where it intersects the *y*-axis.

catenary construction (*ElecEng*) A method of construction used for overhead contact wires of traction systems. A wire is suspended, in the form of a catenary, between two supports, and the contact wire is supported from this by droppers of different lengths, arranged so that the contact wire is horizontal. See COMPOUND CATENARY CONSTRUCTION.

catenation (*BioSci*) The arrangement of chromosomes in chains or in rings.

catenation (*Chem*) See CHAIN.

catenoid (*MathSci*) The surface of revolution generated by the rotation of a catenary about its internal axis.

caterpillar (*BioSci*) The type of larva, found in Lepidoptera, which typically possesses abdominal locomotor appendages (prolegs).

cat flu (*Vet*) See FELINE PNEUMONITIS.

catgut (*Med*) Sterilized strands of sheep's or other animal's intestines formerly used as ligatures.

catharsis (*Psych*) Release of emotions. See ABREACTION.

cathartic (*Pharmacol*) Purgative. A drug which promotes evacuation of the bowel.

cathead (*Eng*) (1) The sheave assembly on the top of a crane jib. (2) A lathe accessory consisting of a turned sleeve having four or more radial screws at each end; used for clamping onto rough work of small diameter and running in the STEADY while centring. Also *spider*.

cathead (*MinExt*) Rotating drum around which a rope can be coiled to provide pull for various operations on a drilling rig.

cathelectrotonus (*Med*) Physiological excitability produced in muscle tissue by passage of electric current.

cathepsins (*BioSci*) A family of intracellular proteolytic enzymes of animal tissues.

Catherine wheel (*Arch*) See ROSE WINDOW.

catheter (*Med*) A rigid or flexible tube for introduction into vessels or organs of the body. Commonly used to drain the urinary bladder, to measure intracardiac pressures or to inject radio-opaque material into the blood vessels (in ANGIOGRAPHY).

cathetometer (*Phys*) An optical instrument for measuring vertical distances not exceeding a few decimetres. A small telescope, held horizontally, can move up and down a vertical pillar. The difference in position of the telescope when the images of the two points whose separation is being measured are lined up with the cross-wires of the telescope is obtained from the difference in vernier readings on a scale marked on the pillar. Also *reading microscope, reading telescope*.

cathexis (*Psych*) A charge of mental energy attached to any particular idea or object.

cathode (*Electronics*) (1) In an electronic tube or valve, an electrode through which a primary stream of electrons enters the inter-electrode space. During conduction, the cathode is negative with respect to the anode. Such a cathode may be cold, electron emission being due to electric fields, photoemission, or impact by other particles, or thermionic, where the cathode is heated by some means. (2) In a semiconductor diode, the electrode to which the forward current flows. (3) In a thyristor, the electrode by which current leaves the thyristor when it is in the ON state. (4) In a light-emitting diode, the electrode to which forward current flows within the device. (5) In electrolytic applications, the electrode at which positive ions are discharged, or negative ions formed.

cathode coating (*Electronics*) A low-work-function surface layer applied to a thermionic or photocathode in order to enhance electron emission or to control spectral characteristics. The cathode coating impedance is between the base metal and this layer.

cathode copper (*Eng*) The product of electrolytic refining, after which the cathodes are melted, oxidized, pooled, and cast into wire-bars, cakes, billets, etc.

cathode efficiency (*Electronics*) The ratio of emission current to energy supplied to cathode. Also *emission efficiency*.

cathode follower (*Electronics*) A valve circuit in which the input is connected between the grid and earth, and the output is taken from between the cathode and earth, the anode being earthed to signal frequencies. It has a high input impedance, low output impedance and unity voltage gain. See COMMON-COLLECTOR CONNECTION.

cathode glow (*Electronics*) Glow near the surface of a cathode, its colour depending on the gas or vapour in the tube. If an arc takes place in a partial vacuum, it may fill the greater part of the discharge tube.

cathode luminous sensitivity (*Electronics*) The ratio of cathode current of photoelectric cell to luminous intensity.

cathode modulation (*Electronics*) Modulation produced by signal applied to cathode of valve through which carrier wave passes.

cathode poisoning (*Electronics*) Reduction of thermionic emission from a cathode as a result of minute traces of adsorbed impurities.

cathode ray (*Electronics*) A stream of negatively charged particles (electrons) emitted normally from the surface of a cathode in a vacuum or low-pressure gas. The velocity of the electrons is proportional to the square root of the accelerating potential, being $6 \times 10^5 \text{ m s}^{-1}$ for 1 volt. They can be deflected and formed into beams by the application of electric or magnetic fields, or a combination of both, and are widely used in oscilloscopes and TV (in cathode-ray tubes), electron microscopes and electron-beam welding, and electron-beam tubes for high-frequency amplifiers and oscillators.

cathode-ray oscillograph (*Electronics*) An oscillograph in which a permanent (photographic or other) record of a transient or time-varying phenomenon is produced by means of an electron beam in a cathode-ray tube. Deprecated term for CATHODE-RAY OSCILLOSCOPE.

cathode-ray oscilloscope (*Electronics*) A device for displaying electronic signals by modulating a beam of electrons before it impinges on a FLUORESCENT SCREEN. Abbrev CRO.

cathode-ray tube (*Electronics*) An electronic tube in which a well-defined and controllable beam of electrons is produced and directed onto a surface to give a visible or otherwise detectable display or effect. Abbrev *CRT*.

cathode spot (*Electronics*) The area on a cathode where electrons are emitted into an arc, the current density being much higher than with simple thermionic emission.

cathodic chalk (*Ships*) A coating of magnesium and calcium compounds formed on a steel surface during CATHODIC PROTECTION in sea water.

cathodic etching (*Electronics*) Erosion of a cathode by a glow discharge through positive-ion bombardment, in order to show microstructure.

cathodic protection (*Electronics, Ships*) The protection of a metal structure against electrolytic corrosion by making it the cathode (electron receiver) in an electrolytic cell, either by means of an impressed emf or by coupling it with a more electronegative metal. In ships and offshore structures, corrosion can be prevented by passing sufficient dc through the sea water to make the metal hull a cathode. See SACRIFICIAL ANODE.

cathodoluminescence (*Phys*) The emission of light, with a possible afterglow, from a material when irradiated by an electron beam, such as occurs in the phosphor of a cathode-ray tube.

cathodophone (*Acous*) Microphone utilizing the silent discharge between a heated oxide-coated filament in air and another electrode. This discharge is modulated directly by the motion of the air particles in a passing sound wave. Also *ionophone*.

catholyte (*Phys*) See CATOLYTE.

cation (*Phys*) The ion in an electrolyte which carries a positive charge and which migrates towards the cathode under the influence of a potential gradient in electrolysis. It is the deposition of the cation in a primary cell which determines the positive terminal. Cf ANION. See panel on BONDING.

cationic detergents (*Chem*) Types of ionic synthetic detergents in which the surface active part of the molecule is the cation, unlike soap and most of the widely used synthetic detergents. Typified by the quaternary ammonium salts such as CETRIMIDE, benzalkonium chloride, domiphen bromide; cetylpyridinium is similar. All have powerful bactericidal activity, and in addition to skin and utensil cleaning are used in lozenges, creams, mouthwashes, etc. Also *invert soaps*.

cationic polymerization (*Chem*) Polymerization using cationic catalyst such as aluminium chloride. Commercial polymers include butyl rubber. See CHAIN POLYMERIZATION.

catkin (*BioSci*) An inflorescence with the flowers sessile on a common axis and typically pendulous, unisexual and wind-pollinated. A common inflorescence of deciduous, north temperate trees.

catolyte (*Electronics*) That portion of the electrolyte of an electrolytic cell which is in the immediate neighbourhood of the cathode. Also *catholyte*.

catophorite (*Min*) See KATOPHORITE.

catoptric element (*Phys*) A component of an optical system that uses reflection, not refraction, in the formation of an image.

catoptric lens (*ImageTech*) See MIRROR LENS.

CAT scanner (*Med*) A device for carrying out *computer-assisted tomography*, a highly advanced X-ray technique that uses a succession of narrow beams passing through the body at different angles. After the beam has encircled the body the computer produces an image resembling a slice through the body

cat's eye (*Glass*) (1) A crescent-shaped blister in glass. (2) A reflecting stud containing glass lenses set in rubber and used for road-marking.

cat's eye (*Min*) A variety of fibrous quartz which shows chatoyancy when suitably cut, as an ornamental stone. The term is also applied to crocidolite when infiltrated with silica (see HAWK'S EYE, TIGER'S EYE). A more valuable form is *chrysoberyl cat's eye* (see CYMOPHANE).

cattle plague (*Vet*) See RINDERPEST.

CATV (*ICT*) Abbrev for COMMUNITY ANTENNA TELEVISION.

catworks (*MinExt*) Assemblage of motors and CATHEADS which provides power for the many secondary activities on drilling platforms like pipe-hoisting.

Cauchy–Riemann equations (*MathSci*) The equations

$$\frac{\partial u}{\partial x} = \frac{\partial v}{\partial y} \text{ and } \frac{\partial u}{\partial y} = -\frac{\partial v}{\partial x}$$

which must be satisfied for a function of a complex variable $f(z) = u(x, y) + iv(x, y)$, u and v being real, to be differentiable. Functions u and v are said to be *conjugate*.

Cauchy's convergence tests (*MathSci*) (1) If

$$\sqrt[n]{u_s} \to l$$

as $n \to \infty$, then the series of positive terms $\sum u_n$ converges if $l < 1$ and diverges if $l > 1$. (2) If $\log(1/u_n)\log^{-1} n \to l$ as $n \to \infty$, then the series of positive terms $\sum u_n$ converges if $l > 1$ and diverges if $l < 1$.

Cauchy's dispersion formula (*Phys*)

$$\mu = A + \frac{B}{\lambda^2} + \frac{C}{\lambda^{21}} + \dots$$

An empirical expression for the relation between the refractive index μ of a medium and the wavelength λ of light; A, B and C are the constants for a given medium.

Cauchy's distribution (*MathSci*) A probability distribution of the form

$$p(x) = \frac{1}{\pi[1 + (x - a)^2]}$$

where the distribution is symmetric about $x = a$.

Cauchy sequence (*MathSci*) A sequence a_r such that for any distance ε, we can choose an N such that the distance between a_m and a_n is less than ε for all m and n greater than N.

Cauchy's inequality (*MathSci*)

$$\left(\sum_{r=1}^{n} a_r b_r\right)^2 \le \left(\sum_{r=1}^{n} a_r^2\right)\left(\sum_{r=1}^{n} b_r^2\right)$$

The equality occurs when $a_r = k b_r$ for all r.

Cauchy's integral formula (*MathSci*) If the function $f(z)$ is analytic within and on a closed contour C, and if a is any point within C, then

$$f(a) = \frac{1}{2\pi i} \oint c \frac{f(z)dz}{z - a_c}$$

Cauchy's theorem (*MathSci*) If a function $f(z)$ is analytic inside and on a contour C, then the integral of $f(z)$ taken round the contour is equal to zero.

cauda (*BioSci*) The tail, or region behind the anus; any tail-like appendage; the posterior part of an organ, such as the *cauda equina*, a bundle of parallel nerves at the posterior end of the spinal cord in vertebrates. Adj *caudal, caudate*.

caudad (*BioSci*) Situated near, facing towards or passing to the tail region.

Caudata (*BioSci*) See URODELA.

caudate (*BioSci*) Bearing a tail-like appendage.

caudate nucleus (*BioSci*) The most frontal of the basal ganglia in the brain. Damage to caudate neurons is characteristic of Huntington's chorea and other motor disorders.

caudex (*BioSci*) A trunk or stock.

caul (*BioSci*) In the higher vertebrates, the AMNION; more generally any enclosing membrane.

caul- (*BioSci*) Prefix meaning stem.

cauldron subsidence (*Geol*) The subsidence of a cylindrical mass of the Earth's crust, bounded by a circular fault up which the lava has commonly risen to fill the cauldron. Good examples have been described from Scotland (Ben Nevis and the Western Isles).

caulescent (*BioSci*) Having a stalk or a stem.

cauliflory (*BioSci*) Production of flowers on trunks, branches and old stems of woody plants rather than near the ends of smaller twigs. Occurs in some trees of tropical forest, eg cocoa.

cauline (*BioSci*) (1) Pertaining to a stem. (2) Leaves borne on an obvious stem, not at the extreme base, and well above soil level, ie not radical.

caulking (*Eng*) The process of closing the spaces between overlapping riveted plates or other joints by hammering the exposed edge of one plate into intimate contact with the other. A filler material is also used esp for closing eg deck planking. Also *calking*.

caulking tool (*Eng*) A tool, similar in form to a cold chisel but having a blunt edge, for deforming the metal rather than cutting it.

caul (*Build*) The rigid sheet used for pressing veneers onto the ground timber.

causalgia (*Med*) Intense burning pain in the skin after injury to the nerve supplying it.

causality (*Phys*) (1) The principle that an event cannot precede its cause. (2) See DETERMINISM.

caustic (*Med*) Destructive or corrosive to living tissue; an agent which burns or destroys living tissue.

caustic curve (*MathSci*) The envelope of the rays from a given point *P* (the *radiant point*) after reflection by a given curve C. Also *catacaustic*. If the rays are refracted, the envelope is called the *diacaustic*.

caustic curve (*Phys*) A curve to which rays of light are tangential after reflection or refraction at another curve.

caustic embrittlement (*Eng*) The intergranular corrosion of steel in hot alkaline solutions, eg in boilers.

caustic lime (*Chem*) The residue of calcium oxide, obtained from freshly calcined calcium carbonate; it reacts with water, evolving much heat and producing slaked lime (calcium hydroxide, hydrate of lime or hydrated lime). Also *quicklime*. See LIME.

caustic paint removers (*Build*) Combinations of strong caustic solutions and a thickening agent such as flour or starch. Surfaces must be thoroughly washed down and neutralized with a mild acid solution before subsequent decoration. Caustic paint removers have an adverse effect on aluminium and zinc substrates.

caustic potash (*Chem*) Potassium hydroxide. The name potash is derived from 'ash' (meaning the ash from wood) and 'pot' from the pots in which the aqueous extract of the ash was formerly evaporated. Highly alkaline.

caustic soda (*Chem*) See SODIUM HYDROXIDE.

caustic surface (*Phys*) A surface to which rays of light are tangential after reflection or refraction at another surface.

CAV (*ImageTech*) Abbrev for CONSTANT ANGULAR VELOCITY.

cavalry twill (*Textiles*) Firm, warp-faced cloth, often wool, with steep double-twill lines, made in different weights for trousers, breeches, rainwear and dresses.

cave (*NucEng*) Well-shielded enclosure in which highly radioactive materials can be kept and manipulated safely. Also *hot cell*.

Cavendish experiment (*Phys*) An experiment, carried out by H Cavendish in 1798, to determine the GRAVITATIONAL CONSTANT. A form of torsion balance was used to measure the very small forces of attraction between lead spheres.

cavern (*Geol*) A chamber in a rock. Caverns are of varying size and are due to several causes, the chief being the solution of calcareous rocks by underground waters, and marine action.

cavernosus (*BioSci*) Honeycombed; hollow; containing cavities, eg CORPORA CAVERNOSA. Also *cavernous*.

cavetto (*Arch*) A hollow moulding, quarter round. See fig. at MOULDINGS.

cavil (*Build*) A small stone axe resembling a JEDDING AXE.

cavilling (*MinExt*) The drawing of lots for working places (usually for three months) in the coal mine.

caving (*MinExt*) Controlled collapse of roof in deep mines to relieve pressure. Same as undercut stoping.

cavitation (*Acous, Eng*) Generation of cavities (eg bubbles) in liquids by rapid pressure changes such as those induced by ultrasound. When cavity bubbles implode, they produce shock waves in the liquid. Components can be damaged by cavitation if it is induced by turbulent flow.

cavitation (*BioSci*) (1) The formation of cavities in any structure of the body, esp the lungs. (2) The sudden development of a gas bubble in a previously sap-filled xylem conduit as a result of excessive tension. See EMBOLISM.

cavity (*Eng*) Hollow space within moulding tool which will form final product.

cavity effect (*Acous*) The enhancement in a microphone due to acoustic resonance in a shallow cavity in front of the diaphragm.

cavity-frequency meter (*Electronics*) One for use in coaxial or waveguide systems. The frequency is related to the wavelength for resonance in the cavity.

cavity magnetron (*ICT*) General term (now deprecated) for MAGNETRONS having cavities cut into the inner faces of a solid cylindrical anode.

cavity modes (*Phys*) Stable electromagnetic or acoustic fields in a cavity which can exhibit resonance. They are DEGENERATE if, having similar frequencies, they have fields which differ in pattern from that of main resonance.

cavity radiation (*Phys*) The radiation emerging from a small hole leading to a constant-temperature enclosure. Such radiation is identical to BLACK-BODY RADIATION at the same temperature, no matter what the nature of the inner surface of the enclosure.

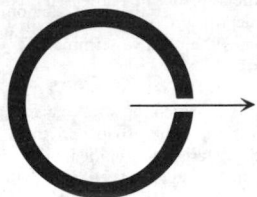

cavity radiation

cavity resonance (*Phys*) The enhancement of airflow for certain frequencies, due to neutralization of the mass (or inertia) reactance with the stiffness reactance of air in a partially enclosed space.

cavity resonator (*Phys*) Any nearly closed section of waveguide or coaxial line in which a pattern of electric and magnetic fields can be established. Also applies to sound fields.

Cavity Transfer Mixing (*Chem*) TN for polymer mixing process using specially designed head fitted to extruder output. The head comprises a cylinder and loose-fitting rotating core (attached to the screw). Each mating surface possesses a number of hemispherical hollows, by means of which the melt is repeatedly turned and thoroughly mixed with pigments etc. See EXTRUSION.

cavity walls (*Build*) Hollow walls, normally built with two $4\frac{1}{2}$ in stretcher bond walls with a 2 in gap between, tied together with wall ties. Cavity walls increase the thermal resistance and prevent rain from reaching the inner face.

cavum (*BioSci*) A hollow or cavity: a division of the concha.

cavus (*Med*) See PES ARCUATUS.

CB (*ElecEng*) Abbrev for *Citizens' Band radio*, used by amateur radio operators.

C-banding (*BioSci*) A method of staining chromosomes. See BANDING TECHNIQUES and panel on CHROMOSOME.

CBT (*ICT*) Abbrev for *computer-based training*.

cc (*Genrl*) Abbrev for *cubic centimetre*, the unit of volume in the CGS metric system. Also cm^3.

CCD (*Geol*) Abbrev for CARBONATE COMPENSATION DEPTH.

CCD (*ICT, Electronics*) Abbrev for CHARGE-COUPLED DEVICE.

CCD array (*Electronics*) An array of many thousands of photodiodes, whose response to an image focused on the surface of the array can be converted into a video signal by employing CCD electronic circuits. An alternative to vacuum tubes in TV cameras.

CC filter (*ImageTech*) Abbrev for *colour correction* or *colour conversion filter*. A gelatine filter used to change the colour temperature of a light source by a specific amount, eg from incandescent 3200 K to daylight 5400 K.

CCIR (*ICT*) Abbrev for *Comité Consultatif International des Radiocommunications*.

CCITT (*ICT*) Abbrev for *Comité Consultatif International Télégraphique et Téléphonique*. United Nations established committee in Geneva for tariffs, technical standards and conformity in order to facilitate international telecommunications.

C-class insulation (*ElecEng*) A class of insulating material which will withstand a temperature of over 180°C.

CCP (*Chem*) Abbrev for CUBIC CLOSE PACKING.

CCT diagram (*Eng*) Abbrev for CONTINUOUS COOLING TRANSFORMATION DIAGRAM.

CCTV (*ImageTech*) Abbrev for *closed-circuit television*, a non-broadcast TV system with the cameras and display receivers directly linked.

CCU (*ImageTech*) Abbrev for CAMERA CONTROL UNIT.

CCV (*Aero*) Abbrev for CONTROL-CONFIGURED VEHICLE.

CD (*Acous*) Abbrev for COMPACT DISC.

Cd (*Chem*) Symbol for CADMIUM.

cd (*Genrl*) Symbol for CANDELA.

CD antigen (*BioSci*) Abbrev for *cluster of differentiation antigen*. An internationally agreed nomenclature for differentiation of antigens on cell surfaces, usually recognized by

specific monoclonal antibodies. By 2005 more than 300 different CD antigens had been described. CD4 and CD8 are eg markers of helper and cytotoxic subsets of T-cells.

cdc gene (*BioSci*) Abbrev for CELL DIVISION CYCLE GENE.

CD-G (*ImageTech*) Abbrev for *CD-graphics*. An audio CD incorporating simple digital graphics which can be decoded and displayed on a TV receiver.

CD-i (*ImageTech*) Abbrev for *CD-interactive*. An established format for CD-size disks carrying digital audio and graphics, and possibly digital still and/or FULL MOTION VIDEO, with a high INTERACTIVE content. See GREEN BOOK, VIDEO CD.

C-display (*Radar*) Display in which bright spot represents the target, with horizontal and vertical displacements representing the bearing and elevation, respectively.

CDK (*BioSci*) Abbrev for CYCLIN-DEPENDENT KINASE. See panel on CELL CYCLE.

CDMA (*ICT*) Abbrev for *code-division multiple access*, a sophisticated digital transmission system used in personal communication devices such as mobile phones.

cDNA (*BioSci*) Abbrev for COMPLEMENTARY DNA.

cDNA cloning (*BioSci*) See COMPLEMENTARY DNA CLONING.

CDR (*BioSci*) Abbrev for COMPLEMENTARITY DETERMINING REGION.

CD-R (*ICT*) Abbrev for *compact disc-recordable*.

CD-ROM (*ICT*) An optical storage disk that stores data in digital form similar in size to a COMPACT DISC. May store in excess of 500 Mbytes of data and are frequently used for encyclopedias, the text of a large set of newspapers or libraries of graphical images. See WORM.

CD-RW (*ICT*) Abbrev for *compact disc-rewritable*.

CdS meter (*ImageTech*) An exposure meter making use of a cadmium sulphide photoresistor powered by a separate battery.

CD-v (*ICT*) Abbrev for *CD-video*. An established standard for storing video data on a CD-ROM. Also *video disk*.

CD video disk (*ICT*) A COMPACT DISC able to store and play synchronized sound and video images that have been compressed to allow about 60 minutes' playing time.

Ce (*Chem*) Symbol for CERIUM.

cecostomy (*Med*) US for CAECOSTOMY.

cecum (*Med*) US for CAECUM.

CED (*Chem*) Abbrev for COHESIVE ENERGY DENSITY.

cedar (*For*) A softwood tree (*Cedrus*), of European (*C. atlantica*) and Asian (*C. deodara*) origin, whose wood is light-brown and straight-grained, with a medium fine and

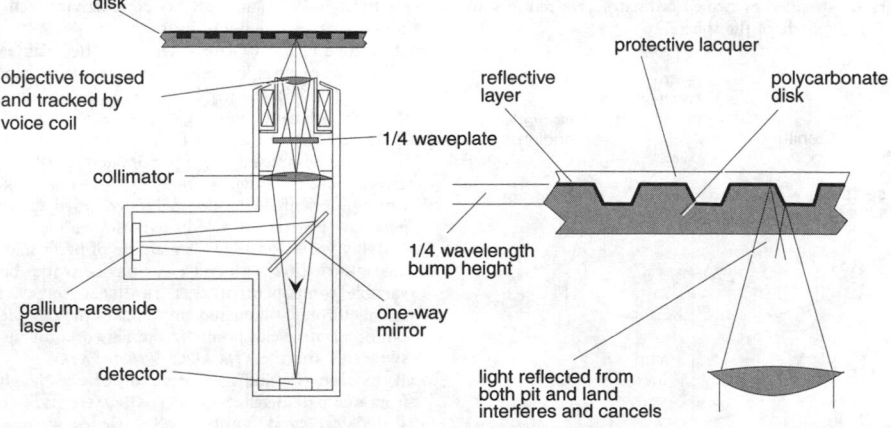

CD-ROM Reading head with enlarged view of disk on right. Signal from transition between pit and land is binary 1.

uniform texture, and a distinctive 'cedary' smell. Also *deodar*.

cedar-tree laccolith (*Geol*) A multiple laccolith, ie a series of laccoliths, one above the other, forming part of a single mass of igneous rock. See DYKES.

CEEB score (*Psych*) A standard score used in the Scholastic Aptitude Test (SAT) that sets the mean to 500 and standard deviation to 100.

Ceefax (*ICT*) See TELETEXT.

ceiling (*Ships*) Heavy planking laid on top of the inner double bottom under hatchways to protect the top of tanks below.

ceiling joist (*Build*) A joist to which a ceiling is fixed.

ceiling plate (*ElecEng*) A plate from which a light fitting may be mechanically supported and the electrical connection made.

ceiling rose (*ElecEng*) A plastic enclosure containing the electrical connections which can support a light fitting.

ceiling switch (*ElecEng*) A switch attached to the ceiling and operated by a cord. Also *pull switch*.

ceiling temperature (*Chem*) Critical temperature above which polymerization cannot occur. Of the order of hundreds of degrees centigrade for most common polymers, it is the result of the high and negative entropy of polymerization competing against the high, negative heat of polymerization. See FLOOR TEMPERATURE.

ceiling voltage (*ElecEng*) A term used to denote the maximum voltage which a machine is capable of giving.

celdecor process (*Paper*) A pulping method for cereals and similar agricultural residues consisting of a mild digestion in caustic soda followed by treatment with gaseous chlorine, caustic soda extraction and bleaching.

-cele (*Med*) Suffix from Gk *kele*, tumour, hernia.

celecoxib (*Pharmacol*) A *non-steroidal anti-inflammatory drug* that acts by selective inhibition of cycloxygenase 2 (COX-2).

celestial equator (*Astron*) The great circle in which the plane of the Earth's equator cuts the celestial sphere; the primary circle to which the co-ordinates right ascension and declination are referred.

celestial mechanics (*Astron*) The study of the motions of celestial objects in gravitational fields. This subject, founded by Newton, classically deals with satellite and planetary motion within the solar system, using Newtonian gravitational theory.

celestial pole (*Astron*) One of the two points at which the Earth's rotation axis, produced indefinitely, cuts the celestial sphere.

celestial sphere (*Astron*) An imaginary sphere, of indeterminate radius, of which the observer is the centre. Positions of stars are specified by two co-ordinates, referred to some chosen great circle of the sphere.

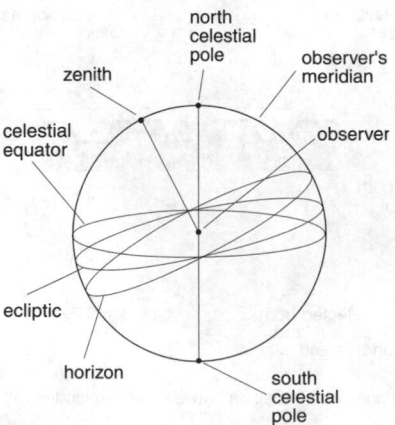

celestial sphere

celestine (*Min*) Strontium sulphate, crystallizing in the orthorhombic system; occurs in association with rock salt and gypsum; also in the sulphur deposits of Sicily, and in nodules of limestone.

Celite (*Chem*) TN for a form of DIATOMITE used as an insulating material and filter aid.

cell (*ElecEng*) (1) Chemical generator of emf. (2) Small item forming part of an assembly, eg Kerr cell, dielectric test cell, etc.

cell (*ICT*) (1) The smallest element in a spreadsheet. Each cell has a unique row and column reference and may be used to store a numerical value, a formula or text. See SPREADSHEET PROGRAM.(2) The coverage area of a BASE STATION in a mobile-telephone network. Cells abut in what is nominally a honeycomb pattern, although in practice they may deviate widely from this. Groups of operating frequencies are arranged so that no two neighbouring cells share the same frequency. Cells may vary in radius from tens of kilometres down to as little as 500 m for the MICROCELLS of a PERSONAL COMMUNICATIONS NETWORK. (3) The basic data packet handled by an ASYNCHRONOUS TRANSFER MODE network. See ATM CELL.

cell (*BioSci*) (1) An autonomous self-replicating unit (in principle) that may constitute an organism (in the case of unicellular organisms, whether prokaryotic or eukaryotic) or be a subunit of multicellular organisms in which individual cells may be more or less specialized (differentiated) for particular functions. All living organisms are composed of one or more cells. (2) One of the spaces into which the wing of an insect is divided by the veins. Fig. ▷

cell (*MathSci*) Small-volume unit in mathematical co-ordinate system.

cell (*NucEng*) (1) Unit of homogeneous reactivity in reactor core. (2) Small storage or workplace for 'hot' radioactive preparations.

cell adhesion molecule (*BioSci*) In principle any molecule on the surface of an animal tissue cell that is involved in intercellular adhesion, but in practice restricted to a subset of molecules that include neural cell adhesion molecule (*NCAM*), liver cell adhesion molecule (*LCAM*), neuroglial cell adhesion molecule (*NgCAM*) and intercellular adhesion molecule (*ICAM*). Abbrev *CAM*. See INTEGRIN, SELECTIN.

cell assembly (*ICT*) In an ASYNCHRONOUS TRANSFER MODE (ATM) network, the process by which a user source, eg a telephone channel, is converted in the ATM ADAPTATION LAYER into the ATM CELL format.

cell cavity (*BioSci*) See LUMEN.

cell constant (*Chem*) The conversion factor relating the electrical conductance of a conductivity cell to the conductivity of the liquid in it.

cell culture (*BioSci*) General term for the maintenance of cell strains or lines in the laboratory by serial subculture.

cell cycle (*BioSci*) The period between one cell division and the next. See panel on CELL CYCLE.

cell death (*BioSci*) Cells die (non-accidentally) either when they have completed an approximately fixed number of division cycles (*Hayflick limit*) or at some earlier stage when programmed to do so. Transformed cells and *cell lines* have generally escaped the limit. See APOPTOSIS.

cell delay variation (*ICT*) A measure of performance in an ASYNCHRONOUS TRANSFER MODE network, being the variable component of CELL TRANSFER DELAY. It arises mainly through filling and emptying of the cell queues that can occur at several points of the network, for instance in SWITCHES and the ATM ADAPTATION LAYER.

cell division (*BioSci*) The formation of two daughter cells from one parent cell. See AMITOSIS, MEIOSIS, MITOSIS.

cell-division cycle gene (*BioSci*) Genes whose mutants block various stages in the cell cycle. Abbrev *cdc gene*. See panel on CELL CYCLE.

cell enlargement (*BioSci*) The growth in volume or length of (generally) a plant cell, produced from a meristem as it

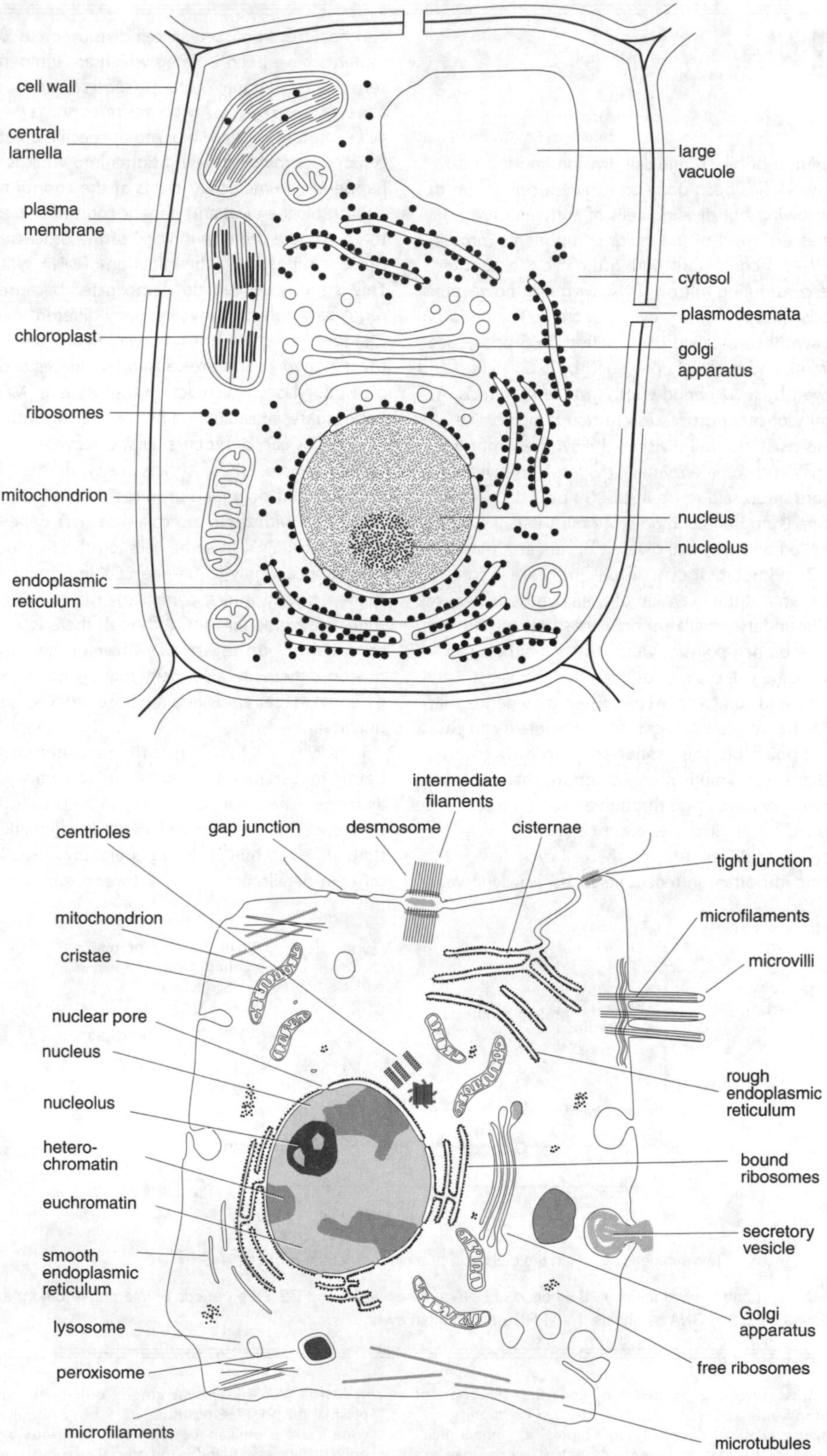

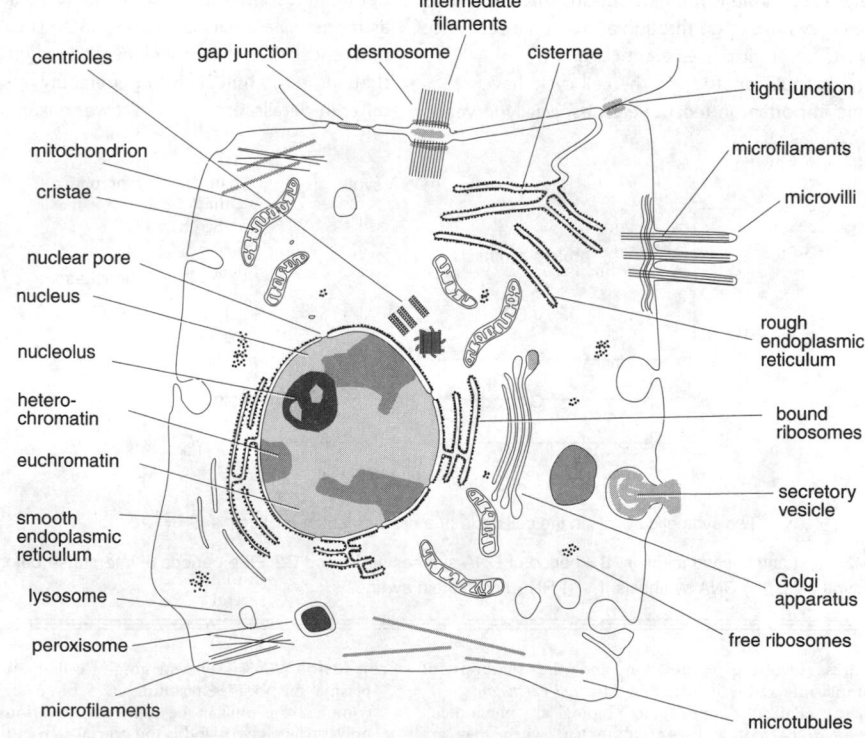

cell wall

central lamella

plasma membrane

chloroplast

ribosomes

mitochondrion

endoplasmic reticulum

large vacuole

cytosol

plasmodesmata

golgi apparatus

nucleus

nucleolus

intermediate filaments

centrioles

gap junction

desmosome

cisternae

tight junction

mitochondrion

microfilaments

cristae

microvilli

nuclear pore

nucleus

nucleolus

rough endoplasmic reticulum

hetero-chromatin

bound ribosomes

euchromatin

smooth endoplasmic reticulum

secretory vesicle

lysosome

peroxisome

Golgi apparatus

free ribosomes

microfilaments

microtubules

cell Generalized plant and animal cell showing common components.

Cell cycle

The period between one cell division and the next. Most work has been done on growing cells or on the non-growing but dividing cells of early embryos. Synthesis of most of the major components (protein and RNA) is continuous during the cycle, except in some cells during mitosis. DNA synthesis, however, is periodic except in fast-growing bacteria. The period of DNA synthesis is called the S period (or S phase, see diagram), and is usually preceded by a G1 period and followed by a G2 period which lasts until mitosis. A minority of other processes are also periodic, like mitosis itself and cell division, the synthesis of nuclear histones and some enzyme activities. Non-dividing cells are often said to be in the G0 period.

There is no general model of how the cell cycle is controlled and why cell division occurs at a particular time. Two important CONTROL POINTS are before mitosis and at the initiation of DNA synthesis near the G1/S boundary. This latter point has been called *start*, and is a branch point at which the cell either goes through the cell cycle or differentiates. In yeast, bacteria and some other cells, there is evidence that the cell has to grow to a critical size before it can pass a control point, but this cannot apply to early embryos. In other cases, time is more important than size since the cell may have to go through a fixed time sequence of events before it passes a critical point.

Genetic tools for studying the cell cycle have become important in recent years, especially in yeast and bacteria. In particular, cell division cycle (*cdc*) mutants have been isolated which are temperature-sensitive conditional mutants (see TEMPERATURE-SENSITIVE MUTANT). At the RESTRICTIVE TEMPERA-TURE, mutant cells are blocked at specific points in the cell cycle, providing new insights into what is happening in molecular terms at the control points.

Both biochemical and genetic tools have been used to investigate the enzymology of the processes which precede mitosis and the initiation of DNA synthesis. This is a very active field, complicated because the regulatory networks involve many different molecules. Changes in protein phosphorylation are key events in these networks and are caused by kinases and phosphorylases which act on the proteins. Many cell cycle kinases are activated by complexing with a group of proteins called CYCLINS. First discovered in sea urchin embryos, these cyclins are synthesized continuously but degraded at mitosis. The kinases associated with them are called CYCLIN-DEPENDENT KINASES or *CDKs*. In some cells, eg the fission yeast, there appears only to be one CDK important for mitosis, but in others, particularly mammalian cells, there are whole families. Although there is some redundancy with several CDKs having the same function, there are also many mammalian CDKs essential for cell survival but whose detailed function is uncertain.

Initially it was hoped that the regulatory networks before mitosis might be the same in all eukaryotic cells as mitosis is a universal process in such cells. The differences that have now emerged may mark the fact that, although mitosis is in general universal, it does differ in details, especially in lower eukaryotic cells.

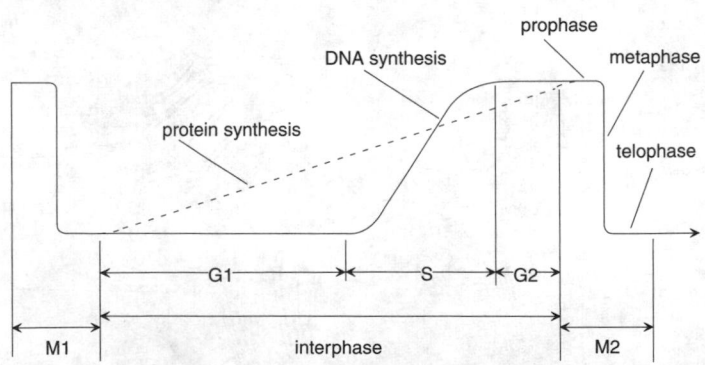

Two synthetic events in the cell cycle of a growing eukaryotic cell

M1, M2 = first and second mitoses, S = period of DNA synthesis, G1 and G2 = the periods of interphase before and after DNA synthesis. (—) DNA synthesis. (- - -) RNA and protein synthesis.

matures, involving vacuolation and the synthesis of protoplasmic and wall materials. Also *cell extension*.

cell-free (*BioSci*) Applied to biological phenomena like TRANSLATION and TRANSCRIPTION when they are made to occur in the laboratory in the absence of intact cells.

cell fusion (*BioSci*) The merging of cells by fusion of their plasma membranes resulting in a bi- or multi-nucleate complex. Fusion can be induced by various agents like polyethylene glycol and is the crucial step in the formation of HYBRIDOMAS during MONOCLONAL ANTIBODY production.

cell genetics (*BioSci*) The study of genetics, particularly the location of genes on chromosomes, by means of cells grown in culture.

Cellidor (*Chem*) European TN for CELLULOSE ACETATE-BUTYRATE.

cell line (*BioSci*) An established cell culture that will proliferate indefinitely given appropriate fresh medium and space. Lines differ from cell strains in that they have escaped the HAYFLICK LIMIT and become immortalized.

cell lineage (*BioSci*) The developmental history of individual cells of an embryo during cell division following fertilization.

cell loss priority (*ICT*) A designated bit in the ASYNCHRONOUS TRANSFER MODE cell header that can be used to indicate that if traffic conditions are causing cell loss then this cell should be among those discarded first. The bit has several applications, eg tagging cells from a user who is exceeding the data rate of his or her connection contract, or in variable bit-rate video, where some elements of the signal can be marked as having lower priority.

cell loss probability (*ICT*) A measure of performance in an ASYNCHRONOUS TRANSFER MODE network. CELLS may be lost through bit errors in the physical layer of the network or through statistically determined buffer overflows in SWITCHES or other elements of the asynchronous transfer mode layer.

cell loss ratio (*ICT*) In an ASYNCHRONOUS TRANSFER MODE network, the ratio of the rate at which CELLS fail to reach their destination, as a proportion of the total rate of cell transfer.

cell-mediated immunity (*BioSci*) Specific immunity that depends on the presence of activated T-lymphocytes. These act as cytotoxic cells and/or release lymphokines that activate monocytes and macrophages. Cell-mediated immunity is responsible for protecting against intracellular microbes, but also for the rejection of allografts and for DELAYED-TYPE HYPERSENSITIVITY reactions.

cell membrane (*BioSci*) A rather imprecise term usually intended to mean PLASMA MEMBRANE.

cellobiose (*Chem*) $C_{12}H_{22}O_{11}$. A disaccharide, obtained by complete hydrolysis of CELLULOSE. G–β1→4–G (G is glucose). It is the repeating unit for cellulose. Also *cellose*.

cello foils (*Print*) See VINYL FOILS.

Cellophane (*Chem*) TN for thin, transparent packaging film (eg on cigarette packets) made from regenerated cellulose by a modified viscose process.

cellose (*Chem*) See CELLOBIOSE.

Cellosolve (*Plastics*) *Hydroxy-ether, 2-ethoxy-ethan-1-ol.* $C_2H_5OCH_2CH_2OH$. A colourless liquid used as a solvent in the plastics industry. It is miscible with water, ethanol and ethoxyethane, and boils at 135·3°C.

cell payload (*ICT*) The user information carried in the last 48 OCTETS of an ASYNCHRONOUS TRANSFER MODE cell.

cell plate (*BioSci*) Characteristic of cell division in larger plants and many algae; a thin partition, bounded by a membrane, growing centrifugally by the coalescence of vesicles across the equatorial plane of the telophase spindle. This effects the division of the cytoplasm and becomes the basis of the middle lamella of the new wall. See CLEAVAGE, PHRAGMOPLAST.

cell splitting (*ICT*) A method of increasing the traffic capacity of a mobile-telephone network by dividing existing CELLS into smaller subcells, thus allowing greater frequency reuse.

cell strain (*BioSci*) Cells adapted to culture, but with finite division potential. See CELL LINE.

cell tester (*ElecEng*) A portable voltmeter for checking the voltage of accumulator cells.

cell transfer delay (*ICT*) A constituent of the overall end-to-end delay in an ASYNCHRONOUS TRANSFER MODE network that excludes CELL ASSEMBLY (and disassembly) time in the ATM ADAPTATION LAYER and delays in the CUSTOMER PREMISES EQUIPMENT but includes delay caused by the physical transmission path and SWITCHES.

cell transformation (*BioSci*) See TRANSFORMATION.

cellular automaton (*ICT*) AUTOMATON consisting of a number of cells, where the state of a given cell depends on the state of adjacent cells.

cellular concrete (*Build*) Concrete of low unit weight in which bubbles of air are induced by chemical or mechanical means during manufacture.

cellular double bottom (*Ships*) A common construction of the bottom of a ship, where an inner bottom extends throughout the length of the ship between the peak bulkheads and over the width of the ship. The space between the outer and inner bottom plating is divided into cells by transverse floors and fore and aft keelsons, some of which are oil- and watertight so that the space may be used for fresh water, water ballast or oil fuel.

cellular fabric (*Textiles*) A woven or knitted fabric featuring an open, or cell-like, structure, used mainly for making shirting, blouses and underwear.

cellular glass (*Glass*) Glass foam often made with H_2S as a blowing agent, hence its dark colour and strong sulphide smell; a very moisture-resistant, low-temperature insulation material.

cellular horn (*Acous*) A horn for a high-frequency loudspeaker (tweeter), in which the path from the throat to the outer air is by a number of expanding channels of equal length, so that marked directivity, arising from the short wavelengths in relation to the width of the total opening, is not apparent.

cellular radio (*ICT*) Radio system linking mobile users to the public telephone system or to each other via the cellular operator's own land network. Each network consists of hundreds of ultrahigh-frequency or microwave transceivers arranged in a cellular pattern, each using a group of frequencies different from those of surrounding cells to prevent interference. Frequencies are reused in more distant cells. See EXTENDED TOTAL ACCESS COMMUNICATIONS SYSTEM, GLOBAL SYSTEM FOR MOBILE COMMUNICATION.

cellular silica (*Chem*) Inorganic silicates containing numerous air cells and low permeability to water vapour.

cellular slime moulds (*BioSci*) See ACRASIOMYCETES.

cellular solid (*Eng*) Generic term for materials comprising an assembly of cells with solid edges or faces, packed together to fill space. Occur naturally as eg wood, cancellous bone, CORK, coral and sponge. Also manufactured as eg FOAM and honeycomb structures.

cellular structure (*Eng*) See EXPANDED PLASTICS, FOAM. Cf NETWORK STRUCTURE.

cellular-type switchboard (*ElecEng*) A switchboard in which each switch with its associated apparatus is contained in a separate cell of fireproof material. Also *cubicle-type switchboard*.

cellulase (*BioSci*) An enzyme or mixture of enzymes capable of catalysing the hydrolysis of cellulose to cellobiose or glucose.

cellulitis (*Med*) A spreading infection of the subcutaneous tissues with pyogenic bacteria, often streptococcal.

celluloid (*Plastics*) Manufactured thermoplastic consisting of nitrocellulose polymer plasticized with camphor. Formerly widely used for moulded products but now restricted to table-tennis balls and spectacle frames owing to competition from cheaper and less flammable synthetic plastics.

cellulose (*BioSci, Chem*) Linear homopolymer of glucose, $(C_6H_{10}O_5)_n$. The most abundant organic polymer on Earth. The glucose repeat units are linked in the β configuration (see panel on BIOLOGICAL ENGINEERING POLYMERS), by contrast with the α configuration in starch. The β configuration allows the chains to crystallize in a linear conformation, so natural cellulose as found in wood and natural fibres such as flax and cotton is highly crystalline (90–95%). The crystallites are organized in fibrils which are aligned along the fibre axis, producing a high tensile modulus of c.100 GPa when stressed. The molecular mass probably exceeds 1 million but natural cellulose is difficult to extract without chain cleavage, whether by natural biodegradation or chemical treatment. Cotton fibre is almost pure cellulose but wood is a natural biocomposite, with 50–60% cellulose content embedded in lignin. Cellulose is the most important constituent of paper and cardboard, as well as providing artificial fibres (principally rayon) and cellulosic thermoplastics (nitrocellulose, celluloid, cellulose acetate). It contains the β1-4 glucoside linkage. See CELLOBIOSE and panels on BIOLOGICAL ENGINEERING POLYMERS, CELL WALL and PAPER AND PAPER-MAKING.

cellulose acetate (*Textiles*) Artificial derivative of CELLULOSE made by treating it with acetic anhydride/methylene chloride/sulphuric acid mixture. Up to three hydroxyl groups in the repeat unit are esterified (eg triacetate), and the product is soluble in acetone or chloroform owing to the breakdown of the crystal structure, lowering of molecular mass and acetylation. Injection moulded products (eg combs) now less popular than extrudates (eg photographic film).

cellulose acetate–butyrate (*Chem*) Random copolymer of cellulose acetate and butyrate. TNs Tenite, Cellidor. Abbrev *CAB*.

cellulose acetates (*Chem*) *Acetylcelluloses*. Ethanoic acid esters of cellulose, obtained by the action of glacial ethanoic acid, ethanoic anhydride and sulphuric acid, upon cellulose. They are considerably less inflammable than cellulose nitrates, and are a raw material for films, windscreens, textile fibres, lacquers, etc.

cellulose esters (*Chem*) Generic term for the manufactured derivatives of CELLULOSE where hydroxyl groups on the repeat unit are esterified, eg cellulose acetate. Other esters include acetate–butyrate (CAB), acetatepropionate and cellulose propionate.

cellulose ethers (*Chem*) Generic term for manufactured derivations of CELLULOSE where hydroxyl groups (—OH) on the molecule are replaced by ether groups (—OR) such as ethyl cellulose (where R is —C_2H_5) used for protective films. Also methyl cellulose (where R is —OCH_3), hydroxyethyl cellulose (where R is —O_2CH_4OH) and hydroxyl propyl cellulose (where R is —OC_3H_6OH), which are water-soluble polymers used as thickening agents, paper sizes, etc.

cellulose lacquers (*Chem*) Lacquers prepared by dissolving nitrocellulose or acetylcellulose in a mixture of suitable solvents, with the admixture of resins, plasticizers and pigments.

cellulose nitrate (*Chem*) Nitro esters of cellulose formed by a mixture of nitric and sulphuric acid acting on cotton fibre. Nitrocellulose with up to four nitro groups per repeat unit is not explosive and is plasticized with camphor to produce thermoplastic products (celluloid). Can also be dissolved in ester solvents for films and coatings eg of playing cards, staples, etc. Much reduced usage due to competition from synthetic plastics, cellulose acetate, etc.

cellulose paints (*Build*) Reversible coatings based on chemically treated cellulose compounds having wide application in the automobile industry but limited use in house painting and decorating. Their storage and use are governed by statutory regulations.

cellulose xanthate (*Chem*) $[C_6H_8O_3(ONa)OCS_2Na]_n$. An acid salt of cellulose–dithiocarbonic acid, obtained by treating cellulose with caustic soda/carbon disulphide solution. The resulting product is called VISCOSE, the intermediate for viscose rayon textile fibres.

cellulosics (*Chem*) General name for useful manufactured materials derived from natural CELLULOSE.

cell unit (*Crystal*) See UNIT CELL.

cell unit (*ElecEng*) A unit which forms the basis for an extended switchboard.

cell wall (*BioSci*) A structure external to the plasmalemma of a plant cell, secreted by the cell and enclosing it. See panel on CELL WALL.

celsian (*Min*) *Barium aluminium silicate*. A barium feldspar found in association with some manganese ores.

Celsius scale (*Phys*) The SI name for *centigrade scale*. The original Celsius scale of 1742 was marked 0 at the boiling point of water and 100 at the freezing point, the scale being inverted by Strömer in 1750. Temperatures on the International Practical Scale of Temperature are expressed in degrees Celsius. See KELVIN THERMODYNAMIC SCALE OF TEMPERATURE.

cement (*BioSci*) In mammalian teeth, a layer resembling bone, covering the dentine behind the enamel.

cement (*Build, Chem*) (1) Generic term for any binding agent, adhesive or glue. (2) Binding agent for concrete, often Portland cement, which hardens as it reacts slowly with water. See panel on CEMENT AND CONCRETE.

cement (*Geol*) The material which binds any loose sediment into a coherent rock. The commonest cements are ferruginous, calcareous and siliceous.

cementation (*CivEng*) See GROUTING.

cementation (*Eng*) Any process in which the surface of a metal is impregnated at high temperature by another substance. Also *carbonization*, *carburization*. See CASE-HARDENING.

cement copper (*Eng*) Impure copper, obtained when the metal is precipitated by means of iron from solutions resulting from leaching. Also *cementation copper*.

cemented carbides (*Eng*) See CARBIDE TOOLS.

cement fillet (*Build*) A substitute for lead flashings in the angles between, eg a chimney stack and roof, weatherproofing being provided by running in a band of cement mortar.

cement grout (*CivEng*) A fluid cement mixture for filling crevices.

cement gun (*CivEng*) Apparatus for spraying fine concrete or cement mortar by pneumatic pressure.

cementite (*Eng*) The iron carbide (Fe_3C) constituent of steel and cast-iron (particularly white cast-iron). Very hard and brittle.

cement joggle (*Build*) A key formed between adjacent stones in parapets etc, by running cement mortar into a square-section channel cut equally into each of the jointing faces, thereby preventing relative movement.

cement mortar (*Build, CivEng*) A hydraulic mortar composed of Portland cement (or other siliceous cement) and sand.

cement paints (*Build*) Powder materials based on Portland cement which are mixed with water immediately before application to eg exterior wall surfaces.

cement rock (*Geol*) Argillaceous limestone containing >18% of clay.

cement–rubber–latex (*Build*) A flooring composed of cement, aggregate and an elastomer to give flexibility; fleximer.

cementstone (*Geol*) Argillaceous limestone suitable for the manufacture of cement.

Cenomanian (*Geol*) The lowest stage in the Upper Cretaceous. See MESOZOIC.

Cenozoic (*Geol*) The youngest era of the Phanerozoic, covering approx 65 million years ago to the present day. It includes the Tertiary and Quaternary periods. Also *Cainozoic*. See appendix on Geological time.

censer mechanism (*BioSci*) A means of seed liberation in which the seeds are shaken out of the fruits as the stem of the plant sways in the wind (eg in the poppy).

censor (*Psych*) In psychoanalytic theory, a powerful unconscious inhibitory mechanism in the mind, which prevents anything painful to the conscious aims of the individual from emerging into consciousness. It is responsible for the distortion, displacement and condensation present in dreams. Also *censorship*.

cent (*NucEng*) Unit of reactivity equal to one-hundredth of a DOLLAR.

Centaurus (Centaur) (*Astron*) A large and rich southern constellation, which contains the closest star to the Sun (PROXIMA CENTAURI) and CENTAURUS A.

Centaurus A (*Astron*) The third most powerful radio source in the sky, also a source of infrared and gamma radiation.

Cell wall

A structure characteristic of plant cells, external to their plasmalemma and secreted by the cell and enclosing it. It is typically tough, sometimes rigid, and protective or skeletal in function but with relatively little effect on solute influx or efflux, which are mainly controlled by the plasmalemma. The properties of cell walls largely determine the shape of cells and, hence, the morphology of plants.

In vascular plants the cell wall of an undifferentiated, parenchyma, cell typically consists of microfibrils of cellulose (say 40% of the dry mass) embedded in a matrix of other substances, rather like a glass-fibre-reinforced plastic. The microfibrils are flattened threads, 5–8 nm wide and relatively long. The matrix includes polysaccharides such as hemicelluloses and pectins together with some protein, especially in growing primary walls. The wall contains much water and is porous, thus allowing free passage to water and to solutes less than about 4 nm in diameter. If, as is nearly always the case, the concentration of solutes in the cell is greater than that in the medium, the cell being *hyperosmotic* to its surroundings, then the wall restrains the tendency of the cell to expand by the osmotic uptake of water. Equilibrium occurs when the excess hydrostatic pressure, the *turgor pressure*, generated by the stretching of the wall, is equal to the difference of osmotic pressure inside and outside the cell. The turgor pressure of a turgid cell might be 5–20 bars.

Cell walls are typically flexible but turgor gives them rigidity and enables non-woody plants to stand erect as in the hydrostatic skeleton of some soft-bodied animals. The cell walls of vascular plants have thin areas of pits and are traversed by plasmadesmata. The walls of some sorts of cells are impregnated with lignin which more or less obliterates the pores and makes the wall incompressible and rigid. The walls of xylem tracheids and vessels, which conduct liquid under tension and must, therefore, resist implosion, are lignified as are those of the secondary xylem or wood cells generally and of most sclerenchyma cells. Woody stems are therefore stiff regardless of water content. Other cell walls contain suberin, cutin, sporopollenin or silica.

The arrangement of the microfibrils appears to relate to the function of the cell and to its manner of growth (see MULTINET GROWTH). Microfibrils are approximately longitudinal in many fibres. Growth of the wall may be controlled by the matrix.

In algae and fungi, cell walls contain other components but there is often a microfibrillar element, eg of cellulose or xylan in some algae or chitin in some fungi, embedded in a matrix of other polysaccharides. Some walls are impregnated with silica, as in diatoms, or calcium carbonate, as in some red and green algae, and may be intrinsically rigid. In some algae, especially motile unicells, the walls are turgor-resisting devices and cell volume may be controlled by the cells being iso-osmotic or having contractile vacuoles as in marine and fresh-water forms, respectively.

The cell walls of prokaryotes are generally turgor-resisting devices but are constructed differently.

Associated with the active galaxy NGC 5128, distance 5 Mpc.

center (*Genrl*) US for CENTRE.

centering (*CivEng*) US for CENTRING.

centi- (*Genrl*) Prefix from Lt *centum*, meaning one-hundredth. Symbol c.

centigrade heat unit (*Genrl*) Unit of heat energy equivalent to 0·01 times the amount of heat required to raise 1 pound of air-free water from 0°C to 100°C at a constant pressure of 1 standard atmosphere. Equal to 1900·44 J. Also *pound-calorie*.

centigrade scale (*Phys*) The most widely used method of graduating a thermometer. The fundamental interval of temperature between the freezing and boiling points of pure water at normal pressure is divided into 100 equal parts, each of which is a *centigrade degree*, and the freezing point is made the zero of the scale. To convert a temperature on this scale to the Fahrenheit scale, multiply by 1·8 and add 32; for the Kelvin-scale equivalent add 273·15. See CELSIUS SCALE.

centile (*MathSci*) Same as PERCENTILE.

centimetre (*Genrl*) A linear measure, the hundredth part of the metre. Abbrev *cm*. US *centimeter*.

centimetre–gram–second unit (*Genrl*) See CGS UNIT.

centimorgan (*BioSci*) Measure of the distance between the loci of two genes on the same chromosome, obtained from the *crossover frequency* (see CROSSOVER). One centimorgan is 1% crossing over; one morgan is equal to

100 centimorgans when summed over short distances between intervening loci.

centipedes (*BioSci*) See CHILOPODA.

centipoise (*Phys*) One-hundredth of a POISE, the CGS unit of viscosity. Symbol cP; 1 cP = 10^{-3} N m^{-2} s.

centistoke (*MinExt*) Unit for measuring the kinematic viscosity of oils. Replaces Redwood and Saybolt seconds, and equivalent to the SI unit, mm^2 s^{-1}.

central angle (*MathSci*) An angle bounded by two radii of a circle.

central dogma (*BioSci*) The postulate that genetic information resides in the nucleic acid and passes to the protein sequence, but cannot flow from protein to nucleic acid.

central force (*Phys*) A force which a particle at an arbitrary point experiences along the line between a certain origin and the point. If the force $f(r)$ is conservative, it can be derived from a potential energy function $V(r)$, a function only of the distance r from the origin: $f(r) = -dV(r)/dr$, where $V(r)$ is the CENTRAL POTENTIAL. See ANGULAR MOMENTUM.

central lubricating system (*Eng*) (1) An oil or grease lubricating system in which the lubricant is distributed by a single pump via pipes to a number of outlets, one actuation of the pump providing a discharge at every outlet. (2) A coolant distribution system in which a battery of machine tools is supplied with cleaned coolant by a single pump/filter unit.

Cement and concrete

Concrete, whether reinforced or not, has become the most widely used civil engineering material. It is made of cement and aggregate, which may have a particular size distribution, mixed with water. Adding water initiates the setting reaction between it and the cement, binding the whole mass together to form concrete. Concrete has the great advantage as an engineering material that during the first few hours of setting the material remains pourable and workable. As a fluid, it can readily be moulded into complex shapes *in situ*, yet subsequently it acquires properties similar to those of natural stone. The most frequently used cement, especially in the UK, is PORTLAND CEMENT, although different materials are used in other parts of the world.

Portland cement is made from a mixture of about 75 wt % limestone, $CaCO_3$, and 25 wt % clay, principally alumino-silicate, but with a significant iron oxide and alkali oxide content. These are ground together and fired with coal in air at up to 1500°C to produce a clinker, which, in turn, is mixed with 3–5 wt % of gypsum and ground again to give cement powder, with a mean particle size of under 10 mm. The particle size is important: the smaller it is, the faster is the setting. The powder is a complex mixture of multicomponent, mineral solid solutions, the principal constituents of which are (approx. wt %): *alite* (tricalcium silicate, $CaOCa_2SiO_4$) 55, *belite* (dicalcium silicate, Ca_2SiO_4) 20, *aluminate* (tricalcium aluminate, $2CaOCa(AlO_2)_2$) 12, *ferrite* (tetracalcium aluminoferrite, $(4CaOAl_2O_3Fe_2O_3)$ 8, *gypsum* (hydrated calcium sulphate, $CaSO_42H_2O$) 3·5 and *oxides* (K_2O, Na_2O, CaO) 1·5.

The sequence and variety of hydration reactions during setting and hardening are also complex, and are still being elucidated. The polymerization of silica, to form an interlocking network of foils and fibres of silica gel around each cement particle, plays an important role, as do the formation and interlocking of needle-like crystals of ettringite, a heavily hydrated calcium aluminosulphate with plate-like crystals of portlandite, $Ca(OH)_2$. The initial setting reaction is very rapid, with a high exotherm, 200 W kg^{-1}, within 30 s. This falls to about 1 W kg^{-1} after an hour, and the cement continues hardening at an increasingly slow rate with changes still detectable after 30 years! Some 40 wt % of water is required for complete hydration.

The tensile strength of hardened cement (about 3–5 MN m^{-2}) is only about 10% of its compressive strength, largely due to porosity, the pores of up to 1 mm in size acting as stress concentrators. Its YOUNG'S MODULUS is about 30 GN m^{-2}, but it shows partially viscoelastic behaviour. Additives, known as *admixtures*, are used to control such properties as viscosity, setting time, freezing temperature, pore size and permeability to water. Macro-defect-free (MDF) cement is made by adding about 5% of polyvinyl alcohol, mixing like a dough and compressing. The material has a maximum pore size of 15 μm and a tensile strength of about 60 MN m^{-2}.

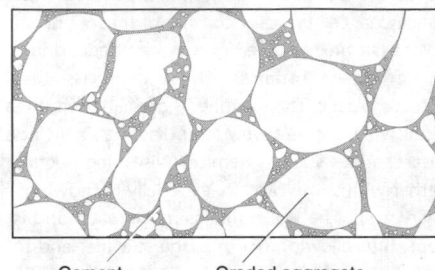

Cement   Graded aggregate

Concrete made with graded aggregate

Concrete is made from about 80 vol % of (graded) aggregate, containing three parts sand to two parts gravel, and 20 vol % cement (see diagram). In reinforced concrete, the concrete is poured around an assemblage of heavily cold-worked, mild-steel reinforcing bars, held together with links made from a soft mild steel. The steel bars are positioned to resist the anticipated tensile stresses in the structure, some of which can be the result of shear and bending. Beams are designed so that the steel carries all the tension, the concrete being allowed to crack in tension, but required to resist loads in compression. *Prestressed concrete beams* are arranged so that all the concrete is under compression. This is applied by high-tensile steel wires which are either held in tension as the concrete is cast around them and then unloaded after the concrete has set (*pre-tensioned*), or the wires are fed into ducts cast in the concrete and tensioned after the concrete has set (*post-tensioned*). For a given bending stiffness, prestressed concrete only requires about half the mass of concrete and a third the mass of steel compared with a reinforced concrete beam.

---

central nervous system (*BioSci*) The main ganglia of the nervous system with their associated nerve cords, consisting usually of a brain or cerebral ganglia and a dorsal or ventral nerve cord which may be double, together with associated ganglia. Abbrev *CNS*.

central office (*ICT*) The local exchange to which CUSTOMER PREMISES EQUIPMENT is connected by the LOCAL LOOP.

central potential (*Phys*) A spherically symmetric potential in which the potential depends only on the distance from some centre; the orbital angular momentum is constant for a single particle moving in such a potential. In quantum mechanics, the SCHRÖDINGER EQUATION can be solved for such a system; the hydrogen atom is an example in which an electron moves in the central coulomb potential provided by the nuclear charge.

central processing unit (*ICT*) See CENTRAL PROCESSOR.

central processor (*ICT*) The main part of a computer consisting of an ARITHMETIC LOGIC UNIT and a CONTROL UNIT. Also *central processing unit*, abbrev *CPU*.

central projection (*MathSci*) The projection of a geometrical configuration in one plane onto another plane by drawing straight lines from a point *O* (not in either plane) through the various points in the configuration.

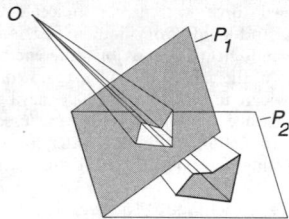

central projection

central tendency (*Genrl*) A statistical measurement attempting to depict the average score in a distribution (see MEAN, MEDIAN, MODE).

centration (*Psych*) The tendency of young children to focus only on their perspective of a specific object and failure to understand that others may see things differently.

centre (*CivEng*) A timber frame built as a temporary support during the construction of an arch or a dome. US *center*.

centre (*Eng*) A conical support for workpieces, used in lathes and some other machine tools. Live centres, for headstocks, are usually left soft and machined to 60° included angle on use in the lathe. Dead centres, also normally at 60°, are heat-treated or carbide-tipped. Anti-friction centres are also used. Running centres rotate with the work. US *center*.

centre (*MathSci*) Of a conic: the point through which all DIAMETERS pass. US *center*.

centre (*Surv*) Accurate vertical alignment of centre of rotation of survey instrument over, or under, a fixed point or station. US *center*.

centre adjustment (*Surv*) On tripod-mounted survey instrument, a device which allows the instrument to be moved horizontally until exactly above (or underground, perhaps beneath) the survey signal point.

centre-bit (*Build*) A wood-boring tool having a projecting central point and two side wings, one of which scribes the boundary of the hole to be cut, while the other removes the material.

centre-contact cap (*ElecEng*) A bayonet cap, fitted to an electric lamp, in which the outer wall forms one of the contacts, the other being the projection.

centre-contact holder (*ElecEng*) A lamp holder with a single centrally located contact, designed to be used with a centre-contact cap.

centre drill (*Eng*) A drill in which a short parallel portion leads to a coned section with an included angle of 60°. Used: (1) to give accurate start to a twist drill; and (2) to provide a conical recess to ends of workpieces to be machined between CENTRES in a lathe or cylindrical grinder. Also *Slocombe drill*.

centre frequency (*ICT*) (1) Geometric or arithmetic mid-point of the cut-off frequencies of a filter. (2) Carrier frequency, when modulated symmetrically.

centre keelson (*Ships*) See KEELSON.

centre lathe (*Eng*) Machine tool for cutting metal in which a workpiece rotates and a rigidly held tool is applied so as to make a circular product. Includes ability to make flat surfaces by facing, and screw cutting, boring and drilling operations. The workpiece may be held in a CHUCK or between CENTRES.

centreless grinding (*Eng*) A method of grinding cylindrical objects. The work is supported on a rest, between a pair of

abrasive wheels revolving at different speeds in opposite directions, instead of between centres as in normal practice.

centre margin ring (*Print*) The bevelled strip around the circumference of the plate cylinder of a letterpress rotary press against which the plates are locked.

centre nailing (*Build*) A method of nailing slates on a roof; the nail is driven in any one slate just above the line of the head of the slate in the course below. Cf HEAD NAILING.

centre note (*Print*) Notes placed between columns, as in bibles, to protect the small type and to economize on space.

centre of action (*EnvSci*) A position occupied, more or less permanently, by an anticyclone or a depression, which largely determines the weather conditions over a wide area. The climate of Europe is dependent on the Siberian anticyclone and the Icelandic depression.

centre of anallatism (*Surv*) In a distance-measuring telescope, the point from which the distance to an observed staff is proportional to the staff intercept as seen between the upper and lower stadia lines of the diaphragm.

centre of buoyancy (*Ships*) The centroid of the immersed portion of a floating body. In ship construction, a ship being symmetrical about a fore and aft plane, the centre of buoyancy lies in that plane. The vertical position is calculated from a succession of water-plane areas and the longitudinal position from a succession of transverse section areas.

centre of curvature (*MathSci*) See CURVATURE.

centre of gravity (*Phys*) The single point through which the resultant of the gravitational forces acting on all the individual particles of a body acts. In a uniform gravitational field, such as that near the Earth's surface, the centre of gravity coincides with the CENTRE OF MASS.

centre of inversion (*MathSci*) See INVERSION.

centre of lens (*Phys*) A point on the principal axis of a lens, through which any ray whose incident and emergent directions are parallel passes. Measurements of object and image distance, focal length, etc, are taken from this point.

centre of mass (*Aero*) Centroid of all distributed masses of a flight vehicle. Important design feature for determination of size of wings, layout, stability and payload. Also important operationally in loading aircraft to safe limits for flight.

centre of mass (*Phys*) The point in an assembly of mass particles where the entire mass of the assembly may be regarded as being concentrated and where the resultant of the external forces may be regarded as acting for considerations not concerned with the rotation of the assembly. The point is defined vectorially by

$$\bar{r} = \frac{\sum mr}{\sum m}$$

where *m* is the mass of a particle at a position *r*, and the summation extends over the whole body.

centre of oscillation (*Phys*) A point in a compound pendulum which, when the pendulum is at rest, is vertically below the point of suspension at a distance equal to the length of the equivalent simple pendulum (ie the simple pendulum having the same period). If the pendulum is suspended at the centre of oscillation, its period is the same as before. Also *centre of percussion*.

centre of percussion (*Phys*) See CENTRE OF OSCILLATION.

centre of pressure (*Aero*) The point at which the resultant of the aerodynamic forces (lift and drag) intersects the chord line of the aerofoil. Its distance behind the leading edge is usually given as a percentage of the chord length. Abbrev *cp*.

centre of pressure (*Phys*) The point in a surface immersed in a fluid at which the resultant pressure over the immersed area may be taken to act.

centre of symmetry (*Crystal*) A point within a crystal such that all straight lines that can be drawn through it pass through a pair of similar points, lying on opposite side of the centre of symmetry and at the same distance from it. Thus, faces and edges of the crystal occur in parallel pairs, on opposite sides of a centre of symmetry.

centre-point steering (*Autos*) The relative positioning of the steered wheel and the swivel axis so as to obtain coincidence between the point of intersection of the swivel axis with the road and the plane of the wheel.

centre punch (*Eng*) A punch with a conical point, used to mark or 'dot' the centres to be drilled etc.

centre section (*Aero*) The central portion of a wing, to which the main planes are attached; in large aircraft it is often built into the fuselage and may incorporate the engine nacelles and the main landing gear.

centre spread (*Print*) A design occupying the area of two pages in the centre opening of a booklet or journal.

centre square (*Eng*) A device for marking the centres of circular objects and bars. The bar is placed in the centre of the square, which is bisected by a blade that serves as a guide for scribing a diametral line.

centrex (*ICT*) A service in which a shared FEATURE SWITCH at the CENTRAL OFFICE provides intelligent PRIVATE BRANCH EXCHANGE facilities for a number of subscribers, for instance direct dialling from the public network to individual extensions.

centre-zero instrument (*ElecEng*) An indicating instrument which has the zero at the centre of the scale and can therefore read both positive and negative values of the quantity indicated.

centric leaves (*BioSci*) Cylindrical leaves, with the palisade tissues arranged uniformly around the periphery of the cylinder, eg onion.

centrifugal (*BioSci*) (1) A term describing a developmental process that starts at the centre and works towards the outside. (2) See EFFERENT.

centrifugal acceleration (*Phys*) The acceleration produced by a centrifugal force. See CENTRIPETAL FORCE.

centrifugal brake (*Eng*) An automatic brake used on cranes etc in which excessive speed of the rope drum is checked by revolving brake shoes forced outwards into contact with a fixed drum by centrifugal force.

centrifugal casting (*Eng*) The casting of large pipes, cylinder liners, etc, in a rotating mould of sand-lined or water-cooled steel.

centrifugal clutch (*Eng*) A type of clutch in which the friction surfaces are engaged automatically at a definite speed of the driving member, and thereafter maintained in contact, by the centrifugal force exerted by the weighted levers.

centrifugal compressor (*Eng*) Compressor which passes entering air through a series of low-pressure, high-volume fans via increasingly restricted chambers so that the emerging air attains the required higher pressure; also, small booster fans (superchargers).

centrifugal elutriator (*PowderTech*) See ELUTRIATOR, CENTRIFUGAL.

centrifugal fan (*Eng*) A fan with an impeller of paddle-wheel form, in which the air enters axially at the centre and is discharged radially by centrifugal force. Also *paddle wheel fan*.

centrifugal-flow compressor (*Aero*) A compressor in which a vaned rotor inspires air near the axis and throws it towards a peripheral diffuser, the pressure rising mainly through centrifugal forces. The maximum pressure ratio for a single stage is about four. Centrifugal compressors are universal for aero-engine superchargers and were widely used in the earlier turbojets.

centrifugal force (*Phys*) See CENTRIPETAL FORCE.

centrifugal pulp cleaner (*Paper*) A cleaning device for pulp stock which produces a high-speed rotation or rotary vortex motion causing dirt particles to separate out.

centrifugal pump (*MinExt*) Continuously acting pump with rotating impeller, used to accelerate water or suspensions through a fixed casing to peripheral discharge at the desired delivery height.

centrifugal starter (*ElecEng*) A device used with small induction motors; it consists of a centrifugally operated switch on the rotor, which automatically cuts out starting resistance or performs some other operation as the motor runs up to speed.

centrifugal tension (*Eng*) The force per unit area of cross-section induced, in consequence of centrifugal force, in the material of a rotating rim, loop or driving belt.

centrifuge (*Eng, MinExt*) Rotating machine which uses centrifugal force to separate molecules from solution, particles and solids from liquids, and immiscible liquids from each other. Depends on differences in the relative densities of the substances to be separated. Used widely in science where forces up to 500 000*g* may be obtained in an ultracentrifuge. In industry they are used in sugar and cream production, separating water from fuel and swarf from cutting oil etc. Also used for separating metals from GANGUE.

centrifuge enrichment (*NucEng*) The enrichment of the ^{235}U isotope by the high-speed centrifugation of uranium hexafluoride gas. It uses the mass difference between ^{235}U and ^{238}U to effect the separation. Also *centrifuge separation*. See panel on URANIUM ISOTOPE ENRICHMENT.

centring (*CivEng*) The general term applied to CENTRES used in constructional work. US *centering*.

centring (*Eng*) (1) The marking of the centres of holes to be drilled in a piece of metal. (2) The adjusting of work in a lathe so that its axis coincides with the lathe axis. US *centering*.

centriole (*BioSci*) A structure virtually identical with the basal body of cilia. In animal cells it is surrounded by the main microtubule-organizing centre of the cell. Prior to mitosis the centriole duplicates and the two daughter centrioles move to lie at the poles of the spindle.

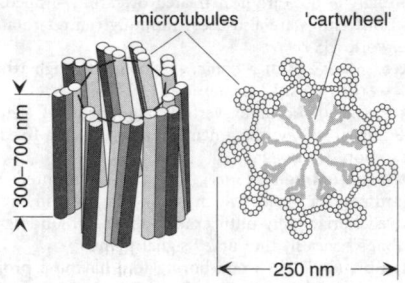

microtubules   'cartwheel'

300–700 nm

250 nm

centriole The right-hand section is of the proximal region.

centripetal (*BioSci*) (1) A term describing a developmental process that starts at the outside and works towards the centre. (2) See AFFERENT.

centripetal acceleration (*Phys*) The acceleration, directed towards the centre of curvature of the path, which is possessed by a body moving along a curved path with constant speed. Its value is v^2/R, where v is the speed and R the radius of curvature.

centripetal force (*Phys*) The force, directed towards the centre of curvature, that constrains a body to move along a curved path. It is equal and opposite to an inertial force directed away from the centre of curvature of the path, the *centrifugal force*. They are both equal in magnitude to the product of the mass and its CENTRIPETAL ACCELERATION.

centroblast (*BioSci*) Stage of B-lymphocyte differentiation after antigen exposure and activation. Centroblasts are found in the dark zone of GERMINAL CENTRES and are rapidly proliferating B-lymphocytes with little or no surface immunoglobulin. These cells undergo SOMATIC MUTATION and class switching of their immunoglobulin genes.

centrocyte (*BioSci*) Non-proliferating progeny of CENTRO-BLAST B-lymphocytes found in the light zone of GERMINAL

CENTRES. These cells re-express surface immunoglobulin and are thought to be positively or negatively selected by their affinity for antigen.

centroid (*MathSci*) (1) The point where the medians of a triangle intersect. (2) The centre of mass of a three-dimensional solid.

centrolecithal (*BioSci*) Having the yolk in the centre.

centromere (*BioSci*) In mitotic metaphase chromosomes, a narrow region in which CHROMATIDS are joined. It has flanking KINETOCHORES to which the microtubules of the spindle attach at mitosis. Also *primary constriction*.

Centronics interface (*ICT*) An international standard INTERFACE and associated protocol commonly used for connecting personal computers to printers.

centrosome (*BioSci*) A self-duplicating *microtubule-organizing centre* adjacent to the interphase nucleus, and from which the fibres of the spindle radiate at mitosis. In animal cells it contains a centriole.

centrum (*BioSci*) The basal portion of a vertebra that partly or entirely replaces the notochord, and from which arise the neural and haemal arches, transverse processes, etc.

cepaceous (*BioSci*) Smelling or tasting like onion or garlic.

cephal-, cephalo- (*Genrl*) Prefixes from Gk *kephale*, head.

cephalad (*BioSci*) Situated near, facing towards or passing to the head region.

cephalic (*BioSci*) Pertaining to or situated on or in the head region.

cephalic index (*Med*) The ratio of maximum breadth of skull divided by maximum length, multiplied by 100.

cephalization (*BioSci*) The specialization of the anterior end of a bilaterally symmetrical animal as the site of the mouth, the principal sense organs and the principal ganglia of the central nervous system; the formation of the head.

cephalo- (*Genrl*) See CEPHAL-.

cephalocele (*Med*) Protrusion of the membranes of the brain, with or without the substance of the brain, through a hole in the skull.

Cephalochordata (*BioSci*) A subphylum of the Chordata. Members have a persistent notochord, metameric muscles and gonads, and a pharynx with a large number of gill slits which are enclosed in an atrial cavity. They lack paired fins, jaws, a brain, and skeletal structures of bone and cartilage; they are marine and sand-living in form. Lancelets.

cephalometry (*Med*) The ultrasound measurement (formerly radiological) of the fetal head dimensions *in utero*.

Cephalopoda (*BioSci*) A class of bilaterally symmetrical Mollusca in which the anterior part of the foot is modified into arms or tentacles, while the posterior part forms a funnel leading out from the mantle cavity. The mantle is undivided, and the shell is a single internal plate, or an external spiral structure, or absent. They range from the Cambrian to the present day. Squids, octopods and pearly nautilus. See AMMONOIDS.

cephaloridine (*Pharmacol*) Cephalosporin C with the acetoxy group replaced by a pyridinium ion. Antibiotic used for treatment of severe sepsis.

cephalosporins (*Pharmacol*) A group of broad-spectrum beta-lactam antibiotics that inhibit bacterial cell-wall synthesis in a similar way to *penicillins*. Various modifications have been made to produce new variants such as *cefalexin* and *cefprozil*.

cephalothorax (*BioSci*) In some Crustacea, a region of the body formed by the fusion of the head and thorax.

Cepheid parallax (*Astron*) See PERIOD–LUMINOSITY LAW.

Cepheid variable (*Astron*) A class of bright periodic variable stars with a period of 1 to 70 days, showing exact correlation between the period and luminosity: the longer the period, the more luminous the star. Measurements of the apparent brightness and the period therefore allow the distance to be estimated. The prototype is DELTA CEPHEI. See panel on REDSHIFT–DISTANCE RELATION.

Cepheus (*Astron*) A northern constellation which includes the star DELTA CEPHEI, the prototype of regular variables used for calibrating the distance scale of the universe.

ceramic (*Eng*) A non-organic and non-metallic substance, often an oxide or a carbide.

ceramic capacitor (*Electronics*) A capacitor using a high-permittivity dielectric such as barium titanate to provide a high capacitance per unit volume.

ceramic filter (*BioSci*) A deep filter, usually finger-like, with fine pores in which small particles or bacteria become trapped. Now largely superseded by MEMBRANE FILTERS. Also *Pasteur filter*.

ceramic filter (*ICT*) A BAND-PASS FILTER constructed from a number of small ceramic PIEZOELECTRIC RESONATORS, commonly used in the INTERMEDIATE-FREQUENCY STRIP of a radio receiver.

ceramic fuel (*NucEng*) Nuclear fuel with high resistance for temperature, eg uranium dioxide, uranium carbide.

ceramic insulator (*ElecEng*) An insulator made of ceramic material, eg porcelain; generally used for outdoor installations.

ceramics (*Chem*) The art and science of non-organic non-metallic materials. The term covers the purification of raw materials, the study and production of the chemical compounds concerned, their formation into components, the study of structure, constitution and properties. See ALUMINA, CARBIDES.

ceramics processing (*Eng*) The methods of making ceramic products before final SINTERING. See panel on CERAMICS PROCESSING.

ceramic transducer (*Eng*) Transducer based on the electrical properties of ceramics such as piezoelectricity.

ceramide (*BioSci*) An N-acyl sphingosine, the lipid moiety of *glycosphingolipids*.

cerargyrite (*Min*) Silver chloride. Also a group name for silver halides. Also *horn silver*.

cerat-, cerato- (*Genrl*) See KERAT-, KERATO-.

cercal (*BioSci*) Pertaining to the tail.

cercaria (*BioSci*) The final larval stage of TREMATODA that develops directly into the adult; usually characterized by the possession of a round or oval body, bearing eyespots, a sucker and a propelling tail.

cercus (*BioSci*) In some ARTHROPODA, a sensory appendage with several joints at the end of the abdomen.

cere (*BioSci*) In birds, the soft skin covering the base of the upper beak. Adj *cereous*. Also *ceroma*.

cereal (*FoodSci*) The fruit of cultivated grasses (*Gramineae*). The main cereals are wheat, barley, oats, rye, maize, rice, sorghum and millets. They are high in soluble carbohydrates, starch and protein, and are the staple diet of well over half the world's population and a major source of animal feed.

cerebellum (*BioSci*) A dorsal thickening of the hind-brain in vertebrates. Adj *cerebellar*. Also *epencephalon*.

cerebr-, cerebro- (*Genrl*) Prefixes from Lt *cerebrum*, brain.

cerebral (*BioSci*) Pertaining to the brain or the CEREBRUM.

cerebral abscess (*Med*) An abscess within the brain.

cerebral blood flow (*Med*) The blood flow to the brain.

cerebral dominance (*Psych*) The phenomenon in which one hemisphere of the human brain, usually the left, is dominant in control over speech, language and movement (handedness), although why the vast majority of people are right-handed is unclear.

cerebral flexure (*BioSci*) The bend that develops between the axis of the fore-brain and that of the hind-brain in adult Craniata.

cerebral fossa (*BioSci*) In mammals, a concavity in the cranium corresponding to the cerebrum.

cerebral haemorrhage (*Med*) Bleeding within the substance of the brain. One of the main causes of a stroke or cerebro-vascular accident.

cerebral hemispheres (*BioSci*) See CEREBRUM.

cerebral palsy (*Med*) Muscular paralysis or other dysfunction resulting from perinatal damage to the motor area of brain.

cerebral thrombosis (*Med*) Thrombosis or clotting within one of the vessels supplying the brain, leading to a stroke or cerebro-vascular accident.

Ceramics processing

The vast majority of ceramics are manufactured from fine powders. Some final grinding, blending or mixing is often carried out before the shaping process itself. Usually the shaping is achieved at low temperatures using one of the variety of methods shown below and then the porous preform, or GREEN COMPACT, is brought to a higher density by the action of heat in the sintering process. Full compaction to the theoretical density is often not achieved and the product then remains porous, which is usually expressed as a percentage of the total volume. The following are the main methods used:

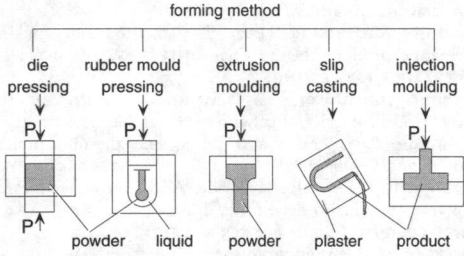

Ceramic processing methods Pressure being applied at P.

Die pressing is by far the most common fabrication process for small ceramic components. High production rates of up to 5000 pieces per minute can be achieved. Sizes can range from about 0·1 to 100 mm with reasonably good dimensional tolerance on the final product (+1%). The main limitations are in the complexity of shape which can be achieved and the uniformity of the product's density.

Isostatic pressing achieves a more uniform density by using a flexible membrane to transmit hydrostatic pressure from a compressed fluid to the piece, thereby ensuring uniform compaction throughout the body. This process is best used for parts with cylindrical symmetry, such as spark plug insulators or tubes, and rates of production of around 1500 pieces per hour are possible. Quite often some final machining will be required because of the imprecise control of dimensions that arises from using a compressible membrane.

Extrusion through a suitable die is an excellent method of producing components with a constant cross-section, provided a batch material with adequate plasticity is available. For clay-based ceramics the plasticity is achieved simply by controlling the amount of water present. For powders, organic plasticizers are needed to provide the correct consistency.

Injection moulding is a technique borrowed from the plastics industry. The batch used contains up to 60–70% by volume of fine ceramic powder in a polymeric matrix. The mix is heated to a plastic state before injecting into a mould under pressure where it cools and sets. The next, delicate step is to eliminate the polymeric material without damaging or disrupting the preform. Long controlled burn-out treatments of, say, 36–48 hours are often required. The method is very good for mass production of complicated shapes, provided the maximum wall thickness is not too great and the initial high capital cost can be justified.

In *slip casting*, the 'slip' is a colloidal suspension of ceramic powder in a liquid which is usually, but not always, water. The slip is poured into a mould which is microporous and slowly draws out the liquid from the slip by capillary action. The result is that a layer of fairly solid material is built up against the mould wall over a period of a few hours. The excess slip can then be poured out to leave a hollow component of uniform wall thickness. The moulds are usually plaster of Paris (hydrated calcium sulphate or gypsum) and are relatively cheap and easy to make. The process is slow, labour-intensive and lacks precision, but used for some engineering components.

All the above processes are followed by drying, and then a sintering process at an appropriate temperature, often around $0.7T_m$, to bring about densification. The temperature required is high because sintering occurs by solid state diffusion, either along the external surfaces or through the crystalline interior. In the latter case, crystal defects such as grain boundaries are preferred paths. Sometimes impurities in the batch or deliberate additions (fluxes) can speed up sintering or allow lower temperatures to be used.

See panels on POLYMERS and POLYMER SYNTHESIS.

---

cerebral tumour (*Med*) A TUMOUR within the cranial cavity.
cerebro- (*Genrl*) See CEREBR-.
cerebroside (*BioSci*) The simplest glycolipid, consisting of N-acyl sphingosine with either a glucose or a galactose residue.
cerebrospinal (*BioSci*) Pertaining to the brain and spinal cord.
cerebrospinal fluid (*Med*) The clear colourless fluid which bathes the surfaces of the brain and spinal cord and fills the ventricles.
cerebro-vascular accident (*Med*) A sudden interruption to the blood supply to the brain, as in cerebral haemorrhage or thrombosis. Also *stroke*.

cerebrum (*BioSci*) A pair of hollow vesicles or hemispheres forming the anterior and largest part of the brain of vertebrates.
Cerenkov counter (*NucEng*) Radiation counter which operates through the detection of Cerenkov radiation.
Cerenkov detector (*Phys*) A device which detects and measures the CERENKOV RADIATION produced as a result of the incidence of high-energy charged particles; from this the speed and charge of the particles may be calculated.
Cerenkov radiation (*Phys*) Radiation emitted when a charged particle travels through a medium at a speed greater than the speed of light in the medium. This occurs

when the refractive index of the medium is high, ie much greater than unity, as for water.

cereous (*BioSci*) See CERE.

Ceres (*Astron*) The first and largest asteroid to be discovered (1801), with a diameter of 1003 km and an orbital period about the Sun of 4·6 years.

ceriferous (*BioSci*) Wax-bearing, wax-producing.

cerium (*Chem*) A steel-grey metallic element, one of the rare earth metals. Symbol Ce, at no 58, ram 140·12, rel.d. at 20°C 6·7, mp 623°C, electrical resistivity 0·78 $\mu\Omega$ m. When alloyed with iron and several rare elements, it is used as the sparking component in automatic lighters and other ignition devices. It is also a constituent (0·15%) in the aluminium base alloy ceralumin and is photosensitive in the ultraviolet region. It is used on tracer bullets, for flashlight powders and in gas mantles. It is a getter for noble gases in vacuum apparatus. Its principal minerals, from which it is extracted, are BASTNAESITE and MONAZITE. There are a number of isotopes which are fission products, ^{144} Ce, with a half-life of 290 days, being a pure electron emitter of importance.

cermet (*Aero*, *Eng*) Ceramic articles bonded with metal. Composite materials combining the hardness and high-temperature characteristics of ceramics with the mechanical properties of metal, eg cemented carbides and certain reactor fuels.

CERN (*Phys*) Byname for *Organisation Européene pour la Recherche Nucléaire*, originally *Conseil Européen pour la Recherche Nucléaire*; the principal European centre for theoretical and experimental research in particle physics, supported by most European countries; located in Geneva. Its facilities include high- and low-energy proton and antiproton accelerators, and an electron–positron collider. See LARGE HADRON COLLIDER.

ceroma (*BioSci*) See CERE.

Certificate of Airworthiness (*Aero*) Certification issued or required by the Civil Aviation Authority confirming that a civil aircraft is airworthy in every respect to fly within the limitations of at least one of six categories. Abbrev *C of A*. US *Approved Type Certificate*.

Certificate of Compliance (*Aero*) Certification that parts of an aircraft have been overhauled, repaired or inspected etc to comply with airworthiness requirements. Abbrev *C of C*.

Certificate of Maintenance (*Aero*) Certification that an aircraft has been inspected and maintained in accordance with its maintenance schedule. Abbrev *C of M*.

ceruminous glands (*BioSci*) Modified sweat glands occurring in the external auditory meatus of mammals and producing a waxy secretion.

cerussite (*Min*) Lead carbonate, crystallizing in the orthorhombic system. A common ore of lead.

cervical ganglia (*BioSci*) Two pairs of sympathetic ganglia, anterior and posterior, situated in the neck of Craniata.

cervical smear (*Med*) The taking of a small sample of cells from the uterine cervix for the detection of cancer or the pre-cancerous stage.

cervicectomy (*Med*) Removal of the cervix uteri.

cervicitis (*Med*) Inflammation of the cervix uteri.

cervicum (*BioSci*) In insects, the neck or flexible intersegmental region between the head and the prothorax; in higher vertebrates, the neck or narrow flexible region between the head and the trunk. Adj *cervical*. Pertaining to the neck or to the cervix uteri.

cervine (*BioSci*) Dark-tawny.

cervix uteri (*BioSci*) The neck of the uterus, situated partly above and partly in the vagina.

Cesarean section (*Med*) US for CAESAREAN SECTION.

cesium (*Chem*) US for CAESIUM.

cespitose (*BioSci*) US for CAESPITOSE.

cesspool (*Build*) (1) A small square wooden box, lined with lead, which serves as a cistern in a parapet gutter at a point where roof water is discharged into a down pipe. (2) Underground pit (also *cesspit*) for the reception of sewage

from houses not connected to a mains or community drainage system.

Cestoda (*BioSci*) A class of PLATYHELMINTHES, all the members of which are endoparasites. There is a tough cuticle and the alimentary canal is lacking. Hooks and suckers for attachment occur at what is considered to be the anterior extremity. Tapeworms.

Cetacea (*BioSci*) An order of large aquatic carnivorous mammals. The forelimbs are fin-like, and hindlimbs are lacking. There is a horizontal flattened tailfin; the skin is thick with little hair. There are two inguinal mammae flanking the vulva and the neck is very short. Whales, dolphins and porpoises.

cetane (*Chem*) Normal (*n*) HEXADECANE.

cetane number (*Chem*) The percentage of cetane in a mixture of cetane and 1-methylnaphthalene that has the same ignition factor as the fuel under test; measure of the ignition value of the fuel. Equivalent for diesel fuels of the octane number for petrol.

cetrimide (*Chem*) One of the quaternary ammonium CATIONIC DETERGENTS, consisting very largely of alkyl-methylammonium bromides. It is a powerful disinfectant for cleansing skin, wounds, etc. TN *Cetavlon*.

Cetus (Whale) (*Astron*) The fourth largest constellation, lying above the equator, but inconspicuous because it has few bright stars.

cetuximab (*Pharmacol*) A *therapeutic antibody* used to treat advanced colorectal cancer.

cetyl alcohol (*Chem*) *Hexadecan-1-ol*. The palmitic ester is the chief constituent of spermaceti.

ceylonite (*Min*) See PLEONASTE.

Ceylon peridot (*Min*) A misnomer for the yellowish-green variety of the mineral tourmaline, approaching olivine in colour; used as a semiprecious gemstone.

Ceylon satinwood (*For*) A lustrous, golden-yellow, HARDWOOD timber with a narrowly interlocked grain, and a fine and uniform texture. From *Chloroxylon swietenia*, a native of India and Sri Lanka (formerly Ceylon). Also *East Indian satinwood*.

Cf (*Chem*) Symbol for CALIFORNIUM.

CFA (*BioSci*) Abbrev for COMPLETE FREUND'S ADJUVANT.

CFA piles (*CivEng*) Abbrev for CONTINUOUS FLIGHT AUGER PILES.

CFC (*EnvSci*) Abbrev for CHLOROFLUOROCARBON.

CFS (*Med*) Abbrev for CHRONIC FATIGUE SYNDROME.

C-format (*ImageTech*) A broadcast standard, composite format using 1 in tape on open reels. Sometimes applied to VHS-C. Also *BVH*.

CFR engine (*Eng*) A specially designed petrol engine, standardized by the Co-operative Fuel Research Committee, in which the knock-proneness or detonating tendency of volatile fuels is determined under controlled conditions and specified as an OCTANE NUMBER. See DETONATION.

CFRP (*Eng*) Abbrev for CARBON-FIBRE-REINFORCED PLASTIC.

CFTR (*BioSci*) Abbrev for CYSTIC FIBROSIS TRANSMEMBRANE CONDUCTANCE REGULATOR.

CGA (*ICT*) Abbrev for *colour graphics adapter*. The IBM PC VIDEO ADAPTER standard capable of supporting text and graphics on a screen 640 pixels wide by 200 pixels high. Cf EGA, VGA.

C-glass (*Glass*) Designation for a chemically resistant grade of glass fibre of composition (percentage by weight): SiO_2 65, CaO 14, Na_2O 8, B_2O_3 6, Al_2O_3 4, MgO 3, Fe_2O_3 0·3.

cg limits (*Aero*) The forward and aft positions within which the resultant centre of gravity of an aircraft must lie if balance and control are to be maintained. Cf LOADING AND CG DIAGRAM.

CGS unit (*Phys*) Abbrev for *centimetre–gram–second unit*, based on the centimetre, the gram and the second as the fundamental units of length, mass and time. For most purposes superseded by SI UNITS.

chabazite (*Min*) A white or colourless hydrated silicate of aluminium and calcium, found in rhombohedral crystals and belonging to the zeolite group.

chaeotropic (*BioSci*) A property of a substance that causes chaos, usually in the sense of disrupting or denaturing macromolecules. Breakdown of the tertiary structure of macromolecules is important for some separation and analytical methods.

chaeta (*BioSci*) In invertebrates, a chitinous bristle, embedded in and secreted by an ectodermal pit.

chaetiferous (*BioSci*) Bearing bristles. Also *chaetigerous, chaetophorous*.

Chaetognatha (*BioSci*) A phylum of hermaphrodite COELOMATA, having the body divided into three distinct regions: head, trunk and tail. The head bears two groups of sickle-shaped setae. Such creatures are small and transparent in form and are of carnivorous habit. They occur in the surface waters of the sea. Arrow worms.

chaetophorous (*BioSci*) See CHAETIFEROUS.

chaetoplankton (*BioSci*) Planktonic organisms bearing bristle-like outgrowths from the cells presumably to increase drag and reduce sedimentation rate.

Chaetopoda (*BioSci*) A group of ANNELIDA, the members of which are distinguished by the possession of conspicuous setae. It includes the POLYCHAETA and the OLIGOCHAETA.

chaff (*Aero*) Radar reflective strip or particles dispensed from aircraft, missiles or guns to confuse radar. First used in World War II and known as 'window', some kinds can be cut to desired length in flight to cope with different radar frequencies.

Chagas' disease (*Med*) See TRYPANOSOMIASIS.

chain (*Chem*) A series of linked atoms, generally in an organic molecule (*catenation*). Chains may consist of one kind of atom only (eg carbon chains), or of several kinds of atoms (eg carbon–nitrogen chains). There are open-chain and closed-chain compounds, the latter being called ring or cyclic compounds.

chain (*Eng*) (1) A series of interconnected links forming a flexible cable, used for sustaining a tensile load. (2) See ROLLER CHAIN.

chain (*MathSci*) Any series, members of which are related in a specific way. See MARKOV CHAIN.

chain (*Surv*) An instrument for the measurement of length. It consists of 100 pieces of straight iron or steel wire, looped together end to end, and fitted with brass swivel handles at both ends, the overall length being 1 chain. See ENGINEER'S CHAIN, GUNTER'S CHAIN, LINK.

chain block (*Eng*) A lifting mechanism comprising chains and sheaves in combination, such as a differential chain block, which allows very heavy loads to be hoisted by hand.

chain branching (*Chem*) Termination reaction which leads to formation of branched polymer (eg low-density polyethylene). See CHAIN POLYMERIZATION.

chain code (*ICT*) A sequence of 2^n or fewer binary digits arranged so that the pattern of n adjacent digits locates their position uniquely.

chain conveyor (*Eng*) Any type of conveyor in which endless chains are used to support slats, apron, pans, buckets, as distinct from the use of a simple band. See APRON CONVEYOR, BUCKET CONVEYOR.

chain coupling (*Eng*) A shaft coupling allowing irregularities in shaft alignment and a small amount of end play, designed for easy disconnection of shafts, in which the torque is transmitted from a chain sprocket keyed to the driving shaft to a similar sprocket keyed to the driven shaft, via a length of duplex roller chain wrapped round the two sprockets.

chain defects (*Chem*) Any kind of deviation from the normal molecular structure of the backbone of a polymer chain, as expressed by the repeat unit. They include chain ends, head-to-head units, oxidized units (eg carbonyl groups), tertiary carbon atoms and itinerant copolymer repeat units.

chain end (*Chem*) Possible weak point in polymer chain to act as a point of initiation of degradation or depolymerization. Some polymers may need capping with a protective end group to inhibit such instability.

chain extending agent (*Chem*) Chemical compound which can react with active chain ends to give much larger polymer molecule.

chain flexibility (*Chem*) Ease or difficulty of movement of a polymer chain, critically dependent on structure (or configuration) of repeat unit and temperature. Polyethylene and silicone polymers eg have very flexible single chains, so have very low T_g and are good low-temperature materials if non-crystalline. Large side groups or aromatic rings in the backbone chain increase rigidity, so increasing T_g and T_m (if chain is regular).

chain-folded crystal (*Chem*) Conformation of polymer chains in lamellar crystals. Chains are oriented perpendicular to main flat surface, where they fold over to maintain continuity. Perfect single crystals can be made by cooling dilute polymer solutions, but are not typical of bulk crystallized polymer, where surface chains may be entangled and connect with adjacent single lamellae (as in the switchboard model). See CRYSTALLIZATION OF POLYMERS and panels on POLYMERS and POLYMER SYNTHESIS.

chain grate stoker (*Eng*) A mechanical stoker for boilers, in which the furnace grate consists of an endless chain built up from steel links.

chaining (*ICT*) A method of processing a large program, segment by segment, often used where the program is too large to fit in the main memory all at once.

chaining (*Psych*) The linking of a series of discrete behaviours.

chain insulator (*ElecEng*) See SUSPENSION INSULATOR.

chain lines (*Paper*) The more widely spaced continuous lines in the watermark of a laid paper.

chain lockers (*Ships*) The subdivisions within a ship's hull for the housing of the anchor cables. They are usually divided on the centre line to form separate compartments for port and starboard cables.

chain polymerization (*Chem, Plastics*) Making high-molecular mass polymer chains by activating a small number of monomer molecules using catalysts such as benzoyl peroxide. Also *chain growth*. See panel on POLYMER SYNTHESIS.

chain printer (*ICT*) LINE PRINTER in which characters are carried on a continuous chain between print hammers and paper.

chain pump (*Eng*) A method of raising water through small lifts by means of disks attached to an endless chain which passes upwards through a tube; a chain alone may be used.

chain reaction (*Chem*) Chemical or atomic process in which the products of the reaction assist in promoting the process itself, such as ordinary fire or combustion, or atomic fission. Often characterized by an 'induction period' of comparatively slow reaction rate, followed, as the chain-promoting products accumulate, by a vastly accelerated reaction rate. See CHAIN POLYMERIZATION.

chain survey (*Surv*) A survey in which lengths only are measured (by means of a chain) and no angular measurements are made.

chain terminator (*BioSci*) See STOP CODON.

chain tongs (*Build*) A pipe grip formed by a chain hooked into a bar with a toothed projection.

chain transfer (*Chem*) Process occurring during free radical polymerization where active chain end terminates by abstracting hydrogen atom from dead chain or solvent or initiator. A new active chain end is thus created, and may lead to graft or branched polymer. See CHAIN POLYMERIZATION.

chain wheel (*Eng*) A toothed disk which meshes with a ROLLER CHAIN to transmit motion. Also *sprocket wheel*.

chair (*Chem*) The conformation of a six-membered ring in which the atoms are alternately above and below their mean plane. It is the stable conformation for cyclohexane and for aldohexoses. Cf BOAT.

chair (*CivEng*) The cast-steel support which is spiked to the sleeper and used to secure BULL-HEADED RAIL in position.

chair (*Glass*) (1) The 'chair' with long arms on which the glass-maker rolls the blowpipe while fashioning the ware. (2) The group of glass-makers who work together in the process of hand fabrication.

chalaza (*BioSci*) (1) The basal portion of the NUCELLUS of an ovule. (2) One of two spirally twisted spindle-like cords of dense albumen that connect the yolk to the shell membrane in a bird's egg.

chalazion (*Med, Vet*) A small nodule on the eyelid due to chronic inflammation of a sebaceous gland. In dogs, a cystic swelling of the Meibomian gland of the eyelid.

chalazogamy (*BioSci*) The entry of the pollen tube through the CHALAZA of the ovule. Cf POROGAMY.

chalcanthite (*Min*) Hydrated copper sulphate. $CuSO_4 \cdot 5H_2O$.

chalcedony (*Min*) A microcrystalline variety of quartz with abundant micropores. It occurs filling cavities in lavas and in some sedimentary rocks; flint is a variety found in the chalk. See panel on SILICON, SILICA, SILICATES.

chalcocite (*Min*) A greyish-black sulphide of copper. Cu_2S. An important ore of copper. Also *copper glance, redruthite.*

chalcogenide (*Chem*) A compound containing an element, or elements, from group VI(B) of the periodic table, ie sulphur, selenium or tellurium.

chalcogenide glass (*Chem*) Generic term for materials made by melt quenching or vapour deposition of chalcogenide compounds. Generally semiconductors, with high transparency in the infrared. Used in XEROGRAPHY, and as solid electrolytes esp in thin-film batteries. Also *chalconide glass.* See panel on GLASSES AND GLASS-MAKING.

chalcophile (*Geol*) Descriptive of elements which have a strong affinity for sulphur and are therefore more abundant in sulphide ore deposits than in other types of rock. Lead is an example.

chalcopyrite (*Min*) Sulphide of copper and iron, crystallizing in the tetragonal system; the commonest ore of copper, occurring in mineral veins. The crystals are brassy yellow, often showing superficial tarnish or iridescence. Also *copper pyrites, cupriferous pyrites.*

chalet (*Arch*) A type of country house, distinguished by having a steeply pitched roof, outside balconies, galleries and staircase; generally built of wood.

chalice (*BioSci*) A flask-shaped gland consisting of a single cell, esp numerous in the epithelia of mucous membranes.

chalk (*Geol*) A white, fine-grained and soft limestone, consisting of finely divided calcium carbonate and minute organic remains. In NW Europe, the chalk forms the upper half of the Cretaceous system.

chalk (*Med*) Compound employed in suitable mixtures as an antacid; also in the treatment of diarrhoea.

chalk gland (*BioSci*) A secreting organ, present in some leaves around which a deposit of calcium carbonate accumulates as in many species of *Saxifraga.*

chalking (*Build*) A paint film defect caused by the disintegration of the binder resulting in a loose powdery deposit.

chalking (*Print*) The failure of a printed ink film to dry and key correctly to the substrate, with the result that it can be removed by light rubbing. Caused by the ink vehicle being rapidly absorbed into the stock and leaving the pigments largely unbound at the surface.

chalk line (*Build, CivEng*) A length of well-chalked string used to mark straight lines on work by holding it taut in position close to the work and plucking it.

chalybite (*Min*) See SIDERITE.

Chamaeleon (*Astron*) A faint southern constellation.

chamaephyte (*BioSci*) Woody or herbaceous plant with perennating buds above but within 25 mm of the soil surface. Includes cushion plants. See RAUNKIAER SYSTEM.

chambered level tube (*Surv*) A level tube fitted at one end with an air chamber. By tilting the tube, air may be added to or taken from the bubble, whose length (which tends to be shortened by rise of temperature) may thus be regulated to maintain the sensitivity of the instrument.

chamber process (*ChemEng*) A process for the manufacture of sulphuric acid, in which the reactions between the air, sulphur dioxide and nitric acid gases necessary to produce the sulphuric acid take place in large lead chambers.

chambray (*Textiles*) Plain-weave cotton cloth with dyed warp and undyed weft producing a speckled effect; used for clothing.

chamfer (*Build*) The surface produced by bevelling an edge or corner.

chamfer plane (*Build*) A plane fitted with adjustable guides to facilitate the cutting of any desired chamfer.

chamomile (*Pharmacol*) Derived from the dried flower-heads of *Anthemis nobilis*; formerly a well-known stomachic and tonic, and sometimes used in powdered form in cosmetic preparations and shampoos. Also *camomile.*

chamosite (*Min*) A hydrous silicate of iron and aluminium occurring in oolitic and other bedded iron ores.

chance (*For*) Strictly, any unit of operation in the woods, with many and varied applications, of which the most familiar is *logging* or *cutting chance.* A logging or pulpwood operating unit.

chancel (*Arch*) The eastern part of a church, originally separated from the NAVE by a screen of latticework, so as to prevent general access thereto, without interrupting either sight or sound.

chancery (*Print*) A style of italic, the original being modelled on the handwriting used in the Papal chancery; Bembo italic is shown below:

RQENbaegn

chancre (*Med*) The hard swelling which constitutes the primary lesion in syphilis.

chancroid (*Med*) Non-syphilitic ulceration of the genital organs due to venereally contracted infection by gram-negative bacillum *Haemophilus ducreyi.*

Chandler wobble (*Astron*) A small periodic variation in the position of the Earth's geographic poles, leading to corresponding changes in latitude of given points on the Earth's surface. Thought to be due mainly to movement of material within the Earth, and seasonal changes in ice and atmospheric mass distribution.

Chandrasekhar limit (*Astron*) The upper mass limit, 1·44 solar masses, for a WHITE DWARF star.

change control (*ICT*) The arrangements by which changes, eg to a set of software components, are managed to ensure consistency and revertibility.

change face (*Surv*) To rotate a theodolite telescope about its horizontal axis so as to change from 'face left' to 'face right', or vice versa. See TRANSIT.

change of state (*Phys*) A change from solid to liquid, solid to gas, liquid to gas, or vice versa.

change-over (*ImageTech*) The transference of projection from one machine to another at the end of one reel and the start of the next, without an apparent break in sequence.

change-over contact (*ICT*) The group of contacts in a relay assembly, so arranged that, on operation, a moving contact separates from a back contact, is free during transit, and then makes contact with a front contact. Cf MAKE-BEFORE-BREAK CONTACT.

change-over switch (*ElecEng*) A switch for changing a circuit from one system of connections to another.

change point (*Surv*) A staff station to which two sights are taken, the first a foresight from one set-up of the level, the second a backsight from the next set-up of the level. Also *turning point*.

change-pole motor (*ElecEng*) An induction motor with a switch for changing the connections of the stator winding to give two alternative numbers of poles, so that the motor can run at either of two speeds.

change-speed motor (*ElecEng*) A motor which can be operated at two or more approximately constant speeds, eg a CHANGE-POLE MOTOR.

change wheels (*Eng*) The gear wheels through which the lead screw of a screw-cutting lathe is driven from the MANDREL, the reduction ratio being varied by changing the wheels. See SCREW-CUTTING LATHE.

changing bag (*ImageTech*) A light-tight bag to accommodate a camera and sensitive photographic material, so that the former can be loaded or unloaded with the latter in daylight.

channel (*Electronics*) The main current path between the SOURCE and DRAIN electrodes in a field-effect transistor.

channel (*Eng*) A standard form of rolled-steel section, consisting of three sides at right angles, in channel form. See ROLLED-STEEL SECTIONS.

channel (*ICT*) (1) Any clear path along which signals and information can be sent. A radio channel may consist of a transmitter and distant receiver, tuned to operate on an assigned frequency or channel; alternatively, a microwave radio system or a long distance underground cable or optical fibre system may have total capacity of several thousand speech channels or several TV channels. Sometimes a single amplifier may be described as a channel, or a stereophonic amplifier has left-hand or right-hand channels; a single TV camera with control equipment is often called a channel. (2) See BUS.

channel (*NucEng*) A passage through the reactor core for coolant, fuel rod or control rod.

channel black (*Chem*) See CARBON BLACK.

channel capacity (*ICT*) The maximum possible information rate for a CHANNEL.

channel effect (*Electronics*) In a transistor, by-passing the base component by leakage due to surface conduction.

channel gulley (*Build*) A channel, about 45 cm long, on the sides of which grease from sink wastes is trapped before reaching the gulley proper. Allows periodic removal.

channelling effect (*NucEng*) The reduced absorption of a radiation-absorbing material with voids relative to similar homogeneous material. Expressed numerically by the ratio of the attenuation coefficients. Particularly important in a moderator which can exist in two phases, eg water and steam. Also used for escape of radiation through flaws in shielding of reactor etc. Also *streaming effect*.

channel map (*ICT*) Specifies the destination channels and devices in a MIDI system.

channel pipe (*Build*) An open drain of half or three-quarter circular section, used in inspection and intercepting chambers.

channel separation (*ICT*) (1) The frequency increment between assigned frequencies for radio communication or broadcasting; chosen with regard to like-frequency variation of the transmitters, bandwidth of the transmissions, and the ability of receivers to discriminate between signals, all with a view to avoiding mutual interference between channels and services. (2) In a stereophonic sound system, the level of interference between left and right channels.

channel width (*ICT*) The width (in kHz, MHz or GHz) of any channel, which could be a radio channel or one within a FREQUENCY-DIVISION MULTIPLEX cable system, that is allotted to any particular form of transmission or service. Speech can be sent over a channel a few kHz wide, but television needs several MHz.

chantlate (*Build*) A projecting strip of wood fixed to the rafters at the eaves and supporting the normal roof covering; it serves to carry the drip clear of the wall.

chaos (*Phys*) A state of disorder and irregularity whose evolution in time, though governed by simple exact laws, is highly sensitive to starting conditions: a small variation in these conditions will produce wildly different results, so that long-term behaviour of chaotic systems cannot be predicted. Some degree of non-linearity is necessary for chaotic behaviour, which is present in most real systems, such as in weather patterns and the motion of planets about the Sun. See FRACTAL.

chapel (*Print*) An association of skilled printers (and, in certain districts, apprentices), who elect a 'father' or 'mother' to safeguard their interests.

chaperone (*BioSci*) Cytoplasmic protein that binds non-covalently to a newly formed polypeptide and ensures its correct folding or transport. The chaperone does not form part of the finished protein.

Chaperon resistor (*ElecEng*) Wirewound resistor of low residual reactance.

chaplet (*Eng*) Iron support for core placed in moulding box in iron foundry.

chaptalization (*FoodSci*) The addition of sugar to grape juice used in wine-making. Only permitted under strict control where the grapes have insufficient natural sugar to produce wine of a given alcoholic strength.

chapter (*Genrl*) One of the Roman figures used on the dials of watches or clocks to mark the hours.

character (*BioSci*) (1) In MENDELIAN GENETICS, the abnormality or variant caused by a gene. (2) In QUANTITATIVE GENETICS, whatever is measured for study, eg weight or yield.

character (*ICT*) Symbol that may be represented in a computer, such as letter, digit, space, punctuation mark. See ALPHANUMERIC, CONTROL CHARACTER.

character code (*ICT*) The BINARY CODE used to represent a character in a computer. There are a number of standard codes: *ASCII*, the American Standard Code for Information Interchange; *ISO7*, the International Organization for Standardization 7 bit code; *EBCDIC*, the Extended Binary-Coded Decimal Interchange Code.

character generator (*ImageTech*) A device for generating alphanumerics for titles etc. See VIDEO TYPEWRITER.

characteristic (*Electronics*) Any measurable property (eg gain, loss, impedance) of a device, active or passive, taken under closely specified operating conditions (supply voltages, temperature, frequency, etc).

characteristic curve (*Electronics*) A graph showing the variation of a particular characteristic of a device with other parameters, eg for a transistor, the collector current against collector–emitter voltage, plotted for different values of base current.

characteristic curve (*ImageTech*) Graph plotting the relationship between the density of a developed emulsion and the logarithm of the exposure. Also *D log E curve, H and D curve, Hurter and Driffield curve*.

characteristic curve (*MathSci*) Given a one-parameter family of surfaces, $\psi(x, y, z, a) = 0$, the limiting position of the curve of intersection of two adjacent surfaces of the family is given by

$$\begin{cases} \psi(x,y,z,a) = 0 \\ \frac{\partial\psi}{\partial a}(x,y,z,a) = 0 \end{cases}$$

for a given value of *a*. This curve is a characteristic curve of the family of surfaces. The locus of all the characteristic curves is the envelope of the family.

characteristic equation of a matrix (*MathSci*) Equation defined only for a square matrix. The equation which results when the determinant of the matrix obtained by replacing every diagonal element a_{ii} of the given square matrix by $a_{ii}-x$ is equated to zero. Its roots are called either *latent roots* or *eigenvalues*.

characteristic equation of an ordinary differential equation (*MathSci*) See AUXILIARY EQUATION.

characteristic function of a set (*MathSci*) A function which is unity for all points of the set, but zero outside the set.

characteristic impedance (*Phys*) (1) For waves propagated through a continuous medium, the ratio Z_0 of exciting force (or voltage) to velocity (or current) at a point. Also *surge impedance*. (2) For acoustical plane waves $Z_0 = \rho v$ where ρ is the density of the medium and v is the velocity of the wave. (3) For waves in a *transmission line*,

$$Z_0 = L_c = \sqrt{L/C}$$

where L and C are the inductance and capacitance per unit length of the line. (4) In free space, $Z_0 = \mu_0 c = 376 \cdot 6 \ \Omega$ where μ_0 is the permeability of free space.

characteristic of a logarithm (*MathSci*) See LOGARITHM.

characteristic points (*MathSci*) (1) Of a one-parameter family of curves in a plane, $\psi(x, y, a) = 0$, the characteristic points or points of contact with the envelope are given by the equations

$$\psi(x, y, a) = 0, \quad \frac{\partial \psi}{\partial a}(x, y, a) = 0$$

for a particular value of the parameter a. The envelope is found by eliminating a from these two equations. (2) Of a two-parameter family of surfaces, $\psi(x, y, z, a, b)$, the characteristic points, as above, are defined by the equations

$$\psi(x, y, z, a, b) = 0,$$
$$\frac{\partial \psi}{\partial a}(x, y, z, a, b) = 0,$$
$$\frac{\partial \psi}{\partial b}(x, y, z, a, b) = 0$$

for fixed a and b. The envelope is found by eliminating a and b from these three equations.

characteristic polynomial (*MathSci*) The polynomial which is equated to zero in the characteristic equation.

characteristic radiation (*Phys*) Radiation from an atom associated with electronic transitions between energy levels; the frequency of the radiation emitted is characteristic of the particular atom.

characteristic spectrum (*Phys*) Ordered arrangement in terms of frequency (or wavelength) of radiation (optical or X-ray) related to the atomic structure of the material giving rise to them.

characteristic velocity (*Space*) The change of velocity and the sum of changes of velocity (accelerations and decelerations) required to execute a space manoeuvre, eg transfer between orbits.

characteristic X-radiation (*Phys*) X-radiation consisting of discrete wavelengths which are characteristic of the emitting element. If arising from the absorption of X- or γ-radiation, it may be called *fluorescence X-radiation*. See CHARACTERISTIC RADIATION.

character recognition (*ICT*) Optical or magnetic reading by a peripheral capable of directly absorbing information from diagrams or the printed or written page.

characters per second (*ICT*) Measure of the speed in which data, in coded form, may be transmitted between two points.

Charadriiformes (*BioSci*) An order of wading and swimming birds found on sea coasts and inland waters. Most feed on animals. Waders, gulls and auks.

Charales (*BioSci*) Small order in the Charophyceae of macroscopic fresh and brackish water algae with a distinct axis; anchored by rhizoids to the bottom, and with whorled branches. One very large cell runs the length of each internode. They are oogamous, eg stoneworts.

charcoal (*Chem, Med*) The residue from the destructive distillation of wood or animal matter with exclusion of air; contains carbon and inorganic matter. Used as an absorbent in its activated form, esp in cases of alkaloid poisoning, and as a palliative for flatulent dyspepsia. Also used in masks to adsorb toxic gases.

charcoal blacking (*Eng*) Powdered charcoal which is dusted over the surface of a mould to improve the smoothness of the casting in fine work.

charcoal iron (*Eng*) Pig iron made in a blast furnace using charcoal instead of coke. Sometimes also wrought-iron made from this. Obsolete.

charge (*Chem*) Electrical energy stored in chemical form in secondary cell.

charge (*Eng*) Amounts (ore, fuel and flux) charged into a furnace at one loading, or proportions of these.

charge (*Glass*) (1) See BATCH. (2) The quantity of batch required to fill a *pot*, the fireclay vessel containing the glass melt in a pot furnace.

charge (*MinExt*) The operation of placing explosive, primer, detonator and fuse in drill hole.

charge (*NucEng*) Fuel material in nuclear reactor.

charge (*Phys*) The quantity of unbalanced electricity in a body, ie excess or deficiency of electrons, giving the body negative or positive electrification, respectively. See NEGATIVE, POSITIVE. The charge of an ion is one or more times that of an electron, of either sign. Unit is the COULOMB.

charge-coupled device (*ICT*) A semiconductor device that relies on the short-term storage of minority carriers in spatially defined depletion zones on its surface. The charges thus stored can be moved about by the application of control voltages via metallic conductors to the storage points, in the manner of a shift register. Single-CHIP memories with a very large storage capacity can be made from charge-coupled devices. Abbrev *CCD*.

charged (*Phys*) The state of: (1) a capacitor when a working potential difference is applied to its electrodes; (2) a secondary cell or battery (accumulator) when it stores the maximum (rated) energy in chemical form, after passing the necessary ampere-hours of charge through it; (3) a conductor when it is held at an operating potential, eg traction conductors or rails, or mains generally.

charge–discharge machine (*NucEng*) A device for inserting or removing fuel in a nuclear reactor without allowing escape of radiation and, in some reactors, without shutting the reactor down. Also *fuelling machine, refuelling machine*.

charged current (*Phys*) Weak interaction with which electrical charges are exchanged; typically involves transmutation of a neutrino into a charged lepton, or the reverse process. Mediated by the W BOSON. Cf NEUTRAL CURRENT.

charged-pair complex (*Chem*) Association of anion and cation, usually in dilute solution.

charged system (*Textiles*) Used in dry-cleaning. By adding a surface active agent a clear dispersion may be obtained when water is added to the organic liquid. The mixture has enhanced cleaning properties.

charge exchange (*Phys*) Exchange of charge between a neutral atom and an ion in a plasma. After their charge is neutralized, high-energy ions can normally escape from plasma – hence this process reduces plasma temperature.

charge face (*NucEng*) In a reactor, that face of the biological shield through which fuel is loaded.

charge-independent (*Phys*) Said of nuclear forces between particles, the magnitude and sign of which do not depend on whether the particles are charged. See NUCLEAR FORCE, SHORT-RANGE FORCE.

charge indicator (*ElecEng*) See POTENTIAL INDICATOR.

charge machine (*NucEng*) See CHARGE–DISCHARGE MACHINE.

charge–mass ratio (*Phys*) The ratio of electric charge to mass of particle, of great importance in physics of all particles and ions.

charging current (*ElecEng*) (1) Current passing through an accumulator during the conversion of electrical energy into stored chemical energy. (2) Impulse of current passing into a capacitor when a steady voltage is suddenly applied, the actual current being limited by the total resistance of the circuit. (3) Alternating current which flows through a capacitor when an alternating voltage is applied.

charging resistor (*ElecEng*) A resistance inserted in series with a switch to limit the rate of rise of current when making the circuit.

charging voltage (*ElecEng*) The emf required to pass the correct charging current through an accumulator, about 2·5 V for each lead–acid cell.

Charles's law (*Phys*) Law stating that the volume of a given mass of gas at constant pressure is directly proportional to the absolute thermodynamic temperature; equivalently, all gases have the same coefficient of expansion at constant pressure. This is approximately true at low pressures and sufficiently high temperatures as the ideal gas behaviour is approached. Also *Gay-Lussac's law*.

charm (*Phys*) One of the six FLAVOURS of QUARKS, with a mass of 1400 MeV and a charge of + 2e/3. Also the property of the charm quark, which is zero for all other particles, 1 for the charm, and −1 for the anticharm quark.

charnockite (*Geol*) A coarse-grained, dark granite rock, consisting of feldspars, blue quartz, and orthopyroxene; it occurs typically in Madras and is named after the founder of Calcutta, Job Charnock.

Charnoid direction (*Geol*) A NW–SE direction of folding in the rocks of central and eastern England. Exemplified by the Precambrian rocks of Charnwood Forest.

Charon (*Astron*) Pluto's only known natural satellite, discovered photographically in 1978. Distance from the planet 20 000 km; diameter 1190 km.

Charophyceae (*BioSci*) A class of the green algae (Chlorophyta) characterized by motile cells (if produced). They are scaly or naked and asymmetric, with two flagella inserted laterally or subapically and associated with a MULTILAYERED STRUCTURE. There is no phycoplast; they are haplontic, the zygote being a resting stage. The class includes unicellular, sarcinoid, filamentous and parenchymatous sorts; they are predominantly fresh-water algae. *Klebsormidium*, *Spirogyra*, the DESMIDS, *Coleochaeta* and *Chara*. (In earlier classification only the Charales were placed in the Charophyceae.)

Charpy test (*Eng*) A flexed beam impact test in which both ends of a notched specimen are supported and a striker carried on a pendulum impacts the specimen centrally on the face opposite the notch; the energy absorbed in fracture is then calculated from the height to which the pendulum rises as it continues its swing. See panel on IMPACT TESTS.

chart comparison unit (*Radar*) A type of display superimposed upon navigational chart. Also *map comparison unit*.

Chartered Engineer (*Eng*) A style or title (UK) which can only be used by persons with acceptable qualifications registered with the Engineering Council, a body set up by Royal Charter in 1981. Abbrev *CEng*.

Chartered Surveyor (*Build*) A style or title which may be used by a fellow or professional associate of the chartered surveyors' professional organization.

chase (*Build*) (1) A trench dug to accommodate a drain pipe. (2) A groove chiselled in the face of a wall to receive pipes etc.

chase (*Print*) An iron or steel frame into which type is locked by means of wooden or mechanical quions.

chase-mortising (*Build*) A method adopted to frame a timber in between two others already fixed, a sloping chase being cut to the bottom of the mortise so that the crosspiece may be placed into position.

chaser (*Eng*) A cutter of which the edge is serrated to the profile of a screw-thread, used for producing or accurately finishing screw-threads. Single chasers may be mounted in lathe tool-posts; sets are used in die-heads, collapsible taps and other screw-cutting tools.

chasmocleistogamous (*BioSci*) Producing both CHASMO-GAMOUS and CLEISTOGAMOUS flowers.

chasmogamous (*BioSci*) Flowers that open normally to expose the reproductive organs, cf CLEISTOGAMOUS flowers.

chassis (*Aero, Electronics*) Rigid base on which electronic units or other electrical components are mounted.

chassis (*Autos*) The main frame of a vehicle (as distinct from the body) to which the engine, steering and transmission gear, wheels, etc, are attached. Vehicles in which a suitably reinforced steel body fulfils this function are sometimes called *chassis-less*, *monocoque* or *unitized*.

chassis (*Eng*) Generally, any major part or framework of an assembly, to which other parts are attached.

Chastek paralysis (*Vet*) Paralytic disease of silver foxes. A disease of fox, mink and ferrets characterized by paralysis, convulsions and death, due to thiamin (vitamin B_1) deficiency. The deficiency is caused by feeding on raw fish, parts of which contain an anti-thiamin factor.

chatoyancy (*Min*) The characteristic optical effect shown by cat's eye and certain other minerals, due to the reflection of light from minute aligned tubular channels, fibres or colloidal particles. When cut *en cabochon* such stones exhibit a narrow silvery band of light which changes its position as the gem is turned.

chat room (*ICT*) A virtual space on the Internet where any number of people can exchange messages.

chatter (*Eng*) Vibration of a cutting tool or of the work in a machine; caused by insufficient rigidity of either, and results in noise and uneven finish.

chatter (*ICT*) Undesired rapid closing and opening of contacts on a relay, reducing their life and making switching uncertain.

Chatterton's compound (*ElecEng*) An adhesive insulating substance consisting largely of gutta percha; used as a cement or filling, esp in cable jointing.

Chattian (*Geol*) A stage in the Oligocene. See TERTIARY.

Chebyshev filter (*ICT*) A constant-*k* filter in which a very rapid cut-off, or increase in attenuation with frequency, is achieved at the expense of the evenness of the INSERTION LOSS within the passband. Named after a Russian mathematician, and can have various anglicized spellings. Cf BUTTERWORTH FILTER.

Chebyshev inequality (*MathSci*) The theorem that if $a_r \geqslant a_{r+1}$ and $b_r \geqslant b_{r+1}$ for all r, then

$$\left(\sum_{r=1}^{n} a_r \right) \left(\sum_{r=1}^{n} b_r \right) < n \sum_{r=1}^{n} a_r b_r$$

Chebyshev polynomials (*MathSci*) The polynomials $T_n(x)$ in the expansion

$$\frac{1 - h^2}{1 - 2xh + h^2} = \sum_{n=0}^{\infty} T_n(x)(2h)^n$$

Under specified conditions these are orthogonal. These polynomials have applications in approximation theory.

checkbox (*ICT*) A small square box displayed in the context of a WIMP environment that can be selected or cleared. When the option is selected a character, such as a cross or a tick, appears in the checkbox.

check digit (*ICT*) Redundant bit in stored word, used in self-checking procedure such as a PARITY CHECK. If there is more than one check digit they form a check number.

checking (*Build*) A minor fault developed by paintwork to relieve stresses in the film, characterized by a network of fine, usually shallow, cracks.

check-lock (*Build*) A device for locking in position the bolt of a door lock.

check nut (*Eng*) See LOCK NUT.

checkpoint (*BioSci*) Any stage in the cell cycle at which the cycle can be halted and entry into the next phase postponed. Two major checkpoints are initiation of DNA synthesis (S-phase) and entry into mitosis.

check rail (*CivEng*) A third rail laid on a curve alongside the inner rail and spaced a little from it, to safeguard rolling stock against derailment due to excessive thrust on the outer rail. Also *guard rail*, *rail guard*, *safeguard*, *safety rail*, *side rail*.

check receiver (*ICT*) Radio or TV receiver for verifying quality or content of a programme. Also *monitoring receiver*.

check sum (*ICT*) A CONTROL TOTAL that may have no external significance but which is used for checking purposes; eg in a banking system, the total number of cash deposits in a file. See HASH TOTAL.

check throat (*Build*) A small groove cut in the face of a short step in the upper surface of a wooden window-sill, just behind the face of the sash. It serves to stop rain from driving up under the sash. Also *anti-capillary groove*.

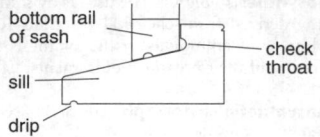

check throat

check valve (*Eng*) A non-return valve, closed automatically by fluid pressure; fitted in a pipe to prevent return flow of the fluid pumped through it. See CLACK, FEED CHECK VALVE.

cheddite (*Chem*) Mixture of castor oil, ammonium perchlorate and 2,4-dinitrotoluene; used as an explosive.

cheek (*BioSci*) In TRILOBITES, the pleural portion of the head; in mammals, the side of the face below the eye, the fleshy lateral wall of the buccal cavity.

cheek (*Build*) One of the solid parts on each side of a mortise, or the removed side of a tenon.

cheese (*FoodSci*) Solids obtained after draining off the liquid (whey) from milk coagulated with rennet and/or lactic acid bacteria. Curds may be salted, then ripened by the actions of moulds and bacteria to bring out the characteristic flavours. An important source of protein and essential nutrients (such as calcium, vitamin A and riboflavin).

cheese (*Textiles*) Densely wound cylindrical package of yarn made on various winding machines for warping and other purposes.

cheese cloth (*Textiles*) A lightweight open cotton fabric used in cheese manufacture but sometimes adapted for shirts and blouses.

cheese-head screw (*Eng*) A screw with a cylindrical head, similar in shape to a round cheese, slotted for a screwdriver.

cheilectropion (*Med*) Turning outwards of the lip.

cheilitis (*Med*) Inflammation of the lip.

cheiropompholyx (*Med*) A skin disease in which vesicles filled with clear fluid suddenly appear on the hands and feet. Sometimes associated with allergic dermatitis.

chela (*BioSci*) In Arthropoda, any chelate appendage. Adjs *cheliferous, cheliform*.

chelate (*BioSci*) In Arthropoda, having the penultimate joint of an appendage enlarged and modified so that it can be opposed to the distal joint like the blades of a pair of scissors to form a prehensile organ. Cf SUBCHELATE.

chelating agent (*Med*) A chemical agent which combines with unwanted metal ions. Used to treat heavy-metal poisoning, thus sodium calcium EDTA is a chelating agent promoting the excretion of lead.

chelicerae (*BioSci*) In Arachnida, a pair of pre-oral appendages, which are usually chelate.

Chelicerata (*BioSci*) An arthropod sub-phylum comprising animals with two major body regions, an anterior prosoma and a posterior opisthosoma, and with the foremost pair of appendages being CHELATE. Spiders, scorpions, horseshoe crabs.

Chelonia (*BioSci*) An order of reptiles in which the body is encased in a horny capsule consisting of a dorsal carapace and a ventral plastron. The jaws are provided with horny beaks in place of teeth, and only the lower temporal row of teeth is present. Tortoises and turtles.

cheluviation (*EnvSci*) Leaching of iron and aluminium oxides from soils after the formation of soluble complexes with esp polyphenols from fresh litter of conifers or heath plants. See PODSOL.

chemical affinity (*Chem*) See AFFINITY.

chemical balance (*Chem*) An instrument used in chemistry for weighing, to a high degree of accuracy, the small amounts of material dealt with.

chemical binding effect (*Phys*) A variation in the cross-section of a nucleus for neutron bombardment depending on how the element is combined with others in a chemical compound.

chemical bond (*Chem*) The electric forces linking atoms in molecules or non-molecular solid phases. Three basic types of bond are usually distinguished: (1) ionic or electrostatic bonding, in which valence electrons are lost or gained, and atoms which are oppositely charged are held together by coulombic forces; (2) covalent bonding, in which valence electrons are associated with two nuclei, the resulting bond being described as polar if the atoms are of differing electronegativity; (3) metallic bonding, in which valence electrons are shared over many nuclei, and electronic conduction occurs. See panel on BONDING.

chemical compound (*Chem*) A substance composed of two or more elements in definite proportions by weight, which are independent of its mode of preparation. Thus the ratio of oxygen to carbon in pure carbon monoxide is the same whether the gas is obtained by the oxidation of carbon, or by the reduction of carbon dioxide.

chemical constitution (*Chem*) See CONSTITUTION.

chemical energy (*Chem*) The energy liberated in a chemical reaction. See AFFINITY.

chemical engineering (*Genrl*) The design, construction and operation of plant and works in which matter undergoes change of state and composition.

chemical equation (*Chem*) A quantitative symbolic representation of the changes occurring in a chemical reaction, based on the requirement that matter is neither added nor removed during the reaction.

chemical finishing (*Textiles*) The final preparation of bleached fabric during which the material is subjected to the action of various chemicals including resin-forming compounds.

chemical fog (*ImageTech*) Overall FOG in a photographic image caused by excessive development or chemical contamination or decomposition.

chemical hygrometer (*EnvSci*) See ABSORPTION HYGRO-METER.

chemical kinetics (*Chem*) The study of the rates of chemical reactions. See ORDER OF REACTION.

chemical lead (*Eng*) Lead of purity exceeding 99·9%; suitable for the lining of vessels used to hold sulphuric acid and other chemicals.

chemically formed rock (*Geol*) Rock formed by precipitation of materials from solution in water, eg calc-tufa and various saline deposits.

chemical oxygen demand (*EnvSci*) See OXYGEN DEMAND. Abbrev *COD*.

chemical potential (*BioSci*) Symbol μ. (1) A measure of the (Gibbs) free energy associated with a given uncharged chemical species under given conditions and hence of its relative ability to perform work. Any non-ionized substance including water moving by diffusion or osmosis will tend to move spontaneously down its own chemical potential gradient. For the jth component:

$$\mu_j = \mu_j^* + RT \ln a_j + PV_j$$

where mu_j^* is the chemical potential of j in some arbitrary standard state, R is the gas constant, T is absolute temperature, ln denotes the natural logarithm, a is the activity ($\approx$ concentration), P is hydrostatic pressure, and V is partial molar volume. The PV term is important for water but not for solutes. Cf ELECTROCHEMICAL POTENTIAL.

chemical precipitation (*Build*) The process of assisting settlement of the solid matter in sewage by adding chemicals before admitting the sewage to the sedimentation tanks.

chemical pulp (*Paper*) Pulp in which the fibres have been resolved by chemical, as distinct from mechanical, means, with the removal of the greater part of the lignin and other non-cellulose material. Cf MECHANICAL WOOD PULP.

chemical reaction (*Chem*) A process in which the structure of a chemical is changed to form another substance which retains the nuclei of the original chemical or chemicals but with a different configuration or content of atoms, energy, etc.

chemical shift (*Phys*) A shift in position of a spectrum peak due to a small change in chemical environment. Observed in the MÖSSBAUER EFFECT and in NUCLEAR MAGNETIC RESONANCE.

chemical shim (*NucEng*) A means of reducing the initial high reactivity at the start-up of a reactor by introducing an element with a high neutron capture cross-section which gradually changes under neutron bombardment to an isotope of lower cross-section. Boron and gadolinium have been used. See BURNABLE POISON.

chemical symbol (*Chem*) A single capital letter, or a combination of a capital letter and a small one, which is used to represent either an atom or a mole of a chemical element; eg the symbol for sodium is Na, for sulphur is S.

chemical synapse (*BioSci*) A nerve–nerve or nerve–muscle junction where the signal is transmitted by release from one membrane of a chemical transmitter that binds to a receptor in the second membrane. Importantly, signalling is unidirectional.

chemical toning (*ImageTech*) The process of converting the silver image into, or replacing it by, a coloured substance by chemicals other than a dye.

chemical vapour deposition (*Chem*) The deposition of solid material, usually as a thin film, from precursors in the gas phase. Abbrev CVD. Cf PHYSICAL VAPOUR DEPOSITION.

chemical warfare (*Genrl*) The use of chemical agents such as WAR GAS and other antipersonnel chemicals as weapons.

chemical wood-pulp (*Paper*) Pulp obtained from wood by the sulphite, sulphate, soda or other chemical process.

chemiluminescence (*BioSci, Chem*) A process in which visible light is produced by a chemical reaction. The technique can be used to measure ATP using the enzyme LUCIFERASE or, with the addition of light-producing substrates, the activation of the MYELOPEROXIDASE–HALIDE SYSTEM in phagocytic cells.

chemiosmosis (*BioSci*) The mechanism underlying the formation of ATP synthesis by oxidative phosphorylation. The energy for ATP synthesis is derived from electrochemical gradients across the inner membrane of the mitochondrion. A similar mechanism operates during PHOTOPHOSPHORYLATION. See OXIDATIVE PHOSPHORYLATION and panel on MITOCHONDRION.

chemisorption (*Chem*) Irreversible adsorption in which the absorbed surface is held on the substance by chemical forces.

chemistry (*Chem*) The study of the composition of substances and of the changes of composition which they undergo. The main branches of the subject are often considered to be inorganic, organic and physical chemistry.

chemoautotroph (*BioSci*) An organism that derives energy from the oxidation of inorganic compounds for the assimilation of simple materials, eg carbon dioxide and ammonia. Also *chemosynthetic autotroph*.

chemokine (*BioSci*) Cytokines that stimulate leucocyte CHEMOTAXIS, eg interleukin-8, which is chemotactic for neutrophils, lymphocytes and basophils.

chemokinesis (*BioSci*) Stimulation of random movement of cells such as leucocytes by substances in the environment.

chemonasty (*BioSci*) A plant movement provoked, but not orientated, by a chemical stimulus. See NASTIC MOVEMENT.

chemoreceptor (*BioSci*) A sensory nerve-ending, receiving chemical stimuli.

chemoreflex (*Med*) A reflex initiated by a chemical stimulus.

chemosis (*Med*) Oedema of the conjunctiva.

chemosphere (*Astron*) The Earth's atmospheric layers between 20 and 200 km.

chemostat (*BioSci*) A culture vessel in which steady-state growth is maintained by appropriate rates of harvest and addition of nutrients.

chemosynthesis (*BioSci*) The use, as by some bacterium, of energy derived from chemical reactions (eg oxidation of sulphur or of ammonia) in the synthesis from inorganic molecules of their organic requirements. Cf PHOTOSYNTHESIS.

chemosynthetic autotroph (*BioSci*) See CHEMOAUTOTROPH.

chemotaxis (*BioSci*) Stimulation of movement by a cell or organism towards or away from substances producing a concentration gradient in the environment.

chemotaxonomy (*BioSci*) The use of chemical evidence (of both primary and secondary metabolism) in taxonomy.

chemotherapy (*Med*) Treatment of disease by chemical compounds selectively directed against abnormal cells and invading organisms.

chemotropism (*BioSci*) A differential growth movement or curvature of part of an organ in relation to a chemical concentration gradient.

chenille (*Textiles*) A type of yarn with fibres projecting all round a central core of threads, and cloth woven with such yarn in the weft.

chenodeoxycholic acid (*Med*) Substance used in certain patients to dissolve gallstones as an alternative to surgery.

Chenopodiaceae (*BioSci*) A family of c.1500 spp of dicotyledonous flowering plants (superorder Caryophyllidae). Mostly herbs, they are temperate and subtropical and are common in desert and saline habitats. The flowers are inconspicuous and wind-pollinated. Many are C4 plants. Includes saltbush, sugar beet, beetroot, leaf beets (all species of *Beta*), spinach and quinoa.

chequer plate (*Eng*) Steel plate used for flooring; provided with a raised chequer pattern to give a secure foothold.

cheralite (*Min*) A radioactive mineral rich in thorium.

Cherenkov radiation (*Phys*) See CERENKOV RADIATION.

Chernikeef log (*Ships*) A device for measuring distance moved through the water. An impeller is lowered through the bottom of the ship to about 40 cm (18 in) below the hull. The rotations of the impeller are transmitted electrically to a distance recorder and combined with time to indicate speed.

Chernobyl (*Genrl*) Location of a major accident to a nuclear reactor on 26 April 1986. The incident led to widespread radioactive contamination locally and over a significant part of N Europe.

chernozem (*EnvSci*) Grassland soil formed in sub-humid cool to temperate areas, with humus at the surface with a blackish layer of mineral soil just beneath which grades downward to a lighter layer where lime has accumulated. Occurs under tall-grass communities such as Russian Steppes, N American Prairies and Argentinian Pampas.

cherry (*For*) A tree of the genus *Prunus*, economically not regarded as a timber tree, though the wood has certain minor uses.

cherry-picker (*CivEng*) A working platform mounted on a two-limbed hydraulically or electrically operated raising mechanism, on the back of a truck; used for access to elevated inaccessible places, eg street-lighting fittings.

chert (*Geol*) A siliceous rock consisting of cryptocrystalline silica, and sometimes including the remains of siliceous organisms such as sponges or radiolaria. It occurs as bedded masses, as well as concretions in limestone. See SILICA.

chestnut (*For*) A light- to dark-brown wood resembling oak; much used for fencing, posts and rails.

cheval-vapeur (*Eng*) The metric unit of horsepower, equivalent to 75 kgf m s^{-1}, 735·5 W or 0·986 hp. Abbrev *CV*. Also *cheval* (*ch*), *Pferdestärke* (*PS*).

Cheyne–Stokes breathing (*Med*) Waxing and waning of respiration, sometimes with periods of apnoea, caused by delay in medullary chemoreceptor response to blood gas changes. Occurs in disease of the heart, kidney or brain.

χ (chi) (*Phys*) Symbol for magnetic susceptibility.

χ² (chi squared) (*MathSci*) See CHI-SQUARED TEST.

chiasma (*BioSci*) (1) Point of contact between chromatids, visible during meiosis and involved in crossover. Pl *chiasmata*. See diagram at MEIOSIS and panel on CHROMOSOME. (2) A structure in the central nervous system, formed by the crossing over of the fibres from the right side to the left side and vice versa.

chiastolite (*Min*) A variety of andalusite characterized by cruciform inclusions of carbonaceous matter.

chicken (*Agri*) An individual of *Gallus domesticus*, the common domesticated fowl, of either sex and any age.

chickenpox (*Med*) A common, mild, acute infectious disease in which papules, vesicles and small pustules appear in successive crops, mainly on the trunk, face, upper arms and thighs. Also *Varicella*.

chicle (*Chem*) The natural LATEX of *Sapota achras* composed of a mixture of *cis-* and *trans*-polyisoprene with natural resins. De-resinated chicle is one base for chewing gum. See JELUTONG.

chiffon (*Textiles*) A very fine, soft, woven dress material of silk or synthetic fibres; the word is also used in other cloth descriptions to indicate lightness, eg chiffon velvet, chiffon taffeta.

chilblains (*Med*) See ERYTHEMA PERNIO.

childhood psychosis (*Psych*) A group of childhood disorders characterized by disturbed social relationships, speech impairment and bizarre motor behaviour. Three subgroups are recognized: infant AUTISM, late-onset psychosis, eg CHILDHOOD SCHIZOPHRENIA, and disorders due to degeneration of the central nervous system.

childhood schizophrenia (*Psych*) A childhood disorder which manifests itself after a period of normal development, often in adolescence, when the child begins to show severe disturbances in social adjustment and reality contact. As a diagnostic category it is not distinguished from adult SCHIZOPHRENIA.

Child–Langmuir equation (*Electronics*) An equation which states that in a thermionic diode, where the anode current is limited by the space charge surrounding the cathode, this current is proportional to the three-halves power of the anode–cathode voltage, ie $I = GV^{3/2}$ where I is the current, V the voltage and G a constant depending on the physical form of the electrodes.

child program (*ICT*) A copy or an adaptation of the PARENT PROGRAM.

Chile nitre (*Chem*) A commercial name for sodium nitrate (V), $NaNO_3$. Also *Chile saltpetre*.

chill (*Eng*) An iron mould, or part of a mould, sometimes water-cooled; used to accelerate cooling and give great hardness and density to the metal which comes into contact during cooling.

chill crystals (*Eng*) Small crystals formed by the rapid freezing of molten metal when it comes into contact with the surface of a cold metal mould.

chilled iron (*Eng*) Cast-iron cast in moulds constructed wholly or partly of metal, so that the surface of the casting is white and hard while the interior is grey.

Chilognatha (*BioSci*) See DIPLOPODA.

Chilopoda (*BioSci*) A class of ARTHROPODA having the trunk composed of numerous segments each bearing one pair of legs. The head bears a pair of uniflagellate antennae, a pair of mandibles and two pairs of maxillae. The first body segment bears a pair of poison-claws; the genital opening is posterior. They comprise active carnivorous forms, some of considerable size and dangerous to humans; some are phosphorescent. Centipedes.

chimaera, chimaeric (*Genrl*) Older spelling of CHIMERA, chimeric.

chimera (*BioSci*) (1) A plant or plant part in which there are two (rarely more) genetically different sorts of cell, a result of mutation or of grafting. The name is not applied to ordinary grafted plants where the stock differs from the scion but rather where cells of one sort have come to form a layer over a core of cells of the other throughout the shoot. See MERICLINAL CHIMERA, PERICLINAL CHIMERA, SECTORIAL CHIMERA. (2) In animals, an individual exhibiting two or more different genotypes in patches derived from two or more different embryos that have become fused, naturally or artificially, at an early stage to make a single embryo. Cf MOSAIC. (3) A DNA or protein molecule with sequences from more than one organism. Adj *chimeric*.

chimeric antibodies (*BioSci*) Genetically engineered monoclonal antibodies that have part of their structure coded for by one species and the rest by a different species. See also HUMANIZED ANTIBODY.

chimney (*Geol*) A volcanic pipe. Chimneys include those through which ocean-floor smokers vent.

chimney (*MinExt*) Either an ore shoot or more usually an esp rich steeply inclined part of the lode.

chimney bar (*Build*) An iron bar supporting an arch over a fireplace opening.

chimney bond (*Build*) A STRETCHING BOND which is generally used for the internal division walls of domestic chimney stacks, as well as for outer walls.

chimneybreast (*Build*) The part of the chimney between the flue and the room.

chimney jambs (*Build*) The upright sides of the fireplace opening.

chimney lining (*Build*) The tile flues, cement rendering or flexible metal tubing within a chimney space.

chimney shaft (*Build*) The part of a chimney projecting above the roof, or a chimney standing isolated like a factory chimney.

chimney stack (*Build*) The unit containing a number of flues together.

china clay (*Geol, Med*) A clay consisting mainly of KAOLINITE, one of the most important raw materials of the ceramics industry. China clay is obtained from kaolinized granite, eg in SW England, and is separated from the other constituents of the granite (quartz and mica) by washing out with high-pressure jets of water. Used in medicine as an internal absorbent (eg of poisons), for poultices and as dusting powder. Also *kaolin, porcelain clay*.

china stone (*Geol*) A kaolinized granitic rock containing unaltered plagioclase. Also applied to certain limestones of exceptionally fine grain and smooth texture.

chine (*Aero*) The extreme outside longitudinal member of the planing bottom of a flying-boat hull, or of a seaplane float; runs approximately parallel with the keel. Also used on sharp fuselage edge of supersonic aircraft, eg SR 71.

Chinese remainder theorem (*MathSci*) A theorem which states that if a_1, a_2, a_3, ..., a_n are pairwise RELATIVELY PRIME, then the system of equations $x \equiv b_i \bmod a_i$ ($i = 1$, 2, ..., n) has a unique solution mod $a_1 a_2 ... a_n$.

chintz (*Textiles*) A plain-woven cotton fabric that has been printed and then glazed by calendering. It may also be stiffened by the addition of starch.

chip (*Eng*) See SWARF.

chip (*ICT*) Popular name for an INTEGRATED CIRCUIT. The term derives from the method of manufacture, as each chip is made as part of a wafer, a flat sheet of silicon, impregnated with impurities in a pattern to form an array of transistors and resistors with the electrical interconnections made by depositing thin layers of gold or aluminium. Many copies of the integrated circuit are formed simultaneously and then have to be broken apart. See SEMICONDUCTOR DEVICE PROCESSING and panel on PRINTED, HYBRID AND INTEGRATED CIRCUITS.

chip (*ImageTech*) Colloq for SOLID-STATE IMAGE SENSOR.

chip & PIN (*ICT*) A system of payment by credit or debit card in which the card-holder keys in a personal identification number to authorize the payment and transactions are verified at the point of sale.

chip-axe (*Build*) A small single-handed tool for chipping timber to rough size.

chipboard (*Build*) A wood-based COMPOSITE MATERIAL made by compression-moulding chips of UREA-FORMALDEHYDE resin-coated waste timber into sheets. May have veneered faces. Also *particle board*.

chip-board (*Paper*) A board, usually made from waste paper, used in box-making.

chip breaker (*Eng*) A groove, step, or similar feature in one face of a (lathe) tool or cutter, designed to break long chips, or swarf, into small pieces which will more easily clear the tool and can be conveniently disposed of.

chip detector (*Aero*) Magnetic device fitted in the lubrication system of aero-engines to collect chips of steel from worn or broken parts.

chip log (*Ships*) Quadranted piece of wood, weighted round its edge, attached to the log line. See NAUTICAL LOG.

chipping (*Eng*) The removing of surface defects from semi-finished metal products by pneumatic chisels. LAPS, ROKES and SEAMS are thus eliminated.

chipping chisel (*Build*) See COLD CHISEL.

chipset (*ICT*) A family of microprocessor and support chips used within a microcomputer, eg Intel 80286, 80386, 80486, etc.

chirality (*Chem*) Absence of symmetry of a molecule with its mirror image, involving reflection or inversion. A molecule is chiral if no single conformation of it may be superimposed on its mirror image. Most chiral organic molecules can be described in terms of so-called chiral centres, in which an atom (usually carbon) has four distinct substituents. See CAHN–INGOLD–PRELOG SYSTEM, ENANTIOMORPHISM.

chirality (*MathSci*) The property of being right-handed or left-handed, usually applied to co-ordinate axes. See HANDEDNESS.

chiropody (*Genrl*) The care and treatment of minor ailments of the feet.

chiropractice (*Med*) A system of healing or alternative medicine, based on manipulation of the spinal column with the intention of restoring normal function to nerves.

Chiroptera (*BioSci*) An order of aerial mammals having the forelimbs specially modified for flight; mainly insect or fruit-eating nocturnal forms. Bats.

chiropterophilous (*BioSci*) Pollinated by bats.

chirp radar (*Radar*) A radar system using linear frequency-modulated pulses.

chisel (*Build*) Steel tool for cutting wood, metal or stone; it consists of a shank whose end is bevelled to a cutting edge.

chi-squared distribution (*MathSci*) The distribution of many quadratic forms in statistics, often encountered as the distribution of the sample variance and of a statistic measuring the agreement of a set of empirically observed frequencies with theoretically derived frequencies. The central chi-squared distribution is indexed by one parameter, the DEGREES OF FREEDOM.

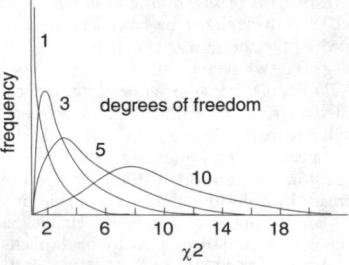

chi-squared distribution

chi-squared test (*MathSci*) A statistical test used to determine the goodness of fit of observed sample data and expected theoretical population. The chi-squared statistic is also used in contingency table testing. Symbol χ^2.

chitin (*BioSci*) A natural nitrogenous polysaccharide with the formula $(C_8H_{13}N_5)_n$ occurring as external skeletal material in many invertebrates, such as insects and arthropods (eg lobster) and also in fungi. Usually occurs in BIOCOMPOSITE form, reinforced by inorganic minerals.

Chladni figure (*Acous*) The visual pattern produced by a fine powder on a vibrating plate. The powder particles accumulate at the nodes where the plate does not move.

Chlamydobacteriales (*BioSci*) Chemosynthetic bacteria-like organisms, probably closely related to the true bacteria. They are filamentous and are characterized by the deposition of ferric hydroxide in or on their sheaths. They occur in fresh water, particularly moorland bogs, where their oxidation of iron (II) to iron (III) results in the deposition of BOG IRON ORE.

chlamydospore (*BioSci*) A hyphal cell that becomes thick-walled, separates from the parent mycelium and functions as a spore.

chloanthite (*Min*) Nickel arsenide occurring in the cubic system. A valuable nickel ore, associated with smaltite. Also *cloanthite*.

chloracne (*Med*) A type of acne caused by exposure to halogenated hydrocarbons by the percutaneous route, ingestion or inhalation. These chemicals are found in industrial oils, herbicides, chemical warfare and contaminated food. The characteristic large comedones (see COMEDO) progress to severe inflammation and scarring.

chloragen (*BioSci*) In Oligochaeta, yellowish flattened cells occurring on the outside of the alimentary canal and concerned with nitrogenous excretion. Also *chloragogen cells*.

chloral (*Chem*) Trichloroethanal. CCl_3CHO, bp 97°C, a viscous liquid, of characteristic odour, obtained by the action of chlorine upon ethanol and subsequent distillation over sulphuric acid.

chloral hydrate (*Chem*) *Trichloroethanal hydrate*. $CCl_3CH(OH)_2$, mp 57°C, bp (with decomposition) 97°C, large colourless crystals, soluble in water, a hypnotic and sedative. Obtained from CHLORAL and water. One of the few compounds having two hydroxy groups attached to the same carbon atom.

chloramines (*Chem*) Compounds obtained by the action of hypochlorite solutions on compounds containing the NH or NH_2 groups. Important as disinfectants. Chloramine T, $CH_3C_6H_4SO_2NClNa$, is the active constituent of an ointment employed as an antidote against *vesicant* war gases.

chloramphenicol (*Pharmacol*) Antibiotic from some species of *Streptomyces*; also obtained synthetically. Of therapeutic value for the treatment of eg typhoid, scrub-typhus, typhus, psittacosis, mycoplasma and chlamydia. Can, rarely, cause aplastic anaemia.

chlorapatite (*Min*) Chlorophosphate of calcium, a member of the apatite group of minerals. See APATITE, FLUORAPATITE.

chlorastrolite (*Min*) A fibrous green variety of *pumpellyite*; it occurs in rounded geodes in basic igneous rocks near Lake Superior. When cut *en cabochon*, it exhibits CHATOYANCY.

chlorates (V) (*Chem*) Salts of chloric acid. Powerful oxidizing agents. Explosive when ground or detonated in contact with organic matter.

chlorazide (*Chem*) N_3Cl; a colourless, highly explosive gas formed by the reaction between sodium chlorate (I) and sodium azide.

chlorazine (*Chem*) *2-Chloro-4,6-bisdiethylamino-1,3,5-triazine*, used as a herbicide.

chlordiazepoxide (*Pharmacol*) A mild tranquillizing drug. TN *Librium*.

Chlorella (*BioSci*) Microscopic unicellular green algae (Chlorophyceae, Chlorococcales) reproducing by autospores, no sexual reproduction; easily grown in laboratory culture and used in biochemical studies.

chlorenchyma (*BioSci*) Tissue composed of cells containing chloroplasts, eg leaf mesophyll.

chlorhexidine (*Med*) An antiseptic and disinfectant used for dressing minor skin wounds or burns.

chloric (V) acid (*Chem*) $HClO_3$. A monobasic acid forming a series of salts, chlorates, where ClO_3 acts as a univalent radical.

chloride of lime (*Chem*) See BLEACHING POWDER.

chloride of silver cell (*ElecEng*) A primary cell having electrodes of zinc and silver and a depolarizer of silver chloride.

chlorides (*Chem*) Salts of hydrochloric acid. Many metals combine directly with chlorine to form chlorides.

chloridizing roasting (*Eng*) The roasting of sulphide ores and concentrates, mixed with sodium chloride, to convert the sulphides to chlorides.

chlorinated hydrocarbons (*Chem*) A major group of compounds used as insecticides. Includes DDT, aldrin, dieldrin. Because of their stability, persistence and toxicity they are major environmental pollutants.

chlorinated polyvinyl chloride (*Chem*) A polymer used in solvent cements. Abbrev CPVC. See SOLVENT WELDING.

chlorinated rubber paints (*Build*) Paints possessing good resistance to chemicals, acids, alkalis, moisture penetration and mould growths. Being derived from rubber they are extremely flexible, but their poor adhesion to bare metal means that a recommended primer must be used.

chlorination (*Build*) The addition of chlorine to sewage as a bactericide in the land treatment method.

chlorination (*Chem*) (1) The substitution or addition of chlorine in organic compounds. (2) The sterilization of water with chlorine, sodium hypochlorite, chloramine or bleaching powder.

chlorine (*Chem*) Element, symbol Cl, at no 17, ram 35·453, valencies 1−, 3+, 5+, 7+, mp −101·6°C, bp −34·6°C. The second halogen, molecular chlorine is a greenish-yellow diatomic gas, with an irritating smell and a destructive effect on the respiratory tract. Its occurrence in sea water is 1·9%, far greater than can be accounted for by the weathering of rocks. Most probably this chlorine comes from volcanic gases (Cl_2 and HCl). In the Earth's crust its abundance is 126 ppm. There are numerous minerals containing the chloride ion, including many in evaporite and volcanic deposits, and in the oxidized zones in metalliferous deposits. In igneous rocks, *chlorapatite* is the most significant chlorine-containing mineral. Produced by the electrolysis of conc brine, or by oxidation of hydrochloric acid. It is a powerful oxidizing agent and is widely used, both pure and as *bleaching powder*, for bleaching and disinfecting. Chlorine is used in the organic chemicals industry to produce tetrachloromethane, trichloromethane, PVC, and many other solvents, plastics, disinfectants and anaesthetics.

chlorine number (*Paper*) An assessment of the bleachability of a pulp by measuring its chlorine demand under specified conditions of test.

chlorine oxides (*Chem*) The (I) *oxide* (Cl_2O) and (IV) *oxide* (ClO_2) are gases similar to chlorine. The (I) oxide (also known as *hypochlorous anhydride*) dissolves in water to give hypochlorous acid. The (IV) oxide is used as a bleach. The (VI) *oxide* (Cl_2O_6) and (VII) *oxide* (Cl_2O_7) are liquids. All are highly unstable, often explosive.

chlorite (*Min*) A group of minerals, typically green, somewhat resembling the micas, composed of hydrated magnesium, iron and aluminosilicates. They occur as alteration products of igneous rocks, chlorite schists and in sediments.

chloritization (*Geol*) The replacement, by alteration, of ferromagnesian minerals by chlorite.

chloritoid (*Min*) Hydrated iron, magnesium, aluminium silicate, crystallizing in the monoclinic or triclinic systems. Characteristic of low- and medium-grade regionally metamorphosed sedimentary rocks.

chloroauric (III) acid (*Chem*) $HAuCl_4$. A complex acid formed when gold (III) oxide (Au_2O_3) dissolves in hydrochloric acid. Forms a series of complex salts called *chloroaurates* (III).

chlorobutadiene (*Chem*) See CHLOROPRENE.

chlorobutyl rubber (*Chem*) Type of butyl rubber used in tyre linings, with enhanced reactivity during VULCANIZATION.

Chlorococcales (*BioSci*) An order of the Chlorophyceae; the members are unicellular, but may form colonies of uninucleate or multinucleate cells, which never divide vegetatively. Asexual reproduction is by zoospores or autospores, and sexual reproduction by biflagellate gametes.

chlorocruorin (*BioSci*) A green respiratory pigment of certain POLYCHAETA. Conjugated protein containing a prosthetic group similar to, but not identical with, reduced haematin.

chlorofibre (*Textiles*) Fibres made from copolymers with vinyl chloride or vinylidene chloride predominating. Other constituent monomers include acrylonitrile. The material has markedly hydrophobic properties.

chlorofluorocarbons (*Aero, EnvSci*) Compounds consisting of ethane or methane with some or all of the hydrogen replaced by fluorine and chlorine. Used as refrigerants, in fire-extinguishers, as aerosol fluids for insecticides, etc, because of their low bp and chemical inertness, and for insulating atmospheres in electrical apparatus because of their high breakdown strengths. Their use is now deprecated because they destroy atmospheric ozone and thus contribute to the GREENHOUSE EFFECT. Abbrev *CFCs*. See panel on ATMOSPHERIC POLLUTION.

chloroform (*Chem*) *Trichloromethane*. $CHCl_3$, bp 62°C, rel.d. 1·49; a colourless liquid of a peculiar ethereal odour, an anaesthetic, solvent for oils, resins, rubber and numerous other substances. It is prepared technically from ethanol and calcium chlorate (I).

chlorohydrins (*Chem*) A group of compounds with −Cl and −OH groups on adjacent carbon atoms, eg *ethylene chlorohydrin* or *1-hydroxy-2-chloroethane*, CH_2ClCH_2OH.

chlorophaeite (*Min*) A mineral closely related to chlorite, dark green when fresh, but rapidly changing to brown, hence the name (Gk *chloros*, yellowish-green; *phaios*, dun). First noted in basic igneous rocks.

chlorophenol red (*Chem*) *Dichlorophenolsulphonphthalein.* An indicator used in acid–alkali titration, having a pH range of 4·8–6·4, over which it changes from yellow to red.

chlorophenols (*Chem*) Class of chlorinated hydrocarbons extensively used as non-agricultural pesticides and biocides, eg as wood preservatives.

Chlorophyceae (*BioSci*) A class of the green algae (CHLOROPHYTA), characterized by the motile cells (if produced) being radially symmetrical with two, four or many flagella inserted apically and with four, cruciate flagellar roots. There is a phycoplast (a set of microtubules oriented parallel to the plane of the new cell wall and involved in wall formation); the life cycle is *haplontic*, the zygote being a resting stage. Such algae live predominantly in fresh water, and include flagellate, coccoid motile and non-motile colonial (COENOBIA), sarcinoid, filamentous and parenchymatous sorts. Examples include *Chlamydomonas, Chlorella, Oedogonium, Volvox.* See CHLOROCOCCALES.

chlorophyll (*BioSci*) Green pigment (variously substituted porphyrin rings with magnesium) involved in photosynthesis. Chlorophyll a is the primary photosynthetic pigment in all those organisms that release oxygen, ie all plants and all algae including the blue-green algae (cyanobacteria). The other chlorophylls are ACCESSORY PIGMENTS; chlorophyll b in vascular plants, bryophytes and green algae; chlorophyll c in brown algae, diatoms, chrysophytes, etc.

Chlorophyta (*BioSci*) The green algae, a division of eukaryotic algae characterized by: chlorophyll a and b; irregularly stacked thylakoids; no chloroplastal endoplasmic reticulum; mitochondrial cristae; flagella not heterokont; storing starch in the chloroplasts. Contains the classes

CHLOROPHYCEAE, Ulvophyceae and Charophyceae, and is obviously ancestral to land plants.

chloroplast (*BioSci*) A *plastid*, one or more in a cell, containing the membranes, pigments and enzymes necessary for photosynthesis in a eukaryotic alga (of any colour) or green plant. A chloroplast in a leaf cell of a typical (C3) vascular plant is an oblate ellipsoid, c.5 × 2 μm bounded by an envelope of two lipoprotein membranes which enclose an aqueous stroma traversed by THYLAKOIDS. The stroma contains enzymes, DNA and, sometimes, granules of starch. The thylakoids (interconnected, flattened, membrane-bound sacs) are organized into GRANA and STROMA LAMELLAE. The thylakoid membranes contain the PHOTO-SYNTHETIC PIGMENTS. The stroma contains the enzymes for the CALVIN CYCLE (panel) and other processes. During photosynthesis, energy from light is used to pump protons from the stroma into the lumen of the thylakoid. This generates both an electrical potential difference (lumen positive) and a pH difference (lumen acid) across the membrane. The resultant 'proton motive force' drives protons back into the stroma through a coupling factor in the membrane, the movement being coupled to the synthesis of ATP from ADP and P_i. NADP is also reduced. The reduced NADP and ATP are used to drive reactions in the stroma, esp the Calvin cycle in which CO_2 is reduced to sugars. See CHEMIOSMOSIS, CHLOROPLAST DNA, PHOTO-PHOSPHORYLATION.

chloroplast DNA (*BioSci*) The chloroplast genome, which is organized like that of prokaryotes and codes for some but not all chloroplast proteins. Genetic recombination within the chloroplast genome has been reported during sexual reproduction of some algae (eg *Chlamydomonas*) and in CYBRIDS. In higher plants, inheritance of chloroplast characters is usually maternal and recombination has not been reported even in the few with biparental inheritance. See RIBULOSE 1,5-BISPHOSPHATE CARBOXYLASE OXYGENASE.

chloroplast ER (*BioSci*) An extension of the ENDOPLASMIC RETICULUM that encloses the chloroplast in some groups of eukaryotic algae, eg the Heterokontophyta.

chloroplatinic (IV) acid (*Chem*) $H_2PtCl_6 \cdot 6H_2O$. Formed when platinum is crystallized from a solution acidified with hydrochloric acid.

chloroprene (*Chem*) 2-*Chlorobutan* 1,2:3,4-*diene*. Starting point for the manufacture of CHLOROPRENE RUBBER.

chloroprene rubber (*Chem*) Chemically resistant elastomer made by polymerizing chloroprene monomer. Also *polychloroprene*. TN Neoprene.

chloroquine (*Pharmacol*) An antimalarial drug, though resistance is now common. Also used for treating rheumatoid arthritis. *Paludrine* is a mixture of chloroquine and proguanil.

chlorosis (*BioSci*) Deficiency of chlorophyll in a normally green part of a plant so that it appears yellow-green, yellow or white, as a result of mineral deficiency, inadequate light or infection.

chlorothiazide (*Pharmacol*) One of the *thiazide* class of diuretics, used in cases of oedema, hypertension and cirrhosis of the liver.

Chloroxone (*Chem*) TN for 2,4-dichlorophenoxyacetic (ethanoic) acid, a selective weedkiller. Also *2,4-D*.

chlorpromazine (*Pharmacol*) A widely used tranquillizing drug, of great value in mental and behavioural disorders, also used as an anti-emetic, analgesic adjuvant and for pre-operational medication.

chlorpropamide (*Pharmacol*) An anti-diabetic drug of the *sulphonylurea* type, for maturity-onset diabetes.

chlortetracycline (*Pharmacol*) A yellow tetracycline antibiotic used to treat a wide variety of bacterial and fungal infections. Proprietary names include *Aureomycin*.

chlorthalidone (*Pharmacol*) *Thiazide*-related diuretic with a long duration of action. Used to treat oedema, hypertension and diabetes insipidus.

choana (*BioSci*) A funnel-shaped aperture; the internal nares of vertebrates. Pl *choanae*

Choanichthyes (*BioSci*) An important, mostly fossil, class of Chordata in which a nostril, connected to the lung, first appeared in evolution. Includes the modern coelacanths and lung fishes (Dipnoi).

choanocyte (*BioSci*) In PORIFERA, a flagellate cell, in which a collar surrounds the base of the flagellum.

Chobert rivet (*Eng*) A hollow rivet, used where only one side of the work is accessible. It comprises a headed tube having a concentric taper of which the smallest diameter is at the shank end. Forcing a wedge-headed mandrel through the bore clinches the rivet by upsetting the shank end on the inaccessible side of the structure. The bore may subsequently be plugged by a parallel sealing pin which also increases the shear strength of the rivet.

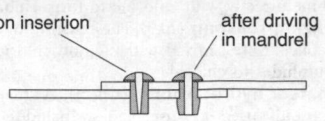

on insertion   after driving in mandrel

Chobert rivet

chock-to-chock (*Aero*) See BLOCK TIME.

chocolate (*FoodSci*) Ground cocoa beans mixed slowly with cocoa butter and sugar often with full-cream milk powder substituted for some of the cocoa and sugar, followed by pressure and temperature treatment to achieve the required flow, setting characteristics and mouthfeel.

chocolate bloom (*FoodSci*) Fat crystals from cocoa butter showing as circular greyish-white marks on the surface of chocolate. Caused usually by incorrect tempering but also by incorrect cooling, storage conditions and, rarely, by micro-organisms. Also *fat bloom*.

chocolate mousse (*MinExt*) Oil (usually crude) and sea-water emulsion often produced following an oil spillage. It floats, is very viscous and contaminates anything it touches.

choice point (*Psych*) The position in a T-maze or other type of maze or in any apparatus involved in discrimination training, when an animal can make only one of two or more alternative responses.

choke (*Autos*) (1) The venturi or throat in the air passage of a CARBURETTOR. (2) A butterfly valve in the carburettor intake which reduces the air supply and so gives a rich mixture for starting purposes. US *strangler*.

choke (*ElecEng*) Colloq for INDUCTOR. Usually refers to coils of high self-inductance used to limit ac flow without power dissipation. Any dc flowing through the choke will be impeded only by the intrinsic resistance of the winding.

choke (*Radar*) In a waveguide, a groove or discontinuity of such a shape and size as to prevent the passage of guided waves within a limited frequency range.

choke coupling (*ElecEng*) (1) Use of the impedance of a choke for coupling the successive stages of a multistage amplifier. (2) In a waveguide, coupling flanges with quarter-wavelength groove which breaks surface continuity at current node.

choke damp (*MinExt*) A term sometimes used for BLACK DAMP (carbon dioxide). More correctly, any mixture of gases which causes choking or suffocation.

choke feed (*ElecEng*) The use of a high-inductance path for the dc component of current to active device in an electrical circuit (eg a transistor). The ac signal is fed through a capacitor so there is separation of ac and dc components.

choke flange (*Radar*) A type of waveguide coupling which obviates the need for metallic contact between the mating flanges, and yet offers no obstruction to the guided waves. One of the waveguide flanges has a slot formed in it of dimensions which prevent energy leakage within the desired frequency range.

choke-jointed (*ElecEng*) See JOINT.

choke line (*MinExt*) See KILL LINE.

cholaemia (*Med*) The presence of bile pigments in the blood. Also *cholemia*.

cholagogue (*Med*) Increasing evacuation of bile; a drug which does this.

cholangitis (*Med*) Inflammation of the bile-duct system. Also *cholangeitis*.

cholecystectomy (*Med*) Excision of the gall bladder.

cholecystenterostomy (*Med*) An artificial opening made between the gall bladder and the upper part of the intestine.

cholecystitis (*Med*) Inflammation of the gall bladder.

cholecystography (*Radiol*) The X-ray investigation of the gall bladder previously filled with a substance opaque to X-rays. A cholecystogram is the X-ray photograph obtained.

cholecystokinin (*Med*) A hormone liberated from the duodenal mucosa, causing contraction of the gall bladder and relaxation of the SPHINCTER OF ODDI.

cholecystostomy (*Med*) The surgical formation of an opening in the wall of the gall bladder.

choledochotomy (*Med*) Incision into one of the main bile ducts.

cholelithiasis (*Med*) Stones in the gall bladder and bile passages.

cholemia (*Med*) See CHOLAEMIA.

cholera (*Med*) An acute bacterial infection by *Vibrio cholerae*; characterized by severe vomiting and diarrhoea, drying of the tissues and painful cramps; spread by infected food and water.

cholera toxin (*BioSci*) A multimeric (AB) protein toxin from *Cholera vibrio* that acts on intestinal cells to cause massive fluid loss and thus diarrhoea.

cholesteatoma (*Med*) A tumour in the brain, or in the middle ear, composed of cells and crystals of cholesterol.

cholesteraemia (*Med*) Also *cholesteremia*. See CHOLESTER-OLAEMIA.

cholesterol (*Chem*, *FoodSci*) $C_{27}H_{45}OH$, a sterol of the alicyclic series, found in nerve tissues, gall stones, and in other tissues of the body. It is a white crystalline solid, mp148·5°C, soluble in organic solvents and fats. There are numerous stereoisomers known. Parent compound for many steroids. Present in meat, fish, poultry and eggs, but only in trace quantities in vegetable fats. Excessive cholesterol in the diet is thought to increase the incidence of heart disease.

cholesterolaemia (*Med*) Excess of cholesterol in the blood. Also *cholesterolemia*.

cholesterosis (*Med*) Diffuse deposits of cholesterol in the lining membranes of the gall bladder.

cholic acid (*Chem*) $C_{23}H_{39}O_3COOH$, the product of hydrolysis of certain bile acids, is conjugated in the body, forming glycocholic acid with glycine and taurocholic acid with taurine.

choline (*Chem*, *BioSci*) *Ethylol-trimethyl-ammonia hydrate*. $OHCH_2CH_2NMe_3OH$, a strong base, present in the bile, brain, yolk of egg, etc, combined with fatty acids or with glyceryl-phosphoric acid (LECITHIN). It is concerned in regulating the deposition of fat in the liver, and its acetyl ester is an important neurotransmitter. See ACETYLCHO-LINE.

cholinergic (*Pharmacol*) Activated by ACETYLCHOLINE, eg transmission at preganglionic synapses of the autonomic system and postganglionic endings of the parasympathetic and some sympathetic nerve fibres. Anti-cholinergic drugs act to block such synapses.

choluria (*Med*) Bile pigments in the urine.

chomophyte (*BioSci*) A plant growing on rock ledges littered with detritus, or in fissures and crevices where root hold is obtainable.

chondr-, chondrio-, chondro- (*Genrl*) Prefixes from Gk *chondros*, cartilage, grain.

chondral (*BioSci*) Pertaining to cartilage.

Chondrichthyes (*BioSci*) A class of cartilaginous fishes in which the skeleton may be calcified but not ossified. The teeth are not fused to the jaw but are serially replaced; fertilization is internal. Rays, dogfish, sharks and chimeras. Cf OSTEICHTHYES. See panel on VERTEBRATE EVOLUTION.

chondrification (*BioSci*) Strictly, the formation of chondrin; hence, the development of cartilage. Also *chondrogenesis*.

chondrin (*BioSci*) Substance obtained by boiling cartilage in water. Probably a degraded form of CHONDROITIN sulphate.

chondrio- (*Genrl*) See CHONDR-.

chondrite (*Min*) A type of stony meteorite containing *chondrules*, nodule-like aggregates of minerals.

chondro- (*Genrl*) See CHONDR-.

chondroblast (*BioSci*) A cell that secretes the extracellular matrix components of CARTILAGE.

chondroclast (*BioSci*) A cartilage cell that destroys the CARTILAGE matrix.

chondrocranium (*BioSci*) The primary cranium of CRA-NIATA, formed by the fusion of the parachordals, auditory capsules and trabeculae.

chondrodite (*Min*) Hydrated magnesium silicate, crystal-lizing in the monoclinic system, and occurring in metamorphosed limestones.

chondroids (*Vet*) Compact lumps of dried pus commonly found in the exudate of inflamed guttural pouches of the horse.

chondroitin (*BioSci*) Natural linear polysaccharide, usually sulphated, that binds with proteins to form proteoglycans. Found in the cornea of the eye as well as structural tissue like cartilage. Derived biomaterial, chondroitin sulphate. See HYALURONIC ACID.

chondroma (*Med*) A tumour composed of cartilage cells.

chondrosamine (*Chem*) *2-aminogalactose*. $C_6H_{13}O_5N$. It is the basis of *chondroitin*, which is the substance of cartilage and similar body tissues.

chondrosarcoma (*Med*) A malignant tumour composed of sarcoma cells and cartilage.

chondroskeleton (*BioSci*) The cartilaginous part of the vertebrate skeleton.

chondrule (*Min*) See CHONDRITE.

chop (*Build*) Movable jaw of a bench vice.

chopped wave (*ElecEng*) A travelling voltage wave which rises to a maximum and then rapidly falls to zero. Such a wave occurs on transmission lines when an ordinary voltage wave has caused an insulator flashover, thereby losing its tail.

chopper (*ElecEng*) (1) Electronic inverter circuit which converts dc to ac. It operates by rapidly switching off and on (chopping) a direct voltage source to produce a square or rectangular waveform which varies between two discrete values, zero and the direct supply level. (2) Light interrupter used to produce ac output from a photocell.

chopper (*NucEng*) A device consisting of a rotating mechanical shutter made of a sandwich of aluminium and cadmium sheets which provides bursts of neutrons for a TIME-OF-FLIGHT SPECTROMETER. See NEUTRON SPEC-TROMETER, NEUTRON VELOCITY SELECTOR.

chopper amplifier (*ElecEng*) See CHOPPER-STABILIZED AMPLIFIER.

chopper disk (*ElecEng*) Rotating toothed or perforated disk which interrupts the light signal falling on a photocell at a desired frequency.

chopper-stabilized amplifier (*ElecEng*) Low-drift dc amplifier in which a semiconductor switch is used to convert a dc or low-frequency ac signal into an ac square wave which can be more readily amplified. A chopper circuit, operating in synchronism with the ac, may be used at the amplifier output as a rectifier. An equivalent input drift voltage of around 0·1 μV per °C can be obtained.

chopping (*NucEng*) Cutting spent fuel elements into lengths suitable for passing into a dissolver cell for extracting uranium and plutonium. Also *shearing*.

chord (*Aero*) See CHORD LINE.

chord (*MathSci*) A straight line drawn between two points on a curve.

chorda (*BioSci*) (1) Any string-like structure, eg the *chordae tendineae*, tendinous chords attaching the valves of the heart. (2) The NOTOCHORD.

chordacentra (*BioSci*) Vertebral centra formed from the notochordal sheaths. Adj *chordacentrous*. Cf ARCHECENTRA.

chordal thickness (*Eng*) Of a gearwheel tooth, the length of the chord subtended by the tooth thickness arc.

Chordata (*BioSci*) A phylum of the METAZOA containing those animals possessing a notochord. See GRAPTOLITES.

chord line (*Aero*) A straight line joining the centres of curvature of the leading and trailing edges of any aerofoil section. The chord (distance) is that between leading and trailing edges measured along this line.

chord of contact (*MathSci*) Of the conic: the line joining the points of contact of the tangents from an external point.

chordoma (*Med*) An invasive tumour arising from the remains of the notochord in the skull and the spinal column.

chordotomy (*Med*) The cutting of the nerve fibres in the spinal cord conveying the sensation of pain; done for the relief of severe pain.

chordotonal organs (*BioSci*) In insects, sense organs consisting of bundles of SCOLOPHORES, sensitive to pressure, vibrations and sound.

chorea (*Med*) Involuntary repetitive jerky movements of the body, particularly limbs and face. A manifestation of a number of neurological diseases including Sydenham's chorea (or *Saint Vitus's dance*) which occurs in the course of acute rheumatic fever and *Huntington's chorea*.

choria (*Psych*) A neurological disorder, characterized by spasmic, jerky and involuntary movements, particularly of the face, tongue, hands and arms.

choriocarcinoma (*Med*) A malignant tumour of the uterus composed of cells derived from the fetal chorion; it appears during or after pregnancy. Also *chorio-epithelioma, hydatidiform mole*.

chorion (*BioSci*) In higher vertebrates, one of the fetal membranes, being the outer layer of the amniotic fold; in insects, the hardened eggshell lying outside the vitelline membrane.

chorionic villus sampling (*Med*) A method for diagnosing human fetal abnormalities in the sixth to tenth week of gestation, in which small pieces of fetally derived chorionic villi are removed through the cervix; chromosomes in the tissue are examined for abnormalities, and DNA may be extracted and probed for disease-associated alleles. Abbrev *CVS*.

chorioptic mange (*Vet*) A mange of horses and cattle due to mites of the genus *Chorioptes*. In horses, known as foot mange or itchy leg.

chorioretinitis (*Med*) Inflammation of the choroid and retina.

Chorleywood bread process (*FoodSci*) The standard method in most plant bakeries in which bread dough is developed by the high-energy input from high-speed mixing. The dough mix needs chemical improvers (eg ascorbic acid) and added fat (up to 0·7% of flour weight) but bulk fermentation is eliminated, flour protein content is less critical, total process time reduced and yield improved.

chorography (*Genrl*) The process or technique of mapping or describing particular districts or regions.

choroid (*BioSci*) The vascular tunic of the vertebrate eye, lying between the retina and the sclera. Adj *choroidal*.

choroiditis (*Med*) Inflammation of the choroid of the eye.

choroid plexus (*BioSci*) In higher CRANIATA, the thickened, vascularized regions of the PIA MATER in immediate contact with the thin epithelial roofs of the DIENCEPHALON and MEDULLA.

choropleth map (*Genrl*) A map in which areas sharing similar geographical, climatic, etc, features are represented by the same colour.

CHP (*EnvSci*) See COMBINED HEAT AND POWER.

CHR (*Chem*) Abbrev for EPICHLORHYDRIN RUBBER.

Christmas tree (*MinExt*) Casinghead assembly which controls oil flow. Also *Xmas tree*. See BLOWOUT PREVENTER.

Christoffel symbols (*MathSci*) The Christoffel symbols of the first kind are defined as

$$\begin{bmatrix} r \\ pq \end{bmatrix} = \frac{1}{2}\left(\frac{\partial g_{pr}}{\partial x^q} + \frac{\partial g_{qr}}{\partial x^p} - \frac{\partial g_{pq}}{\partial x^r}\right)$$

Also denoted by

$$\begin{bmatrix} pq \\ r \end{bmatrix}, [pq, r] C_{pq}^r \text{ and by } \Gamma_{pqr}$$

The Christoffel symbols of the second kind are defined as

$$\left\{\begin{matrix} r \\ pq \end{matrix}\right\} = g^{\alpha r}\begin{bmatrix} \alpha \\ pq \end{bmatrix}$$

Also denoted by

$$\left\{\begin{matrix} pq \\ r \end{matrix}\right\} \text{ and by } \Gamma_{pq}^r$$

g_{pq} and g^{pq} are components of the covariant and contravariant fundamental metric tensors.

chrom-, chromo-, chromat-, chromato- (*Genrl*) Prefixes from Gk *chroma, chromatos*, colour.

chroma (*ImageTech*) See CHROMINANCE.

chroma (*Phys*) In the Munsell colour system a term to indicate degree of saturation; zero represents neutral grey, and, depending on the hue, the numbers 10–18 represent complete saturation.

chroma control (*ImageTech*) Abbrev for *chromaticity control*. Control which adjusts colours in colour TV picture.

chroma-delay (*ImageTech*) A picture defect in which colours smear, caused by timing errors.

chromadizing (*Eng*) Treating aluminium or its alloys with chromic acid to improve paint adhesion.

chromaffinoma (*Med*) A benign or malignant tumour of chromaffin cells usually of the adrenal medulla. Also *phaeochromocytoma*.

chroma-key (*ImageTech*) Video technique for the combination of images from two or more sources. See CSO.

chromat- (*Genrl*) See CHROM-.

chromates (VI) (*Chem*) The salts corresponding to chromium (VI) oxide. The normal chromates, M_2CrO_4, are generally yellow, while the dichromates (VI), $M_2Cr_2O_7$, are generally orange-red. Chromates are used as pigments and in tanning.

chromate treatment (*Eng*) Treating a metal with a hexavalent chromium compound to produce a *conversion coating*, so altering its surface properties.

chromatic aberration (*ImageTech, Phys*) An enlargement of the focal spot caused (1) in a cathode-ray tube, by the differences in the electron velocity distribution through the beam and (2) in an optical lens system using white light, by the refractive index of the glass varying with the wavelength of the light, resulting in coloured fringes surrounding the image.

chromatic adaptation (*BioSci*) Differences in amount or proportion of photosynthetic pigments in response to the amount and colour of light. It can be (1) phenotypic as in many Cyanophyceae or (2) constitutive as in the distribution of littoral algae (red lowest, green at top, brown between).

chromatic colour (*Phys*) A colour which is not grey.

chromatic dispersion (*ICT*) An impairment of OPTICAL FIBRE systems in which light of different frequencies travels at different speeds, causing smearing of data pulses.

chromatic function (*MathSci*) A function defined by the number of ways of COLOURING OF A GRAPH with k colours; it is always a polynomial in k.

chromaticity (*ImageTech, Phys*) Colour quality of light, as defined by its chromaticity co-ordinates, or alternatively by its purity (saturation) and dominant wavelength.

chromaticity diagram (*ImageTech*) Plane diagram in which one of the three chromaticity co-ordinates is plotted against another. Generally applied to the CIE (x, y) diagram in rectangular co-ordinates for colour TV.

chromatic number (*MathSci*) Of a graph, the smallest number k such that the graph may be k-coloured. See COLOURING OF A GRAPH.

chromatics (*Phys*) The science of colours as affected by phenomena determined by their differing wavelengths.

chromatid (*BioSci*) One of the two, thread-like structures joined at the CENTROMERE, which constitute a single METAPHASE chromosome, each containing a single, double-helical DNA molecule with associated protein.

chromatin (*BioSci*) A network of more or less de-condensed DNA, and associated proteins and RNA, in the interphase nucleus forming higher ordered structures in the nucleus of EUKARYOTES.

chromatin bead (*BioSci*) See NUCLEOSOME.

chromato- (*Genrl*) See CHROM-.

chromatography (*Chem*) A method of separating (often complex) mixtures. *Adsorption chromatography* depends on using solid adsorbents which have specific affinities for the adsolved substances. The mixture is introduced onto a column of the adsorbent, eg alumina, and the components eluted with a solvent or series of solvents and detected by physical or chemical methods. *Partition chromatography* applies the principle of COUNTERCURRENT DISTRIBUTION to columns and involves the use of two immiscible solvent systems: one solvent system, the *stationary phase*, is supported on a suitable medium in a column and the mixture introduced in this system at the top of the column; the components are eluted by the other system, the *mobile phase*. See GAS–LIQUID CHROMATOGRAPHY, GEL-PERMEATION CHROMATOGRAPHY, PAPER CHROMATOGRAPHY.

chromatophore (*BioSci*) A cell that contains pigment granules and that may change its shape and colour effect on nervous or hormonal stimulation.

chrome alum (*Chem*) *Potassium chromium sulphate.* $K_2SO_4Cr_2(SO_4)_3 \cdot 24H_2O$, purple octahedral crystals obtained by the reduction of potassium dichromate (VI) solution acidified with sulphuric acid.

chrome brick (*Eng*) Brick incorporating chromite, used as a refractory lining in steel-making furnaces.

chrome iron ore (*Min*) See CHROMITE.

chromel (*Eng*) An alloy used in heating elements, based on nickel with about 10% chromium.

chrome spinel (*Min*) Another name for the mineral *picotite*, a member of the SPINEL group. Cf CHROMITE.

chromic (VI) acid (*Chem*) H_2CrO_4. An aqueous solution of chromic oxide. Often applied as a solution of sodium dichromate in sulphuric acid, used for cleaning glassware and etching plastics etc.

chromic oxide (*Chem*) Chromium (III) oxide, Cr_2O_3, an amphoteric oxide corresponding to chromium (III) salts, CrX_3, and to chromites (chromates (III)), $M_2Cr_2O_4$.

chromic salts (*Chem*) Salts in which chromium is in the (III) oxidation state. They are usually green or violet, and are not readily oxidized or reduced.

chrominance (*ImageTech*) (1) Colorimetric difference between any colour and a reference colour of equal luminance and specified chromaticity. The reference colour is generally a specified white, eg CIE illuminant C for artificial daylight. (2) In TV, the information which defines the colour (hue and saturation) of the image, as distinct from its brightness or LUMINANCE. Also *chroma*.

chrominance channel (*ImageTech*) Circuit carrying the chrominance signal in a colour TV system.

chrominance signal (*ImageTech*) See CHROMINANCE (2).

chrominance subcarrier (*ImageTech*) A subcarrier frequency in a TV signal which is modulated with the chrominance information.

chromite (*Min*) A double oxide of chromium and iron (generally also containing magnesium), used as a source of chromium and as a refractory for resisting high temperatures. It occurs as an accessory in some basic and ultrabasic rocks, and crystallizes in the cubic system as lustrous grey-black octahedra; also massive. Also *chrome iron ore*.

chromium (*Chem*) Metallic element. Symbol Cr, at no 24, ram 51·996, rel.d. at 20°C, 7·138, mp 1900°C. The abundance in the Earth's crust is 122 ppm. The only ore is *chromite*, $FeCr_2O_4$, from which the element is extracted mainly for alloying with nickel in heat-resisting alloys and with iron or iron and nickel in stainless and heat-resisting steels. Also used as a corrosion-resistant plating.

chromium (VI) oxide (*Chem*) CrO_3. Produced by the action of sulphuric acid on a concentrated solution of potassium dichromate (VI). It is deliquescent and a powerful oxidizing agent. See CHROMIC (VI) ACID.

chromo- (*Genrl*) See CHROM-.

chromoblast (*BioSci*) An embryonic cell that will develop into a CHROMATOPHORE.

chromocentre (*BioSci*) Mass of localized interphase CHROMATIN; usually refers to the fused, centromeric regions of dipteran salivary gland polytene chromosomes (see POLYTENY).

chromogen (*Chem*) A coloured compound containing a CHROMOPHORE.

chromolithography (*Print*) The preparation of lithographic printing surfaces, whether for single- or multicolour printing, by hand methods only, without a process camera as in PHOTOLITHOGRAPHY. For multicoloured printing, an outline key drawing is made and transferred to the required number of stones or plates, on each of which the litho artist draws those portions which are required by the colour to be used when printing. See AUTOLITHOGRAPHY, LITHOGRAPHY.

chromomere (*BioSci*) One of the characteristic granules of compacted chromatin in serial array on a metaphase chromosome.

chromonema (*BioSci*) A complete single thread of chromatin.

chromo-optometer (*Phys*) An optometer in which the chromatic aberration of the eye is used to determine its refraction.

chromo paper (*Paper*) Paper which is more heavily coated than art paper; used for chromolithography.

chromophil (*BioSci*) Staining heavily in certain histochemical techniques. Also *chromophilic*.

chromophobe (*BioSci*) Resisting stains, or staining with difficulty, in certain histochemical techniques. Also *chromophobic*.

chromophores (*Chem*) Characteristic groups which are responsible for the colour of dyestuffs. Such groups include: $C = C$, $C = O$, $C = N$, $N = N$, $O \leftarrow N_2O = O$ and others.

chromoplast (*BioSci*) (1) A PLASTID containing a pigment, esp a yellow or orange plastid containing carotenoids. (2) A photosynthetic plastid, now usually called a CHLOROPLAST, of a non-green alga.

chromosomal aberration (*BioSci*) Any visible abnormality in chromosome number or structure, including TRISOMY and TRANSLOCATIONS.

chromosomal chimera (*BioSci*) That in which there are differences in number or morphology of the chromosomes.

chromosomally enriched DNA library (*BioSci*) DNA LIBRARY made from DNA enriched for a specific chromosome; FLOW-SORTED chromosomes are a source.

chromosome (*BioSci*) In eukaryotes the deeply staining rod-like structures seen in the nucleus at cell division. See panel on CHROMOSOME.

chromosome arm (*BioSci*) Part of a chromosome from the CENTROMERE to the end.

Chromosome

In eukaryotes, the deeply staining rod-like structures seen in the nucleus at cell division. Composed of a continuous thread of DNA which with its associated proteins (mainly *histones*) forms higher-order structures called *nucleosomes* and has special regions, *centromere* and *telomere* (see SEX DETERMINATION). Normally constant in number for any species, there are 22 pairs of chromosomes and two sex chromosomes in the human. In micro-organisms the DNA is not associated with histones and does not form visible condensed structures.

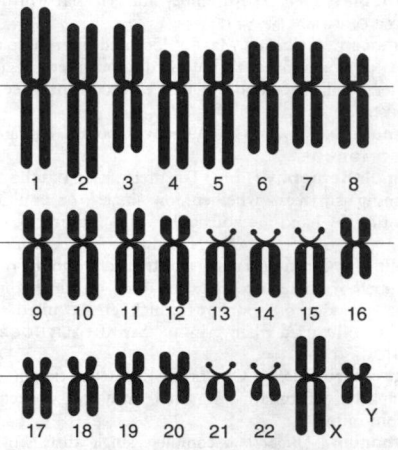

Fig. 1 **Stylized human chromosomes**.

Centromeres are seen in the metaphase chromosome as a narrowed region flanked by the kinetochores to which the microtubules of the spindle are attached. A chromosome arm or chromatid is the part of a chromosome from the centromere to its end at the telomere. Telomeres contain a small number of certain short DNA sequences which have been shown to be progressively lost as the nucleus divides during life, thus providing a mechanism for limiting the longevity of most cells.

Besides carrying genetic information, chromosomes provide the structure for an important method of genetic reassortment or recombination called *crossing over*. This is an exchange of segments of homologous chromosomes primarily during meiosis whereby linked genes become recombined (see Fig. 2). A crossover is the product of such an exchange. The *crossover frequency* is the proportion of gametes bearing a crossover between two specified gene loci. It ranges from 0% for allelic genes to 50% for genes so far apart that there is always a crossover between them. The *crossover site* is the place in the chromosome where breakage and reunion of DNA strands occur during recombination. See CENTIMORGAN, CHIASMA, MEIOSIS.

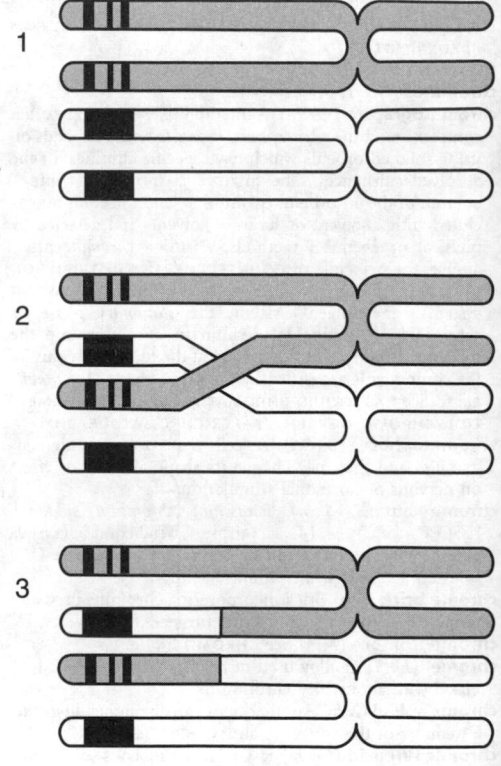

Fig. 2 **Stages in crossover formation**.

chromosome banding (*BioSci*) See BANDING TECHNIQUES and POLYTENE CHROMOSOMES.

chromosome complement (*BioSci*) The set of chromosomes characteristic of the nuclei of any one species of plant or animal. CHROMOSOME SET refers to the haploid set.

chromosome cores (*BioSci*) Non-histone protein network left when HISTONES, DNA and RNA are removed from mammalian metaphase chromosomes.

chromosome mapping (*BioSci*) Assigning GENES or BANDS to specific regions of the haploid chromosome complement. Each chromosome is numbered in order of size, short arms being designated p, long arms q. The bands are numbered consecutively outwards along each arm with the regions within a band also numbered. Thus 11 p2·3 refers to the third region of the second band on the short arm of chromosome 11. The ends are designated *ter*, as in 3 pter, the end of the short arm of chromosome 3.

chromosome-mediated gene transfer (*BioSci*) Transfer of genetic material into a cell by introducing foreign chromosomes. Techniques include CALCIUM PHOSPHATE PRECIPITATION and ELECTROPORATION.

chromosome set (*BioSci*) The whole of the chromosomes present in the nucleus of a gamete, usually consisting of one each of the several kinds that may be present.

Chromosome (Cont.)

Crossing over also occurs in certain somatic cells such as lymphocytes, where it provides one mechanism for the generation of variability in antigen receptor sites.

Identification of chromosomes

Chromosomes differ in shape and size but the detailed identification of parts of chromosomes has been made much easier by the development of special staining or banding techniques to produce patterns of bands characteristic of an individual chromosome (see Fig. 3). All involve staining fixed metaphase chromosomes and fall into five main groups:

(1) Q-banding, staining with quinacrine mustard or 33 258 Hoechst.

(2) G-banding, removal of some protein followed by staining with Giemsa.

(3) R-banding, heat or alkali treatment followed by Giemsa or acridine orange staining (gives reverse pattern of G-banding).

(4) T-banding, variant of R-banding which mainly stains the telomeric regions.

(5) C-banding, treatment which mainly stains constitutive heterochromatin.

In mammals, G- and Q-banding give generally similar patterns, but in plants Q- and C-bands appear to be more closely related.

The bands appear to be made by distinctive interactions between specific types of DNA sequence and proteins. Factors such as base and sequence composition, eg blocks of short repeats, are involved. Polytene chromosomes also exhibit characteristic bands, with or without staining. They are much more numerous than those of metaphase chromosomes and the more stretched state of the polytene chromosomes would be expected to show a greater number of bands. The two kinds of banding patterns may therefore be related. As shown in Fig. 3, the combination of size, shape and banding pattern gives each chromosome a unique appearance, and it is this which has made eukaryotic gene mapping possible.

See panel on DNA AND THE GENETIC CODE.

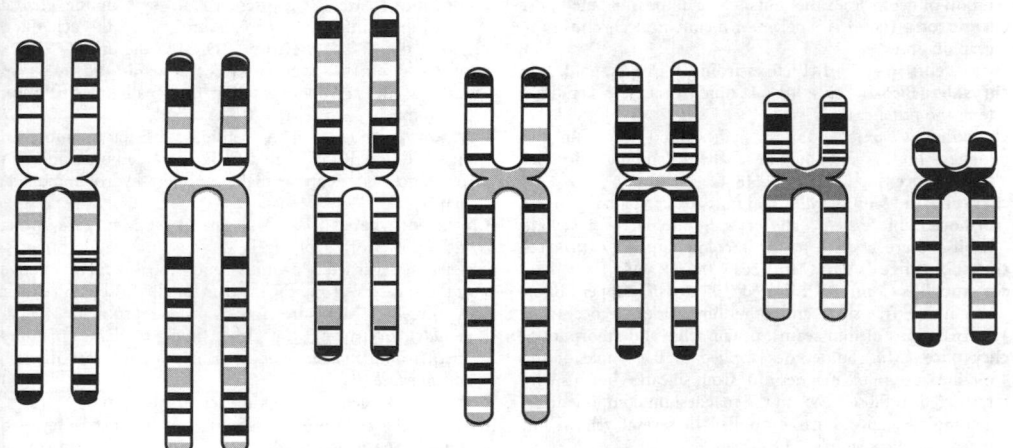

Fig. 3 **Stylized diagram of banded metaphase chromosomes.**

chromosome sorting (*BioSci*) The ability to sort chromosomes by DNA content or size. See FLUORESCENCE-ACTIVATED CELL SORTER.

chromosphere (*Astron*) Part of the outer gaseous layers of the Sun, visible as a thin crescent of pinkish light in the few seconds of a total eclipse of the Sun. Located above the photosphere and below the corona, its temperature rises from a minimum of around 4000 K up to around 50 000 K where it merges with the corona. See panel on SUN AS A STAR.

chromous salts (*Chem*) Salts of chromium in the (II) oxidation state; they yield blue solutions with water and are strong reducing agents.

chromyl chloride (*Chem*) CrO_2Cl_2, a strongly oxidizing red liquid, bp 117°C.

chron (*Geol*) The time span of a CHRONOZONE. See POLARITY CHRON.

chron- (*Genrl*) Prefix from Gk *chronos*, time.

chronaxie (*Med*) A time constant in nervous excitation, equal to the smallest time required for excitation of a nerve when the stimulus is an electrical current of twice the threshold intensity required for excitation when the stimulus is indefinitely prolonged.

chronic (*Med*) Said of a disease which is deep-seated or long-continued. Cf ACUTE.

chronic fatigue syndrome (*Med*) An illness characterized by otherwise unexplained fatigue significantly interfering with normal life. Sometimes associated with transient joint and muscle pains, and various neuropsychological complaints. A complex and controversial condition of unclear incidence and prevalence. Abbrev *CFS*. Also *myalgic encephalomyelitis (ME)*.

chronic granulomatous disease (*Med*) Inherited disease of male children characterized by recurrent abscesses and granuloma formation. The neutrophil leucocytes are deficient in a cytochrome component necessary for generating active oxygen free radicals for intracellular killing of microbes, even though these are ingested normally.

chronic respiratory disease of fowl (*Vet*) An infection of the respiratory tract by organisms of the genus *Mycoplasma*. Abbrev *CRD*.

chronograph (*Genrl*) (1) A watch with a centre seconds hand which can be caused to start, stop and fly back to zero by pressing the button or a push-piece on the side of the case. The chronograph mechanism is independent of the going train, so that the balance is not stopped when the seconds hand is stopped. See STOP WATCH. (2) Any type of mechanism which gives a record of time intervals.

chronometer (*Genrl*) A precision timekeeper. In UK and US the term denotes the very accurate timekeeper kept on board ship for navigational purposes, and fitted with the spring detent mechanism. On the Continent the term is also applied to any very accurate clock or watch which may be fitted with the spring detent or lever escapement.

chronoscope (*Genrl*) Electronic instrument for precision measurement of very short time intervals.

chronostratigraphy (*Geol*) The standard hierarchical definition of geological time units, using all possible methods.

chronozone (*Geol*) A small non-hierarchical chronostratigraphic unit.

chrys-, chryso- (*Genrl*) Prefixes from Gk *chrysos*, gold.

chrysalis (*BioSci*) The PUPA of some insects, esp Lepidoptera; the pupa case.

chrysene (*Chem*) Colourless hydrocarbon found in the highest boiling fractions of coaltar. Exhibits red-violet fluorescence; mp 254°C, bp 448°C.

chrysoberyl (*Min*) Beryllium aluminate, crystallizing in the orthorhombic system. The crystals often have a stellate habit and are green to yellow in colour. See ALEXANDRITE.

chrysoberyl cat's eye (*Min*) See CYMOPHANE.

chrysocolla (*Min*) A hydrated silicate of copper, often containing free silica and other impurities. It occurs in incrustations or thin seams, usually blue and amorphous.

chrysolite (*Min*) Sometimes applied to the whole of the olivine group of magnesium iron silicates but usually restricted to those richer in the magnesium component. In gemmology, an old name applied to several yellow and greenish-yellow stones.

Chrysophyceae (*BioSci*) The golden-brown algae, a class of eukaryotic algae in the division Heterokontophyta comprising c.800 spp. May be naked, scaly (silica) or walled. A diversity of forms occur, including flagellated unicellular, colonial, simple and branched filamentous and a few thalloid types. Mostly found in fresh water and may be prototrophs or heterotrophs.

chrysoprase (*Min*) An apple-green variety of chalcedony; the pigmentation is probably due to the oxide of nickel.

chrysotherapy (*Pharmacol*) Treatment by injections of gold.

chrysotile (*Min*) A fibrous variety of serpentine once forming the most valuable type of the asbestos of commerce but now little used because of its carcinogenicity. Also *Canadian asbestos*.

CHU, Chu (*Genrl*) Abbrevs for CENTIGRADE HEAT UNIT.

chuck (*Eng*) A device attached to the spindle of a machine tool for gripping the revolving work, cutting tool or drill.

chucking machine (*Eng*) A machine tool in which the work is held and driven by a chuck, not supported on centres.

chuffs (*Build*) Bricks rendered useless owing to cracks caused by rain when hot. Also *shuffs*.

chunking (*Psych*) A process of reorganizing materials in memory to reduce the number of items, a form of encoding; eg the four-chunk number 1–2–1–2 can be reduced to the two-chunk 12–12.

churn drill (*MinExt*) See CABLE TOOL DRILLING.

churning loss (*Autos*) In a gearbox, the power wasted in fluid friction through the pumping action of the revolving gears in the oil.

chute (*MinExt*) (1) An inclined trough for the transference of broken coal or ore. (2) An area of rich ore in an inclined vein or lode, generally of much greater vertical than lateral extent.

chute riffler (*PowderTech*) A sampling device consisting of a V-shaped trough from which a series of chutes feed two receiving bins. Alternate chutes feed opposite bins. Particles have in theory an equal chance of being retained in either of the bins.

Chvostek's sign (*Med*) Twitching of the muscles of the face on tapping the facial nerve; a sign of tetany.

chyle (*BioSci*) In vertebrates, lymph containing the results of the digestive processes, and having a milky appearance due to the presence of emulsified fats and oils (*chylomicrons*).

chylification (*BioSci*) Formation of CHYLE. Also *chylifaction*.

chylomicrons (*BioSci, Med*) Minute particles of emulsified fat present in the blood plasma, particularly after digestion of a fatty meal; the plasma lipoprotein with the lowest density. They are the form in which dietary lipids from the intestine are transported to the liver and adipose tissue.

chyloperitoneum (*Med*) The presence of chyle in the peritoneal cavity as a result of obstruction of the abdominal lymphatics.

chylothorax (*Med*) The presence of chyle in the pleural cavity, due to injury to, or pressure on, the thoracic duct.

chyluria (*Med*) The presence of chyle in the urine.

chyme (*BioSci*) In vertebrates, the semi-fluid mass of partially digested food entering the small intestine from the stomach.

chymotrypsin (*BioSci*) A peptidase of the mammalian digestive system that is specific for peptide bonds adjacent to amino acids with aromatic or bulky hydrophobic side chains.

Chytridiomycetes (*BioSci*) A class of the Mastigomycotina with posteriorly uniflagellate zoospores and gametes, generally regarded as primitive and probably representative of the ancestral group for the true fungi. May be unicellular or mycelial. Most are aquatic and saprophytic (*water moulds*), or parasitic on algae, fungi or plants. Includes *Synchytrium endobioticum*, causing potato wart disease, and *Blastocladia*.

Ci (*Phys*) Obsolete unit of radioactivity. See CURIE.

cicatrix (*BioSci, Med*) The scar left after the healing of a wound; one which marks the previous attachment of an organ or structure, particularly in plants.

cicero (*Print*) See DIDOT POINT SYSTEM.

ciclosporin (*Pharmacol*) Immunosuppressive cyclic peptide drug used to prevent transplant rejection. It acts selectively on the production of helper T-cells but can cause renal damage. Formerly *cyclosporin*.

Ciconiiformes (*BioSci*) An order of birds having a DESMOGNATHOUS palate and usually webbed feet. All are long-legged birds of aquatic habit, living mainly in marshes and nesting in colonies. They are powerful flyers and some migrate over long distances. Storks, herons, ibises, spoonbills and flamingoes.

CIE (*Phys*) Abbrev for *Commission Internationale d'Éclairage*. Formed to study problems of illumination.

CIE co-ordinates (*Phys*) Set of colour co-ordinates specifying proportions of three theoretical additive primary colours required to produce any hue. These theoretical primaries were established by the CIE and form the basis of all comparative colour measurement. See CHROMATICITY DIAGRAM.

CI engine (*Eng*) See COMPRESSION–IGNITION ENGINE.

ciguatera (*Med*) A type of food poisoning caused by eating seafood containing natural *ciguatoxins*.

CIIR (*Chem*) Abbrev for *chlorinated isoprene isobutene rubber*. See CHLOROBUTYL RUBBER.

cilia (*BioSci*) (1) Fine hair-like protrusions of the cell surface that beat in unison to create currents of liquid over the cell surface or propel the cell through the medium. Each cilium has a complex and characteristic internal structure built around nine peripheral pairs and one central pair of microtubules. (2) In mammals, the eyelashes; in birds, the barbicels of a feather. Adjs *ciliated, ciliate*.

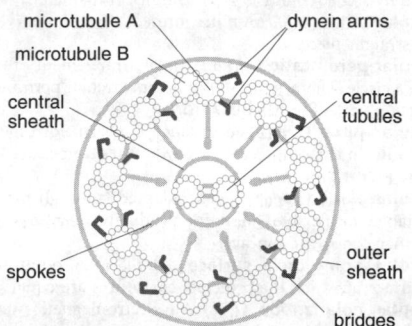

cilia Transverse section showing main parts.

ciliary (*BioSci*) In general, pertaining to or resembling cilia; in vertebrates, used of certain structures in connection with the eye, as the *ciliary ganglion, ciliary muscles, ciliary process*.

ciliate (*BioSci*) (1) Having a fringe of long hairs on the margin. (2) Having flagella. Also *ciliated*.

ciliograde (*BioSci*) Moving by the agency of cilia.

Ciliophora (*BioSci*) A class of Protozoa, comprising forms that always possess cilia at some stage of the life cycle, and usually have a MEGANUCLEUS.

ciliospore (*BioSci*) (1) In Protozoa, a ciliated swarm of spores. (2) In Suctoria, a bud produced by asexual reproduction.

cilium (*BioSci*) Sing of CILIA.

Ciment Fondu (*CivEng*) TN for a type of very-rapid-hardening cement made by heating lime and alumina in an electric furnace to incipient fusion, and afterwards grinding to powder.

cimetidine (*Pharmacol*) A generic drug that heals peptic ulcers by blocking the histamine$_2$ receptors (H_2 *antagonist*), which causes a reduced gastric acid output. Common TN *Tagamet*.

cinching (*ImageTech*) Tightening a roll of film by holding the centre and pulling the edge.

cinch marks (*ImageTech*) Abrasions or scratches along the length of the film caused by movement between surfaces in a roll.

cinchocaine (*Pharmacol*) A long-lasting local anaesthetic. The hydrochloride is used for spinal injection.

cinchona bases (*Chem*) Alkaloids present in cinchona bark, derivatives of quinoline.

cinchonine (*Chem*) $C_{19}H_{22}ON_2$, an alkaloid of the quinoline group, found in cinchona and cuprea barks; crystallizes in rhombic prisms from alcohol, mp 264°C. It behaves as a diacidic base and gives two series of salts.

cinchophen (*Pharmacol*) An analgesic formerly used for gout.

cinder pig (*Eng*) Pig iron made from a charge containing a considerable proportion of slag from puddling or reheating furnaces. Obsolete.

cinders (*Geol*) Volcanic *lapilli* composed mainly of dark glass and containing numerous vesicles (air or gas bubbles).

cine camera (*ImageTech*) A motion picture camera, usually one for narrow-gauge film.

CinemaScope (*ImageTech*) TN for a system of WIDE-SCREEN cinematography using ANAMORPHIC LENSES with a horizontal compression/expansion of 2:1.

cine radiography (*Radiol*) The rapid sequence of X-ray films taken by a camera attached to an image intensifier.

cingulum (*BioSci*) (1) Any girdle-shaped structure. (2) In Annelida, the CLITELLUM. (3) In Rotifera, the outer post-oral ring of cilia. (4) In mammals, a tract of fibres connecting the hippocampal and callosal convolutions of the brain; also in mammals, a ridge surrounding the base of the crown of a tooth and serving to protect the gums from the hard parts of food.

cinnabar (*Min*) Mercury sulphide, HgS, occurring as red acicular crystals, or massive; the ore of mercury, worked extensively at Almadén, Spain, and elsewhere.

cinnamaldehyde (*Chem*) 3-*phenylpropenal*. $C_6H_5CH = CHCHO$, bp 246°C; an oil of aromatic odour; the chief constituent of cinnamon and cassia oils.

cinnamic acid (*Chem*) 3-*Phenylpropenoic acid*. $C_6H_5CH = CHCOOH$, mp 133°C, bp 300°C; an unsaturated monobasic aromatic acid, prepared from benzal chloride by heating with sodium acetate.

cinnamon stone (*Min*) See HESSONITE.

Cinpres (*Eng*) TN for HOLLOW MOULDING process.

cinquefoil (*Arch*) A five-leaved ornament used in panellings etc.

cinture (*Arch*) A plain ring or fillet round a column, generally placed at the top and bottom to separate the shaft from capital and base.

CIP (*MinExt*) Abbrev for CARBON-IN-PULP.

cipher tunnel (*Arch*) A false chimney built on to a house for symmetrical effect.

ciprofloxacin (*Pharmacol*) A broad-spectrum synthetic quinolone antibiotic.

CIPW (*Geol*) A quantitative scheme of rock classification based on the comparison of norms; devised by four US petrologists, Cross, Iddings, Pirsson and Washington.

circadian rhythm (*BioSci*) A cyclical variation in the intensity of a metabolic or physiological process, or of some facet of behaviour, with a periodicity of c.24 hours.

circinate (*BioSci*) Coiled, like a watch spring inwardly from the base towards the apex. The leaf vernation in most ferns, some cycads and some seed ferns is circinate.

circinate (*Med*) Rounded, circular.

Circinus (Compasses) (*Astron*) A small obscure southern constellation.

circle (*MathSci*) A plane curve which is the locus of a point which moves so that it is at a constant distance (the *radius*) from a fixed point (the *centre*). The length of the circumference of a circle is $2\pi r$, and its area πr^2, where r is the radius and π is equal to 3·141 593 (to six places). The cartesian equation of a circle, with centre (a, b) and radius k, is $(x-a)^2 + (y-b)^2 = k^2$.

circle coefficient (*ElecEng*) A term often used to denote the leakage factor of an induction motor.

circle diagram (*ElecEng*) Graphical representation of complex impedances at different points in a transmission system on an orthogonal network. Best-known example is a *Smith chart*.

circle of confusion (*ImageTech*) Strictly *circle of least confusion*, the minimum image area of a point source of light formed by an optical system, the diameter of which determines the effective resolution. In practice, the maximum permissible diameter is affected by viewing conditions; thus it is taken as 0·01 in (0·25 mm) for a photographic print viewed directly, but 0·001 in (0·025 mm) for a motion picture film image greatly magnified on projection.

circle of convergence (*MathSci*) Circle within which a complex power series converges. Its radius is called the *radius of convergence*.

circle of curvature (*MathSci*) See CURVATURE.

circle of inversion (*MathSci*) See INVERSION.

circling disease (*Vet*) See LISTERIOSIS.

circlip (*Eng*) A spring washer in the form of an incomplete circle, usually used as a retaining ring (eg for ball bearings). Internal or external types are fixed in a circular groove in a hole or shaft by temporary distortion (closing or opening the circular shape).

circuit (*ElecEng*) Arrangement of conductors and passive and active components forming a path, or paths, for electric current.

circuit (*ICT*) (1) Complete communication channel. (2) An assembly of electronic (or other) components having some specific function, eg amplifier, oscillator or gate. See panel on PRINTED, HYBRID AND INTEGRATED CIRCUITS.

circuital magnetization (*Phys*) See SOLENOIDAL MAGNE-TIZATION.

circuit breaker (*ElecEng*) A device for opening an electric circuit under abnormal operating conditions, eg excessive current, heat, high ambient radiation level, etc. Also *contact breaker*. See AIR-BLAST SWITCH, OIL SWITCH.

circuit cheater (*ElecEng*) One which, for test purposes, simulates a component or load. Cf DUMMY LOAD.

circuit diagram (*ElecEng*) Conventional representation of wiring system of electrical or electronic equipment.

circuit layer (*ICT*) In a generalized telecommunications network, that part which represents the specific products and services, such as 64 Kbps telephony, offered to customers, rather than physical media or possible routes through the system.

circuit noise (*Electronics*) See THERMAL NOISE.

circuit parameters (*ElecEng*) Relevant values of physical constants associated with circuit elements.

circuit switching (*ICT*) A method of connecting together two users of a transmission service that allocates a circuit for their exclusive use. Once granted the transmission specification should remain stable for its duration. This technique was commonly used in early telephone systems.

circulant (*MathSci*) A determinant in which each row is a cyclic permutation by one position of the previous row.

circular cone (*MathSci*) A CONE (2) with a circular base.

circular dichroism (*Chem*) The differential absorption of left- and right-circularly polarized light by optically active (chiral) substances. Used to study the conformation of proteins in solution. Abbrev *CD*.

circular DNA (*BioSci*) DNA arranged as a closed circle. Replication requires DNA topoisomerase to solve the topological problem that would lead circles to be interlinked. Such DNA is characteristic of prokaryotes but also found in mitochondria, chloroplasts and some viral genomes.

circular form tool (*Eng*) A ring-shaped profile cutter which is gashed to have a substantially radial surface. The curved surface is shaped, across its width, to correspond to the contour of the part to be produced, the cutting edge being formed by the junction of the radial and the curved surfaces.

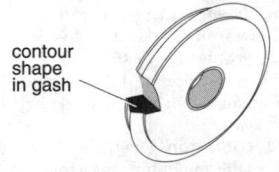

contour shape in gash

circular form tool

circular functions (*MathSci*) The *trigonometrical functions*, more particularly when defined with radian argument. Cf ELLIPTIC FUNCTIONS, HYPERBOLIC FUNCTIONS. All these functions are so named because of their association with the rectification of the similarly named curves.

circular knitting machine (*Textiles*) A WEFT-KNITTING MACHINE that produces fabric of circular cross-section in endless lengths.

circular level (*Surv*) A spirit level with the bubble housed under slightly concave glass.

circularly polarized (*Phys*) A term applied to a particular type of polarized electromagnetic radiation, esp visible light, where the plane of vibration is effectively helical. Produced by circularly polarizing filters, esp POLAROID,

and used for photoelastic analysis of isochromatics. Not to be confused with plane polarized light.

circular magnetization (*Phys*) The magnetization of cylindrical magnetic material in such a way that the lines of force are circumferential.

circular measure (*MathSci*) The expression of an angle in *radians*, 1 radian being the angle subtended at the centre of a circle by an arc of length equal to the radius. There are thus 2π, or approximately 6·283, radians in one complete revolution: 1 radian = 57·2958°; 1° = 0·0174 533 radians.

circular mil (*ElecEng*) US unit for wire sizes, equal to area of wire 1 mil (= 0·001 in = 0·025 mm) in diameter.

circular mitre (*Build*) A mitre formed between a curved and a straight piece.

circular permutation (*MathSci*) An arrangement of objects in a circle. There are $(n-1)!$ different circular permutations of n objects. See PERMUTATIONS.

circular pitch (*Eng*) The distance between corresponding points on adjacent teeth of a GEAR WHEEL, measured along the PITCH CIRCLE.

circular plane (*Build*) A plane adapted (through the use of shaped or flexible soles) for producing curved surfaces, either convex or concave.

circular point on a surface (*MathSci*) A point on the surface at which the principal curvatures are equal.

circular polarization (*ICT*) An electromagnetic wave for which either the electric or magnetic field vector describes a circle at the wave frequency; waves may have left- or right-handed circular polarization. Used widely in satellite communications.

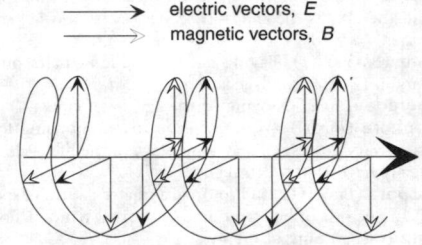

→ electric vectors, *E*

⇢ magnetic vectors, *B*

circular polarization Vectors drawn at 90° intervals.

circular saw (*Eng*) A power-driven steel disk carrying teeth on its periphery, used for sawing wood, metal or other materials.

circular shift (*ICT*) See END-AROUND SHIFT.

circular time base (*Electronics*) Circuit for causing the spot on the screen of a cathode-ray tube to traverse a circular path at constant angular velocity.

circular velocity (*Space*) The horizontal velocity of a body, required to keep it in a circular orbit, at a given altitude, about a planet. For a near-Earth orbit, assuming no air drag, the circular velocity (V_c) is given by

$$V_c = \sqrt{Rg} = 7\cdot91 \text{ km s}^{-1}$$

where R is the radius of the Earth and g the acceleration due to gravity. Also *orbital velocity*.

circular waveguide (*ICT*) A waveguide of circular cross-section (compared with conventional rectangular form). Can be used for very low-loss, high-bandwidth communication links.

circulating current (*ElecEng*) That which flows around the loop of a complete circuit, as contrasted with *longitudinal current*, which flows along the two sides or legs of the same circuit, in parallel.

circulating-current protective system (*ElecEng*) A form of Merz–Price protective system in which the current transformers at the two ends of the circuit to be protected are arranged to circulate current round the pilots, any difference in the currents in the two transformers passing through a relay.

circulating pump (*Eng*) A pump, usually of centrifugal type, used to circulate cooling water or, more generally, any liquid. See CENTRIFUGAL PUMP.

circulation (*Aero*) Used to describe the lift-producing airflow round an aerofoil, but strictly defined as the integral of the component of the fluid velocity along any closed path with respect to the distance round the path. See SUPERCIRCULATION.

circulation (*MathSci*) Of a vector, the line integral of the vector along a closed path in the field of the vector.

circulation (*Med*) The continuous movement of the blood through the heart, arteries, capillaries and veins.

circulation loss (*MinExt*) Loss of mud pumped down a borehole during drilling. It indicates the presence of porous or void conditions *downhole* and is potentially serious.

circulation of electrolyte (*ElecEng*) The process of stirring the electrolyte in an electroplating bath to ensure an even deposit.

circulator (*Radar*) A device having a number of terminals, usually three, with internal circuits which ensure that energy entering one terminal flange flows out through the next in a particular direction. They may appear as waveguide, co-axial or stripline components.

circulatory integral (*MathSci*) The line integral of a function over a closed curve.

circulatory system (*BioSci*) A system of organs through which is maintained a flow of fluid that transports materials between different tissues.

circum- (*Genrl*) Prefix from Lt *circum* denoting around.

circumcentre of a triangle (*MathSci*) The centre of the circumscribed circle. The point at which the perpendicular bisectors intersect.

circumcircle (*MathSci*) See CIRCUMSCRIBED CIRCLE.

circumcision (*Med*) Surgical removal of the prepuce or foreskin of the male or the labia minora in the female.

circumferential register adjustment (*Print*) The maintenance of one operation in relation to others in the direction of the web on rotary presses, manual control being superseded by electronic equipment.

circumnutation (*BioSci*) The rotation of the tip of an elongating stem, so that it traces a helical curve in space.

circumpolar star (*Astron*) A star which, for a given locality on the Earth, does not rise and set but revolves about the elevated celestial pole, always above the horizon. Its declination must exceed the co-latitude of the place in question.

circumscribed circle (*MathSci*) The circle which passes through all the vertices of a polygon. Also CIRCUMCIRCLE.

ciré (*Textiles*) Highly lustrous fabric produced by waxing and mechanical polishing, enhanced if satin weave is used.

cire perdue (*Eng*) See INVESTMENT CASTING.

cirque (*Geol*) A semi-amphitheatre, or 'armchair-shaped' hollow, of large size, excavated in mountain country by, or under the influence of, ice. Also *corrie*.

cirrate (*BioSci*) Bearing cirri. Also *cirriferous*.

cirrhosis (*Med*) A disease of the liver in which there is increase of fibrous tissue and destruction of liver cells. Gk *kirros*, orange-tawny.

cirriferous (*BioSci*) See CIRRATE.

Cirripedia (*BioSci*) A subclass of marine CRUSTACEA, generally of sessile habit when adult. The young are always free-swimming, while the adult possesses an indistinctly segmented body that is partially hidden by a mantle containing calcareous shell plates. There are six pairs of biramous thoracic legs attached by antennules. Many species are parasitic. Barnacles.

cirrocumulus (*EnvSci*) Thin white patch, sheet or layer of cloud without shading, composed of very small elements in the form of grains, ripples, etc, merged or separate, and more or less regularly arranged; most of the elements have an apparent width of less than 1°. Abbrev *Cc*.

cirrose (*BioSci*) Curly, like a waved hair. Consisting of diverging filaments.

cirrostratus (*EnvSci*) Transparent whitish *cloud veil* of fibrous (hair-like) or smooth appearance, totally or partly covering the sky and generally producing halo phenomena. Abbrev *Cs*.

cirrus (*BioSci*) (1) In Protozoa, a stout conical vibratile process, formed by the union of cilia. (2) In some Platyhelminthes, a copulatory organ formed by the protrusible end of the vas deferens. (3) In Annelida, a filamentous tactile and respiratory appendage. (4) In Cirripedia, a ramus of a thoracic appendage. (5) In insects, a hair-like structure on an appendage. (6) In Crinoidea, a slender jointed filament arising from the stalk or from the centrodorsal ossicle and used for temporary attachment. (7) In fish, a barbel.

cirrus (*EnvSci*) Detached *clouds* in the form of white, delicate filaments of white or mostly white patches or narrow bands. These clouds have a fibrous (hair-like) appearance, or silky sheen, or both. Abbrev *Ci*.

cirsoid aneurysm (*Med*) A mass of newly formed, tortuous and dilated arteries.

cis (*MathSci*) The function $\text{cis } x = \cos x + i \sin x$, which equals e^{ix}.

cis- (*Chem*) A prefix indicating that geometrical isomer in which the two radicals are adjacent in a metal complex or on the same side of a double bond or alicyclic ring. Cf *TRANS-*. Also used for control elements in a DNA strand that act downstream of the sequence rather than at a remote site.

CISC (*ICT*) Abbrev for COMPLEX INSTRUCTION SET COMPUTER.

cisplatin (*Pharmacol*) An anti-tumour drug containing platinum, which has an alkylating action and is used in the treatment of ovarian carcinoma and testicular teratoma.

cissing (*Build*) A defect in paint etc due to poor adhesion with pinholes, craters and in serious cases the contraction of the wet paint to form surface blobs.

cissoid (*MathSci*) The inverse of a parabola in respect of its vertex. Its cartesian equation is $y^2(2a-x) = x^3$.

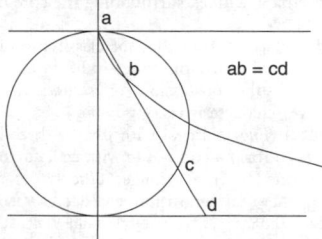

cissoid

cisternum (*BioSci*) A compartment or vesicle, often flattened, formed within the cytoplasm by membranes of the ENDOPLASMIC RETICULUM or GOLGI APPARATUS.

cistron (*BioSci*) A *gene* defined as a stretch of DNA specifying one *polypeptide*.

citral (*Chem*) Geranial, $C_{10}H_{16}O$, an alkene-type terpene, formula $CH_2 = CMeCH_2CH_2CH_2CMe = CHCHO$; bp 110–112°C (12 mm); it occurs in the oil of lemons and oranges and in lemon-grass oil. Used as a flavouring and in perfumery.

citrates (*Chem*) The salts of citric acid.

citrene (*Chem*) See LIMONENE.

citric acid (*Chem*) *2-hydroxypropane-1,2,3-tricarboxylic acid*. $C_6H_8O_7$. An important hydroxy-tricarboxylic acid, occurs in the free state in many fruits, esp lemons, but is now prepared commercially, largely by fermentation with *Aspergillus*. Much used for flavouring effervescent drinks.

citric acid cycle (*BioSci*) See TCA CYCLE.

citrine (*Min*) Not the true TOPAZ of mineralogists but a yellow variety of quartz, which closely resembles it in colour but not in other physical characters; it is of much less value than topaz, and figures under a number of geographical names like *Spanish topaz*. Also *false topaz*. But see BRAZILIAN TOPAZ, the true mineral.

citronellal (*Chem*) $C_{10}H_{18}O$, an aldehyde forming the main constituent of citronella oil and lemon-grass oil. It is used in perfumery. Also *rhodinal*.

civery (*Arch*) One bay of a vaulted ceiling. Also spelt *severy*.

Civil Aviation Authority (*Aero*) An independent body which controls the technical, economic and safety regulations of UK civil aviation. In 1972 it took over the relevant functions of the Department of Trade and Industry, and those of the Air Registration Board and the Air Transport Licensing Board. It is responsible for the civil side of the Joint National Air Traffic Control Services. Abbrev *CAA*.

civil engineer (*CivEng*) Someone engaged in the design, planning and construction of railways, roads, harbours, bridges, etc.

civil twilight (*Astron*) The interval of time during which the Sun is between the horizon and 6° below the horizon, morning and evening.

CJD (*Med*) Abbrev for CREUTZFELDT–JAKOB DISEASE. See panel on TRANSMISSIBLE SPONGIFORM ENCEPHALOPATHY.

Cl (*Chem*) Symbol for CHLORINE.

clack (*Eng*) A check valve admitting water from a feed pump to the boiler of a locomotive. A ball valve is used, the name 'clack' being derived from the characteristic sound of the ball striking its seat.

clacking (*Vet*) An error of gait in the horse in which the toe of a hind foot strikes the sole or shoe of a fore foot. Also *forging*.

clack valve (*Eng*) See CLACK.

cladding (*Build*) The material used for the external covering of a building.

cladding (*MinExt*) The material used for the lining of a mine shaft. Traditionally of timber battens or planks, now frequently of precast reinforced concrete slabs.

cladding (*NucEng*) Thin protective layer, usually metallic, of reactor fuel units to contain fission products and to prevent contact between fuel and coolant. See CAN.

cladding (*Phys*) The homogeneous dielectric material of lower refractive index surrounding the core of an OPTICAL FIBRE.

cladistics (*BioSci*) A method of classifying organisms into groups (taxa) based on 'recency of common descent' as judged by the possession of shared derived (ie not primitive) characteristics.

clad metal (*Eng*) A metal with two or three layers bonded together to form a composite with eg a corrosion-resistant layer formed over a stronger core by co-rolling, heavy plating, chemical deposition, etc. See LAMINATE.

cladode (*BioSci*) (1) Strictly a PHYLLOCLADE of one internode. (2) More commonly, any phylloclade.

cladogram (*BioSci*) A branching diagram (dendrogram) reflecting the relationships between groups of organisms determined by the methods of CLADISTICS.

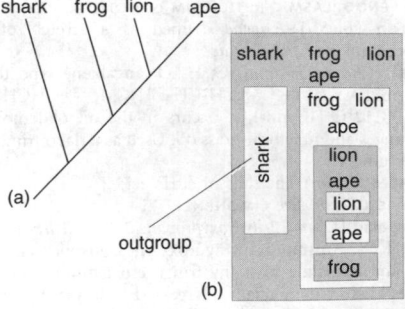

cladogram The diagram for four species (a) and a representative of shared characteristics (b) on which the cladogram is based.

cladophyll (*BioSci*) Same as PHYLLOCLADE.

Clairaut's differential equation (*MathSci*) An equation of the form

$$y = xp + f(p)$$

where

$$p = \frac{dy}{dx}$$

The general solution is $y = cx + f(c)$. There is also a singular solution given parametrically by

$$x = -f^{1}(p),\ y = -pf^{1}(p) + f(p)$$

Claisen condensation (*Chem*) An important synthetic reaction involving condensation between esters, or between esters and ketones, in the presence of sodium ethoxide. Used in the preparation of *ethyl acetoacetate*.

Claisen flask (*Chem*) A distillation flask used for vacuum distillations; it consists of a glass bulb with a neck for a thermometer, to which another neck with outlet tube is attached.

clamp (*Agri*) A shallow pit in which harvested roots and tubers are heaped before covering with hay, straw or turf to provide winter protection from the elements.

clamp (*Build*) See SASH CRAMP.

clamp (*ImageTech, ICT*) Circuit in which a waveform is adjusted and maintained at a definite level.

clamp bar (*Print*) (1) A bar used to lock or unlock plates to a letterpress rotary or offset lithographic machine. (2) A bar to hold blankets in place on the pins of the impression cylinder of a letterpress rotary or the blanket cylinder of an offset press.

clamp connection (*BioSci*) In some Basidiomycetes, a small connecting hypha across the septa between two adjacent cells of dikaryotic hypha. Formed as the two nuclei, lying one behind the other, in the terminal cell divide. Facilitates the maintenance of the dikaryon by allowing the second of the two daughter nuclei to pass the cell wall.

clamp force (*Eng*) See LOCKING FORCE.

clamping diode (*Electronics*) A DIODE used to clamp a voltage at some point in a circuit. See CATCHING DIODE.

clamshell (*Aero*) (1) Cockpit canopy hinged at front and rear. (2) Hinged part of thrust reverser in gas turbine. (3) Hinged door of cargo aircraft.

clamshell phone (*ICT*) A mobile phone consisting of two working parts joined by a hinge so that it can be folded and unfolded.

clan (*Geol*) A suite of igneous rock-types closely related in chemical composition but differing in mode of occurrence, texture and possibly in mineral contents.

clangbox (*Aero*) Deflector fitted to jet engine to divert gas flow for eg V/STOL operation.

clap-board (*Build*) A form of weatherboard which is tongued and rebated and frequently moulded, rather than being feather-edged.

clapper board (*ImageTech*) A board with clapstick and written details of the production (date, scene, take number, etc) photographed at the start of each shot.

clapper box (*Eng*) A slotted tool head carried on the saddle of a planing or a shaping machine. It carries a pivoted block to which the tool is clamped, thus allowing the tool to swing clear of the work on the return stroke of the table or the head.

clappers (*ImageTech*) A pair of hinged arms closed sharply in view of the camera, the action and noise providing clear indications on the separate picture and sound records for subsequent synchronization. Also *clapstick*.

Clapp oscillator (*ICT*) Low-drift COLPITTS OSCILLATOR.

clap-sill (*Build*) See MITRE SILL.

clarain (*Min*) Bands in coal, characterized by bright colour and silky lustre.

clarendon (*Print*) A heavy typeface, eg

RQENbaegn

Clarke (*Min*) The average value of a chemical element in the Earth's crust.

Clark process (*Chem*) A process for effecting the partial softening of water by the addition of sufficient lime water to convert all the acid carbonates of lime and magnesium into the normal carbonates.

claspers (*BioSci*) (1) In insects, an outer pair of GONAPOPHYSES. (2) In male selachian fish, the inner narrow lobe of the pelvic fin, used in copulation. (3) More generally, any organ used by the sexes for clasping one another during copulation.

clasp nail (*Build*) A square-section soft-iron cut nail whose head has two pointed projections that sink into the wood, and the projecting end of the nail is bent over and punched into the back surface.

clasp nut (*Eng*) A nut split diametrally into halves, which may be closed so as to engage with a threaded shaft; used as a clutch between a lathe LEAD SCREW and the SADDLE.

class (*BioSci*) In biometry, a group of organisms all falling within the same range, as indicated by the unit of measurement employed; in zoology, one of the taxonomic groups into which a phylum is divided, ranking next above an order. In botany the taxonomic rank below DIVISION and above ORDER; the names end in -phyceae (algae), -mycetes (fungi) or -opsida (other plants).

class (*ICT*) In object-oriented programming, a structure of data and the functions (or METHODS) by which those data are created and maintained.

class-A amplifier (*ICT*) An amplifier stage (valve or transistor) in which anode or collector current flows throughout the amplitude range of the applied signal.

class-AB amplifier (*ICT*) An amplifier stage (valve or transistor) in which anode or collector current flows for most, but not all, of the amplitude range of the applied signal. For small signals, operation is in CLASS-A.

class-B amplifier (*ICT*) An amplifier stage (valve or transistor) in which anode or collector current flows for approximately half of the amplitude range of the applied signal. Commonly a pair of such amplifiers is used in *push–pull* operation.

class-C amplifier (*ICT*) An amplifier stage (valve or transistor) in which anode or collector current flows during less than half of the amplitude variation of the applied signal.

class frequency (*MathSci*) The number of observations in a set of data in a particular class interval.

Classical (*Arch*) The architecture pertaining to ancient Greece and Rome, the principles of which were revived throughout Europe during the RENAISSANCE.

classical (*Phys*) Said of theories based on concepts established before relativity and quantum mechanics, ie largely conforming with Newtonian mechanics. US *nonquantized*.

classical conditioning (*Psych*) A behavioural technique involving the repeated pairing of two stimuli one of which (the *unconditioned stimulus* or UCS) already elicits a response (the unconditioned response) and the other of which (the *conditioned stimulus* or CS) does not. After one or more pairing procedures the CS comes to elicit a response (the conditioned response) very similar to that initially elicited only by the UCS, even when presented alone.

classical flutter (*Aero*) See FLUTTER.

classical mechanics (*Phys*) See NEWTONIAN MECHANICS.

classical scattering (*Phys*) See THOMSON SCATTERING.

classification (*PowderTech*) Grading in accordance with particle size, shape and density by fluid means.

classification of clouds (*EnvSci*) An agreed classification such as that published by the *World Meteorological Organization*. There are ten *cloud genera*, which can be divided into up to 14 *cloud species*, themselves divisible into one or more out of nine *cloud varieties*. Agreed definitions of the ten genera will be found elsewhere.

classification of ships (*Ships*) Passenger ships are classified according to the nature of the voyages in which they are engaged. Classification societies assign ships to a class in the Register Book so long as they are built, equipped and maintained in accordance with the rules of the society.

classifier (*MinExt*) A machine for separating the product of ore-crushing plant into two portions consisting of particles of different sizes. In general, the finer particles are carried off by a stream of water, while the larger settle. The fine portion is known as the *overflow* or *slime*, the coarse as the *underflow* or *sand*.

class interval (*MathSci*) A subset of the range of values of a variate.

clastic rocks (*Geol*) Rocks formed of fragments of pre-existing rocks.

clathrate (*Chem*) Form of compound in which one component is combined with another by the enclosure of one kind of molecule by the structure of another, eg rare gas in 1,4-dihydroxybenzene. See MOLECULAR SIEVE.

clathrin (*BioSci*) A protein that is the major structural component of the proteinaceous layer of COATED PITS and vesicles. Clathrin takes the form of a triskelion; the association of triskelions into hexagons and pentagons forms the protein lattice of the coated pit and vesicle. See fig. at COATED VESICLE.

Claude process (*Eng*) A method of liquefying air in stages, the expanding gas being cooled by external work on the pistons.

claudication (*Med*) The action of limping. See INTERMITTENT CLAUDICATION.

Clausius–Clapeyron equation (*Chem*) This shows the influence of pressure on the temperature at which a change of state occurs:

$$\frac{dp}{dT} = \frac{\lambda}{T\Delta V}$$

where λ is the heat absorbed (latent heat), T is the absolute temperature, ΔV is the change in volume.

Clausius–Mosotti equation (*Phys*) An equation relating electrical polarizability to permittivity, principally for fluid dielectrics.

Clausius's inequality (*Phys*) For any thermodynamic system undergoing a cyclic process,

$$\oint \frac{dQ}{T} \leq 0$$

where dQ is an infinitesimal quantity of heat absorbed or liberated by the system at the temperature T kelvin. The equality is appropriate to a reversible process.

Clausius virial (*Phys*) A term in an expression for the calculation of the pressure of a gas based on the kinetic theory of gases. The term allows for the inclusion of the effect of forces between the molecules.

Claus process (*ChemEng*) A process for recovering elemental sulphur from H_2S, by desulphurizing. It is a two-stage process as defined in the two chemical equations. The H_2S is divided into two streams. First step: $2H_2S + 3O_2 = 2H_2O + 2SO_2$. Second step: $2H_2S + SO_2 = 2H_2O + 3S$. The second step is carried out at a high enough temperature to produce dry sulphur.

claustra (*Arch*) A panel pierced with geometrical designs giving relief to the concrete structures of the late 19th and early 20th centuries in France.

claustrophobia (*Psych*) Fear of being in a confined space.

clava (*BioSci*) A gradual swelling at the distal end of a structure, resembling a club. Also *clave*. Adj *clavate* as in *clavate antennae*.

clavicle (*BioSci*) In vertebrates, the collar bone, an anterior bone of the pectoral girdle. Adj *clavicular*.

claw (*BioSci*) (1) A curved, sharp-pointed process at the distal extremity of a limb; a nail that tapers to a sharp point. (2) The narrow, elongated lower portion of a petal in some plants.

claw (*Build*) A small tool with a bent and split end, used for extracting tacks.

claw bolt (*Build*) A wrought-iron bolt with a long head flattened in a direction parallel to the bolt, and bent over at right angles near the end.

claw chisel (*Build*) A chisel, having a 2 in (50 mm) long serrated cutting edge, used for rough-dressing building stone.

claw coupling (*Eng*) A shaft coupling in which flanges carried by each shaft engage through teeth cut in their opposing faces, one flange sliding axially to disengage the drive. Also claw clutch.

claw foot (*Med*) See PES ARCUATUS.

claw-hammer (*Build*) A hammer having a bent and split peen used for extracting a nail by levering.

claw hand (*Med*) Claw-like position adopted by the hand when the muscles supplied by the ulnar nerve are paralysed.

claw ill (*Vet*) See FOUL IN THE FOOT.

claws (*ImageTech*) Mechanism for advancing the film frame by frame in a camera or projector by intermittently pulling down the perforation holes.

clay (*Geol*) A fine-textured, sedimentary or residual deposit. It consists of hydrated silicates of aluminium mixed with various impurities. A fine-grained sediment of variable composition having a grain size less than $\frac{1}{256}$ mm (4 μm). Clay for use in the manufacture of pottery and bricks must be fine-grained and sufficiently plastic to be moulded when wet; it must retain its shape when dried, and sinter together, forming a hard coherent mass without losing its original shape, when heated to a sufficiently high temperature. See CLAY MINERAL.

Clayden effect (*ImageTech*) Partial reversal of an image when a very brief exposure to an intensely bright source, such as an electric spark, is followed by a longer exposure at a much lower light level.

clay gun (*Eng*) Arrangement which shoots a ball of fireclay into a blast furnace's taphole.

claying (*CivEng*) The operation of lining a blast-hole with clay to prevent the charge from getting damp.

clay ironstone (*Geol*) Nodular beds of clay and iron minerals, often associated with the Coal Measure rocks.

clay mineral (*Geol*) Any mineral of clay grade but specifically one of a complex group of finely crystalline, metacolloidal or amorphous hydrous aluminosilicates. They have sheet-like lattices (*phyllosilicates*) and most are formed by the weathering of primary silicate minerals. The most common clay minerals belong to the *kaolin*, *montmorillonite* (*smectite*) or *illite* groups. See SILICATES.

clay puddle (*CivEng*) See PUDDLE CLAY.

clay with flints (*Geol*) A stiff clay, containing unworn flints, which occurs as a residual deposit in chalk areas, but which is extensively mixed with other superficial deposits.

clean critical assembly (*NucEng*) A reactor before start-up, when the fuel, moderator and lattice elements have not been irradiated and their compositions are known.

cleaning eye (*Build*) See ACCESS EYE.

clean room (*Space*) A special facility for handling material, destined for space activities, in a sterile and dust-free environment.

clean-up (*ElecEng*) (1) Improvement in vacuum which occurs in an electric discharge tube or vacuum lamp following absorption of the residual gases by the glass. (2) The removal of residual gas by a GETTER.

clear (*ICT*) To remove data from MEMORY so that fresh data can be recorded.

clear-air turbulence (*EnvSci*) Turbulence in the free atmosphere that is not associated with cumulus or cumulonimbus clouds. It occurs mainly in the upper troposphere or lower stratosphere, esp in the vicinity of jet streams. See panels on STRATOSPHERE AND MESOSPHERE and TROPOSPHERE.

clearance (*Eng*) (1) The distance between two objects, or between a moving and stationary part of a machine. (2) The angular backing-off given to a cutting tool in order that the heel shall clear the work.

clearance time (*Agri*) The period that has to elapse between the last pesticide application and the time of consumption of a crop. Also known as *harvest interval* and typically defined by law.

clearance volume (*Eng*) In a reciprocating engine or compressor, the volume enclosed by the piston and the adjacent end of the cylinder, when the crank is on the top dead centre. See COMPRESSION RATIO, CUSHION STEAM.

clearing (*Textiles*) In dyeing or printing, the removal of excess colouring material to improve the appearance of the fabric. Cutting out an imperfection in a yarn and knotting or splicing the resultant ends.

clearing agent (*BioSci*) In microscopical technique, a liquid reagent which has the property of rendering objects immersed in it transparent and so capable of being examined by transmitted light.

clearing hole (*Eng*) A hole drilled slightly larger than the diameter of the bolt or screw which passes through it.

clearing manoeuvre (*Aero*) Alteration of aircraft attitude to give better view of other air traffic.

clear span (*Build*) The horizontal distance between the inner extremities of the two bearings at the ends of a beam.

cleat (*Build*) A strip of wood or metal fixed to another for strengthening purposes, or as a locating piece to ensure that another part shall be in its correct position.

cleats (*MinExt*) The main cleavage planes or joint planes in a seam of coal.

cleavage (*BioSci*) (1) Division of the cytoplasm by infurrowing of the plasmalemma. Cf CELL PLATE. (2) The series of mitotic divisions by which the fertilized ovum is transformed into a multicellular embryo.

cleavage (*Chem*) (1) The splitting of a crystal along certain planes parallel to certain actual or possible crystal faces, when subjected to tension. (2) The splitting up of a complex molecule, such as a protein, into simpler molecules, usually by hydrolysis and mediated by an enzyme.

cleavage (*Geol*) A property of rocks, such as slates, whereby they can be split into thin sheets. Cleavage is produced and oriented by the pressures that have affected rocks during consolidation and earth movements.

cleavage nucleus (*BioSci*) The nucleus of the fertilized ovum produced by the fusion of the male and female pronuclei; in parthenogenetic forms, the nucleus of the ovum.

cleavelandite (*Min*) A platy variety of ALBITE.

cleft palate (*Med*) A gap in the roof of the mouth as a result of congenital maldevelopment, with or without hare-lip.

cleidoic egg (*BioSci*) The egg of a terrestrial animal with a protective shell.

cleidotomy (*Med*) The cutting of the clavicles when the shoulders of the fetus prevent delivery in difficult labour.

cleistocarp (*BioSci*) Same as CLEISTOTHECIUM.

cleistogamous (*BioSci*) A plant that produces flowers, often inconspicuous, that do not open and in which self-pollination occurs. Cf CHASMOGAMOUS.

cleistothecium (*BioSci*) A more or less globose ASCOCARP with no specialized opening.

Clemmensen reduction (*Chem*) Reduction of aldehydes and ketones, by heating with hydrochloric acid and zinc amalgam, to the corresponding hydrocarbons, eg propanone to propane.

clench nailing (*Build*) A method of nailing pieces together in which the end of the nail, after passing right through the last piece, is bent back and driven into this piece, so that it may not be drawn out. Also *clinch nailing*.

clerestorey (*Arch*) The upper stage of the walls of a building, occurring, in the case of a church, above the projecting aisle roofs, and pierced with windows to admit light to the central portion of the building.

clerk of works (*Build, CivEng*) The official appointed by the employer to watch over the progress of any given building works, and to see that contractors comply with requirements for materials and labour.

cleveite (*Chem, Min*) Variety of PITCHBLENDE containing uranium oxide and rare earths; often occluding substantial amounts of helium.

Cleveland Iron Ore (*Min*) An ironstone consisting of iron carbonate, which occurs in the Middle Lias rocks of N Yorkshire near Middlesbrough. The ironstone is oolitic and yields on the average 30% iron.

clevis (*Eng*) U-shaped component with holes drilled through the arms as a means of symmetrical attachment to another component. Used eg for attaching load carriers to overhead conveyors.

clevis joint (*Aero*) Fork and tongue joint secured by metal pin as used in joining solid rocket motor cases.

CLI (*ICT*) Abbrev for COMMAND LINE INTERPRETER.

cliché (*Print*) International name for duplicate printing block; Fr origin.

click (*Acous*) Short impulse of sound, with wide-frequency spectrum, with no perceptible concentration of energy-giving characterization.

clicker (*Print*) A compositor who receives copy and instructions from the overseer and distributes the work among companions.

click stop (*ImageTech*) Aperture control which clicks at each reading, enabling it to be set without looking at the figures.

click-through (*ICT*) An instance of a visitor to a website clicking on an advertisement in order to visit the advertiser's website.

client (*ICT*) An application that retrieves information from a server. Also BROWSER on the World Wide Web. See CLIENT–SERVER SYSTEM.

client application (*ICT*) An APPLICATION PROGRAM whose documents can accept linked or EMBEDDED OBJECTS.

client-centred therapy (*Psych*) A form of psychotherapy based on Carl Roger's beliefs that an individual has an unlimited capacity for psychological growth and is best able to deal with problems if allowed to explore and work them out in a non-directive, non-judgemental environment.

client computer (*ICT*) See CLIENT–SERVER SYSTEM.

client–server system (*ICT*) (1) In the context of FILE systems, a division of labour between two INTELLIGENT computing devices; eg the client computer may provide a WIMP environment for the user whilst the server computer retrieves the required data from a database file and returns the data to the client computer for display. (2) In the context of WINDOWS systems the client is an APPLICATION that accesses services from a *windowing server*. In this model the client and the server can run on the same machine or different machines.

climacteric (*BioSci*) (1) A critical period of change in a living organism, eg the menopause. (2) The period of ripening of some fruit, eg apples, characterized by an increased rate of respiration.

climate (*EnvSci*) The statistical ensemble of atmospheric conditions characteristic of a particular locality over a suitably long period (eg 30 years) including relevant parameters such as mean and extreme values, measures of variability and descriptions of systematic seasonal variations. Aspects considered include temperature, humidity, rainfall, solar radiation, cloud, wind and atmospheric pressure.

climate control (*Genrl*) Air-conditioning.

climate modelling (*EnvSci*) The construction of physical and mathematical models of the Earth's atmosphere, including the influence of seas and oceans, and all relevant aspects of the surface of the land, so that extended

NUMERICAL FORECASTING may be made over periods of simulated time long enough to accumulate sufficient statistical data to define a *model climate*. Climate modelling requires accurate treatment of atmospheric processes such as heat transfer, radiation absorption and reflection, and moisture transfer. Oceanic circulation and changes in continental ice sheets are taken into account in long-term studies.

climatic change (*EnvSci*) Long-term changes of climate, which may be due to variations in the Earth's orbit or modifications in the rate of absorption of carbon dioxide by oceanic phytoplankton (GREENHOUSE EFFECT). See panel on CLIMATIC CHANGE.

climatic factor (*BioSci*) A condition, such as average rainfall, temperature, and so on, that plays a controlling part in determining the features of a plant community and/or the distribution and abundance of animals. Cf BIOTIC FACTOR, EDAPHIC FACTOR.

climatic zones (*EnvSci*) The Earth may be divided into zones, approximating to zones of latitude, such that each zone possesses a distinct type of climate. Eight principal zones may be distinguished: a zone of tropical wet climate near the equator; two subtropical zones of steppe and desert climate; two zones of temperate rain climate; one incomplete zone of boreal climate with a great range of temperature in the northern hemisphere; and two polar caps of arctic snow climate.

climatology (*EnvSci*) The study of climate and its causes.

climax (*BioSci*) The end-point in a succession of vegetation, when the community has reached an approximately steady state, in equilibrium with local conditions. See MONO-CLIMAX THEORY, POLYCLIMAX THEORY, SERE.

climb cutting (*Eng*) The method of machining a surface with a multi-toothed cutter in which the workpiece moves in the same direction as the periphery of the cutter at the line of contact. It produces a better surface finish than can be obtained by cutting upwards but has the disadvantage that the workpiece tends to be drawn towards the cutter.

climbing form (*CivEng*) A type of form sometimes used in the construction of reinforced concrete walls for buildings. The wall is built in horizontal sections, the climbing form being raised, after the pouring of each section, into a position convenient for the pouring of the next section.

clinch (*Eng*) To set or close a fastener, usually a rivet. Also *clench*.

clinch nailing (*Build*) See CLENCH NAILING.

cline (*BioSci*) A quantitative gradation in the characteristics of an animal or plant species across different parts of its range associated with changing ecological, geographic or other factors, eg ECOCLINE, GEOCLINE.

C-line (*Phys*) Fraunhofer line in spectrum of Sun at 656·28 nm, arising from ionized hydrogen in its atmosphere.

clingfilm (*Genrl*) Thin polymer packaging film which quickly bonds to itself or products to be protected.

clinic (*Med*) (1) The instruction of medicine in any of its branches at the bedside or in the presence of hospital patients. (2) An institution, or a department of one, or a group of doctors, for treating patients or for diagnosis. (3) A private hospital: by extension, any similar instructional and/or remedial meeting.

clinical psychology (*Psych*) A branch of psychology concerned with the application of research findings in the field of mental health.

clinical trial (*Med*) The method of testing the efficacy of a treatment or a hypothesis related to the cause of a disease, broadly divisible into two kinds. (1) *Retrospective trial*. This looks at existing records, patient habits, environmental situation, etc, and attempts to make correlations between these and clinical outcome. Because past records may not be standardized and because it may be difficult to obtain a proper control group, this type of trial is at best indicative and is often considered unsatisfactory. (2) *Prospective trial*. In this kind there are two groups of patients, carefully

Climatic change

Change of climate on time-scales longer than that on which the usual working definition of 'climate' is based, ie 30–50 years. Mean values of atmospheric variables such as temperature and rainfall taken over N years fluctuate less and less as N increases from 1 to 30 or so, and a period of 30 years is usually adequate in most regions to provide rough estimates of extremes likely to be encountered over much longer periods. However, it has become increasingly apparent that 30-year means can fluctuate to a degree greater than is to be expected by pure chance, and that even in historical times some periods of 200–300 years may be significantly colder or warmer than others, the best known of these being the 'Little Ice Age' from about 1550 to 1850 which affected much of Europe and the North Atlantic region.

Before the development of thermometers and rain gauges, evidence for weather and climate can be obtained from historical records of harvests, floods, frosts, snowfalls, etc, and for prehistoric times, significant conclusions can often be drawn from *proxy data* including tree-rings, oxygen isotope ratios ($^{18}O/^{16}O$) in layers from borings in polar ice caps and ocean-bed deposits, analysis of pollen and remains of climate-sensitive Coleoptera (beetles) in deep layers of peat and undisturbed soil, varves (year-layers in lake beds), etc. Dating deposits by radiocarbon methods are useful back to about 40 000 years BP (before present), uranium analysis to 300 000 years BP, while for still earlier periods, records of geomagnetic field reversals provide clues to dating.

On the longest time-scale, there is good geological evidence that during the last 10^9 years ice-age epochs, each several million years long, have occurred at intervals of 200 to 300 million years. The last of these epochs began to affect the northern hemisphere about 3 or 4 million years ago (although the Antarctic has been continuously glaciated for 15 million years), possibly as a result of major changes to the oceanic circulation due to continental drift. Since then, the northern hemisphere seems to have experienced glacial or near-glacial conditions for about 90% of the time but with much warmer *interglacials*, each lasting about 10 000 years and occurring at roughly 100 000-year intervals; there have also been *interstadials*, which are periods of a few thousand years when conditions were warmer than those typical of full glaciation but not as consistently warm as in an interglacial.

In the UK, the last ice age (the Devensian) ended about 10 000 years BP, the previous (Ipswichian) interglacial having ended about 110 000 years BP, with interstadials at 60 000 and 45 000 years BP. (There was a further brief warm spell about 12 000 BP.) Between 8000 and 4000 BP the climate in Europe was probably warmer than at present by about 2 and 3°C, while much of what is now the Sahara Desert received generous rainfall and was forested, with animals such as hippopotamus and elephant being depicted in cave paintings; these conditions were most likely caused by the persistence of deep ice sheets over much of North America for much longer than over northern Europe, which caused a persistent deep trough of low pressure over the western Atlantic. For the last 3000 years, variations have not been so extreme but have nevertheless been of considerable social and economic significance. The Little Ice Age, already referred to, had very severe effects on areas with marginal climates such as Greenland, Iceland, Scandinavia and Scotland. It also produced, owing to the strengthening of the thermal gradients in the latitudes of the British Isles, many violent storms over England and the Low Countries with massive flooding as well as the notable winters portrayed in much Flemish painting.

As regards causes, there is increasing evidence that the sequence of glacial, interstadial and interglacial periods, on a time-scale of 10^4 to 10^5 years, is due in some way not yet fully explained to variations in the parameters of the Earth's orbit (MILANKOVITCH THEORY), but very possibly involving modifications of the greenhouse effect by variations in the absorption of atmospheric CO_2 by phytoplankton in the oceans. Suggested causes for variation on the historical scale are still more speculative, including veiling of the Sun's radiation by volcanic dust, and complex feedback processes involving atmospheric and oceanic circulations.

There is good evidence derived from pollen analysis and assemblages of Coleoptera that major climatic changes, eg the end of the Ipswichian interglacial, can take place within two or three centuries, though the development and decay of major continental ice sheets require several millennia. There is increasing speculation at the present time whether human activity associated with rapidly increasing population, industrialization, deforestation and intensive agriculture may affect the climate on a global scale.

See panel on ATMOSPHERIC POLLUTION.

---

matched for age, sex, clinical stage, etc. One group (the control) will receive the current best available treatment or sometimes a PLACEBO, while the other group (the experimental) will receive the new treatment. The clinical outcome is then assessed, preferably without either the patient or the assessor knowing to which group the patient belongs (a *double-blind trial*). Statistical analysis is finally necessary to test whether the outcome in the experimental group differs significantly from that in the control. A phase 1 trial evaluates safety and dose. Effectiveness is evaluated in a phase 2 trial, and a large-scale phase 3 trial confirms effectiveness and safety in preparation for wide-scale use.

clinker (*Eng*) Incombustible residue, consisting of fused ash, raked out from coal- or coke-fired furnaces; used for road-making and as aggregate for concrete. See BREEZE BLOCK.

clinkers (*Build*) See KLINKER BRICK.

clinks (*Eng*) Internal cracks formed in steel by differential expansion of surface and interior during heating. The tendency for these to occur increases with the hardness and mass of the metal, and with the rate of heating.

clink-stone (*Geol*) See PHONOLITE.

clino- (*Genrl*) Prefix from Gk *klinein* (to slant, lean) denoting oblique, reclining.

clinochlore (*Min*) A variety of CHLORITE.

clinograph (*Surv*) A form of adjustable set square, the two sides forming the right angle being fixed, while the third side is adjustable; it differs from the adjustable set square in that no scale is provided to show the angular position of the third side in relation to the other two. Used in surveying deep boreholes to check departure from vertical.

clinohumite (*Min*) One of the four members of the humite group, magnesium silicate with hydroxide, occurring in metamorphosed limestones.

clinoid (*Med*) Bony processes of the sella turcica that encircle the pituitary gland's stalk.

clinometer (*Surv*) A hand instrument for the measurement of angles of slope.

clinopyroxene (*Min*) A general term for the monoclinic members of the pyroxene group of silicates.

clinopyroxenite (*Min*) An ultramafic plutonic rock consisting almost entirely of clinopyroxene.

clino-rhomboidal crystals (*Crystal*) See TRICLINIC SYSTEM.

clinostat (*BioSci*) See KLINOSTAT.

clinozoisite (*Min*) A hydrous calcium aluminium silicate in the epidote group. Differs from zoisite in crystallizing in the monoclinic system.

clintonite (*Min*) Hydrated aluminium, magnesium, calcium silicate. One of the brittle micas; differs optically from xanthophyllite.

clioquinol (*Pharmacol*) Used as an internal amoebicide, also in creams and ointments (sometimes with hydrocortisone) for certain skin infections. Also *Iodochlorhydroxyquinoline*.

clip (*ImageTech*) A short continuous sequence selected from a motion picture or video production. Also *clipping*.

clip art (*ICT*) A library of graphics images that may be imported into design or drawing programs as required or for incorporating into video productions using DESKTOP VIDEO. Such a library is often supplied on CD-ROM because of the large storage requirement.

clipboard (*ICT*) A temporary store for text or graphics that can then be transferred to a different position in the same or another application.

clip gauge (*Eng*) A type of EXTENSOMETER used in TENSILE TESTS; relative movement of two arms clipped to the specimen generates an electrical signal proportional to extension.

clipping (*ICT*) (1) Loss of initial or final speech sounds in telephony due to the operation of voice-operated or other switching devices. (2) Speech distortion caused by over-loading, limiting or amplifier circuits, resulting in the cut-off of amplitude peaks in louder passages. (3) Similar effects on extreme black or white or synchronizing pulse peaks in TV signals. (4) See LIMITER.

clipping circuit (*ICT*) One for removing peaks, tails or high-frequency ripple on leading and trailing edges and within pulses. See LIMITER.

clip screws (*Surv*) The screws by which the two verniers of the vertical circle of a theodolite may be adjusted so as to eliminate index error. Also *antagonizing screws*.

clitellum (*BioSci*) A special glandular region of the epidermis of OLIGOCHAETA that secretes the cocoon and the albuminoid material which nourishes the embryo.

clitoridectomy (*Med*) Surgical removal of the clitoris.

clitoridotomy (*Med*) Circumcision of the female; the surgical removal of the clitoral hood.

clitoris (*BioSci*) In female mammals, a small mass of erectile tissue, homologous with the glans penis of the male, situated just anterior to the vaginal aperture.

cloaca (*BioSci*) (1) Generally, a posterior invagination or chamber into which open the anus, the genital ducts and the urinary ducts. (2) In Urochordata, the median dorsal part of the atrium. (3) In Holothuroidea, the wide posterior terminal part of the alimentary canal into which the respiratory trees open.

cloacitis (*Vet*) Inflammation of the cloaca.

cloanthite (*Min*) See CHLOANTHITE.

clock (*ElecEng*) See COUNTER.

clock (*ICT*) Electronic unit that synchronizes processes within a computer system by generating pulses at a constant rate. Also *master clock*. Cf REAL-TIME CLOCK.

clock-driven behaviour (*Psych*) Rhythmic behaviour under the influence of an endogenous clock, found throughout the animal kingdom and involving many of the annual, lunar and daily rhythms of behaviour found in animals. See ZEITGEBER.

clock gauge (*Eng*) A length-measuring instrument in which the linear displacement of an anvil is magnified and indicated by the deflection of a needle pointer rotating over a dial. Often used as part of a vertical COMPARATOR.

clock gene (*BioSci*) A gene with expression levels that vary in a cyclical fashion and that might therefore be involved in the generation of biological rhythms such as circadian cycles.

clock meter (*ElecEng*) An energy meter in which the current passing causes a change in the rate of a clock.

clock paradox (*Phys*) A phenomenon resulting from an experiment involving two identical clocks, initially together and showing the same time, one of which is carried off on a high-speed round-trip journey. Upon returning, the clock which moved will have lost time relative to the motionless clock by an amount prescribed by special relativity. The principle of relativity which claims that all observers are equal appears violated, but the apparent paradox is resolved by taking into account the acceleration which must occur. Expressed in terms of identical twins, one of whom undertakes a journey and is younger than the other twin upon returning, this is the *twin paradox*.

clock recovery (*ICT*) Any method by which the clock pulses needed in a synchronous transmission system are reconstructed from the data themselves rather than being transmitted independently.

clock speed (*ICT*) The speed of a microprocessor's internal clock, usually measured in MHz.

clock track (*ICT*) A series of optical or magnetic marks on the relevant input medium giving information that locates the read areas.

clomipramine (*Pharmacol*) A tricyclic antidepressant drug used to treat depression and obsessive–compulsive disorder.

clonal deletion (*BioSci*) Negative selection of T- or B-lymphocytes, based on the specificity of their antigen receptors, resulting in the death of cells with a particular antigen receptor. Thought to be one of the main ways in which self-tolerance is achieved by the removal of potentially lethal clones specific for self-antigens. See panel on IMMUNE RESPONSE.

clonal selection (*BioSci*) Hypothesis proposed by Burnet to explain specific antibody production and immunological memory. Immune responses are proposed to occur when antigens selectively stimulate those lymphocytes with receptors capable of recognizing the antigens, from among a large lymphocyte population bearing a great variety of different receptors. The resulting expansion of lymphocyte clones (*clonal expansion*) accounts for the specificity and memory properties of the immune system. See panel on IMMUNE RESPONSE.

clone (*BioSci*) (1) Organisms, cells or micro-organisms all derived from a single progenitor by asexual means. They have therefore an almost identical genotype. In plants it includes those derived by vegetative propagation such as grafting and taking cuttings. (2) Used loosely to describe the procedures by which a VECTOR with an inserted DNA sequence makes multiple copies of itself and thus of the inserted sequence, ie *cloning*. See panels on DNA AND THE GENETIC CODE and GENETIC MANIPULATION.

clone (*ICT*) (1) A computer, software package, etc, that resembles a (usually more costly) product built by another manufacturer. (2) To copy the number (eg of a stolen mobile phone) onto a microchip which is then used in a different phone, so that the owner of the original phone is billed for any calls.

clonic phase (*Psych*) The third phase in a *grand mal seizure* in which the muscles contract and relax rhythmically while the body jerks in violent spasms.

clopidogrel (*Pharmacol*) An antiplatelet drug used to prevent strokes or heart attacks in patients with a history of previous atherosclerotic disease.

cloqué (*Textiles*) Woven or knitted double fabric with a blistered surface pattern.

close annealing (*Eng*) See BOX ANNEALING.

close coupling (*ElecEng*) See TIGHT COUPLING.

closed captions (*ImageTech*) Captions invisibly encoded into tape and disk SOFTWARE for the hearing-impaired, revealed by a special reader.

closed cell (*Eng*) See FOAM.

closed circuit (*ImageTech*) A system of video picture and sound transmission in which the camera and the display, even if remote, are directly linked, in contrast to broadcast methods.

closed circuit (*Phys*) A circuit in which there is zero impedance to the flow of any current, the voltage dropping to zero.

closed-circuit grinding (*ChemEng*) Size reduction of solids done in several stages, the material after each stage being separated into coarser and finer fractions, the coarser being returned for further size reduction and the finer being passed on to a further stage, so that overgrinding is minimized.

closed-coil winding (*ElecEng*) An armature winding in which the complete winding forms a closed circuit.

closed community (*BioSci*) A plant community that occupies the ground without leaving any spaces bare of vegetation.

closed-core transformer (*ElecEng*) A transformer in which the magnetic circuit is entirely of iron, ie with no air-gaps.

closed cycle (*Eng*) Heat engine in which the working substance continuously circulates without replenishment.

closed-cycle control system (*ElecEng*) One in which the controller is worked by a change in the quantity being controlled, eg an automatic voltage regulator in which a field current is actuated by a deviation of the voltage from a desired value, the REFERENCE VOLTAGE.

closed diaphragm (*Acous*) Diaphragm or cone which is not directly open to the air, but communicates with the latter through a horn, which serves to match the high mechanical impedance of the diaphragm with the low radiation impedance of the outer air.

closed herd (*Agri*) A highly inbred group of domesticated animals where reproduction outside the herd is completely prevented.

closed inequality (*MathSci*) An INEQUALITY that defines a CLOSED SET of points, eg $-1 \leqslant x \leqslant +1$.

closed interval (*MathSci*) An INTERVAL, such as $a \leqslant x \leqslant b$, which includes its end-points. Cf OPEN INTERVAL.

closed-jet wind tunnel (*Aero*) Any wind tunnel in which the working section is enclosed by rigid walls.

closed loop (*ICT*) A description of any system in which part of the output is fed back to the input to effect a control or regulatory action; also, the performance of such a system when the feedback path is connected. Cf OPEN LOOP. See FEEDBACK.

closed-loop recycling (*Eng*) Material reclamation within factory or industrial system, eg IN-HOUSE RECYCLING.

closed magnetic circuit (*ElecEng*) The magnetic core of an inductor or transformer without air-gap.

closed mitosis (*BioSci*) Mitosis during which the nuclear envelope remains more or less intact, eg in some algae and fungi.

closed pipe (*Acous*) An organ pipe which is closed at one end. The wavelength of its fundamental resonance is four times the length of the air column inside the pipe. Also *stopped pipe*.

closed pore (*PowderTech*) A cavity within a particle of powder which does not communicate with the surface of the powder.

closed set (*MathSci*) (1) A set of points which includes all its accumulation points. Complement of an OPEN SET. (2) (of a set under an operation) Such that all the results of applying the given operation to members of the set are themselves members of the set.

closed slots (*ElecEng*) In the rotor or stator of an electric machine, slots receiving the armature winding, which are completely closed at the surface and therefore in the form of a tunnel. Also *tunnel slots*.

closed stokehold (*Eng*) A ship's boiler room, closed to allow fans to maintain it at an air pressure slightly above atmospheric, so that forced draft may be provided to the furnaces.

closed subroutine (*ICT*) Separate calls to the SUBROUTINE using the same piece of code. Cf OPEN SUBROUTINE.

closed traverse (*Surv*) A traverse in which the final line links up with the first line.

closed universe (*Astron*) A term used to describe the universe if it has a density greater than a certain critical value. Under this condition space is spherically curved and the universe has a finite volume with no boundary. Eventually gravity would halt the expansion of the universe, which would then contract leading eventually to a BIG CRUNCH. See panels on COSMOLOGY and GRAND UNIFIED THEORIES.

closed user group (*ICT*) (1) A group of users of a public or commercial DATABASE who have exclusive access to certain information stored within it. A subscription fee is usually required. (2) A group of users of a private telecommunications network who, by the use of CALL SCREENING, can talk to each other but prevent other network users from calling them.

closed vascular bundle (*BioSci*) A vascular bundle that does not include any cambium and that will not, therefore, form secondary tissues.

close file (*ICT*) Correct method of giving up access to a file on backing store; this allows operations hidden from a user, such as clearing of buffers, to be successfully completed. Cf OPEN FILE.

close packing of atoms (*Chem*) Metals in the solid state usually pack together regularly to give close-packed crystal structures. See panel on CLOSE PACKING OF ATOMS.

closer (*Build*) Any portion of a brick used, in constructing a wall, to close up the bond next to the brick of a course. Also *bat, glut*.

close sand (*Eng*) See OPEN SAND.

close string (*Build*) See HOUSED STRING.

close timbering (*Build, CivEng*) Trench or excavation lining in which boards have no space between them.

close-up (*ImageTech*) Shot taken from a short distance, typically one showing only the face or head of an actor.

closing error (*Surv*) In a closed traverse, designed to return to the starting point, the discrepancy revealed on plotting, which may be small enough to allow distribution through the series of measurements.

closing layer (*BioSci*) One of the alternating layers of compact and loose suberized tissue in the lenticels of some species of plant.

closing membrane (*BioSci*) Same as PIT MEMBRANE.

closing-up (*Eng*) The operation of forming the head on a projecting rivet shank.

Clos network (*ICT*) A switching network composed of three or more interconnected stages, each comprising a number of CROSSPOINT SWITCHES. Such a network requires a smaller number of crosspoints than one that carries out the same function using a single stage.

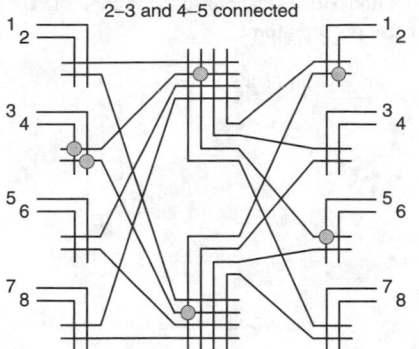

2–3 and 4–5 connected

Clos network This simplified network uses fewer crosspoints than the equivalent single stage switch.

clostridium (*BioSci*) An ovoid or spindle-shaped bacterium, specifically one of the anaerobic genus *Clostridium*, that contains several species pathogenic to humans and animals: *Cl.botulinum* (botulism), *Cl.chauvei* (blackleg), *Cl.tetani* (tetanus), *Cl.welchii* (gas gangrene).

closure (*Eng*) A device used to effect final sealing of containers, eg. screw tops for bottles, lids for boxes. Often polymeric, where design freedom allows discrimination between users, such as child-proof plastic tops for medicine bottles.

clot (*Med*) The semi-solid state of blood or of lymph when they coagulate. See EMBOLISM, THROMBOSIS.

cloth (*Textiles*) Generic term embracing all fabrics woven, felted, knitted or non-woven, etc, using any material or synthetic fibre or continuous filament or blends of these. Including apparel, furnishing or industrial *fabrics*.

clothing (*Build*) The covering (walls, roofing, etc) applied to the structural framework of a steel-framed building.

cloth joints (*Print*) See JOINTS.

clothoid (*MathSci*) See CORNU'S SPIRAL.

cloud (*EnvSci*) A mass of water droplets or ice particles remaining more or less at constant altitude. Cloud is usually formed by condensation brought about by warm moist air which has risen by convection into cooler regions and has been cooled thereby, and, by expansion, below its dew point.

cloud and collision warning system (*Aero*) A primary radar system with forward scanning which gives a visual display of dangerous clouds and high ground at ranges sufficient to allow course to be altered for their avoidance. A second system is usually necessary to give short-range warning of the presence of other aircraft.

cloud chamber (*Phys*) A device for making visible the paths of charged particles moving through a gas. The gas, saturated with a vapour, is suddenly cooled by expansion and the vapour condenses preferentially on the ions produced by the passage of charged particles through the chamber. The tracks of the particles are revealed by the drops of liquid which, if suitably illuminated, can be photographed. See BUBBLE CHAMBER.

cloudiness (*EnvSci*) The amount of sky covered by cloud irrespective of type. Estimated in eighths (*oktas*). Overcast, 8; cloudless, 0.

clouding (*Build*) See BLOOMING.

cloud point (*MinExt*) Standard test on fuel oils, esp *DERV*, for the temperature at which wax begins to appear (as a white cloudiness) in the fuel as it is cooled. Precipitating wax may lead to blocked filters and fuel lines.

cloudy swelling (*Med*) A mild degenerative change in cells in which the swollen cloudy appearance is due to the presence of small vacuoles containing fluid and indicating damage to the 'sodium pump' of the cell membrane.

clouring (*Build*) Chiselling or picking small indentations on a wall surface as a key for the finish.

clout nail (*Build*) A nail having a large, thin, flat head.

clove oil (*Chem*) An oil obtained from the flowers of *Eugenia aromatica*. It is a pale-yellow, volatile liquid, of strong aromatic odour; bp 250–260°C, rel.d. 1·048–1·070. Formerly much used in confectionery, microscopy and dentistry.

cloxacillin (*Pharmacol*) A PENICILLIN which is resistant to penicillinase produced by *Staphylococcus* and therefore used to treat penicillin-resistant staphylococcal infections.

clozapine (*Pharmacol*) An atypical NEUROLEPTIC drug used as a sedative and to treat schizophrenia in patients unresponsive to conventional antipsychotics.

clubbing of the fingers (*Med*) Increased convexity of the nails with loss of nail-bed angle and thickening of soft tissue round the phalanges, associated with a number of diseases (carcinoma of the lung, chronic lung sepsis, chronic cyanotic heart disease, infectious endocarditis and inflammatory bowel disease). Can be familial, cause uncertain.

club line (*Print*) A term used for the last (short) line of a paragraph. It is common practice to avoid the occurrence of a club line at the top of the page. Also *break line*.

club moss (*BioSci*) In various usages, the Lycopsida, the orders Lycopodiales and Selaginellales, or just the Lycopodiaceae.

clunch (*MinExt*) (1) SEAT EARTH below a coal seam of marl or shale; may yield a tough fireclay. (2) Soft limestone or compacted chalky clay formerly used as a building material esp in E England.

Clupeiformes (*BioSci*) An order of Osteichthyes, mostly planktonic feeders containing some economically important species. Herrings and anchovies.

Clusius column (*NucEng*) A device used for isotope separation by method of thermal diffusion, consisting of a long vertical cylinder with a hot wire along the axis.

cluster (*Agri*) A clawpiece, together with four teat cups, applied to a cow's udder for automated milking.

cluster (*Chem*) A molecular compound containing four or more metal atoms bonded to one another without intervening ligands. Clusters have been prepared with more than 40 metal atoms so bonded, and the internal structure of these approximates to that of metals themselves.

cluster (*ICT*) A group of fixed-length sectors defined for storage systems under MS-DOS. Each sector is 512 bytes long and there may be one, two or four sectors per cluster.

cluster (*ImageTech*) On a cinematograph projector, a group of magnetic heads for the reproduction of multichannel sound from striped prints.

cluster (*MathSci*) In statistical sampling, a subset of a population, usually naturally occurring, which has some distinguishing feature.

cluster analysis (*BioSci*) Hierarchical classification technique, often used to reveal patterns of similarity among species lists from many sites.

cluster analysis (*MathSci*) A statistical method of classifying observations into subsets, members of which satisfy some criterion of similarity.

cluster cup (*BioSci*) The popular name for an AECIDIUM.

clustered column (*Arch*) A column formed of several shafts bunched together.

cluster mill (*Eng*) Rolling mill in which two small-diameter working rolls are each backed up by two or more larger rolls.

cluster of differentiation antigen (*BioSci*) See CD ANTIGEN.

cluster variables (*Astron*) Short-period variable stars first observed in globular clusters; they are typical members of Population II. The periods are less than 1 day. Includes RR LYRAE VARIABLE stars.

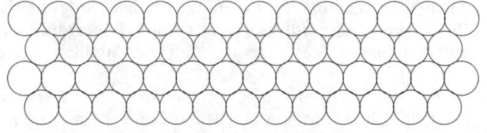

Close packing of atoms

The atoms of the metallic elements of the periodic table can be modelled by perfect spheres. Metals in the solid state usually pack together regularly to give CRYSTAL STRUCTURES, most of which are close-packed. In other words, the spherical atoms arrange themselves so as to fill the available space to the maximum possible extent. On a flat plane, spheres of equal radius pack together most efficiently in a *hexagonal array*, with each sphere surrounded by six *nearest neighbours* (Fig. 1).

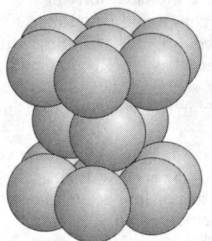

Fig. 1 **A hexagonal array in a flat plane**.

As similar sheets are added successively to form a three-dimensional structure, two types of close-packed systems can be created. The second sheet (B) to be added sits in hollows created by the first layer (A), and if the third sits in the hollows created by B in the same way as the first layer, the *hexagonal close-packed* (hcp) structure is formed (Fig. 2). The sequence of layers is thus ABABAB... and the unit cell in section shows that each atom is surrounded by twelve similar atoms, ie its co-ordination number is 12.

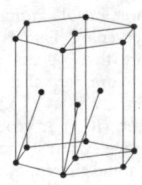

Fig. 2 **Hexagonal close packing**.

Alternatively, the third layer C can be added in a different configuration (C) to either preceding layer (Fig. 3), giving *cubic close packing* (ccp). The sequence of layers is ABCABCABC... and the unit cell is orientated with its diagonal at right angles to the layers of close packed atoms.

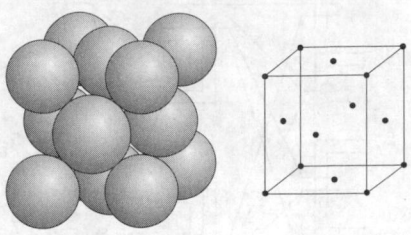

Fig. 3 **Cubic close packing**.

The cubic unit cell possesses a sphere at the centre of each face: hence the alternative term *face-centred cubic* (fcc). Both hcp and fcc structures possess a coordination number of 12 and occupy 74% of the space available.

A third type of close-packed structure is created by square arrays, which can pack in only one way.

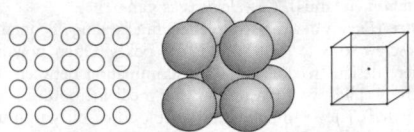

Fig. 4 **Body-centred cubic packing**.

The unit cell possesses a co-ordination number of 8 owing to the less efficiently packed structure (68% of space filled). It comprises a cube with a sphere at the centre and is thus known as *body-centred cubic* (bcc). Other atomic crystals possess unit cells of lower symmetry (eg rhombic, tetragonal) due to distortion of the close-packed structures by bonding effects. The close-packed models of metal structures are particularly useful for explaining alloy formation (eg solid solutions). In addition, the structure of ionic solids (which are the basis of useful ceramics and refractories) can be explained by close-packed lattices of anions into which the smaller cations can be fitted.

See panel on CRYSTAL LATTICE.

---

clutch (*Build, CivEng*) A connecting bar used between adjacent flanges of I-section steel-sheet piles to retain them in position. In section, it consists of a web part running between a pair of adjacent piles, with curved 'flanges' which slide over and secure the flanges of the piles.

clutch (*Eng*) A device by which two shafts or rotating members may be connected or disconnected, either while at rest or in relative motion.

clutch stop (*Autos*) A small brake arranged to act on the driven member of a clutch when it is fully withdrawn, to facilitate an upward gear change. With a modern synchro gearbox this device is only used in heavy gearboxes. Also *inertia brake*.

clutter (*Radar*) Unwanted echoes on a radar display, usually due to terrain in the immediate vicinity of the antenna, or to rain or sea.

CLV (*ImageTech*) Abbrev for CONSTANT LINEAR VELOCITY.

Cm (*Chem*) Symbol for CURIUM.

cm (*Genrl*) Abbrev for CENTIMETRE.

cm³ (*Genrl*) Abbrev for *cubic centimetre*, the unit of volume in the CGS metric system. Also *cc*.

C-MAC (*ImageTech*) See MAC.

CMC (*FoodSci*) Abbrev for CARBOXYMETHYLCELLULOSE.

CMI (*ICT*) Abbrev for COMPUTER-MANAGED INSTRUCTION.

CMOS (*ICT, Electronics*) Abbrev for COMPLEMENTARY METAL–OXIDE SILICON.

C-mount (*ImageTech*) A screw-thread lens mount used on 16 mm cine cameras and also some video cameras, particularly closed-circuit TV.

CMS (*ICT*) Abbrev for *content management system*, a computer program that allows the creation and management of a website without the need for sophisticated technical knowledge or programming ability on the part of the webmaster.

CN (*Chem*) Abbrev for CO-ORDINATION NUMBER.

cn (*MathSci*) Abbrev for ELLIPTICAL FUNCTIONS.

CNC (*ICT*) Abbrev for COMPUTER NUMERICAL CONTROL.

cnemidium (*BioSci*) In birds, the lower part of the leg, bearing usually scales instead of feathers.

cnemis (*BioSci*) The shin or tibia.

CNES (*Space*) Abbrev for *Centre National d'Etudes Spatiales*, the national space agency of France.

C-network (*ICT*) One consisting of three impedance branches in series. The free ends are connected to one pair of terminals and the junctions to another.

Cnidaria (*BioSci*) An aquatic Metazoan phylum with radial or biradial symmetry. They possess a single cavity in the body, the enteron, that has a mouth but no anus. They generally have only two germinal layers, the ectoderm and the endoderm, from one of which the germ cells are always developed. They have specialized stinging cells or CNIDO-BLASTS. They include the important fossil corals which range from the Ordovician to the present. Polyps, corals, sea anemones, jellyfish and hydra. Also *coelenterates*.

cnidoblast (*BioSci*) A thread-cell or stinging-cell, containing a NEMATOCYST; characteristic of the Cnidaria. Also *cnida*.

CNS (*BioSci*) Abbrev for CENTRAL NERVOUS SYSTEM.

C-number (*ICT*) The actual destination number, as opposed to the number dialled by the caller, for a call in an INTELLIGENT NETWORK. See A-NUMBER, B-NUMBER.

Co (*Chem*) Symbol for COBALT.

coacervation (*Chem*) The reversible aggregation of particles of an emulsoid into liquid droplets preceding FLOCCULA-TION.

coach bolt (*Eng*) A bolt having a convex head with a square section beneath it which fits into a corresponding cavity in the material to be bolted and prevents the bolt from turning as the nut is screwed on.

coach screw (*Eng*) A large wood screw with a square or hexagon head which is turned by a spanner; used in heavy timberwork.

co-adaptation (*BioSci*) Correlated adaptation or change in two mutually dependent organisms.

coagulation (*BioSci*) The irreversible setting of protoplasm on exposure to heat, extreme pH or chemicals. Cf DENATURATION.

coagulation (*Chem*) The precipitation of colloids from solutions, particularly of proteins. See panel on GELS.

coagulum (*Med*) A clot.

coal (*Geol*) A general name for firm brittle *carbonaceous rocks*; derived from vegetable debris, but altered, particularly in respect of volatile constituents, by pressure, temperature and a variety of chemical processes. The various types are classified on basis of volatile content, calorific value, caking and coking properties.

coal ball (*BioSci*) A calcareous nodule, usually containing abundant petrified plant remains found in some seams of coal.

coal-cutting machinery (*MinExt*) Mechanized systems used in colliery to detach coal from its face, and perhaps to gather and load it to transporting device.

coalescent (*BioSci*) Grown together, esp by union of walls.

Coal Measures (*Geol*) The uppermost division of the CARBONIFEROUS SYSTEM, consisting of beds of coal interstratified with shales, sandstones, limestones and conglomerates.

coal–oil mixture (*MinExt*) A stabilized suspension of coal in fuel oil that may be transported as a liquid by pumping through a pipeline, or through jets for burning. Abbrev COM.

Coal Sack (*Astron*) A large obscuring dust cloud visible to the naked eye as a dark nebula in the Milky Way near the Southern Cross.

coal sizes (*MinExt*) Sizes officially (UK) recognized are: large coal, over 6 in; large cobbles, 6–3 in; cobbles, 4–2 in; trebles, 3–2 in; doubles, 2–1 in; singles, 1 to $\frac{1}{2}$ in; peas, $\frac{1}{2}$ to $\frac{1}{4}$ in; grains, $\frac{1}{4}$ to $\frac{1}{8}$ in.

coaltar (*Chem*) The distillation products of the high- or low-temperature carbonization of coal. Coaltar consists of hydrocarbon oils (benzene, toluene, xylene and higher homologues), phenols (carbolic acid, cresols, xylenols and higher homologues), and bases, such as pyridine, quinoline, pyrrole and their derivatives.

coaltar paints (*Build*) Surface coatings based on combinations of coaltar pitch and epoxy resins, either single- or two-pack materials, which have excellent resistance to heat, moisture and chemical attack.

coaltar wood preservatives (*Build*) Cheap wood preservatives, eg creosote, derived from coaltar distillation. Normally applied to rough-hewn exterior timber such as garden fencing, telegraph poles, etc. Subsequent decoration using oil paints is not recommended since bleeding will occur.

co-altitude (*Astron*) See ZENITH DISTANCE.

coal washery (*MinExt*) Cleaning plant where run-of-mine coal is processed to remove shale and pyrite, reduce ash and sort in sizes.

Coanda effect (*Aero*) Named after its discoverer, it is the tendency of a fluid jet to attach itself to a downstream surface roughly parallel to the jet axis. If this surface curves away from the jet the attached flow will follow it, deflecting from the original direction.

coarctation (*Med*) A narrowing or constriction esp of the aortic arch causing hypertension in upper part of body.

coarse aggregate (*CivEng*) Gravel or crushed stone (forming a constituent part of concrete) between 4·5 mm ($\frac{3}{16}$ in) and 35 mm ($1\frac{1}{2}$ in) in major diameter. See AGGREGATE.

coarsening (*Eng*) Increase in the grain size of metals usually by heating for a time and at a temperature where grain growth is rapid.

coarse scanning (*Radar*) Rapid scan carried out to determine approximate location of any target.

coarse screen (*ImageTech, Print*) A term applied to halftones for use on rough paper. The *screen* may be up to 33 lines to the centimetre (80 lines to the inch).

coarse stuff (*Build*) A mixture of lime mortar and hair formerly used as a first coat for plastering internal walls.

coastal reflection (*ICT*) Reflection of signal by a land mass so that the resultant received signal consists of direct and reflected waves. This can cause direction-finding errors.

coastal refraction (*ICT*) Refraction, towards the normal, of waves arriving from sea to land at their incidence with the shoreline, resulting in an appreciable error in radio direction-finding for bearings making a small angle with the shoreline.

coastdown (*Eng*) The slowing down of a turbine or other moving part after its energy supply is cut off.

coasting (*ElecEng*) Running with motor supply cut off and the brakes not applied.

coasting of temperature (*Eng*) Rise above correct predetermined operating temperature when the fuel supply has been checked or shut off; due to excessive thermal storage in the furnace brickwork through prior overheating.

coat (*BioSci*) (1) See INTEGUMENT (ovule). (2) See TESTA (seed coat).

coated cathode (*Electronics*) One sprayed or dipped with a compound having a lower WORK FUNCTION than the base metal, in order to enhance electron emission, which may be thermionic or photoemissive.

coated fabric (*Textiles*) A fabric coated on one or both sides with a layer of a coating material such as a plastic, esp polyvinyl chloride or rubber.

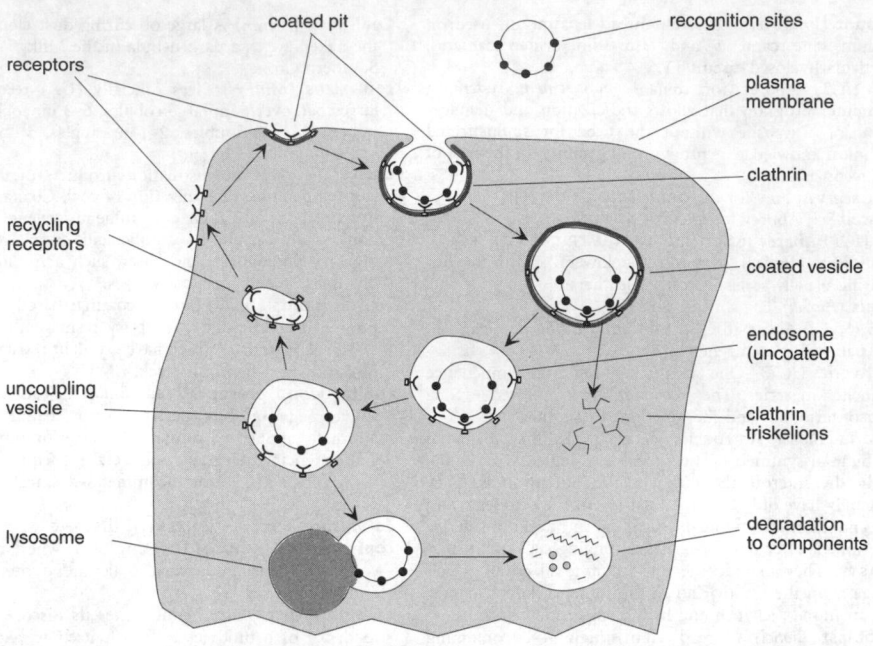

coated vesicle Pathway showing how a protein complex particle is brought into a cell and degraded.

Labels on diagram: coated pit, receptors, recycling receptors, uncoupling vesicle, lysosome, recognition sites, plasma membrane, clathrin, coated vesicle, endosome (uncoated), clathrin triskelions, degradation to components

coated fuel particle (*NucEng*) Fuel in small particles and coated in dense carbon and silicon carbide to minimize the release of fission products.

coated lens (*ImageTech, Phys*) Lens coated with a thin film of transparent material (blooming), to reduce the amount of light reflected from it. The film has a thickness of a quarter of a wavelength and its refractive index is the geometric mean of that of air and glass. Used to increase the light transmission through an optical system by reducing internal reflections, thus also reducing FLARE.

coated pit (*BioSci*) A small surface invagination of the plasma membrane distinguished by a thick proteinaceous layer composed largely of CLATHRIN on its cytoplasmic side. The pits invaginate to form vesicles within the cytoplasm. See RECEPTOR-MEDIATED ENDOCYTOSIS.

coated vesicle (*BioSci*) A cytoplasmic vesicle with a CLATHRIN coat formed by the invagination of a coated pit. Also *spiny vesicle*.

coating (*ElecEng*) The metallic sheets or films forming the plates of a capacitor.

coating (*Eng*) Generic term for protection of steel from corrosion.

coating and wrapping (*MinExt*) The process of covering a pipeline with bitumen and winding protective paper around it. Often done by machine just before the welded pipe is lowered into the trench.

coating machine (*Paper*) Any machine designed to apply a layer of a coating material to the surface(s) of a base paper.

coaxial antenna (*ICT*) Exposed $\lambda/4$ length of coaxial line, with reversed metal cover, acting as a dipole. λ = wavelength.

coaxial cable (*Electronics*) A cable that has a solid inner conductor surrounded by a tubular outer conductor, the dielectric separating them being (in the case of rigid coaxial cables) air or, for flexible cables, solid or expanded plastic, insulating beads or disks. The electromagnetic field associated with the currents in the conductors is confined to the dielectric, and, at high frequencies, a high degree of screening is achieved. A correctly terminated cable, more than a few wavelengths long, will present a characteristic impedance to an imposed ac, its value depending on the

relative dimension of the conductors and the permittivity of the dielectric. Used extensively to feed high-frequency and very-high-frequency antennas, and for multiplex signals in long-distance and submarine telecommunications.

coaxial circles (*MathSci*) Two families of circles having a common radical axis.

coaxial filter (*ICT*) Filter in which a section of coaxial line is fitted with re-entrant elements to provide the inductance and capacitance of a filter section.

coaxial line (*ICT*) A TRANSMISSION LINE coaxial in form, eg COAXIAL CABLE.

coaxial-line resonator (*ICT*) Resonator in which the standing waves are established in a coaxial line, short- or open-circuited at the end remote from the drive. Of very high Q, these are used for stabilizing oscillators, or for selective coupling between amplifying stages.

coaxial propellers (*Aero*) Two propellers mounted on concentric shafts having independent drives and rotating in opposite directions.

coaxial relay (*ICT*) Switching device in which the coaxial circuit on both sides of the contact is maintained at its correct impedance level, thus avoiding wave reflection in the current path.

coaxial stub (*ICT*) Section of coaxial line that is short-circuited at one end and functions as a high impedance at quarter-wave resonance.

cob (*Agri*) (1) A male swan, less commonly a male domesticated bird raised for eating. (2) A short-legged, thickset harness horse below 15 hands.

cob (*Build*) An unburnt brick.

cobalt (*Chem*) Hard, grey metallic element in the eighth group of the periodic system. It is magnetic below 1075°C, and can take a high polish. Symbol Co, at no 27, ram 58·9332, valency 2 or 3, mp 1490°C, electrical resistivity 0·0635 $\mu\Omega$ m at 20°C . There are a number of independent cobalt minerals, eg *smaltite* and *cobaltite*; abundance in the Earth's crust 29 ppm. It is an essential trace element in living systems, but toxic in excess. Tensile strength (commercial grade containing carbon) 450 MN m^{-2}. Similar properties to iron but harder; used extensively in alloys, and as a source for radiography or industrial irradiation. Forms a hexagonal structured compound with samarium ($SmCo_5$) which has high anisotropy and coercivity and is used to make short permanent magnets.

cobaltammines (*Chem*) Complex compounds of cobalt salts with ammonia or its organic derivatives, eg *hexammino (III) compounds*, which contain the group $[Co(NH_3)_6]^{3+}$.

cobalt bloom (*Min*) See ERYTHRITE.

cobalt bomb (*Phys*) (1) Theoretical nuclear weapon loaded with cobalt-59. The long-life radioactive cobalt-60, formed during fission, would make the surrounding area uninhabitable. (2) Radioactive source comprising cobalt-60 in lead shield with shutter.

cobalt carbonyl (*Chem*) $Co_2(CO)_8$, used as a catalyst, esp in the formation of aldehydes and diene polymerization.

cobaltite (*Min*) Sulphide and arsenide of cobalt, crystallizing in the cubic system; usually found massive and compact, with smaltite.

cobalt (II) oxide (*Chem*) CoO. Used to produce a deep-blue colour in glass and, in small quantity, to counteract the green tinge in glass caused by the presence of iron.

cobalt steel (*Eng*) Steel containing 5–12% cobalt, 14–20% tungsten, about 4% chromium, 1–2% vanadium, 0·8% carbon and a trace of molybdenum, used for tools. They have high hardness, and retain this at red heat. See panel on STEELS.

cobalt unit (*Radiol*) Source of radioactive cobalt (half-life 5·3 yr) mounted in a protective capsule.

cobalt vitriol (*Min*) See BIEBERITE.

cobble (*Geol, MinExt*) (1) A rock fragment intermediate in size between a pebble and a boulder, diameter > 64 mm and < 256 mm. (2) See COAL SIZES.

cobblestone (*CivEng*) A smallish roughly squared stone once used for paving purposes; superseded by setts.

cobb sizing test (*Paper*) A method of measuring the amount of water taken up by a test piece in given time under the prescribed conditions of test. Expressed in g m^{-2}.

COBE (*Space*) Abbrev for *cosmic background explorer*. A NASA satellite launched on 18 November 1989 to study the residual radiation from the Big Bang; discovered COSMIC RIPPLES in 1992. See MICROWAVE BACKGROUND.

COBOL (*ICT*) Abbrev for *common business oriented language*. Programming language used for business data processing.

cocaine (*Pharmacol*) A coca-base alkaloid, the methyl ester of benzoyl-L-ecgonine. Used as a local anaesthetic, now mainly for eye, nose and throat surgery, and as a narcotic drug, producing marked addiction.

co-carcinogens (*BioSci*) Substances that, though not carcinogenic in their own right, potentiate the activity of a carcinogen. Strictly speaking they differ from TUMOUR PROMOTERS in requiring to be present concurrently with the carcinogen.

coccidiomycosis (*Med, Vet*) A fungal infection with *Coccidioides immitis* seen in the south-west USA, C and S America. Usually mild upper respiratory tract infection in humans but may cause more severe infestation of lung. Causes a chronic infection of cattle, sheep, dogs, cats and certain rodents. Also *coccidioidomycosis*.

coccidiosis (*Vet*) A contagious infection of animals and birds by protozoa of the genera *Eimeria* and *Isospora*, usually affecting the intestinal epithelium and causing enteritis.

coccoid (*BioSci*) Small, unicellular, walled, spherical and non-motile algae.

coccolith (*BioSci*) Small (say 2–10 μm) calcified scale covering the cells of coccolithophorids (flagellated unicellular algae of the Haptophyceae). Abundant as fossils in chalk.

coccus (*BioSci*) A spherical or near-spherical bacterium with a diameter from 0·5 to 1·25 μm.

coccydynia (*Med*) Severe pain in the coccyx.

coccygectomy (*Med*) Excision of the coccyx.

coccyx (*BioSci*) A bony structure in primates and amphibians, formed by the fusion of the caudal vertebrae; urostyle. Pl *coccyges*.

co-channel interference (*ICT*) In a mobile-telephone network or other system employing FREQUENCY REUSE, interference on one channel from another using the same frequency.

cochineal (*Chem*) The dried bodies of female insects of *Coccus cacti* plus the enclosed larvae. The colouring matter is carminic acid, $C_{17}H_{18}O_{10}$, soluble in water and ethanol.

cochlea (*Acous, BioSci*) In mammals, the complex spirally coiled part of the inner ear which translates mechanical vibrations into nerve impulses.

cochlear potentials (*Acous*) Electrical potentials within the cochlear structures resulting from acoustic situation.

cochleate (*BioSci*) Spirally twisted, like the shell of a snail. Also *cochleariform*.

cock (*Agri*) A male chicken over 1 year of age.

cock (*Build*) See PLUG COCK.

cock-bead (*Build*) A bead projecting from the surface which it is decorating. Also *cocked bead*.

Cockcroft–Walton accelerator (*ElecEng*) High-voltage machine in which rectifiers charge capacitors in series, the discharge of these driving charged particles through an accelerating tube.

cocked bead (*Build*) See COCK-BEAD.

cockerel (*Agri*) A male chicken under 1 year of age; commonly used to describe all intact, male chickens.

cocket centring (*CivEng*) Arch centring which leaves some headroom above the springing lines.

cocking rollers (*Print*) See SLEWING ROLLERS.

cockle-stairs (*Arch*) Winding stairs.

cockling (*Paper*) Local wrinkling of the surface of paper generally due to the release of dried-in strains as the result of moisture take-up.

cockling (*Textiles*) A defect in a fabric which causes wrinkles usually arising from non-uniform shrinking. May be used to produce an interesting irregularity in the appearance of a fabric.

cockpit (*Aero*) The compartment in which the pilot or pilots of an aircraft are seated. It is so called even where it forms a prolongation of the cabin. Also *flight deck*.

cockroach (*Print*) Typography set entirely in lower case, without the use of any capitals.

cockscomb pyrites (*Min*) A twinned form of MARCASITE. Also *cockscomb*.

cockspur fastener (*Build*) A bronze or iron fastener for casement windows, used in conjunction with a stay bar and pin.

cock-up (*Print*) (1) A large initial that extends above the first line and ranges at the foot. (2) In line-casting machines, when the two-letter matrix is in the cock-up position the companion italic or bold letter is cast.

cocoa (*FoodSci*) Beverage obtained from the fruit of the tree, *Theobroma cacao*. After de-husking, cocoa beans with a fat content of 55% are mechanically disintegrated (ground) and a thick liquor (cocoa mass) produced. From this, cocoa butter is extracted by pressing and the remaining part pulverized into cocoa powder.

cocobolo (*For*) A very durable C American hardwood (*Dalbergia*) whose heartwood varies from a rich red to a streaked, mellow orange-red, and has an irregular grain and fine texture.

co-codamol (*Pharmacol*) An analgesic containing a combination of CODEINE and PARACETAMOL.

coconut oil (*Chem*) The oil obtained from the fruit of the coconut palm; a white waxy mass, mp 20–28°C, rel.d. 0·912, saponification value 250–258, iodine value 8·9, acid value 5–50. Chief constituent lauric acid.

cocoon (*BioSci, Textiles*) In insects, a special envelope constructed by the larva for protection during the pupal stage; it consists either of silk or of extraneous matter bound together by silk. If intended for silk manufacture, the cocoon is heated to destroy the pupa and the thread is subsequently reeled.

co-culture (*BioSci*) Growth of distinct cell types in a combined culture. Some cells will only grow at low (clonal) density if grown together with a feeder layer of other cells; the latter are often irradiated to prevent their unwanted proliferation.

co-current contact (*ChemEng*) In chemical engineering processes which involve the transfer of heat or mass between two streams A and B, the opposite of COUNTERCURRENT CONTACT. At the start of the process, fresh A is in contact with fresh B, and at each succeeding stage progressively spent reagents are in contact, so that the effective driving force, and hence the overall economy, is lower.

cocuswood (*For*) A tropical American hardwood of the genus *Brya*, available only in the form of small logs. Also *Jamaica ebony, West Indian ebony.*

COD (*EnvSci*) Abbrev for *chemical oxygen demand*. See OXYGEN DEMAND.

CODAN (*ICT*) Abbrev for *carrier-operated device anti-noise*. Circuit for silencing a receiver in absence of a signal.

Coddington lens (*Phys*) A magnifying lens cut from a spherical piece of glass, having concentric spherical surfaces of equal radius at the ends, and a V-cut round the centre of its length to act as a stop at the centre of the sphere.

code (*ICT*) (1) See BINARY CODE. (2) A set of program instructions. (3) See MACHINE CODE. (4) See BINARY-CODED DECIMAL, UNIVERSAL PRODUCT CODE.

codec (*ICT*) Contraction of coder/decoder. Equipment used to extract from a speech or video signal a set of codes sufficient to reconstruct the signal (coding), or to reconstruct it from these codes (decoding). By eliminating some of the REDUNDANCY of the original signal, codecs are capable of considerably reducing the BANDWIDTH required for its transmission.

code-division multiple access (*ICT*) A SPREAD SPECTRUM technique in which many channels simultaneously occupy the whole of a single wide-band channel. Each channel can be independently recovered by making use of a unique code sequence superimposed on its data stream.

code exchange keying (*ICT*) A SPREAD SPECTRUM technique in which two different code sequences are transmitted, one representing '1' and the other '0'.

codeine (*Pharmacol*) An alkaloid of the morphine group, the methyl derivative of MORPHINE; a widely used analgesic.

co-dependancy (*Psych*) Mutual emotional dependency between two individuals. May be asymmetric in that one may care for or control the other in order to fulfil his or her own emotional needs. US *co-dependency*.

coder (*ICT*) In a pulse-modulation system, the sampler that tests the signal at specified intervals.

codes of construction (*ChemEng*) Written procedures for design, selection of materials, and manufacturing and operating procedures for production of chemical plant items for onerous conditions. Many are internationally acceptable, eg *ASME* (American Society of Mechanical Engineers), *TEMA* (Tubular Exchange Manufacturers Association), *ABCM* (Association of British Chemical Manufacturers).

codes of practice (*Build, CivEng*) Recommendations drawn up by a regulatory authority describing what is regarded as good practice in particular types of work. May not be mandatory.

coding (*ICT*) Process of transforming information into code in accordance with definite rules. See ENCODE, PULSE-AMPLITUDE MODULATION, PULSE-CODE MODULATION, PULSE-WIDTH MODULATION, RUN-LENGTH LIMITED.

coding (*Radar*) Process of subdividing a relatively long pulse of transmitted power into a predetermined pattern of shorter pulses. The receiver uses AUTOCORRELATION or MATCHED FILTER techniques to respond only to echoes bearing the transmitted code. This increases the radar's range, resolution and immunity to interference or jamming.

coding capacity (*BioSci*) The number of different protein molecules that a DNA molecule could specify. Because of the presence of EXONS, reiterated sequences and other non-coding regions, it is not a very helpful concept to apply to whole chromosomes or the genomes of higher organisms.

coding sequence (*BioSci*) That part of a nucleic acid molecule which can be transcribed and translated into polypeptide using the genetic code.

cod-liver oil (*Chem*) Oil obtained from the livers of cod-fish; a yellow or brown liquid, of characteristic odour, rich in vitamins A and D, rel.d. 0·992–0·930, saponification value 182–189, iodine value 141–159, acid value 204–207.

codomain (*MathSci*) Of a function or mapping, a set within which the values of the function or mapping lie, as contrasted with the *range*, which is the set of values the function actually takes, eg for an integral-valued function the codomain may be defined as the integrals or the reals, but the range is by definition the integrals.

codominant (*BioSci*) (1) A term describing a pair of alleles that both show their effects in heterozygotes, eg many blood-group genes. (2) Said of one of two or more species that together dominate a community.

codon (*BioSci*) A triplet of three consecutive bases in the DNA, or in messenger RNA, that specifies (*codes for*) a particular amino acid for incorporation into a polypeptide. See panel on DNA AND THE GENETIC CODE.

coenaesthesis (*Psych*) General consciousness or awareness of one's body. US *cenesthesis*.

coefficient (*MathSci*) In an algebraic expression or equation, a factor multiplying the variable (or quantity) under consideration. It may also be a number or parameter defining a certain characteristic, relationship, etc, eg coefficient of correlation.

coefficient of absorption (*Phys*) See ABSORPTION COEFFICIENT.

coefficient of apparent expansion (*Phys*) The value of the coefficient of expansion of a liquid, which is obtained by means of a dilatometer if the expansion of the dilatometer is neglected. It is equal to the difference between the true coefficient of expansion of the liquid and the coefficient of cubical expansion of the dilatometer.

coefficient of compressibility (*Chem*) A measure of deviation of a gas from BOYLE'S LAW.

coefficient of determination (*Genrl*) The statistic or number determined by squaring the CORRELATION COEFFICIENT. It represents the amount of variance accounted for by that correlation.

coefficient of dispersion (*ElecEng*) See DISPERSION COEFFICIENT.

coefficient of elasticity (*Phys*) See ELASTICITY.

coefficient of equivalence (*Eng*) A factor used in converting amounts of aluminium, iron and manganese into equivalent amounts of zinc, in relation to their effect on the constitution of brass, principally the solubility limit of the alpha solid solution.

coefficient of expansion (*Phys*) The fractional expansion (ie the expansion of the unit length, area or volume) per degree rise in temperature; the value for all gases is very nearly the same, namely 0·003 66 per kelvin when kept at constant pressure. Calling the coefficients of linear, super-ficial and cubical expansion of a substance α, β, and γ respectively, β is approximately twice, and γ three times, α. See THERMAL EXPANSION COEFFICIENT.

coefficient of fineness of water plane (*Ships*) The ratio of the actual area of the water plane to the area of the rectangle having length and breadth the same as the maximum dimensions of the water plane.

coefficient of friction (*Phys*) See FRICTION.

coefficient of perception (*Phys*) A term used in connection with the effect of glare; equal to the reciprocal of Fechner's constant.

coefficient of performance (*Eng*) When applied to a heat pump it is the ratio of high-temperature heat transfer to work input. In a refrigerator it is the ratio of low-temperature heat transfer to work input. It may be greater than unity.

coefficient of reflection (*Phys*) See REFLECTION FACTOR.

coefficient of restitution (*Phys*) The ratio of the relative velocity of two elastic spheres after direct impact to that before impact. If a sphere is dropped from a height onto a fixed horizontal elastic plane, the coefficient of restitution is equal to the square root of the ratio of the height of rebound to the height from which the sphere was dropped. See IMPACT.

coefficient of rigidity (*Phys*) See ELASTICITY OF SHEAR.

coefficient of thermal expansion (*Phys*) See THERMAL EXPANSION COEFFICIENT.

coefficient of utilization (*ElecEng*) A term used in lighting calculations to denote the ratio of the useful light to the total output of the installation.

coefficient of variation (*MathSci*) A dimensionless quantity measuring the relative dispersion of a set of observations, calculated as the ratio of the STANDARD DEVIATION to the mean of the data values (sometimes expressed as a percentage).

coefficient of viscosity (*Phys*) The value of the tangential force per unit area which is necessary to maintain unit relative velocity between two parallel planes unit distance apart in a fluid; symbol η: ie if F is the tangential force on the area A and (dv/dz) is the velocity gradient perpendicular to the direction of flow, then $F = \eta A(dv/dz)$. For normal ranges of temperature, η for a liquid decreases with increase in temperature and is independent of the pressure. Unit of measurement is N m^{-2} s in SI units (= 10 poise or dyne cm^{-2} s in CGS units). See VISCOSITY.

coele-, -coele (*Genrl*) Prefix and suffix from Gk *koilia*, large cavity (of the belly).

coelenterate (*BioSci*) See CNIDARIA.

coeliac (*BioSci*) In vertebrates, pertaining to the belly or abdomen.

coeliac disease (*Med*) A wasting disease in which failure to absorb fat from the intestines is associated with an excess of this substance in the feces, due to a sensitivity to wheat GLUTEN.

coelom (*BioSci*) The secondary body cavity of animals, which is from its inception surrounded and separated from the primary body cavity by mesoderm. Adjs *coelomic, coelomate*.

Coelomata (*BioSci*) A group of Metazoa, all of which possess a COELOM at some stage of their life history.

coelomere (*BioSci*) In metameric animals, the portion of coelom contained within one somite.

coelomocyte (*BioSci*) A cell that circulates within the coelomic fluid of invertebrate animals; usually applied to leucocytes involved in internal defence or wound-healing.

coelomoduct (*BioSci*) A duct of mesodermal origin, opening at one end into the coelom, at the other end to the exterior.

coelomostome (*BioSci*) In vertebrates, the ciliated funnel by which the NEPHROCOEL opens into the SPLANCHNOCOEL.

coelostat (*Astron*) An instrument consisting of a mirror (driven by clockwork) rotating about an axis in its own plane, and pointing to the pole of the heavens. It serves to reflect, continuously, the same region of the sky into the field of view of a fixed telescope.

coelozoic (*BioSci*) Extracellular; living within one of the cavities of the body.

coenobium (*BioSci*) An algal colony in which the number and arrangement of cells are initially determined and that grows only by enlargement of the cells, eg *Pediastrum, Volvox*.

coenocyte (*BioSci*) A multinucleate cell, syncytial tissues formed by the division of the nucleus without division of the cell, as striated muscle fibres and the trophoblast of the placenta. Also *coenocytia*.

coenogamete (*BioSci*) A multinucleate gamete.

coenosarc (*BioSci*) In Hydrozoa, the tubular common stem uniting the individual polyps of a hydroid colony.

coenosteum (*BioSci*) In corals and Hydrocorralinae, the common calcareous skeleton of the whole colony.

coenuriasis (*Vet*) A nervous disease of sheep and goats caused by invasion of the brain and spinal cord by 'coenurus cerebralis', the intermediate stage of the tapeworm *Taenia multiceps*. Controlled by routine worming of canines. Also *gid, sturdy, turnsick*.

coenzyme (*BioSci*) Small molecules that are essential in stoichiometric amounts for the activity of some enzymes. Their loose association with enzymes distinguishes them from prosthetic groups which fulfil a similar role but are tightly bound to the enzyme.

coenzyme A (*BioSci*) The coenzyme that acts as a carrier for acyl groups (A stands for *acetylation*). See ACYL COA.

coenzyme Q (*BioSci*) See UBIQUINONE.

coercimeter (*Phys*) Instrument for measurement of coercive force.

coercive force (*Phys*) Reverse magnetizing force required to bring magnetization to zero after a ferromagnetic material has been left with appreciable residual magnetism.

coercivity (*Phys*) COERCIVE FORCE when the cyclic magnetization reaches saturation.

coesite (*Min*) A high-pressure variety of silica. Found in rocks subjected to the impact of large meteorites; but first made as a synthetic compound.

co-extrusion (*Eng*) A process in which duplex or multiplex film made of two or more polymers is extruded simultaneously.

C of A (*Aero*) Abbrev for CERTIFICATE OF AIRWORTHINESS.

cofactor (*MathSci*) Of an element a_{ij} in the determinant $|A|$ or in the square matrix A: its coefficient A_{ij} in the expansion of $|A|$. It equals $(-1)^{i+j}$ times the minor of a_{ij}. Also *signed minor*.

C of C (*Aero*) Abbrev for CERTIFICATE OF COMPLIANCE.

coffee (*FoodSci*) Beverage produced by infusion of ground roasted coffee beans (berries) from the plants *Coffea arabica* and *C. caneophora*, the coffees being classified as arabica and robusta respectively. The washed berries are allowed to ferment to remove the outer pectinaceous layer by the action of natural enzymes and micro-organisms. After sun-drying the berries are tumbled to remove loose skin and produce the clean 'green beans'. These are roasted to remove moisture and release the aromatic oils responsible for flavour and aroma. Instant coffee is made by infusing the ground coffee in hot water, then concentrating the infusion and drying or freeze-drying.

coffer (*Arch*) A sunk panel in a ceiling or soffit.

coffer (*Build*) A canal lock-chamber.

coffer dam (*CivEng*) A temporary wall or box serving to exclude water from any site normally under water, so as to facilitate the laying of foundations or other similar work; usually formed by driving SHEET PILING.

coffer dam (*Ships*) A short compartment, 1–1·6 m (3–5 ft) in length, separating oil-carrying compartments in an oil tanker from other compartments. Both bulkheads must be oil-tight and the space must be well ventilated.

coffering (*MinExt*) (1) The operation involved in the construction of dams (see COFFER DAM) for impounding water. (2) Shaft lining impervious to normal water pressure.

coffer work (*Build*) Stone-faced rubble-work.

coffin (*NucEng*) See FLASK.

coffinite (*Min*) Black, hydrous uranium silicate, $U(SiO_4)_{1-x}.(OH)_{4x}$. See URANIUM.

C of M (*Aero*) Abbrev for CERTIFICATE OF MAINTENANCE.

cog (*Build*) The solid middle part left between the two notches cut in the lower timber in a COGGING joint.

cog (*Eng*) Any of the wooden teeth along the edge of a gear wheel.

cogging (*Build*) A form of jointing used to connect one beam to another across which it is bearing. Notches as long as the top beam is wide are cut in the top surface of the lower beam opposite one another, so as to leave a solid middle part, and the upper beam has a transverse notch cut in it, to fit over this solid part. Also *caulking, cocking, corking*.

cogging (*Eng*) The operation of rolling or forging an ingot to reduce it to a bloom or billet.

cogging mills (*Eng*) See BLOOMING MILLS.

cognition (*Psych*) Broad general term for mental processes such as thinking, reasoning, insight, imagination, etc.

cognitive behavioural therapy (*Psych*) An approach to psychotherapy based on a combination of behaviourism (based on the theories of learning) and cognitive therapy (based on the theory that thoughts and feelings control much of behaviour).

cognitive dissonance (*Psych*) A situation in which there is recognition of contradiction or inconsistency in one's own beliefs and behaviours. Balance theories, originating with Festinger, assume that reducing the unpleasant state of dissonance is the motivation for reinterpretation of some aspect of experience in order to maximize consistency (*consonance*).

cognitive ethology (*Psych*) A branch of ethology concerned with whether or not conscious awareness, and/or intention, should be taken into account in explanations of animal behaviour.

cognitive map (*Psych*) A mental representation of physical space.

cognitive therapy (*Psych*) An approach to psychotherapy based on the view that emotional disorders are caused primarily by irrational but habitual forms of thinking and that thinking differently will lead to altered emotions and behaviour.

cog-wheel ore (*Min*) See BOURNONITE.

coherence (*Phys*) State existing if two light waves are superposed so as to produce interference effects and there is a constant phase relation maintained between them. Sources producing coherent light are necessary to produce observable interference effects and such sources can be formed by dividing the wave from one source into two parts. Coherence can be thought of in terms of both time and space; lasers are capable of producing light of great time and spatial coherence.

coherent (*BioSci*) United, but so slightly that the coherent organs can be separated without very much tearing.

coherent oscillator (*Radar*) One which is stabilized by being phase locked to the transmitter of a radar for beating with the echo, and used with radar-following circuits.

coherent pulse (*ICT*) One in which individual trains of high-frequency waves are all in the same phase.

coherent sources (*ICT*) Those between which definite phase relationships are maintained, so enabling meaningful interference effects to occur.

coherent unit (*Genrl*) One of a system of units in which no constants appear when units are derived from base units.

cohesion (*BioSci*) The union of plant members of the same kind, as when petals are joined in a sympetalous corolla.

cohesion (*Phys*) The attraction between molecules of a liquid which enables drops and thin films to be formed. In gases the molecules are too far apart for cohesion to be appreciable (but see JOULE–THOMSON EFFECT).

cohesion (*PowderTech*) Attraction forces by which particles are held together to form a body, generally imparted by the introduction of a temporary binder, or by compaction. It is measured as green strength.

cohesionless soils (*CivEng*) Soils such as sand, gravel and ballast which lack cohesion.

cohesion mechanism (*BioSci*) Any mechanism in a plant, in particular one concerned with the dehiscence of a SPORANGIUM, that depends on the cohesive powers of water (ie that a mass of water resists disruption).

cohesion theory (*BioSci*) The generally accepted explanation of water movement through the xylem, that the water is drawn though the vessels or tracheids under tension, the columns of water being maintained by the cohesion of the water molecules and their adhesion to the walls.

cohesive end (*BioSci*) See STICKY END.

cohesive energy density (*Chem*) The measure of the forces of attraction holding molecules together. Defined as

$$\frac{(\lambda - RT)}{M/\rho}$$

where λ is the latent heat of vaporization, R the gas constant, T the absolute temperature, ρ the density and M the molecular mass of the substance. Easy to determine experimentally for small-molecule compounds (eg solvents, plasticizers), more difficult with polymers which cannot vaporize normally. Can, however, be inferred from the structural formula. See STABILITY PARAMETERS.

cohesive soils (*CivEng*) Soils possessing inherent strength due to the surface tension of capillary water, eg saturated clays, silts and some forms of chalk.

cohort (*Psych*, *BioSci*) (1) A group of people or other organisms that share a common characteristic, eg year or location of birth or membership of a functional grouping. Cohort effects, attributable to membership of the cohort rather than other factors, can be important in epidemiological and sociological analyses. (2) A taxonomic group ranking above a superorder.

coil (*ElecEng*) Length of insulated conductor wound around a core which can be eg air (esp at very high frequencies) or iron or ferrite. Because current passing in the coil will create a magnetic field which couples with the winding, the coil can be considered an INDUCTOR.

coil-and-wishbone (*Autos*) A type of independent front-wheel suspension in which a coil spring, usually with a telescopic hydraulic damper, is mounted upon a wishbone-shaped frame, the steering swivels being mounted at the apex and top of the spring taking the weight of one corner of the chassis.

coiled-coil filament (*ElecEng*) A spiral filament for an electric lamp which is coiled into a further helix to reduce radiation losses and enable it to be run at a higher temperature.

coiler (*Textiles*) See CAN COILER.

coil ignition (*Autos*) See BATTERY COIL IGNITION.

coil loading (*ElecEng*) The added inductance, in the form of coils, inserted at intervals along an extended line. Cf CONTINUOUS LOADING.

coil-side (*ElecEng*) That part of an armature coil lying in a single slot.

coil-span (*ElecEng*) The distance, measured round the armature periphery, between one side of an armature coil and the other; usually measured in electrical degrees or slots.

coil-span factor (*ElecEng*) A factor introduced into the equation giving the emf of an electric machine, to allow for the fact that the coils have a fractional pitch and therefore do not embrace the whole flux.

coil-winding machine (*ElecEng*) A machine for automatically, or semi-automatically, winding an electrical conductor coil to a given scheme.

coinage metals (*Chem*) Those TRANSITION METALS renowned for their relative inertness and hence widely used in metal coins. Specifically copper, silver and gold.

coincidence counter (*Phys*) A device which produces an output signal when a prescribed number of two or more input terminals register pulses within a given time interval; used with a series of particle detectors providing the input counts to measure the velocities and directions of particles.

coincidence detection (*Radiol*) The simultaneous detection of two annihilation PHOTONS emitted during positron decay.

coincidence gate (*ICT*) Electronic circuit producing an output pulse only when each of two (or more) input circuits receive pulses at the same instant, or within a prescribed time interval.

coincidence phenomenon (*Acous*) Equality of wavelength of two sound-carrying spheres, eg special form of interaction between the bending wave on a plate and sound waves in the surrounding medium, causing increased sound transmission. See LIMITING FREQUENCY.

coincidence tuning (*ICT*) Tuning all stages to the mid-band frequency. Cf STAGGERED TUNING.

coining (*Eng*) The impressing or repressing of a component in a die and tool set, in which all surfaces of the part are confined, to impart final shape and accuracy of dimensions.

coir (*Textiles*) The reddish-brown coarse fibre obtained from the coconuts of *Cocos nucifera*. The finer fibres are used for making mats, the coarser ones for brushes, and the shortest for filling mattresses and upholstery. Of interest as a substitute for peat-based compost.

coition (*Med*) Sexual intercourse. Also *coitus*.

coitus interruptus (*Med*) Sexual intercourse in which the penis is withdrawn from the vagina before ejaculation occurs.

coke breeze (*Build*) The smaller grades of coke from coke ovens or gasworks, used in the manufacture of breeze concrete.

cokes (*Eng*) Originally, tin plates made from wrought-iron produced in a coke furnace. The term is now applied to plates with a thinner tin coating.

coking coals (*Eng*) Those with more than 15% volatile matter and 80% carbon, which can produce a crush-resistant coke.

col (*EnvSci*) The region between two centres of high pressure or anticyclones.

co-latitude (*Astron*) The complement of the latitude, terrestrial or celestial. On the celestial sphere it is, therefore, the angular distance between the celestial pole and the observer's zenith, and also the meridian altitude of the celestial equator above the observer's horizon.

Colby's bars (*Surv*) Compensated bimetallic bars 10 ft in length, arranged to show an unvarying length despite temperature change; used for baseline measurement in the Ordnance Survey.

colchicine (*BioSci, Chem*) An alkaloid obtained from the root of the autumn crocus, *Colchicum autumnale*, that blocks microtubule assembly by binding to the subunits (tubulin heterodimer) but does not bind once they are assembled into microtubules. As a result of interfering with microtubule reassembly it will block mitosis at metaphase. Inhibition of chromosome separation can double the original number of chromosomes, a technique used in plant breeding for making tetraploids.

cold (*Med*) An acute viral infectious catarrh of the nasal mucous membrane. Also *common cold, coryza*.

cold agglutinin (*BioSci*) Antibody against red cells that causes agglutination at temperatures below body temperature but not at 37°C. Often reactive with the I antigen of red cells and induced by infection with *Mycoplasma pneumoniae*.

cold bend (*Eng*) A test of the ductility of a metal; it consists of bending a bar when cold through a specified angle.

cold-blooded (*BioSci*) Having a bodily temperature that is dependent on the environmental temperature. Also *poikilothermal, poikilothermic*. Cf WARM-BLOODED.

cold boot (*ICT*) See BOOT.

cold casting (*Eng*) A shape made by or the process of pouring a mix of thermosetting resin and metal powder into a mould at ambient temperature and allowing it to cure. Often used for pseudo-bronze sculptures where the product has the appearance of the metal.

cold cathode (*Electronics*) Electrode from which electron emission results from high-potential gradient at the surface at normal temperatures.

cold-cathode discharge lamp (*Phys*) An electric discharge lamp in which the cathode is not heated, the electron emission being produced by a high-voltage gradient at the cathode surface.

cold chisel (*Eng*) A chisel for chipping or cutting away surplus metal; it is used with a hand hammer. Different forms of cutting edges (eg flat, cross-cut, half-round) are used for various purposes. See CROSS-CUT CHISEL, FLAT CHISEL.

cold critical (*NucEng*) The state of a low-power fission reactor in which a chain reaction is sustained without the production of significant heat.

cold drawing (*Eng*) (1) Process of producing bar or wire by drawing through a steel die without heating the material. See WIRE-DRAWING. (2) Tensile deformation of polymer fibre, rod or bar at ambient temperatures (c.25°C) to produce a stiffer product. Caused by CHAIN ORIENTATION along tensile axis.

cold front (*EnvSci*) The leading edge of an advancing mass of cold air, often attended by line squalls and heavy showers.

cold fusion (*Phys*) Nuclear fusion occurring at room temperature; theoretically possible but unlikely, it would allow the production of very cheap electricity.

cold galvanizing (*Eng*) The coating of iron and steel articles with zinc by suspending them in an organic liquid, subsequently evaporated to leave a zinc film on the article. Now also applied to ELECTROPLATING with zinc, alternatively known as *electrogalvanizing*.

cold-heading (*Eng*) The process of forming the heads of bolts or rivets by upsetting the end of the bar without heating the material.

cold insulation mastic (*Build*) A type of coating prepared from bitumen solution and containing fillers, which is used on cold surfaces to prevent entry of moisture and provide protection.

cold junction (*Phys*) The junction of thermocouple wires with conductors leading to a thermoelectric pyrometer or other temperature indicator or recorder.

cold light (*Genrl*) Luminescence.

cold melt (*Print*) An adhesive which does not need heat; used in unsewn binding, eg PVA. Cf HOT MELT.

cold mirror (*ImageTech*) In a lamphouse system, a dichroic mirror reflecting visible light but transmitting the infrared radiations which cause heating.

cold moulding (*Plastics*) The use of resin, filler and accelerator to fill the mould, which then polymerizes to form the component.

cold pinch (*MinExt*) Emergency closing of a ruptured pipeline by flattening it with hydraulic pincers.

cold pool (*EnvSci*) See THICKNESS CHART.

cold riveting (*Eng*) The process of closing a rivet without previous heating; confined to small rivets.

cold roll (*Eng*) Rolling of metal at a temperature close to atmospheric. The cold rolling of metal sheets results in a smooth surface finish.

cold sate (*Eng*) See COLD SETT.

cold saw (*Eng*) A metal-cutting circular saw for cold cutting. The teeth may be either integral with the disk or inserted.

cold set (*Eng*) See COLD SETT.

cold-set ink (*Print*) Printing ink formulated to work in a warmed-up inking system and which sets on the cold stock.

cold sett (*Eng*) A smith's tool similar to a short, stiff, cold chisel; used for cutting bars etc without heating. It is supported by a metal handle and struck with a sledge-hammer. Also *cold sate, cold set*.

cold short (*Eng*) Metals brittle below their recrystallizing temperature.

cold shut (*Eng*) A casting imperfection due to metal entering the mould by different gates or sprues, cooling and failing to unite on meeting.

cold slug well (*Eng*) Cylindrical, deep hole in runner system designed to aid sprue removal in INJECTION MOULDING.

cold start (*ICT*) (1) Restart of a program following a stoppage in which the program has been lost or corrupted and thus must be reloaded. (2) Restart of a computer that involves reloading its operating system. See BOOT, BOOT-STRAP. Cf WARM START.

cold store (*Electronics*) General term for computer memories which, because they depend on superconductivity,

have to be kept at temperatures close to absolute zero. See
SUPERCONDUCTING MEMORY.

cold welding (*Eng*) The forcing together of like or
unlike metals at ambient temperature, often in a shearing
manner, so that normal oxide surface films are ruptured
allowing such intimate metal contact that adhesion takes
place.

cold-working (*Eng*) The operation of shaping metals at
temperatures below their recrystallization temperature (ie
below $0.5T_m$) so as to produce strain-hardening. See WORK
HARDENING.

Cole–Cole plot (*Phys*) A graph of real against imaginary
part of complex permittivity, theoretically a semicircle,
from which the relaxation time of polar dielectrics can be
determined.

colectomy (*Med*) Excision of the colon.

colemanite (*Min*) Hydrated calcium borate, crystallizing in
the monoclinic system; occurs as nodules in clay found in
California and elsewhere.

coleopter (*Aero*) Aircraft having an annular wing, the
fuselage and engine lying on the centre line. Some Fr
designs have vertical take-off and landing capability.

Coleoptera (*BioSci*) An order of Insecta, having the fore-
wings or elytra thickened and chitinized, and meeting in a
straight line. The hind-wings, if present, are membranous
and the mouthparts are adapted for biting. Beetles.

coleoptile (*BioSci*) The sheath, probably a much modified
first leaf, enclosing the epicotyl in the embryo of grasses
and growing up, during germination, as far as the soil
surface to protect the expanding leaves. It is a classical
object for the study of auxin action and of phototropism.

coleorrhiza (*BioSci*) The sheath enclosing the radicle of the
embryo of grasses, through which the radicle grows at
germination.

colibacillosis (*Vet*) Neonatal infection with *Escherichia coli*.
Most common in calves under 1 month old and a
septicaemic form affects calves from 3 to 6 days old. An
enteric toxaemic form also affects calves under a week,
while an enteric form affects those over this age. Infection
may be complicated by a virus. Antisera and vaccines are
available.

colic (*Med, Vet*) Severe cramp-like spasmodic pain due to
intermittent contractions of smooth muscle of the
intestinal, renal or biliary tract.

coliform (*BioSci*) (1) Term used loosely of any rod-shaped
bacterium. (2) Any Gram-negative enteric bacillus. (3)
More specifically, bacteria of the genera *Klebsiella* or
Escherichia.

coliform count (*EnvSci*) A water purity test; the number of
presumptive *coliform* bacteria in 100 ml of water and a
measure of contamination with fecal matter. Since coliform
bacteria vary in virulence it is not directly related to
pathogenicity.

colitis (*Med*) Inflammation of the colon.

collagen (*BioSci*) A family of structural proteins abundant
in the extracellular matrix of animal tissues, especially
BONE (with HYDROXYAPATITE), tendon and skin (with
ELASTIN). Most types, with the exception of Type IV in the
BASAL LAMINA, are fibrillar. Collagen has the general
amino acid sequence –Gly–Pro–Hyp–Gly–*x*– (Hyp =
HYDROXYPROLINE, *x* = any amino acid) arranged in a
crystalline triple α-helix and is resistant to proteases (but
see COLLAGENASE). When degraded by strong alkali,
collagen yields GELATINE. Cross-linking of collagen
increases with chronological age and in mature animals it
contributes to the toughness of the meat.

collagenase (*BioSci*) A proteolytic enzyme capable of
breaking native collagen. Once the initial cleavage is made,
less specific proteases will complete the degradation.
Collagenases from mammalian cells are metallo-enzymes
and are collagen-type specific.

collapse of lung (*Med*) An airless state of the lung,
sometimes caused by obstruction of a bronchus, or by
pneumothorax, or occurring after an abdominal operation.

collapse therapy (*Med*) The treatment of lung disease by
compression of the affected area, eg by injecting air
between the layers of the pleura.

collapsible drum (*Eng*) A device on which a green tyre is
built, needed to make a hollow, cutaway doughnut shape
on the tyre-building machine.

collapsible tap (*Eng*) A screw-cutting TAP in which the
CHASERS can be withdrawn radially, either for the purpose
of producing a tapered thread or, more commonly, to
collapse the tap for quick withdrawal without spoiling the
new thread.

collar (*Arch*) A band, either flat or slightly concave, plain or
decorated, around a column.

collar (*BioSci*) (1) The junction between the stem and root
of a plant, usually situated at soil level. (2) The rim of a
choanocyte. (3) In Hemichordata, a collar-like ridge
posterior to the proboscis. (4) In Gastropoda with a spiral
shell, the collar-like fleshy mantle edge protruding beyond
the lip of the shell. (5) More generally, any collar-like
structure.

collar (*Eng*) A ring of rectangular section secured to a shaft
to provide axial location with respect to a bearing: a similar
ring formed integral with the shaft.

collar (*MinExt*) (1) Concrete platform from which shaft
linings are suspended at top-entry end of drill hole. (2)
Heavy components placed above the drill in a DRILL
STRING to stabilize the drill and to smooth torsional shock.

collar beam (*Build*) The horizontal connecting beam of a
COLLAR-BEAM ROOF.

collar-beam roof (*Arch*) A roof composed of two rafters
tied together by a horizontal beam connecting points about
halfway up the rafters.

collar cell (*BioSci*) See CHOANOCYTE.

collar-head screw (*Eng*) A screw in which the head is
provided with an integral collar; used where fluid leakage
may occur past the threads.

collaring (*Eng*) The term used to indicate that metal passing
through a rolling mill follows one of the rolls so as to
encircle it.

collars (*Eng*) In rolling mills, the sections of larger diameter
separating the grooves in rolls used for the production of
rectangular sections.

collate (*Print*) To check that the sections of a book are in
correct order, after GATHERING.

collateral (*BioSci*) (1) Running parallel or side by side. (2)
Having a common ancestor several generations back.

collateral bud (*BioSci*) An accessory bud located to the side
of an axillary bud.

collateral bundle (*BioSci*) A vascular bundle having phloem
on one side only of the xylem, usually the abaxial side.

collating machine (*Print*) (1) A machine which gathers
sections in sequence prior to binding. (2) A rotary printing
press which collects, in correct register and sequence, the
components of eg multipart stationery, 'collating' being a
misnomer.

collating sequence (*ICT*) Sequence of characters in order
of their character codes.

collecting cell (*BioSci*) A cell of the mesophyll of a leaf lying
below and in contact with cells of the palisade from which
it is presumed to collect photosynthetic products for
transfer to the vascular tissue.

collecting cylinder (*Print*) On rotary presses, the cylinder
for collecting sheets or sections before folding or delivery.

collecting electrode (*ElecEng*) See PASSIVE ELECTRODE.

collecting lens (*ImageTech*) In a multiple condensing lens,
the component lens which is nearest the light source.

collecting power (*Phys*) The property of a lens to render
parallel rays convergent or reduce the divergency of
originally divergent rays.

collection (*ICT*) In object-oriented programming, a group
of objects that are logically related to each other and to
which a logically consistent set of operations are applied.

collection (*Print*) Gathering the sections or pages of a web-
fed rotary press.

collective dose equivalent (*Radiol*) The quantity obtained by multiplying the average effective dose equivalent by the number of persons exposed to a given source of radiation. Expressed as a sievert (*Sv*).

collective drive (*ElecEng*) The drive in an electric locomotive in which all the driving wheels are coupled and powered by a single motor. Cf INDIVIDUAL DRIVE.

collective electron theory (*Phys*) The assumption that ferromagnetism arises from free electrons; FERMI–DIRAC STATISTICS identify the CURIE POINT with the transition from the ferro- to the paramagnetic state.

collective fruit (*BioSci*) A fruit derived from several flowers, as a mulberry.

collective model (*Phys*) Model of the nucleus combining certain features of the SHELL MODEL and LIQUID-DROP MODEL. It assumes that the nucleons move independently in a real potential but the potential is not the spherically symmetric potential of the shell model. It is instead a potential capable of undergoing deformation and this represents the collective motion of the nucleons as in the liquid-drop model.

collective pitch control (*Aero*) A helicopter control by which an equal variation is made in the blades of the rotor(s), independently of their azimuthal position, to give climb and descent.

collective unconscious (*Psych*) According to Jung, those aspects of unconscious mental life that represent the accumulated experiences of the human species, in contrast with the unconscious life of an individual based on personal experience.

collector (*Electronics*) (1) Any electrode which collects electrons which have already completed and fulfilled their function, eg screen grid. (2) Outer section of a transistor which delivers a primary flow of carriers.

collector agent (*MinExt*) In froth flotation, a chemical which is adsorbed by one of the minerals in an ore pulp, causing it to become hydrophobic and removable as a mineralized froth. Also *promoter*.

collector capacitance (*Electronics*) The capacitance of the depletion layer forming the collector–base junction of a bipolar TRANSISTOR.

collector current (*Electronics*) The current which flows at the collector of a transistor on applying a suitable bias.

collector-current runaway (*Electronics*) The continued increase of the collector current arising from an increase of temperature in the collector junction when the current grows. See THERMAL RUNAWAY.

collector efficiency (*Electronics*) The ratio of the useful power output to the dc power input of a transistor.

collector junction (*Electronics*) One biased in the high-resistance direction, current being controlled by MINORITY CARRIERS, eg the semiconductor junction between the collector and base electrodes of a transistor.

collector rings (*ElecEng*) See SLIP RINGS.

collector shoe (*ElecEng*) A metal shoe used on the vehicles of an electric traction system to maintain contact with the conductor rail.

collector strip (*ElecEng*) See CONTACT STRIP.

collect run (*Print*) A method of increasing the number of pages per copy on web-fed presses by COLLECTION before final fold and delivery.

collenchyma (*BioSci*) A mechanical tissue, typical of leaf veins, petioles and the outer cortex of stems, of more or less elongated cells with unevenly thickened non-lignified primary walls.

Colles' fracture (*Med*) Fracture of the lower part of the forearm above the wrist.

collet (*Eng*) An externally coned sleeve, slit in two or more planes for part of its length, and arranged to be closed by being drawn into an internally coned rigid sleeve, for the purpose of accurately gripping articles or material.

collet chuck (*Eng*) A mechanism using COLLETS for holding work or a tool in a lathe, drilling machine, etc.

colleterial glands (*BioSci*) One or two pairs of accessory reproductive glands, present in most female insects. Their secretion forms the OÖTHECA in ORTHOPTERA, and a cement which fastens the eggs to the substratum in many other insects.

colliculus (*BioSci*) A small prominence, as on the surface of the optic lobe of the brain.

colliding-beam experiment (*Phys*) A technique in high-energy physics whereby two beams of particles are made to collide head-on. A greater proportion of the energy of the incident particles is available for the creation of new particles in the collision than in a fixed-target experiment of similar total energy.

colligative properties (*Chem*) Those properties of solutions which depend only on the concentration of dissolved particles, ions and molecules, and not on their nature. They include depression of freezing point, elevation of boiling point and osmotic pressure.

collimation (*Phys*) (1) The process of aligning the various parts of an optical system. (2) The limiting of a beam of radiation to the required dimensions. See COLLIMATOR.

collimation error (*Surv*) An error produced in levelling or in theodolite work when the line of collimation is out of its correct position; the latter, for the level, is parallel to the bubble line and perpendicular to the vertical axis of rotation of the instrument, and for the theodolite it is also perpendicular to the trunnion axis.

collimation system (*Surv*) In levelling along a series of instrument set-ups, transfer of LINE OF COLLIMATION from back to fore sight, by change in reading on graduated staff. See RISE AND FALL SYSTEM.

collimator (*Phys*) A device for obtaining a parallel or near-parallel beam of radiation or particles. An optical collimator consists of a source, usually a fine slit, at the principal focus of a converging lens or mirror. Penetrating radiation such as X-rays or gamma rays is collimated by a series of holes or slits in a highly absorbing material such as lead.

collimator (*Radiol*) The lens of a gamma-camera imaging system which absorbs photons travelling in inappropriate directions and originating from parts of the body other than the region under examination.

collinear array (*ICT*) An antenna array consisting of a number of dipoles connected end to end and operated in phase. Maximum radiation is normal to the line of dipoles (cf END-FIRE ARRAY). Groups of YAGI antennas may also be formed into collinear arrays, achieving very sharp directivity.

collinear transformation of a matrix (*MathSci*) See CONGRUENT TRANSFORMATION.

collinear vectors (*MathSci*) Two vectors which are parallel to the same line.

collineation (*MathSci*) Analytical transformation having a one-to-one correspondence between points, collinear points being projected into collinear points.

Collins process (*Phys*) A process for the liquefaction of helium which combines the JOULE–THOMSON EFFECT with the CLAUDE PROCESS.

collision (*ICT*) (1) In the context of a NETWORK, a collision occurs when two information signals attempt to use the same channel simultaneously. CSMA/CD systems have been developed to prevent the loss of data should a collision occur. (2) In the context of the storage of data RECORDS, a collision occurs when the system attempts to store a record in a LOCATION that is already occupied. The storage ADDRESS may have been generated by means of a HASHING ALGORITHM and an identical address generated.

collision (*Phys*) An interaction between particles in which momentum is conserved. If also the kinetic energy of the particles is conserved, the collision is said to be *elastic*; if not, then the collision is *inelastic*. With particles in nuclear physics, there is no contact unless there is *capture*. Collision then means a nearness of approach such that there is mutual interaction due to the forces associated with the particles. Cf IMPACT.

collisional excitation (*Phys*) The transfer of energy when an atom is raised to an excited state by collision with another particle.

collision bulkhead (*Ships*) A strong watertight bulkhead, not less than 1/20th of the vessel's length from the fore end.

collision diameter (*Chem*) The distance of closest approach of the centres of two identical colliding molecules.

collision number (*Chem*) The frequency of collisions per unit concentration of molecules.

collodion (*Chem*) A cellulose tetranitrate, soluble in a mixture of ethanol and ethoxyethane (1:7); the solution is used for coating materials and, in medicine, for sealing wounds and dressings.

collodion process (*ImageTech*) Early photographic process using glass plates coated with iodized collodion (nitrocellulose) sensitized with silver nitrate solution and exposed in the camera while still wet; hence known as the *wet-plate* process.

colloid (*Chem*) From Gk *kolla*, glue. Name originally given by Graham to amorphous solids, like gelatine and rubber, which spontaneously disperse in suitable solvents to form lyophilic sols. Contrasted with crystalloids on the one hand and with lyophobic sols on the other. The term currently denotes any colloidal system. See panel on GELS.

colloidal electrolyte (*Phys*) Electrolyte formed from long-chain hydrocarbon compounds, end radicals of which can ionize, thus providing some properties of electrolytes.

colloidal fuel (*Eng*) A mixture of fuel oil and finely pulverized coal, which remains homogeneous in storage; calorific value high; used in oil-fired boilers as substitute for fuel oil alone.

colloidal graphite (*Eng*) Extremely fine dispersion of ground graphite in oil. Graphite lowers the surface tension of oil without lowering the viscosity; the oil spreads more easily, taking the graphite to rough surfaces where it can build up a smoothness.

colloidal movement (*Chem*) See BROWNIAN MOVEMENT.

colloidal mud (*MinExt*) Thixotropic mixture of finely divided clays with baryte and/or bentonite, used in the drilling of deep oil bores. See DRILLING MUD and panel on DRILLING RIG.

colloidal state (*Chem*) A state of subdivision of matter in which the particle size varies from that of true 'molecular' solutions to that of coarse suspensions, the diameter of the particles lying between 1 and 100 nm. The particles are charged and can be subjected to electrophoresis, except at the ISO-ELECTRIC POINT. They are subject to Brownian movement and have a large amount of surface activity. See panel on GELS.

colloidal suspension (*Chem*) See panel on GELS.

colloid goitre (*Med*) Abnormal enlargement of the thyroid gland due to accumulation in it of the viscid iodine-containing colloid.

colloid mill (*ChemEng*) A mill with very fine clearance between the grinding components, operating at high speed, and capable of reducing a given product to a particle size of $0 \cdot 1–1 \ \mu m$.

collophane (*Min*) Cryptocrystalline variety of *apatite*.

collotype (*Print*) A PLANOGRAPHIC PROCESS for printing of tone subjects in one or more colours without using the HALF-TONE PROCESS. A film of dichromated gelatine is spread on a glass plate, printed down with a continuous tone negative, and developed with the use of glycerine and water. Dark parts of the subject become hardened and ink-accepting, light parts remain watery and ink-rejecting, and areas of intermediate tone accept ink to correspond with their variations in tone. During preparation, the surface develops a fine grain which provides a bite for the ink. Suitable for runs up to 1000. Used for the finest facsimile reproductions of works of art.

colluviarium (*CivEng*) An access opening in an aqueduct for maintenance and ventilation.

coloboma (*Med*) Congenital defect of development, esp of the lens, the iris or the retina.

coloenteritis (*Med*) Inflammation of the colon and small intestine. Also *enterocolitis*.

colon (*BioSci*) In insects, the wide posterior part of the hind-gut; the large intestine of vertebrates. Adj *colonic*.

colonnade (*Arch*) A row of columns supporting an entablature.

colony (*BioSci*) (1) The vegetative form of many species of algae in which the sister cells are connected in a group to function as a unit. In many sorts (eg *Synura*), the colonies, of no fixed number of cells, grow by division and reproduce asexually by fragmentation. See COENOBIUM. (2) A fungal mycelium grown, eg on an agar plate, from one spore. (3) A bacterial colony similarly initiated and grown. (4) Loosely, a collection of individuals living together and in some degree interdependent, eg a colony of polyps, a colony of social insects. Strictly, the members of such a colony are in organic connection with one another, rather than being a colony.

colony stimulating factors (*BioSci*) Substances that are made by a number of cells and that cause haemopoietic stem cells to proliferate and differentiate into mature forms, appearing as colonies in tissue culture. There are separate factors for granulocytes and macrophages, for eosinophyl leucocytes and for erythrocytes. Abbrev *CSF*.

colopexy (*Med*) The anchoring of part of the colon by sewing it to the abdominal wall. Also *colopexia*.

colophon (*Print*) Originally, a device and/or notice by the scribe or printer, placed at the end of a book before title pages became customary. Replaced today by title page and imprint, although some publishers often append a colophon. In modern practice, a decorative device on the title page or spine of a book.

colophony (*Chem*) Also *colophonium*. See ROSIN.

Colorado beetle (*BioSci*) A black-and-yellow-striped beetle (*Leptinotarsa decemlineata*) that feeds upon potato leaves, causing great destruction.

Colorado ruby (*Min*) An incorrect name for the fiery-red garnet (pyrope) crystals obtained from Colorado and certain other parts of the USA.

Colorado topaz (*Min*) True topaz of a brownish-yellow colour is obtained in Colorado, but quartz similarly coloured is sometimes sold under the same name.

colorimeter (*Phys*) An instrument used for the precise measurement of the hue, purity and brightness of a colour.

colorimetric analysis (*Chem*) Analysis of a solution by comparison of the colour produced by a reagent with that produced in a standard solution.

colorimetric purity (*ImageTech*) The ratio of the luminosity of a dominant hue to the total luminosity of a colour. See SATURATION.

colorimetry (*ImageTech*) The measurement of the spectral transmission and colour density of photographic images.

Color-key (*Print*) TN for a method of producing proofs of colour separations using a film print of each colour mounted in register with the others.

colostomy (*Med*) A hole surgically made into the colon for the escape of feces through the abdominal wall into a colostomy bag when the bowel below is obstructed.

colostrum (*BioSci*) First secretion produced by the mammary gland at the end of pregnancy. Colostrum contains high levels of secretory (IgA) immunoglobulin and antibody-secreting cells which protect the neonate against infection.

colostrum corpuscles (*BioSci*) Large cells that contain fat particles and that appear in the secretion of the mammary glands at the commencement of lactation.

colotomy (*Med*) An incision into the colon; (loosely) colostomy.

colour (*Phys*) A fundamental QUARK property. Each FLAVOUR of quarks comes in three 'colours', red, blue and green, and for antiquarks, three anticolours. All HADRONS are 'colourless', containing either one red plus one quark of each colour, or one quark and its oppositely coloured antiquark. The STRONG INTERACTION couples to the colour of the quarks. *US color*.

colour analyser (*ImageTech*) In film laboratory practice, a calibrated closed-circuit TV system reproducing colour negative as a positive image to assess its required printing levels.

colour balance (*ImageTech*) See BALANCING.

colour bar (*Print*) A standard strip made up of solid and tone printed across the sheet. Used in conjunction with a reflection DENSITOMETER to obtain and maintain a standard ink film thickness and hence consistency of colour during printing.

colour black (*ImageTech*) A signal composed of COLOUR BURST and BLACK LEVEL.

colour bleeding (*ImageTech*) See BLEEDING.

colour blindness (*Med*) The lack of one or more of the spectral colour sensations of the eye. The commonest form, Daltonism, consists of an inability to distinguish between red and green. Even persons of normal sight may be colour blind to the indigo of the spectrum.

colour burst (*ImageTech*) Short sequence of colour subcarrier frequency transmitted as a reference for the chrominance signal at the beginning of each line.

colour cast (*ImageTech*) Predominance of a particular colour affecting the whole image, resulting from unsatisfactory colour BALANCING in reproduction.

colour coder (*ImageTech*) Apparatus in colour TV to generate chrominance subcarrier and composite colour signal from the camera signals.

colour contamination (*ImageTech*) Error in colour reproduction caused by incomplete separation of primaries.

colour contrast (*ImageTech*) Visually, the subjective enhancement of one hue when seen in the surroundings of substantially complementary colours. Photographically, the GAMMA value of one of the component colour images.

colour co-ordinates (*Phys*) Set of numbers representing the location of a hue on a chromaticity diagram.

colour-corrected lens (*ImageTech, Phys*) One in which chromatic aberration is much reduced.

colour correction (*Print*) When making sets of colour printing surfaces, it may be necessary to compensate for inherent faults of the printing inks available. Can be carried out by hand on the separation negatives or positives for any of the printing processes and on the actual plate for relief printing, by COLOUR MASKING or electronically.

colour decoder (*ImageTech*) Circuit in a TV receiver which extracts, decodes and separates the three constituent colours.

colour developer (*ImageTech*) A processing solution in which a colour image is formed in association with the developed silver image.

colour difference signals (*ImageTech*) Signals for colour TV transmission obtained by subtracting the LUMINANCE, Y, from each of the three primary colour signals R, G and B, in the form B−Y, R−Y, which are then coded.

coloured cement (*Build*) Ordinary Portland cement into which selected pigments are introduced in the grinding process.

colour excess (*Astron*) The amount by which the colour index of a star exceeds the accepted value for its spectral class; used as a measure of absorption of starlight.

colour fastness (*Textiles*) The resistance of a dyed textile to change of colour when exposed to specific agents such as water, light or rubbing.

colour fatigue (*Phys*) Changes in the sensation produced by a given colour when the eye is fatigued by another or by the same colour.

colour filter (*Phys*) Film of material selectively absorbing certain wavelengths, and hence changing the spectral distribution of transmitted radiation.

colour gate (*ImageTech*) Circuit in colour TV receiver which allows only primary colour signal, corresponding to excited phosphor, to reach modulation electrode of tube.

colour guides (*Print*) A term sometimes applied to PROGRESSIVE PROOFS.

colour index (*Astron*) The difference between the apparent magnitudes of a star measured at two standard wavelengths,

from which may be deduced the colour and effective temperature.

colour index (*Geol*) A number which represents the percentage of dark-coloured heavy silicates in an igneous rock, and is thus a measure of its leucocratic, mesocratic or melanocratic character.

colour index (*ImageTech*) A systematic arrangement of colours according to their hue, saturation and brightness for identification and reproduction. *The Colour Index* is a publication giving chemical details of commercially available dyestuffs and pigments.

colouring of a graph (*MathSci*) An assignment of colours to each vertex of a graph with the condition that if two vertices are joined by an edge then they have different colours. A *k*-colouring uses *k* distinct colours. See BIPARTITE GRAPH.

colour intermediate (*ImageTech*) A masked INTEGRAL TRIPACK colour film intended for laboratory duplicating purposes, not for projection.

colour killer (*ImageTech*) A circuit rendering the chrominance channel of a colour TV receiver inoperative during the reception of monochrome signals.

colour–luminosity array (*Astron*) A variant of the HERTZSPRUNG–RUSSELL DIAGRAM (panel) in which the absolute magnitudes of stars are plotted as a function of their colours.

colour masking (*ImageTech*) In photographic colour reproduction, the use of additional images to compensate for the deficiencies of the dyes forming the principal colour records. These may be separate black-and-white images but the emulsions of modern colour negative films incorporate additional dye components for this purpose – *integral colour masking*. In the reproduction of colour film on TV, colour masking is carried out electronically by matrixing the three colour signals.

colour masking (*Print*) The use of photographic masks in the process camera to compensate for inherent faults in the inks available for colour printing. Separate masks can be prepared, eg a low-density positive of the cyan printer is used to modify the magenta printer negative, thus reducing the amount of magenta printing in blue areas; or an all-purpose masking colour film can be used, such as Multimask or Trimask. The problem has to some extent been overcome with modern electronic scanning techniques.

colour mixture curve (*Phys*) Representation of the specified three colours which match a given colour.

colour negative (*ImageTech*) A photographic image in which the tonal brightness values of the original subject are inverted, light to dark, and its colours are represented by substantially complementary hues.

colour phase (*ImageTech*) Determines the HUE in a colour TV picture.

colour phase alternation (*ImageTech*) The sequence of the colour signals in the video signal.

colour photographic sensitivity (*ImageTech*) The sensitivity of an emulsion to specified wavelength ranges.

colour picture signal (*ImageTech*) Monochrome video signal, plus a subcarrier conveying the colour information, which is transmitted with synchronizing signals.

colour positive (*ImageTech*) A photographic image in which the tonal values and colours are substantially similar to those of the original subject.

colour primaries (*ImageTech, Print*) Set of colours, usually three, from which a colour picture can be reproduced: red, green and blue for an ADDITIVE PROCESS, cyan, magenta and yellow for a SUBTRACTIVE one.

colour printing (*Print*) The reproduction of an original subject comprising two or more colours. Colour printing is achieved by any of the normal printing processes; each colour is printed separately, in a predetermined order, the superimposed impressions, if accurately registered, building up an image corresponding in colour to the original subject.

colour purity magnet (*ImageTech*) Magnet placed near to neck of colour picture tube to modify path of electron beam and thus improve purity of the displayed colour.

colour pyramid (*ImageTech*) See COLOUR TRIANGLE.

colour reference signal (*ImageTech*) The continuous signal which determines the phase of the burst signal.

colour register (*Print*) The correct superimposing of two or more colours to achieve correct image fit.

colour saturation (*ImageTech*) See SATURATION.

colour screen (*ImageTech*) Either a filter or a mosaic of the primary colours.

colour separation (*ImageTech, Print*) The production in the process camera of three separate negatives of the same subject through green, red and blue filters to record the proportions of the respective colours in the image. The negatives can be used as a basis for colour prints or for half-tone colour blocks. The process can also now be done electronically. See THREE-COLOUR PROCESS.

colour separation overlay (*ImageTech*) A system of video image combination in which the foreground action is shot against a uniform blue backing, these areas being replaced by a picture from another source. Also *chroma-key*.

colour specification (*ImageTech*) The description of a colour in a standard manner, so that it can be duplicated without comparison.

colour standards (*ImageTech*) A standard range of colours for reference purposes in making dyes or filters, or for composing colour patterns for colour photography.

colour subcarrier (*ImageTech*) A signal, conveying the colour information as a modulation, added to the monochrome video and synchronizing signals.

colour temperature (*Phys*) The temperature of a black body which radiates with the same dominant wavelengths as those apparent from a source being described. See PLANCK'S LAW.

colour threshold (*Phys*) The luminance level below which colour differences are indiscernable.

colour transparency (*ImageTech*) A colour photograph to be viewed or projected with transmitted light.

colour triangle (*Phys*) A triangle drawn on a chromaticity diagram to represent the entire range of chromaticities obtainable from additive mixtures of three prescribed primaries, represented by the corners of the triangle.

Colour TrueFinder (*ImageTech*) TN for a COLOUR VIEW-FINDER employing a black-and-white CATHODE-RAY TUBE which sequentially displays RGB SIGNALS at three times the normal field rate in synchronism with a spinning RGB filter wheel. Cf COLOUR VIEWFINDER, ELECTRONIC VIEW-FINDER.

colour under (*ImageTech*) A method of colour video recording in which CHROMA is recorded at a lower carrier frequency than LUMINANCE, thus reducing the bandwidth required but at the sacrifice of some colour quality.

colour viewfinder (*ImageTech*) An ELECTRONIC VIEW-FINDER with usually a SHADOWMASK TUBE or LIQUID CRYSTAL DISPLAY (panel). Some CAMCORDERS have replaced the viewfinder with a larger LCD panel. See COLOUR TRUEFINDER, LCD VIEWFINDER.

colour vision (*BioSci*) The ability of animals to discriminate light of different wavelengths. Depending on the animal there may be specialized receptors (cones in vertebrates) that are preferentially sensitive to two, three or more wavelengths. Trichromatic vision (eg matching any colour by an additive mixture of saturated red, green and blue) is most common. See COLOUR, TRICHROMATIC COEFFI-CIENTS, YOUNG–HELMHOLTZ THEORY.

colpitis (*Med*) Inflammation of the vagina.

Colpitts oscillator (*Electronics*) One in which a parallel tuned circuit is connected between grid and collector (or grid and anode in a valve circuit) with the capacitative part consisting of two series capacitors. The junction of these is at emitter (or cathode) potential and the positive feedback is via the capacitor leading back to the base (or grid).

colpocele (*Med*) A hernia into the vagina.

colpocystocele (*Med*) A hernia formed by protrusion of the bladder into the vagina.

colpocystotomy (*Med*) Incision of the bladder through the wall of the vagina.

colpoperineoplasty (*Med*) Repair of the vagina and perineum by plastic surgery.

colpoperineorrhaphy (*Med*) Sewing up of the torn vagina and perineum.

colpoptosis (*Med*) Prolapse of the vagina.

colporrhaphy (*Med*) Narrowing of the vagina by surgical operation.

colposcope (*Med*) An instrument for inspecting the vagina.

colpospasm (*Med*) Spasm of the vagina.

colpus (*BioSci*) An elongated aperture in the wall of a pollen grain. Cf APERTURATE.

Columba (Dove) (*Astron*) A small southern constellation.

columbite (*Min*) Niobate and tantalate of iron and manganese. When the Nb content exceeds that of Ta, the ore is called columbite. See TANTALITE.

columella (*BioSci*) (1) A small column. (2) The axial part of a root cap in some species, in which the cells are arranged in longitudinal files. (3) The sterile tissue in the centre of the sporangium of bryophytes and some fungi. (4) A radial rod in the wall of a spore or pollen grain. (5) In mammals, the central pillar of the cochlea. (6) In lower vertebrates, the auditory ossicle connecting the tympanum with the inner ear. (7) In some lower tetrapods, the epipterygoid bone that is located above the *pterygoid* bone in the skull. (8) In spirally coiled gastropod shells, the central pillar. (9) In the skeleton of some corals, the central pillar. Adj *columellar*.

column (*BioSci*) (1) Generally, any columnar structure such as the vertebral column. (2) The central portion of the flower of an orchid (probably an outgrowth of the receptacle of the flower), bearing the anther, or anthers and the stigmas. (3) In Crinoidea, the stalk. (4) In vertebrates, a bundle of nerve fibres running longitudinally in the spinal cord, also the edge of the nasal septum.

column (*Chem*) Laboratory or industrial cylindrical vessels of glass or metal, in which solvent extraction or other procedures are carried out.

column (*CivEng, Eng*) A vertical pillar or shaft used to support a compressive load. See STRUT.

column analogy method (*CivEng*) A method of analysing indeterminate frames of non-uniform section by compar-ing equilibrium conditions, at any point, with the stresses in an analogous eccentrically loaded column.

columnar crystals (*Eng*) Elongated crystals formed by growth taking place at right angles to the temperature gradient within a mould, usually at right angles to the mould wall. They initiate from the layer of CHILL CRYSTALS at the surface and the extent to which they grow before solidification is completed is determined by the formation of equi-axed crystals in the interior portions of the casting.

columnar epithelium (*BioSci*) A variety of epithelium consisting of prismatic columnar cells set closely side by side on a basement membrane, generally in a single layer.

columnar structure (*Geol*) A form of regular jointing, produced by contraction following crystallization and cooling in igneous rocks, esp those of basic composition. The columns are generally roughly perpendicular to the cooling surface.

columns across (*Print*) Printing newspapers and magazines with the columns imposed across the plate cylinders.

columns around (*Print*) Printing newspapers and maga-zines with the columns imposed around the plate cylinders.

column vector (*MathSci*) A matrix consisting of a single column. Compare ROW VECTOR.

colures (*Astron*) The great circles passing through: (1) the poles of the celestial equator and ecliptic, and through both solstitial points; or (2) the poles of the celestial equator and both equinoctial points. These two great circles are the solstitial and equinoctial colures respectively.

COM (*ICT*) (1) Abbrev for *component object model*, a proprietary (Microsoft) architectural standard for object-oriented systems design. See also DCOM. (2) Obsolete abbrev for *computer output on microfilm*.

coma (*Astron*) The visible head of a comet.

coma (*BioSci*) A tuft of hairs or leaves.

coma (*Electronics*) Of a cathode-ray tube, plume-like distortion of spot arising from misalignment of focusing elements of gun.

coma (*Med*) A state of complete unconsciousness in which the patient is unable to respond to any external stimulation.

coma (*Phys*) An aberration of a lens or lens system whereby an off-axis point object is imaged as a small pear-shaped blob, due to the power of the zones of the lens varying with distance from the axis.

Coma Berenices (Berenice's Hair) (*Astron*) A faint northern constellation, which includes the COMA CLUSTER.

Coma cluster (*Astron*) A rich cluster of galaxies in the constellation Coma Berenices which contains around 1000 members and has a diameter of around 6 Mpc. Distance approximately 90 Mpc.

comagmatic assemblage (*Geol*) Refers to igneous rocks with a common set of chemical, mineralogical and textural features, suggesting derivation from a common parent magma.

COMAL (*ICT*) Programming language that is an enhanced version of BASIC, with structured programming features.

comatose (*Med*) Being in a state of coma.

comb (*Arch*) The ridge of a roof.

comb (*BioSci*) (1) In CTENOPHORA, a *ctene*. (2) The framework of hexagonal wax cells produced by social bees to shelter the young or for storing food.

comb (*Build*) A flat flexible wire- or rubber-toothed instrument used by the painter for graining surfaces.

comb (*Textiles*) To prepare cotton and wool fibres for spinning by separating and straightening them and removing impurities and fibres below a specified length.

combat rating (*Aero*) See POWER RATING.

comb binding (*Print*) A style of loose-leaf binding in which the leaves are held together by a comb, usually of plastic, being passed through slots in the paper and then curved to form a tube.

combed yarns (*Textiles*) Highest-quality yarns prepared from carded and combed fibres that have been mechanically straightened and freed from NEPS and short fibres.

Combescure transformation of a curve (*MathSci*) A one-to-one transformation which maps one space curve onto another so that the tangents at corresponding points are parallel.

comb filter (*ImageTech, ICT*) Electronic circuit tuned to select a number of specific frequencies for transmission or rejection, leaving unaffected those between them.

combination (*Chem*) Formation of a compound.

combination chuck (*Eng*) A lathe chuck in which the jaws may be operated all together, as in a universal or SELF-CENTRING CHUCK; or each operated separately for holding work of irregular shape, as in an INDEPENDENT CHUCK.

combination cylinder (*Build*) A household hot-water cylinder with a feed tank above it, all within the same outer case.

combination mill (*Eng*) A continuous rolling mill in which the shaping mills follow the roughing mill directly.

combinations (*MathSci*) The different ways of selecting a number of items from a given set without regard to order. There are $n!/r!(n−r)!$ ways of selecting r items from a set of n distinct items. Denoted by

$$^nC_r \text{ or } \left(\begin{smallmatrix} r \\ n \end{smallmatrix}\right)$$

Cf PERMUTATIONS.

combination set (*Eng*) A fitter's instrument comprising a universal protractor, spirit level, rule and straight-edge, centre head and square.

combination therapy (*Pharmacol*) Treatment of a disease with more than one drug, classically used with several antibiotics in tuberculosis and now in cancer and AIDS.

combination tone (*Acous*) An additional tone produced by a non-linear system when two or more tones are applied.

The combination tones have frequencies which are sums (*summation tones*) and differences (*difference tones*) of the frequencies of the applied tones.

combinatorial chemistry (*Chem*) A method by which large numbers of compounds (a 'library') can be made, usually utilizing solid-phase synthesis. The term is often used loosely for any procedure that generates highly diverse sets of compounds; the more recent tendency is to prefer high-speed parallel synthesis in which each reaction chamber contains only one compound. Also *combi-chem*.

combined carbon (*Eng*) In cast-iron, the carbon present as iron carbide as distinct from that present as graphite. See GRAPHITIC CARBON.

combined half-tone and line (*Print*) An illustration in which half-tone and line work are combined.

combined heat and power (*EnvSci*) The use of waste heat from an electricity generating station to provide space and water heating in nearby buildings. Abbrev *CHP*.

combined-impulse turbine (*Eng*) An IMPULSE TURBINE in which the first stage consists of nozzles that direct the steam onto a wheel carrying two rows of moving blades, between which a row of fixed-guide blades is interposed.

combined system (*Build*) A system of sewerage in which only one set of sewers is provided for the removal of both the sewage proper and rain water. Cf SEPARATE SYSTEM.

combine harvester (*Agri*) A self-propelled machine, used to harvest seed crops like wheat, oilseed rape, etc, that separates seeds from straw. The seeds are retained in an on-board hopper and the straw is ejected, commonly chopped or baled.

combing (*Textiles*) The process of further separating and straightening *carded slivers* of fibres and removing impurities and fibres below a specified length.

combining weight (*Chem*) See EQUIVALENT WEIGHT.

Combretastatin (*Pharmacol*) TN of a drug derived from the African bush willow tree, used in cancer treatment.

combustion (*Chem*) Chemical union of oxygen with gas accompanied by the evolution of light and rapid production of (exothermic) heat.

combustion chamber (*Aero*) The chamber in which combustion occurs, ie: (1) the cylinder of a reciprocating engine; (2) the individual chambers or single annular chamber of a gas turbine; (3) the combustion zone of a ramjet duct; or (4) the chamber, with a single venturi outlet, of a rocket.

combustion chamber (*Eng*) (1) In a boiler furnace, the space in which combustion of gaseous products from the fuel takes place. (2) In an internal-combustion engine, the space above the piston (when on its inner dead centre) in which combustion occurs.

combustion control (*Eng*) The control, either by an attendant or by automatic devices, of the rate of combustion in a boiler furnace, in order to adjust it to the demand on the boiler.

combustion noise (*Acous*) Noise caused by combustion. It can be particularly loud if combustion takes place in an acoustic resonator (eg industrial burner, jet aircraft) where a feedback between the released heat and the sound waves can lead to INSTABILITIES.

combustion tube furnace (*Eng*) Laboratory appliance having one or more horizontal refractory tubes heated by gas or electricity; used chiefly for the estimation of carbon content of steels, temperatures 1100 to 1300°C.

come-and-go (*Print*) See FORE-AND-AFT.

comedo (*Med*) A blackhead. A collection of cells, sebum and bacteria, filling the dilated orifice of a sebaceous gland near a hair follicle. Closed comedones are sometimes referred to as whiteheads. Pl *comedones*.

comendite (*Min*) A variety of rhyolite with phenocrysts of quartz, alkali feldspar, sodic pyroxenes and amphiboles.

comet (*Astron*) A member of the solar system, of small mass, becoming visible as it approaches the Sun, partly by reflected sunlight, partly by fluorescence excited by the solar radiation. A bright nucleus is often seen, and sometimes a tail. This points away from the Sun, its gases and fine dust being repelled by radiation pressure and the solar wind.

comfort behaviour (*Psych*) Behaviour that has to do with body care, eg grooming, scratching, preening.

comfort noise (*ICT*) In a digital mobile-telephone system, white noise fed to the receiving earpiece during periods in which either the signal is corrupted or the caller has stopped talking, causing his or her transmitter to switch off with resultant loss of background noise.

comma (*Acous*) The pitch error, not greater than 80:81·1, arising from tuning one note in various ways with natural ratios from a datum note.

commag (*ImageTech*) International code name for a picture film combined with a magnetic sound track.

command (*ICT*) A word, phrase or code typed in by the user to which the machine responds immediately.

command guidance (*Aero*) The guidance of missiles or aircraft by electronic, optical or wire-borne signals from an external source controlled by human operator or automatically.

command language (*ICT*) The LANGUAGE used to communicate with the operating system.

command line (*ICT*) A blank line on the screen to the right of a PROMPT where the user types a COMMAND.

command line interpreter (*ICT*) Part of the OPERATING SYSTEM that analyses a command typed in by the user and executes this immediately; eg in MS-DOS the characters DIR entered at the command line will invoke the appropriate routine within the OPERATING SYSTEM and the user will be presented with the names of the files on the current selected DISK DRIVE.

command module (*Space*) The part of a spacecraft from which operations are directed.

Commelinidae (*BioSci*) A subclass or superorder of monocotyledons. Almost all are terrestrial herbs, often of moist places, with the perianth differentiated into sepals and petals, or reduced and not petaloid. Contains c.19 000 spp in 25 families including Cyperaceae, Gramineae, Bromeliaceae and Zingiberaceae.

commensalism (*BioSci*) An external, mutually beneficial partnership between two organisms (*commensals*); one partner may gain more than the other. Adj *commensal*.

commensurable quantities (*MathSci*) Quantities, each of which is an integral multiple of a common basic quantity or measure, eg the numbers 6 and 15 are commensurable but 6 and $\sqrt{3}$ are not.

commercial sterility (*FoodSci*) Commercial sterility does not imply perfect sterility but indicates that a product will not spoil, or harm the health of the consumer, under normal storage conditions.

comminuted (*Med*) Reduced to small fragments, eg *comminuted* fracture.

comminuted powder (*PowderTech*) Material reduced to a powder by eg attrition, impact, crushing, grinding, abrasion, milling or chemical methods.

comminution (*FoodSci*) Disruption of the normal structure of raw or processed material to produce a more uniform consistency. Used to produce pastes, sausage meats, and to RE-FORM materials to make them suitable for moulding or extrusion. In citrus fruits, the whole fruit can be used rather than just the juice, the product being termed whole orange etc.

comminution (*MinExt*) The process of reducing a material to a powder by eg attrition, impact, crushing, grinding, abrasion, milling or chemical methods.

commissioning (*NucEng*) The process prior to a contractor handing over equipment to a purchaser in which the system is tested to see if it conforms to specification.

commissural bundle (*BioSci*) A small vascular bundle interconnecting larger bundles.

commissure (*BioSci*) A joint, a line of junction between two organs or structures, a bundle of nerve fibres connecting two nerve centres. In plants, esp a surface by which carpels are united.

committed dose equivalent (*Radiol*) The calculated dose equivalent of a given radiation dose integrated over a lifetime, assumed to be 50 years.

commode step (*Build*) A step having a riser curved to present a convex surface; used sometimes at and near the foot of a staircase.

commodity polymer (*Chem*) Commercial polymers produced in high-tonnage quantities, esp polystyrene, low-density polyethylene, high-density polyethylene, polypropylene and polyvinyl chloride.

Common Agricultural Policy (*Agri*) Regulatory measures adopted by the European Union to regulate the agricultural activities of member states. Abbrev CAP.

common air interface (*ICT*) An agreed AIR INTERFACE adhered to by all providers of CT2 cordless telephone services, allowing the same handset to be used at any phonepoint.

common ashlar (*Build*) A block of stone which is pick or hammer dressed.

common-base connection (*Electronics*) The operation of a transistor in which the signal is fed between base and emitter, the output being between collector and base with the latter earthed. Also *grounded base*.

common bricks (*Build*) A class of brick used in ordinary construction (esp in interior work) for filling in and to make up the requisite thickness of heavy walls and piers. They usually have plain sides, are not neatly finished, and are much more absorbent and also much weaker than ENGINEERING BRICKS.

common bundle (*BioSci*) A vascular bundle belonging in part to a stem and in part to a leaf.

common channel signalling (*ICT*) The use of a shared high-speed data link for the signals needed to set up a telephone call, as opposed to sending such signals over the voice link itself. For calls involving several switching centres, call set-up time is considerably reduced.

common-collector connection (*Electronics*) The operation of a transistor in which the collector is earthed, input is between collector and base, and output is between emitter and collector. This circuit provides relatively high input impedance with low output impedance. Voltage gain is unity, while current gain depends on transistor characteristics. Also *emitter follower, grounded collector*.

common command language (*ICT*) A command language used to access a number of different INFORMATION RETRIEVAL SYSTEMS.

common dovetail (*Build*) An angle joint between two members in which both show end grain.

common-emitter connection (*Electronics*) The operation of a transistor in which the signal is fed between base and emitter with the latter earthed. The output is between emitter and collector. Also *grounded emitter*.

common fraction (*MathSci*) See DIVISION.

common-frequency broadcasting (*ICT*) The use of the same carrier frequency by two or more broadcast transmitters, sufficiently separated for their useful service areas not to overlap. Also *shared-channel broadcasting*.

common intermediate format (*ICT*) The picture format required by the CCITT H261 digital video encoding and decoding standard for videoconferencing and videotelephony. A compromise between the US/Japanese NTSC and European PAL standards, frames are presented at 29·97 per second and contain 288×352 luminance elements and 144×176 chrominance elements.

common joist (*Build*) See BRIDGING JOIST.

common lead (*Eng*) Lead of lower purity than chemical or corroding lead (about 99·85%).

common logarithm (*MathSci*) A LOGARITHM to base 10.

common-mode failure (*NucEng*) The failure of two or more supposedly independent parts of a system (eg a reactor) from a common external cause or from interaction between the parts.

common-mode rejection ratio (*Electronics*) For a differential amplifier, the ratio of the gain for a differential input to that for a common-mode input.

common-mode signal (*Electronics*) A signal applied simultaneously to both inputs of a differential amplifier.

common rafter (*Build*) A subsidiary rafter carried on the PURLINS and supporting the roof covering. Also rafters of a common length and bevel in any roof. Also *intermediate rafter*.

common-rail injection (*Autos*) A fuel-injection system for multi-cylinder compression–ignition engines; an untimed pump maintains constant pressure in a pipeline (rail), from which branches deliver the oil to the mechanically operated injection valves.

common return (*ICT*) A single conductor that forms the return circuit for two or more otherwise separate circuits.

communication (*Psych*) The transmission of a message, signal or meaning from one place to another. The sender and receiver must share a common coding system which need not necessarily be language in spoken or written form, eg *non-verbal communication*.

communications satellite (*Space, ICT*) Artificial satellite to aid global communications by the relay of data, voice and TV. See panel on COMMUNICATIONS SATELLITE.

community (*BioSci*) (1) Any group of plants growing together under natural conditions and forming a recognizable sort of vegetation, eg oak wood, blanket bog. (2) The animals inhabiting a restricted area, as field or pond; such a community is not necessarily stable.

community antenna TV (*ICT*) Used in cable TV systems that cover a town or similar area and fed from a single central antenna that may take signals from existing channels or from satellites.

commutating field (*ElecEng*) The magnetic field under the compoles of a dc machine; it induces, in the conductors undergoing commutation, an emf in a direction to assist in the commutation process.

commutating machine (*ElecEng*) An electrical machine provided with a commutator.

commutating pole (*ElecEng*) See COMPOLE.

commutation factor (*Electronics*) The product of rate of current decay and rate of voltage rise after a gas discharge, both expressed per microsecond.

commutation switch (*ICT*) The switch controlling the sequential switching operations required for multichannel pulse communication systems.

commutative (*MathSci*) (of an operation). Such that the order in which it is performed on two quantities does not affect the result, eg the operation of addition in arithmetic is commutative because $a + b = b + a$. Cf ASSOCIATIVE, DISTRIBUTIVE, TRANSITIVE.

commutator (*ElecEng*) The part of a motor or generator armature through which electrical connections are made by rubbing brush contacts.

commutator (*MathSci*) Of two elements a and b of a group: $aba^{-1}b^{-1}$ or $a^{-1}b^{-1}ab$ according as operations are written on the left or the right respectively. Of two functions a and b: $ab-ba$. In either case written $[a, b]$.

commutator bar (*ElecEng*) One of the copper bars forming part of a commutator. Also *commutator segment*.

commutator bush (*ElecEng*) See COMMUTATOR HUB.

commutator face (*ElecEng*) See COMMUTATOR SURFACE.

commutator grinder (*ElecEng*) A portable piece of electric grinding equipment which can be mounted on a commutator machine to grind the commutator surface without removing the armature.

commutator hub (*ElecEng*) A metal structure used for supporting a commutator. Also *commutator bush, commutator shell, commutator sleeve*.

commutator losses (*ElecEng*) Losses occurring at the commutator of an electric machine; they include resistance loss in the segments, in the brushes and at the contact surface, friction loss due to the brushes sliding on the commutator surface, loss due to sparking, and eddy current loss in the segment.

commutator motor (*ElecEng*) An electric motor which embodies a commutator.

commutator ring (*ElecEng*) A ring, usually of cast-iron, made to fit into a dovetail in the commutator segments in order to clamp them firmly in position and including the insulating material between the ring and the segments.

commutator ripple (*ElecEng*) Small periodic variations in the voltage of a dc generator or rotary converter resulting from the fact that there can only be a finite number of commutator segments on the machine.

commutator segment (*ElecEng*) See COMMUTATOR BAR.

commutator surface (*ElecEng*) The smooth portion of a commutator upon which the current-collecting brushes slide. Also *commutator face*.

comopt (*ImageTech*) International code name for a picture film combined with an optical sound track.

comorbid (*Med*) A disease or disorder existing together with another.

compact (*Eng*) The solid produced by confining a powder, with or without a binder, and compressing it in a die. See GREEN COMPACT.

compact disc (*ICT*) Digitally encoded read-only disk, read by laser. Commonly used in domestic hi-fi systems for high-quality sound reproduction. Abbrev *CD*.

compacted graphite cast-iron (*Eng*) Made by a method like that for DUCTILE CAST-IRON but without allowing the formation of completely spherulitic graphite nodules so that the graphite shape is between that of grey and ductile cast-iron. Abbrev *CG iron*. Also *vermicular iron*.

compaction (*Agri*) Reduction of spaces between soil particles to an extent where gas exchange, water-holding capacity and root growth are restricted.

compaction (*Build, CivEng*) The process of consolidating soil or dry concrete by mechanical means or of wet concrete by vibration.

compact space (*MathSci*) Of a topological space, or set within one (compact set), such that every collection of open sets which covers it completely includes a finite subcollection which does so. A closed and bounded set in n-dimensional Euclidean space is compact.

compander (*ICT*) Contraction of *compresser–expander*. Device for compressing the volume range of the transmitted signal and re-expanding it at the receiver, thus increasing the signal-to-noise ratio.

companion cell (*BioSci*) A parenchyma cell, with dense cytoplasm and a conspicuous nucleus, in the phloem of an angiosperm, adjacent to and originating from the same mother cell as a sieve tube member. See fig at SIEVE ELEMENT. Cf ALBUMINOUS CELL.

companion star (*Astron*) The star which accompanies another in a binary system of two stars in orbit.

comparative psychology (*Psych*) A branch of psychology associated with comparing behaviour between different species in order to make generalizations about the mechanisms underlying behaviour.

comparative tracking resistance (*Electronics*) Abbrev *CTR*. See TRACKING RESISTANCE.

comparator (*ICT*) A circuit that compares two sets of impulses, or the magnitudes of two numbers and acts on their matching or otherwise.

comparator (*Phys*) (1) A form of apparatus used for the accurate comparison of standard of length. It has also been used for measuring the coefficients of expansion of metal bars. (2) A form of colorimeter.

comparison lamp (*Phys*) A lamp used, when performing photometric tests, for making successive comparisons between the lamp under test and a standard lamp.

comparison prism (*Phys*) A small right-angled prism placed in front of a portion of the slit of a spectroscope or spectrograph for the purpose of reflecting light from a second source of light into the collimator, so that two spectra may be viewed simultaneously. See COMPARISON SPECTRUM.

comparison spectrum (*Phys*) A spectrum formed alongside the spectrum under investigation, for the purpose of measuring the wavelengths of unknown lines. It is desirable

Communications satellite

An artificial satellite to aid global communications by the relay of data, voice and television. The satellite can be purely reflective, but usually fulfils a repeater role using an on-board TRANSPONDER. The latter amplifies the signal and changes the frequency to avoid interference between the incoming and outgoing electromagnetic signals (see diagram). The system is made up of a space segment, consisting of the satellite and its associated on-board equipment, and the Earth segment, which comprises the ground stations suitably equipped to handle transmission and reception of the signals.

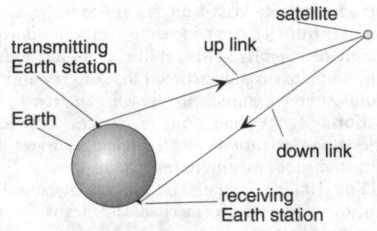

Communications satellite

These satellites may orbit the Earth at any altitude but three regions are either being used or actively developed.

Orbits

(1) Geosynchronous orbits. These are fixed orbits of 35 784 km in which a satellite has a period of 24 hours and will follow the same apparent path across the sky every day. A GEOSTATIONARY orbit is the special instance which lies above the equator and in which the satellite remains over the same spot. Satellites in these orbits have a large FOOTPRINT and only three or four are needed to cover the Earth. When obsolete they can be parked in a higher orbit out of the way. They have the considerable disadvantage for some applications of a long LATENCY of 2·4 seconds for the round trip. This is unacceptable for telephonic communication but acceptable for TV broadcasting. Abbrev GEO.

(2) Medium Earth orbits. These orbits range from 10 000 to 21 000 km and have a latency between 0·06 and 0·14 seconds. About 15 satellites are needed to cover most of the Earth, but like all except the geostationary satellites they spend much of their time over the oceans. Abbrev MEO.

(3) Low Earth orbits. Situated between 800 and 2400 km above the Earth, these orbits have a latency below 0·03 seconds round trip but require a minimum of 50 satellites to give adequate Earth coverage. They will add to the clutter in these near-Earth regions and may well collide with space debris. At the end of their useful life they will have to burn out in the atmosphere. Abbrev LEO.

Frequency bands

Three bands are available for commercial satellites. L-BAND with a frequency range from 1·53 to 2·7 GHz carries least information and needs large antennae (approx 1 m diameter) but can penetrate structures and through rain. Transmitters need to be less powerful for a given range than shorter wavelengths. Ku-band (US) with frequencies of 14·7 to 17·5 GHz uplink and 11·7 to 12·7 downlink can carry considerably more information and penetrate rain and obstacles. It needs more powerful transmitters than L-band and smaller antennae. Ka-band with frequencies from 18 to 31 GHz can carry most information and if used in low Earth orbits will need medium-power transmitters and small receiving antennae. Subject to rain fade, it will also need sophisticated methods of following and switching between the many satellites as they pass overhead.

Uses of communications satellites

DIRECT BROADCAST SATELLITES (DBS), operating at high power from geosynchronous orbits, provide services, such as TV programmes, to geographically remote areas and to their subscribers anywhere. Increasingly, however, commercial operators are setting up broadband services for data, voice and videoconferencing. These will use LEO and MEO orbits and are additional to the international public and private communications services run by such as INTELSAT and EUTELSAT using GEO orbits.

See panel on SPACE and appendix on Radio and radar frequencies.

---

that the comparison spectrum should contain many standard lines of known wavelength. The spectrum of the iron arc is often used for this purpose. See COMPARISON PRISM.
comparison surface (*Phys*) A surface illuminated by a standard lamp or a comparison lamp; used in photometry.
comparison test (*MathSci*) A test for convergence or divergence of the series

$$\sum_{n=0}^{\infty} a_n$$

If $0 \leqslant a_n \leqslant b_n$ for all n then

$$\sum_{n=0}^{\infty} a_n \text{ converges if } \sum_{n=0}^{\infty} b_n \text{ converges.}$$

If $0 \leqslant c_n \leqslant a_n$ for all n then

$$\sum_{n=0}^{\infty} a_n \text{ diverges if } \sum_{n=0}^{\infty} c_n \text{ diverges.}$$

compartment (*BioSci*) (1) Conceptualized part of the body (organs, tissues, cells or fluids) considered as an independent system for purposes of modelling and assessment of distribution and clearance of a substance. (2) Region of an insect embryo within which all cells give rise to the same adult structure, eg leg, fore-wing. Studies with homeotic mutants suggest that cells in different compartments express different sets of genes.

compartment (*For*) Permanent administrative units of a plantation delineated by well-defined natural features.

compartmental analysis (*BioSci*) Mathematical process that models the transport of a substance in terms of compartments and rate constants.

compass (*Surv*) An instrument which indicates (1) magnetic bearings by alignment of its needle on the magnetic poles, or (2) reference bearing by radio signal, or (3) change of orientation in relation to a gyro-maintained line.

compass brick (*Build*) A brick which tapers in at least one direction; specially useful for curved work, in arches, parts of furnaces, etc.

compasses (*Eng*) An instrument for describing arcs, taking or marking distances, etc; it consists essentially of two limbs hinged together at one end.

compass plane (*Build*) A plane used on a curved surface, having a flexible metal sole which can be set concave or convex.

compass roof (*Arch*) A roof with rafters bent to the shape of an arc.

compass-safe distance (*Aero, Ships*) The minimum distance at which equipment may safely be positioned from a direct-reading magnetic compass, or detector unit of a remote-indicating compass, without exceeding the values of maximum compass deviation change.

compass saw (*Build*) A narrow-bladed saw for cutting small-radius curves.

compass traverse (*Surv*) Rapid rough survey method in which the magnetic bearing of each line is measured.

compatibility (*Chem*) (1) General term describing state of mixture of materials whether in liquid or solid state. If two chemically distinct substances mix completely in the liquid state, then they are compatible and miscible, and form a homogeneous fluid. See PHASE DIAGRAMS. If some degree of phase separation occurs, then they are partially miscible, but if totally incompatible, they form immiscible phases. Also a term applied to the solid state, where kinetics of formation may be important. See ISOTHERMAL TRANS-FORMATION DIAGRAMS. Origin of all such effects lies in type and sizes of atoms or molecules and type and magnitude of bonding. (2) Tendency of different materials esp polymers to mix homogeneously at a molecular level. Relatively rare effect in polymers, but see NORYL.

compatible (*ICT*) Computer hardware is compatible if it can use the same software. Software is upwardly compatible if it can continue to be used when a system is improved. See IBM-COMPATIBLE.

compatible colour TV (*ImageTech*) The technique of transmitting TV pictures in colour by a combination of LUMINANCE and CHROMINANCE SIGNALS, compatibility with black-and-white TV being preserved as the chrominance elements are virtually disregarded by the monochrome receiver.

compatible equations (*MathSci*) See CONSISTENT EQUATIONS.

compensated induction motor (*ElecEng*) An induction motor with a commutator winding on the rotor, in addition to the ordinary primary and secondary windings; this winding is connected to the circuit in such a way that the motor operates at unity or at a leading power factor.

compensated pendulum (*Phys*) A pendulum made of two materials which have different coefficients of expansion and are so chosen that the length of the pendulum remains constant when the temperature varies.

compensated scale barometer (*EnvSci*) See KEW-PATTERN BAROMETER.

compensated semiconductor (*Electronics*) A material in which there is a balanced relation between DONORS and ACCEPTORS, by which their opposing electrical effects are partially cancelled.

compensated series motor (*ElecEng*) The usual type of ac series motor, in which a compensating winding is fitted to neutralize the effect of armature reaction and so give a good power factor. Also *neutralized series motor*.

compensated shunt box (*ElecEng*) A shunt box for use with a galvanometer, arranged so that on each step a resistance is put in series with the galvanometer, and the total resistance of galvanometer and shunt is not altered.

compensated voltmeter (*ElecEng*) A voltmeter arranged to indicate the voltage at the remote end of a feeder or other circuit, although connected at the sending end. A special winding compensates for the voltage drop in the feeder.

compensated wattmeter (*ElecEng*) A wattmeter in which there is an additional winding, arranged to compensate for the effect of the current flowing in the voltage circuit.

compensating coils (*ElecEng*) Current-carrying coils to adjust distribution of magnetic flux.

compensating diaphragm (*Surv*) A fitment for a tacheometer which, by an adjustment to the stadia interval determined by the vertical angle, enables the horizontal component of a sloping sight to be deduced from the staff intercept.

compensating error (*Surv*) As opposed to systematic (biased) error in series of observations, an error equally likely to be due to over- or under-measurement, and therefore reasonably likely to be compensated by errors of opposite sign.

compensating field (*ElecEng*) A term sometimes used to indicate the field produced by a compensating winding or, occasionally, by a compole.

compensating filter (*ImageTech*) See CC FILTER.

compensating jet (*Autos*) An auxiliary petrol jet used in some carburettors to supplement the discharge from the main jet at low rates of airflow, and to keep the mixture strength constant. See CARBURETTOR.

compensating pole (*ElecEng*) See COMPOLE.

compensating roller (*Print*) See JOCKEY ROLLER.

compensating winding (*ElecEng*) A winding used on dc or ac commutator machines to neutralize the effect of armature reaction.

compensation (*Acous*) In a sound-reproducing system, adjustment of an actual frequency response to one specified.

compensation (*Psych*) The process of compensating for a sense of failure or inadequacy by concentrating on achievement or superiority, real or fancied, in some other sphere; the defence mechanism involved in this.

compensation point (*BioSci*) The light intensity at which, under specified conditions, photosynthesis and respiration just balance so that there is no net exchange of CO_2 nor O_2. For C3 plants it is 50–70 ppm, for C4 plants 0–10 ppm. Also CO_2 *compensation point*.

compensation point (*Phys*) The temperature (T_{comp}), below the Curie temperature (T_c), at which the magnetization of certain ferrimagnetic materials vanishes. It arises because of differences in the temperature variations of saturation magnetization of the two opposed sublattices in the material. Such ferrimagnetics can be magnetized below T_{comp} and between T_{comp} and T_c.

compensation theorem (*Phys*) A theorem stating that the change in current produced in a network by a small change in any impedance Z carrying a current I is the result of an apparent emf of $-I\delta Z$.

compensation water (*CivEng*) The water which has to be passed downstream from a reservoir to supply users who, prior to the construction of the dam, took their water directly from the stream.

compensator (*ImageTech*) A device such as a graduated filter used with an extreme WIDE-ANGLE LENS to improve uniformity of illumination of the image area.

compensator (*Phys*) (1) A glass plate used in various optical interferometers to achieve equality of optical path length. (2) An apparatus in a polarizing microscope that measures the phase difference between the two components of polarized light, eg a *Berek compensator*. (3) A plate of variable thickness of optically active quartz used to produce elliptically polarized light of a given orientation.

competition (*BioSci*) The struggle between organisms for the necessities of life (water, light, etc).

competitive exclusion principle (*BioSci*) The ecological 'law' that two species cannot occupy the same ecological niche or utilize the same limiting resource. One species always outcompetes the other.

competitive inhibitor (*BioSci*) A molecule that binds to the active site of an enzyme or the binding site of a receptor and thus prevents attachment of the normal substrate or ligand.

compilation error (*ICT*) The error detected during compilation (eg a SYNTAX ERROR).

compiler (*ICT*) A program that translates a HIGH-LEVEL LANGUAGE program into a computer's MACHINE CODE or some other LOW-LEVEL LANGUAGE. Each high-level language instruction is changed into several machine-code instructions. It produces an independent program that is capable of being executed; n *compilation*. Cf ASSEMBLER, INTERPRETER. See EXECUTE.

complanate (*BioSci*) Flattened, compressed.

complement (*BioSci*) (1) A heat-labile cascade of enzymes in plasma associated with response to injury. Activation of the complement cascade occurs through two convergent pathways: the classical pathway, which involves antibody/antigen complexes; and the alternative pathway, which is activated by IgA, endotoxin or polysaccharide-rich surfaces (eg yeast cell wall). The alternative pathway is presumably the ancestral one upon which the sophistication of antibody recognition has been superimposed in the classical pathway. The enzymatic cascade amplifies the response, leads to the activation and recruitment of leucocytes, increases phagocytosis, and induces killing directly. It is subject to various complex feedback controls that terminate the response. (2) See CHROMOSOME COMPLEMENT, CHROMOSOME SET.

complementarity (*Phys*) The quantum mechanical principle that complementary particle and wave aspects occur in nature. There is a correspondence between particles of momentum p and energy E and the associated wavetrain of frequency $v = E/h$ and wavelength $\lambda = h/p$, where h is Planck's constant. A measurement proving the wave character of radiation on matter cannot prove the particle character in the same measurement, and conversely.

complementarity determining region (*BioSci*) Hypervariable region within the antigen binding site of immunoglobulin molecules and T-lymphocyte antigen receptors whose amino acid sequence determines which antigenic EPITOPES bind to the individual immunoglobulins. Also applied to gene sequences encoding the hypervariable regions. Abbrev *CDR*.

complementary (*BioSci*) Relationships between single strands of DNA and RNA are complementary to each other if their sequences are related by the *base-pairing rules*, thus ATCG is complementary to the sequence TAGC, and can pair with it by hydrogen bonding.

complementary after-image (*Phys*) The subjective image, in complementary colours, that is experienced after visual fatigue induced by observation of a brightly coloured object.

complementary angles (*MathSci*) Two angles whose sum is 90°. Each is said to be the *complement* of the other.

complementary colours (*Phys*) Pairs of colours which combine to give spectral white.

complementary DNA (*BioSci*) A DNA sequence complementary to any RNA. Formed naturally in the life cycle of RNA viruses by REVERSE TRANSCRIPTASE, which is widely used in the laboratory to make DNA complements of mRNA. Abbrev *cDNA*.

complementary DNA cloning (*BioSci*) Procedure by which DNA complementary to mRNA is inserted into a VECTOR and propagated. Because mRNA has no INTRONS, such cDNA clones can be made to produce a normal polypeptide product. Also *cDNA cloning*.

complementary function (*MathSci*) If $y = u + v$ is a solution of a differential equation, where u is a particular

integral and v contains the full number of arbitrary constants, v is called the *complementary function*. The complementary function is the general solution of the auxiliary equation.

complementary genes (*BioSci*) Two *non-allelic* genes that must both be present for the manifestation of a particular character.

complementary medicine (*Med*) See ALTERNATIVE MEDICINE.

complementary metal–oxide–silicon (*ICT, Electronics*) A major integrated circuit technology based on combinations of p-channel and n-channel FIELD-EFFECT TRANSISTORS fabricated on the same silicon substrate. Especially attractive in low-power applications since the basic CMOS logic gate only consumes significant power during switching. Abbrev *CMOS*.

complementary symmetry (*Electronics*) Shown by otherwise identical p–n–p and n–p–n transistors.

complementary transistors (*Electronics*) An n–p–n and a p–n–p transistor pair used to produce a push–pull output using a common signal input.

complementation (*BioSci*) The full or partial restoration of normal function when two recessive mutants, both deficient in that function, are combined in a double heterozygote. *Complementing* mutants are *non-allelic, non-complementing* are *allelic*.

complementation (*ICT*) A method of representing negative numbers. See ONE'S COMPLEMENT, TWO'S COMPLEMENT.

complement deficiency (*BioSci*) Hereditary deficiencies of complement components are uncommon in humans although strains of laboratory animals exist lacking C3, C4, C5 or C6. Absence of any single component is compatible with life but persons lacking the early acting components, esp with diminished C3, are unusually liable to bacterial infection and often show signs of immune complex disease. The genes for C4, C2 and Factor B lie within the MAJOR HISTOCOMPATIBILITY COMPLEX.

complement fixation (*BioSci*) A term synonymous with activation but often applied to a system *in vitro* that detects complement by its capacity to cause lysis of red cells with antibody on the surface. Activation of complement by combination of antigen with antibody prior to adding the red cells diminishes the amount of complement available to lyse the red cells. This is a sensitive method for detecting the presence of antigen or antibody but has been superseded by other methods such as ELISA.

complement of a set (*MathSci*) The complement of a given set, A, is the set of all members of the universal set which are not members of A, and is sometimes denoted by A', Ā or $C(A)$.

complement receptor (*BioSci*) Cell surface receptors for various components of COMPLEMENT. At least nine different complement receptors have been described that function at different points in the complement enzyme cascade.

complete combustion (*Eng*) Burning fuel without trace of unburnt gases in the products of combustion, usually accompanied by excess air in the flue products. Cf PERFECT COMBUSTION.

complete differential (*MathSci*) The complete differential df of a function $f(x, y, z)$ is

$$\frac{\partial f}{\partial x}dx + \frac{\partial f}{\partial y}dy + \frac{\partial f}{\partial z}dz$$

Similar expressions apply for more or less variables. Also *total differential*.

complete Freund's adjuvant (*BioSci*) A water-in-oil emulsion with added heat-killed mycobacteria into which is incorporated an antigen for the purpose of immunization against it. This form of adjuvant is very effective for eliciting both T- and B-cell immunity, but it is not used in humans because it is liable to cause suppurating granulomas. Abbrev *CFA*.

complete graph (*MathSci*) A graph in which every two distinct vertices are joined by exactly one edge. The complete graph with *n* vertices is usually called K_n.

complete integral (*MathSci*) The solution of a differential equation containing the full number of arbitrary constants. Also *complete primitive, complete solution, general solution*.

complete metric space (*MathSci*) A metric space in which all CAUCHY SEQUENCES have limits is complete.

complete primitive (*MathSci*) See COMPLETE INTEGRAL.

complete radiator (*Phys*) See BLACK BODY.

complete reaction (*Chem*) A reaction which proceeds until one of the reactants has effectively disappeared.

complete set of functions (*MathSci*) See ORTHOGONAL FUNCTIONS.

complete solution (*MathSci*) See COMPLETE INTEGRAL.

complex (*Psych*) A term introduced by Jung to denote an emotionally toned constellation of mental factors formed by the attachment of instinctive emotions to objects or experiences in the environment, and always containing elements unacceptable to the self. It may be recognized in consciousness, but is usually repressed and unrecognized.

complex amplitude (*Acous*) Complex number with amplitude and phase information of a harmonic signal, eg sound pressure.

complex hyperbolic functions (*Phys*) Hyperbolic functions, with complex quantities as variables, which facilitate calculations of electric waves along transmission lines.

complex instruction set computer (*ICT*) A computer whose central processor INSTRUCTION SET is designed to accept many operands. Typical of most personal computers, each instruction may take many FETCH–EXECUTE CYCLES but a great deal of processing is carried out. Cf RISC.

complex ion (*Chem*) See CO-ORDINATION COMPOUND.

complexity (*ICT*) See COMPUTATIONAL COMPLEXITY.

complexity of DNA (*BioSci*) A measure, obtained from renaturation kinetics, of the number of copies of a given sequence in eg a genome or the mRNA in a cell. Also *complexity of RNA*.

complex modulus (*Eng*) Result from a dynamic mechanical test, where eg tensile modulus E^* is measured. It is related to the real (') and imaginary ('') moduli by the equation: $E^* = E' + iE''$. The ratio E''/E' is $tan\ \delta$ or the LOSS FACTOR.

complex number (*MathSci*) A number *z* of the form $a + ib$, where $i = \sqrt{(-1)}$ and *a* and *b* are real numbers. The number *a* is called the *real part*, written R*z*, and *b* the *imaginary part*, written I*z*: eg if $z = 3 + 4i$, R*z* = 3 and I*z* = 4. Cf ARGUMENT, MODULUS.

complexometric titration (*Chem*) Titration of a metal ion with a reagent, usually *EDTA*, which forms chelate complexes with the metal. The end-point is accompanied by a sharp decrease in the concentration of metal ions, and is observed by a suitable indicator.

complexones (*Chem*) Collective term used to denote chelating reagents (usually organic) used in the analytical determination of metals, eg *EDTA*.

complex Poynting vector (*ElecEng*) See POYNTING VECTOR.

complex tissue (*BioSci*) A plant tissue made up of cells or elements of more than one kind.

complex tone (*Acous*) Strictly, a musical note in which all the separate tones are exact multiples of a fundamental frequency, recognized as the pitch, even when the actual fundamental is absent, as in the lowest octave of the piano. Loosely, a mixed musical chord.

complex wave (*Phys*) A wave with a non-sinusoidal form which can be resolved into a fundamental with superimposed harmonics. See FOURIER PRINCIPLE.

compliance (*Eng*) The ease with which a body can be deformed elastically or the linear displacement produced by unit force. Equals the reciprocal STIFFNESS. It is the mechanical analogue of capacitance.

compliance (*ICT*) The capacity of different pieces of hardware or software to operate together.

compliance (*Psych*) The behaviour of one individual that conforms to the wishes of others; it does not necessarily imply conformity of belief.

complicate (*BioSci*) Folded together.

compo (*Build*) A cement mortar.

compole (*ElecEng*) An auxiliary pole, employed on commutator machines, which is placed between the main poles for the purposes of producing an auxiliary flux to assist commutation. Also *commutating pole, compensating pole, interpole*.

component (*ElecEng*) A term sometimes used to denote one of the component parts into which a vector representing voltage, currents or volt-amperes may be resolved; the component parts are usually in phase with, or in quadrature with, some reference vector. See ACTIVE CURRENT, ACTIVE VOLTAGE, ACTIVE VOLT-AMPERES, REACTIVE COMPONENT OF CURRENT, REACTIVE VOLTAGE, REACTIVE VOLT-AMPERES.

component (*Phys*) The resolved part of a vector quantity such as force, velocity, acceleration or momentum, in any particular direction, eg the component of a force F along a line making an angle θ with the line of action of F is F $\cos\theta$. See RESOLUTION OF FORCES.

component of a vector (*MathSci*) In general, a vector in a given direction, whose sum with one or more vectors in directions perpendicular to the specified direction is the given vector.

components (*Chem*) The individual chemical substances present in a system. See PHASE RULE.

component video (*ImageTech*) A colour TV transmission and recording system with luminance and CHROMINANCE as separate signals. See LUMINANCE SIGNAL.

COM port (*ICT*) See SERIAL PORT.

compose (*Print*) To assemble type matter for printing, either by hand or by typesetting machines.

composing frame (*Print*) A wooden or metal structure at which the compositor works. Originally providing accommodation on top for an upper and lower case with storage below for other cases, it is now made in a large variety of styles to suit the requirements of the work to be done, with a sloping or flat top, and storage for all kinds of typographical material in addition to cases.

composing machines (*Print*) Now almost obsolete, the Monotype composes type matter in separate letters, which may be used for hand-setting or correcting; the Linotype composes in solid lines, or slugs, which must be reset when correcting. Both machines have a keyboard resembling that of a typewriter, but each system has its own layout. See PHOTOTYPESETTING.

composing stick (*Print*) A metal or wooden three-sided box-like receptacle in which the compositor sets the type, letter by letter. The width or measure can be altered as desired.

Compositae (*BioSci*) The daisy family, comprising c.25 000 spp of dicotyledonous flowering plants (superorder Asteridae). It is the largest family of dicotyledons and the plants, mostly herbs and shrubs, are cosmopolitan. The inflorescence is a head (capitulum) made up of many small individual florets or florets surrounded by an involucre of bracts, the whole resembling, and functioning biologically, as a single flower. The florets have a gamopetalous corolla that is usually TUBULAR or LIGULATE, and the ovary is inferior and develops into a one-seeded indehiscent, dry fruit. The modified calyx or pappus is composed of hairs, scales or bristles, and often develops as a feathery parachute aiding wind dispersal. Includes relatively few economic plants, such as sunflower (for oil), lettuce, endive, chicory; also a number of ornamentals, eg chrysanthemum and various daisies. The insecticide pyrethrum comes from the heads of a species of *Tanacetum*. Also *Asteraceae*.

composite (*ImageTech*) A photographic montage of images, eg an audio-visual system in which a number of image areas are presented simultaneously to the viewer.

composite (*Textiles*) A matrix such as cement or plastic reinforced by fibres.

composite beam (*Eng*) A beam composed of two materials properly bonded together and having different moduli of elasticity, eg a reinforced concrete beam, SANDWICH BEAMS.

composite block (*Print*) (1) Combined half-tone and line block. (2) A block made up from two or more originals.

composite cable (*ElecEng*) Cable containing different purpose conductors inside a common sheath.

composite cinematography (*ICT*) Cinematography that involves blending images from two or more sources.

composite compact (*Eng*) A COMPACT made by powder metallurgy which has several adherent layers of different alloys.

composite conductor (*ElecEng*) One in which strands of different metals are used in parallel.

composite material (*Eng*) Structural material made of two or more different materials, eg CERMETS or CARBON- or GLASS-FIBRE-REINFORCED PLASTICS.

composite number (*MathSci*) Any integer which is not PRIME.

composite photography (*ImageTech*) General term for cinematographic SPECIAL EFFECTS in which two or more separate shots are combined to give the effect of a single scene. See BACK PROJECTION, FRONT PROJECTION, TRAVELLING MATTE SHOT.

composite resistor (*ElecEng*) One formed of a solid rod of a carbon compound.

composite structure (*Eng*) Any structure made by bonding two or more different materials, such as metal, plastic, composite material, etc.

composite truss (*Build*) A roof truss formed of timber struts and steel or wrought-iron ties (apart from the main tie, which is usually of timber to simplify fixings).

composite yarn (*Textiles*) Yarn made from a combination of staple fibres and continuous filaments.

composition (*Chem*) The nature of the elements present in a substance and the proportions in which they occur, eg mole fraction.

composition (*MathSci*) The result of applying two or more functions in succession. The composition of f and g is usually written $f \circ g$ and is variously defined as $f(g(x))$ or as $g(f(x))$.

composition fonts (*Print*) The smaller sizes of type, up to 14 point, as used for bookwork.

composition nails (*Build*) Roofing nails made of a cast 60–40 copper–zinc alloy.

composition of atmosphere (*Chem*) Dry atmospheric air contains the following gases in the proportions (by weight) indicated: nitrogen, 75·5; oxygen, 23·14; argon, 1·3; carbon dioxide, 0·05; krypton, 0·028; xenon, 0·005; neon, 0·000 86; helium, 0·000 056. There are variable trace amounts of other gases including hydrogen and ozone. Water content, which varies greatly, is excluded from this analysis.

composition of forces (*Phys*) The process of finding the resultant of a number of forces, ie a single force which can replace the other forces and produce the same effect. See PARALLELOGRAM OF FORCES.

composition rollers (*Print*) (1) For letterpress printing, a mixture of glue, glycerine and molasses. (2) For lithographic printing, vegetable oils and rubber, vulcanized.

compositor (*Print*) A craftworker who sets up type matter by hand, or corrects that set by machine.

compost (*BioSci*) (1) Rotted plant material and/or animal dung, etc, used as a soil conditioner. (2) A medium in which plants (esp plants in pots) are grown, composed of one or more of sand, soil, grit, peat, perlite, vermiculite, etc, with lime and fertilizers as necessary.

compound (*BioSci*) Consisting of several parts, as in a leaf made up of several distinct leaflets, or an inflorescence of which the axis is branched etc. Cf SIMPLE.

compound (*Chem*) See CHEMICAL COMPOUND.

compound arch (*Arch*) An arch having an ARCHIVOLT receding in steps, so as to give the appearance of a succession of receding arches of varying spans and rises.

compound brush (*ElecEng*) A type of brush used for collecting current from the commutator of an electric machine; the brush has alternate layers of copper and carbon so that the conductivity is greater longitudinally (ie in the direction of the main current flow) than laterally.

compound catenary construction (*ElecEng*) A construction used for supporting the overhead contact wire of an electric traction system; the contact wire is supported from an auxiliary catenary which, in turn, is supported from a main catenary, all three wires lying in the same plane.

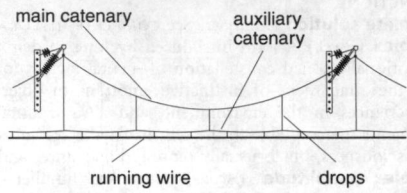

main catenary   auxiliary catenary

running wire   drops

compound catenary construction Longitudinally foreshortened.

compound curve (*Surv*) A curve composed of two arcs of different radii, having their centres on the same side of the curve, connecting two straights.

compound engine (*Eng*) A development of the 'simple' steam engine, the compound engine has two or more cylinders of different size, allowing the steam to expand over several stages and enabling more work to be done per unit mass of steam and thus giving greater efficiency at the cost of increased complexity.

compound eyes (*BioSci*) Paired eyes consisting of many facets or ommatidia, in most adult ARTHROPODA.

compound fault (*Geol*) A series of closely spaced parallel or subparallel faults.

compound fertiliser (*Agri*) A synthetic formulation containing relatively high levels of nitrogen, potassium and phosphorus to improve crop growth and development.

compound-filled apparatus (*ElecEng*) Electrical apparatus (eg bus-bars, potential transformers, switchgear) in which all live parts are enclosed in a metal casing filled with insulating compound.

compound generator (*ElecEng*) A generator which has both series and shunt field windings.

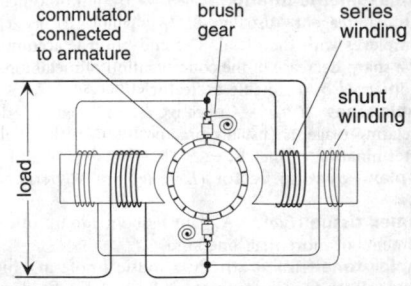

commutator connected to armature   brush gear   series winding

shunt winding

load

compound generator The shunt winding has more turns of lighter wire.

compound girder (*Build*) A rolled-steel joist strengthened by additional plates riveted or welded to the flanges.

compounding (*Chem*) The process of mixing polymer compounds, often using EXTRUSION and usually performed by trade compounders. See MASTERBATCH.

compounding (*Eng*) The principle, or the use of the principle, of expanding steam in two or more stages, either in reciprocating engines or in steam turbines.

compound lever (*Eng*) A series of levers for obtaining a large mechanical advantage, the short arm of one being connected to the long arm of the next; used in large weighing and testing machines.

compound magnet (*ElecEng*) A permanent magnet made up of several laminations.

compound microscope (*Phys*) See MICROSCOPE.

compound modulation (*ICT*) Use of an already modulated wave as a further modulation envelope. Also *double modulation*.

compound motor (*ElecEng*) A motor which has both series and shunt field windings.

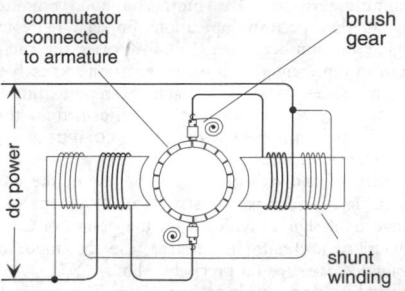

compound motor The shunt winding has more turns of lighter wire.

compound nucleus (*Phys*) A highly excited unstable nucleus formed in certain nuclear reactions, eg that between uranium-235 and a thermal neutron, forming uranium-236. This compound nucleus decays to complete the reaction.

compound pendulum (*Phys*) Any body capable of rotation about a fixed horizontal axis and in stable equilibrium under the action of gravity. If the centre of gravity is a distance h from the axis, and k is the *radius of gyration* about the horizontal axis through the centre of gravity, the period of small oscillations is

$$T = 2\pi \frac{\sqrt{h^2 + k^2}}{hg}$$

compound pillar (*Build*) A pillar formed of a rolled-steel joist or channels strengthened by additional plates riveted or welded to the flanges.

compound press tool (*Eng*) A PRESS TOOL which performs two or more operations at the same station at each stroke of the press.

compound reflex (*BioSci*) A combination of several reflexes to form a definite co-ordination, either simultaneous or successive.

compound slide rest (*Eng*) A device mounted on the upper face of the lathe cross-slide and carrying the tool post. Can be rotated or set over for cutting short internal or external tapers.

compound train (*Eng*) A train of gear wheels in which intermediate shafts carry both large and small wheels, in order to obtain a large speed ratio in a small space.

compressed air (*Eng*) Air at higher than atmospheric pressure. It is used (often at about 600 kN m^{-2}) as a transmitter of energy where the use of electricity or an internal-combustion engine would be hazardous (eg in mining). The exhaust air may be used for cooling or ventilation.

compressed-air capacitor (*ElecEng*) An electric capacitor in which air at several atmospheres' pressure is used as the dielectric, on account of its high dielectric strength at these pressures.

compressed-air disease (*Med*) See CAISSON DISEASE.

compressed-air inspirator (*Eng*) An injector used with pressure-air burners, by which a stream of compressed air

is directed through a venturi throat to inspire additional combustion air.

compressed-air lamp (*MinExt*) An electric lamp for use in fiery mines; it is supplied from a small compressed-air-driven generator incorporated in the lamp-holder.

compressed-air tools (*Eng*) See PNEUMATIC TOOLS.

compressed-air wind tunnel (*Aero*) See VARIABLE-DENSITY WIND TUNNEL.

compressibility (*Phys*) The property of a substance by which it accepts reduction in volume by pressure. It is measured as the ratio of the original volume to the volume of the compact, and is related to the pressure applied. This is sometimes referred to as the *compression ratio*. It is the reciprocal of the bulk modulus. See COEFFICIENT OF COMPRESSIBILITY.

compressibility drag (*Aero*) The sharp increase of drag as airspeed approaches the speed of sound and flow characteristics change from those of a viscous to those of a compressible fluid, causing the generation of shock waves.

compression (*ICT*) See DATA COMPRESSION.

compression cable (*ICT*) See PRESSURE CABLE.

compression fitting (*Build*) A method of joining pipes in which a nut working with the main body of the fitting causes an annular ring with a taper on each side to close onto the pipe and seal the joint. Cf CAPILLARY FITTING.

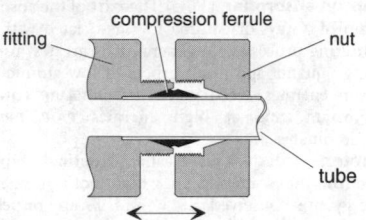

compression fitting Engaging and tightening the nut causes the ferrule to grip the tube.

compression–ignition engine (*Eng*) An internal-combustion engine in which ignition of the liquid fuel injected into the cylinder is performed by the heat of compression of the air charge. See DIESEL ENGINE.

compression moulding (*Chem*) Simple polymer process method where powder, granules or semi-finished product are put directly between heated tool faces, the tool faces brought together under pressure and the material thus shaped. Widely used for products of both simple (gramophone records formerly) and complex shape (tyres). And also for thermoset polymers, but being increasingly replaced by injection moulding etc.

compression plate lock-up (*Print*) A method of locking plates to the cylinder by movable dogs which press the plate from the angled edge at the side towards the centre of the cylinder. Cf TENSION PLATE LOCK-UP.

compression ratio (*Chem*) Injection moulding or EXTRUSION term characterizing the screw dimensions, ratio of volume of one screw flight at entry to that at discharge. See INJECTION MOULDING.

compression ratio (*Eng*) In an internal-combustion engine, the ratio of the total volume enclosed in the cylinder at the outer dead centre to the volume at the end of compression; the ratio of swept volume, plus clearance volume, to clearance volume. See CLEARANCE VOLUME.

compression rib (*Aero*) See RIB.

compression set (*Eng*) Term used to describe permanent CREEP of rubbers under a compressive load. See panel on CREEP AND DEFORMATION.

compression spring (*Eng*) A helical spring with separated coils, or a conical coil spring, with plain, squared or ground ends, made of round, oblong or square-section wire.

compression test (*Eng*) A test in which specimens are subjected to an increasing compressive force, usually until they fail by cracking, buckling or disintegration. A stress–strain curve may be plotted to determine mechanical properties, as in a TENSILE TEST. Compression tests are often applied to materials of high compression but low tensile strength, such as concrete. See STRENGTH MEASURES.

compression waves (*Geol*) See panel on EARTHQUAKE.

compression wood (*For*) Form of REACTION WOOD developed in softwoods with a higher lignin content, hence darker colour, and more brittle, with lower tensile but higher compressive strength than normal. Cf TENSION WOOD.

compressive shrinkage (*Textiles*) A process of forcing a fabric to shrink in length by subjecting it to compressive forces. This makes the fabric less prone to shrink in use.

compressor (*Aero*) Apparatus which compresses the air supply to a GAS TURBINE.

compressor (*BioSci*) A muscle that by its contraction serves to compress some organ or structure.

compressor (*Eng*) A reciprocating or rotary pump for raising the pressure of a gas.

compressor (*ICT*) An electronic amplifier designed to reduce the dynamic range of speech, for transmission at an average higher level in the presence of interference.

compressor drum (*Aero*) A cylinder composed of a series of rings or, more usually, disks wherein the blades of an axial compressor are mounted.

Compton absorption (*Phys*) The part of the absorption of a beam of X-rays or gamma rays associated with Compton scattering processes. In general, it is greatest for medium-energy quanta and in absorbers of low atomic weight. At lower energies PHOTOELECTRIC ABSORPTION is more important, and at high energies PAIR PRODUCTION predominates.

Compton effect (*Phys*) Elastic scattering of photons by electrons, ie scattering in which both momentum and energy are conserved. If λ_s and λ_i are respectively the wavelengths associated with scattered and incident photons, the Compton shift is given by

$$\lambda_s - \lambda_i = \lambda_0(1 - \cos \theta)$$

where θ is the angle between the directions of the incident and scattered photons and λ_0 is the COMPTON WAVELENGTH of the electron ($\lambda_0 = 0.002\,43$ nm). The effect is only significant for incident X-ray and gamma-ray photons.

Compton recoil electron (*Phys*) An electron which has been set in motion following an interaction with a photon (COMPTON EFFECT).

Compton scatter (*Radiol*) A change in the direction of travel of a photon due to the interaction between the photon and the tissue. This is the major cause of loss of resolution in radionuclide imaging.

Compton's rule (*Chem*) An empirical rule that the melting point of an element in kelvins is equal to half the product of the relative atomic mass and the specific latent heat of fusion.

Compton wavelength (*Phys*) Wavelength associated with the mass of any particle, given by $\lambda = h/mc$ where h is PLANCK'S CONSTANT, m is the rest mass of the particle and c is the speed of light.

compulsion (*Psych*) An action which an individual may consider irrational but feels compelled to perform. Usually has implications of repetitive and irrational behaviour.

computability (*ICT, MathSci*) A property of FUNCTIONS. A function is computable if it can be proved that there exists a TURING MACHINE to evaluate it at any given point.

computational complexity (*ICT*) The study of the intrinsic difficulty of computing solutions to different types of mathematically posed problems.

computational fluid dynamics (*Aero*) The calculation of the flow around a surface or in a passage by the solution of mathematical equations over a suitable computing grid and, hence, the calculation of surface pressures, temperatures, overall forces and moments.

computational linguistics (*ICT*) The application of computers to the analysis of natural language, esp *artificial intelligence*.

computed tomography (*Radiol*) See COMPUTER-AIDED TOMOGRAPHY.

computer (*Genrl*) (1) Any device (HARDWARE) which can accept data (INPUT) in a prescribed form, process the data and supply the results (OUTPUT) of the processing in a specified form as information or as signals to control automatically some further machine or process. Processing is done by obeying PROGRAMS (SOFTWARE) composed of a set of arithmetic or logical operations. (2) Any kind of computing device. The most common computers are digital: they perform operations on data represented in digital or number form. In most cases the method of number representation is binary notation and each element in any series must be capable of representing any of the BITS 0 or 1. Operations to be performed are the result of a program. See ANALOGUE COMPUTER, HYBRID COMPUTER.

computer-aided design (*ICT*) The use of the computer particularly with HIGH-RESOLUTION GRAPHICS in a wide range of design activities from the design of cars to the layout of CHIPS. The designs can be modified and evaluated rapidly and precisely. Abbrev *CAD*.

computer-aided engineering (*Eng*) The application of computers to manufacturing processes in which manual control of machine tools is replaced by automatic control resulting in increased accuracy and efficiency. Abbrev *CAE*.

computer-aided learning (*ICT*) A term encompassing the many ways in which a student or teacher in any field may make use of a computer. Also *computer-assisted learning, computer-augmented learning*. Abbrev *CAL*.

computer-aided manufacture (*Eng*) General term used to describe manufacturing processes which are computer-controlled. The data from a COMPUTER-AIDED DESIGN (CAD) system are used directly to produce the program needed for the machine control. Formerly developed as a separate system, it is now often integrated with CAD to form a complete design and manufacturing facility. Abbrev *CAM*. See COMPUTER NUMERICAL CONTROL.

computer-aided tomography (*Radiol*) A method of reconstructing cross-sectional images of the body by using rotating X-ray sources and detectors which move around the body and record the X-ray transmissions throughout the 360° rotation. One detector of a bank is shown below. A computer reconstructs the image in the SLICE. Also *CAT scanner, computed tomography, computerized tomography*.

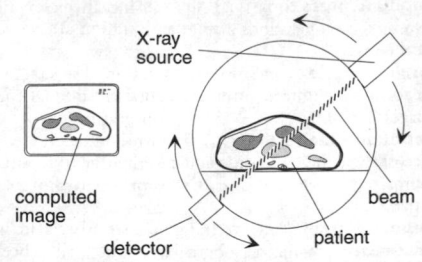

X-ray source

computed image

beam

detector

patient

computer-aided tomography

computer architecture (*ICT*) The structure, behaviour and design of computers. See CENTRAL PROCESSOR, COMPUTER GENERATIONS, DISTRIBUTED COMPUTING.

computer-assisted instruction (*ICT*) Use of the computer to provide educational exercises, eg *PLATO*. Abbrev *CAI*. Cf COMPUTER-AIDED LEARNING, COMPUTER-MANAGED INSTRUCTION.

computer bureau (*ICT*) An organization that sells computer time or computer services to its customers, eg processing a payroll.

computer-friendly (*ICT*) Suitable for use with computers.

computer generations (*ICT*) Convenient means of expressing stages in the advance of digital computer technology. See FIRST-, SECOND-, THIRD-, FOURTH-, FIFTH-GENERATION COMPUTER.

computer graphics (*ICT*) The automatic handling of diagrams, pictures and drawings. The INTERACTIVE input and modification of drawings. See GRAPHICS TABLET, HIGH-RESOLUTION GRAPHICS, JOYSTICK, LIGHT PEN, MOUSE, PLOTTER, RASTER GRAPHICS, VECTOR GRAPHICS.

computer language (*ICT*) The primary symbolic means, comprising meaning, vocabulary and syntactical rules, by which logical instructions are given to a computer.

computer-managed instruction (*ICT*) Computer assistance to teachers for testing and keeping records.

computer numerical control (*ICT*) The control of a machine such as a lathe or milling machine by means of codes sent from the computer. Often these codes will have been generated automatically from the design produced by means of an integrated CAD/CAM system. See also G CODES.

computer science (*ICT*) The study, with the aid of computers, of computable processes and structures. It has many branches, eg ARTIFICIAL INTELLIGENCE, COMMUNICATIONS, COMPUTABILITY, COMPUTATIONAL COMPLEXITY, COMPUTER ARCHITECTURE, CYBERNETICS, DATABASE MANAGEMENT SYSTEMS, FORMAL LANGUAGE THEORY, INFORMATION RETRIEVAL, SOFTWARE ENGINEERING.

computer system (*ICT*) A linked system of PROCESSORS and INPUT/OUTPUT DEVICES with the SOFTWARE necessary to make them operate as a COMPUTER.

computer-to-plate (*Print*) The transfer of information from a PostScript file direct to printer's plate without the intermediate stage of either film or bromide CRC. Abbrev *CTP*.

computer typesetting (*Print*) The use of electronic equipment to process an unjustified input into an output of justified and hyphenated lines. The output can be a new tape, or disk to be used on a phototypesetting machine or it can be the final product in some filmsetting systems. Some equipment can accept a random input of items and produce a classified and alphabetically arranged output.

computer vision (*ICT*) The objective of giving computers the power to 'see' and to interpret what they 'see'.

concanavalin A (*BioSci*) A LECTIN that is derived from jack beans (*Canavalia ensiformis*) and that binds to oligosaccharides present in the membrane glycoproteins on many cells. The lectin has four binding sites and so can cause cross-linking of the glycoproteins. It is a very effective polyclonal mitogen for T-cells, causing them to secrete LYMPHOKINES. Abbrev *ConA*.

concatenate (*ICT*) To join together two STRINGS of characters.

concave brick (*Build*) A COMPASS BRICK.

concave grating (*Phys*) A diffraction grating ruled on the surface of a concave spherical mirror, made usually of speculum metal or glass. Such a grating needs no lenses for collimating or focusing the light. Largely on this account it is the most useful means of producing spectra for precise measurement. See ROWLAND CIRCLE.

concave lens (*Phys*) A divergent lens.

concave mirror (*Phys*) A curved surface, usually a portion of a sphere, the inner surface of which is polished. It is capable of forming real and virtual images, their positions being given by the equation $2/r = 1/l' + 1/l$, where r is the radius of curvature of the surface, l the distance of the object and l' the distance of the image from the mirror (cartesian CONVENTION OF SIGNS).

CONCAWE (*Genrl*) Abbrev for *Conservation Clean Air and Water in Europe*. An association of oil-refining companies operating in Europe that provides a service to international bodies, eg the European Commission, and publishes the results of environmental studies.

conceal/reveal (*ImageTech*) Visual transition effect in which one picture appears to slide across another.

concentrate (*MinExt*) The products of concentration operations in which a relatively high content of mineral has been obtained and which are ready for treatment by chemical methods.

concentrated load (*Build*) A load which is regarded as acting through a point.

concentrates (*Agri*) Processed animal feed containing high levels of nutrient in a limited bulk volume and used to promote growth and development.

concentrating table (*MinExt*) Supported deck, across or along which mineralized sands are washed or moved to produce differentiated products according to the gravitational response of particles of varying size and/or density. Stationary tables include strakes, sluices, buddles; moving tables include shaking tables (eg *Wilfley table*), vanners and rockers.

concentration (*Chem*) (1) The number of molecules or ions of a substance in a given volume, generally expressed as moles per cubic metre or cubic decimetre. (2) The process by which the concentration of a substance is increased, eg the evaporation of the solvent from a solution.

concentration (*Eng*) Production of CONCENTRATE.

concentration cell (*Phys*) A cell with similar electrodes in common electrolyte, the emf arising from differences in concentration at the electrodes.

concentration plant (*MinExt*) *Cleaning plant, concentrator mill, reduction works, washing.* Buildings and installations in which ore is processed by physical, chemical and/or electrical methods to retain its valuable constituents and discard as tailings those of no commercial interest. See MINERAL PROCESSING.

concentration polarization (*Phys*) A form of polarization occurring in an electrolytic cell, due to changes in the concentration of the electrolyte surrounding the electrode.

concentrator (*ICT*) In the context of NETWORKS, a device that channels data from a number of users on to a smaller number of higher-capacity links.

concentric (*ICT*) A term replaced by COAXIAL.

concentric arch (*Arch*) An arch laid in several courses whose curves have a common centre.

concentric chuck (*Eng*) See SELF-CENTRING CHUCK.

concentric plug-and-socket (*ElecEng*) A type of plug-and-socket connection in which one contact is a central pin and the other is a ring concentric with it.

concentric vascular bundle (*BioSci*) A bundle in which a strand of xylem is completely surrounded by a sheath of phloem (amphicribral) or vice versa (amphivasal).

concentric winding (*ElecEng*) An armature winding, used on ac machines, in which groups of concentric coils are used. Also used to denote the type of winding, used on transformers, in which the high-voltage winding is arranged concentrically with the low-voltage winding.

concentric wiring (*ElecEng*) An interior wiring system in which the conductor consists of an insulated central core surrounded by a flexible metal sheath which forms the return lead.

conceptacle (*BioSci*) A flask-shaped cavity in a thallus, opening to the outside by a small pore, and containing reproductive structures, eg in *Fucus*.

conception (*Med*) The fertilization of an ovum with a spermatozoon.

concertina fold (*Print*) A method of folding a leaflet or insert so that it opens out and closes in a zigzag fashion. US *accordion fold*.

concert pitch (*Acous*) The recognized pitch, ie frequency of the generated sound wave, to which musical instruments are tuned, so that they can play together. The exact value has varied considerably during musical history, but it has recently been internationally standardized so that A (above middle C) becomes 440 Hz. Allowance must be made for the rise in temperature experienced in concert halls, which alters the pitch in ways peculiar to the different types of instrument.

concha (*Arch*) The smooth concave surface of a vault.

concha (*BioSci*) In vertebrates, the cavity of the outer ear; the outer or external ear; a shelf projecting inwards from the wall of the nasal cavity to increase the surface of the nasal epithelium.

conchiolin (*BioSci*) A horny substance forming the outer layer of the shell in MOLLUSCA.

conchoid (*MathSci*) If, from a fixed point P, a line is drawn to meet any curve C in the point Q, and if A and B are any two points on the line PQ such that $AQ = BQ$ is a constant, then the locus of the points A and B is a conchoid with respect to the point P. The conchoid of a straight line not passing through P, first discussed by Nicomedes, 250 BC, in connection with the trisection of an angle, and called the *conchoid of Nicomedes*, has a node at P, a cusp at P or no double point at all, depending upon whether the perpendicular length from P to the line is greater than, equal to, or less than AQ respectively.

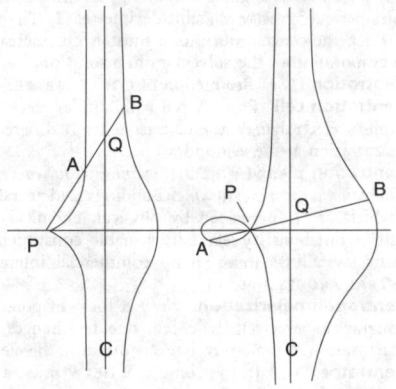

conchoid The line C is straight.

concolor (*BioSci*) Uniform in colour. Also *concolorate*, *concolorous*.

concordant intrusion (*Geol*) An igneous intrusion that lies parallel to the bedding or foliation of the country rock which it intrudes. See SILLS.

concrescence (*BioSci*) Union of originally distinct organs by the growth of the tissue beneath them.

concrete (*Build*, *CivEng*) Artificial stone which can be moulded and then allowed to set. Made from a mixture of cement, aggregate and water. See panel on CEMENT AND CONCRETE.

concrete blocks (*CivEng*) Solid or hollow precast blocks of concrete used in the construction of buildings.

concrete mixer (*Build*, *CivEng*) An appliance in which the constituents of concrete are mixed mechanically.

concrete operational stage (*Psych*) According to Piaget, the period between ages 7 and 11 years when a child acquires the ability to think logically, but only in very concrete terms, and is still deficient in abstract thought.

concrete paving slabs/flags (*Build*) Precast concrete slabs for the top surface of pavements or paths. Often compressed in manufacture.

concrete period (*Psych*) Also *concrete stage*. See CONCRETE OPERATIONAL STAGE.

concrete reinforcement (*CivEng*) The method of overcoming the weakness of concrete in tension by suitable placing of usually steel bars or cables. Fig. ▷

concrete thinking (*Psych*) A form of reasoning that is strongly tied to the immediate situation, or to very tangible and specific information, as opposed to abstract reasoning. See CONCRETE OPERATIONAL STAGE.

concretion (*Geol*) Nodular or irregular concentration of siliceous, calcareous, or other materials, formed by localized deposition from solution in sedimentary rocks.

concretion (*Med*) Collection of organic matter with or without lime salts in bodily organs.

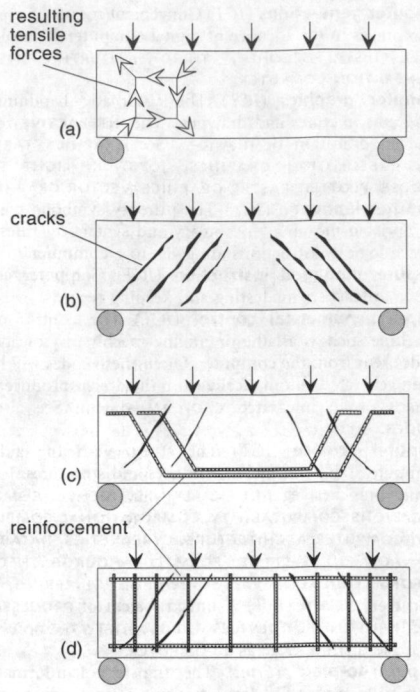

concrete reinforcement Effect of forces (arrows) acting on blocks supported at each end. The resulting cracks (b) and two ways of reinforcing the block (c,d) are shown.

concurrent validity (*Genrl*) A measurement's ability to correlate or vary directly with an accepted measure of the same construct.

concussion (*Med*) A violent shaking or blow (esp of or to the head), or the condition resulting from it.

condensate (*Eng*) The liquid obtained as a result of removing from a vapour such portion of the latent heat of evaporation as it may contain.

condensation (*Chem*) The union of two or more molecules with the elimination of a simpler group, such as H_2O, NH_3, etc.

condensation (*Genrl*) The process of forming a liquid from its vapour.

condensation (*EnvSci*) The formation of liquid water from water vapour when moist air is cooled below its dew point, if there are extended surfaces or nuclei present. These nuclei may be dust particles or ions. Mist, fog and cloud are formed by nuclear condensation.

condensation gutter (*Build*) A small gutter provided at the curb of lantern lights to carry away condensed water formed on the interior surface of the glazing.

condensation sinking (*Build*) A groove cut in the bottom rails of skylights to carry away condensed water formed on the interior surface of the glazing.

condensation trails (*EnvSci*) Artificial clouds caused by the passage of an aircraft due either to condensation following the reduction in pressure above the wing surfaces, or to condensation of water vapour contained in the engine exhaust gases. Also *contrails*.

condensed (*BioSci*) An inflorescence with closely crowded flowers that are short-stalked or sessile.

condensed (*Print*) See ELONGATED.

condensed chromatin (*BioSci*) See HETEROCHROMATIN.

condensed nucleus (*Chem*) A ring system in which two rings have one or more (generally two) atoms in common eg naphthalene, phenanthrene, quinoline.

condensed system (*Chem*) One in which there is no vapour phase. The effect of pressure is then practically negligible, and the *phase rule* may be written $P + F = C + 1$.

condenser (*Chem*) Apparatus used for condensing vapours obtained in distillation. In laboratory practice usually a single tube, either freely exposed to air or contained in a jacket in which water circulates.

condenser (*Electronics*) See CAPACITOR.

condenser (*Eng*) (1) A chamber into which the exhaust steam from a steam engine or turbine is delivered, to be condensed by the circulation or the introduction of cooling water; in it a high degree of vacuum is maintained by an air pump. (2) The part of a refrigeration system in which the refrigerant is liquefied by transferring heat to the cooling medium, usually water or air.

condenser (*Phys*) (1) A large lens or mirror used in an optical projecting system to collect light, radiated from the source, over a large solid angle, and to direct this light onto the object or transparency which is to be focused at a distance by a projection lens. (2) A system of lenses (optical or electrical) below the stage of a microscope arranged to irradiate the object in a manner suitable for the type of observation required. See fig. at MICROSCOPE.

condenser bushing (*ElecEng*) A type of bushing used for terminals of high-voltage apparatus (eg transformers and switchgear) in which alternate layers of insulating material and metal foil form the insulation between the conductor and the outer casing; the metal foil serves to improve the voltage distribution. Also *capacitor bushing, capacitor terminal*.

condenser circulating pump (*Eng*) See CIRCULATING PUMP.

condenser spinning (*Textiles*) A process in which fibres are carded and the resultant web is divided into narrow strips which are rubbed together on oscillating rubber or leather aprons to form twistless SLUBBINGS. These are then spun into CONDENSER YARNS.

condenser terminal (*ElecEng*) Capacitor terminal. See CONDENSER BUSHING.

condenser tissue (*Paper*) Thin rag paper used as a capacitor dielectric. Also *capacitor tissue*.

condenser tubes (*Eng*) The tubes through which the cooling water is circulated in a SURFACE CONDENSER, and on whose outer surfaces the steam is condensed.

condenser yarn (*Textiles*) Yarn spun from clean soft waste material; suitable for cotton blankets, quiltings and towellings.

con–di nozzle (*Aero, Eng*) See CONVERGENT–DIVERGENT NOZZLE.

conditional instability (*EnvSci*) The condition of the atmosphere when the TEMPERATURE LAPSE RATE lies between the *dry* and SATURATED ADIABATIC LAPSE RATES, ie the atmosphere is stable for unsaturated air but unstable for saturated.

conditional instability of the second kind (*EnvSci*) A process whereby low-level CONVERGENCE in the wind field produces convection and cumulus formation thereby releasing latent heat which enhances the convergence and increases convection. This 'positive feedback loop' may lead to the formation of a large-scale disturbance. Abbrev CISK.

conditional jump (*ICT*) See JUMP.

conditionally convergent series (*MathSci*) A series which is convergent but not absolutely convergent, eg $1 - \frac{1}{2} + \frac{1}{3} - \frac{1}{4}\ldots$ which converges to $\ln 2$.

conditionally stable (*ICT*) Said of a system or amplifier that is stable for certain values of input signal and gain, but not others. See NYQUIST CRITERION.

conditional mutation (*BioSci*) A mutation that is only expressed under certain environmental conditions, eg *temperature-sensitive mutants*.

conditional probability (*MathSci*) The probability of occurrence of an event given the occurrence of another *conditioning* event.

conditional probability distribution (*MathSci*) The distribution of a random variable given the value of another (possibly associated) random variable or event.

conditional statement (*MathSci*) A statement of the form 'if p then q'. See MATERIAL IMPLICATION.

condition codes (*ICT*) A set of bits indicating the condition of something within a computer.

conditioned medium (*BioSci*) Cell culture medium that has already been partially used by cells. Although depleted of some components, it is enriched with cell-derived material, probably including small amounts of growth factors; such cell-conditioned medium will support the growth of cells at much lower density.

conditioned reflex (*Psych*) Reflex action by an animal to a previously neutral stimulus as the result of CLASSICAL CONDITIONING.

conditioning (*MinExt*) In froth flotation, the treatment of mineral pulp with small additions of chemicals designed to develop specific aerophilic or aerophobic qualities on surfaces of different mineral species as a prelude to their separation.

conditioning (*Psych*) See CLASSICAL CONDITIONING, OPERANT CONDITIONING.

conditioning (*Textiles*) Allowing materials to reach equilibrium with the surrounding atmosphere. Samples are frequently conditioned in a standard atmosphere before testing. For commercial transactions by weight a percentage moisture content is usually specified. Cf SEASONING.

conditions of severity (*ElecEng*) A term used in connection with the testing of circuit breakers to denote the conditions (eg power factor, rate of rise of restriking voltage, etc) obtaining in the circuit when the test is carried out.

conductance (*Phys*) The ratio of the current in a conductor to the potential difference between its ends; reciprocal of RESISTANCE. SI unit is SIEMENS, symbol S; also reciprocal OHMS (*mhos*).

conductance ratio (*ElecEng*) That between equivalent conductance of a given solution and its value at infinite dilution.

conduct disorders (*Psych*) Childhood disorders involving antisocial behaviour.

conductimetric analysis (*Chem*) Volumetric analysis in which the end-point of a titration is determined by measurements of the conductance of the solution.

conducting tissue (*BioSci*) (1) Xylem and phloem in vascular plants. (2) Leptoids and hydroids in Bryophytes.

conduction (*Phys*) The process by which heat or electricity is transmitted through a material or body without movement of the medium itself. See ELECTRICAL CONDUCTIVITY, THERMAL CONDUCTIVITY.

conduction angle (*ElecEng*) That part of a half cycle (expressed as an angle) for which a rectifier, or controlled rectifier (SILICON-CONTROLLED RECTIFIER or THYRISTOR), conducts. By changing the conduction angle, the power delivered can be controlled.

conduction band (*Electronics*) In the BAND THEORY OF SOLIDS, a band which is only partially filled, so that electrons can move freely in it, hence permitting conduction of current.

conduction by defect (*Electronics*) In a doped semiconductor, conduction by HOLES in the valency electron band.

conduction current (*Electronics*) The current resulting from the flow of charge carriers in a medium in response to a local electric field (same as *drift current*) as distinct from a DISPLACEMENT CURRENT.

conduction electrons (*Electronics*) The electrons situated in the conduction band of a solid, which are free to move under the influence of an electric field.

conduction hole (*Electronics*) In a crystal lattice of a semiconductor, conduction obtained by electrons filling holes in sequence, equivalent to a positive current. See HOLES.

conductivity (*Phys*) The ratio of the CURRENT DENSITY in a conductor to the electric field causing the current to flow.

It is the conductance between opposite faces of a cube of the material of 1 metre edge. Reciprocal of RESISTIVITY. Units $\Omega^{-1}\,m^{-1}$ or $S\,m^{-1}$, symbol σ.

conductivity bridge (*ElecEng*) A form of WHEATSTONE BRIDGE used for the comparison of low resistances.

conductivity cell (*ElecEng*) Any cell with electrodes for measuring conductivity of liquid or molten metals or salts.

conductivity modulation (*Electronics*) That effected in a semiconductor by varying a charge carrier density.

conductivity test (*ElecEng*) See FALL-OF-POTENTIAL TEST.

conductor (*Build*) A pipe for the conveyance of rainwater. Also *leader*.

conductor (*ElecEng*) (1) A material which offers a low resistance to the passage of an electric current. (2) That part of an electric transmission, distribution or wiring system which actually carries the current.

conductor (*MinExt*) See TOP CASING.

conductor (*Phys*) A material which offers a low resistance to the passage of heat. All materials conduct heat to some extent, but it can be channelled (like electricity) using materials of different conductivity. See THERMAL CONDUCTIVITY.

conductor load (*ElecEng*) The total mechanical load to which an overhead electric conductor may be subjected, because of its own weight and that of any adhering matter such as snow or ice.

conductor rail (*ElecEng*) In some electric traction systems a bare rail laid alongside the running rails to conduct the current to or from the train. Also *contact rail*.

conductor-rail insulator (*ElecEng*) An insulator used for supporting a conductor rail and for insulating it from the earth.

conductor-rail system (*ElecEng*) A system of electric traction in which current is collected from a *conductor rail* by *collector shoes*.

conductors and insulators (*ElecEng*) All materials; they have the wide range of RESISTIVITY and CONDUCTIVITY as shown. Fig. ▷

conduit (*BioSci*) Functional element of the conducting system of the xylem: either a whole vessel or a single TRACHEID. Water moves within conduits though cell lumens and perforations in vessels and between conduits through pits (which may resist the spread of an EMBOLISM).

conduit (*Build*) A pipe or channel, usually large, for the conveyance of water.

conduit (*ElecEng*) A trough or pipe containing electric wires or cables, in order to protect them against damage from external causes.

conduit box (*ElecEng*) A box connected to the metal conduit used in some electric wiring schemes. The box forms a base to which fittings (eg switches or ceiling roses) may be attached, or it may take the place of bends, elbows or tees, used to facilitate the installation of the wiring.

conduit fittings (*ElecEng*) A term applied to all the auxiliary items, such as boxes, elbows, etc, needed for the conduit system of wiring.

conduit system (*ElecEng*) (1) A system of wiring, used for industrial and, formerly, domestic premises, in which the conductors are contained in a steel conduit. (2) A system of current collection used on some electric tramway systems; the conductor rail is laid beneath the roadway, and connection is made between it and the vehicle by means of a collector shoe passing through a slot in the road surface.

conduplicate (*BioSci*) Folded longitudinally about the mid-rib so that the two halves of the upper surface are brought together.

condyle (*BioSci*) A smooth rounded protuberance, at the end of a bone, that fits into a socket on an adjacent bone, as the *condyle* of the lower jaw, the occipital *condyles*. Adjs *condylar, condyloid*.

condylomata (*Med*) Inflammatory wart-like papules on the skin round the anus and external genitalia, esp in syphilis. Sing *condyloma*.

cone (*BioSci*) One of a large number of light-sensitive structures in the retina of many vertebrates that respond preferentially to particular wavelengths and thus provide the basis for colour vision. Cf ROD. Also see STROBILUS.

cone (*Eng*) A device used on top of blast furnaces to enable charge to be added without gas escaping. Also *bell*.

cone (*MathSci*) (1) A (conical) surface generated by a line, one point (the *vertex*) of which is fixed and one point of which describes a fixed plane curve. Any line lying in the surface is called a *generator*. (2) A solid bounded by a conical surface and a plane (the *base*) which cuts the surface. If the fixed curve is a circle the cone is called a *circular cone*, and if the vertex is also perpendicularly over the centre of the circle the cone is called a *right circular cone*.

cone (*Textiles*) A conical package of yarn wound onto a conical frame.

cone bearing (*Eng*) A shaft bearing consisting of a conical JOURNAL running in a correspondingly tapered bush, so acting as a combined journal and thrust bearing; used for some lathe spindles.

cone classifier (*MinExt*) Large inverted cone into which ore pulp is fed centrally from above. Coarser material settles to bottom discharge and finer overflows peripherally.

cone clutch (*Eng*) A FRICTION CLUTCH in which the driving and driven members consist of conical frusta with the driven member moved axially to engage the drive.

cone diaphragm (*Acous*) Diaphragm of paper, plastics or metal foil driven by a circular coil carrying speech currents near its apex; widely used for radiating sound in loudspeaking receivers. Also *cone loudspeaker*.

cone drive (*Eng*) A pair of CONE PULLEYS in which a variable speed ratio is obtained by moving the belt mechanically along the pulleys with a striker. Also *cone gear*.

cone-in-cone structure (*Geol*) Cones stacked inside one another, occurring in sedimentary rocks; usually of fibrous calcite, sometimes of other minerals.

cone loudspeaker (*Acous*) See CONE DIAPHRAGM.

cone of silence (*ICT*) Cone-shaped space directly over the antenna of a radio-beacon transmitter in which signals are virtually undetectable.

cone pulley (*Eng*) A belt pulley stepped to give two or more diameters; used in conjunction with a similar pulley to obtain different speed ratios. See CONE DRIVE.

cone sheets (*Geol*) Minor intrusions which occur as inwardly inclined sheets of igneous rock and have the form of segments of concentric cones. See DYKES, SILLS.

confabulation (*Psych*) A tendency to fill in memory gaps with information that fits with what is already known, not necessarily a conscious process.

confectionery (*FoodSci*) Products traditionally made from sugar, intended for eating pleasure rather than necessity.

conference call (*ICT*) A telephone or computer communication in which three or more people can participate simultaneously. With high-bandwidth connectivity simultaneous video transmission is possible (*videoconferencing*). Also *teleconference*.

confervoid (*BioSci*) Consisting of delicate filaments.

confidence head/playback (*ImageTech*) Provides off-tape monitoring during recording.

confidence interval (*MathSci*) An interval so constructed that a statement that the true value of an unknown parameter lies in this interval will be true, in the long run, a proportion of the time that the statement is made, this proportion corresponding to the prescribed level of confidence expressed as a percentage.

config.sys file (*ICT*) In MS-DOS, a text file that contains CONFIGURATION COMMANDS to enable certain system features and establish a chosen configuration when the system is booted.

configuration (*Chem*) The shape of molecules determined by covalent bonds, so invariant unless bonds are broken. In polymers with asymmetric carbon atoms, it is fixed by polymerization, eg atactic or isotactic. See STEREOREGULAR POLYMERS and panel on POLYMERS.

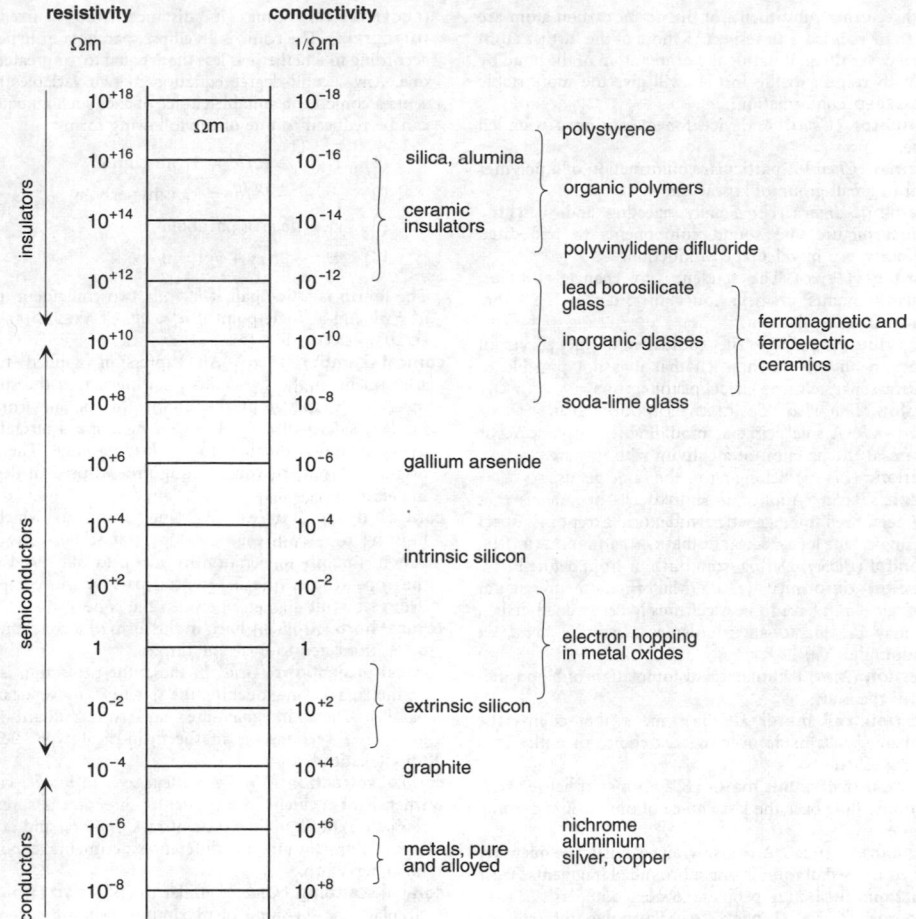

conductors and insulators

configuration (*ICT*) (1) A particular arrangement of HARDWARE and its interconnections. (2) The particular settings chosen for an item of HARDWARE or SOFTWARE.

configuration control (*NucEng*) Control of reactivity of a reactor by alterations to the configuration of the fuel, reflector and moderator assembly.

configuration management (*ICT*) A formal system for the control of changes to a set of important software components or documentation, allowing the state of a system to be known at any time and its previous state accurately recreated.

configure (*ICT*) To set up a computer or application to suit a particular range of PERIPHERALS or working preferences.

confinement (*NucEng*) See CONTAINMENT.

conflict (*Psych, BioSci*) Broad term for any situation in which there are mutually antagonistic or incompatible motives, behaviours, etc. See APPROACH–AVOIDANCE CONFLICT, AVOIDANCE–AVOIDANCE CONFLICT.

conflicted (*Psych*) A term used to describe a person struggling to reconcile contradictory impulses.

confocal (*MathSci*) Having the same foci.

confocal microscope (*BioSci*) A form of light microscope in which an aperture in the illuminating system confines the illumination to a small spot on the specimen and a corresponding aperture in the imaging system (which may be the same aperture in reflecting and fluorescence devices) allows only light transmitted, reflected or emitted by the same spot to contribute to the image. The spots are made, by suitable mechanical or optical means, to scan the specimen as in a TV raster. Compared with conventional microscopy, confocal techniques offer improved resolution (c.0.2 μm in the x- and y-dimensions and 0.7 μm in the z-dimension) and improved rejection of out-of-focus noise.

conformable strata (*Geol*) An unbroken succession of strata. See NON-SEQUENCE, UNCONFORMITY.

conformal-conjugate transformation (*MathSci*) A conformal transformation in which a conjugate system on one surface corresponds to a conjugate system on the other surface.

conformal transformation (*MathSci*) A transformation in which the z-plane is mapped on the w-plane (or vice versa) so that both the magnitude and the sense of the rotation of the angles is the same in either plane. Sometimes used in the sense of *isogonal transformation*.

conformation (*Agri*) The overall physical structure of livestock animals compared with standards for a particular breed or type.

conformation (*Chem*) The shape of molecules determined by rotation about single bonds, esp in polymer chains about carbon–carbon links, eg random coils, oriented chains and zigzag chains. See panel on ELASTOMERS.

conformational analysis (*Chem*) The study of the relative spatial arrangements of atoms in molecules, in particular in saturated organic molecules. Viewed along a C–C bond, the three other substituents of the nearer carbon atom are said to be *eclipsed* with respect to those of the further atom if they cover them. Rotating the nearer atom of the bond by 60° with respect to the further will give the more stable STAGGERED conformation.

conformator (*Genrl*) A device for measuring a rounded shape.

conformer (*Chem*) A particular conformation of a polymer chain or small group of repeat units.

conforming (*ImageTech*) Finally selecting and matching original picture and sound components to an edited continuity, esp in videotape production.

conformity (*Psych*) The tendency to change attitudes, beliefs, thoughts or behaviours in order to be more consistent with others.

confounding (*MathSci*) The confusion of the effects of factors in an experiment such that it is not possible to identify, separately, the effects of the factors.

confusion (*ImageTech*) See CIRCLE OF CONFUSION.

congé (*Arch*) A small circular moulding, either concave or convex, at the junction of a column with its base.

congeneric (*BioSci*) Belonging to the same genus.

congenic (*BioSci*) Applied to inbred cells of animals that have been bred to be genetically identical except in respect to a single gene locus. See RECOMBINANT INBRED STRAINS.

congenital (*BioSci*) Dating from birth or from before birth.

congenital deformity (*Med*) Malformation present at birth. Does not have to be a genetically determined defect, but may be due to environmental factors *in utero*, eg thalidomide.

congestion (*Med*) Pathological accumulation of blood in a part of the body.

congestion call meter (*ICT*) A meter that counts the number of calls made over the last choice of outlet in a grading scheme.

congestion traffic-unit meter (*ICT*) A meter that registers the traffic flow over the last choice of outlets in a grading scheme.

conglomerate (*Geol*) A coarse-grained clastic sedimentary rock composed of rounded or subrounded fragments larger than 2 mm in size (eg pebbles, cobbles, boulders).

Congo red (*Chem*) A benzidene direct dyestuff (scarlet) produced by coupling diphenyl-bis-diazonium chloride with naphthionic acid; first of a long series of derivatives. Used as a chemical indicator for acid solutions in the pH range 3–5 (blue–red).

congruence (*MathSci*) The property of being CONGRUENT.

congruent (*MathSci*) (1) Of geometrical figures, alike in all relevant respects; able to be superimposed on one another. (2) Of two numbers *a* and *b* with respect to a divisor *n*, written $a \equiv b \pmod{n}$, such that the remainders when *a* and *b* are both divided by *n* are equal. More generally, if *M* is a MONOID, then ρ is a congruence on *M* if it is an equivalence relation with the added property that if *a* ρ *b* (*a* is ρ-*equivalent* to *b*) and *a*′ ρ*b*′ then *aa*′ρ*bb*′. (3) Of two matrices, A and B, such that there is a non-singular matrix P with transpose *P*′ such that by $B = P'AP$. The transformation is then also said to be congruent.

congruent melting (*Eng*) Melting at a constant temperature or pressure of a material in which both phases retain the same composition.

Coniacian (*Geol*) A stage in the Upper Cretaceous. See MESOZOIC.

conic (*Genrl*) Used to describe a map projection in which the Earth is projected onto a cone with its apex over one of the poles. Also *conical*.

conic (*MathSci*) The curve in which a plane cuts a circular (but not necessarily right) cone. The curve is a parabola if the cutting plane is parallel to a generator, otherwise it is an ellipse or a hyperbola according to whether a plane parallel to the cutting plane but through the vertex of the cone wholly contains no generators or two generators respectively. Alternatively, the locus of a point which moves so that its distance from a fixed point (FOCUS) is a constant *e*

(ECCENTRICITY) times its distance from a fixed line (DIRECTRIX). The conic is an ellipse, parabola or hyperbola according to whether *e* is less than, equal to, or greater than one. Any second-degree equation in two variables represents a conic. By a suitable choice of axes such an equation can be reduced to one of the following forms:

(1) $x^2/a^2 + y^2/b^2 = 1$, an ellipse

(2) $x^2/a^2 - y^2/b^2 = 1$, a hyperbola

(3) $y^2 = 4ax$, a parabola

(4) $ax^2 + 2hxy + by^2 = 0$

The fourth is a line pair if $h^2 > ab$, two coincident lines if $h^2 = ab$ and a single point if $h^2 < ab$. Cf AXES, DIAMETER, FOCUS, VERTEX. Fig. ▷

conical camber (*Aero*) An expression applied to the adjustment of the CAMBER of a wing across the span to meet the variation of the upflow of the air from the fuselage side to the tip. Used on high-speed aircraft, the 'twist' is applied mainly to the leading edge. The name originates from the conic lofting process used in deriving the aerofoil sections.

conical drum (*MinExt*) Winding drum, to which the hoisting rope of a cage or skip is attached. This shape aids in smooth acceleration from rest to full speed where the rope reaches the flat central part of the compound drum (the full diameter between the cones).

conical horn (*Acous*) A horn in the form of a cone, the apex being truncated to form the throat.

conical projections (*Geol*) In these, the projection is onto an imaginary cone touching the sphere at a given standard parallel. There are equal-area and two standard (*Gall's projection*) versions, as in the cylindrical type. Best for middle latitudes.

conical refraction (*Phys*) An effect seen in *biaxial* crystals when light incident in a particular direction is spread on refraction into a hollow cone of rays. *Internal* and *external* conical refraction require different experimental conditions for observation.

conical scanning (*Radar*) Similar to LOBE SWITCHING, but circular. The direction of maximum response generates a conical surface; used in missile guidance.

conical surface (*MathSci*) See CONE.

conidial (*BioSci*) (1) Referring to, or pertaining to, a CONIDIUM. (2) Producing conidia.

conidiophore (*BioSci*) A simple or branched hypha bearing one or more CONIDIA.

conidiosporangium (*BioSci*) A sporangium capable of direct germination, as well as producing zoospores.

conidium (*BioSci*) An asexual fungal spore. Produced exogenously from a hyphal tip, never within a sporangium.

Coniferales (*BioSci*) The conifers, an order of c.600 spp of gymnosperms (class Coniferopsida) that also includes fossils from the Jurassic. They are mostly trees or shrubs, with simple, usually small, leaves. The reproductive organs are in unisexual cones and fertilization is by means of a pollen tube (siphonogamous). They dominate large areas of the Earth (though reduced since the Cretaceous) and include many important softwood trees for timber and pulp. Includes pines, spruces, cypresses, monkey puzzles.

Coniferopsida (*BioSci*) A class of gymnosperms dating from the Carboniferous. They are mostly substantial trees, with pycnoxylic wood, simple leaves, saccate pollen and flattened (platyspermic) seed. Includes Cordaitales (Carboniferous and Permian), Volztiales (Permian–Jurassic) and the Coniferales, Taxales (yews) and, in some classifications, the Ginkgoales.

coniferous (*BioSci*) Cone-bearing; relating to a cone-bearing plant.

conifuge (*PowderTech*) A conical type of centrifuge for the collection of aerosol particles.

coning and quartering (*MinExt*) Production of a representative sample from a large pile of material such as ore,

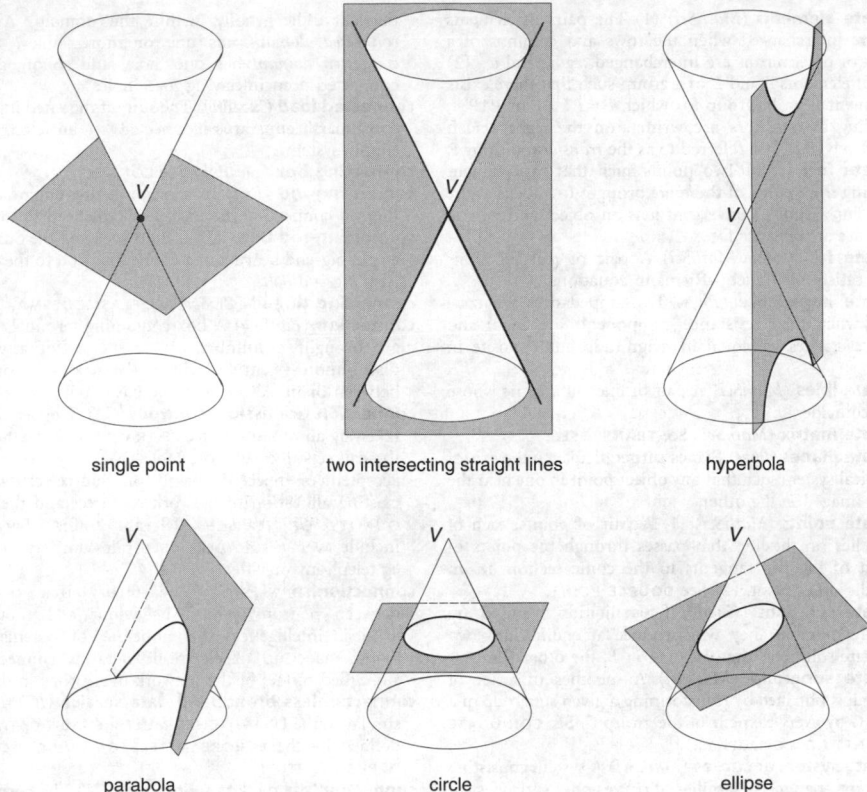

single point   two intersecting straight lines   hyperbola

parabola   circle   ellipse

conic *V* is the vertex of the right cone.

in which it is first formed into a cone by deposition centrally, and then reduced by removal, one shovelful at a time, into four separate piles drawn alternately from four peripheral opposed points. Two are then discarded and the process (perhaps with intermediate-size reduction) is repeated until a manageable hand sample is obtained.

coning angle (*Aero*) The angle between lthe ongitudinal axis of blade of lifting rotor and the tip-path plane in helicopters.

Coniphora puteana (*Build*) Formerly *Coniphora cerebella*. One of several fungi which causes WET ROT in timber.

conjecture (*Genrl*) An unproved theorem, eg GOLDBACH'S CONJECTURE.

conjoined twins (*Med*) Preferred term for Siamese twins.

conjugate (*BioSci*) Molecular species produced by covalently linking two chemical moieties from different sources, eg a conjugate of an antibody with a fluorochrome or enzyme.

conjugate acid and base (*Chem*) These are related, according to the BRÖNSTED–LOWRY THEORY, by the reversible exchange of a proton; thus acid $\rightleftharpoons$ base + H^+.

conjugate algebraic numbers (*MathSci*) Any numbers which are the roots of the same irreducible algebraic equation with rational coefficients.

conjugate angles (*MathSci*) Two angles whose sum is equal to $360°$. Each is said to be the *explement* of the other.

conjugate arcs (*MathSci*) Two arcs whose sum is a complete circle.

conjugate axis of hyperbola (*MathSci*) See AXES.

conjugate Bertrand curves (*MathSci*) Any two curves having the same principal normals. Also *associate Bertrand curves*, *Bertrand curves*.

conjugate branches (*Phys*) Any two branches of an electrical network such that an emf in one branch produces no current in the conjugate branch.

conjugate complex number (*MathSci*) The conjugate $\bar{z}$ of a complex number $z = x + iy$ is $x − iy$.

conjugate deviation (*Med*) The sustained deviation of the eyes in one direction as a result of a lesion in the brain.

conjugate diameters (*MathSci*) Two diameters of a conic which are also conjugate lines with respect to the conic are conjugate diameters. Each bisects chords parallel to the other.

conjugate directions (*MathSci*) At a point *P* on a surface, if *Q* is a point near *P* and *Q* tends to *P*, then the conjugate directions are the directions in the limiting case of: (1) the straight line between *P* and *Q*; and (2) the line of intersection of the tangent planes at *P* and *Q*. If these directions coincide, they are *self-conjugate* or *asymptotic directions*.

conjugate division (*BioSci*) Simultaneous mitosis of a pair of associated nuclei in eg a dikaryotic cell.

conjugate double bonds (*Chem*) Di-alkene compounds with an arrangement of alternate single and double bonds between the carbon atoms, namely RCH=CHCH=CHR. Additive reactions take place, inasmuch as atoms or radicals become attached to the two outside carbon atoms of the chain, thus creating a new alkene linkage in the centre.

conjugate dyadics (*MathSci*) Two dyadics such that each can be obtained from the other by reversing the order of the factors in each dyad. If a dyadic is equal to its conjugate it is said to be *symmetric*; if it is equal to the negative of its conjugate it is *antisymmetric*.

conjugate elements (*MathSci*) (1) The pairs of elements that are interchanged when the rows and columns of a matrix or determinant are interchanged, eg a_{ij} and a_{ji}. (2) Pairs of elements A and B of a group such that there exists an element P in the group for which $B = P^{-1}AP$ or PAP^{-1}, according as operators are written on the right or left respectively. B is then referred to as the *transform* of A by P.

conjugate foci (*Phys*) Two points such that rays of light diverging from either of them are brought to a focus at the other. For a simple convergent lens an object and its real image are at *conjugate foci*.

conjugate functions (*MathSci*) A pair of real functions which satisfy the Cauchy–Riemann equations.

conjugate impedance (*Phys*) Two impedances are conjugate when their resistance components are equal and their reactances are equal in magnitude but opposite in sign.

conjugate lines (*MathSci*) A pair of lines of a conic whose poles coincide.

conjugate matrix (*MathSci*) See TRANSPOSE.

conjugate planes (*Phys*) Planes perpendicular to the axis of an optical system such that any object point in one near the axis is imaged in the other.

conjugate points (*MathSci*) (1) A pair of points, each of which lies on the line that passes through the points of contact of the two tangents to the conic section drawn from the other point. (2) See DOUBLE POINT.

conjugate solutions (*Chem*) If two liquids A and B are partially miscible, they will produce at equilibrium two conjugate solutions, the one of A in B, the other B in A.

conjugate subgroup (*MathSci*) A member of a set of subgroups, obtained by transforming a given subgroup of a group G by every element of the group G. See CONJUGATE ELEMENTS OF A GROUP.

conjugate system of curves (*MathSci*) A system consisting of two one-parameter families of curves on a surface, such that where a member of one family intersects a member of the other family, the directions of the tangents to the curves are conjugate directions.

conjugate triangles (*MathSci*) A pair of triangles such that the vertices of one are the poles of the sides of the other. Also *reciprocal polar triangles*. Cf SELF-POLAR TRIANGLE.

conjugation (*BioSci*) (1) Generally, the process of union between two cells or gametes. (2) In prokaryotes (bacteria), the transfer of genetic material via sex pili. (3) In certain ciliate protozoans, eg *Paramecium*, sexual reproduction and the transfer of micronuclei; see KAPPA PARTICLE. (4) A form of sexual reproduction in some algae and fungi in which there is fusion of two non-flagellated gametes or protoplasts.

conjugation tube (*BioSci*) A tubular outgrowth of a cell through which one non-flagellated gamete moves to fuse with another during conjugation. Also *fertilization tube*.

conjunction (*Astron*) A term signifying that two heavenly bodies have the same apparent geocentric longitude. Applied to Venus and Mercury it is subdivided into *inferior conjunction* and *superior conjunction*, according as the planet is between the Earth and Sun, or the Sun between the Earth and planet, respectively.

conjunctiva (*BioSci*) In vertebrates, the modified epidermis of the front of the eye, covering the cornea externally and the inner side of the eyelid.

conjunctive tissue (*BioSci*) Secondary tissue occupying the space between the vascular bundles where the secondary xylem does not form a solid cylinder.

conjunctivitis (*Med*) Inflammation of the CONJUNCTIVA.

conky (*For*) Applied to a log or tree bearing fruit bodies of wood-rotting fungi.

connate (*BioSci*) Of plant or animal parts that are firmly joined, particularly of like parts. Cf ADNATE.

connate water (*Geol*) Water trapped in a rock at the time of its deposition.

connected domain (*MathSci*) A *simply connected domain* is such that any closed curve in it and the region enclosed by the curve lie wholly within the domain. A *multiply connected domain* has one or more 'holes'. A *doubly connected domain* has one 'hole' and so on, an *n*tuply connected domain having $n-1$ 'holes'.

connected load (*ElecEng*) The sum of the rated inputs of all consumers' apparatus connected to an electric power supply system.

connecting box (*ElecEng*) See CONNECTION.

connecting-rod (*Eng*) In a reciprocating engine or pump, the rod connecting the piston or crosshead to the crank.

connecting-rod bolts (*Eng*) Bolts securing the outer half of a split big-end bearing of a connecting-rod to the rod itself. Also *big-end bolts*.

connecting thread (*BioSci*) See PLASMODESMA.

connection (*ElecEng*) A box containing terminals to which are brought a number of conductors of a wiring or distribution system, to facilitate the making of connections between them. Also *connecting box*.

connection admission control (*ICT*) The set of actions taken by an ASYNCHRONOUS TRANSFER MODE network at the call set-up phase to establish whether a connection is accepted or rejected, based on source characteristics, existing allocation of network resources and the required QUALITY OF SERVICE. Relevant traffic characteristics include average and peak CELL rates and type of source, eg telephony or video.

connectionism (*Psych*) The theory that connections between neurons govern behaviour and thought; (in artificial intelligence) the modelling of systems on networks made up of electronic neurons connected in a simplified model of the network of neurons in the brain.

connectionless broadband data service (*ICT*) A service similar to the US SWITCHED MULTIMEGABIT DATA SERVICE defined by the EUROPEAN TELECOMMUNICATION STANDARDS INSTITUTE.

connectionless packet switching (*ICT*) The normal mode of PACKET SWITCHING, in which each packet or datagram includes the full network destination and source address, and is routed independently over channels that may vary from packet to packet.

connection machine (*ICT*) TN for a massively parallel computer. See PARALLEL PROCESSING, SISMD.

connection-mode packet switching (*ICT*) A form of PACKET SWITCHING in which a fixed route, stored in a data table, is defined for all packets in a given transmission, rather than allowing the route to change from packet to packet. Although the channels of the route are shared by several users, each sees a 'virtual circuit' in which packets arrive with a fixed delay and therefore in the same sequence as they were transmitted.

connective (*BioSci*) A bundle of nerve fibres uniting two nerve centres.

connective tissue (*BioSci*) A group of animal tissues fulfilling mechanical functions, developed from the mesoderm and possessing a large quantity of non-living intercellular matrix, which usually contains fibres; as bone, cartilage and areolar tissue.

connective tissue diseases (*Med*) A term used to cover a number of diseases of uncertain aetiology including disseminated LUPUS ERYTHEMATOSUS, POLYARTERITIS NODOSA and SCLERODERMA.

connectivity (*Chem*) The way in which atoms are bound together in a molecule, ie the configurations of all the atoms. STEREOISOMERS have the same connectivity, while structural isomers do not.

connector (*Build*) See TIMBER CONNECTORS.

connector bar (*ElecEng*) See TERMINAL BAR.

connivent (*BioSci*) Converging and meeting at the tips.

conodont (*Geol*) One of a large number of microscopic phosphatic fossils, of doubtful affinity, having a tooth-like appearance. They ranged from the Cambrian (or earlier) to the Triassic and are useful for stratigraphical correlation.

conoscopic observation (*Crystal*) The investigation of the behaviour of doubly refracting crystal plates under convergent polarized light. Interference pictures obtained are important in explaining crystal optical phenomena.

consanguinity (*Geol*) A term applied to rocks having a similarity or community of origin, which is revealed by common peculiarities of mineral and chemical composition and often also of texture.

conscience (*Psych*) A set of internalized moral principles that are used to evaluate actions and behaviours; according to Freud, the restriction demanded by the SUPEREGO.

consciousness (*Psych*) Several related and general meanings: (1) the state of being alert and capable of action; (2) awareness of the environment, sentience; (3) awareness of a person's thoughts and feelings; (4) the ability of a person to perceive their own mental life; (5) states of mind of which one is aware, and which organize and co-ordinate one's activities, as opposed to subconscious factors organizing and guiding behaviour.

consensus sequence (*BioSci*) A DNA sequence found with minor variations and similar function in widely divergent organisms.

consequent (*MathSci*) (1) In logic, the term of a conditional statement which depends on the other; ie in the material implication 'if p, then q', p is the antecedent, and q is the consequent. (2) (archaic) The DENOMINATOR of a ratio a:b.

consequent drainage (*Geol*) A river system directly related to the geological structure of the area in which it occurs. See DRAINAGE PATTERNS.

consequent pole (*ElecEng*) An effective pole not at the end of ferromagnetic material, eg at the ends of a diameter of a ring magnet; must occur in pairs.

conservancy system (*Build*) Disposing of waste matter from buildings by earth closets and privies, without water.

conservation (*BioSci*) (1) Protection of natural ecosystems from human hands with the intention of preserving them as heritage or as a practical gene-bank. (2) Wise management of ecosystems, allowing exploitation at a level which does not impair the future capacity to produce.

conservation (*Psych*) The understanding, gained during the CONCRETE OPERATIONAL STAGE of development, that quantity, mass and volume remain the same even when the shape changes.

conservation headland (*Agri*) An area between the crop edge and the first set of tractor wheelings, the width of an arm of the sprayer boom, that is left largely untreated so that some weeds and beneficial insects can survive.

conservation laws (*Phys*) Laws concerning certain quantities which remain the same before and after interactions between particles. (1) The classical laws of the separate conservation of mass, of energy and of atomic species, which are sufficiently accurate for most chemical reactions. (2) Intrinsic properties of charge and baryon number are conserved in nuclear interactions. In addition, in strong and electromagnetic interactions between elementary particles, the intrinsic properties of strangeness, charm, topness and bottomness are also conserved but not in weak interactions.

conservation of energy (*Phys*) The constancy of the total energy of an isolated system. Energy may be converted from one form to another, but is not created or destroyed. If the system has only conservative forces, then the total mechanical energy (kinetic and potential) is constant. See MASS–ENERGY EQUATION.

conservation of matter (*Chem*) See LAW OF CONSERVATION OF MATTER.

conservation of momentum (*Phys*) The constancy of the sum of the momenta in a closed system (ie one in which no influences act upon it from outside); it is not affected by processes occurring within the system.

conservative field of force (*Phys*) A field of force such that the work done in moving a particle from A to B is independent of the path followed.

conservative system (*Phys*) A system such that in any cycle of operations where the configuration of the system remains unchanged overall, the work done is zero.

conservator tank (*ElecEng*) A small tank, carried on the top cover of an oil-filled power transformer, to accommodate the change in oil volume with temperature as the load varies, and reduce the oil surface exposed to the air.

conservatory (*Arch*) A glazed building in which plants may be grown under controlled atmospheric conditions.

Considère's construction (*Eng*) A construction on a graph of TRUE STRESS versus ENGINEERING STRAIN which allows the prediction of the onset of NECKING in a material under tensile test. The tangent to the stress–strain curve is drawn that passes through the point (true stress = 0, strain = -1). The point where it touches the curve locates the limit of uniform strain. Important in avoiding necking in metal-forming processes.

consistency (*Build*) The thickness, density or firmness of any thick liquid such as paint or adhesive. Consistency may be measured using a *plastometer*.

consistency (*Paper*) The amount of bone-dry fibre in pulp or stock, expressed as a percentage.

consistent (*MathSci*) (of a set of statements). Capable of all being true at the same time. In particular, of a set of equations: having at least one common solution, eg the pair of equations $x + y = 3$ and $x - y = 1$ are consistent, but the pair $x + y = 3$ and $x + y = 2$ are not. Also *compatible equations*.

console (*Build*) A bracket whose shelf is supported by a spiral scroll or volute.

console (*ElecEng*) A type of CONTROL PANEL: central controlling desk in power station, process plant, computer or reactor, where an operator can supervise operations and give instructions.

console (*ICT*) (1) Special desk equipped with a keyboard and a VDU for communicating with a computer. (2) Any main desk or area from which electronic equipment, distant spacecraft, etc, is controlled and monitored.

consolidation (*Geol*) The drying, compacting and induration of rock strata, as a result of pressures operating after deposition.

consolidation (*ICT*) The process by which VIRTUAL CONTAINERS carrying bursty traffic of a similar kind over several channels of a SYNCHRONOUS DIGITAL HIERARCHY network are concatenated into a single, less bursty channel.

consolidation (*Psych*) Physiological changes in the brain thought to occur following learning and associated with memory storage. The process, possibly involving structural changes, may continue for some time even after the learning phase is complete.

consonance (*Acous*) The condition where two pure tones blend pleasingly.

conspecific (*BioSci*) Relating to the same species. Often used as a noun.

constancy (*BioSci*) The percentage of sample plots in a plant community containing a particular species.

constant (*ICT*) A value that does not change within a program, eg the value of π.

constant (*MathSci*) In an algebraic expression or equation, a quantity (or parameter) which remains unchanged while the variables change in a given context. An *absolute constant*, eg π, has a unique value in all contexts.

constant-amplitude recording (*Acous*) A vinyl-disk recording technique whereby the response of recording, ie amplitude of the track divided by the root of the applied power, is independent of frequency. Cf CONSTANT-VELOCITY RECORDING.

constantan (*ElecEng*) An alloy of about 40% nickel and 60% copper, having a high-volume resistivity and almost negligible temperature coefficient; used as the resistance wire in resistance boxes etc. Also *Eureka*.

constant angular velocity (*ImageTech*) A video disk with a constant rotational speed, and one frame per revolution. Abbrev *CAV*.

constant-boiling mixtures (*Chem*) See AZEOTROPIC MIXTURES.

constant-current characteristic (*Electronics*) A property of a circuit or transformer whereby current supplied to a load is independent of its impedance or any externally applied voltage. A transistor, with base–emitter voltage held constant, can meet this criterion when the load is connected to the collector.

constant-current motor (*ElecEng*) An electric motor designed to operate at a constant current from a constant-current generator.

constant-current system (*ElecEng*) A system of transmitting electric power in which all the equipment is connected in series and a constant current is passed round the circuit. Variations in power result in a variation of the voltage of the system, constant-current generators being used for the supply. See SERIES SYSTEM.

constant-current transformer (*ElecEng*) A transformer designed to maintain a constant secondary current within a specified working range, for all values of secondary impedance and all values of primary voltage.

constant-frequency oscillator (*ElecEng*) One in which special precautions are taken to ensure that the frequency remains constant under varying conditions of load, supply voltage, temperature, etc.

constant-*k* filter (*ICT*) Simple type formed from a CONSTANT-*K* NETWORK only.

constant-*k* network (*ICT*) Iterative network for which the product of series and shunt impedance is frequency independent.

constant-level chart (*EnvSci*) An upper-air chart showing isobars for a particular level. Cf CONSTANT-PRESSURE CHART.

constant-level tube (*Surv*) A special form of level tube in which the volume ratio of bubble to liquid is fixed at such a value that decrease in length of the bubble due to expansion of the liquid is exactly counterbalanced by increase in length of the bubble due to diminished surface tension, so that the length of the bubble – and thus the sensitivity of the level tube – remains unaltered by rise in temperature.

constant linear velocity (*ImageTech*) A video disk whose rotational speed changes as the head tracks across its surface. Abbrev CLV.

constant-mesh gearbox (*Autos*) A gearbox in which the pairs of wheels providing the various speed ratios are always in mesh, the ratio being determined by the particular wheel which is coupled to the mainshaft by sliding dogs working on splines.

constant of integration (*MathSci*) The arbitrary constant resulting from indefinite integration. It arises because the derivative of a constant is zero, so that all functions that differ only by a constant have the same derivative.

constant of inversion (*MathSci*) See INVERSION.

constant-power generator (*ElecEng*) An electric generator which, by variation of the generated voltage, gives a constant-power output at varying currents.

constant-pressure chart (*EnvSci*) An upper-air chart showing contours of the geopotential height above sea level at which a particular pressure occurs. Cf CONSTANT-LEVEL CHART. See THICKNESS CHART.

constant-pressure cycle (*Autos*) See DIESEL CYCLE.

constant proportions (*Chem*) See LAW OF CONSTANT (DEFINITE) PROPORTIONS.

constant region (*BioSci*) The carboxy-terminal half of the light or the heavy chain of an immunoglobulin molecule. Termed constant because the amino acid sequence is the same in all molecules of the same class or subclass.

constant-resistance network (*ICT*) One in which iterative impedance in at least one direction is resistive and independent of the applied signal voltage.

constant-speed propeller (*Aero*) See PROPELLER.

constant time-lag (*ElecEng*) See DEFINITE TIME-LAG.

constant variable transmission (*Autos*) A form of automatic transmission using a belt drive with variable diameter pulleys. The system is becoming more widespread with the development of a segmented steel driving belt.

constant-velocity joint (*Autos*) A joint for transmitting drive, which does not show the cyclical velocity changes of the Hooke-type joint at higher angles. Used for the drive shafts of front-wheel-drive cars.

constant-velocity recording (*Acous*) In vinyl-disk recording the technique whereby the lateral rms velocity of the sinuous track is made proportional to the root of the electrical power applied to the recorder, irrespective of the frequency. Cf CONSTANT-AMPLITUDE RECORDING.

constant-voltage system (*ElecEng*) The usual system of transmission of electric power, in which the voltage between the conductors is maintained approximately constant, and all apparatus is connected to the system in parallel across the conductors.

constant-voltage transformer (*ElecEng*) One with or without extra components which gives a constant voltage with varying load current, or with varying input voltage over a specified range.

constant-volume amplifier (*ICT*) See VOGAD.

constellation (*Astron*) A group of stars, not necessarily connected physically, to which have been given a pictorial configuration and a name (generally of Greek mythological origin) which persist in common use although of no scientific significance.

constipation (*Med*) A condition in which the feces are abnormally dry and hard; retention of feces in the bowel; infrequent evacuation.

constituent (*Chem, Eng*) A component of an alloy or other compound or of a mixture. It may be present as an element or in chemical, physical or intermediate combination.

constituent (*MathSci*) Of a matrix or determinant, one of the numbers in the array which forms the matrix or determinant. Also *element*.

constitution (*Chem, Phys*) Structural distribution of atoms and/or ions composing a regularly co-ordinated substance. Includes percentage of each constituent and its regularity of occurrence through the material.

constitutional ash (*MinExt*) Ash resulting from combustion of coal and derived from siliceous matter in the coal-forming plants. 'Fixed ash' as distinct from entrained impurity of 'free ash'.

constitutional formula (*Chem*) A formula which shows the arrangement of the atoms in a molecule.

constitutional water (*Chem*) See WATER OF CRYSTALLIZATION.

constitution changes (*Eng*) Changes in solid alloys which involve the transformation of one constituent to another (as when pearlite is formed from austenite), or a change in the relative proportions of two constituents.

constitution diagram (*Eng*) See PHASE DIAGRAM.

constitutive enzyme (*BioSci*) An enzyme that is formed under all conditions of growth. Cf INDUCIBLE ENZYME.

constitutive heterochromatin (*BioSci*) HETEROCHROMATIN that is always condensed. SATELLITE DNA is found in these regions and coding sequences are apparently absent.

constraint (*Eng*) The property which distinguishes a mechanism from a KINEMATIC CHAIN. In a mechanism, the motion of one part is followed by a predetermined motion of the remainder of the mechanism.

constriction (*BioSci*) Narrow, localized region in a chromosome, normally found at the centromere (*primary constriction*), and often also at other sites (*secondary constrictions*), including NUCLEOLAR ORGANIZING REGION.

constriction resistance (*Phys*) The resistance across the actual area of contact through which current passes from one metal to another, equal to $(\rho_1 + \rho_2)/4\alpha$, where α is the radius of circular contact area and ρ_1 and ρ_2 are the resistivities of the metals.

constrictive pericarditis (*Med*) A condition in which chronic inflammation encircles the heart and prevents normal pump function.

constrictor (*BioSci*) A muscle that, by its contraction, constricts or compresses a structure or organ.

constringence (*Phys*) The inverse of the dispersive power of a medium. Ratio of the mean refractive index diminished by unity to the difference of the refractive indices for red and violet light.

construct (*Psych*) Any variable that cannot be directly observed and can only be measured through indirect methods, eg intelligence, motivation. The validity of such constructs is often questioned and provides ample opportunity for controversy.

consumer (*BioSci*) In an ecosystem, one of the heterotrophic organisms, chiefly animals, which ingest either other organisms or particulate organic matter. Cf DECOMPOSER, PRODUCER.

consummatory act (*Psych*) Historically in animal behaviour studies, the end phase of goal-oriented behaviour, typified by a series of responses directed at that goal, often of a stereotyped nature; it follows the APPETITIVE PHASE, a goal-seeking phase of behaviour. The rigid distinction between appetitive and consummatory phases of a behaviour sequence has been largely abandoned, although the terms are still used descriptively. Also *consummatory behaviour, consummatory phase*.

contact (*ElecEng*) (1) That part of either of two conductors which is made to touch the other when it is desired to pass current from one to the other, as in a switch for which a suitable minimum pressure and areas of contact are necessary. (2) The juxtaposition of parts, usually of gold, platinum or silver, which, when brought together, provide for the passage of a current as between plugs and sockets or between complex electronic assemblies as in computers.

contact adhesive (*Eng*) A type of polymer glue which only activates when pressed against the surfaces to be joined.

contact angle (*Chem*) The angle between the liquid and the solid at the liquid–solid–gas interface. It is acute for wetting (eg water on glass) and obtuse for non-wetting (eg water on paraffin wax).

contact angle (*MinExt*) The angle (θ) between a bubble of air and the chemically clean, polished and horizontal surface of a specimen of mineral to which it clings, measured between that surface and the side of the bubble. This forms an index to the floatability of the species under prescribed conditions, in which chemicals are added to the water and change in angle observed.

contact aureole (*Geol*) See AUREOLE.

contact bed (*Build*) A tank, filled with material such as broken clinker, used in the final oxidizing stage in sewage treatment, which consists of charging the filtering medium with the liquid sewage, allowing it to stand for a time, draining it off, and finally keeping the tank empty for a time. See PERCOLATING FILTER.

contact bounce (*ICT*) Intermittent opening and closing of relay contacts. Also *chatter*.

contact breaker (*ElecEng*) Circuit breaker.

contact emf (*ElecEng*) ELECTROMOTIVE FORCE which arises at the contact of dissimilar metals at the same temperature, or the same metal at different temperatures.

contact flight (*Aero*) Navigation of an aircraft by the pilot observing the ground only.

contact guidance (*BioSci*) The tendency of some cultured cells to be directed along topographical irregularities in their environment, eg of fibroblasts to align along ridges.

contact herbicide (*BioSci*) A herbicide which kills those plant parts that it contacts, eg ioxynil, paraquat. Cf SOIL-ACTING HERBICIDE, TRANSLOCATED HERBICIDE.

contact hypersensitivity (*BioSci*) Hypersensitivity reaction provoked when substances that act as HAPTENS or as ANTIGENS are applied to the skin. Usually due to prior sensitization by the chemical, and may be of immediate or delayed type.

contact inhibition (*BioSci*) Various behavioural changes that occur when moving or replicating cells come into contact with each other. Thus, *contact inhibition of locomotion* occurs when the direction of movement of a cell alters, or ceases, after collision with another cell; in

contact inhibition of growth (more correctly *density-dependent inhibition of growth*), most normal animal cells in culture stop dividing when a critical cell density is reached, often when all cells are in contact with their neighbours. See TRANSFORMATION.

contact insecticide (*Chem*) One which kills on contact with insect surface (body, legs, etc); used against sucking insects (eg aphids, mosquitoes) which are not affected by insecticides acting only through the alimentary system, eg pyrethrins, rotenone, DDT.

contact ionization (*Electronics*) Loss of electron by an easily ionized atom (eg caesium) when it comes into contact with the surface of a metal with affinity for electrons (eg tungsten).

contact jaw (*ElecEng*) (1) The clamping device of a resistance welding machine, which secures the parts to be welded and also conducts the current to them. (2) The fixed part of a switch, with which the moving blade makes contact in closing the circuit.

contact lens (*Phys*) A lens, usually of plastic material, worn in contact with the cornea instead of spectacles, or for cosmetic purposes.

contact maker (*ElecEng*) Any device used to make an electrical contact, esp a periodical contact, as in an automobile DISTRIBUTOR.

contact metal (*ICT*) That used for contacts on springs of relays, generally silver, platinum, tungsten, etc.

contact metamorphism (*Geol*) The alteration of rocks caused by their contact with, or proximity to, a body of IGNEOUS ROCK.

contact noise (*ElecEng*) Noise voltage arising across a contact, with or without adsorbed gases, arising from differences in work function of contacting metal conductors, one of which may be a semiconductor.

contactor controller (*ElecEng*) A controller in which the various circuits are made and broken by means of contactors.

contactor starter (*ElecEng*) An electric motor starter in which the steps of resistance are cut out, or other operations are performed, by means of contactors.

contactor switching starter (*ElecEng*) A switching starter in which the switching operations are carried out by means of contactors.

contact pad (*Electronics*) An area of metallization or conducting track in eg a transistor to which an external electrical connection is made.

contact potential (*ElecEng*) The electrical potential which arises in equilibrium at the contact of dissimilar metals. A related phenomenon arises when one or both materials are semiconductors, in which case there is a finite region over which the potential is developed. See BUILT-IN VOLTAGE. A contact potential can also be formed at junctions between similar materials at different temperatures.

contact-potential barrier (*Electronics*) Potential barrier formed at the junction between regions with different energy gap or carrier concentration.

contact pressure (*ICT*) The pressure between the contacts on relay springs, a minimum pressure being required for certain contact when the circuit is frequently broken.

contact print (*ImageTech*) A positive photographic or motion picture print made with the sensitive emulsion exposed in physical contact with the negative.

contact process (*ChemEng*) The process of large-scale manufacture of sulphuric acid by the oxidation of sulphur dioxide to sulphur trioxide in the presence of a catalyst.

contact radiation therapy (*Radiol*) Radiation from a very short distance, eg 20 mm, with voltages around 50 kV.

contact rail (*ElecEng*) See CONDUCTOR RAIL.

contact resistance (*ElecEng*) The resistance at the surface of contact between two conductors, influenced by the nature of the materials, the state of the surfaces and the pattern of current flow.

contact scanning (*Eng*) Ultrasonic inspection procedures in which the ultrasonic head is acoustically coupled to the material being scanned.

contact screen (*Print*) A vignetted-dot screen on a film base used in the process camera in close contact with the emulsion, as an alternative to the glass half-tone screen; also for same-size screened work by contact.

contact shoe (*ElecEng*) See COLLECTOR SHOE.

contact spring (*ElecEng*) The flexible metal holder of the contact in a relay. The stiffness of the holder determines the pressure between contacts for a given displacement.

contact strip (*ElecEng*) On a pantograph or bow type of current collector, the renewable metal or carbon strip that actually makes contact with the overhead wire of an electric traction system. Also *bow strip* when used on a bow collector.

contact stud (*ElecEng*) In the surface-contact system of electric traction, the studs in the roadway for making contact with the contact skate on an electric vehicle. The studs are only made alive when the vehicle is actually passing over them.

contact vein (*MinExt*) A vein occurring along the line of contact of two different rock formations, one of which may be an igneous intrusion.

contact wire (*ElecEng*) The overhead conductor from which current is collected, by suitable forms of collector gear, for the vehicles of some electric traction systems.

contagion (*Med*) The communication of disease by direct contact between persons or between an infected object and a person. Adj *contagious*.

contagious bovine pleuro-pneumonia (*Vet*) Lung plague. An acute, subacute or chronic disease of cattle caused by *Mycoplasma mycoides*; characterized by fever, pneumonia and pleurisy. Vaccines have been used experimentally.

contagious catarrh (*Vet*) See INFECTIOUS CORYZA.

contagious distribution (*BioSci*) See AGGREGATION.

contagious equine abortion (*Vet*) A contagious form of abortion in horses due to infection of the placenta by *Salmonella abortivoequina*.

contagious equine metritis (*Vet*) Important cause of equine infertility due to a Gram-negative micro-aerophilic coccobacillus. First reported in 1977, but now rare after intense screening and prohylaxis. Abbrev *CEM*.

contagious ophthalmia (*Vet*) A contagious disease of sheep, characterized by conjunctivitis and keratitis, caused by a rickettsial organism *Colesiota conjunctivae* (*Rickettsia conjunctivae*). Also *heather blindness*.

contagious pustular dermatitis (*Vet*) A contagious virus disease of sheep and goats characterized by vesicle and pustule formation on the skin and mucous membranes, esp on lips, nose and feet. Vaccination available. Also *malignant aphtha, orf*.

containers (*Eng*) (1) Reservoirs for materials, solid or fluid, which must be made of materials themselves inert to both the contents and external environment. Thermoplastics like polyolefins have replaced many traditional materials (eg glass, mild steel) for their ease of shaping, low density and chemical resistance. Some grades are, however, susceptible to ENVIRONMENTAL STRESS CRACKING. (2) Steel boxes, usually 6 or 12 m long by 2·4 m wide and 2·4 m high, used for much present-day freight and handled by special cranes, ships and road and rail vehicles.

containment (*NucEng*) In fusion, the use of shaped magnetic fields or of INERTIAL CONFINEMENT to contain a plasma. Also *confinement*. See MAGNETIC CONFINEMENT.

containment time (*NucEng*) The time for which a given temperature and pressure can be maintained in a fusion experiment.

contaminant (*FoodSci*) Any material or substance that is not a specified part of the food or food product. Includes: foreign bodies (term for non-food material); extraneous vegetable matter (EVM), ie vegetable stalks, leaves, seeds, etc; bones in meat and fish products; biological and microbiological contaminants, ie insects, grubs, yeasts, moulds and bacteria. See ADULTERATION, CROSS-CONTAMINATION, RESIDUES.

contaminated rock (*Geol*) Igneous rock whose composition has been modified by the incorporation of other rock material.

contamination (*Genrl*) The introduction of unwanted material, such as pollution caused by deposits of radio-active materials, or the introduction of impurities or bacteria into a substance.

contamination meter (*Radiol*) A particular design of Geiger–Müller circuit for indicating for civil defence purposes the degree of radioactive contamination in an area, esp for estimating the time for its safe occupation.

content-addressable storage (*ICT*) See ASSOCIATIVE STORAGE.

contention (*ICT*) A conflict in requests for the use of system resources, eg two programs that attempt to print on the same printer simultaneously or two users who attempt to use the same channel simultaneously in a NETWORK.

content validity (*Psych*) A subjective judgement as to whether a test is actually able to measure the variable of interest, based upon a consideration of the content of the test system.

context-dependent memory (*Psych*) The theory that information learned in a particular situation or environment is better remembered when the same conditions exist or are recreated.

context effects (*Psych*) The phenomenon whereby the context in which a stimulus is presented or a behaviour exhibited may well affect the response to the stimulus or the type of behaviour. Recreating the context may or may not be important depending upon the extent of the context effect.

contiguity (*Psych*) The closeness in time of two events which is sometimes regarded as the condition leading to association, esp in CLASSICAL CONDITIONING procedures.

continent (*Geol*) One of the Earth's major land masses, including the dry land and continental shelves.

continental climate (*EnvSci*) A type of climate found in continental areas not subject to maritime influences, and characterized by more pronounced extremes between summer and winter; the winters become colder to a greater degree than the summers become hotter; also relatively small rainfall and low humidities.

continental crust (*Geol*) That part of the Earth's crust which underlies the continents and continental shelves. It is approximately 35 km thick in most regions but is thicker under mountainous areas. Sedimentary rocks predominate in its uppermost part and metamorphic rocks at depth, but the detailed composition of the lower crust is uncertain. Cf OCEANIC CRUST. See panel on EARTH.

continental deposit (*Geol*) A rock formed under subaerial conditions or in water not directly connected with the sea. See AEOLIAN DEPOSITS, GLACIAL DEPOSITS.

continental drift (*Geol*) A hypothesis put forward by Wegener in 1912 to explain the distribution of the continents and oceans and the undoubted structural, geological and physical similarities which exist between continents. The continents were believed to have been formed from one large land mass and to have drifted apart. See panel on PLATE TECTONICS.

continental rise (*Geol*) That part of the continental margin between the CONTINENTAL SLOPE and the ABYSSAL PLAIN. It is characterized by a relatively gentle slope.

continental shelf (*Geol*) The gently sloping offshore zone, extending usually to about 200 m depth.

continental slope (*Geol*) The relatively steep slope between the continental shelf and the more gentle rise from the abyssal plain.

contingency table (*MathSci*) A table giving the frequency of observations cross-classified by variate values.

continued fraction (*MathSci*) A terminating or infinite fraction of the form

$$a + \cfrac{b}{c + \cfrac{d}{e + \cfrac{f}{g +} \dots}}$$

usually written

$$a + \frac{b}{c+} \frac{d}{e+} \frac{f}{g+} \dots$$

The values obtained by ending the fraction at *a*, at *c*, at *e*, etc, are called *convergents*.

continuity (*ImageTech, ICT*) The supervision, coordination and matching in sequence of all the successive scenes and items from various sources in making up a motion picture or broadcast programme, and the records and documents so required.

continuity (*Phys*) The existence of an uninterrupted path for current in a circuit.

continuity-bond (*ElecEng*) A rail-bond used to maintain the continuity of the track- or conductor-rail circuit at junctions and crossings.

continuity-fitting (*ElecEng*) A device used in electric wiring installations for ensuring a continuous electric circuit between adjacent lengths of conduit.

continuous (*MathSci*) (of a function). Informally, able to be graphed without taking the pencil off the paper; formally, $f(x)$ is continuous at a point x_0 if for every neighbourhood U_2 of $f(x_0)$, there exists a neighbourhood U_1 of x_0 such that $f(U_1)$ is contained in U_2.

continuous beam (*Eng*) A beam supported at a number of points and continuous over the supports, as distinct from a series of simple independent beams. Also *continuous girder*.

continuous brake (*Eng*) A brake system used on railway trains, in which operation at one point applies the brakes simultaneously throughout the train. See AIR BRAKE, ELECTROPNEUMATIC BRAKE, PNEUMATIC BRAKE, VACUUM BRAKE.

continuous casting (*Eng*) A method in which the molten metal is added at the top of the mould while the externally solidified material is withdrawn from the bottom, the mould being cooled by water or air jets.

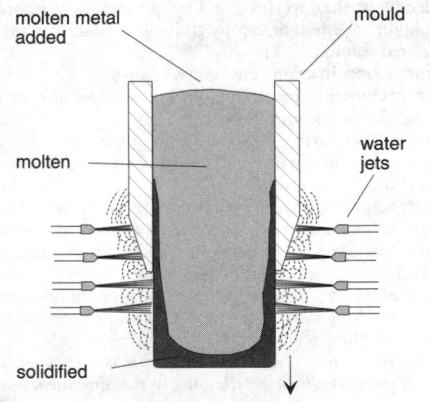

molten metal added — mould
molten — water jets
solidified

continuous casting

continuous control (*ICT*) A system in which the controller is supplied with continuously varying actuating signal. Cf ON–OFF CONTROL.

continuous-cooling transformation diagram (*Eng*) These represent transformation characteristics for materials which undergo solid-state transformations and are very useful for selecting the conditions for achieving the desired mechanical properties for a steel component. They carry time–temperature transformation data but are presented with start- and end-of-transformation lines relating to the cooling of round bars of varying diameter. Abbrev *CCT diagram*. See ISOTHERMAL TRANSFORMATION DIAGRAM.

continuous culture (*BioSci*) A culture maintained at a steady state over a period, usually in a CHEMOSTAT. Cf BATCH CULTURE.

continuous current (*ElecEng*) Earlier name for DIRECT CURRENT. Now obsolete as an electrical distribution system in the UK and USA but retained on the Continent.

continuous diffusion (*Genrl*) Counterflow system of extracting sugar from beet whereby fresh beet slices (cosettes) are extracted by hot dilute sugar solution, and partially extracted slices are finally extracted with fresh hot water.

continuous-disk winding (*ElecEng*) A type of winding used for transformers; the whole winding is made from one continuous length of conductor instead of being split up.

continuous distillation (*ChemEng*) An arrangement by which a fresh distillation charge is continuously fed into the still in the same measure as the still charge is distilled off. The contrary process is known as *batch distillation*.

continuous electrode (*ElecEng*) A type of carbon electrode used in electric furnaces; the electrode is gradually fed forward as the lower part burns away, and the upper part is renewed by adding fresh material. The furnace can thus be worked continuously, without intervals for electrode renewal.

continuous extraction (*ChemEng*) Extraction of solids or liquids by the same solvent, which circulates through the extracted substance, evaporates and is condensed again, and continues the same cycle over again; or by exhaustive extraction with solvents in counter-current arrangement.

continuous feeder (*Print*) An automatic sheet-feeder designed to permit continuous reloading without stopping the machine.

continuous filament yarn (*Textiles*) Yarn composed of one or more unbroken filaments. Synthetic fibres are usually made as continuous filaments although these may be cut or broken into staple fibres for subsequent manufacture into fabric. Silk is a natural continuous filament. Cf STAPLE FIBRES. See panel on FIBRE ASSEMBLIES.

continuous filter (*Build*) See PERCOLATING FILTER.

continuous flight auger piles (*CivEng*) Piles made by augering a hole within a steel tube which is pressed or driven down with the auger. At final depth, high-plasticity concrete is fed to the bottom under pressure through the hollow core of the rotating auger which is then withdrawn with the tube bringing up further excavated material on its flights. Abbrev *CFA piles*.

continuous furnace (*Eng*) A furnace in which the charge enters at one end, moves through continuously, and is discharged at the other.

continuous girder (*Eng*) See CONTINUOUS BEAM.

continuous impost (*Build*) An IMPOST which does not project from the general surfaces of the pier and arch.

continuous loading (*ElecEng*) Inductance loading with series inductions at intervals much less than a wavelength.

continuously regenerating trap (*Eng*) A device for removing particulates, mostly carbon, from the exhaust of, particularly, diesel engines. With the aid of catalysts, nitrogen dioxide oxidizes the particulates which are then emitted as carbon dioxide, so soot does not accumulate.

continuous mill (*Eng*) A rolling mill consisting of a series of pairs of rolls in which the stock undergoes successive reductions as it passes from one end to the other end of the mill. See PULL-OVER MILL, REVERSING MILL, THREE-HIGH MILL.

continuous oscillations (*ICT*) Those which would occur in a tuned circuit of inductance, capacitance but no resistance: in practical circuits, the damping effect of resistance has to be overcome by injecting energy from an external source. Also *undamped oscillations*.

continuous printing (*ImageTech*) Contact printing in which negative film and print stock move continuously past an illuminated exposure aperture.

continuous processing (*Eng*) A method of producing an article continuously and, in theory, indefinitely. See CONTINUOUS CASTING, EXTRUSION.

continuous processing machine (*ImageTech*) Equipment for processing photographic film or paper as a continuous strip passing through the successive chemical solutions and washing baths to the final drier.

continuous projector (*ImageTech*) A motion picture projector in which the film moves continuously and uniformly, the intermittent effect being obtained by optical means, such as a multifaced rotating prism.

continuous rating (*ElecEng*) The maximum power dissipation which could be allowed to continue indefinitely. Cf INTERMITTENT RATING.

continuous reinforcement (*Psych*) *Schedules of reinforcement* in which every correct behaviour is reinforced.

continuous sections (*Print*) The normal arrangement of the sections for bookwork as distinct from INSETTED. Each section is made up of consecutive pages, its last page being followed by the first page of the next section.

continuous spectrum (*Phys*) A spectrum which shows continuous non-discrete changes of intensity with wavelength or particle energy.

continuous stationery (*ICT*) Paper, perforated at page intervals and fan-folded to form a pack, used in DOT MATRIX and LINE PRINTERS.

continuous variation (*BioSci*) Variation of a character whose measurements do not fall into distinct classes, but take any value within certain limits.

continuous vent (*Build*) Extension of a vertical waste pipe above the point of entry of liquid wastes to a point above all windows, to provide ventilation.

continuous vulcanization (*Eng*) A method of processing thermoset rubber product so that it is produced continuously in final cross-linked form. Needs great care to ensure optimum cure, since overcured product is difficult if not impossible to recycle. Simple examples include extruded pipe, more complex than that for making conveyor belting. Abbrev *CV*. See ROTOCURE.

continuous welded rail (*CivEng*) Track which has been prewelded into lengths of up to 366 m (1200 ft) before being transported to the site, where successive lengths are welded together to produce unbroken track many kilometres long. The rail is hydraulically extended to the length which it would take up at a standard temperature (28°C in the UK), so that it is in tension at any lower temperature.

continuum (*BioSci*) The pattern of overlapping populations in a large but definite community with component populations distributed along a gradient of eg altitude.

continuum (*MathSci*) Any compact connected set, in particular the real line. The *cardinality of the continuum*, *C*, is the cardinal number of the points in the real line, which can be proved to be greater than ALEPH-0. The *continuum hypothesis* states that there is no cardinal number greater than aleph-0 and less than *C*, and it has been shown that this hypothesis can be neither proved nor disproved from the basic axioms of set theory.

continuum (*Phys*) See IONIZATION CONTINUUM.

contorted (*BioSci*) Petals in a flower bud that overlap a neighbour to one side and are overlapped on the other, so that the whole appears twisted. See AESTIVATION.

contour (*Surv*) The imaginary intersection line between the ground surface and any given level surface: a line connecting points on the ground surface which are at the same height above DATUM.

contour acuity (*Phys*) The power of the eye to distinguish a displacement between two sections of a line, as in reading a VERNIER.

contour fringes (*Phys*) Interference fringes formed by the reflection of light from the top and bottom surfaces of a thin film or wedge. They correspond to optical thickness. Also *Fizeau fringes*. See NEWTON'S RINGS.

contour gradient (*Surv*) A line on the ground surface having a constant inclination to the horizontal.

contour interval (*Surv*) The vertical distance between adjacent contours in any particular case.

contraception (*Med*) The prevention of conception.

contraceptive (*Med*) Any agent which prevents the fertilization of the ovum with a spermatozoon. See ORAL CONTRACEPTIVES.

contract-demand tariff (*ElecEng*) A form of two-part tariff in which the fixed charge is made proportional to the maximum kilowatt demand likely to be made.

contract growing (*Agri*) Crop production under contract for a specified customer, typically to a detailed specification.

contractile root (*BioSci*) A root, some part of which shortens (by a change in shape of the inner cortical cells) so as to pull eg a herbaceous plant closer to the ground or a bulb or corm deeper into the soil.

contractile tissue (*BioSci*) A group of animal tissues that possess the property of contractility; the most common example is muscle.

contractile vacuole (*BioSci*) In some Protozoa, a cavity, filled with fluid, that periodically collapses and expels its contents into the surrounding medium, so ridding the animal of surplus fluid.

contractility (*BioSci*) The power of becoming reduced in length, exhibited by some cells and tissues, as muscle; the power of changing shape.

contraction (*Eng, PowderTech*) The percentage shrinkage of eg a GREEN COMPACT after processing or heat treatment.

contraction cavities (*Eng*) Porous zones in metal castings, usually in the last portions to solidify, caused by the volume contraction from liquid to solid state not being adequately replaced by fresh liquid from the feeder head. Almost unavoidable in some regions of complex shaped castings. Also *shrinkage cavities*.

contraction coefficient (*ElecEng*) A coefficient used in making calculations on the magnetic circuit of an electric machine, to allow for the effect of the fringing of the flux in the air-gap due to open or semi-closed slots. The actual length of the gap is reduced by the coefficient in order to obtain an effective gap length, which is shorter than the actual value.

contraction in area (*Eng*) See NECKING.

contraction joint (*CivEng*) An interruption of a structure specifically allowing for contraction.

contraction ratio (*Aero*) The ratio of the maximum cross-sectional area of a wind tunnel to that of the working section.

contracture (*Med*) The distortion or shortening of a part, due to spasm or paralysis of muscles, muscular contraction that persists after the stimulus which caused it has ceased, or due to the presence of scar tissue.

contradictory (*MathSci*) Of a set of propositions, not able all to be true at the same time. Of a single proposition, not true in any circumstances.

contrails (*EnvSci*) See CONDENSATION TRAILS.

contralateral (*BioSci*) Pertaining to the opposite side of the body. Cf IPSILATERAL.

contraries (*Paper*) Anything in the pulp or paper stock which is unwanted in the paper.

contrast (*Acous*) The relation, measured in decibels, between the maximum intensity level and the minimum useful intensity level in programme material such as speech or music.

contrast (*ImageTech*) (1) The difference in brightness between the lightest and darkest areas in a subject or its reproduction. (2) In a photographic image, the relation between the maximum and minimum densities. See GAMMA. (3) In lighting a subject, the relation between

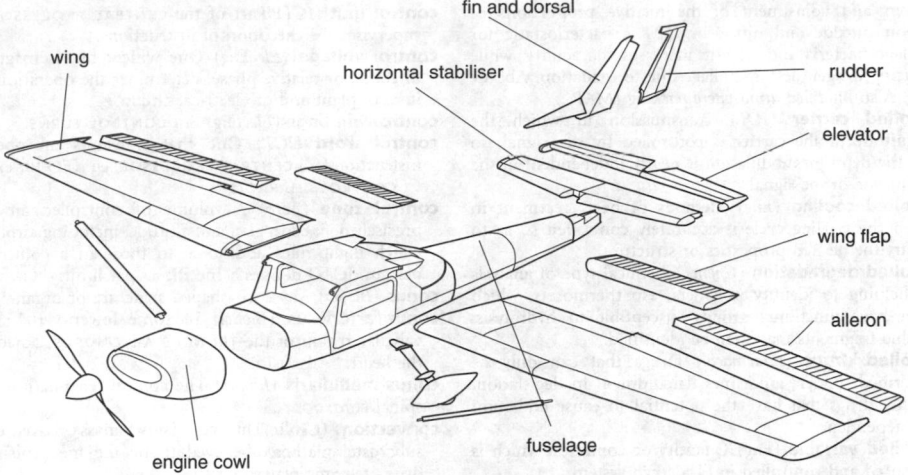

control Exploded view of a light aircraft showing control surfaces.

the KEY illumination and the FILLER LIGHT in the shadow areas.

contrast amplification (*Acous*) Amplification in which dynamic contrast in sound reproduction is increased by electronic means, compensating imprecisely for contrast reduction which is necessary in most communication systems to avoid intrusion of noise.

contrast medium (*Med*) Substance injected into the bloodstream to increase the contrast in X-ray procedures; usually contains iodine. Widely used in diagnostic radiology.

contrast photometer (*Phys*) A class of photometer in which measurement is made by comparing the illumination produced on two adjacent surfaces by the lamp under test and by a standard lamp.

contrast range (*ImageTech*) See DYNAMIC RANGE.

contrate wheel (*Eng*) A toothed wheel, the teeth of which are formed at right angles to the plane of the wheel; a wheel that transmits motion between two arbors at right angles. Useful for light loads.

contra wire (*ElecEng*) The same as *constantan wire*.

control (*Aero*) A device by which the orientation and path of an aircraft can be changed: the more stable an aircraft, the more control force is required for this change. See ARTIFICIAL STABILITY. Control surfaces include *ailerons*, *rudder* and *tailerons*. See MANOEUVRE DEMAND SYSTEM.

control (*ElecEng*) General term for manual or automatic adjustment (usually by potentiometer, fader or attenuator) of power level in a transmission within its dynamic range.

control (*Genrl*) A scientific experiment performed without variables to provide a standard of comparison for other experiments.

control (*ICT*) The selection, interpretation and sequencing of functions to be performed within a computing system. Control is held by the CONTROL UNIT.

control (*NucEng*) Maintenance of power level of a reactor at desired setting by adjustments to the reactivity by control rods or other means. See panel on NUCLEAR REACTOR.

control absorber (*NucEng*) See CONTROL ROD.

control ampere-turns (*ElecEng*) Magnetomotive force applied to a magnetic amplifier.

control board (*ElecEng*) A switchboard on which are mounted the operating handles, push-buttons or other devices for operating switchgear situated remotely from the board. The board usually has mounted on it indicating instruments, key diagrams and other accessory apparatus.

control bus (*ICT*) A BUS that is used to carry control signals between two devices, eg between the PROCESSOR and MAIN

MEMORY or between the PROCESSOR and a PERIPHERAL device.

control channel (*ICT*) See BROADCAST CONTROL CHANNEL.

control character (*ICT*) A non-printing character that is treated as a signal to control operating functions. Cf ALPHANUMERIC.

control characteristic (*ElecEng*) Curve connecting output quantity against control quantity under determined conditions in a MAGNETIC AMPLIFIER.

control chart (*MathSci*) A chart to monitor the behaviour of a process; in particular, to assist in the detection of deviations of the process from a norm.

control circuit (*ElecEng*) A circuit which controls the operation of a piece of equipment or electrical system.

control column (*Aero*) The lever supporting a hand-wheel or hand-grip by which the ailerons and elevator of an aircraft are operated. It may be a simple 'joystick', pivoted at the foot and rocking fore-and-aft and laterally. On military aircraft, usually fighters, it is often hinged halfway up for lateral movement; on transports it is usually either 'spectacle' or 'ram's horn' shape.

control-configured vehicle (*Aero*) One designed with artificial stability giving eg reduced wing size and control surfaces, enhanced manoeuvrability, reduced gust response and flutter suppression.

control current (*ElecEng*) One which, by its magnitude, direction or relative phase, determines the operation of an item of plant and/or electrical circuit.

control electrode (*ElecEng*) One, eg a grid, the primary function of which is to control flow of electrons between two other electrodes, without taking appreciable power itself, control being by voltage which regulates electrostatic fields.

control hysteresis (*ElecEng*) Ambiguous control depending on previous conditions. Jump or snap action arising in electronic or magnetic amplifiers because of excessive positive feedback which occurs under certain conditions of load.

control impedance (*Phys*) The electrical property of a device which controls power in one direction only, such as a gas-filled relay.

controllable-pitch propeller (*Aero*) See PROPELLER.

controlled air space (*Aero*) Areas and lanes clearly defined in three dimensions wherein no aircraft may fly unless it is under radio instructions from AIR-TRAFFIC CONTROL.

controlled atmosphere packing (*FoodSci*) Modifying the atmosphere in a food pack by total or partial removal of oxygen and adjustment of the relative proportions of carbon dioxide and nitrogen. CO_2 is bacteriostatic for spoilage bacteria and slightly increases the acidity, while reducing oxygen delays spoilage due to oxidation. Abbrev CAP. Also *modified atmosphere packing (MAP)*.

controlled carrier (*ICT*) Transmission in which the magnitude of the carrier is controlled by the signal, so that the depth of modulation is nearly independent of the magnitude of the signal.

controlled cooling (*Eng*) Methods of heat treatment in which the cooling cycle is accurately controlled so as to impart the desired properties or structure.

controlled degradation (*Chem*) Chemical type of analysis for helping to identify polymers, esp thermosets, which often have functional groups susceptible to hydrolysis. Soluble fragments may then be identified.

controlled drugs (*Pharmacol*) Drugs that can only be prescribed under guidelines laid down in legislation. Usually drugs that have the potential to cause addiction and dependence.

controlled variable (*Eng*) Quantity or condition which is measured and controlled in eg a servo system.

controller (*ICT*) A device that controls a functional element within a computer system, eg HARD-DISK controller, CACHE CONTROLLER.

controller (*ElecEng*) An assembly of equipment for controlling the operation of electric apparatus.

control limit-switch (*ElecEng*) A limit-switch connected in the control circuit of the motor whose operation is to be limited.

control-line (*ElecEng*) A train-line used on multiple-unit trains for connecting master controllers or contactor gear on the different coaches.

control magnet (*ElecEng*) A magnet used in electric indicating instruments to provide a force for controlling the movement of the moving system.

control measure (*FoodSci*) Action required to eliminate or reduce a hazard to an acceptable level, eg a heat process step at a specified temperature and holding time will eliminate an identified microbiological hazard. See HACCP.

control panel (*ElecEng*) A panel containing a full set of indicating devices and remote-control units required for operation of industrial plant, reactor, chemical works, etc. Cf CONSOLE.

control point (*ElecEng*) The value of a controlled variable, departure from which causes a controller to operate in such a sense as to reduce the ERROR and restore an intended steady state. Also *set point*.

control points (*BioSci*) Places in the cell cycle where the cell's behaviour may be changed. Also *checkpoints*. See panel on CELL CYCLE.

control program (*ICT*) See MONITOR, OPERATING SYSTEM.

control register (*ICT*) Computer REGISTER within the CONTROL UNIT and that stores a single control instruction.

control relay (*ElecEng*) See RELAY.

control reversal (*Aero*) See REVERSAL OF CONTROL.

control rod (*NucEng*) A rod moved in and out of the reactor core to vary reactivity. May be a neutron-absorbing rod, eg boron or cadmium, or, less often, a fuel rod. Also *control absorber*. See REGULATING ROD, SHIM ROD and panel on NUCLEAR REACTORS.

control rod worth (*NucEng*) The change in reactivity of a critical reactor caused by the complete insertion or withdrawal of the control rod.

control total (*ICT*) The sum resulting from the addition of a specified field from each of a group of records, often used for checking purposes.

control track (*ImageTech*) A linear track, outside the video area of the tape, where the control or sync pulses are recorded, one per frame, to ensure accurate tracking during playback.

control turns (*ElecEng*) Those wires on the core of a magnetic amplifier or transductor which carry the control current. Also *control windings*. US *signal windings*.

control unit (*ICT*) Part of the CENTRAL PROCESSOR that supervises the execution of instructions.

control voltage (*ElecEng*) One which, by its magnitude, direction or relative phase, determines the operation of an item of plant and/or electrical circuit.

control windings (*ElecEng*) See CONTROL TURNS.

control word (*ICT*) One that transmits an operating instruction to a CENTRAL PROCESSOR, eg XEQ for execute. Cf CONTROL CHARACTER.

control zone (*Aero*) A volume of controlled air space, precisely defined in plan and altitude, including airports, in which flight rules additional to those in a control area pertain. ICAO defines a specific upper limit.

conus (*BioSci*) Any cone-shaped structure or organ.

conus arteriosus (*BioSci*) In some lower vertebrates, a valvular region of the TRUNCUS ARTERIOSUS, adjacent to the heart.

conus medullaris (*BioSci*) The conical termination of the spinal cord.

convection (*Geol*) The very slow mass movement of subcrustal material; believed to be the mechanism that drives tectonic plates.

convection (*Phys*) The process by which energy or mass is transmitted through a material by bulk motion of the medium itself.

convection current (*Phys*) Current in which the charges are carried by moving masses appreciably heavier than electrons.

convection of heat (*Phys*) The transfer of heat in a fluid by the circulation flow due to temperature differences. The regions of higher temperature, being less dense, rise, while the regions of lower temperature move down to take their place. The convection currents so formed help to keep the temperature more uniform than if the fluid was stagnant.

convective transfer (*Astron*) The transfer of energy from one part of a star to another by convection.

convector (*Eng*) Heater which warms the air passing over it. Cf RADIATOR.

conventional memory (*ICT*) The MAIN MEMORY delivered as standard with an IBM-COMPATIBLE AT computer using MS-DOS. This is usually considered to be the memory between ADDRESS 0 and 640 KBYTES. See fig. at MEMORY MAP.

conventional signs (*CivEng, Surv*) Standard symbols, universally understood, used in the representation on maps and plans of features which would otherwise be difficult or impossible to represent.

convention of signs (*Phys*) Sign convention used in lens calculations to ensure consistency in the derivation and use of lens formulae, in which all distances must be measured from some origin. More than one is in use, but in the *cartesian* convention all distances are measured *from* the reflecting or refracting surface being considered, or *from* the principal planes in the case of a thick lens or lens system. Distances measured in the direction in which the incident light is travelling are given a positive sign, and those in the opposite direction a negative sign. Distances measured perpendicular to the axis, eg size of image or object, are measured from the axis: above is positive, below is negative.

convergence (*EnvSci*) Negative DIVERGENCE.

convergence (*ImageTech*) In a colour TV display, alignment of the three electron or optical beams for correct image registration over the whole picture.

convergence (*MathSci*) (1) An infinite sequence $a_1, a_2, \ldots a_n, \ldots$, is said to be convergent to a if, as n tends to infinity, a_n tends to the limit a, ie

$$\lim_{n \to \infty} a_n = a$$

(2) An infinite series $a_1 + a_2 + \ldots + a_n + \ldots$, is said to be convergent to the sum A, if the sum of the terms tends to the limit A, as the number of terms tends to infinity, ie

$$\lim_{n \to \infty} \sum a_n = A$$

(3) An infinite product, $a_1a_2 \ldots a_n \ldots$ is said to be convergent to α if, as n tends to infinity, the product tends to the limit α, ie

$$\lim_{n \to \infty} \prod a_n = \alpha$$

(4) An infinite integral

$$\int_a^\infty f(t) \, dt$$

is said to converge to I, if $f(t)$ is continuous for $t \geqslant a$ and if

$$\lim_{x \to \infty} \int_a^x f(t) \, dt = I$$

convergence (*Psych*) A binocular cue used in perception of distance: the closer an object, the more inward the eyes need to turn in order to focus.

convergence coils (*ImageTech*) Used to ensure convergence of the electron beams in a triple-gun colour TV tube. Also *convergence magnets*.

convergence surface (*Electronics*) Ideally, the surface generated by the point on which the electron beams in a multibeam cathode-ray tube converge. In practice, this can only be regarded as a point to a first approximation.

convergence zone (*Geol*) In PLATE TECTONICS (panel) the zone where moving plates collide and area is lost by shortening and crustal thickening or by subduction and destruction of the crust. The site of earthquakes, mountain building, trenches and volcanism.

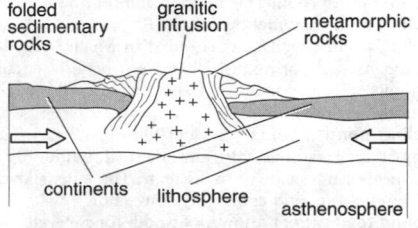

convergence zone

convergent (*MathSci*) See CONTINUED FRACTION.

convergent–divergent nozzle (*Aero, Eng*) A venturi type of nozzle in which the cross-section first decreases to a throat and then increases to the exit, such a form being necessary for efficient expansion of steam, as in the steam turbine, or of other vapours or gases, as in the supersonic aero-engine, ramjet or rocket. Abbrev *con–di nozzle*.

convergent evolution (*BioSci*) The tendency of unrelated species to evolve similar structures, physiology or appearance due to the same selective pressures, eg the succulent CAM plants of deserts. Some apparent convergences, eg the eyes of vertebrates and Cephalopods, prove to be more complex as it becomes clear that the homeobox genes involved are the same.

convergent lens (*Phys*) A lens which increases the convergence or decreases the divergence of a beam of light. A simple lens is convergent if it is double-convex, plano-convex or concavo-convex with the radius of curvature of the concave face greater than that of the convex.

convergent thinking (*Psych*) Logical and conventional thought leading to a single answer in contrast to creative or DIVERGENT THINKING.

converging-beam therapy (*Radiol*) A form of CROSS-FIRE TECHNIQUE in which a number of beams of radiation are used simultaneously to treat a particular region through different entry portals. Fig. ▷

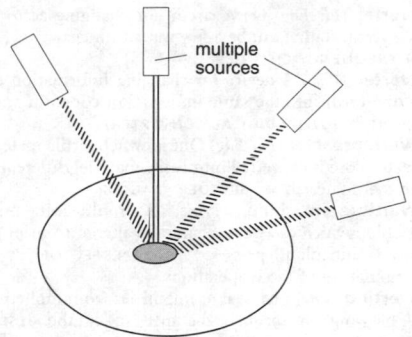

converging-beam therapy

converging-field therapy (*Radiol*) A form of MOVING-FIELD THERAPY in which the source of radiation moves in a spiral path. A mechanical linkage ensures that, whatever the position of the source, the beam is directed at the target.

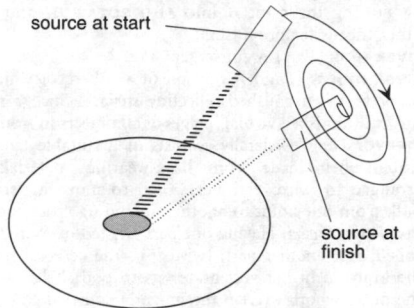

converging-field therapy Shows path of source.

converse magnetostriction (*Phys*) The change of magnetic properties, eg induction, in a ferromagnetic material when subjected to mechanical stress or pressure.

conversion (*Chem*) The fraction of some key reagent which has been used up, hence a measure of the rate or completeness of a chemical reaction.

conversion coating (*Eng*) A coating on a metal produced by chemical treatment, eg chromatization and phosphating, which impart properties like corrosion resistance. See ACTIVATION, PASSIVATION.

conversion coefficient (*Phys*) See INTERNAL CONVERSION.

conversion detector (*ICT*) See MIXER (2).

conversion disorder (*Psych*) Neurosis in which painful emotions are repressed and unconsciously converted into physical symptoms, sometimes with the affected bodily site symbolizing the repressed idea.

conversion efficiency (*Electronics*) See ANODE EFFICIENCY.

conversion electron (*Electronics*) An electron ejected from the inner shell of an atom when its excited nucleus returns to ground state, and the energy released is given to an orbital electron, instead of appearing as quantum.

conversion factor (*Phys*) The factor by which a quantity, expressed in one set of units, must be multiplied to convert it to another.

conversion gain (*ICT*) Effective amplification, or loss, of a conversion detector, measured as the ratio of output voltage at intermediate frequency to input voltage of signal frequency. Commonly expressed in decibels. Also *conversion loss*.

conversion mixer (*ICT*) Same as FREQUENCY CHANGER.

conversion ratio (*NucEng*) The number of fissionable atoms, eg ^{239}Pu, produced per fissionable atom, eg ^{235}U, destroyed in a reactor. Corresponding conversion gain is defined as $R-1$. Symbol R.

converter (*ElecEng*) (1) A circuit for changing ac to dc or vice versa. Rating can be a few watts to megawatts. (2) US for FREQUENCY CHANGER.

converter (*ICT*) A device for changing information coded in one form into the same information coded in another, eg ANALOGUE-TO-DIGITAL CONVERTER.

converter reactor (*NucEng*) One in which fertile material in reactor core is converted into fissile material different from the fuel material. See BREEDER REACTOR.

convertible machine (*Print*) A multi-unit printing machine which can be mechanically altered to print either as a multicolour press or as a PERFECTOR or in a combination of these operations.

converting (*Eng*) Removal of impurities from molten metal by blowing air through the melt in eg the BESSEMER PROCESS.

converting (*Textiles*) Producing a sliver of staple fibres from a continuous filament tow by cutting or breaking.

converting station (*ElecEng*) An electric power system substation containing one or more converters.

convertiplane (*Aero*) A VTOL aircraft which can take off and land like a helicopter, but cruises like an aircraft; by swivelling the rotor(s) and/or wings to act as propellers, or by putting the rotor(s) into AUTOROTATION and using other means for propulsion.

convex lens (*Phys*) A convergent lens.

convex mirror (*Phys*) A portion of a sphere of which the outer face is a polished reflecting surface. Such a mirror forms diminished virtual images of all objects in front of it.

conveyor (*Eng*) Generally consists of a suitable tensioned endless belt made from hard-wearing materials and arranged to run over rollers. Used to move materials in bulk from one point to another, including cross-country.

convolute (*BioSci*) Having one part twisted over, folded, or rolled over another part; twisted; as the cerebral lobes of the brain in higher vertebrates; gastropod shells in which the outer whorls overlap the inner. N *convolution*.

convolution (*Med*) Any elevation of the surface of the brain.

convolution integral (*MathSci*) The integral $\int f(t)g(x-t)dt$, where the limits of the integral are variously defined. Also referred to as the *cross-correlation* between $f(x)$ and $g(x)$. When $f(x) = g(x)$ the integral is sometimes referred to as the *autocorrelation of $f(x)$* or simply the *convolution*.

convulsion (*Med*) Generalized involuntary spasm of the muscles that are normally under control of the will.

cook–chill (*FoodSci*) Fully precooking food, then cooling rapidly and storing at 0–3°C, giving a short-life product (maximum 5 days). The food must be reheated according to the guidelines set by the Department of Health in the UK and consumed within 15 min. Cf COOK–FREEZE.

cook–freeze (*FoodSci*) Fully precooking food, then rapid freezing and storage at below −18°C, to give a storage life, depending on the food and proper temperature maintenance, of between 2 and 9 months. The food must be defrosted, if necessary reheated, and consumed according to the guidelines set by the Department of Health in the UK. Cf COOK–CHILL.

cookie (*ICT*) Computer code that is downloaded to a hard drive when a user visits a website, and will allow the website to identify that computer when the website is visited again. Also *Internet cookie*.

coolant (*Eng*) (1) A mixture of water, soda, oil and soft-soap, used to cool and lubricate the work and cutting tool in machining operations. See CUTTING COMPOUND. (2) A fluid used as the cooling medium in the jackets of liquid-cooled internal combustion engines, eg water, ethylene glycol (ethan 1·2-diol).

coolant, reactor (*NucEng*) The gas, liquid or liquid metal circulated through a reactor core to carry the heat generated in it by fission and radioactive decay to boilers or heat exchangers. In water-cooled reactors, it is often the moderator.

cooled-anode valve (*Electronics*) Large thermionic valve in which special provisions are made for dissipating the heat generated at the anode, effected by circulating water, oil or air around the anode, or by radiation from its surface.

cooling (*NucEng*) The decay of activity of irradiated nuclear fuel or highly radioactive waste before it is processed or disposed of.

cooling analysis (*Eng*) A method of analysing cooling time in moulding of polymers, important because it often forms the larger part of the total cycle time. Uses cooling curves and material data (thermal diffusivity, heat distortion temperature) plus product dimensions (usually greatest thickness) to calculate cooling time from estimated FOURIER NUMBER. Product redesign can then be undertaken to reduce maximum thickness, and so increase productivity. At the same time, care is needed to ensure that product stiffness and strength remain within specification. Also needed for rubber products, eg tyres, where cure kinetics are critical. May be backed up by direct temperature measurement within tyre using thermocouples. See INJECTION MOULDING.

cooling circuit (*Eng*) A system of water tubes within mould tool for maintaining it at a constant preset temperature. It thus ensures product reproducibility. Chilled water is usually used, but some engineering plastics demand high temperatures to minimize orientation. See INJECTION MOULDING.

cooling coil (*Eng*) Tubing which transfers heat from the material or space cooled to the primary refrigerant.

cooling curves (*Eng*) Curves obtained by plotting time against temperature for a metal cooling under constant conditions. The curves show the evolutions of heat which accompany solidification, polymorphic changes in pure metals and various transformations in alloys.

cooling drag (*Aero*) That proportion of the total drag due to the flow of cooling air through and round the engine(s).

cooling duct (*ElecEng*) See DUCT.

cooling pond (*Eng*) An open pond in which water, heated through use in an industrial process, or after circulation through a steam condenser, is, before reuse, allowed to cool through evaporation.

cooling pond (*NucEng*) A water-filled space in which the initial high radioactivity and thermal output of spent elements can be allowed to dissipate. The water allows both safe inspection and cooling by convection.

cooling tower (*Eng*) A tower of wood, concrete, etc, used to cool water after circulation through a condenser. The water is allowed to trickle down over wood slats, thus exposing a large surface to atmospheric cooling. Power station cooling towers are large concrete structures, circular in plan and hyperparabolic in elevation and supported on circular warren trusses. The shape promotes maximum vertical air flow with compressive stress only in the shell which can be made very thin.

Coomassie blue (*BioSci*) Dye used in Bradford method for protein estimation and for detecting proteins on gels. Also *Coomassie Brilliant Blue, Kenacid Blue*.

Coombs test (*Med*) Diagnostic test for determining whether an individual's red blood cells are coated with AUTO-ANTIBODIES or IMMUNE COMPLEXES. Patient's red blood cells are mixed with anti-human immunoglobulin. If antibody is present, the red blood cells will agglutinate.

co-operation (*BioSci*) A category of interaction between two species where each has a beneficial effect on the other, increasing the size or growth rate of the population, but, unlike mutualism, not a necessary relationship. Termed *proto-co-operation* by some, since its basis is neither conscious nor intelligent, as in humans.

co-operativity (*BioSci*) A phenomenon displayed by enzymes or receptors that have multiple binding sites. Binding of one ligand alters the affinity of the other site(s). Both positive and negative co-operativity can occur.

Cooper pair (*Phys*) In a superconducting material below its critical temperature, two weakly bound electrons which do not act independently but as a dynamic pair. The BCS (Bardeen–Cooper–Schrieffer) theory uses this concept to give a detailed microscopic theory of superconductivity. See panel on SUPERCONDUCTORS.

co-ordinate axes (*MathSci*) See AXES.

co-ordinate bond (*Chem*) See COVALENT BOND.

co-ordinated transposition (*ICT*) The reduction of mutual inductive effects in multiline transmission systems (telephony or power) by periodically interchanging positions.

co-ordinate potentiometer (*ElecEng*) One in which two linear potentiometers carry ac currents 90° apart in phase, so that the resultant voltage between tappings can be adjusted in both phase and magnitude.

co-ordinates (*MathSci*) An ordered set of numbers which specify the position or orientation of a point or geometric configuration relative to a set of AXES.

co-ordinating gap (*ElecEng*) A spark gap, used in power transmission schemes, arranged so that it will conduct at a voltage lower than the breakdown voltage of other apparatus in the system ensuring that surge voltages are safely discharged to earth.

co-ordination compound (*Chem*) A compound generally described from the point of view of the central atom to which other atoms are bound or co-ordinated and are called ligands. Normally, the central atom is a (transition) metal ion, and the ligands are negatively charged or strongly polar groups.

co-ordination number (*Chem*) The number of atoms or groups (ligands) surrounding the central (nuclear) atom of a complex ion or molecule. Abbrev *CN*.

co-ordination polymerization (*Chem*) See CHAIN POLYMERIZATION.

cop (*Textiles*) Yarn package built on to mule spindle or ring tube.

copalite (*Min*) Also *copaline*. See HIGHGATE RESIN.

cope (*Eng*) The upper half of a mould or moulding box. See fig. at MOULDING.

Copepoda (*BioSci*) A subclass of CRUSTACEA, mainly of small size. Some are parasitic, others planktonic where they form an important food source for pelagic fish like herring.

Copernican system (*Astron*) The heliocentric theory of planetary motion; called after Copernicus, who introduced it in 1543. It superseded the geocentric or PTOLEMAIC SYSTEM.

co-phenotrope (*Pharmacol*) A mixture of diphenoxylate HCl (an opioid that reduces gut motility) and atropine sulphate (antispasmodic), used to treat diarrhoea. TN *Lomotil*.

coping (*Build*) (1) See COPING BRICK. (2) The operation of splitting stone by driving wedges into it.

coping (*Vet*) The operation of paring or cutting the beak or claws of a bird, particularly of hawks.

coping brick (*Build*) Specially shaped brick used for capping the exposed top of a wall; used sometimes with a CREASING and sometimes without, in which latter case the brick is wider than the wall and has drips under its lower edges.

coping saw (*Build*) Small saw with narrow tensioned blade in a D-shaped bow, for cutting curves in wood up to about 15 mm thick (ie too thick for FRET-SAW).

coplanar vectors (*MathSci*) Vectors are coplanar if they lie in the same plane.

copolymer (*Chem*) Polymer formed from the reaction of more than one species of monomer in order to modify the physical properties of the parent HOMOPOLYMER. See panel on POLYMERS.

copolymer equation (*Chem*) Equation which relates structure of copolymers to propagation rate constants etc. See CHAIN POLYMERIZATION, COPOLYMER.

copolymerization (*Chem*) See COPOLYMER.

copper (*Chem*) Bright, reddish metallic element, symbol Cu, at no 29, ram 63·54, mp 1083°C, resistivity at 0°C, 0·016 mΩ m. Native copper crystallizes in the cubic system. It often occurs in thin sheets or plates, filling narrow cracks or fissures, but it mainly occurs in a large number of other minerals, primarily sulphides, sulphates and other oxidized minerals, and hydrates. Crystallizes in the face-centred cubic

system. There are several grades of commercially pure copper, all of which are ductile, with high electrical and thermal conductivity and good resistance to corrosion; it has many uses, notably as an electrical conductor. Basis of brass, bronze, aluminium bronze and other alloys. Nickel–iron wires with a copper coating are frequently used for *lead-ins* through glass seals, forming vacuum-tight joints. Copper-64 is a mixed radiator, of half-life 13 h. See COPPER ALLOYS.

copper alloys (*Chem*) Alloys with zinc, tin, aluminium, lead, etc, which have been made for different applications and used for over 5000 years. Not surprisingly they have numerous overlapping proprietary and other names. Current international standards adopt a system based on letters for the main alloying base followed by a number designating the particular alloy, eg CZ stands for copper–zinc (brasses), PB for phosphor bronze, LG for leaded gunmetal, CT for copper–tin (bronzes), AB for aluminium–bronze, CN for copper–nickel.

copperas (*Min*) Iron sulphate, $FeSO_4 \cdot 7H_2O$. See MELANTERITE. *White copperas* is goslarite, $ZnSO_4 \cdot 7H_2O$.

copper brushes (*ElecEng*) Brushes occasionally used for electric commutator machines where high conductivity is required; they are made of copper strip, wire or gauze.

copper-clad steel conductor (*ElecEng*) See STEEL-CORED COPPER CONDUCTOR.

copper factor (*ElecEng*) A term used in electric machine design to denote the ratio of the cross-sectional area of the copper in a winding to the total area of the winding, including insulating material and clearance space.

copper glance (*Min*) A popular name for *chalcocite*.

copper glazing (*Build*) Glazing formed of a number of individual panes separated by copper strips on the edges of which small flanges of copper have later been formed by deposition to retain the glass. Also *copperlite glazing*.

copper loss (*ElecEng, ICT*) The loss occurring in electric circuitry according to JOULE'S LAW; it is proportional to the product of I^2R, where I is the current and R is the resistance.

copper nickel (*Min*) See NICCOLITE.

copper nose (*Vet*) See BROWN NOSE DISEASE.

copper pyrites (*Min*) See CHALCOPYRITE.

copper-sheathed cable (*ElecEng*) See MINERAL-INSULATED CABLE.

coppersmith's hammer (*Eng*) A hammer having a long curved ball-pane head, used in dishing copper plates.

copper (II) sulphate (*Chem*) $CuSO_4$. Bluestone; a salt, soluble in water, used in copper-plating baths; formed by the action of sulphuric acid on copper; crystallizes as hydrous copper sulphate, $CuSO_4 \cdot 5H_2O$, in deep-blue triclinic crystals. See BLUE VITRIOL, CHALCANTHITE.

coprecipitation (*Phys*) The precipitation of a radioisotope with a similar substance, which precipitates with the same reagent, and which is added in order to assist the process.

coprime (*MathSci*) See RELATIVELY PRIME.

co-processor (*ICT*) An additional PROCESSOR that performs a special function such as FLOATING-POINT arithmetic.

coprodaeum (*BioSci*) In birds, that part of the cloaca into which the anus opens.

coprolalia (*Psych*) Excessive and often repetitive use of verbal obscenities as in Tourette's syndrome.

coprolite (*Geol*) Fossilized excreta of animals. Generally composed largely of calcium phosphate.

coprophagous (*BioSci*) Dung-eating.

coprophilia (*Psych*) Pleasure or gratification obtained from any dealing with feces.

coprophilous (*BioSci*) Growing on or in dung. Also *coprophilic*.

coprozoic (*BioSci*) Living in dung, as some Protozoa.

copula (*BioSci*) A structure that bridges a gap or joins two other structures, as the series of unpaired cartilages that unite successive gill arches in lower vertebrates.

copulation (*BioSci*) In Protozoa, a type of syngamy in which the gametes fuse completely; in higher animals, union in sexual intercourse.

copulation tube (*BioSci*) See CONJUGATION TUBE.

copy (*ICT*) The transfer of information from one storage medium to another without changing the information in the original.

copy (*Print*) Any matter supplied for setting or for reproduction by any of the printing processes.

copyholder (*Print*) (1) A person who reads aloud from the copy as the proof-corrector follows the reading in the proof. (2) A contrivance for holding up sheets of copy on typesetting machines.

copying machine (*Eng*) A machine for producing numbers of similar objects by an engraving tool or end-cutter, which is guided automatically from a master pattern or template. See DOCUMENT COPYING.

copy lens (*ImageTech*) A lens designed for optimum quality of image formation at comparatively short object distances and small degrees of magnification or reduction.

copy number (*BioSci*) The number of genes or plasmid sequences per genome which a cell contains.

copy-protection (*ICT*) A security precaution to prevent the unauthorized copying of programs and files, eg the use of a special DISK FORMAT to prevent the standard copy COMMAND working properly.

coquille (*Glass*) Glass in thin curved form used in the manufacture of sun glasses. The radius of curvature is traditionally 3·5 in (90 mm). Similar glass of 7 in (180 mm) radius is called *micoquille*.

coquimbite (*Min*) Hydrated ferric sulphate, crystallizing in the hexagonal system, occurring in some ore deposits and also in volcanic fumaroles such as those of Vesuvius.

coquina (*Geol*) A limestone made up of coarse shell fragments, usually of molluscs.

Coraciiformes (*BioSci*) An order of birds, most of which are short-legged arboreal forms that nest in holes and have NIDICOLOUS young. They are mainly tropical and often brightly coloured. Kingfishers, bee-eaters.

coracoid (*BioSci*) In vertebrates, a paired posterior ventral bone of the pectoral girdle, or the cartilage that gives rise to it.

CORAL (*ICT*) Programming language used for ON-LINE REAL-TIME systems.

coral (*BioSci*) The massive calcareous skeleton formed by certain species of ANTHOZOA and some HYDROZOA; the colonies of polyps forming this skeleton. Adjs *corallaceous, coralliferous, coralliform, coralline, coralloid*.

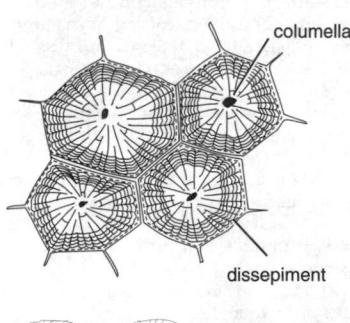

corals Typical fossil forms.

septal view

coral reef (*Geol*) A calcareous bank formed of the calcareous skeletons of corals and algae which live in colonies. The various formations of coral reefs are known as *atolls, barrier reefs* and *fringing reefs*.

coral sand (*Geol*) A sand made up of calcium carbonate grains derived from eroded coral skeletons, often found in deep water on the seaward side of a coral reef.

CORBA (*ICT*) Abbrev for *common object request broker architecture*, a proprietary piece of middleware which enables the control of complex collections of objects by multiple users in an object-oriented system.

corbeille (*Arch*) Carved work representing a basket, used as a form of decoration.

corbel (*Build*) Bricks or stones, frequently moulded, projecting from a wall to support a load.

corbelling (*Build*) Projecting courses of brick or stone forming a ledge used to support a load.

corbel-piece (*Build*) See BOLSTER.

corbicula (*BioSci*) The pollen basket of bees, consisting of the dilated posterior TIBIA with its fringe of long hairs.

corbie-step gable (*Build*) A gable having a series of regular steps up each slope. Also *crow-step gable*.

cord (*For*) Non-metric timber measure, 128 ft³ (8 × 4 × 4 ft, about 3·6 m³).

cordate (*BioSci*) Said of a leaf base that is heart-shaped, with the stem attached to the indentation.

cord blood (*Med*) Blood taken post-partum from the umbilical cord.

corded way (*Build*) A sloping path formed with deep sloping steps separated by timber or stone risers.

cordierite (*Min*) Magnesium aluminium silicate with some iron, typically occurring in thermally metamorphosed rocks and in some gneisses. Often shows cyclic twinning. *Iolite* is a name often used for gem varieties and the mineral is also sometimes called *dichroite* from its strong blue to colourless dichroism.

cordless terminal (*ICT*) Originally a simple telephone handset (CT1) linked by radio to a home or office base station, the term now covers telephones able to make, but not receive, calls via public base stations or TELEPOINTS (CT2).

cords (*Print*) Lengths of hemp across the back of a book, to which the sections are attached by sewing. Used in BOUND BOOKS. See BAND.

Cordtex (*CivEng*) TN for a textile detonating fuse containing a core of pentaerythritol tetranitrate, used in the initiation of large explosive charges. Has a high velocity of detonation and will not inspire detonation unless directly connected to the actual explosive.

corduroy (*Textiles*) Strong, hard-wearing cloth having a rounded or flattened cord or rib of weft pile running longitudinally; made entirely from cotton, or cotton warp and spun-rayon pile.

cordwood (*For*) Tree trunks of medium diameter sawn into uniform lengths.

core (*Aero*) Gas generator portion of a gas-turbine engine which may be developed as a basis of several engines used on different types of aircraft.

core (*Build*) The material removed from a mortise.

core (*CivEng*) A watertight wall built within a dam or embankment as an absolute barrier to the passage of water.

core (*ElecEng*) (1) Region associated with a coil: may be air or a magnetic material to increase inductance. Typical materials are ferrites and punched laminations of soft iron. Construction may be as a complete magnetic circuit (divided to accommodate the coil) or simply a rod inserted into the coil. (2) Assembly consisting of the conductor and insulation of a cable but not including the external protective covering. An arrangement comprising many such assemblies is termed *multicore*.

core (*Eng*) (1) A solid mass of specially prepared sand or loam placed in a mould to provide a hole or cavity in the casting. See fig. at MOULDING. (2) A steel form in an INJECTION MOULDING tool which creates a cavity in the final product. Side cores are retractable to allow removal of the moulding. See FUSIBLE CORE, ROTATING CORE.

core (*Geol*) The central part of the Earth at a depth below 2900 km. It is believed to be composed of nickel and iron.

core (*ImageTech*) (1) A plastic cylinder on which film or magnetic tape is wound. (2) The inner part of the positive carbon of an arc lamp, impregnated with metallic salts to improve colour and brilliance.

core (*MinExt*) Cylindrical rock section cut by rotating hollow drill bit in prospecting, sampling, blasting.

core (*NucEng*) That part of a nuclear reactor which contains the fissile material, either dispersed or in cans.

core (*Phys*) In an atom, the nucleus and all complete shells of electrons. In the atoms of the alkali metals, the nucleus, together with all but the outermost of the planetary electrons, may be considered to be a core, around which the valency electron revolves in a manner analogous to the revolution of the single electron in the hydrogen atom around the nucleus. In this manner, the simple BOHR MODEL may be made to give an approximate representation of the alkali spectra. See panel on ATOMIC STRUCTURE.

core-balance protective system (*ElecEng*) An excess-current protective system for electric power systems, in which any leakage current to earth in a three-phase circuit is made to produce a resultant flux in a magnetic circuit surrounding all three phases; this flux produces a current in a secondary winding on the magnetic circuit, which operates a relay controlling the appropriate circuit breakers.

core bar (*Eng*) (1) An iron bar on which cylindrical loam cores are built up. The bar is supported horizontally and rotated while a loam board is pressed against the core. (2) An iron rod for reinforcing a sand core.

coreboard (*Build*) See BLOCKBOARD.

core box (*Eng*) A wooden box shaped internally for moulding sand cores in the foundry.

core/cladding concentricity (*ICT*) A measure of quality in an OPTICAL FIBRE, expressing how accurately the inner, high refractive index core is centred within the outer layer of lower refractive index. Fibres of high concentricity are easier to join because they can be aligned simply by locating their outer surfaces.

CO₂ recorder (*Eng*) An instrument which analyses automatically the flue gas leaving a furnace, and records the percentage of carbon dioxide (CO_2). See EXHAUST-GAS ANALYSER.

cored electrode (*ElecEng*) A metal electrode provided with a core of flux or other material; used in arc welding.

cored hole (*Eng*) A hole formed in a casting by the use of a core, as distinct from a hole that has been drilled.

cored solder (*Eng*) Hollow solder wire containing a flux paste, which allows flux and solder to be applied to the work simultaneously.

core dump (*ICT*) A dump of the contents of main memory.

coreless induction furnace (*ElecEng*) A high-frequency induction furnace in which there is no iron magnetic circuit other than the charge in the furnace itself.

core losses (*ElecEng*) The losses occurring in electric machinery and equipment owing to hysteresis and eddy-current losses set up in the iron of the magnetic circuit, which are due to an alternating or varying flux.

coremium (*BioSci*) (1) A rope-like strand of anastomosing HYPHAE. (2) A tightly packed group of erect CONIDIO-PHORES, somewhat resembling a sheaf of corn.

core oven (*Eng*) A foundry oven used for drying and baking cores before insertion in a mould.

core plate (*ElecEng*) See LAMINATION.

core plates (*Eng*) Disks attached to a CORE BAR to reinforce large cores.

core prints (*Eng*) Projections attached to a pattern to provide recesses in the mould at points where cores are to be supported.

core register (*Eng*) Corresponding flats or vees formed on cores and core prints, when correct angular location is necessary.

core sample (*MinExt*) Sample from a borehole to give information on the rock formation at side or bottom. Usually a few inches in diameter.

core sand (*Eng*) Moulding sand to which a binding material such as linseed oil has been added to obtain good cohesion and porosity after drying.

core-spun yarn (*Textiles*) A yarn made with a core (usually of continuous filaments) surrounded by a sheath generally made of staple fibres.

core-type induction furnace (*ElecEng*) An induction furnace in which there is an iron core to carry the magnetic flux.

core-type transformer (*ElecEng*) A transformer in which the windings surround the iron core, the former usually being cylindrical in shape.

Corfam (*Chem*) TN for POROMERIC material formerly used in shoe uppers.

coriaceous (*BioSci*) Firm and tough, like leather in texture. Also *corious*.

Coriolis acceleration (*EnvSci*) The apparent tendency of a freely moving particle to swing to one side when its motion is referred to a set of axes that is itself rotating in space. The magnitude of the acceleration for a particle moving horizontally on the surface of the Earth is $2\Omega V \sin \varphi$ where Ω is the angular magnitude of the angular velocity of rotation of the Earth, V is the speed of the particle relative to the Earth's surface, and φ is the latitude. The acceleration is perpendicular to the direction of V and is directed to the right in the northern hemisphere. For general three-dimensional motion the Coriolis acceleration has some other small terms because the Earth's axis does not, normally, lie parallel to the local vertical; for meteorological purposes these additional terms are negligible.

Coriolis effect (*Phys*) In a rotating reference frame, Newton's second law of motion is not valid, but it can be made to apply if, in addition to the real forces acting on a body, a (fictitious) Coriolis force and a *centrifugal force* are introduced. The effect of the Coriolis force is to deviate a moving body perpendicular to its velocity; a body freely falling towards the Earth is slightly deviated from a straight line and will fall to a point east of the point directly below its initial position. Coriolis forces explain the directions of the trade winds in equatorial regions and would affect astronauts in an artificial g environment produced by rotation.

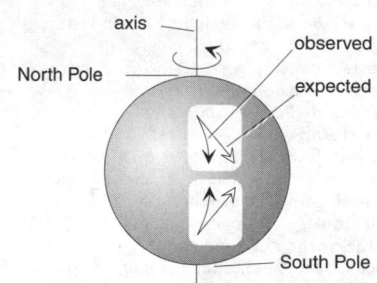

Coriolis effect

Coriolis force (*EnvSci*) The force, acting on a given mass, which would produce the CORIOLIS ACCELERATION.

Coriolis parameter (*EnvSci*) The Coriolis parameter f is defined by $f = 2\Omega \sin \varphi$ where Ω is the angular velocity of rotation of the Earth, and φ is the latitude.

corious (*BioSci*) See CORIACEOUS.

corium (*BioSci*) The dermis of vertebrates.

cork (*BioSci*) The naturally occurring, long exploited, CELLULAR SOLID, density approx $170 \, \text{kg m}^{-3}$, which occurs as a thin layer in the bark of all trees. It is a tissue of dead cells with suberized cells which form a protective layer replacing the epidermis in older stems and roots of many seed plants. The cork oak (*Quercus suber*) is unique in having this layer several cm thick. Used for thermal and acoustic insulation, flooring, packaging, as elastic and chemically inert closures for bottles, and for its buoyancy. Also *phellem*. See CORK CAMBIUM, PERIDERM.

cork cambium (*BioSci*) The layer of meristematic cells lying a little inside the surface of an older root or stem and forming cork on its outer surface and phelloderm internally. Also *phellogen*. See PERIDERM.

cork rubber (*Eng*) Rubber to which cork granules have been added to increase bulk and insulation, much used for gaskets.

corkscrew rule (*ElecEng*) A rule relating the direction of the magnetic field to the current direction in a conductor. The corkscrew is driven in direction of current and sense of rotation of the handle gives direction of magnetic field.

corkscrew staircase (*Arch*) A helical staircase built about a solid central newel.

Corliss valve (*Eng*) A steam engine admission and exhaust valve in the form of a ported cylinder which is given an oscillating rotary motion over the steam port by an eccentric driven wrist-plate.

corm (*BioSci*) Organ of perennation and vegetative propagation in eg *Crocus* consisting of a short, usually erect and tunicated, underground stem of one year's duration, next year's rising on top.

cormophyte (*BioSci*) In former systems of classification, a plant of which the body is differentiated into roots, stems and leaves.

corn (*BioSci*) In the UK, the most important cereal crop of a particular region, especially wheat in England, and oats in Scotland and Ireland. US *maize*.

corn (*Med*) Localized overgrowth of the horny layer of the skin due to local irritation, the overgrowth being accentuated at the centre.

corn (*Vet*) A local inflammation due to bruising or compression of the keratogenous membrane of the posterior portion of the horse's foot; septic corn, an abscess localized to the sole of a bird's foot.

cornea (*BioSci*) In invertebrates, a transparent area of the cuticle covering the eye, or each facet of the eye; in vertebrates, the transparent part of the outer coat of the eyeball in front of the eye. Adj *corneal*.

corned beef (*FoodSci*) Beef ground and cured with sodium nitrite and salt, then canned and thermally processed to COMMERCIAL STERILITY.

cornelian (*Min*) See CARNELIAN.

corneous (*BioSci*) Resembling horn in texture.

corner (*ICT*) See BEND.

corner (*Print*) The piece of leather covering each of the outer corners of a half-bound volume.

corner bead (*Build*) An ANGLE STAFF.

corner cramp (*Build*) See MITRE CRAMP.

corner-lap joint (*Build*) See END-LAP JOINT.

corner reflector (*Radar*) A metal structure of three mutually perpendicular sheets, for returning signals.

corner tool (*Eng*) A sleeking tool for finishing off the internal corners of a mould.

cornice (*Arch*) A projecting moulding decorating the top of a building, window, etc.

corniculate (*BioSci*) (1) Shaped like a horn. (2) Bearing a horn or horn-like outgrowth. Also *cornute*.

Cornish boiler (*Eng*) A horizontal boiler with a cylindrical shell provided with a single longitudinal furnace tube or flue.

corn oil (*Chem*) A pale yellow oil obtained from maize; rel.d. 0·920–0·925, saponification value 188–193, iodine value 111–123, acid value 1·7–20·6. Used as a cooking oil. Also *maize oil*.

cornstone (*Geol*) Concretionary limestone common in the Old Red Sandstone and Permo-Triassic rocks of the UK. Characteristic of soils formed in arid conditions.

cornua (*BioSci*) Horn-like processes; as the posterior *cornua* of the hyoid.

Cornu prism (*Phys*) A 60° prism formed of two 30° prisms cemented together, one being of right-handed and the other left-handed quartz, the optical axes of the two being parallel to the ray passing through the prism at minimum deviation, ie parallel to the base. This device overcomes a defect of using a single 60° quartz prism in a spectrometer.

Cornu's spiral (*MathSci*) A spiral with parametric cartesian equations

$$x = \int_0^s \cos\frac{1}{2}\theta^2 d\theta, \quad y = \int_0^s \sin\frac{1}{2}\theta^2 d\theta$$

It has the property that its radius of curvature $\rho = 1/s$. Also *clothoid*.

cornute (*BioSci*) See CORNICULATE.

corolla (*BioSci*) Inner whorl of the PERIANTH esp if different from the outer and then often brightly coloured, composed of the petals which may be free or fused to one another.

corona (*Arch*) The part of a CORNICE showing a broad projecting face and throated underneath to throw off the water.

corona (*Astron*) The outermost layer of a star's atmosphere consisting of tenuous highly ionized gas. The solar corona, visible as a halo of light during a total eclipse, has a temperature of around 2 000 000 K at a height of 75 000 km, and extends for millions of kilometres into space.

corona (*BioSci*) (1) Generally, the head or upper surface of a structure or organ. (2) A trumpet-like outgrowth from the perianth, as in the daffodil. (3) A ring of small leafy undergrowths from the petals, as in campion. (4) A crown of small cells on the oögonium of Charophyta. (5) In Echinoidea, the shell or test. (6) In Crinoidea, the disk and arms as opposed to the stalk. (7) In Rotifera, the discoidal anterior end of the body. Adj *coronal*.

corona (*ElecEng*) See BRUSH DISCHARGE.

corona (*EnvSci*) A system of coloured rings seen around the Sun or Moon when viewed through very thin cloud. They are caused by diffraction by water droplets. The diameter of the corona is inversely proportional to the size of the droplets.

corona (*Phys*) The phenomenon of air breakdown when electric stress at the surface of a conductor exceeds a certain value. At higher values, stress results in luminous discharge. See CRITICAL VOLTAGE.

Corona Australis (Southern Crown) (*Astron*) A small but prominent southern constellation on the fringes of the Milky Way.

Corona Borealis (Northern Crown) (*Astron*) A small northern constellation.

corona discharge (*Eng*) A method of etching polymer surfaces (esp polyolefins) by electrical discharge. Used for providing chemically active surface ready for laser printing etc. Other thermoplastics such as acrylonitrile–butadiene–styrene (ABS) can be etched chemically with eg chromic acid.

coronagraph (*Astron*) A type of telescope designed by B Lyot in 1930 for observing and photographing the solar corona and prominences at any time.

coronal section (*BioSci*) A cross-section of the brain, taken effectively where the edge of a crown would touch.

coronal suture (*Med*) The serrated line across the skull separating the frontal bone from the parietal bones.

corona radiata (*BioSci*) A layer of cylindrical cells surrounding the developing ovum in mammals.

coronary (*Med*) Relating to the region around an organ, esp the heart.

coronary bypass (*Med*) Surgical bypassing of a blocked or narrow coronary artery.

coronary circulation (*BioSci*) In vertebrates, the system of blood vessels (coronary arteries) that supplies the muscle of the heart wall with blood.

coronary heart disease (*Med*) All forms of heart disorders arising from disease of the coronary arteries. Includes ANGINA and MYOCARDIAL INFARCTION.

coronary sinus (*Med*) A channel opening into the right atrium draining blood from most of the cardiac veins.

coronary thrombosis (*Med*) The formation of a clot in one of the coronary arteries leading to obstruction of the artery and INFARCTION of the area of the heart supplied by it.

Coronaviridae (*BioSci*) A family of single-stranded RNA viruses responsible for respiratory diseases including SARS. The outer viral envelope has club-shaped projections that give a characteristic corona appearance to negatively stained virions.

corona voltmeter (*ElecEng*) An instrument for measuring high voltages by observing the conditions under which a corona discharge takes place on a specially designed wire.

coronene (*Chem*) $C_{24}H_{12}$. A yellow solid hydrocarbon, mp 430°C. Large planar molecule with seven benzene rings.

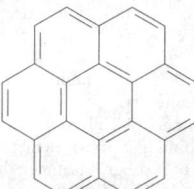

coronene

coronet (*BioSci*) (1) The junction of the skin of the pastern with the horn of the hoof of a horse. (2) The knob at the base of the antler in deer.

coronoid (*BioSci*) (1) In some vertebrates, a membrane bone on the upper side of the lower jaw. (2) More generally, beak-shaped.

co-routine (*ICT*) A kind of SUBROUTINE but whereas a subroutine can only be entered and left in predefined ways, a co-routine can be left and re-entered at any point.

corpora allata (*BioSci*) In insects, endocrine organs behind the brain that secrete neotenin, the JUVENILE HORMONE. In some species they are paired and laterally placed, but in others they fuse during development to form a single median structure, the *corpus allatum*.

corpora bigemina (*BioSci*) In vertebrates, the optic lobes of the brain.

corpora cardiaca (*BioSci*) In insects, paired neurohaemal organs lying behind the brain. They contain the nerve endings of the neurosecretory cells in the brain; these cells release into the blood several hormones including one involved in moulting.

corpora cavernosa (*BioSci*) In mammals, a pair of masses of erectile tissue in the penis.

corpora geniculata (*BioSci*) In the vertebrate brain, paired protuberances lying below and behind the thalamus.

corpora lutea (*BioSci*) See CORPUS LUTEUM.

corpora pedunculata (*BioSci*) In insects, the mushroom or stalked bodies that are the most conspicuous formations in the protocerebral lobes of the brain.

corpora quadrigemina (*BioSci*) The transversely divided optic lobes of the mammalian brain.

cor pulmonale (*Med*) Hypertrophy and failure of the right side of the heart as a consequence of lung disease. In the UK usually as a consequence of chronic bronchitis and emphysema.

corpus (*BioSci*) Inner core of cells, dividing in several planes and distinct from the more superficial TUNICA in a shoot-apical meristem, giving rise to the inner tissues of the shoot. See TUNICA–CORPUS CONCEPT.

corpus adiposum (*BioSci*) See FAT-BODY.

corpus albicans (*BioSci*) See CORPUS MAMILLARE.

corpus callosum (*BioSci*) In the brain of placental mammals, a commissure connecting the cortical layers of the two lobes of the cerebrum.

corpuscle (*BioSci*) A cell that lies freely in a fluid or solid matrix and is not in continuous contact with other cells. But see PACINIAN CORPUSCLE.

corpuscular radiation (*Phys*) A stream of atomic or subatomic particles, which may be charged positively (eg alpha particles), negatively (eg beta particles), or not at all (eg neutrons).

corpuscular theory of light (*Phys*) The view, held by Newton, that the emission of light consisted of the emission of material particles at very high speed. Although this theory was discredited by observations of interference and diffraction phenomena, which could only be explained on the wave theory, the modern concept of wave–particle duality incorporating the photon as the quantum of light maintains aspects of the theory. Cf WAVE THEORY OF LIGHT.

corpus luteum (*BioSci*) The endocrine structure developed in the ovary from a Graafian follicle after extrusion of the ovum, secreting progesterone; the yellow body. Pl *corpora lutea*.

corpus mamillare (*BioSci*) In the brains of higher vertebrates, a protuberance on the floor of the hypothalamic region in which the FORNIX terminates.

corpus spongiosum (*BioSci*) In mammals, one of the masses of erectile tissue composing the penis.

corpus striatum (*BioSci*) In the vertebrate brain, the basal ganglionic part of the wall of each cerebral hemisphere.

corrasion (*Geol*) The work of vertical or lateral cutting performed by a river by virtue of the abrasive power of its load. See RIVERS, GEOLOGICAL WORK OF.

correction for buoyancy (*Phys*) In precision weighing, a correction for the differences in the buoyancy of the air for the body being weighed and for the weights. The correction to be added to the value w of the weights (in grams) is: $1 \cdot 2w(1/D - 1/\delta)$ mg, where D and δ are the densities of the body and of the weights respectively in $g\,cm^{-3}$.

correction of angles (*Surv*) The process of adjusting the observed angles in any triangle so that their sum shall equal 180°.

correlation (*BioSci*) *Mutual relationship*, eg the condition of balance existing between the growth and development of various organs of a plant.

correlation (*Geol*) The linking together of strata of the same age occurring in separate outcrops.

correlation (*MathSci*) (1) The tendency for variation in one variable to be accompanied by linear variation in another. (2) A linear transformation which, in the plane, maps lines into points and points into lines and in space maps points into planes and planes into points.

correlation coefficient (*MathSci*) A dimensionless quantity taking values in the range -1 to 1 measuring the degree of linear association between two variates. A value of -1 indicates a perfect negative linear relationship, 1 a perfect positive relationship.

correlation detection (*ICT*) A method of enhancing weak signals in noise by averaging the product of the received signal and a locally generated signal having some of the known or anticipated properties of the transmitted information. See AUTOCORRELATION, MATCHED FILTER.

correspondence principle (*Phys*) The principle that the predictions of quantum and classical mechanics must correspond in the limit of very large quantum numbers.

corresponding angles (*MathSci*) For a diagram consisting of two straight lines cut by a transversal, the corresponding angles are on the same side of the transversal, each angle being above its appropriate line or each below it. If the two straight lines are parallel then the corresponding angles are equal.

corresponding states (*Phys*) The term applied to substances when their pressures and temperatures are equal fractions of the critical values. A general form of VAN DER WAALS' EQUATION may then be used which is applicable to all gases.

corridor (*Aero*, *Space*) (1) Safe track for intruding aircraft. (2) Path through atmosphere of re-entering aerospace craft above which there is insufficient air density for lift control, and below which kinetic heating is excessive. See panel on ROCKET.

corridor disease (*Vet*) A fatal disease of cattle in Africa due to infection by the protozoon *Theileria lawrencei*; transmitted by ticks.

corrie (*Geol*) See CIRQUE.

corrosion (*Chem*) Chemical degradation of metals and alloys due to reaction with agent(s) in the service environment, eg the rusting of steel in moist air. Eventual failure results from the wasting away of cross-section. May take the form of uniform attack over the whole surface or as highly localized pitting. Sometimes attack may be intergranular and very rapid. See STRESS CORROSION and panel on RUSTING.

corrosion (*Geol*) (1) Petrologically, the modification of crystals formed early in the solidification of an igneous rock by the chemical action of the residual magma. (2) Geomorphologically, erosion by chemical processes.

corrosion embrittlement (*Eng*) Like the effects of severe rusting, the loss of ductility due to corrosion acting between the grains of the material. It is not always readily observed.

corrosion fatigue (*Eng*) Accelerated weakening of a structure exposed to cyclic stress due to both chemical penetration and fatigue. See panel on FATIGUE.

corrosion voltmeter (*ElecEng*) An instrument which locates and estimates corrosion of materials by measuring emf arising from electrochemical action between material and corrosive agent.

corrosive sublimate (*Chem*) Mercuric chloride.

corrugated board (*Paper*) A layered packaging material produced by sticking a suitable liner to both sides of a fluted paper or papers.

corrugated iron (*Build*) Sheet iron, usually *galvanized*, with corrugations for stiffening.

corrupt (*ICT*) To introduce errors into data or programs.

cortex (*BioSci*) (1) Generally, the superficial or outer layers of an organ. Cf MEDULLA. (2) In botany, the tissue (often collenchyma and parenchyma), in a stem or root, between the epidermis and the vascular tissue (ie from hypodermis to endodermis inclusive).

cortical (*BioSci*) (1) Relating to bark. (2) Relating to the CORTEX.

cortical inhibition (*Psych*) The blocking of neural impulses by centres in the cerebral cortex, presumably a higher level control.

cortical microtubules (*BioSci*) Microtubules in the cytoplasm just below the plasmalemma in an interphase plant cell, commonly parallel to and perhaps controlling the shape of the developing cellulose microfibrils in the wall.

corticate (*BioSci*) Possessing or producing a CORTEX.

corticolous (*BioSci*) Living on the surface of bark.

corticosteroids (*Pharmacol*) Steroids secreted by the adrenal cortex or their synthetic analogues. Glucocorticoids, including cortisone and hydrocortisone, act on carbohydrate metabolism; mineralocorticoids such as aldosterone have a primary role in maintaining fluid and electrolyte balance.

corticotrophic (*Med*) Having a stimulatory influence on the adrenal cortex; adrenocorticotrophic. Also *corticotropic*. See ACTH.

corticotrophin (*Med*) See ACTH.

corticotropic (*Med*) See CORTICOTROPHIC.

cortisone (*BioSci*) A crystalline steroid hormone that controls carbohydrate metabolism, isolated from the adrenal cortex. Also *cortisol*. See panel on STEROID HORMONES.

Corti's organ (*BioSci*) In mammals, the modified epithelium forming the auditory apparatus of the ear, in which nerve fibres terminate.

corundum (*Min*) Oxide of aluminium, crystallizing in the trigonal system. It is next to diamond in hardness, and hence is used as an abrasive. The clear-blue variety is *sapphire* and the clear-red *ruby*. See WHITE SAPPHIRE.

Corvus (Crow) (*Astron*) A small northern constellation.

corymb (*BioSci*) A RACEMOSE INFLORESCENCE with the upper flower stalks shorter than the lower so that all the flowers are at approximately the same level.

Corynebacteriaceae (*BioSci*) A family of bacteria belonging to the order *Eubacteriales*, comprising Gram-positive rods. Some pleomorphic species, mainly aerobes, occur in dairy products and the soil. Some species are pathogenic, eg *Erysipelothrix rhusiopathiae* (swine erysipelas), *Corynebacterium diphtheriae* (diphtheria).

coryza (*Med*) See COLD.

coryza (*Vet*) See INFECTIOUS CORYZA, MALIGNANT CATARRHAL FEVER.

cos (*MathSci*) See TRIGONOMETRICAL FUNCTIONS.

cosec, cosecant (*MathSci*) See TRIGONOMETRICAL FUNCTIONS.

cosecant antenna (*ICT*) Antenna comprising a surface so shaped that the radiation pattern is described by a cosecant curve over a wide angle; gives about the same signal strength for near and far sources. Used mainly in navigation radars; another variation is the cosecant-squared antenna.

coset (*MathSci*) A collection of elements in a group formed by combining a fixed element of the group with each element of a subgroup under the group operation.

cosh (*MathSci*) See HYPERBOLIC FUNCTIONS.

cosine law (*Phys*) See LAMBERT'S COSINE LAW.

cosine potentiometer (*ElecEng*) Voltage divider in which the output of an applied direct voltage is proportional to the cosine of the angular displacement of a shaft.

cosine rule (*MathSci*) The theorem that if a, b and c denote the lengths of sides opposite the angles A, B and C of a triangle, then $a^2 = b^2 + c^2 - 2bc \cos A$. When A is a right angle, this reduces to PYTHAGORAS'S THEOREM.

cosmic abundance (*Astron*) The relative proportion of each atomic element found in the universe. The standard values, determined from studies of the solar spectrum and the composition of the Earth, Moon and meteorites, are similar to those found in most stars.

cosmic background radiation (*Astron*) See MICROWAVE BACKGROUND.

cosmic maser (*Astron*) Gas clouds that emit microwaves when excited from radiation from a nearby quasar or bright star. A maser is the microwave equivalent of a laser.

cosmic noise (*ICT*) Interference due to extraterrestrial phenomena, eg sunspots.

cosmic rays (*Astron*) Highly penetrating rays from outer space. See panel on COSMIC RAYS.

cosmic ripple (*Astron*) A slight variation in the temperature of the COSMIC BACKGROUND RADIATION in a certain direction; such variations, discovered in 1992, indicate that the distribution of matter in the early universe on a large scale was slightly uneven. The existence of this 'lumpiness' is thought to have allowed the formation of the galaxies observed today.

cosmic string (*Astron*) A hypothetical supermassive thread-like filament of matter created during the Big Bang. Such strings would have thickness of around 10^{-35} m and a mass density of around 4×10^{22} per metre, and extend infinitely or form closed loops. They may have played a part in the formation of clusters of galaxies during the very early universe.

cosmic year (*Astron*) The orbital period of the Sun around the centre of the Galaxy, equal to around 220 000 000 years.

cosmid (*BioSci*) A type of bacteriophage-lambda CLONING VECTOR that is often used for constructing genomic libraries, because it can carry longer DNA inserts than a PLASMID.

cosmine (*BioSci*) The dentine-like substance forming the outer layer of the COSMOID SCALES of Crossopterygii.

cosmogenic (*Phys*) Said of an isotope capable of being produced by the interaction of cosmic radiation with the atmosphere or the surface of the Earth.

Cosmic rays

Cosmic rays were first detected by V F Hess in 1912 as natural ionizing radiation during a balloon flight. Their astrophysical significance did not become apparent until the 1960s when cosmic-ray detectors could be flown on orbiting satellites. They are extremely energetic particles moving through the universe near to the speed of light, and relativistic effects such as time dilation are therefore important in their study. Their energies are from 10^8 to more than 10^{20} electron-volts (eV) (10^9 eV is the rest mass energy of the proton; around 10^{12} eV is the largest energy attained in particle accelerators).

Cosmic rays are atomic nuclei accelerated to very high energies. Their chemical composition mirrors the cosmic abundance as found in stars like the Sun, although there are small departures at the very highest energies. These observations are very important because cosmic rays are the only particles we can detect that have traversed the Galaxy. Indeed, the ones of the very highest energy may well be coming from quasars and active galactic nuclei, where they were created by unknown processes, probably explosive in nature, that pose a real challenge to modern astrophysics.

Lower-energy cosmic rays (up to 10^{18} eV) are generated by sources within our Galaxy. Almost certainly they originate in supernova explosions, remnants like the Crab Nebula, and pulsars. The energy spectrum of cosmic rays is similar to that of the relativistic electrons that produce the SYN-CHROTRON RADIATION emitted in these objects. These cosmic rays are trapped by the magnetic field of the Galaxy, probably for tens of millions of years. The direction of an incoming cosmic ray, therefore, tells us nothing directly about its origin. Solar flares are a source of the lowest-energy cosmic rays, which increase in intensity at times of solar maximum.

The particles above the atmosphere are known collectively as *primary* cosmic rays and are studied with scintillation counters flown on balloons or spacecraft. The first collision in the atmosphere produces pions which decay rapidly to form gamma rays which in turn produce electrons and positrons via pair production. A single incoming cosmic ray can generate a million secondary particles. This secondary radiation, studied through extensive arrays at ground level, provides most of the knowledge we have about the highest-energy cosmic rays.

See panels on GALAXY, PULSAR, QUASAR, RADIA-TION, SUN AS A STAR; appendices.

---

cosmogonic (*Phys*) Relating to the origin of the universe and therefore to radionuclides surviving from that period. Because of confusion with COSMOGENIC, *primordial* is a better term for such radionuclides.

cosmogony (*Astron*) The science of the origins of stars, planets and satellites. It deals with the genesis of the Galaxy and the solar system.

cosmography (*Genrl*) The science of the constitution of the universe; a description of the world or universe.

cosmoid scale (*BioSci*) In CROSSOPTERYGII, the characteristic type of scale consisting of an outer layer of cosmine, coated externally with vitrodentine, a middle bony vascular layer and an inner isopedin layer. Cf GANOID SCALE.

cosmological constant (*Astron*) A constant introduced by Einstein into his general relativity equations to produce a solution in which the universe would be static; abandoned by him when the universe was later shown to be expanding, but still retained in various theoretical cosmological models with a value close to or equal to zero.

cosmological principle (*Astron*) The postulate, adopted generally in COSMOLOGY (panel), that the universe is uniform, homogeneous and isotropic, ie that it has the same appearance for all observers everywhere in the universe, and there is no preferred position.

cosmological redshift (*Astron*) A REDSHIFT in the spectrum of a galaxy that is caused by the recession velocity associated with the expansion of the universe. See panels on COSMOLOGY and REDSHIFT–DISTANCE RELATION.

cosmology (*Astron*) The study of the universe on the largest scales of length and time, particularly the propounding theories concerning its origin, nature, structure and evolution. See panel on COSMOLOGY.

cosmonaut (*Space*) See ASTRONAUT.

Cosmos (*Space*) General term applied to Soviet satellites used for a variety of missions, eg surveillance, atmospheric research, communications, solar wind studies, testing

propulsion units and military purposes. Cosmos 1 was launched in March 1962 and since then over 2500 Cosmos satellites have been launched; the series continues.

cosmos (*Astron*) The universe as a whole.

cosmotron (*Phys*) Large proton synchrotron using frequency modulation of an electric field; it accelerates protons to energies greater than 3 GeV.

cossyrite (*Min*) See AENIGMATITE.

costa (*BioSci*) (1) In vertebrates, a rib. (2) In insects, one of the primary veins of the wing. (3) In CTENOPHORA, one of the meridional rows of CTENES. (4) More generally, any rib-like structure. Adjs *costal, costate*.

cost–benefit analysis (*BioSci*) An assessment of the relative costs (in terms of the necessary investment of carbohydrate or nitrogen etc) and benefits (in terms of enhanced photosynthesis, reduced losses to herbivores, increased probability of establishment of offspring, etc), and hence the likely selective advantage, of any observed or imagined morphological or physiological variation, such as hairier leaves, synthesis of novel toxin, larger seeds, etc.

costeaning (*MinExt*) Prospecting by shallow pits or trenches designed to expose lode outcrop.

cos θ (*ElecEng*) An expression often used to denote the power factor of a circuit, the power factor being equal to the cosine of the angle (θ) of the phase difference between the current and voltage in the circuit.

cot, cotangent (*MathSci*) See TRIGONOMETRICAL FUNCTIONS.

Cot curve (*BioSci*) A plot of concentration against time for the renaturation of DNA; gives a measure of the number of different sequences present.

cot death (*Med*) See SUDDEN INFANT DEATH SYNDROME.

coterminal angles (*MathSci*) Two angles having the same vertex and the same initial line and whose terminal lines are coincident, eg $60°$ and $420°$.

coterminous (*BioSci*) Of similar distribution.

Cosmology

Cosmology builds models to try and describe what the universe is 'really' like. In historical times, folklore, theology or philosophy provided acceptable cosmologies. Modern cosmology combines astronomical observations with the general theory of relativity to give a mathematical picture of the structure and evolution of the universe. There are two major classes of cosmological model: those in which the universe evolves, so that its appearance will change with time; and those in which the universe remains for ever as a static entity. The first class has given rise to BIG BANG cosmology and the second to steady-state cosmology. Currently almost all astrophysicists choose the Big Bang picture to interpret their results, but steady-state theories were once widely accepted.

Steady-state cosmology

Einstein was the first steady-state cosmologist. Although the general theory of relativity predicted a dynamic universe because of the interplay of gravitational forces, Einstein inserted a completely arbitrary cosmological repulsion factor into the field equations in order to stabilize the universe. Once the REDSHIFT–DISTANCE RELATION was established (1929) this idea was dropped and static models of the universe faded into the background.

In 1948, Herman Bondi and Thomas Gold revived the idea of a steady-state universe by postulating that matter is continuously created, in the expanding universe, to fill the voids left as the galaxies receded from each other. This notion of spontaneous creation was not new: it was suggested by James Jeans (1929), and Pascual Jordan showed (1939) how to modify GENERAL RELATIVITY so that matter could be created. Fred Hoyle extended this work by constructing a model in which the creation rate matches the dilution caused by expansion. The actual form assumed by the newly created matter does not emerge from the theory.

The steady-state theory takes the view that the universe is infinitely old and will continue into the infinite future. The continuous creation of new matter everywhere means that this expanding universe looks more or less the same from all vantage points and at all times. New galaxies form alongside older ones. This has a crucial consequence which is that the universe far away should be more or less like the local universe and thus provides the key to test the theory. The great controversy in the 1960s between Hoyle and radio-astronomer Martin Ryle over the interpretation of the number of galaxies seen at different distances led to considerable public awareness of these rival theories of cosmology. Moreover the discovery of cosmic background radiation or MICROWAVE BACKGROUND, whose presence contradicts the steady-state theory, has not yet eliminated the attractions of the latter in the public mind.

An important outcome of the research on the steady-state theory is frequently overlooked. In the absence of a big bang, the only place where elements heavier than helium can be manufactured is inside stars. Hoyle's brilliant work on NUCLEOSYNTHESIS, motivated by the need to explain the origin of the elements in the steady-state universe, has survived intact as a major contribution to modern astrophysics.

---

coth (*MathSci*) See HYPERBOLIC FUNCTIONS.

co-translational transport (*BioSci*) A process that occurs in eukaryotic cells whereby a protein undergoing synthesis is moved across a cell membrane to be secreted at the same time as it is being translated in the membrane-associated ribosome.

cotransport (*BioSci*) The ACTIVE TRANSPORT of a solute that is driven by a concentration gradient of some other solute, usually an ion; eg the entry of amino acids into animal cells depends upon a sodium ion gradient across the PLASMA MEMBRANE.

co-trimoxazole (*Pharmacol*) A mixture of trimethoprin, which is antibacterial, and sulphamethoxazole, a sulfonamide antibiotic, used to treat *Pneumocystis carinii* pneumonia, toxoplasmosis and nocardiasis.

cotter (*Eng*) A tapered wedge, usually of rectangular section, passing through a slot or cotter way in one member and bearing against the end of a second encircling member whose axial position is to be fixed or adjustable.

cotter pin (*Eng*) A split pin inserted in a hole in a cotter or other part, to prevent loosening under vibration.

cotton (*Textiles*) The seed hairs of the cotton plant of which there are many varieties (*Gossypium* spp). The fibres, consisting of long cellulose chains, vary in length according to variety and country of origin but on average are about 2 to 3 cm. Sea Island and Egyptian cottons are longer and are used for making high-quality fine fabrics and sewing threads.

Cotton balance (*ElecEng*) An instrument for measuring the intensity of a magnetic field by finding the vertical force on a current-carrying wire placed at right angles to the field.

cotton ball (*Min*) See ULEXITE.

cotton linters (*Textiles*) See LINTERS.

Cotton–Mouton effect (*Phys*) The effect occurring when a dielectric becomes double-refracting on being placed in a magnetic field H. The retardation δ of the ordinary over the extraordinary ray in traversing a distance l in the dielectric is given by $\delta = C_m \lambda l H^2$ where λ is the light wavelength and C_m is the Cotton–Mouton constant.

cottonseed oil (*Chem*) Oil from the seeds of *Gossypium heraceum*, a yellow, brown or dark-red liquid, mp 34–40°C, rel.d. 0·922–0·930, saponification value 191–196, iodine value 105–114, acid value 0. Used in manufacture of soaps, fats, margarine, etc.

cottonwood (*For*) A wide-girthed, N American species of POPLAR.

cotton wool (*Med*) Loose cotton or vicose rayon fibres which have been bleached and pressed into a sheet; used as an absorbent or as a protective agent. Medicated cotton wool sometimes has a distinguishing colour to indicate its special property.

Cosmology *(Cont.)*

Big Bang cosmology

Like steady-state cosmology the Big Bang theory is firmly grounded in the general theory of relativity and its account of gravitation and the four-dimensional structure of space–time (cf panel on BLACK HOLE). It is underpinned by two observations. The first, dating from the 1920s, is the HUBBLE RELATION which showed that galaxies are receding from us with velocities that increase with distance. The second is the microwave background, a nearly uniform radio emission with a thermal temperature of 2·7 K that covers the whole sky. Both observations are consistent with an explosive and intensely hot origin for the universe. The galaxies continue to rush away from the explosion into space and the background radiation is a fossil relic of the intense heat.

The *standard model* is extremely detailed and has been thoroughly investigated through physical laws. It is much more than picturesque description. The key feature is that as we extrapolate our understanding of the universe back in time, we encounter an instant in the remote past when all matter was enormously dense. Properly termed a singularity, this is also known as the *primeval atom*. The event was about 10–20 billion years ago and the universe has expanded ever since.

This standard model cannot explain everything. It cannot explain why the universe has the density we observe, nor can it explain how galaxies were formed. It does not account for the extraordinary homogeneity on the large scale which we observe. These difficulties are circumvented in the INFLATIONARY UNIVERSE

model developed since 1980. This remarkable synthesis of elementary particle physics and cosmology pushes the laws of physics back to 10^{-35} seconds and describes a phase change during which the universe expanded by 10^{75} or so. This very successful model implies that we can only ever see a tiny fraction of the physical universe.

In the earliest phase of the Big Bang, after 10^{-35} seconds a slight imbalance between quarks and antiquarks, which together with leptons and antileptons, dominated the universe, resulted in the annihilation of most antiparticles while creating an immense amount of radiation or photons. By 10^{-10} s all four fundamental forces had become distinct: the ELECTROMAGNETIC INTERACTION, GRAVITY, STRONG INTERACTION, WEAK INTERACTION. After a microsecond (10^{-6} s) the universe had cooled sufficiently for the quarks to combine to hadrons, ie protons and neutrons. After about 1 second photons dominated the universe for the next million years, the radiation era, but in the first few minutes of this era, interactions between protons and electrons led to the synthesis of helium nuclei, which accounted for 25% of the matter, with 75% remaining as hydrogen. (All other elements were formed much later in exploding stars.) After this million years, the temperature had fallen sufficiently to allow electrons to combine with nuclei and form stable atoms. During this recombination era the universe became transparent to radiation for the first time. The background radiation comes from this horizon.

See panel on RADIATION; appendices.

---

cotton-wool patches (*Med*) Areas of white exudate in the retina.

Cottrell precipitator (*Eng*) A system used to remove dust from process gases electrostatically.

cotyledon (*BioSci*) The first leaf or one of the first leaves of the embryo of a seed plant; typically one in monocotyledons, two in dicotyledons, two to many in gymnosperms. In non-endospermic seed, eg pea, the cotyledons may act as storage organs. Also *seed leaf*. See ENDOSPERM, EPIGEAL, HYPOGEAL.

cotyledonary placentation (*BioSci*) Having the villi of the placenta in patches, as in ruminants.

cotyloid (*BioSci*) Cup-shaped; pertaining to the acetabular cavity. In some mammals, a small bone bounding part of the acetabular cavity.

cotype (*BioSci*) An additional type specimen, being a brother or sister of the same brood as the type specimen.

couch (*Paper*) To separate the newly formed wet sheet or web from the forming surface and transfer it to a felt.

couching (*Med*) Displacement of the lens in the treatment of cataract.

couch roll (*Paper*) A roll which performs the action of couching.

coudé telescope (*Astron*) An arrangement by which the image in an equatorial telescope is formed, after an extra reflection, at a point on the polar axis. It is then viewed by

a fixed eyepiece looking either down or up the polar axis. This type of mounting is much used for high-dispersion spectroscopy with modern large telescopes. Also *coudé mounting*. See panel on ASTRONOMICAL TELESCOPE.

coulomb (*Phys*) SI unit of electric charge, defined as the charge which is transported when a current of 1 ampere flows for 1 second. Symbol C.

coulomb energy (*Phys*) The fraction of binding energy arising from simple electrostatic forces between electrons and ions.

coulomb force (*Phys*) Electrostatic attraction or repulsion between two charged particles.

coulomb potential (*Phys*) A potential calculated from Coulomb's inverse-square law and from known values of electric charge. The term is used particularly in nuclear physics to indicate the component of the potential energy of a particle which varies with position as a consequence of an inverse-square law of force of YUKAWA POTENTIAL.

coulomb scattering (*Phys*) The scattering of particles by action of coulomb force.

Coulomb's law (*Phys*) Fundamental law which states that the electric force of attraction or repulsion between two point charges is proportional to the product of the charges and inversely proportional to the square of the distance between them. The force also depends on the PERMITTIVITY of the medium in which the charges are placed. In SI units, if Q_1 and Q_2 are point charges a distance d apart, the force is

$$F = \frac{1}{4\pi\varepsilon}\frac{Q_1 Q_2}{d^2}$$

where ε is the permittivity of the medium. The force is attractive for charges of opposite sign and repulsive for charges of the same sign.

Coulomb's law for magnetism (*Phys*) The law which states that the force between two isolated point magnetic poles (theoretical abstractions) is proportional to the product of their strengths and inversely proportional to the square of their distance apart times the permeability of the medium between them, ie F is proportional to

$$F \propto \frac{M_1 M_2}{\mu d^2}$$

where M_1 and M_2 are the strengths of the two poles, d is their distance apart and μ is the relative permeability of the medium.

coulometer (*ElecEng*) Voltameter or electrolytic cell, designed eg for use in measurement of the quantity of electricity passed.

Coulter counter (*BioSci*) TN for an instrument that counts cells by drawing a suspension through an orifice and measuring the change in capacitance as each cell passes.

coumachlor (*Chem*) 3-(a-acetonyl-4-chlorobenzyl)-4-hydroxy-coumarin. A blood anticoagulant type of rodenticide.

coumalic acid (*Chem*) *Pyrone-5-carboxylic acid*, formed by the action of concentrated sulphuric acid on malic acid; mp 206°C. Pyrone is *coumalin*.

coumaric acid (*Chem*) *Hydroxycinnamic acid*. $HOC_6H_4CH{=}CHCOOH$.

coumarin (*Chem*) Odoriferous principle of tonquin beans and woodruff, $C_9H_6O_2$, bp 200°C; used for scenting tobacco.

coumarone (*Chem*) The condensation product of a benzene nucleus with a furan ring. It is a very stable, inert compound, bp 169°C; found in coaltar. Strong acids effect polymerization into *para*-coumarone and coumarone resins.

coumarone resins (*Chem*) Condensation and polymerization products obtained from *coumarone*; used for varnishes, in printing ink, and as plasticizers for moulding powders. They are neutral and acid- and alkali-resisting.

count (*NucEng*) Summation of photons or ionized particles, detected by a counting tube, which passes pulses to counting circuits.

count (*Textiles*) See COUNT OF YARN.

countable set (*MathSci*) See DENUMERABLE SET.

count down (*Space*) A sequence of events in the preparation for the firing of a launch system, denoted by counting time backwards towards zero where zero represents lift-off. The count starts some hours before launch and is finally reckoned in minutes and seconds.

counter (*ElecEng*) (1) Circuit in which a free-running oscillator of known frequency increments a numerical output at regular intervals. Can be used to determine the time between two events by indicating the number of counts which have occurred. (2) Circuit for registering the number of events which have occurred in an external piece of apparatus or circuit. (3) Part of an integrating electricity meter which shows the number of revolutions made by the spindle of the meter which is proportional to the amount of energy which has passed through the circuit. Also *clock*.

counter (*Eng*) An instrument for recording the number of operations performed by a machine, or the revolutions of a shaft.

counter (*NucEng*) Device for detecting ionizing radiation by electric discharge resulting from TOWNSEND AVALANCHE and operating in proportional or GEIGER REGION. Also *counting tube*.

counter (*Ships*) A description applied to a form of ship's stern, implying an overhung portion of deck, abaft the stern post; hence the term 'under the counter'.

counter-arched (*CivEng*) Said of a revetment having arches turned between COUNTERFORTS.

counterboring (*Eng*) The operation of boring the end of a hole to a larger diameter.

counterbracing (*Eng*) The provision of two diagonal tie rods in the panels of a frame girder or other structure. Also *cross-bracing*.

counter-conditioning (*Psych*) A procedure for weakening a classically conditioned response by associating the conditioning stimuli that evoke it to a new and incompatible response.

countercurrent contact (*ChemEng*) In processes involving the transfer of heat or mass between two streams A and B, as in liquid extraction, the arrangement of flow (*countercurrent flow*) so that at all stages, the more spent A contacts the less spent B, thus ensuring a more even distribution and greater economy than with CO-CURRENT CONTACT.

countercurrent distribution (*Chem*) A repetitive distribution of a mixture of solutes between two immiscible solvents in a series of vessels in which the two solvent phases are in contact. The components are distributed in the vessels according to their partition coefficients.

countercurrent treatment (*MinExt*) Arrangement used in chemical extraction of values from ore, in washing rich liquor away from spent sands. The stripping liquid enters 'barren' at one end of a typical layout and the rich ore pulp at the other. They pass countercurrent through a series of vessels, the pulp emerging stripped after its final wash with the new 'barren' liquor and the liquor leaving at the far end, now rich with dissolved values and 'pregnant'.

counter efficiency (*NucEng*) The ratio of counts recorded by a counter to the number of particles or photons reaching the detector. Counts may be lost owing to (1) absorption in the window, (2) passage through the detector without initiating ionization, or (3) passage through the detector during dead time that follows the previous count while the instrument recovers.

counter emf (*ElecEng*) See BACK EMF.

counter-flap hinge (*Build*) A hinge which is arranged, by the provision of separate centres of rotation for each leaf, so that it may fold back to back.

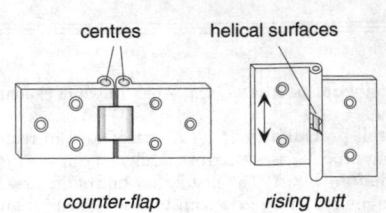

centres   helical surfaces

counter-flap   rising butt

counter-flap hinge And rising butt hinge.

counter-floor (*Build*) An inferior floor laid as a base for a better surface (eg parquet). Also *subfloor*.

counterflow jet condenser (*Eng*) A JET CONDENSER in which the exhaust steam and air flow upwards in the opposite direction to that of the descending spray of cooling water.

counterfort (*CivEng*) A buttress giving lateral support to a retaining wall, to which it is bonded.

counterglow (*Astron*) See GEGENSCHEIN.

counterions (*Chem*) See GEGENIONS.

counter lathing (*Build*) See BRANDERING.

counter life (*NucEng*) The total number of counts a nuclear counter can be expected to make without serious deterioration of efficiency.

counterpoise (*ICT*) A network of conductors placed a short distance above the surface of the ground but insulated from it, and used for the earth connection of an antenna. It wil have a large capacity to earth and serves to reduce greatly the earth current losses that would otherwise take place. Also *artificial earth, capacity earth, counterpoise antenna*.

counter range (*NucEng*) See START-UP PROCEDURE.

countershading (*BioSci*) A type of protective coloration in which animals are darker on their dorsal surface than on their ventral surface, thus ensuring that illumination from above renders them evenly coloured and inconspicuous.

countershaft (*Eng*) An intermediate shaft interposed between driving and driven shafts in a belt or gear drive, either to obtain a larger speed ratio or where direct connection is impossible.

countersinking (*Build, Eng*) The driving of the head of a screw, nail or rivet below the surface, often in a preformed recess. It may be hidden by a plug. See COUNTERBORING.

counterstain (*BioSci*) A stain used in histology to produce a contrasting background in the section or specimen.

counter-stern (*Ships*) A type of ship's stern construction. It is virtually an excrescence to the main hull, and is not waterborne.

countersunk head (*Eng*) A screw or rivet head with a conical base, allowing it to enter the countersunk work-piece so that the top surface of the head is substantially flush with the workpiece surface.

counter-transference (*Psych*) In psychoanalytic theory, the analyst's emotional response to the client, often involving personal and unconscious feeling projected onto the client. See PROJECTION.

counter-vault (*CivEng*) An inverted arch.

countess (*Build*) A roofing slate, 20×10 in (508 × 254 mm).

counting chain (*Phys*) A system for the detection and recording of ionizing radiation. Consists essentially of a detector, linear amplifier, pulse-height analyser and a device to display or record the counts.

counting machine (*Paper*) A piece of equipment that records the number of sheets cut in a sheetcutter or stacked in a pile. Generally also capable of inserting a tab at the chosen sheet count.

counting tube (*NucEng*) See COUNTER.

count of yarn (*Textiles*) A number which designates the size of a yarn. Historically this has been the number of lengths per unit mass with many local variations for the units (see DENIER). Now standardized on SI units of the TEX system which uses the mass per unit length.

count ratemeter (*NucEng*) A ratemeter which gives a continuous indication of the rate of count of ionizing radiation, eg for radiation surveying.

country code (*ICT*) (1) A routing code used in an international VIRTUAL PRIVATE NETWORK which specifies the next country to which a call is to be routed. Unlike the DESTINATION NETWORK IDENTIFIER, it may be changed by an international gateway switch if traffic conditions so require. (2) The country identifier added to a domestic telephone number when it is called from outside that country.

country rock (*MinExt*) The valueless rock forming the walls of a REEF or LODE.

counts (*NucEng, Radiol*) The disintegrations that a radio-nuclide detector records.

coupe (*For*) A felling area, usually one of an annual succession unless otherwise stated.

couple (*Eng*) A system of two equal but oppositely directed parallel forces. The perpendicular distance between the two forces is called its *arm* and any line perpendicular to the plane of the two forces is *axis*. The *moment* of a couple is the product of the magnitude of one of its two forces and its arm. A couple can be regarded as a single statical element (analogous to force); it is then uniquely specified by a vector along its axis having a magnitude equal to its moment. Couples as so specified combine in accordance with the PARALLELOGRAM RULE FOR ADDITION OF VECTORS.

couple-close roof (*Build*) A roof-form derived from the COUPLE ROOF by connecting the lower ends of the two rafters together with a tie, so as to prevent spreading of the roof under load.

coupled control (*NucEng*) The state of a reactor where the output inherently follows the load, as in pressurized-water reactors.

coupled flutter (*Aero*) See FLUTTER.

coupled oscillator (*Electronics*) A circuit in which positive feedback from the output circuit to input circuit of an amplifier by mutual inductance is sufficient to initiate or maintain oscillation.

coupled rangefinder (*ImageTech*) A rangefinder co-ordinated with the focusing mechanism of a camera lens.

coupled switches (*ElecEng*) See LINKED SWITCHES.

coupled vibrations (*Phys*) Vibrations executed by two or more vibrating systems connected in some manner so that energy can be transferred from one to another. The resultant complicated motion can be analysed in terms of a linear combination of the NORMAL MODES of the system.

coupled wheels (*Eng*) The wheels of a locomotive which are connected by coupling rods to distribute the driving effort over more than one pair of wheels.

couple roof (*Build*) A roof composed of two rafters not braced together.

couplers (*ImageTech*) Compounds included in photo-graphic emulsions or processing solutions which form coloured dye images associated with the developed silver image by the oxidation products of the reduction.

coupling (*BioSci*) When two specified *non-allelic* genes are on the same chromosome, having come from the same parent, they are in *coupling*. Cf REPULSION.

coupling (*Build*) (1) A short collar screwed internally at each end to receive the ends of two pipes which are to be joined together. (2) A CAPILLARY FITTING for the same purpose. Also *union*.

coupling (*Eng*) (1) A device for connecting two lengths of hose etc. (2) A device for connecting two vehicles. (3) A connection between two coaxial shafts, conveying a drive from one to the other.

coupling (*ICT*) The interdependence of two or more components (whether of hardware or software) such that one cannot function correctly without the presence of the other(s).

coupling (*MinExt*) A short tube internally threaded at both ends for joining two lengths of drilling tube. See SLIP JOINT.

coupling capacitor (*Electronics*) Any capacitor for coupling two circuits, eg an antenna to a transmitter or receiver, or one amplifying stage to another.

coupling coefficient (*ICT*) Ratio of total effective positive (or negative) impedance common to two resonant circuits to geometric mean of total positive (or negative) reactances of two separate circuits. Also *coupling factor*.

coupling coil (*ICT*) One whose inductance is a small fraction of the total for circuit of which it forms a part; used for inductive transfer of energy to or from the circuit.

coupling element (*ICT*) The component through which energy is transferred in a coupled system.

coupling factor (*ICT*) See COUPLING COEFFICIENT.

coupling factors (*BioSci*) Proteins of the mitochondrial inner membrane, essential for the coupling of the passage of electrons along the electron transport chain with the synthesis of ATP.

coupling loop (*ICT*) A loop placed in a waveguide at a position of maximum magnetic field strength in order to extract energy.

coupling probe (*ICT*) A probe placed in a waveguide at a position of maximum electric field strength in order to extract energy.

coupling resistance (*ICT*) Common resistance between two circuits for transference of energy from one circuit to the other.

coupling transformer (*ElecEng*) A transformer used as a coupling element.

coupon (*Eng*) An extra piece, attached to a forging or casting, from which a test piece can be prepared.

courbaril (*For*) A hardwood (*Hymenaea courbaril*) from the American tropics whose heartwood is reddish- to orange-brown with darker brown and red streaks and a medium texture.

course (*Build*) A horizontal layer of bricks or building stones running throughout the length and breadth of a wall.

course (*Ships*) The angle between some datum line and the direction of the ship's head.

course (*Surv*) The known length and bearing of a survey line.

course (*Textiles*) A row of loops across the width of a knitted fabric.

course correction (*Space*) The firing or burning of a rocket motor, during a coasting flight, in a controlled direction and for a controlled duration to correct an error in course.

course density (*Textiles*) The number of courses per cm measured along a WALE of the knitted fabric.

course length (*Textiles*) The length of yarn in one course of a knitted fabric.

coursing joint (*Build*) The mortar joint between adjacent courses of brick or stone. Also *bed joint*.

Courtelle (*Chem*) TN for synthetic fibres based on POLYACRYLONITRILE.

courtship behaviour (*Psych*) A wide range of behaviours throughout the animal kingdom, often very conspicuous, leading to copulation and rearing of young.

coutil (*Textiles*) Strong, closely woven, herringbone cloth, often of cotton and used for corsetry.

covalency (*Chem*) The union of two or more atoms by the sharing of one or more pairs of electrons.

covalent bond (*Chem*) A chemical bond in which two or more atoms are held together by the interaction of their outer electron clouds. A *co-ordinate bond* involves one atom donating a spare pair of electrons to form a covalent bond. See panels on BONDING and INTRINSIC AND EXTRINSIC SILICON.

covalent radius (*Chem*) Half the internuclear separation of two bonded like atoms. For bonds between unlike atoms, approximate BOND LENGTHS may be derived from the sum of the covalent radii for the two bonded atoms.

covariance (*MathSci*) A measure of the linear dependence between two random variables, equal to the expected value of the product of their deviations from the mean.

cove (*Build*) A hollow cornice, usually large.

coved ceiling (*Arch*) A ceiling which is formed at the edges to give a hollow curve from wall to ceiling, instead of a sharp angle of intersection.

covellite (*Min*) Sulphide of copper, CuS. The colour is indigo-blue or darker. Also *covelline, indigo copper*.

cover (*BioSci*) The percentage of the ground surface covered by a plant species.

cover (*Build*) In coursed work, the hidden or covered width of a slate or tile.

cover (*CivEng*) The thickness of concrete between the outer surface of any reinforcement and the nearest surface of the concrete. See EFFECTIVE DEPTH.

coverage (*ImageTech*) The area over which a lens can give a sharply focused image. Also *covering power*.

covered electrode (*ElecEng*) A metal electrode covered with a coating of flux; used in arc welding.

cover flashing (*Build*) A separate flashing fastened into the upright surface and overlapping the flashing in the angle between the surfaces. See FLASHINGS.

cover glass (*ImageTech*) A square of thin glass mounted to protect a photographic transparency when used as a SLIDE.

covering power (*ImageTech*) See COVERAGE.

cover iron (*Build*) See BACK IRON.

cover paper (*Paper*) A heavy paper or board, generally of distinctive appearance, eg coloured and/or embossed, and intended for use as the cover of booklets, pamphlets, menus, etc.

coverslip (*BioSci*) The thin slip of glass used for covering a specimen that is being observed under a microscope. Essential for all but the lowest magnifications.

coverts (*BioSci*) See TECTRICES.

cover unit (*Print*) A separate printing unit coupled to the main press for producing the cover of a magazine, journal or paperback.

coving (*Build*) (1) The upright splayed side of a fireplace opening. (2) The projection of upper over lower storeys .

co-volume (*Chem*) The correction term b in van der Waals' equation of state, denoting the effective volume of the gas molecules. It is approximately equal to four times the actual volume of the molecules.

cow (*Agri*) The mature female of larger mammals. In cattle, usually indicates a mature animal having had at least one calf.

cow-hocked (*Vet*) Said of horses whose hocks are abnormally close to each other.

cowl (*Build*) A cover, frequently louvred and either fixed or revolving, fitted to the top of a chimney to prevent downdraught.

Cowles dissolver (*Chem*) A type of blender for polymers using a chamber enclosing a rotating impeller blade.

cowl flaps (*Aero*) See GILLS.

cowling (*Aero*) The whole or part of the streamlining covering of any aero-engine; in air-cooled engines, designed to assist cooling airflow.

Cowper's glands (*BioSci*) In male mammals, paired glands whose ducts open into the urethra near the base of the penis.

cow-pox (*Med, Vet*) See VACCINIA.

COX-1, COX-2 (*Pharmacol*) Isoforms of cyclo-oxygenase, a key enzyme in prostaglandin synthesis. Aspirin and many NSAIDS inhibit both isoforms but selective COX-2 inhibitors such as *celecoxib* have the analgesic and anti-inflammatory activity without deleterious effects on the gastric mucosa. The selective COX-2 inhibitors are not, however, entirely without side effects. Can also be used for exacerbations of COPD (chronic obstructive pulmonary disease) and UT (urinary tract) infections, and otitis media in children.

coxa (*BioSci*) In insects, the proximal joint of the leg. Adj *coxal*.

coxalgia (*Med*) Pain in the hip.

coxa valga (*Med*) A deformity of the hip in which the angle between the neck and the shaft of the femur exceeds 140°.

coxa vara (*Med*) A deformity of the hip in which the angle between the neck and the shaft of the femur is less than 120°.

coxib (*Pharmacol*) Any of a group of non-steroidal anti-inflammatory drugs, inhibitors of cyclo-oxygenase 2 (see COX-1, COX-2), used in the treatment of arthritis, eg *celecoxib*.

Coxsackie viruses (*Med*) Species of enteroviruses of the PICORNAVIRIDAE first isolated in Coxsackie, New York. Responsible for various human diseases.

CP (*Chem*) Abbrev for *chemically pure*, a grade of chemical reagent for general laboratory use; less pure than *AR* (analytic reagent).

CP (*Surv*) Abbrev for CHANGE POINT.

cP (*Phys*) Abbrev for CENTIPOISE.

cp (*Aero*) Abbrev for CENTRE OF PRESSURE.

CP filter (*ImageTech*) Abbrev for *colour printing filter*. One of a series of gelatine filters made with specific transmission values for the three primary colours, used for the correction of printing exposure by small increments of colour.

cpi (*ICT*) Abbrev for *characters per inch*.

CPL (*Aero*) Abbrev for *commercial pilot's licence*.

CPM (*Eng*) Abbrev for *critical path method*. See CRITICAL PATH PLANNING.

CP/M (*ICT*) Abbrev for *control program for microcomputers*. TN given to a once commonly used operating system for microcomputers based on the Z80 microprocessor chip. Cf MS-DOS.

cps (*ICT*) Abbrev for CHARACTERS PER SECOND.

cps (*Phys*) Abbrev for *cycles per second*, superseded by hertz (Hz). Also *CPS, c/s*.

CPU (*ICT*) Abbrev for CENTRAL PROCESSOR UNIT.

CPVC (*Chem*) Abbrev for CHLORINATED POLYVINYL CHLORIDE.

CP violation (*Phys*) The violation of a fundamental symmetry principle, observed in the decay of neutral kaons. C represents *charge conjugation* (the operation of turning a particle into its antiparticle) and P represents parity (the operation of changing left- to right-hand co-ordinates).

CR (*Chem*) Abbrev for CHLOROPRENE RUBBER.

CR (*ICT*) Abbrev for CARRIAGE RETURN.

CR1, CR2, CR3 (*BioSci*) Cell surface receptors for COMPLEMENT factor C3b and its decay products.

CR 39 (*Chem*) Abbrev for *Columbia Resin 39*. TN for tough, aliphatic polycarbonate used mainly for spectacle lenses and sunglasses. Thermoset based on allylic resin with high scratch resistance.

crab (*ImageTech*) Movement of a camera sideways, at right angles to its optical axis.

Crab Nebula (*Astron*) An expanding nebulosity in Taurus which represents the remains of the supernova of 1054. It is a powerful source of radio waves and of X-rays. The nebulosity arises from a faint star at the centre, which is a rapid pulsar with a period of 0·033 s. Both the X-ray and optical radiation show the same pulse, the period of which is slowly increasing.

crabwood (*For*) See ANDIROBA.

crack detector (*ElecEng*) An electromagnetic device for detecting flaws by the gathering of fine magnetic powder along the flaw lines in an iron specimen when magnetized, or by the reflection of ultrasonic waves.

crack driving force (*Eng*) Also TOUGHNESS or *critical strain energy release rate*. See STRENGTH MEASURES.

cracked heels (*Vet*) A necrobacillosis of horses' heels due to infection with *Fusiformis necrophorus*, the organism gaining entry through cuts and abrasions. Also *grease*.

cracker (*ICT*) Someone who illegally breaks into a computer system.

cracking (*MinExt*) Controlled breakdown of naphtha to give light olefines such as ethylene, propylene, butylenes by heat and pressure (thermal cracking). Such products can be polymerized after purification to give polymeric materials. Term also used for catalysed process (catalytic cracking), which gives different yield. Ethane, propane, etc, from liquefied petroleum gas and natural gas can be similarly processed to give monomers.

crackled (*Glass*) Said of glassware whose surface has been intentionally cracked by water immersion and partially healed by reheating before final shaping.

crack stopper (*Aero*) In structural design, a means of reducing the progression of potential cracks by placing adjacent components across the likely direction of the crack.

cradle (*Build*) (1) A frame of laths on which scrim is stretched to receive plaster in forming coved or other heavy cornices etc. (2) See CRADLE SCAFFOLD.

cradle (*ElecEng*) An earthed metal net placed below a high-voltage overhead transmission line where it crosses a public highway, railway or telephone circuit; a conductor, if broken, falls on the net and is earthed without further damage. Not now in general use except in construction. Also *catch net*.

cradle (*Vet*) A frame encircling the neck of a horse; used as a means of restraint.

cradle scaffold (*Build*) A form of suspended scaffolding consisting of a strong framework fitted with guard rails and boards for the working platform, and slung from two fixed points or from a wire rope secured between two jibs. Also *boat scaffold*.

cradling (*Build*) Rough timber fixings as grounds around steelwork.

cradling piece (*Build*) (1) A short timber fixed at each side of a fireplace hearth, between a chimneybreast and trimmer, to support the ends of floorboards. (2) See CRADLING.

crag (*Geol*) A rough, steep, precipitous projecting rock. Also *craig* (Scottish).

crag-and-tail (*Geol*) A hill consisting partly of solid rock shaped by ice action, with a *tail* of morainic material banked against it on the lee-side, eg the Castle Rock, Edinburgh.

cramp (*Build*) See SASH CRAMP.

cramp (*Med*) Painful spasm of muscle.

cramp-iron (*Build*) See SASH CRAMP.

crampon (*Build*) An appliance for holding stones or other heavy objects which are to be hoisted by crane. It consists of a pair of bars hinged together like scissors, the points of which are bent inwards for gripping the load, while the handles are connected by short lengths of chain to a common hoist-ring. Also *crampoon*.

Crampton's muscle (*BioSci*) In birds, a muscle of the eye that, by its contraction, decreases the diameter of the eyeball and so aids the eye to focus objects near to it.

crane barge (*MinExt*) In offshore drilling, a special vessel for handling heavy loads in the supply and maintenance of drilling platforms.

crane magnet (*ElecEng*) A magnet (normally electromagnet) used for lifting eg scrap metal.

crane motor (*ElecEng*) A motor specially designed for the operation of a crane or hoist. It should be very robust and have a high starting torque.

crane post (*Eng*) The vertical member of a JIB CRANE to the top of which the jib is connected by a tie rod.

crane rating (*ElecEng*) A term sometimes employed to denote a method of specifying the rating of a motor for intermittent load, such as that of a crane. The maximum power and the load factor are stated.

cranial flexures (*BioSci*) Flexures of the brain in relation to the main axis of the spinal cord, transitory in lower vertebrates, permanent in higher vertebrates. See NUCHAL FLEXURE, PONTAL FLEXURE, PRIMARY FLEXURE.

cranial nerves (*BioSci*) Any of the ten to twelve paired nerves that have their origin in the brain of vertebrates.

Craniata (*BioSci*) See VERTEBRATA.

cranioclasis (*Med*) The crushing of the fetal skull in obstructed labour using a special instrument (a *cranioclast*).

craniosacral system (*BioSci*) See PARASYMPATHETIC NERVOUS SYSTEM.

craniotomy (*Med*) Incision of the fetal skull and removal of its contents in obstructed labour.

cranium (*BioSci*) That part of the skull which encloses and protects the brain; the braincase. Adj *cranial*.

crank (*Eng*) An arm attached to a shaft, carrying at its outer end a pin parallel to the shaft; used either to give reciprocating motion to a member attached to the pin, or in order to transform such motion into rotary motion of the shaft.

crank-brace (*Build*) A brace having a bent handle by which it may be rotated.

crankcase (*Eng*) A box-like casing, usually cast-iron or aluminium, which encloses the crankshaft and connecting-rods of some types of reciprocating engines, air-compressors, etc.

crank effort (*Eng*) The effective force acting on the crank pin of an engine in a direction tangential to the circular path of the pin.

crank pin (*Eng*) The pin which is fitted into the web or arm of a crank, and to which a reciprocating member or connecting-rod is attached.

crankshaft (*Eng*) The main shaft of an engine or other machine which carries a crank or cranks for the attachment of connecting-rods.

crankshaft motion (*Chem*) A form of short-chain movement in polyethylene polymers, where pairs of repeat units collaborate in rotation like the crankshaft of an engine part. Detected as an absorption peak in the infrared.

crank throw (*Eng*) (1) The radial distance from the mainshaft to the pin of a crank, equal to one-half the stroke of a reciprocating member attached to the pin. (2) The web or webs and pin of a crank.

crank web (*Eng*) The arm of a crank, usually of flat rectangular section.

crape (*Textiles*) See CRÊPE.

crash (*ICT*) A failure of a program such that human intervention is required to reset the computer. See HEAD CRASH, SYSTEM CRASH.

crash recorder (*Aero*) See FLIGHT RECORDER.

Crassulaceae (*BioSci*) A family of c.1500 spp of dicotyledonous flowering plants (superorder Rosidae). Most are leaf-succulent perennial plants with CRASSULACEAN ACID METABOLISM; they are widespread but mostly in warm dry temperate regions. They are of little economic importance other than as ornamentals, eg *Sedum, Kalanchoe.*

crassulacean acid metabolism (*BioSci*) A form of PHOTOSYNTHESIS characteristic of desert and some other succulent plants in which CO_2 is taken up (stomata open) during the night and fixed (via PEP CARBOXYLASE) into malic acid from which it is released (stomata shut) during the day and then refixed by RIBULOSE 1,5-BISPHOSPHATE CARBOXYLASE OXYGENASE in the normal way. Abbrev *CAM*. Such CAM plants lose perhaps one-tenth as much water in transpiration than a C3 plant in fixing equivalent amounts of carbon, although the rate at which the carbon is fixed is less. Some plants are facultatively CAM, using C3 when well watered and switching to CAM when water is scarce. CAM has evolved apparently independently in several angiosperm families (eg Crassulaceae, Cactaceae, Bromeliaceae) and in a few ferns. A very few crop plants, eg pineapple and sisal, are CAM plants. A few submerged aquatic plants (without stomata) also operate a form of CAM, presumably taking advantage of the higher concentrations of dissolved CO_2 during the hours of darkness. See C3 PLANT, C4 PLANT.

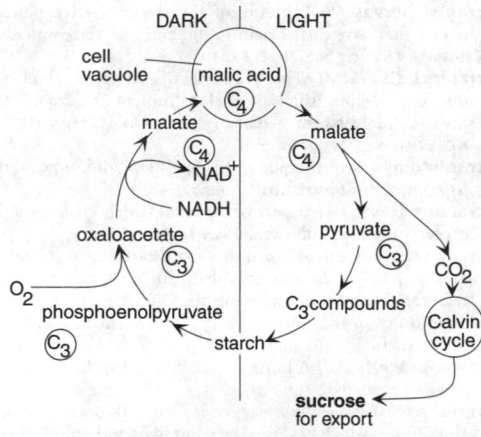

crassulacean acid metabolism Pathway of carbon. NAD^+ = oxidized-, NADH = reduced-nicotinamide adenine dinucleotide. The numbers in circles indicate the number of carbon atoms in the compound.

crater (*Astron, Geol*) Approximately circular depression on the surface of a planetary body caused by volcanic or meteoric activity. Those on Mercury, the Moon, and most of the natural satellites of planets have been formed by impacts with METEORITES in the remote past. The Moon, Mars and Io (one of Jupiter's satellites) also have volcanic craters.

Crater (Cup) (*Astron*) An ancient constellation in the northern sky.

crater lamp (*Electronics*) Discharge tube so designed that a concentrated light source arises in a crater in a solid cathode.

craton (*Geol*) A part of a continent that has attained crustal stability; typically Precambrian or Lower Palaeozoic in age. Also *kraton*.

crawl (*Print*) Continuous running of the press at very low speed. Cf INCHING.

crawler (*ICT*) A computer program that extracts information from sites on the World Wide Web, eg in order to create entries for a search engine index.

crawling (*Build*) A defect in paint or varnish work in which bare patches and paint ridges appear before drying.

crawling (*ElecEng*) A phenomenon sometimes observed with induction motors, the motor running up to about one-seventh of full speed owing to the presence of a pronounced seventh harmonic in the magnetic field, also observed with other harmonics. Also *balking*.

crawl shovel (*Eng*) See POWER SHOVEL.

craze (*Build*) (1) Minute hair cracks which appear on the surface of precast concrete work or artificial stone. (2) Fissuring of faulty coats of paint or varnish in irregular criss-cross cracks. Also *crazing.*

craze (*Eng*) Microfeature associated with fracture and failure of polymers. Consists of voided and oriented material formed at crack tip or rubber particles. Polymers of high molecular mass form stronger crazes than those of low molecular mass, so explaining the strength of eg ultrahigh molecular mass polyethylene (UHMPE). Also gives rise to the STRAIN WHITENING of rubber-toughened polymers. See panel on RUBBER TOUGHENING.

crazy chick disease (*Vet*) See NUTRITIONAL ENCEPHALOMALACIA.

CRC (*Print*) Abbrev for CAMERA-READY COPY.

CRD (*Vet*) Abbrev for CHRONIC RESPIRATORY DISEASE OF FOWL.

C reactive protein (*BioSci*) A plasma protein normally present in low amounts but increased greatly by trauma or infection, ie an *acute phase protein*. It was originally identified by its ability to bind a carbohydrate from *Streptococcus pneumoniae* containing phosphoryl choline groups (C-carbohydrate), but it can bind to nucleic acids, to some lipoproteins and can activate COMPLEMENT. Its biological function is unknown.

cream (*FoodSci*) High-fat (butterfat) fraction that separates from milk after standing; an emulsion stabilized by milk proteins. Artificial cream (non-dairy cream) is made by adding vegetable fat, sugar, suitable emulsifiers and antioxidants to milk or dried milk.

cream-laid (*Paper*) Denoting white writing-paper made with a laid watermark.

cream of tartar (*Chem*) Commercial name for acid potassium tartrate.

cream-wove (*Paper*) Denoting white writing-paper, in the manufacture of which a wove dandy has been used.

crease-resist finish (*Textiles*) A durable finish applied esp to cotton, linen or rayon fabrics to improve the capacity of the materials to resist and recover from creases formed in wear. The fabric is treated with precursors that form a resin within the fibres on curing.

creasing (*Build*) See TILE CREASING.

creatine phosphate (*BioSci*) Phosphate ester of creatine used as a short-term energy reserve in muscle; the breakdown of creatine phosphate is used to convert ADP to ATP. Also *phosphocreatine.*

creatinuria (*Med*) The presence of creatinine in the urine.

creativity (*Psych*) The mental ability to construct original and viable products, ideas, etc, and to go beyond conventional developments. See DIVERGENT THINKING.

creatorrhoea (*Med*) The abnormal presence of muscle fibres in the feces. US *creatorrhea.*

credits (*ImageTech*) Titles at the beginning or end of a film or TV programme listing the cast, technicians and organizations concerned.

creel (*Textiles*) The steel or wooden structure which holds the supply packages at spinning, winding, warping or other machines.

creep (*Build*) (1) The gradual alteration in length or size of a structural member or high-tensile wire due to inherent properties of the materials involved. Not to be confused with shrinkage or expansion. (2) The tendency for lead on a sloping roof to thicken near the lower edge and thin near the top.

creep (*Chem*) (1) The rise of a precipitate on the wet walls of a vessel. (2) The formation of crystals on the sides of a vessel above the surface of an evaporating liquid.

creep (*Eng*) Continuous deformation of metals under steady load. See panel on CREEP AND DEFORMATION.

creep (*MinExt*) Gradual rising of the floor in a coal mine due to pressure. See CRUSH.

creep curve (*Eng*) The presentation of creep data in the form of a strain–time graph, at various different stress levels for a given temperature. Isochronous stress–strain and isometric creep stress curves can be derived by interpolation. Usually commonly used for viscoelastic polymers. See VISCOELASTICITY and panel on CREEP AND DEFORMATION.

creeping weasel (*Eng*) An instrument for measuring the internal details of long, narrow holes, eg the fuel channels in reactor cores.

creep limit (*Eng*) The maximum tensile stress which can be applied to a material at a given temperature without resulting in measurable creep.

creep modulus (*Eng*) Measure of modulus of material, esp polymeric, determined from elongation of specimen (for tensile creep modulus) under a constant applied load. Given by equation $Ec(t) = \sigma/\text{strain}(t)$ where E is the elongation at the constant applied stress σ. Since it is a viscoelastic quantity, the time-scale of measurement must be quoted if it is to be used in QUASI-ELASTIC CALCULATIONS. See VISCOELASTICITY.

creep rupture (*Eng*) A type of failure in materials, esp polymers, where constant applied load causes sample to elongate and finally fail by parting. Often involves slow crack growth. See STATIC FATIGUE.

creep strength (*Eng*) The ability of a material to resist deformation under constant stress, measured as the amount of creep induced by a constant stress acting for a given time and temperature. See panel on CREEP AND DEFORMATION.

creep tests (*Eng*) Methods for measuring the resistance of metals to creep. Time–extension curves under constant loads are determined.

cremaster (*BioSci*) (1) In the pupae of LEPIDOPTERA, an organ of attachment developed from the tenth abdominal somite. (2) In mammals, a muscle of the spermatic cord. (3) In METATHERIA, a muscle whose contraction causes the expression of milk from the mammary gland.

cremorne bolt (*Build*) See ESPAGNOLETTE.

crenate (*BioSci*) Leaf margins etc, having rounded teeth; scalloped.

creosote oil (*Chem*) A coaltar fraction, boiling between 240 and 270°C. The crude creosote oil is used as raw material for producing tar acids etc, or used direct as a germicide, insecticide or disinfectant in various connections (eg soaps, sheep dips, impregnation of railway sleepers, etc).

crêpe (*Textiles*) A woven fabric with a distinctive rough, crinkled appearance because of the special high-twisted yarns from which it is made. Similar fabrics are also made by warp and weft knitting. Also *crape*.

crêpe de chine (*Textiles*) A light crêpe fabric made from continuous filament yarns.

crêpe paper (*Paper*) Crinkled paper produced by doctoring the moist web from a supporting cylinder, so increasing elongation in the machine direction. Used principally for packaging and industrial applications but also for decorative purposes.

Crepe ring (*Astron*) The transparent C ring appearing in Saturn's ring system, with an inner diameter of approx 146 000 km and an outer diameter of approx 183 000 km.

crêpe rubber (*Chem, Eng*) Raw, unvulcanized sheet rubber, not chemically treated in any way.

crepitation (*BioSci*) The explosive discharge of an acrid fluid by certain beetles, which use this as a means of self-defence.

crepitation (*Med*) (1) A crackling sensation felt by the observer on movement of a rheumatic joint. (2) The fine crackling noise made when two ends of a broken bone are rubbed together. (3) The fine crackling sounds heard over the chest in disease of the pleura or of the lungs. Also *crepitus*.

crepuscular (*BioSci*) Active at twilight or in the hours preceding dawn.

crepuscular rays (*EnvSci*) The radiating and coloured rays from the Sun below the horizon, broken up and made apparent by clouds or mountains; also the apparently diverging rays from the Sun passing through irregular spaces between clouds.

crescent wing (*Aero*) A swept-back wing in which the angle of sweep and THICKNESS–CHORD RATIO are progressively reduced from root to tip so as to maintain an approximately constant CRITICAL MACH NUMBER.

cresol red (*Chem*) 2-cresolsulphonphthalein. Indicator used in acid–alkali titrations, having a pH range of 7·2–8·8, over which it changes from yellow to red.

cresol resins (*Plastics*) Resins made from 1-methyl-3-hydroxybenzene and 1-methyl-4-hydroxybenzene and an aldehyde, similar in properties to the phenolics. The 2-compound reacts slowly, and is therefore likely to remain partly unchanged and act as a softener or plasticizer.

cresols (*Chem*) A common name for the hydroxytoluenes, $CH_3C_6H_4OH$, monohydric phenols. There are three isomers: 2-cresol, mp 30°C, bp 191°C; 3-cresol, mp 4°C, bp 203°C; 4-cresol, mp 36°C, bp 202°C. Only 3- and 4-cresol form nitro-compounds with nitric acid, whereas the 2-cresol is oxidized. Important raw materials for plastics, esp the 3-compound; also used for explosives, as intermediates for dyestuffs and as antiseptics.

crest (*BioSci*) A ridge or elongate eminence, esp on a bone.

crest (*Build, CivEng*) The top of a slope or parapet: the ridge of a roof.

crest (*Eng*) Part of screw-thread outline which connects adjacent flanks on the top of the ridge.

crest (*Geol*) The highest part of an anticlinal fold. A line drawn along the highest points of a particular bed is called the *crest line*, and a plane containing the crest lines of successive folds is called the *crest plane*. See FOLDING.

crest factor (*ElecEng*) See PEAK FACTOR.

cresting (*Build*) Ornamental work along a ridge, cornice or coping of a building. Also *bratticing, brattishing*.

crest tile (*Build*) A purpose-made tile having an inverted V-shape for location astride the ridge-line of a roof.

crest value (*ElecEng*) See PEAK VALUE.

crest voltmeter (*ElecEng*) See PEAK VOLTMETER.

Cretaceous (*Geol*) The youngest period of the Mesozoic era covering an approx time span from 145 to 65 million years ago. The corresponding system of rocks; the CHALK is its most striking deposit. See MESOZOIC.

creta preparata (*FoodSci*) Calcium carbonate of BP quality, a dietary calcium supplement for non-self-raising flour with maximum quantities prescribed by regulations. Originally prescribed during World War II as a health measure. Also *creta*.

cretin (*Med*) A person who is affected by CRETINISM. Adjs *cretinoid, cretinous*.

cretinism (*Med*) A congenital condition in which there is failure of mental and physical development, due to absence or insufficiency of the secretion of the thyroid gland.

cretonne (*Textiles*) A heavy printed cotton fabric often used as a furnishing fabric.

Creutzfeldt–Jakob disease (*Med*) A rapidly progressive dementia usually beginning in middle age and now thought to be due to a proteinaceous infective particle or prion, the abnormal form of which arises sporadically. A

Creep and deformation

Continuous deformation of metals under steady load, represented by the curve plotted to display the variation of plastic strain with time at constant temperature and under constant force, usually in tension. The curve has three distinct stages referring to *primary*, *secondary* and *tertiary* creep (see diagram (a)). Families of such curves are often plotted on the same axes in order to show the effect of temperature in a deformation map.

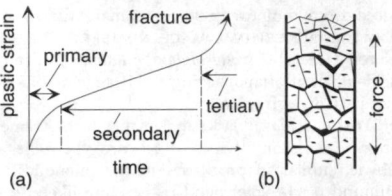

Creep The three stages (a); grain separation and void formation in the tertiary stage (b).

All such creep is a permanent plastic deformation of the material, although the test piece will contract elastically if the load is removed. Stage 1 creep is transient and starts at a high rate that quickly diminishes with time. This leads into the secondary or steady-state creep, where the rate remains constant with time. Stage 2 may be maintained for the remainder of the test but, more usually, the rate increases as stage 3 is reached and then continues to accelerate to the point of fracture.

Each stage of the creep curve is associated with different mechanisms of deformation within the material. Essentially stage 1 is a redistribution of dislocations assisted by thermally activated climb and recovery; stage 2 represents a running balance between work hardening and recovery, with diffusion-controlled creep and GRAIN BOUNDARY slidings involved. In stage 3 grain separation and the formation of voids (see diagram (b)) reduce the effective cross-sectional area supporting the load which then leads to the rate accelerating up to fracture. Unlike a tensile fracture where the cross-section visibly reduces by NECKING prior to the final break, creep fractures show much less change in the original area because the reduction in effective cross-section is accounted for by microscopic internal voids. Oxidation may also play a part in certain types of material.

Increasing temperature speeds up the above mechanisms and thus leads to strain effects presenting at higher rates and in shorter times. At temperatures below $0.3T_m$ (see HOMOLOGOUS TEMPERATURE), stage 3 may never appear, whereas at very high temperatures (above $0.9T_m$) all three stages of the curve may overlap to a degree where none is separately distinguishable.

Creep also occurs in viscoelastic materials (eg polymers) but different mechanisms apply and the creep strains are partially or wholly recoverable on unloading. See VISCOELASTICITY.

---

new variant affects a younger age group with prions identical to those found in cattle infected by BSE. Abbrev *CJD*. See panel on TRANSMISSIBLE SPONGIFORM ENCEPHALOPATHY.

crevasse (*Geol*) A fissure, often deep and wide, in a glacier or ice-sheet.

crevice corrosion (*Eng*) In a liquid-containing system, the acceleration of corrosive attack encountered in crevices and cracks which are partly segregated from the main flow and where build-up of ions and salts or oxygen deficiency may occur. See panel on RUSTING.

CRI (*ImageTech*) Abbrev for *colour reversal intermediate*, a duplicate colour negative printed directly from the original and processed by reversal.

crib (*MinExt*) (1) An interval from work underground for croust, bait, snack, downer, piece, chop, snap, bite or tiffin. (2) A job. (3) A form of timber support.

cribbing (*CivEng*) An interior lining for a shaft, formed of framed timbers backed with boards; it is used to support the sides and keep back water.

crib-biting (*Vet*) A trait acquired by horses, in which some object is grasped with the incisor teeth and air swallowed; may lead to indigestion.

cribellum (*BioSci*) In certain spiders, a perforate oval plate, lying just in front of the anterior spinnerets, which produces a broad strip of silk composed of a number of threads.

cribriform (*BioSci*) Perforate, sieve-like; as the *cribriform plate*, a perforate cartilaginous element of the developing vertebrate skull, which later gives rise to the ECTETHMOID. Also *cribrose*.

cribwork (*CivEng*) Steel or timber cribs or boxes, sometimes filled with concrete and sunk below water level to carry the foundations of bridges.

cricoid (*BioSci*) Ring-shaped; as one of the cartilages of the LARYNX.

cri du chat syndrome (*Med*) A mental-deficiency syndrome due to specific chromosomal aberration. Associated with MICROCEPHALY, widely separated eyes and a characteristic cry resembling that of a cat.

crimp (*Textiles*) The waviness of a fibre, measured as the difference between the straightened and crimped fibre expressed as a percentage of the straightened length. The crimp of a yarn in a woven fabric is measured similarly by comparing the length of the fabric with the length of the yarn removed from it and straightened.

crimping (*Agri*) A procedure for harvesting grass or grain at high moisture content to be broken, flattened and then stored, typically as silage. The product has higher nutritional value than conventionally harvested grass or grain.

crimping (*Eng*) Pressing into small regular folds or ridges in: (1) the reduction of cross-section of a bar material by progressively corrugating it along its surface to give an increase in length; (2) bending or moulding to a required shape; (3) folding or bending sheet metal to provide stiffness.

Crimplene (*Textiles*) TN for polyester fibre with artificial crimp made by heat treating duplex co-extrudate.

crimp stability (*Textiles*) The ability of TEXTURED YARNS to recover after being extended.

crinanite (*Geol*) A basic igneous rock, consisting of intergrown crystals of feldspar, titanaugite, olivine and analcite. Similar to *teschenite*.

crinoid (*BioSci*) See ECHINODERMATA.

Crinoidea (*BioSci*) A class of ECHINODERMATA with branched arms and the oral surface directed upwards. They are attached for part of their life by a stalk that springs from the aboral apex. They have suckerless tube feet and open ambulacral grooves; they have no madreporite, spines or pedicellariae. Most members are extinct.

crippleware (*ICT*) Software that has been partially disabled in order to provide a limited demonstration of its use, often distributed free with magazines.

crisis (*Med*) (1) A painful paroxysm in TABES DORSALIS. (2) The rapid fall of temperature marking the end of a fever.

crispate (*BioSci*) Having a frizzled appearance. Also *crisped*.

crispening (*ImageTech*) The emphasis of edge effects to improve the visibility of video images.

crissum (*BioSci*) In birds, the region surrounding the cloaca or the feathers situated on that area. Adj *crissal*.

crista (*BioSci*) (1) An infolding of the inner membrane of a mitochondrion. Pl *cristae*. (2) A ridge or ridge-like structure; as the projection of the transverse crests of lophodont molars.

crista acustica (*BioSci*) (1) A chordotonal apparatus forming part of the tympanal organ in Tetigoniidae and Gryllidae. (2) A patch of sensory cells in the ampulla, utricle and saccule of the vertebrate ear.

cristate (*BioSci*) Bearing a crest.

cristobalite (*Min*) Usually, the high-temperature form of SiO_2. It is found in volcanic rocks and is stable from 1470 to 1713°C, but exists at lower temperatures. Another, low-temperature form is not stable above atmospheric pressure and is a constituent of *opal*.

crit (*Phys*) Abbrev for CRITICAL MASS (panel).

crith (*Phys*) Unit of mass, that of 1 litre of hydrogen at stp, ie 89·88 mg.

crithidial (*BioSci*) Pertaining to, or resembling, the flagellate genus *Crithidia*; said of a stage in the life cycle of some TRYPANOSOMES.

critical angle (*ICT*) The angle of radiation of a radio wave that will just not be reflected from the ionosphere.

critical angle (*Phys*) See TOTAL INTERNAL REFLECTION.

critical angle

critical coefficient (*Chem*) Additive property, a measure of the molar volume. It is defined as the ratio of the critical temperature to the critical pressure.

critical control point (*FoodSci*) A step or procedure within a food production process where the identified hazard must be controlled (eliminated or reduced to an acceptable level) to ensure food safety. Abbrev CCP. See HACCP.

critical cooling rate (*Eng*) In the heat treatment of steels, the rate of cooling required to prevent nucleation of PEARLITE and to secure the formation of MARTENSITE in steel. With carbon steel, this means cooling in cold water, but the critical rate is reduced by the addition of other elements, hence the use of OIL- or AIR-HARDENING STEEL. Also *critical rate*.

critical corona voltage (*ElecEng*) The voltage at which a corona discharge just begins to take place around an electric conductor.

critical coupling (*ElecEng*) That between two circuits or systems tuned to the same frequency, which gives maximum energy transfer without overcoupling.

critical damping (*Phys*) Damping in an oscillatory electric circuit or in an oscillating mechanical system (such as the movement of an indicating instrument) just sufficient to prevent free oscillations from arising (ringing).

critical dose (*BioSci*) The dose of a substance at and above which adverse changes occur in a cell or an organ.

critical field (*Electronics*) In the case of a MAGNETRON, the smallest steady magnetic flux density (at a constant anode voltage) which would prevent an electron (assumed to be emitted at zero velocity from the cathode) from reaching the anode. It can also mean the magnetic field applied to a conductor below which the superconducting transition occurs at a given temperature. Also *cut-off field*.

critical frequency (*ICT*) The frequency of a radio wave that is just sufficient to penetrate an ionized layer in the upper atmosphere. See MAXIMUM USABLE FREQUENCY.

critical humidity (*Chem*) The humidity at which the equilibrium water vapour pressure of a substance is equal to the partial pressure of the water vapour in the atmosphere, so that it would neither lose nor gain water on exposure.

criticality (*NucEng*) The state in a nuclear reactor when the multiplication factor for neutron flux reaches unity and an external neutron supply is no longer required to maintain power level, ie the chain reaction is self-sustaining. See START-UP PROCEDURE and panel on NUCLEAR REACTOR.

criticality control (*NucEng*) The design procedures required to ensure that a CRITICALITY INCIDENT never occurs. See panel on CRITICAL MASS.

criticality incident (*NucEng*) The accidental accumulation of fissile material in a plant or during handling which leads to criticality and the sudden emission of dangerous amounts of neutrons, gamma rays and heat.

critical limit (*FoodSci*) Known and measurable target that determines if the critical control point is under control or not. See HACCP.

critical load (*Eng*) The acceptable annual dose of sulphur from a thermal power station that a particular area can tolerate. Emissions can then be regulated so that ACID DEPOSITION does not exceed this dose.

critical Mach number (*Aero*) The MACH NUMBER M_{CRIT} at which the airflow over the aircraft first becomes locally supersonic. It can be as low as $M = 0·3$ in leading-edge slat gaps during high-incidence climbing, but more usually is between $M = 0·75$ and $M = 0·95$ on wings of decreasing thickness. It is generally the Mach number above which compressibility effects noticeably alter the handling characteristics.

critical mass (*NucEng*) The minimum mass of fissionable material which can sustain a chain reaction. Abbrev crit. See panel on CRITICAL MASS.

critical path (*ICT*) In systems analysis and project management, the sequence of logically related tasks with the longest overall time-scale, hence the shortest time in which the project can be completed.

critical path planning (*Eng*) A procedure used in planning a large programme of work. Using a digital computer, it determines the particular sequence of operations which must be followed to complete the overall programme in the minimum time and also determines which events have some 'float' or capacity to reprogramme without affecting the whole.

critical period (*Psych*) A time interval deemed highly important for the normal development of a particular emotional, behavioural or cognitive capability or capacity.

critical point (*Eng*) See ARREST POINTS.

critical point (*Phys*) The critical temperature of a substance at which the pressure and volume have their critical values.

Critical mass

The critical mass is the minimum size of fissionable isotope which will sustain a chain reaction. It will depend on the other substances present, particularly non-fissionable isotopes, and on the geometry of the mass in question. Present-day natural uranium has a very low rate of fission producing about 20 neutrons per second per kilogram. All these neutrons are captured or lost, so that a chain reaction can never occur unless the uranium is enriched for the fissile isotope, uranium-235. The figure shows a plot of the *bare sphere* critical mass against the percentage enrichment of uranium-235.

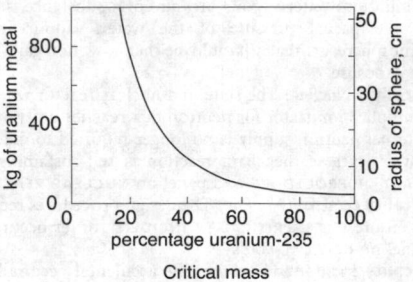

Critical mass

The figure indicates that about 50 kg of 90% enriched uranium in a sphere of about 5 cm radius is

enough to sustain the chain reaction. Various factors alter these numbers: any other shape will increase them as will impurities or the inclusion of uranium oxides or carbides. The critical mass will decrease if neutron reflectors, such as blocks of tungsten carbide, are placed round the sphere to prevent neutron escape. In practice a larger mass of less enriched uranium is cheaper and easier to make.

The effect of becoming critical, by eg carefully adding small pieces of tungsten to form a reflecting wall round a subcritical sphere, is dramatic: a fairly inert material is transformed into a substance emitting over 1000 million neutrons per cubic centimetre and giving off an enormous burst of radiation, becoming red hot and extremely dangerous. Serious accidents of this type occurred in 1945 and 1946. Such a critical mass of uranium will distort and melt before it can explode, and immediately become subcritical again.

One important consequence is the practical one of maintaining CRITICALITY CONTROL during the handling and processing of fissile material. Large-scale chemical processing equipment must be designed so that fissile compounds cannot accumulate in stagnant 'traps' and the total volume of containers may have to be limited so that unscheduled solvent evaporation cannot result in the build-up of a critical mass of the solute. Similarly, enriched fuel elements must be held in containers which cannot inadvertently allow the fuel to aggregate.

At the critical point the densities (and other physical properties) of the liquid and gaseous states are identical.

critical point method (*BioSci*) Technique for preparing tissue or metaphase chromosomes for electron microscopy, by FREEZE-DRYING at the critical point of water. Preserves structural features relatively well.

critical potential (*Phys*) A measure of the energy (in electron-volts) required to ionize a given atom, or raise it to an excited state.

critical pressure (*Phys*) The pressure at which a gas may just be liquefied at its critical temperature.

critical range (*Eng*) The range of temperature in which the reversible change from austenite (stable at high temperature) to ferrite, pearlite and cementite (stable at low temperature) occurs. The upper limit varies with carbon content; the lower limit for slow heating and cooling is about 700°C.

critical rate (*Eng*) See CRITICAL COOLING RATE.

critical size (*NucEng*) The minimum size for a nuclear reactor core of given configuration.

critical solution temperature (*Chem*) The temperature above which two liquids are miscible in all proportions.

critical speed (*Aero*) (1) The speed during take-off at which it has to be abandoned if the aircraft is to stop in the available space. Also *decision speed*. Cf ACCELERATE-STOP DISTANCE. Abbrev V_1. (2) Rotational speed at which resonance or whirling may occur.

critical speed (*Eng*) The rotational speed of a shaft at which some periodic disturbing force coincides with the fundamental or some higher mode of the natural frequency of

torsional or transverse vibration of the shaft and its attached masses.

critical state (*Phys*) The condition of a gas at its critical point, when it appears to hover between the liquid and gaseous states.

critical strain energy release rate (*Eng*) Also TOUGHNESS. See STRENGTH MEASURES.

critical stress intensity factor (*Eng*) Also FRACTURE TOUGHNESS. See STRENGTH MEASURES.

critical temperature (*ElecEng*) The temperature at which magnetic materials lose their magnetic properties; same as *Curie point* (*temperature*).

critical temperature (*Eng*) The temperature at which a sudden transition occurs such as a PHASE CHANGE or the onset of SUPERCONDUCTIVITY. See TRANSFORMATION TEMPERATURE.

critical temperature (*Phys*) The temperature above which a given gas cannot be liquefied. See LIQUEFACTION OF GASES.

critical voltage (*ElecEng*) That which, when applied to a gas-discharge tube, just initiates discharge.

critical volume (*Phys*) The volume of unit mass (usually 1 mole) of a substance under critical conditions of temperature and pressure.

critical wavelength (*Phys*) Free space wavelength corresponding to the CUT-OFF FREQUENCY for a waveguide.

crizzling (*Glass*) Fine cracks in the surface of the glass, occasioned by local chilling during manufacture. See CRAZING.

C–R law (*ElecEng*) Abbrev for CAPACITOR–RESISTANCE LAW.

CRO (*Electronics*) Abbrev for CATHODE-RAY OSCILLOSCOPE.

crochet (*BioSci*) A hook that aids in locomotion, and is associated with the apex of the abdominal legs in insect larvae.

crocidolite (*Min*) The blue asbestos of S Africa, a fibrous variety of the amphibole riebeckite. Long, coarse, flexible spinning fibre with a high resistance to acids. See CAPE ASBESTOS, TIGER'S EYE.

crocin (*FoodSci*) A carotenoid with a natural yellow colour from *Gardenia jasminoides* and *Crocus sativus*; heat stable and water soluble but sensitive to oxidation.

crocking (*Textiles*) Testing the COLOUR FASTNESS of dyes when the fabric is rubbed. The apparatus used for carrying out the tests is called a *crockmeter*.

Crocodilia (*BioSci*) An order of reptiles that have upper and lower temporal arcades, a hard palate, an immovable QUADRATE, loose abdominal ribs and socketed teeth. They are large, powerful and amphibious in form. Crocodiles, alligators, caimans, gavials. Also *Loricata*.

crocodiling (*Build*) A defect on a painted or varnished surface in which ridges or cracks form in irregular patches. Also *alligatoring*.

crocoite (*Min*) Lead chromate, bright orange-red in colour. Also *crocoisite*.

Crohn's disease (*Med*) CHRONIC GRANULOMATOUS DISEASE of unknown aetiology affecting the gut wall.

croissant vitellogène (*BioSci*) In the developing oöcyte, a crescentic area surrounding the archoplasm, in which the mitochondria are grouped.

Cromerian (*Geol*) A temperate stage in the late Pleistocene. See QUATERNARY.

Crookes dark space (*Phys*) A dark region separating the cathode from the luminous 'negative glow' in an electrical discharge in a gas at low pressure. The thickness of the Crookes dark space increases as the pressure is reduced. For air, it is about 1 cm thick at 10 N m^{-2} pressure.

Crookes radiometer (*Phys*) A small mica 'paddlewheel' which rotates when placed in daylight in an evacuated glass vessel. Alternate faces of the mica vanes are blackened and the slight rise of temperature of the blackened surfaces caused by the radiation which they absorb warms the air in contact with them and increases the velocity of rebound of the molecules, the sum of whose impulse constitutes the driving pressure.

Crookes tube (*Phys*) Original gas-discharge tube, illustrating striated positive column, Faraday dark space, negative glow, Crookes dark space and cathode glow.

crop (*BioSci*) See PROVENTRICULUS.

crop (*Geol*) See OUTCROP.

crop (*Print*) To remove portions of copy, eg part of a photograph, so that it better fits the page design or focuses attention on the main subject involved.

crop (*Textiles*) To cut the nap or pile of a fabric to a uniform length.

crop-bound (*Vet*) A term applied to birds suffering from impaction of the crop or ingluvies.

crop diversification (*Agri*) Broadening the range of crops in a rotation to reduce disease build-up, spread economic activity and conserve soil fertility.

crop loss (*Agri*) Reduced quantity or quality of a crop due to pest and disease, environmental conditions or poor practice. Economic loss may also be due to changing market conditions.

crop marks (*Print*) Lines drawn on an OVERLAY or on a photograph to indicate where the image should be trimmed to remove unwanted copy.

cropped (*Print*) Said of the edges of a book which have been cut down to an extent that mars the appearance of the pages.

cropping (*Eng*) The operation of cutting off the end or ends of an ingot to remove the PIPE and other defects.

cropping (*Vet*) The operation of amputating a part of the comb or wattles of birds, or of the ears of dogs.

crop residue (*Agri*) The portion of a crop that is not included in the primary use, eg cereal straw, sugar beet tops.

crop rotation (*Agri*) The regular cycling of a limited number of crops on the same land to improve soil fertility and reduce pests and disease. Typically a legume is included in a rotation to boost soil nitrogen by nitrogen fixation.

crop yield (*Agri*) The actual quantity of harvested product taken off the field, expressed in weight or monetary terms per unit area. The expected yield is the actual yield plus an estimate of known sources of loss during production.

cross (*BioSci*) The mating together of individuals of two different breeds, varieties, strains or genotypes. The progeny are *cross-bred*.

cross (*Build*) A special pipe-fitting having four branches mutually at right angles in one plane.

cross ampere-turns (*ElecEng*) The component of the armature ampere-turns which tends to produce a field at right angles to the main field.

cross-assembler (*ICT*) An ASSEMBLER that runs on one computer but produces CODE for another.

cross-association (*Print*) Placing in position across the machine the required strips or slit webs before finishing operations.

cross-axle (*CivEng*) A driving axle having cranks mutually at right angles.

crossbar exchange (*ICT*) One with the following features: (1) CROSSBAR SWITCHES; (2) common circuits that may be electromechanical or, more commonly, electronic, to select and establish the switched paths and to operate the switch contacts.

crossbar switch (*ICT*) One having multiple vertical and horizontal paths, with electromagnetically operated contacts for connecting any vertical with any horizontal path.

crossbar transformer (*ElecEng*) Coupling device between a coaxial cable and a waveguide, the latter having a short transverse rod, to the centre of which the central conductor is connected.

cross-bearing (*Surv*) Observation on survey point not in the immediate scheme of work, useful for checking purposes.

cross bedding (*Geol*) Internally inclined planes in a rock related to the original direction of current flow. Also *current bedding, false bedding*.

cross-blast explosion pot (*ElecEng*) An explosion pot in which the pressure generated by the arc in the pot causes a stream of oil to be directed across the arc path at right angles to it.

cross-blast oil circuit breaker (*ElecEng*) An oil circuit breaker in which the pressure generated by the arc causes a stream of oil to be forced through ports placed opposite one pair of contacts, thereby cutting across the arc stream.

cross bombardment (*Phys*) A method of identification of radioactive nuclides through their production by differing reactions.

cross-bond (*ElecEng*) A rail-bond for connecting together the two rails of a track or the rails of adjacent tracks.

cross-bonding (*ElecEng*) A process in which the sheath of cable 1 is connected to that of cable 2 and farther on to cable 3. The total induced emf vanishes and there are no sheath-circuit eddies.

cross-bracing (*Eng*) See COUNTERBRACING.

cross-colour noise (*ImageTech*) A picture defect in which spurious colour appears in areas of fine pattern, eg as striped shirts.

cross-compiler (*ICT*) A COMPILER that runs on one computer but produces CODE for another.

cross contamination (*FoodSci*) The transfer of potentially harmful micro-organisms from raw foods to ready-to-eat foods by direct or indirect contact.

cross-correlation (*Acous*) Multiplication of two signals and averaging over a time interval.

cross-correlation (*MathSci*) The comparison of one set of data against another set, usually a time series.

cross-coupling (*ElecEng*) Undesired transfer of interfering power from one circuit to another by induction, leakage, etc.

cross cut (*MinExt*) In metal mining, a level or tunnel driven through the country rock, generally from a shaft, to intersect a vein or lode. See fig. at MINING.

cross-cut chisel (*Eng*) A cold chisel having a narrow cutting edge carried by a stiff shank of rectangular section; used for heavy cuts. See COLD CHISEL.

cross-cut file (*Eng*) A file in which the cutting edges are formed by the intersection of two sets of teeth crossing each other.

cross-cut saw (*Build*) A saw designed for cutting timber across the grain.

cross-direction (*Paper*) In a paper sheet or web, the direction at right angles to the machine direction in the plane of the sheet.

cross-dyeing (*Textiles*) The further dyeing of a fabric made from a blend of fibres after one of the fibres has already been dyed.

crossed-field tube (*ICT*) Any microwave beam tube in which the directions of the static magnetic field, the static electric field and the electron beam are mutually perpendicular, as required for converting the potential energy of the electron beam into radio-frequency energy. Examples include MAGNETRONS, certain backward wave tubes. Cf TRAVELLING-WAVE TUBES in which the kinetic energy of beam electrons is converted to radio-frequency energy.

crossed lens (*Phys*) A simple lens for which the radii of curvature have been chosen to give minimum spherical aberration for parallel incident rays. For a refractive index of 1·5, the radii should be in the ratio 1:6, the surface of smaller radius facing the incident light.

crossed Nicols (*Phys*) Two Nicol prisms arranged with their principal planes at right angles; in this position the plane-polarized light emerging from one Nicol is extinguished by the other. Similarly for *polarizers*.

crossette (*Build*) A projection formed on the flank of an arch stone at the top, giving it a bearing upon the adjacent stone on the side towards the springing. Also *ear*. See fig. at ARCH.

cross-fall (*CivEng*) The difference in vertical height between the highest and lowest points on the cross-section of a road surface. Also, the average rate of fall from one side to the other, or from the crown to a side of a road.

cross-fertilization (*BioSci*) The fertilization of the female gametes of one individual by the male gametes of another individual. Also *allogamy*.

cross-field (*ElecEng*) The component of the flux in an electric machine which is assumed to be produced by the cross ampere-turns.

cross-fire technique (*Radiol*) The irradiation of a deep-seated region in the body from several directions so as to reduce damage to surrounding tissues for a given dose to that region.

cross-frogs (*CivEng*) See CROSSINGS.

cross-front (*ImageTech*) A sliding front carrying the lens in field and technical cameras; used to avoid the consequence of tilting the axis of a camera away from normality with an object.

cross-girders (*Eng*) (1) Short girders acting as ties between two main girders. (2) The members which transmit the weight of the roadway to the main girders of a bridge.

cross-grained float (*Build*) A FLOAT made of cross-grained wood, used in finishing hard-setting plasterwork.

crosshair (*Surv*) A spider's thread fixed across the diaphragm of a level or theodolite.

cross-hatch pattern (*ImageTech*) A test pattern of vertical and horizontal lines used on TV picture tube.

crosshead (*Eng*) A reciprocating block, usually sliding between guides, forming the junction piece between the piston-rod and connecting-rod of an engine.

crosshead (*Print*) A heading or subheading centred on the measure.

crossings (*CivEng*) The cast-steel railway track component which allows passage for the wheel flanges at places where one line crosses another. Also *cross-frogs*. Fig. ▷

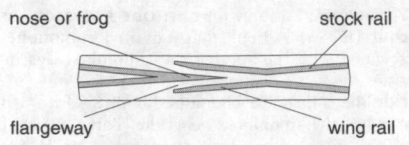

nose or frog   stock rail

flangeway   wing rail

crossings

cross-link density (*Chem*) The density of chemical cross-links in a polymer. In rubbers, the KINETIC THEORY OF ELASTICITY predicts that the SHEAR MODULUS, G, is directly proportional to the cross-link density. Also *degree of cross-linking*.

cross-linking (*Chem*) The formation of side bonds between different chains in a polymer, thus increasing its rigidity. Usually achieved in rubbers with sulphur to give disulphidic or polysulphidic links between vicinal carbon atoms. Physical cross-links are used in THERMOPLASTIC ELASTOMERS to provide elastomeric properties. See panels on ELASTOMERS and THERMOSETS.

cross-magnetizing (*Phys*) The effect of armature reaction on the magnetic field of a current generator.

cross-matching (*BioSci*) A procedure used in selecting blood for transfusion. The red cells to be transfused are mixed with the serum from the patient and if no agglutination occurs the red cells are suitable for transfusion.

cross-modulation (*ICT*) Impression of the envelope of one modulated carrier upon another carrier, due to non-linearity in the medium transmitting both carriers. See LUXEMBURG EFFECT.

cross-neutralization (*Electronics*) A method of neutralization in push–pull amplifiers. Each output is connected by a negative-feedback circuit to the other input.

Crossopterygii (*BioSci*) A subclass of the class OSTEICHTHYES, first known as fossils from the Middle Devonian period, and persisting to the present. They are of interest because the pectoral fins, which are lobed and branched at their tips, are attached to the girdle, an arrangement which could have led to the evolution of the tetrapod limb. Living forms include the coelacanths and Dipnoi (lung-fish).

crossover (*Acous*) In a twin-loudspeaker system, the point in the frequency range above which the amplifier output is fed mainly to the treble speaker and below which mainly to the bass speaker.

crossover (*BioSci*) An exchange of segments of homologous chromosomes during MEIOSIS whereby linked genes become recombined; also the product of such an exchange. The *crossover frequency* is the proportion of gametes bearing a crossover between two specified gene loci. It ranges from 0% for allelic genes to 50% for genes so far apart that there is always a crossover between them. See CENTIMORGAN, CHIASMA.

crossover (*Build*) A special pipe-fitting with its middle length cranked out for use in a 'pass-over offset' when two pipes cross each other in a plane.

crossover (*CivEng*) On railways, a communicating track between two parallel lines, enabling rolling-stock to be transferred from one line to the other.

crossover (*Electronics*) In an electron-lens system, the location where streams of electrons from the object pass through a very small area, substantially a point, before forming an image.

crossover (*ICT*) A filter used in audio systems that uses inductors and capacitors to split the input power into two paths, one containing only frequencies below the crossover frequency and the other containing only frequencies above it. The two paths are then used to feed transducers optimized for these two frequency bands.

crossover area (*Electronics*) The point at which the electron beam comes to a focus inside the accelerating anode of a cathode-ray tube.

crossover frequency (*Acous*) (1) The frequency in a two-channel loudspeaker system at which the high- and low-frequency units deliver equal acoustic power; alternatively it may be more generally applied to electric dividing networks when equal electric powers are delivered to each of adjacent frequency channels. (2) See TURNOVER FREQUENCY.

crossover network (*Acous*) See DIVIDING NETWORK.

crossover site (*BioSci*) The place in the genome where breakage and reunion of DNA strands occur during RECOMBINATION. See panel on CHROMOSOME.

cross-pane hammer (*Build*) A hammer with a pane consisting of a blunt chisel-like edge at right angles to the shaft.

cross-platform (*ICT*) Compatible with different types of computers or software.

cross-ply (*Autos*) A tyre in which reinforcement-belting layers are inclined to one another, and composed only of polymer fibre, eg rayon. Generally of lower life than radial tyre, in which steel belting is also used. See panel on TYRE TECHNOLOGY.

crosspoint switch (*ICT*) A lattice consisting of sets of horizontal and vertical conductors crossing over each other with a switch at each point of crossing, allowing any horizontal conductor to be connected to any vertical conductor.

cross pollination (*BioSci*) The conveyance of pollen from an anther of one flower to the stigma of another, either on the same or on a different plant of the same, or related, species.

cross-product of a vector (*MathSci*) See VECTOR PRODUCT.

cross-protection (*BioSci*) The protection offered by prior, systemic infection by one virus against infection by a second, related virus. Deliberate infection with a symptomless strain of tomato mosaic virus is used commercially to protect tomatoes from infection by other, more damaging strains. The phenomenon is also used experimentally to establish the relatedness of different isolates of viruses.

cross range (*Space*) The distance either side of a nominal re-entry track which may be achieved by using the lifting properties of a re-entering space vehicle.

cross-ratio (*MathSci*) (1) Of four numbers a, b, c, d, taken in the order a, b, c, d: the ratio

$$\frac{a-c}{a-d} / \frac{b-c}{b-d}$$

Denoted by (ab, cd). (2) Of four points on a straight line A, B, C, D: the cross-ratio is

$$(AB, CD) = \frac{AB}{AD} \frac{CD}{CB}$$

(the usual convention as to signs being observed). (3) Of a pencil, consisting of four straight lines OA, OB, OC, OD: the cross-ratio is

$$O(AB, CD) = \frac{\sin AOB}{\sin AOD} \frac{\sin COD}{\sin COB}$$

Any transversal (ie line not through O) cuts the pencil in four points having the same cross-ratio. Some writers associate the above fractions with the order a, c, b, d as opposed to the order a, b, c, d. Also *anharmonic ratio*.

cross-reactivity (*BioSci*) Fortuitous recognition by lymphocyte antigen receptors (B-cell immunoglobulin or T-cell antigen receptors) of EPITOPES on proteins unrelated to those which produced them. Thought to be due to use of similar amino acid sequence or structure by completely different proteins.

cross-section (*Eng*) (1) The section of a body (eg a girder or moulding) at right angles to its length. (2) A drawing showing such a section.

cross-section (*MathSci*) The section of a three-dimensional object made by a plane which cuts the axis of symmetry or the longest axis at right angles.

cross-section (*Phys*) In atomic or nuclear physics, the probability that a particular interaction will take place between particles. The value of the cross-section for any process will depend on the particles under bombardment and upon the nature and energy of the bombarding particles. Suppose I particles per second are incident on a target area A containing N particles and I_r of the incident particles produce a given reaction, then if $I_r < I$,

$$I_r = \frac{IN\sigma}{A}$$

where σ is the cross-section for the reaction; σ can be imagined as a disk of area σ surrounding each target particle. Measured in BARNS.

cross-sectional study (*Psych*) A research study that examines the effects of development (maturation) by examining different subjects of various ages at a single time, as opposed to a longitudinal or follow-up study where the same subjects are studied at different times.

cross-sequential study (*Psych*) A research study that examines the effects of development (maturation) using a combination of cross-sectional and longitudinal studies.

cross-sill (*CivEng*) See SLEEPER.

cross-slide (*Eng*) (1) That part of a planing machine or lathe on which the tool-holder is mounted, and along which it may be moved at right angles to the bed of the machine. (2) The transfer of GLIDE of a DISLOCATION in a material from one slip plane to another during deformation or thermal recovery.

cross-springer (*Build*) In a GROINED ARCH, the rib following the line of a groin.

cross-staff (*Surv*) An instrument for setting out right angles in the field. It consists of a frame or box having two pairs of vertical slits, giving two lines of sight mutually at right angles.

cross-talk (*ICT*) The interference caused by energy from one conversation invading another circuit by electrostatic or electromagnetic coupling. See FAR-END CROSS-TALK, NEAR-END CROSS-TALK.

cross-talk meter (*ICT*) An arrangement for measuring the attenuation between circuits that are liable to permit cross-talk.

crosstie (*CivEng*) See SLEEPER.

cross-tongue (*Build*) A wooden tongue for a PLOUGHED-AND-TONGUED JOINT, cut so that the grain is at right angles to the grooves.

cross-tree (*Ships*) A lateral formation on a ship's mast; its uses are for rigging to top masts, hooks, tackle, etc. The term is derived from antique wooden ships.

cross-wall construction (*Build*) A system where the main supports are the walls running back to front, supporting floors, roof and curtain walls.

crotonaldehyde (*Chem*) *But-2-enal.* $CH_3CH=CHCHO$, a liquid, of pungent odour, bp 105°C; an unsaturated aldehyde, obtained from ethanol by heating with dilute hydrochloric acid or with a solution of sodium acetate. As an intermediate product, ALDOL is formed.

crotonic acid (*Chem*) *But-2-enoic acid.* $CH_3CH=COOH$, an olefinic monocarboxylic acid. There are two stereoisomers: crotonic acid, mp 71°C, bp 180°C; and iso- or allo-crotonic acid, mp 15°C, bp 169°C. The first form is the *cis*-, the latter the *trans*-form. The crotonic acids are also isomers of methacrylic and vinylacetic acids.

croton oil (*BioSci*) A tumour promoter that consists of a purgative and blistering agent extracted from tree seeds and that, when rubbed on skin, causes cell proliferation and significantly increases the effect of a chemical carcinogen given previously. The active ingredient is probably phorbol ester (*phorbol myristate acetate, PMA*).

crotonyl (*Chem*) The group $-CHCH=CHCH_3$ in organic compounds.

crotyl (*Chem*) The group $-CH_2CH=CHCH_3$ in organic compounds.

croup (*Med*) Hoarse croaking cough associated with inflammation of the larynx and trachea in children.

croup (*Vet*) The sacral region of the back of the horse.

crowbar (*Build*) A round iron bar, pointed at one end and flattened to a wedge shape at the other, used as a lever for moving heavy objects.

Crowe process (*MinExt*) A method of removing oxygen from cyanide solution before recovery of dissolved gold, in which liquor is exposed to vacuum as it flows over trays in a tower.

crown (*BioSci*) (1) Generally, crest or head. (2) The part of a polyp bearing the mouth and tentacles. (3) The distal part of a deer's horn. (4) The grinding surface of a tooth. (5) The disk and arms of a Crinoid. (6) A very short rootstock.

crown (*Build, CivEng*) The highest part of an arch. Also *vertex*.

crown (*For*) The upper branchy part of a tree above the bole.

crown (*Paper*) A paper size in use in the UK, 385 × 505 mm. See PAPER SIZES.

crown ether (*Chem*) A cyclic polyether, often of the form $(-O-CH_2-CH_2)_n$. They complex alkali metal ions strongly. See MACROCYCLE.

crown gall (*BioSci*) A disease of dicotyledons, esp of fruit bushes and trees, caused by a soil bacterium, *Agrobacterium tumefaciens* and characterized by the production of large, tumour-like galls. See OPINE, TI PLASMID.

crown-gate (*Build*) A canal-lock headgate.

crown glass (*Glass*) (1) Glass made in disk form by blowing and spinning, having a natural fire-finished surface but varying in thickness with slight convexity, giving a degree of distortion of vision and reflection. Cf BULLION POINT. (2) That class of optical glasses with $v_d > 50$ if $n_d > 1.60$ and $v_d > 55$ if $n_d < 1.60$, where v_d is the Abbé number, or reciprocal of the dispersivity, and n_d is the REFRACTIVE INDEX at 587·6 nm. Cf FLINT GLASS. (3) Loosely used for SODA–LIME–SILICA GLASS.

crown octavo (*Print*) A book size, $7\frac{1}{4} \times 4\frac{7}{8}$ in (186 × 123 mm).

crown of thorns tuning (*Electronics*) Tuning of cavity magnetrons involving changing inductance of cavities by the introduction of conducting rods along their axes.

crown quarto (*Print*) A book size, $9\frac{3}{4} \times 7\frac{3}{8}$ in (246 × 189 mm).

crown-tile (*Build*) An ordinary flat tile. Also *plane-tile*.

crown wheel (*Eng*) The larger wheel of a bevel reduction gear. See BEVEL GEAR.

crow-step gable (*Build*) See CORBIE-STEP GABLE.

crow steps (*Arch*) Steps on the coping of a gable, common in traditional Scottish and Dutch architecture.

croy (*CivEng*) A protective barrier built out into a stream to prevent erosion of the bank.

crozier (*BioSci*) The young ascus when it is bent in the form of a hook.

crozzle (*Build*) An excessively hard and misshapen brick which has been partially melted and overheated.

CRT (*Electronics*) Abbrev for CATHODE-RAY TUBE.

cruciate (*BioSci*) Having the form of, or arranged like, a cross. Also *cruciform*.

cruciate ligament (*Med*) Either of two ligaments (*anterior cruciate ligament* and *posterior cruciate ligament*) that cross each other in the knee.

crucible (*Chem, MinExt*) A refractory vessel or pot in which metals are melted. In chemical analysis, smaller crucibles, made of porcelain, nickel or platinum, are used for igniting precipitates, fusing alkalis, etc.

crucible furnace (*Eng*) A furnace, fired with coal, coke, oil or gas, in which metal contained in crucibles is melted.

crucible steel (*Eng*) Steel made by melting BLISTER BAR or WROUGHT-IRON, charcoal and ferro-alloys in small (45 kg) crucibles. This was the first process to produce steel in a molten condition, hence the product was called *cast steel*. Now obsolete and replaced by electric furnace steel-making.

crucible tongs (*Chem, Eng*) Tongs used for handling crucibles.

Cruciferae (*BioSci*) A family of c.3000 spp of dicotyledonous flowering plants (superorder Dilleniidae). They are cosmopolitan and are mostly herbs or, rarely, shrubs. The flowers characteristically have four sepals, four petals and six stamens, all free, and a superior ovary of two fused carpels. Includes the genus *Brassica* and a number of minor vegetable crops, eg water-cress, and many ornamentals, eg wallflower. Also *Brassicaceae*.

cruciform (*BioSci*) See CRUCIATE.

crude oil (*Chem*) See PETROLEUM.

crude protein (*Agri*) A value that equates to total protein in food products and is calculated as 6.38 times the value for total nitrogen.

cruise control (*Autos*) A system which automatically maintains a selected road speed.

cruise missile (*Aero*) A missile launched from a mobile platform, following a low-altitude course and guided by an inertial guidance system which takes account of minute gravitational anomalies over the terrain on the way to the target.

cruiser stern (*Ships*) Stern construction, integral with the main hull for strength and form and partially waterborne. It assists in manoeuvrability and wave formation and provides underdeck roominess. See COUNTER.

crump (*MinExt*) Rock movement under stress due to underground mining, possibly violent.

crunode (*MathSci*) See DOUBLE POINT.

cruor (*BioSci*) The coagulated blood of vertebrates.

crura (*BioSci*) See CRUS.

crural (*BioSci*) Pertaining to or resembling a leg. See CRUS.

crureus (*BioSci*) A leg muscle of higher vertebrates.

crus (*BioSci*) The *zeugopodium* of the hind-limb in vertebrates; the shank; any organ resembling a leg or shank. Pl *crura*. Adj *crural*.

crush (*MinExt*) The broken condition of pillars of coal in a mine due to pressure of the strata. See panel on CREEP AND DEFORMATION.

crush breccia (*Geol*) A rock consisting of angular fragments, often recemented, which has resulted from the faulting or folding of pre-existing rocks. See CRUSH CONGLOMERATE, FAULT BRECCIA.

crush conglomerate (*Geol*) A rock consisting of crushed and rolled fragments, often recemented; it has resulted from the folding or faulting of pre-existing rocks.

crusher (*MinExt*) A machine used in earlier stages of comminution of hard rock. Typically works on dry feed as it falls between advancing and receding breaking plates. Types include JAW BREAKER or *jaw crusher*, gyrator, ROLLS, STAMPS.

crushing (*ImageTech*) Loss of tonal gradation in the picture through reduced contrast at the extremes of the brightness range: *black crushing* in the shadows, *white crushing* in the highlights.

crushing (*Print*) See SMASHING.

crushing mill (*Agri*) Industrial equipment designed to crush seed to produce a fine product such as soy bean meal. Also used as part of a crop extraction process such as oil and soluble sugar removal.

crushing strip (*Print*) A strip sometimes required on fold rollers to increase the nip.

crushing test (*CivEng*) A test of the suitability of stone to be used for roads or building purposes.

crush syndrome (*Med*) Severe muscle injury resulting in the release of myoglobin and subsequent acute kidney failure.

crust (*CivEng*) See WEARING COURSE.

crust (*Geol*) The outermost layer of the lithosphere consisting of relatively light rocks. Continental crust consists largely of granitic material; oceanic crust is largely basaltic.

Crustacea (*BioSci*) A class of ARTHROPODA, mostly of aquatic habit and mode of respiration. The second and third somites bear antennae and the fourth a pair of mandibles. Shrimps, prawns, barnacles, crabs, lobsters.

crustose (*BioSci*) Forming a crust, esp of lichens, having a crust-like thallus closely attached to, and virtually inseparable from, the surface on which it is growing.

crutching (*Vet*) The operation of removing the wool from around the tail and quarters of sheep as a preventive of MYIASIS.

Crux (Cross) (*Astron*) The smallest constellation in the sky, and one of the most distinctive. Also *Crux Australis, Southern Cross*.

cryo- (*Genrl*) Prefix from Gk *kryos* denoting frost or ice.

cryogenic (*Phys*) The term applied to low-temperature substances and apparatus.

cryogenic freezing (*FoodSci*) Almost instantaneous freezing, by immersion in liquid nitrogen, at $-195°C$, which prevents water from expanding and rupturing the cells, thus minimizing structural breakdown and colour and flavour losses. Also used to preserve living material, eg sperm and ova.

cryogenic gyro (*Phys*) A gyro which depends on electron spin in atoms at very low temperature.

cryogenics (*Phys*) The study of materials at very low temperatures.

cryoglobulin (*BioSci*) Abnormal plasma immunoglobulin (IgG or IgM) that precipitates (*cryoprecipitates*) when serum is cooled.

cryolite (*Min*) Sodium aluminium fluoride, used in the manufacture of aluminium. Also *Greenland spar*.

cryometer (*Phys*) A thermometer for measuring very low temperatures.

cryoprecipitate (*BioSci*) See CRYOGLOBULIN.

cryoprotectant (*BioSci*) A substance that is used to protect from the effects of freezing, largely by preventing large ice-crystals from forming. The two commonly used for freezing cells are dimethylsulphoxide (DMSO) and glycerol.

cryosar (*Electronics*) Low-temperature germanium switch with on–off time of a few nanoseconds.

cryoscope (*Chem*) An instrument for the determination of freezing or melting points.

cryoscopic method (*Chem*) The determination of the relative molecular mass of a substance by observing the lowering of the freezing point of a suitable solvent. See GPC.

cryostat (*Phys*) Low-temperature thermostat.

cryotherapy (*Med*) Medical treatment using the application of extreme cold.

cryotron (*Electronics*) Miniature electronic switch, operating in liquid helium, consisting of a short wire wound with a very fine control wire. When a magnetic field is induced via the control wire, the main wire changes from superconductive to resistive. Used as a memory, characterized by exceedingly short access time. See SUPERCONDUCTIVITY.

crypt (*BioSci*) A small cavity; a simple tubular gland.

cryptic coloration (*BioSci*) Protective resemblance to some part of the environment or camouflage, from simple COUNTERSHADING to more subtle MIMICRY of eg leaves or twigs. Cf APOSEMATIC COLORATION.

crypto- (*Genrl*) Prefix from Gk *kryptos*, hidden.

cryptobiosis (*BioSci*) The state in which an animal's metabolic activities have come effectively, but reversibly, to a standstill. See ANABIOSIS.

cryptocrystalline (*Min*) The texture of a rock in which the crystals are too small to be recognized under a petrological microscope.

cryptogam (*BioSci*) In earlier systems of classification, a plant without flowers or cones in which the method of reproduction was not apparent, ie algae, fungi, bryophytes and pteridophytes.

cryptometer (*Build*) An instrument for determining the opacity of paints and pigments.

Cryptophyceae (*BioSci*) A small class of biflagellate unicellular algae (cryptomonads) with chlorophyll a and c and phycobilins; 4–80 μm long, ovoid or bean-shaped, and dorsoventrally flattened. Found in both fresh-water and marine environments. Protozoologists have considered these as an order, Cryptomonadida, of the class Phytomastigophora.

cryptophyte (*BioSci*) (1) A member of the CRYPTOPHYCEAE. (2) Herb with perennating buds below soil (or water) surface. Includes geophyte, heliophyte, hydrophyte. See RAUNKIAER SYSTEM.

cryptorchid (*Vet, BioSci*) An animal in which one or both testes have not descended from the abdominal cavity to the scrotum within a reasonable time.

cryptorchidectomy (*Vet*) Surgical removal of the testes from a cryptorchid animal.

cryptosystem (*ICT*) Abbrev for *cryptographic system*. That which enables the transmission of secret and authenticated messages.

cryptozoic (*BioSci*) Living in dark places, as in holes, caves or under stones and tree trunks.

crystal (*BioSci*) Crystalline inclusion in a plant cell, usually of calcium oxalate. Types include the DRUSE and the RAPHIDE.

crystal (*Electronics*) Piezoelectric element, shaped from a crystal in relation to crystallographic axes, eg quartz, tourmaline, Rochelle salt, ammonium dihydrogen phosphate, to give FACETS to which electrodes are fixed or deposited for use as transducers or frequency standards.

crystal (*Glass*) See CRYSTAL GLASS.

crystal (*Min*) (1) Homogenous solid body of chemical element, compound or isomorphous mixture having a regular atomic lattice arrangement that may be shown by crystal faces. (2) Old name for QUARTZ. See BRAVAIS LATTICES, UNIT CELL and panel on CRYSTAL LATTICE.

crystal anisotropy (*Crystal*) In general, directional variations of any physical property, eg elasticity, thermal conductivity, etc, in crystalline materials. Leads to existence of favoured directions of magnetization, related to lattice structure in some ferromagnetic crystals.

crystal axes (*Crystal*) The axes of the natural co-ordinate system formed by the crystal lattice. These are perpendicular to the natural faces for many crystals. See BIAXIAL, UNIAXIAL.

crystal boundaries (*Crystal*) The surfaces of contact between adjacent crystals in a metal. Anything not soluble in the crystals tends to be situated at the crystal boundaries, but in the absence of this, the boundary between two similar crystals is simply the region where the orientation changes.

crystal counter (*Electronics*) One in which an operating pulse is obtained from a crystal when made conducting by an ionizing particle or wave.

crystal detector (*Electronics*) A DEMODULATOR, used primarily in microwave applications. The non-linear voltage–current characteristic of a point-contact diode is used to separate the signal information from the carrier frequency.

crystal diamagnetism (*Phys*) Property of negative susceptibility shown by certain substances such as silver and bismuth.

crystal diode (*Electronics*) See DIODE (2).

crystal dislocation (*Crystal*) Imperfect alignment between the lattices at the junctions of small blocks of ions ('mosaics') within the crystal. The resulting mobility and opportunity for realignment of the molecules is of importance in crystal growth, plastic flow, sintering, etc.

crystal drive (*Electronics*) A system in which oscillations of low power are generated in a PIEZOELECTRIC CRYSTAL RESONATOR, being subsequently amplified to a level requisite for transmission.

crystal electrostriction (*Phys*) The dimensional changes of a dielectric crystal under an applied electric field. See ELECTROSTRICTION, MAGNETOSTRICTION.

crystal face (*Crystal*) One of the bounding surfaces of a crystal. In the case of small, undistorted crystals, each face is an optically plane surface. A *cleavage face* is the smooth surface resulting from cleavage; in such minerals as mica, the cleavage face may be almost a plane surface, diverging only by the thickness of a molecule.

crystal filter (*Electronics*) Band-pass filter in which piezoelectric crystals provide very sharp frequency-discriminating elements, esp for group modulation of multichannels over coaxial lines.

crystal-gate receiver (*ICT*) Superhet receiver in which one (or more than one) piezoelectric crystal is included in the intermediate-frequency circuit, to obtain a high degree of selectivity.

crystal glass (*Glass*) A colourless, highly transparent glass of high refractive index, which may be 'lead crystal'. (A somewhat misleading term since it denotes different things in different glass-making districts.) See LEAD GLASS.

crystal goniometer (*Crystal*) An instrument for measuring angles between crystal faces.

crystal growing (*Electronics*) The technique of forming semiconductors by extracting crystal slowly from molten state. Also *crystal pulling*. See CZOCHRALSKI PROCESS.

crystal indices (*Crystal*) See MILLER INDICES.

crystal lattice (*Crystal*) Three-dimensional repeating array of points used to represent the structure of a crystal, and classified into 14 groups by Bravais. See BRAVAIS LATTICES and panel on CRYSTAL LATTICE.

crystalline (*Geol*) Having a crystal structure.

crystalline cone (*BioSci*) The outer refractive body of an OMMATIDIUM which acts as a light guide.

crystalline form (*Crystal*) The external geometrical shape of a crystal.

crystalline lens (*BioSci*) The transparent refractive body of the eye in vertebrates, Cephalopoda, etc. It is compressible by muscles to focus images on the retina.

crystalline overgrowth (*Crystal*) The growth of one crystal round another, frequently observed with isomorphous substances. Cf CUBIC SYSTEM.

crystalline rocks (*Geol*) These consist wholly, or chiefly, of mineral crystals. They are usually formed by the solidification of molten rock, by metamorphic action, or by precipitation from solution.

crystalline solid (*Chem*) A solid in which the atoms or molecules are arranged in a regular manner, the values of certain physical properties depending on the direction in which they are measured. When formed freely, a crystalline mass is bounded by plane surfaces (faces) intersecting at definite angles. See X-RAY CRYSTALLOGRAPHY.

crystalline style (*BioSci*) In BIVALVIA, a transparent rod-shaped mass secreted by a diverticulum of the intestine; composed of protein with an adsorbed amylolytic enzyme.

crystallinity (*Chem, Textiles*) See DEGREE OF CRYSTALLINITY.

crystallins (*BioSci*) Major proteins of the lens of the vertebrate eye.

crystallites (*Chem, Min*) (1) Very small, often imperfectly formed crystals. (2) Minute bodies occurring in glassy igneous rocks, and marking a stage in incipient crystallization.

crystallizable polymers (*Chem*) Those polymers capable of crystallizing owing to stereoregular chains. See panel on POLYMERS.

crystallization (*Chem*) Slow formation of a crystal from melt or solution.

crystallization (*FoodSci*) The formation of crystals as water is evaporated as in manufacture of sugar and crystallized fruits or the solidification of fat and water. Also a form of food spoilage if crystals are unintended. See FAT BLOOM, HONEY.

crystallization of polymers (*Chem*) Polymers are not intrinsically crystallizable because the chain irregularity found in random copolymer units (atacticity) prevents the necessary close approach of chains. See STEREOREGULAR POLYMERS and panel on POLYMERS.

crystallizing rubbers (*Chem*) Those elastomers with stereoregular chains (eg isotactic or *cis*-repeat units) which are capable of crystallizing. They include natural rubber and *cis*-polybutadiene but not styrene butadiene rubber. Crystallization may occur simply in storage (see STARK RUBBER) or by strain-crystallization.

crystalloblastic texture (*Geol*) The texture of metamorphic rocks resulting from the growth of crystals in a solid medium.

crystallographic axes (*Crystal*) Lines of reference intersecting at the centre of a crystal. Crystal (or morphological) axes, usually three in number, by their relative lengths and attitude, determine the system to which a crystal belongs.

crystallographic notation (*Crystal*) Brief method of writing down the relation of any crystal face to certain axes of reference in the crystal.

crystallographic planes (*Crystal*) Any set of parallel and equally spaced planes that may be supposed to pass through the centres of atoms in crystals. As every plane

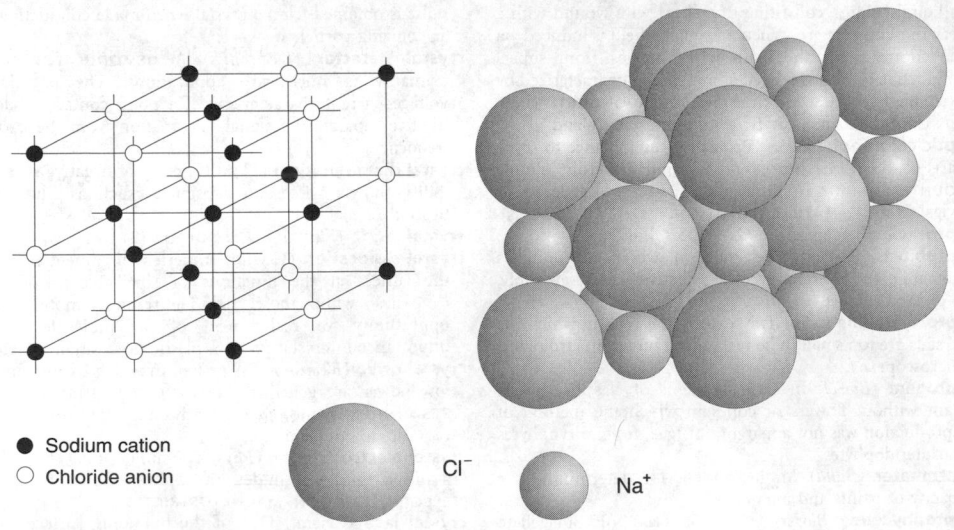

● Sodium cation
○ Chloride anion

Cl⁻   Na⁺

crystal lattice Left, the repeating cubic pattern lattice. Right, a space-filling model of the same.

Crystal lattice

A crystal lattice is the three-dimensional repeating array of points used to represent the structure of a crystal. The lattices were classified by Auguste Bravais in 1845 who showed that, owing to the effects of symmetry, there are only 14 possibly unique crystal lattices. These are shown in Fig. 1.

The Unit Cell
The 14 BRAVAIS LATTICES can be divided into four groups called unit cells which are shown in Fig. 2. Each unit cell packs in a different way in three dimensions.

The primitive crystal lattice has lattice points only at the corners of the unit cell (P-lattice).

The body-centred crystal lattice is one containing two lattice points in which the point at the centre, at the intersection of the four body diagonals, is identical with those at the corners (I-centred lattice).

The face-centred lattice is a lattice in which the unit cell has a lattice point at the centre of each face as well as at each corner (F-centred lattice).

A face-centred lattice may be face-centred on one face only, an A-, B- or C-centred lattice, depending upon the face on which it is centred.

See panel on CLOSE PACKING OF ATOMS.

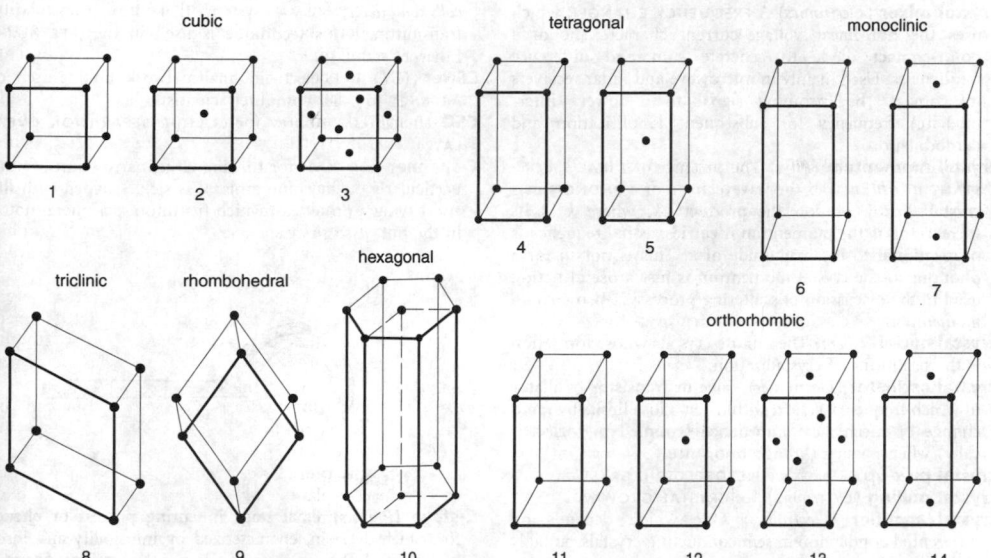

Fig. 1 **Bravais lattices** 1, 4, 6, 8, 9, 10, 11 are primitive; 2, 5, 12 body-centred; 3, 13 face-centred; 7, 14 C-centred.

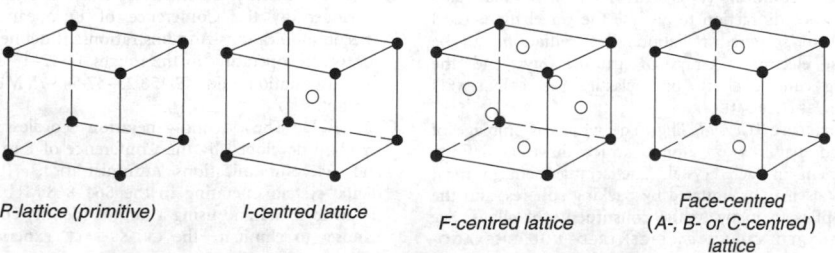

Fig. 2 **The four possible three-dimensional unit cells.**

must pass through atomic centres, and no centres must be situated between planes, the distance between successive planes in a set depends on their direction in relation to the arrangement of atomic centres.

crystallographic system (*Crystal*) Any of the major units of crystal classification embracing one or more symmetry classes.

crystallography (*Genrl*) The study of internal arrangements (ionic and molecular) and external morphology of crystal species, and their classification into types.

crystalloid (*BioSci*) A crystal of protein in eg a cell of a storage organ.

crystal loudspeaker (*Acous*) See PIEZOELECTRIC LOUD-SPEAKER.

crystal melting point (*Chem*) A phase transition where crystals melt to form liquid. For small-molecule substances, usually identical with freezing or crystallization point. For polymers it can be substantially different. Symbol T_m. See CRYSTALLIZATION OF POLYMERS, POLYMER MELTING and panel on POLYMERS.

crystal microphone (*Acous*) See PIEZOELECTRIC MICROPHONE.

crystal mixer (*Electronics*) A FREQUENCY CHANGER which uses the non-linear voltage–current characteristic of a point-contact diode to generate sum and difference frequencies. Used mainly in microwave and radar receivers to convert the incoming signal to a lower (intermediate) frequency for subsequent amplification and demodulation.

crystal momentum (*Phys*) The product of DIRAC'S CONSTANT $\hbar$ ($=h/2\pi$) and the wavevector q of a PHONON in a crystal. For a photon the product $\hbar k$, where k is its wavevector, is the momentum it carries, as its frequency is proportional to the magnitude of k. This is not so for a phonon, so the crystal momentum is just a useful fiction used in the discussion of scattering processes. Also *pseudomomentum*.

crystal nuclei (*Chem*) The minute crystals whose formation is the beginning of crystallization.

crystal oscillator (*Electronics*) Valve or transistor oscillator in which frequency is held within very close limits by rapid change of mechanical impedance (coupled piezoelectrically) when passing through resonance.

crystal pick-up (*Acous*) See PIEZOELECTRIC CRYSTAL.

crystal pulling (*Electronics*) See CRYSTAL GROWING.

crystal rectifier (*Electronics*) One which depends on differential conduction in semiconducting crystals, suitably 'doped', such as Ge or Si.

crystal sac (*BioSci*) A plant cell almost filled with crystals of calcium oxalate.

crystal set (*ICT*) A simple radio receiver using only a crystal detector.

crystal spectrometer (*Phys*) An instrument that uses crystal–lattice diffraction to analyse the wavelengths (and energies) of scattered radiation. The radiation can be neutrons, electrons, X-rays or gamma rays and the scattering can be elastic or inelastic. See TRIPLE-AXIS NEUTRON SPECTROMETER.

crystal structure (*Crystal, Eng*) The whole assemblage of rows and patterns of atoms, which have a definite arrangement in each crystal. The arrangement in most pure metals may be imitated by packing spheres, and the same applies to many of the constituents of alloys. See BODY-CENTRED CUBIC, FACE-CENTRED CUBIC, HEXAGONAL CLOSE PACKING.

crystal systems (*Crystal*) A classification of crystals based on the intercepts made on the crystallographic axes by certain planes. See BRAVAIS LATTICES and CUBIC, HEXAGONAL, MONOCLINIC, ORTHORHOMBIC, TETRAGONAL, TRICLINIC SYSTEMS, and panel on CRYSTAL LATTICE.

crystal texture (*Crystal*) The size and arrangement of the individual crystals in a crystalline mass.

crystal triode (*ICT*) Early name for TRANSISTOR.

crystal tuff (*Min*) A tuff with crystal fragments more abundant than either lithic or vitric fragments.

crystal violet (*Chem*) A dyestuff of the rosaniline series, hexamethyl-4-rosaniline.Used in Gram stain, and as an enhancer for bloody fingerprints. Formerly used topically for skin infections. Also *Gentian violet*.

Cs (*Chem*) Symbol for CAESIUM.

c/s (*Phys*) See CPS.

CSF (*BioSci*) Abbrev for COLONY STIMULATING FACTOR.

CSF (*Paper*) Abbrev for CANADIAN STANDARD FREENESS.

CS gas (*Chem*) *Orthochlorobenzylidene malononitrile.* A potent tear-gas used for crowd dispersal. See WAR GAS.

CSI (*ImageTech*) Abbrev for *compact source iodide.* TN for a type of METAL HALIDE lamp.

CSMA (*ICT*) Abbrev for *carrier sense multiple access.* A method of sharing a communications CHANNEL. Before a device sends information, the sender looks for the presence of a carrier signal indicating that the channel is already in use. If no carrier is detected, the sender transmits. In the context of a NETWORK, many WORKSTATIONS can transmit over one cable without a controlling transmission authority.

CSMA/CD (*ICT*) Abbrev for *carrier sense multiple access with collision detect.* CSMA system that also listens while transmitting. This technique is used in the IEEE 802·3 Ethernet standard.

CSNet (*ICT*) A US electronic mail network, linking users of ARPANET to other computer scientists.

CSO (*ImageTech*) Abbrev for COLOUR SEPARATION OVERLAY.

C-spanner (*Eng*) One for turning large, narrow nuts, found particularly on machine tools. It is sickle-shaped with the end having a projection which fits into a peripheral notch in the nut. Also *sickle spanner*.

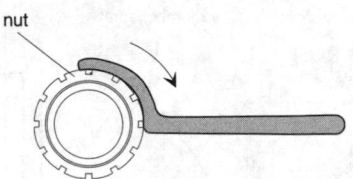

C-spanner

C-stage (*Plastics*) Final stage in curing process of phenol formaldehyde resin, characterized by infusibility and insolubility in alcohol or acetone. See panel on THERMOSETS.

CSS (*ICT*) Abbrev for *cascading stylesheet*, a computer language that describes the basic design characteristics of a document, or a web page, written in HyperText Markup Language (HTML).

CT1 (*ICT*) The first-generation cordless telephone standard developed by the Conference of European Posts and Telecommunications Administrations. It defines an analogue system operating in the ranges 1·642–1·782 MHz for the base station and 47·456 25–47·543 75 MHz for the handset.

CT2 (*ICT*) The second-generation cordless telephone standard developed by the Conference of European Posts and Telecommunications Administrations. It defines a digital system operating in the 864–868 MHz ultrahigh-frequency band and using a unique identity code for each handset to eliminate the CROSS-TALK experienced with CT1 telephones. See TELEPOINT.

ctene (*BioSci*) One of the comb-plates or locomotor organs of CTENOPHORA, consisting of a row of strong cilia of which the bases are fused.

ctenidium (*BioSci*) Generally, any comb-like structure; in aquatic invertebrates a type of gill consisting of a central axis bearing a row of filaments on either side; in insects, a row of spines resembling a comb.

ctenoid (*BioSci*) Said of scales that have a comb-like free border.

Ctenophora (*BioSci*) A phylum of triploblastic animals showing biradial symmetry. They have a system of gastrovascular canals and typically eight meridional rows of swimming plates or ctenes, composed of fused cilia. See acorns, sea gooseberries.

CTP (*Print*) Abbrev for COMPUTER-TO-PLATE.

CTR (*Electronics*) Abbrev for *comparative tracking resistance*. See TRACKING RESISTANCE.

CTR (*NucEng*) Abbrev for *controlled thermonuclear reactor*, or *reaction*. See FUSION REACTOR.

Ctrl/CTRL key (*ICT*) Abbrev for *control key*. A special key on the keyboard that when pressed simultaneously with another will have a special function; eg CTRL and Y will delete a line of text in some EDITORS.

Cu (*Chem*) Symbol for COPPER.

Cuban 8 (*Aero*) Aerobatic manoeuvre in a vertical plane consisting of $\frac{3}{4}$ loop, $\frac{1}{2}$ roll, $\frac{3}{4}$ loop and $\frac{1}{2}$ roll.

Cuban mahogany (*For*) See AMERICAN MAHOGANY.

cube (*MathSci*) (1) A regular three-dimensional solid with six square faces. A square parallelepiped. Its volume is the length of a side raised to third power. (2) The third power of a given quantity: written $a^3 = a \times a \times a$.

cubeoctohedron (*Crystal*) A solid body made up of eight square faces joined at the corners, connected by eight equilateral triangles. It can also be thought of as a cube with each corner chamfered. When twelve equal spheres are packed such that their centres are at the vertices of a cubeoctohedron the internal void will just accommodate one further such sphere.

cube root (*MathSci*) A number or polynomial whose third power is a given number or polynomial.

cubical antenna (*ICT*) One consisting of two or more square loops, with sides one quarter-wavelength long, spaced by one quarter-wavelength; maximum radiation is along the axis of the loops. Also *cubical-quad antenna*.

cubical epithelium (*BioSci*) A form of columnar epithelium in which the cells are short.

cubical-quad antenna (*ICT*) See CUBICAL ANTENNA.

cubic close packing (*Chem*) The stacking of spheres formed by placing close-packed layers in the sequence ABCABC. The unit cell of such an arrangement is a face-centred cube, with four atoms per cell. This structure is adopted by many metals, eg Cu, Ag and Au. See panel on CLOSE PACKING OF ATOMS.

cubic crystal lattice (*Crystal*) See panel on CRYSTAL LATTICE.

cubic equation (*MathSci*) An algebraic equation of the third degree. The usual standard form is

$$x^3 + 3ax + b = 0$$

whose solution can be expressed as

$$x = \sqrt[3]{p} - \frac{a}{\sqrt[3]{p}}$$

where

$$p = -b + \sqrt{b^2 + 4a^3}$$

and where, to obtain all three solutions, all three cubic roots of *p*, including the complex roots, have to be used. Unlike the quadratic, this solution is of little practical value because complex cube roots can only be determined by solving cubic equations. The general equation $y^3 + uy^2 + vy + w = 0$ is reduced to standard form by the substitution $y = x - 1/3u$.

cubicle-type switchboard (*ElecEng*) See CELLULAR-TYPE SWITCHBOARD.

cubic system (*Crystal*) The crystal system which has the highest degree of symmetry; it embraces such forms as the cube and octahedron. Also *isometric system*. See fig. at BRAVAIS LATTICES.

cubic zirconia (*Min*) Artificial diamond simulant; difficult to detect difference.

cubing (*Build*) An approximate method for estimating costs of buildings. The volume of a building is multiplied by a figure known from experience to represent a fair average figure for the cost of unit volume of such building.

cubital (*BioSci*) See SECONDARY.

cubital remiges (*BioSci*) The primary quills connected with the ulna in birds.

cuboid (*MathSci*) A solid with perpendicular rectangular faces, a rectangular parallelepiped.

Cuboni test (*Vet*) A test for pregnancy in the mare, based on the chemical detection of estrogens in the urine.

cucullate (*BioSci*) Hood-shaped.

cud (*Agri*) Ingested food regurgitated from the first stomach of a ruminant back to the mouth for chewing.

cue (*Psych*) Any word or action that serves as a signal or trigger to begin a speech, action, operation, etc; a stimulus that produces a response.

cue marks (*ImageTech*) Dots or circles appearing in the corner of the frame near the end of a reel to warn the projectionist to prepare for a CHANGE-OVER.

cuesta (*Geol*) Common US term for a hill or ridge with steep slope on one side and a gentle dip-slope on the other. Also *escarpment*.

cue track (*ImageTech*) A track carrying coded information to aid in video POST-PRODUCTION (usually a secondary audio track).

cuffing (*Med*) The accumulation of white cells round a blood vessel in certain infections of the nervous system.

cuirasse respirator (*Med*) A respirator which is attached, like armour, to the chest wall only, and assists respiration by fluctuations in its own internal pressure.

cullet (*Glass*) Waste glass added with the raw materials to accelerate the rate of melting of the batch. See panel on GLASSES AND GLASS-MAKING.

culling (*Agri*) Removal and slaughter of livestock from a population as part of a management strategy.

culm (*BioSci*) The stem, esp the flowering stem, of grasses and sedges.

culm (*Geol*) The rocks of Carboniferous age in the south-west of England, consisting of fine-grained sandstones and shales, with occasional thin banks of crushed coal or 'culm'.

culmination (*Astron*) The highest or lowest altitude attained by a heavenly body as it crosses the meridian. *Upper culmination* (or *transit*) indicates its meridian transit above the horizon, *lower culmination* its meridian transit below the horizon or, in the case of a circumpolar, below the elevated pole.

cultivar (*BioSci*) A subspecific rank used in classifying cultivated plants and indicated by the abbreviation cv and/or by placing the name in single quotation marks; defined as an assemblage of cultivated plants that is clearly distinguished by any characters (morphological, physiological, cytological, chemical, etc), and which when reproduced (sexually or asexually, as appropriate) retains its distinguishing characters.

cultivation (*Agri*) Intensive crop production typically involving human intervention in soil structure, nutrient levels and water regimes.

cultivator (*Agri*) A powered but hand-guided tool with rotating tines, used to disturb the soil surface when preparing small areas for cultivation.

culture (*BioSci*) A micro-organism, tissue or organ growing in or on a medium or other support; to cultivate such in this way. See PLANT TISSUE CULTURE.

culvert (*CivEng*) Construction for the total enclosure of a drain or watercourse.

cumarone (*Chem*) See COUMARONE.

cumene (*Chem*) *Isopropyl* (*2-methylethyl*)-*benzene*. $C_6H_5(CH_3)_2$, bp 153°C.

cummingtonite (*Min*) Hydrated magnesium iron silicate, a monoclinic member of the amphibole group, occurring in metamorphic rocks. Differs from *grunerite* in having magnesium in excess of iron.

cumulative distribution (*Chem*) In an assembly of particles, the fraction having less than a certain value of a common property, eg size or energy. Cf FRACTIONAL DISTRIBUTION.

cumulative distribution function (*MathSci*) A function giving the probability that a corresponding continuous random variable takes a value less than or equal to the argument of the function.

cumulative dose (*Radiol*) Integrated radiation dose resulting from repeated exposure.

cumulative errors (*CivEng, MathSci*) See SYSTEMATIC ERRORS.

cumulative excitation (*Electronics*) Successive absorption of energy by electrons in collision, leading to ionization. See AVALANCHE EFFECT.

cumulative median lethal dose (*BioSci*) Estimate of the total administered amount of a substance, given repeatedly in doses that are well below the median lethal dose, that is associated with the death of half a population of animals.

cumulatively compound machine (*ElecEng*) A compound-wound machine in which the series and shunt windings assist each other.

cumulonimbus (*EnvSci*) Heavy and dense *cloud*, with a considerable vertical extent, in the form of a mountain or huge towers. At least part of its upper portion is usually smooth, or fibrous or striated, and nearly always flattened; this part often spreads out in the shape of an anvil or vast plume. Under the base of this cloud, which is often very dark, there are frequently low, ragged clouds either merged with it or not, and precipitation sometimes in the form of VIRGA. Abbrev *Cb*.

cumulus (*BioSci*) The mass of cells surrounding the developing OVUM in mammals.

cumulus (*EnvSci*) Detached *clouds*, generally dense and with sharp outlines, developing vertically in the form of rising mounds, domes or towers, of which the bulging upper part often resembles a cauliflower. The sunlit parts of these clouds are mostly brilliant white; their base is relatively dark and nearly horizontal. Sometimes ragged. Abcumulus are ragged. Abbrev *Cu*.

cumulus oöphorus (*BioSci*) See ZONA GRANULOSA.

cuneate (*BioSci*) Wedge-shaped. Also *cuneal, cuneiform*.

Cunningham correction (*PowderTech*) Modification of STOKES' LAW applied when particles of an aerosol are small compared with the mean free path of the molecules of the gas through which they are falling.

cup-and-cone fracture (*Eng*) A characteristic form of tensile fracture of ductile metals in which one side of the break is flat topped and cone shaped, while the other side is a matching cup. Caused by the fracture starting as a plane crack in the mid-section of the test piece and at 90° to the tensile axis, which then develops as a 45° shear fracture as it approaches the outer surface.

cup chuck (*Eng*) A lathe chuck in the form of a cup or bell screwed to the mandrel nose. The work is gripped by screws in the walls of the chuck. Also *bell chuck*.

cupel (*Eng*) A thick-bottomed shallow dish made of bone ash; used in the cupellation of lead beads containing gold and silver, in the assay of these metals.

cupellation (*Eng*) The operation employed in recovering gold and silver from lead. It involves the melting of the lead containing these metals and its oxidation by means of an air-blast.

cupferron (*Chem*) *Ammonium-nitroso-β-phenylhydrazine*. Reagent used in the colorimetric detection and estimation of copper.

cup flow figure (*Plastics*) The time in seconds taken by a test mould of standard design to close completely under pressure, when loaded with a charge of moulding material, used esp for THERMOSETS (panel).

cup head (*Eng*) A rivet or bolt head shaped like an inverted cup.

cupid's darts (*Min*) See FLÈCHES D'AMOUR.

cup leather (*Eng*) A ring of leather moulded to U-section, used as a seal in hydraulic machinery. Now replaced in many applications by more durable materials like neoprene.

cupola (*Build*) A LANTERN constructed on top of a dome.

cupola (*Geol*) A dome-shaped offshoot rising from the top of a major intrusion.

cupola furnace (*Eng*) A shaft furnace used in melting pig iron (with or without iron or steel scrap) for iron castings. The lining is firebrick. Metal, coke and flux (if used) are charged at the top, and air is blown in near the bottom.

cupped wire (*Eng*) Wire in which internal cavities have been formed during drawings.

cuprammonia (*Chem*) A solvent for cellulose, prepared by adding ammonium chloride and then excess of caustic soda to a solution of a copper (II) salt, washing and pressing the precipitate, and dissolving it in strong ammonia.

cupric (*Chem*) *Copper (II)*. Containing divalent copper. Copper (II) salts are blue or green when hydrated and are stable.

cupriferous pyrites (*Min*) See CHALCOPYRITE.

cuprite (*Min*) Oxide of copper, crystallizing in the cubic system. It is usually red in colour and often occurs associated with native copper; a common ore.

cupro-nickel (*Eng*) An alloy of copper and nickel; usually contains 15, 20 or 30% nickel; is very ductile and has high resistance to corrosion. See COPPER ALLOYS.

cuprous (*Chem*) *Copper (I)*. Containing monovalent copper. Soluble copper (I) salts generally disproportionate to copper (0) and copper (II) salts.

cup shake (*For*) A SHAKE between concentric layers of wood. Also *ring shake*.

cupula (*BioSci*) Any dome-like structure, eg the apex of the lungs, the apex of the cochlea.

cupule (*BioSci*) One of a number of more or less cup-shaped organs, esp the structure that encloses the fruits of oak, beech, chestnut, birch, etc, eg the acorn cup (Cupuliferae).

cup wheel (*Eng*) An abrasive wheel in the form of a cylinder, used mainly for grinding. The cylindrical surfaces may be perpendicular or inclined to the end. Very shallow cup wheels are known as dish wheels or saucers.

curare (*Pharmacol*) S American native poison from the bark of species of *Strychnos* and *Chondodendron*.

curarine (*Pharmacol*) Paralysing toxic alkaloid ($C_{19}H_{26}ON_2$) extracted as *d*-tubocurarine chloride from crude curare; used in anaesthesia as a muscle relaxant.

curb (*Build*) (1) A wall-plate carrying a dome at the springings. Also *curb-plate*. (2) A frame of upstand around an opening in a floor or roof.

curb (*MinExt*) Framework fixed in rock of mine shaft to act as foundation for brick or timber lining.

curb (*Vet*) A swelling occurring just below the point of the hock of the horse, usually due to inflammation caused by sprain of the calcaneo-cuboid ligament.

curb-plate (*Build*) See CURB.

curb-roof (*Build*) See MANSARD ROOF.

curcumin (*FoodSci*) Principal pigment of turmeric with a natural yellow colour. Virtually insoluble in water, unstable to light but affected by metal ions. Stable between pH 1 and 7 with the colour changing to orange above pH 7.

curd (*FoodSci*) (1) Gelled emulsion made with fruit and/or fruit juice, sugar, starch, fat, egg and pectin. A fruit curd must legally contain the named fruit, while in a fruit-flavoured curd it is replaced by flavouring. (2) Starting material in CHEESE production.

cure curve (*Chem*) A plot of viscosity or torque against time for cross-linking polymer system, usually referring to rubbers. Viscosity or torque usually measured with oscillating disk rheometer or Mooney viscometer at standard temperature. Curve usually shows an initial drop, then rises gradually to a plateau where vulcanization is complete. Important for assessing vulcanizing agents etc. See CURE RATE INDEX, OVERCURE, SCORCH, UNDERCURE.

cure cycle (*Chem*) Usually a rubber technology term for the sequence of steps during vulcanization, from initiation of cross-linking to final set of product. Also more widely applied to all THERMOSETS (panel).

cure rate index (*Chem*) A term applied to average gradient of main part of cure curve of VULCANIZING rubber system.

curettage (*Med*) The scraping of the walls of cavities (esp of the uterus) with a curette (or curet), a spoon-shaped instrument.

curie (*Phys*) Unit of radioactivity: 1 curie is defined as $3 \cdot 700 \times 10^{10}$ decays per second, roughly equal to the activity of 1 g of radium-226. Symbol Ci. Now replaced by the BECQUEREL. See panel on RADIATION.

curie balance (*Phys*) A torsion balance for measuring the magnetic properties of non-ferromagnetic materials by the force exerted on the specimen in a non-uniform magnetic field.

Curie constant (*Phys*) See CURIE–WEISS LAW.

Curie point (*Phys*) Also *Curie temperature*. Symbol T_c. (1) Temperature above which a ferromagnetic or ferrimagnetic material becomes paramagnetic. Also *magnetic transition temperature*. (2) Temperature (*upper Curie point*) above which a ferroelectric material loses its polarization. (3) Temperature (*lower Curie point*) below which some ferro-electric materials lose their polarization.

Curies' law (*Phys*) A law stating that, for paramagnetic substances, the magnetic susceptibility is inversely proportional to the absolute temperature.

Curie temperature (*Phys*) See CURIE POINT.

Curie–Weiss law (*Phys*) A relation giving the paramagnetic susceptibility χ of a material well above the Curie temperature (T_c) at which a ferromagnetic material becomes paramagnetic: $\chi = C/(T-T_c)$, where T is the absolute temperature and C is a constant (the *Curie constant*).

curine (*Chem*) $(C_{18}H_{19}O_3N)_2$, an alkaloid of the quinoline group, found in *curare* extract obtained from various *Strychnos* spp.

curing (*Chem*) (1) A term applied to a fermentation or ageing process of natural products, eg rubber, tobacco, etc. (2) The chemical process undergone by a thermosetting plastic by which the liquid resin cross-links to form a solid. This may be initiated, or accelerated, by heat. Curing generally takes place during the moulding operation, and may require from a few seconds to several hours for its completion. Cf C-STAGE, SETTING.

curing (*CivEng*) A method of reducing the rate of contraction of concrete on setting either by spraying water or covering with wet hessian, or, more recently, by covering with a curing membrane (plastic sheet).

curing (*FoodSci*) Preservation of fish and meat by smoking or dry salting with sodium chloride and nitrates and nitrites of sodium or potassium. Sodium chloride alone inhibits microbial growth, but often only at unpalatable levels. Nitrates are reduced to nitrites by bacterial action and prevent the growth of both lactobacilli and the spores of *Clostridium botulinum*. The action of nitrite on meat gives the red colour seen in bacon ham, corned beef and pork-pie fillings. See SMOKING.

curium (*Chem*) Artificially produced radioactive element, symbol Cm, at no 96, derived from americium. There are several long-lived isotopes (up to $1 \cdot 7 \times 10^7$ years half-life), all α-emitters.

curl (*For*) A roughly hewn block of timber cut from a crotch and intended for cutting into veneers.

curl (*MathSci*) For any small vector of any shape at any point in a vector field, the line integral of the vector **v** around its bounding edge will result in an orientation of the area for which the line integral is greatest. The amount of this maximum line integral, expressed per unit area, is called the curl of the vector field at the point and is given the vectorial sense of the positive normal drawn on the small exploring area when in the position giving the greatest integral.

curl (*Paper*) A paper defect caused by unequal dimensions of the top and under sides of the sheet due to changes in the ambient moisture or temperature.

curled toe paralysis (*Vet*) A disease of chicks characterized by leg weakness and inward curling of the toes, associated with degenerative changes in the peripheral nerves; caused by a deficiency of riboflavin in the diet.

curly grain (*For*) A wavy pattern on the surface of worked timber due to the undulate course taken by the vessels and other elements of the wood.

current (*Genrl*) The flow of a substance, such as water or air.

current (*Phys*) The rate of flow of charge in a substance, solid, liquid or gas. Conventionally, it is opposite to the flow of (negative) electrons, this having been fixed before the nature of the electric current had been determined. Practical unit of current is the AMPERE.

current amplification (*ElecEng*) The ratio of output current to input current of an amplifier or photomultiplier, often expressed in decibels.

current antinode (*ElecEng*) A point of maximum current in a standing-wave system along a transmission line or aerial.

current attenuation (*ElecEng*) The ratio of output to input currents of a transducer, expressed in decibels.

current balance (*ElecEng*) A form of balance in which the force required to prevent the movement of one current-carrying coil in the magnetic field of a second coil carrying the same current is measured by means of a balancing mass. Cf MAGNETIC BALANCE.

current bedding (*Geol*) See CROSS-BEDDING.

current-carrying capacity (*ElecEng*) The current which a cable can carry before the temperature rise exceeds a permissible value (usually 40°C). It depends on the size of the conductor, the thermal resistances of the cable and the surrounding medium.

current circuit (*ElecEng*) The electrical circuit associated with the current coil of a measuring instrument or relay.

current coil (*ElecEng*) A term frequently used with wattmeters, energy meters or similar devices, to denote the coil connected in series with the circuit and therefore carrying the main current.

current collector (*ElecEng*) The device used on the vehicles of an electric traction system for making contact with the overhead contact wire or the conductor rail. See BOW, PANTOGRAPH, PLOUGH, TROLLEY SYSTEM.

current density (*ElecEng*) Current flowing per unit cross-sectional area of conductor or plasma, expressed in amperes per square metre.

current efficiency (*ElecEng*) The ratio of the mass of substance liberated in an electrochemical process by a given current, to that which should theoretically be liberated according to FARADAY'S LAW.

current feed (*ICT*) Delivery of radio power to a current maximum (loop or antinode) in a resonating part of an antenna.

current feedback (*ElecEng, Electronics*) In amplifier circuits, a feedback voltage proportional to the load current. It may be applied in series or in shunt with the source of the input signal. See NEGATIVE FEEDBACK, POSITIVE FEEDBACK, VOLTAGE FEEDBACK.

current gain (*Electronics*) In a transistor, the ratio of output current to input current. In common-emitter configuration it may be as high as 100, while in common-base it does not exceed unity.

current generator (*Phys*) Ideally, a current source of infinite impedance such that the current will be unaltered by any further impedance in its circuit. In practice, a generator whose impedance is much higher than that of its load.

current limiter (*ElecEng*) A component which sets an upper limit to the current which can be passed.

current margin (*ElecEng*) In a relay, the difference between the steady-state current needed to just operate the relay and that due to a signal on the winding.

current node (*ElecEng*) A point of zero electric current in a standing-wave system along a transmission line or aerial.

current regulator (*ElecEng*) A circuit employed to control the current supplied to a unit.

current resonance (*ElecEng*) The condition of a circuit when the magnitude of a current passes through the maximum as the frequency is changed. Obsolete terms: *syntony, tuning*.

current saturation (*ElecEng*) The condition when anode current in a triode valve has reached its maximum value.

current sensitivity (*ElecEng*) The effect of a given change in current on the reading of a current-measuring instrument.

current transformer (*ElecEng*) (1) One designed to be connected in series with a circuit, drawing predetermined current. Also *series transformer*. (2) Winding enclosing conductor of heavy ac; steps down the current in known ratio for measurement.

current weigher (*ElecEng*) See CURRENT BALANCE.

cursor (*Eng*) The adjustable fiducial part of a drawing or other instrument, with an engraved line on metal or a transparent window, both placed to reduce parallax error. See VERNIER.

cursor (*ICT*) Character, often flashing on and off, that indicates the current display position on a VIDEO DISPLAY UNIT. See ADDRESSABLE CURSOR.

cursorial (*BioSci*) Adapted for running.

curtail step (*Build*) A step which is not only the lowest step in a flight but is also shaped at its outer end to the form of a scroll in plan.

curtain (*NucEng*) Neutron-absorbing shield, usually made of cadmium.

curtain antenna (*ICT*) A large number of vertical radiators or reflectors in a plane.

curtain wall (*Build*) A thin wall whose weight is carried directly by the structural frame of the building, not by the wall below.

curtain walling (*Build*) Large prefabricated framed sections of lightweight material often predecorated on the exterior surface.

curtate cycloid (*MathSci*) See ROULETTE.

curtate trochoid (*MathSci*) See ROULETTE.

Curtis winding (*ElecEng*) The winding of low-capacitance and low-inductance resistors in which the wire is periodically reversed.

curvature (*MathSci*) (1) Of a plane curve: the rate of change (at the point) in the angle ψ, which the tangent makes with a fixed axis, relative to the arc length s. It is thus defined as

$$K = \frac{d\psi}{ds} = \left(\frac{d^2 y}{dx^2}\right) / \left[1 + \left(\frac{dy}{dx}\right)^2\right]^{\frac{3}{2}}$$

K is the reciprocal of the radius of curvature, p, which is the radius of the circle which touches the curve (on the concave side) at the point in question. The circle is the *circle of curvature*, and its centre is the *centre of curvature* of the curve at the point. The circle of curvature is also called the *osculating* circle. (2) Of a space curve: the rate of change of direction of the tangent with respect to the arc length, ie

$$K = \frac{1}{p} = \frac{d\theta}{ds}$$

where as before K is the curvature, p the radius of curvature, θ represents the change in direction of the tangent, and s is the length. This is called the *first curvature of a space curve*. The second curvature, or *torsion*, of the curve is the corresponding rate of change in the direction of the binormal, ie

$$\lambda = \frac{d\psi}{ds}$$

where ψ is the angle through which the binormal turns. $1/\lambda$ is the radius of torsion. Cf MOVING TRIHEDRAL, OSCULATING SPHERE. (3) Of a surface: at any point P on the surface there is, in general, a single normal line. Planes through this line cut the surface in plane curves called *normal*

curvature at P in that direction. The maximum and minimum values of the normal curvature at P are called the *principal curvatures at P*, and the directions in which they occur, which will be mutually perpendicular, are called the *principal directions at P*. The average of the principal curvatures at P is called the *average, mean* or *mean normal curvature at P*, and their product, ψ, the *total, total normal* or *Gaussian curvature at P*. The reciprocals of the normal and of the principal curvatures at P are called the *normal* and *principal radii of curvature at P* respectively.

curvature correction (*CivEng*) A correction used in the calculation of quantities for earthworks following a curved line in plan; the quantities are taken out as if the line were straight, and a curvature correction made depending on the radius.

curvature of field (*Phys*) For an optical system, the inevitable curvature of the image field of a planar object normal to the axis. Even in the absence of ASTIGMATISM, COMA and SPHERICAL ABERRATION, the image surface will be curved, the PETZVAL CURVATURE.

curvature of spectrum lines (*Phys*) The slightly convex curvature towards the red end of a spectrum produced by a prism. Rays from the ends of the slit are inclined at a small angle to the plane at right angles to the refracting edge of the prism, and so suffer a slightly greater deviation than rays from the centre of the slit, appearing bent towards the violet end of the spectrum.

curve (*Eng*) An instrument used by the technical artist for drawing curves other than circular arcs. It consists of a thin flat piece of transparent plastic or other material, having curved edges which are used as guides for the pencil.

curve (*MathSci*) The locus of a point moving with one DEGREE OF FREEDOM.

curve fitting (*Phys*) The process of finding the best algebraic function to describe a set of experimental measurements. Usually accomplished by using a least-squares process by which the parameters of the function are adjusted to minimize the sum of the squares of the deviations of the observations from the theoretical curve.

curve of pursuit (*Aero*) That path followed by a combat aircraft steering towards the present position of an adversary.

curve ranging (*Surv*) The operation of setting out on the ground points which lie on the line of a curve of given radius.

curvilinear asymptote (*MathSci*) See ASYMPTOTE.

curvilinear co-ordinates (*MathSci*) (1) Of a point in space: the specification of the position of the point as the intersection of three surfaces, one from each of three families of surfaces. If the three families are mutually orthogonal, the co-ordinates are *orthogonal curvilinear co-ordinates*. (2) Of a point on a surface: the values of the parameters that identify that point when the surface is described by parametric equations.

curvilinear distortion (*ImageTech, Phys*) Curvature of lines which should be straight, as seen in the outer portion of the image from a stopped-down simple lens. See BARREL DISTORTION, PINCUSHION (PILLOW) DISTORTION.

Cushing's syndrome (*Med*) The concurrence of obesity, hairiness, linear atrophy of the skin, loss of sexual function, and curvature of the spine, due to a tumour in the pituitary or adrenal gland, causing oversecretion of corticosteroids.

cushion (*Build*) The capping stone of a pier.

cushion craft (*Ships*) The name given to certain types of HOVERCRAFT.

cushion plant (*BioSci*) A plant with many densely crowded upright shoots not more than a few centimetres high, forming a cushion-like mass on the ground; typical of alpine and arctic floras. Also *chamaephyte*.

cushion steam (*Eng*) The steam shut in the cylinder of a steam engine after the closing of the exhaust valve.

Cushyfoot (*Eng*) TN for type of inclined shear mount using a rubber and steel plate laminate.

cusp (*Astron*) One of the horns of the Moon or of an inferior planet in the crescent phase.

cusp (*BioSci*) A sharp-pointed prominence, as on teeth.

cusp (*MathSci*) See DOUBLE POINT. Adj *cuspidate*.

cuspy (*ICT*) Computing slang describing a computer program that is well-written and easy to use.

customer access connection (*ICT*) General term for the final link to a telephone subscriber's premises, used when it is required to leave open whether the link is via wire, fibre or radio.

customer premises equipment (*ICT*) Any part of a telecommunications network that is installed at a customer's premises, eg a telephone or a PRIVATE BRANCH EXCHANGE.

customs plant (*MinExt*) A crushing or concentrating plant serving a group of mines on a contract basis. It buys ore according to valuable content and complexity of treatment, and relies for profit on sale of products.

cusum (*MathSci*) Contraction of *cumulative sum analysis*. The cumulative sum of the deviations of actual versus expected values from the first sampling period up to and including the current period. Extensively used in process monitoring: a change in slope of the cusum chart indicates that there is a change in the way the process is behaving. The absolute value of the cusum is not significant, but the change in value over any period of time represents the cumulative loss or saving.

cut (*Eng*) The thickness of the metal shaving removed by a cutting tool.

cut (*ImageTech*) (1) 'Cut!', instruction by the director to stop shooting. (2) In an edited film, an instantaneous change from one scene to another.

cut (*MinExt*) A petroleum faction.

cut (*NucEng*) The proportion of input material to any stage of an isotope separation plant which forms useful product. Also *splitting ratio*.

cut (*Print*) US for BLOCK.

cut-and-cover (*CivEng*) A method of constructing shallow tunnels by excavating from the surface, forming an invert, walls and roof in the cut and then backfilling the ground around and over the tunnel structure.

cut-and-fill (*CivEng*) A term used to describe any cross-section of highway or railroad earthworks which is partly in cutting and partly in embankment.

cut-and-mitred string (*Build*) A CUT STRING which is mitred at the vertical parts of the notches in the upper surface, so that the end grain of the risers may be concealed. See fig. at STRING.

cut-and-mitred valley (*Build*) A valley formed in a tiled roof by cutting one edge of the tiles on both sides of the valley so that they form a mitre, which is rendered watertight by lead soakers bonded in with the tiles.

cut and paste (*ICT*) A technique of transferring a section of data from one part of a document to another part of the same document or to another document. In some systems a temporary storage area is used to effect this transfer and is known as the CLIPBOARD or NOTEPAD.

cut and sew (*Textiles*) See CUT-UP TRADE.

cutaneous (*BioSci*) Pertaining to the skin.

cutback (*Chem*) BITUMEN that has been diluted with suitable solvents, eg kerosine, to make it liquid and easier to handle.

cut edges (*Print*) Said of the edges of a book when all three are clean cut as distinct from trimmed edges, where only the furthest projecting leaves at the tail are trimmed.

cut flush (*Print*) Cut after limp-binding, so that the cover does not project.

cuticle (*BioSci*) (1) In general, a non-living layer secreted by and overlying the epidermis, as in arthropods. (2) In plants, a layer of cutin on the outside of some plant-cell walls, esp the shoot epidermis. The cuticle forms a continuous layer that, with the epicuticle, has relatively low permeability to water and gases.

cuticle (*Textiles*) Flat overlapping scales that lie on the surface of animal hair and wool. They cover the internal core or cortex.

cuticular transpiration (*BioSci*) The loss of water vapour from a plant through the cuticle.

cuticulin (*BioSci*) The outermost layer of the insect epicuticle, consisting of lipoprotein.

cutin (*BioSci*) A mixture of fatty substances esp of cross-linked polyesters based on mostly C_{16} and C_{18} aliphatic acids and hydroxyacids in the CUTICLE of plants.

cutinization (*BioSci*) The formation of CUTIN; the deposition of cutin in a cell wall to form a cuticle.

cut-in notes (*Print*) Notes occupying a rectangular space, set into the text at the outer edge of a paragraph.

cutis (*BioSci*) The dermis or deeper layer of the vertebrate skin.

cut-off (*Eng*) The point in an engine cycle, expressed as percentage of stroke, at which the supply of steam, fuel oil, etc, is stopped.

cut-off (*Print*) A feature of reel-fed presses; the paper is cut after printing to a size determined by the cylinder periphery; a few models have a selection of cylinder sizes and a consequently variable cut-off.

cut-off current (*Electronics*) The residual current flowing in a valve or transistor when the device is biased off in a specified way.

cut-off field (*Electronics*) See CRITICAL FIELD.

cut-off frequency (*Electronics, ICT*) (1) That above (or below) which gain, efficiency, or other desirable characteristic of a circuit or device is changing so rapidly that it is no longer useful, eg for an amplifier, cut-off frequency is commonly taken as that when gain is 3 dB less than the mid-band value. See ALPHA CUT-OFF, CRITICAL FREQUENCY. (2) For any specified mode of propagation in a lossless waveguide or other structure, the frequency at which the attenuation constantly changes from zero to a positive value or vice versa. See CUT-OFF WAVELENGTH.

cut-off knife (*Print*) A plain or serrated blade which severs each copy on a reel-fed rotary.

cut-off posture (*Psych*) In ethology, a term referring to postures that remove social stimuli (eg a potential mate, or opponent) from sight, and thus may serve to reduce the actor's arousal in a conflict situation.

cut-off rubbers (*Print*) The rubber strip set in a cylinder against which the CUT-OFF KNIFE presses when cutting the product into copies. Also *cutting buffer, cutting strip*.

cut-off wavelength (*ICT*) The free space wavelength corresponding to the lowest frequency at which a waveguide or other propagation structure can support a particular mode or field pattern.

cut-off wheel (*Eng*) Thin abrasive wheel made of flexible material which is used to cut metals, concrete, etc.

cut-out (*ElecEng*) Off-switch operated automatically if safe operating conditions are not maintained, eg water flow cut-out.

cut-out half-tone (*Print*) A half-tone from which the background is removed to give prominence to the subject.

cut-over (*ICT*) The rapid transfer of large numbers of subscribers' lines from one exchange to another, particularly from an electromagnetic (*Strowger*) to an electronic exchange.

cut-stone (*Build*) A stone hewn to shape with a chisel and mallet.

cut string (*Build*) A string whose upper surface is shaped to receive the treads and risers of the steps, while the lower surface is parallel to the slope of the stair. Also *open string*. See fig. at STRING.

cutter (*Acous*) The sapphire or diamond point which removes the thread of lacquer in vinyl disk recording.

cutter (*Eng*) Any tool used for severing, often more specifically a milling cutter. See MILLING MACHINE.

cutter dredge (*CivEng*) Alluvial dredge which loosens material by means of powered cutting ring, draws it to a pump and delivers it for treatment aboard an adjacent plant.

cutter loader (*MinExt*) Coal-cutting machine which both severs the mineral and loads it onto a transporting device such as a face conveyor.

cutters (*Build*) Bricks which are made soft enough to be cut with a trowel to any shape required, and then rubbed to a smooth face and the correct shape. Also *rubbers*.

cutting (*BioSci*) A piece of a plant, usually shoot, root or leaf, that is cut off and induced to form adventitious roots and/or buds as a means of vegetative propagation. See ROOTING COMPOUND.

cutting (*CivEng*) An open excavation through a hill, for carrying a highway or railroad at a lower level than the surrounding ground.

cutting (*ImageTech*) The process of editing a film by the creative assembly of individual scenes of picture and sound to meet the director's intentions, resulting in a *cutting copy* or *work print*. Also, the actual assembly of the original negative to match.

cutting (*Vet*) See BRUSHING.

cutting buffer (*Print*) See CUT-OFF RUBBERS.

cutting compound (*Eng*) A mixture of water, oil, soft soap, etc, used for lubricating and cooling the cutting tool in machining operations. Also *coolant*.

cutting cylinder (*Print*) The cylinder that holds knives to cut the web into separate copies. See CUT-OFF KNIFE, CUT-OFF RUBBERS.

cutting disks (*Print*) See SLITTERS.

cutting gauge (*Build*) A MARKING GAUGE fitted with a bevelled cutter in place of a pin. Used for cutting thin wood and for marking across the grain to obviate tearing.

cutting list (*Build*) A list giving dimensions, sometimes with diagrams of sections, of timber required for any given work.

cutting marks (*Print*) Short lines printed onto the sheet to indicate cutting, slitting or punching positions. Positioned outwith the trimmed size.

cutting speed (*Eng*) The speed of the work relative to the cutting tool in machining operations; usually expressed in feet or metres per minute.

cutting strip (*Print*) See CUT-OFF RUBBERS.

cutting tools (*Eng*) Steel tools used for the machining of metals, eg broach, cutter, lathe tools, milling cutter, planer tools, reamer, screwing die, shaper tools, slotting tools, tap, twist drill.

cuttling (*Textiles*) Operation of folding a fabric to make it convenient to handle; also known as *plaiting*.

cut-up trade (*Textiles*) The section of the knitting industry dealing with fabric made on a circular knitting machine. The material is afterwards cut to shape from patterns, pieces being sewn together to form the final article. Also *cut and sew.*

cut-water (*CivEng*) The angular edge of a bridge pier, shaped to lessen the resistance it offers to the flow of water.

Cuvierian ducts (*BioSci*) In lower vertebrates, a pair of large venous trunks entering the heart from the sides.

CV (*Eng*) Abbrev for: (1) CHEVAL-VAPEUR; (2) CONTINUOUS VULCANIZATION.

cv (*BioSci*) Abbrev for CULTIVAR.

c-value paradox (*BioSci*) The paradox that some very similar animals and plants have unexpectedly large differences in the amount of their genomic DNA, eg amphibian genomes vary by over a hundredfold. Not simply an increase in the number of sequence copies per genome.

CVD (*Chem*) Abbrev for CHEMICAL VAPOUR DEPOSITION.

CVS (*Med*) Abbrev for CHORIONIC VILLUS SAMPLING.

CVT (*Autos*) Abbrev for CONSTANT VARIABLE TRANSMISSION.

CWD (*Med*) Abbrev for *chronic wasting disease.*

CW radar (*Radar*) *Continuous wave radar.* One in which the transmitter emits a continuous radio-frequency signal; the receiving antenna is arranged so that a very small amount of the transmitted power enters it, along with the signal reflected from the target. Movement of the target causes Doppler frequency shift in the reflected signal and this difference can be detected at the output of a mixer. CW radar uses less bandwidth than conventional pulsed radar.

cwt (*Genrl*) Abbrev for HUNDREDWEIGHT.

cyanamide process (*Chem*) The fixation of atmospheric nitrogen by heating calcium carbide (ethynide) in a stream of the gas. Calcium cyanamide, $CaCN_2$, is thus formed, and this, on treatment with water, a little sodium hydroxide and steam under pressure, yields ammonia.

cyanates (*Chem*) Salts containing the monovalent acid radical CNO'.

cyanhydrins (*Chem*) A series of compounds formed by the addition of hydrogen cyanide to aldehydes and ketones. Their general formula is $R'C(OH)(CN)R''$ and they are useful for the preparation of 2-hydroxy-acids. Also *cyanohydrins*. See ACETONE CYANHYDRIN.

cyanicide (*MinExt*) Any constituent in ore or chemical product made during treatment of gold-bearing minerals by cyanidation, which attacks or destroys the sodium or calcium cyanide used in the process.

cyanidation vat (*Eng*) A large tank, with a filter bottom, in which sands are treated with sodium cyanide solution to dissolve out gold.

cyanide-hardening (*Eng*) CASE-HARDENING in which the carbon content of the surface of the steel is increased by heating in a bath of molten sodium cyanide.

cyanides (*Chem*) (1) Salts of hydrocyanic acid. (2) See NITRILES.

cyaniding (*MinExt*) The process of treating finely ground gold and silver ores with a weak solution of sodium cyanide, which readily dissolves these metals. The precious metals were formerly recovered by precipitation from solution with zinc. Currently a number of methods have proved more economical, including adsorption on resins and carbon. See CARBON-IN-PULP, RESIN-IN-PULP.

cyanin (*Chem*) The colouring matter of the cornflower and the rose. It is an anthocyanin, and on hydrolysis yields cyanidin and two molecules of glucose.

cyanite (*Min*) See KYANITE.

cyanoacrylate (*Chem*) Methyl and ethyl 2-cyanoacrylate monomer, well known for its use as an adhesive, with no gap-filling, but excellent penetration of rough surfaces owing to low viscosity. Anionic polymerization is initiated by traces of water on the surfaces to be joined, and occurs rapidly to form a strong bond.

Cyanobacteria (*BioSci*) Modern term for the blue-green algae, prokaryotic cells that use chlorophyll on intracytoplasmic membranes for photosynthesis. According to the endosymbiont hypothesis, Cyanobacteria are the progenitors of chloroplasts. They can be unicellular, colonial or filamentous. Those with heterocysts fix nitrogen. Planktonic kinds may have gas vacuoles. They occur in fresh and salt water (planktonic and benthic), in soils and as nitrogen-fixing symbionts in *Azolla* and the roots of Cycads and some flowering plants and in some lichens. Also *Cyanophyta*.

cyanobiphenyls (*Chem*) Aromatic molecules with linked benzene nuclei, $R–C_6H_5–C_6H_5–CN$, which shows LIQUID CRYSTAL or NEMATIC properties. See LIQUID CRYSTAL PHASES.

cyanogen (*Chem*) A very poisonous, colourless gas with a smell of bitter almonds. It is soluble in four volumes of water, ammonium oxalate (ethandioate) being formed on standing. Its formula is C_2N_2, or $N{\equiv}C–C{\equiv}N$, and it somewhat resembles the halogens in its chemical behaviour.

cyanogenesis (*BioSci*) The release from plant parts, usually after wounding, of hydrogen cyanide by cytoplasmic glycosidase action on a vacuolar glycoside containing eg mandelonitrile. Occurs in leaves of cherry laurel (*Prunus laurocerasus*), seeds of bitter almonds and fronds of bracken. Possibly a deterrent to herbivores.

cyanohydrins (*Chem*) See CYANHYDRINS.

Cyanophyceae (*BioSci*) See CYANOBACTERIA.

cyanosis (*Med*) Blueness of the skin and the mucous membranes due to insufficient oxygenation of the blood.

May be peripheral due to poor circulation or central due to failure of oxygenation.

cyanotype (*ImageTech*) The ferroprussiate process, familiar as blue-printing; it depends on the light reduction of a ferric salt to a ferrous salt, with production of Prussian blue on development.

cyanuric acid (*Chem*) A tribasic, heterocyclic acid, having the formula $H_3C_3N_3O_3$. The trimer of cyanic acid which is too unstable to exist by itself, an aqueous solution being slowly converted to urea.

cyanuric dyes (*Chem*) Relatively new class of dyestuffs based on cyanuric chloride ($C_3N_3Cl_3$). Their importance lies in the ability to link chemically with the fabrics or fibres (particularly cellulosic fibres) being dyed. See REACTIVE DYES.

cybernaut (*ICT*) A SURFER, one who uses the Internet.

cybernetics (*ICT*) The study of control and communications in complex electronic systems and in animals, esp humans.

cyberpet (*Genrl*) A small interactive electronic device with a screen which requires continuous care and slightly simulates a child or pet. Also *virtual pet*.

cyberspace (*ICT*) (1) The three-dimensional environment, or space, of virtual reality, generated by computer. (2) The notional space in which electronic communication takes place over computer networks.

cyberspeak (*ICT*) The jargon of Internet users.

cybersquatting (*ICT*) The purchasing of an Internet domain name, usually one of a famous person or organization, with the intention of selling it on at a profit.

cyberstalking (*ICT*) The use of the Internet, email, etc, to harass, threaten or abuse another person.

cyberterrorism (*ICT*) A person who attempts to cause disruption through the use of computers, especially over the Internet.

cybrid (*BioSci*) A cell, callus, plant, etc, typically resulting from PROTOPLAST FUSION and PROTOPLAST CULTURE, possessing the nuclear genome of one plant with at least some part of the chloroplastal or mitochondrial genome of the other, as opposed to a *hybrid* in which some parts of both parental genomes are present. See CHLOROPLAST, SOMATIC HYBRIDIZATION.

Cycadales (*BioSci*) The cycads, an order of gymnosperms (Cycadopsida) that were widespread in the Mesozoic. There are now c.300 spp in Cl America, S Africa, SE Asia and Australia. The stems are stout, unbranched and MANOXYLIC; the leaves are large and pinnate, with haplocheilic stomata (in which the guard mother cell gives rise to only two guard cells). They are DIOECIOUS; the reproductive organs are in large cones (except for the female *Cycas*), and there are motile flagellated sperm (zooidogamous). The pith of two species is a minor source of sago.

Cycadopsida (*BioSci*) A class of gymnosperms containing the superficially similar Cycadales and Cycadeoidales and a number of other orders. It is probably not a natural group.

cyclamates (*Chem*, *FoodSci*) Derivatives of cyclohexyl-sulphamic acid, having 30 times the sweetening power of sucrose, much used in foods, drinks, and for dietary purposes; banned in some countries including the UK because of supposed health hazards.

cycle (*Genrl*) A series of occurrences in which conditions at the end of the series are the same as they were at the beginning. Usually, but not invariably, a cycle of events is recurrent.

cycle (*MathSci*) For a PERIODIC quantity or function, the set of values that it assumes during a period.

cycle of erosion (*Geol*) The hypothetical course of development followed in landscape evolution; it consists of the major stages of youth, maturity and old age.

cycle time (*Eng*) Overall time to make a shaped product. Applied esp to eg injection moulding process.

cycle-time (*ICT*) The time interval between the start and restart of a particular hardware operation, eg CPU cycle-time, memory cycle-time.

cyclic (*BioSci*) A flower having the parts arranged in whorls, rather than in spirals.

cyclic adenosine monophosphate (*BioSci*) A derivative of adenosine monophosphate in which the phosphate forms a ring involving the 3′ and 5′ hydroxyl groups of ribose. It is of major metabolic importance through its diverse *second messenger* effects on many enzymes. It is also a chemotactic factor for some cellular slime moulds. Abbrev *cAMP*, *cyclic AMP*.

cyclic compounds (*Chem*) Closed-chain or ring compounds consisting either of carbon atoms only (*carbocyclic compounds*), or of carbon atoms linked with one or more other atoms (*heterocyclic compounds*).

cyclic group (*MathSci*) A group in which every element can be expressed as a power of a single element. Cyclic groups are *Abelian*, and those of the same order are *isomorphic*.

cyclic inositol phosphates (*BioSci*) The 1,2-cyclic derivatives of inositol phosphatide that are formed during enzymic hydrolysis of phosphatidyl inositol species in cell membranes. They are important second messengers in various hormone-activated pathways.

cyclic pitch control (*Aero*) Helicopter rotor control in which the blade angle is varied sinusoidally with the blade azimuth position, thereby giving a tilting effect and horizontal translation in any desired direction.

cyclic quadrilateral (*MathSci*) A four-sided polygon whose vertices lie on a circle.

cyclic redundancy check (*ICT*) A form of error detection used in digital systems. The digital information word to be transmitted is clocked through a shift register with specified feedback taps to generate a shorter check word that is transmitted together with the information word. On reception, the information word is clocked through an identical shift register and the result compared with the received check word; any difference indicates a transmission error.

cyclic shift (*ICT*) See END-AROUND SHIFT.

cyclic test (*MinExt*) See LOCKED TEST.

cyclin-dependent kinases (*BioSci*) Enzymes, activated by *cyclins*, that phosphorylate proteins important in regulating the cell cycle. Abbrev *CDK*. See panel on CELL CYCLE.

cyclins (*BioSci*) Family of intracellular proteins whose levels alter cyclically during the phases of the cell cycle. See panel on CELL CYCLE.

cyclitis (*Med*) Inflammation of the ciliary body of the eye.

cyclo- (*Chem*) Containing a closed carbon chain or ring.

cyclo- (*Genrl*) Prefix from Gk *kyklos*, circle.

cycloalkanes (*Chem*) Hydrocarbons containing saturated carbon rings. Also *cyclanes*, *polymethylenes*.

cyclobutane (*Chem*) $(CH_2)_4$. Alicyclic compound; a cycloalkane. Bp 11°C.

cyclodextrin (*Chem*) Any of several water-soluble, cyclic carbohydrates with hydrophobic interiors.

cyclogyro (*Aero*) An aircraft lifted and propelled by pivoted blades rotating parallel to roughly horizontal transverse axes.

cyclohexanamine (*Chem*) See CYCLOHEXYLAMINE.

cyclohexane (*Chem*) C_6H_{11}. Mp 2°C, bp 81°C, rel.d. 0·78. A colourless liquid of mild ethereal odour. Molecules generally adopt a CHAIR conformation.

cyclohexanol (*Chem*) $C_6H_{11}OH$. Mp 15°C, bp 160°C, rel.d. 0·945. An oily, colourless liquid.

cyclohexanone (*Chem*) $C_6H_{10}O$. Bp 154–156°C, rel.d. 0·945. A colourless liquid, of propanone-like odour, solvent for cellulose lacquers.

cycloheximide (*BioSci*) An antibiotic isolated from *Streptomyces griseus*. It blocks eukaryotic, but not prokaryotic, protein synthesis and is a useful experimental tool.

cyclohexylamine (*Chem*) $C_6H_{11}NH_2$. Colourless liquid, bp 134°C. A reduction product of aniline, its derivatives are used in the manufacture of plastics etc. Also *cyclohexanamine*.

cycloid (*MathSci*) An arch-shaped curve with intrinsic equation $s = 4a \sin \psi$. Its parametric cartesian equations are $x = a(\theta + \sin \theta)$ and $y = a(1 - \cos \theta)$. See ROULETTE. Fig. ▷

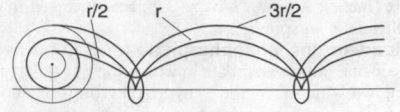

cycloid r = radius of the circle rolled along the line

cycloid (*BioSci*) Evenly curved; said of scales with an evenly curved free border.

cycloidal teeth (*Eng*) Gear teeth whose flank profiles consist of cycloidal curves.

cyclone (*EnvSci*) (1) Same as DEPRESSION. (2) A TROPICAL REVOLVING STORM in the Arabian Sea, Bay of Bengal and S Indian Ocean.

cyclone (*MinExt*) Conical vessel used to classify dry powders or extract dust by centrifugal action. See HYDRO-CYCLONE.

cyclonite (*Chem*) Hexogen, cyclotrimethylene trinitramine $(CH_2)_3(NNO_2)_3$. A colourless, crystalline solid, mp 200–202°C, odourless, tasteless, non-poisonous, soluble in acetone, prepared by oxidative nitration of hexamethylene tetramine. Used, generally with TNT, as an explosive.

cyclo-octadiene (*Chem*) C_8H_{12}. Dimerization product of butadiene obtained by using ZIEGLER CATALYSTS. Used as an intermediate in the preparation of nylon polymers from petrochemical sources.

cyclo-oxygenase (*BioSci*) An enzyme that acts on ARACHIDONIC ACID to produce prostaglandins and thromboxanes, and that can be inhibited by aspirin-like drugs. See COX-1, COX-2.

cycloparaffins (*Chem*) Same as CYCLOALKANES.

cyclopean (*Build*) A name given to ancient dry-masonry works in which the stones are very large and are irregular in size.

cyclopentanal (*Chem*) Cyclic aliphatic alcohol. C_5H_9CHO. Bp 139°C, used as an intermediate in the preparation of perfumery and flavouring esters.

cyclopentane (*Chem*) A cycloalkane with the formula C_5H_{10}. Bp 49°C. The ring of carbon atoms is nearly flat.

cyclopentanone (*Chem*) C_5H_8O. Solvent for a wide range of synthetic polymers, particularly PVC.

cyclophon (*Electronics*) Tube which uses the fundamental principle of electron-beam switching.

cyclophosphamide (*Pharmacol*) A potent alkylating drug that interferes with DNA synthesis and prevents cell replication, used in the treatment of leukaemia and lymphoma.

cycloplegia (*Med*) Paralysis of the ciliary muscle.

cyclopropane (*Chem*) A cycloalkane with the formula C_3H_6. Bp −33°C. As the C–C–C angles are constrained to be 60°C in place of the normal tetrahedral angle, the molecule is very reactive.

cyclosilicates (*Min*) Silicate minerals whose atomic structure contains rings of SiO_4 groups, eg BERYL. See SILICATES.

cyclosis (*BioSci*) See CYTOPLASMIC STREAMING.

cyclospondylous (*BioSci*) Showing partial calcification of cartilaginous vertebral centra in the form of concentric rings.

cyclosporin A (*Pharmacol*) See CICLOSPORIN.

Cyclostomata (*BioSci*) An order of the class AGNATHA, comprising aquatic and gill-breathing organisms with a round jawless suctorial mouth. The buccal cavity contains a muscular tongue bearing horny teeth, which are used to rasp the flesh from the prey. Such organisms have a cartilaginous endoskeleton and lack fins and limb girdles. They have slimy skin with no scales. Lampreys and hagfish.

cyclostrophic wind (*EnvSci*) The theoretical wind that, when blowing round circular isobars, represents a balance between the pressure gradient and the centrifugal force, the CORIOLIS FORCE being neglected; it is a useful approximation only at low latitudes, eg in tropical cyclones.

cyclothem (*Geol*) A series of beds formed during one sedimentary cycle. Particularly associated with coal-bearing rocks.

cyclotron (*Phys*) A machine in which positively charged particles are accelerated in a spiral path within DEES in a vacuum between the poles of a magnet, energy being provided by a high-frequency voltage across the dees. When the radius of the path reaches that of the dees, the particles are electrically deflected out of the cyclotron for use in nuclear experiments. See BETATRON, CYCLOTRON FREQUENCY, SYNCHROCYCLOTRON, SYNCHROTRON.

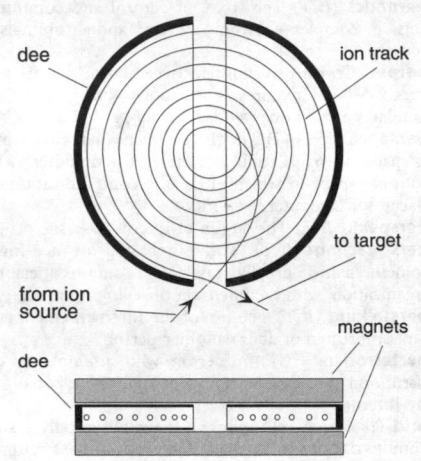

cyclotron Plan and sectional view.

cyclotron frequency (*Phys*) The number of revolutions per second of a particle of charge q moving in a circular path perpendicular to a magnetic field of flux density B, given by

$$f = \frac{Bq}{2\pi m}$$

where m is the mass of the particle. The frequency is independent of the velocity, a result that is used in the CYCLOTRON and the MAGNETRON.

cyclotron resonance (*Phys*) The resonant coupling of electromagnetic power into a system of charged particles undergoing orbital movement in a uniform magnetic field. Used for the quantitative determination of the band parameters in semiconductors. See LANDAU LEVELS.

cyclotron resonance heating (*NucEng*) Mode of heating of a plasma by resonant absorption of energy based on the waves induced in the plasma at the cyclotron frequency of electrons (abbrev *ECRH*) or ions (abbrev *ICRH*).

Cycolac (*Chem*) TN for ACRYLONITRILE–BUTADIENE–STYRENE (ABS) polymers (US).

cyesis (*Med*) Pregnancy. See PSEUDOCYESIS.

Cygnus (Swan) (*Astron*) A large northern constellation, which includes the stars DENEB, CYGNUS A and CYGNUS X-1.

Cygnus A (*Astron*) A powerful radio source in the constellation Cygnus, identified with a distant peculiar galaxy which is also an X-ray source.

Cygnus X-1 (*Astron*) An intense X-ray source in the constellation Cygnus, thought to be a binary star system in which one of the components is a black hole; the X-rays would then be produced as a result of accretion processes.

cylinder (*Eng*) The tubular chamber in which the piston of an engine or pump reciprocates; the internal diameter is called the bore, and the piston-travel the stroke.

cylinder (*ICT*) The name given to the set of tracks on one or more PLATTERS of a HARD DISK that can be read without moving the read head.

cylinder (*MathSci*) (1) A surface generated by a line which moves parallel to a fixed line so as to cut a fixed plane curve. Any line lying in the surface is called a *generator*. (2) A solid bounded by a cylindrical surface and two parallel planes (the *bases*) which cut the surface. A cylindrical surface is named after its normal sections and a solid cylinder after its bases. The axis of a cylinder (if it has one) is its line of symmetry parallel to its generators or the line joining the mid-points of its bases. A circular cylinder is called a *right circular cylinder* if its bases are normal to its axis, otherwise it is called an *oblique circular cylinder*. Adj *cylindrical*.

cylinder barrel (*Eng*) The wall of an engine cylinder, as distinct from the cylinder itself, which term includes the head or covers.

cylinder bearers (*Print*) At each end of impression, blanket and plate cylinders, to provide a datum for packing; may be integral with cylinder or be a removable band (bearer ring).

cylinder bit (*Build*) A steel drill with helical cutting edge, used for precise boring of wood. Also *Forstner bit*.

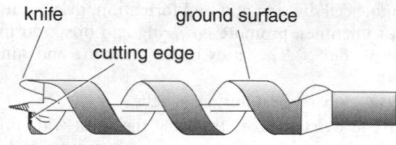

knife   ground surface

cutting edge

cylinder bit

cylinder block (*Eng*) The largest part of an internal-combustion engine which is bored to receive the pistons and contains integrally cast cooling-water channels made from cast-iron or, more recently, from lightweight aluminium alloy.

cylinder bore (*Eng*) See CYLINDER.

cylinder brakes (*Print*) A mechanism on a fast-running printing press which stops it quickly.

cylinder caisson (*CivEng*) A method once widely used for digging foundations under water in which material was excavated from within a stack of cast-iron tubes arranged so that the top was always above water level and the bottom could cut into the silt and sand.

cylinder collection (*Print*) On rotary presses, the gathering of the required number of sections or sheets round a cylinder.

cylinder cover (*Eng*) The end cover of the cylinder of a reciprocating engine or compressor.

cylinder dressing (*Print*) The layers of board, paper or other packing material required to produce the necessary impression.

cylinder-dried (*Paper*) Paper which has been dried by being passed over heated cylinders.

cylinder head (*Eng*) Removable top part of an internal-combustion engine which, when in place, provides a gastight seal for the cylinders. Contains valves, valve ports, combustion chambers and cooling-water channels.

cylinder mould machine (*Paper*) A paper or board machine in which the forming unit comprises an endless wire cloth on the surface of a hollow metal cylinder situated in a vat supplied with stock. Also *vat machine*.

cylinder press (*Print*) A general term used to distinguish cylinder printing machines from hand presses, platens and rotary machines. See SINGLE-REVOLUTION, STOP-CYLIN-DER, TWO-REVOLUTION.

cylinder wrench (*Build*) See PIPE WRENCH.

cylindrical co-ordinates (*MathSci*) Three numbers, r, θ and z, which represent the position of a point in space, the first two numbers r and θ representing, in polar co-ordinates,

the position of the projection of the point on a reference plane, and the third, z, representing the height of the point above the reference plane. Related to rectangular cartesian by the equations $x = r\cos\theta$, $y = r\sin\theta$ and $z = z$.

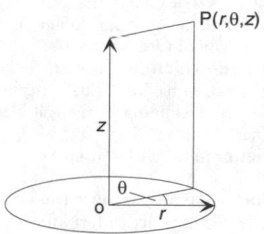

cylindrical co-ordinates

cylindrical gauge (*Eng*) A length gauge of cylindrical form whose length and diameter are made to some standard size. See GAUGE (3).

cylindrical grinding (*Eng*) The operation of accurately finishing cylindrical work by a high-speed abrasive wheel. The work is rotated by the headstock of the machine and the wheel is automatically traversed along it under a copious flow of coolant.

cylindrical lens (*ImageTech*) A lens element having one or both surfaces of cylindrical curvature, used in ANA-MORPHIC systems or to produce a line image.

cylindrical rotor (*ElecEng*) A rotor of an electric machine in which the windings are placed in slots around the periphery, so that the surface is cylindrical.

cylindrical wave (*Phys*) A wave for which equiphase surfaces form coaxial cylinders.

cylindrical winding (*ElecEng*) A type of winding used for core-type transformers; it consists of a single coil of one or more layers wound concentrically with the iron core; it is usually long compared with its diameter.

cyma (*Arch*) A much used moulding showing a reverse curve in profile. Also *OG*, *ogee*. See fig. at MOULDINGS.

cyma reversa (*Arch*) A CYMA which is convex at the top and concave at the bottom. Also *cyma inversa*. See fig. at MOULDINGS.

cymbiform (*Genrl*) Boat-shaped.

cyme (*BioSci*) See CYMOSE INFLORESCENCE.

cymene (*Chem*) $CH_3C_6H_4CH(CH_3)_2$, *1-(1-methylethyl)-4 methylbenzene*, bp 175°C.

cymophane (*Min*) A variety of the gem mineral *chrysoberyl* which exhibits chatoyancy. Also *chrysoberyl cat's eye* or *Oriental cat's eye*.

cymose inflorescence (*BioSci*) An inflorescence in which the main stem and each subsequent branch end in a flower, with any further development of the inflorescence coming from a lateral branch or lateral branches arising below the flower. Also *cyme*. Cf RACEMOSE INFLORESCENCE.

cynopodous (*BioSci*) Having non-retractile claws, as dogs.

Cyperaceae (*BioSci*) The sedge family, comprising c.4000 spp of monocotyledonous flowering plants (superorder Commelinidae). They are mainly rhizomatous, perennial, grass-like herbs. They are cosmopolitan, esp in temperate and arctic regions, and are often found in wet habitats. The aerial stems are typically solid, triangular in section and bear grass-like leaves in three ranks; the flowers are inconspicuous and wind-pollinated. The leaves and stems of some are used for making hats, baskets, mats and paper (papyrus) and for thatching. Includes the large genus *Carex* (1000 spp).

cypress knee (*BioSci*) A vertical upgrowth from the roots of swamp cypress (*Taxidium*), apparently a PNEUMATOPHORE.

Cypriniformes (*BioSci*) An order of OSTEICHTHYES, almost entirely inhabiting fresh water, with over 3000 spp. Characins, loaches and carp.

Cys (*Chem*) Symbol for CYSTEINE.

cyst (*BioSci*) A non-living membrane enclosing a cell or cells; any bladder-like structure, as the gall bladder or the urinary bladder of vertebrates; a sac containing the products of inflammation. Adjs *cystic, cystiform, cystoid*.

cysteine (*Chem*) *2-amino-3-mercaptopropanoic acid*. HSCH$_2$CH(NH$_2$)COOH. The L- or R- form of this amino acid is found in proteins, often in its oxidized form, CYSTINE. Symbol Cys. Abbrev C.

cysteine hydrochloride (*FoodSci*) Reducing agent used as an IMPROVER in baking. Usually *l-cysteine hydrochloride*.

cystic (*BioSci*) Pertaining to the gall bladder; pertaining to the urinary bladder.

cystic adenoma (*Med*) An ADENOMA containing numerous cysts.

cystic duct (*BioSci*) The duct from the gall bladder that meets the hepatic duct to form the common bile duct.

cysticercoid (*BioSci*) A larval stage of some CESTODA that is parasitic within an invertebrate intermediate host and which, when eaten by the vertebrate final host, develops into an adult tapeworm.

cysticercosis (*Med*) Infection with CYSTICERCI of the tapeworm *Taenia solium*.

cysticercus (*BioSci*) A bladderworm, the larval stage in many tapeworms with a fluid-filled sac containing an invaginated SCOLEX.

cystic fibrosis (*Med*) Autosomal recessive genetic disorder causing abnormal viscid mucous production throughout the body but particularly the lungs. Leads to recurrent severe chest infections. Elevated sodium ions in sweat are a diagnostic factor. Also *muco-viscidosis*.

cystic fibrosis transmembrane conductance regulator (*BioSci*) The gene that is defective in cystic fibrosis. Its product, an ABC PROTEIN, is important for chloride ion movement across epithelia, especially in the lung. Abbrev *CFTR*.

cysticolous (*BioSci*) Cyst-inhabiting.

cystidium (*BioSci*) A swollen, elongated, sterile hypha, occurring among the basidia of the HYMENIUM of some Hymenomycetes, usually projecting beyond the surface of the hymenium.

cystine (*Chem*) The dimer resulting from the oxidation of cysteine. The resulting disulphide bridge is an important structural element in proteins, as it often connects groups otherwise distant in the protein chain.

cystitis (*Med*) Inflammation of the bladder.

cystocele (*Med*) Hernia of the bladder.

cystogenous (*BioSci*) Cyst-forming; cyst-secreting.

cystography (*Radiol*) The radiological examination of the urinary bladder following the administration intravenously or through the urethra of a CONTRAST MEDIUM.

cystolith (*BioSci*) The mass of calcium carbonate within a plant cell, on a stalk-like projection from the cell wall.

cystoscope (*Med*) An instrument for inspecting the interior of the bladder.

cystostomy (*Med*) The formation of an opening in the bladder.

cystotomy (*Med*) Incision into the bladder.

cytase (*BioSci*) A general term for an enzyme able to break down the $\beta 1 \rightarrow 4$ link of cellulose.

cyto- (*Genrl*) Prefix from Gk *kytos* denoting cell.

cytochemistry (*BioSci*) Specific staining of cellular components so that they can be localized by optical or electron microscopy. In *immunocytochemistry*, fluorescence- or enzyme-labelled antibodies are used to locate specific antigens.

cytochrome P450 (*BioSci*) A large family of haem-containing microsomal mono-oxygenases catalysing the introduction of oxygen into a substrate, particularly foreign toxins but also therapeutic drugs. The product can then be degraded to an excretable substance. Polymorphism in P450 enzymes often accounts for individual differences in drug metabolism.

cytochromes (*BioSci*) Proteins of the electron transfer chain that can carry electrons by virtue of their haem

PROSTHETIC GROUPS. Cytochromes b, c1 and c have the same prosthetic group as haemoglobin. Cytochromes a and a3 have the related haem A and together form the terminal complex of the chain, cytochrome oxidase. Cytochrome c is one of the most ancient biological molecules known, being found in all animals and plants.

cytogenesis (*BioSci*) The formation and development of cells.

cytogenetic map (*BioSci*) See CHROMOSOME MAPPING and panel on CHROMOSOME.

cytogenetics (*BioSci*) The study of the chromosomal complement of cells, and of chromosomal abnormalities and their inheritance.

cytokines (*BioSci*) Small proteins produced by cells involved in inflammation and immunity that can affect their own growth and behaviour, or that of other cell types. Includes INTERFERONS, INTERLEUKINS, TNF and growth factors.

cytokinesis (*BioSci*) The contraction of an equatorial belt of cytoplasm that brings about the separation of two daughter cells during cell division of animal tissues. In plants the division of the cytoplasm as distinct from the nucleus. See CELL PLATE, CLEAVAGE.

cytokinin (*BioSci*) Any of a group of plant GROWTH SUBSTANCES that are derivatives of adenine, eg zeatin, or their artificial analogues, eg 6-benzylaminopurine (6-benzyladenine). Cytokinins are synthesized esp in roots; promote cell division and bud formation, delay senescence and, sometimes, promote flowering and break dormancy.

cytology (*BioSci*) The study of the structure and functions of cells.

cytolysis (*BioSci*) Dissolution of cells.

Cytomegalovirus (*Med*) Probably the most widespread of the Herpetoviridae group. Infected cells enlarge and have a characteristic inclusion body (composed of virus particles) in the nucleus. The virus causes disease *in utero* (leading to abortion or stillbirth or to various congenital defects), and can be opportunistic in the immuno-compromised host.

cytophilic antibody (*BioSci*) Antibodies that bind to FC RECEPTORS on the cell membrane.

cytoplasm (*BioSci*) That part of the cell outside the nucleus but bounded by the plasma membrane. The cell wall, if it exists, is extracellular.

cytoplasmic inheritance (*BioSci*) Inheritance of traits coded for by the chloroplast or mitochondrial genomes, maternal because of the inheritance of chloroplasts and mitochondria through the egg rather than the sperm or male cell.

cytoplasmic male sterility (*BioSci*) Lack of functional pollen as a maternally inherited trait, resulting from a defective mitochondrial genome. Cf CYTOPLASMIC INHERITANCE. See MALE STERILITY.

cytoplasmic streaming (*BioSci*) Directional flow of cytoplasm within large cells that may facilitate dispersal of metabolites or may occur during cell locomotion, eg in large amoebae. Also *cyclosis*.

cytorrhysis (*BioSci*) Process in which a plant cell wall collapses inwardly, following water loss as a result of the exposure of the cell to a solution of a macromolecular solute to which the cell wall is impermeable, of higher osmotic pressure than that of the cell contents. Turgor will be zero or possibly negative. Cf PLASMOLYSIS.

cytosine (*Chem*) *6-aminopurine-2-one*. One of the five major bases found in nucleic acids. It pairs with guanine in both DNA and RNA. Symbol C. See panel on DNA AND THE GENETIC CODE.

cytosine

cytoskeleton (*BioSci*) The linear multimeric protein assemblages that serve as skeletal elements within the cell, eg MICROFILAMENTS, MICROTUBULES.

cytosol (*BioSci*) The non-particulate components of the cytoplasm.

cytotaxonomy (*BioSci*) The use of studies of chromosome number, morphology and behaviour in taxonomy.

cytotoxic (*BioSci*) Able to kill cells. Applies to cytotoxic T-lymphocytes, to killer cells and NATURAL KILLER CELLS, and also to damage mediated by COMPLEMENT.

cytotoxic antibiotics (*Pharmacol*) Drugs used in the treatment of cancer derived from antibiotics and which mimic radiotherapy (*radiomimetic drugs*). Common examples are *doxorubicin* and *bleomycin*.

cytotoxic drug (*Pharmacol*) A drug that kills cells and is used in the treatment of cancer. They include ALKYLATING DRUGS, ANTIMETABOLITES, CYTOTOXIC ANTIBIOTICS, VINCA ALKALOIDS.

cytotoxin (*BioSci*) A toxin having a destructive action on cells.

cytotrophoblast (*BioSci*) The inner layer of the trophoblast; layer of Langerhans.

Czochralski process (*Electronics*) Single-crystal growth process, esp for semiconductor applications, in which a crystal is grown by slowly withdrawing (pulling) a seed from the melt contained in a crucible. Crystal and melt are continuously counter-rotated to minimize thermal and compositional fluctuation effects. Cf BRIDGMAN PROCESS, FLOAT ZONE.

D

D (*Chem*) Symbol for DEUTERIUM.

D (*Phys*) Symbol for: (1) ANGLE OF DEVIATION; (2) DISPLACEMENT (electric flux density); (3) DIFFUSION COEFFICIENT.

[D] (*Phys*) A group of three FRAUNHOFER LINES in the yellow of the solar spectrum. $[D_1]$ and $[D_2]$, wavelengths 589·6357 and 589·0186 nm, are due to sodium, and $[D_3]$, wavelength 587·5618 nm, to helium.

d (*Genrl*) Symbol for DECI-.

d- (*Chem*) Abbrev for DEXTROROTATORY.

[d] (*Phys*) A line in the blue of the solar spectrum, having a wavelength of 437·8720 nm due to iron.

Δ (*Chem*) Prefixed symbol for a double bond beginning on the carbon atom indicated.

δ- (*Chem*) Substituted on the fourth carbon atom of a chain.

D1 (*ImageTech*) A BROADCAST STANDARD digital COMPONENT VIDEO using $\frac{3}{4}$ in tape in two sizes of cassette.

D2 (*ImageTech*) A BROADCAST STANDARD digital COMPOSITE format using $\frac{3}{4}$ in metal tape in three sizes of cassette.

D3 (*ImageTech*) A BROADCAST STANDARD digital COMPOSITE format using $\frac{1}{2}$ in metal tape in three sizes of cassette.

D5 (*ImageTech*) A BROADCAST STANDARD digital COMPONENT VIDEO using $\frac{1}{2}$ in metal tape in three sizes of cassette. It will also play back tapes recorded on D3.

D-A (*ImageTech*) Abbrev for *digital-to-analogue*, referring to the conversion of signals.

da (*Genrl*) Symbol for DECA-.

DAB (*Genrl*) Abbrev for DIGITAL AUDIO BROADCASTING.

dabbing (*Build*) See DAUBING.

DAC (*ICT*) Abbrev for DIGITAL-TO-ANALOGUE CONVERTER.

dacite (*Geol*) A VOLCANIC ROCK intermediate in composition between andesite and rhyolite, the volcanic equivalent of a granodiorite.

Dacron (*Chem*) TN for a polyester fibre (US).

dacryo-adenitis (*Med*) Inflammation of the lacrimal (tear) gland.

dacryocystitis (*Med*) Inflammation of the lacrimal sac.

dacryocystorhinostomy (*Med*) The formation of a direct opening between the tear sac and the nose.

dactyl (*BioSci*) A digit. Adj *dactylar*.

dactylitis (*Med*) Inflammation of a finger or of a toe.

dado (*Arch*) Panelling applied to the lower half of the walls of a room, or alternatively, decoration to give a similar effect.

dado capping (*Build*) A dado rail when the dado occupies as much as two-thirds of the height of the room.

dado plane (*Build*) A type of grooving plane with two projecting spurs, one on each side at the front, and an adjustable depth stop. Used for making grooves for shelving etc. The spurs cut across the grain and keep the plane in its correct path. Also *trenching plane*.

daft lamb disease (*Vet*) Uncommon disease of sheep and goats caused by a togavirus. Ewe may abort or a poor-viability lamb be born; the latter can have a long coat. Lambs can also show deformity and uncoordinated gait due to chronic contraction of muscles. Also *Border disease, hairy shaker disease, hypomyelinogenesis imperfecta*.

daguerreotype (*ImageTech*) Early process using a silvered copper plate sensitized by fuming with iodine and bromine vapour and developed with mercury vapour.

dahoma (*For*) A durable hardwood (*Piptadeniastrum*) from the African tropical rainforests, whose heartwood is a uniform yellow-orange to golden brown, having interlocked grain with a coarse texture.

daidzein (*Pharmacol*) An estrogen extracted from the root of the kudzu plant, thought to have therapeutic properties.

dailies (*ImageTech*) See RUSHES.

dairy calf (*Agri*) The result of mating a bull and cow from recognized dairy breeds.

dairy cattle (*Agri*) Cattle of a breed selected for milk production, as distinct from beef or dual purpose use. Typically lightly fleshed with prominent bones, eg Friesian, Holstein, Ayrshire, Guernsey and Jersey.

daisy-cutter (*Genrl*) A powerful bomb designed to explode close to the ground, destroying anything within a wide radius.

daisy-wheel printer (*ICT*) Printer in which characters are arranged near the ends of the spokes of a rimless wheel (on the 'petals' of a 'daisy'). Daisy wheels are manually interchangeable to enable alternative character sets to be used.

dalapon (*Chem*) *2,2-dichloropropanoic acid*. Used as a weedkiller.

d'Alembert's principle (*Phys*) The principle that on a body in motion, the external forces are in equilibrium with the inertial forces.

d'Alembert's ratio test (*MathSci*) The theorem that a series of positive terms converges or diverges respectively according to whether the limit of the ratio of a term to its predecessor is less or greater than unity. When the limit is unity the test is inconclusive.

Dalitz pair (*Phys*) Electron–positron pair produced by the decay of a free neutral pion (instead of one of the two gamma quanta normally produced).

DALR (*EnvSci*) See DRY ADIABATIC LAPSE RATE.

Dalradian Series (*Geol*) Very thick and variable succession of sedimentary and volcanic rocks which have suffered regional metamorphism. Occurring in the Scottish Highlands approximately between the Great Glen and the Highland Boundary fault. Referred to the Precambrian to Lower Palaeozoic in age.

dalton (*Chem*) See ATOMIC MASS UNIT.

Daltonism (*Med*) See COLOUR BLINDNESS.

Dalton's atomic theory (*Chem*) States that matter consists ultimately of indivisible, discrete particles (*atoms*) and atoms of the same element are identical; chemical action takes place as a result of attraction between these atoms, which combine in simple proportions. It has since been found that atoms of the chemical elements are not the ultimate particles of matter, and that atoms of different mass can have the same chemical properties (isotopes). Nevertheless, this theory of 1808 is fundamental to chemistry. See panel on ATOMIC STRUCTURE.

Dalton's law (*Chem*) See LAW OF MULTIPLE PROPORTIONS.

Dalton's law of partial pressures (*Chem*) The pressure of a gas in a mixture is equal to the pressure which it would exert if it occupied the same volume alone at the same temperature.

DALY (*Genrl*) Abbrev for DISABILITY-ADJUSTED LIFE YEAR.

dam (*CivEng*) An embankment or other construction made across the current of a stream or river.

dam (*MinExt*) (1) A retaining wall or bank for water or tailings. (2) An airtight barrier to isolate underground workings which are on fire.

damask (*Textiles*) (1) A figured fabric made with satin and sateen weaves, in which background and figure have a contrasting effect; used mainly for furnishing. (2) Linen cloth of damask texture, used for tablecloths and towellings; also a cotton cloth of similar nature, used for tablecloths; both fabrics are reversible.

dam board (*Build*) Steel or timber plate used for temporary stopping of a waterway.

damped balance (*Chem*) Chemical balance using magnetic or air dash pots to bring it quickly to rest.

damped oscillation (*Phys*) Oscillation which dies away from an initial maximum asymptotically to zero amplitude, usually with an exponential envelope; eg the note from a struck tuning fork. See CRITICAL DAMPING, DAMPING, DECAY FACTOR; cf CONTINUOUS OSCILLATIONS.

dampening rollers (*Print*) Rollers on a lithographic printing machine which are kept moist by a water supply, and by means of which the plate is dampened before being rolled by the ink rollers.

damper (*Aero*) Widely used term applied to devices for the suppression of unfavourable characteristics or behaviour: eg *blade damper*, to prevent the hunting of a helicopter rotor; *flame damper*, to prevent visual detection at night of the exhaust of a military aircraft; *shimmy damper*, for the suppression of SHIMMY; a *yaw damper* suppresses directional oscillations in high-speed aircraft, while a *roll damper* does likewise laterally, in both cases the frequency of the disturbances being too high for the pilot to anticipate and correct manually.

damper (*Autos*) Frictional or hydraulic device attached between the chassis and axles to prevent spring rebound and damp out oscillation. Inaccurately referred to as *shock absorber*.

damper (*ElecEng*) Energy-absorbing component often used for reducing the transmission of oscillatory energy from a disturbing source. Also *amortisseur*, *damper winding*, *damping grid*, *damping winding*.

damper (*Eng*) (1) An adjustable iron plate or shutter fitted across a boiler flue to regulate the draught. (2) A device for damping out torsional vibration in an engine crankshaft, the energy of vibration being dissipated frictionally within the damper. See VIBRATION DAMPERS. (3) Device for stiffening the steering of a motor cycle to obviate wheel wobble.

damping (*Acous*) The transfer of sound energy into heat. There are different mechanisms: structure-borne sound is damped eg by molecular displacement processes, and airborne sound by friction on interfaces. See STOKES LAYER.

damping (*Aero*) The capability of an aircraft of suppressing or resisting harmonic excitation and/or flutter. *Internal damping* is intrinsic to the materials, while *structural damping* is the total effect of the built-up structure. See RESONANCE TEST.

damping (*Eng*) Commonly seen in polymeric materials, where viscoelastic effects cause energy dissipation, which appears as heat. Exploited in vibration isolation devices, such as rubber engine mounts etc. Also found in elastic materials. See INTERNAL FRICTION, SNOEK EFFECT.

damping (*Phys*) The extent of reduction of amplitude of oscillation in an oscillatory system, due to energy dissipation, eg friction and viscosity in a mechanical system, and resistance in an electrical system. With no supply of energy, the oscillation dies away at a rate depending on the DEGREE OF DAMPING. The effect of damping is to increase slightly the period of vibrations. It also diminishes the sharpness of resonance for frequencies in the neighbourhood of the natural frequency of the vibrator. See LOGARITHMIC DECREMENT. Fig. ▷

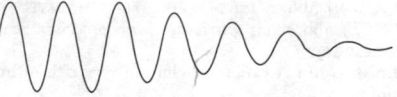

damping A damped sine wave.

damping capacity (*Eng*) The ability to absorb energy from external source, eg sound waves or vibrations. It is a measure of INTERNAL FRICTION associated with the atomic or molecular disturbances induced by the external energy.

damping down (*Eng*) The temporary stopping of a blast furnace by closing all apertures by which air could enter.

damping factor (*Phys*) See DECAY FACTOR.

damping magnet (*ElecEng*) A permanent magnet used to produce damping by inducing eddy currents in a metal disk or other body.

damping-off (*BioSci*) Collapse and death of seedlings around emergence, due usually to fungal attack by *Pythium* and *Fusarium* spp when conditions are unfavourable for the seedlings.

dam plate (*Eng*) Iron vertical plate holding the wall of refractory brick (the dam stone) which forms the fore-hearth of a blast furnace.

damp-proof course (*Build*) A layer of impervious material, as plastic or bituminous sheeting built into a wall 15–25 cm above ground level, to prevent moisture from the foundations rising in the walls by capillary attraction. Vertical damp courses are also used at door and window openings. Also used in chimneys and parapet walls to prevent downward passage of moisture.

damp-proofing (*Build*) The process of coating a wall with a special preparation to prevent ingress of moisture.

DAMPS (*ICT*) Abbrev for DIGITAL ADVANCED MOBILE-PHONE SYSTEM.

danburite (*Min*) A rare accessory mineral, occurring in pegmatites as yellow orthorhombic crystals. Chemically, danburite is a calcium borosilicate, $CaB_2Si_2O_8$.

dancing roller (*Print*) See JOCKEY ROLLER.

dancing step (*Build*) A step intermediate between a rectangular step and a WINDER, having its outer end narrower in plan than its inner end. Helps to form a better shaped handrail. Also *balanced step*.

dandy roll (*Paper*) Hollow cylinder covered with wire cloth situated on top of the paper machine wire so that the surface of the roll makes contact with the upper surface of the wet web. The wire cloth may be such that a wove or laid pattern is imparted to the paper or names or other designs secured to it to produce corresponding WATERMARKS in the paper.

dangerous semicircle (*EnvSci*) The right-hand half of the storm field in the northern hemisphere, the left-hand half in the southern hemisphere, when looking along the path in the direction a TROPICAL REVOLVING STORM is travelling. Cf NAVIGABLE SEMICIRCLE.

Danian (*Geol*) The lowest stage of the Palaeogene (Palaeocene). Some regard it as Cretaceous. See TERTIARY.

Daniell cell (*Chem*) Primary cell with zinc and copper electrodes, the zinc rod being inserted in sulphuric acid contained within a porous pot, which is itself immersed in a copper pot containing copper (II) sulphate solution.

D and K (*Textiles*) *Damaged and kept*, usually by the dyer and finisher of the fabric.

dannemorite (*Min*) A rare manganese-rich monoclinic amphibole, the name being used for the manganese-rich, iron-rich end-member and differing from TIRODITE in having more iron than magnesium.

Dano composting plant (*Build*) A method of composting the organic wastes in domestic refuse with sewage sludge by a process of fermentation and grinding, which reduces the materials to a moist granulated condition. The product is a valuable soil conditioner because of its high humus content.

DAP (*Chem*) Abbrev for DIALPHANYLPHTHALATE.

DAP (*ICT*) Abbrev for *distributed array processor*. See ARRAY PROCESSOR.

daphnite (*Min*) A variety of chlorite very rich in iron and aluminium.

DAP resins (*Chem*) Type of polyester resin made using diallyl phthalate. Compares favourably with phenol formaldehyde resins for electrical insulation. Diallyl isophthalate, abbrev *DAIP*, is also used.

dapsone (*Pharmacol*) A sulphone drug widely used in the treatment of LEPROSY.

daraf (*Phys*) Unit of elastance, the reciprocal of capacitance in farads (FARAD backwards).

darby (*Build*) See DERBY FLOAT.

darcy (*Geol*) A unit used to express the PERMEABILITY COEFFICIENT of a rock, eg in calculating the flow of oil, gas or water. More commonly used is the *millidarcy* (mD), one-thousandth of a darcy (D).

Darcy's law (*PowderTech*) A permeability equation which states that the rate of flow of fluid through a porous medium is directly proportional to the pressure gradient causing the flow.

dark burn fatigue (*Phys*) The decrease of efficiency of a luminescent material during excitation.

dark current (*ImageTech, Phys*) Residual current in a photocell, video camera tube, etc, when there is no incident illumination. The current depends on temperature.

dark fibre (*ICT*) OPTICAL FIBRE that is installed as part of a multifibre cable but not yet in service.

dark ground illumination (*BioSci*) A method for the microscopic examination of living material, eg microorganisms, tissue culture cells, by scattered light. A special condenser with a circular stop illuminates the specimen with a numerical aperture larger than that collected by the objective. Only the scattered rays pass through the objective to reach the eye and the specimen appears luminous against a dark background. It can also be used to detect smoke and other particles too small to be resolved by the light microscope.

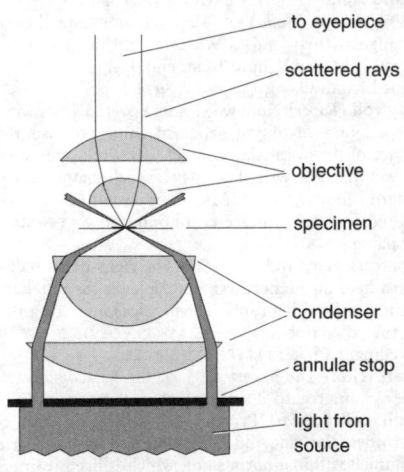

to eyepiece

scattered rays

objective

specimen

condenser

annular stop

light from source

dark ground illumination

dark matter (*Astron*) Material in the universe (over 90% of the total) whose presence is implied by its gravitational effects, but which is not visible through its emission of radiation. The nature of the matter is unknown, and is thought possibly to consist of certain elementary particles, undetectable dim stars or planets. Also *missing mass*.

dark nebula (*Astron*) An obscuring cloud of dust and gas, common throughout the Milky Way, and also observed in other galaxies. Examples include the COAL SACK and the HORSEHEAD NEBULA. Also *absorption nebula*.

darknet (*ICT*) A private network of computer users within which file sharing takes place.

dark reactions (*BioSci*) Those reactions in photosynthesis in which CO_2 is fixed and reduced. They depend on energy and reducing power from the LIGHT REACTIONS. See panel on CALVIN CYCLE.

dark red heat (*Eng*) Glow emitted by metal at temperatures between 550 and 630°C.

dark-red silver ore (*Min*) See PYRARGYRITE.

dark resistance (*Electronics*) The resistance of a photocell in the dark.

dark-room camera (*Print*) A built-in process camera controlled completely from inside the dark room, with the copy board and its illumination equipment outside.

dark slide (*ImageTech*) The carrier for plates to be exposed in cameras, loaded in the dark room and uncovered, after attachment to the camera, by withdrawing a slide.

dark space (*Electronics*) See ANODE DARK SPACE, ASTON DARK SPACE, CROOKES DARK SPACE, FARADAY DARK SPACE.

dark trace screen (*Electronics*) A screen which yields a dark TRACE under electron-beam bombardment.

darmstadtium (*Chem*) An artificially produced radioactive transuranic element (symbol Ds; at no 110), formerly called *ununnilium*.

DARPA (*Aero, ICT*) Abbrev for *Defense Advanced Research Projects Agency*, US. See ARPA.

dart (*BioSci*) Any dart-like structure; eg in certain snails, a small pointed calcareous rod that is used as an incentive to copulation; in certain NEMATODA, a pointed weapon used to obtain entrance to the host.

Darvic (*Plastics*) TN for unplasticized polyvinyl chloride sheet (UK).

Darwinian fitness (*BioSci*) The number of offspring of an individual, or the number relative to the mean (*relative fitness*), that live to reproduce in the next generation; effectively, the number of genes passed on.

Darwinian theory (*BioSci*) See NATURAL SELECTION.

dash pot (*Eng*) A device for damping out vibration or for allowing rapid motion in one direction but only much slower motion in the opposite direction. It consists of a piston attached to the part to be controlled (fitted with a non-return valve if required) sliding in a cylinder containing liquid to impede motion.

dasypaedes (*BioSci*) Birds that, when hatched, have a complete covering of down. Cf ALTRICES.

DAT (*ICT*) Abbrev for *digital audio tape*. A method of recording sound digitally onto magnetic tape. This technology is often used to back up the data from the large-capacity HARD DISK on a FILE SERVER because of the large storage capacity available on the tape.

data (*Genrl*) Facts such as quantities or values from which other information may be inferred. Sing *datum*.

data (*ICT*) All the OPERANDS and results of computer operations directed by the detailed INSTRUCTIONS comprising the program. A program can be *data* for another program, eg a COMPILER takes a program as data.

data back (*ImageTech*) A film back enabling information (time, date, etc) to be imprinted on the film by light-emitting diodes.

data bank (*ICT*) A collection of DATABASES or large files of data.

database (*ICT*) A collection of structured data independent of any particular application.

database management system (*ICT*) Software that handles the storage, retrieval and updating of data in a computer, often integrating data from a number of files. Also *DBMS*. See DATA MODEL.

database typesetting (*Print*) The storing of information in a database for publications such as directories which can be periodically updated by computer processing and prepared for phototypesetting.

data capture (*ICT*) Collecting data for use in a particular computer process, eg for monitoring.

datacasting (*ICT*) A form of broadcasting that enables the rapid dissemination of large amounts of data to a large number of users.

data channel (*ICT*) (1) Alternative term for D-CHANNEL in the INTEGRATED SERVICES DIGITAL NETWORK. (2) Any channel optimized for operation at a variable bit rate, as opposed to the constant bit rate required by services such as telephony.

data compaction (*ICT*) A term often applied to DATA COMPRESSION that involves only the removal of extraneous and unnecessary space and therefore is reversible.

data compression (*ICT*) Altering the form of data to reduce its storage space.

data dictionary (*ICT*) An index of the contents of a set of files or a database. See DIRECTORY.

Data Encryption Standard (*ICT*) An automatic method of data ENCRYPTION designed by IBM and adopted as a standard.

data flow (*ICT*) An approach to the organization of complex algorithms and machines, in which operations are triggered by the arrival of data.

data flowchart (*ICT*) FLOWCHART used to describe a complete processing system, clerical operations and individual programs, but excluding details of such programs. Also *system flowchart*.

dataglove (*ICT*) A glove-like device with sensors that transmits the wearer's movements to a computer.

Datagram (*ICT*) A communication channel that uses information routed through a packet-switching network.

data handling (*Space*) The management of the flow of data to and from a space vehicle; the on-board subsystem might include data buses, commutators, computers, recorders and multiplexers, whereas the ground segment uses equipment like demultiplexers and display units to interpret the transmitted signal which is sent either directly or via a data relay satellite.

data-handling capacity (*ICT*) The maximum amount of information that can be transmitted and received over a given channel or circuit. Also *data-handling capability*.

data link connection identifier (*ICT*) In a FRAME RELAY network, a field of two to four OCTETS placed in the frame header to indicate the PERMANENT VIRTUAL CIRCUIT to which the frame belongs.

data link layer (*ICT*) Level 2 of the OPEN SYSTEMS INTERCONNECTION model that is responsible for the synchronization and handling of errors so that transmission can take place over a physical link.

data logger (*ICT*) See LOGGER.

datamart (*ICT*) A satellite or subdivision of a DATA WAREHOUSE that contains only data relating to a specific operational subject area.

data mining (*Genrl*) Gathering electronically stored information, eg about shopping patterns from loyalty cards.

data model (*ICT*) A structure for the arrangement of data that aids data retrieval. There are three models in general use: a hierarchic model, a network model and one giving a RELATIONAL DATABASE.

data preparation (*ICT*) Translation of data into machine-readable form.

data processing (*ICT*) Traditional name given to business information processing. Abbrev *DP*.

data protection (*ICT*) Safeguards to protect the integrity, privacy and security of data. See DATA PROTECTION ACT.

Data Protection Act (*ICT*) A 1984 Act of Parliament which sets out regulations relating to the use and storage of personal data on computers in the UK. The main provisions are data protection, right of subject access, relevance, accuracy, registration of purpose and disclosure to third parties.

data rate compression (*ICT*) The removal of redundancy from a data stream in order to reduce the BANDWIDTH required for its transmission; eg the data rate of a digital video signal can be greatly reduced by sending only the differences between one frame and the next.

data reduction (*ICT*) The computerized repackaging of observational data to make them more concise and meaningful.

data retrieval (*ICT*) The search for and selection of data from a storage medium.

data signalling rate (*ICT*) The aggregate rate at which binary digits, including any control bits, are transmitted over a channel or circuit, expressed in bps. Cf BAUD.

data storage (*ICT*) See MEMORY CAPACITY.

data stick (*ICT*) (variant of *data pen*). A small removable storage device that plugs into a USB (Universal Serial Bus) port on a computer. Typically contains 126 Mbytes to 2 Gbytes of data.

data structure (*ICT*) Organized form in which grouped data items are held in the computer, such as LIST, STRING, TABLE, TREE.

data type (*ICT*) Most programming languages require a variable to be declared as a data type. Basic restrictions and assumptions will then control the use of the variable. See BOOLEAN, CHARACTER, INTEGER, LIST, QUEUE, REAL, SET, STACK, STRING, TREE.

data warehouse (*ICT*) A form of database design bringing together large volumes of data from a variety of source systems into a heavily *denormalized* structure that allows the efficient retrieval and analysis of those data to answer sophisticated real-world questions.

dating (*Geol*) See panel on RADIOMETRIC DATING.

dative bond (*Chem*) See COVALENT BOND.

datolite (*Min*) Hydrated calcium borosilicate occurring as a secondary product in amygdales and veins, usually as distinct prismatic white or colourless monoclinic crystals.

datum (*Aero*) The *datum level*, or *rigging datum*, is the horizontal plane of reference, in flying attitude, from which all vertical measurements of an aircraft are taken; *cg datum* is the point from which all mass moment arms are measured horizontally when establishing the centre of gravity and loading of an aircraft.

datum (*Eng*) A point, line or surface to which dimensions are referred on engineering drawings and from which measurements are taken in machining or other engineering operations.

datum (*Genrl*) Sing of DATA.

datum (*Surv*) An assumed surface used as a reference surface for the measurement of reduced levels.

daubing (*Build*) (1) The operation of dressing a stone surface with a special hammer in order to cover it with small holes. (2) A rough-stone finish given to a wall by throwing a rough coating of plaster upon it. See ROUGH-CAST.

daughter (*BioSci*) Offspring belonging to the first generation, whether male or female, as *daughter cell*, *daughter nucleus*.

daughter card (*ICT*) A small printed circuit board connected to an ADAPTER CARD or MOTHERBOARD inside a computer.

daughter product (*Phys*) A nuclide that originates from the radioactive disintegration of another *parent* nuclide. Also *decay product*.

Davis apparatus (*Ships*) A respiratory apparatus specially designed to permit escape from a pressure-equalizing chamber in a submarine. Oxygen is breathed from a chamber which, embracing the wearer, gives buoyancy and assists rise to the surface.

Davisson–Germer experiment (*Electronics*) The first demonstration (1927) of wave-like diffraction patterns from electrons by passing them through a nickel crystal.

Davy lamp (*MinExt*) The safety lamp invented by Sir Humphry Davy in 1815.

day (*Astron*) See APPARENT SOLAR DAY, MEAN SOLAR DAY, SIDEREAL DAY.

day for night (*ImageTech*) Creating a night-time effect in daylight by using underexposure and a blue filter.

daylight (*Eng*) The distance between the bed surface and the bottom of the ram of a press.

daylight factor (*ElecEng*) The ratio of the illumination measured on a horizontal surface inside a building to that obtained at the same time from an unobstructed hemisphere of sky. Also *window efficiency ratio*.

daylight lamp (*Phys*) A lamp giving light having a spectral distribution curve similar to that of ordinary daylight.

day-light size (*Build*) The distance between successive mullions in a window and between lintel and sill.

day-neutral plant (*BioSci*) A plant in which flowering is not sensitive to day-length. Cf LONG-DAY PLANT, SHORT-DAY PLANT. See PHOTOPERIODISM.

Db (*Chem*) Symbol for DUBNIUM.

dB (*Acous, ICT*) Abbrev for DECIBEL.

dBA, dBB, dBC (*Acous*) Result of a SOUND PRESSURE LEVEL measurement when the signal has been weighted with a frequency response of the A, B or C curve. The dBA curve approximates the human ear and is therefore used most in noise control regulations.

dBm (*ICT*) A unit for expressing power level in decibels, relative to a reference level of 1 mW.

DBMS (*ICT*) Abbrev for DATABASE MANAGEMENT SYSTEM.

DBS (*ImageTech, ICT*) Abbrev for DIRECT BROADCAST SATELLITE, *direct broadcasting by satellite*.

dc (*ElecEng*) Abbrev for DIRECT CURRENT.

dc (*Print*) Abbrev for *double column, double crown*.

dc amplifier (*ElecEng*) One which uses direct coupling between stages (ie no blocking capacitor) to amplify from zero-frequency (dc) signals to signals of higher frequency.

dc balancer (*ElecEng*) The coupling and connecting of two or more similar dc machines, so that the conductors connected to the junction points of the machines are maintained at constant potentials.

dc bias (*ElecEng*) (1) In an electronic amplifier, the direct signal applied to an active component which sets the quiescent conditions for the device. Thereafter, an ac signal may be applied. (2) In a magnetic tape recorder, the addition of polarizing dc in the signal recording to stabilize magnetic saturation.

dc bridge (*ElecEng*) A four-arm null bridge energized by a dc supply. The prototype is the WHEATSTONE BRIDGE, other examples being the METRE BRIDGE and the POST OFFICE BOX.

dc component (*ImageTech*) That part of the picture signal which determines the average or datum brightness of the reproduced picture.

dc converter (*ElecEng*) A converter which changes dc from one voltage to another.

dc coupling (*Electronics*) See DIRECT COUPLING.

dc/dc converter (*ElecEng*) A dc voltage transformer using an inverter and rectifier. Also *dc transformer*.

DCF (*Build*) Abbrev for *deal-cased frame*.

dc generator (*ElecEng*) A rotary machine to convert mechanical into dc power.

D-channel (*ICT*) A single low-bandwidth (16 Kbps) channel provided as part of the INTEGRATED SERVICES DIGITAL NETWORK. It is intended primarily for signalling to control the 64 Kbps B-CHANNELS.

dc meter (*ElecEng*) One which responds only to the dc component of a signal, eg moving-coil instruments.

DCOM (*ICT*) Abbrev for *distributed component object model*, a proprietary (Microsoft) architectural standard for object-oriented systems design. See also COM.

dc resistance (*ElecEng*) The resistance which a circuit offers to the flow of dc. Also *true (ohmic) resistance*.

dc restoration (*ImageTech*) Reinsertion of a dc or very low-frequency component which has been lost or reduced in transmission; in a TV receiver the use of a CLAMP to hold the level of the dc component.

DCT (*ICT*) Abbrev for DISCRETE COSINE TRANSFORM.

DCT (*ImageTech*) Abbrev for *digital component technology*, TN for a BROADCAST STANDARD digital COMPONENT VIDEO format using $\frac{3}{4}$ in metal tape in three sizes of cassette.

dc testing of cables (*ElecEng*) The application of a dc voltage of five times the rms of the working ac voltage.

Cables which have considerable tracking and are likely to break down in service are broken down by the dc; healthy cables are not affected.

dc transformer (*ElecEng*) (1) Device to measure large dc by means of associated magnetic field. (2) Colloq term for *dc/dc converter*.

dc transmission (*ElecEng*) A method of connecting together different power generating systems for sending and receiving bulk quantities of electricity when ac is not attractive. Losses are lower, insulation is used more effectively, steady-state charging current is zero (important if cables are used for interconnection) and different power systems do not need to be synchronized. Disadvantage is cost of converter equipment at both sending and receiving ends.

dc transmission (*ImageTech*) Inclusion of dc or very low frequency in the transmitted video signal. If omitted, it has to be *restored* in relation to the PEDESTAL in the receiver.

DCW (*Agri*) Abbrev for DRESSED CARCASS WEIGHT.

DDE (*ICT*) (1) Abbrev for DYNAMIC DATA EXCHANGE, a system for linking data in two or more applications. (2) Abbrev for *direct data entry*. The input of data for BATCH PROCESSING using a KEY-TO-DISK UNIT.

DDL (*ICT*) Abbrev for *data description language*. Used in a DATABASE MANAGEMENT SYSTEM.

DDT (*Chem*) Abbrev for a complex chemical mixture, in which *pp*-dichlorodiphenyltrichloroethane predominates; a synthetic insecticide remarkable for high toxicity to insects at low rates of application. A STOMACH INSECTICIDE and CONTACT INSECTICIDE with a very long persistence of activity from residual deposits, which has caused it to be banned in many countries.

DDVP (*Chem*) See DICHLORVOS.

DE (*FoodSci*) Abbrev for DEXTROSE EQUIVALENT.

deactivation (*Chem*) The return of an activated atom, molecule or substance to the normal state. See ACTIVATION (2).

dead (*Acous*) An enclosure which has a period of reverberation much smaller than usual for its size and audition requirements. Applied to sets in film production.

dead (*Build*) Said of materials which have deteriorated.

dead angle (*Eng*) That period of crank angle of a steam engine during which the engine will not start when the stop valve is opened, and which is due to the ports being closed by the slide valve.

dead axle (*Eng*) An axle which does not rotate with the wheels carried by it. Cf LIVE AXLE.

dead bank (*Eng*) A stoker-fired boiler furnace from which the coal feed is shut off, the fire being allowed to burn back as far as possible without going out entirely.

dead-beat compass (*Ships*) A magnetic compass with a short period of oscillation and heavily damped so that it comes to rest very quickly.

dead burnt (*Eng*) Descriptive of such carbonates as limestone, dolomite, magnesite, when they have been so kilned that the associated clay is vitrified, part or all of the volatile matter removed and the slaking quality lowered.

dead centre (*Eng*) (1) Either of the two points in the crankpin path of an engine at which the crank and connecting-rod are in line and the piston exerts no turning effort on the crank. See INNER (TOP) DEAD CENTRE, OUTER (BOTTOM) DEAD CENTRE. (2) A lathe CENTRE. See TAILSTOCK.

dead-centre lathe (*Eng*) A small lathe (used in instrument-making) in which both centres are fixed, the work being revolved by a small pulley mounted on it or by a driving plate running on separate bearings.

dead coil (*ElecEng*) A coil in the winding of a machine which does not contribute any emf to the external circuit, because it is short-circuited or disconnected from the rest of the winding. See DUMMY COIL.

dead crude oil (*MinExt*) Crude oil that has been stabilized so that free gas is absent. Cf LIVE CRUDE OIL.

dead earth (*ElecEng*) A connection between a normally live conductor and earth by means of a path of very low resistance.

dead end (*Build*) The length of pipe between a closed end and the nearest connection to it, forming a 'dead' pocket in which there is no circulation. Also *dead leg*.

dead end (*ICT*) The unused portion of an inductance coil in an oscillatory circuit.

dead-ended feeder (*ElecEng*) See INDEPENDENT FEEDER.

dead-end tower (*ElecEng*) See TERMINAL TOWER.

deadening (*Build*) (1) The operation of dealing with a surface, so as to give it a dead finish. (2) See PUGGING.

dead eye (*Eng*) (1) A sheaveless block used in setting up rigging. (2) A light type of bearing for supporting a spindle; it may consist merely of a hole in a sheet of metal or other material.

dead fingers (*Med*) See VIBRATION WHITE FINGER.

dead finish (*Build*) A dull or rough finish particularly, in painting, a FLAT FINISH.

dead flue (*Build*) A flue which is bricked in at the bottom.

dead ground (*MinExt*) Ground devoid of values: ground not containing veins or lodes of valuable mineral: a barren portion of a coal seam. Also *deads*.

dead knot (*For*) A KNOT which is partially or wholly separated from the surrounding wood.

dead leg (*Build*) See DEAD END.

dead load (*CivEng*) The weight of a structure with finishings, fixtures and partitions. Cf LIVE LOAD.

dead lock (*Build*) A lock the bolt of which is key operated from one or both sides as opposed to spring bolt or latch.

dead-man's handle (*ElecEng*) A form of handle commonly used on the controllers of electric vehicles; designed so that if the driver releases pressure on the handle, owing to sudden illness or other causes, the current is cut off and the brakes applied. *Dead-man's pedal* is a similar foot-operated safety device.

deadmen (*CivEng*) Anchor points for belaying ropes formed by eg driving a short pile.

dead oil (*MinExt*) Crude oil without dissolved gas.

dead points (*Eng*) See DEAD CENTRE (1).

dead rise (*Aero*) At any cross-section of a flying-boat hull or seaplane float, the vertical distance between the keel and the chine.

dead roasting (*Eng*) Roasting carried out under conditions designed to reduce the sulphur content, or that of the other volatile matter, to the lowest possible value. Distinguished from PARTIAL ROASTING and SULPHATING ROASTING.

dead room (*Acous*) See ANECHOIC ROOM.

deads (*MinExt*) (1) Same as DEAD GROUND. (2) Waste (*backfill*) used to support roof.

dead segment (*ElecEng*) A commutator segment which is not connected, either for accidental reasons or for a definite purpose, to the armature winding associated with the commutator.

dead shore (*Build*) A vertical post used to prop up temporarily any part of a building.

dead-smooth file (*Eng*) The smoothest grade of file ordinarily used, having 70–80 teeth to the inch for files of average length; used for finishing surfaces.

dead sounding (*Build*) See PUGGING.

dead spot (*ICT*) A region where the reception of radio transmissions over a particular frequency range is extremely weak.

dead time (*Phys*) Time after ionization during which a detector cannot record another particle. Reduced by a *quench* as in Geiger counters. When the dead time of a detector is variable a fixed electron dead time may be incorporated in subsequent circuits. Also *insensitive time*.

dead-time correction (*NucEng*) Correction applied to the observed rate in a nuclear counter to allow for the probable number of events occurring during the DEAD TIME.

dead water (*Eng*) In a boiler or other plant, water not in proper circulation.

dead weight (*Ships*) The difference, in tons, between a ship's displacement at load draught and light draught. It comprises cargo, bunkers, stores, fresh water, etc.

dead-weight pressure gauge (*Eng*) A device for measuring fluid pressure which is applied to the bottom of a vertical piston, the upward force being balanced by weights on the upper end; used for calibrating BOURDON GAUGES.

dead-weight safety valve (*Eng*) A SAFETY VALVE in which the valve itself is loaded by a heavy metal weight; used for small valves and low pressures.

dead well (*CivEng*) An ABSORBING WELL.

de-aerator (*Eng*) A vessel in which boiler feed-water is heated under reduced pressure in order to remove dissolved air.

deafening (*Build*) See PUGGING.

deafness (*Med*) The lack of sensitivity of hearing in one or both ears, with consequent increase in the threshold of minimum audibility, measurement of which is useful in diagnosis.

deal (*For*) (1) A piece of timber of cross-section roughly 250×75 mm (10×3 in). (2) See RED DEAL, WHITE DEAL.

deaminate (*Chem*) To remove an amino group from a chemical compound.

death (*BioSci*) In a cell or an organism, complete and permanent cessation of the characteristic activities of living matter.

deathnium centre (*Electronics*) Crystal lattice imperfection in semiconductor at which it is believed electron–hole pairs are produced or recombine.

death wish (*Psych*) A wish, conscious or unconscious, for death for oneself or another.

debacle (*EnvSci*) The breaking up of the surface ice of great rivers in spring. Also *débâcle*.

débridement (*Med*) The removal of foreign matter and excision of infected and lacerated tissue from a wound.

debris flow (*Geol*) A mass movement involving rapid flow of various kinds of debris esp in mud. May be associated with earthquakes and volcanic eruptions (eg S America) or with excessive rainfall on unstable material (eg Aberfan disaster). See MUD FLOW.

de Broglie wavelength (*Phys*) The wavelength associated with a particle by virtue of its motion, given by $\lambda = h/p$ where h is PLANCK'S CONSTANT and p is the particle's relativistic momentum. Only for electrons and other elementary particles can the de Broglie wavelength be large enough to produce observable diffraction effects. See ELECTRON DIFFRACTION.

debug (*ICT*) To detect, locate and correct every BUG. See ERROR, DIAGNOSTIC.

debunching (*Electronics*) Tendency for a beam of electrons, or a velocity-modulated beam of electrons, to spread because of their mutual repulsion. See BUNCHER, BUNCHING.

deburring (*Textiles*) Mechanical removal of dirt and vegetable matter from raw wool.

Debye and Scherrer method (*Crystal*) A method of X-RAY CRYSTALLOGRAPHY applicable to powders of crystalline substances or aggregates of crystals.

Debye–Hückel theory (*Chem*) A theory of electrolytes in solution that assumes complete ionization and attributes deviations from ideal behaviour to inter-ionic attraction.

Debye length (*Phys*) Maximum distance at which coulomb fields of charged particles in a plasma may be expected to interact.

Debye temperature (*Phys*) The single fitting parameter required to usefully apply the DEBYE THEORY OF SPECIFIC HEATS OF SOLIDS to many solids.

Debye theory of specific heats of solids (*Phys*) Theory based on the assumption that the thermal vibrations of the atom of a solid can be presented by harmonic oscillators whose energies can be quantized. The oscillator frequencies are distributed, up to a maximum (cut-off) frequency, according to the normal modes of vibration of a *continuous* medium. For many substances the theory gives a satisfactory agreement with experiment over a wide range of temperature. See BLACKMAN THEORY OF SPECIFIC HEATS OF SOLIDS.

Debye unit (*Phys*) Unit of electric dipole moment equal to $3 \cdot 34 \times 10^{-30}$ C m or 10^{-18} esu (electrostatic units).

Debye–Waller factor (*Phys*) The factor by which the intensities of coherently scattered X-rays or neutrons from a crystal are reduced by the thermal vibrations of the atoms, assuming that the thermal vibrations are isotropic. Given by

$$\exp\left(-2B \sin \theta / \lambda^2\right)$$

where θ is the angle of scatter, λ is the wavelength and B is the temperature factor, generally in the range 2–3 K.

deca- (*Genrl*) Prefix from Gk *deka*, signifying ten. Symbol da.

decade (*Phys*) Any ratio of 10:1. Specifically the interval between frequencies of this ratio.

decade box (*ElecEng*) A resistance (capacitance or inductance) box divided into sections so that each section has ten switched positions and ten times the value of the preceding section. The switches can therefore be set to any integral value within the range of the box.

de-caffeination (*FoodSci*) Removal of caffeine from coffee by SOLVENT EXTRACTION. Originally the solvent trichlorethylene was used, but has been largely superseded by liquid carbon dioxide or hot water.

decahydro-naphthalene (*Chem*) See DECALIN.

decalcification (*Med*) The process of absorption of calcium salts from bone.

decalcomania paper (*Paper*) A transfer paper for conveying a design on to pottery etc.

decalescence (*Eng*) The absorption of heat that occurs when iron or steel is heated through the arrest points. See RECALESCENCE.

decalin (*Chem*) Decahydro-naphthalene. $C_{10}H_{18}$. Bp 190°C (103 kN m^{-2}), a product of complete hydrogenation of naphthalene under pressure and in the presence of a catalyst.

decametric waves (*ICT*) Waveband from 10 to 100 m.

decant (*Chem*) To pour off the supernatant liquor when a suspension has settled.

Decapoda (*BioSci*) (1) An order of MALACOSTRACA with three pairs of thoracic limbs modified as maxillipeds, and five as walking legs. Shrimps, prawns, crabs, lobsters, etc. (2) A suborder of CEPHALOPODA having eight normal arms and two longer partially retractile arms. There is a well-developed internal shell, and lateral fins are present; actively swimming forms, usually carnivorous. Squids and cuttlefish.

decapsulation (*Med*) Removal of capsule or covering of an organ, esp of the kidney.

decarboxylase (*BioSci*) An enzyme that catalyses the removal of CO_2 from its substrate.

decarburization (*Eng*) Removal of carbon from the surface of steel by heating in an atmosphere in which the concentration of decarburizing gases exceeds a certain value.

decatizing (*Textiles*) Process for imparting a permanent finish to worsted and woollen fabrics by forcing steam through them while under tension, to improve their appearance and handle.

decay (*Phys*) The process of spontaneous transformation of a radionuclide.

decay (*Psych*) In the study of memory, the theory that memory fades or disappears over time if it is not used or accessed.

decay chain (*Phys*) The series of radionuclides in which one nucleus disintegrates to form another until a stable, non-radioactive isotope is reached.

decay constant (*Phys*) See DISINTEGRATION CONSTANT.

decay factor (*Phys*) The factor which expresses rate of decay of oscillations in a damped oscillatory system, given by the natural logarithm of the ratio of two successive amplitude

maxima divided by the time interval between them. Calculated from the ratio of resistance coefficient to twice mass in a mechanical system, and the ratio of resistance to twice inductance in an electrical system. Also *damping factor*. See LOGARITHMIC DECREMENT.

decay heat (*NucEng*) The heat produced by the radioactive decay of fission products in a reactor core. This continues to be produced even after the reactor is shut down. Also *shutdown heating, shutdown power*.

decay law (*Phys*) An expression describing a physical decay phenomenon. If the rate of decrease of a quantity is proportional to the quantity at that time, then the decay law is exponential, ie

$$N(t) = N_0\, e^{-\lambda t}$$

where $N(t)$ is the quantity at time t, λ is the *decay constant* and N_0 is the value at time $t = 0$. This law holds for the decay of radioactive nuclei.

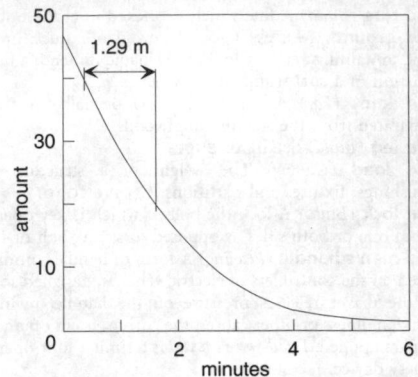

decay law Half-life of 1.29 minutes.

decay product (*Phys*) See DAUGHTER PRODUCT.

decay time (*Phys*) The time in which the amplitude of an exponentially decaying quantity reduces to $\exp(-1)$ ($36 \cdot 8\%$) of its original value.

Decca Navigator (*Aero, Ships*) TN for a navigation system of the radio position-fixing type using continuous waves. The FIX, given by the intersection of two hyperbolic position lines, is indicated on meters and can be plotted on a Decca chart or on a flight log which gives a continuous pictorial presentation of position. See DECTRA.

decelerating electrode (*Electronics*) One which is intended to reduce the velocity of electrons.

deceleration (*Phys*) Slowing, or negative, acceleration. The rate of diminution of velocity with time. Units m s^{-2}.

decerebrate (*BioSci*) Lacking a cerebrum. Also an experimental procedure by which cerebral brain function in an animal is ablated by removing the cerebrum, cutting across the brain stem, or severing certain arteries in the brain stem.

decerebrate rigidity (*Med*) A posture of extensor rigidity of trunk and limbs seen in diseases or lesions of the brain stem.

decerebrate tonus (*BioSci*) A state of reflex tonic contraction of certain skeletal muscles following upon the separation of the cerebral hemispheres from the lower centres.

deci- (*Genrl*) Prefix with physical unit, meaning one-tenth. Symbol d.

deciampere balance (*ElecEng*) An ampere balance having a range $0 \cdot 1$–10 A.

decibel (*Acous*) Ten times the logarithm to base 10 of an energy ratio; eg sound pressure level is measured in decibels and defined as $10 \log p^2/p_0{}^2$, where p is the rms sound pressure and p_0 is a reference sound pressure. For airborne sound, p_0 is the THRESHOLD OF HEARING.

decibel (*ICT*) One-tenth of a *bel*, signifying the ratio of two amounts of power; given by $n = 10 \log_{10} P_1/P_2$, where n is the ratio expressed in decibels, and P_1 and P_2 represent the power levels being compared. Used almost universally as a measure of response or performance in electronic and communication circuits. Under appropriate conditions, usually identical impedance in input and output circuits, the ratio may be expressed as $n = 20 \log_{10} V_1/V_2$ where V_1 and V_2 are the voltage levels of the signals being compared, or as the equivalent current ratio; the latter interpretation is prone to confusion.

decibel meter (*Acous*) A meter which has a scale calibrated in logarithmic steps and labelled with decibel units.

decidua (*BioSci*) In mammals, the modified mucous membrane lining the uterus at the point of contact with the placenta, which is torn away at parturition and then ejected; the afterbirth; the maternal part of the placenta.

deciduate (*BioSci*) Said of mammals in which the maternal part of the placenta comes away at birth. Also *caducous*. Cf INDECIDUATE.

deciduous (*BioSci*) (1) Falling off, usually after a lengthy period of functioning. (2) Said of plants which shed leaves habitually over a cold period, as opposed to evergreen. (3) Said of the first dentition teeth.

decile (*MathSci*) Any of the ten equal groups into which the items in a frequency distribution can be divided; any of the nine values that divide the items in a frequency distribution into ten equal groups, so that the ninth decile is the value below which 90% of the population lie; the fifth decile is the MEDIAN. Compare PERCENTILE.

decimal (*MathSci*) Relating to or using BASE 10. See DENARY.

decimal fraction (*MathSci*) A fraction having a power of ten as denominator. The denominator is not usually written but is indicated by the *decimal point*, a dot (on the Continent, a comma) placed between the whole-part figure and the fractional part. The number of figures after the decimal point is equal to the power of ten of the denominator. Thus 42·017 is equal to

$$42\frac{17}{1000}$$

Also *decimal, decimal number*.

decimal system (*MathSci*) (1) The POSITIONAL number system whose base is ten, so that the digits represent multiples of increasing powers of ten. (2) A measuring system in which each unit is a power of ten times greater than the next smaller one; eg the METRIC SYSTEM and SYSTÉME INTERNATIONAL.

decimetre (*Genrl*) One-tenth of a METRE. US *decimeter*.

decimetric waves (*ICT*) Waveband having a range from 10 cm to 1 m.

decimolar calomel electrode (*Chem*) A calomel electrode containing 0·1 M potassium chloride solution.

decimo octavo (*Print*) See 18 MO.

decimo sexto (*Print*) See 16 MO.

decineper (*Phys*) Unit of voltage and current attenuation in lines and amplifiers, of magnitude one-tenth of the NEPER. Defined by $d = \ln(x_1/x_2)$, where d is number of decinepers, x_1 and x_2 are currents (voltages or acoustic pressures). In properly terminated networks, 1 decineper equals 0·8686 dB. Symbol dN.

decision speed (*Aero*) See CRITICAL SPEED.

decitex (*Textiles*) See TEX.

deck (*ImageTech*) The assembly of transport mechanism and transducer heads for a disk or magnetic tape recording/reproducing system.

deck (*Ships*) A platform which forms the top of one horizontal division of a ship and the bottom of that immediately above.

deck beam (*Ships*) A stiffening member of a deck, which may be either transverse or longitudinal. It is supported at extremities by knee connections to frames or bulkheads or by supporting girders.

deck bridge (*CivEng*) A bridge on which the road or rail track is carried above the support beams or girders. Cf THROUGH BRIDGE.

deck crane (*Ships*) A crane, either fixed or movable, mounted on deck for use in loading or discharging cargo.

deck houses (*Ships*) See TOP HAMPER.

deckle edge (*Paper*) A rough feather edge on the four sides of a hand-made sheet due to stock seeping beneath the DECKLE. Similar effects can be simulated artificially. Also the irregular edges of a web of paper before trimming.

deckle (*Paper*) The width as defined by the web on a FOURDRINIER paper-making machine.

decks (*Print*) On rotary presses, pairs of horizontal printing couples arranged one above the other.

deck stringer (*Ships*) The main strength portion of a ship's deck, being that portion, on both sides, adjacent and attached to the shell plating. It comprises the stringer strake of plating and the stringer angle section forming such attachment.

declaration (*ICT*) A statement in a HIGH-LEVEL LANGUAGE that has the form of descriptive information rather than an explicit instruction. Also *declarative statement*. See DATA TYPE.

declarative memory (*Psych*) Conscious memory that can be communicated (declared) to others. The long-term memory where factual information is stored as opposed to PROCEDURAL MEMORY.

declared efficiency (*ElecEng*) The efficiency which the manufacturers of an electric machine or transformer declare it to have, under certain specified conditions.

declination (*Astron*) The angular distance of a heavenly body from the CELESTIAL EQUATOR measured positively northwards along the HOUR CIRCLE passing through the body. Abbrev *dec*.

declination (*Surv*) See MAGNETIC DECLINATION.

declination circle (*Astron*) A graduated circle on the declination axis of an equatorial telescope which enables the telescope to be set to a given declination or the declination of a given star to be read.

declinimeter (*ElecEng*) An apparatus for determining the direction of a magnetic field with respect to astronomical or survey co-ordinates.

decoction (*Chem*) An extract of a substance or substances obtained by boiling.

decoder (*ICT*) (1) Any circuit that responds to a particular coded signal while rejecting others. (2) A circuit that converts coded information (eg pulse-code-modulated speech) into analogue information. (3) That part of a TV receiver which separates the incoming colour information into the red, green and blue components necessary to operate the cathode-ray tube. (4) In a frequency-modulated broadcast receiver, the circuit that separates the stereophonic sound signal into left and right channels.

decollate (*ICT*) To separate the sheets of multipart continuous stationery.

decollement (*Geol*) Detachment fault of strata due to deformation, with independent styles of deformation above and below a discontinuity, the *plane of decollement*. The rocks above the dislocation commonly show complex folding or thrusting.

decolorize (*Chem*) To remove the coloured material from a liquid by bleaching, precipitation or adsorption.

decolorizers (*Glass*) Materials added to the batch for the express purpose of improving the appearance of the glass by hiding the yellow-green colour due to iron impurities.

decommissioning (*NucEng*) The permanent withdrawal from service of a nuclear facility and the subsequent operations to bring it to a safe and stable condition.

decompensation (*Med*) Failure of an organ to compensate for functional overload produced by a disease.

decomposer (*BioSci*) In an ecosystem, one of the heterotrophic organisms, chiefly bacteria and fungi, which break down the complex compounds of dead protoplasm, absorbing some of the products of decomposition, but also releasing simple substances usable by producers. Cf CONSUMER, PRODUCER.

decomposition (*Chem*) The breaking down of a substance into simpler molecules or atoms.

decomposition voltage (*ElecEng*) The minimum voltage which will cause continuous electrolysis in an electrolytic cell.

decompound (*BioSci*) In botanical anatomy, a COMPOUND structure whose parts are themselves compound. Also *doubly compound*.

decompression (*Med*) Any procedure for relieving pressure or the effects of pressure.

decondensed chromatin (*BioSci*) See EUCHROMATIN.

deconjugation (*BioSci*) The separation of the paired chromosomes before the end of the prophase meiosis.

decontamination (*NucEng*) Removal of radioactivity from area, building, equipment or person to reduce exposure to radiation. More generally, removal or neutralization of bacteriological, chemical or other contamination.

decontamination factor (*NucEng*) The ratio of initial to final level of contamination for a given process.

deconvolution (*ImageTech*) Mathematical processing of a digital image to remove out-of-focus blur. A variety of algorithms are available and have found extensive use in astronomy and biology. The final image can be rotated and viewed from different angles and has usually had noise filtered out so that the image is much clearer and sharper.

decorative laminate (*Plastics*) Laminates with a highly resistant, frequently decorative surface based on MELAMINE RESINS coated on resin impregnated paper. See KRAFT PAPER.

decorticated (*BioSci*) Deprived of bark; devoid of CORTEX.

decortication of the lung (*Med*) Operative removal from the lung of pleura thickened as a result of chronic inflammation.

decoupling (*Electronics*) Reduction of a common impedance between parts of a circuit, eg by using a BYPASS CAPACITOR.

decoupling filter (*Electronics*) Simple resistor–capacitor section(s), which decouple feedback circuits and so prevent oscillation or MOTOR-BOATING (relaxation oscillation).

decreasing (*MathSci*) Of a function or sequence: such that for all $a > b$, $f(a) \leqslant f(b)$. If $f(a) < f(b)$, then the function is said to be *strictly decreasing*. Compare INCREASING.

decrement (*Phys*) The ratio of successive amplitudes in a damped harmonic motion.

decrepitation (*Chem*) (1) The crackling sound made when crystals are heated, caused by internal stresses and cracking. (2) Breakdown in size of the particles of a powder due to internal forces, generally induced by heating.

decryption (*ICT*) The recovery of a plain message from an encrypted one.

DECT (*ICT*) Abbrev for *digital enhanced cordless telecommunications*, a standard for mobile-telephone services.

Dectra (*Aero, Ships*) A radio position-fixing system, based largely on DECCA NAVIGATOR principles, designed to cover specific air route segments and transoceanic crossings. In addition to fix, location along a track and range information are given: hence *Decca, Track and Range*.

decubitus (*Med*) The special or preferred posture in bed of a patient suffering from particular disease states. Patients with pleuritic pain lie on the affected side.

decubitus ulcer (*Med*) An ulcer or bed sore developing from prolonged immobility.

decumbent (*BioSci*) A stem, lying flat, usually with a turned-up tip.

decurrent (*BioSci*) (1) Running down, as when a leaf margin continues down the stem as a wing. (2) With several roughly equal branches, as in shrubs and in the crowns of some trees, esp when old. Cf EXCURRENT.

decussate (*BioSci*) With leaves in opposite pairs, each pair being at right angles to the next. See PHYLLOTAXIS.

decussate texture (*Geol*) The random arrangement of prismatic or tabular crystals in a rock.

decussation (*BioSci*) Crossing over of nerve tracts with interchange of fibres.

Dedekind cut (*MathSci*) A division of the rational numbers into two classes such that all numbers in one are greater than all numbers in the other; used to define irrational numbers.

dedendum (*Eng*) (1) Radial distance from the pitch circle of a gear wheel to the bottom of the spaces between teeth. See fig. at GEAR WHEEL. See INVOLUTE GEAR TEETH, PITCH DIAMETER. (2) Radial distance between the pitch and minor cylinders of an external screw-thread. (3) Radial distance between the major and pitch cylinders of an internal thread.

dedicated computer (*ICT*) One that is permanently assigned to one application.

dedicated protection ring (*ICT*) A network arrangement that provides continuity of service in the event of a link failure. All network nodes are connected in a ring and data are sent both ways round the ring. In the event of failure, each node simply selects from its two links the one that is still working.

dedicated virtual private network (*ICT*) A form of VIRTUAL NETWORK in which hard-wired multiplexers provide continuous shared access to the public network, as opposed to links being set up as and when required. Cf SWITCHED VIRTUAL PRIVATE NETWORK.

dedifferentiation (*BioSci*) Changes in a differentiated tissue, leading to the reversion of cell types to a common undifferentiated form, such as the meristematic state in plants. Although well recognized in plants, it is unclear whether dedifferentiation occurs naturally in animals.

dedolomitization (*Geol*) The recrystallization of a dolomite rock or dolomitic limestone consequent on contact metamorphism; essentially involving the breaking down of the dolomite into its two components, $CaCO_3$ and $MgCO_3$. The former merely recrystallizes into a coarse calcite mosaic; but the latter breaks down further into MgO and CO_2. See FORSTERITE-MARBLE.

deducted spaces (*Ships*) Spaces deducted from the GROSS TONNAGE to obtain the NET REGISTER TONNAGE. In general, deducted spaces are those spaces required to be used in the working of the ship and accommodating the crew.

deductive reasoning (*Psych*) Reasoning that involves a series of logical operations on a set of assumptions, proceeding from the general to the particular, in order to arrive at a conclusion.

de-emphasis (*ICT*) The use of an amplitude–frequency characteristic complementary to the one used for PRE-EMPHASIS prior to transmission and recording.

de-energize (*ElecEng*) To disconnect a circuit from its source of power.

deep drawing (*Eng*) The process of cold-working, or drawing, sheet or strip metal into shapes with dies involving considerable plastic distortion of the metal.

deep-dyeing fibres (*Textiles*) Fibres whose chemical composition has been varied from the normal so that they take up more dyestuff in the dyeing process.

deep etch (*Print*) (1) A lithographic process in which the image is very slightly etched into the surface of the plate, producing more stability than on a SURFACE (or albumen) PLATE. Used for longer print runs. A positive is used when printing down. (2) In letterpress half-tone where highlights have been etched away. See DROPOUT HALF-TONE.

deeply etched (*Print*) A half-tone block given a supplementary deepening etch to improve its printing and duplicating qualities.

deeps (*Surv*) See LEAD LINE.

deep-sea deposits (*Geol*) Pelagic sediments accumulating in depths of more than 2000 m. They include, in order of increasing depth, calcareous oozes, siliceous oozes and red clay. TERRIGENEOUS material is absent.

deep-sea lead (*Surv*) A lead used for attachment to a lead line measuring beyond 100 fathoms.

deep tank (*Ships*) A large tank extending from the bottom up to the first deck and from side to side. May be used for liquid cargo or for water ballast.

deep therapy (*Radiol*) X-ray therapy of underlying tissues by HARD RADIATION (usually produced at more than 180 kVp) passing through superficial layers.

deep-vein thrombosis (*Med*) The formation of a blood clot in a deep vein, sometimes caused by restricted movement in people travelling on long-distance flights with insufficient legroom. Abbrev *DVT*.

deep-well pump (*Eng*) A centrifugal pump, generally electrically driven by a submerged motor built integrally with it, placed at the bottom of a deep borehole for raising water.

deer-fly fever (*Med*) See TULARAEMIA.

deerite (*Min*) A black, monoclinic hydrous silicate of iron and manganese.

dees (*Electronics*) A pair of hollow half-cylinders, ie D-shaped, in the vacuum of a CYCLOTRON for accelerating charged particles in a spiral, a high-frequency voltage being applied to them in anti-phase.

deet or DEET (*Chem*) N,N-diethyl-3-methylbenzamide, a colourless compound used as an insect repellent. Probably from its former chemical name N,N-diethyl-meta-toluamide.

default option (*ICT*) Specific alternative action to be taken automatically by the computer in the event of the omission of a definite instruction or action.

defecation (*Med*) The ejection of feces from the body.

defect (*Crystal*) Crystal lattice imperfection which may be due to the introduction of a minute proportion of a different element into a perfect lattice, eg indium into germanium crystal, to form an intrinsic semiconductor for a transistor. A 'point' defect is a VACANCY or an 'interstitial atom', while a 'line' defect relates to a DISLOCATION in the lattice. Often accounts for the colour of gemstones.

defect (*Eng*) Anything which can cause a product to fail in its specified function. Usefully classified as minor, major, serious and very serious, with the latter two potentially able to cause injury and severe injury and corresponding degrees of economic loss.

defective equation (*MathSci*) An equation derived from another, but with fewer roots than the original.

defective virus (*BioSci*) A virus unable to replicate without a *helper* such as a PLASMID providing a replicative function.

defect sintering (*Eng*) Sintering whereby particles are introduced as a fine dispersion in a sintered body, or by chemical action during sintering. Mobility in heat treatment, with or without working, enables the introduced atoms or molecules to migrate through defects giving marked modification of properties.

defect structure (*Crystal*) Intense localized misalignment or gap in the crystal lattice, due to migration of ions or to slight departures from stoichiometry. The resulting opportunity for mobility is important for semiconductors, catalysis, photography, rectifiers, corrosion, etc. Cf DISLOCATION.

defence mechanism (*Psych*) In psychoanalytic theory, a collective term for a number of unconscious processes that are used to ward off or lessen anxiety by a variety of means that seek to keep the source of anxiety out of consciousness, eg repression.

defensin (*BioSci*) Any of a group of antimicrobial cationic proteins found in vertebrate phagocytes, especially neutrophils, and also identified in insects.

deferent (*Astron*) See EPICYCLE.

defervescence (*Med*) The fall of temperature during the abatement of a fever; the period when this takes place.

defibrillator (*Med*) Electrical apparatus to arrest ventricular FIBRILLATION.

deficiency (*BioSci*) The absence, by loss or inactivation, of a gene or a part of a chromosome, that is normally present.

deficiency (*MathSci*) The difference between the maximum possible number of double points on a curve and the actual number.

deficiency disease (*BioSci, Med*) Any disease caused by a lack of essential minerals, nutrients or vitamins, often due to their unavailability in the soil or diet.

deficient (*Build*) See UNSTABLE.

deficient number (*MathSci*) A natural number for which the sum of the proper factors is less than the number itself; eg 14 is deficient since $1 + 2 + 7 < 14$. Cf ABUNDANT NUMBER, PERFECT NUMBER.

defined medium (*BioSci*) Cell culture medium in which all components are known. In practice this means that the serum (that is normally added to culture medium for animal cells) is replaced by insulin, transferrin and possibly specific growth factors.

definite (*BioSci*) (1) SYMPODIAL GROWTH. (2) A CYMOSE INFLORESCENCE.

definite integral (*MathSci*) If $F(x)$ equals

$$\int f(x)\, dx$$

then

$$\int_a^b f(x)\, dx = F(b) - F(a)$$

is the *definite integral* over the range a to b.

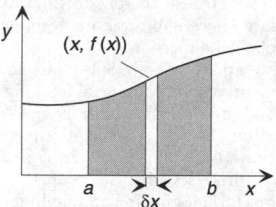

definite integral

definite proportions (*Chem*) See LAW OF CONSTANT (DEFINITE) PROPORTIONS.

definite time-lag (*ElecEng*) A time-lag fitted to relays or circuit breakers to delay their operation; it is quite independent of the magnitude of the current causing that operation. Also *constant time-lag, fixed time-lag, independent time-lag*.

definition (*Acous*) Ill-defined quantity describing one aspect of the quality of concert halls and auditoria. Similar to clarity.

definition (*ImageTech*) The ability of an imaging system to reproduce fine detail, involving both its RESOLUTION and its reproduction of subject contrast. See MTF.

definition (*Phys*) See RESOLVING POWER OF THE EYE.

definitive (*BioSci*) Final, complete: fully developed; defining or limiting.

deflagration (*Chem*) Sudden combustion, generally accompanied by a flame and a crackling sound.

deflation (*Geol*) The winnowing and transport of dry loose material, esp silt and clay, by wind.

deflecting electrodes (*Electronics*) See DEFLECTOR PLATES.

deflection (*Eng*) (1) The amount of bending or twisting of a structure or machine part under load. (2) The movement of the pointer or pen of an indicating or a recording instrument.

deflection angle (*Electronics*) See ANGLE OF DEFLECTION.

deflection angle (*Surv*) The angle between one survey line and the prolongation of another survey line which meets it. See INTERSECTION ANGLE.

deflection coil (*Electronics*) That which, when placed around the neck of a cathode-ray tube and energized with appropriate currents, achieves the desired deflection of the luminous spot on the screen.

deflection defocusing (*Electronics*) Loss of focus of a cathode-ray-tube spot as deflection from the centre of the screen increases.

deflection sensitivity (*Electronics*) (1) The ratio of displacement of the spot or angle of an electron beam to the voltage producing it. (2) The ratio of displacement of the spot or angle of a beam to the magnetic field producing it, or the current in the deflecting coils. Also applies to galvanometer or other instrument for measuring voltage, current or quantity of electricity.

deflection yoke (*Electronics, ImageTech*) An assembly (usually on a moulded plastic former) of specially shaped DEFLECTION COILS placed around the neck of a cathode-ray tube in a TV receiver.

deflectometer (*Eng*) A device for measuring the amount of bending suffered by a beam during a transverse test.

deflector (*Ships*) An instrument used during the adjustment of magnetic compasses. It measures the strength of the field at the compass.

deflector coil (*Electronics*) One of a group of one or more coils so arranged that a current passing through produces a magnetic field which deflects the beam in a cathode-ray tube employing magnetic deflection. Usually applied around the neck of the tube.

deflector plates (*Electronics*) Electrodes so arranged in a cathode-ray tube that the electrostatic field produced by a potential difference deflects the beam. Also *deflecting electrodes*. See ELECTROSTATIC DEFLECTION.

deflexed (*BioSci*) Bent sharply downwards.

deflocculate (*Chem*) To break up agglomerates and form a stable colloidal dispersion.

defoaming agent (*Chem*) Substance added to a boiling liquid to prevent or diminish foaming. Usually hydrophobic and of low surface tension, eg silicone oils.

deforestation (*EnvSci*) Permanent removal of forest.

deformation (*Geol*) A general term used to describe the structural processes that may affect rocks after their formation. Includes folding and faulting.

deformation (*MathSci*) See HOMOTOPIC MAPPING.

deformation map (*Eng*) A two-dimensional representation of temperature and stress which shows the deformation behaviour expected in a given metallic or ceramic material. They are useful in predicting creep behaviour in service at moderately elevated temperatures. Temperature is plotted as homologous temperature, T/T_m, though for practical purposes the Celsius scale may be more useful. The stress axis is normalized by expressing the applied stress as a ratio of the elastic modulus E. See panel on CREEP AND DEFORMATION.

deformation potential (*Electronics*) Potential barrier formed in semiconducting materials by lattice deformation.

deformation ratio (*MathSci*) See MAGNIFICATION.

defragmentation (*ICT*) The rearrangement of BLOCKS of data on a storage device to form contiguous blocks and speed up retrieval.

degassing (*Electronics*) Removal of last traces of gas from valve envelopes, by pumping or *gettering* the gas, with or without heat, with eddy currents, electron bombardment or simple baking.

degassing (*MinExt*) To remove entrained gas from drilling mud; important because the gas may seriously reduce the hydrostatic pressure available in the borehole.

degaussing (*ImageTech*) Removal from colour cathode-ray tube of spurious magnetization which could affect colour purity.

degaussing (*Phys*) Neutralization of the magnetization of a mass of magnetic material, eg a ship, by an encircling current.

degeneracy (*Phys*) A condition which arises if two or more quantum states have the same energy. If the energy level can be realized in n different ways the energy has an n-fold degeneracy.

degenerate (*Phys*) Said of two or more quantum states which have the same energy. The *degeneracy* is the number of states having a given energy.

degenerate gas (*Phys*) (1) That which is so concentrated, eg electrons in the crystal lattice of a conductor, that the MAXWELL–BOLTZMANN DISTRIBUTION LAW is inapplicable. (2) Gas at very high temperature in which most of the electrons are stripped from the atoms. (3) An electron gas which is far below its FERMI LEVEL so that a large fraction of the electrons completely fills the lower energy levels and has to be excited out of these levels in order to take part in any physical processes.

degenerate primer (*BioSci*) A single-stranded synthetic oligonucleotide designed to hybridize to DNA encoding a particular protein sequence but with sufficient variability to hybridize to DNA that differs slightly. Such primers are widely used in screening a genomic library or in degenerate PCR, to identify homologues of already known genes.

degenerate semiconductor (*Electronics*) One in which the conduction approaches that of a simple metal.

degeneration (*BioSci*) Evolutionary retrogression; the process of returning from a higher or more complex state to a lower or simpler state.

degenerative disorders (*Psych*) See DEMENTIA.

deglutition (*BioSci*) The act of swallowing.

degradation (*Chem*) General term for reactions which cause loss of integrity of polymer properties. It covers depolymerization as well as chain oxidation, ozone cracking, ultraviolet degradation, etc.

degradation (*NucEng*) Unintentional loss of energy of neutrons in a nuclear reactor.

degradation (*Phys*) Loss of energy of motion solely by collision. In an isolated system, the ENTROPY increases.

degreasing (*Textiles*) Removal of natural fats, oils and waxes from textiles by extraction, usually with an organic liquid.

degree (*MathSci*) (1) Unit of angle, written °; thus 360° for a revolution, 90° for a right angle. (2) Degree of a polynomial is the highest power of the variable present, so that a quadrate has degree 2, for example. (3) Similar definitions to (2) apply in many areas.

degree (*Phys*) The unit of temperature difference. It is usually defined as a certain fraction of the fundamental interval, which for most thermometers is the difference in temperature between the freezing and boiling points of water. See CELSIUS SCALE, CENTIGRADE SCALE, FAHRENHEIT SCALE, INTERNATIONAL PRACTICAL TEMPERATURE SCALE, KELVIN THERMODYNAMIC SCALE OF TEMPERATURE.

degree of a curve (*Surv*) The angle subtended by a standard chord length of 100 ft at the centre of a curve.

degree of cross-linking (*Chem*) See CROSS-LINK DENSITY.

degree of crystallinity (*Chem, Textiles*) Total crystalline content of a partially crystalline material, esp polymers. Experimentally found using density tests or inferred from wide-angle X-ray scattering measurements. Kinetics of development described by AVRAMI EQUATION. Depends on polymer type and chain regularity as well as rate of cooling from melt. See CRYSTALLIZATION OF POLYMERS and panel on POLYMER SYNTHESIS.

degree of damping (*Phys*) The extent of the damping in an oscillatory system, expressed as a fraction or percentage of that which makes the system critically damped.

degree of dissociation (*Chem*) The fraction of the total number of molecules which are dissociated.

degree of ionization (*Chem*) The proportion of the molecules or 'ion pairs' of a dissolved substance dissociated into charged particles or ions.

degree of polymerization (*Chem*) The number of repeat units in a polymer chain (usually an average). Equals molecular mass of chain divided by molecular mass of repeat unit. Abbrev *DP*.

degree of swelling (*Chem*) Extent of swelling in rubbers, judged either by change in linear dimensions or by volumetric change.

degrees baumé (*FoodSci*) A measure of the potential alcohol by volume in a given must or sugar solution. A hydrometer graduated in degrees baumé sinks to the level which corresponds to the percentage of alcohol which would be produced if all the sugar were to be fermented to alcohol.

degrees of freedom (*Chem*) (1) The number of variables defining the state of a system (eg pressure, temperature) which may be fixed at will. See PHASE RULE. (2) The number of independent capacities of a molecule for holding energy, translational, rotational and vibrational. A molecule may have a total of $3n$ of these, where n is the number of atoms in the molecule.

degrees of freedom (*MathSci*) The minimum number of variables necessary to describe a given system. In statistics, generally one less than the number of random variables.

de Haas–van Alphen effect (*Phys*) Oscillations in the magnetic moment of a metal as a function of $1/B$ where B is the magnetic flux density. Interpretation of this effect gives important information about the shape of the FERMI SURFACE.

dehiscence (*BioSci*) The spontaneous opening at maturity of a fruit, anther, sporangium or other reproductive body, permitting the release of the enclosed seeds, spores, etc. Adj *dehiscent*. Cf INDEHISCENT. More generally the act of splitting open.

dehydration (*Chem*) (1) The removal of H_2O from a molecule by the action of heat, often in the presence of a catalyst, or by the action of a dehydrating agent, eg concentrated sulphuric acid. (2) The removal of water from crystals, tars, oils, etc, by heating, distillation or by chemical action.

dehydration (*FoodSci*) Removing water to produce a dry, shelf-stable material which can be rehydrated to a condition similar to its original form. Traditional methods involved sun drying and air drying. The moisture level is reduced to below that at which spoilage organisms can grow. See DRUM DRYING, FREEZE DRYING, SPRAY DRYING.

dehydration (*Med*) Excessive loss of water from the tissues of the body.

dehydrogenase (*BioSci*) Enzymes that catalyse the oxidation of their substrate with the removal of hydrogen atoms.

de-icing (*Aero*) A method of protecting aircraft against icing by removing built-up ice before it assumes dangerous proportions. It may be based on pulsating pneumatic overshoes, chemical applications or intermittent operation of electrical heating elements. Cf ANTI-ICING.

Deimos (*Astron*) One of the two natural satellites of Mars, discovered in 1877. Distance from the planet 23 460 km; diameter approx 15 km.

de-individuation (*Psych*) A term used by social psychologists to denote a loss of a sense of personal responsibility in conditions of relative anonymity where a person ceases to behave as an individual, acting only as a member of a group. The person engages in activities they would not normally do presumably because of a weakening of internal controls (eg shame or guilt) and loss of personal responsibility.

Deion circuit breaker (*ElecEng*) TN for a circuit breaker fitted with an arc-control device in which the arc takes place within a slot in a stack of insulated plates. The plates contain iron inserts or a magnet coil, so that the arc is blown magnetically towards the closed end of the slot, thereby coming into contact with cool oil which has a de-ionizing action and extinguishes the arc.

de-ionized water (*Chem*) Water from which ionic impurities have been removed by passing it through cation and anion exchange columns.

déjà vu (*Psych*) The illusion that a new scene is familiar, a form of the memory disorder PARAMNESIA.

Deka-ampere balance (*ElecEng*) An ampere balance having a current range from 1 to 100 A.

del (*MathSci*) In cartesian co-ordinates, the vector operator

$$i\frac{\partial}{\partial x} + j\frac{\partial}{\partial y} + k\frac{\partial}{\partial z}$$

Also *nabla*. Symbol ∇.

delamination (*BioSci*) The division of cells in a tissue, leading to the formation of layers.

delamination (*Eng*) (1) Peeling of surface layers in moulded products, a type of mould defect often caused by contamination from mould release agents or foreign polymer. (2) Separation of fabric or reinforced layers in a composite material, caused by water ingress or poor adhesion etc.

De la Rue cell (*ElecEng*) See CHLORIDE OF SILVER CELL.

delay (*Acous*) Time shift which can be introduced into the transmission of a signal, by recording it magnetically on tape, disk or wire, and reproducing it slightly later. Used in public-address systems, to give illusion of distance and to coalesce contributions from original source and reproducers.

delay circuit (*ICT*) One in which the output circuit is delayed by a specified time interval with respect to the input signal; used for phase adjustment or correction in radio-frequency circuits, or in digital (pulse) circuits for making signals from different sources coincide.

delay distortion (*ICT*) Change in waveform during transmission because of non-linearity of delay with frequency, which is $d\beta/d\omega$, where β = phase delay in radians, and $\omega = 2\pi \times$ frequency (Hz). See ENVELOPE DELAY DISTORTION.

delayed action (*ElecEng*) Any arrangement which imposes an arbitrary delay in operation of eg a switch or circuit breaker.

delayed automatic gain control (*ICT*) That which is operative above a threshold voltage signal in a radio receiver. Provides full amplification in radio receivers for very weak signals, and constant output from detector for signals above the threshold. Also *quiet automatic volume control*. See AUTOMATIC GAIN CONTROL.

delayed critical (*Phys*) Assembly of fissile material critical only after release of delayed neutrons. Cf PROMPT CRITICAL.

delayed drop (*Aero*) A live parachute descent in which the parachutist deliberately delays pulling the ripcord until after a descent of several thousand metres.

delayed neutron groups (*NucEng*) Fission products placed into groups with a characteristic decay constant and fractional yield, so that calculations relating to delayed neutrons are made easier. Six groups are commonly used whose parameters vary depending on the fissile nucleus. See INHOUR EQUATION.

delayed neutrons (*NucEng*) Those arising from fission but not released instantaneously. Fission neutrons are always PROMPT NEUTRONS; those apparently delayed (up to seconds) arise from breakdown of fission products, not primary fission. Such delay eases control of reactors.

delayed opening (*Aero*) Delaying the opening of a parachute by an automatic device. In any flight above 40 000 ft (12 500 m), low temperature and pressure require that aircrew must reach lower altitude for survival as rapidly as possible, and it is usual to have a barostatic device to delay opening to a predetermined height, usually 15 000 ft (4500 m).

delayed-type hypersensitivity (*BioSci*) Hypersensitivity state mediated by T-lymphocytes. When the antigen is introduced locally, eg in the skin, a gradual local accumulation of T-cells and monocytes results. The visible reaction is reddening and local swelling, increasing for 24–48 h and then subsiding, sometimes leaving a small scar due to necrosis of blood vessels. Tuberculin testing is a good example. Frequently this and other types of hypersensitivity co-exist, and reactions are not clear cut. Also *Type-4 hypersensitivity*.

delay line (*ICT*) (1) Column of mercury, a quartz plate or length of nickel wire, in which impressed sonic signals travel at a finite speed and which, by the delay in travelling, can act as a STORE, the signals being constantly recirculated and abstracted as required. See FIRST-GENERATION COMPUTER. (2) Real or artificial transmission line or network used to delay propagation of an electrical signal usually by a specified interval.

delay network (*ICT*) Artificial line of electrical networks, designed to give a specified phase delay in the transmission of currents over a specified frequency band.

delay period (*Eng*) The time or crank-angle interval between the passage of the spark and the resulting pressure rise in a petrol or gas engine, or between fuel injection and pressure rise in an oil engine.

Delbrück scattering (*Phys*) Elastic coherent scattering of gamma rays in the coulomb field of a nucleus. The effect is small and so far has not been conclusively detected.

delessite (*Min*) An oxidized variety of chlorite, relatively rich in iron.

deleted neighbourhood (*MathSci*) Of a point z_0; the set of all points z in the domain $|z-z_0|<\alpha$ (α a constant) excluding the point z_0.

deletion mutation (*BioSci*) A mutation in which a base or bases are lost from the DNA. Cf BASE SUBSTITUTION.

Delhi boil (*Med*) Oriental sore resulting from infection of the skin with a protozoal parasite (*Leishmania*). Also *Baghdad boil, tropical sore*.

Delian problem (*MathSci*) One of the three celebrated problems of antiquity: DUPLICATING THE CUBE, SQUARING THE CIRCLE and TRISECTING THE ANGLE.

delimiter (*ICT*) (1) A special character that is used to separate the FIELDS and RECORDS in a DATABASE FILE. A file may be called 'comma delimited' when commas are used in the file to separate the fields. This technique is often used to transfer data files between otherwise incompatible programs. (2) Special characters used by a WORD PROCESSOR PROGRAM that are used to separate words (a space), sentences (a full stop), paragraphs (a hard return) and pages (end of page code).

delinquency (*Psych*) Conduct disorder, usually against the law but including a range of antisocial, deviant or immoral behaviour. The age at which young people cease to be considered juvenile delinquents varies in different legal systems.

deliquescence (*Chem*) The change undergone by certain substances which become damp and finally liquefy when exposed to the air, owing to the very low vapour pressure of their saturated solutions; eg calcium chloride.

delirium (*Med*) A profound disturbance of consciousness occurring in febrile and toxic states; characterized by restlessness, incoherent speech, excitement, delusions, illusions and hallucinations.

delirium tremens (*Med*) An acute delirium in chronic alcoholism, characterized by insomnia, restlessness, terrifying hallucinations and illusions, and loss of orientation to time and place. Abbrev *DT*.

delivery (*Eng*) (1) The discharge from a pump or compressor. (2) The withdrawal of a pattern from a mould.

delivery (*Print*) The mechanical arrangement for delivering sheets after printing, there being several designs.

Dellinger fade-out (*ICT*) Complete fade-out (which may last for minutes or hours) and inhibition of short-wave radio communication because of the formation of a highly absorbing D-layer, lower than the regular E- and F-layers of the ionosphere, on the occasion of a burst of hydrogen particles from an eruption associated with a sunspot.

delph (*Build*) A drain behind a sea embankment, on the land side.

Delphinus (Dolphin) (*Astron*) A small northern constellation.

Delrin (*Plastics*) TN for an acetal resin (US). See POLY-OXYMETHYLENE.

delta (*Geol*) The more or less triangular area of river-borne sediment deposited at the mouth of rivers widely charged with detritus. A delta is formed on a low-lying coastline, particularly in seas of low tidal range and little current action, and in areas where subsidence keeps pace with sediment deposition. The Nile delta is a good example.

delta (*ICT*) The measure of change or difference – usually between the code of two versions of a computer program or the data output from those versions.

Delta Aquarids (*Astron*) See AQUARIDS.

Delta Cephei (*Astron*) A bright variable star in the constellation Cepheus, the prototype of the CEPHEID VARIABLE stars. Distance approx 400 pc.

delta connection (*ElecEng*) The connection of a three-phase electrical system such that the three phasors representing system voltages form a closed triangle.

deltaic deposit (*Geol*) The accumulations of sand, silt, clay and organic matter, deposited as TOPSET, FORESET and BOTTOMSET BEDS. A good active example is the Mississippi delta and, in the geological record, the Millstone grit in England.

delta impulse function (*ICT*) Infinitely narrow pulse of great amplitude, such that the product of its height and duration is unity.

delta iron (*Eng*) The polymorphic form of iron stable between 1403°C and the melting point (about 1532°C). The space lattice is the same as that of α-iron and different from that of γ-iron.

delta-matching transformer (*ICT*) A matching network between two-wire transmission lines and half-wave antennas.

delta modulation (*ICT*) A form of DIFFERENTIAL PULSE-CODE MODULATION in which only 1 bit for each sample is used.

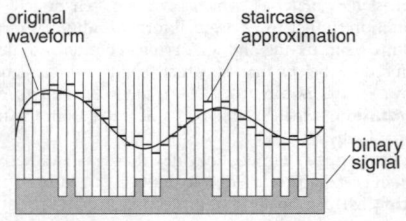

delta modulation

delta network (*ICT*) One with three branches all in series.

delta particle (*Phys*) Very-short-lived HYPERON which decays almost instantaneously through the strong interaction.

delta ray (*Phys*) Any particle ejected by recoil action from passage of ionizing particles through matter.

delta-ray spectrometer (*NucEng*) See SPECTROMETER, DELTA RAY.

Delta rocket (*Space*) NASA satellite launcher series which has been used in over 300 launches between 1960 and present; the series continues to evolve with Delta IV now being developed.

delta V (*Space*) The velocity change required to transform a particular trajectory into another.

delta voltages (*ElecEng*) The voltage between alternate lines in a delta-connected system.

delta waves (*Psych*) (1) Electroencephalographic waves of high amplitude and low frequency (1–3 Hz) that are characteristic of deep sleep. (2) In the heart a delta wave in the ECG is seen in Wolff–Parkinson–White syndrome.

delta wing (*Aero*) A swept-back wing of substantially triangular planform, the trailing edge forming the base. It is longitudinally stable and does not require an auxiliary balancing aerofoil, although tail or nose planes are sometimes fitted to increase pitch control and trim so that landing flaps can be fitted.

deltoid (*BioSci*) Having the form of an equilateral triangle; any triangular structure, as the deltoid muscle of the shoulder.

deltoid (*MathSci*) Recent name for STEINER'S TRI-CUSP hypocycloid.

delusions (*Psych*) An irrational belief which an individual will defend with intensity, despite overwhelming evidence that it has no basis in reality; common among schizophrenic mental disorders.

delustrant (*Textiles*) Dense inorganic material, frequently titanium dioxide, added to a synthetic fibre before it is extruded. In this way a range of fibres may be obtained with different lustres and opacities.

demagnetization (*ElecEng, Phys*) (1) *Removal* of magnetization of ferromagnetic materials by the use of diminishing, saturating and alternating magnetizing forces. (2) *Reduction* of magnetic induction by the internal field of a magnet, arising from the distribution of the primary magnetization of the parts of the magnet. (3) *Removal* by heating above the CURIE POINT. (4) *Reduction* by vibration.

demagnetization (*MinExt*) In DENSE-MEDIA PROCESS, using ferrosilicon, passage of the fluid through an ac field to deflocculate the agglomerated solid.

demagnetization factor (*ElecEng*) Diminution factor (N) applied to the intensity of magnetization (M) of a ferromagnetic material, to obtain the demagnetizing field (ΔH), ie $\Delta H = NM$. N depends primarily on the geometry of the body concerned.

demagnetizing ampere-turns (*ElecEng*) See BACK AMPERE-TURNS.

demagnetizing coil (*Electronics*) One used to eliminate residual magnetization from a record or playback head of a tape recorder; powered by mains frequency ac. Also *degaussing coil*.

demand (*ElecEng*) See MAXIMUM DEMAND.

demand factor (*ElecEng*) The ratio of the maximum demand on a supply system to the total connected load.

demand indicator (*ElecEng*) See MAXIMUM-DEMAND INDICATOR.

demand limiter (*ElecEng*) See CURRENT LIMITER.

demand meter (*ElecEng*) One reading or recording the loading on an electrical system.

demantoid (*Min*) Bright-green variety of the garnet andradite, essentially silicate of calcium and iron.

deme (*BioSci*) A local population of interbreeding organisms.

dementia (*Psych*) Degeneration of various functions governed by the central nervous system, including motor reactions, memory and learning capacity, problem solving, etc. These functions normally decline with age, but several dementia syndromes result from pathological organic deterioration of the brain. See SENILE DEMENTIA.

dementia praecox (*Psych*) Obsolete term for SCHIZOPHRENIA.

demersal (*BioSci*) Found in deep water or on the sea bottom; as fish eggs which sink to the bottom, and midwater and bottom-living fish as opposed to surface fish (eg herring) and shellfish. Cf PELAGIC.

demi- (*Genrl*) Prefix from Fr *demi*, Lt *midium* (the middle), denoting half or half-sized.

demifacet (*BioSci*) One of the two half-facets formed when the articular surface for the reception of the capitular head of a rib is divided between the centra of two adjacent vertebrae.

demijohn (*Glass*) Narrow-necked, glass wine or spirit container of more than 2 l capacity.

de-militarized zone (*ICT*) A layer at the perimeter of a secure computer network that permits outgoing traffic but denies access from external networks. Abbrev DMZ.

demineralization (*ChemEng*) A process for cleaning water in which the anions and cations are removed separately by absorption in synthetic exchange materials, leaving the water free of dissolved salts. The exchange material is regenerated by treatment with alkali and acid.

demister (*Autos*) Ducts arranged so that hot dry air is played on the interior of the windscreen to prevent condensation. Heat source now usually heat dissipated from the engine.

demodectic mange (*Vet*) Mange of animals caused by mites of the genus *Demodex*, which live in the hair follicles and sebaceous glands. Also *Demodectic folliculitis*.

demodulation (*ICT*) The inverse of MODULATION. Applied generally to the process of extracting the original information impressed on a CARRIER. In amplitude modulation the received signal is usually passed through non-linear circuits which generate sum and difference frequencies that allow the modulating signal and the carrier to be separated. In frequency modulation the signal is fed to a frequency-sensitive circuit that generates an output proportional to the frequency shift of the carrier brought about by modulation. See DISCRIMINATOR. In pulse-coded systems the word demodulation is replaced by 'decoding'. See DECODER.

demodulation of an exalted carrier (*ICT*) See HOMODYNE RECEPTION.

demodulator (*ICT*) A circuit or device for DEMODULATION. See DETECTOR.

demographic transition (*EnvSci*) A series of stages, linked to economic development, by which the structure of a human population changes from one characterized by high birth rates and high death rates to one with low birth rates and low death rates.

De Moivre's theorem (*MathSci*) That $(\cos\theta + i\sin\theta)^n = \cos n\theta + i\sin n\theta$. Expressed in terms of the exponential function, the theorem is $(e^{i\theta})^n = e^{ni\theta}$.

demon (*ICT*) A computer process or program, especially in the UNIX operating system, that performs common system tasks independently of a specific user.

demonstrator (*Aero*) A new aircraft, engine or system constructed to prove its novel features prior to embarking on full development.

de Morgan's theorem (*ICT*) A mathematical identity in BOOLEAN ALGEBRA that is used to simplify a logic design, especially to convert a system made of a variety of logic elements to be made from NAND or NOR functions only.

Demospongiae (*BioSci*) A class of PORIFERA usually distinguished by the possession of a skeleton composed of siliceous spicules or, like bath sponges, of spongin.

demoulding temperature (*Eng*) Temperature at which moulding can be safely removed from tool without permanent distortion. It is closely related to the crystallization temperature or glass transition temperature (for non-crystalline polymers). Also *freeze-off temperature*.

demountable (*Phys*) Said of X-ray tubes or thermionic valves when they can be taken apart for cleaning and filament replacement, and are continuously pumped during operation.

demulcent (*Med*) Soothing; allaying irritation.

demulsification number (*MinExt*) The resistance to emulsification by a lubricant when steam is passed through it; indicated in the minutes and half minutes required for the separation of a given volume of oil after emulsification.

demultiplexer (*ICT*) A circuit that enables data from a single source to select one of several possible destinations. Cf MULTIPLEXER.

demy (*Print*) See PAPER SIZES.

demyelinating diseases (*Med*) Diseases, often with an auto-immune aetiology, in which the myelin sheath of nerves is destroyed. Examples are *acute disseminated encephalomyelitis, experimental allergic encephalomyelitis, Guillain–Barre syndrome, multiple sclerosis*.

demy octavo (*Print*) A book size, $8\frac{1}{2} \times 5\frac{1}{2}$ in (216×138 mm).

demy quarto (*Print*) A book size, $11 \times 8\frac{5}{8}$ in (276×219 mm).

denary (*ICT, MathSci*) Using, or relating to, the familiar POSITIONAL NOTATION representing numbers in base 10 with the digits 0, 1, 2, ... , 9.

denaturant (*NucEng*) An isotope added to fissile material to render it unsuitable for military use, eg uranium-238 can be added to uranium-233, but denaturing with fertile uranium-238 necessarily produces plutonium in any reactor and the spent fuel must be either reprocessed or stored.

denaturation (*BioSci*) The destruction of the native conformation or state of a biological molecule by heat, extremes of pH, heavy-metal ions, chaotropic agents, etc, resulting in loss of biological activity. Specifically in DNA, the breakage of the hydrogen bonds maintaining the double-helical structure, a process which can be reversed by renaturation or annealing. In protein it often leads to an irreversible change accompanied by coagulation.

denatured alcohol (*Chem*) Alcohol (ethanol) which according to law has been made unfit for human consumption by the admixture of nauseating or poisonous substances, eg methyl alcohol, pyridine, benzene.

dendr-, dendro- (*Genrl*) Prefixes from Gk *dendron*, tree, tree-like or branching.

dendrimer (*Chem*) A polymer that has a tree-like branched molecular structure.

dendrite (*BioSci*) A branch of a DENDRON. See fig. at NEURON.

dendrite (*Crystal, Eng*) A tree-like crystal formation. Metal crystals grow in the first instance by branches developing in certain directions from the nuclei. Secondary branches are later thrown out at periodic intervals by the primary ones and in this way a skeleton crystal, or 'dendrite', is formed. The interstices between the branches are finally filled with solid which in a pure metal is indistinguishable from the skeleton. In many alloys, however, the final structure consists of skeletons of one composition in a matrix of another.

dendritic cell (*BioSci*) A cell that has branching processes. The term is used in immunology to describe two distinct kinds of cells that have different functions: (1) Antigen-presenting cells with dendritic morphology that occur mainly in the T-cell areas of lymph nodes and the spleen. Very similar cells, probably identical except in location, occur as Langerhans cells in the skin where they have an immunological surveillance role and migrate to lymphoid tissue having encountered antigen. Abbrev *DC*. (2) Follicular dendritic cells. Present in germinal centres where they are in intimate contact with B-cells and retain antigen for long periods. Abbrev *FDC*.

dendritic markings (*Geol*) Tree-like markings, usually quite superficial, occurring on joint faces and other fractures in rocks, frequently consisting of oxide or manganese or of iron. Less frequently the appearance is due to the inclusion of a mineral of dendritic habit in another mineral or rock, eg chlorite in silica as in 'moss agate'.

dendritic ulcer (*Med*) A branching ulcer of the cornea, due to herpes infection of the cornea.

dendro- (*BioSci*) See DENDR-.

dendrochronology (*EnvSci*) The science of dating wooden objects based upon the characteristic pattern of annual radial increments of growth which depends upon climatic factors.

dendroclimatology (*EnvSci*) The science of reconstructing past climates from the information stored in tree trunks as annual radial increments of growth (rings). Also *dendrochronology*, in which timber is dated by matching growth ring patterns to standard climatic periods.

dendrogram (*BioSci*) A branching diagram after the style of a family tree reflecting similarities or affinities of some sort. See CLADOGRAM, PHENOGRAM.

dendrograph (*BioSci*) An instrument that is used to measure the periodic swelling and shrinkage of tree trunks.

dendroid (*BioSci*) (1) Tall, with an erect main trunk, as tree ferns. (2) Freely branched.

dendron (*BioSci*) The afferent or receptor process of a neuron. Often much branched. Cf AXON.

Deneb (*Astron*) A very hot white supergiant star in the constellation Cygnus, the 19th brightest in the sky and the most luminous visible to the naked eye. Distance approx 500 pc. Also *Alpha Cygni*.

denervated (*Med*) Deprived of nerve supply.

dengue (*Med*) A tropical disease in which an influenza-like viral agent is transmitted by mosquito to humans; characterized by severe pain in the joints and a rash. The condition is normally self-limiting, but haemorragic shock can occur and may cause death.

denial (*Psych*) Refusal to acknowledge some unpleasant feature of the external world or some painful aspect of one's own experiences or emotions (a *defence mechanism*).

denial of service (*ICT*) The deliberate swamping of a website or system by a large number of (often automatically generated) user sessions that consume bandwidth and prevent legitimate users from accessing the site.

denier (*Textiles*) A unit used for the thickness of yarns; the mass in grams of 9000 m of yarn. Superseded by TEX. See YARN COUNT.

denim (*Textiles*) A strong cotton twill fabric made from yarn-dyed warp (often blue) and undyed (sometimes even unbleached) weft yarn. Used for the manufacture of overalls, boiler suits and jeans.

denitrification (*BioSci*) See NITRATE-REDUCING BACTERIA.

denominational number system (*MathSci*) A system of representing numbers by a sequence of digits in which each digit contributes an amount dependent upon both its value and its position. Thus with a base of *r*, the *n*-digit number $a_{n-1}\, a_{n-2}\ldots a_2\, a_1\, a_0$ represents

$$\sum_{s=0}^{n-1} a_s r^s$$

where each digit a_s can have any of *r* values; eg in the decimal system $r = 10$, and the number 345 represents $3 \times 10^2 + 4 \times 10 + 5$. See POSITIONAL NOTATION. Cf NON-DENOMINATIONAL NUMBER SYSTEM.

denominator (*MathSci*) See DIVISION.

dense (*Glass*) Of optical glass, having a higher refractive index.

dense-media process (*MinExt*) Dispersion of ferrosilicon or other heavy mineral in water separates lighter (floating) ore from heavier (sinking) ore. Also *heavy-media process*, *sink–float process*.

dens epistrophei (*BioSci*) See ODONTOID PROCESS.

dense set (*MathSci*) A set of points is dense in itself if every point of the set is a limit point. A set is said to be everywhere dense in an interval, if every subinterval (no matter how small) contains points of the set. A subset X of a set Y is said to be dense in Y if the closure of X is Y.

densification (*PowderTech*) All modes of increasing density, including the effect of sintering contraction.

densi-tensimeter (*Chem*) Apparatus for determining both the pressure and the density of a vapour.

Densithene (*Plastics*) TN for polyethylene loaded with lead powder. Used for radioactive shielding.

densitometer (*ImageTech, Phys*) Any instrument for measuring the optical transmission or reflecting properties of a material, in particular the optical density (absorbance) of exposed and processed photographic images.

density (*ICT*) A measure of the PACKING of data onto a BACKING STORAGE medium; eg for a $5\frac{1}{4}$ in floppy disk data may be stored in single density (360 Kbytes), double density (720 Kbytes) or high density (1·2 Mbytes) format.

density (*ImageTech*) A measure of the light-stopping power of a transparent material; it is defined as the logarithm of the OPACITY, which is the reciprocal of the transmission ratio. See REFLECTION DENSITY.

density (*Phys*) The ratio of the mass of a material to its volume. Symbol ρ; units $kg\,m^{-3}$. Its reciprocal is the SPECIFIC VOLUME.

density bottle (*Phys*) A thin glass bottle, accurately calibrated, used for the determination of the density of a liquid.

density change method (*PowderTech*) Particle size analysis technique which measures concentration changes within a sedimenting suspension by measuring the pressure exerted by a column of the suspension.

density dependence (*BioSci*) The phenomenon whereby performance of organisms within a population is dependent on the extent of crowding. See also DENSITY-DEPENDENT INHIBITION OF GROWTH.

density-dependent inhibition of growth (*BioSci*) The phenomenon exhibited by most normal (*anchorage-dependent*) animal cells in culture that stop dividing once a critical cell density is reached, usually, but not necessarily, when all the cells are in contact.

density gradient centrifugation (*BioSci*) The separation of cells, cell organelles or macromolecules according to their density differences by centrifugation to their density equilibrium positions in density gradients established in appropriate solutions. A solution of a highly soluble salt, like CsCl, can be centrifuged to an equilibrium gradient of density in which DNA of different densities will separate.

density of gases (*Chem, Phys*) According to the GAS LAWS, the density of a gas is directly proportional to the pressure and the relative molecular mass and inversely proportional to the absolute temperature. At stp the densities of gases range from $0.0899 \, \text{g dm}^{-3}$ for hydrogen to $9.96 \, \text{g dm}^{-3}$ for radon.

density of states (*Phys*) The number of electronic states per unit volume having energies in the range from E to $E + dE$; an important concept of the BAND THEORY OF SOLIDS.

density range (*Print*) The range of density from shadow to highlight on a film negative or positive or on the printed image. When measured with a DENSITOMETER the range can be expressed numerically.

density test (*Chem*) Analytical method for determining material density. Crudest method involves flotation in fluids of different densities, so only giving density limits to an unknown sample. This test is often combined with the flame test and pencil hardness test to give guide to polymer type. A better method uses calibrated density columns with fluid of regularly varying density, so that the position of the unknown in the tube gives the density precisely. Care is needed in interpretation owing to variation of polymer density with added fillers etc.

dent (*Textiles*) A term denoting one wire in a loom reed; it also refers to the space between two wires, through which warp threads are drawn. The number of dents per cm is the sett of the cloth.

dental arcade (*BioSci*) The row of teeth in the upper or lower jaw that forms an arch or loop that is bilaterally symmetrical.

dental caries (*Med*) A disintegration of enamel and dentine of tooth, common where fluorine content of water is low. Probably due to acids formed by the action of bacteria on dietary carbohydrate.

dental formula (*BioSci*) A formula used in describing the dentition of a mammal to show the number and distribution of the different kinds of teeth in the jaws; thus a bear has in the upper jaw three pairs of incisors, one pair of canines, four pairs of premolars, and two pairs of molars; and in the lower jaw three pairs of incisors, one pair of canines, four pairs of premolars and three pairs of molars. This is expressed by the formula

$$\frac{3142}{3143}$$

dental materials (*Med*) Synthetic materials replacing or restoring function of teeth, eg polymethyl methacrylate for dentures. Other materials include gold, AMALGAMS, IONOMER RESINS, PORCELAIN.

dentary (*BioSci*) In vertebrates, a membrane bone of the lower jaw that usually bears teeth. In mammals, it forms the entire lower jaw.

dentate (*BioSci*) Having a toothed margin.

dentelle (*Print*) A style of decoration of a tooth-like or lace-like character; used in case binding.

denticles (*BioSci*) Any small tooth-like structure; the placoid scales of ELASMOBRANCHII.

denticulated (*Build*) A term applied to mouldings decorated with DENTILS.

dentigerous cyst (*Vet*) A cyst containing teeth; usually a teratomatous cyst on the malar bone of a horse.

dentil (*Build*) A projecting rectangular block forming one of a row of such blocks under the corona of a cornice.

dentine (*BioSci*) Structural BIOCOMPOSITE of very fine (20–40 nm long) needle crystals of HYDROXY APATITE embedded in a COLLAGEN fibre matrix, very similar to compact bone. Main component of teeth, placoid scales and ivory. Adj *dentinal*.

dentirostral (*BioSci*) Having a toothed or notched beak.

dentition (*BioSci*) The kind, arrangement and number of the teeth; the formation and growth of the teeth; a set of teeth, as the milk dentition.

denudation (*Geol*) The laying bare (Lt *nudus*, naked) of the rocks by chemical and mechanical disintegration and the transportation of the resulting rock debris by wind or running water. Ultimately denudation results in the degradation of the hills to the existing base level. The process is complementary to sedimentation, the amount of which in any given period being a measure of the denudation.

denuded quadrat (*BioSci*) A square piece of ground, marked out permanently and cleared of all its vegetation, so that a study may be made of the manner in which the area is reoccupied by plants.

denumerable set (*MathSci*) One that can be put into a one–one correspondence with the positive integers. Also *countable set, enumerable set, numerable set*.

deodar (*For*) See CEDAR.

deodorizing (*ChemEng*) A process extensively used in edible oil and fat refining, in which the oil or fat is held for several hours at high temperatures (varying with product, but 200°C is not uncommon), and low pressure $(10^4 \, \text{N m}^{-2})$, during which time steam is blown through to remove the traces of odour-creating substances (usually free fatty acids of high relative molecular mass).

deoxidant (*Eng*) See DEOXIDIZER.

deoxidation (*Eng*) The process of reduction or elimination of oxygen from molten metal before casting by adding elements with a high oxygen affinity, which form oxides that tend to rise to the surface.

deoxidized copper (*Eng*) Copper from which dissolved oxygen left after refining has been removed by adding deoxidizer. Residual amounts of deoxidizer in solid solution lower the electrical conductivity below that of tough-pitch copper, but the product is more suitable for working operations.

deoxidizer (*Eng*) A substance which will eliminate or modify the effect of the presence of oxygen, particularly in metals. Also *deoxidant*.

deoxyribonuclease (*BioSci*) An endonuclease that preferentially hydrolyses DNA. Abbrevs *DNAase, DNase*.

deoxyribonucleic acid (*BioSci*) See panel on DNA AND THE GENETIC CODE.

deoxyribose (*BioSci*) The sugar that, linked by $3'–5'$ phosphodiester bonds, forms the backbone of DNA.

departure time (*Aero*) The exact time at which an aircraft becomes airborne is an important factor in air-traffic control; estimated time of departure (abbrev *ETD*).

dependent functions (*MathSci*) A set of functions such that one may be expressed in terms of the others.

dependent variable (*MathSci*) A variable whose values are determined by one or more other variables. Cf INDEPENDENT VARIABLE.

depersonalization (*Psych*) A condition in which an individual experiences a range of feelings of unreality or remoteness in relation to the self, to the body or to other people, even extending to the feeling of being dead. A symptom in a range of neurotic syndromes.

depeter (*Build*) Plasterwork finished in imitation of tooled stone, small stones being pressed in with a board before the plaster sets. Also *depreter*.

dephlogisticated air (*Chem*) The name given by Priestley to oxygen. The term is of historic interest only.

dephosphorization (*Eng*) Elimination, partial or complete, of phosphorus from steel, in basic steel-making processes. Accomplished by forming a slag rich in lime. See ACID PROCESS, BASIC PROCESS, BESSEMER PROCESS, OPEN-HEARTH PROCESS.

depilate (*Med*) To remove the hair.

depilatories (*Chem*) Compounds for removing or destroying hair; usually sulphide preparations.

depleted uranium (*Chem*) Natural uranium from which most of the uranium-235 has been removed. It can be used as a heavy metal in ballistic missiles and as a *fertile* material in a fast-breeder reactor. Abbrev *DU*.

depletion (*Electronics*) In semiconductors, the local reduction of the density of charge carriers. Cf ENHANCEMENT.

depletion (*Phys*) Reduction in the proportion of a specific isotope in a given mixture. Cf ENRICHMENT.

depletion allowance (*MinExt*) Analogous to a depreciation allowance, it recognizes that the resources of a well or mine are being depleted as they are produced.

depletion layer (*Electronics*) In semiconductors, a non-neutral region in which the majority carrier density is reduced below that of the local dopant concentration. Such layers arise naturally at the junction between p-type and n-type material and can be induced in the semiconductor adjacent to an insulator in eg METAL–OXIDE–SILICON (MOS) structures.

depletion-mode transistor (*Electronics*) An insulated-gate field effect transistor in which carriers are present in the CHANNEL when the gate–source voltage is zero. Channel conductivity is controlled by changing the magnitude and polarity of this voltage. See ENHANCEMENT-MODE TRANSISTOR, PINCH-OFF.

depluming itch (*Vet*) A skin irritation of fowls due to infestation of the skin around the feather shafts by the mite *Cnemidocoptes gallinae*.

depolarization (*BioSci*) The shift in the negative RESTING POTENTIAL of a cell towards zero, thus reducing its polarity. In an excitable cell such as a neuron, depolarization results from an ACTION POTENTIAL that passes along its membrane.

depolarization (*ElecEng*) Reduction of polarization, usually in electrolytes, but sometimes in dielectrics. In the former it may refer to removal of gas collected at plates of cell during charge or discharge.

depolarizing muscle relaxant (*Pharmacol*) Suxamethonium is the only commonly used drug in this class. It acts by mimicking the action of acetylcholine at the neuro-muscular junction to produce a relatively short period of paralysis and relaxation. It is commonly used to facilitate the passage of an endotracheal tube to maintain airway patency at the start of a surgical operation.

depoling (*MinExt*) See POLING.

depolymerization (*Chem*) The reverse of polymerization which is induced by heat, free radicals, photons, radiation, etc, to produce monomer molecules. Only a few polymers degrade to monomer alone (eg polymethyl methacrylate, polytetrafluoroethylene). Most leave a complex mixture of degradation products.

deposit (*ElecEng*) (1) The coating of metal deposited electrolytically upon any material. (2) The sediment which is sometimes found at the bottom of a secondary cell owing to gradual disintegration of the electrode material.

deposit (*Geol*) See DEPOSITION.

deposit feeder (*BioSci*) An animal that feeds on detritus in or on the substratum. Cf FILTER FEEDER.

deposition (*Geol*) The laying down or placing into position of sheets of sediment (often referred to as *deposits*) or of mineral veins and lodes. Synonymous with *sedimentation* in the former sense.

depreciation factor (*ElecEng*) A term commonly used in the design of floodlighting and similar installations for the ratio of the light output from clean equipment to that after service.

depressant (*Pharmacol*) Lowering functional activity; a medicine which lowers functional activity of the body.

depressed conductor rail (*CivEng*) A section of conductor rail depressed below normal level where contact with the shoes is not desired.

depressing agent (*MinExt*) Agent used in froth flotation to render selected fraction of pulp less likely to respond to aerating treatment. Also *wetting agent*.

depression (*EnvSci*) A cyclone: that distribution of atmospheric pressure in which the pressure decreases to a minimum at the centre. In the northern hemisphere in such a system, the winds circulate in an anticlockwise direction; in the southern hemisphere, in a clockwise direction. A depression usually brings stormy unsettled weather.

depression (*Med*) A state of deep dejection and a pervasive feeling of helplessness, accompanied by apathy and a sense of personal worthlessness, resulting in retardation of thought and bodily functions. Endogenous depression is a spontaneous occurrence. Reactive depression is a response, sometimes exaggerated, to grief or other personal tragedy.

depression of freezing point (*Phys*) Lowering of the freezing point of a solution; it freezes at a lower temperature than the pure solvent by an amount proportional to the concentration of the solution, provided this is not too great. The depression produced by a 1% solution is called the *specific depression* and is inversely proportional to the molecular weight of the solute. Hence the depression is proportional to the number of moles dissolved in unit weight of the solvent and is independent of the particular solute used.

depression of land (*Geol*) Depression relative to sea level may be caused in many ways, including sedimentary consolidation, the superposition of large masses of ice, the migration of magma, or changes of chemical phase at depth. It is generally recognized by the marine transgression produced but is often difficult to distinguish from eustatic changes in sea level. See DROWNED VALLEYS.

depressor (*BioSci*) A muscle that, by its action, lowers a part or organ; a reagent that, when introduced into a metabolic system, slows down the rate of metabolism.

depreter (*Build*) See DEPETER.

depth (*Ships*) The depth measured from the top of the keel to the top of some specified deck.

depth gauge (*Build*) A device clamped to a drill or bit to regulate the depth of the hole bored.

depth multiplex recording (*ImageTech*) A process in which FM audio is recorded into the full depth of the videotape's oxide by audio heads on the DRUM and then shorter-wavelength video is recorded into the surface of the oxide. Cross-talk is largely avoided by frequency differences and AZIMUTH RECORDING.

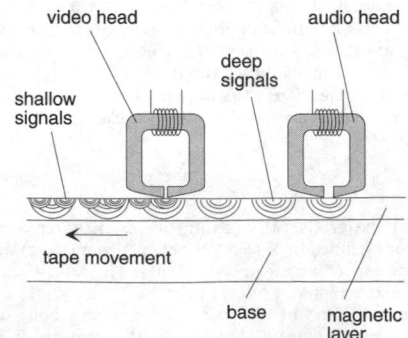

depth multiplex recording The heads are also set at different azimuth angles.

depth of field (*ImageTech*) The range of near and far distances from the camera within which the subject will be in reasonably sharp focus; it is determined by the focal length of the lens used and its STOP as well as the actual focus setting.

depth of focus (*ImageTech*) Within a camera, the range of distances from the lens to the photosensitive surface within which the image is acceptably sharply defined for a given CIRCLE OF CONFUSION.

depth of fusion (*Eng*) The depth to which a new weld has extended into the underlying metal or a previous weld.

depth of modulation (*ICT*) See MODULATION DEPTH, MODULATION INDEX.

depth of penetration (*ElecEng*) Within a plane conductor the magnitude of an electromagnetic field and the associated current falls off exponentially from the surface. The depth of penetration or *skin depth* (δ metres) is normally considered as the depth for which the field magnitude is $1/e$ of its value at the surface. $\delta = (\mu f \sigma)^{-0.5}$ where f is the frequency, μ is the permeability and σ is the conductivity.

deputy (*MinExt*) Position of responsibility in a coal mine. See FIREMAN.

dequeue (*ICT*) To remove (a data processing task) from a list of tasks awaiting processing in a buffer.

dérailleur (*Eng*) Variable-transmission gear mechanism whereby the driving chain may be 'derailed' from one sprocket wheel to another of different size, thus changing the driving ratio. Ten or more ratios are possible. Much used in bicycles. Cf EPICYCLIC GEAR.

de-rating (*ElecEng*) Deliberate reduction in the duty placed on components and equipment by the designer, to maintain an adequate margin of safety and improve reliability.

derby float (*Build*) A large trowel consisting of a flat board with two handles on the back.

Derbyshire neck (*Med*) See GOITRE.

Derbyshire spar (*Min*) A popular name for the mineral *fluorite* or *fluorspar*.

derepress (*BioSci*) To activate a gene by suppressing its repressor.

derivative (*MathSci*) Of a function $f'(x)$, the limit as h tends to zero of

$$\frac{f(x+h) - f(x)}{h}$$

If $y = f(x)$, other notations for $f(x)$ are

$$\frac{dy}{dx}, y', D_x y \text{ and } \frac{d}{dx} f(x)$$

The gradient of the curve $y = f(x)$ at the point (a, b) is $f'(a)$. Also *derived function*, *differential coefficient*. Cf PARTIAL DIFFERENTIAL COEFFICIENT.

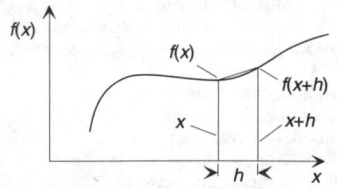

derivative

derivative feedback (*ICT*) Feedback signal in control system proportional to time derivative error.

derived fossils (*Geol*) Fossils eroded from an older sediment and redeposited in a younger sediment.

derived function (*MathSci*) See DERIVATIVE.

derived unit (*Phys*) A unit derived from the fundamental units of a system by consideration of the dimensions of the quantity to be measured. See DIMENSIONS, SI UNITS.

derived working unit (*Radiol*) The upper limit to the concentration of a radioactive substance which can be present continuously without contravening the current dose limitations.

derm (*BioSci*) See DERMIS.

dermal (*BioSci*) Pertaining to the skin; more strictly, pertaining to the dermis. In plants, pertaining to the epidermis or other superficial layer of a plant member. Occasionally *dermic*.

dermal branchiae (*BioSci*) See PAPULAE.

dermatan sulphate (*BioSci*) A GLYCOSAMINOGLYCAN that is found in the extracellular matrix of skin and blood vessels.

dermatitis (*Med*) Inflammation of the surface of the skin or epidermis.

dermatogen (*BioSci*) A HISTOGEN precursor of the epidermis.

dermatographia (*Med*) A sensitive condition of the skin in which light pressure with eg a pencil point will produce a reddish weal. A form of urticaria. Also *dermographia*.

dermatology (*Med*) That branch of medical science which deals with the skin and its diseases.

dermatomyositis (*Med*) A disease, probably of AUTO-IMMUNE origin with inflammation and weakness of muscles, often with a purplish skin rash. Although some patients may have an underlying malignancy the majority respond to corticosteroids and immunosuppressive drugs.

Dermatophagoides pteronyssinus (*BioSci*) A mite present in house dust. Antigens extracted from mites and their feces are a common cause of allergy to house dust in W European countries.

dermatophyte (*BioSci*) Any parasitic fungus that causes a skin disease, eg ring worm or athlete's foot, in animals or humans.

dermatosclerosis (*Med*) See SCLERODERMIA.

dermis (*BioSci*) The inner layers of the integument, lying below the epidermis and consisting of mesodermal connective tissue. Also *derm*.

dermographia (*Med*) See DERMATOGRAPHIA.

dermoid (*Med*) A cyst of congenital origin containing such structures as hair, skin and teeth; occurs usually in the ovary.

dermomuscular layer (*BioSci*) A sheet of muscular tissue underlying the skin in lower Metazoa; it consists of longitudinal and, usually, circular layers.

derrick (*Build, CivEng*) An arrangement for hoisting materials, distinguished by having a boom stayed from a central post, which in turn is usually stayed in position by guys.

derrick (*MinExt*) Steel structure at the well head that supports the drilling assembly, including drill bit and pipe, and the raising and lowering mechanism. See KELLY and panel on DRILLING RIG.

derricking jib crane (*Eng*) A jib crane in which the inclination of the jib, and hence the radius of action, can be varied by shortening or lengthening the derricking ropes between the frame or samson post and jib.

derris (*Chem*) An extract of the root of the *Derris* tree, of which rotenone is the chief toxic constituent; effective contact insecticide.

dertrotheca (*BioSci*) In birds, the horny covering of the maxilla. Also *dertrum*.

DERV (*Genrl*) Diesel fuel (abbrev from *diesel-engined road vehicle*).

DES (*ICT*) Abbrev for DATA ENCRYPTION STANDARD.

desalination (*Chem*) The production of fresh from sea water by one of several processes, including *distillation*, *electrodialysis* and reverse *osmosis*.

desaturation (*ImageTech*) The presence of grey in a colour; sometimes intentionally introduced in a colour reproduction process for artistic effect.

desaturation (*Phys*) Adding white light to a saturated colour (pure spectral wavelength) to produce a desaturated or pale colour. Cf SATURATION.

de-scaling (*Eng*) The process of: (1) removing scale or metallic oxide from metallic surfaces by PICKLING; (2) removing scale from the inner surfaces of boiler plates and water tubes.

Descartes' rule of signs (*MathSci*) The theorem that no algebraic equation $f(x) = 0$ can have more positive or negative roots respectively than there are changes of sign from + to − and from − to + in the polynomial $f(x)$ or $f(-x)$.

descendeur (*Genrl*) A metal device tightened or loosened on a rope by a climber to control speed of descent.

descending (*BioSci*) Running from the anterior part of the body to the posterior part, or from the cephalic to the caudal region.

descending letters (*Print*) Letters the lower part of which is below the baseline; eg *p, q, y*. See fig. at TYPEFACE.

descending node (*Space*) For Earth, the point at which a satellite crosses the equatorial plane travelling north to south.

descloizite (*Min*) Hydrated vanadium–lead–zinc ore with the general formula $PbZn(VO_4)OH$. Important source of vanadium. Occurs in the oxidation zone of lead–zinc deposits.

describer (*CivEng*) Apparatus, either in signal cabins or for public use, which indicates movements, destinations, etc, of trains.

descriptive statistics (*Genrl*) The branch of statistics that focuses on describing in numerical format the status or condition of a population. Descriptive statistics require that all subjects in the population be tested.

de-seaming (*Eng*) The process of removing surface blemishes, or superficial slag inclusions from ingots or blooms.

desensitization (*Med*) A method of reducing or abolishing the effects of a known allergen by injecting it in gradually increasing doses over a period of time until resistance is built up.

desensitization (*Psych*) A decrease in reaction or sensitivity to a stimulus following repeated exposure.

desensitization, systematic (*Psych*) A form of behaviour therapy used esp in the treatment of phobias in which fear is reduced by exposing the individual to the feared object in the presence of a calming stimulus; usually some form of relaxation is involved. See COUNTER-CONDITIONING.

desensitize (*Print*) To treat a lithographic plate with a solution of gum arabic to promote the hydrophilic properties of the plate.

desensitizer (*ImageTech*) A substance which destroys the sensitivity of an emulsion without affecting the latent image.

desertification (*EnvSci*) The formation of deserts from vegetated zones by the action of drought and/or increased populations of humans and herbivores.

desert rose (*Min*) A cluster of platy crystals, often including sand grains, formed in arid climates by evaporation. Typically barytes or gypsum. Also *rock rose*.

desiccants (*Chem*) Substances of a hygroscopic nature, capable of absorbing moisture and therefore used as drying agents; eg anhydrous sodium sulphate, anhydrous calcium chloride, phosphorus pentoxide, etc.

desiccation (*For*) See SEASONING.

desiccator (*Chem*) Laboratory apparatus for drying substances; it consists of a glass bowl with ground-in lid, containing a drying agent, eg concentrated sulphuric acid or anhydrous calcium chloride; a tray for keeping glassware etc in position is also provided, and if desired the desiccator can be evacuated.

design (*Eng*) See ENGINEERING DESIGN, INDUSTRIAL DESIGN, PRODUCT DESIGN.

designation marks (*Print*) An indication of the title of a book, printed in small letters in the same line with the SIGNATURE on the first page of each section.

design engineering (*Eng*) Wide term implying forethought and analysis applied to practical ends. Encompasses all practical methods used in industrial, product and engineering design.

desilverization (*Eng*) The process of removing silver (and gold) from lead after softening. See PARKE'S PROCESS, PATTINSON'S PROCESS.

desizing (*Textiles*) Removal of *size* from woven fabrics.

desktop (*ICT*) The background on which WINDOWS and ICONS will appear on the computer display screen.

desktop conferencing (*ICT*) Remote interworking in which voice, data and video can be exchanged, usually via the INTEGRATED SERVICES DIGITAL NETWORK, by using the software and display capabilities of a personal computer.

desktop publishing (*ICT*) Software for use on a personal computer that allows text to be manipulated with control of eg font, typesize, microjustification, column organization and the incorporation of graphics. This is done with MOUSE and WINDOWS, without the obvious control codes needed in normal typesetting. It will contain a raster image processor producing output compatible with a PAGE DESCRIPTION LANGUAGE. The resulting output should be able to drive a LASER PRINTER for proofing and short runs and a high-resolution typesetter for normal printing and publishing.

desktop video (*ImageTech*) Employing a computer for video POST-PRODUCTION, as an EDIT CONTROLLER and to add titles and graphics. Abbrev *DTV*.

Deslandres equation (*Phys*) An empirical expression for the positions of the origins or heads in a band spectrum: $v = a + bn + cn^2$, where v is the wavenumber of the head, a, b and c are constants, and n takes successive integral values.

de-sliming (*MinExt*) Removal of very fine particles from an ore pulp, or classification of it into relatively coarse and fine fractions.

desmids (*BioSci*) Green algae, two families of the Charophyceae. Unicellular, the cells usually symmetrical about a median constriction, often elaborate in shape, and moving by the secretion of mucilage. Characteristic of rather acid fresh-water habitats.

desmine (*Min*) See STILBITE.

Desmodur (*Plastics*) TN for isocyanates used to produce polyurethanes in conjunction with polyester resins.

desmosomes (*BioSci*) Strong intercellular junctions that bind cells together, either at discrete points at the surface (*spot desmosomes*) or as continuous bands around cells (*belt desmosomes*). The spot desmosome (macula adherens) has intermediate (cytokeratin) filaments attached on the cytoplasmic faces in both cells, while the belt desmosome (zonula adherens) similarly has actin microfilaments. The two membranes of associated cells remain separated in the desmosome by an intercellular space. The hemidesmosome is a similar structure binding an epithelial cell to the basal lamina. Fig. ▷

desmotropism (*Chem*) A special case of TAUTOMERISM which consists of the change of position of a double bond, and in which both series of compounds can exist independently; eg keto and enol form of acetoacetic ester, malonic ester, phenyl-nitromethane, etc.

desorption (*Chem*) Reverse process to ADSORPTION. See OUTGASSING.

de-spun antenna (*Space, ICT*) An antenna used in a *spin-stabilized* spacecraft; the antenna rotates at a speed equal to and in the opposite sense to the body of the satellite, thereby continuing to point in the required direction, usually an antenna on Earth.

desquamation (*Med*) The shedding of the surface layer of the skin.

destination network identifier (*ICT*) A routing code used in an international VIRTUAL PRIVATE NETWORK that specifies the foreign network to which a call is to be routed. Unlike the COUNTRY CODE, it remains unchanged throughout the transit of a call.

Destriau effect (*Phys*) A form of electroluminescence arising from localized regions of very intense electric field associated with impurity centres in the phosphor.

destructive distillation (*Chem*) The distillation of solid substances accompanied by their decomposition. The destructive distillation of coal results in the production of coke, tar products, ammonia, gas, etc.

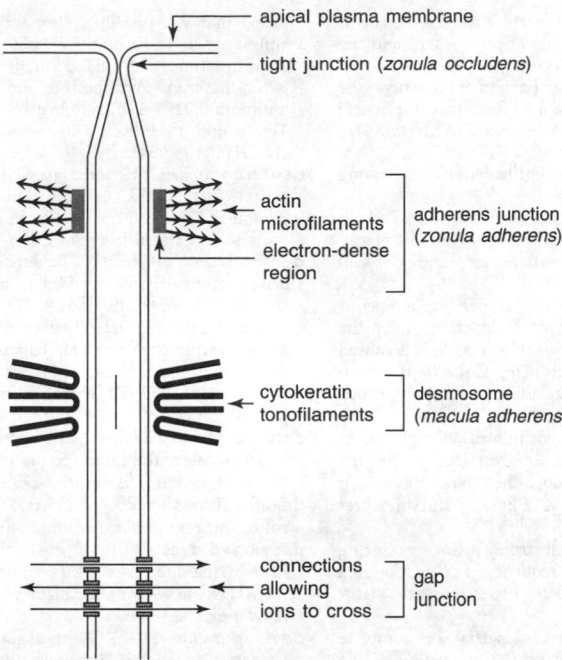

← apical plasma membrane

← tight junction (*zonula occludens*)

actin
microfilaments
electron-dense
region

adherens junction
(*zonula adherens*)

cytokeratin
tonofilaments

desmosome
(*macula adherens*)

connections
allowing
ions to cross

gap
junction

specialized adhesions between cells

destructive read-out (*ICT*) Clearing of a storage location simultaneously with reading its contents. Abbrev *DRO*.

destructor station (*ElecEng*) An electric generating station in which the fuel used consists chiefly of town or other refuse.

desulphurizer (*ChemEng*) Chemical agent added to liquid steel before casting in order to combine with sulphur and transfer to a BASIC SLAG, rich in lime. Similar to *dephosphorizer*.

desuperheater (*Eng*) A vessel in which SUPERHEATED STEAM is brought into contact with a water spray in order to make saturated or less highly superheated steam.

detachable key switch (*ElecEng*) A switch which can be operated only by a special key which is kept under supervisory control.

detail drawing (*Build*) A large-scale working drawing (usually of a part only) giving information which does not appear on small-scale drawings of the whole construction.

detail paper (*Paper*) A translucent tracing paper, usually unoiled.

detector (*ICT*) A circuit that turns any carrier into a form which can be heard or displayed; if the carrier is unmodulated, a dc voltage is the output. A modulated carrier gives the modulating signal as an output; commonly called *demodulation*. See LINEAR DETECTOR, SQUARE-LAW DETECTOR.

detector (*Phys*) A device in which the presence of particles or radiation induces physical change which is observable. See GEIGER COUNTER, GERMANIUM RADIATION DETECTOR, NUCLEAR EMULSION, PROPORTIONAL COUNTER, SCINTILLATION COUNTER.

detector finger (*Print*) See WEB-BREAK DETECTOR.

detent (*Eng*) A catch which, on removal, initiates the motion of a machine. In the chronometer escapement of a clock, the detent carries a stone or jewel for locking the escape wheel.

detergent cracking (*Eng*) The effect of aggressive detergents (eg Teepol) on thermoplastics, esp polyolefins,

causing deep, isolated cracks. Solved by using higher-molecular-mass materials. See ENVIRONMENTAL STRESS CRACKING.

detergents (*Chem*) Cleaning agents (solvents, or mixtures thereof, sulphonated oils, abrasives, etc) for removing dirt, paint, etc. Commonly refers to soapless detergents, containing surfactants which do not precipitate in hard water; sometimes also a protease enzyme to achieve 'biological' cleaning and a whitening agent (see FLUORESCENT WHITENING AGENTS). See CATIONIC DETERGENTS, NON-IONIC DETERGENTS.

determinant (*MathSci*) A square array of numbers representing the algebraic sum of the products of the numbers, one from each row and column, the sign of each product being determined by the number of interchanges required to restore the row suffices to their proper order, eg

$$\begin{vmatrix} a_1\ b_1\ c_1 \\ a_2\ b_2\ c_2 \\ a_3\ b_3\ c_3 \end{vmatrix} = \begin{aligned} & a_1\ b_2\ c_3 - a_1 b_3 c_2 \\ & -a_2\ b_1\ c_3 + a_2 b_3 c_1 \\ & +a_3\ b_1\ c_2 - a_3 b_2 c_1 \end{aligned}$$

determinate (*BioSci*) (1) Growth of a limited extent. (2) SYMPODIAL GROWTH. (3) CYMOSE INFLORESCENCE.

determinate (*Eng*) Said of a structure which is a PERFECT FRAME. Cf INDETERMINATE.

determinate cleavage (*BioSci*) A type of cleavage in which each BLASTOMERE has a predetermined fate in the later embryo.

determination (*BioSci*) The commitment of a ·cell to a particular path of differentiation, though this may not be morphologically apparent.

determinism (*Genrl*) The classical principle that the specification of the initial dynamical variables of a system as well as knowledge of all external forces acting on it allow prediction with certainty of the state of the system at later times. Also *causality*. In psychology this becomes entwined with the question of free will and the extent to which behaviour is preprogrammed, by biology, environment or internal motives.

deterministic (*ICT*) Applies to a machine if the next state of the machine can be predicted from its present state and any new input. Cf NON-DETERMINISTIC.

deterministic effect (*Genrl*) A phenomenon where the particular outcome is determined by fundamental physical principles. Also *deterministic process*. See also STOCHASTIC EFFECT.

de-tinning (*Eng*) Chlorine treatment to remove tin coating from metal scrap.

detonating fuse (*MinExt*) A fuse with a core of TNT which burns at some 6 km s^{-1} and is itself set off by detonator. Varieties include CORDTEX, PRIMACORD and *Cordeau Detonnant*.

detonation (*Autos*) In a petrol engine, spontaneous combustion of part of the compressed charge after the passage of the spark; the accompanying knock. It is caused by the heating effect of the advancing flame front, which raises the gas remote from the plug to its spontaneous IGNITION TEMPERATURE.

detonation (*Chem*) The decomposition of an explosive in which the rate of heat release is great enough for the explosion to be propagated through the explosive as a steep shock front, at velocities above 1 km s^{-1} and pressures above 10^9 N m^{-2}.

detonation meter (*Eng*) An instrument for measuring quantitatively the severity and frequency of detonation in a petrol-engine cylinder. See BOUNCING-PIN DETONATION METER.

detorsion (*BioSci*) In Gastropoda, partial or complete reversal of torsion, manifested by the untwisting of the visceral nerve loop and the altered position of the ctenidium and anus.

detrital mineral (*Geol*) A mineral grain derived mechanically from a parent rock. Typically such minerals are resistant to weathering; eg diamonds, gold and zircon. See HEAVY MINERAL.

detrition (*Geol*) The natural process of rubbing or wearing down strata by wind, running water or glaciers. The product of detrition is *detritus*.

detritovore (*BioSci*) Organism which eats detritus.

detritus (*BioSci*) Organic material formed from decomposing organisms.

detritus (*Geol*) See DETRITION.

detritus chamber (*Build*) A tank through which crude sewage is first passed in order to allow the largest and heaviest of suspended matters to fall to the bottom, from which they can be removed. Also *detritus pit*.

detrusor (*Med*) The muscular coat of the urinary bladder (detrusor urinae); sometimes used for the outer of the three muscular coats of the bladder.

Dettol (*Chem*) TN for non-toxic and non-irritant germicide of which the active principle is chloroxylenol dissolved in a saponified mixture of aromatic oils, eg terpineol.

detumescence (*Med*) The reduction of a swelling.

detuning (*ICT*) Adjustment of a resonant circuit so that its resonant frequency does not coincide with that of the applied emf.

deuteranopic (*Phys*) Colour blind to green.

deuteration (*Chem*) The addition of or replacement by deuterium atoms in molecules.

deuterium (*Chem*) An isotope of element hydrogen having one neutron and one proton in nucleus. Symbol D, when required to be distinguished from natural hydrogen, which is both ^{1}H and ^{2}H because it contains 0·015% of deuterium. Heavy hydrogen is thus twice as heavy as ^{1}H, but similarly ionized in water. Used in isotopic 'labelling' experiments (eg mechanism of esterification). See DEUTERON.

deutero- (*Genrl*) Prefix, in general denoting second in order, derived from Gk *deuteros*, second. Particularly, in chemistry, containing heavy hydrogen (DEUTERIUM).

Deuteromycotina (*BioSci*) The imperfect fungi, for which no sexual reproduction is known. A 'form subdivision' or class of the Eumycota or true fungi (a non-phylogenetic or artificial category based on observable characteristics

rather than relationship). They are typically mycelial with simple septa. The affinities of many appear to be with the Ascomycotina. They include many saprophytes, eg *Aspergillus*, that cause food spoilage, and *Penicillium*, a source of antibiotics. They also include plant parasites, eg *Fusarium*, *Verticillium*, that cause wilt diseases etc. Also *Deuteromycetes*, *Fungi imperfecti*.

deuteron (*Chem*) Charged particle, D$^+$, the nucleus of DEUTERIUM, a stable but lightly bound combination of one proton and one neutron. It is mainly used as a bombarding particle accelerated in cyclotrons.

deuterostoma (*BioSci*) (1) In development, a mouth that arises secondarily, as opposed to a mouth that arises by modification of the blastopore. (2) A taxonomic grouping containing the chordates, hemichordates and echinoderms as well as the chaetognatha. This separates the vertebrate lineage from the majority of invertebrates (protostomates).

deuterotoky (*BioSci*) Parthenogenesis leading to the production of both males and females.

deutocerebron (*BioSci*) In higher Arthropoda, as insects and Crustacea, the fused ganglia of the second somite of the head, forming part of the 'brain'. Also *deutocerebrum*.

developable surface (*MathSci*) A surface that may be rolled out onto a plane without shrinking or stretching.

developed dyes (*Chem*) Dyes which are developed on the fibre by the interaction of the constituents which produce them. Dyeing with aniline black provides an example of a *developed dye*.

developer (*ImageTech*) Chemical solution which converts a LATENT photographic image to a visible one by reducing the exposed silver compounds to metallic silver; in colour photography this metallic silver may be associated with the formation of dyes. In addition to the DEVELOPING AGENT, it generally contains an alkaline ACCELERATOR, a RESTRAINER and a preservative.

developer streaks (*ImageTech*) Streaks, usually following areas of heavy image density, resulting from irregular chemical action by the solution in a processing machine. See DIRECTIONAL EFFECTS.

developing agent (*ImageTech*) A chemical having the property of reducing light-exposed silver halide grains to metallic silver; for black-and-white processing the most common are *hydroquinone, metol* and *phenidone*.

development (*ImageTech*) The conversion of an exposed LATENT IMAGE to a visible one by chemical reaction.

development (*MinExt*) The opening of ore body by ACCESS SHAFTS, CROSSCUTS, DRIVES, RAISES and WINZES for the purpose of proving mineral value (ore reserve) and exploiting it.

developmental psychology (*Psych*) Although originally applied to the study of lifelong changes, the majority of practitioners specialize in the early changes seen during childhood, effectively child psychology.

Devensian (*Geol*) The last glacial stage in the Pleistocene, probable equivalent to the Würm glaciation of the Alps. See QUATERNARY.

deviance (*MathSci*) A statistic measuring the degree of fit of a statistical model by comparison with a more complete model.

deviated drilling (*MinExt*) See ANGLE DRILLING.

deviation (*ElecEng*) The difference between a measured quantity and the 'true' value, eg difference between the setting of a compass needle and the magnetic north, due to a local magnetic disturbance, or between the value of a controlled variable and that to which the controlling mechanism is set in an automatic system. Also *error*.

deviation (*ICT*) In a FREQUENCY MODULATION system, the extent to which the carrier frequency moves from its unmodulated position when the modulating signal is applied.

deviation (*MathSci*) The difference between an observation and a fixed value. Statistical measures of dispersion are often based on functions of the deviation of each value in a set of observations from their common mean.

deviation (*MinExt*) Departure of a drill hole from the vertical either by design or accidentally.

deviation (*Ships*) The angle between the MAGNETIC MERIDIAN and the COMPASS MERIDIAN. A deviation table is normally prepared showing the deviation for various headings.

deviation distortion (*ICT*) The consequence of any restriction of bandwidth or linearity of discrimination in the transmission and reception of a frequency-modulated signal.

deviation IQ (*Psych*) See INTELLIGENCE QUOTIENT.

deviation ratio (*ICT*) In a system using FREQUENCY MODULATION, the ratio of the maximum possible frequency deviation of the carrier to the maximum frequency of the modulating signal.

device driver (*ICT*) See DRIVER.

devil (*EnvSci*) A small whirlwind due to strong convection, which, in the tropics, raises dust or sand in a column.

devil float (*Build*) A square float having four nails projecting from its working face; used to perform the scoring required in DEVILLING.

devilling (*Build*) The operation of scoring the surface of a plaster coat to provide a key for another coat.

devitrification (*Geol*) Deferred crystallization which, in the case of glassy igneous rocks, converts obsidian and pitchstone into dull cryptocrystalline rocks (usually termed FELSITES) consisting of minute grains of quartz and feldspar. Such devitrified glasses give evidence of their originally vitreous nature by traces of perlitic and spherulitic textures.

devitrification (*Glass*) A physical process in silicate glasses which causes a change from the glassy state to a minutely crystalline state. The change can occur over time, and the glass becomes turbid and more brittle. It is employed in a controlled way to produce GLASS-CERAMICS.

Devonian (*Geol*) The oldest period of the Upper Palaeozoic, covering a time span between approx 400 and 360 million years. The corresponding system of rocks. See PALAEO-ZOIC.

dew (*EnvSci*) The deposit of moisture on exposed surfaces which accumulates during calm, clear nights. The surfaces become cooled by radiation to a temperature below the dew point, thus causing condensation from the moist air in contact with them.

Dewar flask (*Chem*) A silvered glass flask with double walls, the space between them being evacuated; used for storing, eg liquid air.

de-watering (*MinExt*) Partial or complete drainage by thickening, sedimentation, filtering or on screen as a process aid, to facilitate shipment or drying of product.

dewaxing (*MinExt*) In the manufacture of lubricating oils, the solvent removal of wax to reduce the POUR POINT.

dew-blown (*Vet*) See BLOAT.

dew claw (*BioSci*) In dogs, the useless claw on the inner side of the limb (esp the hind-limb) that represents the rudimentary first digit.

DEW line (*Radar*) Abbrev for *distant early-warning line*. Line of radar missile warning stations along 70th parallel of latitude from Alaska to Greenland.

dew point (*EnvSci, Phys*) The temperature at which a given sample of moist air will become saturated and deposit dew, if in contact with a colder surface or the ground. Above ground, condensation into water droplets takes place.

dew-point hygrometer (*EnvSci*) A type of hygrometer for determining the dew point, ie the temperature of air when completely saturated. The relative humidity of the air can be ascertained by reference to vapour pressure tables.

dew pond (*EnvSci*) A watertight hollow, usually on elevated land, where the combined effects of rainfall and fog drip exceed that of evaporation. The effect of dew is negligible and the name is the result of an ancient mis-understanding.

dexamethasone (*Pharmacol*) A synthetic steroid analogue (*glucocorticoid*), used as an anti-inflammatory drug.

dexamfetamine (*Pharmacol*) A drug that acts directly on the brain as a stimulant but is a sedative in children and is used for treatment of ATTENTION DEFICIT HYPERACTIVITY DISORDER. Formerly *dexamphetamine*.

dexiocardia (*Med*) See DEXTROCARDIA.

dexiotropic (*BioSci*) Twisting in a spiral from left to right; spiral cleavage.

dextral (*BioSci*) See DEXTRORSE.

dextral fault (*Geol*) A TEAR FAULT in which the rocks on one side of the fault appear to have moved to the right when viewed from across the fault. The opposite of a SINISTRAL FAULT.

dextran (*Chem*) A polyglucose formed by micro-organisms in which the units are joined mainly by $\alpha(1 \to 6)$ links with variable amounts of cross-linking, via $\alpha(1 \to 4)$, $\alpha(1 \to 3)$ or $\alpha(1 \to 2)$. Hydrolysed by dextranases.

dextrin (*Chem, FoodSci*) A term for a group of intermediate products obtained in the transformation of starch into maltose and *d*-glucose. Dextrins are obtained by boiling starch alone or with dilute acids. They do not reduce FEHLING'S SOLUTION. Crystalline dextrins have been obtained by the action of *Bacillus macerans*. Also *starch gum*.

dextro- (*Genrl*) Prefix from Lt *dexter*, Gk *dexious*, denoting to or towards the right.

dextrocardia (*Med*) A congenital anomaly with the heart situated in the right side of the chest. If there is a similar transposition of the abdominal viscera, the condition is termed *situs inversus*.

dextrorotatory (*Phys*) Said of an optically active substance which rotates the plane of polarization in a clockwise direction when looking against the incoming light.

dextrorse (*BioSci*) Helical, twisted or coiled in the sense of a conventional (right-handed) screw-thread or of a Z-helix. Cf SINISTRORSE. Also *dextral*.

dextrose (*Chem*) See GLUCOSE.

dextrose equivalent (*FoodSci*) Classification of glucose syrups according to the proportion of reducing sugars (dextrose) which are present. Different values impart different properties, eg colour, viscosity and crystallization characteristics. Abbrev *DE*.

DFD (*ICT*) Abbrev for *data flow diagram*, a means of representing in graphic form the major flows of information into, out of and within a computer system.

DFS (*ICT*) Abbrev for DISK FILING SYSTEM.

DFVLR (*Space*) Original abbrev for *Deutsche Forschung und Versuchanstalt für Luft und Raumfahrt*, the German centre for aerospace research; now the *Deutsches Zentrum für Luft und Raumfahrt (DLR)*.

d-gluconic acid (*Chem*) $CH_2OH(CHOH)_4COOH$, an oxidation product (at C-1) of D-glucose.

DH (*Build*) Abbrev for *double-hung*.

DHTML (*ICT*) Abbrev for DYNAMIC HYPERTEXT MARKUP LANGUAGE.

di- (*Genrl*) Prefix from Gk *dis* denoting two, twice or double.

dia- (*Genrl*) Prefix from Gk denoting through, across, during or composed of.

diabantite (*Min*) A variety of chlorite relatively rich in iron and poor in aluminium.

diabase (*Geol*) US etc. term for DOLERITE; also used in the UK for altered dolerite.

diabetes insipidus (*Med*) A condition in which there is an abnormal increase in the amount of urine excreted, as a result of disease of, or injury to, the posterior lobe of the pituitary gland, reducing the secretion of anti-diuretic hormone.

diabetes mellitus (*Med*) A complex metabolic disorder, characterized by high blood glucose levels and glucose in the urine, associated with impaired insulin production and/or action. *Type 1 diabetes mellitus* (formerly called *juvenile onset diabetes*) is characterized by excess thirst, weight loss and severe lethargy. It is caused by loss, due to an auto-immune response, of the islet cells of the pancreas (*islets of Langerhans*) leading to an absolute requirement for treatment with insulin. *Type 2 diabetes mellitus* (formerly called *maturity onset diabetes*) is the most common form of the disease and is caused by insulin resistance rather than absolute deficiency. It can be treated by diet alone, oral hypoglycaemic drugs and occasionally with insulin. Diabetes of all types is associated with damage to many organs including retina, nerves, kidney and arteries.

diabetic coma (*Med*) See DIABETIC KETO-ACIDOSIS.

diabetic keto-acidosis (*Med*) When diabetes becomes uncontrolled there is an accumulation of KETONES causing metabolic acidosis. The blood sugar is very high and dehydration and loss of potassium also occur. Also *diabetic coma*. The patient may become confused or stuporous, but this may be a late feature and the term coma is misleading.

diacaustic (*MathSci*) See CAUSTIC CURVE.

diacetone alcohol (*Chem*) 4-hydroxy-2-keto-4-methylpentane. $CH_3CO_2CH_2C(CH_3)_2.OH$. A colourless or light-yellow liquid, mp $-54°C$, bp $130-175°C$, rel.d. $0.915-0.943$. Used as a lacquer solvent.

diacetyl (*Chem*) Butane 2,3-dione. $CH_3COCOCH_3$. A yellow-green liquid, bp $87°C$. It is the simplest diketone, and is obtained by the action of nitrous acid on methyl ethyl ketone. It occurs naturally in butter and is one of the constituents giving butter its characteristic flavour.

diachronism (*Geol*) The transgression across time planes by a geological formation. A bed of sand, when traced over a wide area, may be found to contain fossils of slightly different ages in different places, as, when deposited during a long-continued marine transgression, the bed becomes younger in the direction in which the sea was advancing. Adj *diachronous*.

diacoele (*BioSci*) In Craniata, the third ventricle of the brain.

diacritical current (*ElecEng*) The current in a coil to produce a flux equal to half that required for saturation.

diacritical point (*ElecEng*) The point on the magnetizing curve of a sample of iron at which the intensity of magnetization has half its saturation value.

diacyl glycerol (*BioSci*) A diglyceride that is enzymically released from membrane phospholipids by phospholipase C, activates protein kinase C, and is thus important in the intracellular signalling cascade.

diadelphous (*BioSci*) Stamens of a flower fused by their filaments into two groups.

diadochy (*Geol, Min*) The replacement of one element by another in a crystal structure.

diagenesis (*Geol*) Those changes which take place in a sedimentary rock at low temperatures and pressures after its deposition. These include CEMENTATION, COMPACTION, RECRYSTALLIZATION. Cf METAMORPHISM.

diagnosis (*BioSci*) A formal description of a plant, having special reference to the characters which distinguish it from related species.

diagnosis (*Med*) The determination of the nature of a disordered state of the body or of the mind; the identification of a diseased state.

diagnostic (*ICT*) Aid in the debugging of programs or systems. See SYNTAX ANALYSIS, TEST DATA, TRACE.

diagnostic characters (*BioSci*) Characteristics by which one genus, species, family or group can be differentiated from another.

diagonal (*Eng*) A tie or strut joining opposite corners of a rectangular panel in a framed structure.

diagonal (*MathSci*) (1) Of a polygon: a straight line drawn between two non-adjacent vertices. (2) Of a polyhedron: a straight line drawn between two vertices which do not lie in the same face. (3) Of a square matrix or determinant: the *leading* or *principal diagonal* is that running from the top left-hand corner to the bottom right-hand corner.

diagonal bond (*Build*) See RAKING BOND.

diagonal eyepiece (*Surv*) One incorporating a right-angled prism for convenience in surveying steep lines with telescope or theodolite.

diagonal matrix (*MathSci*) A square matrix in which all the elements except those in the LEADING DIAGONAL are zero.

diagonal symmetry (*MathSci*) Such that a half-turn (180°) about an axis gives the same figure.

diagram (*Eng*) A curve which indicates the sequence of operations in a machine.

diagram (*MathSci*) An outline figure or scheme of lines, points and spaces, designed to: (1) represent an object or area; (2) indicate the relation between parts; (3) show the value of quantities or forces.

Diakon (*Plastics*) TN for polymethyl methacrylate moulding powder (UK).

dial (*ICT*) A calling device operated by the rotation of a disk that, on its release, produces the pulses necessary to establish a connection in an automatic system. Superseded by telephone keypads.

dial (*MinExt*) A large compass mounted on a tripod, used for surveying or mapping workings in coal mines.

dialdehydes (*Chem*) Compounds containing two aldehyde groups. The most important one is GLYOXAL (ethane-1, 2-dial).

dial gauge (*Eng*) A sensitive measuring instrument in which small displacements of a plunger are indicated in $\frac{1}{1000}$ mm, or similar length units, by a pointer moving over a circular scale.

dialkanones (*Chem*) See DIKETONES.

dialkenes (*Chem*) Hydrocarbons containing two double bonds in their molecules. They exist in three forms according to the disposition of the double bonds, ie adjacent as in the allenes; separated by one single bond in the conjugated compounds such as butadiene; or separated by two or more single bonds. Also *diolefins*.

dialkoxyalkanes (*Chem*) See ACETALS.

diallage (*Min*) An ill-defined, altered monoclinic pyroxene, in composition comparable with augite or diopside, with a lamellar structure; occurs typically in basic igneous rocks such as gabbro.

dialling (*MinExt, Surv*) The process of running an underground traverse with a mining DIAL.

dialogite (*Min*) See RHODOCHROSITE.

dialphanyl phthalate (*Chem*) A common plasticizer for polyvinyl chloride. Abbrev *DAP*. See panel on POLYVINYL CHLORIDE.

dialypetalous (*BioSci*) See POLYPETALOUS.

dialyser (*Chem*) Dialysis apparatus.

dialysis (*Chem*) The separation of a colloid from a substance in true solution by allowing the latter to diffuse through a semipermeable membrane. The process is used in the ARTIFICIAL KIDNEY.

diamagnetism (*Phys*) Phenomenon in some materials in which the susceptibility is negative, ie the magnetization opposes the magnetizing force, the permeability being less than unity. It arises from the precession of spinning charges in a magnetic field. The susceptibility is generally one or two orders of magnitude weaker than typical paramagnetic susceptibility.

diameter (*MathSci*) (1) A straight line through the centre of a circle. (2) Straight line bisecting a family of parallel chords of a conic. Cf CONJUGATE DIAMETERS. (3) Straight line which passes through the centre of parallel sections of a quadric surface.

diameter of commutation (*ElecEng*) The diametral plane in which the coils of an armature winding that are undergoing commutation should be situated for perfect commutation.

diametral winding (*ElecEng*) An armature winding in which the number of slots is a multiple of the number of poles.

diametral pitch (*Eng*) Of a gear wheel, the number of teeth divided by the pitch diameter.

diametrical tappings (*ElecEng*) Tappings taken on a closed armature, which are diametrically opposite to each other, ie displaced from each other by 180 electrical degrees.

diametrical voltage (*ElecEng*) The voltage between opposite lines of a symmetrical six-phase system, or the voltage between tappings on an armature winding which are diametrically opposite to each other, ie displaced from each other by 180 electrical degrees.

diametric system (*Crystal*) See TETRAGONAL SYSTEM.

diamines (*Chem*) Compounds containing two amino groups.

diaminoethanetetraacetic acid (*Chem*) $(HOOCCH_2)_2 NCH_2CH_2N(CH_2COOH)_2$, or EDTA as it is usually known. A tetrabasic acid whose anions can function as hexadentate ligands. It is much used in analysis and as a sequestering agent for metal ions, particularly the group IIA ions, Mg^{2+} and Ca^{2+}. Also *ethylenediaminetetraacetic acid*.

diamino-pimelic acid (*BioSci*) A cell-wall constituent of some bacteria and blue-green algae, not known to occur in any other group.

diamond (*Min*) One of the crystalline forms of carbon; it crystallizes in the cubic system, rarely in cubes, commonly in forms resembling an octahedron, and less commonly in the tetrahedron. Curved faces are characteristic. It is the hardest mineral (10 in Mohs' scale); hence valuable as an abrasive, for arming rock-boring tools, etc; its high dispersion and birefringence make it valuable as a gemstone. Occurs in blue ground, in river gravels and in shore sands. See BLACK DIAMOND, BORT, INDUSTRIAL DIAMONDS, KIMBERLITE.

diamond (*Print*) An old type size, approximately $4\frac{1}{2}$ point.

diamond antenna (*ICT*) See *rhombic antenna*.

diamond bit (*MinExt*) Drilling bit in which industrial diamonds are set in the cutting portions of the bit. Cuts hard rock and wears longer, so increasing the time between replacing bits involving a ROUND TRIP. See panel on DRILLING RIG.

Diamond–Blackfan anaemia (*Med*) A rare form of anaemia in which the bone marrow fails to produce red blood cells.

diamond die (*Eng*) A wire-drawing die containing a diamond insert which reduces wear. Cf BUSHING.

diamond dust (*MinExt*) The hardest abrasive agent, used as loose powder or paste in polishing, or embedded in metal tool parts, eg *rock saws* and *drill bits*.

diamond-like carbon (*Eng*) A material containing carbon and hydrogen deposited by plasma-enhanced chemical vapour deposition from hydrocarbon gases. Amorphous but with substantial quantities of sp3-bonded carbon atoms, it has high mechanical hardness (approaching that of diamond), high electrical resistivity and high thermal conductivity. Its optical properties are also similar to those of pure diamond.

diamond mesh (*Build*) A form of EXPANDED METAL with a diamond-shaped network.

diamond ring (*Astron*) An effect seen just before or after a total eclipse of the Sun when a bright flash of light from a BAILY'S BEAD appears briefly.

diamond saw (*Build*) (1) Stone-cutting circular saw used with diamond dust for cutting rock sections. (2) A band or frame saw with diamond-bonded cutting elements, used for cutting stone, concrete (eg in 'bump-cutting' in runways), plastics, etc.

diamond-skin disease (*Vet*) See SWINE ERYSIPELAS.

diamond tool (*Eng*) A diamond of specified shape, mounted in a holder which is used for precision machining of non-ferrous metals and ceramics.

diamond wheel (*Eng*) A rotating wheel for a grinding machine in which small diamonds are embedded, and which is used for grinding and cutting very hard materials.

diamond-work (*Build*) A wall constructed of lozenge-shaped stones laid in courses.

dianysl-trichloroethane (*Chem*) See METHOXYCHLOR.

diapause (*BioSci*) In insects, a state in which development is suspended and cannot be resumed, even in the presence of apparently favourable conditions, unless the diapause is first 'broken' by an appropriate environmental change. May occur at any stage of the life cycle.

diapedesis (*BioSci*) In Porifera, the passage to the exterior of cells primarily occupying the interior of certain types of larva; in vertebrates, the passage of blood leucocytes through the walls of blood vessels into the surrounding tissues.

diaper-work (*Build*) Paving constructed in a chequered pattern, composed of stones or tiles of different colours.

diaphoresis (*Med*) Perspiration.

diaphoretic (*Pharmacol*) Producing perspiration: a medicine which does this.

diaphragm (*Acous*) A vibrating membrane as in a loudspeaker, telephone and similar sound source; also in receivers, eg the human eardrum.

diaphragm (*BioSci*) (1) Generally, a transverse partition subdividing a cavity. (2) In plants, a plate of cells crossing a space, esp a plate one cell thick with many intercellular pores through which air but not water may pass between intercellular air spaces in the submerged stems etc of many aquatic plants. (3) In mammals, the transverse partition of muscle and connective tissue that separates the thoracic cavity from the abdominal cavity. (4) In Anura, a fan-shaped muscle passing from the ilia to the oesophagus and the base of the lungs. (5) In some Arachnida, a transverse septum separating the cavity of the prosoma from that of the abdomen. (6) In certain Polychaeta, a strongly developed transverse partition dividing the body cavity into two regions.

diaphragm (*Build*) A web across a hollow terracotta block, forming compartments.

diaphragm (*ElecEng*) A sheet of perforated or porous material placed between the positive and negative plates of an accumulator cell.

diaphragm (*ImageTech, Phys*) See STOP.

diaphragm (*Surv*) A flanged brass ring which is held in place in a surveying telescope by means of four screws, and which receives the RETICULE.

diaphragm cell (*ChemEng*) An electrolytic cell in which a porous diaphragm is used to separate the electrodes, thus permitting electrolysis of, principally, sodium chloride without recombination in the cell of electrolysis products.

diaphragm plate (*Build*) A connecting stiffener between the webs of a box girder.

diaphragm pump (*Eng*) A pump in which a flexible diaphragm set between two non-return valves replaces a piston or bucket, being clamped round the edge and attached at the centre to a reciprocating rod of short stroke.

diaphysectomy (*Med*) Excision of part of the shaft of a long bone.

diaphysis (*BioSci*) The shaft of a long limb bone.

diaphysitis (*Med*) Inflammation of the diaphysis.

diapir (*Geol*) An intrusion of relatively light material into pre-existing rocks, doming the overlying cover. Applied esp to salt and igneous intrusions. See PETROLEUM RESERVOIRS.

diapiric salt dome (*Geol*) See PIERCEMENT SALT DOME.

diapophyses (*BioSci*) A pair of dorsal transverse processes of a vertebra, arising from the neural arch.

diapositive (*ImageTech*) A positive transparency on glass or film.

diapsid (*BioSci*) Said of skulls in which the supra- and infra-temporal fossae are distinct.

diarrhoea (*Med*) The frequent evacuation of liquid feces. US *diarrhea*.

diarthrosis (*BioSci*) A true (as opposed to a fixed) joint between two bones, in which there is great mobility. A cavity, filled with a fluid, generally exists between the two elements.

diaspore (*Min*) Alumina monohydrate, occurring in bauxite. A dimorph of BOEHMITE.

diastase (*BioSci*) The enzymes α- and β-amylase capable of converting starch into sugar. Produced during the germination of barley in the process of malting.

diastasis (*Med*) The separation, without fracture, of an epiphysis from the bone.

diastatic activity (*FoodSci*) The measure of the ability of diastatic enzymes (α- and β-amylase) to break down starch in flour to form maltose and dextrins.

diastema (*BioSci*) (1) An equatorial modification of protoplasm preceding cell division. (2) A gap in the row of teeth in a jaw, as in herbivorous mammals lacking the canine teeth, eg horses, sheep.

diaster (*BioSci*) In cell division, a stage in which the daughter chromosomes are situated in two groups near the poles of the spindle, ready to form the daughter nuclei.

dia-stereoisomers (*Chem*) Stereoisomers which are not simple mirror images (enantiomers) of one another. For instance, in a molecule with two chiral centres, the L,L- and D,D-forms are enantiomers, while the L,D- and D,L-forms are dia-stereoisomers.

diastole (*BioSci*) Rhythmical expansion, as of the heart, or of a contractile vacuole.

diastolic murmur (*Med*) A murmur heard over the heart during diastole usually indicative of valvular disease.

diastrophism (*Geol*) Large-scale deformation of the Earth's crust.

diathermanous (*Phys*) Relatively transparent to radiant heat.

diathermic surgery (*Med*) The use of an electric arc in preference to a knife. This has the advantage of sealing cuts and reducing bleeding.

diathermy (*Med*) The heating of tissues by high-frequency electric currents. In physiotherapy used to heat structures like joints and muscles under the skin. In surgery to cause coagulation and necrosis by localized application.

diathesis (*Med*) The constitutional state of the body which renders it liable to certain diseases.

diatom (*Geol*) A microscopic single-celled plant of the class Bacillariophyceae growing in both sea water and fresh water, and in soils. It secretes resistant siliceous cell walls called *frustrules* in a large number of forms. Diatoms range from the Cretaceous to the present day but there are diatom-like organisms in the Cambrian and later rocks. See DIATOMITE.

diatomite (*Min*) A siliceous deposit occurring as a whitish powder consisting essentially of the frustules of diatoms. It is resistant to heat and chemical action, and is used in fireproof cements, insulating materials, as an absorbent in the manufacture of explosives and as a filter. Also *diatomaceous earth, infusorial earth, kieselguhr, tripolite.*

diatom ooze (*Min*) A deep-sea deposit consisting essentially of the FRUSTULES of DIATOMS; widely distributed in high latitudes.

diatoms (*BioSci*) See BACILLARIOPHYCEAE.

diatoni (*Build*) QUOINS having two dressed faces projecting from the wall.

diatropism (*BioSci*) A TROPISM in which a plant part becomes aligned at right angles to the source of the orientating stimulus; eg rhizomes are typically 'diagravitropic'. Cf ORTHOTROPIC, PLAGIOTROPIC.

diazepam (*Pharmacol*) A long-acting benzodiazepine used for the treatment of anxiety and insomnia; also used pre-operatively as a relaxant. TN *Valium.*

diazo (*Print*) See DYELINE.

diazoamino compounds (*Chem*) Pale-yellow crystalline substances obtained by the action of a primary or secondary amine on a diazonium salt. Their general formula is RN = NNHR'. They do not form salts and most of them are easily transformed into the isomeric aminoazo compounds.

diazo compounds (*Chem*) Compounds of the general formula RN = NR', obtained by the action of excess nitrous acid on aromatic amines at temperatures below $10°C$. They are important intermediates for dyestuffs.

diazomethane (*Chem*) CH_2N_2, an aliphatic diazo compound, an odourless, yellow, poisonous gas, very reactive, used for introducing a methyl group into a molecule. It is prepared from nitroso-methyl-urethane by decomposition with alcoholic potassium hydroxide.

diazonium salts (*Chem*) The acid salts of diazobenzene of the general formula $(RN{\equiv}N^+)Cl$, important intermediates for azo-dyestuffs. They are usually prepared only in aqueous solution by the action of nitrous acid on an aromatic amine at low temperatures in the presence of excess of acid. The $-N{=}N-$ group can easily be replaced by hydrogen, hydroxyl, halogen, etc, and the diazonium salts can thus be transformed into other benzene derivatives.

diazo process (*ImageTech*) A document-copying process in which diazonium compounds in the paper are destroyed by ultraviolet light and the unexposed areas converted to a coloured dye. See DYELINE.

diazotization (*Chem*) The process of converting amino into diazo compounds.

dibasic acids (*Chem*) Acids containing two replaceable hydrogen atoms in the molecule.

dibenzoyl (*Chem*) See BENZIL.

dibenzoyl-peroxide (*Chem*) $C_6H_5COOOCOC_6H_5$. Relatively stable organic peroxide used mainly as a catalyst in polymerization and polycondensation reactions.

dibenzyl (*Chem*) Symmetrical diphenylethane. C_6H_5 $CH_2CH_2C_6H_5$. Mp $52°C$.

dibenzyl group (*Chem*) A synonym for the stilbene group, comprising compounds containing two benzene nuclei linked together by a chain of two or more carbon atoms.

diborane (*Chem*) The simplest borane, B_2H_6, bp $-92°C$. It is an example of ELECTRON-DEFICIENT bonding.

dibranchiate (*BioSci*) Having two gills or ctenidia.

DIC (*BioSci*) Abbrev for *differential interference contrast.* See DIFFERENTIAL INTERFERENCE CONTRAST MICROSCOPE.

dice (*Electronics*) Small regular pieces of semiconductor material for fabrication of devices.

dicentric (*BioSci*) Having two CENTROMERES.

dicephalus (*Med*) A developmental 'monstrosity' in which a fetus has two heads.

dichasial cyme (*BioSci*) See DICHASIUM.

dichasium (*BioSci*) An inflorescence in which the main stem ends in a flower and bears, from near the flower, two lateral branches also ending in a flower. These branches may in turn bear further similar branches and so on. Also *dichasial cyme.* See CYMOSE INFLORESCENCE, MONOCHASIUM.

dichlamydeous (*BioSci*) Having distinct calyx and corolla. Cf PERIANTH.

dichlorodifluoromethane (*Chem*) Freon 12. CCl_2F_2. Used as a refrigerant, solvent and in fire extinguishers, bp $-30°C$.

dichloroethylenes (*Chem*) Dichloroethenes. $C_2H_2Cl_2$. Exist in *cis*-form, bp $48°C$, and *trans*-form, bp $60°C$. Prepared industrially from tetrachloroethane. Used as source material for vinyl chloride, the monomer of PVC.

dichloromethane (*Chem*) Methylene chloride. CH_2Cl_2. Colourless liquid; bp $41°C$. Used widely as a solvent, eg in manufacture of cellulose acetate.

dichlorophen (*Chem*) Organic compound based on HEXA-CHLOROPHENE widely used as an anti-bacterial agent, particularly for water treatment.

dichlorophenoxyacetic acid (*BioSci*) Synthetic AUXIN used as a selective herbicide and in media for tissue culture. Also 2,4-D.

1,2-dichloropropane (*Chem*) Propylene dichloride. $CH_3CHClCH_2Cl$. Bp $94°C$. Solvent for oils, fats, waxes and resins.

dichocephalous (*BioSci*) Said of ribs that have two heads, a tuberculum and a capitulum. Cf HOLOCEPHALOUS.

dichogamy (*BioSci*) The maturation of the anthers at a different time from the stigmas in the same flower, protandry or protogyny. Cf HOMOGAMY.

dichoptic (*BioSci*) Having the eyes of the two sides distinctly separated.

dichotic (*Genrl*) Involving the simultaneous stimulation of each ear by different sounds.

dichotomy (*Astron*) The half-illuminated phase of a planet, such as the Moon at the quarters, and Mercury and Venus at greatest elongation.

dichotomy (*BioSci*) Bifurcation of an organ, by the division of an apical cell or meristem into two equal parts each growing into a branch. Common in algae and *Selaginella*. Cf FALSE DICHOTOMY.

dichroic (*Chem, Min*) Said of (1) materials, such as a solution of chlorophyll, which exhibit one colour by reflected light and another colour by transmitted light; (2) minerals that display DICHROISM.

dichroic filter (*ImageTech*) A filter whose spectral properties are achieved by coating rather than colouring. It has the advantage of not transmitting infrared. See BLOOMING.

dichroic fog (*ImageTech*) Fog which arises from the formation of an organic compound of silver; so called because of its reddish coloration by transmitted light, and greenish coloration by reflected light.

dichroic mirror (*Phys*) Colour-selective mirror, which reflects a particular band of spectral energy and transmits all others.

dichroism (*Crystal*) A property of certain crystals and other materials that reflect different colours when viewed from different angles, owing to the fact that they have different absorption coefficients for light polarized in different directions.

dichroite (*Min*) CORDIERITE.

dichromates (VI) (*Chem*) See CHROMATES (VI).

dichromatism (*Med*) Colour blindness in which power of accurate differentiation is retained for only two bands of colour in the spectrum.

Dicke radiometer (*ICT*) Sensitive receiving circuit that detects weak signals in noise by modulating them at the input before they reach conventional demodulation circuits. The output of the demodulator is compared with a reference signal from the modulator; when they coincide, this indicates signal presence. Sometimes used for poise measurements in microwave systems.

dickite (*Min*) A clay mineral of the kaolin (or kandite) group, a hydrated aluminosilicate. It generally occurs in hydrothermal deposits.

Dick test (*Med*) A test for immunity against the toxin of *Streptococcus pyogenes* which causes scarlet fever. A small amount of toxin injected into the skin causes an area of redness after six or more hours in persons who do not have antibodies against the toxin.

dicliny (*BioSci*) Having unisexual flowers, either male and female on different individual plants (dioecy) or both on one plant (monoecy). Cf HERMAPHRODITE. Adj *diclinous*.

diclofenac (*Pharmacol*) A non-steroidal anti-inflammatory and analgesic drug.

Diconal (*Pharmacol*) A proprietary combination of cyclizine hydrochloride (an anti-emetic antihistamine) and dipipanone hydrochloride (an opioid analgesic), used for the relief of moderate to severe pain.

dicophane (*Chem*) See DDT.

Dicotyledones (*BioSci*) The dicotyledons, or dicots, the larger of the two classes of angiosperms or flowering plants. They consist of trees, shrubs and herbs of which characteristically the embryo has two cotyledons, the parts of the flowers are in twos or fives or multiples of these numbers, and the leaves are commonly net-veined. The class contains c.165 000 spp in 250 families, which are usually divided among six subclasses or superorders: Magnoliidae, Hamamelidae, Caryophyllidae, Dilleniidae, Rosidae and Asteridae. Also *Magnoliopsida*. Cf MONOCOTYLEDONS. See panel on ANGIOSPERMS (FLOWERING PLANTS).

dicrotic (*Med*) Having a double beat or wave; an acceleration of the normal pulse found in fevers.

dicty-, dictyo- (*Genrl*) Prefixes from Gk *diktyon*, net.

Dictyoptera (*BioSci*) An order of insects with flattened body, long legs and two pairs of wings; it includes cockroaches and mantises.

dictyosome (*BioSci*) The Golgi body. An element of the GOLGI APPARATUS.

dictyostele (*BioSci*) A type of SOLENOSTELE, typical of many ferns, eg *Dryopteris*, in which overlapping leaf gaps

dissect the vascular cylinder into anastomosing strands (meristeles) each with xylem surrounded by phloem and endodermis.

Dictyostelium (*BioSci*) Genus of the cellular slime moulds (*Acrasidae*) much used as an experimental organism.

dicyclic (*BioSci*) Having the perianth in two whorls.

dicyclomine hydrochloride (*Pharmacol*) An *antimuscarinic* drug with antispasmodic properties, used in the treatment of irritable bowel syndrome.

didactyl (*BioSci*) Having all the toes of the hind-feet separate, as in many Marsupialia. Cf SYNDACTYL.

didelphic (*Med*) Pertaining to a double uterus.

dideoxy sequencing (*BioSci*) A standard method of DNA sequencing (Sanger method). A short complementary primer is annealed to single-stranded template DNA, and extended by a DNA polymerase in the presence of low concentrations of fluorescently labelled dideoxy-nucleotides (equivalent to A, T, C and G). Dideoxy nucleotides, when incorporated, block further chain extension and thus a mixture of chains of lengths determined by the template sequence are produced with an identifiable label at the terminus of each. The products of the reaction are separated, now usually by column chromatography in an increasingly automated fashion, and the sequence can be read off.

Didot point system (*Print*) The continental point system based on a 12-point 'cicero', the equivalent of the UK PICA but measuring approximately 12·8 UK points. In the Didot system the point measures 0·376 mm compared with the UK point size of 0·351 mm.

didymous (*BioSci*) Formed of two similar parts, partially attached; twinned.

didynamous (*BioSci*) Having two long and two short stamens.

die (*Build*) (1) The body of a pedestal. (2) The enlarged part at either end of a baluster, where it comes into the coping or the plinth.

die (*Eng*) (1) A metal block used in stamping operations. It is pressed down onto a blank of sheet metal, on which the pattern or contour of the die surface is reproduced. (2) The element complementary to the punch in press tool for piercing, blanking, etc. (3) An internally threaded steel block provided with cutting edges, for producing screw-threads by hand or machine. (4) A tool made of very hard material, often tungsten carbide or diamond, with a (bell-mouthed) hole, usually circular, used to reduce the product cross-section by plastic flow, in wire or tube drawing. (5) The steel tool which shapes a plastic extrudate (see EXTRUSION).

dieback (*BioSci*) Necrosis of a shoot, starting at the apex and progressing proximally.

die box (*Eng*) The holder into which screw dies are fitted in a screwing machine. Also *die head*.

diecasting (*Eng*) The casting of metals or plastics into permanent moulds, made of suitably resistant non-deforming metal. See GRAVITY DIECASTING, PRESSURE DIECASTING.

diecasting alloys (*Eng*) Alloys suitable for diecasting, which can be relied on for accuracy and resistance to corrosion when cast. Aluminium-base, copper-base, tin-base, zinc-base and lead-base alloys are those generally used.

die-fill ratio (*PowderTech*) The ratio of uncompacted powder volume to the volume of the GREEN COMPACT.

die head (*Eng*) See DIE BOX.

diel (*BioSci*) Synonymous with *diurnal* in the sense that it pertains to a period of 24 hours, not in the sense of *by day* rather than *by night*. Also *daily*.

dieldrin (*Chem*) A contact insecticide based on a chlorinated naphthalene derivative. Typical use is mothproofing of carpets and other furnishings. Highly persistent and now widely banned.

dielectric (*Phys*) Substance, solid, liquid or gas, which can sustain a steady electric field, and hence an insulator. It can be used for cables, terminals, capacitors and similar devices. Fig. ▷

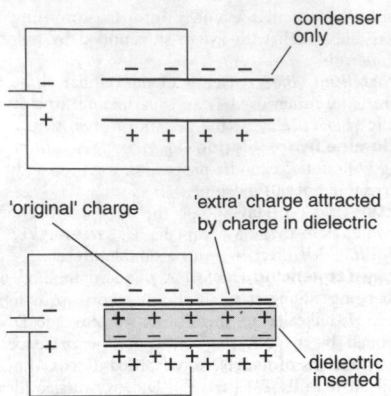

dielectric Effect of dielectric placed between the plates of a condenser.

dielectric absorption (*Phys*) Phenomenon in which the charging or discharging current of a dielectric does not die away according to the normal exponential law, owing to absorbed energy in the medium.

dielectric amplifier (*ElecEng*) One which operates through a capacitor, the capacitance of which varies with applied voltage.

dielectric antenna (*ICT*) One in which required radiation field is principally produced from a non-conducting DIELECTRIC.

dielectric breakdown (*ElecEng*) The effect of a sufficiently intense field strength in causing a normally non-conducting medium to pass current, accompanied by a relative reduction of resistance and, in solids, mechanical damage. See DIELECTRIC STRENGTH.

dielectric constant (*ElecEng*) See DIELECTRIC, PERMITTIVITY.

dielectric current (*Phys*) The displacement current arising due to a changing electric field applied to a perfect dielectric. For a real dielectric there will also be a conduction current or absorption current giving rise to energy loss in the dielectric.

dielectric diode (*ElecEng*) A capacitor whose negative plate can emit electrons into eg CdS crystals, so that current flows in one direction.

dielectric dispersion (*ElecEng*) Variation of permittivity with frequency.

dielectric fatigue (*ElecEng*) Breakdown of a dielectric subjected to a repeatedly applied electric stress, ineffective if applied once or a few times.

dielectric guide (*ElecEng*) Possible transmission of very-high-frequency electromagnetic energy functionally realized in a dielectric channel, the permittivity of which differs from its surroundings.

dielectric heating (*ElecEng*) Radio-frequency heating in which power is dissipated in a non-conducting medium through dielectric hysteresis. It is proportional to $V^2 f S t^{-1}$, where V is the applied voltage, f the frequency, S the area of the heated specimen and t its thickness. This is the principle used in microwave ovens.

dielectric hysteresis (*Phys*) Phenomenon in which the polarization of a dielectric depends not only on the applied electric field but also on its previous variation. This leads to power loss with alternating electric fields.

dielectric lens (*ICT*) A lens made of dielectric material in such a form that it refracts radio waves in much the same way as a glass lens affects light. Used to shape the beam on microwave and radar antennas.

dielectric loss (*Phys*) Dissipation of power in a dielectric under alternating electric stress:

$$W = \omega C V^2 \tan \delta$$

where W is the power loss, V is the rms voltage, C is the capacitance, $\tan \delta$ is the LOSS TANGENT, and $\omega = 2\pi \times$ frequency.

dielectric loss angle (*Phys*) Complement of DIELECTRIC PHASE ANGLE.

dielectric phase angle (*Phys*) The angle between an applied electric field and the corresponding conduction current vector. The cosine of this angle is the power factor of the dielectric.

dielectric polarization (*Phys*) Phenomenon explained by formation of doublets (dipoles) of elements of dielectric under electric stress.

dielectric relaxation (*Phys*) Time delay arising from dipole moments in a dielectric when an applied electric field varies.

dielectric strain (*Phys*) See DISPLACEMENT.

dielectric strength (*ElecEng*) Electric stress necessary to break down a dielectric. It is generally expressed in kV per mm of thickness. The stress, steady or alternating, is normally maintained for 1 minute when testing. See DIELECTRIC.

Diels–Alder reaction (*Chem*) An addition reaction across a pair of conjugated double bonds to form a ring. A typical Diels–Alder reagent is a compound with a double or triple bond activated by electronegative substituents, eg tetra-cyanoethene.

die lubricant (*PowderTech*) Lubricant applied to reduce friction between the powder and the die walls during compaction; it is usually a solid such as a soap. It may be incorporated in powders when interparticle friction is also reduced.

diencephalon (*BioSci*) In vertebrates, the posterior part of the forebrain connecting the cerebral hemispheres with the midbrain.

diene (*Chem*) Organic compound containing two double bonds between carbon atoms in its structure.

diene synthesis (*Chem*) See DIELS–ALDER REACTION.

die nut (*Eng*) Hardened steel nut, usually hexagonal in form with an internal screwed hole. Can be turned by spanner to rectify damage to existing threads.

die pressing (*Eng*) A method of processing ceramics. See panel on CERAMICS PROCESSING.

diesel (*Genrl*) A fuel capable of being burned in engines ignited by compression rather than spark. This oil-based fuel is heavier and oilier than petrol (US gasoline), and has a higher energy density.

diesel cycle (*Autos*) A compression–ignition-engine cycle in which air is compressed, heat added at constant volume by the injection and ignition of fuel in the heated charge, expanded (so doing work on the piston), and the products exhausted, the cycle being completed in either two revolutions (four-stroke) or one (two-stroke). (Rudolph Diesel (1858–1913), German engineer.) See DIESEL ENGINE, FOUR-STROKE CYCLE, TWO-STROKE CYCLE.

diesel–electric locomotive (*Eng*) A locomotive in which a diesel engine is coupled to an electric generator which powers the motors connected to the driving axles.

diesel engine (*Eng*) A compression–ignition engine in which the oil fuel is introduced into the heated compressed-air charge by means of a fuel pump which injects measured quantities of fuel to each cylinder in turn. Earlier models sprayed the fuel by means of an air-blast. See COMPRESSION–IGNITION ENGINE.

diesel-hydraulic locomotive (*Eng*) One whose prime mover is a diesel engine, the power being transmitted through an oil-filled TORQUE CONVERTER.

diesel knock (*Autos*) See KNOCKING.

die set (*Eng*) See SUBPRESS.

die sinking (*Eng*) The engraving of DIES for coining, paper embossing and similar operations.

die stamping (*Print*) An INTAGLIO method of printing requiring a steel die; used mainly for high-class stationery.

die-stock (*Eng*) A hand screw-cutting tool, consisting of a holder with a pair of handles in which screwing dies can be held and rotated.

die swell (*Eng*) The ratio of extrudate diameter to die diameter, caused by elastic recovery of polymer chains after EXTRUSION. It is due to MELT MEMORY and accommodated by careful die design. See MELT ELASTICITY.

diet (*FoodSci*) Collective term for the food supplying the nutrients and energy requirements needed by a person or animal. A balanced diet contains the necessary protein, fat, carbohydrate and essential vitamins and minerals to ensure good health. Special diets are developed for specific needs, eg reduction of weight or cholesterol, or the supply of high energy, or for people with specific allergies or food intolerances.

dietary fibre (*FoodSci*) Carbohydrate or proteinaceous matter which is not completely digested in the alimentary tract eg cellulose, non-cellulosic polysaccharides, lignin. It provides bulk and assists in the removal of body waste and is probably important in the prevention of diverticular disease and bowel cancer.

Dieterici's equation (*Chem*) The VAN DER WAALS' EQUATION modified for the effect of molecules near the boundaries.

diethanolamine (*Chem*) $CH_2OHCH_2NHCH_2CH_2OH$. One of the ETHANOLAMINES used industrially as an intermediate in the preparation of emulsifying agents, corrosion inhibitors, etc. Also used as a stripping agent for CO_2 and H_2S in gas streams.

diethyldithiocarbamic acid (*Chem*) $(C_2H_5)_2NCSSH$. Colourless crystalline solid. Used as a reagent for detecting copper, with which it gives a characteristic brown colour. The zinc salt is used as an accelerator in vulcanization of rubber.

diethylene glycol (*Chem*)
2,2-dihydroxyethoxy-ethane. $(C_2H_4OH)_2O$. Colourless liquid; bp 245°C. Used as a solvent, eg for cellulose nitrate. Its monoethyl ether is known as *carbitol*, also used widely as a solvent. Derivatives (esters and ethers) used as plasticizers.

diethyl ether (*Chem*) *Ethoxyethane*, *ether*. $C_2H_5OC_2H_5$. Mp −113°C, rel.d. 0·72, a mobile, very volatile liquid of ethereal odour, an anaesthetic and a solvent. It is prepared from ethanol and sulphuric acid.

$$H-\underset{\underset{H}{|}}{\overset{\overset{H}{|}}{C}}-\underset{\underset{H}{|}}{\overset{\overset{H}{|}}{C}}-O-\underset{\underset{H}{|}}{\overset{\overset{H}{|}}{C}}-\underset{\underset{H}{|}}{\overset{\overset{H}{|}}{C}}-H$$

diethyl ether

DIF (*ICT*) Abbrev for *data interchange format*, a standard for moving files between different programs.

difference (*MathSci*) See SUBTRACTION.

difference engine (*ICT*) Name given to the very early mechanical computer designed by Charles Babbage and begun in 1823. A complete working replica, based on Babbage's original plans, has subsequently been built and is in the London Science Museum. See ANALYTICAL ENGINE.

difference of phase (*Phys*) See PHASE DIFFERENCE.

difference of potential (*Phys*) See POTENTIAL DIFFERENCE.

difference operators (*MathSci*) An operator used in numerical analysis principally where u_r is a sequence of numbers. (1) $\Delta u_r = u_{r+1} - u_r$ (*descending* or *forward difference*). (2) $\Delta u_{r+1/2} = u_{r+1} - u_r$ (*central difference*). (3) $\nabla u_{r+1} = u_{r+1} - u_r$ (*ascending* or *backward difference*). Higher-order differences are denoted by indices as in the differential calculus. $\Delta_2 u_r$ is a second-order difference.

difference threshold (*Psych*) The amount by which a given stimulus must be increased or decreased in order for a subject to perceive a *just noticeable difference*, JND.

difference tone (*Acous*) See COMBINATION TONE.

differentiable function of a complex variable (*MathSci*) A function that satisfies the Cauchy–Riemann equations, and has each of the four partial derivatives continuous.

differential (*Electronics*) Of a device or circuit whose operation depends on the difference of two opposing effects.

differential (*MathSci*) An arbitrary increment dx of an independent variable x or, if $y = f(x)$, the increment dy of y defined by $dy = f'(x)dx$. See COMPLETE DIFFERENTIAL, DERIVATIVE.

differential absorption ratio (*Radiol*) The ratio of concentration of radioisotope in different tissues or organs at a given time after the active material has been ingested or injected.

differential amplifier (*Electronics*) One whose output is proportional to the difference between two inputs. Usually based on the balanced differential pair. Many linear integrated circuits are of this type.

differential analyser (*ICT*) An obsolete ANALOGUE COMPUTER designed to solve differential equations.

differential anode conductance (*ElecEng*) Reciprocal of DIFFERENTIAL ANODE RESISTANCE.

differential anode resistance (*ElecEng*) The slope of the anode voltage against anode current curve of a multi-electrode valve, when taken with all other electrodes maintained at constant potentials with reference to the cathode. At high frequencies, resistance values are larger than for dc (see SKIN EFFECT) and a separate value may be quoted. Also *ac resistance*, *slope resistance*.

differential booster (*ElecEng*) A booster in which a series winding on the field is connected in opposition to the shunt winding.

differential calculus (*MathSci*) A branch of mathematics dealing with continuously varying quantities; based on the DERIVATIVE or differential coefficient of one quantity with respect to another of which it is a function.

differential capacitor (*ElecEng*) One with one set of moving plates and two sets of fixed plates so arranged that, as the capacitance of the moving plates to one set of fixed plates is increased, that to the other set is decreased.

differential chain block (*Eng*) A lifting tackle in which two connected chain wheels of different diameters carry a continuous chain. Rotating the chain wheels by pulling one loop shortens a second loop which supports the load pulley in such a way as to give a large mechanical advantage. Also *differential pulley block*.

differential coefficient (*MathSci*) See DERIVATIVE.

differential cross-section (*Phys*) For scattering events, the ratio of the number of scattered particles per unit solid angle in a given direction to the number of incoming particles per unit area.

differential dyeing (*Textiles*) Variation in fibres of the same class that leads to their dyeing differently when immersed in a dye-bath.

differential equation (*MathSci*) An equation involving total or partial differential coefficients. Those not involving partial differential coefficients are *ordinary differential equations*.

differential flotation (*Eng*) Production of more than one valuable concentrate by a series of froth flotation treatments of prepared ore pulp.

differential gear (*Eng*) A gear permitting relative rotation of two shafts driven by a third. The driving shaft rotates a cage carrying planetary bevel wheels meshing with two bevel wheels on the driven shafts. The latter are independent, but the sum of their rotation rates is constant.

differential grinding (*MinExt*) Comminution so controlled as to develop differences in grindability of constituents of ore.

differential hardening (*ImageTech*) Hardening of a photographic emulsion in specific areas to form an image, either by the action of light on chemically treated gelatine (eg with bichromate) or by development products associated with the formation of a silver image (TANNING DEVELOPER).

differential interference contrast microscope (*BioSci*) An interference microscope in which the light beam is split by a WOLLASTON PRISM (or a modified version devised by Nomarski) in the condenser, to form slightly divergent beams polarized at right angles. One passes through the specimen (and is retarded if the refractive index is greater), and one through the background nearby: the two are recombined in a second Wollaston or Nomarski prism in the objective and interfere to form a monochromatic image. The image is spuriously 'three-dimensional' – the nucleuseg appears to stand out above the cell because it has a higher refractive index than the cytoplasm. Compared with phase-contrast microscopy, the image is sharper, halo-free and less subject to interference from structures out of the plane of focus. The *Nomarski microscope* is an example. Abbrev *DIC microscope*.

differential ionization chamber (*NucEng*) A two-compartment system in which the resultant ionization current recorded is the difference between the currents in the two chambers. One version (*compensated ion chamber*) may be used to distinguish between neutrons and gamma radiation.

differential iron test (*ElecEng*) Apparatus for testing iron consisting of two magnetic squares, one of the sample to be tested and the other of a standard material. The windings on the squares are connected to a differential wattmeter, so that there will be no deflection when the quality of the two specimens is the same.

differential leakage flux (*ElecEng*) A general term given to the leakage flux occurring in and around the air-gap of an induction motor. See ZIGZAG LEAKAGE.

differentially compound-wound machine (*ElecEng*) A compound-wound dc machine in which the magnetomotive forces of the two windings oppose one another.

differentially wound motor (*ElecEng*) A dc motor with series and shunt windings on the field connected so that the series windings opposes the shunt winding, and therefore causes the speed of the motor to rise as load is put on the machine.

differential-mode signal (*ElecEng*) (1) The difference between two signals, both measured to a common reference, (as opposed to COMMON-MODE SIGNAL). Often used as the input to an amplifier when small signals have to be amplified in a high level of background interference. (2) In a balanced three-terminal system, the signal applied between the two underground terminals.

differential motion (*Eng*) A mechanical movement in which the velocity of a driven part is equal to the difference of the velocities of two parts connected to it.

differential permeability (*ElecEng*) The ratio of a small change in magnetic flux density of magnetic material to change in the magnetizing force producing it, ie slope of the $B(H)$ curve at the point in question, often expressed in dimensionless form as $dB/\mu_0 dH$.

differential pressure gauge (*Eng*) A gauge, commonly of U-tube form, which measures the difference between two fluid pressures applied to it.

differential protective system (*ElecEng*) See BALANCED PROTECTIVE SYSTEM.

differential pulley block (*Eng*) See DIFFERENTIAL CHAIN BLOCK.

differential pulse-code modulation (*ICT*) A version of PULSE-CODE MODULATION in which the difference in value between a sample and its predecessor constitutes the transmitted information. For many types of signal fewer bits are needed; often used in satellite communication.

differential relay (*ElecEng*) See RELAY.

differential resistance (*ElecEng*) The ratio of a small change in the voltage drop across a resistance to the change in current producing the drop, ie the slope of the voltage–current characteristic for the material. See DYNAMIC RESISTANCE.

differential rotation (*Astron*) Rotation of different parts of a system at different angular speeds; occurs eg in the Sun, in which the period of rotation of the polar regions is shorter than the rotation period of the equatorial regions.

differential scanning calorimetry (*Chem*) A type of thermal ANALYSIS similar to DIFFERENTIAL THERMAL ANALYSIS. Abbrev *DSC*.

differential stain (*BioSci*) A stain that picks out details of structure by giving to them different colours or different shades of the same colour.

differential susceptibility (*ElecEng*) The ratio of a small change in the intensity of magnetization of a magnetic material to the change of magnetizing force producing it, ie the slope of the $M(H)$ curve.

differential thermal analysis (*Chem*) The detection and measurement of changes of state and heats of reaction, esp in solids and melts. The sample under investigation together with one of thermally inert material are simultaneously heated side by side and the difference in temperature between them noted. This becomes very marked when one of the two samples passes through a TRANSITION TEMPERATURE with evolution or absorption of heat but the other does not. The technique is particularly used for clay samples. Abbrev *DTA*.

differential titration (*Chem*) Potentiometric titration in which the emf is noted between additions of small increments of titrant, the end-point being where the emf changes most sharply.

differential winding (*ElecEng*) A winding in a compound motor which is in opposition to the action of another winding.

differentiating circuit (*ElecEng*) (1) Amplifier having a combination of resistive input and feedback inductance, or capacitive input and feedback resistance, such that the output is proportional to the rate of change (differential) of the input signal. (2) A passive circuit comprising either R and L or C and R, whose output is proportional to the rate of change of the input signal. This circuit does not produce as accurate a result as the active circuit described above. Used eg to detect sudden changes in otherwise steady wave form and to modify waveforms in digital circuits.

differentiating solvent (*Chem*) See LEVELLING SOLVENT, NON-AQUEOUS SOLVENTS.

differentiation (*BioSci*) (1) The qualitative changes in morphology and physiology occurring in a cell, tissue or organ as it develops from a meristematic, primordial or unspecialized state into the mature or specialized state. (2) Removing excess stain from some of the structures in a microscope preparation so that the whole can be seen more clearly.

differentiation (*Geol*) The process of forming two or more rock types from a common magma.

differentiation (*MathSci*) The operation of finding a DERIVATIVE.

differentiation antigen (*BioSci*) A structural macromolecule that can be detected by immunological reagents and is associated with the differentiation of a particular cell type. See CD ANTIGEN.

differentiator (*ElecEng*) See DIFFERENTIATING CIRCUIT.

diffluence (*EnvSci*) The spreading apart of streamlines.

diffraction (*Phys*) The phenomenon, observed when waves are obstructed by obstacles or apertures, of the disturbance spreading beyond the limits of the geometrical shadow of the object. The effect is marked when the size of the object is of the same order as the wavelength of the waves and accounts for the alternately light and dark bands, diffraction fringes, seen at the edge of the shadow when a point source of light is used. It is one factor that determines the propagation of radio waves over the curved surface of the Earth and it also accounts for the audibility of sound around corners. See FRAUNHOFER DIFFRACTION, FRESNEL DIFFRACTION.

diffraction analysis (*Crystal*) Analysis of the internal structure of crystals by utilizing the diffraction of X-rays caused by the regular atomic or ionic lattice of the crystal.

diffraction angle (*Phys*) The angle between the direction of an incident beam of light, sound or electrons, and the direction of any resulting diffracted beam.

diffraction grating (*Phys*) An optical device for producing spectra. In one form the diffraction grating consists of a flat glass plate with equidistant parallel straight lines ruled in its surface by a diamond. There may be as many as 1000 per mm. If a narrow source of light is viewed through a grating it is seen to be accompanied on each side by one or more spectra. These are produced by diffraction effects from the lines acting as a very large number of equally spaced parallel slits.

diffraction pattern (*Phys*) The pattern formed by equal-intensity contours as a result of diffraction effects, eg in optics or radio transmission.

diffractometer (*Phys*) An instrument used in the analysis of the atomic structure of matter by the diffraction of X-rays, neutrons or electrons by crystalline materials. A mono-chromatic beam of radiation is incident on a crystal mounted on a goniometer. The diffracted beams are detected and their intensities measured by a counting device. The orientation of the crystal and the position of the detector are usually computer-controlled.

diffuse density (*ImageTech*) The density of a photographic image when measured by diffuse light rather than specular. See CALLIER EFFECT.

diffuse growth (*BioSci*) Growth where cells divide through-out the tissue. Cf APICAL GROWTH, INTERCALARY growth, TRICHOTHALLIC GROWTH.

diffuse placentation (*BioSci*) Having the VILLI developed in all parts of the placenta, except the poles, as in lemurs, most ungulates and Cetacea.

diffuse porous (*BioSci*) Wood having the pores distributed evenly throughout a growth ring or changing in diameter gradually across it, eg birch, evergreen oaks. Cf RING POROUS.

diffuser (*Acous*) An irregular structure, eg pyramid or cylinder, to break up sound waves in rooms. See SCATTERER.

diffuser (*Aero*) A means for converting the kinetic energy of a fluid into pressure energy; usually it takes the form of a duct which widens gradually in the direction of flow; also fixed vanes forming expanding passages in a compressor delivery to increase the pressure.

diffuser (*Eng*) A chamber surrounding the impeller of a centrifugal pump or compressor, in which part of the kinetic energy of the fluid is converted to pressure energy by a gradual increase in the cross-sectional area of flow.

diffuser (*ImageTech*) Translucent material in front of studio lamp to diffuse light and soften shadows.

diffuse reflection (*Phys*) See NON-SPECULAR REFLECTION.

diffuse-reflection factor (*Phys*) The ratio of the luminous flux diffusely reflected from a surface to the total luminous flux incident upon the surface.

diffuse series (*Phys*) A series of optical spectrum lines observed in the spectra of alkali metals. Energy levels for which the orbital quantum number is two are designated *d-levels*.

diffuse sound (*Acous*) Sound which is reflected in all directions inside a volume.

diffuse tissue (*BioSci*) A tissue consisting of cells that occur in the plant body singly or in small groups intermingled with tissues of distinct type.

diffuse transmittance (*Phys*) See TRANSMITTANCE.

diffusion (*Chem*) The general transport of matter (atoms, molecules, ions) through thermal agitation. A net flux results from diffusion when there is a concentration gradient. In a crystalline solid, interstitial atom, lattice vacancy and impurity atom diffusion are thermally activated. Diffusion is often used to introduce controlled quantities of impurities in the surfaces of semiconductors (for doping) and metals (for carburizing and nitriding etc). See DIFFUSION COEFFICIENT, FICK'S LAWS OF DIFFUSION.

diffusion activation energy (*Phys*) Activation energy required for temperature-dependent diffusion of inter-stitial atoms, lattice vacancies or impurities in a crystal-line solid. The diffusion coefficient D is given by $D = D_0 \exp(-E/kT)$, where D_0 is a constant, E is the activation energy of the process, T is the temperature and k is Boltzmann's constant.

diffusion area (*NucEng*) A term used in reactor diffusion theory. One-sixth of the mean square displacement (ie direct distance travelled irrespective of route) between point at which neutron becomes thermal and where it is captured.

diffusion attachment (*ImageTech*) Lens accessory for softening the outline of the image in a camera or enlarger, often a disk with a finely etched or engraved surface.

diffusion barrier (*Phys*) Porous partition for gaseous separation according to molecular weight and hydrody-namic velocities, esp for separation of isotopes. A fired but unglazed plate.

diffusion capacitance (*Phys*) The rate of change of injected charge with the applied voltage in a semiconductor diode.

diffusion coating (*Eng*) Methods by which an alloy or metal are allowed to diffuse into the surface of an underlying metal. They can involve heating and exposing the metal to a solution of the coating material.

diffusion coefficient (*Chem*) In the diffusion equation (Fick's first law), the coefficient of proportionality between molecular flux and concentration gradient. Symbol D. Units, $m^2 s^{-1}$, as for THERMAL DIFFUSIVITY. The approx-imate distance, x, moved by a diffusing species in time t, is:

$$x = \sqrt{Dt}$$

Since diffusion is a *thermally activated process*, the Arrhenius rate equation applies,

$$D = D_0 \exp(-E_a/kT)$$

Also *diffusivity*. See FICK'S LAWS OF DIFFUSION.

diffusion constant (*Phys*) The ratio of diffusion current density to the gradient of charge carrier concentration in a semiconductor.

diffusion current (*Chem*) The net flux of particles down a density gradient. In particular, that current resulting from the diffusion of charge carriers in a concentration gradient as distinct from a DRIFT CURRENT. In electrolysis, the maximum current at which a given bulk concentration of ionic species can be discharged, being limited by the rate of migration of the ions through the DIFFUSION LAYER.

diffusion flame (*Eng*) Long luminous gas flame holding practically a constant rate of radiation for its designed length of travel, together with uniform precipitation of free carbon, diffusion occurring between adjacent strata of air and gas.

diffusion law (*Chem*) See FICK'S LAW, GRAHAM'S LAW.

diffusion layer (*Chem*) In electrolysis, the layer of solution adjacent to the electrode, in which the concentration gradient of electrolyte occurs.

diffusion length (*NucEng*) Square root of DIFFUSION AREA.

diffusion length (*Phys*) Average distance travelled by carriers in a semiconductor between generation and recombination.

diffusion of solids (*Phys*) In semiconductors, the migration of atoms into pure elements to form surface alloy for providing minority carriers.

diffusion plant (*NucEng*) Plant used for isotope separation by GASEOUS (MOLECULAR) DIFFUSION or THERMAL DIFFUSION.

diffusion potential (*ElecEng*) The potential difference across the boundary of an electrical double layer in a liquid.

diffusion pump (*ChemEng*) See GAEDE DIFFUSION PUMP.

diffusion theory (*NucEng*) Simplified neutron migration theory based on FICK'S LAW. Less accurate than the more detailed TRANSPORT THEORY.

diffusion transfer reversal (*ImageTech*) A process in which a direct positive image is formed from the material exposed in the camera, hence the basis of many instant photography systems. The exposed emulsion is developed by a viscous solution containing a silver halide solvent while in contact with a receiving layer on paper or transparent base. Unexposed silver halides become transferred to this receptor and form a positive image which can be separated on its new support.

diffusion tube (*EnvSci*) Tube, usually about 70 cm long and 12 mm in diameter, used to obtain a passive sample of the atmosphere. After a predetermined sampling period the tube is closed and the contents analysed.

diffusion welding (*Eng*) A method in which high tempe-rature and pressure cause a permanent bond between two metallic surfaces without melting or large-scale deformation. A solid-metal filler may be sandwiched between the surfaces to aid the process.

diffusivity (*Phys*) See DIFFUSION COEFFICIENT, THERMAL DIFFUSIVITY

difluorophosphoric acid (*Chem*) HPO_2F_2. Formed by partial hydrolysis of phosphoryl fluoride, POF_3, with cold dilute alkali, or preferably by heating phosphoric acid with ammonium fluoride.

digametic (*BioSci*) Having gametes of two different kinds.

digastric (*BioSci*) Of muscles, having fleshy terminal portions joined by a tendinous portion, as the muscles which open the jaws in mammals.

digenesis (*BioSci*) (1) ALTERNATION OF GENERATIONS. (2) The condition of having two hosts; said of parasites. Adj *digenetic*.

digenetic reproduction (*BioSci*) See SEXUAL REPRODUCTION.

digenite (*Min*) A cubic sulphide of copper, usually massive and associated with other copper ores. Composition probably near Cu_2S_5.

di George's syndrome (*Med*) See THYMIC HYPOPLASIA.

digester (*Genrl*) A closed tank used in waste-water (sewage) treatment in which solid waste is broken down by bacterial action.

digester (*Paper*) A vessel in which fibrous raw materials are heated under pressure with chemicals, in the first stages of paper-making.

digestion (*BioSci*) The process by which food material ingested by an organism is rendered soluble and assimilable by the action of enzymes. Adj *digestive*.

digestion (*MinExt*) In chemical extraction of values from ore or concentrate, period during which material is exposed under stated conditions to attacking chemicals.

digestive gland (*BioSci*) Gland(s) present in many invertebrates and Protochordata, in which intracellular ingestion and absorption take place, as opposed to the alimentary canal proper.

Digibox (*ICT*) TN for a brand of set-top box for receiving digital satellite TV.

digicam (*ICT*) A digital camera.

digit (*BioSci*) A finger or toe, one of the free distal segments of a pentadactyl limb.

digit (*ICT*) Discrete sign. See BIT, CHECK DIGIT.

digital (*ICT*) (1) A term used to describe representation of a numerical quantity by a number of discrete signals or by the presence or absence of signals in particular positions. See BIT. (2) Communications circuits in which the information is transmitted in the form of trains of pulses; speech and vision need to be converted into code before such transmission (see PULSE-CODE MODULATION) whereas most data are already in suitable form. Advantages include immunity to noise and the possibility of electronic exchange switching.

digital advanced mobile-phone system (*ICT*) The US equivalent of the European GLOBAL SYSTEM FOR MOBILE COMMUNICATION, with a similar but not identical specification. Abbrev *DAMPS*.

digital audio broadcasting (*Genrl*) The use of digital rather than analogue signals for transmitting sound. It is of higher quality and more channels can occupy a given wavelength band. Abbrev *DAB*.

digital audio player (*ICT*) A device for storing and playing audio data that have been recorded in a digital file format. Abbrev *DAP*.

digital camera (*ImageTech*) Still cameras in which the film is replaced by a charge-coupled device with a resolution between 2 and 6 megapixels in commonly available and affordable types (2006). Images can be stored on various standard memory cards which can then be read in a computer and the images manipulated with standard software. A small screen can be used to view images before exposure and after storage.

digital clock (*Genrl*) A clock which displays the time as a series of numbers instead of the traditional rotating hands or *analogue* manner.

digital computer (*ICT*) See COMPUTER.

digital cross-connect (*ICT*) A form of switch designed to route large groups of digital channels under network management control, as opposed to an exchange, which sets up individual connections in response to circuit signalling. It is used primarily to reconfigure a network in response to demand or fault conditions.

digital differential analyser (*ICT*) An obsolete electronic computer for solution of differential equations by incremental means.

Digital European Cordless Telephone (*ICT*) The pan-European standard for telephones in which the handset is connected to the base by a digital radio link. Operating at around 1·6 GHz, it is designed to support a higher density of traffic than the UK CT2 standard.

digital filter (*ICT*) One that passes, or rejects, pulsed or digital information whose pattern corresponds exactly with that laid down in the design of the filter circuit.

digital imaging back (*ImageTech*) A photographic camera back with a SOLID-STATE IMAGE SENSOR or scanner and a digital storage device or output to a computer.

digitalis (*Pharmacol*) A general term for pharmacologically active compounds from the foxglove (*Digitalis purpurea*). The active substances are the cardiac glycosides, *digoxin*, *digitoxin*, strophanthin and *ouabain*.

digitalization (*Med*) Administration of DIGOXIN to a patient with heart disease, in amounts sufficient to produce full therapeutic effect.

digital local exchange (*ICT*) A CENTRAL OFFICE in which outgoing telephone calls are digitized before switching, and incoming calls are converted from digital to analogue form before delivery over the LOCAL LOOP.

digital meter (*ElecEng*) One displaying the measured quantity as a numerical value.

digital micromirror device (*ImageTech*) TN for a video projector in which plastic micromirrors, each representing 1 PIXEL, twist on a panel in response to an ELECTROSTATIC FIELD produced by a TRANSISTOR beneath and reflect or block light from a PROJECTION LAMP. Abbrev *DMD*.

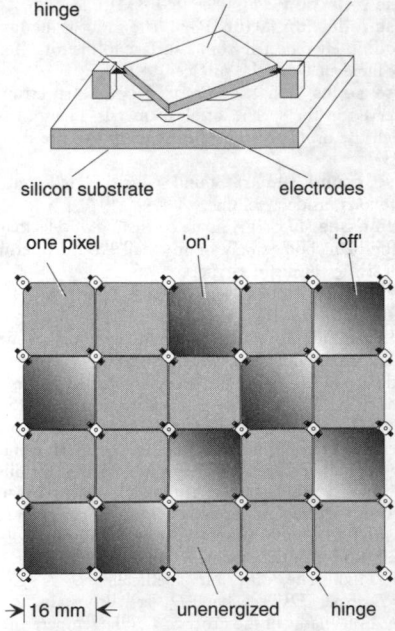

digital micromirror device Each mirror can tilt 10° either way with a 10 ms response.

digital plotter (*ICT*) Graph plotter that receives digital input specifying the co-ordinates of the points to be plotted. Cf INCREMENTAL PLOTTER.

digital radio (*ICT*) A form of radio broadcasting in which the sounds are compressed into and transmitted in digital form.

digital signal processing (*ImageTech*) Conversion of a video signal into digital form so that it may be processed in real time to enhance its quality and, possibly, produce SPECIAL EFFECTS. Abbrev *DSP*.

digital signature (*ICT*) A file that confirms the identity of a person to another user of a computer network.

digital subtraction angiography (*Radiol*) A radiological technique where an initial X-ray image is digitized and subtracted from another taken after the injection of CONTRAST MEDIUM. As only the contrast in the blood vessels is added, high-quality images of these blood vessels can be obtained after a small intravenous injection. Avoids the catheterization of an artery in ANGIOGRAPHY.

digital-to-analogue converter (*ICT*) A device for converting digital signals to analogue ones.

digital tracking (*ImageTech*) An automatic TRACKING CONTROL which applies ideal tracking parameters from a digital store.

digital TV (*ICT*) A form of TV broadcasting in which images and sounds are compressed into and transmitted in digital form.

digital versatile disk (*ImageTech*) A compact disc with compressed digital images and sound recorded on it as from a full-length film. Abbrev *DVD*.

digital video (*ImageTech*) See FULL-MOTION VIDEO.

digital video camera (*ImageTech*) The digital versions of the older analogue cameras in which the image is recorded at TV resolution, 768×568 pixels, although some professional models use 1024×768. The 'domestic' models record on Mini-DV tapes that can play for 30 or 60 minutes. Abbrev *DVC*.

digital video effects (*ImageTech*) Creating SPECIAL EFFECTS from digitized video signals. Abbrev *DVE*.

digital watch (*Genrl*) A watch which displays the time as a series of numbers instead of the traditional rotating hands or *analogue* manner.

digital zoom (*ImageTech*) Enlarges the central portion of pictures taken from a digital video store.

digitate (*BioSci*) See PALMATE.

digitigrade (*BioSci*) Walking on the ventral surfaces of the phalanges only, as terrestrial carnivores.

digitize (*ICT*) To convert an analogue signal to a digital signal. Also *quantize*.

digitizer (*ICT, ElecEng*) See ANALOGUE-TO-DIGITAL CONVERTER.

digit memory span (*Psych*) A test of short-term memory: the number of random digits or items than can be accurately recalled (usually around seven).

digitoxin (*Pharmacol*) A glycoside isolated from the leaves of *Digitalis purpurea*.

digitule (*BioSci*) Any small finger-like process.

digoneutic (*BioSci*) Producing offspring twice a year.

digoxin (*Pharmacol*) A cardiac glycoside used to treat congestive heart failure and supraventricular arrhythmias.

digraph (*MathSci*) A graph whose edges are assigned directions. Applied in many fields including computer science.

dihedral (*MathSci*) The angle between two planes, as measured in the plane normal to their line of intersection.

dihedral angle (*Aero*) Acute angle at which an aerofoil is inclined to the transverse plane of reference. A downward inclination is *negative dihedral*, sometimes *anhedral*.

dihedral group (*MathSci*) The symmetry group of the regular n-sided polygon, normally denoted by D^n, with order $2n$; eg the symmetry group of the square has eight elements: three rotations by $90°$, four reflections and the identity.

diheptal (*Electronics*) Referring to 14 in number, eg pins on the base of a tube or valve.

dihybrid (*BioSci*) The product of a cross between parents differing in two characters determined by single genes; an individual heterozygous at two gene loci.

dihydrofolate reductase (*BioSci*) An enzyme involved in the biosynthesis of tetrahydrofolic acid, an essential vitamin. Abbrev *DHFR*.

dihydroxyacetone (*Chem*) $CH_2OHCOCH_2OH$. The simplest *ketose*, used in suntan lotion.

di-iso-octyl phthalate (*Chem*) Plasticizer used in flexible polyvinyl chloride. Abbrev *DIOP*.

diisopropylnaphthalenes (*Chem*) Compounds used as a solvent for the colour former in carbon-less copy paper. Abbrev *DIPNs*.

dikaryon (*BioSci*) Fungal hypha or mycelium in which two nuclei of different genetic constitution (and different mating type) are present in each cell (or hyphal segment). Adj *dikaryotic*. See DIKARYOPHASE.

dikaryophase (*BioSci*) That part of the life cycle of an ascomycete or basidiomycete in which the cells have two nuclei, ie between PLASMOGAMY and KARYOGAMY. Also *binucleate phase*.

dike (*Build, Geol*) See DYKE.

diketen (*Chem*) Dimer of *ketene*. Bp $127°C$. Useful intermediate in preparative organic chemistry.

diketones (*Chem*) *Dialkanones*. Compounds containing $-CO-$ groups which, according to their position in the molecule, are termed 1,2-diketones $-COCO-$, or 1,3-diketones $-COCH_2CO-$, etc.

dikkop (*Vet*) See AFRICAN HORSE-SICKNESS.

dilambdodont (*BioSci*) Said of teeth in which the paracone and metacone are V-shaped, well separated and placed near the middle of the tooth, as in some INSECTIVORA.

dilapidation (*Build*) The damage done to premises during a period of tenancy.

dilatancy (*Chem*) The behaviour of a fluid when stressed in which it increases its resistance to further stress by increasing its shear rate (eg wet beach sand or polyvinyl chloride plastisol). Pseudo-plastic fluids behave in the opposite way. Cf THIXOTROPY.

dilatometer (*Chem*) Apparatus for the determination of transition points of solids. It consists of a large bulb joined to a graduated capillary tube, and is filled with an inert liquid. The powdered solid is introduced, and the temperature at which there is a considerable change in volume on slow heating or cooling may be noted; alternatively, the temperature at which the volume shows no tendency to change with time may be found.

dilatometry (*Chem*) The determination of transition points by the observation of volume changes.

dilator (*BioSci*) A muscle that, by its contraction, opens or widens an orifice. Cf SPHINCTER.

di-litho (*Print*) See DIRECT LITHOGRAPHY.

Dilleniidae (*BioSci*) A subclass or superorder of dicotyledons consisting of trees, shrubs and herbs. The flowers are polypetalous or sympetalous, stamens (if numerous) develop centrifugally and the ovaries consist of joined carpels (syncarpous), often with *parietal placentation*. Contains c.24 000 spp in 69 families including Malvaceae, Cruciferae and Ericaceae.

diltiazem HCl (*Pharmacol*) A calcium channel blocker used in the treatment of angina.

diluent (*Build*) A volatile liquid able to dilute a genuine paint solvent without disadvantage and used to cheapen the formulation.

diluent air (*Build*) Air admitted or induced into a flue to dilute the noxious effects of combustion.

dilution (*Chem*) (1) Decrease of concentration. (2) The volume of a solution which contains unit quantity of dissolved substance. The reciprocal of *concentration*.

dilution law (*Chem*) See OSTWALD'S DILUTION LAW.

dilution refrigerator (*Phys*) A device for producing very low temperatures, down to 0.004 K, on a small sample. Uses the very low-temperature properties of a mixture of helium-3 and helium-4.

diluvium (*Geol*) An obsolete term for those accumulations of sand, gravel, etc, which, it was thought, could not be accounted for by normal stream and marine action. In this sense, the deposits resulting from the Deluge of Noah would be *diluvial*.

dimension (*ICT*) In data-warehousing the indexes representing significant measures in the real world by which the contents of a fact table may be divided and analysed.

dimensional analysis (*Phys*) (1) Procedure to check that both sides of an equation describing a physical phenomenon have the same dimensions of fundamental quantities such as mass, length, time or current. (2) Analysis of basic dimensions to predict the way in which one quantity depends on others. Dimensional analysis gives no information about dimensionless quantities or pure numbers occurring in an equation.

dimensional stability (*Paper*) The resistance offered by a paper to changes in its dimensions when ambient conditions alter.

dimensioning (*ICT*) The process of designing a telephone network structure and determining the amount of equipment required in each part of the network to satisfy a specific demand with a prescribed QUALITY OF SERVICE.

dimensions (*Phys*) Of a DERIVED UNIT used to express the measurement of a physical quantity, the *powers* to which the fundamental units are involved in the quantity; eg velocity has dimensions +1 in length and −1 in time or $[LT^{-1}]$, force has dimensions +1 in mass, +1 in length and −2 in time, $[MLT^{-2}]$. See appendices on Units of measurement and SI derived units.

dimer (*Chem*) Molecular species formed by the union of two like molecules, the simplest OLIGOMER. Adj *dimeric*.

dimercaprol (*Pharmacol*) A drug used as an antidote to poisoning by heavy metals such as arsenic, lead or mercury.

dimeric (*Chem*) See DIMER.

dimerous (*BioSci*) A flower with two members in a given whorl.

dimethacone (*Pharmacol*) An anti-foaming agent sometimes added to antacids to relieve flatulence.

dimethylformamide (*Chem*) $(CH_3)_2NCHO$. Solvent of wide industrial application, used in plastics manufacture, in gas separation processes, in the artificial fibre industry.

dimethyl glyoxime (*Chem*) Compound used in analysis as a specific and quantitative precipitant for palladium (weakly acid solution) and nickel (ammoniacal solution) with which it gives a brilliant red precipitate.

dimethyl hydrazine (*Chem*) See HYDRAZINE.

dimethyl phthalate (*Chem*) *Dimethyl benzene 1,2-dicarboxylate*. High-boiling-point ester, bp 280°C, widely used as an insect repellent, particularly against midges and mosquitoes.

dimethyl sulphate (*Chem*) $CH_3OSO_2OCH_3$. Bp 187–188°C. Widely used as a methylating (ie introduction of a methyl (CH_3) group) agent in organic preparations. Manufactured from dimethyl ether and sulphur trioxide.

dimethylsulphoxide (*BioSci*) An unpleasant-smelling colourless liquid used experimentally as a solvent for water-insoluble compounds, and as a cryoprotectant when freezing cells for storage. It is used clinically for the treatment of arthritis, although its efficacy is disputed. Abbrev *DMSO*.

diminished stile (*Build*) A door-stile which is narrowed down for a part of its length, as eg in the glazed portion of a sash door. Also *gunstock stile*.

diminishing courses (*Build*) See GRADUATED COURSES.

diminishing pipe (*Build*) A tapered pipe length used to connect pipes of different diameters.

dimity (*Textiles*) Strong cotton fabric that appears striped because of a corded pattern; used chiefly for mattress coverings.

dimorphic (*BioSci*) Organelles, organs or individuals etc existing in two distinct forms. Also *dimorphous*.

dimorphic (*Chem*) Capable of crystallizing in two different forms. Also *dimorphous*.

dimorphism (*BioSci*) The condition of having two different forms, as animals that show marked differences between male and female (sexual dimorphism), animals that have two different kinds of offspring, and colonial animals in which the members of the colony are of two different kinds.

dimorphism (*Min*) Crystallization into two distinct forms of an element or compound, eg carbon as diamond and graphite; FeS_2 as pyrite and *marcasite*.

dimorphous (*BioSci, Chem*) See DIMORPHIC.

DIN (*Genrl, ImageTech*) Abbrev for *Deutsche Industrie Norm*, the German industry standard. (1) An audio connector which can carry video. (2) A photographic film speed rating with logarithmic increments.

Dinantian (*Geol*) The lower Carboniferous rocks of NW Europe, comprising Tournaisian and Viséan Stages. See PALAEOZOIC.

dineutron (*Phys*) A combination of two neutrons, assumed to exist transiently in order to explain certain nuclear reactions.

dingbat (*ICT*) A symbol usually provided in a special font to embellish the text, such as ☛ or ☎.

dinging (*Build*) Rough plastering for walls, a single coat being put on with a trowel and brush.

dinitrocresol (*Chem*) See DNOC.

dinitrogen fixation (*BioSci*) Same as NITROGEN FIXATION.

dinky sheet (*Print*) A web much narrower than the full width of a rotary press.

dinocap (*Chem*) *2-(1-methyl-heptyl)-4,6-dinitrophenyl crotonate*; used as a fungicide.

Dinoflagellata (*BioSci*) See DINOPHYCEAE.

Dinophyceae (*BioSci*) Mesokaryotic algae found in both marine and fresh-water environments. Motile cells have two laterally inserted flagella, one lying in a transverse groove around the cell and helically coiled, the other lying in a longitudinal groove and posteriorly directed. Chloroplasts have chlorophyll a and c, and PERIDININ. Phototrophic species may be flagellate, colonial, coccoid, palmelloid and, in a few cases, filamentous. Phagotrophic, parasitic and various symbiotic sorts occur. RED TIDES are algal blooms of various species. Also *dinoflagellates*. See ACRITARCHS.

dinosaurs (*BioSci*) Any member of a large group of prehistoric reptiles that dominated life on land during the Jurassic and Cretaceous periods, becoming extinct at the end of the Mesozoic era. All species had a specialized type of hip joint, on the basis of which they were divided into two groups: the lizard-hipped dinosaurs (order Saurischia) and the bird-hipped dinosaurs (order Ornithischia). The saurischian dinosaurs included the two-legged carnivores such as *Tyrannosaurus*, and the enormous four-legged, predominantly herbivorous forms such as *Apatosaurus*.

diocoel (*BioSci*) The lumen of the DIENCEPHALON.

dioctylphthalate (*Chem*) A common plasticizer for polyvinyl chloride. Abbrev *DOP*. See panel on POLYVINYL CHLORIDE.

diode (*Electronics*) (1) Simplest electron tube, with a heated cathode and anode; used because of unidirectional and hence rectification properties. (2) Semiconductor device with similar properties, evolved from primitive crystal RECTIFIERS for radio reception.

diode characteristic (*Electronics*) A graph showing the current–voltage characteristics of a vacuum tube or semiconductor diode. In particular it will show marked differences between currents in the forward and reverse directions, and, in the case of AVALANCHE or ZENER DIODES, sudden increases in reverse current when the applied voltage reaches a critical value.

diode clipper (*Electronics*) A limiting circuit using a diode.

diode isolation (*Electronics*) Isolation of the circuit elements in a microelectronic circuit by using the very high resistance of a reverse-biased p–n junction.

diode laser (*Electronics*) See OPTOELECTRONICS.

diode–pentode (*Electronics*) Thermionic diode in same envelope as a pentode.

diode–transistor logic (*Electronics*) Logic circuitry in which arrays of diodes at the inputs perform logic functions prior to controlling the base current of a transistor which subsequently provides power gain for driving additional gates.

diode–triode (*Electronics*) Thermionic diode in same envelope as a triode.

diode voltmeter (*ElecEng*) One in which measured voltage is rectified, amplified and displayed on a moving-coil meter.

dioecious (*BioSci*) Having the sexes in separate individuals; n. *dioecism*.

dioestrus (*BioSci*) In female mammals, the growth period following metoestrus.

diolefin (*Chem*) See DIALKENES.

diols (*Chem*) Dihydric alcohols, chiefly represented by the glycols in which the hydroxyl groups are attached to adjacent carbon atoms.

Dione (*Astron*) The fourth natural satellite of Saturn, discovered in 1684. Distance from the planet 377 000 km; diameter 1120 km. The tiny moon HELENE is associated with it.

DIOP (*Chem*) Abbrev for DI-ISO-OCTYL PHTHALATE.

diophantine equations (*MathSci*) Equations with integral coefficients for which integral or rational solutions are required.

diopside (*Min*) A monoclinic pyroxene, typically occurring in metamorphosed limestones and dolomites, and composed of calcium magnesium silicate, $CaMgSi_2O_6$, usually with a little Fe.

dioptase (*Min*) A rare emerald-green hydrated copper silicate.

dioptre (*Phys*) The unit of power of a lens. A convergent lens of 1 metre focal length is said to have a power of +1 dioptre. Generally, the power of a lens is the reciprocal of its focal length in metres, the power of a divergent lens being given a negative sign.

dioptre lens (*ImageTech*) A supplementary lens used in front of the main camera objective to bring close objects into focus; sometimes covering only part of the field of view to show both close and more distant objects (split dioptre). A *supplementary lens*.

dioptric mechanism (*BioSci*) A mechanism, consisting of the cornea, aqueous humour, lens and vitreous humour, by which the images of external objects may be focused on the retina of the vertebrate eye. OMMATIDIA serve a comparable function in the arthropod eye.

dioptric system (*Phys*) An optical system which contains only refracting components.

diorite (*Geol*) A coarse-grained deep-seated (plutonic) igneous rock of intermediate composition, consisting essentially of plagioclase feldspar (typically near andesine in composition) and hornblende, with or without biotite in addition. Differs from granodiorite in the absence of quartz. See PLUTONIC ROCKS, TONALITE.

dioxan (*Chem*) 1,4-dioxycyclohexane. Colourless liquid; mp 11°C, bp 101°C. Used as a solvent for waxes, resins, viscose, etc.

dioxan

dioxazine dyes (*Chem*) A range of dyestuffs which are relatively complex sulphonated compounds with an affinity for cellulose. Chiefly derived from chloranil by reaction with amines. The chief dyes in the series are light-fast in the blue range.

dioxins (*Chem*) A family of toxic and persistent chlorinated aromatic hydrocarbons of which dioxin (2,3,7,8-tetrachlorodibenzodioxin) was a contaminant of Agent Orange, the defoliant widely used in the Vietnam War. They are lipophilic and are found as trace compounds in the food chain. Fig. ▷

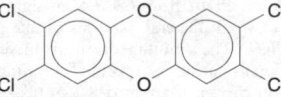

dioxins 2,3,7,8-TCDD or dioxin.

DIP (*Electronics*) Abbrev for DUAL IN-LINE PACKAGE.

dip (*Build*) Any departure from the regular slope at which a pipe is laid, when the slope is increased locally.

dip (*Geol*) The angle that a plane makes with the horizontal. The dip is perpendicular to the STRIKE of the structure. See DRAINAGE PATTERNS.

dip (*Phys*) The angle measured in a vertical plane between the Earth's magnetic field at any point and the horizontal. Also *inclination*.

dip circle (*Phys*) An instrument consisting of a magnetic needle or dip needle pivoted on a horizontal axis; accurate measurements of magnetic DIP can be obtained when it is set in the magnetic north–south plane.

dipentene (*Chem*) *dl*-limonene.

dip fault (*Geol*) A fault parallel to the direction of dip. See STRIKE FAULT.

diphase (*ElecEng*) A term sometimes used in place of *two-phase*.

diphasic (*BioSci*) Of certain parasites, having a life cycle which includes a free active stage. Cf MONOPHASIC.

diphenyl (*Chem*) *Phenylbenzene*. $C_6H_5C_6H_5$. Colourless; mp 71°C; bp 254–255°C; soluble in alcohol and ether. It occurs in coaltar, and is prepared by heating iodobenzene to 220°C with finely divided copper.

diphenyl ether (*Chem*) *Diphenyl oxide*. $C_6H_5OC_6H_5$. A liquid of pleasant odour, mp 28°C, bp 253°C, obtained from phenol by heating with $ZnCl_2$ or $AlCl_3$.

diphenylglyoxal (*Chem*) See BENZIL.

diphenylguanidine (*Chem*) Crystalline solid. Mp 147°C. Used as an accelerator in vulcanization of rubber.

diphenylmethane (*Chem*) $(C_6H_5)_2CH_2$. Colourless needles, mp 26°C, bp 262°C, obtained by the action of chloromethyl benzene on benzene in the presence of aluminium chloride.

diphenyl-methane diisocyanate (*Chem*) A type of aromatic isocyanate monomer used to make polyurethanes. Abbrev *MDI*.

diphtheria (*Med*) Infection, usually airborne, with the bacillus *Corynebacterium diphtheriae*. Bacilli are confined to the throat, producing local necrosis ('pseudo-membrane'), but a powerful EXOTOXIN causes damage esp to heart and nerves.

diphtheria toxin (*BioSci*) An AB TOXIN, made by *Corynebacterium diphtheriae*, that is responsible for the damage caused by clinical infection. The toxin can be neutralized by antitoxin, which is used to treat severe infections in unimmunized persons.

diphtheria toxoid (*Med*) Diphtheria toxin treated with formaldehyde so as to destroy toxicity without altering its capacity to act as an antigen. Used for active immunization against diphtheria. It is usually used after adsorption onto aluminium hydroxide, which acts as an ADJUVANT, and in combination with tetanus toxoid and often with *Bordetella pertussis* vaccine.

diphycercal (*BioSci*) Said of a type of tail-fin (found in lung-fish, adult lampreys, the young of all fish, and many aquatic Urodela) in which the vertebral column runs horizontally, the fin being equally developed above and below it.

diphygenic (*BioSci*) Having two different modes of development.

diphyletic (*BioSci*) Of dual origin: descended from two distant ancestral groups.

diphyodont (*BioSci*) Having two sets of teeth: a deciduous or milk dentition and a permanent dentition, as in mammals.

dipl- (*Genrl*) Prefix from Gk *diploos* signifying double.

diplegia (*Med*) Bilateral paralysis of like parts of the body.

diplex (*ICT*) The simultaneous transmission or reception of two signals using a specified common feature, eg a common carrier or antenna. Cf DUPLEX, MULTIPLEX.

diplexer (*ICT*) A means of coupling that permits two transmitters or receivers to operate with one aerial.

diplobiont (*BioSci*) A plant that alternates between two distinct free-living forms, one haploid (the gametophyte) and one diploid (the sporophyte). Adj *diplobiontic*.

diploblastic (*BioSci*) Having two primary germinal layers, namely ectoderm and endoderm.

diplococcus (*BioSci*) A coccus that divides by fission in one plane, the two individuals formed remaining paired.

diplogangliate (*BioSci*) Having paired ganglia. Also *diplo-ganglionate*.

diplohaplont (*BioSci*) Organism in which there is an *alternation* of haploid and diploid *generations*. Cf DIPLONT, HAPLONT.

diploid (*BioSci*) Possessing two sets of chromosomes, one set coming from each parent. Most organisms are diploid. Cf HAPLOID.

diploidization (*BioSci*) The fusion within the vegetative mycelium of two haploid nuclei to give a diploid nucleus in some fungi.

diplonema (*BioSci*) A stage in the meiotic division (DIPLOTENE stage) at which the chromosomes are clearly double.

diplont (*BioSci*) Organisms in which only the zygote is diploid, meiosis occurring at its germination and the vegetative cells being haploid. Cf DIPLOHAPLONT, HAPLONT.

diplophase (*BioSci*) The period in the life cycle of any organism when the nuclei are diploid. Cf HAPLOPHASE.

diplopia (*Med*) Double vision of objects.

Diplopoda (*BioSci*) A class of Arthropoda having the trunk composed of numerous segments, most formed by fusion of single segments and having two pairs of legs; the head bears a pair of uniflagellate antennae, a pair of mandibles, and a flattened plate as underlip derived from a pair of partially fused maxillae; the genital opening is in the third segment behind the head; vegetarian animals of retiring habits. Millipedes.

diplospondyly (*BioSci*) The condition of having two vertebral centra, or a centrum and an inter-centrum, corresponding to a single myotome. Adjs *diplospondylic, diplospondylous*.

diplostemonous (*BioSci*) Having twice as many stamens as there are petals, with the stamens in two whorls, the members of the outer whorl alternating with the petals.

diplotene (*BioSci*) The fourth stage of meiotic prophase, intervening between pachytene and diakinesis, in which homologous chromosomes come together and there is condensation into tetrads. See fig. at MEIOSIS.

diplozoic (*BioSci*) Bilaterally symmetrical.

dip needle (*Phys, Surv*) Magnetic needle on a horizontal pivot, which swings in vertical plane. When set in magnetic north–south plane, its inclination shows the magnetic DIP. Also *inclinometer*.

Dipnoi (*BioSci*) An order of Sarcopterygii, in which the air bladder is adapted to function as a lung, and the dentition consists of large crushing plates. Also *Dipneusti*. Lung-fish.

dipolar molecule (*Chem*) One which has a permanent moment due to the permanent separation of the effective centres of the positive and negative charges.

dipole (*Acous*) Radiator producing a sound field of two adjacent MONOPOLES in antiphase. A localized fluctuating force is the prototype of a dipole. The directivity of a dipole has the shape of an eight.

dipole (*Phys*) (1) Equal and opposite charges separated by a close distance (*electric dipole*). (2) A bar magnet or a coil carrying a steady current (*magnetic dipole*).

dipole antenna (*ICT*) Wire or rod antenna, half a wavelength long, and split at the centre for connection

with a transmission line. Maximum radiation is at right angles to the axis, and dipoles also have maximum performance with waves polarized in the same plane as the axis. Also *doublet, half-wave dipole*. See FOLDED DIPOLE.

dipole molecule (*Phys*) A molecule which has a permanent moment due to the permanent separation of the effective centres of the positive and negative charges.

dipole moment (*Phys*) See ELECTRIC DIPOLE MOMENT, MAGNETIC DIPOLE MOMENT.

dipping (*Agri*) Total immersion of livestock in liquid pesticide to control pests and disease.

dipping (*Build*) An industrial process of coating articles by temporary immersion in paint.

dipping (*Eng*) The immersion of pieces of material in a liquid bath for surface treatment such as pickling or galvanizing.

dipping needle (*Surv*) See DIP NEEDLE.

dipping refractometer (*Chem*) A type of refractometer which is dipped into the liquid under examination.

diprotodont (*BioSci*) Having the first pair of upper and lower incisor teeth large and adapted for cutting, the remaining incisor teeth being reduced or absent. Cf POLYPROTODONT.

dip slope (*Geol*) A landform developed in regions of gently inclined strata, particularly where hard and soft strata are interbedded. A long gently sloping surface which coincides with the inclination of the strata below ground. See CUESTA.

dip soldering (*Eng*) The method of soldering previously fluxed components by immersing them in a bath of molten solder. Ideal for bulky assemblies with complicated or multiple joints, and for fast automatic operation.

dipsomania (*Med*) The condition in which there is a recurring, temporary and uncontrollable impulse to drink excessively.

DIP switch (*ICT*) A miniature switch in a bank of switches often directly mounted on a printed circuit board. Often used to set the CONFIGURATION of a device, eg a printer or DISPLAY ADAPTER.

Diptera (*BioSci*) An order of insects comprising those that have one pair of wings, the hinder pair being represented by a pair of club-shaped balancing organs or halteres. The mouthparts are suctorial; the larva is legless and sluggish. Flies, gnats and midges.

dip treatment (*Agri*) Brief immersion of all or part of a plant transplant in a required agrochemical before placing it in the field.

dipygus (*Med*) A developmental monstrosity in which a fetus has a double pelvis.

dipyre (*Min*) A member of the SCAPOLITE series, containing 20–50% of the meionite molecule.

dipyridamole (*Pharmacol*) A coronary vasodilator and anti-platelet drug used to treat angina pectoris and prevent blood clotting.

Dirac equation (*Phys*) The basic equation of relativistic quantum mechanics; stated by P Dirac in 1928. It expresses the behaviour of electron waves in a way consistent with special relativity, requiring that electrons have spin $\frac{1}{2}\hbar$. where $\hbar$ is DIRAC'S CONSTANT.

Dirac's constant (*Phys*) PLANCK'S CONSTANT (h) divided by 2π. Usually termed *h-bar*, and written as $\hbar$. See PLANCK'S LAW.

Dirac's theory (*Phys*) Theory using the same postulates as the SCHRÖDINGER EQUATION, plus the requirement that quantum mechanics conforms with the theory of relativity; concluding that an electron must have an inherent angular momentum and magnetic moment. See DIRAC EQUATION.

direct-acting pump (*Eng*) A steam-driven reciprocating pump in which the steam and water pistons are carried on opposite ends of a common rod.

direct-arc furnace (*ElecEng*) An electric-arc furnace in which the arc is drawn between an electrode and the charge in the furnace.

direct broadcast satellite (*ICT*) High power GEOSTA-TIONARY communications satellite, usually having a specially designed antenna so that the FOOTPRINT coincides with the region of the Earth's surface to which TV programmes are to be beamed for direct reception by the viewer, rather than redistribution by cable or other means.

direct capacitance (*Phys*) The capacitance between two conductors, as if no other conductors are present.

direct chill casting (*Eng*) A method like CONTINUOUS CASTING but for larger cross-sections in which the hollow mould is closed at the bottom by a platform. This is gradually lowered as the metal becomes solid on the outside and therefore able to contain the melt, platform, mould and metal being appropriately cooled.

direct-conversion reactor (*ElecEng*) One which converts thermal energy directly into electricity by means of thermoelectric elements (usually of silicon–germanium).

direct cooling (*ElecEng*) The cooling of transformer and machine windings by circulating the coolant through hollow conductors.

direct-coupled exciter (*ElecEng*) An exciter for a synchronous or other electric machine, which is mounted on the same shaft as the machine that it is exciting.

direct-coupled generator (*ElecEng*) A generator which is mechanically coupled to the machine which is driving it, ie not driven through gearing, a belt, etc.

direct coupling (*Electronics*) Interstage coupling without the use of transformers or series capacitors, so that the dc component of the signal is retained. Also *dc coupling*.

direct current (*ElecEng*) Current which flows in one direction only, although it may have appreciable pulsations in its magnitude. Abbrevs *dc, d.c.*

direct-current amplifier (*ElecEng*) See DC AMPLIFIER.

direct-current balancer (*ElecEng*) See DC BALANCER.

direct cycle (*NucEng*) The type of nuclear reactor in which the coolant is allowed to boil and pass directly to the turbines, as in the BOILING-WATER REACTOR.

direct data entry (*ICT*) Input of data directly to the computer using, normally, a KEY-TO-DISK UNIT. The data may be validated while held in a temporary file, before being written to the disk for subsequent processing.

direct-fired (*Eng*) Furnace in which the fuel is delivered into the heating chambers.

directing stimulus (*Psych*) Stimulus which, though not releasing a component of species-specific behaviour, is important in determining the direction of a response.

direct injection (*Aero*) The injection of metered fuel (for a spark-ignition engine) into the supercharger eye of the cylinders, which eliminates the freezing and poor high-altitude behaviour of carburettors.

direct-injection pump (*Aero*) A fuel-metering pump for injecting fuel direct to the individual cylinders.

direct interaction (*Phys*) A mechanism for describing how a nuclear reaction takes place. It assumes that the interaction between bombarding nucleus and target nucleus involves only a few nucleons near the surface of the nuclei. Cf COMPOUND NUCLEUS.

direct inward system access (*ICT*) Access to a VIRTUAL PRIVATE NETWORK from a point outside it, via a PERSONAL IDENTIFICATION NUMBER that gives secure access to the network's facilities.

direction (*MathSci*) The orientation of a straight line in space, sometimes contrasted with SENSE.

directional antenna (*Space, ICT*) Antenna in which the transmitting and/or receiving properties are concentrated along certain directions, used in space over very long distances.

directional circuit breaker (*ElecEng*) A circuit breaker which operates when the current flowing through it is in the direction opposite to normal.

directional coupler (*ICT*) In a transmission line or a waveguide, a device that couples a secondary transmission path to a wave travelling in only one direction on the main path; there is no energy transfer for propagation in the

other direction. The amount of energy coupled is usually only a small proportion, possibly 10–20 dB less than that in the main beam.

directional derivative (*MathSci*) The rate of change of a function with respect to arc length along a given curve (or in a given direction).

directional drilling (*MinExt*) The use of special down-hole drilling assemblies to turn a drill hole in the desired direction.

directional effects (*ImageTech*) Defects of non-conformity in a processed image caused by the action of depleted developer solution where there has been inadequate agitation in a continuous processing machine.

directional filter (*ICT*) A combination of filters, eg a high- and a low-pass filter or two different band-pass filters, to separate the bands of frequencies used for transmission in opposite directions in a CARRIER SYSTEM.

directional gain (*ICT*) The ratio, expressed in decibels, of the response, generally along the axis where it is a maximum, to the mean spherical (or hemispherical with reflector or baffle) response, of an antenna, loudspeaker or microphone. Also *directivity index*.

directional loudspeaker (*Acous*) A loudspeaker which radiates more strongly in one direction than in others. Normally the radiated sound power is directed in a beam. Often a combination of loudspeakers (array) is directional.

directional microphone (*Acous*) Microphone which is directional in response. See ACOUSTIC TELESCOPE.

directional receiver (*ICT*) Receiving system using a directional antenna for discrimination against noise and other transmissions.

directional relay (*ElecEng*) A relay whose operation depends on the direction of the current flowing through it.

directional transmitter (*ICT*) Transmitting system using a directional antenna, to minimize power requirements and to diminish effect of interference.

direction angles (*MathSci*) The three angles which a line in three-dimensional space makes with the positive directions of the co-ordinate axes.

direction components (*MathSci*) See DIRECTION NUMBERS.

direction cosines (*MathSci*) The cosines of the three direction angles of a line.

direction-finding (*ICT*) Using a direction-finder. The principle and practice of determining a bearing by radio means, using a discriminating antenna system and a radio receiver, so that the direction or bearing of a distant transmitter can be determined.

direction numbers (*MathSci*) Of a line, any three numbers, not all of which are zero, which are proportional to the direction cosines of the line. Also *direction components*, *direction ratios*.

direction of a curve (*MathSci*) The direction of the tangent to a curve at the point.

direction of younging (*Geol*) See YOUNGING, DIRECTION OF.

direction ratios (*MathSci*) See DIRECTION NUMBERS.

directive efficiency (*ICT*) The ratio of maximum to average radiation or response of an antenna; the GAIN, in dB, of antenna over a dipole being fed with the same power.

directive force (*ElecEng*) A term used to denote the couple which causes a pivoted magnetic needle to turn into a north and south direction.

directive gain (*ICT*) For a given direction, 4π times the ratio of the radiation intensity in that direction to the total power that is radiated by the aerial.

directivity (*Acous, ICT*) Measurement, in decibels, of the extent to which a directional loudspeaker, microphone or antenna concentrates its radiation or response in specified directions.

directivity angle (*ICT*) Angle of elevation of direction of maximum radiation or reception of electromagnetic wave by an antenna.

directivity factor (*Acous*) Non-dimensional quantity for loudspeakers and microphones which characterizes the strength of the directivity.

directivity index (*ICT*) See DIRECTIONAL GAIN.

direct labour (*Build*) A mode whereby labour is employed directly by the client, as opposed to the usual method of working through independent architect, engineer and surveyor.

direct laying (*Eng*) Cables are laid in a trench and covered with soil and were formerly covered with tiles etc for protection but nowadays a wide brightly coloured plastic tape is often used as a warning only.

direct lighting (*ElecEng*) A system of lighting in which not less than 90% of the total light emitted is directed downwards, ie in the lower hemisphere.

direct lithography (*Print*) Lithographic printing whereby the plate prints directly on the paper, without first offsetting onto a blanket cylinder.

directly heated cathode (*Electronics*) Metallic (coated) wire heated to a temperature such that electrons are freely emitted. Also *filament cathode*.

direct manipulation (*ICT*) An approach to the computer-aided restructuring of data by pointing to and moving the data rather than by entering co-ordinates or descriptions. See MOUSE.

direct memory access (*ICT*) A method for transferring data directly to and from system MEMORY, bypassing the PROCESSOR.

direct metamorphosis (*BioSci*) The incomplete metamorphosis undergone by exopterygote insects, in which a pupal stage is wanting.

direct motion (*Astron*) (1) The apparent eastward motion of a planet viewed from the Earth. Cf RETROGRADE MOTION. (2) The anticlockwise orbital motion of a heavenly body observed from celestial north (ie from west to east). (3) The anticlockwise rotation of a planet viewed from its north pole.

director (*ICT*) Free resonant dipole element in front of antenna array that assists the directivity of the array in the same direction. See YAGI ANTENNA.

director (*Med*) A grooved instrument for guiding a surgical knife.

director circle (*MathSci*) See ORTHOPTIC CIRCLE.

directory (*ICT*) A list of file names, together with information enabling the files to be retrieved from backing store by the operating system. See fig. at HARD DISK. See DATA DICTIONARY.

direct printing (*Print*) A method in which the print is made directly on the paper as in letterpress or photogravure, as distinct from the usual lithographic method of OFFSET PRINTING.

direct process (*Eng*) Now obsolete method for obtaining from ore an iron similar to WROUGHT-IRON without first making pig iron.

direct radiation (*Phys*) See PRIMARY RADIATION.

direct ray (*ICT*) See DIRECT WAVE.

direct-reading instrument (*ElecEng*) An instrument in which the scale is calibrated in the actual quantity measured by the instrument, and which therefore does not require the use of a multiplying constant.

direct-recorded disk (*Acous*) A record produced directly from received signals without subsequent processing.

directrix (*MathSci*) (1) Of a conic or quadric, the POLAR LINE or plane of a focus. See CONIC for alternative definition. (2) A curve of a ruled surface in general through which the generators pass.

direct rope haulage (*MinExt*) Engine plane. An ascending truck is partly balanced by a descending one, motive power being applied to the drum round which the haulage rope passes.

direct sound (*Acous*) The sound intensity arising from a source to a listener, as contrasted with the reverberant sound which has experienced reflections between the source and the listener.

direct stress (*Eng*) The stress produced at a section of a body by a load whose resultant passes through the centre of gravity of the section.

direct stroke (*ElecEng*) When a transmission line or other apparatus is struck by lightning, it is said to receive a *direct stroke*.

direct-suspension construction (*ElecEng*) A form of construction used for the overhead contact wire on electric traction systems; the contact wire is connected directly to the supports without catenary or messenger wires.

direct-switching starter (*ElecEng*) An electric motor starter arranged to switch the motor directly across the supply, without the insertion of any resistance or the performing of any other current-limiting operation.

direct termination overflow (*ICT*) A feature of a VIRTUAL PRIVATE NETWORK whereby a call is routed via the ordinary public network if all dedicated capacity is busy.

direct-trip (*ElecEng*) A term used in connection with circuit breakers, starters, or other similar devices, to indicate that the current which flows in the tripping coil is the main current in the circuit, not an auxiliary current obtained from a battery or other source.

direct vernier (*Surv*) A vernier in which n divisions on the vernier plate correspond in length to $(n-1)$ divisions on the main scale.

direct-vision prism (*Phys*) A compound prism with component prisms of two glasses having different dispersive powers and cemented together so that, in passing through the combination, light suffers dispersion but no deviation.

direct-vision spectroscope (*Phys*) A spectroscope employing a DIRECT-VISION PRISM. Such an instrument is usually in the form of a short straight tube with a slit at one end and an eyepiece at the other; it is used for rough qualitative examination of spectra.

direct vision viewfinder (*ImageTech*) A viewfinder in which the subject is viewed directly, not by reflection.

direct voice input (*Aero*) A means by which a pilot can command an aircraft to respond to spoken instructions for such functions as change of radio frequency, flight performance and possibly weapon aiming and delivery.

direct wave (*ICT*) That portion of the power radiated from an antenna that goes directly to the receiving antenna, without ionospheric reflection. Also *direct ray, ground wave*.

dirigible (*Aero*) A navigable balloon or airship.

dirt (*MinExt*) Broken valueless mineral. Also *gangue*. US *muck*.

DIR technology (*ImageTech*) Abbrev for *developer inhibitor release*. Couplers whose inclusion in photographic emulsions increase BORDER EFFECTS during development, giving improved image sharpness.

dis (*ElecEng*) See DISCONTINUITY.

disability-adjusted life year (*Genrl*) A unit used to assess the loss of economic output in a society due to illness, morbidity, etc. Abbrev *DALY*.

disaccharides (*Chem*) A group of carbohydrates considered to be derived from two molecules of a monosaccharide (either the same or different) by elimination of one molecule of H_2O; eg maltose is $G\alpha-1\rightarrow4G$ (G = glucose).

disadvantage factor (*NucEng*) The ratio of average neutron flux in the reactor lattice to that within the actual fuel element.

disappearing-filament pyrometer (*Eng*) An instrument used for estimating the temperature of a furnace by observing a glowing electric-lamp filament against an image of the interior of the furnace formed in a small telescope. The current in the filament is varied until it is no longer visible against the glowing background. From a previous calibration the required temperature is derived from the current value.

disarticulation (*Med*) Amputation of a bone through a joint.

disassembler (*ICT*) A program that translates from machine code to an assembly language, generally used to decipher existing machine code by generating the equivalent symbolic codes.

disazo dyes (*Chem*) Dyestuffs containing two azo groups of the type $C_6H_5N=NC_6H_4N=NC_6H_4OH$. These dyes are obtained by diazotizing an amino derivative of azobenzene and then coupling it with a tertiary amine or with a phenol, or by coupling a diamine or dihydric phenol with two molecules of a diazonium salt.

disbudding (*Agri*) Removal of small horns, usually by caustic chemicals or thermo-cautery, at the very earliest stage of their development, when an animal is young.

disc (*Genrl*) See MAGNETIC DISK, VIDEO DISK.

discal (*BioSci*) Pertaining to or resembling a disk or disk-like structure; a wing cell of various insects.

discard (*Eng*) A portion of material which has to be rejected by virtue of the nature of a working process, eg in direct extrusion the bulk of the billet can be forced through the die orifice but a small fraction always remains inside the die chamber owing to frictional effects; this portion is separated from the extruded product and becomes the discard.

discard eligibility indicator (*ICT*) A designated FRAME RELAY bit that can be used to indicate that during congestion this frame should be among those discarded first. It can eg be set by the network entry node when too much load is offered or by the source user equipment to discriminate between data and control messages.

discharge (*NucEng*) Unloading of fuel from a reactor.

discharge (*Phys*) (1) The abstraction of energy from a cell by allowing current to flow through a load. (2) Reduction of the potential difference at the terminals (plates) of a capacitor to zero. (3) Flow of electric charge through gas or air due to ionization, eg lightning, or at reduced pressure, as in fluorescent tubes. See FIELD DISCHARGE.

discharge bridge (*ElecEng*) The measurement of the ionization or discharge, in dielectrics or cables, depending on the amplification of the high-frequency components of the discharge.

discharge circuit (*ElecEng*) One arranged to discharge a capacitor or parasitic capacitance, either for circuit operational reasons or for safety.

discharge electrode (*ElecEng*) See ACTIVE ELECTRODE.

discharge head (*MinExt*) Vertical distance between intake and delivery of pump, *plus* allowance for mechanical friction and other retarding resistances requiring provision of extra power.

discharge lamp (*ElecEng*) One in which luminous output arises from ionization in gaseous discharge.

discharge machine (*NucEng*) See CHARGE–DISCHARGE MACHINE.

discharge printing (*Textiles*) The removal of a dye from a fabric to leave a white pattern.

discharger (*ElecEng*) (1) A device, such as a spark gap, which provides a path whereby a piece of electrical apparatus may be discharged. (2) An apparatus containing an electrically heated wire for firing explosives in blasting.

discharge rate (*ElecEng*) A term used in connection with the discharge of accumulators. An accumulator has a certain capacity at, eg a 1 h discharge rate, when that capacity can be obtained if the accumulator is completely discharged in 1 h. If the discharge rate is lower, ie the discharge takes more than 1 h, the capacity obtainable will be higher.

discharge resistance (*ElecEng*) (1) A non-inductive resistance placed in parallel with a circuit of high inductance (eg the field winding of an electric machine) in order to prevent a high voltage appearing across the circuit when the current is switched off. Also *buffer resistance*. (2) Resistance placed in parallel with a capacitance or circuit with parasitic capacitance to provide a discharge path for stored charge, for reasons of circuit operation or for safety.

discharge tube (*Electronics*) Any device in which conduction arises from ionization, initiated by electrons of sufficient energy.

discharge valve (*Eng*) A valve for controlling the rate of discharge of fluid from a pipe or centrifugal pump.

discharging arch (*Build*) An arch built in a wall to protect a space beneath from the weight above, and to allow access or discharge.

discharging tongs (*ElecEng*) A pair of metal tongs used for discharging capacitors before they are touched by hand.

discission (*Med*) An incision into a part, esp needling of a cataract.

disclimax (*BioSci*) A stable community which is not the climatic or edaphic community for a particular place, but is maintained by humans or domestic animals, eg a desert produced by overgrazing, where the natural climax would be grassland. The name derives from *disturbance climax*.

discodermolide (*Pharmacol*) A drug derived from the marine sponge, *Discodermia dissoluta*, used in treating cancer.

Discolichenes (*BioSci*) A group of lichens in which the fungus is a Discomycete.

discomposition effect (*Phys*) See WIGNER EFFECT.

Discomycetes (*BioSci*) A class of fungi in the Ascomycotina in which the fruiting body (ascocarp) is usually an APOTHECIUM. Includes the Lecanorales (lichen-forming fungi), and many saprophytic and mycorrhizal sorts, eg the morels and the truffles.

discone antenna (*ICT*) A biconical antenna, used for short-wave and very-high-frequency communication, having one cone flattened out to form a disk. The transmission line is connected between the centre of the disk and the apex of the cone. Its input impedance and radiation pattern remains constant over a wide frequency range and it gives an omnidirectional pattern in a horizontal plane when the axis of the cone is vertical.

disconformity (*Geol*) A break in the rock sequence in which there is no angular discordance of dip between the two sets of strata involved. Cf UNCONFORMITY.

disconnected set (*MathSci*) One which can be divided into two sets having no points in common and neither containing an ACCUMULATION POINT of the other.

disconnection (*ElecEng*) See DISCONTINUITY.

disconnector (*Build*) See INTERCEPTOR.

discontinued construction (*Build*) See ACOUSTIC CONSTRUCTION.

discontinuity (*ElecEng*) A break, whether intentional or accidental, in the conductivity of an electrical circuit. Colloq *dis*. Also *disconnection*.

discontinuity (*MathSci*) A point at which a function is not CONTINUOUS.

discontinuous distribution (*BioSci*) Isolated distribution of a species, as the tapir, which is found in the Malay Peninsula and Sumatra, and again in Cl and S America.

discontinuous variation (*BioSci*) A sudden change in otherwise smoothly varying characters in a group of organisms over eg a geographical range.

discordant intrusion (*Geol*) An igneous intrusion that cuts across the bedding or foliation of the country rock it intrudes. See DYKE.

discounted cash flow (*Genrl*) An accounting technique used to estimate the relative cost of various schemes. The discounted cash flow takes account of the present value of future expenditures, on the assumption that the money will increase if invested elsewhere, in order to help evaluate the cost of doing things sooner rather than later.

discovery well (*MinExt*) The first well to reveal oil in a new field or at a new level.

discrete cosine transform (*ICT*) A digital video data compression technique that divides the picture into blocks of PIXELS and then reduces the content of each block to a number of coefficients representing the amplitudes of two orthogonal sets of spatial cosine waves. Coefficients representing insignificant image content are discarded. Because of its ease of computation, it has become the most widely used video compression method. Abbrev *DCT*.

discrete mathematics (*MathSci*) The part of mathematics devoted to the study of relationships between sets of discrete objects. Includes graph theory, logic, number theory and abstract algebra.

discretionary hyphen (*Print*) A hyphen inserted by the keyboard operator. If the word needs hyphenation it will override the hyphenation logic of the phototypesetter.

discriminant (*MathSci*) (1) Of a polynomial equation

$$x^n + a_1 x^{n-1} + \ldots + a_n = 0$$

the product of the squares of all the differences of the roots taken in pairs. (2) Of a differential equation, the result of eliminating

$$p \left(= \frac{dy}{dx} \right)$$

between the differential equation $F(x, y, p) = 0$ and

$$\frac{\delta Y}{\delta c}(x, y, c) = 0$$

is the p-discriminant equation, which represents the p-discriminant locus. For a quadratic equation $ax^2 + bx + c = 0$, the discriminant is $b^2 - 4ac$. When a, b, c are all real, the roots are equal if $b^2 - 4ac = 0$, real if $b^2 - 4ac > 0$ and imaginary if $b^2 - 4ac < 0$ (a, b, c being real). If the solution of the differential equation is $u(x, y, c) = 0$ (c an arbitrary constant), the result of eliminating c between $u(x, y, c) = 0$ and

$$\frac{\delta}{\delta p} F(x, y, p) = 0$$

gives the c-discriminant equation.

discriminant analysis (*MathSci*) A method of assigning observations to groups on the basis of values of observations previously obtained from each group.

discriminating circuit breakers (*ElecEng*) A term sometimes used to denote circuit breakers which operate only when the current is in a given direction.

discriminating protective system (*ElecEng*) An excess-current protective system which disconnects only that portion of a power system in which the fault occurred.

discriminating satellite exchange (*ICT*) A small automatic exchange that can decide, without engaging its main exchange, whether or not it can complete a call arising from one of its subscribers.

discrimination (*ICT*) The selection of a signal having a particular characteristic, eg frequency, amplitude, etc, by the elimination of all the other input signals at the discriminator.

discrimination (*Psych*) (1) The capacity to differentiate between two stimuli. (2) In animal behaviour, the ability to respond to different patterns of stimulation, often tested for by using a conditioning procedure (see DISCRIMINATION TRAINING). (3) In social psychology, a term denoting behaviour towards people or groups of people based on their membership of a particular group, eg gender, race.

discrimination training (*Psych*) Learning to respond to certain stimuli that are reinforced, and not to others that are not reinforced.

discriminator (*ICT*) (1) Circuit that rejects pulses below a certain amplitude level, and shapes the remainder to standard amplitude and profile. (2) Circuit used in the DEMODULATION of a frequency-modulated carrier to convert variations in frequency into variations in amplitude.

discs (*Agri*) Disk-shaped ploughshares used on larger ploughs as an alternative to mouldboard-mounted shares.

discus proligerus (*BioSci*) See ZONA GRANULOSA.

dish (*ImageTech, ICT*) Colloq term for parabolic reflector, which may be made of sheet metal or mesh; a form of microwave antenna used for point-to-point and satellite communication and for radio astronomy and satellite broadcast reception.

disharmony (*BioSci*) See HYPERTELY.

dished (*Eng*) Of wheels, esp steering wheels, having the hub inset on a different plane from the rim.

dishing (*NucEng*) Placing depressions at the ends of cylindrical fuel pellets to allow for expansion after irradiation.

dish wheel (*Eng*) See CUP WHEEL.

disincrustant (*Eng*) See ANTI-INCRUSTATOR.

disinfectant (*Chem*) Any preparation that destroys the causes of infection. The most powerful disinfectants are oxidizing agents and chlorinated phenols.

disinfection (*Med*) The destruction of pathogenic bacteria, usually with an antiseptic chemical or DISINFECTANT.

disinfestation (*Med*) The destruction of insects, esp lice.

disinhibition (*Psych*) Loss of an inhibition due to an external factor, either the EXTINCTION of a conditioned reflex or the lowering of social inhibitions as eg a result of drug or alcohol use.

disintegrating mill (*MinExt*) A mill for reducing lump material to a granular product. It consists of fixed and rotating bars in close proximity, crushing being partly by direct impact and partly by interparticulate attrition. See BEATER MILL.

disintegration (*Phys*) A process in which a nucleus ejects one or more particles, esp spontaneous radioactive decay.

disintegration constant (*Phys*) A measure of the probability of radioactive decay of a given unstable nucleus per unit time. Statistically, it is the constant λ, expressing the exponential decay $\exp(-\lambda t)$ of activity of a quantity of this isotope with time. It is also the reciprocal of the mean life of an unstable nucleus. Also *decay constant, transformation constant*. See DECAY LAW.

disintegration energy (*Phys*) See ALPHA DECAY ENERGY.

disintegration of filament (*ElecEng*) The gradual destruction of the filament of a lamp due to the ejection of particles from the filament which adhere to the inner surface of the bulb, causing blackening.

disintegrin (*BioSci*) Any of a group of proteins in snake venom from the viper family that bind to INTEGRINS, thereby blocking platelet aggregation and hence blood-clotting.

disjunct (*BioSci*) Generally, interrupted, disconnected, not continuous. More specifically, having deep constrictions between the different tagmata of the body. See TAGMOSIS.

disjunction (*BioSci*) The separation during meiotic ANAPHASE of the two members of each pair of homologous chromosomes.

disjunctor (*BioSci*) A portion of wall material forming a link between the successive conidia in a chain, and serving as a weak plane where separation may occur.

disk (*BioSci*) (1) Any flattened circular structure. (2) A fleshy outgrowth from the receptacle of a flower, surrounding or surmounting the ovary and often secreting nectar. (3) The central part of a capitulum.

disk (*ICT*) See MAGNETIC DISK, VIDEO DISK.

disk-and-drum turbine (*Eng*) A steam turbine comprising a high-pressure impulse wheel, followed by intermediate and low-pressure reaction blading, mounted on a drum-shaped rotor. Also *combination turbine, impulse-reaction turbine*.

disk area (*Aero*) The area of the circle described by the tips of the blades of a rotorcraft; similarly applied to propellers.

disk armature (*ElecEng*) One for a motor or generator wound to a large diameter on a short axle length.

disk brakes (*Aero, Autos*) A type in which two or more pads close by caliper action onto a disk which is connected rigidly to the landing or car wheel hub; more efficient than drum type, owing to greater heat dissipation.

disk camera (*ImageTech*) A camera in which the images are recorded on a small disk of photographic film, which is rotated through an appropriate angle after each exposure to present a new area.

disk capacitor (*ElecEng*) One in which capacitance is altered by the relative axial motion of disks.

disk centrifuge (*PowderTech*) Apparatus for particle size analysis in which particles are sedimented in a rotating disk.

disk clutch (*Eng*) A friction clutch in which the driving and driven members have flat circular or annular friction surfaces, and consist of one or a number of disks, running either dry or lubricated. Also *plate clutch*. See SINGLE-PLATE CLUTCH, MULTIPLE-DISK CLUTCH.

disk drive (*ICT*) Mechanism that causes magnetic disks to rotate between read/write heads.

diskette (*ICT*) See FLOPPY DISK.

Disk Filing System (*ICT*) Name given to the operating system in the original BBC microcomputer. Abbrev *DFS*.

disk filter (*MinExt*) Continuous heavy-duty vacuum filter in which separating membranes are disks, each revolving slowly through its separate compartment. Also *American filter*.

disk floret (*BioSci*) (1) Usually in the Compositae, one of the florets occupying the central part of the capitulum, whatever its morphology. (2) Sometimes, a tubular floret. Cf RAY FLORET.

disk formatting (*ICT*) The preparation of blank magnetic disk or tape for subsequent writing or reading by adding control information such as track and sector number. See HARD-SECTORED FORMATTING, SOFT-SECTORED FORMATTING.

disk friction (*Eng*) The force resisting the rotation of a disk in a fluid. It is important in the design of centrifugal machinery as it decreases efficiency and causes a rise in the pressure of the fluid being pumped.

disk loading (*Aero*) The lift, or upward thrust, of a rotor divided by the disk area.

disk operating system (*ICT*) See DOS.

disk pack (*ICT*) A set of magnetic disks fitted on a common spindle; each disk has its own set of read/write heads.

disk record (*Acous*) The type of vinyl record, originally made of synthetic thermoplastic resin, and in which the reproducing needle follows a spiral groove while the record is rotated at constant speed; patented by Emile Berliner in 1887.

disk recorder (*ImageTech*) A MAGNETIC DISK recorder used for video editing and INSTANT REPLAY.

disk ruling (*Print*) A method by which ink is applied to the paper by disks instead of pens, permitting higher speeds on the ruling machine.

disk-seal tube (*Electronics*) A valve constructed from metallic disks (which may be the electrodes) sandwiched with glass or ceramic insulating pillars. Characteristics include low interelectrode capacitance, reduced lead inductances and high temperature ratings. Used in high-power applications at frequencies up to 2 GHz.

disk valve (*Eng*) A form of suction and delivery valve used in pumps and compressors; it consists of a light-steel or fabric disk resting on a ported flat seating; steel-valve disks are usually spring-loaded.

disk winding (*ElecEng*) A type of winding for medium and large transformers, in which the turns are made up into a number of annular disks.

dislocation (*Crystal*) A lattice imperfection in a crystal structure which exerts a profound effect on structure-sensitive properties such as strength, hardness, ductility and toughness. Has a configuration of an extra half-plane of atoms inserted in the crystal stacking and the associated structural displacements near the end of the half-plane result in atomic movement, eg slip, at much lower applied stresses than would occur in a perfect crystal. Plays a fundamental role in accounting for deformation and strengthening phenomena in metals. In an annealed crystal the density of dislocation lines is of the order of 10^9 per square mm, which rises to 10^{13} when the material is heavily cold-worked, owing to interactions during deformation. There are two types, edge and screw, both of which are characterized by a BURGERS VECTOR which represents the amount and direction of slip when the dislocation moves. See DEFORMATION MAP and panel on CRYSTAL LATTICE.

dislocation (*Med*) The displacement of one part from another, esp the abnormal separation of two bones at a joint.

dislocation glide (*Crystal*) The movement of dislocations along slip planes during the process of deformation.

disomic (*BioSci*) Relating to two homologous chromosomes or genes.

disorientation (*Psych*) Inability to recognize or be aware of spatial, temporal or contextual cues. Acute disorientation can be pharmacologically induced (eg by alcohol) or be due to trauma; long-term progressive disorientation is symptomatic of psychological or neurological pathology.

dispensable circuit (*ElecEng*) A circuit in a wiring system which allows apparatus to be cut out of the circuit at times of heavy load.

dispensary (*Med*) A place where drugs etc are dispensed (ie are prepared for administration).

dispenser cathode (*Electronics*) One which is not coated, but is continuously supplied with suitable emissive material from a separate cathode element.

dispermy (*BioSci*) Penetration of an ovum by two spermatozoa.

dispersal (*BioSci*) The active or passive movement of individual plants or animals or their disseminules (such as seeds, spores or larvae) into or out of a population or population area. It includes emigration, immigration and migration. Should not be confused with DISPERSION.

dispersed phase (*Chem*) A substance in the colloidal state.

dispersion (*BioSci*) The distribution pattern in an animal or plant population, this being random, uniform (more regular than random) or clumped (see AGGREGATION). Should not be confused with DISPERSAL or with DISTRIBUTION, which refers to the species as a whole, although the dispersion of a population can be described as following a random or POISSON DISTRIBUTION.

dispersion (*Chem*) See MOLECULAR MASS DISTRIBUTION.

dispersion (*MathSci*) The extent to which observations are dissimilar in value, often measured by statistics such as standard deviation, range, etc.

dispersion (*Phys*) The dependence of wave velocity on the frequency of wave motion; a property of the medium in which the wave is propagated. In the visible region of the electromagnetic spectrum, dispersion manifests itself as the variation of refractive index of a substance with wavelength (or colour) of the light. Dispersion enables a prism to form a spectrum. See ANOMALOUS DISPERSION, CAUCHY'S DISPERSION FORMULA, HARTMANN DISPERSION FORMULA.

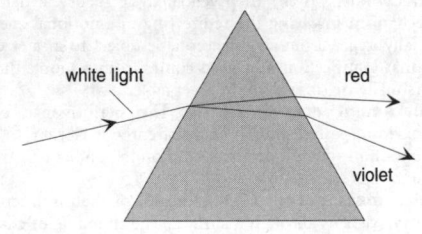

dispersion

dispersion coefficient (*ElecEng*) A term often used to denote the leakage factor of an induction motor.

dispersion curve (*Phys*) A plot of frequency against wavelength for a wave in a dispersive medium. See ACOUSTIC BRANCH, OPTIC BRANCH, PHONON DISPERSION CURVE.

dispersion forces (*Chem*) Old name for weak intermolecular forces. Also *London forces*. See VAN DER WAALS' FORCES.

dispersion hardening (*Eng*) Hardening of a material by introducing a fine dispersion of particles into a ductile matrix which increases the applied stress necessary to move DISLOCATIONS. See PRECIPITATION HARDENING.

dispersion medium (*Chem*) A substance in which another is colloidally dispersed.

dispersion-shifted fibre (*ICT*) A type of OPTICAL FIBRE in which the change in refractive index between core and cladding is graded so as to shift the zero-dispersion wavelength of 1·3 μm given by a normal stepped-index fibre to the 1·5 μm wavelength region at which efficient laser diodes operate.

dispersive medium (*Phys*) The medium in which the phase velocity of a wave is a function of frequency.

dispersive mixing (*Chem*) The process where particle clumps, eg carbon black, are broken down during mixing with polymer by high shear forces. Essential to combine with DISTRIBUTIVE MIXING. Also *intensive mixing*.

dispersive power (*Phys*) See DISPERSIVITY.

dispersivity (*Phys*) The ratio of the difference in the refractive indices (*n*) of a medium for specified wavelengths in the red (R) and violet (V) to the mean refractive index diminished by one. This may be written as follows:

$$v = \frac{n_V - n_R}{\bar{n} - 1}$$

Also *dispersive power*.

dispersivity quotient (*Phys*) The variation of refractive index *n* with wavelength λ, $dn/d\lambda$.

dispersoid (*Chem*) A particle of a second phase material distributed through a host solid by means of eg precipitation.

displaced terrane (*Geol*) Internally consistent rock masses within an orogenic area which are abruptly discontinuous with their surroundings. Also *suspect terrane*. See TERRANE.

displacement (*Aero*) The mass of the air displaced by the volume of gas in any lighter-than-air craft, or water by a seaplane hull or float.

displacement (*Build*) The weight of water displaced by a vessel. It is equal to the total weight of the vessel and contents. See ARCHIMEDES' PRINCIPLE.

displacement (*Eng*) (1) The volume of fluid displaced by a pump plunger per stroke or per unit time. (2) The swept volume of a working cylinder.

displacement (*Phys*) Vector representing the electric flux in a medium and given by D = εE, where ε is the PERMITTIVITY and E is the electric field. Also *dielectric strain*, *electric flux density*. (2) In mechanics, the vector of distance moved by a body.

displacement (*Psych*) In psychoanalytic theory, a defence mechanism involving the redirecting of emotional energy, usually anger, from an unacceptable object to a safer one, so that gratification of a need comes from a source that is personally or socially more acceptable.

displacement activity (*Psych*) The performance of a behaviour pattern which is apparently irrelevant to the situation in which it occurs; common in conflict or anxiety situations.

displacement current (*Phys*) Integral of the displacement current density through a surface. The time rate of change of the electric flux. Current postulated in a dielectric when electric stress or potential gradient is varied. Distinguished from a normal or conduction current in that it is not accompanied by motion of current carriers in the dielectric. Concept introduced by Maxwell for the completion of his electromagnetic equations.

displacement flux (*Phys*) Integral of the normal component of DISPLACEMENT over any surface in a dielectric.

displacement law (*Phys*) The formulation by F Soddy and K Fajans that radiation of an alpha particle *reduces* the atomic number by two and the mass number by four, and that radiation of a beta particle *increases* the atomic number by one, but does not change the mass number. It was later found that emission of a positron *decreases* the atomic number by one, but does not change the mass number. Gamma emission and isomeric transition change neither mass nor atomic number. Displacement laws are summarized as shown below. A change in atomic number means displacement in the periodic classification of the chemical elements; a change in mass number determines the radioactive series. Table ▷

Type of disintegration	Change in atomic number	Change in mass number
alpha emission	−2	−4
beta electron emission	+1	0
beta positron emission	−1	0
beta electron capture	−1	0
isometric transition	0	0
gamma emission	0	0

displacement pump (*MinExt*) A pump with pulsing action, produced by steam, compressed air or a plunger, causing non-return valves to prevent return flow of displaced liquid during the retracting phase of the pump cycle.

displacement series (*Chem*) See ELECTROCHEMICAL POTENTIAL SERIES.

display (*Electronics*) A device for presenting information from an electronic system. See CATHODE-RAY TUBE and panel on LIQUID CRYSTAL DISPLAY.

display adapter (*ICT*) A circuit board containing components which are used to drive the display for a computer. These are usually to a standard specification. See CGA, EGA, VGA.

display behaviour (*Psych*) Species-specific patterns of either sound or movement, often stereotyped in form, and which serve a great variety of communicative functions, eg in courtship or agonistic behaviour.

display work (*Print*) Displayed text setting (such as title pages, jobbing work, advertisements), distinguished from solid text composition.

disposable load (*Aero*) Maximum ramp weight minus *operating weight empty* (OWE); includes crew, fuel, oil and payload (civil) or armament (military).

disposal (*NucEng*) The removal of radioactive waste to a secure place without the intention of recovering it later.

disproportionation (*Chem*) (1) A reaction in which a single compound is simultaneously oxidized and reduced, eg the spontaneous reaction in water of soluble copper (I) salts to form equal amounts of copper (0) and copper (II). (2) A chain termination reaction where two active free-radical chain ends transfer electrons to form two dead chains. See CHAIN POLYMERIZATION and panel on POLYMER SYNTHESIS.

disruptive discharge (*ElecEng*) The discharge arising from the breakdown (puncture) of the dielectric of a CAPACITOR caused by an electric field which it cannot withstand.

disruptive strength (*Eng*) Obsolete term for the stress at which a material fractures under tension. See STRENGTH MEASURES.

disruptive voltage (*ElecEng*) Voltage just sufficient to puncture the dielectric of a capacitor. See BREAKDOWN VOLTAGE.

dissecting (*Print*) Removal of type matter which is to be printed in a second colour, in order to impose it in another chase, position and spacing being carefully regulated.

dissecting aneurysm (*Med*) The leaking of blood through a tear in the inner wall of the AORTA producing a cleavage in the layers of the vessel and tracking of blood in a 'false lumen' along the aorta and its vessel wall. Associated with a high mortality.

disseminated sclerosis (*Med*) See MULTIPLE SCLEROSIS.

disseminated values (*MinExt*) Mode of occurrence in which small specks of concentrate are scattered evenly through the gangue mineral.

dissemination (*BioSci*) The spread or migration of species, usually by means of seeds, spores and larvae.

disseminule (*BioSci*) A propagule.

dissimilar terms (*MathSci*) Terms containing different powers of the same variable(s), or containing different variables, eg xy, $(xy)^2$; $3x$, $3z$.

dissipation (*ICT*) Loss or diminution, usually undesirable, of power, the lost power being converted into heat. Causes power loss in transmission lines etc and can diminish cut-off sharpness in filters. In low-frequency circuits, it is due largely to resistance and eddy-current losses. In high-frequency circuits, resistance, radiation and dielectric losses all contribute. The heat if not removed by heat sinks, air or water cooling can damage components.

dissipation factor (*ElecEng*) The tangent of the phase angle (δ) for an inductor or capacitor.

$$\tan \delta = \frac{\sigma}{\omega\varepsilon}$$

σ is the electrical conductivity, ε the PERMITTIVITY of the medium, and ω is $2\pi \times$ frequency. For low-loss components, the dissipation factor is approximately equal to the power factor. Also *loss tangent*.

dissipationless line (*ICT*) A hypothetical transmission line in which there is no energy loss. Also *lossless line*.

dissipation trails (*EnvSci*) Lanes of clear atmosphere formed by the passage of an aircraft through a cloud. Also *distrails*.

dissipative network (*ICT*) One designed to absorb power, as contrasted with networks that attenuate power by impedance reflection. All networks dissipate to some slight extent, because neither capacitors nor inductors can be made entirely loss-free.

dissipative structure (*Chem*) A system maintained far from chemical/thermodynamic equilibrium, having the potential to form ordered structures.

dissociation (*Chem*) The reversible or temporary breaking down of a molecule into simpler molecules or atoms. See ARRHENIUS THEORY OF DISSOCIATION.

dissociation (*Psych*) An unconscious DEFENCE MECHANISM in which a group of psychological functions are separated from the remainder of the person's activities. In extreme cases this may result in a *dissociative disorder*, eg AMNESIA, FUGUE, MULTIPLE PERSONALITY.

dissociation constant (*Chem*) The equilibrium constant for a process considered to be a dissociation. Commonly it is applied to the dissociation of acids in water.

dissociation of gases (*Eng*) Chemical combustion reaction occurring at the highest temperature of the flame where carbon dioxide and water vapour tend to dissociate into CO, H_2 and O_2.

dissociative disorder (*Psych*) See DISSOCIATION.

dissolution (*Chem*) The taking up of a substance by a liquid, with the formation of a homogeneous solution.

dissolve (*ImageTech*) Transition from one scene to another in which the whole image of the first gradually disappears as it is replaced by the second. Also *lap dissolve*.

dissolving pulp (*Paper*) See ALPHA PULP.

dissonance (*Acous*) The playing of two or more musical terms simultaneously to produce an unpleasant effect on the listener.

dissymmetrical (*Genrl*) See ASYMMETRICAL.

dist (*Build*) Abbrev for (1) *distemper*, (2) *distributed*.

distal (*BioSci*) Far apart, widely spaced; pertaining to or situated at the outer end; farthest from the point of attachment. Cf PROXIMAL.

distance (*MathSci*) (1) Between two points, the length of the straight line joining the points. (2) Between two lines or two planes, the length of the segment of the line perpendicular to the given lines or planes lying between them. (3) Between a line or plane and a point: the length of the perpendicular from the point to the line or plane. See also ANGULAR DISTANCE, METRIC.

distance control (*ElecEng*) See REMOTE CONTROL.

distance mark (*Radar*) Mark on the screen of a cathode-ray tube to denote distance of target.

distance-measuring equipment (*Aero*) Airborne secondary radar which indicates distance from a ground transponder beacon. Abbrev *DME*.

distance meter (*ImageTech*) Same as rangefinder.

distance protection (*ElecEng*) See IMPEDANCE PROTECTIVE SYSTEM.

distance relay (*ElecEng*) See IMPEDANCE RELAY.

distant-reading compass (*Eng*) Gyro flux-gate compass in which the indicator is remote from the sensing device.

distant-reading instrument (*Eng*) A recording or indicating instrument (such as a thermometer or pressure gauge) in which the reading is shown on a scale at some distance from the point of measurement. See REMOTE CONTROL.

distemper (*Build*) A mixture of dry pigment with size, water and sometimes oil, once widely used as a paint for internal walls and ceilings.

disthene (*Min*) A less commonly used name for the mineral KYANITE.

distichiasis (*Med*) A condition in which there are two complete rows of eyelashes in one or both eyelids. Also *distichia*.

distichous (*BioSci*) Leaves on a stem arranged in two diametrically opposite rows.

distillation (*Chem*) A process of evaporation and recondensation used for separating liquids into various fractions according to their boiling points or boiling ranges. See MOLECULAR DISTILLATION.

distillation (*FoodSci*) A process, carried out in a STILL in three distinct stages, to remove the unwanted alcohols (eg methyl, amyl and butyl) which are poisonous and leave only ethyl, propyl and some higher alcohols. It is illegal to operate a still without a licence. Water is distilled to separate out dissolved substances. See WATER PURIFICATION.

distillation flask (*Chem*) Laboratory apparatus, usually made of glass; it consists of a bulb with a neck for the insertion of a thermometer and a side tube attached to the neck, through which the vapours pass to the condenser.

distinct (*BioSci*) Plant members not joined to one another.

distomiasis (*Vet*) Infection of the bile ducts by flukes or trematode worms. Also *distomatosis*.

distorted wave (*ElecEng*) A term often used in electrical engineering to denote a non-sinusoidal waveform of voltage or current.

distorting network (*ICT*) A network altering the response of part of a system, and anticipating the correction of response required to restore a signal waveform before actual distortion has occurred, eg owing to the inevitable frequency distortion in a line, or to minimize noise interference.

distortion (*Eng*) (1) Any departure from an intended or original shape because of internal stress or the release of RESIDUAL STRESS in the material. (2) The permanent change in shape of a moulding or shaped product caused by relief of FROZEN-IN STRAIN, often by a rise in temperature.

distortion (*ICT*) The change of waveform, spectral content or pulse shape of any wave or signal due to any cause.

distortion (*Phys*) An aberration of a lens or lens system whereby a square object is imaged with either concave (BARREL DISTORTION) or convex (PINCUSHION DISTORTION) lines. The type and amount of distortion depend on the position of the lens stop.

distortion factor (*ElecEng*) The ratio of the rms harmonic content to the total rms value of the distorted sine wave.

distortionless line (*Phys*) A transmission line with constants such that the attenuation (a minimum value) and the delay time are constant in magnitude with variation in frequency. The characteristic impedance is purely resistive. For such a line $LG = RC$, where R is resistance, L inductance, G leakage and C capacitance, all being distributed values per unit length. Also *distortionless condition*.

distortion of field (*ElecEng*) A term commonly used in connection with electrical machines to denote the change in the distribution of flux in the air-gap when the machine is put on load.

distraction display (*Psych*) Behaviour well exemplified by female birds that feign injury to lure a predator away from eggs or young.

distrails (*EnvSci*) See DISSIPATION TRAILS.

distributed amplifier (*ICT*) See TRANSMISSION-LINE AMPLIFIER.

distributed capacitance (*Phys*) (1) The capacitance distributed along a transmission line, which, with distributed resistance and/or inductance, reduces the velocity of transmission of signals. (2) The capacitance between the separate parts of a coil, lowering its inductance; represented by an equivalent lumped capacitor across the terminals, giving the same frequency of resonance.

distributed computing (*ICT*) The functional and geographical dispersion of computing power within a fully integrated system of processors and peripherals. It is a more economical and adaptable way to structure a very large computing system than to have one MAINFRAME. See FRONT-END PROCESSOR.

distributed constants (*Phys*) Constants applicable to real or artificial transmission lines and waveguides, because dimensions are comparable with the wavelength of transmitted energy.

distributed database (*ICT*) A DATABASE that is stored on several peripheral storage devices or several computer systems, possibly on separate sites.

distributed inductance (*Phys*) An inductance distributed uniformly along a circuit, eg a power transmission line, a loaded telephone circuit, or a travelling-wave valve or tube.

distributed processing (*ICT*) A system using many PROCESSORS to carry out data processing, possibly on separate sites.

distributed system (*ICT*) Multiple computers connected by a network so that mutual access is allowed.

distributed winding (*ElecEng*) The winding of an electric machine which is spread uniformly over the stator or rotor surface.

distributing centre (*ElecEng*) In an electric power system, a point at which an incoming supply from a feeder is split up amongst a number of other feeders or distributors.

distributing main (*ElecEng*) See DISTRIBUTOR.

distributing point (*ElecEng*) See FEEDING POINT.

distribution (*Autos*) The provision of the same quantity and quality of petrol–air mixture to each of the cylinders of a multicylinder engine by the carburettor and induction manifold.

distribution (*BioSci*) The occurrence of a species, considered from a geographical point of view, or with reference to altitude or other factors. Sometimes used as equivalent to DISPERSAL. Should not be confused with DISPERSION, which refers to individuals.

distribution (*MathSci*) The probabilities of the possible values of a random variable, eg the set of frequencies of observations in a set of intervals. See CUMULATIVE DISTRIBUTION FUNCTION.

distribution board (*ElecEng*) An insulating panel carrying terminals and/or fuses, for the distribution of power supplies to repeaters or telegraph circuits.

distribution cable (*ElecEng*) A communication cable extending from a feeder cable into a defined service area.

distribution coefficients (*NucEng*) In the countercurrent columns of a reprocessing plant, the ratio of the total amount of a substance in the organic phase to that in the aqueous phase. Also *spread factor*.

distribution coefficients (*Phys*) Chromaticity co-ordinates for spectral (monochromatic) radiations of equal power, ie for the component radiations forming an equal energy spectrum.

distribution factor (*ElecEng*) A factor used in the calculation of the emf generated in the winding of an ac machine, to allow for the fact that the emfs in each of the individual coils are not in phase with one another. Also *breadth coefficient, breadth factor*.

distribution factor (*Radiol*) A modifying factor used in calculating biological radiation doses, which allows for the non-uniform distribution of an internally absorbed radioisotope.

distribution frame (*ICT*) A structure with large numbers of terminals, for arranging circuits in specified orders.

distribution-free methods (*MathSci*) Methods of statistical analysis which under certain conditions do not depend on the probability distribution generating the observations.

distribution fuse-board (*ElecEng*) A distribution board having fuses in each of the separate circuits.

distribution law (*Chem*) The total energy in a given assembly of molecules is not distributed equally, but the number of molecules having an energy different from the median decreases as the energy difference increases, according to a statistical law.

distribution pillar (*ElecEng*) A structure in the form of a pillar, containing switches, fuses, etc, for interconnecting the distributing mains of an electric power system.

distribution reservoir (*Build*) See SERVICE RESERVOIR.

distribution switchboard (*ElecEng*) A distribution board having a switch in each of the branch circuits.

distributive (*MathSci*) Of one operation, *, over another, •, such that the result of applying the first to a product under the second equals the product of the results of applying it to the terms individually; ie $a • (b • c) = (a * b) • (a * c)$. Thus in ordinary arithmetic, multiplication *distributes* over addition, but not vice versa.

distributive mixing (*Chem*) Process where additives and fillers are mixed with polymer to produce a homogeneous material. Must be combined with DISPERSIVE MIXING. Also *blending, extensive mixing*.

distributor (*Autos*) A device, geared to the camshaft, whereby high-tension current is transmitted in correct sequence to the sparking plugs.

distributor (*ElecEng*) The cable or overhead line forming that part of an electric distribution system to which the consumers' circuits are connected. Also *distributing main*.

distributor rollers (*Print*) In a printing press, the rollers which distribute the ink, as distinct from the inking rollers which supply ink to the forme or plate.

district (*MinExt*) An underground section of a coal mine serviced by its own roads and ventilation ways: a section of a coal mine.

distrix (*Med*) Splitting of the ends of hairs.

disturbance (*ICT*) Any signal originating from a source other than the wanted transmitter, eg atmospherics, unwanted stations, noise in the receiver.

disulphide bond (*BioSci*) The –SS– linkage that is formed between sulphydryl groups of cysteine and stabilizes the secondary structure of a protein.

ditch canal (*Build*) See LEVEL CANAL.

ditching (*Aero*) Emergency alighting of a landplane on water.

dither (*ElecEng*) Small continuous signal supplied to a servomotor and producing a continuous mechanical vibration of the rotor which prevents sticking.

dither colour (*ICT*) A colour produced by a pattern of coloured dots that simulate the true colour. This technique is used to reduce the required MEMORY capacity. Also *non-solid colour*.

dithionic acid (*Chem*) $H_2S_2O_6$. Its salts are reducing agents and are called dithionates.

dithionous acid (*Chem*) $H_2S_2O_4$. Its salts are strong reducing agents and are called dithionites.

dithiothreitol (*BioSci*) $(CHOHCH_2SH)_2$. A mild reducing agent often used to reduce protein disulphide bonds.

ditrematous (*BioSci*) Of hermaphrodite animals, having the male and female openings separate; of unisexual forms, having the genital opening separate from the anus.

Dittus–Boelter equation (*Eng*) An equation for the transfer of heat from tubes to viscous fluids flowing through them.

$$u \propto c \left(\frac{k\Delta s}{d}\right)^{\frac{1}{3}} \left(\frac{vd}{u/s}\right)^{\frac{1}{12}}$$

where u = film transfer factor, k = thermal conductivity, Δ = logarithmic mean temperature difference between tube and liquid, d = thickness of fluid stream, s = relative density of fluid, c = specific heat capacity of fluid, v = mean velocity of fluid in tube and u/s = kinematic viscosity of the fluid.

diuresis (*Med*) The excretion of urine esp in excess.

diuretics (*Pharmacol*) Producing diuresis. Class of drug which promotes sodium and water loss by the kidneys, usually subdivided into THIAZIDE with related diuretics and the more potent LOOP DIURETICS. See also POTAS-SIUM-SPARING DIURETICS.

diurnal (*Astron, EnvSci*) During a day. The term is used to indicate the variations of an element during an average day.

diurnal libration (*Astron*) The phenomenon by which an observer can see more than half the Moon's surface when daily observations at different times or from different places on the Earth are combined. The effect is one of PARALLAX, the term *libration* being a misnomer in this case.

diurnal parallax (*Astron*) The change in the apparent position of a celestial object which results from the change in the observer's position caused by the Earth's daily rotation. Geometrically it is the angle subtended at the object by the Earth's radius. It is significant only for members of our solar system. Also *geocentric parallax*.

diurnal range (*EnvSci*) The extent of the changes which occur during a day in a meteorological element such as atmospheric pressure or temperature.

diurnal rhythm (*Psych*) Strictly, diurnal is the opposite of nocturnal and thus refers to daylight hours, but commonly used to mean circadian rhythm, where the cycle is of 24 hours. Also *diurnal cycle*.

diurnal variation (*Phys*) A variation of the Earth's magnetic field as observed at a fixed station, which has a period of approximately 24 h.

divalent (*Chem*) Capable of combining with two atoms of hydrogen or their equivalent. Also having an oxidation state of two. Also *bivalent*.

divaricate (*BioSci*) Spreading widely apart, forked, divergent.

dive (*Aero*) A steep descent with or without power. Also *nose dive*.

diver (*MinExt*) Small plummet adjusted to a desired relative density, so that it indicates the density of the fluid in which it is immersed by its up-and-down motion. Also *cartesian diver*.

dive-recovery flap (*Aero*) An AIR BRAKE in the form of a flap to reduce the LIMITING VELOCITY of an aircraft.

divergence (*Aero*) In aircraft stability, a disturbance that increases without oscillation; *lateral divergence* leads to a spin or an accelerating spiral descent; *longitudinal divergence* causes a nose dive or a stall.

divergence (*BioSci*) Evolution of the same basic structure to give organs of different form and function. Also *divergent evolution*. Cf CONVERGENT EVOLUTION. See HOMOLOGY.

divergence (*EnvSci*) If the components of the vector wind are (u, v, w) the divergence is defined as

$$\frac{\partial u}{\partial x} + \frac{\partial v}{\partial y} + \frac{\partial w}{\partial z}$$

The horizontal divergence is defined as

$$\frac{\partial u}{\partial x} + \frac{\partial v}{\partial y}$$

which is usually almost exactly compensated by

$$\frac{\partial w}{\partial z}$$

Furthermore the integrated divergence throughout a column of the atmosphere is almost zero, ie is a small residual of larger positive and negative values; this is known as the *Dines compensation*. Hence strong negative values near the surface are matched by strong positive values at high levels.

divergence (*MathSci*) Of a vector v, its scalar product $\nabla \cdot v$ with the vector operator ∇ (del). Written div v.

divergence (*Phys*) Initiation of a chain reaction in a reactor, in which slightly more neutrons are released than are absorbed and lost. The rate and extent of the divergence are normally controlled by neutron-absorbing rods, eg of boron, cadmium or hafnium.

divergence angle (*Electronics*) Angle of spread of electron beam, arising from mutual repulsion or debunching.

divergence speed (*Aero*) The lowest equivalent AIRSPEED at which AERO-ELASTIC DIVERGENCE can occur.

divergent (*BioSci*) Apices of organs wider apart than their bases.

divergent (*MathSci*) Of a sequence or series, definitions vary. Some writers count anything not CONVERGENT as divergent. Others use it as synonymous with *unbounded*, excluding finitely oscillating sequences such as $u_n = (-1)^n$. Still others confine it to sequences tending to $+\infty$ or $-\infty$, excluding infinitely oscillating sequences such as $u_n = n(-1)^n$.

divergent (*Phys*) A term applied to a reactor or critical experiment when the multiplication constant exceeds unity.

divergent adaptation (*BioSci*) See ADAPTIVE RADIATION.

divergent evolution (*BioSci*) See DIVERGENCE.

divergent junction (*Geol*) A zone where plates move apart and new crust and lithosphere are formed. Characterized by volcanism and earthquakes. eg mid-Atlantic ridge.

divergent lens (*Phys*) A lens which increases the divergence, or diminishes the convergence, of a beam of light passing through it. Such a lens will be double concave, plano-concave, or convexo-concave, the concave surface having the smaller radius of curvature.

divergent nozzle (*Eng*) A nozzle whose cross-section increases continuously from entry to exit; used eg in compound impulse turbines.

divergent strabismus (*Med*) Squint in which the eyes diverge from each other.

divergent thinking (*Psych*) Thinking which is productive and original, involving the creation of a variety of ideas or solutions which tend to go beyond conventional categories. In contrast to CONVERGENT THINKING.

diversion cut (*CivEng*) See BYE CHANNEL.

diversity (*BioSci*) An index of the number of species in a defined area, often represented mathematically. *Alpha diversity* is on a local scale, *beta diversity* on a regional scale. See RICHNESS.

diversity antenna (*ICT*) The antenna system of a diversity receiver.

diversity factor (*ElecEng*) The ratio of the arithmetic sum of the individual maximum demands of a number of consumers connected to an electric supply system, to the simultaneous maximum demand of the group.

diversity reception (*ICT*) System designed to reduce fading; several antennas, each connected to its own receiver, are spaced several wavelengths apart from one another, the demodulated outputs of the receivers being combined. Alternative systems use antennas orientated for oppositely polarized waves (polarized diversity), or independent transmission channels on neighbouring frequencies (frequency diversity).

diver's paralysis (*Med*) See CAISSON DISEASE. Also *diver's palsy*.

diverter (*ElecEng*) A low resistance connected in parallel with the series winding or the compole winding of a dc machine in order to divert some of the current from it, thereby varying the magnetomotive force produced by the winding.

diverter relay (*ElecEng*) A relay employed with certain excess-current protective systems; it increases the stability of the protective system by putting resistance in parallel with the tripping relay in the case of a heavy fault.

diverticulitis (*Med*) Inflammation of diverticula in the colon.

diverticulosis (*Med*) The presence of diverticula in the colon.

diverticulum (*BioSci*) (1) Saccular dilatation of a cavity or channel of the body. (2) Lateral outgrowth of the lumen of an organ. (3) Pouch-like protrusion of the mucous membrane of the colon through the weakened muscular wall. (4) A pouch-like side branch on the mycelium of some fungi. Pl *diverticula*.

divertor (*NucEng*) Trap used in thermonuclear device to divert magnetic impurity atoms from entering plasma, and fusion products from striking walls of chamber. Also *bundle divertor*.

divided bearing (*Eng*) See SPLIT BEARING.

divided pitch (*Eng*) The axial distance between corresponding points on successive threads of a multiple-threaded screw.

divided touch (*Phys*) The magnetizing of a steel bar by stroking it with the opposite poles of two permanent magnets, these being drawn apart from the centre of the bar to the ends.

divided winding (*ElecEng*) A term proposed for that class of windings (for dc machines) usually called multiple or multiplex, in which there are two or more separate windings on the armature, joined in parallel by the brushes.

dividend (*MathSci*) See DIVISION.

divider (*ElecEng*) A circuit which has an output which is a well-defined fraction of a given input; can be constructed using resistors or capacitors. Also *attenuator, voltage divider*.

dividers (*Eng*) Compasses used only for measuring or transferring distances, and not for describing arcs.

dividing box (*ElecEng*) A box for separately bringing out the cores of a multicore cable. The insulation of the cable is hermetically sealed and the cores may be brought out either as bare or insulated conductors.

dividing engine (*Eng*) An instrument for marking or engraving accurate subdivisions on scales; it consists of a carriage adjusted by a micrometer screw and holding a marking tool.

dividing fillet (*ElecEng*) See BARRIER.

dividing head (*Eng*) See INDEXING HEAD.

dividing network (*Acous*) A frequency-selective network which arranges for the input to be fed into the appropriate loudspeakers, usually two, covering high and low frequencies respectively. Also *crossover network, loudspeaker dividing network*.

diving-bell (*CivEng*) A watertight working chamber, open at the bottom, which is lowered into water beneath which excavation or other works are to proceed. The interior is supplied with compressed air to maintain the water level inside at a reasonable height, and thus leave free a space within which people may work.

divinity calf (*Print*) Bindings in dark-brown calfskin, with BLIND TOOLING; used chiefly for theological works.

division (*BioSci*) Highest taxonomic rank used in the classification of plants (equivalent to the zoologist's PHYLUM), ranking above CLASS; the names end in –phyta or, for fungi, –mycota.

division (*MathSci*) (1) For numbers, the operation of ascertaining how many times one number, the *divisor*, is contained in a second, the *dividend*. The result is called the *quotient*, and, if the divisor is not contained an integral number of times in the dividend, any number left over is called the *remainder*. Indicated either by the division sign, ÷, or by a stroke or bar, in which case the expression as a whole is called a *fraction* and the dividend and divisor the *numerator* and *denominator* respectively. Fractions less than one are called *common* or *proper* or *vulgar fractions*, and those greater than one, *improper fractions*. Colloquially, however, a *fraction* is less than one. (2) For complex numbers, the division of $a+ib$ by $c+id$ is given by

$$\frac{a+ib}{c+id} = \frac{ac+bd}{c^2+d^2} + i\frac{bc-ad}{c^2+d^2}$$

(3) For polynomials and other mathematical entities, the inverse operation to multiplication. When appropriate, nomenclature analogous to that outlined above is used.

division plate (*Eng*) A plate used for positioning the plunger of an indexing head; provided with several concentric rings of holes accurately dividing the circumference into various equal subdivisions.

division ring (*MathSci*) A ring which, if zero is removed, is a group under multiplication, ie every non-zero element has an inverse. A commutative division ring is a field.

division wall (*Build*) A wall within a building or serving two houses. Also *party wall*.

divisor (*MathSci*) See DIVISION.

dizygotic twins (*BioSci*) Twins produced from two fertilized eggs. They may be the same or different sexes and are genetically equivalent to full sibs. Also *fraternal twins*. Cf MONOZYGOTIC TWINS.

dl- (*Chem*) Containing equimolecular amounts of the dextrorotatory and the laevorotatory forms of a compound; racemic. Now usually written ±.

D-layer (*ICT*) The lowest region or layer of absorbing ionization, 55–95 km above the Earth. It impedes short-wave communications by absorbing some of the incident power, but it enhances long-wave communication.

DLC (*Chem*) Abbrev for DIAMOND-LIKE CARBON.

d-levels (*Phys*) See DIFFUSE SERIES.

D-lines (*Phys*) See [D].

DLL (*ICT*) Abbrev for DYNAMIC LINK LIBRARY.

D log E curve (*ImageTech*) See CHARACTERISTIC CURVE.

DLR (*Space*) Abbrev for *Deutsches Zentrum für Luft und Raumfahrt*, the German centre for aerospace research; formerly *DFVLR*.

DLVO theory (*BioSci*) A theory of colloid flocculation (formation of a coagulum), advanced independently by Derjaguin and Landau and by Vervey and Overbeek. It was subsequently applied to cell adhesion.

DM (*Build*) Abbrev for *disconnecting manhole*. See INTERCEPTOR.

DMA (*ICT*) Abbrev for DIRECT MEMORY ACCESS.

D MAC, D2 MAC (*ImageTech*) See MAC.

D max, D min (*ImageTech*) Abbrevs for the maximum and minimum densities of a photographic image.

DMC (*Chem*) Abbrev for DOUGH MOULDING COMPOUND.

DMD (*ImageTech*) Abbrev for DIGITAL MICROMIRROR DEVICE.

DMDT (*Chem*) See METHOXYCHLOR.

DME (*Aero*) Abbrev for *distance-measuring equipment*.

D method (*MathSci*) A method used in determining the solution of a linear differential equation with constant coefficients. The operator D represents d/dx, and, under certain conditions, it can be manipulated by some of the procedures of ordinary algebra.

DMF (*Chem*) Abbrev for *dimethylformamide*.

DML (*ICT*) Abbrev for *data manipulation language*. Used in DATABASE MANAGEMENT SYSTEMS. See SQL.

DMOS transistor (*Electronics*) Also *double-diffused transistor*. See DOUBLE-DIFFUSED METAL–OXIDE SEMICONDUCTOR.

DMP (*ICT*) Abbrev for DOT-MATRIX PRINTER.

DMSO (*BioSci*) See DIMETHYL SULPHOXIDE.

DMTA (*Eng*) Abbrev for DYNAMIC MECHANICAL THERMAL ANALYSIS.

DMZ (*ICT*) Abbrev for *de-militarized zone* – in computer security a layer on the perimeter of an organization's network that permits outgoing access to an external network such as the Internet, but prevents incoming access from that external network to the organization's own. (Originally US military term from the 1950s.)

dn (*MathSci*) See ELLIPTIC FUNCTIONS.

DNA (*BioSci*) Abbrev for *deoxyribonucleic acid*. See panel on DNA AND THE GENETIC CODE.

DNA binding proteins (*BioSci*) In prokaryotes, promoters, repressors, etc; in eukaryotes, similar proteins, excluding the histones.

DNA fingerprinting (*BioSci*) See RESTRICTION FRAGMENT LENGTH POLYMORPHISM.

DNA gyrase (*BioSci*) A *topoisomerase* enzyme essential for DNA replication in prokaryotic (circular) DNA. The action of the gyrase maintains a state of negative super-coiling.

DNA library (*BioSci*) A mixture of cloned DNA sequences derived from a single source, like a mouse or a chromosome, and containing ideally all, but in reality most, of the sequences from that source.

DNA ligase (*BioSci*) An enzyme involved in the repair and recombination of DNA. Ligases are used in recombinant DNA technology for joining DNA fragments and preparing radioactive DNA probes by NICK TRANSLATION.

DNA methylation (*BioSci*) The addition of methyl groups to certain nucleotides in genomic DNA of eukaryotes, thereby affecting gene expression; the degree of methylation is passed on to daughter strands at mitosis. In bacteria, methylation plays an important role in the restriction systems.

DNA polymerases (*BioSci*) Enzymes involved in template-directed synthesis of DNA from deoxyribonucleotide triphosphates. Multiple types are found with specific roles. Retroviruses possess a unique DNA polymerase (reverse transcriptase) that uses an RNA template.

DNA repair (*BioSci*) Enzymic correction of errors in DNA structure and sequence arising from environmental damage and replication errors.

DNA transfection (*BioSci*) A method used to introduce genes or gene fragments into cells as DNA.

DNA virus (*BioSci*) A virus in which the nucleic acid is double- or single-stranded DNA (rather than RNA). Major groups of double-stranded DNA viruses are papovaviruses, adenoviruses, herpesviruses, large bacteriophages, and poxviruses: of single-stranded, parvoviruses and coliphages φX174 and M13.

DNC (*Chem*) See DNOC.

DNOC (*Chem*) *2-methyl-4,6-dinitro(1-hydroxybenzene)*. Used as an insecticide and herbicide. Also *dinitrocresol, DNC.*

DNS (*ICT*) Abbrev for DOMAIN NAME SYSTEM.

Doba's network (*ElecEng*) Shaping circuit used in pulse amplifiers where rise times of a few nanoseconds are required.

dobby (*Textiles*) Mechanism over the top or at the side of a loom. Operated by punched cards it lifts and lowers the healds to move the warp threads in timed sequence to form the design in the cloth.

dobby fabric (*Textiles*) A fabric made on a loom fitted with a DOBBY.

Dobson spectrophotometer (*EnvSci*) An instrument used in the routine measurement of atmospheric ozone. It compares, using a photomultiplier and an optical wedge, the intensities of two wavelengths in the solar spectrum in the region of partial ozone absorption (0·30–0·33 lm), and from the result the total amount of ozone in a vertical column can be calculated. The instrument may be used to obtain the vertical distribution of ozone from the UMKEHR EFFECT.

Dobson unit (*EnvSci*) One Dobson unit (DU) is equivalent to 2.7×10^{16} molecules of ozone in the total air column over 1 cm^2 of the Earth's surface.

dockable camera/VTR (*ImageTech*) A camera and VIDEO-TAPE RECORDER that can be attached (docked) to form a CAMCORDER, with all electrical connections made directly instead of by cable.

docking (*Space*) The physical attaching of one space vehicle to another.

doctor (*ElecEng*) A way of further electroplating imperfectly plated parts; it consists of an anode of the metal to be deposited, covered with a spongy material saturated with the plating material.

doctor (*Print*) A blade-like device resting at a shallow angle on the down-running surface of a roll or cylinder to remove unwanted material. Used in intaglio printing and paper-making.

doctor knife (*Textiles*) A metallic blade set near the surface of a printing roller to remove excess colouring matter from the fabric.

doctor test (*Chem*) A test for sulphur in petroleum using a sodium plumbate (II) solution.

doctrine of specific nerve energies (*Psych*) The assertion that qualitative differences in sensory experience depend on which nerve is stimulated and not on the physical attributes of the stimulus.

documentation (*ICT*) Full description in words and diagrams accompanying a package, program or system.

document copying (*Print*) A variety of methods are in use based on DUAL SPECTRUM, DYELINE, electrostatic (XERO-GRAPHY), PHOTOCOPYING, and THERMOGRAPHIC principles, some able to produce lithographic masters for SMALL OFFSET.

document reader (*ICT*) Input device that reads marks and characters made in predetermined positions on special forms.

document retrieval (*ICT*) See INFORMATION RETRIEVAL.

dodeca- (*Genrl*) Prefix from Gk *dodeka* denoting 12.

dodecagon (*MathSci*) A 12-sided polygon.

dodecahedron (*MathSci*) A 12-sided polyhedron.

dodecyl benzene (*Chem*) Important starting material for anionic detergents derived from petroleum. Based on the tetramer of propene.

dodging (*ImageTech*) Manipulation of the light projected through a negative by an enlarger to lighten or darken selected parts of the resultant print.

dodine (*Chem*) Dodecylguanidine; used as a fungicide.

doeskin cloth (*Textiles*) Similar to BEAVER CLOTH but finer and lighter, usually made from merino wool.

doffer (*Textiles*) (1) The operative who removes full cops or other yarn packages from a machine. (2) A fully automatic machine which performs this operation mechanically, eg in spinning.

doffing tube (*Textiles*) The tube of a ROTOR SPINNING unit through which the yarn passes on its way to the rotating collection package.

dog (*Build, Eng*) A steel securing-piece used for fastening together two timbers, as in the process of shoring, for which purpose it is hooked at each end at right angles to the length, so that the hooked ends may be driven into the timbers. The term is also applied to a great variety of gripping implements, including a clutching attachment for withdrawing well-boring tools; a pawl; an adjustable stop used in machine tools; a spike for securing rails to sleepers; a lathe carrier; a circular clawed object to join members of a roof truss.

dog clutch (*Eng*) A clutch consisting of opposed flanges, one with projections and the other with corresponding slots, one of which moves axially to engage the drive. See CLUTCH.

dog down (*Ships*) To secure in position by pieces of bent-round iron, driven through holes in a cast-iron slab in such a manner as to be jammed.

Dogger (*Geol*) The middle epoch of the Jurassic period. See MESOZOIC.

doggers (*Geol*) Flattened ovoid concretions, often of very large size, in some cases calcareous, in others ferruginous, occurring in sands or clays. They may be a metre or more in diameter.

dog-house (*Glass*) A small extension of a glass furnace, into which the BATCH is fed.

dog-legged stair (*Build*) A stair having successive flights rising in opposite directions, and arranged without a well-hole.

dog-nail (*Build*) A large nail having a head projecting over one side.

Dog Star (*Astron*) See SIRIUS.

dog's tooth (*Build*) A string course in which bricks are so laid as to have one corner projecting.

DNA and the genetic code

In its double-stranded form, DNA (deoxyribonucleic acid) is the genetic material of most organisms and organelles, although phage and viral genomes may use single-stranded DNA, single-stranded RNA or double-stranded RNA. The two strands of DNA form a double helix, the strands running in opposite directions, as determined by the sugar-phosphate 'backbone' of the molecule.

The four bases project towards each other like the rungs of a ladder, with a purine always pairing with a pyrimidine, according to the BASE PAIRING RULES, in which thymidine pairs with adenine, and cytosine with guanine. In its B molecular form the helix is 2·0 nm in diameter with a pitch of 3·4 nm (10 base pairs).

Genetic code

The code that relates the four bases of the DNA or RNA to the 20 amino acids found in proteins is shown in the table. There are 64 possible different three-base sequences (triplets) using all permutations of the four bases. One triplet uniquely specifies one amino acid (except for AUG when acting as an initiating codon in bacteria), but each amino acid can be coded by up to six different triplet sequences. The code is therefore *degenerate*. INITIATING CODONS specify the start of a polypeptide chain and the triplets known as *Ochre*, *Amber* and *Opal* are STOP CODONS which terminate the chain. Initiating codons are AUG and GUG in bacteria with the former specifying the amino acid *n*-formylmethionine at the beginning of the chain and methionine within it. In eukaryotes, AUG is the only initiator and always translates as methionine.

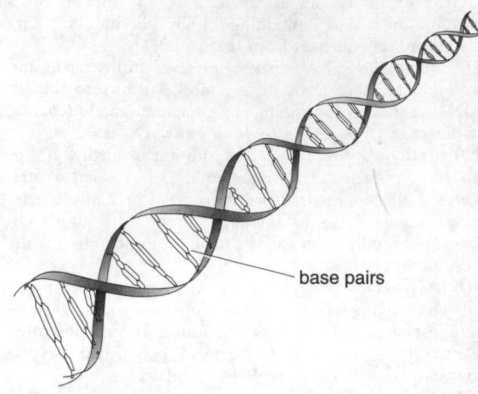

base pairs

DNA The sugar-phosphate backbones displayed as ribbons.

The evidence suggests that the code is universal, applying from the simplest to the most evolutionarily advanced organism, although minor variations have been found, particularly in the mitochondrial DNAs.

In the following table the amino acids and the bases are specified by their three and one letter symbols respectively.

See panels on CHROMOSOME and GENETIC MANIPULATION; appendices.

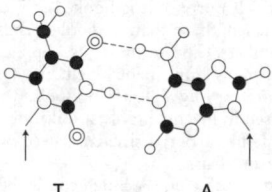

T   A

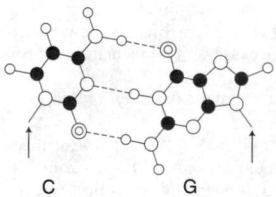

C   G

The DNA base pairs The filled circles are C atoms, the open are N (large) and H (small and the doubled circles are O. The arrows indicate the helix attachment sites which are about 1.1 nm apart. T, A, C and G are the base abbreviations. The dashed lines show the hydrogen bonds.

---

dog-tooth spar (*Min*) A form of calcite in which the scalenohedron is dominant but combined with a prism, giving a sharply pointed crystal like a canine tooth.

Doherty transmitter (*ICT*) One in which high efficiency of amplification of amplitude-modulated wave is obtained by two valves connected to the load, one directly, the other through a 90° retarding network.

Dolby (*Acous, ImageTech*) TN for a noise-reduction system for magnetic and photographic sound recording and reproduction. Also for a system of stereophonic sound presentation in the cinema. Dolby B improves by 10 dB, Dolby C by 15 dB.

doldrums (*EnvSci*) Regions of calm in equatorial oceans. Towards the solstices, these regions move about 5° from their mean positions, towards the north in June and towards the south in December.

dolerite (*Geol*) The general name for basic igneous rocks of medium grain size, occurring as minor intrusions or in the central parts of thick lava flows; much quarried for road metal. Typical dolerite consists of plagioclase near labradorite in composition, pyroxene, usually augite, and iron ore, usually ilmenite, together with their alteration products.

Dolezalek quadrant electrometer (*ElecEng*) The original quadrant electrometer used for measurement of voltages and charges. It consisted of a suspended plate rotating over a metal box divided into four quadrants, two of which are earthed, the other two being charged by the current under

DNA and the genetic code *(Cont.)*

Amino acid	Three-letter	Single-letter	Codons
Alanine	Ala	A	GCU, GCC, GCA, GCG
Arginine	Arg	R	CGU, CGC, CGA, CGG, AGA, AGG
Asparagine	Asn	N	AAU, AAC
Aspartic acid	Asp	D	GAU, GAC
Cysteine	Cys	C	UGU, UGC
Glutamic acid	Glu	E	GAA, GAG
Glutamine	Gln	Q	CAA, CAG
Glycine	Gly	G	GGU, GGC, GGA, GGG
Histidine	His	H	CAU, CAC
Isoleucine	Ile	I	AUU, AUC, AUA
Leucine	Leu	L	UUA, UUG, CUU, CUC, CUA, CUG
Lysine	Lys	K	AAA, AAG
Methionine	Met	M	AUG
Phenylalanine	Phe	F	UUU, UUC
Proline	Pro	P	CCC, CCU, CCA, CCG
Serine	Ser	S	UCU, UCC, UCA, UCG, AGU, AGC
Threonine	Thr	T	AUC, ACC, ACA, ACG
Tryptophan	Trp	W	UGG
Tyrosine	Tyr	Y	UAU, UAC
Valine	Val	V	GUU, GUC, GUA, GUG
Ochre	Stop	–	UAA
Amber	Stop	–	UAG
Opal	Stop	–	UGA

measurement. (This is *heterostatic* operation. *Idiostatic* connection has the plate joined to one pair of quadrants.)

dolichocephalic (*Med*) Long-headed; said of a skull, the breadth of which is less than four-fifths the length.

dolichocolon (*Med*) An excessively long colon.

dolichol phosphate (*BioSci*) A long-chain unsaturated lipid with a terminal pyrophosphate found in the membranes of the ENDOPLASMIC RETICULUM. The core oligosaccharide for N-glycosylation of proteins is constructed on a dolichol phosphate molecule prior to its donation to the nascent polypeptide chain.

doliiform (*BioSci*) Barrel-shaped. Also *dolioform*.

doll (*CivEng*) A small arm or post carrying railway signalling apparatus, mounted on a gantry or bridge.

dollar (*NucEng*) US unit of reactivity for a reactor defined in terms of the effective neutron multiplication factor if prompt neutrons only are assumed to contribute. The CENT is a hundredth of a dollar.

dollar spots (*Vet*) The skin lesions of DOURINE.

dolly (*CivEng*) An object, usually a short length of pile, interposed between the pile hammer and the head of a pile to prevent damage or to help driving under water.

dolly (*ElecEng*) Operating member of a tumbler switch which projects through the outer casing.

dolly (*Eng*) (1) A heavy hammer-shaped tool for supporting the head of a rivet during the forming of a head on the other end. (2) A shaped block of lead used by panel beaters when hammering out dents.

dolly (*ImageTech*) Wheeled mobile mounting for a motion picture or TV camera and its operator, allowing the action to be followed.

dolly truck (*Print*) See REEL BOGIE.

dolly tub (*MinExt*) A large wooden tub used for the final upgrading of valuable minerals separated by water concentration in ore dressing. Also *kieve*. See TOSSING.

dolomite (*Min*) The double carbonate of calcium and magnesium, crystallizing in the rhombohedral class of the trigonal system, occurring as cream-coloured crystals or masses with a distinctive pearly lustre, hence the synonym *pearl spar*. A common mineral of sedimentary rocks and an important gangue mineral.

dolomite limestone (*Eng*) Calcined dolomite is used as a basic refractory for withstanding high temperatures and attack by basic slags in metallurgical furnaces. The name dolomite is also used to describe refractories made from magnesian limestone, which does not necessarily contain the mineral dolomite. See MAGNESIUM OXIDE.

dolomitic limestone (*Geol*) A calcareous sedimentary rock containing calcite or aragonite in addition to the mineral dolomite. Cf DOLOSTONE.

dolomitization (*Geol*) The process of replacement of calcium carbonate in a limestone or ooze by the double carbonate of calcium and magnesium (dolomite).

dolostone (*Geol*) A rock composed entirely of the mineral dolomite. The term is synonymous with *dolomite rock*, but avoids confusion between dolomite rock and the mineral dolomite.

dolphin (*CivEng*) An economic mooring for large vessels in deep waters in which a small number of piles are connected by decking over. Berthing dolphins are fendered while mooring dolphins are placed beyond and behind the berthing face. Originally formed by groups of timber piles bound at the head.

domain (*BioSci*) Part of a molecule that has conserved properties; a consensus sequence in a protein that produces a characteristic structural motif, ligand binding or physical property. Examples include immunoglobulin-like domains, DNA-binding domains, hydrophobic domains. Standard motifs seem to be fairly highly conserved and can be recombined to confer particular properties.

domain (*Chem*) Microstructural unit formed in polymeric materials, esp by segregation of different chain segments in copolymers. In styrene butadiene styrene block copolymers, they may form a regular array, and typically have a diameter of 10 nm. See COPOLYMER. In toughened polymers, they may form within the rubber particles. See panel on RUBBER TOUGHENING.

domain (*MathSci*) (1) A set of points. Also *region*. (2) Of a function or mapping, the set on which the function or mapping is defined, eg for a real function the domain is a subset of the set of real numbers, namely the set of those real numbers for which the function is defined. Compare CODOMAIN, RANGE.

domain (*Phys*) In ferroelectric, ferromagnetic and ferrimagnetic materials, a region where there is saturated polarization, depending only on temperature. The transition layer between adjacent domains is the BLOCH WALL, and the average size of the domain depends on the constituents of the material and its heat treatment. See panel on FERROMAGNETICS AND FERRIMAGNETICS.

domain name system (*ICT*) An Internet naming scheme allowing easy identification of a specific server so that information can be directed to it. Abbrev *DNS*. See panel on INTERNET.

domatium (*BioSci*) A cavity in a plant inhabited by commensal mites or insects.

dome (*Arch*) A vault springing from a circular, or nearly circular, base.

dome (*BioSci*) See APICAL DOME.

dome (*CivEng*) A domed cylinder attached to a locomotive boiler to act as a steam space and to house the regulator valve.

dome (*Crystal*) A crystal form consisting of two similar inclined faces meeting in a horizontal edge, thus resembling the roof of a house. The term is frequently incorrectly applied to a four-faced form which is really a prism lying on an edge.

dome (*Geol*) (1) An igneous intrusion with a dome-like roof. (2) An anticlinal fold with the rock dipping outwards in all directions.

dome nut (*Eng*) See CAP NUT.

dominance (*Crystal*) See HABIT.

dominance hierarchy (*Psych*) An aspect of the social organization of various species, usually referring to aggressive interactions, in which certain individuals predictably dominate, or are dominated by, other members of the group.

dominant (*BioSci*) (1) A version of a gene (*allele*) that shows its effect in those individuals who received it from only one parent, ie in heterozygotes. Also describes a character due to a dominant gene. Cf RECESSIVE. (2) In ecology, the species that, because of its number or size, determines the character of a plant community or vegetation layer. Several species can be co-dominant.

dominant negative (*BioSci*) A mutation that exerts an effect even when only one copy is present, as in a heterozygote.

dominant wavelength (*Phys*) The wavelength of monochromatic light which matches a specific colour when combined in suitable proportions with a reference standard light. On a chromaticity diagram it is determined by a straight line drawn from the achromatic point though the point representing the chromaticity of the colour of interest, to where the line intersects the pure colour perimeter. Also *effective wavelength*.

Domin scale (*EnvSci*) A ten-point non-linear scale used in estimating canopy cover.

donepezil hydrochloride (*Pharmacol*) An *acetylcholine esterase inhibitor* used to treat symptoms of mild to moderate dementia. TN *Aricept*.

dongle (*ICT*) A device used in the illegal cloning of mobile phones; also, formerly a plastic key bundled with software packages and inserted into a port on some early microcomputers to validate the software licence, preventing unlicensed use of the software.

donkey boiler (*Eng*) A small vertical auxiliary boiler for supplying steam-winches and other deck machinery on board ship when the main boilers are not in use.

donkey engine (*Eng*) A small engine used for starting a larger one and sometimes independently.

donkey pump (*Eng*) A small steam reciprocating pump independent of the main propelling machinery of a ship; used for general duty.

Donnan equilibrium (*Chem*) An equilibrium established between a charged, immobile colloid (such as clay, ion exchange resin or cytoplasm) and a solution of electrolyte. The colloid compartment is depleted in ions of like charge and is electrically polarized relative to the solution (a *Donnan potential*).

donor (*Chem*) (1) That reactant in an induced reaction which reacts rapidly with the inductor. (2) That which 'gives', as in *proton-donor* or *electron-pair donor*.

donor (*Electronics*) Impurity atoms which add to the conductivity of a semiconductor by contributing electrons to a nearly empty conduction band, so making n-type conduction possible. See panel on INTRINSIC AND EXTRINSIC SILICON.

donor level (*Electronics*) See ENERGY LEVELS.

donut (*ElecEng*) See DOUGHNUT.

door case (*Build*) The frame into which a door fits to shut an opening.

door check (*Build*) (1) A device fitted to a door to prevent it from being slammed, and yet to ensure that it closes. (2) See DOOR STOPS.

door cheeks (*Build*) The jambs of a door frame. Also *door posts*.

door contact (*ElecEng*) A contact attached to a door or gate, arranged so that it gives an alarm or other signal when the door is opened or closed.

door-knob transformer (*ICT*) A device for coupling a coaxial cable to a waveguide; the inner of the coaxial cable is fed through a hole in the broad wall of the waveguide, and stops at half the depth of the waveguide. This protrusion has a spherical end or 'doorknob' to improve impedance matching.

door posts (*Build*) See DOOR CHEEKS.

door stop (*Build*) (1) A device fitted to a door, or to the floor near to a door, to hold it open. (2) A set of thin timbers round a door opening to stop the door as opposed to rebated jambs. Also *door checks*.

door strip (*Build*) A strip, often of flexible material, attached to a door to cover the space between the bottom of the door and the threshold. Also *draught excluder, weather strip*.

door switch (*ElecEng*) A switch mounted on a door so that the opening or closing of the door operates the switch. Also a *gate switch* when used on electric lifts.

DOP (*Chem*) Abbrev for DIOCTYLPHTHALATE.

DOPA (*Pharmacol*) A precursor of the neurotransmitter DOPAMINE; the L-form (*levodopa*) is used as a treatment for Parkinsonism.

dopamine (*Med*) A SYMPATHOMIMETIC compound that acts on dopaminergic and adrenergic receptors to increase heart rate, cardiac output and blood pressure. Also important as a neural transmitter in the brain. See DOPAMINE HYPOTHESIS.

dopamine hypothesis (*Psych*) The hypothesis that schizophrenia is caused by an excess amount of dopamine in the brain. Some ANTIPSYCHOTIC DRUGS do reduce levels of dopaminergic activity, although others do not.

doped junction (*Electronics*) One with a semiconductor crystal which has had impurity added during a melt.

D operator (*MathSci*) See D METHOD.

doping (*Aero*) A chemical treatment with nitrocellulose or cellulose acetate dissolved in thinners, which is applied to fabric coverings, for the purposes of tautening, strengthening, weatherproofing, etc.

doping (*Electronics*) Addition of known impurities to a semiconductor or other material, to achieve the desired extrinsic properties. See panel on INTRINSIC AND EXTRINSIC SILICON.

Doppler broadening (*Phys*) The frequency spread of radiation in single spectral lines, because of the Maxwellian distribution of velocities in the molecular radiators. This also broadens the resonance absorption curve for atoms or molecules excited by incident radiation.

Doppler effect (*Phys*) The apparent change of frequency (or wavelength) because of the relative motion of the source of radiation and the observer. For instance, the change in frequency of sound heard when a train or aircraft is moving towards or away from an observer. For sound, the observed frequency is related to the true frequency by

$$f_o = \frac{V - V_o(+W)}{V - V_s(+W)}\ f_s$$

where V_s, V_o are the velocities of source and observer, V is the phase velocity of the wave and W is the velocity of the wind. For electromagnetic waves, the LORENTZ TRANSFORMATION is used to give

$$f_o = \frac{\sqrt{1 - v/c}}{1 + v/c}\ f_s$$

where v is the *relative* velocity of the source and observer, and c is the speed of light. In astronomy the measurement of the frequency shift of light received from distant galaxies, the REDSHIFT, enables their recession velocities to be found.

Doppler navigator (*Aero*) Automated dead reckoning by a device which measures true ground speed by the Doppler frequency shift of radio-beam echoes from the ground and computes these with compass readings to give the aircraft's true track and position at any time. Entirely contained in the aircraft, this system cannot be affected by radio interference or hostile jamming.

Doppler radar (*Radar*) Any means of detection by reflection of electromagnetic waves, which depends on measurement of change of frequency of a signal after reflection by a target having relative motion. See CW RADAR, PULSED DOPPLER RADAR.

Doppler spread (*ICT*) A situation that arises when a radio receiver, eg in a mobile-telephone network, receives signals from a moving transmitter via a number of different paths, each giving a different Doppler shift.

Dorado (Swordfish) (*Astron*) A small and inconspicuous southern constellation, which includes the Large MAGELLANIC CLOUD.

DORAN (*Radar*) Doppler ranging system for tracking missiles (*DOppler RANge*).

dore silver (*Eng*) Silver bullion, ie ingots or bars, containing gold.

doric (*Print*) A typeface. The same as sans serif or gothic.

dormancy (*BioSci*) A state of temporarily reduced but detectable metabolic activity, as in seeds.

dormant bolt (*Build*) A hidden bolt sliding in a mortise in a door; operated by turning a knob or by means of a key.

dormer (*Build*) A window projecting from a roof slope.

dormin (*BioSci*) See ABSCISIC ACID.

Dorn effect (*Chem*) The production of a potential difference when particles suspended in a liquid migrate under the influence of mechanical forces, eg gravity; the converse of ELECTROPHORESIS.

dorsal (*Bio*) (1) Generally, that aspect of a bilaterally symmetrical organism that is normally turned away from the ground. (2) The back of any part. (3) In plants, the abaxial surface, which in a leaf is confusingly usually the lower surface.

dorsal fins (*Aero*) Forward extensions along the top of the fuselage to increase effectiveness of the main fin in sideslip, esp in ASYMMETRIC FLIGHT.

dorsalgia (*Med*) Pain in the back.

dorsalis (*BioSci*) An artery supplying the dorsal surface of an organ.

dorsal suture (*BioSci*) The midrib of the carpel in cases where dehiscence occurs along it.

dorsal trace (*BioSci*) The median vascular supply to a carpel.

dorsiferous (*BioSci*) Said of animals that bear their young on their back.

dorsiflexion (*Med*) The bending towards the back of a part, eg flexion of the toes towards the shin.

dorsigrade (*BioSci*) Walking with the backs of the digits on the ground, as sloths.

dor silver (*MinExt*) Silver bullion, ie ingots or bars, containing gold.

dorsiventral (*BioSci*) A flattened plant member having structural differences between its dorsal and ventral sides.

dorsum (*BioSci*) In Anthozoa, the sulcular surface; in Arthropoda, the tergum or notum; in vertebrates, the dorsal surface of the body or back.

Dortmund tank (*Build*) A deep tank, with conical bottom to which liquid sewage is supplied by a pipe reaching down nearly to the bottom. The resulting upward flow assists sedimentation of the sludge.

DOS (*ICT*) Abbrev for *disk operating system*. TN for a microcomputer operating system that handles files and programs stored on disk. Letters are often prefixed, eg *MS-DOS, PC DOS*.

dose (*Chem*) Quantity of material introduced; eg in ion implantation, the number of ions per square metre which are implanted into a surface.

dose (*Med*) The prescribed quantity of a medicine or of a remedial agent.

dose (*Radiol*) General term for quantity of radiation. See ABSORBED DOSE, COLLECTIVE DOSE EQUIVALENT, DOSE EQUIVALENT, EFFECTIVE DOSE EQUIVALENT, GENETICALLY SIGNIFICANT DOSE. See panel on RADIATION.

dose commitment (*Radiol*) See COMMITTED DOSE EQUIVALENT.

dose equivalent (*Radiol*) The quantity obtained by multiplying the absorbed dose by a factor to allow for the different effectiveness of the various ionizing radiations in causing harm to tissues. Unit Sievert (*Sv*). See panel on RADIATION.

dose rate (*Radiol*) The ABSORBED DOSE, or other dose, received per unit time.

dose reduction factor (*Radiol*) A factor giving the reduction in radiation sensitivity for a cell or organism which results from some chemical protective agent.

dosimeter (*Radiol*) Instrument for measuring DOSE and used in radiation surveys, hospitals, laboratories and civil defence. It gives a measure of the radiation field and dosage experienced. Also *dosemeter*.

DOS shell (*ICT*) A technique whereby a user may enter operating system commands whilst the main application program is temporarily suspended.

dot blot (*BioSci*) See BLOTTING.

dot gain (*Print*) An effect whereby HALF-TONE dots are enlarged during image production at plate-making or in the printing operation, thus altering tonal values from the original film element. The amount of dot gain can be measured using a DENSITOMETER to determine whether the dot gain lies within acceptable limits.

dot interference (*ImageTech*) A picture defect in which dot-like interference appears along colour borders.

dotless i (*Print*) A special i sometimes available on phototypesetters to accommodate floating accents or ligatures.

dot-matrix printer (*ICT*) Printer in which characters are formed of ink dots from a rectangular matrix of printing positions, typically 9×7 dots.

DOTS (*Med*) Abbrev for *directly observed therapy, short-course*, a closely supervised drug treatment for tuberculosis.

dot sequential (*ImageTech*) Colour TV system in which the three colour signals are sent in rapid succession for each point in a brief part of the line scan period.

dotting-on (*Build*) The process of adding together similar items when TAKING OFF.

double (*BioSci*) (1) A flower having more than the normal number of petals, commonly by the conversion of stamens (or even carpels) to petals, ie petalody. (2) The capitulum of the Compositae having (in a superficially similar way) some or all of the tubular disk florets converted into ligulate florets.

double (*Build*) A slate size, 13×6 in (330×252 mm).

double-acting engine (*Eng*) Any reciprocating engine in which the working fluid acts on each side of the piston alternately; most steam engines and a few internal-combustion engines are so designed.

double-acting pump (*Eng*) A reciprocating pump in which both sides of the piston act alternately, thus giving two delivery strokes per cycle.

double-action press (*Eng*) A press fitted with two slides, permitting multiple operations, such as blanking and drawing, to be performed.

double amplitude (*Phys*) The sum of the maximum values of the positive and negative half-waves of an alternating quantity. Also *peak-to-peak amplitude*.

double bar and yoke method (*ElecEng*) A ballistic method of magnetic testing, in which two test specimens are arranged parallel to each other and clamped to yokes at the ends to form a closed magnetic circuit. A correction for the effect of the yokes is made by altering their position on the bars and repeating the test.

double bar gauge (*Build*) Marking gauge in which the fence slides on two independently adjustable bars, each with marking pins.

double-bead (*Build*) Two side-by-side beads, separated by a QUIRK.

double-beam cathode-ray tube (*Electronics*) One containing two complete sets of beam-forming and beam-deflecting electrodes operated from the same cathode, thus allowing two separate waveforms to appear on the screen simultaneously (double-gun tube). Also it may use a single gun with a beam-switching circuit, ie two inputs continuously interchanged, or a single-beam tube in which the beam is split, and the two parts are separately deflected.

double-beat valve (*Eng*) A hollow cylindrical valve for controlling high-pressure fluids. The seatings at the two ends exposed to pressure are of only slightly different area, so that the valve is nearly balanced and easily operated.

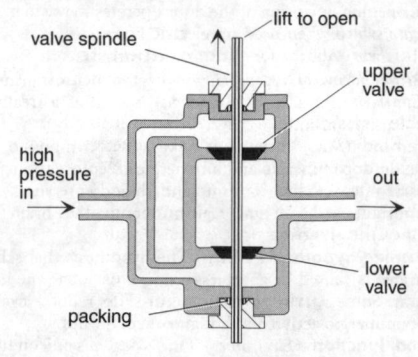

double-beat valve

double-bellied (*Arch*) A term applied to a baluster which has had both ends turned alike.

double beta decay (*Phys*) Energetically possible process involving the emission of two beta particles simultaneously. Not to be confused with DUAL BETA DECAY.

doublebind communication (*Psych*) Mutually contradictory messages which set up a conflict within the individual, who is simultaneously receiving the message that what is meant is the opposite of what is said. Believed by some to be a causative factor in schizophrenia, when a parent is consistently giving doublebind messages to a child.

double bond (*Chem*) Covalent bond involving the sharing of two pairs of electrons.

double-break (*ElecEng*) Said of switches or circuit breakers in which the circuit is made or broken at two points in each pole or phase.

double bridge (*ElecEng*) Network for measuring low resistances. See KELVIN BRIDGE.

double bridging (*Build*) Bridging in which two pairs of diagonal braces are used to connect adjacent floor joists at points dividing their length into equal parts.

double buffering (*ICT*) The use of two BUFFERS so that one may be filled while the other is being emptied.

double-catenary construction (*ElecEng*) A method of supporting the overhead contact wire of an electric traction system; the contact wire is suspended from two parallel catenary wires so that the three wires are in a triangular formation.

double cloth (*Textiles*) Two distinct cloths woven and bound together simultaneously to obtain greater thickness without affecting the face texture, eg heavy overcoatings.

double-coated film (*ImageTech*) A film base with emulsions on both sides.

double-coil loudspeaker (*Acous*) Electrodynamic loudspeaker with two driving coils separated by a compliance, the coils driving one or two open-cone diaphragms, thus operating more effectively over a wide range of frequencies.

double-cone loudspeaker (*Acous*) Large open coil-driven cone loudspeaker with a smaller free-edge cone fixed to the coil former, thus assisting radiation for high audio frequencies.

double contraction (*Eng*) Total shrinkage allowance necessary to add to the dimensions of a finished casting when making a wood pattern from which a metal pattern is to be cast.

double-cover butt joint (*Eng*) A butt joint with a cover plate on both sides of the main plates.

double-current furnace (*ElecEng*) A special form of electric furnace in which dc is used for an electrolytic process and ac for heating, on the principle of the induction furnace.

double-current generator (*ElecEng*) An electric generator which can supply both ac and dc.

double-cylinder knitting machine (*Textiles*) A CIRCULAR KNITTING MACHINE with two cylinders one above the other and having one set of double-ended needles that can knit in either cylinder. Often used for the manufacture of *hosiery*.

double decomposition (*Chem*) A reaction between two substances in which the atoms are rearranged to give two other substances. In general it may be written AB + CD → AC + BD.

double-delta connection (*ElecEng*) A method of connecting windings of a six-phase transformer etc, so that they may be represented diagrammatically by two triangles.

double-density disk (*ICT*) See DENSITY.

double-diffused metal–oxide semiconductor (*Electronics*) A metal–oxide semiconductor manufacturing process involving two stages of diffusion of impurities through a single MASK, enabling depletion-mode or enhancement-mode transistors to be produced on a chip. The technique keeps CHANNELS short so that the devices are suitable for high-speed logic or microwave applications. Abbrev *DMOS*.

double-diffused transistor (*Electronics*) One produced by using DOUBLE-DIFFUSED METAL–OXIDE SEMICONDUCTOR techniques.

double diffusion (*BioSci*) A test for detecting antigens and antibodies in which the two are arranged to diffuse towards one another in a gel (usually agar or agarose). Where they meet lines of precipitation form. Since each antigen forms a separate line the method is used for analysing purity of preparations, and for detecting antigens which share common determinants.

double diode (*Electronics*) Two diodes (usually vacuum tube diodes) in a single package or envelope. May have a common cathode for full-wave rectification.

double-disk winding (*ElecEng*) A form of winding used for transformers, in which two disk coils are wound in such a way that, when placed side by side to form a single coil, the beginning and end of the complete coil are at the outside periphery.

double-dovetail key (*Build*) A piece of wood used to connect together two members lengthwise; shaped and fitted like a SLATE CRAMP.

doubled yarns (*Textiles*) See FOLDED YARNS.

double earth fault (*ElecEng*) A fault on an electric power transmission system, caused by two phases going to earth simultaneously, either at the same point on the system or at different points.

double embedding (*BioSci*) A technique for embedding small objects, otherwise liable to distortion or disorientation, eg the specimen may be first orientated and embedded in celloidin and the small celloidin block embedded in hard paraffin wax.

double-ended boiler (*Eng*) A marine boiler of the shell type provided with a furnace at both ends, each with its independent tubes and uptake.

double-ended bolt (*Eng*) A bar screwed at each end for the reception of a nut.

double-entry compressor (*Aero*) A centrifugal compressor with double-sided impeller so that air enters from both sides.

double exposure (*ImageTech*) The intentional exposure of two (or more) separate images on the same photographic record; they may be superimposed or separated by masking.

double-faced hammer (*Eng*) A hammer provided with a flat face at each end of the head.

double fertilization (*BioSci*) The process, characteristic of angiosperms, in which two male nuclei enter the embryo sac. One fuses with the egg cell nucleus to form the zygote which develops into the embryo; the other typically fuses with the two polar nuclei to form a triploid nucleus from which the ENDOSPERM derives.

double Flemish bond (*Build*) A bond in which both exposed faces of a wall are laid in FLEMISH BOND.

double floor (*Build*) A floor in which the bridging joists are supported at intervals by binding joists. Cf FRAMED FLOOR, SINGLE FLOOR.

double-flow turbine (*Eng*) A turbine in which the working fluid enters at the middle of the length of the casing and flows axially towards each end.

double-glazing (*Build*) Hermetically sealed glazing with two sheets of glass about 22 mm apart, used where insulation and conservation of heat are important. Where sound insulation is the primary function the distance apart should be about 175 mm.

double-helical gears (*Eng*) HELICAL GEARS in which two sets of teeth having opposite inclinations are cut on the same wheel, thus eliminating axial thrust.

double-hump effect (*ICT*) Property of two coupled resonant circuits, each separately resonant to the same frequency, of showing maximum response to two frequencies disposed about the common resonance frequency.

double-hung window (*Build*) A window having top and bottom sashes, each balanced by sash cord and weights so as to be capable of vertical movement in its own grooves.

double-image micrometer (*BioSci, Min*) A microscope attachment for the rapid and precise measurement of small objects, like cells or particles. The principle is that two images of the object are formed by means which allow the separation of the images to be varied and its magnitude read on a scale. By setting the two images edge to edge (a very precise setting) and then interchanging them, the difference of the scale readings is a measure of the diameter of the object. The double image may be formed by birefringent crystals, prisms, an interferometer-like system or a vibrating mirror.

double-insulated (*ElecEng*) Describes portable electric appliances which have two completely separate sets of insulation between the current-carrying parts and any metal accessible to the user. An earth connection is, therefore, unnecessary.

double integral (*MathSci*) A summation of the values of a function $f(x, y)$ over a specified region R of the x, y plane. Written

$$\iint_R f(x, y) \, dx \, dy$$

double jersey (*Textiles*) A range of weft-knitted fabrics (rib or interlock) made on fine-gauge machines. The construction is usually chosen so that the fabric has reduced extensibility.

double junction (*Build*) A drainage or water-pipe fitting made with a branch on each side.

double layer (*Chem*) The zone adjacent to a charged particle in which the potential falls effectively to zero. See ELECTRICAL DOUBLE LAYER.

double-layer winding (*ElecEng*) An armature winding, always used for dc machines and frequently for ac machines, in which the coils are arranged in two layers, one above the other, in the slots.

double-length word (*ICT*) Hardware feature of many computers where two words (of, say, 16 bits) can be joined together and manipulated as a single (32 bit) word in the central processor.

double lock (*Build*) A construction consisting of two side-by-side lock chambers, across the same canal. They are interconnected through a sluice so that the amount of water lost in lockage is only half that which is lost by a single lock.

double magazine (*ImageTech*) A magazine holding two rolls of film feeding a camera or printer; one may be a MASK or MATTE through which the other, the raw stock, is exposed.

double-margined door (*Build*) A door hinged as one leaf but having the appearance of a pair of folding doors.

double moding (*Electronics*) Irregular switches of frequency by magnetron oscillator, due to changing mode.

double modulation (*ICT*) See COMPOUND MODULATION.

double oblique crystals (*Crystal*) See TRICLINIC SYSTEM.

double partition (*Build*) A partition having a cavity in which sliding doors may move.

double pica (*Print*) An old type size, approximately 22 points.

double-pipe exchanger (*ChemEng*) Heat exchanger formed from concentric pipes, the inner frequently having fins on the outside to increase the area. The essential flow pattern is that the two streams are always parallel.

double-pitch skylight (*Build*) A skylight having two differently sloped glazed surfaces.

double plating (*Print*) See TWO SET.

double point (*MathSci*) For the curve $f(x, y) = 0$: one at which

$$\frac{\delta f}{\delta x} = \frac{\delta f}{\delta y} = 0$$

and

$$\frac{\delta^2 f}{\delta x^2}, \frac{\delta^2 f}{\delta x \delta y} \text{ and } \frac{\delta^2 f}{\delta y^2}$$

are not all zero. A double point at which there are two real and distinct tangents, one real tangent or no real tangents, is called a *node*, *cusp* or *isolated point* respectively. A node is sometimes called a *crunode*, a cusp a *spinode* and an isolated point an *acnode* or *conjugate point*.

double-pole (*ElecEng*) Said of switches, circuit breakers, etc, which can make or break a circuit on two poles simultaneously.

double precision (*ICT*) Operation using WORDS of double normal LENGTH. See DOUBLE-LENGTH WORD.

double printing (*ImageTech*) The process of exposing a positive emulsion in a printing machine successively with more than one negative resulting in a superposition of two positive images after development.

double quirk-head (*Build*) A bead sunk into a surface so as to leave a QUIRK on each side. See fig. at BEAD-AND-QUIRK.

doubler (*ICT*) See FREQUENCY DOUBLER.

doubler circuit (*ElecEng*) (1) A form of self-saturating magnetic amplifier having an ac output. (2) One driven so that a frequency can be filtered from the output which is double the frequency of the input.

double reception (*ICT*) Simultaneous reception of two signals on different wavelengths by two receivers connected to the same antenna.

double refraction (*Phys*) The division of an electromagnetic wave in an anisotropic medium into two components propagated with different velocities, depending on the direction of propagation. In uniaxial crystals the components are called the *ordinary* ray where the wavefronts are spherical and the *extraordinary* ray where the wavefronts are ellipsoidal. In BIAXIAL crystals both wavefronts are ellipsoidal. Also *birefringence*.

double roof (*Build*) A roof in which the rafters are supported on purlins between walls.

double-row ball journal bearing (*Eng*) A rolling bearing which has two rows of caged balls running in separate

tracks. There are rigid, self-aligning and angular contact double-row ball journal bearings.

double-row radial engine (*Aero*) A RADIAL ENGINE where the cylinders are arranged in two planes, and operate on two crank pins, 180° apart.

doubles (*MinExt*) See COAL SIZES.

double salts (*Chem*) Compounds having two normal salts crystallizing together in definite molar ratios.

double scalp (*Vet*) A form of OSTEODYSTROPHIA of sheep thought to be caused by phosphorus deficiency; characterized by severe debility, anaemia, and weakening and thinning of the bones, notably the frontal bones.

double seam (*FoodSci*) Seam for sealing the ends of cans which can withstand the internal pressure caused by thermal processing or due to CARBONATION and retain any vacuum needed by a processed product. It is formed by the interlocking of the curled sections formed on the end of the can body (*body hook*) and on the rim of the can end (*cover hook*).

double series (*MathSci*) A series formed from all the terms of a two-dimensional array which extends to infinity in both dimensions, ie

$$\sum_{r,s=1}^{\infty} \alpha_{rs}$$

double shrinkage (*Eng*) See DOUBLE CONTRACTION.

double-sideband system (*ICT*) One that transmits both the sidebands produced in the modulation of a radio-frequency carrier wave. Cf SINGLE-SIDEBAND SYSTEM.

double-six array (*ICT*) A pair of six-element YAGI ANTENNAS placed side by side to improve directivity.

double spread (*Print*) Any pair of facing pages designed as one unit.

double squirrel-cage motor (*ElecEng*) A squirrel-cage motor with two cage windings on its rotor, one of high resistance and low reactance, and the other of high reactance and low resistance. The former carries most of the current at starting, and therefore gives a high starting torque, while the latter carries most of the current when running and results in a high efficiency.

double star (*Astron*) A pair of stars appearing close together as seen by telescope. They may be at different distances (*optical double*) or physically connected as an ECLIPSING BINARY, SPECTROSCOPIC BINARY or VISUAL BINARY.

double-stream amplifier (*Electronics*) A type of travelling-wave tube in which the operation depends upon the interaction of two electron beams of differing velocities.

double superhet receiver (*ICT*) Receiver in which amplification takes place at both high and low intermediate frequencies, thus requiring two frequency-changing stages.

doublet (*ImageTech*) A pair of simple lenses designed to be used together, so that optical distortion in one is balanced by reverse distortion in the other.

doublet (*Phys*) A pair of associated lines in a spectrum, such as the two D-lines of sodium. The arc spectra of the alkali metals consist entirely of series of doublets.

doublet antenna (*ICT*) See DIPOLE ANTENNA.

double tenons (*Build*) Two parallel tenons on the end of thick material.

double-threaded screw (*Eng*) A screw having two threads, whose PITCH is half the LEAD. Also *two-start thread*. See MULTIPLE-THREADED SCREW, DIVIDED PITCH.

double-throw switch (*ElecEng*) One which enables connections to be made with either of two sets of contacts.

double thrust bearing (*Eng*) A thrust bearing for taking axial thrust in either direction.

double-tone ink (*Print*) Ink for half-tone printing, containing a secondary pigment which spreads outwards from each dot while the ink is drying to give a richer effect to the illustration.

double triode (*Electronics*) Thermionic valve with two triode assemblies in the same envelope.

double-trolley system (*ElecEng*) A system of electric traction where, instead of the running rails, a second insulated contact wire is used for the return of negative current. It avoids trouble due to stray earth currents.

double-tuned circuit (*ElecEng*) A circuit containing two elements which may be tuned separately.

double wall (*Acous*) A wall of two plates with a layer of soft material between them. Above a certain frequency, the sound transmission of a double wall is much lower than that of a single wall of the same mass.

double-webbed girder (*Eng*) A built-up box girder in which the top and bottom booms are united by two parallel webs.

double-wedge aerofoils (*Aero*) See WEDGE AEROFOIL.

double-width press (*Print*) A newspaper press with a width of four standard pages.

double window (*Build*) A window arranged with double sashes enclosing air, which acts as a sound and heat insulator. See DOUBLE-GLAZING.

double-wire system (*ElecEng*) The usual system of electric wiring; it employs separate wires for the go and return conductors, instead of using the earth as a return.

doubling course (*Build*) A special course of slates laid at the eaves, to ensure that at the lowest margin there are two thicknesses of slate throughout its depth. Also *double course, eaves course*.

doubling frame (*Textiles*) A machine in which yarns are *folded* or twisted together in a simple manner to give stronger uniform products or to produce a wide variety of fancy effects. Also *twisting frame*.

doubling piece (*Build*) A TILTING FILLET.

doubling time (*NucEng*) (1) Time required for the neutron flux in a reactor to double. (2) In a breeder reactor, time required for the amount of fissile material to double.

doublures (*Print*) The inside of book covers lined with silk or leather and specially decorated.

doubly fed repulsion motor (*ElecEng*) A single-phase repulsion motor in which the armature receives its power partly by conduction and partly by induction.

doubly fed series motor (*ElecEng*) A single-phase series motor in which the armature receives its power partly by conduction and partly by induction.

dough moulding compound (*Plastics*) Unsaturated polyester, thermosetting moulding materials in the form of dough or putty, and usually reinforced with chopped glass fibres. Abbrev *DMC*. US *bulk molding compound* (*BMC*).

doughnut (*ElecEng*) (1) Anchor-ring shape used in circular-path accelerator tubes of glass or metal. (2) Traditional shape of a pile of annular laminations for magnetic testing, since there is no external field. (3) Traditional shape of loading coils on transmission lines, permitting exact balance in addition to inductance. Also *donut, toroid*.

Douglas bag (*Med*) A specially constructed bag for the collection of air expired from the lung, from the analysis of which the oxygen consumption of the body can be determined.

Douglas fir (*For*) A tree, *Pseudotsuga menziesii*, whose wood is the most important timber of the N American continent and one of the best-known softwoods in the world. The wood is light reddish-brown in colour, with prominent growth rings and fairly straight grain; it is moderately durable. Also *British Columbian pine, false hemlock, Oregon pine, red pine*.

Douglas's pouch (*Med*) In the female pelvis, the pouch of peritoneum between the rectum and the posterior wall of the uterus.

dourine (*Vet*) A contagious infection of breeding horses, characterized by inflammation of the external genital organs and paralysis of the hind-limbs; due to *Trypanosoma equiperdum*, which is transmitted through coitus. Also *mal du coit*.

douzième (*Genrl*) A unit of measurement used in the watch trade. See LIGNE.

dovetail (*Build*) A joint formed between a flaring tenon, having a width diminishing towards the root, and a corresponding recess or mortise. Also *swallowtail*.

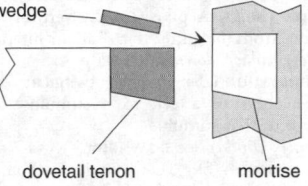

dovetail Showing a blind tenon and wedge.

dovetail halving (*Build*) A form of HALVING in which the mating parts are cut to a dovetail shape.

dovetail hinge (*Build*) A hinge whose leaves increase in width outwards from the hinge joint.

dovetail key (*Build*) A batten, of dovetail-shaped section, which is driven into a corresponding groove cut across the back of adjacent boards in a panel, and serves to prevent warping.

dovetail key (*Eng*) A parallel KEY in which the part sunk in the boss or hub is of dovetail section, the portion on the shaft being of rectangular section.

dovetail mitre (*Build*) See SECRET DOVETAIL.

dovetail saw (*Build*) A saw similar to the tenon saw but of smaller size and having usually twelve teeth to the inch.

dowel (*Build*) A copper, slate, non-ferrous metal or stone pin sunk into opposing holes in the adjacent faces of two stones to give a joint stronger than mortar.

dowel (*Eng*) (1) A pin fixed in one part which, by accurately fitting in a hole in another attached part, locates the two, thus facilitating accurate reassembly. (2) A pin similarly used for locating divided patterns.

dowelling jig (*Build*) A device for directing the bit in drilling holes to receive dowels.

dowel screw (*Build*) See HAND-RAIL SCREW.

down (*ICT*) A term describing a computer when it is not available for normal use.

down (*Phys*) One of the six FLAVOURS of QUARKS with a mass of 10 MeV and a charge of $-e/3$. Protons and neutrons are made up of down and UP quarks. See appendix on Subatomic particles.

downcast (*MinExt*) A contraction for *downcast shaft*, ie the shaft down which fresh air enters a mine. The fresh air may be sucked or forced down the shaft.

downcomer (*Build*) See DOWNPIPE.

downcomer (*Eng*) See DOWNTAKE.

down converter (*ImageTech*) A circuit which converts the high-frequency signals from a DIRECT BROADCAST SATELLITE to a lower, usable frequency.

downdraught (*Autos*) Said of a carburettor in which the mixture is drawn downwards in the direction of gravity.

downdraught (*EnvSci*) The downward draught of air occurring with the approach of a thunderstorm and due to evaporative cooling of descending air by heavy rain.

downer (*Print*) A sudden breaking of the web on a rotary machine, leading to DOWN TIME.

down feathers (*BioSci*) See PLUMULAE.

downhole (*MinExt*) Describes the drills, measuring instruments and equipment used down the borehole.

downlead (*ICT*) See ANTENNA DOWNLEAD.

down link (*ImageTech*) The transmission path from a COMMUNICATIONS SATELLITE to a receiving DISH.

downlink (*ICT*) (1) The radio link from a communications satellite to a ground station. (2) The radio link from a mobile-telephone base station to the mobile telephone.

download (*ICT*) The process of transferring files from one computer to another, usually from a large host computer to a smaller one. The reverse process is upload.

downloadable font (*ICT*) A FONT that is loaded from BACKING STORE when required by the printer or SOFTWARE PACKAGE. Also *soft font*.

down locks (*Aero*) See UP, DOWN LOCKS.

downpipe (*Build*) A pipe (usually vertical) for conveying rainwater from the gutter to a drain or intermediate gulley. Also *downcomer, downspout, fall pipe*.

down-regulation (*BioSci*) The subsequent reduction in the responsiveness of a cell to a stimulus, after the first exposure to that stimulus.

downspout (*Build*) See DOWNPIPE.

Down's process (*Eng*) Electrolytic method of producing sodium metal and chlorine from fused salt at 600°C.

Down's syndrome (*Med*) A form of mental retardation caused by a chromosomal abnormality, trisomy 21; the main features are moderate to severe retardation, a small round head, slanting eyes and minor abnormalities of hands and feet. These children also often have congenital heart lesions. Incidence increases with the mother's age. Formerly *mongolism*.

downstream (*BioSci*) DNA or RNA that is further away from the initiation site on the gene and thus transcribed or translated later.

downtake (*Eng*) The pipe through which the blast furnace gas is taken down outside the furnace from the top of the furnace to the duct catcher. Also *downcomer*.

downthrow (*Geol*) In a FAULT, the vertical displacement of the fractured strata. Indicated on geological maps by a tick attached to the fault line, with (where known) a figure alongside indicating the amount of downthrow.

down time (*Eng*) Time during which a machine, eg a computer or printing press (or series of machines), is idle because of adjustment, cleaning, reloading or other maintenance.

downtime (*ICT*) The percentage of time over a period in which a system is not available for use. As distinct from AVAILABILITY.

Downtonian (*Geol*) The lowest stage of the Old Red Sandstone facies of the Devonian System, named from its typical development around Downton Castle in the Welsh Borderlands.

downward modulation (*ICT*) See NEGATIVE MODULATION.

downwash (*Aero*) The angle through which the airflow is deflected by the passage of an aerofoil measured parallel with the plane of symmetry.

downy mildew (*BioSci*) One of several plant diseases of eg vines, onions, lettuce, caused by biotrophic fungi of the family Peronosporaceae. See OOMYCETES.

Dow process (*Eng*) The extraction of magnesium from sea water by precipitation as hydroxide with lime; also, electrolytic production of magnesium metal from fused chloride.

dowsing (*Genrl*) The process of locating underground water by the twitching of a twig (traditionally hazel or witch-hazel) held in the hand. The phenomenon, if it exists, continues to defy rational explanation, and is accordingly frequently scorned.

doxapram hydrochloride (*Pharmacol*) Centrally acting respiratory stimulant sometimes used in the treatment of severe respiratory failure.

doxazosin (*Pharmacol*) An alpha blocker used to treat hypertension and prostatic hyperplasia.

doxorubicin (*Pharmacol*) A bacterial antibiotic widely used as a cytotoxic drug in the treatment of various forms of cancer.

doxycycline (*Pharmacol*) A broad-spectrum tetracycline antibiotic used for treatment of, among other things, chronic bronchitis, brucellosis, chlamydial infections and rickettsias.

DP (*ICT*) See DATA PROCESSING.

dpi (*ICT*) Abbrev for *dots per inch*. The resolution of a printer is often quoted in dpi. An adequate resolution for general purposes is 300 dpi.

Draco (Dragon) (*Astron*) The eighth largest constellation, a sinuous zone of the northern sky.

dracone (*Ships*) A flexible sausage-shaped envelope of woven nylon fabric coated with synthetic rubber. Floats by reason of buoyancy of its cargo, oil or fresh water. It is towed.

draft (*Genrl*) See DRAUGHT.

draft angle (*Eng*) Slight angle between core and tool to aid the withdrawal of the product from mould at the end of the moulding cycle.

draft ewes (*Agri*) Sexually mature female sheep sold on from an existing flock, commonly from upland to lowland farms.

drafting (*Textiles*) The drawing out or attenuation of the WEB of textile fibres passing through a card, drawframe, speedframe, or spinning machine. Measured by the ratio of the linear density of input to output materials. Cf DRAW RATIO.

draft quality (*ICT*) A low-RESOLUTION printing mode whereby printing may take place at higher speed but at lower quality or whereby the file size and consequent memory capacity may be reduced to increase processing speed.

draftsperson (*Eng*) See DRAUGHTSPERSON.

draft stop (*Build*) See FIRE STOP.

draft tube (*Eng*) A discharge pipe from a water turbine to the tail race. It decreases the pressure at outlet and increases the turbine efficiency.

drag (*Aero*) Resistance to motion through a fluid. As applied to an aircraft in flight it is the component of the resultant force due to relative airflow measured along the drag axis, ie parallel to the direction of motion.

drag (*Build*) (1) A steel-toothed tool for dressing stone surfaces or for keying plasterwork. (2) A three-pronged rake for mixing plaster.

drag (*Eng*) The bottom half of a moulding box or flask. See fig. at MOULDING.

drag (*ICT*) To move an item on the screen using a mouse.

drag angle (*Eng*) The angle between the welding rod and the normal to the surface of the weld.

drag axis (*Aero*) A line through the centre of mass of an aircraft parallel with the relative airflow, the positive direction being downwind.

drag chain conveyor (*Eng*) See DRAG CONVEYOR.

drag classifier (*MinExt*) Endless belt with transverse rakes, which moves upwards through an inclined trough so as to drag settled material up and out while slow-settling sands overflow below as pulp arrives for continuous sorting into coarse and fine sands.

drag conveyor (*Eng*) A conveyor in which an endless chain, having wide links carrying projections or wings, is dragged through a trough into which the material to be conveyed is fed. Also *drag chain conveyor*.

drag-cup generator (*ElecEng*) A servo unit used to generate a feedback signal proportional to the time derivative of the error.

dragged work (*Build*) Stone-dressing done with a DRAG.

dragging-beam (*Build*) The horizontal timber on which the foot of the hip rafter rests. Also *dragon beam*.

drag hinge (*Aero*) The pivot of a rotorcraft's blade which allows limited angular displacement in azimuth.

dragline excavator (*CivEng*) A mechanical excavator comprising a crane with a drag bucket suspended by the hoist rope. The bucket is filled by letting it swing out and then dragging it back by means of a drag rope working from a drum fixed to the front of the machine. The dragline excavator can work at great depth and long radius.

drag link (*Autos*) A link which conveys motion from the drop arm of a steering gear to the steering arm carried by a stub axle, which it connects through ball joints at its ends. Also *steering rod*.

drag link (*Eng*) A rod by which the link motion of a steam engine is moved for varying the cut-off. See LINK MOTION.

dragon-beam (*Build*) See DRAGGING BEAM.

dragon's blood (*Build, Print*) A resinous exudation from the fruit of palm trees and the stems of different species of *Dracaena*. It is a red, amorphous substance, mp 120°C, soluble in organic solvents. By destructive distillation the resin yields methylbenzene and phenylethene. It is used for colouring varnishes and lacquers; also, in photoengraving, to protect parts of a plate in the etching process.

dragon tie (*Build*) See ANGLE BRACE.

drag struts (*Aero*) Structural members designed to brace an aerofoil against air loads in its own plane. Also, landing-gear struts resisting the rearward component of impact loads.

drag wires (*Aero*) Streamlined wires or cables bracing an aerofoil against drag (rearward) loads. Applicable to biplanes and some early monoplanes.

drain (*Electronics*) In a field-effect transistor, the region into which majority CARRIERS flow from the CHANNEL; comparable with the collector in a conventional bipolar transistor.

drain (*Med*) Any piece of material, such as a plastic tube, used for directing away the discharges of a wound.

drainage (*Geol*) The removal of surplus meteoric waters by rivers and streams. The complicated network of rivers is related to the geological structure of a district, being determined by the existing rocks and superficial deposits in the case of youthful drainage systems; but in those that are mature, the courses may have been determined by strata subsequently removed.

drainage (*Med*) The action of draining discharges from wounds or infected areas.

drainage coil (*ElecEng*) A coil bridged between the legs of a communication pair, with its electrical mid-point earthed, in order to prevent the accumulation of static charges on the conductors.

drainage patterns (*Geol*) The complicated network of rivers and streams removing surplus meteoric waters and related to the underlying geological structure. Determined by the existing rocks and superficial deposits in youthful drainage systems, but in more mature ones, often by strata subsequently removed. In youthful systems *consequent streams* run down dip. Later, *subsequent streams* are controlled by the varying resistance to erosion of geological features. A stream or river flowing against the dip of strata is *obsequent*. If the original direction of flow has been reversed, the stream is reversed. See STREAM ORDER.

drain cock (*Eng*) A cock placed (1) at the lowest point of a vessel or space, for draining off liquid; (2) in an engine cylinder, for discharging condensed steam.

drainer (*Paper*) A large compartment with perforated tiles or metal plates in the base for the purpose of allowing water to drain from stuff deposited in it.

drain holes (*MinExt*) Draining oil from strata by boring holes out from the main bore by the use of special DOWNHOLE machinery such as MUD-MOTOR-driven drills.

drain plug (*Build*) A device for closing the outlet of drain pipes. See BAG PLUG, SCREW PLUG.

drain tiles (*CivEng*) Hollow clay pipes or tiles laid end to end without joints, to carry off surface or excess water. Now largely superseded by perforated flexible plastic pipe.

drain-trap (*Build*) See AIR-TRAP.

Dralon (*Chem*) TN for German polyacrylonitrile staple fibre.

DRAM (*ICT*) Abbrev for *dynamic random access memory*. See DYNAMIC MEMORY, RAM.

Draper effect (*Chem*) See PHOTOCHEMICAL INDUCTION.

draught (*Eng*) (1) The flow of air through a boiler furnace. (2) A measure of the degree of vacuum inducing airflow through a boiler furnace.

draught (*Ships*) The depth of water that a ship requires to float freely. More particularly; draught forward and draught aft, the depths of water required at the forward and after perpendiculars respectively. Mean draught is draught at midlength.

draught bead (*Build*) Same as DEEP BEAD.

draught excluder (*Build*) See DOOR STRIP.

draught gauge (*Eng*) A sensitive vacuum gauge for indicating the draught in a boiler furnace or flue.

draughtsperson (*Eng*) One who makes engineering drawings from which prints, at one time blueprints but now usually dye-lines, are made for actual use. See TRACING. The draughtsperson who makes such drawings frequently designs the details, the main design being laid down by an engineer or architect. Also *draftsperson* or technical artist.

dravite (*Min*) Brown tourmaline, sometimes used as a gemstone.

draw (*Eng*) Internal cavity or spongy area occurring in a casting owing to inadequate supplies of molten metal during consolidation. See DRAWING OF PATTERNS.

draw (*MinExt*) (1) To allow ore to run from working places, stopes, through a chute into trucks. (2) To withdraw timber props or sprags from overhanging coal, so that it falls down ready for collection. (3) To collect broken coal in trucks.

draw (*Textiles*) The outward (drafting and twisting) and inward (winding) run of a MULE spinning carriage.

drawability (*Eng*) A measure of the ability of a material to be drawn, as in forming a cup-like object from a flat metal blank.

draw-bar cradle (*Eng*) A closed frame or link for connecting the ends of the draw-bars of railway vehicles, so coupling them together.

draw-bar plate (*Eng*) On a locomotive frame, a heavy transverse plate through which the draw-bar is attached.

draw-bar pull (*Eng*) The tractive effort exerted, in given circumstances, by a locomotive or tractor.

draw-bore (*Build*) A hole drilled transversely through a mortise and tenon so that, when a pin is driven in, it will force the shoulders of the tenon down upon the abutment cheeks of the mortise.

draw-bridge (*CivEng*) A general name for any type of bridge of which the span is capable of being moved bodily to allow the passage of large vessels.

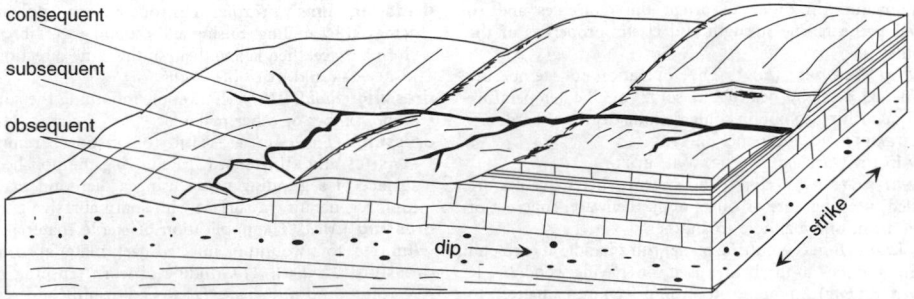

drainage patterns Different types of stream indicated.

drawdown (*MinExt*) The fall of water level in a natural reservoir, eg an AQUIFER, when rate of extraction exceeds rate of replenishment.

drawdown (*Print*) A method of checking the hue of a printing ink sample by scraping it down onto a standard paper alongside a standard sample to produce ink films with graduated density. Comparison of the two can give an indication of the correctness of hue, rate of absorption and drying ability.

drawer-front dovetail (*Build*) See LAP DOVETAIL.

drawer-lock chisel (*Build*) A crank-shaped chisel with two edges, one being parallel to the shank and the other at right angles to it. Used for chopping drawer-lock recesses or other work in restricted positions.

draw-filing (*Eng*) The operation of finishing a filed surface by drawing the file along the work at right angles to the length of the file.

draw-gate (*Build*) A name given to the valve controlling a SLUICE.

draw-in box (*ElecEng*) A box or pit to enable cables to be drawn into, or removed from, a conduit or duct.

drawing (*Eng*) (1) The process of producing wire, or giving rods a good finish and accurate dimensions, by pulling through one or a series of tapered dies. (2) The process of hot or cold plastic stretching of polymeric fibre, tape, rod or sheet to orient the molecules and so increase the tensile strength and elastic modulus in the draw direction. (3) Alternative term for TEMPERING applied to the softening of steels which have been hardened by quenching.

drawing (*Textiles*) Running together and attenuation of a number of slivers or tops of staple fibres, preparatory to making SLUBBINGS and ROVINGS.

drawing-down (*Eng*) The operation of reducing the diameter of a bar, and increasing its length, by forging. Cf COLD DRAWING.

drawing fires (*Eng*) The operation of raking out fires from boiler furnaces when shutting down.

drawing-in (*Textiles*) Drawing the warp yarns through the eyes of the loom heald preparatory to weaving.

drawing of patterns (*Eng*) The removal of a pattern from a mould; also termed lifting of patterns. It is facilitated by the taper or draught of the pattern, and by loosening the pattern by RAPPING.

drawing of tubes (*Eng*) See TUBE DRAWING.

drawing rollers (*Print*) See DRAW ROLLERS.

drawing rollers (*Textiles*) Pairs of steel rollers, each pair running at higher speed to draft the roving passing through. The front pair may comprise a fluted bottom roller with the top one covered with leather, rubber or cork.

drawing staple fibres (*Textiles*) (1) Running together and attenuation of a number of slivers, preparatory to making, slubbing and roving. (2) The operation in which worsted tops are reduced to a roving.

drawing synthetic continuous filaments (*Textiles*) Hot or cold stretching of synthetic filaments during their manufacture in order to orient the molecules and so develop the tensile strength and elastic properties of the filaments.

drawing temper (*Eng*) The operation of tempering hardened steel by heating to some specific temperature and quenching to obtain some definite degree of hardness. See TEMPERING and panel on STEELS.

draw-in pit (*ElecEng*) See DRAW-IN BOX.

draw-in system (*ElecEng*) The system whereby cables are pulled into conduits or ducts of earthenware, concrete or iron, from one manhole to another.

draw knife (*Build*) A cutting blade with a handle at each end at right angles to the blade; used for shaving wood.

draw-nail (*Eng*) A pointed steel rod driven into a pattern to act as a handle for withdrawing it from the mould. See DRAW-SCREW.

drawn on (*Print*) Said of a book cover which is attached by gluing down the back; if pasted down to endpapers it is said to be *drawn on solid*.

drawn-wire filament (*ElecEng*) An incandescent lamp filament, made by a wire-drawing process as opposed to a squirting process.

draw-off valve (*Eng*) A BIB-VALVE.

draw-out metal-clad switchgear (*ElecEng*) Metal-clad switchgear in which the switch itself can be isolated from the bus-bars for inspection and maintenance, by moving it away from the bars along suitable guides.

draw ratio (*Eng*) (1) The ratio of deformed to undeformed length in a polymer sheet during eg vacuum forming. Usually about 2:1 overall, but may reach 5:1 at corners of product. (2) Longitudinal extension of polymer fibre, tape or rod in post-production operation to increase modulus of product. Normally conducted while material is still hot. Draw ratios up to 72 times may be used in gel-spun polymers like Dyneema fibre. See DRAWING.

draw ratio (*Textiles*) The ratio of the linear density of the undrawn yarn to that of the drawn yarn. See DRAWING SYNTHETIC CONTINUOUS FILAMENTS.

draw rollers (*Print*) A pair of rollers, usually both driven, which control the web tension on rotary presses. Also *drawing rollers*.

draw-screw (*Eng*) A screwed rod provided with an eye at the end to act as a handle; screwed into a pattern for lifting it from the mould. See DRAW-NAIL.

draw sheet (*Print*) (1) The sheet drawn over the completed make-ready on a press before proceeding with the printing. (2) The top sheet of a CYLINDER DRESSING. (3) See SHIM.

draw works (*MinExt*) Surface gear of a drilling rig.

dream interpretation (*Psych*) In psychoanalytic theory, a technique for understanding the individual's unconscious life by focusing on dream content and attempting to unravel its hidden meaning, which reflects unconscious wishes and conflicts. See FREUD'S THEORY OF DREAMS.

dredge (*MinExt*) Barge or twin pontoons carrying a chain of digging buckets, or suction pump, with over-gear such as jigs, sluices, trommels, tailing stackers, manoeuvring anchor-lines, and power producer. Used to work alluvial deposits of cassiterite, gold, gemstones, etc.

dredger excavator (*Eng*) See BUCKET-LADDER EXCAVATOR.

dreikanter (*Geol*) A wind-faceted pebble typically having three curved faces (from Ger *drei*, three). Common in desert deposits. See VENTIFACT, zweikanter.

dressed carcass weight (*Agri*) Weight of a prepared carcass ready for sale, typically 50–60% of the live weight. Abbrev DCW.

dressed timber (*Build*) Timber which has been planed more or less to size.

dresser (*Build*) A boxwood tool for straightening lead piping and sheet lead.

dresser (*Eng*) (1) An iron block used in forging bent work on an anvil. (2) A tool for facing and grooving millstones, or for trueing grinding wheels.

dress-face finish (*Textiles*) Lustrous nap on woollen cloths achieved by milling, raising and cropping the fibres at the cloth surface, then laying them in the same direction. Used in BEAVER and DOESKIN cloths.

dressing (*Build*) The operation of smoothing the surface of stone, timber or other material.

dressing (*Eng*) (1) FETTLING of castings, removal of FLASHES and RUNNERS. (2) Removing the polished outer surface of a grinding wheel using a diamond or carbide tool, regenerating a surface with sharp abrasive grains.

dressing (*Med*) The application of sterile material, gauze, lint, etc, to a wound or infected part; material so used.

dressing (*MinExt*) (1) Grinding of worn crushing rolls, to restore cylindrical shape. (2) Rock crushing and screening to required sizes. (3) Preparation of amalgamation plates with liquid mercury for gold recovery.

dressings (*Build*) The mouldings, quoins, strings and like features in a room or building.

drier (*Chem*) A substance accelerating the drying of vegetable oils, eg linseed oil in paints. The most important representatives of this group are the naphthenates, resinates and oleates of lead, manganese, cobalt, calcium and zinc.

drier (*Eng*) Furnace used to de-water ore products without changing their composition.

drift (*Aero*) (1) The motion of an aircraft in a horizontal plane due to crosswind. (2) Slow unidirectional error of instrument or gyroscope.

drift (*CivEng*) The direction in which a tunnel is driven.

drift (*Electronics, ICT*) Slow variation of any performance characteristic (gain, frequency, power output, noise level, etc) of any device or circuit. May be due to gradual self-heating of equipment, ambient temperature or ageing. In particular the tendency of any tuned device (receiver, oscillator, etc) to move slowly away from its intended or selected frequency of operation.

drift (*Eng*) (1) A tapered steel bar used to draw rivet holes into line in 'drifting'. (2) A brass or copper bar used as a punch.

drift (*Genrl*) The rate of flow of a current of water.

drift (*Geol*) A general name for the superficial, as distinct from the solid, formations of the Earth's crust. It includes typically the Glacial Drift, comprising all the varied deposits of boulder clay, outwash gravel, and sand of Quaternary age. Much of the drift is of fluvio-glacial origin.

drift (*MinExt*) (1) A level or tunnel pushed forward underground in a metal mine, for purposes of exploration or exploitation. The inner end of the drift is called a *dead end*. (2) A heading driven obliquely through a coal seam. (3) A heading in a coal mine for exploration or ventilation. (4) An inclined haulage road to the surface. (5) Deviation of borehole from planned course. See fig. at MINING.

drift angle (*Aero, Ships*) The angle between the planned course and the track. Also sometimes used for the angle between heading and track made good.

drift chamber (*Phys*) A particle detector used in high-energy physics experiments to record the tracks of charged particles in interactions.

drift current (*Electronics*) That resulting from the drift of charge carriers in a local electric field as distinct from a DIFFUSION CURRENT.

drift currents (*EnvSci*) Ocean currents produced by prevailing winds.

drifter (*MinExt*) A cradle-mounted compressed-air rock drill, used when excavating tunnels (*drifts* or *cross cuts*).

drifting (*MinExt*) Tunnelling along the strike of a lode, horizontally or at a slight angle.

drifting test (*Eng*) A workshop test for ductility; a hole is drilled near the edge of a plate and opened by a conical DRIFT until cracking occurs.

drift mobility (*Electronics*) See MOBILITY.

drift sight (*Aero*) A navigational instrument for measuring DRIFT ANGLE.

drift space (*Electronics*) In an electron tube which depends on the velocity modulation of electron beams, the space which is free of externally applied alternating fields; here, relative repositioning of the electrodes takes place.

drift transistor (*Electronics*) One in which resistivity increases continuously between emitter and collector junctions, improving high-frequency performance.

drift tube (*Electronics*) A piece of metal tubing, held at a fixed potential, which forms the DRIFT SPACE in a microwave tube or linear accelerator.

drill (*Eng*) (1) A revolving tool with cutting edges at one end, and having flats or flutes for the release of chips; used for making cylindrical holes in metal. (2) Also the DRILLING MACHINE which turns the drill. (3) The operation of making a hole in a workpiece.

drill (*MinExt*) (1) Hand drill, AUGER. (2) A compressed-air-operated rock drill, jackhammer, PNEUMATIC DRILL. (3) Generally, the more elaborate equipment required in *power drilling*. See panel on DRILLING RIG.

drill (*Textiles*) A heavy woven fabric (often cotton) with diagonal lines on the surface.

drill bit (*Build*) Any tool designed for boring a cylindrical hole in wood, metal or masonry. A range of special types with various patterns of flute, designed for use in different materials, with or without percussive action, are available.

drill bit (*MinExt*) The actual cutting or boring tool in a *drill*. In rotary drilling it may be of the *drag* variety, with two or more cutting edges (usually hard-tipped against wear), ROLLER (with rotating hard-toothed rollers) or DIAMOND, with cutting face (annular if core samples are required) containing suitably embedded BORTS (the fig. shows the tricone design). The assembly is attached to the bottom of the DRILL PIPE and rotated with it. DRILLING MUD pumped down the drill pipe cools, lubricates and carries away the debris. See panel on DRILLING RIG.

drill bush (*Eng*) A hard sleeve inserted in a DRILLING JIG to locate and guide a twist drill accurately, in repetition drilling.

drill chuck (*Eng*) A self-centring chuck usually having three jaws which are contracted onto the drill by the rotation of an internally coned sleeve encasing them.

drill collar (*MinExt*) A collar which attaches the drill to the drill string. In deep bores it can weigh 100 tons and is used to damp out torsional stresses and stabilize the drill string. See panel on DRILLING RIG.

drill down (*ICT*) To navigate from a higher to a lower level of a hierarchically structured database.

drilled and strung (*Print*) A method of binding in which holes are drilled close to the back and the leaves then secured by thread or cord; superior to and more expensive than either FLAT STITCHING or STABBING.

drill-extractor (*CivEng*) A tool used to remove from a boring a broken drill or one which has fallen free of the drilling apparatus.

drill feed (*Eng*) The hand- or power-operated mechanism by which a drill is fed into the work in a drilling machine.

drilling (*MinExt*) (1) The operation of boring a hole in the ground, usually for exploration or for the extraction of oil, gas, water or geothermal energy. See panel on DRILLING RIG. (2) The operation of tunnelling or stoping. (3) The operation of making short holes for blasting, or deep holes with a diamond drill for prospecting or exploration. Drilling may be *percussive* (repeated blows on the drilling tool) or *rotary* (circular grinding). Cf BORING.

drilling jig (*Eng*) A device used in repetition drilling, which locates and firmly holds the workpiece accurately in relation to a DRILL BUSH or pattern of drill bushes accurately positioned in the jig.

drilling machine (*Eng*) A machine tool for drilling holes, consisting generally of a vertical standard, carrying a table for supporting the work and an arm provided with bearings for the drilling spindle. See PILLAR DRILL, RADIAL DRILL, SENSITIVE DRILL.

drilling mud (*MinExt*) A mixture of clays, water, density-increasing agents like BARITE and sometimes thixotropic agents like *bentonite* pumped down through the drilling pipe and used to cool, lubricate and flush debris from the drilling assembly. It also helps to seal the bored rock and most importantly provides the hydrostatic pressure to contain the oil and gas when this is reached. MUD MOTORS powered by the mud flow can be used for drilling, particularly in directions away from the main bore axis. See panel on DRILLING RIG.

drilling pipe (*MinExt*) The tubes, joined by screwing together, which connect the KELLY to the drill bit and impart rotary motion to the latter. Also *drill pipe*. See panel on DRILLING RIG.

drilling platform (*MinExt*) Offshore platform which may be floating or fixed to the sea bed and from which over 50 wells may be bored radially and at various angles into the bearing strata. See panel on DRILLING RIG.

drilling rig (*MinExt*) Derrick, surface equipment and related structures used in oilfield exploration. Cf PRODUCTION PLATFORM. See DRILL BIT, DRILLING RIG FLOOR, DRILL STRING and panel on DRILLING RIG.

drilling rig floor (*MinExt*) Horizontal platform on a drill rig where the power-driven rotary table rotates the drill string and drill bit. The rotary table is coupled to the kelly, a pipe of external hexagonal or square section at the top of the drill string, by means of the kelly bushing which has a hole of similar section at its centre. As the drilling proceeds successive lengths of pipe are inserted between the kelly and the top of the drill string. Drilling mud is pumped down the drill pipe to contain the well pressure, cool and lubricate the bit, and bring up the cuttings in the space between the drill pipe and the casing or walls of the hole. See DRILL BIT, DRILL STRING and panel on DRILLING RIG.

drill pipe (*MinExt*) See DRILLING PIPE.

drill string (*MinExt*) The pipe and BOTTOM-HOLE ASSEMBLY in a borehole. See panel on DRILLING RIG.

drill up (*ICT*) To navigate from a lower to a higher level of a hierarchically structured database.

Drinker respirator (*Med*) A respirator in which the whole body, excluding the head, is placed. It assists respiration by moving the chest. Commonly *iron lung*.

driography (*Print*) A planographic printing process akin to lithography whereby the antipathy of the image and non-image areas on the plate is maintained using special plate coatings and inks. Usually relies on the properties of dissimilar silicones.

drip (*Build, CivEng*) A groove in the projecting undersurfaces of a coping brick or stone wider than the wall; designed to prevent water from passing from the coping to the wall. Also *gorge, throat*. See fig. at CHECK THROAT.

drip-dry (*Textiles*) Fabrics and garments that shed creases and wrinkles when hung out wet and allowed to dry. Some light ironing may still be required before wearing. Cotton fabrics require a special treatment with resin-forming compounds to confer drip-dry properties on them.

drip-feed lubricator (*Eng*) A small reservoir from which lubricating oil is supplied in drops to a bearing, sometimes through a glass tube to render the rate of feed visible. See SIGHT-FEED LUBRICATOR.

drip mould (*Build*) A projecting moulding arranged to throw off rainwater from the face of a wall.

dripping eave (*Build*) An eave which is not fitted with a gutter and which therefore allows the rain to flow over to a lower roof or to the ground.

drip-proof (*ElecEng*) Said of an electric machine or other electrical equipment which is protected by an enclosure whose openings for ventilation are covered with suitable cowls, or other devices, to prevent the ingress of moisture or dirt falling vertically.

drip-proof burner (*Eng*) Gas burner designed to prevent the choking of flame ports or nozzles by foreign matter that may drip or fall on to it.

dripstone (*Build*) A projecting moulding built in above a doorway, window opening, etc, to deflect rainwater.

drip tip (*BioSci*) A marked elongation of the tip of the leaf, said to facilitate the shedding of rain from the surface of the leaf.

drive (*ElecEng*) A signal applied to the input of an amplifier, eg current to base of bipolar transistor and voltage to gate of field-effect transistor.

drive (*MinExt*) A tunnel or level driven along or near a lode, vein or massive ore deposit as distinct from country rock. *Driving* is the process.

drive (*ICT*) That which controls a master resonator in an oscillator, eg CAVITY, QUARTZ CRYSTAL, RESONATING LINE. See also DISK DRIVE and LOGICAL DRIVE.

drive (*Psych*) An internal motivation to fulfil a need or avoid the negative aspects of an unpleasant situation. The concept is now considered problematic although needs, eg for food or sleep, certainly exist.

drive-in (*ImageTech*) Open-air cinema where spectators can view the programme from their parked cars, with individual loudspeakers reproducing the sound.

drivenail (*Eng*) See SCREWNAIL.

driven elements (*ElecEng*) Those in an antenna which are fed by the transmitter, as compared with reflector, director or parasitic elements.

driven roller (*Print*) A roller geared to the press drive and used to take the web from one section of the press to another, maintaining tension in the process.

driver (*ICT*) A piece of software that controls a device.

drive-reduction hypothesis (*Psych*) In learning theory, the idea that reinforcing stimuli must reduce some drive or need in an animal if learning is to occur.

driver plate (*Eng*) A disk which is screwed to the mandrel nose of a lathe, and carries a pin which engages with and drives a CARRIER attached to the work. Also *driving plate*.

driver stage (*ICT*) Amplifiers that drive the final stage in a radio transmitter.

driving axle (*Eng*) A vehicle axle through which the driving effort is transmitted to the wheels fixed to it. Also *live axle*.

driving chain (*Eng*) An endless chain consisting of steel links which engage with toothed wheels, so transmitting power from one shaft to another. See ROLLER CHAIN.

driving fit (*Eng*) A degree of fit between two mating pieces such that the inner member, being slightly larger than the outer, must be driven in by a hammer or press.

driving gear (*Eng*) Any system of shafts, gears, belts, chains, links, etc, by which power is transmitted to another system.

driving point impedance (*ElecEng*) The ratio of the emf at a particular point in a system to the current at that point.

driving side (*Eng*) The tension side of a driving belt; the side moving from the follower to the driving pulley.

driving wheel (*Eng*) (1) The first member of a train of gears. (2) The road wheels through which the tractive force is exerted in a locomotive or road vehicle.

drogue (*Aero*) A sea anchor used on seaplanes and flying boats; it is a conical canvas sleeve, open at both ends, like a bottomless bucket. Used to check the way of the aircraft.

drogue parachute (*Aero, Space*) A small parachute used (1) to slow down a descending aircraft or spacecraft, (2) to extract a larger parachute or (3) to extract cargo from a hold or wing mounting.

drone (*Aero*) Pilotless guided aircraft used as a target or for reconnaissance.

drone (*BioSci*) In social bees (Apidae), a male.

droop snoot (*Aero*) Cockpit section hinged onto main fuselage to provide downward visibility at low speeds. Colloq for *droop nose*.

drop (*BioSci*) Premature abscission of fruit esp when half-grown, eg June drop, or when almost mature, eg pre-harvest drop.

drop (*ElecEng*) Common term for VOLTAGE DROP.

drop (*ICT*) Digits are said to 'drop in' if they are recorded without a signal, and 'drop out' if not recorded from a signal.

drop (*MinExt*) The vertical displacement in a downthrow fault: the amount by which the seam is lower on the other side of the fault. See THROW.

drop arm (*Autos*) A lever attached to a horizontal spindle which receives rotary motion from the steering gear; used to transmit linear motion through attached steering rod or drag link to the arms carried by the stub axles.

drop cable (*ICT*) A cable that links a network device to an external TRANSCEIVER attached to a coaxial-cabled LOCAL AREA NETWORK.

drop-down (*ICT*) Denoting a menu on a computer screen that can be accessed by clicking on a button on the toolbar.

drop-down curve (*Build*) The longitudinal profile of the water surface in a channel in which the depth of water has been diminished by a sudden drop in the invert and is therefore not parallel to the invert.

drop elbow (*Build*) A small elbow or tee with ears for fixing to a support.

drop electrode (*Chem*) See DROPPING MERCURY ELECTRODE.

drop foot (*Med*) Dropping of the foot from its normal position, caused by paralysis of the muscles, due to injury or inflammation of the nerves supplying them.

drop forging (*Eng*) The process of shaping metal parts by forging between two dies, one fixed to the hammer and the other to the anvil of a pneumatic or mechanical hammer. The dies are expensive, and the process is used for the mass production of parts such as connecting-rods, crankshafts, etc. Also *drop stamping*.

drop gate (*Eng*) A pouring gate or runner leading directly into the top of a mould.

drop hammer (*Eng*) A gravity-fall hammer or a double-acting stamping hammer used to produce DROP FORGINGS by stamping hot metal between pairs of matching dies secured to the anvil block and to the top of the drop hammer respectively.

drop hammer test (*Paper*) Product impact test of the compression resistance of a paper or cardboard box. See panel on IMPACT TESTS.

dropout (*ICT*) Said of a relay when it de-operates, ie contacts revert to de-energized condition.

dropout (*ImageTech*) A brief loss of signal, esp in magnetic recording.

dropout compensator (*ImageTech*) Obscures any momentary loss of video signal by repeating the previous line. Abbrev *DOC*. See DROPOUT.

dropout half-tone (*Print*) A HALF TONE in which the highlight areas have no screen dot formation. See DEEP ETCH.

dropped beat (*Med*) Intermission of a regular pulse wave at the wrist, due to intermission of the heart beat or to an extrasystole.

dropped elbow (*Vet*) A condition in which there is inability to extend the fore-limb, due to paralysis of the radial nerve.

dropped head (*Print*) The first page of a chapter etc which begins lower down than ordinary pages. As far as possible, the drop should be constant throughout the book.

dropper (*ElecEng*) In catenary constructions for electric traction systems, the fitting used for supporting the contact wire from the catenary wire.

dropping mercury electrode (*Chem*) Polarimeter with a half-element consisting of mercury dropping in a fine stream through a solution. Used in POLAROGRAPHY, a continuously renewed mercury surface being formed at the tip of a glass capillary, the accumulating impurities being swept away with the detaching drops of mercury.

dropping resistor (*ElecEng*) A resistor whose purpose is to reduce a given voltage by the voltage drop across the resistance itself.

drop siding (*Build*) WEATHERBOARDING which is rebated and overlapped.

drop stamping (*Eng*) See DROP FORGING.

dropsy (*Med*) See OEDEMA.

drop tank (*Aero*) A fuel tank designed to be jettisoned in flight. Also *slipper tank*.

drop tee (*Build*) See DROP ELBOW.

drop tracery (*Arch*) Tracery which lies partly below the springing of the arch which it decorates.

drop valve (*Eng*) A conical-seated valve used in some steam engines; rapid operation by a trip gear and return spring reduces wire-drawing losses.

drop-weight impact tester (*Eng*) A method of testing deformation, particularly of plastic pipes, by dropping a known weight from different heights.

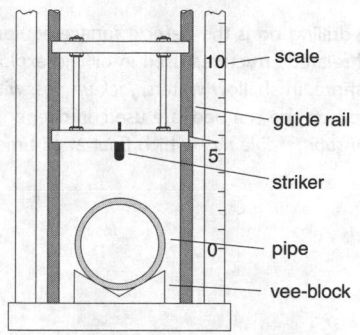

drop-weight impact tester

drop wire (*ICT*) The pair of wires making the final link from a telephone pole to a customer's premises.

drop wrist (*Med*) Limp flexion of the wrist from paralysis of the extensor muscles, as a result of neuritis or of injury to the nerve supplying them, eg in lead poisoning.

drosometer (*EnvSci*) An instrument for measuring the amount of dew deposited.

Drosophila melanogaster (*BioSci*) Common fruitfly, a dipteran used extensively for genetic experiments because its giant SALIVARY GLAND CHROMOSOMES, short generation time and other biological characteristics are very suitable for studies of chromosome organization and GENE MAPPING.

dross (*Eng*) Metallic oxides that rise to the surface of molten metal in metallurgical processes.

dross (*MinExt*) Small coal, inferior or worthless.

drossing (*Eng*) Removal of scums, oxidized films and solidified metals from molten metals.

drought (*EnvSci*) A marked deficiency of rain compared with that usually occurring at the place and season under consideration, sufficiently prolonged for the lack of water to cause serious hydrologic imbalance in the affected area.

drove (*Build*) A broad-edged chisel for dressing stone.

drove (*Genrl*) A narrow channel used for irrigation.

drove work (*Build*) Stone-dressing done with a boaster, leaving rows of parallel chisel marks on the slant across the face. Also *boasted work*.

drowned (*MinExt*) Flooded, eg *drowned workings*.

drowned valleys (*Geol*) Literally, river valleys which have become drowned by a rise of sea level relative to the land. This may be due to actual depression of the land, sea level remaining stationary; or to a eustatic rise in sea level, as during the interglacial periods in the Pleistocene, when melting of the ice caps took place. See RIA, FIORDS.

drowning pipe (*Build*) A cistern inlet pipe, reaching below the water surface and thereby reducing the noise of discharge. Also *silencer pipe*.

Drude law (*Phys*) A law relating the specific rotation for polarized light of an optically active material to the wavelength of the incident light:

$$\alpha = \frac{k}{\lambda^2 - \lambda_0^2}$$

where α is the specific rotation, k is the rotation constant for the material, λ_0^2 is the dispersion constant for material, and λ is the wavelength of incident light. The law does not apply near absorption bands.

Drilling rig

The drilling rig is the derrick, surface equipment and related structures used in oilfield exploration. Offshore, in shallow waters, *jack-up rigs*, with legs based on the sea bed, are used; in deeper waters *semi-submersible rigs* (which float at all times) are employed. In very deep waters drilling ships are used. A PRODUCTION PLATFORM is the offshore platform from which the flow of oil and gas is controlled and usually stored before onward transmission to a refinery.

The drilling rig itself is shown in Figs 1 and 2, and consists of a DERRICK raised above the drill floor where most of the equipment is situated. The derrick supports the weight of the DRILL STRING, and the lengths of CASING when they are being installed. The

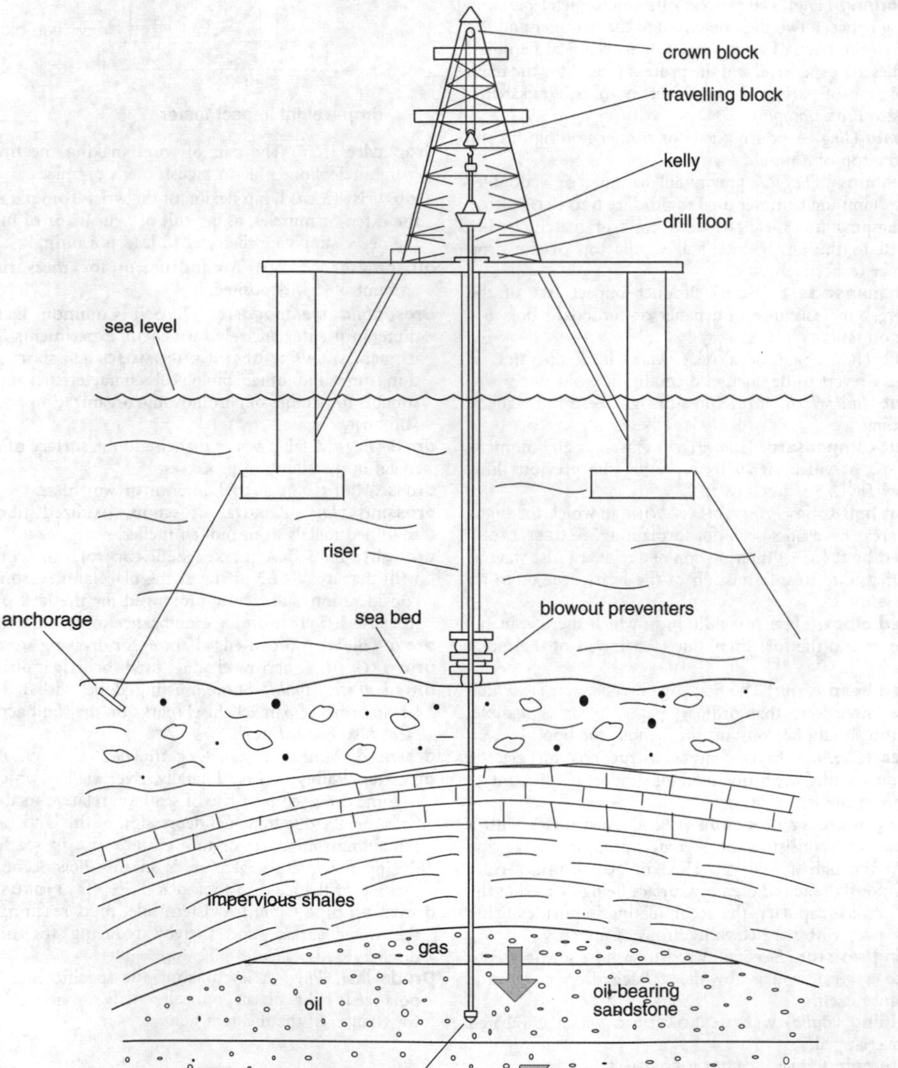

Fig. 1 **Drilling rig** Schematic drawing showing the main components and the oil-bearing strata.

drill string is made of lengths of drill pipe screwed together.

The rotary motion of the drill string and drill bit is imparted by the *power-driven rotary table* (see Fig. 5).

The rotary table is coupled to the KELLY, a pipe of external hexagonal or square section at the top of the drill string by means of the kelly bushing which has a hole of similar section at its centre. As the drilling proceeds successive lengths of pipe are inserted between the kelly and the top of the drill string.

DRILLING MUD is pumped down the drill pipe to contain the well pressure, cool and lubricate the bit, and bring up the cuttings in the space between the drill pipe and the casing or walls of the hole (Fig. 3). Drilling is usually with *tricone bits* to grind up the rock (Fig. 4); diamond-tipped hollow cylindrical bits are used to obtain sample cores.

BLOWOUT PREVENTERS (BOPs) are used to control the pressure in the bore when drilling or *working over*, ie performing remedial operations on a producing well to restore or increase production.

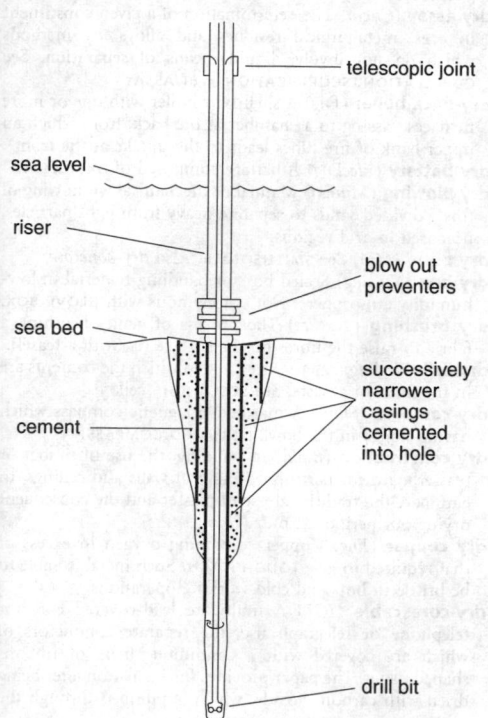

Fig. 2 **Drill string and casing**.

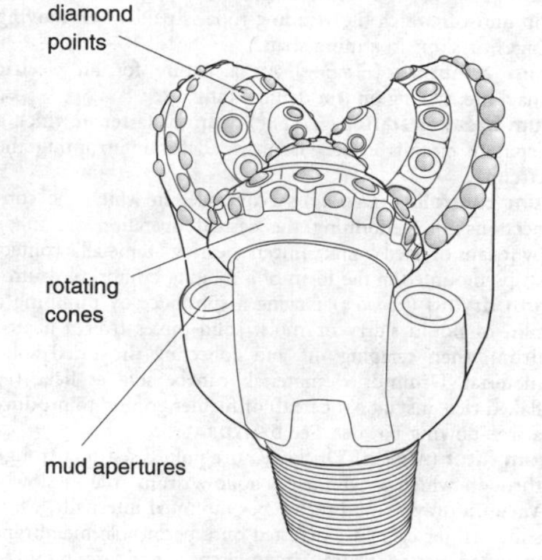

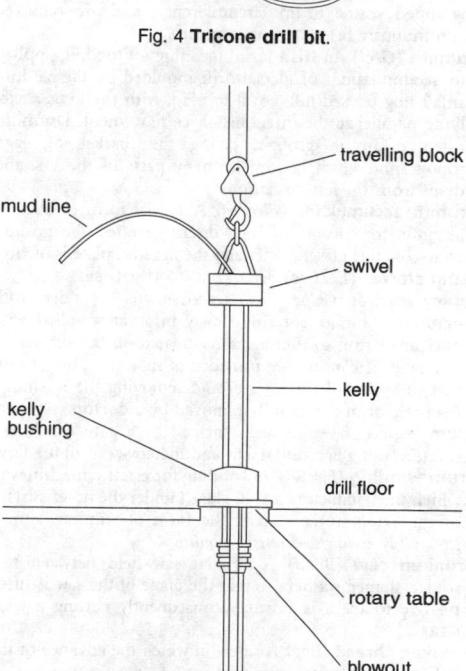

Fig. 4 **Tricone drill bit**.

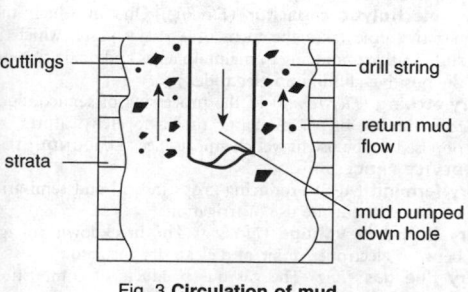

Fig. 3 **Circulation of mud**.

Fig. 5 **Drilling rig floor**.

drug (*Pharmacol*) Any substance, natural or synthetic, which has a physiological action on a living body, when used for the treatment of disease or the alleviation of pain, or for recreation and self-indulgence, leading in some cases to progressive addiction.

drug resistance (*Pharmacol*) The condition in which tissues become resistant after treatment with drugs, commonly found with many anti-tumour treatments. See ANTIBIOTIC RESISTANCE.

drum (*Build*) (1) Any timber structure which is cylindrical in shape. (2) Any cylinder used as a form for bending wood to shape.

drum (*ImageTech*) (1) Heavy cylinder rotating in contact with the film in an optical sound recorder or reproducer to ensure uniform movement. (2) Assembly containing the rotating magnetic heads in a HELICAL-SCAN videotape recorder.

drum (*MinExt*) A cylinder or cone, or compound of these, on and off which the winding rope is paid when moving cages or skips in a mine shaft.

drum armature (*ElecEng*) An armature for an electric machine, having on it a drum winding.

drum breaker starter (*ElecEng*) A drum starter in which a separate circuit breaker is provided for interrupting the circuit.

drum controller (*ElecEng*) A controller in which the connections for performing the desired operation are made by means of fixed contact fingers, bearing on metallic contact strips mounted in the form of a rotating cylindrical drum.

drum drying (*FoodSci*) Drying a substance by pumping a film of liquid slurry or moist solid pieces over a heated drum, then scraping off and collecting the dried solid material. Drum-dried materials can be sold as flake (eg flaked rice, instant oat cereal) or further ground to produce a free-flowing powder. See DEHYDRATION.

drum filter (*MinExt*) Thickened ore pulp is fed to a trough through which a cylindrical hollow drum rotates slowly. Vacuum draws liquid into pipes mounted internally, while solids (filter cake) are arrested on a permeable membrane wrapped round drum circumference, and are removed continuously before re-submergence.

drumlin (*Geol*) An Irish term, meaning a little hill, applied to accumulations of glacial drift moulded by the ice into small hog-backed hills, oval in plan, with the longer axes lying parallel to the direction of ice movement. Drumlins often occur in groups, giving the 'basket of eggs' topography which is seen in many parts of the UK and dates from the last glaciation.

Drumm accumulator (*ElecEng*) A special form of alkaline accumulator capable of high discharge rates; the positive plate contains nickel oxides and the negative plate is of zinc.

drum starter (*ElecEng*) See DRUM CONTROLLER.

drum washer (*Paper*) A large-diameter cylinder with perforated surface rotating slowly in a vat supplied with uncleaned stuff so that a mat is created on the surface by drainage. The means are provided of subjecting the fibrous mat to sprays of clean water and removing the washings. Accepted stuff is generally removed by a doctor.

drum weir (*CivEng*) A weir formed by a gate which can rotate about a horizontal axis and thereby control the flow.

drum winding (*ElecEng*) A winding for electric machines in which the conductors are all placed under the outer surface of the armature core. It is the form of winding almost invariably used. Also *barrel winding*.

drunken saw (*Build*) A circular saw held between two wedge-shaped washers so that the plane of the saw is tilted relative to the axis normal, consequently cutting a wide KERF.

drunken thread (*Eng*) A screw in which the advance of the helix is irregular at every turn.

drupe (*BioSci*) A fleshy fruit with one or more seeds each surrounded by a stony layer (the endocarp), eg plum.

drupel (*BioSci*) A small drupe, usually one of a group forming an aggregate fruit, eg raspberry.

druse (*BioSci*) A globose mass of crystals of calcium oxalate around a central foundation of organic material, in some plant cells.

drusy (*MinExt*) Containing cavities often lined with crystals; said of mineralized lodes or veins. See GEODES.

drusy cavities (*Geol*) See GEODE.

dry adiabatic (*EnvSci*) A curve on an aerological diagram representing the temperature changes of a parcel of dry air subjected to an adiabatic process.

dry adiabatic lapse rate (*EnvSci*) The *temperature lapse rate* of dry air which is subjected to adiabatic ascent or descent. This lapse rate also applies to moist air which remains unsaturated. Its magnitude is $9.76°C\,km^{-1}$. Abbrev *DALR*.

dry area (*Build*) The 2 in or 3 in (50 or 75 mm) cavity in the wall below ground level in basement walls built hollow; the purpose of the cavity is to keep the basement walls dry.

dry assay (*Chem*) The determination of a given constituent in ores, metallurgical residues and alloys, by methods which do not involve liquid means of separation. See CUPELLATION, SCORIFICATION, WET ASSAY.

dry-back boiler (*Eng*) A shell-type boiler with one or more furnaces passing to a chamber at the back, from which an upper bank of fire tubes leads to the uptake at the front.

dry battery (*ElecEng*) A battery composed of DRY CELLS.

dry blowing (*MinExt*) Manual or mechanical winnowing of finely divided sands to separate heavy from light particles, practised in arid regions.

dry bone (*Min*) See SMITHSONITE. Also *dry-bone ore*.

dry box (*NucEng*) Sealed box for handling material in low-humidity atmosphere. Not synonymous with GLOVE BOX.

dry brushing (*Textiles*) The process of gently brushing a fabric to raise the fibres on the surface (eg with a teasel).

dry cell (*ElecEng*) A PRIMARY CELL in which the contents are in the form of a paste. See LECLANCHÉ cell.

dry compass (*Ships*) A mariner's magnetic compass which has no liquid in the bowl. Cf LIQUID COMPASS.

dry construction (*Build*) In building, the use of timber or plasterboard for partitions, lining of walls and ceilings, to eliminate the traditional use of plaster and the consequent drying out period. Also *dry lining*.

dry copper (*Eng*) Copper containing oxygen in excess of that required to give TOUGH PITCH. Such metal is liable to be brittle in hot- and cold-working operations.

dry-core cable (*ICT*) A multicore lead-covered core for telephone or telegraph use, the separate conductors of which are covered with a continuous helix of ribbon-shaped paper. The paper provides the insulation after being dried with carbon dioxide, which is pumped through the cable and kept under pressure.

dry cow (*Agri*) A cow between lactation and a subsequent calving, or one that has failed to lactate for whatever reason.

dry deposition (*EnvSci*) Deposition of gaseous materials esp pollutant gases on natural surfaces. See panel on ATMOSPHERIC POLLUTION.

dry dock (*CivEng*) A dock in which ships are repaired. It is closed by means of gates or caissons and the dock then emptied. Also *graving dock*.

dry electrolytic capacitor (*ElecEng*) One in which the negative pole takes the form of a sticky paste, which is sufficiently conducting to maintain a gas and oxide film on the positive aluminium electrode.

dry etching (*Electronics*) In the processing of semiconductors, the use of gas discharge media to etch features, as opposed to the use of wet chemicals. See SEMICONDUCTOR DEVICE PROCESSING.

dry farming (*Agri*) Producing crops in arid and semi-arid regions without the use of irrigation.

dry flashover voltage (*ElecEng*) The breakdown voltage between electrodes in air of a clean dry insulator.

dry flue gas (*Eng*) The gaseous products of combustion from a boiler furnace, excluding water vapour. See FLUE GAS.

dry fruit (*BioSci*) A fruit in which the pericarp does not become fleshy at maturity.

dry gas (*MinExt*) Petroleum gas in which the lower-boiling-point fractions have been removed or are not naturally present.

dry ice (*Chem*) Solid (frozen) carbon dioxide, used in refrigeration (storage) and engineering. At atmospheric pressures it sublimes slowly.

dry indicator test (*Paper*) One of the methods of determining the resistance of a paper to penetration by aqueous media. A powder that intensifies in colour when wet is sprinkled on a test sample floating on water and the time measured until an agreed degree of colour change has taken place.

drying cabinet (*ImageTech*) A cabinet in which heated air is circulated to dry processed photographic film.

drying cylinder (*Paper*) A hollow metal cylinder, heated internally by live steam to remove moisture from a web of paper in contact with the greater part of its outer surface.

drying cylinder (*Textiles*) A heated rotating hollow cylinder over which wet fabric is passed to dry it. Often a set of cylinders held closely together in a vertical stack is used with the fabric passing under and over each cylinder.

drying off (*Agri*) Allowing cows to cease lactation and achieve optimum condition before a subsequent calving.

drying oils (*Build*) Vegetable or animal oils which harden by oxidation when exposed to air.

drying-out (*ElecEng*) (1) The process of heating the windings of electrical equipment to drive all moisture out of the insulation; usually done by passing current through the windings. (2) In electroplating, the process of removing moisture from a metal by passing it through hot water and then through sawdust or a current of hot air.

drying stove (*Eng*) A large stove or oven in which dry-sand moulds and cores are dried. See DRY SAND.

dry joint (*ElecEng*) A faulty solder joint giving high-resistance contact due to residual oxide film.

dry laying (*Textiles*) The formation of a WEB of fibres by carding or AIR LAYING preparatory to the manufacture of a NON-WOVEN FABRIC.

dry liner (*Eng*) See LINER.

dry mass (*Aero*) The mass of an aero-engine, including all essential accessories for its running and the drives for airframe accessories, without oil, fuel or coolant.

dry moulding (*Eng*) The preparation of moulds in dry sand, as distinct from the use of green sand or LOAM.

dry mounting (*ImageTech*) A method of attaching a photographic print to a mounting card by sandwiching between them a sheet of tissue which becomes adhesive with the application of heat and pressure.

dryness fraction (*Eng*) The proportion, by weight, of dry steam in a mixture of steam and water, ie in wet steam.

dry offset (*Print*) Almost synonymous with LETTERSET PRINTING but chiefly used when relief plates are employed on conventional litho machines, the dampening rollers being unused.

dryout margin (*NucEng*) In a reactor in which steam is directly boiled, the ratio between the heat generated in the fuel at full power to that which would cause steam to boil and cover all the fuel pins and thus markedly reduce the cooling available.

dry pipe (*Eng*) A blanked-off and perforated steam-collecting pipe placed in the steam space of a boiler and leading to the stop valve, for the purpose of excluding water resulting from *priming*. See ANTI-PRIMING PIPE.

dry plate (*ImageTech*) A term used in the early days of photography to describe glass plates coated with a light-sensitive gelatine emulsion, in contrast to the former collodion *wet plates*.

dry-plate rectifier (*ElecEng*) See METAL RECTIFIER.

dry rot (*BioSci, Build*) (1) One of a number of plant diseases, eg of stored potatoes, in which a lesion dries out as it forms. (2) The rotting of timber by the fungus *Serpula* (*Merulius*) *lacrymans*, so that it becomes dry, light and friable, with a cracked appearance. Cf WET ROT.

dry run (*ICT*) Use of test data to check paths through a program or flowchart without the use of a computer. See LOGICAL ERROR.

dry sand (*Eng*) A moulding sand possessing the requisite cohesion and strength when dried. It is moulded in a moist state, then dried in an oven, when a coherent and porous mould results.

Drysdale permeameter (*ElecEng*) An instrument for determining, by a ballistic method, the permeability of a sample of iron; a plug carrying a primary and secondary coil is inserted in an annular hole in the sample of material under test.

Drysdale potentiometer (*ElecEng*) An ac potentiometer of the polar type, comprising a phase-shifting transformer and resistive voltage divider. It is calibrated against a standard cell with dc and ac is then set to the same value using an electrodynamic indicator.

dry spinning (*Textiles*) The production of polymeric filaments by the extrusion of a solution of the polymer in a volatile liquid which is then removed by evaporation.

dry-spun flax (*Textiles*) The coarse flax yarn obtained from a dry roving. See WET SPINNING.

dry steam (*Eng*) Steam free from water, but unsuperheated. Also *dry saturated steam*.

dry sump (*Autos*) An internal-combustion engine lubrication system in which the separate crankcase is kept dry by an oil scavenge pump, which returns the oil to a tank, from which it is delivered to the engine bearings by a pressure pump.

dry-sweating (*Vet*) See ANHIDROSIS.

dry valley (*Geol*) A valley produced at some former period by running water, though at present streamless. This may be due to a fall of the water table, to river capture or to climatic change.

dry weight (*Space*) The weight of a system without fuel and consumables; for a launch vehicle it is the total launch weight minus that of the propellants and pressurizing gases.

Ds (*Chem*) Abbrev for DARMSTADTIUM.

DSC (*Chem*) Abbrev for DIFFERENTIAL SCANNING CALORIMETRY.

DSDM (*ICT*) Abbrev for *dynamic systems development methodology*, a sophisticated form of rapid application development (RAD) which aims to achieve high-quality incremental deliveries by means of prototyping and collaborative working between technologists and users. See PANEL ON RAD/DSDM METHODLOGY.

D segment (*BioSci*) A region that codes for part of the variable region of immunoglobulin heavy chains; heavy-chain genes are assembled from one of several V-region genes, one of several D segments, and one of several J segments. Multiple exons in the segment contribute to antibody diversity (hence the use of *D*).

DSL (*ICT*) Abbrev for *digital subscriber line*, a fast Internet connection over an analogue phone line.

D slide-valve (*Eng*) A simple form of slide-valve, in section like a letter D, sliding on a flat face in which ports are cut. See SLIDE-VALVE.

DSP (*ImageTech*) Abbrev for (1) DIGITAL SIGNAL PROCESSING, (2) *digital sound-field processing*.

DSSS (*ICT*) Abbrev for *direct-sequence spread spectrum*, a technique for reducing interference in radio broadcasting.

DTA (*Chem*) Abbrev for DIFFERENTIAL THERMAL ANALYSIS.

DTF (*ImageTech*) Abbrev for DYNAMIC TRACK FOLLOWING.

DTL (*Electronics*) Abbrev for DIODE−TRANSISTOR LOGIC.

DTP (*ICT*) Abbrev for DESKTOP PUBLISHING.

DTR (*ImageTech*) Abbrev for DIFFUSION-TRANSFER REVERSAL.

DTV (*ImageTech*) Abbrev for: (1) DESKTOP VIDEO; (2) digital TV (see DVB).

DU (*Chem*) Abbrev for DEPLETED URANIUM.

dual (*MathSci*) In projective geometry, the figure obtained by interchanging the lines and points of a given figure. For instance, the dual of the statement 'Any two points are joined by a line' is 'Any two lines meet at a point', which is made true by including points at infinity.

dual-band (*ICT*) Denoting a mobile phone that is capable of operating on two frequency bands.

dual beta decay (*Phys*) BRANCHING where a radioactive nuclide may decay by either electron or positron emission. Not to be confused with DOUBLE BETA DECAY.

dual combustion cycle (*Eng*) An internal-combustion engine cycle sometimes taken as a standard of comparison for the compression-ignition engine, in which combustion occurs in two stages, ie partly at constant volume and partly at constant pressure.

dual-gate MOSFET (*Electronics*) A *metal–oxide semiconductor field-effect transistor* having two separate gate electrodes. This imparts superior performance for applications such as mixers, amplifiers and demodulators.

dual in-line package (*Electronics*) A common integrated-circuit package having two parallel rows of connectors at right angles to the body, as required for insertion into pre-drilled holes in a printed circuit board. Abbrev *DIP*.

dual ion (*Chem*) See ZWITTERION.

dualism (*Psych*) A philosophical position in which it is accepted that two separate states or principles exist. In the context of psychology, mind–body dualism is frequently debated – whether the two are separate but interactive or whether they are parallel manifestations of a complex organism.

duality (*Arch, CivEng*) The repetition of members in the same angular direction in a structure or building.

dualizing (*MathSci*) See POLAR RECIPROCATION.

dual modulation (*ICT*) The modulation of a carrier with two separate types of modulation (eg amplitude and frequency), each carrying different information.

dual spectrum (*Print*) A method of DOCUMENT COPYING in which the document, in contact with the copy paper, is heated by infrared and an image of the document is formed on the copy paper, which is then cascaded with a powder which adheres to the surface, is fused, and can be used for SMALL OFFSET.

dual-tone multifrequency (*ICT*) The system of signalling embodied in telephones with a numeric keypad. Each row and column of the pad are associated with a different audio-frequency. Pressing a key causes the pair of tones corresponding to the row and column intersecting at that key to be sent over the line. Cf LOOP-DISCONNECT PULSING. See MULTIFREQUENCY, TOUCH-TONE.

dual track (*Acous*) Use of two tracks on a magnetic tape, so that recording and subsequent reproduction can proceed along one track and return along the other, thus obviating rewinding.

Duane and Hunt's law (*Phys*) A law stating that the maximum photon energy in an X-ray spectrum is equal to the kinetic energy of the electrons producing the X-rays, so that the maximum frequency, as deduced from quantum mechanics, is eV/h, where V is the applied voltage, e is the electronic charge and h is PLANCK'S CONSTANT.

dubbing (*Build*) The operation of filling in hollows in the surface of a wall with coarse stuff, as a preliminary to plastering.

dubbing (*ImageTech*) Rerecording to combine two or more sound sources including the replacement of original dialogue by another language. Also the transfer of a magnetic sound record to a photographic one or copying video and audio from one videotape recorder to another.

dubnium (*Chem*) An artificially manufactured radioactive metallic element, symbol Db, at no 104, of the transactinide series, formed by bombarding CALIFORNIUM with carbon nuclei. It has ten isotopes with half-lives of up to 70 seconds. Also *unnilquadium*.

du Bois balance (*ElecEng*) An instrument used for measuring the permeability of iron or steel rods. The magnetic attraction across an air-gap in a magnetic circuit, of which the sample forms a part, is balanced against the gravitational force due to a sliding weight on a beam.

Duchemin's formula (*Aero*) An expression giving the normal wind pressure on an inclined area in terms of that on a vertical area. It states that:

$$N = F \frac{2 \sin \alpha}{1 + \sin^2 \alpha}$$

where F = pressure of wind in N m^{-2} of vertical surface; α = angle of the inclined surface with the horizontal; N = normal pressure in N m^{-2} of inclined surface.

Duchenne–Erb paralysis (*Med*) A form of paralysis in which the arm can be neither abducted nor turned outwards nor raised nor flexed at the elbow, as a result of a lesion of the fifth and sixth cervical nerves in the brachial plexus.

Duchenne muscular dystrophy (*Med*) A common form of the inherited MUSCULAR DYSTROPHIES, affecting only male children and leading to progressive muscular weakness.

duchess (*Build*) A slate, 24×12 in $(610 \times 305$ mm).

duck (*Textiles*) A plain, bleached cotton or linen cloth, used for tropical suitings. Heavier makes are used for sails, tents and conveyor belting.

duckboard (*Build*) A board which has slats nailed across it at intervals and is used as steps in repair work on roofs, or for walking in excavations or valley gutters.

duck cholera (*Vet*) An infection of ducks by the bacterium responsible for FOWL CHOLERA.

duckfoot bend (*Build*) See REST BEND.

duckfoot quotes (*Print*) The chevron-shaped quotation marks used by continental printers.

duck viral hepatitis (*Vet*) An acute and highly fatal virus disease of ducklings under 3 weeks old, characterized by liver cell necrosis and sudden death. At least three different viruses have been identified. Notifiable disease in the UK for which an experimental live vaccine has been used.

duct (*BioSci*) A tube formed of cells; a tubular aperture in a non-living substance, through which gases and liquids or other substances (such as spermatozoa, ova, spores) may pass. Also *ductus*.

duct (*ElecEng*) An air passage in the core or other parts of an electric machine along which cooling air may pass. Also *cooling duct, ventilating duct*.

duct (*Eng*) (1) A hole, pipe or channel carrying a fluid, eg for lubricating, heating or cooling. (2) A large sheet-metal tube or casing through which air is passed for forced-draught, ventilating or conditioning purposes.

duct (*Print*) A reservoir holding the ink in a printing machine. Usually the supply is regulated by a number of screws and by a ratchet.

ducted cooling (*Aero*) A system in which air is constrained in ducts that convert its kinetic energy into pressure for more efficient cooling of an aero-engine or of its radiator.

ducted fan (*Aero*) A gas turbine aero-engine in which part of the power developed is harnessed to a fan mounted inside a duct. Also *turbofan*.

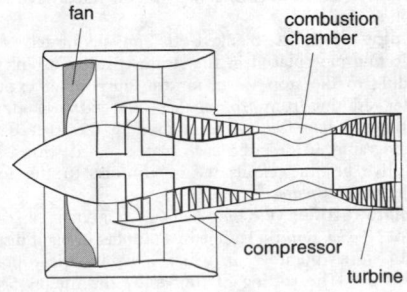

ducted fan

duct height (*EnvSci*) The height above the Earth's surface of the lower effective boundary of a tropospheric radio duct.

ductile–brittle transition temperature (*Eng*) That at which the failure mode of a material, esp metals and plastics, changes from ductile, higher energy, to brittle, lower energy, as the temperature is reduced. Symbol T_b. Transition often mapped by impact tests. Also *brittle temperature*. See panel on IMPACT TESTS.

ductile cast-iron (*Eng*) Cast-iron in which the free graphite has been induced to form as nodules by adding cerium or magnesium in the molten state which gives a marked increase in ductility. Abbrev *SG iron* for SPHERULITIC GRAPHITE CAST-IRON.

ductile fracture (*Eng*) A type of fracture in any material where substantial deformation has occurred away from fracture surfaces. Usually associated with yielding in materials.

ductility (*Eng*) The ability of metals and alloys to retain strength and freedom from cracks when shape is altered. See WORK HARDENING.

ductless glands (*BioSci*) Masses of glandular tissue that lack ducts and discharge their products directly into the blood; as the lymph glands and the endocrine glands.

ductule (*BioSci*) A duct with a very narrow lumen: a small duct; the fine terminal portion of a duct.

ductus (*BioSci*) See DUCT.

ductus arteriosus (*Med*) A blood vessel important in fetal development linking the pulmonary artery to the aorta. It closes at birth. In some cases there is abnormal persistence known as *patent ductus arteriosus* where blood flows from the aorta to the pulmonary artery creating an abnormal shunt.

ductus caroticus (*BioSci*) In some vertebrates, a persistent connection between the systemic and carotid arches.

ductus Cuvieri (*BioSci*) See CUVIERIAN DUCTS.

ductus ejaculatorius (*BioSci*) In many invertebrates, as the platyhelminthes, a narrow muscular tube forming the lower part of the vas deferens and leading into the copulatory organ.

ductus endolymphaticus (*BioSci*) In lower vertebrates and the embryos of higher vertebrates, the tube by which the internal ear communicates with the surrounding medium.

ductus pneumaticus (*BioSci*) A duct that connects the gullet with the air bladder in *physostomous* fish, eg trout.

duct thickness (*EnvSci*) See DUCT WIDTH.

duct waveguide (*Phys*) A layer in the atmosphere which, because of its refractive properties, keeps electromagnetic radiated (or acoustic) energy within its confines. It is *surface* or *ground-based* when the surface of the Earth is one confining plane.

duct width (*EnvSci*) The difference in height between the upper and lower boundaries of a tropospheric radio duct. Also *duct thickness*.

duff (*MinExt*) Fine coal too low in calorific value for direct sale.

duffel (*Textiles*) Heavy, low-quality, woven woollen fabric raised on both sides. Short coats made from this fabric. Also *duffle*.

Dühring's rule (*ChemEng*) If the temperatures at which two chemically similar liquids that have the same vapour pressure are plotted against each other, a straight line results, ie $t' = \alpha + \beta t$, where t' and t are the boiling points of the two liquids on the same scale of temperature and at the same pressure, and α and β are constants.

duke (*Paper*) A notepaper size, 178×143 mm ($7 \times 5\frac{3}{8}$ in).

dulcin (*Chem*) Sucrol, 4-ethoxyphenylurea. Colourless crystalline substance which is about 200 times as sweet as sugar.

dull (*Med*) Not resonant to percussion; said of certain regions of the body, esp the chest.

dull-emitter cathode (*Electronics*) One from which electrons are emitted in large quantities at temperatures at which incandescence is barely visible. The emitting surface is the oxide of an alkaline earth metal.

Dulong and Petit's law (*Chem*) See LAW OF DULONG AND PETIT.

dulosis (*BioSci*) Among ants (Formicoidea), an extreme form of social parasitism in which the work of the colony of one species is done by captured 'slaves' of another species called *amazons*. Also *helotism, slavery*.

Dumbbell nebula (*Astron*) A famous planetary nebula in the constellation Vulpecula with a distinctive hour-glass shape. Distance approx 220 pc.

dumb buddle (*MinExt*) A buddle without revolving arms or sweeps, for concentrating tin ores.

dumb compass (*Ships*) See PELORUS.

dumb iron (*Autos*) Forgings attached to the front of the side members of the frame, to carry the spring shackles and front cross-member, now found only on commercial vehicles.

dumb terminal (*ICT*) A TERMINAL that has no inherent PROCESSOR or processing power with all processing done by the central HOST computer. Cf INTELLIGENT TERMINAL.

dummy (*Print*) An unprinted volume, made up for the use of publishers in estimating their requirements. Thickness should be measured at the fore-edge and tail.

dummy antenna (*ICT*) See ARTIFICIAL ANTENNA.

dummy cell (*ICT*) In an ASYNCHRONOUS TRANSFER MODE network, empty CELLS inserted by the receiving ATM ADAPTATION LAYER to replace any lost in transmission. This maintains isochronicity for services that require it, eg video.

dummy coil (*ElecEng*) A coil put onto an armature to preserve mechanical balance and symmetry, but not electrically connected to the rest of the winding.

dummying (*Eng*) The preliminary rough shaping of the heated metal before placing between the dies for drop forging.

dummy load (*ElecEng*) One which matches a feeder or transmitter; so designed to absorb the full load without radiation, particularly for testing. Also *antenna load*.

dummy piston (*Eng*) A disk placed on the shaft of a reaction turbine; to one side of it steam pressure is applied to balance the end thrust. Also *balance piston*.

dummy variable (*ICT*) Identifier used for syntax reasons in a program, but that will be replaced by some other variable identifier when the program is executed; a common example is the use of a dummy variable in the definition of a subroutine parameter.

dump (*ICT*) To copy the contents of a file, or the contents of the immediate access store, to backing store or to an output device. The output is known as the dump, and may be used to ensure the INTEGRITY of the data or to assist in program error detection.

dump (*MinExt*) The heap of accumulated waste material from a metal mine, or of treated tailings from a mill or ore-dressing plant. Also *tip*.

dump condenser (*NucEng*) Condenser or water-filled tank into which steam destined for the turbines in a nuclear power station can be diverted if the electrical load is suddenly removed.

dumper (*CivEng*) A wagon used, in the construction of earthworks, for conveying excavated material on site and dumping it where required.

dumpling (*CivEng*) The soil remaining in the centre of an open excavation which is commenced by sinking a trench around the site; the dumpling is removed later.

dump valve (*Aero*) (1) An automatic safety valve which drains the fuel manifold of a gas turbine when it stops, or when the fuel pressure fails. (2) A large-capacity valve to release residual pressure in any fluid system for emergency or operational reasons after landing, or to release all cabin pressure in an in-flight emergency.

dumpy level (*Surv*) A type of level in which the essential characteristic is the rigid connection of the telescope to the vertical spindle.

dune (*Geol*) An accumulation of sand formed in an area with a prevailing wind. The principal types are: *barchans*, crescent-shaped dunes which migrate in the direction of the point of the crescent; *seifs*, elongated ridges of sand aligned in the wind direction; *transverse dunes*, at right angles to the wind; and *whaleback dunes*, very large elongated dunes. Fossil sand dunes can be recognized, and they indicate desert conditions and wind conditions during past geological periods.

dune bedding (*Geol*) A large-scale form of cross-bedding characteristic of wind-blown sand dunes.

dungannonite (*Geol*) A corundum-bearing diorite containing nepheline, originally described from Dungannon, Ontario.

dungaree (*Textiles*) A cotton cloth, with a twill weave, made from coloured warp and weft yarns, generally used for overalls.

dunite (*Geol*) A coarse-grained, deep-seated igneous rock, almost monominerallic, consisting essentially of olivine, though chromite is an almost constant accessory. In several parts of the world (eg Bushveld Complex, S Africa) it contains native platinum and related metals. Named from Mt Dun, New Zealand.

dunkop (*Vet*) See AFRICAN HORSE-SICKNESS.

dunnage (*Ships*) Loose wood laid in the hold to keep the cargo out of the bilge water, or wedged between parts of the cargo to keep them steady.

duode (*Acous*) Electrodynamic open-diaphragm loudspeaker driven by eddy currents in a metal former, the VOICE COIL being wound over a rubber compliance; the arrangement gives enhanced width of response with damping of diaphragm resonances.

duodecal (*Electronics*) Twelve-contact tube base.

duodecimal system (*MathSci*) The positional number system with base 12.

duodecimo (*Print*) The twelfth of a sheet, or a sheet folded to make twelve leaves or 24 pages. Also *12mo*.

duodenal ulcer (*Med*) A PEPTIC ULCER occurring in the first part of the duodenum.

duodenectomy (*Med*) Excision of the duodenum.

duodenitis (*Med*) Inflammation of the duodenum.

duodenocholecystostomy (*Med*) A communication, made surgically between the duodenum and the gall bladder.

duodenojejunostomy (*Med*) A communication, made surgically between the duodenum and the jejunum.

duodenum (*BioSci*) In vertebrates, the region of the small intestine immediately following the PYLORUS of the stomach, distinguished usually by the structure of its walls. Adj *duodenal*.

duolateral coil (*ElecEng*) See BASKET COIL.

duophase (*ElecEng*) Use of an inductor in the output circuit of the active device in an amplifier to obtain a reversed-phase voltage for driving a push–pull stage.

duotone (*Print*) Two half-tone plates made from monochrome copy, the key plate at 45° and the tint at 75°, to produce a two-tone effect.

dupe (*ImageTech*) Contraction of *duplicate negative*, prepared from the original camera negative for protection or to incorporate visual effects not originally photographed.

Duperry's lines (*Phys*) Lines on a magnetic map showing the direction in which a compass needle points, ie the direction of the magnetic meridian.

duplex (*BioSci*) A double-stranded part of a nucleic acid molecule.

duplex (*ICT*) Data transmission in both directions simultaneously over the same channel, as in a normal telephone link. Also *full duplex*. Cf HALF DUPLEX.

duplex balance (*ICT*) Telegraph name for line balance.

duplex burner (*Aero*) A gas turbine fuel injector with alternative fuel inlets, but a single outlet nozzle.

duplex chain (*Eng*) A roller chain construction using two sets of rollers and three sets of link plates, used where the chain tension exceeds that which can be transmitted by a simple chain. It avoids matching two simple chains run side by side.

duplexer (*Radar, ICT*) (1) In radar, a system that takes advantage of the time delay between transmission of a

pulse and reception of its echo to allow the use of the same aerial for transmission and reception. *Transmit–Receive* (TR) or sometimes TR and *Anti-Transmit–Receive* switches are used to isolate the delicate receiver during the high-power pulse transmission. (2) More generally in radio, any system or network allowing simultaneous transmission and reception on a single aerial, although separation is normally achieved by using different frequencies for transmission and reception.

duplex lathe (*Eng*) A lathe in which two cutting tools are used, one on each side of the work, either to avoid it springing or to increase the rate of working. See MULTIPLE-TOOL LATHE.

duplex paper (*Paper*) Deprecated term for a paper or board with two noticeably different layers, eg by reason of colour or finish.

duplex processes (*Eng*) The combination of two alternative methods in performing one operation; as when steel-making is carried out in two stages, first in the open hearth and second in the electric furnace.

duplex pump (*Eng*) A pump with two working cylinders side by side.

duplex set (*Print*) A combination of BARRING MOTOR and main drive motor used to provide all speeds from INCHING and CRAWL to full speed.

duplex winding (*ElecEng*) A winding for dc machines in which there are two separate and distinct windings on the machine, the two being connected in parallel by the brushes.

duplicate feeder (*ElecEng*) A feeder forming an alternative path to that normally in use.

duplicate plates (*Print*) Stereotypes, electrotypes, and rubber and plastic plates made from original plates or from type formes.

duplicating (*Eng*) Use of special equipment for machining or forming an object which is a copy of a two- or three-dimensional master.

duplicating the cube (*MathSci*) One of the classical problems of antiquity: to construct with straight edge and compasses a cube with volume double that of a given cube. This was proved to be impossible in 1837 by P Wantzell. Cf DELIAN PROBLEM.

duplication (*BioSci*) Doubling of a gene or a larger segment of a chromosome, by a variety of mechanisms including unequal crossing over and fusion between a chromosomal fragment and a whole chromosome of the same sort.

duplicator paper (*Paper*) Paper intended for use in stencil duplicator machines.

duplicident (*BioSci*) Having two pairs of incisor teeth in the upper jaw, as hares and rabbits.

Dupuit relation (*PowderTech*) A relationship between the apparent linear velocity of a fluid flow through an isotropic powder bed and the actual velocity of flow which depends on the porosity of the powder bed.

Dupuytren's contraction (*Med*) Thickening and contraction of the fascia of the palm of the hand, with resulting flexion of the fingers, esp of the ring and little fingers.

durable press (*Textiles*) A treatment applied to fabrics or garments to make them retain desired creases in normal wearing and washing. Compounds similar to those for *crease-resist finishes* are used.

durain (*Geol*) A separable constituent of dull coal; of firm, rather granular structure, sometimes containing many spores.

Dural (*Eng*) TN for a precipitation hardenable aluminium alloy containing nominally 4% copper and 0·5% manganese. Also *Duralumin*. See ALUMINIUM ALLOYS.

dura mater (*BioSci*) A tough membrane lining the cerebrospinal cavity in vertebrates.

duramen (*BioSci*) See HEARTWOOD.

Duranol dyes (*Chem*) Acetate (ethanoate) silk (viscose) dyes, derived from amino-anthraquinones.

durene (*Chem*) 1:2:4:5-tetramethylbenzene. Mp 79·3°C, bp 196°C.

duricrust (*Geol*) A hard crust formed in or on soil in a semi-arid environment. It is formed by the precipitation of soluble minerals from mineral waters, particularly during the dry season.

Durosier's murmur (*Med*) A murmur heard over the femoral artery during diastole of the heart; indicative of aortic valve incompetence.

durum wheat (*FoodSci*) Hard-grained wheat milled to produce semolina and pasta. The best quality from *Triticum durum* is grown mainly in N America and has a 10–16·5% protein content.

dust (*PowderTech*) Particulate material which is or has been airborne and which passes a 200 mesh BS rest sieve (76 μm).

dust chamber (*Eng*) Fume chamber in which dust is arrested as dry furnace gases are filtered, cycloned or baffled.

dust core (*Phys*) Magnetic circuit embracing or threading a high-frequency coil, made of ferromagnetic particles compressed into an insulating matrix binder, thus obviating losses at high frequency because of eddy currents.

dust counter (*EnvSci*) An instrument for counting the dust particles in a known volume of air.

dust counter (*MinExt*) Apparatus, usually portable, for collecting dust in mines, to display and check on working conditions underground.

dust cover (*Print*) See JACKET.

dust explosion (*Eng*) An explosion resulting from the ignition of small concentrations of flammable dust (eg finely divided metal particles, coal dust, sugar or flour) in the air.

dust figure (*ElecEng*) See LICHTENBERG FIGURE.

dust monitor (*NucEng*) An instrument which separates airborne dust and tests for radioactive contamination.

dust-proof (*ElecEng*) Said of a piece of electrical apparatus which is constructed so as to exclude dust or textile flyings.

Dutch barn (*Arch*) A simple open structure, generally of mild steel, the roof support being formed of braced trusses, arch-shaped, and carried by slender columns spaced normally at about 4 m intervals. The roof covering may be of corrugated iron or asbestos cement.

Dutch bond (*Build*) A bond differing from English bond only in the angle detail, the vertical joints of one stretching course being in line with the centre of the stretchers in the next stretching course.

Dutch elm disease (*For*) See ELM.

Dutch gold (*Eng*) A cheap alternative to GOLD LEAF, consisting of copper leaf, which, by exposure to the fumes from molten zinc, acquires a yellow colour.

dutchman (*Build*) A piece of wood driven into a gap left in a joint which has been badly cut.

Dutch process (*Chem*) A process of making WHITE LEAD by corroding metallic lead in stacks where fermentation of tan or bark is taking place, in the presence of dilute ethanoic acid.

Dutch roll (*Aero*) Lateral oscillation of an aircraft, in particular an oscillation having a high ratio of rolling to yawing motion. Dutch roll can be countered by yaw dampers.

duty (*ElecEng*) The cycle of operations which an apparatus is called upon to perform whenever it is used, eg with a motor it is the starting, running for a given period, and stopping; or with a circuit breaker, it may be closing and opening for a given number of times with given time intervals between; or the prescription of a process timer.

duty factor (*ElecEng*) (1) The ratio of the equivalent current taken by a motor or other apparatus running on a variable load to the full-load current of the motor (continuous rating). (2) The ratio of pulse duration to space interval in a train of pulses. Also *mark/space ratio*.

D-value (*FoodSci*) Decimal reduction time. The time required, in minutes at a specified temperature, to reduce the number of a given micro-organism by 90%.

DVB (*ImageTech*) Abbrev for the European Commission's *Digital Video Broadcasting* project to oversee the development of digital cable, satellite and terrestrial broadcasting in Europe.

DVC (*ImageTech*) Abbrev for DIGITAL VIDEO CAMERA.

DVD (*ImageTech*) Abbrev for DIGITAL VERSATILE DISK.

DVE (*ImageTech*) Abbrev for DIGITAL VIDEO EFFECTS.

DVORAK keyboard (*ICT*) One laid out to minimize finger movement.

DVT (*Med*) Abbrev for DEEP VEIN THROMBOSIS.

DVTR (*ImageTech*) Abbrev for *digital videotape recorder*. See BETACAM, D1, D2, D3, D5, DCT.

dwang (*Build*) (1) See BRIDGING. (2) A mason's term for a crowbar.

dwarfism (*ImageTech*) Negative size distortion in stereoscopic film caused by the camera lens separation being in excess of the normal interocular distance.

dwarf male (*BioSci*) A male animal greatly reduced in size, and usually in complexity of internal structure also, in comparison with the female of the same species; such males may be free-living but are more usually carried by the female, to which they may be attached by a vascular connection in extreme cases, as some kinds of deep-sea angler fish.

dwarf rafter (*Build*) A JACK RAFTER.

dwarf shoot (*BioSci*) See SHORT SHOOT.

dwarf star (*Astron*) A low-luminosity star of the MAIN SEQUENCE.

dwell (*Eng*) Of a cam, the angular period during which the cam follower is allowed to remain at a constant lift.

dwell (*Plastics*) A term used to describe the pause in the application of pressure in a moulding press, which allows the escape of gas from the moulding material.

dwell (*Print*) The slight pause in the motion of a hand press or platen when the impression is being made.

Dwight Lloyd machine (*MinExt*) A type of continuous *sintering* machine characterized by having air drawn down through the burning bed on a travelling grate into 'wind boxes' and used in roasting pyritic ore so that by segregating and recycling the gas streams, concentrations of sulphur dioxide sufficiently high for conversion to sulphuric acid are readily obtained.

DX film speed coding (*ImageTech*) Automatic film speed setting for 35 mm cameras, with the ISO SPEED coded as a pattern of printed and bare metal on the cassette side which is read by electrical contacts in the film chamber.

Dy (*Chem*) Symbol for DYSPROSIUM.

dyad (*BioSci*) Half of a tetrad group of chromosomes passing to one pole at meiosis.

dyad (*MathSci*) A TENSOR formed from the product of two vectors. For the vectors A and B it is written AB where, unlike scalar and vector multiplication, no multiplication symbol separates the factors.

dyadic (*MathSci*) The sum of two or more dyads. See CONJUGATE DYADICS.

Dycril plates (*Print*) See PHOTOPOLYMER PLATES.

dye (*Textiles*) A coloured compound that has SUBSTANTIVITY for the textile.

dye-coupling process (*ImageTech*) One using COUPLERS.

dye-destruction (*ImageTech*) Colour-printing process in which the tripack layers incorporate dyes of the respective primary colours and the image is differentially bleached in development.

dye diffusion printer (*ICT*) See DYE SUBLIMATION PRINTER.

dye laser (*Phys*) A laser using an organic dye and excited by a separate laser. It can be tuned over a significant fraction of the visible spectrum by using a reflection grating as one of the cavity mirrors.

dyeline (*Print*) A DOCUMENT COPYING method using DYELINE BASE PAPER, printing down with ultraviolet light which neutralizes diazonium salts in non-image areas, and processing with ammonium which develops the image. Sometimes called *diazo*.

dyeline base paper (*Paper*) Paper with controlled chemical and physical properties to enable it to be coated satisfactorily with a diazo compound and thereafter used to make a dyeline print.

dyestuffs (*Chem*) Groups of aromatic compounds having the property of dyeing textile fibres, and containing characteristic groups essential to their qualification as dyes. The more important dyestuffs are classified as follows: (1) nitroso- and nitro-dyestuffs, (2) azo-dyes, (3) stilbene, pyrazole and thiazole dyestuffs, (4) di- and triphenylmethane dyes, (5) xanthene dyestuffs, (6) acridine and quinoline dyestuffs, (7) indamine and indophenol dyestuffs, (8) azines, oxazines and thiazines, (9) hydroxyketone dyestuffs, (10) sulphide dyes, (11) vat dyestuffs, indigo and indanthrenes, (12) reactive dyestuffs, (13) ACID DYES, used in photography.

dye sublimation printer (*ImageTech*) A colour printer which produces hard copy from captured or computer-generated digital or video images using dye transfer sheets (yellow, magenta and cyan). The image is colour separated and the signals fed to a thermal head which evaporates the dyes in sequence onto printing paper. Also *dye diffusion printer, thermal dye transfer printer*.

dye toning (*ImageTech*) The chemical process whereby a dye is made to replace the silver in a normal photographic image, or to adhere to it by mordanting.

dye transfer (*ImageTech*) Colour print process using three gelatine matrices taken from separation negatives and processed with a tanning developer. These are treated with dye of the appropriate colour and the images transferred in register onto paper.

dying shift (*MinExt*) The night (graveyard) shift.

dyke (*Build*) A wall or embankment of timber, stone, concrete, fascines or other material, built as training works for rivers, to confine flow rigidly within definite limits over the length treated.

dyke (*Geol*) Discordant igneous intrusion of a tabular nature, usually nearly vertical, cutting the bedding or foliation of the country rock. Usually a metre to a few metres in width but may extend laterally for hundreds of km. Dykes are often grouped with SILLS as minor intrusions or hypabyssal rocks; they are commonly dolerites and basalts. Also *dike*.

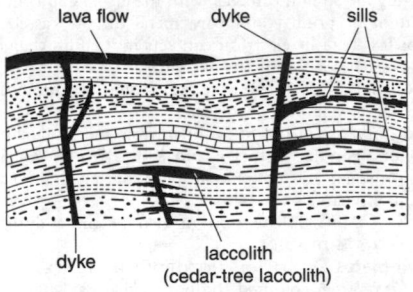

lava flow   dyke   sills

dyke   laccolith (cedar-tree laccolith)

dykes

dyke phase (*Geol*) That episode in a volcanic cycle characterized by the injection of minor intrusions, esp dykes. The dyke phase usually comes after the major intrusions, and is the last event in the cycle. See DYKE.

dyke swarm (*Geol*) A series of dykes of the same age, usually trending in a constant direction over a wide area. Occasionally, dykes may radiate outwards from a volcanic centre, as the Tertiary dyke swarm in Rum, Scotland; but usually they are parallel; eg the Old Red Sandstone dyke swarm of SW Scotland, of which the trend is north-east to south-west. See DYKE.

dynamical stability (*Ships*) The work done, usually in ft-tons or megajoules, when a vessel is heeled to a particular angle. See HEEL.

dynamical theory of X-ray and electron diffraction (*Phys*) A theoretical approach that takes into account the dynamical equilibrium between the incident and diffracted beams in a crystal, eg the effect of interference between the incident beam and multiply diffracted beams.

dynamic balance (*Aero*) The condition wherein centrifugal forces due to any rotating mass, eg a propeller, produce neither couple nor resultant force in the shaft and hence a reduction of vibration.

dynamic balancing (*Acous*) The technique of balancing the centrifugal forces in rotating machines so that there is no residual unbalance, and consequent vibration, to give rise to noise.

dynamic characteristic (*ElecEng*) Any characteristic curve obtained under normal working conditions for the device under question, eg the collector current (collector–emitter voltage relationship for a bipolar transistor when the effect of load impedance is included).

dynamic damper (*Eng*) Supplementary rotating mass driven through springs attached to a crankshaft at a point remote from the node to eliminate a troublesome critical speed. Also *detuner*.

dynamic data exchange (*ICT*) A method that links data in two or more applications so that changes in one are immediately reflected in the other, as between a database and a report program. Abbrev *DDE*.

dynamic electricity (*ElecEng*) A term sometimes used to denote electric currents, ie electric charges in motion, as opposed to static electricity, in which the charges are normally stationary. Also *current electricity*.

dynamic friction (*Eng*) See FRICTION.

dynamic heating (*Aero*) Heat generated at the surface of a fast-moving body by the bringing to rest of the air molecules in the boundary layer either by direct impact or by viscosity.

dynamic hypertext markup language (*ICT*) Text formatting instructions specifically designed to aid interaction between the viewer and a web page and to facilitate repositioning material. See DHTML.

dynamic isomerism (*Chem*) See TAUTOMERISM.

dynamic link library (*ICT*) A set of programming routines in the Windows (PC) operating system that can be called by other routines. Similar to *extensions* in the Macintosh operating system. Abbrev *DLL*.

dynamic loudspeaker (*Acous*) Open diaphragm, driven by a VOICE COIL on a former; intended to be used in a plane (Rice–Kellog) baffle, from the side of a box (box baffle) or at the neck of a large horn (flare).

dynamic mechanical test (*Eng*) A type of test which seeks to measure mechanical properties, eg tensile modulus E, under dynamic conditions, such as regular vibrations. A common apparatus for polymeric materials is the TORSION PENDULUM. Similar tests give a COMPLEX MODULUS, which can be divided into real and imaginary parts.

dynamic mechanical thermal analysis (*Eng*) A method of measuring complex moduli of materials as a function of temperature, using a torsion pendulum, vibrating reed, etc. Abbrev *DMTA*.

dynamic memory (*ICT*) One that needs to be periodically refreshed, ie read and rewritten every 2 ms or so, as stored charge tends to leak. Cf STATIC MEMORY. Also *volatile memory*.

dynamic model (*Aero*) A free-flight aircraft model in which the dimensions, inertia and masses are such as to duplicate full-scale behaviour.

dynamic noise suppressor (*Acous*) A device which automatically reduces the effective audio bandwidth, depending on the level of the required signal to that of the noise.

dynamic pressure (*Aero*) The pressure resulting from the instantaneous arresting of a fluid stream, the difference between total and static pressure.

dynamic psychology (*Psych*) A school of thought that assumes the importance of inner and subjective mental and emotional events in the explanation of behaviour. See PSYCHODYNAMICS.

dynamic range (*ICT*) The full range of signal levels, from highest to lowest, contained in any signal, transmission or recording; normally expressed in DECIBELS. Needs to be assessed in most forms of electronic design.

dynamic range (*ImageTech*) The range between the maximum signal which can be satisfactorily handled and the minimum inherent noise level of the system.

dynamic resistance (*ElecEng*) The relationship between voltage and current at a given position on the non-linear static characteristic of a device in an electrical circuit, eg a diode. It may be regarded as the tangent to the characteristic curve at that point and is often assumed to be constant over a small range of voltage and current. See DIFFERENTIAL RESISTANCE.

dynamic routing (*ICT*) A process for selecting the most appropriate path for a PACKET to travel across a NETWORK.

dynamics (*MathSci*) That branch of applied mathematics which studies the way in which force produces motion.

dynamic satellite constellation (*ICT*) A group of LOW EARTH ORBIT SATELLITES that between them guarantee full Earth coverage for a mobile communication system.

dynamic sensitivity (*ElecEng*) The alternating component of the output of a photoelectric device divided by the alternating component of incident radiant flux.

dynamic track following (*ImageTech*) A video tracking system which uses control signals in the video tracks and video heads on PIEZOELECTRIC mounts. See AUTOMATIC TRACKING.

dynamic viscosity (*Phys, Eng*) See COEFFICIENT OF VISCOSITY.

dynamite (*Chem*) A mixture of nitroglycerine with kieselguhr, wood pulp, starch flour, etc, making the nitroglycerine safe to handle until detonated. The most common industrial high explosive. Also *giant powder*.

dynamo (*ElecEng*) Electromagnetic machine which converts mechanical energy into ac or dc electrical supply. See ALTERNATOR.

dynamo-electric amplifier (*ElecEng*) Low- and zero-frequency mechanically rotating armature in a controlled magnetic field; used in servo systems.

dynamometer (*ElecEng*) An instrument for measurement of supply torque. *Electric dynamometers* can be used for measurement of act, voltage or power. See SIEMENS DYNAMOMETER.

dynamometer (*Eng*) A machine for measuring the brake horsepower of a prime mover or electric motor. See ABSORPTION DYNAMOMETER, ELECTRIC DYNAMOMETER, FROUDE BRAKE, ROPE BRAKE, TRANSMISSION DYNAMOMETER.

dynamometer ammeter (*ElecEng*) An ammeter operating on the dynamometer principle, the fixed and moving coils being connected in series and carrying the current to be measured.

dynamometer voltmeter (*ElecEng*) A voltmeter operating on the dynamometer principle, the fixed and moving coils being connected in series, and in series with a high resistance across the voltage to be measured.

dynamometer wattmeter (*ElecEng*) A commonly used type of wattmeter operating on the dynamometer principle, the fixed coil being usually in the current circuit and the moving coil in the voltage circuit.

dynamothermal metamorphism (*Geol*) A regional metamorphism involving both heat and pressure.

dynamotor (*ElecEng*) An electric machine having two armature windings, one acting as a generator and the other as a motor, but only a single magnet frame. Also *rotary transformer*.

dynatron (*Electronics*) A circuit in which steady-state oscillations are set up in a tuned circuit between screen and anode of a tetrode, the latter exhibiting negative resistance when the anode potential is below the potential of the screen.

dyne (*Phys*) The unit of force in the CGS system of units. A force of 1 dyne acting on a mass of 1 g imparts to it an acceleration of 1 cm s^{-2}. 10^5 dynes = 1 N.

Dyneema (*Eng*) TN for gel-spun polyethylene fibre (Europe). See panel on HIGH-PERFORMANCE POLYMERS.

dynein (*BioSci*) Large protein which forms columns of side arms along the peripheral MICROTUBULES of cilia and mediates the sliding of the microtubules relative to each other, causing the cilia to bend. It has ATPase activity. Cytoplasmic isoforms are known and are responsible for some intracellular movement of materials.

dynode (*Electronics*) Intermediate electrode (between cathode and final anode) in photomultiplier or electron multiplier tube. Dynode electrons are those which emit secondary electrons and provide the amplification. See fig. at PHOTOMULTIPLIER.

dynode chain (*Electronics*) Resistance potential divider employed to supply increasing potentials to successive dynodes of an electron multiplier.

dys- (*Genrl*) Prefix from Gk *dys-* denoting mis-, un-.

dysadaptation (*Med*) Marked reduction in rapidity of adaptation of the eye to suddenly reduced illumination, as in vitamin A deficiency.

dysarthria (*Med*) Difficult articulation of speech, due to a lesion in the brain.

dysbasia angiosclerotica (*Med*) Pain in the legs on walking, due to thickening of the arteries. See INTERMITTENT CLAUDICATION.

dyscalculia (*Psych*) Same as ACALCULIA.

dyschezia (*Med*) A form of constipation in which the feces are retained in the rectum, as a result of blunting of a normal reflex, due to faulty habits.

dyscrasia (*Med*) Any disordered condition of the body, esp of the body fluids.

dyscrasite (*Min*) Silver ore consisting mainly of a silver antimonide, Ag_3Sb.

dysdiadokokinesia (*Med*) Inability to perform rapid alternate movements as a result of a lesion in the cerebellum.

dysentery (*Med*) A term formerly applied to any condition in which inflammation of the colon was associated with the frequent passage of bloody stools. Now confined to (1) bacillary dysentery, due to infection with *Bacterium dysenteriae*; (2) amoebic dysentery, the result of infection with the *Entamoeba histolytica*.

dysgenic (*BioSci*) Causing, or tending towards, racial degeneration.

dysgraphia (*Med*) Inability to write, as a result of brain damage or other cause.

dyskinesia (*Med*) A term for a number of conditions in which involuntary movements follow a definite pattern, eg tics.

dyslalia (*Med*) Articulation difficulty due to defects in speech organs.

dyslexia (*Med*) Great difficulty in learning to read, write or spell, which is unrelated to intellectual competence and of unknown cause. Also *word blindness*.

dysmelia (*Med*) Misshapen limbs.

dysmenorrhoea (*Med*) Painful and difficult menstruation. US *dysmenorrhea*.

dysmetria (*Med*) Faulty estimation of distance in the performance of muscular movements, due to a lesion in the cerebellum.

dysmorphia (*Psych*) Any psychological condition involving a distorted perception of one's own body.

dysostosis (*Med*) Defect in the normal ossification of cartilage.

dyspareunia (*Med*) Painful or difficult coitus.

dyspepsia (*Med*) Indigestion: any disturbance of digestion.

dysphagia (*Med*) Difficulty in swallowing.

dysphasia (*Med*) Disturbed utterance of speech due to a lesion in the brain.

dysphonia (*Med*) Difficulty of speaking, due to any affection of the vocal cords.

dysphoria (*Med*) Unease; absence of feeling of well-being.

dysplasia (*Med*) Abnormality of development.

dyspnoea (*Med*) Laboured or difficult respiration. US *dyspnea*.

dyspraxia (*Med*) Difficulty with carrying out tasks involving co-ordinated motor activities, particularly complex ones, although there is no apparent physical impairment. In some cases may affect speech (*speech dyspraxia*).

dysprosium (*Chem*) A metallic element, a member of the rare earth group. Symbol Dy, at no 66, ram 162·5.

dyssynergia (*Med*) Incoordination of muscular movements, due to disease of the cerebellum.

dystectic mixture (*Chem*) A mixture with a maximum melting point. Cf EUTECTIC.

dystocia (*Med*) Painful or difficult childbirth. Also *dystokia*.

dystrophia adiposogenitalis (*Med*) A condition characterized by obesity, hairlessness of the body and underdeveloped genital organs, due to disordered function of the pituitary gland. Also *Fröhlich's syndrome*.

dystrophia myotonica (*Med*) See MUSCULAR DYSTROPHY.

dystrophic (*EnvSci*) Said of a lake in which the water is rich in organic matter, such as humic acid; this consists mainly of undecomposed plant fragments, and nutrient salts are sparse.

dystrophin (*BioSci*) A protein found in skeletal muscle, but that is missing or abnormal in patients with DUCHENNE MUSCULAR DYSTROPHY.

dystrophy (*Med*) A condition of impaired or imperfect nutrition, as in MUSCULAR DYSTROPHY.

dysuria (*Med*) Painful or difficult passage of urine. Also *dysury*.

E

E (*BioSci*) Abbrev for ERYTHROCYTE. See EA, EAC.

E (*ChemEng*) Symbol for EDDY DIFFUSIVITY.

E (*Genrl*) Symbol for EXA-.

E (*Phys*) Symbol for: (1) ELECTROMOTIVE FORCE; (2) ELECTRIC FIELD STRENGTH; (3) energy; (4) (with subscript) single electrode potential, thus E_H, on the hydrogen scale; E_O, standard electrode potential; (5) INTERNAL ENERGY (US); (6) YOUNG'S MODULUS.

E- (*Chem*) Prefix denoting 'on the opposite side' (Ger *entgegen*), and roughly equivalent to *trans-*. See CAHN–INGOLD–PRELOG SYSTEM.

[E] (*Phys*) One of the FRAUNHOFER LINES in the green of the solar spectrum. Its wavelength is 526·9723 nm, and it is due to iron.

e (*Build*) Symbol for eccentricity of a load.

e (*MathSci*) Symbol for: (1) the base of natural logarithms (2·718 281 828 5…), defined as the limiting value of

$$\left(1+\frac{1}{m}\right)^m$$

as *m* approaches infinity; (2) the eccentricity of a CONIC.

e (*Phys*) Symbol for: (1) the ELECTRON (e^-); (2) the POSITRON (e^+).

e (*Phys*) Symbol for the charge on the electron, $1·6022 \times 10^{-19}$ C.

e-, ex- (*Genrl*) Prefixes from Lt *ex*, *e*, out of.

ε (*Chem*) Symbol for MOLAR EXTINCTION COEFFICIENT.

ε (*Phys*) Symbol for: (1) EMISSIVITY; (2) linear STRAIN; (3) PERMITTIVITY.

ε- (*Chem*) (1) Substituted on the fifth carbon atom. (2) *epi-*, ie containing a condensed double aromatic nucleus substituted in the 1·6 positions. (3) *epi-*, ie containing an intramolecular bridge.

η (*Chem, ElecEng*) Symbol for: (1) ELECTROLYTIC POLARIZATION; (2) OVERVOLTAGE.

η (*Phys*) Symbol for COEFFICIENT OF VISCOSITY.

EA, EAC (*BioSci*) Abbrevs for *Erythrocyte with Antibody on its surface*, and *Erythrocyte with Antibody and Complement*, the latter describing components that have become attached following activation (eg EAC 1423). Such red cells are used to detect Fc receptors or complement receptors on other cells.

Eagle mounting (*Phys*) A compact mounting of a concave diffraction grating based on the principle of the ROWLAND CIRCLE. The mounting suffers from less astigmatism than either the Rowland or ABNEY MOUNTINGS, and is useful for studying higher-order spectra.

EAN (*ICT*) Abbrev for *European academic network*. A communications network for the European research community.

ear (*BioSci*) (1) Strictly, the sense organ that receives auditory impressions. (2) In insects, various tympanic structures on the thorax or forelegs. (3) In some birds and mammals, a prominent tuft of feathers or hair close to the opening of the external auditory meatus. (4) In mammals, the pinna. (5) More generally, any ear-like structure. (6) The seed-holding inflorescence of a cereal plant.

ear (*Build*) A CROSSETTE.

ear (*Eng*) A projection integral with, or attached to, an object, for supporting it, or attaching another part to it pivotally. Also *finplate*, *lug*.

ear defender (*Acous*) A plug or muff for insertion into, or fitting over, the ear to reduce reception of noise.

ear drum (*Med*) The outer termination of the aural mechanism of the ear, consisting of a membrane, in tension, for transferring the acoustic pressures applied from without to the ossicles for transmission to the inner ear.

ear emergence (*Agri*) The date when the first spikelet of the inflorescence is visible on 50% of the plants in a crop.

earing (*Eng*) Excessive elongation along edges and folds of metals being shaped by deep drawing or rolling.

Early Bird (*ICT*) The first (1964) communications satellite providing regular commercial telecommunications. See panel on COMMUNICATIONS SATELLITE.

Early effect (*Electronics*) The variation of junction capacitance and effective base thickness of a transistor with the supply potentials.

early replicating regions (*BioSci*) Parts of chromosomes that are replicated early in S-PHASE.

early-warning radar (*Radar*) A system for the detection of approaching aircraft or missiles at greatest possible distances. See BMEWS.

early wood (*BioSci*) The wood formed in the first part of a growth layer during the spring, typically with larger cells with thinner cell walls than the LATE WOOD. Also *spring wood*.

ear muff (*Acous*) (1) A pad of rubber or similar material placed on a telephone receiver to minimize discomfort during long use. (2) An EAR DEFENDER.

EARN (*ICT*) Abbrev for *European academic research network*. A European-wide network that forms part of BITNET. It was supported by funding from IBM Corporation.

earphone coupler (*Acous*) A suitably shaped cavity with an incorporated microphone used for acoustical testing of earphones.

earphones (*Acous*) Electro-acoustic transducers which transform electric signals into acoustic signals and are worn over the ear.

ear tag (*Agri*) A plastic tag clipped into the ear of livestock for identification. A legal requirement in many countries.

Earth (*Astron*) Third planet from the Sun and largest of the inner planets, with a mean equatorial radius of 6378·17 km, mass $5·977 \times 10^{24}$ kg and mean density 5·517 times that of water. See panel on EARTH.

earth (*ICT*) A system of plates or wires buried in the ground to allow a path to earth for currents flowing in an antenna, used largely to improve the efficiency of broadcast and short-wave antenna. US *ground*.

earth (*Phys*) Connection to main mass of earth by means of a conductor having a very low impedance.

earth capacitance (*Phys*) The capacitance between an electrical circuit and earth (or conducting body connected to earth). Also *ground capacitance*.

earth coil (*ElecEng*) A pivoted coil of large diameter for measuring the strength of the Earth's magnetic field; this is done by suddenly changing the position of the coil in this field and observing the throw of a ballistic galvanometer connected to it. Also *earth indicator*.

earth continuity conductor (*ElecEng*) A third conductor (with line and neutral) in a mains distribution system, bonded to earth, provided for connection to any metal component not in the electrical circuit; for safety purposes. Abbrev ECC.

Earth

The third planet in order from the Sun with a mean equatorial radius of 6378·17 km, mass $5·977 \times 10^{24}$ kg and mean density 5·517. From the astronomical perspective, Earth belongs to the group of rocky terrestrial planets which also includes Mercury, Venus and Mars. It is with this group, and also the Moon, that its origin, structure and evolution are often compared. Earth has an atmosphere intermediate in thickness between those of Venus and Mars. It is unique in possessing vast oceans of liquid water. The complex interaction between the oceans, the atmosphere and the continental surfaces determines the energy balance, the temperature regime and hence the climate. Cloud cover is typically 50% and heat trapped within the atmosphere by the GREENHOUSE EFFECT raises the average temperature by more than 30°C above that expected for the Earth's distance from the Sun.

The present composition of the atmosphere is 77% molecular nitrogen, 21% molecular oxygen, 1% water vapour and 0·9% argon, with the balance in trace components. The high concentration of oxygen, which dates from 2000 million years ago, is a direct result of the presence of plants. The presence of oxygen allowed the formation of the high-level ozone layer, which shields the surface from solar ultraviolet radiation that damages higher forms of life.

Earth is the most geologically active of the major planets in this solar system. Its large-scale features have all been determined by the creation, destruction, relative movement and interaction of a dozen or so crustal plates – the LITHOSPHERE – which slide over the less rigid ASTHENOSPHERE, as shown below. Collisions between the plates produce folded mountains, and zones of seismic activity are concentrated along the plate boundaries.

Seismic waves, such as are generated during earthquakes, reveal the internal structure of the Earth. At the centre, there is a molten metallic core of iron and nickel, possibly with a solid core at the very centre at a temperature around 4000°C. The silicate mantle overlies the core. The outermost crust is about 10 km thick under the oceans and 30 km thick where there are continents.

In planetary terms, the surface rocks of the Earth are very young, with the basaltic rocks forming the ocean floors among the youngest. The Precambrian shields, which occupy about 10% of the surface, are the oldest and the nearest approximation to the cratered terrain that forms a large part of other planetary surfaces. Weathering and erosion have removed all but a few traces of any impact craters that now remain.

The molten metallic core gives rise to the Earth's magnetic field and MAGNETOSPHERE. A layer of electrically charged particles (from the Sun) at a height of 200–300 km forms the IONOSPHERE. The funnelling of charged particles by the magnetic field to regions between latitudes of 60° and 75° creates the phenomenon of the aurora. Satellite measurements have shown that the Earth is also an intense source of radio waves at kilometric wavelengths.

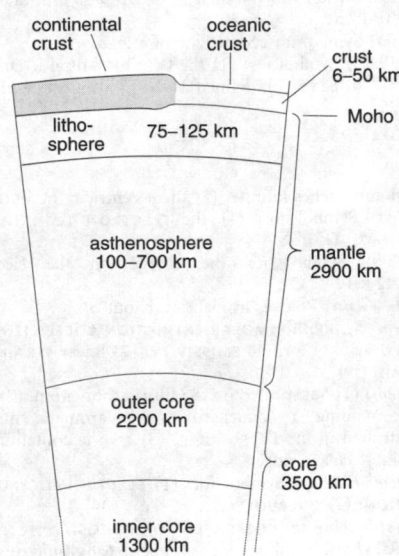

Earth Sectional drawing (not to scale)

See panels on ATMOSPHERIC BOUNDARY LAYER, GEOLOGICAL COLUMN, PLATE TECTONICS, STRATOSPHERE AND MESOSPHERE, TROPOSPHERE and appendices.

---

earth currents (*ElecEng*) (1) Currents in the Earth which, by electromagnetic induction, cause irregular currents to flow in submarine cables and so interfere with the reception of the transmitted signals. (2) Direct currents in the Earth, which are liable to cause corrosion of the lead sheaths of cables; they can be the earth-return currents of power systems. (3) Any current flowing in an item of equipment through that which is grounded or at earth potential.

earth dam (*CivEng*) A dam built of local gravels, earth, etc. Usually provided with impervious clay core which must be taken down in a trench to an impervious layer or otherwise protected from passage of water under it.

earth detector (*ElecEng*) See LEAKAGE INDICATOR.

earthed aerial (*ICT*) Marconi aerial, in which an elevated wire is earthed at its lower end.

earthed circuit (*ElecEng*) An electric circuit connected to earth at one or more points.

earthed concentric wiring system (*ElecEng*) A two-wire system for wiring or general distribution, which uses twin concentric conductors, the outer conductor being earthed.

earthed neutral (*ElecEng*) A neutral point of a polyphase system or piece of electrical apparatus which is connected to earth, either directly or through a low impedance.

earthed pole (*ElecEng*) The pole or line of an earthed circuit which is connected to earth.

earthed switch (*ElecEng*) A switch in which all exposed metal parts can be earthed.

earthed system (*ElecEng*) A system of electric supply in which one pole or the neutral point is earthed, either directly or through a low impedance, the former being known as a solidly earthed system.

earth electrode system (*ElecEng*) The totality of conductors, conduits, shields and screens which are earthed by low-impedance conductors.

earthenware (*Glass*) General term for vessels and other products made from opaque, permeable ceramic white-ware, which is fired at a lower temperature (1100–1150°C), and so has a lower proportion of glass phase, and is more porous and less strong than eg PORCELAIN. Must be glazed to make it impermeable. Cf BONE CHINA, FAIENCE, STONEWARE, TERRACOTTA.

earth fault (*ElecEng*) An accidental connection between a live part of an electrical system and earth.

earth impedance (*Phys*) The impedance as normally measured, with all extraneous emfs reduced to zero, between any point in a communicating system or a measuring circuit and earth.

earth inductor (*ElecEng*) See EARTH COIL.

earthing autotransformer (*ElecEng*) See EARTH REACTOR.

earthing resistor (*ElecEng*) A resistance through which the neutral point of a supply system is earthed, in order to limit the current which flows on the occurrence of an earth fault. Also *earthing resistance.*

earthing tyres (*Aero*) Tyres for aircraft having an electrically conductive surface in order to discharge static electricity upon landing.

earth lead (*ICT*) Connection between a radio transmitting or receiving apparatus and its earth.

earth leakage protection (*ElecEng*) Protection system, suitable for domestic use, in which imbalance between live and neutral currents is used to trip a circuit breaker, thus isolating the supply after an earth fault. See BALANCED PROTECTIVE SYSTEM.

Earth observation (*Space*) Observation by remote sensing techniques from satellites. Provides detailed information for military purposes, and for the oil and mining industries, fisheries, weather forecasting, climate studies, forestry and agriculture. Movement of ice and locust swarms, location of mineral deposits, and distribution of pollutants can all be monitored. Sensing may be: (1) *passive*, eg by RADIOMETER sensitive to visible and infrared radiation, and to microwaves which can pass cloud cover to provide all-weather information; or (2) *active*, using SYNTHETIC APERTURE RADAR (where reflections of a signal generated on board the satellite are received and analysed), LIDAR, and *acoustic sounding* which uses bursts of high-frequency sound for wind and atmospheric temperature studies. GEOSYNCHRONOUS and SUN-SYNCHRONOUS ORBITS may be used.

earth-pillars (*Geol*) These occur where sediments consisting of relatively large and preferably flat stones, embedded in a soft, finer-grained matrix, are undergoing erosion, esp in regions of heavy rainfall. As the ground is progressively lowered the flat stones protect the softer material beneath them and are thus left standing on tall, acutely conical pillars.

earth plate (*ElecEng*) A metal plate buried in the earth for the purpose of providing an electrical connection between an electrical system and the earth.

earth potential (*Phys*) The electric potential of the Earth; usually regarded as zero, so that all other potentials are referred to it.

earth pressure (*CivEng*) The pressure exerted on a wall by earth which is retained, ie supported laterally by the wall.

earthquake (*Geol*) A shaking of the Earth's crust, usually caused by displacement along a fault. See SEISMOLOGY, panel on EARTHQUAKE and appendices.

earthquake intensity (*Geol*) The measurement of the effects of an earthquake at a particular place. One earthquake has one magnitude but different intensities. See EARTHQUAKE MAGNITUDE, MERCALLI SCALE, panel on EARTHQUAKE and appendices.

earthquake magnitude (*Geol*) A measure of the strength of an earthquake as determined by instruments, expressed on a scale called the *Richter scale*. See EARTHQUAKE INTENSITY and appendices.

earthquake mount (*Build*) Laminated bearing of steel and rubber for buildings, designed to resist vibrations from earthquakes. Widely used in California and Japan.

earth reactor (*ElecEng*) (1) A REACTOR connected between the neutral point of an ac supply system and earth, in order to limit the earth current which flows on the occurrence of an earth fault. (2) An arrangement of reactors or transformers, so connected to a polyphase system that a neutral point is artificially obtained. Also *earthing autotransformer, neutralator, neutral autotransformer, neutral compensator.*

earth resistance (*ElecEng*) The resistance offered by the earth between two points of connection, and therefore forming a coupling between all circuits making use of the same current path in the earth.

earth-return circuit (*ElecEng*) One which comprises an insulated conductor between two points, the circuit being completed through the earth.

Earth's atmosphere (*Astron*) The gaseous envelope surrounding the Earth. Various layers are recognized: the troposphere, the lowest part, comprising 80% of atmospheric mass; the TROPOPAUSE, the upper limit of the troposphere; the stratosphere, which accounts for most of the remaining atmospheric mass; the STRATOPAUSE; the IONOSPHERE, where ionized gases reflect radio waves (TV wavelengths penetrate and escape), which includes the mesosphere, where ozone is produced, and the MESOPAUSE, the mesosphere boundary; and the outermost EXOSPHERE above the ionosphere. See panels on ATMOSPHERIC BOUNDARY LAYER, STRATOSPHERE AND MESOSPHERE and TROPOSPHERE.

earth science (*Geol*) A term frequently used as a synonym for geology. Has also been used to include sciences which fall outside the scope of geology, eg meteorology (although meteorology is included in 'EnvSci' in this dictionary).

earthshine (*Astron*) The faint light which often covers the entire disk of the Moon close to the new Moon, due to the reflection of sunlight from the Earth. Also *ashen light* or, picturesquely, *the old Moon in the new Moon's arms.*

Earth's magnetic field (*ElecEng*) See TERRESTRIAL MAGNETISM.

Earth station (*ICT*) A single transmitter and steerable antenna, or a site housing several of these, with the sole purpose of transmitting to or receiving from communication and direct broadcast satellites, and linking them to terrestrial communications networks.

earth system (*ICT*) See panel on EARTH.

earth terminal (*ElecEng*) A terminal provided on the frame of a machine or piece of apparatus to make a connection to earth.

earth thermometer (*EnvSci*) A thermometer used for measuring the temperature of the ground at depths up to a few metres. *Symon's earth thermometer* (the most commonly used) consists of a mercury thermometer, with its bulb embedded in paraffin wax, suspended in a steel tube.

earth trip (*ElecEng*) See BALANCED PROTECTIVE SYSTEM.

earthware duct (*ElecEng*) A conduit made of earthenware, for carrying underground cable.

earthwork (*CivEng*) A bank or cutting.

earthy (*ElecEng*) Said of (1) circuits when they are connected to earth, either directly (as for dc) or through a capacitor (with ac); (2) any point in a communicating

Earthquake

This is a shaking of the Earth's crust caused, in most cases, by displacement along a fault. The place of maximum displacement is the *focus*, as in Fig. 1(a). The *epicentre* of an earthquake is the point on the surface of the Earth lying immediately above its focus. On the Earth's surface, lines of equal intensity are *isoseismal* lines. Although the amount of displacement may be small, a matter of centimetres only, the destruction wrought at the surface may be very great, owing in part to secondary causes, eg the severance of gas and water mains, as in the great San Francisco earthquake in 1906.

Most earthquakes take place at the boundaries of the major tectonic plates of the Earth's crust (see panel on PLATE TECTONICS). They are measured around the world by *seismographs* which record the types of wave received (Fig. 1(b)). The earthquake releases energy as *shear waves* (S-wave, Secondary) and *compressional waves* (P-waves, Compression, Primary), as well as *surface waves* (L-waves).

A typical seismic wave pattern is shown in Fig. 2. P-waves can travel through any material including liquids and are faster than the other waves. S-waves will only pass through solids. The surface or L-waves are the slowest and are confined to the Earth's crustal layers. The worldwide seismographic network enables the location, strength and depth of focus of earthquakes to be determined and sheds much light on the structure of the Earth (see panel on EARTH). Since S-waves cannot traverse the core there is a shadow zone on the far side of the globe from the epicentre.

Earthquake intensities are expressed on a scale known as the Modified Mercalli Scale which expresses the degree of shaking as in the table. This scale is empirical and in 1935 C F Richter devised a *scale of magnitude* which is logarithmic, ie the difference

between magnitude 4 and 5 is one-tenth of that between 5 and 6. One of the strongest earthquakes of this century, in Alaska in 1964, probably had a magnitude of 8·6. Submarine earthquakes may give rise to a *tsunami*, a wave produced in the oceans. It can be very destructive and travels great distances at very high speeds, up to 950 km h^{-1}.

See appendix on Earthquake severity measurement scales.

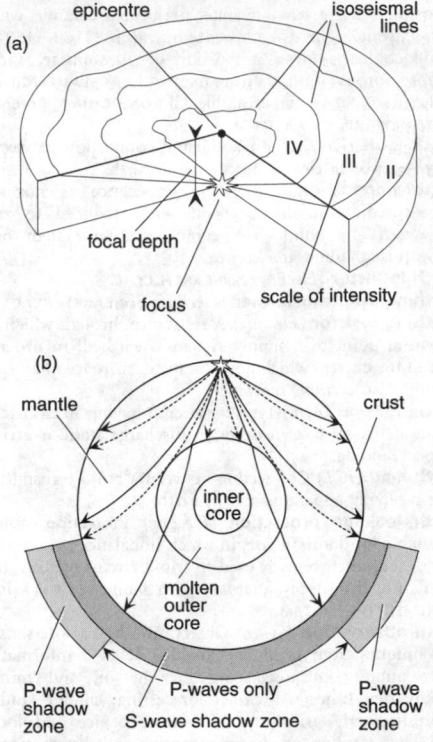

Fig. 1 **Earthquake** (a) Slice from Earth's crust. (b) Section through centre; P-waves are solid and S-waves dashed.

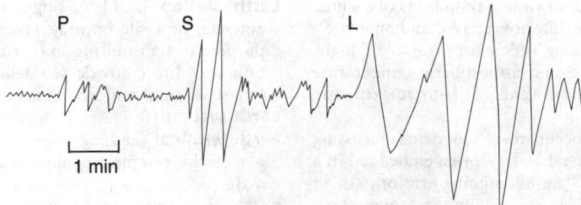

Fig. 2 **Seismogram**, showing arrival of P-, S- and L-waves from an epicentre 1400 km away.

system (eg the mid-point of a shunting resistance across a balanced line) which is at earth potential, although not actually connected to earth through zero impedance; (3) points of a bridge when reduced to earth potential by a WAGNER EARTH.
earthy cobalt (*Min*) A variety of wad containing up to about 40% of cobalt oxide. Also *asbolite*.

EAS (*Aero*) Abbrev for EQUIVALENT AIRSPEED.
easement curve (*Surv*) See TRANSITION CURVE.
easer (*Print*) A general term for additives mixed with printing ink to produce some particular result, eg quicker drying, reduction of tack.
easing (*Build*) The shaping of a curve so that there is no abrupt change in curvature, particularly in hand rails.

easing centres (*Build, CivEng*) The process of gradually removing the centring from beneath a newly completed arch, thereby transferring its weight slowly to the arch abutments.

easing wedges (*CivEng*) Same as STRIKING WEDGES.

East African olive (*For*) See OLIVE.

East Coast fever (*Vet*) A disease of cattle in Africa due to infection by the protozoan *Theileria parva*; characterized by fever, enlarged lymph glands, respiratory distress, diarrhoea and loss of condition. Transmitted by the tick *Rhipicephalus appendiculatus*.

Easter egg (*ICT*) An undocumented sequence of code in a computer program activated by a specific set of keystrokes. Often intended as a joke or as the signature of the author.

East Indian satinwood (*For*) See CEYLON SATINWOOD.

easting (*Surv*) Departure, or displacement, along an E–W line of a point with reference to a true N–S datum.

easy axis (*Phys*) Crystal axis in magnetic materials along which magnetization is energetically preferred in the absence of external fields. See MAGNETOCRYSTALLINE ANISOTROPY.

easy-care (*Textiles*) See DRIP-DRY.

eau de Javelle (*Chem*) See JAVEL WATER.

eave (*Build*) The lower part of a roof which projects beyond the face of the walls.

eave-board (*Build*) See TILTING FILLET.

eave-lead (*Build*) A lead gutter behind a parapet, around the edge of a roof.

eaves course (*Build*) See DOUBLING COURSE.

eaves fascia (*Build*) See FASCIA (2).

eaves gutter (*Build*) A trough fixed beneath an eave to catch and carry away the rain flow from the roof. Also *shuting*.

eaves plate (*Build*) A beam carried on piers or posts and supporting the feet of roof rafters in cases where there is no wall beneath.

eaves soffit (*Build*) The horizontal surface beneath a projecting eave.

e-banking (*ICT*) Banking carried out though electronic communications.

EBCDIC (*ICT*) See CHARACTER CODE.

EBD (*Psych*) Abbrev for *emotional and behavioural difficulties* (or *disorder*).

E-bend (*ICT*) A smooth bend in the axis direction of a waveguide, the axis being maintained in a plane parallel to the polarization direction.

Eberhard effect (*ImageTech*) Border effect in the developing of a heavy photographic image, showing higher density at the edges than in the centre.

EBM (*Aero, Eng*) Abbrev for ELECTRON BEAM MACHINING.

EBNA (*BioSci*) Abbrev for *Epstein–Barr virus nuclear antigen*. Antigen detected in the nuclei of B-cells and tumour cells in conditions associated with infection by the Epstein–Barr virus, such as *infectious mononucleosis, Burkitt's lymphoma* and *nasopharyngeal carcinoma*.

Ebola disease (*Med*) A severe and often fatal viral disease causing one form of African haemorrhagic fever. The other is caused by the Marburg virus, which is closely related.

ebonite (*Chem*) Highly cross-linked and filled natural or synthetic rubber, vulcanized with up to about 40% sulphur. A hard brittle material, it was formerly much used for battery cases etc but has now been largely displaced by polypropylene, polycarbonate, etc. Also *hard rubber, vulcanite*.

ebony (*For*) Hardwood from a tree of the genus *Diospyros*, a native of W Africa (*D. crassiflora*), India and Sulawesi (*D. celebica*). It is very dense, dark brown to black, straight grained and fine textured.

e-book or ebook (*ICT*) A publication in electronic form that can be stored and read on a computer (in full, *electronic book*).

EBU (*Genrl*) Abbrev for *European Broadcasting Union*.

ebullator (*Phys*) A heated surface used to impart heat to a fluid in contact.

ebulliometer (*Phys*) A device which enables the true boiling point of a solution to be determined.

ebullioscopy (*Chem*) The determination of the molecular weight of a substance by observing the elevation of the boiling point of a suitable solvent.

ebullition (*Phys*) See BOILING.

eburnation (*Med*) Ivory-like hardening of bone which occurs at the joint surfaces in osteoarthritis.

EBV (*Agri*) Abbrev for ESTIMATED BREEDING VALUE.

EBV (*BioSci*) Abbrev for EPSTEIN–BARR VIRUS.

EBW (*Aero, Eng*) Abbrev for ELECTRON BEAM WELDING.

ECAC (*Aero*) Abbrev for EUROPEAN CIVIL AVIATION CONFERENCE.

ecad (*BioSci*) A species with distinctive forms that depend simply on the environment rather than on genotypic differences. Cf ECOTYPE.

ECC (*ElecEng*) Abbrev for EARTH CONTINUITY CONDUCTOR.

eccentric (*Eng*) (1) Displaced with reference to a centre; not concentric. (2) A crank in which the pin diameter exceeds the stroke, resulting in a disk eccentric to the shaft; used particularly for operating steam-engine valves, pump plungers, etc.

eccentric angle (*MathSci*) Of a point P on an ellipse, the angle θ, where ($a \cos \theta$, $b \sin \theta$) are the parametric co-ordinates of P referred to the axes of the ellipse.

eccentric fitting (*Build*) A fitting in which the centre line is offset.

eccentric groove (*Acous*) See LOCKED GROOVE.

eccentricity (*Astron*) The extent to which an elliptical orbit deviates from circularity, given by the distance between the two focal points of the ellipse divided by twice the length of the major axis.

eccentricity (*Build*) The distance from the centre of application of a load or system of loads to the centroid of the section of the structural member to which it or they are connected. Symbol e.

eccentricity (*MathSci*) See CONIC.

eccentric load (*Build*) A load which is carried by a structural member at a point other than the centroid of the section.

eccentric pole (*ElecEng*) A pole on an electric machine in which the pole face is not concentric with the armature but has a greater air-gap at one pole tip than at the other to assist in neutralizing the effect of armature reaction.

eccentric sheave (*Eng*) The disk of an ECCENTRIC, often formed integral with the shaft.

eccentric station (*Surv*) One not physically occupied during triangulation etc but serving as a fixation point.

eccentric strap (*Eng*) A narrow split bearing, fitting onto an eccentric sheave and bolted to the end of a valve rod etc; corresponds to the 'big end' of a connecting-rod.

eccentric throw-out (*Eng*) A device for engaging the back gear of a lathe. The back-gear shaft runs in eccentric-bored bearings, which are rotated to bring the gears in and out of mesh with those on the mandrel. See BACK GEAR.

ecchondroma (*Med*) A tumour composed of cartilage and growing from the surface of bone.

ecchondrosis (*Med*) An abnormal outgrowth of the joint cartilage, in chronic arthritis.

ecchymosis (*Med*) A large discoloured patch due to extravasation of blood under the skin.

Eccles–Jordan circuit (*Electronics*) Original bistable multi-vibrator using two triodes or transistors. See FLIP-FLOP.

eccrine (*BioSci*) Said of a gland whose product is excreted from its cells.

ecdemic (*BioSci*) Foreign; not indigenous or ENDEMIC. Also *exotic*.

ecdysis (*BioSci*) The act of casting off the outer layers of the integument, as in ARTHROPODA.

ecdysone (*BioSci*) A steroid hormone found in arthropods and plants (*phytoecdysone*). In arthropods α-ecdysone stimulates moulting of the cuticle.

ECFA (*BioSci*) Abbrev for *eosinophil chemotactic factor of anaphylaxis*, a peptide that is released from mast cells and causes eosinophils to move into the site from the bloodstream. Local accumulation of eosinophil granulocytes takes place where type-I allergic reactions occur (see HYPERSENSITIVITY REACTIONS).

ECG (*Med*) Abbrev for ELECTROCARDIOGRAM.

echelon grating (*Phys*) A form of interferometer resembling a flight of glass steps, light travelling through the instrument in a direction parallel to the treads of the steps. The number of interfering beams is therefore equal to the number of steps. Owing to the large path difference, $t(\mu-1)$, where t is the thickness of a step and μ is the REFRACTIVE INDEX, the order of interference and therefore the resolving power are high, making the instrument suitable for studying the fine structure of spectral lines.

echin-, echino- (*Genrl*) Prefixes from Gk *echinos*, hedgehog; meaning spiny.

echinococcosis (*Vet*) An infection of sheep, pigs and cattle, and sometimes humans, by the intermediate hydatid stage of the tapeworm *Echinococcus granulosus*. The adult worm occurs in dogs and other carnivora.

echinococcus (*BioSci*) A CYSTICERCUS or bladderworm possessing a well-developed bladder containing daughter bladders, each with numerous SCOLICES.

Echinodermata (*BioSci*) A phylum of radially symmetrical marine animals that have a body wall strengthened by calcareous plates. There is a complex coelom; locomotion is usually carried out by the tube feet, which are distensible finger-like protrusions of a part of the coelom known as the water vascular system. The larva is bilaterally symmetrical and shows traces of metamerism. Starfish, sea urchins, brittle stars, sea cucumbers and sea lilies.

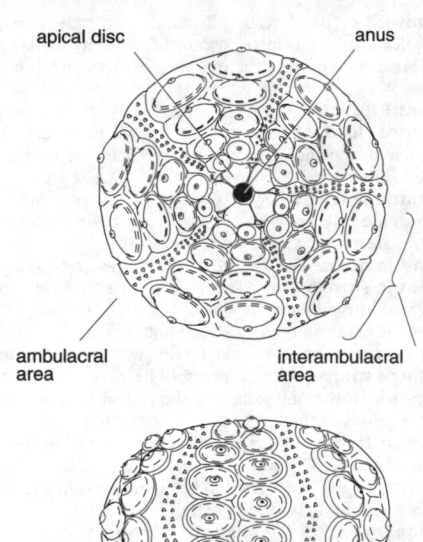

Echinodermata Upper and lateral views of a sea urchin shell or test.

echinoderms (*BioSci*) See ECHINODERMATA.

Echinoidea (*BioSci, Geol*) A class of ECHINODERMATA having a globular, ovoid or heart-shaped body that is rarely flattened. There are no arms; the tube feet possess ampullae and occur on all surfaces, but not in grooves. The anus is aboral or lateral and the madreporite aboral. There is a well-developed skeleton and free-living forms exist. Fossil echinoids are found in strata ranging from the Lower Palaeozoic to the present. They are particularly important in the Jurassic (Clypeus Grit etc) and Cretaceous, where, in the Chalk, they have proved useful indices of horizon, esp the various species of *Micraster* and *Holaster*. Sea urchins.

echinus (*Arch*) An ornament in the shape of an egg carved on a moulding etc.

Echiuroidea (*BioSci*) A phylum of sedentary marine worm-like animals, in which nearly all trace of metamerism has been lost in the adult. The body is sac-shaped, and feeding is effected by an anterior non-retractile proboscis, bearing a ciliated groove leading to the mouth.

echo (*Acous*) Received acoustic wave, distinct from a directly received wave, because it has travelled a greater distance due to reflection.

echo (*ICT*) (1) Data transmission in which data are returned to the point of origin for comparison with the original data. (2) The reception of a signal additional to, and later than, the desired signal; caused by reflection from hills etc, or travel completely round the Earth.

echo (*Radar*) Return signal in radar, whether from wanted object, or from side or back lobe radiation.

echo box (*Radar*) Adjustable test resonator of high Q for returning a signal to the receiver from the transmitter.

echocardiography (*Med*) Examination of the structure and function of the heart using reflected pulsed ultrasound.

echo chamber (*Acous*) See REVERBERATION CHAMBER.

echo flutter (*Radar*) A rapid sequence of reflected radar (or sound) pulses arising from one initial pulse.

echograph (*Genrl*) A device used for echo sounding.

echographia (*Med*) Ability to copy writing, associated with inability to express ideas in writing, due to a lesion in the brain.

echoic memory (*Psych*) The brief retention of auditory information, in an unprocessed or *echo* form; fades within 2–6 seconds. Cf ICONIC MEMORY.

echolalia (*Psych, Med*) (1) A stage in language development, typically at between 6 and 9 months of age, involving repetition of syllables, eg dada. (2) Aimless repetition of words heard without regard for their meaning, occurring in disease of the brain or in mental illness; often seen in catatonic schizophrenia and autistic children.

echolocation (*BioSci*) Means of locating objects in conditions of poor visibility; involves the production of high-frequency sounds and the detection of their echoes.

echopraxia (*Med*) Imitation by an insane person of postures or of movements of those near them; commonly present in the catatonic type of schizophrenia. Also *echopraxis*.

echo-ranging sonar (*Acous*) Determination of distance and direction of objects, such as submarines, by the reception of the reflection of a sound pulse under water. See ASDIC.

echo sounding (*Acous*) Use of echoes of pressure waves sent down to the bottom of the sea and reflected, the delay between sending and receiving times giving a measure of the depth; used also to detect wrecks and shoals of fish.

echo studio (*Acous*) An enclosure of long reverberation time, used for the artificial introduction of an adjustable degree of reverberation in the main channel of a broadcast programme.

echo suppression (*ICT*) In telephone two-way circuits, the attenuation of echo currents in one direction which is due to telephone currents in the other direction.

ECHO virus (*Med*) Abbrev for *Enteric Cytopathic Human Orphan* virus. A group of small RNA viruses not linked to any human disease but a common cause of enteric infections.

eckermannite (*Min*) A rare monoclinic alkali amphibole; a hydrous sodium lithium aluminium magnesium silicate.

eclampsia (*Med*) A term now restricted to the acute toxaemia occurring in pregnancy, parturition or in the puerperium, associated with hypertension, convulsions and loss of consciousness. Adj *eclamptic*. See PRE-ECLAMPSIA.

E classification (*BioSci*) Enzyme classification based on the recommendations of the Committee on Enzyme Nomenclature of the International Union of Biochemistry. The first number indicates the broad type of enzyme

(1 = oxidoreductase; 2 = transferase; 3 = hydrolase; 4 = lyase; 5 = isomerase; 6 = ligase (synthetase)). The second and third numbers indicate subsidiary groupings, and the last number, which is unique, is assigned arbitrarily in numerical order by the Committee.

E-class insulation (*ElecEng*) Insulating material which can withstand a temperature of 120°C. See CLASS-A INSULATING MATERIAL etc.

eclipse (*Astron*) (1) The total or partial disappearance from view of an astronomical object when it passes directly behind another object. (2) The passage of a planet or satellite through the shadow cast by another body so that it is unable to shine as it normally does by reflected sunlight. An eclipse of the Sun (*solar eclipse*) can only occur at new Moon, when the Moon is directly between the Earth and the Sun; a coincidence of nature makes them appear nearly the same size in our sky. A *total eclipse*, when the whole disk is obscured, can last no more than 7·5 minutes. A *partial eclipse* of much longer duration occurs before, after and to each side of the *path of totality*. If the apparent size of the lunar disk is too small for a total eclipse, there is an *annular eclipse* in which a bright ring of sunlight surrounding the Moon is seen. A *lunar eclipse* occurs when the full Moon passes into the shadow of the Earth. Moons and satellites of other bodies in the solar system are eclipsed when they pass through the shadow of their primary bodies.

eclipse year (*Astron*) The interval of time between two successive passages of the Sun through the same NODE of the Moon's orbit; it amounts to 346·620 03 days.

eclipsing binary (*Astron*) A BINARY STAR system whose orbital plane lies so nearly in the line of sight that the components pass in front of each other in the course of their mutual revolution.

ecliptic (*Astron*) The great circle in which the plane containing the centres of the Earth and Sun cuts the celestial sphere; hence the apparent path of the Sun's annual motion through the fixed stars. See OBLIQUITY OF THE ECLIPTIC.

eclogite (*Geol*) A coarse-grained, deep-seated metamorphic rock, consisting essentially of pink garnet, green pyroxene (some of which is often chrome-diopside) and (rarely) kyanite.

eclosion (*BioSci*) The act of emergence from an egg or pupa case.

ECM (*BioSci*) Abbrev for EXTRACELLULAR MATRIX.

ECM (*Eng*) Abbrev for ELECTROCHEMICAL MACHINING.

ECM (*ICT*) Abbrev for ELECTRONIC COUNTERMEASURES.

ecobalance (*EnvSci*) A description of an industrial process in terms of the materials and energy inputs and the outputs of solid, liquid and gaseous wastes. Also *ecoprofile*. See LIFE-CYCLE ANALYSIS, LIFE-CYCLE INVENTORY.

ecoclimate (*EnvSci*) A local, or micro-, climate regarded as an ecological factor.

ecocline (*BioSci*) A CLINE occurring across successive zones of an organism's habit.

E. coli (*BioSci*) See ESCHERICHIA COLI.

ecological efficiency (*BioSci*) Ratios between the amount of energy flow at different points along a food chain, eg the *primary* or *photosynthetic efficiency* is the percentage of the total energy falling on the Earth which is fixed by plants, this being approximately 1%.

ecological factor (*BioSci*) Anything in the environment that affects the growth, development and distribution of plants, and therefore aids in determining the characters of a plant community.

ecological footprint (*EnvSci*) The amount of renewable and non-renewable ecologically productive land area required to support the resource demands and absorb the wastes of an individual, a population or specific activities. If everybody had an equal share then it is calculated that there are about 1.89 hectares available for each person; in Europe the individual average is nearer 6.3 ha and in the US around 10 ha.

ecological indicator (*BioSci*) An organism whose presence in a particular area indicates the occurrence of a particular set of conditions, eg water and soil conditions, temperature zones. Large species with relatively specific requirements are most useful in this way, and numerical relationships between species, populations or whole communities are more reliable than a single species.

ecological niche (*BioSci*) The position or status of an organism within its community or ecosystem. This results from the organism's structural adaptations, physiological responses and innate or learned behaviour. An organism's niche depends not only on where it lives but also on what it does.

ecological pyramid (*EnvSci*) A diagram in which the producer level forms the base and successive trophic levels the remaining tiers. They include the BIOMASS, PYRAMID OF NUMBERS and energy.

ecological succession (*BioSci*) See SUCCESSION.

ecology (*Genrl*) The scientific study of: (1) the interrelations between living organisms and their environment, including both the physical and biotic factors, and emphasizing both interspecific and intraspecific relations; (2) the distribution and abundance of living organisms (ie exactly where they occur and precisely how many there are), and any regular or irregular variations in distribution and abundance, followed by explanation of these phenomena in terms of the physical and biotic factors of the environment.

Econet (*ICT*) TN for a LOCAL AREA NETWORK.

econometrics (*MathSci*) The application of statistical methods to economic phenomena.

economic geology (*Geol*) The study of geological bodies and materials that can be used profitably by humans.

economic ratio (*CivEng*) In reinforced concrete work, the ratio between steel reinforcement and concrete which allows the full strength of both to be developed after solidification.

economizer (*Eng*) A bank of tubes, placed across a boiler flue, through which the feed-water is pumped, being heated by the otherwise waste heat of the flue gases.

Economo's disease (*Med*) Also *von Economo's disease*. See ENCEPHALITIS LETHARGICA.

economy resistance (*ElecEng*) A resistance inserted into the circuit of a contactor coil or other electromagnetic device after its initial operation, in order to reduce the current to a value just sufficient to hold the device closed.

ecophysiology (*BioSci*) The branch of physiology concerned with how organisms are adapted to their natural environment.

ecoprofile (*EnvSci*) See ECOBALANCE.

ecospecies (*BioSci*) An ECOTYPE sufficiently distinct to be given a subspecific name.

ecosystem (*BioSci*) Conceptual view of a plant and animal community, emphasizing the interactions between living and non-living parts, and the flow of materials and energy between these parts. Ecosystems are usually represented as flow diagrams, showing the path of these flows between producers, consumers and decomposers.

ecoterrorism (*Genrl*) Violence carried out to draw attention to environmental issues.

ecotone (*EnvSci*) A transitional zone between two habitats.

ecotype (*BioSci*) A form or variety of any species possessing special inherited characteristics enabling it to succeed in a particular habitat. Cf ECAD.

ECRH (*NucEng*) Abbrev for *electron cyclotron resonance heating*. See CYCLOTRON RESONANCE HEATING.

ecru (*Textiles*) Unbleached knitted fabrics and their colour.

Ecstasy (*Pharmacol*) Slang for 3,4-methylenedioxy-methamphetamine, (MDMA), a drug of abuse.

ECT (*Med*) Abbrev for ELECTROCONVULSIVE THERAPY.

ectasia (*Med*) Pathological dilation or distension of any structure of the body. Also *ectasis*. Adj *ectatic*.

ectethmoid (*BioSci*) One of a pair of cartilage bones of the vertebrate skull, formed by ossification of the ethmoid plate.

ecthoraeum (*BioSci*) The thread of a NEMATOCYST.

ecthyma (*Med*) Local gangrene and ulceration of the skin as a result of infection, the ulcer being covered by a crust and the skin round it being inflamed.

ectoblast (*BioSci*) See EPIBLAST.

ectoderm (*BioSci*) The outer layer of cells forming the wall of a GASTRULA; the tissues directly derived from this layer.

ectoenzyme (*BioSci*) An enzyme secreted from a cell or located on the outer surface of the plasma membrane and thus able to act on extracellular substrates.

ectogenesis (*BioSci*) Development outside the body.

ectogenous (*BioSci*) Independent; self-supporting.

ectolecithal (*BioSci*) Said of ova in which the yolk is deposited peripherally.

ectomorph (*Psych*) One of Sheldon's somatotyping classifications; ectomorphs are delicately built, not very muscular, and are withdrawn and intellectual. See SOMATOTYPE THEORY.

-ectomy (*Genrl*) Suffix from Gk *ektome*, cutting out, used esp in surgical terms.

ectomycorrhiza (*BioSci*) See ECTOTROPHIC MYCORRHIZA.

ectoparasite (*BioSci*) A parasite living on the surface of the host; may feed on the internal tissues of the host, but has all, or the greater part, of its body and its reproductive structures on the surface. Also *epiparasite*, and, in the case of parasitic animals, *ectozoon*.

ectophloedal (*BioSci*) Living on the outside of bark.

ectophloic (*BioSci*) A stele or stem having phloem on the side of the xylem nearer the periphery of the organ, but not on the other side.

ectopia cordis (*Med*) Congenital displacement of the heart outside the thoracic cavity.

ectopia (*Med*) Displacement from normal position. Also *ectopy*. Adj *ectopic*.

ectopia vesicae (*Med*) A congenital abnormality in which the anterior wall of the bladder is absent and the posterior wall opens on to the surface of the abdomen, the lower abdominal wall being also absent.

ectopic gestation (*Med*) Fertilization of the ovum and growth of the fetus outside the uterus.

ectoplasm (*BioSci*) A layer of clear non-granular cytoplasm at the periphery of the cell adjacent to the PLASMA MEMBRANE.

Ectoprocta (*BioSci*) A phylum of Metazoa with the anus outside the LOPHOPHORE. There is a coelomic body cavity and the lophophore is retractable into a tentacle sheath. There are no excretory organs. Bryozoa. Also *Polyzoa*.

ectopy (*Med*) See ECTOPIA.

ectotrophic mycorrhiza (*BioSci*) A mycorrhiza in which the root is surrounded by a well-developed layer of fungal mycelium the hyphae of which interconnect with hyphae both within the root cortex and also ramifying through the soil. Most trees form an ectotrophic mycorrhizal association often with a basidiomycete.

ectozoon (*BioSci*) See ECTOPARASITE.

ectromelia (*BioSci*) An infectious disease of mice, due to a virus.

ectropion (*Med*) Eversion of the eyelid. Also *ectropium*.

eczema (*Med*) Itching, inflammatory skin condition in which papules, vesicles and pustules may be present together with oedema, scaling or exudation. Although the immediate cause may not be known, underlying hypersensitivity to food (eg milk proteins) or an environmental allergen is often detectable in *atopic* persons. Allergens include chemical agents, plant poisons and materials used in trades.

eczematous conjunctivitis (*Med*) See PHLYCTENULAR CONJUNCTIVITIS.

edaphic climax (*BioSci*) A climax community of which the existence is determined by some property of the soil.

edaphic factor (*BioSci*) Any property of the soil, physical or chemical, that influences plants growing on that soil.

EDAX (*Eng*) Abbrev for ENERGY DISPERSIVE ANALYSIS OF X-RAYS.

eddy (*Aero*) An interruption in the steady flow of a fluid, caused by an obstacle situated in the line of flow; the VORTEX so formed.

eddy current brake (*ElecEng*) (1) A form of brake for the loading of motors during testing; it consists of a mass of metal rotating in front of permanent magnets so that heavy eddy currents are set up in it. (2) A form of brake, used on tramways, in which the retarding force is produced by the induction of eddy currents in the rail by an electromagnet on the vehicle.

eddy current heating (*ElecEng*) See INDUCTION HEATING.

eddy currents (*ElecEng*) Those resulting from the changing emfs induced by varying magnetic fields, resulting in losses from the latter and dissipation of power. They are one of the main causes of heating in motors, transformers, etc (known also as *iron loss*, cf COPPER LOSS). To minimize them the iron cores of such machines are composed of many layers of thin iron sheet (laminations) which are insulated from one another to reduce the currents' ability to flow in the direction in which the field tries to induce them. This in turn slightly increases the reluctance of the magnetic circuit, thereby reducing the overall efficiency of the machine, hence a compromise must be struck. Used for mechanical damping and braking (as in electricity meters) and for induction heating in eg case-hardening. Also *Foucault current*.

eddy current speed indicator (*ElecEng*) A speed indicator consisting of a rotating disk and a spring-controlled magnetic needle; the latter is deflected as a result of eddy currents induced in the disk.

eddy current testing (*Eng*) NON-DESTRUCTIVE TEST using an electromagnetic field to induce eddy currents in a component. Discontinuities, flaws, internal cracks and changes in shape in a metal component are detected by variations in the signal produced in pickup coils located nearby.

eddy diffusion (*ChemEng*) The migration and interchange of portions of a fluid as a result of their turbulent motion. Cf DIFFUSION.

eddy diffusion (*EnvSci*) The transport of quantities such as heat and momentum by eddies in regions of the atmosphere which are in a state of turbulent motion. Eddy diffusion is roughly 10^5 times as effective as molecular diffusion which for meteorological purposes can normally be ignored.

eddy diffusivity (*ChemEng*) Exactly analogous, for eddy diffusion, to the DIFFUSIVITY for molecular diffusion. Symbol *E*.

eddy flow (*Phys*) See TURBULENT FLOW.

edelopal (*Min*) A variety of *opal* with an exceptionally brilliant play of colours.

edema (*Med*) Also *edematous*. See OEDEMA, OEDEMATOUS.

edenite (*Min*) An end-member compositional variety in the hornblende group of monoclinic amphiboles: hydrated magnesium, calcium and aluminium silicate.

Edentata (*BioSci*) An order of primitive terrestrial mammals characterized by the incomplete character of the dentition. The testes are abdominal. Such animals may be phytophagous or insectivorous in form. Sloths, ant-eaters, armadillos.

edentulous (*BioSci*) Without teeth. Also *edentate*.

edge (*Aero*) See LEADING EDGE, TRAILING EDGE.

edge coal (*MinExt*) Highly inclined coal seams.

edge dislocation (*Crystal*) Line defect within a crystal in which the BURGERS VECTOR is perpendicular to the line of the DISLOCATION.

edge effect (*Acous*) In acoustic absorption measurements, the variations which arise from the size, shape or division of the areas of material being tested in a reverberation room.

edge effect (*ElecEng*) The deviation from parallelism in fields at the edge of parallel-plate capacitors, or between poles of permanent or electromagnets, thus leading to non-uniformity of the field at the edges.

edge effect (*ImageTech*) See BORDER EFFECT.

edge-emitting diode (*ICT*) A LIGHT-EMITTING DIODE in which the radiation is parallel to the surface and perpendicular to the current flow; used as a light source in optical fibre communications.

edge filter (*ChemEng*) A type of filter in which a large number of disks are clamped on a perforated hollow shaft joining a cylinder. This is contained in another cylindrical vessel into which liquid is pumped and flows through the narrow spaces between the disks, the solids being trapped on the disk edges. Filtrate leaves via the perforations in the hollow shaft.

edge numbers (*ImageTech*) A series of numerals and letters printed or photographically exposed along the edge of a strip of motion picture film, usually at intervals of 1 foot (16 frames on 35 mm film), allowing individual frames to be identified and located.

edge plane (*Build*) One with the cutter at the extreme front for working corners.

edge planing (*Print*) Trimming and squaring the edges of printing plates, either original or duplicate, by hand- or power-operated tools.

edge-rolled (*Print*) Said of a pattern on the edges of leather bindings, either blind or decorated, applied by a ROLL.

edge-sealed (*Textiles*) Descriptive of the edge of a fabric that has been made without the usual SELVEDGE and has been cut and sealed (usually by the heating and melting of a thermoplastic fibre) to prevent fraying.

edge tone (*Acous*) A tone produced by the impact of an air jet on a sharply edged dividing surface, such as in an organ-pipe mouth. See WHISTLE.

edge tool (*Eng*) A hand-worked, mallet-struck or machine-operated cutting tool with one or a regular pattern of cutting edges, ie excluding grinding wheels.

edge-trimming plane (*Build*) One with a rebated sole, for squaring up edges.

edge water (*Geol*) That pressing inward upon the gas or oil in a natural reservoir.

edge winding (*ElecEng*) A form of winding frequently used for the field windings of salient-pole synchronous machines; it consists of copper strip wound on edge around the pole. Such a winding has good heat-dissipating properties.

edgewise instrument (*ElecEng*) A switchboard indicating instrument in which the pointer moves in a plane at right angles to the face of the switchboard. The end of the pointer is bent and moves over a narrow scale.

EDI (*ICT*) Abbrev for *electronic data interchange*. The practice of transferring large volumes of data between systems automatically.

EDIFACT (*ICT*) Abbrev for *electronic data interchange for administration, commerce and transport*.

Edison effect (*Electronics*) The phenomenon of electrical conduction between an incandescent filament and an in-dependent cold electrode contained in the same envelope, when the second electrode is made positive with reference to a part of the filament. Precursor of FLEMING DIODE. Also *Richardson effect*.

Edison screw-cap (*ElecEng*) A lamp cap in which the outer wall forms one of the contacts, and which is in the form of a coarse screw for inserting into a corresponding socket. A central pin forms the other contact. Sizes in descending order include goliath, large, medium, small, miniature and lilliput.

Edison screw-holder (*ElecEng*) A holder for electric lamps with EDISON SCREW-CAPS.

E-display (*Radar*) Display in which target range and elevation are plotted as horizontal and vertical co-ordinates of the blip.

edit controller (*ImageTech*) A device used remotely to control the functions of a video EDIT RECORDER and one or more SOURCE PLAYER(S).

edit decision list (*ImageTech*) A list containing the locations of all wanted shots on the video master tape(s),

together with any instructions, eg fade to black. In a computerized EDIT SUITE the list will run the editing process (auto CONFORMING).

editing (*ImageTech*) The selection and arrangement in sequence of the individual scenes of a film or video production in accordance with the director's interpretation of the script. In film editing, the chosen sections are physically cut from *rush prints* and joined together but in video editing they are rerecorded in the required order.

editing terminal (*Print*) In phototypesetting the video display unit (VDU) on which is displayed the results of keyboarding. Corrections and other editorial functions can be carried out before final phototypesetting.

edition (*Print*) A number of copies printed at one time, either as the original issue (first edition), or when the text has undergone some change, or the text has been partly or entirely reset, or the format has been altered.

edition binding (*Print*) The normal style of binding for hardcover books, highly mechanized, only occasionally sewn on tapes, with case usually cloth-covered and lettered on spine only. Also *publisher's binding*. Cf BOUND BOOK, LIBRARY BINDING.

editor (*ICT*) Program that enables the user to inspect and alter a program or data. See SCREEN EDITOR.

editor (*ImageTech*) (1) The creative craftworker who collaborates with the director to achieve the final assembly of a film or video production. (2) Equipment for the detailed examination of the action in a motion picture film on a small illuminated screen; an editing table may also include paths for the synchronization and reproduction of the associated magnetic sound records.

edit recorder (*ImageTech*) A videotape recorder designed with the necessary accuracy and facilities to perform edits when connected to a SOURCE PLAYER.

edit suite (*ImageTech*) The video EDIT RECORDER and one or more SOURCE PLAYERS together with ancillary equipment.

EDL (*ImageTech*) Abbrev for EDIT DECISION LIST.

edriophthalmic (*BioSci*) Having sessile eyes, as some Crustacea.

Edser and Butler's bands (*Phys*) Dark bands, having a constant-frequency separation, which are seen in the spectrum of white light which has traversed a thin parallel-sided plate of a transparent material, or a thin parallel-sided film of air between glass plates.

EDTA (*BioSci, Chem*) Abbrev for *ethylene diamine tetra-acetic (ethanoic) acid, diamino ethane tetraacetic acid*. $(HOOCCH_2)_2NCH_2CH_2N(CH_2COOH)_2$. A CHELATING AGENT frequently used to protect enzymes from inhibition by traces of metal ions and as an inhibitor of metal-dependent proteases because of its ability to combine with metals. It is also used in special soaps to remove metallic contamination. Also *complexone*.

Edwards' roaster (*MinExt*) Long horizontal furnace through which sulphide minerals are rabbled counterwise to hot air, to remove part or all of the sulphur by ignition.

E-E (*ImageTech*) Abbrev for *electronics to electronics*. A signal passed through a videotape recorder and not coming off-tape.

EEG (*Med*) Abbrev for ELECTROENCEPHALOGRAPH (or *-gram*).

eel (*BioSci*) Any fish of the Anguillidae, Muraenidae or other family of the Anguilliformes. The name is extended to other fish of similar form, eg sand eel.

eel-grass (*BioSci*) Species of *Zostera*, grass-like monocoty-ledons which grow in the sea mostly around or below the low-water mark, historically used for sound insulation and the correction of acoustic defects.

eelworms (*BioSci*) See NEMATODA.

effective address (*ICT*) One that is obtained by modifying the address part of an INSTRUCTION during processing.

effective antenna height (*ICT*) Height (in metres) that, when multiplied by the field strength (in volts per metre) incident upon the antenna, gives the emf (in volts) induced therein. It is less than the physical height and differs from the equivalent height in that it is also a function of the direction of arrival of the incident wave.

effective bandwidth (*ICT*) The bandwidth of an ideal (rectangular) band-pass filter that would pass the same proportion of the signal energy as the actual filter.

effective column length (*Build*) The column length which is used in finding the *ratio of slenderness* after taking into account the rigidity of the end fixings.

effective depth (*CivEng*) The depth of a reinforced concrete beam as measured from the surface of the concrete on the compression side to the centre of gravity of the tensile reinforcement. See COVER.

effective dose equivalent (*Radiol*) The quantity obtained by multiplying the dose equivalents to various tissues and organs by the risk weighting factor appropriate to each organ and summing the products. Unit sievert (*Sv*).

effective energy (*Radiol*) The quantum energy (or wavelength) of a monochromatic beam of X-rays or γ-rays with the same penetrating power as a given heterogeneous beam. Its value depends upon the nature of the absorbing medium. Also *effective wavelength*.

effective half-life (*Phys*) The time required for the activity of a radioactive nuclide in the body to fall to half its original value as a result of both biological elimination and radioactive decay. Its value is given by

$$\frac{\tau(b\frac{1}{2})\tau(r\frac{1}{2})}{\tau(b\frac{1}{2}) + \tau(r\frac{1}{2})}$$

where $\tau(b\frac{1}{2})$ and $\tau(r\frac{1}{2})$ are the biological and radioactive half-lives respectively.

effective heating surface (*Eng*) The total area of a boiler surface in contact with water on one side and with hot gases on the other.

effective isotropic radiated power (*ImageTech*) A satellite's signal-transmission power.

effective mass (*Electronics*) For electrons and/or holes in a semiconductor, effective mass is a parameter which may differ appreciably from the mass of a free electron, and which depends to some extent on the position of the particle in its energy band. This modifies the mobility and hence the resulting current.

effective particle density (*PowderTech*) The mass of a particle divided by its volume, including opened and closed pores.

effective pixels (*ImageTech*) The image-forming PIXELS in a SOLID-STATE IMAGE SENSOR. Others provide housekeeping functions and OPTICAL BLACK.

effective porosity (*PowderTech*) That portion of the powder porosity which is readily accessible to a fluid moving through a powder compact.

effective radiated power (*ICT*) Actual maximum or unmodulated power delivered to a transmitting antenna multiplied by the factor of gain in a specified direction in the horizontal plane.

effective range (*Eng*) That part of the scale of an indicating instrument over which a reasonable precision may be expected.

effective resistance (*ElecEng*) The total ac resistance caused by eddy current losses, iron losses, dielectric and corona losses, transformed power, as well as conductor loss. For a sinusoidal current it is the component of the voltage in phase with current divided by that current. Measured in ohms.

effective sieve aperture size (*PowderTech*) To allow for the size aperture distribution in a real sieve for accurate particle size analysis, sieves have to be calibrated. This is sometimes done by analysing a powder of known distribution. If *A* percentage by weight should be retained on a sieve of nominal aperture size *a*, then *B* percentage is retained on the actual sieve; from the particle size distribution of the powder *B*, the size at which *B* percentage of the powder is greater than this size is read off. The

effective size of the sieve aperture is then stated to be *B*. The value of *B* will depend upon the size distribution of the powder used to calibrate the sieve. The magnitude of the quantity $a-B$ is termed the *aperture error of a sieve*.

effective span (*Build*) The horizontal distance between the centres of the two bearings at the ends of a beam.

effective temperature (*Astron*) The temperature which a given star would have if it were a perfect radiator, or a BLACK BODY, with the same distribution of energy among the different wavelengths as the star itself has.

effective value (*Genrl*) A simple parameter value which has the same effect as a more complex one, eg rms value of ac=dc value for many purposes.

effective wavelength (*Phys*) See DOMINANT WAVELENGTH.

effective wavelength (*Radiol*) See EFFECTIVE ENERGY.

effector (*BioSci*) A tissue complex capable of effective response to the stimulus of a nervous impulse, eg a muscle or gland.

effector neuron (*BioSci*) A motor neuron.

effects track (*ImageTech*) A sound track containing only the sound effects and noises required in a production, excluding speech and music. Abbrev FX.

effect threads (*Textiles*) Yarns of striking appearance that are included in a fabric to attract attention.

efferent (*BioSci*) Carrying outwards or away from; as the *efferent branchial vessels* in a fish, which carry blood away from the gills, and *efferent nerves*, which carry impulses away from the central nervous system. Cf AFFERENT.

effervescence (*Chem*) The vigorous escape of small gas bubbles from a liquid, esp as a result of chemical action.

efficiency (*Genrl*) A dimensionless measure of the performance of a piece of apparatus, eg an engine, obtained from the ratio of the output of a quantity, eg power, energy, to its input, often expressed as a percentage. The power efficiency of an internal-combustion engine is the ratio of the shaft or brake horsepower to the rate of intake of fuel, expressed in units of energy content per unit time. It must always be less than 100% which would imply perpetual motion. Not to be confused with *efficacy*, which takes account only of the output of the apparatus, and is not given an exact quantitative definition.

efficiency of impaction (*PowderTech*) The ratio of the cross-sectional area of the stream, from which particles of a given size are removed, to the total cross-sectional area of the jet stream, both areas measured at the mouth of the jet.

efficiency of screening, numerical index of (*MinExt*) A quantity used to assess the efficiency of industrial screening procedures, ie sieving procedures. It is defined by the equation $E = F(D/B–C/A)$, where E = the numerical index of efficiency of screening, F = the percentage of the powder supply passed by the screen, A = the percentage of difficult oversize in the powder supply, B = the percentage of difficult undersize in the powder supply, C = the percentage of the difficult oversize passed by the screen, D = the percentage of the difficult undersize passed by the screen.

efficiency ratio (*Eng*) Of a heat engine, the ratio of the actual thermal efficiency to the ideal thermal efficiency corresponding to the cycle on which the engine is operating.

effleurage (*Med*) The action of lightly stroking in massage.

efflorescence (*Build*) The formation of a white crystalline deposit on the face of a wall; due to the drying out of salts in the mortar or stone.

efflorescence (*Chem*) The loss of water from a crystalline hydrate on exposure to air, shown by the formation of a powder on the crystal surface.

efflorescence (*Min*) A fine-grained crystalline deposit on the surface of a mineral or rock.

effluent (*Agri*) Unwanted liquid runoff containing potentially environmentally damaging compounds if there is uncontrolled release into the environment, eg drainage from a cattle yard, manure heap or silage clamp.

effluent (*Build*) Liquid sewage after having passed through any stage in its purification.

effluent (*Eng*) The liquid or gaseous waste from a chemical or other plant.

effluent monitor (*NucEng*) Instrument for measuring level of radioactivity in fluid effluent.

efflux (*Aero*) The mixture of combustion products and cooling air which forms the propulsive medium of any jet or rocket engine.

effort syndrome (*Med*) A condition in which the subject complains of palpitations, breathlessness and chest pain, often after exercise, in the absence of heart disease. Thought to be due to psychological factors. Also *soldier's heart*.

effusiometer (*Chem*) Apparatus for comparing the relative molecular masses of gases by observing the relative times taken to stream out through a small hole.

effusion (*Chem*) The flow of gases through holes larger than those to which diffusion is strictly applicable; see GRAHAM'S LAW. The rate of flow is approximately proportional to the square root of the pressure difference.

effusion (*Med*) An abnormal outpouring of fluid into the tissues or cavities of the body, as a result of infection, or of obstruction to blood vessels or lymphatics.

effusive (*Geol*) Extrusive, volcanic.

eflornithine (*Pharmacol*) A drug that affects cell division through enzyme inhibition, used to slow hair growth and also used in the treatment of sleeping sickness.

EFP (*ImageTech*) Abbrev for *electronic field production*, video programme production shooting outside a studio, usually with lightweight portable cameras.

EFT (*ICT*) Abbrev for ELECTRONIC FUNDS TRANSFER.

EFTPOS (*ICT*) Abbrev for *electronic funds transfer at point of sale*. The term is usually used in the context of retailing whereby money is transferred directly from the customer's bank account to the shop's bank account directly by electronic means, ie without using a cheque.

EGA (*ICT*) Abbrev for *enhanced graphics adapter*. The IBM PC video ADAPTER standard capable of supporting text and graphics on a screen consisting of 640 PIXELS wide by 350 PIXELS high. See CGA, VGA.

egest (*BioSci*) To throw out, to expel; to defecate, to excrete. The egested materials are *egesta*.

EGF (*BioSci*) Abbrev for EPIDERMAL GROWTH FACTOR.

egg (*BioSci*) See OVUM.

egg and anchor (*Arch*) An ornament carved on a moulding, resembling eggs separated by vertical anchors.

egg and dart (*Arch*) Similar to EGG AND ANCHOR, arrows taking the place of anchors.

egg apparatus (*BioSci*) The egg and the two synergidae in the embryo sac of an angiosperm.

egg-bound (*Vet*) Said of the oviduct of birds when obstructed by an egg.

egg-box lens (*ICT*) See SLATTED LENS.

egg-cell (*BioSci*) The ovum, as distinct from any other cells associated with it.

egg-eating (*Vet*) A vice developed by individual birds, characterized by the eating of their own eggs or of those of other birds.

egg glair (*Build*) A substance produced by mixing the white of an egg with half a litre of lukewarm water. Egg glair is applied to painted surfaces and allowed to dry prior to gilding work. The glair, which removes any tack from the painted surface, is destroyed by the gold size applied to areas to be gilded. The surface must be washed with lukewarm water to remove excess glair when gilding work is complete.

egg nucleus (*BioSci*) The female pronucleus.

egg-peritonitis (*Vet*) Septic peritonitis in birds, extending from an infected and obstructed oviduct.

egg-shaped sewer (*Build*) A type of sewer section much used for fluctuating flows; the section resembles the longitudinal profile of an egg placed with the smaller radius at the bottom, thus increasing the velocity of flow at the bottom.

eggshell finish (*Build*) Paint or varnish which dries with a degree of sheen or lustre between matt and semi-gloss.

eggshell finish (*Paper*) A soft dull finish on paper.

egg sleeker (*Eng*) A moulder's sleeker with a spoon-shaped end; used for smoothing rounded corners in a mould.

egg tooth (*BioSci*) A sharp projection at the tip of the upper beak of young birds and monotremes, by means of which they break open the eggshell.

E-glass (*Glass*) Designation for a glass of composition (percentage by weight): SiO_2 52·9, CaO 17·4, Al_2O_3 14·5, B_2O_3 9·2, MgO 4·4, Na_2O 1·0, K_2O 1·0. It is the most widely used glass fibre for the reinforcement of COMPOSITE MATERIALS, with a chemical resistance between those of A- and C-GLASS.

ego (*Psych*) Originally, a term in philosophy, denoting the existence of a sense of self; in psychoanalytic theory, the rational, reality-oriented level of personality which develops in childhood, as the child gathers awareness of and comes to terms with the nature of the social and physical environment; represents reason and common sense, in contrast to the ID.

egocentrism (*Psych*) In Piaget's theory, a form of thinking most typically found in young children, in which the individual perceives and comprehends the world from a totally subjective point of view, being unable to differentiate between the objective and subjective components of experience.

ego ideal (*Psych*) A component of the SUPEREGO; offering reward in the form of pride in good behaviour.

egomania (*Psych*) A pathological preoccupation with self.

egophony (*Med*) See AEGOPHONY.

ego psychology (*Psych*) The school of psychology represented by those Freudian theorists who emphasize ego processes, such as reality perception, learning and conscious control of behaviour, and who argue that the ego has its own energy and autonomous functions, not derived from the ID; in contrast to *instinct theory*, which states that all mental energy is ultimately derived from the id.

egressive (*Genrl*) In phonetics, descriptive of speech sounds pronounced with exhalation rather than inhalation of breath.

EGT (*Aero*) Abbrev for *exhaust gas temperature*.

EGTA (*BioSci*) Abbrev for *ethyleneglycol-aminoethyl-tetra-acetic acid*. A CHELATING AGENT for divalent cations, with a higher affinity for calcium than magnesium. See EDTA.

Egyptian (*Print*) See SLAB SERIF.

Egyptian blue (*Min*) An artificial mineral, calcium copper silicate, $CaCuSi_4O_{12}$, thought to be the pigment of ancient Egypt.

Egyptian jasper (*Min*) A variety of jasper occurring in rounded pieces scattered over the surface of the desert, chiefly between Cairo and the Red Sea; used as a broochstone and for other ornamental purposes. Typically shows colour zoning.

Eh (*Chem*) Symbol for OXIDATION POTENTIAL.

EHF (*ICT*) Abbrev for EXTREMELY HIGH FREQUENCY.

ehp (*Aero*) Abbrev for TOTAL EQUIVALENT BRAKE HORSE-POWER. Also *tehp*.

EI (*ImageTech*) Abbrev for *exposure index*. A photographic film speed rating which takes into account the effects on a film of different developers and development times.

EIA (*EnvSci*) Abbrev for ENVIRONMENTAL IMPACT ASSESSMENT.

eicosanoid (*BioSci*) A term used for compounds such as LEUKOTRIENES and PROSTAGLANDINS that are derived from arachidonic acid.

eicosapentaenoic acid (*Chem*) A fatty acid found in fish oils. Abbrev *EPA*.

EIDE (*ICT*) Abbrev for ENHANCED INTEGRATED DRIVE ELECTRONICS.

eidetic imagery (*Psych*) The ability to generate a vividly clear and detailed picture or visual memory image of previously seen objects. Commonly present in children up to 14 years, and occasionally persisting into adult life.

eidograph (*Surv*) An instrument for reducing and enlarging plans.

Eifelian (*Geol*) A stage in the Middle Devonian. See PALAEOZOIC.

Eiffel wind tunnel (*Aero*) An open-jet, non-return-flow wind tunnel.

eigen- (*Genrl*) Prefix from Ger *eigen* signifying proper, own.

eigenfrequency (*Acous*) Frequency of vibration of a system which vibrates freely. See ACOUSTIC RESONANCE.

eigenfunction (*Phys*) A solution of a wave equation which satisfies a set of boundary conditions. In quantum mechanics, a possible solution for the SCHRÖDINGER EQUATION for a given system.

eigentone (*Phys*) One of the natural frequencies of vibration of a system.

eigenvalue (*Phys*) A possible value for a parameter of an equation for which the solution will be compatible with the boundary conditions. In quantum mechanics, the energy eigenvalues for the SCHRÖDINGER EQUATION are possible ENERGY LEVELS for the system. See ENERGY BAND.

eigenvalues (*MathSci*) See CHARACTERISTIC EQUATION OF A MATRIX.

eigenvector (*MathSci*) Any non-zero vector v which satisfies $Av = \lambda v$, where λ is an eigenvalue obtained from the CHARACTERISTIC EQUATION of A.

Einstein–de Haas effect (*Phys*) Effect which occurs when a magnetic field is applied to a body; the precessional motion of the electrons produces a mechanical moment that is transferred to the body as a whole. See BARNETT EFFECT.

Einstein–de Sitter universe (*Astron*) Cosmological model in which space has no curvature and the matter distribution extends to infinity at all times.

Einstein diffusion equation (*Chem*) As a result of BROWNIAN MOTION, molecules or colloidal particles migrate an average distance, δ, in each small time interval, τ. The equation for the diffusion of a spherical particle of radius, r, through a fluid of viscosity, η is:

$$D = \frac{\delta^2}{2\tau} = \frac{RT}{6\pi\eta r N_A}$$

where D is the diffusion coefficient, N_A is AVOGADRO'S NUMBER, R is the GAS CONSTANT and T the absolute temperature.

Einstein energy (*Phys*) See MASS–ENERGY EQUATION.

Einstein equation for the specific heat of a solid (*Chem*) One mole of the solid consists of N_A atoms, each vibrating with frequency, v, in three dimensions. From quantum theory, the molar specific heat is:

$$C_v = \frac{3Rx^2 e^x}{(e^x - 1)^2}$$

where $x = hv(kT)^{-1}$, R is the gas constant, h is PLANCK'S CONSTANT, k is BOLTZMANN'S CONSTANT and T is the absolute temperature.

einsteinium (*Chem*) Artificial element, symbol Es, at no 99, produced by bombardment in a cyclotron, but also recognized in H-bomb debris, its having been produced by beta decay of uranium which had captured a large number of neutrons. The longest-lived isotope is ^{254}Es, with a half-life of greater than 2 years.

Einstein law of photochemical equivalence (*Chem*) Each quantum of radiation absorbed in a photochemical process causes the decomposition of one molecule.

Einstein photoelectric equation (*Chem*) That which gives the energy of an electron, just ejected photoelectrically from a surface by a photon, ie $E = hv - \varphi$, where E = kinetic energy, h = PLANCK'S CONSTANT, v = frequency of photon and φ = WORK FUNCTION.

Einstein shift (*Astron*) See GRAVITATIONAL REDSHIFT.

Einstein theory of specific heats of solids (*Phys*) A theory based on the assumption that the thermal vibrations of the atoms in a solid can be represented by harmonic oscillators of one frequency, whose energy is quantized. See BLACKMAN THEORY, DEBYE THEORY.

Einthoven galvanometer (*ElecEng*) A galvanometer in which the current is carried by a single current-carrying filament in a strong magnetic field, the deflection usually being magnified by a microscope. Also *string galvanometer*.

EIRP (*ImageTech*) Abbrev for EFFECTIVE ISOTROPIC RADIATED POWER.

EIS (*EnvSci*) Abbrev for ENVIRONMENTAL IMPACT STATEMENT.

EIS (*ICT*) Abbrev for *executive information system*. A specialized form of computer system, similar to a MANAGEMENT INFORMATION SYSTEM, that presents complex data in a highly refined, easy-to-read form for the benefit of time-poor high-status users.

EISA (*ICT*) Abbrev for EXTENDED INDUSTRY STANDARD ARCHITECTURE.

ejaculatory duct (*BioSci*) See DUCTUS EJACULATORIUS.

ejecta (*Geol*) Solid material thrown from a volcano or an impact crater. Also *ejectamenta*.

ejection capsule (*Aero*) (1) A cockpit, cabin, or portion of either, in a high-altitude and/or high-speed military aircraft which can be fired clear in emergency and which, after being slowed down, descends by parachute. (2) Container of recording instruments ejected and recovered by parachute.

ejective (*Genrl*) In phonetics, of or relating to ejection, or of a type of consonant in some languages that is pronounced with a glottal stop.

ejector (*Build*) An appliance used for raising sewage from a low-level sewer to a sewer at a higher elevation; worked by compressed air.

ejector (*Eng*) (1) A device for exhausting a fluid by entraining it by a high-velocity steam or air jet, eg AIR EJECTOR. (2) A mechanism for removing a part or assembly from a machine at the end of an automatic sequence of operations, eg in assembling or machining.

ejector pin (*Eng*) Steel pin which is activated at end of moulding cycle to push moulding from tool cavity.

ejector seat (*Aero*) A crew seat for high-speed aircraft which can be fired, usually by slow-burning cartridge, clear of the structure in emergency. Automatic releases for the occupant's safety harness and for parachute opening are usually incorporated. Also *ejection seat*.

eka- (*Chem*) A prefix, coined by Mendeleyev, denoting the element occupying the next lower position in the same group in the periodic system; used in the naming of new elements and unstable radioelements.

ekistics (*Genrl*) The science or study of human settlements.

Ekman spiral (*Aero*) The theoretical path traced by the end of the horizontal wind velocity vector, plotted with a fixed origin on a wind HODOGRAPH, as the wind varies with height up through the ATMOSPHERIC BOUNDARY LAYER (panel) on the assumption that density, pressure gradient and eddy viscosity are constant. It shows the approach of the wind velocity from zero near the ground to the *free atmosphere* value at a height of about 1 km and often gives a good approximation to reality. Wind-driven currents in the surface layers of the ocean have a similar variation with depth.

ektrodactylia (*Med*) Congenital absence of one or more fingers or toes.

elaeodochon (*BioSci*) See OIL GLAND.

elaeolite (*Min*) A massive form of the mineral nepheline, greenish grey or (when weathered) red in colour.

elaiosome (*BioSci*) An outgrowth from the surface of a seed, containing fatty or oily material (often attractive to ants) and serving in seed dispersal.

Elara (*Astron*) The seventh natural satellite of Jupiter, discovered in 1905. Distance from the planet 11 740 000 km; diameter 80 km.

Elasmobranchii (*BioSci*) A subclass of GNATHOSTOMATA comprising highly developed, usually predacious fishes with a cartilaginous skeleton and plate-like gills. Includes sharks, skates, rays.

elastance (*Phys*) The reciprocal of the capacitance of a capacitor, so termed because of electromechanical analogy with a spring. Unit is the DARAF.

elastase (*BioSci*) The proteolytic enzyme (peptidase) secreted by the pancreas which digests *elastin*; a similar enzyme is also found in neutrophil leucocytes.

elastic bitumen (*Min*) See ELATERITE.

elastic collision (*Phys*) A collision between two bodies in which, in addition to the total momentum being conserved, the total kinetic energy of the bodies is conserved.

elastic constant (*Phys*) A property of materials; the ratio of stress to strain for different modes of ELASTIC DEFORMATION. In linear, ISOTROPIC, elastic materials, only four constants are needed to characterize fully the elastic behaviour of the material: YOUNG'S MODULUS, E, SHEAR (or *rigidity*) MODULUS, G, BULK MODULUS, K, which have the units of stress (N m^{-2}) and Poisson's ratio, v, which is dimensionless. These are related through

$$E = 2G(1 + v) = 3K(1 - 2v)$$

so that either E and v, or K and G, are sufficient to allow the other pair to be determined. In anisotropic elastic materials, up to 36 elastic constants are required. Also *elastic modulus, modulus of elasticity*. See LAMÉ CONSTANTS, SPECIFIC MODULUS, TENSILE MODULUS.

elastic deformation (*Eng*) Any change in shape in response to an applied force in which the initial shape is recoverable with no sensible time delay when the applied force is removed.

elastic fabric (*Textiles*) Fabric containing rubber or ELASTOMERIC YARNS that give it good elastic recovery and shape-retaining properties.

elastic fatigue (*Eng, Phys*) Obsolete and confusing term for time-dependent recovery from 'elastic' deformation. Now recognized as an aspect of VISCOELASTICITY.

elastic fibres (*BioSci*) See YELLOW FIBRES.

elastic fibrocartilage (*BioSci*) See YELLOW FIBROCARTILAGE.

elasticity (*Phys*) The tendency of matter, whether gaseous, liquid or solid, to return to its original size or shape, after having been stretched, compressed or otherwise deformed. See ELASTIC CONSTANT, HOOKE'S LAW. Cf PLASTICITY, VISCOELASTICITY.

elasticity of bulk (*Phys*) The ELASTICITY for changes in the volume of a body caused by changes in the pressure acting on it. The bulk modulus is the ratio of the change in pressure to the fractional change in volume.

elasticity of elongation (*Phys*) The ELASTICITY where the stress is a stretching force per unit area of cross-section and the strain is the elongation per unit length. The modulus of elasticity of elongation is known as YOUNG'S MODULUS.

elasticity of gases (*Phys*) If the volume V of a gas is changed by δV when the pressure is changed by δp, the modulus of elasticity is given by

$$-V \frac{\delta p}{\delta V}$$

This may be shown to be numerically equal to the pressure p for isothermal changes, and equal to γp for adiabatic changes, γ being the ratio of the specific heats of the gas.

elasticity of shear (*Phys*) The elasticity of a body which has been pulled out of shape by a shearing force. The stress is equal to the tangential shearing force per unit area, and the strain is equal to the angle of shear, ie the angle turned through by a straight line originally at right angles to the direction of the shearing force. See ELASTICITY, POISSON'S RATIO.

elastic limit (*Eng, Phys*) The highest stress or strain that can be applied to a material without producing a measurable amount of plastic (ie permanent) deformation. Usually assumed to coincide with the PROPORTIONAL LIMIT. See STRENGTH MEASURES.

elastic liquid (*Eng*) Polymer melt or concentrated solution exhibiting memory of previous condition, usually behaving like an uncrosslinked elastomer. In one demonstration, a viscous fluid snaps back into its container after cutting the pouring liquid with scissors! See DIE SWELL, KAYE EFFECT, MELT FRACTURE, WEISSENBERG EFFECT.

elastic medium (*Phys*) A medium which obeys HOOKE'S LAW. No medium is perfectly elastic, but many are sufficiently so to justify the making of calculations which assume perfect elasticity.

elastic modulus (*Eng*) See BULK MODULUS, COMPLEX MODULUS, ELASTIC CONSTANT, POISSON'S RATIO, SHEAR MODULUS, TENSILE MODULUS, YOUNG'S MODULUS.

elastic scattering (*Phys*) See SCATTERING.

elastic strain (*Eng*) The recoverable strain or fractional deformation undergone by a material, ie that which disappears as the straining force is removed.

elastic tissue (*BioSci*) A form of connective tissue in which elastic fibres predominate.

elastin (*BioSci*) A structural protein abundant in elastic body tissues such as skin and internal membranes. A heteropolymer of some 16 different amino acids, it is largely non-crystalline and exhibits long-range reversible elasticity like natural rubber. Permanent cross-links occur through the lysine residues in the main chain. See RESILIN.

elastivity (*Phys*) The reciprocal of the permittivity of a dielectric.

elastomer (*Chem, Plastics*) A material, usually synthetic, having elastic properties akin to those of rubber. See panels on ELASTOMERS and RUBBER TOUGHENING.

elastomeric state (*Chem*) General condition of amorphous thermoplastic polymers above glass transition temperature. See VISCOELASTICITY.

elastomeric yarns (*Textiles*) Yarns comprising filaments or staple fibres (eg polyurethane), which have good elastic properties.

elaterite (*Min*) A solid bitumen resembling dark-brown rubber. Elastic when fresh. Also *mineral caoutchouc*.

E-layer (*Phys*) The most regular of the ionized regions in the ionosphere, which reflects waves from a transmitter back to Earth. Its effective maximum density increases from zero before dawn to its greatest at noon, and decreases to zero after sunset, at heights varying between 110 and 120 km. There are at least two such layers. Also *Heaviside layer, Kennelly–Heaviside layer.*

elbaite (*Min*) One of the three chief compositional varieties of tourmaline; a complex hydrated borosilicate of lithium, aluminium and sodium. Most of the gem varieties of tourmaline are elbaites.

elbow (*Build*) An arch stone whose lower bed is horizontal, while its upper bed is inclined towards the centre of the arch, to correspond with those of the voussoirs.

elbow (*Eng*) A short right-angle pipe connection, as distinct from a bend, which is curved, not angular.

elbow-board (*Build*) The window-board beneath a window, in the interior.

elbow linings (*Build*) The panelling at the sides of a window recess, running from the floor to the level of the window-board. Cf JAMB LININGS.

Electra complex (*Psych*) See OEDIPUS COMPLEX.

electret (*ElecEng*) Permanently polarized dielectric material, formed by cooling a FERROELECTRIC (eg barium titanate) from above a CURIE POINT or waxes (eg carnauba) in a strong electric field.

electret transmitter (*ICT*) A type of telephone transmitter in which the active element is a capacitor, one of whose plates carries a permanent electrostatic charge. This form of transmitter is lighter than the carbon granule type it replaces, but unlike it requires a transistor amplifier.

electric (*Phys*) Said of any phenomena which depend essentially on electric charges or currents. Commonly used for ELECTRICAL.

Elastomers

Any material in the elastomeric state, specifically polymers which are elastomeric at ambient temperatures and are lightly cross-linked for stability. Most linear polymers become elastomeric as the temperature rises above T_g (see VISCOELASTICITY) but it is those polymers which are elastomeric at ambient temperature or below which have become important for engineering applications, whether *general purpose rubbers* (GP) or *speciality materials*. Historically, the most important elastomer was natural rubber cross-linked with sulphur (VULCANIZATION) and often compounded with carbon black to reinforce and toughen the material. A wide range of synthetic rubbers is now available, many of which compete directly with the natural material.

All these materials possess a double bond in the main chain, a feature which reduces steric hindrance, making the polymer chains highly flexible and able to adopt a large number of different conformations at ambient temperature (see ROTATIONAL ISOMERISM). This is the origin of both long-range elasticity and the retractive force. Provided chain slippage (ie viscous flow) is prevented by a small degree of crosslinking (1–2% typically), then elastomers will extend by several hundred per cent, when stressed, and return to their original shape when released.

Natural rubber and butadiene polymers are cross-linked (vulcanized) with sulphur, the chains being cross-linked between sites adjacent to the double bond (see diagram). Accelerators are normally used to improve the efficiency of the reaction. The relatively small stress needed to extend elastomers is entropic rather than energetic in origin, as it is in most rigid materials like glass and steel, reflecting the tendency of chains to return to their equilibrium RANDOM COIL conformation. It also helps to explain why the modulus increases with rising temperature, rather than falling, as in glass and steel. Thermal vibration increases with temperature, so increasing the number of possible conformations of the random coils. See KINETIC THEORY OF ELASTICITY.

Another phenomenon shown by elastomers is their rise in temperature when strained quickly, a direct analogue of the JOULE–THOMSON EFFECT in gases. The temperature rise in natural rubber, eg common elastic bands when stressed on the lips, is even greater due to strain crystallization. The release of the heat of fusion (ΔH_f) heats the sample up by about 1°C, but on release, the crystallites melt and the rubber cools by roughly the same amount.

For most engineering applications (bushes, engine mounts, seals, bearings) the high extensibility is not needed and it is usually desirable to increase the modulus. Although increasing the degree of cross-linking will help, the material loses its toughness (eg ebonite). The preferred alternative is to reinforce with carbon black, as in the ubiquitous car tyre.

Crosslinking of rubbers by S–S bridges

---

electrical (*Phys*) Descriptive of means related to, pertaining to or associated with electricity, but not inherently functional. Commonly used for ELECTRIC.

electrical absorption (*Phys*) An effect in a dielectric whereby, after an initially charged capacitor has been once discharged, it is possible after a few minutes to obtain from it another discharge, usually smaller than the first.

electrical analogy (*Acous*) The correspondence between electrical and acoustical systems, which assists in applying to the latter procedures familiar in the former.

electrical bias (*ICT*) The use of a polarizing winding on a relay core, for adjusting the sensitivity of the relay to signal currents.

electrical conductivity (*Phys*) The ratio of current density to applied electric field. Expressed in siemens per metre (S m^{-1}) or Ω^{-1} m^{-1} in SI units. Conductivity of metals at high temperatures varies as T^{-1}, where T is absolute temperature. At very low temperatures, variation is complicated but it increases rapidly (at one stage proportional to T^{-5}), until it is finally limited by material defects of structure.

electrical degrees (*ElecEng*) Angle, expressed in degrees, of phase difference of vectors, representing currents or voltages, arising in different parts of a circuit.

electrical discharge machining (*Eng*) SPARK EROSION technique in which metal is removed as sparks pass between a shaped electrode and the work. Can be used for machining irregular holes etc. Abbrev *EDM*.

electrical discharger (*Aero*) A device for discharging static electricity, eg earthing tyres, static wick dischargers.

electrical double layer (*Chem*) The layer of adsorbed ions at the surface of a dispersed phase which gives rise to the ELECTROKINETIC EFFECTS.

electrical engineering (*Genrl*) The branch of engineering chiefly concerned with the design and construction of all electrical machinery and devices, as well as eg power transmission.

electrical power distribution (*Space*) The provision, conditioning and supply of electrical power to satisfy the needs of a spacecraft and its payload. Continuous sources of power may be the Sun (eg SOLAR CELLS, thermal devices) or carried on board (eg FUEL CELLS, radioisotopes). During certain mission phases of space-flight, such as launch and re-entry, an auxiliary power source must be used.

electrical prospecting (*MinExt*) A form of GEOPHYSICAL PROSPECTING which identifies anomalous electrical effects associated with buried ore bodies. Most important techniques utilize resistivity or inverse conductivity and the inductive properties of ore minerals.

electrical reset (*ElecEng*) The restoration of a magnetic device, eg relay or circuit breaker, by auxiliary coils or relays.

electrical resistivity (*Phys*) The reciprocal of ELECTRICAL CONDUCTIVITY. Units Ω m. See RESISTIVITY.

electricTRS resonance (*ElecEng*) Condition arising when a maximum of current or voltage occurs as the frequency of the electric source is varied; also when the length of a transmission line approximates to multiples of a quarter wavelength and the current or voltage becomes abnormally large.

electrical synapse (*BioSci*) A connection between two electrically excitable cells, such as neurons or muscle cells, via arrays of gap junctions.

electrical thread (*ElecEng*) The threadform used on screwed steel conduit for electrical installation work.

electric-arc furnace (*ElecEng*) See ARC FURNACE.

electric-arc welding (*ElecEng*) See ARC WELDING.

electric axis (*ICT*) Direction in a crystal that gives the maximum conductivity to the passage of an electric current. The x-axis of a piezoelectric crystal.

electric balance (*ElecEng*) A name sometimes applied to a type of electrometer, to a current weigher (which establishes the absolute ampere) and to a Wheatstone bridge.

electric bell (*ElecEng*) A bell with a solenoid-operated hammer giving a rapid succession of strokes by means of a make-and-break contact on the solenoid.

electric braking (*ElecEng*) A method of braking for electrically driven vehicles; the motors are used as generators to return the braking energy to the supply, or to dissipate it as heat in resistances.

electric calamine (*Min*) US for HEMIMORPHITE.

electric cautery (*Med*) Burning of parts of the human body for surgical purposes by means of electrically heated instruments. Also *electrocautery*.

electric charge (*Phys*) See CHARGE.

electric circuit (*Phys*) Series of conductors forming a partial, branched or complete path around which either dc or ac can flow.

electric component (*ElecEng*) That component of an electromagnetic wave which produces a force on an electric charge, and along the direction of which currents in a conductor in the field are urged to flow. Also *electrostatic component*.

electric conduction (*ElecEng*) Transmission of energy by flow of charge along a conductor.

electric current (*Phys*) See CURRENT.

electric dipole (*Phys*) See ELECTRIC DOUBLET.

electric dipole moment (*Phys*) The product of the magnitude of the electric charges and the distance between it and its opposite charge in an electric dipole.

electric discharge (*Phys*) See FIELD DISCHARGE.

electric-discharge lamp (*ElecEng*) A lamp in which the light is obtained from the discharge between two electrodes in an evacuated glass tube. Also *gas-discharge lamp*.

electric double layer (*ElecEng*) A distribution of positive and negative layers of electric charge very close together so that effectively the total charge is zero but the two layers form an assembly of dipoles, thus giving rise to an electric field.

electric doublet (*Phys*) A system with a definite electric moment, mathematically equivalent to two equal charges of opposite sign at a very small distance apart.

electric dynamometer (*Eng*) An electric generator which is used for measuring power. The stator frame is capable of partial rotation in bearings concentric with those of the armature, and the torque is balanced and measured by hanging weights on an arm projecting from the frame.

electric eye (*ICT*) Miniature cathode-ray tube used, eg in a radio receiver, to exhibit a pattern determined by the rectified output voltage obtained from the received carrier, thus assisting in tuning the receiver. Also used for balancing ac bridges. Also *cathode-ray tuning indicator, electron-ray indicator tube, magic eye*.

electric field (*Phys*) A region in which forces are exerted on any electric charge present.

electric field strength (*Phys*) The strength of an electric field is measured by the force exerted on a unit charge at a given point. Expressed in volts per metre ($V\,m^{-1}$). Symbol *E*.

electric flux (*Phys*) Surface integral of the electric field intensity normal to the surface. The electric flux is conceived as emanating from a positive charge and ending on a negative charge without loss. Symbol Ψ.

electric flux density (*Phys*) See DISPLACEMENT.

electric generator (*ElecEng*) See GENERATOR.

electric harmonic analyser (*ElecEng*) An electrical device for determining the magnitude of the harmonics in the waveshape of an ac or voltage. Also *spectrum analyser*.

electricity (*ElecEng*) The form of energy associated with static or dynamic electric charges.

electricity meter (*ElecEng*) See INTEGRATING METER.

electric-light ophthalmia (*Med*) See PHOTOPHTHALMIA.

electric locomotive (*ElecEng*) A locomotive in which the motive power is by electric motor, supplied from batteries (battery locomotive), from a diesel engine/electrical generator set on the locomotive (diesel electric), from an overhead contact wire or from a contact rail.

electric machine (*ElecEng*) See ELECTRIC MOTOR, ELECTROMAGNETIC GENERATOR, ELECTROSTATIC GENERATOR.

electric moment (*ElecEng*) The product of the magnitude of either of two equal electric charges and the distance between their centres, with axis direction from the negative to the positive charge. See MAGNETIC MOMENT.

electric motor (*ElecEng*) Any device for converting electrical energy into mechanical torque. Also *electromotor*.

electric organ (*BioSci*) A mass of muscular or nervous tissue, modified for the production, storage and discharge of electric energy; occurring in fish.

electric oscillations (*ICT*) Electric currents that periodically reverse their direction of flow, at a frequency determined by the constants of a resonant circuit. See CONTINUOUS OSCILLATIONS, ELECTRONIC OSCILLATIONS.

electric polarization (*ElecEng*) The dipole moment per unit volume of a dielectric.

electric potential (*ElecEng*) That measured by the energy of a unit positive charge at a point, expressed relative to ZERO POTENTIAL.

electric propulsion (*Space*) The use of electrostatic or electromagnetic fields to accelerate ions or plasma, thereby producing propulsive thrust. See ION PROPULSION.

electric resistance welded tube (*ChemEng*) Much used in heat exchangers, it is made continuously by forming an accurately rolled strip over a mandrel and welding the edges electrically. Abbrev *ERW tube*.

electric shielding (*ElecEng*) See FARADAY CAGE.

electric storm (*EnvSci*) A meteorological disturbance in which the air becomes highly charged with static electricity. In the presence of clouds this leads to thunderstorms.

electric strength (*ElecEng*) Maximum voltage which can be applied to an insulator or insulating material without sparkover or breakdown taking place. The latter arises when the applied voltage gradient coincides with a breakdown strength at a temperature which is attained through normal heat dissipation.

electric susceptibility (*ElecEng*) The amount by which the relative permittivity of a dielectric exceeds unity, or the ratio of the polarization produced by unit field to the permittivity of free space. See DIELECTRICS, FERROELECTRICS.

electric traction (*ElecEng*) The operation of a railway or road vehicle by means of electric motors, which obtain their power from an overhead contact wire or from batteries mounted on the vehicle.

electric wind (*ElecEng*) Stream of air caused by the repulsion of charged particles from a sharply pointed portion of a charged conductor.

electrification (*ElecEng*) (1) Charging a network to a high potential. (2) Conversion of any plant to operation by electricity, eg changeover of steam-driven locomotives to electricity. (3) Charging a conductor by electric induction from another charged conductor. (4) Provision of a supply of electrical energy where none existed, eg rural electrification is the provision of electricity to consumers in country areas.

electro-, electr- (*Genrl*) Prefix from Gk *elektron* (amber), denoting electric or electrolytic.

electro-acoustics (*Acous*) The branch of technology dealing with the interchange of electric and acoustic energy, eg as in a transducer.

electroactive polymer composites (*Eng*) Electroactive composites are an attempt to combine the desirable properties of piezoelectric ceramics and polymers without suffering their intrinsic limitations. They consist of a mixture of processed piezoelectric ceramic and an electrically inert polymer such as polypropylene. The non-piezoelectric polymer gives the composite the desired low permittivity and density whilst its compliant nature enables it to be shaped into specialized geometries and to resist thermal shock. Examples are lead titanate zirconate (PZT) and polyvinylidene fluoride (PVDF).

electroanalysis (*Chem*) Electrodeposition of an element or compound to determine its concentration in the electrolysed solution. See CONDUCTIMETRIC ANALYSIS, POLAROGRAPH, POTENTIOMETER, VOLTAMETER.

electroarteriograph (*Med*) An instrument for recording blood flow rates.

electrobrightening (*Eng*) See ELECTROLYTIC POLISHING.

electrocapillary effect (*Chem*) The decrease in interfacial tension, usually of mercury, caused by the mutual repulsion of adsorbed ions opposing the attractive force of interfacial tension. See CAPILLARY ELECTROMETER, ELECTRICAL DOUBLE LAYER.

electrocapillary maximum (*Chem*) The potential at which a mercury surface in an electrolyte is charge-free and consequently has the maximum interfacial tension; about -0.28 volts.

electrocardiogram (*Med*) A record of the electrical activity of the heart. Abbrev *ECG*.

electrocardiography (*Med*) The study of electric currents produced in cardiac muscular activity.

electrocataphoresis (*Phys*) See ELECTROPHORESIS.

electrocautery (*Med*) See ELECTRIC CAUTERY.

electroceramic processing (*Eng*) See panel on ELECTRO-CERAMIC PROCESSING.

electrochemical constant (*Chem*) See FARADAY.

electrochemical equivalent (*Phys*) The mass of a substance deposited at the cathode of a voltameter per coulomb of electricity passing through it.

electrochemical machining (*Eng*) Removing material from a metal by anodic dissolution in a bath in which electrolyte is pumped rapidly through the gap between the shaped electrode and the stock. Abbrev *ECM*.

electrochemical potential (*Chem, BioSci*) (1) Difference in voltage between anode and cathode in an electrochemical cell. (2) A thermodynamic measure that combines the energy stored in the form of chemical potential (concentration gradient) and electrostatics. Important in analysing the movement of ions across eg a membrane; if movement is up a gradient of electrochemical potential then it is an ACTIVE TRANSPORT process of some sort.

electrochemical potential series (*Chem*) The classification of redox half-reactions, written as reductions, in order of decreasing reducing strength. Thus the combination of any half-reaction with the reverse of one further down the series will give a spontaneous reaction. For reference, the half-reaction $2H^+ + e^- = H_2$ is taken as having an energy of zero. Also *electrochemical series, electrode potential series, electromotive series*.

electrochemistry (*Chem*) That branch of chemistry which deals with the electronic and electrical aspects of processes in a liquid or solid phase.

electrochronograph (*ElecEng*) The combination of an electrically driven clock and an electromagnetic recorder for recording short time intervals.

electrocoagulation (*Med*) The coagulation of bodily tissues by high-frequency electric current.

electroconvulsive therapy (*Med*) A form of therapy in which an electric current is passed through the brain (usually under anaesthetic), resulting in convulsive seizures, used primarily in the treatment of depression. Abbrev *ECT*.

electro copper glazing (*Build*) See COPPER GLAZING.

electrocution (*Med*) Death caused by an electric shock.

electrocyte (*BioSci*) A cell, usually muscle but sometimes nerve, which is specialized to generate an electric discharge.

electrode (*ElecEng*) (1) Conductor whereby an electric current is lead into or out of a liquid (as in an electrolytic cell), or a gas (as in an electric-discharge lamp or gas tube), or a vacuum (as in a photomultiplier).

electrode (*Electronics*) In a semiconductor, the emitter or collector of electrons or holes.

electrode boiler (*ElecEng*) A boiler in which heat is produced by the passage of an electric current through the liquid to be heated.

electrode characteristic (*ElecEng*) The graph relating current in electronic device to potential of one electrode, that of all others being maintained constant.

electrode current (*Electronics*) The net current flowing in a valve (or tube) from an electrode into the surrounding space.

electrode dark current (*Electronics*) The current which flows in a camera tube or phototube when there is no radiation incident on the photocathode, given certain specified conditions of temperature and shielding from radiation. It limits the sensitivity of the device.

electrode dissipation (*Electronics*) The power released at an electrode, usually an anode, because of electron ion impact. In large valves the temperature is held down by radiating fins, graphiting to increase radiation, or by water or oil cooling.

electrode efficiency (*ElecEng*) The ratio of the quantity of metal deposited in an electrolytic cell to the quantity which should theoretically be deposited according to Faraday's laws.

electrode holder (*ElecEng*) In electric-arc welding, a device used for holding the electrode and leading the current to it.

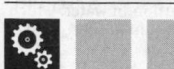

Electroceramic processing

Ceramic is the term given to a broad class of materials that are composed of inorganic, polycrystalline compounds. They are commonly complex oxides or nitrides of metals or semimetals and are characteristically refractory in nature; ceramics are brittle and difficult to form into special shapes.

Ceramics play an important role in a number of electronic devices such as those used in piezoelectric, pyroelectric, electro-optic and electromagnetic applications. Their most common use is as dielectric fillings for capacitors in passive components although more specialized applications, such as ferroelectric memories and transducers, are becoming increasingly common.

Bulk ceramics for electronic applications are formulated to a specific composition by a number of distinct chemical and mechanical processes. Typical examples are the high-temperature superconductor containing yttrium, barium and copper, $YBa_2Cu_3O_{7-x}$ (*YBCO*), where x denotes a departure from stoichiometry, and the ferroelectric, barium titanate, $BaTiO_3$ (*BT*). Powders of the oxides of the various metal, or cation, components are mixed thoroughly in the appropriate proportions for the composition. Alternatively carbonates or nitrates, which decompose readily to oxides, may be used: eg $BaCO_3$ and TiO_2 are mixed in a 1:1 cation ratio in the fabrication of BT. The oxide/carbonate/nitrate mix is heated to a temperature at which the individual powders partially react to form a single compound. This process is called *calcination* and is typically performed in the range 500–800°C. The calcined powder is then finely ground, or milled, before being compacted, typically in cylindrical dies, in pressures up to 80 MPa.

In a further heat treatment, the pressed pellet is sintered for several hours at a temperature usually a few hundred degrees above that needed for calcination. This process completes the reaction of the constituent cations and transforms the compacted powder into a single ceramic body with grains typically 5–20 mm in diameter. The sintered block is usually sliced into wafers for further processing such as lapping, polishing, electroding and, for ferroelectric devices, poling by the application of an electric field. These steps are summarized in the diagram.

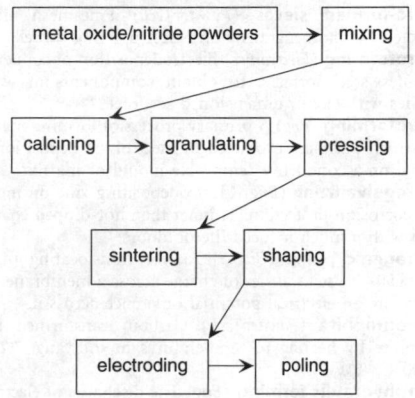

Electroceramic processing Main steps.

A variety of methods have been developed to improve the chemical uniformity and mechanical properties of the processed ceramics. A homogeneous mixture of the component cations, for instance, can be achieved by dissolving soluble forms of the base compounds (usually nitrates) in deionized water and spray-drying them at 200°C. Again, sintering under pressure (HOT ISOSTATIC PRESSING) in a controlled oxidizing environment may give fully dense, pore-free ceramic. This process is aided significantly for most ceramics if the pre-sintered powder consists of fine, spherical particles at around 1 mm diameter.

See panel on CERAMICS PROCESSING.

---

electrodeposition (*Chem*) Deposition electrolytically of a substance on an electrode, as in ELECTROFORMING or ELECTROPLATING.

electrode potential (*Chem*) The voltage between the electrode and electrolyte in an electrochemical cell.

electrode potential series (*Chem*) See ELECTROCHEMICAL POTENTIAL SERIES.

electrodermal effect (*Med*) The change in skin resistance following an emotional reaction.

electrodialysis (*Chem*) The removal of electrolytes from a colloidal solution by an electric field between electrodes in pure water outside the two dialysing membranes between which is contained the colloidal solution.

electrodisintegration (*Phys*) The disintegration of a nucleus under electron bombardment.

electrodissolution (*ElecEng*) Dissolving a substance from an electrode by electrolysis.

electrodynamic instrument (*ElecEng*) An electrical measuring instrument which depends for its action on the electromagnetic force between two or more current-carrying coils.

electrodynamic loudspeaker (*Acous*) A loudspeaker in which the radiating cone is driven by current in a coil which moves in a constant magnetic field.

electrodynamic microphone (*Acous*) The inverse of an ELECTRODYNAMIC LOUDSPEAKER. Also *moving-coil microphone*.

electrodynamics (*Phys*) The study of the motion of electric charges caused by electric and magnetic fields.

electrodynamic wattmeter (*ElecEng*) One for low-frequency measurements. It depends on the torque exerted between currents carried by fixed and movable coils.

electroencephalogram (*Med*) A record of the electrical activity of the brain. Abbrev *EEG*.

electroencephalograph (*Med*) An instrument for the study of voltage waves associated with the brain; effectively comprises a sensitive detector (voltage or current), a stable dc amplifier and a recording system. Abbrev *EEG*.

electro-endosmosis (*Chem*) The movement of liquid, under an applied electric field, through a fine tube or membrane. Also *electro-osmosis, electrosmosis*.

electroextraction (*Eng*) The recovery of a metal from a solution of its salts by electrolysis, the metal depositing on the cathode. Also *electrowinning*.

electrofacing (*ElecEng*) The process of coating, by electrodeposition, a metal surface with a harder metal to render it more durable.

electrofluor (*Phys*) A transparent material which has the property of storing electrical energy and releasing it as visible light.

electrofluorescence (*Phys*) See ELECTROLUMINESCENCE.

electro-formed sieves (*PowderTech*) Fine-mesh sieves formed by photoengraving and electroplating nickel.

electroforming (*Electronics*) Electrodeposition of copper on stainless-steel formers, to obtain components of waveguides with closely dimensioned sections.

electroforming (*Eng*) A primary process of forming metals, in which parts are produced by electrolytic deposition of metal on a conductive removable mould or matrix.

electrogalvanizing (*Eng*) Electrodepositing zinc on metals for corrosion protection. Thinner than hot-dipped coating and with a much reduced life outdoors.

electrogenic pump (*BioSci*) An ion-translocating pump that causes a net transfer of charge across a membrane and therefore an electrical potential difference across it.

electrographite (*PowderTech*) Carbon transformed into graphite by heating to temperatures in the range 2200–2800°C.

electrohydraulic forming (*Eng*) The discharge of electrical energy across a small gap between electrodes immersed in water, which produces a shock wave. The wave travels through the water, until it hits and forces a metal workpiece into the particular shape of the die which surrounds it, thus eg piercing holes, crushing rock, etc. Also *explosive forming*.

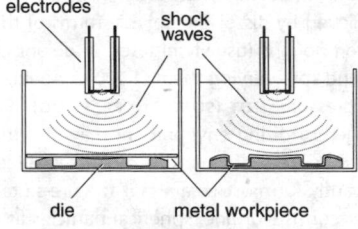

electrodes

shock waves

die   metal workpiece

electrohydraulic forming

electrokinetic effects (*Chem*) Phenomena due to the interaction of the relative motion with the potential between the two phases in a dispersed system. There are four: ELECTRO-ENDOSMOSIS, ELECTROPHORESIS, SEDIMENTATION POTENTIAL and STREAMING POTENTIAL.

electrokinetic potential (*Chem*) Potential difference between the surface of a solid particle immersed in aqueous or conducting liquid and the fully dissociated ionic concentration in the body of the liquid. Concept important in froth flotation, electrophoresis, etc. Also *zeta potential*.

electrokinetics (*Phys*) The science of electric charges in motion, without reference to the accompanying magnetic field.

electroluminescence (*Phys*) The luminescence produced by the application of an electric field to a dielectric PHOSPHOR (such as manganese-doped zinc sulphide). Also *electrofluorescence*. See GUDDEN–POHL EFFECT.

electrolysis (*Chem*) Chemical change, generally decomposition, effected by a flow of current through a solution of the chemical, or its molten state, based on ionization.

electrolysis (*Med*) The removal of hair by applying an electrically charged needle to the follicle.

electrolyte (*Chem*) Chemical, or its solution in water, which conducts current through ionization. Molten salts are also electrolytes.

electrolyte (*Med*) A compound that in solution dissociates into ions. In clinical practice sodium, potassium and bicarbonate are the electrolytes of everyday concern.

electrolyte strength (*Chem*) The extent towards complete ionization in a dilute solution. When concentrated, the ions join in groups, as indicated by lowered MOBILITY.

electrolytic capacitor (*ElecEng*) An electrolytic cell in which a very thin layer of non-conducting material has been deposited on one of the electrodes by an electric current. This is known as 'forming' the capacitor, the deposited layer providing the dielectric. Because of its thinness a larger capacitance is achieved in a smaller volume than in the normal construction of a capacitor. In the so-called *dry electrolytic capacitor* the dielectric layer is a gas which, however, is actually 'formed' from a moist paste within the capacitor.

electrolytic cell (*Chem*) An electrochemical cell in which an externally applied voltage causes a non-spontaneous change to occur, such as the breakdown of water into hydrogen and oxygen. Opposite of GALVANIC CELL.

electrolytic copper (*Eng*) Copper refined by electrolysis. This gives metal of high purity (over 99·94% copper), and enables precious metals, such as gold and silver, to be recovered.

electrolytic corrosion (*Eng*) Corrosion produced by contact of two different metals when an electrolyte is present and current flows.

electrolytic depolarization (*ElecEng*) See DEPOLARIZATION.

electrolytic dissociation (*Chem*) The reversible splitting of substances into oppositely charged ions.

electrolytic grinding (*Eng*) A metal-bonded and diamond-impregnated grinding wheel removes metal from the stock but in addition the abradant is an insulator between the wheel acting as anode and the stock which has electrolyte flowing over it. This allows ELECTROCHEMICAL MACHINING to occur.

electrolytic instrument (*ElecEng*) An instrument depending for its operation upon electrolytic action, eg an ELECTROLYTIC METER.

electrolytic lead (*Eng*) Lead refined by the Betts process; has a purity of about 99·995–99·998% lead.

electrolytic lightning arrester (*ElecEng*) A lightning arrester consisting of a number of electrolytic cells in series; it breaks down, allowing the lightning stroke to discharge to earth, when the voltage across it exceeds about 400 V per cell.

electrolytic machining (*Eng*) An electrochemical process based on the same principles as electroplating, except that the workpiece is the anode and the tool is the cathode, resulting in a deplating operation. Sometimes, eg in ELECTROLYTIC GRINDING, combined with some abrasive action.

electrolytic meter (*ElecEng*) An integrating meter whose operation depends on electrolytic action, eg water breakdown.

electrolytic polarization (*ElecEng*) The change in the potential of an electrode when a current is passed through it. As current rises, polarization reduces the potential difference between the two electrodes of the system.

electrolytic polishing (*Eng*) By making the metal surface an anode and passing a current under certain conditions, there is a preferential solution so the microscopic irregularities vanish, leaving a smoother surface. Also *anode brightening, anode polishing, electrobrightening, electropolishing*.

electrolytic refining (*Eng*) The method of producing pure metals, by making the impure metal the anode in an electrolytic cell and depositing a pure cathode. The impurities either remain undissolved at the anode or pass into solution in the electrolyte. Also *electrofining*.

electrolytic tank (*ElecEng*) A device used to simulate field systems, eg the electrostatic field around the electrodes in a cathode-ray tube. The tank is filled with a poorly conducting fluid in which is immersed a scale model in metal of the desired system. Appropriate voltages are applied between the various parts and the equipotentials are traced out using a suitable probe.

electrolytic zinc (*Eng*) Zinc produced from its ores by roasting (to convert sulphide to oxide), solution of oxide in sulphuric acid, precipitation of impurities by adding zinc dust, and final electrolytic deposition of zinc on aluminium cathodes. Product has purity over 99·9%.

electromagnet (*ElecEng*) Soft-iron core, embraced by a current-carrying coil, which exhibits appreciable magnetic effects only when current passes.

electromagnetic brake (*ElecEng*) A brake in which the braking force is produced by the friction between two surfaces pressed together by the action of a solenoid, or by magnetic attraction, the necessary flux being produced by an electromagnet.

electromagnetic clutch (*Autos, Eng*) A friction clutch without pressure springs, operated by a solenoid. A switch in the base of the gear lever breaks the circuit and releases the pressure for gear changes.

electromagnetic component (*ICT*) Strictly, that component of the combined field surrounding a transmitting antenna which represents the radiated energy.

electromagnetic control (*ElecEng*) A form of remote control for switchgear etc in which operation is effected by means of a solenoid.

electromagnetic damping (*ElecEng*) See MAGNETIC DAMPING.

electromagnetic deflection (*Electronics*) Deflection of the beam in a cathode-ray tube by a magnetic field produced by a system of coils carrying currents, eg for scanning a TV image or for providing a time-base deflection.

electromagnetic field theory (*Phys*) Theory based on MAXWELL'S EQUATIONS describing ELECTROMAGNETIC WAVES (including visible light), interdependent transverse waves of electric and magnetic fields.

electromagnetic focusing (*Electronics*) The focusing of a beam of charged particles by magnetic fields associated with current-carrying coils. Used in the cathode-ray tube and electron microscope. See ELECTRON LENS and panel on ELECTRON MICROSCOPE.

electromagnetic generator (*ElecEng*) An electric generator which depends for its action on the induction of emfs in a circuit by a change in the magnetic flux linking with that circuit.

electromagnetic horn (*ICT*) See HORN ANTENNA.

electromagnetic induction (*ElecEng*) The transfer of electrical power from one circuit to another by varying the magnetic linkage. See FARADAY'S LAW OF INDUCTION.

electromagnetic inertia (*ElecEng*) The energy required to stop or start a current in an inductive circuit. Inductance is analogous to mass in a mechanical system.

electromagnetic instruments (*ElecEng*) Electrical measuring instruments whose action depends on the electromagnetic forces set up between a current-carrying conductor and a magnetic field. See DYNAMOMETER, MOVING-COIL INSTRUMENT.

electromagnetic interaction (*Phys*) An interaction between charged elementary particles and mediated by PHOTONS; completed in about 10^{-18} s. Intermediate in strength between strong and weak interactions. See panel on GRAND UNIFIED THEORIES.

electromagnetic lens (*Electronics*) One using current-carrying coils to focus electron beams.

electromagnetic loudspeaker (*Acous*) A device where a fluctuating magnetic field, induced by an electric current, excites a force in magnetic material connected to the loudspeaker cone. Also *inductor loudspeaker*.

electromagnetic microphone (*Acous*) The inverse of an ELECTROMAGNETIC LOUDSPEAKER.

electromagnetic mirror (*ElecEng*) A reflecting surface for electromagnetic waves.

electromagnetic moment (*Phys*) The magnetic moment of a current-carrying coil; given by the product of the current, coil area and number of turns.

electromagnetic pickup (*Acous*) Pickup in which the motion of the stylus, in following the recorded track, causes a fluctuation in the magnetic flux carried in any part of a magnetic circuit, with consequent emfs in any coil embracing such magnetic circuit.

electromagnetic pole piece (*ElecEng*) In a U-shaped core, the pole pieces are attached at the free end and are often conical in shape to concentrate the magnetic field in the air-gap.

electromagnetic prospecting (*Geol*) A method in which distortions produced in electric waves are observed and lead to pinpointing of geological anomalies.

electromagnetic pump (*ElecEng*) A pump designed to conduct fluids, eg liquid metals, and to maintain circulation without moving parts.

electromagnetic radiation (*Phys*) The emission and propagation of ELECTROMAGNETIC WAVES.

electromagnetic reaction (*ElecEng*) The reaction between the anode and grid circuits of a valve obtained by electromagnetic coupling; also *inductive reaction, magnetic reaction*.

electromagnetic separation (*MinExt*) Removal of ferromagnetic objects from town refuse, or TRAMP IRON from bulk materials, as they travel along a conveyor, over a drum or into a revolving screen, by setting up a magnetic field which diverts the ferromagnetic material from the rest. Concentration of ferromagnetic minerals from gangue.

electromagnetic separation (*NucEng*) Isotope separation by ELECTROMAGNETIC FOCUSING, as in a mass spectrometer.

electromagnetic spectrum (*Phys*) See ELECTROMAGNETIC WAVE and appendices.

electromagnetic switch (*ElecEng*) A switch moved by electromagnets or solenoids.

electromagnetic theory (*Phys*) See ELECTROMAGNETIC FIELD THEORY.

electromagnetic units (*Phys*) Any system of units based on assigning an arbitrary value to μ_0, the permeability of free space.

$$\mu_0 = 4\pi \times 10^{-7} \text{ H m}^{-1}$$

in the SI system; μ_0 is unity in the CGS electromagnetic system.

electromagnetic wave (*Phys*) A wave comprising two interdependent mutually perpendicular transverse waves of electric and magnetic fields. The speed of propagation in free space for all such waves is that of the speed of light, $2·997\,924\,58 \times 10^8$ m s^{-1}. The electromagnetic spectrum ranges from wavelengths of 10^{-15} m to 10^5 m, ie including gamma rays, X-rays, ultraviolet radiation, visible light, infrared radiation, microwaves and radio waves. Electromagnetic waves undergo reflection and refraction, exhibit interference and diffraction effects, and can be polarized. The waves can be channelled, eg by waveguides for microwaves or fibre optics for visible light.

electromagnetism (*ElecEng*) The science of the properties of, and relations between, magnetism and electric currents. Also *electromagnetics*.

electromechanical brake (*ElecEng*) An electric brake in which the braking force is due partly to the attraction of two magnetized surfaces and partly to friction of the clamping action caused by the solenoid.

electromechanical counter (*ElecEng*) One which records mechanically the number of electric pulses fed to a *solenoid*.

electromechanical recorder (*ElecEng*) One which changes electrical signals into a mechanical motion of a similar form, and cuts or records the shape of the motion in an appropriate medium.

electrometallization (*Eng*) The electrodeposition of a metal on a non-conducting base, either for decorative purposes or to give a protective covering.

electrometallurgy (*Eng*) A term covering the various electrical processes for the industrial working of metals, eg electrodeposition, electrorefining and operations in electric furnaces.

electrometer (*ElecEng*) Fundamental instrument for measuring potential difference, depending on the attraction or repulsion of charges on plates or wires.

electrometer gauge (*ElecEng*) A small attracted-disk electrometer sometimes attached to the needle of a quadrant electrometer to determine whether the needle is sufficiently charged.

electrometric titration (*Chem*) See POTENTIOMETRIC TITRATION.

electromotive force (*ElecEng*) The difference of potential produced by sources of electrical energy which can be used to drive currents through external circuits. Abbrev *emf*. Unit *volt*.

electromotive series (*Chem*) See ELECTROCHEMICAL POTENTIAL SERIES.

electromyography (*Med*) The study of electric currents set up in muscle fibres by bodily movement.

electron (*Phys*) A fundamental particle with negative electric charge of $1\cdot602 \times 10^{-19}$ C and mass $9\cdot109 \times 10^{-31}$ kg. Electrons are a basic constituent of the atom; they are distributed around the nucleus in *shells* and the electronic structure is responsible for the chemical properties of the atom. Electrons also exist independently and are responsible for many electric effects in materials. Owing to their small mass, the wave properties and relativistic effects of electrons are marked. The *positron*, the antiparticle of the electron, is an equivalent particle but with a positive charge. Either electrons or positrons may be emitted in beta decay. Electrons, muons and neutrinos form a group of fundamental particles called LEPTONS.

electron affinity (*Chem*) The energy required to remove an electron from a negatively charged ion to form a neutral atom.

electron affinity (*Electronics*) (1) The tendency of certain substances, notably oxidizing agents, to capture an electron. (2) See WORK FUNCTION.

electron attachment (*Phys*) The formation of a negative ion by attachment of a free electron to a neutral atom or molecule.

electron beam (*Electronics*) A stream of electrons moving with the same velocity and direction in neighbouring paths and usually emitted from a single source such as a cathode.

electron-beam analysis (*Eng*) Scanning a microbeam of electrons over a surface *in vacuo* and analysing the secondary emissions to determine the distribution of selected elements.

electron-beam tube (*Electronics*) One in which several electrodes control one or more electron beams. Also *electron-beam valve*.

electron-beam welding (*Eng*) Heating components to be welded by a concentrated beam of high-velocity electrons *in vacuo*.

electron binding energy (*Phys*) See IONIZATION POTENTIAL.

electron camera (*ImageTech*) Generic term for a device which converts an optical image into a corresponding electric current directly by electronic means, without the intervention of mechanical scanning.

electron capture (*Phys*) The capture of shell electron (K or L) by the nucleus of its own atom, decreasing the atomic number of the atom without change of mass. The capture is accompanied by the emission of a neutrino.

electron charge/mass ratio (*Electronics*) The ratio

$$e/m = 1\cdot759 \times 10^{11} \text{ C kg}^{-1}$$

Fundamental physical constant, the mass being the rest mass of the electron.

electron cloud (*Chem*) (1) The density of electrons in a volume of space, as the position and velocity of an electron cannot be simultaneously specified. (2) The nature of the valence electrons in a metal, where their non-attachment to specific nuclei gives rise to electronic conduction.

electron compound (*Chem*) A solid phase appearing in metallic systems whose composition is largely determined by a particular ratio of valency electrons to number of atoms, eg CuZn occurring at a ratio of 3:2. Usually associated with extensive solubility on either side of the ideal ratio.

electron conduction (*Electronics*) That which arises from the drift of free electrons in metallic conductors when an electric field is applied. See N-TYPE and P-TYPE SEMICONDUCTOR.

electron coupling (*Electronics*) That between two circuits, due to an electron stream controlled by the one circuit influencing the other circuit. Such coupling tends to be unidirectional, the second circuit having little influence on the first.

electron-deficient (*Chem*) A substance which does not have enough valence electrons to form 'normal' chemical bonds. Usually, the term is used for the compounds of boron, but all metallic bonding is of this type.

electron density (*Phys*) The number of electrons per gram of a material. Approx 3×10^{23} for most light elements. In an ionized gas the equivalent electron density is the product of the ionic density and the ratio of the mass of an electron to that of a gas ion.

electron device (*Electronics*) One which depends on the conduction of electrons through a vacuum, gas or semiconductor.

electron diffraction (*Electronics*) The investigation of crystal structure by the patterns obtained on a screen from electrons diffracted from the surface of crystals or as a result of transmission through thin metal films.

electron discharge (*Electronics*) The current produced by the passage of electrons through otherwise empty space.

electron-discharge tube (*Electronics*) Highly evacuated tube containing two or more electrodes between which electrons pass.

electron dispersion curve (*Phys*) A curve showing the electron energy as a function of the wavevector under the influence of the periodic potential of a crystal lattice. Experiments and calculations which determine such curves give important information about the energy gaps, the electron velocities and the DENSITY OF STATES.

electron drift (*Electronics*) The actual transfer of electrons in a conductor as distinct from energy transfer arising from encounters between neighbouring electrons.

electronegative (*Phys*) Carrying a negative charge of electricity. Tending to form negative ions, ie having a relatively positive electrode potential.

electronegativity (*Chem*) The relative ability of an atom to retain or gain electrons. There are several definitions, and the term is not quantitative. It is, however, useful in predicting the strengths and the polarities of bonds, both of which are greater when there is a significant electronegativity difference between the atoms forming the bond. On the commonly used scale of Pauling, the range of values is from less than one (alkali metals) to four (fluorine). See panel on BONDING.

electron–electron scattering (*Phys*) A possible process that contributes to the electrical resistivity of metals. Important at low temperatures in transition metals.

electron emission (*Electronics*) The liberation of electrons from a surface.

electron gas (*Electronics*) The 'atmosphere' of free electrons *in vacuo*, in a gas or in a conducting solid. The laws obeyed by an electron gas are governed by FERMI–DIRAC STATISTICS, unlike ordinary gases to which MAXWELL-BOLTZMANN STATISTICS apply.

electron gun (*Electronics*) An assembly of electrodes in a cathode-ray tube which produces the electron beam, comprising a cathode from which electrons are emitted, an apertured anode, and one or more focusing diaphragms and cylinders.

electronic (*Genrl*) Pertaining to devices or systems which depend on the flow of electrons; the term covers most branches of electrical science other than electric power generation and distribution. Telecommunications, radar and computers all use electronic components and techniques. Electronic engineering is a field which encompasses the application of electronic devices, as opposed to physical electronics which is the study of electronic phenomena in vacuum, in gases or in solids.

electronic charge (*Phys*) The magnitude of the charge on the electron and the unit in which nuclear charges are expressed, equal to 1.602×10^{-19} C; that on the electron is negative.

electronic configuration (*Chem*) The descriptions of the electrons of an atom or a molecule in terms of orbitals.

electronic control (*ElecEng*) A method of control which is based on the use of electronic circuits, suitable transducers and actuators where necessary. Modern examples often incorporate programmable units for both accuracy and flexibility. Applications are widespread, eg other circuits, instruments, machinery, etc.

electronic countermeasures (*ICT*) An offensive or defensive tactic using electronic systems and reflectors to impair the effectiveness of enemy guidance, surveillance or navigational equipment that depend on electromagnetic signals. Also *electronic warfare*, *EW*. Abbrev *ECM*.

electronic engineering (*Electronics*) See ELECTRONIC.

electronic engraving (*Print*) The making of plates direct from the copy without the use of the camera or the etching bath. As the copy is scanned the plates are engraved. Some models produce sets of colour plates, applying the necessary COLOUR CORRECTION electronically.

electronic flash (*Electronics, ImageTech*) A battery or mains device which charges a capacitor, the latter discharging through a tube containing neon or xenon when triggered and producing a very brief source of illumination. Used for photography and stroboscopy.

electronic funds transfer (*ICT*) The use of computer systems to transfer credits and debits between co-operating organizations, such as banks or large companies. Abbrev *EFT*. See EFTPOS.

electronic ignition (*Autos*) Generic term for various systems which employ electronic circuitry, rather than the traditional coil and circuit breaker, to provide a timed high-voltage pulse to the sparking plugs. Cf COIL IGNITION.

electronic intelligence (*ICT*) Using airborne equipment, ground stations and surveillance satellites to monitor and record enemy electromagnetic emanations and to reveal the nature and deployment of guidance, navigational and communication systems.

electronic keying (*ICT*) The production of telegraphic signals by an all-electronic system.

electronic mail (*ICT*) The sending of text messages via computer systems. Messages are usually stored centrally until they are accessed by the recipient. Abbrev *E-mail*, *email*.

electronic microphone (*Acous*) A microphone in which the acoustic pressure is applied to an electrode of a valve.

electronic music (*Acous*) See ELECTROSONIC MUSIC.

electronic news gathering (*ICT*) The production and delivery of TV images, usually news, to a central point by wholly electronic means rather than by physically trans-porting photographic film or videotape.

electronic oscillations (*ICT*) Those generated by electrons moving between electrodes in an amplifying device, the frequency being determined by the transit time.

electronic pickup (*Electronics*) One in which external vibration affects the grid of a valve and thereby modulates the anode current. See MICROPHONIC.

electronic purse (*ICT*) A portable device, such as a swipecard, that can be loaded with monetary value for future redemption against currency, goods or services, eg from a vending machine or ATM.

electronic rectifier (*ElecEng*) See RECTIFIER.

electronic register control (*Print*) Equipment to maintain lateral and lengthwise register on web-fed printing and paper-converting machines.

electronic retouching (*ImageTech*) Digitizing a photo-graphic image and altering it in a computer.

electronics (*Genrl*) The science and technology of the conduction of electricity in a vacuum, a gas or a semi-conductor, and the devices based on this.

electronic shutter (*ImageTech*) (1) A mechanical shutter in which the rate of opening and closing is controlled electronically. (2) An electronic shutter using a polarizing device or LIQUID CRYSTAL DISPLAY (panel) to control the amount of light reaching the focal plane. (3) A video FAST SHUTTER.

electronic still camera (*ImageTech*) Any electronic camera that records still images, digital or video, stored internally or output to an external device. Abbrev *ESC*. See DIGITAL CAMERA, STILL VIDEO BACK, STILL VIDEO CAMERA.

electronic switch (*ElecEng*) A device for opening or closing an electric circuit by electronic means.

electronic theory of valency (*Chem*) Valency forces arise from the transfer or sharing of the electrons in the outer shells of the atoms in a molecule. The two extremes are complete transfer (IONIC BOND) and close sharing (COVALENT BOND), but there are intermediate degrees of bond strength and distance.

electronic traction control (*Autos*) A system which adjusts the power provided to the driven wheels to prevent wheelspin by sensing excessive acceleration of the wheels.

electronic tuning (*Electronics*) (1) Changing the operating frequency of an amplifier, oscillator, receiver, or other device, by altering a control voltage or signal rather than by mechanically changing any physical properties of tuned circuits. (2) Changing the operating frequency of a high-frequency valve, which depends on an electron beam for its operation, by altering the characteristics of the beam.

electronic viewfinder (*ImageTech*) A video camera viewfinder, normally containing a black-and-white CATHODE-RAY TUBE, which displays an image from the PICKUP DEVICE or off-tape from the CONFIDENCE HEADS. Abbrev *EVF*. See COLOUR TRUEFINDER, COLOUR VIEWFINDER, LCD VIEWFINDER.

electronic voltmeter (*ElecEng*) One whose operation depends on the detection, measurement and amplifying properties of electronics. Normally the display is digital. See DIGITAL METER.

electronic white board (*ICT*) Communication device for TELEPRESENCE whereby anything written or drawn on a wall-mounted board is instantly reproduced on similar boards in remote offices.

electron lens (*Electronics*) A composite arrangement of magnetic coil and charged electrodes, to focus or divert electron beams in the manner of an optical lens.

electron mass (*Phys*) A result of relativity theory, that mass can be ascribed to kinetic energy, is that the effective mass (m) of the electron should vary with its velocity according to the experimentally confirmed expression:

$$m = \frac{m_0}{\sqrt{1 - (v/c)^2}}$$

where m_0 is the mass for small velocities, c is the velocity of light, and v that of the electron.

electron micrograph (*BioSci*) A photomicrograph of the image of an object, taken by substituting a photographic plate for the fluorescent viewing screen of an electron microscope. See panel on ELECTRON MICROSCOPE.

electron microprobe (*BioSci*) Use of ELECTRON PROBE ANALYSIS in the electron microscope. Also *electron-beam analysis*.

electron microscope (*BioSci*) A tube in which electrons emitted from the cathode are focused, by suitable magnetic

Electron microscope

Abbrev *EM*. Any of a number of devices, consisting essentially of an evacuated tube in which a beam of electrons is caused to interact with a specimen. The electron beam is made by accelerating electrons through a potential difference of from 1 to 1500 kV between an *electron gun* and an anode and focused by electrical or magnetic fields produced by *electron lenses*.

An image is formed either directly, by focusing the electrons that pass through the specimen onto a fluorescent screen or photographic emulsion, or indirectly, by using the information carried by eg the X-rays or secondary electrons which are emitted during the interaction of the primary electrons with the specimen. A number of electron microscopes have been developed for specific purposes.

Transmission electron microscope

Abbrev *TEM*. A form of electron microscope in which the specimen is usually either a thin (570 nm) section of fixed, embedded material, often stained with heavy metals, or virus particles or macromolecules, often negatively stained or shadowed with heavy metals and supported on a thin film. The specimen is evenly illuminated by a broad beam of electrons at 40–100 kV and the image is formed directly by focusing those electrons which pass more or less unscattered through the specimen (the *transmitted electrons*) on a fluorescent screen for viewing and selecting the field, or on a photographic film for recording at high resolution (see Fig. 1). The screen image is darker where the specimen is denser. The main technical problems are in

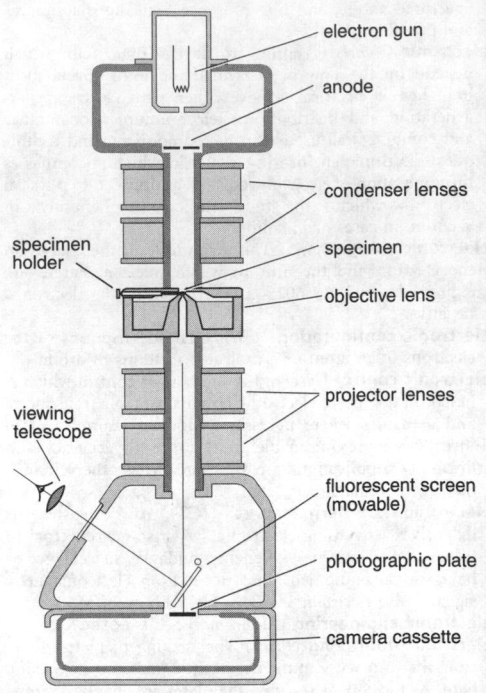

Fig. 1 **Schematic diagram of a transmission electron microscope** The 'lenses' are magnetic coils and only the objective has a representation of the pole pieces.

the design of the electron lenses so as to reduce aberrations.

The TEM has the merit of much finer resolution (0·3 nm or less) than the light microscope but is usually not suitable for living specimens for several reasons. Energy absorption by the specimen can also damage it, particularly at high magnifications, and high resolutions need very thin specimens and no water.

and electrostatic fields, and pass through a specimen; an image is formed on a fluorescent screen. See panel on ELECTRON MICROSCOPE.

electron mirror (*Electronics*) A 'reflecting' electrode in an electron tube, eg reflex klystron.

electron mobility (*Electronics*) See MOBILITY.

electron multiplier (*Electronics*) Electron tube in which the anode is replaced by a series of auxiliary electrodes, maintained at successively increasing positive potentials up to the final anode. Electrons emitted from the cathode impinge on the first of the auxiliary electrodes, from which secondary electrons are ejected and travel to the next electrode, where the process is repeated. With suitable materials for the auxiliary electrodes, the number of secondary electrons emitted at each stage is greater than the number of incident electrons, so that very high overall amplification of the original tube current results. See DYNODE, PHOTOMULTIPLIER.

electron octet (*Chem, Phys*) The (up to) eight valency electrons in an outer shell of an atom or molecule. Characterized by great stability, in so far as the complete shell round an atom makes it chemically inert, and round a molecule (by sharing) makes a stable chemical compound.

electron optics (*Phys*) Control of free electrons by curved electric and magnetic fields, leading to focusing and formation of images.

electron pair (*Chem*) Two valence electrons shared by adjacent nuclei, so forming a bond.

electron paramagnetic resonance (*Chem*) See ELECTRON SPIN RESONANCE.

electron–phonon scattering (*Phys*) An important process that contributes to the cause of electrical resistivity; the electrons are scattered by the thermal vibrations of the crystal lattice.

electron probe analysis (*Chem*) A beam of electrons is focused onto a point on the surface of the sample, the elements being detected both qualitatively and quantitatively by their resultant X-ray spectra. An accurate (1%) non-destructive method needing only small quantities (micrometre size) of sample. See ENERGY DISPERSIVE ANALYSIS OF X-RAYS.

electron radius (*Phys*) The classical theoretical value for the radius of the electron, $2·82 \times 10^{-15}$ m. Experimentally the effective value varies greatly with the interaction concerned.

electron runaway (*Electronics*) The condition in a plasma when the electric fields are sufficiently large for an electron

In passing through the specimen some electrons are scattered and lose amounts of energy characteristic of the atoms with which they have interacted. More sophisticated instruments can select electrons of a particular energy band and form images showing eg the distribution of a chemical element.

Some TEMs are equipped for X-ray analysis and in others, *high-voltage TEMs*, the electron beam is accelerated through 500 to 1500 kV; such beams can penetrate relatively thick, even hydrated and living, specimens.

Scanning electron microscope
This microscope (abbrev *SEM*) uses a very fine beam of electrons at 5 to 100 kV which is made to scan the specimen as a raster of parallel contiguous lines. The specimen is usually a solid object, not a section, and the number of secondary electrons emitted by the surface, or just below it, will depend on its topology and nature. These are collected, amplified and analysed before modulating the beam of a cathode-ray tube scanned in synchrony with the scanning beam (see Fig. 2).

The images resemble those seen through a hand lens but at a much higher resolution (say 5 nm). Useful magnifications up to 10^5 are possible and there is a much greater depth of focus than in the light microscope. These are the typical high-contrast images of insect eyes or cuticle.

Another type is the scanning transmission electron microscope (abbrev *STEM*). This uses the field emission from a very fine tungsten point as the source of electrons. The electrons transmitted through the specimen are either unscattered, or elastically or inelastically scattered. They are then collected, separated and analysed to produce an image on a screen. The STEM is widely used for producing detailed images of images just below the resolution of the light microscope. A refinement of the principle usable at

Electron microscope *(Cont.)*

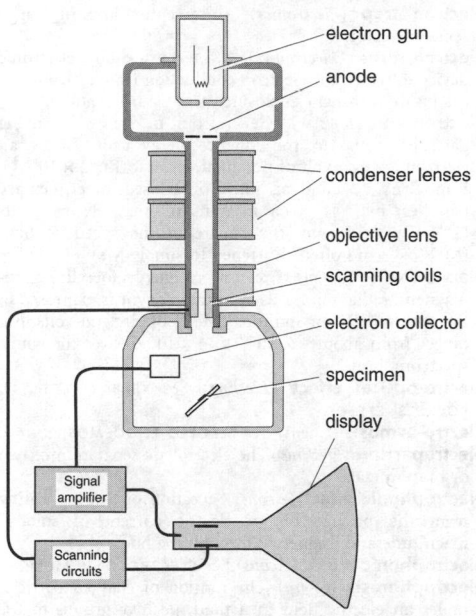

Fig. 2 **Schematic diagram of a scanning electron microscope**

much higher resolutions is the scanning tunnelling electron microscope.

Here the probe is an 'atomic microtip' floated, using SUPERCONDUCTING LEVITATION, over the surface being scanned. The tip–surface distance is of the order of atomic diameters so that the electron current obtained is through quantum mechanical tunnelling. A horizontal resolution of ~ 0.2 nm and a vertical resolution of ~ 0.01 nm can be achieved and in favourable instances individual atoms visualized.

to gain more energy from the field than it loses, on average, in a collision. See AVALANCHE EFFECT.

electron scanning (*ImageTech*) Scanning or establishing a TV image by an electron beam in a TV camera tube or a cathode-ray tube, normally using a rectangular raster, with horizontal lines.

electron sheath (*Electronics*) Electron space charge around an anode, when the supply of electrons is greater than demanded by the anode circuit. See ELECTRON CLOUD.

electron shell (*Phys*) A grouping of electrons surrounding the nucleus of an atom. The *Pauli exclusion principle* governs the way in which electrons can fill the available *orbitals*. For a given principal quantum number n, there are n allowed values of l, the orbital angular momentum quantum number; for each value of l, there are $(2l+1)$ allowed values of m_l, the magnetic angular momentum quantum number; for each value of m_l, there are two values of m_s, the magnetic spin number. This makes a total of $2n^2$ orbitals for a given value of n, and as the Pauli principle allows only one electron for each set of four quantum numbers n, l, m_l, m_s, this limited number of allowed orbitals makes up the electron shell for a given n. Starting with the innermost shell, the shells are called K, L, M, ..., Q corresponding to the principal quantum numbers n of 1, 2, 3, ..., 7. Inner filled shells are relatively inert and the chemical properties of the atom are determined by the electron arrangement in the outermost shell. See K-SHELL, L-SHELL, M-SHELL, ..., Q-SHELL and panel on ATOMIC STRUCTURE.

electron spectroscopy (*Chem*) See PHOTOELECTRON SPECTROSCOPY.

electron spin (*Electronics*) See SPIN.

electron spin resonance (*Phys*) A branch of microwave spectroscopy in which there is resonant absorption of radiation by a paramagnetic substance, possessing unpaired electrons, when the energy levels are split by the application of a strong magnetic field. The difference in energy levels is modified by the environment of the atoms. Information on impurity centres in crystals, the nature of the chemical bond and the effect of radiation damage can be found. Also *electron paramagnetic resonance*. Abbrev *ESR*.

electron transfer chain (*BioSci*) A series of chemical components, starting with glucose, each successively oxidized in AEROBIC RESPIRATION to release energy and synthesize ATP by the stepwise transfer of electrons to their ultimate acceptor, oxygen (reduced to water). The process involves three large multiprotein complexes (*flavoproteins*, *cytochromes* and other metaloproteins), with the smaller components of *ubiquinone* (also *coenzyme Q*) and *cytochrome c*. See panel on MITOCHONDRION.

electron trap (*Electronics*) An acceptor impurity in a semiconductor.

electron tube (*Electronics*) US term for any electronic device in which the electron conduction is in a vacuum or gas inside a gastight enclosure. Also *electron valve*.

electron-volt (*Phys*) (1) General unit of energy of moving particles, equal to the kinetic energy acquired by an electron losing 1 volt of potential, equal to 1.602×10^{-19} J. Symbol eV. (2) Unit of 'mass' often used for elementary particles; mass is given in units of mega-electron-volts (10^6 eV) divided by the square of the speed of light (MeV c^{-2}) and often shortened to simply MeV.

electron-volt (*Radiol*) The unit of energy for all electromagnetic radiation. As used in radiology it is expressed in thousands (keV) or millions (MeV) of electron-volts, but ranges from about 4×10^{-9} to 4×10^9 eV over the whole spectrum.

electro-optical effect (*ElecEng*) See KERR EFFECT (1), POCKELS' EFFECT.

electro-osmosis (*Chem*) See ELECTRO-ENDOSMOSIS.

electroparting (*ElecEng*) The electrolytic separation of two or more metals.

electrophonic effect (*Acous*) Sensation of hearing arising from the passage of an electric current of suitable magnitude and frequency through the body.

electrophonic music (*Acous*) See ELECTROSONIC MUSIC.

electrophoresis (*Chem*) The motion of charged particles under an electric field in a fluid, positive groups to the cathode and negative groups to the anode.

electrophorus (*ElecEng*) A simple electrostatic machine for repeatedly generating charges. A resinous plate, after rubbing, exhibits a positive charge, which displaces a charge through an insulated metal plate placed in partial contact. Earthing the upper surface of this plate leaves a net negative charge on the metal plate when it is removed, a process which can be repeated indefinitely.

electrophysiology (*BioSci*) The study of electrical phenomena associated with living organisms, particularly nervous conduction.

electroplaque (*BioSci*) Large, flat, disk-shaped ELECTROCYTE. Usually stacked in series to produce a substantial voltage pulse.

electroplating (*ElecEng*) The deposition of one metal on another by electrolytic action on passing a current through a cell, for decoration or for protection from corrosion etc. Metal is taken from the anode and deposited on the cathode, through a solution containing the metal as an ion.

electroplating bath (*ElecEng*) A tank in which objects to be electroplated are hung. It is filled with electrolyte at the correct temperature, with anodes of the metal to be deposited on articles which are made cathodes.

electroplating generator (*ElecEng*) A dc electric generator, specially designed for electroplating work; it gives a heavy current at a low voltage.

electropneumatic (*ElecEng*) Said of a control system using both electrical and pneumatic elements.

electropneumatic brake (*CivEng*) A type of brake which can be applied simultaneously throughout the length of the train.

electropneumatic contactor (*ElecEng*) A contactor operated by compressed air but controlled by electrically operated valves.

electropneumatic control (*ElecEng*) A form of remote control in which switches or other apparatus are operated by compressed air controlled by electrically operated valves; commonly used on electric trains.

electropneumatic signalling (*ElecEng*) A signalling system operated by compressed air, the valves which control the latter being operated electrically.

electropolishing (*Eng*) See ELECTROLYTIC POLISHING.

electroporation (*BioSci*) A method of introducing foreign DNA or chromosomes into cells by subjecting them to a brief voltage pulse, which transiently increases membrane permeability, allowing uptake of DNA or chromosomes from the surrounding buffer.

electropositive (*Phys*) Carrying a positive charge of electricity. Tending to form positive ions, ie having a relatively negative electrode potential.

electroradiescence (*Phys*) The emission of ultraviolet or infrared radiation from dielectric PHOSPHORS on the application of an electric field.

electroreceptor (*BioSci*) Sense organ specialized for the detection of electric discharges. Found in a variety of fish, particularly mormyrids, gymnotids and some elasmobranchs.

electrorefining (*Eng*) See ELECTROLYTIC REFINING.

electroremediation (*EnvSci*) A method for removing environmental pollutants using electrodes placed in the contaminated site to accumulate heavy metals etc.

electroscope (*ElecEng*) Indicator and measurer of small electric charges, usually two gold leaves which diverge because of repulsion of like charges; with one gold leaf and a rigid brass plate, indication is more precise.

electroscopic powder (*ElecEng*) A mixture of finely divided materials which can acquire charges by rubbing together, so that, if dusted onto a plate, the different materials adhere to differently charged portions of the plate, forming a figure.

electrosmosis (*Chem*) See ELECTRO-ENDOSMOSIS.

electrosonic music (*Acous*) Music or other sounds produced by electronic means (eg by oscillators, photocells, generators or microprocessors), then combined electrically and reproduced through loudspeakers. Also *electronic music, electrophonic music*.

electrospray mass spectroscopy (*Chem*) A method of mass spectroscopy in which the sample is introduced as a fine spray from a highly charged needle so that each droplet has a strong charge. Solvent rapidly evaporates from the droplets leaving the free macromolecule.

electrostatic accelerator (*ElecEng*) One which depends on the electrostatic field due to large dc potentials. Used eg for particle accelerators in high-energy physics.

electrostatic actuator (*Acous*) Apparatus used for absolute calibration of microphone through application of known electrostatic force.

electrostatic adhesion (*ElecEng*) That between two substances, or surfaces, due to electrostatic attraction between opposite charges.

electrostatic application (*Agri*) A technique that applies an electrostatic charge to spray droplets to ensure more effective coverage and retention on the target surfaces.

electrostatic bonding (*Chem*) See ELECTROVALENCE.

electrostatic charge (*ElecEng*) Electric charge at rest on the surface of an insulator or insulated body, and consequently leading to the establishment of an adjacent electrostatic field system.

electrostatic component (*ElecEng*) See ELECTRIC COMPONENT.

electrostatic coupling (*Electronics*) That between circuit components, one applying a signal to the next through a capacitor.

electrostatic deflection (*Electronics*) Deflection of the beam of a cathode-ray tube by an electrostatic field produced by two plates between which the beam passes on its way to the fluorescent screen.

electrostatic field (*ElecEng*) Electric field associated with stationary electric charges.

electrostatic flocking (*Textiles*) The application of a coloured flock directed by an electrostatic field onto a fabric pretreated with an adhesive. The fibres of the flock protrude from the surface of the fabric giving it a characteristically prickly feel. The products are often used as wall-hangings.

electrostatic focusing (*Electronics*) Focusing in high-vacuum cathode-ray tubes by the electrostatic field produced by two or more electrodes maintained at suitable potentials.

electrostatic generator (*ElecEng*) One operating by electrostatic induction, eg WIMSHURST MACHINE and VAN DE GRAAFF GENERATOR. Also *frictional machine, induction machine, influence machine, static machine.*

electrostatic induction (*ElecEng*) Movement and manifestation of charges in a conducting body by the proximity of charges in another body. Also the separation of charges in a dielectric by an electric field.

electrostatic instrument (*ElecEng*) A measuring instrument depending on electrostatic forces set up between charged bodies. See ELECTROMETER, ELECTROSCOPE.

electrostatic Kerr effect (*ElecEng*) See KERR EFFECTS (2).

electrostatic lens (*Electronics*) An arrangement of tubes and diaphragms at different electric potentials.

electrostatic loudspeaker (*Acous*) The inverse of an ELECTROSTATIC MICROPHONE.

electrostatic machine (*ElecEng*) See ELECTROSTATIC GENERATOR.

electrostatic memory (*Electronics*) A device in which the information is stored as electrostatic energy, eg storage tube. Also *electrostatic storage.*

electrostatic microphone (*Acous*) Microphone with a stretched or slack foil diaphragm, which is polarized at steady potential, or at a high-frequency voltage, both of which are modulated by variations in capacitance due to varying pressure.

electrostatic precipitation (*ElecEng*) Use of an electrostatic field to precipitate solid (or liquid) particles in a gas, eg dust removal in power stations.

electrostatic precipitator (*Eng*) A device for negatively charging coal particles or ash in flue gas which then precipitate onto earthed plates. Highly efficient method for removing most forms of small particles in power stations.

electrostatic printing (*Print*) See XEROGRAPHY.

electrostatics (*Genrl*) The section of the science of electricity which deals with the phenomena of electric charges substantially at rest.

electrostatic separator (*ElecEng*) Apparatus in which materials having different permittivities are deflected by different amounts when falling between charged electrodes, and therefore fall into different receptacles.

electrostatic shield (*ElecEng, ICT*) Conducting shield surrounding instruments or other apparatus to protect them from the effects of external electric fields, or between two circuits to prevent unwanted capacitance coupling.

electrostatic spraying (*Build*) Spray application in which paint particles are charged by spraying through a high-voltage electric field while the article to be painted is earthed, attracting the paint particles. Overspray is eliminated since paint is attracted to the sides of the article not facing the gun.

electrostatic storage (*Electronics*) See ELECTROSTATIC MEMORY.

electrostatic units (*Phys*) Units for electric and magnetic measurements in which the PERMITTIVITY of free space is taken as unity, with no dimensions in the CGS system.

electrostatic voltmeter (*ElecEng*) One indicating the attraction or repulsion between charged bodies. Usual unit of calibration is *kilovolt.* See ELECTROMETER.

electrostatic wattmeter (*ElecEng*) One which utilizes electrostatic forces to measure ac power at high voltages.

electrostriction (*Phys*) The change in the dimensions of a dielectric accompanying the application of an electric field.

electrosyntonic switch (*ElecEng*) A switch remotely controlled by means of a high-frequency current superimposed on the main circuit.

electrotaxis (*BioSci*) See GALVANOTAXIS.

electrotellurograph (*ElecEng*) Apparatus for the study of earth currents.

electrotherapy (*Med*) The treatment of diseases involving the use of electricity. Also *electrotherapeutics.*

electrothermal instrument (*ElecEng*) One depending on Joule heating of a current for its operation. Also *electrothermic instrument.* See BOLOMETER.

electrothermoluminescence (*Phys*) Changes in electroluminescent radiation resulting from changes of dielectric temperature. (Some dielectrics show a series of maxima and minima when heated.) The complementary arrangement of observing changes in thermoluminescent radiation when an electric field is applied is termed THERMOELECTROLUMINESCENCE.

electrotint (*Print*) A printing block produced by drawing with varnish on a metal plate, and depositing metal electrically on the parts not covered by the varnish.

electrotonus (*Med*) The state of a nerve which is being subjected to a steady discharge of electricity.

electrotropism (*BioSci*) See GALVANOTAXIS.

electrotype (*Print*) A hard-wearing printing plate made by depositing a film of copper electrolytically on a mould taken from type or an original plate. The copper shell is backed with a lead alloy. Abbrev *electro.*

electrovalence (*Chem*) Chemical bond in which an electron is transferred from one atom to another, the resulting ions being held together by electrostatic attraction.

electroviscosity (*Chem*) Minor change of viscosity when an electric field is applied to certain polar liquids.

electroweak theory (*Phys*) The WEINBERG–SALAM THEORY unifying electromagnetic and weak interactions between particles. See LEPTON–QUARK SYMMETRY.

electrowinning (*Eng*) See ELECTROEXTRACTION.

electrum (*Eng*) (1) An alloy of gold and silver (55–88% of gold) used for jewellery and ornaments. (2) Nickel–silver (copper 52%, nickel 26% and zinc 22%); it has the same uses as other nickel–silvers.

electuary (*Pharmacol*) A medicine consisting of the medicinal agent mixed with honey, syrup or jam.

Elektron alloys (*Eng*) TN for magnesium-based light alloys with up to 4·5% copper, up to 12% aluminium and perhaps some manganese and zinc.

element (*Chem*) A simple substance which cannot be resolved into simpler substances by normal chemical means. Because of the existence of ISOTOPES of elements, an element cannot be regarded as a substance which has identical atoms, but as one which has atoms of the same ATOMIC NUMBER.

element (*ElecEng*) A term often used to denote the resistance wire and former of a resistance type of electric heater.

element (*Electronics*) Unit of an assembly, esp the detailed parts of electron tubes which affect operation or performance.

element (*ImageTech, Phys*) The component of a lens, eg six-element lens.

element (*MathSci*) (1) A basic object of a particular mathematical system, such as a number in arithmetic. (2) A member of a SET. (3) A CONSTITUENT of an ARRAY.

elemental analysis (*Chem*) Quantitative analysis of a substance to determine the relative amounts of the elements that make it up.

elementary bodies (*Med*) Particles present in cells of the body in virus infections.

elementary colours (*Phys*) See PRIMARY COLOURS.

elementary matrix (*MathSci*) One which differs from the identity matrix by a single ELEMENTARY MATRIX OPERATION.

elementary matrix operation (*MathSci*) A matrix operation which consists of: (1) interchanging two rows; (2) multiplying a row by a non-zero constant; or (3) adding a multiple of one row to another row. Similar operations may be defined for the columns.

elementary particle (*Phys*) A particle believed to be incapable of subdivision; the term FUNDAMENTAL PARTICLE is now more generally used.

element-former (*ElecEng*) A refractory substance upon which the heated wire of a resistance type of electric heater is wound. Also *element-carrier.*

elements of an orbit (*Astron*) The six data mathematically necessary to determine completely a planet's orbit and its position in it: (1) longitude of the ascending NODE; (2) inclination of the orbit; (3) longitude of perihelion; (4) semi-axis major; (5) eccentricity; (6) epoch, or date of planet's passing perihelion. Analogous elements are used in satellite and double-star orbits.

elephant (*Paper*) UK size of paper (23 × 28 in). Cf ISO sizes.

elephantiasis (*Med*) Enlargement of the limbs, or of the scrotum, from thickening of the skin and stasis of lymph; due to obstruction of lymphatic channels, esp by filarial worms.

Elephantide pressboard (*ElecEng*) TN for pressboard with a large cotton content, used for insulation of transformers, armature and stator coils. It is especially suitable for use under oil.

eleutherodactyl (*BioSci*) Having the hind-toe free, as in some passerine birds and some tropical frogs.

elevated duct (*EnvSci*) Tropospheric radio duct which has both upper and lower effective boundaries elevated.

elevation (*Arch, Build*) The view or representation of any given side of a building.

elevation (*Eng*) A view (eg side or end elevation) of a component or assembly drawn in projection on a vertical plane.

elevation (*Surv*) US for REDUCED LEVEL.

elevation of boiling point (*Chem*) The raising of the boiling point of a liquid by substances in solution. May be used to determine molecular weights of solutes.

elevator (*Aero*) An aerodynamic surface, operated by fore-and-aft movement of the pilot's control column, governing motion in pitch. See fig. at CONTROL.

elevator (*BioSci*) A muscle that, by its contraction, raises a part of the body. Cf DEPRESSOR.

elevator (*Eng*) (1) A type of CONVEYOR for raising or lowering material which is temporarily carried in buckets or fingers attached to an endless chain or belt. (2) US for LIFT.

elevator chain (*Eng*) A chain used to carry a series of buckets or slats and to which the elevator drive is applied. Cast chain is used for slow to medium speeds, precision (roller) chain for higher speeds.

elevons (*Aero*) Hinged control surfaces on the wing trailing edge of tailless or delta aircraft which are moved in unison to act as elevators and differentially as ailerons.

elfin forest (*BioSci*) See KRUMMHOLZ.

ELFO (*MinExt*) Abbrev for EXTRA LIGHT FUEL OIL.

elimination (*Chem*) Removal of a simple molecule (eg of water, ammonia, etc) from two or more molecules, or from different parts of the same molecule. See CONDENSATION (2).

elimination filter (*ICT*) See BAND-STOP FILTER.

ELINT (*ICT*) Abbrev for ELECTRONIC INTELLIGENCE.

Elinvar (*Eng*) A nickel–chromium steel alloy with variable proportions of manganese and tungsten. Used for watch hairsprings because of its constant elasticity at different temperatures.

ELISA (*BioSci*) Abbrev for ENZYME-LINKED IMMUNOSORBENT ASSAY.

elixir (*Pharmacol*) A strong extract or tincture.

ell (*Build*) A short L-shaped connecting pipe.

Ellingham diagram (*Chem*) The graph of standard free energy of formation of esp metallic oxides plotted against temperature. Enables prediction of the stability of oxides and the ease of reduction to metal using carbon, hydrogen or light metals such as aluminium in the ALUMINOTHERMIC PROCESS.

ellipse (*MathSci*) A plane figure that is the locus of a point that moves so that its distance from two given points is constant. It is the intersection of a right cone with a plane placed obliquely to its base. See fig. at CONIC.

ellipsoid (*MathSci*) See QUADRIC.

elliptical arch (*Build*) The arch formed to an elliptical curve, or sometimes to a curve which is not a true ellipse

but a combination of circular arcs from three or five centres.

elliptical galaxy (*Astron*) A common type of galaxy of symmetrical form but having no spiral arms; the nearer elliptical galaxies have been resolved into stars, but contain no dust or gas. See panel on GALAXY.

elliptical orbit (*Space*) The orbit of a space vehicle about a primary body in the shape of an ellipse. The primary centre of mass is one of the foci and the nearest and farthest points from it are the *pericentre* and *apocentre* respectively.

elliptical point on a surface (*MathSci*) One at which the curvatures of all normal sections are of the same sign. Cf HYPERBOLIC, PARABOLIC, UMBILICAL POINT ON A SURFACE.

elliptical polarization (*Phys*) A type of polarization produced by two mutually perpendicular plane-polarized components which are not in phase. The vector representing the wave varies as the radius of an ellipse while the vector rotates about a point. See CIRCULAR POLARIZATION.

elliptic functions (*MathSci*) In general, $f(z)$ is an elliptic function if it is doubly periodic, ie if there exist two numbers w_1 and w_2, $w_1 \neq w_2$, such that $f(z) = f(z + w_1) = f(z + w_2)$. The simplest elliptic functions are the Jacobian sn, cn and dn functions of u which are defined as follows:

$$\operatorname{sn} u = x$$
$$\operatorname{cn} u = \sqrt{1 - x^2}$$
$$\operatorname{dn} u = \sqrt{1 - k^2 x^2}$$

where u is the elliptic integral of the first class. If x is replaced by $\sin \varphi$, the following equations can be written from which the basic properties of the Jacobian sn, cn and dn functions follow:

$$\operatorname{sn} u = \sin \varphi$$
$$\operatorname{cn} u = \cos \varphi$$
$$\operatorname{dn} u = \sqrt{1 - k^2 \sin^2 \varphi} = \delta \varphi \text{ (say)}$$
$$\varphi = \operatorname{am} u \text{ (say)}$$

elliptic geometry (*MathSci*) See ABSOLUTE.

elliptic integral (*MathSci*) All integrals of the type $\int \sqrt{P} dx$, where P is a fourth-degree polynomial, can be expressed in one of the following forms called respectively elliptic integrals of the first, second and third class:

$$F(k, x) = \int_0^x \frac{dt}{\sqrt{(1 - t^2)(1 - k^2 t^2)}}$$

$$E(k, x) = \int_0^x \sqrt{\frac{(1 - k^2 t^2)}{(1 - t^2)}}\, dt$$

$$\Pi(n, k, x) = \int_0^x \frac{dt}{(1 + nt^2)\sqrt{(1 - t^2)(1 - k^2 t^2)}}$$

k is called the modulus and may be taken as $0 < k < 1$. Cf ELLIPTIC FUNCTIONS.

ellipticity (*Phys*) See AXIAL RATIO.

elliptic trammel (*Eng*) An instrument for drawing ellipses, consisting of a straight arm having a pencil point at one end and two adjustable studs (all three of which project at right angles to the arm), and a frame with two grooves crossing one another at right angles. If a stud is placed in each of the grooves, then, as the arm is rotated, the pencil point describes an ellipse.

elm (*For*) Trees of the genus *Ulmus*, native to N Europe (*U. hollandica*) and N America (*U. americana*). The wood is coarse textured and light to reddish brown, and is noted for its durability under water. The trees are susceptible to Dutch elm disease, caused by a fungus *Ceratocystis ulmi* that, spread by bark beetles carrying infected spores, defoliates and kills the tree within weeks. Also *American white elm, Dutch elm, English elm, smooth-leaved elm.*

Elmendorf tear tester (*Paper*) A paper test apparatus utilizing the acceleration of a falling quadrant-shaped pendulum, mounted on a frictionless bearing, to measure the force necessary to continue a tear, already made in the test piece, through a given length.

El Niño (*EnvSci*) El Niño – The Child – is the name originally given locally to a weak warm ocean current flowing south along the coast of Ecuador at Christmas time. The El Niño southern oscillation is the term now applied to a more intense, extensive and prolonged warming of the eastern tropical Pacific occurring every few years which is associated with major meteorological anomalies. Extreme cases have serious effects on fisheries, bird life and mainland weather. See SOUTHERN OSCILLATION.

elongated (*Print*) A narrow form of type, often used in display work. It is commonly known as CONDENSED.

elongation (*Astron*) The angular distance between the Moon or planets and the Sun. The planets Mercury and Venus have maximum elongations of around 28° and 48° respectively.

elongation (*Eng*) The percentage extension produced in a TENSILE TEST.

elongation factor (*BioSci*) An enzyme associated with ribosomes that adds an incoming amino acid onto the existing peptide.

elongation ratio (*PowderTech*) (1) The ratio of the length of a powder particle to its breadth. See ASPECT RATIO. (2) See EXTENSION RATIO.

Elrod (*Print*) A machine for producing lead-alloy strip material, spacing and rule, cut to size or in long lengths.

ELT (*ICT*) Abbrev for *extract, load & transform.* In data warehousing, the process of extracting data from a source database, transforming its structure (generally by optimizing it for query) and then loading it in its transformed state into the warehouse schema.

elution (*MinExt*) Washing of loaded ion-exchange resins to remove captured uranium ions or other seized elements using washing liquor. This liquor is the *eluent* and the enriched solution it becomes is the *eluate.* The resin (or ZEOLITE) is regenerated (like rinsing water softener with brine).

elutriation (*ChemEng*) The process for separating into sized fractions finely divided particles in accordance with their rate of gravitation relative to a rising stream of fluid. Used for biological molecules in a centrifugal *elutriator.*

elutriator, centrifugal (*PowderTech*) A device for fractionating powder particles into sized fractions by means of a suspension undergoing centrifugal motion.

eluvial gravels (*Geol*) Those gravels formed by the disintegration *in situ* of the rocks which contributed to their formation. Also *eluvium.* Cf ALLUVIUM.

ELV (*Space*) Abbrev for EXPENDABLE LAUNCH VEHICLE.

elvan (*Geol*) A term applied by Cornish miners to the dyke rocks associated with the Armorican granites of that county. Elvans are actually quartz-porphyries, microgranites, and other medium- to fine-textured dyke rocks of granitic composition.

elytra (*BioSci*) In COLEOPTERA, the hardened, chitinized fore-wings that form horny sheaths to protect the hind-wings when the latter are not in use; in certain POLYCHAETA, plate-like modifications of the dorsal CIRRI, possibly for respiration. Adjs *elytriform, elytroid.*

em (*Print*) The square of the body of any size of type; the 12-point em is the unit of measurement for spacing material and the dimensions of pages. Six ems of 12 points = approx 1 in (25 mm). See EM QUAD.

e/m (*Phys*) See SPECIFIC CHARGE.

emaciation (*Med*) Extreme wasting.

emagram (*EnvSci*) An aerological diagram, the name being derived from *energy per unit mass diagram.* The axes, rectangular or oblique, are temperature and log(pressure).

E-mail (*ICT*) Also *email.* Abbrev for ELECTRONIC MAIL.

emanations (*Chem*) Heavy isotopic inert gases resulting from decay of natural radioactive elements. They are radioisotopes 222, 220, 219 of element 86 or RADON; these gases are short-lived and decay to other radioactive elements.

emanometer (*Radiol*) Meter for measuring RADON.

emarginate (*BioSci*) Notched; esp of a petal, with a small indentation at the apex.

emasculation (*BioSci*) (1) In plant breeding, the surgical removal of the stamens from a flower, usually before pollen is shed, to prevent self-pollination when cross-pollination is planned. (2) In animals, removal of the testes or of the testes and penis.

emasculator (*Vet*) An instrument for castrating horses and bulls by crushing the spermatic cord.

embankment wall (*CivEng*) A retaining wall from the top of which the supported earth normally rises at a slope.

embattlemented (*Arch*) A term applied to a building feature (such as a parapet) which is indented along the top like a battlement.

Embden–Meyerhof pathway (*BioSci*) The main pathway for anaerobic degradation of carbohydrate (glycolysis).

embed (*ICT*) To insert information (an object) that was created in one document into another document. The EMBEDDED OBJECT may be edited directly from within the destination document.

embedded column (*Arch*) A column which is partly built into the face of a wall.

embedded object (*ICT*) Information created in one document and inserted into another document. See EMBED.

embedded temperature detector (*ElecEng*) A resistance thermometer or thermocouple built into a machine or other piece of equipment during its construction, in order to be able to ascertain the temperature of a part which is inaccessible under working conditions.

embedding (*BioSci*) The technique of embedding biological specimens in a supporting medium, such as paraffin wax or plastics like epoxy resin, in preparation for sectioning with a *microtome.*

embellishment (*Arch*) Ornamentation applied to building features.

embolectomy (*Med*) Removal of an embolus.

embolic gastrulation (*BioSci*) Gastrulation by invagination.

embolism (*Med, BioSci*) (1) The blocking of a blood vessel usually by a blood clot or thrombus from a remote part of the circulation. (2) In plants, the blockage of a xylem conduit by a bubble of air as a result of damage or following cavitation. See TYLOSIS.

embolus (*Med*) A clot or mass formed in one part of the circulation and impacted in another, to which it is carried by the blood stream.

emboly (*BioSci*) Invagination; the condition of pushing in or growing in. Adj *embolic.*

embossed paper (*Paper*) Paper, to the surface of which a pattern has been imparted by passing the sheet or web through the nip of suitable rolls in an embossing CALENDER. The upper roll is of steel engraved with the appropriate design and the other, backing, roll is of compressible fibrous material.

embossing (*Print*) Producing a raised design on paper or board by the use of a die; if the design is unprinted it is called *blind embossing.*

embrasure (*Arch*) The splayed reveal of a window opening.

embrittlement (*Eng*) The reduction in TOUGHNESS developing after heat treatment or over a period of service. Some metals and plastics exhibit reduced impact toughness at subzero temperatures or may degrade at ordinary temperatures in ways which reduce their ability to absorb energy when stressed to the point of fracture.

embrocation (*Med*) The action of applying or rubbing a medicated liquid into an injured part; the liquid so used.

embroidery (*Textiles*) (1) Lace work consisting of a ground of net on which an ornamental design has been stitched. (2) Ornamental work done by needle or machine on a cloth, canvas, or other ground.

embryo (*BioSci*) (1) An immature organism in the early stages of its development, before it emerges from the egg or from the uterus of the mother. In humans the term is restricted to the stages between 2 and 8 weeks after conception. (2) A plant at an early stage of development, eg within a seed. Adj *embryonic*.

embryo culture (*BioSci*) The aseptic culture on a suitable medium of an embryo excised at an early stage. The technique is useful in plant breeding in cases where, as in some hybrids, the embryos abort if left in the ovule. Cf OVULE CULTURE.

embryogenesis (*BioSci*) (1) The processes leading to the formation of the embryo. (2) Also production of EMBRYOIDS in PLANT CELL CULTURE.

embryogeny (*BioSci*) See EMBRYOGENESIS.

embryoid (*BioSci*) In plants, an embryo-like structure that may subsequently grow into a plantlet. See ANTHER CULTURE, MICROPROPAGATION, PLANT CELL CULTURE, TOTIPOTENCY.

embryology (*BioSci*) The study of the formation and development of embryos.

embryoma (*Med*) A tumour formed of embryonic or fetal elements.

embryonic fission (*BioSci*) See POLYEMBRYONY.

embryonic stem cell (*BioSci*) A totipotent cell cultured from an early embryo that can be used to produce chimeric embryos and thus transgenic animals. Their potential for treatment of degenerative diseases is being actively researched.

embryonic tissue (*BioSci*) Meristematic tissue in plants.

embryophyte (*BioSci*) A member of those plant groups in which an embryo, dependent at least at first on the parent plant, is formed, ie the Bryophytes and vascular plants.

embryo sac (*BioSci*) The female gametophyte in angiosperms, formed within the ovule by the enlargement of the functional megaspore and containing usually eight nuclei.

embryotomy (*Med*) Removal of the viscera or of the head of a fetus, in obstructed labour.

emerald (*Min*) Brilliant green gemstone, a form of beryl; silicate of beryllium and aluminium, crystallizing in hexagonal prismatic forms, occurring in mica-schists, calcite veins and rarely in pegmatites. See GEMS AND GEMSTONES.

emerald copper (*Min*) See DIOPTASE.

emergence (*BioSci*) (1) In plants, an outgrowth that is derived from epidermal and cortical tissues and that does not contain vascular tissue or develop into a stem or leaf. (2) In animals, an epidermal or subepidermal outgrowth. (3) The appearance above ground of germinating seedlings, eg in pre-emergence, post-emergence. (4) In insects, the appearance of the imago from the cocoon, pupa-case or pupal integument.

emergence (*Geol*) The uplift of land relative to the sea.

emergency diesel supply (*NucEng*) To provide essential electrical power to a nuclear power plant in the event of loss of grid.

emergency shutdown (*NucEng*) Rapid shutdown of a reactor to forestall or remedy a dangerous situation.

emergency shutdown system (*NucEng*) A system of shutting down the reactor if other methods fail, eg injection of boron spheres which absorb neutrons strongly and quickly make the reactor subcritical. See SECONDARY SHUTDOWN SYSTEM.

emergency switch (*ElecEng*) A switch placed in a convenient position for cutting off the supply of electricity to a piece of apparatus or to a building, in case of emergency.

emergent (*Psych*) The description of novel unpredicted properties that appear in a complex system as a result of interactions. Consciousness is often considered to be an emergent property of brain neurophysiology.

emergent ray point (*Radiol*) See EXIT PORTAL.

emersed (*BioSci*) Raised above or rising out of a surface, esp growing up out of water.

emersion (*Astron*) The exit of the Moon, or other body, from the shadow which causes its eclipse.

Emerson enhancement effect (*BioSci*) The more than additive effect on the rate of photosynthesis (in plants and algae) of illuminating simultaneously with far-red light ($\lambda > 680$ nm) and light of shorter wavelength ($\lambda < 680$ nm), indicating the existence of the two PHOTOSYSTEMS, I and II.

emery (*Min*) An abrasive powder consisting of a mixture of CARBORUNDUM with either MAGNETITE or HAEMATITE, occurring naturally in Greece and localities in Asia Minor etc. Paper (*emery paper*), or more often cloth (*emery cloth*), can be surfaced with emery powder, held on by an adhesive solution, and is used for polishing and cleaning metal.

emery wheel (*Eng*) A GRINDING WHEEL in which the abrasive grain consists of emery powder, held by a suitable bonding material.

emesis (*Med*) The act of vomiting.

emetic (*Pharmacol*) Having the power to cause vomiting; a medicament which has this power.

emetine (*Chem*) $C_{29}H_{40}O_4N_2$. An alkaloid obtained from the roots of Brazilian ipecacuanha. It forms a white amorphous powder, mp $74°C$, soluble in ethanol, ether or trichloromethane, slightly in water. It is used in medicine as an emetic; its principal use, however, is in the form of emetine bismuthous iodine or emetine hydrochloride, remedies for amoebic dysentery.

emf, e.m.f. (*ElecEng*) Abbrevs for ELECTROMOTIVE FORCE.

EMI (*ICT*) Abbrev for *electromagnetic interference*, particularly from computers for which strict standards are now commonly set.

emigration (*BioSci*) A category of population dispersal covering one-way movement out of the population area. Cf IMMIGRATION, MIGRATION.

emissary (*BioSci*) Passing out, as certain veins in vertebrates that pass out through the cranial wall.

emission (*Phys*) The release of electrons from parent atoms on absorption of energy in excess of normal average. This can arise from: (1) *thermal* (thermionic) agitation, as in valves, Coolidge X-ray tubes, cathode-ray tubes; (2) *secondary* emission of electrons, which are ejected by impact of higher-energy primary electrons; (3) *photoelectric* release on absorption of quanta above a certain energy level; (4) *field* emission by actual stripping from parent atoms by a strong electric field.

emission current (*Electronics*) The total electron flow from an emitting source.

emission efficiency (*Electronics*) See CATHODE EFFICIENCY.

emission nebula (*Astron*) A region of hot ionized interstellar gas and dust which shines owing to excitation by a nearby hot star.

emission spectrum (*Phys*) Wavelength distribution of electromagnetic radiation emitted by self-luminous source.

emission tax (*EnvSci*) Fiscal tax, proportional to the quantity and toxicity of the pollutant emitted, levied to encourage environmentally sounder industrial practices.

emission tomography (*Radiol*) Transverse section reconstruction of the radionuclide distribution within the body, obtained by acquiring images or SLICES of the head or body. This may be done by using COINCIDENCE DETECTION (positrons) or single-photon detection from gamma-ray emitters. Also *body-section radiography, cross-section radiography*. See COMPUTER-AIDED TOMOGRAPHY.

emissive power (*Phys*) The energy radiated at all wavelengths per unit area per unit time from a surface. It depends on the nature of the surface and on its temperature. See EMISSIVITY, EMISSION.

emissivity (*Phys*) The ratio of emissive power of a surface at a given temperature to that of a black body at the same temperature and with the same surroundings. Values range from 1·0 for lampblack down to 0·02 for polished silver. See STEFAN–BOLTZMANN LAW.

emitter (*Electronics*) In a transistor, the region from which charge carriers, that are minority carriers in the base, are injected into the base.

emitter follower (*Electronics*) See COMMON-COLLECTOR CONNECTION.

emitter junction (*Electronics*) One biased in the low-resistance direction, so as to inject minority carriers into the base region.

emmetropia (*Med*) The normal condition of the refractive system of the eye, in which parallel rays of light come to a focus on the retina, the eyes being at rest. Adj *emmetropic*.

emotion (*Psych*) Rather loose term for short-term states of arousal and desire such as fear, pity, love, etc, generally taken to be more extreme than feelings. Affective disorders are ones in which there is an inappropriate expression or experience of emotion.

empathy (*Psych*) The ability to share the emotions of another person, to be able to imagine oneself in their place.

empennage (*Aero*) See TAIL UNIT.

emphasizer (*ICT*) An audio-frequency circuit that selects and amplifies specific frequencies or frequency bands.

emphysema (*Med*) (1) The presence of air in the connective tissues. (2) The formation in the lung of bullae or spaces containing air, as a result of destruction of alveoli and rupture of weakened alveolar walls.

emphysematous chest (*Med*) The barrel-shaped, immobile chest which is the result of chronic bronchitis and emphysema.

empirical formula (*Chem*) The formula deduced from the results of analysis which is merely the simplest expression of the ratio of the atoms in a substance. In molecular materials it may, or may not, show how many atoms of each element the molecule contains: eg methanal, CH_2O, ethanoic acid, $C_2H_4O_2$, and lactic acid, $C_3H_6O_3$, have the same percentage composition, and consequently, on analysis, they would all be found to have the same empirical formula.

empirical formula (*Genrl*) A relationship founded on experience or experimental data only, not deduced in form from purely theoretical considerations.

empiricism (*Genrl*) The regular scientific procedure whereby scientific laws are induced by inductive reasoning from relevant observations. Critical phenomena are deduced from such laws for experimental observation, as a check on the assumptions or hypotheses inherent in the theory correlating such laws. Scientific procedure, described by empiricism, is not complete without the experimental checking of deductions from theory.

empiricism (*Psych*) A philosophical position based on the view that all knowledge comes from experience and thus that experiment is more important than theory.

emplastrum (*Med*) A medicated plaster for external application.

emplectum (*Build*) An ancient form of masonry, showing a squared stone face, sometimes interrupted by courses of tiles at intervals.

empress (*Build*) A slate size, 26×16 in (660×406 mm).

emprosthotonos (*Med*) Bending of the body forwards caused by spasm of the abdominal muscles, as in tetanus.

empty band (*Phys*) See ENERGY BAND.

empty set (*MathSci*) The set with no elements denoted by $\varnothing$ or {}.

empyema (*Med*) Accumulation of pus in any cavity of the body, esp the pleural cavity.

em quad (*Print*) A square quadrat of any size of type. Less than type height, it is used for spacing. Usually called a 'mutton' to distinguish it clearly from an en (or nut quad).

em rule (*Print*) The dash (—). A thin horizontal line 1 em of the type body in width. Apart from its uses as a mark of punctuation, it is often used to build up rules in tabular work. Also *em score*, *metal rule*, *mutton rule*.

EMS (*ICT*) (1) Abbrev for *enhanced messaging (or message) service* for mobile phones. (2) Abbrev for EXPANDED MEMORY specification.

Emsian (*Geol*) A stage in the Lower Devonian. See PALAEOZOIC.

emulation (*ICT*) A mode in which a device may emulate the operational characteristics of another device, eg one printer may behave like another type of printer.

emulator (*ICT*) A program that causes one COMPUTER to behave as if were another type of computer; eg a computer may emulate an IBM-compatible one.

emulsification (*Print*) A fault found in lithographic printing whereby fine droplets of water become dispersed in the ink on the press rollers during printing.

emulsified coolant (*Eng*) These are used as cutting media in (metal) machining. The three main types are: (1) an emulsion of water, a thick soap solution and a mineral oil, (2) an emulsion of a mineral oil and soft soap or some other alkaline soap solution, and (3) an emulsion of a sulphurized or sulphonated oil neutralized and blended with a soluble oil.

emulsifier (*Chem*) Apparatus with a rotating, stirring or other device used for making emulsions.

emulsifying agent (*Chem*) A substance whose presence in small quantities stabilizes an emulsion, eg ammonium linoleate, certain benzene-sulphonic acids, etc.

emulsion (*Chem*) A colloidal suspension of one liquid in another, eg milk. See panel on GELS.

emulsion (*FoodSci*) A colloid in which the dispersed and continuous phases are two immiscible liquids. Emulsions can be oil in water, eg cream, or water in oil, eg butter. Since they tend to be unstable, an emulsifying agent may be required to stabilize the droplets in the disperse phase.

emulsion (*ImageTech*) A suspension of finely divided silver halide crystals in a medium such as gelatine which provides the light-sensitive coating on film, glass plates and paper and plastic supports. It may also contain sensitizing dyes and colour-forming COUPLERS.

emulsion paint (*Build*) A water-thinnable paint made from a pigmented emulsion or dispersion of a resin (generally synthetic) in water. The resin may be polyvinyl acetate, polyvinyl chloride, an acrylic resin or the like.

emulsion polymerization (*Chem*) See CHAIN POLYMERIZATION and panel on POLYMER SYNTHESIS.

emulsion technique (*Phys*) A method used to study subatomic particles by means of tracks formed in photographic emulsion.

emunctory (*Med*) Conveying waste matter from the body; any organ or canal which does this.

en (*Chem*) Abbrev for *1,2-diamino ethane, ethylene diamine*, $(CH_2NH_2)_2$, in complexes.

en (*Print*) A unit of measurement used in reckoning up composition. It is assumed that the average letter has the width of 1 en and that the average word, including the space following, has the width of 6 ens. See EN QUAD.

enabling pulse (*Electronics*) One which opens a GATE which is normally closed.

enalapril (*Pharmacol*) An *ACE inhibitor* used to treat hypertension and, together with diuretics, to treat heart failure.

enamel (*BioSci*) The hard external layer of a tooth consisting almost entirely of large elongated apatite (calcium phosphate) crystals set vertically on the surface of the underlying dentine; enamel also occurs in certain scales.

enamel (*ElecEng*) General term for hard surface coatings, eg oil-based paints containing resin and used for wire insulation and in building.

enamel (*Glass*) Surface coating of opaque glass fused onto metal articles for decoration or to provide a hard, inert and impermeable layer on eg cooking vessels. Also *vitreous enamel*. See FRIT.

enamel cell (*BioSci*) See AMELOBLAST.

enamel-insulated wire (*ElecEng*) Wire having an insulating covering of ENAMEL used for winding small magnet coils etc.

enamelled brick (*Build*) A brick having a glazed surface.

enamel paint (*Build*) A high-grade paint prepared by careful grinding of pigment in an oily medium containing a proportion of resin, and the usual lesser ingredients. It may be anything from glossy to flat.

enanthema (*Med*) An eruption on a mucous membrane. Also *enanthem*.

enantiomer (*Chem*) Either of a pair of STEREOISOMERS of a compound that has chirality.

enantiomerism (*Chem*) See ENANTIOMORPHISM.

enantiomorphism (*Chem*) Mirror-image isomerism. A classical example is that of the crystals of sodium ammonium tartrates which Pasteur showed to exist in mirror-image forms. Also *enantiomerism*. Adj *enantiomorphous*.

enargite (*Min*) Sulpharsenide of copper, often containing a little antimony.

enarthrosis (*BioSci*) A ball-and-socket joint.

enation (*BioSci*) (1) Generally an outgrowth, eg those on leaves caused by some viruses. (2) A non-vascularized, spine-like outgrowth from the axis of some primitive vascular plants, eg *Zosterophyllum*. See ENATION THEORY.

enation theory (*BioSci*) A theory that regards the MICRO-PHYLLS of the Lycopodiales etc as simple enations that have become vascularized and therefore different from MEGA-PHYLLS which are regarded as having evolved from branch systems.

encallow (*Build*) The material which is first removed from the surface of the site where clay for brick-making is to be obtained, and which is stored in a spoil bank for resurfacing later.

encapsulation (*Eng*) Provision of a tightly fitted envelope of material to protect eg a metal during treatment or use in an environment with which it would otherwise react in a detrimental manner. Similarly, the coating of eg an electronic component in a resin to protect it against the environment.

encapsulation (*ICT*) In object-oriented programming the approach of hiding data structures or logic (which may be subject to change) within a wrapper or layer, thus protecting the system of which it is a part from the effects of change within the encapsulated object.

encase (*Build*) To surround or enclose with linings or other material.

encastré (*Build*) A term applied to a beam when the end of it is fixed. Also *encastered*.

encaustic painting (*Build*) A process of painting in which hot wax is used as a medium.

encaustic tile (*Build*) An ornamental tile whose colours are produced by substances added to the clay before firing.

Enceladus (*Astron*) The second natural satellite of Saturn, discovered in 1789. Distance from the planet 238 000 km; diameter 500 km.

enceph-, encephalo- (*Genrl*) Prefixes from Gk *enkephalos*, brain.

encephalalgia (*Med*) Pain inside the head.

encephalitis (*Med*) Inflammation of the brain substance usually by a viral agent. Adj *encephalitic*.

encephalitis lethargica (*Med*) An acute inflammation of the brain caused by a virus; characterized by fever and disturbances of sleep, and followed by various persisting forms of nervous disorder (eg Parkinsonism), or by changes in character. Last UK outbreak in 1920s. Also *sleepy sickness*, *von Economo's disease*.

encephalitis periaxialis (*Med*) A disease characterized by progressive destruction of the nerve fibres composing the central white matter of the brain, causing blindness, deafness, paralysis and amentia. Also *Schilder's disease* and *encephalitis periaxialis diffusa* .

encephalitogen (*BioSci*) Substances present in extracts of brain which when administered with eg complete Freund's adjuvant, cause experimental allergic encephalitis. The active material is myelin basic protein.

encephalocele (*Med*) Hernial protrusion of brain substance through a defect in the skull.

encephalogram (*Radiol*) X-ray plate produced in ENCE-PHALOGRAPHY.

encephalography (*Radiol*) Radiography of the brain after its cavities and spaces have been filled with air, or dye, previously injected into the space round the spinal cord.

encephalomalacia (*Med*) Pathological softening of the brain.

encephalomyelitis (*Med*) Diffuse inflammation of the brain and the spinal cord.

encephalon (*BioSci*) The brain.

encephalopathy (*Med*) A generalized disease of the brain often associated with toxic poisoning, eg lead encephalopathy is the brain disorder caused by lead poisoning; hepatic encephalopathy is the confusional state associated with severe liver disease.

encephalospinal (*BioSci*) See CEREBROSPINAL.

enchondroma (*Med*) A tumour, often multiple, composed of cartilage and occurring in bones.

Encke's Comet (*Astron*) A well-known regular comet with period 3·3 yr.

Encke's Division (*Astron*) The principal division in the A ring of Saturn, around 200 km wide, containing very little orbiting material. Also *A-Ring Gap*.

enclitic (*Med*) Having the planes of the fetal head inclined to those of the maternal pelvis.

enclosed fuse (*ElecEng*) A fuse in which the fuse wire is enclosed in a tube or other covering.

enclosed self-cooled machine (*ElecEng*) An electric machine which is enclosed in such a way as to prevent the circulation of air between the inside and the outside and needing special provision for additional cooling.

enclosed–ventilated (*ElecEng*) Said of electrical apparatus which is protected from ordinary mechanical damage by an enclosure, with openings for ventilation.

encode (*ICT*) (1) Conversion of information, by means of a code, in such a way that it can be subsequently reconverted to its original form. (2) Conversion of one system of communication to another, as from amplitude to pulse-code modulation.

encoding (*Psych*) The processing and transformation of information to be stored in memory. See CHUNKING.

encoding altimeter (*Aero*) An ALTIMETER designed for automatic reporting of altitude to AIR-TRAFFIC CONTROL. A special encoding disk within the instrument rotates in response to movement of the aneroid capsules, and transmits a signal which is amplified, fed to the aircraft's transponder and thence automatically to ATC.

encopresis (*Med*) Involuntary defecation.

encounter group (*Psych*) A general term for a range of group therapies with the general aim of increasing personal awareness and encouraging creative and open relations with others. Procedures vary widely, though all tend to emphasize free and candid expression of feeling and thought within the group.

encrinal limestone (*Geol*) A crinoidal limestone. Also *encrinital limestone*.

encryption (*ICT*) Encoding data to make them incomprehensible to those without a decoder. See DECRYPTION.

encysted (*Med*) Enclosed in a cyst or a sac.

encystment (*BioSci*) (1) The formation by an organism of a protective capsule surrounding itself. (2) The formation of a walled non-motile body from a swimming spore. Also *encystation*. V *encyst*.

end (*MinExt*) Solid rock at end of underground passage.

end (*Textiles*) (1) In spinning, an individual strand. (2) In weaving, a warp thread. (3) In finishing, each passage of a fibre through a machine. (4) A length of finished fabric shorter than the normal PIECE.

endangered species (*BioSci*) A species of animal or plant that is in danger of extinction.

endarch (*BioSci*) A xylem strand having the protoxylem on the side nearer to the centre of the axis. Cf EXARCH, MESARCH.

end-around shift (*ICT*) One in which digits drop off one end of a word and return at the other. Also *circular shift*, *cyclic shift*, *ring shift*, *rotation shift*.

endarterectomy (*Med*) Surgical removal of material obstructing blood flow in an artery.

endarteritis (*Med*) Inflammation of the inner lining (*intima*) of an artery.

endarteritis obliterans (*Med*) Obliteration of the lumen of an artery, as a result of inflammatory thickening of the intima.

end-artery (*Med*) A terminal artery which does not anastomose with itself or with others.

end bell (*ElecEng*) A strong metal cover placed over the end-windings of the rotor of a high-speed machine, eg a turbo-alternator, to prevent their displacement by centrifugal forces.

end bracket (*ElecEng*) An open structure fitted at the end of an electrical machine, for the purpose of carrying a bearing.

end cell (*ElecEng*) See REGULATOR CELLS.

end connections (*ElecEng*) That part of an armature winding which does not lie in the slots, and which serves to join the ends of the active or slot portions of the conductors.

end correction (*Acous*) The ratio between two lengths in an open-ended resonating tube, where the pressure node is just outside the tube. The numerator is the distance between the tube end and the pressure node, and the denominator is the tube radius.

end-down (*Textiles*) Broken yarn twisting on the rotating spindles in spinning. A warp thread broken during weaving.

endellionite (*Min*) Another name for BOURNONITE.

endellite (*Min*) US for HALLOYSITE.

endemic (*BioSci, Med*) (1) A species or family confined to a particular region and thought to have originated there, eg *Primula scotica*, native only in the north of Scotland. (2) A disease permanently established in moderate or severe form in a defined area. Also *indigenous*.

enderbite (*Min*) A member of the charnockitic group of rocks, consisting of quartz, antiperthite, hypersthene and magnetite; the equivalent of a hypersthene tonalite.

endergonic (*BioSci*) An energy-requiring biochemical reaction which cannot therefore proceed spontaneously. Cf EXERGONIC.

end-feed magazine (*Eng*) One of several types of magazine used for the automatic loading of centreless grinding machines. It is suitable for short, conical components.

end-fire array (*Radar, ICT*) A linear array of radiators in which the maximum radiation is along the axis of the array; the antenna may be uni- or bi-directional. A YAGI array is an example, though in most end-fire arrays, each radiator is fed from a transmission line, the relative phase of each element being determined by its position along the line. Also *end-fire aerial array*.

end fixing (*Build*) A term used in referring to the condition of the ends of a column, whether fixed or only partially so. See EFFECTIVE LENGTH.

end gauge (*Eng*) A gauge consisting of a metal block or cylinder the ends of which are made parallel within very small limits, the distance between such ends defining a specified dimension. See LIMIT GAUGE.

end-group analysis (*Chem*) A method of determining molecular mass of polymers by titration of eg acidic groups at chain ends. Accuracy limited to relatively low-mass polymers such as prepolymers.

ending (*Textiles*) A dyeing fault that results in the end of the fabric being a different shade from the other end or the bulk of the fabric.

end-labelling (*BioSci*) The enzymatic attachment of a label (eg radioactive nucleotide or fluorescent conjugate) to the end of a DNA or RNA molecule.

end-lap joint (*Build*) A halving joint formed between the ends of two pieces of timber intersecting at an angle.

end leakage flux (*ElecEng*) Leakage flux associated with the end connections of an electric machine.

endless rope haulage (*MinExt*) A method of hauling trucks underground by attachment to a long loop of rope, guided by many pulleys along the roads or haulage ways, and actuated by a power-driven drum.

endless saw (*Eng*) See BANDSAW.

end links (*Eng*) The links at either end of a chain; they are made slightly stronger than the remainder to allow for wear when attached to hooks, couplings, etc.

end matter (*Print*) The items which follow the main text of a book, ie appendices, notes, glossary, bibliography, index. US *backmatter*.

end measuring instruments (*Eng*) Measuring instruments (eg micrometers, callipers, gauges) which measure length by making contact with the ends of an object.

end-member (*Min*) Any of two or more chemical compounds that make up a solid solution series of minerals of similar crystal structure.

end mill (*Eng*) A MILLING CUTTER having radially disposed teeth on its circular cutting face; often small in diameter and used for facing and grooving operations.

endo- (*Genrl*) Prefix from Gk *endon*, within.

endo-aneurysmorrhaphy (*Med*) (1) Obliteration, by suture at either end, of an aneurysm of an artery. (2) Obliteration, by suture, of the aneurysmal sac, with reconstruction of the original arterial channel.

endobiotic (*BioSci*) (1) Growing inside another plant. (2) Formed inside the host cell.

endoblast (*BioSci*) See HYPOBLAST.

endocardiac (*BioSci*) Within the heart.

endocarditis (*Med*) Inflammation of the lining membrane of the heart, esp of that part covering the valves.

endocardium (*BioSci*) In vertebrates, the layer of endothelium lining the cavities of the heart.

endocarp (*BioSci*) A differentiated innermost layer of a pericarp, usually woody in texture like the hard outside of a peach stone.

endocervicitis (*Med*) Inflammation of the mucous membrane lining the cervix uteri.

endochondral (*BioSci*) Situated within or taking place within cartilage; as *endochondral ossification*, which begins within the cartilage and works outwards.

endocoelar (*BioSci*) See SPLANCHNOPLEURE.

endocranium (*BioSci*) In insects, internal processes of the skeleton of the head, serving for muscle attachment.

endocrine (*BioSci*) Internally secreting; said of certain glands, principally in vertebrates, that pour their secretion into the blood, and so can affect distant organs or parts of the body. See HORMONE.

endocrinology (*Med*) The study of the internal secretory glands and the effects of the hormones that they secrete.

endocrinopathy (*Med*) Any disease due to disordered function of the endocrine glands.

endocuticle (*BioSci*) The laminated inner layer of the insect cuticle.

endocytobiosis (*BioSci*) See ENDOSYMBIOSIS.

endocytosis (*BioSci*) The entry of material into the cell by the invagination of the PLASMA MEMBRANE. Material can enter in the fluid phase of the resulting vesicle or attached to its membrane.

endoderm (*BioSci*) The inner layer of cells forming the wall of a gastrula and lining the archenteron; the tissues directly derived from this layer. Also *entoderm*.

endodermis (*BioSci*) (1) The innermost layer of the cortex in roots and stems, sheathing the stele, one cell thick. In roots typically and sometimes in stems there are no radial intercellular spaces and all anticlinal walls have a CASPARIAN STRIP. In some roots the cell walls become thickened later. See STARCH SHEATH. (2) A similar layer, with casparian strip or the like, elsewhere, eg surrounding the vascular tissues in the pine leaf.

endodontics (*Med*) The branch of dentistry concerned with disorders of the tooth pulp.

endodyne (*ICT*) See AUTODYNE.

endo-ergic process (*Phys*) Nuclear process in which energy is consumed. Also *endo-energetic*. The equivalent thermodynamic term is *endothermic*.

endo-exo configuration (*Chem*) Special case of *cis–trans* isomerism in ring compounds, eg borneol shows the *trans-* or *endo*-configuration while *iso*-borneol shows the *cis-* or *exo*-configuration.

end-of-file marker (*ICT*) Marker signalling the end of a file. Abbrev *EOF*.

endogamy (*BioSci*) The production of a zygote by the fusion of gametes from two closely related parents, or from the same individual. Cf EXOGAMY. See AUTOGAMY, INBREEDING.

endogenetic (*Geol*) Describes (1) the processes and resultant products that originate within the Earth, eg volcanism, extrusive rocks; (2) the processes leading to *endomorphism*, modification of igneous rocks by assimilation of country rocks. Also *endogenous*. Cf EXOGENETIC.

endogenous (*BioSci*) Processes or organs that originate, develop or occur within the organism. In higher animals, said of metabolism that leads to the building of tissue, or to the replacement of loss by wear and tear. In the case of psychological conditions the root cause may be physiological (*somatogenic*) or mental (*psychogenic*). Cf EXOGENOUS.

endogenous antigen (*BioSci*) Antigenic determinants produced inside a cell and processed and presented to T-lymphocytes in conjunction with MAJOR HISTOCOMPAT-IBILITY COMPLEX class I proteins. Normally applies to foreign antigens produced by cells infected with virus.

endogenous pyrogen (*BioSci*) Substances produced as a consequence of inflammation which induce fever, primarily INTERLEUKIN 1.

endogenous rhythm (*BioSci*) A rhythm in movement or in a physiological process that depends on internal rather than external stimuli. It will often persist under constant environmental conditions and may show ENTRAINMENT. See CIRCADIAN RHYTHM.

endoglycosidase (*BioSci*) An enzyme that hydrolyses the internal glycosidic bonds of polysaccharides or oligosaccharides, converting them to disaccharides and terminal dextrins.

endolithic (*BioSci*) Growing within rock as some algae do in limestone.

endolymph (*BioSci*) In vertebrates, the fluid that fills the cavity of the auditory vesicle and its outgrowths (semicircular canals etc). Adj *endolymphatic*.

endolymphangial (*BioSci*) Situated within a lymphatic vessel.

endolytic insecticides (*Agri*) Systemic insecticides that remain in their original form until metabolized by insects that have absorbed them.

endometatoxic insecticides (*Agri*) Systemic pesticides that are taken into the insect and subsequently metabolized into an active, secondary toxic compound.

endometrial (*Med*) Pertaining to, or having the character of, endometrium.

endometrioma (*Med*) An endometrial tumour consisting of glandular elements and a cellular connective tissue, occurring in the regions where endometrium is normally absent. Also *adenomyoma*.

endometriosis (*Med*) A condition in which fragments of tissue resembling ENDOMETRIUM occur in other tissue or organs. It is subject to the same menstrual changes as normal endometrium and causes severe pain and discomfort during menstruation.

endometritis (*Med*) Inflammation of the endometrium.

endometrium (*Med*) The mucous membrane lining the cavity of the uterus.

endomitosis (*BioSci*) The process, occurring naturally in some differentiating cells or when induced by, eg colchicine, in which the chromosomes divide without cell division giving double the original number of chromosomes in the cell. See RESTITUTION NUCLEUS.

endomorph (*Psych*) One of Sheldon's somatotyping classifications; endomorphs are soft and round; they are described as loving comfort, affection and are sociable. See SOMATOTYPE THEORY.

endomorphy (*Psych*) The physical and emotional make-up associated with Sheldon's ENDOMORPH somatotype.

endomyocardial fibrosis (*Med*) A form of CARDIOMYO-PATHY common in E Africa which causes fibrosis of endocardium and myocardium leading to severely restricted function.

endomyocarditis (*Med*) Inflammation of the heart muscle and of the membrane lining the cavity of the heart.

endomysium (*BioSci*) The intramuscular connective tissue that unites the fibres into bundles.

endoneurium (*BioSci*) Delicate connective tissue between the nerve fibres of a FUNICULUS.

endonuclease (*BioSci*) An enzyme that cuts a polynucleotide chain internally.

endoparasite (*BioSci*) A parasite living inside the body of its host.

endopeptidase (*BioSci*) An enzyme that cuts a polypeptide chain internally, not just removing terminal peptides.

endophlebitis (*Med*) Inflammation of the intima of a vein.

endophyte (*BioSci*) A plant living inside another plant, but not necessarily parasitic on it, eg the hyphae of a symbiotic fungus. Adj *endophytic*.

endophytic mycorrhiza (*BioSci*) Same as ENDOTROPHIC MYCORRHIZA.

endoplasmic reticulum (*BioSci*) A series of flattened membranous tubules and *cisternae* in the cytoplasm of eukaryotic cells, which can either show a smooth profile (*smooth endoplasmic reticulum*; SER) or be decorated with ribosomes (*rough endoplasmic reticulum*; RER). Smooth ER is the site of synthesis of lipids, and rough ER of proteins destined for secretion. The membrane is continuous with that of the nuclear envelope. Abbrev *ER*.

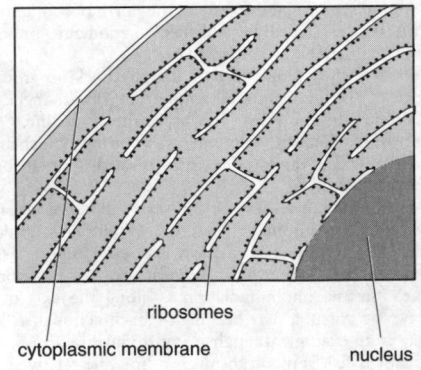

ribosomes

cytoplasmic membrane　　　　　　nucleus

endoplasmic reticulum

endopodite (*BioSci*) The inner ramus of a biramous arthropod appendage.

endopolyploid (*BioSci*) The product of ENDOMITOSIS.

Endoprocta (*BioSci*) See ENTOPROCTA.

Endopterygota (*BioSci*) A subclass of the Insecta with complete metamorphosis and a larval form in which the wings develop internally. Also HOLOMETABOLA. Cf EXOPTERYGOTA.

endoradiosonde (*Electronics, Med*) A miniature, battery-powered transmitter, encapsulated like a pill; designed to be swallowed by patients in order to transmit physiological data from their gastro-intestinal tracts.

endorhachis (*BioSci*) In vertebrates, a layer of connective tissue that lines the canal of the vertebral column and the cavity of the skull.

endorphins (*Pharmacol*) Peptides that are synthesized in the pituitary gland and that have analgesic properties associated with their affinity for opiate receptors in the brain.

endoscope (*Eng, Med*) An instrument for inspecting and photographing (1) internal cavities of the body in medicine or (2) remote and inaccessible sites in industry. Flexible fibre optics are normally used both to illuminate and inspect the remote site from outside and there may be facilities for eg biopsy.

endoscopic embryology (*BioSci*) The condition when the apex of the developing plant embryo points towards the base of the ARCHEGONIUM.

endoscopy (*Med*) Any technique using an ENDOSCOPE.

endoskeleton (*BioSci*) In Craniata, the internal skeleton, formed of cartilage or cartilage bone. In Arthropoda, the endophragmal skeleton, ie hardened invaginations of the integument forming rigid processes for the attachment of muscles and the support of certain other organs.

endosome (*BioSci*) A cytoplasmic vesicle derived from a COATED VESICLE by removal of the protein coat. See RECEPTOR-MEDIATED ENDOCYTOSIS.

endosperm (*BioSci*) Tissue formed within the embryo sac, usually triploid (see DOUBLE FERTILIZATION), serving in the nutrition of the embryo and often increasing to form a storage tissue in the mature seed, eg cereals.

endospermic (*BioSci*) Said of a mature seed having endosperm. Also *albuminous*, or *endospermous*.

endospore (*BioSci*) (1) The innermost layer of the wall of a spore or intine. (2) A very resistant thick-walled spore formed within a bacterial cell. (3) A spore produced within or by the division of the contents of the parent cell.

endosporic (*BioSci*) A GAMETOPHYTE developing within the spore, as in seed plants.

endostyle (*BioSci*) In some Protochordata and in the larvae of Cyclostomata, a longitudinal ventral groove of the pharyngeal wall, lined by ciliated and glandular epithelium. Adj *endostylar*.

endosulphan (*EnvSci*) An organochlorine insecticide used in agriculture. Highly toxic to fish.

endosymbiont hypothesis (*BioSci*) Generally accepted hypothesis that the chloroplasts, mitochondria and other plastids of most eukaryotes evolved from endosymbiotic prokaryotes, which were able to photosynthesize or respire aerobically. Some subsequent transfer of genes from endosymbiont to host must also have occurred. See panel on MITOCHONDRION.

endothecium (*BioSci*) (1) The fibrous layer in the wall of an anther. (2) The inner tissues in the young sporophyte of bryophytes, giving rise to the sporogenous tissue and/or the columella. Cf AMPHITHECIUM.

endotheliochorial placenta (*BioSci*) See VASOCHORIAL PLACENTA.

endothelioma (*Med*) A tumour arising from the lining membrane of blood vessels or lymph channels. Also a tumour arising in the pleura, in the peritoneum or in the meninges.

endothelium (*BioSci*) Pavement epithelium occurring on internal surfaces, such as those of the SEROUS MEMBRANES, blood vessels and lymphatics.

endotherm (*BioSci*) An animal that is able to generate heat internally and maintain its body temperature above ambient levels.

endothermic (*Chem*) An endothermic chemical reaction is one that absorbs thermal energy (heat), ie ΔH positive. Without an input of thermal energy the reaction will not proceed. Cf EXOTHERMIC.

endotoxin (*BioSci*) Generally, any heat-stable polysaccharide-like toxin bound to a bacterial cell. The term is used more specifically to refer to lipopolysaccharide (LPS) of the outer membrane of Gram-negative bacteria.

endotoxin shock (*Med*) Syndrome following administration of endotoxin or systemic infection with endotoxin-producing bacteria. Characterized by prostration, hypotension, fever and leucopenia.

endotrophic mycorrhiza (*BioSci*) A mycorrhiza in which the fungal hyphae grow between and within the cells of the root cortex and connect with hyphae ramifying though the soil but which do not form, as in ECTOTROPHIC MYCORRHIZA, a thick mantle on the surface of the root. Vesicular–arbuscular MYCORRHIZAS and the mycorrhizas of orchids and of the Ericaceae are endotrophic.

endozoic (*BioSci*) (1) Living inside an animal. (2) Said of the method of seed dispersal in which seeds are swallowed by an animal and voided after having been carried for some distance.

end-papers (*Print*) Stout papers formed from a folded sheet which is firmly attached to the first and last sections of a volume at the fold. One-half of each sheet is securely pasted to the inner side of the front and the back cover, the other half forming a fly leaf.

end plate (*BioSci*) A form of motor nerve ending in muscle.

end-plate fins (*Aero*) Fins mounted at or near the tips of the tailplane or wing to increase its efficiency.

end-point (*Chem*) The point in a volumetric titration at which the amount of added reagent is equivalent to the solution titrated.

end product (*Phys*) The stable nuclide forming the final member of a radioactive decay series.

end-quench test (*Eng*) See JOMINY TEST.

endrin (*BioSci*) An organic pesticide isomeric with dieldrin.

end shield (*ElecEng*) A cover which wholly or partially encloses the end of an electric machine.

end speed (*Aero*) Naval term for the speed of an aircraft relative to its aircraft carrier at the moment of release from catapult or accelerator.

end-to-end delay (*ICT*) The total delay accumulating in a network between a given source and destination.

end-to-end performance (*ICT*) The performance of a network in terms of such parameters as CELL LOSS PROBABILITY and END-TO-END DELAY as measured between a given source and destination.

end-to-end testing (*ICT*) The phase of a testing cycle in which components of a system are brought together and an entire process, usually involving the transfer of data between systems, is evaluated for suitability for purpose.

endurance (*Aero*) The maximum time that an aircraft can continue to fly without refuelling, under any agreed conditions.

endurance limit (*Eng*) In FATIGUE TESTING, the number of cycles which may be withstood without failure at a particular level of stress. See FATIGUE STRENGTH and panel on FATIGUE.

endurance range (*Eng*) See LIMITING RANGE OF STRESS.

end user (*ICT*) The 'everyday' user of a computer system or software.

end-windings (*ElecEng*) See ARMATURE END CONNECTIONS.

end-window counter (*Phys*) Geiger counter designed so that radiation of low penetrating power can enter one end, usually through a thin mica sheet.

endysis (*BioSci*) The formation of new layers of the integument following ECDYSIS.

enema (*Med*) Injection of fluid into the rectum to promote evacuation of the bowels.

energetics (*Chem*) The abstract study of the energy relations of physical and chemical changes. See THERMODYNAMICS.

energy (*Phys*) The capacity of a body for doing work. Mechanical energy may be of two kinds: *potential energy*, by virtue of the position of the body, and *kinetic energy*, by virtue of its motion. Both *mechanical* and *electrical* energy can be converted into *heat* which is itself a form of energy. Electrical energy can be stored in a capacitor to be recovered at discharge of the capacitor. *Elastic potential energy* is stored when a body is deformed or changes its configuration, eg in a compressed spring. All forms of wave motion have energy; in electromagnetic waves it is stored in the electric and magnetic fields. In any closed system, the total energy is constant – the *conservation of energy*. SI unit is the JOULE; CGS unit is the ERG. The foot-pound force (ft-lbf) of the UK system equals 1·356 J. See BRITISH THERMAL UNIT, ELECTRON-VOLT, KILOWATT-HOUR, KINETIC ENERGY, MECHANICAL EQUIVALENT OF HEAT, POTENTIAL ENERGY.

energy balance (*ChemEng*) See HEAT BALANCE.

energy balance (*Eng*) A balance sheet drawn up from an engine test, expressing the various forms of energy produced by the engine (eg BHP or output power, heat to cooling water, heat carried away in exhaust gases, and heat unaccounted for) as percentages of the heat supplied from the calorific value of the fuel.

energy balance (*Genrl*) (1) The detailed study of the energy flow in an industrial process. (2) The collection and publication of detailed information on the energy production and demand of a country, as compiled by eg the International Energy Agency. (3) A quantitative account of the exchanges of energy between organisms and their surroundings.

energy band (*Phys*) One of the bands of allowed energies separated by forbidden regions which arise in a solid when the energy levels of the individual atoms combine. The individual electrons are considered to belong to the crystal as a whole rather than to a particular atom. The energy bands are the consequence of the motion of the electron in the periodic potential of the crystal lattice. A solid for which a number of the bands are completely filled and the others empty is an insulator provided the energy gaps are large. If one band is incompletely filled or the bands overlap then metallic conduction is possible. For semiconductors, there is a small energy gap between the filled and empty band and *intrinsic conduction* occurs when some electrons acquire sufficient energy to surmount the gap.

energy barrier (*Chem*) The minimum amount of free energy which must be attained by a chemical entity in order to undergo a given reaction. See ACTIVATION ENERGY.

energy component (*ElecEng*) See ACTIVE COMPONENT.

energy confinement time (*NucEng*) In fusion, the ratio of the total energy of a confined plasma to the rate of energy loss from it.

energy curve (*ImageTech*) The spectral distribution of radiated energy in a source of light, eg an arc, the ordinate being proportional to the energy contained in a specified small band of wavelengths.

energy density of sound (*Acous*) The sum of potential and kinetic sound energy per unit volume.

energy dispersive analysis of X-rays (*Chem*) A method of elemental analysis of materials by scanning back-scattered X-rays from high-voltage electron bombardment, usually in a SCANNING ELECTRON MICROSCOPE. Characteristic emission peaks enable identification of most elements. Useful for analysing inclusions, impurities, etc. Abbrev EDAX.

energy-equivalent sound-pressure level (*Acous*) Sound-pressure level, where the squared sound pressure is averaged over a long time, typically more than 15 minutes. Used to characterize strongly fluctuating sound levels such as those of traffic noise.

energy fluence (*Phys*) The radiation intensity integrated over a short pulse.

energy gap (*Phys*) The range of forbidden energy levels between two permitted bands. See ENERGY BAND.

energy level (*Phys*) See LEVEL.

energy levels (*Electronics*) (1) In semiconductors, a donor level is an intermediate level close to the conduction band; being filled at absolute zero, electrons in this level can acquire energies corresponding to the conduction level at other temperatures. An acceptor level is an intermediate level close to the normal band, but empty at absolute zero; electrons corresponding to the normal band can acquire energies corresponding to the intermediate level at other temperatures. See FERMI LEVEL. (2) Electron energies in atoms are limited to a fixed range of values termed permitted energy levels, and represented by horizontal lines drawn against a vertical energy scale. Cf BAND THEORY OF SOLIDS. See EIGENVALUES, SCHRÖDINGER EQUATION.

energy management (*Aero*) Operational technique of minimizing energy loss by advanced automatic flight and engine monitoring and control systems. The actual method takes several forms but includes eg the measurement of an individual aircraft and engine performance in flight, and

adjustment of height and Mach number to suit the monitored conditions.

energy–mass equation (*Phys*) See MASS–ENERGY EQUATION.

energy policy (*Genrl*) A statement of a country's energy production and demand and future intentions.

energy product (*Phys*) The maximum product BH_{max} of magnetizing field H and magnetic flux density B associated with the upper left quadrant of a hysteresis loop; its value reflects a material's ability to remain magnetized in the presence of a demagnetizing field.

energy values (*FoodSci*) The measure of energy supplied by a given food or raw material. The basic unit of energy is the joule (4·184 joules = 1 calorie). In human nutrition kilojoule (kJ) and kilocalorie (kcal) are preferred because energy values are calculated per 100 g or 100 ml of food. To calculate the energy content simply use: energy in kilojoules = protein × 17 (4) + carbohydrate × 17 (4) + fat × 37 (9) + alcohol × 29 (7) + organic acid × 13 (3) + POLYOLS × 10(2·4) where the figures in parentheses are the energy yield in kilocalories.

enfleurage (*Chem*) The process of cold extraction with fat, eg of essential oils from flowers. Used in perfumery.

ENG (*ImageTech*) Abbrev for *electronic news gathering*, recording and sometimes transmitting TV news items direct from their location using very lightweight video equipment.

eng (*For*) See KERUING.

engaging speed (*Aero*) The relative speed of a carrier-borne aircraft to its ship at the moment when the ARRESTER GEAR is engaged.

Engel process (*Plastics*) Patented process for the production of large vessels, as in ROTOMOULDING.

engine (*Eng*) Generally, a machine in which energy is applied to do work: particularly, a machine for converting heat energy into mechanical work; loosely, a locomotive.

engine (*ICT*) A software system underlying a class of computer programs.

engine cylinder (*Eng*) That part of an engine in which the heat and pressure energy of the working fluid do work on the piston, and so are converted into mechanical energy. See CYLINDER, CYLINDER BARREL.

engineer (*Eng*) A person engaged in the science and art of engineering practice. The term is a wide one, but it is properly confined to someone qualified to design and supervise the execution of mechanical, electrical, hydraulic and other devices, public works, etc. See CHARTERED ENGINEER.

engineered safeguards (*NucEng*) The special features built into a nuclear reactor to cope with accidents and malfunctions.

engineered storage (*NucEng*) Facilities specially designed to store highly radioactive waste, eg the dry storage bunkers for spent fuel rods, which must be cooled and kept safe until their contents can be disposed of or processed.

engineering (*Genrl*) The design and putting to practical use of engines or machinery of any type; the design and construction of public works such as roads, railways and harbours.

engineering bricks (*CivEng*) Bricks made of semi-vitreous materials. They are dense and of high strength and low porosity. Manufactured chiefly in Staffordshire, Lancashire, Sussex, N Wales and Yorkshire. Used where severe loading conditions exist or exposure to damp, frost, acid or acid fumes, etc.

engineering design (*Eng*) Methods used to plan, test and produce engineering devices and machines, such as CAD, failure modes and effects analysis, materials selection. See INDUSTRIAL DESIGN, PRODUCT DESIGN.

engineering geology (*Eng, Geol*) Geology applied to engineering practice, esp in mining and civil engineering.

engineering polymer (*Eng*) A term rather loosely applied to speciality polymers offering improved properties over commodity polymers, eg polysulphones for heat resistance or polycarbonate for toughness.

engineering strain (*Eng*) See STRAIN.

engineers' bending theory (*Eng*) Simple relationship between the parameters in the bending of beams, encapsulated in the double equality:

$$\frac{M}{I} = \frac{\sigma}{y} = \frac{E}{R}$$

where M is the BENDING MOMENT, I is the second moment of area, σ is the STRESS developed a distance y from the NEUTRAL AXIS, E is YOUNG'S MODULUS of the beam material and R is the radius of curvature to which it is bent.

engineer's chain (*Surv*) A 100 ft measuring chain of 100 1 ft links.

engine friction (*Eng*) The frictional resistance to motion offered by the various working parts of an engine. See FRICTION HORSEPOWER.

engine pit (*Eng*) (1) A hole in the floor of a garage to enable a mechanic to work on the underside of a motor vehicle. (2) An engine sump or crank pit; the box-like lower part of the crankcase. (3) A large pit for giving clearance to the flywheel of any large gas engine or winding engine.

engine plane (*MinExt*) In a coal mine, a roadway on which tubs, trucks or trains are hauled by means of a rope worked by an engine.

engine pod (*Aero*) A complete turbojet power unit, including cowlings, supported on a pylon, usually under the wings of an aircraft: an installation method commonly adopted on most types of multi-engined high-subsonic-speed aircraft.

engine sized paper (*Paper*) Paper made from stock to which appropriate chemicals have been added to confer resistance to penetration by aqueous liquids.

engine speed (*Eng*) In a turbine engine, the revolutions per minute of the main rotor assembly; in a reciprocating engine, those of the crankshaft.

englacial (*EnvSci*) Occurring or situated in a glacier, eg an *englacial stream*

Engler distillation (*Chem*) The determination of the boiling range of petroleum distillates, carried out in a definite prescribed manner by distilling 100 cm^3 of the substance and taking the temperature after every 5 or 10 cm^3 of distillate has collected. The initial and final boiling points are also measured.

Engler flask (*Chem*) A 100 cm^3 distillation flask of definite prescribed proportions used for carrying out an ENGLER DISTILLATION.

English (*Print*) An old type size, approximately 14 point.

English bond (*Build*) The form of bond in which each course is alternately composed entirely of headers or of stretchers.

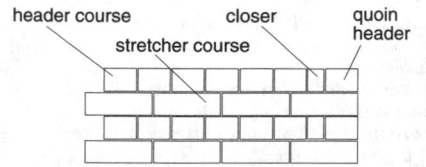

header course   closer   quoin header
stretcher course

English bond

English cross bond (*Build*) A DUTCH BOND.

English roof truss (*Build*) A truss for roofs of large span in which the sloping upper and lower chords are symmetrical about the central vertical, and are connected by vertical and diagonal members.

engorgement (*Med*) Congestion of a tissue or organ with blood.

engram (*Psych*) A permanent physical change in the brain brought about by a stimulus or experience; such memory traces have proved elusive as discrete entities, possibly because they are actually diffuse. Also *engramma*.

enhanced integrated drive electronics (*ICT*) Disk controller components on a special board, or integrated into the motherboard, which allow up to four hard disks to be controlled, with high-speed data transfer for some of them. Abbrev *EIDE*.

enhancement (*Electronics*) In metal–oxide–silicon devices, the creation of a conducting channel as a consequence of an externally applied gate–substrate bias voltage. Semiconductor beneath the gate accumulates an excess of charges which are nominally the minority carriers and is said to have undergone inversion.

enhancement effect (*BioSci*) Same as EMERSON ENHANCEMENT EFFECT.

enhancement-mode transistor (*Electronics*) A field-effect transistor in which there are no charge carriers in the channel when the gate–source voltage is zero. Channel conductivity increases when a gate–source voltage of appropriate polarity is applied.

enhancer (*BioSci*) A DNA sequence that can stimulate transcription of a gene while being at a distance from it.

enkephalins (*Pharmacol*) Pentapeptides, isolated from the brain, that have the same N-terminal amino acid sequences as ENDORPHINS and share their pain-relieving properties.

enlarger (*ImageTech*) Apparatus for making photographic prints of larger size than the original camera negative by projecting an illuminated image through a lens onto the sensitized paper.

enlarging lens (*ImageTech*) Lens designed for use in a photographic enlarger, required to project at comparatively short distances and provide a flat distortion-free image from a flat original, the negative.

enneagon (*MathSci*) A more correct but less common name for *nonagon*.

enneagram (*Psych*) A nine-sided diagram, now often taken to represent the nine supposed different types of personality.

enol form (*Chem*) Unsaturated alcohol form of a substance exhibiting keto-enol tautomerism, ie that form in which the mobile hydrogen atom is attached to the oxygen atom, and therefore has acidic properties, eg *ethyl aceto-acetate*.

enophthalmos (*Med*) Abnormal retraction of the eye within the orbit. Also *enophthalmus*.

enostosis (*Med*) A bony growth formed on the internal surface of a bone.

enprint (*ImageTech*) Standard size enlargement produced by D and P firms in lieu of contact print.

en quad (*Print*) A type space half an em wide. Usually called a 'nut' to distinguish it clearly from an EM QUAD (or mutton).

enqueue (*ICT*) To add a data processing task to a list of tasks awaiting processing in a buffer.

enriched uranium (*Phys*) Uranium in which the proportion of the fissile isotope uranium-235 has been increased above its natural abundance.

enrichment (*Arch, Build*) Ornamentation applied to building features.

enrichment (*BioSci*) A method of increasing the proportion of cells with a mutation that cannot be selected directly. Mutants that are unable to grow in a given medium are tolerant of agents such as penicillin that only kill growing cells.

enrichment (*MinExt*) The effect of superficial leaching of lode, whereby part of value is dissolved and redeposited in a lower enriched zone. See SECONDARY ENRICHMENT.

enrichment (*NucEng*) (1) Raising the proportion of uranium-235 fissile nuclei above that for natural uranium in reactor fuel. (2) Raising the proportion of the desired isotope in an element above that present initially by isotope separation.

enrichment factor (*Phys*) (1) In the UK, the abundance ratio of a product divided by that of the raw material. The *enrichment* is the enrichment factor less unity. (2) In the US, SEPARATION FACTOR.

en rule (*Print*) A dash (–) cast on an en body. Half the width of an EM rule, it is often used to divide dates, express range, etc. Also *en score*.

ensemble (*Phys*) In statistical mechanics, a set of a very large number of systems, all dynamically identical to the system under consideration and differing in the initial condition.

ensiform (*Genrl*) Shaped like the blade of a sword.

ensiform process (*BioSci*) See XIPHISTERNUM.

ensilage (*Genrl*) See SILAGE.

enstatite (*Min*) An orthorhombic pyroxene, chemically magnesium silicate, $MgSiO_3$; it occurs as a rock-forming mineral and in meteorites.

enstatitite (*Min*) A variety of orthopyroxenite composed almost entirely of ENSTATITE.

enstrophy (*EnvSci*) Half the square of the vorticity. It is conserved in two-dimensional, adiabatic, non-dissipative flow.

ENT (*Med*) Abbrev for *ear, nose and throat* as a department of medical practice.

entablature (*Arch, Eng*) (1) The whole of the parts immediately supported upon columns, consisting of an architrave, a frieze and a cornice. (2) The framework of a stationary steam engine supported on columns.

entamoebiasis (*Med*) Infection with *Entamoeba histolytica*; amoebic dysentery.

entanglement molecular mass (*Chem*) See MOLECULAR MASS DISTRIBUTION.

entanglements (*Chem*) Knots and loops formed between polymer chains above a critical molecular mass. See MOLECULAR MASS DISTRIBUTION.

entasis (*Arch*) The slight swelling towards the middle of the length of a column to correct for the appearance of concavity in the outline of the column if it had a straight taper.

enter-, entero- (*Genrl*) Prefixes from Gk *enteron*, intestine.

enteral (*Med*) Within, or by way of, the intestine.

enteral (*BioSci*) Parasympathetic.

enterclose (*Arch*) A corridor separating two rooms.

enterectomy (*Med*) Removal of part of the bowel.

enteric (*Med*) (1) Pertaining to the intestines. (2) Synonym for 'enteric fever' (see TYPHOID FEVER).

enteric-coated (*Pharmacol*) Of a medicinal capsule, coated in a substance that prevents it from releasing its contents until it has passed through the stomach and into the intestine.

entering edge (*ElecEng*) The edge of the brush of an electric machine which is first passed by a point on the commutator or slip ring. Also *leading edge, toe of the brush*. Cf LEAVING EDGE.

entering tap (*Eng*) See TAPER TAP.

enteritis (*Med*) Inflammation of the small intestine.

entero- (*Genrl*) See ENTER-.

entero-anastomosis (*Med*) The surgical union of two parts of the intestine; the operation for doing this.

Enterobacteriaceae (*BioSci*) A family of bacteria belonging to the order Eubacteriales, comprising aerobic Gram-negative rods. They are carbohydrate fermenters, and many are saprophytic. They include some gut parasites of animals and some blights and soft rots of plants, eg *Salmonella typhi* (typhoid), *Escherichia coli* (some strains cause enteritis in calves and infants).

enterobiasis (*Med*) Infection by the thread- or pin-worm, *Enterobius vermicularis*. Commonest intestinal parasitic infection in the UK, causing anal irritation in children.

enterocele (*Med*) A hernia containing intestine.

enterocentesis (*Med, Vet*) Surgical puncture of the intestine. In veterinary practice, puncturing the distended intestine of a horse suffering from colic, in order to liberate the gas.

enterocoel (*BioSci*) A coelom formed by the fusion of coelomic sacs that have separated off from the ARCH-ENTERON. In Chaetognatha, Echinodermata and Cephalochordata.

enterocolitis (*Med*) Inflammation of the small intestine and of the colon.

enterocolostomy (*Med*) A surgical communication between the small intestine and colon.

enterocystocele (*Med*) Hernia containing intestine and bladder.

entero-enterostomy (*Med*) The surgical formation of a communication between two separate parts of the small intestine; the operation for doing this.

enterogenous cyanosis (*Med*) A disorder characterized by chronic cyanosis and by the presence of methaemoglobin or sulphaemoglobin in the blood; due usually to taking drugs, esp aniline derivatives.

enterolith (*Med*) A concretion of organic matter and lime, bismuth or magnesium salts, formed in the intestine.

enteron (*BioSci*) The single body cavity of Cnidaria; it corresponds to the archenteron of a GASTRULA; in higher forms, the alimentary canal. Adj *enteric*. See ARCHENTERON.

Enteropneusta (*BioSci*) See HEMICHORDATA.

enteropathy (*Med*) Any intestinal disorder.

enteroproctia (*Med*) The condition in which an artificial anus is formed from intestine.

enterostomy (*Med*) The surgical formation of an opening in the intestine for draining intestinal contents.

enterosympathetic (*BioSci*) Said of that part of the AUTONOMIC NERVOUS SYSTEM which supplies the alimentary tract.

enterotomy (*Med*) Incision of the intestinal wall.

enterotoxaemias of sheep (*Vet*) A group of fatal toxaemic diseases of sheep due to alimentary infection and toxin production by the bacteria *Clostridium perfringens* (*Cl. welchii*), types B, C or D. Vaccines widely used.

enterotoxins (*BioSci*) A group of exotoxins produced by bacteria in the intestine and that act on the intestinal mucosa. Cholera toxin is the best known example.

enthalpy (*Phys*) Thermodynamic property of a working substance defined as $H = U + PV$ where U is the internal energy, P the pressure and V the volume of a system. Associated with the study of heat of reaction, heat capacity and flow processes. SI unit is the JOULE.

enthalpy heat drop (*Eng*) The enthalpy drop which occurs during the adiabatic expansion of unit mass of steam or other vapour and is capable of transformation into work.

entire (*BioSci*) (1) Said of the margin of a flattened organ when it is continuous, being neither toothed nor lobed. (2) A male animal that has not been castrated.

entire function (*MathSci*) See ANALYTIC FUNCTION.

ento- (*Genrl*) Prefix from Gk *entos*, within.

entoderm (*BioSci*) See ENDODERM.

entogastric (*BioSci*) Within the stomach or enteron.

entomology (*BioSci*) The branch of zoology that deals with the study of insects.

entomophagous (*BioSci*) Feeding on insects.

entomophily (*BioSci*) Pollination by insects. Adj *entomophilous*.

Entoprocta (*BioSci*) A phylum of Metazoa in which the anus is inside the circlet of tentacles; mainly marine forms. Also *Endoprocta*.

entovarial (*BioSci*) Within the ovary.

entozoic (*BioSci*) Living inside an animal.

entozoon (*BioSci*) An animal parasite living within the body of the host. Adj *entozoic*.

entrain (*Eng*) To suspend bubbles or particles in a moving fluid.

entrainer (*ChemEng*) See AZEOTROPIC DISTILLATION.

entrainment (*Chem*) Transport of small liquid particles in vapour, eg when drops of water are carried over in steam.

entrainment (*Psych*) The process whereby an endogenous clock-driven rhythm is synchronized to the rhythm of environmental events. See ZEITGEBER.

entrance lock (*Build*) A lock through which vessels must pass in entering or leaving a dock because of tides outside.

entrance pupil (*Phys*) The image of the aperture stop as formed by that part of an optical system nearer the object. It defines the cone of rays entering the system.

entresol (*Arch*) A MEZZANINE floor usually between the ground and first floors.

entropy (*Phys*) In thermal processes, a quantity which measures the extent to which the energy of a system is available for conversion to work. If a system undergoing an infinitesimal reversible change takes in a quantity of heat dQ at absolute temperature T, its entropy is increased by $dS = dQ/T$. The area under the absolute temperature–entropy graph for a reversible process represents the heat transferred in the process. For an adiabatic process, there is no heat transfer and the temperature–entropy graph is a straight line, the entropy remaining constant during the process. When a thermodynamic system is considered on the microscopic scale, equilibrium is associated with the distribution of molecules that has the greatest probability of occurring, ie the state with the greatest degree of disorder. Statistical mechanics interprets the increase in entropy in a closed system to a maximum at equilibrium as the consequence of the trend from a less probable to a more probable state. Any process in which no change in entropy occurs is said to be *isentropic*.

entropy of fusion (*Phys*) A measure of the increased randomness that accompanies the transition from solid to liquid or liquid to gas; equal to the latent heat divided by the absolute temperature.

entropy of polymerization (*Chem*) Change of entropy during polymerization. Large and negative due to ordering of monomer molecules into chains. See CEILING TEMPERATURE.

entry (*Space*) See RE-ENTRY.

entry corridor (*Space*) See panel on ROCKET.

entry portal (*Radiol*) Area through which a beam of radiation enters the body.

enucleate (*BioSci*) (1) Lacking a nucleus. (2) To remove, eg by microdissection, the nucleus from a cell.

enucleation (*BioSci*) The removal of the nucleus from a cell, eg by microdissection. Adj and v *enucleate*.

enucleation (*Med*) Removal of any tumour or globular swelling so that it comes out whole.

E number (*FoodSci*) All additives approved for food use in the European Union are assigned an E number. Some additives, only approved in the UK, are not prefixed with an 'E'.

enumerable set (*MathSci*) See DENUMERABLE SET.

enuresis (*Med*) A lack of bladder control past the age when such control is normally achieved.

envelope (*Aero*) The gasbag of a balloon, or of a non-rigid or semi-rigid airship.

envelope (*ICT*) (1) A term used to describe how a musical sound varies in loudness. In some implementations of HIGH-LEVEL LANGUAGES, envelope commands are included to enable the COMPUTER's sound facilities to be programmed. (2) The modulation waveform within which the carrier of an amplitude-modulated signal is contained, ie the curve connecting the peaks of successive cycles of the carrier wave.

envelope (*MathSci*) Of a family of plane curves: the curve which touches every curve of the family.

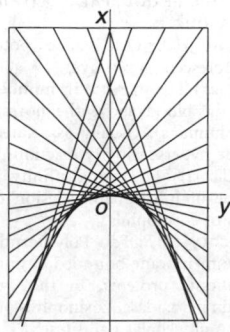

envelope Family of straight lines with an envelope of a parabola.

envelope (*Paper*) A container for a letter or similar flat document, generally made from paper by cutting a suitable blank shape which is then folded and glued. A flap is usually provided by which the contents may be sealed, either by tucking in or by use of an applied adhesive.

envelope delay (*ICT*) The time taken for the envelope of a signal to travel through a transmission system, without reference to the time taken by the individual components. Cf GROUP DELAY.

envelope-delay distortion (*ICT*) Distortion arising when the rate of change of phase shift with frequency for a transmission system is variable over the required frequency range.

envelope velocity (*ICT*) See GROUP VELOCITY.

environment (*Genrl, EnvSci*) (1) The physical and chemical surroundings of an object, eg the temperature and humidity, the physical structures, the gases. (2) When applied to human societies, the cultural, aesthetic and any other factors which contribute to the quality of life are included in the definition.

environmental archaeology (*EnvSci*) The study of past environments in which humans lived, using the techniques and methods of archaeology.

environmental audit (*EnvSci*) An investigation of the extent to which an organization's activities pollute the environment.

environmental burden (*EnvSci*) An approach to ranking the impact of the emission of different substances based upon known detrimental effects, potency of impact and the combined impact of the amount released and its potency. The calculated burden will vary according to the method of release (eg whether by dispersion in the atmosphere or by landfill).

environmental control (*Space*) The provision and control of the environment of a space vehicle, or part of it, so that its payload (including crew) can operate efficiently. It can involve control of temperature, humidity, atmosphere and contamination.

environmental geology (*Geol*) The application of geological knowledge to problems created by humans in the physical environment and by the use of physical resources. The term includes problems of a global scale.

environmental impact assessment (*Genrl*) The European Economic Community's equivalent of the US ENVIRONMENTAL IMPACT STATEMENT. Abbrev *EIA*.

environmental impact statement (*Genrl*) In the US, a detailed analysis, required of any agency undertaking a major federal project, of its effects on the environment. Abbrev *EIS*.

environmentally sensitive area (*EnvSci*) An area recognized by the government as having a landscape, wildlife or historic feature of national importance. Abbrev *ESA*.

environmental monitoring (*EnvSci*) Continuous or repeated measurement of agents in the environment to evaluate environmental exposure and possible damage by comparison with appropriate reference values based on knowledge of the probable relationship between ambient exposure and resultant adverse effects.

environmental pathway (*NucEng*) The route by which a radionuclide in the environment can reach humans, eg from radioactivity in rain to grass, to cows, to milk, to people.

environmental-stress cracking (*Eng*) Describes a variety of phenomena and mechanisms in which the initiation and propagation of cracks in materials subject to stress is accelerated by environmental chemicals. Their FRACTURE TOUGHNESS is thereby reduced and therefore their structural integrity. In polymers, the RESIDUAL STRESSES (or frozen-in strains) resulting from processing can be sufficient to cause cracking in the presence of many organic liquids (cf CRAZING) and the term *environmental-stress cracking* (abbrev *ESC*) tends to be used. This also affects the response of polymers to external loading, eg CREEP RUPTURE and fatigue. In other materials (metals, ceramics, glasses, etc) the term *environmental-stress corrosion cracking* (abbrev *ESCC*)

Enzyme

Most chemical reactions in living organisms only proceed sufficiently fast if mediated by catalytic proteins known as enzymes. Metabolism therefore depends upon the balanced and coordinated effects of these proteins controlling the rates of both synthetic and degradative reactions.

The catalytic effect depends on a restricted region of the protein molecule, known as the *active centre* of the enzyme, which consists of a specific and unique configuration of a few amino acid side chains brought into confluence by the *secondary* and *tertiary folding* of the protein into its native conformation. The *substrate* of the enzyme is bound at this site when the catalytic events take place. As the integrity of the active centre depends upon the stability of the native conformation, enzymes will only retain activity when this conformation is stable. Hence extremes of temperature and pH, organic solvents and heavy metals will all inhibit activity by denaturing the protein. Enzymes may also be inhibited by molecules which react specifically with the active centre. Such poisons may bind to the active centre because they are structurally related to the proper substrate and can be displaced by high concentrations of the substrate. Thus the dicarboxylic acid (malonic acid) will inhibit succinate oxidase, which normally converts another dicarboxylic acid (succinic acid) to fumaric acid and the inhibition will be reversed by an excess of succinate. This reversible inhibition is termed *competitive inhibition*. In other instances, eg most nerve gases, the poison is unrelated to the substrate and the inhibition is termed *non-competitive*.

An enzyme's activity is restricted to a specific set of metabolic reactions depending on the nature of the chemical bond to be modified and the ability of the enzyme to bind to the substrate. For instance, enzymes which catalyse the hydrolysis of peptide bonds, *peptidases*, usually have their activity restricted to peptide bonds adjacent to certain amino acids in the polypeptide sequence of the substrate protein.

---

is preferred and the phenomenon is more usually associated with external loading, eg fatigue and STATIC FATIGUE. Cf STRESS CORROSION. See panel on FATIGUE.

environmental variance (*BioSci*) Variation of quantitative character due to non-genetic causes.

environment variable (*ICT*) A VARIABLE that can be used to specify a particular parameter such as path, drive or file name. This variable may then be used by the OPERATING SYSTEM as required: eg in MS-DOS, setting PATH to a:\system would cause the operating system to search in the DIRECTORY called 'system' on the FLOPPY DISK drive a: for files by default.

enzootic (*Vet*) Said of a disease prevalent in, and confined to, animals in a certain area; corresponding to ENDEMIC in man.

enzootic ataxia (*Vet*) See SWAYBACK.

enzootic bovine leukosis (*Vet*) Caused by an oncogenic RNA virus, imported by infected animals. Tumour masses are found in lymph nodes and some other organs, eg spinal cord, heart and abomasum. Notifiable in the UK where a slaughter policy exists. Abbrev EBL.

enzootic haematuria (*Vet*) See BOVINE CYSTIC HAEMATURIA.

enzootic marasmus (*Vet*) See PINE.

enzootic ovine abortion (*Vet*) A contagious form of abortion in sheep caused by a chlamydial infection. Most common cause of ovine abortion in the UK, but vaccine is available. Also *kebbing*.

enzyme (*BioSci*) A protein with catalytic activity. The activity is restricted to a limited set of reactions, defining the specificity of the enzyme. See panel on ENZYME.

enzyme-linked immunosorbent assay (*BioSci*) An assay method in which antigen or antibody is detected by means of an enzyme chemically coupled either to antibody specific for the antigen or to anti-Ig which in turn will bind to the specific antibody. Either the antigen or the antibody to be detected is attached to the surface of a small container or to plastic beads, and the specific antibody allowed to bind in turn. The amount bound is subsequently measured by addition of a substrate for an enzyme which develops a colour when hydrolysed. Commonly used enzymes are *horseradish peroxidase* or *alkaline phosphatase*. Abbrev *ELISA*.

Eocambrian (*Geol*) A poorly fossiliferous sequence of Late Precambrian rocks, approx the equivalent of Riphean. See PRECAMBRIAN.

Eocene (*Geol*) The oldest division of Tertiary rocks, now regarded as an epoch within the Palaeogene system.

EOF (*ICT*) Abbrev for END-OF-FILE. See END-OF-FILE MARKER.

eolian deposits (*Geol*) See AEOLIAN DEPOSITS.

eolith (*Geol*) Literally 'dawn stone'; a term applied to the oldest-known stone implements used by early humans which occur in the Stone Bed at the base of the Crag in E Anglia and in high-level gravel deposits elsewhere. The construction is crude, and some authorities question the human origin, thinking it likely that the chipping has been produced by natural causes.

Eolithic period (*Geol*) The time of the primitive peoples who manufactured and used eoliths: the dawn of the Stone Age. Cf NEOLITHIC PERIOD, PALAEOLITHIC PERIOD.

eon (*Geol*) See AEON.

eosin (*Chem*) $C_{20}H_6Br_4O_5K_2$. The potassium salt of tetrabromo-fluorescein, a red dye.

eosinophilia (*Med*) The increase in numbers of eosinophil leucocytes in the blood above the normal levels (up to $400\ mm^{-3}$ in humans). Usually associated with repeated immediate-type hypersensitivity reactions.

eosinophilic (*BioSci*) (1) Having affinity for the red dye eosin. (2) Type of inflammatory lesion characterized by large numbers of eosinophils.

eosinophil leucocyte (*BioSci*) Polymorphonuclear leucocyte with large membrane-bounded eosinophilic granules, containing cationic proteins, in the cytoplasm. Other granules contain peroxidase. Eosinophils play an important role in killing multicellular parasites.

Eötvös balance (*MinExt*) A torsion balance sensitive to minute gravitational differences in land masses.

Eötvös equation (*Chem*) The molecular surface energy of a substance decreases linearly with temperature, becoming zero about 60°C below the critical point.

Eozoic (*Geol*) A term suggested for the Precambrian system, but little used. It means the 'dawn of life', and is comparable with *Cenozoic, Mesozoic* and *Palaeozoic*.

eozoon (*Geol*) A banded structure found originally in certain Canadian Precambrian rocks and thought to be of organic origin; now known to be inorganic and a product of dedolomitization, consisting of alternating bands of calcite and serpentine replacing forsterite.

EPA (*Chem*) Abbrev for *eicosapentaenoic acid*.

epapophysis (*BioSci*) A median process of a vertebral neural arch.

eparterial (*Med*) Situated over an artery, as *Eparterial bronchus*, the first division of the right bronchus, which passes to the upper lobe of the right lung.

epaxial (*BioSci*) Above the axis, esp above the vertebral column, therefore dorsal; as the upper of two blocks into which the myotomes of fish embryos become divided. Adj *epaxonic*. Cf HYPAXIAL.

epaxonic (*BioSci*) See EPAXIAL.

EPDM (*Chem*) Abbrev for ETHYLENE–PROPYLENE–DIENE RUBBER.

epeirogenic earth movements (*Geol*) Continent-building movements, as distinct from mountain-building movements, involving the coastal plain and just-submerged 'continental platform' of the great land areas. Such movements include gentle uplift or depression, with gentle folding and the development of normal tensional faults.

epencephalon (*BioSci*) See CEREBELLUM.

ependyma (*BioSci*) In vertebrates, the layer of columnar ciliated epithelium, backed by neuroglia, that lines the central canal of the spinal cord and the ventricles of the brain. Adj *ependymal*.

ependymitis (*Med*) Inflammation of the ependyma.

ependymoma (*Med*) A tumour within the brain arising in or near the ventricles and containing ependyma-like cells.

ep-, eph-, epi- (*Genrl*) Prefixes from Gk *epi* signifying above, over, upon, on, in addition or after.

ephedrine (*Pharmacol*) An alkaloid originally isolated from *Ephedra vulgaris*. Used in medicine as a *sympathomimetic* agent for many conditions (asthma, allergies), and as an surgical adjuvant.

ephemeral (*BioSci*) Descriptive of an organism that completes its life cycle in a short period, weeks rather than months.

ephemeral fever (*Vet*) See THREE-DAY SICKNESS.

ephemeris (*Astron*) A compilation, published at regular intervals, in which are tabulated the daily positions of the Sun, Moon, planets and certain stars, with other data necessary for the navigator and observational astronomer. See ASTRONOMICAL EPHEMERIS, NAUTICAL ALMANAC.

ephemeris time (*Astron*) Uniform or Newtonian time, as used in the calculation of future positions of the Sun and planets. The normal measurement of time by observations of stars includes the irregularities due to the changes in the Earth's rate of rotation. The difference between ephemeris time and universal time is adjusted to zero at an epoch in 1900; it amounted to around 40 s in 1970. The *ephemeris second* is the fundamental unit of time adopted by the International Committee of Weights and Measures, its defined value being 1/315 569 25·9747 of the tropical year for 1900 January, 0 d at 12 h ET. Ephemeris time has been superceded by *dynamical time*. See TIME and appendices. Abbrev *ET*.

Ephemeroptera (*BioSci*) An order of insects, in which the adults have large membranous fore-wings and reduced hind-wings, and the abdomen bears two or three long caudal filaments. The adult life is very short and the mouthparts are reduced and functionless; the immature stages are active aquatic forms. Mayflies.

epi (*Electronics*) Abbrev for an epitaxially grown layer, especially silicon semiconductor.

epi- (*Chem*) (1) Containing a condensed double aromatic nucleus substituted in the 1,6-positions. (2) Containing an intramolecular bridge.

epibasal half (*BioSci*) The anterior portion of an embryo.

epibenthos (*BioSci*) The organisms living on the floor of a sea or lake.

epibiosis (*BioSci*) Relationship in which one organism lives on the surface of another without causing it harm. Plant epibionts are *epiphytes*, animal epibionts, *epizoites*.

epiblast (*BioSci*) The outer germinal layer in the embryo of a metazoan animal, which gives rise to the ectoderm. Cf HYPOBLAST.

epiblem rhizodermis (*BioSci*) The outermost cell layer (epidermis) of a root.

epiboly (*BioSci*) Overgrowth; growth of one part or layer so as to cover another. Adj *epibolic*.

epicalyx (*BioSci*) An extra, calyx-like structure below and close to the real calyx in some flowers, eg many Rosaceae, including strawberry, and many Malvaceae.

epicanthus (*Med*) A semi-lunar fold of skin above, and sometimes covering, the inner angle of the eye; a normal feature of the Mongolian races.

epicardium (*BioSci*) In vertebrates, the serous membrane covering the heart; in UROCHORDATA, diverticula of the pharynx, which grow out and surround the digestive viscera like a perivisceral cavity. Adj *epicardial*.

epicarp (*BioSci*) The superficial layer of the PERICARP, esp when it can be stripped off as a skin. Also *exocarp*.

epicentre (*Geol*) That point on the surface of the Earth lying immediately above the focus of an earthquake or nuclear explosion. See panel on EARTHQUAKE.

epichlorhydrin (*Chem*) 1-chloro-2,3-epoxypropane. C_3H_5ClO. Bp 117°C. A liquid derivative of glycerol formed by reaction with hydrogen chloride to give *dichlorohydrin*, which in turn is treated with concentrated potassium hydroxide solution. Used in the production of EPOXY RESINS. See panel on THERMOSETS.

epichlorhydrin rubber (*Chem*) A speciality rubber made with ethylene oxide as co-monomer. Abbrev *CHR*.

epicoele (*BioSci*) In Craniata, the cerebellar ventricle or cavity of the cerebellum.

epicondyle (*BioSci*) The proximal part of the condyle of the humerus or femur.

epicontinental (*Geol*) Within the limits of a continental mass, including the continental shelf.

epicontinental sea (*Geol*) Sea covering the continental shelf or part of a continent.

epicormic shoot (*BioSci*) A shoot growing out, adventitiously or from a dormant bud, from a tree trunk or substantial woody branch.

epicotyl (*BioSci*) Either (1) the part of the shoot of an embryo or seedling above the cotyledon(s), ie the whole of the plumule, or (2) the stem between the cotyledon(s) and the first leaf or leaves, ie the first internode of the plumule.

epicritic (*BioSci*) Pertaining to sensitivity to slight tactile stimuli.

epicuticle (*BioSci*) (1) Layer of waxes including long-chain (>C20) alkanes, alcohols, acids and esters on the surface of the plant cuticle. Cf BLOOM. (2) The thin outermost layer of the insect cuticle consisting of lipid and protein. See CUTICULIN.

epicycle (*Astron*) The term applied in early descriptions of the solar system (eg the Ptolemaic and Copernican systems) to a small circle, described uniformly by the Sun, Moon or planet, the centre of that circle itself describing uniformly a larger orbit.

epicyclic gear train (*Eng*) A system of gears, in which at least one wheel axis itself revolves about another fixed axis either inside or outside it; used for giving a large reduction ratio in small compass. Fig. ▷

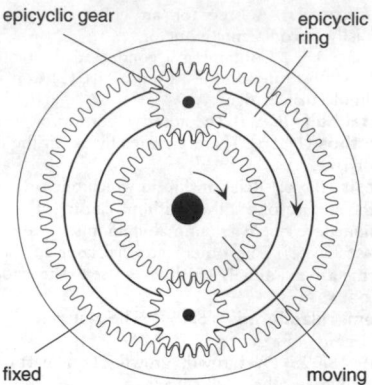

epicyclic gear   epicyclic ring

fixed   moving

epicyclic gear train One member is fixed and the epicyclic ring moves through half the angle moved by the central gear in this example.

epicycloid (*MathSci*) See ROULETTE.

epidemic (*Med*) An outbreak of an infectious disease spreading widely among people at the same time in any region. Also as adj.

epidemic parotitis (*Med*) See MUMPS.

epidemic tremor (*Vet*) See INFECTIOUS AVIAN ENCEPHA-LOMYELITIS.

epidemiology (*Med*) The study of disease in the population, defining its incidence and prevalence, examining the role of external influences such as infection, diet or toxic substances, and examining appropriate preventive or curative measures.

epidermal (*BioSci*) Relating to the epidermis. Also *epidermatic*.

epidermal growth factor (*BioSci*) A polypeptide growth factor that stimulates division of a variety of cell types, not just epidermal cells. Abbrev *EGF*.

epidermis (*BioSci*) (1) In plants, a layer, usually one cell thick, forming a skin over the young shoots and roots of plants, continuous except where perforated by stomata (on shoots) or over lateral or adventitious roots. It often carries hairs. It is eventually replaced by the periderm in woody plants. (2) Those layers of the animal integument that are of ectodermal origin; the epithelium covering the body.

epidermoid cyst (*Med*) A cyst lined with epithelium, occurring in the scalp. Also *wen*.

epidermolysis bullosa (*Med*) A congenital defect of the skin, in which the slightest blow produces a blister.

epidiascope (*Phys*) A projection lantern which may be used for transparencies or for opaque objects. See EPISCOPE.

epididymectomy (*Med*) Removal of the epididymis.

epididymis (*BioSci*) In the male of Elasmobranchii, some amphibians and Amniota, the greatly coiled anterior end of the Wolffian duct, which serves as an outlet for the spermatozoa.

epididymitis (*Med*) Inflammation of the epididymis.

epididymo-orchitis (*Med*) Inflammation of the epididymis and the testes.

epididymotomy (*Med*) Incision into the epididymis.

epidiorite (*Geol*) A metamorphosed gabbro or dolerite in which the original pyroxene has been replaced by fibrous amphibole. Other mineral changes have also taken step in the conversion, by dynamothermal metamorphism, of a basic igneous rock into a green schist.

epidosites (*Geol*) Metamorphic rocks composed of epidote and quartz. See EPIDOTIZATION.

epidote (*Min*) A hydrated aluminium iron silicate, occurring in many metamorphic rocks in lustrous yellow-green crystals, and as an alteration product in igneous rocks. Also *pistacite*.

epidotization (*Geol*) A process of alteration, esp of basic igneous rocks in which the feldspar is albitized with the separation of EPIDOTE and ZOISITE representing the anorthite molecule of the original plagioclase.

epidural anaesthesia (*Med*) Loss of painful sensation in the lower part of the body produced by injecting an anaesthetic into the epidural space surrounding the spinal canal.

epifauna (*BioSci*) Animals that inhabit the surfaces of water and sediment, or submerged rocks.

epifluorescence (*BioSci*) A method of fluorescence microscopy in which the excitatory light is transmitted through the objective onto the specimen rather than through the specimen; reflected excitatory light, much less intense than transmitted light, is filtered out.

epigaeous (*BioSci*) See EPIGEAL.

epigamic (*BioSci*) Attractive to the opposite sex, as *epigamic colours*.

epigastric (*BioSci*) Above or in front of the stomach; said of a vein in birds which represents the anterior part of the anterior abdominal vein of lower vertebrates.

epigastrium (*Med*) The abdominal region between the umbilicus and sternum.

epigeal (*BioSci*) (1) Germinating with the cotyledons appearing above the surface of the ground. (2) Living on the soil surface. Also *epigaeous*.

epigenesis (*BioSci*) The theory, now universally accepted, that the development of an embryo consists of the gradual production and organization of parts: as opposed to the theory of *preformation*, which supposed that the future animal or plant was already present complete, although in miniature, in the fertilized egg.

epigenetic (*Min*) Descriptive of ore deposits formed later than the rocks enclosing them. See SYNGENETIC.

epigenetics (*BioSci*) The study of mechanisms involved in the production of phenotypic complexity in morphogenesis. The pattern of selective gene expression brought about by eg DNA methylation can have important consequences; study of the epigenome, rather than the full unmodified genome, is expected to be informative.

epiglottidectomy (*Med*) Excision of the epiglottis.

epiglottis (*BioSci*) In insects, the epipharynx; in mammals, a cartilaginous flap which protects the glottis.

epignathous (*BioSci*) Having the upper jaw longer than the lower jaw, as in sperm whales.

epignathus (*Med*) A fetal malformation in which the deformed remnants of one twin project through the mouth of the more developed twin.

epigynous (*BioSci*) Having the calyx, corolla and stamens inserted on the top of the inferior ovary.

Epikote (*Plastics*) TN for a range of EPOXY RESINS, used for castings, encapsulation (potting) and surface coatings.

epilation (*Med*) Removal of the hair by the roots.

epilepsy (*Med*) A general term for a sudden disturbance of cerebral function accompanied by loss of consciousness, with or without convulsion. See GRAND MAL, JACKSONIAN EPILEPSY, PETIT MAL.

epileptiform (*Med*) Resembling epilepsy.

epileptogenic (*Med*) Exciting an attack of epilepsy.

epilimnion (*EnvSci*) The warm upper layer of water in a lake. Cf HYPOLIMNION.

epilithic (*BioSci*) Growing on a rock surface.

epiloia (*Med*) A condition characterized by feeble-mindedness, epileptic fits, sclerosis of the brain, and tumours in the skin and viscera; due to a defect in development. See TUBEROUS SCLEROSIS.

epimenorrhoea (*Med*) Too frequent occurrence of menstrual periods. US *epimenorrhea*.

epimerization (*Chem*) A type of asymmetric transformation in organic molecules, eg shown by the change of D-gluconic acid into an isomeric mixture of D-gluconic and D-mannonic acids.

Epimetheus (*Astron*) The eleventh natural satellite of Saturn, discovered in 1980. Distance from the planet 151 000 km; diameter 140 km.

epimorph (*Min*) A natural cast of a crystal.

epimysium (*BioSci*) The investing connective-tissue coat of a muscle.

epinasty (*BioSci*) NASTIC MOVEMENT in which the upper side of the base of an organ grows more than the lower, resulting in a downward bending of the organ as in the petals of an opening flower or the leaves of a tomato plant with waterlogged roots. Cf HYPONASTY.

epinephrine (*Pharmacol*) Synonym (US) for ADRENALINE.

epinephros (*BioSci*) See SUPRARENAL BODY.

epineural (*BioSci*) In Echinodermata, lying above the radial nerve; in vertebrates, lying above or arising from the neural arch of a vertebra.

epineurium (*BioSci*) The connective tissue that invests a nerve trunk, uniting the different funiculi and joining the nerve to the surrounding and related structures.

epiparasite (*BioSci*) See ECTOPARASITE.

epipetalous (*BioSci*) Attached to or inserted upon the petals, as stamens are in many flowers.

epipharynx (*BioSci*) In insects, the membranous roof of the mouth which in some forms is produced into a chitinized median fold, and in Diptera is associated with the labrum, to form a piercing organ; in Acarina, a forward projection of the anterior face of the pharynx. Adj *epipharyngeal*.

epiphenomenon (*Psych*) An event that accompanies, but is not causally related, to another event or process.

epiphloeodal (*BioSci*) Growing on the surface of bark. Also *epiphloeodic*.

epiphora (*Med*) An overflow of the lacrimal secretion due to obstruction of the channels which normally drain it.

epiphragm (*BioSci*) In Gastropoda, a plate, mostly composed of calcium phosphate, with which the aperture of the shell is sealed during periods of dormancy.

epiphyllous (*BioSci*) Growing upon, or attached to, the upper surface of a leaf; sometimes, growing on any part of a leaf.

epiphysis (*BioSci*) A separate terminal ossification of some bones, which only becomes united with the main bone at the attainment of maturity; the pineal body; in Echinoidea, one of the ossicles of Aristotle's lantern. Adj *epiphysial*.

epiphysis cerebri (*BioSci*) See PINEAL ORGAN.

epiphysitis (*Med*) Inflammation of the epiphysis of a bone.

epiphyte (*BioSci*) A plant that is attached to another with benefit to the former but not to the latter. Also *air plant*. See NEST EPIPHYTE.

epiphytotic (*BioSci*) A widespread outbreak of disease among plants, by analogy with epidemic.

epipleura (*BioSci*) (1) In COLEOPTERA, the reflexed sides of the ELYTRA. (2) In bony fish, upper ribs formed from membrane bone. (3) In CEPHALOCHORDATA, horizontal shelves of membrane arising from the inner sides of the metapleural folds, and forming the floor of the atrial cavity.

epiplocele (*Med*) A hernia containing the EPIPLOON.

epiploic foramen (*BioSci*) See WINSLOW'S FORAMEN.

epiploon (*BioSci*) In mammals, a double fold of serous membrane connecting the colon and the stomach (the great omentum). Adj *epiploic*.

epipubic (*BioSci*) In front of or above the pubis; pertaining to the epipubis.

episclera (*Med*) The connective tissue between the conjunctiva and the sclera.

episcleritis (*Med*) Inflammation of the episclera, sometimes involving the sclera.

episcope (*Phys*) A projection lantern which throws an enlarged image of a brilliantly illuminated opaque object onto a screen.

episematic (*BioSci*) Serving for recognition; as *episematic colours*.

episepalous (*BioSci*) (1) Borne on the sepals. (2) Placed opposite to the sepals.

episiostenosis (*Med*) Narrowing of the vulvar orifice.

episiotomy (*Med*) Cutting the vulvar orifice to facilitate delivery of the fetus.

episodic memory (*Psych*) Personal memory of an event tied to a particular time and place, a subset of declarative memory. Cf SEMANTIC MEMORY.

episome (*BioSci*) A self-replicating element able to grow independently of the host's chromosome, but also to integrate into it. Often termed a *plasmid* in prokaryotes.

epispadias (*Med*) Congenital defect in the anterior or dorsal wall of the urethra, commoner in the male than in the female.

epispore (*BioSci*) The outermost layer of a spore wall, often consisting of a deposit forming ridges, spines or other irregularities of the surface.

epistatic (*BioSci*) Describes a gene or character, whose effect overrides that of another gene with which it is not allelic; analogous to *dominant* applied to genes at different loci. More generally, *epistasis* exists when the effect of two or more non-allelic genes in combination is not the sum of their separate effects.

epistaxis (*Med*) Bleeding from the nose.

epistemics (*Psych*) The scientific study of the perceptual, intellectual and linguistic processes by which knowledge and understanding are acquired and communicated.

epistemology (*Psych, Genrl*) Study of the origin, nature, methods and limits of human knowledge.

epistilbite (*Min*) A white or colourless zeolite; hydrated calcium aluminium silicate, crystallizing in the monoclinic system.

epistomatal (*BioSci*) Said of a leaf having stomata on the upper surface only. Also *epistomatic*. Cf AMPHISTOMATAL, HYPOSTOMATAL.

epistropheus (*BioSci*) The axis vertebra.

epistyle (*Arch*) See ARCHITRAVE.

epitaxial (*Electronics*) Having the same crystal axes. Used to describe extensions grown or deposited onto a single crystal substrate; the material in the epitaxial layer must have a lattice spacing and structure close to that of the substrate. Abbrev *epi*.

epitaxial growth of semiconductors (*Electronics*) The deposition of single-crystal material, often with added dopants, onto a single-crystal substrate. In this way the crystalline orientation of the deposited layer is determined by the substrate but its purity or composition is independently controlled. See panel on SEMICONDUCTOR FABRICATION.

epitaxial transistor (*Electronics*) One in which the collector consists of a high-resistivity epitaxial layer deposited on a low-resistivity substrate, the emitter and base regions being formed in or on this layer by diffusion techniques. This type of construction results in a very thin base region, important for effective high-frequency operation, and a relatively large collector, which ensures good heat dissipation.

epitaxy (*Crystal*) Unified crystal growth or deposition of one crystal layer on another. See EPITAXIAL GROWTH OF SEMICONDUCTORS.

epithalamus (*BioSci*) In the vertebrate brain, a dorsal zone of the DIENCEPHALON.

epithelioid (*Med*) Resembling epithelium.

epithelioma (*Med*) A malignant growth derived from epithelium.

epithelioma contagiosa (*Vet*) See FOWL POX.

epithelium (*BioSci*) A sheet of cells, one or several layers thick, organized above a basal lamina (sometimes referred to as a basement membrane), and often specialized for mechanical protection, active transport, or secretion. Examples include skin, and the lining of lungs, gut and blood vessels. Adj *epithelial*.

epithermal neutrons (*NucEng*) Neutrons having energies just above thermal, comparable with chemical bond energies. See NEUTRON.

epithermal reactor (*NucEng*) See INTERMEDIATE REACTOR.

epitokous (*BioSci*) Said of a free-swimming stage of some POLYCHAETA.

epitope (*BioSci*) A term used to describe an antigenic determinant in a molecule which is specifically recognized by an antibody combining site or by the antigen receptor of a T-cell.

epitope tag (*BioSci*) A short peptide sequence, an epitope for an existing antibody, used in molecular biology to label transgenic proteins in order to follow their expression and fate by immunocytochemistry or Western blotting.

epitrichium (*BioSci*) A superficial layer of the epidermis in mammals, which consists of greatly swollen cells and is found on parts of the body devoid of hair. Adj *epitrichial*.

epitrochlea (*Med*) The inner condyle, or bony eminence, on the inner aspect of the lower end of the humerus.

epitrochoid (*MathSci*) See ROULETTE.

epitropic fibre (*Textiles*) A synthetic fibre whose surface contains particles (eg of carbon) that are electrically conducting.

epituberculosis (*Med*) The congestion and inflammation of the area surrounding a tuberculous focus, esp in the lung.

epixylous (*BioSci*) Growing on wood.

epizoic (*BioSci*) (1) Growing on a living animal, eg epibiosis. (2) Having the seeds or fruits dispersed by animals.

epizoon (*BioSci*) An animal that lives on the skin of some other animal; it may be an ECTOPARASITE or a COMMENSAL. Adj *epizoan*.

epizootic (*Vet*) Applied to a disease affecting a large number of animals simultaneously throughout a large area and spreading with great speed.

epizootic catarrhal fever (*Vet*) See EQUINE INFLUENZA.

epizootic lymphangitis (*Vet*) A chronic contagious lymphangitis of horses, due to infection by *Histoplasma farciminosus*, a yeast. Also *African glanders*.

epoch (*Astron*) The precise instant to which the data of an astronomical problem are referred; thus the elements of an orbit when referred to a specific epoch also implicitly define the obliquity of the ecliptic, the rate of precession and other conditions obtaining only at that instant.

epoch (*Geol*) A subdivision of a geological period. eg the WENLOCK epoch of the SILURIAN period.

EPOS (*Chem*) TN for database of properties of polymers supplied by large UK manufacturer.

epoxy resins (*Chem*) Polymers derived from epichlorhydrin and bisphenol-A. Widely used as structural plastics, surface coatings and adhesives, and for encapsulating and embedding electronic components. Characterized by low shrinkage on polymerization, good adhesion, mechanical and electrical strength and chemical resistance. Also *epoxide resins*. See panel on THERMOSETS.

EPR (*Chem*) Abbrev for ELECTRON PARAMAGNETIC RESONANCE.

EPROM (*ICT*) See ERASABLE PROGRAMMABLE READ-ONLY MEMORY.

EPS file (*ICT*) Abbrev for *encapsulated PostScript file*. Format used for transferring text and graphics between APPLICATION PROGRAMS. See POSTSCRIPT.

Epsilon Canis Majoris (*Astron*) See ADHARA.

Epsom salts (*Min*) Hydrated magnesium sulphate, $MgSO_4 \cdot 7H_2O$. Occurs as incrustations in mines, in colourless acicular to prismatic crystals. Many chemical and other uses. Also *epsomite*.

Epstein–Barr virus (*Med*) A human herpes virus that can infect B-lymphocytes, causing rapid proliferation and accumulation leading to INFECTIOUS MONONUCLEOSIS (*glandular fever*). The virus is also implicated in some B-cell tumours including the endemic Burkitt's lymphoma found in tropical countries and B-cell lymphomas of immunocompromised individuals. Abbrev *EBV*.

epulis (*Med*) A tumour, innocent or malignant, of the gums, growing from the periosteum of the jaw.

EQ gate (*ICT*) See EQUIVALENCE GATE.

equal-area criterion (*ElecEng*) A term used in connection with the stability of electric power systems. The stability limit occurs when two areas on the power–angle diagram, governed by the load conditions obtaining, are equal.

equal-energy source (*Phys*) An electromagnetic or acoustic source whose radiated energy is distributed equally over its whole frequency spectrum.

equal falling particles (*MinExt*) Particles possessing equal TERMINAL VELOCITIES; the underflow, oversize product of a classifier.

equalization (*ICT*) Electronically, the reduction of distortion by compensating networks that allow for the particular type of the distortion over the requisite band.

equalization of boundaries (*Surv*) See GIVE-AND-TAKE LINES.

equalizing bed (*CivEng*) Usually a bed or bedding of fine ballast or concrete laid immediately underneath a pipeline, eg in a trench, in cases where the bottom of the excavation is sound but uneven (as in rock, hard chalk, etc).

equalizing network (*ICT*) (1) Network, incorporating any inductance, capacitance or resistance, that is deliberately introduced into a transmission circuit to alter the response of the circuit in a desired way; particularly to equalize a response over a frequency range. (2) A similar arrangement incorporated in the coupling between stages in an amplifier.

equalizing pulse (*ImageTech*) A pulse used in TV at twice the line frequency, which is applied immediately before and after the vertical synchronizing pulse. This is done to reduce any effect of line-frequency pulses on the interlace.

equalizing signals (*ImageTech*) Those added to ensure triggering at the exact time in a frame cycle.

equal-signal system (*Aero, ICT*) One in which two signals are emitted for radio range, an aircraft receiving equal signals only when on the indicated course.

equal-tempered scale (*Acous*) See TEMPERED SCALE.

equation (*Chem*) See CHEMICAL EQUATION.

equation (*MathSci*) (1) A sentence in which the verb is 'is equal to', such as $2 + 2 = 4$ or $x^2 + 6 = 5x$, where x is a variable. Those values of x which turn the equation into a true statement constitute the *solution set* of the equation. For the equation $x^2 + 6 = 5x$ the solution set is [2,3]. If the equation is satisfied by every element of the set E, then the equation is said to be an *identity*; eg $\sin^2 x + \cos^2 x = 1$, where x is a variable on the set of real numbers, is an identity, for it is satisfied by every real number. (2) The equation of a curve or surface is an equation satisfied by the co-ordinates of every point on the curve or surface, but not satisfied by the co-ordinates of any point not on the curve or surface.

equation of maximum work (*Chem*) See GIBBS–HELMHOLTZ EQUATION.

equation of motion (*Phys*) One of the relations between distance s, time t, initial velocity u, final velocity v and acceleration a for a moving body:

$$v = u + at$$
$$v^2 = u^2 + 2as$$
$$s = \frac{1}{2}(u+v)t$$
$$s = ut + \frac{1}{2}at^2$$

equation of state (*Chem*) An equation relating the volume, pressure and temperature of a given system, eg van der Waals' equation.

equation of state (*Phys*) Any equation relating thermodynamic state functions, ie those system variables whose value is independent of the path taken to reach a particular state of the system. Examples include the GIBBS–HELMHOLTZ EQUATION and the IDEAL GAS equation. See THERMODYNAMICS.

equation of time (*Astron*) The difference between the right ascensions of the true and mean Sun, and hence the difference between apparent and mean time. In the sense mean time minus apparent time, it has a maximum positive value of nearly $14\frac{1}{2}$ min in February, and a negative maximum of nearly $16\frac{1}{2}$ min in November, and vanishes four times a year.

equator (*Astron*) See CELESTIAL EQUATOR, TERRESTRIAL EQUATOR.

equatorial (*Astron*) An astronomical telescope which is so mounted that it revolves about a polar axis parallel to the Earth's axis; when set on a star it will keep that star in the field of view continuously, without adjustment. It has two graduated circles reading RIGHT ASCENSION and DECLINATION respectively. See panel on ASTRONOMICAL TELESCOPE.

equatorial (*BioSci*) Situated or taking place in the equatorial plane; as the *equatorial furrow* that precedes division of an ovum into upper and lower blastomeres, and the *equatorial plate* that, during mitosis, is the assembly of chromosomes on the spindle in the *equatorial plane*.

equatorial horizontal parallax (*Astron*) See HORIZONTAL PARALLAX.

equi- (*Genrl*) Prefix from Lt *aequus*, equal.

equiangular spiral (*MathSci*) A spiral in which the angle between the tangent and the radius vector is constant. Also a LOGARITHMIC SPIRAL. Its polar equation is $\log r = \alpha\theta$.

equilateral arch (*Build*) An arch in which the two springing points and the crown of the intrados form an equilateral triangle.

equilateral roof (*Build*) A pitched roof having rafters of a length equal to the span.

equilateral triangle (*MathSci*) A triangle having three equal sides.

equilibration (*Psych*) In Piaget's theory, the motivational mechanism underlying intellectual development; the process of balancing incoming information with existing knowledge; contradictory explanations of perceived events produce a state of disequilibrium and this acts as a motivation to reorganize thinking on a higher cognitive level.

equilibrium (*Chem*) The state reached in a reversible reaction when the reaction velocities in the two opposing directions are equal, so that the system has no further tendency to change.

equilibrium (*Phys*) (1) The state of a body at rest or moving with constant velocity. A body on which forces are acting can be in equilibrium only if the resultant force is zero and the resultant torque is zero. (2) Thermal state of a system at which no further heat flow occurs and all components of the system are at the same temperature.

equilibrium constant (*Chem*) The ratio, at equilibrium, of the product of the active masses of the molecules on the right side of the equation representing a reversible reaction to that of the active masses of the molecules on the left side. See LAW OF MASS ACTION.

equilibrium diagram (*Eng*) See PHASE DIAGRAM.

equilibrium moisture (*ChemEng*) The percentage water content of a solid material, when the vapour pressure of that water is equal to the partial pressure of the water vapour in the surrounding atmosphere.

equilibrium of floating bodies (*Phys*) The balance which occurs for a body which floats, partly immersed in liquid; the weight of the body is equal to the weight of the fluid it displaces. The ratio of the volume of the body to the volume immersed is the ratio of the density of the fluid to that of the body. See ARCHIMEDES' PRINCIPLE, CENTRE OF BUOYANCY, METACENTRE.

equilibrium relative humidity (*FoodSci*) The relative humidity of the surrounding atmosphere when a substance neither gains nor loses moisture. Abbrev *ERH*. See RELATIVE HUMIDITY.

equilibrium still (*ChemEng*) One designed to produce a boiling liquid mixture in complete phase equilibrium with its vapour, for the purposes of physicochemical measurement.

equilux spheres (*Phys*) Spherical surfaces which are concentric with a source of light so that the illumination is constant.

equine contagious catarrh (*Vet*) See STRANGLES.

equine encephalomyelitis (*Vet*) An acute disease of horses, mules and donkeys, due to a virus which causes an encephalomyelitis; characterized by fever, nervous symptoms and often death. Transmitted by mosquitoes, ticks and mites. Several strains of virus occur: eastern American, western American, Venezuelan and Russian. Another form, BORNA DISEASE, is caused by an unrelated virus.

equine infectious anaemia (*Vet*) See INFECTIOUS ANAE-MIA OF HORSES.

equine influenza (*Vet*) 'The cough'. Caused by myxoviruses, *A Equi 1* (Prague), *A Equi 2* (Miami). Highly contagious disease occurring as epidemics. Symptoms include fever, cough, nasal discharge and depression. Vaccine available. Also *epizootic catarrhal fever, pink eye, shipping fever, stable pneumonia*.

equine thrush (*Vet*) Inflammation of the frog of the horse's foot, attended with a fetid discharge.

equine viral arteritis (*Vet*) An acute, respiratory viral disease of horses, characterized by degenerative and inflammatory changes in the small arteries; the main symptoms are fever, respiratory difficulty, oedema of the legs, diarrhoea, and in pregnant mares, abortion. Modified live vaccines available in certain places.

equine virus abortion (*Vet*) See EQUINE VIRUS RHINOP-NEUMONITIS.

equine virus rhinopneumonitis (*Vet*) *Equine herpes virus one infection, EHV1*. Two subtypes. Subtype one strains associated with respiratory, neurological and neonatal disease, and abortion. Subtype two with mainly respiratory disease. Vaccination available.

equinoctial points (*Astron*) The two points, diametrically opposite each other, in which the celestial equator is cut by the ecliptic; called respectively the *First Point of Aries* (the origin from which both right ascension and celestial longitude are measured) and the *First Point of Libra*, from the signs of the Zodiac of which they are the beginning.

equinox (*Astron*) (1) Either of the two points on the CELESTIAL SPHERE where the ECLIPTIC intersects the CELESTIAL EQUATOR. Physically they are the points at which the Sun, in its annual motion, crosses the celestial equator: the *vernal equinox* as the Sun crosses from south to north, and the *autumnal equinox* as it crosses from north to south. The vernal equinox is the zero point in celestial co-ordinate systems. (2) Either of the two instants of time at which the Sun crosses the celestial equator, being around 21 March and 23 September.

equipartition of energy (*Chem*) The MAXWELL–BOLTZMANN LAW, which states that the available energy in a closed system eventually distributes itself equally among the DEGREES OF FREEDOM present.

equipotent (*BioSci*) See TOTIPOTENT.

equipotential surface (*ElecEng*) One where there is no difference of potential and hence no electric field. See FARADAY CAGE. Also *equipotential region*.

Equisetales (*BioSci*) The horsetails, an order of c.20 spp of the genus *Equisetum*. They are cosmopolitan except for Australia and New Zealand; the order also includes fossils from the Upper Carboniferous onwards. The sporophyte has roots and rhizomes, and an aerial stem which bears whorls of very small fused leaves and of branches. The stems are photosynthetic. The gametophytes are thalloid and photosynthetic.

equitant (*BioSci*) DISTICHOUS leaves folded longitudinally and each overlapping the next. See VERNATION.

equi-tempered scale (*Acous*) See TEMPERED SCALE.

equitonic scale (*Acous*) The musical scale in which the main notes progress by whole tones, as contrasted with the Pythagorean diatonic, which uses both whole tones and half-tones.

equity theory (*Psych*) General class of theory that seeks to explain behaviour on the basis of equity, that rewards for individuals in a group or relationship are in proportion to their contribution. Proponents argue that altruism, co-operation, power and aggression can be analysed using this framework.

equivalence class (*MathSci*) The set of elements of a domain that are related by an EQUIVALENCE RELATION.

equivalence gate (*ICT*) A GATE with two input signals. If both are 0, output is 1. If both are 1, output is 1. When incoming signals differ, output is 0. Abbrev *EQ gate*. Also *equivalence element*.

equivalence, photochemical (*Chem*) See EINSTEIN LAW OF PHOTOCHEMICAL EQUIVALENCE.

equivalence principle (*Genrl*) See PRINCIPLE OF EQUIVALENCE.

equivalence relation (*MathSci*) Relation which is reflexive, symmetric and transitive. It partitions a set into EQUIVALENCE CLASSES. For instance, *having the same father* is an equivalence relation, but *having the same sister* is not.

equivalent (*Chem*) See EQUIVALENT WEIGHT.

equivalent airspeed (*Aero*) INDICATED AIRSPEED corrected for position error (angle of incidence) and air compressibility. Abbrev *EAS*.

equivalent circuit (*ElecEng*) One consisting of passive components (*R*, *L* and *C*) and ideal current (or voltage) sources, which behaves, as far as current and voltage at its terminals are concerned, exactly as some other circuit or component, eg a bipolar transistor may be represented by a combination of resistors, capacitors and a current generator.

equivalent conductance (*Chem*) Electrical conductance of a solution which contains one gram-equivalent weight of solute at a specified concentration, measured when placed between two plane-parallel electrodes, 1 m apart. *Molar conductance* is more often used.

equivalent electrons (*Electronics*) Those which occupy the same orbit in an atom, hence have the same principal and orbital quantum numbers.

equivalent focal length (*Phys*) The focal length of a thin lens which is equivalent to a thick lens in respect of the size of image it produces.

equivalent free-falling diameter (*PowderTech*) The diameter of a sphere which has the same density and the same free-falling velocity in a given fluid as an observed particle of powder.

equivalent height (*ICT*) That of a perfect antenna, erected over a perfectly conducting ground, which, when carrying a uniformly distributed current equal to the maximum current in the actual antenna, radiates the same amount of power.

equivalent hiding power (*PowderTech*) Of particles interrupting a light beam, in photo-sedimentation analysis, the weight of a given size, expressed as a fraction or a percentage of the weight of a standard size, which has the same hiding power as unit weight of the standard of the size when in suspension. The standard size is usually the largest particle present.

equivalent lens (*Phys*) A simple lens of the same equivalent focal length as an optical system which, when placed at the first principal plane of the system, produces an image identical to that of the system except for a shift along the axis of magnitude equal to the distance between the principal planes of the system.

equivalent network (*ICT*) One identical to another network either in general or at some specified frequency. The same input applied to each would produce outputs identical in both magnitude and phase generated across the same internal impedance.

equivalent points (*Phys*) The principal points of a lens that is used with the same medium on both sides. See CARDINAL POINTS.

equivalent potential temperature (*EnvSci*) The equivalent potential temperature of an air sample is the EQUIVALENT TEMPERATURE of the sample when brought adiabatically to a pressure of 1000 mb. It is a conservative property for both dry and saturated adiabatic processes.

equivalent proportions (*Chem*) See LAW OF EQUIVALENT PROPORTIONS.

equivalent reactance (*ElecEng*) The value of the reactance of an equivalent circuit which allows it to represent the system of magnetic or dielectric linkages present in the actual circuit.

equivalent resistance (*ElecEng*) The value of the resistance of an equivalent circuit which allows its loss to represent the total loss occurring in the actual circuit.

equivalent simple pendulum (*Phys*) See CENTRE OF OSCILLATION.

equivalent sine wave (*ElecEng*) One which has the same frequency and the same rms value as a given non-sinusoidal wave.

equivalent surface diameter (*PowderTech*) The diameter of a sphere which has the same effective surface as that of an observed particle when determined under stated conditions.

equivalent telephony erlang (*ICT*) A measure of traffic in a network offering several different types of service that takes into account the ratio between their various bandwidths and that of the basic voice service. See ERLANG.

equivalent temperature (*EnvSci*) The equivalent temperature of a sample of moist air is the temperature which would be attained by condensing all the water vapour in the sample and using the latent heat thus released to raise the temperature of the sample.

equivalent T-networks (*ICT*) T- or pi-networks equivalent in electrical properties to sections of transmission line, provided these are short in comparison with the wavelength.

equivalent volume diameter (*PowderTech*) The diameter of a sphere which has the same effective volume as that of an observed particle when determined under the same conditions.

equivalent weight (*Chem*) That quantity of one substance which reacts chemically with a given amount of a standard. In particular, the equivalent weight of an acid will react with 1 mole of hydroxide ions, while the equivalent weight of an oxidizing agent will react with 1 mole of electrons. Also *equivalent*.

equivalve (*BioSci*) Said of bivalves that have the two halves of the shell of equal size.

Equuleus (Little Horse) (*Astron*) The second-smallest constellation in the northern sky.

ER (*BioSci*) Abbrev for *endoplasmic reticulum*.

Er (*Chem*) Symbol for ERBIUM.

era (*Geol*) A geological time unit within an eon, the formal chronostratigraphic unit above a SYSTEM, eg *Mesozoic* era. See appendix on Geological time.

erasable programmable read-only memory (*Electronics*) A read-only memory in which stored data can be erased, by ultraviolet light or other means, and reprogrammed bit by bit with appropriate voltage pulses. Abbrev *EPROM*.

erase head (*Acous*) In magnetic tape recording, the head which saturates the tape with high-frequency magnetization, in order to remove any previous recording.

erasion (*Med*) Removal of all diseased structures from a joint by cutting and scraping.

erbium (*Chem*) Symbol Er, at no 68, ram 167·26. A metallic element, a member of the group of RARE EARTH ELEMENTS. Found in the same minerals as dysprosium (gadolinite, fergusonite, xenotime), and in euxenite.

erbium-doped fibre amplifier (*ICT*) An OPTICAL FIBRE that by the addition of a small proportion of erbium is made to act as a laser and so to amplify optical signals travelling along it.

erbium laser (*Phys*) A laser using erbium in YAG (*yttrium–aluminium–garnet*) glass. It has the advantage of operating between 1·53 and 1·64 m, a range in which there is a high attenuation in water. This feature is of particular importance in laser applications to eye investigations, since a great deal of energy absorption will now occur in the cornea and aqueous humour before reaching the delicate retina.

ERC (*ImageTech*) Abbrev for *ever-ready case*.

erect (*BioSci*) Set at approximately right angles to the part from which it grows or emerges.

erectile tissue (*BioSci*) Tissue which contains extensive blood-spaces, which when distended with blood becomes turgid.

erecting prism (*Phys*) A right-angled prism used for erecting the image formed by an inverting projection system. The prism is used with its hypotenuse parallel to the beam of light incident on one of the other faces, which is totally reflected at the hypotenuse and emerges from the third face parallel to its original direction.

erecting shop (*Eng*) That part of an engineering works where finished parts are assembled or fitted together.

erection (*BioSci*) The turgid condition of certain animal tissues when distended with blood; an upright or raised condition of an organ or part. Adj *erect*.

erection (*Build*) The assembly of the parts of a structure into their final positions.

erections (*Ships*) See TOP HAMPER.

erector (*ImageTech*) A lens added to an optical viewing system to provide an image the right way up, rather than inverted.

erector (*Med*) A muscle which, by its contraction, assists in raising or erecting a part or organ.

erg (*Phys*) Unit of work or energy in CGS system: 1 erg of work is done when a force of 1 dyne moves its point of application 1 cm in the direction of the force. See ENERGY, FUNDAMENTAL DYNAMICAL UNITS.

ergastic substances (*BioSci*) Non-protoplasmic substances; storage and waste products; starch, oil, crystals, tannins.

ergate (*BioSci*) A sterile female ant or worker.

ergatogyne (*BioSci*) A wingless queen ant.

ergatoid (*BioSci*) Resembling a worker; said of sexually perfect but wingless adults of certain social insects.

ergonomics (*Psych, Eng*) Methods intended to improve product performance by matching its design and materials of construction to the user. It includes study of human physiology and biomechanics (eg muscle strength), variability in human form (eg hand size) on the one side and product properties on the other (eg mechanical strength and stiffness). Hand tool design, eg, has improved dramatically through use of lightweight plastics carefully shaped to fit the user's limbs. US *human factors*.

ergonomy (*Med*) The physiological differentiation of functions.

ergosphere (*Astron*) The area around a black hole.

ergosterol (*Chem*) A STEROL which occurs in ergot, yeast and moulds. Traces of it are associated with cholesterol in animal tissues. On irradiation with ultraviolet light, vitamin D_2 is produced. See panel on VITAMINS.

ergot (*FoodSci*) Toxic contaminant in flour from rye and, to a lesser extent, other cereals, due to the growth of the fungus *Claviceps purpurea*. Rarely found today exceeding the maximum limit of 0·05% set in 1980–1 by the European Commission, although higher levels can be tolerated in countries which have a high consumption of rye bread.

ergotamine tartrate (*Pharmacol*) Ergot preparation which relieves migraine by constricting cranial arteries.

ergotism (*Med*) A condition characterized by extreme vasoconstriction leading to gangrene and convulsions; due to eating the grains of cereals which are infected by the ergot fungus *Claviceps purpurea*. Formerly known as *St Anthony's Fire*.

ERH (*FoodSci*) Abbrev for EQUILIBRIUM RELATIVE HUMIDITY.

Ericaceae (*BioSci*) A family of c.3000 spp of dicotyledonous flowering plants (superorder Dilleniidae). They are mostly shrubs, mostly calcifuge, and are cosmopolitan. The family includes the heaths (*Erica* spp) and heathers (*Calluna* spp) and a number of ornamental plants, eg rhododendron.

ericaceous (*BioSci*) Heather-like.

ericeticolous (*BioSci*) Growing on a heath.

Erichsen test (*Eng*) A test in which a piece of metal sheet is pressed into a cup by means of a plunger; used to estimate the suitability of sheet for pressing or drawing operations.

ericoid (*BioSci*) Having very small tough leaves like those of heather.

Eridanus (*Astron*) An inconspicuous southern constellation, but the sixth largest. Its brightest star is ACHERNAR. Eridanus is the name of a river, probably mythical, possibly the one into which Phaeton fell when Zeus toppled him from the chariot of the Sun with a thunderbolt.

erionite (*Min*) One of the less common zeolites; a hydrated aluminium silicate of sodium, potassium and calcium.

eriophorous (*BioSci*) Having a thick cottony covering of hairs.

erlang (*ICT*) International unit of traffic flow in telephone calls.

Erlangen program (*MathSci*) A program which classifies geometries according to properties left invariant under groups of transformations. Instituted in 1872 at the University of Erlangen by F Klein.

Erlenmeyer flask (*Chem*) A conical glass flask with a flat bottom, widely used for titrating, as it is easily cleaned, stood, stoppered and swirled.

ERM (*ICT*) Abbrev for *entity–relationship model*. A graphic representation of the logical relations between the sets and items of data operated on by a system.

Ernie (*ICT*) Abbrev for *electronic random number indicating equipment*. A so-called 'computer' used to select winning numbers in the UK Premium Bond lottery. Similar random-number generators are used in the MONTE CARLO METHOD.

erogenous zones (*Psych*) Sensitive areas of the body, the stimulation of which arouses sexual feelings and responses. They function as substitutes for the genital organs and are associated with stages of development in childhood.

eros (*Psych*) In psychoanalytic theory, the constructive life instinct, the urge for survival and procreation.

erosion (*Eng*) The removal of metal from components subject to fluid flow, particularly when the liquid contains solid particles.

erosion (*Geol*) The removal of the land surface by weathering, corrasion, corrosion and transportation, under the influence of gravity, wind and running water. Also, the eating away of the coastline by the action of the sea. Soil erosion may result from factors such as bad agricultural methods, excessive deforestation, overgrazing.

erratics (*Geol*) Stones, ranging in size from pebbles to large boulders, which were transported by ice, which, on melting, left them stranded far from their original source. They furnish valuable evidence of the former extent and movements of ice sheets.

error (*ElecEng*) See DEVIATION.

error (*Eng*) In servo or other control systems, the difference between the actual value of a quantity arising from the process and the desired value stored in the controller.

error (*ICT*) Fault or mistake causing the failure of a computer program or system to produce expected results. See COMPILATION ERROR, EXECUTION ERROR, LOGICAL ERROR, SYNTAX ERROR.

error-correcting code (*ICT*) A code that by including extra PARITY BITS can detect and correct certain types of errors that may occur during reading, writing and transmission of data.

error handling (*ICT*) Same as EXCEPTION HANDLING.

error in indication (*ElecEng*) The difference between the indication of an instrument and the true value of the quantity being measured. It may be expressed as a percentage of the true value when a positive value of the error means that the instrument indicates a greater than the true value.

error message (*ICT*) Indication that an error has been detected.

error of closure (*Surv*) See CLOSING ERROR.

error of the first kind (*MathSci*) The incorrect rejection of the status quo or a working (null) HYPOTHESIS in favour of an alternative hypothesis.

error of the second kind (*MathSci*) The incorrect failure to reject the status quo or a working (null) HYPOTHESIS in favour of an alternative hypothesis.

error signal (*ICT*) Feedback signal in control system representing deviation of controlled variable from set value.

Ertel potential vorticity (*EnvSci*) A rigorous formulation by Ertel of the idea of POTENTIAL VORTICITY for any compressible, thermodynamically active, inviscid fluid in adiabatic flow. If S is some conservative thermodynamic property of the fluid (eg the potential temperature), Ω is the angular velocity of the co-ordinate system, ρ is the density, and V is the velocity of the fluid relative to the co-ordinate system, then the Ertel potential vorticity Π is defined by

$$\Pi = \nabla S \left(\frac{2\Omega + \text{curl } V}{\rho} \right)$$

Π is a conservative property for all individual fluid particles. Approximations to the Ertel potential vorticity are useful in dynamical studies of the general circulation of the atmosphere and in NUMERICAL FORECASTING.

erubescite (*Min*) See BORNITE.

erucic acid (*Chem, FoodSci*) $CH_3[CH_2]_7CH=CH[CH_2]_{11}COOH$. Mp $33°C$. Mono-saturated fatty acid in the oleic acid series. Occurs in rapeseed oil and mustard seed oil, and implicated in health problems. Regulated in Europe to levels below 5%. New rape cultivars have no erucic acid, and are now used to produce culinary rapeseed oil.

eructation (*Med*) A belching of gas from the stomach through the mouth.

erumpent (*BioSci*) Developing at first beneath the surface of the substratum, then bursting out through the substratum and spreading somewhat.

eruption (*Med*) A breaking out of a rash on the skin or on the mucous membranes; a rash.

eruptive rocks (*Geol*) A term sometimes used for all igneous rocks; but usually applied to VOLCANIC or EXTRUSIVE ROCKS.

erysipelas (*Med*) A diffuse and spreading inflammation of skin and subcutaneous cellular tissue esp of face, neck, forearm and hands. The inflamed area being red, shiny and sharply demarcated; caused by *Streptococcus pyogenes*.

erythema (*Med*) A superficial redness of the skin, due to dilation of the capillaries.

erythema multiforme (*Med*) A skin disease in which raised red patches appear and reappear, esp on the upper part of the body, associated in severe cases with extensive skin necrosis and involvement of kidneys, lungs and gastrointestinal system.

erythema nodosum (*Med*) A skin disease in which red, painful, oval swellings appear, usually on the shins, associated with fever, joint pains and sore throat; now believed to be due to a hypersensitive phase in certain infections, such as tuberculosis or sarcoidosis, or certain streptococcal infections, or to drug hypersensitivity.

erythema pernio (*Med*) Painful red swellings of the extremities, esp on hands and feet, caused by constriction of small blood vessels in cold weather. Also *chilblains*.

erythraemia (*Med*) US *erythremia*. See *polycythaemia*.

erythrasma (*Med*) Infection of the horny layer of the skin with the fungus *Microsporon minutissimum*, giving rise to superficial, reddish-yellow patches.

erythremia (*Med*) US for ERYTHRAEMIA.

erythrite (*Min*) Hydrated cobalt arsenate, occurring as pale reddish crystals or incrustations. Also *cobalt bloom*.

erythritol (*Pharmacol*) A sugar alcohol extracted from certain algae and lichens and used medicinally to dilate blood vessels.

erythroblast (*BioSci*) A nucleated mesodermal embryonic cell, the cytoplasm of which contains haemoglobin, and which will later give rise to an erythrocyte. See MEGALOBLAST.

erythroblastosis (*Med*) The abnormal presence of erythroblasts in the blood.

erythroblastosis fetalis (*Med*) Disease of the human fetus due to immunization of the mother. Escape of fetal red cells into the maternal circulation during pregnancy (or during a previous pregnancy) elicits antibodies in the mother. If these are IgG, and can cross the placental membranes into the fetal circulation, they cause haemolysis of the fetal red cells. This may be sufficient to cause stillbirth or anaemia and severe jaundice with brain damage. *Rhesus antigens* are the most common cause. Also *erythroblastosis foetalis, haemolytic disease of the newborn*. See RHESUS BLOOD GROUP SYSTEM.

erythrocyte (*BioSci*) One of the red blood corpuscles of vertebrates; flattened oval or circular disk-like cells (lacking a nucleus in mammals), whose purpose is to carry oxygen in combination with the pigment haemoglobin in them, and to remove carbon dioxide.

erythrocytosis (*Med*) Excess in the number of red cells in the blood.

erythromelalgia (*Med*) A condition characterized by pain, redness and swelling of the toes, feet and hands, often associated with vascular disease.

erythromycin (*Pharmacol*) A broad spectrum *macrolide* antibiotic often used as an alternative to *penicillins*.

erythropenia (*Med*) Diminution, below normal, of the number of red cells in the blood.

erythrophore (*BioSci*) A chromatophore containing a reddish pigment.

erythropoiesis (*Med*) The formation of red blood cells.

erythropoietin (*Med*) A substance produced by the juxtaglomerular cells of the kidneys which stimulates the production of red blood cells by the bone marrow.

erythropsia (*Med*) The state in which objects appear red to the observer, eg snow blindness.

erythropterin (*BioSci*) A red heterocyclic compound deposited as a pigment in the epidermal cells or the cavities of the scales and setae of many insects.

erythrose (*Chem*) A TETROSE.

Es (*Chem*) Symbol for EINSTEINIUM.

ESA (*Space*) Abbrev for *European Space Agency*, formed in 1975 combining the activities of the European Space Research Organisation (ESRO) and the European Launcher Development Organisation (ELDO). ESA, an intergovernmental agency, co-ordinates European space activities and related technologies; in particular, it instigates and manages international space programmes on behalf of 17 member states.

Esaki diode (*Electronics*) See TUNNEL DIODE.

ESC (*Eng*) Abbrev for ENVIRONMENTAL STRESS CRACKING.

ESC (*ImageTech*) Abbrev for ELECTRONIC STILL CAMERA.

escape (*BioSci*) A plant growing outside a garden and derived from cultivated specimens, surviving but not well naturalized.

escape behaviour (*Psych*) (1) Defensive behaviour against a predator, often involving specialized evasive manoeuvres or structures. (2) a mental process by which a person avoids something unpleasant (*escape mechanism*).

escape boom (*MinExt*) Means of escape from an offshore platform. Pivoted chutes with a buoyant outer end are normally secured inboard. In emergency they are released and rotate outwards until the outer end floats and personnel can slide to safety.

escape character (*ICT*) A keyboard character that terminates an action or restores a previous state.

escape conditioning (*Psych*) A learning or conditioning procedure in which escape behaviour is acquired as a response to a negative reinforcement or aversive stimulus. Also *avoidance learning*.

escape key (*ICT*) A special key on a keyboard that has the effect of terminating the execution of a program or leaving the currently active MENU and returning to a previous one.

escapement (*Build*) The cut-away part above the mouth of a plane through which the shavings are voided.

escapement (*Eng*) (1) A mechanism allowing one component at a time to move automatically from one machine to the next in a line. (2) That portion of the mechanism of a clock that regulates the gradual release of the potential energy stored in the spring or in the weight system. The *escape wheel* allows a small amount of energy delivered by the GOING TRAIN to pass to the MOTION WORK with every swing of the pendulum or rotation of the balance wheel.

escape sequence (*ICT*) A sequence of escape characters.

escape velocity (*Astron, Space*) The minimum velocity v_e necessary for an object to travel in a parabolic orbit about a massive primary body, and thus to escape its gravitational attraction. An object which attains this or any greater velocity will coast away from the primary. For the surface of the Earth this velocity is $11 \cdot 2$ km s^{-1}, for the Moon $2 \cdot 4$ km s^{-1}, and for the Sun $617 \cdot 7$ km s^{-1}. The formal relation is

$$v_e = \sqrt{2GM/r}$$

where G is the Newtonian constant of gravitation, M is the mass of the object and r is its radius. The escape velocity of a BLACK HOLE (panel) exceeds the speed of light.

escarpment (*Geol*) A long cliff-like ridge developed by denudation where hard and soft inclined strata are interbedded, the outcrop of each hard rock forming an *escarpment*, such as those of the Chalk (Chiltern Hills, N and S Downs) and the Jurassic limestones (Cotswold Hills). Generally an escarpment consists of a short steep rise (the *scarp face*) and a long gentle slope (the *dip slope*). Cf CUESTA.

ESCC (*Eng*) Abbrev for *environmental-stress corrosion cracking*. See ENVIRONMENTAL-STRESS CRACKING.

eschar (*Med*) A dry slough produced by burning or by corrosives.

Escherichia coli (*BioSci*) The archetypal bacterium for biochemists, used very extensively in experimental work. A rod-shaped Gram-negative bacillus ($0.5 \times 2.0 \ \mu$m) abundant in the large intestine (colon) of mammals. Most strains are non-pathogenic, but the *E. coli* O157 strain, common in the intestines of cattle, has caused a number of human deaths.

Eschka's reagent (*Chem*) A mixture of MgO and Na$_2$CO$_3$ (2:1); used for estimation of the sulphur content of fuels.

escribed circle of a triangle (*MathSci*) A circle that touches one side of a triangle externally and the extensions of the other two sides.

escutcheon (*Build*) A perforated plate around an opening, such as a keyhole plate or the plate to which a doorknob is attached.

Esda (*Genrl*) Abbrev for *electrostatic document analysis* (or *apparatus*), a forensic test (or the equipment for this) used to reveal impressions on paper or evidence of amendments to documents.

eserine (*Pharmacol*) An alkaloid *acetylcholine esterase* inhibitor isolated from the Calabar bean, used in the treatment of glaucoma. Also *physostigmine*.

esker (*Geol*) A long, narrow, steep-sided, sinuous ridge of poorly stratified sand and gravel deposited by a subglacial or englacial stream. Found in glaciated areas.

Esmarch's bandage (*Med*) A rubber bandage which, applied to a limb from below upwards, expels blood from the part. Used to make it easier to operate surgically on a limb.

esophageal valve (*BioSci*) US for *oesophageal valve*. See CARDIAC VALVE.

esophagus (*BioSci*) US for OESOPHAGUS.

esophoria (*Med*) Latent internal squint, revealed in an apparently normal person by passing a screen before the eye.

espagnolette (*Build*) Fastening for a French window, having two long bolts which operate in slots at the top and bottom when the door handle is turned. Also *cremorne bolt*.

esparto (*Paper*) A rush (*Macrochloa tenacissima*) native to N Africa and S Spain; the main raw material for paper pulp before wood pulp, and yielding paper with excellent printing qualities. Also *esparto grass, Spanish grass*.

ESPRIT (*ICT*) Abbrev for *European Strategic Programme for Research in Information Technology*. European Economic Community initiative, begun 1982, to fund and encourage collaborative research in computing and information technology.

espundia (*Med*) An ulcerative infection of the skin, and of the mucous membranes of the nose and mouth, by the protozoal parasite *Leishmania brasiliensis*; occurs in S America.

ESR (*Chem*) Abbrev for ELECTRON SPIN RESONANCE.

essential amino acids (*BioSci*) Amino acids that cannot be synthesized by an organism and must therefore be present in the diet. The term is often applied anthropocentrically to those amino acids required by humans (Ileu, Leu, Lys, Met, Phe, Thr, Tyr and Val).

essential element (*BioSci*) An element that is necessary for the growth and reproduction of an organism and that cannot be replaced by another element. See DEFICIENCY DISEASE, MACRONUTRIENT, MICRONUTRIENT.

essential fatty acids (*BioSci*) The three fatty acids required for growth in mammals: arachidonic, linolenic and linoleic acids. Only linoleic acid needs to be supplied in the diet; the other two can be made from it.

essential minerals (*Geol*) Components present, by definition, in a rock, the absence of which would automatically change the name and classification of the rock. Cf ACCESSORY MINERALS.

essential oils (*BioSci*) Volatile SECONDARY METABOLITES formed mainly in oil glands, rarely in ducts; mostly terpenoids, also aliphatic and aromatic esters, phenolics and substituted benzene hydrocarbons, responsible for the odours of many aromatic plants (and steam distilled as perfumes from some). Some appear to deter insects or herbivores, others to be ALLELOPATHIC.

essential organs (*BioSci*) The stamens and carpels of a flower.

Essex board (*Build*) A building board made of layers of compressed wood-fibre material cemented with a fire-resisting cement.

essexite (*Geol*) A coarse-grained deep-seated igneous rock, a nepheline monzogabbro or nepheline monzodiorite containing labradorite, alkali-feldspar, titanaugite, kaersutite and biotite. Named from Essex County, Massachusetts.

Esson coefficient (*ElecEng*) See SPECIFIC TORQUE COEFFICIENT.

essonite (*Min*) Original spelling of HESSONITE.

EST (*BioSci*) Abbrev for EXPRESSED SEQUENCE TAG.

establishment charges (*Build, Eng*) See ON-COSTS.

ester (*Chem*) Esters are derivatives of acids obtained by the exchange of the replaceable hydrogen for alkyl radicals. Many esters have a fruity smell and are used in artificial fruit essences; also used as solvents. See POLYESTERS.

esterase (*BioSci*) An enzyme that catalyses the hydrolysis of ester bonds.

esterification (*Chem*) The direct action of an acid on an alcohol, resulting in the formation of esters. Catalysed by hydrogen ions. See STEP POLYMERIZATION and panel on POLYMER SYNTHESIS.

ester value (*Chem*) The number of milligrams of potassium hydroxide required to saponify the fatty acid esters in 1 gram of a fat, wax, oil, etc, equal to the saponification value minus the acid value.

esthiomene (*Med*) A condition in which there is chronic hypertrophy and destructive ulceration of the external genitals of the female.

estimate (*MathSci*) A function of a set of observations which is used as a value of an unknown parameter.

estimated breeding value (*Agri*) A value calculated from recorded reproductive history of the individual animal and its parents.

estradiol (*Med, Pharmacol*) The US name and now British Approved Name (BAN) for *oestradiol*. Female sex hormone, secreted by the ovarian follicle (follicular hormone) and responsible for the development of the sexual characteristics of the female. Estradiol ($C_{18}H_{24}O_2$) is a sterol. The synthetic form is used in hormone replacement therapy.

estriol (*Pharmacol*) A female sex hormone (related to the sterols) produced in the ovaries and found in the urine in pregnancy. Also *oestriol*.

estrogen (*Med*) The generic term for a group of female sex hormones which induce estrus. Estradiol (ethinylestradiol) and mestranol are the common estrogen components of some oral contraceptives. Also *oestrogen*. Cf ANDROGEN. See panel on STEROID HORMONES.

estrone (*Pharmacol*) An estrogen produced by the ovaries, weaker than estradiol. Also *oestrone*.

estrous cycle (*BioSci*) In female mammals, the succession of changes in the genitalia commencing with one estrous period and finishing the next. Also UK *oestrous cycle*.

estrus (*BioSci*) In female mammals, the period of sexual excitement and acceptance of the male occurring between proestrum and metoestrum; more generally, the period of sexual excitement. Adj *estral, oestral*. UK (becoming obsolete) *oestrus, oestrum*.

estuarine deposition (*Geol*) Sedimentation in the environment of an estuary. The deposits differ from those which form in a deltaic environment, chiefly in their closer relationship to the strata of the adjacent land, and are usually of finer grain and of more uniform composition. Both are characterized by brackish water and sediments which contain land-derived animal and plant remains.

estuarine muds (*Geol*) So-called *estuarine muds* are, in many cases, silts admixed with sufficient true clay to give them some degree of plasticity; they are characterized by a high content of decomposed organic matter.

estuary (*Geol*) An inlet of the sea at the mouth of a river; developed esp in areas which have recently been submerged by the sea, the lower end of the valley having been thus drowned. Cf DELTA. See FIORDS.

Et (*Chem*) Symbol for the ethyl radical C_2H_5–.

ETA (*Aero*) Abbrev for *estimated time of arrival*, as forecast on a FLIGHT PLAN, for the destination time of a civil aircraft, or the time of arrival over a target for a military aircraft.

Eta (*Genrl*) A correlational technique used primarily for non-linear relationships. For instance, income and age are positively correlated until older age at which point the correlation reverses itself to some extent.

eta (*NucEng*) (η) In reactor theory, one of the factors in the FOUR-FACTOR FORMULA which represents average number of fission neutrons produced per neutron absorbed in the fuel.

Eta Aquarids (*Astron*) See AQUARIDS.

ETACS (*ICT*) Abbrev for EXTENDED TOTAL ACCESS COMMUNICATIONS SYSTEM.

etalon (*Phys*) An interferometer consisting of an air film enclosed between half-silvered plane-parallel plates of glass or quartz having a fixed separation. It is used for studying the fine structure of spectral lines. See FABRY–PÉROT INTERFEROMETER.

etanercept (*Pharmacol*) A *cytokine* inhibitor, used in the treatment of rheumatoid arthritis.

Etard's reaction (*Chem*) The formation of aromatic aldehydes by oxidizing methylated derivatives and homologues of benzene with chromyl chloride, CrO_2Cl_2.

etchant (*Chem*) Chemical for removing copper from laminate during production of printed circuits.

etched figures (*Crystal*) Small pits or depressions of geometrical design in the faces of crystals, due to the action of some solvent. The actual form of the figure depends upon the symmetry of the face concerned, and hence they provide invaluable evidence of the true symmetry in distorted crystals. Also *etch-figures*.

etching (*Eng*) (1) A method of showing the structure of metals and alloys by attacking a highly polished surface with a reagent that has a differential effect on different crystals or different constituents. (2) Removing films from the surface of materials to facilitate the subsequent deposition of another coating, eg paint. See DRY ETCHING, SEMICONDUCTOR DEVICE PROCESSING.

etching (*ImageTech*) The process of: (1) dissolving, with an acid, portions of a surface, such as copper or zinc sheet, where it is not protected with a resist; (2) soaking away gelatine differentially to form a relief image.

etching pits (*Eng*) Small cavities formed on the surface of metals during etching.

etching test (*Chem*) Test for the detection of fluorides. The substance under examination is heated with sulphuric acid in a lead vessel covered with a glass lid. If fluorides are present the glass will be etched owing to the action of hydrogen fluoride produced by the action of the acid on the fluoride.

ETD (*Aero*) Abbrev for *estimated time of departure*. See DEPARTURE TIME.

ethacrynic acid (*Pharmacol*) A LOOP DIURETIC used in the treatment of oedema and oliguria due to renal failure.

ethanal (*Chem*) See ACETALDEHYDE.

ethanamide (*Chem*) See ACETAMIDE.

ethandiamide (*Chem*) See OXAMIDE.

ethandioates (*Chem*) See OXALATES.

ethane (*Chem*) H_3CCH_3. Bp $-84°C$. A colourless, odourless gas of the alkane series; the critical temperature is $+34°C$, the critical pressure is $50·2$ atm. The second member of the alkane series of hydrocarbons. Chemical properties similar to those of METHANE.

ethanoates (*Chem*) See ACETATES.

ethanoic acid (*Chem*) See ACETIC ACID.

ethanol (*Chem*) C_2H_5OH. Mp $-114°C$, bp $78·4°C$. The IUPAC name for *ethyl alcohol*, a colourless liquid, of vinous odour, miscible with water and most organic solvents, rel.d. $0·789$; formed by the hydrolysis of ethyl chloride or of ethyl hydrogen sulphate; it may be obtained by absorption of ethylene in fuming sulphuric acid at 160°C, followed by hydrolysis with water, by reduction of acetaldehyde, or by direct synthesis from ethylene and water at high temperatures in the presence of a catalyst. It is prepared technically by the alcoholic fermentation of sugar. It forms alcoholates with sodium and potassium.

ethanol

ethanolamines (*Chem*) Amino derivatives of ethyl alcohol, *monoethanolamine*, $CH_2OHCH_2NH_2$; *diethanolamine*, $(CH_2CH_2OH)_2NH$; *triethanolamine*, $(CH_2CH_2OH)_3N$; hygroscopic solids with strong ammoniacal smell. Used, in combination with fatty acids, to produce detergents and cosmetic products.

ethanoyl (*Chem*) Same as ACETYL.

ethanoylation (*Chem*) See ACETYLATION.

ethanoyl chloride (*Chem*) See ACETYL CHLORIDE.

ethene (*Chem*) $H_2C=CH_2$. Mp$-169°C$, bp$-103°C$. The IUPAC name for *ethylene*. A gas of the alkene series, contained in illuminating gas and in gases obtained from the cracking of petroleum. Used for making polyethylene and polyethylene oxide, and for maturing fruit in storage.

ethene

ethenoid resins (*Plastics*) Resins made from compounds containing a double bond between two carbon atoms, ie the acrylic, vinyl and styrene groups of plastics.

ethephon (*BioSci*) *2-chloroethylphosphoric acid*, which breaks down rapidly in water to yield ethylene and is used to promote controlled ripening of fruit.

ether (*Chem*) *Alkoxyalkane*. (1) Any compound of the type R–O–R′, containing two identical, or different, alkyl groups united to an oxygen atom; they form a homologous series $C_nH_{2n+2}O$. (2) Specifically, DIETHYL ETHER.

ether (*Phys*) A hypothetical medium supposed (mainly during the 19th century) to fill all space whether 'empty' or occupied by matter; light was the result of undulations of the ether. The theory that electromagnetic waves need such a medium for propagation is no longer tenable. Also *aether*.

Ethernet (*ICT*) A LOCAL AREA NETWORK designed on the principle that one computer wishing to communicate with another broadcasts onto the network. Acknowledgement establishes the link.

ethidium bromide (*BioSci*) *3,8-diamino-5-ethyl-6-phenyl-phenanthridinium bromide*. A fluorescent reagent that binds to double-stranded DNA and RNA and is used for their detection after fractionation in a gel matrix. A MUTAGEN.

Ethiopian region (*BioSci*) One of the primary faunal regions into which the surface of the globe is divided; it includes all of Africa and Arabia south of the Tropic of Cancer.

ethmo- (*Genrl*) Prefix from Gk *ethmos*, sieve.

ethmohyostylic (*BioSci*) In some vertebrates, having the lower jaw suspended from the ethmoid region and the hyoid bar.

ethmoidalia (*BioSci*) A set of cartilage bones (ethmoids) forming the anterior part of the brain case in the vertebrate skull.

ethmoidectomy (*Med*) Surgical removal of the ethmoid cells or of part of the ethmoid bone.

ethmoiditis (*Med*) Inflammation of the ethmoid cells.

ethmoturbinal (*BioSci*) In mammals, a paired bone or cartilage of the nose, on which are supported the folds of the olfactory mucous membrane.

ethogram (*Psych*) A complete behavioural inventory for a particular species; the term has its origin in early instinct theories of animal behaviour and is rare in modern usage, because of the methodological complexities involved in categorizing units of behaviour.

ethology (*Psych*) Describes an approach to the study of animal behaviour in which attempts to explain behaviour combine questions about its immediate causation, development, function and evolution.

ethoxyl group (*Chem*) The group $-OC_3H_5$.

ethyl acetate (*Chem*) *Ethanoate*. $CH_3COOC_2H_5$. Mp $-82°C$, bp $77°C$. Colourless liquid of fruity odour, used as a lacquer solvent and in medicine.

ethyl aceto-acetate (*Chem*) $CH_3COCH_2COOC_2H_5$. One of the best-known examples of organic compounds existing in keto and enol forms. Widely used as a chemical intermediate, including the manufacture of pyrazolone dyes and mepacrine. Also *aceto-acetic ester*.

ethyl acrylate (*Chem*) $CH_2 = CCOOC_2H_5$. Bp $101°C$. Colourless liquid, used in the manufacture of plastics.

ethyl alcohol (*Chem*) See ETHANOL.

ethylamine (*Chem*) $C_2H_5NH_2$. Bp $19°C$. A liquid or gas of ammoniacal odour, which dissolves in water, and forms salts; it dissolves $Al(OH)_3$.

ethylene (*BioSci*) A gas that is a plant GROWTH SUBSTANCE and is produced esp in wounded, diseased, ripening and senescent tissues, interacting with auxin and promoting eg fruit ripening, leaf abscission and epinasty. Ethylene and substances that release it (see ETHEPHON) are used commercially to regulate the ripening of fruit, esp stored fruit. Also *ethene*.

ethylene (*Chem*) See ETHENE.

ethylene diamine tartrate (*Electronics*) Chemical in crystal form, exhibiting marked piezoelectric phenomena; used in narrow-band carrier filters. Abbrev *EDT*.

ethylene diamine tetra-acetic acid (*Chem*) See EDTA.

ethylene dichloride (*Chem*) 1,2-dichloroethane.

ethylene glycol (*Chem*) *Glycol*. $HOCH_2CH_2OH$. Bp $197.5°C$, rel.d. 1.125. A colourless, syrupy, hygroscopic liquid, miscible with water and ethanol. Prepared from ethylene dibromide or ethylene chlorhydrin by hydrolysing with caustic soda. Intermediate for glycol esters, which are solvents and plasticizers for lacquers; used in the textile industry, for printing inks, foodstuffs, antifreezing mixtures, and for de-icing aeroplane wings.

ethylene oxide (*Chem*) C_2H_4O. Bp $13.5°C$. A mobile colourless liquid of ethereal odour, obtained by distilling 2-chloroethanol with concentrated potassium hydroxide. Manufactured directly from ethylene and oxygen in the presence of a catalyst. Very useful organic intermediate in the manufacture of solvents, detergents, etc. Can also be used as a sterilizing medium in the gaseous state.

ethylene polymers (*Plastics*) A group of common polymers made from ethylene monomer. Includes low-density polyethylene, high-density polyethylene and medium-density polyethylene plus copolymers like ETHYLENE-PROPYLENE–DIENE RUBBER. See CHAIN POLYMERIZATION and panel on POLYMER SYNTHESIS.

ethylene–propylene–diene rubber (*Plastics*) A TER-POLY-MER which is elastomeric owing to low crystallinity from random copolymer structure. A small amount of diene is added to aid VULCANIZATION. Abbrev *EPDM*. Also *ethylene–proplyene rubber, EPM*. See panel on ELASTOMERS.

ethylene–propylene rubber (*Plastics*) See ETHYLENE-PROPYLENE–DIENE RUBBER.

ethylene-vinylacetate copolymer (*Plastics*) The relatively low-cost thermoplastic elastomer used as a HOT-MELT ADHESIVE and for small injection mouldings (eg stapler base-pads). Abbrev *EVA*.

ethyl group (*Chem*) The monovalent radical $-C_2H_5$.

ethylidene (*Chem*) The organic group CH_3CH.

ethyl mercaptan (*Chem*) *Ethane thiol*. C_2H_5SH. Bp $37°C$. A liquid of vile odour.

ethyne (*Chem*) See ACETYLENE.

ethynide (*Chem*) See ACETYLIDE.

etiolation (*BioSci*) The condition of a green plant that has not received sufficient light; the stems are weak, with abnormally long internodes, the leaves are small, yellowish or whitish, and the vascular strands are deficient in xylem. Adj *etiolated*.

etiology (*Med*) US for AETIOLOGY.

etioplast (*BioSci*) The form of PLASTID that develops in plants grown in the dark; colourless and containing a PROLAMELLAR BODY, rapidly converted to a chloroplast on illumination.

E-transformer (*ElecEng*) An electric sensing device in an automatic control system which gives an error voltage in response to linear motion. It consists of coils for detecting small displacements of magnetic armature, which affects balance of currents when off-centre.

Etruscan (*Arch*) An ancient Italian civilization which immediately preceded and influenced that of Ancient Rome. It is noted for the massive scale of its fortifications, built from hewn rock, many of which still stand today.

Ettinghausen effect (*Phys*) A difference in temperature established between the edges of a metal strip carrying a current longitudinally, when a magnetic field is applied perpendicular to the plane of the strip. Effect is very small and is analogous to HALL EFFECT.

Eu (*Chem*) Symbol for EUROPIUM.

eu- (*Genrl*) Prefix from Gk *eu*, well, good.

Eubacteriales (*BioSci*) One of the two main orders of the true bacteria, distinguished from the *Pseudomonadales* by the peritrichous flagella of the motile members. They are spherical or rod-shaped cells that have no photosynthetic pigments. They are not acid fast, and are readily stained by aniline dyes.

euchromatin (*BioSci*) Chromatin in the nucleus of interphase eukaryotic cells that appears light after staining (unlike HETEROCHROMATIN) because it is not heavily complexed with protein and is transcriptionally active.

euclase (*Min*) A monoclinic mineral, occurring as prismatic, usually colourless, crystals; hydrated beryllium aluminium silicate.

Euclidean construction (*MathSci*) An angle or figure constructed using only a straight edge and compasses.

Euclidean geometry (*MathSci*) The geometry of the plane, in which the axioms stated by Euclid are satisfied, so that there is one and only one line through any given point parallel to a given plane. Compare NON-EUCLIDEAN GEOMETRY.

Euclidean space (*MathSci*) A vector space consisting of n-tuples of numbers, on which a distance function is defined. With suitable interpretation three-dimensional Euclidean space (*Euclidean n-space*) describes the physical world.

Euclid's algorithm (*MathSci*) A procedure to find the GREATEST COMMON DIVISOR (gcd) of two positive integers. For $a \geqslant b$ the method is as follows: divide a by b, then b by the remainder, the first remainder by the second, and so on. When a remainder of zero is obtained, the last divisor is the greatest common divisor: to find the gcd of 28 and 12, $28/12 = 2$ remainder 4 and $12/4 = 3$ remainder 0, so 4 is the gcd.

eucrite (*Geol*) A coarse-grained, usually ophitic, deep-seated basic gabbro containing plagioclase near bytownite in composition, both ortho- and clino-pyroxenes, together with olivine. Eucrite is an important rock type in the Tertiary complexes of Scotland.

eucryptite (*Min*) A hexagonal lithium aluminium silicate, commonly found as an alteration product of spodumene.

eudialyte (*Min*) A pinkish-red complex hydrated sodium calcium iron zirconosilicate. It crystallizes in the rhombohedral system and occurs in some alkaline igneous rocks. Also *eudialite*.

eudiometer (*Chem*) (1) A voltameter-like instrument in which quantity of electricity passing can be found from volume of gas produced by electrolysis. (2) Similar system to determine volume changes in a gas mixture due to combustion.

eugamic (*BioSci*) Pertaining to the period of maturity.

eugenics (*BioSci*) The study of the means whereby the characteristics of human populations might be improved by the application of genetics.

eugenol (*Chem*) $C_6H_3(OH)(OCH_3)(CH_2CH=CH_2)$. Bp 252°C, rel.d. 1·07. A phenol homologue, the chief constituent of oil of cloves and cinnamon leaf oil; used for manufacturing VANILLIN.

eugeosyncline (*Geol*) A geosyncline, characterized by intermittent volcanic activity.

euglenoid movement (*BioSci*) A type of movement undergone by many Protozoa, which possess a definite body form, by means of contractions of the protoplasm stretching the pellicle. Also *metaboly*.

Euglenophyceae (*BioSci*) Small class of unicellular eukaryotic algae, flagellated, usually with two flagella, sometimes palmelloid. Chloroplasts, if present, have chlorophyll a and b. They may be phototrophs (often auxotrophic) or heterotrophs (both osmotrophic and phagotrophic) and are found mostly in fresh water, though a few species are found in marine environments. See PELLICLE.

euhedral crystals (*Geol*) See IDIOMORPHIC.

eukaryote (*BioSci*) A higher organism; literally those that have 'good nuclei' in their cells; ie animals, plants and fungi in contrast to the prokaryotic cells of bacteria and blue-green algae. Eukaryotes possess a nucleus bounded by a membranous nuclear envelope and have many cytoplasmic organelles. Their nuclear DNA is linear and complexed with proteins to form chromosomes. Adj *eukaryotic*. Cf PROKARYOTE.

Euler buckling limit (*Eng*) Criterion for the elastic BUCKLING of a structural member under compressive loading where the loading is along the centroid of the member's cross-section. The buckling load, P, is given by:

$$P = \frac{n^2\pi^2 E I}{L^2}$$

where E is YOUNG'S MODULUS, I is the minimum SECOND MOMENT OF AREA of the section, L is the loaded length of the member and n is the half wavelength of the buckled shape.

Eulerian angles (*MathSci*) Three independent angles φ, θ and ψ which specify the direction of a line in space.

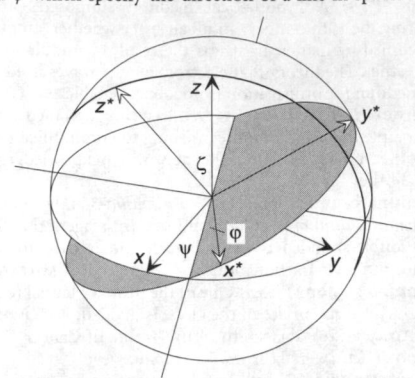

Eulerian angles The angles ψ, φ, ζ by which the sphere x,y,z has turned to its new position x*,y*,z*.

Eulerian path (*MathSci*) Of a graph, a closed path which includes every edge.

Euler's constant (*MathSci*) The limit of

$$\sum_{r=1}^{n} \frac{1}{r} - \ln n, \text{ as } n \to \infty$$

Denoted by γ or C. Its value is 0·577 215 66.....

Euler's formula (*CivEng*) A formula giving the collapsing load for a long, thin column of given sizes. It only applies where the load passes down the centroid of the column. It states that

$$P = \frac{\pi^2 E I \text{ (min)}}{L^2}$$

where $P =$ the collapsing load, $E =$ Young's modulus, $I =$ the least moment of inertia, and $L =$ the length of the pin-jointed column.

Euler's formula (*MathSci*) The formula stating that if v, e and f denote the number of vertices, edges and faces of a planar graph or polyhedron then $v - e + f = 2$.

Euler's theorem (*MathSci*) $K = A\cos^2\theta + B\sin^2\theta$, where K is the curvature of a normal section of a surface displaced by an angle θ from a principal direction, A is the curvature in that principal direction, and B is the other principal curvature.

EUMETSAT (*Space*) Abbrev for *European Meteorological Satellite organization*, an intergovernmental agency which provides operational meteorological data for its member states.

Eumycota (*BioSci*) A division containing the true fungi. These are eukaryotic, heterotrophic, walled organisms, typically with hyphae, or single cells (eg yeasts) or chains of cells. The walls usually contain CHITIN (chitosan in Zygomycotina) as a major constituent. The division comprises the subdivisions Mastigomycotina, Zygomycotina, Ascomycotina and Basidiomycotina.

eunuch (*Med*) A male who has no testes.

eunuchoid (*Med*) A male with testes, but no secondary male characteristics.

Euphausiacea (*BioSci*) An order of shrimp-like Malacostraca that are filter-feeding and pelagic. They are a major source of food for some whales. Krill.

euphenics (*BioSci*) The science concerned with the physical improvement of human beings by modifying their development after birth.

Euphorbiaceae (*BioSci*) A family of c.7000 spp of dicotyledonous flowering plants (superorder Rosidae), comprising trees, shrubs and herbs, mostly tropical, with unisexual flowers, and usually with latex. It includes *Hevea*, the source of most natural rubber, manioc (*Manihot utilissima*, cassava or tapioca) and the very large genus *Euphorbia*, the spurges (1600 spp), the members of which include C3 plants, C4 plants and leafless, stem-succulent CAM plants. See CYATHIUM.

euphoria (*Med*) A feeling of well-being, not necessarily indicative of good health. Also *euphory*.

euphotic zone (*BioSci*) See PHOTIC ZONE.

euploidy (*BioSci*) Having a chromosome complement consisting of one or more whole *chromosome sets* and, therefore, haploid, diploid or polyploid. Cf ANEUPLOID. Adj *euploid*.

eupyrene (*BioSci*) Spermatozoa that are normal, typical.

Euratom (*NucEng*) Abbrev for EUROPEAN ATOMIC ENERGY COMMUNITY.

Eureka (*ElecEng*) See CONSTANTAN.

eureka (*Radar*) See REBECCA-EUREKA.

euro-connector (*ImageTech*) See SCART.

Eurokom (*ICT*) A *teleconferencing* system for the participants of the ESPRIT program.

Euronet (*ICT*) A European information exchange NET-WORK.

Europa (*Astron*) The second natural satellite of Jupiter, discovered by Galileo in 1610, and encased in a mantle of ice. Distance from the planet 671 000 km; diameter 3140 km.

European plane (*For*) A hardwood (*Platanus acerifolia*) that yields a useful general purpose, but perishable, timber. It is light reddish-brown with characteristic fleck markings, straight-grained and fine-textured. Selected logs are cut at about 45° to the growth rings (quarter-sawn) to accentuate the flecked appearance. This is LACEWOOD and used for decorative panelling etc. The related *P. occidentalis* is known as sycamore in the USA, but should not be confused with the true SYCAMORE, *Acer pseudoplatanus*.

European Space Agency (*Space*) See ESA.

European Telecommunications Standards Institute (*ICT*) A body founded in 1988 with the aim of accelerating technical harmonization in telecommunications throughout the European Community by developing and promulgating a range of standards covering all developing areas of the technology, including their impact on related subjects such as broadcasting and information technology. Based in France and holding its General Assemblies in Nice.

europium (*Chem*) Symbol Eu, at no 63, ram 151·96, valency 2 and 3. Metallic element of the rare earth group. Contained in black monazite, gadolinite, samarskite, xenotime.

Eurovision (*ImageTech*) Exchange of TV programmes between member countries of the European Broadcasting Union.

euryhaline (*BioSci*) Descriptive of marine animals that will tolerate a wide variation in salinity.

Eurypterida (*Geol*) An order of Crustaceans ranging from the Ordovician to Permian and represented by such types as *Eurypterus* and *Stylonurus*, the latter reaching almost 2 metres in length with a scorpion-like appearance. They were entirely aquatic. Also *sea scorpions*.

eusporangium (*BioSci*) A sporangium of a vascular plant in which the wall develops from superficial cells, and the sporogenous tissue from internal cells of the sporophyll or sporangiophore. The wall is usually more than one cell thick at maturity. Cf LEPTOSPORANGIATE.

Eustachian tube (*BioSci*) In land vertebrates, a slender duct connecting the tympanic cavity with the pharynx.

Eustachian valve (*BioSci*) In mammals, a rudimentary valve separating the opening of the SUPERIOR VENA CAVA from that of the INFERIOR VENA CAVA.

eustatic movements (*Geol*) Changes of sea level, constant over wide areas, due probably to alterations in the volumes of the seas. These may be due to variations in the extent of the polar ice caps, large-scale crustal movements in ocean basins, or submarine volcanic activity.

eustele (*BioSci*) A stele in which the primary vascular tissue is organized into discrete vascular bundles surrounding a pith; typical of dicotyledons and gymnosperms.

eustomatous (*BioSci*) With a well-defined mouth or opening.

eustyle (*Arch*) A colonnade in which the space between the columns is equal to 21/4 times the lower diameter of the columns.

eutaxitic (*Geol*) Descriptive of the streaky banded structure of certain pyroclastic rocks, as contrasted with the smooth layered structure of flow-banded lavas.

eutectic (*Chem*) The isothermal transformation of a liquid solution simultaneously to different solid phases. It represents the lowest temperature for solidification of any mixture in that part of the system and involves simultaneous crystallization of two constituents in a binary system (or of three in a ternary system, four in a quaternary, and so on) and occurs at a unique temperature. Eutectic liquids also exhibit high fluidity compared with compositions on either side, which freeze over a range of temperature and pass through a pasty region during solidification. See fig. at PHASE DIAGRAM.

eutectic change (*Eng*) The transformation from the liquid to the solid state in a eutectic alloy. It involves the simultaneous crystallization of two constituents in a binary system and of three in a ternary system.

eutectic point (*Eng, Min*) The point in the binary or ternary constitutional diagram indicating the composition of the eutectic alloy, or mixture of minerals, and the temperature at which it solidifies.

eutectic structure (*Eng*) The particular arrangement of the constituents in a eutectic alloy which arises from their simultaneous crystallization from the melt.

eutectic system (*Eng*) A binary or ternary alloy system in which one particular alloy solidifies at a constant temperature which is lower than the beginning of solidification in any other alloy.

eutectic welding (*Eng*) A metal welding process carried out at relatively low temperature, using the eutectic properties of the metals involved.

eutectoid (*Eng*) Similar to a eutectic except that it involves the simultaneous formation of two or three constituents from another solid constituent instead of from a melt. Eutectoid point and eutectoid structure have similar meanings to those given for EUTECTIC POINT and structure.

eutectoid steel (*Eng*) Steel having the same composition as the eutectoid point in the iron–carbon system (0·87%C), and which therefore consists entirely of the eutectoid at temperatures below 723°C. See PEARLITE.

EUTELSAT (*Space, ICT*) The European Telecommunications Satellite organization, an intergovernmental agency that provides satellite communications for its participating countries. See panel on COMMUNICATIONS SATELLITE.

eutexia (*Eng*) The property of being easily melted, ie at a minimum melting point.

euthanasia (*Med*) Easy (good) or painless death; the action of procuring this.

euthenics (*BioSci*) The science concerned with the improvement of living conditions.

Eutheria (*BioSci*) An infraclass of viviparous mammals in which the young are born in an advanced stage of development. There is no marsupial pouch, and an allantoic placenta occurs. The scrotal sac is behind the penis, the angle of the lower jaw is not inflexed and the palate is imperforate. The higher mammals.

euthyroid (*Med*) Having a normal level of thyroid activity.

eutrophic (*EnvSci*) Said of a type of lake in which the HYPOLIMNION becomes depleted of oxygen during the summer by the decay of organic matter falling from the

EPILIMNION. A eutrophic lake is usually shallow, with much primary productivity. Cf OLIGOTROPHIC.

eutrophication (*EnvSci*) The enrichment of water with plant nutrients, which enter the water by leaching from the land drained by the water body; can be accelerated by the leaching of agricultural fertilizers and the discharge of effluents containing nitrogen and phosphorus.

euxenite (*Min*) Niobate, tantalate and titanate of yttrium, erbium, cerium, thorium and uranium, and valuable on this account. Commonly massive and brownish black in colour.

EV (*ImageTech*) Abbrev for EXPOSURE VALUE.

eV (*Phys*) Abbrev for ELECTRON-VOLT.

EVA (*Chem*) Abbrev for ETHYLENE–VINYLACETATE COPO-LYMER.

EVA (*Space*) Abbrev for *extra-vehicular activity*, ie operations performed outside the 'living environment' of a space vehicle. To accomplish EVA, it is necessary to wear a space (or pressure) suit provided with pressure control and life support systems.

evaginate (*BioSci*) Not having a sheath.

evagination (*BioSci*) (1) Withdrawal from a sheath. (2) The development of an outgrowth. (3) Eversion of a hollow ingrowth. (4) Turning inside out of an organ. (5) An out-growth or an everted hollow ingrowth. Cf INVAGINATION.

evanescent mode (*ICT*) A waveguide propagation mode at a frequency below the CUT-OFF WAVELENGTH. In this mode, the amplitude of the wave diminishes rapidly along the waveguide, but the phase does not change; applied in certain special waveguide filter designs.

evanescent wave (*Acous*) A non-decaying surface wave.

evaporation (*Phys*) The conversion of a liquid into vapour, at temperatures below the boiling point. The rate of evaporation increases with rise of temperature, since it depends on the saturated vapour pressure of the liquid, which rises until it is equal to the atmospheric pressure at the boiling point. Evaporation is used to concentrate a solution.

evaporative capacity (*Eng*) The mass of steam at a stated temperature and pressure produced in 1 hour from unit area of heating surface from feed-water at a stated temperature, when steaming at the most economical rate. Measured in $lb\,ft^{-2}\,h$ or $kg\,m^{-2}\,h$. Also *evaporation rate*.

evaporative cooling (*Aero, ChemEng*) The process of evaporating part of a liquid by supplying the necessary latent heat from the main bulk of liquid which is thus cooled. Used for some piston aero-engines in the 1930s, some current types of turbine and rocket components, and also for cooling purposes in cabin air-conditioning systems. Cf SWEAT COOLING.

evaporator (*Chem*) A still designed to evaporate moisture or solvents to obtain the concentrate.

evaporimeter (*EnvSci*) An instrument used for measuring the rate of natural evaporation.

evaporite (*Geol*) Sedimentary deposit of material previously in aqueous solution and concentrated by the evapora-tion of the solvent. Normally found as the result of evaporation in lagoons or shallow enclosed seas, eg salt and anhydrite.

evapotranspiration (*EnvSci*) The total water loss from a particular area, being the sum of evaporation from the soil and transpiration from vegetation.

evection (*Astron*) The largest of the four principal periodic inequalities in the mathematical expression of the Moon's orbital motion; due to the variable eccentricity of the Moon's orbit, with a longitude displacement maximum value of $1°16'$ and a period of $31\cdot81$ days.

even–even nuclei (*Phys*) Nuclei with even numbers of both protons and neutrons; normally stable.

even function (*MathSci*) f is an even function if $f(-x) = f(x)$. Cf ODD FUNCTION.

evening primrose oil (*BioSci*) Oil obtained from the seeds of *Oenothera biennis*, particularly rich in essential fatty acids; said to have remarkable medicinal properties.

evening star (*Astron*) The name given in popular language to a planet, usually Venus and sometimes Mercury, seen in the western sky at or just after sunset. Also used loosely to describe any planet which transits before midnight.

even number (*MathSci*) An integer divisible by two without remainder.

even–odd nuclei (*Phys*) Nuclei with an even number of protons and an odd number of neutrons.

even parity (*ICT*) Binary representation in which the number of ones is even. See PARITY CHECK.

even pitch (*Eng*) In screw-cutting in the lathe, the state when the PITCH of the thread is equal to, or an integral multiple of, the pitch of the LEAD SCREW.

even small caps (*Print*) Small capitals set up without capitals. Also *level small caps*.

event horizon (*Astron*) The boundary of a black hole: no light can escape from inside this boundary. See panel on BLACK HOLE.

eventration (*Med*) Protrusion of the abdominal contents outside the abdomen, eg through the diaphragm into the thorax.

event tree (*NucEng*) A method of investigating a real or simulated accident which, starting from an initial event, plots the alternative ways in which the accident can proceed depending on whether or not particular safety features function.

even working (*Print*) The situation when the pages of a book occupy a number of complete SECTIONS (usually of 16 or 32 pages). If an oddment of four or eight pages is required it is an *uneven working*.

Everest theodolite (*Surv*) A form of theodolite differing from the transit in that reversal of the line of sight is effected by removing the telescope from its trunnion supports and turning it.

ever-ready case (*ImageTech*) A case from which the camera can operate without being removed. Usually the camera is screwed into the case, whose top and front can hinge down in one piece to expose lens and controls. Abbrev ERC

eversafe (*NucEng*) Used to describe a nuclear processing plant which is designed so that a critical amount of fissile material can never accumulate.

EVF (*ImageTech*) Abbrev for ELECTRONIC VIEWFINDER.

evisceration (*Med*) Surgical removal of thoracic and abdominal contents from the fetus in obstructed labour; surgical removal of a structure (eg the eye) from its cavity. Also *disembowelment*.

evocation (*BioSci*) The initial event at the root apex in response to the arrival of the floral stimulus which commits the apex to the subsequent formation of flower primordia. Also *floral evocation*.

evolute (*BioSci*) Having the margins rolled outwards.

evolute (*MathSci*) Of a curve: locus of centre of curvature or envelope of normals. The original curve is the INVOLUTE of the evolute.

evolution (*BioSci*) Changes in the genetic composition of a population during successive generations. The gradual development of more complex organisms from simpler ones.

evolution (*MathSci*) Raising a number to a power $1/n$, where n is a positive integer, ie finding a root. Cf INVOLUTION (1).

evolutionarily stable strategy (*BioSci*) A strategy such that, if most members of a population adopt it, would give a reproductive fitness higher than any mutant strategy.

evolutionary computation (*MathSci*) The branch of artificial intelligence concerned with solving optimization problems using ALGORITHMS modelled on the process of evolution in nature.

evolutionary operation (*MinExt*) Introduction of section-ally controlled variants into a commercial process, during transfer of laboratory research into better production.

evulsion (*Med*) Plucking out by force.

EW (*ICT*) Abbrev for *electronic warfare*. See ELECTRONIC COUNTERMEASURES.

E-wave (*ICT*) See TM-WAVE.

ewe (*Agri*) A sexually mature female sheep, maidens not yet bred and ewe lambs up to 1 year old.

Ewing permeability bridge (*ElecEng*) A measuring device in which the flux produced in a sample of iron is balanced against that from a standard bar of the same dimensions. The magnetizing force on the bar under test is varied until balance is obtained when the permeability can be estimated.

EWT (*Build*) Abbrev for ELSEWHERE TAKEN.

ex- (*Genrl*) See E-.

ex. & ct. (*Build*) Abbrev for *excavate and cart away*.

exa- (*Genrl*) Prefix denoting 1 million million million, or 10^{18}. Symbol E.

exacerbation (*Med*) An increase in the severity of a disease or of its manifestations.

exact equation (*MathSci*) The differential equation

$$P(x,y)dx + Q(x,y)dy = 0, \text{ where } \frac{\partial P}{\partial y} = \frac{\partial Q}{\partial x}$$

The primitive is $u(x, y) = c$, c being a constant, where

$$\frac{\partial u}{\partial x} = P \text{ and } \frac{\partial u}{\partial y} = Q$$

exalbuminous (*BioSci*) A seed, lacking endosperm when mature; non-endospermic.

exalted carrier (*ICT*) The addition of a synchronized carrier before demodulation, to improve linearity and to mitigate the effects of fading during transmission.

exanthema (*Med*) An eruption on the surface of the body. Used specifically for infectious diseases characterized by an exanthem. Also *exanthem*. Pl *exanthemata*.

exarch (*BioSci*) Of a xylem strand, having the protoxylem to the side towards the outside of the axis (the normal condition in eg angiosperm roots).

excavation (*Med*) The process of hollowing out; a part hollowed out.

exception dictionary (*Print*) Words excluded from the hyphenation logic of the phototypesetter, often added by the operator.

exception handling (*ICT*) Programming routines written to cope with, or trap, errors, as when a missing file is called. Also *error handling*.

excess air (*Eng*) The proportion of air that has to be supplied in excess of that theoretically required for complete combustion of a fuel, because of the imperfect conditions under which combustion takes place in practice.

excess code (*ICT*) One that increases decimal digits before conversion to binary digits.

excess-3 code (*ICT*) Binary coding of a decimal number, to which has been added three. Complements are formed by 1 changed to 0, and 0 changed to 1, in all numerics.

excess conduction (*Electronics*) That arising from excess electrons provided by donor impurities.

excess electron (*Electronics*) One added to a semiconductor by eg a donor impurity.

excess feed (*Print*) The drawing through of too much paper when the peripheral speed of a web-driving component is greater than the next ahead. Also *making paper*. Cf INSUFFICIENT FEED.

excess voltage (*ElecEng*) See OVERVOLTAGE PROTECTIVE DEVICE.

exchange (*Phys*) (1) The interchange of one particle between two others (eg a pion between two nucleons), leading to establishment of exchange forces. (2) Possible interchange of state between two indistinguishable particles, involving no change in the wavefunction of the system.

exchangeable disk pack (*ICT*) A DISK PACK that can be removed from the disk unit for storage and later use on the same or a compatible machine.

exchange energy (*Phys*) A term, of quantum mechanical origin, in the energy balance of magnetic domains which

accounts for the interaction between neighbouring dipole moments. Cf SUPER EXCHANGE ENERGY.

exchange force (*Phys*) A force acting between particles due to the exchange of some property. In quantum mechanics, such forces can arise when two particles interact. In the ion of the hydrogen molecule, the forces responsible for the binding can be regarded as the continual exchange of the single electron between the two protons. The strong force between nucleons is the exchange of a pion (π-meson) between the two interacting nucleons. See PARTICLE EXCHANGE.

excimer (*Chem*) An excited dimer in which one of the two bound atoms is in a higher energy state.

excipient (*Pharmacol*) The inert ingredient in a medicine which takes up and holds together the other ingredients.

excision (*Med*) The action of cutting a part out or off; the surgical removal of a part.

excision repair (*BioSci*) Enzymatic DNA repair process in which a mismatching DNA sequence is removed and the gap filled by synthesis of a new sequence complementary to the remaining strand.

excitable tissue (*BioSci*) That which responds to stimulation, eg muscle or nervous tissue.

excitant (*ElecEng*) A term occasionally used to denote the electrolyte in a primary cell.

excitation (*ElecEng*) (1) Current in a coil which gives rise to a magnetomotive force (mmf) in a magnetic circuit, esp in a generator or motor. (2) The mmf itself. (3) The magnetizing current of a transformer.

excitation (*ICT*) A signal that drives any amplifier stage in a transmitter or receiver.

excitation (*Phys*) The addition of energy to a system, such as an atom or nucleus, raising it above the GROUND STATE.

excitation anode (*Electronics*) An auxiliary anode used to maintain the cathode spot of a mercury-pool cathode valve.

excitation–contraction coupling (*BioSci*) The process of coupling the arrival of an excitatory nerve impulse at the neuromuscular junction (motor end plate) with the DEPOLARIZATION of the muscle membrane and subsequent contraction. See STIMULUS–SECRETION COUPLING.

excitation loss (*ElecEng*) The ohmic loss (I^2R) in the field or exciting windings of an electric machine excited by dc.

excitation potential (*Electronics*) Potential required to raise orbital electron in atom from one energy level to another. Also *resonance potential*. See IONIZATION POTENTIAL.

excitatory amino acid (*BioSci*) An amino acid that acts as an excitatory neurotransmitter, eg glutamate.

excited atom (*Phys*) An atom which has more energy than in the normal or ground state. The excess may be associated with the nucleus or an orbital electron.

excited ion (*Phys*) Ion resulting from the loss of a valence electron, and the transition of another valence electron to a higher energy level.

excited nucleus (*Phys*) A nucleus raised to an excited state with an excess of energy over its ground state. Nuclear reactions frequently leave the product nucleus in an excited state. It returns to its ground state with the emission of gamma rays.

exciter (*ElecEng*) A small machine for producing current, usually dc, necessary for supplying the exciting winding of a larger machine. It is frequently mounted on the shaft of the machine which it is exciting.

exciter field rheostat (*ElecEng*) A rheostat in the field of an EXCITER whereby the voltage of the exciter, and, therefore, the excitation on the main machine, can be controlled.

exciter lamp (*ImageTech*) In optical sound reproduction, a small incandescent lamp whose light is modulated by the passage of the photographic sound track on the film.

exciter set (*ElecEng*) An assembly of one or more exciters driven by an electric motor.

exciting circuit (*ElecEng*) The complete circuit through which flows the current for exciting an electric machine. It comprises the exciter, the windings of the main machine, and possibly a field rheostat and measuring instruments.

exciting coil (*ElecEng*) A coil on a field magnet, or any other electromagnet, which carries the current for producing the magnetic field.

exciting current (*ElecEng*) That drawn by a transformer, magnetic amplifier, or other electric machine under no load conditions.

exciting winding (*ElecEng*) The winding which produces the emf to set up the flux in an electric machine or other apparatus.

exciton (*Electronics*) Bound hole–electron pair in semiconductor. These have a definite half-life during which they migrate through the crystal. Their eventual recombination energy releases a photon or, less often, several photons.

excitron (*Electronics*) A single-anode steel-tank mercury-arc rectifier, with a means for maintaining a continuous cathode spot, even when current is not flowing to the anode. This improves the turn-on characteristics of the rectifier.

excluded volume (*Chem*) A property of real polymer chains, where they cannot pass through one another. Leads directly to problem of entanglements in chains. See STATISTICAL CHAIN.

exclusion principle (*Phys*) See PAULI EXCLUSION PRINCIPLE and panel on ATOMIC STRUCTURE.

excoriation (*Med*) Superficial loss of skin, eg through scratching.

excrescence (*Med*) Any abnormal outgrowth of tissue.

excreta (*BioSci*) Poisonous or waste substances eliminated from a cell, tissue or organism. Adj *excreted*, n *excretion*.

excurrent (*BioSci*) (1) Running out, as when the midrib of a leaf is prolonged into a point. (2) With a single main axis or trunk and subordinate laterals, as in trees, eg pine, esp when young. (3) Of ducts etc carrying an outgoing current.

excursion (*NucEng*) Rapid increase of reactor power above set operation level, either deliberately caused for experimental reasons or accidental.

execute (*ICT*) Carry out instructions specified by the machine-code version of a program. See COMPILER, MACHINE-CODE INSTRUCTION.

execution error (*ICT*) Error detected during program execution (eg overflow, division by zero). Also *run-time error*.

executive program (*ICT*) See COMPILER, EDITOR, OPERATING SYSTEM.

exempted spaces (*Ships*) Certain spaces in the ship which are not included in the GROSS TONNAGE, eg double bottoms used for water ballast, wheelhouse, galley, water closets, etc.

exenteration (*Med*) Complete removal of the contents of a cavity, eg disembowelment.

exergonic (*BioSci*) A biochemical reaction accompanied by the release of energy (strictly, with negative ΔG) and capable, therefore, of proceeding spontaneously.

Exeter hammer (*Build*) See LONDON HAMMER.

exfoliation (*BioSci*) The process of falling away in flakes, layers or scales, as some bark in plants.

exfoliation (*Eng*) Lifting away the surface of a metal due to the formation of corrosion products beneath the surface, a common result of rusting. Also *spalling*.

exfoliation (*Geol*) The splitting off of thin folia or sheets of rock from surfaces exposed to the atmosphere, particularly in regions of wide temperature variation. It is one of the processes involved in spheroidal weathering.

exhalant (*BioSci*) Emitting or carrying outwards a gas or fluid; as the *exhalant siphon* in some Mollusca.

exhaust (*Eng*) (1) The working fluid discharged from an engine cylinder after expansion. (2) That period of the cycle occupied by the discharge of the used fluid.

exhaust cone (*Aero*) In a turbojet or turboprop, the duct immediately behind the turbine and leading to the *jet pipe*, consisting of an inner conical unit behind the *turbine disk* and an outer unit of frustum form connecting the *turbine shroud* to the jet pipe.

exhaust-driven supercharger (*Aero*) A piston-engine supercharger driven by a turbine motivated by the exhaust gases. Also *turbo-supercharger*.

exhaust fan (*Eng*) A fan used in artificial draught systems; placed in the smoke uptake of a boiler to draw air through the furnace and exhaust the flue gases.

exhaust gas (*Eng*) The gaseous exhaust products of an internal-combustion engine, containing in general CO_2, CO, O_2, N_2 and water vapour.

exhaust-gas analyser (*Autos*) An instrument which records the mixture strength supplied to a petrol engine by automatic electrical measurement of the thermal conductivity of the exhaust gas. See CO_2 RECORDER.

exhausting (*FoodSci*) Removing air from a can before closing and seaming by hot-filling and then steam-purging the HEADSPACE. The advantage is that oxygen content is minimized and spoilage due to oxidation and vitamin loss is prevented or delayed.

exhaustion dyeing (*Textiles*) The process of dyeing in which the textile takes up substantially all the available dyestuff.

exhaustive methylation (*Chem*) The process of converting bases into their quaternary ammonium salts and subsequent distillation with alkalis, resulting in the formation of simpler unsaturated compounds which can be reduced to known saturated compounds. This method was of particular value for investigating the constitution of alkaloids and other complicated ring systems.

exhaust lap (*Eng*) Of a slide-valve, the distance moved by the valve from mid-position on the port face, before uncovering the steam port to exhaust. Also *inside lap*.

exhaust line (*Eng*) The lower line of the enclosed area of an INDICATOR DIAGRAM showing the back pressure on the piston during the exhaust stroke of an engine.

exhaust manifold (*Autos*) Branched pipe which channels burnt gases from the combustion chambers to the exhaust system.

exhaust pipe (*Eng*) The pipe through which the exhaust products of an engine are discharged.

exhaust port (*Eng*) In an engine cylinder, the port or opening through which a valve allows egress of the exhaust.

exhaust stator blades (*Aero*) An assembly of stator blades, usually in sections to allow for thermal expansion, mounted behind the turbine to remove residual swirl from the gases.

exhaust steam (*Eng*) See LIVE STEAM.

exhaust stroke (*Eng*) In a reciprocating engine, the piston stroke during which the EXHAUST is ejected from the cylinder. Sometimes *scavenging stroke*.

exhaust valve (*Eng*) The valve controlling the discharge of the exhaust gas in an internal-combustion engine. Risk of overheating presents a design problem, particularly in aircraft engines, which is met by the use of heat-resisting steels, facing with Stellite or a similar hard deposit.

exhaust velocity (*Space*) The velocity at which a propellant gas leaves a rocket motor. It is related to the SPECIFIC IMPULSE, I_{sp}, by the expression $v_e = I_{sp}g$, where v_e is the exhaust velocity and g is the acceleration due to gravity at the Earth's surface.

exhibitionism (*Psych*) Describes behaviour in which sexual gratification is obtained through displaying the genitals to members of the opposite sex; by extension all behaviour motivated by the pleasure of self-display.

exine (*BioSci*) The outer part of the wall of a pollen grain or embryophyte spore, from the patterns of the surface of which it is often possible to identify the genus or even species of plant from which the pollen has come. Cf INTINE. See POLLEN ANALYSIS, SPOROPOLLENIN.

exinguinal (*BioSci*) In land vertebrates, outside the groin.

exinite (*Geol*) A hydrogen-rich MACERAL which is found in coal.

existential psychology (*Psych*) A philosophical movement that forms the basis of some psychological approaches to therapy that emphasise free will, individuality and the subjective nature of experience.

exit (*ICT*) To transfer control from a SUBPROGRAM back to the calling program or from the program entirely.

There can be more than one exit in a program or sub-program.

exit portal (*Radiol*) The area through which a beam of radiation leaves the body. The centre of the exit portal is sometimes called the emergent ray point.

exit pupil (*Phys*) The image of the aperture stop as formed by that part of an optical system on the image side of the aperture. It defines the emergent cone of rays from the system. In a microscope or telescope it is usually the image of the objective formed by the eyepiece, and it is the position which should be occupied by the eye of the observer.

Exner function (*EnvSci*) If p is the atmospheric pressure, p_0 a reference pressure, and γ is the ratio of the specific heats of a perfect gas, then the Exner function P of p is given by

$$P = (p/p_0)^{(\gamma-1)/\gamma}$$

It is useful in studies of compressible adiabatic flow.

exo- (*Genrl*) Prefix from Gk *exo* signifying outside.

exobiology (*BioSci*) The study of putative living systems that probably exist elsewhere in the universe.

exocardiac (*BioSci*) Outside the heart.

exocarp (*BioSci*) See EPICARP.

exoccipital (*BioSci*) A paired lateral cartilage bone of the vertebrate skull, forming the side wall of the braincase posteriorly.

exocoelar (*BioSci*) See SOMATOPLEURE.

exocoelom (*BioSci*) The extraembryonic coelom of a developing bird, reptile or elasmobranch.

exocrine (*BioSci*) A gland that secretes its hormone into a duct rather than directly into the blood stream as do endocrine glands.

exocuticle (*BioSci*) The layer of the insect cuticle, between the epicuticle and endocuticle, which becomes hardened and darkened in most insects.

exocytosis (*BioSci*) The exit of material from the cell by fusion of internal vesicles with the PLASMA MEMBRANE. The vesicles either void their contents to the exterior or introduce new surface material into the plasma membrane.

exodermis (*BioSci*) The outermost layer or layers of the cortex of some roots with more or less thickened and/or suberized cell walls; a specialized hypodermis.

exo-electron (*Electronics*) One, emitted from the surface of a metal or semiconductor, which comes from a metastable trap with very low binding energy under conditions such that electrons in their ground state could not be emitted.

exo-ergic process (*Phys*) Nuclear process in which energy is liberated. Also *exo-energetic*. The equivalent thermodynamic term is *exothermic*.

exogamete (*BioSci*) A gamete that unites with one from another parent.

exogamy (*BioSci*) The production of a zygote by the fusion of gametes from two unrelated parents. Cf ALLOGAMY, ENDOGAMY, OUTBREEDING.

exogenetic (*Geol*) Describes (1) processes originating at or near the surface of the Earth, eg weathering, denudation, and the rocks (esp sedimentary rocks), ore deposits and landforms to which they give rise; (2) the processes leading to *exomorphism*, modification of the country rocks by igneous rocks intruding them. Also *exogenous*. Cf ENDOGENETIC.

exogenous (*BioSci*) (1) Resulting from causes external to an organism. (2) Produced on the outside of another organ or developed from tissues at or near the surface (as leaf primordia are). (3) Growing by the addition of new layers at or near the surface. (4) In higher animals, said of metabolism which leads to the production of energy for activity. Cf ENDOGENOUS.

exogenous antigen (*BioSci*) Antigenic determinants taken up, rather than produced by a cell and processed and presented to T-lymphocytes in conjunction with MAJOR HISTOCOMPATIBILITY COMPLEX class II proteins.

exomphalos (*Med*) A hernia formed by the protrusion of abdominal contents into the umbilicus.

exon (*BioSci*) That part of the transcribed nuclear RNA of eukaryotes which forms the mRNA after the excision of the INTRONS.

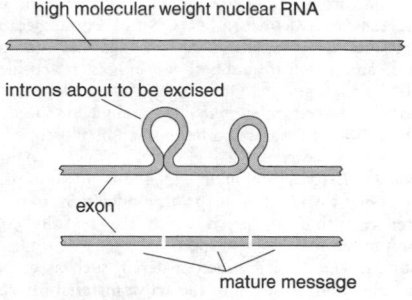

exon Stages in the maturation of messenger RNA.

exonuclease (*BioSci*) An enzyme that cleaves nucleic acids from a free end. It can thus digest a linear molecule by steps. Cf ENDONUCLEASE.

exophoria (*Med*) Latent external squint revealed in an apparently normal person by passing a screen before the eye.

exophthalmic goitre (*Med*) The protrusion of the eyes, or eye-muscle disorder, associated with hyperthyroidism. Also *Graves' disease*.

exoplanet (*Astron*) A planet that orbits a star other than the Sun.

exopodite (*BioSci*) The outer ramus of a crustacean appendage.

Exopterygota (*BioSci*) A subclass of insects in which wings occur, although they are sometimes secondarily lost. The change from young form to adult is gradual, and the wings develop externally. The young form is usually a NYMPH.

exoscopic embryology (*BioSci*) The condition when the apex of the embryo is turned towards the neck of the archegonium.

exoskeleton (*BioSci*) Hard supporting or protective structures that are external to and secreted by the ectoderm, eg in vertebrates, scales, scutes, nails and feathers; in invertebrates, the carapace, sclerites, etc.

exosphere (*Astron*) Outermost region of the Earth's atmosphere, beyond around 500 km from the surface.

exospore (*BioSci*) (1) The outermost layer of the wall of a spore or exine. (2) A spore formed by the extrusion of material from the parent cell.

exostosis (*Med*) A bony tumour growing outwards from a bone.

exothermic (*Chem*) Descriptive of a chemical reaction which evolves thermal energy (heat), ie ΔH negative. Cf ENDOTHERMIC.

exotic (*BioSci*) Not native. Also *ecdemic*.

exotoxin (*BioSci*) A toxin released by a bacterium into the medium in which it grows. Frequently very toxic, eg neurotoxins which destroy cells of the nervous system, haemolytic toxins which lyse red blood cells.

expanded (*Build*) Of cellular structure and therefore light in weight. Thus expanded concrete is *lightweight concrete*.

expanded memory (*ICT*) In a computer with 80286 or higher CHIPSETS using MS-DOS, special additional memory up to 32 Mbytes in size beyond the normal limit of 640 Kbytes of MAIN MEMORY. This memory is accessed using the technique of BANK SWITCHING. See fig. at MEMORY MAP.

expanded metal (*Build*) A metal network formed by suitably stamping or cutting sheet metal and stretching it to form open meshes. It is used as a reinforcing medium in concrete construction, as lathing for plasterwork and various other purposes. Cf BRC FABRIC.

expanded plastics (*Plastics*) Foamed plastic materials, eg polyvinyl chloride, polystyrene, polyurethane, polythene, etc, created by the introduction of pockets or cells of inert gas (air, carbon dioxide, nitrogen, etc) at some stage of manufacture. Used for heat insulation purposes or as core materials in SANDWICH BEAM construction, because of their low densities; also, because of lightness, for packaging, and (as foam rubber) for upholstery cushioning, artificial sponges, etc. Often highly flammable. See FOAM.

expanded sweep (*Electronics*) A technique for speeding up the motion of the electron beam in an oscilloscope during a part of the sweep.

expander (*Acous*) Amplifying apparatus for automatically increasing the contrast in speech modulation, particularly after reception of speech which has had its contrast compressed by a COMPRESSOR.

expander (*ElecEng*) An inert material, such as carbon or barium sulphate, added to the active material in accumulator plates to prevent shrinkage of the mixture.

expanding arbor (*Eng*) A lathe arbor expandable by blades or keys sliding in taper grooves, which allows work of various bore diameters to be supported and located.

expanding bit (*Build*) A boring bit carrying a cutter on a radial arm, the position of the cutter being adjustable so that holes of different sizes may be cut.

expanding brake (*Eng*) A brake in which internal shoes are expanded to press against the drum, usually by a cam or toggle mechanism.

expanding cement (*Build, CivEng*) Cement containing a chemical agent which induces predetermined expansion to minimize the normal shrinkage which occurs in concrete during the setting and drying-out process.

expanding mandrel (*Eng*) See MANDREL (1).

expanding metals (*Eng*) Metals or alloys, eg two parts antimony to one part bismuth, which expand in final stage of cooling from liquid; used in type-founding.

expanding plug (*Build*) A BAG PLUG.

expanding reamer (*Eng*) A REAMER, (1) partially slit longitudinally, and capable of slight adjustment in diameter by a coned internal plug, (2) machined with six or more grooves cut at a small angle to the axis along which hard steel blades, correspondingly tapered, can be moved by nuts at either end. The effective circumference of the blades is thus adjustable over a range of about 1·5 mm.

expanding universe (*Astron*) The view, based on the evidence of the REDSHIFT of galaxies, that the whole universe is expanding. See panels on COSMOLOGY and REDSHIFT–DISTANCE RELATION.

expansion (*Electronics*) A technique by which the effective amplification applied to a signal depends on the magnitude of the signal being larger for the bigger signals. See AUTOMATIC VOLUME EXPANSION.

expansion (*Eng*) (1) Increase in volume of working fluid in an engine cylinder. (2) Piston stroke during which such expansion occurs.

expansion (*Phys*) See ADIABATIC CHANGE.

expansion board (*ICT*) Also *expansion card*. See ADAPTER CARD.

expansion circuit breaker (*ElecEng*) A circuit breaker in which arc extinction occurs as a result of the rapid cooling produced by the expansion of steam or of gases; these result from the arc which arises between the contacts in water or in a small quantity of oil.

expansion curve (*Eng*) The line on an INDICATOR DIAGRAM which shows the pressure of the working fluid during the expansion stroke.

expansion engine (*Eng*) An engine which utilizes the working fluid expansively.

expansion gear (*Eng*) That part of a steam-engine valve gear through which the degree of expansion can be varied.

expansion joint (*CivEng*) A joint arranged between two parts to allow them to expand with temperature rise,

without distorting laterally, eg the gap left between successive lengths of rail or the joint made between successive sections of carriageway in road construction. See CONTINUOUS WELDED RAIL.

expansion joint (*Eng*) A special pipe joint used in long pipelines to allow for expansion, eg a horseshoe bend, a corrugated pipe acting as a bellows, a sliding socket joint with a stuffing box.

expansion line (*Eng*) See EXPANSION CURVE.

expansion of gases (*Phys*) See ABSOLUTE TEMPERATURE, COEFFICIENT OF EXPANSION, GAS LAWS.

expansion pipe (*Build*) In a domestic system of heating, a pipe carried up from the hot-water tank to a point above the level of the cold-water tank and bent over, so that boiling water or steam is carried away.

expansion ratio (*Aero*) The ratio between the gas pressure in a rocket combustion chamber, or a jet pipe, and that at the outlet of the propelling nozzle.

expansion rollers (*Eng*) Rollers on which one end of a large girder or bridge is often carried, to allow for movement resulting from thermal expansion; the other end of the girder etc is fixed. See BRIDGE BEARING.

expansion slot (*ICT*) A set of connectors joined electrically to the MOTHERBOARD into which ADAPTER CARDS may be inserted.

expansion tank (*Build*) In a hot-water system, the tank connected to, and above, the hot-water cylinder to allow expansion of the water on heating; often the cold-water feed tank is so used. See EXPANSION PIPE.

expansion valve (*Eng*) An auxiliary valve working on the back of the main slide-valve of some steam engines to provide an independent control of the point of cut-off.

expansive working (*Eng*) The use of a working fluid expansively in an engine; an essential feature of every efficient working cycle.

expectancy theory (*Psych*) A cognitive theory of animal and human learning which posits that the anticipation of events, and esp of rewards, is an important aspect of many learning events.

expectation (*MathSci*) The average value of a random variable in an infinite series of repetitions of an experiment or observations.

expectoration (*Med*) The coughing up of mucous or sputum from the air passages.

expedor phase advancer (*ElecEng*) A phase advancer which injects into the secondary circuit of an induction motor an emf which is a function of the secondary current. Cf SUSCEPTOR PHASE ADVANCER.

expendable launch vehicle (*Space*) A launch system which is made up of throwaway stages and has no recoverable parts. Abbrev ELV.

experiment (*Genrl*) A controlled trial to test a theory.

experimental allergic encephalomyelitis (*BioSci*) An auto-immune disease produced by injections into mice of proteins present in brain and spinal cord together with complete Freund's adjuvant. After a few days acute encephalomyelitis develops, accompanied by demyelination and progressive paralysis. The main factor is T-cell sensitization against myelin basic protein but antibodies may also be involved. Used as an experimental model for multiple sclerosis.

experimental embryology (*BioSci*) The experimental study of the physiology and mechanics of development.

experimental mean pitch (*Aero*) The distance of travel of a propeller along its own axis, while making one complete revolution, assuming the condition of its giving no thrust.

experimental petrology (*Geol*) A branch of petrology concerned with the laboratory study of rocks and minerals under different physical and chemical conditions.

experimental psychology (*Psych*) An area of research psychology that depends upon experimental rather than observational methods; has close links with fields of sensory physiology, neurobiology and ethology.

expert system (*ICT*) Software designed to function like a specialist consultant. It usually has two parts, a base of organized expert knowledge that can be easily expanded, and a set of rules for reaching conclusions. See KNOWL-EDGE-BASED SYSTEMS.

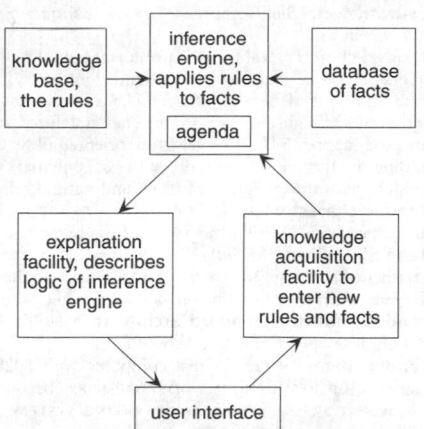

expert system The agenda consists of the ordered set of rules which satisfy the facts.

expiration (*BioSci*) The expulsion of air or water from the respiratory organs.

explant (*BioSci*) A piece of tissue or an organ removed for experimental purposes, from a plant esp to start a PLANT CELL CULTURE.

explantation (*BioSci*) In experimental zoology, the culture, in an artificial medium, of a part or organ removed from a living individual, tissue culture. N *explant*.

explement (*MathSci*) See CONJUGATE ANGLES.

expletive (*Build*) A stone used as a filling for a cavity.

explicit function (*MathSci*) The expression of the values of one variable directly in terms of the values of another variable, such as $y = x^2$. Cf IMPLICIT FUNCTION.

exploding star (*Astron*) See NOVA.

exploitable girth (*For*) (1) The minimum girth at breast height at or above which trees are considered suitable for felling or for the purpose specified. (2) The girth down to which all portions of a bole or tree must be exploited as timber or fuel under a permit licence.

exploitation well (*MinExt*) Well drilled in a proved deposit.

exploration well (*MinExt*) Well drilled in unproven ground to test for oil after a positive seismic or other survey.

exploratory behaviour (*Psych*) A form of APPETITIVE BEHAVIOUR, by which an animal gains information about its environment; some forms of exploratory behaviour are goal-linked; others seem motivated by a general curiosity mingled with mild fear and are not terminated by any apparent end goal.

Explorer (*Space*) A series of US artificial satellites used to study the physics of space cosmic rays, temperatures and meteorites; responsible for the discovery of the VAN ALLEN RADIATION BELTS. Explorer 1 was the first Earth satellite launched by the US (January 1958); Explorer 83 was launched in 2004.

exploring brush (*ElecEng*) A small brush fitted to a dc machine for experimental purposes; it can be moved in order to investigate the distribution of potential round the commutator.

exploring coil (*ElecEng*) A way of measuring magnetic field strengths by recording the emfs from within the field and from positions remote to the field. Also *search coil*.

explosion (*Chem*) A rapid increase of pressure in a confined space. Explosions are generally caused by the occurrence of exothermic chemical reactions in which gases are produced

in relatively large amounts. For nuclear explosion see ATOMIC BOMB, HYDROGEN BOMB.

explosion pot (*ElecEng*) A strong metal container surrounding the contacts of an oil circuit breaker; the high pressure set up inside the pot when an arc occurs assists in the extinction of the arc.

explosion-proof (*ElecEng*) Said of electrical apparatus so designed that an explosion of flammable gas inside the enclosure will not ignite flammable gas outside. Such apparatus is used in mines or other places having an explosive atmosphere. Also *flame-proof*.

explosion welding (*Eng*) Welding of two components made to hit each other at high speed due to a controlled explosion. Also *explosive welding*.

explosive (*MinExt*) There are two main classes, 'permitted' and 'non-permitted', ie those which are safe for use in coal mines and those which are not. Ammonium nitrate mixtures are mostly used in coal mines; nitroglycerine derivatives in metal mines. ANFO (*ammonium nitrate and fuel oil*) is now widely used in hard-rock mining. Explosives are used as propellants (*low explosives*) and for blasting (*high explosives*), in both civil and military applications.

explosive decompression (*FoodSci*) The explosive expansion of water vapour from a wet starchy feedstock kept at a high temperature and pressure when the pressure is suddenly released, as in making puffed cereals and snack products.

explosive forming (*Eng*) One of a range of high-energy rate-forming processes by which parts are formed at a rapid rate by extremely high pressures. Low and high explosives are used in variations of the explosive forming process: with the former, known as the cartridge system, the expanding gas is confined; with the latter, the gas need not be confined and pressure of up to 1 million atmospheres may be attained. See ELECTROHYDRAULIC FORMING.

explosive fracturing (*MinExt*) The use of an explosive charge to crack strata and increase oil flow round a borehole. Invented by Col. Roberts in Titusville, Pennsylvania, as the *Roberts torpedo*.

explosive limits (*Eng*) The upper and lower limits of the ratio of an inflammable gas to air within which the mixture can explode. Petrol vapour will only explode between 7·6 and 1·4% vapour to air by volume.

explosive rivet (*Eng*) A type of BLIND RIVET which is clinched or set by exploding, electrically, a small charge placed in the hollow end of the shank.

exponent (*MathSci*) See INDEX.

exponential baffle (*Acous*) A baffle approximating to a short section of an exponential horn.

exponential distribution (*MathSci*) A probability distribution often applied to represent the distribution of the time that elapses before the occurrence of an event.

exponential function (*MathSci*) The function $\exp(x)$ or e^x. See EXPONENTIAL SERIES.

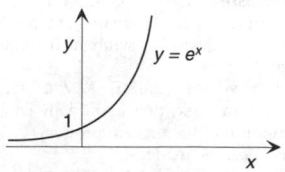

exponential function

exponential growth (*BioSci*) A stage of growth occurring in populations of unicellular micro-organisms when the logarithm of the cell number increases linearly with time.

exponential horn (*Acous*) A horn for which the taper or flare follows an exponential law.

exponential reactor (*NucEng*) A reactor with insufficient fuel to make it diverge; it needs excitation by an external source of neutrons for the determination of its properties. See DIVERGENCE.

exponential series (*MathSci*) The series

$$1 + \frac{x}{1!} + \frac{x^2}{2!} + \frac{x^3}{3!} + \cdots + \frac{x^n}{n!} + \cdots$$

which converges to the value e^x for all values of x.

export (*BioSci*) The process of transferring proteins across a membrane either into the medium or into another cellular compartment.

expose (*ICT*) In object-oriented programming, the approach of selectively revealing to other parts of the system data or logic which has been encapsulated within an object.

exposure (*EnvSci*) The method by which an instrument is exposed to the elements. The exposure in meteorological stations is standardized so that records from different stations are comparable.

exposure (*ImageTech*) (1) The process of allowing light to fall on a photosensitive surface; numerically specified by the intensity of light I and the duration of the exposure, time T, $E = IT$, in candela-metre-seconds or lux-seconds. (2) In practice, exposure in a system is determined by the illumination and reflectivity of the subject, the light transmission of the lens, its APERTURE, and the length of time the shutter is open; each of these factors must be adjusted to suit the sensitivity of the material being exposed.

exposure dose (*Radiol*) See DOSE.

exposure learning (*Psych*) Changes in behaviour that result from an individual being exposed to an object or situation, under circumstances in which no consistent response apart from investigatory or exploratory behaviour is elicited by that situation, and with no obvious reward. See OBSERVATIONAL LEARNING.

exposure meter (*ImageTech*) Instrument, a combination of photocell and current meter, for measuring the light reflected from or falling on a subject (incident light) and expressing it in terms of exposure for a given stop or as an EXPOSURE VALUE. Now often incorporated in the camera.

exposure value (*ImageTech*) An index combining both aperture and shutter speed, expressed as a single figure which remains constant when both are altered, eg to change depth of focus, in such a way that the amount of light reaching the sensitized emulsion is unchanged. Some shutters have stop and speed rings coupled to give a constant exposure value when either is altered. Abbrev *EV*.

expressed sequence tag (*BioSci*) A DNA sequence from the end of a random cDNA clone from an expression library. Such tags provide a way of identifying cDNAs of interest, which in turn indicates which proteins are being produced. Abbrev *EST*.

expression (*Textiles*) Residual liquid left in a fabric after squeezing (eg on a mangle), calculated as a percentage of the dry fabric.

Expressionism (*Arch*) A concept of architectural design which prevailed in Germany at the beginning of the 20th century, based on symbolism frequently inspired by biological forms.

expression vector (*BioSci*) A VECTOR, used in genetic manipulation work, that is specially constructed so that a large amount of the protein product, coded by an inserted sequence, can be made.

expulsion fuse (*ElecEng*) An enclosed fuse-link in which the arc occurring when the link melts is extinguished by the lengthening of the break due to expulsion of part of the fusible material through a vent in the container.

expulsion gap (*ElecEng*) A special form of expulsion fuse connected in series with a gap and placed across insulator strings on an overhead transmission line; a voltage surge breaks down the gap and the resulting arc is quickly broken by the fuse, so that no interruption to the supply need take place.

exsanguination (*Med*) Severe loss of blood. Exsanguination is often used as a method for killing animals, although it is not very humane.

exserted (*BioSci*) Protruding.

exstrophy (*Med*) A turning inside out of a hollow organ (esp the bladder). Also *extraversion*, *extrophy*.

ex. sur. tr. & ct. (*Build*) Abbrev for *excavate surface trenches and cart away*.

extended-chain crystal (*Chem*) Conformation of polymers where individual chains are fully extended in the crystalline state. In polyethylene, the extended chain is a linear zigzag and occurs in high-density polyethylene crystallized at high pressure (approx 5 kbar) or gel-spun, oriented fibre. Chain folding is the most common state of polymer single crystals, with the exception of PTFE and natural cellulose.

extended character set (*ICT*) A set of CHARACTERS larger than that defined within the ASCII set of characters.

extended delivery (*Print*) The delivery of a machine lengthened by design to give the printed image on the sheet the maximum time to set before being covered by the next.

extended industry standard architecture (*ICT*) A bus system developed from ISA. Abbrev *EISA*.

extended memory (*ICT*) In a computer with 80286 or higher CHIPSETS using MS-DOS, MEMORY beyond the 1 Mbyte ADDRESS limit of the OPERATING SYSTEM which will require special programs, eg Microsoft Windows, to make use of it. Also *XMS*. See fig. at MEMORY MAP.

extended total access communications system (*ICT*) Enhanced UK analogue cellular mobile-telephone system with specification identical to the former TOTAL ACCESS COMMUNICATIONS SYSTEM (TACS) but having an additional 720 channels provided in the bands 872–890 MHz (mobile) and 917–935 MHz (base station). Abbrev *ETACS*.

extender (*Build, Plastics*) (1) A substance added to paint as an adulterant or to give it body, eg barytes, china clay, French chalk, gypsum, whiting. (2) In synthetic resin adhesives, a substance (eg rye flour) added to reduce the cost of gluing or to adjust viscosity. (3) A non-compatible plasticizer used as an additive to increase the effectiveness of the compatible plasticizer in the manufacture of elastomers. (4) Low-molecular-mass substance which may be used to replace plasticizer, esp in polyvinyl chloride compounds. See panel on POLYVINYL CHLORIDE.

extender (*ImageTech*) See TELECONVERTER.

extenders (*Print*) The parts of lower case type characters which extend above or below the X-HEIGHT, ie the ascending and descending parts in eg b, f, p. See fig. at TYPEFACE.

extensible markup language (*ICT*) Text formatting instructions specifically designed to aid data searching and the formatting of the results. An extension of HYPERTEXT MARKUP LANGUAGE. Abbrev *XML*.

extensin (*BioSci*) The major structural protein of the primary cell walls of higher plants, rich in hydroxyproline and highly insoluble owing to dimerization of tyrosine residues.

extension (*ICT*) See DYNAMIC LINK LIBRARY and FILE NAME EXTENSION.

extension flap (*Aero*) A landing flap which moves rearward as it is lowered so as to increase the wing area. See FOWLER FLAP.

extension ore (*MinExt*) In assessment of reserves, ore which has not been measured and sampled, but is inferred as existing from geological reasoning supported by proved facts regarding adjacent ore.

extension ratio (*Eng*) The ratio of extended length to original length of rubber samples. Symbol λ (= new length/original length). Also *elongation ratio*. See STRAIN.

extension spring (*Eng*) A helical spring with looped ends which stores energy when in the stretched condition.

extension tubes (*ImageTech*) Tubes which increase the distance between the lens and the focal plane of a camera for close-up work. Made in sets to give different degrees of enlargement.

extensive farming (*Agri*) Production with low levels of input per unit area, including energy and labour. Characteristic of large, upland farms in the UK.

extensometer (*Eng*) An instrument for measuring dimensional changes of a material esp during a mechanical test such as the TENSILE TEST. Many different principles of operation, eg optical lever, MOIRÉ FRINGES, optical interferometry, various transducers, etc. See CLIP GAUGE.

extensor (*BioSci*) A muscle which, by contracting, straightens a limb or other part of the body. Cf FLEXOR.

extensor–plantar response (*Med*) The extension of big toe and fanning of other toes on stimulation of sole of foot; a sign of organic disease of the nervous system. It is normal in infants. Modern name for *Babinski's sign*.

exterior angle (*MathSci*) The angle between any side produced and the adjacent side (not produced) of a polygon.

external angle (*Build*) A vertical or horizontal angle forming part of the projecting portion of a wall or feature of a building. Also *salient junction*. See ARRIS.

external characteristic (*ElecEng*) A curve showing the relation between the terminal voltage of an electric generator and the current delivered by it.

external circuit (*ElecEng*) The circuit to which current is supplied from a generator, a battery, or other source of electrical energy.

external compensation (*Chem*) Neutralization of optical activity by the mixture or loose molecular combination of equal quantities of two enantiomorphous molecules.

external conductor (*ElecEng*) The outer earthed conductor of an earthed concentric wiring system.

external digestion (*BioSci*) A method of feeding, adopted by some Cnidaria, Turbellaria, Oligochaeta, Echinodermata, Insecta and Araneida, in which digestive juices are poured onto food outside the body and imbibed when they have dissolved some or all of the food.

external firing (*Eng*) The practice of heating a boiler or pan by a furnace outside the shell; all modern boilers have internal furnaces and flues.

external indicator (*Chem*) An indicator to which are added drops of the solution in which the main reaction is taking place, away from the bulk of the solution.

externalization (*Psych*) (1) In learning, the way in which external stimuli activate drives. (2) In developmental psychology, the process by which a child differentiates between self and the outside world. (3) In social psychology, the attribution of behavioural cause to external factors; by extension, the process of PROJECTION.

external respiration (*BioSci*) See RESPIRATION.

external screw-thread (*Eng*) A screw-thread cut on the outside of a cylindrical bar. Also *male thread*.

external secretion (*BioSci*) A secretion that is discharged to the exterior, or to some cavity of the body communicating with the exterior. Cf INTERNAL SECRETION.

exteroceptor (*BioSci*) A sensory nerve ending, receiving impressions from outside the body. Cf INTEROCEPTOR.

extinction (*Astron*) The reduction of radiation received on Earth from celestial bodies owing to absorption and scattering by intervening dust particles in interstellar space and in the Earth's atmosphere.

extinction (*BioSci*) The total elimination or dying out of any plant or animal species, or a whole group of species, worldwide. Present-day extinctions are usually due to human activity, eg hunting, pollution, or destruction of natural habitats such as rainforests. The disappearance of the dinosaurs (and various marine invertebrates, including ammonites) was associated with a mass extinction about 65 million years ago.

extinction (*Psych*) The weakening and eventual disappearance of a learned or conditioned response after exposure to unreinforced presentations of the conditioned stimulus.

extinction coefficient (*Chem*) Spectroscopic term applied to molecular group, determining absorption at a particular wavelength. See BEER'S LAW, MOLAR ABSORBANCE.

extinction frequency (*ImageTech*) The frequency at which the wavelength during playback is equal to the width of the HEAD GAP and the signal is extinguished.

extinction potential (*Electronics*) See EXTINCTION VOLTAGE.

extinction ratio (*ICT*) The ratio, usually expressed in DECIBELS, between the powers of an optical signal when gated 'on' and 'off' by an ACOUSTO-OPTIC MODULATOR.

extinction voltage (*Electronics*) (1) Lowest anode potential which sustains a discharge in a gas at low pressure. (2) In a gas-filled tube, the potential difference across the tube which will extinguish the arc. Also *extinction potential*.

extra- (*Genrl*) Prefix from Lt *extra*, beyond, outside of, outwith.

extra bound (*Print*) A book completely bound by hand, expensive, using carefully chosen materials, sewn headbands, special end-papers, gold tooling, etc.

extracellular (*BioSci*) Located or taking place outside a cell. Cf INTERCELLULAR, INTRACELLULAR.

extracellular enzyme (*BioSci*) One secreted out of the cell, eg into the intercellular space or the lumen of the gut.

extracellular matrix (*BioSci*) A non-cellular matrix of proteins and glycoproteins surrounding cells in some tissues. It can be extensive as in cartilage and connective tissue and calcified as in bone. Abbrev *ECM*.

extra-chromosomal DNA (*BioSci*) (1) Non-integrated viral DNA and other EPISOMES. (2) DNA of cytoplasmic organelles, ie mitochondrial and chloroplastal DNA.

extra-chromosomal inheritance (*BioSci*) See CYTOPLASMIC INHERITANCE.

extraction (*Chem*) A process for dissolving certain constituents of a mixture by means of a liquid with solvent properties for one of the components only. Substances can be extracted from solids, eg grease from fabrics with petrol; or from liquids, eg extraction of an aqueous solution with ethoxyethane, the efficiency depending on the partition coefficient of the particular substance between the two solvents.

extraction column (*NucEng*) In the nuclear industry, the large vertical columns in which an organic solvent is used to extract uranium and plutonium from nitric acid solution.

extraction fan (*Eng*) A fan used to extract foul air, fumes, suspended paint particles, etc, from a working area.

extraction metallurgy (*MinExt*) The first stage or stages of ore treatment, in which gangue minerals are discarded and valuable ones separated and prepared for working up into finished metals, rare earths or other saleable products. Characteristically, the methods used do not change the physical structure of these products save by comminution.

extraction rate (*FoodSci*) The percentage of grain made into flour or meal during milling, important with white flour because of the increased chance of particles of bran being present to discolour the flour. It may be as low as 50% for very white flour.

extraction thimble (*Chem*) A porous cylindrical cup containing the solid material to be extracted, usually by placing it under the hot reflux in a still.

extraction turbine (*Eng*) A steam turbine from which steam for process work is tapped at a suitable stage in the expansion, the remainder expanding down to condenser pressure.

extractive distillation (*ChemEng*) A technique for improving, or achieving in cases impossible without it, distillation separation processes by the introduction of an additional substance which changes the system equilibrium. It is essentially different from AZEOTROPIC DISTILLATION in that the added substance is not distilled itself but is added as a liquid at some point, usually at or near the top in the distillation column, and leaves at the base as liquid.

extrados (*Build, CivEng*) The back or top surface of an arch. See fig. at ARCH. See INTRADOS.

extradural (*Med*) Situated outside the DURA MATER. Also *epidural*.

extra-embryonic (*BioSci*) In embryos developed from eggs containing a great deal of yolk, eg birds, pertaining to that part of the germinal area outside the embryo.

extra-floral nectary (*BioSci*) A nectary occurring on or in some part of a plant other than a flower. Also *extra-nuptial nectary.*

extragalactic (*Astron*) Used to describe an object which is outside our Galaxy.

extra-high voltages (*ElecEng*) A term used in official regulations for voltages above 3·3 kV; but more commonly employed to denote voltages of the order of 100 kV or more.

extra-lateral rights (*MinExt*) See APEX LAW.

extra light fuel oil (*MinExt*) A heating oil with a Redwood viscosity of 32″. Abbrev *ELFO.*

extraneous ash (*MinExt*) In raw coal, the so-called 'free dirt' or associated shale and enclosing beds.

extra-nuptial nectary (*BioSci*) See EXTRA-FLORAL NEC-TARY.

extraordinary ray (*Phys*) See DOUBLE REFRACTION.

extrapolation (*MathSci*) The estimation of the value of a function at a particular point from values of the function on one side only of the point. Cf INTERPOLATION.

extras (*Build, CivEng*) All work the inclusion of which is not expressed or implied in the original contract price. Also *variations.*

extrasensory perception (*Psych*) The perception of pheno-mena without the use of the ordinary senses. Abbrev *ESP.*

extrasystole (*Med*) A premature contraction of the heart interrupting the normal rhythm, the origin of the impulse to contraction being abnormally situated either in the ventricles or in the atrium.

extraterrestrial (*Astron*) Used to describe a phenomenon or object beyond the Earth, in particular (hypothetical) intelligent life elsewhere in the universe.

extra thirds (*Paper*) A former size of cut card, 44·5 × 76 mm ($1\frac{3}{4}$ × 3 in).

extra-uterine (*Med*) Situated or happening outside the uterus.

extravasation (*BioSci*) The abnormal escape of fluids, such as blood or lymph, from the vessels which contain them. V. *extravasate.*

extravascular (*Med*) Placed or happening outside a blood vessel.

extraversion (*Med*) See EXSTROPHY.

extreme breadth (*Ships*) The greatest breadth measured to the farthest-out part of the structure on each side, including any rubbing strakes or other permanent attach-ments to the hull or SUPERSTRUCTURE but not to the TOP HAMPER.

extreme dimensions (*Ships*) These dimensions are pro-vided for general information, mainly for docking pur-poses. They are EXTREME BREADTH, LENGTH OVERALL and DEPTH or SUMMER DRAUGHT.

extremely low-frequency radiation (*Radiol*) Frequencies below 3 kHz. See panel on NON-IONIZING FIELDS AND RADIATION.

extreme pressure lubricant (*Eng*) A solid lubricant, such as graphite, or a liquid lubricant with additives which form oxide or sulphide coatings on metal surfaces exposed where, under very heavy loading, the liquid film is interrupted, thus mitigating the effects of dry friction.

extreme programming (*ICT*) A paradigm for software development involving the close collaboration of two or more developers on small, low-level, rapidly delivered components of code in which the design may be subject to sudden evolutionary change. Often characterized by the self-conscious use of terminology derived from the martial arts.

extremophile (*BioSci*) An organism, usually a micro-organism, that is capable of living in hostile conditions or an extreme environment.

extrinsic (*BioSci*) Said of appendicular muscles of verte-brates that run from the trunk to a girdle or the base of a limb. Cf INTRINSIC.

extrinsic (*Crystal*) Said of electrical conduction properties arising from impurities in the crystal.

extrinsic motivation (*Psych*) Motivation that derives from factors outside the individual, driven by potential external rewards or punishments.

extrinsic semiconductor (*Electronics*) A semiconductor made of extrinsic silicon. See panel on INTRINSIC AND EXTRINSIC SILICON.

extrinsic silicon (*Electronics*) Silicon owing its conduc-tivity to charge carriers deliberately introduced through the controlled addition of dopant atoms. See panel on INTRINSIC AND EXTRINSIC SILICON.

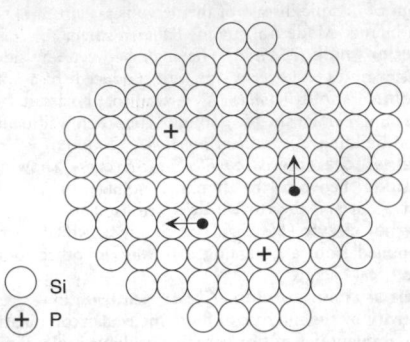

extrinsic silicon

extrophy (*Med*) See EXSTROPHY.

extrorse (*BioSci*) Directed or bent outwards; facing away from the axis, esp of stamens opening towards the outside of the flower. Cf INTRORSE.

extrovert (*BioSci*) An extrusible proboscis, found in certain aquatic animals. See LOPHOPHORE.

extrovert (*Psych*) In Jungian theory, a person who tends to focus on the external and objective aspects of experience rather than on internal and subjective ideas. Unduly simplistic, but in popular usage, extroverts are considered outgoing and introverts are seen as withdrawn.

extruded foods (*FoodSci*) Foods forced through an orifice; the simplest are cold extrusions like sausage meat and the oldest pasta and noodles. In modern continuous cooker-extruders starchy raw materials are mixed and cooked by a steam-heated screw which also provides the pressure to extrude the mixture. On release into the atmosphere, water trapped in the extrudate expands and puffs the product to produce a crisp open texture. See EXPLOSIVE DECOMPRESSION.

extrusion (*Eng*) The operation of producing rods, tubes and various solid and hollow sections, by forcing suit-able material through a die by means of a ram. Applied to numerous non-ferrous metals, alloys and other sub-stances, notably plastics (for which a screw drive is frequently used). In addition to rods and tubes, extruded plastics include sheets, film and wire coating. See panel on CERAMICS PROCESSING.

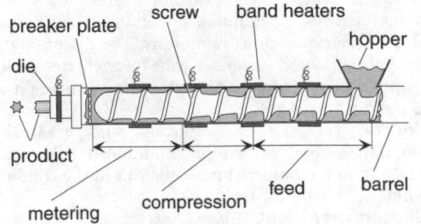

extrusion Barrel cut to show screw.

extrusion (*Textiles*) The process used in the manufacture of synthetic fibres in which a viscous solution of the polymer or the molten polymer is forced through the fine holes of a SPINNERET.

extrusion blow moulding (*Eng*) Conventional route for making plastic bottles etc. Cf INJECTION BLOW MOULDING. See BLOW MOULDING.

extrusion zones (*Eng*) Parts of barrel of an extrusion-moulding machine where different processes occur, eg feed section. See EXTRUSION.

extrusive rocks (*Geol*) Rocks formed by the consolidation of magma on the surface of the ground as distinct from *intrusive rocks*, which consolidate below ground. Commonly referred to as *lava flows*; normally of fine grain or even glassy.

exudate (*Med*) The fluid which has formed in the tissues or the cavities of the body as a result of inflammation; it contains protein and many cells, and clots outside the body. N *exudation*.

exudation pressure (*BioSci*) See ROOT PRESSURE.

exudative diathesis of chicks (*Vet*) Subcutaneous oedema in chicks associated with excessive capillary permeability due to vitamin E deficiency.

exumbrella (*BioSci*) The upper convex surface of a medusa. Adj *exumbrellar*.

exuviae (*BioSci*) The layers of the arthropod integument cast off in ECDYSIS.

exuvial (*BioSci*) Pertaining to, or facilitating, ECDYSIS.

eye (*Arch*) (1) The circular opening in the top of a dome. (2) A circular or oval window.

eye (*BioSci*) The sense organ that receives visual impressions.

eye (*Build*) Of an axe or other tool, the hole or socket in the head for receiving the handle.

eye (*Eng*) (1) A loop formed at the end of a steel wire or bolt. See EYE BOLT. (2) The central inlet passage of the impeller of a centrifugal compressor or pump.

eye (*EnvSci*) The central calm area of a cyclone or hurricane, which advances as an integral part of the disturbed system.

eye (*Glass*) The hole in the centre (or elsewhere) of the floor of a pot furnace up which the combustible gases rise as flame to heat the furnace.

eye-and-object correction (*Surv*) A correction applied in precise work to the average angle of elevation read on the vertical circle, in order to compensate for the vertical axis of the theodolite not being truly vertical. The correction is

$$+\left(\frac{\Sigma_0 - \Sigma_e}{4}\right)\theta$$

where Σ_0 = object-end reading of the altitude level, Σ_e = eye-end reading of the altitude level, and θ = angular value of one division of altitude level.

eye bolt (*Eng*) A bolt carrying an eye instead of the normal head; fitted to heavy machines and other parts for lifting purposes.

eye-cup (*ImageTech*) A soft rubber fitting for the eyepiece of a camera viewfinder.

eye diagram (*ICT*) The pattern formed by an OSCILLO-SCOPE when fed with a binary signal and sweeping once per symbol interval. The pattern resembles a human eye, the extent of whose opening indicates the amount of INTERSYMBOL INTERFERENCE present, as well as the optimum sampling point and noise margin of the signal.

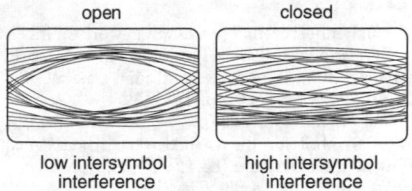

open closed

low intersymbol high intersymbol
interference interference

eye diagram

eye-ground (*Med*) The fundus; that part of the cavity of the eyeball which can be seen through the pupil with an ophthalmoscope.

eyelids (*Aero*) Jet engine thrust-reverser nozzle deflectors, shaped and closed like eyelids.

eyepiece (*Phys*) In an optical instrument, the lens or lens system to which the observer's eye is applied in using the instrument.

eyepiece graticule (*BioSci*) Grid or scale incorporated in the eyepiece for measuring objects under the microscope. A special type used in particle size analysis consists of a rectangular grid for selecting the particles and a series of graded circles for use in sizing the particles. Also *micrometer eyepiece*. US *ocular micrometer*.

eye spot (*BioSci*) Usually an orange or red spot composed of droplets containing carotenoids. Found near photo-receptive areas in motile cells of many algae, phytoflagel-lates and lower animals, and presumed to help detect the direction of light in phototaxis. Also *stigma*.

eye stalk (*BioSci*) A paired stalk arising close to the median line on the dorsal surface of the head of many Crustacea, bearing an eye.

eyot (*Genrl*) A small island (pronounced *ait*).

Eyring formula (*Acous*) A formula proposed for the period of reverberation of an enclosure, taking into account that the sound waves lose energy every time they are reflected. The results are not identical to those of the SABINE REVERBERATION FORMULA.

F

F (*BioSci*) Symbol for *filial generation* in work on inheritance; usually distinguished by a subscript, either F_1 or F1, first filial generation; F_2 or F2, second filial generation. Cf P.

F (*Chem*) Symbol for FLUORINE.

F (*ElecEng*) Symbol for FARAD.

F (*Phys*) Symbol for the FAHRENHEIT SCALE when used following a temperature (eg 41°F).

F (*Build*) Abbrev for *face* or *flat*.

F (*Phys*) Symbol for: (1) FARADAY; (2) FORCE; (3) HELMHOLTZ FREE ENERGY (US).

[F] (*Phys*) A Fraunhofer line in the blue of the solar spectrum of wavelength 486·1527 nm. It is the second line in the Balmer hydrogen series, known also as $H\beta$.

f (*Genrl*) Symbol for FEMTO-.

f (*Chem*) Symbol for: (1) activity coefficient, for molar concentration; (2) partition function.

f_α (*Electronics*) (1) In transistors, symbol for common-emitter forward-current gain cut-off frequency. Symbols f_{hfe}, f_{hfb}, f_T are more commonly used. (2) Abbrev for *cut-off frequency*.

f_{hfb} (*Electronics*) Symbol for common-base-mode forward-current gain cut-off frequency.

f_{hfe} (*Electronics*) Symbol for common-emitter-mode forward-current gain cut-off frequency.

f_τ (*Electronics*) In transistors, symbol for transition frequency; that at which the common-emitter forward-current gain is reduced to unity.

F1 hybrid (*BioSci*) Crop or strain variety, characterized by unusual vigour and uniformity, produced by crossing two selected inbred lines. Cf CYTOPLASMIC MALE STERILITY, HETEROSIS, NICK.

FAA (*Aero*) Abbrev for *Federal Aviation Administration*, a US government agency responsible for all aspects of US civil aviation. Cf CAA.

Fabaceae (*BioSci*) See LEGUMINOSAE.

Fab fragment (*BioSci*) Fragment of immunoglobulin molecule obtained by hydrolysis with papain. It consists of one light chain linked to the N-terminal part of the adjacent heavy chain. Two Fab fragments are obtained from each molecule. Each Fab contains one antigen-binding site, but none of the heavy-chain Fc.

F(ab')₂ fragment (*BioSci*) Fragment of immunoglobulin obtained by pepsin digestion. It contains both Fab fragments plus a short section of the hinge region of the heavy chain Fc. It behaves as a bivalent antibody but lacks properties associated with the Fc fragment (eg complement activation, placental transmission).

fabric (*Build*) Walls, floors and roof of building.

fabric (*Geol*) The sum of all the textural and structural features of a rock.

fabric (*Textiles*) A coherent assembly of fibres and/or yarns that is long and wide but relatively thin and strong; a *cloth*. See panel on FIBRE ASSEMBLIES.

Fabry–Pérot interferometer (*Phys*) An instrument in which multiple-beam circular HAIDINGER FRINGES are produced by the passage of monochromatic light through a pair of plane-parallel half-silvered glass plates, one of which is fixed while the other can be moved by an accurately calibrated screw. In transmission, the fringes appear as sharp bright fringes on a dark background. By observing the fringes as the separation of the plates is changed, the wavelength of the light can be determined.

façade (*Arch*) The front elevation of a building.

face (*Build*) (1) The front of a wall or building. (2) The exposed vertical surface of an arch.

face (*Crystal*) See CRYSTAL and panel on CRYSTAL LATTICE.

face (*Eng*) The working surface of any part; as the sole of a carpenter's plane, the striking surface of a hammer, the surface of a slide-valve, or the surface of the steam chest on which it slides, the seating surface of a valve, the flank of a gear-tooth, etc.

face (*MinExt*) The exposed surface of coal or other mineral deposit in the working place where mining, winning or getting is proceeding.

face-airing (*MinExt*) The operation of directing a ventilating current along the face of a working place; also *flushing*.

face-centred cubic (*Chem*) A crystal lattice with a cubic unit cell, the centre of each face of which is identical in environment and orientation to its vertices. Specifically a common structure of metals, in which the unit cell contains four atoms, based on this lattice. Abbrev *FCC*. See panel on CLOSE PACKING OF ATOMS.

face chuck (*Eng*) A large disk fixed to the mandrel of a lathe and provided with slots and holes for securing work of a flat or irregular shape. Also *face plate*.

faced cloth (*Textiles*) General term for fabrics whose surfaces have been treated to give a rich, luxuriant effect by laying the pile, eg BEAVER and doeskin cloths. See DRESS-FACED CLOTH.

face edge (*Build*) See WORKING EDGE.

face hammer (*Build*) A hammer having a peen which is flat rather than pointed or edged.

face lathe (*Eng*) A lathe designed for work of large diameter but short length (eg large wheels or disks), often with a vertical spindle.

face left (*Surv*) Expression referring to the pointing of a theodolite telescope when the vertical circle is left of the telescope, as seen from the eyepiece end.

face mark (*Build*) A mark distinguishing that face of wood etc to which the other faces have been trued.

face mix (*Build*) A mixture of cement and stone dust used for facing concrete blocks in imitation of real stone.

face mould (*Build*) A templet used as a reference for shaping the face of wood, stone, etc.

face plate (*Electronics*) That part of a cathode-ray tube which carries the phosphor screen.

face plate (*Eng*) (1) See FACE CHUCK. (2) A SURFACE PLATE.

face-plate breaker controller (*ElecEng*) A face-plate controller having a separate contactor for breaking the circuit.

face-plate breaker starter (*ElecEng*) A face-plate starter having a separate interlocked contactor for breaking the circuit.

face-plate controller (*ElecEng*) See FACE-PLATE STARTER.

face-plate coupling (*Eng*) See FLANGE COUPLING.

face-plate starter (*ElecEng*) An electric motor starter in which a contact lever moves over a number of contacts arranged upon a plane surface. Also *face-plate controller*.

face right (*Surv*) Expression referring to the pointing of a theodolite telescope when the vertical circle is right of the telescope, as seen from the eyepiece end.

face shovel (*CivEng*) A mechanical mobile device for cutting into the vertical face of an excavation and depositing soil into vehicles for transporting it elsewhere.

facet (*Arch*) See FACETTE.

facet (*BioSci*) One of the corneal elements of a compound eye; a small articulatory surface.

facet (*Crystal*) A flat polished face on a gemstone.

facette (*Arch*) A projecting flat surface between adjacent flutes in a column. Also *facet, listel*. See fig. at MOULDINGS.

face-wall (*Build*) The front wall.

facial (*BioSci*) Pertaining to or situated on the face; the seventh cranial nerve of vertebrates, supplying the facial muscles and tongue of higher forms, the neuromast organs of the head and snout in lower forms, and the palate in both.

facies (*Geol*) The appearance or aspect of any rock; the sum total of its characteristics. Used of igneous, metamorphic and esp sedimentary rocks.

facies, stratigraphic (*Geol*) The sum of the rock and fossil features of a sedimentary rock. They include *lithofacies*: mineral composition, grain size, texture, colour, cross-bedding and other sedimentary features; and *biofacies*: the fossil plant and animal characteristics of the rock.

facilitated diffusion (*BioSci*) The rapid permeation of solutes into cells by interaction of the solutes with specific carriers which facilitate their entry. Facilitated diffusion is distinguished from ACTIVE TRANSPORT (panel) because it does not allow the entry of solutes against their concentration gradients.

facilitation (*BioSci*) The augmented response of a nerve due to pre-stimulation; the activation of physiological and behavioural response resulting from non-specific stimulation from a conspecific. See SOCIAL FACILITATION.

facilities house (*ImageTech*) A company that provides video production and/or POST-PRODUCTION facilities.

facing (*Eng*) (1) The operation of turning a flat face on the work in the lathe. (2) A raised machined surface to which another part is attached.

facing bond (*Build*) A general term for any bond consisting mainly of stretchers.

facing bricks (*Build*) Bricks of better quality and appearance used for facing work but not made to withstand heavy loads.

facing paviours (*Build*) A class of hard-burnt bricks used as facing bricks in high-class work.

facing points (*CivEng*) See POINTS.

facing sand (*Eng*) Moulding sand with admixed coal dust, used near pattern in foundry flask to give the casting a smooth surface.

faconné (*Textiles*) Figured fabric with a DOBBY PATTERN, JACQUARD.

FACS (*BioSci*) See FLUORESCENCE-ACTIVATED CELL SORTER.

facsimile (*ICT*) The scanning of any still graphic material to convert the image into electrical signals, for subsequent reconversion into a likeness of the original. Abbrev *fax*.

facsimile bandwidth (*ICT*) The frequency difference between the highest and lowest components necessary for the adequate transmission of the facsimile signals.

facsimile receiver (*ICT*) One for translating signals from a communication channel into a facsimile record of the original copy.

facsimile transmitter (*ICT*) Means for translating text, lines and half-tone copy into signals suitable for a communication channel.

factor (*BioSci*) Obsolete term for *gene*.

factor analysis (*MathSci*) Any statistical method for analysing given data in terms of a number of variables which are then postulated as the causal explanation of the observed data.

factor B, factor D (*BioSci*) Components involved in the alternative pathway of COMPLEMENT activation.

factor group (*MathSci*) See QUOTIENT GROUP.

factor H, factor I (*BioSci*) A glycoprotein and an enzyme respectively that together inactivate C3b and act as a damper on activated COMPLEMENT.

factorial (*MathSci*) The product of all whole numbers from a given number (*n*) down to 1; it is equal to the number of PERMUTATIONS of *n* objects in order. Written

$$n! = n(n-1)(n-2)(n-3) \ldots 3.2.1.$$

Cf GAMMA FUNCTION, STIRLING'S APPROXIMATION, SUBFACTORIAL *N*.

factor of merit (*ElecEng*) Of reflecting galvanometers, the deflection in millimetres, produced on a scale at a distance of 1 m by a current of 1 μA, the deflection being corrected for coil resistance and time of swing.

factor of safety (*Build, Eng*) The ratio, allowed for in design between the ultimate stress in a member or structure and the safe permissible stress in it. Abbrev *FS*.

factor P (*BioSci*) See PROPERDIN.

Factors I–XIII (*BioSci*) Blood-clotting factors, especially from humans.

factory fitting (*ElecEng*) An electric-light fitting in which the lamp is housed in a strong protecting enclosure. Also *mill fitting*.

fact table (*ICT*) In data-warehousing, the heavily denormalized area of a database in which 'facts' extracted from source systems data are stored for future analysis according to a number of significant dimensions.

faculae (*Astron*) Large bright areas of the photosphere of the Sun. They can be seen most easily near sunspots and at the edge of the Sun's disk; they are at a higher temperature than the average for the Sun's surface.

facultative (*BioSci*) Able but not obliged to function in the way specified; a facultative anaerobe can grow in and perhaps use free oxygen, but will also survive and perhaps grow in its absence. Cf OBLIGATE.

facultative heterochromatin (*BioSci*) Chromatin that is condensed in some cell types but not in others and that is not expressed when condensed, despite containing CODING SEQUENCES. Cf CONSTITUTIVE HETEROCHROMATIN.

facultative parasite (*BioSci*) A parasite able to live saprophytically and be cultured on laboratory media. Cf OBLIGATE PARASITE. See NECROTROPH.

fade (*ICT*) Variation of strength, sometimes periodic, of the signal received from a distant transmitter. In short-wave communication fading may be due to interference between the reflected and direct waves or to change in the properties of the ionosphere during the transmission. In microwave links, it may be caused by atmospheric absorption of the transmitted signal, or by refraction of the highly directional beams causing them to *miss* the receiving antenna. Also *fading*.

fade-in (*ImageTech*) (1) Gradual appearance of a picture from uniform black. (2) Gradual increase in the level of a sound signal from silence or a very low intensity.

fade-out (*ImageTech*) (1) Gradual disappearance of a picture to uniform black. (2) Gradual decrease in the level of a sound signal to silence or a very low intensity.

fader (*ImageTech*) A control to vary audio or video signal level to produce fade effects.

fade shutter (*ImageTech*) Variable shutter on a motion picture camera or printer whose opening can be continuously altered while running.

fading (*Autos*) See BRAKE-FADE.

fading (*Build*) The loss of colour in paint due to ageing, exposure to sunlight, etc.

fading (*ICT*) See FADE.

fading (*ImageTech*) Weakening of a photographic image as a result of age, exposure to light or chemical reactions. Black-and-white images may become pale brown or yellow, while in colour images the components may often fade to different extents, resulting in gross distortion of colour balance.

fading (*Textiles*) The change in colour that sometimes takes place when a material is exposed to light or atmospheric fumes.

fading area (*ICT*) That in which fading is experienced in night reception of radio waves, between the primary and secondary areas surrounding a station transmitting on medium and long waves.

fadometer (*Chem*) An instrument used to determine the resistance of a dye or pigment to fading.

faeces (*BioSci*) The US term, FECES, has become so commonplace that it has been used throughout this dictionary.

Fagaceae (*BioSci*) A family of c.100 spp of dicotyledonous flowering plants (superorder Hamamelidae). They are mostly trees, and are often dominant in broad-leaved forests of esp the northern hemisphere. The flowers are typically in catkins. Includes oak, beech, chestnut and southern beech.

faggot (*CivEng*) A bundle of brushwood. See FASCINE.

fagopyrism (*Vet*) A form of photosensitization affecting animals with white or lightly pigmented skin; due to ingestion of a fluorescent substance occurring in buckwheat (*Fagopyrum sagittatum*), with subsequent exposure to strong sunlight. Also *buckwheat rash*.

fagot (*Genrl*) Also *fagoted*. See FAGGOT.

Fahrenheit scale (*Phys*) The method of graduating a thermometer in which freezing point of water is marked 32° and boiling point 212°, the fundamental interval being therefore 180°. Fahrenheit has been largely replaced by the Celsius (centigrade) and Kelvin scales. To convert °F to °C subtract 32 and multiply by 5/9. For the Rankine scale equivalent add 459·67 to °F; this total multiplied by 5/9 gives the Kelvin equivalent.

FAI (*Build*) Abbrev for FRESH-AIR INLET.

faïence (*Build*) Originally the 16th-century Fr name for EARTHENWARE glazed with an opaque, tin-based glaze, from Faenza, Italy. Now used in the UK for large glazed blocks and slabs used in buildings, in France for any glazed, porous earthenware, and in the US for a decorated earthenware with a transparent glaze. Cf BONE CHINA, EARTHENWARE, PORCELAIN, STONEWARE, TERRACOTTA.

fail-operational (*Aero*) System designed so that it can continue to function after a single failure, warning being indicated.

failover (*ICT*) The orderly failure of part of a computer system in which the processes, data and users are reassigned to another part of the system in a seamless manner without abridgement of service.

fail safe (*NucEng*) A design in which power supply, control or structure is able to return to a safe condition in the event of failure or maloperation, by automatic operation of protective devices or otherwise.

failure modes and effects analysis (*Aero, Eng*) The design activity for process and product which aims to eliminate defects before production or launch of product in marketplace. Involves a committee of designers and engineers who review systematically potential defects in terms of ease of detection, seriousness and ease of correction. Each defect is assigned a number (scale of 1–10) by mutual agreement, and the product number matched against the RPN (risk priority number). If above a critical value, the defect is corrected. Of particular importance in product liability. Abbrevs *FMEA*, *FMECA*.

failure of materials (*Eng*) Loss of structural integrity in products or samples by some form of change in the material, eg chemical (see CORROSION AND DEGRADATION), mechanical (see FATIGUE (panel), STRENGTH MEASURES), dimensional (see TOLERANCE), physical (see DEMAGNETIZATION), electrical (see DIELECTRIC FATIGUE), or any combination of such kinds of change. Studied systematically in products using FAILURE MODES AND EFFECTS ANALYSIS or FAULT TREE ANALYSIS, or in materials using FRACTOGRAPHY, FRACTURE MECHANICS, etc.

fair cutting (*Build*) The operation of cutting brickwork to the finished face of the work. Abbrev FC.

fair ends (*Build*) Projecting masonry ends requiring to be dressed to a finished surface.

fairfieldite (*Min*) A hydrated phosphate of calcium and manganese, crystallizing in the triclinic system as prismatic crystals or fibrous aggregates.

fairing (*Aero*) A secondary structure added to any part of an aircraft to reduce drag by improving the streamlining.

fairing (*Ships*) The process of ensuring that the lines of intersection of all planes with a true ship form are 'fair'; the resulting lines are known as QUARTER LINES.

Fair's graticules (*PowderTech*) Types of eyepiece graticule marked with rectangles and circles for use in microscope methods of particle size analysis.

fairy ring (*BioSci*) A ring, usually in grass, in which the plants near the periphery are green and healthy and those near the centre less so. It will persist and expand for many years and is associated with the mycelium of the fungus which forms fruiting bodies at the ring's periphery.

Fajans rule for ionic bonding (*Chem*) The conditions which favour the formation of ionic (as opposed to covalent) bonds are (1) large cation, (2) small anion, (3) small ionic charge, (4) the possession by the cation of an inert gas electronic structure.

Fajans–Soddy law of radioactive displacement (*Chem*) The atomic number of an element decreases by two upon emission of an α-particle, and increases by one upon emission of a β-particle.

falcate (*BioSci*) Sickle-shaped. Also *falciform*.

falciform ligament (*BioSci*) In higher vertebrates, a peritoneal fold attaching the liver to the diaphragm.

falciparum malaria (*Med*) A severe form of *malaria* caused by the parasite *Plasmodium falciparum*.

Falconbridge process (*Eng*) A method of separating copper from nickel, in which matte is acid-leached to dissolve copper, after which residue is melted and refined electrolytically.

Falconiformes (*BioSci*) One of the two orders of diurnal birds of prey, the falcons. Cf Accipitriformes, Strigiformes.

falcula (*BioSci*) A sharp curved claw. Adj *falculate*.

fall (*CivEng*) The inclination of rivers, streams, ditches, drains, etc, quoted as a fall of so much in a given distance. See CROSS-FALL.

fall (*Eng*) A hoisting rope.

fall (*MinExt*) (1) The collapse of the roof of a level or tunnel, or of a flat working place or stall; the collapse of the hanging wall of an inclined working place or stope. (2) A mass of stone which has fallen from the roof or sides of an underground roadway, or from the roof of a working place.

fall bar (*Build*) The part of a latch which pivots on a plate screwed to the inner face of a door, and drops into a hook on the frame.

falling mould (*Build*) The development in elevation of the centre line of a hand rail.

fall-of-potential test (*ElecEng*) A test for locating a fault in an insulated conductor; the voltage drop along a known length of the conductor is compared with the voltage drop between one end of the conductor and the fault. Also CONDUCTIVITY TEST, DROP TEST.

Fallopian tube (*BioSci*) In mammals, the anterior portion of the Müllerian duct; the oviduct.

Fallot's tetralogy (*Med*) A relatively common type of cyanotic congenital heart disease (blue baby) where there is an incomplete septum between left and right ventricles (hole in the heart), the pulmonary valve is small and stenotic, the aorta lies over both left and right ventricles leading to right ventricular overload and hypertrophy.

fallout (*EnvSci, Phys*) Particulate matter in the atmosphere, which is transported by natural turbulence, but which will eventually reach the ground by sedimentation or dry or wet deposition. Applied esp to airborne radioactive contamination resulting from eg a nuclear explosion, inadequately filtered reactor coolant, or the failure of reactor containment after an accident.

fallow (*Agri*) Arable land that has not been seeded and from which a crop will not be taken. Ploughing and tilling during the fallow period is a method for reducing troublesome weeds.

fall pipe (*Build*) See DOWNPIPE.

fall time (*ICT*) The decaying portion of a wave pocket or pulse; usually the time taken for the amplitude to decrease from 90% to 10% of the peak amplitude.

false amethyst (*Min*) An incorrect name given to a purple gemstone. Applied (wrongly) to purple fluorite, and sometimes to purple corundum. See ORIENTAL AMETHYST.

false amnion (*BioSci*) See CHORION.

false annual ring (*BioSci*) A second ring of xylem formed in one season, following the defoliation of the tree by the attacks of insects or other accident; oaks are liable to this, as they may be completely stripped of leaves by the oak tortrix.

false bands (*Print*) Strips of board or leather glued across the spine of hollow-backed books before covering with leather. See RAISED BAND.

false bearing (*Build*) A beam, such as a sill, when not supported under its entire length, is said to have a false bearing.

false bedding (*Geol*) See CROSS BEDDING.

false body (*Build*) An apparent defect in which a full-bodied paint undergoes a sharp but temporary drop in viscosity under agitation or brushing. This may prevent brush marks flowing out but is now considered a useful property of thixotropic paints.

false bottom (*Eng*) (1) A removable bottom placed in a vessel to facilitate cleaning; a casting placed in a grate to raise the fire bars and reduce the size of the fire. (2) Any secondary bottom plate or member used to reduce the volume of a container or to create a secondary container.

false colour (*ImageTech*) A system of colour photography used for camouflage detection and aerial survey; a multi-layer colour film is used in which one layer is sensitized to infrared, so that natural green subjects such as grass and foliage are reproduced in magenta.

false curvature (*Phys*) The curvature of particle tracks (eg in cloud chambers, bubble chambers, spark chambers or photographic emulsions) which results from undetected interactions and not from an applied magnetic field.

false diamond (*Min*) Several natural minerals are sometimes completely colourless and, when cut and polished, make brilliant gems. These include zircon, white sapphire and white topaz. All three, however, are birefringent and can be easily distinguished from true diamond by optical and other tests. Many artificial diamond *simulants* are now made, some difficult to detect. See CUBIC ZIRCONIA, YAG.

false ellipse (*Build*) An approximate ellipse, composed of circular arcs.

false fruit (*BioSci*) A fruit formed from other parts of the flower in addition to the carpels, eg the strawberry where the receptacle becomes fleshy.

false header (*Build*) A half-length brick, sometimes used in Flemish bond.

false hemlock (*For*) See DOUGLAS FIR.

false key (*Eng*) A circular key for attaching a hub to a shaft; it is driven into a hole which is parallel with the shaft axis and has been drilled half in the hub and half in the shaft. A similar tapped hole can be used with a screw.

false memory syndrome (*Psych*) A condition of erroneous memory, sometimes attributed to adults who recall childhood experiences, esp of sexual abuse, under hypnosis or other medical inducement. Abbrev FMS.

false pregnancy (*Med, BioSci*) See PSEUDOCYESIS.

false ribs (*BioSci*) In higher vertebrates, ribs that do not reach the sternum. Also *floating ribs*.

false ruby (*Min*) Some species of garnet (*Cape ruby*) and some species of spinel (*balas ruby* and *ruby spinel*) possess the colour of ruby, but have neither the chemical composition nor the physical attributes of true ruby.

false septum (*BioSci*) A REPLUM.

false tissue (*BioSci*) See PSEUDOPARENCHYMA.

false topaz (*Min*) A name wrongly applied to yellow quartz. See CITRINE.

false-twist (*Textiles*) A method of producing textured, continuous filament yarn from thermoplastic fibres, in which it is twisted, heated, allowed to cool and set, and then twisted in the opposite direction. See TEXTURED YARN.

falsework (*CivEng*) Any temporary works used to support a structure during construction including any plant required and any formwork for concrete.

false-zero test (*ElecEng*) A test, made on a bridge or potentiometer, in which a balance is obtained, not with zero galvanometer reading, but with some definite value caused by a constant extraneous current.

falx (*BioSci*) Any sickle-shaped structure. Adjs *falcate, falciform.*

falx cerebri (*BioSci*) A strong fold of the DURA MATER, lying in the longitudinal fissure between the two cerebral hemispheres.

Famennian (*Geol*) The highest stage in the Devonian. See PALAEOZOIC.

familial hypercholesterolaemia (*Med*) An excess of cholesterol in plasma as a result of defects in the recycling process that lead to reduced uptake of LDL (low-density lipoprotein) into coated vesicles. US *familial hypercholesterolemia*.

family (*BioSci*) A group of similar genera of taxonomic rank below ORDER and above GENUS; with plants, the names usually end in -aceae.

family (*Phys*) The group of radioactive nuclides which form a decay series.

family therapy (*Psych*) Psychotherapy which regards the family as a unit and as the object of therapy, rather than its individual members, so that roles and attitudes within the family can be explored and changed.

family tool (*Eng*) Injection moulding tool possessing cavities of different shape, so that several parts can be made in one operation. They are usually subcomponents of one final product, such as a telephone handset.

famotidine (*Pharmacol*) A *histamine H2-receptor antagonist* used to treat gastric and duodenal ulcers.

fan (*Aero*) Rotating bladed device for moving air, eg in ducts or in wind tunnels. Cf PROPELLER. See AERO-ENGINE.

fan (*Eng*) (1) A device for delivering or exhausting large volumes of air or gas with only a low pressure increase. It consists either of a rotating paddle-wheel or an airscrew. (2) A small vane to keep the wheel of a wind pump at right angles to the wind.

fan (*Geol*) (1) A detrital cone found at the foot of mountains and also in the deep sea (submarine fan). (2) Fan cleavage; an axial-plane cleavage in which the cleavage planes fan out.

fan antenna (*ICT*) Antenna in which a number of vertically inclined wires are arranged in a fanwise formation, the apex being at the bottom.

fan characteristic (*Eng*) A graph showing the relation between pressure and delivery, used as a basis for fan selection. The characteristic is determined by the shape of the fan blades.

Fanconi's anaemia (*Med*) An inherited form of anaemia that leads to aplastic anaemia and may increase susceptibility to cancer (also *Fanconi anaemia*).

Fanconi's syndrome (*Med*) A kidney disease where the renal tubules are unable to conserve amino acids, phosphates, glucose, bicarbonate and water. Patients present with ACIDOSIS or with RICKETS. It may occur as a genetically determined disease or may complicate AMYLOIDOSIS, multiple myeloma or poisoning with heavy metals.

fan cooling (*Autos*) The use of an engine-driven fan to induce a greater airflow through the radiator at low speeds than would result from the forward motion of the vehicle.

fancy yarns (*Textiles*) Yarns made for decorative purposes. The ornamentation of the thread may be due to a variety of reasons, such as (1) colour; (2) the combination of threads of different types; (3) the production of thick and thin places; (4) the production of curls, loops, slubs, etc, at suitable intervals. The majority of these fancy yarns are *folded yarns*, two or more threads being combined in some special way to produce the desired effect.

fan dipole (*ICT*) A DIPOLE antenna consisting of two triangular sheets of metal, with the feeder connected between them. Used for very-high frequency and ultra-high frequency, with the advantage of a broad operating wavelength band.

fan drift (*MinExt*) Ventilating passage along which air is moved by means of a fan.

fang (*Build*) The part of an iron railing which is embedded in the wall.

fang (*BioSci*) The grooved or perforated poison-tooth of a venomous serpent; one of the cuspidate teeth of carnivorous animals, esp the CANINE or CARNASSIAL.

fang bolt (*Eng*) A bolt having a nut which carries pointed teeth for gripping the wood through which the bolt passes, so preventing the nut from rotating when the bolt is tightened.

fanglomerate (*Geol*) A conglomerate formed by lithification of a fan.

fan-guard (*Build*) A protective parapet formed of boarding secured around the platforms of builders' stagings or gantries, when the platforms are to be used for receiving and distributing materials.

fan marker beacon (*Aero*) A form of marker beacon radiating a vertical fan-shaped pattern.

fan shaft (*MinExt*) Mine shaft or pit at the top of which a ventilating fan is placed.

fantail burner (*Eng*) A pulverized-coal burner which discharges the fuel and primary air vertically downwards into the furnace in a thin flat stream to meet heated secondary air which is discharged horizontally from the walls. Also *streamline burner*.

fantasy (*Psych*) Generally, sequences of private mental images, sometimes in anticipation of possible events, but sometimes irrational and referring to extremely unlikely possibilities. In *psychoanalytic theory*, an imaginary episode, operating on either a conscious or unconscious level, in which the subject is a central figure and which fulfils a conscious or unconscious wish. Also *phantasy*.

fan vaulting (*Arch*) Tracing rising from a capital or a corbel, and diverging like the folds of a fan on the surface of a vault.

farad (*ElecEng*) The practical and absolute SI unit of electrostatic capacitance, defined as that which, when charged by a potential difference of 1 volt, carries a charge of 1 coulomb. Equal to 10^{-9} electromagnetic units and 9×10^{11} electrostatic units. Symbol F. This unit is in practice too large, and the subdivisions, *microfarad* (μF), *nanofarad* (nF) and *picofarad* (pF), are in more general use.

faraday (*Phys*) The quantity of electric charge carried by 1 mole of singly charged ions, ie 9.6487×10^4 C. Symbol F.

Faraday cage (*ElecEng*) An arrangement of conductors, or conducting sheet or conducting mesh, bonded together so that they provide an ELECTROSTATIC SHIELD, but connected in such a way that induced currents cannot circulate. Used, for example, to provide an equipotential screen around equipment and/or personnel to enable them to work on live high-voltage equipment. Also *Faraday shield*.

Faraday dark space (*Phys*) Dark region in a gas-discharge column between the negative glow and the positive column.

Faraday disk (*ElecEng*) Rotating disk in the gap of an electromagnet, so that a low emf is generated across a radius. Used eg to generate a calculated emf to balance against the drop across a resistance due to a steady current, thus establishing the latter in absolute terms from the calculation of the mutual inductance between the disk and the exciting air-cored coil.

Faraday effect (*Phys*) Rotation of the plane of polarization of plane-polarized light when passed through an isotropic transparent medium placed in a strong magnetic field, the light being passed in a direction in which the field has a

component. If l is the length of path traversed, H the strength of the field in the direction of propagation and θ is the angle of rotation, then $\theta = ClH$ where C is VERDET'S CONSTANT. The effect is also exhibited by a plane-polarized microwave passing through a ferrite.

Faraday shield (*ElecEng*) See FARADAY CAGE.

Faraday's ice-pail experiment (*Phys*) Classical experiment which consists of lowering a charged body into a metal pail connected to an electroscope, in order to show that charges reside only on the outside surface of conductors.

Faraday's law of induction (*Phys*) The law stating that the emf induced in any circuit is proportional to the rate of change of magnetic flux linked with the circuit. Principle used in every practical electrical machine. MAXWELL'S EQUATIONS involve a more general mathematical statement of this law.

Faraday's laws of electrolysis (*Phys*) (1) The principle that the amount of chemical change produced by a current is proportional to the quantity of electricity passed. (2) The pinciple that the amounts of different substances liberated or deposited by a given quantity of electricity are proportional to the chemical equivalent weights of those substances.

faradmeter (*ElecEng*) Generic name for direct-reading capacitance meters. Typically, they use the mains voltage in series with an ac milliammeter.

farcy (*Med, Vet*) See GLANDERS.

far-end cross-talk (*ICT*) Cross-talk heard by a listener, and caused by a speaker at the distant end of the parallelism.

far field (*Acous*) Sound field a long distance from the source. Every sound source has a *far field* and a *near field*, eg a monopole source has a far field which decays as r^{-1} (r is the distance from the source) and a near field which decays as r^{-2} so that the far field dominates at a large distance.

farina (*Genrl*) A floury powder, generally ground corn, meal or starch.

farinose (*BioSci*) Covered with a mealy powder.

-farious (*Genrl*) Suffix meaning *arranged in so many rows*.

farmer's lung (*Med*) A respiratory disease due to hypersensitivity to spores of a thermophilic bacterium, *Micropolyspora faeni*, present in the dust of mouldy hay. It mainly occurs among farmers working in damp areas. It is characterized by attacks of breathlessness coming on some hours after inhaling the dust. A combination of type 3 and type 4 *hypersensitivity reactions* are probably involved.

Farmer's reducer (*ImageTech*) A reducing bath for photographic images made by the addition of potassium ferricyanide to hypo.

farmyard manure (*Agri*) Livestock feces (typically from cattle) mixed with straw or other bedding material from stables, barns or night yards. It is composted before use and applied to soil to increase crop yield and water-retaining capacity. Abbrev *FYM*.

farnesylation (*BioSci*) A post-translational modification of protein by the addition of the farnesyl group (three isoprene units), probably acting as a membrane attachment device.

far point (*Phys*) Object point conjugate to the eye's retina when accommodation is completely relaxed; at infinity in emmetropia, between infinity and the eye in myopia, and behind the eye in hyperopia.

far-red light (*BioSci*) Light of wavelength around 730 nm, effective in plant responses mediated by PHYTOCHROME.

Farror's process (*Eng*) CASE-HARDENING by a mixture of ammonium chloride, manganese dioxide and potassium ferrocyanide.

farrowing (*Agri*) The production of a litter of pigs.

fascia (*Arch, Build*) (1) A wide flat member in an entablature. (2) A board embellishing a gutter around a building. (3) The broad flat surface over a shop front or below a cornice.

fascia (*Autos*) The instrument board of an automobile.

fascia (*BioSci*) Any band-like structure, esp the fibrous connective tissue bands that envelope, separate or bind together muscles, organs and other soft structures of the body. Adj *fascial*.

fasciation (*BioSci*) An abnormal condition, resulting from damage, infection or mutation, in which a shoot (or other organ) grows broad and flattened, resembling several shoots fused laterally.

fascicle (*BioSci*) A VASCULAR BUNDLE.

fascicular cambium (*BioSci*) The flat strand of cambium between xylem and phloem in a VASCULAR BUNDLE.

fasciculus (*BioSci*) A small bundle, as of muscle or nerve fibres.

fasciitis (*Med*) Inflammation of fascia. Also *fascitis*.

fascine (*CivEng*) A bundle of brushwood used to help make a foundation on marshy ground, or to make a wall to protect a shore against erosion by sea or river, or to accumulate sand and silt on the bed of an estuary.

fascine buildings (*Build*) A building constructed with logs and boards.

fasciola (*BioSci*) A narrow band of colour; a delicate lamina in the vertebrate brain.

fascioliasis (*Med, Vet*) Infection of humans and other animals with the liver fluke *Fasciola hepatica*.

fasciotomy (*Med*) Surgical incision of FASCIA.

fascitis (*Med*) See FASCIITIS.

fassaite (*Min*) A monoclinic pyroxene rich in aluminium, calcium and magnesium, and poor in sodium; found in metamorphosed limestones and dolomites.

fast (*Acous*) Measuring mode of a SOUND-LEVEL METER with a time constant of 0.125 s.

fast (*Build*) Said of colours which are not affected by the conditions of their use. N *fastness*.

fast (*ImageTech*) Contributing to reduction of time of exposure; said of an emulsion or lens.

fast (*MinExt*) (1) A heading or working place which is driven in the solid coal, in advance of the open places, said to be in the *fast*. (2) A hole in coal which has had insufficient explosive used in it, or which has required undercutting. (3) In shaft sinking, a hard stratum under poorly consolidated ground, on which a WEDGING CRIB can be laid.

fast-acting relay (*ICT*) A relay designed to act with minimum delay after the application of voltage, usually by increasing the resistance of the circuit in comparison with the inductance, and by minimizing moving masses.

fast coupling (*Eng*) A coupling which permanently connects two shafts.

fast effect (*Phys*) See FAST FISSION.

fastener (*Eng*) An article designed to fasten together two or more other articles, usually in the form of a shaft passing through the articles to be fastened, eg nails, screws, rivets, pins.

fastext (*ImageTech*) A development of TELETEXT in which related pages are grabbed and held by the TV to speed up access.

fast fission (*Phys*) Fission by fast neutrons. Uranium-238 has a fission threshold for neutrons of energy about 1 MeV, and the fission cross-section increases rapidly with energy. Fission of this isotope by fast neutrons may cause a substantial increase in the reactivity of a thermal reactor (*fast effect*). See panel on NUCLEAR REACTOR.

fast fission factor (*NucEng*) Ratio of the total number of fast neutrons produced by fissions due to neutrons of all energies (fast and thermal) to the number resulting from thermal/neutron fissions. Symbol ε.

fast Fourier transform (*MathSci*) A particularly rapid numerical method for calculating Fourier transforms. Abbrev *FFT*.

fast head (*Eng*) The fixed headstock of a lathe.

fastigiate (*BioSci*) Having the branches more or less erect and parallel, eg Lombardy poplar.

fastigium (*Med*) The highest point of temperature in a fever.

fast-needle surveying (*Surv*) See FIXED-NEEDLE SURVEY-ING.

fastness (*Textiles*) The ability of a colour to remain unchanged when exposed to a specified agency including light, rubbing and washing.

fast neutron (*Phys*) See NEUTRON.

fast pulley (*Eng*) A pulley fixed to a shaft by a key or set bolt, as distinct from a LOOSE PULLEY which can revolve freely on the shaft.

fast reaction (*Phys*) Nuclear reaction involving the strong interaction and occurring in a time of the order of 10^{-23} s.

fast reactor (*NucEng*) One without a moderator in which a chain reaction is maintained almost entirely by fast fission. It may also be a BREEDER REACTOR. See panel on NUCLEAR REACTOR.

fast resource management (*ICT*) A facility that may be provided in an ASYNCHRONOUS TRANSFER MODE network, in which a user who requires to send a burst of data is allocated capacity such as BANDWIDTH or buffer space for the duration of the burst.

fast sheet (*Build*) See STAND SHEET.

fast shutter (*ImageTech*) A feature built into many SOLID-STATE IMAGE SENSORS in which the early charge is drained away and further charge retained to give the brief duration required. Also *electronic shutter, high-speed shutter, variable-speed shutter*. See SLOW SHUTTER, STROBE EFFECT.

fast store (*ICT*) Computer memory with a very fast ACCESS TIME. See MAIN MEMORY.

fast-time constant circuits (*ICT*) Those for which the circuit parameters (particularly resistance and capacitance) permit a very rapid response to a step signal.

FAT (*ICT*) Abbrev for FILE ALLOCATION TABLE.

fat (*BioSci*) See ADIPOSE TISSUE.

fat (*Build*) Part of a cement mortar mix containing a higher proportion of cement than the rest. This comes to the surface before the mixture has set.

fat (*Chem*) See FATS.

fata morgana (*EnvSci*) A complicated mirage caused by the existence of several layers of varying refractive index, resulting in multiple images, possibly elongated. Especially characteristic of the Strait of Messina and arctic regions.

fat bloom (*FoodSci*) See CHOCOLATE BLOOM.

fat board (*Build*) A board on which the bricklayer collects the fat during the process of pointing.

fat-body (*BioSci*) In insects, a mesodermal tissue of fatty appearance, the cells of which contain reserves of fat and other materials and play an important part in the metabolism of the animal; in amphibians, highly vascular masses of fatty tissue associated with the gonads.

fat coals (*MinExt*) Coals which contain plenty of volatile matter (gas-forming constituents).

fat edges (*Build*) A defect in paintwork in which ripples form at edges and in angles, due to excess of paint.

fat face (*Print*) Heavy types with hairline serifs and main strokes at least half as wide as the height of the letters, Ultra Bodoni being a well-known example.

father file (*ICT*) See GRANDFATHER FILE.

father or mother of the chapel (*Print*) A person elected by the associated employees of a printing department to represent them and to safeguard their interests.

fathom (*For*) A cubic timber measure, $(6 \text{ ft})^3$ or 216 ft^3.

fathom (*Genrl*) A unit of measurement. Generally, a nautical measurement of depth = 6 ft (1·8 m).

fathom (*MinExt*) In general mining, the volume of a 6 ft cube; in gold mining, often a volume 6 ft by 6 ft by the thickness of the reef; in lead mining, sometimes a volume 6 ft by 6 ft by 2 ft. It is the unit of performance of a rock drill – 'fathoms per shift'.

fathometer (*Acous*) An ultrasonic depth-finding device.

fatigue (*BioSci*) The state of an excitable cell or tissue that responds less readily to an immediate second stimulation than to one occurring later.

fatigue (*Eng*) A phenomenon which results in the sudden fracture of a component after a period of cyclic loading in the elastic regime. See panel on FATIGUE.

fatigue limit (*Eng*) The upper limit of the range of stress that a metal can withstand indefinitely. If this limit is exceeded, failure will eventually occur.

Fatigue

Fatigue causes the sudden fracture of a component after a period of cyclic loading. It involves the initiation and growth of a crack at a site of STRESS CONCEN-TRATION usually on, but sometimes below, the surface. The crack reduces the effective cross-sectional area so that the component ruptures under a normal service load which had been satisfactorily withstood before the propagation of the crack. Depending on the material, the final fracture may occur in a *ductile* or *brittle* mode.

The initiation site and progressive development of the crack front culminating in final fracture give the site a characteristic appearance. See diagram (left), which illustrates fatigue failure in a circular shaft. The initiation site is shown and the shell-like markings, often referred to as beach markings because of their resemblance to the ridges left in the sand by retreating waves, are caused by arrests in the propagation of the crack front. The hatched region on the opposite side to the initiation site is the final region of ductile fracture.

If there is more than one initiation point, two or more cracks may propagate as shown in the diagram (right) with the final area of ductile fracture being a band across the middle. Such fractures occur when a component is continuously bent in opposite directions or when it is bent towards a crack that has already formed on one side.

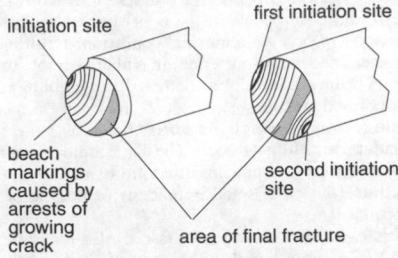

Fatigue Fracture from single and double initiation sites

Fatigue strength

This is determined by applying different levels of cyclic stress to individual test specimens and measuring the number of cycles to failure. Standard laboratory tests use various methods for applying the cyclic loading, eg rotating bend, cantilever bend, axial push-pull and torsion. The data are plotted in the form of a *stress–number of cycles to failure (S/N) curve*. Owing to the statistical nature of the failure, several specimens have to be tested at each stress level. Some materials, notably low-carbon steels, show a flattening off at a particular stress level which is referred to as the fatigue limit. In principle, components designed so that the applied stresses do not exceed this level should not fail in service. The difficulty is that a localized stress concentration may be present or introduced during service which leads to failure, despite the design stress being nominally below the 'safe' limit.

See STRESS CONCENTRATION, STRESS–STRAIN CURVE.

---

fatigue of metals (*Eng*) The phenomenon of the failure of metals under the repeated application of a cycle of stress. Factors involved include amplitude, average severity, rate of cyclic stress and temperature effect. NOTCH BRITTLENESS commences at a scratch or blemish.

fatigue test (*Eng*) A test made on a material to determine the range of alternating stress to which it may be subjected without risk of ultimate failure. By subjecting a series of specimens to different ranges of stress, while the mean stress is constant, a stress–number curve is obtained.

fatigue-testing machine (*Eng*) A machine for subjecting a test piece to rapidly alternating or fluctuating stress, in order to determine its FATIGUE LIMIT.

fat lime (*Build*) LIME made by burning a pure, or very nearly pure, limestone, such as chalk.

fat matter (*Print*) A composing-room term for easily set portions of the work in hand.

fat-necrosis (*Med*) The splitting of fat, due to the escape of a fat-splitting enzyme from the pancreas into the abdominal cavity, with death of the fat-containing cells so affected.

fats (*Chem, FoodSci*) An important group of naturally occurring substances consisting of the glycerides of higher fatty acids (eg palmitic acid, stearic acid, oleic acid) which are solid at room temperature (by contrast with OILS which are liquid or semi-solid). Essential component of the human diet.

fat splitting (*Chem*) The term used to describe the hydrolysis of animal and vegetable fats into glycerol and

fatty acids. Can be effected in a number of ways but chiefly by using strong alkalis (as in soap-making) or inorganic acids.

fat stock (*Agri*) See FINISHED STOCK.

fattening (*Build*) Thickening of varnish in the can, esp the appearance of gelatinous bits. Also *curdling, livering*.

fatty acids (*Chem*) A term for the whole group of saturated and unsaturated monobasic aliphatic carboxylic acids. The lower members of the series are liquids of pungent odour and corrosive action, soluble in water; the intermediate members are oily liquids of unpleasant smell, slightly soluble in water. The higher members from C_{10} upwards are mainly solids, insoluble in water, but soluble in ethanol and in ethoxyethane. Saturated fatty acids with no double bonds are linked with the development of ATHEROMA. In contrast the polyunsaturated fatty acids, linoleic, linolenic and arachidonic, are termed essential fatty acids as they must be included in the diet and may have a preventive role against atheroma and are required for the synthesis of prostaglandins.

fatty degeneration (*Med*) Degeneration of the cell substance, accompanied by the appearance in it of droplets of fat, due to the action of poisons or to lack of oxygen.

faucet (*Build*) (1) A small tap or cock. (2) The enlarged or socket end of a pipe at a *spigot-and-socket joint*.

faucet ear (*Build*) See CROISETTE.

faujasite (*Min*) One of the less common zeolites; a hydrated silicate of sodium, calcium and aluminium. It exhibits a wider range of molecular absorption than any other zeolite.

fault (*Geol, MinExt*) A fracture in rocks along which some displacement (the *throw* of the fault) has taken place. The displacement may vary from a few millimetres to thousands of metres. Movement along faults is the common cause of earthquakes.

fault breccia (*Geol*) A fragmental rock of breccia type resulting from shattering during the development of a fault.

fault condition (*Eng*) A departure from normal operating conditions which might lead directly or indirectly to an accident, damage or a shutdown.

fault current (*ElecEng*) That caused by defects in electrical circuit or device, such as short circuit in system. The peak value of current is the accepted measure.

fault-finding (*ElecEng*) General description of locating and diagnosing faults, according to a prearranged schedule, generally arranged in a chart or table, with or without special instruments. US *trouble-shooting*.

fault rate (*ElecEng*) See RELIABILITY.

fault resistance (*ElecEng*) A term sometimes used to denote insulation resistance, but more commonly the resistance of an actual fault, eg an arc between a conductor and earth.

fault tolerance (*ICT*) The ability of a system to execute specific tasks correctly regardless of failures and errors. See REDUNDANCY.

fault tolerant (*ElecEng*) The method of design and construction of an electrical circuit to make it highly reliable. The circuit configuration is arranged in such a way that strategic-components failure does not mean circuit operation is completely lost.

fault trap (*Geol*) A petroleum reservoir in which oil in a pervious bed cannot escape because of an impervious bed across the fault.

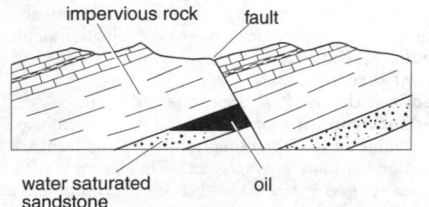

fault trap Any gas not shown.

fault tree (*Eng*) A representation of the different initial events and the possible successive malfunctions which would lead to an accident.

fault tree analysis (*Eng*) Design method which aims to track a specific product failure back to the original cause(s) using a network. Complementary to FAILURE MODES AND EFFECTS ANALYSIS. Abbrev *FTA*.

fauna (*BioSci*) A collective term denoting the animals occurring in a particular region or period. Pl *faunas* or *faunae*. Adj *faunal*.

faunal region (*BioSci*) An area of the Earth's surface characterized by the presence of certain species of animals.

faveolate (*BioSci*) Resembling a honeycomb in appearance. *Favous*, said of a surface pitted like a honeycomb. *Favus*, a hexagonal pit or plate. Also *favose*.

favism (*Med*) Haemolytic red cell destruction, in people of Mediterranean origin with glucose–6–phosphate dehydrogenase deficiency, after ingesting the broad bean, *Vicia faba*.

favose (*BioSci*) See FAVEOLATE.

favourite (*ICT*) A website that has been BOOKMARKED in a browser.

favus (*Med*) A contagious skin disease, esp of the scalp, due to infection with the fungus *Achorion schönleinii*.

fax (*ICT*) Abbrev for FACSIMILE.

fayalite (*Min*) A silicate of iron, Fe_2SiO_4, crystallizing in the orthorhombic system; a common constituent of slags but occurring also in igneous rocks, chiefly of acid composition, including pitchstone, obsidian, quartz-porphyry, rhyolite, and also in ferrogabbro.

faying face (*Eng*) That part of a surface of wood or metal specially prepared to fit an adjoining part. Also *faying surface*.

FBR (*NucEng*) Abbrev for *fast-breeder reactor*. See BREEDER REACTOR, FAST REACTOR.

FC (*Build*) Abbrev for FAIR CUTTING.

FCB (*ICT*) Abbrev for FILE CONTROL BLOCK.

FCC (*Chem*) Abbrev for FACE-CENTRED CUBIC.

F-centre (*Phys*) An electron trapped at a negative ion vacancy in an ionic crystal. *F*-centres can be formed by the release of electrons by irradiation with X-rays or by producing stoichiometric excess of anions in the crystal, and give rise to broad optical absorption bands. F_1-centres consist of *F*-centres with a further electron trapped in the same vacancy. F_A-centres are *F*-centres modified by one neighbouring cation being different from the cations of the lattice. *F*-centre aggregates can be formed by arrays of nearest-neighbour centres.

F-centred lattice (*Phys*) See *F*-CENTRE. See fig. at UNIT CELL. Also *face-centred crystal lattice*.

Fc fragment (*BioSci*) Fragment of immunoglobulin obtained by papain hydrolysis representing the C-terminal halves of the two heavy chains linked by disulphide bonds. It has no antigen-binding activity but contains the sites involved in complement activation and placental transmission, and some of the Gm allotype markers.

F-class insulation (*ElecEng*) A class of insulating material which will withstand a temperature of 155°C.

Fc receptor (*BioSci*) Any of a number of receptors that are present on the plasma membranes of cells and that bind the *Fc fragment* of immunoglobulin. Neutrophils, mononuclear phagocytes, eosinophils and B-lymphocytes have receptors for the Fc of IgG; different receptors have varying affinities for different IgG subclasses. Mast cells, basophil leucocytes and eosinophils have receptors for Fc of IgE.

fd (*Build*) Abbrev for *framed*.

FDDI (*ICT*) Abbrev for FIBRE DISTRIBUTED DATA INTERFACE.

F-diagram (*ChemEng*) The cumulative residence time distribution in a continuous flow system, plotted in dimensionless co-ordinates. Important in assessing the performance of chemical reactors, kilns, etc.

F-display (*Radar*) A type of radar display, used with directional antenna, in which the target appears as a bright spot which is off-centre when the aim is incorrect.

F-distribution (*MathSci*) The distribution of the random variable formed by taking the ratio of two chi-squared random variables, each of which is divided by its degrees of freedom. The central F-distribution is indexed by two parameters, the degrees of freedom associated with the chi-squared variables which form the numerator and denominator of the ratio.

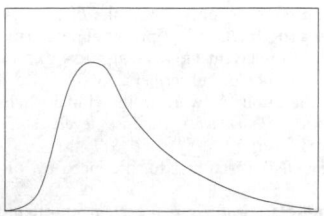

F-distribution

FDM (*ICT*) Abbrev for FREQUENCY-DIVISION MULTIPLEXING.

FDS law (*Phys*) Fermi–Dirac–Sommerfeld law, which gives the algebraic number of a quantized system of particles which have velocities within a small range.

Fe (*Chem*) Symbol for IRON.

fear (*Psych*) An emotional state aroused by the presence or anticipation of dangerous or noxious stimuli and which normally gives rise to avoidance, defensive or escape behaviour.

feasibility diagram (*Eng*) A plot of flow length against cavity thickness in an injection moulding tool, with curves for shot volume, clamp force and melt pressure. Together, they define a feasibility envelope, which with mouldability index curves defines the window within which moulding can occur.

feather (*Build*) A thin strip of wood fitted into a groove on each edge of adjacent butted boards.

feather (*Eng*) (1) A rectangular key sunk into a shaft to permit a wheel to slide axially, while preventing relative rotation. (2) Iron slips for reducing the friction between a wedge and an object to be split.

Feather analysis (*Phys*) An approximate method of determining the range of beta rays forming part of a combined beta–gamma spectrum, by comparison of the absorption curve with that for a pure beta emitter.

feather eating (*Vet*) A habit, acquired by birds, characterized by pecking, plucking or eating their own plumage or that of other birds. The habit may develop into CANNIBALISM.

feather-edge brick (*Build*) A brick similar to a compass brick, used esp for arches.

feather-edged coping (*Build*) A coping stone sloping in one direction on its top surface. Also *splayed coping*.

feathering (*Paper*) The irregular edge of an ink line due to the paper being insufficiently sized.

feathering hinge (*Aero*) A pivot for a rotorcraft blade which allows the angle of incidence to change during rotation.

feathering paddles (*Eng*) Paddle-wheels so controlled that the floats enter and leave the water at right angles to the surface.

feathering pitch (*Aero*) The blade angle of a propeller giving minimum drag when the engine is stopped.

feathering propeller (*Aero*) See PROPELLER.

feathering pump (*Aero*) A pump for supplying the necessary hydraulic pressure to turn the blades of a *feathering propeller* to and from the feather position.

feather joint (*Build*) See PLOUGHED-AND-TONGUED JOINT.

feather ore (*Min*) A plumose or acicular form of the sulphide of lead and antimony, Also *jamesonite, stibnite*.

feathers (*BioSci*) Epidermal outgrowths forming the body covering of birds; distinguished from scales and hair, to which they are closely allied, by their complex structure, and by the possession of a vascular core which at first projects from the surface.

feather tongue (*Build*) A wooden tongue with its grain across the grooves for a PLOUGHED-AND-TONGUED JOINT.

feature interaction (*ICT*) A problem that can arise in an INTELLIGENT NETWORK when a service called up by the user in turn calls up another service whose characteristics or requirements conflict with the first.

feature switch (*ICT*) Equipment at a CENTRAL OFFICE that provides intelligent PRIVATE BRANCH EXCHANGE facilities for a number of subscribers, for instance direct dialling from the public network to individual extensions.

febrifuge (*Pharmacol*) Against fever; a remedy which reduces fever.

febrile (*Med*) Pertaining to, produced by, or affected with fever.

fecal pellets (*Geol*) Animal excrement, often in the form of rods or ovoid pellets, found in sedimentary rocks. See COPROLITE.

feces (*BioSci*) The indigestible residues remaining in the alimentary canal after digestion and absorption of food. Formerly (UK) *faeces*.

Fechner colours (*Phys*) The visual sensations of colour which are induced by intermittent achromatic stimuli.

Fechner's law (*Psych*) A law stating that logarithmic changes in stimulus intensity produce linear changes in

subjective sensations. See WEBER'S LAW, with which it is often linked.

fecundity (*BioSci*) The number of young produced by a species or individual.

feebly hydraulic lime (*Build*) LIME made by burning a limestone containing 5–12% clay.

feed (*Eng*) (1) The rate at which the cutting tool is advanced. (2) Fluid pumped into a vessel, eg feed-water to a boiler. (3) Mechanism for advancing material or components into a machine for processing.

feed (*ICT*) To offer a programme or signal at some point in a communication network.

feed (*MinExt*) Forward motion of drill or cutter.

feed (*NucEng*) The loaded solution introduced into the next stage of a reprocessing plant.

feedback (*Acous*) Phenomenon in which part of an output signal is fed back into the input of the system. If the feedback signal is in phase with the primary input signal, the system can become unstable (positive feedback). This often occurs in electro-acoustic systems in which microphone and loudspeaker are in the same room. Negative feedback occurs if the feedback signal decreases the input signal.

feedback (*BioSci*) A mechanism of control that uses the consequences of a process to regulate the rate at which the process occurs. It may be slowed down or inhibited (NEGATIVE FEEDBACK), as in many metabolic pathways, or may be enhanced (POSITIVE FEEDBACK).

feedback (*ICT*) The transfer of some output energy of an amplifier to its input, so as to modify its characteristics. *Current* (or *voltage*) *feedback* is when feedback signal depends on current (or voltage) in output respectively. See NEGATIVE FEEDBACK, POSITIVE FEEDBACK.

feedback admittance (*ICT*) The short-circuit transfer admittance from the output to the input terminals of any circuit, filter, electronic device or combination of all three.

feedback characteristic (*Electronics*) See TRANSISTOR CHARACTERISTICS.

feedback circuit (*ICT*) See FEEDBACK PATH.

feedback control loop (*ICT*) A closed transmission path including an active transducer, forward and feedback paths and one or more mixing points. The system is such that a given relation is maintained between the input and output signals of the loop.

feedback control system (*ICT*) US for CLOSED LOOP SYSTEM.

feedback factor (*ICT*) Defined as $m = 1 - \beta A$; m is the feedback factor, β is the FEEDBACK RATIO and A is the open-loop gain of the amplifier.

feedback path (*ICT*) That from the loop output signal to the feedback signal in a feedback control loop.

feedback ratio (*ICT*) A property of the feedback path that determines the amount of feedback applied to the input of an amplifier. Defined by $\beta = e_f/e_o$ where β is the feedback ratio, e_o is the signal voltage at the amplifier output and e_f is the signal fed back to the input via the feedback path.

feedback signal (*ICT*) That which is responsive in an automatic controller to the value of the controlled variable.

feedback transducer (*ICT*) One that generates a signal, generally electrical, depending on quantity to be controlled, eg for rotation potentiometer, synchro or tacho, giving proportional derivative or integral signals respectively.

feedback windings (*ElecEng*) Those control windings in a saturable reactor to which are made the feedback connections.

feed check valve (*Eng*) Non-return valve in the delivery pipe between feed-water pump and boiler.

feeder (*Build*) A natural or artificial channel supplying water to a reservoir or canal.

feeder (*ElecEng*) (1) An overhead or underground cable, of large current-carrying capacity, used in the transmission of electric power; it serves to interconnect generating stations, substations and feeding points, without intermediate connections. (2) In electrical circuits, the lines running from the main switchboard to the branch panels in an installation.

feeder (*Eng*) The runner or riser hole of a mould, containing sufficient molten metal to feed the casting and so compensate for contraction of the solidifying metal.

feeder (*ICT*) Conductor, or system of conductors, connecting the radiating portion of an antenna to the transmitter or receiver. It may be a balanced pair, a quad, coaxial or waveguide.

feeder (*MinExt*) A mechanical appliance for supplying broken rock or crushed ore, at a predetermined rate, to some form of crusher or concentrator.

feeder bus-bars (*ElecEng*) In a generating station or main substation, bus-bars to which the outgoing feeders are connected.

feeder cell (*BioSci*) A cell type that is added to another cell population in culture, to facilitate growth of the second type by secreting growth factors or compensating for a genetic deficiency etc. For example, peritoneal macrophages are used as feeder cells for HYBRIDOMA cells. Collectively *feeder layer*.

feeder ear (*ElecEng*) A type of ear for attaching an overhead contact wire of a tramway system to the supporting wire; it serves also to lead current to the contact wire.

feeder head (*Eng*) See HOT TOP.

feeder layer (*BioSci*) See FEEDER CELL.

feeder mains (*ElecEng*) See FEEDER (1).

feeder panel (*ElecEng*) A switchboard panel on which are mounted the switchgear and instruments for controlling one or more feeders.

feeder pillar (*ElecEng*) A pillar containing switches, links and fuses, for connecting the feeders of an electric power distributing system with the distributors.

feed finger (*Eng*) A rod used in association with a collet to push or pull the bar of material forward in a capstan or automatic lathe until the bar touches a stop, thus governing the feed length.

feedhorn (*ImageTech*) A receiving device at the focal point of a DISH which collects the focused signals and directs them to the LOW-NOISE BLOCK CONVERTER.

feeding (*Build*) The thickening of paint, often due to a reaction between non-compatible solvents etc.

feeding head (*Eng*) See FEEDER.

feeding point (*ElecEng*) The junction point between a feeder and a distribution system. Also *distributing point*.

feeding rod (*Eng*) A heated iron rod inserted in the feeder of a mould, and worked with a pumping motion to assist feeding during the cooling of the molten metal.

feeding-up (*Build*) A thickening of paints and varnishes in the can, making them unsatisfactory.

feed pipe (*Eng*) The pipe carrying feed-water from the feed pump to a boiler.

feed reel (*ImageTech*) The reel of film which is being unwound as the film is taken off to pass through the gate in a camera, printer or projector.

feed rollers (*Print*) Driven rollers which convey a web into a printing couple. Cf IDLER, IDLING ROLLER.

feed screw (*Eng*) A screw used for supplying motion to the feed mechanism of a machine tool.

feedstock energy (*Chem*) Energy of oil, gas or coal used as feedstock for polymer production. Part of the feedstock energy can be recouped by incineration of polymer products following use.

feedthrough (*ElecEng*) A conductor used to connect patterns on opposite sides of the board of a printed circuit, or an insulated conductor for connection between two sides of a metal earthing screen.

feed-water (*Eng*) The de-aerated and chemically treated water fed into a boiler for evaporation.

feed-water heater (*Eng*) An arrangement for heating boiler feed-water by means of steam which has done work in an engine or turbine. It is similar in principle to a steam condenser of either the surface or the jet type.

feel (*Textiles*) A term describing the physical character of a cloth when handled.

feeler (*Textiles*) Mechanical or electrical device, used with automatic weft replenishing motion on a loom to determine when the pirn change is necessary.

feeler gauge (*Eng*) A thin strip of metal of known and accurate thickness, usually one of a set, used to measure the distance between surfaces or temporarily placed between working parts while setting them an accurate distance apart.

feeler switch (*ElecEng*) A switch sometimes forming part of the equipment of an auto-reclose circuit breaker; it determines whether the fault has cleared before allowing the circuit breaker to reclose.

feet-switch (*ElecEng*) See TROPICAL SWITCH.

Fehling's solution (*Chem*) A solution of cupric sulphate and potassium sodium tartrate (Rochelle salt) in alkali, used as an oxidizing agent. It is an important analytical reagent for aldehydes, glucose, fructose, etc, which reduce it to cuprous oxide.

feint (*Print*) The pale, edge-to-edge, horizontal ruling in account books and notebooks.

feldspar (*Min*) A most important group of rock-forming silicates of aluminium, together with sodium, potassium, calcium, or (rarely) barium, crystallizing in closely similar forms in the monoclinic and triclinic systems. The chief members are *orthoclase* and *microcline* (potassium feldspar); *albite* (sodium feldspar); and the *plagioclases* (sodium–calcium feldspar). The form *felspar*, though still commonly used, perpetuates a false derivation from the Ger *fels* (rock); actually it is from the Swedish *feldt* (field).

feldspathic sandstone (*Geol*) See ARKOSE.

feldspathoids (*Min*) A group of rock-forming minerals chemically related to the feldspars, but undersaturated with regard to silica content, and therefore incapable of free existence in the presence of magmatic silica. The chief members of the group are HAÜYNE, LEUCITE, NEPHELINE, NOSEAN and SODALITE. Colloq *foids*.

Felici balance (*ElecEng*) An ac electrical measuring bridge for determining mutual inductance between windings.

Felici generator (*ElecEng*) A modern form of electrostatic high-voltage generator developed in France and comparable with a van de Graaff generator.

feline distemper (*Vet*) See FELINE ENTERITIS.

feline enteritis (*Vet*) Highly contagious and often fatal parvovirus disease of cats characterized by vomiting, dehydration, abdominal pain, anorexia and death. Abortion and fetal resorption can occur in queens. Kittens can be born with cerebellar ataxia. Vaccination widely used. Also *feline distemper, panleucopenia*.

feline immunodeficiency virus (*Vet*) Widespread *lentivirus* (retrovirus) that causes an immunodeficiency in domestic cats. Abbrev FIV.

feline infectious anaemia (*Vet*) Caused by *Haemobartonella felis*. Symptoms include inappetance, weakness, anaemia, jaundice, wasting and splenomegaly. Parasite is associated with red blood cells.

feline influenza (*Vet*) A complex of conditions with several pathogens involved and numerous secondary invaders. The two most commonly encountered agents are feline rhinotracheitis virus (FVR) and feline calcivirus (FCV). Vaccines available. Colloq *cat flu*.

feline leukaemia (*Vet*) Characterized by fever, weakness, inappetance, wasting and anaemia. Caused by leukaemia virus. Mainly abdominal and thymic forms. Experimental vaccines have been used.

feline pneumonitis (*Vet*) Caused by a cat-adapted *Chlamydia psittaci*. The symptoms are mucopurulent conjunctivitis, sneezing and a mild nasal discharge. No vaccine is yet available. Also *cat flu*.

feline viral rhinotracheitis (*Vet*) An acute viral infection of the upper respiratory tract of the cat. See FELINE INFLUENZA. Vaccines available.

fell (*Textiles*) The edge of the cloth in a loom, where the PICKS of weft are beaten up by the reed.

fellmongering (*Textiles*) Obtaining wool from the skins of slaughtered sheep. Cf PULLING, SHEARING, SKIN WOOL.

felloe (*Build*) (1) The outer part of the framing for a centre. (2) A segment of the rim of a wooden wheel, about which a tyre is usually shrunk. The term is sometimes applied to the whole rim.

felon (*Vet*) Suppurative arthritis of cattle; commonly associated with mastitis. Also *fellon*.

felsic (*Min*) Mnemonic for FELdspars, FELspathoids and quartz (SIliCa) actually present as mineral constituents in a rock. Also applied to rocks largely composed of feldspars and quartz. See MAFIC.

felsite (*Min*) Fine-grained igneous rocks of acid composition, occurring as lavas or minor intrusions, and characterized by felsitic texture; a fine patchy mosaic of quartz and feldspar, resulting from the devitrification of an originally glassy matrix.

felspar (*Min*) See FELDSPAR.

felt (*Build*) A fibrous material, treated so as to be rendered watertight, used as underlining for roofs; also as an overlining for roofs when underlaid with asphalt or an asphalt compound.

felt (*Paper*) A woven blanket or a synthetic fabric in the form of an endless band to give support to the web at various points on the paper machine, eg at a wet press, MG DRYER or drying cylinders. See panel on PAPER AND PAPERMAKING.

felt (*Textiles*) (1) A densely matted non-woven fabric containing wool or hair that has passed through a felting process of heat, steam and pressure. (2) Heavily milled woven fabric with a matted fibrous surface.

felting (*Paper*) The natural action by which fibres adhere in paper-making.

felting (*Textiles*) The formation of a felt during processing (eg by MILLING) or during wear; a property esp of woollen fabrics resulting from the structure of the cuticle of the fibres.

female (*BioSci*) (1) An individual whose gonads produce ova. (2) The larger and less motile gamete, the egg. (3) Gametophytes and their reproductive structures that produce eggs but not male gametes. (4) Sporophytes and their reproductive structures that produce the megaspores and hence seeds. (5) Individual seed plants or flowers that have functional carpels but not functional stamens. Cf HERMAPHRODITE, MALE.

female (*Eng*) See MALE AND FEMALE.

female gauge (*Eng*) See RING GAUGE.

female pronucleus (*BioSci*) The nucleus remaining in the ovum after maturation.

female thread (*Eng*) See INTERNAL SCREW-THREAD.

femerell (*Build*) A roof lantern having louvres for ventilation.

femic constituents (*Geol*) Those minerals which are contrasted with the salic constituents in determining the systematic position of a rock in the *CIPW* scheme of classification. Note that these are the *calculated* components of the 'norm'; the corresponding *actual* minerals in the 'mode' are said to be *mafic*, ie rich in magnesium and iron.

femto- (*Genrl*) Prefix denoting a thousand million millionth (10^{-15}). Symbol f.

femtometre (*Phys*) See FERMI.

femur (*Arch*) See MEROS.

femur (*BioSci*) The proximal region of the hind-limb in land vertebrates; the bone supporting that region; the third joint of the leg in insects, Myriapoda and some Arachnida. Adj *femoral*.

fen (*EnvSci*) Vegetation developed naturally on waterlogged land, forming a peat that is neutral or alkaline from the dead parts of tall grasses, sedges and herbs. See CARR. Cf BOG.

fence (*Build, Eng*) (1) An adjustable plate directing or limiting the movement of one piece with respect to another; a guard or stop to limit motion. (2) An attachment to a plane (cf FILLISTER) or hand circular saw which keeps the blade at a fixed distance from the edge of the work.

Fenchel wet expansion tester (*Paper*) Test apparatus in which a strip of paper is held vertically between two clamps under low tension and immersed in water. The resultant increase in length (wet expansion) is indicated by a needle connected to one of the clamps.

fenchone (*Chem*) Crystalline solid. Mp 5°C, bp 192°C. A dicyclic ketone. Optical isomers occur in fennel and lavender oils (dextrorotatory form) and in thuja oil (laevorotatory).

fender (*ElecEng*) A metal cover attached to the end of the frame of an electric machine in such a way as to prevent accidental contact with live or moving parts. It does not carry a bearing. Also a *protection cap*.

fender wall (*Build*) A dwarf brick wall supporting the hearth of a ground-floor fireplace.

fenestra (*Arch*) A window or other opening in the outer walls of a building. Pl *fenestrae*.

fenestra (*BioSci*) An aperture in a bone or cartilage, or an opening between two or more bones.

fenestra ovalis (*BioSci*) The upper of the two membrane-covered openings between the middle ear and the inner ear. It transmits auditory vibrations from the TYMPANUM, conveyed via the ossicles of the middle ear, to the COCHLEA of the inner ear. See AUDITORY OSSICLES.

fenestra rotunda (*BioSci*) The lower of the two membrane-covered openings between the middle ear and the inner ear that vibrates in response to changes in pressure in the PERILYMPH surrounding the COCHLEA.

fenestrate (*BioSci*) With window-like perforations; perforated or having translucent spots. Also *fenestrated*.

fenestration (*Build*) (1) The arrangement of window and other openings in the outer walls of a building. (2) The controlling of light emission into a room or building.

fenestration (*Med*) A surgical operation to improve hearing which involves a new 'window' being opened to the inner ear.

fenfluramine (*Pharmacol*) A drug, now withdrawn, formerly used in conjunction with phentermine to treat obesity.

fenite (*Min*) A metasomatic leucocratic alkalisyenite usually associated with cabonatites.

fenitization (*Min*) The process of conversion of granite rocks into fenite by alkali metasomatism.

fenoprop (*Chem*) 2-(2,4,5-trichlorophenoxy)propanoic acid. Used as a weedkiller. Also *silvex, 2,4,5-TP*.

fent (*Textiles*) Damaged pieces of cloth or off-cuts sold by weight to wholesalers or market traders for sale to consumer; if suitable, recycled within factory or system. See WASTE.

fentanyl (*Pharmacol*) A powerful analgesic resembling morphine in its action.

FEP (*Chem*) Abbrev for FLUORINATED ETHYLENE PROPYLENE.

feral (*BioSci*) Of a domesticated animal that has reverted to the wild.

ferberite (*Min*) Iron tungstate, the end-member of the wolframite group of minerals, the series from $FeWO_4$ to $MnWO_4$.

fergusite (*Min*) An alkaline SYENITE containing large crystals of pseudoleucite in a matrix of aegirine-augite, olivine, apatite, sanidine and iron oxides.

fergusonite (*Min*) A rare mineral occurring in pegmatites; it is a niobate and tantalate of yttrium, with small amounts of other elements.

Fermat prime (*MathSci*) Any prime of the form $2^{2^n}+1$. Cf MERSENNE PRIME.

Fermat's last theorem (*MathSci*) The conjecture that no integral values of x, y and z can be found to satisfy the equation $x^n + y^n = z^n$ if n is an integer greater than two. A general proof has been sought for over 350 years; one proposed in 1993 is now generally accepted.

Fermat's principle of least time (*Phys*) A principle stating that the path of a ray of light from one point to another (including refractions and reflections) will be that taking the least time.

Fermat's spiral (*MathSci*) See PARABOLIC SPIRAL.

fermentation (*BioSci, Chem*) The biochemical pathway whereby organic compounds, especially carbohydrates, are broken down enzymatically in the absence of oxygen. It is a form of anaerobic respiration, and is used industrially for the manufacture of antibiotics and certain other important drugs by the action of bacteria, yeasts, moulds, or other micro-organisms. Two basic types of fermentation are important in the food industry. (1) In *alcoholic fermentation* the yeast *Saccharomyces* is grown in a sugar-rich solution and produces carbon dioxide and ethyl alcohol. (2) In *lactic acid fermentation* enzymes from *Lactobacillus* convert sugar into lactic acid, as in yoghurt, cheese, etc.

fermi (*Phys*) A length unit equal to 10^{-15} m; also *femtometre* (fm). Used in nuclear physics, being of the order of the radius of the proton, 1·2 fm.

Fermi age (*NucEng*) Slowing-down area for neutrons calculated from *Fermi age theory* which assumes that neutrons, on being slowed down, lose energy continuously in an infinite homogeneous medium. Has the dimensions of length squared. Also *neutron age*. See AGE THEORY.

Fermi characteristic energy level (*Electronics*) See FERMI LEVEL.

Fermi constant (*Phys*) A universal constant which indicates the coupling between a nucleon and a lepton field. Its value is $1·4 \times 10^{-50}$ J m^{-3}, and it is important in beta decay theory.

Fermi decay (*Electronics*) The theory of ejection of electrons as β-particles.

Fermi–Dirac distribution curve (*Electronics*) A function, ranging from unity to zero, specifying the probability that an electron in a semiconductor will occupy certain quantum states when thermal equilibrium exists. The energy level at which the value of the function is 0·5 is called the FERMI LEVEL.

Fermi–Dirac gas (*Phys*) An assembly of particles which obey FERMI–DIRAC STATISTICS and the PAULI EXCLUSION PRINCIPLE. For an extremely dense Fermi gas, such as electrons in a metal, all energy levels up to a value E_F, the Fermi energy, are occupied at absolute zero.

Fermi–Dirac–Sommerfeld law (*Phys*) See FDS LAW.

Fermi–Dirac statistics (*Phys*) Statistical mechanics laws obeyed by system of particles whose wavefunction changes sign when two particles are interchanged, ie the PAULI EXCLUSION PRINCIPLE applies.

Fermi level (*Electronics*) The energy level at which there is a 0·5 probability of finding an electron; it depends on the distribution of energy levels and the number of electrons available. In semiconductors, the number of electrons is relatively small and the Fermi level is affected by DONOR and ACCEPTOR impurities.

fermion (*Phys*) A particle which obeys FERMI–DIRAC STATISTICS. Fermions have total spin angular momentum of $(n+\frac{1}{2})\hbar$ where $n = 0,1,2,\ldots$ and $\hbar$ is DIRAC'S CONSTANT. BARYONS and LEPTONS are fermions and are subject to the PAULI EXCLUSION PRINCIPLE. See panel on ATOMIC STRUCTURE.

Fermi plot (*Phys*) See KURIE PLOT.

Fermi potential (*Phys*) The equivalence of the energy of the Fermi level as an electric potential.

Fermi selection rules (*Phys*) See NUCLEAR SELECTION RULES.

Fermi surface (*Phys*) A constant energy surface in k-space which encloses all occupied electron states at absolute zero in a crystal.

Fermi temperature (*Phys*) The degeneracy temperature of a FERMI–DIRAC GAS which is defined by E_F/k, where E_F is the energy of the FERMI LEVEL and k is BOLTZMANN'S CONSTANT. This temperature is of the order of tens of thousands of degrees kelvin for the free electrons in a metal.

fermium (*Chem*) Artificial element, at no 100. Principal isotope ^{257}Fm, half-life 95 days. Symbol Fm.

ferns (*BioSci*) Pteridophytes of the class Filicopsida.

ferractor (*Phys*) A MAGNETIC AMPLIFIER with a FERRITE CORE.

Ferranti effect (*ElecEng*) The rise in voltage which takes place at the end of a long transmission line when the load is thrown off; it is due to the charging current flowing through the inductance of the line.

Ferranti–Hawkins protective system (*ElecEng*) A discriminative protective system for feeders; core balance transformers are placed at each end, with their secondary windings connected to each other through pilot wires.

Ferranti meter (*ElecEng*) A name often given to the mercury-motor type of supply meter invented by Ferranti.

ferredoxin (*BioSci*) Non-haem iron–sulphur protein, a component of the electron transport system in PHOTO-SYNTHESIS.

Ferrel cell (*EnvSci*) A mid-latitude mean atmospheric circulation cell proposed by Ferrel in the 19th century, in which air flows polewards and eastwards near the surface and equatorwards and westwards at higher levels. This disagrees with reality. The term is now sometimes used to describe a mid-latitude circulation identifiable in mean meridional wind patterns.

ferri-, ferro- (*Chem*) Denoting trivalent and divalent iron respectively. Prefixes from Lt *ferrum*, iron.

ferric iron (III) chloride (*Chem*) FeCl$_3$. Brown solid. Deliquescent, soluble in ethanol. Uses: coagulant in sewage and industrial wastes, mordant, photo-engraving etching of copper, chlorination and condensation catalyst, disinfectant, pharmaceutical, analytical reagent.

ferric iron (III) oxide (*Chem*) Fe$_2$O$_3$. The common red oxide of iron. Used in metallurgy, pigments, polishing and theatrical rouge, gas purification, and as a catalyst.

ferric iron (III) sulphate (*Chem*) Fe$_2$(SO$_4$)$_3$. A yellowish-white powder which dissolves slowly in water. Uses: pigments, water purification, dyeing, disinfectant, medicine.

ferricyanide (*Chem*) The complex ion Fe(CN)$_6{}^{3-}$. Also *hexacyano ferrate* (III).

ferrimagnetic (*Phys*) A material, or of a material, which exhibits FERRIMAGNETISM. See panel on FERROMAG-NETICS AND FERRIMAGNETICS.

ferrimagnetism (*Phys*) Phenomenon in some magnetically ordered materials in which there is incomplete cancellation of the *antiferromagnetically* arranged spins giving a net magnetic moment; observed in ferrites and similar materials.

ferrimolybdite (*Min*) Hydrated molybdate of iron. Most so-called molybdite is ferrimolybdite. It occurs as a yellowish alteration product of molybdenite.

ferrite (*Chem*) (1) The BODY-CENTRED CUBIC form of iron and of solid solutions based on it. In pure iron, α-ferrite is stable up to 1183 K, whilst δ-ferrite occurs between 1663 K and the melting point (1811 K). (2) A ceramic iron oxide compound having FERRIMAGNETIC properties. Those with INVERSE SPINEL (cubic) structure and the general formula MOFe$_2$O$_3$, where M is a divalent TRANSITION METAL (such as Mn, Fe, Co, Ni, Cu), tend to have high permeability and low coercivity. Being ceramics they have high resistivity and are much used in high-frequency applications where eddy current losses are critical. The hexagonal structured ferrites such as barium and strontium ferrites, MO(Fe$_2$O$_3$)$_6$, have a very high coercivity. Useful as short, permanent magnets. (3) tetracalcium aluminoferrite, 4CaOAl$_2$O$_3$Fe$_2$O$_3$; a constituent ($\approx 8\%$ by weight) of Portland cement, it hydrates at a slow-to-medium rate during the setting reaction. See panels on CEMENT AND CONCRETE and FERROMAGNETICS AND FERRIMAGNETICS.

ferrite bead (*ICT*) A small element of ferrite material used particularly for threading onto wire of transmission line to increase the series inductance, but also once used for computer memory as ferrite-bead memory.

ferrite core (*Phys*) A magnetic core, usually in the form of a small toroid, made of FERRITE material such as nickel ferrite, nickel–cobalt ferrite, manganese–magnesium ferrite or yttrium–iron garnet. These materials have high resistance and make eddy current losses very low at high frequencies.

ferrite-core memory (*ICT*) Memory widely used in SECOND-GENERATION COMPUTERS. It consisted of tiny rings of magnetic material through which several wires were threaded. Each core could be magnetized in either direction to store a BIT.

ferrite-rod antenna (*ICT*) A small reception antenna, using a ferrite rod to accept electromagnetic energy, output being from an embracing coil. Also *loopstick antenna*.

ferritin (*BioSci*) A protein that functions as an iron store in the liver. As the central iron core is visible in the electron microscope, ferritin can be used as a tag for the localization of proteins in electron microscopy.

ferro- (*Genrl*) Prefix. See FERRI-.

ferro-actinolite (*Min*) An end-member compositional variety in the monoclinic amphiboles; essentially $Ca_2Fe_5Si_8O_{22}(OHF)_2$, but the name is applied to a member of the actinolite series with more than 80% of this molecule.

ferrochromium (*Eng*) A MASTER ALLOY of iron and chromium (60–72% chromium) used in making additions of chromium to steel and cast-iron.

ferrocyanide (*Chem*) The complex ion $Fe(CN)_6^{4-}$. Also *hexacyano ferrate* (II).

ferrodynamometer (*ElecEng*) Any DYNAMOMETER incorporating ferromagnetic material to enhance the torque.

ferro-edenite (*Min*) An end-member compositional variety in the monoclinic amphiboles; a hydrous sodium, iron, calcium and aluminium silicate.

ferroelectric material (*Phys*) A dielectric material (usually ceramic) with domain structure which exhibits spontaneous electric polarization. Analogous to ferromagnetic material. Such materials have relative permittivities of up to 10^5, and show dielectric hysteresis. See panel on FERROMAGNETICS AND FERRIMAGNETICS.

ferrogedrite (*Min*) See GEDRITE.

ferrohastingsite (*Min*) A compositional variety in the hornblende group of monoclinic amphiboles.

ferromagnesian (*Min*) Containing a relatively large proportion of iron and magnesium, as in minerals, eg hypersthene, or rocks, eg peridotite, theralite.

ferromagnetic (*Phys*) A material, or of a material, which exhibits FERROMAGNETISM. See panel on FERROMAGNETICS AND FERRIMAGNETICS.

ferromagnetic amplifier (*ElecEng*) A paramagnetic amplifier which depends on the non-linearity in ferroresonance phenomena at high radio-frequency power levels.

ferromagnetic resonance (*ElecEng*) A special case of PARAMAGNETIC RESONANCE, exhibited by ferromagnetic materials and often termed *ferroresonance*. Explained by simultaneous existence of two different pseudo-stable states for the magnetic material *B–H* curve, each associated with a different magnetization current for the material. Oscillation between these two states leads to large currents in associated circuitry.

ferromagnetism (*Phys*) A phenomenon in some magnetically ordered materials in which there is a bulk magnetic moment and the magnetization is large. The electron spins of the atoms in microscopic regions, *domains*, are aligned. In the presence of an external magnetic field the domains oriented favourably with respect to the field grow at the expense of the others and the magnetization of the domains tends to align with the field. Above the *Curie temperature*, the thermal motion is sufficient to offset the aligning force and the material becomes *paramagnetic*. Certain elements (iron, nickel, cobalt), and alloys with other elements (titanium, aluminium) exhibit permeabilities up to 10^4 (*ferromagnetic materials*). Some show marked hysteresis and are used eg for permanent magnets and magnetic amplifiers. See panel on FERROMAGNETICS AND FERRIMAGNETICS.

ferromanganese (*Eng*) A MASTER ALLOY of iron and manganese, used in making additions of manganese to steel or cast-iron.

ferromolybdenum (*Eng*) A MASTER ALLOY of iron and molybdenum (55–65% molybdenum) used in adding molybdenum to steel and cast-iron.

ferronickel (*Eng*) An alloy of iron and nickel containing more than 30% nickel. Lower-nickel alloys are known as nickel steel. See ELINVAR, INVAR, MUMETAL, PERMALLOY.

ferroprussiate paper (*Paper*) See BLUEPRINT PAPER.

ferroresonance (*ElecEng*) See FERROMAGNETIC RESONANCE.

ferrosilicon (*Eng*) A MASTER ALLOY of iron and silicon, used in making additions of silicon to steel and cast-iron. When containing 15% silicon, widely used in DENSE-MEDIA PROCESSES.

ferrospinel (*Eng*) A crystalline material which has the equivalent function to that of the divalent transition metal in a ferrite. See FERRITE (2).

ferrotype (*ImageTech*) A wet-collodion positive on a plate of darkened metal. Formerly favoured by street photographers owing to immediate availability of the print. Also *melanotype, tin-type*.

ferrous iron (II) oxide (*Chem*) FeO. Black oxide of iron.

ferrous iron (II) sulphate (*Chem*) $FeSO_4 \cdot 7H_2O$. Also the mineral MELANTERITE.

ferroxyl indicator (*Chem*) A little potassium hexacyanoferrate (III) and phenolphthalein, together with a corroding solution, eg of sodium chloride, made into a jelly with agar. It is used to show the presence of anodic and cathodic areas in an apparently uniform piece of iron, by turning blue and pink respectively.

ferruginous deposits (*Geol*) Sedimentary rocks containing sufficient iron to justify exploitation as iron ore. The iron is present, in different cases, in silicate, carbonate or oxide form, occurring as the minerals chamosite, thuringite, siderite, haematite, limonite, etc. The ferruginous material may have formed contemporaneously with the accompanying sediment, if any, or may have been introduced later.

ferrule (*Build*) The brass ring round the handle of a chisel or similar tool, at the end where the tang enters or at the striking end to prevent splitting.

ferrule (*Eng*) (1) A short length of tube. (2) A circular gland nut used for making a joint. (3) A slotted metal tube into the ends of which the conductors of a joint are inserted. The whole is soldered solid. When the conductors are oval, the ferrule is in two parts to allow for the fact that the major axes of the oval sections may not coincide.

fertile (*BioSci*) Able to produce asexual spores and/or sexual gametes.

fertile (*Phys*) The term describing an isotope in a nuclear reactor which can be converted by the capture of a neutron into a fissile isotope, eg uranium-238 which can be converted by series of reactions into plutonium-239.

fertile flower (*BioSci*) (1) A flower with functional carpels and/or stamens. (2) Sometimes a female flower.

fertilisin (*BioSci*) A substance that is present in the cortex of an ovum and that increases sperm motility.

fertility (*BioSci*) The number or percentage of eggs produced by a species or individual which develop into living young. Cf FECUNDITY.

fertilization (*BioSci*) The union of two sexually differentiated gametes to form a zygote.

fertilization cone (*BioSci*) A conical projection of protoplasm arising from the surface of an ovum containing many microtubules which are thought to facilitate the entry of the sperm.

fertilization tube (*BioSci*) See CONJUGATION TUBE.

fertilizer (*Agri*) A product containing nutrient supplements to enhance plant growth and development. It may be natural or synthetic and applied to soil or foliage.

Fery spectrograph (*Phys*) A spectrograph in which the only optical element is a back-reflecting prism with cylindrically curved faces. Considerable astigmatism is experienced with this instrument.

Fessenden oscillator (*Acous*) A low-frequency underwater sound source of the moving-coil type.

Ferromagnetics and ferrimagnetics

Ferromagnetic and ferrimagnetic materials are able to interact strongly with external magnetizing fields (H) because of the electric charge and the arrangement of their atoms. The effects are described in terms of the magnetization (M). Thus the magnetism of individual atoms (their MAGNETIC MOMENT) is associated with electron orbits and electron spin, but in solids the nature of atomic bonding causes the cancellation of the basic orbital effects.

Solids with atoms without unpaired electrons exhibit DIAMAGNETISM, a small effect arising from the reaction of electron orbits to the application of an external magnetic field (LENZ'S LAW). The diamagnetic susceptibility (M/H) is small and negative. Transition elements with d electron shells with unpaired spins have a paramagnetic response. Such atoms behave like tiny, independent 'bar magnets', aligning with and enhancing an external magnetizing field. Paramagnetic susceptibilities are small, positive and diminished by increasing temperature.

Technologically important magnetic materials have a long-range magnetic order in which the atomic 'bar magnets' co-operate to form large groups or *domains* with a common orientation of their magnetism. Only a very few elements, notably iron, cobalt and nickel, show ferromagnetism or spontaneous magnetization. The ordered state is destroyed by thermal agitation above a characteristic, critical temperature known as the CURIE POINT or temperature (T_c), above which paramagnetic behaviour occurs. Many alloys containing ferromagnetic elements are also ferromagnetic.

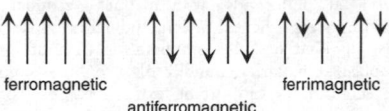

ferromagnetic   ferrimagnetic
antiferromagnetic

Magnetic order

In antiferromagnetic materials the opposite alignment of atomic 'bar magnets' is favoured so no net magnetization is exhibited. A similar situation exists in ferrimagnetic materials but the oppositely aligned atoms have unequal magnetic moments and a net magnetization persists (see diagram). In ferrites the magnetic moments are due to unpaired spins of electrons in the d electron shell of transition metal ions. Both ferromagnetics and ferrimagnetics are generally referred to simply as magnetic materials.

---

festination (*Med*) Involuntary quick walking with short steps, occurring in certain diseases of the nervous system, eg in PARKINSON'S DISEASE.

festoon dryer (*Paper*) A means of drying coated paper by the circulation of hot air while the web is in the form of slowly travelling loops carried on wooden rods.

FET (*Electronics*) Abbrev for FIELD-EFFECT TRANSISTOR.

fetal membranes (*BioSci*) In reptiles, birds and mammals, outgrowths from the embryo, or the extraembryonic tissue, that surround and protect the fetus and facilitate respiration. See ALLANTOIS, AMNION, CHORION.

fetch (*EnvSci*) The length of the traverse of an airstream of fairly uniform direction across a sea or ocean area.

fetch/execute cycle (*ICT*) See INSTRUCTION CYCLE.

fetishism (*Psych*) Sexual gratification in which an object, or some part of the body, is the main source of sexual arousal, to the exclusion of the person as a whole.

fetlock (*Vet*) The metacarpophalangeal and metatarsophalangeal regions of horses.

fettler (*Textiles*) An operative who clears away the fibrous waste and dirt from card cylinders, doffers and rollers etc.

fettling (*Eng*) 'Making good', eg: (1) of hearth and walls of a furnace, where erosion has damaged the refractory lining; (2) the trimming of feeders and excess material from a moulding or casting

fetus (*BioSci*) A young mammal within the uterus of the mother, or in oviparous animals the young within the egg, from the beginning of organ development until birth. Also *foetus*. Adjs *fetal*, *foetal*.

Feulgen reaction (*BioSci*) Specific staining procedure for DNA: mild acid hydrolysis makes the aldehyde group of deoxyribose available to react with Schiff's reagent to give a purple colour.

fever (*Med*) The complex reaction of the body to infection, associated with a rise in temperature. Less accurately, a rise of the temperature of the body above normal.

fexofenadine (*Pharmacol*) An antihistamine used to treat hay fever.

Feynman diagram (*Phys*) A diagram which shows the contributions to the rate of an elementary particle reaction. A powerful method of finding the physical properties of a system of interacting particles.

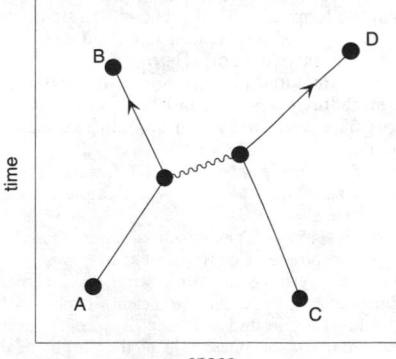

Feynman diagram AB and CD are the paths of two particles which interact.

FF (*ICT*) Abbrev for FORM FEED.

f-factor (*Radiol*) The ratio of absorbed dose to exposure dose for a given material and X-ray energy.

FFT (*MathSci*) Abbrev for FAST FOURIER TRANSFORM.

FGF (*BioSci*) See FIBROBLAST GROWTH FACTOR.

FHP (*Eng*) Abbrev for FRICTION HORSEPOWER.

FHSS (*ICT*) Abbrev for *frequency-hopping spread spectrum*, a technique for reducing interference in radio broadcasting.

Fibonacci numbers (*MathSci*) The sequence of integers, of which each is the sum of the two preceding it: 1,1,2,3,5,8,13,21,.... Also *Fibonacci sequence, Fibonacci series.*

fibrates (*Pharmacol*) A group of lipid-lowering drugs that reduce plasma triglycerides and increase breakdown of LDL cholesterol.

fibre (*BioSci*) An elongated sclerenchyma cell, typically tapering at both ends, with a thick secondary wall containing steeply helical or longitudinal cellulose microfibrils, lignified or not, and with or without a living protoplast at maturity. Bundles of such fibres constitute some economically important textile fibres, eg flax, jute, hemp, sisal.

fibre (*Eng*) The term for arrangement of the constituents of metals parallel to the direction of working. It is applied to the elongation of the crystals in severely cold-worked metals, to the elongation and stringing out of the inclusions in hot-worked metal, and to preferred orientations.

fibre (*Textiles*) Any type of vegetable, animal, regenerated, synthetic or mineral filament which is long in relation to its thickness and is fine and flexible. Yarns and fabrics are manufactured from them, by spinning and weaving, or knitting, felting, bonding, etc, and they are also used directly as reinforcement in COMPOSITE MATERIALS.

fibre assemblies (*Textiles*) The way in which textiles may be grouped. See panel on FIBRE ASSEMBLIES.

fibreboard (*Build*) Building or insulating board made from fibrous material such as wood pulp, waste paper and other waste vegetable fibre. May be homogeneous, bitumen-bonded or laminated. See HARDBOARD, INSULATING BOARD.

fibre brushes (*Build*) Cheap paint brushes in which the filling consists of vegetable fibre. Normally limited to work with alkaline materials such as paint removers, cement paints, etc.

fibre bundle (*Phys*) A bundle of optical fibres used in endoscopes for the inspection of body cavities. An incoherent bundle is used for illumination. A coherent bundle, in which the relative positions of the individual fibres are maintained, is used for the transmission of images. See FIBRE OPTICS.

fibre camera (*Phys*) An instrument for measuring the X-ray diffraction pattern of fibrous materials.

fibre distributed data interface (*ICT*) A LOCAL AREA NETWORK using TOKEN RING topology in which data are conveyed by means of OPTICAL FIBRES.

fibre in the loop (*ICT*) The use of OPTICAL FIBRES for some part of the connection between a CENTRAL OFFICE and the CUSTOMER PREMISES EQUIPMENT.

fibre, manufactured (*Textiles*) See MANUFACTURED FIBRE.

fibre metallurgy (*Eng*) The metallurgy of the manufacture of the fibres and products made from metallic fibres by sintering.

fibre optic inter-repeater link (*ICT*) A high-capacity link between REPEATERS on a NETWORK connected via FIBRE OPTIC CABLE.

fibre optics (*Phys*) The branch of optics based on the properties and use of OPTICAL FIBRES.

fibre-optics gyro (*Aero*) An instrument for measuring angular rotation by passing two beams of coherent light in both directions round a closed loop of optical fibre. Rotation affects the phase shift at the output of the two beams. The loops are often of triangular shape, each side being between 20 and 200 mm long. The 'gyro' is fixed to the aircraft, has no rotating parts and is strictly not a gyro, but is so called because it provides equivalent information.

fibre optic transceiver (*ICT*) An electronic device that converts electrical data signals into light signals suitable for transmission down a FIBRE OPTIC CABLE and vice versa.

fibre, synthetic (*Textiles*) See SYNTHETIC FIBRE.

fibre tracheid (*BioSci*) A cell of the secondary xylem intermediate between a LIBRIFORM FIBRE and a TRACHEID.

fibril (*BioSci*) Any minute thread-like structure, such as the longitudinal contractile elements of a muscle fibre. Also *fibrilla*. Adjs *fibrillar, fibrillate.*

fibril (*Chem*) (1) Bundle of aligned, crystalline polymer chains, as in cellulose. (2) Form of growth of aligned polymer lamellae. See CRYSTALLIZATION OF POLYMERS and panel on POLYMERS.

fibrillation (*Med*) (1) Twitching of individual muscle fibres, or bundles of fibres, in certain nervous diseases. (2) Incoordinate contraction of individual muscle fibres of the heart, giving rise to an irregular and inefficient action of the heart in eg ATRIAL FIBRILLATION, VENTRICULAR FIBRILLATION.

fibrillation (*Textiles*) The process by which a film (eg of polypropylene) is converted into fibres. The film is deliberately stretched in order to orientate the molecules. Further manipulation such as twisting splits the film longitudinally into a yarn comprising an interconnected mass of fibres. See panel on PAPER AND PAPER-MAKING.

fibrin (*BioSci*) An insoluble protein precipitated from blood which forms a network of fibres during the process of clotting. Its immediate precursor is the soluble protein FIBRINOGEN.

fibrinogen (*BioSci*) The soluble precursor of FIBRIN. It is converted to fibrin under the influence of thrombin by the proteolytic cleavage of terminal peptides.

fibrinolysis (*Med*) Enzymatic breakdown of FIBRIN. A variety of compounds can be used as fibrinolytic drugs in the treatment of disease where THROMBUS or clotting takes place.

fibrino-purulent (*Med*) Containing fibrin and pus.

fibro-adenoma (*Med*) An adenoma in which there is an overgrowth of fibrous tissue.

fibroblast growth factor (*BioSci*) Acidic FGF (a-FGF) and basic FGF (b-FGF) are the original members of a family of structurally related growth factors for mesodermal and neuroectodermal cells.

fibroblasts (*BioSci*) Flattened connective tissue cells of irregular ('fibroblastic') form, easily able to grow in tissue culture.

fibrocartilage (*BioSci*) A form of cartilage that has white or yellow fibres embedded in the matrix.

fibroid (*Med*) Resembling fibrous tissue; a fibromyoma, usually applied to muscular and fibrous growths in the uterus.

fibroin (*Textiles*) The structural protein of silk fibre. See panel on BIOLOGICAL ENGINEERING POLYMERS.

fibrolite (*Min*) A variety of the aluminium silicate SILLIMANITE occurring as felted aggregates of exceedingly thin fibrous crystals; also used when the mineral is cut as a gemstone.

fibroma (*Med*) A tumour composed of fibrous tissue.

fibromyectomy (*Med*) Removal of a FIBROMYOMA.

fibromyositis (*Med*) Inflammation of fibrous tissue in muscle and in the muscle fibres adjacent to it.

fibronectin (*BioSci*) A large glycoprotein, found in fibrous form in the extracellular matrix and in soluble form in plasma (cold-insoluble globulin), that interacts with extracellular substances such as collagen, fibrin and heparin, and also with receptors (INTEGRINS) on responsive cells such as fibroblasts.

fibrosarcoma (*Med*) A malignant tumour derived from the fibroblasts of the connective tissue.

fibrose (*Med*) To form fibrous tissue.

fibrosis (*Med*) The formation of fibrous tissue as a result of injury or inflammation of a part, or of interference with its blood supply. Adj *fibrotic.*

fibrositis (*Med*) Inflammation (often presumed rheumatic) of fibrous tissue.

fibrous concrete (*Build*) Concrete in which fibrous aggregate, such as asbestos, sawdust, etc, is incorporated either as alternative or additional to the sand and gravel.

Fibre assemblies

Fibre assemblies can be made from either short staple or continuous fibres in several distinctly different ways as shown in Fig. 1. Both can be twisted together to form yarn, but staple fibres can be used directly for felt and non-woven fabrics.

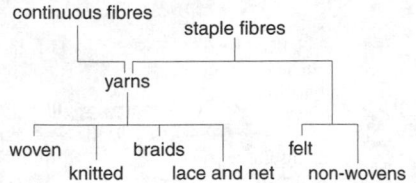

Fig. 1 **Fibre assemblies** Classification of fabric types.

Woollen felts for hats, billiard tables, blankets, carpet underlay, etc, are formed by repeatedly compressing woollen fibres together so that the scales on the fibre surface can interlock and bind the material. Smoother fibres, such as the synthetics, have to be entangled by repeatedly punching with barbed needles to create a felt. Such materials are used as geotextiles to provide a porous medium which will stabilize underlying soil, sand, etc. Another familiar form of non-woven fabric is formed by a more or less random, planar array of cellulose fibres, hydrogen bonded together at their points of contact, namely paper. See panel on PAPER AND PAPER-MAKING.

Spinning fibres to make yarn is the basis of one of our oldest technologies. The main variables governing the integrity of the yarn are the type of fibre, which affects the frictional force between fibres, the number of fibres in the cross-section (a minimum of 30 is usually required) and the degree of twist. Twist direction is described as S-twist or Z-twist (Fig. 2). The maximum strength of a spun staple fibre is half that of the individual fibres, while spun continuous fibres have a maximum strength the same as in the untwisted state.

Fig. 2 **Twist direction**.

Weaving is the process of creating a fabric by building up a parallel array of yarns (the *weft*) that pass over and under a perpendicular array (the *warp*) in a repeating pattern. Different repeat sequences plus the use of different yarn materials, colours, textures and sizes lead to the huge variety of woven cloth types (eg Fig. 3), a selection of which is listed in the dictionary.

In contrast to weaving, knitting forms a fabric by the intermeshing of loops of yarn in repeated sequences, in effect creating a series of knots (Fig. 4). Each horizontal row of loops is called a *course* and each vertical line a *wale*. In a way analogous to weaving, a large variety of different effects can be obtained.

Another important class of fibre assembly is represented by rope and the related string and cable. Rope is distinguished from string in having a minimum diameter of 4 mm. The traditional construction is known as *hawser-laid*, in which three strands are twisted together in a helical array. Each strand consists

---

fibrous layer (*BioSci*) In the wall of the anther in angiosperms, a layer of cells below the epidermis with uneven cell walls probably responsible, because of the way they shrink on drying, for the opening of the mature anther. Also ENDOTHECIUM.

fibrous plaster (*Build*) (1) Prepared plaster slabs formed of canvas stretched across a wooden frame and coated with a thin layer of gypsum plaster. (2) Any decorative plasterwork reinforced with hessian or similar.

fibrous root system (*BioSci*) Root system composed of many roots of roughly equal thickness and length, as in grasses. Cf TAPROOT SYSTEM.

fibrous tissue (*BioSci*) A form of connective tissue in animals consisting mainly of bundles of white (collagen) fibres; any tissue containing a large number of fibres.

fibrovascular bundle (*BioSci*) A vascular bundle accompanied, usually on its outer side, by a strand of SCLERENCHYMA.

fibula (*BioSci*) In Tetrapoda, the posterior of the two bones in the middle division of the hind-limb.

fibula (*Build*) A bent iron bar, used to fasten together adjacent stones.

fibulare (*BioSci*) In Tetrapoda, a bone of the proximal row of the tarsus, in line with the fibula.

Fick principle (*Med*) A method introduced by the German physiologist A E Fick for the measurement of cardiac output. Flow or cardiac output is the oxygen uptake of either an organ or the whole body, divided by the oxygen extraction of the tissue being examined.

Fick's laws of diffusion (*Chem*) Model of the DIFFUSION process expressed as two laws. The first states that the rate of diffusion of a species, or molar flux, J, in a given direction is proportional to the concentration gradient in that direction, ie

$$J = -D\frac{\partial C}{\partial x}$$

where J is expressed as (number) $m^{-2}\,s^{-1}$, C as (number) m^{-3} and D is the DIFFUSION COEFFICIENT ($m^2\,s^{-1}$). The negative sign indicates flow down the gradient. The second law combines the first with a continuity equation and the assumption that D is a constant, and states that the rate of change of concentration with time, t, is proportional to the

Fibre assemblies (Cont.)

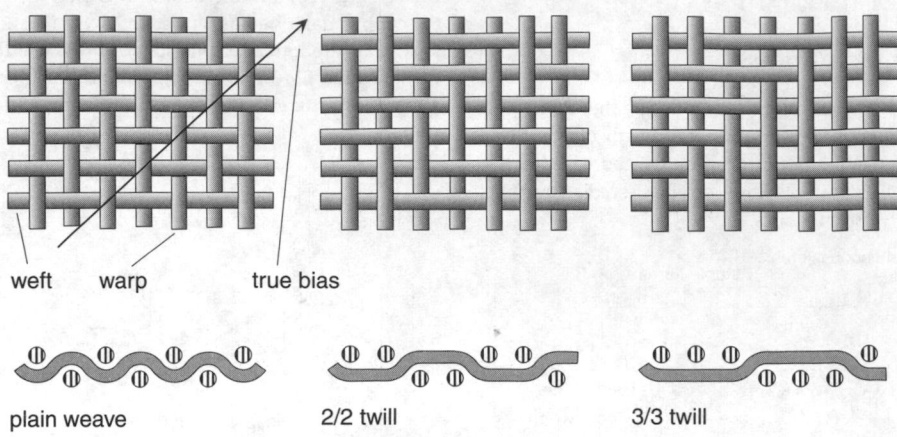

weft   warp   true bias

plain weave   2/2 twill   3/3 twill

Fig. 3 **Weaving** Simple patterns.

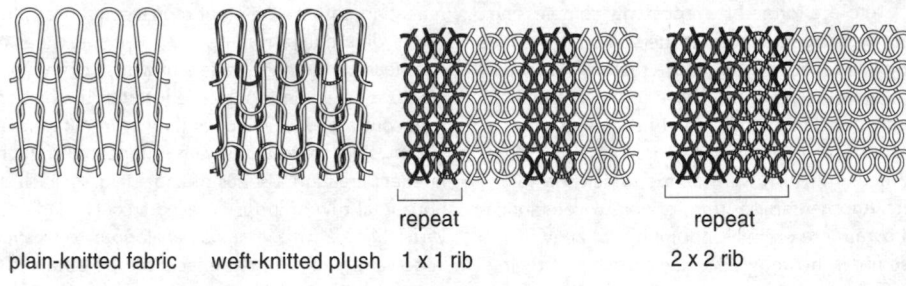

plain-knitted fabric   weft-knitted plush   1 x 1 rib   2 x 2 rib

repeat   repeat

Fig. 4 **Knitting sequences**.

of *base yarns* twisted into a *primary yarn*, which in turn is spun with other primaries into a *roping yarn*. Roping yarns are then spun to form the strand. A structure

more resistant to untwisting is the *plaited* rope in which the strands are woven together rather than just twisted.

change in concentration gradient with distance in a given direction, ie in one dimension:

$$\frac{\partial C}{\partial t} = D \frac{\partial^2 C}{\partial x^2}$$

ficoll-hypaque (*BioSci*) A proprietary mixture of a large polysaccharide and a dense synthetic organic molecule used in radiography, but the density can also be adjusted to allow separation of different cells by centrifugation. Frequently used for separating mononuclear cells and granulocytes from blood.

fiddleback (*For*) See SYCAMORE.

fidelity (*BioSci*) The degree of restriction of a species to a particular situation, a species with high fidelity having a strong preference for a particular type of community.

fidelity (*ICT*) The exactness of reproduction of the input signal at the output end of a system transmitting information.

fiducial (*Surv*) Said of a line or point established accurately as a basis of reference.

fiducial points (*ElecEng*) Points on the scale of an indicating instrument located by direct calibration, as contrasted with

the intervening points, which are inserted by interpolation or subdivision.

fiducial temperature (*EnvSci*) The temperature at which a sensitive barometer reads correctly, the maker's calibration holding for latitude 45° at the temperature 285 K (12°C) only.

field (*ICT*) Predetermined section of a record.

field (*ImageTech*) A single complete scanning of the picture from top to bottom; in an INTERLACED SCANNING system two successive fields complete a FRAME.

field (*MathSci*) See RING.

field (*Phys*) (1) Concept used in explanations of the interaction between bodies or particles. For instance, the *potential energy* of a body may depend on its position and then is represented by a *scalar field* with magnitude only. Other physical quantities carry direction as well as magnitude and they are represented by *vector fields*, eg electric, magnetic or gravitational fields. (2) Space in which there are electromagnetic oscillations associated with a radiator. The *induction* field which represents the interchange of energy between the radiator and space is within a few wavelengths of the radiator; *radiation field* represents the energy lost from the radiator to space. The region where components radiated by antenna elements are parallel is the *Fraunhofer region*; that where they are not parallel is the *Fresnel region*. The latter will exist between the antenna and the Fraunhofer region and is usually taken to extend a distance $2D^2\lambda^{-1}$, where λ is the wavelength of the radiation and D is the aerial aperture in a given aspect.

field ampere-turns (*ElecEng*) The ampere-turns producing the magnetic field of an electric machine.

field blanking (*ImageTech*) The period between successive FIELDS during which the picture information is suppressed and SYNC PULSES transmitted. Also *vertical blanking interval*.

field-breaking resistance (*ElecEng*) See FIELD-DISCHARGE RESISTANCE.

field-breaking switch (*ElecEng*) See FIELD-DISCHARGE SWITCH.

field capacity (*EnvSci*) The amount of water held in a given soil by capillarity against drainage by gravity to a water potential or suction of about −0·05 bar.

field coil (*ElecEng*) The coil which carries the current for producing the mmf to set up the flux in an electric machine. Also *field spool*. See MAGNETIZING COIL.

field control (*ElecEng*) The adjustment of the field current of a generator or motor to control the voltage or speed respectively.

field copper (*ElecEng*) A term used in the design of electrical machines to denote the total quantity of copper used in the field windings of a machine.

field current (*ElecEng*) The current in the field winding of an electric machine.

field density (*Phys*) The number of lines of force passing normally through unit area of an electric or magnetic field.

field discharge (*Phys*) The passage of electricity through a gas as a result of ionization of the gas; it takes the form of a brush discharge, an arc or a spark. Also *electric discharge*.

field-discharge resistance (*ElecEng*) A discharge resistance connected across the shunt or separately excited field winding of an electric machine, to prevent high induced voltages when the field circuit is interrupted. Also *field-breaking resistance*.

field-discharge switch (*ElecEng*) A switch for controlling the field circuit of a generator which connects a discharge resistance across the winding as the field circuit is broken. Also *field-breaking switch*.

field-diverter rheostat (*ElecEng*) A rheostat connected in parallel with the series field winding or the compole winding of a dc machine to give control of the mmf produced by the winding, independently of the current flowing through the main circuit.

field drain (*CivEng*) See SUBSOIL DRAIN.

fielded panel (*Build*) A panel which is moulded, sunk, raised or divided into smaller panels.

field-effect transistor (*Electronics*) A transistor in which the field associated with the voltage applied to a gate electrode creates/destroys/modifies a conducting channel between source and drain electrodes. Abbrev *FET*.

field emission (*Electronics*) That arising, at normal temperature, through a high-voltage gradient causing an intense electric field at a metallic surface and stripping electrons from surface atoms.

field-emission microscope (*Phys*) A microscope in which the positions of the atoms in a surface are made visible by means of the electric field emitted on making the surface the positive electrode in a high-voltage discharge tube containing argon at very low pressure. When an argon atom passes over a charged surface atom, it is stripped of an electron, and thus is drawn towards the negative electrode, where it hits a fluorescent screen in a position corresponding to that of the surface atom.

field enhancement (*Phys*) Local increase in electrical field strength due to convex curvature of an electrode, or proximity of another electrode.

field frequency (*ImageTech*) The number of fields scanned per second, currently 60 Hz for US broadcast practice and 50 Hz for European.

field intensity (*Phys*) See FIELD STRENGTH.

field-ion microscope (*Phys*) A modification, with greater resolving power, of the FIELD-EMISSION MICROSCOPE, in which ions of a gas (usually helium) which is adsorbing on the metal point are repelled and produce the image on the screen; used in conjunction with an instrument such as a mass spectrometer to identify the atoms, it is known as an *atom-probe field-ion microscope*.

field lens (*Phys*) A lens placed in or near the plane of an image to ensure that the light to the outer parts of the image is directed into the subsequent lenses of the system, and thus uniform illumination over the field of view is ensured. See fig. at MICROSCOPE.

field magnet (*ElecEng*) The permanent or electromagnet which provides mmf for setting up the flux in an electric machine. See ROTATING-FIELD MAGNET.

field of force (*Phys*) The concept arising from the principle of *action at a distance*: a region in which mechanical forces are experienced by an electric charge, a magnet or a mass, at a distance from an independent electric charge, magnet or mass, because of fields established by these and described by uniform laws.

field of view (*Phys*) The area over which the image is visible in the eyepiece of an optical instrument. It is usually limited by a circular stop in the focal plane of the eye-lens (*eyepiece*). See SAGITTAL FIELD, TANGENTIAL FIELD.

field oscillator (*Electronics*) See FRAME OSCILLATOR.

field oxide (*Electronics*) Silicon dioxide grown on a wafer for the purpose of isolating the active regions where devices are fabricated.

field resistance (*BioSci*) Resistance that is shown by a plant to natural infection in the field and that is more dependent on the environment and the nature of pathogen and vector than on inoculation in laboratory or greenhouse.

field rheostat (*ElecEng*) A variable resistance connected in series or parallel with the field winding of an electrical machine to vary the current in the winding. Also *field regulator*.

field rivet (*Eng*) A rivet which is put in when the work is on the site. Also *site rivet*.

field sequential (*ImageTech*) A colour TV system in which successive fields are scanned in the three primary colours.

Field's siphon flush tank (*Build*) See FLUSHING TANK.

field star (*Astron*) An individual star which is not a member of any star cluster or association. Field stars are numerous on all astronomical photographs.

field strength (*ICT*) The measure of the intensity of an electric, magnetic or electromagnetic field, most commonly expressed in $V\,m^{-1}$, $mV\,m^{-1}$ or $\mu V\,m^{-1}$. In microwave engineering, field strength is often identified with the power flux density, measured in $W\,m^{-2}$. Field strength is inversely proportional to distance from an antenna.

field strength (*Phys*) Vector representing the quotient of a force and the charge (or *pole*) in an electric (or magnetic) field, with the direction of the force. Also *field intensity*.

field strength meter (*ICT*) A calibrated radio receiver and antenna system for measuring FIELD STRENGTH.

field suppressor (*ElecEng*) An arrangement for automatically reducing the field current of a generator when a short circuit or other fault occurs on the machine or its adjacent connections.

field sync pulse (*ImageTech*) A signal transmitted during the FIELD BLANKING interval to synchronize the time base of the receiver with the transmitter.

field theory (*Phys*) A theory in which the basic quantities are FIELDS. Classically the equations describing them are given; quantum field theories include QUANTUM CHROMODYNAMICS, QUANTUM ELECTRODYNAMICS and the WEINBERG–SALAM THEORY.

field theory (*Psych*) A development of Gestalt psychology that emphasizes the totality of environmental influences and their interactions rather than reductionist explanations.

field tube (*Eng*) A special form of boiler tube, consisting of an outer tube which is closed at its lower end and contains a second concentric tube down which the water passes to return up the annular space between the two.

field winding (*ElecEng*) The winding placed on the field magnets of an electric machine and producing the mmf necessary to set up the exciting flux.

fiery mine (*MinExt*) One in which there is a possibility of explosion from gas or coal dust.

FIFO (*ICT*) Abbrev for *first in, first out*. See QUEUE.

fifth freedom traffic (*Aero*) Passengers or freight carried between two countries by an airline of a third country.

fifth-generation computer (*ICT*) A forward-looking generic term to apply to knowledge-based computer systems predicted for the near future. Will have very fast processing with VLSI based on LOGIC PROGRAMMING, providing access to large knowledge bases through novel human/machine interfaces. Such a computer would also use artificial intelligence techniques to learn and process natural human languages. See COMPUTER GENERATIONS.

figured fabric (*Textiles*) Any fabric with a complex woven pattern produced by the DOBBY or JACQUARD mechanisms.

figure of loss (*ElecEng*) A term occasionally used in connection with transformers to denote the energy loss per unit mass of material (iron or copper).

figure of merit (*ElecEng*) General parameter which describes the quality of performance of an instrument, a circuit or other item associated with a given system, eg the voltage gain of an amplifier or the bandwidth of a filter.

filament (*BioSci*) (1) Generally, any fine thread-like structure. (2) A chain (unbranched or branched) of cells joined end on end. (3) A fungal hypha. (4) The stalk of a stamen in angiosperms. (5) The axis of a down-feather. (6) See INTERMEDIATE FILAMENT, MICROFILAMENT.

filament (*ElecEng*) A fine wire of high resistance, which is heated to incandescence by the passage of an electric current. In an electric filament lamp it acts as the source of light, and in thermionic tubes it acts as an emitter of electrons.

filament (*Electronics*) Historically, this term remains from when the source of electrons was simply a fine wire heated to incandescence by a current. See HEATER.

filament (*Textiles*) See CONTINUOUS FILAMENT YARN.

filamentation (*Textiles*) Mechanical damage to continuous filaments in a yarn or fabric giving the material a fuzzy appearance.

filament getter (*Electronics*) One which adsorbs gas readily when hot, and is used for this purpose in a high-vacuum assembly.

filament lamp (*Phys*) An electric lamp in which a filament in a glass bulb, evacuated or filled with inert gas, is raised to incandescence by the passage through it of an electric current.

filariasis (*Med, Vet*) Infestation with nematode worms of the family Filariidae, which inhabit the lymphatic channels, often causing elephantiasis.

filar micrometer (*BioSci, Min*) Modified eyepiece graticule with movable scale or movable cross-hair.

file (*ICT*) General term for named set of data items stored in machine-readable form. See MASTER FILE, TRANSACTION FILE.

file allocation table (*ICT*) A list that maps a file to BLOCKS of data storage on a DISK. This list is stored on the disk and as FILES are added to or deleted from the disk this table is updated as necessary. Particularly used in the operating system MS-DOS. See fig. at HARD DISK.

file attribute (*ICT*) See ATTRIBUTE.

file card (*Eng*) A short wire brush used to clean SWARF from files.

file name extension (*ICT*) Under eg MS-DOS, the name given to a FILE may have up to eight characters followed by a full stop followed by an optional file name extension of up to three characters. The extensions are often chosen to help identify the type of file, eg DOC for word processed files. Under MS-DOS some file extensions have special

significance, eg BAT for BATCH FILES, COM for COMMAND files and EXE for files capable of direct EXECUTION.

file server (*ICT*) The file store in a distributed computing environment.

file sharing (*ICT*) The activity of making files freely available for other users on a computer network while also downloading files made available by others.

filet net (*Textiles*) Any type of square ground mesh of lace; a woven net.

file transfer and access method (*ICT*) An OPEN SYSTEMS INTERCONNECTION application program that provides access to files on a network irrespective of the operating system being used by the host computer.

file transfer protocol (*ICT*) Computing language that enables files to be remotely transferred between computers. Abbrev *FTP*. Cf ANONYMOUS FTP. See panel on INTERNET.

Filicales (*BioSci*) The leptosporangiate ferns, an order of c.9000 extant spp of Filicopsida that also includes fossils from the Carboniferous onwards. The sporophyte may have long, creeping, horizontal rhizomes with fronds at intervals, short, more or less upright stems with a rosette of fronds, or grow into tree ferns. The leaves are CIRCINATE in bud and usually pinnately compound. Many are poisonous and some are carcinogenic. The expanding fronds (croziers) of a few spp are eaten; some are cultivated as ornamentals. Also *Polypodiales*.

Filicopsida (*BioSci*) The ferns, a class of Pteridophytes existing from the Devonian onwards. The sporophyte usually has roots, a rhizome or stem and spirally arranged, often pinnately compound, leaves (fronds, megaphylls) with circinate venation. (The palaeozoic sorts show little differentiation between stem and leaf.) The sporangia are typically borne marginally or abaxially on leaves; most ferns are homosporous. The spermatozoids are multiflagellate. The class includes Marattiales (eusporangiate), Filicales and Salviniales (floating, heterosporous aquatica). Also *Polypodiopsida*, *Pteropsida*.

filiform (*BioSci*) Thread-like, eg *filiform antenna*.

fill (*Build, CivEng*) Rock or soil dumped to bring a site to the required level. After dumping, the fill has to be progressively consolidated.

filled band (*Chem*) An energy-level band in which there are no vacancies. Its electrons do not contribute to valence or conduction processes.

filler (*Build*) (1) A material used to fill in the pores of, or any holes in, wood, plaster, etc, which is to be painted, varnished or otherwise decorated. Supplied in several forms: (a) powder to be mixed with water; (b) ready mixed, oil, water or resin based; (c) two-pack materials; (d) cellulose materials. (2) An EXTENDER.

filler (*CivEng*) A finely divided substance added to bituminous material for road surfacing, in order to reduce it to a suitable consistency.

filler (*Paper*) White mineral matter in finely divided form added to the stock to improve opacity, dimensional stability and reduce costs of paper. In some circumstances it may assist in achieving a smooth finish. Certain loadings such as titanium dioxide are expensive but have special properties such as preventing undue loss of opacity when paper is waxed. Also *loading*.

filler (*Plastics*) Any inert solid substance added to plastics either for economy (eg wood flour, clay) or to modify its properties (eg mica, aluminium or other metal powder).

filler joist floor (*Build*) A type of floor in which the principal supporting members are steel joists, the gaps between the joists being filled with concrete.

filler light (*ImageTech*) Lighting directed into the shadow areas of a scene in order to avoid excessive contrast.

filler metal (*Eng*) The metal required to be added at the weld in welding processes in which the fusion temperature is reached. In metal-electrode welding, the electrode, usually fluxed and coated, is melted down to provide the filler metal.

filler rod (*ElecEng*) See WELDING ROD.

fillet (*Aero*) A fairing at the intersection of two surfaces intended to improve the airflow by reducing breakaway and turbulence.

fillet (*Arch*) (1) A flat and narrow surface separating or strengthening curved mouldings. (2) A LISTEL. See FACETTE.

fillet (*Build*) A thin strip of wood fixed into the angle between two surfaces, to a wall as a shelf support, to a floor as a door stop, etc.

fillet (*Eng*) (1) A narrow strip of metal raised above the general level of a surface. (2) A radius provided, for increased strength, at the intersection of two surfaces, particularly in a casting. See STRESS CONCENTRATION.

fillet (*Plastics*) A rounded internal corner in a plastic article, to avoid a possible weakness arising from an abrupt change in cross-section. See STRESS CONCENTRATION.

fillet (*Print*) A band or line of gold leaf, or a plain band or line, on a cover.

filleting (*Build*) See CEMENT FILLET, FILLET.

fillet weld (*Eng*) A weld at the junction of two parts, eg plates, at right angles to each other, in which a fillet of welding metal is laid down in the angle created by the intersection of the surfaces of the parts.

filling (*MinExt*) (1) The loading of tubs or trucks with coal, ore or waste. (2) The filling up of worked-out areas in a metal mine.

filling (*Textiles*) (1) Insoluble materials such as China clay added with starch or gum to increase the weight of fabrics or to alter their appearance. Also *loading*. (2) Used in Canada and the US for WEFT YARNS.

filling-in (*Print*) The spreading of ink on the printing image causing half-tone dots to join up or the bowls of letters to be filled with ink.

filling post (*Build*) A middle post in a timber frame.

fillister (*Build*) (1) A rebate in the edge of a sash bar, to receive the glass and putty. (2) A rabbeting plane having a movable stop to regulate depth or cut.

fillister-head screw (*Eng*) A CHEESE-HEAD SCREW with a slightly convex upper head surface.

filly (*Agri*) A sexually mature female horse yet to produce a foal, usually less than 4 years old.

film (*Chem*) (1) Any thin layer of substance, eg that which carries a light-sensitive emulsion for photography, that which carries iron (III) oxide particles in a matrix for sound recording. (2) Thin layer of material deposited, formed or adsorbed on another, down to monomolecular dimensions, eg electroplated films, oxide on aluminium, sputtered depositions on glass or microcomponents. (3) Packaging material, esp from a thermoplastic polymer.

film (*ImageTech*) (1) A photographic emulsion carried on a thin flexible transparent support; motion picture film is manufactured in long continuous strips with perforation holes along one or both edges. (2) The craft and techniques of cinematography. (3) A completed motion picture production.

film badge (*Radiol*) Small photographic film used as radiation monitor and dosimeter. Normally worn on lapel, wrist or finger and sometimes partly covered by cadmium and tin screens so that exposure to neutrons, and to beta and gamma rays, can be estimated separately.

film circuits (*Electronics*) See THIN-FILM CIRCUIT.

film coefficient (*ChemEng*) The constant of proportionality between the flux (eg of mass or heat) and the difference in driving force (eg concentration or temperature) across a film. Symbol h.

film foils (*Print*) Blocking foils with a viscose film substrate. These give better release and finer detail than PAPER FOILS.

film glue (*Build*) Resin-impregnated paper, or phenolic resin film, used in making resin-bonded plywood.

film pack (*ImageTech*) A package of cut sheets of photographic film which can be successively exposed in a camera and individually removed for processing.

film recorder (*ImageTech*) An apparatus for recording electronic images onto motion picture or photographic film, using a scanning spot generated by a CATHODE-RAY TUBE, high-power lamp or laser. See panel on LASER.

film recording (*Acous*) The process of recording sound on a sound track on the edge of cinematograph film, for synchronous reproduction with the picture.

film scanner (*ImageTech*) Apparatus for converting motion picture film images into digital signals. The film is illuminated by a high-power lamp and read by a line-array CHARGE-COUPLED DEVICE one line at a time. See TELECINE.

filmsetting (*Print*) See PHOTOTYPESETTING.

film sizing (*MinExt*) The concentration of finely divided heavy mineral by gently sloped surfaces which may be plane, riffled or vibrated. See BUDDLE, WILFLEY TABLE.

film speed (*ImageTech*) (1) The sensitivity of a photographic material as rated by a standard method of exposure determination, such as DIN and ISO SPEED. Also *speed*. (2) The rate at which film passes through a motion picture camera or projector, in frames per second (fps) or feet or metres per minute.

film strip (*ImageTech*) A strip of film, usually 35 mm gauge, containing a series of still pictures for individual projection, often used for educational illustration and demonstrations.

film theory (*ChemEng*) Mass transfer between two well-stirred fluid phases, or between a solid and a well-stirred fluid, takes place by eddy diffusion in the bulk fluid phase, combined with molecular diffusion through a stagnant film at the interface, the latter being the slower process.

filoplumes (*BioSci*) Delicate hair-like contour feathers with a long axis and few barbs, devoid of locking apparatus at the distal end.

filopodium (*BioSci*) A long filamentous spike containing actin filaments found at the surface of some cells.

Filoviridae (*Med*) A family of single-stranded RNA viruses, similar in some respect to rhabdoviruses. Marburg and

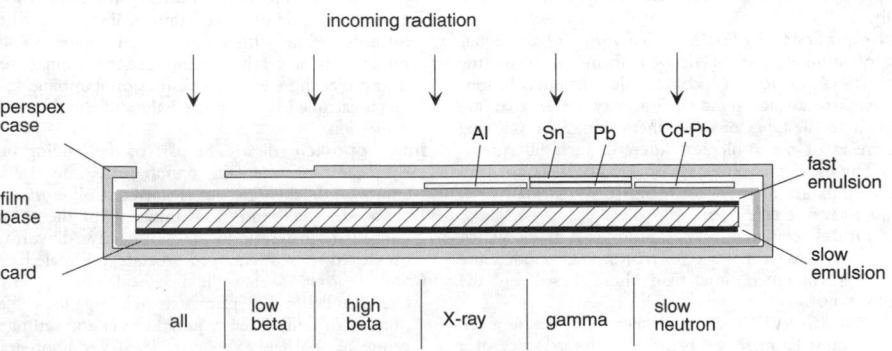

film badge Showing radiation types which can be differentially detected.

Ebola viruses, both of which cause severe haemorrhagic fevers, are the only two members of the family at present.

filter (*Chem*) A sheet of material with pore sizes within a defined range and used to separate particles or macromolecules from a suspension or solution. See FILTER PAPER, MEMBRANE FILTER.

filter (*ChemEng*) Equipment used to separate liquids and suspended solids, either for recovery of solid and classification of liquid, or both simultaneously. The liquid flows through the pores in a cloth, wire mesh or granular bed, and thus the solid is sieved out.

filter (*Electronics*) A transmission network used in electronic circuits, or an optical device in optical communication systems, for the selective enhancement or reduction of specified components of an input signal. Filtering is achieved by selectively attenuating those components of the input signal which are undesired, relative to those which it is desired to enhance. A filter may consist of inductances and capacitances, resistor and capacitors and gyrators (to transform specified capacitors into inductances); it may contain amplifying stages (see ACTIVE FILTER), or it may rely on resonances in piezoelectric, ceramic or magnetic materials. Digital filters depend on the action of externally manipulated gates to block or pass selectively certain of the pulses making up a digital signal.

filter (*ICT*) (1) A program that reads information from a standard input, transforms it in some way (usually by eliminating extraneous data) and writes it to a standard output. The term is mostly used in the context of the UNIX operating system. (2) A method of selecting or listing certain types of files, eg to list only text files the FILTER may be set to *.TXT in MS-DOS.

filter (*ImageTech*) A device, usually consisting of a glass plate or a sheet of gelatine, interposed across a beam of light for the purpose of altering the relative intensity of the different component wavelengths in the beam.

filter aid (*ChemEng*) A material such as diatomaceous earth that is added to solutions, or precoated onto the filter before use, to aid separation of difficult solids, usually colloidal, which would otherwise pass through or choke the filter.

filter attenuation (*ICT*) The loss of signal power in its passage through a FILTER due to absorption, reflection or radiation. It is usually given in decibels.

filter attenuation band (*ICT*) See STOP BAND.

filter bed (*Build*) A CONTACT BED or any similar bed used for filtering purposes.

filter cake (*Chem*) The layer of precipitate which builds up on the cloth of a filter press.

filter circuit (*Phys*) A circuit, usually composed of REACTORS arranged in resonant circuits, designed to accept certain desired frequencies and to reject all others. Often used to remove noise from a signal, eg between a radio transmitter and its antenna. Also *filter network*.

filter cut (*ImageTech*) The wavelength at which the relative absorption for light of different wavelengths changes rapidly.

filtered equations (*EnvSci*) Modified forms of the equations of motion, esp of derived forms such as the VORTICITY EQUATION, which exclude certain solutions such as fast-moving sound and gravity waves that are irrelevant to the types of atmospheric motion producing phenomena of meteorological interest. Such filtering is effected by use of the HYDROSTATIC APPROXIMATION, and the judicious use of the GEOSTROPHIC APPROXIMATION and the BALANCE EQUATION.

filtered model (*EnvSci*) A numerical forecast model which makes use of the FILTERED EQUATIONS. Such models use much less computer time than those based on the PRIMITIVE EQUATIONS.

filter factor (*ImageTech*) The number of times a given exposure must be increased because of the presence of a filter which absorbs light and reduces the effective exposure of a lens system.

filter feeder (*BioSci*) In benthic and planktonic communities, detritus feeders which remove particles from the water. In benthic communities they usually predominate on sandy bottoms. Cf DEPOSIT FEEDER.

filtering basin (*Build*) A tank through which water passes on its way from the reservoir to the mains, and in which it is subjected to a process of filtration.

filter network (*Phys*) See FILTER CIRCUIT.

filter overlap (*ImageTech*) The band of wavelengths transmitted in common by filters of different (usually broad) spectral absorption.

filter paper (*Chem*) Paper, consisting of pure cellulose, which is used for separating solids from liquids by filtration. Filter paper for quantitative purposes is treated with acids to remove all or most inorganic substances, and has a definite ash content.

filter press (*Chem*) Apparatus used for filtrations; it consists of a set of frames covered with filter cloths into which the mixture which is to be filtered is pumped.

filter-press action (*Geol*) A differentiation process involving the mechanical separation of the still liquid portion of a magma from the crystal mesh. The effective agent is pressure operating during crystallization.

filter pulp (*Paper*) Rag fibre made into convenient slubs or cakes at the paper mill.

filter transmission band (*ICT*) See PASS BAND.

filtrate (*Chem*) The liquid freed from solid matter after having passed through a filter.

filtration (*Chem*) The separation of solids from liquids by passing the mixture through a suitable medium, eg filter paper, cloth, glass wool, which retains the solid matter on its surface and allows the liquid to pass through. Related methods extend the range of retained particles to submicrometre regions. Fig. ▷

filtration (*Radiol*) Removal of longer wavelengths in a composite beam of X-rays by the interposition of thin metal, eg copper or aluminium. Similarly for energy of other wavelengths.

fimbria (*BioSci*) (1) Generally, any fringing or fringe-like structure. (2) The delicate processes fringing the internal opening of the oviduct in mammals. (3) The ridge of fibres running along the anterior edge of the hippocampus in mammals. (4) The processes fringing the openings of the siphons in molluscs. (5) Projections from the surface of some bacteria, see PILI. Pl *fimbriae*.

fimbriate (*BioSci*) Having a fringed margin. Also *fimbriated*.

fimbriocele (*Med*) Hernia containing the fimbriae of the Fallopian tube.

fimicolous (*BioSci*) Growing on or in dung.

fin (*Aero*) A fixed vertical surface, usually at the tail, which gives directional stability to a fixed-wing aircraft in motion and to which a rudder is usually attached. In an airship, any fixed stabilizing surface. See STABILIZER.

fin (*BioSci*) In fish, some Cephalopoda and other aquatic forms, a muscular fold of integument used for locomotion or balancing, supported in fish by internal FIN RAYS.

fin (*Eng*) (1) One of several thin projecting strips of metal formed integral with an air-cooled engine cylinder or a pump body or gearbox to increase the cooling area. (2) A thin projecting edge on a casting or stamping, formed by metal extruded between the halves of the die; any similar projection.

final approach (*Aero*) The part of the landing procedure from the time when the aircraft turns into line with the runway until the flare-out is started. Colloq *finals*.

finasteride (*Pharmacol*) An inhibitor of the enzyme that converts testosterone to the more active dihydrotestosterone. Used to treat enlarged prostate and male baldness.

finder (*Astron*) A small auxiliary telescope of low power, fixed parallel to the optical axis of a large telescope for the purpose of finding the required object and setting it in the centre of the field; also used in stellar photography for guiding during an exposure.

finder (*ImageTech*) See VIEWFINDER.

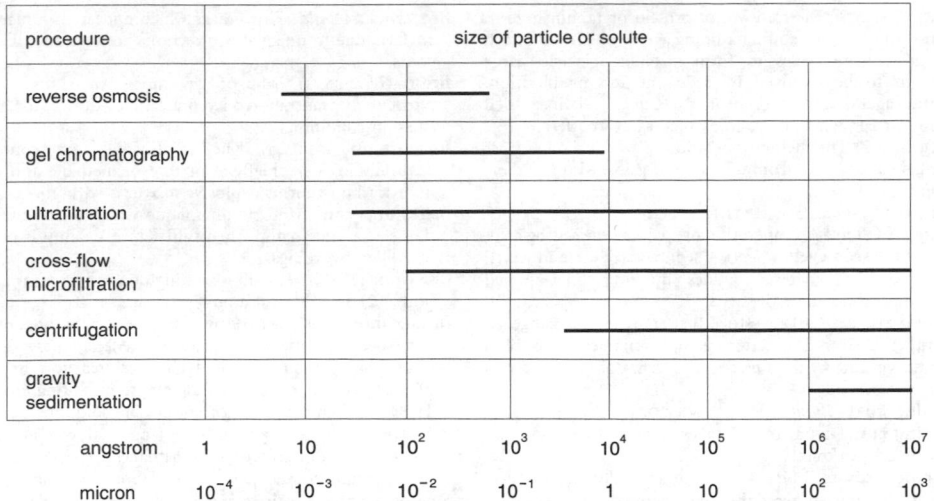

procedure	size of particle or solute							
reverse osmosis								
gel chromatography								
ultrafiltration								
cross-flow microfiltration								
centrifugation								
gravity sedimentation								
angstrom	1	10	10^2	10^3	10^4	10^5	10^6	10^7
micron	10^{-4}	10^{-3}	10^{-2}	10^{-1}	1	10	10^2	10^3

filtration The relation of forms of filtration to other separative procedures.

fine aggregate (*CivEng*) Sand or the screenings of gravel or crushed stone. For concrete the material should pass a 5 mm (3/16 in) sieve.

fine boring (*Eng*) A high-precision final machining process for bores of internal diameters between 5 mm and 750 mm, using a single-point cutter in a very accurate and rugged machine tool in which the cutting stresses, and consequently the clamping distortion, are kept particularly low.

fine etching (*Print*) The finishing stages in the making of a half-tone block to achieve the required contrast and range of tones.

fine gold (*Eng*) Pure 24-carat gold.

fine-grain developer (*ImageTech*) Developer solution which minimizes clumping of the silver grains by partial development only, but at the cost of a lower image density.

fine-grained (*EnvSci*) Used esp of a small-scale pattern of environmental or resource heterogeneity, often in relation to the foraging activities of animals.

fine machining (*Eng*) A family of precision finishing processes, including fine boring, milling, grinding, honing, lapping, diamond turning and superfinishing, using particularly accurate machine tools with special provision to eliminate vibration and, often, cemented carbide or diamond cutters to remove small amounts of material in order to attain high accuracy and excellent surface finish.

fineness (*Chem*) The state of subdivision of a substance.

fineness (*Eng*) The purity of a gold or silver alloy; stated as the number of parts per thousand that are gold (or silver).

fineness modulus (*CivEng*) In US practice a number indicating the grading of an aggregate, obtained by finding the percentage retained by weight on a set of nine standard sieves with apertures ranging from 37·5 mm (1·5 in) to 0·15 mm (0·06 in), summing and dividing by 100. UK practice is to use grading curves.

fineness-of-grind gauge (*PowderTech*) A device to provide control tests for the presence of large particles of powder in a slurry. It generally comprises a stainless-steel block with parallel grooves of gradual decrease of depth from an inch to zero, along which the slurry is distributed by means of a smooth stroke with a rigid blade. When the groove depth is equal to the largest particle present, the smooth surface of the slurry film is marked by score lines.

fineness ratio (*Aero*) The ratio of the length to the maximum diameter of a streamlined body, or flying boat planing bottom.

fine papers (*Paper*) Papers of high quality, used for graphic purposes.

fines (*PowderTech*) That portion of a powder composed of particles under a specified size.

fine screen (*ImageTech, Print*) The term for half-tones suitable for art paper, usually 52 but also 59 lines per centimetre.

fine silt (*Geol*) See PARTICLE SIZE.

fine structure (*Electronics*) The splitting of optical spectrum lines into multiplets due to interaction between spin and orbital angular momenta of electrons in the emitting atoms.

fine structure analysis (*NucEng*) The detailed analysis of the neutron flux through the moderator, coolant, structures and fuel in the design of a nuclear reactor.

fine-structure constant (*Phys*) A dimensionless constant, equal to $7·297\,351 \times 10^{-3}$ (approximately $\frac{1}{137}$), given by 2π times the square of the electron charge, divided by the product of the speed of light and Planck's constant.

fine stuff (*Build*) Fine type of plaster, usually composed of lime and plaster of Paris, used for the finishing coat.

FINGAL (*NucEng*) Abbrev for *fixation in glass of active liquid*. A method for the long-term storage of active wastes.

fingering (*Textiles*) Combed, soft-twisted, worsted yarn of the type generally used for hand knitting; usually 2, 3 or 4 ply.

finger plate (*Build*) A plate fixed on the side of the meeting stile of a door, near the lock, to prevent damage to the paintwork by fingermarks.

fingerprinting (*BioSci*) By analogy with the eponymous forensic tool, application of analytical methods (esp SPECTROSCOPY) to materials which allow identification of an unknown sample by comparison with standards. Macromolecules like DNA can be cut specifically to give a characteristic pattern of oligomers of different lengths. See MATERIALS MATCHING, RESTRICTION FRAGMENT LENGTH POLYMORPHISM.

finger stop (*Eng*) A sliding stop in a press tool, used to locate and position the material to be processed in relation to the tool, usually at the commencement of a production run.

finger-type contact (*ElecEng*) A type of contact which, as usually fitted to drum-type controllers, is in the form of a finger which is pressed against the contact surface by means of a spring.

finial (*Build*) A term applied to an ornament placed at the summit of a gable, pillar or spire.

fining (*FoodSci*) The removal of protein or tannin haze in wines and beers by adding a fining agent, which will attach to protein molecules to form insoluble particles large enough to be removed by sedimentation or filtration. Fining agents are either proteins such as gelatine, dried blood or egg white, or colloids such as BENTONITE.

fining (*Glass*) The melting operation in which molten glass is made almost free from undissolved gases. Also *founding*, *plaining*, *refining*.

fining coat (*Build*) See SETTING COAT.

fining-off (*Build*) The operation of applying the setting coat.

fining-upwards cycle (*Geol*) A sedimentary cycle in which coarse-grained material grades up into finer-grained material, eg turbidite deposits.

finished stock (*Agri*) Livestock that are ready for slaughter.

finishing (*Print*) The lettering and ornamentation of a bound volume by the finisher; the term does not apply to 'cased' work.

finishing coat (*Build*) See SETTING COAT.

finishing cut (*Eng*) A fine cut taken to finish the surface of a machined workpiece.

finishing stove (*Print*) A small gas or electric stove on which the finisher heats the tools required for work.

finishing system (*Agri*) The strategy and practice of feeding livestock to reach the condition required for commercial slaughter.

finishing tool (*Eng*) A lathe or planer tool, generally square-ended and cutting on a wide face. Used for taking the final or finishing cut.

fink truss (*Eng*) See FRENCH TRUSS.

finplate (*Eng*) See EAR.

fin rays (*BioSci*) In fish, the distal skeletal elements which support the fins; in CEPHALOCHORDATA, the rods of connective tissue supporting the dorsal and ventral fins.

fiord (*Geol*) Narrow winding inlet of the sea bounded by mountain slopes; formed by the drowning of steep-sided valleys, deeply excavated by glacial action; in many cases a rock bar partially blocks the entrance and impedes navigation. Also *fjord*.

FIR (*Aero*) Abbrev for FLIGHT INFORMATION REGION.

fir (*For*) Trees of the genus *Abies*, giving a valuable structural softwood. The species *A. balsamea*, which is found in eastern N America, yields CANADA BALSAM and is used for PULP. See WHITEWOOD.

fire (*Min*) Flashes of different spectral colours seen in diamond and other gemstones as a result of *dispersion*. See GEM.

fireball (*Astron*) See BOLIDE.

fire bank (*MinExt*) A slack or rubbish heap or dump, at surface on a colliery, which becomes fired by spontaneous combustion.

fire barriers (*Build*) Fire-resisting doors, enclosed staircases and similar obstructions to the spread of fire. See FIRE STOP.

fire bars (*Eng*) Cast-iron bars forming a grate on which fuel is burnt, as in domestic fires, boiler furnaces, etc.

fire-box (*Eng*) That part of a locomotive-type boiler containing the fire; the grate is at the bottom, the walls and top being surrounded by water. See LOCOMOTIVE BOILER.

firebrick arch (*Eng*) An arch built at the end of a boiler furnace, either to deflect the burning gases or to assist the combustion of volatile products. Also *flue bridge*.

fire cement (*Build*) See REFRACTORY CEMENT.

fireclay (*Geol, Eng*) Clay consisting of minerals predominantly of SiO_2 and Al_2O_3, low in Fe_2O_3, CaO, MgO, etc. Those clays which soften only at high temperatures are used widely as refractories in metallurgical and other furnaces. Fireclays occur abundantly in the Carboniferous system, as 'seat earths' underneath the coal seams.

fire climax (*EnvSci*) A vegetation climax, the composition of which is determined by fires that occur repeatedly; over time, fires favour those species which can regenerate quickly in the ashes or that are fire-resistant, and these become dominant. Also *pyroclimax*.

fire cracks (*Build*) Fine cracks which appear in a plastered surface, due to unequal contractions between the different coats.

fired (*Electronics*) Said of gas tubes when discharging, particularly transmit–receive pulse tubes during the transmission condition.

fire damp (*MinExt*) The combustible gas contained naturally in coal; chiefly a mixture of methane and other hydrocarbons; forms explosive mixtures with air.

fire-damp cap (*MinExt*) Blue flame which forms over the flame of a safety lamp when sufficient fire damp is present in colliery workings.

fire door (*Build*) (1) A fire-resisting door of wood, metal or both. (2) The door of a boiler furnace.

fire extinguisher (*Genrl*) Any of several portable apparatuses for emergency use against fire. In general, these depend upon the ejection (by rapid chemical reaction or compressed gas, CO_2 or nitrogen) of the fire-inhibiting medium. The latter may be: water or an alkaline solution; a rapidly evolved chemical foam; CO_2 gas and 'snow'; tetrachloromethane, CCl_4; chlorobromomethane, CH_3ClBr, or bromochlorodifluoromethane, CF_3BrCl, both heavy smothering gases; a suitable hydrogen carbonate in dry-powder form.

firefly (*BioSci*) Beetles, family Lampyridae, with light-producing organs whose main function is in attracting a mate.

fire foam (*Genrl*) A mixture of foaming but non-flammable substances used to seal off oxygen and to extinguish fire without use of water.

fire load (*Build*) The heat in $MJ\,m^{-2}$ or $Btu\,ft^{-2}$ of floor area of a building if destroyed by fire (including contents).

fireman (*MinExt*) (1) In a metal mine, a miner whose duty it is to explode the charges of explosive used in headings and working places. (2) In a coal mine, an official responsible for safety conditions underground.

fire opal (*Min*) A variety of opal (cryptocrystalline silica) characterized by a brilliant orange-flame colour. Particularly good specimens, prized as gemstones, are of Mexican origin.

fire polishing (*Glass*) The polishing of silicate glass and glassware by localized melting of the surface in a flame.

fireproof aggregates (*Build*) Materials such as crushed firebricks, fused clinkers, slag, etc, incorporated in concrete to render it fire-resisting.

fire refining (*Eng*) The refining of blister copper by oxidizing the impurities in a reverberatory furnace and removing the excess oxygen by POLING. May be used as an alternative to electrolytic refining, and is in any case carried out as a preliminary to this.

fire-retardant adhesive (*Eng*) Heavy-duty adhesives intended to fasten lagging around hot surfaces. These are usually alkyd based, but may contain chlorinated paraffins and antimony oxide to render the film non-combustible.

fire-retardant paints (*Build*) Surface coatings which reduce the degree of flame spread over a surface and protect the underlying substrate. These coatings act as a buffer coat between a flammable substrate and any source of ignition. Some rely solely upon their incombustible nature while those of the INTUMESCENT type swell on heating to form an insulating barrier.

fire ring (*Eng*) A top piston ring of a special heat-resisting design, used in some two-stroke oil engines.

fire sand (*Eng*) Refractory oxide or carbide suitable for lining furnaces.

fire stink (*MinExt*) The smell given off underground when a fire is imminent, eg in the GOB; also, smell of sulphuretted hydrogen from decomposing pyrite.

firestone (*Geol*) A stone or rock capable of withstanding a considerable amount of heat without injury. The term has been used with reference to certain Cretaceous and Jurassic sandstones employed in the manufacture of glass furnaces.

fire stop (*Build*) An obstruction across an air passage in a building to prevent flames from spreading further. Also *draft stop*.

fire top-centre (*Autos*) The top dead centre of an internal-combustion engine, when the piston is about to make its power stroke.

fire-tube boiler (*Eng*) A boiler in which the hot furnace gases, on their way to the chimney, pass through tubes in the water space, as opposed to a WATER-TUBE BOILER. See LOCOMOTIVE BOILER, MARINE BOILER.

firewall (*Aero*) See BULKHEAD.

firewall (*ICT*) A program or device that filters information coming through an Internet connection and blocks packets that the user has determined may be threatening or unwanted while allowing traffic from trusted sources.

firing (*Electronics*) Establishment of discharge through gas tube.

firing (*Eng*) (1) The process of adding fuel to a boiler furnace. (2) The ignition of an explosive mixture, as in a petrol or gas engine cylinder. (3) Excessive heating of a bearing.

firing (*Vet*) The application of thermocautery to the tissues of animals.

firing angle (*Electronics*) Phase angle (of an applied voltage) at which a thyristor (or a gas-filled valve, eg thyratron) starts to conduct.

firing key (*ElecEng*) A key which fires a charge of explosive by completing the electric circuit to a fuse.

firing order (*Autos*) The sequence in which the cylinders of a multicylinder internal-combustion engine fire, eg 1, 3, 4, 2 for a four-cylinder engine.

firing power (*Electronics*) The minimum radio-frequency power required to start a discharge in a switching tube for a specified ignitor current.

firing stroke (*Autos*) The power or expansion stroke of an internal-combustion engine.

firing time (*Electronics, Radar*) (1) The interval between applying a dc voltage to the trigger electrode of a thyristor or switching tube and the beginning of conduction. (2) In radar, the time required to establish a radio-frequency discharge in a switching tube (transmit–receive or anti-transmit–receive) after the application of radio-frequency power.

firing tools (*Eng*) Implements (eg shovels, rakes, and slicers or slicing bars) used in firing a boiler furnace by hand.

firmer chisel (*Build*) A woodcutting chisel, usually thin in relation to its width ($\frac{1}{16}$–2 in or 3–50 mm). Stouter than a paring chisel but less robust than a mortise chisel.

firmer gouge (*Build*) Standard type of GOUGE with the bevel on the outside. Used for cutting grooves and recesses. Cf SCRIBING GOUGE.

firmware (*ICT*) See HARDWIRED LOGIC.

firn (*Geol*) See NÉVÉ.

firring (*Build*) Timber strips of constant width but varying depth, which are nailed to the wood bearers of flat roofs as a basis for roof boarding, to which they give a suitable fall. Also *furring*.

first-angle projection (*Eng*) A system of projection used in engineering drawing, in which each view shows what would be seen by looking on the far side of an adjacent view.

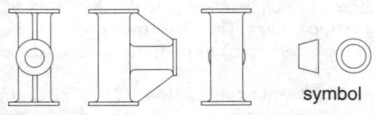

symbol

first-angle projection Three of possible five views are shown with the conventional symbol.

first detector (*ICT*) See MIXER.

first-generation computer (*ICT*) An early machine built around 1951 that used electronic VALVES and MERCURY DELAY LINES in the processor and electrostatic and MAGNETIC DRUMS and MAGNETIC TAPE for on-line storage. In 1949 the first operational STORED PROGRAM computer, the Manchester Mark 1, was completed. Other important development machines were the Cambridge EDSAC, NPL Pilot ACE, Washington SEAC, BINAC, the very large commercial UNIVAC, EDVAC, Princeton's IAS, WHIRLWIND 1. Many of these were logically and functionally equivalent to a present-day MICROCOMPUTER. See COMPUTER GENERATIONS.

first-in, first-out (*ICT*) Abbrev FIFO. See QUEUE.

first isomorphism theorem (*MathSci*) A theorem stating that if G and H are groups with $\varphi{:}G{\to}H$ a HOMOMORPHISM between them, then the QUOTIENT GROUP $G/\ker \phi$ is isomorphic to the image of φ.

first-order reaction (*Chem*) One in which the rate of reaction is proportional to the concentration of a single reactant, ie

$$\frac{dc}{dt} = -kc$$

where c is the concentration of the reagent.

First Point of Aries (*Astron*) See EQUINOCTIAL POINTS.

First Point of Libra (*Astron*) See EQUINOCTIAL POINTS.

first runnings (*Chem*) The first fraction collected from a FRACTIONAL DISTILLATION process, usually containing low-boiling impurities.

first ventricle (*BioSci*) In vertebrates, the cavity of the left lobe of the cerebrum.

first weight (*MinExt*) The first indications of roof pressure which occur after the removal of coal from a seam.

fir tree roots (*Aero*) A certain type of fixing adopted for turbine blades, the outline form of the root resembling that of a fir tree.

FIS (*Aero*) Abbrev for FLIGHT INFORMATION SERVICE.

Fischer–Tropsch process (*Chem*) The method of obtaining fuel oil from coal, natural gas, etc. The 'synthesis gas', hydrogen and carbon monoxide in proportional volumes 2:1, is passed, at atmospheric or slightly higher pressure and temperature up to 200°C, through contact ovens containing circulating water with an iron or cobalt catalyst. The gases are washed out and the resultant oil contains alkanes and alkenes, from the lower members up to solid waxes; fractionation yields petrols, diesel oils, etc. Cf HYDROGENATION.

FISH (*BioSci*) Abbrev for FLUORESCENCE *IN SITU* HYBRIDIZATION.

fish-beam (*CivEng*) A beam which is fish-bellied, ie it curves along its bottom edge to be convex downwards.

fish bellied (*Eng*) Said of (1) steel girders with a convex lower edge; (2) long straight-edges, which are convex upward. Such a form results in greater resistance to bending.

fishbone antenna (*ICT*) An END-FIRE ARRAY consisting of dipoles spaced along a twin transmission line, in the same pattern as the bone structure of many fishes.

fished joint (*CivEng*) A BUTT JOINT between rails or beams, in which fish-plated or cover straps are fitted on both sides of the joint and bolted together.

fish-eye lens (*ImageTech*) Camera lens covering an extremely wide angle, up to at least 180°, but with considerable barrel distortion and foreshortening of perspective. Also *bug-eye lens*.

fish glue (*Genrl*) (1) ISINGLASS. (2) Any glue prepared from the skins of fish (esp sole, plaice), fish bladders and offal.

fishing (*Eng*) Recovering tools dropped from drilling tackle during deep rock drilling operations.

fishing tool (*MinExt*) A tool attached to a drill string and designed to catch and retrieve components lost down-hole. Also *overshot tool*.

fish louse (*BioSci*) Crustacean ectoparasites (*Argulus*), 5–19 mm long, of fresh-water and marine fish; a major problem in fish farms causing tissue damage and increasing the likelihood of secondary infection.

fish-plate (*CivEng*) A steel or wood cover-plate, fitted one to each side of a fished joint between successive lengths of beam or rail. Also *fish-bar, fish piece, shin, splice piece*.

fish-wire (*ElecEng*) A thin wire drawn into a conduit for electric cables or wires during construction, and subsequently used for drawing in the cables or wires themselves.

fisk (*ICT*) To post a worthless or insubstantial message to a BLOG. Originally associated with the UK journalist Robert Fisk.

fissile (*Phys*) Capable of nuclear fission, ie breakdown into lighter elements of certain heavy isotopes (uranium-232, uranium-235, plutonium-239) when these capture neutrons of suitable energy. Also *fissionable*. See REACTOR.

fission (*BioSci*) Asexual reproduction of some unicellular organisms in which the cell divides into two more or less equal parts as in the fission yeast *Schizosaccharomyces*. Also *binary fission*. Cf BUDDING.

fission (*Phys*) Spontaneous or induced fragmentation of heavy elements into two or more light atoms of comparable mass together with energetic neutrons. Used as the basis for nuclear power generation through release of nuclear binding energy. See panel on BINDING ENERGY OF THE NUCLEUS.

fissionable (*Phys*) See FISSILE.

fission bomb (*Phys*) See ATOMIC BOMB.

fission chain (*Phys*) The short radioactive decay chain which occurs because atoms formed by uranium or plutonium fission have too high a neutron–proton ratio for stability. This is corrected either by neutron emission (the delayed neutrons) or more usually by the emission of a series of beta particles.

fission chamber (*NucEng*) Ionization chamber lined with a thin layer of uranium. This can experience fission by slow neutrons, which are thereby counted by the consequent ionization. Also *fission counter*.

fission counter (*NucEng*) See FISSION CHAMBER.

fission neutrons (*Phys*) Neutrons released by nuclear fission, having a continuous spectrum of energy with a maximum of about 10^6 eV.

fission parameter (*Phys*) The square root of the atomic number of a fissile element divided by its relative atomic mass.

fission poisons (*NucEng*) Fission products with abnormally high thermal neutron absorption cross-sections, which reduce the reactivity of nuclear reactors. Principally ^{135} Xe and ^{149} Sm.

fission products (*Phys*) Atoms, often radioactive, resulting from nuclear fission. They have masses of roughly half of that of the fissioning nucleus, eg strontium-90, a major contributor to radiation in fallout from nuclear explosions.

fission spectrum (*Phys*) The energy distribution of neutrons released by nuclear fission.

fission-track analysis (*Geol*) The examination of defects caused in solids, including minerals, by the spontaneous fission of the heavy nuclide uranium-238. Fission tracks are principally found in micas, sphene, zircon and apatite, and are mainly used for age determination.

fission-track dating (*Min*) Certain glassy minerals may contain uranium-238 which undergoes spontaneous fission at a known rate. Each fission results in the production of two heavy nuclei which travel through and disrupt the molecular lattice of the material, forming microscopically visible tracks. If the amount of uranium is known and the tracks counted per unit area, an estimate can be made of the time since the mineral was last heated to the temperature needed to eliminate any previous tracks. See RADIOMETRIC DATING.

fission yeast (*BioSci*) See *SCHIZOSACCHAROMYCES POMBE*.

fission yield (*Phys*) The percentage of fissions for which one of the products has a specific mass number. Fission yield curves show two peaks of approximately 6% for mass numbers of about 97 and 138. The probability of fission dividing into equal mass products falls to about 0·01%.

fissiped (*BioSci*) Having free digits. Cf PINNATIPED.

fissure (*Geol, MinExt*) A cleft in rock determined in the first instance by a fracture, a joint plane, or fault, subsequently widened by solution or erosion; may be open, or filled in with superficial deposits, often minerals of pneumatolytic and hydatogenetic provenance. See GRIKE.

fissure (*Med*) (1) Any normal cleft or groove in organs of the body. (2) Linear ulceration of the anus, usually the result of constipation.

fissure eruption (*Geol*) Throwing-out of lava and (rarely) volcanic 'ashes' from a fissure, which may be many kilometres in length. Typically there is no explosive violence, but a quiet welling-out of very fluid lava. Recent examples are known from Iceland.

fistula (*Build*) An ancient name for a water pipe.

fistula (*Med*) An epithelial lined track connecting two hollow viscera. Adj *fistulous*.

fistulous withers (*Vet*) Abscess and fistula formation in the withers of the horse; *Brucella abortus* infection has been found in some cases, and in others a nematode worm, *Onchocerca cervicalis*, has been found in the affected region.

fit (*Eng*) The dimensional relationship between mating parts. Limits of tolerances for shafts and holes to result in fits of various qualities, eg clearance, transition and interference fits, are laid down by national standard specifications.

fit (*Med*) A sudden attack of disturbed function of the sensory or of the motor parts of the brain, with or without loss of consciousness. See EPILEPSY.

FITC (*BioSci*) Abbrev for FLUORESCEIN ISOTHIOCYANATE.

FIT CCD (*ImageTech*) Abbrev for *frame interline transfer charge-coupled device*. A SOLID-STATE IMAGE SENSOR in which the charge is transferred from the PIXELS during FIELD BLANKING to vertical columns of adjacent opaque cells and thence to an opaque storage area on the other half of the chip for reading out. See CCD ARRAY, FT CCD, IT CCD.

fitch (*Build*) A small, long-handled, hog's-hair brush, used for fine finishing work. Made with chisel, filibert or round tapered tips.

fitness (*BioSci*) (1) In natural history, the degree to which an organism is adapted to its environment and can therefore survive the struggle for existence. (2) In evolutionary ecology, the extent to which an individual passes on its genes to the next generation. Cf DARWINIAN FITNESS, INCLUSIVE FITNESS.

fitness, Darwinian (*BioSci*) See DARWINIAN FITNESS.

fitness, inclusive (*BioSci*) See INCLUSIVE FITNESS.

fitter (*Eng*) A mechanic who assembles finished parts in an engineering workshop.

fitter's bench (*Eng*) A heavy wooden bench provided with a vice and a drawer for tools.

fitter's hammer (*Eng*) A hand hammer having a flat striking face, and a straight, cross or ball pane.

Fittig's synthesis (*Chem*) The synthesis of benzene hydrocarbon homologues by the action of metallic sodium on a mixture of a brominated benzene hydrocarbon and bromo- or iodo-alkane in a solution of dry ether.

fitting (*ElecEng*) A device used for supporting or containing a lamp.

fitting (*Eng*) Hand or bench work involved in the assembly of finished parts by a fitter.

fittings (*Eng*) (1) Small auxiliary parts of an engine or machine. (2) Boiler accessories, as valves, gauges, etc.

fitting shop (*Eng*) The department of an engineering workshop where finished parts are assembled. See ERECTING SHOP.

FitzGerald–Lorentz contraction (*Phys*) The contraction in dimensions of a body in the direction of its motion when its speed is comparable with that of light, relative to the frame of reference (Lorentz frame) from which measurements are made. Also *Lorentz contraction*.

five-centred arch (*Build*) An arch having the form of a false ellipse struck from five centres.

five-unit code (*ICT*) The Baudot code, as used for machine transmission of telegraphic signals in synchronous and start–stop systems.

fix (*Aero*) The exact geographical position of an aircraft, as determined by terrestrial or celestial observation, or by radio cross-bearing. Cf PINPOINT.

fix (*Surv*) Point accurately established on plan, perhaps by observations for latitude and longitude.

fixation (*BioSci*) (1) See CARBON FIXATION, NITROGEN FIXATION, PHOSPHATE FIXATION. (2) In microscopy, preparative treatment, esp before embedding and sectioning, aimed to stabilize (or fix) structures against subsequent treatments; eg for electron microscopy, with glutaraldehyde that cross-links proteins. (3) The action of certain muscles that prevent disturbance of the equilibrium or position of the body or limbs. (4) The process of attachment of a free-swimming animal to a substratum, on the commencement of a temporary or permanent sessile existence.

fixation (*Psych*) In psychoanalytic theory, a defence mechanism caused by acute anxiety and frustration during development; the individual is temporarily or permanently halted at a particular psychosexual stage of growth and this is reflected in an unevenness of personality development (eg fixation at the oral stage may result in a compulsive eating or talking disorder).

fixation of nitrogen (*Chem*) For commercial purposes (fertilizers etc) the HABER PROCESS is now the most important method of nitrogen fixation.

fixed action pattern (*Psych*) In ethology, species-specific movements recognized by their relatively high degree of stereotyping. Originally all fixed action patterns were assumed to be innate, but it is now recognized that such patterns are often influenced by environmental factors during development. Abbrev *FAP*.

fixed beam (*Build*) A beam with fixed ends.

fixed carbon (*Eng*) Residual carbon in coke after removal of hydrocarbons by distillation in inert atmosphere.

fixed carbon (*EnvSci*) Carbon dioxide fixed by photosynthesis and thus immobilized in biomass. The amount fixed per unit area is a measure of the primary productivity of an ecosystem.

fixed contact (*ElecEng*) The contact of a switch or fuse which is permanently fixed to the circuit terminal.

fixed earth station (*ICT*) In a satellite mobile-telephone system, the high-power/high-sensitivity ground station that sends signals to and receives signals from a mobile telephone (UNIVERSAL TERMINAL) via a satellite visible to both of them.

fixed eccentric (*Eng*) An eccentric which is permanently keyed to a shaft, not capable of angular movement, unlike a LOOSE ECCENTRIC.

fixed end (*Build*) The end of a beam when it is built in or otherwise secured. In theory the tangent to the end of the curve taken up by the beam, deflecting under applied load, should remain fixed.

fixed expansion (*Eng*) A steam engine in which the CUT-OFF cannot be altered and which thus works with a constant expansion ratio.

fixed head disk unit (*ICT*) One where a separate read–write head is positioned over each track on each surface; this reduces access time, at increased cost.

fixed interval schedule (*Psych*) See INTERVAL SCHEDULE OF REINFORCEMENT.

fixed-length record (*ICT*) One where the number of available bits (or characters) is predetermined. Cf VARIABLE-LENGTH RECORDS.

fixed lens (*ImageTech*) A term used to describe lenses of single FOCAL LENGTH, eg 50 mm.

fixed-loop aerial (*Aero*) A loop aerial, used with a homing receiver, which is fixed in relation to the aircraft's centre line.

fixed-needle surveying (*Surv*) A traverse with magnetic compass locked, the instrument being used like a theodolite for measuring azimuth angles, except where there is no nearby iron or steel to affect bearing, when compass needle may be used to give a check reading.

fixed-pitch propeller (*Aero*) See PROPELLER.

fixed point (*Phys*) A temperature which can be accurately reproduced, and used to define a temperature scale and for the calibration of thermometers. The temperature of pure

melting ice and that of steam from pure boiling water at 1 atmosphere pressure define the Celsius and Fahrenheit scales. The *International Practical Temperature Scale* defined ten fixed points ranging from the triple point of hydrogen (13·81 K) to the freezing point of gold (1337·58 K). See KELVIN THERMODYNAMIC SCALE OF TEMPERATURE, TRIPLE POINT.

fixed-point notation (*ICT*) Numbers are expressed as a set of digits with the decimal point in position (eg 142·687). The position of the decimal point is generally maintained by the program or programmer. The magnitude of fixed-point numbers is limited by the construction of the computer, but operations are generally very fast and preferred for most commercial data processing work. See FLOATING-POINT NOTATION.

fixed pulley (*Eng*) A pulley keyed to its shaft. See FAST PULLEY.

fixed ratio schedule (*Psych*) See RATIO SCHEDULE OF REINFORCEMENT.

fixed sash (*Build*) (1) A STAND SHEET. (2) A sash permanently fixed in a solid frame. Also *dead light*.

fixed time-lag (*ElecEng*) See DEFINITE TIME-LAG.

fixed-trip (*ElecEng*) A term for those forms of circuit breaker or motor starter in which the tripping mechanism cannot operate while the breaker or starter is being closed. Also *fixed-handle circuit breaker*. Cf FREE-TRIP.

fixed-type metal-clad switchgear (*ElecEng*) Metal-clad switchgear in which all parts are permanently fixed, no provision being made for easy removal of any part for inspection or maintenance purposes.

fixing (*ImageTech*) A process for removing unreduced silver halides after development of an emulsion.

fixing block (*Build*) Brick-shaped material built into the surface of a wall to provide a substance to which joinery may be nailed. Fixing blocks are made of porous concrete, of coke breeze or of a special brick made with a mixture of sawdust which burns away in the kiln to leave a porous material. See WOOD BRICK.

fixing fillet (*Build*) (1) A SLIP. (2) A strip fixed to a surface to support something such as a shelf. Also *fixing pad*.

fixing plugs (*Build*) Plugs made of plastic, metal or other material used in conjunction with screws or bolts to provide fixings to walls or concrete surfaces.

fixings (*Build*) Supports, such as grounds and plugs, for securing joinery in position.

fixture (*Build*) An attachment to a building.

fixture (*Eng*) A device used in the manufacture of (interchangeable) parts to locate and hold the work without guiding the cutting tool. Cf JIG.

Fizeau fringes (*Phys*) See CONTOUR FRINGES.

fjord (*Geol*) See FIORD.

flabellate (*BioSci*) Shaped like a fan. Also *flabelliform*.

flaccid (*BioSci*) A cell or cells in a tissue that are not turgid but limp. Turgor pressure is zero as a result of water loss. See PLASMOLYSIS.

flag (*Geol*) Natural flagstones are sedimentary rocks of any composition which can be readily separated, on account of their distinct stratification, into large slabs. They are often fine-grained sandstones interbedded with shaly or micaceous partings along which they can be split.

flag (*ICT*) Indicator that can be set or unset to indicate a condition in a program or a set of data. See CARRY FLAG, EOF, OVERFLOW FLAG, SENTINEL, SEPARATOR.

flag (*ImageTech*) A small opaque shield used in studio lighting to provide a local shadow. Also *french flag*.

flag (*Paper*) A coloured slip of paper inserted in a reel to indicate the position of a join or break in the web.

flag alarm (*ElecEng*) See FLAG INDICATOR.

flagellar root (*BioSci*) A group of microtubular structures lying under the PLASMALEMMA close to the basal body in flagellated cells.

Flagellata (*BioSci*) See MASTIGOPHORA.

flagellate (*BioSci*) (1) Having flagella. (2) An organism in which the body form is a unicell or colony of cells with flagella. (3) Bearing a long thread-like appendage. (4) A member of the Mastigophora.

flagellin (*BioSci*) The protein monomer that is assembled into a helical array to form the filament of a bacterial flagellum.

flagellum (*BioSci*) (1) Long hair-like extension from the cell surface whose undulation or beating is used for locomotion. In eukaryotes, the central *axoneme* is of a very standard '9+2' arrangement of microtubules; *cilia* are basically the same but shorter (see diagram at CILIA). (2) Prokaryotic (bacterial) flagella are made of a single protein, flagellin, and rotate. (3) In Arthropoda, a filiform extension to an appendage, eg a crustacean limb, an insect antenna. (4) In Gastropoda, one of the male genitalia.

flag indicator (*ElecEng*) In navigational and other instruments, a semaphore-type signal which warns the viewer when instrument readings are unreliable. Also *alarm flag, flag alarm.*

flag leaf (*BioSci*) The topmost leaf on a culm, just below the ear, in such cereal grasses as wheat and barley.

flag stage (*Agri*) The stage in cereal development where the last leaf, the FLAG LEAF, becomes visible.

flail chest (*Med*) A term used to describe paradoxical movement of the chest wall during respiration which can occur after major injury to the chest with multiple fractured ribs.

flail joint (*Med*) A joint in which there is, as a result of disease or of operation, excessive mobility.

flaking (*Build*) (1) The breaking away of surface plaster due to non-adhesion with the undercoat or to free lime or impurities in the basic surface. (2) A defect in paintwork, the paint film breaking away from the surface it was covering.

flame (*Chem*) A region in which chemical interaction between gases occurs, accompanied by the evolution of light and heat.

flame (*ICT*) To email or post an angry, intemperate or abusive message to an individual or website.

flame blow-off factor (*Eng*) The relation between the velocity of combustible mixture and the rate of flame propagation, the latter varying appreciably with gases of different composition. See BURNER FIRING BLOCK, FLAME RETENTION, PILOTED HEAD, TUNNEL BURNERS.

flame cell (*BioSci*) See SOLENOCYTE.

flame cleaning (*Build*) A method of preparing steelwork in which an oxyacetylene flame is passed over the surface to loosen mill scale and rust. The rust and scale are then removed by wire brushing and the surface primed while still warm.

flame cutting (*Eng*) Cutting of ferrous metals by oxidation, using a stream of oxygen from a blow pipe or torch on metal preheated to about 800°C by fuel gas jets in the cutting torch.

flame damper (*Aero*) See DAMPER.

flame failure (*Eng*) The accidental extinction of a burner flame, eg in an oil-fired boiler, which has to be detected in order to stop or otherwise control the supply of further fuel, to prevent explosion or damage.

flame-failure control (*Eng*) Direct-flame thermostat with interconnected relay valve, which provides a constantly burning pilot flame for igniting the main gas burners, and automatically shuts off the gas supply to the main burner in the event of the pilot flame becoming extinguished.

flame hardening (*Eng*) The use of an intense flame from eg an oxyacetylene burner for local heating of the surface of a hardenable ferrous alloy which is then immediately cooled.

flame ionization gauge (*Chem*) Used in gas chromatography, as a very sensitive detector for the separate fractions in the effluent carrier gas, by burning it in a hydrogen flame, the electrical conductivity of which is measured by inserting two electrodes in the flame with an applied voltage of several hundred volts.

flame out (*Aero*) The loss of power from a jet engine when the fuel stops burning.

flame photometry (*Chem*) Like ATOMIC ABSORPTION SPECTROSCOPY, except that the flame spectrum is viewed or measured directly.

flame plates (*Eng*) Those plates of a boiler firebox subjected to the maximum furnace temperature.

flame-proof (*ElecEng*) See EXPLOSION-PROOF.

flame retarders (*Eng*) Speciality chemicals added to normally flammable materials, esp polymers and fibres, to increase resistance to fire and flame. Solids include $CaCO_3$ (powdered chalk, whiting, etc), antimony oxide (Sb_2O_3) and hydrated alumina. Fluids include tritolylphosphate and trixylylphosphate plasticizers, esp for polyvinyl chloride conveyor belting.

flame retention (*Eng*) The ability to retain a stable flame with gas burners at all rates of gas flow, irrespective of adverse combustion conditions. See BURNER FIRING BLOCK, FLAME BLOW-OFF FACTOR, PILOTED HEAD, TUNNEL BURNERS.

flame spectrum (*Phys*) The spectrum emitted from a flame which is characteristic of substances volatilized in it.

flame speed (*MinExt*) The rate at which a flame front will travel through an inflammable mixture of gas and air. Highly dependent on the fuel and the physical conditions.

flame temperature (*Eng*) The temperature at the hottest spot of a flame.

flame test (*Chem*) (1) The detection of the presence of an element in a substance by the coloration imparted to eg a Bunsen flame. (2) Simple way of identifying polymers by smelling (with care) odour emitted by burning material. Systematic methods like infrared spectroscopy are much more accurate and safer. Such tests are usually combined with pencil hardness and density tests etc to give a rough guide to polymer identity.

flame trap (*Autos*) A gauze or grid of wire, or coiled corrugated sheet, placed in the air intake to a carburettor to prevent the emission of flame from a 'pop-back'.

flame trap (*Eng*) A device inserted in pipelines carrying air–gas mixture of a combustible or self-burning nature to arrest the flame in the event of a flash-back (or backfire) occurring at the burner.

flame tube (*Aero*) The perforated inner tubular 'can' of a gas turbine combustion chamber in which the actual burning occurs. Cf COMBUSTION CHAMBER.

flame war (*ICT*) Computing slang for an acrimonious exchange of electronic-mail messages.

flammability (*Chem*) The tendency of a material to ignite and continue to burn. Since most polymers contain carbon and hydrogen, their heats of combustion are large, so will tend to be flammable (esp if finely divided, as in fibres or particles). Burning behaviour can be measured in several ways, as in ASTM D635-74, where a rod is held horizontally in a flame. See OXYGEN INDEX, UL 94 FLAMMABILITY.

Flandrian (*Geol*) The final post-glacial, temperate stage of the Quaternary, by definition from 10 000 years ago to the present. Equivalent to the Holocene. See QUATERNARY.

flange (*Eng*) (1) A projecting rim, as the rim of a wheel which runs on rails. (2) The top or bottom members of a rolled I-beam. (3) A disk-shaped rim formed on the ends of pipes and shafts, for coupling them together; or on an engine cylinder, for attaching the covers. See FLANGED RAIL.

flange coupling (*Eng*) A shaft coupling consisting of two accurately faced flanges keyed to their respective shafts and bolted together.

flanged beam (*Eng*) A rolled-steel joist of I-section. Also *flanged girder.*

flanged nut (*Eng*) A nut having a flange or washer formed integral with it. See COLLAR-HEAD SCREW.

flanged pipes (*Eng*) Pipes provided either with integral or attached flanges for connecting them together by means of bolts.

flanged rail (*CivEng*) A rail section of inverted-T shape, now almost universally used in which the flange is at the bottom and the end of the stem of the T at the top. The latter part is enlarged locally to form the head of the rail. Also *flat-bottomed rail*.

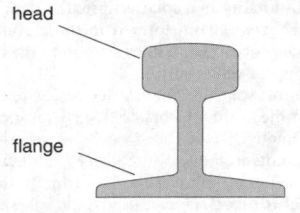

flanged rail Cross section.

flanged seam (*Eng*) A joint made by flanging the ends of furnace tubes and bolting them together between a pair of steel rings.

flange joint (*Eng*) Any joint between pipes, made by bolting together a pair of flanged ends.

flange protection (*ElecEng*) The rendering of electrical apparatus flame-proof by providing all joints with very wide flanges.

flank (*Eng*) (1) That part of a gear-tooth profile which lies inside the pitch line or circle. (2) The working face of a cam. (3) The straight side connecting the crest and the root of a screw-thread.

flank angle (*Eng*) An angle between the flanks of a thread and a plane perpendicular to the axis, measured in an axial plane.

flank dispersion (*ElecEng*) See END LEAKAGE FLUX.

flanking transmission (*Acous*) In architectural acoustics, sound transmission by the flanking walls of two rooms. When the wall between the two rooms has a high transmission loss (eg double wall), flanking transmission limits the sound isolation of the two rooms.

flanking window (*Build*) A window located beside an external door.

flanks (*Build, CivEng*) (1) The parts of the intrados of an arch near to the abutments. Also *haunches*. (2) The side surfaces of a building stone or ashlar, when it is built into a wall.

flank wall (*Build*) A side wall.

flannel (*Textiles*) A soft all-wool fabric, the weave being either plain or twill. The cloth is pre-shrunk and lightly raised.

flannelette (*Textiles*) A cotton fabric of plain or twill weave, raised on both sides and used for pyjamas, nightdresses, sheets and working shirts.

flanning (*Build*) The internal splay of a window jamb or of a fireplace.

flap (*Aero*) Any surface attached to the wing, usually to the trailing edge, which can be adjusted in flight, either automatically or through controls, to alter the lift as a whole; primarily on fixed-wing aircraft, but occasionally on rotor systems.

flap (*Med*) An area of tissue partly separated by the knife from the surface of the body, in connection with amputation of a limb or for the purpose of grafting skin.

flap angle (*Aero*) The angle between the chord of the flap, when lowered or extended, and the wing chord.

flap attenuator (*ICT*) One consisting of a strip of absorbing material which is introduced through a non-radiating slot in a waveguide.

flapping angle (*Aero*) The angle between the tip-path plane of a helicopter rotor and the plane normal to the hub axis.

flapping hinge (*Aero*) The pivot which permits the blade of a helicopter to rise and fall within limits, ie variation of zenithal angle in relation to the rotor head.

flap setting (*Aero*) The flap angle for a particular condition of flight, eg take-off, approach, landing.

flap tile (*Build*) A purpose-made tile, shaped so as to fit over a hip or valley line or to catch water.

flap trap (*Build*) A type of anti-flood valve, in which back flow is prevented by a hinged metal flap fitted in an intercepting chamber, to allow flow in one direction only.

flap valve (*Build, Eng*) (1) A sheet of flexible material like polythene hinged about one edge, allowing one-way flow of air through ventilators. Formerly made of mica. See FLAP TRAP. (2) A similar non-return valve for liquids, made of metal, faced with rubber or an O-ring, used for low pressures.

flare (*Acous*) The prominent part of the opening of a horn, bell or trumpet attached to a loudspeaking unit. Also *mouth*.

flare (*Astron*) An energetic outburst in the upper chromosphere/inner corona of the Sun associated with sunspots, causing intense radio and particle emission.

flare (*Genrl*) A flame from an oil well or landfill site in which excess gas is burnt. See FLARE STACK.

flare (*ImageTech*) The scattering of light within an optical system, such as a camera lens; it produces unwanted exposure, unrelated to the required image, and may appear as a patch of light or as an overall haze, reducing the effective tonal range. See FLARE FACTOR.

flare factor (*ImageTech*) The ratio between the luminance scale of the subject and that of the camera image.

flare gas (*MinExt*) See FLARE STACK.

flare header (*Build*) A brick which has been burnt to a darker colour at one end, so that it may be used with others in facing-work to vary the effect.

flare-out (*Aero*) Controlled approach path of aircraft immediately prior to landing.

flare stack (*MinExt*) Vertical pipe for the safe dispersal of hydrocarbon vapours from an oil rig or refinery. Steam is injected into the gas flow to ensure complete combustion.

flaring (*Acous*) A term applied to the end of eg a pipe when it is shaped out so as to be of increasing diameter towards the end. See FLARE.

flaser structure (*Geol*) A streaky, patchy structure in a dynamically metamorphosed rock. Flaser gabbro, eg shows an apparent crude flow structure round unaltered granular lenses.

flash (*Eng*) A thin fin of metal formed at the sides of a forging where some of the metal is forced between the faces of the forging dies. By extension, a similar extrusion in other (eg moulded) materials.

flash (*Plastics*) The excess material forced out of a mould during moulding. Sometimes referred to as a *fin*, if the mould is in two halves.

flashback voltage (*Electronics*) The inverse peak voltage in a gas tube at which ionization occurs.

flash boiler (*Eng*) A steam boiler consisting of a long coil of steel tube, usually heated by oil burners, in which water is evaporated as it is pumped through by the feed pump. See STEAM CAR.

flashbulb (*ImageTech*) A source of very brief illumination by the electrical ignition of metallic wire (magnesium or zirconium) in an oxygen-filled bulb; it can only be used once but multi-bulb arrays are made for operation in rapid succession.

flashbulb memory (*Psych*) Clear recollections that people sometimes have of what they were doing and experiencing at the time of some emotionally charged and dramatic event.

flash-butt welding (*ElecEng*) See RESISTANCE-FLASH WELDING.

flash colours (*BioSci*) See STARTLE COLOURS.

flash distillation (*Chem*) The spraying of a liquid mixture into a heated chamber of lower pressure, in order to drive off some of the more volatile constituents. Used in the petroleum industry.

flash drive (*ICT*) A portable data-storage device that can be connected to a computer through a serial port.

flash drying (*Eng*) Removal of moisture as stream of small particles falls through current of hot gas.

flashed glass (*Glass*) A term sometimes applied to glass coloured by the application of a thin layer of densely coloured glass to a thicker, colourless, base layer.

flasher (*ElecEng*) (1) An oscillator, or thermally operated switch and heater, arranged to switch a lamp on and off repetitively. (2) The lamp itself.

flash gun (*ImageTech*) An assembly of flashbulb or flash tube in a reflector complete with its power source, mounted on or connected to the camera for synchronization.

flash hook (*ICT*) A signal sent by a caller to a telephone CENTRAL OFFICE by moving the handset rest or SWITCH HOOK to interrupt the LOCAL LOOP several times in quick succession, usually to indicate that a follow-on call is required from a payphone or when using a CALLING CARD.

flashing (*Build*) (1) The method of brick burning in which the air supply is periodically stopped in order that the colouring of the bricks shall be irregular. (2) Glossy patches or streaks on flat-finished surfaces, usually attributable to poor technique. See FLASHINGS.

flashing (*Genrl*) The process of passing a boat across any point in a river where there is a sudden fall; effected by constructing a convergent passage from the high to the low level, shutting it by a sluice gate to allow the water to pond up, and then opening the gate and allowing the boat to be carried through the sluiceway by the artificially deepened water.

flashing board (*Build*) A board to which FLASHINGS are secured.

flashing compound (*Build*) Thick non-drying materials used for filling in crevices, eg between insulation blocks. Essential properties are impermeability and elasticity.

flashing light (*Ships*) A navigation mark identified during darkness by a distinctive pattern of flashes of light. See ALTERNATING LIGHT, OCCULTING LIGHT.

flashing-off (*Eng*) The production of steam by reducing the pressure on superheated water.

flashings (*Build*) Lead, copper or other sheeting used for waterproofing eg the gap between a wall and a lower roof. The flashing is cemented into a groove or RAGGLE in the wall and dressed over the tiles or slates below.

flashlight photography (*ImageTech*) Photography where rapid subject movement must be frozen, or where inadequate available light is supplemented, using a very brief source of illumination. Formerly, magnesium powder was ignited but current practice employs ELECTRONIC FLASH, usually synchronized with the camera shutter.

flash memory (*ICT*) A form of peripheral data storage device, typically a rewritable memory chip, such as a DATA STICK, with fast access times and non-volatile storage.

flashover (*ElecEng*) An electric discharge over the surface of an insulator.

flashover test (*ElecEng*) A test applied to electrical apparatus to determine the voltage at which a flashover occurs between any two parts, or between a part and earth. Also *sparkover test*.

flashover voltage (*ElecEng*) The highest value of a voltage impulse which just produces flashover.

flash pasteurization (*FoodSci*) Preserving a liquid product by rapid heating then rapid cooling in order to destroy micro-organisms and deactivate enzymes while minimizing physical and chemical changes.

flash photolysis (*Chem*) PHOTOLYSIS induced by light flashes of short duration but high intensity, eg from a laser.

flash point (*Build*) The temperature at which material gives off a vapour which will ignite on exposure to a flame.

flash point (*MinExt*) The temperature at which a liquid, heated in a Cleveland cup (open test) or in a Pensky–Martens apparatus (closed test), gives off sufficient vapour to flash momentarily on the application of a small flame. The *fire point* is ascertained by continuing the test.

flash radiography (*Radiol*) High-intensity, short-duration X-ray exposure.

flash roasting (*Eng*) The roasting of finely ground concentrates by introducing them into a large combustion chamber in which the sulphur is burned off as they fall.

flash spectrum (*Astron*) A phenomenon seen at the first instant of totality in a solar eclipse: the dark FRAUNHOFER LINES of the spectrum formed in the chromosphere flash out into bright emission lines as soon as the central light of the Sun is cut off.

flash suppressor (*ElecEng*) A device for preventing flash-overs on the commutators of dc generators; it consists of an automatically operated switch for short-circuiting certain points in the winding, thereby reducing the voltage to zero before a flashover has had time to develop.

flash-synchronized (*ImageTech*) Said of shutters which, when released, can close an electric circuit for firing a flash gun.

flash test (*ElecEng*) A test applied to electrical equipment for testing its insulation strength; it consists of the application of a voltage of about twice the working voltage, for a period of not more than about 1 minute.

flash tube (*ImageTech*) Discharge tube used for ELECTRO-NIC FLASH.

flash welding (*Eng*) An electric welding process similar to butt welding, in which the parts are first brought into very light contact. A high voltage starts a flashing action between the two surfaces, which continues while sufficient forging pressure is applied to the parts to complete the weld.

flask (*Eng*) A moulding box of wood, cast-iron or pressed steel for holding the sand mould in which a casting is made; it may be in several sections. See COPE, DRAG.

flask (*NucEng*) Lead case for storing or transporting multicurie radioactive sources or a container for the transport of irradiated nuclear fuel. Also *cask, casket, coffin*.

flat (*ElecEng*) A term used to denote a point on the surface of a commutator where the bars are lower than normal, owing to wear or displacement.

flat (*ImageTech*) The unit panel from which studio sets are constructed.

flat arch (*Build*) An arch whose intrados has no curvature and whose voussoirs (laid in parallel courses) are arranged to radiate to a centre. It is used over a doorway, fireplace, and window openings to relieve the pressure on the beam or lintel below it. Also *jack arch, straight arch*.

flatback stope (*MinExt*) A stope, in overhand stoping, worked upwards into a lode more or less parallel to level. See OVERHAND STOPES.

flat band (*Build*) A square and plain impost stone.

flat bed (*Print*) A general name distinguishing a cylinder machine with a horizontal flat bed from a ROTARY MACHINE or from a *platen machine*.

flatbed scanner (*ICT*) A SCANNER that has a flat sheet of glass onto which documents are placed for scanning by the light beam. Hand-held scanners are also available which are moved across the document by hand.

flat-bottomed rail (*CivEng*) See FLANGED RAIL.

flat chisel (*Eng*) A cold chisel having a relatively broad cutting edge, used in chipping flat surfaces.

flat-compounded (*ElecEng*) Said of a compound-wound generator whose series winding has been designed so that the output voltage remains constant at all loads. Also *level-compounded*.

flat file (*ICT*) See RELATIONAL DATABASE.

flat finish (*Build*) Surface paint coating showing no gloss, sheen or lustre.

flat foot (*Med*) See PES PLANUS.

flat four (*Aero*) Four-cylinder, horizontally opposed piston engine.

flat glass (*Glass*) Generic term for FLOAT, PLATE and SHEET GLASS suitable for eg windows. Made from soda–lime–silica glass of typical composition (percentage by weight): SiO_2 72, Na_2O 14, CaO 8, MgO 3, Al_2O_3 2. See panel on GLASSES AND GLASS-MAKING.

flat gouge (*Build*) A gouge having a cutting edge shaped to a large radius of curvature.

flat joint (*Build*) The type of mortar joint made in FLAT POINTING.

flat-joint jointed (*Build*) A flat joint which has had a narrow groove struck along the middle of its face by means of a jointer.

flat keel (*Ships*) See KEELSON.

flat knitting machine (*Textiles*) A WEFT-KNITTING MACHINE with straight needle beds.

flat lead (*Build*) Sheet lead.

flat lighting (*ImageTech*) General diffuse lighting, minimizing the contrast between high-light and shadow areas of the scene.

flat of keel (*Ships*) The portion of a ship's form actually coinciding with the baseline in a transverse plane.

flat pointing (*Build*) The method of pointing, used for uncovered internal wall surfaces, in which the stopping is formed into a smooth flat joint in the plane of the wall.

flat random noise (*Acous*) See WHITE NOISE.

flat region (*NucEng*) Portion of reactor core over which neutron flux (and hence power level) is approximately uniform.

flat roof (*Build*) A roof surface laid nearly horizontal, ie having a fall of only about 1 in 80, provided for drainage.

flats (*Eng*) (1) Iron or steel bars of rectangular section. (2) The sides of a (hexagonal) unit.

flat-screen TV (*ImageTech*) Any screen with minimal depth, employing technologies such as LIGHT-EMITTING DIODE, LIQUID CRYSTAL DISPLAY (panel) and plasma display, leading ultimately to the wall screen.

flat spot (*Autos*) In a carburettor, a point during increase of airflow (resulting from increased throttle opening or speed) at which the air–fuel ratio becomes so weak as to prevent good acceleration.

flat stitching (*Print*) The stitching of a book close to the back with wire which passes through from the first page to the last, as distinct from SADDLE STITCHING. Cf DRILLED AND STRUNG, STABBING.

flattener (*Glass*) One who takes a cylindrical piece of glass like a wide tube, cracked longitudinally, and, after heating it to softening in a furnace, flattens it out to form a sheet. An old process only used for making special types of sheet.

flattening material (*NucEng*) Neutron absorber or depleted fuel rod used in centre of reactor core to give larger flat region.

flatter (*Eng*) (1) Smith's tool resembling a flat-faced hammer, which is placed on forged work and struck by a sledge hammer. (2) Draw plate for producing flat wire such as hair springs.

flatting mill (*Eng*) Rolling mill which produces strip metal or sheet.

flatting varnish (*Build*) An oil varnish containing resin used as a basis for the final coat of varnish after having been rubbed down with abrasives to give a flatter surface.

flat-top antenna (*ICT*) One in which the uppermost wires are horizontal. Also *roof antenna*.

flat tuning (*ICT*) A broadband receiver or amplifier, unable to discriminate between closely spaced signals having different frequencies.

flat twin cable (*ElecEng*) Cable for wiring work in which two insulated conductors are laid side by side (not twisted) and surrounded by a suitable protective sheath. Often includes separate uninsulated earth conductor.

flatulence (*Med*) Air or gas in the intestinal tract. Also *flatus*.

flat universe (*Astron*) Cosmological model in which space is not curved (ie can be described by Euclidean geometry); such a universe is characterized by a certain critical density, is always infinite and will continue to expand for an infinite time.

flat yarn (*Textiles*) (1) Manufactured continuous filaments that have not been twisted or *textured*. (2) A straw-like filament.

flaunching (*Build*) The slope given to the top surface of a chimney to throw off the rain.

flavescent (*BioSci*) Becoming yellow; yellowish.

Flaviviridae (*BioSci*) The family of enveloped RNA viruses with spherical particles 40–50 nm in diameter. Cause dengue haemorrhagic fever, Japanese encephalitis, tick-borne encephalitis and yellow fever (the last being the source of the name).

flavone (*Chem*) A yellow plant pigment; the phenyl derivative of chromone, parent substance of a number of natural vegetable dyes.

flavone

flavonoids (*BioSci*) A large group of SECONDARY METABO-LITES of bryophytes and vascular plants, based on 2-phenylbenzopyran, often as glycosides. Some sorts are pigments and others may be PHYTOALEXINS, insecticides or of no known function. Many are significant in chemotaxonomy.

flavoproteins (*BioSci*) Proteins that serve as electron carriers in the electron transfer chain by virtue of their prosthetic group, *flavine adenine dinucleotide*.

flavor (*Phys*) US for FLAVOUR.

flavour (*Phys*) A label which denotes different types of quarks. The six types are: up (u), down (d), strange (s), charm (c), bottom (or beauty) (b) and top (or truth) (t). US *flavor*.

flavour enhancer (*FoodSci*) A substance with little flavour of its own, which enhances a food's basic flavour, eg common salt, monosodium glutamate, RNA hydrolysates.

flavour profile (*FoodSci*) A semi-quantitive means of describing a flavour in which a food is tasted by a range of people and scored on a number of different stimuli, eg sweetness, bitterness (astringency), acidity (sourness) and often aroma. Average scores are plotted for each stimulus to give a multidimensional diagram. Computers can then process this ORGANOLEPTIC data to produce profiles of natural substances or competitors' products so that acceptable replicates can be produced.

flax (*Textiles*) Plants of *Linum usitatissimum* grown in temperate regions and used for the fibres obtained from their stems and for their seeds (linseed oil and animal food). The cellulosic fibres are prepared by RETTING followed by cleaning (including HACKLING and SCUTCHING).

flax tow (*Textiles*) The short flax fibres removed during cleaning.

F-layer (*Phys*) Upper ionized layer in the ionosphere resulting from ultraviolet radiation from the Sun and capable of reflecting radio waves back to Earth at frequencies up to 50 MHz. At a regular height of 300 km during the night, it falls to about 200 km during the day. During some seasons, this remains as the F_1 layer while an extra F_2 layer rises to a maximum of 400 km at noon. Considerable variations are possible during particle bombardment from the Sun, the layer rising to great heights or vanishing. Also *Appleton layer*.

fleaking (*Build*) Thatching a roof with reeds.

fleam (*Build*) Angle of rake between the cutting edge of a sawtooth and the plane of the blade.

fleam-tooth (*Build*) A sawtooth having the shape of an isosceles triangle.

flèche (*Arch*) A slender spire, particularly a timber one, springing from a roof ridge.

flèches d'amour (*Min*) Acicular, hair-like crystals of rutile, a crystalline form of the oxide of titanium, TiO_2, embedded in quartz. Used as a semiprecious gemstone.

Also *love arrows* (the literal translation), *cupid's darts* or *Venus's hair stone*.

fleece wool (*Textiles*) Wool obtained by shearing the living sheep.

fleecy fabric (*Textiles*) (1) The term in the hosiery trade for fabrics having at the back a thick yarn which is brushed to raise a pile. (2) Any woven apparel cloth raised on one or both sides and made to resemble a fleece.

fleeting tetanus (*Vet*) See TRANSIT TETANY.

flehmen (*Psych*) Lip-curling behaviour; a common facial expression in mammals probably associated with olfactory sensing.

Fleming diode (*Electronics*) The 1904 detector of radio signals, with an incandescent filament and a separate anode.

Flemish bond (*Build*) A bond consisting of alternate headers and stretchers in every course, each header being placed in the middle of the stretchers in the courses above and below.

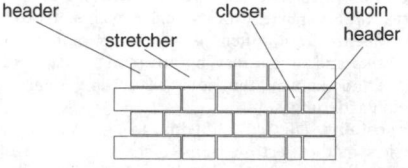

Flemish bond

Flemish garden-wall bond (*Build*) A bond in which each course consists of three stretchers alternating with a header, each header being placed in the middle of the stretchers in the courses above and below.

Fletcher–Munson curves (*Acous*) Equal-loudness curves for aural perception, measured just outside the ear, extending from 20 to 20 000 Hz, and from the threshold of hearing to the threshold of pain. They are the basis of the PHON scale.

F-levels (*Phys*) See FUNDAMENTAL SERIES.

flexible band (*Print*) One used in attaching the sections of a volume together where the bands are not let into the back of the sections, the sewing thread passing completely round each band. See RAISED BAND.

flexible cable (*ElecEng*) A cable containing one or more cores of such cross-section and fine stranding as to make the whole quite flexible.

flexible cord (*ElecEng*) A flexible cable of small circular cross-section protective plastic, each conductor having a large number of fine wire strands insulated by plastic. Used for connections to portable domestic apparatus, pendant lamps, etc. See TWIN FLEXIBLE CORD.

flexible coupling (*Eng*) A coupling used to connect two shafts in which perfectly rigid alignment is impossible; the drive is commonly transmitted from one flange to another through a resilient member, such as a steel spring, or a rubber disk or bushes.

flexible manufacturing system (*Eng*) An arrangement of computer-controlled machines which can easily be adapted and used in different sequence to modify the manufacturing route for a particular component or to restructure operations for a different product. Abbrev *FMS*.

flexible resistor (*ElecEng*) One resembling a flexible cable.

flexible roller bearing (*Eng*) An anti-friction bearing containing hollow cylindrical rollers made by winding strip steel into helical form. The hollow construction permits greater deflection under load. Also *Hyatt roller bearing*.

flexible support (*ElecEng*) A support for an overhead transmission line, which is designed to be flexible in a direction along the line, but rigid in a direction at right angles to the line.

flexible suspension (*ElecEng*) A method of suspending the contact wire of a traction system so that it has a certain amount of lateral and vertical movement relative to the fixed supports.

fleximer (*Build*) See CEMENT–RUBBER–LATEX.

flexographic printing (*Print*) Usually reel-fed rotary, using relief rubber plates and spirit- or water-based inks which dry quickly, leaving no odour; much used for food wrappers.

flexor (*BioSci*) A muscle that, by contracting, bends a limb or other part of the body. Cf EXTENSOR.

flexuose (*BioSci*) A zigzag, wavy stem. Also *flexuous*.

flexural rigidity (*Phys*) A measure of the resistance of a beam to bending.

flexural strength (*Eng*) To avoid problems of failure at the grips with brittle materials, their strength is often measured in bending rather than in the tensile test; the flexural strength tends to be somewhat higher than the tensile strength. See STRENGTH MEASURES.

flexural test (*Eng*) See BENDING TEST.

flexure (*Build*) The bending of a member, eg under load.

flex-wing (*Aero*) A collapsible single-surface fabric wing of delta planform investigated first for the return of space vehicles as gliders; later applied to army low-performance tactical aircraft of collapsible type; now for MICROLIGHTS and hang gliders.

flicker (*Phys*) Visual perception of fluctuation of brightness at frequencies lower than that covered by persistence of vision. Threshold of flicker depends on brightness and angle from optical axis. *Chromaticity flicker* arises from variations in CHROMATICITY only.

flicker effect (*Electronics*) Irregular emission of electrons from a thermionic cathode due to spontaneous changes in condition of emitting surface; resulting in an electronic noise.

flicker-fusion frequency (*Phys*) The rate of successive intermittent light stimuli just necessary to produce complete fusion, and thereby the same effect as continuous lighting.

flicker noise (*Electronics*) Noise with an associated power spectrum that is inversely proportional to frequency, associated with a variety of phenomena such as fluctuations in the value of the resistivity of a resistor material.

flicker photometer (*Phys*) A photometer in which a screen is illuminated alternately and in quick succession by the lamp under test and a standard lamp, thus producing a flickering effect. When the illumination from the two sources of light is equal, the flickering effect disappears.

flicker shutter (*ImageTech*) Rotating shutter in a motion picture projector interrupting the light at least 48 times a second to minimize the perception of flicker in the projected picture.

flick roll (*Aero*) See ROLL.

flier (*Build*) See FLYING SCAFFOLD.

flight (*Eng*) (1) The helical element in a worm or screw conveyor, usually fabricated from sheet metal, which may be hollow to allow a heating or cooling liquid to pass through it. (2) The helical element in a vibratory bowl feeder over which components pass while being unscrambled and orientated.

flight angle (*Eng*) The property of a screw helix in injection moulding or extrusion. Defined as the angle subtended by screw flight to the line orthogonal to screw axis. Optimal value of 30° for conveying powders and 10–15° for granules.

flight control (*Aero*) Control of vehicle, eg spacecraft, missile or module, so that it attains its target, taking all conditions and corrections into account. Generally done by computer, controlled by signals representing actual and intended path.

flight deck (*Aero*) (1) Upper part of an aircraft carrier. (2) Crew compartment of a large aircraft.

flight director (*Aero*) An aircraft instrument (ie blind) flying system in which the dials indicate what the pilot must do to achieve the correct flight path as well as the actual attitude of the aircraft. The dial display is usually a GYRO HORIZON with a spot or pointer which must be centred. The equipment can be coupled to a radio or GYROCOMPASS to bring the aircraft onto a desired heading, preset on the compass, or it can be coupled to the INSTRUMENT LANDING SYSTEM to receive these signals. In all cases, the flight director can be made automatic by switching its signals into the autopilot.

flight engineer (*Aero*) A member of the flying crew of an aircraft responsible for engineering duties, ie management of the engines, fuel consumption, power systems, etc.

flight envelope (*Aero*) The plot of altitude against speed defining performance limits within which an aircraft and/ or its equipment can operate.

flight fine pitch (*Aero*) The minimum blade angle, held by a removable stop, which the propeller of a turboprop engine can reach while in the air, and which provides braking drag for the landing approach.

flight information (*Aero*) A *flight-information centre* provides a FLIGHT-INFORMATION SERVICE of weather and navigational information within a specified FLIGHT-INFORMATION REGION.

flight-information region (*Aero*) An air space of defined dimensions within which information on air-traffic flow is provided according to the types of air space therein. Abbrev *FIR*.

flight-information service (*Aero*) One giving advice and information to assist in the safe, efficient conduct of flights. Abbrev *FIS*.

flight level (*Aero*) AIR-TRAFFIC CONTROL instructions specify heights at which controlled aircraft must fly and these are given in units of 100 ft (30 m) for altitudes of 3000 or 5000 ft (900 or 1500 m) and above.

flight Mach number (*Aero*) The ratio of true airspeed of an aircraft to speed of sound under identical atmospheric conditions.

flight management system (*Aero*) Computer-controlled automatic flight control system allowing the pilot to select specific modes of operation. These could include: standard instrument departure; autothrottle; standard terminal arrival system; Mach hold.

flight path (*Aero*) The path in space of the centre of mass of an aircraft or projectile. (Its *track* is the horizontal projection of this path.)

flight plan (*Aero*) A legal document filed with AIR-TRAFFIC CONTROL by a pilot before or during a flight (by radio), which states the destination, proposed course, altitude, speed, ETA and alternative airfield(s) in the event of bad weather, fuel shortage, etc.

flight recorder (*Aero*) A device which records data on the functioning of an aircraft and its systems on tape or wire. (1) The *flight data recorder* (FDR) should be in a crashproof, floatable box which may be ejected in case of an accident, and is usually fitted with a homing radio beacon and flashlight. (2) The *cockpit voice recorder* (CVR) stores all speech between crew on the flight deck, and between crew and ground AIR-TRAFFIC CONTROL. (3) The *maintenance data recorder* receives data from hundreds of inputs from engineering systems. Colloq *black box* (frequently a misnomer).

flight time (*Aero*) See BLOCK TIME.

Flinders bar (*Ships*) A bar of soft iron attached to the magnetic compass binnacle to correct for disturbing effects of that part of the ship's field due to vertical soft iron in the fore and aft line.

flint (*Geol*) Concretions of silica, sometimes tabular, but usually irregular in form, particularly abundant on the bedding planes of the Upper Chalk. See CHERT, PARAMUDRAS, SILICA.

flint glass (*Glass*) (1) Originally LEAD GLASS; the good-quality silica needed to ensure freedom from colour was obtained from crushed flint. (2) Now that class of optical glasses with $v_d < 50$ if $n_d < 1·60$ and $v_{nd} < 55$ if $n_d > 1·60$,

where v_d is the *Abbé number*, or reciprocal of the DISPERSIVITY, and n_d is the REFRACTIVE INDEX at 587·6 nm. Cf CROWN GLASS. (3) Also loosely used for all colourless glass other than FLAT GLASS.

flint gravel (*Geol*) A deposit of gravel in which the component pebbles are dominantly of flint, eg the Tertiary and fluvioglacial gravels in SE England.

flip chip (*Electronics*) Mounting method to maximize heat transfer and minimize lead inductance in which CHIPS are inverted and directly bonded to contacts in a HYBRID or PRINTED CIRCUIT. See panel on PRINTED, HYBRID AND INTEGRATED CIRCUITS.

flip-flop (*ICT*) Circuit that can be in either of two states, reversed by a pulse. It is used to construct DELAY ELEMENTS and as a 1 bit storage device. Also *bistable multivibrator, trigger circuit*.

flip phone (*ICT*) A mobile phone with a hinged cover that protects it when it is not in use.

FLIR (*Aero*) Abbrev for FORWARD-LOOKING INFRARED.

flirt (*Eng*) A device for bringing about the sudden movement of a mechanism.

flitch (*For*) A piece of timber of greater size than 4×12 in, intended for reconversion.

flitch beam (*Build*) A built-up beam formed with an iron plate between two timber beams.

flit-plug (*ElecEng*) A detachable connecting-box for coupling cables.

float (*Aero*) (1) The distance travelled by an aircraft between flattening-out and landing. (2) A watertight buoyancy unit which is of combined streamline and hydrodynamic form to reduce air and water resistance; *main floats* are the principal hydrodynamic support of float planes, while *wing-tip floats*, often retractable, give lateral stability to flying boats.

float (*Autos*) A small buoyant cylinder of thin brass, steel or proofed cork, placed in the float chamber of a CARBURETTOR for actuating a valve controlling the petrol supply from the main tank.

float (*Build*) (1) A plasterer's trowel. (2) A polishing block used by marble-workers. Cf FLOAT STONE. (3) A flat piece of wood with a handle on one side used for FLOATING.

float (*Eng*) A single-cut file (or float-cut file), ie a file having only one set of parallel teeth, as distinct from a CROSS-CUT FILE.

float (*For*) A measure of timber, equalling 18 loads.

float (*MinExt*) (1) Values so fine that they float on the surface of the water when crushed or washed, eg *float gold*. (2) Surficial deposit of rock or mineral detached from the main dyke or vein. (3) The term for blocks of bedrock within soil or superficial deposits encountered during prospecting or drilling, eg ERRATICS.

float (*Textiles*) (1) In a woven fabric the length of yarn between adjacent intersections. (2) A defect in a fabric caused by a thread passing over other threads with which it is designed to interweave. (3) A pattern thread in a lace.

float bowl (*Autos*) US for FLOAT CHAMBER.

float chamber (*Autos*) In a CARBURETTOR, the petrol reservoir from which the jets are supplied, and in which the fuel level is maintained constant by means of a float-controlled valve. US FLOAT BOWL.

float-cut file (*Eng*) See FLOAT.

floated coat (*Build*) A coat of plaster smoothed with a float.

floated rate-integrating gyro (*Aero, Space*) An electrically driven single-degree-of-freedom gyro whose cylindrical or spherical case floats with neutral buoyancy in a fluid within an outer casing. Precession of the gyro is detected electrically and these signals combined with the viscous torque set up by relative motion between rotor and case are used to measure the integral of the angular motion. Cf TUNED ROTOR GYRO. Abbrev *RIG, MIG* if miniaturized.

float glass (*Glass*) The vast majority of FLAT GLASS is now produced by the float process. A broad stream of molten glass at about 1050°C is spread across the surface of a bath of molten tin (in a reducing atmosphere to avoid oxidation of the tin) and is drawn off at the other end when rigid, at about 600°C. The two glass surfaces emerge flat, parallel and fire-polished, so that they give clear, undistorted vision without requiring any further working. Cf PLATE GLASS, SHEET GLASS. See panel on GLASSES AND GLASS-MAKING.

floating (*Build*) (1) The second of three coats applied in plastering, applied with a FLOAT to bring the coat level with the SCREEDS. (2) A paint film defect caused when pigments separate and rise to the surface during drying, resulting in a patchy finish.

floating accents (*Print*) Free-standing accents which can be placed at will over individual characters.

floating address (*ICT*) See SYMBOLIC ADDRESS.

floating anchor (*Ships*) See SEA ANCHOR.

floating battery (*ElecEng*) An electrical supply system in which a storage battery and electric generator are connected in parallel to share load, so that the former carries the whole load if the generator fails.

floating bay (*Build*) An area between screeds, which is to be filled in with plaster.

floating-carrier wave (*ICT*) Transmission in which the carrier is reduced or even suppressed to zero during intervals of no modulation by signals, to economize in power or to avoid interference with reception.

floating compass (*Genrl*) An early type of primitive compass, probably Chinese. The needle was attached to a float of wood or straw and floated in a bowl of water. The directions were marked inside the bottom of the bowl. See LIQUID COMPASS.

floating dam (*Build*) A CAISSON.

floating floor (*Acous*) Common type of floor to reduce sound transmission. The load-carrying floor is covered with a layer of fibrous material or soft foam, and a hard plate of about 40 mm thickness is put over the foam.

floating gudgeon pin (*Eng*) One free to revolve in both the connecting-rod and the piston bosses.

floating harbour (*Build*) (1) A breakwater formed of booms fastened together and anchored, to afford protection to vessels behind it. (2) Sometimes used to distinguish the locked from the tidal part of a harbour; ie it is that part in which vessels can always float.

floating kidney (*Med*) See NEPHROPTOSIS.

floating lens element (*ImageTech*) An ELEMENT which moves independently of others during focusing to maximize optical quality.

floating-point notation (*ICT*) Numbers are expressed as a fractional value (mantissa) followed by an integer exponent of the base, eg $0.347\,91 \times 10^3$. This enables a large range of numbers to be represented with a fixed number of digits. The advantage of the extra range is gained at the expense of processing time and precision. Cf FIXED-POINT NOTATION.

floating potential (*Electronics*) That appearing on an isolated electrode when all other potentials on electrodes are held constant.

floating ribs (*BioSci*) See FALSE RIBS.

floating roof (*MinExt*) A tightly fitting but free cover which floats on an otherwise open tank of fuel. There is no space for the build-up of inflammable vapour.

floating rule (*Build*) A long straight-edge used to form flat surfaces in plaster or cement work.

floating temperature control (*Eng*) The use, in a furnace, of an automatic temperature controller which functions by electrically operated valves.

float seaplane (*Aero*) An aircraft of the seaplane type, in which the water support consists of floats in place of the main undercarriage, and sometimes at the tail and wing tips. It may be of the *single-* or *twin-float type*.

float stone (*Build*) A shaped iron block which is rubbed over curved brickwork, such as cylindrical backs, in order to remove marks left on the surface by rough dressing.

float stone (*Min*) A coarse, porous, friable variety of impure silica, which, on account of its porosity, floats on water until saturated.

float switch (*ElecEng*) A switch operated by a float in a tank or reservoir, and usually controlling the motor of a pump.

float zone (*Chem*) The process technique for growing single crystals without the use of a crucible. A seed is fused to a bar of purified stock material. Stock and seed ends are counter-rotated about a vertical axis while a local molten zone (created by a radio-frequency coil around the rod) is swept slowly away from the seed. Cf BRIDGMAN PROCESS, CZOCHRALSKI PROCESS.

flocculation (*BioSci*) The formation of floccules in a precipitin test or of agglutinated bacteria in an agglutination test for flagellar antigens of *Salmonella* species.

flocculation (*Chem*) The coalescence of a finely divided precipitate into larger particles. See panel on GELS.

flocculation (*MinExt*) The coagulation of ore particles by use of reagents which promote formation of flocs, as a preliminary to settlement and removal of excess water by thickening and/or filtration.

flocculent (*Chem*) Existing in the form of cloud-like tufts or flocs.

flocculi (*Astron*) See PLAGE.

floccus (*BioSci*) In birds, the downy covering of the young forms of certain species; in mammals, the tuft of hair at the end of the tail; more generally, a tuft.

flock (*Psych*) A group of birds that remain together because of social attraction between individuals, rather than because of eg a shared interest in some environmental feature (an *aggregation*). See HERDING, SCHOOLING.

flock (*Textiles*) (1) Waste fibres produced in the processes of finishing woollen cloths, used for bedding and upholstery purposes. (2) Short-cut or ground wool, cotton or manufactured fibres for spraying onto adhesive-coated backings for furniture and upholstery purposes. See ELECTROSTATIC FLOCKING.

flogger (*Build*) A brush with a long bristle filling used in graining to create the effect of pores in certain timbers.

flogging (*Build*) The operation of rough-dressing a timber to shape, when the material is removed in large pieces.

flong (*Paper, Print*) A board made of PAPIER MÂCHÉ used for making moulds from an original typesetting for casting duplicate metal *stereotype* printing plates.

flood (*ImageTech*) A studio light source to give general illumination over a wide angle. Also *floodlight*.

floodable length (*Ships*) The maximum length of that portion of a ship, centred at a given point, which can be flooded without submerging the MARGIN OF SAFETY LINE.

flood basalts (*Geol*) Widespread plateau basalts originating from fissure eruptions.

flood flanking (*Build*) The constructing of an embankment by depositing stiff moist clay in separate small loads, so that each shall unite so far as is possible with the others, while the crevices left when the clay has dried out are filled with SLUDGING.

flooding (*Chem*) The condition of a fractionating column in which the upward flow of vapour has become great enough to prevent the downward flow of liquid, giving poor performance.

flooding (*Med*) Copious bleeding from the uterus.

flooding (*Psych*) A therapeutic approach to treating phobias in which repeated confrontation with the anxiety-producing stimulus is used to extinguish the avoidance and fear responses.

floodlighting (*ElecEng*) The lighting of a large area or surface by means of light from projectors situated at some distance from the surface.

floodlight projector (*ElecEng*) The housing and support for a lamp used in a floodlighting scheme; it is designed with a reflector which directs the light from the lamp into a suitable beam.

flood plain (*Geol*) A plain of stratified alluvium bordering a stream and covered when the stream floods.

flood track (*ImageTech*) A photographic sound track with full uniform exposure across its complete width.

floor (*Eng*) The bed of sand constituting the floor of a foundry; in it large castings are often made.

floor (*MinExt*) The upper surface of the stratum underlying a coal seam.

floor (*Ships*) Transverse vertical plate connected to the shell plating and to the inner bottom and extending from side to side.

floor cramp (*Build*) A cramp for closing up the joints of floorboarding as it is nailed.

flooring saw (*Build*) One with a curve towards the toe and extra teeth on the back above the toe. Used for cutting floorboards in order to raise them.

floor joist (*Build*) A BRIDGING JOIST.

floor line (*Build*) A mark made at the lower end of a door post, or other finishing, to indicate the level of the floor when the finishing is in position.

floor plan (*Arch*) A plan drawn for any given floor of a building, normally showing the dimensions of the rooms etc.

floor sand (*Eng*) Foundry sand in which new and used moulding sand and coal dust are mixed.

floor stop (*Build*) A door stop projecting from the floor near a door.

floor temperature (*Chem*) Critical temperature below which polymerization cannot occur. Unlike ceiling temperature, a relatively rare phenomenon but encountered in plastic sulphur.

flop damper (*Build*) A damper which stays under its own weight in the open or shut position.

flopgate (*MinExt*) Diverting gate which directs moving material into alternative routes. Can be worked by remote control.

flop-over (*ImageTech*) A visual effect in which the picture is reversed from left to right.

floppy disk (*ICT*) Lightweight, flexible magnetic disk that behaves as if rigid when rotated rapidly. It is robust and light enough to send through the post. Also *diskette, floppy*.

FLOPS (*ICT*) Abbrev for *floating-point operations per second*. A measure of speed of a processor. Also *megaflops*. Cf LIPS, MIPS.

floptical (*ICT*) A floppy disk that uses an optical system to locate the tracks so that more information can be recorded.

flora (*BioSci*) (1) The plants of a particular region habitat or epoch. (2) A catalogue or description of such plants.

floral diagram (*BioSci*) A conventional plan of the arrangement of the parts of a flower as seen in cross-section.

floral envelope (*BioSci*) The calyx and corolla, or perianth.

floral formula (*BioSci*) A representation of the structure of a flower by means of the letters P (perianth) or K (calyx) and C (corolla), A (androecium) and G (gynoecium), of figures (numbers of parts), of parentheses (connation), of horizontal brackets (adnation) and of a line above or below the G (superior or inferior ovary); thus P 3+3 A 3+3 G(3) is the tulip.

floral leaf (*BioSci*) (1) A bract or bracteole. (2) A sepal or petal.

floral mechanism (*BioSci*) The arrangement of the flower parts to ensure either cross- or self-pollination.

Florence flask (*Chem*) A long-necked round flask with a flat bottom.

Florentine arch (*Arch*) An arch having a semicircular intrados and a pointed extrados, giving greater strength at the crown.

flore pleno (*BioSci*) With double flowers. Abbrev *fl pl*.

floret (*BioSci*) An individual flower in a crowded inflorescence. In grasses, typically consists of an ovary (with two styles) and three stamens enclosed by two bracts (lemma and palea).

floriated (*Arch*) Said of an elaborately ornamented building style, particularly with flowers and leaves featured.

florigen (*BioSci*) A hypothetical plant hormone, possibly a protein, inferred to be induced in leaves, to move to the shoot apices and there to cause the initiation of flowers.

flos ferri (*Min*) A stalactite form of the orthorhombic carbonate of calcium, *aragonite*, some of the masses resembling delicate coralline growths; deposited from hot springs.

flotation (*MinExt*) See FROTH FLOTATION.

flour (*Build, CivEng*) The fine dust incidentally formed in crushing material to be used as an aggregate.

flour (*FoodSci*) The fine meal produced when the starchy endosperm is milled or ground. In isolation 'flour' generally means wheat flour but it can also be made from potato and starch. The interaction of starch and protein following hydration and heating gives flour with its useful properties in baking etc. See EXTRACTION RATE, GLUTEN, HIGH-RATIO FLOUR, MILLING.

flow box (*Paper*) A compartment immediately before the machine wire or other forming unit, supplied with stock and designed to ensure uniform mixing within the stock and the means to control its flow onto the wire and even distribution. Also *head box*.

flowchart (*ICT*) Diagrammatic representation of the operations involved in an ALGORITHM or automated system. Flow lines indicate the sequence of operations or the flow of data, and special standard symbols are used to represent particular operations. See DATA FLOWCHART and PROGRAM FLOWCHART.

flow counter (*NucEng*) See GAS-FLOW COUNTER.

flow cytometry (*BioSci*) See FLUORESCENCE ACTIVATED CELL SORTER.

flower (*BioSci*) The reproductive structure in angiosperms consisting of a shoot axis bearing, as lateral members traditionally interpreted as modified leaves, one or more of sepals and petals, or tepals, stamens and carpels.

flowering plants (*BioSci*) The angiosperms, comprising the monocotyledons and the dicotyledons. See panel on ANGIOSPERMS (FLOWERING PLANTS).

flowers (*Print*) Small type ornaments, copied from early designs, used for building up fancy borders etc.

flowers of sulphur (*Chem*) Finely divided sulphur obtained by allowing sulphur vapour to fall on a cold surface.

flow forming (*Eng*) A metal-forming operation, in which thick blanks of aluminium, copper, brass, mild steel or titanium are made to flow plastically by rolling them under pressure to form components, often conical, having a wall thickness much less than the original blank thickness. Also *flospinning*.

flow improver (*MinExt*) Chemicals added to oil passing through a pipeline which reduce frictional losses.

flow length ratio (*Eng*) The ratio of greatest distance from gate to end of cavity, to wall thickness in an injection moulding tool. Used to choose best grade of polymer for mouldfill.

flow-line production (*Eng*) A system of mass production, and certain types of batch production, in which machines are arranged in flow lines to enable components to progress from operation to operation in correct sequence.

flow lines (*Eng*) (1) Witness marking or lines which appear on the surface of manufactured components to reveal the direction of material flow in the shaping operation. (2) Bands or structural features within a sectioned component which reveal the direction of material flow during working. (3) Lines found on exterior of moulded polymer products, due to the poor mixing of polymer melt.

flowmeter (*Phys*) Device for measuring, or giving an output signal proportional to, the rate of flow of a fluid in a pipe.

flow noise (*Acous*) Acoustic signal caused by a flow process, eg siren, ventilator noise, jet noise. Also *aerodynamic sound*.

flow-off (*Eng*) A channel cut from a riser to allow metal to escape when it has reached a predetermined height.

flow process (*Glass*) See GOB PROCESS.

flow sheet (*MinExt*) A diagram showing the sequence of operations used in a process of production with a given plan, eg the extraction and refining of metals.

flow-structure (*Geol*) A banding, often contorted, resulting from flow movements in a viscous magma, adjacent bands differing in colour and/or degree of crystallization. It is also shown by the alignment of phenocrysts or of minute crystals and crystallites, in the groundmass of lavas and, more rarely, minor intrusions. Also loosely used of metamorphic rocks.

fl oz (*Genrl*) Abbrev for FLUID OUNCE.

fl pl (*BioSci*) Abbrev for FLORE PLENO.

flu (*Med, Vet*) See EQUINE INFLUENZA, INFLUENZA, SWINE INFLUENZA.

fluconazole (*Pharmacol*) A triazole antifungal drug used for treatment of candidiasis, athlete's foot and cryptococcal meningitis.

fluctuation (*Med*) The palpable undulation of fluid in any cavity or abnormal swelling of the body.

fluctuation noise (*Acous*) Noise produced in the output circuit of an amplifier by shot and flicker effects.

fludrocortisone acetate (*Pharmacol*) A *mineralocorticoid* used in replacement therapy for adrenocortical insufficiency.

flue (*Eng*) A passage or channel through which the products of combustion of a boiler etc are taken to the chimney.

flue bridge (*Eng*) See FIREBRICK ARCH.

flue gas (*Eng*) The gaseous products of combustion from a boiler furnace, consisting chiefly of CO_2, CO, O_2, N_2 and water vapour, whose analysis is used as a check on the furnace efficiency. See CO_2 recorder.

flue gas desulphurization (*Eng*) Methods of removing sulphur dioxide from the flue gas of a plant burning sulphur-containing fuel by either absorbing onto sodium sulphite solution or washing in a limestone solution. The by-products often have little value and the lower the sulphur dioxide content required, the more expensive the process becomes.

flue-gas temperature (*Eng*) Temperature of flue gases at the point in the flue where it leaves the furnace.

flue gathering (*Build*) See GATHERING.

flueing soffit (*Build*) A flush soffit under a geometrical stair.

flue lining (*Build*) Flexible stainless-steel, fireclay or fire-resistant concrete pipe, arranged within a flue passage to protect the walls.

fluff (*Paper*) Fibres or other debris removed from the surface of the paper during printing and converting operations.

fluffing (*Print*) A tendency with soft-sized paper for fibres to be detached from the surface and edges, producing difficulties in printing, particularly if by lithography.

fluid (*Genrl*) Any substance such as a gas, liquid or powder which flows. It differs from a solid in that it can offer no permanent resistance to change of shape.

fluid bearing motor (*ICT, Eng*) A motor in which ball bearings are replaced by films of an oil whose properties are stable over a wide temperature range. Used in gyroscopes and other precision machinery, it was developed by Seagate for hard-disk drives. Ball bearings introduce non-repeatable runout (NNR) which causes eccentric tracks.

fluid bilayer model (*BioSci*) Generally accepted model for membranes in cells, basically a bilayer of phospholipids with a central hydrophobic domain. Membrane proteins, that may traverse the membrane or be limited to the inner or outer leaflet, are free to move in the plane of the membrane. Also *fluid mosaic model*.

fluid flywheel (*Eng*) A device for transmitting power through the medium of the change in momentum of a fluid, usually oil. Similar in principle to a FROUDE BRAKE in which the stator is released and forms the driven member.

fluidics (*Eng, Phys*) The science and technology of using a flow of liquid or gas to simulate electron flow in conductors and conducting plasma. The interaction of streams of fluid can thus be used for the control of instruments or industrial processes without the use of moving parts.

fluid inclusion (*Min*) See INCLUSION.

fluidity (*Phys*) The inverse of VISCOSITY.

fluidization (*ChemEng*) A technique whereby gas or vapour is passed through solids so that the mixture behaves as a liquid, and of special significance when the solid is a catalyst to induce reactions in the fluidizing medium.

fluidized bed (*ChemEng*) If a fluid is passed upwards through a bed of solids with a velocity high enough for the particles to separate and become freely supported in the fluid, the bed is said to be fluidized. The total fluid frictional force on the particles is then equal to the effective weight of the bed. Fluidized beds are used in the chemical industry because of the intimate contact between solid and gas, the high rates of heat transfer and the uniform temperatures within the bed, and the high heat transfer coefficients from the bed to the walls of the containing vessel. They are also used to ensure close contact between catalyst and vaporized feedstock in refining oil.

fluidized-bed combustion (*Eng*) The use of a FLUIDIZED BED as a furnace with small coal or crude oil as the fuel. Lower temperatures and the possible inclusion of limestone to remove sulphur oxides might make it

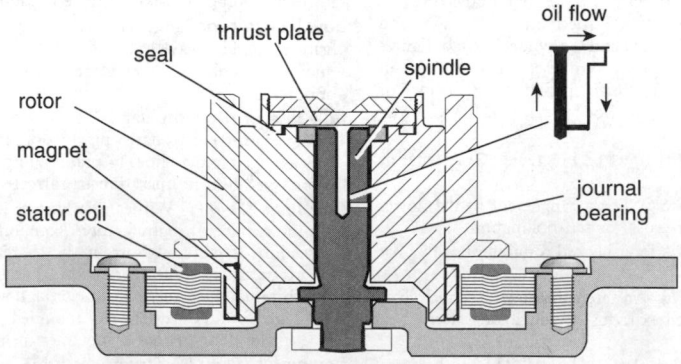

fluid bearing motor All rotating parts are hatched and the inset (top right) shows the direction of oil flow through the bearing.

attractive as a less polluting alternative to open flame furnaces.

fluidized-bed processing (*FoodSci*) Blowing hot or very cold air through discrete and fairly small food particles to dry or quick-freeze them. Used to produce 'free-flow' and 'individually quick frozen' products.

fluidized-bed reactor (*NucEng*) A reactor in which the active material is supported in a finely divided form by an upwardly moving gas or liquid, as in certain designs of nuclear reactor.

fluid lubrication (*Eng*) A state of perfect lubrication in which the bearing surfaces are completely separated by a fluid or viscous oil film which is induced and sustained by the relative motion of the surfaces.

fluid-mosaic model (*BioSci*) See FLUID BILAYER MODEL.

fluid needle (*Build*) The fluid control in a spray gun. Operated by the trigger, it seats into the fluid tip which meters and directs the fluid into the airstream.

fluid ounce (*Genrl*) Volume measure equal to 0·0284 litres ($\frac{1}{20}$ pints). US 0·0295 litres ($\frac{1}{16}$ US pints).

fluing (*Build*) A term applied to window jambs which are splayed. See SPLAYED JAMBS.

fluing arch (*Build*) See SPLAYING ARCH.

fluke (*BioSci*) (1) A semi-popular name for worms belonging to the group Trematoda. Blood flukes are responsible for SCHISTOSOMIASIS and liver flukes for FASCIOLIASIS. (2) The tail of a whale.

flukes (*Ships*) The flattened and curving points terminating the arms of an anchor.

flume (*Build, CivEng*) A metal chute used for the distribution of concrete from a placing plant or mobile concrete mixer.

flume (*MinExt*) A flat-bottomed timber trough, or other open channel, generally nowadays formed in concrete, for the conveyance of water, eg to ore-washing plant, or as a bypass.

flume stabilizer (*Ships*) A roll stabilization system using passive fluid tanks fitted athwartships, and having specifically designed nozzles to cause a phase difference of 90° between the movement of the liquid in the tank and the roll of the ship.

Fluon (*Chem*) TN for POLYTETRAFLUOROETHYLENE.

fluorapatite (*Min*) $Ca_5(PO_4)_3F$. The commonest form of APATITE.

fluorene (*Chem*) *Diphenylenemethane.* $(C_6H_4)_2CH_2$. Mp 113°C, bp 295°C. Colourless fluorescent plates; contained in coaltar; produced by leading diphenylmethane through red-hot tubes.

fluorescein (*Chem*) *1,3-dihydroxybenzene phthalein.* $C_{20}H_{12}O_5$. Red crystals which dissolve in alkalis with a red colour and green fluorescence.

fluorescein isothiocyanate (*BioSci*) Fluorescein-derivative commonly used to conjugate fluorescein with proteins or with antibodies for use in INDIRECT IMMUNOFLUORESCENCE. Abbrev FITC.

fluorescence (*Phys*) The emission of radiation, generally light, from a material during illumination by radiation of usually higher frequency, or from the impact of electrons. See PHOSPHOR. Cf PHOSPHORESCENCE.

fluorescence activated cell sorter (*BioSci*) An instrument in which cells or chromosomes in a suitable medium have their fluorescence measured as they pass down a fine tube. They are then ejected from an aperture that causes the stream to break up into fine droplets which pass between electrostatic deflector plates. Depending on the amount or nature of the fluorescence, the droplet containing the cell or chromosome is given an electrical charge which causes it to be diverted into an appropriate reservoir. Analysers measure the fluorescence but do not separate the cells and are extensively used to measure eg DNA content or antigen expression in or on the cells. Abbrev FACS. Fig. ▷

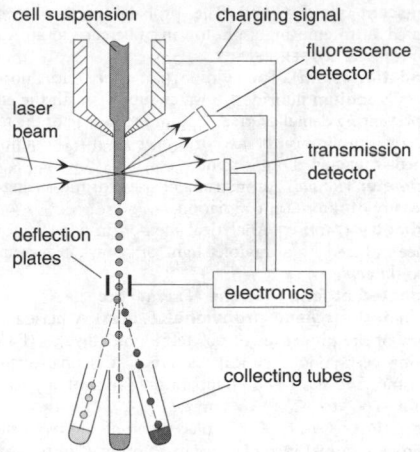

fluorescence activated cell sorter   Each droplet should contain a single cell or chromosome.

fluorescence energy transfer (*BioSci*) The transfer of energy from one fluorochrome to another. If light at the emission wavelength of the recipient fluorochrome is detected, it implies that the two fluorochromes were physically very close and the techniques is used to probe protein–protein interactions. Also *fluorescence resonant energy transfer*, abbrev *FRET*.

fluorescence in situ hybridization (*BioSci*) A technique of directly mapping the position of a gene or DNA clone within a genome by *in situ* hybridization to metaphase spreads. The DNA probe is labelled with a fluorophore, and the hybridization sites visualized by epifluorescence microscopy. Several probes can be used at one time, to mark specific chromosomes with different coloured fluorophores (*chromosome painting*). Abbrev *FISH*.

fluorescence microscopy (*BioSci*) Light microscopy in which the specimen is irradiated at wavelengths which will excite the natural or artificially introduced fluorochromes. An optical filter absorbs the exciting wavelengths but transmits the fluorescent image which can be studied normally.

fluorescence recovery (*BioSci*) The method for measuring the fluidity or viscosity of cell surfaces. If an area of fluorochrome-labelled cell or organism is photobleached by laser illumination, the bleached (dark) patch regains fluorescence when unbleached, labelled molecules move sideways in the cell membrane and repopulate the area. Also *fluorescence recovery after photobleaching*, abbrev *FRAP*.

fluorescent brightener (*Textiles*) See BRIGHTENING AGENT.

fluorescent lamp (*ElecEng*) A mercury-vapour electric-discharge lamp having the inside of the bulb or tube coated with fluorescent material so that ultraviolet radiation from the discharge is converted to light of an acceptable colour.

fluorescent penetrant inspection (*Eng*) PENETRANT FLAW DETECTION with a fluorescent dye.

fluorescent screen (*Electronics*) One coated with a layer of luminescent material so that it fluoresces when excited, eg by X-rays or cathode rays. See FLUOROSCOPY, PHOSPHOR.

fluorescent whitening agents (*Chem*) Special dyes widely used to 'whiten' textiles, paper, etc, and sometimes incorporated in detergents. Their effect is based on ability to convert invisible ultraviolet light into visible blue light, giving fabrics greater uniformity of reflectance over the visible part of the spectrum. Also *optical bleaches*, *optical whites*.

fluorescent yield (*Phys*) The probability of a specific excited atom emitting a photon in preference to an Auger electron. See AUGER EFFECT.

fluoridation (*Med*) The addition of inorganic fluorides (usually sodium fluoride) to water supplies with the effect of preventing dental CARIES. It is said to be one of the most effective public health measures ever taken. The amount added is usually about 1 ppm.

fluorimeter (*Radiol*) An instrument used for measuring the intensity of fluorescent radiation.

fluorimetry (*MinExt*) Analytical method in which fluorescence induced by ultraviolet light or X-rays is measured. Also *fluoremetry*.

fluorinated ethene propene (*Plastics*) See FEP.

fluorinated ethylene propylene (*Plastics*) A plastic with many of the properties of polytetrafluoroethylene (PTFE), having very good chemical resistance; it is unaffected by moisture and has a wide temperature range of application from $-260°$ to $+200°C$. Abbrev *FEP*.

fluorination (*Chem*) The replacement of atoms (usually hydrogen atoms) in molecules by fluorine. Can be carried out catalytically using fluorine in the vapour phase or by using fluorine 'carriers' such as cobaltic fluoride or silver fluoride in solution with increased temperature.

fluorine (*Chem*) Symbol F, at no 9, ram 18·9984, valency 1, mp $-223°C$, bp $-187°C$, density $1·696 \, g \, dm^{-3}$ at stp. The lightest halogen and the most electronegative (non-metallic) element, molecular fluorine is a pale greenish-yellow diatomic gas. Discovered by Scheele in 1771 and isolated by Moissan in 1886, fluorine is chemically highly reactive and thus is never found as the free element. Its abundance in the Earth's crust is 544 ppm and in sea water 1·3 ppm. Its ionic size (1·33 Å) is almost identical with those of OH^- and O^{2-}, and it enters into many silicate minerals having OH groups in their structures. There are many independent fluorine minerals of which the most important is fluorite, CaF_2, and it is also present in apatite. Fluorine has an essential biological role, toxic in excess. Used in separating the isotopes of uranium (see URANIUM HEXAFLUORIDE). Combines with carbon to form inert polymers with low coefficient of friction, eg PTFE (Teflon). See FLUOROCARBONS.

fluorite (*Min*) Calcium fluoride, CaF_2, crystallizing in the cubic system, commonly in simple cubes. Occasionally colourless, yellow, green, but typically purple; the coloured varieties may fluoresce strongly in ultraviolet light. Also *fluorspar*. See BLUE JOHN and panel on TWINNED CRYSTALS.

fluoroboric acid (*Chem*) A complex monobasic acid formed by the combination of hydrogen fluoride and boron (III) fluoride. Salts called *fluoroborates*.

fluorocarbons (*Chem*) Hydrocarbons in which some or all of the hydrogen atoms have been replaced by fluorine. The fluorinated derivatives of methane are widely used as refrigerating agents and propellants for aerosols. See FREONS, POLYTETRAFLUOROETHENE.

fluorochrome (*Chem*) A molecule or chemical group which fluoresces on irradiation. Can be made to bind to a specific site and thus localize it.

fluorography (*Radiol*) The photography of fluoroscopic images.

fluorophore (*Chem*) A group of atoms which give a molecule fluorescent properties.

fluoropolymers (*Chem*) All those organic polymers containing fluorine atoms and hence showing some degree of heat and solvent resistance, polytrichlorofluoroethylene, polytetrafluoroethylene, Viton rubber, etc.

fluoroscope (*Radiol*) Measurement system for examining fluorescence optically. Fluorescent screen assembly used in FLUOROSCOPY.

fluoroscopy (*Eng, Radiol*) The examination of objects by observing their X-ray shadow on a fluorescent screen; used to examine contents of luggage, packages without unwrapping, quality of welding, etc.

fluorosis (*Med*) Chronic poisoning with fluorine.

fluorouracil (*Pharmacol*) 5-*fluorouracil*. A cytotoxic antimetabolite used in the treatment of solid tumours, especially colorectal and breast carcinoma.

fluorspar (*Min*) See FLUORITE.

Fluothane (*Chem*) See HALOTHANE.

fluoxetine (*Pharmacol*) An antidepressant drug of the *SSRI* class best known by its proprietary name *Prozac*.

flush (*BioSci*) Area of land, fed by groundwater. If the water is eutrophic it may enrich the soil locally, but not if the water is oligotrophic.

flush (*Build*) In the same plane.

flush bead (*Build*) A sunk bead, finished so as to be level with the surface which it decorates.

flush boards (*Print*) A method of binding in which boards are drawn on and trimmed with the book. The covers are then flush with the page edges.

flush-bolt (*Build*) A sliding bolt sunk into the side or edge of a door so as to be flush with the surface.

flushing (*Build*) A crushing of the edges of a stone at a HOLLOW BED, due to excessive pressure upon them.

flushing (*MinExt*) The operation of clearing off accumulation of fire damp or noxious gases underground by means of air currents. See FACE-AIRING.

flushing of ewes (*Agri*) The practice of increasing the nutritional plane of ewes a few weeks before they are served by the ram, with the object of increasing fertility. It is done by sharply reducing food intake for a brief period and then restoring to better grazing and supplementing with concentrates.

flushing tank (*Build*) A tank used to accumulate the water for flushing a drain or sewer which is not laid at a self-cleansing gradient. The discharge is often effected automatically by a siphoning device.

flush joint (*Build*) The type of mortar joint made in FLAT POINTING.

flush panel (*Build*) A panel whose surface is in line with the faces of the stiles.

flush plate (*ElecEng*) See SWITCH PLATE.

flush soffit (*Build*) The continuous surface formed under any ceiling or stair etc.

flush switch (*ElecEng*) A switch which can be mounted flush with the wall. Also *panel switch, recessed switch*.

flute (*Build*) A long vertical groove, usually circular in form, in the surface of a column etc. Also *fluting*.

flute cast (*Geol*) A hollow eroded by turbulent flow and subsequently filled by sediment. More properly called a *mould*, it is common in TURBIDITE deposits, and can be used to determine the direction of flow of the depositing currents.

fluticasone proprionate (*Pharmacol*) A synthetic *glucocorticoid* used by inhalation for treatment of asthma.

fluting (*Build*) See FLUTE.

fluting (*Eng*) Parallel channels or grooves, longitudinal or helical, cut in a cylindrical object such as a tap or reamer.

fluting plane (*Build*) A plane for cutting round-bottomed grooves.

flutter (*Acous*) Rapid fluctuation of frequency or amplitude.

flutter (*Aero*) Sustained oscillation, usually on wing, fin or tail, caused by interaction of aerodynamic forces, elastic reactions and inertia, which rapidly break the structure. *Asymmetrical flutter* occurs where the port and starboard sides of the aircraft simultaneously undergo unequal displacements in opposite directions, as opposed to *symmetrical flutter*, where the displacements and their direction are the same; *classical* or *coupled flutter* is due solely to the inertial, aerodynamic or elastic coupling of two or more degrees of freedom.

flutter (*ImageTech*) Variation in brightness of a reproduced picture, arising from additional radio reflection from a moving object, eg an aircraft.

flutter (*Med*) An abnormality of cardiac rhythm, in which the atrium of the heart contracts regularly at a greatly increased frequency (between 180 and 400 beats per minute), the ventricles contracting at a slower rate.

flutter echo (*Acous*) See MULTIPLE ECHO.

flutter speed (*Aero*) The lowest EQUIVALENT AIRSPEED at which flutter can occur.

fluvial (*BioSci, Geol*) Of, occurring in or formed by the flow of a river or stream. Also *fluviatile*.

fluviatile deposits (*Geol*) Sand and gravel deposited in the bed of a river.

fluvioglacial (*Geol*) Relating to the meltwater streams flowing from a glacier; the deposits and landforms produced by such streams, eg KAMES.

fluviomarine (*BioSci*) Able to live in rivers and in the sea, as the salmon.

fluvioterrestrial (*BioSci*) Found in rivers and on their banks, as the otter.

flux (*Chem*) A substance added to a solid to increase its fusibility, eg that added to a joint prior to welding, soldering or brazing which improves wetting by the filler, prevents oxidation of the heated surfaces and dissolves existing infusible oxide films.

flux (*Eng*) The material added to a furnace charge, which combines with those constituents not wanted in the final product and issues as a separate slag.

flux (*MathSci*) Through a surface in a vector field: the integral over the surface of the product of the elementary area by the normal component of the vector, ie $\int F \cdot ds$.

flux (*Phys*) The rate of flow of mass, volume or energy per unit cross-section normal to the direction of flow and therefore a vector field quantity, eg ELECTRIC FLUX, MAGNETIC FLUX.

flux density (*Phys*) (1) The number of photons (or particles) passing through unit area normal to the beam, or the energy of the radiation passing through this area. (2) See DISPLACEMENT. (3) See MAGNETIC FLUX DENSITY.

flux gate (*ElecEng*) Magnetic reproducing head in which magnetic flux (due to flux leakage from signals recorded on magnetic tape) is modulated by high-frequency saturating magnetic flux in another part of the magnetic circuit.

flux-gate compass (*Eng*) A device in which the balance of currents in windings is affected by the Earth's magnetic field.

flux guidance (*ElecEng*) Directing the electric or magnetic flux in high-frequency heating by shaped electrodes or magnetic materials, respectively.

flux link (*ElecEng*) Conservative flux across a surface bounded by a conducting turn. For a coil, *flux linkage* is the integration of the flux with individual turns.

fluxmeter (*ElecEng*) Electrical instrument for measuring total quantity of magnetic flux linked with a circuit; it consists of a search coil placed in the magnetic field under investigation, and a ballistic galvanometer, or an uncontrolled moving-coil element (Grassot) or a semiconductor probe generating Hall voltage. See GAUSSMETER.

flux quantization (*Phys*) A magnetic effect in superconductors. A ring of material is placed in a uniform magnetic field and then cooled below its critical temperature so that it becomes superconducting. When the external field is removed it is found that the ring has trapped the field in its hole. If the flux of the trapped field is measured as a function of the strength of the applied field, it is found to be quantized in steps of $(h/2e)$ where h is PLANCK'S CONSTANT and e is the electronic charge. This shows superconductivity to be a quantum effect.

fly (*Textiles*) Fine short fibres that escape into the atmosphere esp in the carding of cotton.

fly ash (*Build*) A fine ash from the pulverized fuel burned in power stations, used in brick-making and as a partial substitute for cement in concrete.

flyback (*ImageTech, Radar*) The return of the scanning beam to its starting point at the completion of a radar trace or a line of a TV picture, the line being blanked out during the process. Also *retrace*.

fly before buy (*Aero*) The process of procuring new military aircraft by flying a prototype prior to giving the production order. An alternative is to order 'off the drawing board',

which shortens delivery time at the risk of inadequate performance or delays in fixing inadequacies found during flight test.

fly-blown (*Vet*) Affected by MYIASIS.

fly-by (*Space*) A type of space mission where the spacecraft passes close to the target but does not rendezvous with it, orbit around it or land on it.

fly-by-light (*Aero*) Flight control system in which the signalling is performed by coherent light beams travelling in optical fibres.

fly-by-wire (*Aero*) Flight control system using electric/ electronic signalling.

fly cutter (*Eng*) A single-point tool used in a milling machine to produce a flat surface.

fly disease (*Vet*) See NAGANA.

flyer (*Build*) See FLYING SCAFFOLD.

fly hand (*Print*) A rotary press assistant who removes the printed product from the delivery or the conveyor.

flying boat (*Aero*) A seaplane wherein the main body or hull provides water support.

flying bond (*Build*) See MONK BOND.

flying buttress (*Arch*) An arched buttress giving support to the foot of another arch. Also *arc-boutant, arch(ed) buttress*.

flying deck (*Ships*) See HURRICANE DECK.

flying erase head (*ImageTech*) A head mounted on the DRUM which erases the video track immediately before the video record head. Also *rotary erase head*.

flying levels (*Surv*) Back-sight and fore-sight readings taken between any two points, without reference to bench marks, when only the difference of level of the points is required.

flying paster (*Print*) See AUTOMATIC REEL CHANGE.

flying scaffold (*Build*) A support, independent from the ground, between two buildings, used temporarily after the removal of one building from a row. Also *flier, flyer, flying shore, suspended scaffold*.

flying shore (*Build*) See FLYING SCAFFOLD.

flying speed (*Aero*) The *maximum flying speed* is the highest attainable speed in level flight, under specified conditions and corrected to standard atmosphere. The *minimum flying speed* is the lowest speed at which level flight can be maintained.

flying-spot microscope (*BioSci*) A type of light microscope in which the object is scanned in two dimensions by a light spot formed by the diminished image of a cathode-ray tube placed at the eyepiece plane of a compound microscope. All the transmitted energy can be collected by a photomultiplier and an image, suitable for electronic analysis, reconstructed using the timing circuits driving the cathode-ray tube. The SCANNING ELECTRON MICROSCOPE is an analogous instrument.

flying-spot scanner (*ImageTech*) A device used in TELECINE equipment which scans film with a spot of light generated on a cathode-ray tube with the modulated transmitted light received by a line-array CHARGE-COUPLED DEVICE or photosensor. See FILM SCANNER.

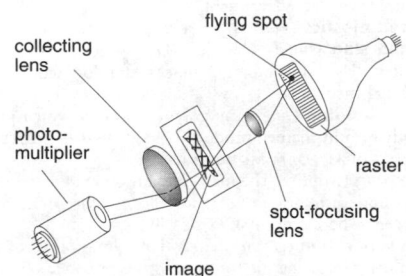

flying-spot scanner The collecting lens focuses the spot-focusing lens aperture onto the photocathode.

flying tail (*Aero*) See ALL-MOVING TAIL.

flying tuck (*Print*) A gear-driven rotating folding blade mounted on a cylinder on web-fed presses.

fly leaf (*Print*) A blank leaf at the beginning and at the end of a bound volume. It may be part of an endpaper.

fly nut (*Eng*) See WING NUT.

fly press (*Eng*) A hand-operated press for punching holes, making driving fits, etc; it consists of a bed supporting a vertical frame through which a square-threaded screw is fitted. The screw is turned by a cross-piece terminating in one or two heavy steel balls, for giving additional impetus to the descent of the die attached to the bottom end.

flysch (*Geol*) Sediments derived from the Alpine orogeny. More generally applied to almost any TURBIDITE deposits, derived from developing, large-scale, fold structures.

fly shuttle (*Textiles*) A mechanism, invented by John Kay (Bury, Lancashire) in 1733, for propelling the shuttle across the loom. It superseded hand-shuttling.

flywheel (*Eng*) A heavy wheel attached to a shaft (eg an engine crankshaft) either to reduce the speed fluctuation resulting from uneven torque, or to store up kinetic energy to be used in driving a punch, shears, etc, during a short interval.

flywheel effect (*ImageTech*) Maintenance of synchronization in a TV receiver by internal circuits which can operate for a short time in the absence of transmitted sychronizing signals, to assist under adverse reception conditions.

FM (*Phys*) Abbrev for FREQUENCY MODULATION.

Fm (*Chem*) Symbol for FERMIUM.

FMEA (*Aero*) Abbrev for FAILURE MODES AND EFFECTS ANALYSIS.

fMRI (*Med*) Abbrev for *functional magnetic resonance imaging*, a technique used to determine which parts of the brain are activated by particular stimuli or processes by using MRI to identify areas of increased blood flow.

FMS (*Eng*) Abbrev for FLEXIBLE MANUFACTURING SYSTEM.

FMS (*Psych*) Abbrev for FALSE MEMORY SYNDROME.

FMV (*ImageTech*) Abbrev for FULL-MOTION VIDEO.

f-number (*ImageTech*) The relative APERTURE of a lens, particularly when stopped down, representing its light transmission; it expresses the diameter of the lens diaphragm as a fraction of its focal length, eg f/8, also written f:8 or f8. Also *aperture number*. See APERTURE, STOP.

foam (*Chem*) A class of CELLULAR SOLID which can be thought of as a type of COMPOSITE material whose matrix incorporates a gas rather than a solid. Polymers (see EXPANDED PLASTICS), metals, ceramics and glass (see CELLULAR GLASS) can all be foamed, variously by blowing or beating the gas into the melt, by chemical reactions evolving a gas, by using hollow fillers (*syntactic foam*) or by dissolving or melting out a particulate second phase. Used for eg buoyancy, shock absorption, cushioning, thermal insulation and, in sandwich panels, for enhanced flexural rigidity with small weight penalty. See RETICULATED FOAM.

foambacked fabric (*Textiles*) Dress and furnishing fabrics bonded on the back to polyether or polyester foams by adhesive or flame treatment.

foamed plastics (*Plastics*) See EXPANDED PLASTICS.

foamed slag (*Build*) Blast furnace slag aerated while still molten. Used for building blocks and for acoustic and thermal insulation.

foam moulding (*Eng*) Any polymer process which gives a partly or fully foamed product, esp SANDWICH MOULDING, STRUCTURAL FOAM MOULDING.

foam plug (*MinExt*) The mass of foam generated and blown into underground workings to seal off a fire or keep out oxygen, where a fire risk exists.

foam separation (*Chem*) Removal of solutes or ions from a liquid by bubbling air through in the presence of surface active agents which tend to be adsorbed on to the bubbles. Cf FROTH FLOTATION, for larger particles.

focal adhesions (*BioSci*) Areas of close apposition, anchorage points, of the plasma membrane of a cell to the substratum over which it is moving. Usually $1 \times 0.2 \ \mu m$ with the long axis parallel to the direction of movement; always associated with a cytoplasmic microfilament bundle.

focal length (*Phys*) For a lens, the distance measured along the principal axis, between the principal focus and the second principal point. In a thin lens both principal points may be taken to coincide with the centre of the lens. See BACK FOCUS, CONVENTION OF SIGNS, EQUIVALENT FOCAL LENGTH.

focal plane (*ImageTech, Phys*) The plane, at right angles to the principal axis of a lens or lens system, in which the image of a particular object is formed. The principal focal plane passes through the principal focus, and contains the images of objects at infinity. It is the normal location of a film or plate, a ground-glass focusing screen or a PICKUP DEVICE.

focal-plane shutter (*ImageTech*) Camera shutter in the form of a blind with a slot, which is pulled rapidly across, and as close as practicable to, the film or plate, exposure time being varied by adjusting the width of the slot. Called *self-capping* because the slot is closed during retensioning.

focal point (*Phys*) The focal spot formed on the axis of a lens or curved mirror by a parallel beam of incident radiation. In its general form, this definition includes acoustic lenses, electron lenses, and lenses or mirrors designed for use with radio waves, infrared or ultraviolet radiation.

focal spot (*Phys*) A spot onto which a beam of light or charged particles converges. See X-RAY FOCAL SPOT.

focometer (*Phys*) An instrument for measuring the focal length of a lens.

focus (*Geol*) See panel on EARTHQUAKE.

focus (*MathSci*) Of a conic: a point on the convex side of a conic in terms of which the locus of points constituting the conic is defined. The two lines of every pair of conjugate lines through it are mutually perpendicular. The ellipse and hyperbola each have two real and two imaginary foci, and the parabola has one real focus. See CONIC for alternative definition. Of a quadric: a point not on the quadric, such that the three planes of every set of three mutually conjugate planes through it are mutually perpendicular.

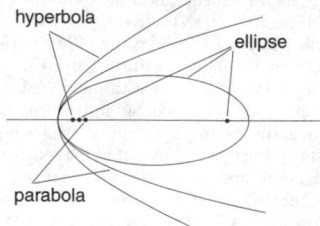

focus Indicated with the appropriate curve.

focus (*Phys*) A point to which rays converge after having passed through an optical system, or a point from which such rays appear to diverge. In the first case the focus is said to be *real*; in the second case, *virtual*. The *principal focus* is the focus for a beam of light rays parallel to the principal axis of a lens or spherical mirror.

focusing (*ImageTech*) Adjustment of the axial position of a camera lens to form a sharp image in the required plane, such as the film or plate in photography or the camera tube target in TV.

focusing (*Phys*) The convergence to a point of: (1) beams of electromagnetic radiation; (2) charged particle beams; or (3) sound or ultrasonic beams.

focusing coil (*ElecEng*) One used to focus a charged particle beam by a magnetic or electrostatic field. Also *focusing electrode*.

focusing screen (*ImageTech*) A screen, usually of ground glass, located in the place of a film or plate, or on the top of a reflex camera, for adjusting the focusing of the lens before exposure.

focus skin distance (*Radiol*) The distance from the focus of an X-ray tube to the surface of incidence on a patient, usually measured along the beam axis. Abbrev *FSD*.

fodder (*Agri*) A crop presented as animal food, either fresh or after storage.

foetus (*BioSci*) Obsolete spelling of FETUS.

fog (*EnvSci*) Minute water droplets with radii in the range 1 to 10 μm suspended in the atmosphere and reducing visibility to below 1 km (1100 yd in UK).

fog (*ImageTech*) An overall density in a photographic record not related to the exposed image. It may be caused by an unwanted exposure to light or radiation such as X-rays, by incorrect chemical processing or by protracted and unsuitable storage.

fogbow (*EnvSci*) A bow seen opposite the Sun in fog. The bow is similar to the rainbow, but the colours are faint, or even absent, owing to the smallness of the drops, which causes diffraction scattering of the light.

fog fever (*Vet*) An acute respiratory distress syndrome of cattle which usually comes on within 2 weeks of introduction to lush pasture. Most common in suckler herds where morbidity may reach 50%. Also *Atypical interstitial pneumonia*.

fogging (*Agri*) The application of water or pesticide solution to plants as a controlled, fine mist.

fog levels (*ImageTech*) The minimum density of a processed photographic image in an unexposed area.

fog signal (*CivEng*) A detonating cap which is placed on a rail before the passage of a train, so that the detonation occurring when a wheel passes over it will serve as a signal to the driver in bad visibility.

fog-type insulator (*ElecEng*) A type of overhead-line insulator having long leakage distances; specially designed for areas in which fog is prevalent.

föhn wind (*EnvSci*) A warm dry wind which blows to the lee of a mountain range. It is prevalent on the northern slopes of the Alps.

foid (*Min*) A term meaning FELDSPATHOID used by international agreement on rock classification.

foidite (*Min*) Internationally ('IUGS') agreed usage for volcanic rocks containing more than 60% feldspathoids ('foid') by volume among light-coloured constituents. The most abundant feldspathoid name should be used if possible, eg nephelinite, leucitite, etc. See VOLCANIC ROCKS.

foidolite (*Min*) Internationally ('IUGS') agreed usage for plutonic rocks containing more than 60% feldspathoids ('foid') by volume among light-coloured constituents. The most abundant feldspathoidal name should be used if possible, eg nephelinolite, leucitolite, etc. See PLUTONIC ROCKS.

FOIRL (*ICT*) Abbrev for FIBRE OPTIC INTER-REPEATER LINK.

foldback DNA (*BioSci*) Sequence complementarity that allows a single-stranded molecule to form secondary structure. HAIRPIN DNA is one form with a minimal loop at its end.

folded dipole (*ICT*) A DIPOLE antenna with a separate, parallel rod or wire connected from one tip to another of the original dipole; done to increase the impedance at the feed point, for matching purposes.

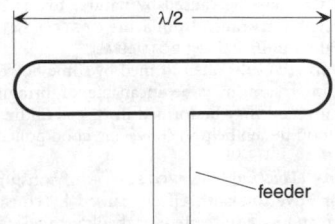

folded dipole Size determined by wavelength (λ).

folded horn (*Acous*) An acoustic horn which is turned back in itself to reduce necessary space.

folded yarns (*Textiles*) Yarns formed from two or more single yarns twisted together for strength or special appearance. The products are known as two-, three-, four-, etc, ply yarns. See DOUBLING FRAME.

folder (*Print*) The section of a web-fed press where the webs are associated, folded, cut, delivered and sometimes stitched. See ASSOCIATION.

folding (*Geol*) Folding (or bending) of strata is usually the result of compression that causes the formation of the geological structures known as anticlines, synclines, monoclines, isoclines, etc. The amplitude (ie vertical distance from crest to trough) of a fold ranges from a centimetre to thousands of metres.

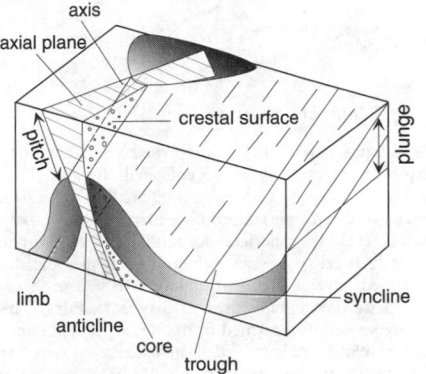

folding Showing main terms.

folding blade (*Print*) A metal strip which thrusts the web or webs into jaws or rollers to produce a fold. Also *tucker, tucking blade*.

folding cylinder (*Print*) The cylinder which holds the folder blade or jaw for making a fold on web-fed presses.

folding jaws (*Print*) The gripping section on a NIP AND TUCK FOLDER.

folding plates (*Print*) (1) Adjustable parts on a folding machine into which the paper travels and, being stopped, receives a BUCKLE FOLD. (2) Large-sized illustrations which require folding before inclusion in the book.

folding rollers (*Print*) Driven rollers on folding machines, the paper being folded as it passes between them.

folding strength (*Paper*) The number of double folds needed to break a test strip of the paper, under prescribed conditions.

folding wedges (*Build, CivEng*) STRIKING WEDGES used for tightening and easing shoring and centring, and in some joint construction in joinery.

fold-out (*Print*) A large leaf in a book which must be *folded out* when the book is being used.

foliaceous (*BioSci*) (1) Flat and leaf-like. (2) Bearing leaves. Also *foliose*. Cf CRUSTOSE, FRUTICOSE.

foliar feeding (*BioSci*) Supplying mineral elements in solution by watering them onto the foliage; useful where soil conditions prevent uptake through the roots, as in LIME-INDUCED CHOROSIS.

foliar gap (*BioSci*) See LEAF GAP.

foliar trace (*BioSci*) See LEAF TRACE.

foliation (*Geol*) The arrangement of minerals normally possessing a platy habit (such as the micas, chlorites and talc) in folia and leaves, lying with their principal faces and cleavages in parallel planes; such arrangement is due to development under great pressure during regional metamorphism.

folic acid (*Chem*) See vitamin B complex in panel on VITAMINS.

folio (*Print*) (1) A sheet of paper folded in half. (2) A book made up of sheets folded once, so having four pages to the sheet. (3) The number of a page.

foliose (*BioSci*) See FOLIACEOUS.

folium of Descartes (*MathSci*) The curve defined by the cartesian equation, $x^3 + y^3 = 3axy$.

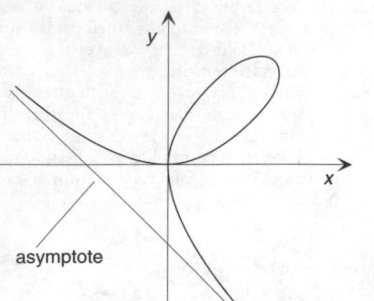

folium of Descartes

folk weave (*Textiles*) A loosely woven rough fabric made from coarse yarns including coloured ones.

follicle (*BioSci*) (1) Generally, any small sac-like structure, such as the pit surrounding a hair root. See GRAAFIAN FOLLICLE. (2) Spherical accumulations of lymphocytes (largely B-cells) present in lymphoid tissues that become enlarged after antigenic stimulation to become secondary follicles or germinal centres. (3) In plants, a dry, dehiscent, many-seeded fruit formed from one carpel, dehiscing along one side only. Adjs *follicular, folliculose*.

follicle-stimulating hormone (*Med*) A gonadotrophic mucoprotein hormone secreted by the basophil cells of the pars anterior of the pituitary gland, which stimulates growth of the Graafian follicles of the ovary and spermatogenesis. Abbrev *FSH*.

follicular dendritic cells (*BioSci*) Cells of uncertain (mesenchymal or haematopoietic) lineage found in germinal centres. These cells are very important in the selection of functional B-lymphocytes by their AFFINITY for antigen.

follicular mange (*Vet*) See DEMODECTIC MANGE.

folliculitis (*Med*) Inflammation of a follicle, esp of the ovary.

folliculoma (*Med*) A tumour arising from cells in the GRAAFIAN FOLLICLE of the ovary.

follower (*CivEng*) An intermediate length of timber which transmits the blow from the pile hammer to the PILE; used when driving below water level. Also *long dolly*. See DOLLY.

follower (*Eng*) (1) That which follows the profile of a CAM. (2) The driven wheel of a pair of gears engaging with each other.

follower (*Surv*) A chain assistant who has charge of the rear end of a chain and is responsible for lining-in the leader at each chain's length.

followers (*Agri*) Immature livestock mixed in with the commercial herd or flock. Specifically, young cattle being reared as replacements to dairy or beef cow herds.

follow focus (*ImageTech*) Adjusting lens focus during action at varying distances to maintain a sharp image.

following dirt (*MinExt*) A thin bed of loose shale above coal; a parting between the top of a coal seam and the roof. Also *following stone*. See PUG.

following response (*Psych*) The response in which newly hatched presocial birds follow a moving object during a fairly brief period soon after hatching; under natural conditions this would be the parents or siblings. See IMPRINTING.

follow-me (*ICT*) A service provided by an INTELLIGENT NETWORK in which a caller can send instructions from a telephone which he or she is visiting, telling the network to contact his or her home exchange and instruct it to divert calls to the visited telephone.

follow-rest (*Eng*) A supporting member attached to the rear of the saddle of a lathe to steady the workpiece or bar material against the cutting pressure. See STEADY.

folly (*Arch*) A structure, usually a tower or sham ruin built in a GOTHIC or CLASSICAL style, the purpose of which was to enhance a view. Follies were built generally towards the end of the eighteenth century, when landscaping became very popular. See PICTURESQUE.

Fomalhaut (*Astron*) A bright white star in the constellation Piscis Austrinus, the 18th brightest in the sky. Distance 7·0 pc. Also *Alpha Piscis Austrini*.

fomes (*Med*) Any infected inanimate object other than food. Pl *fomites*. Fomites such as clothing, bedding, etc, may convey infection from one person to another.

font (*ICT, Print*) A set of printing or display characters in a particular TYPEFACE, size (POINT size) and weight, ie bold, normal or italic; eg Times 18-pt bold is a common font for column headings in a newspaper. In modern electronic typesetting in which any size of type can be scaled from an outline usually held at 1-pt size, the term is increasingly applied to a particular design and style irrespective of size. Also *fount*.

fontanelle (*BioSci*) In Craniata, a gap or space in the roof of the cranium.

font cartridge (*ICT*) A piece of hardware containing an EPROM that is plugged into a printer to supply it with a set of FONTS for printing purposes. The cartridge may be changed by the user to provide different sets of fonts as required.

font management system (*Print*) Character-generating software which controls the format of typefaces produced by a non-impact printer.

food allergy (*FoodSci, Med*) The presence of ALLERGENS in food that provoke immunological, ie hypersensitivity, reactions, most commonly characterized by skin rashes, swelling or respiratory problems, eg infants may be allergic to bovine milk.

food body (*BioSci*) A soft mass of cells, containing oil and other nutrient substances, attached to the outside of the seed coat; it is eaten by ants, which drag the seed along, leave it when they have eaten the food body, and so assist dispersal.

food chain (*BioSci*) A sequence of organisms dependent on each other for food. The number of links in the chain is usually only four or five. See FOOD WEB.

food colouring (*FoodSci*) An ADDITIVE used in a food, particularly sugar and flour confectionery and soft drinks, to enhance its colour. In the past, AZO DYES have been extensively used, but the modern trend is to encourage the use of NATURAL COLOURS.

food groove (*BioSci*) A groove along which food is passed to the mouth; median and ventral in Branchiopoda; along the edge of each ctenidium in Bivalvia.

food intolerance (*FoodSci*) The inability to digest certain foods, often causing nausea, vomiting or diarrhoea.

food irradiation (*FoodSci*) The use of ionizing radiation to destroy micro-organisms in food products. See panel on FOOD IRRADIATION.

food poisoning (*FoodSci*) Harmful effect caused by bacterial or chemical contamination of food. Toxins may be formed in food by the growth of micro-organisms, or be present as chemical residues or contaminants (eg heavy metals). Can also be caused by natural toxins present in foods. The effects range from acute GASTRO-ENTERITIS to life-threatening illness (eg BOTULISM).

food pollen (*BioSci*) Pollen formed by some flowers, which attracts insects; it may be incapable of bringing about fertilization and may be formed in special anthers. Insects seeking food pollen help in conveying good pollen to other flowers.

food safety (*FoodSci*) The PROTECTION of consumers from injury or adverse health effects caused by consuming or handling spoilt, adulterated or badly stored foods. See GOOD MANUFACTURING PRACTICE.

food spoilage (*FoodSci*) Changes leading to food becoming unpalatable, indigestible or unsafe for consumption, brought about by chemical reactions, enzyme activity or

Food irradiation

Gamma rays are able to kill bacteria and yeast and high-energy sources, using cobalt-60, have been used for a long time to sterilize medical apparatus like syringes, tubing and dialysis equipment. It was therefore natural to consider whether similar methods could be used to treat certain foodstuffs like spices, which have in the past been disinfested with ethylene oxide gas, particularly because gamma-ray radiation cannot induce radioactivity in the food. The radiation dosage required depends on the desired effect and is expressed in *kilograys* in the following table; 1 gray is the *absorbed dose* and is equivalent to 1 joule absorbed by 1 kilogram of tissue.

A major problem for the public acceptability of food irradiation has been the difficulty of finding a reliable test to distinguish irradiated food. It has proved very difficult to detect consistent changes due to irradiation, which lasts the lifetime of the food. However, many foods are covered with minute quantities of mineral matter deposited from the soil or by the wind, and these have proved very sensitive indicators of any previous irradiation.

A few micrograms of common feldspars and clay minerals are, after irradiation, sufficient to contribute reproducible signals when tested by thermoluminescence and photoluminescence. Radiation, absorbed by these minerals, causes electrons and ions to become trapped in the mineral structure, where they remain in a METASTABLE STATE until released either by heat or by a strong beam of light at defined wavelengths. The stored energy in the crystals is emitted as light of a wavelength different from the incident light, in a process called LUMINESCENCE. In the laboratory the mineral deposits are carefully removed from the sample and then tested for luminescence. A high signal indicates irradiation but a low signal might mean that the mineral was insensitive. It is then necessary to irradiate part of the sample specially kept back to see if this shows luminescence. If it does, the foodstuff was definitely not irradiated previously.

An objective method of detecting any past history of irradiation should allow a more direct comparison of radiation compared with chemical, heat or other methods of preserving and decontaminating food.

Purpose of irradiation	Dose in kilograys
Inhibition of sprouting	0.05–0.15
Delay in ripening	0.5–1
Insect disinfestation	0.15–0.5
Extension of shelf life	1–3
Eliminating spoilage and pathogens	1–7
Improving appearance and odour	1–7

microbial growth and prevented or retarded by good hygiene or food processing such as heating, chilling, effective barrier packaging and aseptic packing.

food vacuole (*BioSci*) In the cytoplasm of some Protozoa, a space surrounding a food particle and filled with fluid.

food web (*BioSci*) An interlocking pattern formed by a series of interconnected food chains.

foolscap (*Paper*) A superseded size of writing and printing paper 13 × 17 in (US 13 × 16 in).

foolscap octavo (*Print*) A book size, $6\frac{1}{2} \times 4\frac{1}{8}$ in (159 × 105 mm).

fool's gold (*Min*) See PYRITE.

foot (*BioSci*) (1) Generally in animals, a locomotor appendage. (2) In Crustacea, any appendage used for swimming or walking. (3) In Arachnida, Myriapoda and Insecta, the tarsus. (4) In Echinodermata, the podia (see PODIUM). (5) In Mollusca, a median ventral muscular mass, used for fixation or locomotion. (6) In land vertebrates, the podium of the hind-limb, or of all limbs in Tetrapoda. (7) In plants, the basal part of an embryo, a developing sporophyte or a spore-producing body, embedded in the parental tissues and serving to absorb nutrients.

foot (*Genrl*) The unit of linear measure equal to 0·3048 m. Symbol ft.

foot-and-mouth disease (*Vet*) An acute febrile contagious disease of cloven-footed animals, due to infection by a virus; characterized by a vesicular eruption on the mucous membrane and skin, esp in the mouth and in the clefts of the feet. Controlled by slaughter in the UK and by vaccines where it is endemic. Also *aphthous fever*.

foot block (*Build*) An ARCHITRAVE BLOCK.

foot-board (*CivEng*) See FOOT-PLATE.

foot bolt (*Build*) A robust form of tower bolt, fixed vertically near the foot of a door.

footer (*ICT*) Text that appears at the bottom of every page of a document, eg page number.

footing (*Build, CivEng*) The lower part of a column or wall, standing immediately upon the foundation; usually enlarged locally in order to distribute the load.

footing (*ElecEng*) The foundation which is set in the ground to support a tower of an overhead transmission line.

footing resistance (*ElecEng*) The ohmic resistance between a transmission-line tower and the earth.

foot irons (*Build*) Shaped iron bars which can be built into the joints of a manhole wall, leaving projecting steps for use by workers descending the manhole. Also *step irons*.

foot mange (*Vet*) See CHORIOPTIC MANGE.

Footner process (*Build*) A process which uses phosphate pickling to remove scale from steelwork prior to painting.

foot-plate (*Build*) A HAMMER-BEAM.

foot-plate (*CivEng*) The platform on which the driver and fireman of a steam locomotive stand.

foot-pound (*Phys*) Unit of work in the foot–pound–second system of units. The work done in raising a mass of 1 pound through a vertical distance of 1 foot against gravity, ie 1·3558 J.Also *foot-pound force*. Symbol lbf. See FUNDAMENTAL DYNAMICAL UNITS.

footprint (*ICT*) (1) A measure of the space that the equipment will occupy on a desktop or table. (2) The service area, or the outline of the area on the Earth's surface within which a communications or direct broadcast satellite gives satisfactory results.

footprinting (*BioSci*) The method for identifying the site where a protein binds a nucleic acid. By comparing the pattern of the fragments after limited hydrolysis, the binding position can be determined because the protein will have masked a cleavage site.

foot rot (*BioSci*) (1) One of a number of plant diseases, caused by a variety of fungi, in which the primary symptom is the death of the roots and stems around or below soil level.

foot rot (*Vet*) A suppurative infection between the horn and the sensitive corium of the hoof of the sheep, caused by *Fusiformis nodosus* and possibly other bacteria; causes lameness. Vaccines available.

foot run (*Build*) A term meaning foot of length, as in speaking of a loading or of a price *per foot run*.

foot-stall (*Build*) The base of a pillar.

footstone (*Build*) The lowest coping stone over a gable.

foot super (*For*) Namely, 1 ft^2 or 144 in^2 of timber.

foot switch (*ElecEng*) A switch arranged for operation by the foot.

foot thumper (*Aero*) Stall warning device that vibrates the rudder pedals when the detector senses that a stall is imminent.

foot-ton (*Phys*) Namely, 2240 lbf. Also *foot-ton force*. See FOOT-POUND.

foot valve (*Eng*) (1) The non-return or suction valve fitted at the bottom of a pump barrel, or in the valve chest of a pump. (2) A non-return valve at the inlet end of a suction pipe.

footwall (*Geol, MinExt*) The lower wall of country rock in contact with a vein or lode. The upper wall is the *hanging wall*.

footway (*MinExt*) A colliery shaft in which ladders are used for descending and ascending.

forage (*Agri*) Crops grown as feed to be grazed or fed after harvest.

forage harvester (*Agri*) A trailed or self-propelled machine used to gather and chop forage crops for silage production.

forage mites (*Vet*) Acari of the family Acaridae, which commonly infest the skin of animals and birds.

foraging (*Psych*) Behaviour that involves searching for, capturing and consuming food.

foramen (*BioSci*) An opening or perforation, esp in a chitinous, cartilaginous or bony skeletal structure. The *foramen magnum* is the main opening at the back of the vertebrate skull, by which the spinal cord issues from the braincase.

Foraminifera (*BioSci*) An order of SARCODINA, the members of which have numerous fine anastomosing pseudopodia and a shell which is usually calcareous; the ectoplasm is sometimes vacuolated.

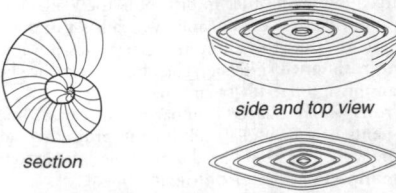

side and top view

section

Foraminifera Two typical fossil shells.

forb (*BioSci*) Any herb other than a grass.

forbidden band (*Phys*) The gap between two bands of allowed energy levels in a crystalline solid. See ENERGY BAND, ENERGY GAP.

forbidden clone (*BioSci*) Clones of lymphocytes reactive with 'self' antigens. According to the clonal selection theory they should have been eliminated during the early maturation of lymphocytes, and hence forbidden. The term is of historical interest, and is sometimes used, but the actual mechanisms by which 'self-reactivity' is avoided are more complicated.

forbidden lines (*Astron, Phys*) Spectral lines which cannot be reproduced under laboratory conditions. Such lines correspond to transitions from a metastable state, and occur in extremely rarefied gases, eg in the solar corona and in gaseous nebulae.

forbidden transition (*Phys*) A transition of electrons between energy states which has a very low probability; that which has a high probability is an *allowed transition*.

force (*Phys*) The influence which, when acting on a body which is free to move, produces an acceleration in the motion of the body, measured by rate of change of momentum of body. Extended to denote loosely any operating agency. Electromotive force, magnetomotive force and magnetizing force are strictly misnomers. The unit of force is that which produces unit acceleration in unit mass. See DYNE, NEWTON, POUNDAL.

force constant (*Chem*) Bond stiffness defined as force needed to deform a specific covalent bond divided by the deformation produced by that force (k_1 for stretching and k_2 for bending). For the single carbon–carbon bond, $k_1 = 436\,\text{N m}^{-1}$ and $k_2 = 0.35\,\text{N m}^{-1}$. See THEORETICAL STIFFNESS.

forced choice (*Psych*) A procedure in which the test subject must select from a list of alternatives thereby giving the experimenter greater control over the assessment process.

forced-circulation boiler (*Eng*) Steam boiler in which water and steam are continuously circulated over the heating surface by pumps (as opposed to natural circulation systems) in order to increase the steaming capacity. Also *forced-flow boiler*. See LÖFFLER BOILER.

forced commutation (*ElecEng*) (1) The usual process of commutation in which the change of direction of the current in the coil actually undergoing commutation is assisted by flux from a commutating pole. (2) The method of commutating conducting devices in an electronic converter whereby circuits containing inductance and capacitance are arranged to drive the current through the conducting devices to zero. Cf NATURAL COMMUTATION.

forced development (*ImageTech*) Processing with increased time or temperature to increase the image density for underexposed film.

forced draught (*ElecEng*) Said of electrical apparatus cooled by ventilating air supplied under pressure from some external source.

forced draught (*Eng*) An air supply to a furnace driven or induced by fans or steam jets (as opposed to the natural draught created by a chimney) in order to obtain a high rate of combustion. See BALANCED DRAUGHT, CLOSED STOKEHOLD, INDUCED DRAUGHT.

forced-draught furnace (*Eng*) A furnace, but more particularly a boiler furnace, arranged to work under FORCED DRAUGHT.

forced-flow boiler (*Eng*) See FORCED-CIRCULATION BOILER.

force diagram (*Eng*) A diagram in which the internal forces in a framed structure, assumed pin-jointed, are shown to scale by lines drawn parallel to the members themselves. Also *reciprocal* or *stress diagram*.

forced lubrication (*Eng*) The lubrication of an engine or machine by oil under pressure. See FORCE FEED, FULL FORCE FEED.

forced oscillation (*Phys*) An oscillatory current whose frequency is determined by factors other than the constants of the circuit in which it is flowing, eg those flowing in a resonant circuit coupled to a fixed-frequency oscillator. See FORCED VIBRATION.

force drying (*Build*) Accelerated paint drying through moderate heat, generally below about 82°C.

forced vibration (*Phys*) A vibration which results from the application of a periodic force to a body capable of vibrating. The amplitude of forced vibrations becomes very great when resonance occurs, ie when the frequency of the applied force equals the natural frequency of the vibrator, particularly if the damping is small.

force feed (*Eng*) Lubrication of an engine by forcing oil to main bearings and through the hollow crankshaft to the big-end bearings.

force on a moving charge (*Phys*) If a charge q is moving with a velocity **v** in a magnetic field **B**, then the force on the charge is

$$F = q\,(\mathbf{v} \times \mathbf{B})$$

If the moving charges are within a conductor, the force of a short length **l** is

$$F = i\,(\mathbf{l} \times \mathbf{B})$$

where i is the current. See LORENTZ FORCE.

forceps (*BioSci*) (1) In Dermaptera, the pincer-shaped cerci. (2) In Arachnida and Crustacea, the opposable distal joints of the chelae. (3) In Echinodermata, the distal opposable jaws of pedicellariae. Adj *forcipulate*.

forceps (*Med*) (1) A pincer-like instrument with two blades, for holding, seizing or extracting objects. *Obstetrical forceps* have large blades which, applied to the fetal head, aid delivery. (2) That part of either of the two ends of the CORPUS CALLOSUM of the brain which diverges into the adjacent brain tissue on each side.

force pump (*Build*) An air pump used to clean out gas and other service pipes by blowing air through them.

force pump (*Eng*) A pump which delivers liquid under a pressure greater than its suction pressure. It consists of a barrel fitted with a solid plunger, and a valve chest with suction and delivery valve.

Ford cup (*Build*) A device used to ascertain the viscosity of paints. The time taken for paint to flow from the cup though an accurately machined orifice in its base is measured in seconds.

fore-and-aft (*Print*) The first and last sections of the book are printed and folded in one sheet, the first page being head to head with the last page; the second section is similarly combined with the second last section; and so on. Used for the production of large paperback editions; there are fewer sheets to handle and the economy of two-on working is maintained throughout all printing and binding operations. Also *come-and-go*.

forearc basin (*Geol*) A sedimentary basin developed in the gap between a volcanic arc and its SUBDUCTION ZONE.

forebody strake (*Aero*) Low-aspect-ratio extension of the wing at the root along sides of the forebody. These create powerful vortices during high-incidence flight, thereby improving handling and increasing lift.

fore-brain (*BioSci*) In vertebrates, that part of the brain which is derived from the first or anterior brain vesicle of the embryo, comprising the olfactory lobes, the cerebral hemispheres and the DIENCEPHALON; the first or anterior brain-vesicle itself.

forecast (*EnvSci*) A statement of the anticipated weather conditions in a given region, for periods of from 1 hour to 30 days in advance, the longer term being less reliable; made from a study of current *synoptic charts*, or by carrying out a NUMERICAL FORECAST.

fore-edge (*Print*) The outside margin of a book page; the edge opposite to the back; the outer edge of a volume. Cf HEAD, TAIL.

fore-edge painting (*Print*) Pictures painted on the fanned-out edges of a book, unseen when the book is closed, as an additional decoration to an already elaborate binding.

foreground/background processing (*ICT*) The method of organizing a TIME-SHARING computer system so that while main tasks may claim the use of the computer when required, other less pressing tasks utilize the remaining time.

fore-gut (*BioSci*) That part of the alimentary canal of an animal which is derived from the anterior ectodermal invagination or stomodaeum of the embryo.

forehand welding (*Eng*) That in which the palm of the torch or electrode hand faces the direction of travel so that the metal ahead of the weld position is preheated.

forehearth (*Eng*) Bay in front of a furnace into which molten products can be run.

foreign body (*FoodSci*) See CONTAMINATION.

foremilk (*Agri*) The volume of milk in the teat before milking begins. It is most likely to contain contaminating bacteria and is discarded. Clotted foremilk is an indicator of mastitis.

forensic engineering (*Eng*) The application of engineering methods to help determine facts at issue in civil and criminal cases. Typical civil cases involve personal injury, product liability, contract and intellectual property. Methods include those of forensic sciences as applied to practical situations, fractography applied to broken products, design analysis in patent cases, stress analysis of products, etc.

forensic medicine (*Med*) Any aspect of medicine which has civil or statutory legal implications. Most simply *legal medicine*.

forensic pathology (*Psych*) The application of pathology for legal purposes, particularly in investigation of unnatural deaths.

forensic science (*Genrl*) The application of scientific methods to legal problems, in order to determine facts at issue in both criminal and civil cases. Methods include materials identification and materials matching using chemical (eg elemental analysis) or physical (eg X-ray analysis) techniques.

fore peak (*Ships*) The spaces forward of the COLLISION BULKHEAD. Lower part frequently used as fresh-water tank and upper part may be used as storeroom.

foreplane (*Aero*) Horizontal aerofoil mounted on front fuselage for pitch control. In a tail-first or *canard* configuration it replaces the function of the tailplane; in a delta-wing design it may assist slow-speed behaviour. Never used for roll control, a foreplane may be fixed or retractable and have ELEVATORS, FLAPS or SLATS.

forepoling (*MinExt*) A method of progressing through loosely consolidated ground by driving poles forward over frames.

foreset beds (*Geol*) Gently inclined cross-bedded units progressively covering BOTTOMSET BEDS and covered in turn by TOPSET BEDS. Foreset beds form the major bulk of a delta.

fore sight (*Surv*) The levelling staff reading as taken forward to a station which has not been passed by the instrument. It transfers collimation line from back-sight station to fore-sight station. See BACK OBSERVATION, INTERMEDIATE SIGHT.

forest (*EnvSci*) (1) Natural vegetation in which the dominant species are trees, with crowns that touch each other to form a continuous canopy. (2) Almost any vegetation with trees, including plantations. (3) In the UK, areas which were used for hunting large game and may have been relatively treeless.

forest (*MathSci*) In graph theory, a family of TREES.

forfex (*BioSci*) A pair of pincers, as of an earwig.

forge (*Eng*) Open hearth or furnace with forced draught; place where metal is heated and shaped by hammering.

forge pigs (*Eng*) Pig iron suitable for the manufacture of wrought-iron.

forge tests (*Eng*) Rough workshop tests made to check the malleability and ductility of iron and steel.

forging (*Eng*) The operation of shaping malleable materials by means of hammers or presses.

forging (*Vet*) See CLACKING.

forging machines (*Eng*) Power hammers and presses used for forging and drop forging. See DROP FORGING.

forked lightning (*EnvSci*) A popular name given to a lightning stroke; the name derives from the branching of the stroke channel which is commonly observed.

forked tenon (*Build*) A joint formed by a slot mortise astride a tenon cut across the length of a member.

fork-lift truck (*Genrl*) A vehicle with power-operated prongs which can be raised or lowered at will, for loading, transporting and unloading goods; chiefly used in factories and warehouses. Loads are usually stacked upon stands or pallets with sufficient ground clearance for the prongs to be inserted beneath them.

form (*CivEng*) Preferred term for shuttering in which wet concrete is shaped for its ultimate purpose. Also *formwork*.

form (*Crystal*) A complete assemblage of crystal faces similar in all respects as determined by the symmetry of a particular class of crystal structure. Thus the *cube*, consisting of six similar square faces, and the *octahedron*, consisting of eight faces, each an equilateral triangle, are crystal *forms*. The number of faces in a form ranges from 1 (the pedion) to 48 (the hexakis-octahedron). A natural crystal may consist of one form or many.

formability (*Eng*) See DRAWABILITY.

formaldehyde (*Chem*) Methanal. HCHO. Bp $-21°$C. A gas of pungent odour, readily soluble in water, and usually used in aqueous solution. Formaldehyde easily polymerizes to give POLYOXYMETHYLENE or metaformaldehyde. It is produced by oxidation of methanol, or by the oxidation of ethene in the presence of a catalyst. It forms with ammonia *hexamethylene-tetramine*. It is a disinfectant, tissue *fixative* and is of great importance in plastics manufacture. See ACETAL RESIN.

formaldehyde resins (*Chem, Plastics*) Synthetic resins which are condensation products of methanal with hydroxybenzenes, urea, etc. Also *methanal resins*. See panel on THERMOSETS.

formalin (*Chem*) TN for a commercial 40% aqueous methanal (FORMALDEHYDE) solution.

formal language theory (*ICT*) A subject in computer science concerned with the specification and use of artificial languages.

formal operational stage (*Psych*) In Piagetian theory, the fourth and final major stage of cognitive development, occurring during adolescence; it reflects a transition from logic bound to the real and concrete to a logic capable of dealing with abstract events. See CONCRETE OPERATIONAL STAGE.

formal specification (*ICT*) (1) The subject concerned with techniques and methods of specifying a program using a FORMAL LANGUAGE. (2) A program specification in a formal language.

formant (*Acous*) An envelope of the frequency pattern arising in the vocal cords, which determines the distribution of energy in unvoiced sounds and the reinforcement of harmonics in voiced sounds. It is the formant which leads to recognition of speech sounds by aural perception.

format (*ICT*) See DISK FORMATTING.

format (*Print*) The general size, quality of paper, typeface and binding of a book.

formates (*Chem*) The salts of FORMIC ACID. Also *methanoates*.

formation (*Geol*) A stratigraphical rock unit used as a basis for rock mapping. A formation has a recognizable lithological identity and is divisible into *members* or combined into *groups*. It has been casually used in the UK but more precisely in N America.

formation (*Paper*) The pattern of the fibres in the paper when viewed by transmitted light.

formative time (*Electronics*) The time interval between the first Townsend discharge in a given gap and the formation there of a self-maintaining glow discharge.

formatting (*Print*) The term used to describe the translating of type specifications into format or command codes prior to phototypesetting.

form drag (*Aero*) The difference obtained when the *induced drag*, ie the fraction of the total drag induced by lift, is subtracted from the PRESSURE DRAG. See DRAG.

forme (*Print*) Type matter assembled and locked up in a chase ready for printing.

formed plate (*ElecEng*) A type of plate used in lead–acid accumulators; made by electrolytically converting the substance of which the plate is made into active material.

former (*Aero*) A structural member of a fuselage, nacelle, hull or float to which the skin is attached, and having the primary purpose of preserving form or shape. It generally carries structural loads. Cf FRAME.

former (*ElecEng*) A tool for giving a coil or winding the correct shape; it sometimes consists of a frame upon which the wire can be wound, the frame afterwards being removed.

forme rollers (*Print*) Rollers which supply ink to a forme or plate, as distinct from DISTRIBUTOR ROLLERS.

former-wound coil (*ElecEng*) An armature coil built to the correct shape by means of a FORMER, it being then dropped into the slots on the armature.

form factor (*Phys*) The ratio of the effective value of an alternating quantity to its average value over a half-period.

form-feed (*ICT*) Instruction to a printer to start a new page or sheet by sending a special CONTROL CHARACTER from the computer.

form genus (*BioSci*) See FORM TAXON.

form grinding (*Eng*) See PROFILE GRINDING.

formic acid (*Chem*) HCOOH. Mp 9°C, bp 101°C. A colourless liquid, of pungent odour; corrosive. Prepared by passing carbon monoxide and steam at 200–300°C under pressure over a catalyst. Also *methanoic acid*.

forming (*Eng*) Changing the shape of a metal component without in general altering its thickness. Cf DRAWING.

forming cutter (*Eng*) See FORM TOOL.

forming rollers (*Print*) See BENDING ROLLERS.

formol titration (*Chem*) A method of estimating volumetrically the amount of amino acids present in a solution. It is based upon the fact that amino acids and their derivatives possess both a carboxyl and an amino group which neutralize each other, and that by the addition of methanal the amino group is converted into a methylene derivative without basic properties, by which reaction it becomes possible to titrate subsequently the carboxyl in the usual manner.

formol toxoid (*BioSci*) Any toxoid prepared by formaldehyde treatment of a toxin.

formoxyl (*Chem*) The organic radical OCHO–.

form taxon (*BioSci*) An artificial taxon, eg form genus, intended to provide a name for morphologically similar but possibly unrelated organisms or parts of organisms when it is not possible to determine their correct taxonomic position. Thus an organ genus in palaeobotany provides names for fossil leaves, spores, pieces of cuticle, etc, where it is not (yet) possible to reconstruct the whole plant. Again in Deuteromycetes, it provides names for fungi with no, or no known, sexual reproduction.

form tool (*Eng*) Any cutting tool which produces a desired contour on the workpiece by being merely fed into the work, the cutting edge having a profile similar to, but not necessarily identical with, the shape produced. Also *forming cutter*. See CHASER.

formula (*Chem*) The representation of the types and relative numbers of atoms in a compound, or the actual number of atoms in a molecule of a compound. It uses chemical symbols and subscripts, eg H_2SO_4 or C_6H_6.

formula (*Genrl*) A fixed rule or set form.

formula (*MathSci*) Any expression using algebraic symbols.

formula (*Med*) A prescription or specification.

formwork (*CivEng*) See FORM.

formyl (*Chem*) The organic radical OCH–.

fornacite (*Min*) A basic chromarsenate of copper and lead, crystallizing in the monoclinic system. Also *furnacite*.

Fornax (Furnace) (*Astron*) A faint southern constellation.

fornix (*BioSci*) In the brains of higher vertebrates, a tract of fibres connecting the posterior part of the cerebrum with the hypothalamus.

Forssman antigen (*BioSci*) A glycolipid antigen present on tissue cells of many species, and on the red cells of some species such as sheep. Sheep red cells coated with antibody directed against the antigen (*Forssman antibody*, often found naturally in humans) are used in complement fixation tests.

forsterite (*Min*) An end-member of the olivine group of minerals, crystallizing in the orthorhombic system. Chemically forsterite is a silicate of magnesium, Mg_2SiO_4. It typically occurs in metamorphosed impure dolomitic limestones.

forsterite-marble (*Geol*) A characteristic product of the contact metamorphism of magnesian (dolomitic) limestones containing silica of organic or inorganic origin. Also *ophicalcite*.

Forstner bit (*Build*) An accurate centre-bit for sinking blind holes in timber which may overlap the edge or adjacent holes. It has a short cylindrical part attached to the shank with a small centre bit and integral cutters.

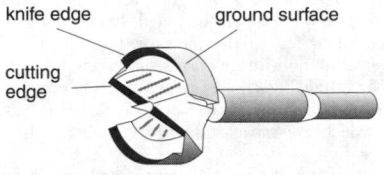

knife edge   ground surface

cutting edge

Forstner bit

FORTH (*ICT*) Programming language using REVERSE POLISH NOTATION, with applications in control.

Fortin's barometer (*EnvSci*) A type of mercury barometer suited for accurate readings of the pressure of the atmosphere. The zero of the scale is indicated by a pointer inside the mercury cistern, the bottom of which is flexible and may be moved by an adjusting screw until the mercury surface just touches the pointer.

FORTRAN (*ICT*) Abbrev for *formula translation*. A programming language widely used for scientific work.

forward-bias (*Electronics*) Said of a semiconductor diode, p–n junction, or the emitter–base junction of a transistor, when the polarity of the applied emf is such as to allow substantial conduction to take place.

forward chaining (*ICT*) An INFERENCE METHOD used in EXPERT SYSTEMS where the IF-portion of rules is matched against facts to establish new facts, ie by using known facts and existing rules, further facts can be deduced: eg the rule '*If it is raining then carry an umbrella*' along with the fact '*it is raining*', will allow the system to deduce the fact that '*an umbrella is carried*' Cf BACKWARD CHAINING.

forward eccentric (*Eng*) On a steam engine having link motion reverse gear, the eccentric which drives the valve when the engine is going ahead. See LINK MOTION.

forward error correction (*ICT*) The use by a digital coding scheme of REDUNDANCY in such a form that it allows a terminal receiving a corrupted signal to determine the actual signal sent. Such codes are important where the need for isochronicity makes delays due to retransmission unacceptable, eg in video links.

forward explicit congestion notification (*ICT*) A method of dealing with overloading in a FRAME RELAY or ASYNCHRONOUS TRANSFER MODE network by setting a specific bit in the header of a FRAME or CELL sent to the destination from a source of traffic which is experiencing congestion. On receiving this, the destination reduces its demand for data.

forwarding (*Print*) The operations entailed in hand bookbinding after sewing but before FINISHING.

forward lead (*ElecEng*) See FORWARD SHIFT.

forward-looking infrared (*Aero*) Sensor systems in the nose of aircraft or guided weapons for target detection and vehicle guidance.

forward path (*ICT*) The transmission path from the loop-activating signal to the loop-output signal in a feedback control loop.

forward perpendicular (*Ships*) The forward side of a ship's stem post when this is truly perpendicular to the longitudinal baseline; but in cases when the stem post is 'raked', ie angled to the baseline, it is the perpendicular intersecting the forward side of the stem post at the summer load water line. Abbrev *FP*.

forward scatter (*ICT*) Multiple transmission on centimetric waves, using reflection downwards and forwards from ionization in troposphere; range about 150 km. See TROPOSPHERIC SCATTER.

forward scatter (*Phys*) Scattering of particles through an angle of less than 90° to the original direction of the beam. Cf BACK-SCATTER.

forward shift (*ElecEng*) A movement of the brushes of a commutator machine around the commutator, from the neutral position, and in the same direction as that of rotation. Also *forward lead*.

forward shovel (*Eng*) See POWER SHOVEL.

forward sweep (*Aero*) See SWEEP.

forward transfer function (*ICT*) That of the forward path of a feedback control loop.

forward voltage (*ElecEng*) The voltage of the polarity which produces the larger current in an electrical system.

forward wave (*Electronics*) One whose group velocity is in the same direction as the motion of an electron stream in a travelling-wave tube. See BACKWARD WAVE.

fossa (*BioSci*) A ditch-like or pit-like depression, as the *glenoid fossa*.

fossette (*BioSci*) In general, a small pit or depression: in some Arthropoda, the socket which receives the base of the antennule.

fossil (*Geol*) The relic or trace of some plant or animal which has been preserved by natural processes in rocks of the past.

fossil fuel (*MinExt*) Coal, oil or natural gas, which is derived from fossilized organic matter.

fossil meal (*Build*) A diatomaceous earth used in the manufacture of lightweight and heat-resistant, often hollow, blocks.

fossil zone (*Geol*) The stratigraphical horizon characterized by a *zone fossil*. See ZONE.

fossorial (*BioSci*) Adapted for digging.

Foster–Seeley discriminator (*ElecEng*) One for demodulating FM transmission using a balanced pair of diodes. To these are applied voltages which are sum and difference of limiter signal voltage and half transformer-coupled voltage, diode outputs being differenced.

FOT (*ICT*) Abbrev for FIBRE OPTIC TRANSCEIVER.

Föttinger coupling (*Eng*) A hydraulic coupling, gear or clutch for transmitting power from eg an engine to a ship's propeller; it consists essentially of an outward-flow water turbine driving an inward-flow turbine, within a common casing. Similar device formerly used for speed reduction. Also *Föttinger transmitter*.

Foucault currents (*ElecEng*) See EDDY CURRENTS.

Foucault knife-edge test (*Phys*) A method of testing for the form and optical quality of a lens or mirror by placing a knife edge in the focus of a point source and observing the pattern of light and shade as seen in the lens from immediately behind the knife edge.

Foucault's measurement of the speed of light (*Phys*) One of the first successful attempts to obtain an accurate result for the constant value of the speed of light. Foucault, in 1862, made use of a rapidly rotating mirror sending light to a distant fixed concave mirror which reflected it back. Measurement of the displacement of the reflected image gave a value of $2 \cdot 986 \times 10^8$ m s^{-1} for the speed of light in a vacuum.

Foucault's pendulum (*Astron*) An instrument devised by Foucault in 1851 to demonstrate the rotation of the Earth; it consists of a heavy metal ball suspended by a very long fine wire. The plane of oscillation slowly changes through 15° sin(latitude) per sidereal hour.

foul anchor (*Ships*) An anchor which has become entangled in its own cable or with some obstruction on the bottom.

foulard (*Textiles*) A lightweight dress fabric with a printed pattern, made either of silk or synthetic filament yarns or of high-quality Sea Island or Egyptian cotton.

foul berth (*Ships*) A situation where anchored ships may collide when swinging round with the change of tidal stream.

foul clay (*Build*) A brick earth composed of silica and alumina combined with only a small percentage of lime, magnesia, soda or other salts. Such a clay lacks sufficient fluxing material to fuse its constituents at furnace temperature, and is improved by the addition of sand or loam, lime or ashes. Also *plastic clay, pure clay, strong clay.*

fouling (*Eng*) (1) Coming into accidental contact with. (2) Deposition or incrustation of foreign matter on a surface, as of carbon in an engine cylinder, or marine growth on the bottom of a ship, or on structures subject to the action of sea water.

fouling point (*CivEng*) The location before the meeting of two tracks where the loading-gauge outlines come into contact.

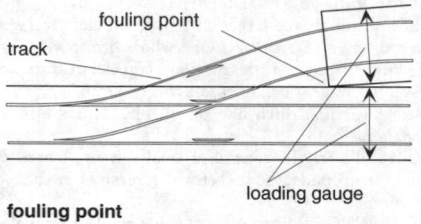

fouling point

foul in the foot (*Vet*) Suppuration and necrosis of the interdigital tissues of the foot in cattle due to infection by the bacterium *Sphaerophorus necrophorus* (*Fusiformis necrophorus*); causes lameness. Also *claw ill, infectious pododermatitis.*

fouls (*MinExt*) The cutting-out of portions of the coal seam by 'wash outs'; barren ground.

foul solution (*MinExt*) In cyanide process, one contaminated by salts or metals other than gold and silver, and which is not fit to be recirculated in the process.

foul water (*Build*) Water contaminated by soil, waste or trade effluent.

foundation cylinder (*Eng*) A large steel or iron cylinder sunk into the ground to provide a solid foundation for bridge piers etc in soft ground.

foundation piles (*CivEng*) Piles supporting a structure.

foundation ring (*Eng*) In a locomotive boiler, a rectangular iron ring of rectangular section, to which the lower edges of the inner and outer plates of the fire-box are secured.

founded (*CivEng*) Said of a caisson which has been sunk to a firm level.

founders' type (*Print*) Fonts of type produced by the specialist typefoundries, in hard type metal as distinct from that produced by the printer; used mainly for display types.

founding (*Glass*) See FINING.

foundry (*Eng*) A workshop in which metal objects are made by casting in moulds.

foundry (*Print*) That department of a printing establishment where work in connection with duplicate plates is carried out.

foundry chase (*Print*) A chase with type-high rims and sometimes with built-in locking-up device, used when taking a mould of type matter.

foundry ladle (*Eng*) A steel ladle lined with fireclay; used for transporting molten metal from a foundry cupola to the

moulds. Small ladles are carried by hand, large ones by a truck or crane. See HAND SHANK.

foundry pig iron (*Eng*) Bars of cast-iron up to 1 m in length and 1·5 cm in diameter, as bought by an iron foundry.

foundry pit (*Eng*) A large hole in the floor of a foundry, which serves the purpose of a moulding box for very large or deep castings.

foundry sand (*Eng*) Silica-based sand with either enough natural clay or special additions (oil, molasses, fuller's earth, etc) to give it some cohesive strength when used in moulding.

foundry stove (*Eng*) A large oven for drying moulds and cores, heated either externally by hot gases or internally by a fire basket.

fount (*Print*) See FONT.

fountain (*Print*) The ink reservoir on a printing press. In LITHOGRAPHY that part of the machine which holds the dampening or fountain solution.

fountain effect (*Phys*) Elegant experiment in superfluidity demonstrating the behaviour of liquid helium II: if two containers of superfluid helium are connected by a narrow tube and one is heated, helium flows out from the colder container into the hotter one.

fountain solution (*Print*) In lithographic printing a solution of water or alcohol, acid, buffer and gum arabic which when fed to the plate by the machine dampers prevents the non-image areas from taking ink.

four-bar chain (*Eng*) A common and simple versatile mechanism comprising an assemblage of four rigid members, which are capable of relative motion. Also *four-bar linkage.*

four-centred arch (*Arch*) A pointed arch struck from four centres.

fourchette (*Med*) The posterior or junction of the labia minora.

four-colour press (*Print*) Sheet-fed and web-fed models are available for the main printing processes. Four colours are the minimum for full-colour reproduction but extra units are often added for additional printing or varnishing.

four-colour reproduction (*Print*) Printing by any of the main printing processes with standard yellow, magenta, cyan and black inks, not always in that order, from sets of colour plates or cylinders, to produce a reproduction of a coloured subject. See COLOUR CORRECTION, THREE-COLOUR PROCESS.

four-colour theorem (*MathSci*) A theorem stating that it is possible to colour any map using no more than four colours in such a way that countries which share a border have different colours. Equivalently, every planar graph has a four colouring. See COLOURING OF A GRAPH.

fourdrinier (*Paper*) The standard type of paper-making machine characterized by a wire part of which (the upper, forming, surface) is horizontal or nearly so.

four-factor formula (*NucEng*) That giving the multiplication factor of an infinite thermal reactor as the product of FAST FISSION FACTOR, RESONANCE ESCAPE PROBABILITY, THERMAL UTILIZATION FACTOR and the number of neutrons absorbed per neutron absorbed in the fuel (η).

Fourier analysis (*MathSci, Phys*) The determination of the harmonic components of a complex waveform (ie the terms of a Fourier series that represents the waveform) either mathematically or by a wave-analyser device.

Fourier half-range series (*MathSci*) Fourier series with only sine or only cosine terms. Valid for x between 0 and π.

Fourier integral (*Phys*) The limiting form of the FOURIER SERIES when the period of the waveform becomes infinitely long. It is the Fourier representation of a non-repeated waveform, ie pulses, wave packets and wavetrains of limited extent.

Fourier number (*Phys*) A dimensionless parameter used for studying heat-flow problems. It is defined by $\lambda t / C_p \rho l^2$ where λ is the thermal conductivity, t is the time, C_p the specific heat at constant pressure, ρ the density and l a linear dimension.

Fourier optics (*Phys*) The application of Fourier analysis and the use of Fourier transforms to problems in optics, in particular to image formation. The *Fraunhofer diffraction* pattern is the Fourier transform of the distribution of amplitude of light across the diffracting object. The distribution of amplitude in the Fraunhofer pattern is modified by the optical system and the image formed is the transform of this modified distribution. The same principle is used in X-ray crystal structure analysis where an 'image' of the atomic arrangement is constructed mathematically from the X-ray diffraction pattern.

Fourier principle (*Phys*) The principle that all repeating waveforms can be resolved into sine wave components consisting of a fundamental and a series of harmonics at multiples of this frequency. It can be extended to prove that non-repeating waveforms occupy a continuous frequency spectrum.

Fourier series (*MathSci*) The series

$$\frac{1}{2}a_0 + \sum_{n=1}^{\infty}(a_n \cos nx + b_n \sin nx)$$

where

$$a_n = \frac{1}{\pi}\int_{-\pi}^{+\pi} f(x)\cos nx\, dx$$

and

$$b_n = \frac{1}{\pi}\int_{-\pi}^{+\pi} f(x)\sin nx\, dx$$

which, subject only to certain very general restrictions upon $f(x)$, converges to $f(x)$ in the interval $(-\pi, +\pi)$. Fourier series are used in the determination of the component frequencies of vibrating systems. See FOURIER ANALYSIS.

Fourier transform (*MathSci*, *Phys*) A mathematical relation between the energy in a transient and that in a continuous energy spectrum of adjacent component frequencies. The Fourier transform $F(u)$ of the function $f(x)$ is defined by

$$F(u) = \int_{-\infty}^{+\infty} e^{-iut} f(t)\, dt$$

Sometimes $\exp(iut)$ is used instead of $\exp(-iut)$.

Fourier transform infrared (*Phys*) A form of infrared spectroscopy involving interferometric methods to give enhanced resolution; spectra are produced by Fourier transformation of the interferometer output data. Abbrev FTIR.

Fourier transform spectroscopy (*Phys*) The production of a spectrum by taking the FOURIER TRANSFORM of a two-beam interference pattern. See FOURIER TRANSFORM INFRARED.

four-jaw chuck (*Eng*) A chuck used in a lathe or in certain other types of machine tools, comprising four jaws disposed at right angles to each other. Usually, the jaws act independently of each other. Used for holding rectangular or irregularly contoured workpieces, or for revolving circular workpieces eccentrically.

four-part vault (*Arch*) A vault formed at the intersection of two barrel vaults.

four-phase system (*ElecEng*) A name sometimes given to a two-phase system in which the mid-points of the two phases are connected to form a neutral point.

four-squares theorem (*MathSci*) Theorem stating that any positive integer may be expressed as the sum of squares of four integers.

four-stroke cycle (*Autos*) An engine cycle completed in four piston strokes (ie in two crankshaft revolutions), consisting of suction or induction, compression, expansion or power stroke and exhaust. See DIESEL CYCLE, OTTO CYCLE.

four-terminal resistor (*ElecEng*) Laboratory standard fitted with current and potential terminals, arranged so that measurement of the potential drop across the resistor is not affected by contact resistances at the terminals.

fourth-generation computer (*ICT*) A term somewhat imprecisely applied to a computer system in use in 1987; physically small and neat using low-cost CHIP technology with LSI. Processors and computers were being linked up to form systems and NETWORKS rather than a computer. COMPUTER GRAPHICS was blossoming. Typical machines were Amdahl 470, Intel 8748. See COMPUTER GENERATIONS, DISTRIBUTED COMPUTING, MICROCOMPUTER, WORKSTATION.

fourth-generation language (*ICT*) A language specific to a particular class of application. These languages were designed to be easy to use by non-specialists but were not particularly efficient as regards speed of EXECUTION; eg Filetab and RPG were used for producing management reports. Many PACKAGES had such languages included within them, such as dBase IV for DATABASE applications.

fourth rail (*ElecEng*) A conductor rail on an electric traction system. When there are two running rails and two conductor rails, the fourth rail generally carries the return current, instead of its being allowed to return along the running rails.

fourth ventricle (*BioSci*) In vertebrates, the cavity of the hind-brain.

fourth wire (*ElecEng*) A name sometimes given to the neutral wire in a three-phase, four-wire distribution system.

four-wave mixing (*ICT*) An effect due to non-linearity in an OPTICAL FIBRE that gives rise to unwanted products at frequencies equal to the sum and difference of signal and noise frequencies or, in a multichannel regime, the sum and difference of the signal frequencies.

four-way tool post (*Eng*) A supporting and clamping mechanism for cutting tools used in a lathe or certain other machine tools, in which up to four tools may be mounted simultaneously but in various positions, the required tool being brought into the operative position by indexing the tool post appropriately.

four-wire circuit (*ICT*) One in which information is transmitted in one direction on one path and in the opposite direction on the other path. Not necessarily with four separate wires as the effect of a physical four-wire circuit may be reproduced electronically.

four-wire repeater (*ICT*) A repeater for insertion into a four-wire telephone circuit, in which the two amplifiers, one for amplifying in each direction, are kept separate.

four-wire system (*ElecEng*) A system of distribution of electric power requiring four wires. In a three-phase system, the four wires are connected to the three line terminals of the supply transformer and the neutral point; and in the two-phase system, the wires are connected to the ends of the two transformer windings.

fovea (*BioSci*) A small pit or depression. Adjs *foveate*, *foveolar*, *foveolate*. Also *foveola*.

fovea centralis (*BioSci*) In the vertebrate eye, the areas of greatest visual resolution, seen as a small depression at the centre of the MACULA LUTEA.

foveola (*BioSci*) See FOVEA.

fowl cholera (*Vet*) An acute and usually fatal septicaemia of domestic fowl and other birds, caused by the bacterium *Pasteurella multocida* (*P. aviseptica*). Characterized by dejection, loss of appetite, raised temperature and diarrhoea. Mild and chronic forms of the disease also occur.

fowl coryza (*Vet*) See INFECTIOUS CORYZA.

Fowler flap (*Aero*) A high-lift trailing-edge flap that slides backwards as it moves downwards, thereby increasing the wing area, also leaving a slot between its leading edge and the wing when fully extended.

Fowler position (*Med*) The semi-sitting position in which the patient may be placed in bed after an abdominal operation with head elevated and knees drawn up so that the pelvis is the lowest part of the body.

fowl paralysis (*Vet*) Chickens are the only important natural host. It is one of the most ubiquitous avian infections, caused by a herpes virus with three serotypes. Paralysis is sometimes seen but the bird usually shows depression prior to death. Various peripheral nerves become enlarged and lymphoid tumours are observed in various organs. Controlled by vaccination. Also *Marek's disease, neuro-lymphomatosis.*

fowl pest (*Vet*) A term usually embracing both FOWL PLAGUE and NEWCASTLE DISEASE.

fowl plague (*Vet*) An acute, contagious virus disease of chickens and other domestic and wild birds; the main symptoms are high temperature, oedema of the head, nasal discharge and rapid death.

fowl pox (*Vet*) Caused by large DNA virus. A worldwide and relatively slow-spreading infection of chickens and turkeys. Lesions consist of nodules in the skin which progress to scabs, and of diphtheria membranes in the respiratory and upper digestive tracts. Also *avian diphtheria, epithelioma contagiosa.*

fowl typhoid (*Vet*) A contagious septicaemic disease of chickens and other domesticated birds caused by *Salmonella gallinarum.* Vaccination using a *rough strain* of *S. gallinarum.*

fox encephalitis (*Vet*) An acute infection characterized by severe nervous symptoms, caused by the virus of INFECTIOUS CANINE HEPATITIS. Also *epizootic fox encephalitis, infectious encephalitis of foxes.*

fox marks (*Print*) Brown stains on the pages of books that have been allowed to get damp.

foxtail wedging (*Build*) The tightening up of a tenon in a blind mortise by inserting small wedges in saw-cuts in the end of the tenon before inserting the latter in the mortise. The operation of driving the tenon into position then forces the wedges into the saw-cuts and spreads the fibres of the tenon, giving a secure hold resisting withdrawal. Also *fox wedging.*

fox wedging (*Build*) See FOXTAIL WEDGING.

foyaite (*Geol*) A widely distributed variety of nepheline-syenite, described originally from the Foya Hills in Portugal. Typically it contains about equal amounts of nepheline and potassium feldspar, associated with a subordinate amount of coloured mineral such as aegirine.

FP (*Ships*) Abbrev for FORWARD PERPENDICULAR.

fp (*Phys*) Abbrev for FREEZING POINT.

FPS (*Genrl*) The system of measuring in *feet, pounds* and *seconds.*

Fr (*Chem*) Symbol for FRANCIUM.

fracking (*MinExt*) Forcing liquid out into the strata round a well bottom to increase the permeability of the petroleum formation. The liquid contains sand or other material which remains in the fissures and prevents them closing. Also *fracturing.* See HOT DRY ROCK, PROPPANTS, TERTIARY PRODUCTION.

fractal (*MathSci*) A set that has a non-integral dimension, usually characterized by self-similarity. For instance, an irregular coastline under increasing magnification shows the same degree of irregularity each time, and is estimated to have a dimension of 1·25.

fractals (*ICT*) Geometrical entities characterized by basic patterns that are repeated at ever-decreasing sizes. For instance, trees describe an approximate fractal pattern, as the trunk divides into branches that further subdivide into smaller branches which ultimately subdivide into twigs; at each stage of division the pattern is a smaller version of the original. Fractals are not able to fill spaces, and hence are described as having fractional dimensions. They were devised in 1967 by French mathematician Benoît Mandelbrot (1924–86), during a study of the length of the coastline of Great Britain. They are relevant to any system involving self-similarity repeated on diminishing scales, such as in the study of CHAOS, forked lightning or the movement of oil through porous rock. They are also used in computer graphics and in lossy IMAGE COMPRESSION.

fraction (*MathSci*) See DIVISION.

fractional crystallization (*Chem*) The separation of substances by the repeated partial crystallization of a solution.

fractional crystallization (*Geol*) The separation of a cooling magma into parts by crystallization of different minerals at successively lower temperatures.

fractional distillation (*Chem*) Distillation process for the separation of the various components of liquid mixtures. An effective separation can only be achieved by the use of FRACTIONATING COLUMNS attached to the still. Also *fractionation.* See PETROLEUM.

fractional distribution (*Chem*) In an assembly of particles having different values of some common property such as size or energy, the fraction of particles in each range of values is called the fractional distribution in that range. Cf CUMULATIVE DISTRIBUTION.

fractional pitch (*Eng*) In screw-cutting in the lathe, the state when the pitch is not an integral multiple or submultiple of the pitch of the LEAD SCREW. See EVEN PITCH.

fractionating column (*Chem*) A vertical tube or column attached to a still and usually filled with packing, eg Raschig rings, or intersected with bubble plates. An internal reflux takes place, resulting in a gradual separation between the high- and the low-boiling fractions inside the column, whereby the fractions with the lowest boiling point distil over. The efficiency of the column depends on its length or on the number of bubble plates used, and also on the ability of the packing to promote contact between the vapour and liquid phases.

fractionation (*BioSci*) A general term used for methods of separating or purifying materials, either molecules, eg by CHROMATOGRAPHY, or subcellular components, eg by differential CENTRIFUGATION.

fractionation (*Chem*) See FRACTIONAL DISTILLATION.

fractionation (*Radiol*) A system of treatment commonly used in radiotherapy in which doses are given daily or at longer intervals over a period of 3 to 6 weeks.

fraction one protein (*BioSci*) See RIBULOSE 1,5-BISPHOSPHATE CARBOXYLASE OXYGENASE.

fractography (*Eng*) The study of fracture surfaces of materials to determine nature and origin of product failure, eg whether brittle or ductile, single or multiple origins, association with stress concentrations, nature of crack propagation. For instance, stop–start mechanism may be indicated by beach marks on surface, so indicating FATIGUE (panel). Analysis usually starts with macrography, then micrography, often using SCANNING ELECTRON MICROSCOPY. Important investigative method in FORENSIC ENGINEERING. See BEACH MARKS, FRACTURE LANCES, HACKLE, MIRROR, MIST, STRIATIONS, WALLNER LINES.

fracture (*Med*) Breaking of a bone. Fractures may be *simple* (broken bone only), *compound* (external wound communicating with fracture), *complicated* (additional injury, eg to internal organs, blood vessels, etc), *comminuted* (bone broken in several or many parts), *fissured* (bone cracked, eg skull), IMPACTED, GREENSTICK.

fracture (*Min*) The broken surface of a mineral as distinct from its cleavage. The fracture is described, in different cases, as conchoidal (shell-like), platy or flat, smooth, hackly (like that of cast-iron) or earthy. Thus calcite has a perfect rhombohedral cleavage, but conchoidal fracture.

fracture cleavage (*Geol*) A set of closely spaced, parallel joints. Common in shallowly deformed metamorphic rocks.

fracture energy (*Eng*) The energy required to form unit area of fractured surface, numerically equal to half the TOUGHNESS. See STRENGTH MEASURES.

fracture lances (*Eng*) Characteristic, spear-like, fracture surface markings arising from fracture under torsional loading.

fracture mechanics (*Eng*) Stress analysis of the conditions and criteria for crack propagation in materials in terms of such properties as TOUGHNESS and FRACTURE TOUGHNESS. See FRACTURE MODE, GRIFFITH EQUATION, STRENGTH MEASURES, STRESS-INTENSITY FACTOR.

fracture mode (*Eng*) Three fundamental ways of loading a crack in a body: the opening mode, forward shear (or sliding) mode and transverse shear (or tearing) mode, designated modes I, II and III respectively.

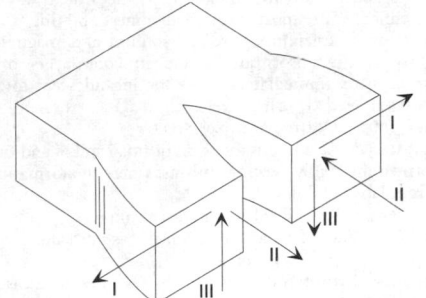

fracture mode I, opening; II, sliding; III, tearing.

fracture of materials (*Eng*) Loss of structural integrity by propagation of cracks in products or test samples. Often analysed using FRACTURE MECHANICS and FRACTOGRAPHY. May be brittle or ductile, depending on state of material, stress concentrations, rate of test, etc.

fracture stress (*Eng*) The tensile load at failure in a tensile test divided by the final cross-sectional area. In brittle materials equal to TENSILE STRENGTH. See STRENGTH MEASURES.

fracture test (*Eng*) Any test to determine strength or crack propagation resistance of a material. Additional information is provided by FRACTOGRAPHY of the fracture surfaces. See STRENGTH MEASURES.

fracture toughness (*Eng*) A materials property derived from FRACTURE MECHANICS and providing a measure of the material's resistance to crack propagation. It is the critical value of K_{IC}, the mode I STRESS INTENSITY FACTOR, for a crack in a material to propagate to failure. It is related to the TOUGHNESS G_C by

$$K_{IC} = \sqrt{E' G_C}$$

For PLANE STRAIN, $E'=E$, the YOUNG'S MODULUS of the material, while for PLANE STRESS

$$E' = \frac{E}{1 - v^2}$$

where v is POISSON'S RATIO. The plane strain value for K_{IC} is lower than the plane stress value, so is normally taken as the conservative measure of resistance to cracking. See FRACTURE MODE, STRENGTH MEASURES.

fracturing (*MinExt*) See FRACKING.

fragile-X syndrome (*Med*) Hereditary human condition involving mental subnormality in which a portion of the X chromosome tends to show breaks in appropriately prepared lymphocytes.

fragmental (*Min*) Applied to rocks that are formed of fragments of pre-existing rocks. Also *clastic*.

fragmentation (*BioSci*) (1) See AMITOSIS. (2) The breaking off of a portion from the main body of a chromosome. (3) A form of vegetative or asexual reproduction in which pieces of the parent become detached and grow into new individuals, esp in many filamentous algae.

fragmentation (*ICT*) A situation that occurs on a DISK when a FILE is stored in non-consecutive BLOCKS. This will increase the time taken to read or write to the file. See DEFRAGMENTATION.

fraktur (*Print*) A BLACK LETTER type peculiar to Germany.

framboesia (*Med*) See YAWS. Also *frambesia*.

frame (*Aero*) A transverse structural member of a fuselage, hull, nacelle or float, which follows the periphery and supports the skin or the skin-stiffening structure. Cf FORMER. See SPAR.

frame (*Build*) See FRAMEWORK.

frame (*ICT*) (1) One complete picture on a TV or VDU screen, consisting of 625 lines; a new frame is transmitted every 1/25th second, as two interlaced fields: the odd-numbered lines followed by the even-numbered lines. (2) A fixed-length sequence of bits in a digital communication system that repeats at regular intervals and normally consists of a fixed number of bits identifying the frame and a larger number of bits conveying transmitted data belonging to one user.

frame (*ImageTech*) The individual unit picture on a strip of motion picture film.

frame (*Print*) The compositor's workplace, providing storage for cases of type and provision for mounting them on top.

frame antenna (*ICT*) One comprising a loop of one or more turns of conductor wound on a frame, its plane being oriented in the direction of the incoming waves or, in transmission, for the direction of maximum radiation. Also *coil antenna*, *loop antenna*.

frame counter (*ImageTech*) (1) Automatic indicator in a camera showing the number of frames of film which have been exposed. (2) A measurer for processed motion picture film, showing the length in feet and frames which has passed over a SPROCKET.

framed (*Build*) Said of work assembled with mortise and tenon joints.

framed and braced door (*Build*) A boarded door secured in a frame consisting of two stiles, and top, middle and bottom rails, with diagonal braces between.

framed floor (*Build*) A floor in which the bridging joists are supported at intervals by binding joists, which in turn are supported at intervals by girders. Cf DOUBLE FLOOR.

framed grounds (*Build*) Grounds used in good work around openings such as door openings, the head being tenoned into the posts on each side.

frame direction-finding system (*ICT*) Simple type of direction-finder using a loop, preferably screened to obviate THE ANTENNA EFFECT, the polar response of which is a figure of eight; the loop is rotated until the received signal vanishes, when the axis of frame is in line with direction of arrival of wave. Also *loop direction-finding system*.

framed, ledged and braced door (*Build*) A boarded door secured in a frame consisting of two stiles and a top rail, and braced on one side with middle and bottom rails and diagonal braces.

frame finder (*ImageTech*) Viewfinder comprising a wire frame and a peep hole separated by a distance equivalent to the focal length of the lens.

frame frequency (*ImageTech*) The number of complete interlaced pictures, FRAMES, scanned per second; currently 30 Hz for US and Japanese broadcast practice and 25 Hz for Europe.

frame grabber (*ImageTech*) A device which captures a single FRAME from an off-air or video source and holds it in a FRAME STORE.

frame grid (*Electronics*) Said of a rugged high-performance thermionic valve in which the grid is held very rigid and close to the cathode surface by winding it on stiff rods.

frame level (*Build*) A mason's level.

frame line (*ImageTech*) The thin black line dividing the frames in the positive projection print of a motion picture.

frame noise (*ImageTech*) The noise caused in an optical sound reproducer if the slit is displaced so as to scan the edge of the picture area, including the frame lines.

frame of reference (*Phys*) See REFERENCE FRAME.

frame oscillator (*Electronics*) That which generates frame-scanning voltage or current. Also *field oscillator*.

frame relay (*ICT*) A basic service provided via INTEGRATED SERVICES DIGITAL NETWORK or other network whereby FRAMES are simply transferred between terminals without acknowledgement or error checking.

frames (*Build, CivEng*) (1) The centring used in concrete construction. (2) The built-up superstructure in any suitable structural material to form the skeleton of a building. (3) The surrounding timber or metal members of a door or window.

frame saw (*Build*) A thin-bladed saw, which is held taut in a special frame. Also *span saw*.

frame shift mutation (*BioSci*) The DNA base sequence is read in blocks of three so that insertion or deletion of a number of bases, not divisible by three, causes a drastic change and usually results in the generation, downstream, of a nonsense, chain-termination codon.

frame slip (*ImageTech*) The lack of exact synchronization of the vertical scanning and the incoming signal whereby the reproduced video picture progresses vertically.

frame store (*ImageTech*) A device for storing the information of two successive interlaced FIELDS; it can then release this in sequential (non-interlaced) form.

frame-sync pulse (*ImageTech*) A pulse transmitted at the end of each complete frame-scanning operation, to synchronize the framing oscillator at the receiver with that at the transmitter.

frame turner (*Ships*) A skilled worker engaged in turning and bevelling ships' frames, when red hot, to the shape of the ship's form.

frame-type switchboard (*ElecEng*) See SKELETON-TYPE SWITCHBOARD.

frame weir (*CivEng*) A type of movable weir consisting of a wooden barrier supported against iron frames placed at intervals across a river, and capable of being lowered onto the bed of the river in flood-time, or of being entirely removed.

framework (*Build*) The supporting skeleton of a structure.

framing (*Build*) The operation of assembling into final position the members of a structure.

framing (*ImageTech*) Vertical adjustment of film in the gate of a projector to centre the picture correctly in the aperture.

framing (*Psych*) The context or framework in which information is presented and that can affect, either positively or negatively, the influence it has on an individual or group.

francium (*Chem*) Symbol Fr, at no 87. The heaviest alkali metal. No stable isotopes exist; ^{223}Fr of half-life 22 min is the most important.

Franck–Condon principle (*Phys*) The principle that an electronic transition takes place so fast that a vibrating molecule does not change its internuclear distance appreciably during the transition. Applied to the interpretation of molecular spectra.

Franck–Hertz experiment (*Phys*) An early experiment that showed the existence of energy levels in the atom by measuring the kinetic energy lost by electrons in inelastic collisions with atoms in a discharge tube; used to determine excitation and ionization potentials.

Frankenstein food (*BioSci*) A foodstuff made or derived from plants or animals that have been genetically modified by other than conventional breeding techniques. Also *Frankenfood*.

franking (*Build*) The operation of notching a sash bar to make a mitre joint with a transverse bar.

franklinite (*Min*) Iron–zinc spinel, occurring rarely, as at the type locality, Franklin, New Jersey.

FRAP (*BioSci*) Abbrev for *fluorescence recovery after photobleaching*. See FLUORESCENCE RECOVERY.

Frasch process (*MinExt*) The method of mining elemental sulphur, by drilling into deposit and flushing it out by hot compressed air as a foam of low density. The sulphur is melted by superheated water.

Frasnian (*Geol*) The lower stage of the Upper Devonian rocks of Europe. See PALAEOZOIC.

frass (*BioSci*) Feces or excrement of wood-boring larvae.

frater (*Arch*) A refectory; sometimes applied in error to a monastic common room or to a chapter house. Also *frater house, fratry*.

Fraunhofer diffraction (*Phys*) Diffraction of a parallel light beam observed at an effectively infinite distance from the diffracting aperture; usually achieved with lenses to produce a parallel beam before diffraction and to focus the pattern onto an image plane. Cf FRESNEL DIFFRACTION.

Fraunhofer lines (*Astron*) Sharp narrow absorption lines in the spectrum of the Sun, 25 000 of which are now identified. The most prominent lines are due to the presence of calcium, hydrogen, sodium and magnesium. Most of the absorption occurs in cool layers of the atmosphere, immediately above the incandescent PHOTOSPHERE. See [A], [B], [C], etc.

Fraunhofer region (*Phys*) See FIELD.

frazil ice (*EnvSci*) Ice, in the form of small spikes and plates, formed in rapidly flowing streams, where the formation of large slabs is inhibited.

free (*BioSci*) Not joined to another member.

free (*Phys*) Said of a transducer when it is not *loaded*, eg the input has a *free impedance*.

free acceleration test (*EnvSci*) The test used to measure a vehicle's exhaust emissions.

free-air anomaly (*Geol*) A gravity anomaly which had been corrected for the height of the measured station above datum but without allowance for the attractive effect of topography. Cf BOUGUER ANOMALY.

free-air dose (*Radiol*) A dose of radiation, measured in air, from which secondary radiation (apart from that arising from air, or associated with the source) is excluded.

free association (*Psych*) A psychoanalytic technique for probing unconscious ideas and feelings; the individual is given a single word or concept and then verbalizes whatever thought comes to mind without structuring or censoring the content.

free atmosphere (*EnvSci*) The atmosphere above the friction layer where motion is determined primarily by the large-scale pressure field.

free atom (*Chem*) Unattached atom assumed to exist during reactions. See FREE RADICALS.

free balloon (*Aero*) Any balloon floating freely in the air, not propelled or guided by any power or mechanism, either within itself or from the ground.

free beaten stuff (*Paper*) Lightly beaten stuff with minimum hydration. The resultant paper is low in strength, bulky, porous and opaque. Such stock permits easy drainage of water on the machine wire.

freeboard (*Ships*) An assignment made by law to prevent overloading of a ship; calculated from statutory tables based on the vessel's form. Permanent markings are made on the ship's side to indicate the depth to which a ship may be loaded, and severe penalties are incurred by any overloading.

freeboard deck (*Ships*) Uppermost complete deck having permanent means of closing all openings.

free cell formation (*BioSci*) Free nuclear division followed by delimitation of separated cells without all parental cytoplasm being used, as in ascospore formation.

free cementite (*Eng*) Iron carbide in cast-iron or steel not associated with the ferrite in PEARLITE.

free central placentation (*BioSci*) Placentation in plants where the placentas develop on a central column or projection that arises from the base of a unilocular ovary and is not connected by septa to the ovary wall. Seen in many Caryophyllaceae, Primulaceae.

free charge (*ElecEng*) An electrostatic charge which is not bound by an equal or greater charge of opposite polarity.

free charge (*Phys*) A charge which is not bound to an atomic nucleus.

free cutting (*Eng*) See FREE MACHINING.

free-cutting brass (*Eng*) α-brass containing about 2–3% of lead, to improve the machining properties. Used for engraving and screw machine work.

freedom (*Eng*) A body free to move in space is said to have six degrees of freedom, three of translation and three of rotation.

free electron theory (*Electronics*) Early theory of metallic conduction based on concept that outer valence electrons, which do not form crystal bonds, are free to migrate through crystal, so forming ELECTRON GAS. Now superseded by ENERGY BAND theory.

free end (*Build*) The end of a beam which is not fixed or built in.

free energy (*Chem*) The capacity of a system to perform work, a change in free energy being measured by the maximum work obtainable from a given process. See GIBBS FREE ENERGY, HELMHOLTZ FREE ENERGY.

free energy equation (*Chem*) See GIBBS–HELMHOLTZ EQUATION.

free fall (*Space*) The motion of an unpropelled body in a gravitational field. In orbit beyond the Earth's atmosphere, free fall produces near-weightlessness. See MICROGRAVITY.

free-falling velocity (*PowderTech*) The velocity of fall of a particle of powder through a still fluid, at which the effective weight of the particle is balanced by the drag exerted by the fluid on the particle.

free ferrite (*Eng*) Ferrite in steel or cast-iron not associated with the cementite in pearlite.

free field (*Acous*) Sound field which is radiated directly from a source, without being reflected elsewhere.

free-field emission (*Phys*) That from an emitter when the electric gradient at a surface is zero.

free-flight wind tunnel (*Aero*) (1) A wind tunnel, usually of up-draught type, wherein the model is not mounted on a support, but flies freely. (2) One in which three pilots each control one axis of a free-flying model, as used by NASA in the large Ames 80 ft tunnel. (3) Ballistic shape fired into airflow of a wind tunnel for shock wave or re-entry experiments.

free floating anxiety (*Psych*) See GENERALIZED ANXIETY DISORDER.

free-form database (*ICT*) A DATABASE system that does not have strict layout constraints as regards RECORD formats.

free-handle (*ElecEng*) See FREE-TRIP.

free-hearth electric furnace (*ElecEng*) A direct-arc furnace in which one electrode forms a part of the bottom of the hearth.

free impedance (*Phys*) The impedance at input terminals of a transducer when its load impedance is zero.

freeing port (*Ships*) An opening in the bulwark to free the deck from large quantities of water which may come on board in heavy weather. Usually with a hinged plate fitted to prevent water coming in.

free machining (*Eng*) An alloy with additions to make it easier to machine and so reduce machining time and power consumption, usually effected by causing chips to break up rather than absorb energy by plastically deforming, eg the addition of lead to steels and copper alloys.

freemartin (*Vet*) In cattle, a sterile female intersex occurring as co-twin with a normal bull-calf. Sex hormones from the male sterilize the female calf.

free milling (*MinExt*) Descriptive term for gold or silver ore, which contains its metal in amalgamable state.

free-needle surveying (*Surv*) Traverse work done with a compass, the magnetic bearing of each line being read in turn. Cf FIXED-NEEDLE SURVEYING.

free oscillations (*ICT*) (1) Oscillatory currents whose frequency is determined by constants of the circuit in which they are flowing, eg those resulting from the discharge of a capacitance through an inductance. (2) Mechanical oscillations governed solely by the natural elastic properties of a vibrating body.

free path (*Phys*) See MEAN FREE PATH.

freephone (*ICT*) An INTELLIGENT NETWORK service that, usually via a fixed 'freephone' prefix, allows telephone callers to contact a subscribing organization, which then pays the call charges.

free-piston engine (*Eng*) A prime mover of the internal-combustion type, in which a power piston acts directly on a compressor piston working in the opposite direction. May be compounded, as an alternative to a rotary compressor, with a gas turbine.

free pole (*ElecEng*) A magnet pole which is imagined, for theoretical purposes, to exist separately from its corresponding opposite pole.

free-radical polymerization (*Chem*) Polymerization of doubly bonded monomers using a free-radical catalyst such as benzoyl peroxide. See CHAIN POLYMERIZATION and panel on POLYMER SYNTHESIS.

free radicals (*Chem*) Groups of atoms in particular combinations capable of free existence under special conditions, usually for only very short periods (sometimes only microseconds). Because they contain unpaired electrons, they are paramagnetic, and this fact has been used in determining the degree of dissociation of compounds into free radicals. The existence of free radicals such as methyl (CH_3.) and ethyl (C_2H_5.) has been known for many years. See CHAIN POLYMERIZATION and panels on POLYMERS and POLYMER SYNTHESIS.

free range (*Agri*) A poultry-rearing system where birds can range at will over relatively large areas with no restriction on behaviour.

free-range grazing (*Agri*) Grazing that allows livestock unrestricted access to all areas of the pasture.

free recall (*Psych*) A test of memory in which the subject is required to produce as many learned items of the learned task as possible, without regard for the order in which it was first learned.

free-running (*Psych*) Experimental condition used in testing for circadian or other rhythms, in which the environmental conditions are kept constant. Animals kept in complete darkness will often continue to exhibit an approximately 24-hour cycle of activity and rest.

free-running circuit (*Electronics*) Running without external synchronization.

free-running frequency (*ICT*) That of an oscillator, otherwise uncontrolled, when not locked to a synchronizing signal, eg a time-base generator in a TV receiver.

free-running speed (*ElecEng*) The speed which a vehicle or train will attain when propelled by a constant tractive effort, ie the speed at which the applied tractive effort exactly equals the forces resisting motion. Also *balancing speed*.

free-settling (*MinExt*) In classification of finely ground mineral into equal-settling fractions, use of conditions in which particles can fall freely through the environmental fluid, as opposed to the packed conditions of mineral settling.

free space (*BioSci*) The APOPLAST, esp in estimates of the fraction it forms (the apparent free space) of a tissue.

free space impedance (*Phys*) For electromagnetic waves the characteristic impedance of a medium is given by the square root of the ratio of permeability to permittivity (or 4π times this in unrationalized systems of units). This gives a free space value of $376 \cdot 6 \ \Omega$.

freestone (*Build*) A building-stone which can be worked with a chisel without tending to split into definite layers.

free sulphur dioxide (*FoodSci*) Dissolved but uncombined sulphur dioxide together with sulphite and bisulphite ions, present in food preserved by adding SULPHUR DIOXIDE.

free surface (*Ships*) A free surface exists when any compartment in the ship is partly filled with liquid.

free-trip (*ElecEng*) Said of certain types of circuit breaker or motor starter in which the tripping mechanism is independent of the closing mechanism, and will therefore allow the switch to trip while the latter is being operated. Also *free-handle*. Cf FIXED-TRIP.

free turbine (*Aero*) A power take-off turbine mounted behind the main turbine/compressor rotor assembly and either driving a long shaft inside the main rotor to a gearbox at the front of the engine, or a short shaft to a gearbox at the rear of the engine. It can also be a separate unit fed by a remotely produced gas supply.

free vibrations (*Phys*) The vibrations which occur at the natural frequency of a body when it has been displaced from its position of rest and allowed to vibrate freely without the application of any periodic force.

free volume (*Phys*) The term for the empty space surrounding atoms and molecules in solid or fluid materials. Symbol $V_f = V - V_0$ where V is the SPECIFIC VOLUME (reciprocal of density) and V_0 the specific volume extrapolated to 0 K. Used in molecular theories of the glass transition and of viscosity.

free vortex flow (*Aero*) A vortex persisting away from a solid surface as in a natural tornado. A bound vortex is one attached to the body creating eg a wing-tip vortex.

freeware (*ICT*) A program that can be copied without charge but not sold.

free-wheel (*Autos*) A one-way clutch, usually depending on the wedging action of rollers, placed in the transmission line, so as to transmit torque only when the engine is driving.

freeze branding (*Agri*) Livestock branding of animals, esp horses, using an ultra-cold iron. The regrown hair has no pigment so the technique is effective for pigmented animals. It is said to cause less hide damage and distress to the branded animal.

freeze-drying (*BioSci*) A method of fixing tissues or complex macromolecules (*lyophilization*) by freezing sufficiently rapidly as to inhibit the formation of ice crystals, eg by immersion in isopentane cooled to −190°C, followed by dehydration *in vacuo*. Also used as a method of food preservation in which the moisture is removed after freezing by sublimation under vacuum. See ACCELERATED FREEZE-DRYING.

freeze-etch (*BioSci*) A technique for specimen preparation for electron microscopy in which tissue is frozen (strictly vitrified) by very rapid cooling to below −100°C, fractured and a surface replica made (for microscopical examination) either immediately (known as *freeze-fracture*) or after allowing some water to sublime (freeze-etch). See CARBON REPLICA TECHNIQUE.

freeze frame (*ImageTech*) Repetition of a single frame to hold the action stationary.

freeze line (*Plastics*) A line formed on BLOWN FILM showing the onset of crystallization, so transparent below but translucent above.

freezer burn (*FoodSci*) FOOD SPOILAGE in which white patches occur on the surface of frozen food. Caused by localized moisture loss which is sometimes reversible by rehydration on defrosting, but in meat and fish is due to irreversible protein denaturation resulting in toughness and unacceptable texture.

freeze-sinking (*MinExt*) Shaft sinking, or penetration of waterlogged strata, in which refrigerated brine is circulated to freeze the ground and make establishment of an imperviously lined shaft possible.

freeze-substitution (*BioSci*) Specimen preparation for electron microscopy in which tissue is rapidly 'frozen' (see FREEZE-ETCH) and then the ice dissolved out at low temperatures with suitable solvents (often containing a fixative) before embedding in the normal way.

freeze–thaw stability (*FoodSci*) The ability of a frozen food to retain shape and texture without fluid loss after defrosting, related to the rate of freezing and method of defrosting and also to the amount of water in the cells or the emulsion.

freezing (*FoodSci*) Reducing the temperature of a food product to below −18°C, often resulting in ice crystals puncturing the cell membranes which causes water losses on defrosting. The quicker the freezing, the smaller the crystals and the less the effect. Quick-freezing techniques include blast freezing in air cooled below −25°C and contact freezing on cold metal plates. See CRYOGENIC FREEZING.

freezing (*Phys*) The conversion of a liquid into the solid form. This process occurs at a definite temperature for each substance, this temperature being known as the *freezing point*. The freezing of a liquid invariably involves the extraction of heat from it, known as *latent heat of fusion*. See DEPRESSION OF FREEZING POINT, SPECIFIC LATENT HEAT.

freezing mixture (*Chem*) A mixture of two substances, generally of ice and a salt, or solid carbon dioxide with an alcohol, used to produce a temperature below 0°C.

freezing point (*Phys*) The temperature at which a substance solidifies; the same as that at which the resulting solid melts (the *melting point*). The freezing point of water is used as the lower fixed point in graduating a thermometer. Its temperature is defined as 0°C (273·15 K). Pure materials, eutectics and some intermediate constituents freeze at constant temperature; alloys generally solidify over a range. Abbrev *fp*. See DEPRESSION OF FREEZING POINT, PHASE DIAGRAM, TRIPLE POINT, WATER.

freezing point method (*Chem*) See CRYOSCOPIC METHOD.

freight car (*CivEng*) US for a goods wagon.

fremitus (*Med*) Palpable vibration, esp of the chest wall during speech or coughing; variations in intensity are of diagnostic value.

Fremont test (*Eng*) A type of impact test in which a beam specimen notched with a rectangular groove is broken by a falling weight.

Frémy's salt (*Chem*) Potassium hydrogen fluoride, acid potassium fluoride (KHF_2).

French bit (*Build*) A boring tool having a flat blade, shaped at the two cutting edges in continuous curves, from the point to and beyond a place of maximum diameter; it is used in a lathe-head for drilling hard wood.

French chalk (*Min*) The mineral, talc, ground into a state of fine subdivision, its softness and its perfect cleavage contributing to its special properties when used as a filler or dry lubricant.

French curve (*Eng*) A drawing instrument used to guide the pen or pencil in drawing curved lines. It consists of a thin flat sheet of transparent plastic or other material cut to curved profiles at the edges.

French door (*Arch*) FRENCH WINDOW.

French flag (*ImageTech*) See FLAG.

French fliers (*Build*) Steps in an OPEN NEWEL STAIR with QUARTER-SPACE LANDINGS.

French fold (*Print*) A sheet of paper folded twice at right angles to make eight pages, but left uncut, and printed only on pages 1, 4, 5 and 8.

French gold (*Eng*) Oroide. A copper-based alloy with about 16·5% zinc, 0·5% tin and 0·3% iron.

French groove (*Print*) See FRENCH JOINT.

French joint (*Print*) One in which the book cover is given a wide space between board and spine to enable the board to hinge back freely. Also *French groove*.

frenchman (*Build*) A joint-trimming tool, used for pointing.

French moult (*Vet*) A defective development of the first plumage of birds leading to the shedding of the wing and tail primaries; particularly observed in aviary-bred budgerigars.

French pitch (*Phys*) A standard pitch for musical instruments. A is 435 Hz at 15°C. See CONCERT PITCH.

French polish (*Build*) A solution of shellac dissolved in methylated spirit. Applied to wood surfaces to produce a high polish.

French roof (*Build*) A MANSARD ROOF.

French sewing (*Print*) (1) The normal method of present-day machine sewing, ie without tapes. (2) A method of hand sewing at the edge of the bench without using a frame.

French stuc (*Build*) Plasterwork finished to present a surface resembling that of stonework.

French truss (*Eng*) A symmetrical roof truss for large spans, composed of a pair of braced isosceles triangles based on the sloping sides of the upper chord, their apices being jointed by a horizontal tie. Also *Belgian truss*, *Fink truss*.

French window (*Arch*) A glazed casement, serving as both window and door.

Frenet's formulae (*MathSci*)

$$\mathbf{t}' = +\frac{1}{\rho}\mathbf{n}$$
$$\mathbf{n}' = -\frac{1}{\rho}\mathbf{t} \quad +\frac{1}{\tau}\mathbf{b}$$
$$\mathbf{b}' = -\frac{1}{\tau}\mathbf{n}$$

where $\mathbf{t}$, $\mathbf{n}$ and $\mathbf{b}$ are unit vectors along the tangent, principal normal and binormal respectively at a point on a space curve, ρ and τ are the radii of curvature and torsion respectively, and a dash (') denotes differentiation with respect to arc length. Also *Serret–Frenet formulae*.

Frenkel defect (*Crystal*) Disorder in the crystal lattice, due to some of the ions (usually the cations) having entered interstitial positions, leaving a corresponding number of normal lattice sites vacant. Likely to occur if one ion (in practice the cation) is much smaller than the other, eg in silver chloride and bromide.

frenotomy (*Med*) Cutting of the frenum (the small band of tissue that connects the tongue to the floor of the mouth) for tongue-tie. Also *fraenotomy*.

Freons (*Chem*) TN for compounds consisting of ethane or methane with some or all of the hydrogen substituted by fluorine, or by fluorine and chlorine, ie CFCs. Principal ones are: Freon 11, trichlorofluoromethane (CCl_3F); Freon 12, dichlorodifluoromethane (CCl_2F_2); Freon 21, dichlorofluoromethane ($CHCl_2F$); Freon 114, dichlorotetrafluoroethane ($CClF_2CClF_2$); Freon 142, 1-chloro-1:1-difluoroethane ($CH_2 CClF_2$). Use now deprecated. See CHLOROFLUOROCARBONS.

frequency (*BioSci*) The number of any given species in an area, or percentage of sample squares containing the species.

frequency (*Phys*) The rate of repetition of a periodic disturbance, measured in hertz (cycles per second). Also *periodicity*.

frequency allocation (*ICT*) The frequency on which a transmitter has to operate, within specified tolerance. Bands of frequencies for specified services are allocated by international agreement.

frequency band (*ICT*) The interval in the frequency spectrum occupied by a modulated signal. In sinusoidal amplitude modulation, it is twice the maximum modulation frequency, but it is much greater in frequency or pulse modulation. See FREQUENCY ALLOCATION.

frequency bridge (*ElecEng*) An ac measuring bridge whose balancing condition is a function of the supply frequency, eg *Robinson bridge*.

frequency changer (*ElecEng*) A circuit (or machine) designed to receive an input at one frequency and to convert this to an output at a different frequency. US *converter*.

frequency changer (*ICT*) Mixer or frequency converter. See SUPERHET RECEIVER.

frequency converter (*ICT*) Same as frequency changer. See SUPERHET RECEIVER.

frequency counter (*Electronics*) A device which produces an alphanumeric read-out of the frequency of an incoming signal. Frequencies up to 100 MHz can be measured, the accuracy being limited only by the internal reference oscillator or frequency source. Microwave frequencies can be accommodated by means of external frequency converters and oscillators. See FREQUENCY DIVIDER.

frequency demultiplication (*ICT*) See FREQUENCY DIVISION.

frequency departure (*ICT*) Discrepancy between actual and nominal carrier frequencies of a transmitter. Formerly termed *frequency deviation*.

frequency deviation (*ICT*) (1) In frequency modulation, maximum departure of the radiated frequency from mean quiescent frequency of the carrier. (2) Greatest deviation allowable in operation of frequency modulation. In broadcast systems within the range 88–108 MHz, the maximum deviation is ± 75 kHz. (3) See FREQUENCY DEPARTURE.

frequency-discriminating filter (*ICT*) See FILTER.

frequency discriminator (*ICT*) A circuit, the output from which is proportional to frequency or phase change in a carrier from condition of no frequency or phase modulation.

frequency distortion (*Acous*) In sound reproduction, variation in the response to different notes solely because of frequency discrimination in the circuit or channel. Generally plotted as a decibel response on a logarithmic frequency base.

frequency distribution (*MathSci*) See DISTRIBUTION.

frequency diversity (*ICT*) See DIVERSITY RECEPTION.

frequency divider (*Electronics*) A digital circuit, an essential part of a FREQUENCY COUNTER, which by responding to individual cycles of an incoming signal, and feeding the

resultant pulses to an array of gating circuits, can produce subharmonics of the input frequency.

frequency division (*ICT*) Dividing a frequency by harmonic locking oscillators or stepping pulse circuits. Specific integral division alone can be obtained; an arbitrary collection of frequencies cannot be divided, except by recording and reproducing at a lower speed. Also *frequency demultiplication*.

frequency-division multiple access (*ICT*) The creation of several channels within an allocated frequency band by assigning a fixed sub-band to each channel, eg to each of the mobile telephones served by a particular CELL.

frequency-division multiplex (*ICT*) A method of multiplex transmission in which individual speech or information channels are modulated to separate channels and then transmitted simultaneously over a cable or microwave link. Single-sideband suppressed carrier methods are normally employed. See GROUP.

frequency-division multiplexing (*ICT*) A technique by which each CHANNEL is transmitted at a different CARRIER FREQUENCY.

frequency doubler (*ICT*) Frequency multiplier in which the output current or voltage has twice the frequency of the input. Achieved by a combination of non-linear and tuned circuits.

frequency doubling (*ICT*) The introduction of marked double-frequency components through lack of polarization in an electromagnetic or electrostatic transducer, in which the operating forces are proportional to the square of the operating currents and voltages respectively.

frequency drift (*ICT*) The change in frequency of oscillation because of internal (ageing, change of characteristic) or external (variation in supply voltages or ambient temperature) causes. Also *oscillator drift*.

frequency factor (*Chem*) The pre-exponential factor in ARRHENIUS'S EQUATION when applied to chemical reactions, expressing the frequency of successful collisions between the reactant molecules.

frequency hopping (*ICT*) A radio security technique in which the CARRIER frequency of a transmission is rapidly and pseudo-randomly changed in order to defeat unauthorized access. The authorized receiver reconstructs the signal by hopping in step with the transmitter, using a matching pseudo-random-number generator.

frequency meter (*Electronics*) A device which compares an unknown frequency with known standard, which may be a standard-frequency transmission derived from an ATOMIC CLOCK, the output of a crystal-controlled or other precision oscillator, or the comparison may be made with the resonant frequency of a tuned circuit or resonant line or structure. See ABSOLUTE WAVEMETER, FREQUENCY COUNTER, HETERODYNE WAVEMETER, WAVEMETER.

frequency-modulated cyclotron (*ElecEng*) One in which frequency of the voltage applied to the DEES is varied, so as to keep synchronous orbiting of accelerated particles when their mass increases through relativity effect at the high velocities attained.

frequency modulation (*ICT*) The method of impressing a signal on a carrier wave in which the frequency of the carrier is made to vary in proportion to the instantaneous value of the signal, the amplitude remaining constant. The spectrum occupied is theoretically infinite but in practice can be taken as approximately the band occupied by the carrier between its maximum and minimum frequencies plus twice the bandwidth of the modulating signal. Abbrev *FM*.

frequency-modulation receiver (*Electronics, ICT*) One incorporating a frequency demodulator. See FOSTER-SEELEY DISCRIMINATOR, FREQUENCY DISCRIMINATOR, RATIO DETECTOR.

frequency monitor (*ElecEng*) Nationally or internationally operated equipment to ascertain whether or not a transmitter is operating within its assigned channel.

frequency multiplier (*Electronics*) Non-linear circuit in which the output circuit is tuned to select a harmonic of the input signal. Transistor or varactor diode circuits may be used.

frequency of gyration (*Electronics*) That of electrons about a line indicating direction of magnetic field in ionosphere.

frequency of infinite attenuation (*ICT*) A frequency at which a filter inserted in a communication channel provides a maximum attenuation theoretically infinite with loss-free inductances and capacitances. Such large attenuation is generally provided by an anti-resonant series arm, or by an acceptance resonant shunt arm.

frequency of penetration (*Phys*) The frequency of a wave which just fails to be reflected by ionospheric layer.

frequency overlap (*ICT*) Common parts of frequency bands used eg for the regular video signal and the chrominance signal in colour TV.

frequency pulling (*ICT*) The change in oscillator frequency resulting from variation of load impedance.

frequency relay (*ElecEng*) See RELAY.

frequency response (*Phys*) For constant input power applied to a TRANSDUCER through a range of discrete frequencies, the ENVELOPE of the output powers at each of those frequencies over the given range. The response may be measured absolutely in watts against hertz; by implication, eg volts or intensity against frequency; or proportionately, eg DECIBELS below peak output response against frequency. A *flat* or *level response* therefore indicates equal response to all frequencies within the stated range, eg for audio equipment an equal response to within, say, 1 dB for the range 20 Hz to 20 kHz.

frequency selectivity (*ICT*) See SELECTIVITY.

frequency-shift keying (*ICT*) In radiotelegraphy, altering the carrier by mark-and-space keying.

frequency-shift transmission (*ICT*) A form of modulation used in communication systems in which the carrier is caused to shift between two frequencies denoting respectively on and off pulses.

frequency stabilization (*ICT*) The prevention of changes produced in frequency of oscillation of a self-oscillating circuit by changes in supply voltage, load impedance, valve parameters, etc. Achieved by resonating crystals, tuned cavities or transmission lines.

frequency standard (*ICT*) (1) Reference oscillator of very-high-frequency stability; may be quartz-crystal controlled, although atomic beam standards provide the ultimate reference. (2) Special transmissions, often with precision time codes added, that can be received worldwide and used as standards.

frequency swing (*ICT*) Extreme difference between maximum and minimum instantaneous frequencies radiated by a transmitter.

frequency synthesizer (*ICT*) A source of signals of precisely defined frequency; they may be sine, square or pulse waveform and frequencies may range from zero (dc) to microwave. The output is derived from one or more precision crystal-controlled oscillators, working with multipliers, dividers and mixers.

frequency table (*MathSci*) A table classifying a set of observations by the number of occurrences of particular values or types.

frequency tolerance (*ICT*) The extent to which frequency of the carrier of a transmission is permitted to deviate from its allocation.

frequency transformer (*ElecEng*) See FREQUENCY CHANGER.

frequency translation (*ICT*) Shifting all frequencies in a transmission by the same amount (not through zero).

frequency tripler (*Electronics*) See FREQUENCY MULTIPLIER.

fresco (*Build*) A method of painting on plastered walls with lime-fast colours while the plaster is still wet.

fresh-water allowance (*Ships*) The difference between the FREEBOARD in sea water of 1025 and in fresh water of 1000 oz ft^{-3} or g dm^{-3}.

fresh-water sediments (*Geol*) Sediments which are accumulating or have accumulated in fresh water, ie river, lake or glaciofluvial environments.

fresnel (*Phys*) A unit of optical frequency, equal to 10^{12} Hz = 1 THz (terahertz).

Fresnel–Arago laws (*Phys*) Laws concerning the condition of interference of polarized light. (1) Two rays of light emanating from the same polarized beam, and polarized in the same plane, interfere in the same way as ordinary light. (2) Two rays of light emanating from the same polarized beam and polarized at right angles to each other will interfere only if they are brought into the same plane of polarization. (3) Two rays of light polarized at right angles and emanating from ordinary light will not interfere if brought into the same plane of polarization.

Fresnel diffraction (*Phys*) The study of the diffracted field at a distance from an aperture in an absorbing screen, the distance being large compared with the wavelength but not so large that the curvature of the wavefronts can be neglected. See FIELD. Cf FRAUNHOFER DIFFRACTION.

Fresnel ellipsoid (*Phys*) A method of representing the doubly refracting properties of a crystal, used in crystal optics.

Fresnel lens (*ImageTech*) A lens having a surface of stepped concentric circles, thinner and flatter than a conventional lens of equivalent focal length; used in viewfinders and as a condenser in studio SPOT LIGHTS.

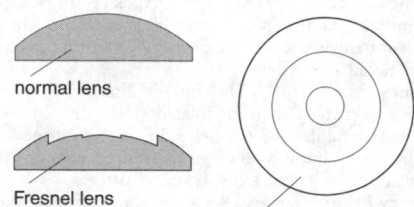

normal lens

Fresnel lens

Fresnel lens in plan view

Fresnel lens Annuli in practice have conical surfaces and are closer together.

Fresnel region (*Phys*) See FIELD.

Fresnel's biprism (*Phys*) An isosceles prism having an angle of nearly 180°, used for producing interference fringes from the two refracted images of an illuminated slit.

Fresnel's mirrors (*Phys*) Two plane mirrors inclined at an angle of a little less than 180°, used for producing interference fringes from the two reflected images of an illuminated slit.

Fresnel's reflection formula (*Phys*) A formula giving the fraction of the intensity of unpolarized incident light reflected at the surface of a transparent medium. The fraction equals

$$\frac{1}{2}\left(\frac{\sin^2(i-r)}{\sin^2(i+r)} + \frac{\tan^2(i-r)}{\tan^2(i+r)}\right)$$

where i and r are the angles of incidence and refraction respectively.

Fresnel's rhomb (*Phys*) A glass rhomb which is used for obtaining circularly polarized light from plane-polarized light by total internal reflection. The rhomb is so constructed that two such reflections at an angle of 54° are obtained, each of which introduces a phase difference of one-eighth of a period between the two components obtained from the incident plane-polarized light.

Fresnel zone (*Phys*) One of the zones into which a wavefront is divided, according to the phase of the radiation reaching any point from it. The radiation from any one zone will reach the given point half a period out of phase with the adjacent zones.

FRET (*BioSci*) Abbrev for fluorescence resonance energy transfer; see FLUORESCENCE ENERGY TRANSFER.

fret (*Print*) A continuous border design of interlaced bands or fillets, tooled on the case of a book or used for typographic decoration.

fret-saw (*Build*) One with a very thin and narrow blade (usually 5 in or 125 mm long) kept in tension by a metal bow, elongated 12–20 in (300–500 mm). Used for cutting narrow curves in thin wood. There is also a mechanical type.

fretted lead (*Build*) Lead section used for joining glass panes. See CAME.

fretting corrosion (*Eng*) Corrosion due to slight movements of unprotected metal surfaces, left in contact either in a corroding atmosphere or under heavy stress.

fret-work (*Build*) (1) A mode of glazing in which diamond-shaped panes (quarrels) are connected together by leaden canes to form a window. (2) Panels with holes cut through to form designs.

Freudian slip (*Psych*) An error in speech, which Freud believed revealed unconscious ideas or wishes.

Freud's theory of dreams (*Psych*) Holds that the dream is a disguised transformation of unconscious wishes. The *manifest content* of the dream refers to the images we remember upon waking, and is a transformation of repressed wishes and ideas, the *latent dream content*; dream interpretation consists of working out the nature and meaning of the transformation.

Freundlich's adsorption isotherm (*Chem*) The concentration c of adsorbate is related to the equilibrium partial pressure p as $c = ap^b$, where a and b are empirical constants.

Freund's adjuvant (*BioSci*) A water-in-oil emulsion that is used to stimulate a vigorous response to a soluble antigen. Complete adjuvant contains heat-killed tubercle bacteria, but these are omitted from incomplete adjuvant.

Freysinnet (*CivEng*) A method used for the stretching and anchoring of high-tensile wires in a post-tensioned reinforced concrete system, after the French engineer (1879–1962) who developed prestressed concrete.

friable (*MinExt*) Of ore, easily fractured or crumbled during transport or comminution.

friar (*Print*) An area of print with too little ink. See MONK.

friction (*Med*) (1) The sound produced by the rubbing together of two inflamed surfaces, as in pleurisy or pericarditis. (2) Rubbing of a part, as in massage.

friction (*Phys*) The resistance to sliding motion between two surfaces in contact. The frictional force opposing the motion is equal to the applied force up to the onset of motion when its value is known as the *limiting friction*. Any increase in the applied force will then cause slipping. *Static friction* is the value of the limiting friction just before slipping occurs. *Dynamic friction* is the value of the frictional force after slipping has occurred, is smaller than the static friction, and can depend on sliding speed. The *coefficient of friction*, symbol μ, is the ratio of the limiting friction to the normal reaction between the sliding surfaces. Under normal conditions μ is a constant between a given pair of materials of specified surface quality and lubrication conditions. With polymers, μ depends on the normal reaction and on the duration of loading or the sliding speed.

frictional damper (*Eng*) A device consisting of a supplementary mass frictionally driven from a crankshaft at a point remote from a node, which dissipates vibrational energy in heat.

frictional electricity (*Phys*) Static electricity produced by rubbing bodies or materials together, eg an ebonite rod with fur. See TRIBOELECTRIFICATION.

frictional machine (*ElecEng*) See ELECTROSTATIC GENERATOR.

friction and windage loss (*ElecEng*) Losses in an electrical machine due to friction of sliding parts (see FRICTION LOSS) and also to air resistance. These losses are frequently considered together in designing and testing electrical machinery.

friction calendering (*Textiles*) Passing fabric between two rolls of a CALENDER designed so that one of the rolls is highly polished and rotates faster that the other. This produces a glaze on the fabric surfaces.

friction clutch (*Eng*) A device for connecting or disconnecting two shafts which are in line, in any relative position, through the friction of two surfaces in contact. It consists of a pair of opposed members, between which the drive is transmitted

through the friction of their contact surfaces, and which may be separated by a lever system. Also *friction drive*.

friction compensation (*ElecEng*) A small torque, additional to the main torque, provided in a motor-type integrating meter to compensate for the effect of friction of the moving parts.

friction drive (*Eng*) See FRICTION CLUTCH.

friction feed (*ICT*) A mechanism for advancing paper in a printer by gripping it between rollers.

friction gear (*Eng*) A gear in which power is transmitted from one shaft to another through the tangential friction set up between a pair of wheels pressed into rolling contact. One of the contacting surfaces is usually fabric-faced. Suitable only for small powers.

friction glazing (*Paper*) A method of glazing paper in which one of the CALENDER rolls revolves at a greater peripheral speed than that of the others. A very high polish is obtained.

friction horsepower (*Eng*) That part of the gross or indicated horsepower developed in an engine cylinder which is absorbed in frictional losses; the difference between the indicated and the brake horsepower.

friction layer (*EnvSci*) The atmospheric layer, extending to a height of about 600 m, in which the influence of surface friction is appreciable. Also *planetary boundary layer*.

friction loss (*ElecEng*) The power absorbed in the bearings, commutator, or slip-ring surfaces, or at any other sliding contacts of an electric machine.

friction pile (*Build*) A pile which supports its load only by the friction on its sides.

friction rollers (*Eng*) See ANTI-FRICTION BEARING.

friction spinning (*Textiles*) A method for converting staple fibres into a yarn by feeding a sliver onto a rotating perforating roller through which air is being sucked. Another roller is set near so that the yarn is formed from the rapidly twisting fibres at the nip of the rollers.

friction twisting (*Textiles*) A method of texturing yarns.

friction welding (*Eng*) Welding in which the necessary heat is produced frictionally, eg by rotation, and forcing the parts together.

Friedel–Crafts polymers (*Chem*) Heat-resistant aromatic polymers made by alkylation of phenols, structurally intermediate between PF RESINS and polyphenylenes.

Friedel–Crafts synthesis (*Chem*) The synthesis of alkyl-substituted benzene hydrocarbons and aromatic ketones, by the action of halogenoalkanes or acyl halides on aromatic hydrocarbons in the presence of anhydrous aluminium chloride.

Friedmann universe (*Astron*) Cosmological model in which the universe is homogeneous, isotropic and non-static, and whose dynamics are determined by the gravitational effects of its non-zero matter density.

Friedreich's ataxia (*Med*) An autosomal recessive hereditary disease of the central nervous system which usually occurs in childhood. It is often associated with skeletal discontinuities and about half have cardiomyopathy.

friendly numbers (*MathSci*) Pairs of numbers each of which is the sum of the factors of the other, including unity, eg 220 and 284. Also *amicable numbers*. Cf PERFECT NUMBER.

frieze (*Arch, Build*) (1) The middle part of an entablature, between the architrave and the cornice. (2) The decorated upper part of a wall, below the cornice.

frieze (*Textiles*) Woven woollen overcoating fabric whose surface has been heavily milled and raised.

frieze panel (*Build*) An upper panel in a six-panel door.

frieze rail (*Build*) The rail next to the top rail in a six-panel door.

frilling (*ImageTech*) Crinkling and separation of the emulsion layer from the glass of a photographic plate, usually as a result of too high a temperature in processing.

fringe area (*ImageTech*) In radio or TV broadcasting, regions remote from a transmitter in which good reception is uncertain.

fringed micelle model (*Chem*) A model of polymer crystallites, where chains wander between crystallites rather than folding into lamellae. See CRYSTALLIZATION OF POLYMERS and panel on POLYMERS.

fringe effect (*ImageTech*) A faint light line on the low-density side of the boundary between a lightly and a heavily exposed area on a developed emulsion. Also *edge effect*. See MACKIE LINES.

fringe medicine (*Med*) See ALTERNATIVE MEDICINE.

fringes (*ImageTech*) Coloured edges visible in an image caused by CHROMATIC ABERRATION or as a result of imperfect registration of the colour components.

fringing (*ElecEng*) The spreading of flux lines in a given field configuration (conduction, electric or magnetic), eg at the edges of a parallel-plate capacitor or the edges of an air-gap in a magnetic circuit.

fringing coefficient (*ElecEng*) A coefficient used in making magnetic circuit calculations to allow for the effect of fringing of the flux.

fringing reef (*Geol*) A coral reef directly attached to or bordering the shore of an island or continent, having a rough table-like surface exposed at low tide. Cf ATOLL, BARRIER REEF.

frisket (*Print*) (1) On a hand press, a thin iron frame covered with paper, to hold the sheet to the tympan, prevent it from being soiled, and strip it from the forme after printing. (2) On a platen machine, the adjustable metal fingers which strip the sheet from the forme after printing.

frisking (*NucEng*) Searching for radioactive radiation by contamination meter, usually a portable ionization chamber.

frit (*Glass*) The pulverized GLAZE used esp for making ENAMEL ware.

fritting (*Eng*) A condition in fire assaying in which the powdered ore, flux and other reagents are in a pasty condition a little below melting point.

frog (*Build*) Of a plane, the surface against which the blade rests. It determines the PITCH.

frog (*CivEng*) The point of intersection of the inner rails, where a train or tram crosses from one set of rails to another. The frog is in the form of an X. See CROSSING, TURNOUT.

frog (*Vet*) A V-shaped band of horn passing from each heel to the centre of the sole of a horse's foot.

Fröhlich's syndrome (*Med*) See DYSTROPHIA ADIPOSO-GENITALIS.

Froin's syndrome (*Med*) The presence of yellow cerebrospinal fluid, which has a high content of protein but no cells, below the site of obstruction (eg by a tumour) of the spinal cord.

frond (*BioSci*) (1) A large compound leaf, esp of a fern, cycad or palm. (2) A more or less leaf-like thallus of lichen, liverwort or alga. (The term is imprecise in either usage.)

frons (*BioSci*) In insects, an unpaired sclerite of the front of the head; in higher vertebrates, the front of the head above the eyes. Adj *frontal*.

front (*Build*) The sole face of a plane.

front (*EnvSci*) (1) The surface of separation of two air masses. (2) The line of intersection of the surface of separation of two air masses with another surface or with the ground. If warm air replaces cold, it is a warm front; if cold air replaces warm, it is a cold front.

frontage line (*Build*) The BUILDING LINE.

frontal (*BioSci*) (1) A paired dorsal membrane bone of the vertebrate skull, lying between the orbits. (2) Pertaining to the FRONS.

frontal lobes (*BioSci*) The front part of the cortex. It is thought to be involved in immediate memory.

frontal plane (*BioSci*) The median horizontal longitudinal plane of an animal.

frontal sinuses (*BioSci*) Air cavities, connected with the nasal chambers, extending into the frontal bones, in mammals.

frontal zone (*EnvSci*) The three-dimensional transition zone of large horizontal density gradient that separates adjacent air masses.

front clearance (*Eng*) In a single-point cutting tool, the angle of the edge below the point to the perpendicular to the tool shank base.

front end (*ICT*) The client part of a CLIENT–SERVER APPLICATION that requests a service from the SERVER.

front end (*ImageTech*) General term for all the operations from original photography through editing and sound mixing up to the first show print.

front-end processor (*ICT*) A small computer that receives data from a number of input devices, then organizes and transmits the data to a more powerful computer for processing.

front-end shovel (*Eng*) See POWER SHOVEL.

front-end system (*Print*) The parts of a phototypesetting system which control all typesetting operations prior to processing through an imagesetter or phototypesetting unit.

front hearth (*Build*) The part of the hearth extending beyond the chimneybreast.

frontispiece (*Print*) An illustration facing the title page of a book.

frontogenesis (*EnvSci*) The intensification or realization of a *front*; and its weakening or disappearance (*frontolysis*).

front porch (*ImageTech*) A short period of BLACK-LEVEL signal transmitted between the end of the picture information and the horizontal SYNC PULSE.

front projection (*ImageTech*) The projection of a picture from the same side of the screen as the audience. In particular, the system of composite cinematography also known as REFLEX PROJECTION.

front-to-back ratio (*ICT*) The ratio of effectiveness of a directional antenna or microphone, etc in the forward and reverse directions.

frontwall cell (*Electronics*) Semiconductor cell in which light passes through a conducting layer to the active layer, which is separated from the base metal by a semiconductor.

frost (*EnvSci*) A frost is said to occur when the air temperature falls below the freezing point of water (0°C or 32°F). See AIR FROST, GROUND FROST, HOAR FROST.

frost point (*EnvSci*) The temperature at which air becomes saturated with respect to ice if cooled at constant pressure.

froth (*Chem, MinExt*) Liquid foam. A gas–liquid continuum in which bubbles of gas are contained in a much smaller volume of liquid, which is expanded to form bubble walls. The system is stabilized by oil, soaps or emulsifying agents which form a binding network in the bubble walls.

frother (*Eng*) A substance used to promote the formation of a foam in the FROTH FLOTATION process.

froth flotation (*MinExt*) A process in dominant use for concentrating values from low-grade ores. After fine grinding, chemicals are added to a pulp (ore and water) to develop differences in surface tension between the various mineral species present. The pulp is then copiously aerated, and the preferred (*aerophilic*) species clings to bubbles and floats as a mineralized froth, which is skimmed off.

Froude brake (*Eng*) An absorption dynamometer consisting of a rotor inside a casing, itself free to rotate, the space between the two being filled with water. The energy is dissipated in eddy formation and heat, the torque absorbed being measured by the torque necessary to prevent rotation of the casing.

Froude's transition curve (*Surv*) A transition curve the equation to which is that of a cubic parabola, the offset y from the straight produced being given by

$$y = \frac{x^3}{6lr}$$

where x = distance from tangent point, l = length of transition, r = radius of the circular arc.

frozen bearing (*Eng*) A seized bearing. See SEIZURE.

frozen equilibrium (*Chem*) The state of a solid at low temperature, which is prevented from attaining the theoretically possible thermodynamic equilibrium, because its molecular motion has become too slow. Cf NERNST THEORY.

frozen-in strain (*Chem*) Non-equilibrium state of polymer moulding where chain orientation is high. Can be relieved by annealing, which may result in distortion of shape.

Detected using polarized light in transparent polymers. Also *residual strain*.

frozen-in stress (*Chem*) See RESIDUAL STRESS.

fructification (*BioSci*) Any seed- or spore-bearing structure like the large spore-bearing structures of many fungi. Also *fruit body, fruiting body*.

L-fructose (*Chem, FoodSci*) Fruit-sugar, laevulose. $C_6H_{12}O_6$. Mp 95°C. A KETOHEXOSE which is prepared by heating inulin with dilute acids, and is always found together with D-glucose in sweet fruit juices. See SUCROSE.

frue vanner (*MinExt*) Endless rubber belt which is driven slowly upslope while finely ground ore is washed gently downslope. Belt is given a sideways shake to aid distribution, and wash water is so adjusted that heavy material stays on belt, while light gangue is washed down to bottom end of pulley system round which belt circulates.

frugivorous (*BioSci*) Fruit-eating.

fruit (*BioSci*) (1) The structure that develops from the ovary of an angiosperm as the seeds mature, with (false fruit) or without (true fruit) associated structures. (2) Sometimes, any of the various structures associated with the mature seed of a gymnosperm, esp eg the fleshy cone scales of the juniper 'berry'. (3) A FRUCTIFICATION.

fruiting body (*BioSci*) Also *fruit body*. See FRUCTIFICATION.

frusemide (*Med*) See FUROSEMIDE.

frustration (*Psych*) (1) *Animal behaviour*: a motivational state assumed to occur when an animal's actions do not lead to an expected outcome (eg a food reward). Often the initial response to a frustrating situation is to increase the persistence and intensity of behaviour; continued frustration often leads to REDIRECTED BEHAVIOUR. (2) *Human behaviour*: refers to either the prevention of activity that is directed towards a goal, or to the psychological state that results from being prevented from reaching a goal.

frustule (*BioSci*) The silicified cell wall of a diatom (Bacillariophyceae) consisting of two halves which fit one (the *hypotheca*) inside the other (the *epitheca*) like the two halves of a petri-dish (together, in some usages, with the cell itself).

frustum of a cone (*MathSci*) A part of a cone lying between its base and a plane parallel to its base.

frutescent (*BioSci*) Shrubby.

fruticose (*BioSci*) Shrubby.

frying (*FoodSci*) The relatively fast cooking or part cooking of a food in oil or fat which produces rapid and desirable MAILLARD REACTIONS and textural changes (browning and crispness) because of the high boiling point of the fat or oil.

frying arc (*ElecEng*) An arc which hisses.

Fsc (*ImageTech*) Abbrev for *frequency of subcarrier*. The sampling frequency of a digital COMPOSITE signal based on multiples of the COLOUR SUBCARRIER frequency (4·43 MHz PAL). See 4:2:2.

F-scale (*Psych*) A scale constructed to measure the authoritarian personality; contains statements, with which an individual agrees or disagrees on a numerical scale, designed to measure proposed aspects of authoritarianism, eg submission to authority, admiration for power, etc.

FSD (*Radiol*) Abbrev for FOCUS SKIN DISTANCE.

FSH (*Med*) Abbrev for FOLLICLE-STIMULATING HORMONE.

FSK (*ICT*) Abbrev for FREQUENCY-SHIFT KEYING.

f state (*Electronics*) That of an orbital electron when the orbit has angular momentum of three DIRAC UNITS.

FTA (*Eng*) Abbrev for FAULT TREE ANALYSIS.

FTAM (*ICT*) Abbrev for FILE TRANSFER AND ACCESS METHOD.

FT CCD (*ImageTech*) Abbrev for *frame transfer charge-coupled device*. A SOLID-STATE IMAGE SENSOR in which the charge from the image area is transferred through other pixels during FIELD BLANKING to an opaque storage area on the other half of the chip for read-out. See CCD ARRAY, FIT CCD, IT CCD.

F-test (*MathSci*) A statistical test used to test a hypothesis concerning the equality of two population variances based on the variances of two samples, one from each population.

FTIR (*Phys*) Abbrev for FOURIER TRANSFORM INFRARED.

ftp (*ICT*) Abbrev for FILE TRANSFER PROTOCOL. Cf ANONYMOUS FTP.

fuchsin (*Chem*) Magenta, the hydrochloride of rosaniline, a basic triphenylmethane dyestuff, dark green crystals, dissolving in water to form a purple-red solution. Used as a disinfectant, esp in certain skin infections.

fuchsite (*Min*) A green variety of muscovite (white mica) in which chromium replaces some of the aluminium.

fucivorous (*BioSci*) Seaweed-eating.

fucoxanthin (*BioSci, Chem*) $C_{40}H_{56}O_6$ or $C_{40}H_{60}O_6$; a XANTHOPHYLL, a major ACCESSORY PIGMENT in most members of the Heterokontophyta and responsible for their brownish colours. The main carotenoid found in brown algae.

fudge (*Print*) A space reserved in a newspaper for late news. Also *stop press*.

fuel (*NucEng*) Fissile material inserted in or passed through a reactor; the source of the chain reaction of neutrons, and so of the energy released. See REACTOR.

fuel accumulator (*Aero*) A reservoir which augments the fuel supply when the critical fuel pressure is reached during the starting cycle of a gas turbine.

fuel assembly (*NucEng*) A group of nuclear fuel elements forming a single unit for purposes of charging or discharging a reactor. The term includes bundles, clusters, stringers, etc.

fuel cell (*Chem*) A galvanic cell in which the oxidation of a fuel (eg methanol) is utilized to produce electricity.

fuel-cooled oil cooler (*Aero*) A compact oil cooler for high-performance gas turbines in which heat is transferred to fuel passing in the counter bores of the device, instead of to air.

fuel cut-off (*Aero*) A device which shuts off the fuel supply to an aero-engine. Also *slow-running cut-out*.

fuel cycle (*NucEng*) The stages involved in the supply and use of fuel in nuclear power generation. The main steps are mining, milling, extraction, purification, enrichment (if required), fuel fabrication, irradiation in the reactor, cooling, reprocessing, recycling, and waste management and disposal.

fuel element (*NucEng*) A unit of nuclear fuel which may consist of a single cartridge, or a cluster of fuel PINS.

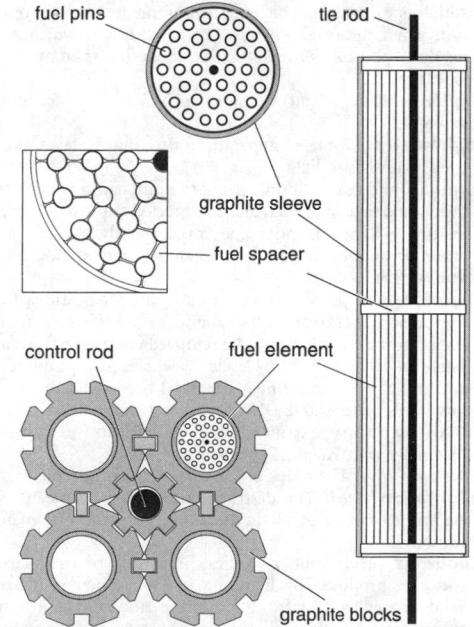

fuel element Advanced gas-cooled reactor.

fuel grade (*Aero*) The quality of piston aero-engine fuel as expressed by its KNOCK RATING.

fuel injection (*Autos*) A method of operating a spark-ignition engine by injecting liquid fuel directly into the induction pipe or cylinder during the suction stroke, thus dispensing with a carburettor; the standard method in diesels, it avoids freezing in aero-engines.

fuel injector (*Autos*) See INJECTOR.

fuel jettison (*Aero*) Apparatus for the rapid emergency discharge of fuel in flight.

fuelling machine (*NucEng*) See CHARGE–DISCHARGE MACHINE.

fuel manifold (*Aero*) The main pipe, or gallery, with a series of branch pipes, which distributes fuel to the burners of a gas turbine.

fuel oils (*Chem*) Oils obtained as residues in the distillation of petroleum; used, either alone or mixed with other oils, for domestic heating and for furnace firing (particularly marine furnaces); also as fuel for internal-combustion engines.

fuel rating (*NucEng*) The ratio of total energy released to initial weight of heavy atoms (U, Th, Pu) for reactor fuel. Usually expressed in megawatts per tonne. US *specific power*.

fuel reprocessing (*NucEng*) The processing of nuclear fuel after use to remove fission products etc and to recover fissile and fertile materials for further use.

fuel rod (*NucEng*) A unit of nuclear fuel in rod form for use in a reactor. Short rods are sometimes termed *slugs*.

fuel tanks (*Aero*) These may be of many forms, for which the names vary. The *main tanks* are normally all those carried permanently, and are usually either flexible self-sealing bags or cells in wing or fuselage, or are integral with the wing structure; *auxiliary tanks* can be mounted additionally to increase range. See DROP TANK.

fuel trimmer (*Aero*) A variable-datum device for resetting in flight the automatic fuel regulation, by *barostat*, of a gas turbine to meet changes in ambient temperature.

fugacity (*Chem*) The tendency of a gas to expand or escape; substituted for pressure in the thermodynamic equations of a real gas. Analogous to ACTIVITY. See IDEAL GAS.

fugitive (*Geol*) Descriptive of the dissolved volatile constituents of magma, which are commonly lost by evaporation when the magma is erupted as lava, and which are partly responsible for metasomatic alteration when magma is intruded.

fugitive colours (*Build*) Colours which fade on exposure to sunlight.

fugitometer (*Chem*) An apparatus for testing the fastness of dyed materials to light.

fugue (*Psych*) A condition, related to amnesia, in which the individual takes a sudden and unexpected trip from home, assuming a new identity and forgetting the past in an attempt to escape from overwhelming stress. See DISSOCIATION.

Fugu rubripes (*BioSci*) Japanese puffer fish. Notorious for the poison (TETRODOTOXIN) found in lethal amounts in the poison gland, that must be removed before the fish can safely be eaten, and at low levels elsewhere. Also of interest and utility as an experimental animal because of the very low levels of repetitive DNA found in the genome.

Fujian flu (*Med*) A virulent strain of influenza associated originally with Fujian, a province in China.

fulcrum (*Phys*) The point of support, or pivot, of a lever.

fulguration (*Med*) The destruction of tissue by means of electric sparks. Used in the treatment of some malignant tumours.

fulgurites (*Min*) Tubular bodies, branching or irregularly rod-like, produced by lightning in loose unconsolidated sand; caused by the vitrification of the sand grains forming silica glass. Although of very narrow cross-section, some specimens have been found to exceed 6 m in length. Also *lechatelierite, lightning tubes*.

fuliginous (*BioSci*) Soot-coloured.

full (*Eng*) A term signifying slightly larger than the specified dimension. Cf BARE.

full adder (*ICT*) Logic circuit which adds a pair of corresponding bits of two numbers expressed in binary form and any carry from a previous stage, producing a sum and a new carry. Also *three-input adder*. Cf HALF ADDER.

full annealing (*Eng*) Of steel, heating above the critical range, followed by slow cooling, as distinguished from (1) annealing below the critical range, and (2) normalizing, which involves air cooling. See ANNEALING.

full aperture (*ImageTech*) A frame size occupying the full width of 35 mm film between the perforations.

full-bound (*Print*) Said of a volume the sides and back of which are covered with one-piece leather or cloth. Also *whole-bound*.

full-centre arch (*Arch*) A semicircular arch or vault.

full coverage spray (*Agri*) A spray applied in quantity to the point of visible runoff from the crop.

full duplex (*ICT*) See DUPLEX.

fullerene (*Chem*) The compound formed when a large number of carbon atoms form ball-shaped molecules. See BUCKMINSTERFULLERENE.

fuller's earth (*Geol*) A non-plastic clay consisting essentially of the mineral montmorillonite, and similar in this respect to bentonite. Used originally in 'fulling', ie absorbing fats from wool, hence the name. The Fuller's Earth of English stratigraphy is a small division of the Jurassic system in the S Cotswolds.

full-force feed (*Eng*) An engine lubrication system in which oil is forced to main bearings, connecting-rod big-end bearings, and thence, by drilled holes or attached pipes, to the gudgeon pins and cylinder walls.

full gear (*Eng*) Of a steam-engine valve gear, the position giving maximum valve travel and cut-off for full power.

full hard (*Eng*) The stage in the tempering of some ferrous and non-ferrous alloys just below that at which the metal cannot be formed by bending.

fulling (*Textiles*) See MILLING.

full-load (*ElecEng*) The normal rated output of an electric machine or transformer.

full load (*Eng*) The normal maximum load under which an engine or machine is designed to operate.

full mass (*Aero*) See WEIGHT.

full Moon (*Astron*) The instant when the geocentric longitudes of the Sun and Moon differ by 180°; the Moon is then opposite the Sun, and therefore fully illuminated.

full-motion video (*ImageTech*) Moving video images in digital, compressed form. Also digital video (abbrev *DV*) when it conforms to the MPEG-1 standard. Abbrev *FMV*. See VIDEO CD.

full out (*Print*) An instruction to set the matter with no indents.

full-pitch winding (*ElecEng*) An armature winding in which the span of the coils is equal to a pole pitch.

full radiator (*Phys*) See BLACK BODY.

full-satellite exchange (*ICT*) A small automatic telephone exchange that is entirely dependent for completion of calls on its parent or main exchange.

full shroud (*Eng*) A gearwheel in which the shrouding extends up to the tips of the teeth.

full thread (*Eng*) A screw-thread cut to the theoretically correct depth.

full-wave rectification (*ElecEng*) That in which current flows, during both half-cycles of the alternating voltage, through similar rectifying devices alternately, eg in a double diode or bridge rectifier.

fully fashioned (*Textiles*) Knitted fabrics and garments that are shaped by increasing or decreasing the number of wales. This ensures that the garment fits more closely.

fulminates (*Chem*) Compounds containing the ion CNO^-, which explode under slight shock or on heat detonators, eg mercury fulminate, $Hg(CNO)^-$.

fumaric acid (*Chem*) *Ethane 1,2-dicarboxylic acid.* HOOCCH=CHOOH. Small prisms which do not melt, but sublime at about 200°C, with the formation of maleic anhydride. Fumaric acid is the *trans*-form. Used in polyester resins.

fumaroles (*Geol*) Small vents on the flanks of a volcanic cone, or in the crater itself, from which gaseous products emanate.

fume (*Chem, PowderTech*) A cloud of airborne particles, generally visible, of low volatility and less than a micrometre in size, arising from condensation of vapours or from chemical reaction.

fume cupboard (*Chem*) A glass chamber or cupboard where laboratory operations involving obnoxious fumes are carried out under forced ventilation.

fumigants (*Chem*) Substances which, when volatilized, are capable of destroying vermin, insects, bacteria, moulds, or which act as disinfectants. Examples are hydrogen cyanide and ethylene oxide for vermin and insects, and formaldehyde (disinfectant).

fumigation (*BioSci*) Disinfection by means of gas or vapour. Killing infesting insects by placing a raw material or product in a controlled atmosphere containing a gaseous or aerosol insecticide or similarly treating an enclosed room or storage container.

fuming liquids (*Chem*) Liquids which give off vapours which unite with water to form a mixture or compound with a lower vapour pressure than water.

fuming sulphuric acid (*Chem*) A solution of sulphur (VI) oxide in concentrated sulphuric acid.

function (*BioSci*) The normal activity of a biological structure, as *digestive function* or *ribosomal function*.

function (*ICT*) See OPERATION, SUBROUTINE.

function (*MathSci*) A relation between two sets (the *domain* and the *codomain*) that associates a unique member of the second with each member of the first. The function (or *mapping*), f, from a set X to a set Y, written f: $X \rightarrow Y$, associates the element x of X with the element $y = f(x)$ of Y.

functional (*BioSci*) Carrying out normal activities; active (as opposed to *passive*).

functional disease (*Med*) The term used to describe symptoms for which it is believed there is no organic basis.

functional group (*Chem*) A small cluster of linked atoms with chemically active bonds, of esp interest in step polymerization. Examples in monomers include hydroxyl (–OH), carboxyl (–COOH), amine (–NH$_2$) and isocyanate (–NCO). Examples in polymers include amide (–CO–NH–), ester (–COO–), carbonate (–OCOO–) and urethane (–NH–COO–). See STEP POLYMERIZATION.

functionalism (*Psych*) The school of thought that analyses mind and behaviour in terms of functions or utilities rather than contents.

functionality (*ICT*) A specific subset of an application program.

functional psychosis (*Psych*) A severe mental disorder which cannot be attributed to any certain physical pathology.

function chamber (*Build*) A closed chamber, generally of brick and concrete, inserted in a sewer system for accepting the inflow of one or more sewers and allowing for the discharge thereof.

function generator (*ICT*) (1) An element in an analogue computer capable of generating a voltage wave approximately following any desired, single-valued, continuous algebraic function of one variable. (2) A signal generator with a range of alternative non-sinusoidal output waveforms.

function key (*ICT*) One of a set of special keys on a keyboard that performs different tasks in different programs; eg F10 may be used to save a file or to call a menu depending on the program.

function switch (*ICT*) A network with a number of input and output lines, connected so that the output signals give the information in a different code from that of the input.

functor (*MathSci*) A mapping between categories which preserves their structure.

fundamental colours (*Phys*) See PRIMARY ADDITIVE COLOURS.

fundamental component (*ICT*) The harmonic component of an alternating wave that has the lowest frequency and that usually represents the major portion of the wave.

fundamental crystal (*Electronics*) A crystal which is designed to vibrate at the lowest order of a given mode.

fundamental dynamical units (*Phys*) A set of units for which the basic equations of dynamics are the same. Unit force acting on unit mass produces unit acceleration; unit force moved through unit distance does unit work; unit work done in unit time is unit power. Four systems are, or have been, in general use, the SI system now being the only one employed in scientific work.

fundamental fracture mode (*Eng*) See FRACTURE MODE.

fundamental frequency (*ICT*) In a steady periodic oscillation, a frequency that divides into all components in the waveform. See HARMONIC.

fundamental frequency of antenna (*ICT*) Lowest frequency at which antenna is resonant when not loaded.

fundamental interval (*Phys*) The number of degrees between the two fixed points on a thermometer scale.

fundamental metric tensor (*MathSci*) See METRIC (1).

fundamental mode (*ICT*) (1) In an antenna, the pattern of current and voltage distribution at the fundamental frequency, usually showing a maximum at one end of a radiator and a minimum at the other. (2) In a waveguide or resonant cavity, the field distribution at frequencies (or corresponding free-space wavelengths) between the lowest which that structure can support and the first harmonic of that frequency.

	Fundamental dynamical units				Dimensions
	ft lb sec	Gravitational	CGS	SI (MKSA)	
length	foot (ft)	foot	centimetre (cm)	metre (m)	L
mass	pound (lb)	slug	gram (g)	kilogram (kg)	M
time	second (s)	second	second	second	T
velocity	ft s^{-1}	ft s^{-1}	cm s^{-1}	m s^{-1}	LT^{-1}
acceleration	ft s^{-2}	ft s^{-2}	cm s^{-2}	m s^{-2}	LT^{-2}
force	poundal (pdl)	pound force (lb f)	dyne	newton (N)	MLT^{-2}
work	ft pdl	ft lb f	erg	joule (J)	ML^{-2}T^{-2}
power	ft pdl s^{-1}	ft lb f s^{-1}	erg s^{-1}	watt (W)	ML^{-2}T^{-3}

Notes: (1) There is no name for the unit of power except in the SI system. It is possible to express power in the ft lb sec system and in the gravitational system by the horse power (550 ft lb f s^{-1}) and in the CGS system by the watt (10^7 erg s^{-1}).
(2) The unit of force (lbf) in the gravitational system is also known as the pound weight (lbwt).

fundamental particle (*Phys*) A particle that is incapable of subdivision. There are believed to be three kinds of such particle: GAUGE BOSONS, LEPTONS, QUARKS.

fundamental series (*Phys*) Series of optical spectrum lines observed in the spectra of alkali metals. Energy levels for which the orbital quantum number is three are designated *f-levels*.

fundamental theorem of algebra (*MathSci*) The theorem that every *n*-degree polynomial has exactly *n* roots.

fundamental theorem of arithmetic (*MathSci*) The theorem that every positive integer has a unique factorization into prime numbers.

fundamental unit (*Phys*) Any of a number of arbitrarily defined units in a system of measurement, such as metre, second or candela, from which the other quantities in the system are derived.

fundamental wavelength (*ICT*) Wavelength in free space corresponding to a fundamental frequency.

fundoplication (*Med*) A surgical procedure in which the fundus of the stomach is gathered, wrapped and sutured around the lower end of the oesophagus in order to alleviate reflux oesophagitis.

fundus (*Med*) That part of the cavity of the eyeball which can be seen through the pupil with an ophthalmoscope.

fungi (*BioSci*) Heterotrophic, eukaryotic organisms reproducing by spores, and their allies. See EUMYCOTA, MYXOMYCOTA.

fungible (*MinExt*) Oil products which are interchangeable and can therefore be mixed during transport. Makes it difficult to trace the origins of a given sample.

fungicidal paints (*Build*) Various forms of liquid materials from antiseptic washes to wood preservatives and stains which contain substances which destroy fungi.

fungicide (*BioSci*) A substance that kills fungal spores and/ or mycelium.

Fungi Imperfecti (*BioSci*) Older name for DEUTEROMYCOTINA.

fungistatic (*BioSci*) Preventing the growth of a fungus without killing it.

funicle (*BioSci*) The stalk by which the ovule (and seed) is attached to the placenta in angiosperms.

funicular railway (*CivEng*) A form of CABLE RAILWAY.

funiculus (*BioSci*) In some invertebrates (as Bryozoa), thickened strands of mesoderm attaching the digestive organs to the body wall; more generally, any small cord, as a tract of nerve fibres in the central nervous system. Adj *funicular*.

funnel (*BioSci*) A modified part of the foot in Cephalopoda, protruding from the mantle cavity and acting as an exhalant channel. In Annelida, a NEPHROSTOME.

funnelling (*EnvSci*) The strengthening of a wind blowing along a valley, esp when the valley narrows.

fur (*BioSci*) In mammals, the thick undercoat of short, soft, silky hairs.

fural (*Chem*) See FURFURAL.

furan (*Chem*) A heterocyclic ring compound, like pyrrole but with oxygen in place of –NH.

furan

furan group (*Chem*) A group of heterocyclic compounds derived from furan, C_4H_4O, a compound containing a ring of four carbon atoms and one oxygen atom.

furan resins (*Chem*) A group of plastics derived from the partial polymerization of furfuryl alcohol, or from condensation of furfuryl alcohol with either furfural or methanal, or of furfural with ketones, and used widely as baked plastic coatings on metal, as adhesives and as resin binders for stoneware.

furca (*BioSci*) Any forked structure; in vertebrates, a divergence of nerve fibres; in some arthropods, a pair of divergent processes at the end of the abdomen.

furcula (*BioSci*) (1) In Collembola, the leaping apparatus, consisting of a pair of partially fused appendages arising from the fourth abdominal somite. (2) In birds, the partially fused clavicles, the wishbone. (3) More generally, any forked structure.

furfural (*Chem*) C_4H_3OCHO. Bp 162°C. A colourless liquid, obtained by distilling pentoses with diluted hydrochloric acid. Used as a solvent, particularly for the selective extraction of crude rosin, also as raw material for synthetic resins. Used in petroleum refining for the selective extraction of aromatics and naphthenes and to allow their subsequent recovery. Also *fural, furfuraldehyde*.

6-furfurylaminopurine (*BioSci*) See KINETIN.

furlong (*Genrl*) A distance of 10 Gunter's chains, ie 220 yd or one-eighth of a mile.

furnace atmosphere (*Eng*) Three main classes: (1) *oxidizing*, produced when air volumes are in excess of fuel requirements; (2) *neutral*, when air-to-fuel ratios are perfectly proportioned; (3) *reducing*, due to deficiency of combustion air. See PROTECTIVE FURNACE ATMOSPHERE.

furnace brazing (*Eng*) A high-production method of copper-brazing steel, without flux, in a reducing atmosphere, or of brazing steels, copper and copper alloys with brasses or silver-brazing alloys, in continuous or in batch furnaces.

furnace clinker (*Build*) The final residue from the combustion of coke or coal which has been burnt and reburnt in order to consume the maximum of combustible matter in it. It is useful as an aggregate in the manufacture of concrete.

furnace linings (*Eng*) The interior portions of metallurgical furnaces which are in contact with hot gases and the charge, and must therefore be constructed of materials resistant to heat, abrasion, chemical action, etc. See REFRACTORIES.

furnacite (*Min*) See FORNACITE.

furnish (*Paper*) The ingredients from which paper is manufactured.

furniture (*Build*) A general name for all metal fittings for doors, windows, etc.

furniture (*Print*) Lengths of wood, plastic or metal, less than type height, used in a forme for making margins etc. They are made to standard widths and lengths.

furosemide (*Pharmacol*) A *loop diuretic* used for treatment of oedema associated with heart failure or liver disease. Formerly *frusemide*.

furring (*Build*) See FIRRING.

furrow application (*Agri*) Delivery of pesticide into the furrow as seeds are sown.

furrowed (*Build*) A term applied to margin-drafted ashlars having parallel vertical grooves cut in the face.

furrowing (*BioSci*) Cell division by intucking the plasmalemma to pinch the cell into two. Cf CELL PLATE.

furuncle (*Med*) See BOIL.

furunculosis (*Med, Vet*) The condition of having several boils. A major problem in fish-farming, caused by infection with *Aeromonas salmonicida*.

fusain (*Min*) Mineral charcoal, the soiling constituent of coal, occurring chiefly as patches or wedges. It consists of plant remains from which the volatiles have been eliminated.

fusarium (*BioSci*) Any fungus of the genus *Fusarium*, esp any of several species that cause serious disease in plants.

fuse (*ElecEng*) A device for protecting electrical apparatus against the effect of excess current; it consists of a piece of fusible metal, which is connected in the circuit to be protected, and which melts and interrupts the circuit when an excess current flows. The term fuse also includes the necessary mounting and cover (if any). Cf CIRCUIT BREAKER.

fuse (*MinExt*) A thin waterproof canvas length of tube containing gunpowder arranged to burn at a given speed for setting off charges of explosive.

fuse-board (*ElecEng*) See DISTRIBUTION FUSE-BOARD.

fuse-box (*ElecEng*) The term sometimes used for a distribution fuse-board enclosed in a box.

fuse-carrier (*ElecEng*) A carrier for holding a fuse-link; arranged to be easily inserted between fixed contacts, so that a replacement of the fuse-link can be quickly carried out. Also *fuse-holder*.

fused junction (*Electronics*) See ALLOY JUNCTION.

fused ring (*Chem*) See CONDENSED NUCLEUS.

fused silica (*Glass*) See VITREOUS SILICA.

fusee (*Eng*) (1) A spirally grooved pulley of gradually increasing diameter, used to equalize the pull of the mainspring of a clock as it runs down by increasing the leverage on a chain or gut-line wound round the fusee and the mainspring barrel. (2) A match with large head for igniting the fuse of an explosive charge.

fuse-element (*ElecEng*) The essential part of a fusible cut-out. Also *fuse, fuse-link*.

fuse-holder (*ElecEng*) See FUSE-CARRIER.

fuselage (*Aero*) The main structural body of a heavier-than-air craft, other than the hull of a flying boat or amphibian.

fuse-link (*ElecEng*) See FUSE-ELEMENT.

fusel oil (*Chem*) Mainly optically inactive 3-methyl-butan-1-ol, $(CH_3)_2CHCH_2CH_2OH$, accompanied by active amyl alcohol, usually occurring in the products of alcoholic fermentation.

fuse rating (*ElecEng*) The maximum current a fuse will carry continuously and/or (less frequently) the minimum current at which it can be relied upon to blow.

fuse-switch (*ElecEng*) A SWITCH-FUSE.

fuse tongs (*ElecEng*) Tongs with insulating handles, used for withdrawing or replacing fuses on high-voltage circuits.

fusible alloys (*Eng*) Alloys of bismuth, lead and tin (and sometimes cadmium or mercury) which melt in the 47–248°C temperature range; used as solders and for safety devices in fire extinguishers, boilers, etc.

fusible core (*Eng*) Low-melting-alloy core in injection moulding tool. Used to create products with complex shapes and many re-entrant angles, and removed at end of cycle by melting out (eg sports racquet, inlet manifolds).

fusible cut-out (*ElecEng*) See FUSE.

fusible metals (*Eng*) See FUSIBLE ALLOYS.

fusible plug (*Eng*) A plug containing a metal of low melting point used eg in the crown of a boiler fire-box to prevent serious overheating of the plates if the water level falls below them.

fusiform (*BioSci*) Elongated and tapering towards each end; shaped like a spindle.

fusiform initials (*BioSci*) More or less elongated initial cells, in the cambium, giving rise to all the cells of the secondary xylem and phloem except for the ray cells. Cf RAY INITIALS.

fusing factor (*ElecEng*) The minimum current required to blow a fuse, expressed as a ratio to the rated current.

fusing point (*Eng*) See MELTING POINT.

fusion (*Phys*) (1) The solid to liquid phase change; the reverse of freezing. Fusion of a substance takes place at a definite temperature, the melting point, and is accompanied by the absorption of latent heat of fusion. (2) The process of forming new atomic nuclei by the fusion of lighter ones; principally the formation of helium nuclei by the fusion of hydrogen and its isotopes. The energy released in the process is referred to as *nuclear energy* or *fusion energy*. The controlled release of energy from the thermonuclear fusion reaction between deuterium and tritium nuclei (requiring very high temperature and pressure) has been attained for short periods in experimental reactors and has been proposed as the basis for power generation. See panel on BINDING ENERGY OF THE NUCLEUS.

fusion bomb (*Phys*) See HYDROGEN BOMB.

fusion cones (*Eng*) See SEGER CONES.

fusion drilling (*MinExt*) The method of hard-rock boring with a paraffin–oxygen jet which melts the rock, the slag being decrepitated and flushed out by a water spray.

fusion energy (*Phys*) The energy released by the process of nuclear FUSION, usually in the formation of helium from lighter nuclei. The energy released in stars is by fusion processes. Fusion is the source of energy in the hydrogen bomb. See CARBON CYCLE, PROTON–PROTON CHAIN.

fusion–fission hybrid reactor (*NucEng*) Proposed reactor system in which neutrons from fusion produce fissile material from a U or Th blanket and electricity. See BLANKET, BREEDER REACTOR.

fusion protein (*BioSci*) Protein produced from a hybrid gene that combines two gene sequences. Usual rationale is to add a sequence with convenient experimental properties – for purification or to localize the protein. Also *recombinant protein*.

fusion reactor (*NucEng*) Reactor in which NUCLEAR FUSION is used to produce useful energy. A field of active current research.

fusion splicing (*ICT*) The method of joining OPTICAL FIBRES in which, after aligning the fibres with a small gap between them, an electric arc across the interface melts the glass to unite the fibres.

fusion welding (*ElecEng*) A process of welding metals in which the weld is carried out solely by the melting of the metals to be joined, without any mechanical pressure.

fust (*Arch*) The shaft of a column.

fustian (*Textiles*) A term including a number of hard-wearing fabrics usually of cotton but differing widely in structure and appearance, but all heavily wefted; they are used for clothing and furnishings. See CORDUROY, MOLESKIN, VELVETEEN.

fuzzy logic (*ICT*) An approach to reasoning in which truth values carry probabilities or labels such as 'true', 'very true', etc. The rules of inference are thus approximate. Often used in the context of EXPERT SYSTEMS.

fuzzy logic control (*ImageTech*) A system which attempts to provide human-like control over variables such as EXPOSURE and WHITE BALANCE by responding smoothly to changes rather than with discrete steps.

F_0 value (*FoodSci*) See PROCESS VALUE.

FYM (*Agri*) See FARMYARD MANURE.

G

G (*BioSci*) Symbol for GUANINE.

G (*Build*) Abbrev for GULLEY.

G (*ElecEng*) See G-VALUE.

G (*Genrl*) Symbol for GIGA-.

G (*Phys*) Symbol for GAUSS.

G (*Chem*) Symbol for: (1) thermodynamic potential; (2) GIBBS FUNCTION; (3) FREE ENERGY.

G (*Eng*) Symbol for TOUGHNESS or strain energy release rate.

G (*Phys*) Symbol for: (1) CONDUCTANCE; (2) GRAVITATIONAL CONSTANT; (3) SHEAR MODULUS.

[G] (*Phys*) A pair of FRAUNHOFER LINES in the deep blue part of the solar spectrum. One, of wavelength 430·8081 nm, is due to iron; the other, of wavelength 430·7907 nm, is due to calcium.

Γ (*Chem*) Symbol for SURFACE CONCENTRATION EXCESS.

Γ (*MathSci*) Symbol for gamma function.

Γ–way (*ICT*) Abbrev for INFORMATION (SUPER)HIGHWAY.

g (*Aero*) See LOAD FACTOR.

g (*Chem*) Abbrev for GRAM.

g (*Chem*) Symbol for OSMOTIC COEFFICIENT.

g (*Phys*) Symbol for ACCELERATION DUE TO GRAVITY.

γ (*Chem*) Symbol for: (1) substituted on the carbon atom of a chain next but two to the functional group; (2) substituted on one of the central carbon atoms of an anthracene nucleus; (3) substituted on the carbon atom next but two to the hetero-atom in a heterocyclic compound; (4) a stereoisomer of a sugar.

γ (*MathSci*) See EULER'S CONSTANT.

γ (*Phys*) Symbol for: (1) ratio of specific heats of a gas; (2) SURFACE TENSION; (3) propagation coefficient; (4) Grüneisen's constant; (5) molar ACTIVITY COEFFICIENT; (6) coefficient of cubic thermal expansion; (7) the greatest refractive index in a biaxial crystal; (8) ELECTRICAL CONDUCTIVITY.

G0 (*BioSci*) The period after the end of mitosis in a cell that has not been stimulated into a further division. See panel on CELL CYCLE.

G1 (*BioSci*) The period of growth in the cell cycle between the end of cell division and the beginning of DNA synthesis. See panel on CELL CYCLE.

G2 (*BioSci*) As for G1 but between the end of DNA synthesis and the beginning of mitosis. See panel on CELL CYCLE.

G-11 (*Chem*) See HEXACHLOROPHENE.

Ga (*Chem*) Symbol for GALLIUM.

gab (*Build*) A pointed tool for working hard stone.

GABA (*BioSci*) Abbrev for GAMMA-AMINO BUTYRIC ACID.

gabapentin (*Pharmacol*) An anticonvulsant drug used in the treatment of epilepsy.

gabbart scaffold (*Build*) Scaffolding in which sawn timbers are used instead of round poles. Also *gabers scaffold*.

gabbro (*Geol*) The name of a rock clan, and also of a specific igneous rock type. The rock gabbro is a coarse-grained plutonite, consisting essentially of plagioclase, near labradorite in composition, and clinopyroxene, with or without olivine in addition. The gabbro clan includes also norite, eucrite, troctolite, kentallenite, etc. See PLUTONIC ROCKS.

gaberdine (*Textiles*) A firm twill fabric (eg with worsted warp and cotton weft), with the warp predominating on the surface; used for dress and suiting cloths and light showerproof overcoatings. All-cotton gaberdine is also used for similar purposes.

gabers scaffold (*Build*) See GABBART SCAFFOLD.

gabion (*CivEng*) A wicker or steel mesh basket, containing earth or stones, deposited with others for revetting river walls.

gable (*Build*) The triangular part of an external wall at the end of a ridged roof.

gable board (*Build*) A BARGE BOARD.

gable shoulder (*Build*) The projecting masonry or brickwork supporting the foot of a gable.

gable springer (*Build*) The concrete, brick or tile corbel supporting the gable shoulder.

gable tiles (*Build*) Purpose-made arris tiles to cover the intersection between gable and roof.

gaboon (*For*) Mahogany-like wood from a HARDWOOD tree (*Aucoumea*), found in parts of C and W Africa.

Gabriel synthesis (*Chem*) A reaction used in organic syntheses for the preparation of pure primary amines. Based on the use of potassium phthalimide (benzene 1,2-dicarboximide) and halogenoalkanes.

gadding (*Vet*) The excited behaviour of cattle when irritated by gad-flies.

gad-fly (*Vet*) A fly of the genus *Hypoderma*, the larva of which parasitizes cattle and is known as a *warble*.

Gadiformes (*BioSci*) An order of mainly marine and often deep-water OSTEICHTHYES with elongated body and long dorsal and anal fins. Some are economically important. Cod, haddock and grenadiers.

gadolinite (*Min*) Silicate of beryllium, iron and yttrium, often with cerium; occurs in pegmatite.

gadolinium (*Chem*) Symbol Gd, at no 64, ram 157·25. A rare metallic element; trivalent; a member of the rare earth group. Only known in combination; obtained from the same sources as europium. Natural gadolinium has a high neutron capture cross-section of 49 000 barns and is expensive, but gadolinium oxide has been used as a BURNABLE POISON during the start-up of reactors and in nitrate form as a deliberate poison during the shutdown of CANDU reactors.

gadolinium gallium garnet (*Chem*) See GGG.

Gaede diffusion pump (*ChemEng*) A pump using mercury vapour, which entrains molecules of gas from a low pressure established by a backing pump. Oil of low vapour pressure (apiezon) is a modern alternative.

Gaede molecular pump (*ChemEng*) A rotary pump which ejects molecules of gas by imparting a drift velocity to their random motion.

gaffer (*ImageTech*) Senior lighting electrician on a film or TV unit.

GAG (*BioSci*) Abbrev for GLYCOSAMINOGLYCAN.

gag (*Med*) To retch; also a device for keeping the mouth open for surgical procedures.

gagger (*Eng*) See LIFTER.

gahnite (*Min*) A mineral belonging to the spinel group; occurs as grey octahedral cubic crystals. Also *zinc-spinel* (see SPINEL) the composition being zinc aluminate, $ZnAl_2O_4$.

Gaia hypothesis (*EnvSci*) Hypothesis proposed by J E Lovelock in 1979 that suggests that the Earth can be considered as a single organism and that there are complex interrelationships that tend to maintain homeostasis. It is not synonymous with the biosphere but includes earth, water and air.

gain (*ICT*) (1) In electric systems, generally provided by insertion of an amplifier into a transmission circuit, or by matching impedances by a loss-free transformer. Measured in decibels or nepers, and defined as the increase in power level in the load, ie the ratio of the actual power delivered to that which would be delivered if the source were correctly matched, without loss, to the load, on the absence of the amplifier. (2) In a directional antenna, the ratio (expressed in decibels) of voltage produced at the receiver terminals by a signal arriving from the direction of maximum sensitivity of the antenna, to that produced by the same signal in an omnidirectional reference antenna (generally a half-wave dipole). In a transmitting antenna, the ratio of the field strength produced at a point along the line of maximum radiation by a given power radiated from antenna, to that produced at the same point by the same power from an omnidirectional antenna.

gain–bandwidth product (*ICT*) A FIGURE OF MERIT rating for amplifiers, or transmission paths incorporating amplifiers, based on the product of gain and bandwidth as measured under specified conditions.

gain control (*ICT*) Means for varying the degree of amplification of an amplifier, often a simple potentiometer. See AUTOMATIC GAIN CONTROL.

gain-up (*ImageTech*) Increasing video amplification to enable a camera to operate in low light, at the expense of increased NOISE and reduced colour SATURATION.

gaiting (*Textiles*) The operation of preparing a loom for weaving by placing the warp in position.

Gal (*Chem*) Abbrev for GALACTOSE.

gal (*Phys*) Unit of acceleration used in gravity measurements: 1 cm s^{-2}. In honour of Galileo. Often used as *milligal*.

galactagogue (*Pharmacol*) Promoting the secretion of breast milk (Gk *gala*, gen *galaktos*); any medicine which does this.

galactans (*Chem*) The anhydrides of galactose. They comprise many gums, agar and fruit pectins, and occur in algae, lichens and mosses.

galactic circle (*Astron*) The great circle of the celestial sphere in which the latter is cut by the galactic plane: hence the primary circle to which the galactic co-ordinates are referred.

galactic cluster (*Astron*) See OPEN CLUSTER.

galactic co-ordinate (*Astron*) Two spherical co-ordinates referred to the galactic plane; the origin of galactic longitude lies at RA 17 h 42·4 min, dec $-28°55'$ (1950), approximately at the galactic centre; galactic latitude is measured positively from the galactic plane towards the north galactic pole.

galactic equator (*Astron*) The great circle which cuts the centre of our galaxy and has the Earth at its centre. It is inclined at approximately 62° to the celestial equator.

galactic halo (*Astron*) An almost spherical aggregation of stars, gas and dust, which is concentric with our galaxy. It contains stars of Population II, and is responsible for much of the background of radio emission from the sky. Similar halos surround other galaxies.

galactic noise (*ICT*) That arriving from outer space, similar to electronic circuit noise, but arising from sources in galaxies.

galactic plane (*Astron*) The disk-like plane of our Galaxy which is densely populated with stars; its cross-section is observed in the sky as the Milky Way.

galactic pole (*Astron*) Either of the two points on the celestial sphere which lie at 90°N and 90°S of the galactic equator.

galactic rotation (*Astron*) The rotation of our galaxy about its centre. The velocity of rotation increases sharply to approx 150 km s^{-1} at 1 kpc from the centre, increases more gradually to a peak near the Sun's position, and is roughly constant further out. In the Sun's vicinity the velocity due to galactic rotation is about 250 km s^{-1}. The Sun takes around 220 million years to complete one orbit round our galaxy.

galactobolic (*BioSci*) Refers to the action of neurohypophysial peptides that contract mammary myoepithelium and so eject milk.

galactocele (*Med*) A cystic swelling in the breast, due to retention of milk as the result of a blockage of a milk duct.

galactophoritis (*Med*) Inflammation of the milk ducts.

galactophorous (*BioSci*) See LACTIFEROUS.

galactopoiesis (*Med*) Increase in milk secretion.

galactorrhoea (*Med*) Excessive secretion of milk by the breast, causing it to overflow through the nipple. US *galactorrhea*.

galactosaemia (*Med*) An inborn error of metabolism. Infants with the defect are unable to metabolize GALACTOSE to glucose because of the absence of the enzyme galactose-1-phosphate uridyl transferase. Untreated survivors are mentally retarded. Also *galactosemia*.

galactose (*Chem*) $CH_2OH(CHOH)_4CHO$. A hexose. Thin needles. Mp 166°C. Dextrorotatory. It is formed together with D-glucose by the hydrolysis of milk-sugar with dilute acids. Stereoisomeric with glucose, which it strongly resembles in properties. Present in certain gums and seaweeds as a polysaccharide *galactan* and as a normal constituent of milk.

galactosis (*BioSci*) See LACTATION.

galactotrophic (*Med*) Stimulating the secretion of milk by the mammary gland. Also *galactotropic*.

galantamine (*Pharmacol*) A tertiary amine compound originally derived from flowers (daffodils and snowdrops), now synthesized. A specific, competitive and reversible *acetylcholine esterase inhibitor* used in treating dementia. Also *galanthamine*.

galatin dynamite (*MinExt*) High explosive containing nitroglycerine, sodium nitrate, collodion cotton, and such inert fillers as wood meal and sodium carbonate.

galaxite (*Min*) A rare manganese aluminium spinel, formula $MnAl_2O_4$.

Galaxy (*Astron*) (1) Our own Galaxy; the entire system of dust, gases and stars within which the Sun moves; now known to have the typical spiral structure. (2) An automatic star-plate analyser for measuring brightness and position of high-density photographic images of portions of our galaxy (*General Automatic Luminosity And XY measuring engine*).

galaxy (*Astron*) A large-scale system of stars, dust and gas, containing typically 10^7–10^{12} solar masses. See panel on GALAXY.

gale (*EnvSci*) A wind having a speed of 34 knots (63 km h^{-1}) or more, at a height of 33 ft (10 m) above the ground. On the Beaufort scale, a gale is a wind of force 8.

galeate (*BioSci*) Shaped like a helmet or a hood. Also *galeiform*.

galena (*Min*) Lead sulphide, PbS; commonest ore of lead, occurring as grey cubic crystals, often associated with zinc blende, in mineralized veins. Also *lead glance*.

galet (*Build*) See SPALL.

Galilean binoculars (*Phys*) Binoculars in which the objectives are doublet telescope objectives and the eyepieces are negative lenses.

Galilean moons (*Astron*) The four moons of Jupiter (Io, Europa, Ganymede and Callisto) discovered by Galileo in 1610.

Galilean telescope (*Astron*) A telescope consisting of a single long-focus objective lens and a powerful diverging lens eyepiece; introduced by Galileo in 1609.

Galilean transformation (*Phys*) In classical kinematics, the space and time co-ordinates of an event in one frame of reference as seen by an observer in another frame moving with a constant velocity relative to it. See LORENTZ TRANSFORMATION.

Galileo project (*Space*) US spacecraft launched from the Space Shuttle *Atlantis* in October 1989 bound for Jupiter. Found evidence for water on Europa; crashed into Jupiter in September 2003.

Galaxy

A galaxy is a large family of stars, interstellar gas and dust, held together by its mutual gravitational force, and generally isolated by almost empty space from its neighbour galaxies. Galaxies are the basic large-scale components from which our universe is constructed. Their masses range from 10^7 to 10^{12} solar masses. Our own Milky Way Galaxy has about 10^{11} solar masses all together. Since galaxies come in a bewildering variety of forms, it is useful to distinguish three major classes: spiral, elliptical and irregular.

The *spiral* galaxies are flat disks of stars with two spiral arms emerging from the nuclear region, and account for 25% of all galaxies. In 30% of spirals a central bar of stars links the arms. Diameters are 25–800 kiloparsecs (kpc). The arms are rich in interstellar matter, and they play a key role in star formation. The spiral pattern does not rotate rigidly, like a wheel. Theorists believe a compression wave propagates through the galaxy, and the spiral pattern delineates its location. At the leading edge of the arm, compression of the interstellar gas triggers vigorous star formation.

Most galaxies are *elliptical*, with diameters up to 100 kpc. They range from almost spherical up to a ratio of about three for length relative to diameter. They contain hardly any dust or gas, and so there is very little star formation. Elliptical galaxies are thus characterized by an ageing population of stars. *Irregular* galaxies have no clear morphology and they embrace many of the more active subtypes.

There are no clear explanations as to why galaxies come in spiral and elliptical shapes. It appears rather unlikely that spirals change into ellipticals, but beyond that not much is known. Dust may hold the key: in a collapsing protogalaxy with dust, the formation of a flat disk is inevitable. Most galaxies are extremely old: the Milky Way is at least 12 billion years old. The epoch of galaxy formation was perhaps 1 billion years after the onset of the BIG BANG.

Distances to the galaxies are mainly derived from the Hubble law and they are very large. The Magellanic Clouds, satellites of the Milky Way, are 55 kpc away. Our own LOCAL GROUP, which includes the Andromeda galaxy, will fit in a sphere a few Mpc across. The next significant group of galaxies is tens of Mpc away. The Virgo cluster, for instance, is 2 Mpc across, 20 Mpc distant, and it has hundreds of members. Clusters of galaxies sit at the apex of the hierarchical structure of the universe: as the universe expands, clusters and the galaxies within them are unchanged, but the spacing between clusters increases monotonically with time. Very remote galaxies are of considerable importance in cosmology. At distances of eg 5000 Mpc, they bring information on the universe as it was some 15 billion years ago. The HUBBLE SPACE TELESCOPE has been investigating very distant galaxies as one of its major tasks.

Active galaxies are numerically quite rare, but have been much observed on account of the exotic phenomena within them. About one galaxy per million is a giant radio galaxy. These have two clouds of radio emission extending to 1 Mpc from the nucleus, and their total luminosity (10^{38} W) is around a million times brighter than a normal galaxy. This suggests that stars cannot be the source of their energy. QUASARS (see panel) are similar to radio galaxies as regards their radio emissions, but they have higher REDSHIFTS (up to $z = 5$) and much smaller angular sizes than galaxies. Possibly they are extremely active galaxies that are so far away that we can only see their brilliant nuclei. If the Hubble law holds out to large distances, these are the remotest objects known to us.

Individual astronomers have studied and classified particular sorts of galaxies that have taken the name of their investigator: eg Seyfert galaxies are spirals with brilliant nuclei and faint arms; Markarian galaxies have unusual blue spectra; the Maffei galaxies are two under-luminous members of our local group; and Zwicky galaxies are very compact, only just distinguishable from stars on photographs.

An important goal of theoretical work is to provide a unified picture of galactic activity on all scales. The commonest model invokes a central black hole of up to a billion solar masses. Matter falling into this black hole behaves in bizarre ways: a gigantic *accretion disc* is formed and relativistic jets can emerge along its rotation axis. In a qualitative way many of the observed features of galactic nuclei can be explained. Alternatives to such models include the *starburst galaxy* in which it is envisaged that a temporarily enhanced rate of star formation causes much higher luminosity for a while. A few astronomers have chosen to challenge the Hubble law, and argue that the active galaxies are much closer than supposed, in which cases their energy requirements would not be as extreme.

See panels on BLACK HOLE, COSMOLOGY and REDSHIFT–DISTANCE RELATION.

---

gall (*BioSci*) (1) An abnormal localized swelling or outgrowth, usually of characteristic shape, that follows an attack by a parasite or pest. (2) An injury of the skin of animals due to the pressure of harness. (3) See BILE.

gall (*Med*) See BILE.

gallamine (*Pharmacol*) A muscle relaxant used in anaesthesia.

gall bladder (*BioSci*) In vertebrates, a lateral diverticulum of the bile duct, in which the bile is stored.

gallery (*Arch*) An elevated floor projecting beyond the walls of a building and supported on pillars, brackets or otherwise, so as to command a view upon the main floor, as at a theatre etc. Sometimes cantilevered to eliminate obstruction of view by pillars.

gallery (*MinExt*) A tunnel or passage in a mine.

gallery furnace (*Eng*) Furnace type used in distillation of mercury from its ores.

gallet (*Build*) A splinter of stone.

galleting (*Build*) See GARRETING.

galley (*Print*) A steel tray open at one end, in which type matter is held after setting. Corrections and deletions are more easily made to type in galley form than in page form, and are traditionally marked on the galley proof or slip, itself referred to as a *galley*.

galley press (*Print*) A proofing press which allows for the thickness of the galley.

galley proof (*Print*) A proof taken after text has been typeset (traditionally from type on a GALLEY) and before it is made up into pages.

gallic acid (*Chem*) 3,4,5-trihydroxybenzoic acid. $C_6H_2(OH)_3COOH$. Crystallizes with one H_2O; thin needles; decomposes at about 200°C into CO_2 and pyrogallol (3,4,5-trihydroxybenzene). Occurs in nut-galls, tea, divi-divi and other plants; it is obtained from tannin by hydrolysis.

Galliformes (*BioSci*) An order of ground-living birds with feet well adapted for running. Game birds that seek their food (berries, seeds, buds and insects) on the ground. Brush-turkeys, curassows, turkeys, pheasants, partridges, grouse and quail.

gallium (*Chem*) Symbol Ga, at no 31, ram 69·72, rel.d. 5·9, oxidation state + 3, mp 30·15°C, bp 2000°C. A metallic element in the third group of the periodic system. Used in fusible alloys and high-temperature thermometry. Gallium arsenide is an important semiconductor.

gallium arsenide (*Chem*) GaAs. Compound semiconductor in near-stoichiometric proportions that is more difficult to process than silicon but with a higher band gap and higher electron mobility. It has SPHALERITE (zinc blende) structure.

gallon (*Genrl*) Liquid measure. One imperial gallon is the volume occupied by 10 lb avoirdupois of water. One imp gallon = 4·546 09 litres = $\frac{6}{5}$ US gallons. One US gallon = 3·785 43 litres = $\frac{5}{6}$ imperial gallons.

galloon (*Arch*) Decorated work for a band or moulding, to which is applied a row of small round balls.

gallop rhythm (*Med*) Heard when listening to the heart beat when there is an added heart sound. Usually indicates heart failure.

Galloway boiler (*Eng*) A cylindrical boiler of the Lancashire type, in which the two furnace tubes unite, at a short distance from the grates, into a single arched oval flue, crossed by inclined water tubes (Galloway tubes).

gallows bracket (*Eng*) See ANGLE BRACKET.

gall-sickness (*Vet*) See ANAPLASMOSIS.

gallstones (*Med*) Pathological concretions in the gall bladder and bile passages. They do not have a uniform composition but some constituents may be preponderant, eg cholesterol, or calcium carbonate and phosphate. Also *biliary calculi*.

GALT (*BioSci*) Abbrev for GUT ASSOCIATED LYMPHOID TISSUE.

galvanic cell (*Chem*) An electrochemical cell from which energy is drawn. Cf ELECTROLYTIC CELL.

galvanic corrosion (*Eng*) Corrosion resulting from the current flow between two dissimilar metals in contact with an electrolyte. For instance, if zinc and copper are in electrical contact in a damp atmosphere, a current will flow under a potential of 1·14 V due to the zinc becoming anodic and corroding away while the copper will form the cathode of the cell and remain unaffected.

galvanic current (*Med*) Steady unidirectional current for therapeutic use.

galvanic series (*Chem*) Electrochemical series for different metals and alloys in specific electrolytes, eg sea water.

galvanic skin response (*Psych*) A change in the electrical resistance of the skin, recorded by a polygraph; used as an indicator of physiological arousal, also, with less justification, of lie detection.

galvanized iron (*Build*) Iron which has been subjected to GALVANIZING, eg zinc coating, widely used, esp as corrugated iron, for minor roofing purposes, eg on wooden buildings etc. Also for nails, bolts, etc, where moisture may produce corrosion.

galvanizing (*Eng*) The coating of steel or iron with zinc, generally by immersion in a bath of zinc, covered with a flux, at a temperature of 425–500°C. The zinc may alternatively be electrodeposited from cold sulphate solutions. The zinc is capable of protecting the iron from atmospheric corrosion even when the coating is scratched, since the zinc is preferentially attacked by carbonic acid, forming a protective coating of basic zinc carbonates. Also *hot-dip galvanizing*. See ELECTROGALVANIZING.

galvanoluminescence (*ElecEng*) Feeble light emitted from the anode in some electrolytic cells.

galvanomagnetic effect (*ElecEng*) See HALL EFFECT.

galvanometer (*ElecEng*) Current-measuring device depending on forces on the sides of a current-carrying coil normal to magnetic fields in gaps. In a moving-magnet instrument, the suspended coil is replaced by an astatic magnet system which is magnetically shielded for very sensitive work.

galvanometer constant (*ElecEng*) A number by which the scale reading of a galvanometer must be multiplied in order to give a reading of current in amperes or other suitable units.

galvanometer shunt (*ElecEng*) A shunt connected in parallel with a galvanometer to reduce its sensitivity. See UNIVERSAL SHUNT.

galvanotaxis (*BioSci*) The tendency of organisms to grow or move into a particular orientation relative to an electric current passing through the surrounding medium. Also *electrotaxis, electrotropism, galvanotropism*.

gam-, gamo- (*Genrl*) Prefixes from Gk *gamos*, marriage, union.

gambrel roof (*Build*) See MANSARD ROOF.

games paddle (*ICT*) The name for a hand-held device used for computer games.

gametangium (*BioSci*) Any cell or organ within which gametes are formed, eg antheridium, archegonium, oögonium.

gametes (*BioSci*) Reproductive cells that will unite in pairs to produce zygotes; germ cells. Adj *gametal*.

game theory (*Psych*) A mathematical formalization of decision and strategic processes involving the probabilities and values of various outcomes of action choices for the decision-makers.

gametocyte (*BioSci*) A cell that divides to produce gametes.

gametogenesis (*BioSci*) The formation of gametes from gametocytes: *gametogeny*.

gametophore (*BioSci*) The branch of a moss that bears the sex organs.

gametophyte (*BioSci*) The characteristically haploid generation in the life cycle of a plant showing ALTERNATION OF GENERATIONS; it produces the gametes to form the zygote, which germinates to give the SPOROPHYTE.

gamma (*ImageTech*) A measure of the contrast in image reproduction. In a photographic system it is the increment of image density produced by a given increment of log exposure, $\gamma = \Delta D/\Delta \log E$, generally derived from the CHARACTERISTIC CURVE. In TV, overall gamma similarly relates the logarithmic increments of receiver screen luminance and those of the brightness of the original scene.

gamma-amino butyric acid (*BioSci*) An inhibitory NEUROTRANSMITTER in the mammalian brain. It also inhibits the peripheral nervous system in Crustacea and in the leech. Abbrev *GABA*.

gamma-BHC (*Chem*) See GAMMEXANE.

gamma brass (*Eng*) The γ-constituent in brass is hard and brittle and is stable between 60 and 68% of zinc at room temperature. γ-Brass is an alloy consisting of this constituent.

gamma camera (*Radiol*) See SCINTILLATION CAMERA.

gamma correction (*ImageTech*) Non-linear amplification of the picture signal applied on transmission to compensate for preceding tonal distortions and to obtain the desired contrast in the received image.

gamma detector (*Radiol*) A radiation detector specially designed to record or monitor gamma radiation.

gamma function (*MathSci*) The function $\Gamma(x)$ defined by Euler as

$$\int_0^\infty e^{-t} t^{x-1} dt$$

and by Weierstrass by the equation

$$\frac{1}{\Gamma(x)} = x e^{\gamma x} \prod_{n=1}^\infty \left(1 + \frac{x}{n}\right) e^{-x/n}$$

where γ is Euler's constant. Its main properties are

$$\Gamma(1 + x) = x\Gamma(x)$$

$$2^{2x-1}\Gamma(x)\Gamma\left(x + \frac{1}{2}\right) = \sqrt{\pi}\ \Gamma(2x)$$

$$\Gamma(x)\Gamma(1 - x) = \frac{\pi}{\sin \pi x}$$

$$\Gamma\left(\frac{1}{2}\right) = \sqrt{\pi}$$

$$\Gamma(n + 1) = n!$$

gamma globulin (*BioSci*) Describes the serum proteins that on electrophoresis have the lowest anodic mobility at neutral pH: these are mainly immunoglobulins. The term was used to describe immunoglobulins until more specific means of distinguishing them were developed.

gamma infinity (*ImageTech*) The maximum γ obtainable with prolonged development.

gamma iron (*Eng*) The polymorphic form of iron stable between 906 and 1403°C. It has a FACE-CENTRED CUBIC lattice and is non-magnetic. Its range of stability is lowered by carbon, nickel and manganese, and it is the basis of the solid solutions known as AUSTENITE.

Gamma Orionis (*Astron*) See BELLATRIX.

gamma radiation (*Phys*) Electromagnetic radiation of high quantum energy emitted after nuclear reactions or by radioactive atoms when nucleus is left in excited state after emission of an alpha or beta particle. In medicine this is the commonest form of radionuclide emission in patient imaging.

gamma-ray astronomy (*Astron*) The study of radiation from celestial sources at wavelengths shorter than 0·01 nm. Gamma rays have been detected from the gamma-ray background, a few energetic galaxies and quasars, and from certain highly evolved stars. See panels on ASTRONOMICAL TELESCOPE, GALAXY and QUASAR.

gamma-ray burst (*Astron*) An intense burst of high-energy X-rays or gamma rays over a period of tenths to tens of seconds from an astronomical source; can occur several times a year and the widely distributed sources, though unknown, are thought to be galactic in origin.

gamma-ray capsule (*NucEng*) Usually metal, sealed and of sufficient thickness to reduce γ-ray transmission to a safe value.

gamma-ray energy (*Phys*) The energy of a gamma-ray photon, given by hv where v is the frequency and h is PLANCK'S CONSTANT. The energy may be determined by diffraction by a crystal or by the maximum energy of photoelectrons ejected by the gamma rays. The depth of penetration into a material is determined by the energy.

gamma-ray imaging (*Radiol*) In medical diagnosis, the commonest way of showing the distribution of radioactive isotopes in the body.

gamma-ray photon (*Phys*) A quantum of gamma-radiation energy given by hv where v is the frequency and h is PLANCK'S CONSTANT.

gamma-ray source (*Radiol*) A quantity of matter emitting γ-radiation in a form convenient for radiology.

gamma-ray spectrometer (*NucEng*) Instrument for investigation of energy distribution of γ-ray quanta. Usually a scintillation or germanium counter followed by a PULSE-HEIGHT ANALYSER.

Gammexane (*Chem*) TN applied to the γ isomer of BENZENE HEXACHLORIDE; a synthetic stomach and contact insecticide of great toxicity to a wide range of pests. Also *gamma-BHC*.

gamo- (*Genrl*) See GAM-.

gamocyte (*BioSci*) In Protozoa, a phase developing from a TROPHOZOITE and giving rise to gametes.

gamone (*BioSci*) Any chemical substance released by a gamete or hypha that is attractive to another appropriate gamete or hypha in sexual reproduction, eg malic acid in ferns, called sirenin.

gamopetalous (*BioSci*) Having a corolla consisting of a number of petals united by their edges. Cf POLYPETALOUS.

gamophyllous (*BioSci*) Having the perianth members united by their edges. Cf POLYPHYLLOUS.

Gamow–Teller selection rules (*Phys*) See NUCLEAR SELECTION RULES.

gamut (*Phys*) The range of chromaticities available through the addition of three colours.

ganciclovir (*Pharmacol*) An anti-herpes pro-drug used in the treatment of cytomegalovirus infections in AIDS patients. Formerly *gancyclovir*.

G and M codes (*ICT*) These codes are used within the programming language for COMPUTER NUMERICAL CONTROL machines such as lathes and milling machines; eg G00 X30 Y20 M03 would cause rapid tool movement followed by the spindle motor being turned on.

gang (*MinExt*) A train or *journey* of tubs or trucks.

gang boarding (*Build*) A board with battens nailed across to form steps; used as a gangway during building operations.

ganged capacitor (*ICT*) Assemblage of two or more variable capacitors mechanically coupled to the same control mechanism.

ganging (*ElecEng*) Mechanical coupling of the movements of several circuit elements for simultaneous control.

ganging oscillator (*ElecEng*) One giving a constant output, whose frequency can be rapidly varied over a wide range; used for testing accuracy of adjustment of ganged circuits over their tuning range.

gangliectomy (*Med*) See GANGLIONECTOMY.

ganglion (*BioSci*) A plexiform collection of nerve fibre terminations and nerve cells. Pl *ganglia*.

ganglion (*Med*) (1) An aggregation of nerve cells. (2) A type of cyst (a fluid-filled sac) attached to a tendon sheath often in the wrist. Also *Bible cyst, dorsal tendon cyst, wrist cyst*.

ganglionectomy (*Med*) Surgical removal of a nerve ganglion or of a ganglion arising from a tendon sheath. Also *gangliectomy*.

ganglioneuroma (*Med*) A tumour composed of ganglion nerve cells, nerve fibre and fine fibrous tissue, usually arising in connection with sympathetic nerves, eg in the medulla of the suprarenal gland.

ganglionitis (*Med*) Inflammation of a nerve ganglion.

ganglioside (*BioSci*) Glycolipids derived from CEREBROSIDE by the addition of complex oligosaccharide chains.

gang milling (*Eng*) The use of several milling cutters on one spindle to produce a surface with a required profile or to mill the face and sides of the work at one operation.

gang mould (*Build, CivEng*) A mould in which a number of similar concrete units may be cast simultaneously.

gangrene (*Med*) Death of a part of the body, associated with putrefaction; due to infection or to cutting off of the blood supply.

gangrenous coryza (*Vet*) See MALIGNANT CATARRHAL FEVER.

gang saw (*For*) An arrangement of parallel saws secured in a frame to operate simultaneously in sawing a log into strips.

gang switches (*ElecEng*) A number of switches mechanically connected together so that they can all be operated simultaneously.

gang tool (*Eng*) A tool holder having a number of cutters; used in lathes and planers, each tool cutting a little deeper than the one ahead of it.

gangue (*Eng*) Valueless rock or mineral aggregates in an ore.

gangway (*MinExt*) Main haulage road, or level underground.

gannister (*Geol*) A particularly pure and even-grained siliceous grit or loosely cemented quartzite, occurring in the Upper Carboniferous of northern England, and used in the manufacture of silica bricks. Also *ganister*.

ganoid (*BioSci*) Formed of, or containing, GANOIN. Said of fish scales of rhomboidal form, composed of an outer layer of ganoin and an inner layer of isopedin; (fish) having these scales.

ganoin (*BioSci*) A calcareous substance secreted by the dermis and forming the superficial layer of certain fish scales; it was formerly supposed to be homologous with enamel.

gantry (*Build*) A temporary erection having a working platform used as a base for building operations or for the support of cranes, scaffolding or materials.

gantry (*Space*) The servicing tower beside a rocket on its launching pad.

Gantt chart (*Eng*) Graphic construction chart which shows each operation and its connection and timing as part of the overall scheme.

Ganymede (*Astron*) The third natural satellite of Jupiter, discovered by Galileo in 1610. Distance from the planet 1 070 000 km; diameter 5260 km. It is the largest moon in the solar system, larger than Mercury, and the brightest of the Galilean satellites.

gap (*Acous*) Air-gap in magnetic circuit of recording or erasing head in tape recorder; allows signal to interact with oxide film.

gap (*Aero*) The distance from the leading edge of a biplane's upper wing to the point of its projection onto the chord line of a lower wing.

gap (*Electronics*) The range of energy levels between the lowest of conduction electrons and the highest of valence electrons.

gap (*ICT*) Digits that separate signals for data or program. Also *space*.

gap (*Phys*) The space between discharge electrodes.

GAPAN (*Aero*) Abbrev for *Guild of Air Pilots and Air Navigators*.

gap arrester (*ElecEng*) A lightning arrester consisting essentially of a small air-gap connected between the circuit to be protected and earth; the gap breaks down on the occurrence of a lightning surge, and discharges the surge to earth. See MULTIGAP ARRESTER.

gap bed (*Eng*) A lathe bed having a gap near the headstock, to permit turning of large flat work of greater radius than the centre height.

gap breakdown (*Phys*) The cumulative ionization of the gas between electrodes, leading to a breakdown of insulation and a TOWNSEND AVALANCHE.

gap bridge (*Eng*) A bridge casting of the same cross-section as the bed in a gap-bed lathe, and used to close the gap.

gape (*BioSci*) The width of the mouth when the jaws are open.

gape (*MinExt*) Aperture below which a crushing machine can receive and work on entering ore.

gapes (*Vet*) See SYNGAMIASIS.

gap factor (*ElecEng*) The ratio of energy gain in electron-volts traversing a gap across which an accelerating field acts, to the actual voltage across the gap.

gap filler (*Aero*) Radar to supplement long-range surveillance radar.

gap-filling (*Chem*) A term applied to the property of adhesives in adequately packing the free space in a joint;

implies a high-viscosity material such as epoxy resin rather than monomeric cyanoacrylate.

gap junction (*BioSci*) A junction between cells that allows direct communication between cells by molecules that diffuse through pores in the junction. The flow is controlled by the opening or closing of these pores. See DESMOSOMES.

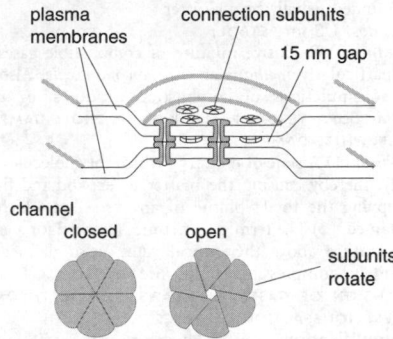

gap junction Showing channels between two adjacent cells.

gap lathe (*Eng*) A lathe with a GAP BED.

gap length (*Acous*) The distance between adjacent surfaces of the poles in a longitudinal magnetic recording system.

gap window (*Arch*) A long and narrow one.

garboard strake (*Ships*) The first strake or line of plating attached to the keel on either side.

garden city (*Arch*) An independent community established on the outskirts of cities from the end of the 19th century, in an attempt to integrate industry with the pastoral life of the village.

garden-wall bonds (*Build*) Forms of bond with an increased number of stretchers used largely for building low boundary walls of single brick thickness, when the load to be carried is that of the wall only and it is desired to show a fair face on both sides of the wall.

gargle (*Acous*) A WOW which has fluctuation changes ranging between about 30 and 200 Hz.

gargoyle (*Build*) A grotesquely shaped spout projecting from the upper part of a building, to carry away the rainwater.

gargoylism (*Med*) Former name for HURLER'S SYNDROME.

garnet (*Min*) A group of cubic minerals including $Mn_3Al_2Si_3O_2$, in which the aluminium ion is tetrahedrally co-ordinated and the silicon is octahedrally co-ordinated. They occur typically in metamorphic rocks, eg garnetiferous schists. Some species are of value as gems, rivalling ruby in colour. See ANDRADITE, GROSSULAR, MELANITE, PYROPE, SPESSARTITE, UVAROVITE. Also a related class of FERRIMAGNETIC MATERIALS, having a similar basic structure, but containing three trivalent cations. See panel on FERROMAGNETICS AND FERRIMAGNETICS.

garnet hinge (*Build*) A form of STRAP HINGE.

garneting (*Build*) See GARRETING.

garnet paper (*Build*) A type of sandpaper having powdered garnet as the abrasive coating.

garnett machine (*Textiles*) A strong carding machine consisting of rollers fitted with saw teeth; used for breaking down waste materials to fibres for reuse.

garnierite (*Min*) A bright-green nickeliferous serpentine, hydrated nickel magnesium silicate. It occurs in serpentinite as a decomposition product of olivine, and in other deposits, and is an important ore of nickel.

garnish bolt (*Build*) A bolt whose head is chamfered or faceted.

garreting (*Build*) A term applied to inserting small stone splinters in the joints of coarse masonry. Also *galleting*, *garneting*.

garret window (*Build*) A skylight of which the glazing lies along the slope of the roof.

garter spring (*Eng*) An endless band formed by connecting the two ends of a long helical spring; used to exert a uniform radial force on any circular piece round which it is stretched, as in a CARBON GLAND.

garth (*Arch*) An enclosed area attached to a building and surrounded usually by a cloister.

gas (*Autos*) US for PETROL.

gas (*MinExt*) Explosive mixture of combustible gases with air, particularly *methane* and *carbon monoxide*. Also used for accumulations of combustion products, eg *carbon dioxide*. See AFTERDAMP, BLACK DAMP, CHOKE DAMP, FIRE DAMP, WHITE DAMP.

gas (*Phys*) (1) A state of matter in which the molecules move freely, thereby causing the matter to expand indefinitely, occupying the total volume of any vessel in which it is contained. (2) The term is sometimes reserved for a gas at a temperature above the critical value. Also defined as a definitely compressible fluid. See DENSITY OF GASES, EXPANSION OF GASES, GAS LAWS, KINETIC THEORY OF GASES, LIQUEFACTION OF GASES.

gas amplification (*Phys*) The increase in sensitivity of a Geiger or proportional counter compared with a corresponding ionization chamber.

gas analysis (*Chem*) The quantitative analysis of gases by absorption. A measured quantity of gas, $100 \, cm^3$, is brought into intimate contact with the various reagents, and the reduction in volume is measured after each absorption process. Carbon dioxide is absorbed in a concentrated potassium hydroxide solution, oxygen in alkaline pyrogallol solution, carbon monoxide in ammoniacal cuprous chloride solution, unsaturated compounds by absorption in bromine water etc, hydrogen by combustion with a measured quantity of air over palladium asbestos. Nitrogen is estimated by difference.

gas-and-pressure-air burner (*Eng*) Industrial gas burner designed to operate with low-pressure gas and with air under pressure from fans and compressors.

gas-bag (*Aero*) Any separate gas-containing unit of a rigid airship.

gas black (*Chem*) See CARBON BLACK.

gas-blast circuit breaker (*ElecEng*) A high-power circuit breaker in which a blast of gas is directed across the contacts at the instant of separation in order to extinguish the arc. See AIR-BLAST SWITCH.

gas cap (*MinExt*) The free gas phase overlying liquid hydrocarbon in a reservoir.

gas carbon (*Chem*) A hard dense deposit of almost pure carbon which slowly collects on the inside of a coal-gas retort.

gas carburizing (*Eng*) The introduction of carbon into the surface layers of mild steel by heating in a reducing atmosphere of gas high in carbon, usually hydrocarbons or hydrocarbons and carbon monoxide.

gas cell (*Chem*) A galvanic cell in which at least one of the reactants is a gas.

gas chromatography (*Chem*) See GAS–LIQUID CHROMATOGRAPHY.

gas colic (*Vet*) Colic associated with TYMPANITES.

gas concrete (*CivEng*) Lightweight concrete in which bubbles of gas are generated when a metallic additive (powdered aluminium or zinc) reacts chemically with the water and cement in the concrete.

gas conditioning (*Eng*) See PROTECTIVE FURNACE ATMOSPHERE.

gas constant (*Phys*) The constant of proportionality R in the equation of state for 1 mole of an ideal gas, $pV = RT$, where p is the pressure, V the volume and T the absolute temperature. $R = 8 \cdot 314 \, J \, K^{-1} \, mol^{-1}$. Also *molar gas constant*.

gas-cooled reactor (*NucEng*) One in which the cooling medium is gaseous, usually carbon dioxide, air or helium. See figs at FUEL ELEMENT, REACTOR.

gas counter (*NucEng*) (1) A gas-filled counter operating in the proportional or Geiger–Müller modes. (2) Geiger counter into which radioactive gases can be introduced.

gas counting (*NucEng*) That of radioactive materials in gaseous form. The natural radioactive gases (radon isotopes) and carbon dioxide ($^{14}CO_2$) are common examples. See GAS-FLOW COUNTER.

gas-discharge lamp (*Phys*) See ELECTRIC-DISCHARGE LAMP.

gas-discharge tube (*Electronics*) Generally, any tube in which an electric discharge takes place through a gas. Specially, a tube comprising a hot or cold cathode, with or without a control electrode (grid) for initiating the discharge, and with gas at an appreciable pressure. See GAS-FILLED RELAY, GLOW TUBE, GRID-GLOW TUBE, IGNITRON, MERCURY-ARC RECTIFIER, THYRATRON.

gas drain (*MinExt*) A tunnel or borehole for conducting gas away from old workings.

gas drilling (*MinExt*) See AIR DRILLING.

gas electrode (*Phys*) An electrode which holds gas by adsorption or absorption, so that it becomes effective as an electrode in contact with an electrolyte.

gas engine (*Autos*) One in which gaseous fuel is mixed with air to form a combustible mixture in the cylinder and fired by spark ignition. Used for stationary power and may operate on two- or four-stroke cycle.

gaseous diffusion (*NucEng*) Isotope separation process based on principle of molecular diffusion.

gaseous diffusion enrichment (*Phys*) The enrichment of uranium isotopes using gaseous UF_6 passing through a porous barrier. See panel on URANIUM ISOTOPE ENRICHMENT.

gaseous discharge (*Phys*) The flow of charge arising from ionization of low-pressure gas between electrodes, initiation being by electrons of sufficient energy released from hot or cold cathodes. Various gases give characteristic spectral colours, eg mercury vapour, sodium vapour, neon, hydrogen.

gas evolution (*Eng*) The liberation of gas bubbles during the solidification of metals. It may be due to the solubility of the gas being less in the solid than in the molten metal, as when hydrogen is evolved by aluminium and its alloys, or to the promotion of a gas-forming reaction, as when iron oxide and carbon in molten steel react to form carbon monoxide. See BLOWHOLE, UNSOUNDNESS.

gas exchange (*BioSci*) The uptake and output of gases, esp of carbon dioxide and oxygen in photosynthesis and RESPIRATION. Also *gaseous exchange*.

gas-filled cable (*ICT*) An impregnated paper-insulated power cable in which gas (nitrogen) at a high pressure is admitted within the lead sheath to minimize ionization.

gas-filled photocell (*Electronics*) One in which anode and photocathode are enclosed in atmosphere of gas at low pressure. It is more sensitive than the corresponding high-vacuum cell because of the formation of positive ions by collision of the photoelectrons with the gas molecules.

gas-filled relay (*Electronics*) Thermionic tube, usually of the mercury-vapour type, when used as a relay; a thyratron.

gas-flow counter (*NucEng*) (1) Counter tube through which gas is passed to measure its radioactivity. (2) One used to measure low intensity α or β sources. These are introduced into the interior of the counter, and to prevent the ingress of air; the counting gas flows through it at a pressure slightly above atmospheric. Also *flow counter*.

gas gangrene (*Med*) Rapidly spreading infection of a wound with gas-forming anaerobic bacteria, causing progressive gangrene of the infected part.

gas generator (*Aero*) The high-pressure compressor/combustion/turbine section of a gas turbine which supplies a high-energy gas flow for turbines which drive propellers, fans or compressors.

gas generator (*Chem*) Chemical plant for producing gas from coal, eg water gas, by alternating combustion of coal and reduction of steam.

gas gland (*BioSci*) A structure in the wall of the air bladder in certain fish that is capable of secreting gas into the bladder. See RETE MIRABILE.

gas governor (*Eng*) A device in which diaphragms are used to maintain a constant-pressure supply to an appliance.

gash (*ElecEng*) Abbrev for *guanidine aluminium sulphate hexahydrate*, a ferroelectric compound with an almost square hysteresis loop suitable for constructing a binary cell.

gasification (*EnvSci*) The production of gaseous fuel by treating solid carbon-rich material at high temperature in the presence of air and water vapour. In some cases the feedstock is coal, in other cases complex mixtures of waste.

gas-impregnated cable (*ICT*) See GAS-FILLED CABLE.

gasket (*Build*) Hemp or cotton yarn wound round the spigot end of a pipe at a joint, and rammed into the socket of the mating pipe to form a tight joint. Also *gaskin*.

gasket (*Eng*) (1) A layer of packing material placed between contact surfaces or parts needing a sealed joint. It can consist of thin copper sheets, compressed rubber bonding, asbestos, etc. Use of asbestos has, however, gone out of favour. Used between cylinder blocks and heads etc. (2) Jointing or packing material, such as cotton rope impregnated with graphite grease; used for packing stuffing-boxes on pumps etc.

gas laws (*Phys*) Boyle's law, Charles's law and the pressure law which are combined in the equation $pV = RT$, where p is the pressure, V the volume, T the absolute temperature and R the GAS CONSTANT for 1 mole. A gas which obeys the gas laws perfectly is known as an *ideal* or *perfect* gas. Cf VAN DER WAALS' EQUATION.

gas lift (*MinExt*) A method of pumping oil from the bottom of a well by releasing compressed liquid gas there. On vaporization it lifts and entrains the oil.

gas lime (*Chem*) The spent lime from gasworks after it has been used for the absorption of hydrogen sulphide and carbon dioxide in the gas purifiers.

gas–liquid chromatography (*Chem*) A form of partition or adsorption chromatography in which the mobile phase is a gas and the stationary phase a liquid. Solid and liquid samples are vaporized before introduction onto the column. The use of very sensitive detectors has enabled this form of chromatography to be applied to submicrogram amounts of material. Abbrev *GLC*.

gas mantle (*Eng*) A small dome-shaped structure of knitted or woven ramie or rayon, impregnated with a solution of the nitrates of cerium and thorium, then dried, and the textile fabric burned off. Also *incandescent mantle*.

gas maser (*Electronics*) One in which the interaction takes place between molecules of gas and the microwave signal.

gas mask (*Chem*) A device for protection against poisonous gases, which are absorbed by activated charcoal or by other reactive substances, eg soda–lime. The choice of the absorbing material depends on the nature of the gas to be counteracted.

gas moulding (*Eng*) See HOLLOW MOULDING.

gasoline (*Eng*) US for PETROL. Often colloq *gas*.

gas-pipe tongs (*Eng*) A wrench used for turning pipes when screwing them into, or out of, coupling pieces.

gas pliers (*Eng*) Stout pliers with narrow jaws, the gripping faces of which are concave and serrated, to provide a secure grip.

gas porosity (*Eng*) Small voids within the body of a moulding or casting caused by the evolution and/or entrapment of gas.

gas-pressure cable (*ElecEng*) See PRESSURE CABLE.

gas-pressure regulator (*Phys*) Diaphragm-operated valve or other device actuated by gas pressure and balanced to produce a constant outlet pressure irrespective of fluctuating initial pressures.

gas producer (*Aero*) A turbo-compressor of which the power output is in the form of the gas energy in the efflux, sometimes mixed with air from an auxiliary compressor.

Essentially, the gas producer is mounted remotely from the point of utilization of its energy, eg helicopter ROTOR-TIP JETS or a FREE TURBINE.

gas pump (*Autos*) US for PETROL PUMP.

gas pump (*Eng*) See HUMPHREY GAS PUMP.

gas regulator (*Eng*) (1) Also GAS-PRESSURE REGULATOR. (2) The throttle valve of a gas engine.

gas tar (*Chem*) Coaltar condensed from coal gas, consisting mainly of hydrocarbons. Distillation of tar provides many substances, eg ammoniacal liquor, 'benzole', naphtha and creosote oils, with a residue of pitch. Dehydrated, it is known as 'road tar', and used as a binder in road-making.

gas temperature (*Aero*) The temperature of the gas stream resulting from the combustion of fuel and air within a turbine engine. For engine performance monitoring, the temperature may be measured at either of two points signified by the abbrevs *JPT* and *TGT*. *JPT* (jet pipe temperature) is the measured temperature of the gas stream in the exhaust system, usually at a point behind the turbine. *TGT* (turbine gas temperature) is the measured temperature of the gas stream between turbine stages. *EGT* (exhaust gas temperature), frequently used, is synonymous with *JPT*.

gaster (*BioSci*) The abdomen proper in Hymenoptera, being the region posterior to the first abdominal segment which, in many members, is constricted.

gastero-, gastr-, gastro- (*Genrl*) Prefixes from Gk *gaster*, gen *gastros*, stomach.

Gasteromycetes (*BioSci*) The puffballs, earth-stars and stinkhorn fungi, a class of Basidiomycotina in which the hymenium is enclosed until after the spores have matured. Most are saprophytic in soil.

gas thread (*Eng*) See BRITISH STANDARD PIPE THREAD.

gastr- (*Genrl*) See GASTERO-.

gas transport (*BioSci*) The transport by the blood of oxygen from the site of external respiration to cells where it is needed for aerobic respiration (see RESPIRATION, AEROBIC), this frequently involving a RESPIRATORY PIGMENT, and the transport away from the respiring cells of any carbon dioxide produced.

gastrectomy (*Med*) Removal of the whole, or part, of the stomach.

gastric (*BioSci*) Pertaining to, or in the region of, the stomach.

gastric juice (*Med*) Human gastric juice consists principally of water (99·44%), free HCl (0·02%) and small quantities of NaCl, KCl, CaCl$_2$, Ca$_3$(PO$_4$)$_2$, FePO$_4$, Mg$_3$(PO$_4$)$_2$ and organic matter including various digestive enzymes.

gastrin (*BioSci*) A polypeptide hormone, secreted by specialized cells in the stomach mucosa, that stimulates the secretion of acid and pepsin by other cells in the mucosa.

gastritis (*Med*) Inflammation of the mucous membrane of the stomach.

gastro- (*Genrl*) See GASTERO-.

gastrocele (*Med*) Hernia of the stomach.

gastrocentrous (*BioSci*) Said of vertebrae in which the centrum is composed largely of the pleurocentrum and the intercentrum is reduced or absent; in all Amniota.

gastrocnemius (*BioSci*) In land vertebrates, a muscle of the lower leg.

gastrocoel (*BioSci*) See ARCHENTERON.

gastrocolic (*Med*) Pertaining to, or connected with, the stomach and the colon.

gastroduodenal (*Med*) Pertaining to, or connected with, the stomach and the duodenum.

gastroduodenitis (*Med*) Inflammation of the mucous membrane of the stomach and the duodenum.

gastroduodenostomy (*Med*) A surgical communication between the stomach and the duodenum.

gastro-enteritis (*Med*) Inflammation of the mucous membrane of the stomach and the intestines.

gastroenterostomy (*Med*) A surgical communication between the stomach and the small intestine.

gastrogastrostomy (*Med*) A surgical communication between the upper and lower parts of the stomach after these are pathologically separated by a stricture.

gastrojejunal (*Med*) Pertaining to, or connected with, the stomach and the jejunum.

gastrojejunostomy (*Med*) A surgical communication between the stomach and the jejunum.

gastrolienal (*Med*) Pertaining to the stomach and the spleen.

gastromyotomy (*Med*) Incision of the muscle of the stomach round a gastric ulcer.

gastroparesis (*Med*) A disorder, often associated with diabetes, in which paralysis of the stomach muscles delays the passage of food through the stomach.

gastropexy (*Med*) The operation of suturing the stomach to the abdominal wall for the treatment of gastroptosis.

Gastropoda (*BioSci*) A class of MOLLUSCA with a distinct head bearing tentacles and eyes, a flattened foot, and a visceral hump that undergoes torsion to various degrees and is often coiled. Such organisms are always bilaterally symmetrical to some extent, with the shell usually in one piece. Snails, slugs, whelks, etc. Also *Gasteropoda*, *gastropods*.

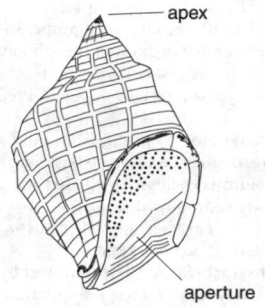

— apex

— aperture

Gastropoda Typical shell.

gastroptosis (*Med*) Abnormal downward displacement of the stomach in the abdominal cavity.

gastroscope (*Med*) An ENDOSCOPE used to inspect the interior of the stomach.

gastrostaxis (*Med*) Bleeding or oozing of blood from the stomach, the mucous membrane of which is intact.

gastrostomy (*Med*) The surgical formation of an opening into the stomach, through which food may be passed when the normal channels are obstructed.

gastrotomy (*Med*) Incision of the stomach wall.

gastrovascular (*BioSci*) Combining digestive and circulatory functions, as the canal system of Ctenophora and the coelenteron of Cnidaria.

gastrula (*BioSci*) In development, the double-walled stage of the embryo which succeeds the blastula.

gastrulation (*BioSci*) The process of formation of a gastrula from a blastula during development.

gas tube (*Phys*) A tube in which the pressure of residual gas is sufficiently high to influence the operation. Since it is so far impossible to obtain a perfect vacuum or even one approaching outer space, all tubes and valves are gas tubes. See GAS-FILLED PHOTOCELL, THYRATRON.

gas tungsten-arc welding (*Eng*) See TUNGSTEN INERT GAS WELDING.

gas turbine (*Eng*) A simple, high-speed machine used for converting heat energy into mechanical work in which stationary nozzles discharge jets of expanded gas (usually products of combustion) against the blades of a turbine wheel. Used in stationary power and other plants, locomotives, marine (esp naval) craft, jet aero-engines, and experimentally in road vehicles. See BYPASS TURBOJET, DUCTED FAN, GAS PRODUCER, SHAFT TURBINE, TURBOJET, TURBOPROP.

gas turbine plant (*Eng*) A power plant comprising at least a gas turbine, an air compressor driven by the gas turbine, and a combustion chamber in which liquid (usually) fuel is burnt to form the working medium of the turbine. Intercooling or other thermodynamic modifications may be added to increase the thermal efficiency of the plant.

gas vacuole (*BioSci*) Structure in the cells of some planktonic cyanobacteria that provides buoyancy and is composed of many small, more or less cylindrical, gas-filled vesicles.

gas welding (*Eng*) Any metal welding process in which gases are used in a combination to obtain a hot flame. The most commonly used gas welding process employs the oxyacetylene combination which develops a flame temperature of 3200°C. Some plastics, esp polythene, may be fusion jointed by a form of low-temperature gas welding without flame.

gas well (*Geol*) A deep boring, generally in an oilfield, which yields natural gas rather than oil. See NATURAL GAS.

gate (*Build*) A movable barrier for (1) closing an entrance and (2) stopping or regulating the flow in a channel, eg a LOCK GATE.

gate (*Electronics*) The control electrode in a FIELD-EFFECT TRANSISTOR which electrically modifies the nature of a conducting channel between source and drain electrodes.

gate (*Eng*) In a mould, the channel or channels through which the molten metal is led from the runner, down gate or pouring gate to the mould cavity. Also *geat, git, sprue*. See fig. at MOULDING.

gate (*ICT*) A circuit that controls the flow of binary signals. The components of a computer processor and memory can all be constructed from combinations of very simple gates, each with a combination of input signals and a single output. See AND GATE, EQUIVALENCE GATE, NAND GATE, NOR GATE, NOT GATE, OR GATE, XOR GATE.

gate (*ImageTech*) The aperture and associated mechanism at which film is exposed in a motion picture camera, printer or projector.

gate (*Plastics*) A term used to denote the small restricted space in an injection mould between the mould cavity and the passage carrying the plastic moulding material. Also *diaphragm, fan, submarine gates, tab*, by reference to their shape. See INJECTION MOULDING.

gate array (*Electronics*) An integrated logic circuit consisting of a two-dimensional array of logic cells, each consisting of one or more GATES. The final details of metallization and interconnection determine the performance of the chip, according to the required application.

gate-chambers (*Build*) Recesses in the side walls of a lock to accommodate the gates when open.

gate-change gear (*Autos*) A multi-speed gearbox in which the control lever moves sideways as well as backwards and forwards in a *gate*. This may be in the form of a simple H, or a more complex pattern depending on the number of ratios. Moving the lever into each arm of the gate selects one ratio.

gate current (*ElecEng*) (1) Current flowing in the gate–cathode circuit of a THYRISTOR. (2) Current flowing in the gate–source circuit of a field-effect transistor (normally very small). (3) The ac or dc pulse which saturates the core of a reactor. Also *gate drive* in (1) and (2).

gated-beam tube (*Electronics*) A valve in which a flat electron beam passes through a slotted plate; beam deflection causes very rapid cut-off of current. Used in some switching and FM detector circuits.

gate detector (*Electronics*) One whose operation is controlled by an external gating signal.

gated throttle (*Aero*) A supercharged aero-engine throttle quadrant with restricting stop(s) to prevent the throttle being wrongly used. See BOOST CONTROL.

gate leakage current (*Electronics*) The dc gate current which flows in a field-effect transistor under normal operating conditions.

gate stick (*Eng*) A stick placed vertically in the cope while it is rammed up; on removal it provides the gate or runner passage into the mould. Also *runner stick*.

gate valve (*Eng*) One which can be moved across the line of flow in a pipeline.

gate voltage (*Electronics*) The control voltage for an electronic GATE. The voltage applied to the 'gate' electrode of a field-effect transistor.

gateway (*ICT*) A device connecting one communicating NETWORK with another.

gateway (*MinExt*) A road through the worked-out area (goaf) for haulage in longwall working of coal. Road connecting coal working with main haulage. Also *gate road*.

gate winding (*ElecEng*) That used to obtain gating action in a magnetic amplifier.

gather (*Glass*) A charge of glass picked up by the gatherer.

gatherer (*Glass*) A person who gathers a charge of glass (a *gather*) on a blowpipe or gathering iron for the purpose of forming it into ware or feeding a charge to a machine for that purpose.

gathering (*Build*) The contracting portion of the chimney passage to the flue, situated a short distance above the source of heat.

gathering (*Print*) Collecting and arranging in proper sequence the folded sections forming a volume.

gathering line (*MinExt*) Small-bore pipes which collect oil or gas from peripheral wells and take them to a central distributing station.

gathering motor (*MinExt*) Light electric loco used to move loaded coal trucks from filling points to main haulage system.

gating (*Phys*) Selection of part of a wave on account of time or magnitude. Operation of a circuit when one wave allows another to pass during specific intervals.

gating (*Psych*) Selective damping or inhibition of sensory input.

gating (*Textiles*) See GAITING.

Gatso (*Genrl*) TN of an automatic photographic device used to identify vehicles exceeding the speed limit.

Gattermann reactions (*Chem*) Reactions used in organic syntheses for preparing aromatic aldehydes from hydrocarbons. Based on passing a mixture of carbon monoxide and hydrogen chloride into the hydrocarbon in the presence of cuprous chloride and aluminium chloride.

gauche (*Chem*) A form of staggered molecular conformation in which the substituents being considered make a dihedral angle of 60° to one another. The most stable conformation of hydrogen peroxide has the hydrogen atoms in this relationship.

Gaucher's disease (*Med*) A rare familial disorder of lipid metabolism resulting in the accumulation of abnormal GLUCOCEREBROSIDES in the RETICULOENDOTHELIAL SYSTEM. The condition usually presents in childhood with anaemia and enlargement of liver and spleen.

gauge (*Build*) A device for marking lines parallel with an edge. See MARKING AND CUTTING GAUGE.

gauge (*CivEng*) The distance between the inside edges of the rails of a permanent way.

gauge (*Eng*) (1) An object or instrument for the measurement of dimensions, pressure, volume, etc. See PRESSURE GAUGE, WATER GAUGE. (2) An accurately dimensioned piece of metal for checking the dimensions of work or less precisely made gauges. See LIMIT GAUGE, MASTER GAUGE, PLUG GAUGE, RING GAUGE. (3) A tool used for measuring lengths, as a MICROMETER GAUGE. (4) The diameter of wires and rods. See BIRMINGHAM WIRE GAUGE, BROWN & SHARPE WIRE GAUGE. (5) That portion of a test piece over which some property such as strain or elongation is to be measured.

gauge (*ImageTech*) The width of motion picture film or magnetic tape; for film standard sizes are 70 mm, 35 mm, 16 mm and 8 mm, while for tape they are 2 in, 1 in, 3/4 in, 1/2 in, 1/4 in and 8 mm.

gauge (*Textiles*) Relates to the fineness of a knitted fabric and generally indicates the number of needles per cm in warp or weft knitting machines.

gauge boson (*Phys*) A particle that mediates the interaction between two fundamental particles. There are four types: PHOTONS for electromagnetic interactions, GLUONS for strong interactions, INTERMEDIATE VECTOR BOSONS for weak interactions and GRAVITONS for gravitational interactions.

gauge box (*CivEng*) A box which measures a known quantity of material, such as cement, sand or coarse aggregate or similar substance, for testing or making mixtures. Also *batch box*.

gauge cocks (*Eng*) Small cocks fitted to pressure vessels to which pressure gauges are attached or which carry liquid-level gauge glasses.

gauge-concussion (*CivEng*) The lateral outward impact of the wheel flanges against the rails due to centrifugal force.

gauged arch (*Build*) An arch built from special bricks cut with a bricklayer's saw and rubbed to exact shape on a stone.

gauged mortar (*Build*) A mortar made of cement, lime and sand, to proportions suitable for the bricks, blocks or other material used.

gauge door (*MinExt*) A door underground for controlling the supply of air to part of the mine.

gauged stuff (*Build*) A stiff plaster used for cornices, mouldings, etc, made with lime putty to which plaster of Paris is added to hasten setting. Also *putty and plaster*.

gauge glass (*Eng*) The tube fitted vertically between a pair of gauge fittings and used to indicate the liquid level in a tank or boiler. Such glasses are usually protected by flat glass panels or perforated metal.

gauge number (*Eng*) An (arbitrary) number denoting the gauge or thickness of sheet metal or the diameter of wire, rod or twist drills in one of many gauge number systems, eg BIRMINGHAM WIRE GAUGE (BWG), BRITISH STANDARD WIRE GAUGE (SWG), BROWN & SHARPE WIRE GAUGE in the USA. See LETTER SIZES.

gauge pressure (*Eng*) Of a fluid, the pressure as shown by a pressure gauge, ie the amount by which the pressure exceeds the atmospheric pressure, the sum of the two giving the absolute pressure.

gauge rod (*Build*) A rod used in laying graduated courses of slates.

gauge theory (*Phys*) Theories in particle physics which attempt to describe the various types of interaction between fundamental particles. QUANTUM ELECTRODYNAMICS describes relativistic quantum fields. The WEINBERG-SALAM THEORY unifies the weak and electromagnetic interactions. QUANTUM CHROMODYNAMICS is designed to explain the binding together of QUARKS to form HADRONS through the STRONG INTERACTION. Grand unified theories are gauge theories which set out to unify all four basic interactions (the strong, ELECTROMAGNETIC and WEAK INTERACTIONS and GRAVITY). See panels on GRAND UNIFIED THEORIES and QUANTUM THEORY.

gauging-board (*Build*) A platform on which mortar or concrete may be mixed.

gaul (*Build*) A hollow spot formed in the setting of plaster etc.

Gault (*Geol*) A blue-grey clayey formation in the Cretaceous (Albian) of England. Also *gault clay*.

Gause's principle (*BioSci*) The idea that closely related organisms do not co-exist in the same niche, except briefly. See COMPETITIVE EXCLUSION PRINCIPLE.

gauss (*Phys*) CGS electromagnetic unit of magnetic flux density; equal to 1 maxwell cm^{-2}, each unit magnetic pole terminating 4π lines. Now replaced by the SI unit of magnetic flux density, the tesla (T); $1\ T = 10^4$ gauss.

Gauss eyepiece (*Phys*) A form of eyepiece used in optical instruments, such as spectrometers and refractometers, to facilitate setting the axis of the telescope at right angles to a plane-reflecting surface. Light enters the side of the eyepiece and is reflected down the telescope tube by a piece of unsilvered glass, being then reflected back into the eyepiece by the plane surface.

Gaussian curvature (*MathSci*) See CURVATURE (3).
Gaussian distribution (*MathSci*) See NORMAL DISTRIBUTION.
Gaussian noise (*Acous*) Any noise whose frequency shows a Gaussian distribution.
Gaussian noise (*ICT*) An unwanted, randomly fluctuating electrical signal, due for example to thermal motion of electrons in a resistor, whose probability density function is given by the Gaussian or normal distribution. For a stationary system in which the average noise voltage is zero, the average noise power is proportional to the VARIANCE of the signal.
Gaussian optics (*Phys*) Simple optical theory which does not consider the aberration of lenses. Practically, it applies only to paraxial rays.
Gaussian points (*Phys*) See PRINCIPAL POINTS OF A LENS.
Gaussian response (*Phys*) The response, eg of an amplifier, for a transient impulse, which, when differentiated, matches the Gaussian distribution curve.
Gaussian units (*Phys*) Formerly widely used system of electric units where quantities associated with electric field are measured in ELECTROSTATIC UNITS and those associated with magnetic field in ELECTROMAGNETIC UNITS. This involves introducing a constant c (the free space velocity of electromagnetic waves) into MAXWELL'S EQUATIONS.
Gaussian well (*Phys*) A particular form of POTENTIAL WELL used to describe the distribution of potential energy of a nuclear particle in the field of the nucleus or other nuclear particle.
gaussmeter (*ElecEng*) An instrument measuring magnetic flux density. This term is most widely used in the USA.
Gauss's convergence test (*MathSci*) The theorem that if, for a series of positive terms $\sum a_n$

$$n\left(\frac{a_n}{a_{n+1}} - 1\right) = \sigma + \left(\frac{1}{n^\delta}\right), \quad \delta > 0$$

then $\sum a_n$ is convergent if $\sigma > 1$ and divergent if $\sigma \leqslant 1$. This test is an extension of Raabe's test.
Gauss's differential equation (*MathSci*) The equation

$$x(1-x)\frac{d^2y}{dx^2} + [c - (a+b+1)x]\frac{dy}{dx} - aby = 0$$

It is satisfied by the hypergeometric function $F(a;b;c;x)$. Also *hypergeometric equation*.
Gauss's laws (*ElecEng*) Laws concerning electrostatics and magnetostatics. The surface integral of the normal component of electric displacement (or magnetic flux) over any closed surface in a dielectric is equal to the total electric charge enclosed (or to zero in the magnetic case). Differential forms of these laws constitute two of MAXWELL'S FIELD EQUATIONS. See POISSON'S EQUATION.
Gauss's theorem (*MathSci*)

$$\int_S \frac{\partial \varphi}{\partial n} dS = \int_V \nabla^2 \varphi dV$$

where the surface S is the boundary of the volume V and n is the normal direction to S. Cf GREEN'S THEOREM.
gauze (*Textiles*) A lightweight woven fabric of open texture.
gavage (*Agri*) Forced feeding of birds being fattened for meat production.
gavel (*Build*) A mallet used for setting stones.
gavelock (*Build*) An iron crowbar. Also *gablock*.
Gay-Lussac's law (*Chem*) (1) Of volumes: when gases react, they do so in volumes which bear a simple ratio to one another and to the volumes of the resulting substances in the gaseous state, all volumes being measured at the same temperature and pressure. (2) See CHARLES'S LAW.
gay-lussite (*Min*) A rare grey hydrated carbonate of sodium and calcium, occurring in lacustrine deposits.
gazebo (*Arch*) A summerhouse resembling a temple in form and commanding a wide view.

Gb (*ICT*) Abbrev for GIGABYTE.
G-banding (*BioSci*) See BANDING TECHNIQUES and panel on CHROMOSOME.
GBL (*Chem*) Abbrev for *gamma-butyrolactone*, a colourless liquid used as a solvent and sometimes as a recreational drug.
GCA (*Aero, Radar*) Abbrev for GROUND-CONTROLLED APPROACH.
GCI (*Aero, Radar*) Abbrev for GROUND-CONTROLLED INTERCEPTION.
G cramp (*Build*) One in the shape of a G, with a screw passing through one end. The shoe is sometimes swivelled to enable the cramp to be used on tapered surfaces. Also *G clamp*.
Gd (*Chem*) Symbol for GADOLINIUM.
G-display (*Radar*) Similar to F-DISPLAY but indicating increasing or diminishing range of target by increasing or diminishing lateral extension of the spot.
Ge (*Chem*) Symbol for GERMANIUM.
geanticline (*Geol*) A regional upwarping of the crust of the Earth. Cf GEOSYNCLINE.
gear (*Eng*) (1) Any system of moving parts transmitting motion, eg levers, gear wheels, etc. (2) A set of tools for performing some particular work. (3) A mechanism built to perform some special purpose, eg steering gear, valve gear. (4) The position of the links of a steam-engine valve motion, as astern gear, mid-gear, etc. (5) The actual gear ratio in use, or the gear wheels involved in transmitting that ratio, in an automobile gearbox, as 'first gear', 'third gear', etc.
gearbox (*Eng*) Casing containing a GEAR TRAIN. The term commonly stands also for the casing including its gear train, particularly when applied to gearboxes used with engines or with machine tools.
gear cluster (*Eng*) A set of gear wheels integral with, or permanently attached to, a shaft, as on the lay shaft of an automobile gearbox.
gear cutters (*Eng*) Milling cutters, hobs, etc, having the requisite tooth form for cutting teeth on gear wheels.
geared lathe (*Eng*) A lathe provided with a BACK GEAR or a multi-speed gearbox between the driving motor and the head.
geared turbogenerator (*ElecEng*) An electric generator driven through a reduction gear from a steam turbine, the object being to enable both machines to operate at their most economical speeds.
gearing (*Eng*) Any set of gear wheels transmitting motion. See GEAR.
gearing-down (*Eng*) A reduction in speed between a driving and a driven wheel or unit, eg between the engine of an automobile and the road wheels.
gearing-up (*Eng*) Raising the speed of a driven unit above that of its driver by the use of gears.
gearless motor (*ElecEng*) A traction motor mounted directly on the driving axle of an electric locomotive.
gear lever (*Eng*) A lever used to move gear wheels relative to each other to change gear. In a motor car, this lever acts on the gear wheels indirectly through SELECTOR FORKS.
gear marks (*Print*) Slurred streaks or bands across the printed sheet or web caused by uneven rotation of cylinder.
gear pump (*Eng*) A small pump consisting of a pair of gear wheels in mesh, enclosed in a casing, the fluid being carried round from the suction to the delivery side in the tooth spaces; used for lubrication systems etc.
gear-tooth forming (*Eng*) A family of engineering processes, including casting, plastic moulding, stamping from sheet metal for watch and clock gears, form cutting, gear shaping, hobbing and other methods of gear-tooth generating.
gear train (*Eng*) Two or more GEAR WHEELS, transmitting motion from one shaft to another. With external spur or bevel gears, the velocity ratio is inversely proportional to the number of gear teeth.

gear wheel (*Eng*) A toothed wheel used in conjunction with another, or with a rack, to transmit motion.

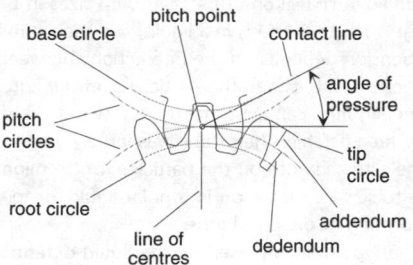

base circle   pitch point   contact line

pitch circles   angle of pressure

tip circle

root circle   addendum

line of centres   dedendum

gear wheel

Gedinnian (*Geol*) The oldest stage in the Devonian period. See PALAEOZOIC.

gedrite (*Min*) An orthorhombic amphibole, containing more aluminium and less silicon than anthophyllite. The iron–aluminium end-member has been called *ferrogedrite*. Gedrite occurrences are restricted to metamorphic and metasomatic rocks.

gegenions (*Chem*) The simple ions, of opposite sign to the colloidal ions, produced by the dissociation of a colloidal electrolyte. Also *counterions*.

gegenschein (*Astron*) Ger for *counter-glow*. A term applied to a faint illumination of the sky sometimes seen in the ecliptic, diametrically opposite the Sun, and connected with the zodiacal light.

gehlenite (*Min*) The calcium aluminium end-member of the melilite group of minerals, $Ca_2Al_2SiO_7$.

Geiger characteristic (*Phys*) The plot of recorded count rate against operating potential for a Geiger or proportional counter detecting a beam of radiation of constant intensity.

Geiger counter (*NucEng*) An instrument for measuring ionizing radiation, with a tube carrying a high-voltage wire in an atmosphere containing argon plus halogen or organic vapour at low pressure, and an electronic circuit which quenches the discharge and passes on an impulse to record the event. Also *Geiger–Müller counter*, *G–M counter*. See TOWNSEND AVALANCHE.

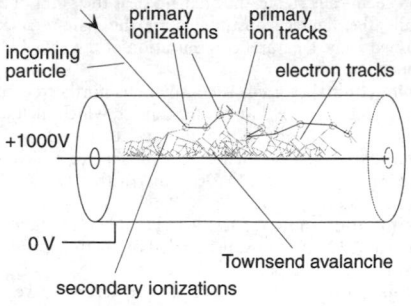

incoming particle   primary ionizations   primary ion tracks   electron tracks

+1000V

0 V   Townsend avalanche

secondary ionizations

Geiger counter

Geiger–Müller tube (*NucEng*) The detector of a Geiger counter, ie without associated electronic circuits.

Geiger–Nuttall relationship (*Phys*) An empirical rule relating the half-life *T* of radioactive materials emitting alpha particles to the range *R* of the particles emitted:

$$\log(1/T) = a\log R + b$$

where *a* and *b* are constants.

Geiger region (*NucEng*) That part of the characteristic of a counting tube, where the charge becomes independent of the nature of the ray intercepted. Also *Geiger plateau*, since in this region the efficiency of counting varies only slowly with voltage on the tube.

Geiger threshold (*NucEng*) Lowest applied potential for which Geiger tube will operate in Geiger region.

Geissler pump (*Chem*) A glass vacuum pump which operates from the water supply.

geitonogamy (*BioSci*) Fertilization involving pollen and ovules from different flowers on the same individual plant (ramet) or from the same clone (genet); a form of ALLOGAMY (1). See INBREEDING, SELF-FERTILIZATION, SELF-POLLINATION.

gel (*BioSci*) A colloid in which the disperse phase is solid and the continuous phase liquid. In most food gels the liquid phase is water. Gels are used as an inert matrix for applications such as separation of polynucleotides and polypeptides by electrophoresis (eg polyacrylamide or starch), or within which cells can be cultured (eg agarose or collagen). See GEL ELECTROPHORESIS and GEL FILTRATION.

gel (*Chem*) A substance with properties intermediate between the liquid and the solid states. See panel on GELS.

gel (*ImageTech*) A coloured gelatine filter for light sources.

gelatin (*Chem*, *FoodSci*, *Build*) The product of partial hydrolysis of COLLAGEN from the tissues and bones of slaughtered animals with different iso-electric points depending on the hydrolysis conditions. It is colourless, odourless and tasteless and used as a thickener and gelling agent in many areas of food production and for photographic films, glues, etc. Used as a mordant in glass gilding. Also *gelatine*. See BLOOM GELOMETER.

gelatin filter (*ImageTech*) Also *gelatine filter*. See FILTER.

gelation (*Plastics*) The process whereby plasticized polyvinyl chloride compounds by the application of heat undergo an irreversible change to soft, rubbery thermoplastic materials. See PLASTISOLS.

gel diffusion tests (*BioSci*) Precipitin tests in which antigen and antibody are placed separately in a gel medium (commonly agar) and allowed to diffuse towards one another and to form lines of precipitate where they meet in suitable proportions.

gelding (*Agri*) A castrated male horse.

gel electrophoresis (*BioSci*) A method of separating large molecules on the basis of their charge, which affects their rate of migration in an electric field and their size, which affects their movement through the gel supporting phase. Polyacrylamide gels are used extensively but agar and starch gels can also be used. See ELECTROPHORESIS.

gel filtration (*BioSci*) A type of column CHROMATOGRAPHY that separates molecules in solution according to size; smaller molecules enter the porous beads of eg polyacrylamide or sugar polymer (Sepharose, Sephadex) and their flow is retarded relative to larger molecules. Cf AFFINITY CHROMATOGRAPHY.

gelignite (*Chem*, *MinExt*) Explosive used for blasting, composed of a mixture of nitroglycerine (60%), guncotton (5%), wood pulp (10%) and potassium nitrate (25%).

gelling (*FoodSci*) Forming a gel by adding a GELLING AGENT to a liquid, first changing its viscosity and then causing it to set to form a stable elastic solid. The process is temperature and pH dependent, usually requiring heat to melt a gel and cooling to allow the gel to set.

gelling agent (*FoodSci*) A substance which forms a gel when added to and mixed with a liquid, commonly gelatine, alginate and gums. See GEL.

gel-permeation chromatography (*Chem*) A method for analysing molecular mass distribution of non-cross-linked polymers, by selective elution of polymer solution through a microporous gel (cross-linked polystyrene). Common solvent is tetrahydrofuran. For insoluble polymers, high-temperature gel-permeation chromatography in decalin etc is available. Abbrev GPC. See MOLECULAR MASS DISTRIBUTION and panel on GELS.

gel spinning (*Chem*, *Plastics*) Polymer fibre-making process that involves creating a dry gel from THETA SOLVENTS. See panels on GELS and HIGH-PERFORMANCE POLYMERS.

Gels

Gels have properties intermediate between the liquid and the solid states. They deform elastically and recover, yet can often be induced to flow at higher stresses. They are extended three-dimensional network structures based on polymeric molecules and can be permanent or temporary. They are highly porous and many gels contain a very high proportion of liquid to solid.

Gels occur in many fields. Examples include: vulcanized rubber when it absorbs a liquid and swells; plasticized PVC, thixotropic paint, photographic emulsion, jam and mayonnaise; GEL PERMEATION CHROMATOGRAPHY (GPC), for measuring the molecular mass distribution in polymers; finings for beers and wines (eg isinglass and gelatin); the drying agent, silica gel; gel spinning to produce highly oriented polymer fibres. They form the basis of the sol-gel process for making inorganic glasses and ceramics in which the gel network is formed by polymerization in solution. The interstitial liquid is then eliminated and the residue collapsed by sintering to produce a glass without melting.

The classical starting point for gels is a suspension of a colloid (a collection of particles with sizes in the range 1 nm to 0·1 mm) in a liquid, and their stability in suspension depends on the interactions between the particles and between the particles and the liquid. Some, eg inks, can exist indefinitely, whilst others, eg milk, are unstable. Their stability is largely governed by the electric charges on the particles, which might be due to surface ionization in a polar liquid, or to absorption of dissolved ions.

The type of ions present in the liquid determine whether the particles are positively or negatively charged; in a colloidal suspension in water, the charges on the particles change from positive to negative as the concentration of OH^- ions is increased (ie the pH rises). The ISO-ELECTRIC POINT is the pH at which the changeover occurs and as the particles are then uncharged, there is no electrostatic repulsion between them. Brownian motion can bring them close enough to combine through eg van der Waals forces, forming larger and larger particles which fall out of the suspension, a process known as *coagulation*. If polymeric molecules are present, these can provide bridges between the particles, creating the open three-dimensional network required of a gel; this is known as FLOCCULATION.

See panel on HIGH-PERFORMANCE POLYMERS.

---

gem (*Min*) A mineral or other natural material (eg amber, coral, pearl and shells), which when cut and polished possesses the qualities of beauty and durability that make it suitable for personal adornment or as ornaments. Rarity is also an essential element. Precious stones are generally taken to be diamond, ruby, sapphire, emerald and opal (and somewhat differently, pearl). Semiprecious stones are arbitrarily and less clearly defined but usually include beryls, chrysoberyl, cordeirite, garnets, olivine, sphene, spinels, topaz, tourmaline, zircon, jade, lapis lazuli and turquoise. Gemstones are cut to enhance their beauty.

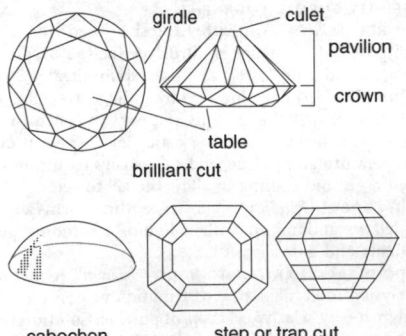

brilliant cut

cabochon   step or trap cut

gems and gemstones Typical cuts.

gemel window (*Arch*) A two-bay window.
gemfibrosil (*Pharmacol*) A *fibrate* used for treatment of hyperlipidaemias.
gem gravels (*Geol*) Sediments of the gravel grade containing appreciable amounts of gem minerals, and formed by the disintegration and transportation of pre-existing rocks, in which the gem minerals originated. They are really placers of a special type, in which the heavy minerals are not gold or tin, but such minerals as garnets, rubies, sapphires, etc. As most of the gem minerals are heavy and chemically stable, they remain near the point of origin, while the lighter constituents of the parent rocks are washed away, a natural concentration of the valuable components resulting.

Gemini (Twins) (*Astron*) A conspicuous northern constellation, lying between Taurus and Cancer, which includes the bright pair of stars CASTOR and POLLUX.
Geminids (*Astron*) A major meteor shower which shows maximum activity on 13 December with a rate of approx 58 per hour.
Gemini programme (*Space*) A series of two unmanned and 10 manned NASA missions carried out between April 1964 and November 1966.
gemma (*BioSci*) (1) A multicellular structure for vegetative reproduction in algae, pteridophytes and esp bryophytes. (2) Same as CHLAMYDOSPORE. (3) A bud that will give rise to a new individual. Pl *gemmae*.
gemma cup (*BioSci*) Cup-like structure in which gemmae are borne in some mosses and liverworts.
gemmation (*BioSci*) Budding; gemma-formation.
gemmiferous (*BioSci*) Producing GEMMAE. Also *gemmiparous*.
gemmule (*BioSci*) In fresh-water Porifera, an aggregation of embryonic cells within a resistant case, which is formed at the onset of hard conditions when the rest of the colony dies down, and which gives rise to a new colony when conditions have once more become favourable.
gemstone (*Min*) See GEM.
gen-, geno- (*Genrl*) Prefixes from Gk *genos*, race, descent.
-gen, -gene (*Genrl*) Suffixes meaning *generating, producing*.

gender identity (*Psych*) Internal sense of being either male or female: when this is in conflict with biological sex, the result is gender identity disorder.

gender role (*Psych*) The set of attitudes and behaviours a given culture considers appropriate for each sex.

gene (*BioSci*) The hereditary determinant of a specified difference between individuals. Can refer either to a particular *allele* or to all alleles at a particular *locus*. Molecular analysis has shown that a specific sequence or parts of a sequence of DNA can be identified with the classical gene. See CISTRON.

genealogy (*BioSci*) The study of the development of plants and animals from earlier forms.

gene amplification (*BioSci*) Selective replication of a DNA sequence to produce multiple extra copies of that sequence in the cell. The best-known example is in the oöcyte of *Xenopus*, where there are about 2 million copies of the ribosomal RNA genes.

gene bank (*BioSci*) A collection of plants or, more often, of seeds, cell cultures, frozen pollen, etc, of species of known or potential use to humans, and esp of LANDRACES that may contain genes of use in crop breeding.

gene chip (*BioSci*) An array of oligonucleotides immobilized on a surface that can be used to screen an RNA sample (after reverse transcription) and thus a method for rapidly determining which genes are being expressed in the cell or tissue from which the RNA was extracted. The immobilized oligonucleotides may be either a random set of defined sequences or probes for known genes.

gene cloning (*BioSci*) The insertion of a DNA sequence into a vector that then multiplies in a host cell or organism and produces a large number of copies of the sequence. The sequence need not necessarily be for a complete gene.

genecology (*BioSci*) The branch of ecology which seeks genetic explanations of the patterns of distribution of plants and animals in time and space.

gene dosage (*BioSci*) The number of copies of a gene in a given cell or individual organism.

gene expression (*BioSci*) The full use of the information in a gene, via transcription, leading to production of RNA, and to a protein if the RNA is translated, and hence the appearance of the phenotype determined by that gene.

gene-for-gene concept (*BioSci*) The concept that there are corresponding genes for virulence and resistance in pathogen and host respectively.

gene frequency (*BioSci*) The proportion of all representatives of a particular gene in a population that contain the specified allele.

gene mapping (*BioSci*) The determination of the positions and relative distances of genes on chromosomes by means of their *genetic linkage*.

gene number (*BioSci*) The total number of different CODING SEQUENCES that are transcribed into RNA in an individual or species.

general adaptation syndrome (*Psych*) Physiological characterization of the biological response to severe stress: the first alarm stage is subdivided into shock and counter-shock phases and is followed by the second resistance stage during which recuperation, begun during counter-shock, continues. If the stress exceeds tolerable limits then the third stage, exhaustion, ensues and can lead to death. Abbrev *GAS*.

general aviation (*Aero*) Private, agricultural and survey aviation.

general circulation of the atmosphere (*EnvSci*) Caused by the way excess solar heat energy received in equatorial regions is distributed upwards and polewards by the wind, until lost to space by radiation. Cooled polar air returning towards the equator travels at the surface of the Earth. Winds are affected by the CORIOLIS FORCE as the Earth rotates, and by land masses, mountain ranges and the seasons. See panels on STRATOSPHERE AND MESOSPHERE and TROPOSPHERE.

general integral (*MathSci*) See COMPLETE INTEGRAL.

generalist (*BioSci*) An organism with many food sources, or able to live in many habitats. Cf SPECIALIST.

generalization (*Genrl, Psych*) (1) Forming a judgement or decision that is applicable to a class or category of objects, ideas, etc, on the basis of a limited sample. (2) Extending a principle to new situations. (3) Scientifically, any broad principle that encompasses a number of observations. (4) In psychology, the tendency to respond similarly to stimuli that have properties in common or are grouped, eg semantically. (5) In classical conditioning, a conditioned response elicited by a stimulus similar to the conditioned stimulus.

generalized anxiety disorder (*Psych*) A chronic state of diffuse, unfocused anxiety, in the absence of specific symptoms such as are found in phobic reactions, and without any associated specific stimuli or objects.

general lighting (*ElecEng*) A system of lighting employing fittings which emit the light in approximately equal amounts in an upward and a downward direction.

general linear group (*MathSci*) The group containing all the $n \times n$ invertible matrices with entries from the field F. Usually denoted by $GL_n(F)$. Cf SPECIAL LINEAR GROUP.

general paresis (*Psych*) A psychosis produced by syphilitic infection in which there is a progressive deterioration of cognitive or motor functions, culminating in death.

general purpose foils (*Print*) Blocking foils suitable for marking paper, bookcloths, etc, but not normally used on thermoplastics.

general relativity (*Phys*) Generalization of Einstein's SPECIAL RELATIVITY theory to accelerating frames of reference (1915), which replaces the Newtonian notion of instantaneous action at a distance via the gravitational field with a distortion of space–time due to the presence of mass. Supported by experiments which measure the bending of starlight due to the presence of the Sun's mass, and also the precession of Mercury's orbit.

general sexual dysfunction (*Psych*) In women, the absence or weakness of the physiological changes normally accompanying the excitement phase of sexual response.

general solution (*MathSci*) See COMPLETE INTEGRAL.

general theory of relativity (*Phys*) See GENERAL RELATIVITY.

generating circle (*Eng*) Any circle in which a point on the circumference is used to trace out a curve when the circle rolls along a straight line or curve.

generating function (*MathSci*) A function which can be regarded as summarizing a sequence of functions which become apparent when the generating function is expanded as a series; eg (1) the function $(q + pt)^n$ generates the binomial probability distribution when expanded in powers of t. The coefficient of t^x in the expansion of $(q + pt)^n$ is the probability $\binom{n}{x}p^x q^{n-x}$ of x successes out of n trials. (2) The function $(1 - 2xh + h^2)^{-1/2}$, when expanded in powers of h, generates the Legendre polynomials $P_n(x)$.

generating line (*Eng*) A straight or curved line rotated about some axis to generate a surface.

generating set (*ElecEng*) An electric generator, together with the prime mover which drives it.

generating station (*ElecEng*) A building containing the necessary equipment for generating electrical energy.

generation (*BioSci*) The individuals of a species that are separated from a common ancestor by the same number of broods in the direct line of descent.

generation (*ImageTech*) Whether a recorded tape is a master (first generation), or a copy (second generation, or first-generation copy), etc.

generation (*Phys*) Two leptons and two quarks forming groups which interact. First generation is (up quark, down quark) and (electron, electron neutrino); second generation is (charmed quark, strange quark) and (muon, muon neutrino); third generation is (top quark, bottom quark) and (tau lepton, tau neutrino). See LEPTON–QUARK SYMMETRY.

Genetic manipulation

The term for the procedures with which it is now possible to combine DNA sequences from widely different organisms *in vitro*, often with great precision. Two major advances have made these procedures possible: the discovery of RESTRICTION ENZYMES and the ability to construct suitable vectors (such as plasmids) into which a DNA sequence can be inserted to form a hybrid molecule. Such hybrid molecules are able to multiply in a rapidly growing host (bacterium or yeast). A necessary further part of the procedure is the selection of those host cells which contain hybrid molecules. One way to do this is to arrange for the inserted DNA to disrupt a vector sequence which normally prevents growth (*positive selection*).

Restriction enzymes which cut eg eukaryote DNA at specific sites enable a sequence such as part of a gene to be identified and purified. They are also the main way in which vectors can be constructed so that they are stripped of unnecessary functions while still being able to replicate after the donor DNA has been incorporated.

'Gene manipulation' has the following legal definition in the UK: 'The formation of new combinations of heritable material by the insertion of nucleic acid molecules, produced by whatever means outside the cell, into any virus, bacterial plasmid or other vector system so as to allow their incorporation into a host organism in which they do not naturally occur but in which they are capable of continued propagation.' See POLYMERASE CHAIN REACTION for a method of amplifying a DNA sequence without using living organisms.

Clones

Clones are exact genetic copies of an individual animal, plant or bacterium. In other words, within the limits set by SOMATIC MUTATION and by gene reassortment in eg the production of lymphocytes, their DNA sequences should be identical. Clones have been propagated by procedures long practised in horticulture to provide exact copies of a plant for sale by nurseries.

In 1997 it was announced that the first true clone of an adult mammal had been made. What distinguishes this event was that a nucleus from a differentiated adult tissue cell had been injected into an egg from which the nucleus had been removed. Electrical stimulation caused the transplanted nucleus to divide and the egg was implanted into a parent animal and a healthy sheep (Dolly) was born. Further research showed in 1998 that Dolly was a true clone of the adult donor. Further, Dolly has a natural lamb, Bonnie. The successful outcome followed many attempts and stems from work already undertaken at the Roslin Institute, Edinburgh, which has allowed the production of human blood factors in sheep's milk.

In 1998 similar experiments were repeated in mice which, because of their shorter reproductive life and cheaper maintenance, will allow this important discovery to be more easily investigated. It already seems that the mechanisms thought to determine the lifespan of somatic cells can be reset.

---

generation of diversity (*BioSci*) The process by which T- and B-lymphocyte antigen receptors are generated from the available genes by recombination and, in B-lymphocytes, SOMATIC MUTATION.

generation rate (*Electronics*) The rate of production of ELECTRON–HOLE PAIRS in semiconductors.

generation time (*Phys*) Average life of a fission neutron before absorption by a fissile nucleus.

generative cell (*BioSci*) A cell of the male gametophyte in pollen grain or pollen tube. In gymnosperms it divides to give the sterile (stalk) cell and the spermatogenous (body) cell. In angiosperms it divides to give the two male gametes. See DOUBLE FERTILIZATION.

generator (*ElecEng*) Electrostatic or electromagnetic device for conversion of mechanical into electrical energy. Also *electric generator*.

generator (*MathSci*) See CONE (1), CYLINDER (1), RULED SURFACE.

generator bus-bars (*ElecEng*) Bus-bars in a generating station to which all the generators can be connected.

generator-field control (*ElecEng*) See VARIABLE VOLTAGE CONTROL.

generator panel (*ElecEng*) A panel of a switchboard upon which are mounted all the switches, instruments, and other apparatus necessary for controlling a generator.

generator potential (*BioSci*) An electrical potential arising in a sensory neuron as a result of a sensory stimulus.

genesis (*BioSci*) The origin, formation or development of some biological entity. Adj *genetic*.

genet (*BioSci*) A genetic individual, the product of one zygote; either an individual plant grown from a sexually produced seed or all the individuals produced by vegetative reproduction from one such plant. Cf RAMET.

gene therapy (*BioSci*) Colloq term for the substitution of a functional for a defective gene as a treatment for a *genetic disease*.

genetically significant dose (*Radiol*) The dose that, if given to every member of a population prior to conception of children, would produce the same genetic or hereditary harm as the actual doses received by the various individuals.

genetic code (*BioSci*) The rules that relate the four bases of DNA (adenine, guanine, cytosine, thymine) or RNA (adenine, guanine, cytosine, uracil) with the 20 amino acids found in proteins. See panel on DNA AND THE GENETIC CODE.

genetic correlation (*BioSci*) The measure of the extent to which the variation of two different quantitative characters is caused by the same genes.

genetic drift (*BioSci*) The process by which *gene frequencies* are changed by the chances of random sampling in small populations.

genetic engineering (*BioSci*) Colloq term for GENETIC MANIPULATION (panel).

Genetic manipulation *(Cont.)*

Plant genetic manipulation
The use of various techniques, except in most usages of conventional breeding, to produce plants containing foreign DNA as part of their genomes. The techniques include both the isolation of DNA from another organism and its transfer to a plant, and also PROTOPLAST FUSION and culture. *Genetic engineering* and *genetic manipulation* usually refer to the former rather than the latter techniques. Both, however, make it possible to transfer genes from much less closely related organisms than does conventional breeding, even from bacteria with DNA transfer. Plants in which foreign DNA is incorporated and expressed are called *transformed* or *transgenic plants*. See PLANT CELL CULTURE for additional methods.

DNA containing a desired gene may be prepared as cDNA or be selected from a DNA library. One method of transferring the DNA to the plant is to use a natural vector, most successfully the Ti plasmid of *Agrobacterium tumefaciens*, which inserts DNA into the nuclear genome; plants may then be regenerated from the transformed cells by plant cell culture. In a second method, protoplasts are stimulated to take up the DNA directly by polyethylene glycol or by ELECTROPORATION, the plants being regenerated from protoplasts by protoplast culture. Both methods have been more successful with dicotyledons than with cereals because *Agrobacterium* naturally infects dicotyledons only and current methods have rarely achieved regeneration from the protoplasts of cereals. In a third method, successful with at least one cereal, DNA is injected directly into the bud which will give rise to the ear.

In protoplast fusion, protoplasts are prepared from cells of two plants and induced to fuse by polyethylene glycol or electric shock (*electrofusion*). The fused protoplast initially contains the complete nuclear, chloroplast and mitochondrial genomes of both plants. During protoplast culture, individual chromosomes, chloroplasts or mitochondria may be lost which, together with recombination in both mitochondria and chloroplasts, may result in cells with almost any combination of parental genomes. Some of these cells may be capable of regenerating hybrid or cybrid plants.

Genetic manipulation is important in research into the functioning of the genome and has considerable promise in the production of new cultivars. Useful traits which may be introduced into crop plants include: herbicide resistance (to increase the range of herbicides that can be used with a given crop), disease resistance (plants able to make the coat protein of a virus may be resistant to that virus), pest resistance (plants making the toxin of *Bacillus thuringiensis*, toxic to caterpillars) and male sterility (to facilitate the production of F_1 hybrids). Protoplast fusion has made possible the transfer of cytoplasmic male sterility and some cytoplasmically inherited herbicide resistance between crops. Once made, transgenic plants may be used in conventional breeding programmes.

Plant products which have been produced by these kinds of procedures include rot-resistant tomatoes, soya flour, Golden rice and maize.

---

genetic equilibrium (*BioSci*) The situation reached when the frequencies of genes and of genotypes in a population remain constant from generation to generation.

genetic linkage (*BioSci*) Genes located close together on the same chromosome tend to be inherited together unless there is a recombination event between them; the closer they are, the rarer the event. Linkage is a measure of the percentage recombination between loci, unlinked genes showing 50% recombination. See LINKAGE DISEQUILIBRIUM, LINKAGE EQUILIBRIUM.

genetic load (*BioSci*) Generally the decrease in fitness of a population due to deleterious mutations in the population gene pool. More specifically, the average number of recessive lethal mutations estimated to be present in the genome of a heterozygous individual in a population.

genetic manipulation (*BioSci*) A term for the procedures with which it is now possible to combine DNA sequences from widely different organisms *in vitro*, often with great precision. Cf CLONE. See PLANT GENETIC MANIPULATION, PLASMID, RESTRICTION ENDONUCLEASE ENZYME, VECTOR and panels on DNA AND THE GENETIC CODE and GENETIC MANIPULATION.

genetic recombination (*BioSci*) The formation of new combinations of alleles in offspring as a result of exchange of DNA sequences between molecules. It occurs naturally, in crossing over between homologous chromosomes in meiosis, or experimentally, as a result of genetic engineering techniques.

genetics (*BioSci*) The study of heredity; of how differences between individuals are transmitted from one generation to the next; and of how the information in the genes is used in the development and functioning of the adult organism.

genetic spiral (*BioSci*) In phyllotaxis, a hypothetical line through the centres of successive leaf primordia at the shoot apex.

genetic variance (*BioSci*) The measure of the variation between individuals of a population due to differences between their *genotypes*.

Geneva movement (*Eng*) Intermittent movement of a wheel with slots from the continuous motion of a driver.

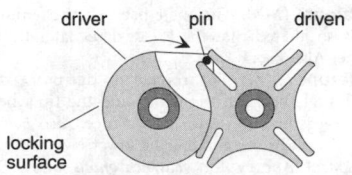

Geneva movement Three or more slots can be used.

genial (*BioSci*) Pertaining to the chin.

genicular (*BioSci*) Pertaining to, or situated in, the region of the knee.

geniculate (*BioSci*) Bent rather suddenly, like the leg at the knee, as *geniculate antennae*.

geniohyoglossus (*BioSci*) The muscle that moves the tongue in vertebrates.

genioplasty (*Med*) Plastic surgery of the chin.

Genistein (*Pharmacol*) TN for an isoflavone plant estrogen, found in soya, that may have a chemopreventive role in cancer.

genital atrium (*BioSci*) In Platyhelminthes and some Mollusca, a cavity into which open the male and female genital ducts.

genitalia (*BioSci*) The gonads and their ducts and all associated accessory organs. Also *genitals*.

genital stage (*Psych*) In psychoanalytic theory, the final phase of psychosexual development in which sexual pleasure involves not only gratification of one's own impulses but attention to and pleasure in, the social and physical pleasure of one's mate. Freud was referring to heterosexual patterns, and to sexual pleasure derived mainly through the genitalia.

genito-urinary (*Med*) Pertaining to the genital and the urinary organs.

genlock (*ImageTech*) A device for synchronizing locally generated pulse signals with those transmitted from a distant source.

Gennari's band (*Med*) A layer of nerve fibres in the cerebral cortex. Also *Gennari's fibres, Gennari's line*.

geno- (*Genrl*) See GEN-.

genome (*BioSci*) The totality of the DNA sequences of an organism or organelle.

genomic DNA (*BioSci*) Nuclear chromosomal DNA.

genomic imprinting (*BioSci*) The phenomenon in which copies of paternally derived chromosomes are in some way different from the copies of maternally derived chromosomes (thus 'imprinted' with their origin), so that the two copies are not equally active and cannot substitute for each other. This has important implications for attempts to correct genetic disorders by genetic engineering. See DNA METHYLATION.

genomic library (*BioSci*) DNA LIBRARY derived from a whole, single GENOME. See panel on DNA AND THE GENETIC CODE.

genotype (*BioSci*) The particular alleles at specified loci present in an individual; the genetic constitution. Adj *genotypic*. Cf PHENOTYPE.

gentian violet (*Chem*) A mixture of three dyes, methyl rosaniline, methyl violet and crystal violet, which is antiseptic and bactericidal. Used as a fungicide and anthelminthic.

gentiobiose (*Chem*) $C_{12}H_{22}O_{11}$. A disaccharide based on glucose. A reducing sugar which occurs in combination in AMYGDALIN. Contains $G\beta1 \rightarrow 6G$ (G = glucose unit).

genu (*BioSci*) A knee-like structure, ie a bend in a nerve tract; more particularly, part of the CORPUS CALLOSUM in mammals.

genu recurvatum (*Med*) The condition in which there is hyperextension of the knee joint.

genus (*BioSci*) A taxonomic rank of closely related forms that is further subdivided into species and therefore below FAMILY and above SPECIES. Pl *genera*. Adj *generic*.

genu valgum (*Med*) The angle between the femur and the tibia is so altered that the leg deviates laterally from the midline. Also *knock knee*.

genu varum (*Med*) The reverse of GENU VALGUM, the altered angle between the femur and the tibia being such that the legs bow outwards at the knee. Also *bow leg*.

genys (*BioSci*) In vertebrates, the lower jaw.

GEO (*Space*) Abbrev for *geosynchronous orbit*. See GEO-SYNCHRONOUS.

geo- (*Genrl*) Prefix from Gk *ge* denoting the Earth.

geobiotic (*BioSci*) Terrestrial; living on dry land.

geobotanical indicator (*MinExt*) See GEOBOTANICAL SURVEYING.

geobotanical surveying (*MinExt*) A form of GEOCHEMICAL PROSPECTING. (1) Identification and systematic surveying of distribution of metallophile plant species, eg *calamine violet* associated with zinc anomalies in C Europe (*geobotanical indicators*). (2) Identification and systematic surveying of pathological conditions in plants caused by metal toxaemia. (3) Systematic sampling of vegetation to identify anomalous concentrations of metals in plant tissues.

geocarpy (*BioSci*) The ripening of fruits underground, the young fruits being pushed into the soil by a post-fertilization curvature of the stalk.

geocentric (*Astron*) The term applied to any system or mathematical construction which has as its point of reference the centre of the Earth.

geocentric altitude (*Surv*) The TRUE ALTITUDE of a heavenly body as corrected for GEOCENTRIC PARALLAX.

geocentric latitude (*Astron*) See LATITUDE AND LONGITUDE, CELESTIAL.

geocentric parallax (*Astron*) See DIURNAL PARALLAX.

geocentric parallax (*Surv*) The correction which must be applied to the altitude of a heavenly body in the solar system as observed, in order to give the altitude corrected to the Earth's centre. Its value is given by $p = + P \cos \alpha$, where p = geocentric parallax, P = HORIZONTAL PARALLAX, and α = observed altitude corrected for refraction.

geochemical prospecting (*MinExt*) Application of GEOCHEMISTRY to mineral exploration by the systematic analysis of bedrock, soil, stream, river and groundwater, stream gravels and vegetation for the purposes of identifying anomalous concentrations of particular elements of economic interest, or elements commonly associated with such ore bodies. See GEOBOTANICAL SURVEYING, SOIL SAMPLING, STREAM SAMPLING, TRACERS.

geochemistry (*Chem*) The study of the chemical composition of the Earth's crust.

geochronology (*Geol*) The study of time with respect to the history of the Earth, primarily through the use of *absolute*, or *isotopic*, and *relative age-dating* methods.

geocline (*BioSci*) A CLINE occurring across topographic or spatial features of an organism's range.

geode (*Geol, MinExt*) Hollow, rounded rock, mineral nodule or concretion, often lined with crystals which have grown inwards. Also *drusy cavity*.

geodemographics (*Genrl*) The study of demographic data in terms of geographical area.

geodesic (*MathSci*) The shortest path between two points on any surface.

geodesic structures (*CivEng*) Structures consisting of a large number of identical parts and therefore simple to erect; and whose pressure is load-shared throughout the structure, so that the larger it is, the greater its strength. Fig. ▷

geodesy (*Surv*) The branch of surveying concerned with extensive areas, in which, to obtain accuracy, allowance must be made for the curvature of the Earth's surface. Also *geodetic surveying*.

geodetic construction (*Aero*) A redundant space frame whose members follow diagonal geodesic curves to form a lattice structure, such that compression loads induced in any member are braced by tension loads in crossing members. Does not need stress-carrying covering.

geodetic surveying (*Surv*) See GEODESY.

geodynamics (*Genrl*) The study of the dynamic processes and forces within the Earth.

geognosy (*Geol*) An old term for absolute knowledge of the Earth, as distinct from geology, which includes various theoretical aspects.

geographical latitude (*Astron*) See LATITUDE AND LONGITUDE, TERRESTRIAL.

geographical mile (*Genrl*) The length of 1 minute of latitude, a distance varying with the latitude, and having a mean value of 6076·8 ft (1852·2 m). US, 1 minute of longitude at the equator, ie 6087·1 ft (1855·3 m).

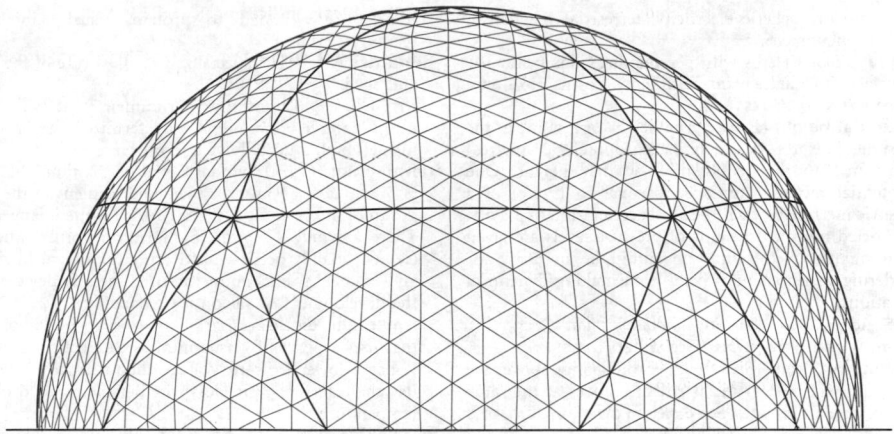

geodesic structures Buckminster Fuller dome.

geographical race (*BioSci*) A collection of individuals within a species that differ constantly in some slight respects from the normal characters of the species, but not sufficiently to cause them to be classified as a separate species, and that are peculiar to a particular area.

geography (*Genrl*) The science of the surface of the Earth and its inhabitants.

geoid (*Surv*) The figure of the Earth, considered as a smooth oblate spheroid or ellipsoid, and taken as the reference for GEODETIC LEVELLING.

geoisotherms (*Geol*) Lines or surfaces of equal temperature within the Earth. Also called *isogeotherms*.

geological column (*Geol*) A diagram that shows the subdivisions of part or all of geological time or the stratigraphical sequence in a particular area. See appendix on Geological time.

geological time (*Geol*) The time extending from the end of the Formative period of Earth history to the beginning of the Historical period. It is conveniently divided into periods, each being the time of formation of one of the systems into which the geological column is divided. See appendix on Geological time.

geological work of rivers (*Geol*) See RIVERS, GEOLOGICAL WORK OF.

geology (*Geol*) The study of the planet Earth. It embraces mineralogy, petrology, geophysics, geochemistry, physical geology, palaeontology and stratigraphy. It increasingly involves the use of the chemical, physical, mathematical and biological sciences. See EARTH SCIENCE and panel on EARTH.

geomagnetic effect (*Phys*) The effect of the Earth's magnetic field on cosmic rays by which positively charged particles are deflected towards the east.

geomagnetism (*Geol*) The Earth's magnetic field and the study of this.

geometrical attenuation (*Phys*) The reduction in intensity of radiation on account of the distribution of energy in space, eg due to the inverse-square law, or, in acoustics, progression area along the axis of a horn.

geometrical cross-section (*Phys*) The area subtended by a particle or nucleus. This does not usually resemble the interaction cross-section.

geometrical isomerism (*Chem*) A form of stereoisomerism in which the difference arises because of hindered rotation about a double bond or a bond that is part of a ring. Thus but-2-ene has two isomers, depending on whether the methyl groups are on the same (cis) or opposite (trans) sides of the double bond. In the CAHN–INGOLD–PRELOG SYSTEM, these are called Z- and E- respectively. See panel on POLYMERS.

geometrical optics (*Phys*) The study of optical problems based on the conception of light rays. See PHYSICAL OPTICS.

geometrical stair (*Build*) A stair arranged about a well-hole and curved between the successive flights.

geometric capacitance (*Phys*) The capacitance of an isolated conductor in a vacuum, uninfluenced by dielectric material, and depending only on shape.

geometric distortion (*ImageTech*) Any departure in an image from the representation of the correct form and perspective of the original subject. See BARREL DISTORTION, KEYSTONE DISTORTION, PINCUSHION DISTORTION.

geometric distribution (*MathSci*) The probability distribution of the total number of consecutive outcomes of a particular kind in a sequence of trials before the occurrence of an outcome of another kind, given that the probability of the first kind of outcome is constant at each trial and the different trials being statistically independent.

geometric mean (*MathSci*) Of n positive numbers a_r: the nth root of their product, ie

$$\left(\prod a_r\right)^{1/n}$$

geometric sequence (*MathSci*) The sequence of numbers $a, ar, ar^2, ar^3, \ldots$, in which each term is obtained from the preceding one by multiplication by the common ratio r. The sum of the first n terms of this sequence $(a + ar + ar^2 + ar^3 + \ldots + ar^{n-1})$ is called the geometric series of order n and is equal to $[a(1-r^n)/(1-r)]$.

geometry (*MathSci*) A branch of mathematics concerned generally with the properties of lines, curves and surfaces. Usually divided into *pure*, *algebraic* and *differential geometry* in accordance with the mathematical techniques utilized.

geometry factor (*Phys*) The factor $1/(4\pi)$ of the solid angle subtended by the window or sensitive volume of a radiation detector at the source.

geomorphology (*Geol*) The study of landforms and their relationship to the underlying geological structure.

geophagous (*BioSci*) Earth-eating.

geophilous (*BioSci*) Living on, or in, the soil.

geophones (*MinExt*) An array of sound detectors used to collect information in seismic surveys from planned explosions.

geophysical prospecting (*MinExt*) Prospecting by using quantitative physical measurements directly or indirectly, including magnetic, gravitational, electrical, electromagnetic, seismic and radioactive methods.

geophysics (*Genrl*) The study of physical properties of the Earth; it makes use of the data available in geodesy, seismology, meteorology and oceanography, as well as that relating to atmospheric electricity, terrestrial magnetism and tidal phenomena.

geophyte (*BioSci*) Herbs with perennating buds below the soil surface. It includes plants with bulbs, corms, rhizomes, etc. See RAUNKIAER SYSTEM.

geopotential height (*EnvSci*) The height of a point in the atmosphere expressed in units (geopotential metres) proportional to the geopotential at that height. One geopotential metre is numerically equal to (9·8/g) of a geometric metre, where g is the local acceleration of gravity. All reported measured heights of pressure levels are given in geopotential metres thus obviating the necessity for considering variations of gravity when making dynamical calculations.

George (*Aero*) Colloq term for *automatic pilot*.

Georgian wired glass (*Glass*) See WIRED GLASS.

geostationary orbit (*Space*) An orbit lying above the equator, in which an artificial satellite moves at the same speed as the Earth rotates. See panel on COMMUNICATIONS SATELLITE.

geostrophic approximation (*EnvSci*) Use of the GEOSTROPHIC WIND as an approximation to the actual wind, either in operational forecasting or as a replacement for certain terms in the equations of motion.

geostrophic force (*EnvSci*) A virtual force used to account for the change in direction of the wind relative to the Earth's surface, arising from the Earth's rotation and the CORIOLIS FORCE.

geostrophic wind (*EnvSci*) The theoretical wind arising from the PRESSURE GRADIENT force and the GEOSTROPHIC FORCE.

geosynchronous orbit (*Space*) An orbit of a satellite which has a period of 24 hours and so follows the same apparent path across the sky every day. See panel on COMMUNICATIONS SATELLITE.

geosyncline (*Geol*) A major elongated downwarp of the Earth's crust, usually hundreds of kilometres long and filled with sediments and lavas many kilometres in thickness. The rocks are generally deformed and metamorphosed later.

geotaxis (*BioSci*) The locomotory response of a motile organism or cell to the stimulus of gravity. Adj *geotactic*. Cf GEOTROPISM.

geotechnical process (*CivEng*) Means employed to alter the properties of soil for construction purposes. These include compaction and vibration, soil stabilization by the addition of cement, bitumen emulsions, etc; the latter sometimes carried out by injection.

geotechnics (*Genrl*) The application of scientific and engineering principles to the solution of civil engineering and other problems created by the nature and constitution of the Earth's crust.

geotextile (*Textiles*) A textile material used in civil engineering. Strong fabrics made of synthetic fibres are frequently used in road-making and for the stabilization of embankments.

geothermal gradient (*Geol*) The rate at which the temperature of the Earth's crust increases with depth.

geothermal power (*Geol*) Power generated by using the heat energy of crustal rocks. Active volcanic areas have traditionally been a source, but more recently deep boreholes in areas with a high geothermal gradient have shown economic potential.

geotropism (*BioSci*) The reaction of a plant member or sessile animal to the stimulus of gravity, shown by a growth curvature; cells become more elongated on one side than the other, tending to bring the axis of the affected part into a particular relation to the force of gravity. Adj *geotropic*. Cf GEOTAXIS.

geranial (*Chem*) See CITRAL.

geraniol (*Chem*) $C_{10}H_{18}O$. A terpene alcohol forming a constituent of many of the esters used in perfumery.

geranyl (*BioSci*) Prenyl group ((2E)-3,7-dimethyl-2,6-octadien-1-yl). Intermediate in cholesterol synthesis and in production of geranyl–geranyl group. Can be post-

translationally added to proteins usually in tandem (*geranyl–geranyl*).

geriatrics (*Med*) The specialized medical care of the elderly and aged.

germ (*BioSci*) The primitive rudiment that will develop into a complete individual, as a fertilized egg or a newly formed bud; a unicellular micro-organism.

germanium (*Chem*) Symbol Ge, at no 32, ram 72·59, rel.d. 5·47, mp about 958°C. A metalloid element in the fourth group of the periodic system. Greyish white in appearance. There are only 1·5 ppm in the Earth's crust, where the element mainly occurs substituting for silicon in silicates, and in coal. A semiconductor, used in early devices before the development of silicon processing.

germanium diode (*Electronics*) A semiconductor diode that uses a pellet of germanium as the rectifying element. It has a lower forward voltage than a silicon diode, but its characteristics are less predictable and drift with temperature.

germanium radiation detector (*NucEng*) A semiconductor detector with relatively large sensitive volume for γ-ray spectrometry. It has a much higher resolution but (in general) less sensitivity than a scintillation spectrometer.

germanium rectifier (*ElecEng, Electronics*) A p–n junction diode. It requires a lower forward voltage than a silicon diode, but it has higher reverse leakage current.

German lapis (*Min*) See SWISS LAPIS.

German measles (*Med*) See RUBELLA.

German siding (*Build*) Weatherboards finished with a hollow curve along the outside of the top edge, and rebated along the inside of the lower edge.

German silver (*Eng*) A series of alloys containing copper, zinc and nickel within the limits: copper, 25–50%; zinc, 10–35%; and nickel 5–35%. Also *nickel silver*. See COPPER ALLOYS.

germ band (*BioSci*) In insects, a ventral plate of cells, produced in the egg by cleavage, which later gives rise to the embryo.

germ cells (*BioSci*) In Metazoa, the reproductive cells. Gametes; spermatozoa and ova, or the cells that give rise to them. The germ cell line is formed very early in embryonic development. Also *germinal cells*.

germinal aperture (*BioSci*) See GERM PORE.

germinal cells (*BioSci*) See GERM CELLS.

germinal centre (*BioSci*) An aggregation of lymphocytes, mainly B-cells with numerous blast forms undergoing cell division, that develops from a primary follicle in response to antigenic stimulation. Follicular dendritic cells presenting antigen and macrophages are also conspicuously present. Thought to be sites at which memory B-cells are produced with receptors that recognize antigens in the complexes.

germinal disk (*BioSci*) The flattened circular region at the top of an egg with large quantities of yolk in which cleavage takes place.

germinal epithelium (*BioSci*) A layer of columnar epithelium that covers the stroma of the ovary in vertebrates.

germinal layers (*BioSci*) See GERM LAYERS.

germinal pore (*BioSci*) See GERM PORE.

germinal vesicle (*BioSci*) See BLASTOCYST.

germination (*BioSci*) The beginning of growth in a spore, seed, zygote, etc, esp following a dormant period.

germ layers (*BioSci*) The three primary cell layers in the development of Metazoa, ie ectoderm, mesoderm and endoderm. Also *germinal layers*.

germ line (*BioSci*) The cells whose descendants give rise to the *gametes*.

germ-line therapy (*BioSci*) Gene therapy in which genes are introduced into the germ cells and the change would therefore be hereditable.

germ nucleus (*BioSci*) See PRONUCLEUS.

germ pore (*BioSci*) A thin area in the wall of a spore or pollen grain through which the germ or pollen tube emerges at germination. Also *germinal aperture*.

germ tube (*BioSci*) The hypha or other tubular outgrowth that emerges from a spore at germination (eg pollen tube).

gerontic (*BioSci*) Pertaining to the senescent period in the life history of an individual.

gerontology (*Med*) The scientific study of the processes of ageing.

gersdorffite (*Min*) Metallic grey sulphide–arsenide of nickel, occurring as cubic crystals or in granular or massive forms.

gesso (*Build*) A pasty mixture of whiting, prepared with size or glue, applied to a surface as a basis for painting or gilding.

Gestalt (*Psych*) Ger word that has been incorporated into general usage because there is no exact English equivalent, but has the general sense of referring to 'the unified whole'. Gestalt psychology is an approach to perception and problem solving that stresses the need to understand the underlying organization of these functions and believes that to dissect experience into constituent parts is to lose its essential meaning.

Gestalt therapy (*Psych*) A therapeutic approach that focuses on the present manifestations of past conflict often using role playing and other acting-out techniques in order to help individuals gain insight into themselves and their behaviour. Not to be confused with the work of the Gestalt school of psychology. See GESTALT.

gestation (*BioSci*) In mammals, the act of retaining and nourishing the young in the uterus; pregnancy.

get (*MinExt*) To win or mine.

getter (*Electronics*) Material (K, Na, Mg, Ca, Sr or Ba), used, when evaporated by high-frequency induction currents, for 'cleaning' the vacuum of valves, after sealing onto the pump line during manufacture.

gettering (*Electronics*) The removal of harmful impurities (or defects in a crystalline solid) by the scavenging action of other impurities (or defects). Strategy used in semiconductor technology to improve crystal purity in active regions.

gettering discharge (*Electronics*) That used to assist the GETTER in the clean-up of vacuum in valves, through ionization of remaining gas molecules.

GeV (*Phys*) Abbrev for *giga-electron-volt*; unit of particle energy, 10^9 electron-volts, $1{\cdot}602 \times 10^{-10}$ J. US sometimes *BeV* (billion-electron-volt).

geyser (*Geol*) A volcano in miniature, from which hot water and steam are erupted periodically instead of lava and ashes, during the waning phase of volcanic activity. Named from the Great Geyser in Iceland, though the most familiar example is probably 'Old Faithful' in Yellowstone Park, Wyoming. The eruptive force is the sudden expansion which takes place when locally heated water, raised to a temperature above boiling point, flashes into steam. Until the moment of eruption, this had been prevented by the pressure of the overlying column of water in the pipe of the geyser, which is usually terminated upwards by a sinter crater. Also *gusher*.

geyserite (*Min*) See SINTER.

g factor (*Psych*) General factor. A single factor proposed as a measure of general intelligence. Originally identified on the basis of factor analysis of the various components of an intelligence test, it is a second-order factor and its value will depend upon the method of factor analysis used.

GFRP (*Plastics*) Abbrev for GLASS-FIBRE-REINFORCED PLASTIC.

G-gas (*Phys*) Gaseous mixture (based on helium and isobutane) used in low-energy beta counting (eg of tritium) by a gas-flow proportional counter. See GAS-FLOW COUNTER.

GGG (*Min*) Abbrev for *gadolinium gallium garnet*. A simulant of diamond.

ghost (*Eng*) In steel, a band in which the carbon content is less than that in the adjacent metal and which therefore consists mainly of FERRITE. Also *ghost line*.

ghost (*ImageTech*) Vertical streaks on highlights in a projected picture, arising from incorrect phasing of the rotary shutter with respect to the moving film.

ghost crystal (*Min*) A crystal within which may be seen an early stage of growth, outlined by a thin deposit of dust or other mineral deposit.

ghost image (*Phys*) The image arising from a mirror when the rays have experienced reflection within the glass between the surface and the silvering.

ghost line (*Eng*) See GHOST.

ghrelin (*BioSci*) An appetite-stimulating hormone produced in the stomach wall (*growth-hormone-releasing peptide*).

Ghyben–Herzberg principle (*EnvSci*) The principle that determines the amount of fresh water than can be abstracted from a source which is open to the sea before it becomes contaminated by salt water.

GI (*BioSci*) Abbrev for GLYCAEMIC INDEX.

giant (*MinExt*) *Hydraulic giant*. See MONITOR.

giant cells (*BioSci*) Cells of unusual size, as the MYELOPLAXES of bone marrow; certain cells of the excitable region of the cerebrum; certain cells sometimes found in lymph glands; large multinucleate cells of the thymus and of the spleen.

giant fibres (*BioSci*) In many invertebrates (eg Annelida, Cephalopoda and some Arthropoda) enlarged motor axons in the ventral nerve cord that transmit impulses very rapidly and initiate escape behaviour.

giantism (*BioSci*) Abnormal increase in size of the body, due to overactivity of the anterior lobe of the pituitary gland. Often associated with polyploidy in plants. Also *gigantism*.

giant planet (*Astron*) One of the planets Jupiter, Saturn, Uranus and Neptune.

giant powder (*MinExt*) Dynamite.

giant source (*Med*) Large source of radioactivity, eg 150 000 curies of ^{60}Co, used for industrial sterilization of packed food or chemical processing (eg cross-linking of polymers).

giant star (*Astron*) A star which is more luminous than the main sequence stars of the same spectral class. Smaller groups of subgiants and supergiants are recognized.

giardiasis (*Med*) Infestation of the intestinal tract with the flagellate protozoon *Giardia lamblia*, sometimes causing severe diarrhoea.

gib (*Eng*) (1) Metal piece used to transmit the thrust of wedge or cotter, as in some connecting-rod bearings. (2) A brass working surface let in to the working surface of a steam-engine cross-head. (3) Tapered or parallel strip in bearings for reciprocating slides, used to fit or to clamp the slide in the guide.

gibberellic acid (*BioSci*) A GIBBERELLIN obtained from cultures of the fungus *Gibberella fujikuroi*, used commercially eg to promote rapid, even malting of barley.

gibberellin (*BioSci*) Any of a large group of terpenoid plant GROWTH SUBSTANCES, synthesized mainly in young leaves and promoting stem elongation and, sometimes, flowering, germination and the utilization of reserves as in germinating barley grains. See GIBBERELLIC ACID.

gibbous (*Astron*) The word applied to the phase of the Moon, or of a planet, when it is between either quadrature and opposition, and appears less than a circular disk but greater than a half disk.

gibbous (*BioSci*) (1) Swollen, esp if swollen more on one side than another. (2) With a pouch.

Gibbs' adsorption theorem (*Chem*) Solutes which lower the surface tension of a solvent tend to be concentrated at the surface, and conversely.

Gibbs–Duhem equation (*Chem*) For binary solutions at constant pressure and temperature, the chemical potentials (μ_1, μ_2) vary with the mole fractions (x_1, x_2) of the two components as follows:

$$\frac{\partial \mu_1}{\partial \ln x_1} = \frac{\partial \mu_2}{\partial \ln x_2}$$

Gibbs' free energy (*Chem*) The difference of the enthalpy and the product of the entropy and the temperature of a system. Usual symbol G ($G = H - TS$). A calculated negative change in G indicates a spontaneous process in a closed system at constant pressure.

Gibbs–Helmholtz equation (*Chem*) An equation of THERMODYNAMICS,

$$\triangle G = \triangle H - T \triangle S$$

where $\triangle G$ = change in free energy, $\triangle H$ = change in enthalpy, $\triangle S$ = entropy change and T = the absolute temperature. It is applied to electrochemical cells in the form

$$\triangle G = -zFE$$

where $\triangle G$ = the free energy change in the cell reaction, z = the number of electrons transferred, $F = 1$ faraday (96 496 coulombs) and E = the reversible emf of the cell. See NERNST EQUATION.

gibbsite (*Min*) Hydroxide of aluminium, $Al(OH)_3$, occurring as minute mica-like crystals, concretional masses, or incrustations. An important constituent of bauxite. Also *hydrargillite*.

Gibbs–Konowalow rule (*Chem*) For the phase equilibrium of binary solutions. At constant pressure the equilibrium temperature is a maximum or minimum when the compositions of the two phases are identical, and vice versa, eg in eutectics or azeotropes. The corresponding statements hold for pressure at constant temperature.

Gibbs' phase rule (*Chem*) See PHASE RULE.

gib-headed key (*Eng*) A key for securing a wheel etc to a shaft, having a head formed at right angles to its length.

giblet check (*Build*) An exterior rebate for a door which opens outwards.

gid (*Vet*) See COENURIASIS.

Giemsa stain (*BioSci*) A stain containing eosin and methylene blue used to stain blood films, haemoprotozoan parasites and chromosomes.

Gies' biuret reagent (*Chem*) A reagent for testing the presence of proteins by the biuret reaction; it consists of a solution of 10% KOH and 0·075% copper (II) sulphate.

GIF (*ICT*) Abbrev for *graphic interchange format*. A standard image file format.

Giffard's injector (*Eng*) The original steam INJECTOR.

giga- (*Genrl*) Prefix used to denote 10^9 times, eg a gigawatt is 10^9 watts. Symbol G.

gigabyte (*ICT*) Usually 10^9 but sometimes 2^{30} bytes. Abbrev *Gb*.

giga-electron-volt (*Phys*) See GEV.

gigaflop (*ICT*) A unit of processor speed equal to 2^{30} floating-point operations per second.

giggering (*Print*) A method of producing lines on the back of a volume by means of a catgut cord.

gig stick (*Build*) See RADIUS ROD.

gilder's cushion (*Build*) A pad comprising a flat wooden board padded with felt covered with tightly stretched chamois leather having a parchment screen at one end. A gilder uses the pad as a base on which to manipulate and cut loose-leaf metals when gilding.

gilder's knife (*Build*) A long, thin, blunt-edged knife used for cutting loose-leaf gold on a gilder's cushion.

gilder's mop (*Build*) A soft brush with a round, bushy camel-hair filling used to press gold leaf into recesses and enrichments.

gilder's tip (*Build*) A small brush comprising a single line of badger hair set between two sheets of thin card. It is used to pick up loose gold leaf from the gilder's cushion and transfer it to the surface being gilded.

gilder's wheel (*Build*) A device used to apply lines from a roll of ribbon gold to a prepared surface.

gilding (*Build*) The application of gold leaf by means of an adhesive, normally gold size.

gilding metal (*Eng*) Copper–zinc alloy containing zinc up to 15%. See COPPER ALLOYS.

gill (*BioSci*) (1) One of the vertical plates, bearing the hymenium, on the undersurface of the cap of the fruiting body of the mushroom and other agarics. (2) A membranous respiratory outgrowth of aquatic animals, usually in the form of thin lamellae or branched filamentous structures.

gill arch (*BioSci*) In fish, the incomplete jointed skeletal ring supporting a single pair of gill slits; one segment of the branchial basket.

gill bars (*BioSci*) See GILL RODS.

gill basket (*BioSci*) In fish and cyclostomes, the skeletal ring supporting a single pair of gill slits; one segment of the branchial basket.

gill book (*BioSci*) The book-like respiratory lamellae of Xiphosura borne by the opisthosoma, of which they represent the appendages. Also *book gill*.

gill clefts (*BioSci*) See GILL SLITS.

gill cover (*BioSci*) See OPERCULUM.

gillion (*Genrl*) Rarely used for 10^9; preferred term is *giga-* or *G* (US) and now, more generally, a billion.

gill pouch (*BioSci*) One of the pouch-like gill slits of cyclostomes and fish.

gill rakers (*BioSci*) In some fish, small processes of the branchial arches, which strain the water passing out via the gill slits and prevent the escape of food particles.

gill rods (*BioSci*) In CEPHALOCHORDATA, skeletal bars that support the pharynx. Also *gill bars*.

gills (*Aero*) Controllable flaps which vary the outlet area of an air-cooled engine cowling or of a radiator. Also *cooling gills, cowl flaps, radiator flaps*.

gill slits (*BioSci*) In Chordata, the openings leading from the pharynx to the exterior, on the walls of which the gills are situated. Also *brachial clefts, gill clefts*.

gilsonite (*Min*) See UINTAITE.

gilt (*Agri*) A female pig yet to produce a first litter.

gimbal mount (*Eng*) One giving rotational freedom about two perpendicular axes, as used for gyroscope and nautical compass.

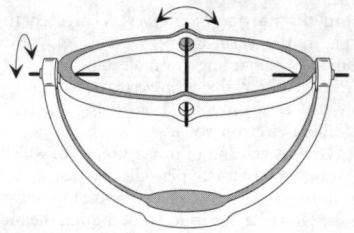

gimbal mount Shows the two axes.

gimbals (*Eng*) Self-aligning bearings for supporting a chronometer or other instrument and used to ensure that it is kept level, irrespective of eg ship's motion. Also *gymbals*.

gimmick (*ElecEng*) Colloq a small capacitor formed by twisting insulated wires.

gin (*Eng, MinExt*) (1) A hoist which consists of a chain or rope barrel supported in bearings and turned by a crank. Often horse-powered (*horse-gin*). (2) A portable tripod carrying lifting tackle.

gin (*Med*) A distilled grain spirit flavoured with juniper berries, often mixed with tonic water for medicinal purposes.

gin (*Textiles*) A machine (*cotton gin*) that is used to remove cotton fibres from the seeds: the process is known as *ginning*.

ginger-beer plant (*BioSci*) A symbiotic association of a yeast and a bacterium that ferments a sugary liquid containing oil of ginger, giving ginger beer. Often known popularly as *Californian bees*, and by similar names.

gingham (*Textiles*) Lightweight, plain weave cotton cloth woven with coloured yarns to produce a check pattern.

gingival (*BioSci*) In mammals, pertaining to the gums.

gingivectomy (*Med*) The cutting back of inflamed or excess gum.

gingivitis (*Med*) Inflammation of the gums.

ginglymus (*BioSci*) An articulation that allows motion to take place in one plane only; a hinge joint. Adj *ginglymoid*.

Ginkgoales (*BioSci*) An order of zooidogamous gymnosperms, abundant worldwide in the Mesozoic. There is one extant species, *Ginkgo biloba*, a Chinese tree with small, fan-shaped leaves.

Giorgi system (*Phys*) A system of units proposed in 1904 and later adopted as the MKSA system, the common PRACTICAL UNITS of eg ohm, volt and ampere becoming identified with ABSOLUTE UNITS of the same entities. See SI UNITS.

Giotto (*Space*) The ESA spacecraft which flew through the coma of Halley's comet in March 1986 and approached within 600 km of the nucleus.

GIP (*Paper*) See GLAZED IMITATION PARCHMENT.

girder (*BioSci*) Usually longitudinal strand of mechanical tissue, often T- or I-shaped in cross-section, giving strength and esp stiffness to a plant part as in a grassleaf.

girder (*Eng*) A beam, usually steel, to bridge an open space. Girders may be rolled sections, built up from plates or of lattice construction. See CONTINUOUS BEAM.

girder bridge (*CivEng*) A bridge in which the loads are sustained by beams (generally compound) resting across the bridge supports.

girdle (*BioSci*) In vertebrates, the internal skeleton to which the paired appendages are attached, consisting typically of a U-shaped structure of cartilage or bone with the free ends facing dorsally. In *Amphineura*, the part of the mantle which surrounds the shells.

girdle (*For*) A continuous incision which is made all round a bole, cutting through at least bark and cambium, generally with the object of killing the tree.

girth (*For*) See EXPLOITABLE GIRTH, QUARTER GIRTH, TOP GIRTH.

girt strip (*Build*) See RIBBON STRIP.

GIS (*Genrl*) Abbrev for *geographical information system*, a computer system used to process geographically referenced information.

gismondine (*Min*) A rare zeolite; a hydrated calcium aluminium silicate which occurs in cavities in basaltic lavas.

give-and-take lines (*Surv*) Straight lines drawn on a plan of any area having irregular boundaries, each line following the trend of a part of the boundary so that any small piece that it cuts off the area is balanced by an equal piece added by it.

Givetian (*Geol*) A stage in the Middle Devonian. See PALAEOZOIC.

gizzard (*BioSci*) See PROVENTRICULUS.

glabrescent (*BioSci*) (1) Almost hairless. (2) Becoming hairless. Also *glabrate*.

glabrous (*BioSci*) Having a smooth hairless surface.

glacial acetic acid (*Chem*) Pure concentrated acetic (ethanoic) acid. Owing to its comparatively high mp (16·6°C) it solidifies easily, forming ice-like crystals.

glacial action (*Geol*) All processes relating to the action of glacier ice, comprising: (1) the grinding, scouring, plucking and polishing effected by ice, armed with rock fragments frozen into it; and (2) the accumulation of rock debris resulting from these processes.

glacial deposits (*Geol*) These include spreads of boulder clay, sheets of sand and gravel occurring as outwash DELTAS, outwash FANS and KAMES; also deposits of special topographical form, such as DRUMLINS and ESKERS.

glacial erosion (*Geol*) The removal of rock materials by the action of glaciers and associated meltwater streams. Includes grinding, scouring, plucking, grooving and polishing by rock fragments contained in the ice.

glacial phosphoric acid (*Chem*) See METAPHOSPHORIC ACID.

glacial sands (*Geol*) These cover extensive areas in advance of sheets of boulder clay, and together with glacial (largely

fluvioglacial) gravels, represent the outwash from ice sheets.

glaciation (*Geol*) Embraces both the processes and products arising from the presence of ice masses on the Earth. The effects are most obvious on land but there is increasing evidence that the shallow sea floors too were affected. Glaciation, traditionally connected with the Pleistocene period, is now known from older geological periods including the Permo-Carboniferous and the Precambrian.

glacier (*Geol*) A large mass of ice. Three kinds can be recognized: (1) valley glaciers, (2) Piedmont glaciers which overflow from valleys and coalesce on the lower ground, and (3) large continental ice sheets (eg Greenland) and smaller ice caps (eg Iceland).

glacier lake (*Geol*) See LAKE.

glaciology (*Genrl*) The study of the geological nature, distribution and action of ice.

glacis (*CivEng*) An inclined bank.

Gladstone and Dale law (*Phys*) Law relating the refractive index of a gas and its density as it is changed by variations in pressure and temperature:

$$\frac{n-1}{\rho} = \text{constant}$$

where n is the refractive index and ρ the density.

glair (*Print*) A preparation made from white of egg and vinegar, used as the adhesive for gold leaf in gold finishing and blocking.

glance (*Min*) Opaque mineral with a resinous or shining lustre.

glancing angle (*Phys*) The complement of the ANGLE OF INCIDENCE.

gland (*BioSci*) A structure at or near the surface of a plant or a single or an aggregate of epithelial cells in animals, specialized for the elaboration of a secretion useful to the organism or of an excretory product. Adj *glandular*.

gland (*Eng*) (1) A device for preventing leakage at a point where a rotating or reciprocating shaft emerges from a vessel containing a fluid under pressure. (2) A sleeve or nut used to compress the packing in a STUFFING-BOX.

gland bolts (*Eng*) Bolts for holding and tightening down a gland.

gland cell (*BioSci*) A unicellular gland, consisting of a single goblet-shaped epithelial cell producing a secretion, usually mucus.

glanders (*Vet*) A contagious bacterial disease of horses, mules and asses, due to infection by *Actinobacillus mallei* (*Malleomyces mallei*); inflammatory nodules occur in the respiratory passages and lungs and also in other parts of the body. Infection of the lymphatics under the skin is known as *farcy*. Glanders is communicable to humans.

glandular epithelium (*BioSci*) Epithelial tissue specialized for the production of secretions.

glandular fever (*Med*) See INFECTIOUS MONONUCLEOSIS.

glandular tissue (*BioSci*) See GLANDULAR EPITHELIUM.

glans (*BioSci*) (1) The highly vascularized apex of the penis, and the extremity of the clitoris. (2) A fruit in which the receptacle, pedicel or peduncle is enlarged, eg the acorn.

glans penis (*BioSci*) A dilatation of the extremity of the mammalian penis.

glare (*Phys*) The visual discomfort experienced by observers in the presence of a visible source of light. The term can also refer to the visual disability produced by the presence of visible sources in the field of view when these sources do not assist the viewing process.

glarimeter (*Paper*) An instrument for measuring the gloss of a paper surface based on light reflectance.

glass (*Genrl*) A hard, amorphous, brittle substance, made by fusing together one or more of the oxides of silicon, boron or phosphorus, with certain basic oxides (eg sodium, magnesium, calcium, potassium), and cooling the product rapidly to prevent crystallization or devitrification. The melting point varies between 800 and 950°C but it is

Glasses and glass-making

The word 'glass' is most commonly associated with the family of hard, brittle, non-crystalline solids based on silica (see panel on SILICON, SILICA AND SILICATES), fused together with various other oxides (eg those of calcium, sodium, boron, phosphorous, magnesium and potassium). Polymeric glasses such as acrylic sheet are also well-known; less so are those based on the chalcogenides (sulphur, selenium and tellurium), the halides, certain organic compounds and on carbon. More recently, metallic glasses have been produced. Their common characteristic is that their structures are amorphous, ie with none of the regularity of crystals.

Glasses are sometimes defined as those non-crystalline solids which exhibit a transition in behaviour, the GLASS TRANSITION, with temperature; the remaining non-crystalline solids are classified as amorphous. The location of the glass transition temperature, T_g, is not as well-defined as the crystalline melting temperature, T_m, and depends on the rate of cooling. The most common route to a glass is the cooling of a liquid at a sufficiently high rate to avoid nucleation and growth of crystals (in metals this requires cooling rates of around 10^6 K s^{-1}!). The resulting solid retains the structural disorder of the liquid. Other methods exist, including condensing a gas, creating disorder in solids by shock waves or radiation damage, and chemical reactions in solution to produce a gel precursor (the SOL-GEL PROCESS).

The manufacture of silicate glasses is based on the readily available ingredients, silica and sodium carbonate. Glasses made from these alone are vulnerable to attack by water. To reduce this effect, some of the sodium carbonate is replaced by calcium carbonate to make soda–lime–silica glass as used for such products as windows, bottles and jars, light bulbs and spectacle lenses. Since the carbonates decompose to form oxides and CO_2 during glass-making, the glass is formed from a mixture of oxides (as its name suggests).

The first stage in making the glass is to melt the pulverized raw materials in a furnace. One of these (SiO_2) has a very high melting point ($\sim$2000 K). To avoid the need (and expense) of heating to this temperature, a quantity of powdered scrap glass, known as *cullet*, is added to the raw materials. This makes up 15 to 30% of the total, and by melting before any of the raw constituents it conducts heat and dissolves the other ingredients together which reduces the time needed to melt the mixture. Shown in Fig. 2 are a type of continuous glass-melting furnace, and a sketch of the temperature profile along its length. The processes taking place are complex and interdependent and can be divided into melting, refining (or fining) and homogenization. The furnace is either gas- or oil-fired and is usually arranged so that the maximum temperature (about 1750 K for soda–lime–

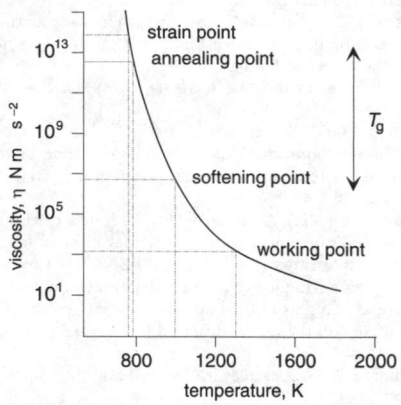

Fig. 1 **Glass** Viscosity plotted against temperature.

---

worked at higher temperatures. The tensile strength of glass resides almost entirely in the outer skin; if this is scratched or corroded, the glass is much more easily broken. See NATURAL GLASS.

glass blocks (*Arch, Build*) Hollow blocks made of glass usually with a patterned surface. Used where translucence and decorative effect are required. They also provide insulation against heat and sound. Also *glass bricks*.

glass-bulb rectifier (*ElecEng*) A mercury-arc rectifier in which the arc takes place within a glass bulb. Cf STEEL-TANK RECTIFIER.

glass–ceramic (*Glass*) Ceramic processed to final shape as a glass and then induced to crystallize by controlled heat treatment. Has improved thermal and mechanical properties over parent silicate glass. Used in eg cooker hobs. See SILICA, SILICATES.

glassed (*Build*) A term applied to stones such as granite and marble which are highly polished by being held against a revolving disk covered with felt.

Glasser's disease (*Vet*) Disease of pigs in which *Haemophilus suis*, *H. parasuis* and *Mycoplasma* spp are all implicated. Sudden-onset fever often with difficult and rapid breathing, anorexia and lameness with all joints swollen and painful. Also *infectious polyarthritis of pigs*.

glasses (*Glass*) Non-crystalline solids which exhibit a transition in behaviour with temperature. See panel on GLASSES AND GLASS-MAKING.

glass fibre (*Textiles*) Glass melted and then drawn out through platinum bushings (almost always in multiples of 204) into fine fibres (diameters range from 3 to 25 μm) which may be spun into threads and woven into tapes and cloths. The fibres may also be formed into pads and quiltings for thermal insulation. Plastics reinforced with glass fibres are used for making some car and boat bodies. See A-, C-, E-, S- and Z-GLASS and panel on FIBRE ASSEMBLIES.

glass-fibre paper (*Paper*) That formed by using glass fibres as part or the whole of the FURNISH.

silica glass) is reached approximately one-third of the way along the melting chamber where the refining process takes place. Refining clears the CO_2 bubbles formed during the reaction of silica with sodium and calcium carbonates and fining agents such as sodium sulphate are sometimes used.

Beyond the throat of the furnace is the working end where the fining and homogenization occur, as the temperature is reduced to the working range of the glass. Since there are no abrupt changes in the viscosity of glass as a function of temperature, glasses are frequently characterized by a set of reference temperatures at which their viscosities have defined values. In the melting chamber the viscosity Z is of the order of $10 \ N \ m^{-2} \ s$ (similar to that of thick treacle). When the glass is being worked (either moulded or drawn) Z is between 10^2 and $10^5 \ N \ m^{-2} \ s$, so the temperature at which Z is $10^3 \ N \ m^{-2} \ s$ is arbitrarily called the *working point*.

The *softening point* is defined to be the temperature at which $Z = 10^{6.6} \ N \ m^{-2} \ s$, which is a viscosity associated with the CREEP at a prescribed rate of a given length of a glass rod or fibre under its own weight. The *annealing point* ($Z = 10^{12.4} \ N \ m^{-2} \ s$) is the temperature at which any internal stresses are relieved after several minutes. The *glass transition temperature*, T_g, lies somewhere between the softening and annealing points. Finally, the *strain point* ($Z = 10^{13.5} \ N \ m^{-2} \ s$) is the highest temperature from which the glass can be cooled rapidly without causing significant levels of internal stress to be developed.

The structure of the resulting material is a random network of predominantly covalently bonded SiO_4 tetrahedra, in which each oxygen atom is shared between two tetrahedra (*bridging*). The sodium and calcium oxides ionize and donate oxide ions to the network, partially breaking up the network and forming gaps in which sodium and calcium ions sit and provide ionic bonds between unbridged tetrahedra.

Glasses and glass-making *(Cont.)*

The viscosity of the melt is reduced owing to the lower energy and non-directionality of these ionic bonds so that processing can occur at lower temperatures. Ions such as Na^+ and Ca^{2+} are known as NETWORK MODIFIERS.

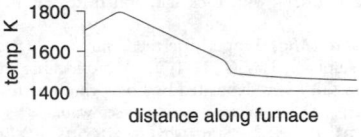

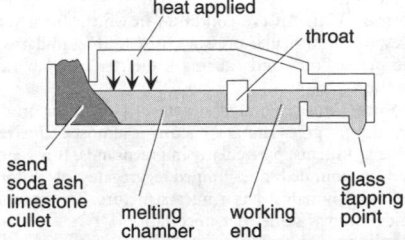

Fig. 2 **Glass-making** A furnace showing temperature of melt in top graph.

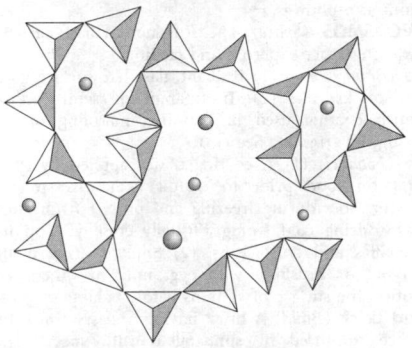

⊙ network modifier, eg sodium or calcium ions

Fig. 3 **Glass structure** The drawing shows SiO_4 tetrahedra and modifiers.

---

glass-fibre-reinforced plastic (*Eng*) Class of COMPOSITE MATERIALS comprising a polymeric MATRIX reinforced with glass fibre. The matrix can be thermoplastic (used chiefly with chopped fibres) or thermosetting (allowing use with continuous fibres and/or glass textiles). Widely used as a structural material in eg aircraft, boats and ships, road and rail vehicles, sports equipment, printed circuit boards. Abbrev *GFRP*.

glass gilding (*Build*) The application of loose-leaf gold to the reverse side of glass using isinglass or gelatine as a mordant.

glassification (*NucEng*) See VITRIFICATION.

glassine (*Paper*) Transparent glazed greaseproof paper. Produced by long beating of the stock and high glazing of the finished paper.

glass-making (*Glass*) The manufacture of silicate glass, based on the readily available ingredients, silica and sodium carbonate. See panel on GLASSES AND GLASS-MAKING.

glasspaper (*Paper*) Paper coated with glue on which is sprinkled broken glass of a definite grain size; used for rubbing down surfaces. Cf EMERY PAPER, SANDPAPER.

glass shot (*ImageTech*) A composite shot in which live action is photographed through the clear portions of a glass plate on which the remainder of the scene has been painted.

glass structure (*Chem*) The network of silicate tetrahedra modified by sodium, calcium or other ions.

glass tile (*Build*) A small glass sheet in a roof, bonded in with slates, plain tiles or pantiles, to admit light within the roof space.

glass transition (*Chem*) Characterized by a change in slope of the volume against temperature and enthalpy against temperature curves as glass-forming materials are cooled or heated, the transition is general to all glass materials whether inorganic, polymeric or metallic. Its location, the *glass transition temperature*, T_g, varies with the rate of cooling or heating. Theories and models of the transition abound, but none is generally applicable or accepted. See VISCOELASTICITY and panels on GLASSES AND GLASS-MAKING and POLYMERS.

glass transition temperature (*Chem*) See GLASS TRANSITION.

glass wool (*Build, Chem*) A felt of fine glass fibres; used as a relatively inert filter, packing or insulation.

glassy state (*Glass*) See VITREOUS STATE.

glauberite (*Min*) Monoclinic sulphate of sodium and calcium, occurring with rock salt, anhydrite, etc, in saline deposits.

Glauber salt (*Min*) Properly termed *mirabilite* (hydrated sodium sulphate, $Na_2SO_4.10H_2O$). A monoclinic mineral formed in salt lakes, deposited by hot springs, or resulting from the action of volcanic gases on sea water.

glaucescent (*BioSci*) (1) Somewhat GLAUCOUS. (2) Becoming glaucous.

glaucoma (*Med*) An eye condition in which, from various causes, the intra-ocular pressure rises, leading to damage to optic nerve fibres and causing a quarter of all blindness after the age of 45.

glauconite (*Min*) Hydrated silicate of potassium, iron and aluminium, a green mineral occurring almost exclusively in marine sediments, particularly in greensands. It is generally found in rounded fine-grained aggregates of ill-formed platelets, although it has a mica structure. The manner of its formation is somewhat uncertain.

glaucophane (*Min*) A blue monoclinic amphibole occurring in crystalline schists formed at high pressures. A hydrated sodium iron magnesium silicate, the name is used for an end-member compositional variety of amphibole.

glaucous (*BioSci*) Greyish or bluish green; covered with a bloom as a plum is.

GLAVCOSMOS (*Space*) The Russian civilian agency for marketing space expertise and facilities.

glaze (*Build, Glass*) (1) Brilliant glass-like surface given to tiles, bricks, etc. (2) Transparent or semi-transparent varnish coating used in graining, marbling or colour glazing. Cf ENAMEL. See FRIT.

glaze (*FoodSci*) (1) A coating of ice applied by spraying water on food prior to QUICK FREEZING, to reduce moisture loss during freezing and protect from FREEZER BURN during cold storage, usually employed on frozen vegetables and crustaceans. (2) Shiny coating made by spraying or brushing with egg, milk or other protein solution the surface of eg buns before baking.

glazed brick (*Build*) A brick having a glassy finish to the surface produced by spraying it with special surface preparations before firing.

glazed door (*Build*) A door fitted with glass panels.

glazed frost (*EnvSci*) A smooth layer of ice which is occasionally formed when rain falls and the temperature of the air and the ground is below freezing point.

glazed imitation parchment (*Paper*) A highly glazed packaging paper, generally bleached and frequently opacified, suitable for waxing or coating for use in food-wrapping applications. Abbrev GIP.

glazed morocco (*Print*) Goatskin crushed and polished by rolling.

glazier (*Build*) A worker who cuts panes of glass to size and fits them in position.

glazier's putty (*Build*) A mixture of whiting and linseed oil, forming a plastic substance for sealing panes of glass into frames.

glazing (*Build*) The operation of fitting panes of glass into sashes.

glazing (*ImageTech*) Forming a shiny surface on photographic prints by heating them in contact with a highly polished surface, eg chromium plate.

glazing bead (*Build*) A bead nailed, instead of putty, to secure a pane.

gleba (*BioSci*) The spore-bearing tissue enclosed within the peridium of the fructification of Gasteromycetes, and in truffles.

gleet (*Med*) Chronic discharge from the urethra as a result of gonococcal infection.

glei soil (*EnvSci*) See GLEY SOIL.

glenoid (*BioSci*) Socket-shaped; any socket-shaped structure; as the cavity of the pectoral girdle which receives the basal element of the skeleton of the fore-limb.

glenoid fossa (*BioSci*) In mammals, a hollow beneath the zygomatic process.

gley soil (*EnvSci*) A soil that is permanently or periodically waterlogged and therefore anaerobic, characterized by its blue-grey colours (due to ferrous iron) often mottled with orange-red (ferric iron). Also *glei soil*.

glia (*BioSci*) Same as NEUROGLIA, from Gk *glia*, meaning glue.

glibenclamide (*Pharmacol*) A *sulphonylurea* used in the treatment of maturity onset diabetes.

glide (*Min*) The movement of dislocations along slip planes. See DISLOCATION.

glide path (*Aero*) The approach slope (usually $3\frac{1}{2}°$ or $5°$) along which large aircraft are assumed to come in for a landing. The term is imprecise because such aircraft do not glide, but are brought in with a considerable amount of power.

glide-path beacon (*Aero*) A directional radio beacon, associated with an ILS, which provides an aircraft, during approach and landing, with indications of its vertical position relative to the desired approach path.

glide-path landing beam (*Aero*) Radio signal pattern from a radio beacon, which aids the landing of an aircraft during bad visibility.

glider (*Aero*) A heavier-than-air craft not power driven within itself, although it may be towed by a power-driven aircraft. Cf SAILPLANE.

gliding (*Aero*) (1) Flying a heavier-than-air craft without assistance from its engine, either in a spiral or as an approach glide before flattening out antecedent to landing. (2) The sport of flying *gliders* which are catapulted into the air, or launched by accelerating with a winch, or towed by car, or towed to height by an aircraft.

gliding (*BioSci*) A form of movement in algae in which cells or filaments, without flagella, move against a solid surface.

gliding angle (*Aero*) The angle between the flight path of an aircraft in a glide and the horizontal.

gliding growth (*BioSci*) Same as SLIDING GROWTH.

gliding planes (*Crystal*) In minerals, planes of molecular weakness along which movement can take place without actual fracture. Thus calcite crystals or cleavage masses can be distorted by pressure and pressed into quite thin plates without actual breakage. See SLIP PLANES.

glimmerite (*Geol*) An ultrabasic rock composed mainly of mica, usually of the biotite variety.

glint (*Radar*) Pulse-to-pulse variation in the amplitude of a reflected signal; may be due to reflection from different surfaces of a rapidly moving target, or from propellers or rotor blades.

glioblastoma multiforme (*Med*) A very malignant tumour of the central nervous system.

glioma (*Med*) A general term applied to tumours arising from neuroglial nervous tissue in the brain and, more rarely, in the spinal cord.

gliomatosis (*Med*) Diffuse overgrowth of neuroglia in the brain or in the spinal cord.

glipizide (*Pharmacol*) A *sulphonylurea* used in the treatment of non-insulin-dependent diabetes mellitus.

glissette (*MathSci*) The locus of any point, or envelope of any line, moving with a first curve which slides against two fixed curves. Many curves can be regarded as either glissettes or roulettes, eg the astroid.

glitch (*ICT*) Colloq engineering term for any short-lived disturbance that causes mal-operation of electronic equipment, typically noise spikes on power supplies or momentary false logic states in digital circuitry arising from poor design. Fig. ▷

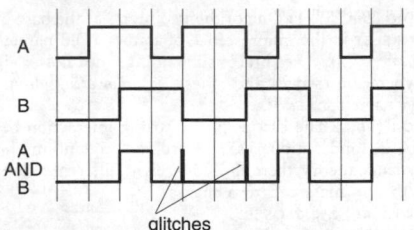

glitch Caused by inputs of an AND gate changing simultaneously.

glitch (*ImageTech*) A burst of NOISE, usually marking a bad video edit.

Glivec(*Pharmacol*) TN for imatinib mesylate, a *protein tyrosine kinase* inhibitor used to treat some forms of chronic myeloid leukaemia. US *Gleevec*.

Gln (*Chem*) Symbol for GLUTAMINE.

global positioning system (*ICT*) A navigation system that uses data from several satellites (at least four) to fix the geographical position of the hand-held receiver, and its altitude. Current systems are accurate to within 20 m in three dimensions. Also *Navstar*. Abbrev *GPS*.

global system for mobile communication (*ICT*) A digital mobile-telephone standard developed by the Groupe Spéciale Mobile of the EUROPEAN TELECOMMUNICATION STANDARDS INSTITUTE to replace Europe's many different analogue systems. It allows pan-European roaming and is used across many countries and territories.

global variable (*ICT*) One that is available for the whole of a program, including any subprograms.

globe photometer (*Phys*) See ULBRICHT SPHERE PHOTOMETER.

globigerina ooze (*Geol*) A deep-sea deposit covering a large part of the ocean floor (one-quarter of the surface of the globe); it consists chiefly of the minute calcareous shells of the foraminifer, *Globigerina*.

globin (*BioSci*) The polypeptide moiety of haemoglobin. In the adult human the haemoglobin molecule has two α- and two β-globin chains.

globoid (*BioSci*) A rounded inclusion of phytin in a protein body.

globular cementite (*Eng*) In steel, cementite occurring in the form of globules instead of in lamellae (as in PEARLITE) or as envelopes round the crystal boundaries (as in HYPEREUTECTOID STEEL). Produced by very slow cooling, or by heating to between 600 and 700°C.

globular cluster (*Astron*) Densely packed family of stars arranged characteristically as a compact sphere of stars. They contain tens of thousands to millions of stars, all thought to have been formed at the same time. Over 100 are known as members of our Galaxy, distributed through the GALACTIC HALO. See panels on GRAND UNIFIED THEORIES and HERTZSPRUNG–RUSSELL DIAGRAM.

globular pearlite (*Eng*) See GRANULAR PEARLITE.

globule (*Astron*) A small dark nebula composed of opaque gas and dust, representing an early stage of star formation.

globulites (*Geol*) Crystallites (ie incipient crystals) of minute size and spherical shape occurring in natural glasses such as pitchstones.

globus (*BioSci*) Any globe-shaped structure; as the *globus pallidus* of the mammalian brain. Adj *globate*.

globus hystericus (*Med*) The sensation as of a lump in the throat experienced in hysteria.

glochidiate (*BioSci*) Bearing barbed hairs or bristles.

glomera carotica (*Med*) See CAROTID BODIES.

glomerate (*BioSci*) Collected into heads.

Glomeromycota (*BioSci*) See ZYGOMYCOTA.

glomerulitis (*Med*) Inflammation of the glomeruli of the kidney.

glomerulonephritis (*Med*) A kidney disease in which the major lesion is in the glomeruli. The capillary walls contain deposits of IgG and often of activated complement components. These are due to deposition of immune complexes from the circulation, or to antibodies against some component present in the glomerular capillary wall. The glomeruli are infiltrated with inflammatory cells.

glomerulus (*BioSci*) A capillary blood-plexus, as in the vertebrate kidney; a nest-like mass of interlacing nerve fibrils in the olfactory lobe of the brain. Adj *glomerular*.

GLORIA (*Geol*) Abbrev for *geological long-range inclined asdic*. Long-range side-scan sonar by which very large areas of the ocean floor have been surveyed.

glory (*EnvSci*) A small system of coloured rings surrounding the shadow of the observer's head, cast by the Sun on a bank of mist, as in the BROCKEN SPECTRE. The *glory* is produced by diffraction caused by the water droplets in the mist.

glory-hole (*Eng, Glass*) (1) A subsidiary furnace, in which articles may be reheated during manufacture. (2) An opening exposing the hot interior of a furnace and used for eg fire polishing of glass.

glory-hole (*MinExt*) A combination of open-pit mining with underground tunnel through which spoil is removed after gravitating down.

gloss-, glosso- (*Genrl*) Prefixes from Gk *glossa*, tongue.

glossa (*BioSci*) In vertebrates, the tongue; any tongue-like structure. Adjs *glossate, glossal*.

glossectomy (*Med*) Removal of the tongue.

glossitis (*Med*) Inflammation of the tongue.

glosso- (*Genrl*) See GLOSS-.

glossodynia (*Med*) Pain in the tongue.

glossopharyngeal (*BioSci*) Pertaining to the tongue and the pharynx; the ninth cranial nerve of vertebrates, running to the first gill-cleft in lower forms, to the tongue and the gullet in higher forms.

glossoplegia (*Med*) Paralysis of the tongue.

glossospasm (*Med*) Spasm of the muscles of the tongue.

gloss paint (*Build*) Paint to which varnish is added as an ingredient in the manufacturing process; characterized by a glossy finish.

glottis (*BioSci*) In higher vertebrates, the opening from the pharynx into the trachea.

glove box (*NucEng*) An enclosure in which radioactive or toxic material can be handled by use of special rubber gloves attached to the sides of the box thus preventing contamination of the operator. Normally operated at a slightly reduced pressure so that any leakage is inward.

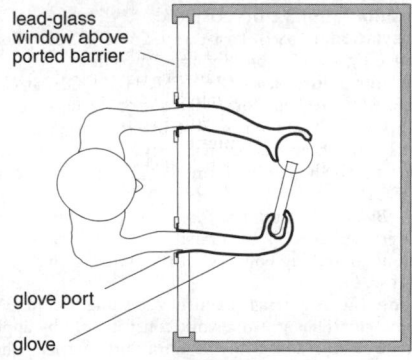

glove box

Glover tower (*ChemEng*) A tower of a sulphuric acid plant used to recover the nitrogen oxides from the Gay-Lussac tower, to cool the gases from the burners, to concentrate the acid trickling down the tower, to partly oxidize the gases from the sulphur burners, and to introduce the necessary nitric acid into the chambers by running nitric acid down the tower along with the nitrated acid from the Gay-Lussac tower.

glow discharge (*Electronics*) An electrical discharge through a low-pressure gas, producing a PLASMA. The ionized media of glow discharges are much used in surface engineering and microfabrication.

glow plug (*Aero*) In a gas turbine, an electrical igniting plug which can be switched on to ensure automatic relighting when the flame is unstable, eg under icing conditions. Also used in diesel engines.

glow potential (*Electronics*) The potential which initiates sufficient ionization to produce a gas discharge between electrodes, but is below sparking potential.

glow switch (*Electronics*) A tube in which a glow discharge thermally closes a contact, starting fluorescent tubes.

glow tube (*Electronics*) Cold-cathode, gas-filled diode, with no space-current control, the colour of glow depending on contained gas.

Glu (*Chem*) Abbrev, depending on the context, for GLUCOSE or GLUTAMIC ACID.

glucagon (*Med*) A hormone released by the alpha cells of the ISLETS OF LANGERHANS which raises the blood sugar by stimulating the breakdown of glycogen in the liver to glucose.

glucans (*Chem*) The condensation polymers of glucose, eg cellulose, starch, dextrin, glycogen, etc.

glucocorticoid (*Pharmacol*) Any of the CORTICOSTEROIDS which act on carbohydrate, fat and protein metabolism.

gluconic acid (*Chem*) See D-GLUCONIC ACID.

glucophore (*Chem*) A group of atoms which causes sweetness of taste.

glucosamine (*BioSci*) An amino-sugar (*2-amino-2-deoxy-glucose*); component of *chitin, heparan sulphate, chondroitin sulphate*, and many complex polysaccharides. Usually found as β-D-N-acetylglucosamine. Used as a dietary supplement to alleviate arthritic pain.

glucosaminoglycan (*BioSci*) See GLYCOSAMINOGLYCAN.

glucose (*BioSci, Chem*) $C_6H_{12}O_6$. The commonest aldo-hexose, and the major source of energy in animals. Starch and cellulose are condensation polymers of glucose. Also *dextrose, grape sugar*.

glucose-6-phosphate dehydrogenase (*Med*) An enzyme that catalyses the dehydrogenation of glucose-6-phosphate to 6-phosphogluconolactone, an important pathway in carbohydrate metabolism. Deficiency may cause the development of HAEMOLYTIC ANAEMIA.

glucosinolates (*BioSci*) A class of naturally occurring pesticides found particularly in brassicas such as broccoli and Brussels sprouts.

glucosuria (*Med*) See GLYCOSURIA.

glucosylation (*BioSci*) Post-translational modification of protein by the addition of glucose moieties.

glucuronic acid (*Chem*) $CHO(CHOH)_4COOH$. It can be prepared by reduction of the lactone of saccharic acid, and occurs in small amounts in the urine. It forms glycosides or esters with phenols or aromatic acids which are removed from the body in the form ($6\text{-}CH_2OH$ of glucose → $6\text{-}COOH$).

glue (*Build*) A substance used as an adhesive agent between surfaces to be united. Glue is obtained from various sources, eg bones, gelatine, starch, resins, etc. See ADHESIVE.

glueline (*ElecEng*) High-frequency heating technique for drying glue films in woodwork construction, by applying electric field in line with the film, with specially shaped electrodes. The film should have a high loss factor compared with medium to be 'glued'.

glue size (*Build*) Refined animal material available in granular or powder form which is soaked in cold water before melting with boiling water and allowed to cool and gel. Used to pretreat the surface to equalize porosity before paperhanging.

glue sniffing (*Med*) A form of drug abuse in which solvents are sniffed to produce intoxication.

Glufosinate (*BioSci*) TN for a non-selective chemical herbicide that can be used only on tolerant crops.

glume (*BioSci*) (1) One of the two bracts at the base of each SPIKELET in the inflorescence of a grass. (Old name: *sterile glume*.) (2) The bract subtending the flower in the Cyperaceae (sedges and allies). (3) *Flowering glume* is an old name for LEMMA.

gluon (*Phys*) The carrier of the strong interaction between quarks (and antiquarks). According to quantum chromo-dynamic theory, there should be eight different gluons each with zero mass and zero charge.

glut (*Build*) See CLOSER.

glutamate (*BioSci*) The anion of GLUTAMIC ACID, important in protein metabolism, the major EXCITATORY AMINO ACID in the mammalian central nervous system, and a neurotransmitter in arthropod skeletal muscle.

glutamate receptor (*BioSci*) A protein that is a LIGAND-GATED ION CHANNEL specific for the amino acid GLUTAMATE, and is important in brain function.

glutamic acid (*Chem*) *2-amino-pentane-1,6-dioic acid.* $HOOCCH_2CH_2CH(NH_2)COOH$. The *L*- or *S*-form is a constituent of proteins, and is classed as acidic as it has two carboxylic acid groups. Symbol Glu, short form E.

glutamine (*Chem*) The 6-amide of glutamic acid. Symbol Gln, short form Q.

glutaraldehyde (*Chem*) $CHO(CH_2)_3CHO$. An oil, soluble in water and volatile in steam; used in tanning leather, esp for clothing, as it imparts resistance to perspiration. Used as a *fixative* in preparing biological specimens for electron microscopy and as a disinfectant.

glutathione (*BioSci*) The tripeptide γ-glutamylcysteinylglycine. Present in animal cells at millimolar concentration where it acts as a sulphydryl and redox buffer. An important cofactor for some enzymes

gluteal (*BioSci*) Pertaining to the buttocks.

gluten (*FoodSci*) Hydrolysis product of two protein chains GLIADIN and GLUTENIN present in wheat flour. Forms the elastic network which allows CO_2 retention during yeast fermentation in the mixing of a bread dough. Among cereal flours only rye flour gives a similar product. Produces severe illness in people with COELIAC DISEASE. Dried gluten can be used as a supplement for natural protein in high-fibre, high-protein bread.

gluteus (*BioSci*) In land vertebrates, a retractor and elevator muscle of the hind-limb.

Gly (*Chem*) Symbol for GLYCINE.

glycaemic (*Med*) Relating to the concentration of glucose in the blood. US *glycemic*.

glycaemic index (*BioSci*) A system for categorizing foods according to how quickly their carbohydrates are broken down into blood sugar. Abbrev GI.

glycerides (*Chem*) A term for glycerine esters, the most important of which are the FATS.

glycerination (*BioSci*) Permeabilization of the cell membrane of cells by incubating in aqueous glycerol at low temperature. Glycerinated muscle will contract if exogenous ATP and calcium are added.

glycerine (*Chem*) *Glycerol, propan-1,2,3-triol.* $CH_2OHCHOHCH_2OH$. Mp 17°C, bp 290°C. A syrupy hygroscopic liquid, obtained by the hydrolysis of oils and fats, or by the alcoholic fermentation of glucose in the presence of sodium sulphite solution, which reacts with the aldehydes formed, thus liberating a larger amount of glycerine. It is also prepared synthetically from propylene by chlorination and hydrolysis. Glycerine is a trihydric alcohol, forming alcoholates, esters and numerous derivatives. Colourless. It is a raw material for ALKYD RESINS, NITROGLYCERINE, printing inks, foodstuff preparations, etc. Also *glycerin*.

glycerine litharge cement (*Chem*) A mixture of litharge (lead (II) oxide) and glycerine which rapidly sets to a hard mass.

glycerol (*Chem*) See GLYCERINE.

glycerol-phthalic resins (*Plastics*) See ALKYD RESINS.

glyceryl monostearate (*FoodSci*) A widely used emulsifying agent which is water soluble (hydrophilic) at the glycerol end and fat soluble (lyophilic) at the stearate end.

glyceryl trinitrate (*Pharmacol*) A short-acting vasodilator that reduces venous return to the heart, heart work and oxygen consumption. Its main use is in the treatment of ANGINA PECTORIS.

glycine (*Chem*) *Aminoethanoic acid*. The simplest naturally occurring amino acid. Symbol Gly, short form G.

glycocalyx (*BioSci*) The carbohydrate layer on the outer surface of animal cells, composed of the oligosaccharide termini of the integral membrane glycoproteins and glycolipids.

glycoconjugate (*BioSci*) Any biological macromolecule containing a carbohydrate moiety, thus a generic term to cover glycolipids, glycoproteins and proteoglycans.

glycogen (*BioSci*) Storage polysaccharide, α1-4-linked with frequent α1-6 branches, existing as small granules in blue-green algae and bacteria and in the cytoplasm of eumycete fungi and animals but not in plants or eukaryotic algae. Glycogen in the liver and muscle is an important energy store broken down to glucose on demand.

glycol (*Chem*) See ETHYLENE GLYCOL.

glycolipid (*BioSci*) A glycosylated lipid. See CEREBROSIDE, GANGLIOSIDE.

glycols (*Chem*) Dihydric alcohols, of the general formula $C_nH_{2n}(OH)_2$, viscous liquids with a sweet taste or crystalline substances. They give all the alcohol reactions and, having two hydroxyl groups in the molecule, they can also form mixed compounds, eg ester-alcohols.

glycolysis (*BioSci*) The sequence of reactions that converts glucose to pyruvate with the concomitant net synthesis of two molecules of ATP. Under aerobic conditions it is the prelude to the complete oxidation of glucose via the TCA CYCLE.

glycophorin (*BioSci*) A glycoprotein that spans the human erythrocyte membrane and acts as a chloride ion channel.

glycophyte (*BioSci*) A plant that will grow only in soils containing little sodium chloride or other sodium salt. Most plants are glycophytes. Cf HALOPHYTE.

glycoprotein (*BioSci*) A protein with covalently linked sugar residues. The sugars may be bound to OH side chains of the polypeptide (*O-linked*) or to the amide nitrogen of asparagine side chains (*N-linked*).

glycosaminoglycan (*BioSci*) A large polysaccharide of repeating disaccharide units of amino sugars. GAGs are attached as side chains to a protein core to make a PROTEO-GLYCAN. Common GAGs are hylauronate, and chondroitin, dermatan, heparan and keratan sulphates. Abbrev GAG.

glycosidase (*BioSci*) General and imprecise term for an enzyme that degrades linkage between sugar subunits of a polysaccharide, thus any of the EC 3.2 class of hydrolases that cleave glycosidic bonds.

glycosides (*Chem*) A group of compounds, derived from the other monosaccharides in the same way as glucosides are derived from glucose, including glucosides as a subclass.

glycosome (*BioSci*) A membrane-bound organelle that contains glycolytic enzymes, found within kinetoplastid Protozoa.

glycosphingolipids (*BioSci*) Ceramide derivatives containing more than one sugar residue. If sialic acid is present these are called *gangliosides*.

glycosuria (*Med*) The presence of sugar in the urine. Also *glucosuria*.

glycosylation (*BioSci*) The addition of sugar units, as in the addition of glycan chains to proteins.

glycosyltransferase (*BioSci*) An enzyme that catalyses the transfer of sugars onto a protein or another sugar side chain. These enzymes bring about the glycosylation of glycoproteins.

glycuronic acid (*Chem*) See GLUCURONIC ACID.

glyoxal (*Chem*) *Ethan 1,2-dial*. CHOCHO. Mp 15°C, bp 51°C. A dialdehyde. A yellow liquid forming green vapours, but it is not stable, and polymerizes to insoluble paraglyoxal, $(CHOCHO)_3$.

glyoxalic acid (*Chem*) *Ethan-1-al-2-oic acid*. CHO-COOH + H_2O or $CH(OH)_2COOH$. An aldehyde monobasic acid occurring in unripe fruit; rhombic prisms, soluble in water, volatile in steam, which can be obtained by the oxidation of ethanol with nitric acid, or by the hydrolysis of dichloroacetic or dibromoacetic acid.

glyoxalines (*Chem*) See IMIDAZOLES.

glyoxisome (*BioSci*) See GLYOXYSOME.

glyoxylate cycle (*BioSci*) A series of reactions resulting in the formation of succinate from acetyl CoA, enabling carbohydrates to be made from fatty acids as in germinating oil-storing seeds. See GLYOXYSOME.

glyoxylic acid (*Chem*) See GLYOXALIC ACID.

glyoxysome (*BioSci*) Cytoplasmic organelles of plant cells similar to PEROXISOMES but also containing enzymes of the *glyoxylate cycle*, a cyclic metabolic pathway that generates succinate from acetate. They are abundantly present in eg the endosperm or cotyledons of oil-rich seeds. Also *glyoxisome*.

glyph (*Arch*) A short upright flute. See TRIGLYPH.

glyptal resins (*Plastics*) See ALKYD RESINS.

Gm allotype (*BioSci*) An allotypic determinant (recognizable by specific antibodies) due to amino acid substitutions at various positions in the constant region of human IgG heavy chains. At least 25 different allotypes are known; useful in genetic studies.

G–M counter (*NucEng*) See GEIGER COUNTER.

gmelinite (*Min*) A pseudohexagonal zeolite, white or pink in colour and rhombohedral in form, resembling chabazite. Chemically, it is hydrated silicate of aluminium, sodium and calcium.

Gmelin test (*Chem*) A test for the presence of bile pigments; based upon the formation of various coloured oxidation products in treatment with concentrated nitric acid.

GMO (*BioSci*) Abbrev for GENETICALLY MODIFIED ORGANISM.

GMP (*FoodSci*) Abbrev for GOOD MANUFACTURING PRACTICE.

GMS (*FoodSci*) Abbrev for GLYCERYL MONOSTEARATE.

GMT (*Astron*) Abbrev for GREENWICH MEAN TIME.

gnathic (*BioSci*) Pertaining to the jaws.

gnathites (*BioSci*) Mouthparts, esp those of insects.

gnathobase (*BioSci*) In Arthropoda, a masticatory process on the inner side of the first joint of an appendage.

gnathopod (*BioSci*) In Arthropoda, any appendage modified to assist in mastication.

Gnathostomata (*BioSci*) A superclass of Chordata, including all vertebrate animals with upper and lower jaws. It comprises a wide range of animals, from fish to tetrapods. Cf AGNATHA.

gnathostomatous (*BioSci*) Having the mouth provided with jaws.

gnathotheca (*BioSci*) The horny part of the lower beak of birds.

gneiss (*Geol*) A metamorphic rock of coarse grain size, characterized by a mineral banding, in which the light minerals (quartz and feldspar) are separated from the dark ones (mica and/or hornblende). The layers of dark minerals are foliated, while the light bands are granulitic. See METAMORPHISM.

gneissose texture (*Geol*) A rock texture in which foliated and granulose (granulitic) bands alternate.

Gnetopsida (*BioSci*) A class of the gymnosperms with c.80 spp. Mostly trees, shrubs, lianas and switch plants. Leaves are reticulately veined or scaled. As in angiosperms, the xylem has vessels. Reproductive structures organized into compound strobili. There are only three genera: *Gnetum*, *Ephedra* and *Welwitchia* (the last a curious turnip-like plant with two strap-shaped leaves, of the Namibian desert).

gnomon (*MathSci*) The remainder of a parallelogram after a similar parallelogram has been removed from one corner.

gnomon (*Surv*) An early instrument for determination of time and latitude, involving the measurement of the shadow of an upright rod as cast by the Sun. The pointer of a sundial.

gnotobiotic (*BioSci*) Of a known (defined) environment for living organisms, either *in vitro* as a sterile culture inoculated with one or a few strains of bacteria, or as an environment in which animals can be reared and in which all the living microbes are known. This avoids antigenic stimulation by casual infections. If all living microbes are absent the environment is said to be *germ-free*.

go (*Build*) The GOING.

goaf (*MinExt*) See GOB.

goal-directed behaviour (*Psych*) Implies that the individual has some model of the goal situation, and that discrepancies between the current and goal situation are used to guide behaviour; conscious awareness of the goal or complex cognitive abilities are not necessarily associated with goal-directed behaviour, which probably occurs in many species.

goat pox (*Vet*) An epidemic disease of goats due to infection by a virus; characterized by fever and a papulo-vesicular eruption of the skin and mucous membranes.

gob (*Glass*) (1) A measured portion of molten glass as fed to machines making glass articles. (2) A lump of hot glass gathered on a punty or blowing iron.

gob (*MinExt*) The space left by the extraction of a coal seam, into which waste is packed, also loose waste. Also *goaf*.

gob fire (*MinExt*) A fire occurring in a worked-out area, due to ignition of timber or broken coal left in the gob.

gob heading (*MinExt*) A roadway driven through the gob after the filling has settled. Also *gob road*.

goblet cell (*BioSci*) A goblet- or flask-shaped epithelial gland cell, occurring usually in columnar epithelium.

gobo (*ImageTech*) An opaque mask on a stand used in the studio to provide local shadow areas or shield the camera lens.

gob process (*Glass*) One for making hollow ware, in which glass is delivered by an automatic feeder in the form of soft lumps of suitable shape to a forming unit. Also *flow process*, *gravity process*.

gob road (*MinExt*) See GOB HEADING.

gob stink (*MinExt*) A smell indicating spontaneous combustion or a fire in the gob.

godcast (*ICT*) A PODCAST containing exclusively religious subject matter.

godets (*Textiles*) (1) Small rollers used to regulate the speed of extruded filaments during synthetic fibre production. (2) An insert of material used to shape a garment being assembled from fabric.

go-devil (*MinExt*) Cylindrical plug with brushes, scrapers and rollers on its periphery and able to move under the oil pressure through a pipeline and clean it. Also *pig*, *rabbit*.

goethite (*Min*) Orthorhombic hydrated oxide of iron with composition FeOOH. Dimorphous with lepidocrocite.

Goetz size separator (*PowderTech*) An instrument for the size classification of airborne particles, comprising a high-speed rotor with a helical channel along which the particles move in laminar flow and are deposited by centrifugal force on a removable envelope surrounding the rotor.

going (*Build*) The horizontal interval between consecutive risers in a stair. See fig. at STRING.

going light (*Vet*) A term popularly applied to emaciation of animals.

going line (*Build*) See WALKING LINE.

going rod (*Build*) A rod used for setting out the going of the steps in a flight.

going train (*Eng*) The set of gears in a clock that transmit the power from the mainspring or weight system to the ESCAPEMENT. This in turn drives the MOTION WORK.

goitre (*Med*) Morbid enlargement of the thyroid gland as in BASEDOW'S DISEASE. Adj *goitrous*.

goitrogenous (*Med*) Producing, or tending to produce, GOITRE.

Golay cell (*Phys*) Pneumatic cell used as detector of heat radiation, eg in an infrared spectrometer.

gold (*Chem*) Symbol Au (from *aurum*), at no 79, ram 196·967, rel.d. at 20°C 19·3, mp 1062°C, electrical resistivity about 0·02 $\mu\Omega$ m. A heavy, yellow, metallic element, occurring in the free state in nature. Most of the metal is retained in gold reserves but some is used in jewellery, dentistry, and for decorating pottery and china. In coinage and jewellery, the gold is alloyed with varying amounts of copper and silver. *White gold* is usually an alloy with nickel, but as used in dentistry this alloy contains platinum or palladium.

gold amalgam (*Min*) A variety of native gold containing approximately 60% of mercury.

Goldbach's conjecture (*MathSci*) The unproven hypothesis that every even number greater than two can be written as the sum of two prime numbers.

Goldberg–Hogness box (*BioSci*) See TATA BOX.

gold blocking (*Print*) The process of pressing a design upon gold leaf spread out on the cover of a book, the tools or dies, which are heated, leaving the desired impression. Also carried out by machine, gold foil being fed from a spool.

gold cushion (*Print*) A small board, covered with rough calfskin, which is padded with a soft material. The gold leaf required for gold blocking is placed on the cushion ready for use.

golden beryl (*Min*) A clear yellow variety of the mineral beryl, prized as a gemstone. Heliodor is a variety from SW Africa. Also *chrysoberyl*.

golden mean (*MathSci*) The number obtained when the line AB is divided by a point P such that the ratios AB:AP and AP:PB are equal. It follows that AB/AP = $(1 + \sqrt{5})/2$ ≈ 1·62. This proportion has been used in architecture and in the construction of many classical works of art and culture.

golden number (*Astron*) A term derived originally from medieval church calendars, still used to signify the place of a given year in the METONIC CYCLE of 19 years.

Golden rice (*BioSci*) Rice that has been genetically engineered with genes that produce high levels of β-carotene, that is converted to vitamin A within the body, and should help to prevent blindness due to vitamin deficiency. The carotene makes the rice yellow in colour, hence the name.

golden section (*Arch*) A proportion thought to be ideal by Renaissance theorists. It is defined as a line cut in such a way that the smaller part is to the greater part as the greater part is to the whole (See GOLDEN NUMBER.)

gold filled (*Eng*) The agreed term for a coating of gold over a base metal which should be stamped with a fraction giving the proportion by weight of the gold present and the karat fineness of the gold alloy.

gold grains (*Radiol*) Small lengths of activated gold wire (half-life 2·70 days), used in a similar manner to RADON SEEDS.

gold leaf (*Build, Eng*) Leaf of 23–24 carat gold which is beaten into extremely thin sheets, so that it may be applied to surfaces which are to be gilded; available in books of 25 sheets 82 mm^2 in either loose or transfer form.

gold-leaf electrometer (*ElecEng*) GOLD-LEAF ELECTRO-SCOPE, modified for the measurement of very small currents by observing the rate of movement of the gold leaf through a microscope. The PERSONAL DOSIMETER is a development of this instrument.

gold-leaf electroscope (*ElecEng*) A device for detecting small electric charges which are applied to a piece of thin gold foil usually attached at upper end to a metal electrode. Mutual repulsion between the foil and the similarly charged plate electrode leads to the former being displaced.

Goldmann equation (*BioSci*) An equation that predicts the electrical potential across a membrane on the basis of the distributions and relative permeabilities of the main permeant ions (typically sodium, potassium and chloride). Assumes constancy of the electric field across the membrane, and the absence of active ion transport, but gives a reasonable approximation to the observed values.

gold paints (*Build*) Paints made of bronze powders mixed with varnish.

gold rug (*Print*) A piece of flannel cloth or soft rubber (*gold rubber*) used for wiping off surplus gold after gilding. It is sold to a refiner after a period of use.

Goldschmidt process (*Chem*) (1) See ALUMINOTHERMIC PROCESS. (2) Detinning of coated iron by use of chlorine.

gold size (*Build*) A type of SIZE used as a basis to secure gold leaf onto surfaces which are to be gilded, and for other purposes. Formulated with specific drying times, eg 1 h, 4 h and 24 h.

Gold slide (*EnvSci*) A slide, named after its designer, which is attached to a marine mercury barometer to make the corrections for index error, latitude, height and temperature mechanically.

gold toning (*ImageTech*) A metallic toning process in which the original silver image becomes associated with a gold deposit.

gold tooling (*Print*) The decorating of book covers by hand, using gold leaf and heated tools.

golf-ball printer (*ICT*) Uses a moving spherical print head that is easily removable, enabling many different character sets to be used.

Golgi apparatus (*BioSci*) A cytoplasmic organelle consisting of a stack of plate-like CISTERNAE often close to the nucleus. It is the site of protein glycosylation. In plants, the *dictyosome*.

Golgi body (*BioSci*) See GOLGI APPARATUS.

Golgi's organs (*BioSci*) In vertebrates, a type of sensory nerve ending occurring in tendons near the point of attachment of muscle fibres, stretch receptors.

goliath Edison screw-cap (*ElecEng*) An Edison screw-cap having a screw-thread about 1·5 in diameter, with 4 threads per inch; used with large metal-filament lamps.

gomphosis (*BioSci*) A type of articulation in which a conical process fits into a cavity; as the roots of teeth into their sockets.

gon-, gono- (*Genrl*) Prefixes from Gk *gonos*, seed, offspring.

gonad (*BioSci*) A mass of tissue arising from the primordial germ cells and within which the spermatozoa or ova are formed; a sex gland, ovary or testis. Adj *gonadal*.

gonadotrophic (*Med*) Pertaining to gonadotrophins, gonad-stimulating hormones. Also *gonadotropic*.

gonal (*BioSci*) Forming or giving rise to a GONAD, as the *gonal ridge*.

gonapophysis (*BioSci*) Tubular outgrowth from the medial borders of the COXITES of the GONOPODS in insects; in females sometimes modified to form an ovipositor or sting. Also loosely any insect genital appendage.

Gondwanaland (*Geol*) The hypothetical continent of the southern hemisphere which broke up and drifted apart to form bits of the present continents of S America, Africa, India, Australia and Antarctica.

gone to bed (*Print*) A newspaper term meaning that the newspaper has been made up and sent to press.

goni- (*Genrl*) Prefix from Gk *gonia*, angle; *gony*, knee. Not to be confused with prefixes GON-, GONO-.

goniatite (*BioSci*) Any cephalopod belonging to the order Goniatitida. See AMMONOIDS.

gonidial layer (*BioSci*) See ALGAL LAYER.

gonidium (*BioSci*) (1) A cell that gives rise to an asexual daughter colony in *Volvox*. (2) Any algal cell in the thallus of a lichen.

gonimic layer (*BioSci*) See ALGAL LAYER.

goniometer (*Eng*) A device for measuring angles or for direction finding.

gonioscope (*Med*) An instrument used in ophthalmology to measure and inspect the anterior chamber of the eye (ie the region between cornea and iris).

gonitis (*Vet*) Inflammation of the stifle joint of animals.

gonnardite (*Min*) A rare zeolite; a hydrated sodium, calcium, aluminium silicate.

gono- (*Genrl*) See GON-.

gonoblast (*BioSci*) A reproductive cell.

gonochorism (*BioSci*) Sex determination.

gonochoristic (*BioSci*) Having separate sexes.

gonococcus (*Med*) A Gram-negative diplococcus, the causative agent of GONORRHOEA.

gonocoel (*BioSci*) That portion of the coelom, the walls of which give rise to the gonads: hence, the cavity of the gonads.

gonoduct (*BioSci*) A duct conveying genital products to the exterior; a duct leading from a gonad to the exterior.

gonopods (*BioSci*) The external organs of reproduction in insects, associated with the eighth and ninth abdominal segments in females and with the ninth only in males. In Chilopoda, a pair of modified appendages borne on the 17th (genital) body segment.

gonopore (*BioSci*) The aperture by which the reproductive elements leave the body.

gonorrhoea (*Med*) A contagious infection of the mucous membrane of the genital tract with the gonococcus, contracted usually through sexual intercourse. US *gonorrhea*.

gonosome (*BioSci*) In colonial animals, all the individuals concerned with reproduction.

go, not-go gauge (*Eng*) A LIMIT GAUGE of which one section is above and another below the specified dimensional limits of the part to be gauged. If the 'go' section accepts the part and the 'not-go' section interferes with the part, the latter is accurate within the limits specified.

Gooch crucible (*Chem*) A filter used in laboratories, which consists usually of a small porcelain cup, the bottom of which is perforated with numerous small holes and covered with a thin layer of washed asbestos fibres, which act as a filtering medium. Now mainly superseded by sintered glass crucibles. Also *Gooch filter*.

good colour (*Print*) A term which indicates consistent distribution of ink throughout a print job.

good manufacturing practice (*FoodSci*) The application of responsible practice in all aspects of food production so that manufacturing specifications and statutory health, safety and compositional requirements are always met. It requires measures for proper training of personnel, maintenance of equipment and processing environment, and the prevention of foreign body contamination. Comparable specifications for good practice exist for the manufacture of pharmaceuticals. Abbrev *GMP*. See HAZARD ANALYSIS.

goodness of fit (*MathSci*) The extent to which observed data match the values predicted by a theory.

Goodpasture's syndrome (*Med*) A VASCULITIS of the small blood vessels of the kidney and lung. Probably an abnormal immune reaction.

Google (*ICT*) TN of a widely used Internet search engine, characterized by the use of sophisticated algorithms to return popular matches for search terms. (Originally derived from the invented arithmetic term 'googol', a number 1 followed by 100 zeros.)

google (*ICT*) To attempt to find out about someone or something by entering their name into an Internet search engine (usually Google).

googol (*MathSci*) The number 10^{100}.

gooseberry stone (*Min*) The garnet GROSSULAR, which was named for the resemblance of this green variety, in both form and colour, to the gooseberry, which has the botanical name *Ribes grossularia*.

goose flesh (*Med*) See HORRIPILATION.

go-out (*Build*) See TIDAL FLAP.

gopher (*ICT*) A menu-based information retrieval system that allows access to, usually, plain text information held by institutions. Their addresses on the Internet start with gopher:\\. See panel on INTERNET.

Gordon–Hauss jitter (*ICT*) An effect produced during SOLITON PROPAGATION in an OPTICAL FIBRE, in which AMPLIFIED SPONTANEOUS EMISSION induces a random shift in the carrier frequency of the soliton, which interacts with CHROMATIC DISPERSION to produce a random timing jitter, causing an increased BIT ERROR RATIO.

Gordon's formula (*CivEng*) An empirical formula giving the collapsing load for a given column. It states that

$$P = \frac{fc.A}{1 + c(l/d)^2}$$

where P = the collapsing load, fc = safe compressive stresses for very short lengths of the material, A = area of cross-section, l = length of the pin-jointed column, d = the least breadth or diameter of the cross-section, c = a constant for the material and the shape of cross-section.

gore (*Aero*) One of the sector-like sections of the canopy of a parachute.

Gore-tex (*Textiles*) TN for microporous polytetrafluoroethylene, made by stretching polymer in a controlled way so as to create fine pores, which allow diffusion of air and water vapour but prevent liquid water ingress. Used in waterproof clothing.

gorge (*Build*) A DRIP.

gorge (*Geol*) A general term for all steep-sided, relatively narrow valleys, eg canyons, overflow channels, etc.

goslarite (*Min*) Hydrated zinc sulphate, a rare mineral precipitated from water seeping through the walls of lead mines; formed by the decomposition of sphalerite. See WHITE VITRIOL.

gossan (*MinExt*) The leached, oxidized material found in surface exposures of an ore deposit; represents the residue left after SECONDARY ENRICHMENT of a mineral vein or lode. Often stained brown by iron oxides and rich in quartz, gossans are an indication of mineral deposits below the surface, although they may not be of any value themselves.

Gothic (*Arch*) The style of architecture, mainly ecclesiastical, prevalent in W Europe from the late 12th century. It is characterized by pointed arches, ribbed vaults, cruciform plans, flying buttresses and glazed CLERESTOREYS.

Gothic (*Print*) Originally a term applied to black letter or Old English type, it is now used to include bold sans serif faces.

Gothic Revival (*Arch*) An architectural movement of the late 18th and the 19th centuries to revive the GOTHIC style. Since it had originated as a Christian form of architecture it was appropriate that it should be adopted in the construction of the many new parish churches being built at that time.

Gothic wing (*Aero*) A very low ASPECT RATIO wing with the double curvature plan form, similar to a Gothic arch of the perpendicular period, used for supersonic aircraft to combine low WAVE DRAG with SEPARATED LIFT. See OGEE WING.

Gothlandian (*Geol*) Obsolete name used in Europe for SILURIAN.

Göttingen wind tunnel (*Aero*) A return-flow wind tunnel in which the working section is open. See OPEN-JET WIND TUNNEL.

Gott's method (*ElecEng*) A bridge method of finding the capacitance of a cable. The cable is made to form one arm of the bridge, the others being a standard capacitor and two sections of a slide wire.

gouache (*Build*) Opaque colours mixed with water, honey and gum, applied in impasto style.

gouge (*Build*) Similar to a chisel, but having a curved blade and a cutting edge capable of forming a rounded groove.

gouge (*Med*) A hollow chisel for removing and cutting bone.

gouge (*Print*) A hand tool used to form curved lines.

gouge slip (*Build*) A shaped piece of oilstone on which the concave side of the cutting edge of a gouge may be rubbed to sharpen it. Also *oilstone slip, slip stone*.

goundou (*Med*) Symmetrical bony overgrowth at the sides of the nose, thought to be a late sequel of yaws. Also *dog nose, gros nez*.

gout (*Med*) A disorder of metabolism in which there is an excess of uric acid in the blood; this is deposited, as sodium biurate, in the joints, bones, ligaments and cartilages. In acute gout there is a sudden very painful swelling of the joint, usually of the big toe. Also *podagra*.

Gouy layer (*MinExt*) Diffuse layer of counterions surrounding charged lattices at surface of particle immersed in liquid.

government rubber-styrene (*Eng*) The chief synthetic rubber developed in the USA during World War II. Based on styrene and butadiene. Abbrev *GR-S*. See SBR.

governor (*ElecEng*) Speed regulator on rotating machine, eg turbo-alternator set.

governor (*Eng*) A device for controlling the fuel or steam supply to an engine in accordance with the power demand, so that the speed is kept constant under all conditions of loading.

GPC (*Chem*) Abbrev for GEL PERMEATION CHROMATOGRAPHY.

GPI (*Med*) Abbrev for GENERAL PARALYSIS OF THE INSANE. See GENERAL PARESIS.

G-protein (*BioSci*) See GTP-BINDING PROTEIN.

G-protein-coupled receptor (*BioSci*) Major and very diverse class of cell surface receptors that are coupled to *GTP-binding proteins*. All have seven membrane-spanning domains. Examples include many hormone receptors, olfactory receptors, photoreceptors: GPCRs are the target for many drugs. Abbrev *GPCR*.

GPRS (*ICT*) Abbrev for *general packet radio service*. An advanced packet-switching system.

GPS (*ICT*) Abbrev for GLOBAL POSITIONING SYSTEM.

G–P zones (*Eng*) Abbrev for GUINIER–PRESTON ZONES.

Graafian follicle (*BioSci*) A vesicle, containing an ovum surrounded by a layer of epithelial tissue, that occurs in the ovary of mammals.

grab (*CivEng*) A steel bucket or cage made of two halves hinged together, so that they dig out and enclose part of the material on which they rest; used in mechanical excavators and dredgers. Also *grab-bucket*. See GRABBING CRANE, GRAB-DREDGER.

grabbing crane (*CivEng*) An excavator consisting of a crane carrying a large GRAB or bucket.

grab-bucket (*CivEng*) See GRAB.

grab-dredger (*CivEng*) A dredging appliance consisting of a GRAB or grab-bucket suspended from the jib-head of a crane, which does the necessary raising and lowering. Also *grapple dredger*.

graben (*Geol*) An elongated downthrown block bounded by faults along its length. Cf HORST.

gracilis (*BioSci*) A thigh muscle of land vertebrates.

grade (*Chem*) Individual specification of a particular polymer. Includes details of composition (additives, fillers, etc), polymer type (homopolymer, copolymer, etc) and molecular mass distribution. Usually given a proprietary code by the manufacturer. Applies *mutatis mutandis* to other materials.

grade (*CivEng*) US for GRADIENT.

grade (*Eng*) A measure (expressed by a letter of the alphabet) of the hardness of a grinding wheel as a whole, due to the strength of the bond holding the abrasive particles in place.

graded bedding (*Geol*) Bedding which shows a sorting effect with the coarser material at the base progressively changing upwards to finer sediment at the top. Occasionally such grading may be *reversed*.

graded brush (*ElecEng*) A brush for collecting current from the commutator of an electrical machine; made up of layers of different materials, or of material which has different values of lateral and longitudinal resistance.

graded filter (*ImageTech*) See GRADUATED FILTER.

graded-index fibre (*ICT*) A fibre in an OPTICAL FIBRE cable, in which the refractive index of the glass decreases gradually or in several small steps from the central core to the periphery. See MONOMODE FIBRE, MULTIMODE FIBRE, STEPPED-INDEX FIBRE FIG. ▷.

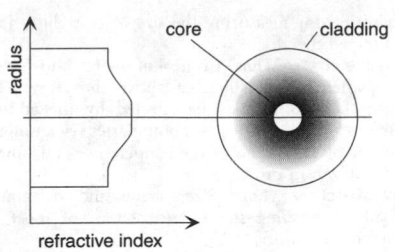

graded-index fibre

grade of service (*ICT*) The proportion of calls in the busy hour that must fail to be completed through insufficiency of apparatus.

grade pegs (*Surv*) Pegs driven into the ground as references, to establish gradients in constructional work. Also *gradient pegs*.

grader (*CivEng*) A power-operated machine provided with a blade for shaping excavated surfaces to the desired shape of slope.

gradient (*CivEng*) The degree of slope, eg of a highway or railway. US *grade*.

gradient (*MathSci*) Of a plane curve at a given point: the slope of the tangent at the point. Of a scalar function $f(x,y,z)$: the vector $\nabla f(x,y,z)$ where ∇ is the vector operator nabla (see DEL). Written grad f.

gradient (*Phys*) The rate of change of a quantity with distance, eg the temperature gradient in a metal bar is the rate of change of temperature along the bar.

gradient (*Surv*) The ratio of the difference in elevation between two given points and the horizontal distance between them, or the distance for unit rise or fall. Also *incline*.

gradient analysis (*BioSci*) A method of ordination in which vegetation samples or stands are plotted on axes representing environmental variables.

gradient of reinforcement (*Psych*) The curve which describes the declining effectiveness of reinforcement as the interval between the response and reinforcement increases.

gradient pegs (posts) (*Surv*) See GRADE PEGS.

gradient post (*CivEng*) A short upright post fixed at the side of a railway at a point of change of gradient. An arm on each side of the post indicates whether the track rises or falls, the figure on it giving the distance to be travelled to rise or fall one of the same unit of distance.

gradient wind (*EnvSci*) A theoretical wind which, when blowing along curved isobars with no tangential acceleration, represents a balance between the pressure gradient, the CORIOLIS force and the centrifugal force. It is less than the GEOSTROPHIC WIND round a DEPRESSION, and greater round an ANTICYCLONE.

grading (*Build*) (1) The proportions of the different sizes of stone or sand used in mixing concrete. (2) The selection of these proportions.

grading (*CivEng*) The operation of preparing a surface to follow a given gradient.

grading (*ICT*) (1) The scheme of connecting trunks or outlets so that a group of selectors is given access to individual trunks, while larger groups of selectors share trunks when all the individual trunks are found to be in use. (2) An arrangement of trunks connected to the banks of selectors by the method of grading. See GRADE OF SERVICE.

grading (*ImageTech*) The process of selecting optimum printing values of colour and intensity for the successive scenes of an assembled motion picture negative.

grading coefficient (*ElecEng*) A figure denoting the ratio of the lower to the upper limit of current for motor starters.

grading group (*ICT*) The group of selectors that are concerned in one grading scheme.

grading instrument (*Surv*) A general name for any instrument of the GRADIOMETER class.

grading shield (*ElecEng*) A circular conductor placed concentric with a string of suspension insulators on an overhead transmission line to equalize the potential across the individual insulator units. Also *arcing shield*.

grading, size (*Chem*) See SIZE-GRADING.

gradiometer (*Phys*) Magnetometer for measurement of the gradient of a magnetic field.

gradiometer (*Surv*) An instrument for setting out long uniform gradients; it consists of a level that may be elevated or depressed, by known amounts, by means of a vertical tangent screw.

graduated circle (*Surv*) A circular plate, marked off in degrees and parts of degrees, used on surveying instruments as a basis for the measurement of horizontal or vertical angles. See HORIZONTAL CIRCLE, VERTICAL CIRCLE.

graduated courses (*Build*) Courses of slates laid so that the gauge diminishes from eaves to ridge.

graduated filter (*ImageTech*) A camera FILTER having variations of density, colour or diffusion across its surface to modify selected areas of the resultant image(s). Also *graded filter*.

graduated vessels (*Chem*) Vessels which are used for measuring liquids and are adapted to measure or deliver definite volumes of liquid.

Graff's 'C' stain (*Paper*) A microscopical stain for assisting the identification of paper, prepared from aluminium, calcium and zinc chlorides, together with iodine and potassium iodide.

Grafil (*Chem*) TN for carbon fibre (Japan).

graft (*Med*) Any transplanted organ or tissue.

graft chimera (*BioSci*) A more or less stable PERICLINAL CHIMERA in which the skin and core (typically) derive from different species. Graft chimeras arise, rarely, from the junction of stock and scion in interspecific grafts, can be propagated only vegetatively and exhibit characteristics intermediate between the two species. Graft chimeras are indicated by a + sign before the specific or generic name (intra- and intergeneric chimeras respectively). The most famous is probably + *Laburnocytisus adami* which consists of a skin of *Cytisus purpureus* over a core of *Laburnum anagyroides*.

graft hybrid (*BioSci*) In some usages, GRAFT CHIMERA, in others, BURDO.

grafting (*BioSci*) The placing for propagation or experiment of a piece of one plant (scion) onto a piece, usually with roots, of another (stock) so that the tissues may unite and growth follow.

grafting (*Build*) The operation of lengthening a timber by jointing another piece onto it.

graft polymer (*Chem*) Polymer where chain branches occur, esp where branches are different to main chain. See panel on POLYMERS.

graft rejection (*BioSci*) The process by which the immune system recognizes and ultimately destroys non-tissue-matched organ grafts. Tissues that are not well matched for the major histocompatibility complex antigens rapidly become invaded by lymphocytes and the cells of the graft are killed. Patients who have received organ transplants require constant immunosuppressive therapy to prevent rejection.

graft-versus-host disease (*Med*) A GRAFT-VERSUS-HOST REACTION of transfused or transplanted lymphocytes against host antigens. Major complication of BONE MARROW GRAFTING. Abbrev *GVH*.

graft-versus-host reaction (*BioSci*) Occurs when a tissue graft (notably bone marrow) contains T-lymphocytes that can respond to antigens present in the recipient that are not identical with those of the donor. If the recipient is unable to suppress the donor lymphocytes, because of immunological immaturity or immunosuppression due to X-irradiation or to drugs, they cause severe damage to the recipient that begins in the skin and gut and may progress to death.

grahamite (*Min*) A member of the asphaltite group.

Graham's law (*Chem*) The velocity of effusion of a gas is inversely proportional to the square root of its density.

grain (*Agri*) Seeds harvested from cereal plants.

grain (*BioSci*) See CARYOPSIS.

grain (*Chem*) Unit of mass. One grain = 15·4 g. See APOTHECARIES' WEIGHT.

grain (*Eng*) Small particles of a crystalline substance. May be discrete as in a single grain of mineral or one in a polycrystalline array, as in a phase mixture or a recrystallized metal (see ANNEALING). The crystal structure is continuous across each individual grain and boundaries are formed where one grain meets its neighbours because the orientation of the crystal lattice is different. Grains may sometimes be visible to the unaided eye without any preparation, but more usually a section needs to be polished and etched to reveal their size, form and distribution. Cf BANDING in metals.

grain (*For*) A feature of the texture of wood determined by the direction of the long axes of the TRACHEID cells (ie parallel to the long axes of trunk and branches) and showing the marked ANISOTROPIC nature of wood.

grain (*Geol, Min*) (1) See RIFT AND GRAIN. (2) Average size of mineral crystals composing a rock. Direction in which it tends to split.

grain (*ImageTech*) The component particles of black silver which make up a photographic image after development and their aggregation into clumps; colour images from which silver has been removed may still show something of the original grouping. *Graininess* is a subjective term describing the visual appearance of a picture, whereas *granularity* is an objective measurement using a microdensitometer to examine the image structure.

grain boundary (*Eng*) Zone formed at the junction of single crystals in a polycrystalline material. Impurities tend to accumulate here by being excluded from normal growth of each crystal.

grain growth (*Eng*) A stage in the annealing process of cold-worked metals, in which holding the metal at above about $0·4$–$0·5T_m$ (melting temperature), after recrystallization has taken place, allows the average grain size of the metal crystals to increase. See ANNEALING.

graining (*Build*) The operation of brushing, combing or otherwise marking a painted surface while the paint is still wet, in order to produce an imitation of the grain of wood.

graining boards (*Print*) Boards or metal plates with parallel lines in relief, running diagonally. Used to produce a diced effect on covers.

graining comb (*Build*) See COMB.

graining plates (*Print*) The aluminium or zinc plates for lithography require a grained surface, which will retain moisture and also hold the image areas or 'work'. Plates can be used several times and regrained after each use in a graining machine, where the plate is covered with balls of porcelain, glass or steel, and, after adding water and an abrasive such as carborundum, is vibrated mechanically to produce the required grain.

grain refining (*Eng*) Production of small closely knit grains, resulting in improved mechanical properties. Particularly with aluminium alloys it is achieved by small additions to melts of substances such as boron which cause fine nucleation on casting.

grain legume (*Agri*) See PULSE.

grains (*MinExt*) See COAL SIZES.

grain size (*Geol, Eng*) The average diameter or expressed dimension of the grains or crystals in a sample of metal or rock. Also *particle size*.

grain-size analysis (*PowderTech*) US term for particle size analysis. It is also the literal translation for the Ger phrase used to specify particle size analysis. Confusion arises because grain-size analysis is also used to describe

procedures for measuring the size of crystallites in a cast or sintered metal.

grain-size control (*Eng*) Control of the rate at which grains grow when metal is heated above the recrystallization range; the control may be effected by the addition of elements which anchor grain boundaries (eg aluminium to steel) or by regulation of the temperature and time of the recrystallization process.

grain structure (*Eng*) Size, shape and orientation of crystallites forming the microstructure of most metals, alloys and ceramics.

gram (*Phys*) The unit of mass in the CGS system. It was originally intended to be the mass of 1 cm^3 of water at 4°C but was later defined as one-thousandth of the mass of the International Prototype Kilogram, a cylinder of platinum–iridium kept at Sèvres. Also *gramme*.

gram-atom (*Chem*) The quantity of an element whose mass in grams is equal to its relative atomic mass. A MOLE of atoms.

gram-calorie (*Genrl*) See CALORIE.

gramicidin (*BioSci*) A heterogeneous group of ionophores. Thus gramicidin A is an open-chain polypeptide while gramicidin S is a cyclic peptide.

graminacious (*BioSci*) Relating to grasses. Also *gramineous*.

Gramineae (*BioSci*) The grass family, comprising c.9000 spp of monocotyledonous flowering plants (superorder Commelinidae). They are annual or perennial herbs, sometimes woody (the bamboos), and often tufted or rhizomatous. They are cosmopolitan and are represented in most habitats on Earth. The aerial stems are usually hollow and bear leaves; the leaves sheathe the stem at the base and have flat, long and narrow blades, in two ranks. The flowers are relatively inconspicuous and wind pollinated (see FLORET and SPIKELET). Many tropical grasses are C4 plants, including maize. The grasses are extremely important economically for food (all the cereals, eg rice, wheat, oats, maize, etc, and sugar cane), for fodder, and also for some constructional and furniture-making materials (bamboos) and in lawns, sportsfields, etc. Also *Poaceae*.

gramineous (*BioSci*) See GRAMINACIOUS.

graminicolous (*BioSci*) Living on grasses, esp of parasitic fungi.

graminivorous (*BioSci*) Grass-eating.

gram-ion (*Phys*) The mass in grams of an ion, numerically equal to that of the molecules or atoms constituting the ion.

grammage (*Paper*) Preferred term for the mass in grams of 1 square metre of the paper or board under the prescribed conditions of test. See BASIS WEIGHT, SUBSTANCE.

grammatite (*Min*) Synonym for TREMOLITE.

gramme (*Phys*) See GRAM.

gram-molecular volume (*Chem*) See MOLAR VOLUME.

gram-molecule (*Chem*) See MOLE.

Gram-negative bacteria (*BioSci*) Those bacteria that fail to stain with Gram's reaction. The reaction depends on the complexity of the cell wall and has for long determined a major division between bacterial species. Cf GRAM-POSITIVE BACTERIA. See panel on BACTERIA.

gramophone (*Acous*) Archaic term for an instrument for reproducing sound, using a stylus in contact with a spiral groove on a revolving vinyl disk. In US this term is proprietary: the general equivalent is *phonograph*.

gramophone record (*Acous*) See DISK RECORD.

Gram-positive bacteria (*BioSci*) The comparative simplicity of the cell wall of some bacterial species allows them to be stained by Gram's procedure. Cf GRAM-NEGATIVE BACTERIA. See panel on BACTERIA.

gram-roentgen (*Phys*) The real conversion of energy when 1 roentgen is delivered to 1 g of air (approx 8·4 μJ). A convenient multiple is the MEGA-GRAM-ROENTGEN (10^6 gram-roentgens) or meg.

grams per square metre (*Paper*) The mass in grams of 1 square metre of the paper or board under the prescribed

Grand unified theories

Modern theoretical physics aims to describe the nature of the physical universe with as few assumptions and laws as possible. In a complete framework of physical theory, those parameters, which we now measure experimentally like the speed of light or the fine-structure constant, would emerge naturally from the equations. Just as Newton was able to explain Kepler's laws of planetary motion through his more powerful theory of gravitation, so in modern physics there is a desire to find all-embracing physical laws.

Physics recognizes four distinct interactions that affect matter. These are the *electromagnetic* interaction, the *strong* and *weak* interactions affecting particles and nuclei, and the *gravitational* interaction. Although gravitation is by far the weakest of these, it is always an attractive force and it acts over immense distances; that is why it is so important in COSMOLOGY (panel). Each of these interactions has its own theoretical formalism. Gravitation is now described through eg the general theory of relativity. Electromagnetic, weak and strong interactions are also described through different GAUGE THEORIES. (Gauge group theory is a powerful branch of mathematics.)

In the 1960s and 1970s, physicists such as S Weinberg, A Salam and S Glashow found ways of achieving partial unification of these different classes of interaction. *Electroweak theory* could explain most features of electromagnetism and the weak interaction. The more elaborate quantum chromodynamics described the interactions between quarks and heavy particles such as neutrons. The unified theories achieved well-publicized successes at the time by predicting new elementary particles that were eventually found by the CERN accelerator.

Grand unified theories (or GUTs) are particularly associated with H M Georgi and they aim to fold the electromagnetic, weak and strong interactions into a single gauge group. This theory predicts that the proton is not stable, although the half-life is extremely long: 10^{29} years. Experiments to measure proton decay are being attempted as an important test of the theory. The theory also attempts to describe how elementary particles with energies of 10^{21} electron-volts will behave. This energy is so colossal, with enough to run a light bulb for a minute residing on a single particle, that they will never be made in accelerators. They were, however, abundant in the first 10^{-32} seconds of the BIG BANG and it is possible that they could be found in COSMIC RAYS (panel). It is because of these astrophysical connections that astronomers, as well as physicists, are interested in GUTs. A more distant goal is the unification of gravitation also, to produce a viable theory of quantum gravity.

See panel on COSMOLOGY.

---

conditions of test. An expression of the basis weight or substance (now preferred term) of paper.

grandfather file (*ICT*) Along with the *father file* and the *son file*, one of the three most recent versions of a file that is periodically updated, retained for security purposes.

grand mal (*Med*) General convulsive epileptic seizure, with loss of consciousness. See EPILEPSY, PETIT MAL.

grand period of growth (*BioSci*) The period in the life of a plant, or of any of its parts, during which growth begins slowly, gradually rises to a maximum, gradually falls off, and comes to an end, even if external conditions remain constant.

grand swell (*Acous*) A swell-like balanced pedal for bringing in, as it is depressed, all the stops in an organ in a graded series.

grand unified theory (*Phys*) A theory which simultaneously describes the four forces of nature: strong nuclear, weak nuclear, electromagnetic and gravitational. Although such a unification has not yet been achieved, the WEINBERG–SALAM THEORY (or electroweak theory) has successfully unified the electromagnetic and weak interactions. Abbrev *GUT*. See panel on GRAND UNIFIED THEORIES.

granite (*Geol*) A coarse-grained igneous rock containing megascopic quartz, averaging 25%, much feldspar (orthoclase, microcline, sodic plagioclase), and mica or other coloured minerals. In the wide sense, granite includes alkali-granites, adamellites and granodiorites, while the *granite clan* includes the medium- and fine-grained equivalents of these rock types. See PLUTONIC ROCKS.

granite-aplite (*Geol*) See APLITE.

granite-porphyry (*Geol*) Porphyritic microgranite, a rock of granitic composition but with a groundmass of medium grain size in which larger crystals (phenocrysts) are embedded.

granite series (*Geol*) A series relating the different types of granitic rock with respect to their time and place of formation in an orogenic belt, starting with early formed, deep-seated autochthonous granites and ending with post-tectonic high-level plutons.

granitic finish (*Build*) A surface finish, resembling granite, given to cement work by the use of a suitable face mix.

granitic texture (*Geol*) See GRANITOID TEXTURE.

granitization (*Geol*) A metamorphic process by which rocks can be changed into granite *in situ*.

granitoid texture (*Geol*) A rock fabric in which the minerals have no crystal form and occur in shapeless interlocking grains. Such rocks are in the coarse grain-size group. Also *xenomorphic granular texture*.

granoblastic texture (*Geol*) An arrangement of equigranular mineral grains in a rock of metamorphic origin similar to that of a normal granite but produced by recrystallization in the solid and not by crystallization from a molten condition. The grains show no preferred orientation.

granodiorite (*Geol*) An igneous rock of coarse grain size, containing abundant quartz and at least twice the amount of plagioclase over orthoclase, in addition to coloured minerals such as hornblende and biotite. See PLUTONIC ROCKS.

granolithic (*Build*) A rendering of cement and fine granite chippings, used as a covering for concrete floors, on which it is floated in a layer 1–2 in (25–50 mm) in thickness. Used because of its hard-wearing properties.

granophyre (*Geol*) An igneous rock of medium grain size, in which quartz and feldspar are intergrown as in graphic granite.

granular (*PowderTech*) Said of particles having an approximately equidimensional but irregular shape.

granular pearlite (*Eng*) PEARLITE in which the CEMENTITE occurs as globules instead of as lamellae. Produced by very slow cooling through the critical range, or by subsequent heating just below the critical range. Also *globular pearlite*.

granularity (*ICT*) The fineness with which available bandwidth can be subdivided in a given transmission system.

granularity (*ImageTech*) See GRAIN.

granulation (*Astron*) The mottled appearance of the Sun's PHOTOSPHERE when viewed at very high resolution; small, bright, rapidly evolving granules (around 1000 km in diameter) appear against the darker background.

granulation tissue (*Med*) A new formation of vascular connective tissue which grows to fill up the gap of a wound or ulcer; when healing is completed, a white scar is left.

granule (*Geol*) A rock or mineral with a grain size between 2 and 4 mm.

granules (*Chem*) Main form in which a compounded polymer is supplied for further processing to shape.

granulite (*Geol*) A granular-textured metamorphic rock, a product of regional metamorphism.

granulitic texture (*Geol*) The texture of a granulite, sometimes referred to as *granulose* or *granoblastic*, is an arrangement of shapeless interlocking mineral grains resembling the granitic texture but developed in metamorphic rocks. Fewer than 10% of the grains have a preferred orientation.

granulitization (*Geol*) The process in regional metamorphism of reducing the components of a solid rock to grains. If the reduction of the size of the particles goes farther, mylonite is produced.

granulocyte (*BioSci*) A general term describing polymorphonuclear leucocytes.

granulocytopenia (*Med*) An abnormal diminution in the number of granulocytes in the blood.

granuloma (*BioSci*) A localized accumulation of macrophages around the site of some continuing stimulus, such as a persisting antigen that causes delayed-type hypersensitivity and the release of lymphokines chemotactic to monocytes (which turn into macrophages). The macrophages may fuse together so as to form multinucleated giant cells. Adj *granulomatous*.

granuloma annulare (*Med*) A condition in which rings of white cellular nodules appear on the back of the hands, and occasionally elsewhere.

granuloma inguinale (*Med*) A chronic disease occurring in the tropics, in which ulcerating nodules appear on the genital organs, the perineum and the groin. Also *granuloma venereum, ulcerating granuloma*.

granulomatous disease (*Med*) (1) Chronic X-linked genetic disease in which phagocytic cells lack the ability to kill invading bacteria. Patients suffer from repeated bacterial infection and abscesses in many tissue sites. (2) One of a group of diseases characterized by granuloma formation in tissues. See CROHN'S DISEASE, SARCOIDOSIS.

granum (*BioSci*) A stack, rather like a pile of coins, of c.5–50 THYLAKOIDS (or disks) forming one of say 40–60 such bodies in the STROMA of a chloroplast in vascular plants, bryophytes and some green algae.

grape-sugar (*Chem*) GLUCOSE.

graph (*MathSci*) A representation, generally a drawing depicting the relationship between two or more variables. The relationship may be a mathematical function, eg the graph of the equation $x^2 + y^2 = 1$ is a circle, or it may be a physical phenomenon, eg the graph of a hospital patient's temperature against time.

graphecon (*Radar*) A double-ended storage tube used for the integration and storage of radar information, and as a translating medium.

graphene (*Chem*) A single planar sheet of carbon atoms, packed in an hexagonal array, the two-dimensional counterparts of three-dimensional graphite or equivalent to unrolled carbon nanotube. They appear to have interesting electrical properties.

graphical display unit (*ICT*) Output device, incorporating a cathode-ray tube, on which line drawings and text can be displayed. Used in conjunction with a LIGHT PEN to input or reposition data.

graphical methods (*Eng*) Methods in which items, such as forces in structures, are determined by drawing diagrams to scale.

graphical user interface (*ICT*) A method of interacting with a computer and its software by means of selecting ICONS rather than issuing COMMANDS; the advantage is that the interface is more intuitive. It will often include WINDOWS and PULL-DOWN MENUS and some sort of pointer, often controlled by a MOUSE. Taken together this is called a WIMP environment. This approach was pioneered by Apple Computers Inc. in the early 1980s based on work from Xerox Park Laboratories. Abbrev *GUI*. See fig. at WINDOWS.

graphic formula (*Chem*) A formula in which every atom is represented by the appropriate symbol, valency bonds being indicated by dashes, eg H–O–H, the graphic formula for water.

graphic granite (*Geol*) Granite of pegmatitic facies, in which quartz and alkali feldspar are intergrown in such a manner that the quartz simulates runic characters. Also *runite*.

graphics (*ICT*) See COMPUTER GRAPHICS, GRAPHICS ADAPTER, RASTER GRAPHICS, VECTOR GRAPHICS.

graphics accelerator (*ICT*) A hardware device generally on a circuit board that is dedicated to increasing the speed and performance of GRAPHICS. Graphics accelerators calculate PIXEL VALUES and write them into a BUFFER, thus freeing the CPU for other operations.

graphics adapter (*ICT*) A CIRCUIT BOARD that is used to drive a particular standard VIDEO DISPLAY UNIT. Common standards in IBM-COMPATIBLE computers are CGA, EGA and VGA.

graphics card (*ICT*) A printed circuit board that stores visual data (in digital form) and conveys them to the display screen.

graphics mode (*ICT*) A mode of operation for VIDEO DISPLAY UNITS whereby graphics images consisting of sets of PIXELS may be displayed. See CGA.

graphics resolution (*ICT*) The number of PIXELS making up a graphics image as in the VGA standard with up to 640 pixels wide by 480 pixels high. See RESOLUTION.

graphics tablet (*ICT*) An input device where the movement of a pen over a sensitive pad is translated into digital signals giving the pen's position.

graphic statics (*Eng*) A method of finding the stresses in a framed structure, in which the magnitude and direction of the forces are represented by lines drawn to a common scale.

graphic texture (*Geol*) A rock texture in which one mineral intimately intergrown with another occurs in a form simulating ancient writing, esp runic characters; produced by simultaneous crystallization of two minerals present in eutectic proportions. See GRAPHIC GRANITE.

graphite (*Chem, Min*) One of the two naturally occurring forms of crystalline carbon, the other being diamond. It occurs as black, soft masses and, rarely, as shiny crystals (of flaky structure and apparently hexagonal) in igneous rocks; in larger quantities in schists, particularly in metamorphosed carbonaceous clays and shales, and in marbles; also in contact-metamorphosed coals and in meteorites. Graphite has numerous applications in trade and industry now much overshadowing its use in 'lead' pencils; in particular, ultrapure graphite is used as a MODERATOR in nuclear reactors. Much graphite is now produced artificially in electric furnaces using petroleum as a starting material. Also *black lead, plumbago*. See COLLOIDAL GRAPHITE, GRAPHENE.

graphite brush (*ElecEng*) A brush, made of graphite, for collecting the current from the commutator of an electric machine. It has a higher conductivity and better lubricating properties than an ordinary carbon brush.

graphite paint (*Build*) Paint consisting of fine graphite and bitumen solution, with or without oil, and supplied chiefly for the protection of hot wet surfaces such as the insides of industrial boilers.

graphite reactor (*NucEng*) One in which fission is produced principally or substantially by slow neutrons moderated by graphite. See figs at FUEL ELEMENT, REACTOR. See ADVANCED GAS-COOLED REACTOR, MODERATOR.

graphite resistance (*ElecEng*) A resistance unit consisting of a rod of graphite, which has a high ohmic value; also a variable resistance made up of piles of graphitized disks of cloth under a variable pressure.

graphitic acid (*Chem*) Graphite which has been treated with nitric acid and potassium (V) chlorate for a prolonged period. It is an *intercalation compound*.

graphitic carbon (*Eng*) In CAST-IRON, carbon occurring as GRAPHITE instead of cementite.

graphitization (*Chem*) The transformation of amorphous carbon to graphite brought about by heat. It results in a volume change due to the alteration in atomic lattice layer spacing. It is reversible under bombardment by high-energy neutrons and other particles. See WIGNER EFFECT.

graph theory (*MathSci*) The study of discrete structures consisting of vertices and edges which connect them. See fig. at BIPARTITE GRAPH. See COLOURING OF A GRAPH.

grapnel (*Eng*) Any device used to grapple with an object which is obscured to view, eg underwater. Grapnels generally take the form of grapnel hooks having several flukes. Also *grappel*.

grapnel (*MinExt*) An extracting tool used in boring operations. Also *grappel*.

graptolite (*Geol*) An extinct group of animals, represented by their abundant fossil remains in Palaeozoic rocks, and used for dating Ordovician and Silurian sediments. Generally thought to be CHORDATA.

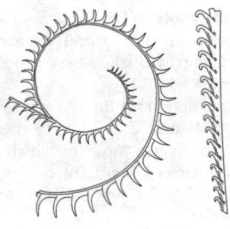

graptolite Typical fossils from the Silurian.

grass (*BioSci*) See GRAMINEAE.

grass (*Radar*) Irregular deflection from the time base of a radar display, arising from electrical interference or noise.

grass disease (*Vet*) See BOVINE HYPOMAGNESAEMIA.

grassland (*EnvSci*) Natural, semi-natural or farm vegetation in which the dominant species are grasses. Major grasslands of the world include pampas, prairies and savannas.

Grassot fluxmeter (*ElecEng*) See FLUXMETER.

grass sickness (*Vet*) A disease of unknown cause affecting the horse, often occurring soon after the horse is put onto grass, associated with dysfunction of the bowels. Acute cases are characterized by colic and more chronic cases by emaciation. The essential pathological change is believed to be degeneration of the neurons in the ganglia of the autonomic nervous system.

grass staggers (*Vet*) See BOVINE HYPOMAGNESAEMIA.

grass table (*Build*) A GROUND TABLE.

grass tetany (*Vet*) See BOVINE HYPOMAGNESAEMIA.

grate (*Eng*) That part of a furnace which supports the fuel. It consists of FIRE BARS or bricks so spaced as to admit the necessary air.

grate area (*Eng*) The area of the grate in a furnace burning solid fuel; for a boiler furnace, a measure of the evaporative capacity of the boiler.

graticule (*Surv*) See RETICULE.

grating (*Phys*) An arrangement of alternate reflecting and non-reflecting elements, eg wire screens or closely spaced lines ruled on a flat (or concave) reflecting surface, which, through diffraction of the incident radiation, analyses this into its frequency spectrum. An *optical grating* can contain a thousand lines or more per centimetre. A STANDING-WAVE system of high-frequency sound waves with their alternate compressive and rarefied regions can give rise to a diffraction grating in liquids and solids. With a criss-cross system of waves, a three-dimensional grating is obtainable.

grating spectrum (*Phys*) An optical spectrum produced by a DIFFRACTION GRATING. Cf PRISMATIC SPECTRUM.

Grätz rectifier (*ElecEng*) A type of bridge rectifier using six rectifying elements for three-phase supply.

gravel (*Build*) A natural mixture of rounded rock fragments and generally sand used in the manufacture of concrete.

gravel (*Geol*) An aggregate consisting dominantly of pebbles, though usually a considerable amount of sand is intercalated. The grain size variously defined as 2–20 mm. In the Stratigraphical Column, gravels of different ages and origins occur abundantly, eg in SE England, where they consist chiefly of well-rounded flint pebbles originally derived from the Chalk. These gravels are mainly of fluviatile and fluvioglacial origin, but marine gravels are also common in the littoral zone. The indurated equivalent of gravel is conglomerate. See PARTICLE SIZE, WENTWORTH SCALE.

gravel (*Med, Vet*) Small calculi in the kidneys, ureters or urinary bladder.

gravel board (*Build*) A long board standing on its edge at the bottom of a wooden fence, so that the upright boards of the fence do not have to reach down to the ground. Also *gravel plank*.

graveolent (*BioSci*) Having a strong rank odour.

Grave's disease (*Med*) An auto-immune disease of the thyroid gland. Auto-antibodies are produced that act against receptors for thyroid-stimulating hormone, resulting in thyroid hyperactivity. This results in excessive secretion of thyroid hormone, causing THYROTOXICOSIS with loss of weight, sweating, rapid pulse and prominent eyes (exophthalmic goitre).

graveyard (*NucEng*) See BURIAL SITE.

gravid (*BioSci*) Pregnant; carrying eggs or young.

gravimeter (*Phys*) An instrument for measuring variations in gravity at points on the Earth's surface.

gravimetric analysis (*Chem*) The chemical analysis of materials by the separation of the constituents and their estimation by weight.

gravimetry (*Phys*) The measurement of weight or density.

graving dock (*CivEng*) See DRY DOCK.

gravipause (*Space*) A point or border in space in which the gravitational force of one body is matched by the counter-gravity of another. Also *neutral point*.

graviperception (*BioSci*) The perception of gravity by plants.

gravitation (*Phys*) The force of nature which manifests itself as a mutual attraction between masses, and whose mathematical expression was first given by Newton, in the law which states: 'Any two particles of matter attract one another with a force directly proportional to the product of their masses and inversely proportional to the square of the distance between them.' This may be expressed by the equation

$$F = G\frac{m_1 m_2}{d^2}$$

where F is the force of gravitational attraction between bodies of mass m_1 and m_2, separated by a distance d. G is the constant of gravitation, equal to $6\cdot670 \times 10^{-11}$ N m^2 kg^{-2}.

gravitational astronomy (*Astron*) See CELESTIAL MECHANICS.

gravitational collapse (*Astron*) The contraction of a celestial body due to its own gravitational attraction.

gravitational constant (*Phys*) The constant of proportionality in Newton's law of gravitation, equal to $6.670 \times 10^{-11}\ \mathrm{N\ m^2\ kg^{-2}}$.

gravitational differentiation (*Geol*) The production of igneous rocks of contrasted types by the early separation of denser crystals such as olivine, pyroxenes, etc, which become concentrated in the basal parts of intrusions. The ultramafic rocks such as peridotites and picrites may have been formed in this way.

gravitational field (*Phys*) A region of space in which at all points a gravitational force would be exerted on a test particle.

gravitational lensing (*Astron*) The effect whereby the image of a distant quasar or other astronomical body can be distorted when the gravity of an extremely massive intervening object bends the light rays, sometimes producing multiple images. See panel on QUASAR.

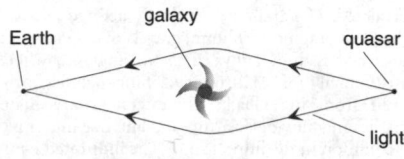

Earth   galaxy   quasar

light

gravitational lensing

gravitational radiation (*Astron*) Very weak gravity waves produced when a massive body is disturbed or accelerated. The phenomenon is predicted by the general theory of relativity, but not yet observed with certainty.

gravitational redshift (*Astron*) The effect whereby light loses energy and increases its wavelength when it is emitted from a region containing a strong gravitational field. Also *Einstein shift*.

gravitational system of units (*Phys*) A system of units in which time, length and force form the fundamental quantities; the basic unit of force is defined as the gravitational force on a standard object at a certain location on Earth's surface.

gravitational wave (*Phys*) A hypothetical progressive energy-carrying wave whose existence was postulated by Einstein in 1916. Also *gravity wave*.

graviton (*Phys*) The hypothesized carrier of gravitational interactions between particles. The graviton has not been detected but is believed to have zero mass and zero charge. It bears the same relationship to the gravitational field as the photon does to the electromagnetic field. Gravitational interactions between particles are so weak that no quantified effects have been observed.

gravity (*Phys*) The force of attraction between bodies as a result of their mass.

gravity assist (*Space*) Use of a planet to provide an energy boost to a spacecraft on a trajectory planned to pass close to the body for this purpose.

gravity-controlled instruments (*ElecEng*) Electrical measuring instruments in which the controlling torque is provided solely by the action of gravity.

gravity conveyor (*Eng*) A conveyor for articles with weight enough to allow them to move unassisted down eg an inclined runway.

gravity dam (*CivEng*) A dam which is prevented by its own weight from overturning.

gravity diecasting (*Eng*) A process by means of which castings of various alloys are made in steel or cast-iron moulds, the molten metal being poured by hand. See DIECASTING, PRESSURE DIECASTING.

gravity drop hammer (*Eng*) A type of machine hammer used for drop forging, in which the impact pressure is

obtained from the kinetic energy of the falling ram and die. Also *board hammer*.

gravity feed gun (*Build*) A spray gun for use with gravity feed containers having the fluid connection on the upper gun body.

gravity feed tanks (*Aero, Autos*) Fuel tanks, situated above the point of delivery to the engine, which feed the engine solely by gravity.

gravity plane (*MinExt*) An inclined plane on which the descending full trucks pull up the ascending empty ones.

gravity process (*Glass*) See GOB PROCESS.

gravity roller conveyor (*Eng*) A fabricated framework, usually in longitudinal sections and slightly inclined, carrying freely rotating tubular rollers on which a load will move under gravity.

gravity scale (*Eng*) A measure of certain characteristics of diesel fuel oil, related to the density of the oil by an arbitrary formula.

gravity separation (*MinExt*) The use of differences between relative densities of roughly sized grains of mineral to promote settlement of denser species while less dense grains are washed away. See BUDDLES, DENSE MEDIA PROCESS, HUMPHREY SPIRALS, HYDROCYCLONES, JIG, SHAKING TABLE, SLUICE, VANNER.

gravity stabilization (*Space*) A way of stabilizing the orientation of a spacecraft with reference to a primary body (such as Earth) by using the gravity gradient, in which the long axis of the spacecraft is directed towards the centre of the body.

gravity stamp (*MinExt*) Set of five heavy pestles, lifting about 10 in at 90 times a minute by a cam and allowed to fall on ore spread in a mortar box, for crushing purposes. Also *Californian stamp*.

gravity tectonics (*Geol*) Processes of rock deformation and folding which are activated by gravity applied over considerable periods of geological time.

gravity transport (*Geol*) The movement of material under the influence of gravity. It includes downhill movement of weathering products, and movement of unweathered material in landslides.

gravity water system (*Phys, NucEng*) A system in which flow occurs under the natural pressure due to gravity, incorporated as part of the safety procedures for flooding nuclear reactors.

gravity wave (*Phys*) (1) A liquid surface wave controlled by gravity and not surface tension. In meteorology, they are atmospheric transverse waves in which the restoring force is due to the effect of gravity on pressure and density fluctuations. Examples are LEE WAVES. (2) See GRAVITATIONAL WAVE.

gravure (*Print*) Abbrev for PHOTOGRAVURE.

Grawitz's tumour (*Med*) See HYPERNEPHROMA.

gray (*Radiol*) SI unit of absorbed dose; $1\ \mathrm{J\ kg^{-1}}$. Symbol Gy. See panel on RADIATION.

Gray code (*ICT*) A modification of a number in binary code in which consecutive integers are represented by binary numbers differing in only one digit, the absolute difference being the value 1.

Gray–King Test (*MinExt*) A test of coking quality of coal under prescribed conditions of heating to 600°C.

graywacke (*Geol*) See GREYWACKE.

grazing angle (*Phys*) A very small GLANCING ANGLE.

grease (*Vet*) A chronic seborrhoeic dermatitis affecting horses, associated with the caudal aspect of the pastern and fetlock. Usually associated with poor management. Also *greasy-heels, sore-heels.*

grease cup (*Eng*) A lubricating device consisting of an externally threaded cylindrical cup. Screwing down an internally threaded cap forces grease into the bearing on which the grease cup is fixed.

grease gun (*Eng*) A device for forcing grease into bearings under high pressure. It consists of a cylinder from which the grease is delivered by hand pressure on the piston, intensified by a second plunger which forms the delivery pipe and which is pressed against a nipple screwed into the bearing.

greaseproof (*Paper*) A quality of paper possessing grease-resistant characteristics brought about by heavy beating of a suitable fibrous furnish.

greaseproof paper (*Paper*) Any paper that in its natural state or as the result of coating or other treatment resists penetration by oils or greases.

grease-spot photometer (*Phys*) A simple means of comparing the intensities of two light sources. A screen of white paper, rendered partially translucent by a spot of grease, is illuminated normally by the two sources, one on each side. The position of the screen is adjusted until the grease spot is indistinguishable from its surround, when the illuminations on the two sides may be assumed to be equal. Also *Bunsen photometer*. See PHOTOMETER.

grease table (*MinExt*) Sloping table anointed with petroleum jelly, over which diamondiferous concentrate is washed, the diamond adhering strongly while the gangue is worked away.

grease trap (*Build*) A trapped gulley receiving sink wastes, and specially designed to prevent obstruction of gulley or drain by congealed fatty matter.

greasy pig disease (*Vet*) Infection of piglets with *Staphylococcus hyicus*. An exudative epidermitis in which the skin becomes covered with a greasy brown exudate which dries and then peels off. Mortality high if left untreated. Also *marmite disease*.

great circle (*MathSci*) The intersection of a sphere with a plane passing through its centre. The shortest distance between two points on the surface of a sphere is along the great circle passing them.

greatest common divisor (*MathSci*) The largest integer which exactly divides given integers a and b, which may be calculated using EUCLID'S ALGORITHM. Often written $\gcd(a, b)$, $\mathrm{hcf}(a, b)$ or in short form (a, b). Also *greatest common factor, highest common factor*. Abbrev: *gcd, gcf, hcf*.

Great Ice Age (*Geol*) See PLEISTOCENE.

great primer (*Print*) An old type size, approximately 18-point.

Great Red Spot (*Astron*) See JUPITER.

Great Wall (*Astron*) A giant wall of galaxies, discovered during a sky survey, which forms a structure of larger scale than a supercluster.

greeked text (*ICT*) A method of displaying text common in DESKTOP PUBLISHING programs whereby a simple symbol is substituted for the actual text for the purpose of reducing the time taken to display the image while retaining the appearance of the layout.

green (*CivEng*) Colloq term for concrete in the hardening stage, after pouring and before setting.

green (*PowderTech*) Describes the unheat-treated condition of preforms held together by the cohesive forces resulting from compaction alone or with the assistance of a temporary binder.

green algae (*BioSci*) The CHLOROPHYTA.

greenalite (*Min*) A septechlorite related to the chlorites chemically and to the serpentines structurally. A hydrous, iron silicate of composition $Fe_6Si_4O_{10}(OH)_3$.

Green Book (*ImageTech*) The technical specifications that define the CD-I standard.

green bricks (*Build*) Moulded clay shapes which after undergoing a burning process will become bricks.

Greenburg–Smith impinger (*PowderTech*) Device for sampling gas streams, consisting of a tall, flat-bottomed flask containing a vertical tube narrowing to a jet at the lower end, which is immersed in dust-free water. The gas stream is sucked through the jet and any dust particles are collected in the water.

green carbonate of copper (*Min*) See MALACHITE.

green compact (*Eng*) Ceramic *preforms* before sintering, which are held together by the cohesive forces resulting from compaction alone or with the assistance of a temporary binder.

green flash (*Astron*) A phenomenon sometimes seen in clear atmospheres at the instant when the upper rim of the Sun finally disappears below the horizon as a bright blob of light; green is the last apparent colour from the Sun, because the more greatly refracted blue is dispersed.

green fluorescent protein (*BioSci*) The protein from luminous jellyfish *Aequorea victoria* that, when excited by blue light, emits green light. The gene has been extensively used as a reporter or in recombinant proteins as a marker. Abbrev *GFP*.

green glands (*BioSci*) The antennal excretory glands of decapod Crustacea.

greenheart (*For*) Very strong, dense, yellowish-green to black timber of an evergreen HARDWOOD tree (*Ocotea*) from S America

greenhouse effect (*Astron, EnvSci*) Phenomenon by which thermal radiation from the Sun is trapped by water vapour and carbon dioxide on a planet's surface, thus preventing its re-emission as long-wave radiation. This leads to the temperature at the planet's surface being considerably higher than would otherwise be the case. The effect is most pronounced for Venus and the Earth.

greenhouse gases (*EnvSci*) Collective term for gases that contribute to the greenhouse effect. The main gases involved (with their relative contribution in parentheses) are carbon dioxide (50%), chlorofluorocarbons (15%), methane (20%), ozone (10%), nitrous oxide and water vapour (5%), some being more damaging on a per molecule basis than others.

Greenland spar (*Min*) See CRYOLITE.

green manure (*Agric*) A method of increasing soil organic matter by planting a crop on temporarily free land and ploughing it in while still green.

greenockite (*Min*) Cadmium sulphide occurring as small, yellow hexagonal crystals in cavities in altered basic lavas.

green sand (*Eng*) Highly siliceous sand, with some alumina, dampened and used with admixed coal or charcoal in moulding.

greensand (*Geol*) A sand or sandstone with a greenish colour due to the presence of the mineral glauconite.

green sand casting (*Eng*) One made in foundry sand not strengthened by kiln drying before pouring.

Green's theorem (*MathSci*) (1) If C is the boundary curve of plane region R, then

$$\int_C (Pd_x + \varphi dy) = \int_R \left(\frac{\partial \varphi}{\partial x} - \frac{\partial P}{\partial y} \right) dx\, dy$$

Also known as *Gauss's theorem* or *Ostrogradski's theorem*. It is a case of *Stokes's theorem*. (2) If S is the boundary surface of volume V, then

$$\int_S F \cdot dS = \int_V \operatorname{div} F\, dV$$

Also known as *Green's lemma, Gauss's theorem, Ostrogradski's theorem* or the *divergence theorem*. (3) With S and V as above,

$$\int_V (P \nabla^2 \varphi + \varphi \nabla^2 P)\, dV = \int_S \left(P \frac{\partial \varphi}{\partial n} - \varphi \frac{\partial P}{\partial n} \right) dS$$

where n is in the direction normal to S. Cf GAUSS'S THEOREM, STOKES'S THEOREM.

greenstick fracture (*Med*) Fracture of immature bones, which break like a green stick with one side of the bone broken and the other bent.

greenstone (*Geol*) An omnibus term lacking precision and applied indiscriminately to basic and intermediate igneous rocks in which much chlorite has been produced as a result of metamorphism.

green strength (*Eng*) See STRENGTH MEASURES.

green tyre (*Eng*) Tyre in unvulcanized state ready for final shaping. See panel on TYRE TECHNOLOGY.

green vitriol (*Min*) See MELANTERITE.

Greenwich Mean Time (*Astron*) MEAN SOLAR TIME referred to the zero meridian of longitude, ie that through Greenwich; the basis for scientific and navigational purposes. Abbrev *GMT*. See UNIVERSAL TIME.

gregaria phase (*BioSci*) In locusts (Orthoptera), a phase characterized by high activity and gregarious tendencies, differing morphologically from the SOLITARIA PHASE with which, under natural conditions, it alternates.

Gregorian calendar (*Astron*) The civil calendar now used in most countries of the world. It is the Julian calendar as reformed by decree of Pope Gregory XIII in 1582. The Gregorian reform omitted certain leap years, and brought the length of the year on which the calendar is based nearer to the true astronomical value.

Gregorian telescope (*Astron*) A form of reflecting telescope, very similar in principal to the Cassegrain telescope, in which the large mirror is pierced at the centre and the light is reflected back into an eyepiece in the centre by a small concave mirror on the principal axis and a little outside the focus.

Gregory formula (*MathSci*) Obtained from the Newton interpolation formula, used for numerical integration. It takes the form

$$\int_a^b f(x)dx = \frac{h}{2}(y_0 + 2y_1 + 2y_2 + \cdots + 2y_{n-1} + y_n)$$
$$- \frac{h}{12}(\Delta y_{n-1} - \Delta y_0) - \frac{h}{24}(\Delta^2 y_{n-2} + \Delta^2 y_0)$$
$$- \frac{19h}{720}(\Delta^3 y_{n-3} - \Delta^3 y_0) - \frac{3h}{160}(\Delta^4 y_{n-4} - \Delta^4 y_0) - \cdots$$

where h is the common interval between successive values of x, and $\Delta^m y_{n-m}$ represent finite differences.

greige (*Textiles*) See GREY.

greisen (*Geol*) A rock composed essentially of mica and quartz, resulting from the alteration of a granite by percolating solutions. Greisens often contain small amounts of fluorite, topaz, tourmaline, cassiterite, and other relatively uncommon minerals, and may be associated with mineral deposits, as in Cornwall. See PNEUMATOLYSIS.

greisenization (*Geol*) The process by which granite is converted to GREISEN. Greisenization is a common type of wall rock alteration in areas where granite is traversed by hydrothermal veins. Also *greisening*.

grenz rays (*Radiol*) X-rays produced by electron beams accelerated through potentials of 25 kV or less. These are generated in many types of electronic equipment; low penetrating power.

grey (*ImageTech*) Neutral in colour, having no perceptible hue and zero saturation; in colour reproduction, a balance of the component colours to give this subjective impression.

grey (*Textiles*) Fabrics in the state they leave the loom or knitting machine before any scouring or bleaching has been carried out.

grey body (*Phys*) A body which behaves as a black body in respect of the absorbed fraction of the incident radiation, with $E_a = E_i(1-A)$ where E_a is the absorbed fraction, E_i incident energy, and A albedo, and complies with the STEFAN–BOLTZMANN LAW.

grey copper ore (*Min*) See TETRAHEDRITE.

grey iron (*Eng*) Pig or cast-iron in which nearly all the carbon not included in PEARLITE is present as graphite carbon. See MOTTLED IRON, WHITE IRON.

grey key image (*ImageTech*) A neutral image printed in shades of black and white used to supplement the primary colour images in colour reproduction, extending the density range and adding visibility.

grey matter (*BioSci*) The centrally situated area of the central nervous system, mainly composed of cell bodies. Cf WHITE MATTER.

greyscale (*ImageTech*) A series of shades of grey ranging from white to black used as a test object in reproduction

processes as a measure of greyness in a monochrome image.

greywacke (*Geol*) A sandstone containing silt, clay and rock fragments in addition to quartz grains. It is much more poorly sorted than other types of sandstone, and often occurs on beds which show gradation in grain size from fine at the top to coarse at the bottom. See TURBIDITE.

greywethers (*Geol*) Grey-coloured rounded blocks of sandstone or quartzite left as residual boulders on the surface of the ground when less resistant material was denuded. From a distance they resemble sheep grazing. See SARSEN.

grid (*Build*) The plan layout of the structure of a given building.

grid (*ElecEng*) The electrical power generation, transmission and, to a lesser extent, distribution system. See ACCUMULATOR GRID, DAMPER RESISTANCE GRID, SUPERGRID.

grid (*Electronics*) Control electrode having an open structure (eg mesh) allowing the passage of electrons; in an electron gun it may be a hole in a plate.

grid (*Eng*) A grating made up of a number of parallel bars, such as that required to prevent foreign matter from entering a pump intake.

grid (*Surv*) A network of lines superimposed upon a map and forming squares for referencing, the basis of the network being that each line in it is at a known distance either east or north of a selected origin. In ordnance surveying, triangulation system covering large area.

grid bias (*Electronics*) The dc negative voltage applied to the control grid of a thermionic valve.

grid-bias battery (*ElecEng*) Once widely used to provide a potential difference for the grid polarization of valves. US C-battery.

grid capacitor (*ElecEng*) A capacitor between the grid and the remainder of the grid circuit.

grid circuit (*ElecEng*) The circuit connected between the grid and the cathode of a thermionic valve.

grid conductance (*Phys*) The in-phase component of the grid input admittance, due eg to grid current, Miller effect.

grid control (*Electronics*) That provided by voltage on grid of a thyratron or mercury-arc rectifier; at a sufficient positive voltage, anode current flows and grid loses control until deionization is effected after loss of anode voltage.

grid-controlled mercury-arc rectifier (*ElecEng*) One in which initiation of arc in each cycle is determined by phase of voltage on a grid for each anode.

grid drive (*ElecEng*) Voltage or power required to drive a valve when delivering a specified load.

grid-glow tube (*Electronics*) Cold-cathode gas-discharge tube in which glow is triggered by a grid.

grid modulation (*Electronics*) See SUPPRESSOR-GRID MODULATION.

grid navigation (*Aero, Ships*) A navigational system in which a grid instead of true north is used for the measurement of angles and for heading reference. The grid is of parallel lines referenced to the 180° meridian which is taken as Grid North. Used principally in polar route flying and, as the system is independent of convergence of meridians, headings remain the same from departure to destination.

grid neutralization (*ElecEng*) The method of neutralization of an amplifier through an inverting network in the grid circuit, which provides the requisite phase shift of 180°. Also *phase inverter*.

grid pool tube (*Electronics*) A tube having a mercury-pool cathode, one or more anodes, and a control electrode or *grid* which controls the onset of current flow in each cycle. The *excitron* and *ignitron* are examples.

grid ratio (*Radiol*) In grid therapy, the ratio of the total area of holes to that of the grid. Also ratio of lead strips to radiolucent plastic used in radiography to reduce scattered radiation effect on film.

grid reference (*Genrl*) A series of numbers and letters used to indicate the precise location of a place on a map.

grid resistance (*ElecEng*) A resistance unit for heavy-current work, eg starters for railway motors; it is made up of a number of resistance grids placed side by side and mounted in a metal frame.

grid therapy (*Radiol*) A method of treatment in radiotherapy, in which radiation is given through a grid of holes in a suitable absorber, eg lead, rubber. By careful positioning, a greater depth dose can be given, and skin breakdown reduced to a minimum.

grief (*Psych*) A deep sense of sorrow and suffering caused by bereavement. In excessive prolongation it may develop into overt depressive illness.

grief stem (*MinExt*) See KELLY.

Griffin mill (*MinExt*) A pendulum mill in which a vertically suspended roller bears against a stationary bowl as it rotates, crushing passing ore.

Griffith equation (*Eng*) Relates the TENSILE STRENGTH, σ_t, of a material to the critical flaw size, or crack length, a_c for an edge crack or $2a_c$ for an internal crack, required for fracture, ie failure by crack propagation:

$$\sigma_t = \sqrt{E'G_c/\pi a_c}$$

where G_c is the TOUGHNESS of the material. For PLANE STRAIN, $E' = E$, the YOUNG'S MODULUS of the material, while for PLANE STRESS,

$$E' = \frac{E}{1 - v^2}$$

where v is POISSON'S RATIO. See STRENGTH MEASURES.

Grignard reagents (*Chem*) Mixed organometallic compounds prepared by dissolving dry magnesium ribbon or filings in an absolutely dry ethereal solution of an alkyl bromide or iodide. The value of Grignard reagents in the preparation of secondary or tertiary alcohols, and the alkenes, so easily produced therefrom, and in the synthesis of alkyl and aryl derivatives from many halogen compounds, can hardly be overestimated.

grike (*Geol*) A fissure in limestone rock caused by the solvent action of rainwater. See KARST.

grill (*Build*) A layer of joists in a GRILLAGE FOUNDATION.

grillage foundation (*Build*) A type of foundation often used at the base of a column. It consists of one, two or more tiers of steel beams superimposed on a layer of concrete, adjacent tiers being placed at right angles to each other, while all tiers are encased in concrete.

grille (*Build*) A plain or ornamental openwork of wood or metal, used as a protecting screen or grating.

grindability (*MinExt*) Empirical assessment of response of ore to pulverizing forces, applied under specified conditions.

grinder (*Paper*) Equipment for producing mechanical wood pulp by holding logs of wood against the surface of a revolving natural or synthetic grindstone kept wet by water showers.

grinder's rot (*Med*) Lung disease caused by inhalation of metallic particles by steel-grinders.

grinding (*MinExt*) Comminution of minerals by dry or, more usually, wet methods, mainly in rod, ball or pebble mills.

grinding-in (*Eng*) The process of obtaining a pressure-tight seal between a conical-faced valve and its seating by grinding the two together with an abrasive mixture such as silicon carbide and oil.

grinding machine (*Eng*) A machine tool in which flat, cylindrical or other surfaces are produced by the abrasive action of a high-speed grinding wheel. See CENTRELESS GRINDING, CYLINDRICAL GRINDING, PROFILE GRINDING, SURFACE GRINDING MACHINE, THREAD GRINDING.

grinding medium (*Eng*) The solid charges (balls, pebbles, rods, etc) used in suitable mills for grinding certain materials, eg cement, pigments, etc, to a fine powder.

grinding wheel (*Eng*) An abrasive wheel for cutting and finishing metal and other materials. It is composed of abrasive particles, such as silicon carbide (carborundum) or emery, held together by a bonding or binding agent, which may be either vitrified or a softer material, eg shellac or rubber.

GRIN glass (*ImageTech*) Abbrev for *gradient refractive index glass*. Used in lens ELEMENTS to vary the REFRACTIVE INDEX from edge to centre to improve their optical properties.

grinning through (*Build*) A paintwork defect characterized by the fact that the HIDING POWER of the paint is insufficient to obscure completely the surface underneath.

grip (*Build*) (1) A small channel cut to carry away rainwater during construction of foundations. (2) Small channel across the roadside to conduct surface water to a drain. Also *grippers, offlet*.

grip (*ImageTech*) A member of a camera crew moving equipment or camera mountings.

gripper edge (*Print*) See GRIPPERS.

gripper-feed mechanism (*Eng*) A type of strip-feeding mechanism used in presswork, which advances the strip material during its forward stroke, and then returns to its initial position with grippers open, and thus inoperative.

grippers (*Print*) Attachments which grip the edge (gripper edge) of a sheet of paper when it is fed into the printing machine.

gripper-shuttle (*Textiles*) A projectile that grips the weft thread and inserts it into the fabric being manufactured on a weaving machine.

grisaille (*Arch*) Decorative painting in monochromatic colours to give the impression of design in relief.

griseofulvin (*Pharmacol*) An oral antibiotic which is excreted in the cells of the skin, thus destroying any cutaneous fungus infections.

grit (*Eng*) A measure indicating the sizes of the abrasive particles in a grinding wheel, usually expressed by a figure denoting the number of meshes per linear inch in a sieve through which the particles will pass completely.

grit (*Geol*) Siliceous sediment, loose or indurated, the component grains being angular. Sometimes applied to a hard coarse-grained sandstone.

grit (*PowderTech*) Hard particles, usually mineral, of natural or industrial origin, retained on a 200 mesh BS test sieve (76 μm).

grit blasting (*Eng*) A process used in the preparation for metal spraying or other coating, which cleans the surface and gives it the roughness required to retain the coating.

grit cell (*BioSci*) A STONE CELL occurring in a leaf or in the flesh of a fruit, eg pear.

grit chamber (*Build*) A DETRITUS CHAMBER.

grizzle bricks (*Build*) Bricks which are underburnt and of bad shape. They are soft inside and unsuitable for good work but are often used for the inside of walls. Also *place bricks, samel bricks*.

grizzly (*MinExt*) The set of parallel bars or grating used for the coarse screening of ores, rocks, etc.

grog (*Build*) Bricks or waste from a clayworks broken down and added to clay to be used for brick manufacture.

groin (*Arch*) The line of junction of the two constituent arches in a GROINED ARCH.

groined arch (*Arch*) An arch which is intersected by other arches cutting across it transversely.

groin rib (*Arch*) A projecting member following the line of a groin.

grommet (*Eng*) A ring or collar used to line a sharp-edged hole through which a cable or similar material passes.

grooming (*ICT*) A process in which the signals on a number of channels, each carrying mixed traffic, eg video, data and telephony, are interchanged so that each channel carries only one type of traffic, with a consequent increase in efficiency.

grooming (*Psych*) Maintenance of and attention to all aspects of the body surface; it can be performed by individuals to their own bodies, or between members of a dyad or larger social group.

groove (*Print*) The separation between the board and the spine of the book cover, permitting free opening.

grooving (*Eng*) (1) Cracking of the plates of steam boilers at points where stresses are set up by the differential expansion of hot and colder parts. (2) Producing a rectangular, V-shaped or similar groove or channel by milling, shaping, grinding, etc, to provide eg a location for a reciprocating slide or as a lubricant reservoir.

grooving plane (*Build*) One with a narrow, interchangeable blade and adjustable fence, for cutting grooves.

grooving saw (*Build*) A circular saw which may be of the drunken type, used for cutting grooves.

gros nez (*Med*) See GOUNDOU.

gross energy requirement (*Eng*) Total energy expended in manufacture of different materials. Highly electropositive metals (eg aluminium, titanium) possess high-energy contents and recycling after use is widely practised. Polymers possess intermediate-energy contents (process energy + feedstock energy), which are greater than that of glass or paper.

gross register tonnage (*Ships*) See GROSS TONNAGE.

gross tonnage (*Ships*) The cubic capacity (100 ft^3 = 1 ton) of all spaces below the freeboard deck and permanently closed-in spaces above that deck, except EXEMPTED SPACES.

grossular (*Min*) An end-member of the garnet group, the composition being represented by $Ca_3Al_2Si_3O_{12}$; formed in the contact metamorphism of impure limestone. Commonly contains some iron, and is greenish, brownish or pinkish. Also *gooseberry stone*. See HYDROGROSSULAR.

gross weight (*Aero*) The BASIC WEIGHT of an aircraft plus the weight of additional loads such as fuel and crew.

grosswetterlage (*EnvSci*) The sea-level pressure distribution averaged over a period during which the essential characteristics of the atmospheric circulation over a large region remain nearly unchanged.

gross wing area (*Aero*) The full area of the wing, including that covered by the fuselage and any nacelles.

grotesque (*Print*) The name given to early sans serif types and still used, esp the contraction *grot*.

Grotthus–Draper law (*Chem*) Only such energy of electromagnetic radiation as is absorbed by tissue is effective in chemical action following ionization in that tissue.

ground (*ICT*) US for EARTH.

ground (*MinExt*) The mineralized deposit and rocks in which it occurs, eg payground, payable reef; barren ground, rock without values.

ground (*Textiles*) (1) In LACE manufacture, the mesh which forms a foundation for a pattern. (2) The base fabric to which is secured the figuring, threads, eg pile or loops in carpet or terry cloths.

ground absorption (*ICT*) The energy loss in radio-wave propagation due to absorption in the ground.

ground air (*Build*) In land used for sewage treatment, the air contained in the upper layers in the subsoil; it has a variable composition, including carbon dioxide, ammonia and other gases resulting from oxidation of organic matters, and may be noxious.

ground auger (*Build*) An auger specially adapted for boring holes in the ground, eg for artesian wells.

ground capacitance (*Phys*) See EARTH CAPACITANCE.

ground clutter (*Radar*) The effect of unwanted ground-return signals on the screen pattern of a radar indicator.

ground coat (*Build*) A base coat of paint on which further treatments such as glazing, graining or marbling will be worked.

ground control (*Radar*) Control of an aircraft or guided missile in flight from the ground.

ground-controlled approach (*Aero, Radar*) Aircraft landing system in which information is transmitted by a *ground controller* from a ground radar installation at end of runway to a pilot intending to land. Also *talk-down*. Abbrev *GCA*.

ground-controlled interception (*Aero, Radar*) Radar system whereby aircraft are directed onto an interception course by a station on the ground. Abbrev *GCI*.

grounded base (*Electronics*) See COMMON-BASE CONNECTION.

grounded-cathode amplifier (*Electronics*) One with the cathode at zero alternating potential, drive on the grid and power taken from the anode; the normal and original use of a triode valve.

grounded circuit (*ElecEng*) A circuit which is deliberately connected to earth at one point or more, for safety or testing.

grounded collector (*Electronics*) See COMMON-COLLECTOR CONNECTION.

grounded emitter (*Electronics*) See COMMON-EMITTER CONNECTION.

grounded-grid amplifier (*Electronics*) One with the grid at zero alternating voltage, drive between cathode and earth, output being taken from the anode; there is no anode–grid feedback.

ground effect (*Genrl*) The result of interaction between directly transmitted sound and that which has been reflected from the ground. Depending upon the nature of the ground surface, there can be attenuation of particular frequencies.

ground engineer (*Aero*) An individual, selected by the licensing authorities, who has power to certify the safety for flight of an aircraft, or certain specified parts of it. Term now superseded by *licensed aircraft engineer*.

ground fine pitch (*Aero*) A very flat blade angle on a turboprop propeller, which gives extra braking drag and low propeller resistance when starting the engine. Colloq *disking*.

ground frost (*EnvSci*) A temperature of 0°C (32°F) or below, on a horizontal thermometer in contact with the shorn grass tips on a turf surface. See AIR FROST.

ground joist (*Build*) A horizontal timber supported off the ground at a basement or ground-floor level.

ground-level (*Surv*) The REDUCED LEVEL of a survey station with reference to an official bench mark.

ground loop (*Aero*) An uncontrollable and violent swerve or turn by an aircraft while taxiing, landing or taking off.

groundmass (*Geol*) In igneous rocks which have crystallized in two stages, the groundmass is the finer-grained portion, in which the phenocrysts are embedded. It may consist wholly of minute crystals, wholly of glass, or partly of both.

ground meristem (*BioSci*) Partly differentiated meristematic tissue (PRIMARY MERISTEM) derived from the apical meristem and giving rise to ground tissue.

ground mould (*CivEng*) A timber piece or frame used as a template to bring earthworks such as embankments to the required form.

ground noise (*Acous*) See BACKGROUND NOISE.

ground plan (*Build, CivEng*) A drawing showing a plan view of the foundations for a building or of the layout of rooms etc on the ground floor.

ground plate (*Build*) The bottom horizontal timber to which the frame of a building is secured.

groundplot (*Aero*) A method of calculating the position of an aircraft by relating groundspeed and time on course to starting position.

ground-position indicator (*Aero*) An instrument which continuously displays the dead-reckoning position of an aircraft.

ground ray (*ICT*) See DIRECT RAY, WAVE.

ground reflection (*Radar*) The wave in radar transmission which strikes the target after reflection from the Earth.

ground resonance (*Aero*) A sympathetic response between the dynamic frequency of a rotorcraft's rotor and the natural frequency of the alighting gear which causes rapidly increasing oscillations.

ground return (*Radar*) The aggregate sum of the radar echoes received after reflection from the Earth's surface. May include CLUTTER.

ground run (*Aero*) The distance that an aircraft travels down the runway before lift-off or after touching down before reaching a stop.

grounds (*Build*) Strips of wood which are nailed to a wall or partition (fixing plugs being used when necessary) as a basis for the direct attachment of joinery.

ground safety lock (*Aero*) See RETRACTION LOCK.

ground sill (*Build*) A SLEEPER.

ground sills (*Build*) Underwater walls built at intervals across the bed of a channel to prevent excessive scour of the bed or to increase the width of flow.

ground sluicing (*MinExt*) Bulk concentration of heavy minerals *in situ*, by causing a stream of water to flow over unconsolidated alluvial ground with just enough force to flush away the lighter, less valuable sands leaving the heavier ones to be removed for further treatment.

groundspeed (*Aero*) The speed of an aircraft or missile relative to the ground and not to the surrounding medium. Cf AIRSPEED.

ground state (*Phys*) The state of a system of particles such as a nucleus or atom when at its lowest energy, ie not excited. Also *normal state*.

ground support equipment (*Aero*) All the handling facilities employed to service aircraft on the airport, eg tractors, steps, fuelling tankers, food and cleaning supplies. Also weapon trolleys and installation check-out gear for military aircraft.

ground system (*ICT*) See EARTH SYSTEM.

ground table (*Build*) The course of stones at the foundation of a building. Also *grass table*.

ground tissue (*BioSci*) Tissue other than vascular tissue, epidermis and periderm, mostly parenchyma and collenchyma of eg pith and cortex.

ground water (*Build*) Water naturally contained in, and saturating, the subsoil.

ground water (*Geol*) Water occupying space in rocks. It may be *juvenile*, having arisen from a deep magmatic source, or *meteoric*, the result of rain percolating into the ground. Also *groundwater*.

ground wave (*Phys*) See WAVE.

groundwood (*Paper*) Pulp produced in a grinder.

group (*Chem*) (1) A vertical column of the periodic system, containing elements of similar properties. (2) Metallic radicals which are precipitated together during the initial separation in qualitative analysis. (3) A number of atoms which occur together in several compounds.

group (*Geol*) A stratigraphic rock unit consisting of two or more FORMATIONS.

group (*ICT*) (1) A collection of documents, APPLICATION PROGRAMS or FILES represented as ICONS within a WINDOW in a graphical environment. (2) An assembly of telephone speech channels, 48 kHz wide, comprising 12 channels of 4 kHz each, in a FREQUENCY-DIVISION MULTIPLEX system.

group (*MathSci*) A set together with an operation * that satisfies the following four conditions: (1) The set G is *closed* with respect to *; ie for all elements x,y in G, $x*y$ is an element of G. (2) The operation * in G is *associative*; ie for all elements x, y, z in G, $(x*y)*z = x*(y*z)$. (3) The set G has an *identity element* with respect to *; ie there is an element in G such that, for all elements x in G, $x*e = x = e*x$. (4) Each element in G has an *inverse* with respect to *; ie for each element x in G there is an element x' in G such that $x*x' = e = x'*x$. A group (G;*) is said to be an *abelian group* if the operation * in G is *commutative*; ie if, for all elements x, y in G, $x*y = y*x$. The real numbers with addition, the non-zero real numbers with multiplication, and the three-dimensional vectors with vector addition are abelian groups.

group delay (*ICT*) For any signal travelling through a network or along a transmission line, the derivative of phase with respect to frequency. Where the medium is non-dispersive, group delay is equal to envelope display; in real situations, the phase response of a device or transmission path nearly always varies with frequency, so that group delay becomes a complex quantity.

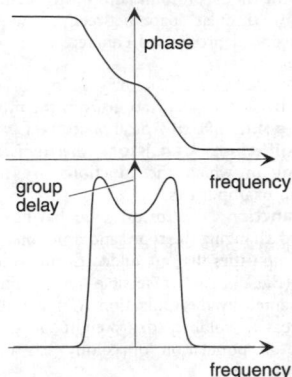

group delay Typical bandpass filter, shown relative to its phase response.

group dynamics (*Psych*) The collective interactions that occur within a small social group.

group icon (*ICT*) The ICON that represents a GROUP when the WINDOW is itself minimized to a single icon.

group mixer (*ICT*) See MIXER. Also *group fader*.

group modulation (*ICT*) Use of one carrier for transmitting a group of communication channels, with demodulation on reception and ultimate separation. Carrier side frequencies may represent different groups.

group polarization (*Psych*) The tendency for a cohesive group to make more extreme decisions than its members would as individuals.

group reaction (*Chem*) The reaction by which members of a GROUP are precipitated.

group selection (*BioSci*) A form of natural selection proposed to explain the evolution of behaviour which appears to be for the long-term good of a group or species, rather than for the immediate advantage of the individual.

group theory (*NucEng*) Approximate method for the study of neutron diffusion in a reactor core, in which neutrons are divided into a number of velocity groups in which they are retained before transfer to the next group. See ONE-group, TWO-GROUP, MULTIGROUP THEORY.

group therapy (*Psych*) Any form of psychotherapy involving several persons at the same time, such as a small group of patients with similar psychological or physical problems who discuss their difficulties under the chairmanship of eg a doctor. They thus learn from the experiences of others and teach by their own.

group think (*Psych*) A breakdown in rational decision making when members of a cohesive group try to reach consensus decisions; the desire for unanimity is greater than the concern for a good solution.

group velocity (*Phys*) The velocity of energy propagation for a wave in a dispersive medium. Given by

$$v_g = \left(\frac{d\omega}{dk} \right) = v - \lambda \frac{d\chi}{d\lambda}$$

where $k = 2\pi/\lambda$, $\omega = 2\pi \times$ frequency, v is the phase velocity and λ is the wavelength.

groupware (*ICT*) Computer systems or SOFTWARE that are specifically designed to be used by a number of USERS simultaneously all carrying out related tasks.

grouse disease (*Vet*) A popular term for the specific infection of the intestines of grouse by the nematode worm *Trichostrongylus pergracilis*.

grout (*CivEng*) See CEMENT GROUT.

grouting (*Build, CivEng*) The process of injecting cement grout for strengthening purposes. Also *cementation*.

grouting (*Eng*) (1) Setting a machine foot or base to a required level by filling the space between it and the supporting floor or foundation with cement grout. (2) Filling the annular space between a stressing tendon in post-tensioned prestressed concrete work with cement or other grout.

growing (*Electronics*) The production of semiconductor crystals by slow crystallization from the molten state.

growing point (*BioSci*) Apical MERISTEM.

grown diffusion transistor (*Electronics*) A junction transistor in which the junctions are formed by the diffusion of impurities.

grown junction (*Electronics*) One having junctions produced by changing the types and amounts of acceptor and donor impurities that are added during GROWING.

growth (*BioSci*) An irreversible change in an organism accompanied by the utilization of material, and resulting in increased volume, dry weight or protein content; increase in population or colony size of a culture of micro-organisms.

growth (*Eng*) Permanent increase in size of a component, leading to distortion, which occurs when materials are in service over a long period. For example, common cast-iron fire bars, subjected to thermal cycling in the range 700–900°C or held for long periods above 480°C, suffer extensive distortion and cracking due to an increase in SPECIFIC VOLUME resulting from the breakdown of iron carbides to graphite in the microstructure. The cracking leads to internal oxidation which further adds to the damage.

growth (*Phys, NucEng*) (1) Elongation of fuel rods in reactor under irradiation. (2) Build-up of artificial radioactivity in a material under irradiation, or of activity of a daughter product as a result of decay of the parent.

growth cone (*BioSci*) A specialized region at the tip of a growing nerve process, whose outward growth towards a target cell is influenced or inhibited by various physical or molecular factors.

growth curvature (*BioSci*) A curvature in an elongated plant organ, caused by one side growing more than the other.

growth factor (*BioSci*) Any of a group of polypeptide hormones that regulate the division of cells, eg PDGF (platelet-derived growth factor), NGF (NERVE GROWTH FACTOR), TGF (transforming growth factor).

growth form (*BioSci*) See LIFE FORM.

growth hormones (*BioSci*) Substances having growth-promoting properties, eg pituitary growth hormone; in plants, *auxins*.

growth inhibitor (*BioSci*) A substance that inhibits plant growth at low concentrations, esp an endogenous substance, of which the best characterized is ABSCISIC ACID.

growth in soft agar (*BioSci*) The ability of cells in culture to grow in a low-concentration gel of agar. It is one of the properties that distinguish a TRANSFORMED CELL from normal cells that can only grow in culture in contact with a substratum upon which they can spread.

growth movement (*BioSci*) Movement of a plant part brought about by differential growth. Cf TURGOR MOVEMENT.

growth regulator (*BioSci*) A GROWTH SUBSTANCE, esp one of the synthetic types.

growth retardant (*BioSci*) A synthetic substance used to retard the growth of a plant, eg to stop the sprouting of stored onions or to restrict the height of grain crops, eg maleic hydrazide.

growth ring (*BioSci*) A recognizable increment of wood (secondary xylem) in a cross-section of a stem; most commonly an ANNUAL RING, but under some conditions more than one (or no) growth ring is formed within one year. See EARLY WOOD, LATE WOOD.

growth room (*BioSci*) A room in which plants are grown under controlled artificial lighting, photoperiod, temperature, etc.

growth substance (*BioSci*) One of a number of substances (sometimes called *hormones*) formed in plants, or a synthetic analogue thereof, that have specific effects on plant growth or development at low concentrations (say >10 μM). See ABSCISIC ACID, AUXIN, CYTOKININ, ETHYLENE, GIBBERELLIN.

groyne (*Build*) Low wall, usually of timber, steel sheet piles or concrete running out from high-water springs to just below low water to prevent or reduce lateral drift of beach material.

GR-S (*Plastics*) Abbrev for GOVERNMENT RUBBER-STYRENE.

grub saw (*Build*) A handsaw for cutting marble, having a steel blade stiffened along the back with wooden strips.

grub screw (*Eng*) A small headless screw used without a nut to secure a collar or similar part to a shaft.

grummet (*Build*) Hemp and red-lead putty mixed as a jointing material for watertightness.

Grüneisen's relation (*Phys*) The coefficient of volume expansion of a solid β, is given by $\beta = \gamma\kappa C$, where κ is the compressibility, C is the specific heat capacity of unit volume and γ is Grüneisen's constant which for most materials lies between 1 and 2 and is practically independent of temperature.

grunerite (*Min*) A monoclinic calcium-free amphibole; a hydrated silicate of iron and magnesium, differing from cummingtonite in having Fe > Mg. Typically found in metamorphosed iron-rich sediments.

Grus (Crane) (*Astron*) A southern constellation.

gryke (*Geol*) See GRIKE.

GSE (*Aero*) Abbrev for GROUND SUPPORT EQUIPMENT.

gsm (*Print*) Abbrev for *grams per square metre*.

GSR (*Psych*) Abbrev for GALVANIC SKIN RESPONSE.

G-string (*Phys*) Colloq a single-wire transmission line loaded with dielectric so that surface-wave propagation can be employed (named after Dr Groubau).

g-suit (*Aero*) See ANTI-G SUIT.

g tolerance (*Space*) The tolerance of an object or person to a given value of *g* force.

GTP (*BioSci*) Guanosine 5′-triphosphate. Like ATP a source of phosphorylating potential, but particularly important for the activation of regulatory G-proteins (GTP-BINDING PROTEINS).

GTP-binding protein (*BioSci*) Heterotrimeric G-proteins associated with receptors of the seven transmembrane domain superfamily that are involved in signal transduction; small cytoplasmic G-proteins play an important part in regulating many intracellular processes including cytoskeletal organization and secretion. Also *G-proteins*.

guag (*MinExt*) The space left after the mineral has been extracted. Also *gunis, gunnice, gunnies, gunnis*.

guaiacol (*Chem*) $HOC_6H_4OCH_3$. The monomethyl ether of *catechol* (*1-methoxybenzene 2-hydroxybenzene*), found in beechwood tar; a very unstable compound, with strong reducing properties. Used in medicine and in veterinary practice as an expectorant.

guanidine (*Chem*) $HNC(NH_2)_2$. Imido-urea or the amidine of amidocarbonic acid; a crystalline compound, easily soluble in water, strongly basic.

guanine (*Chem*) Purine base which occurs in DNA and RNA, pairing with cytosine. See panel on DNA AND THE GENETIC CODE.

guanophore (*BioSci*) See XANTHOPHORE.

guanosine (*BioSci*) 9-β-D-ribofuranosyl guanine. The nucleoside formed by linking ribose to guanine.

guard (*Build, Eng*) A protection on a scaffold to prevent persons from falling; a fence or other safety device on a machine, to prevent injury to the operator or others from gears, cutting tools, etc.

guard (*ICT*) (1) Signal that prevents accidental operation by spurious signals or avoids ambiguity. (2) The wire accompanying a speaking pair through an automatic telephone exchange; it is earthed while the speaking pair is being used by subscribers, thus indicating that the speaking pair cannot be engaged by any other circuit. Also *guard wire*.

guard band (*ImageTech, ICT*) (1) In radio and TV transmission, an additional frequency band on each side of the allocated band to reduce interference from adjacent channels. (2) In magnetic recording, an unused space between tracks to avoid cross-talk.

guard cell (*BioSci*) One of a pair of specialized epidermal cells that surround and, by increase and decrease in their turgor, open and close a stoma, thus regulating GAS EXCHANGE.

guard circle (*Acous*) Inner groove on vinyl disk recording which protects stylus from being carried to centre of turntable.

guard cradle (*ElecEng*) A network of wires serving the same purpose as a GUARD WIRE. Also *guard net*.

guard magnet (*MinExt*) A strong magnet, usually suspended above moving stream of lump ore, to remove steel from broken drills etc (TRAMP IRON), which might damage the crushing machine.

guard net (*ElecEng*) See GUARD CRADLE.

guard rail (*CivEng*) See CHECK RAIL.

guard ring (*ElecEng*) Auxiliary electrode used to avoid distortion of electric (or heat) field pattern in working part of a system as a result of the EDGE EFFECT, or to bypass leakage current through insulator to earth in an ionization chamber.

guard-ring capacitor (*ElecEng*) A standard capacitor consisting of circular parallel plates with a concentric ring maintained at the same potential as one of the plates to minimize the edge effect.

guards (*Print*) (1) Narrow strips of paper or linen projecting between sections in a book, for the attachment of plates, maps, etc. (2) Strips of paper sewed into a book to compensate for the thickness of inserted plates.

guard wire (*ElecEng*) An earth wire used on an over-head transmission line; it is arranged in such a position that, should a conductor break, it will immediately be earthed by contact with the wire. See PRICE'S GUARD WIRE.

guard wire (*ICT*) See GUARD.

guayule (*For*) A bush which occurs in the US and Mexico and is a potential source of natural rubber.

gubernaculum (*BioSci*) (1) In mammals, the cord support-ing the testes, in the scrotal sac. (2) In Hydrozoa, an ectodermal strand supporting the gonophore in the gonotheca. (3) In Mastigophora, a posterior flagellum used in steering. Adj *gubernacular*.

Gudden–Pohl effect (*Phys*) A form of electroluminescence which follows metastable excitation of a PHOSPHOR by ultraviolet light.

Guddermanian (*MathSci*) A curious function defined by

$$\mathrm{gd}\,x = \int_0^x \mathrm{sech}\,\theta\,d\theta$$

It then follows that if $u = \mathrm{gd}\,x$, then $\sin u = \tanh x$ and $\cos u = \mathrm{sech}\,x$, and that

$$x = \int_0^u \sec\theta\,d\theta$$

gudgeon (*Build*) The pin fastened to a gatepost or door frame about which the leaf of a strap hinge turns.

gudgeon (*Eng*) A pivot at the end of a beam or axle on which a bell, wheel, etc, works.

gudgeon pin (*Autos*) The pin connecting the piston with the bearing of the little end of the connecting-rod. Also *piston pin*. See FLOATING GUDGEON PIN.

Guerin process (*Eng*) Used in presswork to cut or form sheet metal by placing it between a die made of a cheap material and a thick rubber pad which adapts itself to the die while under pressure Fig. ▷.

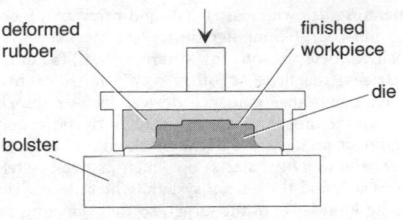

Guerin process

guest (*BioSci*) An animal living and/or breeding in the nest of another animal, as a Myrmecophile in an ant's nest.

GUI (*ICT*) Abbrev for GRAPHICAL USER INTERFACE.

guidance (*Aero*) See GUIDED MISSILE.

guide bars (*Eng*) (1) Bars with flat or cylindrical surfaces provided to guide the crosshead of a steam engine, and so avoid lateral thrust on the piston rod. Also *motion bars*, *slide bars*. (2) Any bars used as guides to control one machine element in relation to another.

guide bead (*Build*) A bead fixed to the inside of a cased frame as a guide for the sliding sash. Also *inner bead*, *parting bead*.

guided atmospheric flight (*Aero*) Refers to any unpiloted and *aeronautical* flight. Includes: (1) guided missiles, eg the unpowered *guided bomb*, which has aerodynamic steering; the unpowered *smart bomb*, with terminal guidance; the low-flying CRUISE MISSILE; the *ground to air* ground-launched anti-aircraft missile, which includes the anti-missile missile; the *ground to ground* missile, which includes the intercontinental ballistic missile (eg Atlas), for attacking ground targets; the *air to sea* air-launched missile for attacking shipping (eg Exocet); the *air to ground* missile for anti-tank, anti-radar installation or other ground target attack; the *underwater to ground* submar-ine-launched ballistic missile (eg Polaris, Trident); (2) remotely piloted military aircraft (RPV), which are small and propeller-driven for photoreconnaissance, returning to catch-nets for reuse; (3) drones, which include expendable radio-controlled unmanned target vehicles used for the development of experimental aircraft or weapon systems, or for training operational crews; and (4) radio-controlled model aircraft used for leisure or aerial photography. *Propulsion* may be by RAMJET, TURBOJET or solid- or liquid-fuelled ROCKET, or combinations of these. The missile body contains fuel, warhead, control and guidance systems, and self-destruct and other safety systems similar to those for a GUIDED MISSILE. See INERTIAL GUIDANCE, ROCKET PROPULSION, STAND-OFF BOMB, STRATEGIC DEFENSE INITIATIVE.

guided missile (*Aero*) *Guided weapon*, UK. Strategic and tactical unmanned weapon which is guided to its target; propulsion is usually by RAMJET, ROCKET or simplified short-life TURBOJET. Guided missiles are broadly divided into categories using the initials of the launch and target media (Air, Surface or Underwater) followed by M for missile, eg ASM or SSM. In addition there are the ICBM (*Inter-Continental Ballistic Missile*) and IRBM (*Intermedi-ate Range Ballistic Missile*) which, as their names imply, are guided on only part of their journeys and end in a ballistic descent through the atmosphere. Guidance systems vary greatly: *direct command guidance* is control entirely from the launcher by radio or wire signals; *radar command guidance* is radio-signal guidance from a lock-on radar/computer system on the launcher; *beam rider* is a missile which follows a radar beam directly from launcher to target; *semi-active homing* is the radar 'illumination' of the target, on the reflection from which the missile 'homes'; *collision course homing* is similar, but employs an offset missile aerial to bring it onto a converging course with the target; *fully-active homing* is self-contained, the missile generating its own radar signals and carrying a lock-on or collision-course computer device; *passive homing* is by a sensitive detector on to infrared, heat, sound, static electricity, magnetic or other wave emissions from the target. Long-range guidance devices include the *celestial*, wherein the missile's automatic astronavigation equipment is given a preflight programme relating its course to the target and to a fixed star(s) on which its tracking telescope is focused; and the *inertial* system, which depends upon a precise knowledge of the target, so that the course can be planned by the variation of inertial forces, which are followed by the internal device which can measure and correct minute variations in gravitational, and other, forces. The latter is the only known system completely independent of jamming or other external interference. See CRUISE MISSILE, STAND-OFF BOMB.

guided wave (*Phys*) Electromagnetic or acoustic wave which is constrained within certain boundaries as in a waveguide.

guide field (*ElecEng*) That component of field particles in a cyclotron or betatron which maintains particles in their intended path.

guide mill (*Eng*) A rolling mill equipped with guides to ensure that the stock enters the mill at the correct point and angle.

guide pulley (*Eng*) A loose pulley used to guide a driving belt past an obstruction or to divert its direction. Also *idler pulley*.

guide rail (*CivEng*) See CHECK RAIL.

guides (*MinExt*) Timbers, ropes or steel rails at sides of shaft used to steady the cage or skip.

guide track (*ImageTech*) Sound track, usually speech or dialogue, recorded during picture shooting under adverse recording conditions, to serve as a guide for subsequent replacement in the studio.

guide-vanes (*Aero*) A general term for aerofoils which guide the airflow in a duct. Also *cascades*. See IMPELLER-INTAKE VANES, NOZZLE VANES, TOROIDAL-INTAKE VANES.

guide wavelength (*Phys*) The wavelength in a guide, operated above the cut-off frequency: $1/\lambda_g^2 = 1/\lambda_0^2 - 1/\lambda_{0c}^2$, where λ_g is the guide wavelength, λ_0 the wavelength in the unbounded medium at the same frequency, and λ_{0c} the wavelength in the unbounded medium at the cut-off frequency for the mode in question.

guild (*BioSci*) (1) A division or category of a plant species from one area, made on the basis of similar PHENOLOGY and morphology. See SYNUSIA. (2) Of an animal community, a group of species within the same trophic level which exploits a common resource, in a similar manner.

Guillain–Barre syndrome (*Med*) An acute or subacute disorder of peripheral nerves causing progressive muscular weakness which may even paralyse the muscles of respiration. Often thought to be related to a recent viral illness but AETIOLOGY is not certain. Provided the patient is kept alive during acute phase full recovery is usual.

Guillemin effect (*Phys*) The tendency of a bent magnetostrictive rod to straighten in a longitudinal magnetic field.

Guillemin line (*ICT*) A network designed to produce a nearly square pulse with a steep rise and fall.

guillotine (*Med*) An instrument for cutting off tonsils.

guillotine (*Paper*) A machine having a heavy steel blade, used for cutting or trimming stacks of paper, or for trimming books.

guinea-pig paralysis (*Vet*) A viral infection of cavies characterized by a diffuse meningomyeloencephalitis.

Guinier–Preston zones (*Eng*) A clustering of solute atoms on certain crystallographic planes in supersaturated solid solutions which occurs during precipitation hardening. Two types are found, referred to as GP I in the initial stage followed by GP II at an intermediate stage of the hardening process. See PRECIPITATION HARDENING.

gula (*BioSci*) In vertebrates, the upper part of the throat.

gular (*BioSci*) In some fish, a bone developed between the rami of the lower jaw; in Chelonia, an anterior unpaired element of the plastron.

gulching (*MinExt*) The noise which generally precedes a fall or settlement of overlying strata.

Guldberg and Waage's law (*Chem*) See LAW OF MASS ACTION.

Guldin's theorem (*MathSci*) See PAPPUS'S THEOREM.

gullet (*BioSci*) The oesophagus; in Protozoa the cytopharynx.

gullet (*Build*) A depression cut in the face of a saw in front of each tooth, alternately on each side of the blade.

gulleting (*CivEng*) The process of excavating road or railway cuttings on a series of steps worked simultaneously. See GULLET.

gullet saw (*Build*) A saw with gullets cut in front of each tooth. Also *brier-tooth saw*.

gulley (*Build*) A fitting installed at the upper end of a drain, to receive the discharge from rainwater or waste pipes. Abbrev G.

gulley trap (*Build*) A device installed at a GULLEY to imprison foul air within the drain pipe. Also *yard trap*.

gulose (*Chem*) A monosaccharide belonging to the group of aldohexoses.

gum (*BioSci*) Viscid plant secretion exuding naturally or on wounding, soluble or swelling in water. Mostly complex, branching polysaccharides. Some, eg gum arabic, alginic acid, agar, are economically important.

gum (*For*) See AMERICAN RED GUM, RED RIVER GUM.

gum arabic (*Chem*) A fine, yellow or white powder, soluble in water, rel.d. 1·355. It is obtained from certain varieties of acacia, the world's main supply coming from Sudan and Senegal. Used in pharmacy for making emulsions and pills; also in glues and pastes. Also *acacia gum, Senegal gum*.

gum-boil (*Med*) A small abscess on the palate, associated with a carious tooth, or the result of infection following upon local injury.

gum dichromate process (*ImageTech*) The use of gum as a vehicle for pigments and dichromate on printing papers; the exposed image being developed by water.

gum-lac (*BioSci*) An inferior type of lac, containing much wax, produced by some lac insects (Hemiptera) in Madagascar.

gumma (*Med*) A mass of cellular granulation tissue, due to syphilitic infection in the late or tertiary stage. Adj *gummatous*.

gummed paper (*Paper*) Paper coated with a moisture-activated adhesive, dextrin, gum arabic, etc.

gumming up (*Print*) Application of film of gum arabic to the surface of a lithographic plate, thinning down and drying it evenly, to protect and enhance the hydrophilic properties on the non-image areas.

gummosis (*BioSci*) A pathological condition of plants characterized by the conspicuous secretion of gums.

gums (*BioSci*) In higher vertebrates, the thick tissue masses surrounding the bases of the teeth.

gums (*Chem*) Non-volatile, colloidal plant products which either dissolve or swell up in contact with water. On hydrolysis, they yield certain complex organic acids in addition to pentoses and hexoses.

gum streaks (*Print*) Streaks across a lithographic plate caused through improper gumming-up of the plate. If the gum is too thick and not thinned down, it can attack the plate image and cause it to lose ink receptivity.

gun (*Build*) See SPRAY GUN.

gun (*Electronics, ImageTech*) Assemblage of electrodes, comprising cathode, anode, focusing and modulating electrodes, from which the electron beam is emitted before being subjected to deflecting fields in a cathode-ray or TV tube.

guncotton (*Chem*) A NITROCELLULOSE (cellulose hexanitrate) with a high nitrogen content. It burns readily and explodes when struck or strongly heated. Used for explosives.

gun current (*Electronics*) In electron beam tubes, the total electron current flowing to the anode, only part of which forms the beam current.

gun drill (*Eng*) A trepanning drill or a centre-cut drill used for deep-hole drilling.

gunis (*MinExt*) See GUAG.

gunite (*CivEng*) A finely graded cement concrete (a mixture of cement and sand), which is sprayed into position under air pressure to produce a dense, impervious adherent layer; used to line or repair existing works, or to build up dense concrete in inaccessible areas of operation.

gunmetal (*Eng*) A copper–tin alloy (ie bronze) either Admiralty gunmetal (copper 88%, tin 10%, zinc 2%) or copper 88%, tin 8% and zinc 4%. Lead and nickel are frequently added, and the alloys are used as cast where resistance to corrosion or wear is required, eg in bearings, steam-pipe fittings, gears. See COPPER ALLOYS.

Gunn effect (*Electronics*) A pair of electrodes connected across a small piece of n-type gallium arsenide can, when the applied voltage exceeds a certain level, demonstrate a negative resistance effect, due to the electron velocity actually reducing when the electric field exceeds about 3 kV cm^{-1}. Installed in a suitable tuned circuit or cavity, a Gunn-effect diode can operate as an oscillator from 500 Hz to well over 50 GHz.

gunnice (*MinExt*) See GUAG. Also *gunnies*.

gunning (*Build*) The forcible application of refractory, sound insulating or corrosion-resistant linings with a gun, usually operated by compressed air.

gunnis (*MinExt*) See GUAG.

gunstock stile (*Build*) See DIMINISHED STILE.

Gunter's chain (*Surv*) A chain having overall length of 66 ft (= 20·116 m): 1 acre equals 10 square chains; and 1 chain is 1/10th of a furlong.

Günz (*Geol*) The name of an early glacial stage of the Pleistocene epoch in the Alps. See QUATERNARY.

Guppy (*Aero*) Aircraft modified by substituting a very-large-diameter section for a major part of the fuselage. Nose and tail hinge sideways for carrying large indivisible loads as in Boeing 377. Airbus major assemblies are delivered this way. Also *Pregnant Guppy, Super Guppy*.

gurley (*Paper*) The reading, generally expressed in seconds, obtained from testing a paper for air resistance utilizing the Gurley densometer.

Gurley densometer (*Paper*) A test instrument for measuring the air resistance of a sample of paper. It consists of two concentric upright cylinders, the outer being closed at the bottom and containing oil, the inner open at both ends and having a clamping device at the top. After raising the inner cylinder and placing the sample in position it is allowed to fall under its own weight, its progress being retarded by the entrapped air which passes through the sample at a rate dependent on the resistance of the paper. The time in seconds is recorded for the passage of a given volume of air, eg 100 ml is called the *gurley*.

gusher (*Geol*) See GEYSER.

gusset (*Eng*) A bracket or stay, cast or built up from plate and angle, used to strengthen a joint between two plates which meet at a joint, as the junction of a boiler shell with the front and back plates, or between connecting members of a structure.

gusset piece (*Build*) A piece of timber covering the triangular end-gap between the roof slope and the horizontal gutter boarding behind a chimney stack.

gusset plate (*Eng*) See GUSSET.

gustation (*Psych*) The sense of taste.

GUT (*Astron, Phys*) Abbrev for *grand unified theory*. See panel on GRAND UNIFIED THEORIES.

gut (*BioSci*) The alimentary canal.

gut associated lymphoid tissue (*BioSci*) Lymphoid tissues found in the digestive tract. Includes PEYER'S PATCHES, the appendix and the diffuse lymphoid tissue of the lamina propria. This tissue is well situated to deal with gut pathogens. Abbrev *GALT*.

Gutenberg discontinuity (*Geol*) The seismic-velocity discontinuity separating the mantle of the Earth from the core at a depth of approximately 2900 km.

gut oedema (*Vet*) See BOWEL OEDEMA DISEASE.

gutta (*BioSci*) A patch of colour or other marking, resembling a small drop, on the surface of an animal. Pl *guttae*. Adj *guttulate*.

guttae (*Arch*) An ornament in the form of a line of truncated cones used to decorate entablatures or hollow mouldings.

gutta percha (*Chem*) The coagulated LATEX of *Palaquium oblongifolium*. Consists mainly of 60% crystalline trans-1,2-polyisoprene compared with 0–10% crystallinity for natural rubber (the *cis* isomer). A hard inelastic solid ($T_m = 56$–$65°C$), it was formerly used for submarine cables before being displaced by polyethylene etc. See NATURAL RUBBER.

guttation (*BioSci*) The exudation of drops of water (containing some solutes) from an uninjured part of a plant, commonly under conditions of high humidity, typically from HYDATHODES at tips or margins of leaves.

gutter (*Build, CivEng*) A channel along the side of a road, or around the eaves of a building, to collect and carry away rainwater.

gutter (*Print*) (1) The spacing material in a forme between the fore-edges of the pages. (2) The space between two fore-edges on the printed sheet and including any trims.

gutter bearer (*Build*) A timber about $2 \times 1\frac{1}{2}$ in (50×38 mm), carrying gutter boarding.

gutter bed (*Build*) A lead sheet fixed behind the eaves gutter and over the tilting fillet to prevent overflow from the gutter from soaking into the wall.

gutter boards (*Build*) See SNOW BOARDS.

gutter bolt (*Build*) A securing bolt between the spigot and the socket ends at a joint in a gutter.

gut-tie (*Vet*) Strangulation of a loop of intestine which has herniated through a rupture in the peritoneal covering of the right spermatic cord of castrated cattle.

guttural (*BioSci*) Pertaining to the throat.

guttural pouch (*Vet*) A diverticulum of the Eustachian tube of the horse.

gutturoliths (*Vet*) See CHONDROIDS.

Gutzeit test (*Chem*) A method of determining arsenic, by adding metallic zinc and hydrochloric acid. The evolved gases darken mercury (II) salts.

guy (*CivEng*) A rope holding a structure in a desired position.

guy (*ICT*) Thin tension support for antenna mast or similar structure.

guy derrick (*Build, CivEng*) A crane operating from a mast held upright by guy-ropes.

guying (*CivEng*) The operation of holding or adjusting a structure in position by means of guy-ropes.

guyot (*Geol*) A flat-topped seamount, a topographic feature of the ocean floor.

G-value (*Chem*) A constant in radiation chemistry denoting the number of molecules reacting as a result of the absorption of 100 eV radiation energy.

GVH (*BioSci*) Abbrev for GRAFT-VERSUS-HOST REACTION.

gymbals (*Eng*) See GIMBALS.

gymno- (*Genrl*) Prefix from Gk *gymnos*, naked.

gymnocyte (*BioSci*) A rarely used term for a cell without a cell wall.

Gymnomycota (*BioSci*) See MYXOMYCOTA.

gymnosperms (*BioSci*) A group (classified as a division Gymnospermophyta or a class Gymnospermae, or regarded as polyphyletic) containing those SEED PLANTS in which the ovules are not enclosed in carpels, the pollen typically germinating on the surface of the ovule. There is no double fertilization and the xylem is vessel-less (except in the Gnetales). There are c.700 extant spp, often classified as three classes: Cycadopsida, Coniferopsida and Gnetopsida. See panel on ANGIOSPERMS (FLOWERING PLANTS) and appendix on Classification of the plant kingdom.

gyn-, gyno-, gynaeco- (*Genrl*) Prefixes from Gk *gyne*, gen *gynaikos*, woman. US *gyneco-*.

gynaecium (*BioSci*) See GYNOECIUM.

gynaeco- (*Genrl*) See GYN-.

gynaecology (*Med*) That branch of medical science which deals with the functions and diseases peculiar to women. US *gynecology*.

gynaecomastia (*Med*) Abnormal enlargement of the male breast. US *gynecomastia*.

gynandrism (*BioSci*) See HERMAPHRODITE.

gynandromorph (*BioSci*) An animal exhibiting male and female characters.

gynandromorphism (*BioSci*) The occurrence of secondary sexual character of both sexes in the same individual.

gynandrous (*BioSci*) Having the stamens and style united.

gyne- (*Med*) See GYNAE-.

gyno- (*Genrl*) See GYN-.

gynobasic (*BioSci*) A style that appears, because of the infolding of the ovary wall, to be inserted at the base of the ovary, eg Labiatae.

gynodioecious (*BioSci*) A species having some individual plants with hermaphrodite flowers only and others with female flowers only. Cf DIOECIOUS.

gynoecium (*BioSci*) The female part of a flower, consisting of one or more carpels. Also *gynaecium*. Cf ANDROECIUM.

gynomonoecius (*BioSci*) Species having all the plants bearing both female and hermaphrodite flowers. Cf MONOECIOUS.

gynophore (*BioSci*) An elongation of the receptacle between ANDROECIUM and GYNOECIUM.

gynospore (*BioSci*) Same as MEGASPORE in heterosporous plants. Cf ANDROSPORE.

gypsum (*Min*) Crystalline mineral of hydrated calcium sulphate, $CaSO_4.2H_2O$. Occurs in bulk form as ALABASTER, in fibrous form as satin spar, and as clear, colourless, monoclinic crystals of selenite. Used in making PLASTER OF PARIS, plaster and plasterboard, and is an important constituent ($\approx 3.5\%$ by weight) of PORTLAND CEMENT. It hydrates very rapidly during the setting reaction, and helps to control the initial setting rate. See panel on CEMENT AND CONCRETE.

gypsum plate (*Phys*) A thin plate of gypsum used in the determination of the sign of the birefringence of crystals in a polarizing microscope.

gyrator (*ElecEng*) Electronic component which does not obey reciprocity law. Frequently based on the FARADAY EFFECT in ferrites.

gyrator (*Electronics*) A circuit which can perform impedance inversion; the equivalent impedance at one port is inversely proportional to the impedance connected at the other port. With suitable phase alteration, a capacitance connected at the output can be made to appear as an inductance at the input. Used in integrated circuits where inductance cannot be realized on account of bulk.

gyratory breaker (*MinExt*) A widely used form of rock-breaker, in which an inner cone gyrates in a larger outer hollow cone.

gyro (*Eng*) See GYROSCOPE.

gyrocompass (*Eng*) A gyroscope, electrically rotated, controlled and damped either by gravity or electrically so that the spin axis settles in the meridian.

gyrodyne (*Aero*) A form of rotorcraft in which the rotor is power-driven for take-off, climb, hovering and landing, but is in AUTOROTATION for cruising flight, there usually being small wings further to unload the rotor.

gyro frequency (*Electronics*) See FREQUENCY OF GYRATION.

gyro horizon (*Aero*) An instrument which employs a gyro with a vertical spin axis, so arranged that it displays the attitude of the aircraft about its pitch and bank axes, referenced against an artificial horizon. It is normally electrically, but sometimes pneumatically, operated.

gyrolite (*Min*) A hydrated calcium silicate, formula $Ca_2Si_3O_7(OH)_2.H_2O$. Often occurs in amygdales with apophyllite.

gyromagnetic compass (*Aero*) A magnetic compass in which direction is measured by gyroscopic stabilization.

gyromagnetic effect (*Phys*) See BARNETT EFFECT.

gyromagnetic ratio (*Phys*) The ratio γ of the magnetic moment of a system to its angular momentum. For orbiting electrons $\gamma = e/2m$ where e is the electronic charge and m is the mass of the electron. γ for electron spin is twice this value.

gyroplane (*Aero*) Rotorcraft with unpowered rotor(s) on a vertical axis. Also *autogyro*.

gyroscope (*Eng*) Spinning body in a GIMBAL MOUNT or similar, which resists torques altering the alignment of the spin axis, and in which PRECESSION or NUTATION replace the direct response of static bodies to such applied torques. Used as a compass, or as a controlling device for servos which reduce the misalignment and thus correct the course or eg stabilize a ship. Also *gyrostat*. Abbrev *gyro*. See FIBRE OPTIC GYRO, FLOATED RATE-INTEGRATING GYRO, LASER GYRO, RATE GYRO, TUNED-RATE GYRO.

gyrostat (*Eng*) See GYROSCOPE.

Gyrosyn (*Aero, Ships*) TN for a remote-indicating compass system employing a directional GYROSCOPE which is monitored by and synchronized with signals from an element fixed in azimuth and designed to sense its angular displacement from the Earth's magnetic meridian by FLUX GATE. The element is located at some remote point, eg wing tips, away from extraneous magnetic influences.

gyrotron (*NucEng*) In fusion studies, very-high-frequency power generator for microwave heating at the electron cyclotron resonance.

gyrus (*BioSci*) A ridge between two grooves; a convolution of the surface of the cerebrum.

G-Y signal (*ImageTech*) Component of colour TV signal which, when combined with the Y SIGNAL, produces the green chrominance signal.

Gzelian (*Geol*) The youngest epoch of the Pennsylvanian period.

H

H (*Chem*) Symbol for HYDROGEN.

H (*Phys*) Symbol for HENRY.

H (*ElecEng*) Symbol for MAGNETIC FIELD STRENGTH.

H (*Phys*) Symbol for ENTHALPY.

[H] (*Phys*) One member of the strongest pair (H and K) of FRAUNHOFER LINES in the solar spectrum, almost at the limit of visibility in the extreme violet. Their wavelengths are [H], 396·8625 nm; [K], 393·3825 nm; and the lines are due to ionized calcium.

h (*Genrl*) Symbol for HECTO-.

h (*Genrl*) Symbol for: (1) height; (2) PLANCK'S CONSTANT.

ℏ (*Phys*) H-bar. Symbol for DIRAC'S CONSTANT. Equal to PLANCK'S CONSTANT h divided by 2π.

H_V (*Eng*) Symbol for Vickers hardness. Also *VHN, VPN*. See VICKERS HARDNESS TEST.

h_{fb} (*Electronics*) In transistors, the symbol for common-base mode current gain. See TRANSISTOR PARAMETERS.

h_{fe} (*Electronics*) In transistors, the symbol for common-emitter mode current gain. See TRANSISTOR PARAMETERS.

$H\alpha, H\beta, H\gamma, H\delta$ (*Phys*) The lines of the BALMER SERIES in the hydrogen spectrum. Their wavelengths are: $H\alpha$ 656·299 nm; $H\beta$ 486·152 nm; $H\gamma$ 434·067 nm; $H\delta$ 410·194 nm. The series continues into the ultraviolet, where about 20 more lines are observable.

haar (*EnvSci*) Local term for a wet sea-fog advancing in summer from the North Sea upon the shores of England and Scotland.

Haas effect (*Acous*) Phenomenon associated with a long-delayed echo which has been applied to reinforcement systems in auditoria.

habenula (*BioSci*) A strap-like structure; in particular, a nerve centre of the diencephalon.

Haber process (*Chem*) Currently the most important process of fixing nitrogen, in which the nitrogen is made to combine with hydrogen under influence of high temperatures (400–500°C), high pressure (2×10^7 N m^{-2}), and catalyst of finely divided iron from iron (III) oxides, in large continuous enclaves. Many variants operate at different pressures according to the catalyst. The product ammonia may be dissolved in water or condensed, and unreacted gases recycled.

habit (*BioSci*) The established normal behaviour of an animal species.

habit (*Crystal*) A term used to cover the varying development of the crystal forms possessed by any one mineral. Thus calcite may occur as crystals showing the faces of the hexagonal prism, basal pinacoid, scalenohedron and rhombohedron. According to the relative development or *dominance* of one or other of these forms, the habit may be prismatic, tabular, scalenohedral or rhombohedral.

habit (*Psych*) As a general term, and in learning theory, refers to learned patterns of behaviour which are very consistent and predictable in particular situations. Its specific usage varies in different branches of psychology, eg referring to compulsive behaviour in clinical psychology, to a particular cognitive style in cognitive psychology.

habitat (*BioSci*) The normal locality inhabited by a plant or animal, particularly in relation to the effect of its environmental factors.

habit spasm (*Psych*) A repeated, rapidly performed, involuntary and uncoordinated movement, occurring in a nervous person. Also *tic*.

habituated culture (*BioSci*) A plant tissue culture that has developed an ability to synthesize auxin since its isolation and can now grow in its absence.

habituation (*Psych*) An aspect of learning in which there is a decrease in responsiveness as a consequence of repeated stimulation; the habituated response reappears if the stimulus is withheld for a long time.

HAC (*Build*) Abbrev for HIGH ALUMINA CEMENT.

HACCP (*FoodSci*) Abbrev for HAZARD ANALYSIS AND CRITICAL CONTROL POINT.

hachure (*Surv*) The use of lines to shade a plan and indicate hills and valleys. Also *hatching*.

hack (*Build*) A long parallel bank, about 6 in (150 mm) high, of brick, rubbish and ashes, on which bricks are laid in the course of manufacture, when it is intended to dry them in the open.

hack-barrow (*Build*) A barrow used to carry green bricks to the HACK for drying.

hack-cap (*Build*) A small timber structure erected to provide cover for a HACK.

hacker (*ICT*) Colloq term for (1) a person who writes programs of a poorly designed or unmaintainable nature; (2) a person who obtains unauthorized access through information NETWORKS to computer systems.

hacking (*Build*) (1) The operation of piling up green bricks on a HACK to dry. (2) The process of making a surface rough, in order to provide a key for plasterwork. (3) A course of stones in a rubble wall, the course being composed partly of single stones of the full height of the course and partly of shallower stones arranged two to the height of the course.

hackle (*Eng*) Rough, ridged region of tensile fracture surface in brittle materials which follows the MIRROR and MIST regions. The surface in the hackle region makes an angle to the previous plane of crack propagation, because the crack has branched while running at maximum fracture velocity, frequently ejecting a piece of material. See FRACTOGRAPHY.

hackling (*Textiles*) The process of combing scutched flax in the hackling machine, in order to divide and parallelize the long fibres, and to remove the short ones and impurities.

hackmanite (*Min*) A fluorescent variety of sodalite, showing on freshly fractured surfaces a pink colour which fades on exposure to light, but which returns if kept in the dark or subjected to X-rays or ultraviolet light. See TENEBRESCENCE.

hacksaw (*Eng*) (1) A mechanic's hand-saw used for cutting metal. It consists of a steel frame, across which is stretched a narrow saw blade of hardened steel. (2) A larger saw, similar to the above, but power-driven.

Hackworth valve gear (*Eng*) A radial gear in which an eccentric opposite the crank operates a link whose other end slides along an inclined guide, the valve rod being pivoted to a point on the link.

Hadar (*Astron*) A bright white star in the constellation Centaurus which forms a visual binary system. Distance approx 120 pc. Also *Beta Centauri*.

hade (*Geol*) The angle of inclination of a fault plane, measured from the vertical.

Hadfield's steel (*Eng*) See MANGANESE STEEL.

Hadley cell (*EnvSci*) A meridional circulation of the atmosphere consisting of low-level equatorward movement of air from about 30° to the equator, rising air near the equator, poleward flow aloft, and descending motion near 30°. It was suggested by Hadley in the 18th century, and is at least partially confirmed by observation.

hadrom (*BioSci*) (1) The conducting elements and associated parenchyma in xylem tissue. Cf LEPTOM. (2) The HYDROIDS of mosses. Also *hadrome*.

hadron (*Phys*) An elementary particle that interacts with other particles via the STRONG INTERACTION. Hadrons include BARYONS and MESONS.

Haeckel's law (*BioSci*) See RECAPITULATION THEORY.

haem-, haemat-, haemato-, haemo- (*Genrl*) Prefixes from Gk *haima, -atos* denoting blood. US *hem-, hemat-, hemato-, hemo-*. All 'haem-' compound words have the alternate US spelling 'heme-'.

haem (*BioSci*) Compounds of iron complexed in a porphyrin (*tetrapyrrole*) ring that differ in side chain composition. *Haems* are the prosthetic groups of cytochromes and are found in most oxygen carrier proteins. HAEMOGLOBIN is *globin* with haem as a prosthetic group. US *heme*.

haemad (*BioSci*) Situated on the same side of the vertebral column as the heart.

haemagglutinin (*BioSci*) A substance that agglutinates red blood cells. This may be a specific antibody, or a LECTIN, or a component of certain viruses (eg influenza or measles) by which they bind to cell surfaces.

haemal (*BioSci*) Pertaining to the blood or to blood vessels. Also *haematal, haemic*.

haemal arch (*BioSci*) A skeletal structure arising ventrally from a vertebral centrum, which encloses the caudal blood vessels.

haemal canal (*BioSci*) The space enclosed by the centrum and the haemal arch of a vertebra, through which pass the caudal blood vessels.

haemal ridges (*BioSci*) See HAEMAPOPHYSES.

haemal spine (*BioSci*) The median ventral vertebral spine formed by the fusion of the haemapophyses, below the haemal canal.

haemal system (*BioSci*) The system of vessels and channels in which the blood circulates.

haemangioma (*Med*) A tumour composed of blood vessels irregularly disposed and of varying size. Also *angioma*.

haemapoiesis (*Med*) See HAEMATOGENESIS.

haemapophyses (*BioSci*) A pair of plates arising ventrally from the vertebral centrum, and meeting below the haemal canal to form the haemal arch and spine. Also *ridges*.

haemarthrosis (*Med*) A joint containing blood which has effused into it.

haemat- (*Genrl*) See HAEM-.

haematal (*BioSci*) See HAEMAL.

haematemesis (*Med*) The vomiting of blood, or of blood-stained contents of the stomach.

haematinic (*Med*) Pertaining to the blood.

haematite (*Min*) Oxide of iron, Fe_2O_3, crystallizing in the trigonal system. It occurs in a number of different forms: kidney iron ore massive; specular iron ore in groups of beautiful, lustrous, rhombohedral crystals as eg from Elba; bedded ores of sedimentary origin, as in the Precambrian throughout the world; and as a cement and pigment in sandstones. US *hematite*.

haemato- (*Genrl*) See HAEM-.

haematobium (*BioSci*) An organism living in blood. Also *hematobium*. Adjs *haematobic, hematobic*.

haematoblast (*BioSci*) A primitive blood cell that may develop into an erythrocyte or a leucocyte.

haematocele (*Med*) An effusion of blood localized in the form of a cyst in a cavity of the body.

haematochrome (*BioSci*) A red or orange pigment accumulated in the cells of some green algae usually when nitrogen starved, as *Chlamydomonas nivalis*, responsible for 'red snow'.

haematocolops (*Med*) Accumulation of menstrual blood in the vagina, due to an imperforate hymen.

haematocolpometra (*Med*) Accumulation of menstrual blood in the vagina and uterine cavity.

haematocrit (*BioSci*) A graduated capillary tube of uniform bore in which whole blood is centrifuged, to determine the ratio, by volume, of blood cells to plasma. Often used to mean the value of the ratio determined in this way.

haematogenesis (*BioSci*) The differentiation process by which new blood cells are made. Also *haemapoiesis, haematopoiesis, haematosis, haemopoiesis*.

haematogenous (*BioSci*) Having origin in the blood.

haematologist (*Med*) One who specializes in the study of the blood and its diseases.

haematoma (*Med*) A swelling composed of blood effused into connective tissue.

haematophagous (*BioSci*) Feeding on blood.

haematopoiesis (*BioSci*) See HAEMATOGENESIS.

haematopoietic stem cell (*BioSci*) Stem cell found mostly in bone marrow that produces red and white cells of the blood.

haematosis (*BioSci*) See HAEMATOGENESIS.

haematoxylin (*BioSci*) Basophilic stain that gives a blue colour (eg to the nucleus of a cell), commonly used in conjunction with eosin that counterstains the cytoplasm pink/red (the histopathologist's '*H and E*' stain).

haematozoon (*BioSci*) An animal living parasitically in the blood.

haematuria (*Med*) Presence of blood in the urine.

haemic (*BioSci*) See HAEMAL.

haemin (*Chem*) $C_{34}H_{32}N_4O_4FeCl$. The hydrochloride of haematin, brown crystals. Its molecule contains four pyrrole radicals.

haemochromatosis (*Med*) A disease in which the iron-containing pigment haemosiderin is deposited in excess in the organs of the body, giving rise to cirrhosis of liver, enlargement of spleen, diabetes, skin pigmentation. Also *bronzed diabetes*.

haemocoel (*BioSci*) A secondarily formed body cavity derived from the blood vessels that replaces the coelom in arthropods and molluscs.

haemocyanin (*BioSci*) A blue respiratory pigment, containing copper, in the blood of Crustacea and Mollusca. It has respiratory functions similar to haemoglobin.

haemocytes (*BioSci*) The leucocytes of HAEMOLYMPH, involved predominantly in defence and blood-clotting.

haemocytoblast (*BioSci*) A stem cell in bone marrow or in other haemopoietic tissues.

haemocytolysis (*BioSci*) See HAEMOLYSIS.

haemocytometer (*BioSci*) An apparatus consisting of a special glass slide with a grid of lines engraved on the bottom of a shallow rectangular trough so that if a coverslip is placed over the trough the grid demarcates known volumes. Cells from a well-mixed suspension are introduced into the space and the number in the grid squares counted under the microscope. Used for blood counts, mitotic counts, etc.

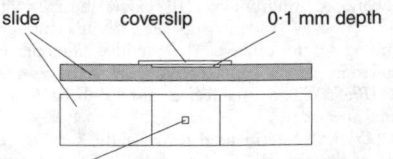

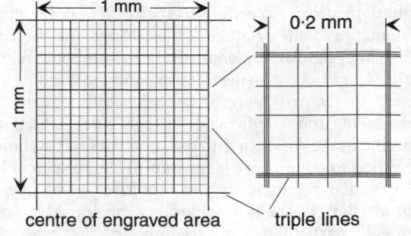

haemocytometer Neubauer type illustrated.

haemodialysis (*Med*) The restoration of diffusable chemical constituents of the blood towards normal, by passing blood across a cellulose membrane which has on the other side a fluid containing ELECTROLYTES at the desired concentration. Principally used to restore body chemistry in patients with kidney failure using an ARTIFICIAL KIDNEY.

haemoglobin (*BioSci*) The red pigment of blood whose major function is the transport of oxygen from the lungs to the tissues. It is a protein of four polypeptide chains each bearing a haem prosthetic group which serves as an oxygen binding site.

haemoglobinaemia (*Med*) The abnormal presence of haemoglobin in the blood, as a result of destruction of red blood cells.

haemoglobinometer (*Med*) An instrument for measuring the percentage of haemoglobin in the blood.

haemoglobinopathies (*Med*) Disorders due to abnormalities in the haemoglobin molecule, the best known being sickle-cell anaemia in which there is a single amino acid substitution (valine for glutamate) in position 6 of the β-globin chain. See also THALASSAEMIA.

haemoglobinuria (*Med*) The presence of haemoglobin in the urine, as a result of excessive destruction of red blood cells.

haemolymph (*BioSci*) The watery fluid containing leucocytes found in the haemocoelic body cavity of certain invertebrates.

haemolysin (*BioSci*) (1) Antibody capable of lysing red cells in the presence of complement. (2) A bacterial toxin which lyses red cells.

haemolysis (*BioSci*) The lysis of red blood cells. Also *haemocytolysis*.

haemolytic anaemia (*Med*) Anaemia due to an abnormal increase in the rate of destruction of circulating erythrocytes. This can result from the presence of antibodies against the erythrocytes (eg against the Rhesus antigen as in ERYTHROBLASTOSIS FOETALIS, or to auto-antibodies); or from overactivity of mononuclear phagocytes in association with grossly enlarged spleen (hypersplenism); or from metabolic abnormalities of the erythrocytes such as glucose-6-phosphatase deficiency aggravated by some drugs.

haemolytic disease of newborn animals (*Vet*) Seen mainly in fowl and piglets, it is the result of an immune reaction between antigens on the RED BLOOD CELLS of the neonate and colostral antibody from the dam. Intravascular haemolysis occurs giving rise to weakness, dyspnoea, anaemia and jaundice. Cf HAEMOLYTIC DISEASE OF THE NEWBORN in humans.

haemolytic disease of the newborn (*Med*) Anaemia and jaundice that can occur when a *Rhesus-negative* mother carries a *Rhesus-positive* child. It can be prevented by immunizing the mother so that maternal antibodies are not produced. See ERYTHROBLASTOSIS FOETALIS.

haemolytic plaque assay (*BioSci*) A method used to detect and enumerate individual cells secreting antibody *in vitro*. Sheep red cells (treated when necessary so as to bind the antibody) are mixed with a cell suspension to be assayed in a thin layer of agarose, and incubated. Antibody secreted by test cells diffuses outwards and binds to adjacent red cells; when complement is added cells secreting antibody are revealed by the presence of an area of haemolysis around them.

haemopericardium (*Med*) The presence of blood in the pericardial sac.

haemophilia (*Med*) A hereditary bleeding tendency, with deficiency of a normal blood protein (factor VIII, antihaemophilic globulin) preventing normal clotting. The classic SEX-LINKED disorder. Clinically indistinguishable from Christmas disease, in which there is a factor-IX deficiency and which is sometimes called haemophilia B.

haemopneumothorax (*Med*) The presence of blood and air in the pleural cavity.

haemopoiesis (*BioSci*) See HAEMATOGENESIS.

haemoptysis (*Med*) The expectoration of blood, or of blood-stained sputum.

haemorrhage (*Med*) Bleeding; escape of blood from a ruptured blood vessel.

haemorrhagic disease (*Vet*) See REDWATER.

haemorrhagic septicaemia (*Vet*) A bacterial disease of cattle caused by *Pasteurella multocida* and characterized by high fever, pneumonia and oedematous swelling of the skin. A similar disease occurs in sheep. *Pasteurella* vaccines available. Also *bovine pasteurellosis, shipping fever, shipping pneumonia*.

haemorrhoid (*Med*) Varicose dilation of the haemorrhoidal veins at the lower end of the rectum and the anus. Also *pile* (usually pl).

haemosiderosis (*Med*) Deposition, in the tissues of the body, of the iron-containing pigment haemosiderin, after excessive destruction of red blood cells. See HAEMOCHROMATOSIS.

haemostasis (*Med*) The arrest of bleeding. Adj *haemostatic*.

haemotropic (*BioSci*) Affecting blood.

hafnium (*Chem*) Symbol Hf, at no 72, ram 178·49 rel.d. 12·1, mp about 2000°C. A metallic element in the fourth group of the periodic system. It occurs in minerals containing zirconium, to which it is chemically similar, but with a much higher neutron capture cross-section. This makes it a troublesome impurity in the zirconium alloys used as fuel cladding. See ZIRCALOY. Used to prevent recrystallization of tungsten filaments.

haft (*Build*) A tool handle.

hagatalite (*Min*) A variety of ZIRCON containing an appreciable quantity of the rare earth elements.

Hagberg falling number (*Agri*) A measure of the viscosity of a slurry of starch produced from cereal grains. Lower numbers indicate a lower starch content, commonly attributed to early sprouting among the grains, which is undesirable.

ha-ha (*Build*) A fence sunk in a ditch below ground level so as to give an uninterrupted view.

hahnium (*Chem*) An artificially manufactured radioactive chemical element, symbol Hn, at no 108, of the transactinide series. Also *unniloctium*.

Haidinger fringes (*Phys*) Optical interference fringes produced by transmission and reflection from two parallel, partly reflecting surfaces, eg a plate of optical glass. The fringes are produced by division of amplitude of the wavefront and are circular fringes formed at infinity (cf CONTOUR FRINGES). Used extensively in INTERFEROMETERS, eg FABRY–PÉROT INTERFEROMETER, MICHELSON INTERFEROMETER.

hail (*EnvSci*) Precipitation in the form of hard pellets of ice (hailstones), which often fall from cumulonimbus clouds and accompany thunderstorms. They are formed when raindrops are swept up by strong air currents into regions where the temperature is below freezing point. In falling, the hailstone grows by condensation from the warm moist air which it encounters.

hail stage (*EnvSci*) That part of the condensation process taking place at a temperature of 0°C so that water vapour condenses to water liquid which then freezes.

hailstone (*EnvSci*) See HAIL.

hair (*BioSci*) (1) In animals, any thread-like growth of the epidermis, and in mammals, a slender, elongate structure, mostly composed of KERATINS, arising by proliferation of cells from the Malpighian layer of the epidermis. (2) In plants, see TRICHOME.

hair (*Textiles*) Animal fibre of variable diameter and length. Formerly used for rope, fabric, etc, or to reinforce building materials such as plaster and brick clay in composite form. Excludes sheep's wool and the invertebrate product, silk, but the soft shorter fibres from certain animals may also be called wool but qualified by the animal's name, eg angora wool.

hair cloth (*Textiles*) A material generally composed of coarse hair and cotton yarn; used as a stiffening for coats and in upholstery.

haircord (*Textiles*) Cotton fabrics of light weight, in which cords are produced by running two warp threads together at frequent intervals.

haircord carpet (*Textiles*) A hard carpet made from animal hair.

hair follicle (*BioSci*) See FOLLICLE.

hair hygrometer (*EnvSci*) A form of HYGROMETER which is controlled by the varying length of a human hair with humidity. It is not an absolute instrument, but it can be used at temperatures below freezing point, and it can be made self-recording.

hairline (*Phys*) A fine straight line in the optical system of an instrument; used for the positive location of an image or correlation onto a measuring scale.

hairpin loop (*BioSci*) A short double-stranded region made possible in a single-stranded nucleic acid molecule because of the complementarity of neighbouring sequences. Common in tRNA.

hair plates (*BioSci*) Groups of articulated sensory hairs occurring near the joints of the appendages in insects and acting as proprioceptors.

hair space (*Print*) The thinnest space between words, about 1 point wide, $\frac{1}{10}$ of an em of the body size.

hair stippler (*Build*) A decorative brush having a bristle filling set in rows in ebonite secured to a rectangular wooden base. Used in colour blending and colour glazing to equalize colour distribution and eliminate brush marks.

hake (*Build*) A HACK built to dry tiles in the course of their manufacture.

halation (*Electronics*) In a cathode-ray tube, glow surrounding spot on phosphor arising from internal reflection within the thickness of the glass. Also *halo*.

halation (*ImageTech*) Unwanted exposure surrounding the image of a bright object caused by light reflected from the rear surface of the film base or plate; it is reduced by a light-absorbing backing layer. See ANTI-HALATION.

Haldane apparatus (*Chem*) Apparatus for the analysis of air; used for the analysis of mine gases.

half adder (*ICT*) Logic circuit that adds a pair of binary bits producing a sum and a carry. Also *two-input adder*. Cf FULL ADDER.

half-bed (*Build*) In pricing the labour charge for stonework, each horizontal surface on a stone is spoken of as a *half-bed*, as it contributes one-half to the cost of preparing each bed joint; similarly, *half-joint* refers to the vertical jointing surfaces.

half-blind dovetail (*Build*) See LAP DOVETAIL.

half-bound (*Print*) Said of a book having its back, a portion of the sides, and the corners bound in one material (originally leather) and the remainder of the sides in some other material (eg cloth or paper).

half-brick wall (*Build*) A wall built entirely of stretchers and therefore 4·5 in thick.

half-case (*Print*) A type case of the usual width, but half the length, for holding display type or special letters.

half-cell (*ElecEng*) See SINGLE-ELECTRODE SYSTEM.

half-closed slot (*ElecEng*) See SEMICLOSED SLOT.

half-column (*Arch*) An EMBEDDED COLUMN of which half projects.

half-deflection method (*ElecEng*) A method of finding the internal resistance of a cell when the value is known to be high. A second cell, a galvanometer and a resistance are connected in series with the cell under test, and the value of the resistance required to give a galvanometer deflection of half the value obtained with the cell alone is found.

half duplex (*ICT*) Data transmission in both directions but in only one direction at any one time, as in eg a simple radio link.

half-element (*ElecEng*) See SINGLE-ELECTRODE SYSTEM.

half-frame camera (*ImageTech*) A camera which takes pictures half the normal size for the film used, in particular, half the size of the standard 36 × 24 mm still picture on 35 mm film.

half-hour rating (*ElecEng*) A form of rating for electric machinery supplying an intermittent load. It indicates that the machine delivers the specified rating for a period of half an hour without exceeding the specified temperature rises. Cf ONE-HOUR RATING.

half-joint (*Build*) See HALF-BED.

half-lap joint (*Build*) The joint formed by the process of HALVING.

half-lattice girder (*Build, CivEng*) See WARREN GIRDER.

half life (*BioSci*) Time required for the concentration of a reactant to reach a value that is the arithmetic mean of its initial and final (equilibrium) values. For a reactant that is entirely consumed, it is the time taken for the reactant concentration to fall to one-half its initial value. Also $t_{1/2}$.

half-life (*Phys*) (1) Time in which half of the atoms of a given quantity of radioactive nuclide undergo at least one disintegration. The half-life T is related to the *decay constant* λ by

$$T_{\frac{1}{2}} = \frac{0·693}{\lambda}$$

Also *half-period, half-value period*. (2) See BIOLOGICAL HALF-LIFE, EFFECTIVE HALF-LIFE.

half-line block (*Print*) One in which the strength of the lines is reduced to produce a pleasing effect by using in the process camera a coarse cross-line screen or a single-line screen vertical, horizontal or diagonal.

half-measure (*Print*) Type matter set to half the usual page or column width to accommodate an illustration.

half-period (*Phys*) See HALF-LIFE.

half-period zones (*Phys*) A conception, due to Huygens, whereby an optical wavefront is considered to be divided into a number of concentric annular zones, so that, at a given point in front of the wave, the illumination from each zone is half a period out of phase with that from its neighbour. The use of half-period zones facilitates the study of diffraction problems.

half plate (*ImageTech*) A standard format, $4\frac{3}{4} \times 6\frac{1}{2}$ in.

half-power (*Acous*) The condition of a resonant system (electrical, mechanical or acoustical) when amplitude response is reduced to $1/\sqrt{2}$ of maximum, ie by 3 dB.

half-power point (*ICT*) Useful reference point in the performance characteristics of any device, eg amplifier, antenna or transistor, whose performance varies with frequency; it is the frequency or frequencies at which the gain or response falls 3 dB below a mean value.

half-principal (*Build*) A short rafter which does not reach the ridge of a roof.

half-residence time (*Phys*) Time in which half the radioactive debris deposited in the stratosphere by a nuclear explosion would be carried down to the troposphere.

half-roll (*Aero*) See ROLL.

half-round chisel (*Eng*) A COLD CHISEL having a small half-round cutting edge; used for chipping semicircular grooves such as oil-ways.

half-round file (*Eng*) A file whose cross-section has one flat and one convex face.

half-round screws (*Eng*) See BUTTON-HEADED SCREWS.

half-sawn (*Build*) Said of a stone face as left from the saw.

half-secret dovetail (*Build*) See LAP DOVETAIL.

half-section (*Eng*) On an engineering drawing, a sectional view of a symmetrical workpiece or assembly, terminating at an axis of symmetry.

half-section (*ICT*) See SECTION.

half-sheet work (*Print*) The method in which one forme or plate is used to print on each side of the paper which is split after printing, thus producing two identical copies, each requiring half the sheet of paper. See WORK-AND-TURN.

half-shroud (*Eng*) A gearwheel shroud extending only up to half the tooth height. See FULL SHROUD, SHROUD.

half-silvered (*Phys*) A metallic film deposited, eg by anode sputtering, onto glass or other surface which reflects a substantial proportion of incident light.

half-socket pipe (*Build*) A drainpipe having a socket for the lower half only.

half space (*Build*) (1) A landing at the end of a flight of steps. (2) A raised floor in a window bay.

half-stuff (*Paper*) Paper raw materials which have been converted into pulp but not yet beaten.

half-supply voltage principle (*ElecEng*) If the collector–emitter voltage of a power stage transistor is less than half the supply voltage, the circuit will be inherently safe from thermal runaway, as the thermal loop gain will be less than unity.

half-thickness (*Phys*) See HALF-VALUE THICKNESS.

half-timbering (*Build*) An early mode of house-building in which the foundations and principal members were of stout timber, and walls were formed by filling the spaces between members with plaster.

half-title (*Print*) Widely used to describe the title of a book printed on the leaf preceding the title page. More accurately, section titles set within the body of the book. See BASTARD TITLE.

half-tone (*ICT, Print*) An image that simulates a GREYSCALE by varying the sizes of the dots making up the image. See HALF-TONE PROCESS.

half-tone block (*Print*) A screened plate mounted on a base to type height and used for the reproduction of continuous tone copy, such as photographs.

half-tone process (*ImageTech, Print*) The reproduction of a subject containing a range of tones by photographing through a cross-line screen which translates these tones into dots of varying size. Used for all the usual printing methods, relief (letterpress), planographic (lithography), intaglio (gravure) and stencil (screen process).

half-tone screen (*Print*) A cross-ruled glass used in the production of half-tone illustrations. The transparent squares break down the continuous tone original into a series of various-sized dots.

half-twist bit (*Build*) A bit shaped like a gimlet, used for making screw-holes.

half-uncials (*Print*) An early style of lettering in which square forms of letter are mixed with rounded forms; a few typefaces are available which copy this style.

half-value layer (*Phys*) See HALF-VALUE THICKNESS.

half-value period (*Phys*) See HALF-LIFE.

half-value thickness (*Phys*) The thickness of a specified substance which must be placed in the path of a beam of radiation in order to reduce the transmitted intensity by one-half. Also *half-thickness, half-value layer*.

half-wave antenna (*ICT*) An antenna whose overall length is one half-wavelength. The voltage distribution is from a maximum at one end to a minimum in the middle and a maximum at the other.

half-wave plate (*Phys*) A plate of doubly refracting, uniaxial crystal cut parallel to the optical axis, of such thickness that, if light is transmitted normally through it, a phase difference of half a period is introduced between the ordinary and extraordinary waves.

half-wave rectification (*ElecEng, ICT*) Rectification in which current flows only during the positive (or negative) half cycles of the alternating voltage. The commonest form of detection of amplitude-modulated radio signals.

half-wave rectifier (*ElecEng*) One which conducts for part of half of the alternating voltage cycle.

half-wave suppressor coil (*ICT*) An inductance coil inserted at half-wavelength intervals along an antenna wire; used in some forms of directional antenna to suppress radiation in reverse phase from alternate half-wavelength sections of the wire.

half-wave transmission (*ElecEng*) A method of transmission of electrical energy in which the natural period of oscillation of the transmission line equals four times the frequency of the transmitted current.

half-width (*Phys*) A measure of sharpness on any function $y = f(x)$ which has a maximum value y_m at x_0 and also falls off steeply on either side of the maximum. The half-width is the difference between x_0 and the value of x for which $y = y_m/2$. Used particularly to measure the width of spectral lines or of a response curve.

hali-, halo- (*Genrl*) Prefixes from Gk *hals*, salt.

halides (*Chem*) Fluorides, chlorides, bromides, iodides and astatides.

haliplankton (*BioSci*) The plankton of the seas.

halite (*Min*) Common or rock salt. The naturally occurring form of sodium chloride, crystallizing in the cubic system; forming deposits of considerable thickness in close association with anhydrite and gypsum esp in the Permian and Triassic rocks, the salt is pumped out as brine or mined. Salt domes, with which oil or gas may be associated, occur in many parts of the world.

halitosis (*Med*) The condition of having bad breath.

Hallade recorder (*Eng*) An instrument for recording vibration of rolling stock due to track irregularities etc in planes parallel, transverse and perpendicular relative to the track.

Hall coefficient (*Electronics*) The coefficient of proportionality (R_H) in the HALL EFFECT relation $E_H = R_H jB$ where E_H is the resulting transverse electric field, j is the current density and B is the magnetic flux density.

Hall effect (*Electronics*) The generation of a transverse electric field in a conductor or semiconductor when carrying current across a magnetic field.

Halley's comet (*Astron*) A famous regular comet, period 76 yr, whose 1758 return was successfully predicted by Edmond Halley. Its most recent close approach to the Sun was in 1986.

Hall mobility (*Electronics*) Mobility (mean drift velocity in unit field) of current carriers in a semiconductor as calculated from the product of the Hall coefficient and the conductivity.

halloysite (*Min*) One of the clay minerals, a hydrated form of kaolinite and member of the kandite group; consists of hydrated aluminium silicate.

Hall–Petch equation (*Eng*) An empirical relation between the yield stress σ_Y of a metal or alloy and its average grain size d which states that

$$\sigma_Y = \sigma_0 + kd^{-\frac{1}{2}}$$

where σ_0 and k are experimentally determined constants. The effect is due to the pinning of dislocations by grain boundaries. The equation shows why grain size should be kept as small as possible to achieve highest strength. Also *Hall–Petch relation*. See STRENGTH MEASURES.

Hall probe (*Phys*) A small probe which uses the HALL EFFECT to compare magnetic fields.

Hall process (*Eng*) A process for the extraction of aluminium by the electrolysis of a fused solution of alumina in cryolite at a temperature of approximately 1000°C. The aluminium is molten at this temperature and settles to the bottom of the bath, from which it is drawn off. Contains up to 99·8% aluminium.

hallucination (*Psych*) A perceptual experience that occurs in the absence of any appropriate external stimulus.

hallucinogen (*Pharmacol*) A drug or chemical which induces HALLUCINATIONS (eg lysergic acid (LSD) or mescaline).

hallux (*BioSci*) In land vertebrates, the first digit of the hind-limb.

hallux flexus (*Med*) A late stage of HALLUX RIGIDUS, the big toe being rigidly flexed on the sole.

hallux rigidus (*Med*) Rigid stiffness of the big toe, due to osteoarthritis of the joint between the toe and the foot.

hallux valgus (*Med*) A deformity of the big toe, in which it turns towards and comes to lie above the toe next to it; usually associated with bunion.

hallux varus (*Med*) A rare deformity of the big toe, in which it diverges from the toe next to it.

halo (*Astron*) (1) An arc of light which can appear to surround the Sun or Moon owing to refraction by ice crystals in the Earth's atmosphere. The *large halo*, of radius 22°, is due to light refracted at minimum deviation by ice crystals in high cirrostratus clouds. (2) See GALACTIC HALO.

halo (*Electronics*) See HALATION.

halo (*EnvSci*) See CORONA.

halo (*MinExt*) The product of diffusion in rock surrounding an ore deposit of traces of mineral or element being sought, identified by geochemical tests. The halo may be the only surface indication of an ore deposit at depth. See GEOCHEMICAL PROSPECTING.

halo- (*Genrl*) See HALI-.

halobacteria (*BioSci*) See HALOPHILIC BACTERIA.

Halobacterium halobium (*BioSci*) Photosynthetic (halo-philic) bacterium that has patches of purple membrane containing the pigment bacteriorhodopsin.

halobiotic (*BioSci*) Living in salt water, esp in the sea.

halochromism (*Chem*) The formation of coloured salts from colourless organic bases by the addition of acids.

halo effect (*Psych*) In the perception of other people, the tendency to generalize an impression of one characteristic of a personality to other, unrelated aspects of the personality.

halogen (*Chem*) One of the seventh group of elements in the periodic table, for which there is one electron vacancy in the outer energy level, namely fluorine (F), chlorine (Cl), bromine (Br), iodine (I), astatine (At). The main oxidation state is −1.

halogenation (*Chem*) The introduction of HALOGEN atoms into an organic molecule by substitution or addition.

halogen-quench Geiger tube (*NucEng*) Low-voltage tube for which halogen gas (normally bromine) absorbs residual electrons after a current pulse, and so quenches the discharge in preparation for a subsequent count.

haloid acids (*Chem*) A group consisting of hydrogen fluoride, hydrogen chloride, hydrogen bromide and hydrogen iodide.

halolimnic (*BioSci*) Originally marine but secondarily adapted to fresh water.

halons (*Chem*) Analogues of CHLOROFLUOROCARBONS, but containing bromine.

haloperidol (*Pharmacol*) A drug belonging to the butyr-ophenones which as a class all have anti-psychotic actions. Haloperidol is used in the treatment of schizophrenia.

halophile (*BioSci*) A freshwater species capable of surviving in salt water.

halophilic bacteria (*BioSci*) Salt-tolerant bacteria occurring in the surface layers of the sea, where they are important in the nitrogen, carbon, sulphur and phosphorus cycles. Many are pigmented or phosphorescent and if present in quantity may colour the surface water. Some halophilic bacteria, eg halobacterium, are able to live in salted meats.

halophyte (*BioSci*) A plant able to grow where the soil is rich in sodium chloride or other sodium salts. Cf GLYCOPHYTE.

halophytic vegetation (*BioSci*) A population of halo-phytes, eg on mangrove swamp, salt marsh or alkaline soil.

halosere (*BioSci*) A SERE that starts on land emerging from the sea, eg a salt marsh.

Halothane (*Chem*) *1-chloro-1-bromo-2,2,2 trifluoroethane*. $CF_3CHClBr$. TN for a non-explosive anaesthetic, now widely used, containing one atom of chlorine in its molecular structure. Also *Fluothane*.

halotrichite (*Min*) Hydrated sulphate of iron and alumi-nium, occurring rarely as yellowish, fibrous, silky colour-less crystals. Also *iron alum*.

halteres (*BioSci*) A pair of capitate threads that take the place of the hind-wings in DIPTERA, and that assist the insect to maintain its equilibrium while flying; balancers.

halving (*Build*) A method of jointing (eg two timbers) in which half the thickness is cut from the face of one, and half from the back of the other, so that when the two pieces are put together the outer surfaces are flush. Cf COMBED JOINT.

Hamamelidae (*BioSci*) A subclass or superorder of dicoty-ledons, comprising mostly woody plants. The perianth is poorly developed or absent; the flowers are often unisexual, often borne in catkins and often wind pollinated. There are c.3400 spp in 23 families including Betulaceae and Fagaceae (includes the Amentiferae).

hamartoma (*Med*) A benign tumour due to excess growth of one of the cellular elements, often vascular, in normal tissue.

hambergite (*Min*) Basic beryllium borate, crystallizing in the orthorhombic system.

Hamiltonian path (*MathSci*) A closed path of a graph which includes every vertex.

Hamilton's principle (*MathSci*) The principle that the motion of a system from time t_1 to t_2 is such that the line integral

$$\int_{t_1}^{t_2} (T + W)\, dt$$

is stationary, where T is the kinetic energy of the system and

$$W = \sum_i F_i r_i$$

where F_i are the external forces acting at points r_i. This principle can be taken as a basic postulate in place of Newton's laws. For systems to which Newton's laws can be applied it gives the same results, but it is important because it can also be applied to other systems.

hammer-axe (*Build*) A tool with a double head, axe at one side and hammer at the other.

hammer beam (*Build*) A short cantilever beam projecting into a room or hall from the springing level of the roof, strengthened by a curved strut underneath, and carrying a HAMMER-BEAM ROOF.

hammer-beam roof (*Build*) A type of timber roof existing in various forms, all affording good headroom beneath. It consists essentially of arched ribs, supported on HAMMER BEAMS at their feet, and carrying the principal rafters, strengthened sometimes by a collar beam and/or struts.

hammer blow (*Eng*) (1) The alternating force between a steam locomotive's driving wheels and the rails, due to centrifugal force of the balance weights used to balance the reciprocating masses. (2) The noise due to a pressure wave travelling along a pipe in which the flow of a liquid has been suddenly impeded.

hammer-dressed (*Build*) A term applied to stone surfaces left with a rough finish produced by the hammer.

hammer-drill (*MinExt*) A compressed-air rock drill in which the piston is not attached to the steel or borer but moves freely.

hammer-drive screw (*Eng*) A self-tapping screw with a very long pitch, akin to a nail, which can be driven by a hammer or inserted by a press into a plain hole in a relatively soft material.

hammerhead crane (*Eng*) See TOWER CRANE.

hammer-headed (*Build*) A term applied to masons' chisels intended to be struck by a hammer rather than by a mallet.

hammerman (*Eng*) (1) The operator of a power-driven hammer. (2) A smith's mate.

hammer mill (*MinExt*) See BEATER MILL, IMPACT CRUSHER.

hammer scale (*Eng*) The scale of iron oxide which forms on work when it is heated for forging.

hammer test (*Eng*) Drop test for impact strength of large metal parts, eg rails. Weight is dropped from increasing heights until specified deflection is produced.

hammer toe (*Med*) A deformity of any toe, esp the second, in which the toe, flexed on itself, is, at its junction with the foot, bent towards the instep.

hammer track (*Phys*) Highly characteristic track resembling a hammer, formed by decay of a lithium-8 nucleus into two alpha particles emitted in opposite directions at right angles to the lithium track.

hance (*Arch*) That part of the intrados, close to the springing of an elliptical or many-centred arch, which forms the arc of smaller radius.

Hancock jig (*MinExt*) One in which ore is jigged up and down with some thrown forward in a tank of water, the heavy mineral stratifying down and being separately removed.

hand (*Textiles*) US for HANDLE.

hand brace (*Build*) See BRACE.

H and D curve (*ImageTech*) See CHARACTERISTIC CURVE.

hand-cut overlay (*Print*) Plies of paper cut out in the highlights and built up in the dark tones and included in the MAKE-READY to improve the printing of half-tone blocks. Cf MECHANICAL OVERLAY.

hand feed (*Eng*) The hand operation of the feed mechanism of a machine tool. See FEED.

hand hole (*Eng*) A small hole, closed by a removable cover, in the side of a pressure vessel or tank; it provides means of access for the hand to the inside of the vessel.

hand ladle (*Eng*) A small foundry casting ladle supported by a long handle of steel bar. See HAND SHANK.

handle (*Textiles*) The subjective reaction obtained by feeling a fabric and assessing its roughness, harshness, flexibility, softness, etc.

hand lead (*Surv*) Lead plummet used at sea, by attachment to a lead line measuring within 100 fathoms (183 m).

hand letters (*Print*) Letters formed of brass and mounted on a handle, with which the finisher impresses the title on the back or side of a bound volume.

handle-type fuse (*ElecEng*) A fuse in which the carrier containing the fuse-link is provided with a handle to facilitate withdrawal and replacement.

hand level (*Surv*) Small and light sighting tube above which a spirit level and mirror are arranged, so that the bubble can be seen while sighting on a station.

hand-made paper (*Paper*) Paper made in single sheets by dipping a mould into a vat containing stock so that the requisite amount of stock is picked up and, by skilful shaking, is disturbed and formed into the sheet. The wet sheets are couched onto felts, pressed to remove water, dried and, if necessary, tub sized. Hand-made papers are characterized by their permanence and durability, appearance of quality and excellent properties for watercolour painting.

hand monitor (*Radiol*) Radiation monitor designed to measure radioactive contamination on the hands of an operator, or to be held in the hand.

hand mould (*Paper*) Wooden frame, accompanied by a pair of deckles, covered with a wove or laid pattern woven wire on which a sheet of hand-made paper is formed.

hand-off (*Electronics*) See HANDOVER.

handover (*ICT*) The process by which a mobile telephone that is getting out of range of its BASE STATION is assigned a new one. Measurement of its signals by neighbouring stations leads to allocation of a new frequency, which is communicated to it by the current station. While the mobile retunes, the switching centre diverts the call to the new base station for transmission at the new frequency. Also *hand-off*.

handover hysteresis (*ICT*) Avoidance of rapidly repeated HANDOVERS in a mobile-telephone network by allowing the received signal from a mobile to fall below the threshold that would normally trigger handover if a handover has just taken place.

hand press (*Print*) A press on which the paper is fed by hand; sometimes powered by a treadle, or hand-operated lever.

hand-rail bolt (*Build*) A rod which is threaded and fitted with a nut at both ends; used to draw together the mating surfaces of a butt joint, such as that between adjacent lengths of hand-rail. The rod passes though holes in the members, and at one end the nut, a square one, is housed in a square mortise which prevents it from turning. At the other end the nut is circular, with notches cut or holes drilled in its periphery, so it can be turned within its mortise.

hand-rail plane (*Build*) A plane having a specially shaped sole and cutting iron, adapting it to the finishing of the top surface of a hand-rail.

hand-rail screw (*Build*) A small rod, with opposite-handed screw-threads at each end, used to connect adjacent lengths of hand-rail as an alternative to the HAND-RAIL BOLT. Also *dowel screw*.

hand reset (*ElecEng*) Restoration of a magnetic device, eg relay or circuit breaker, by a manual operation.

hand rest (*Eng*) A support, shaped like a letter T, on which a turner rests a hand tool, during wood-turning or metal-spinning in a lathe.

hand roller (*Print*) A roller used for inking type matters on the hand-operated press, preparatory to pulling a proof.

Hand–Schuller–Christian disease (*Med*) A lipoid granuloma of unknown aetiology, affecting mainly the skull, which may lead to exophthalmus and diabetes insipidus. The granulomas contain cholesterol.

handscrew (*Build*) A wooden clamp consisting of two parallel bars connected by two tightening screws; used to hold parts together while a glued joint dries, or to secure work in process of being formed.

handset (*ICT*) (1) A combined telephone transmitter (microphone) and receiver in a form convenient for holding simultaneously to mouth and ear. (2) The portable part of a cordless telephone or car-mounted mobile telephone.

handshaking (*ICT*) The PROTOCOL that enables two devices to establish (or break) communications.

hand shank (*Eng*) A FOUNDRY LADLE supported at the centre of a long iron bar, formed into a pair of handles at one end for control during pouring; carried by two people.

hand specimen (*Min*) A piece of rock or mineral of a size suitable for megascopic study, further investigation or preserving in a collection.

hand tools (*Eng*) All tools used by fitters when doing hand work at the bench, as hammers, files, scrapers, etc.

hand winding (*ElecEng*) Winding a machine by placing the coils, turn by turn, into the slots, used when it is impractical to use former-wound coils.

hangar (*Aero*) A special construction for the accommodation of aircraft.

hanger (*Build*) A bracket for the support of a gutter at the eaves.

hanger (*ElecEng*) (1) Plates of glass or other material standing on edge in an accumulator cell, and supporting the accumulator plates by means of their lugs. (2) A fitting used for supporting the overhead contact wire of a traction system from a transverse wire or structure.

hanger (*Eng*) A bracket, usually of cast-iron, bolted to a wall or to the underside of a girder, to hold a bearing for supporting overhead shafting.

hangfire (*MinExt*) Unexpected delay or failure of explosive charge to detonate, thus creating a dangerous situation.

hang glider (*Aero*) Original manned glider as used by Lilienthal in Germany in the 1890s. Revived in the 1960s as the ROGALLO WING, employing flexible wing surfaces and now a major class of ultra-light aircraft. Both flexible and fixed hard wings are now used.

hanging (*Eng*) In the blast furnace, adhesion of partly melted charge to walls, thus upsetting smooth working. Also *hang-up*.

hanging buttress (*Build*) A buttress carried upon a corbel at its base.

hanging-drop preparation (*BioSci*) A preparation for the microscope in which the specimen, in a drop of medium on the undersurface of a coverslip, is suspended over a hollow-ground slide, to which it is sealed to prevent evaporation.

hanging figures (*Print*) Figures normally supplied with OLD STYLE founts; 1, 2, 0 conform to the X-HEIGHT, 6 and 8 have the height of ascenders, and 3, 4, 5, 7 and 9 hang below the BASELINE. Also *old style figures*.

hanging indentation (*Print*) Layout in which the first line of the paragraph is set full out and the succeeding lines are indented one em or more, as employed in this dictionary.

hanging post (*Build*) See HINGING POST.

hanging sash (*Build*) A sash arranged to slide in vertical grooves, and counterweighted so as to be balanced in all positions.

hanging steps (*Build*) Steps which are built into a wall at one end and are unsupported at the other end. Also *cantilevered steps*.

hanging stile (*Build*) That stile of a door to which the hinges are secured.

hanging valley (*Geol*) A tributary valley not graded to the main valley. It is a product of large-scale glaciation and due to the glacial overdeepening of the main valley relative to the hanging valley. There may be rapids or waterfalls from the tributary to the main valley. See VALLEYS.

hanging wall (*MinExt*) Rock above the miner's head, usually the country rock above the deposit being worked.

hang-over (*ICT*) (1) The delay in restoration of speech-activated switches, as in the VODAS, to ensure the non-clipping of weak final consonants of words. (2) Excessive prolongation of any *on–off* type of signal or current or voltage pulse.

hang-up (*Eng*) See HANGING.

hank (*Textiles*) A general term for a reeled length of yarn.

Hankel functions (*MathSci*) See BESSEL FUNCTIONS.

Hansen's disease (*Med*) Leprosy.

hantavirus (*Med*) A virulent and often fatal virus with flu-like early symptoms. First identified near the Hantaan River in Korea.

hapanthous (*BioSci*) See HAPAXANTHIC.

hapaxanthic (*BioSci*) Flowering and fruiting once and then dying. Also *hapanthous*. See MONOCARPIC.

haplo- (*Genrl*) Prefix from Gk *haploos*, single, simple.

haplobiont (*BioSci*) A plant that has only one kind of individual or form in its life history. Adj *haplobiontic*.

haplodiploid (*BioSci*) Species in which one sex has haploid cells and the other has diploid cells, as eg honeybees.

haplodiploidy (*BioSci*) A means of sex determination where females develop from fertilized eggs and are therefore diploid and the males from unfertilized eggs and are haploid, eg honeybees.

haplodont (*BioSci*) Having molars with simple crowns.

haploid (*BioSci*) Of the reduced number of chromosomes characteristic of the germ cells of a species, equal to half the number in the somatic cells. Cf DIPLOID.

haploidization (*BioSci*) In the PARASEXUAL CYCLE of fungi, the progressive loss of chromosomes from the diploid set by occasional non-disjunction until stable haploid nuclei are formed.

haplont (*BioSci*) Organism in which only the gametes are haploid, meiosis occurring at their formation and the vegetative cells being diploid. Cf DIPLONT.

haplophase (*BioSci*) The period in the life cycle of any organism when the nuclei are haploid. Cf DIPLOPHASE.

haplostele (*BioSci*) A protostele in which the solid central core of xylem is circular in cross-section.

haplostemonous (*BioSci*) Having a single whorl of stamens. Cf DIPLOSTEMONOUS.

haplotype (*BioSci*) A particular set of *alleles* at several very closely linked loci.

haploxylic (*BioSci*) Said of a leaf containing one vascular strand.

hapten (*BioSci*) A substance that can combine with antibody but cannot itself initiate an immune response unless it is attached to a carrier molecule. Most haptens are small molecules (eg dinitrophenyl). Haptens are often conjugated chemically to carrier proteins for experimental purposes, since they provide easily recognized antigenic determinants.

hapteron (*BioSci*) A holdfast, ie a unicellular or multi-cellular organ attaching a plant to the substrate.

haptic (*Psych*) Relating to the sense of touch.

haptonema (*BioSci*) Appendage arising between the flagella of the motile cells of the Haptophyceae sometimes serving for temporary attachment to a surface.

Haptophyceae (*BioSci*) A class of eukaryotic planktonic algae, mostly marine. There are flagellated and palmelloid sorts (often interconvertible) and motile cells usually have two equal, smooth, flagella and a HAPTONEMA . Some sorts or stages bear COCCOLITHS, which are frequently found as fossils and are the major component of chalk. Most are phototrophic although some are heterotrophic. Also *Prymnesiophyceae*.

haptotaxis (*BioSci*) Strictly speaking, a directed response of cells in a gradient of adhesion, but often loosely applied to situations where an adhesion gradient is thought to exist and local trapping of cells seems to occur.

haptotropism (*BioSci*) A TROPISM like that of a tendril coiling round its support in which differential growth is determined by touch. Also *thigmotropism*.

hard (*Build*) A layer of gravel or similar materials put down on swampy or sodden ground to provide a way for passage on foot.

hard (*Electronics*) Adjective, synonymous with high-vacuum, which differentiates thermionic-vacuum valves from gas-discharge tubes.

hard (*Eng*) Said of a bulk magnetic material which retains its magnetization. See MAGNETIC MATERIALS.

hard (*Glass*) Having a relatively high softening point.

hard acids and bases (*Chem*) Terminology used with LEWIS ACIDS AND BASES to indicate non-polarizable; polarizable acids and bases are termed *soft*. In general, hard–hard and soft–soft interactions are stronger than hard–soft ones.

hard bast (*BioSci*) SCLERENCHYMA present in phloem.

hardboard (*Build*) Fibreboard that has been compressed in drying, giving a material of greater density than INSULATING BOARD.

hard bronze (*Eng*) Copper-based alloy used for tough or dense castings; based of 88% copper plus tin with either some lead or zinc. See COPPER ALLOYS.

hard copy (*ICT*) Computer output printed on paper.

hard core (*Build, CivEng*) Lumps of broken brick, hard natural stone, etc, used to form the basis of a foundation for road or paving or floors to a building.

hard disk (*ICT*) Rigid magnetic disk. It has a higher recording density and rotational speed than a floppy disk and provides more storage for the same physical dimensions. May be stacked as platters. Fig. ▷

hard drawn (*Eng*) A term applied to wire or tube which has been greatly reduced in cross-section without annealing.

hardenability (*Eng*) The propensity of a steel to transform to a hard MARTENSITE when cooled from the AUSTENITIC state. A steel with low hardenabilty will only form martensite when cooled rapidly (eg water quenched in thin section) whereas one possessing a high hardenability may be cooled slowly in air and will still transform to a hard martensite. Improvement of hardenability is one of the prime reasons for alloying medium carbon steels, since it allows components of large cross-section to be hardened prior to tempering without risk of cracking. Commonly assessed by the JOMINY (END QUENCH) TEST. See panel on STEELS.

hardener (*ImageTech*) Chemical (formalin, acrolein, chrome alum, etc) added to the fixing bath to toughen the emulsion of a film.

hardener (*Plastics*) An ACCELERATOR.

hardening (*BioSci*) Increasing resistance to cold as temperatures are gradually lowered either naturally or as the result of horticultural practice. Analogous hardening to drought, heat, wind, etc, occurs.

hardening (*Eng*) The process of making steel hard by cooling from above the critical range at a rate that prevents the formation of ferrite and pearlite and results in the formation of martensite. May involve cooling in water, oil or air, according to composition and size of article and the HARDENABILITY of the steel. The steel must contain sufficient carbon (above about 0·3%) to achieve a useful hardening response.

hardening media (*Eng*) Liquids into which steel components are plunged for hardening. They include cold water and brine to increase the cooling power for the fastest quench rates and special mineral oils and polymers for slower and intermediate rates.

hardening of oils (*Chem*) The hydrogenation of oils in the presence of a catalyst, usually finely divided nickel, in which the unsaturated acids are transformed into saturated acids, with the result that the glycerides of the unsaturated acids become hard. This process is of great importance for the foodstuffs industries, eg margarine is prepared in this way.

Harder's glands (*BioSci*) In most of the higher vertebrates, an accumulation of small glands near the inner angle of the eye, closely resembling the lacrimal gland.

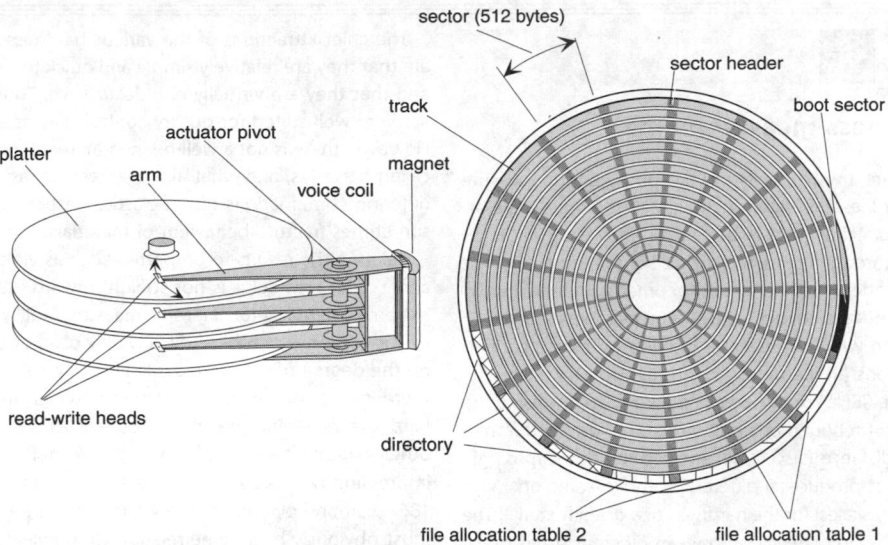

hard disk Schematic drawing of a three platter drive (left) and a plan of the domains on platter (right) showing many fewer tracks than would be normal. A cylinder is all the tracks on the same radius.

hard-facing (*Eng*) (1) The application of a surface layer of hard material to impart, in particular, wear resistance. (2) A surface so formed. The composition is generally of high-melting-point metals, carbides, etc, applied by powder, wire or PLASMA-ARC SPRAYING, or by welding.

hard glass (*Chem*) BOROSILICATE GLASS, whose hardness is principally due to boron compounds. Resistant to heat and to chemical action.

hard-gloss paint (*Build*) A popular class of paint that dries hard with a high gloss. It always contains some hard resin in the medium.

hard head (*Eng*) Alloy of tin with iron and arsenic left after refining of tin.

hard heading (*MinExt*) Sandstone or other hard rock encountered in making headings or tunnels in a coal mine.

hard hyphen (*Print*) A hyphen which is essential to the correct spelling of a word.

Hardinge mill (*MinExt*) Widely used grinding mill, made in three sections: a flattish cone at the feed end, a cylindrical drum centrally and a steep cone; the assembly is hung horizontally between trunnions.

hard lead (*Eng*) All antimonial lead; metal in which the high degree of malleability characteristic of pure lead is destroyed by the presence of impurities, of which antimony is the most common.

hardmetal (*Eng*) Sintered tungsten carbide. Used for the working tip of high-speed cutting tools. See SINTERING.

hard metals (*Eng*) Metallic compounds with high melting points; typified by refractory carbides of the TRANSITION METALS (fourth to sixth groups of the periodic system), notably tungsten, tantalum, titanium and niobium carbides.

hardness (*Electronics*) The degree of vacuum in an evacuated space, esp of a thermionic valve or X-ray tube. Also penetrating power of X-rays, which is proportional to frequency.

hardness (*Eng*) Signifies, in general, resistance to cutting, indentation and/or abrasion. It is actually measured by determining the resistance to indentation, as in BRINELL, ROCKWELL, SCLEROSCOPE and VICKERS HARDNESS TESTS. The values of hardness obtained by the different methods are to some extent related to each other and to the ultimate tensile stress of non-brittle metals. See panel on HARDNESS MEASUREMENTS.

hardness (*Min*) The resistance which a mineral offers to abrasion. The absolute hardness is measured with the aid of a SCLEROMETER. The comparative hardness is expressed in terms of Mohs' scale, and is determined by testing against 10 standard minerals: (1) talc, (2) gypsum, (3) calcite, (4) fluorite, (5) apatite, (6) orthoclase, (7) quartz, (8) topaz, (9) corundum, (10) diamond. Thus a mineral with 'hardness 5' will scratch or abrade fluorite, but will be scratched by orthoclase. Hardness varies on different faces of a crystal, and in some cases (eg kyanite) in different directions on any one face.

hardness measurements (*Eng*) A variety of relatively simple tests purporting to measure the hardness of materials. See panel on HARDNESS MEASUREMENTS.

hardness (water) (*Chem*) See HARD WATER.

hard packing (*Print*) Hard paper employed to cover the cylinder of a printing press when printing on hard, smooth papers from engravings etc; used in order to obtain a sharp impression. Also necessary when printing from plastic or rubber plates.

hard pad (*Vet*) A name for the HYPERKERATOSIS of dog's pads due to infection of the animal with CANINE DISTEMPER.

hard palate (*BioSci*) In mammals, the anterior part of the roof of the buccal cavity, consisting of the horizontal palatine plates of the maxillary and palatine bones covered with mucous membrane.

hard pan (*Min*) (1) A hardened impervious layer of soil cemented by iron oxides and hydroxides (*iron pan*), silica, carbonates, etc, sometimes clayey. (2) A layer of partly cemented gravel below the surface in a gold placer.

hard plaster (*Build*) A term usually applied to hard-setting forms of gypsum plaster, eg KEENE'S CEMENT.

hard plating (*Eng*) Chromium plating deposited in appreciable thickness directly onto the base metal, ie without a preliminary deposit of copper or nickel. The coating is porous, but offers resistance to corrosion and to wear.

hard radiation (*Radiol*) Qualitatively, the more penetrating types of X-, beta and gamma rays.

hard return (*ICT*) A LINE FEED typed in by the USER when editing a text document. Unlike a SOFT RETURN, the WORD PROCESSOR will not remove this line feed when the text is JUSTIFIED.

Hardness measurements

There are a wide variety of tests purporting to measure the hardness of materials. These range from a scale of what scratches what (MOHS' SCALE, PENCIL HARDNESS TEST), through measuring the size of the impression left by an indenter of prescribed geometry under a known load (BRINELL, KNOOP and Vickers hardness), or the depth to which an indenter penetrates under specified conditions (INTERNATIONAL RUBBER HARDNESS DEGREE (IRHD) or ROCKWELL B AND C, SHORE A), to the height of rebound of a ball or hammer dropped from a given distance (SHORE SCLEROSCOPE). Not surprisingly each test produces a different number (some on arbitrary scales) for the hardness of a given material. The approximate correlation between different scales of hardness is shown in the diagram.

The chief attractions of the various hardness tests are that they are relatively simple and quick to perform, and that they are virtually non-destructive. Thus they are very well suited for quality control purposes. However, there is not a well-defined materials property called hardness, and what all these tests measure is differing combinations of the elastic, plastic, and sometimes fracture behaviour of materials.

Relating the results to properties such as yield stress and Young's modulus is not straightforward. At one extreme, the IRHD (or Shore A) measures solely the elastic response of rubbers. It is widely used as a check on the degree of cure or cross-linking of a rubber, and there is an approximate correlation between the hardness value and the shear modulus. At the other extreme, the size of the plastically deformed impression produced by indenters such as the 136° diamond pyramid in the VICKERS HARDNESS TEST must obviously bear some relation to the yield stress.

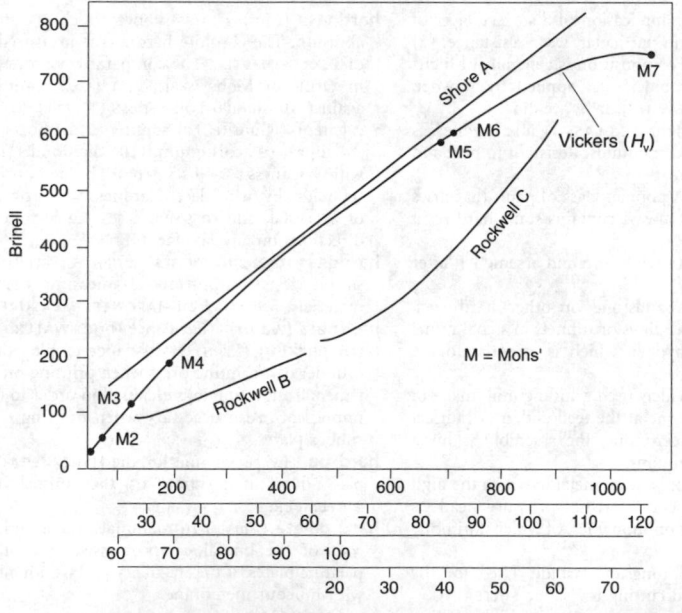

Hardness measurements Relations between the various kinds.

hard-rock geology (*Geol*) An informal term for the geology of igneous and metamorphic rocks. Cf SOFT-ROCK GEOLOGY.

hard-rock mining (*MinExt*) A term used to distinguish between deposits soft enough to be detached by mechanical excavator and those which must first be loosened by blasting.

hard-rock phosphate (*Geol*) A phosphatic deposit resulting from the leaching of calcium carbonate out of a phosphatic limestone, leaving a phosphatic residue. Applied specifically to the phosphate deposits of Florida which have this origin.

hard-sectored disk (*ICT*) One that is formatted partially or wholly by markers which are put on the disk when it is made.

hard segment (*Chem*) A term used for rigid parts of block copolymer polymers, esp polyurethanes and block polyesters, formed from isocyanate or polyester groups respectively. Such rigid chain segments often co-crystallize, so reinforcing the final material.

hard soaps (*Chem*) See SOAPS.

hard solder (*Eng*) One containing more than 0·3% of carbon.

hard solders (*Eng*) For joining metals, usually copper–silver–zinc alloys, which melt at temperatures between 600 and 850°C; they have greater strength than those based on lead–tin alloys. See SILVER SOLDER.

hard space (*ICT*) A space character that is put in by the user, eg between words. This space will not be removed by the program if the text is reformatted. A soft space will be inserted by the computer to format the text if required.

hard stocks (*Build*) Bricks which are sound but have been overburnt and are not of good shape and colour.

hard twist (*Textiles*) A yarn with more than the standard amount of twist, inserted to secure the desired effects in particular fabrics, eg crêpes, voiles, gaberdines.

hardware (*ICT*) General term for all mechanical, electrical and electronic components of a COMPUTER SYSTEM including PERIPHERALS. Cf SOFTWARE.

hardware (*Space*) The physical material (eg structure, electrical harness, computers) produced for space systems as opposed to non-tangible aspects such as computer software and operating procedures.

hard waste (*Textiles*) Waste from single or folded twisted yarns from cop ends and waste made during winding, warping, reeling and weaving.

hard water (*Chem*) Water having magnesium and calcium ions in solution; the hardness can be either temporary, due to calcium or magnesium carbonates, or permanent, due to sulphates and other salts and that cannot be removed by boiling. Hard water resists lather formation by soaps. Water with less than the equivalent of 50 ppm of calcium carbonate is classed as being soft. See PERMANENT HARDNESS, TEMPORARY HARDNESS.

hardwired logic (*ICT*) Permanent circuitry, often integrated circuit elements and their interconnections. Also *firmware*. See LOGIC ARRAY.

hardwood (*For*) Dense, close-grained wood from deciduous dicotyledonous trees (oak, beech, ash, teak, etc). Cf SOFTWOOD.

Hardy and Schulze law (*MinExt*) The observation that the efficiency of an ion used as a coagulating agent is roughly proportional to its state of oxidation.

Hardy–Weinberg law (*BioSci*) The gene frequencies in a large population remain constant from generation to generation if mating is at random and there is no selection, migration or mutation. If two alleles A and a are segregating at a locus, and each has a frequency of p and q respectively, then the frequencies of the genotypes AA, Aa and aa are p^2, $2pq$ and q^2 respectively.

harelip (*Med*) A congenital cleft in the upper lip, often associated with cleft palate.

Hare's apparatus (*Phys*) A simple apparatus for comparing the densities of two liquids. An inverted U-tube has its limbs dipping each into one of the liquids. By suction at the top of the U-tube, the liquids rise in the limbs. The densities are inversely proportional to the heights the liquids rise above their reservoirs.

Harker diagram (*Geol*) A variation diagram in which chemical analyses of rocks are plotted to show their relationships. The constituents are plotted as ordinates against the silica content as abscissa.

H-armature (*ElecEng*) See SHUTTLE ARMATURE.

harmonic (*Phys*) Sinusoidal component of repetitive complex waveform with frequency which is an exact multiple of basic repetition frequency (the fundamental). The full set of harmonics forms a Fourier series which completely represents the original complex wave. In acoustics, harmonics are often termed overtones, and these are counted in order of frequency above, but excluding, the lowest of the detectable frequencies in the note; the label of the harmonic is always its frequency divided by the fundamental. The nth overtone is the $(n + 1)$th harmonic. Fig. ▷

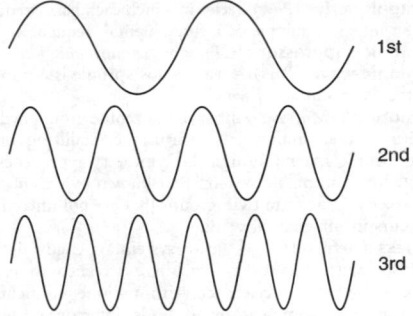

harmonic

harmonic absorber (*Phys*) An arrangement for removing harmonics in current or voltage waveforms, using tuned circuits or a wave filter.

harmonic analysis (*Phys*) The process of measuring or calculating the relative amplitudes of all the significant harmonic components present in a given complex waveform. The result is frequently presented in the form of a Fourier series, eg $A = A_0 \sin \omega_t + A_1 \sin (2\omega_t + \varphi_1) + A_2 \sin (3\omega t + \varphi_2) + \ldots$, where ω is the pulsatance and φ is the phase angle.

harmonic antenna (*ICT*) One whose overall length is an integral number (greater than one) of quarter-wavelengths.

harmonic components (*Phys*) Any term (except the first) in a Fourier series which represents a complex wave.

harmonic conjugate (*MathSci*) Two of four collinear points are harmonic conjugates of the other two if the CROSS-RATIO of the four points has the value −1. Also used of a pencil of four concurrent lines.

harmonic distortion (*Phys*) The production of harmonic components from a pure sine wave signal as a result of non-linearity in the response of a transducer or amplifier.

harmonic drive (*ICT*) See HARMONIC EXCITATION.

harmonic excitation (*ICT*) (1) Excitation of an antenna on one of its harmonic modes. (2) Excitation of a transmitter from a harmonic of the master oscillator. Also *harmonic drive*.

harmonic filter (*ICT*) One that separates harmonics from fundamental in the feed to an antenna. Also *harmonic suppressor*.

harmonic function (*MathSci*) A function which satisfies Laplace's equation.

harmonic generator (*Phys*) Waveform generator with controlled fundamental frequency, producing a very large number of appreciable-amplitude odd and even harmonic components which provide a series of reference frequencies for measurement or calibration. See MULTIVIBRATOR.

harmonic interference (*ICT*) That caused by harmonic radiation from a transmitter and outside the specified channel of radio communication.

harmonic mean (*MathSci*) (1) A number inserted between two numbers such that the resulting sequence is a harmonic progression. (2) The *harmonic mean* of n numbers a_r is the reciprocal of the arithmetic mean of their reciprocals, ie

$$n / \sum_{r=1}^{n} \frac{1}{a_r}$$

harmonic motion (*Phys*) See SIMPLE HARMONIC MOTION.

harmonic progression (*MathSci*) The numbers $a, b, c, d \ldots$ form a harmonic progression if their respective reciprocals form an arithmetic progression.

harmonic ratio (*MathSci*) The relationship between four collinear points when their CROSS-RATIO equals −1.

harmonic series (*MathSci*) The series $1 + \frac{1}{2} + \frac{1}{3} + \frac{1}{4} + \ldots$, which can be shown to diverge.

harmonic series (*Phys*) Series in which each basic frequency is an integral multiple of a fundamental frequency.

harmonic suppressor (*ICT*) Same as *harmonic filter*.

harmonic wave (*Phys*) A wave whose profile is a pure sine curve. Also *sinusoidal wave*.

harmotome (*Min*) A member of the zeolite group, hydrated silicate of aluminium and barium, crystallizing on the monoclinic system, though the symmetry approaches that of the orthorhombic system. Best known by reason of the distinctive cruciform twin groups that are not uncommon. Occurs in mineralized veins.

harness (*Aero*) (1) The entire system of engine ignition leads, particularly those which are screened to prevent electromagnetic interference with radio equipment. (2) The parallel combination of leads interconnecting the thermocouple probes of a turbine engine exhaust gas temperature-indicating system. See GAS TEMPERATURE. (3) Prefabricated electrical connections for any electrical or electronic system. (4) Straps by which aircrew are held in their seats.

Harris process (*Eng*) A method of softening lead. Arsenic, antimony and tin are oxidized by adding sodium nitrate and lead oxide, and the oxides formed are caused to react with sodium hydroxide and chloride to form arsenates, antimonates and stannates.

harrow (*Agri*) Trailed equipment with upright teeth or discs, used to break up and level previously cultivated land.

hartite (*Min*) A naturally occurring hydrocarbon compound, $C_{20}H_{34}$, crystallizing in the triclinic system.

hartley (*ICT*) Unit of INFORMATION CONTENT.

Hartley oscillator (*Electronics*) An oscillator circuit incorporating a parallel-tuned resonant circuit with a tapping point on the inductance. Feedback is provided by the section of inductance connected between grid and cathode (in valve circuit) or between source and gate (in a field-effect transistor) or between emitter and base (in a transistor).

Hartley principle (*ICT*) General statement that amount of information that can be transmitted through a channel is the product of frequency bandwidth and time during which it is open, whether time division is used or not. See INFORMATION CONTENT.

Hartmann dispersion formula (*Phys*) An empirical expression for the variation of refractive index n of material with the wavelength of light λ:

$$n = n_0 + \frac{c}{(\lambda - \lambda_0)^a}$$

where n_0, c and λ_0 are constant for a given material. For glass a is about 1·2.

Hartman oscillator (*Acous*) A device, consisting basically of a conical nozzle, supersonic gas jet and cylindrical cavity (resonator), used for generating high-intensity ultrasound in fluids or gases.

Hartman test (*Phys*) A test for aberration of a lens, in which a diaphragm containing a number of small apertures is placed in front of the lens and the course of the rays is recorded by photographing the pencils of light in planes on either side of the focus.

Hartnell governor (*Eng*) An engine governor in which the vertical arms of two or more bell-crank levers support heavy balls, the horizontal arms carrying rollers which abut against the central spring-loaded sleeve operating the engine-governing mechanism.

Hartree equation (*Electronics*) One relating flux density in a travelling-wave magnetron to minimum anode potential required for oscillation in any given mode. Graphs representing this relationship for different mode numbers form a *Hartree diagram*.

Harvard classification (*Astron*) A method of classifying stellar spectra, employed by the compilers of the Draper Catalogue of the Harvard Observatory and now in universal use. See SPECTRAL TYPES.

harvest interval (*Agri*) See CLEARANCE TIME.

harvest mite (*Vet*) The popular name of the parasitic larval stage of the mite *Trombicula autumnalis*, which occurs in the skin of animals and humans. The nymphal and adult stages of the mite are free-living.

harvest Moon (*Astron*) The name given in popular language to the full Moon occurring nearest to the autumnal equinox, at which time the Moon rises on several successive nights at almost the same hour. This retarded rising, due to the small inclination of the Moon's path to the horizon, is most noticeable at the time of the full Moon, although it occurs for some phase of the Moon each month.

harvest spider (*BioSci*) Common name for arachnids of the order OPILIONES. Also *harvestmen*.

Harvey process (*Eng*) Toughening treatment for alloy steels, involving superficial carburization, followed by heating to a high temperature and quenching with water.

harzburgite (*Geol*) An ultrabasic igneous rock belonging to the peridotite group. It consists almost entirely of olivine and orthopyroxene, usually with a little chromite, magnetite and diopside.

Harz jig (*MinExt*) Concentrating appliance in which water is pulsed through a submerged fixed screen, across which suitably sized ore moves. Heaviest particles gravitate down and through and lighter ones overflow.

Hashimoto thyroiditis (*Med*) A disease of the thyroid gland, characterized by chronic inflammatory changes due to infiltration by lymphocytes, plasma cells and macrophages. The gland becomes enlarged and hard. Autoantibodies against thyroglobulin and other thyroid antigens are usually present in the blood.

hashing (*ICT*) Generating HASH TOTALS to associate with items of data. The numbers are used as index numbers or storage addresses for the data.

hashish (*BioSci*) See CANNABIS.

hash total (*ICT*) A meaningless number generated from a coded data item or its key. See CHECK SUM, HASHING.

hasp (*Build*) A fastening device in which a slotted plate fits over a staple and is secured to it by means of a padlock or peg.

Hassal's corpuscles (*BioSci*) Characteristic cellular structures of unknown function found in the medulla of the thymus. They consist of flattened spheres of epithelial cells.

hastate (*BioSci*) Shaped like an arrow head with narrow basal lobes pointing outwards (ie halberd-shaped).

hastingsite (*Min*) A monoclinic amphibole, hydrated sodium calcium iron aluminosilicate. The name is used for an end-member compositional variety of amphibole.

hatch coaming (*Ships*) The strengthened frame surrounding a HATCHWAY. Its functions are to replace the strength lost by cutting the beams and deck plating, protect the opening and provide support for the hatch cover.

hatchet (*Build*) A small axe used for splitting or rough-dressing timber.

Hatch–Slack–Kortshak pathway (*BioSci*) Metabolic pathway responsible for primary carbon fixation in C4 plants, most of which are found in hot climates with high light intensity.

Hatch–Slack pathway (*BioSci*) A method of CO_2 fixation in C4 plants. See PEP CARBOXYLASE and panel on C4 PHOTOSYNTHETIC PATHWAY.

hatchway (*Ships*) An opening in the deck to allow cargo to be shipped into and removed from the hold.

HAT medium (*BioSci*) A growth medium for animal cells in culture that contains hypoxanthine, aminopterin and thymidine as sources of purines and pyrimidines. Cells that can utilize these exogenous sources are selected for (eg HYBRIDOMA cells in the production of monoclonal antibodies) because parental cells that lack the necessary salvage enzymes are eliminated.

haulage level (*MinExt*) Underground tramming road in, or parallel to, the strike of the ore deposit, usually in footwall. Broken ore gravitates, or is moved, to ore chutes and drawn to trucks in this level.

haul distance (*CivEng*) The distance, at any particular time, that excavated material from eg a cutting has to be carried before deposition in order to form an embankment. The 'haul' is the sum of the products of each load by its haul distance.

haulm (*Agri*) Residual foliage of a crop left after harvest.

haunch (*Build*) The part forming a stub tenon, left near the root of a HAUNCHED TENON. Also *hauncheon*.

haunched tenon (*Build*) A tenon from the width of which a part has been cut away, leaving a haunch near its root.

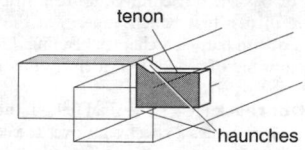

haunched tenon Shown in a stopped version with a blind mortise.

hauncheon (*Build*) See HAUNCH.

haunches (*Build*, *CivEng*) See FLANKS.

haunching (*Build*) A mortise cut to receive the HAUNCH of a HAUNCHED TENON.

Hausdorff space (*MathSci*) A topological space in which any two points belong to disjoint open sets.

hausmannite (*Min*) A blackish-brown crystalline form of manganese oxide, occurring with other manganese ores. Crystallizes in the tetragonal system but is often found massive.

haustellate (*BioSci*) Mouthparts modified for sucking as in many insects.

haustellum (*BioSci*) In Diptera, the distal expanded portion of the proboscis. Adj *haustellate*.

haustorium (*BioSci*) An outgrowth from a parasite that penetrates a tissue or cell of its host and acts as an organ for absorbing nutrients.

Hauterivian (*Geol*) A stage in the Lower Cretaceous. See MESOZOIC.

haüyne (*Min*) A feldspathoid, crystallizing in the cubic system, consisting essentially of silicate of aluminium, sodium and calcium, with sodium sulphate; occurs as small blue crystals chiefly in phonolites and related rock types.

Haversian canals (*BioSci*) Small channels pervading compact bone and containing blood vessels.

Haversian lamellae (*BioSci*) In compact bone, the concentrically arranged lamellae which surround a Haversian canal.

Haversian spaces (*BioSci*) In the development of bone, irregular spaces formed by the internal resorption of the original cartilage.

Haversian system (*BioSci*) In compact bone, a Haversian canal with surrounding lamellae.

haversine (*MathSci*) Half of the VERSINE, ie $\frac{1}{2}(1-\cos\theta)$.

Hawaiian-type eruption (*Geol*) An eruption characteristic of a shield volcano with large quantities of fluid lava, mainly as lava fountains from fissures, and only rare explosive phenomena.

hawaiite (*Geol*) An olivine-bearing oligoclase or andesine trachyandesite.

hawk (*Build*) A small square board, with handle underneath, used to carry plaster or mortar.

Hawking radiation (*Astron*) A type of radiation predicted by S Hawking (1974) to emerge continuously from black holes. Of pairs of particles produced by quantum effects in space near a black hole, one is absorbed by the black hole whilst the other is radiated; the black hole slowly evaporates into photons and other particles, finally expiring in a huge burst of gamma rays. See panel on BLACK HOLE.

hawk's eye (*Min*) A dark-blue form of silicified crocidolite found in Griqualand West; when cut EN CABOCHON, it is used as a semiprecious gemstone. Cf TIGER'S EYE.

hawser (*Ships*) A tubular casting fitted to the bows of a ship, through which the anchor chain or cable passes. Also *hawse-pipe*.

Hawthorne effect (*Psych*) The observation that experimental subjects who are aware that they are part of an experiment often perform better than totally naive subjects, and that innovations produce positive effects independently of the nature of the change.

hay (*Agri*) Plants, commonly grasses, dried for fodder.

Hay bridge (*ElecEng*) An ac bridge quite widely used for the measurement of inductance.

hay fever (*Med*) Acute nasal catarrh and conjunctivitis in atopic subjects, caused by inhalation of allergens (usually pollens). Its occurrence is due to an immediate type 1 *hypersensitivity reaction* resulting from combination of cell–IgE antibody with the causative allergen. There is a tendency for this type of sensitization to be inherited. Also *seasonal allergic rhinitis*.

Hayflick limit (*BioSci*) The limit to the number of division cycles that animal tissue cells will undergo in culture (about 50). Cells that surpass the limit are often said to be transformed.

hay tedder (*Agri*) A machine that turns cut hay left on the field, to aid drying before gathering.

hazard (*FoodSci*) Any physical, chemical or microbiological material within food that may cause harm to the consumer. See HACCP.

hazard analysis (*FoodSci*) The identification of all potential hazards at each stage in a food production process. See HACCP.

hazard analysis and critical control point (*FoodSci*) A systematic study of all stages of food production, from raw material supply through to storage and distribution, to identify all potential hazards, assess the risk of likely occurrence, determine the critical control points, establish appropriate control measures and their critical limits. Usually done by a team of food production and technical specialists. A HACCP approach to food production ensures food safety and is a legal requirement in the UK. Abbrev *HACCP*.

hazardous waste (*Genrl*) Any waste material, usually generated by industry, that is potentially damaging to the environment or human health, eg radioactive waste, arsenic, cyanide compounds, heavy metals and other toxic compounds, corrosive agents, organic solvents, pesticides and explosives. Also *hazardous materials*.

haze (*EnvSci*) A suspension of solid particles of dust and smoke etc, reducing visibility above 1 km.

Hazen and Williams' formula (*Eng*) An empirical formula relating the flow in pipes and channels to the hydraulic radius and the slope, using a coefficient which varies with the surface roughness.

Hazen number (*EnvSci*) A unit of measurement of coloration of water based upon the colour produced by 1 mg of platinum per litre in the presence of a cobalt chloride.

Hazen's uniformity coefficient (*PowderTech*) A measure of the range of particle sizes present in a given powder. It is defined by the equation $H = A/B$, where A is the 60th percentile of the cumulative percentage undersize by weight of the powder, and B is the 10th.

HAZMAT (*Genrl*) Abbrev for *hazardous materials*.

Hb (*Chem*) Symbol for a molecule of haemoglobin minus the iron atom.

H-bar (*h̄*) (*Phys*) Symbol for DIRAC'S CONSTANT. Equal to PLANCK'S CONSTANT *h* divided by 2π.

HCI (*ICT*) Abbrev for *human–computer interface*. See USER INTERFACE.

H-class insulation (*ElecEng*) A class of insulating material which can withstand a temperature of 180°C.

HCP (*Chem*) Abbrev for HEXAGONAL CLOSE PACKING.

HDD (*Aero*) See HEAD-DOWN DISPLAY.

HDI (*Chem*) Abbrev for HEXAMETHYLENE DIISOCYANATE.

H-display (*Radar*) Modified B-DISPLAY to include angle of elevation. The target appears as two adjacent bright spots and the slope of the line joining these is proportional to the sine of the angle of elevation.

HD-MAC (*ImageTech*) A WIDE-SCREEN, HIGH-DEFINITION version of the MAC colour TV system with 1250 lines.

HDPE (*Chem*) Abbrev for HIGH-DENSITY POLYETHYLENE.

HDTV (*ImageTech*) Abbrev for *high-definition television*, in general systems using 1000 or more scanning lines per frame.

HD video disk (*ImageTech*) A VIDEO DISK capable of carrying HIGH-DEFINITION signals in compressed analogue or digital form.

HD VTR (*ImageTech*) A videotape recorder able to record HIGH-DEFINITION signals. See UNIHI.

He (*Chem*) Symbol for HELIUM.

head (*Arch*) The capital of a column.

head (*BioSci*) A dense inflorescence of sessile flowers, usually a CAPITULUM.

head (*Build*) See LINTEL.

head (*ElecEng*) Recording and reproducing unit for magnetic tape, containing exciting coils, a laminated core, in ring form with a minute gap. Flux leakage across this gap enters the tape and magnetizes it longitudinally. See FLUX GATE.

head (*Eng*) (1) Any part having the shape or position of a head, eg the head of a bolt. (2) Any part or principal part analogous to a head, eg the head of a hammer or a lathe. (3) A distance representing the height above a datum which would give unit mass of a fluid in a conduit the potential energy equal to the sum of its actual potential energy, its kinetic energy and its pressure energy. See BERNOULLI'S LAW.

head (*Geol*) A superficial deposit consisting of angular fragments of rock, originating from the breaking up of rock by alternate freezing and thawing of its contained water, followed by downhill movement. Head is found in valleys in periglacial regions, ie those formerly near the edge of an ice sheet. In SE England, also *coombe rock*.

head (*ImageTech*) (1) General term for the central mechanism of a motion picture projector or printer. See SOUND HEAD. (2) The mounting for a camera on a tripod. (3) The start of a roll of film or magnetic tape; *head end*. (4) The top of a motion picture frame.

head (*MinExt*) (1) An advance main roadway driven in solid coal. (2) The top portion of a seam in the coal face. (3) The difference in air pressure producing ventilation. (4) The whole falling unit in a stamp battery, or merely the weight at the end of the stem.

head (*Print*) The top edge of a volume: the top margin of a page. See FORE-EDGE, TAIL.

head amplifier (*Electronics*) A pre-amplifier which has to be placed physically adjacent to the signal source in order to provide gain before the signal is lost though transmission lines or cables or before it is swamped by interference. Used esp with receiving aerials and sensitive microphones.

headband (*Print*) A decorative band of silk or other material at the head of a book, between the back and the case. See TAILBAND.

head bay (*Build*) The part of a canal lock immediately above the headgates.

head-box (*Paper*) See FLOW BOX.

head cap (*Print*) The leather at the head and tail of the spine folded over in a curve, the HEADBAND if present, being left visible.

head clog (*ImageTech*) Clogging of the HEAD GAP with oxide particles shed by the tape, or dust, largely preventing recording and playback.

head crash (*ICT*) Dramatic and expensive descent of the read/write head on to the surface of a magnetic disk, caused by mechanical malfunction.

head-down display (*Aero*) Usually a visual display mounted inside the cockpit to supplement the HEAD-UP DISPLAY.

head end (*ICT*) The part of a community antenna TV system that includes antenna(s), receivers and signal distribution equipment.

head end (*NucEng*) That part of a reprocessing plant which precedes solvent extraction; it therefore includes facilities for storing and handling fuel assemblies, breaking down and dissolving them. See THORP.

header (*Build*) A whole brick which has been laid so that its length is at right angles to the face of the wall. See fig. at ENGLISH BOND.

header (*ElecEng, Electronics*) The base for a relay, or transistor, can with hermetically sealed insulated leads.

header (*Eng*) A box or manifold supplying fluid to a number of tubes or passages, or connecting them in parallel.

header (*ICT*) (1) Text which appears at the top of every page of a document, eg a chapter heading. (2) Data placed at the beginning of a FILE that identify the file and which may describe its structure.

header error check (*ICT*) The final field in the header of an ATM CELL. It provides a check sum over the contents of the header, allowing detection and sometimes correction of errors, and can also be used to find the start of unaligned cells arriving at a SWITCH or terminal.

header joist (*Build*) See TRIMMER.

headframe (*MinExt*) The steel or timber frame at the top of a shaft, which carries the sheave or pulley for the hoisting rope, and serves such other purposes as, eg acting as transfer station for hoisted ore, or as loading station for man and materials. Also *headgear*.

head gap (*ImageTech*) The space between the pole pieces of a HEAD in which the magnetic field is produced or induced during recording or playback.

headgates (*Build*) The gates at the high-level end of a lock.

headgear (*MinExt*) See HEADFRAME.

head grit (*Vet*) See YELLOWSES.

heading (*Build*) See HEADING COURSE.

heading (*CivEng*) A relatively small passage driven in the line of an intended tunnel, the latter being afterwards formed by enlarging the former.

heading (*MinExt*) Passageway through solid coal.

heading bond (*Build*) The form of bond in which every brick is laid as a header, each 4·5 in (112 mm) face breaking joint above and below; used for footings and corbellings but not for walling.

heading chisel (*Build*) See MORTISE CHISEL.

heading course (*Build*) An external or visible course of bricks which is made up entirely of headers. Also *heading*.

heading date (*Agri*) The date on which 50% of the flower spikes of a crop are completely emerged above the FLAG LEAF.

heading indicator (*Aero*) The development of supersonic fighters, which climb at extremely steep angles, made it essential to have an even more comprehensive instrument than the *attitude indicator*. A sphere enables compass heading to be included, thereby giving complete 360° presentation about all three axes.

heading joint (*Build*) A joint between the ends of boards abutting against each other.

headings (*MinExt*) Concentrate settling nearest to the entry point of a concentrating device such as a SLUICE or BUDDLE. Also *head tin*.

heading stage (*Agri*) The stage of grain crop development when the seed head first emerges from the sheath.

headland (*Agri*) The unploughed area at the end of field where the plough is turned.

headline (*Print*) The line of type placed at the top of the page, giving either the title of the book or the chapter heading.

head motion (*MinExt*) Vibrator, a sturdy device which gives reciprocating movement to shaking tables, used in gravity methods of concentration.

head moulding (*Build*) A moulding situated above an aperture.

head nailing (*Build*) The method of nailing slates on a roof in which nails are driven in the slates near their heads or higher edges. Cf CENTRE NAILING.

head race (*Build*) A channel conveying water to a hydraulically operated machine.

head-rail (*Build*) The horizontal member of a door case.

headroom (*Build*) The uninterrupted height within a building on any floor, or within a staircase, tunnel, doorway, etc. Also *headway*.

heads (*Build*) The tiles forming a course around the eaves.

headspace (*FoodSci*) The space between the surface of a food and the top of its packaging which needs to be large enough to allow sufficient air to be withdrawn for a good vacuum to be obtained in vacuum packaging and to allow the product to be sufficiently agitated by shaking to ensure uniformity in some dry goods and liquids.

headstock (*Eng*) In general, a device for supporting the end or head of a member or part; eg (1) the part of a lathe that carries the spindle, (2) the part of a planing machine that supports the cutter or cutters, (3) the supports for the gudgeons of a wheel, (4) the movable head of some measuring machines.

head tip (*ImageTech*) The end of the HEAD which extends beyond the circumference of the DRUM in a videotape recorder. Also *pole tip*. See TIP PENETRATION.

head-to-head unit (*Chem*) A pair of asymmetric repeat units of a polymer fused at identical ends. Normal structure of chain is head-to-tail. A kind of chain defect.

head tree (*Build*) A timber block placed on the top of a post, so as to provide increased bearing surface.

head-up display (*Aero*) The projection of instrument information onto the windscreen or a sloping glass screen in the manner of the REFLECTOR SIGHT, so that the pilot can keep a continual lookout and receive flight data at the same time. Used originally with fighter interception and attack radar, later adapted for blind approach, landing and general flight data. Abbrev *HUD*.

head valve (*Eng*) The delivery valve of a pump, as distinct from the suction or FOOT VALVE.

head wall (*Build*) A wall built in the same plane as the face of a bridge arch.

headway (*Build*) See HEADROOM.

headworks (*Eng*) In a hydroelectric scheme, a dam forming a reservoir or a low weir across a river or stream, which possesses the necessary intakes and control gear to divert the water into an aqueduct.

Heaf test (*Med*) A commonly used tuberculin test in which tuberculin is injected intradermally with a multiple puncture apparatus. A positive reaction indicates the presence of cellular immunity to tuberculosis.

heald (*Textiles*) Part of the loom mechanism used to raise and lower the warp. Comprises eyes formed of coated twine or wire through which warp ends are drawn.

healing (*Build*) The operation of covering a roof with tiles, lead, etc.

health physics (*Radiol*) The branch of radiology concerned with health hazards associated with ionizing radiations, and protection measures required to minimize these. Personnel employed for this work are health physicists or radiological safety officers.

health psychology (*Psych*) The speciality within applied psychology that is concerned with the use of psychology to promote personal and public health.

heap leaching (*MinExt*) This, perhaps aided by heap roasting, is the dissolution of copper from oxidized ore by solvation with sulphuric acid. The resulting liquor is run over scrap iron to precipitate out its copper.

heap sampling (*MinExt*) See QUARTERING.

hearing (*Acous*) The subjective appreciation of externally applied sounds.

hearing aid (*Acous*) A device used by a person with hearing loss to improve audition of external sounds, either in the form of an acoustic amplifier (collector), or in the form of a microphone–receiver combination, with or without amplifier.

hearing loss (*Acous*) Of a partially deaf ear, the difference in decibels between the threshold of hearing and that of a normal ear at any frequency.

heart (*BioSci*) A hollow organ, with muscular walls, that by its rhythmic contractions pumps the blood through the vessels and cavities of the circulatory system.

heart attack (*Med*) The common term for a MYOCARDIAL INFARCTION.

heart block (*Med*) The condition in which a lesion of the special tissue that conducts the contraction impulse from the atrium to the ventricle prevents the spread of the wave of contraction, usually leads to a slow pulse.

heart bond (*Build*) A form of bond having no through-stones, headers consisting of a pair of stones meeting in the middle of the wall, the joint between them being covered by another header stone.

heartburn (*Med*) A burning sensation in the midline of the chest, usually associated with dyspepsia.

heart cam (*Eng*) A cam in the form of a heart, used in stopwatches and chronographs to bring the recording hand instantly back to zero on pressing a button.

heart failure (*Med*) A condition in which the heart fails to maintain an adequate flow of blood to all the body tissues, for whatever reason.

hearth (*Eng*) The floor of a reverberatory, open hearth, cupola or blast furnace, made of refractory material able to support the charge and collect molten products for periodic removal.

hearting (*Build*) The operation of building the inner part of a wall, between its facings.

heart sounds (*Med*) Sounds generated by the heart and heard by AUSCULTATION over the left side of the chest. The first sound coincides with the closure of the mitral and TRICUSPID valves, the second sound with closure of the aortic and pulmonary valves. Added sounds may be heard in diseases of the heart.

heart transplant (*Med*) When a human donor heart is implanted into a patient whose heart disease can be controlled by no other means. The donor heart takes on the pumping function for the recipient.

heart valve implant (*Med*) Artificial valve grafted to heart tissue to replace diseased valves; usually composed of metal frame with synthetic fabric covering (eg polyethylene terephthalate) and polymer (eg polypropylene) ball or disk.

heartwater (*Vet*) A disease of cattle, sheep and goats, occurring in parts of Africa, caused by *Cowdria ruminantium* (*Rickettsia ruminantium*), and transmitted by ticks of the genus *Amblyomma*. Characterized by fever, nervous symptoms and often death.

heartwood (*For*) The older, inner core of wood in tree trunks and branches, which no longer functions in storage and conduction; in many species it is impregnated with gums and resins which help it to resist decay and make it denser and darker than the surrounding sapwood. Also *duramen*. Cf SAPWOOD. See panel on WOOD.

heat (*BioSci*) The period of sexual excitation.

heat (*Phys*) Energy in the process of transfer between a system and its surroundings as a result of temperature differences. However, the term is still used also to refer to the energy contained in a sample of matter. Also for TEMPERATURE, eg forging or welding heat. For some of the chief branches in the study of heat, see CALORIMETRY, HEAT UNITS, INTERNAL ENERGY, SPECIFIC LATENT HEAT, MECHANICAL EQUIVALENT OF HEAT, RADIANT HEAT, SPECIFIC HEAT CAPACITY, TEMPERATURE, THERMAL CONDUCTIVITY, THERMOMETRY and panel on RADIATION.

heat balance (*ChemEng*) A statement which relates for a chemical process or factory all sources of heat and all uses. The allowance of heats of reaction, solution and dilution must be included as well as conventional sources such as fuel, steam, etc. If done graphically, it is usually called a *Sankey diagram*. Also *energy balance*.

heat balance (*Eng*) Evaluation of operating efficiency of a furnace or other appliance, the total heat input being apportioned as to heat in the work, heat stored in brickwork, or refractory materials, loss by conduction, radiation, unburnt gases in waste products, sensible heat in dry flue gases, and latent heat of water vapour, thus determining the quantity and percentage of heat usefully applied and the sources of heat losses.

heat capacity (*Phys*) See SPECIFIC HEAT CAPACITY.

heat density (*Eng*) Weight and pressure of live gases in heating chambers of industrial furnaces, upon which the rate of heat transfer depends.

heat detector (*Eng*) An indirect-acting thermostat for operation in conjunction with a gas-flow control valve, and for controlling working temperatures in furnaces and heating appliances up to about 1000°C.

heat-distortion temperature (*Chem*) The point at which solid polymers sag or cannot maintain structural integrity as temperature is raised, so lies near their T_g. It is important in injection moulding, determining the point below which the product can be removed safely from the machine.

heat drop (*Eng*) Colloq term for transfer of heat energy, but often used for ENTHALPY HEAT DROP.

heater (*Electronics*) Conductor carrying current for heating a cathode, generally enclosed by the latter.

heater box (*Print*) The heated part of a blocking press. This is usually electrically heated and thermostatically controlled.

heater platen (*Print*) A grooved platen sliding into the heater box of a blocking press and holding the die.

heater transformer (*Electronics*) In equipment using thermionic valves, a mains transformer giving a low-voltage output for the valve heaters.

heat exchange (*ChemEng*) The process of using two streams of fluid for heating or cooling one or the other either for conservation of heat or for the purpose of adjusting process streams to correct processing temperatures.

heat exchanger (*NucEng*) Any device for the exchange of heat between two substances without intermixing. Typically those in which the gaseous or liquid coolant from a nuclear reactor heats water to provide steam for the turbines in an indirect-cycle reactor.

heat exhaustion (*Med*) Less severe than HEAT STROKE and usually occurring after several days of salt and water depletion due to heat exposure.

heat filter (*ImageTech*) In an illuminating system, a filter which reduces or reflects infrared radiation in the beam and hence its heating effect.

heat-flow measurement (*Geol*) The measurement of the amount of heat leaving the Earth.

heat flux (*ChemEng*) The total flow of heat in HEAT EXCHANGE, in appropriate units of time and area.

heath (*EnvSci*) Vegetation type consisting of evergreen woody shrubs growing on acid soil. In N Europe the species are largely members of the *Ericaceae*, but the term heath is often used more widely to cover dwarf shrub communities in other parts of the world.

heather blindness (*Vet*) See CONTAGIOUS OPHTHALMIA.

heating curve (*Eng*) A curve obtained by plotting time against temperature for a metal heating under constant conditions. The curve shows the absorption of heat which accompanies melting and arrest points marking polymorphic changes.

heating depth (*ElecEng*) Thickness of skin of material which is effectively heated by dielectric or eddy current induction heating, or radiation.

heating element (*ElecEng*) The heating resistor, together with its former, of an electric heater, electric oven, or other device, in which heat is produced by the passage of an electric current through a resistance.

heating inductor (*ElecEng*) Conductor, usually water-cooled, for inducing eddy currents in a charge, workpiece or load. Also *applicator*, *work coil*.

heating limit (*ElecEng*) See THERMAL LIMIT.

heating muff (*Aero*) A device for providing hot air, consisting of a chamber surrounding an exhaust pipe or jet pipe.

heating resistor (*ElecEng*) The wire or other suitable material used as the source of heat in an electric heater.

heating time (*Electronics*) The time required after switching on, before a valve cathode, etc, reaches normal operating temperature.

heat-insulating concrete (*Eng*) High-alumina cement and a lightweight aggregate, eg kieselguhr, diatomic earth or vermiculite, to reduce heat transfer through furnace walls etc.

heat liberation rate (*Eng*) A measure of steam boiler performance, expressed in kJ l^{-1} h^{-1} (= 26·8 Btu ft^{-3} h^{-1}).

heat of formation (*Chem*) Strictly the *enthalpy of formation*. The net quantity of heat evolved or absorbed during the formation of 1 mole of a substance from its component elements in their standard states. Symbol ΔH_f. See STANDARD HEAT OF FORMATION.

heat of polymerization (*Chem*) Strictly, the *enthalpy of polymerization*. Heat given out during polymerization, caused by bond formation. Strict control is necessary in large-scale commercial operations. Symbol ΔDH_p. See CHAIN POLYMERIZATION.

heat of solution (*Chem*) The quantity of heat evolved or absorbed when 1 mole of a substance is dissolved in a large volume of a solvent. Symbol ΔH_s.

heat pipe (*Space*) A means of cooling where heat is transferred along a tube from a heat source to a heat sink of small temperature difference. Heat transfer is effected by a liquid which vaporizes at a desired temperature.

heat pump (*Eng*) Machine operating on a reversed heat engine cycle to produce a heating effect. Energy from a low-temperature source, eg earth, lake or river, is absorbed by the working fluid, which is mechanically compressed, resulting in a temperature increase. The high-temperature energy is transferred in a heat exchanger.

heat regenerator (*ChemEng*) A matrix of metal which is alternately heated by the waste gases from an industrial process and used to heat up the incoming gases.

heat-resisting alloy (*Eng*) Alloys developed to withstand high stresses at very high temperatures as in the fan blades of aero-engines.

heat-resisting paints (*Build*) Paints able to withstand high temperatures using synthetic resins with pigments unaffected by heat. Classified as withstanding temperatures: (1) up to 93°C (200°F) includes alkyd resin paints pigmented with titanium oxide, synthetic reds, monastral blue, yellow oxides, etc; (2) between 93 and 260°C (500°F) includes acrylic resin based paints pigmented with titanium oxide and suitable colours; (3) between 260 and 540°C (1000°F) based on silicone resins.

heat run (*ElecEng*) A test in which an electric machine or other apparatus is operated at a specified load for a long period to ascertain the temperature which it reaches.

heat-set ink (*Print*) Printing ink formulated to set by heating the surface of the stock.

heat setting (*Textiles*) Stabilizing fibres, yarns or fabrics by means of heating under controlled conditions. Thus a fabric may be heat set when it is held flat under tension while being extended in length and breadth. The resultant fabric will tend to retain its flatness in use. Pleats may also be heat set so that they remain clearly defined in a garment.

heat shock protein (*BioSci*) A conserved class of proteins found in both pro- and eukaryotic organisms, generally produced in excess when the organism or culture is subject to an elevated temperature or other environmental stress. Some act as chaperones to prevent denaturation of proteins.

heat sink (*ElecEng*) Usually metal plate designed eg with matt black fins, to conduct and radiate heat from an electrical component, eg a transistor or microprocessor.

heat spot (*BioSci*) An area of the skin sensitive to heat owing to the presence of certain nerve endings beneath the skin.

heat stroke (*Med*) The combination of coma, convulsions, a high temperature and other symptoms, as a result of exposure to excessive heat. Also *heat hyperpyrexia*.

heat transfer (*Phys*) See CONDUCTION, CONVECTION OF HEAT and panel on RADIATION.

heat transfer coefficient (*ChemEng*) The rate of heat transfer q between two phases may be expressed as

$$q = h\,A(T_1 - T_2)$$

where A is the area of the phase boundary, $(T_1 - T_2)$ is their difference in temperature, and the heat transfer coefficient *h* depends on the physical properties and relative motions of the two phases.

heat transfer salt (*ChemEng*) Molten salts used as a heating or quenching medium. Usually mixtures of sodium or potassium nitrate or nitrite, range 200–600°C. Abbrev HTS.

heat treatment (*Eng*) Generally, any heating operation performed on a solid metal, eg heating for hot-working, or annealing after cold-working. Particularly, the thermal treatment of steel by normalizing, hardening, tempering, etc, used also in connection with precipitation-hardening alloys, such as those of aluminium.

heat units (*Phys*) See BRITISH THERMAL UNIT, CALORIE, CALORIFIC VALUE, JOULE.

heave (*Geol, MinExt*) (1) The horizontal distance separating parts of a faulted seam, bed, vein or lode, measured normal to the fault plane. (2) Rising of the floor of a mine.

heavier-than-air aircraft (*Aero*) See AERODYNE.

Heaviside–Campbell bridge (*ElecEng*) An electrical network for comparing mutual and self-inductances.

Heaviside layer (*Phys*) See E-LAYER.

Heaviside–Lorentz units (*Phys*) Rationalized CGS Gaussian system of units, for which corresponding electric and magnetic laws are always similar in form. See RATIONALIZED UNITS.

Heaviside unit function (*ICT*) Step change in a magnitude, with an infinite rate of change, required in pulse analysis and transient response of circuits.

heavy-aggregate concrete (*NucEng*) That containing very dense aggregate material such as lead, barytes or iron nodules in place of some or all of the usual gravel, so increasing its gamma-ray absorption coefficient. Also used for appropriate general purposes (eg sea walls). See LOADED CONCRETE.

heavy chain (*BioSci*) A polypeptide chain in immunoglobulins that, together with the light chain, makes up the complete molecule. Each heavy chain consists of a variable region and a constant region composed of three or four domains, depending on the class. See IMMUNOGLOBULIN.

heavy chemicals (*Chem*) Those basic chemicals which are manufactured in large quantities, eg sodium hydroxide, chlorine, nitric acid, sulphuric acid.

heavy ground (*MinExt*) Unstable roof rock requiring special care and support.

heavy hydrogen (*Chem*) See DEUTERIUM.

heavy liquids (*MinExt*) Liquids, organic or solutions of heavy salts, of relative densities adjustable in the range 1·0–4·1, used to separate ore constituents into relatively heavy (sink) and light (float) fractions with fair precision, and to carry out specific gravity tests on minerals including gemstones. They include carbon tetrachloride, bromoform, methylene iodide and Clerici's solution.

heavy media separation (*MinExt*) A method of upgrading ore by feeding it into liquid slurry of intermediate density, the heavier fraction sinking to one discharge arrangement and the light ore overflowing. Used to remove shale from coal (*sink–float process*) and to reject waste from ore. Also *dense media separation*.

heavy metal (*BioSci*) (1) In electron microscopy, metal of high atomic number used to introduce electron density into a biological specimen by staining, negative staining or shadowing. (2) In plant nutrition, metals of moderate to high atomic number, eg Cu, Zn, Ni, Pb, present in soils due to an outcrop or mine spoil, preventing growth except for a few tolerant species and ecotypes. (3) In food technology, natural or consequential contaminants like Hg, Pb, Cd and As which are toxic. Their maximum concentration is set by law.

heavy metal (*EnvSci*) Loose term for polluting metal ions that are persistent and potentially toxic in the environment; includes arsenic, cadmium, lead, mercury, vanadium and zinc.

heavy-metal replacement (*Crystal*) See ISOMORPHOUS REPLACEMENT.

heavy mineral (*Geol*) A detrital mineral from a sedimentary rock having a higher than normal specific gravity. Commonly applied to minerals which sink in bromoform (density 2·9).

heavy particle (*Phys*) See HYPERON.

heavy spar (*Min*) See BARYTES.

heavy water (*Chem*) Deuterium oxide, or water containing a substantial proportion of deuterium atoms (D_2O or HDO).

heavy-water reactor (*NucEng*) One using HEAVY WATER as moderator, eg CANDU and SGHWR types. See figs at CANDU, REACTOR.

hebephrenia (*Psych*) A type of schizophrenia characterized by incoherence of thought, odd and child-like behaviour, and inappropriate emotional expression, occurring about puberty. Also *disorganized schizophrenia, hebephrenic schizophrenia*.

Heberden's nodes (*Med*) Small bony knobs occurring on the bones of fingers of the old.

hebetude (*Med*) Lethargy and mental dullness, with impairment of the special senses.

hectare (*Surv*) Metric unit of area, equal to 100 *ares*. Abbrev *ha*; 1 ha = 2·47 acres; 100 ha = 1 km^2, = 0·386 sq miles.

hecto- (*Genrl*) Metric prefix meaning × 100. Symbol h.

hectobar (*Eng*) See BAR.

hectocotylized arm (*BioSci*) In some male Cephalopoda, one of the tentacles modified for the purpose of transferring sperm to the female.

hectometric waves (*ICT*) See MEDIUM FREQUENCY.

hectorite (*Min*) A rare lithium-bearing mineral in the montmorillonite group of clay minerals.

heddle (*Textiles*) See HEALD.

hedenbergite (*Min*) An important calcium–iron pyroxene, $CaFeSi_2O_6$, occurring as black crystals, and also as a component molecule in many of the rock-forming clinopyroxenes.

Hedley's dial (*Surv*) A form of compass adapted for taking inclined sights; it consists of a pair of sighting vanes (or a telescope capable of rotation about a horizontal axis), carrying with them a vertical graduated arc moving over a fixed reference mark.

hedonism (*Psych*) Psychological theory that behaviour is motivated by the search for pleasure and the avoidance of pain.

-hedron (*Genrl*) Denotes a geometric solid figure or body with a specified number of plane faces, as in tetrahedron.

heel (*Build*) The back end of a plane.

heel (*Ships*) An angle of transverse inclination arising from a force external to the ship, eg wind. Cf LIST.

heeling error (*Ships*) DEVIATION at a magnetic compass occurring when a vessel HEELS. It can be eliminated by a group of vertical permanent magnets.

heel post (*Build*) The hinge post of a mitre lock gate. Also *quoin post*.

heel strap (*Build*) A strap fastening the foot of a principal rafter to the tie-beam on a timber truss. The strap is threaded at the ends to take a cover plate so that the joint is completely encircled. Nowadays such joints are commonly made by toothed plates hydraulically pressed into both sides of the timbers.

Heenan dynamatic dynamometer (*Eng*) An eddy current dynamometer in which the steel rotor has the shape of a gearwheel and rotates in a cylindrical stator, with smooth bore, surrounded by dc excited field coils.

Heenan hydraulic torque meter (*Eng*) An arrangement of input and output shafts with vanes forming a concentric annular working compartment through which oil is forced to transmit torque, the differential pressure across the vanes being a measure of the torque.

hefted sheep (*Agri*) Upland sheep grazed in unfenced areas and not moved but, instinctively, do not stray. A hefted flock is usually sold with the farm.

Hegman gauge (*PowderTech*) A FINENESS-OF-GRIND GAUGE used widely in the UK paint industry. The groove depth from four-thousandths of an inch (0·1 mm) to zero is marked off arbitrarily into a linear scale of eight units. The fitness of a paint pigment is often quoted by reference to the *Hegman gauge number* at which scoring occurs.

Hehner's test (*Chem*) A test for the presence of formaldehyde in milk. It is based upon the appearance of a blue or violet ring when the milk is mixed with a dilute ferric chloride solution, and concentrated sulphuric acid is added to form a layer beneath the milk.

heifer (*Agri*) A young cow up to the birth of its first calf or in first lactation following the first calving.

height board (*Build*) A gauge for the treads and risers of a timber staircase.

height control (*ImageTech*) The means of varying the amplitude of the vertical deflection of a cathode-ray oscilloscope.

height of a transfer unit (*ChemEng*) A measure of the separating efficiency of packed columns for mass transfer operations. It is the height of packed column used for liquid–liquid extraction in which one theoretical transfer unit takes place. Abbrev *HTU*.

height of instrument (*Surv*) (1) In levelling or trigonometrical survey work, the vertical distance of the plane of collimation of the level, or the horizontal axis of the theodolite above datum. (2) In tacheometry, the vertical distance of the horizontal axis of the instrument above ground level. Abbrev *HI*.

Heimlich manoeuvre (*Med*) An emergency method of dislodging an obstruction from a choking person's windpipe by applying a sharp thrust below the breastbone. Also *Heimlich procedure*.

Heisenberg uncertainty principle (*Phys*) See UNCERTAINTY PRINCIPLE.

Heising modulation (*Electronics*) Constant-current modulation, arising from one valve driven by signal and another valve driven by carrier, having their anodes fed through the same inductor; the modulated carrier is taken from the anode circuit by capacitive or inductive coupling.

held water (*Geol*) That kept above the natural water table through capillary force. US *free water*.

Helene (*Astron*) The 12th natural satellite of Saturn, a tiny object just 35 km in diameter. It is orbitally associated with the much larger DIONE.

helianthine (*Chem*) 4-*dimethylamino-azobenzene-4-sulphonic acid*. $(CH_3)_2NC_6H_4N=NC_6H_4SO_3H$. A chrysoidine dye. The sodium salt of this acid is methyl orange, used as an indicator.

helical antenna (*ICT*) An antenna in the form of a helix; when the circumference is one wavelength; maximum radiation is along its axis. Used in very-high-frequency and ultrahigh-frequency bands.

helical coil model (*BioSci*) The model of METAPHASE chromosome organization that envisages that the primary DNA helix is packed by secondary and higher orders of coiling.

helical gears (*Eng*) Gearwheels in which the teeth are not parallel with the wheel axis, but helical (ie parts of a helix described on the wheel face), being therefore set at an angle with the axis. See DOUBLE-HELICAL GEARS.

helical hinge (*Build*) A type of hinge used for hanging swing-doors which have to open both ways.

helical rising or setting (*Astron*) The rising or setting of a star or planet, simultaneously with the rising or setting of the Sun. It was much observed in ancient times as a basis for a solar calendar for agricultural purposes.

helical scan (*ImageTech*) Describing videotape recording and reproducing systems in which the tape path follows a helix around a rotating drum containing the magnetic heads. Fig. ▷

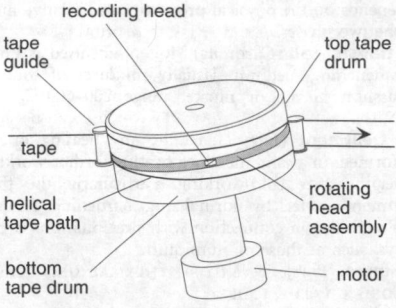

helical scan The head assembly rotates at 1500 rpm while each tape drum moves with the tape.

helical spring (*Eng*) A spring formed by winding wire into a helix along the surface of a cylinder; sometimes erroneously termed a *spiral spring*.

helical thickening (*BioSci*) A secondary wall deposited in the form of a helix in the tracheids and vessel elements of xylem.

helical tip speed (*Aero*) The resultant velocity of the tip of a propeller blade, which is a combination of the linear speed of rotation and the flight speed.

helical waveguide (*ICT*) A CIRCULAR WAVEGUIDE consisting of closely spaced turns of fine copper wire, clad in a jacket that absorbs any radiation which might otherwise escape.

helicase (*BioSci*) An enzyme that uses energy from ATP to unwind the DNA helix at the replication fork so that the single strands can be copied.

Helicobacter pylori (*Med*) A helix-shaped Gram-negative bacterium that is found in the gastric mucous layer or adherent to the epithelial lining of the stomach. Infection with *H. pylori* causes more than 90% of duodenal ulcers and up to 80% of gastric ulcers although many more people are infected than are affected. Treatment is with antibiotics in conjunction with *H2-blockers* and/or *proton pump inhibitors*.

helicoid (*BioSci*) Coiled like a flat spring.

helicoid (*MathSci*) A surface generated by a screw motion of a curve about a fixed line (axis); ie the curve rotates about the axis and moves in a direction of the axis in such a way that the ratio of the rate of rotation and rate of translation is constant. This constant multiplied by 2π is called the *pitch* of the helicoid. See RIGHT HELICOID.

helicopter (*Aero*) A ROTORCRAFT whose main rotor(s) are power driven and rotate about a vertical axis, and which is thus capable of vertical take-off and landing.

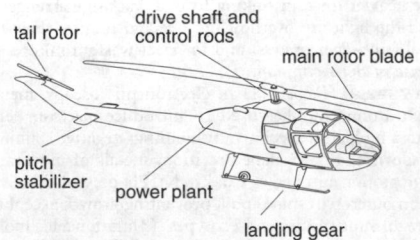

helicopter

helictic (*Geol*) Descriptive of the S-shaped trails of inclusions found in the minerals of some metamorphic rocks, esp abundant in garnet and staurolite. These inclusion trails are often continuous with the mineral alignments of surrounding crystals and may help to indicate the crystallization history of the rock.

helio- (*Genrl*) Prefix from Gk *helios*, Sun.

heliocentric parallax (*Astron*) See ANNUAL PARALLAX.

heliocentric system (*Astron*) A system centred on the Sun.

heliodor (*Min*) A beautiful variety of clear yellow beryl.

heliograph (*Surv*) An instrument similar to the heliostat but fitted with a spring device by which it can be made to flash long or short flashes.

heliometer (*Astron*) An instrument for determining the Sun's diameter and for measuring the angular distance between two celestial objects in close proximity. It consists of a telescope with its object glass divided along a diameter, the two halves being movable, so that a superposition of the images enables a value of the angular separation to be deduced from a reading of the micrometer.

heliophyte (*BioSci*) See SUN PLANT.

heliostat (*Astron*) An instrument designed on the same principle as the COELOSTAT, but with certain modifications that make it more suitable for reflecting the image of the Sun than for use on a larger region of the sky; hence used, in conjunction with a fixed instrument, esp for photographic and spectroscopic study of the Sun.

heliostat (*Surv*) An instrument used to reflect the Sun's rays in a continuous beam, so as to serve as a signal enabling a station to be sighted over long distances.

heliotaxis (*BioSci*) The response or reaction of an organism to the stimulus of the Sun's rays. Adj *heliotactic*. Also *heliotropism*.

heliotherapy (*Med*) The treatment of disease by the exposure of the body to the Sun's rays.

heliotrope (*Min*) See BLOODSTONE.

heliotrope (*Surv*) A form of HELIOGRAPH.

heliotropin (*Chem*) See PIPERONAL.

heliotropism (*BioSci*) See HELIOTAXIS. More particularly response by growth curvature. Adj *heliotropic*.

Heliozoa (*BioSci*) Protozoa of the Order Heliozoida. They are generally free-floating, spherical cells with many straight, slender microtubule-supported pseudopods (*axopodia*) radiating from the cell body like a sunburst.

Helipot (*ElecEng*) TN for precision multi-turn potential divider, resettable to within 0·1%.

helium (*Chem*) Chemically inert element, symbol He, at no 2, ram 4·0026. The gas is monatomic, liquefies at temperatures below 4 K, and undergoes a phase change to a form known as *liquid helium II* at 2·2 K. The latter form has many unusual properties believed to be due to a substantial proportion of the molecules existing in the lowest possible quantum energy state; see SUPERFLUID. The abundance of helium in the Earth's crust is 0·003 ppm and in the atmosphere 5·2 ppm (vol). Almost all of it is of RADIOGENIC origin and obtained from gas wells. Liquid helium is the standard coolant for devices working at cryogenic temperatures. It has also been used as a coolant in GAS-COOLED REACTORS where its chemical stability and very low neutron capture cross-sections are advantageous, but it is expensive and only finds wide use in low-volume applications like surrounding the fuel in FUEL PINS. The atom has an extremely stable nucleus identical to an α-particle. See NOBLE GASES.

helium-arc welding (*Eng*) A welding process in which helium is used to shield the weld area from contamination by atmospheric oxygen and nitrogen.

helium diving-bell (*CivEng*) A DIVING-BELL in which the nitrogen in the compressed air is replaced by helium, thus reducing tendency to the *bends* and permitting effective operation at greater depths than with normal air. The helium content also gives the divers high squeaky voices.

helium flash (*Astron*) The sudden onset of helium nuclear reactions in the cores of giant stars, resulting in core expansion and loss of degeneracy.

helium–neon laser (*Phys*) A laser using a mixture of helium and neon, energized electrically. Its output is in the visible region at 632·8 nm and can have a continuous power of 1 W or a pulsed output with peak powers up to 100 W. Also *He–Ne laser*. See panel on LASER.

helium star (*Astron*) A star of spectral type B in the Harvard classification whose spectrum shows only dark lines, with those due to the element helium predominating.

helix (*Genrl*) A line, thread, wire, or other structure curved into a shape such as it would assume if wound in a single layer round a cylinder or cone; a form like a screw-thread which is very common in biological macromolecules, eg DNA, and the path followed by a charged particle in a magnetic field.

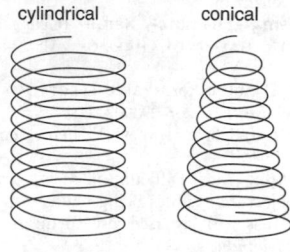

helix

helix (*ICT*) (1) See HELICAL ANTENNA. (2) The SLOW-WAVE STRUCTURE in a travelling-wave tube.

Helmert's formula (*Phys*) An empirical formula giving the value of *g*, the acceleration due to terrestrial gravity, for a given latitude and altitude: $g = 9.806\ 16 - 0.025\ 928 \cos 2\lambda + 0.000\ 069 \cos^2 2\lambda - 0.000\ 003\ 086H$, where λ is the latitude and H is the height in metres above sea level; g is in m s^{-2}.

helmet (*CivEng*) A cast-iron DOLLY used at the head of a reinforced concrete pile, the two being separated by a resilient cushion.

Helmholtz coils (*ElecEng*) A pair of identical compact coaxial coils separated by a distance equal to their radius. These give a uniform magnetic field over a relatively large volume at a position midway between them.

Helmholtz double layer (*Chem*) Electrical double layer. This assumes an interphase between a relatively insoluble solid and an ambient ionized liquid, in which oppositely charged ions tend to concentrate in layers. See ELECTRO-KINETIC POTENTIAL.

Helmholtz free energy (*Chem*) Similar to Gibbs free energy but with internal energy substituted for enthalpy. Symbol *A* (or *F* in US). A negative change in *A* is indicative of a spontaneous change in a closed system at constant volume.

Helmholtz galvanometer (*ElecEng*) A type of TANGENT GALVANOMETER in which an approximately uniform field is produced by having two coils parallel to each other, a few centimetres apart.

Helmholtz resonance (*Acous*) The type of acoustic resonance arising in a HELMHOLTZ RESONATOR.

Helmholtz resonator (*Acous*) An air-filled cavity with an opening. The resonance frequency depends on the stiffness of the cavity and the mass of air which oscillates in the opening.

Helmholtz's theorem (*ElecEng*) See THÉVENIN'S THEOREM.

Helminthes (*BioSci*) A name formerly used in classification to denote a large group of worm-like invertebrates now split up into PLATYHELMINTHES, NEMATODA and smaller groups.

helminthiasis (*Med*) Infestation of the body with parasitic worms.

helm wind (*EnvSci*) Local Cumbrian name for a cold north-easterly wind blowing into the Eden Valley from the western slopes of Crossfell: a line of cloud (the *helm*) forms along the crest of the ridge. See ROTOR CLOUD.

helophyte (*BioSci*) Marsh plant with perennating buds below the surface of the marsh. See RAUNKIAER SYSTEM.

helper T-lymphocyte (*BioSci*) Often termed *helper T-cell*. A thymus-derived lymphocyte that co-operates with B-lymphocytes to enable them to produce antibodies when stimulated by antigen or by some polyclonal mitogens.

Helper T-cells release lymphokines causing differentiation and growth of B-cells. They also influence the generation of cytotoxic and suppressor T-cells. See panel on IMMUNE RESPONSE.

helve (*Build*) The handle of an axe, hatchet or similar chopping tool.

helve hammer (*Eng*) A power hammer used on plating etc in which the head is secured to one end of a helve or beam, which is supported on a pivot and actuated by a crank or cam mechanism.

hem-, hemat-, hemato-, hemo- (*Genrl*) US for HAEM-, HAEMAT-, HAEMATO-, HAEMO-. All entries are under haem-.

HEM (*Phys*) Abbrev for HYBRID ELECTROMAGNETIC WAVE.

hematite (*Min*) US for HAEMATITE.

heme (*Genrl*) US for HAEM. All heme- entries are under haem-.

hemeralopia (*Med*) Difficulty in seeing clearly in a bright light, objects sometimes being better seen in a dull light. The term is wrongly used also to mean night-blindness. Also *day-blindness*.

hemi- (*Genrl*) Prefix from Gk *hemi*, half.

hemianaesthesia (*Med*) Loss of sensibility to touch on one side of the body; usually connotes also loss of sensibility to pain and temperature. US *hemianesthesia*.

hemianalgesia (*Med*) Loss of sensibility to pain on one side of the body.

hemianesthesia (*Med*) US for HEMIANAESTHESIA.

hemianopia (*Med*) Loss of half the field of vision. Also *hemianopsia*.

Hemiascomycetes (*BioSci*) A class of fungi in the Ascomycotina in which no ascocarps are formed. They are mostly unicellular or have a poorly developed mycelium. Includes the yeasts, *Saccharomyces* and *Schizosaccharomyces*, and some plant parasites causing eg peach leaf curl.

hemiataxy (*Med*) Loss of co-ordination of the muscles of one side of the body. Also *hemiataxia*.

hemiatrophy (*Med*) Wasting of muscles of one side of the body, or of one-half of a part of the body.

hemiballism (*Med*) Involuntary violent twitching and jerking affecting one side of the body. Condition due to a hypothalamic lesion.

hemibranch (*BioSci*) The single row of gill lamellae or filaments, borne by each face of a gill arch in fish; a gill arch with respiratory lamellae or filaments on one face only.

hemicelluloses (*BioSci*) A group of polysaccharides in the matrix of plant cell walls; homo- and hetero-polymers, linear and branched, of xylose, glucose and other sugars.

Hemichordata (*BioSci*) A subphylum of Chordata, lacking any bony or cartilaginous skeletal structures. They lack a tail or atrium and have a reduced notochord in the pre-oral region. The central nervous system is simple and the three primary coelomic cavities persist in the adult. Also *Protochordata*.

hemichorea (*Med*) CHOREA on one side of the body.

hemicolloid (*Chem*) A particle up to $2 \cdot 5 \times 10^{-8}$ m in length; 20–100 molecules.

hemicrania (*Med*) See MIGRAINE.

hemicryptophyte (*BioSci*) Plant with buds at soil level. See RAUNKIAER SYSTEM.

hemicrystalline rocks (*Geol*) Those rocks of igneous origin which contain some interstitial glass, in addition to crystalline minerals. Cf HOLOCRYSTALLINE ROCKS.

hemicyclic (*BioSci*) A flower with some parts inserted in helices and some in whorls.

hemignathous (*BioSci*) Having jaws of unequal length.

hemihydrate plaster (*Build*) See PLASTER OF PARIS.

Hemimetabola (*BioSci*) See EXOPTERYGOTA.

hemimetabolous (*BioSci*) Insect species that do not undergo any marked change in body-plan from larval to adult, apart from the development of wings, eg grasshoppers and crickets. Cf HOLOMETABOLOUS. Also *hemimetabolic*.

hemimorphism (*Min*) The development of polar symmetry in minerals, in consequence of which different forms are exhibited at the ends of bi-terminated crystals. HEMIMORPHITE shows this character in a marked degree.

hemimorphite (*Min*) An orthorhombic hydrated silicate of zinc; one of the best minerals for demonstrating polar symmetry, the two ends being distinctly dissimilar. US *calamine, electric calamine*.

hemiparasite (*BioSci*) See PARTIAL PARASITE.

hemipenes (*BioSci*) In snakes and lizards, SQUAMATA, the paired eversible copulatory organs.

hemiplegia (*Med*) Paralysis of one side of the body.

Hemiptera (*BioSci*) An order of Insecta, comprising insects that have two pairs of wings of variable character. The mouthparts are symmetrical and adapted for piercing and sucking. In some forms, the females are wingless. Many of these organisms are ectoparasitic; others feed on plant juices. Bugs, cicadas, aphids, plant lice, scale insects, leaf hoppers, white flies, black flies, green flies, cochineal insects.

hemisection (*Med*) The cutting through of half of a part, eg of the spinal cord.

hemisphere (*BioSci*) One of the cerebral hemispheres (see CEREBRUM).

hemisphere (*Genrl*) The half of a sphere, obtained by cutting it by a plane passing through the centre. As applied to the Earth, the term usually refers to the *northern* or the *southern hemisphere*, the division being by the equatorial plane.

hemithyroidectomy (*Med*) Removal of one-half of the thyroid gland.

hemizygous (*BioSci*) Having only one representative of a gene or a chromosome, as are male mammals which have only one X-CHROMOSOME.

hemlock (*For*) A N American hardwood (*Tsuga*) whose heartwood is a pale brown, with a straight grain and fine texture.

hemo- (*Genrl*) See HAEM-.

hemoglobin (*BioSci*) US for HAEMOGLOBIN.

hemp (*Textiles*) The BAST FIBRE of the hemp plant, *Cannabis sativa*, generally used for making string and ropes. Certain other fibres such as manila and sisal are sometimes incorrectly called hemps.

Hempel burette (*Chem*) Used for measuring the volume of a gas, eg in gas analysis.

HEMT (*Electronics*) Abbrev for HIGH ELECTRON MOBILITY TRANSISTOR.

Henderson–Hasselbach equation (*BioSci*) Equation of the form $pH = pK_a - \log([HA]/[A^-])$ for the calculation of the pH of solutions where the ratio $[HA]/[A^-]$ is known, $[HA]$ and $[A^-]$ are the concentrations of protonated and deprotonated forms of an acid, respectively, and K_a is the acid dissociation constant.

He–Ne laser (*Phys*) See HELIUM–NEON LASER.

henequen (*Textiles*) Bast fibre obtained from *Agava fourcroydes* and is similar to SISAL.

Henle's loop (*BioSci*) See LOOP OF HENLE.

Henoch–Schönlein purpura (*Med*) A disease characterized by purpura, urticaria, swollen joints and abdominal pain, thought to be caused by an immune mechanism. In adults GLOMERULONEPHRITIS may predominate.

henry (*Phys*) SI unit of mutual and self-inductance. (1) A circuit has an inductance of 1 henry if an emf of 1 volt is induced in the circuit by a current variation of 1 ampere per second. (2) A coil has a self-inductance of 1 henry when the magnetic flux linked with it is 1 weber per ampere. (3) The mutual inductance of two circuits is 1 henry when the flux linked with one circuit is 1 weber per ampere of current in the other. Symbol H.

Henry's law (*Chem*) The amount of a gas absorbed by a given volume of a liquid at a given temperature is directly proportional to the pressure of the gas.

Henry Williams fishplate (*Eng*) A drop-forged fishplate for insulated rail joints in which the insulating fibre is not subject to mechanical wear, the load on the joint being transmitted by side flanges secured together by bolts.

Henschel mixer (*Plastics*) A type of dry blender of polymer powders using a high-speed rotating disk fitted with sharp blades, which creates circulating vortex.

heparan sulphate (*BioSci*) A GLYCOSAMINOGLYCAN, a constituent of membrane-associated proteoglycans.

heparin (*Pharmacol*) A short-acting anticoagulant used to prevent thrombus and clot formation. Can only be given parenterally and usually to initiate treatment of deep venous thrombosis and pulmonary embolism until anticoagulation by oral drugs can be established. Also used in low doses to prevent deep-vein thrombosis formation during surgery.

hepat-, hepato- (*Genrl*) Prefixes from Gk *hepar*, gen *hepatos*, the liver.

hepatectomy (*Med*) Removal of part of the liver.

hepatic (*BioSci*) (1) Associated with the liver. (2) A liverwort. See HEPATICOPSIDA.

Hepaticae (*BioSci*) See HEPATICOPSIDA.

hepatic artery (*BioSci*) In Craniata, a branch of the coeliac artery that conveys arterial blood to the liver.

hepatic duct (*BioSci*) In Craniata, a duct conveying the bile from the liver and discharging into the intestine as the common bile duct.

Hepaticopsida (*BioSci*) A class of the Bryophyta containing c.10 000 spp. The gametophyte is thalloid or leafy with unicellular rhizoids and the capsule (sporophyte) is without a columella. Liverworts. Also *Hepaticae*.

hepatic portal system (*BioSci*) In Craniata, the part of the vascular system that conveys blood to the liver; it consists of the HEPATIC ARTERY and the HEPATIC PORTAL VEIN.

hepatic portal vein (*BioSci*) In Craniata, the vein that conveys blood from the alimentary canal to the liver.

hepatitis (*Med*) Inflammation of the liver. See INFECTIOUS HEPATITIS.

hepatitis contagiosa canis (*Vet*) See INFECTIOUS CANINE HEPATITIS.

hepatitis virus (*Med*) Hepatitis A virus is a small single-stranded RNA virus that causes 'infectious hepatitis'. Hepatitis B is a DNA virus responsible for 'serum hepatitis'. Hepatitis C virus is becoming increasingly important and is associated with hepatocarcinoma. Other forms (D and E) are also known.

hepatization (*Med*) Pathological change of tissue so that it becomes liver-like in consistency; as of the lung in pneumonia.

hepato- (*Genrl*) See HEPAT-.

hepatocyte (*BioSci*) Epithelial cell of liver. Often considered the paradigm for an unspecialized animal cell.

hepatogenous (*Med*) Having origin in the liver.

hepatolenticular degeneration (*Med*) Degeneration of the liver and the lenticular nucleus (part of the brain). See WILSON'S DISEASE.

hepatolith (*Med*) A gallstone present in the liver.

hepatoma (*Med*) A tumour composed of liver tissue.

hepatomegaly (*Med*) Enlargement of the liver. Also *hepatomegalia*.

hepatopancreas (*BioSci*) In many invertebrates (as Mollusca, Arthropoda, Brachiopoda), a glandular diverticulum of the mid-gut, frequently paired, consisting of a mass of branching tubules, and believed to carry out the functions proper to the liver and pancreas of higher vertebrates.

hepatopexy (*Med*) Fixation of the liver by suturing it to the abdominal wall.

hepatoportal system (*BioSci*) See HEPATIC PORTAL SYSTEM.

hepatorrhexis (*Med*) Rupture of the liver.

hepatotomy (*Med*) Incision of the liver.

hept-, hepta- (*Genrl*) Prefixes from Gk *hepta*, seven.

hepta- (*Chem*) Containing seven atoms, groups, etc.

heptagon (*MathSci*) A seven-sided polygon.

heptane (*Chem*) C_7H_{16}. Bp 98°C, rel.d. 0·68. There are nine isomers. The foregoing properties relate to *normal* heptane, an alkane hydrocarbon, a colourless liquid, which is a constituent of petrol and resembles hexane in its chemical behaviour.

heptavalent (*Chem*) Capable of combining with seven hydrogen atoms or their equivalent. Having an oxidation or co-ordination number of seven. Also *septavalent*.

heptoses (*Chem*) A subgroup of the monosaccharides containing seven carbon atoms, of the general formula $HOCH_2(CHOH)_5CHO$.

herb (*BioSci*) (1) A plant that does not develop persistent woody tissues above ground. (2) A plant that is used for medicinal purposes or for flavour.

herbaceous (*BioSci*) A soft and green plant organ or a plant without persistent woody tissues above ground.

herbaceous perennial (*BioSci*) A perennial plant with a perennating structure at or below ground level and producing aerial shoots that die at the end of the growing season.

herbarium (*BioSci*) A collection of dried plants: by extension, the place where such a collection is kept.

herbicide (*Agri*) An agrochemical employed to kill plants.

Herbig–Haro object (*Astron*) A small highly luminous nebula containing several star-like objects. Associated with T TAURI STARS.

Herceptin (*Pharmacol*) A proprietary name for TRASTUZU-MAB.

hercogamy (*BioSci*) Physical arrangement of anthers and stigma so that pollen is not transferred from one to the other in the absence of an insect visit.

Hercules (*Astron*) A faint constellation, the fifth largest in the northern sky.

Hercynian orogeny (*Geol*) The late Palaeozoic orogeny in Europe. Formerly *Armorican orogeny*.

hercynite (*Min*) See SPINEL.

herdbook (*Agri*) The pedigree, reproductive and performance records of a livestock herd.

herding (*Psych*) Gregariousness, the tendency to live in herds or flocks, found in many species.

hereditary (*BioSci*) Inherited; capable of being inherited; passed on or capable of being passed on from one generation to another.

hereditary angioneurotic oedema (*Med*) A disease characterized by recurrent episodes of transient oedema of skin and mucous membranes, and due to absence or functional inactivity of C1a-esterase inhibitor. Inherited as a Mendelian dominant. See C1A-INHIBITOR. Also *Milroy's disease*.

hereditary ataxia (*Med*) A group of inherited central nervous system diseases causing ATAXIA. FRIEDREICH'S ATAXIA is the best known, presenting in late childhood with poor co-ordination and the development of sensory and motor loss. About half of those affected have CARDIOMYOPATHY.

hereditary haemorrhagic telangiectasia (*Med*) A genetic disease with multiple vascular anomalies (telangiectasia) of the skin and mucous membranes. The condition is often complicated by internal bleeding. Also *Osler–Rendu–Weber syndrome*.

heredity (*BioSci*) The relation between successive generations, by which characters persist.

Hereford disease (*Vet*) See BOVINE HYPOMAGNESAEMIA.

HERF (*Eng*) Abbrev for HIGH-ENERGY RATE FORGING or FORMING.

heritability (*BioSci*) Measure of the degree to which the variation of a character is inherited. Ranges from 0 to 100%. It is the proportion of *additive genetic variance* in the total *phenotypic variance*. Usually symbolized by H_2.

hermaphrodite (*BioSci*) (1) An individual organism that has both male and female reproductive organs and produces both male and female gametes. (2) Flowers with male and female organs functional in the one flower. Also *monoclinous*. Cf DIOECIOUS, UNISEXUAL.

hermeneutics (*Psych, Genrl*) The science of interpretative procedures.

hernia (*Med*) Protrusion of a viscus, or part of a viscus, through an opening or weak spot or defective area in the cavity containing it; esp of an abdominal viscus.

herniorrhaphy (*Med*) Surgical repair of a hernia.

herniotomy (*Med*) Cutting operation for hernia.

Hero's formula (*MathSci*) The formula for the area of a triangle with sides a, b and c, is given by

$$\sqrt{s(s-a)(s-b)(s-c)}$$

where $2s = a + b + c$. Also *Heron's formula*.

Héroult process (*Eng*) An electrolytic process for the manufacture of aluminium from a solution of bauxite in fused cryolite.

herpes simplex (*Med*) A virus causing a number of clinical disorders ranging from the common eruption of vesicles round the mouth in a febrile illness, through the disabling genital herpes, to the life-threatening infection of the brain, encephalitis.

herpes zoster (*Med*) After chicken pox and variella, the variella-zoster virus becomes latent in nerve cells. Later in life, often at a time of decreased immunity, there is reactivation of the virus with a belt of blisters (dermasomes) over the skin supplied by that nerve, the posterior root ganglia of which are inflamed. Almost always on one side of the body only. After the blisters heal, pain can be severe and prolonged in post-herpetic neuralgia. Also *shingles*.

herpetology (*Genrl*) The study of reptiles and amphibians.

herringbone (*Textiles*) Any cloth made from the eponymous weave.

herringbone ashlar (*Build*) Blocks of stone tooled in grooves of herringbone design.

herringbone bond (*Build*) A form of RAKING BOND in which the bricks are laid with rake in opposite directions from the centre of the wall, to form a herringbone pattern. This bond is also used for brick pavings, and has the advantage of making an effective bond in the middle.

herringbone gear (*Eng*) See DOUBLE-HELICAL GEARS.

herringbone parlour (*Agri*) A common layout of a milking parlour where the cows stand in two rows, overlapping each other to form the herringbone pattern.

Herschel formula (*Eng*) A hydraulic formula used to calculate the rate of flow over a drowned nappe (dam with smoothed sides) or weir.

hertz (*Phys*) SI unit of frequency, indicating number of cycles per second. Symbol Hz.

Hertz antenna (*ICT*) Original half-wave dipole, fed at the centre.

Hertzian dipole (*ElecEng*) A pair of opposite and varying charges, close together, with an electric moment. Also *Hertzian doublet*.

Hertzian oscillator (*ICT*) Idealized system envisaged by Hertz, comprising two point charges of opposite sign and separated by an infinitesimal distance, whose electric moment varies harmonically with time.

Hertzian waves (*Phys*) Electromagnetic waves in the approximate frequency range 10^4–10^{10} Hz, used for communication through space, covering the range from very low to ultrahigh frequencies, ie from audio reproduction, through radio broadcasting and TV to radar.

Hertzsprung–Russell diagram (*Astron*) A graphical representation of the correlation between the SPECTRAL TYPE and luminosity for a sample of stars, useful in the study of stellar evolution. See panel on HERTZSPRUNG–RUSSELL DIAGRAM.

Herzberg stain (*Paper*) A general purpose stain for identifying paper fibres under the microscope, consisting of a mixture of zinc chloride, iodine and potassium iodide.

hesitation line (*Plastics*) Visual defect formed on surface of hollow moulded plastic product where extra sharp impression of tool cavity surface is formed by onset of gas pressurization.

Hesperus (*Astron*) An ancient name for Venus when it appears as an 'evening star'.

hessian (*MathSci*) The determinant obtained by replacing the n functions u_i in the Jacobian by the n partial derivatives of a single function u.

hessian (*Textiles*) A strong plain-weave jute fabric, used for packing material, sacks, in tarpaulin manufacture, and as a furnishing fabric and wall covering.

hessite (*Min*) Silver telluride, a metallic-grey pseudo-cubic mineral occurring in silver ores in various parts of the world.

hessonite (*Min*) A variety of garnet containing a preponderance of the grossular molecule, and characterized by a pleasing reddish-brown colour. Also *Cinnamon stone*.

Hess's law (*Chem*) The net heat evolved or absorbed in any chemical change depends only on the initial and final states, being independent of the stages by which the final state is reached.

heter-, hetero- (*Genrl*) Prefixes from Gk *heteros*, other, different.

hetero-agglutination (*BioSci*) (1) The adhesion of spermatozoa to one another by the action of a substance produced by the ova of another species. (2) The adhesion of erythrocytes to one another when blood of different groups is mixed. Cf ISO-AGGLUTINATION. See AGGLUTININ.

hetero-auxin (*BioSci*) Old name for indole acetic acid; auxin.

heteroblastic (*BioSci*) (1) Animals showing *indirect development* in which embryonic cell lineages of fixed fate and limited division capacity give rise to the larval structures and the adult arises from cells in the larva that are 'set-aside' and retain extensive proliferative capacity. (2) Having a marked morphological difference between the first-formed structures, like leaves on a seedling, and those formed later. Cf HOMOBLASTIC. See JUVENILE PHASE. (3) A plant whose seeds vary in the conditions they require for germination.

heterocercal (*BioSci*) Said of a type of tail fin, found in adult sharks, rays, sturgeons and many other primitive fish, in which the vertebral column bends abruptly upwards and enters the epichordal lobe, which is larger than the hypochordal lobe.

heterochlamydeous (*BioSci*) Having a distinct calyx and corolla.

heterochromatin (*BioSci*) Relatively dense chromatin visible by microscopy in eukaryotic cell nuclei. Generally contains DNA sequences inactive in transcription. See CONSTITUTIVE HETEROCHROMATIN, FACULTATIVE HETEROCHROMATIN.

heterocoelous (*BioSci*) Said of vertebral centra in which the anterior end is convex in vertical section, concave in horizontal section, while the posterior end has these outlines reversed.

heterocotylized arm (*BioSci*) See HECTOCOTYLIZED ARM.

heterocyclic compounds (*Chem*) Cyclic or ring compounds containing carbon atoms and other atoms, eg O, N, S, as part of the ring.

heterodactylous (*BioSci*) Of birds, having the first and second toes directed backwards, the third and fourth forwards, as the trogons.

heterodesmic structure (*Crystal*) A structure which includes two or more types of crystal bonding.

heterodimer (*BioSci*) A dimer in which the two subunits are different, though often similar. An example is tubulin that is found in cells as a heterodimer of α- and β-tubulin.

heterodont (*BioSci*) Of teeth, having different forms adapted to different functions.

heterodromous (*BioSci*) Having two kinds of asymmetric flowers, one the mirror image of the other and sometimes associated in pairs.

heteroduplex DNA (*BioSci*) A duplex (double-stranded) DNA made by renaturing (or *annealing*) single DNA strands from different sources. By varying the conditions, the sequence similarity required to get duplex formation can be adjusted. Under conditions of high stringency a good match is required, which indicates similarity and possibly evolutionary relatedness. See panel on DNA AND THE GENETIC CODE.

Hertzsprung–Russell diagram

Spectral type and luminosity can be measured with relative ease for a great many stars. The first tells us the type of star and its surface temperature, and the second is a measure of the absolute quantity of energy being radiated by the star. The Hertzsprung–Russell diagram (or HR diagram) is named for E Hertzsprung and H N Russell who independently (1911 and 1913) discovered that the two quantities are correlated.

This correlation is very striking. About 90% of all stars lie on a narrow diagonal band known as the main sequence. Stars on the left side of this diagram are hot, those to the right cool. There is also a line of cool luminous stars. These are giants on the giant red branch. WHITE DWARF stars form a separate population at the lower left: they are hot but dim. The fact that not all combinations of temperature and luminosity are found means that the study of stars is drastically simplified compared with the study of galaxies or the interstellar medium. In particular, this branch of astrophysics is amenable to investigation via rigorous mathematical physics.

The explanation of the HR diagram has been one of the great triumphs of stellar evolution theory. Essentially, when new stars form, the only parameter that distinguishes one from another is mass. For most of its life a star burns hydrogen in the core to form helium. The more massive stars have higher central temperatures and burn hydrogen faster, have hotter surfaces and are more luminous. Those stars burning hydrogen in the core are all found on the main sequence. The Sun is one such star. While on the main sequence the temperature and luminosity scarcely change, so a given star will be at more or less the same place on the main sequence.

Eventually hydrogen burning in the core stops. Structural changes occur and the star moves off the main sequence onto the giant branch. Although the surface temperature falls, the surface area expands greatly as the star swells. The luminosity therefore goes up temporarily. As the star nears the end of its life it contracts again, crosses the main sequence from right to left and eventually settles as a white dwarf.

The HR diagram is a valuable tool for investigating discrete populations of stars, such as those found in star clusters. All stars in a cluster are at more or less the same distance. For such a population the HR diagram can be plotted in relative terms without knowing the exact distance. The point at which the main sequence breaks over to the RED GIANT branch is an indication of cluster age. These vary from 20 million years for a young OPEN CLUSTER such as the Pleiades, to 10 000 million years for highly evolved GLOBULAR CLUSTERS. The HR diagram can also be used to gauge a star's distance. The spectral type is a proxy indicator of absolute magnitude, which can be compared with the directly measured relative magnitude to yield a distance.

Related to the HR diagram are the COLOUR–LUMINOSITY ARRAY and *colour–colour* diagrams. In these plots the colour is the difference in magnitude when the star is examined through two or more colour filters. The advantage of this cruder measure of star type is that large numbers of colours can be measured from a series of survey photographs taken through filters. This technique can be automated. Traditional spectroscopy is, by contrast, very much slower.

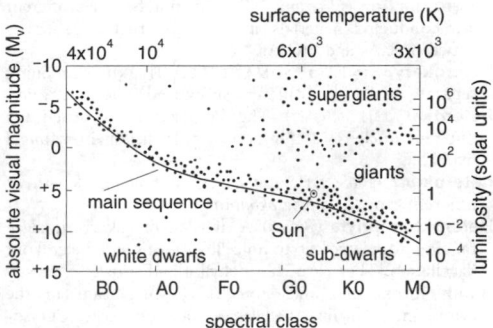

Hertzsprung–Russell diagram The dots are schematic but represent the scatter observed.

See panels on COSMOLOGY and SUN AS A STAR.

---

heterodyne (*ICT*) Combination of two sinusoidal radio-frequency waves in a non-linear device resulting in sum and difference frequencies. The latter is the heterodyne frequency, and will produce an audio-frequency beat note when the original two sine waves are sufficiently close in frequency.

heterodyne conversion (*ICT*) Change in the frequency of a modulated carrier wave produced by heterodyning it with a second unmodulated signal. The sum and difference frequencies will carry the original modulation signal and either of these can be isolated for subsequent amplification. The frequency-changing stage of a superhet radio receiver employs this principle, an oscillator being tuned to a fixed amount away from the signal frequency so that the difference frequency (intermediate-frequency signal) remains constant for all incoming signals.

heterodyne-frequency meter (*ICT*) See HETERODYNE WAVEMETER.

heterodyne interference (*ICT*) That arising from simultaneous reception of two stations the difference between whose carrier frequencies is an audible frequency.

heterodyne wavemeter (*ICT*) One in which a continuously variable oscillator is adjusted to give zero beat with an unknown frequency, the value of which then coincides with the calibration of the oscillator. Also *beat-frequency wavemeter*.

heterodyne whistle (*ICT*) See HETERODYNE INTERFERENCE.

heteroecious (*BioSci*) A parasite that requires two, usually unrelated, host species to complete different stages of its life cycle. Also *metoecious, metoxenous*. Cf AUTOECIOUS.

heterogamete (*BioSci*) See ANISOGAMETE.

heterogametic sex (*BioSci*) The sex that is heterozygous for the *sex-determining chromosome*, the male in mammals, the female in birds; its gametes determine the sex of the progeny. Cf HOMOGAMETIC.

heterogamous (*BioSci*) Producing unlike gametes or flowers.

heterogeneous (*Chem*) Said of a system consisting of more than one phase.

heterogeneous nuclear RNA (*BioSci*) The population of RNA molecules in the nucleus including the precursors of mature messenger RNA, which are eventually found in the cytoplasm. Abbrev *hnRNA*.

heterogeneous radiation (*Phys*) Radiation comprising a range of wavelengths or particle energies.

heterogeneous reactor (*NucEng*) One in which the fuel is present as rods spread in an array or lattice within (but separate from) the moderator. Cf HOMOGENEOUS REACTOR.

heterogeneous summation (*Psych*) Phenomenon that where a response is influenced by stimulus characters acting through more than one sensory modality, their effects may supplement each other, eg parts of the stimulus presented successively may produce the same response as when they are presented simultaneously.

heterogenesis (*BioSci*) Adj *heterogenetic*. See ABIOGENESIS, ALTERNATION OF GENERATIONS.

heterogeny (*BioSci*) Cyclic reproduction in which several broods of parthenogenetic individuals alternate with one or more broods of sexual forms.

heterogony (*BioSci*) Reproduction by both parthenogenesis and amphigony.

heterojunction (*Electronics*) The junction between different semiconductor materials in a heterostructure, such as between GaAs and AlGaAs.

heterokaryon (*BioSci*) SOMATIC CELL HYBRID containing separate nuclei from different species. Also *heterokaryote*.

heterokaryosis (*BioSci*) The coexistence of genetically different nuclei in a common cytoplasm, esp in a fungal hypha.

heterokont (*BioSci*) Having flagella of unequal length or different type. Also *heterokontan*.

Heterokontophyta (*BioSci*) A division of eukaryotic algae ranging from large kelp to unicellular forms with heterokont flagellation. They have chlorophyll a and usually c, mostly with fucoxanthin and store β, $1 \rightarrow 3$ glucans in the cytoplasm. Contains the classes Xanthophyceae, Chrysophyceae, Bacillariophyceae, Raphidophyceae and Paeophyceae.

heterolecithal (*BioSci*) With unequally distributed yolk.

heteromerous (*BioSci*) A lichen thallus with the algal cells confined to a distinct layer, with pure fungus above and below. Cf HOMOIOMEROUS.

heterometabolic (*BioSci*) Having incomplete metamorphosis.

heterometry (*Chem*) A process of titration in which precipitation is plotted as an optical density curve.

heteromorphic (*BioSci*) (1) Having more than one form. (2) A species with ALTERNATION OF GENERATIONS in which the latter differ, often considerably, in form. Also *heteromorphous*.

heteromorphous rocks (*Geol*) Rocks of closely similar chemical composition, but containing different mineral assemblages.

heteronomous (*BioSci*) Subject to different laws, esp of growth and specialization. Cf AUTONOMOUS.

heterophil antigens (*BioSci*) Antigens that occur on the cells of many different species of animals, plants and bacteria, and that are sufficiently similar to elicit antibodies which cross-react extensively, eg *Forssman antigen*.

heterophoria (*Med*) Latent squint revealed by passing a screen before each eye. See ESOPHORIA, EXOPHORIA.

heterophylly (*BioSci*) An individual plant with two or more different forms of leaf as in the submerged, floating and aerial leaves of many water plants.

heteroplasma (*BioSci*) In TISSUE CULTURE, a medium prepared with plasma from an animal of a different species from that from which the tissue was taken. Cf AUTOPLASMA, HOMOPLASMA.

heteroplastic (*BioSci*) In experimental zoology, said of a graft that is transplanted to a site different from its point of origin, eg epithelial cells of cornea to a skin site.

heteroplasty (*Med*) The operation of grafting on one person body-tissue removed from another.

heteropolar (*Chem*) Having an unequal distribution of charge, as in covalent bonds between unlike atoms.

heteropolar generator (*ElecEng*) An electromagnetic generator of the usual type, ie one in which the conductors pass alternate north and south poles, or in which alternate poles pass the conductors.

heteropolar liquids (*Chem*) Compounds such as alcohols, amines and organic acids which contain molecules that have localized associated polar groups.

heteropolymers (*Chem*) See panel on POLYMERS.

heteropycnosis (*BioSci*) Differential stainability of bands of chromosomes.

heteroscedastic (*MathSci*) Having unequal variances; applied to sets of observations.

heterosexuality (*Psych*) Sexual interest directed at members of the opposite sex.

heterosis (*BioSci*) The difference between the mean of a quantitative character in a crossbred generation and the mean of the two parental strains. Also *heterozygous advantage*. See HYBRID VIGOUR.

heterospory (*BioSci*) The production of more than one type of spore, typically megaspores and microspores. Adj *heterosporous*. Cf HOMOSPORY.

heterostructure (*Electronics*) In semiconductor technology, certain materials grown epitaxially in layers having different composition and properties offering extra degrees of freedom in device engineering. See EPITAXIAL GROWTH OF SEMICONDUCTORS.

heterostyly (*BioSci*) The condition in which individuals of a species have style and thus stigma lengths falling into two or more distinct classes. This causes the anthers to be placed low in flowers with high stigmas and vice versa. Heterostyly appears to promote cross pollination. Cf HOMOSTYLY, ILLEGITIMATE POLLINATION, PIN, THRUM.

heterothallism (*BioSci*) The condition in which there are two (or more) mating types with sexual reproduction only successful between individuals of a different type. Also *self-incompatibility*. Adj *heterothallic*. Cf HOMOTHALLISM.

heterotopia (*Med*) Displacement of a group of cells of an organ from their normal position during the course of development.

heterotrichous (*BioSci*) Having cilia or flagella of two or more different kinds.

heterotrophic (*BioSci*) Organisms that require carbon in organic form, as do all animals, fungi, some algae, parasitic plants and most bacteria. Cf AUTOTROPHIC.

heterotypic (*BioSci*) Differing from the normal condition. Cf HOMOTYPIC.

heterotypic division (*BioSci*) The first (reductional) of two nuclear divisions in meiosis in which the number of chromosomes is halved. The second (equational) is a *homotypic division*.

heterozygosis (*BioSci*) The condition of being *heterozygous*.

heterozygosity (*BioSci*) The proportion of individuals in a population that are *heterozygous* at a specified locus, or at a number of loci averaged.

heterozygote (*BioSci*) An individual with two different alleles at a particular locus, the individual having been formed from the union of gametes carrying different alleles. Adj *heterozygous*. Cf HOMOZYGOTE.

HETP (*ChemEng*) Abbrev for *height equivalent to a theoretical plate*. A measure of the separation efficiency of a distillation column, ie the height of packing in a packed distillation column which behaves as a THEORETICAL PLATE.

Hettangian (*Geol*) The oldest stage in the Jurassic. See MESOZOIC.

heulandite (*Min*) One of the best-known zeolites, often beautifully crystalline, occurring as coffin-shaped monoclinic crystals in cavities in basic igneous rocks. In composition similar to hydrated calcium sodium aluminium silicate.

heuristic (*Genrl*) Describes an approach based on common-sense rules and trial and error rather than on comprehensive theory.

heuristic (*Psych*) A rule of thumb, based on experience, used to make decisions and in problem solving by helping to reduce the range of options.

heuristic program (*ICT*) One that attempts to improve its own performance as a result of learning from previous actions within the program.

Hewlett disk insulator (*ElecEng*) A disk form of suspension-type insulator.

hewn stone (*Build*) Blocks of hammer-dressed stone.

hex (*ICT*) See HEXADECIMAL NOTATION.

hex- (*Chem*) Colloq *uranium* (VI) *fluoride*, the compound used in the separation of uranium isotopes by gaseous diffusion.

hex- (*Genrl*) Prefix from Gk *hex*, six.

hexa- (*Chem*) Containing six atoms, groups, etc.

hexachlorophene (*Chem*) *2,2′-methylene bis-(3,4,6 trichloro-hydroxybenzene*). Mp approx 160°C. White powder. Widely used bactericide in soaps, deodorants and other toilet products. Also *G-11*.

Hexactinellida (*BioSci*) A class of Porifera, usually distinguished by the possession of a siliceous skeleton composed of triaxial spicules, and large thimble-shaped flagellated chambers.

hexadecane (*Chem*) $C_{10}H_{34}$. An alkane hydrocarbon found in petroleum, esp that showing a straight chain structure.

hexadecimal notation (*ICT, MathSci*) The positional number system with base 16 generally written using the digits $0,1,2,\ldots,9,A,B,C,D,E,F$.

hexafluorophosphoric acid (*Chem*) HPF_6. Produced by the action of strong hydrofluoric acid on difluorophosphoric acid. Also *phosphorofluoric acid*.

hexagon (*MathSci*) A six-sided polygon.

hexagonal close packing (*Chem*) The stacking of close-packed layers of spheres in an ABAB sequence. Several metals have this structure, including Mg. See panel on CLOSE PACKING OF ATOMS.

hexagonal packing (*Crystal*) System in which many metals crystallize, thus achieving minimum volume. Each lattice point has 12 equidistant neighbours in such a cell construction.

hexagonal system (*Crystal*) A crystal system in which three equal coplanar axes intersect at an angle of 60°, and a fourth, perpendicular to the others, is of a different length. See fig. at BRAVAIS LATTICES.

hexagon dresser (*Eng*) A metal disk tool used for *dressing* grinding wheels.

hexagon voltage (*ElecEng*) The voltage between two lines, adjacent as regards phase sequence, of a six-phase system.

hexahydrobenzene (*Chem*) See CYCLOHEXANE.

hexahydrocresol (*Chem*) See METHYLCYCLOHEXANOL.

hexahydrophenol (*Chem*) See CYCLOHEXANOL.

hexahydropyridine (*Chem*) See PIPERIDINE.

hexamerous (*BioSci*) Having parts in sixes.

hexametaphosphates (*Chem*) Salts of hexametaphosphoric acid, $H_6(PO_3)_6$, a polymer of metaphosphoric acid. Cf CALGON.

hexamethylene (*Chem*) See CYCLOHEXANE.

hexamethylenediamine (*Chem*) *1,6-diamino-hexane*. $H_2N(CH_2)_6NH_2$. Important as a constituent material of nylon 6·6, which is a step-growth polymer formed from nylon salt. See STEP POLYMERIZATION.

hexamethylene diisocyanate (*Chem*) A type of aliphatic isocyanate monomer used to make polyurethanes. Abbrev *HDI*.

hexamethylenetetramine (*Chem*) *Hexamine*. $(CH_2)_6N_4$. A condensation product of methanal with ammonia, a crystalline substance with antiseptic and diuretic properties. Used in the production of *cyclonite*, a highly efficient explosive.

hexamitiasis (*Vet*) A disease of turkeys due to infection by the flagellate protozoon *Hexamita meleagridis*, which causes enteritis.

hexane (*Chem*) C_6H_{14}. There are five compounds with this formula: normal hexane, a colourless liquid, of ethereal odour, bp 69°C, rel.d. 0·66, is an important constituent of petrol and of solvent petroleum ether or ligroin.

hexapod (*BioSci*) Having six legs.

Hexapoda (*BioSci*) See INSECTA.

hexarch (*BioSci*) Having six strands of protoxylem.

hexastyle (*Arch*) A portico formed of six columns in front.

hexavalent (*Chem*) Capable of combining with six hydrogen atoms or their equivalent. Having an oxidation or co-ordination number of six. Also *sexavalent*.

hexobarbitone sodium (*Pharmacol*) The monosodium derivative of 5-Δ′-cyclohexenyl- 5-methyl-*N*-methyl-barbituric acid ($C_{12}H_{15}O_3N_2Na$), used intravenously or intramuscularly as a basic anaesthetic. TN *Evipan*.

hexogen (*Chem*) See CYCLONITE.

hexoses (*Chem, FoodSci*) A subgroup of the monosaccharides containing six carbon atoms, of the general formulae $HOCH_2(CHOH)_4CHO$ and $HOCH_2(CHOH)_3COCH_2OH$. The first formula signifies an *aldohexose*, the second a *ketohexose*.

hexphase (*ElecEng*) A term sometimes used instead of *six-phase*.

Heyland diagram (*ElecEng*) A particular application of the circle diagram of an ac circuit to represent the behaviour of an induction motor.

HF (*Aero*) Abbrev for *high frequency*. Radio transmissions between 3000 and 30 000 kHz.

Hf (*Chem*) Symbol for HAFNIUM.

HFC (*Chem*) Abbrev for HYDROFLUOROCARBON.

Hg (*Chem*) Symbol for MERCURY.

H-girder (*Build*) See BEAM.

H-hinge (*Build*) A hinge which when opened has the shape of the letter H. Also *parliament hinge*.

H-2 histocompatibility system (*BioSci*) The major histocompatibility system in the mouse. H-2 genes determine the major histocompatibility antigens on the surface of somatic cells and also the immune response (Ir) genes. The antigens in a given strain of mice are controlled by alleles within the H-2 locus. The H-2 system corresponds closely to the HLA SYSTEM of humans.

HI (*Surv*) Abbrev for HEIGHT OF INSTRUMENT.

Hi8 (*ImageTech*) A HIGH-BAND version of the 8 mm format, using a Y/C signal. Both consumer and industrial equipment is available.

HI Arc (*ImageTech*) Abbrev for *high-intensity arc*, a high-current carbon arc.

hiatus (*Geol*) A break or gap in the stratigraphical record, because of non-deposition or erosion.

Hib (*Med*) Abbrev for *Haemophilus influenzae* type b, a bacterium that can cause meningitis and other serious illnesses in young children.

Hibbert standard (*ElecEng*) A standard of magnetic flux linkage suitable for fluxmeter or galvanometer calibration. It comprises a stabilized magnet producing a radial field in an annular gap, through which a cylinder carrying a multiturn coil can be dropped.

hibernation (*BioSci*) The condition of partial or complete torpor into which some animals relapse during the winter season. V *hibernate*.

hiccup (*Med*) Sudden spasm of the diaphragm followed immediately by closure of the glottis.

hickie (*Print*) A blemish on a printed image which appears as a spot surrounded by a halo, caused by a small fragment of paper adhering to the plate as a result of FLUFFING or PICKING.

hick joint (*Build*) A FLAT JOINT formed in fine mortar when pointing, after the old mortar has been raked out of the joints.

hickory (*For*) The product of a hardwood tree (*Carya*) common to eastern N America. The heartwood is reddish-brown to brown (red hickory) while the sapwood is much lighter (white hickory). Fairly straight-grained with a coarse texture.

Hicks hydrometer (*ElecEng*) A form of hydrometer used for finding the relative density of the electrolyte in an accumulator, to determine the state of its charge; the hydrometer consists of a glass tube containing a number of coloured beads, which float at different relative densities.

hidden file (*ICT*) A file whose name is hidden from users for security or convenience by altering a file attribute. OPERATING SYSTEM files are often stored in this way.

hiddenite (*Min*) See SPODUMENE.

hidden Markov models (*MathSci*) Graphical models, originally developed by computer scientists studying machine learning and speech recognition but now more widely applied to analysis of gene sequences and phylogenetics. They describe a probability distribution over an infinite number of sequences.

hiding power (*Build*) The power of a paint to obscure a black-and-white contrast; generally expressed as the number of square feet per gallon or square metres per litre of paint. Also *obliterating power*.

hidrosis (*BioSci*) Formation and excretion of sweat.

hierarchical database (*ICT*) A DATABASE where the data are held in the form of a TREE structure.

hierarchy of needs (*Psych*) Maslow's theory that motives are based upon a hierarchy of needs with physiological needs (food, water, etc.) at the top level followed by safety, affiliation to a group, esteem, knowledge, aesthetic needs and at the lowest level the need for self-fulfilment. Until the high level needs are satisfied, the lower-ranking ones assume lesser importance.

hi-fi (*Acous*) Abbrev for HIGH FIDELITY.

Higgs' boson (*Phys*) A massive zero-spin meson whose existence is predicted by unified theories of the weak and electromagnetic forces.

high-alumina cement (*Eng*) Cement containing a higher proportion of alumina (30–50%) than ordinary Portland cement, it is faster setting, less affected by low temperature during setting, and more resistant to sea water and acids when set. But it has been found to degrade in warm, humid environments, eg swimming pools. Made by fusing a mixture of bauxite and chalk or limestone and grinding the resultant clinker. Abbrev *HAC*.

high aspect ratios (*Aero*) See ASPECT RATIO.

high-band (*ImageTech*) A format variant in which the FM carrier frequency is higher than in LOW-BAND, allowing room for a greater LUMINANCE SIGNAL bandwidth to increase the horizontal resolution, and greater frequency deviation to improve the signal-to-noise ratio and tonal gradation. Recordings are incompatible with low-band. See BETA, BETACAM, HI8, S-VHS, U-MATIC.

high brass (*Eng*) Common brass of 65/35 copper–zinc alloying, as distinct from deep-drawing brass with 66–70% copper. See COPPER ALLOYS.

high bypass ratio (*Aero*) Applied to a TURBOFAN in which the AIR MASS FLOW ejected directly as propulsive thrust by the fan is more than twice the quantity passed internally through the *gas generator* section.

high-carbon steel (*Eng*) Hypereutectoid steels containing more than 0·8% carbon. Such steels consist of iron carbide (*cementite*) and PEARLITE when slow cooled. They are capable of being heat treated to high hardness but tend to be brittle. Used for metal-working formers and fine-edge cutting tools, eg files. See panel on STEELS.

high-conductivity copper (*Eng*) Metal of high purity, having an electrical conductivity not much below that of the international standard, which is a resistance of 0·153 028 ohms for a wire 10 m in length and weighing 10 g.

high-definition (*ImageTech*) In current TV practice, the term is applied to systems using 1000 or more scanning lines to make up the picture.

high-definition developer (*ImageTech*) One which increases the contrast at boundaries between light and dark tones, the light boundary being enhanced by bromide from the heavily exposed part of the image and the dark boundary by comparatively fresh developer from the lightly exposed part.

high-density lipoproteins (*BioSci*) Subclass of lipoproteins involved in cholesterol transport in blood. Abbrev *HDL*.

high-density polyethylene (*Chem*) Highly crystalline ethylene polymer made at low pressure using Ziegler–Natta-type catalysts. Linear chains with little branching, with high T_m (approx 140°C) and high density (approx 960 kg m^{-3}). Competes with low-density polyethylene for packaging. Abbrev *HDPE*.

high electron mobility transistor (*Electronics*) A HETEROJUNCTION device in which electron current is confined to an undoped, high-mobility region. Abbrev *HEMT*.

high endothelial venule (*BioSci*) Specialized venules in the thymus-dependent area of lymph nodes characterized by prominent cuboidal endothelial lining cells rather than the normal squamous form. Recirculation of lymphocytes from blood to lymph takes place through the walls of these venules.

high-energy ignition (*Aero*) A gas turbine ignition system using a very high-voltage discharge.

high-energy phosphate compounds (*BioSci*) Phosphate compounds with a high negative free energy of hydrolysis. Endergonic metabolic reactions are driven by coupling them with the exergonic hydrolysis of these phosphate esters, the most common example being the hydrolysis of ATP.

high-energy physics (*Phys*) See PARTICLE PHYSICS.

high-energy rate forging (*Eng*) Methods in which the ram is accelerated to very high velocities by the release of compressed gas, usually to complete an operation in one blow. Abbrev *HERF*. Also *high-velocity forging*.

high-energy rate forming (*Eng*) Any of a recently developed family of processes, in which metal parts are rapidly compacted, forged and extruded, by the application of extremely high pressures.

higher-order conditioning (*Psych*) A form of conditioning in which a conditioned stimulus from earlier training serves as an unconditioned stimulus.

highest common factor (*MathSci*) See GREATEST COMMON DIVISOR.

high explosive (*MinExt*) One in which the active agent is in chemical combination and is readily detonated by application of shock. Nitrated cotton, nitroglycerine and ammonium nitrate are widely used, diluted to required explosive strength by inert fillers such as kieselguhr or wood pulp. See GELIGNITE, TRINITROTOLUENE.

high-fidelity (*Acous*) Said of high-quality sound reproduction. Abbrev *hi-fi*.

high-fidelity amplifier (*Acous*) Amplifier in which the input signal is reproduced with a very high degree of accuracy.

Highfield booster (*ElecEng*) An automatic battery booster consisting of a generator, a motor and an exciter. Automatic regulation is carried out by balancing the exciter voltage against that of the battery.

high-flux reactor (*NucEng*) One designed to operate with a greater neutron flux than normal for testing materials for radiation effects and experiments requiring intense beams of neutrons. Also *materials testing reactor*.

high-frequency amplification (*ICT*) That at frequencies used for radio transmission. In a receiver, any amplification which takes place before detection, frequency conversion, or demodulation.

high-frequency capacitance microphone (*Acous*) Microphone which uses audio variation of capacitance to vary the frequency of an oscillator, or response of a tuned circuit.

high-frequency heating (*Phys*) Heating (induction or dielectric) in which the frequency of the current is above mains frequency; from rotary generators up to ≈3000 Hz and from electronic generators 1–100 MHz. Also *radio heating*. See MICROWAVE HEATING.

high-frequency induction furnace (*Eng*) Essentially an air transformer, in which the primary is a water-cooled spiral of copper tubing, and the secondary the metal being melted. Currents at a frequency above about 500 Hz are used to induce eddy currents in the charge, thereby setting up enough heat in it to cause melting. Used in melting steel and other metals. Lower frequency (50 Hz) is used for melting non-ferrous metals, where a loop of liquid forms the secondary of the transformer and the furnace is never emptied completely in order to preserve this loop.

high-frequency resistance (*ICT*) That of a conductor or circuit as measured at high frequency, greater than that measured with dc because of the SKIN EFFECT.

high-frequency transformer (*ICT*) One designed to operate at high frequencies, taking into account self-capacitance, usually with band-pass response.

high-frequency welding (*ElecEng*) Welding by radio-frequency heating. See SEAM WELDING (2).

Highgate resin (*Min*) A popular name for the fossil gum-resin occurring in the Tertiary London Clay at Highgate in North London. Also *copaline, copalite*.

high grading (*MinExt*) (1) Selective mining, in which subgrade ore is abandoned unworked. (2) Theft of valuable concentrates or specimens such as nuggets of gold.

high-intensity separation (*MinExt*) Dry concentration of small particles of mineral in accordance with their relative ability to retain ionic charge after passing through an ionizing field.

high key (*ImageTech*) Describing a scene containing mainly light tones well illuminated without large areas of strong shadow.

Highland Boundary Fault (*Geol*) One of the most important dislocations in the UK, extending from the Clyde to Stonehaven and separating the Highlands of Scotland from the Midland Valley.

high-lead bronze (*Eng*) Soft matrix metal used for bearings, of copper/tin/lead alloys in approximate proportions 80, 10 and 10.

high-level language (*ICT*) Problem-orientated programming language in which each instruction may be equivalent to several MACHINE-CODE INSTRUCTIONS.

high-level modulation (*ICT*) Conditions where modulation of a carrier for transmission takes place at high level for direct coupling to the radiating system; *low-level modulation* requires subsequent push–pull or straight amplification. Also *high-power modulation*.

high-level radio-frequency signal (*Radar*) A signal having sufficient power to fire a switching tube.

high-level waste (*NucEng*) Nuclear waste requiring continuous cooling to remove the heat produced by radioactive decay.

highlight (*ImageTech*) An area of maximum brightness in a scene and its reproduction in a photographic or TV image.

highlighted (*ICT*) Indicates that an object or text is selected and will be affected by the user's next action. Highlighted text will usually appear in REVERSE VIDEO or as a different colour.

high-memory area (*ICT*) In a computer with 80286 or higher CHIPSETS using MS-DOS operating system, the first 64 Kbytes of EXTENDED MEMORY. See fig. at MEMORY MAP.

high-opacity foils (*Print*) A type of blocking foil esp suitable for marking undressed book cloths and deep-grained materials. These are only made in white and pastel

shades and have a considerable weight of pigment to obliterate the surface completely.

high-pass filter (*ICT*) One that freely passes signals of all frequencies above a reference value known as the cut-off frequency, f_c. (NB Beyond f_c attenuation only rises slowly and seldom approaches complete cut-off as implied by this name.)

high-performance polymers (*Chem*) Polymers whose orientation and crystallinity allow their strength to approach that of the carbon–carbon bond. See panel on HIGH-PERFORMANCE POLYMERS.

high-power modulation (*ICT*) See HIGH-LEVEL MODULATION.

high-pressure compressor (*Aero*) In a gas turbine engine with two or more compressors in series, the last is the high-pressure one. In a dual-flow turbojet this feeds the combustion chamber(s) only. Abbrev *HP compressor*.

high-pressure cylinder (*Eng*) The cylinder of a compound or multiple-expansion steam engine in which the steam is first expanded.

high-pressure hose (*MinExt*) Armoured hose, reinforced with circumferentially embedded wire, and hence able to withstand moderately high pressure and rough usage.

high-pressure turbine (*Aero*) The first turbine after the combustion chamber in a gas turbine engine with two or more turbines in series. Abbrev *HP turbine*.

high-pressure turbine stage (*Aero*) The first stage in a MULTISTAGE TURBINE. Abbrev *HP stage*.

high recombination rate contact (*Electronics*) The contact region between a metal and semiconductor (or between semiconductors) in which the densities of charge carriers are maintained effectively independent of the current density.

high-resistance joint (*ElecEng*) See DRY JOINT.

high-resistance voltmeter (*ElecEng*) One drawing negligible current and typically having a resistance in excess of 1000 ohms per volt.

high-resolution graphics (*ICT*) A term generally applied to graphical display units capable of fine definition by plotting around 600 or more distinct points in the width of a video display unit.

high spaces (*Print*) The normal height of spacing is about 0·75 in (19 mm) but higher spaces, cast to the shoulder height of the type, are more convenient when pages are to be stereotyped.

high-speed circuit breaker (*ElecEng*) A circuit breaker in which special devices are used to ensure very rapid operation; used particularly on dc traction systems.

high-speed shutter (*ImageTech*) (1) Shutter using mechanical or electrical means to provide short exposures, eg 0·001 s. (2) See FAST SHUTTER.

high-speed steam engine (*Eng*) A vertical steam engine, generally compound, using a piston valve, or valves, whose moving parts are totally enclosed and pressure-lubricated.

high-speed steel (*Eng*) A range of high-alloy steels used for metal-cutting tools. They are formulated to retain their hardness at a low red heat, and hence tend not to soften when used at high rates of machining, as would lower-alloy steels of similar hardness. High-speed steels usually contain 12–18% tungsten or 6–8% tungsten plus 5–8% molybdenum, with up to 5% chromium and 5% cobalt. Carbon is in the range 0·7–1·2% and small amounts of other elements, eg vanadium, are usually included. Abbrev *HS steel*. See panel on STEELS.

high-speed videotape duplicator (*ImageTech*) An apparatus for duplicating videotape SOFTWARE in a fraction of the programme's duration by running master and copy in contact and transferring the magnetic signal by applying heat or a magnetic field. See SLAVE VIDEOTAPE RECORDER.

high-speed wind tunnel (*Aero*) A high SUBSONIC wind tunnel in which compressibility effects can be studied.

high spot (*Radiol*) A small volume so situated that the dose therein is significantly above the general dose level in the region treated.

High-performance polymers

The structure of polymer chains is specified in two ways: by the shape of individual or groups of chains in space, CONFORMATION, and by the way each chain is constructed from its covalently bonded atoms, CON-FIGURATION. The chain configuration is determined mainly during polymerization, when the monomer units are linked together. Monomer configuration is often very similar to that of the repeat unit, but there are subtleties of chain structure which are not present in the monomer unit. The molecular size of the polymer helps to determine the mechanical properties, especially strength.

The intrinsic strength of the carbon–carbon bond (eg as in diamond) suggests that organic polymers should be among the strongest materials known. But the opposite seems to be true: their tensile moduli are among the lowest of engineering materials. In fact, a high-performance organic fibre has been known since the late 1950s, ie carbon fibre made by controlled pyrolysis and orientation of polyacrylonitrile (PAN).

The sheets of carbon atoms bonded are aligned along the fibre axis and laterally interleaved like a rolled-up newspaper. When strained along the fibre axis, carbon–carbon bonds are stressed directly, resulting in a maximum TENSILE MODULUS of 520 GN m^{-2} (over 2·5 times that of steel). With a specific gravity of 1·96, its SPECIFIC MODULUS ($E\rho^{-1}$) is almost ten times that of the best steel wire. When woven into cloth and impregnated with epoxy or polyester resin, carbon fibres are therefore ideal materials for aerospace applications.

A similar conversion of linear polymers into high-modulus fibres was not achieved until 1968 when a way of spinning liquid crystal aramid oligomers into ARAMID FIBRE (TNs Kevlar, Twaron) was discovered. The material is nearly 100% crystalline, unlike most polymers, and all the chains are aligned along the fibre axis.

The material is stabilized laterally by hydrogen bonds (cf PROTEIN STRUCTURE) and the material is also very stable thermally, showing no T_g or T_m but degrading at temperatures in excess of 450°C. Like carbon fibre, several grades of aramid fibre are available commercially for application in composites,

but unlike carbon fibre, aramid can be made into rope for engineering applications. Such cables (rated loads 10–200 tonnes) have been used for tethering oil rigs as well as reinforcing cracked structures like concrete cooling towers.

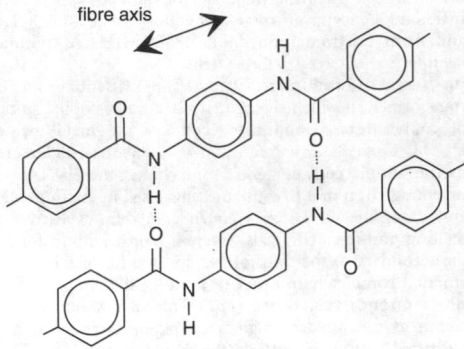

aramid fibres Showing hydrogen bonds between chains.

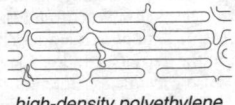

high-density polyethylene

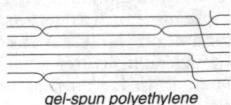

gel-spun polyethylene

Gel spinning

More recently a method of spinning high-perfor-mance fibres from any linear polymer has been developed. The process, known as GEL SPINNING, involves creating a dry gel from organic solvents (THETA SOLVENTS) in which each ultrahigh molecular mass polymer chain is a random coil, not entangled with neighbouring coils, thus eliminating persistent chain-folded crystals normally present. When hot stretched and drawn some 72 times, the gel-spun chains crystallize and orient along the fibre axis. Although this material is not as stable thermally as either aramid or carbon fibre, it possesses a specific gravity of less than one, so finds application for yacht rope etc. Like the other fibres, it is prepared with a diameter of about 10 μm and is routinely available as continuous fibre.

---

high-stop filter (*ICT*) See LOW-PASS FILTER.

high-strength brass (*Eng*) A type of brass based on the 60% copper–40% zinc composition, to which manga-nese, iron and aluminium are added to increase strength. MANGANESE BRONZE denotes a variety in which manganese is the principal addition, but most varieties now contain all three elements. See COPPER ALLOYS.

high-temperature reactor (*NucEng*) One designed to attain core temperatures above 660°C. Usually requires coated uranium dioxide or carbide pellets, cooled by helium gas.

high-temperature superconductors (*Phys*) Ceramics of perovskite crystal structure able to maintain superconduc-tion well above 77 K, the boiling point of liquid nitrogen. See panel on SUPERCONDUCTORS.

high-tension (*ElecEng*) See HIGH-VOLTAGE.

high-tension battery (*ElecEng*) Battery once widely used for supplying power for the anode current of valves. Also *anode battery*. US *B-battery*.

high-tension ignition (*ElecEng*) An ignition system for internal-combustion engines which employs a spark from a high-tension magneto or an induction coil.

high-tension magneto (*ElecEng*) The form of magneto once used for producing the high-voltage spark for internal-combustion engines.

high-tension separation (*MinExt*) Electrostatic separation, in which small particles of dry ore fall through a high-voltage dc field, and are deflected from gravitational drop or otherwise separated in accordance with the electric charge they gather and retain.

high-vacuum (*Electronics*) A system so completely evacuated that the effect of ionization on its subsequent operation may be neglected. See HARD.

high voltage (*ElecEng*) Legally, any voltage above 650 volts. In batteries etc, often called *high-tension*. See HIGH-TENSION BATTERY.

high-voltage electron microscope (*BioSci*) An accelerating voltage in the range of 10^6 volts shortens the wavelength of the electrons, and thus increases resolution and penetration compared with the normal transmission electron microscope, so that thicker specimens can be examined. Abbrev *HVEM*. See panel on ELECTRON MICROSCOPE.

high-voltage test (*ElecEng*) The application of a voltage greater than working voltage to a machine, transformer, or other piece of electrical apparatus to test the adequacy of the insulation.

highway (*ICT*) (1) UK term, now dying out, for BUS. (2) A high-capacity data link.

high-wing monoplane (*Aero*) An aircraft with the wing mounted on or near the top of the fuselage.

Hi-k capacitor (*ElecEng*) One in which the dielectric of barium and strontium titanates has permittivities above 1000.

Hilbert transformer (*ElecEng*) A device for obtaining a phase shift of 90°. It consists of a delay line, fed from a travelling-wave source and terminated by a negligibly small resistor, the potential difference across this forming the 0° output signal. The 90° signal is obtained by integrating the voltage along the line with a weighting function inversely proportional to the distance from the termination.

Hildebrand electrode (*Chem*) See HYDROGEN ELECTRODE.

hill-climbing (*ICT*) Continuous or periodic adjustment of self-regulating adaptive control systems to achieve an optimum result.

Hill coefficient (*BioSci*) A measure of co-operativity in a binding process. A Hill coefficient of 1 indicates independent binding, a value of greater than 1 shows positive co-operativity.

hillebrandite (*Min*) Dicalcium silicate hydrate. Occurs as white fibrous aggregates in impure thermally metamorphosed limestones and in boiler scale.

Hill reaction (*BioSci*) The light-driven transport of electrons from water to some acceptor other than CO_2 (eg ferricyanide) with the production of oxygen, by isolated chloroplasts or chloroplast-containing cells.

hi/lo (*Aero*) Refers to the high- and low-compression stages of the compressor of a gas turbine. See AERO-ENGINE.

Hilt's law (*Geol*) An expression of the observation that the more deeply buried a coal seam, the higher is the rank of its coal.

hilum (*BioSci*) (1) A scar or mark, esp on the testa where the stalk was attached to the seed. (2) The central part of a starch grain around which the starch is deposited. (3) A small depression in the surface of an organ that usually marks the point of entry or exit of blood vessels, lymphatics or an efferent duct. Also *hilus*.

Himalia (*Astron*) The sixth natural satellite of Jupiter, discovered in 1904. Distance from the planet 11 480 000 km; diameter approx 180 km.

HiMAT (*Aero*) Abbrev for *Highly Manoeuvrable Aircraft Technology*.

hind-brain (*BioSci*) In vertebrates, that part of the brain which is derived from the third or posterior brain vesicle of the embryo, comprising the CEREBELLUM and the MEDULLA OBLONGATA, the posterior brain vesicle itself.

hindcast (*Genrl*) A test of the accuracy of a predictive model by checking whether it can predict a known historical outcome from the events known to have preceded it.

hindered settling (*MinExt*) Hydraulic classification of sand-sized particles in accordance with their ability to gravitate through a column of similar material expanded by a rising current of water.

hindered settling (*PowderTech*) The settling of solids in a suspension of a concentration greater than 15% by volume, in which the predominant physical process is the draining of the fluid out of a thick slurry. No particle segregation by size occurs, a clear boundary being formed between the supernatant fluid and the settling suspension.

hind-gut (*BioSci*) That part of the alimentary canal of an animal which is derived from the posterior ectodermal invagination or proctodaeum of the embryo.

hindrance (*Phys*) Impedances (0 for zero, and 1 for infinite) used in theoretical manipulation of switching.

hinge (*BioSci*) The flexible joint between the two valves of the shell in a bivalve invertebrate, such as a pelecypod mollusc or a brachiopod; any similar structure; a joint permitting of movement in one plane only.

hinge-bound door (*Build*) A door which will not close easily or fully owing to the hinges being too deeply sunk.

hinge fault (*Geol*) A fault along which the displacement increases from zero at one end to a maximum at the other end.

hinge ligament (*BioSci*) The tough uncalcified elastic membrane that connects the two valves of a bivalve shell.

hinge moment (*Aero*) The moment of the aerodynamic forces about the hinge axis of a control surface, which increases with speed, necessitating AERODYNAMIC BALANCE.

hinge region (*BioSci*) A flexible region of immunoglobulin heavy chains near the junction of the Fab and Fc portions. This flexibility allows the angle between the arms bearing the antigen-combining sites to vary widely and so accommodate different dispositions of the antigen.

hinging post (*Build*) The post from which a gate is hung. Also *swinging post*.

HIP (*Eng*) Abbrev for HOT ISOSTATIC PRESSING.

hip (*Arch*) The salient angle formed by the intersection of two inclined roof slopes.

hip hook (*Build*) A strap of wrought-iron fixed at the foot of a hip rafter and bent into the form of a scroll, as a support for the hip tiles.

hip iron (*Build*) See HIP HOOK.

hip joint implant (*Med*) An artificial composite joint used to replace a diseased hip joint. Typically, the socket is constructed from ultrahigh-molecular-mass polyethylene; a ceramic ball is attached to a metal alloy stem that is adhesively bonded to the thigh bone using polymethyl methacrylate. The technology has been extended to most joints of the human body.

hip knob (*Build*) A finial surmounting the peak of a gable or a hipped roof.

hipped end (*Build*) The triangular portion of roof covering the sloping end of a hipped roof.

hipped roof (*Build*) A pitched roof having sloping ends at the gable ends.

hippocampus (*BioSci*) In the vertebrate brain, a tract of nervous matter running back from the olfactory lobe to the posterior end of the cerebrum. It forms part of the limbic system and is involved in memory and the transfer of information from short-term to long-term memory. See panel on BRAIN STRUCTURE. Adj *hippocampal*.

hippuric acid (*Chem*) Benzoyl-aminoethanoic acid. $C_6H_5CONHCH_2COOH$. Mp 187°C. Rhombic crystals, occurring in the urine of many animals, particularly herbivores.

hippus (*Med*) Rhythmical alternate contraction and dilatation of the pupil of the eye.

hip rafter (*Build*) The rafter at the hip of a roof where the two slopes meet. It supports the top ends of the jack rafters.

hip replacement (*Med*) Operative replacement of hip joint particularly for severe osteoarthritis.

hip roll (*Build*) A timber of circular section with a vee cut out along its length, so as to adapt it for sitting astride the hip of a roof.

HIPS (*Chem*) Abbrev for HIGH-IMPACT POLYSTYRENE. See COPOLYMER.

hip tile (*Build*) A form of arris tile laid across the hip of a roof.

HI-PVC (*Chem*) See panel on POLYVINYL CHLORIDE.

Hirschsprung's disease (*Med*) A condition occurring in children in which there is great hypertrophy and dilatation of the colon. Also *megacolon*.

hirsute (*BioSci*) Hairy; having a covering of stiffish hair or hair-like feathers.

hirsuties (*Med*) Excessive hairiness.

hirudin (*BioSci*) An anticoagulant, present in the salivary secretion of the leech, that prevents blood clotting by inhibiting the action of thrombin on fibrinogen.

Hirudinea (*BioSci*) A class of Annelida, the members of which are ectoparasitic on a great variety of aquatic and terrestrial animals. They possess anterior and posterior suckers, and most of them lack setae. They are hermaphrodite animals with median genital openings; the development is direct. Leeches.

His (*Chem*) Symbol for HISTIDINE.

hispid (*BioSci*) Coarsely and stiffly hairy as in many Boraginaceae; having a covering of stiffish hair or hair-like feathers.

hiss (*ICT*) See NOISE.

His's bundle (*Med*) See BUNDLE OF HIS.

hist-, histo- (*Genrl*) Prefixes from Gk *histos* (a web) denoting animal or plant tissue.

histamine (*Med*) 2-*imidazolyl*-4 (*or* 5) *ethylamine*. A base, $C_5H_9N_3$. Formed *in vivo* by the decarboxylation of HISTIDINE. Released during allergic reactions, eg hay fever; large releases cause the contraction of nearly all smooth muscle and dilatation of capillaries, which cause a fall of arterial blood pressure and shock.

histamine H2-receptor antagonists (*Pharmacol*) A class of drugs that block the type of histamine receptors found in the stomach and thereby inhibit production of gastric acid. Used for treatment of gastric and duodenal ulcers. Examples: *cimetidine* (Tagamet), *famotidine* (Pepcid), *ranitidine* (Zantac).

histamine receptors (*Pharmacol*) There are two main classes of receptor, H1 and H2; the former are mostly in skin, nose and airways and are targeted by antihistamine drugs to inhibit allergic responses. *H2-receptor antagonists* block receptors found mostly in the stomach.

histidine (*Chem*) 2-*amino*-3-*imidazolepropanoic acid*. The *L*- or *S*-isomer is a constituent of proteins and a precursor of HISTAMINE. Symbol His, short form H.

histiocyte (*BioSci*) A macrophage found within the tissues in contrast to those found in the blood (monocytes).

histioma (*Med*) Any tumour derived from fully developed tissue, such as fibrous tissue, cartilage, muscle, blood vessels. Also *histoma*.

histo- (*Genrl*) See HIST-.

histochemistry (*BioSci*) The chemistry of living tissues. Mainly used in the context of staining of specimens for microscopic examination and the use of dyes that selectively bind to materials of different composition.

histocompatibility antigen (*BioSci*) Genetically determined antigens present on the surface of nucleated cells, including blood leucocytes. Coded for by *MHC genes*. They are responsible for the differences between genetically non-identical individuals which cause rejection of homografts. See MAJOR HISTOCOMPATIBILITY COMPLEX.

histocompatibility testing (*BioSci*) Tests whereby donor and recipient are matched as closely as possible prior to tissue grafting in humans.

histogen (*BioSci*) One of three meristems (DERMATOGEN, PERIBLEM, PLEROME) at the shoot or root tip that give rise exclusively to particular tissues in that organ (epidermis, cortex, stele + pith, respectively). The concept of discrete histogens has now been replaced by the TUNICA-CORPUS CONCEPT.

histogenesis (*BioSci*) The formation of new tissues.

histogram (*MathSci*) A graphical representation of class frequencies as rectangles whose base is the class interval, the value of frequency being proportional to the area of the corresponding rectangle.

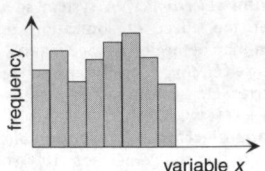

histogram

histology (*BioSci*) The study of the minute structure of tissues and organs.

histolysis (*BioSci*) The breakdown, and sometimes liquefaction, of tissues.

histones (*BioSci*) Basic proteins involved in the packaging of DNA in the eukaryotic nucleus to form CHROMATIN, which is folded into NUCLEOSOMES, the first level of chromosome organization above the DNA helix. There are five types of histone molecule, four of which have been highly conserved in sequence throughout eukaryotic evolution.

histoplasmosis (*Med, Vet*) A disease of animals and humans due to infection by the fungal organism *Histoplasma capsulatum*. Affects the lungs in humans, being relatively common in the USA.

historical geology (*Geol*) The major branch of geology that is concerned with the evolution of the Earth and its environment from its origins to the present day.

histozoic (*BioSci*) Living in the tissues of the body, amongst the cells.

histrionic personality disorder (*Psych*) A disorder characterized by behaviour that is self-centred, attention-seeking, overly dramatic and manipulative. Formerly referred to as *hysterical personality*.

hit (*ICT*) A request to a web server, from a client browser or a request broker, for a single file.

hitch (*MinExt*) (1) A fault of minor importance, usually not exceeding the thickness of a seam. (2) A ledge cut in the rock face to hold mine timber in place.

hit rate (*ICT*) The proportion of RECORDS in a FILE that will be selected or modified when the file is edited or updated.

Hittorf dark space (*Phys*) See CROOKES DARK SPACE.

HIV (*BioSci*) Abbrev for HUMAN IMMUNODEFICIENCY VIRUS.

Hi-vision (*ImageTech*) A wide-screen, high-definition, colour TV system with 1125 lines. See MUSE, UNIHI.

HLA-A, HLA-B, HLA-C (*BioSci*) Histocompatibility antigens, each coded by different loci in the MHC genes, and for which there are numerous allelic products at each locus. These antigens belong to Class I and have similar structures, consisting of a membrane-bound polypeptide to which is attached non-covalently $\beta2$ microglobulin. Class I antigens are present on almost all nucleated cells. Cytotoxic T-cells recognize antigens in association with Class I molecules.

HLA-D (*BioSci*) Histocompatibility antigens coded for by separate loci in the MHC gene complex. There are three distinct loci, DP, DR and DQ, each of which has several alleles. These antigens belong to Class II, and are dimers composed of two different membrane-bound polypeptide chains. HLA-D antigens are normally absent from most nucleated cells but present on B-lymphocytes, dendritic (interdigitating) cells, and macrophages stimulated by INTERFERON. T-helper cells recognize antigens in association with Class II molecules.

HLB (*Chem*) Abbrev for HYDROPHILIC–LIPOPHILIC BALANCE referring to emulsifying agents.

HMI (*ImageTech*) TN for a type of metal halide lamp.

Hn (*Chem*) Symbol for HAHNIUM.

H-network (*ICT*) Symmetrical section of circuit, with one shunt branch and four series branches.

HnRNA (*BioSci*) Abbrev for HETEROGENEOUS NUCLEAR RNA.

HNW (*Build*) Abbrev for *head, nut and washer*.

Ho (*Chem*) Symbol for HOLMIUM.

hoarding (*Psych*) The storing of food or other items in the animal's home or territory; occurs in small mammals and in some bird species.

hoar frost (*EnvSci*) A deposit of ice crystals formed on objects, esp during cold clear nights when the dew point is below freezing point. The conditions favouring the formation of hoar frost are similar to those which produce *dew*.

hob (*Eng*) (1) A hardened master punch, used in die sinking, which is a duplicate of the part to be produced by the die. It is pressed into an unheated die blank to make the die impression. (2) A gear-cutting tool resembling a milling cutter or a worm gear, whose thread is interrupted by grooves so as to form cutting faces. Also *hobbing cutter*.

hobbing machine (*Eng*) A machine for cutting teeth on gear blanks, for the production of spur, helical and worm gears by means of a hobbing cutter.

hobbles (*Vet*) An apparatus applied to the legs of a horse for casting.

Höchstäter cable (*ElecEng*) A high-voltage multicore cable in which a thin metallized sheath is placed over the insulation of each core, in order to control the distribution of electric stress in the dielectric and ensure that it is purely radial.

hock (*BioSci*) The tarsal joint of a mammal.

hod (*Build*) A three-sided container, supported on a long handle, used for carrying bricks and mortar on the site.

Hodgkin's disease (*Med*) One of a group of diseases named lymphomas which involve lymphoid tissues. It is characterized by destruction of the normal architecture of lymph nodes and replacement with reticular cells, lymphocytes, neutrophil and eosinophil leucocytes, and an unusual kind of giant cell with two nuclei. This is accompanied by deficient cell-mediated immunity although antibody formation is normal.

hodograph (*Phys*) A curve used to determine the acceleration of a particle moving with known velocity along a curved path. The hodograph is drawn through the ends of vectors drawn from a point to represent the velocity of the particle at successive instants.

hodoscope (*NucEng*) Apparatus (eg an array of radiation detectors) which is used for tracing paths of charged particles in a magnetic field.

Hoechst 33258 (*BioSci*) TN for a DNA-specific stain, used in chromosome BANDING TECHNIQUES.

Hofmann degradation (*Chem*) A process used in organic chemistry to determine the structure of amines. It involves exhaustive methylation, ie the use of excess of a methylating agent, such as methyl iodide. Good for alkaloids.

Hofmann's reaction (*Chem*) A method of preparing primary amines from the amides of acids by the action of bromine and then of caustic soda. The number of carbon atoms in the chain should not be more than six, and the resulting amine has one carbon atom less then the amide from which it has been prepared.

Hofmeister series (*Chem*) The simple anions and cations arranged in the order of their ability to coagulate solutions of lyophilic colloids.

hogback (*Geol*) A ridge with a sharp summit and steep slopes on both sides, usually 20°.

hogback girder (*Build, CivEng*) A girder which curves along its top edge to be convex upwards.

hog cholera (*Vet*) See SWINE FEVER.

hog-frame (*Build*) A term applied to some forms of truss which are shaped so as to bulge on the upper side.

hogget (*Agri*) (1) A castrated male sheep in its first year. (2) An uncastrated male pig.

hogging (*Build*) A mixture of gravel and clay, used for paving. Also *hoggin*.

hogging (*Ships*) This occurs when the middle of the ship is supported in the crest of a wave while the ends are in troughs or when the ends are more heavily loaded than the middle. If the ship actually bends it is said to be hogged. Cf SAGGING.

hog-pit (*Paper*) The pit below the couch of the paper-making machine, equipped with an agitator, in which backwater, edge trims (and the whole web at times of a break) are collected for pumping to an earlier part of the system for reuse.

hohlraum (*Acous*) Ger for *cavity*.

Hohmann orbit (*Space*) A space trajectory tangential to, or osculating, two co-planar planetary orbits at its PERIHELION and APHELION respectively: it is also the most energetically economical transfer orbit.

hoist (*MinExt*) An engine with a drum, used for winding up a load from a shaft or in an underground passage such as a winze.

hol-, holo- (*Genrl*) Prefixes from Gk *holos*, whole.

holandry (*BioSci*) Inheritance of characters specified by genes on the male chromosome and therefore only expressed in the male.

Holarctic region (*BioSci*) One of the primary faunal regions into which the surface of the globe is divided. It includes N America to the edge of the Mexican plateau, Europe, Asia (except Iran, Afghanistan, India south of the Himalayas, the Malay Peninsula), Africa north of the Sahara and the Arctic islands.

hold (*Electronics*) The maintenance, in charge-storage tubes, of the equilibrium potential by means of electron bombardment.

hold (*ImageTech, ICT*) Synchronization control, in which oscillator frequency of the receiver is adjusted to that of incoming synchronizing pulses.

hold (*Ships*) A compartment within a ship's hull for the carriage of cargo. Below the lowermost deck it is termed hold; above this, 'tween decks. For identification, the holds are numbered from the fore end of the ship.

holdback (*NucEng*) An agent for reducing an effect, eg a large quantity of inactive isotope reduces the coprecipitation or absorption of a radioactive isotope of the same element.

hold-down roller (*ImageTech*) A small roller pressing the edges of motion picture film against a sprocket to ensure that the teeth fully engage the perforations. Also *pad roller*.

Holden permeability bridge (*ElecEng*) A permeability bridge in which the standard bar and the bar under test carry magnetizing coils, and are connected by yokes to form a closed magnetic circuit. The magnetizing currents are varied until there is no magnetic leakage between the yokes.

holderbat (*Build*) A metal collar formed in two half-round parts, capable of being clamped together around a rainwater, soil or waste pipe, and having a projecting leg on one part for fixing to a wall.

Holder's inequality (*MathSci*) If a_r and b_r are positive, non-proportional, sets, and if $\alpha + \beta = 1$, then (1)

$$\sum_1^n a_r^\alpha b_r^\beta < \left(\sum_1^n a_r\right)^\alpha \left(\sum_1^n b_r\right)^\beta$$

if α and β are positive, and (2)

$$\sum_1^n a_r^\alpha b_r^\beta > \left(\sum_1^n a_r\right)^\alpha \left(\sum_1^n b_r\right)^\beta$$

if $\alpha > 1$ or if $\alpha < 0$.

holdfast (*BioSci*) Any single-celled or multicellular organ other than a root, which attaches a plant (esp an alga) to the substrate.

holdfast (*Build*) A device for holding down work on a bench, comprising a main pillar which passes through a hole in the bench, and an adjustable clamp with a shoe for placing on the work.

hold frame (*ImageTech*) See FREEZE FRAME.

holding altitude (*Aero*) The height at which a controlled aircraft may be required to remain at a HOLDING POINT.

holding anode (*Electronics*) Auxiliary dc anode in a mercury-arc rectifier for maintaining an arc.

holding beam (*Electronics*) Widely spread beam of electrons used to regenerate charges retained on the dielectric surface of a storage tube or electrostatic memory.

holding pattern (*Aero*) A specified flight track, eg *orbit* or figure of eight, which an aircraft may be required to maintain about a holding point.

holding point (*Aero*) An identifiable point, such as a radio beacon, in the vicinity of which an aircraft under AIR-TRAFFIC CONTROL may be instructed to remain.

holding-up (*Eng*) The action of pressing a heavy hammer against the head of a rivet while closing or forming the head on the shank.

hold-on coil (*ElecEng*) An electromagnet holding the moving arm of a switch in the 'on' position against a spring which causes the arm to return to the 'off' position if the current in the coil is reduced or interrupted.

hold-up (*ChemEng*) In any process plant, the amount of material which must always be present in the various reactors etc to ensure satisfactory operation.

hole (*Electronics*) A vacancy in a normally filled energy band, either as result of the electron being elevated by thermal energy to the conduction band, and so producing a hole–electron pair; or as a result of one of the crystal lattice sites being occupied by an acceptor impurity atom. Such vacancies are mobile and contribute to electric current in the same manner as positive carriers and are equivalent to positrons. Analogously, in oxide ceramics, a singly charged oxygen ion can be viewed as a doubly charged ion plus a hole. Also *negative-ion vacancy*.

hole control (*MinExt*) Altering the composition of the DRILLING MUD, drill pressure and rate to accommodate changes in the rock formation in which the hole is being drilled. See panel on DRILLING RIG.

hole current (*Electronics*) That part of the current in a semiconductor due to the migration of HOLES.

hole density (*Electronics*) The density of the holes in a semiconductor in a band which is otherwise full.

hole injection (*Electronics*) Holes can be emitted in n-type semiconductor by applying a metallic point to its surface.

hole mobility (*Electronics*) See MOBILITY.

hole theory of liquids (*Chem*) Interpretation of the fluidity of liquids by regarding them as disordered crystal lattices with mobile vacancies or holes.

hole trap (*Electronics*) An impurity in a semiconductor which can release electrons to the conduction or valence bands and so trap a 'hole'.

holiday (*Build*) A greater or lesser part of the surface accidentally missed during painting.

holland (*Textiles*) A glazed cotton or linen fabric used principally for window-blinds and INTERLININGS.

hollander beater (*Paper*) A horizontal trough with a dividing wall parallel to its longer side which stops short of the ends to provide a continuous channel around which stuff may circulate under the propulsion of the rotating beater roll. The surface of the latter is fitted with metal bars parallel to the roll shaft which impart a rubbing and cutting action on the fibres in the narrow gap between them and similar fixed bars beneath the roll. The extent of this action largely controls the properties of the finished paper. Now generally superseded by the REFINER, which performs a similar function.

Hollofil (*Textiles*) TN (US) for hollow polyester fibre, with lowered thermal conductivity due to entrapped air, used therefore for insulation in sleeping bags etc.

hollow back (*Print*) A type of binding in which the spine of the book is not pasted or glued down, leaving a space between the back of the sections and the leather or cloth when the book is open. Also *open back*.

hollow bed (*Build*) A bed joint in which, owing to the surfaces of the stones not being plane, there is contact only at the outer edges.

hollow-cathode tube (*Electronics*) A gas-discharge lamp in which the glowing plasma forms only inside a small tubular cathode, giving, under appropriate current and gas pressure conditions, an intense light with high spectral purity.

hollow glass microspheres (*Glass*) Glass bubbles with diameters in the region 10–150 μm and typical density of 280 kg m^{-3}. Used as fillers for plastics, chiefly for the low density of the resulting composite material. Cf BALLOTINI.

hollow mandrel lathes (*Eng*) The term formerly applied to lathes capable of having bar stock fed through the mandrel for repetition work.

hollow moulding (*Eng*) A type of INJECTION MOULDING process where nitrogen gas is injected into the molten polymer product to create a void. Also *gas-melt process*.

hollow newel (*Build*) The well-hole of a winding stair.

hollow punch (*Eng*) A hollow cylindrical tool tapered on its outside diameter to form a cutting edge and used to punch circular washers from sheets of soft materials, eg rubber, leather, etc.

hollow quoin (*Build*) A quoin accommodating the heel post of a lock gate in a vertical recess.

hollow roll (*Build*) A joint between the edges of two lead sheets on the flat, made by turning up each edge at right angles to the flat surface, bringing the two turned-up parts together, and shaping them over to form a roll.

hollows (*Eng*) Fillets, or curves of small radius, uniting two surfaces intersecting at an angle; added to a pattern to give strength to the casting and facilitate withdrawal of the pattern from the mould.

hollows (*Print*) Strips of strong paper etc attached to the case spine and to the back of a book to strengthen the spine. Also *back lining*.

hollow spindle spinning (*Textiles*) A specialized method of yarn formation in which a core of untwisted staple fibres is wrapped with a binder thread as it passes through a rotating hollow spindle.

hollow walls (*Build*) Same as CAVITY WALLS.

holly (*For*) A widespread hardwood (*Ilex*) whose heartwood is fine-textured, irregularly grained and creamy white.

holmium (*Chem*) Symbol Ho, at no 67, ram 164·930. A metallic element, a member of the rare earth group. It occurs in euxenite, samarskite, gadolinite and xenotime.

holmquistite (*Min*) A rare lithium-bearing calcium-free variety of orthorhombic amphibole.

holo- (*Genrl*) See HOL-.

holoaxial (*Crystal*) A term applied to those classes of crystals characterized by axes of symmetry only; such crystals are not symmetrical about planes of symmetry.

holobenthic (*BioSci*) Passing the whole of the life cycle in the depths of the sea.

holoblastic (*BioSci*) Said of ova which exhibit total cleavage.

holobranch (*BioSci*) In fish, a branchial arch carrying two rows of respiratory lamellae or filaments, one on the posterior and one on the anterior face.

holocarpic (*BioSci*) Having the whole thallus transformed at maturity into a sporangium or a sorus of sporangia.

Holocene (*Geol*) The younger, temperate, epoch of the Quaternary period. Its base is taken as 10 000 years before the present (*BP*). Also *Recent*. See QUATERNARY.

holocentric chromosome (*BioSci*) One lacking a localized CENTROMERE, and along whose full length spindle MICROTUBULES attach. Found in some plants, protozoa and certain classes of insect.

holocephalous (*BioSci*) Said of single-headed ribs.

holocrine (*BioSci*) A form of secretion in which the whole cell is shed from the gland, usually after becoming packed with the main secretory substance. In mammals, sebaceous glands are one of the few examples.

holocrystalline rocks (*Geol*) Those igneous rocks in which all the components are crystalline; glass is absent. Cf HEMICRYSTALLINE ROCKS.

holoenzyme (*BioSci*) The complete enzyme complex composed of the protein portion (*apoenzyme*) and cofactor or coenzyme.

hologamy (*BioSci*) The condition of having gametes that resemble the ordinary cells of the species in size and form; union of such cells.

hologram (*Phys*) Image produced by HOLOGRAPHY.

holographic interferometry (*Eng*) The superimposition of two holograms of the same object produces a pattern of interference fringes if any changes in surface displacement have occurred. Used as a *non-destructive testing method* for measuring deformations (eg STRESS ANALYSIS), vibration analysis and detection and monitoring of cracks.

holography (*Phys*) Imaging technique which records and reconstructs the wavefront emanating from an illuminated object. Coherent light from a laser is split in two: one is a reference beam and the other illuminates the object. The waves scattered by the object and from the reference beam are recombined to form an interference pattern on a photographic plate, the *hologram*; this records both the amplitude and phase of the scattered light. When the hologram is itself illuminated by light from a laser or other point source, two images are produced: one is virtual but the other is real and can be viewed directly, so a three-dimensional image of the object can be produced.

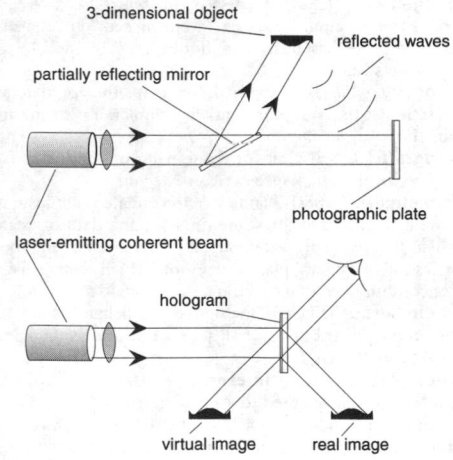

holography

holohedral (*Crystal*) A term applied when a crystal is complete, showing all possible faces and angles.

holomastigote (*BioSci*) Having numerous flagella scattered evenly over the body.

Holometabola (*BioSci*) See ENDOPTERYGOTA.

holometabolic (*BioSci*) Showing a complete metamorphosis, as members of the ENDOPTERYGOTA. N *holometabolism*. Cf HEMIMETABOLIC.

holomorphic function (*MathSci*) See ANALYTIC FUNCTION.

holophrastic stage (*Psych*) Early stage of language development in which mostly single words are uttered, often forcefully. Despite the absence of complex syntax, the meaning is generally quite clear.

holophytic (*BioSci*) Living by photosynthesis, like most plants that are not parasitic, saprophytic, phagocytic, etc. Cf HOLOZOIC.

holoplankton (*BioSci*) Organisms which remain as plankton throughout their life cycles.

holostyly (*BioSci*) A type of jaw suspension in which the upper jaw fuses with the cranium, the hyoid arch playing no part in the suspension, found in some, largely extinct, cartilaginous fish.

Holothuroidea (*BioSci*) A class of ECHINODERMATA having a sausage-shaped body without arms; the tube feet possess ampullae and may occur on all surfaces; the anus is aboral, the madreporite internal and the skeleton is reduced to small ossicles embedded in the soft integument. Generally free-living mud feeders. Sea cucumbers.

holotrichous (*BioSci*) Bearing cilia of uniform length over the whole surface of the body.

holotype (*BioSci*) The one specimen designated as the nomenclatural type in a published description of the species.

holozoic (*BioSci*) Living by ingestion or phagocytosis and digestion, like most animals; not on the uptake of soluble materials. N *holozoon*. Cf HOLOPHYTIC.

home cinema (*ImageTech*) Recreating the cinema experience in the home with a large-screen TV or PROJECTION TV and SURROUND SOUND.

home location register (*ICT*) The database in a PUBLIC LAND MOBILE NETWORK that stores details of each mobile, such as class of service and, if switched on, the VISITOR LOCATION REGISTER on which it is currently registered.

homeobox (*BioSci*) A conserved sequence of about 180 base pairs, originally detected in genes that gave rise to segmentation and homeotic mutants in *Drosophila*. The homeobox product, the *homeodomain*, confers specific DNA-binding properties on the protein in which it is located. Homeoboxes fall into groups and each group appears to be particularly involved in the development of a particular complex structure, such as an eye; remarkably, the homeobox controlling vertebrate eye development (*Pax6*) is highly homologous with that for the insect compound eye, forcing some reconsideration of concepts of anatomical homology and analogy.

homeodomain (*BioSci*) Conserved 60-amino-acid sequence, the product of a particular HOMEOBOX.

homeomorphic (*MathSci*) Topological spaces T_1 and T_2 are said to be homeomorphic, or topologically equivalent, if there is a mapping f from T_1 onto T_2 which is one-to-one and such that both f and its inverse f^{-1} are continuous. Such a mapping f is said to be a *homeomorphism*. In effect, two geometrical figures are homeomorphic if each can be transformed into the other by a continuous deformation, eg the circle, the square and the triangle are homeomorphic, but a circle and a figure of eight are not.

homeopathy (*Med*) A system of medicine, founded by Dr Samuel Hahnemann (1755–1843), the basic principle of which is *Similia similibus curantur* (let likes be cured by likes). Experimental observations led Hahnemann to conclude that (1) a disease is characterized by a definite symptom complex, (2) it can be effectively treated by the drug which produces in a healthy individual the most similar symptom complex, and (3) for a given disease, the proven drug is best administered in extreme dilution. Also *homoeopathy*.

homeostasis (*BioSci*) (1) The tendency for the internal environment of an organism to be maintained constant. (2) The tendency for plant and animal populations to remain constant as a result of density-dependent mechanisms operating on birth rate, survival or death rate.

homeotic mutants (*BioSci*) Mutants in *Drosophila* that effect large-scale changes in development, eg the substitution of a leg for a wing.

homeotypic division (*BioSci*) See MITOSIS and panel on CELL CYCLE.

home page (*ICT*) The introductory first page of an Internet website. See panel on INTERNET.

home public land mobile network (*ICT*) In a mobile-telephone network that permits ROAMING, the PUBLIC

LAND MOBILE NETWORK containing a mobile's HOME LOCATION REGISTER.

homer (*ICT*) Any arrangement that provides signals or fields which can be used to guide a vehicle to a specified destination, eg a homing aid for aircraft, a leader cable for ships or a guidance system for missiles.

home range (*BioSci*) A definite area to which individuals, pairs or family groups of many types of animals restrict their activities.

home shopping (*ICT*) Buying products and services by means of telecommunication, eg over the Internet or cable TV. See panel on INTERNET.

homespun (*Textiles*) A coarse tweed hand-woven from hand-spun wool.

homing (*ICT*) The operation of a selector in returning to a predetermined normal position following the release of the connection.

homing aid (*Aero*) Any system designed to guide an aircraft to an airfield or aircraft carrier.

homing behaviour (*Psych*) Navigational behaviour in a number of species, ranging from returning home after a daily foraging or other excursion, to the more complex navigational task involved in long migrations.

hominin (*BioSci*) Current new name for humans and their ancestors (*Homo sapiens*, *H. ergaster*, *H. rudolfensis*, the Australopithecines, *Australopithicus africanus*, *A. boisei*, etc., and other ancient forms like Paranthropus and Ardipithecus). This nomenclature recognizes the evolutionary divergence from chimps and gorillas that, together with hominids, are in the subfamily Homininae (within the Hominoid Family).

hominoid (*BioSci*) Primate Family that is now generally subdivided into two subfamilies: Ponginae (orang-utans) and Homininae (humans and their ancestors, and chimps and gorillas). See HOMININ.

homo-, homeo-, homoeo-, homoio- (*Genrl*) Prefixes from Gk *homos*, *homoios*, same.

homoblastic (*BioSci*) (1) Animals showing direct embryonic development from similar cells and without intervening larval forms. (2) Term describing a species in which the first-formed leaves in a seedling or shoot are very like those formed later. Cf HETEROBLASTIC.

homocentric (*Phys*) A term applied when rays are either parallel or pass through one focus.

homocercal (*BioSci*) Said of a type of tail fin, found in all the adults of the higher fish, in which the vertebral column bends abruptly upwards and enters the epichordal lobe, which is equal in size to the hypochordal lobe.

homochlamydeous (*BioSci*) Having a perianth consisting of members all of the same kind, not distinguishable into sepals and petals.

homocyclic (*Chem*) Containing a ring composed entirely of atoms of the same kind.

homocysteine (*BioSci*) An oxidized form of cysteine that occurs as a by-product of the metabolism of protein, thought to be an indicator of heart disease when present in high levels.

homocystinuria (*Med*) A metabolic defect resulting in the excretion of homocystine (the oxidized form of cysteine, an essential amino acid) in the urine.

homodesmic structure (*Crystal*) Crystal form with only one type of bond (either ionic or covalent).

homodont (*BioSci*) Said of an animal whose teeth all have the same characteristics.

homodyne reception (*ICT*) That using an oscillating valve adjusted to, or locked with, an incoming carrier, to enhance its magnitude and improve demodulation. Also *demodulation of an exalted carrier*.

homoeomerism (*BioSci*) In metameric animals, the condition of having all the somites alike. Adj *homoeomeric*. Cf HETEROMERISM.

homoeopathy (*Med*) See HOMEOPATHY.

homogametic (*BioSci*) Having all the gametes alike.

homogametic sex (*BioSci*) The sex that is *homozygous* for the *sex-determining chromosomes*. Cf HETEROGAMETIC SEX.

homogamy (*BioSci*) (1) Inbreeding, usually due to isolation. (2) The simultaneous maturation of the anthers and stigmas in a flower. Cf DICHOGAMY, PROTANDRY, PROTOGYNY.

homogeneous (*Chem*) Said of a system consisting of only one phase, ie a system in which the chemical composition and physical state of any physically small portion are the same as those of any other portion.

homogeneous co-ordinates (*MathSci*) A system of co-ordinates in which any multiple of the co-ordinates of a point or line represents the same point or line, eg the line (*a*,*b*,*c*) is the same as the line (*ka*,*kb*,*kc*).

homogeneous function (*MathSci*) An algebraic function such that the sum of the indices occurring in each term is constant, eg $x^3 + 2x^2y + y^3$.

homogeneous ionization chamber (*NucEng*) One in which both walls and gas have similar atomic composition, and hence similar energy absorption per unit mass.

homogeneous light (*Phys*) See MONOCHROMATIC LIGHT.

homogeneous radiation (*Phys*) Radiation of constant wavelength (monochromatic), or constant particle energy.

homogeneous reactor (*NucEng*) One in which the fuel and moderator are finely divided and mixed (or the fuel may be dissolved in a liquid moderator) so as to produce an effectively homogeneous core material. See SLURRY REACTOR.

homogenesis (*BioSci*) The type of reproduction in which the offspring resemble the parents.

homogenisation (*FoodSci*) Reducing fat particles in an emulsion (eg milk, ice cream) to a common size to improve stability and texture.

homogenization (*Glass*) See panel on GLASSES AND GLASSMAKING.

homogenizer (*Phys*) A device in which coarse and polydisperse emulsions are transformed into nearly monodisperse systems. The liquid is subjected to an energetic shear.

homogeny (*BioSci*) Individuals or parts thereof that are HOMOLOGOUS and substantially similar in form and function. Adj *homogenous*.

homograft (*BioSci*) Graft of tissue from one individual to another of the same species. Also *allograft*.

homoiohydric (*BioSci*) Plants able to regulate water loss and to remain hydrated for some time (hours, days or years) when the external water supply is restricted, eg most terrestrial vascular plants, few of which can survive desiccation. Cf POIKILOHYDRIC.

homoiomerous (*BioSci*) Descriptive of a lichen thallus that has an even distribution of algal cells through its thickness. Cf HETEROMEROUS.

homoioplastic (*BioSci*) In experimental zoology, said of a graft that is transplanted to a site identical with its point of origin, eg a skin graft to a skin site. Also *homoplastic*.

homoiosmotic (*BioSci*) Of an aquatic animal, maintaining a relatively constant internal osmotic pressure. Cf POIKILOSMOTIC.

homoiothermal (*BioSci*) See WARM-BLOODED.

homokaryon (*BioSci*) SOMATIC CELL HYBRID containing separate nuclei from the same species.

homologous (*BioSci*) Of the same essential nature and of common descent although the functions of homologous organ may differ; eg pentadactyl limbs are homologous but may be used in very different modes of locomotion.

homologous alternation of generations (*BioSci*) See HOMOLOGOUS THEORY OF ALTERNATION, ISOMORPHIC ALTERNATION OF GENERATIONS. Cf ANTITHETIC THEORY OF ALTERNATION.

homologous chromosomes (*BioSci*) Chromosomes that pair with each other during synapsis at *meiosis*, so that one member of each pair is carried by every gamete.

homologous organs (*BioSci*) Organs that are equivalent morphologically and of common evolutionary origin but that may be similar or dissimilar in appearance or function.

homologous recombination (*BioSci*) Genetic recombination involving exchange of homologous loci during meiosis. Also *crossing over*.

homologous series (*Chem*) A series of organic compounds, each member of which differs from the next by the insertion of a –CH$_2$– group in the molecule. Such a series may be represented by a general formula and shows a gradual and regular change of properties with increasing molecular weight.

homologous theory of alternation (*BioSci*) The hypothesis that the sporophyte is of a similar nature to the gametophyte and thus that vascular plants evolved from algae with an ISOMORPHIC ALTERNATION OF GENERATION. Also *homologous alternation of generation*. Cf ANTITHETIC THEORY OF ALTERNATION.

homologous variation (*BioSci*) The occurrence of similar variations in related species.

homology (*BioSci*) (1) Morphological equivalence, common evolutionary origin. N *homologue*. Cf ANALOGY. (2) Of DNA sequences or peptide sequences, the degree of similarity.

homology group (*MathSci*) One of a class of groups used to classify topological spaces.

homomorphic (*BioSci*) Said of chromosome pairs that have the same form and size.

homomorphism (*MathSci*) A structure preserving mapping from one group to another. If $(G, \bigcirc)$ and $(H, \bigstar)$ are groups with group operators $\bigcirc$ and $\bigstar$ respectively, then a homomorphism from G to H is a mapping $\varphi : G \to H$ which satisfies the condition $\varphi(x \bigcirc y) = \varphi(x) \bigstar \varphi(y)$ for all x, y in G.

homomorphous (*BioSci*) Alike in form.

homoplasma (*BioSci*) In TISSUE CULTURE, a medium prepared with plasma from another animal of the same species as that from which the tissue was taken. Cf AUTOPLASMA, HETEROPLASMA.

homoplastic (*BioSci*) (1) Of the same structure and manner of development but not descended from a recent common source. (2) See HOMOIOPLASTIC. N *homoplasty*.

homopolar (*Chem*) Having an equal distribution of charge, as in a covalent bond between like atoms.

homopolar generator (*ElecEng*) Low-voltage dc generator based on Faraday disk principle which produces ripple-free output without commutation.

homopolar magnet (*ElecEng*) One with concentric pole pieces.

homopolar molecule (*ElecEng*) One without effective electric dipole moment.

homopolymer (*BioSci*) A polymer made up of identical subunits; commonly used of DNA or RNA strands whose nucleotides are all of the same kind. Usually made enzymatically from a single nucleotide precursor.

homopolymer (*Chem*) A polymer in which all repeat units are identical. See panel on POLYMERS.

homoscedastic (*MathSci*) Having the same variance (applied to sets of observations).

homosexuality (*Psych*) Sexual interest directed at members of one's own sex.

homospory (*BioSci*) A species that produces only one type of a spore. Adj *homosporous*. Cf HETEROSPORY.

homostyly (*BioSci*) The condition in which all the styles are the same length. Cf HETEROSTYLY.

homotaxis (*Geol*) A term introduced by T H Huxley in 1862 to indicate that strata or sequences of strata in different areas sharing the same fossil characteristics are not necessarily the same age. A faunal assemblage may originate in locality A, be gradually dispersed or migrate to locality B and eventually reach locality C. The strata accumulating at these three localities are *homotaxia* although not necessarily contemporanous.

homothallism (*BioSci*) The condition in which successful fertilization can take place between any two gametes from the same organism. It is analogous to self-compatibility in flowering plants. Also *self-compatibility*. Adj *homothallic*. Cf HETEROTHALLIC.

homothermous (*BioSci*) See WARM-BLOODED.

homotopic mapping (*MathSci*) Two continuous mappings f and g of a topological space A into a topological space B are said to be homotopic if there is a function $F(x,t)$, representing a continuous mapping of $A \times I$ into B (I being the unit interval), for which $F(x,0) = f(x)$, $F(x,1) = g(x)$, for all x in A and for $0 \leqslant t \leqslant 1$. f and g are also said to be continuously deformed into each other.

homotypic (*BioSci*) Conforming to the normal condition. Cf HETEROTYPIC.

homozygosis (*BioSci*) The condition of being *homozygous*.

homozygote (*BioSci*) An individual whose two genes at a particular locus are the same allele, the individual having been formed by the union of gametes carrying the same allele. Adj *homozygous*. Cf HETEROZYGOTE.

homunculus (*BioSci*) A dwarf of normal proportions; a mannikin or little man created by the imagination; a miniature human form believed by animalculists to exist in the spermatozoon.

Honduras mahogany (*For*) See AMERICAN MAHOGANY.

hone (*Geol*) A term applied to fine-textured even-grained indurated sedimentary rocks which may be used as oilstones for imparting a keen edge to cutting tools. Honestone has been largely replaced now by emery and silicon carbide products. Also *honestone, whetstone*.

honey (*FoodSci*) A natural sweet substance made by bees from nectar and stored in honeycomb made from beeswax. Flavour depends on the predominant flower frequented by the bees but the basic composition of honey is fairly constant at 77% sugar, 17·5% water giving a supersaturated solution that sets readily. Clear honey is produced by heating and stirring to reduce the crystal size and is fairly stable unless the temperature falls too low, when crystallization occurs.

honeycomb (*Aero*) A gridwork across the duct of a wind tunnel to straighten the airflow. Also *straighteners*.

honeycomb (*BioSci*) A mass of hexagonal wax cells built by honeybees in their nests and used to contain their larvae or to store honey and pollen.

honeycomb (*Eng*) A cellular solid, a structural material made by bending and bonding together thin sheets of eg aluminium or paper to give an array of channels of hexagonal cross-section like its eponym. Used as the filling in SANDWICH BEAMS to give lightweight, rigid products such as aircraft flooring.

honeycomb (*Textiles*) Fabric with the threads forming ridges and hollows to give a cell-like appearance. Generally woven from coarse soft yarns in compact structures and used for towels and bedspreads.

honeycomb bag (*BioSci*) See RETICULUM.

honeycomb base (*Print*) A metal base for printing plates which both secures and moves them into register. It is drilled with a pattern of holes in which register hooks are placed, there being suitable holes for any position of the plate.

honeycomb coil (*ICT*) One in which wire is wound in a zigzag formation around a circular former. The adjacent layers are staggered, so that the wires cross each other obliquely to reduce capacitance effects between turns.

honeycomb structure (*Aero*) Lightweight, very rigid, material for aircraft skin or floors, usually made from thin light-alloy plates with a bonded foil interlayer of generally honeycomb-like form. For high supersonic speeds, a heat-resistant material is made by brazing together stainless-steel or nickel-alloy skins and honeycomb core.

honeycomb wall (*Build*) A wall built so as to leave regular spaces, usually entirely with stretchers. See SLEEPER WALL.

honey dew (*BioSci*) A sweet substance secreted by certain HEMIPTERA; emitted through the anus.

honey guide (*BioSci*) Lines or dots (sometimes visible only in UV photographs), on the perianth, that direct a pollinating insect to the nectar in a flower. Also *nectar guide*.

honing (*Eng*) The process of finishing cylinder bores etc to a very high degree of accuracy by the abrasive action of stone or silicon carbide slips held in a head having both a rotatory and axial motion.

honing machine (*Eng*) A partly hand-operated or wholly automatic machine for HONING.

hood jettison (*Aero*) A mechanism, often operated by explosive bolts or cartridge, for releasing the pilot's canopy in flight. Confined to military aircraft.

hood mould (*Build*) A projecting moulding above a door or window opening.

hoof (*BioSci*) In ungulates, a horny proliferation of the epidermis, enclosing the ends of the digits.

hook-down (*Print*) See HOOK-UP.

hooked joint (*Build*) A form of joint used between the meeting edges of a door and its case when an airtight or dustproof joint is necessary; the meeting edges on the door have a projection on them fitting into a corresponding recess in the case.

hooked plates (*Print*) Single-leaf illustrations sewn in with the *section*, having been printed on paper large enough to allow a narrow fold down the back edge.

Hooke's joint (*Eng*) A common form of UNIVERSAL JOINT, comprising two forks arranged at right angles and coupled by a cross-piece.

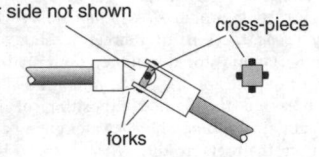

Hooke's joint

Hooke's law (*Phys*) The basic statement of linear elasticity, originally formulated by Robert Hooke in 1676; it was Thomas Young who realized over 100 years later that the proportionality was between STRESS, σ, and strain, ε, ie $\sigma = E\varepsilon$, where the proportionality constant is YOUNG'S MODULUS. This is the constitutive equation for elastic deformation in tension, and analogous equations apply for other deformation modes. Although materials are intrinsically non-linear, the law is a good approximation to the behaviour of most types within the range of recoverable, small strains. The main exceptions are polymers, which are not only viscoelastic, but also significantly non-linear, and elastomers, where strains can reach several hundred per cent.

hook-up (*ICT*) A temporary communication channel.

hook-up (*MinExt*) In pilot plant testing, flexible assembly of machines into a continuous flow line before final treatment.

hook-up (*Print*) The end of a line turned over and bracketed in the line above or below (*hook-down*). Often used in setting up poetry.

hookworm disease (*Med*) See ANKYLOSTOMIASIS.

hookworms (*BioSci*) Parasitic strongyloid nematodes with hook-like organs on the mouth for attachment to the host. Humans are attacked chiefly by the genus *Ankylostoma*, which penetrates the bare feet and induces a form of anaemia. See ANKYLOSTOMIASIS.

Hoopes process (*Eng*) A process for the refining of aluminium electrolytically to a purity of 99·99%. Metal made by the HALL PROCESS is alloyed with 33% of copper and made the anode in a non-aqueous electrolytic bath composed of alumina and fluorides of barium, sodium and aluminium. When current is passed between the hearth and carbon electrodes on top of the bath, aluminium dissolves from the anode alloy and pure metal accumulates at the cathodes.

hooping (*CivEng*) Reinforcing bars for concrete, bent to either a circular or helical shape.

hoop iron (*Build*) Thin strip-iron for securing barrels, and also for various purposes in the building trades, eg as reinforcement in brick walls, also for the packaging of bricks or building blocks to enable them to be lifted in quantity.

hoop stress (*Eng*) The largest of the three stresses in the wall of a tube under pressure, acting around the circumference of the tube; usual symbol, $\sigma\theta$. The other two are the axial stress, which is one-half of the hoop stress, and the radial stress through the thickness of the tube wall, which is negligible when the diameter is greater than 20 times the wall thickness.

hoose (*Vet*) See HUSK.

hop (*ICT*) The distance along the Earth's surface between successive reflections of a radio wave from an ionized region. Also *skip*.

Hope sapphire (*Min*) Synthetic stone having the composition of spinel and blue colour which turns purple in artificial light. First produced in the attempts to synthesize sapphire.

Hopkinson test (*ElecEng*) A method of testing two similar dc machines on full load without requiring a large consumption of power from the supply; one machine fed from the supply drives the other as a generator, which returns power to the supply.

Hopkinson–Thring torsion meter (*Eng*) A torsion meter in which two mirrors are mounted on a flange fixed to one end of the shaft under test, one mirror being fixed and the other so mounted that it tilts as the shaft twists. The separation of light beams reflected by the mirrors is a measure of the shaft twist.

hopper (*Build*) A draught-preventer at the side of a HOPPER LIGHT.

hopper (*MinExt*) A container or surge bin for broken ore, used to hold small amounts.

hopper crystal (*Crystal, Min*) A crystal which has grown faster along its edges than in the centres of its faces, so that the faces appear to be recessed. This type of skeletal crystallization is often shown by rock salt.

hopper dredger (*CivEng*) A dredger which not only dredges material from below but has hopper compartments fitted with flap-doors which allow the material to be discharged after the vessel has moved to the place of deposit.

hopperfeed (*Eng*) A machine used for unscrambling identical components from bulk, orientating them, and delivering them in an orderly arrangement to an assembling machine or other mass-production process.

hopper light (*Build*) A window sash arranged to open inwards about hinges on its lower edge.

hopper window (*Build*) A HOPPER LIGHT fitted at the sides with HOPPERS.

Hoppus foot (*For*) A largely obsolete unit of volume obtained by using the square of the quarter girth for trees or logs instead of the true sectional area; only applied to timber in the round.

hops (*FoodSci*) The coniform blossom of a perennial climbing plant *Humus lupulus*, used in brewing to provide flavour, aroma and bitterness. Now mainly dried and pelletized but flavour and aroma can be improved by adding volatile extracts near the end of brewing.

hopsacking (*Textiles*) A development of plain weave in which two or more ENDS and PICKS weave as one.

hordeolum (*Med*) See STYE.

Hordeum vulgare (*BioSci*) Cultivated barley, very important for the brewing industry.

horizon (*Astron*) The great circle, of which the zenith and the nadir are the poles, in which the plane tangent to the Earth's surface, considered spherical at the point where the observer stands, cuts the celestial sphere.

horizon (*EnvSci*) A layer in a soil distinguishable from others by colour, hardness, inclusions or other visible or tangible properties.

horizon (*Geol*) The surface separating two beds of rock. It has no thickness, and is more frequently used in the sense of a thin bed or time-plane with a characteristic lithofacies or biofacies, persistent over a wide area.

horizon (*Psych*) The more or less coloured visual impression experienced subjectively by blind persons.

horizon (*Surv*) A plane perpendicular to the direction of gravity shown by a plumbline at the point of observation.
horizon glass (*Surv*) See SEXTANT.
horizon sensor (*Electronics*) Sensor providing a stable vertical reference level for missiles and depending on the use of a thermistor to detect the thermal discontinuity between Earth and space.
horizontal antenna (*ICT*) One comprising a system of one or more horizontal conductors, radiating or responding to horizontally polarized waves.
horizontal axis (*Surv*) See TRUNNION AXIS.
horizontal blanking (*ImageTech*) The elimination of the horizontal trace in a cathode-ray tube during flyback.
horizontal circle (*Surv*) The graduated circular plate used for the measurement of horizontal angles by theodolite.
horizontal component (*ElecEng*) The component of Earth's magnetic field which acts on a unit pole in a horizontal direction.
horizontal engine (*Eng*) Any engine in which the cylinders are horizontal.
horizontal flash tool (*Eng*) The tool normally used in injection moulding, where mating parts meet at right angles to main injection direction. Flash polymer may be extruded here during moulding, and must be removed in a post-moulding operation. See VERTICAL FLASH TOOL.
horizontal parallax (*Astron*) The value of the DIURNAL PARALLAX for a heavenly body in the solar system when the body is on the observer's horizon. In astronomy, the *equatorial horizontal parallax* of a planet or of the Moon is the angle subtended at the centre of that body by the equatorial radius of the Earth. See SOLAR PARALLAX.
horizontal polarization (*ICT*) The transmission of radio waves in such a way that the electric lines of force are horizontal and the magnetic vertical; transmitting and receiving DIPOLES are mounted horizontally to handle signals polarized in this way. Cf VERTICAL POLARIZATION.
horizontal polarization (*Phys*) The polarization of an electromagnetic wave when the alternating electric field is horizontal.
horizontal resolution (*ImageTech*) The number of separate picture elements which can be resolved along each horizontal scanning line in a TV picture or facsimile reproduction.
horizontal stabilizer (*Aero*) Tailplane. See STABILIZER.
horme (*Psych*) Goal-directed or purposive behaviour.
hormesis (*Pharmacol*) A phenomenon whereby substances that are toxic in large doses have a beneficial effect when absorbed in very small doses. Might apply to radiation by eg stimulating the production of repair enzymes.
hormone (*BioSci*) Generally, a signalling substance that is released by cells and acts remotely on other cells, which have specific surface or intracellular receptors for the hormone. Plant hormones, eg abscisic acid, auxins, cytokinins, ethylene, gibberellins, are more akin to growth substances; animal hormones are released into the circulation by endocrine glands and tend to have longer-lasting and more widespread activity than neuronal signals.
hormone replacement therapy (*Med*) The use of female hormones (estrogens, progestogen) to relieve symptoms that occur once the ovaries cease secreting hormones or are removed. A wide range of regimens and preparations are available although there are both advantages and disadvantages for post-menopausal women.
horn (*Acous*) A tube of continuously varying cross-section used in launching or receiving of radiation, eg acoustic horn, electromagnetic horn. Horns are best classified by their geometric or other shapes which include: *compound, conical, corner, exponential, folded, logarithmic, pyramidal, re-entrant, sectoral, tractrix*.
horn (*BioSci*) (1) One of the pointed or branched hard projections borne on the head in many mammals, composed largely of the protein KERATIN often together with inorganic minerals in a biocomposite. (2) Any conical or cylindrical projection of the head resembling a horn. (3)

In some birds, a tuft of feathers on the head. (4) In some GASTROPODA, a tentacle. (5) In some fish, a spine. Adjs *horned, horny*.
horn (*Eng*) (1) Any projecting part, such as the two jaws of a horn plate carrying an axle-box. (2) The material derived from natural hard body parts, shaped by cutting and often also thermoformed to make thin sheet (formerly for lantern glazing) and cups etc.
horn (*Geol*) A steep-sided mountain peak formed by the coalescence of three or more cirques.
horn antenna (*ICT*) A microwave antenna consisting of a flared-out section of a waveguide that may be rectangular, square or circular. Maximum radiation is along the central axis of a straight horn; other types may be curved, folded or bifurcated.
horn arrester (*ElecEng*) A lightning arrester with a HORN GAP which rapidly extinguishes an arc from a lightening surge owing to the special shape of the electrodes.
horn balance (*Aero*) An AERODYNAMIC BALANCE consisting of an extension forward of the hinge line at the tip of a control surface; it may be *shielded* (ie screened by a surface in front) or *unshielded*.
hornbeam (*For*) A European hardwood tree of the genus *Carpinus*, yielding a fairly perishable, dull-white wood.
hornblende (*Min*) Important members of the amphibole group of rock-forming minerals. They are of complex composition, essentially silicate of calcium, magnesium and iron, with smaller amounts of sodium, potassium, hydroxyl and fluorine; crystallize in the monoclinic system; occur as black crystals in many different types of igneous and metamorphic rocks, including hornblende-granite, syenite, diorite, andesite, etc and hornblende-schist and amphibolite.
hornblende-gneiss (*Geol*) A coarse-grained metamorphic rock, containing hornblende as the dominant coloured constituent, together with feldspar and quartz, the texture being that typical of the gneisses. Differs from hornblende-schist in grain size and texture only.
hornblende-schist (*Geol*) A type of green schist, formed from basic igneous rocks by regional metamorphism, and consisting essentially of sodic plagioclase, hornblende and sphene, frequently with magnetite and epidote. See GLAUCOPHANE.
hornblendite (*Geol*) An igneous rock composed almost entirely of hornblende.
horn-break fuse (*ElecEng*) A fuse fitted with arcing horns to assist in the rapid extinction of any arc which may be formed.
horn centre (*Eng*) A small transparent disk, originally of horn, used to protect the drawing paper at frequently used centres.
Horner's syndrome (*Med*) The combination of small pupil (miosis), sunken eye (enophthalmus) and drooping of upper eyelid (ptosis), due to paralysis of the sympathetic nerve in the region of the neck.
horn feed (*Radar*) The feed to an antenna in the form of a horn.
hornfels (*Geol*) A fine-grained rock which has been partly or completely recrystallized by contact metamorphism.
horn gap (*ElecEng*) A spark gap of gradually increasing length, such that an arc struck across it gets longer and finally extinguishes itself.
horn gate (*Eng*) Horn-shaped ingates or sprues, radiating from the bottom of a runner, which supply several small moulds made in the same moulding box. See INGATE.
horn lead (*Min*) Sometimes applied to the mineral PHOSGENITE.
horn loudspeaker (*Acous*) A loudspeaker in which the radiating device is acoustically coupled to the air by means of a horn.
horns (*Build*) The ends of the head in a door or window frame when these project beyond the outer surfaces of the posts.
horn silver (*Min*) See CERARGYRITE.

hornwort (*BioSci*) See ANTHOCEROTOPSIDA.

Horologium (Clock) (*Astron*) A faint southern constellation.

horology (*Genrl*) The science of time measurement, or of the construction of timepieces.

horripilation (*Med*) Erection of the hairs on the skin, giving rise to goose flesh.

horse (*Build*) (1) One of the STRINGS supporting the treads and risers of a stair. (2) A trestle for supporting a board or timber while it is being sawn.

horsehead (*MinExt*) The curved part at the end of the arm of an oil-well pump; it keeps the cable attached between the arm and the pump rods running at a fixed radius.

Horsehead Nebula (*Astron*) A famous dark nebula in the constellation Orion; it bears a plausible resemblance to the silhouette of a horse's head.

horse latitudes (*EnvSci*) See TRADE WINDS.

horse path (*Build*) A canal towing path.

horsepower (*Eng*) The mechanical engineering unit of power, equivalent to working at the rate of 33 000 ft lb min^{-1}, or 42·41 Btu min^{-1}, or 745·70 W. Abbrev *h.p.* See CHEVAL-VAPEUR for the metric equivalent. Replaced in the current SI system of units by the watt.

horse pox (*Vet*) A contagious viral infection of equines. It is characterized by a papulo-vesicular eruption of the skin and mucous membranes. Also *thrush*.

horseradish peroxidase (*BioSci*) An enzyme commonly conjugated with antibodies for use in *immunoassays*. Catalyses deposit of dye at the site of binding. Abbrev *HRP*.

horseshoe curve (*Surv*) A curve whose arc subtends an angle of more than 180° at the centre, so that the INTERSECTION POINT lies on the same side of the curve as the centre.

horseshoe kidney (*Med*) Congenital fusion of corresponding poles of two kidneys.

horseshoe magnet (*ElecEng*) Traditional form of an electro- or permanent magnet, as used in many instruments, eg meters, magnetrons.

Horseshoe Nebula (*Astron*) See OMEGA NEBULA.

horsing up (*Build*) The building-up of the mould used in running cornices etc.

horst (*Geol*) An elongated uplifted block bounded by faults along its length. Cf GRABEN.

horst faults (*Geol*) Two parallel normal faults heading outwards and throwing in opposite directions, the resulting structure being termed a *horst*.

horticulture (*Agri*) The intensive cultivation of fruit, flowers, vegetables and medicinal plants.

Horton sphere (*ChemEng*) A strong, spherical tank for storing liquefied propane or butane gas under pressure to prevent losses due to vaporization.

hose (*Textiles*) A tubular woven fabric made from flax or hemp for conveying liquids under pressure (eg a fire hose). May also be a fibre-reinforced plastic tube.

hose bit (*Build*) A type of SHELL BIT with a projecting tip which withdraws the waste.

hosiery (*Textiles*) Knitted articles for covering the feet and legs. Formerly used to include all types of knitted fabrics and garments.

hospital bus-bars (*ElecEng*) A set of bus-bars provided in a power station or substation for temporary or emergency purposes.

hospital switch (*ElecEng*) (1) A switch used on tramway or railway controllers to cut a faulty motor out of circuit. (2) Any switch for changing a circuit over to an emergency supply in case of failure of the main supply.

host (*BioSci*) An organism that supports another organism (parasite) at its own expense, in molecular biology that in which a plasmid or virus can replicate.

host (*ICT*) A (large) computer system on which applications may be executed and which provides a service to USERS over a NETWORK.

host (*Phys*) Essential crystal, base material, or matrix of a luminescent material.

host range (*BioSci*) The range of species or strains of bacteria that will support the replication of a plasmid or virus.

host rock (*MinExt*) See COUNTRY ROCK.

host-versus-graft reaction (*BioSci*) Cell-mediated reactions of a host against allogeneic or xenogeneic cells or tissues acquired as a graft or otherwise that leads to damage or/and destruction (rejection) of the grafted cells.

hot (*Phys*) Charged to a dangerously high potential.

hot (*Radiol*) Colloq reference to high levels of radioactivity; hence 'hot laboratory', a designated area for handling radioactive substances with extra precautions against irradiation of staff.

hot-air engine (*Eng*) One in which the working fluid, air, is alternately heated and cooled by a furnace and regenerator. See STIRLING ENGINE.

hot-air gun (*Build*) An electric power tool which generates a stream of hot air which, when directed onto a paint film, softens it sufficiently to facilitate its removal.

hot-blast stoves (*Eng*) Large stoves, filled with a brick chequerwork, used for preheating the air blown into the blast furnace. Also *Cowper stoves*.

hot cathode (*Electronics*) One in which the electrons are produced thermionically.

hot-cathode discharge lamp (*Phys*) A discharge lamp employing a heated cathode to increase its efficiency, improve the starting, and reduce the voltage drop across the tube.

hot-cathode rectifier (*Electronics*) One with emitting cathode heated independently of the rectified current, eg a mercury thyratron.

hot cell (*NucEng*) See CAVE.

hot crack (*Eng*) Crack formed during cooling by the stresses set up by the solidification of one or other of the components of an alloy or metal.

hot-die steel (*Eng*) A shock- and temperature-resistant alloy used in high-temperature forging. See panel on STEELS.

hot-dip galvanizing (*Eng*) See GALVANIZING.

hot-drawn (*Eng*) A term describing metal wire, rod or tubing which has been produced by pulling it heated through a constricting orifice.

hot dry rock (*Geol*) Potential source of heat energy from hot underground rocks. Water is pumped down an injection well into artificially induced fractures, and recovered, hot, in a second well drilled into the fracture system. The fracturing is initially caused by injecting water under very high pressure (*hydrofracking, hydrofracting*). Heat is extracted at the surface in a heat exchanger and the water recycled. Favourable conditions are provided by radiothermal granites.

hot electron (*Electronics*) An electron in excess of the thermal-equilibrium number and, for metals, having energy greater than the FERMI LEVEL. In semiconductors, the energy must be a definite amount above the edge of the conduction band. Hot electrons (or holes) can be generated by photoexcitation, tunnelling, minority-carrier injection or Schottky emission over a forward-biased p–n junction, or by abnormal electric fields in non-conductors.

hot-fluid injection (*MinExt*) Pumping steam, hot water or gas into the formation to increase flow of low-specific-gravity oil. Also *hot footing*.

hot galvanizing (*Eng*) See GALVANIZING.

hot-ground pulp (*Paper*) Mechanically ground wood pulp in which the minimum of water is used, so allowing the temperature to rise by friction.

hot insulation mastics (*Build*) 'Breathing' mastics prepared by emulsifying bitumen in water and appropriate filler. Although the film formed is black and waterproof, it permits the passage of water vapour and allows the lagging around a hot surface to dry out.

hot isostatic pressing (*Eng*) Making a compact by application of heat as well as pressure. The combined effect is general softening and/or the liquefaction of a phase to allow sintering. Far greater amounts of pressure and heat would be required separately. Also *hot pressing*. See panel on CERAMICS PROCESSING.

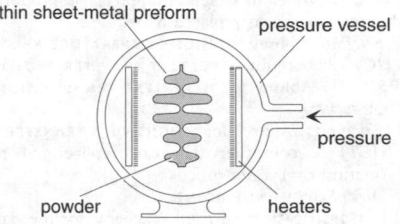

thin sheet-metal preform

pressure vessel

← pressure

powder

heaters

hot isostatic pressing

hot key (*ICT*) A key on the keyboard that has a special effect such as calling up a MENU or executing another program. The user may use this key as a shortcut.

hot melt (*Print*) An adhesive which requires to be heated prior to application during unsewn binding. Normally sets very quickly during binding.

hot-melt adhesive (*Eng*) Polymeric equivalent of solder. A thermoplastic applied in molten state by glue gun directly onto surfaces to be joined, which bond together as the melt solidifies. Low-molecular-mass ethylene vinylacetate is a popular hot-melt adhesive.

hot-plate welding (*Eng*) A method of joining similar thermoplastics by heating to above T_g or T_m with a steel platen, and bringing together under load. Used for joining medium-density polyethylene gas pipes, battery cases, etc. Care needed in control of temperature and load to ensure good fusion at the weld line. SACRIFICIAL-TAPE WELDING is used for critical products.

hot press (*Print*) See BLOCKING PRESS.

hot-pressed (*Paper*) Descriptive of paper finished by glazing between hot-metal plates.

hot pressing (*PowderTech*) See HOT ISOSTATIC PRESSING.

hot-press stamping (*Print*) See BLOCKING.

hot runner (*Eng*) RUNNER into injection-moulding tool cavity which is heated to conserve polymer. See INJECTION MOULDING.

hot shoe (*ImageTech*) Camera ACCESSORY SHOE with electrical connections for flash sychronization.

hot-short (*Eng*) Said of metals that tend to be brittle at temperatures at which hot-working operations are performed. Also *red-short*.

hot spot (*Autos*) Part of the wall surface of the induction manifold of a petrol engine on which the mixture impinges; heated by exhaust gases or coolant water to assist vaporization and distribution.

hot spot (*BioSci*) The region of a chromosome peculiarly susceptible to mutation or recombination.

hot spot (*Electronics*) A small region of an electrode in a valve (or CRO screen) which has a temperature above the average.

hot spot (*Geol*) A region of high thermal activity deep within the Earth's mantle, associated with an area of high volcanic activity, and thought to be the surface manifestation of rising convection currents of molten mantle material.

hot spot (*ImageTech*) (1) Local area of excessive brightness in a projected image. (2) A marked area in an interactive video image which can be clicked on with a pointing device to access further information etc.

hot spot (*NucEng*) (1) The position of highest temperature in a reactor fuel pin. (2) A highly radioactive region in a plant or reactor, needing protective screening.

hot top (*Eng*) Feeder head, often heated by exothermic reaction compounds, containing a reservoir of molten metal drawn on by a cooling ingot as it solidifies, thus avoiding porosity. See INGOT, INGOT MOULD.

hot well (*Eng*) The tank or pipes into which the condensate from a steam engine or turbine condenser is pumped, and from which it is returned by the feed pump to the boiler.

hot-wire (*ElecEng*) Said of an indicating instrument which depends on the thermal expansion or change in resistance of a wire or strip when it carries a current.

hot-wire ammeter (*ElecEng*) An ammeter operating on the HOT-WIRE principle; of use chiefly for very high frequencies.

hot-wire anemometer (*EnvSci*) An instrument which measures wind velocities by using their cooling effect on a wire carrying an electric current, the resistance of the wire being used as an indication of the velocity.

hot-wire detector (*ICT*) A fine wire that is heated by the passage of high-frequency currents, producing a change in its dc resistance. See BOLOMETER.

hot-wire microphone (*Acous*) A microphone in which a dc signal through a hot wire is modulated by resistance variations consequent upon the cooling effect of an incident sound wave. Also *thermal microphone*.

hot-wire voltmeter (*ElecEng*) A voltmeter operating on the *hot-wire* principle, the current heating the wire which expands and moves the needle.

hot-working (*Eng*) The process of shaping metals by rolling, extrusion, forging, etc, at temperatures above about $0.6T_m$. The hot-working range varies from metal to metal, but it is, in general, a range in which recrystallization proceeds concurrently with the working, so that no strain hardening occurs.

Houdry process (*Chem*) Catalytic cracking of petroleum, using activated aluminium hydrosilicate.

hour angle (*Astron*) The angle, generally measured in hours, minutes and seconds of time, which the hour circle of a heavenly body makes with the observer's meridian at the celestial pole; it is measured positively westwards from the meridian from 0 to 24 h.

hour circle (*Astron*) (1) The great circle passing through the celestial poles and a heavenly body, cutting the celestial equator at 90°. (2) The graduated circle of an equatorial telescope which reads sidereal time and right ascension.

hour counter (*ElecEng*) See TIME-METER.

hour-glass stomach (*Med*) Constriction of the middle part of the stomach, due either to spasm of stomach muscle or to the formation of scar tissue in connection with a gastric ulcer, the constriction in the latter case being permanent.

hour meter (*ElecEng*) See TIME-METER.

house corrections (*Print*) The correcting of mistakes made by the compositor when setting by hand or machine, as distinct from author's corrections made by him or her after checking the proofs.

housed joint (*Build*) A fitted joint, such as a tenon in its mortise.

housed string (*Build*) A string which has its upper and lower edges parallel to the slope of the stair, and which houses, in grooves specially cut in the inner side, the ends of the steps. Also *close string*.

housekeeping (*ICT*) Colloq term for allocation decisions and record keeping within a computer system.

housekeeping data (*Space*) A term used to denote information on the working of a space system, subsystem or component (as opposed to scientific data per se).

housekeeping gene (*BioSci*) Genes that code for proteins or RNAs that are important for all cells and are thus constitutively active. Term used by contrast with 'luxury' proteins, those that are only produced by differentiated cells.

housemaid's knee (*Med*) Inflammation of the bursa in front of the patella of the knee.

house mites (*BioSci*) Various species of arthropods, *Dermatophagoides*, thought to play an important allergenic role in house-dust-induced ASTHMA.

house service meter (*ElecEng*) An integrating meter for measuring the electrical energy consumption of a domestic installation.

house style (*Print*) See STYLE OF THE HOUSE.

housing (*Build*) A method of jointing two timbers in which the whole of the end of one is fitted into a corresponding blind mortise cut in the other.

housing (*ElecEng*) Containment of apparatus to prevent damage in handling or operation.

hoven (*Vet*) See BLOAT.

hovercraft (*Ships*) A craft which can hover over or move across water or land surfaces while being held off the surfaces by a cushion of air. The cushion is produced either by pumping air into a plenum chamber under the craft or by ejecting air downwards and inwards through a peripheral ring of nozzles. Propulsion can be by tilting the craft, or by jet, or air propeller, or, over water, by water propeller, or, over land, by low-pressure tyres or tracks.

hovership (*Ships*) A HOVERCRAFT intended for use over water only but capable of coming in over a beach to land.

Hovmüller diagram (*EnvSci*) A diagram, with one axis representing time and the other longitude, on which are shown isopleths of an atmospheric variable such as pressure or *thickness*, usually averaged over a band of latitude. It demonstrates, very effectively, the movement over weeks or months of large-scale atmospheric features.

Howard protective system (*ElecEng*) A form of earth-leakage protection sometimes applied to ac machines and equipment. It consists of a current transformer connected between the frame of the machine and earth; if a current flows through the transformer a relay is operated, which opens the main circuit breaker.

howieite (*Min*) A black triclinic hydrated silicate of sodium, manganese and iron.

howl (*Acous*) A high-pitched audio tone due to unwanted acoustic (or electrical) feedback.

howl (*Aero*) See SCREECHING.

howler (*ICT*) A device that uses acoustic feedback between a telephone transmitter and a telephone receiver to maintain a continuous oscillation and so provides suitable currents for testing telephonic apparatus.

Hoxnian (*Geol*) An interglacial stage of the Late Pleistocene. See QUATERNARY.

Hoyt's metal (*Eng*) A tin base (91·5%) WHITE METAL, containing also antimony (3·4%), lead (0·25%), copper (4·3%) and nickel (0·55%).

h.p. (*Eng*) Abbrev for HORSEPOWER.

h-parameters (*Electronics*) See TRANSISTOR PARAMETERS.

H-plane (*Phys*) The plane containing the magnetic vector H in electromagnetic waves, and containing the direction of maximum radiation. The electric vector is normal to it.

HPLC (*Chem*) Abbrev for *high-pressure liquid chromatography*. Chromatographic method in which the sample is forced at high pressure through a tightly packed column of finely divided particles that present a very large surface area. HPLC gives good separation very rapidly. Sometimes, esp in adverts, spoken of as *high-performance liquid chromatography*.

HPV (*BioSci*) Abbrev for *human papilloma virus*.

HQ (*ImageTech*) Abbrev for *high quality*. An upgrade of the original VHS specification which increases the WHITE CLIP level and adds a detail enhancer (and possibly LUMINANCE SIGNAL noise reduction).

H-radar (*Radar*) Navigation system in which an aircraft interrogates two ground stations for distance.

HRP (*BioSci*) Abbrev for HORSERADISH PEROXIDASE.

H-section (*ICT*) An electrical network derived from the T-section, in which half of each series arm is placed in the other leg of the circuit, making the section balanced.

HS steel (*Eng*) Abbrev for HIGH-SPEED STEEL.

HTLV-1 (*Med*) Abbrev for *human T-cell leukaemia virus type 1*. The first human retrovirus to be discovered. HTLV-1 is the causative agent of adult T-cell leukaemia (ATL). HTLV-1 infects CD4-positive T-cells and in ATL causes uncontrollable cell proliferation. The virus is endemic in SW Japan, C Africa and the Caribbean islands. A small proportion of infected individuals develop ATL after very

long latent periods (> 30 years). The virus is also associated with a variety of non-malignant disorders.

html (*ICT*) Abbrev for HYPERTEXT MARKUP LANGUAGE.

HTO (*Chem*) Water in which an appreciable proportion of ordinary hydrogen is replaced by tritium.

HTR (*NucEng*) Abbrev for HIGH-TEMPERATURE REACTOR.

http (*ICT*) Abbrev for HYPERTEXT TRANSFER PROTOCOL.

HTTPS (*ICT*) Abbrev for HYPERTEXT TRANSMISSION PROTOCOL *secure*.

HTU (*ChemEng*) Abbrev for HEIGHT OF A TRANSFER UNIT.

hub (*ICT*) A computer that can receive and redirect information over a network.

hub (*Surv*) See CHANGE POINT.

Hubble classification (*Astron*) A scheme for the classification of galaxies according to morphology, the principal types being BARRED SPIRAL, ELLIPTICAL and SPIRAL. See panel on GALAXY.

Hubble constant (*Astron*) See HUBBLE PARAMETER.

Hubble diagram (*Astron*) A plot of redshifts of galaxies against their inferred distances. See HUBBLE RELATION and panel on REDSHIFT–DISTANCE RELATION.

Hubble parameter (*Astron*) Originally known as the *Hubble constant* of proportionality, symbol H_0. It is a measure of the rate at which the recession speed of galaxies (due to the expansion of the universe) varies with distance, and has the value 50–100 km s^{-1} Mpc^{-1}. It is determined from the observed REDSHIFTS of distant galaxies.

Hubble relation (*Astron*) The relationship first described by E Hubble in 1925, which states that the recession velocity of a distant galaxy is directly proportional to the distance from the observer; specifically, $v = H_0d$, where v is the recession velocity of a galaxy, d is its distance and H_0 is the HUBBLE PARAMETER. Also *Hubble law*. See panel on COSMOLOGY.

Hubble Space Telescope (*Space*) A 2·4 m diameter optical telescope deployed in orbit by the Space Shuttle in 1990, expected to operate for 15 years. After transmitted data showed that the main mirror was manufactured to the wrong shape, corrective optics were installed during a repair mission in 1993; now performing as originally intended, it will continue visible and ultraviolet observations of faint objects. A further repair mission is planned for late 2007 that should extend the working life until 2013. See panel on ASTRONOMICAL TELESCOPE.

hübnerite (*Min*) Manganese tungstate, the end-member of the wolframite group of minerals, the series from $MnWO_4$ to $FeWO_4$. Also *huebnerite*.

huckaback (*Textiles*) A woven linen or cotton cloth with a rough surface, used for towels and glass-cloths. Different from *terry-towelling fabric* which is a pile fabric. See TERRY FABRIC.

HUD (*Aero*) Abbrev for HEAD-UP DISPLAY. Cf HDD.

hue (*Phys*) The perception of colour which discriminates different colours as a result of their wavelengths. Hue, chroma (degree of saturation) and value (brightness) specify a colour on the Munsell scale.

huebnerite (*Min*) See HÜBNERITE.

hue sensibility (*Phys*) The ability of the eye to distinguish small differences of colour.

Huff separator (*MinExt*) High-tension or electrostatic separator used to concentrate small particles of dry ore.

hull (*Aero*) The main body (structural, flotation and cargo-carrying) of a flying boat or boat amphibian.

hull (*Ships*) A term used in its widest sense to signify the ship itself exclusive of masts, funnels and TOP HAMPER, but including SUPERSTRUCTURE. In a more restricted sense it means the shell of the ship. Also used to distinguish between ship and machinery.

hullite (*Min*) See CHLOROPHAEITE.

hulls (*NucEng*) Small pieces of fuel-rod cladding left after fuel pins have been dissolved in a reprocessing plant.

hum (*Acous*) Objectionable low-frequency components induced from power mains into sound reproduction, caused by inadequate smoothing of rectified power supplies, induction into transformers and chokes, unbalanced capacitances, or leakages from cathode heaters.

human–computer interface (*ICT*) See USER INTERFACE.

human factors (*Genrl*) US for ERGONOMICS.

Human Genome Diversity Project (*BioSci*) The plan to analyse and compare variations in DNA samples from hundreds of different ethnic groups, in order to understand the origins and migrations of human populations and the genetic basis of differing susceptibility to disease.

Human Genome Project (*BioSci*) The international project that led to the mapping and sequencing of the 3 billion bases of the human genome.

human immunodeficiency virus (*Med*) The retrovirus responsible for AIDS, several types of which have been recognized. Abbrev *HIV*, with a number added to denote the type.

humanistic psychology (*Psych*) An approach to psychology that stresses a positive view of human nature, the importance of subjective experience, free will and the uniqueness of the individual.

humanized antibody (*BioSci*) An engineered antibody for human therapeutic use in which the antigen recognition site is from mouse (usually), but a substantial proportion of the molecule has human sequence and is therefore non-antigenic.

hum bars (*ImageTech*) A horizontal band in the picture, often moving vertically, caused by ac interference, usually at mains frequency.

hum-bucking coil (*ElecEng*) See BUCKING COIL.

humectant (*FoodSci*) Any hygroscopic substance added to food products to keep them moist. The most common humectants used as food additives are the polyhydric alcohols, mannitol (E421), sorbitol (E420), glycerol (E422) and xylitol.

humeral (*BioSci*) In vertebrates, pertaining to the region of the shoulder; in insects, pertaining to the anterior basal angle of the wing.

humerus (*BioSci*) The bone supporting the proximal region of the fore-limb in land vertebrates. Adj *humeral*.

humic acids (*Chem*) Complex organic acids occurring in the soil and in bituminous substances formed by the decomposition of dead vegetable matter.

humicole (*BioSci*) Growing on soil or on humus. Also *humicolous*.

humidity (*EnvSci*) See RELATIVE HUMIDITY, SPECIFIC HUMIDITY, VAPOUR CONCENTRATION.

humidity mixing ratio (*EnvSci*) The ratio of the mass of water vapour in a sample of moist air to the mass of dry air with which it is associated.

humification (*BioSci*) The formation of humus during the decomposition of organic matter in soils.

humite (*Min*) An orthorhombic magnesium silicate, also containing magnesium hydroxide. Found in impure marbles. The humite group also includes chondrodite, clinohumite and norbergite.

hummer screen (*MinExt*) A type used to grade smallish minerals, using ac to provide vibration by solenoid action.

hum note (*Acous*) The pitch of the note of the sound from a bell which persists after the strike note has died away.

humor (*BioSci*) US for HUMOUR.

humoral immunity (*BioSci*) Specific immunity attributable to antibodies as opposed to cell-mediated immunity.

humoralism (*Med*) The doctrine that diseases arise from some change in the humours or fluids of the body. Also *humorism*.

humour (*BioSci*) A fluid; as the *aqueous humour* of the eye. US *humor*.

Humphrey gas pump (*Eng*) A large water pump in which periodic gas explosions are made to act directly on an oscillating column of water, thereby effecting a pumping cycle; used in waterworks.

Humphreys spiral (*MinExt*) A spiral sluice which combines separation of mineral sands by simple gravitational drag with mild centrifugal action as the pulp cycles downward.

hump speed (*Aero*) The speed, on the water, at which the water resistance of the floats or boat body of a seaplane or flying boat is a maximum. After this is past the craft begins to plane over the water.

humulene (*Chem*) $C_{15}H_{24}$. A sesquiterpene found in oil of hops.

humus (*BioSci*) Amorphous, black or dark-brown material, a mixture of macromolecules based on benzene carboxylic and phenolic acids, often complexed with clays, which results from the decomposition of organic matter in soils. See MODER, MOR, MULL.

humus plant (*BioSci*) A flowering plant, often with little chlorophyll, depending for much of its nutrition on a mycorrhizal association with a fungus growing in a rich humus.

hundredweight (*Genrl*) (1) 50 kg, known as metric hundredweight. (2) $\frac{1}{20}$ of a ton (50·80 kg), known as a long hundredweight. (3) In US 100 lb (45·3 kg), known as a short hundredweight.

Hund's rule (*Chem*) For electrons of otherwise equal energy, spins are aligned parallel as much as possible. This results in the strongly paramagnetic properties of many transition metal compounds.

Hungarian cat's-eye (*Min*) An inferior greenish cat's-eye obtained in the Fichtelgebirge in Bavaria. No such stone occurs in Hungary.

hung sash (*Build*) See HANGING SASH.

hunting (*Aero*) (1) An uncontrolled oscillation, of approximately constant amplitude, about the flight path of an aircraft. (2) The angular oscillation of a rotorcraft's blade about its drag hinge. (3) The oscillation of instrument needles.

hunting (*Autos*) Irregular running of a gasoline engine at idling speed, usually caused by overlean mixture.

hunting (*Eng*) The tendency of rotating mechanisms which should run at constant speed to pulsate above and below that speed, due to shortcomings in the governor or other speed control systems. Also *cycling*, *oscillation*.

hunting (*ICT*) The operation of a selector, or other similar device, to establish connection with an idle circuit of a group.

hunting (*ImageTech*) The tendency of an AUTOMATIC FOCUSING system to oscillate about the plane of focus, usually due to low light or contrast. See AUTOFOCUS ASSIST.

huntingtin (*BioSci*) Protein product of the gene that is defective in HUNTINGTON'S CHOREA. The normal gene is widely expressed and required for normal development; the product of the abnormal gene has multiple polyglutamine repeats and altered interactions with other proteins.

Huntington mill (*MinExt*) Wet-grinding mill in which four cylindrical mullers, hung inside a steel tub, bear outwards as they rotate, thus grinding the passing ore.

Huntington's chorea (*Med*) Hereditary CHOREA occurring in adults; associated with progressive mental deterioration and involuntary movements. See HUNTINGTIN.

hunting tooth (*Eng*) An extra tooth added to a gearwheel in order that its teeth shall not be an integral multiple of those in the pinion.

Hurler's syndrome (*Med*) A congenital disorder of lipid metabolism showing abnormal development of many parts of the body. Formerly *gargoylism*.

Huronian (*Geol*) A major division of the Proterozoic of the Canadian Shield, typically exposed on the northern shores of Lake Huron.

hurricane (*EnvSci*) (1) A wind of force 12 on the BEAUFORT SCALE. A mean wind speed of 75 mph (120 km h^{-1}). (2) A *tropical revolving storm* in the N Atlantic, eastern N Pacific and the western S Pacific.

hurricane deck (*Ships*) A term, not normally in use, for a superstructure deck. Sometimes termed flying deck. It is independent of the ship from the point of view of strength.

hurter (*Build*) A cast-iron, timber, stone or concrete block, which is so placed as to protect a quoin from damage from passing vehicles.

Hurter and Driffield curve (*ImageTech*) See CHARACTERISTIC CURVE.

hushing (*MinExt*) A washing away of the surface soil to lay bare the rock formation for prospecting. Also *hush*.

husk (*Vet*) An infection of cattle with *Dictyocaulus viviparus*. Affects young stock. Adult worms infest the bronchi and bronchioles resulting in varying degrees of respiratory disease. Vaccine available. Also *Dictyocauliasis, hoose, lungworm disease, parasitic bronchitis*.

hutch (*MinExt*) (1) A small train or wagon. (2) A basket for coal. (3) A compartment of a jig used for washing ores. The concentrate which passes through a jig screen is called the *hutchwork*.

Hutchinson's teeth (*Med*) Narrowing and notching of the permanent incisor teeth, occurring in congenital syphilis.

Huygens' eyepiece (*Phys*) A combination of two plano-convex lenses placed with their plane sides towards the observer, at a distance apart equal to half the sum of their focal lengths, which are in the ratio of three to one, the shorter-focus lens being nearer the observer. Huygens' eyepiece is often used in microscopes, but is not suited for use with cross-wires or an eyepiece scale.

Huygens' principle (*Phys*) The assumption that every element of a wavefront acts as a source of so-called secondary waves. The principle can be applied to the propagation of any wave motion, but it is more frequently used in optics than in acoustics.

HVEM (*BioSci*) Abbrev for HIGH-VOLTAGE ELECTRON MICROSCOPE. See panel on ELECTRON MICROSCOPE.

HVP (*FoodSci*) Abbrev for HYDROLYSED VEGETABLE PROTEIN.

H-wave (*ICT*) See TE-WAVE.

hyacinth (*Min*) The reddish-brown variety of transparent ZIRCON, used as a gemstone. The name has also been used for a brownish GROSSULAR from Sri Lanka. Also *jacinth*.

Hyades (*Astron*) An open star cluster in the constellation Taurus which is visible to the naked eye.

hyal-, hyalo- (*Genrl*) Prefixes from Gk *hyalos*, clearstone, glass.

hyaline (*BioSci*) Translucent and colourless; without fibres or granules, eg *hyaline cartilage*.

hyaline membrane disease (*Med*) One of the causes of the respiratory distress syndrome of the newborn where SURFACTANT has not yet developed sufficiently to maintain normal ALVEOLAR opening. Cyanosis and difficulty in breathing result.

hyalite (*Min*) A colourless transparent variety of *opal*, occurring as globular concretions and crusts. Also *Müller's glass*.

hyalo- (*Genrl*) See HYAL-.

hyaloid (*BioSci*) Clear, transparent; as the *hyaloid membrane* of the eye which envelops the vitreous humour.

hyalophane (*Min*) Feldspar containing barium, with up to 30% $BaAl_2Si_2O_8$.

hyalopilitic texture (*Geol*) A texture of andesitic volcanic rocks in which the groundmass consists of small microlites of feldspar embedded in glass.

hyaloplasm (*BioSci*) (1) The ground substance or matrix of the cytoplasm, between the organelles and filamentous components. (2) Obsolete term for cytoplasm.

hyaluronic acid (*BioSci*) Natural polysaccharide that exists in body tissues (eg joint synovial fluid) and often binds with proteins. See CHONDROITIN.

hyaluronidase (*BioSci*) Lysosomal enzyme that degrades hyaluronic acid.

Hyatt roller bearing (*Eng*) See FLEXIBLE ROLLER BEARING.

hybrid (*BioSci*) Offspring of a *cross* between two different strains, varieties, races or species. (Adj or n.)

hybrid antibody (*BioSci*) (1) An antibody molecule in which the two antigen-combining sites are of different specificities. This can be achieved by recombining half molecules of two specific antibodies *in vitro*, or may also occur when cells from two different antibody-producing cell lines are fused to make a hybridoma. (2) An antibody with the variable (antigen-specific) region of a monoclonal antibody from one species joined to the constant region of immunoglobulin of another species. An example is a *humanized antibody*.

hybrid circuit (*Electronics*) An electronic subassembly formed by combining different types of individual integrated circuits with some discrete components. See panel on PRINTED, HYBRID AND INTEGRATED CIRCUITS.

hybrid coil (*ICT*) In telephone circuits a coil, comprising four equal windings and an additional winding, used for the separation of incoming and outgoing currents in a two-wire repeater. Also *bridge transformer, hybrid transformer*.

hybrid computer (*ICT*) One that combines a digital processor with a number of analogue units.

hybrid electromagnetic wave (*Phys*) A wave which has longitudinal components of both the electric and magnetic field vectors. Abbrev *HEM*.

hybrid integrated circuit (*Electronics*) A complete circuit formed by combining different types of individual integrated circuits and, in some cases, a number of discrete components. See panel on PRINTED, HYBRID AND INTEGRATED CIRCUITS.

hybridization (*BioSci*) (1) The formation of a new organism by normal sexual processes or more recently by protoplast fusion. (2) When DNA or RNA molecules from two different sources are annealed or renatured together, they will form hybrid molecules, whose stability is a measure of their sequence relatedness.

hybrid junction (*ICT*) A waveguide transducer that is connected to four branches of a circuit and designed so that these branches conjugate in pairs.

hybridoma (*BioSci*) (1) B-cell hybridoma. A cell line obtained by the fusion of a myeloma cell line that is able to grow indefinitely in culture, with a normal antibody-secreting B-cell. By choosing a myeloma that has ceased to make its own immunoglobulin product, the hybridoma secretes only the normal B-cell antibody and, since the cell line is cloned, the antibody is monoclonal. (2) T-cell hybridoma. A cell line obtained by fusion of a T-lymphoma cell line with a normal T-lymphocyte. Such hybridomas provide a source of homogeneous T-cell antigen receptors.

hybrid-p (*Electronics*) The most complex EQUIVALENT CIRCUIT used for analysis of the characteristics of transistors operating at high frequencies on the common-emitter configuration.

hybrid ring (*ICT*) A microwave junction, realized in waveguide, coaxial or stripline form, that has the properties of a HYBRID JUNCTION. Also *rat race*.

hybrid rocks (*Geol*) Rocks which originate by interaction between a body of magma and its wall rock or roof rock, which may be another igneous, or sedimentary, or metamorphic rock.

hybrid set (*ICT*) In telephony, two or more coils or transformers connected together to form a hybrid junction. See HYBRID.

hybrid sterility (*BioSci*) The lack of fertility of some interspecific hybrids that results from lack of homology between chromosome sets and thus abnormal segregation. It may be overcome by chromosome doubling. See AMPHIDIPLOID.

hybrid tee (*ICT*) A four-way waveguide junction arranged so that the power entering one port always divides into the two adjacent ports, but never reaches an opposite port. Sometimes used as a waveguide BALANCED MIXER.

hybrid transformer (*ICT*) See HYBRID COIL.

hybrid vehicle (*Eng*) A vehicle able to be powered by more than one type of energy, esp one that can run using internal combustion and electricity.

hybrid vigour (*BioSci*) The increase in desirable qualities such as growth rate, yield, fertility, often exhibited by HYBRIDS, ie favourable *heterosis*.

hydathode (*BioSci*) Structure through which water is exuded from uninjured plants.

hydatid cyst (*BioSci*) A large sac or vesicle containing a clear watery fluid and encysted immature larval CESTODA.

hydatidiform mole (*Med*) An affection of the chorionic villi (vascular tufts of the fetal part of the placenta) whereby they become greatly enlarged, resembling a bunch of grapes.

hydr-, hydro- (*Genrl*) Prefixes from Gk *hydor*, gen *hydatos*, water.

Hydra (Sea Serpent) (*Astron*) The largest constellation in the southern hemisphere, containing only one bright star.

hydragogue (*Pharmacol*) (1) Having the property of removing water. (2) A purgative drug which produces watery evacuations.

hydralazine (*Pharmacol*) Vasodilator drug used to treat hypertension.

hydramnios (*Med*) Excess of fluid in the amniotic sac of the fetus.

hydranth (*BioSci*) In HYDROZOA, a nutritive polyp of a hydroid colony.

hydrargillite (*Min*) See GIBBSITE.

hydrargyrism (*Med*) The state of being poisoned by mercury and its compounds.

hydrarthrosis (*Med*) Swelling of the joint, due to the accumulation in it of clear fluid.

hydrated electron (*Chem*) Very reactive free electron released in aqueous solutions by the action of ionizing radiation.

hydrated ion (*Chem*) An ion surrounded by molecules of water which it holds in a degree of orientation. Hydronium ion. Solvated H-ion of formula $[H_2O \rightarrow H]^+$ or H_3O^+.

hydrate of lime (*Build*) Also *hydrated lime*. See CAUSTIC LIME.

hydrates (*Chem*) Salts which contain water of crystallization. See WATER OF HYDRATION.

hydration (*Geol*) The addition of water to anhydrous minerals, the water being of atmospheric or magmatic origin. Thus anhydrite, by hydration, is converted into gypsum, and feldspars into zeolites.

hydration (*Paper*) The result in the fibre of prolonged or heavy beating/refining whereby the stuff is said to be 'wet', ie water does not drain very readily from it on the machine wire.

hydraulic (*Genrl*) Operated by the action of water or similar fluid.

hydraulic accumulator (*Aero, Eng*) A weight-loaded or pneumatic device for storing liquid at constant pressure, to steady the pump load in a system in which the demand is intermittent. In aircraft hydraulic systems an accumulator also provides fluid under pressure for operating components in an emergency, eg failure of an engine-driven pump.

hydraulic air compressor (*MinExt*) Arrangement in which water falling to the bottom of a shaft entrains air which is released in a tunnel at depth, while the water rises to a lower discharge level.

hydraulic amplifier (*Eng*) A power amplifier employed in some servomechanisms and control systems, in which power amplification is obtained by the control of the flow of a high-pressure liquid by a valve mechanism.

hydraulic blasting (*MinExt*) In fiery mines, rock breaking by means of a ram-operated device acting on a HYDRAULIC CARTRIDGE.

hydraulic brake (*Eng*) (1) An absorption dynamometer. See FROUDE BRAKE. (2) A motor-vehicle brake applied by small pistons operated by oil under pressure supplied from a pedal-operated master cylinder.

hydraulic burster (*CivEng*) See HYDRAULIC CARTRIDGE.

hydraulic capacity (*EnvSci*) The amount of water held by a soil between FIELD CAPACITY and the PERMANENT WILTING POINT, a measure of available water.

hydraulic cartridge (*CivEng*) An apparatus for splitting rock, mass concrete, etc, it consists of a long cylindrical body which has numerous pistons projecting from one side and moving in a direction at right angles to the body (under hydraulic pressure from within the body), which is placed in a hole drilled to take it. Also *hydraulic burster*.

hydraulic cement (*Build, CivEng*) A cement which will harden under water.

hydraulic classifier (*MinExt*) A device in which a vertically flowing column of water is used to carry up and on light and small mineral particles, while heavy and large ones sink.

hydraulic conductivity (*EnvSci*) A measure of the ease with which water moves through an aquifer, the rate of flow of groundwater across a unit cross-section of aquifer under unit hydraulic gradient. Also *hydraulic permeability*.

hydraulic control (*MinExt*) The use of fluid in a sensing mechanism to actuate a signalling or a correcting device, in response to pressure changes.

hydraulic coupling (*Eng*) A traction coupling used for automobiles or for diesel engines and electric motors, in which each half-coupling contains a number of radial vanes and rotation causes a vortex in the hydraulic fluid between them, power being dissipated by the slip.

hydraulic cyclone elutriator (*PowderTech*) A *hydrocyclone* fitted with an apex container which serves as a return-flow device for recycling the fine particles in suspension to give very efficient fractionation.

hydraulic engineering (*Eng*) That branch of engineering chiefly concerned in the design and production of hydraulic machinery, pumping plants, pipelines, etc.

hydraulic fill (*Eng*) An embankment or other fill in which the materials are deposited in place by a flowing stream of water, gravity and velocity control being used to bring about selected deposition.

hydraulic fracturing (*MinExt*) A method of increasing oil flow from less permeable strata by forcing liquid into them under very high pressure. See FRACKING.

hydraulic glue (*Chem*) A glue which is partially able to resist the action of moisture.

hydraulic gradient (*EnvSci*) The difference in hydraulic head along the fluid flow path.

hydraulic intensifier (*Eng*) A device for obtaining a supply of high-pressure liquid from a larger flow at a lower pressure.

hydraulicity (*Build, CivEng*) The property of a lime, cement or mortar which enables it to set under water or in situations where access of air is not possible.

hydraulic jack (*Eng*) A JACK in which the lifting head is carried on a plunger working in a cylinder, to which oil or water is supplied under pressure from a pump.

hydraulicking (*MinExt*) See HYDRAULIC MINING.

hydraulic leather (*Genrl*) A flexible leather prepared by being heavily treated, after tanning, with cod oil and then stoved; while hot, it may be shaped to requirements.

hydraulic lift (*Eng*) A lift or elevator operated either directly by a long vertical ram, working in a cylinder to which liquid is admitted under pressure, or by a shorter ram through ropes. See JIGGER.

hydraulic mining (*MinExt*) The operation of breaking down and working a bank of gravel, alluvium, poorly consolidated or decomposed bedrock by high-pressure water jets as in mining gold, tin PLACERS or china clay deposits. Also *hydraulicking*.

hydraulic mortar (*Build, CivEng*) A mortar which will harden under water.

hydraulic motor (*Eng*) A multicylinder reciprocating machine, generally of radial type, driven by water or oil.

hydraulic packing (*Eng*) L- or U-section rings providing a self-tightening packing under fluid pressure; used on rams and piston rods of hydraulic machines. See U-LEATHER.

hydraulic press (*Eng*) An upstroke, downstroke or horizontal press, with one or more rams, working at approximately constant pressure for deep drawing and extruding operations or at progressively increasing pressure for plastics moulding etc.

hydraulic ram (*Eng*) (1) The plunger of a hydraulic press. (2) A device whereby the energy provided by stopping the flow of a moving column of water is used to deliver some of the water under pressure.

hydraulic reservoir (*Aero*) In an aircraft hydraulic system, the header tank which holds the fluid; not to be confused with the *accumulator*, which is a pressure vessel wherein hydraulic energy is stored.

hydraulic riveter (*Eng*) A machine for closing rivets by hydraulic power.

hydraulics (*Genrl*) The science relating to the flow of fluids.

hydraulic squeezer (*Eng*) See SQUEEZER.

hydraulic stowing (*MinExt*) The filling of worked-out portions of a mine with water-borne waste material. The water drains off and is pumped to surface.

hydraulic test (*Eng*) A test for pressure-tightness and strength applied to pressure vessels, pipelines, instruments, etc.

hydraulic torque converter (*Eng*) A variable-speed mechanism similar to a hydraulic coupling, in which a decrease in secondary speed is accompanied by an increase in torque.

hydrazides (*Chem*) The mono-acyl derivatives of hydrazine.

hydrazine (*Chem*) A fuming strongly basic liquid. H_2NNH_2. Bp 113°C. A powerful reducing agent, it (also its derivatives, in particular DIMETHYL HYDRAZINE) is used as a high-energy propellant in rockets.

hydrazine hydrate (*Chem*) $N_2H_4.H_2O$. Diacid base. Attacks glass, rubber and cork. Used in some liquid rocket fuels and as an oxygen scavenger in boiler water treatment. Derivatives are widely used as blowing agents in the production of expanded plastics and rubbers.

hydrazo compounds (*Chem*) Symmetric derivatives of hydrazine, colourless, crystalline, neutral substances, obtained by the reduction of azo compounds.

hydrazoic acid (*Chem*) N_3H, HN^- $N^+{\equiv}N$. The aqueous solution is a strong monobasic acid and forms azides with many common metals. Used in the Schmidt reaction for converting aromatic carboxylic acids into primary amines.

hydrazones (*Chem*) The condensation products of aldehydes and ketones with hydrazine, water being eliminated from the two molecules.

hydrides (*Chem*) Compounds formed by the union of hydrogen with other elements. Those of the non-metals are generally molecular liquids or gases, certain of which dissolve in water (oxygen hydride) to form acid (eg hydrogen chloride) or alkaline (eg ammonia) solutions. The alkali and alkaline earth hydrides are crystalline, salt-like compounds, in which hydrogen behaves as the electronegative element. They contain H^+ ions and, when electrolysed, give hydrogen at the anode. Transition elements give alloy or interstitial hybrids.

hydriodic acid (*Chem*) HI. An aqueous solution of hydrogen iodide. Forms salts called *iodides*. Easily oxidized.

hydro- (*Genrl*) See HYDR-.

hydroa (*Med*) A skin disease in which groups of vesicles appear on reddened patches in the skin, associated with intense itching.

hydro-acoustics (*Acous*) The branch of acoustics concerned with radiation and propagation of sound in water.

hydroborons (*Chem*) See BORANE.

hydrobromic acid (*Chem*) HBr. An aqueous solution of hydrogen bromide.

hydrocarbons (*Chem*) A general term for organic compounds which contain only carbon and hydrogen. They are divided into saturated and unsaturated hydrocarbons, aliphatic (alkane or fatty) and aromatic (benzene) hydrocarbons. Crude oil is essentially a complex mixture of hydrocarbons.

hydrocele (*Med*) A swelling in the scrotum due to an effusion of fluid into the sac (tunica vaginalis) which invests the testis.

hydrocelluloses (*Chem*) Products obtained from cellulose by treatment with cold concentrated acids. They still retain the fibrous structure of cellulose, but are less hygroscopic.

hydrocephalus (*Med*) An abnormal accumulation of cerebrospinal fluid in the cavities (ventricles) of the brain, distending them and stretching and thinning the brain tissue over them.

hydrocerussite (*Min*) A colourless hydrated basic carbonate of lead occurring as an encrustation on native lead, galena, cerussite and other lead minerals.

hydrochloric acid (*Chem*) HCl. An aqueous solution of hydrogen chloride gas. Dissolves many metals, forming chlorides and liberating hydrogen. Used extensively in industry for numerous purposes, eg for the manufacture of chlorine, pickling, tinning, soldering, etc.

hydrochlorofluorocarbons (*EnvSci*) Family of compounds used as a replacement for chlorofluorocarbons in refrigeration etc because they have lower ozone-lowering potential. One example is CF_3CHCl_2. Abbrev *HCFCs*.

hydrocodone (*Pharmacol*) An opioid analgesic and antitussive. Also *hydrocodone bitartrate*.

hydrocoel (*BioSci*) In ECHINODERMATA, the WATER-VASCULAR SYSTEM.

hydrocooling (*FoodSci*) Cooling of vegetables or aseptically packed foods by immersion in potable water.

hydrocortisone (*Pharmacol*) A powerful corticosteroid, with both mineralocorticoid and glucocorticoid activity, produced by cells of the zona reticularis in the adrenal gland. It has potent anti-inflammatory effects. Also *cortisol*; *17-hydroxy-corticosterone*.

hydrocyanic acid (*Chem*) An aqueous solution of HYDROGEN CYANIDE. Dilute solution called *prussic acid*. Monobasic. Forms cyanides. Very poisonous.

hydrocyclone (*MinExt*) A small cyclone extractor for removing suspended matter from a flowing liquid by means of the centrifugal forces set up when the liquid is made to flow through a tight conical vortex. Used to separate solids in mineral pulp into coarse and fine fractions. Fluent stream enters tangentially to cylindrical section and coarser sands gravitate down steep-sided conical section to controlled apical discharge. Bulk of pulp, containing finer particles, overflows through a pipe inserted in the central vortex. Classification into fractions is aided by the centrifugal force with which the pulp is delivered to the appliance, which can handle large tonnages.

hydrodynamic lubrication (*Eng*) Thick-film lubrication in which the relatively moving surfaces are separated by a substantial distance and the load is supported by the hydrodynamic film pressure.

hydrodynamic power transmission (*Eng*) A power-transmission system which, in general, employs a hydraulic coupling or a torque converter or a reaction coupling.

hydrodynamic process (*Eng*) A process for shallow forming and embossing operations, in which high-pressure water presses the blank against a female die, there being no solid punch to conform to the die contour. Similar processes involve explosive charges.

hydrodynamics (*Eng*) That branch of dynamics which studies the motion produced in fluids by applied forces.

hydroelectric generating power station (*ElecEng*) An electric generating station in which the generators are driven by water turbines.

hydroelectric generating set (*ElecEng*) An electric generator driven by a water turbine.

hydrofining (*Chem*) The process of removing undesirable impurities, particularly sulphur and unsaturated compounds, from petroleum fractions using hydrogen at high temperature and pressure over a platinum catalyst. Also *hydrotreating*. Obsolete term *hydroforming*.

hydrofluoric acid (*Chem*) Aqueous solution of hydrogen fluoride. Dissolves many metals, with evolution of hydrogen. Etches glass owing to combination with the silica of the glass to form silicon fluoride, hence it is stored in eg polyethylene vessels. See ETCHING TEST.

hydrofoil (*Aero*) An immersed aerofoil-like surface to facilitate the take-off of a seaplane by increasing the hydrodynamic lift.

hydrofoil (*Ships*) A fast, light craft fitted with wing-like structures (foils) on struts under the hull. These may be extendable and adjustable. Propelled by propeller in water or air, or by jet. Foils may act entirely or partly submerged with hull lifted clear of the water at speed. Steered by water rudder. Cf HYDROPLANE.

hydroforming (*Chem*) See HYDROFINING.

hydroforming (*Eng*) A hydraulic forming process in which the shape is produced by forcing the material by means of a punch against a flexible bag partly filled with hydraulic fluid and acting as a die.

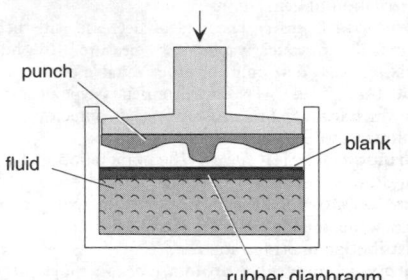

hydroforming Blank before punching.

hydrofuge (*BioSci*) Water-repelling; said of certain hairs possessed by some aquatic insects and used for retaining a film of air.

hydrogel (*Chem*) A gel the liquid constituent of which is water.

hydrogen (*Chem*) Symbol H, at no 1, ram 1·007 97, valency 1. The element with the lightest atoms, forming diatomic molecules H_2. Molecular hydrogen is a colourless, odourless, diatomic gas, water being formed when it is burnt; mp $-259\cdot14°C$, bp $-252\cdot7°C$, density $0\cdot089\,88\,g\,dm^{-3}$ at stp. The element is widely distributed as part of the water molecule, occurs in many minerals, eg petroleum, and in living matter. It is cosmically the most abundant of all elements but in its elemental form is not of major importance in the Earth's crust (abundance 1520 ppm; 0·53 ppm by volume in the atmosphere). Hydrogen is manufactured chiefly as a by-product of electrolysing caustic soda, water gas and gas cracking. It is used in the Haber process for the fixation of nitrogen, and in the hardening of fats (eg in the manufacture of margarine); also for hydrogenation of oils, manufacture of hydrochloric acid, filling small balloons, and as a reducing agent for organic synthesis and metallurgy, oxyhydrogen and atomic hydrogen welding flames. As a component of water, hydrogen is of great importance in the moderation (slowing down) of neutrons as hydrogen atoms are the only ones of similar mass to a neutron and are therefore capable of absorbing an appreciable proportion of the neutron energy on collision. Isotopes, with one proton and one electron are shown in the table.

Name	Atomic symbol	Relative atomic mass	Atomic number	Neutrons
protium	^1H	1·007 825	1	0
deuterium	D, ^2H	2·014 10	2	1
tritium	T, ^3H	3·0221	3	2

hydrogen I, II (*Astron*) The hydrogen of interstellar space, known in two clearly defined states: HI (neutral hydrogen) and HII (ionized hydrogen). The HI regions, confined mainly to the arms of spiral galaxies, emit no visible light, and are detected solely by their emission of the 21 cm radio line. The HII regions, found in the gaseous nebulae, emit both visible light and radio waves.

hydrogenation (*Chem*) Chemical reactions involving the addition of hydrogen, present as a gas, to a substance, in the presence of a catalyst. Important processes are: the hydrogenation of coal; the hydrogenation of fats and oils; the hydrogenation of naphthalene and other substances. See FATTY ACID, FISCHER–TROPSCH PROCESS.

hydrogen bacteria (*BioSci*) Chemosynthetic bacteria that obtain the energy required for carbon dioxide assimilation from the oxidation of hydrogen to water. Organic compounds may also be oxidized by these species under suitable conditions.

hydrogen bomb (*Phys*) A bomb which uses the nuclear fusion process to release vast amounts of energy. Extremely high temperatures are required for the process to occur and these temperatures are obtained by a fission bomb around which the fusion material is arranged. Lithium deuteride can initiate a number of fusion processes involving the hydrogen isotopes, deuterium and tritium. These reactions also produce high-energy neutrons capable of causing fission in a surrounding layer of the most abundant isotope of uranium, uranium-238, so that further energy is released.

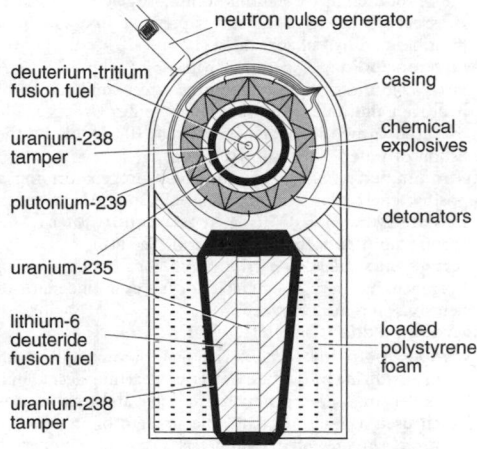

hydrogen bomb Probable arrangement.

hydrogen bond (*Chem*) A weak inter- or intramolecular force resulting from the interaction of a hydrogen atom bonding with an electronegative atom with a lone pair of electrons, eg O or N. Hydrogen bonding is important in associated liquids, particularly water, and in polyamides. It is responsible for much of the secondary and tertiary structure of proteins. See ALPHA HELIX and panels on BONDING and PROTEIN.

hydrogen bonding (*BioSci*) Hydrogen atoms in the groups –O–H and –N–H can form non-covalent bonds with a nitrogen or oxygen atom. The bonding is ionic and weak, but hydrogen bonding between purines and pyrimidines contributes to the stability of the DNA double helix, the planar stacking of the bases and co-operation between adjacent hydrogen bonds making for a very stable and rigid structure.

hydrogen bromide (*Chem*) HBr. Hydrogen bromide gas. Can be made by direct combination of the two elements, particularly in the presence of a catalyst. Closely resembles hydrogen chloride, but is less stable.

hydrogen chloride (*Chem*) HCl. A colourless gas which dissolves in water to form hydrochloric acid. Produced by the action of concentrated sulphuric acid on chlorides, on the industrial scale as a by-product of the chlorination of hydrocarbons, and directly from the elements H_2 and Cl_2, both of which are formed as a by-product of caustic soda manufacture. Uses: to produce hydrochloric acid; chlorination of unsaturated organic compounds, eg chloroethene in polymerization, isomerization and alkylation, as a catalyst.

hydrogen cooling (*ElecEng*) A method of cooling rotating electric machines; the machine is totally enclosed and runs in an atmosphere of hydrogen.

hydrogen cyanide (*Chem*) HCN. Mp $-13.4°C$, bp $26°C$, permittivity 95, dissociation constant 1.3×10^{-9} at $18°C$. Highly poisonous liquid. Dissolves in water to form hydrocyanic or prussic acid. Faint odour of bitter almonds, undetectable by some people. Uses: fumigant; medicine; organic synthesis, eg acrylic resins; analytical reagent.

hydrogen electrode (*Chem*) For pH measurement, a platinum-black electrode covered with hydrogen bubbles. Although rarely used in practice, it defines the HYDROGEN SCALE of electrode potentials. Also *Hildebrand electrode*. See ELECTROCHEMICAL SERIES.

hydrogen embrittlement (*Eng*) The effect produced on metal by sorption of hydrogen during pickling or electroplating operations.

hydrogen fluoride (*Chem*) HF. A liquid which fumes strongly in air. Dissolves in water to form hydrofluoric acid. Produced by the action of sulphuric acid on fluorides. Uses: catalyst in organic reactions; preparation of uranium; fluorides and hydrofluoric acid.

hydrogen iodide (*Chem*) HI. Mp $-50°C$, bp $-35°C$. A heavy colourless gas, formed by the direct combination of hydrogen and iodine; fumes strongly in air. Usually made by the decomposition of phosphorus (III) iodide by the action of water. See HYDRIODIC ACID.

hydrogen ion (*Chem*) An atom of hydrogen carrying a positive charge, ie a proton; in aqueous solution, hydrogen ions are hydrated, H_3O^+ (the HYDROXONIUM ION).

hydrogen ion concentration (*Chem*) See PH.

hydrogenous (*Phys*) Descriptive of a substance rich in hydrogen and therefore suitable for use as a moderator of neutrons in a nuclear reactor.

hydrogen oxide (*Chem*) H_2O. Water.

hydrogen peroxide (*Chem*) H_2O_2. A viscous liquid with strong oxidizing properties. Powerful bleaching agent; and, as its decomposition products are water and oxygen, it is much used as a disinfectant. The strength of an aqueous solution is represented commercially by the number of volumes of oxygen which $100\ cm^3$ of the solution will give on decomposition. Recent interest in high-concentration (approx 90%) hydrogen peroxide has centred on its use as both a technical and a military oxidizing agent, esp in rocket fuels as an *oxidant*. Prepared via the anodic oxidation of hydrogen sulphates to peroxydisulphate followed by hydrolysis and steam distillation.

hydrogen phosphide (*Chem*) PH_3. See PHOSPHINE (1).

hydrogen scale (*Chem*) A system of relative values of electrode potentials, based on that for hydrogen gas, at standard pressure, against hydrogen ions at unit ACTIVITY (2), as zero. See HYDROGEN ELECTRODE.

hydrogen sulphide (*Chem*) H_2S. May be prepared by direct combination of the two elements or by the action of dilute hydrochloric or sulphuric acid on iron sulphide. It is readily decomposed. Reacts with bases forming sulphides and with some metals to produce metal sulphides and liberate hydrogen. Poisonous, with a characteristic smell of rotten eggs.

hydrogen thyratron (*Electronics*) A THYRATRON containing hydrogen, used in radar transmitters to provide high peak currents at high voltage with very rapid switch-on and recovery time.

hydrogeology (*Geol*) The study of the geological aspects of the Earth's water.

hydrographical surveying (*Surv*) A branch of surveying dealing with bodies of water at the coastline and in harbours, estuaries and rivers.

hydrography (*Genrl*) The study, determination and publication of the conditions of seas, rivers, lakes and groundwater. Includes surveying and charting of coasts, rivers, estuaries and harbours, and supplying particulars of eg depth, bottom, tides and currents.

hydrogrossular (*Min*) A hydrated variety of garnet, close to GROSSULAR in composition but with some of the silica replaced by water. A massive variety from S Africa is sometimes called 'Transvaal Jade'.

hydrohaematite (*Min*) $Fe_2O_3.nH_2O$. Probably a mixture of the two minerals *haematite* and *goethite*, the former being in excess. It is fibrous and red in the mass, with an orange tint when powdered. Also *turgite*. US *hydrohematite*.

hydroid (*BioSci*) (1) Water-conducting cell in the stems of some moss gametophytes, eg *Polytrichum*. (2) An individual of the asexual stage in those HYDROZOA that show alternation of generations.

hydrolapse (*EnvSci*) The rate of decrease with height of atmospheric water vapour as measured by humidity mixing ratio, dew point or other suitable quantity.

Hydrolastic (*Autos*) TN for a proprietary type of suspension in which front and rear suspensions are interconnected and compensated hydraulically.

hydrological cycle (*EnvSci*) The evaporation and condensation of water on a world scale.

hydrology (*Genrl*) The study of water, including rain, snow and water on the Earth's surface, covering its properties, distribution and utilization.

hydrolysed vegetable protein (*FoodSci*) Soya protein etc chemically degraded to amino acids. Used to accentuate natural savoury flavours in sauces, soups and meat products.

hydrolysis (*Chem*) (1) The formation of an acid and a base from a salt by interaction with water; it is caused by the ionic dissociation of water. (2) The decomposition of organic compounds by interaction with water, either in the cold or on heating, alone or in the presence of acids or alkalis; eg esters form alcohols and acids; oligo- and polysaccharides on boiling with dilute acids yield monosaccharides.

hydrolysis of polymers (*Chem*) Degradation of stepgrowth polymers by water. Must be guarded against during heat treatment or moulding. See STEP POLYMERIZATION.

hydromagnesite (*Min*) Hydrated magnesium hydroxide and carbonate, occurring as whitish amorphous masses, or rarely as monoclinic crystals in serpentines.

hydromagnetic (*ElecEng*) Pertaining to the behaviour of a plasma in a magnetic field. See MAGNETOHYDRODYNAMICS.

Hydromedusae (*BioSci*) See HYDROZOA.

hydromelia (*Med*) Dilatation of the central canal of the spinal cord.

hydrometallurgy (*MinExt*) Extraction of metals or their salts from crude or partly concentrated ores by means of aqueous chemical solutions; also electrochemical treatment including electrolysis or ion exchange.

hydrometeor (*EnvSci*) A generic term for all products of the condensation or sublimation of atmospheric water vapour, including: ensembles of falling particles which may either reach the Earth's surface (rain and snow) or evaporate during their fall (virga); ensembles of particles suspended in the air (cloud and fog); particles lifted from the Earth's surface (blowing or drifting snow and spray); particles deposited on the ground or on exposed objects (dew, hoar frost, etc).

hydrometer (*Phys*) An instrument by which the relative density of a liquid may be determined by measuring the length of the stem of the hydrometer immersed, when it floats in the liquid with its stem vertical.

hydronephrosis (*Med*) Distension of the kidney with urine held up as a result of obstruction elsewhere in the urinary tract.

hydronium ion (*Chem*) See HYDROXONIUM ION.

hydropericardium (*Med*) The collection of clear fluid in the pericardial sac.

hydroperitoneum (*Med*) Accumulation of clear fluid in the abdominal cavity. Also *ascites*.

hydroperoxides (*Chem*) Intermediate compounds formed during the oxidation of unsaturated organic substances, eg fatty oils, such as linseed oil. They contain the group –OOH.

hydrophane (*Min*) A variety of opal which, when dry, is almost opaque, with a pearly lustre, but becomes transparent when soaked with water, as implied in the name.

hydrophilic colloid (*Chem*) A colloid which readily forms a solution in water.

hydrophilic group (*Chem*) A polar group or one that can take part in hydrogen bond formation, eg OH, COOH, NH$_2$. Confers water solubility or, in lipids and macromolecules, causes part of the structure to make close contact with the aqueous phase.

hydrophily (*BioSci*) (1) Living in water. (2) Pollination by water.

hydrophobia (*Med*) Literally fear of water, a symptom of rabies. Used synonymously with *rabies*.

hydrophobic bonding (*Chem*) An important determinant of protein conformation and of lipid structures, and is considered to be a consequence of maximizing polar interactions and reducing exposure of non-polar residues to an aqueous (*hydrophilic*) environment, rather than a positive interaction between apolar residues.

hydrophobic cement (*Build, CivEng*) Cement into which a special agent is introduced during the grinding process to reduce the deterioration which occurs in other cements when exposed to damp.

hydrophobic colloid (*Chem*) A colloid which forms a solution in water only with difficulty.

hydrophone (*Acous*) Electro-acoustic transducer used to detect sounds or ultrasonic waves transmitted through water. Also *subaqueous microphone*.

hydrophyte (*BioSci*) (1) A plant with leaves partly or wholly submerged in water. (2) Water plant with perennating buds at the bottom of the water (a sort of CRYPTOPHYTE). See RAUNKIAER SYSTEM.

hydroplane (*Ships*) (1) A powered boat which skims the surface of the water; cf HYDROFOIL. (2) A planing surface which enables a submarine to submerge.

hydroponics (*BioSci*) A technique of growing plants without soil for experimental purposes and sometimes crops. The roots can be in either a nutrient solution or in an inert medium percolated by such a solution as in water culture and sand culture. See NUTRIENT FILM TECHNIQUE.

hydropote (*BioSci*) A gland-like structure found on the submerged surfaces of the leaves of many water plants.

hydrops folliculi (*Med*) An ovarian cyst formed by the accumulation of clear fluid in a Graafian follicle.

hydropyle (*BioSci*) A modified area of the serosal cuticle of the developing egg of some ORTHOPTERA for the uptake or loss of water.

hydroquinone (*Chem, ImageTech*) 1,4-dihydroxybenzene. A strong reducing agent extensively used as a developer in photography. Also *quinol*.

hydrosalpinx (*Med*) Accumulation of clear fluid in a Fallopian tube which has become shut off as a result of inflammation.

hydrosere (*BioSci*) A SERE beginning with submerged soil, as at the margin of a lake.

hydroskis (*Aero*) Hydrofoils, proportioned like skis and usually retractable, fitted to seaplanes without a PLANING BOTTOM as the sole source of hydrodynamic lift. They are also fitted to aircraft landing gear to make them amphibious (*pantobase*), in which case a minimum taxiing speed is necessary to keep the aircraft above water.

hydrosol (*Chem*) A colloidal solution in water.

hydrosphere (*EnvSci*) The water on or near the surface of the Earth as opposed to the solid crust and the gaseous atmosphere.

hydrostatic approximation (*EnvSci*) The assumption that the atmosphere is in hydrostatic equilibrium in the vertical, which is equivalent to ignoring the vertical components of acceleration and CORIOLIS FORCE in the equations of motion. The approximation is valid for atmospheric

disturbances on horizontal scales of not less than about 10 km.

hydrostatic extrusion (*Eng*) A form of extrusion in which material is preshaped to occlude a die which forms the lower end of a high-pressure container filled with liquid. The pressure in the liquid is increased until the material is forced through the die orifice. See EXTRUSION.

hydrostatic instability (*EnvSci*) See STATIC INSTABILITY.

hydrostatic pressing (*PowderTech*) The application of liquid pressure directly to a preform which has been sealed, eg in a plastic bag. It is characterized by equal pressure in all directions maintaining preform shape to reduced scale. Equivalent to isostatic pressing.

hydrostatics (*Genrl*) The branch of statics which studies the forces arising from the presence of fluids.

hydrostatic skeleton (*BioSci*) A body form maintained by muscles acting upon a fluid-filled cavity, usually the coelom as in many soft-bodied invertebrates.

hydrostatic test (*Build*) A test to detect leakage in a drain by the fall in water level when the outlet end is temporarily plugged.

hydrostatic valve (*Eng*) Apparatus which tends to maintain an underwater body (eg a moving torpedo) at the desired depth.

hydrosulphides (*Chem*) Salts formed by the action of hydrogen sulphide on some of the hydroxides. They contain the ion HSa.

hydrosulphuric acid (*Chem*) An aqueous solution of hydrogen sulphide.

hydrotaxis (*BioSci*) The response or reaction of an organism to the stimulus of moisture. Adj *hydrotactic*.

hydrotherapy (*Med*) The treatment of disease by water, externally or internally.

hydrothermal (*Min*) Relating to the action of hot water, and mineral deposits formed by such processes. The term is sometimes restricted to water of magmatic origin.

hydrothermal metamorphism (*Geol, Min*) That kind of change in the mineral composition and texture of a rock which was effected by heated water.

hydrothermal mineralization (*Min*) Processes involving hot water, usually of magmatic origin, by which mineral deposits may be formed.

hydrothorax (*Med*) Clear fluid in the pleural cavity formed by transudation from blood vessels.

hydrotropism (*BioSci*) A tropism in which the orientating stimulus is water.

hydrovane (*Aero*) See HYDROFOIL, HYDROSKIS, SPONSON.

hydroxides (*Chem*) Compounds of the basic oxides with water. The term *hydroxide* (a contraction of *hydrated oxide*) is applied to compounds that contain the –OH or hydroxyl group.

hydroxonium ion (*Chem*) The hydrogen ion, normally present in hydrated form as H$_3$O$^+$. Also *hydronium ion*.

hydroxyapatite (*Min*) Hydrated calcium phosphate occurring widely in natural biomaterials such as ENAMEL, bone, etc. Also contains fluoride, chloride or carbonate ions in place of the phosphate (PO$_4^-$) group.

hydroxyl (*Chem*) –OH. A monovalent group consisting of a hydrogen atom and an oxygen atom linked together.

hydroxylamine (*Chem*) *Hydroxyammonia*. NH$_2$OH. Mp 33°C, bp 58°C at 3 kN m^{-2}. Rather explosive, deliquescent colourless crystals which may be obtained by the reduction of nitric oxide, ethyl nitrate or nitric acid under suitable conditions. Its aqueous solution is alkaline, its salts are powerful reducing agents.

hydroxylamines (*Chem*) Derivatives of hydroxylamine, NH$_2$OH, in which the hydrogen has been exchanged for alkyl radicals.

hydroxyproline (*BioSci, Chem*) 4-hydroxypyrrole-2-carboxylic acid. An *imino* acid formed by the post-translational hydroxylation of proline residues within protein molecules, common in cell-wall proteins and in collagen.

hydrozincite (*Min*) A monoclinic hydroxide and carbonate of zinc. It is an uncommon ore, occurring with smithsonite as an alteration product of sphalerite in the oxide zone of some lodes. Also *zinc bloom.*

Hydrozoa (*BioSci*) A class of CNIDARIA, in which alternation of generations typically occurs; the hydroid phase is usually colonial, and gives rise to the medusoid phase by budding.

Hydrus (Water Snake) (*Astron*) An inconspicuous southern constellation.

hyetograph (*EnvSci*) An instrument which collects, measures and records the fall of rain (Gk *hyetos*, rain).

hygristor (*ElecEng*) A resistance element sensitive to ambient humidity.

hygro- (*Genrl*) Prefix from Gk *hygros*, wet, moist.

hygrochastic (*BioSci*) A plant movement caused by the absorption of water.

hygroma (*Vet*) A fluid-filled swelling, usually associated with a joint, due to distension of a synovial sac.

hygrometer (*Phys*) An instrument for measuring or giving output signal proportional to atmospheric humidity. Electrical hygrometers make use of HYGRISTORS.

hygrometry (*EnvSci*) The measurement of the hygrometric state, or RELATIVE HUMIDITY, of the atmosphere.

hygrophyte (*BioSci*) A plant of wet or waterlogged soils.

hygroscopic (*Chem*) Tending to absorb moisture; in the case of solids, without liquefaction.

hygroscopic movement (*BioSci*) That caused by changes in moisture content of unevenly thickened cell walls, eg of *elators* or the awns of the grains of some grasses. Also *hygrometric movement, imbibitional movement.*

hygrostat (*Chem*) Apparatus which produces constant humidity.

hylophagous (*BioSci*) Wood-eating.

hymen (*BioSci*) In mammals, a fold of mucous membrane that partly occludes the opening of the vagina in young forms.

hymenitis (*Med*) Inflammation of the hymen.

hymenium (*BioSci*) The fertile layer containing the asci or basidia in Ascomycetes and Basidiomycetes. Also *sporogenous layer.*

Hymenomycetes (*BioSci*) The mushrooms, toadstools, agarics and bracket fungi, a class of the Basidiomycotina in which the HYMENIUM is exposed at the time of spore formation. Many are saprophytic in soil, dung, leaf litter, etc; some are parasitic, eg honey fungus *Armillaria mellea.* Many form ECTOTROPHIC MYCORRHIZAS with trees.

hymenophore (*BioSci*) Any fungal structure that bears a HYMENIUM.

Hymenoptera (*BioSci*) An order of ENDOPTERYGOTA having usually two almost-equal pairs of transparent wings, which are frequently connected during flight by a series of hooks on the hind-wing. Mandibles always occur but the mouthparts may be suctorial. Adults are usually of diurnal habit but the larvae show great variation of form and habit. Saw-flies, gall flies, ichneumons, ants, bees, wasps.

hymenotomy (*Med*) Cutting of the hymen.

hyo- (*Genrl*) Prefix from Gk *hyoeides*, U-shaped.

hyoid (*BioSci*) In higher vertebrates, a skeletal apparatus lying at the base of the tongue, derived from the hyoid arch of the embryo.

hyoid arch (*BioSci*) The second pair of visceral arches in lower vertebrates, lying between the mandibular arch and the first branchial arch.

hyoideus (*BioSci*) In vertebrates, the post-trematic branch of the facial (seventh cranial) nerve. It runs to the mucosa of the mouth and to the muscles of the hyoid region, and, in aquatic forms, to the neuromast organs of the region below and behind the orbit.

hyomandibular (*BioSci*) In Craniata, the dorsal element of a hyoid arch.

hyomandibular nerve (*BioSci*) In Craniata, the post-trematic branch of the seventh cranial nerve; it divides into anterior and posterior branches.

hyoscine (*Pharmacol*) A coca-base alkaloid, present in *Datura meteloides.* It is the scopoline ester of tropic acid. It has a sedative effect on the central nervous system and the hydrobromide is used to induce twilight sleep; it was once used to obtain criminal confessions. Also *scopolamine.*

hyoscyamine (*Chem*) $C_{17}H_{23}O_3N$. Mp 109°C. A coca-base alkaloid, optically active, stereoisomeric with *atropine*, forming colourless needles or plates. It can be prepared from *Datura stramonium.*

hyoscyamine sulphate (*Pharmacol*) An *anti-muscarinic* drug that is used for relief of gut spasm. It is the active component of belladonna extract (atropine is a racemic mixture of D- and L-hyoscyamine).

hyostyly (*BioSci*) A type of jaw suspension, found in most fish, in which the upper jaw is attached to the cranium anteriorly by a ligament, posteriorly by the hyomandibular. Adj *hyostylic.*

hypabyssal rocks (*Geol*) Literally, igneous rocks that are not quite abyssal (ie deep-seated, plutonic), occurring as minor intrusions. The three main divisions, based on mode of occurrence, are plutonic, hypabyssal and volcanic. See DYKES, SILLS.

hypaesthesia (*Med*) Diminished susceptibility to physical stimuli. US *hypesthesia.*

hypaethral (*Arch*) Said of a building without a roof, or with an opening in its roof.

hypalgesia (*Med*) Diminished sensitivity to pain.

Hypalon (*Eng*) TN for chlorosulphonated polyethylene rubber (US).

hypanthium (*BioSci*) The flat or concave receptacle of a perigynous flower.

hypapophyses (*BioSci*) Paired ventral processes of the vertebrae of many higher Craniata.

hypaxial (*BioSci*) Below the axis, esp below the vertebral column, therefore ventral; as the lower of two blocks into which the myotomes of fish embryos become divided.

hyper- (*Genrl*) Prefix from Gk *hyper*, above.

hyperaccumulator (*BioSci*) An organism, usually a plant, that absorbs large amounts of heavy metals from the soil.

hyperacidity (*Med*) Excessive acidity, esp of the stomach juices.

hyperactive state (*Psych*) See HYPERKINETIC STATE.

hyperactivity (*Psych*) Behavioural state in children, characterized by motor restlessness, poor attention and excitability; often associated with learning difficulties. Also *attention deficit hyperactivity disorder, ADHD.*

hyperacusis (*Med*) Abnormally increased acuity of hearing.

hyperadrenalism (*Med*) Abnormally increased activity of the adrenal gland.

hyperaemia (*Med*) Congestion, or excess of blood, in a part of the body. US *hyperemia.*

hyperaesthesia (*Med*) Lowered threshold to a given stimulus, eg excessive sensitivity to painful or tactile stimuli. US *hyperesthesia.*

hyperalgesia (*Med*) Heightened sensitivity to painful stimuli.

hyperaphia (*Med*) Excessive sensitivity to touch.

hyperbaric chamber (*Med*) A chamber containing oxygen at high pressures, in which patients undergo radiotherapy or treatment for forms of poisoning.

hyperbarism (*Space*) An agitated bodily condition when the pressure within the body tissues, fluids and cavities is countered by a greater external pressure, such as may happen in a sudden fall from a high altitude.

hyperbilirubinaemia (*Med*) Excess of the bile pigment BILIRUBIN in the blood. US *hyperbilirubinemia.*

hyperbola (*MathSci*) The intersection of a cone with a plane which cuts both branches of the cone. See also fig. at CONIC. Fig. ▷

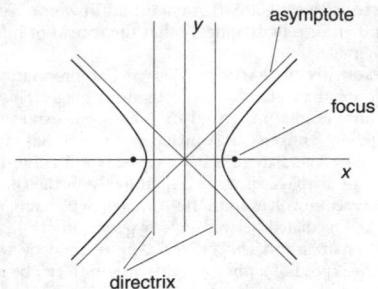

hyperbola The rectangular type is drawn.

hyperbolic (*ICT*) Said of any system of navigation that depends on difference in cycles and fractions between locked waves from two or more stations, eg DECCA.

hyperbolic functions (*MathSci*) A set of six functions, analogous to the trigonometric functions. The *hyperbolic sine*, sinh x, is defined as $1/2(e^x - e^{-x})$; the *hyperbolic cosine*, cosh x, as $1/2(e^x + e^{-x})$; and the *hyperbolic tangent*, tanh x, as sinh x/cosh x. The *hyperbolic secant*, sech x, the *hyperbolic cosecant*, cosech x, and the *hyperbolic cotangent*, coth x, are defined as the inverses of cosh x, sinh x, and tanh x respectively. In electrical engineering, these functions are useful in calculations involving the transmission of currents along wires and in filters.

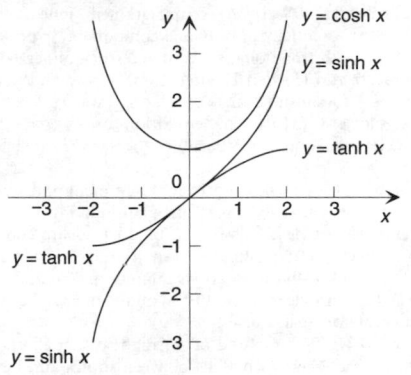

hyperbolic functions

hyperbolic geometry (*MathSci*) See ABSOLUTE.

hyperbolic paraboloid (*MathSci*) A saddle-shaped surface generated by the equation $x^2/a - y^2/b = z$.

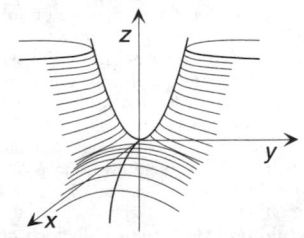

hyperbolic paraboloid

hyperbolic point on a surface (*MathSci*) One at which the curvatures of the normal sections are not all of the same sign. The directions which separate normal sections with positive curvature from those with negative curvature are called *asymptotic directions*. Cf ELLIPTICAL POINT ON A SURFACE, PARABOLIC POINT ON A SURFACE, UMBILICAL POINT ON A SURFACE.

hyperbolic spiral (*MathSci*) A spiral with polar equation $r\theta = a^2$.

hyperboloid (*MathSci*) See QUADRIC.

hypercalcaemia (*Med*) Rise in the calcium content of the blood beyond normal limits. US *hypercalcemia*.

hypercapnia (*Med*) Excess of carbon dioxide in the lungs or the blood.

hypercharge (*Phys*) See STRANGENESS.

hyperchlorhydria (*Med*) Increased secretion of hydrochloric acid by the acid-secreting cells of the stomach.

hypercholesterolaemia (*Med*) Increase of cholesterol in the blood beyond normal limits. US *hypercholesterolemia*.

hypercube (*ICT*) A MULTIDIMENSIONAL DATABASE, often used in OLAP systems to bring together complex arrays of data that may then be analysed in a variety of ways determined by the user.

hyperdactyly (*BioSci*) The condition of having more than the normal number (five) of digits, as in Cetacea.

hyperdiploidy (*BioSci*) The condition where the full chromosome complement is present, as well as a portion of one chromosome which has been translocated.

hyperemesis (*Med*) Excessive vomiting.

hyperemesis gravidarum (*Med*) Continued vomiting during pregnancy.

hyperemia (*Med*) US for HYPERAEMIA.

hyperesthesia (*Med*) US for HYPERAESTHESIA.

hypereutectoid steel (*Eng*) Steel containing more carbon than the eutectoid *pearlite*; ie in carbon steels, one containing more than 0·8% carbon. See panel on STEELS.

hyperfine structure (*Phys*) (1) Splitting of spectral lines into two or more closely spaced components due to the interaction between the orbital magnetic moment and the nuclear magnetic moment of the atom. (2) The effect of nuclear mass on the energy levels in isotopically related atoms. Both hyperfine effects are of the same order of magnitude.

hyperfocal distance (*ImageTech*) The distance in front of a lens focused at infinity beyond which all objects are acceptably in focus. If a lens is focused on the hyperfocal distance, the DEPTH OF FIELD extends from half that distance to infinity.

hypergammaglobulinaemia (*Med*) A term used to describe clinical conditions in which the concentration of immunoglobulins in the blood exceeds normal limits. May result from continuous antigenic stimulation in chronic infections, from auto-immune diseases, or from abnormal proliferation of B-cells as occurs in *Waldenstrom macroglobulinaemia* or in *myelomatosis*. US *hypergammaglobulinemia*.

hypergeometric distribution (*MathSci*) The probability distribution of the random variable corresponding to the number of outcomes of a particular kind in a sampling procedure in which sampling is without replacement from a population of finite size and in which, for each draw from the population, the outcome can be one of only two kinds. The numbers of members of each kind in the population, and hence the total population size, and the size of the sample are fixed in advance.

hypergeometric equation (*MathSci*) See GAUSS'S DIFFERENTIAL EQUATION.

hypergeometric function (*MathSci*) The function $F(a;b;c;x)$ which is the sum of the *hypergeometric series*

$$1 + \frac{ab}{1c}x + \frac{a(a+1)b(b+1)}{1.2.c(c+1)}x^2$$
$$+ \frac{a(a+1)(a+2)b(b+1)(b+2)}{1.2.3.c(c+1)(c+2)}x^3 + \dots$$

which converges either if $|x| < 1$ or if $x = 1$ and $a + b < c$.

hypergeometric series (*MathSci*) See HYPERGEOMETRIC FUNCTION.

hyperglycaemia (*Med*) An increase in the sugar content of the blood beyond normal limits. US *hyperglycemia*. See DIABETES MELLITUS.

hypergolic (*Space*) A rocket-propellant mixture (fuel and oxidizer) which ignites spontaneously upon mixing.

Hyper HAD (*ImageTech*) TN for a SOLID-STATE IMAGE SENSOR of the frame interline transfer type (see FIT CCD). The hole accumulated diode uses transparent and buried electrodes to increase the light-sensitive area and a microlens over each PIXEL to capture more light.

hyperhidrosis (*Med*) Excessive perspiration. Also *hyperidrosis*.

hypericism (*Vet*) A form of photosensitization occurring in sheep and cattle following the ingestion of the St John's Wort plant (*Hypericum perforatum*).

hyperidrosis (*Med*) See HYPERHIDROSIS.

hyperinosis (*Med*) Excess of fibrin in the blood; the opposite of *hypinosis*.

hyperinsulinism (*Med*) A condition in which the blood sugar falls below normal limits, due to oversecretion of insulin by the pancreas; usually associated with pancreatic tumours, which produce the excess insulin.

Hyperion (*Astron*) The seventh natural satellite of Saturn, discovered in 1848. Distance from the planet 1 481 000 km; diameter 300 km.

hyperkeratosis (*Med*) Overgrowth of the horny layer of the skin.

hyperkinesia (*Med*) Excessive motility of a person, or of muscles.

hyperkinetic state (*Psych*) See HYPERACTIVITY.

hyperlink (*ICT*) A link between documents, or items within a document, created using HTML.

hyperlipidaemia (*Med*) Elevated levels of low-density lipoprotein in blood, correlated with increased risk of cardiovascular disease. See HYPERCHOLESTEROLAEMIA.

hypermetamorphic (*BioSci*) Of insects, passing through two or more sharply distinct larval instars. N *hypermetamorphosis*.

hypermetropia (*Med*) An abnormal condition of the eyes in which parallel rays of light come to a focus behind the retina instead of on it, the eyes being at rest. Also *hyperopia*, *long-sightedness*.

hypermnesia (*Med*) Exceptional power of memory.

hypernephroma (*Med*) A tumour occurring in the kidney, resembling adrenal tissue but now known to be a carcinoma of the renal tubular cells.

hypernova (*Astron*) A star that collapses to form a black hole, emitting very large quantities of light and gamma radiation.

hyperon (*Phys*) Elementary particles with masses greater than that of a neutron and less that of a deuteron and having a lifetime of the order of 10^{-10} s.

hyperopia (*Med*) See HYPERMETROPIA.

hyperosmotic (*BioSci*) See HYPERTONIC.

hyperparasite (*BioSci*) An organism parasitic on a parasite.

hyperparasitism (*BioSci*) The condition of being parasitic on a parasite. N *hyperparasite*.

hyperphalangy (*BioSci*) The condition of having more than the normal number of phalanges, as in whales.

hyperpharyngeal (*BioSci*) Above the pharynx, as the *hyperpharyngeal band* in TUNICATA.

hyperpituitarism (*Med*) Overactivity of the pituitary gland; any condition due to overactivity of the pituitary gland, eg acromegaly, gigantism.

hyperplasia (*BioSci*) Abnormal, usually pathological, enlargement of an organ or tissue by an increase in the numbers of cells as a result of proliferation. Adj *hyperplastic*. Cf HYPERTROPHY, HYPOPLASIA.

hyperploid (*BioSci*) Having a chromosome number slightly exceeding an exact multiple of the haploid number.

hyperpnoea (*Med*) Increase in the depth and frequency of respiration; over-ventilation of the lungs. US *hyperpnea*.

hyperpolarization (*BioSci*) A shift towards greater negativity, and thus greater polarization, of a cell's RESTING POTENTIAL. Cf DEPOLARIZATION.

hyperpyrexia (*Med*) A degree of body temperature greatly above normal (eg > 41°C or 105°F). See HEAT STROKE.

hypersensitivity (*BioSci*) A condition in which a pathogen kills its host cells so quickly that the spread of infection is prevented.

hypersensitivity reactions (*BioSci*) Inappropriate immune responses that actually cause tissue damage. This type of response occurs on second or subsequent exposure to the triggering antigen and includes the various forms of ALLERGY. There are four basic types: type 1 is IgE-mediated atopy or anaphylaxis, type 2 is antibody-mediated cytotoxicity, type 3 involves immune-complex deposition, and type 4 is cell-mediated delayed-type hypersensitivity.

hypersensitization (*ImageTech*) Any method by which the effective speed of a photographic emulsion can be increased before exposure, eg by chemical solutions or vapours or by pre-exposure uniformly to very low light levels.

hypersonic (*Aero, Space*) Velocities of MACH NUMBER 5 or more. *Hypersonic flow* is the behaviour of a fluid at such speeds, eg in a shock tube. See panel on AERODYNAMICS.

hypersound (*Acous*) Sound of a frequency over 10^9 Hz, far above the audible frequency range.

hypersthene (*Min*) An important rock-forming silicate of magnesium and iron, $(Mg, Fe)SiO_3$, crystallizing in the orthorhombic system; an essential constituent of norite, hypersthene-pyroxenite, hypersthenite, hypersthene-andesite and charnockite. Strictly, an ortho-pyroxene containing 50–70% of the enstatite molecule.

hypersthenic (*Med*) Having increased strength or tonicity as in *hypersthenic gastric diathesis*, the constitutional disposition in which the stomach is short and overactive in secretion and movement.

hypersthenite (*Geol*) A coarse-grained igneous rock, consisting essentially of only one component, hypersthene, together with small quantities of accessory minerals.

hyperstomatal (*BioSci*) Having stomata only on the upper surface. Cf AMPHISTOMATAL, HYPOSTOMATAL.

hypertelorism (*Med*) The condition of excessive width between two organs or parts, particularly related to the eyes.

hypertely (*BioSci*) The progressive attainment of disproportionate size, either by a part or by an individual.

hypertension (*Med*) Increase in blood pressure above the normal which, if prolonged, can lead to heart or renal failure, strokes and myocardial infarction. The majority of cases have no clear cause but a minority may be due to endocrine or renal causes.

hypertext (*ICT*) A method of storing text that allows users to construct their own links between topics and between chosen parts of the text. This may be achieved by means of browsing or searching for specific words or terms – the system will store the route taken. Preprogrammed links between topics may also be used.

hypertext markup language (*ICT*) A simplified set of text formatting instructions based on the STANDARD GENERALIZED MARKUP LANGUAGE (SGML) that allows HYPERTEXT links, and is used for designing pages on the WORLD WIDE WEB. Abbrev *HTML*. See panel on INTERNET.

hypertext transfer protocol (*ICT*) The system by which HYPERTEXT links in documents are accessed over the Internet. Constitutes the first part of an Internet address. Abbrev *http*. See panel on INTERNET.

hyperthermophile (*BioSci*) Members of the Archaea that live and thrive in temperatures above 60°C, sometimes above 100°C (cf. *thermophiles* that have a tolerance ceiling of about 60°C).

hyperthyroidism (*Med*) Abnormally high rate of secretion of thyroid hormone by the thyroid gland, resulting in an increase in the basal metabolic rate, with symptoms of tiredness, anxiety, heat intolerance, palpitations and weight loss.

hypertonic (*BioSci*) Having a higher osmotic pressure than a standard, eg that of blood, or of the sap of cells which are being tested for their osmotic properties. Also *hyperosmotic*.

hypertonic solution (*BioSci*) A solution with a water potential lower than that of a cell suspended in it; water passes into the cell until the water potentials are equalized. See ISO-OSMOTIC SOLUTION.

hypertonus (*Med*) A state of excessive muscular tone; the opposite of *hypotonus*.

hypertrichiasis (*Med*) Abnormal overgrowth of hair; excessive hairiness. Also *hypertrichosis*.

hypertrophic obstructive cardiomyopathy (*Med*) A relatively common disease of the heart where there is abnormal thickening of the muscle of the left ventricle below the aortic valve, often causing obstruction to ejection of blood during SYSTOLE.

hypertrophic pyloric stenosis (*Med*) A disorder in children in which there is hypertrophy of the muscle in the pyloric region of the stomach, leading to obstruction to the passage of food into the small intestine, and vomiting.

hypertrophy (*BioSci*) Abnormal, usually pathological, enlargement of an organ or cell. Cf HYPERPLASIA.

hypervariable region (*BioSci*) (1) Any region of a DNA or protein sequence that shows very high level of variation between individuals or species. (2) More specifically, those regions of the heavy or light chains of immunoglobulins in which there is considerable sequence diversity and that confer the antigen-binding specificity of the antibody.

hypervitaminosis (*Med*) The condition arising when too much of any vitamin (esp vitamin D) has been taken.

hypha (*BioSci*) (1) The mycelium of a fungus which is a branching, filamentous structure with apical growth; the tubular cytoplasm contains the nuclei and may be divided by septa. (2) Elongated tubular cell in the thallus of some algae.

hyphopodium (*BioSci*) A more or less lobed outgrowth from a hypha, often serving to attach an epiphytic fungus to a leaf.

hypidiomorphic (*Geol*) A term referring to the texture of igneous rocks in which some of the component minerals show crystal faces, the others occurring in irregular grains. Also *subhedral*. Cf IDIOMORPHIC.

hypinosis (*Med*) See HYPERINOSIS.

hypnagogic images (*Psych*) Hallucination-like images seen during the transitional state of consciousness while falling asleep.

hypnagogic state (*Psych*) The transitional state of consciousness while falling asleep.

hypnophrenosis (*Med*) Any type of disturbance of sleep.

hypnosis (*Psych*) A temporary, trance-like state, induced by certain verbal or non-verbal procedures (hypnotism), characterized by heightened suggestibility both during and after hypnosis. See POST-HYPNOTIC SUGGESTION.

hypnospore (*BioSci*) A thick-walled, non-motile, resting spore. See APLANOSPORE.

hypnotherapy (*Psych*) Any psychotherapeutic method that involves the use of hypnosis.

hypnotic (*Med*) Of the nature of, or pertaining to, hypnosis; a drug which induces sleep.

hypo (*Chem, ImageTech*) Collog abbrev for *sodium thiosulphate*, the normal fixing solution for silver halide emulsions, the unreduced silver being removed by the hypo, which forms complexes with Ag^+.

hypo- (*Genrl*) Prefix from Gk *hypo*, under.

hypoacidity (*Med*) A deficiency of acid, esp in the gastric juice.

hypoadrenalism (*Med*) The condition in which the activity of the adrenal glands is below normal.

hypoblast (*BioSci*) The innermost germinal layer in the embryo of a metazoan animal, giving rise to the endoderm and sometimes also to the mesoderm. Cf EPIBLAST.

hypobranchial (*BioSci*) The lowermost element of a branchial arch.

hypobranchial space (*BioSci*) The space below the gills in DECAPODA.

hypocalcaemia (*Med*) A calcium content of the blood below normal limits. US *hypocalcemia*

hypocaust (*Build*) A hollow space beneath the floor of a room or bath, serving as a flue for the hot gases from a furnace which, in circulating, give warmth to the room or bath.

hypocercal (*BioSci*) Said of a type of caudal fin, found in Anaspida and Pteraspida, in which the vertebral column bends downwards and enters the hypochordal lobe which is larger than the epichordal lobe.

hypochlorhydria (*Med*) Diminished secretion of hydrochloric acid by the acid-secreting cells of the mucous membrane of the stomach.

hypochlorites (*Chem*) See HYPOCHLOROUS ACID.

hypochlorous acid (*Chem*) HClO. An aqueous solution of chlorine (I) oxide. Monobasic acid which forms salts called *hypochlorites (chlorates* (I)). Weak acid, easily decomposed.

hypochlorous anhydride (*Chem*) See CHLORINE OXIDES.

hypochondria (*Psych*) A disorder characterized by excessive preoccupation with bodily functions and sensations with the false belief that the latter indicate bodily disease; associated with a number of psychiatric syndromes. Also *hypochondriasis*.

hypocone (*BioSci*) A fourth cusp arising on the cingulum on the postero-internal side of an upper molar tooth, producing a quadritubercular pattern.

hypocotyl (*BioSci*) The part of the axis of the embryo between the radicle and the cotyledon(s), and the region of the seedling that derives from it.

hypocretin (*BioSci*) Synonym for OREXIN.

hypocycloid (*MathSci*) See ROULETTE.

hypoderm (*BioSci*) See HYPODERMIS.

hypodermic (*Med*) (1) Under the skin. (2) A medical agent injected under the skin.

hypodermis (*BioSci*) (1) In plants, a layer, one or more cells thick, immediately below the epidermis and differing morphologically from the underlying ground tissue. (2) In Arthropoda and other invertebrates with a distinct cuticle, the epithelial cell layer underlying the cuticle, by which the cuticle is secreted. Also *hypoderm*. Adj *hypodermal*.

hypodermoclysis (*Med*) The injection of fluid (eg salt solution) under the skin.

hypo eliminator (*ImageTech*) The solution used for removing all trace of fixative after washing prints, thus ensuring permanence. Usually hydrogen peroxide and ammonia.

hypoeutectoid steel (*Eng*) Steel with less carbon than is contained in PEARLITE, ie the iron–cementite eutectoid. In plain carbon steels, one containing less than 0·8% carbon. See panel on STEELS.

hypogaeous (*BioSci*) See HYPOGEAL.

hypogammaglobulinaemia (*Med*) A condition in which the concentration of immunoglobulins in the blood is much lower than normal. In the infantile sex-linked form there is a maturation defect in B-cells. A late acquired form has various causes, one of which is excessive activity of suppressor T-cells. There is greatly increased susceptibility to bacterial, but not to viral infections. US *hypogammaglobulinemia*.

hypogastrium (*Med*) The lower median part of the abdomen. Adj *hypogastric*. Cf EPIGASTRIUM.

hypogeal (*BioSci*) (1) Living beneath the surface of the ground. (2) Germinating with the cotyledons remaining in the soil. Also *hypogaeous*.

hypogene (*Geol*) Said of rocks formed, or agencies at work, under the Earth's surface.

hypoglossal (*BioSci*) (1) Underneath the tongue. (2) The twelfth cranial nerve of vertebrates, running to the muscles of the tongue.

hypoglottis (*BioSci*) In vertebrates, the underpart of the tongue; in beetles, part of the labium.

hypoglycaemia (*Med*) A reduction in the level of sugar (glucose), in the blood. US *hypoglycemia*.

hypognathous (*BioSci*) Having the under jaw protruding beyond the upper jaw; having the mouthparts directed downwards.

hypogonadism (*Med*) The condition in which there is a deficiency of the internal secretion of the gonads.

hypogynous (*BioSci*) A flower having the stamens inserted on the convex receptacles close beside or beneath the base of the ovary, eg buttercup (*Ranunculus*). Cf EPIGYNOUS, PERIGYNOUS. See SUPERIOR.

hypohidrosis (*Med*) Abnormal diminution in the secretion of sweat.

hypohyal (*BioSci*) The lowermost element of a hyoid arch.

hypoid bevel gear (*Eng*) A bevel gear in which the axes of the driving and driven shafts are at right angles but not in the same plane, resulting in some sliding action between the teeth; used in the back-axle drive of some automobiles.

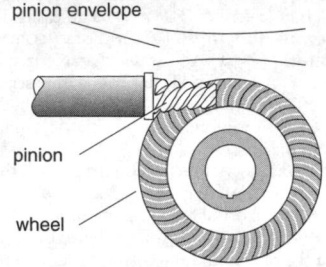

hypoid bevel gear Note hyperbolic shape of the pinion envelope.

hypolimnion (*EnvSci*) The cold lower layer of water in a lake. Cf EPILIMNION.

hypomania (*Med*) A condition characterized by mental excitement in the absence of mental confusion or of symptoms of insanity. Also *Simple mania*.

hypomenorrhoea (*Med*) The condition in which the interval between two menstrual periods is increased to between 35 and 42 days. US *hypomenorrhea*.

hyponasty (*BioSci*) The greater growth of the lower side which causes the upward bending of an organ. Cf EPINASTY. See NASTIC MOVEMENT.

hyponome (*BioSci*) In CEPHALOPODA, the funnel by which water escapes from the mantle cavity.

hypo-osmotic (*BioSci*) See HYPOTONIC.

hypopharyngeal (*BioSci*) Below the pharynx, as the *hypopharyngeal groove* of CEPHALOCHORDATA.

hypophloedal (*BioSci*) Growing just within the surface of bark.

hypophosphoric acid (*Chem*) H_2PO_3, or $H_4P_2O_6$. Obtained by the slow oxidation of phosphorus in moist air. Stable at ordinary temperatures. Hydrolysed by mineral acids, forming a mixture of phosphoric and phosphorous acids.

hypophosphorous acid (*Chem*) H_3PO_2. Feeble monobasic acid, which forms a series of salts called *hypophosphites* (phosphates (I)), oxidized to phosphates by oxidizing agents.

hypophysectomy (*Med*) Removal of the pituitary gland.

hypophysis (*BioSci*) A downwardly growing structure. (1) In Cephalochordata, the olfactory pit. (2) In vertebrates, the pituitary body. (3) In plants, the suspensor cell closest to the embryo. Adj *hypophysial*.

hypopituitarism (*Med*) A general term for any condition caused by diminished activity of the pituitary gland; characterized usually by obesity and imperfect sexual development.

hypoplasia (*BioSci*) Abnormal, usually pathological, under-development of a tissue because the cells are smaller or fewer. Adj *hypoplastic*. Cf HYPERPLASIA, HYPERTROPHY.

hypoploid (*BioSci*) Having a chromosome number a little less than some exact multiple of the haploid number.

hypopteronosis cystica (*Vet*) An inherited disease of budgerigars and canaries characterized by the formation of dermal cysts containing immature feathers.

hypopyon (*Med*) A collection of pus in the anterior chamber of the eye, between the iris and the cornea.

hyposensitization (*BioSci*) The administration of a graded series of doses of an allergen to atopic subjects suffering from immediate-type hypersensitivity to it. The aim is to reduce the severity of response to the allergen.

hypospadias (*Med*) A congenital deficiency in the floor of the urethra.

hypostasis (*Med*) Sediment or deposit. Passive hyperaemia in a dependent part owing to sluggishness of the circulation.

hypostatic (*BioSci*) The opposite of *epistatic*; analogous to *recessive* applied to genes at different loci.

hypostoma (*BioSci*) (1) In some Cnidaria, the raised oral cone. (2) In insects, the labrum. (3) In Crustacea, the lower lip or fold forming the posterior margin of the mouth. (4) In some Acarina, the lower lip formed by the fusion of the pedipalpal coxae. Also *hypostome*.

hypostomatal (*BioSci*) Leaf, etc, with stomata only on the lower surface. Cf AMPHISTOMATAL, EPISTOMATAL.

hypostomatous (*BioSci*) Having the mouth placed on the lower side of the head, as sharks.

hypostome (*BioSci*) See HYPOSTOMA.

hypostyle hall (*Arch*) A hall having columns to support the roof.

hyposulphuric acid (*Chem*) Dithionic acid. $H_2S_2O_6$.

hyposulphurous acid (*Chem*) Dithionous acid. $H_2S_2O_4$.

hypotarsus (*BioSci*) In birds, the FIBULARE of tetrapods.

hypotension (*Med*) Low blood pressure.

hypotenusal allowance (*Surv*) The distance added to each chain length, when chaining along sloping ground, in order to give a length whose horizontal projection shall be exactly 1 chain. For the 100-link chain the hypotenusal allowance is $100(\sec\theta - 1)$, where θ is the angle of slope of the ground from the horizontal.

hypotenuse (*MathSci*) The side opposite the right angle of a right-angled triangle.

hypothalamus (*BioSci*) In the vertebrate brain, the ventral zone of the diencephalon. In humans, the part of the brain that makes up the floor and part of the lateral walls of the third ventricle. The mammillary bodies, tuber cinerium, infundibulum, neurohypophysis and the optic chiasma are also part of the hypothalamus. The hypothalamus controls the autonomic nervous system and therefore maintains the body's homeostasis (controls body temperature, metabolism and appetite). It also translates extreme emotions into physical responses. See panel on BRAIN STRUCTURE.

hypothermia (*Med*) Subnormal body temperature; occurs in the very young, the aged or in coma; used therapeutically for heart and other surgery.

hypothesis (*Genrl*) A prediction based on theory, an educated guess derived from various assumptions, which can be tested using a range of methods, but is most often associated with experimental procedure; a proposition put forward for proof or discussion.

hypothesis (*MathSci*) Any unproven assertion or theory from which consequences are derived, or for which a proof is sought.

hypothesis testing (*MathSci*) A procedure for testing the compatibility of a set of data with a particular hypothesis, known as the status quo or working or null hypothesis, regarding the probability model which generated the data, in particular when there is a clearly specified alternative hypothesis.

hypothetical exchange (*ICT*) A telephone exchange that, until a new exchange is constructed, is made up from parts of existing exchanges, the subscribers being numbered according to the system of the new exchange.

hypothyroidism (*Med*) The condition accompanying the diminished secretion of the thyroid gland. See CRETINISM, MYXOEDEMA.

hypotonic (*BioSci*) Having a lower osmotic pressure than a standard, eg that of blood, or of the sap of cells which are being tested for their osmotic properties. Also *hypoosmotic*.

hypotonic solution (*BioSci*) A solution with a water potential greater than that of a cell suspended in it; water will flow out of the cell until the water potentials are equalized. See ISO-OSMOTIC SOLUTION.

hypotonus (*Med*) See HYPERTONUS.

hypotrichous (*BioSci*) Having cilia, principally on the lower surface of the body.

hypotrochoid (*MathSci*) See ROULETTE.

hypovitaminosis (*Med*) The condition resulting from deficiency of a vitamin in the diet.

hypovolaemia (*Med*) An abnormal reduction in the volume of blood or blood plasma. Also (US) *hypovolemia*. Adj *hypovolaemic* or (US) *hypovolemic*.

hypoxanthine (*Chem*) 6-*hydroxypurine*. Formed by the breakdown of nucleoproteins by enzymatic action.

hypoxia (*Med*) Lack of oxygen supply.

hypso- (*Genrl*) Prefix from Gk *hypsos*, height.

hypsochrome (*Chem*) A radical which shifts the absorption spectrum of a compound towards the violet end of the spectrum. A *bathochrome* shifts it the other way.

hypsochromic (*Chem*) Changing to a shorter wavelength (blueshift) in the absorption spectrum of a compound; the opposite of bathochromic.

hypsodont (*BioSci*) Of a mammalian tooth with a high crown and deep socket. Cf BRACHYODONT.

hypsoflore (*Chem*) A radical which tends to shift the fluorescent spectrum of a compound towards shorter wavelengths. A BATHOFLARE shifts it the other way.

hypsometer (*Phys*) An instrument used for determining the boiling point of water, with a view to ascertaining altitude, by calculating the pressure, or for correcting the upper fixed point of the thermometer used.

hypsophyll (*BioSci*) A non-foliage leaf inserted high on a shoot, eg a floral bract. Cf CATAPHYLL.

Hyracoidea (*BioSci*) An order of small terrestrial eutherian mammals that have four digits on the fore-limb and three on the hind-limb. They have pointed incisor teeth with persistent pulps, and lophodont grinding teeth. The males have no scrotal sac, the females have six mammae. Confined to Africa. Dassies, hyraxes.

hyster-, hystero- (*Genrl*) Prefixes from Gk *hysteria*, womb, or *hysteros*, later.

hysteranthous (*BioSci*) Said of leaves that develop after the plant has flowered.

hysterectomy (*Med*) Surgical removal of the uterus.

hysteresis (*Phys*) The retardation or lagging of an effect behind the cause of the effect, eg DIELECTRIC HYSTERESIS or magnetic hysteresis. A property commonly seen in the stress–strain curves of polymeric materials where unloading gives a curve which does not coincide with the original. See DAMPING, HYSTERESIS LOOP.

hysteresis coefficient (*ElecEng*) See STEINMETZ COEFFICIENT.

hysteresis error (*ElecEng*) Instruments or control systems may show non-reversibility similar to hysteresis. The maximum difference between the readings or settings obtainable for a given value of the independent variable is the *hysteresis error*.

hysteresis heat (*ElecEng*) That arising from HYSTERESIS LOSS, in contrast with that from ohmic loss associated with eddy currents.

hysteresis loop (*ElecEng*) A closed figure formed by plotting magnetic flux density B in a magnetizable material against magnetizing field H when the latter is taken through a complete cycle of increasing and decreasing values. The area of this loop measures the energy dissipated during a cycle of magnetization. Also *B/H loop*. Similar effects occur with applied mechanical or electrical stresses, and this is by analogy known as *mechanical* or *electrical hysteresis*. Dielectric materials with appreciable hysteresis loss are termed *ferroelectric materials*.

hysteresis loss (*ElecEng*) Energy loss in taking unit quantity of material once round a hysteresis loop. It can arise in a polymeric material subjected to a varying stress, in a dielectric material subjected to a varying electric field or in a magnetic material in a varying magnetic field. See DIELECTRIC, MAGNETIC MATERIALS.

hysteresis motor (*ElecEng*) A synchronous motor which starts by reason of the hysteresis losses induced in its steel secondary by the revolving field of the primary.

hysteresis tester (*ElecEng*) A device, invented by J A Ewing, for making a direct measurement of magnetic hysteresis in samples of iron or steel.

hysteria (*Psych, Med*) A state characterized by hallucinations, functional anaesthesia, paralysis and dissociation: no longer considered to be a genuine syndrome.

hystero- (*Genrl*) See HYSTER-.

hysterocolpectomy (*Med*) Removal of the vagina (or part of it) and the uterus.

hysteropexy (*Med*) The fixation of a displaced uterus surgically.

hysterotomy (*Med*) Incision of the uterus.

Hytrel (*Eng*) TN for polyester rubber, a kind of thermoplastic elastomer (US).

Hz (*Phys*) Symbol for HERTZ, the SI unit of frequency.

I

I (*Chem*) Symbol for IODINE.

I (*Chem*) Symbol for IONIC STRENGTH.

I (*Eng*) Symbol for SECOND MOMENT OF AREA.

I (*Phys*) Symbol for: (1) electric current; (2) LUMINOUS INTENSITY.

I$_{sp}$ (*Aero*) See SPECIFIC IMPULSE.

i (*Chem*) Symbol for VAN'T HOFF'S FACTOR.

i (*MathSci*) The imaginary number whose square equals minus one, ie $i^2 = -1$. See *j*.

i- (*Chem*) Abbrev for: (1) OPTICALLY INACTIVE; (2) *iso*-, ie containing a branched hydrocarbon chain.

IÅ (*Phys*) Abbrev for INTERNATIONAL ANGSTROM.

IAA (*BioSci*) Abbrev for INDOLE-3-ACETIC ACID.

Ia antigens (*BioSci*) Class II histocompatibility antigens with functions similar to human HLA-D ANTIGENS on mouse B-cells, macrophages and accessory cells.

IACS (*Eng*) Abbrev for *International Annealed Copper Standard*, relating the electrical conductivity of a metal or alloy to that of copper in percentage terms.

IAEA (*NucEng*) Abbrev for *International Atomic Energy Agency*. Autonomous intergovernmental body concerned with the promotion of nuclear energy and to ensure as far as possible that it is not used to further military objectives.

IAP (*ICT*) Abbrev for INTERNET ACCESS PROVIDER.

Iapetus (*Astron*) The eighth natural satellite of Saturn, discovered in 1671. Distance from the planet 3 560 000 km; diameter 1460 km.

Iapetus Ocean (*Geol*) An ocean which is thought to have existed from Late Precambrian to Lower Palaeozoic times in the general position of the present Atlantic Ocean. The *Iapetus suture* along which the ocean opened, separating northern and southern faunal provinces, and eventually closed, is thought to have traversed Ireland and the Solway Firth. The name is from *Iapetus*, father of *Atlas*, after whom the Atlantic Ocean was named.

IAR (*ICT*) See PROGRAM COUNTER.

IAS (*Aero*) Abbrev for INDICATED AIR SPEED.

IAS (*Space*) Abbrev for IMAGE ANALYSIS SYSTEM.

IATA (*Aero*) Abbrev for INTERNATIONAL AIR TRANSPORT ASSOCIATION.

iatrochemistry (*Pharmacol*) The study of chemical phenomena in order to obtain results of medical value; practised in the 16th century. Modern equivalents: CHEMOTHERAPY, PHARMACOLOGY.

iatrogenic disease (*Med*) A disease produced by a doctor, usually occurring as a side effect of pharmacological agents.

IBA (*BioSci*) Abbrev for INDOLE-3-BUTYRIC ACID.

I-beam (*Build, CivEng*) See BEAM.

IBM-compatible (*ICT*) A personal computer which operates to the same specification as that developed from the original PC specified by IBM although the internal ARCHITECTURE of the machine may be different. See EISA, ISA, MCA.

ibuprofen (*Pharmacol*) A non-steroidal anti-inflammatory drug with fewer gastric side effects than many of this class of drugs.

IC (*ICT*) Abbrev for INTEGRATED CIRCUIT.

ICAM (*BioSci*) Abbrev for *intercellular adhesion molecule*, found on the luminal surface of endothelial cells, the ligand for *integrins* on circulating leucocytes. It is upregulated in response to inflammatory cytokines and is also the site to which rhinovirus binds and to which erythrocytes infected with *Plasmodium falciparum* adhere.

ICAO (*Aero*) Abbrev for INTERNATIONAL CIVIL AVIATION ORGANIZATION.

ICE (*ICT*) Abbrev for INTRUSION COUNTERMEASURE ELECTRONICS.

ice (*EnvSci*) Ice is formed when water is cooled below its freezing point. It is a transparent crystalline solid of rel.d. 0·916 and specific heat capacity 0·50. Because water attains its maximum density at 4°C, ice is formed on the surface of ponds and lakes during frosts, and thickens downwards.

ice action (*Geol*) The work and effects of ice on the Earth's surface. See GLACIAL EROSION, GLACIATION, GLACIER.

ice age (*Geol*) A period when glacial ice spread over regions which were normally ice-free. *The Ice Age* is a synonym of the Pleistocene epoch.

ice beer (*FoodSci*) Beer brewed using a process that freezes the beer and then removes some of the ice, so increasing the alcoholic content.

iceberg (*EnvSci*) A large mass of ice, floating in the sea, which has broken away from a glacier or ice barrier. Icebergs are carried by ocean currents for great distances, often reaching latitudes of 40°–50° before having completely melted. Approximately one-tenth of an iceberg shows above the surface.

iceblink (*EnvSci*) A whitish glare in the sky over ice which is too distant to be visible.

ice-breaker (*CivEng*) (1) Protection on the upstream side of a bridge pier. (2) A projecting pier so arranged in relation to a harbour entrance that floating ice is kept outside. (3) A vessel specially equipped for clearing a passage through ice-bound waters.

ice colours (*Chem*) Dyestuffs produced on the cotton fibre direct, by the interaction of a second component with a solution of a diazo-salt cooled with ice.

ice contact slope (*Geol*) The steep slope of material originally deposited at an ice front and in contact with it.

ice guard (*Aero*) A wire-mesh screen fitted to a piston aero-engine intake so that ice will form on it and not inside the intake; a *gapped ice guard* is mounted ahead of the intake so that air can pass round it, while a *gapless ice guard* is inside the intake and an alternative air path comes into use when it ices up.

Iceland agate (*Min*) A name quite erroneously applied to the natural glass OBSIDIAN.

Iceland spar (*Min*) A very pure transparent and crystalline form of calcite, first brought from Iceland. It has perfect cleavage, is noted for its double refraction, and hence is used in construction of the Nicol prism.

IC engine (*Autos*) Abbrev for INTERNAL-COMBUSTION ENGINE.

I-centred lattice (*Crystal*) Body-centred crystal lattice. See fig. at UNIT CELL.

ice pellets (*EnvSci*) Precipitation of transparent or translucent pellets of ice with diameters of 5 mm or less.

ice shelf (*EnvSci*) A floating sheet of ice permanently attached to a land mass and projecting into the sea.

ice storm (*EnvSci*) A storm of freezing rain.

ichor (*Geol*) The name applied by Sederholm to highly penetrating granitic liquids, charged with magmatic vapours (emanations), which he believed to operate in palingenesis.

ichor (*Med*) A thin, watery discharge from a wound or a sore. Adj *ichorous*.

ichthy-, ichthyo- (*Genrl*) Prefixes from Gk *ichthys*, fish.

ichthyic (*BioSci*) Pertaining to, or resembling, fish.

ichthyo- (*Genrl*) See ICHTHY-.

ichthyopterygium (*BioSci*) A paddle-like fin, or limb, used for swimming, eg pectoral or pelvic fin of fish.

ichthyosis (*Med*) A disease characterized by dryness and roughness of the skin, resembling fish scales, due to lack of secretion of the sweat and the sebaceous glands. Also *xeroderma, xerodermia*.

ichthyosis (*Vet*) Hardening of the skin which develops cracks which become filled with dirt and thereby suppurate. Seen congenitally in calves, also over the elbows and hocks of dogs.

ICM (*Agri*) See INTEGRATED CROP MANAGEMENT.

icon (*ICT*) A small on-screen symbol used to represent a function of a program etc. By moving a pointer to the icon (often by using a MOUSE) the user selects the function by 'clicking' on it, ie by pressing a button on the mouse or on the keyboard. See fig. at WINDOWS. See WIMP.

iconic memory (*Psych*) A form of sensory memory; a transient visual trace that fades rapidly after removal of the stimulus. Cf ECHOIC MEMORY.

icosahedron (*MathSci*) A 20-faced POLYHEDRON. The faces of a *regular icosahedron* are identical regular equilateral triangles.

icositetrahedron (*Min*) A solid figure having 24 trapezoidal faces, and belonging to the cubic system. Exemplified by some garnets.

ICRH (*NucEng*) Abbrev for *ion cyclotron resonance heating*. See CYCLOTRON RESONANCE HEATING.

ICSH (*Med*) Abbrev for *interstitial cell stimulating hormone*. See LUTEINIZING HORMONE.

ICSI (*BioSci*) Abbrev for *intracytoplasmic sperm injection*, a method of *in vitro* fertilization.

ICT (*Genrl*) Abbrev for *information and communications technology*, used in this dictionary to cover computing and telecommunications matters. Sometimes *information and computing technology*.

icterus (*Med*) Adj *icteric*. See JAUNDICE.

ictus (*Med*) A stroke or sudden attack.

ICW (*ICT*) Abbrev for INTERRUPTED CONTINUOUS WAVES.

id (*Psych*) In the Freudian model, that part of the personality which contains primitive animalistic impulses such as sex, anger and hunger. The irrational, demanding part of the personality that operates according to the pleasure principle.

iddingsite (*Min*) An alteration product of olivine consisting of goethite, quartz, montmorillonite group clay materials, and chlorite.

IDE (*ICT*) Abbrev for INTEGRATED DRIVE ELECTRONICS and INTEGRATED DEVELOPMENT ENVIRONMENT.

ideal (*MathSci*) A subset of a RING which is a SUBGROUP with respect to addition and which contains all products of its elements with any element of the ring.

ideal crystal (*Crystal*) One in which there are no imperfections or alien atoms.

ideal gas (*Chem*) A gas with molecules of negligible size and exerting no intermolecular forces. Such a gas is a theoretical abstraction which would obey the ideal gas law under all conditions:

$$pV = nRT$$

where p = pressure, V = volume, n = number of moles, R = GAS CONSTANT and T = absolute temperature. The behaviour of real gases becomes increasingly close to that of an ideal gas as their pressure is reduced. Also *perfect gas*.

ideal self (*Psych*) Humanistic term representing the characteristics, behaviours, emotions and thoughts to which a person aspires.

ideal transducer (*ElecEng*) Any transducer which converts without loss all the power supplied to it.

ideal transformer (*ElecEng*) A hypothetical transformer corresponding to one with a coefficient of coupling of unity.

ideas of reference (*Psych*) A characteristic of some mental disorders, notably schizophrenia, in which the individual perceives irrelevant and independent environmental and social events as relating to him- or herself ('people are looking at me').

idempotent (*MathSci*) An element, *e*, of a set on which an operation * is defined, such that $e^*e = e$. For the usual addition of real numbers, zero is the only idempotent; for the usual multiplication of real numbers, zero and one are the only idempotents. A GROUP has one and only one idempotent, namely its identity element. There is no restriction on the number of idempotents to be found in a semi-group.

identification (*Psych*) In psychoanalytic theory, the way in which an individual incorporates (introjects) the values, standards, sexual orientation and mannerisms of the same-sex parent, as part of the development of the SUPEREGO. It can also be used to describe the influence of any relevant and powerful figure for the internalization of external norms.

identification dimensions (*Ships*) See REGISTERED DIMENSIONS.

identifier (*ICT*) Name or label chosen by the programmer.

identity (*MathSci*) (1) See EQUATION. (2) An element, *e*, of a set on which an operation * is defined, such that, for all elements *x* in the set S we have $x^*e = x = e^*x$, eg in the set of real numbers, zero is an identity element with respect to addition, and one is an identity element with respect to multiplication. Also *neutral element*.

identity mapping (*MathSci*) The identity mapping, or identity function, on a set S is the mapping i_s from S onto S defined by $i_s(x) = x$, for all elements *x* in S.

idio- (*Genrl*) Prefix from Gk *idios*, peculiar, distinct.

idioblast (*BioSci*) A cell of clearly different properties to the others in the tissue, as a stone cell in pear fruit.

idioblast (*Geol*) A crystal which grew in metamorphic rock and is bounded by its own crystal faces. Adj *idioblastic*. Cf IDIOMORPHIC. See PORPHYROBLASTIC.

idioglossia (*Med*) The wrong use of consonants by a child, making speech unintelligible.

idiogram (*BioSci*) A diagram (or photomontage) of the chromosome complement of a cell, conventionally arranged to show the general morphology including relative sizes, positions of centromeres, etc. Also *karyogram*.

idiographic (*Psych*) Used to describe any research, system or philosophy that focuses on the individual and does not attempt to formulate general laws. See NOMOTHETIC.

idiomorphic (*Geol*) A term used for igneous rock minerals which are bounded by the crystal faces peculiar to the species. Cf ALLOTRIOMORPHIC (anhedral), HYPIDIOMORPHIC (subhedral).

idiopathy (*Med*) Any morbid condition arising spontaneously, having no known origin. Adj *idiopathic*.

idiosyncrasy (*Med*) Individual hypersensitivity to a drug or food but not explained by altered immunity.

idiot (*Med*) A term no longer used but it formerly described a person so defective in mind from birth as to be unable to protect him- or/ herself against ordinary physical dangers. Now defined as 'mentally severely handicapped'.

idiothermous (*BioSci*) See WARM-BLOODED.

idiotope (*BioSci*) Antigenic determinant on immunoglobulin molecules characteristic of the product of a single clone or a small minority of clones, and associated with or part of the antigen binding site.

idiot savant (*Psych*) A child who, despite generally diminished skills, shows astonishing proficiency in one isolated skill ('foolish wise one').

idiot tape (*Print*) A continuous unjustified tape, containing only signals for new paragraphs, which must be processed into a new justified tape before it can control a typesetting or filmsetting machine. See COMPUTER TYPESETTING.

idiotype (*BioSci*) Set of one or more *idiotopes* by which a clone of immunoglobulin-forming cells can be distinguished from other clones. Some idiotypes appear to be unique to an individual animal; others are found in many members of the same animal species.

idioventricular (*Med*) Pertaining to the ventricle of the heart alone.

I-display (*Radar*) A display representing the target as a full circle when the antenna is pointed directly at it, the radius being in proportion to the range.

idler (*Eng*) See IDLE WHEEL.

idler (*Print*) A free-running roller on a web-fed press. Also *idling roller*.

idler pulley (*Eng*) See GUIDE PULLEY.

idle wheel (*Eng*) A wheel interposed in a gear train, to reverse the direction of rotation, modify the spacing of centres or reduce the size of driver and driven gears, without affecting the ratio of the drive. Also *idler*, *intermediate wheel*.

idle wire (*ElecEng*) The part of the armature winding of an electric machine which does not actually cut the lines of force, ie that part comprising the end connections.

idling (*Autos*) The slow rate of revolution of an automobile or aero-engine, when the throttle pedal or lever is in the closed position.

idling adjustment (*Autos*) A setting of the slow-running jet and throttle position of a carburettor, so as to give regular IDLING.

idling roller (*Print*) See IDLER.

idocrase (*Min*) See VESUVIANITE.

idose (*Chem*) A monosaccharide belonging to the group of ALDOHEXOSES.

idoxuridine (*Pharmacol*) Antiviral agent effective against herpes viruses. Used topically in the treatment of herpetic lesions.

IEC (*Electronics*) Abbrev for *International Electrotechnical Commission*. Main standards setting body at international level for electronic materials, devices, etc.

IEEE (*ICT*) Abbrev for *Institute of Electrical and Electronics Engineers*. A US society that amongst other things establishes international standards in the computing, electronic and telecommunications fields.

IEEE 802 (*ICT*) A set of standards established by the IEEE relating to NETWORKS. In particular, 802·2 is the data-link layer; 802·3 is a physical network layer using CSMA/CD PROTOCOLS; 802·4 is a physical layer using TOKEN- passing PROTOCOLS on a BUS TOPOLOGY LAN; 802·5 is a physical layer that uses a TOKEN-passing protocol over a RING TOPOLOGY LAN.

IF (*ICT*) Abbrev for INTERMEDIATE FREQUENCY.

IFF (*Radar*) Abbrev for World War II system whereby vessels and aircraft carried TRANSPONDERS capable of indicating to a 'friendly' radar system that they were not hostile. *IFF* is used now as a general description of such radar identification systems. See ATCRBS.

iff (*MathSci*) Contraction of 'if and only if'.

IFIP (*ICT*) Abbrev for INTERNATIONAL FEDERATION FOR INFORMATION PROCESSING.

IFR (*Aero*) Abbrev for INSTRUMENT FLIGHT RULES.

IFRB (*ICT*) Abbrev for INTERNATIONAL FREQUENCY REGISTRATION BOARD.

Ig (*BioSci*) General abbrev for IMMUNOGLOBULIN.

IgA (*BioSci*) The major immunoglobulin present in mucosal secretions, where it constitutes the main humoral defence mechanism, apparently by preventing microbial adhesion. Usually found as dimers or trimers.

IgD (*BioSci*) Immunoglobulin present only in very low concentrations in blood, but present at the surface of B-lymphocytes; probably an antigen receptor.

IgE (*BioSci*) Immunoglobulin the Fc region of which binds very strongly to a receptor on the surface of mast cells and basophils. Normally present in blood only in very low concentrations, but the concentration is increased in atopic subjects, and in infection by several helminth parasites.

IgG (*BioSci*) The major immunoglobulin in humans and most species from amphibians upwards (but not in fish). IgG fixes complement and crosses the placenta (ie can be passed from mother to fetus). In humans there are four varieties (subclasses) termed IgG1, IgG2, IgG3 and IgG4. Also *7S antibody*.

IgM (*BioSci*) High-molecular-weight immunoglobulin, consisting in humans of five basic units (of two light and two heavy chains) arranged as a pentamer and joined together by disulphide bonds and a small link peptide (J chain). IgM antibodies are the first to be synthesized and released after a primary antigenic stimulation, and are the predominant form made in response to many bacterial capsular polysaccharides. Also *19S antibody*.

igneous complex (*Geol*) A group of rocks, occurring within a comparatively small area, which differ in type but are related by similar chemical or mineralogical peculiarities. This indicates derivation from a common source.

igneous cycle (*Geol*) The sequence of events usually followed in igneous activity; it consists of an eruptive phase, a plutonic phase and a phase of minor intrusion. Also *magmatic cycle*.

igneous intrusion (*Geol*) A mass of igneous rock which crystallized before the magma reached the Earth's surface, including BATHOLITHS, BOSSES, DYKES, SILLS and STOCKS.

igneous rocks (*Geol*) Rock masses generally accepted as being formed by the solidification of magma injected into the Earth's crust, or extruded on its surface.

ignimbrite (*Geol*) A pyroclastic rock consisting originally of lava droplets and glass fragments which were so hot at the time of deposition that they were welded together. Also *welded tuffs*.

ignite (*Chem*) To heat a gaseous mixture to the temperature at which combustion occurs, eg by means of an electric spark.

igniter (*CivEng*) A blasting fuse or other contrivance used to fire an explosive charge.

igniter plug (*Aero*) An electrical-discharge unit for lighting up gas turbines.

ignition (*Autos, ElecEng*) The firing of an explosive mixture of gases, vapours or other substances by means of eg an electric spark.

ignition advance (*Autos*) The crank angle before top dead centre, at which the spark is timed to pass in a petrol or gas engine. See ANGLE OF ADVANCE, IGNITION TIMING.

ignition coil (*Autos*) An induction coil for converting the low-tension current supplied by the battery into the high-tension current required by the sparking plugs.

ignition lag (*Autos*) Of a combustible mixture in an engine cylinder, the time interval between the passage of the spark and the resulting pressure rise due to combustion.

ignition rating (*ElecEng*) A special rating, in *ampere-hours* employed for accumulators used for supplying ignition systems; it is generally twice the continuous rating at a low discharge rate.

ignition system (*Autos*) The arrangement for providing the high-tension voltage required for ignition. See ELECTRONIC IGNITION, IGNITION COIL, MAGNETO IGNITION.

ignition temperature (*Eng*) In the combustion of gases, the temperature at which the heat loss due to conduction, radiation, etc, is more than counterbalanced by the rate at which heat is developed by the combustion reaction.

ignition temperature (*NucEng*) In fusion, the point at which alpha-particle heating can sustain the fusion reaction.

ignition timing (*Autos*) The crank angle relative to top dead centre at which the spark passes in a petrol or gas engine. See ANGLE OF ADVANCE.

ignition voltage (*ElecEng*) That required to start discharge in a gas tube.

ignitor (*Electronics*) See PILOT ELECTRODE.

ignitor drop (*Electronics*) The voltage drop between cathode and anode of the ignitor discharge in a switching tube.

ignitron (*Electronics*) Mercury-arc rectifier with ignitor, which is an electrode which can dip into the cool mercury pool and draw an arc to start the ionization.

IGV (*Aero*) Abbrev for INLET GUIDE VANES.

IHP (*Eng*) Abbrev for INDICATED HORSEPOWER.

IIR (*Eng*) Abbrev for ISOPRENE–ISOBUTENE RUBBER.

ijolite (*Geol*) A coarse-grained igneous rock, consisting of nepheline, aegirine-augite, with usually melanite garnet as a prominent accessory, occurring in nepheline–syenite complexes in the Kola Peninsula, the Transvaal and elsewhere.

IKBS (*ICT*) Abbrev for *intelligent knowledge-based system*. See KNOWLEDGE-BASED SYSTEM.

IL-1, IL-2, etc (*BioSci*) See INTERLEUKIN.

Ile (*Chem*) Symbol for ISOLEUCINE.

ileitis (*Med*) Inflammation of the ileum.

ileocolitis (*Med*) Inflammation of the ileum and the colon.

ileocolostomy (*Med*) The making of a communication between the ileum and the colon by operation.

ileostomy (*Med*) An artificial opening in the ileum, made surgically.

ileum (*BioSci*) In vertebrates, the posterior part of the small intestine.

ileus (*Med*) Colic due to obstruction in the intestine; obstruction of the intestine. See PARALYTIC ILEUS.

Ilgner system (*ElecEng*) See WARD–LEONARD–ILGNER SYSTEM.

iliac region (*BioSci*) The dorsal region of the pelvic girdle in vertebrates.

iliac veins (*BioSci*) In fish, the paired veins from the pelvic fins, draining into the lateral veins.

ilio- (*Genrl*) Prefix referring to the ilium, used in the construction of compound terms, eg *iliofemoral*, pertaining to the ilium and the femur.

ilium (*BioSci*) A dorsal cartilage bone of the pelvic girdle in vertebrates. Adj *iliac*.

Ilkovic equation (*Chem*) In polarography, the equation expresses the current to the DROPPING-MERCURY ELECTRODE as $i = AnCD^{1/2}m^{2/3}\tau^{1/6}$, where i = diffusion current, A = numerical constant, n = ionic charge, D = diffusivity, m = rate of flow of mercury, τ = lifetime of a drop.

ill-conditioned (*Surv*) A term used in TRIANGULATION to describe triangles of such a shape that the distortion resulting from errors made in measurement and in plotting may be great, the criterion often used being that no angle in a triangle should be less than 30°.

illegitimate pollination (*BioSci*) The transfer of pollen in a way the floral structure appears to discourage, eg from the anther of one PIN flower to the stigma of another.

illegitimate recombination (*BioSci*) Recombination between species whose DNA shows little or no homology, facilitated by duplicate sequences casually present.

illite (*Min*) A monoclinic clay mineral, a hydrated silicate of potassium and aluminium. It is the dominant clay mineral in shales and mudstones. The illite group somewhat resembles muscovite.

illuminance (*Phys*) See ILLUMINATION.

illuminated diagram (*ElecEng*) A circuit diagram on a switchboard, or a track diagram in a railway signal box, so arranged that lamps behind the diagram illuminate any part of the circuit which is alive or any part of the track upon which a train is standing.

illumination (*Phys*) The quantity of light or luminous flux falling on unit area of a surface. Illumination is inversely proportional to the square of the distance of the surface from the source of light, and proportional to the cosine of the angle made by the normal to the surface with the direction of the light rays. The unit of illumination is the *lux*, which is an illumination of 1 lumen per square metre. Symbol E. Also *illuminance*.

illusion (*Psych*) A surprising perceptual experience, due to the fact that some aspect of the relation between the physical stimulus and the individual's perception of it violates normal expectation.

ilmenite (*Min*) An oxide of iron and titanium, crystallizing in the trigonal system; a widespread accessory mineral in igneous and metamorphic rocks, esp in those of basic composition. A common mineral in detrital sediments, often becoming concentrated in beach sand.

ILS (*Aero*) Abbrev for INSTRUMENT LANDING SYSTEM.

ILT (*Vet*) Abbrev for INFECTIOUS LARYNGOTRACHEITIS.

ilvaite (*Min*) A silicate of iron, calcium and manganese. It crystallizes in the orthorhombic system.

image (*Phys*) The figure of an object formed by reflected or refracted rays of light. A *real image* is one which is formed by the convergence of rays which have passed through the image-forming device (usually a lens) and can be thrown onto a screen, as in the camera and the optical projector. A *virtual image* is one from which rays appear to diverge. It cannot be projected onto a screen or a sensitive emulsion.

image (*Print*) The general term to describe the subject on negatives, positives and plates, at any stage of preparation.

image admittance (*Phys*) The reciprocal of IMAGE IMPEDANCE.

image analysis system (*Space*) A computer system, initially developed for planetary missions and now applied extensively in geology. A range of spatially related data (eg geological, geochemical or geophysical), including remotely sensed data, are analysed interactively. Abbrev *IAS*.

image charge (*Phys*) Hypothetical charge used in electrostatic theory as a substitute for a conducting (equipotential) surface. The charge must not modify the field distribution at any point outside this surface.

image converter tube (*Phys*) A tube in which an optical image applied to a photoemissive surface produces a corresponding image on a luminescent surface.

image curvature (*Phys*) In electron microscopes, image curvature due to aberrations. Provided curvature of the field is the only aberration present, a sharp image is formed on a curved surface tangential to the image plane on the axis.

image-dissection camera (*ImageTech*) High-speed camera having, in place of a normal lens, an array of parallel lightguides made of thin glass fibres embedded in an opaque matrix, which transmit light by internal reflection onto successive portions of a moving plate.

image enhancement (*ICT*) A method of enhancing the definition of an image using a computer program to convert shades of grey into either black or white.

image force (*Phys*) The force on an electric (or magnetic) charge between itself and its image induced in a neighbouring body.

image frequency (*ICT*) In a SUPERHET receiver, that which differs from the local oscillator frequency by an amount equal to the intermediate frequency, and is on the opposite side of the local oscillator frequency from that of the desired signal; a direct result of a mixer's ability to produce sum and difference frequencies. Unless the mixer is preceded by tuned stages to provide image rejection, images can cause serious interference with reception.

image impedance (*Phys*) The quantity Z_0 given by $Z_0^2 = Z_{sc}Z_{oc}$, where Z_{sc} is the short-circuit impedance and Z_{oc} the open-circuit impedance for a network. The network is image operated when the input generator impedance equals the network input impedance, and the load impedance equals the output impedance of the network. See ITERATIVE IMPEDANCE.

image intensifier (*ImageTech*) A device which amplifies light using a PHOTOCATHODE to convert photons into electrons, which are then accelerated by a high voltage or multiplied by a channel plate on their way to a fluorescent screen. This may be viewed directly, or passed to film or a pickup device.

image intensifier (*Radiol*) An electronic device screen for enhancing brightness of an image in fluoroscopy, at the same time reducing patient dose. See INTENSIFYING SCREEN. Fig. ▷

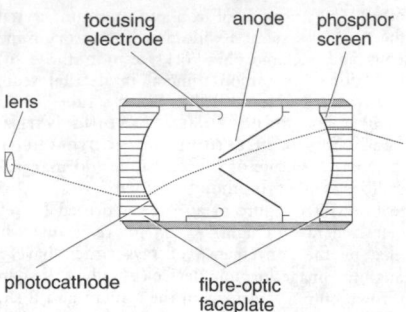

image intensifier Electrostatic type; can be cascaded.

image interference (*ICT*) That produced by any signal received with a frequency at or near the image frequency.

image phase constant (*ICT*) In a filter section, terminated in both directions, unreal or imaginary part of the (image) transfer constant; phase delay of the section in radians.

image processing (*ICT*) Techniques for filtering, storing and retrieving images.

imager (*ICT*) A device that records images.

image rejection ratio (*ICT*) The ratio of the image frequency signal input at the aerial to the desired signal input, in a superhet receiver, for identical outputs.

image response (*ICT*) Unwanted response of a superhet receiver to the image frequency.

imagery (*Psych*) Forming a mental picture of a sensory experience, although not necessarily a visual representation.

image stabilizer (*ImageTech*) A means of reducing the effects of camera shake, compensating for the movement either optically by shifting the image or electronically by passing the signal from a SOLID-STATE IMAGE SENSOR through a digital store from where it is retimed.

image transfer (*ImageTech*) A process in which the unexposed areas of a photographic emulsion are used to form an image in another layer; used in INSTANT PHOTOGRAPHY and document copying. See DIFFUSION TRANSFER REVERSAL.

image tube (*Electronics*) One in which an optically focused image on a photoemissive plate releases electrons which are focused on a phosphor by electric or magnetic means. Used in X-ray intensifiers, infrared telescopes, electron telescopes and microscopes.

imaginal bud (*BioSci*) One of a number of masses of formative cells which are the principal agents in the development of the external organs of the imago, during the metamorphosis of the ENDOPTERYGOTA. Also *imaginal disc.*

imaginary axis (*MathSci*) See ARGAND DIAGRAM.

imaginary circle (*MathSci*) A circle defined analytically so as to have an imaginary radius, eg the circle defined by $(x-a)^2 + (y-b)^2 + r^2 = 0$.

imaginary modulus (*Eng*) Also *loss modulus.* See COMPLEX MODULUS.

imaginary number (*MathSci*) The product of a real number x and i, where $i^2 = -1$. A complex number in which the real part is zero.

imaginary part (*MathSci*) See COMPLEX NUMBER.

imaging system (*Space*) An instrument with its supporting HARDWARE and SOFTWARE for obtaining remote images of the Earth and objects in space. The types of image may be optical, microwave or obtained by using some other selected part of the electromagnetic spectrum. The images may be recorded, eg on film or digitally, and recovered later or transmitted in real time to an Earth receiving station.

imago (*BioSci*) The form assumed by an insect after its last ECDYSIS, when it has become fully mature, the final INSTAR. Adj *imaginal.*

IMAX (*ImageTech*) A wide-screen projector for 70 mm film in which the height of the image is set by the width of the film. Used to fill a large screen that surrounds the audience's field of vision.

imbalance (*Med*) A lack of balance, as between the ocular muscles, or between the activities of endocrine hormones, or between parts of the involuntary nervous system.

imbecile (*Med*) A term formerly used for a person with moderately severe mental subnormality.

imbibition (*BioSci*) (1) Uptake of water where the driving force is a difference of MATRIC POTENTIAL rather than osmotic. (2) Uptake of water and swelling by seeds, the first step in germination.

imbibition (*Chem*) The absorption or adsorption of a liquid by a solid or a gel, accompanied by swelling of the latter.

imbibition (*ImageTech*) The transfer of dye from a matrix to an absorbing surface, such as a gelatine layer, to produce a dye image. Also *dye transfer.*

imbibition matrix (*ImageTech*) A relief or differentially hardened image capable of selectively absorbing dye and transferring it to another receptive layer.

imbricate (*BioSci*) Organs that overlap like the tiles on a roof, such as scales, leaves or petals in buds. Cf VALVATE. See AESTIVATION, VERNATION.

imbricate structure (*Geol*) (1) A structure produced by thrust faulting leading to the development of numerous small faults and rock slices arranged in parallel like a pack of fallen cards. (2) A sedimentary structure in which pebbles with a flat surface are stacked in the same direction, dipping upcurrent.

IMC (*Aero*) Abbrev for *instrument meteorological conditions,* wherein aircraft must conform to INSTRUMENT FLIGHT RULES.

IMEP (*Eng*) Abbrev for INDICATED MEAN EFFECTIVE PRESSURE.

Imhoff cone (*EnvSci*) A large graduated glass cone used for determining the amount of solid material in eg sewage sludge that is capable of settling.

Imhoff tank (*Build*) A form of settling tank to which sewage is passed, the solid matter being exposed to a fermentation process, with the production of methane gas and an inoffensive sludge which can be easily dried.

imidazoles (*Chem*) *Glyoxalines*; heterocyclic compounds produced by substitution in a five-membered ring containing two nitrogen atoms on either side of a carbon atom. Benzimidazoles are formed by the condensation of *ortho*-diamines with organic acids, and contain a condensed benzene nucleus.

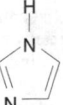

imidazole

imides (*Chem*) Organic compounds containing the group –CONHCO–, derived from acid anhydrides. See POLYIMIDES.

imino group (*Chem*) The group RR′NH. Imino compounds are secondary amines obtained by the substitution of two hydrogen atoms in ammonia by alkyl radicals. PROLINE is actually an imino acid, although generally considered within the amino acids since it is a constituent of proteins.

imitation (*Psych*) See OBSERVATIONAL LEARNING.

imitation parchment (*Paper*) A wood-pulp paper to which some degree of transparency, and grease resistance, has been imparted by prolonged beating of the pulp.

immature cotton (*Textiles*) Cotton picked before it is fully mature. The fibres are not properly formed and the yarn made from them is generally weaker and inferior.

immediate access store (*ICT*) See MAIN MEMORY.

immediate address (*ICT*) The address field itself is used to hold data which are needed for the operation.

immediate hypersensitivity (*BioSci*) Hypersensitivity response, mediated by IgE, that is characteristically due to release of histamine and other vasoactive substances. Also *type-1 hypersensitivity reaction*.

immediate mode (*ICT*) The use, for immediate execution, of a program language statement outside the program.

immersed liquid-quenched fuse (*ElecEng*) A fuse in which liquid is used for extinguishing the arc, the fuse-link being totally immersed in the liquid.

immersed pump (*Aero*) An electrical pump mounted inside a fuel tank.

immersible apparatus (*ElecEng*) Electrical apparatus designed to operate continuously under water.

immersion (*Astron*) The entry of the Moon, or other body, into the shadow which causes its eclipse.

immersion foot (*Med*) A cold injury to the foot produced by cold wet exposure causing vasoconstriction resulting in anoxic tissue damage. Also *trench foot*.

immersion heater (*ElecEng*) An electric heater designed for heating water or other liquids by direct immersion in the liquid.

immigration (*BioSci*) A category of population dispersal covering one-way movement into the population area. Cf EMIGRATION, MIGRATION.

immiscibility (*Chem*) The property of two or more liquids of not mixing and of forming more than one phase when brought together.

immittance (*Phys*) Combined term covering IMPEDANCE and ADMITTANCE.

immobilized culture (*BioSci*) The use of plastic foam, beads or sheets to let cells grow on the surfaces and interstices in close contact with the circulating medium. Combines the advantages of surface and suspension culture.

immortalization (*BioSci*) Escape from the normal limitation on population growth of a finite number of division cycles (the *Hayflick limit*), by variants in animal cell cultures, and cells in some tumours. Immortalization in culture may be spontaneous, induced by mutagens, or by transfection of certain oncogenes.

immune (*BioSci*) Not susceptible to infection with a particular infectious agent; an organism in this state has immunity. See panel on IMMUNE RESPONSE.

immune adherence (*BioSci*) Adherence of antigen–antibody complexes or antibody-coated microbes that contain bound C3b or C4b to complement receptors on erythrocytes or platelets of some species and on macrophages and polymorphs.

immune complex (*BioSci*) A type of macromolecular complex consisting of antigen and antibody linked by their combining sites. The size of the complex depends upon the ratio of antigen to antibody. Such complexes, if present in the blood stream, may be deposited in the walls of small blood vessels or in renal glomeruli, leading to *immune complex diseases*. Immune complex deposition is a type 3 hypersensitivity response.

immunization (*BioSci*) Administration either of antigen to produce active immunity or of antibody to confer passive immunity, and thereby to protect against the harmful effects of antigenic substances or microbes.

immunoassay (*BioSci*) A large group of procedures which exploit the ability of antibodies to recognize specific antigens. Either the antibody or antigen can be identified with radioactive, fluorescent or chemiluminescent labels and often the reaction is linked to an enzyme system to amplify or modify the effect.

immunoblot (*BioSci*) See BLOTTING.

immunocytochemistry (*BioSci*) Laboratory techniques involving the staining of tissue sections using antibodies that specifically bind to particular cell proteins. The antibodies are then detected using secondary reagents that are coupled to enzymes. When the enzymes are reacted with their substrates, colour develops. This allows the identification of individual cells within a tissue that bear the protein of interest. Also *immunohistochemistry*.

immunodeficiency disease (*Med*) Heterogeneous group of diseases resulting in lack or impaired efficiency of the immune response. The impaired immune response may be inherited (genetic/primary immune deficiency) or be acquired as a result of disease, eg AIDS. Genetic defects may be in a single arm of the immune response, eg lack of a particular complement component, or may affect the whole system, eg severe combined immune deficiency.

immunofluorescence (*BioSci*) A technique in which antigen or antibody is conjugated to a fluorescent dye and then allowed to react with the corresponding antibody or antigen in a tissue section or a cell suspension. This enables the location of antibodies or antigens in or on cells to be determined by fluorescence microscopy.

immunogen (*BioSci*) A substance that stimulates humoral and/or cell-mediated specific immunity when introduced into the body.

immunoglobulin (*BioSci*) A family of proteins all of which have a similar basic structure, made up of LIGHT CHAINS and HEAVY CHAINS linked together by disulphide bonds so as to form a Y-shaped molecule with two flexible arms. See panel on IMMUNE RESPONSE.

immunoglobulin genes (*BioSci*) Genes that code for immunoglobulins. See panel on IMMUNE RESPONSE.

immunoglobulin superfamily (*BioSci*) A large group of proteins with immunoglobulin-like domains. Most are involved with cell surface recognition events.

immunohistochemistry (*BioSci*) See IMMUNOCYTOCHEMISTRY.

immunological memory (*BioSci*) A term describing the fact that antibody and cell-mediated responses occur more rapidly and are quantitatively greater after a second exposure to antigen, provided that the interval is more than a few days and less than several years. This phenomenon depends upon resting long-lived 'memory cells' that respond on second exposure to the antigen.

immunological tolerance (*BioSci*) A state in which an animal fails to respond to an antigen normally capable of inducing humoral or cell-mediated immunity. This can result from administration of antigens in very early life, before the immune system is fully developed.

immunology (*Genrl*) The scientific study of immunity and the defence mechanisms of the body. See panel on IMMUNE RESPONSE.

immunopathology (*BioSci*) Disease caused as a result of tissue destruction by the immune response or disease of the immune system itself.

immunophilin (*BioSci*) A general term for intracellular proteins that bind immunosuppressive drugs such as CICLOSPORIN.

immunoprecipitation (*BioSci*) The precipitation of a multivalent antigen by a bivalent antibody, resulting in the formation of a large complex. See IMMUNE COMPLEX.

immunosorbent (*BioSci*) The use of an insoluble preparation of an antigen to bind specific antibodies from a mixture, so that the antibodies can later be eluted in pure form.

immunostimulatory complex (*BioSci*) A method for presenting viral proteins to the immune system. Viral protein is mixed with a compound called Quill A and detergent to form a small basketwork structure. Abbrev ISCOM.

immunosuppression (*BioSci*) The artificial suppression of immune responses by the use of drugs that interfere with lymphocyte growth (anti-metabolites), or by irradiation or by antibodies against lymphocytes. A state of immunosuppression can also exist as a result of infections (such as by HUMAN IMMUNODEFICIENCY VIRUS, HIV, or cytomegalovirus) which damage lymphocytes.

immunosuppressive (*Med*) Applied to drugs which lessen the body's rejection of eg transplanted tissue and organs, or decrease the response to AUTO-IMMUNE DISEASE. See panel on AUTO-IMMUNITY.

immunotoxin (*BioSci*) A complex consisting of a cytotoxic molecule coupled to an antibody. The antibody is chosen

Immune response

Animals are continuously invaded by all kinds of foreign bodies ranging from inanimate material to parasitic worms, bacteria and viruses. They have evolved, therefore, a series of primary defences like the skin, the acidity of the stomach and mucus in the lungs. Nevertheless, if invaders succeed in penetrating further, highly sophisticated cellular mechanisms are called into play. Some cellular defences, like the ability of PHAGOCYTES to ingest foreign bodies or of LYSOZYME to dissolve their cell walls, are always present, but vertebrates have evolved cellular and molecular systems which greatly enhance both the specificity and the effectiveness of these secondary defences. These are the immune responses, which make use of a network of special cells, particularly lymphocytes, and an exquisite mechanism at the genetic level, for producing an enormous range of molecules able to bind to chemical groups whatever their origin.

All macromolecules, and therefore the surfaces of all cells, have sites capable of provoking an immune response, and by definition act as ANTIGENS. What the system has to do is to recognize the antigens of harmful cells or molecules and to trigger a response which neutralizes the harmful cell by eg causing phagocytic cells to ingest them. This recognition system involves a number of cell types and some of them are listed below.

B-cells

In mammals, this class of lymphocytes is derived from precursors in the bone marrow, but they do not undergo differentiation in the thymus and circulate freely within the lymphoid system. Each individual cell is programmed to place on its surface about 10^5 molecules of one, and only one, immunoglobulin, and broadly, every cell makes a different immunoglobulin at this stage. Any antigen molecule is faced, therefore, with a vast assortment of recognition sites and will bind to those with the best fit. Antigen binding, then, triggers (via associated proteins) the multiplication of B-cells to form a clone of plasma cells with the same specificity as the original B-cell. This is called CLONAL SELECTION and it results in the eventual production of a large amount of a specific antibody, which may be enough to suppress an infection.

Antibodies circulate in body fluids, where they bind to certain bacteria and viruses; thus, they neutralize bacterial toxins and promote uptake by macrophages. They also lyse their targets and promote inflammation by fixing COMPLEMENT. However, other viruses and bacteria remain hidden within the host's own cells. A second system involving different lymphocytes has evolved to cope with them. These are killer T-lymphocytes, which like other T-cells recognize short foreign peptides bound to proteins of the major histocompatibility complex (see later) on the surface of host cells.

T-cells

These lymphocytes are also derived from cells produced in the bone marrow but they mature and differentiate in the thymus, which is of major importance for the development of the immune response. Lymphocyte precursors differentiate and divide several times in the cortex of the thymus and it is here that the genes controlling the T-cell antigen receptor become reas-sorted from their germ line configuration. This antigen receptor consists of two POLYPEPTIDE chains, called α and β, each of which also has a constant and variable region. These two chains, together with associated proteins, sit across the membrane of the lymphocyte. The diversity of these receptors is generated by a mechanism rather like, but not the same as, that which produces variability in immunoglobulins.

In the cortex only those T-cells which recognize self-MHC molecules initially survive and multiply, and later those among them which recognize 'self-antigenic' determinants are eliminated by APOPTOSIS and give rise to self-tolerance (but see panel on AUTO-IMMUNITY). The remainder undergo further differen-tiation in the medulla, becoming separated into mature helper and down-regulatory T-lymphocytes, which recognize antigenic determinants associated with Class II molecules of the major histocompatibility complex (MHC) and cytotoxic T-lymphocytes, which recognize antigenic determinants associated with Class I major histocompatibility complex molecules.

Host cells have specialized machinery for chopping up the proteins of any foreign organisms which they happen to contain, and then making the resulting

---

to recognize an antigen that is specifically expressed by a particular cell type, allowing the cytotoxic molecule to be selectively targeted.

iMode (*ICT*) TN of a technology enabling the Internet to be accessed from cellular phones.

impact (*Eng, Phys*) Elastic or inelastic collision between bodies during which the rate of change of momentum is high, so that large contact forces are generated. The duration of an impact is frequently of the same order or less than the time for elastic stress waves to propagate through the body,

'dynamic loading', so that static analyses of the deformations of the body do not apply. For the direct impact of two elastic bodies, the ratio of the relative velocity after impact to that before impact is constant and is called the *coefficient of restitution* for the materials of which the bodies are composed. This constant has the value 0·95 for glass–glass and 0·2 for lead–lead, the values for most other solids lying between these two figures. See panel on IMPACT TESTS.

impact accelerometer (*Aero*) An ACCELEROMETER which measures the deceleration of an aircraft while landing.

Immune response (Cont.)

peptides available to Class I MHC molecules; thus, killer T-cells recognize antigens from the cell's interior. On the other hand Class II take up peptides derived from exterior antigens, fitting them for their major task of directing helper T-cells to assist B-cells make antibodies. Helper T-cells act principally through the secretion of CYTOKINES.

It is common experience that childhood illnesses such as measles or mumps do not recur, although the flood of specific plasma cells and cytotoxic T-cells subsides. Long-lasting protection is due to immunological memory, which has three components: (1) long-lasting plasma cells located in bone marrow which produce small amounts of high-affinity antibody (see below); (2) expanded numbers of specific T- and B-cells; (3) the heightened reactivity of memory T-cells. This is the basis of protective vaccination which uses the strategy of administering antigen in a harmless form to invoke a heightened response to later virulent infections. The relative importance of B-cells and the two types of T-cells is best revealed by the IMMUNO-DEFICIENCY DISEASES. Children lacking molecules essential for thymus function suffer from severe combined immunodeficiency and are at grave risk from many infections unless transplanted with bone marrow. Children lacking antibodies are at lesser risk, mainly from pus-forming bacteria.

The major histocompatibility complex
Abbrev *MHC*. The MHC consists of a cluster of genes linked together on a single chromosome. The genes vary in number from one species to another, with three each of Class I and Class II genes in humans. Having so many genes expressed at the same time provides an individual with a large array of molecules and thus enables him or her to present a large assortment of antigenic peptides, or determinants, to his or her T-cells. The array is further enlarged by the likelihood that the ALLELES expressed at each gene in that individual will differ, as very many alleles are present in the human population. The MHC proteins are strong antigens, and an individual will react vigorously against the proteins which the individual does not express. This is how the MHC was discovered, as it is the main cause of rejection of organ transplants.

Antibodies, T-cell receptors and MHC proteins all belong to the immunoglobulin superfamily, composed of characteristically folded domains that occur repeatedly in the genome; domains are often assembled into series making up a single polypeptide chain. They probably all arose by gene duplication from single-domain genes present in invertebrates. All these members of the family occurring in the immune system have domains which handle foreign material, and which therefore vary for the reasons explained above for the MHC. Antibodies are an extreme instance of this variability, as they assemble their variable regions from up to three sets of germline-encoded minigenes (see the figure); later, the diversity is further increased by somatic mutation, which enables the high-affinity antibodies mentioned above to form.

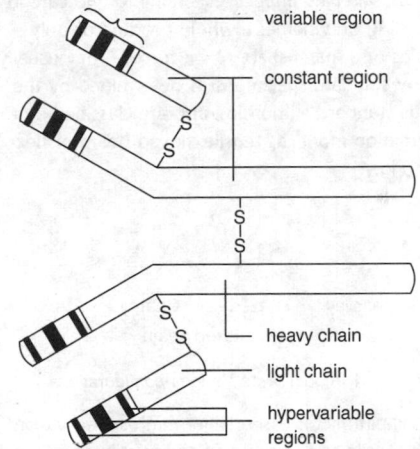

Immunoglobulin Schematic diagram.

In antibodies, the specific antigen-binding site is formed by the variable regions of the heavy and the light chains. Other sites on the constant regions have the biological functions of stimulating phagocytes and of activating the multistage complement pathway which results in the production of an enzyme complex able to attack bacterial membranes. Essentially, the circulating antibody forms a bridge between the antigens on, say, a bacterium and the surface of the phagocyte, and stimulates the phagocyte to engulf the bacterium. The more antibodies there are, the more effective this stimulation is.

See COMPLEMENT, INTERLEUKINS.

---

impact crusher (*MinExt*) A machine in which soft rock is crushed by swift blows struck by rotating bars or plates. The material may break against other pieces of rock or against casing plates surrounding the rotating hammers. Also *hammer mill*.

impacted (*Med*) Firmly fixed, pressed closely in; said of a tooth which has failed to erupt, or of a fracture in which the broken bones are firmly wedged together.

impacter forging hammer (*Eng*) A horizontal forging machine in which two opposed cylinders propel the dies until they collide on the forging, which is worked equally on both sides.

impact extrusion (*Eng*) A fast, cold-working process for producing tubular components by one blow with a punch on a slug of material placed on the bottom of a die, so that the material squirts up around the punch into the die clearance.

Impact tests

Under rapid LOADING, a material may respond very differently compared with slower rates of loading. If the loading rate is comparable with the velocity of elastic waves in the material, interactions can occur which greatly increase the stress locally. Time-dependent materials can undergo a transition from ductile to brittle behaviour as the loading rate increases.

Testing the response of materials under impact loading, together with the TENSILE TEST and hardness tests, are among the most widely used mechanical tests on materials. Impact tests fall into two categories depending on whether a whole product or only samples of a material are to be tested. They usually employ standard specimens as determined by the various standard authorities, and are classified by their deformation mode, as tensile, flexed beam or flexed plate (Fig. 1).

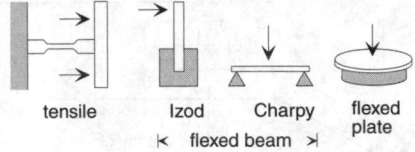

Fig. 1 **Impact tests** Various configurations.

Tensile impact is essentially a high-speed variant of the tensile test. Loading in the tensile and flexed

beam cases is conventionally achieved by a falling pendulum (Fig. 2), hence PENDULUM IMPACT TEST, whilst the flexed plate is usually loaded by a dropping weight. Impact speeds range from 1 to 6 m s^{-1}.

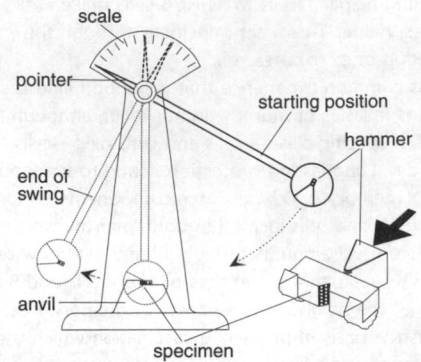

Fig. 2 **Pendulum impact tester**.

Despite many of these tests being specified in standards, correlations between different tests are, at best, qualitative. Their utility in predicting the impact performance of products in service is, therefore, not high. They do, however, provide a quick and simple method of quality control, and are used to determine the DUCTILE–BRITTLE TRANSITION TEMPERATURE in metals and plastics.

See panel on HARDNESS MEASUREMENTS.

---

impact ionization (*Electronics*) The loss of orbital electrons by an atom of a crystal lattice which has experienced a high-energy collision.

impaction sampler (*PowderTech*) A device for sampling particles of dust and spray droplets from a gaseous stream, which is forced through a jet and directed to a horizontal surface, usually a microscope slide smeared with grease, on which particles are collected by inertial impaction. Also *jet impactor*.

impact modifier (*Chem*) Specific polymer added esp to polyvinyl chloride to increase impact strength, eg ABS, MBS. Copolymers now preferred, eg HI-PVC.

impact parameter (*Phys*) The distance at which two particles which interact would have passed if no interaction had occurred between them.

impact printer (*ICT*) Any printer where the standard typewriter action of 'character pressing on paper through inked ribbon' is used, eg a daisy-wheel or dot-matrix printer.

impact strength (*Eng*) A measure of the resistance of materials to impact loading applied in an impact test; it is not a true strength but the energy absorbed per unit area of fractured material. See STRENGTH MEASURES and panel on IMPACT TESTS.

impact tests (*Eng*) Tests designed to determine the response of a material under very rapid loading. See panel on IMPACT TESTS.

IMPATT diode (*Electronics*) Abbrev for *impact avalanche transit-time diode*, a microwave diode which exhibits a

negative resistance characteristic due to avalanche breakdown and charge carrier transit time in chips made of gallium arsenide or silicon. When linked to a waveguide or resonant cavity or similar structure it can be used as a microwave oscillator or amplifier.

impedance (*Phys*) (1) Complex ratio of sinusoidal voltage to current in an electric circuit or component. Its real part is the RESISTANCE (dissipative or wattful impedance) and its imaginary part REACTANCE (non-dissipative or wattless), which may be positive or negative according to whether the phase of the current lags or leads that of the voltage. Resistance, reactance and impedance are all measured in ohms. Expressed symbolically, the impedance Z is given by $R + jX$ where R is resistance, X is reactance and $j = \sqrt{-1}$. Also *apparent resistance*. (2) Component offering electrical impedance. Also *impedor*.

impedance bond (*ElecEng*) A special rail bond of high reactance and low resistance designed to allow the passage of dc traction current but not the ac used for signalling purposes.

impedance circle (*ElecEng*) The locus of the end of the impedance vector in an Argand (R,X) diagram of a system, eg drawn to show the variation of input impedance of an improperly terminated line with frequency.

impedance coupling (*ElecEng*) The coupling of two circuits by means of a tuned circuit or an impedance.

impedance drop (rise) (*ElecEng*) A drop or rise in the voltage at the terminals of a circuit, caused by current passing through the impedance of the circuit.

impedance factor (*ElecEng*) The ratio of the impedance of a circuit to its resistance.

impedance matching (*ElecEng*) See MATCHING.

impedance matching stub (*ElecEng*) See MATCHING STUB.

impedance matching transformer (*ElecEng*) See MATCHING TRANSFORMER.

impedance protective system (*ElecEng*) Discriminative protective equipment in which discrimination is secured by a measurement of the impedance between the point of installation of the relays (impedance relays) and the point of fault. Also *distance protection*.

impedance relay (*ElecEng*) A relay, used in discriminative protective gear, whose operation depends on a measurement of the impedance of the circuit beyond the point of installation of the relay; if this falls below a certain value, when a fault occurs, the relay operates. Also *distance relay*.

impedance rise (*ElecEng*) See IMPEDANCE DROP.

impedance transforming filter (*ElecEng*) A filter network which has differing image impedances, and which can therefore act as a transformer over a band of frequencies.

impedance triangle (*ElecEng*) The right-angled triangle formed by the vectors representing the resistance drop, the reactance drop and the impedance drop of a circuit carrying an alternating current.

impedance voltage (*ElecEng*) The voltage produced as a result of a current flowing through an impedance.

impedometer (*ElecEng*) A device for measuring impedances in waveguides.

impedor (*Phys*) Physical realization of an impedance. An inductor, capacitor or resistor, or any combination of these.

impeller (*Aero*) The rotating member of a CENTRIFUGAL-FLOW COMPRESSOR or SUPERCHARGER.

impeller (*Eng*) The rotating member of a centrifugal pump or blower, which imparts kinetic energy to the fluid.

impeller-intake guide-vanes (*Aero*) The curved extension of the vanes of a centrifugal impeller which extend into the intake eye or throat, and which thereby give the airflow initial rotation.

imperfect (*Build*) Said of a structural framework which has either more or fewer members than it would require to be determinate.

imperfect dielectric (*Phys*) A dielectric in which there is a loss element resulting in part of the electric energy of the applied field being used to heat the medium.

imperfect flower (*BioSci*) A flower in which either the stamens or the carpels are lacking or, if present, non-functional.

imperfect fungi (*BioSci*) The DEUTEROMYCOTINA.

imperfect stage (*BioSci*) A stage in the life cycle of eg a fungus in which the organism can only reproduce asexually if at all.

imperforate (*BioSci*) Lacking apertures, esp of shells; said of gastropod shells which have a solid columella.

imperforate (*Med*) Not perforated; closed abnormally.

imperial (*Build*) (1) A slate size, 33×24 in. (2) A domed roof shaped to a point at the top.

imperial measure (*Genrl*) Non-metric standard of measure or weight (eg imperial gallon, imperial yard and imperial pound) as fixed by parliament for the UK (final act 1963).

Imperial Standard Wire Gauge (*Eng*) See STANDARD WIRE GAUGE.

impermeable (*Chem, Geol*) Not permitting the passage of liquids or gases etc.

impervious (*Build*) Said of materials which prevent the passage of water.

impetigo (*Med*) A contagious skin disease, chiefly of the face and hands, due to infection with pus-forming bacteria (*Staphylococcus aureus*). Adj *impetiginous*.

implant (*Med, BioSci*) (1) A graft of an organ or tissue to an abnormal position. (2) Engineered devices constructed of artificial materials replacing and restoring living tissues or organs within the body and exposed to body fluids, eg hip joint implant, corneal lens implant, heart valve implant.

implant (*Radiol*) The radioactive material, in an appropriate container, which is to be imbedded in a tissue for therapeutic use, eg needle or seed.

implantation (*BioSci*) In mammals, the process by which the blastocyst becomes attached to the wall of the uterus.

implementation (*ICT*) The various steps involved in producing a functioning system from a design.

implicit function (*MathSci*) The expression of the relationship between the values of two variables indirectly rather than by an EXPLICIT FUNCTION, eg $x^2 + xy - y^2 = 5$.

imploding linear system (*NucEng*) Fusion device in which a cylindrical plasma is formed by the implosion of material lining the reactor vessel.

implosion (*Eng*) Mechanical collapse of a hollow structure, eg a cathode-ray tube.

implosive therapy (*Psych*) A behaviour therapy in which the patient imagines exposure to the feared stimulus with the aim of extinguishing a phobia. See FLOODING.

import (*ICT*) To bring a file or document created using one piece of software into another, eg a spreadsheet file into a word processor application.

imposex (*BioSci*) The superimposition of male sexual characteristics onto female gastropods, caused by pollutants such as TRIBUTYLTIN.

imposing stone (*Print*) A heavy iron-topped table on which the type matter is locked up preparatory to printing. In the early days of printing, level stone-topped tables were used.

imposition (*Print*) (1) The process of assembling letterpress pages of type in their proper order on the stone, arranging appropriate furniture or spacing material, and locking the whole into a chase. After imposition the unit is known as a FORME, and from it a book section or signature is printed. (2) Assembly of film elements for plate-making in a predetermined pattern so that when the work is printed and folded the pages will run in the correct sequence. (3) A plan showing the arrangement of pages to suit the printing and folding operations and to give the desired page sequence.

impost (*Build*) The top member of a pier or pillar from which an arch springs. See fig. at ARCH.

impregnation (*BioSci*) The passage of spermatozoa from the body of the male into the body of the female.

impregnation (*Eng*) Strengthening of porous material such as wood, cement or plaster by exposure to low-molecular-mass monomer or oligomer. They diffuse rapidly into the porous solid and are polymerized catalytically.

impregnation (*PowderTech*) The partial or complete filling of the pores of a powder product with an organic material, glass, salt or metal, to make it impervious or impart to it secondary properties. Vacuum, pressure and capillary forces may be employed. It may include stoving to produce setting. See INFILTRATION.

impression (*Eng*) One of several similar cavities in a single mould tool.

impression (*Print*) (1) All copies of a book printed at one time from the same type or plates. (2) The pressure applied to a type forme by the cylinder or platen.

impression cylinder (*Print*) The cylinder which presses the stock against the printing surface, which may be flat or cylindrical, and letterpress, gravure or lithographic. See fig. at LITHOGRAPHY.

impression formation (*Psych*) A traditional area of research in social psychology, referring to the issue of how information about other individuals is integrated into a unified impression, often on the basis of very little information.

imprint (*Print*) The name of the publisher and/or printer which must appear on certain items, particularly books, periodicals and election literature.

imprinting (*Psych*) An aspect of learning in some species, through which attachment to the important parental figure develops; best-known example is the tendency for newly hatched birds to follow the first large moving object they see. But see GENOMIC IMPRINTING.

improper fraction (*MathSci*) See DIVISION.

improver (*FoodSci*) Substance added to flour (usually for bread-making) to produce 'better' products. Used for oxidizing and bleaching, crumb softening, modification of gluten elasticity and retarding staling, eg sulphur dioxide, ascorbic acid, soya flour, l-cysteine.

improving (*Eng*) See SOFTENING.

impsonite (*Min*) A member of the asphaltite group.

impulse (*ICT*) Obsolete term for PULSE.

impulse (*Phys*) The change of momentum produced in either body when two bodies collide; given by the time integral $\int F dt$, the impulse of the force.

impulse circuit (*ICT*) In an automatic switching exchange, a source of machine-generated impulse trains for operating step-by-step switches, controlled by relays.

impulse circuit breaker (*ElecEng*) A circuit breaker, requiring only a small quantity of oil, in which the arc is extinguished by a mechanically produced flow of oil across the contacts.

impulse excitation (*Electronics*) Maintenance of oscillatory current in a tuned circuit by pulses synchronous with free oscillations, or at a submultiple frequency.

impulse flashover voltage (*ElecEng*) The value of the impulse voltage which just causes flashover of an insulator or other apparatus.

impulse frequency (*ICT*) The number of impulses per second in the impulse trains used in dialling and operating selectors.

impulse function (*ICT*) See DELTA IMPULSE FUNCTION.

impulse generator (*ElecEng*) A circuit providing a single or continuous series of pulses, generally by capacitor discharge and shaping, eg by the charging of capacitors in parallel and the discharging of them in series. Also *surge generator*.

impulse inertia (*ElecEng*) That property of an insulator by which the voltage required to cause disruptive discharge varies inversely with its time of application.

impulse machine (*ICT*) A machine that generates accurately timed pulses for operating selector switches.

impulse period (*ICT*) The time between identical phases of a train of impulses: the time between the start of one impulse and the start of the next.

impulse ratio (*ElecEng*) The ratio between the breakdown voltage of an insulator or piece of insulating material when subjected to an impulse voltage to the breakdown when subjected to a normal-frequency (50 or 60 Hz) voltage.

impulse ratio (*ICT*) The ratio of the time during an impulse to the total time of impulse plus interval before another impulse.

impulse–reaction turbine (*Eng*) See DISK-AND-DRUM TURBINE.

impulse repeater (*ICT*) A relay mechanism for repeating impulses from one circuit to another.

impulse starter (*Aero*) A mechanism in a magneto which delays the rotor against a spring so that, when released, there is a strong and retarded spark to help starting.

impulse turbine (*Eng*) A steam turbine in which steam is expanded in nozzles and directed onto blades carried by a rotor, in one or more stages, there being no change in pressure as the steam passes the blade ring.

impulse voltage (*ElecEng*) A transient voltage lasting only for a few microseconds; very frequently used in high-voltage testing of electrical apparatus in order to simulate voltage due to lightning strikes or other similar causes.

impulse wheel (*Eng*) The wheel of an IMPULSE TURBINE. Also used to denote one of the two principal types of turbines, in which the whole available head is transformed into kinetic energy before reaching the wheel. See PELTON WHEEL.

impulsive sound (*Acous*) Short sharp sound, the energy spectrum of which spreads over a wide frequency range.

impurity (*Electronics*) Small proportion of foreign matter, eg arsenic, boron, phosphorus, etc, added to a pure semiconductor, eg silicon, to obtain the required type of conduction and conductivity for solid-state devices. The impurity in the crystal lattice may add to or subtract from the average densities of free electrons and holes in the semiconductor. See ACCEPTOR, CARRIER, DONOR.

impurity levels (*Electronics*) In the band theory of solids, localized energy levels in the band gap introduced by the presence of impurities in a crystal lattice.

In (*Chem*) Symbol for INDIUM.

in (*For*) See KERUING.

in- (*Genrl*) Prefix from Lt *in*, in(to), not.

inactivation (*Chem*) The destruction of the activity of a catalyst, serum, etc.

inanition (*Med*) Exhaustion and wasting of the body from lack of food.

inappropriate affect (*Psych*) Exhibiting contradictory or abnormal behaviour when describing or experiencing an emotion, eg laughing when describing sorrowful matters.

inband (*Build*) A header stone.

inband rybat (*Build*) A header stone laid to form the jamb of an opening.

inbetweening (*ImageTech*) Animation technique in which beginnings and ends of short sequences are drawn leaving the intermediate drawings, 'inbetweens', to be produced by a computer or novice animator, 'inbetweener'.

in-box (*ICT*) A file for storing incoming electronic mail.

inbred (*BioSci*) The condition of the offspring produced by inbreeding. See INBREEDING COEFFICIENT.

inbred line (*BioSci*) A strain that has been inbred over many generations and whose INBREEDING COEFFICIENT is nearly 100%. All members are genetically identical and *homozygous* at all loci, or very nearly so. Cf ISOGENIC.

inbreeding (*BioSci*) The mating together of individuals that are related by descent. The offspring are inbred to an extent depending on the degree of relationship. Inbreeding produces HOMOZYGOSIS. See INBREEDING COEFFICIENT.

inbreeding coefficient (*BioSci*) A measure of the degree to which an individual is inbred. Ranges from zero when the parents are unrelated to 100% when the parents for many generations back have been related. Usually symbolized by *F*.

inbreeding depression (*BioSci*) The reduction of desirable characters such as growth rate, yield, fertility, consequent on the homozygosis produced by inbreeding, esp in those which normally outbreed.

inbye (*MinExt*) The direction from a haulage way to a working face.

incandescence (*Phys*) The emission of light by a substance because of its high temperature, eg a glowing electric-lamp filament. In the case of solids and liquids, there is a relation between the colour of the light and the temperature. Cf LUMINESCENCE.

incandescent lamp (*Phys*) A lamp in which light is produced by heating some substance to a white or red heat, eg a filament lamp.

incandescent mantle (*Eng*) See GAS MANTLE.

incentive learning (*Psych*) A motivational concept that refers to the expectation of rewards or punishments from the environment. The high or low incentive value of a goal is reflected in the amount of energy the organism will expend to obtain it.

incept (*BioSci*) The rudiment of an organ.

incertae sedis (*BioSci*) Of uncertain taxonomic position.

incertum (*Build*) An early form of masonry work in which squared stones were used as a facing, with rubble filling as a backing.

incest (*Med*) Sexual intercourse between close relatives, eg brother and sister, father and daughter.

incest taboo (*Psych*) A strong negative social sanction which forbids sexual relations between members of the same immediate family, found in all human societies.

inch (*Genrl*) One-twelfth of a foot, equal to 2·54 cm.

inching (*Eng*) Very slow, closely controlled step-by-step movement of a usually fast-moving machine, eg a printing press for adjustment, using an electrical inching button switch. Cf CRAWL.

inching starter (*ElecEng*) An electric-motor starter in which provision is made for inching the motor, ie running it very slowly for such purposes as the threading of the paper in a printing press.

inch-penny weight (*MinExt*) In valuation of gold ore, the width of the lode or reef measured normal to the enclosing rock, multiplied by the assay value in penny weights per ton.

inch-tool (*Build*) A steel chisel having a cutting edge 1 in (25 mm) wide, used by the mason for dressing stone.

inch trim moment (*Ships*) See MOMENT TO CHANGE TRIM ONE INCH. Abbrev *ITM*.

incidence (*Med*) The frequency with which new cases of a given disease presents in a particular period for a given population. Cf PREVALENCE.

incidence, angle of (*Aero, Phys*) See ANGLE OF INCIDENCE.

incidental learning (*Psych*) Learning without trying to learn.

incident beam (*Phys*) Any wave or particle beam the path of which intercepts a surface of discontinuity.

incineration (*Eng*) One route to disposal of combustible materials following use and loss of function, particularly where costs of separation from domestic waste are prohibitive. Probably best route for tyre disposal, where the thermoset nature of several rubbers used in construction precludes solvent extraction, melt processing, etc.

incipient plasmolysis (*BioSci*) The state of a plant cell in which the TURGOR pressure is zero but the protoplast is in contact with the cell wall all round. Loss of water will result in PLASMOLYSIS; uptake of water will generate turgor.

incise (*Arch*) To cut in; to carve.

incised meander (*Geol*) An entrenched bend of a river, which results from renewed downcutting at a period of rejuvenation.

incisiform (*BioSci*) Shaped like an incisor tooth.

incision (*Med*) A surgical cut.

incisors (*BioSci*) The front teeth of mammals; they have a single root, are adapted for cutting, and are the only teeth borne by the premaxillae in the upper jaw.

incisura (*Med*) A cut or notch. Various notches in the body are thus designated.

inclination (*Phys*) See DIP.

inclination factor (*Phys*) The angle-dependent term used in Fresnel's theory of the propagation of light waves, where the disturbance at a point, due to the contributions from the secondary waves, is assumed to depend on the angle θ between the normal to the primary wavefront and the direction to the point. Also *obliquity factor*.

incline (*Surv*) See GRADIENT.

inclined-catenary construction (*ElecEng*) A catenary construction for the overhead contact wire of an electric traction system; in it, the catenary wire is not placed vertically above the contact wire.

inclined plane (*Phys*) A smooth plane inclined at an angle θ to the horizontal, used to examine simple principles in mechanics. The force parallel to the plane required just to move a mass up the plane is $mg \sin \theta$. The inclined plane may therefore be regarded as a machine having a velocity ratio of cosec θ.

inclined shear mount (*Eng*) Type of laminated bearing for controlling stiffness along several axes.

inclined shore (*Build*) See RAKING SHORE.

inclining experiment (*Eng*) A practical method of determining the metacentric height and the height of the centre of gravity of a floating vessel; accomplished by observing the angle of heel of the vessel resulting from a measured transverse movement of a known weight across the deck.

inclinometer (*Phys, Surv*) See DIP NEEDLE.

inclusion (*Eng*) A particle of alien material retained in a solid material. In metals such inclusions are generally oxides, sulphides or silicates of one or other of the component metals of the alloy, but may also be particles of refractory materials picked up from the furnaces or ladle

lining. In oxide glasses such particles are called 'stones' and act as deleterious stress concentrators.

inclusion (*Min*) A foreign body (gas, liquid, glass or mineral) enclosed by a mineral. Fluid inclusions (eg liquid carbon dioxide) may be used to study the genesis of the minerals in which they occur. See XENOLITH.

inclusion bodies (*Med*) Particulate bodies found in the cells of tissue infected with a virus.

inclusive fitness (*BioSci*) Same as DARWINIAN FITNESS, but including those genes that the individual shares with its relatives and are passed on by them.

inclusive page depth (*Print*) The number of lines of text on the page.

incoherent (*Phys*) Said of radiation of the same frequency emitted from discrete sources with random phase relationships. All visible light sources except the laser emit incoherent radiation. See panel on LASER.

Incoloys (*Eng*) Proprietary range of corrosion-resistant and high-temperature alloys containing 30% nickel, 20% chromium and 48% iron with small amounts of carbon, aluminium and titanium.

incoming feeder (*ElecEng*) A feeder in a substation through which power is received.

incompatibility (*BioSci*) (1) Mismatch that may be due to immunological, chemical or physical factors. Commonly applied to blood transfusion of mismatched or incompatible blood groups. (2) The consistent failure of fertilization or hyphal fusion between particular combinations of individual plants, algae, fungi, etc. See SELF-INCOMPATIBILITY. (3) In horticultural practice, interaction between stock and scion resulting in the failure of the graft either immediately or after some years of apparently successful growth. (4) The relationship between a plant and a pathogen to which the plant is not susceptible.

incompatibility (*Chem*) The tendency of different polymers to form separate phases when mixed together. See COMPATIBILITY.

incompatible behaviours (*Psych*) Behaviour patterns which cannot occur simultaneously, because of reciprocal inhibition in eg the case of reflexes, but also due to psychological factors, such as the limits of attention.

incompetence (*Med*) The inability to perform proper function; said esp of diseased valves of the heart which allow the blood to pass in the wrong direction, eg *aortic incompetence, mitral incompetence.*

incompetent rock (*Geol*) A rock that yields by plastic flow, folding or shearing during deformation. Similarly *incompetent bed.*

incomplete flower (*BioSci*) A flower in which the calyx and corolla (or one of these) are lacking.

incomplete metamorphosis (*BioSci*) In insects, a more or less gradual change from the immature to the mature state, a pupal stage being absent and the young forms resembling the parents, except in the absence of wings and mature sexual organs.

incomplete reaction (*Chem*) A reversible reaction which is allowed to reach equilibrium, a mixture of reactants and reaction products being obtained.

incompressible volume (*Chem*) See CO-VOLUME.

Inconels (*Eng*) Nickel-based heat-resistant alloys containing some 13% of chromium, 6% iron, and a little manganese, silicon or copper. See NICKEL ALLOYS.

incontinence (*Med*) Inability to retain voluntarily natural excretions of the body (eg feces and urine); lack of self-control.

inco-ordination (*Med*) Inability to combine muscular movements in the proper performance of an action, the component muscle groups working independently instead of together.

increasing (*MathSci*) Of a function or sequence: such that for all $a > b$, $f(a) \geq f(b)$. If $f(a) > f(b)$, then the function is said to be *strictly increasing*. Compare DECREASING.

incremental backup (*ICT*) A copy containing files altered since the last back-up.

incremental delivery (*ICT*) The practice of making regular, frequent deliveries of small portions of functionality to a system each of which extends that previously delivered.

incremental hysteresis loss (*ElecEng*) A small pulsation of the magnetic field about a fixed value leading to a small hysteresis loop on the boundary of a full loop.

incremental induction (*ElecEng*) The difference between the maximum and minimum value of a magnetic induction at a point in the polarized material, when subjected to a small cycle of magnetization. Cf INCREMENTAL HYSTERESIS LOSS.

incremental iron losses (*ElecEng*) A term sometimes used to denote iron losses occurring in an ac machine owing to frequencies higher than the fundamental, eg tooth pulsation losses.

incremental permeability (*ElecEng*) The gradient of the curve relating flux density to magnetizing force (the *B/H* curve). This represents the effective permeability for a small alternating field superimposed on a larger steady field. See MAGNETIC MATERIALS.

incremental plotter (*ICT*) Graph plotter that receives input data specifying increments to its current position, rather than data specifying co-ordinates. Cf DIGITAL PLOTTER.

incremental resistance (*ElecEng*) The small signal resistance for a component or network, $r = \Delta V / \Delta I$.

incrustation (*Build*) A term applied to a wall facing which is of different material from that forming the rest of the wall.

incubation (*Med*) The period intervening between the infection of a host by bacteria or viruses and the appearance of the first symptoms.

incubation (*Psych*) Behaviour which maintains the eggs of birds and other species in a fairly stable thermal and gaseous environment; most bird species accomplish it by sitting on the eggs, but other methods are also used by various species of birds and insects, eg mound-building, sunning.

incubation time (*FoodSci*) When applied to food-related disease, the period between consuming contaminated food and the onset of symptoms.

incubous (*BioSci*) Said of the leaf of a liverwort when its upper border (the border towards the apex of the stem) overlaps the lower border of the next leaf above it and on the same side of the stem.

incudectomy (*Med*) Removal of the incus by operation.

incurrent (*BioSci*) Carrying an ingoing current, said of ducts as in some sponges.

incus (*BioSci*) In mammals, an ear ossicle, derived from the quadrate; more generally, any anvil-shaped structure. Pl *incudes*.

indanthrene (*Chem*) *N-dihydro-1,2,2′,1′-anthraquinone-azine.* $C_{28}H_{14}O_4N_2$. An anthraquinone vat dyestuff, a dark-blue powder, practically insoluble in water and organic solvents. For dyeing purposes, indanthrene is reduced by sodium hydrosulphite to the water-soluble salt of the dihydro derivative, and re-oxidized to indanthrene by exposure to air. TNs *Indanthrone, Caledon blue* and *Duranthrene*.

indeciduate (*BioSci*) Said of mammals in which the maternal part of the placenta does not come away at birth. Also *non-caducous*.

indefinite (*BioSci*) (1) Numerous but not fixed in number. (2) MONOPODIAL GROWTH. (3) A RACEMOSE INFLORESCENCE.

indefinite integral (*MathSci*) See INTEGRAL.

indehiscent (*BioSci*) An organ that does not open spontaneously to release the seeds, spores, etc. Cf DEHISCENCE.

indene (*Chem*) C_9H_8. Bp 182°C. An aromatic double-ring liquid hydrocarbon occurring in coaltar. Usually contains coumarone.

indent (*Build*) A notch made in a timber.

indent (*Print*) To commence a line with a blank space, which in bookwork paragraphs may be 1, 1·5 or 2 ems, according to the width of the line.

indentation test (*Build, CivEng*) A test for a paving, roofing or road-making asphalt, in which a steady load is applied, under constant temperature conditions, to the asphalt surface, through the sector of a wheel resting upon it, the amount of indentation being measured after a fixed time.

indented bar (*CivEng*) A special type of reinforcing bar used in reinforced concrete work to provide a mechanical bond and having for its full length a series of depressions and ridges all round.

indenter (*CivEng*) A roller having projections from its curved surface, so that, when it is rolled over newly laid asphalt paving, indentations shall be left in the latter surface to render it non-skid. Also *branding iron, crimper.*

independent axle-drive (*ElecEng*) See INDIVIDUAL AXLE-DRIVE.

independent chuck (*Eng*) A lathe chuck in which each of the jaws is moved independently by a key; used for work of irregular shape, or when very accurate centring is needed.

independent equations (*MathSci*) A set of equations none of which can be deduced from a combination of any of the others.

independent feeder (*ElecEng*) A feeder in an electric-power distribution system which is used solely for supply to a substation or a feeding point, and not as an interconnector. Also *dead-ended feeder, radial feeder.*

independent particle model (*Phys*) A model of the nucleus in which each nucleon is assumed to act quite separately in a common field to which they all contribute.

independent suspension (*Autos*) A springing system in which the wheels are not connected by an axle beam, but are mounted separately on the chassis through the medium of springs and guide links, so as to be capable of independent vertical movement.

independent time-lag (*ElecEng*) See DEFINITE TIME-LAG.

independent trip (*ElecEng*) A tripping device for a circuit breaker, starter, or similar apparatus, in which the current operating the device is independent of the current flowing in the circuit to which the device is connected.

independent variable (*MathSci*) See DEPENDENT VARIABLE.

indestructibility of matter (*Chem*) See LAW OF CONSERVATION OF MATTER.

indeterminacy principle (*Phys*) See UNCERTAINTY PRINCIPLE.

indeterminate (*Eng*) Said of a structure which is REDUNDANT. Cf DETERMINATE.

indeterminate equations (*MathSci*) Equations that do not have a unique solution.

index (*MathSci*) (1) The small number written to the right of and above a number or term to indicate how many times that number or term has to be multiplied by itself, eg 3 in $a^3 = a \times a \times a$, or x in $(a + b)^x$. Note:

$$a^0 = 1, \quad a^{\frac{1}{2}} = \sqrt{a}, \quad a^{x/y} = \sqrt[y]{a^x},$$

$$a^{-1} = \frac{1}{a}, \quad a^x = e^{x \log a}.$$

Also *exponent*. (2) Of a subgroup, the order of the group divided by the order of the subgroup.

index case (*Med*) The first or original case of a disease. A term used in the EPIDEMIOLOGY of infectious disease. In genetics synonymous with the proband or propositus.

indexed address (*ICT*) The DIRECT or INDIRECT ADDRESS as modified by the addition of a number held in an INDEX REGISTER.

indexed sequential access (*ICT*) The process of storing or retrieving data directly, but only after reading an index to locate the address of that item. See DIRECT ACCESS.

index error (*Surv*) The difference between the horizontal or vertical angular reading of a theodolite and the true line of collimation, with regard to concentric centring of the azimuth and plate circle and accurate engraving of the reading lines.

index fossil (*Geol*) A fossil species which characterizes a particular geological HORIZON. It tends to be abundant, with a narrow time range and a wide geographical spread.

indexing head (*Eng*) A machine-tool attachment for rotating the work through any required angle, so that faces can be machined, holes drilled, etc, in definite angular relationship.

index mineral (*Geol*) One whose appearance marks a particular grade of metamorphism in progressive regional metamorphism.

index of refraction (*Phys*) See REFRACTIVE INDEX.

index search (*ImageTech*) A means of locating places on videotape by coded signals recorded in the CONTROL TRACK. Also *VISS* (*video/VHS index search system*).

Indian hemp (*BioSci*) See CANNABIS.

Indian ink (*Genrl*) Ink in a solid form made from lampblack mixed with parchment size or fish glue. Rubbed down in water it produces an intensely black permanent ink, used for line and wash drawings.

Indian topaz (*Min*) See CITRINE; also a misnomer for yellow corundum.

India paper (*Paper*) A thin, strong, opaque rag paper, made for bibles and other books where many pages are required in a small compass.

india-rubber (*Chem*) See RUBBER.

indicated airspeed (*Aero*) The reading of an airspeed indicator which, when corrected for instrument errors, reads low by a factor equal to the square root of the relative air density as the latter falls with altitude. Abbrev *IAS*.

indicated horsepower (*Eng*) Of a reciprocating engine, the horsepower developed by the pressure–volume changes of the working agent within the cylinder; it exceeds the useful or brake horsepower at the crankshaft by the power lost in friction and pumping. Abbrev *IHP*.

indicated mean effective pressure (*Eng*) The average pressure exerted by the working fluid in an engine cylinder throughout the cycle, equal to the mean height of the indicator diagram in $kN\ m^{-2}$ or $lbf\ in^{-2}$. Abbrev *IMEP*.

indicated ore (*MinExt*) Proved limits of deposit, in the light of known geology of mine and economic factors.

indicated power (*Eng*) See INDICATED HORSEPOWER.

indicated thermal efficiency (*Eng*) The ratio between the indicated power output of an engine and the rate of supply of energy in the steam or fuel.

indicating instrument (*Eng*) One in which the immediate value only of the measured quantity is visually indicated.

indication (*Eng*) A sign on inspection which indicates an imperfection of the material.

indicator (*BioSci*) (1) The presence of a species that gives an indication of features of the habitat, or method of land management by growing well or badly, eg the stinging nettle which indicates a high level of available phosphorus in the soil. (2) Plants that react to a particular pathogen or environmental factor with obvious symptoms and may, therefore, be used to identify that pathogen or factor.

indicator (*Chem*) (1) A substance whose colour varies with the acidity or alkalinity of the solution in which it is dissolved. (2) Any substance used to indicate the completion of a chemical reaction, generally by a change in colour.

indicator (*ElecEng*) See ANNUNCIATOR.

indicator (*Eng*) An instrument for obtaining a diagram of the pressure–volume or pressure–time changes in an engine or compressor cylinder during the working cycle.

indicator card (*Eng*) A chart on which the trace of an INDICATOR is recorded, producing an INDICATOR DIAGRAM.

indicator diagram (*Eng*) A graphical representation of the pressure and volume changes undergone by a fluid, while performing a work cycle in the cylinder of an engine or compressor, the area representing, to scale, the work done during the cycle. See INDICATED MEAN EFFECTIVE PRESSURE, LIGHT-SPRING DIAGRAM.

indicator gate (*Electronics*) A step or pulse signal applied to a cathode-ray tube to control its sensitivity in order to highlight a certain part of the display.

indicator range (*Chem*) The range of pH values within which an indicator (1) changes colour.

indicator species (*BioSci*) A species whose presence or absence indicates particular conditions in a habitat or that are particularly susceptible to the effects of some environmental factor.

indicator species analysis (*BioSci*) Multivariate statistical technique to enable classification of vegetation on the basis of the presence or absence of key species (INDICATOR SPECIES).

indicator tube (*Electronics*) Miniature cathode-ray tube in which size or shape of target glow varies with input signal.

indicator vein (*MinExt*) In prospecting, one associated with the lode or vein being traced, thus guiding the search.

indices of crystal faces (*Crystal*) See MILLER INDICES.

indicial admittance (*ICT*) Transient current response of a circuit to the application of a STEP FUNCTION of 1 volt, using Heaviside operational calculus.

indicial response (*ICT*) Output waveform from a system when a step pulse of unit magnitude is applied to the output.

indicolite (*Min*) A blue (either pale or bluish-black) variety of tourmaline. Also *indigolite*.

indigenous (*BioSci*) Native; not imported.

indigestion (*Med*) A condition, marked by pain and discomfort, in which the normal digestive functions are impeded. Also *dyspepsia*.

indigo (*Chem*) $C_{16}H_{10}N_2O_2$. A dye occurring in a number of plants, esp in species of *Indigofera*, in the form of a glucoside. It is an indole derivative. Indigo is a very important blue vat dyestuff, and can be synthesized in various ways.

indigo

indigo copper (*Min*) See COVELLITE.

indigolite (*Min*) See INDICOLITE.

indinavir (*Pharmacol*) A *protease inhibitor* used as a drug in the treatment of HIV.

indirect address (*ICT*) The address specified in the instruction is that of a location which in turn contains the required address.

indirect-arc furnace (*ElecEng*) An electric-arc furnace in which the arc is struck between two electrodes mounted above the charge, the latter being heated chiefly by radiation.

indirect cycle (*NucEng*) Nuclear power plant in which the core coolant passes through a heat exchanger in which the secondary circuit of water produces steam for the turbines, as in a pressurized water reactor.

indirect-fired furnace (*Eng*) One in which the combustion chamber is separate from the one in which the charge is heated.

indirect heating (*Eng*) A system of heating by convection, as opposed to *direct heating* by radiation.

indirect immunofluorescence (*BioSci*) A technique in which a specific antibody is first bound to its antigen. A fluorochrome visible under FLUORESCENCE MICROSCOPY and conjugated to a second antibody specific to the first is then used to detect the presence of the first antibody and therefore the original antigen.

indirect lighting (*ElecEng*) A system of lighting in which more than 90% of the total light flux from the fittings is emitted in the upper hemisphere.

indirectly heated cathode (*Electronics*) One with an internal heater, highly insulated from the cathode on a surrounding ceramic cylinder.

indirectly heated valve (*Electronics*) A valve using an INDIRECTLY HEATED CATHODE.

indirect metamorphosis (*BioSci*) The complex change characterizing the life cycles of ENDOPTERYGOTA; the young are larvae and the imago is preceded by a pupal instar.

indirect wave (*ICT*) Also *indirect ray*. See IONOSPHERIC WAVE.

indium (*Chem*) Symbol In, at no 49, ram 114·82, mp 155°C, bp 2100°C, rel.d. 7·28 at 13°C, electrical resistivity 9×10^{-8} Ω m. A silvery metallic element in the third group of the periodic system. Found in traces in zinc ores. The metal is soft and marks paper like lead; it forms compounds with carbon compounds. It has a large cross-section for slow neutrons and so is readily activated and it has eight isotopes. Also used in the manufacture of transistors and as bonding material for acoustic transducers.

indium-doped tin oxide (*Electronics*) Used in thin-film form as a transparent electrode material in electronic display devices. Abbrev ITO.

individual (*BioSci*) A single member of a species; a single zooid of a colony of Cnidaria or Polyzoa; a single unit or specimen.

individual axle-drive (*ElecEng*) A term applied to the arrangement of an electric locomotive in which each driving axle is driven by a separate motor. Also *independent axle-drive*.

individual distance (*Psych*) A spatial relationship between members of a flock (or other social group, eg a school or herd) which is maintained through the two conflicting tendencies of social attraction and aggression or avoidance.

individual risk (*Radiol*) The probability of radiation damage to an individual.

indole (*Chem*) *Benzpyrrole*. C_8H_7N. Mp 52°C, bp (decomposition) 245°C, volatile in steam. Colourless plates. Indole forms the basis of the indigo molecule and results from the condensation of a benzene nucleus with a pyrrole ring.

indole

indole-3-acetic acid (*BioSci*) The commonest naturally occurring plant growth substance of the AUXIN type. Also *auxin, heteroauxin*. Abbrev IAA.

indole-3-butyric acid (*BioSci*) A synthetic plant growth regulator with auxin-like activity used esp in ROOTING COMPOUNDS. Abbrev IBA.

indolent (*Med*) Causing little or no pain, eg *indolent* ulcer.

indometacin (*Pharmacol*) A non-steroidal anti-inflammatory drug. Formerly *indomethacin*.

induced charge (*Phys*) Charge produced on a conductor as a result of a charge on a neighbouring conductor.

induced current (*Phys*) The current which flows in a circuit as a result of induced emf.

induced dipole moment (*Phys*) The induced moment of an atom or molecule which results from the application of an electric or magnetic field.

induced drag (*Aero*) The portion of the DRAG of an aircraft attributable to the derivation of lift.

induced draught (*Eng*) A forced draught system used for boiler furnaces, in which a fan placed in the uptake induces an airflow through the furnace. See BALANCED DRAUGHT, EXTRACTION FAN, FAN.

induced emf (*Phys*) The emf which appears in a circuit as a result of changes in the interlinkages of magnetic flux with part of the circuit; the emf in the secondary of a

transformer. Discovered by Faraday in 1831. See FARADAY'S LAW OF INDUCTION.

induced moving-magnet instrument (*ElecEng*) An instrument whose operation depends on the force exerted by the resultant of the fields produced by a fixed coil carrying a current, and a permanent magnet fixed at an angle thereto, on a movable piece of magnetic material.

induced polarization (*Phys*) Polarization which is not permanent in a dielectric, but arises from applied fields.

induced radioactivity (*Phys*) Radioactivity induced in non-radioactive elements by neutrons in a reactor, or protons or deuterons in a cyclotron or linear accelerator. X-rays or gamma rays do not induce radioactivity unless the gamma-ray energy is exceptionally high.

induced reaction (*Chem*) A chemical reaction which is accelerated by the simultaneous occurrence in the same system of a second, rapid reaction.

inducer (*BioSci*) An agent that increases the transcription of specific genes.

inducible enzyme (*BioSci*) An enzyme that is formed only in response to an inducing agent, often its substrate. Cf CONSTITUTIVE ENZYME.

inductance (*Phys*) (1) The property of a circuit element which, when carrying a current, is characterized by the formation of a magnetic field and the storage of magnetic energy. (2) The magnitude of such capability. Symbol L. See MUTUAL INDUCTANCE, SELF-INDUCTANCE.

inductance factor (*ElecEng*) A term sometimes used to denote the ratio of the reactive current to the total current in an ac circuit, ie the *sine* of the angle of lag.

induction (*BioSci*) The production of a definite condition or differentiated characteristic by the action of an external factor, eg in early development.

induction (*Chem*) A change in the electronic configuration and hence reactivity of one group in a molecule upon addition of a neighbouring polar group.

induction (*ElecEng*) The driving of electric current by time-varying magnetic fields.

induction (*MathSci*) See MATHEMATICAL INDUCTION.

induction (*Psych*) (1) A reasoning process in which general principles are derived from consideration of specific instances. (2) The transfer of effects from one thing to another, eg emotions from person to person by sympathetic induction.

induction balance (*ElecEng*) An electrical network, ie a bridge, to measure inductance.

induction coil (*ElecEng*) A transformer for producing high-voltage pulses in the secondary winding, obtained from interrupted dc in the primary, as for a petrol engine. The original Ruhmkorff induction coil was magnetically open-circuit and self-interrupting, like a buzzer or relay; used for early discharges in gas tubes.

induction compass (*Phys*) A compass which indicates the direction of the Earth's magnetic field by a rotating coil, in which an emf is induced.

induction field (*Phys*) See FIELD.

induction flame damper (*Autos*) See FLAME TRAP.

induction furnace (*ElecEng*) Application of induction heating in which the metal to be melted forms the secondary of a transformer.

induction generator (*ElecEng*) An electric generator similar in construction and operation to an induction motor; in order to generate, it must be driven above synchronous speed and must be excited from the ac supply into which it is delivering power.

induction hardening (*Eng*) Using high-frequency induction to heat a metal part for surface hardening. The heating is rapid and lends itself to control of the thermal gradient and hence the depth of hardening, since the penetration is inversely proportional to the frequency.

induction heating (*ElecEng*) That arising from eddy currents in conducting material, eg solder, profiles of gear wheels, etc. Generated with a high-frequency source,

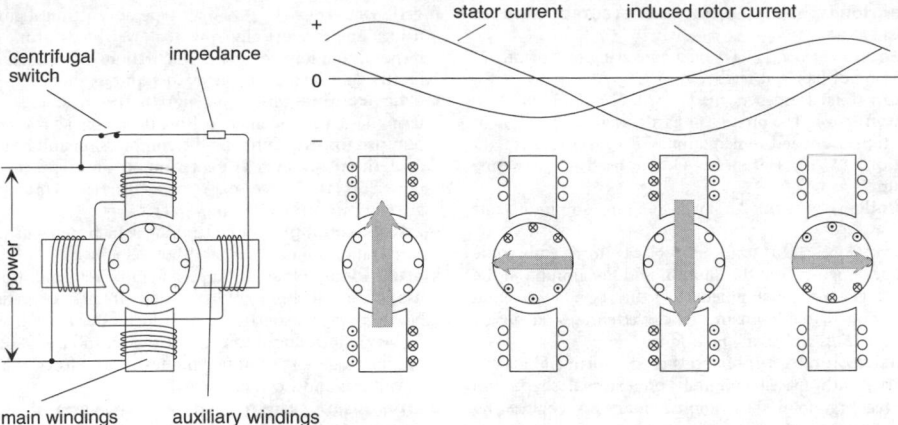

induction motor Single phase. The four drawings to the right show two poles with main windings. The large arrows indicate the size and direction of the resultant flux which causes the rotor to turn.

usually oscillators of high power, operating at 10^6–10^7 Hz. Also *eddy current heating*.

induction lamp (*ElecEng*) See NEON INDUCTION LAMP.

induction machine (*ElecEng*) See INDUCTION GENERATOR, INDUCTION MOTOR.

induction manifold (*Autos*) In a multicylinder petrol engine, the branched pipe which leads the mixture from the carburettor to the combustion chambers. Also *inlet manifold*.

induction meter (*ElecEng*) The most common type of ac integrating meter; it is a motor meter in which the torque is produced as the result of the interaction between an alternating flux and currents induced in a disk by this flux.

induction motor (*ElecEng*) An ac motor in which currents in the primary winding (connected to the supply) set up a flux which causes currents to be induced in the secondary winding (usually the rotor); these currents interact with the flux to produce rotation. Also *asynchronous motor*, *non-synchronous motor*.

induction motor generator (*ElecEng*) A motor-generator set driven by an induction motor.

induction period (*Chem*) The interval of time between the initiation of a chemical reaction and its observable occurrence.

induction port (*Autos*) The port through which the charge is induced into the cylinder during the suction stroke. Also *inlet port*.

induction regulator (*ElecEng*) A voltage regulator having a winding connected in series with the supply; voltages are induced in this winding from a primary winding connected across the supply, and regulation of the voltage is carried out by varying the relative position of the two windings.

induction relay (*ElecEng*) A relay, for use in an electrical circuit, in which the contacts are closed as the result of the interaction between an alternating flux and currents induced in a disk by this flux.

induction stroke (*Autos*) The suction stroke, charging stroke or intake stroke, during which the working charge or air is induced into the cylinder of an engine.

induction valve (*Autos*) The valve through which the charge is induced into the cylinder during the suction stroke. Also *inlet valve*.

inductive (*ElecEng*) Said of an electric circuit or piece of apparatus which possesses mutual or self-inductance, which tends to prevent current changes. Always present to some extent, but may often be neglected.

inductive circuit (*ElecEng*) One in which effects arising from inductances are not negligible, the back emf tending to oppose a change in current, leading to sparking or arcing at contacts which attempt to open the circuit.

inductive drop (*ElecEng*) Voltage drop produced in an ac circuit owing to its mutual or self-inductance.

inductive load (*ElecEng*) Terminating impedance which is markedly inductive, taking current lagging in phase on the source emf, eg electrodynamic loudspeaker or motor. Also *lagging load*.

inductively coupled plasma-mass spectrometry (*Chem, Phys*) The plasma torch provides a rich source of free atoms and ions from the sample elements. Part of the sample stream from the centre of the fireball is directed to a mass spectrometer which identifies the elements from the mass number of the ion peaks and measures their amounts from the peak size. Used in ecological surveys for minor or trace elements, it can give the isotopic composition of certain elements like lead in environmental contamination. The radiogenic source of lead means that different deposits may have different isotopic composition which can be used to distinguish between lead from paint or water pipes, and from anti-knocking compounds. In geochemistry it is a useful detection method for isotopic ratios of eg strontium (Sr) and in biology certain studies can be made with non-radioactive tracers. Abbrev *ICP-MS*.

inductive neutralization (*ElecEng*) An amplifier in which the feedback of the self-capacitance of the circuit elements is balanced by the equal and opposite susceptance of an inductor.

inductive pick-off (*ElecEng*) One in which changes in RELUCTANCE of a laminated path alter a current or generate an emf in a winding.

inductive reactance (*ElecEng*) That which relates the current through an inductance to the voltage appearing across it. Of magnitude *WL*, where *W* is angular frequency of supply and *L* is inductance in HENRYS and associated phase angle such that current lags voltage by 90°.

inductive reaction (*ElecEng*) See ELECTROMAGNETIC REACTION.

inductive resistor (*ElecEng*) Wirewound resistor having appreciable inductance at frequencies in use.

inductor (*Chem*) A substance which accelerates a slow reaction between two or more substances by reacting rapidly with one of the reactants.

inductor (*ElecEng*) (1) Any circuit component whose inductance cannot be treated as negligible. (2) A coil designed to exhibit INDUCTANCE.

inductor generator (*ElecEng*) An electric generator in which the field and armature windings are fixed relative to each other, the necessary changes of flux to produce the emf being produced by rotating masses of magnetic material.

inductor loudspeaker (*Acous*) See ELECTROMAGNETIC LOUDSPEAKER.

indumentum (*BioSci*) (1) The hairy covering of a plant. (2) A covering of hair and feathers.

indurated (*Eng*) Hardened, made hard. N *induration*.

induration (*Geol*) The process of hardening a soft sediment by heat, pressure and cementation. Cf DIAGENESIS.

induration (*Med*) Hardness. Often used to describe tumours.

Indus (Indian) (*Astron*) An inconspicuous southern constellation.

indusium (*BioSci*) In some insects, a third embryonic envelope lying between the chorion and the amnion in the early stages of development of the egg; a cerebral convolution of the brain in higher vertebrates; an insect larva case. Adjs *indusiate, indusiform*.

industrial design (*Genrl*) Methods used to plan and market products, with special emphasis on external shape and form (eg ergonomics). Complementary to engineering design. See PRODUCT DESIGN.

industrial diamond (*Min*) Small diamonds, not of gemstone quality, eg BLACK DIAMOND and BORT; used to cut rock in borehole drilling, and in abrasive grinding. Now synthesized on a considerable scale by subjecting carbon to ultrahigh pressures and temperature.

industrial frequency (*ElecEng*) A term used to denote the frequency of the ac used for ordinary industrial and domestic purposes, usually 50 or 60 Hz. Also *mains* or *power frequency*.

industrial melanism (*BioSci*) MELANISM which has developed as a response to blackening of trees and other habitats by industrial pollution. This favours melanic forms, esp among moths which rest on trees during the day.

industrial/organizational psychology (*Psych*) The branch of psychology that is concerned with the study of organizations and organizational behaviour. It includes the study of things such as personnel practices, reward systems and market research, and is also applied to non-industrial organizations such as hospitals and universities. Abbrev *I/O psychology*.

industry standard architecture (*ICT*) A standard ARCHITECTURE for IBM-COMPATIBLE computers.

inelastic collision (*Phys*) In atomic or nuclear physics, a collision in which there is a change in the total energies of the particles concerned resulting from the excitation or de-excitation of one or both of the particles. See COLLISION.

inelastic scattering (*Phys*) See SCATTERING.

inequality (*Astron*) The term used to signify any departure from uniformity in orbital motion; it may be *periodic*, ie completing a full cycle within a specific time and then repeating it, or *secular*, ie increasing steadily in magnitude with time.

inequality (*MathSci*) A statement as to which is the larger or smaller of two quantities. The *strict inequality* that a is greater than b is written $a > b$, and the equivalent statement that b is less than a, $b < a$. The derived *weak inequalities* that a is not less than b, and that b is not greater than a, are written $a \geq b$ and $b \leq a$.

inequipotent (*BioSci*) Possessing different potentialities for development and differentiation.

inequivalve (*BioSci*) Having the two valves of the shell unequal.

inert (*Chem*) Not readily changed by chemical means.

inert anode (*Ships*) An anode of platinized titanium, used in CATHODIC PROTECTION. Requires an impressed direct current. Long-lasting. Cf VIRTUALLY INERT ANODE.

inert gases (*Chem*) See NOBLE GASES.

inertia (*ImageTech*) A factor used in some systems of photographic speed rating, obtained by extrapolating the linear portion of the CHARACTERISTIC CURVE to indicate the nominal exposure for zero density.

inertia (*Phys*) The property of a body, proportional to its mass, which opposes a change in the motion of the body. See INERTIAL FORCE, INERTIAL REFERENCE FRAME.

inertia governor (*Eng*) A shaft type of centrifugal governor using an eccentrically pivoted weighted arm, which responds rapidly to speed fluctuations by reason of its inertia, and in such a way as to suppress them.

inertial confinement (*NucEng*) In fusion studies, short-term plasma confinement arising from inertial resistance to outward forces (mainly by the compression and heating of deuterium or mixed deuterium–tritium pellets by a powerful laser). See CONTAINMENT, INERTIAL FUSION SYSTEM, MAGNETIC CONFINEMENT.

inertial damping (*Phys*) Damping which depends on the acceleration of a system, and not velocity.

inertial force (*Phys*) An apparent force resulting from the use of accelerating and rotating frames of reference. Newton's laws may still be used in these frames if inertial forces are introduced to preserve the second law of motion. In the case of rotating frames these forces are the centrifugal and Coriolis forces.

inertial fusion system (*NucEng*) A system in which small capsules (*pellets*) containing deuterium and tritium are injected into a reaction chamber and ignited by high-energy laser or ion beams. See MAGNETIC CONFINEMENT FUSION SYSTEM.

inertial guidance (*Aero*) Navigation of a aircraft, spacecraft or missile by measuring the inertial forces during flight and comparing them with a program held on board. It is not subject to outside interference.

inertial impaction (*PowderTech*) A method of collecting small particles of dust and droplets from a fluid stream by allowing them to impinge upon an interposed deflecting surface.

inertial navigation system (*Aero*) An assembly of highly accurate gyros to stabilize a platform supported on gimbals on which are mounted a group of similarly accurate accelerometers (typically one for each of the three rectilinear axes) which measure all accelerations imparted. With one automatic time integration, these measurements give a continuous read-out of velocity and with another a read-out of present position related to the start. Accuracy is typically 1 in 10^9. See FIBRE OPTIC GYRO.

inertial reference frame (*Phys*) In mechanics, a reference frame in which Newton's first law of motion is valid.

inertia switch (*ElecEng*) One operated by an abrupt change in its velocity, as for some meters, to avoid overloading.

inertia wheel (*Space*) See MOMENTUM WHEEL.

inertinite (*Geol*) A carbon-rich MACERAL found in coal.

inert metal (*Phys*) Alloy (usually Ti–Zr) for which scattering of neutrons by nuclei is negligible.

I neutron (*Phys*) A neutron possessing such energy as to undergo resonance absorption by iodine.

infanticide (*Med*) Killing of an infant, particularly the killing of a newborn infant by its mother.

infantile paralysis (*Med*) See POLIOMYELITIS.

infantilism (*Med*) A disturbance of growth, the persistence of infantile characters being associated with general retardation of development.

infarct (*Med*) That part of an organ which has had its blood supply cut off, the area so deprived undergoing aseptic necrosis.

infarction (*Med*) The formation of an INFARCT; the infarct itself.

infection (*Med*) The invasion of body tissue by living micro-organisms, with the consequent production in it of morbid change; a diseased condition caused by such invasion; the infecting micro-organism itself.

infectious anaemia of horses (*Vet*) An acute or chronic RNA viral disease of equines spread by biting insects and other mechanical vectors. Symptoms include pyrexia, depression, oedema and anaemia. Virus found in all tissues and persist in WHITE BLOOD CELLS for life. Vaccines available. Notifiable in the UK. Also *swamp fever*.

infectious avian bronchitis (*Vet*) An acute, highly contagious respiratory disease of chickens, caused by a virus and associated with inflammation of the respiratory tract, esp the trachea and bronchi; the main symptoms are nasal discharge, gasping, rales and coughing.

infectious avian encephalomyelitis (*Vet*) An encephalomyelitis of young chicks, caused by a virus, and characterized by muscular inco-ordination, muscular tremor and death. Also *epidemic tremor*.

infectious bovine rhinotracheitis (*Vet*) Common herpes virus infection of cattle. Most common signs are associated with the upper respiratory tract, with reproductive, nervous and alimentary symptoms also. Vaccines available.

infectious bulbar paralysis (*Vet*) See AUJESKY'S DISEASE.

infectious canine hepatitis (*Vet*) Caused by canine adenovirus, type 1. Acute, contagious and often fatal disease of the dog characterized by fever, diarrhoea, vomiting melaena, abdominal pain, jaundice and nervous signs. Corneal oedema (blue eye) sometimes during recovery. Vaccines available. Also *Rubarth's disease*.

infectious coryza (*Vet*) An acute, contagious bacterial infection of the upper respiratory tract of domestic fowl. Also *contagious catarrh, fowl coryza, roup*.

infectious endocarditis (*Med*) A serious infection of the endocardium overlying the heart valves, particularly if they are diseased or altered. Formerly termed *sub-acute bacterial endocarditis* as the majority were caused by *Streptococcus viridans*. Because normal valves can be attacked by a wide variety of micro-organisms, the term infectious endocarditis has been adopted.

infectious hepatitis (*Med*) A term applied to viral infection of the liver, causing jaundice. Hepatitis A is usually foodborne with an incubation period of 1 month with recovery the rule. Hepatitis B is more severe and is usually transmitted by infected blood or needles or instruments contaminated with blood. It is now recognized to be sexually transmitted also. The incubation period is 3–6 months with a mortality rate of 5–20%. The virus may remain in the blood in the so-called 'carrier state'. Variants 'C', 'D' and 'E' have also been described.

infectious icterohaemoglobinuria (*Vet*) See REDWATER.

infectious jaundice (*Med*) See LEPTOSPIROSIS.

infectious keratitis (*Vet*) See INFECTIOUS OPHTHALMIA.

infectious laryngotracheitis (*Vet*) A highly contagious, often fatal, virus infection of the respiratory tract of the chicken. Signs include coughing and sneezing. Vaccination available. Abbrev *ILT*.

infectious mononucleosis (*Med*) An acute viral infectious disease characterized by slight sore throat, enlargement of glands in the neck, and an increase in the white (mononuclear) cells of the blood. The causative virus is the EPSTEIN–BARR VIRUS. Common in adolescents and usually mild, but jaundice may occur in some and convalescence may also be greatly prolonged. Also *glandular fever*.

infectious ophthalmia (*Vet*) A contagious form of conjunctivitis and keratitis occurring in cattle. *Moraxella bovis* with or without *Neisseria* spp are implicated. Also *infectious bovine kerato-conjunctivitis, infectious keratitis, New Forest disease, pink eye*.

infectious parotitis (*Med*) See MUMPS.

infectious pig paralysis (*Vet*) See TESCHEN DISEASE.

infectious pododermatitis (*Vet*) See FOUL IN THE FOOT.

infectious sinusitis of turkeys (*Vet*) A disease of turkeys caused by infection of the infraorbital sinuses of the head by organisms of the genus *Mycoplasma*; characterized by swelling of the face and a discharge from the eyes and nostrils. Also *big head disease of turkeys*.

infectious synovitis (*Vet*) A disease of chickens caused by infection by organisms of the genus *Mycoplasma* resulting in exudative synovitis.

inference engine (*ICT*) Part of an EXPERT SYSTEM that contains the general problem-solving knowledge or ALGORITHMS.

inferior (*BioSci*) (1) Lower, under, situated beneath, eg the inferior rectus muscle of the eyeball. (2) In plants, an ovary having perianth and stamens inserted round the top, ie epigynous. The ovary appears to be sunk into and fused with the receptacle. Cf SUPERIOR.

inferior (*Print*) A term used to describe small figures or letters set below the general level of the line; as in chemical formulae, eg C_6H_3.

inferior conjunction (*Astron*) See CONJUNCTION.

inferiority complex (*Psych*) A concept, first proposed by Adler, referring to the repressed and powerful conviction of inferiority, whose basis lies in the universal experience of infantile helplessness and dependency; these feelings become repressed during development, and are a powerful dynamic force in determining adult personality and, often, character disorders.

inferior planet (*Astron*) See PLANET.

inferior vena cava (*BioSci*) See POSTCAVAL VEIN.

inferred-zero instrument (*ElecEng*) See SUPPRESSED-ZERO INSTRUMENT.

infertility (*Med*) Inability to produce offspring.

infestation (*Med, Food*) (1) The condition of being occupied or invaded by parasites, usually parasites other than bacteria. (2) The presence of animals in eg storage premises. Evidence can be seen in foods or raw materials as live animals or their remains, eg excrement, hair, feathers.

infilling (*Build*) Material, such as hard core, used for making up levels, eg under floors.

infiltration (*Med*) (1) The accumulation of abnormal substances (or of normal constituents in excess) in cells of the body. (2) The gradual spread of infection in an organ (eg tuberculous infiltration of the lung).

infiltration (*PowderTech*) Impregnation using capillary forces to soak up the impregnant.

infimum (*MathSci*) Greatest lower bound. See BOUNDS OF A FUNCTION.

infinite attenuation (*Phys*) The property of some filters of providing a theoretically infinite attenuation for one or more specified frequencies against which strong discrimination is required.

infinite line (*Phys*) A transmission line which is infinitely long, or finite but terminated with its characteristic impedance, and along which there is uniform attenuation and phase delay.

infinite loop (*ICT*) A LOOP from which there is no exit, other than by terminating the run.

infinite set (*MathSci*) A set that can be put into a one–one correspondence with part of itself, eg the positive integers, which can be put into a one–one correspondence with the positive even integers.

infinitesimal (*MathSci*) A vanishingly small part of a quantity, which, although it retains the dimensions or other qualities of the quantity, is negligible in magnitude compared with the quantity.

infinity (*MathSci*) Loosely, a quantity larger than any quantified value. When conceived of as a number larger than any natural number, for some purposes it may be considered as the reciprocal of zero, and minus infinity equated to plus infinity. Formally an infinite sequence is one such that every term has a successor, no matter how far the sequence is extended, and an infinite set is one that can be put in one-to-one correspondence with a proper subset. There are in fact infinitely many distinct infinite CARDINAL NUMBERS. Symbol ∞. See also POINTS AT INFINITY.

infinity plug (*ElecEng*) A plug in a resistance box which, when withdrawn, breaks the circuit, ie introduces an infinite resistance.

infix (*ICT*) A form of algebraic notation in which the operators are placed between the operands (eg A + B, X and Y). Cf POSTFIX.

inflammation (*Med*) The reaction of living tissue to injury or to infection, characterized by increased blood supply to the affected area, which becomes red, hot, painful and swollen, with exudation of lymph and escape into the tissue of blood cells.

inflatable aircraft (*Aero*) A small low-performance aircraft for military use, in which the aerofoil surfaces (and sometimes the fuselage) are inflated so that it can be compactly packed for transport.

inflation (*Aero*) The process of filling an airship or balloon with gas. Also *gassing*.

inflationary universe (*Astron*) A model of the very early universe (10^{-44} seconds after the Big Bang) in which the universe expands momentarily much faster than the speed of light. This model is able to account for the flatness of space–time and the isotropy of the observed universe. See panel on COSMOLOGY.

inflected arch (*CivEng*) See INVERTED ARCH.

inflexion (*MathSci*) See POINT OF INFLEXION.

infliximab (*Pharmacol*) A therapeutic antibody used to treat Crohn's disease and some types of arthritis.

inflorescence (*BioSci*) (1) Flowering branch (or portion of the shoot above the last stem leaves) including its branches, bracts and flowers. See CAPITULUM, CYMOSE, MIXED INFLORESCENCE, PANICLE, RACEMOSE, UMBEL. (2) In bryophytes, a group of antheridia or archegonia and associated structures.

influence line (*Build, Eng*) A curve whose ordinate represents the value of some variable (such as the bending moment) at particular points in the structure, due to the application of a unit load at that point. Also *load curve*.

influence machine (*ElecEng*) See ELECTROSTATIC GENERATOR.

influenza (*Med, Vet*) An airborne respiratory virus infection, causing epidemics which are often worldwide, but whose severity varies with the (constantly changing) virus type. Abbrev *flu*. See EQUINE INFLUENZA, SWINE INFLUENZA.

information (*Genrl*) Meaning given to data through analysis and interpretation.

information (*ICT*) Any intelligence (code, speech, images, etc) that can be communicated to a remote destination by electrical or electromagnetic means.

information content (*ICT*) In information theory, a measure of the information conveyed by the occurrence of an allowable symbol in a transmitted message; defined as the negative of the probability that this particular symbol might be sent. If the base of the logarithm chosen is 2, then information content is measured in hartleys: 1 hartley equals $\log_2 10$, or $3 \cdot 323$ BITS.

information processing (*ICT*) The organization, manipulation and distribution of information. Central to almost every use of the computer and almost synonymous with computing.

information rate (*ICT*) The number of symbols transmitted per second multiplied by the average INFORMATION CONTENT per symbol. See BIT.

information retrieval (*ICT*) Abbreviated expression for *information storage and retrieval*. A major branch of COMPUTER SCIENCE concerned with the use of computers to structure and retrieve information from large information stores. Data retrieval is the special case where the nature of the information stored is fully expressed and retrieval involves perfect matching. The more difficult situation occurs when, as in document retrieval, the nature of the stored information cannot be fully summarized and the request for information does not fully anticipate the nature of the response. Abbrev *IR*. See KEYWORDS.

information storage and retrieval (*ICT*) See INFORMATION RETRIEVAL.

information superhighway or highway (*ICT*) Digital electronic telecommunication systems, including computer networks (esp the Internet), cable and satellite TV, and telephone links. See panel on INTERNET.

information technology (*ICT*) Abbrev *IT*. The application to INFORMATION PROCESSING of current technologies from computing, MICROELECTRONICS and TELECOMMUNICATIONS. Now more commonly *information and communications technology*, abbrev *ICT*.

information theory (*ICT*) Mathematical study of the information rate, channel capacity, noise and other factors affecting information transmission and reception. Initially applied to electrical communications, now applied universally to business systems and other areas concerned with information in its broadest sense and its flow through networks. See SHANNON'S THEOREM.

infra- (*Genrl*) Prefix from Lt *infra*, below.

infrablack (*ImageTech*) The amplitude in a TV signal beyond the black level of the picture.

infracostal (*Med*) Beneath the ribs.

infradyne (*ICT*) Supersonic heterodyne receiver in which the intermediate frequency is higher than that of the incoming signal.

inframarginal (*BioSci*) Below the margin; a marginal structure; in tortoises and turtles, one of certain plates of the carapace lying below the marginals; in star fish, one series of ossicles situated on the lower margin of each ray.

infraorbital foramen (*BioSci*) In mammals, a foramen on the outer surface of each maxilla for the passage of the second division of the fifth cranial nerve.

infraorbital glands (*BioSci*) In mammals, one of the four pairs of salivary glands.

infrared astronomy (*Astron*) The study of radiation from celestial objects in the wavelength range 800 nm–1 mm. Absorption by water vapour in our atmosphere poses severe difficulties, some of which are overcome at high-altitude observatories such as Mauna Kea in Hawaii at 4000 m. There are many thousands of infrared sources in our Galaxy, principally cool giant stars, nascent stars and the galactic centre itself.

infrared countermeasures (*Aero*) Means of deceiving missiles guided by infrared sensors by (1) flares deployed as decoys from aircraft or (2) in design, by reducing heat output from jet exhaust or shielding hot parts.

infrared detection (*Phys*) The detection and photographic registration of infrared rays with special dyes: photosensitively with a special Cs–O–Ag surface; by photoconduction of lead sulphide and telluride; and, in absolute terms, by bolometer, thermistor, thermocouple or Golay detector.

infrared maser (*Phys*) A maser which radiates or detects signals of millimetre wavelengths. See panel on LASER.

infrared photography (*ImageTech*) Photography using materials specially sensitized to infrared radiation; applications include photography without visible light, haze penetration, camouflage and forgery detection and medical records. See FALSE COLOUR.

infrared radiation (*Phys*) Electromagnetic radiation in the wavelength range of 0·75–1000 μm approximately, ie between the visible and microwave regions of the spectrum. The *near* infrared is from 0·75 to 1·5 μm, the *intermediate* from 1·5 to 20 μm and the *far* from 20 to 1000 μm. See fig. at appendix on Electromagnetic spectrum.

infrared spectrometer (*Phys*) An instrument similar to an optical spectrometer but employing non-visual detection and designed for use with infrared radiation. The infrared spectrum of a molecule gives information as to the functional groups present in the molecule and is very useful in the identification of unknown compounds.

infrared spectroscopy (*Phys*) Routine analytical tool for detection of functional groups by infrared absorption in molecules, esp polymers, which can be easily examined in thin-film form. Can also use fluid smear on sodium chloride disks. Absolute structure determination is more difficult than matching unknown spectrum with standard spectra, although additives cause problems by obscuring the spectrum of the matrix polymer. Analysts then turn to NUCLEAR MAGNETIC RESONANCE SPECTROSCOPY or ULTRAVIOLET SPECTROSCOPY if a polymer solution can be made.

infrasound (*Acous*) The sound of frequencies below the usual audible limit of 20 Hz.

infundibulum (*BioSci*) (1) Generally, a funnel-shaped structure. (2) In vertebrates, a ventral outgrowth of the brain or a pulmonary vesicle. (3) In CEPHALOPODA, the siphon. (4) In CTENOPHORA, the flattened gastric cavity. Adj *infundibular*.

infusible (*Eng*) Not rendered liquid under specified conditions of pressure, temperature or chemical attack. Also *refractory*.

infusion (*Chem*) A solution of the soluble constituents of vegetable matter, obtained by steeping the vegetable matter in liquid, often hot and sometimes under pressure.

infusorial earth (*Min*) See TRIPOLITE.

ingate (*Eng*) The channel or channels, by which the molten metal is led from the runner hole into the interior of a mould.

ingestion (*BioSci*) The act of swallowing or engulfing food material (ingesta) so that it passes into the body. V *ingest*.

ingluvies (*BioSci*) An oesophageal dilatation of birds, the crop.

ingluvitis (*Vet*) Inflammation of the crop, or ingluvies, of birds.

ingo (*Build*) See REVEAL. Also *ingoing*.

ingot (*Eng*) A metal casting of a shape suitable for subsequent hot-working, eg for rolling or forging.

ingot iron (*Eng*) Iron of comparatively high purity, produced in the same way as steel, but under conditions that keep down the carbon, manganese and silicon content.

ingot mould (*Eng*) The mould or container in which molten metal is cast and allowed to solidify to form an ingot.

ingot stripper (*Eng*) Mechanism for extracting ingots from ingot moulds.

ingravescent (*Med*) Gradually increasing in severity.

ingroup, outgroup (*Psych*) A set of related concepts used in studies of intergroup relations; a person's ingroup is the group the person perceives as the group they belong to; the outgroup is any other identifiable group.

inguinal (*BioSci*) Pertaining to, or in the region of, the groin, eg the *inguinal canal* through which the testes descend in male mammals.

inguinodynia (*Med*) Pain in the groin.

inhalant (*BioSci*) Pertaining to, or adapted for, the action of drawing in a gas or liquid, eg the *inhalant siphon* in some MOLLUSCA.

inhalation (*Med*) The act of breathing in, or taking into the lungs; any medicinal agent breathed into the lungs.

inherent ash (*MinExt*) Non-combustible material intimately bound in the original coal-forming vegetation, as distinct from 'dirt' from extraneous sources.

inherent filtration (*Radiol*) That introduced by the wall of the X-ray tube as distinct from added primary or secondary filters.

inherent floatability (*MinExt*) The natural tendency of some mineral species to repel water and to become part of the 'float' without preliminary conditioning in the froth-flotation process.

inherent regulation (*ElecEng*) The change in voltage at the output terminals of an electric machine (ac or dc generator, or a converter) when the load is applied or removed, all other conditions remaining constant. The change in secondary voltage of a transformer when the load is changed between zero and full load.

inherent viscosity (*Chem*) The natural logarithm of relative viscosity divided by polymer concentration in dilute solution. Also *log viscosity number*.

inheritance (*ICT*) In object-oriented programming, the process by which a new class takes on the data, methods and attributes of a previously existing class. This enables the reuse of data structures and code.

inherited error (*ICT*) That in one stage of a multistage calculation which is carried over as an initial condition to a subsequent stage.

inhibited oil (*ElecEng*) Transformer or switch oil which includes an anti-oxidant to delay the onset of sludge and acid formation.

inhibition (*BioSci*) The stopping or deceleration of a metabolic process. Adj *inhibitory*. Cf EXCITATION.

inhibitor (*BioSci*) A substance that reduces or prevents some metabolic or physiological process.

inhibitor (*Chem*) An additive which retards or prevents an undesirable reaction: eg (1) phosphates, which prevent corrosion by the glycols in ANTIFREEZE solutions; (2) anti-oxidants in rubber; (3) aromatic amines added to light petroleum products to prevent gum formation by the polymerization of unsaturated components.

inhibitory (*BioSci*) Said of a nerve, whose action upon other nerves tends to render them less liable to stimulation.

inhibitory phase (*Chem*) The protective colloid in a lyophobic sol.

inhour (*NucEng*) The unit of reactivity equal to the reciprocal of period of nuclear reactor in hours, eg a reactivity of 2 inhours will result in a PERIOD of half an hour.

inhour equation (*NucEng*) An equation relating the decay constant, ω, to the excess multiplication δk of the overall neutron flux in a reactor.

$$\delta k = l\omega + k\omega \sum_{i=1}^{I} \frac{\beta i}{\omega + \lambda_i}$$

where l is the neutron lifetime (typically 10^{-3} s), k is the multiplication constant of the reactor, i refers to the DELAYED NEUTRON GROUPS (conventionally six), λ is the decay constant of group i and β is the fractional yield of the group. For small values of δk, when ω can be neglected, the right-hand term in the equation is 0·083 s for uranium-235 fissions, which gives a reactor period of 850 s. This shows the important effect of the delayed neutrons on the stability of a reactor.

in-house recycling (*Eng*) Material reclamation within the factory, eg granulated thermoplastic runners and scrap mouldings fed back into injection-moulding process.

ininquiline (*BioSci*) A guest animal living in the nest of another animal, or making use of the food provided for itself or its offspring by another animal.

inion (*Med*) The external bony protuberance on the occiput, at the back of the skull.

initial (*BioSci*) (1) A dividing cell in a meristem; one of the daughters or its progeny adding to the tissues of the plant, the other remaining in the meristem and repeating the process. (2) A cell in the earliest stages of specialization.

initial conditions (*ICT, MathSci*) Starting values of the parameters of an algorithm that are used in calculating numerical solutions.

initial consonant articulation (*ICT*) See ARTICULATION.

initialization files (*ICT*) FILES that contain information necessary to start up a computer or start up a particular APPLICATION PROGRAM. Within WINDOWS these files are given a .INI FILE EXTENSION.

initialize (*ICT*) To set counters or variables to some starting value (often zero).

initial permeability (*ElecEng*) Limiting value of differential permeability when magnetization tends to zero. Applicable to small alternations of magnetization when there is no magnetic bias.

initial stability (*Ships*) The moment of the couple tending to return the vessel to its equilibrium position when the ANGLE OF HEEL is small. Depends directly on META-CENTRIC HEIGHT and is frequently expressed as such.

initiation (*Chem*) The start of polymerization, often catalytically induced. See CHAIN POLYMERIZATION and panel on POLYMER SYNTHESIS.

initiator (*Chem*) The substance or molecule which starts a chain reaction. See CHAIN POLYMERIZATION and panel on POLYMER SYNTHESIS.

initiator codon (*BioSci*) The sequence AUG or, rarely in prokaryotes, GUG or UUG, that always specifies the first amino acid of a protein. See panel on DNA AND THE GENETIC CODE.

injected (*BioSci*) Having the intercellular spaces filled with water or other fluid.

injection (*Autos*) The process of spraying fuel into the inlet manifold or cylinder of an internal-combustion engine by means of an injection pump.

injection (*Geol*) The emplacement of fluid rock matter in crevices, joints or fissures found in rocks. See INTRUSION.

injection blow moulding (*Plastics*) The process for making plastic bottles, by injection moulding of PARISON followed by blow moulding of reheated parison. Widely used for polyethylene terephthalate (PET) beverage bottles.

injection carburettor (*Aero*) A pressure carburettor in which the fuel delivery to the jets is maintained by pressure instead of by a float chamber. It is unaffected by negative *g* in aerobatics or severe atmospheric turbulence.

injection complex (*Geol*) An assemblage of rocks, partly igneous, partly sedimentary or metamorphic, the former in intricate intrusive relationship to the latter, occurring in zones of intense regional metamorphism.

injection condenser (*Eng*) See JET CONDENSER.

injection efficiency (*Electronics*) The fraction of the current flowing across the emitter junction in a transistor which is due to the minority carriers.

injection lag (*Autos*) In a compression–ignition engine, the time interval between the beginning of the delivery stroke of the fuel injection pump and the beginning of injection into the engine cylinder.

injection moulding (*Plastics*) A method for the fabrication of thermoplastic materials. The viscous resin is squirted, by means of a plunger, out of a heated cylinder into a water-chilled mould, where it is cooled before removal. Method used also with thermosetting moulding powders. See MOULD CYCLE and panel on CERAMICS PROCESSING.

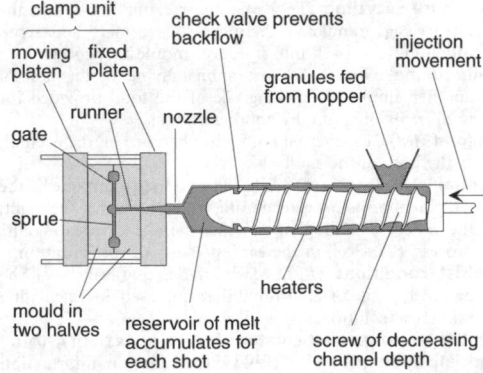

injection moulding

injection pump (*Autos*) A unit of a diesel engine, or petrol-injection system, which injects measured quantities of fuel to each cylinder in turn on the compression stroke. See COMMON-RAIL INJECTION, FUEL INJECTION, JERK-PUMP.

injection string (*MinExt*) Pipe run in addition to the *production string* in the borehole to allow the passage of additives.

injection valve (*Autos*) See ATOMIZER, INJECTOR.

injector (*Autos*) A plug with a valved nozzle through which fuel is metered to the combustion chambers in diesel- or fuel-injection engines. Connected with the fuel pump by a capillary tube. In diesels, the injectors are generally screwed into the cylinder head (corresponding to the sparking plug in a conventional petrol engine); in petrol-injection engines, they are located beside the inlet valves.

injector (*Eng*) A device by which a stream of fluid, as steam, is expanded to increase its kinetic energy, and caused to entrain a current of a second fluid, as water, so delivering it against a pressure equal to, or greater than, that of the steam; the steam injector is commonly used for feeding boilers. SEE GIFFARD'S INJECTOR.

ink (*BioSci*) See INK SAC.

ink (*Genrl*) A dispersion of a pigment or a dye solution produced as a fluid, paste or powder. Writing ink usually consists of a fluid extract of tannin with the addition of solutions of iron salts. The semi-solid ink used in ball-point pens consists of highly concentrated dyes in a non-volatile solvent. Coloured writing inks are prepared by dissolving suitable dyes in water. Marking inks are made from solutions of silver or copper compounds, aniline being sometimes added. See INDIAN INK, PRINTING INK.

inkblot test (*Psych*) See RORSCHACH INKBLOT TEST.

inkjet printer (*ICT*) A printer that produces characters by using a fine jet of quick-drying ink.

inkjet printing (*Print*) A process whereby charged ink particles are projected in a continuous stream (or singly) and then deflected by an electrostatic field using computer control to create text and graphics on the printing substrate.

ink misting (*Print*) Ink is sometimes reduced to a fine mist, particularly on fast-running rotary presses, by the friction between two surfaces rotating in opposite directions, LONG INKS being more susceptible than SHORT INKS.

ink pump (*Print*) On rotary presses: (1) a special type of pump-operated ink supply system; (2) a circulating pump to agitate liquid inks, particularly for photogravure; and (3) a pump to maintain ink supply to the ink ducts.

ink rail (*Print*) Part of a pump-operated ink duct which extends across the machine to apply ink to the system.

ink sac (*BioSci*) In some CEPHALOPODA, a large gland, opening into the alimentary canal near the anus, which secretes a dark-brown pigment (sepia).

ink table (*Print*) A flat surface on a printing machine, part of the ink distribution system; also a flat surface used with a hand roller. Also *ink slab*.

inlay (*Build*) Pieces of metal, ivory, etc, inserted ornamentally (inlaid) in furniture etc.

inlay (*ImageTech*) A combination of two video images using a separate signal to KEY one into the other.

inlet guide vanes (*Aero*) Radially positioned aerofoils in the annular air intake of axial-flow compressors which direct the airflow onto the first stage at the most efficient angle. The vanes are often rotatable about their mounting axes so that different entry airspeeds can be accommodated; they are then called variable inlet guide vanes. Abbrev *IGV*.

inlet manifold (*Autos*) See INDUCTION MANIFOLD.

inlet port (*Autos*) See INDUCTION PORT.

inlet valve (*Autos*) See INDUCTION VALVE.

inlier (*Geol*) An outcrop of older rocks surrounded by those of younger age.

in-line assembly (*Electronics*) Assembly by machine, in which an array of heads inserts components into a printed circuit board.

in-line engine (*Autos*) A multicylinder engine, consisting of a bank of cylinders mounted in line along a common crankcase.

innate (*BioSci*) In plants: (1) Sunken into or originating within the THALLUS. (2) An anther joined by its base to the filament.

innate (*Psych*) Said of behaviour which normally occurs in all members of a species despite natural variation in environmental influences. The development of innate behaviour is complex, and often involves both endogenous and environmental factors. In psychology, however, the term is used simply to mean inherited behaviour, not necessarily common to all member of the species.

innate capacity for increase (*BioSci*) See R.

innate immunity (*BioSci*) Immunity conferred before birth and not acquired subsequently by exposure to antigens from the environment. Also *natural immunity*.

innate releasing mechanism (*Psych*) In classical ethological models of motivation, a postulated mechanism in which instinctive acts are inhibited until the appropriate stimulus, the sign stimulus, is presented.

inner bead (*Build*) See GUIDE BEAD.

inner cell mass (*BioSci*) A group of cells in the mammalian blastocyst that give rise to the embryo and are potentially capable of forming all tissues, embryonic and extra-embryonic, except the trophoblast. Abbrev *ICM*.

inner conductor (*ElecEng*) (1) See INTERNAL CONDUCTOR. (2) The neutral conductor of a three-wire system.

inner dead centre (*Eng*) Of a reciprocating engine or pump, the piston position at the beginning of the outstroke, ie when the crank pin is nearest to the cylinder. Also *top dead centre*.

inner-directed (*Psych*) A tendency to be guided by one's own principles rather than by external influences.

inner ear (*Med*) Structure encased in bone and filled with fluid, including the cochlea, the semicircular balancing canals and the vestibule.

inner forme (*Print*) In SHEET-WORK printing the imposition containing the second and second-last pages of the printed sheet, the *outer forme* containing the first and last pages and being printed on the other side.

inner glume (*BioSci*) Also PALEA.

inner marker beacon (*Aero*) A vertically directed radio beam which marks the airport boundary in a beam-approach landing system, such as the INSTRUMENT LANDING SYSTEM.

innervation (*BioSci*) The distribution of nerves to an organ.

innings (*CivEng*) Lands reclaimed from the sea. See POLDER.

innocent (*Med*) (1) Not malignant; not cancerous. See TUMOUR. (2) Type of heart murmur of no pathological significance.

innominate (*BioSci*) Without a name, eg the *innominate artery* of some mammals, which leads from the aortic arch to give rise to the carotid artery and the subclavian artery; the *innominate vein* of Cetacea, Edentata, Carnivora and primates, which leads across from the jugular–subclavian trunk of one side to that of the other; the *innominate bone*, which is the lateral half of the pelvic girdle.

inoculation (*BioSci*) (1) The introduction into an experimental animal, by various routes, of infected material or of pathogenic bacteria. (2) The injection of a vaccine into a person for protection against subsequent infection by the organisms contained in the vaccine, but also used loosely in immunology for the introduction of a substance into the body, usually by parenteral injection. (3) The placing of cells, spores, etc, on or in a potential host, soil or culture medium.

inoculation (*Chem*) The introduction of a small crystal into a supersaturated solution or supercooled liquid in order to initiate crystallization.

inoculation (*Eng*) The modification of crystallizing habit or grain refinement for imparting alloy qualities to molten metal in furnace or ladle, by addition of small quantities of other metals, deoxidants, etc.

inoculum (*BioSci*) The material used in inoculation.

inorganic chemistry (*Chem*) The study of the chemical elements and their compounds, other than the compounds of carbon; however, the oxides and sulphides of carbon and the metallic carbides are generally included in inorganic chemistry.

inorganic polymers (*Chem*) Polymers whose chains are composed of atoms other than carbon. Many common materials contain such chains, eg silicate minerals, cement, where they perform an important reinforcement role. Chain bonds are Si–O (silicates), P–O (polyphosphates), P–N (phosphonitrilics), B–O (polyborates), etc. Silicone rubbers are of mixed lineage.

inosilicates (*Min*) Those silicate minerals which have an atomic structure in which the SiO_4 groups are linked together in chains, eg PYROXENE GROUP. See SILICATES.

inositol (*Chem*) Cyclohexanehexol, hexahydroxycyclohexane. $C_6H_6(OH)_6$.

inositol trisphosphate (*BioSci*) Inositol 1,4,5 trisphosphate is an important second messenger released from phosphatidyl inositol bisphosphate in the cell membrane by the action of a specific phospholipase C enzyme; binds to and activates a calcium channel in the endoplasmic reticulum.

inotropic (*BioSci*) Anything that alters the rate of heartbeat. Example: adrenaline has a positive inotropic effect and increases rate of beating.

in parallel (*ElecEng*) See PARALLEL.

in phase (*ElecEng*) See PHASE.

in-phase component (*ElecEng*) See ACTIVE COMPONENT.

in-phase loss (*ElecEng*) See OHMIC LOSS.

in-pile test (*NucEng*) One in which the effects of irradiation are measured while the specimen is subjected to radiation and neutrons in a reactor.

input (*ICT*) (1) Information entering a device, data about to be processed. (2) To enter information, to read in data.

input capacitance (*Phys*) The effective capacitance between the input terminals of a network. See MILLER EFFECT.

input characteristic (*Electronics*) See TRANSISTOR CHARACTERISTICS.

input device (*ICT*) PERIPHERAL that can accept data, presented in the appropriate machine-readable form, decode them and transmit them as electrical pulses to the CENTRAL PROCESSING UNIT.

input gap (*Electronics*) See BUNCHER.

input impedance (*ElecEng, Electronics*) The small signal impedance measured between the input terminals of a network.

input/output processor (*ICT*) Processor that supervises input–output operations. See FRONT-END PROCESSING.

input signal (*ElecEng*) That connected to the input terminals of any instrument or system (usually electronic).

input transformer (*ElecEng*) One used for isolating a circuit from any dc voltage in the applied signal and/or to provide a change in voltage. Also used to match the impedance of an input signal to that of the circuit to maximize power transfer.

input voltage (*ElecEng, Electronics*) The voltage applied between the input terminals of a network.

inquartation (*Eng*) Removal of silver from gold–silver bullion. The proportion of silver to gold must be raised by fusion to at least three to one, the silver being then dissolved in nitric acid. The silver can also be dissolved in concentrated sulphuric acid or converted to chloride by bubbling chlorine through the molten bullion. Also *parting of bullion, quartation*.

insanity (*Psych*) A medical legal term used in the defence of individuals who have committed crimes while their capacity for rational thought and behaviour was seriously impaired.

inscribed circle of a triangle (*MathSci*) The circle touching all three sides of a triangle internally. Its centre is the intersection of the bisectors of the angles of the triangle.

Insecta (*BioSci*) A class of mainly terrestrial mandibulate Arthropoda that breathe by tracheae. They possess uniramous appendages; the head is distinct from the thorax and bears one pair of antennae. There are three pairs of similar legs attached to the thorax, which may also bear wings. The body is sharply divided into head, thorax and abdomen. Also *Hexapoda*.

insecticides (*Chem*) Natural (eg derris, pyrethrins) or synthetic substances for destroying insects. The widely used synthetic compounds are broadly classified according to chemical composition as chlorinated (eg DDT), organophosphates (eg MALATHION), carbamates, dinitrophenols. Some of these have properties now considered undesirable, like persistence in *food chains*. Newer compounds like the modified pyrethrins have greater stability than the natural substance and still retain their effectiveness. See CONTACT INSECTICIDE, FUMIGANTS, STOMACH INSECTICIDE

Insectivora (*BioSci*) An order of small, mainly terrestrial insectivorous mammals that have numerous sharp teeth, tuberculate molars, well-developed collar bones, and plantigrade unguiculate pentadactyl feet. Shrews, moles and hedgehogs. Sometimes divided into two orders: Lipotyphla and Menotyphla.

insectivorous plant (*BioSci*) See CARNIVOROUS PLANT.

inselberg (*Geol*) A steep-sided knob or hill arising from a plain; often found in the semi-arid regions of tropical countries.

insemination (*BioSci*) The approach and entry of the spermatozoon to the ovum, followed by the fusion of the male and female pronuclei.

insensitive time (*Electronics*) See DEAD TIME.

insert (*Eng*) (1) In casting, a small metal part which is fitted into the mould or die in such a manner that the material flows around it to cast it in position. It may provide a hard metal wearing surface, sintered metal oil-retaining bearing surface, etc. (2) A metal core inserted into a plastic article during the (injection-)moulding process. (3) Special metal parts placed in a tool cavity which are designed to be incorporated in the final product (eg screw-thread) after injection moulding.

insert (*ImageTech*) A single shot, such as the close-up of a newspaper headline, photographed separately and inserted in a sequence.

insert (*Print*) See INSET.

insert edit (*ImageTech*) Inserting new material into an existing recording with no disturbance at the edit in and out points because the existing CONTROL TRACK is used. Professionals also ASSEMBLE EDIT in insert mode. See BLACKING A TAPE, FLYING ERASE HEAD.

inserted tooth cutter (*Eng*) A milling cutter in which replaceable teeth, of expensive material, are inserted into slots in the cutter body.

insertion (*BioSci*) (1) The point or area of attachment of a muscle, mesentery or other organ. (2) The place where one plant member grows out of another, or is attached to another. (3) A stretch of chromosome that has been inserted into another.

insertional mutagenesis (*BioSci*) Mutagenesis of DNA by the insertion of one or more bases. An extreme case is the insertion of a retrovirus adjacent to a cellular proto-oncogene, causing tumorigenesis.

insertion element (*BioSci*) A DNA sequence a few hundred bases long that can be naturally inserted into genomic DNA. Such an element does not code for mRNA but can inactivate a coding sequence by its presence.

insertion gain (*ElecEng*) The gain in decibels when a transformer, filter or impedance matching transducer or network is inserted into a circuit.

insertion head (*Eng*) An automatic mechanism, which may incorporate cutting, forming or fastening tools, for placing components into a partly completed assembly.

insertion loss (*ElecEng*) The loss in decibels when a transformer, filter or impedance matching transducer or network is inserted into a circuit.

insertion point (*ICT*) The place where text will be inserted into a document when the user types. At the insertion point a CURSOR will usually appear as a flashing line or vertical bar on the screen.

insert mode (*ICT*) A mode of operation within an EDITOR whereby text is inserted at the CURSOR as the user types; the existing text is pushed along to the right. Cf OVERTYPE MODE.

insessorial (*BioSci*) Adapted for perching.

inset (*Print*) One folded sheet put inside another and bound in; an *insetted* book is one having its section placed one within the other. Also *insert*.

insetter (*Print*) Electronically controlled equipment for registering pre-printed reels into rotary presses to be incorporated in the final product, overprinted if required; with controlled AUTOMATIC REEL CHANGE; half-width or full-width reels.

inside callipers (*Eng*) CALLIPERS with the points turned outwards, used for taking inside dimensions.

inside crank (*Eng*) A crank, with two webs, placed between bearings, as distinct from an outside or overhung crank.

inside cylinders (*Eng*) In a steam locomotive, those cylinders which are fixed inside the frame.

inside lap (*Eng*) See EXHAUST LAP.

inside-out vesicle (*BioSci*) A small vesicle surrounded by a bilayer membrane, produced when a cell membrane is mechanically disrupted. The vesicles may be right-side-out (ROV) or inside-out, and are a useful experimental tool for studying transmembrane fluxes. Abbrev *IOV*.

inside spider (*Acous*) A flexible device inserted within a voice coil so as to centre the coil accurately between the pole pieces of a dynamic loudspeaker.

insight (*Psych*) (1) An understanding of the inner nature of something. (2) Self-knowledge. (3) In psychotherapy, understanding one's mental condition, esp in situations where this understanding has been absent. (4) In Gestalt psychology, a process by which problems are solved suddenly, apparently without conscious effort.

insight learning (*Psych*) A form of intelligent activity that involves the apprehension, often suddenly, of relations between the elements of a problem; it is often contrasted with a more passive trial and error mode of learning or problem solving.

in silico (*BioSci*) A term used (rather colloq) for bioscientific experiments done using a computer (ie on a silicon chip). Bioinformatics has become crucial in handling the data from genomics and array studies, and increasingly sophisticated models of cellular systems are being developed.

in situ hybridization (*BioSci*) The identification by AUTORADIOGRAPHY in a fixed cell or section of the place where a radioactive or similar PROBE binds. Can identify a complementary sequence or an antigen.

insolation (*EnvSci*) The radiation received from the Sun. This depends on the position of the Earth in its orbit, the thickness and transparency of the atmosphere, the inclination of the intercepting surface to the Sun's rays, and the SOLAR CONSTANT.

insolation (*Med*) See SUNSTROKE.

insoluble (*Chem*) Incapable of being dissolved. Most 'insoluble' salts have a definite, though very limited, solubility.

insoluble electrode (*ElecEng*) Non-ionizing cathode or anode.

insomnia (*Med*) A persistent inability to sleep.

inspection (*ICT*) A technique for reviewing the quality of code or system documentation.

inspection chamber (*Build*) A pit formed at regular points in the length of a drain or sewer to give access for purposes of inspection, testing and the removal of obstructions.

inspection fitting (*Eng*) A bend, elbow or tee used in a conduit wiring system, which is fitted with a removable cover to facilitate the drawing in of the wires and subsequent inspection.

inspection gauges (*Eng*) Gauges used by the inspection department of a works for testing the accuracy of finished parts.

inspection junction (*Build*) A drainpipe with a vertical branch to ground level to allow access for inspection.

inspiration (*BioSci*) The drawing in of air or water to the respiratory organs.

inspirator (*Eng*) The injector of a pressure gas burner, combined with a venturi mixing tube, primary combustion air being entrained by the projection of a gas stream into the injector throat.

inspissation (*Med*) The thickening of pus as a result of the removal of fluid.

instability (*Acous*) Vibration with an exponentially growing amplitude. It occurs if there is some sort of positive FEEDBACK, eg thermoacoustic instability, screech.

instability (*Aero*) An aircraft possesses *instability* when any disturbance of its steady motion tends to increase, unless it is overcome by a movement of the controls by the pilot.

instability (*ICT*) The tendency for a circuit to break into unwanted oscillation.

instability (*NucEng*) In a *plasma*, various instabilities of which the principal ones are: bending of plasma, *kink* *instability*, and bead-like instability, *sausage instability*, in both of which the magnetic field change magnifies the variations; acceleration-driven *Rayleigh–Taylor* instability when the magnetic fields and plasma behave like a light fluid in contact with and accelerated against a heavier one, causing spiky irregularities; *flute instability* in mirror machines.

instability (*Phys*) See UNSTABLE EQUILIBRIUM.

install shield (*ICT*) A common name for a program enabling the process of installation.

install/uninstall/deinstall (*ICT*) The process of loading software onto a computer in an orderly fashion, identifying its presence to the operating system and ensuring that the program is configured to run correctly. Conversely, the orderly removal from the computer of installed software such that the remaining programs and system function correctly.

instantaneous automatic gain control (*Radar*) A rapid action system in radar for reducing the clutter.

instantaneous carrying-current (*ElecEng*) Of switches, circuit breakers and similar apparatus, the maximum value of current which the apparatus can carry instantaneously.

instantaneous centre (*Eng*) The imaginary point, not necessarily in the body, about which a body having general motion may be considered to be rotating at a given instant.

instantaneous frequency (*ICT*) That calculated from the instantaneous rate of change of angle of a waveform on a time base. In any oscillation, the rate of change of phase divided by 2π.

instantaneous fuse (*MinExt*) Rapid burning, as distinct from slow fuse. Ignition proceeds at a few kilometres per second, but is slower than that of DETONATING FUSE.

instantaneous power (*ElecEng*) For a circuit or component, the product of the instantaneous voltage and the instantaneous current. This may not be zero even for a non-dissipative (wattless) system on account of stored energy, although in this case its time integral must be zero.

instantaneous specific heat capacity (*Phys*) Specific heat capacity at any one temperature level; the *true* as opposed to the *mean* specific heat capacity.

instantaneous value (*Genrl*) A term used to indicate the value of a varying quantity at a particular instant. More correctly it is the average value of that quantity over an infinitesimally small time interval.

instantaneous velocity of reaction (*Chem*) Rate of reaction, measured by the change in concentration of some key reagent. For instance, in the first-order reaction, $A \rightarrow B$, the instantaneous velocity of reaction at any time t is given by $dC_A/dt = -dC_B/dt = -kC_A$, where C_A, C_B are the concentrations of reagent and product, respectively, and k is the 'velocity constant' of the reaction. Similarly, for the second-order reaction, $A + B \rightarrow C$, the instantaneous velocity of reaction is $dC_A/dt = dC_B/dt = -dC_C/dt = -kC_A C_B$. Abbrev *ivr*.

instantizing (*FoodSci*) See AGGLOMERATION.

instant messaging (*ICT*) The practice of conversing by text in real time with one or more other persons over a network of computers.

instant photography (*ImageTech*) A system which rapidly produces a positive picture from the material originally exposed, sometimes by viscous processing within the camera itself.

instant replay (*ImageTech*) Using a magnetic disk which records inwards and outwards continuously so that approximately 18 seconds of video is always available for replay (more disks can be added), or a video disk recorder (VDR) with a separate playback head to replay any part of the recording while recording continues.

instant video rushes (*ImageTech*) Using the recorded output from VIDEO ASSIST to view the scenes just shot, rather than waiting for film RUSHES.

instar (*BioSci*) The form assumed by an insect during a particular stage or stadium between two ECDYSES.

in step (*ElecEng*) See STEP.

instinct (*Psych*) A generally discredited term for innate, unlearned behaviour.

institution (*Psych*) In sociology, an organized system of relationships and roles characteristic of a given society. Thus marriage may be regarded as an institution currently under threat.

institutionalization (*Psych*) A syndrome of apathy and withdrawal resulting from long periods in any under-stimulating institution, eg a mental hospital.

instron tensile strength tester (*Paper*) A precise instrument for measuring tensile strength, in which the rate of strain on the paper sample is constant.

instruction (*ICT*) (1) A MACHINE-CODE INSTRUCTION. (2) A statement in a HIGH-LEVEL LANGUAGE. See MACRO-INSTRUCTION, MICRO-INSTRUCTION.

instruction address register (*ICT*) See PROGRAM COUNTER.

instruction cycle (*ICT*) Complete process of fetching one INSTRUCTION from memory, decoding and executing it by the CENTRAL PROCESSOR. Also *fetch/execute cycle*. See MACHINE-CODE INSTRUCTION.

instruction decoder (*ICT*) Part of the CENTRAL PROCESSOR that decodes MACHINE-CODE INSTRUCTIONS.

instruction number (*ICT*) One used to label a specific statement in a program so that the user and the machine can refer back to it subsequently. Such numbers do not indicate the order in which instructions must be carried out, or the number of the address at which the instruction will be stored. Also *instruction number, line number*.

instruction set (*ICT*) Complete collection of instruments available in a particular MACHINE CODE or ASSEMBLY LANGUAGE.

instrument (*ElecEng*) The term for any item of electrical or electronic equipment which is designed to carry out a specific function or set of functions. Examples are found in a wide range of applications and include: in an electrical network, eg a voltmeter; in an engineering system, eg a displacement measuring unit; or in medicine, eg an ECG unit.

instrument (*Psych*) Any psychological test, eg an IQ test or a survey.

instrumental behaviour (*Psych*) Behaviour that is goal-directed.

instrumental conditioning (*Psych*) Experimental protocol in which reinforcement is only given once a proper response has been made by the subject, eg by selecting the correct arm of a maze.

instrumental sensitivity (*ElecEng*) The ratio of the magnitude of the response to that of the quantity being measured, eg divisions per milliamp.

instrument approach (*Aero*) Aircraft approach made using INSTRUMENT LANDING SYSTEM.

instrument flight rules (*Aero*) The regulations governing flying in bad visibility under strict flight control. Abbrev IFR.

instrument landing system (*Aero*) A localizer for direction in the horizontal plane and glide slope in the vertical plane, usually inclined at 3° to the horizontal. Two markers are used for linear guidance. Abbrev ILS.

instrument range (*NucEng*) The intermediate range of reaction rate in a nuclear reactor, when the neutron flux can be measured by permanently installed control instruments, eg ion chambers. See START-UP PROCEDURE.

instrument shunt (*ElecEng*) A resistance of appropriate value connected across the terminals of a measuring instrument to extend its range.

instrument transformer (*ElecEng*) One specially designed to maintain a certain relationship in phase and magnitude between the primary and secondary voltages or currents.

insufficient feed (*Print*) When too little paper is drawn through the press, there is an increase in tension between units or components. Cf EXCESS FEED.

insufflation (*Med*) (1) The action of blowing gas, air, vapour or powder into a cavity of the body, eg the lungs. (2) The powder etc used in insufflation.

insufflator (*Med*) An instrument used for INSUFFLATION.

insulance (*ElecEng*) See INSULATION RESISTANCE.

insulant (*Build*) See INSULATION (2).

insulated (*Arch*) Said of any building or column which stands detached from other buildings.

insulated bolt (*ElecEng*) A bolt having a layer of insulating material around its shank.

insulated clip (*ElecEng*) A clip, incorporating an INSULATED EYE, used for supporting flexible electrical connections.

insulated eye (*ElecEng*) An eye for supporting flexible connections; it has an insulating bush to prevent these from making contact with the metal.

insulated hanger (*ElecEng*) A hanger for the contact wire of an electric traction system, which insulates the contact wire from the main supporting system.

insulated hook (*ElecEng*) A hook terminating in an INSULATED EYE through which flexible electric connections may be passed and supported.

insulated metal roofing (*Build*) Roofing panels made up of light-gauge aluminium or sheet steel, corrugated for strength, and with a top covering of insulation board and bituminous felt. The ceiling underlining is also frequently pre-fixed.

insulated neutral (*ElecEng*) A term used to denote: (1) the neutral point of a star-connected generator or transformer when it is not connected to earth directly or through a low impedance; (2) the middle wire of a three-wire distribution system when the wire is an insulated cable.

insulated-return system (*ElecEng*) A system of supply for an electric traction system, in which both the outgoing and the return conductors are insulated from earth.

insulated screw-eye (*ElecEng*) A screw terminating in an IINSULATED EYE through which flexible electric connections may be passed.

insulated system (*ElecEng*) A system of electrical supply in which each of the conductors is insulated from earth for its normal voltage.

insulated wire (*ElecEng*) A solid conductor insulated throughout its length.

insulating beads (*ElecEng*) Beads of glass or similar material strung over a bare conductor to provide an insulating and heat-resisting covering which is also flexible.

insulating board (*Build*) FIBREBOARD of low density (300 kg m^{-3}, 20 lb ft^{-3}) used for thermal insulation and acoustical control.

insulating compound (*ElecEng*) An insulating material which is liquid at fairly low temperatures so that it can be poured into joint boxes of cables and other similar pieces of apparatus and then allowed to solidify.

insulating oils (*ElecEng*) Special types of oil having good insulating properties; used for oil-immersed transformers, circuit breakers, etc.

insulating tape (*ElecEng*) Tape impregnated with insulating compounds, frequently adhesive; used for covering joints in wires etc.

insulation (*Genrl*) (1) Any means for confining as far as possible a transmissible phenomenon (eg electricity, heat, sound, vibration) to a particular channel or location in order to obviate or minimize loss, damage or annoyance. (2) Any material (also *insulant*) or means suitable for such a purpose in given conditions, eg dry air suitably enclosed, polystyrene and polyurethane foam slab, glass fibre, rubber, porcelain, mica, asbestos, hydrated magnesium carbonate, cork, kapok, crumpled aluminium foil. Insulating materials are classified as class A, class B, etc, according to the temperature which they may be expected to withstand. See LAGGING, SHIELDING.

insulation resistance (*ElecEng*) The resistance between two conductors, or between a conductor and earth, when they are separated only by insulating material. Also *insulance*. See FAULT RESISTANCE.

insulation test (*ElecEng*) A test made to determine the insulation resistance of a piece of apparatus or of a system of electric conductors.

insulation tester (*ElecEng*) An instrument for measuring INSULATION RESISTANCE.

insulator (*EleeEng*) A component of insulating material specifically designed to give insulation and mechanical support, eg for telegraph lines, overhead traction and transmission lines, plates and terminals of capacitors.

insulator (*Electronics*) That part of an electrical or electronic device which is intended to block the passage of current.

insulator (*Phys*) (1) A material having a high electrical resistivity in the range $10^6–10^{15} \Omega$ m. The electrons in the material are not free to move under the influence of an electric field. (2) A material which offers a high resistance to the flow of heat.

insulator arcing horn (*ElecEng*) A metal projection placed at the upper and lower ends of a suspension-type or other insulator, in order to deflect an arc away from the insulator surface.

insulator cap (*ElecEng*) A metal cap placed over the top of a suspension insulator, which serves to attach it to the next insulator.

insulator pin (*ElecEng*) The central metal support of a pin insulator, or the metal projection on the under side of a suspension insulator, serving to attach it to the cap of the next unit in the string.

insulator strength (*ElecEng*) The maximum mechanical and/or the maximum electrical stress, which can be applied.

insulin (*BioSci*) A protein hormone that is produced by the islets of Langerhans of the pancreas. Widely used in the treatment of DIABETES MELLITUS, insulin injection results in a prompt decline in blood glucose concentration and an increase in formation of products derived from glucose. The first protein to have its structure completely established, it consists of two peptide chains joined by two disulphide bridges. Insulin was formerly entirely produced from pancreatic extracts from pig and oxen, but can now be made from a GENETIC ENGINEERED strain of the bacterium *Escherichia coli*.

insulin-dependent diabetes (*BioSci*) See DIABETES MELLITUS.

intaglio (*Glass*) A form of decoration in which the depth of cut is intermediate between deep cutting and engraving.

intaglio (*Print*) A printing process, such as photogravure, in which the printing image is cut or etched into the surface of the plate or cylinder, normally in a characteristic cell formation. The thickness of ink transferred to the paper during printing varies with the depth of the cell.

intarsia (*Textiles*) A weft-knitted fabric having designs in two or more colours.

integer (*ICT, MathSci*) Any of the set of positive, zero or negative whole numbers; ie the set $(...-4,-3,-2,-1,0,1,2,3,4,...)$. Cf RATIONAL NUMBER, REAL NUMBER. See DATA TYPE.

Integra (*Pharmacol*) TN for an artificial skin consisting of pure bovine collagen and glycosaminoglycan made from shark cartilage, with an outer layer of silicone. Laid on injured tissue to encourage cell growth.

integral (*MathSci*) (1) Relating to the integers or having a value which is an integer. (2) The function which, when differentiated, equals the given function. If $\varphi(x)$ is the derivative of $f(x)$, then $f(x)$ is the integral of $\varphi(x)$. More precisely, $f(x)$ is the *indefinite integral* of $\varphi(x)$. Also used generally as the adjective. Also *antiderivative*.

integral calculus (*MathSci*) The inverse of the DIFFERENTIAL CALCULUS, its chief concern being to find the value of a function of a variable when its differential coefficient is known. This process, termed *integration*, is used in the solution of such problems as finding the area enclosed by a given curve, the length of a curve, or the volume enclosed by a given surface.

integral colour masking (*ImageTech*) See COLOUR MASKING.

integral convergence test (*MathSci*) The series of positive terms

$$\sum_1^n f(r)$$

where $f(r)$ is a monotonic decreasing function, and the integral

$$\int_1^n f(x)\,dx$$

converge and diverge together.

integral domain (*MathSci*) A ring in which the product of two non-zero elements is non-zero.

integral dose (*Radiol*) See DOSE.

integral function (*MathSci*) See ANALYTIC FUNCTION.

integral membrane protein (*BioSci*) A protein that is firmly anchored in the lipid bilayer of a cell membrane; those that cross the entire membrane are referred to as *transmembrane proteins*.

integral stiffeners (*Aero*) The stiffening ridges left when an aircraft skin panel is machined from a solid billet.

integral tripack (*ImageTech*) Colour film or paper containing three distinct emulsion layers, each sensitized to a specific spectral range, which can be processed to produce superimposed colour separation images. Also termed *monopack*.

integrand (*MathSci*) A function upon which integration is to be effected.

integrated circuit (*Electronics*) An electronic subassembly in which several components are fabricated on the same semiconductor substrate. See panel on PRINTED, HYBRID AND INTEGRATED CIRCUITS.

integrated control (*Agri*) The combined use of both biological and chemical methods against pests and diseases.

integrated crop management (*Agri*) The combined use of a range of techniques to produce and protect crops whilst attempting to conserve the environment.

integrated development environment (*ICT*) A software user interface. Abbrev *IDE*.

integrated drive electronics (*ICT*) A control system that allows communication between a computer and hard disks (cf. ENHANCED INTEGRATED DRIVE ELECTRONICS). Abbrev *IDE*.

integrated pest management (*Agri*) A system integrating a range of methods of pest control to produce healthy crops economically and to reduce or minimize risks to human health and the environment.

integrated services digital network (*ICT*) A digital network designed to deliver a wide range of services including telephony, data transfer and video over the same bearer by means of a number of 64 Kbps bearer channels plus a 16 Kbps channel for signalling. Abbrev *ISDN*. See B-CHANNEL, BROADBAND INTEGRATED SERVICES DIGITAL NETWORK, D-CHANNEL, NARROW-BAND INTEGRATED SERVICES DIGITAL NETWORK.

integrated services digital network user part (*ICT*) A COMMON CHANNEL SIGNALLING protocol of the CCITT that supports network services including CALLING LINE IDENTIFICATION and CALL FORWARDING.

integrated virus (*BioSci*) A viral DNA sequence that has become integrated into the host's chromosomes at one or more sites.

integrating ammeter (*ElecEng*) Instrument which measures the time integral of the named quantity. Also *integrating voltmeter*, *integrating wattmeter*, etc.

integrating circuit (*ElecEng*) (1) An amplifier with feedback capacitance and input resistance whose output is proportional to the integral of the input signal. (2) A network comprising passive components, which will perform the same mathematical function. This technique is not as accurate as that described in (1).

integrating factor (*MathSci*) A multiplying factor which enables a differential equation to be transformed into an EXACT EQUATION.

integrating frequency meter (*ElecEng*) A meter which sums the total number of cycles of an ac supply in a given time. Also *master frequency meter*.

integrating meter (*ElecEng*) An electrical instrument which sums up the value of the quantity measured with respect to time.

integrating motor (*ElecEng*) A permanent magnet dc motor whose angular rotation records the integration of current in its armature.

integrating network (*ElecEng*) See INTEGRATING CIRCUIT.

integrating photometer (*Phys*) A photometer which measures the total luminous radiation emitted in all directions.

integration (*BioSci*) The insertion of one DNA sequence into another by recombination.

integration (*MathSci*) See INTEGRAL CALCULUS.

integration (*Psych*) The fusion of disparate elements into a unified whole, eg the formation of a unified personality or a social group; alternatively, adopting the mores of an integrated group so as to be accepted.

integration by parts (*MathSci*) The rule used in integrating the product of two differentiable functions, which states that

$$\int f(x)\,g'(x)\,dx = f(x)\,g(x) - \int f'(x)\,g(x)\,dx$$

where $f'(x)$ and $g'(x)$ are the first derivatives of $f(x)$ and $g(x)$.

integration testing (*ICT*) The phase of a testing cycle in which individual components of software or hardware are brought together into a complete or partial model of the system and tested together to establish their fitness for purpose.

integrator (*ICT*) (1) A device that performs the mathematical operation of integration, usually with reference to time. (2) Any device that integrates a signal over a period of time.

integrins (*BioSci*) A family of heterodimeric proteins found in the surface membranes of a wide variety of cells. These proteins are involved in cell–cell adhesion and in the adhesion of cells to extracellular matrix components. Cells of the immune system express a subset of integrins, the leucocyte integrins. See CELL ADHESION MOLECULES.

integrity (*ICT*) A property of data that have retained their accuracy and correctness during processing and transmission. Cf CORRUPT.

integument (*BioSci*) (1) A covering layer of tissue; in animals, esp the skin and its derivatives. (2) In plants, one or several layers of tissue that develop into the seed coat (*testa*). Adj *integumented*.

intellectual property (*Genrl*) Legal term covering patents, registered designs, design right, copyright, confidential information and moral rights.

intelligence (*Psych*) A complex and much disputed concept. One view is that intelligence is the ability to profit by experience and therefore adapt to, and succeed within, a particular environment. Whether this is measured by intelligence tests is far from obvious.

intelligence quotient (*Psych*) The score on an intelligence test. In early versions it was obtained by multiplying by 100 the ratio of *mental age* to *chronological age* to find out whether a child's mental age was ahead or behind its chronological age; in newer versions it is read off a table constructed from norms for individuals of the same age. Also *deviation IQ*. Abbrev *IQ*.

intelligent design (*Genrl*) The creation of the universe by a rational agent, rather than by random processes and NATURAL SELECTION. A belief, not a scientific hypothesis.

intelligent materials (*Eng*) See SMART MATERIALS.

intelligent network (*ICT*) A telephone network in which switching systems have access to external computers, databases and software that allow them to offer the caller a wider range of services than a simple exchange, eg the ability to dial numbers for which the called party has agreed to pay the call charges.

intelligent peripheral (*ICT*) A unit in an ADVANCED INTELLIGENT NETWORK that, under control of a SERVICE CONTROL POINT, provides voice and data services to users via a SERVICE SWITCHING POINT.

intelligent terminal (*ICT*) A terminal within which a certain amount of computing can be done without contacting a central computer, eg a PC connected to a host computer.

intelligibility (*Acous, ICT*) The percentage of correctly understood syllables, words or sentences. Important for speech transmission.

INTELSAT (*Space, ICT*) The International Telecommunications Satellite organization. Body responsible for the design, construction, development, operations and maintenance of the worldwide satellite communication system. Originally an intergovernmental organization, now a private company. See panel on COMMUNICATIONS SATELLITE.

intensification (*ImageTech*) Increasing the density of a processed image, usually by chemical addition of another metal, such as chromium or mercury, to the developed silver.

intensifier electrode (*Electronics*) One which provides post-deflection acceleration in the electron beam of a cathode-ray tube.

intensifying screen (*Radiol*) (1) Layer or screen of fluorescent material adjacent to a photographic surface, so that registration by incident X-rays is augmented by local fluorescence. (2) Thin layer of lead which performs a similar function for high-energy X-rays or gamma rays, as a result of ionization produced by secondary electrons.

intensitometer (*Radiol*) An instrument for measuring intensities of X-rays during exposures.

intensity level (*Acous*) The level of power per unit area as expressed in decibels above an arbitrary zero level, eg sound intensity level.

intensity modulation (*ImageTech*) The modulation of the beam current of a cathode-ray tube to vary the spot luminosity. Also *Z-modulation*.

intensity of field (*ElecEng*) The vector quantity by which an electric or magnetic field at a point is measured. Precisely, it is the same number of units of intensity as the force in newtons on a unit charge (unit electric charge or unit fictitious pole) placed at the point in the electric or magnetic field respectively. Also *field strength*.

intensity of magnetization (*Phys*) The vector of magnetic moment of an element of a substance divided by the volume of that element.

intensity of radiation (*Phys*) The energy flux, ie of photons or particles, per unit area normal to the direction of propagation.

intensity of sound (*Acous*) The magnitude of a sound wave, measured in terms of the power transmitted, in watts, through unit area normal to the direction of propagation.

intensity of wave (*ElecEng*) The energy carried by any sinusoidal disturbance is proportional to the square of the wave amplitude. See RMS VALUE.

intensive farming (*Agri*) (1) Production with high levels of input per unit area, including energy and labour. Characteristic of large, lowland farms in the UK. (2) Producing the maximum possible output from crops and livestock on the available land.

intentional learning (*Psych*) Learning when informed that there will be a later test of learning.

intention movement (*Psych*) The incomplete phases of behaviour patterns that often occur in conflict situations; many are assumed to provide the phylogenetic basis for the evolution of threat and courtship displays.

intention tremor (*Med*) Tremor of the arms on carrying out a voluntary movement, indicative of disease of the nervous system.

interaction (*Phys*) (1) Transfer of energy between two particles. (2) Interchange of energy between particles

and a wave motion. (3) Superposition of waves. See INTERFERENCE.

interaction gap (*Electronics*) The space in a microwave tube in which the electron beam interacts with the wave system.

interaction space (*Electronics*) That in a vacuum tube where an electron beam interacts with an alternating electromagnetic field.

interactive (*ICT*) Describes a program or system that allows two-way communication between user and computer, or computer and computer.

interactive computing (*ICT*) A conversational mode of communication between computer and user. Input is commonly via a KEYBOARD or a MOUSE and both input and output may be shown on a video display. See LIGHT PEN, LOG IN/OUT, MULTI-ACCESS, PROMPT, TELETYPE-WRITER.

interactive TV/cable (*ImageTech*) Using the TV and telephone for home shopping, calling up video films and games, etc. Cable offers the further possibility of complete two-way interaction, enabling programmed viewing, access to databases, etc.

interambulacrum (*BioSci*) In Echinodermata, esp Echinoidea, the region between two ambulacral areas.

interatomic forces (*Chem*) Interactions between atoms in molecules or materials. Generally very strong covalent, ionic or metallic bonds. See panel on BONDING.

interatomic potential energy curve (*Chem*) The diagram showing the relation between potential energy and interatomic distance for various bond types. See panel on BONDING.

interbase current (*Electronics*) The current flowing between the two base connectors in junction-type tetrode transistors.

interblock gap (*ICT*) The space separating two recorded portions of data on backing store (eg to allow a tape to stop or to accelerate to reading speed).

interbranchial septa (*BioSci*) The stout fibrous partitions separating the branchial chambers in fish.

intercalare (*BioSci*) A cartilage or ossification lying between the basiventrals, or between the basidorsals of the vertebral column.

intercalary (*BioSci*) Placed between other bodies or the ends of a stem, filament, hypha, etc. See INTERCALATE.

intercalary meristem (*BioSci*) A meristem located somewhere along the length of a plant member and which divides to give INTERCALARY growth, as at the base of a grass leaf.

intercalate (*BioSci*) To add, to insert, as in *intercalated somite*. Adj *intercalary*.

intercalating dyes (*BioSci*) Chemical compounds with a high affinity for DNA whose molecules intercalate between the planar base pairs of the DNA helix. They are used for visualizing DNA, changing its density and to induce breakage. Often mutagenic and carcinogenic.

intercarrier spacing (*ImageTech*) The frequency separating video and audio in a broadcast TV signal to minimize mutual interference, ie 6 MHz for UK PAL.

intercavitary X-ray therapy (*Radiol*) X-ray therapy in which the appropriate part of suitable X-ray apparatus is placed in a body cavity.

intercellular (*BioSci*) Between cells. Cf EXTRACELLULAR, INTRACELLULAR.

intercellular spaces (*BioSci*) The interconnecting spaces between cells in a tissue, air-filled in vascular plants, and providing for gas exchange.

intercentra (*BioSci*) See HYPAPOPHYSES.

interceptor (*Build*) A trap fitted in the length of a house drain, close to its connection to the sewer, which provides a water seal against foul gases rising up into the drain. Also *disconnector, intercepting trap*.

interchange (*BioSci*) The mutual transfer of parts between two chromosomes.

interchondral (*BioSci*) Said of certain ligaments and articulations between the cartilages of the rib.

interclavicle (*BioSci*) In vertebrates, a bone lying between the clavicles, forming part of the pectoral girdle.

intercolumniation (*Arch*) The distance between the columns in a colonnade, in terms of the lower diameter of the columns as a unit.

interconnected star connection (*ElecEng*) See ZIGZAG CONNECTION.

interconnecting (*ICT*) The commoning of outlets for the bank multiples of selectors on different shelves, when there is an insufficiency of outlets for full availability. See GRADING.

interconnecting feeder (*ElecEng*) A feeder which serves to interconnect two substations or generating stations, and along which energy may flow in either direction. Also *interconnector*.

interconnectivity (*ICT*) The ability of different parts of a system to operate together.

intercooler (*Aero*) (1) A heat exchanger on the delivery side of a supercharger which cools the charge heated by compression. (2) A secondary heat exchanger used in a cabin air-conditioning system for cooling the charge air from the compressor to the turbine of a BOOTSTRAP COLD-AIR UNIT.

intercooler (*Autos*) A form of heat exchanger which lowers the temperature of inlet air to improve volumetric efficiency in an engine with a TURBOCHARGER.

intercooler (*Eng*) A cooler, generally consisting of water-cooled tubes, interposed between successive cylinders or stages of a multistage compressor or blower, to reduce the work of compression.

intercostal (*BioSci*) Between the ribs.

intercrystalline failure (*Eng*) Refers to metal fractures that follow the crystal boundaries instead of passing through the crystals, as in the usual transcrystalline fracture. It is frequently due to the combined effect of stress and chemical action, but may be produced by stress alone when the conditions permit a certain amount of recrystallization under working conditions.

interdentil (*Build*) The space between successive DENTILS.

interdependent functions (*MathSci*) See DEPENDENT FUNCTIONS.

interdigital cyst (*Vet*) An abscess occurring between the digits of the paw of the dog due to bacterial infection.

interdigital structure (*Electronics*) One in which the effective path length between a pair of terminal electrodes is increased by an interlocking-finger design, which may be three-dimensional or formed by metallization of an insulating surface. Used for transistors, capacitors and integrated-circuit devices, for converting microwave signals into surface acoustic waves and as a slow-wave structure or part of a filter.

interdigitating cells (*BioSci*) Bone-marrow-derived cells found in lymphoid tissues where they interdigitate with T-lymphocytes. Interdigitating cells express high amounts of major histocompatibility Class II proteins and are involved in the presentation of antigen to CD4-positive T-lymphocytes.

interdorsal (*BioSci*) An intercalary element lying between adjacent basidorsals of the vertebral column.

interelectrode capacitance (*Electronics*) That of any pair of electrodes in a valve, other electrodes being earthed. Also *internal capacitance*.

interface (*Aero, Space*) The relationship between parts of a system or subsystem which ensures that their eventual meeting will be harmonious; the interface may be physical (eg mechanical, thermal, electrical) or non-physical (eg software, organizational) and all conditions of the eventual union are controlled as part of the system documentation.

interface (*Electronics*) A shared boundary. It may be a piece of hardware used between two pieces of equipment, a portion of computer storage accessed by two or more programs, or simply a surface that forms the boundary between two different materials.

interface (*ICT*) Hardware and associated software needed to enable one device to communicate with another.

interface (*MinExt*) A sharp contact boundary between two phases, either or both of which may be solid, liquid or gaseous. Differs from interphase in lacking a diffuse transition zone.

interfacial film (*MinExt*) A special state developed in emulsification, in which oriented molecules of the emulsifying agent are loosely aggregated in such a way as to surround and enclose droplets of one phase of the emulsified mixture.

interfacial layer (*Genrl*) An inhomogeneous region intermediate between two bulk phases in contact, and where properties are significantly different from, but related to, the properties of the bulk phases.

interfacial surface tension (*Phys*) The SURFACE TENSION at the surface separating two non-miscible liquids.

interfasicular cambium (*BioSci*) Vascular cambium between the vascular bundles of the stem and joined to the cambium in the bundles to make a complete cylinder.

interfasicular region (*BioSci*) Tissue between the vascular bundles of a stem. Also *medullary ray, pith ray*.

interference (*Aero*) Mutual aerodynamic interactions between solid bodies in airflow, the drag of the combined bodies exceeding that of their separate drags by the interference drag. Thus the lift of the lower wing of a biplane is reduced by the flow under the upper wing.

interference (*BioSci*) The usually negative effect that the presence of one CHIASMA has on the probability of a second occurring in its vicinity.

interference (*ICT*) Any detectable energy that tends to interfere with the reception of desired signals. May be artificial, eg from electrical machinery, power lines or radio-frequency heating systems, or due to natural phenomena, esp atmospheric electricity. In crowded wavebands or abnormal propagation conditions interference may be from other transmitters.

interference (*Phys*) The interaction between two or more waves of the same frequency emitted from coherent sources. The wavefronts are combined, according to the PRINCIPLE OF SUPERPOSITION, and the resulting variation in the disturbances produced by the waves is the interference pattern. See INTERFERENCE FRINGES.

interference factor (*ICT*) See TELEPHONE INFLUENCE FACTOR.

interference fading (*ICT*) The fading of signals because of interference among the components of the signals that have taken slightly different paths to the receiver.

interference figure (*Crystal*) More or less symmetrical pattern of concentric rings or lemniscates, cut by a black cross or hyperbola, exhibited by a section of anisotropic mineral when viewed in convergent light between crossed Nicol prisms or polarizers in a polarizing microscope. See BIAXIAL, UNIAXIAL.

interference filter (*ICT*) Means of reducing interference, eg a tuned rejector circuit for a single steady transmission, or a band-pass filter to reduce the accepted band of frequencies to the minimum. Also *interference trap*.

interference filter (*Phys*) Light filter which uses interference principles to select a range of wavelengths for transmission. If a beam of light is incident normally on a Fabry–Pérot etalon with a plate spacing d of between 2×10^{-7} m and 6×10^{-7} m, then there will be only one wavelength in the visible region for which there is an interference maximum ($n\lambda = 2d$, where λ is the wavelength and $n = 0,1,2,...$). The highly reflecting surfaces of the etalon ensure that only a narrow band of wavelengths around λ is transmitted. Interference filters can also be made by successively evaporating dielectric and silvered films of suitable thickness on a glass plate.

interference fit (*Eng*) A negative fit, necessitating force sufficient to cause expansion in one mating part, or contraction in the other, during assembly.

interference fringes (*Phys*) Alternate light and dark bands formed when two beams of monochromatic light having a constant phase relation overlap and illuminate the same portion of a screen. The method of producing fringes is either by division of wavefront (see FRESNEL'S BIPRISM, LLOYD'S MIRROR) or by division of amplitude (see FABRY–PÉROT INTERFEROMETER, FIZEAU FRINGES, HAIDINGER FRINGES, NEWTON'S RINGS).

interference microscope (*BioSci*) A microscope in which the phase changes caused by differences in optical path (refractive index times thickness for transmitted light) within the specimen can be measured or made visible as differences in brightness or colour in the image. The light is split into two beams, one passing through the specimen, the other, the *reference beam*, ideally through empty medium near the specimen. The two beams, with suitably manipulated phase, are then made to interfere at the image plane. Areas of the specimen which have similar optical paths appear in the image similarly bright or coloured. Because the refractive index of an aqueous solution is nearly proportional to concentration of solutes, the microscope can be used to estimate the dry mass (to 'weigh') microscopic objects. Cf DIFFERENTIAL INTERFERENCE CONTRAST MICROSCOPE. See PHASE-CONTRAST MICROSCOPY.

interference microscopy (*Eng*) Special optical microscopic method for examining polished material surfaces. Utilizes monochromatic light shone vertically onto surface through angled glass plate. Interference of reflected beam with incident beam gives fringe map of surface microtopography. Used to examine cracks, wear and scratches, etc. See NOMARSKI INTERFERENCE which uses white light.

interference pattern (*Phys*) See INTERFERENCE.

interference reflection microscopy (*BioSci*) An optical technique for detecting the topography of the surface of a cell in contact with the substratum by imaging the interference of light reflected from the medium–substratum and immediately adjacent cell–medium interfaces.

interference trap (*ICT*) See *interference filter*.

interfering (*Vet*) An injury inflicted by a horse's foot on the opposite leg during progression.

interferometer (*Eng*) An instrument in which an acoustic, optical or microwave interference pattern of *fringes* is formed and used to make precision measurements, mainly of wavelength.

interferon (*BioSci, Med*) A group of proteins with antiviral activity. IFN-α is made by leucocytes and IFN-β by fibroblasts after viral infection. IFN-γ is produced by immune cells after antigen stimulation. IFN-α and IFN-β are also known as Type-I interferons. IFN-γ, a Type-II interferon, is more usually classed as a *cytokine*.

interflow (*EnvSci*) Lateral movement of water at a soil–rock interface.

interfluve (*Geol*) A ridge separating two parallel valleys.

intergalactic medium (*Astron*) General term for any material which might exist isolated in space far from any galaxy.

interglacial stage (*Geol*) A period of milder climate between two glacial stages.

intergranular corrosion (*Eng*) Corrosion in a polycrystalline mass of metal, taking place preferentially at the boundaries between the crystal grains. This leads to disintegration of the metallic mass before the bulk of the metal has been attacked by the corrosive agent.

intergranular texture (*Geol*) A texture characteristic of holocrystalline basalts and doleritic rocks, due to the aggregation of augite grains between feldspar laths arranged in a network.

interhalogen compound (*Chem*) A compound of two members of the halogen family, eg iodine monochloride ICl, iodine heptafluoride IF$_7$.

interkinesis (*BioSci*) See INTERPHASE.

interlaced scanning (*ImageTech*) A scanning system in which the lines of successive FIELDS are displaced to form an interlaced pattern of alternate lines.

interlay (*Print*) Paper inserted between a printing plate and its mount in order to raise the plate to type height. Its thickness may be varied to produce increased or decreased impression where necessary.

interleukin (*BioSci*) A term originally used to describe products of macrophages and T-lymphocytes that influence the differentiation of themselves or of other cells. The numbers of interleukins described has progressively grown and by November 2006 there were 33. Only a few are described below.

interleukin-1 (*BioSci*) One of the major inflammatory cytokines that activates a cascade of changes in various cells. It acts on the thermoregulation centre in the brain to cause fever (IL-1 is *endogenous pyrogen*) and stimulates the synthesis and release of acute-phase proteins by liver cells. It is also required for activation of T-lymphocytes to develop receptors for interleukin-2 and to secrete interleukin-2. It has a direct mitogenic activity for thymocytes. Abbrev *IL-1*.

interleukin-2 (*BioSci*) A growth factor for T-lymphocytes, made by T-lymphocytes that have been activated by IL-1 and a mitogenic agent such as concanavalin A or a specific antigen. Abbrev *IL-2*.

interleukin-3 (*BioSci*) A factor made by activated T-lymphocytes that acts to stimulate growth and differentiation of the progenitors of all haematopoietic cells. Abbrev *IL-3*.

interleukin-4 (*BioSci*) A factor made by activated T-lymphocytes that causes resting B-lymphocytes to divide. Abbrev *IL-4*.

interleukin-5 (*BioSci*) A factor produced by T-lymphocytes and mast cells; it causes B-lymphocyte activation and proliferation. It is chemotactic for eosinophils. Abbrev *IL-5*.

interleukin-6 (*BioSci*) A factor produced by T-lymphocytes, monocytes/macrophages, fibroblasts, hepatocytes. It induces growth and differentiation of T- and B-lymphocytes, hepatocytes, keratinocytes and nerve cells, and stimulates the production of ACUTE-PHASE PROTEINS by the liver. Abbrev *IL-6*.

interleukin-7 (*BioSci*) A factor produced by bone marrow stromal cells and fetal liver cells. It supports the growth of B-lymphocyte precursor cells and T-lymphocytes. Abbrev *IL-7*

interleukin-8 (*BioSci*) A factor produced by a wide variety of cells including lymphocytes, monocytes, fibroblasts, keratinocytes, neutrophils and epithelial cells. It is chemotactic for neutrophil granulocytes. Abbrev *IL-8*.

interleukin-9 (*BioSci*) A factor produced by T-lymphocytes. It enhances T-lymphocyte growth and mast cell activity. Abbrev *IL-9*.

interleukin-10 (*BioSci*) A factor produced by lymphocytes, keratinocytes and macrophages. It suppresses macrophage activation. Abbrev *IL-10*.

interleukin-12 (*BioSci*) A factor produced by B-lymphocytes and macrophages. It induces the differentiation of T-HELPER CELLS and NATURAL KILLER CELLS. Abbrev *IL-12*.

interlining (*Textiles*) A fabric placed in a garment between the lining and outer layer to act as a stiffener or to add bulk.

interlobar (*Med*) Situated or happening between two lobes, esp between two lobes of the lung.

interlock (*ElecEng*) An arrangement of controls in which those intended to be operated later are disconnected until the preliminary settings are correct; eg it is normally impossible to apply the anode voltage to an X-ray tube before the cooling water is flowing.

interlock (*ImageTech*) Maintaining synchronism between a motion picture camera and a sound recorder by electrical control.

interlock (*MinExt*) An arrangement of switchgear by which the controlling source prevents premature loading, starting or continuance during partial malfunction where a series of operations is so interlinked as to require smooth on-line operation.

interlock (*NucEng*) A mechanical and/or electrical device to prevent hazardous operation of a reactor, eg to prevent withdrawal of control rods before coolant flow has been established.

interlock (*Textiles*) A double-faced weft-knitted fabric made of two rib fabrics joined by interlocking loops. Although originally made from cotton for underwear the fabrics are now knitted from various fibres and also used for outerwear.

intermediate (*Chem*) (1) Starting point for manufacture of materials or products, but usually excluding raw material. Normally a chemical compound or mixture of compounds. (2) A short-lived species in a chemical reaction.

intermediate (*ImageTech*) General term for a print of a motion picture film made on colour duplicating stock having integral masking. The detailed usage is: interdupe, a duplicate colour negative derived from an interpositive; interneg, a duplicate colour negative from a reversal original; interpositive, interpos, a colour master positive printed from an original negative.

intermediate constituent (*Eng*) A constituent of alloys that is formed when atoms of two metals combine in certain proportions to form crystals with a different structure from that of either of the metals. The proportions of the two kinds of atoms may be indicated by formulae, eg CuZn; hence these constituents are also known as *intermetallic compounds*.

intermediate coupling (*Phys*) Coupling between spin and orbital angular momenta of valence electrons with characteristics between those of J–J COUPLING and those of RUSSELL–SAUNDERS COUPLING.

intermediate filaments (*BioSci*) Cytoplasmic filaments of diameter intermediate between those of thick and thin filaments, ie from 7 to 11 nm. There are several different subclasses, eg *desmin*, *cytokeratin*, *lamin*, *vimentin*, each with a characteristic cellular distribution. They are all constructed of proteins possessing a rod-like structure built from four α-helical domains.

intermediate frequency (*ICT*) Output carrier frequency of a frequency changer (first detector) in a superhet receiver, adjusted to coincide with the centre of the frequency pass band of the intermediate amplifier. Abbrev *IF*.

intermediate-frequency amplifier (*ICT*) That which, in a superhet receiver, which is between the first (frequency changer) and second (final) demodulator, and which provides the main gain and band pass of the receiver.

intermediate-frequency oscillator (*ICT*) That which, in heterodyne reception, is combined with the output of the intermediate amplifier for demodulation in the second (final) detector. See HETERODYNE CONVERSION.

intermediate-frequency response ratio (*ICT*) The ratio in decibels of an input signal at the intermediate frequency to one at the required signal frequency that would produce a corresponding output.

intermediate-frequency strip (*ICT*) An intermediate-frequency amplifier. Abbrev *IF strip*.

intermediate-frequency transformer (*ICT*) One specially designed for coupling components and providing selectivity in an intermediate-frequency amplifier, the pass band being determined by antiresonance of the windings and coefficient of coupling.

intermediate host (*BioSci*) In the life history of a parasite, a secondary host that is occupied by the young forms, or by a resting stage between the adult stages in the primary host.

intermediate igneous rocks (*Geol*) Igneous rocks containing from 55% to 66% silica, and essentially intermediate in composition between the acid (granitic) and basic (gabbroic or basaltic) rocks. See DIORITE, SYENITE, SYENODIORITE.

intermediate-level waste (*NucEng*) Nuclear waste not included in the categories HIGH-LEVEL WASTE or LOW-LEVEL WASTE.

intermediate neutrons (*Phys*) See NEUTRON.

intermediate phase (*Eng*) A homogeneous phase in an alloy with a composition range different from the pure components of the system.

intermediate-pressure compressor (*Aero*) The section of the axial compressor of a turbofan or bypass turbojet between the *LP* and *HP* sections. It may be on a shaft of its own with a separate turbine, or it may be mounted on either of the other shafts.

intermediate rafter (*Build, CivEng*) See COMMON RAFTER.

intermediate reactor (*NucEng*) One designed so that the majority of fissions will be produced by the absorption of intermediate NEUTRONS. Also *epithermal reactor*.

intermediate sight (*Surv*) Reading on levelling staff, held at point not to be occupied by level other than back- or foresight observation from a levelling point between these last.

intermediate switch (*ElecEng*) A switch for controlling a circuit where more than two positions of control are required; it is connected between the two-way switches which must also be used in such a scheme.

intermediate vector boson (*Phys*) The carrier of weak interactions between particles. There are three intermediate vector bosons, W^+, W^- and Z^0, and all have been observed. See WEINBERG–SALAM THEORY.

intermediate wheel (*Eng*) See IDLE WHEEL.

intermedium (*BioSci*) A small bone of the proximal row of the basipodium, lying between the tibiale and fibulare, or between the radiate and ulnare.

intermenstrual (*Med*) Occurring between two menstrual periods.

intermetallic compounds (*Eng*) See INTERMEDIATE CONSTITUENTS.

intermingled yarn (*Textiles*) A continuous-filament yarn in which the constituent filaments are entangled by passing a turbulent airstream through the yarn.

intermission (*Med*) Temporary cessation, as of fever or of the normal pulse.

intermittent (*ImageTech*) The mechanism of a motion picture camera or projector by which the film is advanced frame by frame.

intermittent claudication (*Med*) Pain in the legs on walking caused by inadequate blood supply to the leg muscles; ATHEROSCLEROSIS is responsible for arterial narrowing.

intermittent control (*ICT*) Control system in which the controlled variable is monitored periodically, an intermittent correcting signal thus being supplied to the controller.

intermittent duty (*ICT*) The conditions of use for a component operated at its intermittent rating.

intermittent earth (*ElecEng*) An accidental earth connection which is present occasionally.

intermittent filtration (*Build*) The LAND TREATMENT process of sewage purification, in which the land is drained artificially by ordinary earthenware or perforated plastic pipes. Cf BROAD IRRIGATION.

intermittent jet (*Aero*) See PULSE JET.

intermittent rating (*ElecEng*) The specified power handling capacity of a component or instrument under specified condition of continuous usage. See CONTINUOUS RATING, HALF-HOUR RATING, ONE-HOUR RATING.

intermittent reinforcement (*Psych*) Schedules of reinforcement in which only some responses are reinforced; it can be based on *ratio* or *interval reinforcement*.

Intermix (*Plastics*) TN for type of internal mixer for polymers using intermeshing rotors.

intermodulation (*ICT*) Undesired modulation of all frequencies with each other in passing through a non-linear element in a transmission path.

intermodulation distortion (*Acous*) Amplitude distortion in which the intermodulation products are of greater importance than the harmonic products, as in AF amplifiers for high-quality speech or music.

intermolecular forces (*Chem*) Interactions between molecules generally involving van der Waals' bonds and hydrogen bonds. Relatively weak compared with covalent, ionic or metallic bonds, but in the case of hydrogen bonds, critical for many biomaterials (eg COLLAGEN, DNA, KERATIN) and some synthetic polymers. See VAN DER WAALS' FORCES and panels on BONDING and DNA AND THE GENETIC CODE.

intermontane basin (*Geol*) A basin between mountain ranges often associated with a graben, eg the Midland Valley of Scotland.

internal capacitance (*Phys*) See INTERELECTRODE CAPACITANCE.

internal characteristic (*ElecEng*) A curve showing the relation between the load on an electric generator and the internal emf.

internal-combustion engine (*Eng*) An engine in which combustion of a fuel takes place within the cylinder, and the products of combustion form the working medium during the power stroke. See COMPRESSION–IGNITION ENGINE, DIESEL ENGINE, GAS ENGINE, PETROL ENGINE.

internal compensation (*Chem*) Neutralization of optical activity within the molecule by the combination of two enantiomorphous groups, eg in meso-tartaric acid.

internal conductor (*ElecEng*) The inner conductor of a concentric cable. Also *inner conductor*.

internal consistency (*Psych, Maths*) An estimate of whether subsections of a test are measuring the same variable. See TEST–RETEST RELIABILITY.

internal conversion (*Phys*) Nuclear transition where energy released is given to orbital electron which is usually ejected from the atom (conversion electron), instead of appearing in the form of a gamma-ray photon. The *conversion coefficient* for a given transition is given by the ratio of conversion electrons to photons.

internal emf (*ElecEng*) The emf generated in an electric machine; the voltage appearing at the terminals is the internal emf minus any voltage drop caused by the current in the machine.

internal energy (*Chem, Phys*) The kinetic energy of the constituent molecules of a system and their potential energies due to the molecular interactions. The internal energy is manifest as the temperature of the system, latent heat, as shown by a change of state, or the repulsive forces between molecules, seen as expansion. For a thermodynamic system, the difference between the heat absorbed by the system and the external work done by the system is the change in its internal energy (the *first law of thermodynamics*); this depends only upon the initial and final conditions, and is therefore independent of the paths of change. Symbol U. SI unit is the JOULE.

internal-expanding brake (*Eng*) See EXPANDING BRAKE.

internal flue (*Eng*) A furnace tube, or fire tube, running through the water space of a boiler.

internal focusing telescope (*Surv*) One in which focusing is effected by the movement of an internal concave lens fitted between the object glass and the eyepiece, both of which are fixed.

internal friction (*Eng*) That which gives rise to HYSTERESIS and DAMPING in elastic bodies. Also used for that which opposes flow in liquids and gases giving rise to viscosity.

internal gear (*Eng*) A spur gear in which teeth, formed on the inner circumference of an annular wheel, mesh with the external teeth of a smaller pinion. Both wheels revolve in the same direction.

internal grinding (*Eng*) The grinding of internal cylindrical surfaces by a rotating abrasive wheel, which is either traversed along the revolving work or, the work being fixed, given a planetary motion.

internal impedance (*Phys*) The OUTPUT IMPEDANCE of an amplifier or generator.

internal indicator (*Chem*) An indicator which is dissolved in the solution in which the main reaction takes place.

internally fired boiler (*Eng*) A boiler whose fire box or furnace is inside the boiler and surrounded by water, as in the Lancashire, marine and locomotive types.

internal mixer (*Eng*) A type of mixer for polymers using rolls within a closed chamber, tending to replace two-roll mill. See BANBURY, INTERMIX.

internal pair production (*Phys*) The production of an electron–positron pair in the coulomb field of a nucleus. For transitions where the excitation energy released exceeds 1·02 MeV, this process will be competitive with both internal conversion and gamma-ray emission. It occurs most readily in nuclei of low atomic number when the excitation energy is several mega-electron-volts.

internal phloem (*BioSci*) Primary phloem on the centric side of the xylem as in the Solanaceae. Also *intraxylary phloem*.

internal resistance (*ElecEng*) That of any voltage source resulting in a drop in terminal voltage when current is drawn.

internal respiration (*BioSci*) See RESPIRATION.

internal screw-thread (*Eng*) A screw-thread cut on the inside of a cylindrical surface, as distinct from an EXTERNAL SCREW-THREAD. Also *female thread*.

internal secretion (*BioSci*) A secretion that is poured into the blood vessels, or into the canal of the spinal cord; a hormone.

internal store (*ICT*) See MAIN MEMORY.

internal stress (*Eng*) RESIDUAL STRESS in a material due to differential effects of heating, cooling or working operations or to constitutional (eg phase) changes in a solid. To satisfy equilibrium, the net force on the body due to internal stresses must equal zero.

internal validity (*Genrl*) A measure of the trustworthiness of a sample of data. Internal validity looks at the subject testing and environment in which the data collection took place.

internal voltages (*Electronics*) Those, such as CONTACT POTENTIAL or WORK FUNCTION, which add an effect to the external voltages applied to an active device.

internasal septum (*BioSci*) See MESETHMOID.

International Air Transport Association (*Aero*) Association founded in Havana (1945) for the promotion of safe, regular and economic air transport, to foster air commerce and collaboration among operators, and to co-operate with ICAO and other international organizations. Abbrev *IATA*.

international angstrom (*Phys*) A unit which, although very nearly equal to the angstrom unit (10^{-10} m), is defined in a different way. It is such that the red cadmium line at 15°C and 760 mm Hg pressure would have a wavelength of 6438·4696 IÅ. Formerly the reference standard for metrology and spectroscopy. Symbol IÅ.

International Annealed Copper Standard (*Eng*) Standard reference for the conductivity of copper and its alloys: 100%IACS represents a conductivity of 58 megasiemens per metre (MS^{-1}). Abbrev *IACS*.

International Atomic Energy Agency (*NucEng*) Organization for promoting the peaceful uses of nuclear power. Abbrev *IAEA*.

international candle (*Phys*) Former unit of luminous intensity stated in terms of a point source, which was expressed in an equivalence of a standard lamp burning under specified conditions. Replaced by CANDELA.

International Civil Aviation Organization (*Aero*) An intergovernmental co-ordinating body, which has its headquarters at Montreal, Canada, for the regulation and control of civil aviation and the co-ordination of AIRWORTHINESS requirements and other safety measures on a worldwide basis. Abbrev *ICAO*.

international electrical units (*Phys*) Units (amp, volt, ohm, watt) for expressing magnitudes of electrical quantities, adopted internationally until 1947; replaced first by MKSA, later by SI UNITS.

International Frequency Registration Board (*ICT*) The committee responsible for reviewing and allocating radio frequencies used throughout the world, in order to ensure efficient use of channel space and avoid interference. Abbrev *IFRB*.

international gateway (*ICT*) A switching centre designed to route traffic between a national network and the BEARERS that will convey it to the international gateway of the destination country.

International Organization for Standardization (*Genrl*) Main international body setting standards by agreement with national standards bodies, eg BSI, DIN. Abbrev *ISO*.

international paper sizes (*Paper*) See ISO PAPER SIZES.

international practical temperature scale (*Phys*) A practical scale of temperature which is defined to conform as closely as possible with the thermodynamic scale. Various fixed points were defined initially using the gas thermometer, and intermediate temperatures are measured with a stated form of thermometer according to the temperature range involved. The majority of temperature measurements of this scale are now made with platinum resistance thermometers.

International Rubber Hardness Degree (*Eng*) A measure of the depth of penetration of an indenter into an elastomer, used to monitor the degree of cure. Abbrev *IRHD*. See RUBBER HARDNESS and panel on HARDNESS MEASUREMENTS.

international screw-thread (*Eng*) A metric system on which the pitch of the thread is related to the diameter, the thread having a rounded root and flat crest and a 60° included angle.

International Space Station (*Space*) The name for the space station jointly being developed by NASA, ESA, Japan, Canada and Russia. See panel on SPACE STATION.

International Standard Atmosphere (*Aero*) A standard fixed atmosphere, adopted internationally, used for comparing aircraft performance; mean sea-level temperature 15°C at 1013·2 millibars, lapse rate 6·5°C km^{-1} altitude up to 11 km (ISA tropopause), above which the temperature is assumed constant at −56·5°C in the stratosphere. Abbrev *ISA*.

International Style (*Arch*) US term, originally introduced as the title of a book to complement the first International Exhibition of Modern Architecture which took place in New York in 1932, but now widely accepted as a name for the architectural style which has prevailed since the 1920s and which has formed the basis of architectural development to the present day.

international system of units (*Genrl*) See SI UNITS.

International Telecommunication Union (*ICT*) Specialized agency of the United Nations, established to promote international collaboration in telecommunications with a view to improving the efficiency of worldwide services. It has three permanent committees: the CCIR, the CCITT and the IFRB. Abbrev *ITU*.

International Union of Pure and Applied Chemistry (*Chem*) A body responsible, among other things, for the standardization of chemical nomenclature, which it alters frequently. Abbrev *IUPAC*.

Internet (*ICT*) A linkage of worldwide networks designed to store its information at many widely separated sites. See panel on INTERNET.

Internet access provider (*ICT*) Same as INTERNET SERVICE PROVIDER. Abbrev *IAP*.

Internet protocol (*ICT*) A communication protocol used to transfer packets of data across a network of computers.

Internet service provider (*ICT*) A commercial organization that provides access to the Internet. Abbrev *ISP*. See panel on INTERNET.

interneuron (*BioSci*) A nerve cell within the central nervous system that is neither sensory nor motor but communicates between other nerve cells. Also *interneurone*.

internode (*BioSci*) (1) In plants, a region between two successive nodes of a stem or hypha. (2) In a nerve, the myelinated region between adjacent *nodes of Ranvier*.

internuncial (*BioSci*) Interconnecting, eg a neuron of the central nervous system interposed between afferent and efferent neurons of a reflex arc.

interoceptor (*BioSci*) A sensory nerve ending, specialized for the reception of impressions from within the body. Cf EXTEROCEPTOR.

interoperable (*ICT*) A term applied to hardware or software systems that are able to exchange and use information from different computer systems.

interopercular (*BioSci*) In fish, a ventral membrane bone supporting the operculum.

interparietal (*BioSci*) A median dorsal membrane bone of the vertebrate skull, situated between the parietals and the supraoccipital.

interpenetration twins (*Min*) Two or more crystals united in a regular fashion which appear to have grown through one another. Cf JUXTAPOSITION TWINS. See panel on TWINNED CRYSTALS.

interphase (*BioSci*) The period of the cell cycle between mitoses. See panel on CELL CYCLE.

interphase (*MinExt*) Transition zone between two phases in a system (solid–liquid, liquid–liquid, liquid–gas). In solid–fluid system, zone of shear through which the physical or chemical qualities of the contacting surfaces migrate towards one another.

interphase transformer (*ElecEng*) A centre-tapped, iron-cored reactor, connected between the two star joints of the supply transformer for a double three-phase mercury-arc rectifier. The centre tap forms the dc negative terminal. Also *absorption inductor, interphase reactor, phase equalizer*.

interplane struts (*Aero*) In a multiplane structure, those struts, either vertical or inclined, connecting the spars of any pair of planes, one above the other.

interplanetary matter (*Astron*) Material in the solar system other than the planets and their satellites. Includes the COMETS, dust, INTERPLANETARY MEDIUM and METEOROIDS.

interplanetary medium (*Astron*) The tenuous ionized gas found between the planets due to the SOLAR WIND.

interplanetary space (*Space*) The region of space generally outside the influence of planetary bodies which is occupied by a tenuous gas. It consists of charged and neutral particles of gas (mainly hydrogen), dust and streams and clouds of particles (eg protons or electrons) ejected by the Sun. A weak magnetic field pervades the whole region.

interpolation (*MathSci*) The estimation of the value of a function at a particular point from values of the function on either side of the point. Cf EXTRAPOLATION.

interpolation of contours (*Surv*) The joining of points derived from spot levelling or in conformity with the existing mapped contour lines, assuming that the recorded data disclose all abrupt changes in ground level and configuration.

interpole (*ElecEng*) See COMPOLE.

interpole motor (*ElecEng*) An electric motor fitted with COMPOLES.

interpolymerization (*Chem*) A mixture of two or more individual homopolymers at a molecular scale. Made by polymerizing a monomer-swollen gel. Not to be confused with mixture of compatible polymers (eg Noryl).

interpositional growth (*BioSci*) See INTRUSIVE GROWTH.

interpreter (*ICT*) A program that translates and executes a source program one statement at a time. Cf COMPILER.

interquartile range (*MathSci*) The difference between the lower and upper QUARTILES.

interrenal body (*BioSci*) In selachian fish, a ductless gland that lies between the kidneys and corresponds to the cortex of the suprarenal gland of higher vertebrates.

interrogation (*ICT*) Transmission of a pulse to a TRANSPONDER (pulse repeater).

interrupt (*ICT*) A signal that causes a break in the execution of the current routine. Control passes to another routine in such a way that later the original routine may be resumed. See PACKET SWITCHING, TIME SHARING.

interrupted continuous waves (*ICT*) Electromagnetic waves generated by an oscillator the output from which is interrupted periodically at an audible frequency, so that

Internet

A vast, uncontrolled network of computers connected by data links which anyone can use and for which anyone can provide information. The only constraint is that every computer must use common software systems known as PROTOCOLS, the most important of which is called TCP/IP for the transmission control protocol and Internet protocol.

Originally developed by the US Defense Department to counter the vulnerability of a command structure dependent on central computers, it was set up by distributing the computers around the country. The system had many links to nodes operated by the military and by universities which had the necessary mainframe computers and the expertise to develop the protocols needed. It was organized so that it could operate despite the destruction of a proportion of its links and nodes and of necessity the network extended to Europe and the Far East. Eventually, the academic part of the network was taken over by the US National Science Foundation and gradually became more widely accessible both in the USA and elsewhere.

Nowadays any user with a modem or other terminal adapter can be linked to a server by telephone line or radio. The server is one of a number of computers operated by an Internet service provider (ISP) who usually has a local telephone access number and connections through a GATEWAY to the Internet. Service providers often allow the user to have a number of WORLD WIDE WEB (or web) pages, limited by the storage space (measured in megabytes) on the host's server. They must also store incoming messages which can then be downloaded at leisure and route outgoing messages towards the recipient. They handle all the search and downloading traffic originated by a user. For these facilities they charge a fee which often allows unlimited access. In return they must maintain sufficient computer capacity and often duplicate popular and distant sites.

The modern Internet
The modern Internet is an all-inclusive system, encompassing much more than just the websites. These include GOPHER sites which predate the web and are a menu-based search system mainly for information peculiar to a given institution, eg a university; FILE TRANSFER PROTOCOL (ftp) sites which contain many hundreds of thousands of files which can be downloaded; *searchable database* sites (eg WAIS, or wide area information server); and many more.

Long messages are split into separate packets of information, each with a header, containing the address and other information. These packets can travel over different links as determined by the different routers and are eventually reassembled at the recipient's server. All addresses conform to a convention called the universal resource locator (URL) which is unique for each user.

Browsers and search engines
The software involved includes a client application, or browser in World Wide Web parlance, which manages communication between the user and service provider. The struggle for market share between Netscape and Microsoft's Internet Explorer has effectively been won by the latter. Apple's Safari and open-source products such as Firefox/Mozilla occupy niche positions. All provide access to as many of the Internet's functions as possible in an easy and intuitive fashion. The growth in downloadable applets and controls using the Java virtual machine technology has enabled the

---

the received signal is directly audible after detection. Abbrev *ICW*. Also *chopped continuous waves*.

interruptedly pinnate (*BioSci*) A pinnate leaf having pairs of large and small leaflets alternating along the rachis.

interrupter (*ElecEng*) A device which interrupts periodically the flow of a continuous current, such as the mechanical 'make and break' of an induction coil.

interrupt request lines (*ICT*) Hardware lines over which devices can send INTERRUPT signals. Abbrev *IRQ*.

intersatellite link (*ICT*) A radio or optical link between communications satellites, allowing some traffic to be rerouted without involvement of ground stations.

interscapular (*Med*) Between the two shoulder blades.

intersection (*MathSci*) Of sets A and B: the set of elements *x* such that *x* is in A and *x* is also in B.

intersection (*Surv*) Plane-table surveying in which the plane table is set up consecutively at each end of a measured baseline. Rays are drawn on paper at each set-up to show the direction of the point that it is required to fix on the plan, the intersection of these rays giving the position of the point.

intersection angle (*Surv*) The angle of deflection as measured at the intersection point between the straights of a railway or highway curve.

intersection point (*Surv*) The point at which the straights of a railway or highway curve would meet if produced.

intersegmental membrane (*BioSci*) The flexible infolded portion of the cuticle between adjacent definitive segments to allow the freedom of movement of the body in ARTHROPODA.

intersertal texture (*Geol*) The texture characterized by the occurrence of interstitial glass between divergent laths of feldspar in basaltic rocks.

intersetting (*Print*) The introduction on a rotary press of a pre-printed web in register.

intersex (*BioSci*) An individual that exhibits characters intermediate between those of the male and female of the same species. Often due to a chromosomal abnormality.

intersheath (*ElecEng*) Cylindrical electrodes in the interior of a cable dielectric, used for the purpose of keeping the variation of stress at a minimum. The intersheaths must be kept at certain potentials in order to achieve this purpose. See STRESS.

interspecific (*BioSci*) Said of an event such as a cross between individuals from separate species. Cf INTRA-SPECIFIC.

Internet *(Cont.)*

widespread use of multimedia embellishments independent of client operating system and browser. The other area of notable change has been in the explosive growth of search engines, websites accessed via a browser which employ sophisticated algorithms to return lists of content matching user-input search terms. Their ability to do this depends upon the use of 'crawler' or 'spider' programs which continuously scan web pages, generating indexes of keywords and content. Probably the most widely used of these is Google, which has added a variety of further searchable facilities, among them images of the Earth and, controversially, scanned text from conventionally published material.

Usergroups and Usenet
A network of tens of thousands of discussion groups, open to anyone and on any subject to which participants contribute hundreds of thousands of pages every day.

Online services
If this account gives the impression of a somewhat disorganized system, it is, and some specialized service providers tried to impose some order, initially by being independent of the Internet while using its methods. They provided an easy interface and access to a very wide range of facilities at their own nodes worldwide. Many of these developed into ISPs offering access via their portal sites to the web as well as facilities such as e-mail, chatrooms and newsgroups. Some large ISPs merged with conventional media and communications companies to provide increasingly seamless access to content across electronic media and telecommunications. The growth in e-commerce, pioneered by companies such as Amazon which exploited the moribund mail-order business model, has led on the one hand to conventional businesses adapting to new media with concepts such as ticketless air travel, and on the other hand to the emergence of unprecedented services such as eBay, a democratic worldwide marketplace in which individuals offer and bid for goods.

The World Wide Web
The web is, as we have seen, a large part but certainly not the whole of the Internet. It is distinguished by the ability to use hypertext links to other sites and other documents on the web. A single click on a highlighted word or phrase sends the user to a new page which may be on a site in another part of the world. The enormous strength of this approach has ensured that the web has grown exponentially in use over the last few years.

Problems and limitations
The Internet is not without problems, which in the public's mind are those associated with its anarchic and uncontrolled nature, often highlighted by lurid stories in the conventional media. There are technical problems as well. Speed of access has been improved by the spread of broadband technology and reliance on metered dial-up telephony is becoming less common, at least in countries with technologically advanced infrastructures. The more subtle problem of the fragility of web content remains – hypertext links are easily broken and cannot respond when a website is discontinued. More seriously, the question of content authority is pervasive – users have no easy means of judging the accuracy or provenance of web-based information, and a generation of surfers has now arisen with an implicit and often unwarranted trust in Internet content. Many ISPs and search engines are opaque with regard to the commercial and political screening of content.

---

interstation interference (*ICT*) That from another transmitter on the same or an adjacent wavelength, as distinct from atmospheric interference.

interstation noise suppression (*ICT*) Suppression of radio receiver output when no carrier is received.

interstellar hydrogen (*Astron*) The hydrogen gas which dominates the INTERSTELLAR MEDIUM. Cool diffuse clouds of atomic hydrogen (*HI regions*) are detected by FORBIDDEN LINE radio emission at a frequency of 1420·4 MHz (the 21 CENTIMETRE LINE). Ionized hydrogen (*HII regions*) exists as discrete clouds often associated with star formation or other sources of ultraviolet radiation.

interstellar medium (*Astron*) The gaseous and dusty matter pervading interstellar space and amounting to one-tenth of the mass of our galaxy. Where cold gas and dust conglomerate, regions of star formation occur. The medium comprises hydrogen, helium, INTERSTELLAR MOLECULES and dust. It is replenished by stellar winds and the ejecta of NOVA outbursts and SUPERNOVA explosions.

interstellar molecule (*Astron*) One of more than 50 species of molecule found within the gases of the INTERSTELLAR MEDIUM, particularly in cold dense clouds. The most common molecules include: CO, H_2O, NH_3, $HCHO$, CH_4 and CH_3OH.

interstice (*Chem*) The space between atoms in a lattice where other atoms can be located, eg in close-packed metallic lattices.

interstitial (*BioSci*) Occurring in the interstices between other structures.

interstitial cell stimulating hormone (*Med*) Abbrev ICSH. See LUTEINIZING HORMONE.

interstitial compounds (*Eng*) Metalloids in which small atoms of non-metallic elements (H, B, C, N) occupy positions in the interstices of metal lattices. In general, the structure of the metal is preserved, though somewhat distorted. They are often (esp those of group IV and V metals) characterized by exceptionally high melting points and hardness, by chemical inertness and by metallic lustre and conductivity.

interstitial solid solution (*Chem*) A type of solid solution formed when there is a large difference in relative atomic sizes, usually the solute being less than 0·59 that of the solution. This enables the solute atoms to take up positions within the interstices of the crystal lattice of the solvent. The commonest example is that of carbon in iron.

intersymbol interference (*ICT*) A source of errors in digital transmissions systems, in which departure of pulses from their ideal shape and timing causes the probability of their being recognized as '1' or '0' to depend on the values of their neighbours. See EYE DIAGRAM.

intertrack bond (*ElecEng*) A conductor for connecting electrically the rails of separate tracks on electric railways or tramways to reduce the total resistance of the return path.

intertrigo (*Med*) Excoriation of the skin in the skin folds, usually due to excessive moisture.

intertrochanteric (*Med*) Situated between the two trochanters of the upper part of the femur.

intertropical convergence zone (*EnvSci*) A narrow, low-latitude zone in which air masses originating in the northern and southern hemispheres converge. Over the Atlantic and Pacific Oceans it is the boundary between the NE and SE trade winds. The mean position is somewhat north of the equator but over the continents the range of movement is considerable. It is a zone of generally cloudy, showery weather. Abbrev *ITCZ*. Also *intertropical front* (*ITF*).

interval (*Acous*) The ratio of the frequencies of two sounds; or in some cases its logarithm. See SEMITONE.

interval (*MathSci*) See CLOSED INTERVAL, OPEN INTERVAL.

interval mathematics (*MathSci*) A branch of mathematics used to obtain algorithms with computationally rigorous bounds for the solution of numerical problems.

intervalometer (*ImageTech*) A device for operating a camera shutter at set intervals.

interval scale (*MathSci*) Any scale of measurement with elements at equal intervals, but with no fixed zero, eg temperatures, in which it makes no sense to say 100°C (212°F) is twice as hot as 50°C (122°F). Compare ORDINAL SCALE, RATIO SCALE.

interval schedule of reinforcement (*Psych*) A schedule in which reinforcement is given for the first response made after a certain period of time has passed. On fixed-interval schedules, the period of time is always the same; on variable-interval schedules the period of time fluctuates. Both types are forms of *partial* or *intermittent* reinforcement. Cf CONTINUOUS REINFORCEMENT.

intervening sequence (*BioSci*) See INTRON.

interventional radiology (*Radiol*) A term used to describe radiological procedures undertaken for therapeutic rather than diagnostic purposes, eg ANGIOPLASTY, where a balloon on a catheter is dilated in a narrowed artery.

intervertebral (*BioSci*) Between the vertebrae; said of the fibro-cartilage disks between the vertebral centra in crocodiles, birds and mammals.

intestine (*BioSci*) In vertebrates, that part of the alimentary canal leading from the stomach to the anus; in invertebrates, that part of the alimentary canal which was thought by early investigators to correspond to the vertebrate intestine. Adj *intestinal*.

intima (*BioSci*) The inner layer of a blood vessel, comprising an endothelial monolayer in contact with blood and a subcellular elastic layer with smooth muscle cells. Below the intima is the media, then the adventitia. The term may be applied to other organs.

intine (*BioSci*) The inner part of the wall of a pollen grain or vascular plant spore. Cf EXINE.

into (*MathSci*) See ONE-TO-ONE.

intoxication (*Med*) The state of being poisoned. Colloq implies the toxin is alcohol.

intracapsular (*Med*) Situated within a capsule, esp within the ligamentous joint capsule enveloping the head and neck of the femur.

intracavitary therapy (*Radiol*) Treatment applied within cavities of the body; said eg of radium placed in the cavity of the uterus, also of irradiation of part of the body through natural or artificial body cavities.

intracavitary X-ray therapy (*Radiol*) X-ray therapy in which the appropriate part of suitable X-ray apparatus is placed in a body cavity.

intracellular (*BioSci*) Within the cell.

intracellular enzyme (*BioSci*) One that functions within the cell. Cf EXTRACELLULAR ENZYME.

intracerebral (*Med*) Situated in the substance of the brain.

intracervical (*Med*) Situated in, or applied to, the canal of the cervix uteri (the lowest part or neck of the uterus).

intracranial (*Med*) Situated within the skull.

intradermal (*Med*) Situated in, or introduced into, the skin.

intrados (*Build*) The under surface of an arch. Also *soffit*. See fig. at ARCH. See EXTRADOS.

intrafusal (*BioSci*) A muscle fibre contained within a muscle spindle. Concerned with proprioceptive reflexes.

intrahepatic (*Med*) Situated or occurring in the substance of the liver.

intramammary (*Med*) Situated or occurring in the breast.

intramedullary (*Med*) In the substance of the medulla oblongata (the brain stem); situated or occurring in the substance of the spinal cord.

intramolecular forces (*Chem*) Interactions within a single molecule, generally involving van der Waals' and hydrogen bonds. Although relatively weak compared with the covalent bonds of the backbone chain, they are important in polymers for determining chain flexibility, and hence a wide range of polymer properties. See ROTATIONAL ISOMERISM.

intranuclear forces (*Phys*) Forces operative between nucleons at close range comprising COULOMB FORCES and SHORT-RANGE FORCES. According to the hypothesis of charge independence, the former are always the same for two nucleons of corresponding angular momentum and spin regardless of whether these are protons or neutrons. See ISOTOPIC SPIN.

intra-ocular (*Med*) Situated within the eyeball.

intraperitoneal (*Med*) Situated in, or introduced into, the peritoneal cavity.

intrapleural (*BioSci*) Within the thoracic cavity.

intraspecific (*BioSci*) Pertaining to interactions within members of a species. Cf INTERSPECIFIC.

intrathecal (*Med*) Within the sheath of membranes investing the spinal cord.

intratracheal (*Med*) Within, or introduced into, the trachea.

intratubal (*Med*) Situated within a Fallopian tube.

intra-uterine (*Med*) Situated within, or developing within, the uterus.

intra-uterine device (*Med*) Contraceptive device of relatively high efficacy, inserted in the uterus; consists usually of a spiral or a loop-shaped copper-coated filament. Its action, though not completely understood, may be physiological. Abbrevs *IUD*, *IUCD*.

intravenous (*Med*) Within, or introduced into, a vein.

intravitam staining (*BioSci*) The artificial staining of living cells. Usually by injection of the stain into a circulatory system. Also *vital staining*.

intraxylary phloem (*BioSci*) See INTERNAL PHLOEM.

intrazonal soil (*EnvSci*) Well-developed soil in which local conditions have modified the influence of the regional climate. Cf AZONAL SOIL.

intrinsic (*BioSci*) Said of appendicular muscles of vertebrates that lie within the limb itself, and originate either from the girdle or from the limb bones. Cf EXTRINSIC.

intrinsic angular momentum (*Phys*) The total spin of an atom, nucleus or particle as an idealized point, or arising from orbital motion. Spin and orbital angular momentum are quantized in integral multiples of

$$\frac{1}{2}\frac{h}{2\pi} \text{ and } \frac{h}{2\pi}$$

respectively, where h is PLANCK'S CONSTANT.

intrinsic conduction (*Electronics*) That in a semiconductor when electrons are raised from a filled band into the conduction band by thermal energy, so producing hole–electron pairs. It increases rapidly with rising temperature. See panel on INTRINSIC AND EXTRINSIC SILICON.

intrinsic crystal (*Crystal*) A crystal, the photoelectric properties of which do not depend on impurities.

intrinsic equation (*MathSci*) Of a curve: an equation connecting the arc length s of any point on the curve with the gradient ψ at the point. Such an equation is independent of any co-ordinate system and is an inherent property of the curve.

intrinsic factor (*Med*) A glycoprotein secreted from the parietal cells of the gastric mucosa which is essential for the absorption of cyanocobalamin. Absence of intrinsic factor causes PERNICIOUS ANAEMIA.

intrinsic impedance (*Phys*) Wave impedance depending on the medium alone, ie $Z_0^2 = \mu/\varepsilon$, where μ is the absolute permeability and ε the absolute permittivity of the medium.

intrinsic induction (*Phys*) Equivalent to INTENSITY OF MAGNETIZATION, although in unrationalized units, it is 4π times this quantity.

intrinsic mobility (*Electronics*) The mobility of electrons in an intrinsic semiconductor. See panel on INTRINSIC AND EXTRINSIC SILICON.

intrinsic motivation (*Psych*) The motivation or desire to do something based on internal satisfaction or fulfilment rather than for external reward (reinforcement).

intrinsic semiconductor (*Electronics*) A semiconductor made of intrinsic silicon. See panel on INTRINSIC AND EXTRINSIC SILICON.

intrinsic silicon (*Electronics*) Pure crystalline silicon which thus owes its conductivity to its intrinsic properties. See panel on INTRINSIC AND EXTRINSIC SILICON.

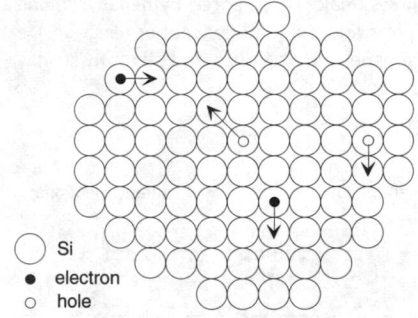

○ Si
● electron
○ hole

intrinsic silicon

intrinsic viscosity (*Phys*) Polymer viscosity as determined from dilute solution. Found by extrapolating curves of reduced viscosity and inherent viscosity to zero concentration. Gives a measure of molecular mass when inserted into the MARK–HOUWINK EQUATION. Symbol η. Also *limiting viscosity number*.

introgression (*BioSci*) Incorporation of genes of one species into the gene pool of another via an interspecific hybrid. Also *introgressive hybridization*.

introitus (*Med*) Entry to a cavity, eg vagina.

introjection (*Psych*) In social psychology, the internalization of the norms and values of one's social group, so that the individual comes to be guided by a sense of what is appropriate or inappropriate, rather than by external rewards and punishments. In psychoanalytic theory, the term is synonymous with IDENTIFICATION.

intromission (*Med*) The insertion of one part into another, esp of the penis into the vagina.

intromittent (*BioSci*) Adapted for insertion, as the copulatory organs of some male animals.

intron (*BioSci*) Genes in eukaryotes are organized in such a way that while the whole sequence is transcribed, only part of it forms the messenger RNA. *Introns* or *intervening sequences*, which are often long, are excised during the maturation of the RNA. Cf EXONS, the parts which are expressed in the protein product.

introrse (*BioSci*) Directed or bent inwards. Facing towards the axis, esp stamens opening towards the centre of the flower. Cf EXTRORSE.

introvert (*BioSci*) A structure or part of the body that may be turned inside out or involuted, as the proboscis of a nemertinean worm.

introvert (*Psych*) In Jungian theory, a person who is inwardly directed, resulting in the shunning of interpersonal relations and absorption in their own thoughts. Generally, a person more interested in their own thoughts and feelings than in the external world and social activity. N *introversion*. See EXTROVERT.

intrusion (*Geol*) The process of emplacement of a magma into pre-existing rock. Also used in the sense of injection of a plastic sediment, eg a salt dome.

intrusion countermeasure electronics (*ICT*) Software designed to prevent unauthorized access to data. Abbrev ICE.

intrusive growth (*BioSci*) A type of tissue growth in which an elongating cell grows by insertion between other cells. Cf SYMPLASTIC GROWTH.

intubation (*Med*) The introduction of a tube, esp through the larynx into the trachea.

intumescence (*Chem*) The swelling of material on heating, often with the violent escape of moisture.

intumescence (*Med*) The process of swelling; a swelling.

intumescent paints (*Build*) Fire-retardant paints which upon exposure to heat swell to form an insulating barrier which protects the underlying surface.

intussusception (*BioSci*) New material inserted into the thickness of an existing cell wall. Cf APPOSITION.

intussusception (*Med*) The pushing down, or invagination, of one part of the intestine into the part below it. Usually occurs in children causing intestinal obstruction.

inulin (*Med*) A polysaccharide obtained from the tubers of the dahlia, used in assessing renal function.

invagination (*BioSci*) (1) Insertion into a sheath. (2) The development of a hollow ingrowth. (3) The pushing in of one side of the blastula in embolic gastrulation. Adj *invaginate*. Cf EVAGINATION.

inv allotypes (*BioSci*) Allotypic antigenic determinants on the constant region of the kappa chain of human immunoglobulins.

Invar (*Eng*) TN for iron–nickel alloy. Composed of 36% nickel, 63·8% iron and 0·2% carbon. COEFFICIENT OF THERMAL EXPANSION is very small. Used for measuring tapes, tuning forks, pendulums and in instruments.

invariable plane (*Astron*) A certain plane which remains absolutely unchanged by any mutual action between the planets in the solar system; defined by the condition that the total angular momentum of the system about the normal to this plane is constant. The plane is inclined at 1°35′ to the ecliptic.

invariant (*Chem*) Possessing no DEGREES OF FREEDOM (1).

invariant (*MathSci*) Any characteristic of a system which is unchanged by transformations of the type under consideration.

invariant chain (*BioSci*) Polypeptide chain of invariant sequence associated with MAJOR HISTOCOMPATIBILITY COMPLEX CLASS II molecules inside ANTIGEN PRESENTING CELLS. Functions by covering the peptide binding groove of the MHC II molecules until they encounter foreign peptide in the lysosomal compartment of the cell.

invar shadowmask (*ImageTech*) A TV shadowmask which is more resistant to thermal distortion than a conventional type, allowing a brighter picture. See INVAR, SHADOW MASK TUBE.

inventory (*NucEng*) Total quantity of fissile material in a reactor.

inverse (*MathSci*) Of a given element, x, of a set with respect to an operation *, an element x' such that $x*x' = e = x'x$, where e is the IDENTITY element of the set. In particular: (1) For the usual addition of real numbers, the inverse of x is written $-x$ and is called the *negative* of x. (2) For the

Intrinsic and extrinsic silicon

Pure crystalline silicon is an *intrinsic semiconductor*. At absolute zero all four outer electrons of each atom in the bulk are involved in COVALENT BONDS with surrounding atoms. The structure is like that of diamond. As the temperature is raised above absolute zero, an increasing fraction of bonding electrons is able to escape the covalent partnership and move almost freely from atom to atom. Such electrons are then termed CONDUCTION ELECTRONS. At the same time, the holes left behind in the network of bonds can be filled either by an electron from a neighbouring bond or else by recombination with one of the conduction electrons: in this way the hole apparently moves or vanishes. A hole is effectively a carrier of positive charge since it is drawn to regions of negative potential. Electrical conductivity arises from the presence and mobility of both conduction electrons and holes (see diagram).

At room temperature, on average about 3 in 10^{13} covalent bonds in silicon are incomplete – a tiny fraction, but sufficient to endow the material with a conductivity of about $4 \times 10^{-4} \, \Omega^{-1} \, m^{-1}$. An increase of 100 K in the ambient temperature raises the conductivity to about $0.3 \, \Omega^{-1} \, m^{-1}$.

Extrinsic silicon owes its conductivity to charge carriers introduced by the careful addition of dopant atoms. When a group V element, like phosphorus or arsenic, is substituted into a silicon (group IV) lattice site, it acts as an electron donor as only four of its five valence electrons are required for bonding. The fifth is weakly bound and easily dislodged thermally. In the same way a group III element, aluminium or gallium, introduces a hole as it can only provide three of the four required bonding electrons. A hole is launched into the valence band when the so-called acceptor dopant attracts an electron from a neighbouring silicon–silicon bond to make good its own deficit.

Doping with donors tends to make n-type semiconductors, introducing, almost pro rata, additional conduction electrons. The electrons are the majority charge carrier, being mostly balanced by an immobile positive charge associated with donor atoms which have been ionized. Holes form a group of minority carriers with a reduced lifetime owing to the increased risk of recombination posed by the larger electron population. The net charge is zero. Doping with acceptors tends to make p-type semiconductors, with the holes (majority) balanced by negatively ionized acceptors and the minority species (electrons).

See panels on BONDING and SILICON, SILICA, SILICATES.

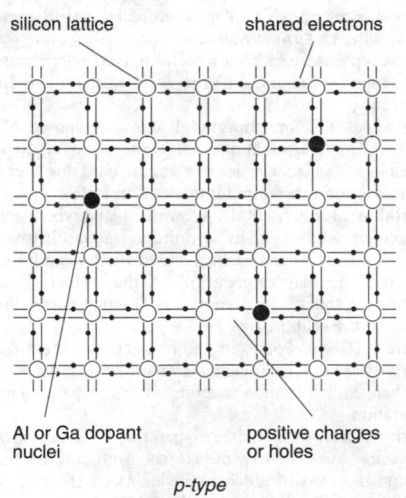

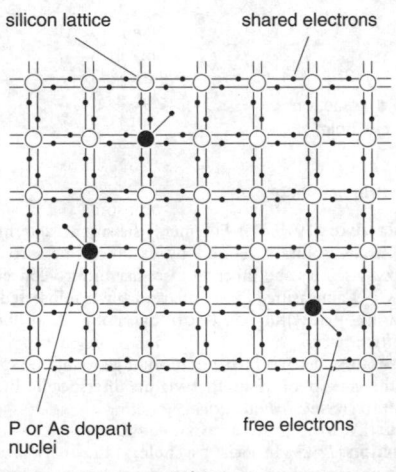

silicon lattice — shared electrons

Al or Ga dopant nuclei — positive charges or holes

p-type

P or As dopant nuclei — free electrons

n-type

Semiconductor The two types.

usual multiplication of real numbers, the inverse of the non-zero real number x is written $1/x$ or x^{-1} and is called the *reciprocal* of x. Note that

$$x + (-x) = 0 = (-x) + x,$$
$$xx^{-1} = 1 = x^{-1}x$$

(3) For a square matrix, the inverse is its adjoint divided by its determinant provided the latter is not zero. The inverse

of the square matrix A has the same order as A and is denoted by

$$A A^{-1} = I = A^{-1}A$$

where I is the unit matrix having the same order as A. (4) For a function or one-to-one mapping, f, from set S onto set T, the *inverse function* is the mapping g from T to S, written f^{-1}, such that $g[f(x)] = x$.

inverse agonist (*BioSci*) Any ligand that binds to receptors and reduces the proportion of receptors in the active state. Has the opposite effects to an agonist and may actually reduce the background level of activity. Not the same as a *partial agonist*.

inverse current (*Electronics*) Current in the reverse direction.

inverse hyperbolic functions (*MathSci*) The INVERSE FUNCTION of any of the HYPERBOLIC FUNCTIONS: eg the inverse hyperbolic sine function is $x = $ arc sinh y or $x = \sinh^{-1} y$, where arc sinh y or $\sinh^{-1} y$ is the number whose sinh is x. Similarly for arc cosh y ($\cosh^{-1} y$), arc tanh y ($\tanh^{-1} y$), etc.

inverse networks (*ElecEng*) A pair of two-terminal networks whose impedances are such that their product is independent of frequency.

inverse power factor (*ElecEng*) A term sometimes used to denote secφ ($= 1/\cos \varphi$).

inverse segregation (*Eng*) A type of segregation in which the content of impurities, inclusions and low-melting-point constituents in cast metals tends to be higher at the surface than in the axial regions. See NORMAL SEGREGATION, SEGREGATION.

inverse-speed motor (*ElecEng*) See SERIES-CHARACTERISTIC MOTOR.

inverse spinel (*Crystal*) The crystal structure of MAGNETITE and the other ferrites which are ferrimagnetic, similar to that of the mineral SPINEL, but with the divalent ions and half of the trivalent ions interchanged. See panel on FERROMAGNETICS AND FERRIMAGNETICS.

inverse-square law (*Phys*) A law stating that the intensity of a field of radiation is inversely proportional to the square of the distance from the source. Applies to any system with spherical wavefront and negligible energy absorption.

inverse time-lag (*ElecEng*) A time-lag which is approximately proportional to the inverse of the current causing its operation. Also *inverse time element, inverse time limit*.

inverse trigonometrical functions (*MathSci*) The inverse function of any of the trigonometric functions. For example, the inverse sine function is $x = $ arc sin y or $x = \sin^{-1} y$, where arc sin y or $\sin^{-1} y$ is the number whose sin is x. Similarly for arc cos y ($\cos^{-1} y$), arc tan y ($\tan^{-1} y$), etc.

inverse video (*ICT*) Also REVERSE VIDEO.

inverse voltage (*ElecEng*) That generated across a rectifying element during the half-cycle when no current flows. Its maximum value is the *peak inverse voltage*.

inversion (*BioSci*) A stretch of chromosome that has been turned round so that the order of the nucleotides in the DNA is reversed.

inversion (*Chem*) The formation of a laevorotatory solution of fructose and glucose by the hydrolysis of a dextrorotatory solution of sucrose (cane sugar).

inversion (*Electronics*) In semiconductor devices, the local accumulation of nominally minority carriers to such an extent that they are actually in the majority. Such an effect occurs in the formation of a conducting channel between source and drain in METAL–OXIDE–SILICON FIELD-EFFECT TRANSISTORS (MOSFETs).

inversion (*EnvSci*) Reversal of the usual temperature gradient in the atmosphere, the temperature increasing with height. Inversions are of frequent occurrence near the ground on clear nights and in anticyclones; a layer of warm air lying over a layer of cooler air prevents water vapour and pollutants from rising and leads to mist or fog being trapped.

inversion (*MathSci*) The operation of deriving from a first set of points P a second set of points P' by the rule $OP.OP' = a^2$ where O is a fixed point (the *centre of inversion*), a is a constant (*the radius of inversion*), and P and P' lie on a straight line through O and on the same side of it.

inversion of relief (*Geol*) A condition whereby synclinal ridges are separated by anticlinal depressions.

inversion temperature (*Phys*) The temperature of a gas above which the temperature rises when it is passed through a nozzle and allowed to expand.

invert (*Build, CivEng*) The lowest part of the inner surface of the cross-section of a non-vertical drain or sewer at any given point, and related to the datum level.

Invertebrata (*BioSci*) A collective name for all animals other than those in the phylum Chordata; ie all those animals that do not exhibit the characteristics of vertebrates, namely possession of a notochord or vertebral column, ventral heart, eyes, etc. Some animals near the chordate boundary line, eg Hemichorda and Urochorda, are sometimes regarded as invertebrates. The term is not used as a scientific classification.

inverted arch (*CivEng*) An arch having the crown below the line of the springings, eg the floor of a tunnel, in order to distribute the pressure on the walls over a greater area. Also *inflected arch*.

inverted-brush contact (*ElecEng*) A laminated switch contact in which the laminations are carried on the fixed, instead of on the moving, contact.

inverted engine (*Aero*) An in-line engine having its cylinders below the crankshaft. Adopted in certain types of aircraft to improve the forward view of the pilot.

inverted-L antenna (*ICT*) One comprising a vertical wire joined to one end of a horizontal conductor.

inverted loop (*Aero*) An *aerobatic* manoeuvre consisting of a complete revolution in the vertical plane with the upper surface of the aircraft outside, which is started from the inverted position. Also *outside loop*.

inverted machine (*ElecEng*) Any electric machine in which the usual arrangement of the stator and rotor windings is inverted, eg an induction motor in which power is supplied to the rotor, the stator winding being short-circuited.

inverted rectifier (*ElecEng*) (1) Circuit arranged to convert dc to ac. (2) Amplifier which inverts polarity. Also *inverter*.

inverted rotary converter (*ElecEng*) A rotary converter which is used to convert from dc to ac.

inverted speech (*ICT*) Inversion of the order of speech frequencies before modulation for privacy; effected by modulating with a carrier just above the maximum speech frequency, then discarding this carrier and its upper sideband.

inverted telephoto lens (*ImageTech*) Wide-angle lens having a short focal length and comparatively long back focus, thus allowing space for a prism system between the rear of the lens and the sensitive surface.

inverted-V antenna (*ICT*) Two wires, several quarter-wavelengths long, joined at the top, one lower end being terminated by a resistance, the other end by the transmitter or receiver; the direction of maximum response is horizontal and in the plane of the wires.

inverter (*ElecEng*) A dc–ac converter using eg a transistor or thyristor oscillator.

inverter (*ICT*) An arrangement of modulators and filters for inverting speech or music for privacy. Also *speech inverter*.

invertible (*MathSci*) Having an inverse.

invert soap (*Chem*) A term sometimes used to denote certain CATIONIC DETERGENTS, eg cetyl pyridinium bromide.

invert sugar (*Chem, FoodSci*) The product obtained by the hydrolysis of cane sugar with acids; it is a mixture of equal parts of d-fructose and d-glucose, and is a useful HUMECTANT. Most fruits contain invert sugar, and honey averages over 70%; also obtained from starch.

investment (*BioSci*) The outer layers of an organ or part, or of an animal.

investment casting (*Eng*) Forming a mould round a pattern whose shape can be destroyed to allow its removal. Patterns of complex shapes can be used which would otherwise be impossible to withdraw from the mould. Patterns may be of wax, plastic, frozen mercury, etc. The ancient lost wax or 'cire perdue' process is a good example. Cf HOLLOW MOULDING. See FUSIBLE CORE.

invisible glass (*Phys*) Glass which has its surface curved, or has been coated with molecular thickness material, to eliminate surface reflections.

in vitro (*BioSci*) Literally 'in glass'. Used to describe the experimental reproduction of biological processes in the more easily defined environment of the culture vessel or plate. Cf IN VIVO.

in vitro fertilization (*BioSci*) A technique in which an ovum is fertilized by sperm outside the body. The resulting zygote is allowed to undergo several cell divisions *in vitro* and the embryo is then implanted into the uterus to undergo gestation. Abbrev *IVF*.

in vitro transcription (*BioSci*) Use of a laboratory medium without the presence of cells to obtain specific mRNA production from a DNA sequence. Also *cell-free transcription*.

in vitro translation (*BioSci*) The use of pure mRNA, ribosomes, factors, enzymes and precursors to obtain a specific protein product without the presence of cells. The procedure can be coupled to IN VITRO TRANSCRIPTION. Also *cell-free translation*.

in vivo (*BioSci*) Used to describe biological processes occurring within the living organism or cell. Cf IN VITRO.

involucre (*BioSci*) (1) Bracts forming a calyx-like structure close to the base of a usually condensed inflorescence, as in the Compositae. (2) A sheath or ring of leaves surrounding a group of ARCHEGONIA or ANTHERIDIA in bryophytes.

involucrum (*Med*) Sheath of new bone formed round bone which has died as the result of infection of the bone.

involuntary muscle (*BioSci*) Muscle that is not under conscious control; eg heart muscle and most of the smooth muscle that surrounds internal organs of the body.

involute (*BioSci*) Tightly coiled; said of gastropod shells, and of the margins of a leaf rolled towards the adaxial surface. Cf REVOLUTE. See VERNATION.

involute (*MathSci*) A line traced out by a point on a piece of string unwinding from the curve. See EVOLUTE.

involute gear teeth (*Eng*) Gear teeth whose flank profile consists of an involute curve given by the locus of the end of a taut string uncoiled from a base circle; the commonest form of tooth.

involution (*MathSci*) (1) Raising a number to a positive integral power. Cf EVOLUTION. (2) An AUTOMORPHISM, which, when applied twice, gives the identity mapping.

Io (*Astron*) The first natural satellite of Jupiter, discovered by Galileo in 1610. It is characterized by very active volcanoes, and stimulates radio emission in the ionosphere of Jupiter. Distance from the planet 422 000 km; diameter 3630 km.

I/O addresses (*ICT*) Locations in MAIN MEMORY used to communicate with a peripheral device, eg a printer.

iodates (V) (*Chem*) HIO_3. Salts of iodic acid.

iodazide (*Chem*) N_3I. An iodine azide.

IO device (*ICT*) Abbrev for *input/output device*. Peripheral that may be used as an INPUT DEVICE or as an OUTPUT DEVICE.

iodic (V) acid (*Chem*) HIO_3. Formed by the direct oxidation of iodine with nitric acid. White crystalline solid. Soluble in water. Forms iodates (V).

iodic anhydride (*Chem*) See IODINE OXIDES.

iodides (*Chem*) Salts of hydriodic acid. Most metallic iodides liberate free iodine and leave behind the metal or metallic oxide when heated. See HYDRIODIC ACID.

iodination (*Chem*) (1) The substitution or addition of iodine atoms in or to organic compounds. (2) The addition of iodine to table salt in districts where thyroid deficiencies are prevalent. See GOITRE.

iodine (*Chem*) Symbol I, at no 53, ram 126·9044, valencies 1, 3, 5, 7. Mp 113·5°C, bp 184·35°C, rel.d. 4·95 at 20°C. Non-metallic element in the seventh group of the periodic system, one of the halogens. It forms blackish scales with a violet lustre and a characteristic smell. Concentrated by biological processes in marine plants and animals. It has eleven isotopes of which ^{135}I with its 6·7 h half-life is an important fission product in nuclear reactors because it is the major source, by β-decay, of the nuclear poison, ^{135}Xe. It is therefore the reason for the long delay before xenon reaches its maximum after power is reduced in a reactor. It is widely but sparingly distributed as iodides, and is a constituent of the thyroid gland. The important commercial sources are crude Chile saltpetre (caliche) and certain seaweeds; it is used in organic synthesis. Also a powerful disinfectant once widely used and a constituent, as iodide, of CONTRAST MEDIA. See RADIO-IODINE.

iodine monochloride (*Chem*) ICl. Mp 13·9°C. Brown-red crystals. Prepared by reaction of chlorine with iodine. Used as an iodinating agent in analytical and organic chemistry.

iodine oxides (*Chem*) Iodine has four oxides with the empirical formulae I_2O_4, I_4O_9, I_2O_5, IO_4. They differ in marked degree from those of the other halogens. The (V) oxide, I_2O_5 (IODIC ANHYDRIDE) has acidic properties; stable, with water yields iodic acid.

iodine pentafluoride (*Chem*) IF_5. Mp -8°C, bp 97°C. Colourless liquid, formed with incandescence by the direct combination of fluorine and iodine.

iodine trichloride (*Chem*) ICl_3. Yellow powder, a powerful disinfectant.

iodine value (*Chem*) The number of grams of iodine absorbed by 100 g of a fat or oil. It gives an indication of the amount of unsaturated acids present in fats and oils.

iodism (*Med*) The condition resulting from overdosage of, or sensitivity to, iodine; characterized by running at the eyes and the nose, salivation and skin eruptions.

iodo compounds (*Chem*) Compounds containing the covalent iodine radical –I, eg iodobenzene.

iodoform (*Chem*) Tri-iodomethane. CHI_3. Mp 119°C. Yellow hexagonal plates, of peculiar, saffron odour, volatile in steam, mildly antiseptic. It is prepared by warming alcohol with iodine and alkali; or by an electrolytic method, in which a current is passed through a solution containing KI, Na_2CO_3 and ethanol, the temperature being about 65°C. This reaction is the basis of the *iodoform test* for compounds containing the group $-COCH_3$ or $-CHOHCH_3$.

iodogorgoic acid (*Chem*) 3,5-di-iodotyrosine. A constituent of the thyroid hormone.

iodometric (*Chem*) Measured by iodine, eg in a titration.

iodophilic bacteria (*BioSci*) Bacteria that stain blue with iodine, revealing the presence of starch-like compounds. They occur in large numbers in the rumen where they are associated with cellulose decomposition.

iodopsin (*Phys*) See VISUAL VIOLET.

iodoso compound (*Chem*) Compound containing the iodoso radical –IO.

iodoxy compounds (*Chem*) Compound containing the radical $-IO_2$.

iodyl ion (*Chem*) The monoxide of trivalent electropositive iodine: $(IO)^+$. Forms salts, eg iodyl sulphate $(IO)_2SO_4$, which are hydrolysed by water to give iodine, iodic (V) acid, and the acid corresponding to the anion.

iolite (*Min*) See CORDIERITE.

ion (*Phys*) Strictly, any atom or molecule which has resultant electric charge due to loss or gain of valency electrons. Free electrons are sometimes loosely classified as *negative ions*. Ionic crystals are formed of ionized atoms and in solution exhibit ionic conduction. In gases, ions are normally molecular and cases of double or treble ionization may be encountered. When almost completely ionized, gases form a fourth state of matter, known as a PLASMA. Since matter is electrically neutral, ions are normally produced in pairs. See IONIC RADIUS.

ion beam (*Phys*) A beam of ions moving in the same direction with similar speeds, esp when produced by some form of accelerating machine or mass spectrograph.

ion bombardment (*Phys*) The impact on the cathode of a gas-filled electron tube of the positive ions created by ionization of the gas. The bombardment may cause electrons to be ejected from the cathode.

ion burn (*Electronics*) Damage to the phosphor of a magnetic-deflection cathode-ray tube because of relatively minute deflection of heavy negative ions by the magnetic control which deflects the electron beam.

ion channel (*BioSci*) A hydrophilic pore that allows ions to cross a cell membrane.

ion cluster (*Chem*) A group of molecules loosely bound (by electrostatic forces) to a charged ion in a gas.

ion concentration (*Chem*) That expressed in moles per unit volume for a particular ion. Also *ionic concentration*. See PH.

ion concentration (*Phys*) The number of ions of either sign, or of ion pairs, per unit volume. Also *ionization density*.

ion cyclotron frequency (*Phys*) The frequency of circular motion, similar to that in a cyclotron, followed by an ion in a uniform magnetic field perpendicular to its motion. The critical frequency depends on the magnitude of the field and on the charge and mass of the ion.

ion engine (*Electronics*) A device for propulsion or attitude control of spacecraft or satellites, esp in outer space. Ions and electrons are expelled at high velocity from a combustion chamber. They recombine outside, thus preventing any space-charge effect which would counteract thrust.

ion-exchange capacity (*MinExt*) Electrical charge surplus to that uniting the framework of the ion-exchange resin or other exchange vehicle.

ion-exchange chromatography (*Chem, Med*) The separation of molecules by absorption and desorption from charged polymers, usually packed into a column down which the mixture is passed, separate effluent fractions being collected. A diverse range of polymers have been developed for specific purification processes, particularly for proteins. Ion-exchange resins may be administered orally or rectally to patients with renal failure as a way of reducing potassium levels.

ion-exchange liquids (*MinExt*) Those immiscible with water (eg kerosine) which are rendered active by addition of suitable chemicals, such as tri-lauryl amine. These liquids can be used in solvent extraction processes in place of solid ion-exchange resins.

ion-exchange resins (*Chem*) A term applied to a variety of materials, usually organic, which have the capacity of exchanging the ions in solutions passed through them. Different varieties of resin are used dependent on the nature (cationic or anionic) of the ions to be exchanged. Many of the resins in present-day use are based on polystyrene networks cross-linked with divinyl benzene. See GEL-PERMEATION CHROMATOGRAPHY.

ion flotation (*Chem*) Removal of ions or gels from water by adding a surface active agent which forms complexes. These are floated as a scum by the use of air bubbles. See FROTH FLOTATION, GIBBS' ADSORPTION THEOREM.

ionic (*Phys*) Appertaining to or associated with gaseous or electrolytic ions. *Ion* is frequently used interchangeably with *ionic* as an adjective, eg in *ion(ic)* conduction.

ionic beam (*Phys*) See ION BEAM.

ionic bombardment (*Electronics*) The impact on the cathode of a gas-filled electron tube of the positive ions created by ionization of the gas. The bombardment may cause electrons to be ejected from the cathode.

ionic bond (*Chem*) Coulomb force between ion pairs in molecule or ionic crystal. These bonds usually dissociate in solution.

ionic concentration (*Chem*) See ION CONCENTRATION.

ionic conduction (*Chem*) That which arises from the movement of ions in a gas or electrolytic solution. In solids, it refers to electrical conductivity of an ionic crystal, arising from movement of positive and negative ions under an applied electric field. It is a diffusion process and hence very temperature dependent.

ionic conductivity (*Chem*) An additive property, symbol l_i; the equivalent CONDUCTIVITY of the ion i at infinite dilution. Thus for KCl, $\Lambda_\infty = l_{K^+} + l_{Cl^-}$.

ionic conductor (*Chem*) One in which conduction is predominantly by ions, rather than by electrons and holes.

ionic crystal (*Crystal*) A lattice held together by the electric forces between ions, as in a crystalline chemical compound.

ionic current (*Electronics*) The current carried by positively charged ions in a gas at low pressure.

ionic-heated cathode (*Electronics*) Hot cathode that is heated primarily by ionic bombardment of the emitting surface.

ionic materials (*Chem*) Solids with structure composed partly or wholly of charged species, anions and cations. Includes simple salts like magnesia (Mg^{2+}, O^{2-}) which has a crystal lattice of anions in which cations sit at interstitial sites, ceramics made of crystalline silicate sheets, chains or networks like mica, and glasses where the silicate chains form an amorphous network. See panel on SILICON, SILICA, SILICATES.

ionic migration (*Chem*) The transport of ion-bearing particle to an electrode oppositely charged with electricity.

ionic potential (*Electronics*) The ratio of an ionic charge to its effective radius, regarded as a capacitor.

ionic product (*Chem*) The product of the activities (see ACTIVITY (2)) of the ions into which a pure liquid dissociates. For water, these ions are H_3O^+ and OH^-.

ionic radius (*Crystal*) Approximate limiting radius of ions in crystals, ranging, for common metals (including carbon), from a fraction to about 1 nanometre. Fig. ▷

ionic strength (*Chem*) Half the sum of the terms obtained by multiplying the ACTIVITY (2) of each ion in a solution by the square of its valency; it is a measure of the intensity of the screening of electrostatic potential in solutions of electrolytes. See DEBYE–HÜCKEL THEORY.

ionic theory (*Chem*) The theory that substances whose solution conducts an electric current undergo electrolytic dissociation on dissolution. This assumption explains both the laws of electrolysis and the abnormal COLLIGATIVE PROPERTIES, such as osmotic pressure, of electrolyte solutions.

ion implantation (*Electronics*) A technique used to produce semiconductor devices. Impurities are introduced by firing high-energy ions at the substrate material.

ionization (*Phys*) The formation of ions by separating atoms, molecules or radicals, or by adding or subtracting electrons from atoms by strong electric fields in a gas, or by weakening the electric attractions in a liquid, particularly water.

ionization by collision (*Phys*) The removal of one or more electrons from an atom by its collision with another particle such as another electron or an alpha particle. Very prominent in electrical discharge through a rarefied gas.

ionization chamber (*NucEng*) An instrument used in the study of ionized gases and/or ionizing radiations. It comprises a gas-filled enclosure with parallel-plate or coaxial electrodes, in which ionization of the gas occurs. For fairly large applied voltages, the current through the chamber is dependent only upon the rate of ion production. For very large voltages, additional ionization by collision enhances the current. The system is then known as a GEIGER–MÜLLER COUNTER or PROPORTIONAL COUNTER.

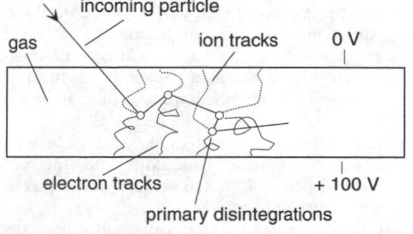

ionization chamber

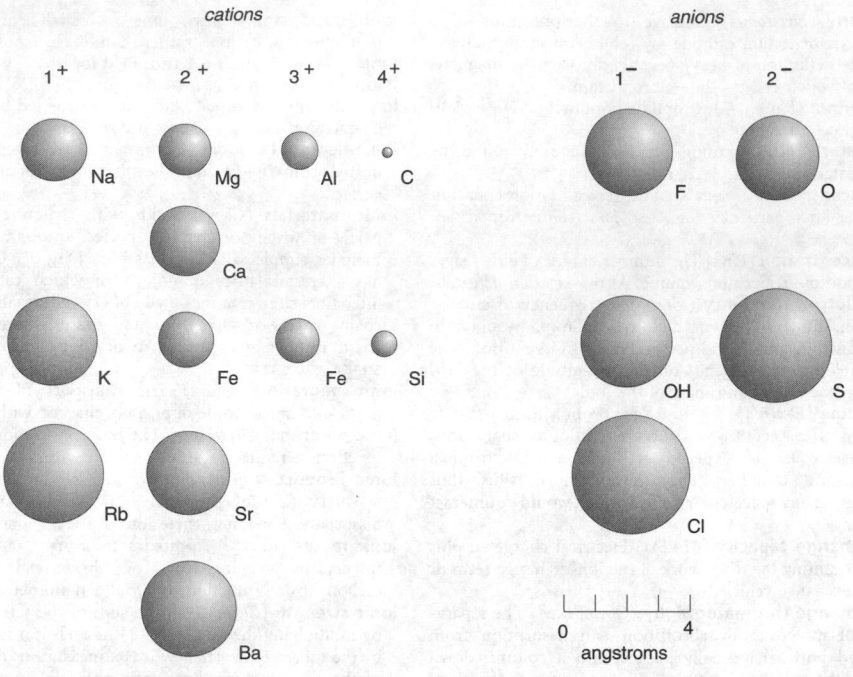

ionic radius

ionization constant (*Chem*) The ratio of the product of the activities of the ions produced from a given substance to the activity of the undissociated molecules of that substance. See ACTIVITY (2), DISSOCIATION CONSTANT.

ionization continuum (*Phys*) The energy spectrum above the ionization threshold of an atom.

ionization cross-section (*Phys*) The effective geometrical cross-section offered by an atom or molecule to an ionizing collision.

ionization current (*Phys*) (1) The current passing through an ionization chamber. (2) The current passed by an ionization gauge, when used for measuring low gas pressures.

ionization density (*Phys*) See ION CONCENTRATION.

ionization gauge (*Electronics*) Vacuum gauge formed by small thermionic triode attached to a chamber in which it is desired to measure residual gas pressure. Current is passed from anode to cathode, the grid being made negative; grid current is measured, giving degree of vacuum.

ionization manometer (*NucEng*) See IONIZATION GAUGE.

ionization potential (*Phys*) The energy in electron-volts required to detach an electron from a neutral atom. For hydrogen, the value is 13·6 eV. Atoms, other than hydrogen, may lose more than one electron and can be multiply ionized. Also *electron binding energy, radiation potential*. See PHOTOELECTRON SPECTROSCOPY.

ionization temperature (*Astron, Phys*) A critical temperature, different for different elements, at which the constituent electrons of an atom will become dissociated from the nucleus; hence a factor in deducing stellar temperatures from observed spectral lines indicating any known stage of ionization.

ionization time (*Electronics*) The delay between the application of ionizing conditions and the onset of ionization, depending on temperature and other factors, eg in a mercury-pool rectifier.

ionized (*Chem, Phys*) (1) Electrolytically dissociated. (2) Converted into an ion by the loss or gain of an electron.

ionized atom (*Phys*) An atom with a resultant charge arising from the capture or loss of electrons; an ION.

ionizing collision (*Phys*) The interaction between atoms or elementary particles in which an ion pair is produced.

ionizing energy (*Phys*) The energy required to produce an ion pair in a gas under specified conditions. Measured in electron-volts. For air it is about 32 eV.

ionizing event (*Phys*) Any interaction which leads to the production of ions.

ionizing particle (*Phys*) Charged particle which produces considerable ionization on passing through a medium. Neutrons, neutrinos and photons are not ionizing particles although they may produce some ions.

ionizing radiation (*Phys*) Any electromagnetic or particulate radiation which produces ion pairs when passing through a medium. See panels on NON-IONIZING FIELDS AND RADIATION and RADIATION.

ionizing voltage (*Electronics*) See STARTING VOLTAGE.

ion migration (*Chem*) The movement of ions in an electrolyte or semiconductor due to applying a voltage across electrodes.

ion mobility (*Chem*) ION VELOCITY in unit electric field (1 volt per metre).

ionogenic (*Chem*) Forming ions, eg electrolytes.

ionomer resins (*Chem*) Ethylene copolymerized with small amount of acrylic acid and treated with zinc or sodium salt so that ion-acid groups act as physical cross-links, stabilizing the material. A type of POLYELECTROLYTE.

ionophone (*Acous*) See CATHODOPHONE.

ionophore (*BioSci*) A compound that enhances the permeation of biological membranes by specific ions, acting either as specific ion carriers or by creating ion channels across the membrane, eg *gramicidin, valinomycin*.

ionosphere (*EnvSci*) That part of the Earth's atmosphere (the upper atmosphere) in which an appreciable concentration of ions and free electrons normally exist. This shows daily and seasonal variations. See E-LAYER, F-LAYER, HOP.

ionospheric control points (*ICT*) Points in the ionosphere 2000 km and 1000 km distant from each ground terminal, used respectively to control transmission by way of the F_2- and E-layers.

ionospheric disturbance (*EnvSci*) An abnormal variation of the ion density in part of the ionosphere, commonly produced by solar flares. Has a marked effect on radio communication. See IONOSPHERIC STORM.

ionospheric forecast (*EnvSci*) The forecasting of ionospheric conditions relevant to communication. Also *ionospheric prediction*.

ionospheric ray (*ICT*) See IONOSPHERIC WAVE.

ionospheric regions (*EnvSci*) These are: D region, between 90 and 150 km; F region, over 150 km; all above surface of Earth. Internal effective layers are labelled E, *sporadic E*, E_2, F, F_1, $F_{1.5}$, F_2.

ionospheric storm (*ICT*) Turbulence in parts of the ionosphere, probably connected with sunspot activity, causing dramatic changes in its reflective properties and sometimes totally disrupting short-wave communications.

ionospheric wave (*ICT*) Radiation reflected from an upper ionized region between transmitter and receiver. Also *indirect wave, reflected wave* or *sky wave*.

ionotropic (*BioSci*) Denoting a neurotransmitter receptor which affects cell activity by regulating the cell's ion channels (cf METABOTROPIC).

ionotropy (*Chem*) The reversible interconversion of certain organic isomers by migration of part of the molecules as an ion; eg hydrogen ion (PROTOTROPY).

ion pair (*Phys*) A pair of positive and negative ions produced together by transfer of an electron from one atom or molecule to another.

ion propulsion (*Space*) A method of rocket propulsion in which charged particles (eg lithium or caesium ions) are accelerated by an electrostatic field, giving a small thrust but a high specific impulse; the thrust-to-weight ratio is large so that the main use is for station keeping.

ion-selective electrode (*BioSci*) An electrode half-cell, with a semipermeable membrane that only allows the passage of certain selected ion species. The electrical potential measured between this and a reference half-cell (eg a calomel electrode) is thus the Nernst potential for the ion and the activity of the ion in the unknown solution can be measured.

ion source (*Phys*) A device for producing ions for ion implantation or in a particle accelerator. Various configurations exist, eg deriving ions directly from ionized gases (PLASMAS), from liquid metals (by FIELD EMISSION from protuberances) or from solids (by surface ionization). In a particle accelerator a minute jet of gas or vapour of the required compound is ionized by heating at a filament or with an electron beam. The water-cooled magnet concentrates the ions near the filament and focuses them as they are accelerated into the main beam by the high-voltage electrode.

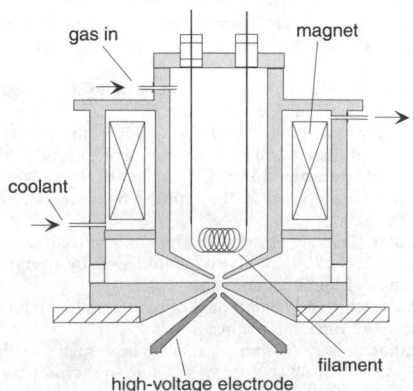

ion source

ion spot (*Electronics*) Deformation of target, cathode or screen by ion bombardment in a camera tube or cathode-ray tube. In a camera, a spurious signal results; in a CRT, ION BURN becomes apparent.

iontophoresis (*Chem*) Migration of ions into body tissue through electric currents. See ELECTROPHORESIS.

ion trap (*Electronics*) Means of preventing ION BURN in a magnetic-deflection cathode-ray tube. The electron beam is deflected through a large angle to reach the screen, so that the heavier negative ions fall elsewhere.

ion velocity (*Chem*) The velocity of translation of drift of ions under the influence of an electric field, in a gas or electrolyte. See ION MOBILITY.

ion yield (*Phys*) The average number of ion pairs produced by each incident particle or photon.

IOP (*ICT*) Abbrev for INPUT/OUTPUT PROCESSOR.

IOV (*BioSci*) Abbrev for INSIDE-OUT VESICLE.

IPA (*BioSci*) Abbrev for ISOPENTENYL ADENINE.

IP address (*ICT*) Abbrev for *Internet protocol address*. A numerical series that uniquely identifies a computer on a network such as the Internet.

iph (*Print*) Abbrev for *impressions per hour*.

IPM (*Agri*) See INTEGRATED PEST MANAGEMENT.

iPod (*ICT*) A proprietary name (Apple) for a personal digital audio player.

IPPC (*EnvSci*) Abbrev for *integrated pollution prevention and control*, a European Union directive requiring major industrial installations to be licensed and to maintain defined standards with regard to emissions, waste disposal, noise levels, etc.

ipratropium bromide (*Pharmacol*) An *antimuscarinic* drug used as a bronchodilator in chronic bronchitis.

ipsilateral (*BioSci*) Pertaining to the same side of the body. Cf CONTRALATERAL.

ipso- (*Chem*) Prefix meaning 'itself'. Used to describe the atom of a substituent group through which that group is attached, eg the carbon atom of a phenyl ring attached to phosphorus in triphenylphosphine.

Ipswichian (*Geol*) A temperate stage of the Late Pleistocene. See QUATERNARY.

IPT thermometers (*Chem*) Thermometers conforming to the standards laid down by the *Institute of Petroleum Technologists*.

IPX (*ICT*) Abbrev for INTERNETWORK PACKET EXCHANGE.

IR (*Chem*) Abbrev for ISOPRENE RUBBER.

IR (*ICT*) Abbrev for INFORMATION RETRIEVAL.

IR (*Phys*) Abbrev for INFRARED.

IRCM (*Aero*) Abbrev for INFRARED COUNTERMEASURES.

IR drop (*ElecEng*) The voltage drop due to a current flowing through a resistance.

I region (*BioSci*) The region in the murine major histocompatibility complex that contains genes coding for Class II histocompatibility antigens.

Ir gene (*BioSci*) Abbrev for *immune response gene*. Found in the I region, and so called because in inbred strains of mice the ability to respond to certain simple peptides depends upon which Class II antigens are expressed on their cells.

IRHD (*Eng*) Abbrev for INTERNATIONAL RUBBER HARDNESS DEGREE.

irid-, irido- (*Genrl*) Prefixes from Gk *iris*, gen *iridos*, rainbow.

iridalgia (*Med*) Pain in the iris of the eye.

iridectomy (*Med*) Excision of part of the iris of the eye.

iridescence (*Phys*) The production of fine colours on a surface; due to the interference of light reflected from the front and back of a very thin film. Also *irisation*.

iridescent clouds (*EnvSci*) High clouds which show colours, generally delicate pink and green, in irregular patches. It is thought that the effect is caused by the diffraction of sunlight by supercooled water droplets.

iridium (*Chem*) Symbol Ir, at no 77, ram 192·2, rel.d. at 20°C 22·4, mp 2410°C, electrical resistivity 6×10^{-8} Ω m. A brittle, steel-grey metallic element of the platinum family. Alloyed with platinum or osmium to form hard, corrosion-resistant alloys, used for pen points, watch and compass bearings, crucibles, standards of length. The radioactive isotope ^{192}Ir is a medium-energy gamma emitter used for industrial radiography. High concentrations of iridium in clay bands near the Cretaceous–Tertiary boundary worldwide have been ascribed to an impact by an iridium-rich extraterrestrial object like an asteroid.

irido- (*Genrl*) See IRID-.

iridochoroiditis (*Med*) Inflammation of the iris and of the choroid of the eye.

iridocoloboma (*Med*) Congenital absence of part of the iris, a gap or fissure being present in it.

iridocyclitis (*Med*) Inflammation of the iris and of the ciliary body of the eye.

iridocyte (*BioSci*) A reflecting cell in certain eyes giving an iridescent appearance.

iridodialysis (*Med*) Separation of the iris from its attachment to the ciliary body of the eye.

iridokeratitis (*Med*) Inflammation of the iris and of the cornea.

iridoplegia (*Med*) Paralysis of the sphincter, or circular muscle, of the iris.

iridosmine (*Min*) An ore of iridium and osmium, a natural alloy, with Os greater than 35%. Crystallizes in the hexagonal system.

iridotomy (*Med*) Surgical cutting of the iris.

iridovirus (*BioSci*) A non-occluded virus of insects; the crystalline array of the intracytoplasmic virus particles makes the insect appear iridescent, often (as in the case of infected leatherjackets) blue.

irinotecan (*Pharmacol*) A topoisomerase inhibitor used for treatment of colonic and rectal cancer.

iris (*BioSci*) In the vertebrate eye, that part of the choroid, lying in front of the lens, which takes the form of a circular curtain with a central opening. Adj *iridial*.

iris (*ElecEng*) Apertured diaphragm across a waveguide, for introducing specific impedances.

iris (*ImageTech*) Adjustable circular aperture used in conjunction with camera lens for control of exposure.

iris (*Min*) A form of quartz showing chromatic reflections of light from fractures, often produced artificially by suddenly cooling a heated crystal. Also *rainbow quartz*.

irisation (*Phys*) See IRIDESCENCE.

Irish moss (*FoodSci*) See CARRAGHEENIN.

iris recognition (*ICT*) A security system that uses a digital camera to detect the unique marks on a person's iris and map these to information on a database in order to confirm the person's identity. Also *iris scanning*.

iris wipe (*ImageTech*) A transitional wipe effect in which the image boundary is in the form of a circle, increasing (*iris-in*) or decreasing (*iris-out*).

iritis (*Med*) Inflammation of the iris.

I^2R loss (*ElecEng*) The power loss caused by the flow of a current I through a resistance R.

IRM (*Psych*) Abbrev for INNATE RELEASING MECHANISM.

iroko (*For*) A general utility timber from *Chlorophora excelsa*, a tropical African hardwood tree, it is golden-orange to brown, with interlocked grain and coarse but even texture.

iron (*Chem*) Symbol Fe, at no 26, ram 55·847, rel.d. at 20°C 7·86, mp 1525°C, bp 2800°C, electrical resistivity $9 \cdot 8 \times 10^{-8}$ Ω m. A metallic element in the eighth group of the periodic system. It exists in three forms: ALPHA-, DELTA-, GAMMA-. As basis metal in steel and cast-iron, it is the most widely used of all metals. It is the fourth commonest element of the Earth's crust, with an abundance of 6·2%, and it is thought to make up 80% of the core of the Earth. It has twelve isotopes. See IRON ORES.

iron alum (*Min*) See HALOTRICHITE.

iron arc (*Phys*) An arc between iron electrodes, used for obtaining light containing standardized lines, for spectrometer and spectrograph calibrations.

iron bacteria (*BioSci, Min*) Filamentous bacteria that can convert iron oxide to iron hydroxide, which is deposited on their sheaths. They are important in the formation of BOG IRON ORE.

ironbark (*For*) Wood from *Eucalyptus leucoxylon*, which varies in colour from greyish- to reddish-brown.

iron-clad electromagnet (*ElecEng*) One in which the return path for the flux is formed by an iron covering surrounding the winding.

iron-clad switchgear (*ElecEng*) See METAL-CLAD SWITCHGEAR.

iron-deficiency anaemia (*Med*) ANAEMIA due to either poor intake or loss of iron. Common in menstruating women or when there is blood loss from gastro-intestinal disease.

iron dust core (*ElecEng*) One used in a high-frequency transformer or inductor to minimize eddy current losses. It consists of minute magnetic particles bonded in an insulating matrix.

iron-glance (*Min*) From the German *Eisenglanz*, a name sometimes applied to specular iron ore (haematite).

iron loss (*ElecEng*) The power loss due to HYSTERESIS and EDDY CURRENTS in the iron of magnetic material in transformers or electrical machinery.

iron meteorites (*Geol*) One of the two main categories of meteorites, the other being the *stony meteorites*. They are composed of iron and of iron–nickel alloy, with only a small proportion of silicate or sulphide minerals.

iron–nickel accumulator (*ElecEng*) See NICKEL–IRON–ALKALINE ACCUMULATOR.

iron-olivine (*Min*) See FAYALITE.

iron ores (*Geol*) Rocks or deposits containing iron-rich compounds in workable amounts; they may be primary or secondary; they may occur as irregular masses, as lodes or veins, or interbedded with sedimentary strata. See CHAMOSITE, GOETHITE, HAEMATITE, LIMONITE, MAGNETITE, SIDERITE.

iron (II) oxide (*Chem*) See FERROUS OXIDE.

iron (III) oxide (*Chem*) See FERRIC OXIDE.

iron pan (*Geol*) A hard layer often found in sands and gravels; caused by the precipitation of iron salts from percolating waters. It is formed a short distance below the soil surface. See HARD PAN.

iron pattern (*Eng*) See METAL PATTERN.

iron pentacarbonyl (*Chem*) $Fe(CO)_5$. Formed at ordinary temperatures when carbon monoxide is passed over finely divided iron. A liquid which readily decomposes.

iron pyrites (*Min*) See PYRITE.

iron spinel (*Min*) See SPINEL.

ironstone (*Geol*) An iron-rich sedimentary rock, found in nodules, layers or beds.

ironwood (*For*) See LIGNUM VITAE.

irradiance (*Phys*) See RADIANT-FLUX DENSITY.

irradiation (*FoodSci*) See panel on FOOD IRRADIATION.

irradiation (*ImageTech*) Image spread, arising from scatter within the emulsion from the silver halide grains. See HALATION.

irradiation (*Radiol*) Exposure of a body to X-rays, gamma rays or other ionizing radiations.

irradiation swelling (*NucEng*) Changes in density and volume of materials due to neutron irradiation.

irrational number (*MathSci*) A REAL NUMBER which cannot be expressed as the ratio of two integers, eg $\sqrt{2}$. Cf SURD.

irregular (*BioSci*) (1) Asymmetric, not arranged in an even line or circle. (2) Not divisible into halves by an indefinite number of longitudinal planes.

irregular-coursed (*Build*) Said of rubble walling built up in courses of different heights.

irregular galaxy (*Astron*) A small galaxy, such as either of the MAGELLANIC CLOUDS, showing no symmetry and containing little dust or gas. See panel on GALAXY.

irregular variable (*Astron*) See VARIABLE STAR.

irreversibility (*Phys*) A feature of a physical system which has the common tendency to change spontaneously from one state to another but not to change in the reverse direction. ENTROPY provides an indication of irreversibility.

irreversible colloid (*Chem*) See LYOPHOBIC COLLOID.

irreversible controls (*Aero*) A flying control system, hydraulically or electrically operated, wherein there is no feedback of aerodynamic forces from the control surfaces.

irreversible reaction (*Chem*) A reaction which takes place in one direction only, and therefore proceeds to completion.

irrigation (*Build*) The method of sewage disposal by LAND TREATMENT.

irritability (*BioSci*) A property of living matter, namely the ability to receive and respond to external stimuli.

irritant (*BioSci*) Any external stimulus that produces an active response in a living organism.

irrotational field (*ElecEng*) A field in which the *field circulation* is everywhere zero.

ISA (*Aero*) Abbrev for INTERNATIONAL STANDARD ATMOSPHERE.

ISA (*ICT*) Abbrev for INDUSTRY STANDARD ARCHITECTURE.

isallobar (*EnvSci*) The contour line on a weather chart, signifying the location of equal changes of pressure over a specified period.

isallobaric high and low (*EnvSci*) Centres, respectively, of rising and falling BAROMETRIC TENDENCY.

isallobaric wind (*EnvSci*) Theoretical component of the wind arising from the spatial non-uniformity of local rates of change of pressure.

ISAS (*Space*) Abbrev for *Institute of Space and Aeronautical Science* of the University of Tokyo, mainly responsible for the Japanese scientific satellites.

ischaemia (*Med*) Deficiency of blood flow to part of the body, causing inadequate tissue perfusion with oxygen. Adj *ischaemic*. US *ischemia*.

ischium (*BioSci*) A posterior bone of the pelvic girdle in tetrapods. Adjs *ischiadic* ,*ischial*.

ISCOM (*BioSci*) Abbrev for IMMUNOSTIMULATORY COMPLEX.

ISDN (*ICT*) Abbrev for INTEGRATED SERVICES DIGITAL NETWORK.

isenthalpic (*Phys*) Of a process carried out at constant ENTHALPY, or heat function H.

isentropic (*Phys*) See ENTROPY.

Isherwood system (*Ships*) A method of ship construction of which the dominant feature is longitudinal framing. Named after its originator.

I signal (*ImageTech*) In the NTSC colour TV system, that corresponding to the wide-band axis of the chrominance signal.

isinglass (*Chem*) Fish glue. A white solid amorphous mass, prepared from fish bladders; chief constituent, GELATINE. It has strong adhesive properties. Used eg in various food preparations, as an adhesive, as a MORDANT in gilding on glass and in the fining of beers and wines.

island arc (*Geol*) A chain of volcanic islands formed at a convergent plate boundary. Deep ocean trenches occur on the convex side and deep basins on the concave side, eg Japan.

island biogeography (*BioSci*) The study of the number and distribution of plant and animal species on islands, or in areas completely surrounded by ecosystems of an entirely different type which resemble islands.

islet cells (*BioSci*) Insulin-secreting cells of the *Islets of Langerhans* within the pancreas.

ISO (*Genrl*) Abbrev for INTERNATIONAL ORGANIZATION FOR STANDARDIZATION. See ISO SIZES, ISO SPEED.

ISO7 (*ICT*) See CHARACTER CODE.

iso- (*Genrl*) A prefix indicating: (1) Having the same value, identical, equal. (2) In chemical nomenclature,

the presence of a branched carbon chain in the molecule.

iso-agglutination (*BioSci*) (1) The adhesion of spermatozoa to one another by the action of some substance produced by the ova of the same species. (2) The adhesion of erythrocytes to one another within the same blood group. Cf HETERO-AGGLUTINATION.

isobar (*Chem*) A curve relating qualities measured at the same pressure.

isobar (*EnvSci*) A line drawn on a map through places having the same atmospheric pressure at a given time.

isobar (*Phys*) One of a set of nuclides having the same total of protons and neutrons with the same mass number and approximately the same atomic mass; eg the isotopes of hydrogen and helium, hydrogen-3 and helium-3.

isobaric spin (*Phys*) See ISOTOPIC SPIN.

isobarometric charts (*EnvSci*) Maps on which isobars are drawn. See ISOBAR. Also *isobaric charts*.

isobases (*Geol*) Lines drawn through places where equal depression of the land mass took place in Glacial times, as a result of the weight of the ice load.

isobilateral (*BioSci*) Divisible into symmetrical halves along two distinct planes; esp of a leaf which has palisade towards both faces.

isobutane (*Chem*) An isomeric form of BUTANE having the structure $(CH_3)_2CHCH_3$, as compared with $CH_3CH_2CH_2CH_3$ for normal butane. Found in natural gas, and produced by cracking petroleum. Used as a refrigerant.

isobutyl alcohol (*Chem*) *2-methylpropan-1-ol*. $(CH_3)_2CHCH_2OH$. Bp 107°C. Partly miscible with water; formed during sugar fermentation.

isocercal (*BioSci*) Said of a type of secondarily symmetrical tail fin in fish in which the areas of the fin above and below the vertebral column are equal.

isochore (*Chem*) A curve relating quantities measured under conditions in which the volume remains constant.

isochromatic (*Phys*) Interference fringe of uniform hue observed with a white light source, esp in PHOTOELASTIC ANALYSIS where it joins points of equal phase retardation.

isochrone (*ICT*) Hyperbola (or ellipse) on a chart or map, along which there is a constant difference (or sum) in time of arrival of signals from two stations at the ends of a baseline.

isochronism (*Phys*) Regular periodicity, as the swinging of a pendulum. Adj *isochronous*.

isochronous data (*Eng*) Measurements taken at constant times; eg an isochronous stress–strain curve for a polymer is constructed by measuring constant-time tensile creep strains at different stress levels, and replotting data in stress–strain form.

isoclinal fold (*Geol*) A fold in which both limbs dip in the same direction. See FOLDING.

isocline (*Phys*) A line on a map, joining points where the angle of DIP (or inclination) of the Earth's magnetic field is the same.

isoclinic (*Aero*) A wing designed to maintain a constant angle of incidence even when subject to dynamic loads.

isoclinic (*Eng*) Loci of points at which directions of principal axes of stress are parallel to the axes of the crossed plane polarizers in photoelasticity. Appear as black bands in white light, not observed in circularly polarized light. Cf ISOCHROMATIC.

isocoria (*Med*) Equality in the size of the pupils of the eye.

isocount contours (*Radiol*) The curves formed by the intersection of a series of ISOCOUNT SURFACES with a specified surface.

isocount surface (*Radiol*) A surface on which the counting rate is everywhere the same.

isocyanates (*Chem*) Compounds with isocyanate group –NCO. Di-isocyanates such as TDI and MDI are used for making polyurethanes. They are highly reactive compounds, forming amines (with water), which can react with further isocyanate to give urea (–NH–CO–NH–) groups.

isocyanides (*Chem*) Isonitriles or carbylamines. Isocyano-compounds, $R-N\equiv C$. They are colourless liquids, only slightly soluble in water with a feebly alkaline reaction, having a nauseous odour, obtained by the action of trichloromethane and alcoholic potash on primary amines. They are very stable towards alkali, form additive compounds with halogens, HCl, H_2S, etc, and can be hydrolysed into methanoic acid and a primary amine, having one carbon atom less than the original compound.

isocyclic compounds (*Chem*) Also CARBOCYCLIC COMPOUNDS.

isodactylous (*BioSci*) Having all the digits of a limb the same size.

isodesmic structure (*Crystal*) Crystal structure with equal lattice bonding in all directions, and no distinct internal groups.

isodiametric (*Genrl*) Of the same length vertically and horizontally. Also *isodiametrical*.

isodiapheres (*Phys*) Two or more nuclides having the same difference between the number of neutrons and the number of protons.

isodimorphous (*Chem*) Existing in two isomorphous crystalline forms.

isodisperse (*Chem*) Dispersible in solutions having the same PH value.

isodomon (*Build*) An ancient form of masonry in which the facing consisted of squared stones laid in courses of equal height, and the filling of coursed stones of smaller size.

isodont (*BioSci*) Having all the teeth similar in size and form.

isodose chart (*Radiol*) A graphical representation of a number of isodose contours in a given plane, which usually contains the central ray.

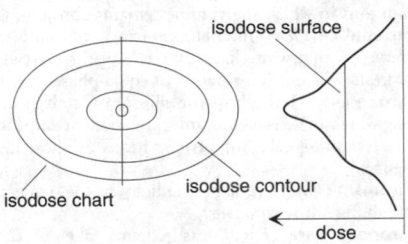

isodose surface

isodose contour

isodose chart

dose

isodose chart

isodose curve (*Radiol*) The curve obtained at the intersection of a particular ISODOSE SURFACE with a given plane. Also *isodose contour*.

isodose surface (*Radiol*) A surface on which the dose received is everywhere the same.

isodulcite (*Chem*) See RHAMNOSE.

isodynamic lines (*Phys*) Lines on a magnetic map which pass through points having equal strengths of the Earth's field.

iso-electric focusing (*BioSci*) A technique for separating proteins according to their ISO-ELECTRIC POINTS. The mixture of proteins is placed in a pH gradient established by AMPHOLINES in an electric field across a liquid or a gel matrix; the proteins move in the electric field until they reach their iso-electric points in the pH gradient where, having lost net charge, they are focused.

iso-electric point (*Chem*) Hydrogen ion concentration in solutions, at which dipolar ions are at a maximum. The point also coincides with minimum viscosity and conductivity. At this pH value, the charge on a colloid is zero and the ionization of an ampholyte is at a minimum. It has a definite value for each amino acid and protein. See panel on GELS.

iso-electronic (*Electronics*) Said of similar electron patterns, as in valency electrons of atoms.

iso-enzymes (*BioSci*) See ISOZYME.

isoform (*BioSci*) A protein having the same function and similar (or identical sequence), but the product of a different gene and (usually) tissue-specific. Rather stronger in implication than 'homologous'.

isogamy (*BioSci*) Sexual fusion of similar gametes. Cf ANISOGAMY, OÖGAMY.

isogeneic (*BioSci*) Genetically identical. Also *syngeneic*.

isogenetic (*BioSci*) Having a similar origin.

isogenic (*BioSci*) Describes a strain in which all individuals are genetically identical but not necessarily homozygous. Cf INBRED LINE.

isogeotherms (*Geol*) See GEOISOTHERMS.

isogonal transformation (*MathSci*) A transformation from the z plane to the w plane, in which corresponding curves intersect at the same angle in each plane.

isogonic line (*Phys*) Line on a map joining points of equal magnetic declination, ie corresponding variations from true north.

isograd (*Geol*) A line joining points where metamorphic rocks have attained the same facies, by being subjected to the same temperature and pressure.

isogrivs (*Aero, Ships*) Lines drawn on a GRID NAVIGATION system chart joining points at which convergency of meridians and magnetic variation are equal.

isohel (*EnvSci*) A line drawn on a map through places having equal amounts of sunshine.

isohydric (*Chem*) Having the same PH value, or concentration of hydrogen ions.

isohyet (*EnvSci*) A line drawn on a map through places having equal amounts of rainfall.

isokinetic sample (*PowderTech*) Particles from a fluid stream or aerosol entering the mouth of the sampling device at the linear velocity at which the stream was flowing at that point prior to the insertion of the device.

isokont (*BioSci*) Having two or more flagella of equal length or identical morphology as the whiplash flagella of *Chlamydomonas*. Cf HETEROKONT. Also *isokonton*.

isolate (*BioSci*) To establish a pure culture of a microorganism; such a culture.

isolated essential singularity (*MathSci*) See POLE.

isolated phase switchgear (*ElecEng*) Switchgear in which all the apparatus associated with each phase is segregated in separate cubicles or on separate floors of the switch house.

isolated point (*MathSci*) (1) A point of a set in the neighbourhood of which there is no other point. (2) See DOUBLE POINT.

isolating mechanism (*BioSci*) Anything that prevents the exchange of genetic material between two populations. It can include geography, physiology or behaviour.

isolation (*Acous*) Prevention of sound transmission by eg walls, mufflers or resilient mounts.

isolation box (*Agri*) Housing for individual animals suspected of having an infectious disease, or under treatment.

isolation diode (*ICT*) One used to block signals in one direction but to pass them in the other.

isolation transformer (*ElecEng*) One used to isolate electrical equipment from its power supply.

isolator (*ICT*) Passive device for insertion into very-high-frequency or microwave transmission lines (waveguide, coaxial or stripline) with low loss in the forward direction and high attenuation in the reverse. Used to isolate oscillators, transmitters, etc, from a mismatched or reflective load. See GYRATOR.

isolecithal (*BioSci*) Said of ova that have yolk distributed evenly through the protoplasm.

isoleucine (*Chem*) $CH_3CH_2CH(CH_3)CH(NH_2)COOH$. An essential amino acid. Symbol Ile, short form I.

isolux (*Phys*) Locus, line or surface where the light intensity is constant. Also *isophot*.

isomagnetic lines (*Phys*) Lines connecting places at which a property of the Earth's magnetic field is a constant.

isomastigote (*BioSci*) Having two or four flagella of equal length.

isomerism (*Chem*) The existence of more than one substance having a given molecular composition and relative molecular mass but differing in constitution or structure. (See OPTICAL ISOMERISM.) The compounds themselves are called *isomers* or *isomerides* (Gk 'composed of equal parts'). Isobutane and butane have the same formula, C_4H_{10}, but their atoms are placed differently; one type of alkane molecule, $C_{40}H_{82}$, has over 50^{12} possible isomers. Isomerism is common in POLYMERS (panel), among organic compounds and complex inorganic salts.

isomerism (*Phys*) The existence of nuclides which have the same atomic number and the same mass number, but are distinguishable by their energy states: that having the lowest energy is stable, the others having varying lifetimes. If the lifetimes are measurable the nuclides are said to be in *isomeric states* and undergo *isomeric transitions* to the ground state.

isomerization (*MinExt*) A petroleum refinery process to improve the quality of straight-run components from crude oil. Thus for motor fuels, straight-chain components, eg *n*-butane, are converted to the more desirable branched-chain components, eg *iso*-butane, over a catalyst of aluminium chloride.

isomerized rubber (*Chem*) Rubber in which the molecules have been rearranged by heating in solution in the presence of suitable catalysts.

isomerous (*BioSci*) Equal numbers as in the parts in two whorls of a flower.

isomers (*Chem*) See ISOMERISM.

isomer separation (*Phys*) The chemical separation of the lower-energy member of a pair of nuclear isomers.

isometric contraction (*BioSci*) The type of contraction involved when a muscle produces tension but is held so that it cannot change its length.

isometric data (*Eng*) Measurements taken with constant dimensions; eg isometric stress against time creep curves for a polymer are obtained at constant strain, and can be derived from a creep curve or an isochronous stress–strain curve.

isometric projection (*Arch*) A type of axonometric drawing in which all horizontal lines are drawn at 30° to the horizontal plane of projection; the result is a three-dimensional drawing which gives equal emphasis to all three planes.

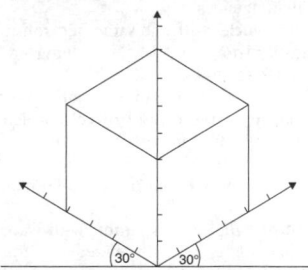

Isometric projection   A cube is shown.

isometric system (*Crystal*) The cubic system.

isometry (*MathSci*) A geometrical transformation which preserves distance.

isomorphic (*BioSci*) Morphologically similar.

isomorphic alternation of generations (*BioSci*) Alternation of generations that are morphologically alike. Also *homologous alternation of generations*. Cf HETEROMORPHIC ALTERNATION OF GENERATIONS.

isomorphic groups (*MathSci*) The groups $(G_1;*)$ and $(G_2;\bigcirc)$ are said to be isomorphic to one another if there exists a one-to-one mapping f from G_1 onto G_2, and for all elements x, y in G_1 we have $f(x*y) = f(x)\bigcirc f(y)$.

isomorphism (*Crystal*) The phenomenon whereby two or more minerals, which are closely similar in their chemical constitution, crystallize in the same class of the same system of symmetry, and develop very similar forms. Adjs *isomorphic, isomorphous*.

isomorphism (*MathSci*) A one-to-one mapping from one algebraic system onto another which shows the systems to have the same abstract structure.

isomorphism theorem (*MathSci*) One of the results concerning the quotient groups of a particular parent group and the ISOMORPHISMS between them. See FIRST, SECOND and THIRD ISOMORPHISM THEOREMS.

isomorphous replacement (*Crystal*) Replacing atoms at a given position in a crystal structure by others, usually those of a heavy metal. The determination of crystal structure of particularly complex molecules is made much more difficult by the investigator's inability to determine the phase relations of the diffraction pattern. Heavy-metal replacement is an important method of overcoming the problem.

isoniazid (*Pharmacol*) A bactericidal antibiotic used in combination therapy for tuberculosis.

isonitriles (*Chem*) See ISOCYANIDES.

isonome (*BioSci*) A line on a map joining points of equal abundance of a given species of plant.

isopach (*Geol*) A line drawn through points of equal thickness on a rock unit.

isopentenyladenine (*BioSci*) A natural CYTOKININ.

isophot (*Phys*) See ISOLUX.

isopiestic (*Chem*) Having equal pressure in the system or conditions described.

isopleth (*Chem*) A line of constant composition on a PHASE DIAGRAM.

isopleth (*MathSci*) See NOMOGRAM.

Isopoda (*BioSci*) An order of Malacostraca, in which the carapace is absent, the eyes are sessile or borne on immovable stalks, the body is depressed and the legs used for walking. Such organisms show great variety of form, size and habit; some are terrestrial, plant-feeders or ant-guests, others are marine, free-living, and feeding on seaweeds or ectoparasitic on fish. Woodlice etc.

isopodous (*BioSci*) Having the legs all alike.

isoprenaline (*Pharmacol*) A drug mimicking the action of sympathetic stimulation by adrenaline to produce tachycardia and increase the output of the heart. See SYMPATHOMIMETICS.

isoprene (*Chem*) $CH_2 = C(CH_3)CH = CH_2$. Bp 37°C. A diene, colourless liquid, obtained by dehydrogenation of 2-methylbutane, from propylene, or by several other methods. It is the monomer for synthesis of isoprene rubber and a co-monomer for isoprene–isobutene rubber.

isoprene–isobutene rubber (*Chem*) A butyl rubber. See panel on ELASTOMERS.

isopropyl alcohol (*Chem*) Propan-2-ol. $(CH_3)_2CHOH$. Bp 81°C. A colourless liquid, miscible with water.

isopropyl benzene (*Chem*) See CUMENE.

isopropyl group (*Chem*) The monovalent radical, $(CH_3)_2CH-$.

Isoptera (*BioSci*) An order of social EXOPTERYGOTA living in large communities that occupy nests excavated in the soil or built up from mud and wood. Different polymorphic forms or castes occur in each species. The mouthparts are adapted for biting; both pairs of wings, if present, are membranous and can be shed by means of a basal suture. Such organisms are exclusively herbivorous. White ants, termites.

isopycnic (*EnvSci*) A line on a chart joining points of equal atmospheric density.

isoquinoline (*Chem*) Occurs in coaltar. Mp 23°C, bp 240°C. It is an isomer of quinoline and a condensation product of a benzene ring with a pyridine ring. It forms colourless crystals.

isosbestic (*BioSci*) Wavelength at which the absorption coefficients of equimolar solutions of two different substances are identical.

isosceles triangle (*MathSci*) A triangle having two equal sides.

isoseismal line (*Geol*) A line drawn on a map through places recording the same intensity of earthquake shocks. See panel on EARTHQUAKE.

ISO sizes (*Paper*) A series of trimmed, international, metric paper sizes based on a width-to-length ratio of 1:1·414 (ie 1:$\sqrt{2}$). The next smaller size in the series is produced by halving the longer dimension. The range comprises the A-, B- and C-series of sizes, based on basic sheets of 1 m^2 (= 2^0 m^2), $2^{1/2}$ m^2 and $2^{\frac{1}{4}}$ m^2 respectively. Thus A0 is 841 mm × 1189 mm (= 1 m^2); A1, 594 mm × 841 mm; A2, 420 mm × 594 mm, etc. B0 is 1000 mm × 1414 mm, and C0 is 917 mm × 1297 mm.

isosmotic (*BioSci*) Having the same OSMOTIC PRESSURE.

isosorbide mononitrate (*Pharmacol*) A *nitrate* drug used in treatment of angina and with other drugs in treatment of congestive heart failure.

ISO speed (*ImageTech*) Abbrev for *International Organization for Standardization speed*. A photographic film speed rating, expressed on an arithmetic scale; equivalent to former ASA rating.

isospin (*Phys*) Abbrev for ISOTOPIC SPIN.

isostasy (*Geol*) The process whereby areas of crust tend to float in conditions of near equilibrium on the plastic mantle.

isostatic pressing (*Eng*) See HOT ISOSTATIC PRESSING.

isostemonous (*BioSci*) With stamens in one whorl and equal in number to petals.

isostere (*EnvSci*) A line on a chart joining points of equal atmospheric SPECIFIC VOLUME.

isosteric (*Chem*) Consisting of molecules of similar size and shape.

isotach (*EnvSci*) A line on a chart joining points of equal wind speed.

isotactic (*Plastics*) A term denoting linear-substituted hydrocarbon polymers in which the substituent groups all lie on the same side of the carbon chain. See ATACTIC, STEREOREGULAR POLYMERS, SYNDIOTACTIC.

isotaxy (*Chem*) Polymerization in which the monomers show stereochemical regularity of structure. Adj *isotactic*. See fig. at STEREOREGULAR POLYMERS. Cf SYNDIOTAXY.

isoteniscope (*Chem*) An instrument for the static measurement of vapour pressure by observing the change of level of a liquid in a U-tube.

isotherm (*EnvSci*) A line drawn on a chart joining points of equal temperature.

isothermal (*Phys*) (1) Occurring at constant temperature. (2) A curve relating quantities measured at constant temperature.

isothermal change (*Phys*) A change in the volume and pressure of a substance which takes place at constant temperature. For gases, BOYLE'S LAW applies to isothermal changes.

isothermal curve (*Phys*) A curve obtained by plotting pressure against volume for a gas kept at constant temperature. For a gas sufficiently above its critical temperature for BOYLE'S LAW to be obeyed, such curves are rectangular hyperbolas.

isothermal efficiency (*Eng*) Of a compressor, the ratio of the work required to compress a gas isothermally to the work actually done by the compressor.

isothermal process (*Eng*) A physical process, particularly one involving the compression and expansion of a gas, which takes place without temperature change.

isothermal temperature coefficient (*NucEng*) In a reactor in which the temperature is uniform over the core, the change in reactivity for a given temperature alteration.

isothermal transformation (*Eng*) The change in phase which occurs in a metal or alloy at constant temperature after cooling or heating through the equilibrium temperature. Fig. ▷

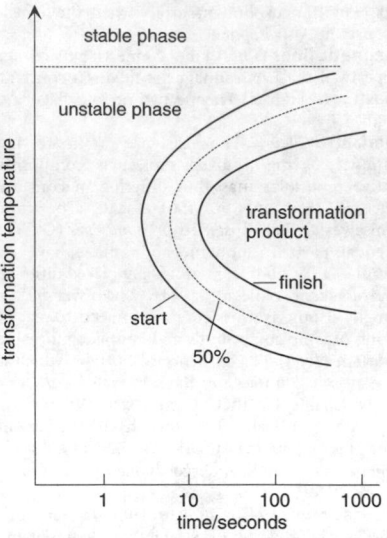

isothermal transformation diagram

isothermal transformation diagram (*Eng*) A diagram consisting of single or overlapping C-shaped curves, displayed on logarithmic time and linear temperature axes which depict the beginning and end of a solid-state transformation. They represent a balance between the nucleation and growth of a new phase during the transformation of a metastable phase to an equilibrium state. The curves are determined by cooling specimens of the material from a temperature where the transforming phase is stable and holding it isothermally at a lower temperature where it is unstable, then measuring some physical property or characteristic which allows the time to be determined for transformation to begin and end. Conducting such experiments at different temperatures produces start–finish curves displaced on the time axis. Also *TTT diagram* (time–temperature–transformation diagram). See also CONTINUOUS COOLING TRANSFORMATION diagram.

isothiocyanates (*BioSci*) A group of sulphur-containing compounds, some of which are produced by cabbages, cress and other cruciferous vegetables and that act as herbicides or fungicides.

isotones (*Phys*) Nuclei with the same neutron number but different atomic numbers, ie those lying in a vertical column of a SEGRÈ CHART.

isotonic (*Chem*) Having the same osmotic pressure, eg as that of blood, or of the cytosol of cells which are being tested for their osmotic properties.

isotonic contraction (*BioSci*) The type of contraction involved when a muscle shortens while maintaining a constant tension.

isotonic solution (*BioSci*) A solution with water potential equal to that of a cell suspended in the solution; consequently there is no water flux between cell and solution. In animal cells these conditions are established for non-permeating solutes at iso-osmolarity but in plant cells the contribution of the hydrostatic pressure of the cell wall must be taken into account.

isotope (*Phys*) One of a set of chemically identical species of atom which have the same ATOMIC NUMBER but different MASS NUMBERS. A few elements have only one natural isotope, but all elements have artificially produced radioisotopes.

isotope geology (*Geol*) The study of the relative abundances of radioactive and stable isotopes in rocks to determine radiometric ages and conditions of formation.

isotope separation (*Phys*) The process of altering the relative abundance of isotopes in a mixture. The separation may be virtually complete as in a mass spectrograph, or may give slight enrichment only as in each stage of a diffusion plant. See panel on URANIUM ISOTOPE ENRICH-MENT.

isotope structure (*Phys*) Hyperfine structure of spectral lines resulting from mixture of isotopes in the source material. The wavelength difference is termed the *isotope shift*.

isotope therapy (*Radiol*) Radiotherapy by means of radioisotopes.

isotopic abundance (*Phys*) The proportion of one isotope to the total amount of an element, as it occurs in nature. See ABUNDANCE.

isotopic age (*Geol*) See ABSOLUTE AGE.

isotopic dating (*Phys*) The calculation of the age in years for geological (or archaeological) materials using the known radioactive decay rates from parent to daughter isotopes. See POTASSIUM–ARGON DATING, RADIOCARBON DATING, RUBIDIUM–STRONTIUM DATING, URANIUM–LEAD DATING.

isotopic dilution (*Radiol*) The mixing of a particular nuclide with one or more of its isotopes.

isotopic dilution analysis (*Phys*) A method of determining the amount of an element in a specimen by observing the change in isotopic composition produced by the addition of a known amount of radioactive allobar.

isotopic number (*Phys*) See NEUTRON EXCESS.

isotopic spin (*Phys*) A quantum number assigned to members of a group of elementary particles differing only in electric charge; the particle groups are known as *multiplets*. Thus it is convenient to regard protons and neutrons as two manifestations of the nucleon, with isotopic spin either parallel or antiparallel to some preferred direction, ie they have isotopic spin $+\frac{1}{2}$ and $-\frac{1}{2}$. The nucleon is then a doublet. This can be extended to all baryons and mesons; eg the triplet π-meson consists of three pions. The small mass differences between the members of a multiplet are associated with their differing charges. The number of members of a multiplet set is $2I + 1$ where I is the isotopic spin, 0 for a singlet, $\frac{1}{2}$ for a doublet, 1 for a triplet, etc. The justification for the classification of particles is that all the members of a multiplet respond identically to strong nuclear interactions, the charges affecting only electromagnetic interactions. Isotopic spin is conserved in all strong interactions and never changes by more than one in a weak interaction. This classification is introduced by analogy with the spin or intrinsic angular momentum of atomic spectroscopy; isotopic spin has nothing to do with the nuclear spin of the particles. Also *isobaric spin*, *isospin*, *i-spin*.

isotopic symbols (*Chem*) Numerals attached to the symbol for a chemical element, with the following meanings: upper left, mass number of atom; lower left, nuclear charge of atom; lower right, number of atoms in molecule, eg

$$_{1}^{2}H_2, \quad _{12}^{24}Mg$$

isotron (*Phys*) A device for the separation of isotopes. Pulses from a source of ions are synchronized with a deflecting field. The ions undergo deflections according to their mass.

isotropic (*Phys*) Having properties which do not vary with direction. Cf ANISOTROPIC.

isotropic dielectric (*ElecEng*) One in which the electrical properties are independent of the direction of the applied electric field.

isotropic etching (*Electronics*) Describes an etching process which proceeds equally in all directions such as in semiconductor processing when etching is accomplished with wet chemicals or by dry etching without substantial ion bombardment. Cf ANISOTROPIC ETCHING.

isotropic radiator (*ICT*) An idealized antenna which sends out energy equally in all directions; virtually impossible to realize in practice. Cf OMNIDIRECTIONAL ANTENNA.

isotropic source (*Electronics*) Theoretical source which radiates all its electromagnetic energy equally in all directions.

isozyme (*BioSci*) Electrophoretically distinct forms of an enzyme with identical activities, usually coded by different genes. Also *isoenzyme*.

ISP (*ICT*) Abbrev for INTERNET SERVICE PROVIDER.

i-spin (*Phys*) See ISOTOPIC SPIN.

ISRO (*Space*) Abbrev for *Indian Space Research Organisation*, which oversees all Indian space activities.

ISS (*Space*) Abbrev for INTERNATIONAL SPACE STATION.

isthmus (*BioSci*) A neck connecting two expanded portions of an organ; as the constriction connecting the midbrain and the hind-brain of vertebrates.

IT (*ICT*) See INFORMATION TECHNOLOGY.

itacolumite (*Geol*) A micaceous sandstone with loosely interlocking grains, which enable the rock to bend when cut into thin slabs.

Italian asbestos (*Min*) A name often given to tremolite asbestos to distinguish it from Canadian or chrysotile asbestos.

Italian roof (*Arch*) See HIPPED ROOF.

italic, italics (*Print*) A sloping style of type, thus *italic*.

italite (*Geol*) A rare coarsely granular plutonic rock composed of leucite and a little glass, a *leucitolite*.

IT CCD (*ImageTech*) Abbrev for *interline transfer charge-coupled device*. A SOLID-STATE IMAGE SENSOR in which the charge is transferred almost continuously from the PIXELS to vertical columns of adjacent opaque cells for reading out. See CCD ARRAY, FIT CCD, FT CCD.

itchy leg (*Vet*) See CHORIOPTIC MANGE.

ITCZ (*EnvSci*) Abbrev for INTERTROPICAL CONVERGENCE ZONE.

iter (*BioSci*) A canal or duct, as the reduced ventricle of the midbrain in higher vertebrates.

iterated fission expectation (*Phys*) Limiting value, after a long time, of the number of fissions per generation in the chain reaction initiated by a specified neutron to which this term applies.

iteration (*ICT*) To obtain a result by repeatedly performing the same sequence of steps until a specified condition is satisfied. See LOOP.

iterative impedance (*Phys*) The INPUT IMPEDANCE of a four-terminal network or transducer when the output is terminated with the same impedance, or when an infinite series of identical such networks are cascaded. See IMAGE IMPEDANCE.

iteroparous (*BioSci*) Reproducing on two or more occasions during a lifetime.

ITF (*EnvSci*) Abbrev for *intertropical front*. See INTERTROPI-CAL CONVERGENCE ZONE.

ITM (*Ships*) Abbrev for *inch trim moment*. Also MOMENT TO CHANGE TRIM ONE INCH.

ITO (*Eng*) Abbrev for INDIUM-DOPED TIN OXIDE.

ITU (*ICT*) Abbrev for INTERNATIONAL TELECOMMUNICA-TIONS UNION.

IUGS (*Geol*) Abbrev for *International Union of Geological Sciences*. Under its aegis a classification of igneous rocks has been agreed internationally.

IUPAC (*Chem*) Abbrev for the *International Union of Pure and Applied Chemistry*.

ivermectin (*Pharmacol, Vet*) A semi-synthetic anthelminthic used in the treatment of onchocerciasis and extensively in veterinary practice. Ivermectin is derived from the *avermectins*, a class of highly active broad-spectrum anti-parasitic agents isolated from the fermentation products of *Streptomyces avermitilis*.

IVF (*BioSci*) Abbrev for IN VITRO FERTILIZATION.

ivory (*BioSci*) The dentine of teeth esp the massive type occurring in elephants, mammoths, etc. Formerly used for tools (eg harpoon tips) and still used for decorative products despite limitation attempts.

ivory board (*Paper*) Genuine ivory board is formed from high-quality papers by starch-pasting two or more together.

ivorywood (*For*) Rare Australian hardwood from *Siphonadendron*, prized for engraving, turnery, mirror frames, inlaying, etc.

IVP, IVU (*Med*) Abbrev for *intravenous pelography, intravenous urography*, ie the demonstration by X-ray of the renal tract after the intravenous injection of a radio-opaque contrast medium.

IW (*Chem*) Abbrev for ISOTOPIC WEIGHT.

IX (*Chem, MinExt*) Abbrev for ION EXCHANGE.

Izod test (*Eng*) A flexed cantilever-beam, notched-specimen, impact test in which one end of a notched specimen is held in a vice while the other end is struck by a striker carried on a pendulum; the energy absorbed in fracture is then calculated from the height to which the pendulum rises as it continues its swing. See panel on IMPACT TESTS.

Izod value (*Eng*) The energy absorbed in fracturing a standard specimen in an Izod pendulum impact-testing machine. See panel on IMPACT TESTS.

J

J (*Chem*) In names of dyestuffs, a symbol for yellow.

J (*Phys*) Symbol for JOULE.

J (*Eng*) Symbol for polar moment of inertia.

J (*Phys*) Symbol for: (1) electric current density; (2) MAGNETIC POLARIZATION.

j (*ElecEng*) The symbol *j* is used by electrical engineers in place of the mathematician's *i*. Its main use is that in circuits carrying sinusoidal current of angular frequency ω, any inductance *L* and any capacitance *C* can be replaced by reactances $j\omega L$ and

$$\frac{1}{j\omega C}$$

respectively, and Ohm's law can then be used. Also referred to as the $90°$ operator.

jacaranda (*For*) See BRAZILIAN ROSEWOOD.

jacaranda rosa (*For*) BRAZILIAN TULIPWOOD.

jacinth (*Min*) See HYACINTH.

jack (*Eng*) A portable lifting machine for raising heavy weights through a short distance, consisting either of a screw raised by a nut rotated by hand gear and a long lever, or of a small hydraulic ram. See HYDRAULIC JACK.

jack (*ICT*) Socket whose connections are short-circuited until a jack plug is inserted. A break jack is one which breaks the normal circuit on inserting plug, while a branch jack is one which does not.

jack arch (*Build*) See FLAT ARCH.

jackbit (*MinExt*) Detachable cutting end fitted to shank of miner's rock drill, used to drill short blast holes. Also *rip-bit*.

jackblock (*Build*) A method of system building in which the roof and floor slabs are cast on top of each other and hydraulically jacked up to their respective levels, walls being built as required.

jack box (*ElecEng*) One containing switches or connections for changing circuits.

jacket (*Eng*) An outer casing or cover constructed round a cylinder or pipe, the space being filled with a fluid for either cooling or heating the contents, or with insulating material for keeping the contents at substantially constant temperature, eg the water jackets of an internal-combustion engine.

jacket (*MinExt*) Erected around an offshore oil well and resting on the sea bed, it supports the platform carrying the drilling derrick, operating equipment and living accommodation.

jacket (*NucEng*) See CAN.

jacket (*Print*) The wrapper, or dust cover, in which a book is enclosed. Book jackets are usually artistically designed and executed in colour, as their purpose is to enhance the appeal of the volume as well as to protect it.

jackhammer (*MinExt*) A hand-held compressed-air hammer drill for rock drilling.

jack plane (*Build*) A bench plane about 16 in long, used for bringing the work down to approximate size, prior to finishing with a trying or smoothing plane.

jack rafter (*Build*) A short rafter connecting a HIP RAFTER and the eaves, or a VALLEY RAFTER and the ridge.

jack shaft (*ElecEng*) An intermediate shaft used in locomotives having collective drive; the jack shaft is geared to

the motor shaft and carries cranks which drive the coupling rods on the driving wheels.

Jacksonian epilepsy (*Med*) A convulsion of a limited group of muscles spreading gradually from one group to the other, usually without loss of consciousness; the result of a lesion (eg tumour) of the brain.

Jackson structured programming (*ICT*) A method of top-down programming developed by Michael Jackson during the 1970s. The method stresses stepwise refinement, breaking down the problem to smaller and smaller units until each can be programmed directly; the structure has three basic 'building blocks' and an emphasis on DATA STRUCTURES.

jack-up rig (*MinExt*) A prefabricated well-drilling assembly mounted on a barge towed to the drilling site in moderately shallow water (usually less than 100 m). Three or more legs are flooded and lowered to the sea bed, and the barge and superstructure are jacked up out of the water and clear of wave action.

Jacobian (*MathSci*) Of *n* functions u_i each of *n* variables x_j, the determinant whose *i*, *j*th element is $\partial u_i/\partial x_j$. Written

$$\frac{\partial(u_1, u_2, u_3, \dots, u_n)}{\partial(x_1, x_2, x_3, \dots, x_n)}$$

Jacobian elliptic functions (*MathSci*) See ELLIPTIC FUNCTIONS.

jacobsite (*Min*) An oxide of manganese and iron, often with considerable replacement of manganese by magnesium; crystallizes in the cubic system (usually in the form of distorted octahedra). A spinel.

jacob's ladder (*Eng*) Vertical belt conveyor with cups or buckets.

Jacobson's glands (*BioSci*) In some vertebrates, nasal glands, the secretion of which moistens the olfactory epithelium.

Jacobson's organ (*BioSci*) In some vertebrates, an accessory olfactory organ developed in connection with the roof of the mouth.

jacquard (*Textiles*) A device, frequently incorporating punched cards or punched continuous strip, used to produce patterned fabrics during weaving, warp knitting, weft knitting and lace-making. Named after the French inventor, Joseph-Marie Jacquard, 1752–1834. Also applied to the fabrics so produced.

Jacquet's method (*Eng*) Final polishing of metal surfaces by ELECTROLYSIS.

jactitation (*Med*) Restless tossing of a severely ill patient; a twitching or convulsion of muscle or of a limb.

jacupirangite (*Geol*) A nepheline-bearing pyroxenite consisting of titanaugite, biotite, iron ores, and nepheline, the last being subordinate to the mafic minerals.

jad (*MinExt*) A deep groove cut into the bed to detach a block of natural stone. To undercut. A *jadder* is a stonecutter and the working tool is a *jadding pick*. Also *jud*.

jade (*Min*) A general term loosely used to include various mineral substances of tough texture and green colour used for ornamental purposes. It properly embraces NEPHRITE and JADEITE but is sometimes misapplied to green varieties of minerals such as AMAZONSTONE, BOWENITE, HYDRO-GROSSULAR, QUARTZ and VESUVIANITE.

jadeite (*Min*) A monoclinic member of the pyroxene group; sodium aluminium silicate. Usually white, grey or mauve, it occurs only in metamorphic rocks, and is the rare form of jade (*Chinese jade*).

jag-bolt (*Eng*) See RAG-BOLT.

jail fever (*Med*) See TYPHUS.

jalousies (*Build*) Hanging or sliding wooden sun-shutters giving external protection to a window, and allowing for ventilation through louvres or holes cut in the shutters themselves. Also *Venetian shutters*.

jam (*FoodSci*) Gel made from fruit, fruit juice, sugar and natural or added pectin with the aroma, taste and colour of the parent fruit. A shelf-stable preserve achieved by the low water activity after the soluble solids are increased by boiling to around 67%.

jamaicin (*Chem*) See BERBERINE.

jamb (*Build*) The side of an aperture.

jamb linings (*Build*) The panelling at the sides of a window recess, running from the floor to the level of the window head. Cf ELBOW LININGS.

jamb post (*Build*) An upright member on one side of a doorway opening.

jamb stone (*Build*) A stone forming one of the upright sides of an aperture in a wall.

James–Lange theory of emotions (*Psych*) A theory that emotion is the subjective experience of one's own bodily reactions in the presence of certain arousing stimuli; the stimuli cause certain physiological responses, and the awareness of these responses causes emotion.

jamesonite (*Min*) See FEATHER ORE.

Jamin interferometer (*Phys*) A form of interferometer in which two interfering beams of light pursue parallel paths a few centimetres apart. The instrument is used to measure the refractive index of a gas, by observing the fringe shift when one of the light beams traverses a tube filled with the gas, while the other traverses a vacuum.

Jamin–Lebedeff system (*BioSci*) A system of interference microscopy in which object and reference beams are split and later recombined by birefringent calcite plates, but pass through the same optical components (in contrast to the MACH–ZEHNDER SYSTEM).

jamming (*ICT*) Deliberate interference of transmission on one carrier by transmission on or near the same frequency, with wobble or noise modulation.

JANET (*ICT*) Abbrev for *Joint Academic Network*. Dating back to the 1970s, it links universities and research institutions in the UK.

jansky (*Astron*) A unit in radio astronomy (symbol Jy) to measure the power received at the telescope from a cosmic radio source: $1 \, \text{Jy} = 10^{-26} \, \text{W m}^{-2} \, \text{Hz}^{-1}$.

J-antenna (*ICT*) Half-wave antenna fed and matched at the end by a quarter-wavelength line.

Janus (*Astron*) The tenth natural satellite of Saturn, discovered in 1980. Distance from the planet 151 000 km; diameter 200 km.

Janus (*ICT*) Transmitting or receiving antenna which can be switched between opposite directions. Used for airborne Doppler navigation systems.

japan (*Build*) A black, glossy, paint based on asphaltum and drying oils. Also *black japan*.

Japan camphor (*Chem*) See CAMPHOR.

Japanese paper (*Paper*) Japanese hand-made paper prepared from mulberry bark. The surface is similar to that of JAPANESE VELLUM.

Japanese planes (*Build*) Planes with a STOCK made of rectangular hardwood with a steel blade and CAP IRON situated near one end. They are designed for pulling on the cutting stroke with the work supported on an inclined beam. Fig. ▷

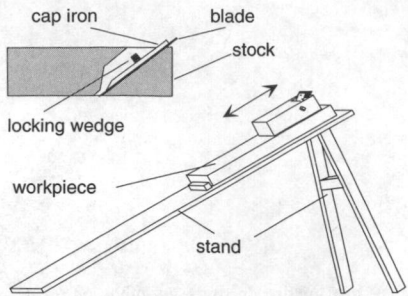

Japanese planes Inset shows plane in section.

Japanese vellum (*Paper*) An expensive hand-made paper. Prepared from the inner bark of the mulberry tree, thicker than JAPANESE PAPER.

japanners' gold size (*Build*) See GOLD SIZE.

japanning (*Build*) The process of finishing an article with japans, esp the STOVING of japans.

Japan wax (*Chem*) A natural wax obtained from sumach, mp 50°C. It has a high content of palmitin.

jappe (*Textiles*) Lightweight, fine, plain-weave cloth usually made from silk, used for linings and dresses.

jargon aphasia (*Med*) Rapid unintelligible utterance, due to a lesion in the brain.

jargons (*Min*) A name given in the gem trade to the zircons (chiefly colourless, smoky or of golden-yellow colour) from Sri Lanka. They resemble diamonds in lustre but are less valuable. See HYACINTH. Also *jargoons*.

jarosite (*Min*) A hydrous sulphate of iron and potassium crystallizing in the trigonal system; a secondary mineral in ferruginous ores.

jarrah (*For*) A dense hardwood (*Eucalyptus marginata*) from Australia, with a rich deep-red colour, usually straight-grained and even but medium- to coarse-textured.

jar-ramming machine (*Eng*) See JOLT-RAMMING MACHINE.

jasmonate (*BioSci*) Any of several organic compounds that occur in plants and are thought to control processes such as growth and fruit ripening and to aid the plant's defences against disease and insect attack.

jaspé (*Textiles*) (1) Plain woven fabric with a shaded appearance resulting from a warp-thread colour pattern. Used mainly for bedspreads and curtains. (2) Yarn made from two chemically different continuous filament yarns (eg nylon and polyester) *textured* together and then dyed in such a way that only one component is coloured.

jasper (*Min*) An impure opaque chalcedonic silica, commonly red owing to the presence of iron oxides.

JATO (*Aero*) Abbrev for *jet-assisted take-off*. See TAKE-OFF ROCKET.

jaundice (*Med*) Yellow coloration of the skin and other tissues of the body, by excess of bile pigment present in the blood and the lymph. May be caused by excessive breakdown of blood (haemolytic jaundice), by failure of the liver to transport the pigments (hepatic jaundice) or failure to excrete the pigment through the biliary system (obstructive jaundice). Also *icterus*.

Java (*ICT*) Programming language designed specifically to be used on the Internet, independent of an operating system. It also automatically reclaims memory no longer being used and forces the programmer to write EXCEPTION HANDLING routines. See panel on INTERNET.

JavaBeans (*ICT*) TN for small applications (applets) developed by Sun Microsystems as an extension of the JAVA language for downloading over the Internet. Similar to ACTIVEX CONTROLS. See panel on INTERNET.

JavaScript (*ICT*) A scripting programming language with a C-like syntax, used in web-based systems in particular to provide access to objects within other systems.

Javel water (*Chem*) A mixture of potassium chloride and hypochlorite in solution. Chiefly used for bleaching and disinfecting. Also *eau de Javelle*.

jaw (*Eng*) (1) One of a pair or group of members between which an object is held, crushed or cut, as the jaws of a vice or chuck. (2) One of several members attached to an object, to locate it by embracing another object.

jaw breaker (*MinExt*) Heavy-duty rock-breaking machine with fixed vertical and inclined swing jaw, between which large lumps of ore are crushed. Also *alligator*, *Blake crusher*, *jaw crusher*.

jaws (*BioSci*) In gnathostomatous vertebrates, the skeletal framework of the mouth enclosed by flesh or horny sheaths, assisting in the opening and closing of the mouth, and usually furnished with teeth or horny plates to facilitate seizure of the prey or mastication; in invertebrates, any similar structures placed at the anterior end of the alimentary tract.

jaws (*Print*) See FOLDING JAWS.

J chain (*BioSci*) A polypeptide chain with a high content of the amino acid cysteine, which enables it to form disulphide bonds. It helps to link together IgA molecules into polymeric forms and to hold IgM in pentameric form.

JCL (*ICT*) See JOB CONTROL LANGUAGE.

J-display (*Radar*) A modified *A-display* with circular time base. See R-DISPLAY.

jean (*Textiles*) Strong woven twilled fabric, used for overalls or casual wear. See DENIM.

jedding axe (*Build*) An axe having one flat face and one pointed peen.

jejunectomy (*Med*) Excision of part of the jejunum.

jejunitis (*Med*) Inflammation of the jejunum.

jejunocolostomy (*Med*) The formation, by operation, of a communication between the jejunum and the colon.

jejunoctomy (*Med*) Incision of the jejunum.

jejunojejunostomy (*Med*) The formation, by anastomosis, of a communication between two parts of the jejunum, thus short-circuiting the part in between.

jejunostomy (*Med*) The operative formation of an opening into the jejunum.

jejunum (*BioSci*) In mammals, that part of the small intestine which intervenes between the duodenum and the ileum.

jelly (*ImageTech*) See GEL.

jelutong (*For*) Malayan hardwood (*Dyera costulata*) that is almost white in colour and is straight-grained with a fine, even texture, but is non-durable. The tree also yields a LATEX that was once an important substitute for natural rubber, and is now used to make chewing gum.

jemmy (*Build*) A small crowbar. Also *jimmy*.

Jeppesen chart (*Aero*) Airway charts, airport maps and information, named after Ebroy Jeppesen, who built up the basic format from 1926 to 1940.

jerk (*Med*) A sudden and brief contraction of a group of muscles. Often used to test a REFLEX, eg knee jerk.

jerkin head (*Build*) The end of a pitched roof which is hipped, but not down to the level of the feet of the main rafters, thus leaving a half-gable. Also *shread head*.

jerk-pump (*Autos*) A timed fuel-injection pump in which a cam-driven plunger overruns a spill port, thus causing the abrupt pressure rise necessary to initiate injection through the atomizer.

jerks (*Print*) Violent intermittent pulls of paper through a web-fed printing press or folder due usually to incorrect or worn drives.

jersey fabric (*Textiles*) The general name for knitted fabrics supplied in lengths.

JET (*NucEng*) Abbrev for JOINT EUROPEAN TORUS.

jet (*Genrl*) (1) A fluid stream issuing from an orifice or nozzle. (2) A small nozzle, such as the jet of a carburettor.

jet (*Min*) A hard coal-black variety of lignite, sometimes exhibiting the structure of coniferous wood; worked for jewellery in the last century.

jet coefficient (*Aero*) The basic non-dimensional thrust–lift relationship of the JET FLAP;

$$C_j = \frac{J}{\frac{1}{2}e v^2 s}$$

where J = jet thrust, e = air density, v = speed and S = wing area.

jet condenser (*Eng*) One in which exhaust steam is condensed by jets of cooling water introduced into the steam space.

jet deflection (*Aero*) A jet-propulsion system in which the thrust can be directed downwards to assist take-off and landing.

jet drilling (*MinExt*) See FUSION DRILLING.

jet dyeing (*Textiles*) A machine for dyeing delicate fabrics, or garments, in which the material is gently circulated by the dye liquor being pumped at high velocity through jets or a narrow throat.

jet flap (*Aero*) A high-lift flight system in which (1) the whole efflux of turbojet engines is ejected downward from a spanwise slot at the wing trailing edge or (2) a large surplus efflux from turboprop engines is so ejected, with the propellers providing a relative airflow over the wing. The downward ejection of the jet forms a barrier to the passage of air under the wing and induces more air to flow across the upper surface, so giving very high lift coefficients of the order of ten and even higher. See NGTE RIGID ROTOR.

jet impactor (*PowderTech*) See IMPACTION SAMPLER.

jet lag (*Aero*) Delayed bodily effects felt after long flight by fast jet aircraft and arriving in a different time zone.

jet loom (*Textiles*) A high-speed machine in which the weft is propelled through the SHED by a high-pressure jet of air or water.

jet mill (*ChemEng*) A mill in which particles are pulverized to micrometre size by the collisions occurring among them when they are swept into a small jet of gas at sonic velocity.

jet noise (*Aero*) The noise of jet efflux, which varies as the eighth power of its velocity.

jet nozzle process (*NucEng*) A process whereby isotope separation, based on the mass dependence of centrifugal force, is obtained by the fast flow of uranium hexafluoride in a curved duct.

jet pipe shroud (*Aero*) A covering of heat-insulating material, usually layers of bright foil, round a jet pipe.

jet propulsion (*Aero*) Propulsion by reaction from the expulsion of a high-velocity jet of the fluid in which the machine is moving. It has been used for the propulsion of small ships by pumping in water and ejecting it at increased velocity, but the principal application is to aircraft. See PULSE JET, RAMJET, REACTION PROPULSION, ROCKET PROPULSION, TURBOJET.

jet pump (*MinExt*) Hydraulic elevator, in which a jet of high-pressure water rises in a pipe immersed in a sump containing the water, sands or gravels which are to be entrained and pumped. Inefficient but cheap and effective where surplus hydraulic power exists.

jet shales (*Geol*) Shales containing 'jet rock', found in the Upper Lias of the Whitby district of England.

jet spinning (*Textiles*) See AIR JET SPINNING.

jet stream (*EnvSci*) A fairly well-defined core of strong wind, perhaps 200–300 miles (320–480 km) wide with wind speeds up to perhaps 200 mph (320 km h^{-1}) occurring in the vicinity of the TROPOPAUSE.

jet-textured yarn (*Textiles*) See TEXTURED YARN.

jetting (*Eng*) An injection moulding defect where a thin stream of polymer is forced into the furthest part of the tool cavity.

jetting-out (*Arch*) The projection of eg a corbel from the face of a wall.

jetty (*Build*) A structure jutting out from shore into navigable water to provide a vertical berthing face for mooring vessels.

jewel (*Eng*) Natural ruby or sapphire, or synthetic stone, used for bearings, particularly in clocks. Owing to the high polish and surface hardness obtainable they provide wearing surfaces which have a long life with little friction.

Jewel Box (*Astron*) An open cluster in the constellation Crux which contains over 100 stars of many different colours.

J exon (*BioSci*) See J GENE.

JFET (*Electronics*) See JUNCTION FIELD-EFFECT TRANSISTOR.

jib (*Eng*) The boom of a crane or DERRICK.

jib barrow (*Eng*) A wheelbarrow consisting of a platform without sides; used in foundries and workshops.

jib crane (*Eng*) An inclined arm or jib attached to the foot of a rotatable vertical post and supported by a tie rod connecting the upper ends of the two. The load rope or chain runs from a winch on the post, and over a pulley at the end of the jib. There are many modern variants including A-frame designs and means of altering the radius and therefore the reach of the jib.

jib door (*Build*) A door which carries and continues the general decoration of the wall.

jig (*Eng*) A device used in the manufacture of (interchangeable) parts to locate and hold the work and to guide the cutting tool. Cf FIXTURE.

jig (*MinExt*) A device for concentrating ore according to the relative density of its constituent minerals.

jig (*Textiles*) Machines with two rollers commonly used for dyeing. The open-width fabric passes repeatedly from one roller to the other and back again while immersed in a bath of the appropriate solution. The machines are also used for scouring and bleaching fabrics. Also *jigger*.

jig borer (*Eng*) A vertical-spindle machine for accurately boring and locating holes, having a horizontal table which can be precisely positioned by transverse and longitudinal feed motions.

jigger (*Eng*) A hydraulic lift or elevator in which a short-stroke hydraulic ram operates the lift through a system of ropes and pulleys in order to increase the travel.

jig saw (*Build*) A mechanical saw with a short narrow reciprocating blade which cuts on the up stroke; used for curved as well as straight cuts.

jim crow (*Eng*) (1) A rail-bending device, operated by hand or by hydraulic power. (2) A swivelling tool-head used on a planing machine, cutting during each stroke of the table.

jimmy (*Build*) See JEMMY.

JISC (*ICT*) Abbrev for *Joint Information Systems Committee*, a UK institution co-ordinating the use of information technology in further and higher education.

JIT (*Eng*) Abbrev for JUST IN TIME.

jitter (*ICT*) (1) Any small rapid irregularities in a waveform arising from fluctuations in supply voltages, components, etc. (2) Short-term variations of the significant instants of a digital signal from their ideal positions in time.

jitter (*ImageTech*) Picture unsteadiness caused by TIME-BASE errors. See TBC.

j–j coupling (*Phys*) Extreme form of coupling between the orbital electrons of atoms. Electrons showing individual spin–orbit coupling also interact with each other. See INTERMEDIATE COUPLING, RUSSELL–SAUNDERS COUPLING.

Jl (*Chem*) Symbol for JOLIOTIUM.

JND (*Psych*) Abbrev for JUST NOTICEABLE DIFFERENCE.

job (*ICT*) Normal term for a complete item of work performed by a computer system.

jobbing fonts (*Print*) Fonts of type used for display purposes. Often decorative.

jobbing machines (*Print*) The class of machines, usually platens, used for printing commercial or jobbing work.

jobbing work (*Print*) Small printed matter such as handbills, billheads, cards, etc.

job control (*ICT*) One function of the OPERATING SYSTEM as it controls the provision of internal resources and the flow of each job through the computer system.

job control language (*ICT*) Special language used to identify a JOB and describe its requirements to the OPERATING SYSTEM. Abbrev *JCL*.

job queue (*ICT*) A queue of jobs waiting to be serviced within a MULTIPROGRAMMING or BATCH PROCESSING system. See BACKGROUND JOB.

jockey roller (*Print*) On web-fed machines, a roller, usually the first to be traversed by the web, arranged to compensate any uneven tension as the reel unwinds.

jog (*Phys*) A discontinuity in an edge DISLOCATION in a crystal. The dislocation will be made up of many sections of varying lengths lying on neighbouring slip planes and joined together by jogs.

jogger (*Print*) (1) A rapidly vibrating inclined tray in which sheets of paper are placed to be jogged up to two adjacent edges prior to cutting. (2) An adjustable fitment at the delivery of the printing machine which straightens the sheets as they are delivered, giving a neat delivery pile.

joggle (*Build*) A short stub tenon for fitting into a mortice.

joggle (*Eng*) (1) A small projection on a piece of metal fitting into a corresponding recess in another piece, to prevent lateral movement. (2) A lap joint in which one plate is slightly cranked so as to allow the inner edges of the two plates to form a continuous surface.

joggle (*Ships*) A sharp distortion in a plate, angle or other section, made purposely to permit overriding of contacting members. It reduces the amount of steel packing.

joggle joint (*Build*) A connection between adjacent ashlars in which JOGGLES are used.

joggle work (*Build*) Coursed masonry in which slipping between the stones is prevented by the insertion of JOGGLES.

jog/shuttle (*ImageTech*) A videotape recorder playback control with an inner (jog) dial to advance or reverse one FRAME at a time, and an outer (shuttle) ring to select speeds from slow to fast motion in either direction.

johannsenite (*Min*) A silicate of calcium, manganese and iron. It is a member of the pyroxene group, crystallizing in the monoclinic system.

Johne's disease (*Vet*) A chronic disease of cattle, sheep and goats caused by infection by *Mycobacterium paratuberculosis* (*M. johnei*) which causes a chronic enteritis affecting the small intestine, caecum and colon; the main symptoms are diarrhoea and emaciation. Also *paratuberculosis*.

Johnson concentrator (*MinExt*) Machine used to arrest heavy auriferous material flowing in ore pulp. An inclined cylindrical shell rotates slowly, metallic particles being caught in rubber-grooved linings at periphery, lifted and separately discharged.

Johnson noise (*Electronics*) Noise in resistors, thermally generated and having a flat power spectrum. Associated with the random motion of charge carriers within the material. Cf FLICKER NOISE, SHOT NOISE. See THERMAL NOISE.

joiner (*ImageTech*) See SPLICER.

joiner's chisel (*Build*) See PARING CHISEL.

joinery (*Genrl*) (1) The craft of working timber to form the finishings of a building, as distinct from carpentry. (2) The material worked in this way.

joint (*ElecEng*) (1) A permanent connection between two lengths of cable or waveguides which may be *butt-jointed* (intermetallic contact maintained) or *choke-jointed* (when a half-wavelength short-circuited line is used to provide an effective contact at the guide walls). (2) The contact formed when either two conductors, or a conductor and a device, are connected together.

joint (*Eng*) The parting plane in the sand round a rammed mould, to enable the pattern to be withdrawn. It is covered with PARTING SAND before the cope or top half is rammed.

joint (*Geol*) An actual or potential fracture in a rock, in which there is no displacement. See COLUMNAR STRUCTURE, RIFT AND GRAIN.

joint (*MinExt*) A length of drilling pipe or casing, usually 20–30 feet long.

joint (*Radar*) Permanent or semi-permanent connection between two lengths of waveguide. It may consist of plain flanges with direct metallic contact and no discontinuity in the waveguide walls or it may be a CHOKE FLANGE.

joint chair (*CivEng*) A type of chair used at the joint between successive lengths of rail and providing support for the ends of both lengths.

joint efficiency (*Eng*) The ratio, expressed in per cent, of the strength of an analogous section of solid plate to that of the joint.

jointer (*Build*) A tool used by bricklayers for pointing the mortar joint between courses of bricks.

jointer plane (*Build*) See JOINTING PLANE.

Joint European Torus (*Phys*) A research facility in nuclear fusion comprising a large tokamak-type experimental fusion reactor, which became operational in 1983 at Culham, Oxfordshire, UK. It is funded by the countries of the European Union plus Switzerland.

joint fastening (*CivEng*) A fish-plate or other means of fastening together the adjacent ends of successive lengths of rail.

joint hinge (*Build*) See STRAP HINGE.

joint-ill (*Vet*) A disease of young foals, calves, lambs and piglets, caused by a variety of bacteria, and characterized by abscess formation in the umbilicus, pyaemia, abscesses in various organs, and arthritis affecting notably the leg joints. Also *navel-ill, pyosepticaemia*.

jointing (*Build*) The operation of making and/or finishing the joints between bricks, stones, timbers, pipes, etc.

jointing (*Eng*) Material used for making a pressure-tight joint between two surfaces, eg asbestos sheet, corrugated steel rings, vulcanized rubber, etc. See GASKET.

jointing (*Geol*) See JOINT.

jointing plane (*Build*) A bench plane, similar to the jack plane but larger (30 in, 750 mm long), used for truing the edges of timbers which are to be accurately fitted together. Also *jointer plane, shooting plane*.

jointing rule (*Build*) A straightedge about 6 ft (2 m) long, used as a guide when pencilling, ie painting the mortar joints of bricks to accentuate them.

jointing stage (*Agri*) The stage of grain crop development when the stem internodes are elongating.

jointless flooring (*Build*) See MAGNESITE FLOORING.

joint-mouse (*Med*) A hard body, often a piece of cartilage, loose in the joint cavity; found esp in the joints of those suffering from osteoarthritis.

joints (*Print*) The lateral projections formed on each side of the spine of a volume in the process of backing; the case hinges along it. A linen strip pasted down the fold of the endpaper is a *cloth joint*.

Joint Tactical Information Distribution System (*Aero*) A full-scale tactical, jam-resistant, command and control system developed by the US Air Force. It integrates the hitherto separate functions of communication, navigation and identification. Abbrev *JTIDS*.

joist (*Build*) A horizontal beam of timber, reinforced or prestressed concrete, or steel, used with others as a support for a floor and/or ceiling.

joliotium (*Chem*) An artificially manufactured metallic element, symbol Jl, at no 105, of the transactinide series that has six isotopes, all with half-lives of a fraction of a second; sometimes referred to as NIELSBOHRIUM. Also *unnilpentium*.

Jolly balance (*Chem*) A spring balance used to measure density by weighing in air and water.

Jolly's apparatus (*Chem*) Apparatus for the volumetric analysis of air.

jolt-ramming machine (*Eng*) A moulding machine (see MACHINE MOULDING) in which the box, pattern and sand are repeatedly lifted by a table operated by air pressure and allowed to drop by gravity, the resulting jolt or jar packing or ramming the sand in an efficient manner. Also *jar-ramming machine, jolt-ram machine*.

jolt-squeeze machine (*Eng*) A moulding machine (see MACHINE MOULDING) used for deep patterns; in it, jolting is used to pack the sand onto the pattern followed by squeezing from the top to complete the ramming.

Jominy test (*Eng*) A test for determining the relative HARDENABILITY of steels, in which one end of a heated cylindrical specimen is quenched from the austenitic region and the longitudinal hardness gradient along a ground flat measured. The hardness decreases from the martensitic level at the quenched end towards that at the air-cooled end; the greater the hardenability of the steel the farther the distance before a significant reduction appears. Also *end-quench test*.

Jones–Mote hypersensitivity (*Med*) A form of HYPER-SENSITIVITY REACTION characterized by infiltration of the skin by basophil granulocytes.

Joosten process (*MinExt*) The use of chemical reaction between solutions of calcium chloride and sodium silicate to consolidate running soils or gravels when tunnelling. A water-resistant gel is formed.

Jordanon species (*BioSci*) One of a number of true-breeding, morphologically slightly different, lines within a complex of largely inbreeding plants. Also *microspecies*. Cf LINNAEAN SPECIES.

jordan refiner (*Paper*) A type of conical refiner in which a tapering rotor revolves in a hollow shell, both being equipped with longitudinal metal bars so that stuff passing through the gap between them is abraded and cut.

Josephson effects (*Electronics, Phys*) Two effects which can occur when, at very low temperatures, two superconductors are separated by a narrow insulating gap. By TUNNELLING through the gap a direct current can pass from one superconductor to another without an applied potential. Also, when a potential difference V is established between the superconductors there is an alternating current across the gap of frequency $v = 2Ve/h$, where e is the charge on the electron and h is PLANCK'S CONSTANT. Applications include ultrahigh-speed switching of logic circuits, memory cells and parametric amplifiers operating up to 300 GHz; the Josephson effect is being widely adopted as the basis of the STANDARD VOLT. See panel on SUPERCONDUCTORS.

Joshi effect (*Phys*) The change of current in a gas because of light irradiation.

joule (*Phys*) SI unit of work, energy and heat: 1 joule is the work done when a force of 1 newton moves its point of application 1 metre in the direction of the force; symbol J; 1 erg = 10^{-7} J, 1 kW h = 3.6×10^6 J, 1 eV = 1.602×10^{-19} J, 1 calorie = 4.18 J, 1 Btu = 1055 J. See SI UNITS and appendices on SI conversion factors and SI derived units.

Joule effect (*ElecEng*) The production of heat solely arising from current flow in a conductor. See JOULE'S LAW.

Joule effect (*Phys*) The slight increase in the length of an iron core when longitudinally magnetized. See MAGNE-TOSTRICTION.

Joule–Kelvin effect (*Phys*) See JOULE–THOMSON EFFECT.

Joule magnetostriction (*Phys*) Magnetostriction for which length increases with increasing longitudinal magnetic field. Also *positive magnetostriction*.

Joule meter (*ElecEng*) An integrating wattmeter whose scale is calibrated in joules.

Joule's equivalent (*Phys*) See MECHANICAL EQUIVALENT OF HEAT.

Joule's law (*Chem*) (1) The internal energy of a given mass of gas is a function of temperature alone; it is independent of the pressure and volume of the gas. (2) The molar heat capacity of a solid compound is equal to the sum of the atomic heat capacities of its component elements in the solid state.

Joule's law (*ElecEng*) A law giving the heat H liberated by the flow of current I in a conductor with resistance R for a time t: $H = I^2Rt$. This is the basis of all electrical heating, wanted or unwanted. With high-frequency ac, R is an

EFFECTIVE RESISTANCE and *I* may be confined to a thin skin of the conductor.

Joule–Thomson effect (*Phys*) (1) The effect in which the temperature of a gas generally increases when subjected to an adiabatic expansion through a porous plug or similar device. This is due to energy being used to overcome the cohesion of the molecules of the gas. The liquefaction of gases by the LINDÉ PROCESS depends on this effect. (2) Thermodynamic heating when a rubber is stretched quickly under adiabatic conditions. Also *Joule–Kelvin effect*.

journal (*Eng*) That part of a shaft which is in contact with, and supported by, a bearing.

journal file (*ICT*) A permanent record of every interaction with the computer OPERATING SYSTEM.

Jovian planets (*Astron*) The planets Jupiter, Saturn, Uranus and Neptune.

joypad (*ICT*) Hand-held peripheral device consisting of a pad with buttons on it used for controlling the motion of objects in a computer game.

joystick (*Aero*) Colloq term for CONTROL COLUMN.

joystick (*ICT*) A control lever that can be moved in two dimensions, widely used in computer games and as an input device for computer-aided design systems.

Joy's valve gear (*Eng*) A steam-engine valve gear of the radial type used on some locomotives; in it, motion is taken entirely from a point on the connecting-rod.

JP- (*Aero*) Nomenclature for jet fuels: JP-1, *Avtur*, original jet fuel, now Jet A-1, NATO F35; JP-4, *Avtag*, wide-range distillate, now Jet-B, NATO F40; JP-5, *Avcat*, dense, high-flash kerosine, NATO F44; JP-7 is for high-speed aircraft, eg SR-71; JP-10 is a special high-density fuel for missiles.

J/ψ particle (*Phys*) A MESON with zero STRANGENESS and zero CHARM, but with a very high mass, 3·2 GeV, and an exceptionally long lifetime, 10^{-20} s. Discovered independently in 1974 at Brookhaven (J-particle) and Stanford (ψ-particle). The existence of this particle necessitated the postulation of the charm and the anticharm QUARKS; it is composed of one charm and one anticharm. Also *J particle* or *ψ particle*.

JPEG (*ICT*) Abbrev for *Joint Photographic Experts Group*, a standard image file format.

JPT (*Aero*) Abbrev for *jet pipe temperature*. See GAS TEMPERATURE.

J segment (*BioSci*) A short sequence of DNA coding for part of the hypervariable region of immunoglobulin light or heavy chains near to the site of joining to the constant region. There are several possible J exons of which any may be used. The gene for the beta chain of the T-cell antigen receptor also includes a different J exon which has similar variability.

JSP (*ICT*) Abbrev for JACKSON STRUCTURED PROGRAMMING.

JT60 (*NucEng*) Large tokamak experiment at the Japan Atomic Energy Research Institute. See TOKAMAK.

JTIDS (*Aero*) Abbrev for JOINT TACTICAL INFORMATION DISTRIBUTION SYSTEM.

jud (*MinExt*) See JAD.

jugal (*BioSci*) A paired membrane bone of the zygoma of the vertebrate skull, lying between the squamosal and the maxilla.

jugular (*BioSci*) Pertaining to the throat or neck region, eg a *jugular vein*.

jugular nerve (*BioSci*) The posterior branch of the hyomandibular component of the seventh cranial nerve in vertebrates; it carries the visceromotor component of the hyomandibular branch.

Julian calendar (*Astron*) The system of reckoning years and months for civil purposes, based on a tropical year of 365·25 days; instituted by Julius Caesar in 45 BC and still the basis of our calendar, although modified and improved by the Gregorian reform.

Julian date (*Astron*) The number of days which have elapsed since 12·00 GMT on 1 January 4713 BC. This consecutive numbering of days gives a calendar independent of month

and year which is used for analysing periodic phenomena, esp in astronomy. This system devised in 1582 by J. Julius Scaliger has no connection whatsoever with the Julian calendar.

Julia set (*MathSci*) Any set defined as the boundary between the values of a parameter for which a given complex function is bounded and those for which it is not, that produces a convoluted FRACTAL curve when plotted on a graph.

jump (*ICT*) Departure from the normal sequential execution of program steps. May be conditional on the result of a test or unconditional. Also *branch*.

jump cut (*ImageTech*) In film editing, the intentional deletion of part of the continuous action within a scene.

jumper (*Build*) See THROUGH-STONE.

jumper (*CivEng*) A pointed steel rod which is repeatedly dropped on the same spot from a suitable height (being turned slightly between blows), and which, by pulverizing the earth, forms a borehole.

jumper (*ElecEng*) (1) A short section of overhead transmission line conductor serving to form an electrical connection between two sections of a line. (2) Multicore flexible cable making connection between the coaches of a multiple-unit train.

jumper (*MinExt*) The borer, steel or bit for a compressed-air rock drill.

jumper (*ICT*) Length of wire used in telephony to rearrange permanent circuit connections.

jumper cable (*ElecEng*) A cable for making electrical connection between two sections of conductor-rail in an electric traction system.

jumper field (*ICT*) The cross-connection or translation field in a director. Any space devoted to jumpers, ie temporary connections, esp within a distribution frame. Also *cross-connection field*.

jumper-top blast pipe (*Eng*) A locomotive BLAST PIPE in which the back pressure, and hence the draught, is automatically limited by the lifting of an annular valve, which increases the nozzle area.

jumper wire (*ICT*) See JUMPER.

jumping-up (*Eng*) The operation of thickening the end of a metal rod by heating and hammering it in an endwise direction. Also *upsetting*.

jump joint (*Eng*) A butt joint made by JUMPING-UP the ends of the two pieces before welding them together.

junction (*Electronics*) The area of contact between semiconductor material having different electrical properties.

junction (*ICT*) Union or division of waveguides in either H- or E-planes, tee or wye, tapered to broaden the frequency response, or *hybrid* to direct flow of wave energy.

junction capacitance (*Electronics*) Capacitance associated with the effective storage of charge (arising from the ionized, fixed dopant ions) in the depletion region of a p–n junction. Cf DIFFUSION CAPACITANCE.

junction chamber (*Build*) A closed chamber, generally of brick and concrete, inserted in a sewer system for accepting the inflow of one or more sewers and allowing for the discharge thereof.

junction circuit (*ICT*) One directly connecting two exchanges situated at a distance apart less than that specified for a trunk circuit.

junction coupling (*ElecEng*) In coaxial line cavity resonators, coupling by direct connection to the coaxial conductor.

junction diode (*Electronics*) One formed by the junction of n- and p-type semiconductors, which exhibits rectifying properties as a result of the potential barrier built up across the junction by the diffusion of electrons from the n-type material to the p-type. Applied voltages, in the sense that they neutralize this potential barrier, produce much larger currents than those that accentuate it.

junction field-effect transistor (*Electronics*) FIELD-EFFECT TRANSISTOR in which the conducting channel is in effect actively controlled by a p–n junction which bounds it. Cf METAL–OXIDE–SILICON FIELD-EFFECT TRANSISTOR.

junction potential (*BioSci*) Potential difference at the boundary between dissimilar solutions; arises from differences in diffusion constants of ions.

junction rectifier (*Electronics*) One formed by a p–n junction by HOLES being carried into the n-type semiconductor.

junction transistor (*Electronics*) See BIPOLAR TRANSISTOR.

juniper (*For*) A conifer, therefore a softwood, *Juniperus virginiana* yielding an essential oil used medicinally; its fruits are used to flavour gin, and its fragrant, aromatic wood is the standard material for pencils. Also *Virginian pencil cedar*, though not a true cedar.

junk DNA (*BioSci*) Genomic DNA that serves, as yet, no known function.

junk ring (*Eng*) A metal ring attached to a steam-engine piston for confining soft packing materials; or for similarly holding a cast-iron piston ring in position.

Juno (*Astron*) The third asteroid to be discovered (1804), with a diameter of around 247 km.

Jupiter (*Astron*) The fifth planet from the Sun, orbiting at a mean distance of 7.783×10^8 km and a sidereal period of 11.86 years. Its mass, 1.90×10^{27} kg is 318 times that of the Earth. The relative density is 1.33. Composition is 82% hydrogen, 17% helium and 1% all heavier elements. Rotation period is 9 h 50 min at the equator, but five times longer at the poles. This differential rotation of Jupiter's cloudy atmosphere leads to a richly coloured banded structure, the most prominent feature of which is the *Great Red Spot*, a long-lived storm in the atmosphere. Owing to rapid rotation Jupiter is oblate, the equatorial diameter being 6.4% larger than the polar diameter. There are 16 natural satellites, the four largest of which (CALLISTO, EUROPA, GANYMEDE and IO) are visible with the simplest telescope. Beneath the atmosphere, the hydrogen is compressed so enormously that it becomes a metallic conductor, and this is the source of Jupiter's very strong magnetic field. Within the associated MAGNETO-SPHERE intense bursts of radio emission are detected. See appendix on Planets.

Jurassic (*Geol*) The middle period of the Mesozoic era covering an approx time span from 215 to 145 million years ago. Named after the type area, the Jura Mountains. The corresponding system of rocks. See MESOZOIC.

jury strut (*Aero*) A strut giving temporary support to a structure. Usually required for folding-wing biplanes, sometimes for naval monoplanes with folding parts.

justification (*Print*) The correct spacing of words to a given measure of line.

just in time (*Genrl*) Manufacturing philosophy with the aim of reducing stock levels of parts needed for final product by minimizing delivery time to factory or production line. It may also involve tighter quality control and replacing external by internal suppliers (eg in-house injection moulding replacing trade moulding) as well as parts rationalization. Abbrev *JIT*.

just noticeable difference (*Psych*) See DIFFERENCE THRESHOLD. Abbrev *JND*.

just scale (*Acous*) See NATURAL SCALE. Also *just temperament*.

jute (*Textiles*) Strong, brownish, bast fibre from the Asian plants *Corchorus olitorius* and *C. capsularus*. Used in cordage, canvas, hessian and carpet backings.

juvenile (*BioSci*) A structure characteristic of the JUVENILE PHASE.

juvenile hormone (*BioSci*) See NEOTENIN.

juvenile phase (*BioSci*) The phase before flowering in woody plants; it differs in many attributes including the ease of rooting from cuttings; usually transient but can be maintained as in some cultivated ornamental plants including many conifers.

juvenile water (*Geol*) Water derived from magma, as opposed to meteoric water (derived from rain or snow), or connate water (trapped in sediments at the time of deposition).

juxtaglomerular (*Med*) Close to the renal glomerulus, eg *juxtaglomerular apparatus*, a small group of cells located on the afferent arteriole of the glomerulus of the kidney, secreting RENIN.

juxtaposition twins (*Min*) Two (or more) crystals united regularly in accordance with a 'twin law', on a plane (the 'composition plane') which is a possible crystal face of the mineral. Cf INTERPENETRATION TWINS. See panel on TWINNED CRYSTALS.

Jy (*Phys*) Symbol for JANSKY.

K

K (*Chem*) Symbol for POTASSIUM.

K (*Phys*) Symbol for: (1) KELVIN; (2) THERMAL DIFFUSIVITY.

K (*Chem*) Symbol for: (1) EQUILIBRIUM CONSTANT; (2) K_s, SOLUBILITY PRODUCT.

K (*ChemEng*) See κ.

K (*Eng*) Symbol for: (1) STRESS INTENSITY FACTOR; (2) BULK MODULUS.

[K] (*Phys*) A very strong FRAUNHOFER LINE in the extreme violet of the solar spectrum. See [H].

K_c (*Eng*) Critical STRESS INTENSITY FACTOR at which a sharp crack will propagate catastrophically in a strained material. See FRACTURE TOUGHNESS.

K_{cat} (*BioSci*) Catalytic constant of an enzyme, also referred to as the *turnover number*. Represents the number of reactions catalysed per unit time by each active site.

K_{eq} (*BioSci*) The equilibrium constant for a reversible reaction. $K_{eq} = [AB]/[A][B]$ where the square brackets [ ] signify concentration.

K_m (*BioSci*) The *Michaelis constant*. A kinetic parameter used to characterize an enzyme; defined as the concentration of substrate that permits half-maximal rate of reaction. An analogous constant K_a is used to describe binding reactions, in which case it is the concentration at which half the receptors are occupied.

k (*Genrl*) Symbol for KILO-.

k (*Chem*) Symbol for the VELOCITY CONSTANT of a chemical reaction.

k (*ChemEng*) Symbol for MASS TRANSFER COEFFICIENT.

k (*Phys*) Symbol for: (1) BOLTZMANN'S CONSTANT; (2) RADIUS OF GYRATION; (3) THERMAL CONDUCTIVITY.

κ- (*Chem*) Symbol for: (1) *cata-*, ie containing a condensed double aromatic nucleus substituted in the 1,7 positions; (2) substitution on the tenth carbon atom; (3) ELECTROLYTIC CONDUCTIVITY.

κ- (*Phys*) Symbol for: (1) COMPRESSIBILITY; (2) MAGNETIC SUSCEPTIBILITY.

K-acid (*Chem*) *1,8-aminonaphthol-4,6-disulphonic acid*. An intermediate for dyestuffs.

kaersutite (*Min*) A hydrated silicate of calcium, sodium, magnesium, iron, titanium and aluminium; a member of the amphibole group crystallizing in the monoclinic system. It occurs in somewhat alkaline igneous rocks.

kainite (*Min*) Hydrated sulphate of magnesium, with potassium chloride, which crystallizes in the monoclinic system. It usually occurs in salt deposits.

Kainozoic (*Geol*) See CENOZOIC.

kairomone (*BioSci*) A chemical signal that is produced by one species of animal and that influences the behaviour of members of another species, to the benefit of the recipient. Cf ALLOMONE.

kaiwekite (*Geol*) A volcanic rock containing phenocrysts of olivine, titanaugite, barkevikite and anorthoclase, probably a hybrid between basalt and trachyte.

kala-azar (*Med*) A disease due to infection with the protozoan *Leishmania donovani*; characterized by enlargement of the liver and spleen, anaemia, wasting and fever. Also *leishmaniasis*.

Kalanite (*ICT*) TN for a hard insulating material, not affected by oil, used for spreaders in cable joints.

kaliophilite (*Min*) A silicate of potassium and aluminium, which crystallizes in the hexagonal system. It has a similar composition to kalsilite and is probably metastable at all temperatures at atmospheric pressure.

kallitron (*ICT*) A periodic combination of two triode valves for obtaining negative resistance.

kalsilite (*Min*) A potassium–aluminium silicate, $KAlSiO_4$; it is related to nepheline and crystallizes in the hexagonal system.

kamacite (*Min*) A variety of nickeliferous iron, found in meteorites; it usually contains about 5·5% nickel. Metallurgically, alpha-iron.

kame (*Geol*) A mound of gravel and sand which was formed by the deposition of the sediment from a stream as it ran from beneath a glacier. Kames are thus often found on the outwash plain of glaciers.

kanamycin (*Pharmacol*) An *aminoglycoside* antibiotic obtained from a bacterium (*Streptomyces kanamyceticus*) found in soil and used in the treatment of Gram-negative bacterial infections.

kandite (*Min*) A collective term for the KAOLIN minerals or members of the kaolinite group. These include kaolinite, dickite, nacrite, anauxite, halloysite, meta-halloysite and allophane.

kanthals (*Eng*) Alloys with high electrical resistivity, used as heating elements in furnaces. General composition iron 67%, chromium 25%, aluminium 5%, cobalt 3%.

kaolin (*Geol*) The main constituents are *kaolinite*, a hydrated aluminium silicate, and *illite*, a mica-like alumino-silicate together with potassium, magnesium and iron. Basis of PORCELAIN and BONE CHINA and used as a filler for paper, rubber and toothpaste. Also *china clay*.

kaolinite (*Min*) A finely crystalline form of hydrated aluminium silicate, $(OH)_4Al_2Si_2O_5$, occurring as minute monoclinic flaky crystals with a perfect basal cleavage; resulting chiefly from the alteration of feldspars under conditions of hydrothermal or pneumatolytic metamorphism, or by weathering. The kaolinite group of clay minerals includes the polymorphs DICKITE and NACRITE. See KANDITE.

kaolinization (*Geol*) The process by which the feldspars in a rock such as granite are converted to kaolinite.

kaon (*Phys*) See MESON.

Kaplan water turbine (*Eng*) A propeller-type water turbine in which the pitch of the blades can be varied in accordance with the load, resulting in high efficiency over a large load range.

kapok (*Textiles*) The seed hairs of the kapok tree, *Ceiba pentranda*. They are light and fluffy and in loose form are used as an insulating or flotation material eg in life jackets. They are not spun or converted into fabrics commercially.

Kaposi's sarcoma (*Med*) A malignant vascular skin tumour that appears in two forms. The first is a slowly progressive lesion on toes or legs of elderly men of Mediterranean origin. The second is a much more invasive disseminated form and is common in children and young men in C Africa. It is now seen as a common feature in AIDS.

kappa chain (*BioSci*) One of the two types of LIGHT CHAIN of immunoglobulins, the other being the lambda chain. An individual immunoglobulin molecule bears either two kappa or two lambda chains. The proportion of molecules bearing each chain varies in different species. In humans about 60% are of the kappa and 40% of the lambda type.

kappa number (*Paper*) A bleachability test for pulps measured by the number of ml of 0·1N potassium permanganate solution consumed by 1 g of dry pulp.

kappa particle (*BioSci*) An endosymbiotic Gram-negative bacterium, found in the cytoplasm of some strains of the ciliate protozoan *Paramecium*, that produces a 'killer' factor toxic to strains that lack the endosymbiont. The factor is transferred during CONJUGATION.

Kapp phase advancer (*ElecEng*) A form of phase advancer for use with slip-ring induction motors. It consists of a small armature connected to each phase of the rotor circuit and allowed to oscillate freely in a dc field so that it has a leading emf induced in it. Also *Kapp vibrator*.

Kapton (*Plastics*) TN for polyimide film (US).

kapur (*For*) SE Asian hardwood (*Dryobalanops*), which also yields BORNEOL. A uniform light- to deep-red brown, straight-grained with a coarse but even texture, and a camphor-like odour.

karat (*Genrl*) See CARAT.

Karbate (*ChemEng*) TN for a form of extremely dense carbon having high enough strength to permit its being made into special shapes, eg tubes, pumps, and possessing corrosion resistance and good heat conductivity.

Karman vortex street (*Acous*) Regular vortex pattern behind an obstacle in a flow where vortices are generated and travel away from the object. The frequency of vortex generation is determined by the STROUHAL NUMBER.

karri (*For*) An Australian hardwood (*Eucalyptus*) with a dense deep-red wood similar to, but not so durable as, JARRAH, and also used as a structural timber.

karst (*Geol*) Any uneven limestone topography, characterized by joints enlarged into criss-cross fissures (*grikes*) and pitted with depressions resulting from the collapse of roofs of underground caverns. It is formed by the action of percolating waters and underground streams.

kary-, karyo- (*Genrl*) Prefixes from Gk *karyon*, nucleus. Also *cary-, caryo-*.

karyogamy (*BioSci*) The fusion of the two gametes to form a zygote. It usually follows cytoplasmic fusion but may as in some fungi be followed by a prolonged binucleate stage, *dikaryophase*. See PLASMOGAMY.

karyogram (*BioSci*) See IDIOGRAM.

karyon (*BioSci*) The cell nucleus, only used in compound words. From Gk for nut or kernel.

karyoplast (*BioSci*) A nucleus that has been removed from a eukaryotic cell and is surrounded by a thin layer of cytoplasm and a plasma membrane.

karyotype (*BioSci*) The appearance, number and arrangement of the chromosomes in the cells of an individual.

Kasimovian (*Geol*) An epoch of the Pennsylvanian period.

Kaspar–Hauser experiments (*Psych*) Experiments in which animals are reared in complete isolation from other animals of their own or other species.

kata- (*Genrl*) Prefix from Gk *kata*, down. Also *cata-*.

katabatic wind (*EnvSci*) A local wind flowing down a slope cooled by loss of heat through radiation at night. It is caused by the difference in density between cold air in contact with the ground and the warmer air at corresponding levels in the *free atmosphere*. It is often quite strong, esp when channelled down a narrow valley, and can reach gale force at the edges of the Antarctic and Greenland ice caps. Cf *anabatic wind*.

katadromous (*BioSci*) Of fish, migrating to water of greater density than that of the normal habitat to spawn, as the freshwater eel which migrates from fresh to salt water to spawn. Also *catadromous*. Cf ANADROMOUS.

kata-front (*EnvSci*) A situation at a front, warm or cold, where the warm air is sinking relative to the FRONTAL ZONE.

kataklasis (*Geol*) See CATACLASIS.

kataphorite (*Min*) See KATOPHORITE.

kataplexy (*BioSci*) The state of imitation of death, adopted by some animals when alarmed.

katathermometer (*MinExt*) An instrument used in mine ventilation survey to assess the cooling effect of air current.

The thermometer bulb is first exposed dry, then when covered with wetted gauze, and time taken for temperature to fall from 100 to 95°F (38 to 35°C) is observed.

Katayama disease (*Med*) A disease due to invasion of the body by the blood fluke *Schistosoma japonicum*, characterized by urticaria, painful enlargement of liver and spleen, bronchitis, diarrhoea, loss of appetite and fever lasting a few days to several weeks.

katharometer (*Chem*) An instrument for the analysis of gases by means of measurements of thermal conductivity.

katophorite (*Min*) Hydrated silicate of sodium, calcium, magnesium, iron and aluminium; a member of the amphibole group. It crystallizes in the monoclinic system and occurs in basic alkaline igneous rocks. Also *cataphorite, kataphorite*.

kauri-butanol value (*Chem*) A term applied to solvents to indicate their dissolving powers for resins. Based on natural KAURI GUM as the standards.

kauri gum (*Chem*) A gum found in New Zealand, used for varnishes and linoleum cements. It is the resinous exudation of the kauri pine (*Agathis australis*), a tree whose timber is of value for general joinery and decorative purposes.

kautschuk (*Chem*) Ger for natural rubber. See CAOUTCHOUC.

Kawasaki's disease (*Med*) A disease occurring in children under 5 years which causes a skin rash and glandular enlargement. In some it produces severe dilatation of the coronary arteries with thrombus formation leading to occlusion. Also *mucocutaneous lymph node syndrome*.

Kaye disk centrifuge (*PowderTech*) A transparent disk centrifuge with an entry port coaxial with the axis of revolution. The suspension of the powder under test is injected into the centrifuge while it is running at the final speed. Concentration changes are measured optically.

Kaye effect (*Chem*) Phenomenon shown by concentrated polymer solutions when poured onto a liquid surface, where the impinging jet rebounds to form a second, rising jet. See ELASTIC LIQUIDS.

Kayser–Fleischer ring (*Med*) A ring of brownish-yellow pigmentation found in the cornea of patients suffering from WILSON'S DISEASE.

Kazanian (*Geol*) A stratigraphical stage in the Permian rocks of Russia and E Europe.

kb, kbp (*BioSci*) See KILOBASE.

K-band (*ICT, Radar*) US designation of microwave band between 12 and 40 GHz, widely adopted in the absence of internationally agreed standards. Subdivided into Ku-band (K-under), 12–19 GHz, K-band, 18–26 GHz and Ka (*K-above*), 26–40 GHz; replacing approximately the UK designations of J-, K- and Q-bands.

K-capture (*Phys*) Absorption of an electron from the innermost (K) shell of an atom into its nucleus. An alternative to ejection of a positron from the nucleus of a radioisotope. Also *K-electron capture*. See X-RAY.

K-cell (*BioSci*) A cell, generally resembling a lymphocyte, that bears Fc receptors by which it can bind to other cells to which are attached antibodies against antigens on the cell surface (thus exposing the Fc portions). K-cells induce lesions in the membrane of the target cell which kill it. This is an example of antibody-dependent cell-mediated cytotoxicity.

kCi (*Phys*) Abbrev for *kilocurie*, an obsolete unit of radioactivity equivalent to 1000 CURIES.

kDa (*Chem*) Abbrev for KILODALTON. Often, though erroneously, kD.

K-display (*Radar*) A form of *A-display* produced with a lobe-switching antenna. Each lobe produces its own peak and the antenna is directly on target when both parts of the resulting double peak have the same height. See R-DISPLAY.

keatite (*Min*) A high-pressure synthetic form of silica.

kebbing (*Vet*) See ENZOOTIC OVINE ABORTION.

kedge anchor (*Ships*) A small anchor used for steadying and warping purposes.

keel block (*Eng*) A standard casting shaped like a ship's keel which is used to provide a test specimen for steel or other alloys subject to high shrinkage.

keelson (*Ships*) (1) A term describing the longitudinal strengthening members of a ship or flying boat, which form the shell-plating stiffeners. A *flat keel* is the lower horizontal member of the ship's backbone; a *centre keelson* is the vertical member thereof; a *bar keel* is similar to the latter, external to the hull; *side keelsons* are vertical members, off the ship's centre line. (2) The wrought-iron saddles or standards which support cylindrical boilers of the Scotch marine type. Sometimes *boiler cradles*.

Keene's cement (*Build*) A quick-setting hard plaster, made by soaking plaster of Paris in a solution of alum or borax and cream of tartar.

keep (*Eng*) See KEEPER.

keep down (*Print*) A typographic instruction to avoid the use of capital initials so far as possible.

keeper (*Build*) (1) The part of a Norfolk latch limiting the travel of the fall bar. (2) The socket fitted on a door jamb to house the bolt of the lock in the shut position.

keeper (*Eng*) The lower part of the bearing in a railway-truck axle-box, which limits the downward movement of the box due to track irregularities.

keeper (*Phys*) Soft-iron bar (or similar) used to close the magnetic circuit of a permanent magnet when not in use, thereby conserving its strength. Also *armature*.

keep standing (*Print*) An instruction to keep the printing surface intact, particularly to keep the type locked up in chase, usually in anticipation of a reprint.

keep up (*Print*) (1) A typographic instruction to use capital initials in preference to small letters. (2) See KEEP STANDING.

Keewatin Group (*Geol*) A series of basic pillow lavas associated with sedimentary iron ores (worked in the 'Iron Ranges'); forms part of the Precambrian succession in the Canadian Shield.

kefir (*Chem*) A fermentation product of milk in which the lactose has undergone both alcoholic and lactic fermentation simultaneously.

keilhauite (*Min*) A variety of sphene containing more than 10% of the rare earths.

K-electron capture (*Phys*) See K-CAPTURE.

Kel-F (*Plastics*) TN for POLYTRIFLUOROCHLOROETHYLENE (US).

Kelling's test (*Chem*) A test for the detection of lactic acid in gastric juice, based upon the colouring effect produced by the addition of a few drops of a very dilute neutral iron (III) chloride solution.

kelly (*MinExt*) The topmost JOINT of a drill string, attached below to the next drill joint and above to the swivel and mud-hose connection. It is a heavy tubular part of square or hexagonal cross-section externally which fits into the corresponding hole of the KELLY BUSHING, itself fixed to the ROTARY TABLE. To add a fresh pipe to the drill string, it has to be raised by the DRAW WORKS until the kelly and bushing are clear of the rotary table. The string is then locked by wedge-shaped *slips* to the table and the kelly joint unscrewed and parked in the *rat hole*. After a new pipe has been attached the kelly is rescrewed and the whole string lowered so that drilling can be restarted. See panel on DRILLING RIG.

kelly bushing (*MinExt*) The replaceable bearing with a square hole in which the KELLY slides and which is attached to the ROTARY TABLE during drilling.

keloid (*Med*) A dense new growth of skin occurring in skin that has been injured.

kelp (*BioSci*) (1) A general name for large seaweeds, esp *Laminaria* and allies. (2) The ashes from burning such seaweed, formerly a source of soda and iodine.

kelvin (*Phys*) Symbol K. (1) The SI unit of temperature. It is defined by fixing the triple point of water (ie the temperature and pressure at which pure ice, water and water vapour can coexist at equilibrium) as exactly 273·16 K above ABSOLUTE ZERO. Note that K does not take the degree ° sign. (2) The SI unit of temperature interval: 1 K is 1/273·16 of the interval between absolute zero and the triple point of water; the interval equals the interval 1°C. See KELVIN THERMODYNAMIC SCALE OF TEMPERATURE.

Kelvin ampere-balance (*ElecEng*) Laboratory instrument for measuring current; in it the calculated force between two coils carrying the current to be measured is balanced by the force of gravity on a weight sliding along a beam.

Kelvin bridge (*ElecEng*) An electrical network involving two sets of ratio arms used for accurate measurement of low resistances.

Kelvin compass (*Ships*) A form of ship's compass having a very light card and a number of short parallel needles held by silk cords, as well as other special features. Also *Thomson's compass*.

Kelvin effect (*ElecEng*) Also SKIN EFFECT.

Kelvin electrometer (*ElecEng*) Also QUADRANT ELECTRO-METER.

Kelvin's law (*ElecEng*) A principle regarding the transmission of electrical energy. It states that the most economical size of conductor to use for a line is that for which the annual cost of the losses is equal to the annual interest and depreciation on that part of the capital cost of the conductor which is proportional to its cross-sectional area.

Kelvin thermodynamic scale of temperature (*Phys*) A scale of temperature based on the thermodynamic principle of the performance of a reversible heat engine. The scale cannot have negative values so ABSOLUTE ZERO is a well-defined thermodynamic temperature. The temperature of the TRIPLE POINT of water is assigned the value 273·16 K. The temperature interval corresponds to that of the Celsius scale so that the freezing point of water (0°C) is 273·15 K. Unit is the KELVIN. Symbol K.

Kelvin–Varley slide (*ElecEng*) Constant-resistance decade voltage divider of the type used in vernier potentiometers. It consists of a resistor of 2r ohms shunting two adjacent units in a series array of eleven resistors, each r ohms. The 2r resistor may similarly be divided into eleven equal parts, and shunted if additional decade(s) are required.

kelyphitic rim (*Geol*) A shell of one mineral enclosing another in an igneous rock, produced by the reaction of the enclosed mineral with the other constituents of the rock. Kelyphitic rims are most common in basic and ultrabasic rocks.

Kematal (*Eng*) TN for acetal resin (UK).

kenacid blue (*BioSci*) See COOMASSIE BLUE.

Kennelly–Heaviside layer (*Phys*) See E-LAYER.

kentallenite (*Geol*) A coarse-grained, basic igneous rock, named for the type locality, Kentallen, Argyllshire, Scotland; it consists essentially of olivine, augite and biotite, with subordinate quantities of plagioclase and orthoclase in approximately equal amounts.

Kent claw hammer (*Build*) One with a thick, slightly curved claw and hexagonal head.

kentledge (*CivEng*) Scrap iron, rails, heavy stones, etc, used as loading on a structure (eg upon the top section in sinking a cylinder caisson), or as a counterbalance for a crane.

kenyte (*Geol*) A fine-grained igneous rock, occurring as lava flows on Mt Kenya, E Africa, and in the Antarctic; essentially an olivine-bearing phonolite with phenocrysts of anorthoclase.

Kepler's laws (*Astron*) Three laws describing planetary motion. (1) The planets describe ellipses with the Sun at a focus. (2) The line from the Sun to any planet sweeps across equal areas in equal times. (3) The squares of the periodic times of the planets are proportional to the cubes of their mean distances from the Sun.

kerat-, kerato- (*Genrl*) Prefixes from Gk *keras, -atos*, horn. Also *cerat-, cerato-*.

keratan sulphate (*BioSci*) See GLYCOSAMINOGLYCAN.

keratectasia (*Med*) Large bulging of part of the cornea.

keratectomy (*Med*) Excision of part of the cornea.

keratin (*BioSci*) A structural protein of hair and wool; the class of INTERMEDIATE FILAMENT that is characteristic of epithelial cells (*cytokeratin*). α-keratin is a major component of skin; β-keratin exists as a β-sheet and is a major component of silk.

keratinizing epithelium (*BioSci*) An epithelium, such as vertebrate epidermis, in which a keratin-rich layer that is mechanically protective is formed from residual intracellular cytokeratins as the outermost keratinocytes die.

keratitis (*Med*) Inflammation of the cornea.

kerato- (*Genrl*) See KERAT-.

keratocele (*Med*) Protrusion of the innermost layer (Descemet's membrane) of the cornea through a corneal ulcer.

keratoconus (*Med*) Conical cornea. Cone-shaped deformity of the cornea due to a weakness and thinness of the centre.

keratodermia blennorrhagica (*Med*) Red patches on the skin which become hard, dry, yellow and raised above the skin, occurring in gonorrhoea.

keratogenous (*BioSci*) See KERATIN.

keratoma (*Med*) A tumour of the skin in which overgrowth of the horny layer predominates.

keratomalacia (*Med*) A disease in which the cornea first becomes dry and lustreless and then softens; associated with deficiency of vitamin A in the diet.

keratomileusis (*Med*) The surgical reshaping of the cornea to correct defective vision.

keratophyre (*Geol*) A fine-grained igneous rock of intermediate composition. It is essentially a soda-trachyte, containing albite-oligoclase or anorthoclase in a cryptocrystalline groundmass. The pyroxenes, when present, are often altered to chlorite or epidote.

keratoplasty (*Med*) The grafting of a new cornea onto an eye whose cornea has become opaque.

keratoscleritis (*Med*) Inflammation of the cornea and of the sclera of the eye.

keratosis (*Med*) Overgrowth of the horny layer of the skin.

keratotomy (*Med*) Incision of the cornea.

kerf (*Build*) The cut made by a saw.

kerf (*Eng*) The part of the original material which was removed by cutting, pressing, etc.

kerf (*MinExt*) In coal winning, undercut made by coal-cutting machine to depth of 1 m or more. Also *kirve*.

kerma (*Radiol*) The initial energy of all the charged ionizing particles released by an uncharged ionizing particle in a given mass of tissue. Measured in grays (J kg^{-1}). See panel on RADIATION.

kermesite (*Min*) Oxysulphide of antimony, which crystallizes in the monoclinic system. It is a secondary mineral occurring as the alteration product of stibnite. Also *pyrostibnite*.

kern (*Print*) The portion of some type letters which projects beyond the body and rests on the body of the preceding or following letter, eg the tail of an italic *f*.

kernel (*ICT*) The minimum set of instructions in an operating system needed to run the microprocessor, sometimes as little as 15 calls in a microkernel. All other processes, eg file and device management, run in independent sections of the code.

kernel (*MathSci*) The kernel of a mapping is the set of elements mapped to the identity (0 or 1 according to context).

kernicterus (*Med*) Damage to the brain in infants caused by the passage of unconjugated bilirubin across the blood–brain barrier. Usually a complication of HAEMOLYTIC DISEASE OF THE NEWBORN.

Kernig's sign (*Med*) A sign of meningitis. When the patient is lying on his or her back with the thigh bent at right angles to the body, the leg cannot be straightened at the knee.

kerning (*ICT*) The process of adjusting the spacing between two characters such that the letter spacing will be visually attractive and easy to read.

kernite (*Min*) Hydrated sodium borate. $Na_2B_4O_7.4H_2O$.

kerogen (*Geol*) Fossilized, insoluble, organic material present in a sediment, that yields petroleum products on distillation.

Kerr cell (*ElecEng*) A light modulator consisting of a liquid cell between crossed polaroids. The light transmission is modulated by an applied electric field. See KERR EFFECT.

Kerr coefficient (*ICT*) Coefficient expressing the extent of non-linearity in an OPTICAL FIBRE caused by the dependence of its refractive index n on the intensity of the optical signal I: $n = n_1 + n_2I$, where n_2 is the Kerr coefficient, with a typical value of 2.7×10^{-20} m^2 W^{-1}.

Kerr effect (*Phys*) (1) Double refraction produced in certain transparent dielectrics by the application of an electric field. Also *electro-optical effect*. (2) Dispersion of the plane of polarization experienced by a beam of plane-polarized light on its passage through a transparent medium subjected to an electrostatic strain. Also *electrostatic Kerr effect*. (3) Modification of the state of polarization of light on reflection from the polished surface of a magnetized material. Also *magneto-optical effect*. See FARADAY EFFECT.

kersantite (*Geol*) A mica-lamprophyre, named from the type locality Kersanton, near Brest; it consists essentially of biotite and plagioclase feldspar. See MINETTE.

kersey (*Textiles*) A heavy woollen cloth, milled and raised giving a lustrous nap, similar to melton and used in eg overcoats.

keruing (*For*) A moderately durable, dark reddish-brown wood from a number of related species of SE Asian evergreen, hardwood trees (*Dipterocarpus*). Also *eng*, *in*.

ketenes (*Chem*) Compounds of the general formula $R_2C=C=O$. The ketene series may be considered homologues of carbon monoxide. The first member of the series is ketene, $CH_2=CO$, which is readily obtainable by passing propanone vapours through a red-hot glass tube filled with broken tile. Propanone is then decomposed into ketene and methane. Ketenes form acids or acid derivatives on adding water, alcohols, ammonia, amines, etc, to one of their double bonds. They are liable to auto-oxidation; they react with other unsaturated compounds, forming four-membered rings. They are very unstable and polymerize easily.

ketoacidosis (*Med*) The excessive formation in the body (due to incomplete oxidation of fats) of ketone or acetone bodies (aceto-acetic acid and β-hydroxybutyric acid) which are accompanied by ketonaemia and ketonuria; occurs eg in diabetes. Also *ketosis*.

ketoconazole (*Pharmacol*) A powerful imidazole antifungal drug used in treating resistant candidiasis, gastrointestinal infections and infections of skin and nails.

keto–enolic tautomerism (*Chem*) The formation by certain compounds of two series of derivatives, based upon their ketonic or enolic constitution. The enol-form is produced from the keto-form by the migration of a hydrogen atom, which forms a hydroxyl group with the ketone oxygen, accompanied by a change in the position of the double bond. Thus

$$-CH_2-CO- \rightleftharpoons -CH=C(OH)-$$

keto-form   enol-form

keto form (*Chem*) That form of a substance exhibiting keto–enolic tautomerism which has the properties of a ketone, eg ethyl aceto-acetate.

ketogenic (*Med*) Capable of producing ketone bodies, eg *ketogenic diet*.

ketohexoses (*Chem*) $HOCH_2(CHOH)_3COCH_2OH$, a group of carbohydrates, isomers of *aldohexoses*. There are three asymmetric carbon atoms and thus four pairs of stereoisomers: the D- and L-forms of fructose, sorbose, tagatose and allulose. They reduce an alkaline copper solution. Ketohexoses can be oxidized to acids containing fewer carbon atoms in the molecule.

ketonaemia (*Med*) the presence of ketone bodies in the blood. See KETOACIDOSIS. Also *ketonemia*.

ketone body (*Med*) Any of three compounds that are produced when fatty acids are broken down in the liver. Not a body in any conventional sense.

ketones (*Chem*) Compounds containing a carbonyl group, –CO–, in the molecule attached to two hydrocarbon radicals. The simplest ketone is acetone (propanone), CH_3COCH_3, but more generally either methyl group can be replaced by alkyl or other organic groups.

acetone

ketone Acetone, the simplest ketone, is shown.

ketonuria (*Med*) The presence of KETONE BODIES in the urine. Also *acetonuria*. See KETOACIDOSIS.

ketoses (*Chem*) A general term for monosaccharides with a ketonic constitution. They always form mixtures of acids on oxidation, containing a smaller number of carbon atoms than the original ketose.

ketosis (*Med, Vet*) See KETOACIDOSIS and PREGNANCY TOXAEMIA.

ketoximes (*Chem*) The reaction products of ketones with hydroxylamine, containing the oximino group =NOH attached to the carbon atom.

kettle (*Eng*) An open-top vessel used in carrying out metallurgical operations on low-melting-point metals, eg in drossing and desilverizing lead.

kettle (*Geol*) A steep-sided basin in glacial drift, often a lake or swamp, and derived from the melting of a block of stagnant ice.

kettlestitch (*Print*) The stitch which is made at the head and tail of each section on a book to interlock the sections.

Keuper (*Geol*) The upper series of rocks assigned to the Triassic system in NW Europe, lying above the Muschelkalk.

keV (*Phys*) Abbrev for kilo-electron-volt; unit of particle energy equal to 10^3 ELECTRON-VOLTS.

kevel (*Build*) A hammer, edged at one end and pointed at the other, used for breaking and rough-hewing stone.

Kevlar (*Eng*) TN for ARAMID FIBRE (US).

Kew Certificates (*Genrl*) Certificates of performance of watches and chronometers issued by the National Physical Laboratory, Teddington. (This work was originally undertaken at Kew, hence the name.) For an 'absolutely perfect' watch, 100 marks are awarded, made up as follows: 40 for a complete absence of variation of daily rate; 40 for absolute freedom from change of rate with change of position; and 20 for perfect compensation for effects of temperature.

Keweenawan (*Geol*) Conglomerates, arkoses, red sandstones and shales of desert origin, associated with great thicknesses of basic lavas and intrusives. This is the youngest of the Precambrian divisions in the Canadian Shield.

Kew-pattern barometer (*EnvSci*) The Adie barometer. Specially graduated so that error arising from changes in the free level in the cistern is obviated. Also *compensated scale barometer*.

Kew-pattern magnetometer (*ElecEng*) A delicate type of reflecting magnetometer which records changes in the Earth's field photographically.

key (*BioSci*) A set of instructions devised to enable an unknown organism to be identified on the basis of critical characteristics.

key (*Build*) In any surface to be plastered, the roughening, lathing, or other preliminary process undertaken in order to give a grip to the coat of plaster and so enable it to adhere more satisfactorily.

key (*Eng*) A piece inserted between a shaft and a hub to prevent relative rotation. It fits in a key-way, parallel with the shaft axis, in one or both members, the commonest form being the parallel key, of rectangular section.

key (*For*) In tree felling, the portion of a tree bole deliberately left uncut until the last moment to lessen the risk of splitting. Also *hinge*.

key (*ICT*) Part of the data description used to determine their location in memory. See ASSOCIATIVE MEMORY, HASHING.

key (*ImageTech*) By analogy with music, the character of the tonal range of subject matter and brightness in a picture as a subjective whole. See HIGH KEY, LOW KEY.

key (*Print*) A small tool used for securing the bands while a book is being sewn.

keyboard (*ICT*) INPUT DEVICE with an array of keys to be pressed. See ASCII keyboard, DVORAK KEYBOARD, NUMERIC KEYBOARD, QWERTY KEYBOARD, TELETYPEWRITER.

keyboard buffer (*ICT*) A temporary storage area in memory that stores each key pressed even though the computer may not immediately respond.

key boss (*Eng*) A local thickening up of a boss or hub at the point at which a key-way is cut, to compensate for loss of strength due to the cut.

key card (*ICT*) A small usually plastic card that is punched, has a magnetically coded strip or incorporates a microchip, and that electronically operates a mechanism eg a lock or a cash dispenser. Also *swipecard*.

key chuck (*Eng*) A jaw chuck whose jaws are adjusted by screws turned by a key or spanner. See SELF-CENTRING CHUCK.

key click (*ICT*) The click produced by RINGING in a key circuit as contact is opened or closed. Normally minimized by use of key-click filter. Also *key chirp*.

key course (*CivEng*) The course of stones in an arch corresponding to the keystone.

key drawing (*Print*) In lithography and line-colour blockmaking, an outline drawing which serves as a guide in the making of the separate colour plates.

key drop (*Build*) A guard plate covering a keyhole and falling into position by its own weight.

keyed pointing (*Build*) Pointing which is finished with lines or grooves struck on the flat joint. See FLAT POINTING.

keyer (*ICT*) A device for changing the output of a transmitter from one frequency (or amplitude) to another according to the intelligence transmitted.

key fossil (*Geol*) See INDEX FOSSIL.

Key–Gaskel syndrome (*Vet*) Feline dysautonomia of unknown aetiology. Symptoms include dilated pupils, dehydration, constipation, regurgitation and sometimes loss of anal sphincter tone.

keyhole limpet haemocyanin (*BioSci*) A large coppercontaining protein from the keyhole limpet. Haemocyanins normally function as oxygen-carrying molecules, but are widely used as immunogens in immunology since they are likely to be completely foreign to mammals. Also used as a marker in electron microscopy, being large and of distinctive (cubic) form. Abbrev *KLH*.

keyhole saw (*Build*) One with a stiff, narrow blade 6–10 in (150–250 mm) long, for internal, curved and small cuts. Also *padsaw*.

keying (*Eng*) The process of fitting a key to the key-ways in a shaft and boss.

keying (*ImageTech*) A video switching effect which creates a space within a picture into which another image is inserted.

keying wave (*ICT*) See MARKING WAVE.

key light (*ImageTech*) The principal lighting of the main subject in a scene.

keymap (*ICT*) A MIDI PATCH-MAP entry that translates key values for certain MIDI messages.

key plan (*Build, Eng*) A small-scale plan showing the relative disposition of a number of items in a scheme.

key plate (*Build*) An ESCUTCHEON.

key print (*ImageTech*) See GREY KEY IMAGE.

key seating (*Eng*) A key-way, or the surface onto which a key is bedded.

key-seating machine (*Eng*) A machine tool for milling key-ways in shafts etc by means of an end mill, the work being supported on a table at right angles to the axis of the spindle. Feed is obtained by an automatic traverse of either the tool or the table.

keystone (*CivEng*) The central voussoir at the crown of an arch.

keystone distortion (*ImageTech*) The distortion of an optical or electronic image in which a rectangle is reproduced as a trapezium with the vertical sides converging. Generally a result of the beam axis not meeting the screen at right angles.

key-to-disk (*ICT*) Input device for accepting data from a keyboard and writing directly to magnetic disk.

key-to-tape unit (*ICT*) Input device for accepting data from a keyboard and writing it directly onto tape.

key-way (*Eng*) A longitudinal slot cut in a shaft or hub to receive a KEY.

key-way tool (*Eng*) A slotting machine tool used for the vertical cutting of key-ways, equal in width to that of the key-way. See SLOTTING MACHINE.

keywords (*ICT*) The most informative or significant words in a piece of text. These are some of the elements stored in most INFORMATION RETRIEVAL systems. Also *index terms*.

kibble (*MinExt*) See BOWK.

kick (*MinExt*) Sudden increase in pressure during well drilling, which if not controlled quickly can cause a BLOWOUT. See KILL LINE.

kick-back (*Autos*) Shocks felt at the steering wheel, due to the *reversibility* of many steering devices.

kick copy (*Print*) On a web-fed printing press the printed and folded product is counted off in batches, each batch indicated by a displaced copy.

kicker (*Print*) The mechanism on a web-fed printing press which displaces one copy of the product to indicate the intervals between batches. See QUIRE SPACING.

Kick's law (*MinExt*) A law assuming that the energy required for subdivision of a definite amount of material is the same for the same fractional reduction in average size of the individual particles, ie $E = k_k \ln d_1/d_2$, where E is the energy used in crushing, k_k is a constant, depending on the characteristics of the material and method of operation of the crusher, and d_1 and d_2 are the average linear dimensions before and after crushing.

kicksorter (*ICT*) See PULSE-HEIGHT ANALYSER (1).

kick stage (*Space*) A propulsive stage used to provide an additional velocity increment required to put a spacecraft on a given trajectory.

kidney (*BioSci*) A paired organ for the excretion of nitrogenous waste products in vertebrates.

kidney machine (*Med*) See ARTIFICIAL KIDNEY.

kidney ore (*Min*) A form of the mineral haematite, oxide of iron, Fe_2O_3, which occurs in reniform masses, hence the name (Lt *ren*, kidney).

kidney stone (*Min*) (1) A pebble or nodule of limestone, resembling a kidney and found in Jurassic rocks. (2) A misnomer for NEPHRITE, which was once supposed to be efficacious in diseases of the kidney (Gk *nephros*, kidney).

kidney stones (*Med*) Hard deposits formed in the kidney. The composition varies, and kidney stones have been found to consist of uric acid and urates, calcium oxalate, calcium and magnesium phosphate, silica and alumina, cystine, xanthine, fibrin, cholesterol and fatty acids. Passage of the stones down the ureter may cause severe pain (renal colic).

kidney worm disease (*Vet*) In pigs the causative parasite is *Stephanurus dentalus* with infestation mainly in the kidney with occasional spinal canal involvement. The intermediate host is the earthworm. *Dictyophyma renale* is the parasite in dogs.

kier (*Textiles*) A large steel vessel in which yarn and cloth are boiled with alkaline liquors for scouring and bleaching. Now frequently replaced by continuous processing machinery.

kieselguhr (*Min*) See DIATOMITE.

kieserite (*Min*) Hydrated magnesium sulphate which crystallizes in the monoclinic system; found in large amounts in some salt deposits.

kieve (*MinExt*) See DOLLY TUB.

kieving (*MinExt*) See TOSSING.

kill (*Print*) In printing, an editorial instruction to delete entirely some item in preparation, derived from the use of the word as an instruction to distribute type.

killas (*Geol*) A name used in SW England for Palaeozoic slates or phyllites metamorphosed in contact with granite or mineral veins.

killed steel (*Eng*) Steel that has been 'killed', ie fully deoxidized before casting, by the addition of manganese, silicon and sometimes aluminium. There is practically no evolution of gas from the reaction between carbon and iron oxide during solidification. Sound ingots are obtained. See RIMMED STEEL, RIMMING.

killer (*Phys*) See POISON.

killer cells (*BioSci*) (1) Mammalian cells, probably of the mononuclear phagocyte lineage, which can lyse antibody-coated target cells. Not to be confused with *cytotoxic T-cells* (CTLs) which recognize targets by other means. (2) *Natural killer cells* (NK cells) are CD3-negative large granular lymphocytes that kill cells that do not express Class I or II major histocompatibility antigens. (3) *Lymphokine-activated killer cells* (LAK cells) are NK cells activated by interleukin-2.

kill line (*MinExt*) Small-bore pipelines connected through the BLOWOUT PREVENTER stack; they allow denser mud to be pumped into a borehole which has been shut out because of the danger of a BLOWOUT. Also *choke line*.

kill string (*MinExt*) See INJECTION STRING.

kiln (*MinExt*) A furnace used for: drying ore; driving off carbon dioxide from limestone; roasting sulphide ores or concentrates to remove sulphur as dioxide; reducing iron (II) ores to magnetic state in reducing atmosphere.

kiln drying (*For*) Accelerated seasoning of timber (typically 2–5 days as opposed to 2–10 years) in a temperature- and humidity-controlled chamber.

kilo- (*Genrl*) Prefix denoting 1000; used in the metric system, eg 1 kilogram = 1000 grams. Symbol k.

kilobase (*BioSci*) A thousand base pairs of DNA. Strictly should be kbp (kilobase pairs) but usually truncated. Abbrev kb (kbp).

kilobit (*ICT*) Commonly, a thousand BITS. Sometimes, 2^{10} or 1024 bits.

kilobyte (*ICT*) Commonly, a thousand BYTES. Sometimes, 2^{10} or 1024 bytes.

kilocalorie (*Phys*) See CALORIE.

kilocurie source (*Radiol*) Powerful radioactive source, usually in form of ^{60}Co.

kilocycles per second (*Phys*) See KILOHERTZ.

kilodalton (*Chem*) Unit of molecular weight; a thousand times the ATOMIC MASS UNIT. Abbrev kDa, kd.

kilo-electron-volt (*Phys*) A thousand ELECTRON-VOLTS.

kilogram (*Genrl*) Unit of mass in the MKSA (SI) system, being the mass of the International Prototype Kilogram, a cylinder of platinum–iridium alloy kept at Sèvres, Paris. Also *kilogramme*.

kilohertz (*Phys*) A thousand HERTZ or cycles per second. Symbol kHz.

kilometre (*Genrl*) A thousand metres. Abbrev km. US *kilometer*.

kilometric waves (*ICT*) Those with wavelengths between 1000 and 10 000 m.

kiloparsec (*Astron*) See PARSEC.

kilotex (*Textiles*) See TEX.

kiloton (*Phys*) Unit of explosive power for nuclear weapons equal to that of 10^3 tons of TNT.

kilovar (*ElecEng*) A unit of reactive volt-amperes equal to 1000 V Ar. Abbrev KV AR.

kilovolt-ampere (*ElecEng*) A commonly used unit for expressing the rating of ac electrical machinery and for other purposes; it is equal to 1000 volt-amperes. Abbrev KV A.

kilowatt (*ElecEng*) A unit of power equal to 1000 watts and approximately equal to 1·34 h.p. Abbrev KW.

kilowatt-hour (*ElecEng*) The commonly used unit of electrical energy, equal to 1000 watt-hours or 3·6 MJ. Often called simply a *unit*; abbrev *kWh*. See BOARD OF TRADE UNIT.

Kimball tag (*ICT*) Small punched card attached to merchandise that is detached when goods are sold, to provide machine-readable sales data.

Kimberley horse disease (*Vet*) A disease of horses in Australia due to poisoning caused by eating whitewood (*Atalaya hemiglauca*). Also *walk-about disease*.

kimberlite (*Geol*) A type of mica-peridotite, occurring in volcanic pipes in S Africa and elsewhere, and containing xenoliths of many types of ultramafic rocks, and diamonds.

Kimmeridgian (*Geol*) A stage in the Upper Jurassic. See MESOZOIC.

kinaesthesis (*Psych*) A general term for sensory feedback from muscles, tendons and joints, which informs the individual of the movements of the body or limbs, and the position of the body in space. US *kinesthesis*.

kinaesthetic (*BioSci*) Pertaining to the perception of muscular effort. US *kinesthetic*.

kinase (*BioSci*) An enzyme that catalyses the phosphorylation of its substrate by ATP. Thus protein kinases phosphorylate proteins and hexose kinases phosphorylate hexoses.

kinematical theory of X-ray diffraction (*Phys*) A treatment which does not take account of the attenuation of the incident beam as it passes through the crystal nor the interference between the incident beam and multiply diffracted beams; can be applied to very thin or very small crystals.

kinematic chain (*Eng*) A number of links connected to one another to allow motion to take place in combination. It becomes a mechanism when so constructed as to allow constrained relative motion between its links.

kinematics (*MathSci*) That branch of applied mathematics which studies the way in which velocities and accelerations of various parts of a moving system are related.

kinematic viscosity (*Phys, Eng*) The COEFFICIENT OF VISCOSITY of a fluid divided by its density. Symbol v. Thus $v = \eta/\rho$. Unit is $m^2 s^{-1}$ in the SI system and the stoke ($cm^2 s^{-1}$) in the CGS system.

kinesalgia (*Med*) Feeling of pain on movement.

kinescoping (*ImageTech*) US term for film recording of a TV programme. Also *telerecording*.

kinesin (*BioSci*) An intracellular protein of wide distribution in eukaryotes. It is responsible for the movement of organelles to which it is attached along microtubules using energy from the hydrolysis of ATP.

kinesis (*BioSci*) A simple response to environmental stimuli in which the response is proportional to the intensity of stimulation; it involves a change in speed of movement (positive or negative orthokinesis) or rate (probability) of turning (klinokinesis). Unlike a TAXIS, the animal or cell is not oriented to the stimulus although the effected movements often produce a change of position relative to it. Cf TAXIS.

kinetic energy (*Phys*) Energy arising from motion. For a particle of mass m moving with a velocity v it is

$$\tfrac{1}{2}mv^2$$

and for a body of mass M, moment of inertia I_g, velocity of centre of gravity v_g and angular velocity ω, it is

$$\tfrac{1}{2}Mv_g^2 + \tfrac{1}{2}I_g\omega^2$$

kinetic friction (*Phys*) See FRICTION.

kinetic heating (*Aero*) See DYNAMIC HEATING.

kinetic pressure (*Aero*) See DYNAMIC PRESSURE.

kinetic theory of elasticity (*Chem*) The explanation of elastomeric properties in molecular terms using the STATISTICAL CHAIN MODEL. Chains are random coils in an elastomer at rest, but, when strained, deform so that chains orient and elongate. The possible conformations are thereby reduced, so the entropy of the system decreases. When released, the elastomer snaps back to its original equilibrium state under the driving force to increase entropy. The theory accounts for the surprising increase in retractive force in elastomers as the temperature is increased. The retractive force in strained metals, ceramics and glasses is, by contrast, energetic in origin. Also *statistical theory of rubber elasticity*. See panel on ELASTOMERS.

kinetic theory of gases (*Phys*) A theory which accounts for the bulk properties of gases in terms of the motion of the molecules of the gas. In its simplest form the gas molecules are conceived as elastic spheres whose bombardment of the walls of the containing vessel causes the pressure exerted by the gas. If it is assumed that the size of the molecules is small compared with their mean spacing and that the molecules do not exert forces on each other except on collision, then the theory gives a simple explanation of the GAS LAWS and yields useful results concerning gaseous viscosity and thermal conductivity.

kinetin (*BioSci*) 6-*furfurylaminopurine*. Synthetic plant growth regulator of the cytokinin type.

kinetochore (*BioSci*) Paired structures within the CENTROMERIC region of metaphase chromosomes, to which spindle microtubules attach. They lie on each side of the PRIMARY CONSTRICTION and, when viewed with the electron microscope, appear as a trilaminar plate with microtubules entering at regular intervals.

kinetodesma (*BioSci*) See KINETY.

kinetoplast (*BioSci*) A complex of interlocked circles of mitochondrial DNA that is located near the base of the flagellum in flagellate Protozoa.

kinetosome (*BioSci*) See BASAL BODY.

kinetosomes (*BioSci*) See KINETY.

kinety (*BioSci*) A unit of structure in the Protozoa comprising the kinetosomes (the basal granules of the cilia and flagella) and the kinetodesma (a fine strand running from the kinetosomes).

king closer (*Build*) A three-quarter brick used to maintain the bond of the surface.

kingdom (*BioSci*) In current taxonomy, any of the six major groupings in the classification of living organisms: Archaea, Eubacteria, Protista, Plantae, Animalia, Fungi. Many alternative schemata have been proposed and quite commonly Monera are considered as including Eubacteria and Archaea, making only five kingdoms.

king pile (*CivEng*) A pile driven down the centre of a wide trench to enable two short struts (butting on opposite sides of the pile) to be used, instead of one long one, for keeping the poling boards of opposite sides of the trench in position.

king pin (*Autos*) The pin by which a stub axle is articulated to an axle beam or steering head; it is inclined to the vertical to provide caster action. For light vehicles, the king pin is now usually replaced by a pair of ball joints. Also *swivel pin*.

king post (*Build*) A vertical timber tie connecting the ridge and the tie-beam of a roof, shaped at its lower end to afford bearing to two struts supporting the middle points of the rafters. Also *broach post, joggle piece, joggle post, king piece, middle post*. See fig. at ROOF TRUSS.

king rod (*Build*) A vertical steel rod connecting the ridge and tie-beam of a couple-close roof, to prevent sagging of the tie-beam when it is required to support ceiling loads.

Kingston valve (*Eng*) A sea valve fitted to a ship's side for the purpose of admitting water to circulating pumps, or flooding or blowing out ballast tanks.

kingswood (*For*) See BRAZILIAN KINGSWOOD.

kingwood (*For*) Dense Brazilian hardwood (*Dalbergia*) with a characteristic multicoloured heartwood on a rich violet-brown background, a straight grain and lustrous, fine, even texture. Not the same as Brazilian kingswood.

kinin (*BioSci*) See CYTOKININ.

kinins (*Pharmacol*) A class of vasoactive peptides that are associated with local regulation of blood flow, eg bradykinin.

kink (*Electronics*) An abrupt change or reversal in the slope of a characteristic curve: eg the change from forward to reverse bias in a semiconductor diode, the sudden increase of reverse current at a certain voltage in a Zener diode, or the region where a negative-resistance device shows increasing current with decreasing applied voltage.

kink instability (*Phys*) See INSTABILITY.

kino (*For*) A red resinous exudate, rich in tannins, from various trees, notably *Pterocarpus* spp. Used in tanning. Also *gum kino*.

kin selection (*BioSci*) Natural selection for behaviour that lowers an individual's own chance of survival but raises that of a relative.

Kipp's apparatus (*Chem*) A generator for hydrogen sulphide, carbon dioxide, or other gases, by reacting a solid with a liquid. By opening or closing the gas outlet, gas pressure automatically fills or empties the solids container with liquid.

Kirchhoff's diffraction theory (*Phys*) A mathematical description of diffraction based on HUYGENS' PRINCIPLE.

Kirchhoff's equation (*Phys*) An expression for the rate of change of the heat ΔH of a process with temperature, carried out at constant pressure:

$$\left(\frac{\partial \Delta H}{\partial T}\right)_p = \Delta\left(\frac{\partial H}{\partial T}\right)_p = \Delta C_p$$

where ΔC_p is the change in the heat capacity at constant pressure for the same process. Similarly, for a process carried out at constant volume:

$$\left(\frac{\partial \Delta U}{\partial T}\right)_v = \Delta\left(\frac{\partial U}{\partial T}\right)_v = \Delta C_v$$

Kirchhoff's law (*Phys*) A law stating that the ratio of the ABSORPTION COEFFICIENT to the coefficient of emission for radiation is the same for all substances and depends only on the temperature. The law holds for the total emission and also for the emission of any particular frequency.

Kirchhoff's laws (*ElecEng*) Generalized extensions of *Ohm's law* employed in network analysis. They may be summarized as: (1) $\sum i = 0$ at any junction (2) $\sum E = \sigma iZ$ round any closed path. E = emf, i = current, Z = complex impedance.

Kirkendall effect (*Eng*) If a piece of a pure metal is placed in contact with a piece of an alloy of that metal and the whole heated, the constituent elements from the alloy will diffuse into the pure metal, causing a shift of the original interface.

Kirkwood gaps (*Astron*) Regions of very few asteroid orbits within the main asteroid belt of the solar system which exist owing to the gravitational force of Jupiter.

kish (*Eng*) Solid graphite which has separated from, and floats on the top of, a molten bath of cast-iron or pig iron which is high in carbon.

kiss impression (*Print*) The lightest possible impression on paper so that the type or blocks do not press into the paper, requiring careful MAKE-READY; also required when printing rubber or plastic plates to avoid distortion.

kiss of life (*Med*) The 'mouth-to-mouth' method of resuscitation in which the rescuer places his or her mouth over the patient's and inflates the latter's lungs by breathing into them.

kitchen midden (*Geol*) The dump of waste material, largely shells and bones associated with ashes, marking the site of a kitchen in a settlement of early humans.

kite (*Aero*) Any aerodyne anchored or towed by a line, not mechanically or power driven. Derives its lift from the aerodynamic forces of the relative wind.

kite winder (*Build*) A WINDER used at the angle of a change of direction in a stair, and therefore kite shape in plan.

Kjeldahl flask (*Chem*) A glass flask with a round bottom and a long wide neck. Used in Kjeldahl's method for the estimation of nitrogen.

Kjeldahl's method (*Chem*) A method for the quantitative estimation of nitrogen in organic compounds, based on the conversion of the organic nitrogen into ammonium sulphate, and subsequent distillation of ammonia after the solution has been made alkaline. The ammonia which distils over can be titrated.

Klebs–Löffler bacillus (*Med*) A rod-shaped Gram-positive bacterium (*Corynebacterium diphtheriae*) that causes diphtheria in humans and similar diseases in other animals.

Klein bottle (*MathSci*) A three-dimensional surface with only one side obtained by pulling the narrow end of a tapering cylinder through the wall of the cylinder and then stretching the narrow end and joining it to the larger end. Cf MÖBIUS STRIP.

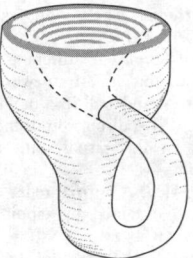

Klein bottle

Kleine–Levin syndrome (*Med*) A rare disorder of young men with drowsiness, excessive eating and perverted sexual behaviour.

Klein–Gordon equation (*Phys*) An equation describing the motion of a spinless charged particle in an electromagnetic field.

Klein–Nishina formula (*Phys*) Theoretical expression for the cross-section of free electrons for the scattering of photons. See COMPTON EFFECT.

kleptomania (*Psych*) Morbid condition where there is an uncontrollable desire to steal.

KLH (*BioSci*) Abbrev for *keyhole limpet haemocyanin*.

Klinefelter's syndrome (*Med*) Impaired gonadal development in males with one or more extra X chromosomes; it results in underdeveloped testes, mixed (male and female) secondary sex characteristics, sterility and, sometimes, mild mental retardation.

K-lines (*Phys*) Characteristic X-ray frequencies from atoms due to excitation of electrons in the K-shell. Denoted by Kα, Kβ, Kγ; the lines are all doublets.

klinker brick (*Build*) A very hard type of brick much used in the Netherlands and Germany, principally for paving purposes.

klinokinesis (*BioSci*) A KINESIS in which the rate or probability of turning is affected, either positively or negatively.

klinostat (*BioSci*) Apparatus to rotate a plant, eg slowly about a horizontal axis, in order to cancel out at least some of the effects of gravity in investigations of gravitropism etc. Also *clinostat*.

klippe (*Geol*) An erosional remnant of a thrust sheet. It is essentially an outlier lying on a thrust plane. Pl *klippen*.

Klippel–Feil syndrome (*Med*) A congenital shortening of the neck that may be associated with SYRINGOMYELIA.

klystron (*Electronics*) General name for class of ultrahigh-frequency and microwave electron tubes (amplifiers, oscillators, frequency multipliers, cascade amplifiers, etc), in which electrons in a stream have their velocities varied (velocity modulation) by a high-frequency field; and subsequently impart energy to it or to other high-frequency fields, contained in a resonator. See REFLEX KLYSTRON, RHUMBATRON.

klystron frequency multiplier (*Electronics*) A two-cavity klystron whose output cavity is tuned to an integral multiple of the fundamental frequency.

km (*Genrl*) Abbrev for KILOMETRE.

knapping hammer (*Build*) A hammer used for breaking and shaping stones and flints.

knapsack abutment (*CivEng*) An abutment provided with an additional integral slab behind it to reduce the overturning pressure.

knebelite (*Min*) A silicate of manganese and ferrous iron, in the olivine series. It crystallizes in the orthorhombic system and occurs as the result of metamorphism and metasomatism.

knee (*BioSci*) In land vertebrates, the joint between the femur and the crus.

knee (*Build*) A sudden rise in a hand rail when it is convex upwards. Cf RAMP.

knee (*Build, Eng*) An elbow pipe. See ELBOW.

knee (*Electronics*) The region of maximum curvature on a characteristic curve.

knee brace (*Build*) A stiffening member fixed across the inside of an angle in a framework, particularly at the angle between roof and wall in a building frame.

knee gall (*Vet*) A distension of the carpal synovial sheath at the back of the carpal joint of the horse. Also *knee throughpin*.

knee jerk (*Med*) A normal reflex extension of the leg at the knee joint, obtained by tapping the tendon below the patella.

kneeler (*Build*) The return of the dripstone at the spring of an arch.

knee rafter (*Build*) A rafter having its lower end bent downwards.

knee roof (*Arch*) See MANSARD ROOF.

knee throughpin (*Vet*) See KNEE GALL.

knickpoint (*Geol*) A change of slope in the longitudinal profile of a stream as a result of REJUVENATION.

knife cheeks (*Print*) Spring-loaded gripping edges to hold the web while it is being severed.

knife coating (*Textiles*) A method of coating fabric with flexible layer (eg plasticized polyvinyl chloride) by passing fabric between roller and knife blade. Produces flat surface finish after curing, unlike reverse roll coating, which gives even finish over an often rough surface.

knife edge (*Eng*) Support for a balance beam or similar instrument member, usually in the form of a hardened steel wedge, the apex of which gives line support.

knife fold (*Print*) A fold made by pushing the paper into the folding rollers with a blunt knife, a necessary arrangement when the fold is at right angles to the leading edge of the paper, or to the previous fold. Cf BUCKLE FOLD.

knife switch (*ElecEng*) An electric circuit switch in which the moving element consists of a flat blade which engages with fixed contacts.

knife tool (*Eng*) A lathe tool having a straight lateral cutting edge, used for turning right up to a shoulder or corner.

Knight shift (*Phys*) A shift in NUCLEAR MAGNETIC RESONANCE frequency in metals from that of the same isotope in chemical compounds in the same magnetic field. It is due to the paramagnetism of the conduction electrons.

knitting (*Textiles*) The process of making a fabric from yarn by the formation of intermeshing loops. See WARP KNITTING, WEFT KNITTING.

knitting machine (*Textiles*) A machine for making fabrics by WARP KNITTING or WEFT KNITTING.

knitwear (*Textiles*) A general term used for all knitted outer garments except stockings and socks.

knobbing (*Build*) The operation of breaking projecting pieces off stones in the quarry.

knock (*Autos, Eng*) See KNOCKING.

knocked up toe (*Vet*) Synovitis and fibrositis affecting the first or second interphalangeal joint of the toe of the dog; caused by sprain and usually met with in racing greyhounds.

knocker-out (*Eng*) The horns of a planing machine against which the tappets strike to reverse the motion of the table.

knocking (*Autos*) The characteristic metallic noise, often called 'pinking', resulting from DETONATION in a petrol engine, or 'diesel knock', due to the rapid pressure rise during combustion in a diesel engine. Also *knock*.

knocking (*Eng*) A periodic noise made by a worn bearing in a reciprocating engine, due to reversal of the load on the shaft or pin. Also *knock*.

knocking-down iron (*Print*) A piece of iron, fixed to the lying press, on which joints are beaten out and lacings flattened.

knockings (*Build*) The stone pieces, smaller than spalls, knocked off in the process of chiselling or hammering.

knockings (*MinExt*) See RIDDLE.

knock knee (*Med*) See GENU VALGUM.

knockout (*BioSci*) Informal term for a mutant organism in which the expression or function of a particular gene has been completely eliminated. The effects may differ remarkably according to the genomic background.

knockout mouse (*BioSci*) Transgenic mouse in which a particular gene has been deleted. Often shows disappointingly little phenotypic change – possibly because there are alternative mechanisms or because some genes are probably unnecessary for the survival of a well-fed laboratory mouse in very well-regulated surroundings.

knock rating (*Autos*) The measurement of the ANTIKNOCK VALUE of a volatile liquid fuel.

knoll (*Geol*) See REEF KNOLLS.

Knoop hardness (*Eng*) Hardness measured with a Knoop diamond, an elongated rhombus pyramid with angles of 172·5° and 130° at the apex, giving a hardness value, $H_K = 14.233Fd^{-2} \, \text{MN m}^{-2}$, where F is the load in newtons and d is the longer diagonal in mm. It is a shallower diamond than the VICKERS, and so is better suited to measuring surface layers. See panel on HARDNESS MEASUREMENTS.

knot (*Aero, Ships*) Speed of 1 nautical mph (1·15 mph or 1·85 km h^{-1}) used in navigation and meteorology.

knot (*BioSci*) (1) A node in a grass stem. (2) A hard and often resinous inclusion in timber, formed from the base of a branch which became buried in secondary wood as the trunk thickened.

knot (*MathSci*) A closed curve in space.

knotter (*Paper*) An appliance for the removal of knots, undigested particles or similar unwanted matter from paper pulp.

knotting (*Build*) A solution of shellac in spirit used for covering knots in wood, prior to painting, to prevent exudation of resin.

knowledge-based system (*ICT*) Any software system that aims to store and effectively utilize large amounts of specialist knowledge. See EXPERT SYSTEM.

knowledge engineer (*ICT*) A person who designs and builds EXPERT SYSTEMS.

knuckle (*Build*) The parts of a hinge receiving the pin.

knuckle joint (*Eng*) A hinged joint between two rods, in which the ends are formed into an eye and a fork respectively and united by a pin.

knuckle-joint press (*Eng*) A heavy power press incorporating a knuckle-joint mechanism for actuating the slide, It has a high mechanical advantage near the bottom end of its stroke, which is relatively short, and is therefore used for coining, sizing and embossing work.

knuckling (*Vet*) A state of abnormal flexion of the fetlock joint of the horse, due to a variety of causes, and associated with an inability or lack of desire to extend the joint fully.

Knudsen flow (*ChemEng*) The flow of low-pressure gas through a tube whose diameter is much smaller than the mean free path of the molecules of the gas. Under these conditions, the macroscopic concept of viscosity needs to be modified since the resistance to motion is due primarily to molecular collisions with the passage walls. Also *molecular streaming*.

knurling tools (*Eng*) Small, hard steel rollers, serrated or cross-hatched on their peripheries, mounted on axes carried by a holder. They are pressed against circular work in the lathe, to knurl or roughen a surface required to give a grip to the fingers, as the head of a 'milled' screw.

kobellite (*Min*) A lead, bismuth, antimony sulphide, crystallizing in the orthorhombic system.

Koch resistance (*Chem*) The resistance of a vacuum photocell or phototube when its active surface is irradiated with light.

Koch's postulates (*Med*) Four criteria laid down by Koch to show that a disease was caused by a micro-organism: (1) the organism must be observed in all cases of the disease; (2) it must be isolated and grown in pure culture; (3) the culture must be capable of reproducing the disease when inoculated into a suitable experimental animal; (4) the organism must be recovered from the experimental disease.

Kodak relief plate (*Print*) TN for an easily processed relief plate suitable for either direct or 'dry offset' printing with considerable wearing capacity.

Kohler's disease (*Med*) Aseptic necrosis of the navicular bone in the tarsus; a self-limiting disease.

Kohlrausch's law (*Chem*) A law stating that the contribution from each ion of an electrolytic solution to the total electrical conductance is independent of the nature of the other ion.

koilonychia (*Med*) Spoon-shaped depression of the finger nails.

konimeter (*PowderTech*) A type of impaction sampler in which dirty air, sucked through a round orifice, impinges upon a circular glass plate smeared with grease, which collects the particles. In eg mining, it is used to check dust content of mine atmosphere.

Kooman's array (*ICT*) Original name for the category of BROADSIDE ANTENNAS consisting of rows and columns of dipoles. The rows are a half-wavelength apart and fed by twin-wire feeders, twisted between each row. This ensures that the dipoles are all fed in the same phase.

Koplik's spots (*Med*) Bluish-white specks on the inner side of the cheek, occurring in measles 2 to 3 days before the rash appears.

Köppen classification (*EnvSci*) A system of classifying climate by groups of letters. The general climatic type is shown by a letter or letter-pair in capitals, and further detail is given by additional lower case letters, eg the climate of the UK is coded as Cf where C indicates 'warm or temperate rainy' and f is 'moist, no marked dry season'.

Kopp's law (*Chem*) See JOULE'S LAW.

Korndorfer starter (*ElecEng*) A variant of the AUTOTRANSFORMER STARTER for three-phase induction motors which involves no interruption of the supply of power after the starting cycle has been initiated. Used with high-powered and high-voltage machines.

kornerupine (*Min*) A magnesium aluminium borosilicate, crystallizing in the orthorhombic system. Rare, but collected and cut as a gemstone.

Korotkoff sounds (*Med*) Sounds heard on auscultation over the brachial artery when a pneumatic cuff is deflated. Korotkoff's first sound corresponds to SYSTOLIC blood pressure and his fifth sound when the sounds disappear is taken as the DIASTOLIC blood pressure.

Korsakoff's syndrome (*Psych*) An irreversible nutritional deficiency, caused by chronic alcoholism and related malnutrition; characterized by severe ANTEROGRADE AMNESIA and CONFABULATION. Also *Korsakoff's psychosis*.

Kovar (*Eng*) TN for an alloy of Ni–Co–Fe for glass-to-metal seals over working ranges of temperature, when temperature coefficients of expansion coincide.

Kozeny–Carman equation (*PowderTech*) A permeability equation used to calculate the surface area of a powder from the permeability of a packed powder bed to the flow of fluid through it.

Kr (*Chem*) Symbol for KRYPTON.

Kraemer–Sarnow test (*Build, CivEng*) A test for the determination of the melting point of a bitumen for use in building or road-making. It is the temperature at which mercury flows through a sample of bitumen under gravity.

kraft paper (*Paper*) Strong paper made from a sulphate pulp (kraft pulp), used in packaging, and in laminated plastics, eg melamine-formaldehyde laminates. From Ger for *strength*.

Kramer control (*ElecEng*) A form of speed and power factor control for large induction motors in which the slip energy is supplied to a rotary converter, the resulting dc power being used to drive a motor either mounted on the main motor shaft or driving an alternator and returning power to the supply.

Krantz anatomy (*BioSci*) The leaf anatomy of most C4 plants, characterized by the bundle sheath cells being large and having conspicuous chloroplasts. The mesophyll cells have less conspicuous choloroplasts.

Kraton (*Chem*) TN for styrene–butadiene–styrene thermoplastic elastomer.

kraton (*Geol*) See CRATON.

kraurosis (*Med*) Atrophy of the vulva, associated with narrowing of the vaginal orifice.

Krebs' cycle (*BioSci*) Also *citric acid cycle*. See TCA CYCLE.

kriging (*EnvSci*) Ordinary kriging (OK) is a geostatistical approach to modelling. Instead of weighting nearby data points by some power of their inverted distance, OK relies on the spatial correlation structure of the data to determine the weighting values. Effectively a method of interpolation which predicts unknown values from data observed at known locations.

kriging (*MinExt*) A statistical technique used in calculating grade and tonnage of ore reserves.

Kroll's process (*Eng*) The reduction to metal of tetrachloride of titanium or zirconium *in vacuo* or by reaction with magnesium in a neutral atmosphere.

Kronecker delta (*MathSci*) A function δ_{ij} of two variables i and j, defined to have the value zero unless $i = j$ when its value is unity.

Kronig–Penny model (*Phys*) A relatively simple model for a one-dimensional lattice from which the essential features of the behaviour of electrons in a periodic potential may be illustrated. See BAND THEORY OF SOLIDS.

Kruger flap (*Aero*) Leading-edge flap which is hinged at its top edge so it normally lies flush with the underwing surface, and swings down and forward to increase circulation and lift for low-speed flight.

Krukenberg's tumour (*Med*) A solid tumour appearing in each ovary and believed to be always secondary to cancer of the stomach.

Krummholz (*BioSci*) Stunted, wind-trimmed trees between the timber and tree lines on mountains. Also *elfin forest*.

krypton (*Chem*) Symbol Kr, at no 36, ram 83·80, mp $-169°C$, bp $-151·7°C$, density $3·743 \, g \, dm^{-3}$ at stp. A zero valent element, one of the noble gases. It is a colourless and odourless monatomic gas, and constitutes about one-millionth by volume of the atmosphere, from which it is obtained by liquefaction. It is used in certain gas-filled electric lamps, forms few compounds, eg KrF_4, and has 15 isotopes. Krypton-85 is discharged into the atmosphere by nuclear plants, but not at a level considered to be a problem at present.

Ks (*Chem*) Symbol for SOLUBILITY PRODUCT.

K-shell (*Phys*) The innermost ELECTRON SHELL in an atom corresponding to a principal quantum number of unity. The shell can contain two electrons. See panel on ATOMIC STRUCTURE.

k-space (*Phys*) Symbol for *momentum space* or *wavevector space*. This is an important concept in semiconductor energy band theory. See BRILLOUIN ZONE.

K-strategist (*BioSci*) An organism that assigns relatively little of its resources to reproduction. Such organisms are characteristic of stable, saturated communities, and have low fecundity and high competitive ability. Cf R-STRATEGIST.

Kuchemann tip (*Aero*) Low-drag wing-tip shape developed by Kuchemann at RAE for transonic aircraft. It has a large-radius curve in plan finishing with a corner at the trailing edge.

Kuiper belt (*Astron*) A large ring of icy bodies orbiting the Sun just beyond Neptune.

kulaite (*Geol*) An amphibole-bearing nepheline basalt.

Kümmell's disease (*Med*) Delayed collapse or crumbling of one or more spinal vertebrae after injury to the spine.

Kummer's convergence test (*MathSci*) If, for a divergent series of positive terms $\sum (v_n)^{-1}$ and a series of positive terms $\sum u_n$, the expression

$$v_n \frac{u_n}{u_{n+1}} - v_{n+1} \to l$$

then the series $\sum u_n$ converges if $l > 0$, and diverges if $l < 0$.

Kundt constant (*Phys*) The ratio of VERDET'S CONSTANT to magnetic susceptibility.

Kundt's rule (*Phys*) A rule that the refractive index of a medium on the shorter-wavelength side of an absorption band is abnormally low, and on the other side abnormally high. This gives rise to ANOMALOUS DISPERSION.

Kundt's tube (*Acous*) Transparent tube in which standing sound waves are established, indicated by lycopodium powder, which accumulates at the nodes. Used for measuring sound velocities. See ACOUSTIC STREAMING.

kunzite (*Min*) See SPODUMENE.

kupfernickel (*Min*) See NICCOLITE.

Kupffer cell (*BioSci*) The form of macrophages that line the blood sinuses of the liver. They are phagocytic and remove foreign particles from the blood. They have Fc receptors and can engulf blood cells that have become coated with antibody.

Kurie plot (*Phys*) A plot used for determining the energy limit of a beta-ray spectrum from the intercept of a straight-line graph. Prepared by plotting a function of the observed intensity against energy, the intercept on the axis being the energy limit for the spectrum. Also *Fermi plot*.

kurtosis (*MathSci*) A measure of the extent to which a distribution is concentrated around its MEAN. See LEPTO-KURTIC, MESOKURTIC, PLATYKURTIC.

kuru (*Med*) A rapidly fatal brain degeneration formerly found in people living in the Fore region of Papua New Guinea. Shown to be caused by the ritual custom of eating human brain, thus transmitting infectious particles identical to those responsible for the sporadic and late-onset form of CREUTZFELDT–JAKOB DISEASE. See panel on TRANSMISSIBLE SPONGIFORM ENCEPHALO-PATHY.

Kussmaul breathing (*Med*) Deep-sighing breathing often known as air-hunger seen in severe metabolic ACIDOSIS.

kV (*Phys*) Abbrev for *kilovolt*.

kV A (*ElecEng*) Abbrev for KILOVOLT-AMPERE.

K value (*Chem*) Practical measure of molecular mass of polyvinyl chloride, obtained from intrinsic viscosity experiments. A K value of 69 corresponds to a molecular mass of 240 000.

kV Ar (*ElecEng*) Abbrev for KILOVAR.

kVp (*Radiol*) Abbrev for *kilovolts, peak*. Voltage applied across an X-ray tube, hence designating the maximum energy of emitted X-ray photons.

kW (*ElecEng*) Abbrev for KILOWATT.

kwashiorkor (*Med*) A disease of the tropics due to protein and calorie deficiency in children. There is gross loss of weight, oedema and pigmentation. Also known as *protein-energy malnutrition*.

kWh (*ElecEng*) Abbrev for KILOWATT-HOUR.

Kwok's disease (*FoodSci*) Food intolerance caused by excessive consumption of MONOSODIUM GLUTAMATE.

kX unit (*Radiol*) See X-RAY UNIT.

kyanite (*Min*) A silicate of aluminium which crystallizes in the triclinic system. It usually occurs as long-bladed crystals, blue in colour, in metamorphic rocks. Also *cyanite*. See DISTHENE.

kylite (*Geol*) An olivine-rich theralite.

kymography (*Radiol*) A method of recording in a single radiograph the excursions of moving organs in the body, by use of a kymograph, which records physiological muscular waves, pulse beats or respirations.

Kyoto Agreement (*EnvSci*) An international meeting of heads of state held in Kyoto in 1997 at which various promises were made about reductions in greenhouse gas emission, notably that levels by 2010 would be 12.5% below 1990 levels. There is some scepticism as whether these laudable goals will be realized.

kyphoscoliosis (*Med*) A deformity of the spine in which dorsal convexity is increased, the spine being also bent laterally.

kyphosis (*Med*) A deformity of the spine in which the dorsal convexity is increased.

L

L (*Chem*) Symbol for MOLAR LATENT HEAT.

L (*Phys*) Symbol for: (1) ANGULAR MOMENTUM; (2) LUMINANCE; (3) INDUCTANCE.

L- (*Genrl*) Prefix denoting LAEVO-, LAEVOROTATORY.

Λ (*Chem*) Symbol for MOLAR CONDUCTANCE; Λ_0, at infinite dilution.

L_{10}, L_{90} (*EnvSci*) Noise indices that are exceeded for 10% and 90% of the day, respectively.

l (*Chem*) Abbrev for LITRE ($1\ dm^3$ is now used officially).

l (*Chem*) Symbol for: (1) specific latent heat per gram; (2) mean free path of molecules; (3) (with subscript) equivalent ionic conductance, 'mobility'.

l- (*Chem*) Abbrev for LAEVOROTATORY.

λ (*Phys*) Symbol for: (1) DISINTEGRATION CONSTANT; (2) linear coefficient of thermal expansion; (3) MEAN FREE PATH; (4) THERMAL CONDUCTIVITY; (5) WAVELENGTH.

La (*Chem*) Symbol for LANTHANUM.

label (*Arch*) A projecting moulding above a door or window opening.

label (*ICT*) Used in a PROGRAMMING LANGUAGE to identify a particular statement, eg a line number in BASIC.

label-corbel table (*Build*) A DRIPSTONE supported by a corbel.

labelled atom (*Phys*) An atom which has been made radioactive, in a compound introduced into a flowline or laboratory procedure in order to trace progress. Also *tagged atom*. See RADIOACTIVE TRACER.

labelled molecule (*Phys*) A molecule containing a radioactive isotopic tracer.

labelling theory (*Psych*) The view that no actions are inherently deviant or abnormal, and that the label *mental illness*, applied to certain behaviours and to individuals, acts as a self-fulfilling prophecy; once applied, the individual's self-expectations, and the expectations of others, result in behaviour associated with the label, and so perpetuate the condition.

labellum (*BioSci*) (1) The posterior petal of the flower of an orchid and often the most conspicuous part of the flower. (2) The lower lip of the corolla forming a landing platform for pollinating insects in a number of families, eg Labiatae. (3) A spoon-shaped lobe at the apex of the glossa in bees. (4) (pl) In certain Diptera, a pair of fleshy lobes into which the proboscis (*labium*) is expanded distally. Pl *labella*.

labia (*BioSci*) (1) Any structures resembling lips. (2) The lesser and greater pudendal lips of female external genitalia in primates. See sing form LABIUM. Adj *labial*.

labia majora (*BioSci*) In mammals, the two prominent folds that form the outer lips of the vulva.

labia minora (*BioSci*) In mammals, the small inner lips of the vulva; nymphae.

labiate (*BioSci*) (1) With a lip or lips, as the corolla of various Labiatae. (2) A species of the Labiatae.

labile (*Chem*) Unstable, liable to change. Usually a kinetic, not a thermodynamic, term.

labile (*Electronics*) Said of an oscillator which can be synchronized from a remote source.

labioglossopharyngeal (*Med*) Pertaining to the lips, tongue and pharynx.

labiovelar (*Genrl*) In phonetics, a sound produced by the lips and soft palate together, eg 'w'.

labium (*BioSci*) In insects, the lower lip, formed by the fusion of the second maxillae; in the shells of Gastropoda, the inner or columellar lip of the margin of the aperture. See pl form LABIA.

laboratory sand-bath (*Phys*) Gas-heated steel bath containing sand in which crucibles or other vessels are indirectly heated.

labradorescence (*Min*) A brilliant play of colours shown by some labradorite feldspars. Probably due to a fine intergrowth of two phases.

labradorite (*Min*) A plagioclase feldspar containing 50–70% of the anorthite molecule; occurs in basic igneous rocks; characterized by a beautiful play of colours in some specimens, due to schiller structure.

labrum (*BioSci*) In insects and crustacea, the plate-like upper lip; in the shells of Gastropoda, the outer lip or right side of the margin of the aperture. Adj *labral*.

labyrinth (*Acous*) (1) Folded path for loading the rear of loudspeaker diaphragms and microphone ribbons. (2) Inner ear containing cochlea and semicircular canals.

labyrinth (*BioSci*) Any convolute tubular structure; esp the bony tubular cavity of the internal ear in vertebrates, or the membranous tube lying within it.

labyrinth (*NucEng*) See RADIATION TRAP.

labyrinthectomy (*Med*) Surgical removal of the labyrinth of the ear.

labyrinthitis (*Med*) Inflammation of the labyrinth of the ear.

labyrinthodont (*BioSci*) Having the dentine of the teeth folded in a complex manner.

labyrinth packing (*Eng*) A type of gland, with radial or axial clearance, used eg in steam engines and turbines.

lac (*BioSci*) A resinous substance, an excretion product of some coccid insects (Gascardia, Tachardia, etc), in certain jungle trees; used in the manufacture of shellac. Chief source, India. See SHELLAC, for which the name is frequently loosely used.

laccolith (*Geol*) A concordant intrusion with domed top and flat base, the magma having been instrumental in causing the up-arching of the 'roof'. Cf PHACOLITH. See fig. at DYKES.

lace (*Plastics*) Long, thin extrudate which, when cut by a rotating blade fitted after the die, gives polymer granules.

lace (*Textiles*) A fine open-work decorative fabric comprising a ground net on which patterns are formed by looping, twisting or knitting. The process may be done by hand using a needle or bobbin or by machine.

laced valley (*Build*) A valley formed in a tiled roof by interlacing tile-and-a-half tiles across a valley board.

lace machine (*Textiles*) There are several different machines developed for the manufacture of lace but particularly important are the Leavers machines. In these the threads are in narrow brass bobbins mounted in carriages and the pattern threads and the warp threads are manipulated by guide bars controlled by a jacquard to produce the desired pattern.

Lacerta (Lizard) (*Astron*) A fairly small, faint northern constellation. It includes BL Lacertae, the prototype of a class of active galaxy.

lacewood (*For*) See EUROPEAN PLANE.

lachrymal (*BioSci*) See LACRIMAL.

lacing course (*Build*) A brickwork bond-course built into rubble or flint walls.

lacing-in (*Print*) The operation in hand binding of attaching the boards to a book by lacing the ends of the bands through holes made in the boards to receive them.

laciniate (*BioSci*) Deeply cut into narrow segments; fringed.

lac operon (*BioSci*) See LACTOSE OPERON.

lacquer (*Build*) A solution of film-forming substances in volatile solvents, eg a spirit lacquer or varnish consists of shellac or other gums dissolved in methylated spirit. Lacquers may be pigmented or clear, and dry by the evaporation of solvents only, not by oxidation. (In the US lacquers include those drying by oxidation.) The most important lacquers are the CELLULOSE LACQUERS.

lacquering (*ImageTech*) Coating-processed motion picture film with a transparent protective layer to minimize scratches and abrasions in subsequent handling.

lacrimal (*BioSci*) Pertaining to, or situated near, the tear gland; in some vertebrates, a paired lateral membrane bone of the orbital region of the skull, in close proximity to the tear gland.

lacrimal duct (*BioSci*) In most of the higher vertebrates, a duct leading from the inner angle of the eye into the cavity of the nose; it serves to drain off the secretion of the lacrimal gland from the surface of the eye.

lacrimal gland (*BioSci*) In most of the higher vertebrates, a gland situated at the outer angle of the eye; it secretes a watery substance which washes the surface of the eye and keeps it free from dust.

lacrimation (*Med*) Shedding of tears.

lactalbumin (*BioSci*) Milk protein fraction containing β-lactoglobulin and α-lactalbumin.

lactam (*Chem*) A cyclic amide compound (lactone + amide).

lactation (*BioSci*) The production of milk from the mammary glands of mammals.

lactation tetany (*Vet*) See BOVINE HYPOMAGNESAEMIA.

lacteals (*BioSci*) In vertebrates, lymphatics, in the region of the alimentary canal, in which the lymph has a milky appearance, due to minute fat globules in suspension.

lactic (*BioSci*) Pertaining to milk.

lactic acid (*Med*) The dextrorotatory isomer D-lactic acid is produced by anaerobic metabolism of glucose. Pathological accumulations of lactic acid occur where there is poor tissue perfusion or where the liver is unable to metabolize lactic acid.

lactic acid bacteria (*FoodSci*) See LACTOBACILLACEAE.

lactic acids (*Chem, FoodSci*) *Hydroxy-propanoic acids.* There are two isomers: 2-hydroxy-propanoic acid, $CH_3CH(OH)$-COOH, and 3-hydroxy-propanoic acid, $CH_2(OH)CH_2$-COOH; only the former, known as *lactic acid* or *fermentation lactic acid*, is of importance. Lactic acid is a synonym for racemic 2-hydroxy-propanoic acid, hygroscopic crystals, mp 18°C, bp 83°C at 1 mm, rel.d. 1·248. It is produced commercially by the lactic fermentation of sugars with *Bacillus acidi lactici* or by heating glucose with caustic potash solution of a certain concentration.

lactiferous (*BioSci*) (1) Milk-producing, milk-carrying. (2) In plants, producing or containing *latex*.

Lactobacillaceae (*BioSci*) A family of bacteria belonging to the order EUBACTERIALES, comprising Gram-positive cocci or rods that are carbohydrate fermenters. They are anaerobic or grow best at low oxygen tensions. The family includes several pathogenic organisms, eg *Streptococcus pyogenes* (scarlet fever, tonsillitis, erysipelas, puerperal fever), *Diplococcus pneumoniae* (one cause of pneumonia), but also those (*lactic acid bacteria*) that cause desirable flavour and aroma changes when used in the production of fermented foods such as yoghurt, cheese and sauerkraut. Can also cause food spoilage, eg souring.

lactoferrin (*BioSci*) Iron-binding protein of very high affinity found in milk and in the specific granules of neutrophil leucocytes.

lactogenic hormone (*Med*) See PROLACTIN.

lactones (*Chem*) Intramolecular esters of hydroxycarboxylic acids. Most lactones have five- or six-membered rings (γ- or δ-lactones) but others are known.

lactose (*Chem*) $C_{12}H_{22}O_{11}.H_2O$, rhombic prisms which become anhydrous at 140°C, mp 201–202°C with decomposition; when hydrolysed it forms D-galactose and D-glucose. It reduces Fehling's solution and shows MUTAROTATION. It occurs in milk and is not so sweet as glucose.

lactose intolerance (*FoodSci*) With the exception of Caucasians, a common condition in adult humans due to the absence of the enzyme lactase which hydrolyses lactose to glucose in the digestive system.

lactosuria (*Med*) The presence of lactose in the urine.

lacuna (*BioSci*) A cavity, gap, space or depression such as a space in a plant tissue caused by the breakdown of the protoxylem, or sometimes an intercellular space such as the cell-containing cavities in bone. Adjs *lacunar, lacunose*.

lacustrine (*Geol*) Related to a LAKE.

LAD (*Psych*) Abbrev for *language acquisition device*.

ladder (*Textiles*) In a knitted structure, esp stockings, a defect caused by the breaking of stitches resulting in a thread reverting in long runs in the wale direction to its original linear form.

ladder network (*ICT*) One that comprises a number of filter sections, all alike, in series, acting as a transmission line, with attenuation and delay properties.

ladder polymers (*Chem*) Polymers consisting of two chains covalently bonded together, such as Black Orlon. See panel on HIGH-PERFORMANCE POLYMERS.

ladies (*Build*) Slates 16×8 in (406×203 mm).

ladle (*Eng*) An open-topped vessel lined with refractory material; used for conveying molten metal from the furnace to the mould or from one furnace to another.

ladle addition (*Eng*) The addition of alloying metal to molten metal in a ladle before casting.

Laënnec's cirrhosis (*Med*) A term formerly used for multilobular cirrhosis of the liver, in which degeneration of liver cells is associated with areas of fibrosis enclosing many regenerating lobules of the liver. See CIRRHOSIS.

laevo- (*Genrl*) Prefix from Lt *laevus*, left. US *levo-*.

laevorotatory (*Phys*) Said of an optically active substance which rotates the plane of polarization in an anticlockwise direction when looking against the oncoming light.

laevulose (*Chem*) $C_6H_{12}O_6$. Fruit sugar or L-*fructose*. Also *levulose*.

lag (*ICT*) Any time delay between an initiating action and a desired effect; eg a trigger pulse to a circuit may be followed by a lag before the circuit operates, or the image on a TV camera tube may persist for a number of frames, impairing the depiction of rapid movement.

lag (*MinExt*) To protect a shaft or level from falling rock by lining it with timber (*lagging*).

lagena (*BioSci*) In higher vertebrates, a pocket lined by sensory epithelium and developed from the posterior side of the sacculus. It becomes transformed in mammals into the scala media or canal of the cochlea.

lager (*FoodSci*) Traditional term for beers bulk stored in chilled conditions underground. Today refers to any brewing process employing bottom-fermenting yeasts fermented between 5 and 12°C and matured at around 0°C. From Ger *Lager*, a storehouse. Also *lagering*.

lagging (*BioSci*) Slow movement towards the poles of the spindle by one or more chromosomes in a dividing nucleus, with the result that these chromosomes do not become incorporated into a daughter nucleus.

lagging (*Build, CivEng*) (1) The wooden boards nailed across the framework of a centre, to form the immediate supporting surface for the arch. (2) Material generally used in floors and roofs to eliminate sound and loss of heat. See INSULATION.

lagging (*Eng*) (1) The process of covering a vessel, pipe or boiler with a non-conducting material to prevent either loss or gain of heat. (2) The non-conducting material itself. See INSULATION.

lagging current (*ElecEng*) An ac which reaches its maximum value later in the cycle than the voltage which is producing it.

lagging load (*ElecEng*) See INDUCTIVE LOAD.

lagging phase (*ElecEng*) A term used in connection with measuring equipment on three-phase circuits to denote a phase whose voltage is lagging behind that of one of the other phases by approximately 120°. Also used, particularly in connection with the two-wattmeter method of three-phase power measurement, to denote the phase in which the current at unity power factor lags behind the voltage applied to the meter in which that current is flowing.

Lagomorpha (*BioSci*) An order of Eutheria with two pairs of upper, and one of lower, incisor teeth which grow throughout life; there are no canines, and there is a wide diastema between the incisors and the cheek teeth into which the cheeks can be tucked to separate the front part of the mouth from the hind during gnawing. Rabbits and hares.

lagoon (*Geol*) A shallow stretch of sea water close to the sea but partly or completely separated from it by a low strip of land. Often associated with coral reefs.

lagopodous (*BioSci*) Having feet covered by hairs or feathers.

Lagrange interpolation (*MathSci*) A method to approximate a function when values are known at n points: eg when $f(x)$ is known at three points x_1, x_2 and x_3 then

$$f(x) = f(x_1)\frac{(x-x_2)(x-x_3)}{(x_1-x_2)(x_1-x_3)}$$
$$+f(x_2)\frac{(x-x_1)(x-x_3)}{(x_2-x_1)(x_2-x_3)}$$
$$+f(x_3)\frac{(x-x_1)(x-x_2)}{(x_3-x_1)(x_3-x_2)}$$

Lagrange's dynamical equations (*MathSci*) The equations

$$\frac{d}{dt}\left(\frac{\delta T}{\delta q_i}\right) - \frac{\delta T}{\delta q_i} = Q_i, \quad i = 1,2,3,\cdots$$

where T is the kinetic energy of a system, q_i the various co-ordinates which define the position of the system and Q_i the so-called generalized forces; ie $Q_i\delta q_i$ is the work done by the external forces during a small displacement δq_i. These equations are characterized by being independent of any reactions internal to the system.

Lagrange's theorem (*MathSci*) If H is a subgroup of a finite group G, then the order of H divides the order of G and $|G| = |H|.|G/H|$.

Lagrangian point (*Astron*, *Space*) One of the five points associated with a binary system (particularly Earth–Moon), where a small third body can maintain an orbit. Owing to perturbations, three of these are in practice unstable. Also *libration point*.

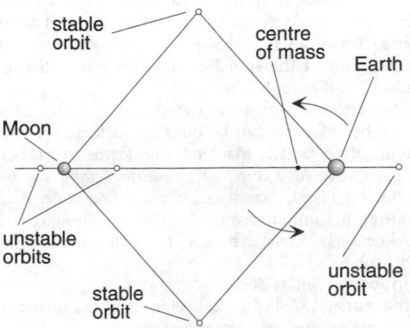

Lagrangian point The stable and unstable orbits at the Lagrangian points of the Earth–Moon system.

lahar (*Geol*) A mud flow of volcanoclastic material on the slopes of a volcano.

LAI (*BioSci*) Abbrev for LEAF AREA INDEX.

laid dry (*Build*) Said of bricks or blocks which have been laid without mortar.

laid-in moulding (*Build*) A moulding cut out of a separate strip of wood of the required section, and sunk in a special groove in the surface which it is intended to decorate.

laid-on moulding (*Build*) A PLANTED MOULDING.

laid paper (*Paper*) Writing or printing paper watermarked with a pattern of spaced parallel lines (*chain lines*) generally disposed in the machine direction and usually accompanied by more closely spaced parallel lines at right angles (*laid lines*).

laitance (*Build*) (1) The milky scum from grout or mortar, squeezed out when tesserae or tiles are pressed into place. (2) The milky scum formed on trowelled cement, concrete or rendering.

lake (*Geol*) A body of water lying on the surface of a continent, and unconnected (except indirectly by rivers) with the ocean. Lakes may be *freshwater lakes*, provided with an outlet to the sea: or *salt lakes*, occurring in the lowest parts of basins of inland drainage, with no connection with the sea. Lakes act as natural settling tanks, in which sediment carried down by rivers is deposited, containing the shells of molluscs etc. The lakes of former geological periods may thus be recognized by the nature of the sediments deposited in them and the fossils they contain. Lakes occur plentifully in glaciated areas, occupying hollows scooped out by the ice, and depressions lying behind barriers of morainic material.

lakes (*Chem*) Pigments formed by the interaction of dyestuffs and 'bases' or 'carriers', which are generally metallic salts, oxides or hydroxides. The formation of insoluble lakes in fibres, which are being dyed, is known as *mordanting*, the hydroxides of aluminium, chromium and iron generally being employed as *mordants*.

lalling (*Med*) Babbling speech of infants; lack of precision in the articulatory mechanism of the mouth. Also *lallation*.

Lamarckism (*BioSci*) A theory, now discredited, that evolutionary change takes place by the inheritance of acquired characters, ie that characters acquired during the lifetime of an individual (eg an athlete's strong muscles) are transmitted to its offspring.

lamb (*Agri*) A young sheep still with the ewe or up to 5 months old.

lambda chain (*BioSci*) One of the two types of light chain of immunoglobulins, the other being the KAPPA CHAIN.

lambda leak (*Phys*) The leakage of liquid helium II through holes so small that normal liquids cannot pass.

lambda particle (*Phys*) HYPERON with hypercharge 0 and isotopic spin 1.

lambda phage (*BioSci*) A well-studied temperate phage that can either grow in synchrony with its host (*E. coli*) in its LYSOGENIC phase or go into a *lytic* phase, when its genome is replicated many times by a *rolling circle* mechanism. An important vector in genetic manipulation procedures.

lambda point (*Phys*) (1) Transition temperature of helium I to helium II. (2) A temperature characteristic of a second-order phase change, eg ferromagnetic Curie point.

lamb dysentery (*Vet*) An acute and fatal toxaemic disease of newborn lambs caused by an intestinal infection by the bacterium *Clostridium perfringens* (*Cl. welchii*), Type B. The main symptom is diarrhoea due to enteritis. Vaccination widely used.

lambert (*Phys*) Unit of luminance or surface brightness of a diffuse reflector emitting 1 lm cm^{-2} or $10^4/\pi$ cd m^{-2}. The millilambert is used for low illuminations. See LUMEN.

Lambert's cosine law (*Phys*) A law stating that the energy emitted from a perfectly diffusing surface in any direction is proportional to the cosine of the angle which that direction makes with the normal.

Lambert's law (*Phys*) A law stating that the illumination of a surface on which the light falls normally from a point source is inversely proportional to the square of the distance of the surface from the source.

lambing sickness (*Vet*) See MILK FEVER.

Lamb shift (*Phys*) A small difference in the 2s and 2p energy levels in the hydrogen atom, which, according to the Dirac theory, should be the same. The effect is explained by the theory of QUANTUM ELECTRODYNAMICS.

lambswool roller (*Build*) A tool consisting of handle, frame and sheepskin fabric sleeve used for application of paints. The sleeves are made in short, medium and long pile to suit the appropriate paint and surface.

lamé (*Textiles*) Fabric with conspicuous decorative metallic threads.

Lamé constants (*Phys*) Constants which appear in the three-dimensional form of HOOKE'S LAW. For co-ordinate axes x, y and z and with the stresses (σ) expressed in terms of the strains (ε), the law is given by

$$\sigma_x = (\lambda + 2G)\,\varepsilon_x + \lambda\varepsilon_y + \lambda\varepsilon_z$$

where

$$\lambda = \frac{\nu E}{(1 + \nu)(1 - 2\nu)}$$

and the SHEAR MODULUS, G, are the Lamé constants, ν is POISSON'S RATIO and E is YOUNG'S MODULUS.

Lamé formula (*Eng*) A formula for calculating the stresses in thick (hydraulic) cylinders under elastic deformation.

lamell-, lamelli- (*Genrl*) Prefixes from Lt *lamella*, thin plate.

lamella (*BioSci*) (1) A thin plate or layer of cells. (2) A gill of an agaric. Adjs *lamellar, lamellate*.

lamellae (*Chem*) Plate-like microscopic crystals found in partially crystalline polymers. They usually consist of chain-folded molecules, and may possess regular edges (eg truncated lozenges) as well as having hollow centres. See CRYSTALLIZATION OF POLYMERS and panel on POLYMERS.

lamellar magnetization (*ElecEng*) Magnetization of a sheet or plate distributed in such a way that the whole of the front of the sheet forms one pole and the whole of the back forms the other.

lamellar phase (*BioSci*) See PHOSPHOLIPID BILAYER.

lamelli- (*Genrl*) See LAMELL-.

lamellibranch (*BioSci*) Having plate-like gills, as members of the group, Bivalvia.

lamina (*BioSci*) (1) Any flattened plate-like structure. (2) Expanded blade-like part of a leaf, as distinct from petiole and leafbase. See also NUCLEAR LAMINA.

lamina (*ElecEng*) Thin sheet steel.

lamina (*Geol*) See LAMINATION.

lamina propria (*BioSci*) A layer of connective tissue that supports the epithelium of the digestive tract and with it forms the mucous membrane. The lamina propria is the site at which lymphocytes, plasma cells, mast cells and accessory cells accumulate in immunological reactions involving the gut.

laminar flow (*Phys*) A type of fluid flow in which adjacent layers do not mix except on the molecular scale. Also *streamline flow*. See BOUNDARY LAYER, VISCOUS FLOW.

laminarin (*BioSci*) The storage polysaccharide of the brown algae, a β 1→3 glucan.

laminate (*Eng*) A structural member made from two or more components bonded together, eg two sheets separated by a honeycomb.

laminate (*Plastics*) A structural sheet material made from two or more dissimilar layers (laminae) bonded together, eg laminated glass, paper, plastics. See ANGLE-PLY LAMINATE, BALANCED LAMINATE, LAMINATED PLASTICS.

laminated arch (*Build*) An arch formed of successive thicknesses of planking, which are bent into shape and secured together with fastenings or glue.

laminated bearing (*Eng*) A type of bridge and building support for absorption of movement and vibration. Comprises alternate layers of steel and rubber sheet

(natural rubber or chloroprene rubber), laid perpendicular to main downward thrust of structure. See SHAPE FACTOR. Similar principle used in helicopter rotor bearings. See BRIDGE BEARING.

laminated bending (*Eng*) The practice of bending several layers of material and at the same time joining them along their surfaces in contact to form a unit.

laminated brush (*ElecEng*) A brush for an electric machine, made up of a number of layers insulated from one another, so that the resistance is greater across the brush than along its length (ie in the direction of normal current flow).

laminated-brush switch (*ElecEng*) A switch in which one or both of the contacts are laminated. See LAMINATED CONTACT.

laminated conductor (*ElecEng*) A conductor commonly used for armature windings of large machines or for heavy-current bus-bars; it is made up of a number of thin strips, in order to reduce eddy currents in the conductor or to make it more flexible.

laminated contact (*ElecEng*) A switch contact made up of a number of laminations arranged so that each lamination can be pressed into contact with the opposite surface, thereby giving a large area of contact and also a wiping action. Also *brush contact*.

laminated core (*ElecEng*) A core for a transformer or electric machine made up from insulated laminations for the purpose of reducing losses associated with EDDY CURRENTS.

laminated glass (*Glass*) See SAFETY GLASS.

laminated magnet (*ElecEng*) (1) A permanent magnet built up of magnetized strips to obtain a high intensity of magnetization. (2) An electromagnet for ac circuits, having a laminated core to reduce EDDY CURRENTS.

laminated paper (*Paper*) A product formed by bonding the whole of the surface of a sheet of paper to another paper or sheet material such as metal foil, plastics film, etc.

laminated plastics (*Plastics*) Superimposed layers of a synthetic resin-impregnated or coated filler (eg KRAFT PAPER or fibre reinforcement) which have been bonded together, usually by means of heat and pressure, to form a single piece. Also *laminate*.

laminated pole (*ElecEng*) A pole for the field windings of an electric machine, having the core built up of laminations to reduce EDDY CURRENTS caused by flux pulsations in the air-gap.

laminated record (*Acous*) A vinyl disk record in which the surface material differs from that in the inside, or core, in being finer grained and therefore freer from surface noise. The superior material is carried on a fine sheet of paper, being pressed onto the hot core in the press.

laminated spring (*Eng*) A flat or curved spring consisting of thin plates or leaves superimposed, acting independently, and forming a beam or cantilever of uniform strength. Also *carriage spring*.

laminated yoke (*ElecEng*) A yoke for an electric machine, built up of laminations; used in some forms of ac motors.

lamina terminalis (*BioSci*) In Craniata, the anterior termination of the spinal cord which lies at the anterior end of the diencephalon.

lamination (*ElecEng*) A sheet-steel stamping shaped so that a number of them can be built up to form the magnetic circuit of an electric machine, transformer, or other piece of apparatus. Also *core plate, punching, stamping*.

lamination (*Geol*) Stratification on a fine scale, each thin stratum, or *lamina*, frequently being a millimetre or less in thickness. Typically exhibited by shales and fine-grained sandstones.

laminboard (*Build*) See BLOCKBOARD.

laminectomy (*Med*) Surgical removal of the posterior arch or arches of one or more spinal vertebrae.

laminin (*BioSci*) A large fibrous protein that is a major component of the basal lamina.

laminitis (*Vet*) Inflammation of the sensitive laminae of the hoof of the horse and ox.

laminography (*Radiol*) See EMISSION TOMOGRAPHY.

lamins (*BioSci*) A group of three proteins, lamin A, B and C, of the INTERMEDIATE FILAMENT type. These proteins form the nuclear lamina.

Lamont's law (*ElecEng*) A law stating that the permeability of steel, at any flux density, is proportional to the difference between that flux density and the saturation value.

lamotrigine (*Pharmacol*) An anticonvulsant drug used to treat partial seizures.

lampas (*Vet*) A swelling of the palatal mucous membrane behind the upper incisor teeth of horses.

lampblack (*Chem*) The soot (and resulting pigment) obtained when substances rich in carbon (eg mineral oil, turpentine, tar, etc) are burnt in a limited supply of air so as to burn with a smoky flame. The pigment is black with a blue undertone, containing 80–85% carbon and a small percentage of oily material. See CARBON BLACK.

lampbrush chromosome (*BioSci*) A chromosome of the first meiotic prophase, observed in many EUKARYOTES, that has a characteristic appearance due to the orderly series of lateral loops of chromatin, arranged in pairs on either side of the chromosome axis.

lamphouse (*ImageTech*) That part of a projector, printer or enlarger which contains the source of illumination.

lamping (*MinExt*) Use of ultraviolet light to detect fluorescent minerals either when prospecting or in checking concentrates during ore treatment.

lamp man (*MinExt*) Colliery surface worker in charge of miners' lamps, working in lamp room or cabin and controlling repairs, recharge, issue, lighting, etc, of portable lamps.

lamp resistance (*ElecEng*) A resistance consisting of one or more electric filament lamps.

lamproite (*Geol*) A variety of lamprophyric extrusive rock rich in potassium and magnesium. In the Argyle mine in W Australia it has become a major source of the world's diamonds.

lamprophyres (*Geol*) Igneous rocks usually occurring as dykes intimately related to larger intrusive bodies; characterized by abnormally high contents of coloured silicates, such as biotite, hornblende and augite, and a correspondingly small amount of feldspar, some being feldspar-free. See ALNÖITE, CAMPTONITE, KERSANTITE, MINETTE, MONCHIQUITE, SPESSARTITE, VOGESITE.

lamp working (*Glass*) Making articles, usually from glass tubing or rod, with the aid of an oxy-gas or air–gas flame.

lamziekte (*Vet*) A form of botulism occurring in cattle in S Africa due to the ingestion, in phosphorus-deficient areas, of bones contaminated with *Clostridium botulinum*, Type D.

LAN (*ICT*) Abbrev for LOCAL AREA NETWORK.

lanarkite (*Min*) A rare monoclinic sulphate of lead, occurring with anglesite and leadhillite (into which it easily alters), as at Leadhills, Lanarkshire, Scotland.

lanate (*BioSci*) Woolly.

LANC (*ImageTech*) Abbrev for *local area network control bus*. See SYNCHRO-EDIT.

Lancashire boiler (*Eng*) A cylindrical steam boiler having two longitudinal furnace tubes containing internal grates at the front. After leaving the tubes the gases pass to the front along a bottom flue, and return to the chimney along side or wing flues.

lance (*Build*) A sharp scribing part of a wood-cutting tool, serving to cut through the grain in advance and on each side of the cutting tool proper.

lanceolate (*BioSci*) Lance-shaped; much longer than broad and tapering at both ends.

lancet arch (*Arch*) A sharply pointed arch, of greater rise than an equilateral arch of the same span.

lancet window (*Arch*) A tall narrow window surmounted by a lancet arch.

lancewood (*For*) Durable straight-grained wood chiefly from *Oxandra lanceolata*, a native of tropical America.

lancinating (*Med*) Of pain, acute, shooting, piercing, cutting.

lancing (*Eng*) A line cut made in a press which does not remove metal but only separates it.

land and sea breezes (*EnvSci*) Winds occurring at the coast during fine summer weather, esp when the general pressure gradient is small. During the day, when the land is warmer than the sea, the air over the land becomes warmer and less dense than that over the sea and a local circulation is created with air flowing from land to sea at high levels and from sea to land near the surface. At night, conditions are reversed. The sea breeze can penetrate many miles inland and can become strong at low latitudes.

Landau levels (*Phys*) The allowed energy levels of CONDUCTION ELECTRONS of a solid in a magnetic field. The electrons will describe complete orbits if $\omega\tau > 1$ where ω is the CYCLOTRON FREQUENCY and τ is the time between scattering events. The electron density of states will be altered and the Landau levels will differ by $\hbar\omega$ where $\hbar$ is Dirac's constant, equal to PLANCK'S CONSTANT h divided by 2π. See DE HAAS–VAN ALPHEN EFFECT.

Landau theory (*Phys*) (1) Theory for calculating diamagnetic susceptibility produced by free conduction electrons. (2) That explaining the anomalous properties of liquid helium II in terms of a mixture of normal and superfluids. See HELIUM, SUPERFLUIDITY.

Land camera (*ImageTech*) See POLAROID CAMERA.

Landé splitting factor (*Phys*) The factor employed in the calculation of the splitting of atomic energy levels by a magnetic field. See ZEEMAN EFFECT.

landfill (*Eng*) The disposal of solid refuse by depositing it in relatively thin layers in trenches prepared on low-lying, flat or reclaimed land. Each successive layer of refuse is compacted and covered with a thin layer of earth, chalk, furnace ash, gravel, etc, so that the level of the ground is gradually raised.

Landfill Directive (*EnvSci*) A European directive (1998) covering waste disposal. It sets targets for reduction, pretreatment and the avoidance of mixing of hazardous and non-hazardous waste.

landfill gas (*EnvSci*) Gas generated in landfill sites by the decomposition of organic material. It is mostly methane and carbon dioxide but with minor traces of odiferous compounds.

landfill site (*EnvSci*) A site at which waste is buried. In a well-regulated site the burial site is lined and once the filling and compaction phases are complete the site is capped with clay and topsoil with provision for drainage and for safe venting of landfill gas.

landing (*MinExt*) A stage in the hoisting shaft at which cages are loaded or discharged.

landing area (*Aero*) That part of the movement area of an unpaved airfield intended primarily for take-off and landing.

landing beacon (*Aero, ICT*) A transmitter used to produce a LANDING BEAM.

landing beam (*Aero, ICT*) A beam from a transmitter along which an aircraft approaches a landing field during blind landing. See INSTRUMENT LANDING SYSTEM.

landing direction indicator (*Aero*) A device indicating the direction in which landings and take-offs are required to be made, usually a T, towards the cross bar of which the aircraft is headed.

landing gear (*Aero*) That part of an aircraft which provides for its support and movement on the ground, and also for absorbing the shock on landing. It comprises main support assemblies incorporating single or multi-wheel arrangements, and also auxiliary supporting assemblies such as nose wheels, tail wheels or skids. See BOGIE-, DRAG STRUTS, OLEO.

landing ground (*Aero*) In air transport, any piece of ground that has been prepared for landing of aircraft as required; not necessarily a fully equipped airfield. An *emergency landing ground* is any area of land that has been surveyed and indicated to pilots as being suitable for forced or emergency landings. See AIRSTRIP.

landing parachute (*Aero*) See BRAKE PARACHUTE.

landing procedure (*Aero*) The final approach manoeuvres, beginning when the aircraft is in line with the axis of the runway, either for the landing, or upon the reciprocal for the purpose of a procedural turn, until the actual landing is made, or *overshoot* action has to be taken.

landing speed (*Aero*) The minimum airspeed, with or without engine power, at which an aircraft normally alights.

landing wires (*Aero*) Wires or cables which support the wing structure of a biplane when on the ground and also the negative loads in flight. Also *anti-lift wires*.

landrace (*BioSci*) Ancient or primitive cultivar of a crop plant.

Landsat (*Space*) One of a successful continuing series of Earth observation satellites developed by NASA, operating from July 1972; originally named *Earth Resources Technology Satellites* (ERTS). The current version, Landsat 7, should function until 2007.

Landsberger apparatus (*Chem*) An apparatus for the determination of the boiling point of a solution by using the vapour of the solvent to heat the solution. Prevents errors due to superheating.

landscape (*Print*) A term which indicates that the breadth of the page is greater than the depth: eg 138 × 216 mm, the convention being to state the upright edge first. Also applied to a similarly shaped full-page book illustration or map imposed at right angles to the main text. US *horizontal format*.

landscape lens (*ImageTech*) Photographic objective of meniscus form, generally with the concave surface towards the object and with the diaphragm on the object side of the lens.

landscape marble (*Geol*) A type of limestone containing markings resembling miniature trees etc when polished; the surface has the appearance of a sepia drawing.

landslip (*Geol*) The sudden sliding of masses of rock, soil, or other superficial deposits from higher to lower levels, on steep slopes. Landslips on a very large scale occur in mountainous districts as a consequence of earthquake shocks, stripping the valley sides bare of all loose material. In other regions landslips occur particularly where permeable rocks, lying on impermeable shales or clays, dip seawards or towards deep valleys. The clays hold up water, becoming lubricated thereby, and the overlying strata, fractured by joints, tend to slip downhill, a movement that is facilitated on the coast by marine erosion.

land treatment (*Build*) The final or oxidizing stage in sewage treatment, in which the liquid sewage is distributed over an area of land, through which it filters to underdrains. If the land will not permit of easy filtering, the sewage is applied to one plot of land by irrigation, and is then passed on to a second, third and fourth plot, before final discharge into a stream.

Langerhans cell (*BioSci*) (1) A type of DENDRITIC CELL present in the epidermis, very important in antigen presentation to T-lymphocytes. MHC Class II antigens are strongly expressed. (2) Spindle-shaped cells in the centre of each acinus of the pancreas. See ISLET CELLS.

Langerhans islets (*BioSci*) See ISLET CELLS, LANGERHANS CELL (2).

Langevin equation (*Phys*) (1) A classical expression for diamagnetic susceptibility produced by the orbital electrons of atoms. (The quantum mechanical equivalent of this was derived subsequently by W Pauli.) (2) An expression for the resultant effect of atomic magnetic moments which enters into explanations of both FERROMAGNETISM and PARAMAGNETISM. See panel on FERROMAGNETICS AND FERRIMAGNETICS.

Langhian (*Geol*) A stage in the Miocene. See TERTIARY.

langite (*Min*) A very rare ore of copper occurring in Cornwall, blue to greenish-blue in colour; essentially hydrated copper sulphate, crystallizing in the orthorhombic system.

lang lay (*Eng*) A method of making wire ropes in which the wires composing the strands, and the strands themselves are laid in the same direction of twist. Cf CABLE-LAID ROPE.

Langmuir adsorption isotherm (*Chem*) The fraction of the adsorbent surface which is covered by molecules of adsorbed gas is given by $\theta = bp(1 + bp)$ where p is the gas pressure and b is a constant.

Langmuir–Blodgett film (*Chem*) Monomolecular (organic) assemblies on a substrate. An integral part of MOLECULAR ELECTRONICS.

Langmuir dark space (*Electronics*) Non-glowing region surrounding a negative electrode placed in the luminous positive column of a gas discharge.

Langmuir frequency (*Electronics*) The natural frequency of oscillation for electrons in a plasma.

Langmuir law (*Electronics*) See CHILD–LANGMUIR EQUATION.

Langmuir probe (*Electronics*) Electrode(s) introduced into gas-discharge tube to study potential distribution along the discharge.

Langmuir's theory (*Chem*) (1) The assumption that the extranuclear electrons in an atom are arranged in shells corresponding to the periods of the periodic system. The chemical properties of the elements are explained by supposing that a complete shell is the most stable structure. (2) The theory that adsorbed atoms and molecules are held to a surface by residual forces of a chemical nature.

Langmuir trough (*MinExt*) Apparatus in which a rectangular tank is used to measure surface tension of a liquid.

language (*ICT*) See ASSEMBLY LANGUAGE, FORMAL LANGUAGE, HIGH-LEVEL LANGUAGE, LOW-LEVEL LANGUAGE, NATURAL LANGUAGE, PROGRAMMING LANGUAGE.

language acquisition device (*Psych*) According to Chomsky, a hypothetical brain structure that enables an individual to learn the rules of grammar on exposure to spoken language. Abbrev *LAD*.

laniary (*BioSci*) Adapted for tearing, as a canine tooth.

La Niña (*EnvSci*) A meteorological phenomenon in which unusually cold ocean temperatures in the tropical Pacific cause extreme weather conditions.

lanolin (*Chem*) Fat from wool (*adeps lanae*), a yellowish viscous mass of wax-like consistency, very resistant to acids and alkalis; it emulsifies easily with water and is used for making ointments. It consists of the palmitate, oleate and stearate of cholesterol.

lansfordite (*Min*) Hydrated magnesium carbonate, crystallizing in the monoclinic system. Lansfordite occurs in some coal mines but is not stable on exposure to the atmosphere and becomes dehydrated.

lansoprazole (*Pharmacol*) A *proton-pump inhibitor* drug used in the treatment of gastric ulcers.

lantern (*Arch*) An erection on the top of a roof, projecting above the general roof level, and usually having glazed sides to admit light, as well as openings for ventilation.

lanthanide contraction (*Chem*) The peculiar characteristic of the LANTHANIDE SERIES that the ionic radius decreases as the atomic number increases, because of the increasing pull of the nuclear charge on the unchanging number of electrons in the two outer shells. Thus the elements after lanthanum, eg platinum, are very dense and have chemical properties very similar to their higher homologues, eg palladium.

lanthanides (*Chem*) The rare earth elements of at nos 57–71, after lanthanum, the first of the series. Cf ACTINIDES. In both series an incomplete *f*-shell is filling. Also *lanthanide series*.

lanthanum (*Chem*) Symbol La, at no 57, ram 138·91, mp 921°C. A metallic element in the third group of the periodic system, belonging to the rare earth group.

lanthanum glass (*Glass*) Optical glass used for high-quality photographic lenses etc with high REFRACTIVE INDEX and LOW DISPERSION.

lanuginose (*BioSci*) Woolly.

lanugo (*BioSci*) In mammals, prenatal hair.
lap (*Build*) The length of overlap (6–10 cm, $2\frac{1}{2}$–4 in) of a slate over the slate next but one below it, in centre-nailed work; or that of a slate over the nail securing the slate next but one below it, in head-nailed work.
lap (*Eng*) (1) A surface defect on rolled or forged steel. It is caused by folding a fin to the surface and squeezing it in; as welding does not occur, a seam appears on the surface. (2) Soft metal embedded with abrasive powder used in LAPPING. (3) Polishing cloth impregnated with diamond dust or other abrasive, used in polishing rock specimens, gemmology, etc. (4) In mining, one coil of rope on the mine hoisting drum.
lap (*Glass*) (1) A rotating disk or other tool for grinding or polishing glass. (2) A square piece of material, usually rubber, to protect the hands when handling glass.
lap (*Textiles*) (1) A sheet of fibres or fabric wrapped round a core (eg of wood) for transfer to the next process, eg the rolled sheet of fibres collected from the opening machine ready for carding. (2) The undesired build-up of fibres on a roller, eg after a yarn break in spinning. (3) The length of fabric between folds when it is being plaited into a vertical pile.
laparoscopy (*Med*) The insertion of a rigid or flexible device to inspect the abdominal cavity. Simple surgical operations, eg TUBAL LIGATION, may be carried out using a laparoscope.
laparotomy (*Med*) Cutting into the abdominal cavity. *Exploratory laparotomy*, the operation of cutting into the abdominal cavity so that direct examination of abdominal organs may be made.
lap dissolve (*ImageTech*) A technique by which one scene in a film fades out while the next scene fades in, so that the two images momentarily overlap.
lap dovetail (*Build*) An angle joint between two members, in which only one shows end grain, a sufficient thickness of wood having been left on this member, in cutting the joint, to cover the end grain of the other member. Also *drawer front dovetail*, from one of its common uses.
lapel microphone (*Acous*) Small microphone which does not impede vision of the speaker.
lapidicolous (*BioSci*) Living under stones.
lapilli (*Geol*) Small rounded pieces of lava whirled from a volcanic vent during explosive eruptions; lapilli are thus similar to volcanic bombs but smaller in size, with a mean diameter 2–64 mm. Sing *lapillus*.
lapilli tuff (*Geol*) A compact pyroclastic rock composed of lapilli.
lapis lazuli (*Min*) Original sapphire of ancients, a beautiful blue stone used extensively for ornamental purposes. It consists of the deep-blue feldspathoid LAZURITE, usually together with calcite and spangles of pyrite.
lap joint (*Build*) A joint between two pieces of timber, formed by laying one over the other for a certain length and fastening the two together with metal straps passing around the timbers, or with bolts passing through them.
lap joint (*Eng*) A plate joint in which one member overlaps the other, the two being riveted or welded along the seam.
Laplace linear equation (*MathSci*) A differential equation of the form

$$(a_0 + b_0 x)\frac{d^n y}{dx^n} + (a_1 + b_1 x)\frac{d^{n-1}y}{dx^{n-1}} + \cdots$$
$$+ (a_n + b_n x)y = 0$$

Laplace's equation (*MathSci*) The equation $\nabla^2 V = 0$, where V is the vector operator nabla, ie

$$\frac{\partial^2 V}{\partial x^2} + \frac{\partial^2 V}{\partial y^2} + \frac{\partial^2 V}{\partial z^2} = 0$$

It is satisfied by electric and gravitational potential functions. Cf POISSON'S EQUATION.

Laplace transform (*MathSci*) The Laplace transform $F(p)$ of the function $f(t)$ is defined by

$$F(p) = \int_0^\infty e^{-pt} f(t)\, dt$$

In the *two-sided* Laplace transform the range of integration is from $-\infty$ to $+\infty$.
La Pointe picker (*MinExt*) Small belt conveyor on which ore is so displayed that radioactive pieces are removed as they pass a Geiger–Müller counter.
lapping (*ElecEng*) The final abrasive polishing of a quartz crystal to adjust its operating frequency. Also, smoothing of surface of crystalline semiconductors.
lapping (*Eng*) The finishing of spindles, bored holes, parallel pieces, etc, to fine limits, by the use of laps of lead, brass, etc, impregnated with abrasive paste.
lapping (*ImageTech*) Rubbing one surface against another, generally with an abrasive such as rouge, so that the softer takes up the contour of the harder, eg in polishing a lens or optical flat.
lapping machine (*Eng*) A machine tool for finishing the bores of cylinders etc to fine limits by the use of revolving circular laps supplied with an abrasive powder suspended in the coolant.
lapse rate (*EnvSci*) The rate of fall of a quantity with increase in height.
lap-shear test (*Eng*) A method of testing adhesives using a LAP JOINT, which is then tensioned, putting the joint into a state of shear.
laptop (computer) (*ICT*) A portable computer with its own power supplies and hard disk with a solid-state flat screen and a keyboard.
lap winding (*ElecEng*) A form of two-layer winding for electric machines in which each coil is connected in series with the one adjacent to it. Cf WAVE WINDING.
Laramide orogeny (*Geol*) In the narrow sense, that mountain-building movement associated with the production of the Rocky Mountains between Late Cretaceous and Palaeocene times. More broadly sometimes embraces all orogenies that took place during that span of time.
lardalite (*Geol*) See LAURDALITE.
large-area foils (*Print*) Blocking foils for display card printing. These are capable of filling in very large areas and yet of giving fine detail for half-tone work.
large calorie (*Phys*) See CALORIE.
large crown octavo (*Print*) The book size $7\frac{7}{8} \times 5\frac{1}{2}$ in (198 × 129 mm).
Large Hadron Collider (*Phys*) A 27 km ring particle accelerator to be completed at CERN in Geneva scheduled (currently) to start operation in November 2007. Expected to collide protons at energies of around 13 TeV. Abbrev LHC.
large intestine (*BioSci*) See COLON.
large-scale integration (*Electronics*) Fabricating a very large number of electron devices on a single chip. Abbrev LSI.
Lariam (*Pharmacol*) TN for *mefloquine*, an antimalarial drug.
larmier (*Build*) A corona placed over a door or window opening to serve as a dripstone.
Larmor frequency (*Electronics*) The angular frequency of precession for the spin vector of an electron acted on by an external magnetic field.
Larmor precession (*Electronics*) The precessional motion of the orbit of a charged particle when subjected to a magnetic field. Precession occurs about the direction of the field.
Larmor radius (*Phys*) The radius of the circular or helical path followed by a charged particle in a uniform magnetic field.
larnite (*Min*) A rare calcium silicate, Ca_2SiO_4, formed at very high temperatures in contact-metamorphosed limestone.

larry (*Build*) A tool having a curved steel blade fixed to the end of a long handle, to which it is bent normally; used for mixing mortar, or for mixing hair with coarse stuff to form a plaster.

larrying (*Build*) The process of pouring a mass of mortar upon the wall and working it into the joints; sometimes used in building large masses of brickwork.

larva (*BioSci*) The young stage of an animal if it differs appreciably in form from the adult; a free-living embryo.

larvikite (*Geol*) See LAURVIKITE.

larviparous (*BioSci*) Giving birth to offspring that have already reached the larva stage.

larvivorous (*BioSci*) Larva-eating.

laryngectomy (*Med*) Surgical removal of the larynx.

laryngismus (*Med*) Spasm of the larynx.

laryngitis (*Med*) Inflammation of the larynx.

laryngofissure (*Med*) Surgical exposure of the larynx by dividing the thyroid cartilage (Adam's apple) in the midline. Also *thyrotomy*.

laryngology (*Med*) That branch of medical science which treats diseases of the larynx and adjacent parts of the upper respiratory tract. N *laryngologist*.

laryngopharyngitis (*Med*) Inflammation of the larynx and the pharynx.

laryngophone (*Acous*) See THROAT MICROPHONE.

laryngoscope (*Med*) An instrument used for viewing the larynx.

laryngostenosis (*Med*) Pathological narrowing of the larynx.

laryngostomy (*Med*) The surgical formation of an opening in the larynx.

laryngotomy (*Med*) The operation of cutting into the larynx.

laryngotracheal chamber (*BioSci*) In amphibians, a small chamber into which the lungs open anteriorly, and which communicates with the buccal cavity by the glottis.

larynx (*BioSci*) An organ in the throat of humans and vertebrates which with the lungs forms the voice.

LASEK (*Med*) Abbrev for *laser-assisted epithelial keratomileusis* (or *keratectomy*), a surgical procedure in which tissue on the surface of the cornea is reshaped with a laser.

laser (*Phys*) Abbrev for *light amplification by stimulated emission of radiation*. See panel on LASER.

laser-beam cutting (*Eng*) Using the intense narrow beam of radiation from a laser to cut often complex shapes in sheet or plate. Good finish and high precision can be achieved.

laser-beam machining (*Eng*) The use of a focused beam of high-intensity radiation from a laser to vaporize and so machine material at the point of focus. Cf LASER-BEAM CUTTING.

laser card (*ICT*) A plastic card with digital information stored in the same way as on a LASER DISK.

laser compression (*NucEng*) See INERTIAL CONFINEMENT, LASER FUSION.

laser diode (*Electronics*) See OPTOELECTRONICS

laser disk (*ImageTech*) A play-only disk system carrying analogue video and digital audio, represented by microscopic pits of varying lengths in its surface which alter the brightness of a laser beam as it is reflected to a photosensor. LaserActive (TN) is an interactive variant. Abbrev *LD*. See CONSTANT ANGULAR VELOCITY, CONSTANT LINEAR VELOCITY, MASTER.

laser enrichment (*Phys*) The enrichment of uranium isotopes using a powerful laser to ionize atoms of the selected isotope. See panel on URANIUM ISOTOPE ENRICHMENT.

laser fusion (*Phys*) The initiation of the nuclear fusion process by directing energy from a laser beam onto the fusion fuel contained in pellets. The laser both heats and compresses the material. Fig. ▷

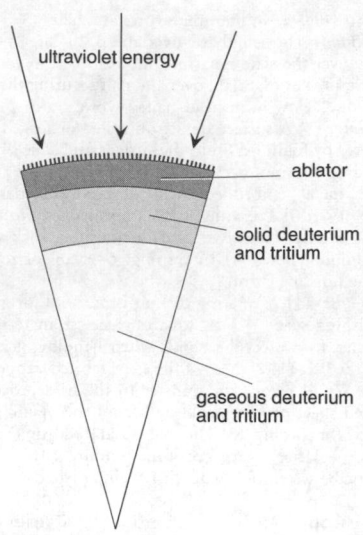

laser fusion Segment of spherical pellet.

laser fusion reactor (*NucEng*) A proposed type of fusion reactor in which pellets of deuterium and tritium are contained in a small target sphere and bombarded by a pulsed laser beam. Also *inertial fusion system*. See MAGNETIC CONFINEMENT FUSION SYSTEM.

laser gyro (*Eng*) An integrating rate gyroscope which combines the properties of the optical oscillator, the laser and general relativity. It differs from conventional gyros in the absence of a spinning mass and so its performance is not affected by accelerations. It is an instrument, usually triangular with internally reflecting prisms at the corners, which senses angular rotation in its plane by measuring the frequency shift of laser energy passing around the circuit. It has a low power consumption and is not subject to prolonged starting time; used in missile guidance systems. Modern versions can measure rates as low as $0 \cdot 1\underline{o}$ per hour.

laser level (*Surv*) An instrument for accurate single-handed levelling. A laser beam, narrowly confined in the vertical dimension, is continuously rotated around an accurately vertical axis. The staff has a cursor which is able to sense the laser beam and indicate its height which is then read.

laser pointer (*Genrl*) A small hand-held device that projects a laser beam, usually red, as a directional pointer for slide shows etc.

laser printer (*ICT*) A printer that uses a laser beam to form characters on paper. Fig. ▷

laser threshold (*Phys*) The minimum pumping power (or energy) required to operate a laser.

lashing (*MinExt*) A S African term for removing broken rock after blasting. Canadian term, *mucking, mucking out*.

lash-up (*ICT*) The temporary connection of apparatus for experimental or emergency use.

LASIK (*Med*) Abbrev for *laser-assisted in situ keratomileusis*, a surgical procedure in which a flap is created in the cornea and the underlying tissue is reshaped with a laser.

Lassa fever (*Med*) A viral haemorrhagic fever which occurs sporadically in rural W Africa. The animal reservoir is a wild rodent, *Mastomys natalensis*. Causes a severe illness with shock and has a high mortality.

Lassaigne's test (*Chem*) A test for the presence of nitrogen in an organic substance. The sample is heated with metallic sodium in a test tube; the product is placed in water and filtered. Iron (II) sulphate is added and the mixture boiled, cooled, and sodium cyanide and iron (III) chloride added. If nitrogen is present the characteristic *Prussian blue* colour is observed.

Laser

Abbrev for *light amplification by stimulated emission of radiation*. It is a source of intense monochromatic light in the ultraviolet, visible or infrared region of the spectrum, and operates by producing a large population of atoms with their electrons in a certain high energy level. By STIMULATED EMISSION, transitions to a lower level are induced, the emitted photons travelling in the same direction as the stimulating photons. If the beam of inducing light is produced by reflection from mirrors or BREWSTER WINDOWS at the ends of a resonant cavity, the emitted radiation from all stimulated atoms is in phase or *coherent*, and the output is a very narrow beam of coherent monochromatic light. Solids, liquids and gases have been used as lasing materials. They can vary in size from those used in printers and compact disc readers, or even smaller, to powerful lasers which can cut thick metal, separate uranium isotopes or implode fusion devices.

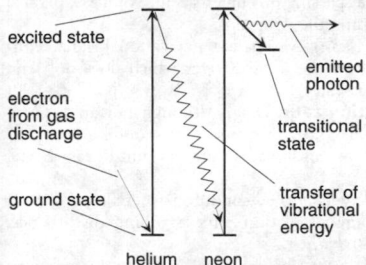

Laser Energy states between the initial gas discharge electron and the final monoenergetic photon.

The quantum photoelectric theory shows how a photon with the right energy can cause an electron to jump to a more energetic orbit, but, in addition, an electron already in a more energetic orbit will be forced to return to its original orbit if it collides with an electron with the appropriate energy. In doing so it will emit an additional photon with the same energy and the same phase, and travelling in the same direction. This is called STIMULATED EMISSION. Normally most electrons will be in their least energetic state, but in special conditions it is possible to force a high proportion into a more energetic state so that a POPULATION INVERSION occurs. An incoming electron can then cause a burst of coherent and monochromatic light.

There are many kinds of laser depending on the output and wavelength required, but the helium–neon laser can illustrate the principle. If helium at low pressure is placed in a gas-discharge tube, the electrons passing from the cathode to the anode will excite the helium atoms into a high-energy state (see diagram) in which the excited atoms can collide with any other atoms that are present. It so happens that neon also has an excited state at the same energy level as helium, and if excited helium atoms collide with rest-state neon atoms, there is a good chance of the neon atoms moving into their excited state. By adjusting the proportions of helium and neon in the tube, a high proportion of the neon atoms can be maintained in the high-energy state and a population inversion will occur. An appropriate incoming photon can now force a neon electron into a lower-energy state and emit another photon which then stimulates another neon electron, producing a pulse of highly coherent light. The output can be further enhanced by having a fully reflecting mirror at one end and a semi-reflecting mirror at the other end of the tube and adjusting its length (see MODE-LOCKING). This will cause the photons to traverse the tube many times, stimulating further emissions before they finally escape.

Powerful lasers are often arranged in tandem so that the stimulated emission of one laser can cause the population inversion in a second and so enhance the power of the whole system. See CARBON DIOXIDE LASER, MASER, Q-SWITCHING.

---

last-in, first-out (*ICT*) See STACK.
last-subscriber release (*ICT*) The release of automatic switching plant when the last of both subscribers has replaced the receiver and opened the loop. Also *last-party release*.
latanoprost (*Pharmacol*) A prostaglandin analogue used to reduce pressure in the eye caused by glaucoma. It acts by increasing the efflux of aqueous fluid.
latching (*ICT*) Arrangement whereby a circuit is held in position, eg in read-out equipment, until previous operating circuits are ready to change this circuit. Also *locking*.
latching (*Surv*) See DIALLING.
latch needle (*Textiles*) See NEEDLE.
late-choice call meter (*ICT*) A traffic meter so connected as to record the number of calls carried by a late-choice trunk of a grading.

latency (*ICT*) (1) General term for the time that elapses between the issuing of an instruction or the presentation of data for transmission and the start of a system's response. (2) The period between transmitting and receiving a signal via a COMMUNICATIONS SATELLITE (panel), about 0·24 seconds round trip for one in geostationary orbit. (3) Delay, in digital computers, between the initiation of the call for data and the start of the transfer. Latency forms part of the total ACCESS TIME.
latency (*Psych*) General term for the interval before some reaction; also, the dormancy of a particular behaviour or response.
latency period (*Psych*) According to Freud, the fourth psychosexual stage of development, in which sexuality lies essentially dormant, occurring roughly between the age of 6 and the onset of puberty. Also *latency stage*.

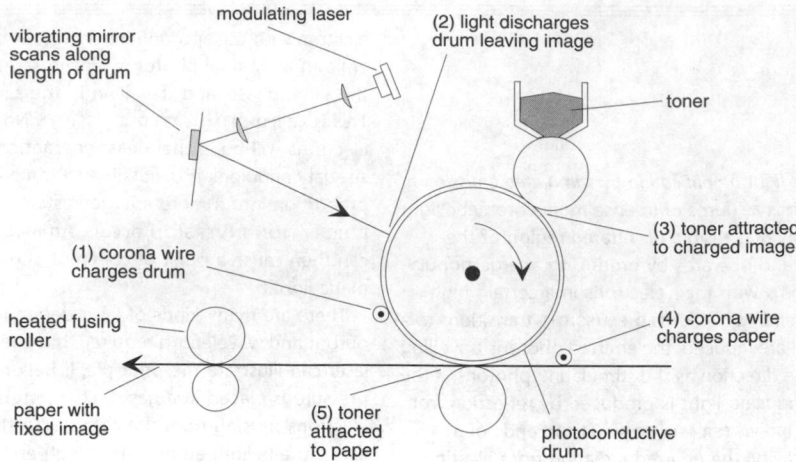

modulating laser

vibrating mirror
scans along
length of drum

(2) light discharges
drum leaving image

toner

(3) toner attracted
to charged image

(1) corona wire
charges drum

heated fusing
roller

(4) corona wire
charges paper

paper with
fixed image

(5) toner
attracted
to paper

photoconductive
drum

laser printer The numbers refer to the main sequence of events.

latensification (*ImageTech*) Increasing the LATENT IMAGE before development, usually by a long uniform exposure to light of very low intensity, but chemical methods can be used.

latent (*BioSci*) In a resting condition or state of arrested development, but capable of becoming active or undergoing further development when conditions become suitable; said also of hidden characteristics that may become evident under the right circumstances. Latent viruses are integrated into the host genome but are inactive unless triggered eg by stress.

latent content (*Psych*) In psychoanalytic theory, the unconscious material or hidden meaning of a dream that is being expressed in a disguised fashion through symbols contained in the dream. See MANIFEST DREAM CONTENT.

latent heat (*Phys*) See SPECIFIC LATENT HEAT.

latent image (*ImageTech*) The invisible image formed in a sensitive emulsion by molecular ionization in the silver halide grains affected by exposure to radiation; it is made visible by the process of DEVELOPMENT.

latent instability (*EnvSci*) A type of CONDITIONAL INSTABILITY of the atmosphere which exists only if a rising parcel of air reaches a critical level.

latent learning (*Psych*) Learning the characteristics of a situation which is not manifested by any immediate overt behaviour, but which is revealed when the individual is later placed in a situation requiring previously acquired information.

latent magnetization (*Phys*) The property possessed by certain feebly magnetic metals (eg manganese and chromium) of forming strongly magnetic alloys or compounds.

latent neutrons (*Phys*) In reactor theory, the delayed neutrons due from (but not yet emitted by) fission products.

latent period (*BioSci*) (1) The period of time between stimulation and the first signs of a response. (2) The period of time between infection and the appearance of the symptoms of disease.

latent period (*Radiol*) That between exposure to radiation and its effect.

latent roots (*MathSci*) See CHARACTERISTIC EQUATION OF A MATRIX.

latent virus (*BioSci*) Virus integrated within host genome but inactive; may be reactivated by stress such as ultraviolet irradiation.

later (*Build*) A brick or tile.

lateral (*BioSci*) Arising from a parent axis or the structure so formed. Cf LEADER.

lateral (*Genrl*) Situated on or at, or pertaining to, a side.

lateral axis (*Aero*) The cross-wise axis of an aircraft, particularly that passing through its centre of mass, parallel to the line joining the wing tips.

lateral canal (*Build*) A separate navigational canal constructed to follow the lie of a river which does not lend itself to canalization.

lateral contraction ratio (*Eng*) Although not an ELASTIC CONSTANT, it is the parameter corresponding to POISSON'S RATIO in anisotropic and/or non-linear elastic materials.

lateral instability (*Aero*) A condition wherein an aircraft suffers increasing oscillation after a rolling disturbance. Also *rolling instability*.

lateralization (*Psych*) The organization of brain functions such that each of the cerebral hemispheres controls different psychological functions, particularly in verbal and visual spatial skills. In most right-handers, the left hemisphere is specialized for language functions, while the right hemisphere is better at various visual and spatial tasks. Also *laterality*.

lateral line (*BioSci*) (1) In fish, a line of neuromast organs running along the side of the body. (2) In Nematoda, a paired lateral concentration of hypodermis containing the excretory canal.

lateral load (*Eng*) A force acting on a structure or a structural member in a transverse direction, eg wind forces on a bridge or building at right angles to its length, which trusses and girders are not primarily designed to withstand.

lateral meristem (*BioSci*) A MERISTEM lying parallel to the sides of the axis, eg cambium.

lateral moraine (*Geol*) A low ridge formed at the side of a valley glacier. It is composed of material derived from glacial plucking and abrasion or material that has fallen onto the ice.

lateral plate (*BioSci*) (1) In Craniata, a ventral portion of each mesoderm band that surrounds the mesenteron in embryo. (2) In insects, the paired lateral region of the germ band of the embryo that becomes separated from the opposite member of its pair by the middle plate.

lateral recording (*Acous*) A recording in which the cutting stylus removes a thread (*swarf*) from the surface of a blank disk, the modulation being realized as a lateral (radial) deviation as the spiral is transversed, in contrast with the obsolete method of *vertical recording*. Also *radial recording*. Cf STEREOPHONIC RECORDING.

lateral shift (*Geol*) The displacement of outcrops in a horizontal sense, as a consequence of faulting. Cf THROW.

lateral stability (*Aero*) The stability of an aircraft's motions out of the plane of symmetry, ROLLING, SIDESLIP, YAW.

lateral traverse (*Eng*) The longitudinal play given to locomotive trailing axles to permit taking sharp curves.

lateral velocity (*Aero*) The rate of SIDESLIP of an aircraft, ie the component of velocity which is resolved in the direction of the lateral axis.

laterigrade (*BioSci*) Moving sideways, as some crabs.

laterite (*Geol*) A residual clay formed under tropical climatic conditions by the weathering of igneous rocks, usually of basic composition. Consists chiefly of hydroxides of iron and aluminium. See BAUXITE.

laterization (*Geol*) The process whereby rocks are converted into laterite. Essentially, the process involves the abstraction of silica from the silicates. See LATERITE.

laterosphenoid (*BioSci*) The so-called 'alisphenoid' of fish, reptiles and birds (representing an ossification of the wall of the chondrocranium), as distinct from the alisphenoid of mammals (developed from the splanchnocranium).

late wood (*BioSci*) The wood formed in the later part of a growth ring, usually having smaller cells with thicker walls than the EARLY WOOD.

latex (*BioSci*) Fluid, often milky, exuded from cells and vessels (laticifers) when many plants are cut. It consists of a watery solution containing many different substances including terpenoids which form rubber, alkaloids such as the opium alkaloids, sugar, starch, etc.

latex (*Chem*) (1) A milky viscous fluid extruded when a rubber tree (eg *Hevea brasiliensis*) is tapped. It is a colloidal system of CAOUTCHOUC dispersed in an aqueous medium, density $990 \, \text{kg m}^{-3}$, which forms rubber by coagulation. The coagulation of latex can be prevented by the addition of ammonia or formaldehyde. Latex may be vulcanized directly, the product being known as *vultex*. (2) In synthetic rubber manufacture, the process stream in which the polymerized product is produced.

latex (*Paper*) An aqueous emulsion of synthetic rubber-like compounds used to increase the flexibility and durability of paper, eg for bookbinding papers or base material for imitation leather. It may be added to the stock or used as an impregnant. Also extensively employed as the binder in mineral coating applications.

lath (*Build*) See LATHING.

lath (*Min*) A term commonly applied to a lath-like (long and thin) crystal.

lathe (*Eng*) A machine tool for producing cylindrical work, facing, boring and screw cutting. It consists generally of a bed carrying a headstock and tailstock, by which the work is driven and supported, and a saddle carrying the slide rest by which the tool is held and traversed.

lathe bed (*Eng*) That part of a lathe forming the support for the headstock, tailstock and carriage. It consists of a rigid-cast box-section girder, supported on legs, its upper face being planed and scraped to provide true working surfaces, or 'ways'.

lathe carrier (*Eng*) A clamp consisting of a shank which is formed into an eye at one end and provided with a set screw. It is attached to work supported between centres and driven by the engagement of the driver plate pin with the shank or 'tail' of the carrier, which may be straight or bent.

lathe tools (*Eng*) Turning tools with edges of various shapes (round-nosed, side-cutting, etc) and cutting angles, varying with the material worked on, formed by giving clearance to the front of the cutting edge and rake to the top of the tool. See FINISHING TOOL, KNIFE TOOL, ROUGHING TOOL, SIDE TOOL. Fig. ▷

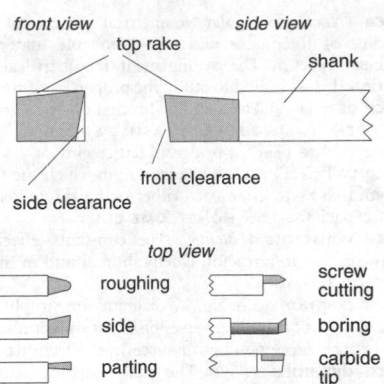

lathe tools Representative types below.

lathe work (*Eng*) Any work ordinarily performed in the lathe, such as turning, boring and screw cutting.

lathing (*Build*) Material fixed to surfaces to provide a basis for plaster. Formerly softwood strips (laths) 3–4 ft long were used; nowadays steel meshing or perforated steel sheet is sometimes used as an alternative to plasterboard.

lathyrism (*Med*) A disease characterized by stiffness and paralysis of the legs, due to poisoning with certain kinds of chick pea.

lati- (*Genrl*) Prefix from Lt *latus*, wide, broad.

laticifer (*BioSci*) A cell or vessel containing LATEX, present in tissues of many plants.

Latin square (*MathSci*) A square array in which no letter or figure occurs more than once in the same row or column. A modern example is sudoku.

latite (*Min*) A volcanic rock containing approximately equal amounts of alkali feldspar and sodic plagioclase, ie the approx equivalent of monzonite. See VOLCANIC ROCKS.

latitude (*ImageTech*) The range of exposure permissible, or range of density usefully obtainable, in a photographic emulsion. The range of exposure obtainable with the linear portion of the gamma curve of an emulsion.

latitude (*Surv*) The distance of a point north (if positive) or south (negative) from the point of origin of survey. This, together with departure (easting), locates the point in a rectangular grid orientated on true meridian. Also *northing*.

latitude and longitude, celestial (*Astron*) Spherical co-ordinates referred to the ecliptic and its poles. *Celestial latitude* is the angular distance of a body north or south of the ecliptic. *Celestial longitude* is the arc of the ecliptic intercepted between the latitude circle and the First Point of Aries, and is measured positively eastwards from 0° to 360°.

latitude and longitude, terrestrial (*Genrl*) Spherical co-ordinates referred to the Earth's equator and its poles; used to specify a point on the Earth's surface. The angular elevation of the celestial pole above a plane tangential to the Earth at a given place is known as the *geographical latitude*; the *geocentric latitude* is the angle made with the equatorial plane by the radius of the Earth through the given point. The latter is slightly less than the former owing to the oblate form of the Earth. *Terrestrial longitude* is the arc of the equator between the meridian through the point and the meridian at Greenwich, England; generally measured from 0° to 180° east or west of Greenwich.

latterkin (*Build*) Hardwood suitably shaped for clearing grooves in CAMES.

lattice (*Crystal*) A regular spaced arrangement of points as for the sites of atoms in a crystal.

lattice (*MathSci*) A set with a partial order in which any two elements have a least UPPER BOUND and a greatest LOWER BOUND.

lattice (*NucEng*) Regular geometrical pattern of discrete bodies of fissionable and non-fissionable material in a nuclear reactor. The arrangement is subcritical or just critical if it is desired to study the properties of the system.

lattice bars (*Eng*) The diagonal bracing of struts and ties in an OPEN-FRAME GIRDER or LATTICE GIRDER.

lattice bridge (*Eng*) A bridge of lattice girders.

lattice coil (*ICT*) An inductance coil in which the turns are wound so as to cross each other obliquely, to reduce the self-capacitance. See HONEYCOMB COIL.

lattice constants (*NucEng*) The constants effecting the neutron flux in a reactor, chiefly those found in the FOUR-FACTOR FORMULA.

lattice diagram (*ElecEng*) A diagram for simplifying the calculation of travelling waves on a transmission line when there are a large number of successive reflections.

lattice dynamics (*Phys*) The study of the excitations a crystal lattice can experience and their consequences for the thermal, optical and electrical properties of solids.

lattice energy (*Crystal*) The energy required to separate the ions of a crystal from each other to an infinite distance.

lattice filter (*ICT*) One or more lattice networks acting as a wave filter.

lattice girder (*Eng*) A girder formed of upper and lower horizontal members connected by an open web of diagonal crossing members, used in structures such as bridges and large cranes.

lattice hypothesis (*BioSci*) A hypothesis proposed to explain how aggregates are formed when antibodies combine with a soluble polyvalent antigen, the size and composition of which varies with the ratio of the two components.

lattice network (*ICT*) One formed by two pairs of identical arms on opposite sides of a square, the input terminals being across one diagonal and the output terminals across the other. Also *bridge network, lattice section*.

lattice structure (*Crystal*) One of three types of crystal structure: (1) ionic, with symmetrically arranged ions and good conducting power; (2) molecular, covalent, usually volatile and non-conducting; (3) layer, with large ions each associated with two small ones, forming laminae weakly held by non-polar forces.

lattice vibration (*Crystal*) The vibration of atoms or molecules in a crystal due to thermal energy.

lattice water (*Chem*) WATER OF CRYSTALLIZATION, which is present in stoichiometric proportions and occupies definite lattice positions, but is in excess of that with which the ions could be co-ordinated. This water apparently fills in holes in the crystal lattice, as with CLATHRATE compounds.

lattice winding (*ElecEng*) A winding, made up of LATTICE COILS, for electric machines; always used for dc machines and frequently for ac machines.

lattice window (*Arch*) A window in which diamond-shaped panes are supported in a leaden frame consisting of diagonally intersecting CAMES, the longer axes of the diamonds being vertical.

laudanum (*Pharmacol*) Tincture of opium.

Laue pattern (*Crystal*) A pattern of spots produced on photographic film when a heterogeneous X-ray beam is passed through a thin crystal, which acts like an optical grating. Used in the analysis of crystal structure.

laughing gas (*Med*) Colloq term for anaesthetic gas, nitrous oxide.

laumontite (*Min*) A zeolite consisting essentially of hydrated silicate of calcium and aluminium, crystallizing in the monoclinic system; occurs in cavities in igneous rocks and in veins in schists and slates.

launcher (*Space*) See LAUNCH SYSTEM.

launching (*ICT*) Said of the operation of transmitting a signal from a conducting circuit into a waveguide.

launch pad (*Space*) Special area from which a launch system is fired; it contains all the necessary support facilities such as a servicing tower, cooling water, safety equipment and flame deflectors.

launch system (*Space*) An assemblage of propulsive devices (stages) capable of accelerating a space vehicle to a velocity needed to achieve a particular space trajectory. Examples are: Titan and Space Shuttle (US); Proton (USSR); Ariane (Europe); H-1 (Japan) and Long March (China). Also *launcher, launch vehicle*.

launch window (*Space*) The time slot within which a spacecraft must be launched to best achieve its given mission trajectory; a launch window may become available a number of times during a particular launch opportunity.

launder (*MinExt*) Inclined trough for conveying water or crushed ore and water (pulp).

Lauraceae (*BioSci*) A family of c.2500 spp of dicotyledonous flowering plants (superorder Magnoliidae) that are mostly woody, tropical and subtropical. It includes avocado, cinnamon, camphor, bay laurel and some timber trees.

Laurasia (*Geol*) The palaeogeographic supercontinent of the northern hemisphere corresponding to Gondwanaland in the southern hemisphere.

laurdalite (*Geol*) A coarse-grained soda-syenite from S Norway; it resembles laurvikite but contains nepheline (elaeolite) as an essential constituent. Also *lardalite*.

laurel (*For*) See STINKWOOD.

Laurentian Granites (*Geol*) Precambrian granitic intrusives in the Canadian Shield.

Laurent's expansion (*MathSci*) If the function $f(z)$ is analytic in the annulus formed by two circles c_1 and c_2, of radii r_1 and r_2, respectively, then within this annulus:

$$f(z) = \sum_{n=1}^{\infty} a_n(z-z_0)^n + \sum_{n=1}^{\infty} b_n(z-z_0)^{-n}$$

where

$$a_n = \frac{1}{2\pi i}\int_{c_1} \frac{f(z)}{(z-z_0)^{n+1}}\,dz$$

$$b_n = \frac{1}{2\pi i}\int_{c_2} (z-z_0)^{n-1} f(z)\,dz$$

lauric acid (*Chem*) Dodecanoic acid. $CH_3(CH_3)_{10}COOH$. Mp 44°C. Crystalline solid. Occurs as glycerides in milk, laurel oil, palm oil, etc.

laurvikite (*Geol*) A soda syenite from S Norway, very popular as an ornamental stone when cut and polished; widely used for facing buildings, the distinctive feature being a fine blue colour, produced by schiller structure in the anorthoclase feldspars. Also *blue granite, larvikite*.

lauryl alcohol (*Chem*) Dodecan-1-ol. $CH_3(CH_2)_{10}CH_3OH$. Mp 24°C. Insoluble in water, crystalline solid. Used in manufacture of detergents.

lautarite (*Min*) Monoclinic iodate of calcium, occurring rarely in caliche in Chile.

lava (*Geol*) The molten rock material that issues from a volcanic vent or fissure and consolidates on the surface of the ground (*subaerial lava*), or on the floor of the sea (*submarine lava*). Chemically, lava varies widely in composition; it may be in the condition of glass, or a holocrystalline rock. Also *basalt, obsidian, pillow structure, pumice*. See VOLCANO.

lava flows (*Geol*) See EXTRUSIVE ROCKS.

lavage (*Med*) Irrigation or washing out of a cavity, such as the stomach or the bowel, eg gastric lavage.

Lavalier microphone (*Acous*) A microphone worn on a cord round the neck. It has a frequency response such as to compensate for the change in directivity produced by that of the mouth.

law (*Genrl*) In science, a rule or generalization which describes specified natural phenomena within the limits of experimental observation. An apparent exception to a law tests the validity of the law under the specified conditions. A true scientific law admits of no exception but is of no scientific value unless it can be related to other laws comprehending relevant phenomena.

law calf (*Print*) Calf leather with a rough surface and light in colour; used for account book bindings etc.

lawn (*Textiles*) Fine, lightweight, plain-weave cloth made of flax or cotton yarns. See ORGANDIE.

law of conservation of matter (*Chem*) A law stating that matter is neither created nor destroyed during any physical or chemical change. See RELATIVITY.

law of constant proportions (*Chem*) Every pure substance always contains the same elements combined in the same proportions by weight. Also *law of definite proportions*.

law of Dulong and Petit (*Chem*) The law that the atomic heat capacities of solid elements are constant and approximately equal to 25 (when the specific heat capacity is in $J\,mol^{-1}\,K^{-1}$). Certain elements of low atomic mass and high melting point have, however, much lower atomic heat capacities at ordinary temperatures.

law of effect (*Psych*) Thorndike's formulation of the importance of reward in learning, which states that the tendency of a stimulus to evoke a response is strengthened if the response is followed by a satisfactory or pleasant consequence, and is weakened if the response is followed by an annoying or unpleasant consequence.

law of equilibrium (*Chem*) See LAW OF MASS ACTION.

law of equivalent proportions (*Chem*) The proportions in which two elements separately combine with the same weight of a third element are also the proportions in which the first two elements combine together. Also *law of reciprocal proportions*.

law of Guldberg and Waage (*Chem*) See LAW OF MASS ACTION.

law of isomorphism (*Chem*) See MITSCHERLICH'S LAW OF ISOMORPHISM.

law of mass action (*Chem*) Fundamental law applying to equilibrium chemical reactions of the general form

$$aA + bB + \cdots \rightarrow cC + dD + \cdots$$

The law states that concentrations of reactants and products, shown thus [..], are related by the equation

$$K_c = \frac{[C]^c\,[D]^d}{[A]^a\,[B]^b}$$

where K_c is known as the EQUILIBRIUM CONSTANT. When expressed in terms of activity, it is K_a, and in partial pressures, K_p. The concentration equilibrium constant is also the ratio of the forward and reverse reaction rate constants:

$$K_c = \frac{k_f}{k_r}$$

It thus obeys ARRHENIUS'S RATE EQUATION. Also *law of equilibrium*. See ORDER OF REACTION.

law of mixtures (*Chem*) See RULE OF MIXTURES.

law of multiple proportions (*Chem*) When two elements combine to form more than one compound, the amounts of one of them which combine with a fixed amount of the other exhibit a simple multiple relation. Also *Dalton's law*.

law of octaves (*Chem*) The relationship observed by Newlands (1863) which arranges the elements in order of atomic weight and in groups of eight (octaves) with recurring similarity of properties. See PERIODIC TABLE.

law of partial pressures (*Chem, Phys*) See DALTON'S LAW OF PARTIAL PRESSURES.

law of photochemical equivalence (*Chem*) See EINSTEIN LAW OF PHOTOCHEMICAL EQUIVALENCE.

law of rational indices (*Crystal*) A fundamental law of crystallography which states, in the simplest terms, that in any natural crystal the indices may be expressed as small whole numbers.

law of reciprocal proportions (*Chem*) See LAW OF EQUIVALENT PROPORTIONS.

law of superposition (*Geol*) See SUPERPOSITION, LAW OF.

law of the minimum (*BioSci*) In its original form, the law which states that the rate at which a plant grows, the size it attains and its overall health all depend on the amount available to it of the scarcest of its essential nutrients. More recently it has been expanded into a general principle covering the factors limiting the growth and health of all organisms.

law of volumes (*Chem*) See GAY-LUSSAC'S LAW.

lawrencium (*Chem*) Transuranic element, symbol Lr, at no 103. Its only known isotope has a short half-life of 8 seconds.

laws of reflection (*Phys*) Laws stating that: (1) when a ray of light is reflected at a surface, the reflected ray is found to lie in the plane containing the incident ray and the normal to the surface at the point of incidence; (2) the angle of reflection equals the angle of incidence.

Lawson criterion (*NucEng*) The minimum physical conditions of plasma temperature (T), plasma density (n) and confinement time (τ) needed for the production of net power in fusion. It is expressed as $n\tau f(T)$ where $f(T)$ has a minimum value of around $2 \times 10^{20}\,s\,m^{-3}$ for the deuterium–tritium reaction.

lawsonite (*Min*) An orthorhombic hydrated silicate of aluminium and calcium. It occurs in low-grade regionally metamorphosed schists, particularly in glaucophane schists.

laxative (*Pharmacol*) An agent to promote evacuation of the bowel.

LAXS (*Phys*) Abbrev for LOW-ANGLE X-RAY SCATTERING.

lay (*ICT*) The axial length of one turn of the helix formed by the core (in a telephone cable) or a strand of a conductor (in a power cable). See LAY RATIO.

lay (*Print*) See LAY EDGES.

lay (*Textiles*) Fabrics set out in piles of identical length ready for cutting out for making into garments.

lay barge (*MinExt*) A vessel for the storage, welding and laying of pipelines underwater.

lay boy (*Paper*) Equipment at the end of a sheet cutter to collect the cut sheets and stack them in neat piles with aligned edges.

lay cords (*Print*) Cords with which a book is tied, to prevent the covers from warping while drying.

lay edges (*Print*) The edges of a sheet of paper which are laid against the guides or lays in a printing or folding machine. The front edge (or *gripper edge*) is laid to the *front lays* and the side of the sheet to the *side lay*.

layer (*Build*) (1) A course in a wall. (2) A bed of mortar between courses.

layer (*Phys*) An ionized region in space; these vary vertically and affect radio propagation. See E-LAYER, F-LAYER.

layer board (*Build*) See LEAR BOARD.

layered igneous rocks (*Geol*) Igneous rocks which display layers of differing mineral and chemical composition, eg the Bushveld complex in S Africa, the Skaergaard complex of E Greenland.

layered map (*Surv*) See RELIEF MAP.

layering (*BioSci*) (1) See STRATIFICATION. (2) A method of artificial propagation in which stems are pegged down and covered with soil until they root, when they can be detached from the parent plant.

layering (*Geol*) The high-temperature sedimentation feature of igneous rocks.

layer lattice (*Crystal*) The concentration of bonded atoms in parallel planes, with weaker and non-polar bonding between successive planes. This gives marked cleavage, eg in graphite, mica.

layer line (*Crystal*) One joining a series of spots on an X-ray rotating-crystal diffraction photograph. Its position enables the crystal lattice spacing parallel to the axis of rotation to be determined.

laying (*Build*) The first coat of plaster in two-coat work.

laying-in (*Build*) A term generally meant to mean the covering of a surface with a material, eg paint stripper, texture, goldsize, etc, before further working.

laying-off (*Build*) The finishing strokes in the brush painting process. These strokes should eliminate or minimize brushmarks.

laying-off (*Ships*) The process of transferring the design form to full scale, for the purpose of FAIRING and ultimately of fabrication of details.

laying-out (*Eng*) The marking-out of material, esp plate work, full size, for cutting and drilling.

laying press (*Print*) See LYING PRESS.

laying the bearings (*Acous*) In tuning fixed-pitch instruments, such as the piano or organ, the technique of tuning the twelve semitones of a central octave. All other notes are then tuned by unison or octaves.

lay light (*Build*) A window or sash, fixed horizontally in a ceiling, to admit light to a room.

lay marks (*Print*) Marks printed on the sheet, outwith the trimmed size, to indicate the LAY EDGES of the sheet during printing. Used as a guide for subsequent operations to ensure uniformity.

layout (*Print*) (1) The general appearance of a printed page. (2) The art and practice of arranging display (eg advertising) matter to the best advantage. (3) An outline and accurate drawing showing all the information required to assemble a printed job, including the relative positions of the text and illustrations, dimensions, folds, trims, etc.

lay panel (*Build*) A long panel of small height formed in a panelled wall above a doorway, or all round the room immediately below the cornice.

lay ratio (*ICT*) The ratio of the LAY to the mean diameter of the helix.

Layrub universal joint (*Eng*) A flexible joint depending on the deflection of a group of rubber bushes. Frequently used where the angle between shafts is not great, say up to 10° on either side of the straight line.

lay shaft (*Eng*) An auxiliary geared shaft, eg a secondary shaft running alongside the main shaft of an automobile gearbox, to and from which the drive is transferred by gear wheels of varying ratio.

lazulite (*Min*) A deep sky blue, strongly pleochroic mineral, crystallizing in the monoclinic system. In composition essentially a hydrated phosphate of aluminium, magnesium and iron, with a little calcium. Found in aluminous high-grade metamorphic rocks and in granite pegmatites.

lazurite (*Min*) An ultramarine blue mineral occurring in cubic crystals or shapeless masses; it consists of silicate of sodium and aluminium, with some calcium and sulphur, and is considered to be a sulphide-bearing variety of haüyne. A constituent of LAPIS LAZULI.

lb (*Genrl*) Abbrev for POUND.

L-band (*ICT*) A radar and microwave frequency band, generally accepted as being 1–2 GHz; not frequently referred to because of lack of international conformity in its use.

lb.s.t. (*Aero*) Abbrev for STATIC THRUST in pounds.

LC$_{50}$ (*BioSci*) Concentration of substance that is lethal to 50% of individuals tested.

lc (*Print*) Abbrev for LOWER CASE.

LCA (*Eng*) Abbrev for LIFE-CYCLE ANALYSIS.

L-capture (*Phys*) Absorption of an electron from the L-SHELL into the nucleus of an atom, giving rise to X-rays of characteristic wavelength depending on atomic number of the element. See K-CAPTURE, X-RAY.

LC coupling (*Electronics*) Inductor output load of an amplifier circuit is connected through a capacitor to the input of another circuit.

LCD (*Electronics*) Abbrev for LIQUID CRYSTAL DISPLAY (panel).

LCD overhead projector (*ImageTech*) An overhead projector having a liquid crystal display panel, accepting inputs from computer and video hardware. See panel on LIQUID CRYSTAL DISPLAYS.

LCD projector (*ImageTech*) A video projection system using either a single-colour liquid crystal display panel or three filtered black-and-white panels for red, green and blue signals transilluminated by the projection lamp(s). See panel on LIQUID CRYSTAL DISPLAYS.

LCD viewfinder (*ImageTech*) An ELECTRONIC VIEWFINDER using a back-illuminated, colour liquid crystal display panel. See COLOUR VIEWFINDER and panel on LIQUID CRYSTAL DISPLAYS.

L-C filter (*ElecEng*) Property of a component (inductor) in a circuit whereby back emf arises because of rate of change of current. It is 1 *henry* when 1 volt is generated by a rate of change of 1 ampere per second. Often abbreviated to *inductance*.

LCI (*NucEng*) Abbrev for LIFE-CYCLE INVENTORY.

LCLV projector (*ImageTech*) Abbrev for LIQUID CRYSTAL LIGHT VALVE PROJECTOR.

LCN (*Aero*) Abbrev for LOAD CLASSIFICATION NUMBER.

LD$_{50}$ (*BioSci*) The dose of a toxic substance that, administered in a named way, will kill 50% of a large number of individuals of a given species.

L-display (*Radar*) A radar display in which the target appears as two horizontal pulses, left and right from a central vertical time base, varying in amplitude according to accuracy of aim.

L $_+$, L$_0$, L$_R$ dose of toxin (*BioSci*) Used in standardizing diptheria toxin and antitoxin. Describes the quantity of toxin that, when mixed with one standard unit of antitoxin, respectively kills a guinea pig under standardized conditions, or when injected subcutaneously produces no observable reaction, or when inoculated into the skin produces a minimal lesion.

LDPE (*Chem*) Abbrev for LOW-DENSITY POLYETHYLENE.

L/D ratio (*Eng*) Property of an injection moulding screw or extrusion screw, defined by ratio of screw length to flight diameter. It is typically 20–30 to 1 for thermoplastics and 5–10 to 1 for rubbers. See EXTRUSION.

leachate (*EnvSci*) Fluid seeping from a landfill site or spoil heap.

leached zone (*MinExt*) See GOSSAN, SECONDARY ENRICHMENT.

leaching (*EnvSci*) The removal of substances from soils by percolating water. Cf CHELUVIATION, FLUSH, LESSIVAGE, PODSOL. See LEACHATE.

leaching (*MinExt*) The extraction of a soluble metallic compound from an ore by dissolving it in a solvent, eg cyanide, sulphuric acid. The metal is subsequently precipitated or adsorbed from the solution.

leaching (*NucEng*) The removal, usually by water, of a substance from a solid as in the removal of fission products from the vitrified blocks designed to contain them.

lead (*Build*) The leaden came of a lattice window.

lead (*Chem*) Symbol Pb, at no 82, ram 207·19, valency 2 or 4, mp 327·4°C, bp 1750°C, rel.d. at 20°C 11·35. Specific electrical resistivity $20·65 \times 10^{-8}$ Ω m. A metallic element in the fourth group of the periodic system. Abundance in the Earth's crust 13 ppm. Occurs chiefly as GALENA. The naturally occurring stable element consists of four isotopes: ^{204}Pb (1·5%), ^{206}Pb (23·6%), ^{207}Pb (22·6%) and ^{208}Pb (52·3%). It is used as shielding in X-ray and nuclear work because of its relative cheapness, high density and nuclear properties. Other principal uses: in storage batteries, ammunition, foil and as a constituent of bearing metals, solder and type metal. Lead can be hardened by the addition of arsenic or antimony. It occurs very rarely in the native form, and then appears to have been formed by fusion of some simple lead ore accidentally incorporated in lava. Toxic.

lead (*CivEng*) (1) See HAUL DISTANCE. (2) In railway track, the distance from the nose of the CROSSING to the nose of the SWITCH.

lead (*ElecEng*) (1) A term often used to denote an electric wire or cable. (2) See BACKWARD SHIFT, BRUSH SHIFT, FORWARD SHIFT.

lead (*Eng*) The distance between consecutive contours, on the same helix of a screw-thread measured parallel to the axis of the screw; it is the axial distance a nut would advance in one complete revolution, equal to PITCH in single-start threads.

lead (*Surv*) The leaden sinker secured at one end of a LEAD LINE. See LEADS.

lead–acid accumulator (*ElecEng*) A *secondary cell* consisting of lead electrodes, the positive one covered with lead dioxide, dipping into sulphuric acid solution. Its emf is about 2 volts; very widely used. Also *lead accumulator*. See ACCUMULATOR (2).

lead age (*Min*) The age of a mineral or rock calculated from the ratios of its radiogenic and non-radiogenic lead isotopes.

lead- and leave-edges (*Print*) On printing presses the first and last edges of the product to be printed.

lead azide (*Chem*) PbN_6. Explosive like most azides. Sometimes used, instead of mercury fulminate, as a detonator for TNT.

lead burning (*Build*) The process of welding together two pieces of lead, thus forming a joint without the use of solder.

lead (II) carbonate (*Chem*) $PbCO_3$. Occurs in nature as cerussite. At about 200°C it decomposes into the (II) oxide and carbon dioxide. Readily reduced to metal by CO.

lead (II) chromate (IV) (*Chem*) $PbCrO_4$. Precipitated when potassium chromate (VI) is added to the solution of a lead salt. Used as pigments, called *chrome yellows*. The colour may be varied by varying the conditions under which the precipitation is made.

lead disilicate (*Chem*) Obtained by fusing lead (II) oxide and silica together. As lead FRIT, it is used as a ready means of incorporating lead oxide in the making of lead GLAZES.

lead dot (*Build*) A lead peg or dowel used to fasten sheet lead to the upper surface of a coping or cornice, for which purpose it is run into a mortise in the stone.

leaded (*Eng*) A term descriptive of copper, bronze, brass, steel, nickel and phosphor alloys to which from 1% to 4% of lead has been added, mainly to improve machinability.

leaded lights (*Build*) A window formed of (usually) diamond-shaped panes of glass connected together by leaden cames.

lead equivalent (*Radiol*) Absorbing power of a radiation screen expressed in terms of the thickness of lead which would be equally effective for the same radiation.

leader (*BioSci*) The younger part of the main stem or a main branch of woody plants. Its branches are called *laterals*. See SPUR.

leader (*Build*) See CONDUCTOR.

leader (*ImageTech*) A strip at the beginning (head leader) or end (tail leader) of a reel of film or tape for protection and identification.

leader (*MinExt*) A thin mineralized vein parallel to or otherwise related to the main ore-carrying vein, so aiding its discovery.

leader (*Print*) A group of dots (...) cast on a 1 em body set together to form dotted lines, or spaced at intervals to guide the eye on contents pages etc.

leader (*Surv*) A surveyor's assistant who has charge of the forward end of a chain. He is directed into line by the follower.

leader-hook (*Build*) A device, such as a HOLDERBAT, for securing a rainwater pipe to a wall.

lead frit (*Chem*) See LEAD DISILICATE.

lead glance (*Min*) See GALENA.

lead glass (*Glass*) Glass containing lead oxide. The amount may vary from 3–4% to 50% or more in special cases. *English Lead Crystal*, used for tableware because of its high refractive index, has the composition (percentage by weight): SiO_2 56, PbO 29, K_2O 13, Na_2O 2. A similar glass is used in the *pinch* of incandescent light bulbs. Lead glass is also extensively used as a transparent radiation shield, esp for X-rays.

lead grip (*ElecEng*) A bonding device for providing continuity of a lead-sheathed cable.

leadhillite (*Min*) Hydrated carbonate and sulphate of lead, so called from its occurrence with other ores of lead at Leadhills (Scotland).

lead (II) hydroxide (*Chem*) $Pb(OH)_2$. Dissolves in excess of alkali hydroxides to form plumbites (plumbates (II)).

lead-in (*Acous*) Unmodulated groove at the start of a recording on a vinyl disk, so that the stylus falls into the groove correctly before the start of the modulation. The corresponding final groove after modulation ends is the *lead-out*.

leading (*ICT*) The spacing between lines of type.

leading current (*ElecEng*) An ac whose phase is in advance of that of the applied emf creating the current. For a pure capacitance in circuit, the phase of the current is $\pi/2$ radians in advance of that of the applied voltage. See ANGLE OF LEAD, PHASE ANGLE.

leading diagonal (*MathSci*) Of an array, the sequence of elements beginning with the first element of the first line, and continuing with the nth element of the nth line.

leading edge (*Aero*) The edge of a streamline body or aerofoil which is forward in normal motion; structurally, the member that constitutes that part of the body or aerofoil.

leading edge (*ElecEng*) A term used in connection with the brushes of electric machines. See ENTERING EDGE.

leading edge (*ICT*) Rising amplitude portion of a pulse signal. See RISE TIME, TRAILING EDGE.

leading-edge flap (*Aero*) A hinged portion of the wing leading edge, usually on fast aircraft, which can be lowered to increase the camber and so reduce the stalling speed. Colloq *droops*.

leading load (*ElecEng*) See CAPACITATIVE LOAD.

leading note (*Acous*) The note one semitone below the tonic or key note of the normal musical scale; essential in combining harmonic frequencies.

leading-out (*Print*) The process of inserting LEADS between lines of type matter, in order to open them out, thus presenting more white space between the printed lines. Also, in PHOTOTYPESETTING, adjusting the space between lines.

leading-out wire (*ElecEng*) The flexible insulated wire which is attached to the more delicate insulated wire used for the windings of transformers etc. It is sufficiently robust for connecting to terminals.

leading phase (*ElecEng*) (1) A term used in connection with measuring equipment on three-phase circuits to denote a phase whose voltage is leading upon that of one of the other phases by approximately 120°. (2) Particularly in connection with the two-wattmeter method of three-phase power measurement, the phase in which the current at unity power factor leads upon the voltage applied to the meter in which that current is flowing.

leading pole horn (*ElecEng*) The portion of the pole shoe of an electric machine which is first met by a point on the armature or stator surface as the machine revolves. Hence also *leading pole tip*.

leading ramp (*ElecEng*) The sloped end of a conductor rail at which the collector shoe of an electric train first makes contact.

lead line (*Surv*) A line with which soundings are taken. The depth of water is indicated on the line by 'marks', or by knots in the line, indicating fathoms; fathoms not indicated on the line (eg between 7 and 10) are DEEPS.

lead monoxide (*Chem*) A bright-yellow solid compound used in pigments and paints. See LITHARGE.

lead network (*Electronics*) One which provides a signal proportional to rate of change, ie time differential or derivative, of error signal.

lead of the web (*Print*) The reel of paper is torn diagonally to thread it through the press; when the press is started a wrongly threaded path is called a *lost lead* or *wrong lead*.

lead-out (*Acous*) See LEAD-IN.

lead (II) oxide (*Chem*) PbO. An oxide of lead, varying in colour from pale yellow to brown depending on the method of manufacture. An intermediate product in the manufacture of red lead. See LITHARGE, MASSICOT.

lead (IV) oxide (*Chem*) PbO_2. A strong oxidizing agent. Industrial application very limited. Present, in certain conditions, in accumulators or electrical storage batteries as a chocolate-brown powder.

lead oxychloride (*Chem*) $PbCl_2PbO$. See CASSEL'S YELLOW.

lead paint (*Build*) Paints containing lead. Because of toxicity their preparation, storage and application is usually governed by regulation.

lead paint regulations (*Build*) Detailed regulations which govern the use of lead paints, which are toxic.

lead plug (*Build*) A cast lead connecting piece binding together adjacent stones in a course; formed by running molten lead into suitably cut channels in the jointing faces.

lead poisoning (*Med*) Chronic poisoning due to inhalation, ingestion and skin absorption of lead. Recognized as a hazard, both for young children (formerly through sucking lead or lead-painted articles or toys), and also industrially. Characterized by anaemia, constipation, severe abdominal pain and perhaps ultimately renal damage. Lesser degrees now recognized and soft water delivered by lead pipes is a potential hazard.

lead protection (*Radiol*) Protection provided by metallic lead against ionizing radiation. Joined with other substances to provide further protection, eg lead glass, lead rubber.

lead response (*Autos*) The increase in octane number of a given motor or aviation fuel mixture for a particular concentration of added lead, eg LEAD TETRAETHYL.

lead rubber (*Radiol*) Rubber containing high properties of lead compounds. Used as flexible protective material for eg gloves and aprons.

leads (*Eng*) Lengths of thin lead wire inserted between a very large JOURNAL and the bearing cap during assembly to test the clearance.

leads (*Print*) Strips of lead placed between lines of type. Usual thicknesses: 1-, $1\frac{1}{2}$-, 2- and 3-point (*thick*), 6-point (*nonpareil*), 12-point (*pica*).

lead sulphate (*Chem*) $PbSO_4$. Formed as a white precipitate when sulphuric acid is added to a solution of a lead salt.

lead (II) sulphide (*Chem*) Found in nature as GALENA. Black precipitate formed when lead sulphate is reduced by carbon and when hydrogen sulphide is passed through a solution of a lead salt.

lead tetraethyl (*Chem*) $Pb(C_2H_5)_4$. A colourless liquid, obtained by the action of a zinc or magnesium ethyl halide on lead chloride. Used in motor spirit to increase the ANTIKNOCK VALUE.

lead tree (*Chem*) The form, in the shape of a tree, which lead takes after electrodeposition from simple salts.

lead zirconate titanate (*Chem*) A piezoelectric ceramic with a higher CURIE POINT than barium titanate (IV). Abbrev *PZT*. See DIELECTRIC AND FERROELECTRIC MATERIALS.

leaf (*BioSci*) (1) In modern vascular plants a lateral organ of limited growth that develops from a primordium at a shoot apex. In angiosperms a leaf typically has a bud in its axil. Most leaves are more or less flat and green and photosynthetic in function; modified leaves include bud scales, bulb scales, many sorts of spine and tendril, bracts and probably sepals and petals, and possibly stamens and carpels. Cf SPOROPHYLL. (2) In bryophytes, similar but usually smaller and thinner (mostly one cell thick) structures called phyllids.

leaf area index (*BioSci*) The ratio of the total area of leaves of a plant or crop to the area of soil available to it.

leaf filter (*ChemEng*) A type of filter in which large thin frames with a pipe connection through the frame are covered in cloth or metal gauze and placed inside a closed cylindrical vessel. Liquid to be filtered is forced under pressure into cylinder and clear filtrate leaves via frame's pipe connections (Kelly and Sweetland are this type). Similarly ore pulp is filtered by use of vacuum or pressure.

leaf gap (*BioSci*) A region of parenchyma in the vascular cylinder of a stem above a leaf trace.

leafing (*Build*) The property of the small metallic flakes used in aluminium, bronze and similar paints, of floating flat at the surface of the paint, thus increasing the metallic lustre. The overlapping effect of the flakes forms a barrier coat resistant to permeation.

leaflet (*BioSci*) One of the leaf-like units that together make up the lamina of a compound leaf.

leaf mosaic (*BioSci*) See MOSAIC.

leaf scar (*BioSci*) The scar, usually covered by a thin protective layer, left on a stem following the abscission of a leaf.

leaf sheath (*BioSci*) The sheath surrounding the stem at the base of a leaf in grasses.

leaf spot (*BioSci*) Any of several plant diseases characterized by the appearance of dark spots on the leaves.

leaf spring (*Eng*) A machine component comprising one or a group of relatively thin, flexible or resilient strips, reacting as a spring to forces applied to a main surface.

leaf succulent (*BioSci*) A plant with succulent photosynthetic leaves. Many are CAM plants, eg *Aloe*. Cf STEM SUCCULENT.

leaf trace (*BioSci*) A vascular bundle in a stem from its junction with another bundle of the stele to the base of the leaf; a leaf may have one trace or more.

leaf types (*BioSci*) The arrangement of the parts of a leaf as in the illustration.

leak (*ElecEng*) (1) A path between electrically isolated parts of a circuit, or of a component, which has reduced resistance and can cause small unwanted currents to flow. (2) A high-valued resistor deliberately placed in a circuit to permit the controlled discharge of electrically charged components, usually associated with the discharge of a capacitor.

leakage (*NucEng*) Net loss of particles from a region or across a boundary, eg neutrons from the core of a reactor, often split up for calculation purposes into *fast non-leakage probability* (P_f); *resonance non-leakage probability* (P_r); *thermal non-leakage probability* (P_t).

leakage coefficient (*ElecEng*) The ratio of the total flux in the magnetic circuit of an electric machine or transformer to the useful flux which actually links with the armature of secondary winding. Also *leakage factor*.

leakage conductance (*ElecEng*) In electrical circuits, the leakage current expressed by the reciprocal of insulation resistance of the circuit.

leakage current (*ElecEng*) Small unwanted current flowing through a component such as a capacitor or a reverse-biased diode.

leakage factor (*ElecEng*) See LEAKAGE COEFFICIENT.

leakage flux (*ElecEng*) (1) That which, in any type of electric machine or transformer, does not intercept all the turns of the winding intended to enclose it. (2) That which crosses the air-gap of a magnet other than through the intended pole faces. Also *lost flux*.

leakage indicator (*ElecEng*) An instrument for measuring or detecting a leakage of current from an electric system to earth. Also *earth detector*.

leakage protective system (*ElecEng*) A protective system which operates as a result of leakage of current from electrical apparatus to earth.

leakage reactance (*ElecEng*) That, in a transformer, which arises from flux in one winding not entirely linking another winding; measured by the inductance of one winding when the other is short-circuited. Leakage inductance, taken with the effective self-capacitance of a winding, determines the range of frequencies effectively passed by the transformer.

leak detector (*NucEng*) See BURST-CAN DETECTOR.

leak detector (*Phys*) A device for indicating points where gases leak into a high-vacuum system.

leaky bucket (*ICT*) A method of controlling peak or mean CELL rate in an ASYNCHRONOUS TRANSFER MODE network by counting cell arrivals using a counter that is decremented periodically.

leaky mutation (*BioSci*) Mutation in which subnormal function exists, eg if a mutation leads to instability in a protein rather than its complete absence, or there is reduced expression of a gene.

lean-burn engine (*Autos*) Engine employing STRATIFIED CHARGE COMBUSTION to burn unusually weak (non-stoichiometric) mixtures to reduce unwanted exhaust emissions.

lean ore (*MinExt*) Ore of marginal value; low-grade ore.

lean-to roof (*Arch*) A roof having only one slope.

leap-frog test (*ICT*) A program used to test the internal operation of a computer.

leaping weir (*CivEng*) A special arrangement whereby flood flows may be diverted from a channel into which normal flows would ordinarily pass, the water having to go over a weir set at such a height that flood flows leap beyond the channel to an overflow. Also *separating weir*.

leap second (*Astron*) A periodic adjustment of time signal emissions to maintain synchronism with co-ordinated UNIVERSAL TIME (UTC). A positive leap second may be inserted, or a negative one omitted, at the end of December or June.

leap year (*Astron*) One of the years in which an extra day (29 February) is added to the civil calendar to allow for the fractional part of a tropical year of 365·2422 days. Since the Gregorian reform of the Julian calendar, the leap years are those whose number is divisible by four, except centennial years unless these are divisible by 400.

lear board (*Build*) A board carrying a lead gutter. Also *layer board*.

learned helplessness (*Psych*) A condition characterized by a general sense of powerlessness, which has its origins in a traumatic and inescapable event, but which persists in situations where escape or avoidance is possible. It specifically refers to effects in laboratory animals exposed to uncontrollable events but is speculated by some to be an underlying factor in human depression.

learning set (*Psych*) The increased ability to solve a particular kind of problem as a consequence of previous experience with similar kinds of problems.

learning theory (*Psych*) A range of theories that attempt to explain behaviour in terms of learning and conditioning. Largely the offspring of behaviourism, it takes as its starting point the association of various stimuli and responses. See BEHAVIOUR THERAPY, SOCIAL LEARNING THEORY, S–R THEORY.

lease rods (*Textiles*) Two rods, across a warp sheet, which separate the yarns, equalize the tension and keep them at the correct height.

least action, principle of (*Phys*) See PRINCIPLE OF LEAST ACTION.

least distance of distinct vision (*BioSci*) For a normal eye it is assumed that nothing is gained by bringing an object to be inspected nearer than 25 cm, owing to the strain imposed on the ciliary muscles if the eye attempts to focus for a shorter distance.

least energy principle (*Phys*) The principle that a system is only in stable equilibrium under those conditions for which its potential energy is a minimum.

least significant bit (*ICT*) The BIT in a binary number or binary WORD which has the least value.

least time, principle of (*Phys*) See FERMAT'S PRINCIPLE OF LEAST TIME.

leat (*MinExt*) A ditch following a contour line used to conduct water to working place.

leather (*Genrl*) A material made by TANNING and other treatment of the hides or skins of a great variety of creatures (mostly domesticated animals, but including also the whale, seal, shark, crocodile, snake, kangaroo, camel and ostrich). Artificial or imitation leathers are based on plasticized polyvinyl chloride or polyurethene, suitably filled with pigment and with an embossed surface to simulate the natural product.

leatherboard (*Paper*) Board made largely or wholly from leather scraps, generally on an intermittent board machine and containing LATEX as a binder and to impart flexibility.

leathercloth (*Textiles*) A woven or knitted fabric coated on one side with a polymer (eg rubber, cellulose derivative, polyvinyl chloride) which is embossed to simulate leather.

leather hollows (*Eng*) Strips of leather used by pattern makers to form the fillets in wood patterning.

leathery region (*Chem*) Glass-transition region of master curve of polymers, where material behaves like very stiff rubber. See VISCOELASTICITY.

leavening agent (*FoodSci*) Substance added to flour which when hydrated and warmed reacts to evolve carbon dioxide. The simplest is BAKING POWDER which contains sodium bicarbonate $Na(HCO_3)_2$ and an acidic substance, eg cream of tartar, gluconodeltalactone, acid calcium phosphate, acid sodium pyrophosphate. Yeast fermentation is also a form of leavening.

Leavers machine (*Textiles*) See LACE MACHINE.

leaving edge (*ElecEng*) The edge of the brush of an electric machine which is last met during revolution by a point on the commutator or slip ring. Also *back, heel, trailing edge*.

Leber's disease (*Med*) Hereditary optic atrophy; hereditary optic neuritis. A hereditary SEX-LINKED disease in which there is gradual loss of sight due to an affection of the optic nerve behind the eyeball.

Lebesgue integral (*MathSci*) An advanced concept based upon the theory of sets whereby functions, not ordinarily integrable, can be said to be integrated, eg the function which is one when x is rational and zero elsewhere is not ordinarily integrable but it does have a Lebesgue integral. Where both an ordinary integral (ie that of Riemann) and a Lebesgue integral exist, they are equal.

Leblanc connection (*ElecEng*) A method of connecting transformers for linking a three-phase to a two-phase system.

Leblanc phase advancer (*ElecEng*) A dc armature with three sets of brushes per pole pair on the commutator; the brushes are connected to the three slip rings of the induction motor, and the advancer is driven from the motor shaft at an appropriate speed, causing it to take a leading current and improve the motor power factor.

Leblanc process (*Chem*) A process, formerly of great importance but now obsolete, for the manufacture of sodium carbonate and intermediate products from common salt, coal, limestone and sulphuric acid.

LE cell (*BioSci*) See LUPUS ERYTHEMATOSUS CELL.

Le Chatelier–Braun principle (*Chem*) See LE CHATELIER'S PRINCIPLE.

lechatelierite (*Min*) A name sometimes applied to naturally fused vitreous silica, such as that which occurs as fulgurites (*lightning tubes*).

Le Chatelier's principle (*Chem*) A principle stating that if any change of conditions is imposed on a system at equilibrium, then the system will alter in such a way as to counteract the imposed change. This principle is of extremely wide application. It is a statement of the LAW OF MASS ACTION. Also *Le Chatelier–Braun principle*.

Le Chatelier test (*Build*) A simple method for checking the soundness of cement by checking its expansion.

Lecher wires (*ElecEng*) Two insulated parallel stretched wires tunable by means of sliding short-circuiting copper strip. The wires form a microwave electromagnetic transmission line which may be used as a tuned circuit, as an impedance matching device or for the measurement of wavelengths. Also *parallel-wire resonator*.

lecith-, lecitho- (*Genrl*) Prefixes from Gk *lekithos*, yolk of egg.

lecithin (*BioSci*) See PHOSPHATIDYLCHOLINE.

lecitho- (*Genrl*) See LECITH-.

lecithocoel (*BioSci*) The segmentation cavity of a holoblastic egg.

Leclanché cell (*ElecEng*) A PRIMARY CELL, good for intermittent use, which has a positive electrode of carbon surrounded by a mixture of manganese dioxide and powdered carbon in a porous pot. The pot and the negative zinc electrode stand in a jar containing ammonium chloride solution. The emf is approximately 1·4 V. The *dry cell* is a particular form of Leclanché.

lectin (*BioSci*) Protein, usually from plants, that binds specifically to sugars or to oligosaccharides. Since similar sugar residues are present in glycoproteins or glycolipids on the surface of many cells, multivalent lectins can cause agglutination of these cells. Some lectins bind specifically to certain cell types, or to cells at a particular stage of differentiation; several act as polyclonal mitogens for lymphocytes and some can cause illness if not completely denatured when present in foods. Examples include concanavalin A (ConA) from Jack beans, phytohaemagglutinin (PHA) from red kidney beans and pokeweed mitogen from *Phytolacca americana*.

LED (*Electronics*) Abbrev for LIGHT-EMITTING DIODE. See OPTOELECTRONICS.

Leda (*Astron*) The 13th natural satellite of Jupiter, discovered in 1974. Distance from the planet 11 100 000 km; diameter 20 km.

ledge (*Build*) One of the battens across the back of a batten door.

ledged-and-braced door (*Build*) A door similar to a batten door, but framed diagonally with braces across the back, between the battens.

ledged door (*Build*) See BATTEN.

ledgement (*Build*) A horizontal line of mouldings or a string course.

ledger (*Build*) A horizontal pole or member, lashed or otherwise fastened across the standards in a scaffold or in a trench.

ledger board (*Build*) See RIBBON STRIP.

Leduc effect (*Electronics*) A magnetic field, applied at right angles to the direction of a temperature gradient in an electrical conductor, will produce a temperature difference at right angles to the direction of both the temperature gradient and the magnetic field.

LEED (*Phys*) Abbrev for *low-energy electron diffraction*. Used to study the structure of surfaces using electrons of energy in the range 10–500 eV.

lee wave (*EnvSci*) Stationary wave set up in an airstream to the lee of a hill or range over which air is flowing. Special conditions of atmospheric stability and vertical wind structure are required. Lee waves are sometimes of large amplitude and can be dangerous to aircraft.

LEFM (*Eng*) Abbrev for LINEAR ELASTIC-FRACTURE MECHANICS.

left-handed engine (*Aero*) An aero-engine in which the propeller shaft rotates counterclockwise, viewed with the engine between the observer and the propeller.

left-hand propeller (*Aero*) See PROPELLER.

left-hand rule (*Phys*) (1) The rule that if the fingers of the left hand are placed around a current-carrying wire with the thumb pointing in the direction of electron flow, the position of the fingers gives the direction of the magnetic field produced by the current. (2) The rule that if the thumb, first finger and second finger of the left hand are extended at right angles to each other, with the second finger representing the direction of current flow in a wire and the first finger pointing in the direction of a magnetic field in which it lies, the thumb will point in the direction of force on the wire.

left-hand thread (*Eng*) A screw-thread cut in the opposite direction to the normal right hand. Viewed in elevation, the external thread is inclined upwards from left to right. Used when a normal thread would tend to unscrew.

left-hand tools (*Eng*) Lathe side tools with the cutting edge on the right, thus cutting from left to right.

leg (*BioSci*) In horticulture a single short trunk from which branches arise, as of a fruit bush that is not managed as a STOOL.

leg (*ICT*) One side of a loop circuit, ie either the *go* or *return* of an electric circuit.

legacy code (*ICT*) Code retained, particularly in operating systems, to allow older application programs to run.

legacy system (*ICT*) Denoting software, hardware, etc, that is outdated or discontinued, or no longer supported by its manufacturer.

legend (*Print*) See CAPTION.

Legendre's differential equation (*MathSci*) The equation

$$(1 - x^2)\frac{d^2 y}{dx^2} - 2x\frac{dy}{dx} + n(n+1)y = 0$$

It is satisfied by the Legendre polynomial $P_n(x)$.

Legendre's polynomials (*MathSci*) The polynomials $P_n(x)$ in the expansion

$$(1 - 2x + h^2)^{-\frac{1}{2}} = \sum_{n=0}^{\infty} P_n(x)\, h^n$$

In particular, $P_0(x) = 1$, $P_1(x) = x$, $P_2(x) = 1/2(3x^2 - 1)$, and $P_3(x) = 1/2(5x^3 - 3x)$. Cf CHEBYSHEV POLYNOMIALS.

leghaemoglobin (*BioSci*) A protein, similar to haemoglobin, in the nitrogen-fixing root nodules of leguminous plants, where it is involved in the maintenance of a low concentration of oxygen. See NITROGENASE.

Legionnaire's disease (*Med*) An atypical pneumonia caused by *Legionella pneumophila*. Infection is by droplet inhalation from air-conditioners and showers. In severe cases there is also renal, hepatic and neurological involvement.

legume (*BioSci*, *FoodSci*) (1) A fruit from one carpel, dehiscent along top and bottom, eg pea pod. (2) A member of the Leguminosae. Adj *leguminous*.

Leguminosae (*BioSci*) The pea family, c.17 000 spp of dicotyledonous flowering plants (superorder Rosidae). Trees, shrubs and herbs, cosmopolitan. The flowers have five free petals and a superior ovary of one carpel. The fruit is characteristically a legume or pod. There are three subfamilies (sometimes ranked as families): Mimosoideae, Caesalpinioideae and Papilionoideae, the last with more or less butterfly-shaped or 'pea' flowers. They form root nodules with symbiotic, nitrogen-fixing bacteria, *Rhizobium* spp. Although the seeds of many are poisonous (eg uncooked), the Papilionoideae include extremely important crops, eg the protein-rich seeds and pods of many sorts of peas and beans, lentils and groundnuts and forage crops, eg alfalfa. Also *Fabaceae*.

lehr (*Glass*) An enclosed, tunnel-like oven or furnace used for annealing glass or other forms of heat treatment. Hot glass from the forming process is passed through it to cool slowly, so that strain is removed, and cooling takes place without additional strain being introduced. Lehrs may be of the open type (in which the flame comes in contact with the ware) or of the muffle type. Also *lear*, *leer*, *lier*.

Leica (*ImageTech*) TN of the first miniature still camera using 35 mm film, giving its name to the corresponding standard frame size of 36 × 24 mm.

leiomyoma (*Med*) A tumour composed of non-striated (smooth) muscle fibres.

leishmaniasis (*Med*) A term applied to a group of diseases caused by infection with protozoal parasites of the genus *Leishmania*. See KALA-AZAR. Also *leishmaniosis*.

Leishman's stain (*BioSci*) A mixture of basic and acid dyes that, like GIEMSA STAIN, is used to stain blood smears and haemoprotozoa.

Lemberg's stain test (*Geol*) A black iron sulphide stain, used to distinguish between calcite and dolomite in limestones.

lemma (*BioSci*) The lower of the two bracts enclosing a floret of a grass. Cf PALEA.

lemma (*MathSci*) A subsidiary theorem. The proof of a complicated theorem is generally made more clear by proving any subsidiary facts in a series of lemmas.

lemniscate of Bernoulli (*MathSci*) A figure-of-eight-shaped curve with polar equation $r^2 = a^2 \cos 2\theta$. Cf OVALS OF CASSINI.

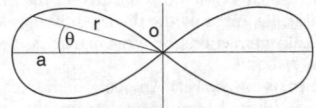

lemniscate of Bernoulli

length (*ICT*) The number of BITS in a WORD, or the number of columns or spaces in a FIELD.

length between perpendiculars (*Ships*) The length on the summer load waterline from the fore side of the stem to the after side of the stern post or, in a vessel without a stern post, to the centre of the rudder stock.

length contraction (*Phys*) See FITZGERALD–LORENTZ CONTRACTION.

length-fast (*Min*) In optically birefringent minerals, term used to denote that the fast vibration direction is aligned parallel or nearly parallel to the length of prismatic crystals.

length fold collection (*Print*) Bringing together, on rotary presses, folded products from a series of formers sometimes mounted one above another. See BALLOON FORMER.

length of lay (*ICT*) See LAY.

length overall (*Ships*) The length from the foremost point of the structure, including any figurehead or bowsprit, to the aftermost point of the ship.

length-slow (*Min*) In optically birefringent minerals, term used to denote that the slow vibration direction is aligned parallel or nearly parallel to the length of prismatic crystals.

Lennard-Jones potential (*Chem*) Potential due to molecular interaction: also called the *6–12 potential* because the attractive potential varies as the inverse sixth power of the intermolecular distance, and the repulsive potential as the inverse twelfth.

leno fabric (*Textiles*) A woven fabric in which the warp threads are made to cross one another between PICKS. Lightweight fabrics of this kind are known as gauzes; heavier qualities of cotton are used as blankets.

lens (*BioSci*) In Arthropoda, the cornea of an ocellus or compound eye; in Craniata and Cephalopoda, a structure immediately behind the iris which serves to focus light onto the retina. See CRYSTALLINE LENS.

lens (*Electronics*) (1) Device for focusing radiation. See LENS ANTENNA. (2) An arrangement of coils, permanent magnets or electrodes which can produce magnetic electrical fields capable of focusing electron beams. Used particularly in electron microscopes and accelerators.

lens (*Phys*) A portion of a homogeneous transparent medium bounded by usually spherical surfaces. Each of these surfaces may be convex, concave or plane. If in passing through the lens a beam of light becomes more convergent or less divergent, the lens is said to be *convergent* or *convex*. If the opposite happens, the lens is said to be *divergent* or *concave*. See CHROMATIC ABERRATION, FOCAL LENGTH, IMAGE, LENS FORMULA, SPHERICAL ABERRATION, THICK LENS.

lens antenna (*ICT*) A microwave antenna in which a focusing arrangement is placed in front of the radiator of energy in order to concentrate it into a beam of predetermined form and direction. The lens may consist of a system of metal slats or pieces of shaped dielectric material, or a lattice of metal spheres; the intention of all designs is to introduce selective phase delays over different paths through the lens in order to shape the beam.

lens barrel (*ImageTech*) The metal tube in which one or more lenses are mounted.

lens cap (*ImageTech*) A temporary light-tight protective covering for the external end of a camera lens; removed during exposure and focusing.

lens formula (*Phys*) An equation giving the relation between the image and object distances, l' and l, and the focal length f of a lens

$$\frac{1}{f} = \frac{1}{l'} - \frac{1}{l}$$

in the cartesian CONVENTION OF SIGNS for a thin lens in air.

lens grinding (*Glass*) The process of grinding pieces of flat sheet glass (or pressed blanks) to the correct form of the lens. Cast-iron 'tools' of the correct curvature, supplied with a slurry of abrasive and water, are used.

lens hood (*ImageTech*) Tubular or rectangular shade in front of the camera to exclude stray light from entering the lens; it may also include mountings for filters and fixed MATTES. See MATTE BOX.

lens mount (*ImageTech*) The metal tubular fitting housing the elements of a camera lens, incorporating the focus adjustment and sometimes the iris diaphragm; it is usually attached to the camera body by a standard screw or bayonet lock.

lentic (*BioSci*) Associated with standing water; inhabiting eg ponds or swamps.

lenticel (*BioSci*) A small patch of the periderm in which intercellular spaces are present allowing some gas exchange between the internal tissues and the atmosphere.

lenticle (*Geol*) A mass of lens-like (lenticular) form. The term may refer to masses of clay in sand, or vice versa, and, in metamorphic rocks, to enclosures of one rock type in another.

lenticonus (*Med*) Abnormal curvature of the lens of the eye, in which the surface becomes conical instead of spherical.

lenticular (*BioSci*) Shaped like a double convex lens.

lenticular (*Min*) Said of a mineral or rock of convex-lens shape, embedded in a matrix of a different kind.

lenticular girder bridge (*Eng*) A type of girder bridge composed of an arch whose thrust is taken by a suspension system hanging below it, the anchorages of the latter system being provided by the arch thrust.

lenticular process (*ImageTech*) Obsolete process of colour photography and cinematography in which a black-and-white emulsion was exposed through a base embossed with minute lenses or ribs, using a banded tri-colour filter on the camera lens. After reversal processing, the image was projected through a similar filter to produce an *additive* colour picture.

lentigo (*Med*) Freckles. Adjs *lentiginose*, *lentiginous*.

Lentivirinae (*Med*) Subfamily of the Retroviridae, non-oncogenic retroviruses that cause 'slow diseases'. These diseases are characterized by horizontal transmission, long incubation periods and chronic progressive phases. Visna virus is in this group, and there are similarities between visna, equine infectious anaemia virus and HIV.

Lentz valve gear (*Eng*) A locomotive valve gear in which the steam is admitted and exhausted through two pairs of *poppet valves*, spring controlled and operated from a camshaft rotating at engine speed.

Lenz's law (*ElecEng*) An induced emf will tend to cause a current to flow in such a direction as to oppose the cause of the induced emf

LEO (*Space*) Abbrev for LOW EARTH ORBIT. See panel on COMMUNICATIONS SATELLITE.

Leo (Lion) (*Astron*) A northern constellation, lying between Cancer and Virgo. Leo Minor is a faint constellation near Leo.

Leonids (*Astron*) A periodic meteor shower which is usually minor (maximum hourly rate of ten around 17 November) but increases activity dramatically every 33 yr, corresponding to the orbital period of the comet Tempel–Tuttle whose dust debris causes meteor storms when the Earth crosses its path close behind it.

leontiasis ossea (*Med*) A rare condition characterized by diffuse hypertrophy of the bones of the skull.

lepido- (*Genrl*) Prefix from Gk *lepis*, gen *lepidos*, a scale.

lepidocrocite (*Min*) An orthorhombic hydrated oxide of iron (FeOOH) occurring as scaly brownish-red crystals in iron ores. It is often one of the constituents of limonite, together with its dimorph goethite.

lepidolite (*Min*) A lithium-bearing mica crystallizing in the monoclinic system as pink to purple scaly aggregates. It is essentially a hydrated silicate of potassium, lithium and aluminium, often also containing fluorine, and is the most common lithium-bearing mineral. It occurs almost entirely in pegmatites. Also *lithia mica*.

lepidomelane (*Min*) A variety of biotite, rich in iron, which occurs commonly in igneous rocks.

Lepidoptera (*BioSci*) An order of Endopterygota, having two pairs of large and nearly equal wings densely clothed in scales. The mouth parts are suctorial, the mandibles are absent and the maxillae form a tubular proboscis. The larva or caterpillar is active and usually herbivorous, with biting mouth parts. Butterflies and moths.

lepidote (*BioSci*) With a covering of scale-like hairs.

lepospondylous (*BioSci*) Said of vertebral centra in which there is a skeletal ring constricting the notochord in the intervertebral region, with an expansion between each pair of adjacent centra.

leproma (*Med*) A nodular lesion of leprosy.

lepromin test (*Med*) A test in which killed *Mycobacterium leprae* organisms are injected into the skin of subjects with leprosy. Those with lepromatous leprosy do not react, whereas those with the tuberculoid form of leprosy show a tuberculin-type response. Not diagnostic of leprosy but an aid to classification and prognosis.

leprosy (*Med*) Infection with the bacillus *Mycobacterium leprae*; a highly chronic and moderately contagious disease. Lepromatous and neural forms are recognized, the one with nodular and destructive lesions chiefly of the face and nose, the other, even more chronic, involving chiefly the nerves. Treatable with DAPSONE. Also *Hansen's disease*.

leptin (*BioSci*) Protein product of the *ob* (obesity) gene. Found in plasma of mouse and humans: reduces food uptake and increases energy expenditure.

leptite (*Geol*) An even-grained metamorphic rock composed mainly of quartz and feldspars. Approximately synonymous with *granulite* and *leptynite*.

lepto- (*Genrl*) Prefix from Gk *leptos*, slender.

leptocercal (*BioSci*) Having a long slender tail. Also *leptocercous*.

leptodactylous (*BioSci*) Having slender digits.

leptodermatous (*BioSci*) Thin-skinned.

leptokurtic (*MathSci*) Of a curve or distribution, concentrated around the MEAN to a greater extent than a normal curve. Compare MESOKURTIC, PLATYKURTIC. See KURTOSIS.

leptom (*BioSci*) The sieve elements and associated parenchyma cells (but not sclerenchyma cells) of the phloem. Cf HADROM. Also *leptome*.

leptomeninges (*Med*) The two innermost membranes, the arachnoid and the pia mater, investing the brain and the spinal cord.

leptomeningitis (*Med*) Inflammation of the leptomeninges.

lepton (*Phys*) Fundamental particles with spin $\frac{1}{2}$ which are not affected by the STRONG INTERACTION. The lepton generations are electron, electron neutrino; muon, muon neutrino; tau, tau neutrino; with their ANTILEPTONS. The charge of the electron, muon and tau is $-e$ while the charge of the neutrinos is zero. See LEPTON NUMBER and appendix on Subatomic particles.

leptonema (*BioSci*) See LEPTOTENE.

lepton number (*Phys*) An intrinsic property of leptons. Each of the LEPTON generations has its own lepton number, L_e, L_μ and L_τ which is 1 for the respective particles, -1 for the antiparticles. It is zero for the leptons of the other generations and for all QUARKS and GAUGE BOSONS. The lepton number of the charged leptons appears to be conserved, but there is recent evidence that the lepton number conservation of the neutrinos may be violated in a process called lepton oscillation.

lepton–quark symmetry (*Phys*) Observation that the number of pairs of leptons is equal to the number of pairs of quarks. Required by the theory unifying electromagnetic and weak interactions between particles. See WEINBERG–SALAM THEORY.

leptospirosis (*Med, Vet*) An acute disease characterized by fever, jaundice, haemorrhages from the mucous membranes, enlargement of the liver, and nephritis; due to infection with *Leptospira icterohaemorrhagiae*, conveyed to humans by rats, which excrete the organism in their urine. Vaccines can be used. Two other species are involved in related diseases. Also *Leptospirosis icterohaemorrhagica* and *infectious jaundice*, *spirochaetosis icterohaemorrhagica*, *Weil's disease*.

leptosporangium (*BioSci*) A sporangium characteristic of the Filicales, the Order to which most modern ferns belong. Cf EUSPORANGIUM.

leptotene (*BioSci*) The first stage of meiotic prophase, in which the chromatin thread acquires definite polarity. See fig. at MEIOSIS.

leptynite (*Geol*) See LEPTITE.

Lepus (Hare) (*Astron*) A constellation in the southern hemisphere.

lesbianism (*Med*) Homosexuality between females.

lesion (*Med*) Any wound or morbid change anywhere in the body.

Leslie matrix model (*BioSci*) Specific deterministic model to predict the age structure of a population given the age structure at some past time and the age-specific survival and fecundity rates. Proposed by P H Leslie in 1945.

lessivage (*EnvSci*) The process in which clay particles are washed downwards in a soil by percolating water. Cf CHELUVIATION, LEACHING.

lethal (*BioSci*) Causing death. (1) Of an environmental factor, fatal to an organism. (2) Of a genetic factor causing death, often, as in bacteria, only in a particular medium or environment, then a *conditional lethal*.

lethal value (*FoodSci*) See Z-VALUE.

lethargic encephalitis (*Med*) See ENCEPHALITIS LETHARGICA.

lethargy of neutrons (*NucEng*) The natural logarithm of the ratio of initial and actual energies of neutrons during the moderation process. The lethargy change per collision is defined similarly in terms of the energy values before and after the collision. Also *logarithmic energy decrement*.

let-off motion (*Textiles*) The mechanism behind the machine to regulate the tension and delivery of the sheet of warp threads coming from the beam to the loom or warp-knitting machine.

Letraset (*Print*) TN for a TRANSFER LETTERING SYSTEM.

letterbox (*ImageTech*) A WIDE-SCREEN image that is less than the height of a TV screen, leaving black at top and bottom. See ANAMORPHIC, PALPLUS.

letterpress (*Print*) (1) A term applied to printing from relief surfaces (type or blocks), as distinct from lithography, intaglio, etc. See PRINTING. (2) The reading matter in a book, apart from illustrations etc.

letterset printing (*Print*) See OFFSET PRINTING.

letter sizes (*Eng*) A series of drill sizes in which each letter of the alphabet represents a diameter, increasing in irregular increments from A = 0·234 in (5·94 mm) to Z = 0·413 in (10·49 mm). Below A, gauge numbers, from 1 (0·228 in, 5·77 mm) to 80 (0·0135 in, 0·343 mm), are used in the reverse direction.

letting down (*Aero*) The reduction of altitude from cruising height to that required for the approach to landing.

letting down (*Eng*) The process of tempering hardened steel by heating until the desired temperature, as indicated by colour, is reached, and then quenching.

Leu (*Chem*) Symbol for LEUCINE.

leucine (*Chem*) 2-amino-4-methylpentanoic acid. $(CH_3)_2CHCH_2CH(NH_2)COOH$. The *L*- or *S*-isomer is a constituent of proteins. Symbol Leu, short form L.

leucite (*Min*) A silicate of potassium and aluminium, related in chemical composition to orthoclase, but containing less silica. At ordinary temperatures, it is tetragonal (pseudo-cubic), but on heating to about 625°C it becomes cubic. Occurs in igneous rocks, particularly potassium-rich, silica-poor lavas of Tertiary and Recent age, as eg at Vesuvius.

leucitite (*Geol*) A volcanic rock composed of leucite and clinopyroxene.

leucitophyre (*Geol*) A fine-grained igneous rock, commonly occurring as a lava, carrying phenocrysts of leucite and other minerals in a matrix essentially trachytic; a well-known example comes from Rieden in the Eifel region of Germany.

leucistic (*BioSci*) Lacking pigmentation in the skin, but differing from an albino in having blue eyes.

leuco-, leuko- (*Genrl*) Prefixes from Gk *leukos*, white.

leucobases (*Chem*) Colourless compounds formed by the reduction of dyes, which when oxidized are converted back into dyes.

leucoblast (*BioSci*) A cell that will give rise to a leucocyte.

leucocompounds (*Chem*) See LEUCOBASES.

leucocratic (*Geol*) A term used to denote a light colour in igneous rocks, due to a high content of felsic minerals, and a correspondingly small amount of dark, heavy silicates.

leucocyte (*BioSci*) A white blood corpuscle; one of the colourless amoeboid cells occurring in suspension in the blood plasma of many animals. Also *leukocyte*.

leucodermia (*Med*) A condition in which white patches, surrounded by a pigmented area, appear in the skin. Also *leucoderma, vitiligo, melanodermia*.

leuco-erythroblastic anaemia (*Med*) A form of leukaemia in which immature forms of red and white cell series are found in the peripheral blood.

leucopenia (*Med*) Abnormal diminution in the numbers of white cells in the blood. Also *leucocytopenia*.

leucoplakia (*Med*) The stage of a chronically inflamed area at which the surface becomes hard, white and smooth. On the tongue, it is usually due to syphilitic infection.

leucoplast (*BioSci*) A colourless PLASTID.

leucopoiesis (*Med*) The production of white cells of the blood.

leucorrhoea (*Med*) A whitish discharge from the vagina. US *leucorrhea*.

leucosapphire (*Min*) See WHITE SAPPHIRE.

leucotomy (*Med*) Surgical scission of the association fibres between the frontal lobes of the brain and the thalamus, in order to reduce the effects of emotions on intellectual processes (the premises for this supposition are dubious). Occasionally performed to relieve cases of severe schizophrenia and manic depression. Also *pre-frontal lobotomy*.

leucoxene (*Min*) An opaque whitish mineral formed as a decomposition product of ilmenite. Normally consists of finely crystalline rutile, but may be composed of finely divided brookite.

leukaemia (*Med*) Progressive uncontrolled overproduction of any one of the types of white cell of the blood, with suppression of other blood cells and infiltration of organs such as the spleen and liver. Cancers of the blood-cell-forming tissues. Occurs in acute and chronic LYMPHATIC, MONOCYTIC and MYELOID varieties. Also (esp US) *leukemia*.

leuko- (*Genrl*) See LEUCO-.

leukocytosis (*Med*) An increase in the number of leucocytes in the blood.

leukotrienes (*Pharmacol*) Pharmacologically active substances related to PROSTAGLANDINS and generated from arachidonic acid by the action of lipoxygenases. A series of hydroxyicosatetraenoic acids, of which leukotriene B_4 is chemotactic for granulocytes; leukotrienes C_4, D_4 and E_4 (which contain cysteine) have the properties of a 'slow-reacting substance of anaphylaxis' (SRS-A), ie they cause contraction of some types of smooth muscle, esp bronchial muscle, and increase vascular permeability. Leukotrienes may be released from platelets and various leucocytes when damaged, and SRS-A activity is generated as a result of combination of antigen with IgE antibody on mast cells.

levator (*BioSci*) See ELEVATOR.

levee (*Geol*) A long low ridge built up on either side of a stream on its flood plain. It consists of relatively coarse sand and silt deposited by the stream when it overflows its banks. Now more commonly the raising of a river bank to prevent flooding. In the USA, a landing place.

level (*Acous*) Logarithmic ratio of two energies or two field quantities, where the nominator is the measured quantity and the denominator a reference quantity.

level (*Build*) See LEVEL TUBE.

level (*CivEng*) (1) To reduce a cut or fill surface to an approximately horizontal plane. (2) The state whereby any given surface is at a tangent to the perfect Earth's perimeter at any given point. (3) A ditch or channel for drainage, esp in flat country.

level (*ICT*) (1) The difference between a measured and an arbitrarily defined reference quantity, usually expressed as the logarithm of the ratio of the quantities. In telecommunications, a common reference level is 1 mW, and power levels are expressed in decibels relative to this (dBm). (2) A specified point on an amplitude scale applied to a signal waveform, eg reference white or black in a TV transmission. (3) A single bank of contacts, eg in a stepping relay.

level (*MinExt*) An approximately horizontal tunnel in a mine, generally marking a working horizon or level of exploitation, and either in or parallel to the ore body.

level (*Phys*) Possible energy value of electron or nuclear particle.

level (*Surv*) An instrument for determining the difference in height between two points. See LEVEL TUBE.

level canal (*Build*) A canal which is level throughout. Also *ditch canal*.

level-compounded (*ElecEng*) See FLAT-COMPOUNDED.

level indicator (*ICT*) A voltage indicator on a transmission line, calibrated to indicate decibels in relation to a zero power level; also *power-level indicator*. See VOLUME UNIT.

levelling (*Surv*) The operation of finding the difference of elevation between two points.

levelling (*Textiles*) A process that allows a freshly dyed material to become uniformly coloured.

levelling agent (*Chem*) Agent added to a dye bath to produce uniform precipitation of the dye onto the fibre.

levelling bulb (*Chem*) In a gas pipette, the pressure may be equalized to that of the atmosphere by connecting with flexible tubing the liquid in the pipette to that in an open bulb, the level of which is adjusted till the heights of the two liquid interfaces (and hence their hydrostatic pressures) are equal.

levelling solvent (*Chem*) One which has a high dielectric constant and high polarity, so that most electrolytes appear strong in solution.

levelling staff (*Surv*) Light extensible system originally of wooden rods, now often telescopic, graduated in feet and tenths or metrically, used to transfer LINE OF COLLIMATION of surveyor's level from back sight, on which it is first held vertically, to fore sight, its second vertical station. As the line of collimation is horizontal, change of height is thus measured. See TARGET ROD.

level-luffing crane (*Eng*) A jib crane in which, during derricking or luffing, the load is caused to move radially in a horizontal path, with consequent power saving.

level setting (*ICT*) Provision for adjusting the base voltage for an irregular waveform, eg in TV scanning circuit voltages and signals. See CLAMP.

level small caps (*Print*) See EVEN SMALL CAPS.

level trier (*Surv*) Apparatus for measuring the angular value of a division on a level tube; it consists of a beam, hinged about a horizontal axis at one end and capable of being moved up or down at the other end by means of a micrometer screw, which records the inclination corresponding to a given number of divisions of movement of the bubble.

level tube (*Surv*) A specially shaped glass tube nearly filled with spirit, so as to leave a 'bubble' of air and spirit vapour, which always rises to the highest part of the tube. The level tube is used to test whether a surface to which it is applied is horizontal. It is an essential feature of many forms of surveying instrument. Also *spirit level*. See DUMPY LEVEL, TILTING LEVEL.

lever (*Phys*) One of the simplest machines. It may be considered as a rigid beam pivoted at a point called the *fulcrum*, a load being applied at one point in the beam and an effort, sufficient to balance the load, at another. Three classes of lever may be distinguished: (1) fulcrum between effort and load; (2) effort between fulcrum and load; and (3) load between fulcrum and effort. See MACHINE, MECHANICAL ADVANTAGE, VELOCITY RATIO.

lever key (*ICT*) A hand-operated key for telephone switchboards; operated by a small lever, which opens and closes one or more spring contacts. May be locking or non-locking.

lever rule (*Chem*) A method for estimating the proportions of phases present in a mixture. See PHASE DIAGRAM.

lever safety valve (*Eng*) A safety valve in which the valve is held on its seating by a long lever, loaded by a weight at the other end; a form of DEAD-WEIGHT SAFETY VALVE.

lever-type brush-holder (*ElecEng*) A type of brush-holder in which the brush is held at the end of an arm pivoted about the brush spindle.

lever-type starter (*ElecEng*) See FACE-PLATE STARTER.

lever-wind (*ImageTech*) A device on a camera which transports the film and sets the shutter in one movement.

levigation (*MinExt*) The use of sedimentation or elutriation to separate finely ground particles into fractions according to their movement through a separating fluid. Also used in connection with wet grinding.

levitation melting (*Eng*) A process in which the melt is suspended in an electromagnetic field while being heated by INDUCTION *in vacuo*. It is therefore not contaminated by the material of a container while molten. Only suitable for quantities of a few grams, depending on the density.

levitron (*NucEng*) Toroidal fusion system in which the poloidal magnetic field is provided by a current flowing in a solid ring situated at the circular axis of the torus. The ring may be supercooled so can be levitated indefinitely, avoiding the use of material supports.

levo- (*Genrl*) US for LAEVO-.

levodopa (*Pharmacol*) The amino acid precursor of DOPAMINE; by replenishing dopamine in the basal ganglia it improves mobility in PARKINSONISM. Usually administered with an extra-cerebral dopa-decarboxylase inhibitor (carbidopa) to prevent peripheral effects of dopamine.

levonorgestrel (*Pharmacol*) A progestogen-type hormone used in oral contraceptives, esp as an post-coital contraceptive.

Levothyroxine (*Pharmacol*) A preparation of thyroid hormone (thyroxine) that is given when normal levels are deficient. Also *L-Thyroxine sodium*.

levyne (*Min*) A zeolite; hydrated silicate of calcium and aluminium crystallizing in the trigonal system. Also *levynite*.

lew (*Build*) A light covering or roof of straw used to protect bricks on HACKS during the drying period.

lewis (*Build*) A truncated steel wedge or dovetail made in three pieces, with the larger end downwards and fitting into a similarly shaped hole in the top of a block of masonry; it then provides, by its attached hoist ring, a means of lifting the stone. Also *lewisson*. Fig. ▷

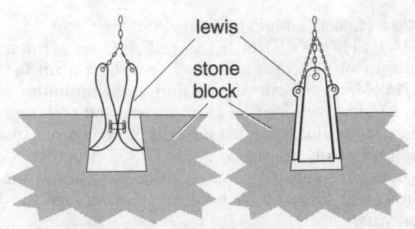

lewis Two kinds shown.

Lewis acids and bases (*Chem*) The concept that defines any substance donating an electron pair as a base, and any substance accepting an electron pair as an acid. Conventional acids and bases fit this definition, as do complex-forming reactions.

lewis bolt (*CivEng*) A foundation bolt with a tapered and jagged head, which is securely fixed into a hole in the anchoring masonry by having molten lead run round it; also used in concrete foundations.

Lewis formula (*Eng*) A formula used to calculate the strength of gear teeth.

Lewisian complex (*Geol*) The oldest rocks in the UK, a division of the Scottish Precambrian. Much of the complex is of Archaean origin, subsequently affected by later Precambrian events.

lewisite (*Chem*) 2-chloro-ethenyl dichloro arsine. $ClCH = CHAsCl_2$. Dark-coloured oily liquid (colourless when pure) with a strong smell of geraniums. Vesicant having lachrymatory and nose-irritant action. Used as poison gas.

lewisson (*Build*) See LEWIS.

Lewis's theory (*Chem*) The assumption that atoms can combine by sharing electrons, thus completing their shells without ionization.

Lewis structure (*Chem*) A possible structure for a molecule in which all electrons are specifically associated with one or two atoms. Many structures can only be described by a mixture of two or more Lewis structures, the best known example being benzene. Also *resonance structure*.

Lexan (*Chem*) TN for polycarbonate (US).

lexical analysis (*ICT*) The stage during the COMPILATION of a program, in which standard components of a statement, such as PRINT, IF, etc, are replaced by internal codes (tokens) which identify their meaning.

ley (*Agri*) Arable field sown with grass seed for one or more years to be used as relatively short-term pasture, as compared with permanent grass.

Leyden jar (*Phys*) The earliest device for storing electric charge (capacitor), named after the University of Leyden (1746). A glass jar was coated inside and outside with metal foils, which were connected by a rod passing the insulating stopper.

Leydig's duct (*BioSci*) See WOLFFIAN DUCT.

L-forms (*BioSci*) Morphological variants developed from some bacteria following prolonged culture; do not have normal cell walls and frequently revert to the normal form. If blood or serum is supplied in their nutrient medium, may breed true for some generations.

LH (*Med*) Abbrev for LUTEINIZING HORMONE.

LHC (*Phys*) Abbrev for LARGE HADRON COLLIDER.

L-head (*Autos*) A petrol-engine cylinder head carrying the inlet and exhaust valves in a pocket at one side; resembles an inverted L.

lherzolite (*Geol*) An ultramafic plutonic rock, a peridotite, consisting essentially of olivine, with both ortho- and clino-pyroxene; named from Lake Lherz in the Pyrenees.

L'Hospital's rule (*MathSci*) A rule for evaluating certain limits. If $f(a) = 0$ and $g(a) = 0$, and if n is the lowest integer for which $f_n(a)$ and $g_n(a)$ are not both zero, then

$$\lim_{x \to a} \frac{f(x)}{g(x)} = \frac{f^n(a)}{g^n(a)}$$

Li (*Chem*) Symbol for LITHIUM.

liana (*BioSci*) A climbing plant, esp a woody climber of tropical forests. Also *liane*.

Lias (*Geol*) The oldest epoch of the Jurassic period.

Lias Clay (*Geol*) A thick bed of clay found in the Lower Jurassic rocks of the UK. See MESOZOIC.

liberation (*MinExt*) The first stage in ore treatment, in which comminution is used to detach valuable minerals from gangue.

libethenite (*Min*) An orthorhombic hydrated phosphate of copper, occurring rarely as olive-green crystals in the oxide zone of metalliferous lodes.

libido (*Psych*) According to Freud, a motivating force associated with instinctual drives towards survival, pleasure and the avoidance of pain, and which manifests itself in wish fulfilment.

libollite (*Min*) A pitch-like member of the asphaltite group.

Libra (Scales) (*Astron*) An inconspicuous southern constellation, lying between Virgo and Scorpius.

library (*BioSci*) See DNA LIBRARY.

library (*ICT*) A collection of programs, MACROINSTRUCTIONS, SUBROUTINES and sections of code available in a computer system for a user to CALL or incorporate in a program.

library binding (*Print*) Stronger than the usual EDITION BINDING; sewn on tapes which are inserted in SPLIT BOARDS.

library program (*ICT*) One of a collection of programs on backing store available to computer users. Also *library routine*, *library subroutine*.

libration (*Astron*) An apparent oscillation of the Moon (or other body). The actual physical librations (due to changes in the Moon's rate of rotation) are very small; the other librations are librations in latitude or in longitude, and diurnal libration.

libration in latitude (*Astron*) A phenomenon by which an observer on the Earth sees alternately more of the north and south regions of the lunar surface, and so, in a complete period, more than a hemisphere. This is due to the Moon's rotation axis not being perpendicular to its orbital plane.

libration in longitude (*Astron*) A phenomenon by which, owing to the uniform rotation of the Moon on its axis combined with its non-uniform orbital motion, an observer on the Earth sees more, sometimes on the east and sometimes on the west, of the lunar surface than an exact hemisphere.

libration point (*Astron*) See LAGRANGIAN POINT.

libriform fibre (*BioSci*) Fibre in the XYLEM, usually the longest cell type in the tissue with thick walls and slit-like pits.

licensed aircraft engineer (*Aero*) An engineer licensed by the airworthiness authority (in the UK the CIVIL AVIATION AUTHORITY) to certify that an aircraft and/or component complies with current regulations.

lichen (*BioSci*) A symbiotic association of a fungus with an alga forming a macroscopic body, a thallus. Lichen-forming fungi are mostly Ascomycotina (see DISCOMYCETES); a few are Basidiomycotina. The alga is usually Chlorophyceae, occasionally Cyanophyceae, rarely one of each. Lichens are mostly very sensitive to air pollution (esp by SO_2) and the species present can be used as an index of pollution. Estimates suggest 20 000 spp.

lichen (*Med*) A term for papular skin eruption. *Lichen planus* has a papular patchy eruption with itch and a blue-violet colour.

lich gate (*Arch*) See LYCH GATE.

Lichtenberg figure (*ElecEng*) A figure appearing on a photographic plate or on a plate coated with fine dust when the plate is placed between electrodes and a high voltage is applied between them.

lid (*EnvSci*) Temperature inversion in the atmosphere which prevents the mixing of the air above and below the inversion region.

lidar (*EnvSci*) Abbrev for *light detection and ranging*. Device for detection and observation of distant cloud patterns by measuring the degree of back-scatter in a pulsed laser beam.

lidocaine (*Pharmacol*) Local anaesthetic agent which is also used to treat ventricular arrhythmias of the heart. Also *lignocaine*.

Lie algebra (*Phys*) Algebra dealing with groups of quantities subject to relationships which reduce the number that are independent. These are known as *special unitary groups*, and groups SU(2) and SU(3) have proved very valuable in elucidating relationships between fundamental particles.

Lieberkühn's crypts (*BioSci*) Simple tubular glands occurring in the mucous membrane of the small intestine in vertebrates.

Liebermann–Storch test (*Paper*) A test for the detection of rosin in paper by extracting a sample with acetic anhydride, cooling and adding a drop of concentrated sulphuric acid which produces a red colour if rosin is present.

Liebermann test for phenols (*Chem*) Colour changes observed on treating the phenol with sulphuric acid and sodium nitrite (nitrate (III)), then pouring the solution into excess of aqueous alkali.

Liebig condenser (*Chem*) The ordinary water-cooled glass condenser used in laboratory distillations.

lien (*BioSci*) See SPLEEN.

lienal (*BioSci*) Pertaining to the spleen.

lienogastric (*BioSci*) Pertaining to, or leading to, the spleen and the stomach, as the *lienogastric artery* and *vein* in vertebrates.

lier (*Glass*) See LEHR.

Liesegang rings (*Chem*) The stratification, under certain conditions, of precipitates formed in gels by allowing one reactant to diffuse into the other.

life cycle (*BioSci*) The various stages through which an organism passes, from fertilized ovum to the fertilized ovum of the next generation.

life-cycle analysis (*Genrl*) The flow of energy and materials through a manufacturing system from raw material in ground, through processing to shape, assembly of finished product and disposal following use. Consists of three stages: an inventory of inputs and outputs; assessment of the impact; and formulation of solution. Also *womb-to-tomb* (US), *cradle-to-grave* (Europe) analysis. Not to be confused with PRODUCT LIFE CYCLE. Abbrev *LCA*.

life-cycle inventory (*Genrl*) The measurement or calculation of all material and fuel inputs and ouputs from a process starting with raw materials in the ground and ending with waste disposal back into the earth or atmosphere. Abbrev *LCI*.

life form (*BioSci*) The overall morphology of a plant categorizing it as an annual herb, a shrub, a succulent, etc. See RAUNKIAER SYSTEM.

life support (*Space*) The provision of the necessary conditions of health and comfort to support a human in space, either during the occupation of a space vehicle or during extra-vehicular activity (EVA).

life table (*BioSci*) Tabulated display of the mortality schedule of a population, first devised by insurance companies and now used by ecologists in the study of wild populations.

lifetime (*Phys*) The mean period between the generation and recombination of a charge carrier in a semiconductor. See HALF-LIFE, MEAN LIFE.

Lifetime Study (*Radiol*) The long-term study of the survivors of the Hiroshima and Nagasaki bombs. See panel on LIFETIME STUDY OF THE NUCLEAR BOMB SURVIVORS.

LIFO (*ICT*) Abbrev for *last-in, first-out*. See STACK.

lift (*Aero*) *Aerodynamic lift* is the component of the aerodynamic forces supporting an aircraft in flight, along the lift axis, due solely to relative airflow. Lift force acts at right angles to the direction of the undisturbed airflow relative to the aircraft.

Lifetime study of the nuclear bomb survivors

The long-term study of the survivors of the Hiroshima and Nagasaki bombs, detonated on 6 and 10 August 1945, is now supported by the Radiation Effects Research Foundation, and has been carried out by Japanese and US scientists who have published a large number of reports over the years. It is by far the largest study of the biological effects of a single acute radiation dose.

Two methods have been used. For 80 000 of the survivors it has been possible to calculate from position, shielding and symptoms, dose rates varying from nothing to over 4 grays. Their death rates from cancer were then compared with that of the total Japanese population. In addition, about 68 000 children were divided into three groups depending on their parents' history: those irradiated within 2000 metres of the two epicentres; those placed at more than 2500 metres from them; and those who returned to the cities after the bombs. The children were further classified according to the gonadal dose to both their parents.

The absolute risk for different cancers in the survivors is shown in the diagram (the scale is the excess death per million person-years for a dose of 10 milligrays). It shows the expected rise in excess deaths from leukaemia comparatively soon after irradiation, followed by a fall 30 years later. Other cancers, with their longer latent periods, become more significant after 20 years. In general, those who received more than 2 grays had about twice as many cancers during the 33 years following the bomb, except for leukaemias and multiple myeloma, where there were just over ten and five times as many cases.

The 68 000 children have been investigated in many ways, new methods being added as they have become available. The number of cancers was recorded, but also the number of congenital abnormalities, deaths other than cancer, chromosomal rearrangements, abnormal number of sex chromosomes, sex ratio, abnormal proteins as evidence for genetic recessives, and general growth and development. But there is no statistically significant evidence for any difference between the children of irradiated parents and the matched controls. Despite this, and because such a difference has been found in mice, the investigators have tried to derive a figure for the genetic doubling dose. In other words, what is the dose necessary to double the number of mutations compared with the background? They have calculated a figure between 2 and 3 sieverts for gonadal exposure compared with 0·3 to 0·4 sieverts for male mice.

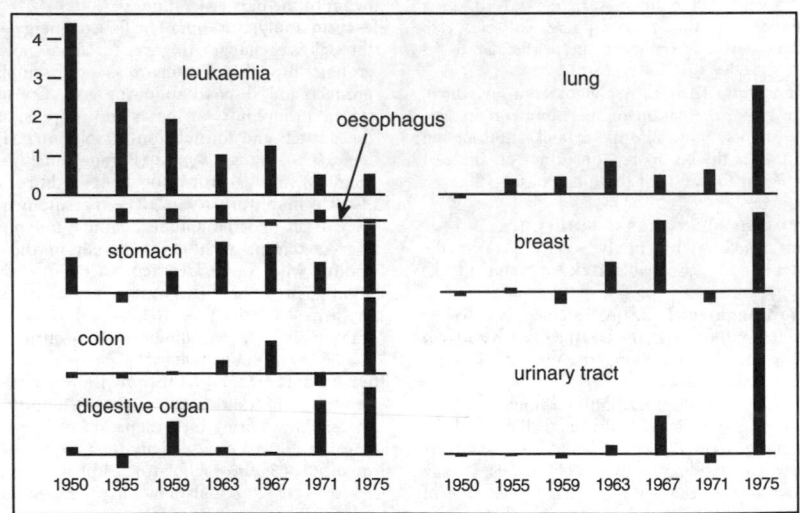

Cancer incidence after the Hiroshima/Nagasaki bombs

---

lift (*Build*) An enclosed platform or car moving in a well to carry persons or goods up and down. US *elevator*. Cf PATERNOSTER.

lift (*ImageTech*) See PEDESTAL.

lift axis (*Aero*) See AXIS.

lift bridge (*CivEng*) A type of movable bridge which is capable of being lifted bodily through a sufficient vertical distance to allow the passage of a vessel beneath.

lift coefficient (*Aero*) A non-dimensional number representing the aerodynamic lift on a body.

$$C_{\mathrm{L}} = \frac{L}{\frac{1}{2}\rho V^2 S}$$

where L = lift, ρ = air density, V = airspeed, S = wing area.

lift engine (*Aero*) An engine used on VTOL or STOL aircraft, having the primary purpose of providing lifting force. It is shut down in normal flight and the intake usually closed by a sealing flap.

lifter (*Eng*) An L- or Z-shaped bar of cast- or wrought-iron, used for supporting the sand in a cope, the upper end being hooked on to a box bar.

lifter (*MinExt*) (1) Projecting rib or wave in lining plates of ball or rod mill. (2) Perforated plate in drum washer or heavy media machine which aids tumbling action or removes floating fraction of ore being treated.

lifting (*Build*) The result of applying a paint on top of a coating which is not fully dry. This can also happen when subsequent coats of paint contain excessively strong solvents.

lifting blocks (*Eng*) A lifting machine consisting of a continuous rope passing round pulleys mounted in blocks, whereby an effort applied at the free end of the rope lifts a larger weight attached to the lower block. See MECHANICAL ADVANTAGE.

lifting magnet (*ElecEng*) A large electromagnet used, instead of a hook, on cranes or hoists, when lifting iron and steel.

lifting of patterns (*Eng*) See DRAWING OF PATTERNS.

lifting screw (*Eng*) An iron rod screwed into a pattern to withdraw it from the mould.

lifting the offsets (*Ships*) The process of measuring the ship's form as 'laid off' on buttocks, waterlines and sections. These offsets are the permanent record and are used to reproduce the ship's form, initially, for design work and ordering material and, subsequently, in cases of repair or alterations.

lifting truck (*Eng*) A truck with three or four wheels, drawn by a handle which can be raised and lowered to lift a loaded platform standing on feet. The lift is effected either by leverage or by hydraulic mechanism.

lift-lock (*Build*) A canal lock serving to lift a vessel from one reach of water to another, the lock, water and vessel being raised mechanically.

lift-off (*Aero*) The speed at which a pilot pulls back on the control column to make an aircraft leave the ground. It is a carefully defined value for large aircraft depending upon the weight, runway surface, gradient, altitude and ambient temperature. It is one of the functions established from WAT CURVES. Colloq *unstick*.

lift-off (*Space*) The point at which the vertical thrust of a spacecraft exceeds the force due to local gravity and the spacecraft begins to leave the ground.

lift-valve (*Eng*) Any valve consisting of a disk, ball, plate, etc, which lifts or is lifted vertically to allow the passage of a fluid.

ligament (*BioSci*) A bundle of fibrous tissue joining two or more bones or cartilages.

ligand field theory (*Chem*) An essentially ionic interpretation of bonding in transition metal compounds, in which the spectroscopic and magnetic properties are rationalized in terms of the distortion of the non-bonding d-electrons of the metal by the electric field of the ligands.

ligand-gated ion channel (*BioSci*) A transmembrane ION CHANNEL whose permeability to certain ions is increased by the binding of a specific ligand, eg a neurotransmitter such as GAMMA AMINO BUTYRIC ACID or ACETYLCHOLINE.

ligands (*Chem*) (1) In a complex ion, the ions, atoms or molecules surrounding the central (nuclear) atom, eg $(CN)^-$ in $Fe(CN)_6^{4-}$. (2) More generally, esp in bioscience, small molecules that bind to receptors.

ligase (*BioSci*) An enzyme that seals nicks in one strand of a duplex DNA, much used to seal the gaps formed when one DNA sequence is artificially inserted into another.

ligasoid (*Phys*) A colloidal system in which the continuous phase is gaseous and the dispersed phase is liquid.

ligate (*Med*) To tie with a ligature. N *ligation*.

ligation (*BioSci*) Joining two linear nucleic acid molecules together by a phosphodiester bond. Such linear molecules often have STICKY ENDS which facilitate ligation.

ligature (*Med*) A piece of thread, silk, wire, catgut or any other material, for tying round blood vessels etc.

ligature (*Print*) Two or more letters, originally cast together on one body, eg

ct ff ffi ffl fl st

light (*Build*) (1) A term applied to any glazed opening admitting light to a building. (2) A single division of a window.

light (*Glass*) Of optical glass, having a lower refractive index.

light (*Phys*) (1) Electromagnetic radiation capable of inducing visual sensation, with wavelengths between about 380 and 780 nm. (2) More generally, electromagnetic waves of any wavelength, including eg ultraviolet and infrared radiation. See ILLUMINATION, SPEED OF LIGHT.

light-adapted (*Phys*) See ADAPTATION OF THE EYE.

light air (*EnvSci*) A wind of between about 1 and 3 mph, force 1 on the Beaufort scale.

light aircraft (*Aero*) One having a maximum take-off weight less than 12 500 lb (5670 kg).

light alloys (*Eng*) Alloys based on the low-density metals aluminium and magnesium. See ALUMINIUM ALLOYS.

light breeze (*EnvSci*) A wind of between about 4 and 7 mph, force 2 on the Beaufort scale.

light-centre length (*ElecEng*) The distance from the geometrical centre of the filament of an electric filament lamp to the contact plate or plates at the end of the lamp cap remote from the bulb. With automobile headlight lamps the measurement is taken from the bulb side instead of from the remote side of the pin.

light chain (*BioSci*) A polypeptide chain present in all immunoglobulin molecules. In most molecules two identical light chains are linked to two heavy chains by disulphide bonds. There are either of two types of light chain, KAPPA and LAMBDA, but never both on the same molecule. Light chains have an N-terminal variable region that forms part of the antigen combining site and a C-terminal constant region.

light change points (*ImageTech*) The series of preset steps of exposure used in a motion picture film printer; abbrev *LCP*. (US *printer lights*).

light-curve (*Astron*) The line obtained by plotting, on a graph, the apparent change of brightness of an object (such as a variable star or supernova) against time.

light distribution curve (*Phys*) A graph showing the relation between the luminous intensity of a light source and the angle of emission.

light efficiency (*Phys*) The ratio of total luminous flux over total power input, expressed in lumens per watt for eg an electric lamp.

light-emitting diode (*Electronics*) Semiconductor diode which radiates in the visible region. Used in alphanumeric displays and as an indicator or warning lamp. Abbrev *LED*. See OPTOELECTRONICS.

lighter-than-air craft (*Aero*) See AEROSTAT.

light face (*Print*) A lighter weight of typeface, 'medium' being the usual normal in the TYPE FAMILY.

light fastness (*Build*) The ability of paint pigments or dyes to retain their colour when exposed to light.

light filter (*ImageTech*) A filter used in photography etc to change or control the total (or relative) energy distribution from the source. It consists of a homogeneous optical medium (sometimes of a specific thickness as in interference filters) with characteristic light absorption regions. Can be graded for special effects.

light flux (*Phys*) The measure of the quantity of light passing through an area, eg through a lens system. Light flux is measured in lumens. Illumination is light flux per unit area.

light fog (*ImageTech*) See FOG.

lighting contrast (*ImageTech*) The ratio between the level of the KEY LIGHT plus filler to the FILLER LIGHT alone.

light meter (*ImageTech*) A device for measuring the intensity of incident or reflected light, usually comprising a photosensitive cell and a meter; when this measurement is expressed as the corresponding camera lens aperture and exposure time for a given photographic sensitivity, it becomes an EXPOSURE METER.

lightness (*ImageTech*) The degree of illumination of a surface, measured in lumens per square metre.

lightning (*EnvSci*) The luminous discharge of electric charges between clouds, and between cloud and earth (or sea). A path is found by the *leader stroke*, the main discharge following along this ionized path, with possible repetition. See THUNDERSTORM.

lightning arrester (*ElecEng*) A device for the protection of apparatus from damage by a lightning discharge or other accidental electrical surge. A surge arrester is effectively a shunt device whose impedance is sensitive to voltage. At normal operating voltage, impedance is high, but, under conditions of overvoltage, the impedance reduces and the device protects by providing a bypass path for the surge.

lightning conductor (*ElecEng*) A metal strip connected to earth at its lower end, and its upper end terminated in one or more sharp points where it is attached to the highest part of a building. By electrostatic induction it will tend to neutralize a charged cloud in its neighbourhood and the discharge will pass indirectly to earth through the conductor. Also *lightning rod*.

lightning protector (*ElecEng*) See LIGHTNING ARRESTER.

lightning rod (*ElecEng*) See LIGHTNING CONDUCTOR.

lightning tubes (*Min*) See FULGURITES.

light oils (*Chem*) A term for oils with a boiling range of about 100–210°C, obtained from the distillation of coaltar.

light painting (*ImageTech*) A photographic technique in which the camera shutter is opened and a fibre optic light hose or portable light is used to selectively light the subject in total darkness.

light pen (*ICT*) INPUT DEVICE used in conjunction with a GRAPHICAL DISPLAY UNIT. The graphical display unit's hardware, together with special software, senses the position of the pen and relays this information to the CENTRAL PROCESSING UNIT.

light quanta (*Phys*) When light interacts with matter, the energy appears to be concentrated in discrete packets called *photons*. The energy of each photon is $E = h\nu$ where ν is the frequency and h is PLANCK'S CONSTANT. See QUANTUM and panel on QUANTUM THEORY.

light railway (*CivEng*) One of standard or narrow gauge, subject to severe restrictions on speed, load, etc, and thereby relieved of many of the operating requirements (signalling etc) applied to mainline railways.

light reactions (*BioSci*) Those reactions in PHOTOSYNTHESIS in which light is absorbed, transferred between pigments, and used to generate the ATP and reduced NADP used in the DARK REACTIONS for CO_2 fixation. See PHOTOSYNTHETIC PIGMENTS, PHOTOSYSTEMS I AND II.

light-red silver ore (*Min*) See PROUSTITE.

light relay (*ICT*) A relay operating circuit using a photoelectric detector.

light resistance (*Electronics*) The resistance, when exposed to light, of a photocell of the photoconductive type. Cf DARK RESISTANCE.

light scattering (*Chem*) A method for determining molecular mass of molecules, esp soluble polymers. Gives absolute measure of M_w. See MOLECULAR MASS DISTRIBUTION

light-sensitive (*Phys*) Said of a thin surface for which the electrical resistance, emission of electrons or generation of a current depends on the incidence of light.

light-spring diagram (*Eng*) An indicator diagram taken by a piston or diaphragm-type engine indicator, using a specially weak control spring or diaphragm in order to reproduce the low-pressure part of the diagram to a large scale.

light table (*Print*) A ground-glass surface, illuminated from below, there being a variety of sizes and styles, for use in the graphic reproduction processes when working with negatives, positives and proofs. Also *shiner*.

light trap (*ImageTech*) General term for any mechanical arrangement which permits movement while excluding light, such as the doors and partitions for operational access to a dark room (light-lock) or the rollers and lips over which film passes from and into a magazine.

light-up (*Aero*) The period during the starting of a *turbojet* or *turboprop* engine when the fuel/air mixture has been ignited.

light valve (*ImageTech*) A device for rapidly varying light transmission and hence the exposure of motion picture film, as in photographic sound recording or in printing from an assembled negative.

light valve projector (*ImageTech*) A video projector using a schlieren optical system and a light valve, consisting of a transparent liquid film whose surface is optically distorted by an ELECTRON BEAM modulated by the picture signal, to produce an image in the light. See DIGITAL MICROMIRROR DEVICE, LIQUID CRYSTAL LIGHT VALVE.

light water (*Chem*) Normal water (H_2O) as distinct from HEAVY WATER (D_2O).

light-water reactor (*NucEng*) A reactor using ordinary water as moderator and coolant. See figs at BOILING-WATER REACTOR, PRESSURIZED-WATER REACTOR, REACTOR.

light watt (*Phys*) The photometric radiation equivalent, the radiation being measured in lumens or watts. The ratio of lumens to watts depends on the wavelength. At the wavelength of greatest sensitivity of the eye ($\lambda = 555$ nm), the value of the light watt is 682 lm.

lightweight aggregate (*Build, CivEng*) Aggregate which is used in the manufacture of lightweight concrete. Normally clinker ash or clays or other materials which have been fired to reduce their carbon content and weight.

lightweight concrete (*Build*) Concrete of low density ($20–90$ lb ft^{-3} or $300–1400$ kg m^{-3}) made by using special aggregates or air entraining processes. Used particularly where lightness and insulation are required rather than load-resisting properties.

lightwood (*For*) Coniferous wood having an abnormally high resin content.

light-year (*Astron*) An astronomical measure of distance, being the distance travelled by light in space during a year, which is approximately $9 \cdot 46 \times 10^{12}$ km.

ligne (*Genrl*) A unit used in the measurement of watch movements. It is equal to $2 \cdot 256$ mm. The *twelfth*, or *douzième* ($0 \cdot 188$ mm), is the unit used for the height or thickness of a movement. There are twelve twelfths in a ligne. Women's wristwatches vary from 3 to 8 lignes; men's wristwatches from 10 to 13, pocket watches from 17 to 19, and deck watches from 20 to 32. Symbol *'''*. Also *line*.

lignicole (*BioSci*) Growing or living on or in wood, or on trees, eg certain species of termites. Adj *lignicolous*.

lignin (*BioSci*) A complex cross-linked polymer, based on variously substituted p-hydroxyphenyl propane ($HOC_3H_6–$) units; a constituent with cellulose etc of the cell walls of xylem tracheids and vessels and of many sclerenchyma and some parenchyma cells etc where its major function appears to be to impart rigidity to an otherwise flexible wall. See panel on WOOD.

lignite (*Geol*) A brownish, black coal intermediate between peat and sub-bituminous coal. It is commoner than coal in Mesozoic and Tertiary deposits. Also *brown coal*.

lignivorous (*BioSci*) Wood-eating.

lignocaine (*Med*) See LIDOCAINE.

ligno-celluloses (*Chem*) Compounds of lignin and cellulose found in wood and other fibrous materials.

lignum vitae (*For*) One of the hardest and densest commercial timbers; some 25% of its dry weight is guaiac resin which makes it self-lubricating. A hardwood (*Guaiacum*) from the W Indies and tropical America, its heartwood is greenish-brown to black, with a heavily interlocked grain and fine, uniform texture.

ligroin (*Chem*) A term for a petroleum fraction with a boiling range of from about 90 to 120°C.

ligulate (*BioSci*) Strap-shaped.

likelihood function (*MathSci*) The expression, as a function of the parameter, of the joint probability density function or probability function of observations in a random sample from a distribution which is dependent on a parameter.

Liliaceae (*BioSci*) A family of c.3500 spp of monocotyledonous flowering plants (superorder Liliidae). They are mostly herbs, often with bulbs, corms or rhizomes, and are cosmopolitan. The flowers are usually showy, with six perianth members, six stamens and a superior ovary of three carpels. Includes *Allium*, the onions, garlic, leeks, *Colchicum*, the source of COLCHICINE, and many ornamental plants.

Liliidae (*BioSci*) A subclass or superorder of monocotyledons. Mostly herbs with both sepals and petals usually petaloid. Syncarpous, with the ovary normally inferior. Contains c.28 000 spp in 17 families including Liliaceae, Amaryllidaceae, Iridaceae and Orchidaceae.

Liliopsida (*BioSci*) See MONOCOTYLEDONES.

lim (*MathSci*) Abbrev for *limit*.

limaciform (*BioSci*) Slug-like.

limaçon (*MathSci*) A heart-shaped curve with an inner loop at its vertex. An epitrochoid in which the rolling circle equals the fixed circle. Polar equation $r = 2a(k + \cos\theta)$.

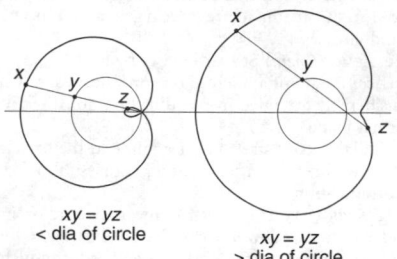

$xy = yz$
< dia of circle

$xy = yz$
> dia of circle

limaçon

limb (*Astron*) The edge or rim of a heavenly body having a visible disk; used specially of the Sun and Moon.

limb (*BioSci*) (1) A jointed appendage, eg a leg. (2) The lamina of a leaf. (3) The widened upper part of a petal. (4) The upper, often spreading, part of a sympetalous corolla.

limb (*Geol*) The side of a fold.

limb (*Surv*) Lower horizontal plate of theodolite.

limba (*For*) See AFARA.

limb darkening (*Astron*) The apparent darkening of the limb of the Sun due to the absorption of light in the deeper layers of the solar atmosphere near the edge of the disk.

limberneck (*Vet*) BOTULISM of birds.

limbic (*Med*) Marginal, bordering; eg lobe of cerebral cortex.

limbic system (*Psych, BioSci*) A complex set of evolutionarily ancient structures of the fore-brain, including the hippocampus, anterior thalamus, amygdala and parts of the hypothalamus, that play a role in emotional expression, particularly in the emotional component of behaviour, memory and motivation. See panel on BRAIN STRUCTURE.

limbous (*BioSci*) Overlapping.

limbric (*Textiles*) Plain-weave cotton cloth of light to medium weight in which the weft predominates.

limburgite (*Geol*) An ultramafic, fine-grained igneous rock occurring in lava flows, similar to the dyke rock monchiquite, but having interstitial glass between the dominant olivine and augite crystals. Typically, limburgite is feldspar-free; but it does occur.

limbus (*BioSci*) A distinct edge or border, eg that between cornea and sclera in the eye.

lime (*Chem*) A substance produced by heating limestone to 825°C or more, as a result of which the carbonic acid and moisture are driven off. Lime, which is much used in the building, chemical, metallurgical, agricultural and other industries, may be classified as *high-calcium, magnesian* or *dolomitic* depending on the composition. Unslaked lime is commonly known as *caustic lime*, also *anhydrous lime, burnt lime, quicklime*.

lime (*For*) W European hardwood (*Tilia*), creamy-yellow, straight-grained and with a fine, even texture.

lime bag (*Eng*) A bag of powdered lime used for testing the fit of joints. Lime is sprinkled on the parting face, and the cope is lowered and lifted; if the lime adheres to the top face, the joint is good.

lime-induced chlorosis (*BioSci*) Chlorosis of young leaves induced in (relatively) calcifuge plants by growth on calcareous or over-limed soils (ie soils of too high pH), caused apparently by a disturbance in iron metabolism and curable horticulturally by the application of eg chelated iron, SEQUESTRENE, to leaves or roots.

lime light (*Eng*) Intense white light obtained by heating a cylinder of lime in an oxyhydrogen flame.

lime mortar (*Build, CivEng*) A mortar composed of lime and sand, with the addition sometimes of other material, such as crushed bricks, ground slag or coke. It is not suitable for use under water.

limen (*Psych*) The threshold of consciousness; the limit below which a stimulus is not perceived.

lime paste (*Build*) Slaked lime.

lime powder (*Build*) The material produced as a result of subjecting quicklime to the process of air-slaking.

lime-silicate rocks (*Geol*) These result from the contact (high-temperature) metamorphism of limestones containing silica in detrital grains, nodules of flint or chert, or siliceous skeletons, the silica combining with the lime to form such silicates as lime-garnet, anorthite, wollastonite.

lime slurry (*MinExt*) Thick aqueous suspension of finely ground slaked lime, used to control alkalinity of ore pulps in flotation and cyanide process.

lime–soda process (*Chem*) Standard water-softening process, carried out either hot or cold; uses lime and soda ash to reduce the hardness of the treated water by precipitating the dissolved calcium and magnesium salts as insoluble calcium carbonate and magnesium hydroxide respectively. Often used in conjunction with ZEOLITE PROCESS. Also used in the preparation of caustic soda, by mixing slaked lime with soda and filtering off the precipitated calcium carbonate. $Ca(OH)_2 + Na_2CO_3 = 2NaOH + CaCO_3$.

limestone (*Geol*) A sedimentary rock consisting of more than 50% by weight of calcium carbonate. Mineralogically, limestone consists of either aragonite or calcite, although the former is abundant only in Tertiary and Recent limestones. Dolomite may also be present, in which case the rock is called *dolostone*. Limestones may be *organic*, formed from the calcareous skeletal remains of living organisms, *chemically precipitated* or *detrital*, formed of fragments from pre-existing limestones. The majority are organic, and can be classified according to their texture and the nature of the organisms whose skeletons are incorporated, eg oölitic limestone, shelly limestone, algal limestone, crinoidal limestone.

limewash (*Build*) A mixture which is prepared by slaking lump lime with about one-third of its weight of water, and then adding sufficient water to make a 'milk'; used as wall covering in cases where a frequent application is necessary. An improved, more water-resistant type embodies tallow.

lime water (*Chem*) Saturated calcium hydroxide solution.

limicolous (*BioSci*) Living in mud.

limit (*MathSci*) A number L such that a given sequence or function approaches L infinitesimally closely; formally u_r has a *limit* L if for every NEIGHBOURHOOD U of L there exists m such that for all $n \geqslant m$, u_n is in U.

limited stability (*ICT*) A property of a servomechanism or a communication system that is characterized by stability only when the input signal falls within a particular range.

limiter (*ICT*) Any transducer in which the output, above a threshold or critical value of the input, does not vary, eg a shunt-polarized diode between resistors. Particularly applied to the circuit in a frequency-modulation receiver in which all traces of amplitude modulation in the signal have to be removed before final demodulation.

limiter (*NucEng*) An aperture which defines the boundary of a plasma and protects the vacuum vessel from damage by contact with hot plasma.

limit gauge (*Eng*) A gauge used for verifying that a part has been made to within specified dimensional limits. Limit gauges consist eg of a pair of plug gauges on the same bar, one of which should just enter a hole ('go') and the other just not enter ('not go').

limit gauging (*Eng*) A method of measurement which ensures that pieces intended to fit together shall do so within certain specified limits of clearance, and that similar pieces shall be interchangeable.

limiting conductivity (*Chem*) The molar conductivity of a substance at infinite dilution, ie when completely ionized.

limiting density (*Chem*) The relative density of a gas at vanishingly low pressures.

limiting factor (*BioSci*) The variable, or combination of variables, which limits the rate of growth of an organism or a population, or which limits the rate of a physiological process.

limiting frequency (*Acous*) (1) The frequency at which the wavelength of an airborne sound wave coincides with the wavelength of a bending wave of a wall or plate. Important in sound transmission. (2) The name for the highest or lowest frequency transmitted by an electro-acoustic system or by a waveguide.

limiting friction (*Phys*) See FRICTION.

limiting gradient (*CivEng*) Same as RULING GRADIENT.

limiting Mach number (*Aero*) The maximum permissible *flight Mach number* at which any particular aircraft may be flown, either because of the BUFFET BOUNDARY or for structural strength limitations.

limiting range of stress (*Eng*) The greatest range of stress (mean stress zero) that a metal can withstand for an indefinite number of cycles without failure. If exceeded, the metal fractures after a certain number of cycles, which decreases as the range of stress increases. See FATIGUE STRENGTH.

limiting velocity (*Aero*) The steady speed reached by an aircraft when flown straight, the angle to the horizontal, power output, altitude and atmospheric conditions all specified. See MAX LEVEL SPEED, TERMINAL VELOCITY.

limiting viscosity number (*Phys*) See INTRINSIC VISCOSITY.

limit load (*Aero*) The maximum load anticipated from a particular condition of flight and used as a basis when designing an aircraft structure.

limit of proportionality (*Eng*) See PROPORTIONAL LIMIT.

limit point (*MathSci*) See ACCUMULATION POINT.

limit state (*CivEng*) The state at which a structure has become unfit for use. The main limit states are *limit state of collapse* and *serviceability limit state*.

limit switch (*ElecEng*) A switch fitted particularly to automatic equipment which cuts off the power supply when some part of the equipment travels beyond a specified limit.

limit values (*Genrl*) A European Community term specifying environmental quality standards and emission standards, which defines the limits that member states must set, eg on concentrations of smoke and sulphur dioxide at ground level.

limivorous (*BioSci*) Mud-eating, like certain aquatic invertebrates that swallow mud for its organic constituents.

limnobiotic (*BioSci*) Living in fresh water.

limnology (*Geol*) The study of lakes.

limnophilous (*BioSci*) Living in marshes, esp freshwater marshes.

limonene (*Chem*) *Hesperidene, citrene, carvene*. The oil of the orange peel consists almost entirely of (+)-limonene, bp 175°C;. (−)-limonene is present in the oil of fir cones. These compounds, which are monocyclic terpenes, are also called *1,8-menthadiene*.

limonite (*Min*) Although originally thought to be a definite hydrated oxide of iron, now known to consist mainly of cryptocrystalline goethite or lepidocrocite along with adsorbed water; some haematite may also be present. It is a common alteration product of most iron-bearing minerals and also the chief constituent of bog iron ore.

limp (*Print*) Said of a book having non-rigid sides; described as *limp cloth*, *limp leather*, according to the covering material.

limpet washer (*Build*) A form of washer used in fixing corrugated sheeting, for which purpose it is shaped on one side to conform to the curve of a corrugation.

linarite (*Min*) Hydrated sulphate of lead and copper, found in the oxide zone of metalliferous lodes; a deep-blue mineral resembling azurite, also crystallizing in the monoclinic system.

linch pin (*Eng*) A pin placed in a transverse hole on the outside of the axles of a vehicle, to retain a wheel; at the top it has a projection on one side, to prevent it from passing through the hole.

Lincoln index (*BioSci*) See MARK AND RECAPTURE.

Lindane (*BioSci*) TN for gamma-hexachlorocyclohexane (*benzene hexachloride*). An organochlorine pesticide also used for treating head lice and scabies. Banned in many countries because of its toxic side effects.

Lindeck potentiometer (*ElecEng*) One which differs from most potentiometers in using a fixed resistance and variable current to obtain balance.

Lindé process (*Chem*) A process for the liquefaction of air and for the manufacture of oxygen and nitrogen from liquid air.

Lindé sieve (*Chem*) See MOLECULAR SIEVE.

line (*Acous*) Said of a microphone or loudspeaker when it is considerably extended in one direction, for directivity in a normal plane.

line (*Build*) A cord stretched as a guide to the bricklayer for level and direction of succeeding courses; also for setting out foundations.

line (*ElecEng*) (1) The direction of an electric or magnetic field, as a line of flow or force. See LINE OF FLUX (or *force*). (2) The term used to describe overhead conductor in a power transmission system, normally of steel-core aluminium.

line (*Genrl*) (1) The twelfth part of an inch. (2) See LIGNE.

line (*ICT*) Transmission line, coaxial, balanced pair, or earth return, for electric power, signals or modulation currents.

line (*ImageTech*) Single scan in a facsimile or TV picture transmission system.

line (*MathSci*) An infinitely long straight one-dimensional figure with no thickness. This is an undefined primitive term of Euclidean geometry. On a plane, a line is the shortest distance between two points.

line (*Radiol*) Single frequency of radiation as in a luminous, X-ray or neutron spectrum.

linea alba (*Med*) The tendinous line which extends down the front of the belly, from the lower end of the chest to the pubic bone, and gives attachment to abdominal muscles.

lineament (*Geol*) A long feature on the Earth's surface, often more clearly visible by satellite photography. It may be structural or volcanic and can be related to PLATE TECTONICS (panel).

line amplifier (*ICT*) In broadcasting, that which supplies power, at a specified level to the line, either to a control centre or to a transmitter.

line amplitude (*ImageTech*) The amplitude of the voltage generated by the line-scanning generator, or the length of the line on the screen produced thereby.

linea nigra (*Med*) Pigmented linea alba, occurring in pregnant women.

linear (*BioSci*) (1) A leaf having parallel sides, and at least four to five times as long as broad. (2) A tetrad of pollen grains in a single row.

linear (*Phys, Eng*) Said of any device or motion where the effect is exactly proportional to the cause, such as rotation and progression of a screw, current and voltage in a wire resistor (Ohm's law) at constant temperature, output versus input of a modulator (or demodulator). See ULTRALINEAR.

linear absorption coefficient (*Phys*) See ABSORPTION COEFFICIENT.

linear accelerator (*Electronics*) A large device for accelerating electrons or positive ions up to nearly the speed of light. The particles are accelerated through loaded waveguides by high-frequency pulses or oscillations of the correct phase.

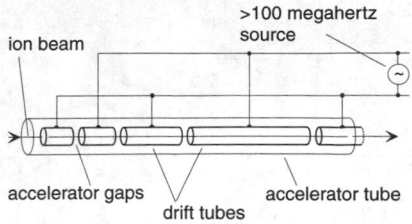

linear accelerator Only first stages shown.

linear amplifier (*Electronics*) One for which the output power is directly proportional to the input level.

linear array (*ICT*) See COLLINEAR ARRAY.

linear-ball bearing (*Eng*) The linear analogue of the radial ball bearing, in which a column of balls rolls in a straight track between the bearing and a hardened and ground shaft, recirculating in a space formed parallel to the track but out of contact with the shaft. Commonly five or six tracks are arranged round the bearing, constraining it radially. Cf RECIRCULATING-BALL THREAD.

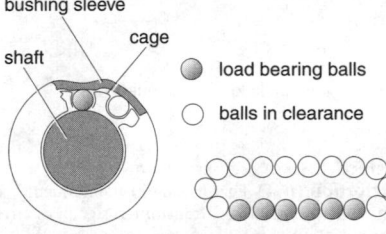

linear-ball bearing At right the pattern of one of the five ball runs in the bearing.

linear density (*Textiles*) The weight of a fixed length of textile yarn, eg DENIER, TEX; the smaller the number, the finer the thread. Cf YARN COUNT.

linear detector (*ICT*) One in which the demodulated signal voltage is directly proportional to the changes in input carrier amplitude or frequency, for amplitude or frequency modulation respectively.

linear differential equation (*MathSci*) An equation of the form

$$q_0(x)\frac{d^n y}{dx^n} + q_1(x)\frac{d^{n-1}y}{dx^{n-1}} + \cdots$$
$$+ q_{n-1}(x)\frac{dy}{dx} + q_n(x)y = f(x)$$

where n is the *order* of the equation.

linear distortion (*ICT*) That which results in the non-linear response of a system, such as an amplifier, to the envelope of a varying signal, such as speech, without distorting (within acoustic perception) the detailed waveform.

linear elastic-fracture mechanics (*Eng*) Abbrev *LEFM*. See FRACTURE MECHANICS.

linear energy transfer (*Phys*) The linear rate of energy dissipation by particulate or electromagnetic radiation while penetrating absorbing media. Abbrev *LET*.

linear irrigation (*Agri*) A continuous length of horizontal water pipe mounted at intervals on motorized trucks. The

system moves forward across the area to be irrigated, spraying water as it goes.

linearity (*ICT*) The condition in which the change in value of one quantity is directly proportional to that of another quantity; said esp of amplifiers, operating in the range where this gives minimum distortion, before the process of limiting sets in as the output approaches the maximum available.

linearity control (*Electronics*) See STROBE (2).

linear low-density polyethylene (*Chem*) A grade of polymer made by the low-pressure catalytic route with controlled level of short-chain branches. Abbrev *LLDPE*.

linearly dependent (*MathSci*) Of a sequence of n quantities $u_1, u_2, \ldots, u_n$ such that it is possible to find n numbers $\lambda_1, \lambda_2, \ldots, \lambda_n$ (at least one of which is not zero) such that $\lambda_1 u_1 + \lambda_2 u_2 + \ldots + \lambda_n u_n = 0$. Otherwise the u_n are *linearly independent*.

linear modulation (*ICT*) Modulation in which the change in the modulated characteristic of the carrier signal is proportional to the level of the modulating signal.

linear momentum (*Phys*) See MOMENTUM.

linear motor (*ElecEng*) A form of induction motor in which the stator and rotor are linear and parallel instead of cylindrical and coaxial. Has been used in railroad traction. Fig. ▷

linear network (*ICT*) One with electrical elements that are constant in magnitude with varying current.

linear phase condition (*ICT*) A term describing the state of a transmission line that exhibits no frequency dispersion. Because such a line delays all frequencies by the same amount, the phase shift it produces is directly proportional to frequency.

linear predictive coding (*ICT*) A method of coding and resynthesizing sampled speech widely used to obtain data reduction in transmission and storage. The speech signal is modelled as the output of a digital filter excited by a pulse or noise source and is transmitted or stored as a sequence of coefficients characterizing these two elements.

linear programming (*Eng*) A general term implying that, in a complicated process of operations on an article, the various steps have been mutually planned for maximum economy.

linear rectifier (*ElecEng*) One in which the output current is strictly proportional to the envelope of the applied alternating voltage.

linear resistor (*ElecEng*) One which 'obeys' Ohm's law, ie under certain conditions the current is always proportional to voltage. Also *ohmic resistor*.

linear scan (*Electronics*) The sweeping of a cathode spot across a screen at constant velocity using a deflecting sawtooth waveform.

linear stopping power (*Phys*) See STOPPING POWER.

linear superpolymer (*Chem*) Polymer in which the molecules are essentially in the form of long chains with an average molecular weight greater than 10 000.

linear sweep (*Electronics*) The use of a sawtooth waveform to obtain a linear scan.

linear time-base generator (*Electronics*) The circuit for producing the deflecting voltage (or current) of the linear time base. See LINEAR SCAN, SWEEP CIRCUIT. Also *linear time-base oscillator*.

linear transformation (*MathSci*) (1) A mapping from a vector space to itself which satisfies: (a) if V is a real vector space with X and Y arbitrary vectors from V, then $f(X + Y) = f(X) + f(Y)$; and (b) for any real number λ, $f(\lambda X) = \lambda f(X)$. (2) A transformation between two sets of n variables represented by n equations in which the variables of both sets occur linearly. Thus when $n = 1$, the equation would be $ax + bx' + cx'd = 0$. Cf BILINEAR TRANSFORMATION.

linear video (*ImageTech*) Any video programme which is viewed from beginning to end, rather than interactively.

line at infinity (*MathSci*) The imaginary line joining POINTS AT INFINITY.

lineation (*Geol*) Any one-dimensional structure in a rock.

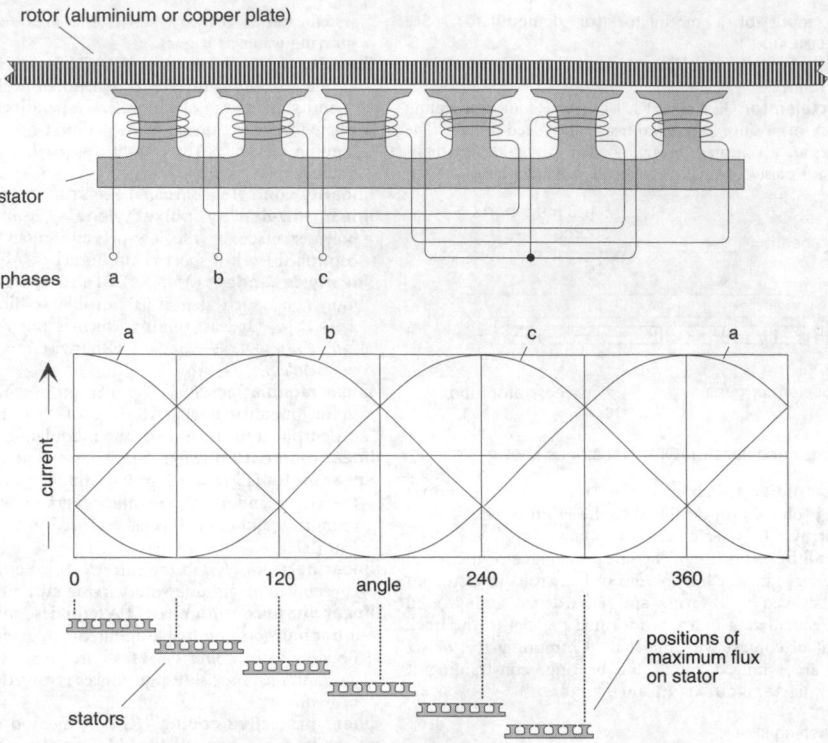

linear motor Windings of six-pole stator (top) and the movement of the flux field through one cycle of alternating current (bottom).

line-blanking (*ImageTech*) The reduction of the amplitude of the TV video signal to below the black level at the end of each line period. This allows the transmission of the line-synchronizing pulses.

line block (*Eng*) A sliding block pivoted to the end of the valve rod, and working in the slotted link of a LINK MOTION.

line block (*Print*) Relief block surface produced from a line drawing and incapable of printing with tonal gradation.

line break (*ICT*) The place where a line of text ends. WORD PROCESSORS normally only break lines between words.

line-breaker (*ElecEng*) A contactor on an electric vehicle, arranged for closing or interrupting the main current circuit.

line broadening (*Astron*) A term used for the increase in width of the lines of a stellar spectrum due to rotation of the star, turbulence in the stellar atmosphere, or the STARK EFFECT or ZEEMAN EFFECT.

line choking coil (*ElecEng*) An inductor included in an electric power supply circuit in order to protect plant connected to the line from the effect of high-frequency or steep-fronted surges. Also *screening protector.*

line colour (*Print*) A colour picture produced by super-imposed impressions of two or more line illustrations printed in different colours. Varied tones are obtainable by the use of stipples.

line co-ordinates (*MathSci*) A co-ordinate system in which each set of co-ordinates specifies a line, usually by reference to cartesian axes. Thus the line co-ordinates (a,b,c) could specify a line whose ordinary cartesian equation is $ax + by + c = 0$. In a line co-ordinate system a point is specified by a linear equation.

line coupling (*ICT*) The transfer of energy between resonant (tank) circuits in a transmitter, using a short length of line with small inductive coupling at each end. Also *link coupling.*

line defect (*Crystal*) See DEFECT.

line distortion (*ICT*) That arising in the frequency content or phase distribution in a transmitted signal, as a result of the propagation constant of the line.

line doubling (*ImageTech*) Inserting lines averaged from those before and after to double the definition of normal programming for showing on a HIGH-DEFINITION receiver.

line drop (*ICT*) The potential drop between any two points on a transmission line due to resistance, leakage or reactance.

line equalizer (*ICT*) A device that compensates for attenuation and/or phase delay for transmission of signals along a line over a band of frequencies. Also *lumped loading, phase compensation, phase equalization.*

line feed (*ICT*) A CONTROL CHARACTER that has the effect of moving the current character printing position to the next line down. See CARRIAGE RETURN.

line flyback (*Electronics*) (1) The time interval correspond-ing to the steeper portion of a sawtooth wave. (2) The return time of the image spot from its deflected position to its starting point.

line focus (*Electronics*) Cathode-ray tube in which the electron beam meets the screen along a line and not at a point, due to astigmatism when unintended.

line frequency (*ImageTech*) The number of lines scanned per second; eg with 625-line pictures at 25 frames per second, the line frequency is 15 625 Hz.

line-frequency generator (*ImageTech*) The generator of the voltage or current which causes the scanning spot to traverse each line of the image.

line hold (*ImageTech*) Time-based control in a TV receiver to synchronize with the incoming signal.

line impedance (*ICT*) The impedance measured across the terminals of a transmission line; when the line is correctly matched to its load, this equals the *characteristic impedance* of that particular type of line.

line integral (*MathSci*) Of a function along a path: the limiting sum along the path of the product of the value of the function and an element of length of the path.

linen (*Textiles*) Yarns, fabrics and articles made from FLAX fibres.

line of action (*Phys*) The line along which a force acts.

line of apsides (*Astron*) See APSE LINE.

line of collimation (*Surv*) In a surveying telescope, the imaginary line passing through the optical centre of the object glass and the intersection of the cross-hairs in the diaphragm. Also *line of sight*.

line of flux (*ElecEng, Phys*) A line drawn in a magnetic (or electric) field so that its direction at every point gives the direction of magnetic (or electric) flux (or force) at that point. Also *line of force*.

line of sight (*Astron*) The imaginary line between an observer and an astronomical object under observation.

line of sight (*ICT*) Said of a transmission system when there has to be a straight line between transmitting and receiving antennas, as in ultra-high frequency and radar.

line of sight (*Surv*) See LINE OF COLLIMATION.

line-of-sight velocity (*Astron*) The velocity at which a celestial body approaches, or recedes from, the Earth. It is measured by the Doppler shift of spectral lines emitted by the body as observed on the Earth. Also *radial velocity*.

line oscillator (*Electronics*) One which has its frequency stabilized either by a resonant low-loss (high-Q) coaxial line, or by a resistance–capacitance ladder which gives the necessary delay (phase shift) in a feedback loop. Also *phase-shift oscillator*.

line pad (*ICT*) A resistance–attenuation network that is inserted between the programme amplifier and the transmission line to the broadcasting transmitter. Its purpose is to isolate electrically the amplifier output from the variations of impedance of the line.

line printer (*ICT*) Prints a complete line of characters at one time and hence is generally faster than a character printer, such as a DAISY-WHEEL PRINTER. Abbrev *LPT*.

line profile (*Phys*) A graph showing the fine structure of a spectral line, the intensity of the line, measured with a microphotometer, being plotted as a function of wavelength or frequency.

liner (*Eng*) A separate sleeve placed within an engine cylinder to form a renewable and more durable rubbing surface; in internal-combustion engines, termed 'dry' if in continuous contact with the cylinder wall, and 'wet' if supported only at the ends and surrounded by cooling water.

line ranger (*Surv*) An instrument for locating an intermediate point in line with two distant signals. It consists of two reflecting surfaces so arranged as to bring images of the two signals into coincidence when the instrument is in line with the signals.

line reflection (*ICT*) The reflection of some signal energy at a discontinuity in a transmission line.

liner-off (*Ships*) A tradesperson engaged in shipbuilding, whose function it is to 'mark off' by fair lines, using battens to enable plate workers and others to prepare material to fit *in situ*.

line scanning (*ImageTech*) A method of scanning in which the scanning spot repeatedly traverses the field of the image in a series of straight lines.

line screen process (*ImageTech*) A colour photographic process in which the screen takes the form of lines ruled on the emulsion.

line-sequential (*ImageTech*) Said of colour TV systems in which successive scanning lines generate images in each of the three primary colours.

line shafting (*Eng*) Overhead shafting formerly used in factories to transmit power from a prime mover to many individual machines by pulley wheels and belts.

line slip (*ImageTech*) An apparent horizontal movement of part (or all) of a reproduced screen picture due to lack of synchronism between the line frequencies of the signal and the scanning system.

lines of curvature (*MathSci*) Curves on a surface that are tangential to the principal directions at every point of the surface.

line spectrum (*Phys*) A spectrum consisting of relatively sharp lines, as distinct from a BAND SPECTRUM or a CONTINUOUS SPECTRUM. Line spectra originate in the atoms of incandescent gases or vapours. See BOHR MODEL, SPECTRUM.

line squall (*EnvSci*) A system of squalls occurring simultaneously along a line, sometimes hundreds of km long, which advances across the country. It is characterized by an arch or line of low dark cloud, a sudden drop in temperature and rise in pressure. Thunderstorms and heavy rain or hail often accompany these phenomena.

line stabilization (*ICT*) The dependence of an oscillator on a section of transmission line for stabilization of its frequency of oscillation; eg a quarter-wave line acts as a rejector circuit of very high Q, thus giving a highly critical change in phase at the resonant frequency.

line standard (*Eng*) A standard of length consisting of a metal bar near whose extremities are engraved fine lines, the standard length being the distance between these lines measured under specified conditions.

line stretcher (*ICT*) A section of waveguide or coaxial line tuned by adjusting its physical length.

line synchronization (*ImageTech*) Synchronization of the line-scanning generator at the receiver with that at the transmitter so that the scanning spots at the two ends keep in step throughout each line.

line termination unit (*ICT*) CENTRAL OFFICE telephone equipment dedicated to a single subscriber line.

line testing (*ImageTech*) Checking whether animated images flow smoothly.

line-up (*ICT*) Adjustment of a number of circuits in series so that they function in the desired manner.

line voltage (*ElecEng*) See VOLTAGE BETWEEN LINES.

Lineweaver–Burke plot (*BioSci*) A plot of $1/v$ against $1/S$ for an enzyme-catalysed reaction, where v is the initial rate and S the substrate concentration. From the equation $1/v = 1/V_{max}(1 + K_m/S)$ the parameters V_{max} (maximal reaction velocity) and K_m (Michaelis constant) can be determined.

line width (*ICT*) The wavelength spread (or energy spread) of radiation that is normally characterized by a single value. The spread is defined by the separation between the points having half the maximum intensity of the line. In TV, the reciprocal of the number of lines per unit length in the direction of line progression.

lingering period (*Phys*) The time interval during which an electron remains in its orbit of highest excitation before jumping to the energy level of a lower orbit.

lingua (*BioSci*) Any tongue-like structure; in insects, the hypopharynx; in ACARINA, the floor of the mouth.

lingual (*BioSci*) Pertaining in arthropods to the lingua, in molluscs to the radula and in vertebrates to the tongue.

linguidental (*Med*) Of the tongue and teeth.

lingulate (*BioSci*) Tongue-shaped.

linguogingival (*Med*) Of the tongue and the gums.

lining (*Build*) A layer of clay puddle covering the sides of a canal, making them watertight.

lining (*Print*) (1) The operation of pasting a strip of strong paper down the back of a book after backing. (2) A strip of linen fixed down the middle of a section for strengthening purposes.

lining (*Textiles*) A fabric attached to the inside of a garment to make it more acceptable to the wearer (eg more comfortable, more durable).

lining figures (*Print*) Said of figures which have the height of the capital letters of a font, as distinct from *hanging figures*; eg 1,6,7. Also *ranging figures*.

lining fitch (*Build*) A hog-hair brush cut at a slant and used in conjunction with a wooden straight edge for the purpose of painting lines; made in various sizes.

lining-papers (*Print*) See END-PAPERS.

lining-up (*Eng*) The operation of arranging the bearings of an engine crankshaft etc in perfect alignment.

lining wheel (*Build*) A small device consisting of a metal cylinder and a range of serrated wheels used to paint lines of various widths. An alternative to a lining fitch.

linisher (*Eng*) A machine with a moving belt covered with an abrasive for giving a smooth or clean surface to metal etc.

link (*Eng*) (1) Any connecting piece in a machine, pivoted at the ends. (2) The curved slotted member of a LINK MOTION.

link (*ICT*) (1) A connection. (2) A circuit or outlet between one rank of selectors and the next in order of operation, or between such selectors and a manual position. US *trunk*. See HYPERLINK.

link (*Surv*) The one-hundredth part of a chain. In the Gunter's chain, 1 link = 7·92 in; in the engineer's chain, 1 link = 1 ft.

linkage (*BioSci*) The tendency of genes, or characters, to be inherited together because the genes are on the same chromosome.

linkage (*Chem*) A chemical bond, particularly a covalent bond in an organic molecule.

linkage (*ElecEng*) The product of the total number of lines of magnetic flux and the number of turns in the coil (or circuit) through which they pass.

linkage equilibrium (*BioSci*) The situation that should exist in a population undisturbed by selection, migration, etc, in which all possible combinations of linked genes should be present at equal frequency. The situation is no more common than are such undisturbed populations.

linkage group (*BioSci*) All the genes known from their linkage to be on the same chromosome.

linkage map (*BioSci*) A diagram showing the positions of genes on a chromosome or set of chromosomes.

link coupling (*ICT*) See LINE COUPLING.

linked list (*ICT*) A list where each item contains both data and a pointer to the next item.

linked numbering scheme (*ICT*) One in which a range of numbers (including the code numbers in a director automatic area) are distributed between the subscribers on several exchanges in a given area.

linked object (*ICT*) A visual representation, eg an ICON, of an OBJECT in a destination document.

linked switches (*ElecEng*) Switches mechanically linked, so that they operate together or in a definite sequence. Also *coupled switches*.

linker (*ICT*) A program that links programs together.

link mechanism (*Eng*) A system of rigid members joined together with constraints so that motion can be amplified or can be changed in direction.

link motion (*Eng*) A valve motion, invented by Stephenson, for reversing and controlling the cut-off of a steam engine. It consists of a pair of eccentrics, set for ahead and reverse rotation, connected to the ends of a slotted link carrying a block attached to the valve rod. Variation of the link position (known as 'linking up') makes either eccentric effective, and also varies the cut-off.

link resonance (*ICT*) That resulting in a repeated network when the sections are coupled together.

lin–log receiver (*ICT*) A radio receiver that gives a linear amplitude response for small signals but logarithmic for larger signals.

Linnaean species (*BioSci*) A species defined broadly, often including many varieties. Also *Linnean species*. Cf JORDANON.

Linnaean system (*BioSci*) The system of classification and of BINOMIAL NOMENCLATURE established by the Swedish naturalist Linnaeus. Also *Linnean system*.

Linnean (*BioSci*) See LINNAEAN SYSTEM.

linocut (*Print*) A relief printing surface cut by hand in linoleum, giving a bold broad effect; not suitable for fine detail or long runs.

linoleate (*Chem*) A salt or ester of linoleic acid.

linoleic acid (*Chem*) $CH_3(CH_2CHCH)_3(CH_2)_7COOH$. Unsaturated fatty acid. Occurs as glycerides in various vegetable oils, esp linseed oil to which, by forming solid oxidation products on exposure to air, it imparts the 'drying' quality responsible for its utility in paints etc.

linoleum (*Build*) A floor-covering material, now largely superseded by *vinyl floor coverings*, made by impregnating a foundation of hessian fabric or waterproofed felt with a mixture of oxidized linseed oil (linoxyn), resins (eg kauri gum) and fillers. Abbrev *lino*.

linseed oil (*Chem*) An oil obtained from the seeds of FLAX (*Linum usitatissimum*). It contains solid and liquid glycerides of oleic and other unsaturated acids. Its iodine value is 160–200, which puts it into the class of drying oils. It oxidizes and polymerizes in air to form a brittle polymer. Once much used for the mixing of paints and varnishes and for the manufacture of linoleum. Still used in PUTTY for glazing windows etc.

lint (*Textiles*) (1) The main seed hairs of the cotton plant. (2) A plain-woven cotton fabric raised on one side to make it highly absorbent; used, after sterilization, as a wound dressing.

lintel (*Build*) A beam across the top of an aperture. Also *head*.

linters (*Textiles*) Short fuzzy fibres remaining on cotton seeds after substantial removal of the longer fibres (*lint*); removed before the seeds are crushed. Linters are used extensively in the manufacture of cellulose nitrate and acetate etc.

Linux (*ICT*) A widely used open-source computer operating system, the kernel of which was developed by Linus Torvalds.

Linville truss (*Eng*) See WHIPPLE–MURPHY TRUSS.

lionism (*Med*) A lion-like appearance, often characteristic of lepromatous leprosy.

lip-, lipo- (*Genrl*) Prefixes from Gk *lipos* denoting fat.

lipaemia (*Med*) Excess of fat in the blood. US *lipemia*.

liparite (*Geol*) See RHYOLITE.

lipase (*BioSci*) An enzyme that cleaves the hydrocarbon chains from lipids.

lipectomy (*Med*) Surgical removal of fatty tissue.

lipid body (*BioSci*) A cytoplasmic inclusion, esp in the storage tissue of oil- and fat-rich seeds, typically spherical, $0·1–5$ μm diameter, bounded by what appears to be half a unit membrane, and containing lipids.

lipids (*Chem*) Generic terms for oils, fats, waxes and related products found in living tissues. Also *lipoids*.

lip microphone (*Acous*) Microphone constructed in the form of a box shaped to fit the face around the mouth. This reduces extraneous noises as at eg sporting events.

lipo- (*Genrl*) See LIP-.

lipochromes (*Chem*) Pigments of butter fat.

lipocortin (*BioSci*) Any of the annexin family of proteins that inhibit phospholipase A and thus inhibit release of ARACHIDONIC ACID.

lipofuscin (*BioSci*) A brown pigment, characteristic of ageing, that is deposited in the skin, eg 'liver spots' on the face and hands.

lipogenous (*BioSci*) Fat-producing.

lipoids (*Chem*) See LIPIDS.

lipolysis (*BioSci*) The breaking down of fat into fatty acids and glycerol.

lipoma (*Med*) A tumour composed of cells containing fat.

lipophorin (*BioSci*) A high-density LIPOPROTEIN found in insect haemolymph that is involved in haemolymph clotting and the transport of diacyl glycerols.

lipoplast (*BioSci*) A lipid body.

lipopolysaccharide (*BioSci*) A term commonly used to refer to bacterial lipopolysaccharides that constitute the O-antigens of Gram-negative bacilli, esp enterobacteria.

Chemically they consist of various antigenically specific polysaccharides linked to lipid A. The latter contains rhamnose linked to galactosamine diphosphate which is esterified with myristic acid, and is responsible for the endotoxin properties common to all lipopolysaccharides. They are also polyclonal mitogens for B-lymphocytes. Abbrev *LPS*.

lipoprotein (*BioSci*) A complex of protein and lipid in varying proportions, classified according to their increasing density into CHYLOMICRONS, *high-density lipoproteins (HDL), low-density lipoproteins (LDL)* and *very-low-density lipoproteins (VLDL)*. They transport lipids in the plasma between various organs in the body, particularly the gut, liver and adipose tissues.

liposome (*BioSci*) Spherical shell formed when mixtures of phospholipids, with or without cholesterol, are dispersed in aqueous solutions. Liposomes are made up of one or several concentric phospholipid bilayers within which other molecules can be incorporated. They simulate many permeability properties of membranes and are used for the administration of certain drugs.

liposuction (*Med*) The removal of subcutaneous fat by suction after injection of normal saline. It is usually performed for cosmetic reasons but has been tried (unsuccessfully) as a treatment for the metabolic complications of obesity.

lipotropic (*Med*) Having an effect on fat metabolism by accelerating fat removal or decreasing fat deposition, eg of liver fat.

Lippmann process (*ImageTech*) Early method of colour photography using a transparent silver halide emulsion backed with mercury. When viewed at an angle the reflected image appears in natural colours.

LIPS (*ICT*) Abbrev for *logical inferences per second*. A measure of speed of a processor. Cf FLOPS, MIPS.

lip seal (*Eng*) An oil-retaining shaft seal, often used adjacent to a bearing, in which an annular rubber sealing element having an aperture slightly smaller than the shaft diameter is deformed into the shape of a sealing lip, the sealing pressure being due partly to the deformation stress and partly to the force supplied by an additional GARTER SPRING.

lip sync (*ImageTech, ICT*) Precise synchronization in simultaneous picture and sound recording, eg in a video telephone or conferencing system, in contrast to WILD SHOOTING. Because VIDEO CODECS introduce longer delays than AUDIO CODECS, this may require the audio channel to be buffered and played out in accordance with the timestamps present in the two data streams.

lipuria (*Med*) The presence of fat in the urine.

liquation (*Eng*) Partial melting of an alloy due to heterogeneity of composition.

liquefaction (*Geol*) The change in packing of the grains of a water-filled sediment, turning it into a fluid mass which can then flow.

liquefaction of gases (*Phys*) Converting a gas to a liquid by cooling below its *critical temperature* and, in some cases, compressing it. For the so-called 'permanent' gases, oxygen, nitrogen, hydrogen and helium, having very low critical temperatures, the problem of liquefaction becomes one of obtaining low temperatures. This is done mainly by allowing the compressed gas to expand through a nozzle, cooling occurring by the JOULE–THOMSON EFFECT.

liquefaction temperature (*Phys*) The temperature at which a gas changes state to liquid. Physically the same as the boiling point.

liquefied petroleum gases (*Chem*) Gases such as propane and butane, used for fuels. Abbrev *LPG*.

liquid (*Genrl*) A state of matter between a solid and a gas, in which the shape of a given mass depends on the containing vessel, the volume being independent. Liquids are almost as incompressible as solids.

liquid compass (*Ships*) A magnetic compass fitted in a bowl containing a suitable liquid. The weight of the card

on the pivot is reduced and oscillations of the card are damped out.

liquid counter (*Phys*) A counter for measuring the radioactivity of a liquid, usually designed to measure beta as well as gamma rays. Also *liquid scintillation counter*.

liquid crystal (*Chem*) One of a number of pure liquids which are turbid and, like crystals, anisotropic over a definite range of temperature above their freezing points. See panel on LIQUID CRYSTAL DISPLAYS.

liquid crystal display (*Electronics*) A digital or alphanumeric display consisting of two sheets of glass separated by a sealed-in liquid crystal material. See panel on LIQUID CRYSTAL DISPLAYS.

liquid crystal light valve (*ImageTech*) A projection device consisting of three layers – amorphous silicon, liquid crystal and DIELECTRIC mirror – which produces a polarized replica of an image projected onto it from a small CATHODE-RAY TUBE, which is picked up by the projection beam and passed through a polarizing filter. See DIGITAL MICROMIRROR DEVICE, LIGHT VALVE PROJECTOR.

liquid crystal phases (*Chem*) Liquid crystals are mesophases arising between the normal solid and liquid phases of certain organic substances. Can be described with reference to a phase diagram in terms of either the composition range (lyotropic) or temperature range (thermotropic). They are anisotropic owing to their organic constituents and their long-range order. Thermotropic mesophases are found in various display devices and considerable effort has gone into extending and controlling the temperature range over which stable liquid crystal phases can be formed.

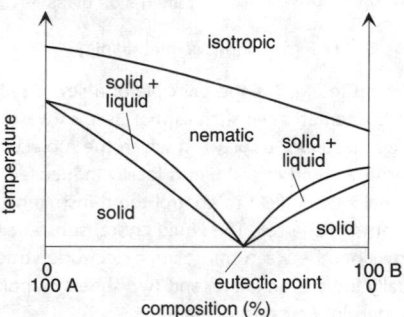

liquid crystal phases Different mixtures of A and B.

liquid crystal polymers (*Chem*) Chains or precursors aligned along the flow direction in processing or polymerization, similar to nematic liquid crystals. They exhibit high stiffness along this axis. Moulding materials include copolymers of polyethylene terephthalate and PHB (para-hydroxy benzoic acid) and aramid fibres, Kevlar and Twaron. See panel on HIGH-PERFORMANCE POLYMERS.

liquid-drop model (*Phys*) A model of the atomic nucleus using the analogy of a liquid drop in which various concepts of eg surface tension and heat of evaporation are employed. A semi-empirical mass formula can be developed from the model which describes the masses of many stable and unstable nuclei in terms of a few parameters.

liquid-flow counter (*Phys*) A counter used for continuous monitoring of radioactivity in flowing liquids.

liquid helium II (*Chem*) See HELIUM.

liquid honing (*Eng*) Honing by a jet of liquid containing abrasives.

liquid–liquid extraction (*ChemEng*) A process, both batch and continuous, whereby two non-mixing liquids are brought together to transfer soluble substance from one to the other for useful recovery of these soluble substances.

liquid-metal reactor (*NucEng*) (1) Normally used of a reactor designed for liquid-metal (usually sodium) cooling.

Liquid crystal displays

Liquid crystals are fluids containing stiff, rod-like organic molecules which have a tendency to form ordered structures. They exhibit some of the properties of crystalline solids. In particular, the elastic, optical and dielectric properties of liquid crystals are highly ANISOTROPIC and these are exploited in the construction of display devices.

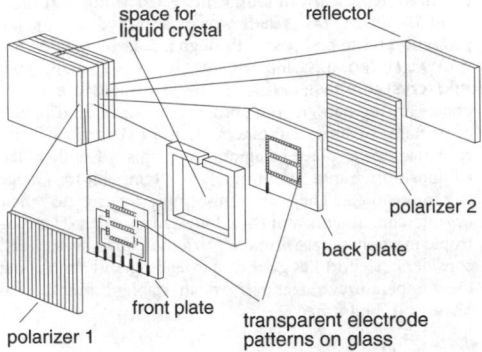

Fig. 1 **Liquid crystal display**.

Shown in Fig. 1 is the principle of a seven-segment 'twisted nematic' cell such as that used in watch and pocket calculator displays, in which the molecular order in a liquid crystal film is locally influenced by an external electric field to control the transparency of elements of the cell. The liquid crystal substance is sandwiched between transparent electrode structures (usually indium tin oxide) and two sheets of polarizer arranged in a crossed orientation.

Electrically, a segment of the display looks like a parallel-plate capacitor with a liquid crystal DIELECTRIC. With no field across a segment, the molecular ordering is locally determined by a combination of molecule–

molecule and molecule–surface interactions. The device illustrated uses a NEMATIC phase. The inner surfaces of the cell are coated with a thin polymeric layer in which are fine parallel grooves; the rod-like molecules preferentially align themselves with these grooves. On opposite sides of the cell, the grooves are at right angles so that the molecular orientation through the bulk twists smoothly through 90° (see left side of Fig. 2). The properties of the nematic phase ensure that the plane of polarization of light passing between the two polarizers also twists smoothly through 90°, allowing it to escape despite the crossed orientation of the polarizers. In practice the beam is reflected back through the cell and out again, so that in this region the cell appears relatively bright.

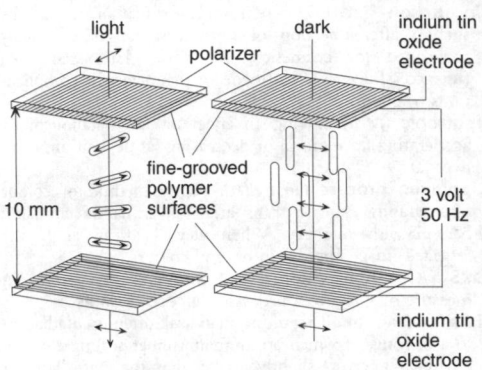

Fig. 2 **Twisted nematic elements of a liquid crystal display**.

A relatively strong electric field (typically around 300 kV m^{-1}) across a segment causes the molecules locally to reorient along the field, destroying the smooth rotation of the plane of polarization. Now light passing is absorbed by the crossed polarizers and the segment appears dark (see Fig. 2). When the field is removed, thermal agitation allows the molecules to jostle each other back into the twisted orientation.

---

(2) Occasionally used to indicate a liquid-metal-fuelled reactor. Abbrev *LMR*.

liquid oxygen (*Chem*) See OXYGEN.

liquid paraffin (*Chem*) A liquid form of PETROLEUM JELLY, colourless and tasteless, used as a mild laxative. A mixture of alkanes with more than twelve carbon atoms to the molecule.

liquid-penetrant inspection (*Eng*) See FLUORESCENT PENETRANT INSPECTION.

liquid-phase epitaxy (*Electronics*) Layers grown from a molten source on a crystalline substrate. See panel on SEMICONDUCTOR FABRICATION.

liquid-phase sintering (*Eng*) Sintering in which a small proportion of the material becomes liquid. It may or may not speed up the sintering process by solution transfer of the phases forming the matrix.

liquid-quenched fuse (*ElecEng*) A fuse in which a liquid is used for quenching the arc. See IMMERSED LIQUID-QUENCHED FUSE, SEMI-IMMERSED LIQUID-QUENCHED FUSE.

liquid resistance (*ElecEng*) A resistance consisting of a liquid of low conductivity, the current being led to and from the liquid by means of suitable electrodes.

liquid rheostat (*ElecEng*) One in which a liquid column is used as the resistive element, the terminals being attached to suitable metal plates, one of which is usually movable, thus providing a continuous variation. Only used where current control need not be too precise.

liquid scintillation counter (*Phys*) See LIQUID COUNTER.

liquid starter (*ElecEng*) A liquid rheostat arranged to operate as a motor starter.

liquidus (*Chem*) A line in a phase diagram indicating the temperatures at which solidification of one phase or constituent begins or melting is completed. See PHASE DIAGRAM, SOLIDUS.

liquor amnii (*Med*) The clear fluid in the amniotic cavity, in which the embryo is suspended.

liquor ratio (*Textiles*) Used in dyeing or finishing to express the ratio of the weight of liquor used to the weight of material being treated.

L-iron (*Eng*) See ANGLE IRON.

lisinopril (*Pharmacol*) An *ACE inhibitor* used in the treatment of heart failure and hypertension.

lisle (*Textiles*) Long-staple, highly twisted, folded cotton hosiery yarn, gassed (ie passed through gas flame or over hot element to remove protruding fibre ends) and often mercerized to produce a lustrous effect. See MERCERIZATION.

LISP (*ICT*) Abbrev for *list processing*. A general purpose programming language in which the expressions are represented as lists.

Lissajous' figures (*MathSci, Phys*) Plane curves formed by the composition of two sinusoidal waveforms in perpendicular directions. The parametric cartesian equations are: $x = a \sin(\omega t + \alpha)$, $y = b \sin(\omega t + \beta)$. The curves embrace a great variety of forms. If the frequencies are commensurable they consist, in general, of a number of loops determined by the ratio of the frequencies, and this property is extensively exploited to compare the frequencies of two sinusoidal voltages by applying them to the plates of a cathode-ray tube. Also *Lissajous' curves*.

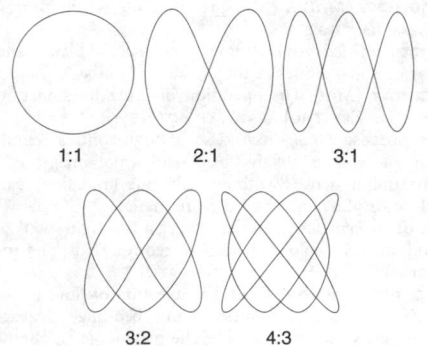

1:1   2:1   3:1

3:2   4:3

Lissajous' figures Various frequency ratios shown.

lissencephalous (*BioSci*) Having smooth cerebral hemispheres.

list (*ICT*) (1) A linearly ordered data structure. See ARRAY, LINKED LIST, QUEUE, STACK. (2) A DATA TYPE, a list of ATOMS or smaller lists. See LISP. (3) To print or display a LISTING.

list (*Ships*) An angle of transverse inclination arising from unsymmetrical distribution of internal weight. Cf HEEL.

listel (*Arch*) See FACETTE and fig. at MOULDINGS.

listening key (*ICT*) The lever key that the operator throws, to put the head-set onto a cord circuit and speak to a subscriber.

listerellosis (*Vet*) See LISTERIOSIS.

Listeria monocytogenes (*BioSci, FoodSci*) A genus of small Gram-positive mobile rod-like bacteria, found in human and animal feces, that can be transmitted to humans via certain foods, eg contaminated soft cheese, causing the disease LISTERIOSIS.

listeriosis (*Med, Vet*) A disease, caused by infection with the bacterium *Listeria monocytogenes* (*Listerella monocytogenes*), which has symptoms resembling those of meningitis or encephalitis, and may cause miscarriage or fetal damage if contracted by pregnant women. Infection can also occur in cattle and sheep (circling disease) and more rarely in pigs. In poultry, and sometimes in young animals, the disease occurs as a septicaemia. Also *listerellosis*.

listing (*Build*) (1) A narrow edge of a board. (2) The operation of removing the sappy edge of a board.

listing (*ICT*) The sequences of program statements or data in printed form or on a screen.

listing (*Textiles*) (1) In a dyed fabric, a variation in shade between the selvedges and the centre. (2) SELVEDGE.

liter (*Genrl*) US for LITRE.

literal (*Print*) Casual error of composition, such as one character substituted for another, worn letters, transpositions, turned letters, etc.

lith-, litho- (*Genrl*) Prefixes from Gk *lithos* denoting stone or (in medicine) calculus.

litharge (*Build, Chem*) Lead (II) oxide, used in paint-mixing as a drier and in the rubber and electrical accumulator industries.

litharge cement (*Chem*) See GLYCERINE LITHARGE CEMENT.

lithia (*Chem*) Lithium oxide. Li_2O.

lithia emerald (*Min*) Misnomer for HIDDENITE.

lithia mica (*Min*) See LEPIDOLITE.

lithiasis (*Med*) The formation of calculi in the body. The condition in which an excess of uric acid and urates is excreted in the urine; the gouty diathesis.

lithic arenite (*Geol*) A sediment of sand-sized particles many of which are rock fragments rather than mineral grains.

lithic tuff (*Geol*) A volcanic tuff in which rock fragments are more abundant than crystals or vitric fragments.

lithification (*Geol*) Processes which convert an unconsolidated sediment into a SEDIMENTARY ROCK.

lithiophilite (*Min*) Orthorhombic phosphate of lithium and manganese, forming with triphylite a continuously variable series.

lithite (*BioSci*) See STATOLITH.

lithium (*Chem*) An element, symbol Li, ram 6·939, at no 3, mp 186°C, bp 1360°C, rel.d. 0·585. It is the least dense solid, chemically resembling sodium but less active. Abundance in the Earth's crust 20 ppm. There are numerous minerals mainly occurring in pegmatites associated with granitic rocks. *Spodumene* ($LiAlSi_2O_6$), lithium-bearing micas, *lepidolite* and *zinnwaldite*, *amblygonite* [$LiAlPO_4(OH,F)$] and *petalite* ($LiAlSi_4O_{10}$) are the commonest lithium minerals. It is used in alloys and in the production of tritium; also as a basis for lubricant grease with high resistance to moisture and extremes of temperature; and as an ingredient of high-energy fuels. In nuclear engineering used as the *chemical blanket* for extracting heat from neutron capture in fusion reactors. Its salts are used medically as an anti-depressant.

lithium aluminium hydride (*Chem*) *Lithium tetrahydroaluminate*. $LiAlH_4$. Important reducing agent widely used in organic chemistry because of its effectiveness in 'difficult' reductions, eg the reduction of carboxylic acid groups ($COOH$) to primary alcohol groups (CH_2OH).

lithium-drifted silicon detector (*Phys*) An energy-sensitive solid-state detector for ionizing radiation. Lithium is thermally diffused into almost pure but p-type silicon crystal. Used for low-energy X-ray and gamma-ray spectroscopy. The crystal is kept at liquid-nitrogen temperature to reduce thermal noise and to avoid the redistribution of lithium ions.

lithium-drifted germanium detector (*NucEng*) One type of GERMANIUM RADIATION DETECTOR. There is also an *intrinsic germanium detector* which does not contain lithium.

lithium hydride (*Chem*) LiH. Formed when lithium unites with hydrogen at a red heat, it is a strong reducing agent, used as a hydrogen carrier, eg for small balloons at sea.

lithium hydroxide (*Chem*) $LiOH$. A strong base, but less hygroscopic than sodium hydroxide. Used in the form of a coarse, free-flowing powder as a carbon dioxide absorber, eg in submarines. Also to make lithium stearates etc in driers and lubricants.

lithium 12-hydroxy stearate (*Chem*) A lithium 'soap' widely use in high-performance greases as the main thickening agent. Helps to confer high water resistance and good low-temperature performance.

lithium niobate (*Chem*) Abbrev *LN*. See DIELECTRIC, FERROELECTRIC.

lithium salts (*Pharmacol*) Lithium salts (carbonate, citrate) are used for treatment and prevention of mania, manic-depressive illness and recurrent depression; lithium succinate has anti-inflammatory and antifungal activity.

lithium tantalate (*Chem*) Abbrevs *LT, LTO*. See DIELECTRIC, FERROELECTRIC.

litho (*Print*) Abbrev for LITHOGRAPHY.

litho- (*Genrl*) See LITH-.

lithocyst (*BioSci*) See STATOCYST.

lithodomous (*BioSci*) Living in rocks.

lithogenous (*BioSci*) Rock-building, as certain corals.

lithographic oil (*Print*) See LITHOGRAPHIC VARNISH.

lithographic paper (*Paper*) High machine-finished or supercalendered paper, made so that any stretch occurs the narrow way of the sheet.

lithographic stone (*Geol*) A compact, porous fine-grained limestone, often dolomitic, employed in lithography. Pale creamy-yellow in colour, but occasionally grey. Fair samples may be obtained from the Jurassic rocks in the UK, but the finest material comes from Solenhofen and Pappenheim in Bavaria. See SOLENHOFEN STONE.

lithographic varnish (*Print*) *Heat-bodied* oil used as binder for lithographic inks. Also *lithographic oil, litho oil*.

lithography (*Electronics*) Pattern transfer process used to define tracks on printed circuit boards and to create electronic device features. In semiconductor technology, optical (ultraviolet), electron-beam or X-ray irradiation through a mask transfers the pattern to a thin coating of appropriate resist on the wafer surface. Subsequent etching is able to proceed through open areas in the developed resist layer. See SEMICONDUCTOR DEVICE PROCESSING.

lithography (*Geol*) The systematic description of rocks, more esp sedimentary rocks. See PETROLOGY.

lithography (*Print*) Originally the art of printing from stone (*lithographic stone*), but now applied to planographic printing processes depending on the mutual repulsion of water and greasy ink. Damp rollers pass over the surface, followed by inking rollers. The design, which is greasy, repels the water but retains the ink (also greasy), which is transferred to the paper. In modern practice, a metal or paper plate is commonly used in place of stone, the plate being attached to a cylinder for rotary printing, either sheet-fed or web-fed. Most lithography is offset, the impression from the plate being transferred to a rubber-covered blanket cylinder and thence to the stock which may be paper or board or other flexible sheets including metal. See CHROMOLITHOGRAPHY, OFFSET PRINTING, PHOTOLITHOGRAPHY, SMALL OFFSET.

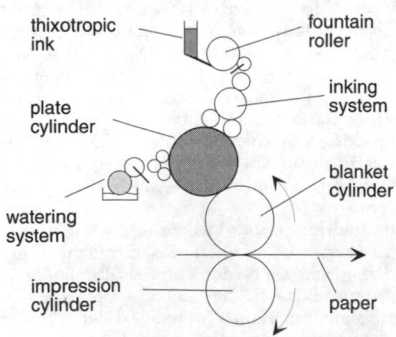

lithography Much simplified drawing of an offset rotary machine.

litholapaxy (*Med*) The operation of crushing a stone in the bladder, followed by the washing out of the crushed fragments.

lithology (*Geol*) The character of a rock expressed in terms of its mineral composition, its structure, the grain size and arrangement of its component parts; ie all those visible characters that in the aggregate impart individuality to the rock. The term is most commonly applied to sedimentary rocks in hand specimen and outcrop.

litho oil (*Print*) See LITHOGRAPHIC VARNISH.

lithophagous (*BioSci*) Stone-eating, like graminivorous birds that take small stones into the gizzard to aid mastication, and certain molluscs that tunnel in rock.

lithophile (*Geol*) An element that is concentrated in the silicate minerals rather than in sulphide or metal phases, and occurs in stony meteorites or the Earth's crust. Also *oxyphile*.

lithophyte (*BioSci*) A plant growing on rocks or stones, esp an alga attached to a rock or stone.

lithosphere (*Geol*) The outer, rigid shell of the Earth lying over the ASTHENOSPHERE. It includes the *continents, crust* and *plates*. See CONTINENTAL CRUST, OCEANIC CRUST and panel on PLATE TECTONICS.

lithostratigraphy (*Geol*) Stratigraphy based on the observable petrographical characters of the rock successions.

lithotomous (*BioSci*) Stone-boring, as certain molluscs.

lithotomy (*Med*) Cutting into the bladder or ureter for the removal of a stone or calculus.

lithotripsy (*Med*) A technique using ultrasound for destroying kidney stones.

lithotrite (*Med*) An instrument, with special blades, adapted for crushing stones in the bladder.

lithotrity (*Med*) The operation of crushing stones in the bladder. Cf LITHOLAPAXY, LITHOTRIPSY.

lith process (*ImageTech*) Use of high-contrast emulsions and developers to give black-and-white images of high maximum density without half-tone gradation, particularly for photomechanical reproduction.

litmus (*Chem*) A material of organic origin used as an indicator; its colour changes to red for acids, and to blue for alkalis at pH > 7. See INDICATOR.

lit-par-lit injection (*Geol*) The injection of fluid or molten material, mostly granitic, along bedding, cleavage or schistosity planes in a rock. The product is an alternation of apparently igneous and non-igneous material, known as a *migmatite*. Some of the rocks previously considered to have been formed by the process are now thought to be the products of partial melting, the igneous layers representing the fraction of the rock which became liquid.

litre (*Genrl*) Unit of volume equal to 1 cubic decimetre or 10^{-3} m^3. US *liter*.

litre-atmosphere (*Chem*) A former unit of work equal to 101·325 joules.

litter (*BioSci*) More or less undecomposed fallen leaves and other plant residues at the soil surface.

little-end (*Autos*) That part of the connecting rod which is attached to the gudgeon pin.

Little Ice Age (*Geol*) Variously applied to cool periods in otherwise warm stages, including periods in the 16th and 18th centuries.

Little's disease (*Med*) Spastic paralysis mainly affecting the lower half of the body, producing the classical 'spasm' child owing to aortic infarction of the white matter of brain. Strongly associated with prematurity, better management of which is reducing this form of cerebral palsy.

littoral (*BioSci*) (1) The area between high- and low-tide marks. Cf NERITIC ZONE. (2) The shallower water of lakes where light reaches the bottom and where rooted plants may grow.

littoral zone (*BioSci*) Faunal zone bounded by the continental shelf, ie down to approximately 200 m.

Littrow grating mounting (*Phys*) The mounting of a plane diffraction grating such that a plane mirror reflects back light transmitted by the grating so that it passes through the grating a second time. By rotating the mirror the spectrum can be scanned.

Littrow prism spectrograph (*Phys*) A spectrograph in which the same lens serves both to collimate the light and to focus the spectrum on the photographic plate, the light being reflected back through the prism by a plane mirror behind the prism.

litz wire (*ElecEng*) Multiple-stranded wire, each strand being separately insulated so as to reduce the relative weighting of the SKIN EFFECT, ie concentration of high-frequency currents in the surface. Much used in compact low-loss coils in filters and in high-frequency tuning circuits.

live (*Acous*) Said of an enclosure, or filmset, which is not rendered dead by the presence of sound-absorbing areas, and in which the reverberation is normal or above normal.

live (*ElecEng*) Connected to a voltage source.

live (*ImageTech*) Direct transmission of sound or TV without recording. A *live insert* is that part of an otherwise recorded transmission which is live.

live axle (*Eng*) A revolving axle to which the road wheels are rigidly attached, as distinct from a fixed or DEAD AXLE.

live centre (*Eng*) In machine tools, a centre which rotates with the workpiece.

live crude oil (*MinExt*) That arriving at the well head and containing gas. Cf DEAD CRUDE OIL.

livedo (*Med*) Reddish mottling of the skin.

live load (*Eng*) A moving load or a variable force on a structure; eg that imposed by traffic movement over a bridge, as distinct from dead weight or load due to the weight of the bridge.

liver (*BioSci*) In invertebrates, the digestive gland or HEPATOPANCREAS; in vertebrates, a large mass of glandular tissue arising as a diverticulum of the gut, which secretes the bile and plays an important part in the storage and synthesis of metabolites.

live rail (*ElecEng*) The supply rail of a three- or four-rail electric traction system, usually a few hundred volts dc above or below earth potential.

liver flukes (*BioSci*) A group of trematode parasites, esp *Fasciola hepatica* and *Clonorchis sinensis*, infecting humans via various species of water snail as intermediate hosts, causing damage to the liver and surrounding organs.

livering (*Build*) The condition when a paint in bulk becomes jelly-like or tough. Also *curdling, livering up*.

live ring (*Eng*) A large roller bearing, used for supporting turntable and revolving cranes.

live room (*Acous*) A room which has a longer period of reverberation than the optimum for the conditions of performance and listening.

liver opal (*Min*) A form of opaline silica, in colour resembling liver.

liver rot (*Vet*) See DISTOMIASIS.

liverworts (*BioSci*) See HEPATICOPSIDA.

live steam (*Eng*) Steam supplied direct from a boiler, as distinct from exhaust steam or steam which has been partly expanded.

livestock unit (*Agri*) An arbitrary score given to each category of herd or flock member, eg cow in milk, yearling, to allow an indicative calculation of grazing pressure for the herd to be made.

living polymers (*Chem*) Anionic polymers which still have active chain ends and have not been terminated. Active ends are charged-pair complexes, which often produce highly coloured solutions. Useful for making block copolymers (see CHAIN POLYMERIZATION). Not to be confused with polymers from living organisms, eg biomaterials.

lixiviation (*Eng*) See LEACHING.

lizardite (*Min*) A mineral of the SERPENTINE group.

Ljungstrom regenerative air heater (*Eng*) A heat exchanger largely used in power stations. It consists essentially of a large, vertical, slowly rotating drum filled with a honeycomb arrangement of very thin steel sheet, the flue gases passing upwards through one half of the honeycomb while the air passes downwards through the other.

llama fibre (*Textiles*) The hair of the llama (*Lama glama*) of S America.

Llandeilo (*Geol*) An epoch of the Ordovician period. See PALAEOZOIC.

Llandovery (*Geol*) The oldest epoch of the Silurian period. See PALAEOZOIC.

Llanvirn (*Geol*) An epoch of the Ordovician period. See PALAEOZOIC.

LLDPE (*Chem*) Abbrev for LINEAR LOW-DENSITY POLY-ETHYLENE.

L-lines (*Phys*) Characteristic X-ray frequencies from atoms due to the excitation of electrons from the L-shell. Denoted by $L\alpha$, $L\beta$, …; the lines are multiplets.

Lloyd Morgan's canon (*Psych*) The proposal that one should never explain behaviour in terms of a higher mental function if a more simple process will explain it.

Lloyd's mirror (*Phys*) A device for producing interference fringes. A slit, illuminated by monochromatic light, is placed parallel to and just in front of the plane of a plane mirror or piece of unsilvered glass. Interference occurs between direct light from the slit and that reflected from the mirror. See INTERFERENCE FRINGES.

Lloyd's rules (*Ships*) A set of rules laid down by the Classification Society, Lloyd's Register of Shipping, governing the construction of steel ships and their machinery.

LLTV (*Aero*) Abbrev for LOW-LIGHT TELEVISION.

lm (*Phys*) Abbrev for LUMEN.

LMFBR (*NucEng*) Abbrev for *liquid-metal-cooled fast-breeder reactor*. See BREEDER REACTOR, FAST REACTOR.

LMR (*NucEng*) Abbrev for LIQUID-METAL REACTOR.

LN (*Chem*) Abbrev for *lithium niobate*. See DIELECTRIC AND FERROELECTRIC MATERIALS.

LNB (*ImageTech*) Abbrev for LOW-NOISE BLOCK CONVERTER.

L network, attenuator, filter (*ICT*) Half an unbalanced T network. See L SECTION.

lo (*Aero*) Low-level military flight, usually at less than 200 ft.

load (*ElecEng*) (1) Electrical impedance to which the output of source is connected. The source may be eg an electrical generator, a transformer, an amplifier or even a single transistor. (2) Total demand for electrical power on a supply, which may be eg an electrical generator or an amplifier. (3) Material placed between electrodes for the inductance of heat-through dielectric loss by means of high-frequency electric fields.

load (*For*) The measure for the equivalent area or volume of timber that will weigh approx 1 ton, ie 600 ft^2 of 1-inch planks, 40 ft^3 of unhewn timber or 50 ft^3 of hewn timber.

load (*ICT*) (1) Termination of an amplifier or line that absorbs the transmitted power, which is a maximum when this load MATCHES the output impedance. (2) The actual power drawn by or received by a terminating impedance. (3) In computing, to transfer a program and/or data into MAIN MEMORY from a BACKING STORE.

load (*Phys*) (1) The weight supported by a structure. (2) Synonym for force applied to a body.

load balancing (*ICT*) Arrangements by which the number of requests to a service or subsystem are automatically distributed between several instances of that service, or across several pieces of hardware, thus ensuring optimal performance of the system.

load capacitor (*ElecEng*) That which tunes and maximizes the power to a load in induction or dielectric heating.

load cast (*Geol*) A *sole mark* appearing as an irregular bulge on the underside of a sedimentary rock. It lacks evidence of current direction. See FLUTE CAST.

load cell (*Eng*) A load-detecting and measuring element utilizing electrical or hydraulic effects which are remotely indicated or recorded.

load characteristic (*ElecEng*) *Instantaneous* voltage–current characteristic at the output of a generator or amplifier under loading conditions. Also *operating characteristic*.

load classification number (*Aero*) A number defining the load-carrying capacity of the paved areas of an airport without cracking or permanent deflection. Abbrev *LCN*.

load coil (*ElecEng*) The coil in an induction heater used to carry the ac which induces the heating current in the specimen or charge.

load curve (*ElecEng*) A curve whose ordinates represent the load on a system or piece of apparatus, and whose abscissae represent time of day, month or year, so that the curve indicates the value of the load at any time.

load curve (*Eng*) See INFLUENCE LINE.

load despatcher (*ElecEng*) An engineer who is responsible for the distribution of load over a large interconnected power system.

load displacement (*Ships*) A ship's displacement at load draught, ie the draught to the centre of the freeboard disk marking, which is set off to the summer freeboard.

load draught (*Ships*) The draught when loaded to the minimum FREEBOARD permitted for the place and season.

load dumping (*ElecEng*) The automatic disconnection of circuits connected to an electrical supply. This is done either to maintain system frequency, if there is a loss in supply, or to prevent overloading of the supply equipment. See LOAD GOVERNING, LOAD MANAGEMENT.

loaded antenna (*ICT*) An antenna in which series inductance has been added to increase its natural wavelength.

loaded concrete (*NucEng*) Concrete used for shielding nuclear reactors, loaded with elements of high atomic number, eg lead, iron shot, barytes.

loaded impedance (*ICT*) That of the input of a transducer or other device when output load is connected.

load efficiency (*Electronics*) The ratio of the useful power delivered by the output stage to a specified load, and the dc input power to the stage.

loader (*ICT*) A program for controlling the loading of other programs.

loader (*MinExt*) A mechanical shovel or other device for loading trucks underground.

load–extension curve (*Eng*) A curve, plotted from the results obtained in a TENSILE TEST, showing the relations between the applied load and the extension produced.

load factor (*Aero*) (1) In relation to the structure, the ratio of an external load to the weight of an aircraft. Loads may be centrifugal and aerodynamic owing to manoeuvring, to gravity, to ground or water reaction; usually expressed as g, eg $7g$ is a load seven times the weight of the aircraft. (2) In aircraft operations, the actual PAYLOAD on a particular flight as a percentage of the maximum permissible payload.

load factor (*ElecEng*) The ratio of the average load to peak load over a period. See PLANT LOAD FACTOR.

load following (*NucEng*) See SELF-REGULATING.

load governing (*ElecEng*) The automatic maintenance of supply frequency from a rotating machine by connecting and disconnecting load as the input power increases and decreases respectively. This is often preferred to conventional governing in small plant because the equipment is cheaper.

load impedance (*ICT*) The impedance of any device that accepts power from a source, eg amplifiers, antenna, loudspeakers, magnetizing coils.

loading (*Build*) See ON-COSTS.

loading (*Chem*) The addition of fillers to polymers, eg carbon black in rubbers. See panel on TYRE TECHNOLOGY.

loading (*ElecEng*) An inductance added to a line for the purpose of improving its transmission characteristics throughout a specified frequency band. See COIL LOADING, CONTINUOUS LOADING.

loading (*MinExt*) Adsorption by resins of dissolved ionized substances in ion-exchange process. In uranium technology, the loading factor is the mass of U_3O_8 adsorbed by unit volume of the resin.

loading (*NucEng*) The introduction of fuel into a reactor.

loading (*Paper*) Adding filler to paper.

loading (*Textiles*) Increasing the weight of fabrics by use of starch, size, china clay, etc. Also *filling*.

loading and cg diagram (*Aero*) A diagram, usually comprising a side elevation of the aircraft concerned, with a scale and the location of all items of removable equipment, payload and fuel, which is used to adjust the weight of the aircraft so that its resultant lies within the forward and aft CG LIMITS.

loading and unloading machine (*NucEng*) A system for introducing and withdrawing fuel elements from a reactor, with safety provision for personnel. See CHARGE–DISCHARGE MACHINE.

loading capacity (*MinExt*) In ion exchange, saturation limit of resin.

loading coil (*ElecEng*) The coil inserted, in series with a line's conductors, at regular intervals. Also *Pupin coil*.

loading gauge (*CivEng*) (1) The limiting dimensions governing height, width, etc, of rolling stock to ensure that adequate clearance is obtained for passage under bridges and through tunnels. (2) A shaped bar suspended over a railway track, at the correct height and position, to check compliance of rolling stock passing underneath with the above limiting dimensions.

load leads (*ElecEng*) The connections or transmission lines between the power source of an induction (or dielectric) heater and the load coil or applicator.

load-levelling relay (*ElecEng*) A relay used in connection with certain, temporarily dispensable, forms of apparatus such as storage water heaters. It automatically switches them off when the demand on the system exceeds a certain value.

load line (*Electronics*) On a set of output characteristic curves for an amplifying device, a line representing the load, straight if entirely resistive, elliptical if reactive.

load lines (*Ships*) A group of lines marked on the outside of both sides of a ship to mark the minimum FREEBOARD permitted in different parts of the world and seasons. Also *plimsoll mark*.

load management (*ElecEng*) The connection and disconnection of circuits fed from an electrical supply to maintain the average load constant, or within preset limits. This is done for reasons of stable operation of a power system and for economy.

load matching (*ICT*) Adjusting circuit conditions to meet requirements for maximum energy transfer to load.

load on top (*Ships*) At one time, tankers would discharge their tank water washings and water ballast overboard before taking on a new cargo of crude oil, contributing to marine pollution. The practice was voluntarily replaced in the 1960s so that new cargoes are now loaded on top of all contaminated ballast and washings which are later treated ashore at special installations. New tankers must be fitted with separate, clean ballast tanks.

load-rate prepayment meter (*ElecEng*) A form of prepayment meter in which the charge per unit is changed whenever the load exceeds a certain predetermined value.

loadstone (*Min*) See LODESTONE.

loam (*Eng*) A clayey sand milled with water to a thin plastic paste, from which moulds are built up on a backing of soft brick; generally swept or strickled to shape without the use of a pattern.

loam (*EnvSci*) A rich friable soil that is a mixture of sand, silt, clay and usually organic matter, a desirable texture for horticultural and agricultural uses.

loam board (*Eng*) See STRICKLE BOARD.

loam bricks (*Eng*) Cakes of loam, or soft building bricks built up with loam, which form a solid but porous support for the loam forming the wall of the mould. See LOAM.

loan (*Paper*) High-quality bond paper, originally and occasionally made from a rag furnish and tub sized, of durable character intended for documents which resist repeated handling and are required to last.

lobar pneumonia (*Med*) Inflammation of one or more lobes of the lung, the affected lobes becoming solid; due usually to infection with the pneumococcus. See PNEUMONIA.

lobe (*Autos*) A rounded projection, usually on a cam.

lobe (*BioSci*) (1) A rounded or flap-like projection. (2) A natural division, formed by a fissure, of an organ, eg the brain, liver, lungs. See OLFACTORY LOBES, OPTIC LOBES. (3) The soft lower part of the external ear. (4) Curved or rounded part of a leaf, petal, etc, connected to others by an undivided centre. Adjs *lobate*, *lobed*, *lobose*, *lobulate*.

lobe (*ICT*) Enhanced response of an antenna in the horizontal or vertical plane, as indicated by a lobe or loop in its radiation pattern. The BEAM effect arises from a major lobe, generally intended to be along the forward axis.

lobeline (*Chem*) $C_{22}H_{27}O_2N$. Mp 130–131°C. A piperidine alkaloid obtained from *Lobelia inflata*. Forms broad needles. It is monoacidic, and is remarkable for yielding acetophenone when heated with water. Used as a smoking deterrent.

lobe switching (*Radar*) A method of determining the precise direction of a target without resorting to impossibly narrow antenna beams. While the antenna is turning, its beam is switched periodically to the left and right of the dead-ahead position; when equal signals are received in both positions, the antenna is accurately aimed.

lobotomy (*Med*) See LEUCOTOMY.

lobule (*BioSci*) A small lobe; one of the polyhedral cell masses forming the liver in vertebrates. Adjs *lobular, lobulate*. Also *lobulus*.

LOCA (*NucEng*) Abbrev for *loss-of-coolant accident*. The conditions which might arise in the event of loss of primary or secondary coolant in a reactor.

local action (*ElecEng*) Deterioration of battery due to currents flowing to and from the same electrode.

local area network (*ICT*) High-bandwidth (allowing high data transfer rate) computer network, operating over a small area, such as an office or group of offices. Abbrev *LAN*. See METROPOLITAN AREA NETWORK, NETWORK, WIDE AREA NETWORK.

local attraction (*Phys*) See MAGNETIC ANOMALY.

local bus video (*ICT*) A technique of connecting the video ADAPTER CARD via a high-speed DATA BUS directly to the PROCESSOR rather than using the conventional bus that connects the processor to the peripheral devices. VESA is one such standard.

local carrier (*ICT*) In suppressed carrier systems, demodulation with an adequate carrier wave inserted before demodulation.

local exchange (*ICT*) The exchange to which a given subscriber has a direct line. Sometimes *local central office*.

local group (*Astron*) The family (or cluster) of galaxies to which our own Galaxy belongs. There are two dozen members within 5 million light-years or so. The ANDROMEDA GALAXY is another prominent member of the group.

localization (*Psych*) An ability of the sense organs of an animal to determine the source of a stimulus in space. Many sense organs can detect direction, but probably only the eyes and ears or lateral line system, can judge distance. See ECHOLOCATION and STEREOPHONY.

localized vector (*MathSci*) See BOUND VECTOR.

localizer beacon (*Aero*) A directional radio beacon associated with the ILS, which provides an aircraft during approach and landing with an indication of its lateral position relative to the runway in use.

local junction circuit (*ICT*) A junction between exchanges in the local call area, ie where connections are made by local codes as opposed to SUBSCRIBER DIRECT DIALLING.

local loop (*ICT*) In a telephone system, the connection between the CUSTOMER PREMISES EQUIPMENT and the local exchange or CENTRAL OFFICE that serves it.

local Mach number (*Aero*) The ratio of the velocity of the airflow over a part of a body in flight to the local speed of sound. Usually it is concerned with a part of greater curvature, where the airflow accelerates momentarily, thereby increasing the Mach number above that of the body as a whole, eg over the wing or canopy.

local oscillations (*ICT*) Those generated within the apparatus which uses them.

local oscillator (*ICT*) That in a SUPERHET receiver which is mixed with the incoming signal, in the mixer or frequency changer to produce an output of sum and difference frequencies. The difference signal will generally be quite low and it is this which is the intermediate frequency.

local printer (*ICT*) A printer directly connected to one of the PORTS on the computer.

local standard of rest (*Astron*) The frame of reference in which the peculiar motions of local stars are averaged to zero, hence representing the general motion of the local stars around the centre of the galaxy.

local time (*Astron*) Applied to any of the three systems of time reckoning – sidereal, mean solar or apparent solar time – it signifies the hour angle of the point of reference in question measured from the local meridian of the observer. The local times of a given instant at two places differ by the amount of their difference in longitude expressed in time, the local time at a place east of another being the greater.

local variable (*ICT*) One whose use is restricted to a particular subprogram.

local vent (*Build*) A connection enabling foul air in a room or plumbing fixture to escape to the outer air.

location (*Eng*) A geometrical feature, eg a projection or a recess, by which one machine component or article may be correctly positioned by engagement in or with a corresponding feature in another, for the correct spatial relationship of machine or other components.

location (*ICT*) A position in a MEMORY that holds a WORD or part of a word.

location (*ImageTech*) Place of motion picture or video production other than a permanently established studio.

locator (*Acous*) See SOUND LOCATOR.

locator beacon (*Aero*) The 'homing' beacon on an airfield used by the pilot until he or she picks up the localizer signals of the INSTRUMENT LANDING SYSTEM.

lochia (*Med*) The normal discharge from the vagina during the first week or two after childbirth.

lock (*Build*) A communicating channel, having gates at each end, between the higher and lower reaches of a canal. It is used to transfer a vessel from one REACH to the other.

lock (*ICT*) An ATTRIBUTE that may be set as a security measure to prevent the accidental erasure of a FILE. The lock has to be 'turned off' before the file can be deleted, which reduces the chance of unintentional erasure.

lock (*Print*) A safety device in most control boxes to prevent the press being moved under power; the press can be made free only by cancelling the lock; may also be used during the run to prevent increase of speed.

lockage (*Build*) Water lost, ie transferred from a higher to a lower level, in the operation of passing a vessel through a lock.

lock bay (*Build*) The water space enclosed in a *lock chamber*, between the headgates and tailgates.

locked (*ElecEng*) Said of an oscillator when it is held to a specific frequency by an external source.

locked-coil conductor (*ElecEng*) A form of stranded conductor in which the outer wires are so shaped that they are prevented from having any radial movement.

locked groove (*Acous*) Finishing groove on the surface of a vinyl disk record, motion of the needle in this groove operating a stopping mechanism. Also *eccentric groove*.

locked-in syndrome (*Med*) A neurological condition resulting from brain-stem damage in which the subject is conscious and aware but unable to move or communicate other than sometimes through blinking.

locked oscillator detector (*ICT*) A form of frequency-modulated (FM) detector using a pentode valve that incorporates a self-oscillating tuned circuit connected to the suppressor grid. The FM input signal is applied to a resonant tuned circuit that is connected to the control grid. The tuned circuit of the suppressor grid locks to the frequency of the incoming signal and leads to an AF anode current proportional to the FM input; this circuit does not respond to AM signals, including noise, so that a separate limiting stage is unnecessary.

locked test (*MinExt*) In preliminary tests on unknown ores, retention of a potentially troublesome fraction of the test product for addition to a new batch of ore, to ascertain whether in continuous work such a build-up would be upsetting. Also *cyclic test*.

lock-fit (*Eng*) Close-tolerance mechanical joint between two components. Used for joints not intended to be disassembled.

lock gate (*Build*) A pair of doors at one end of a lock, serving in conjunction with gates at the other end to enclose water within the lock chamber. Many different kinds of lock gate are now used.

lock-in (*Electronics*) Generally, to synchronize one oscillator with another, as in a homodyne or frequency doubler. One oscillator must be free-running and capable of being pulled. See PHASE-LOCKED LOOP.

lock-in amplifier (*ICT*) Synchronous amplifier, sensitive to variation of signal of its own frequency.

locking (*ImageTech, ICT*) (1) See LATCHING. (2) The control of frequency of an oscillating circuit by means of an applied signal of constant frequency, eg controlling the time base of a TV receiver by the incoming signal.

locking force (*Eng*) See INJECTION MOULDING.

locking key (*ICT*) A hand-operated telephone key that, when operated, remains in its operated position until released by the hand.

locking relay (*ICT*) A telephone relay that, when operated, remains in its operated condition when the operating current ceases, either by closing a winding that carries a sustaining current or, more rarely, by mechanical means.

lockjaw (*Med*) See TETANUS.

locknit (*Textiles*) A fabric produced by WARP KNITTING.

lock nut (*Eng*) (1) An auxiliary nut used in conjunction with another, to prevent it from loosening under vibration. (2) Any special type of nut designed to prevent accidental loosening. Also *check nut*.

lock paddle (*Build*) A sluice through which water is passed to fill an empty lock chamber.

lock rail (*Build*) (1) The door rail which is level with the lock. (2) The front member of a piano, running along the keyboard, which usually contains the lock.

lockrand (*Build*) A course of stones laid as bondstones.

lock seam (*Eng*) A type of joint, not absolutely tight, produced in the manufacture of sheet-metal drums, cans, etc, by simply folding, interlocking and pressing together the longitudinal edge areas of the product.

lock sill (*Build*) See MITRE SILL.

lock stile (*Build*) That stile of a door in or on which the lock is fastened.

locks, up and down (*Aero*) See UP, DOWN LOCKS.

lock-woven mesh (*CivEng*) A mechanically woven fabric, made of steel wires crossing at right angles and secured at the intersections, used in reinforced concrete construction.

locomotive (*Eng*) A vehicle driven by oil, electricity or coal, for moving rolling stock on a railway.

locomotive boiler (*Eng*) The type of boiler used on steam locomotives; it consists of an internal fire-box at one end of the horizontal cylindrical shell, from which the hot gases are led through fire-tubes passing through the water space into the smoke-box at the front of the boiler. See FIRE-BOX, FIRE-TUBE BOILER.

locomotor ataxia (*Med*) Also *locomotor ataxy*. See TABES DORSALIS.

Loctite (*Eng*) TN for cyanoacrylate monomer adhesives.

locular (*BioSci*) Divided into compartments by septa. Also *loculatous*.

locule (*BioSci*) A cavity, esp one within a sporangium or an ovary, containing the spores or ovules respectively. Also *loculus*. Adj *loculicidal*.

locus (*BioSci*) The position on a chromosome occupied by a specified gene and its alleles. Pl *loci*.

locus of control (*Psych*) In social psychology, the perceived source of control over an individual's behaviour.

locust (*BioSci*) One of several kinds of winged insects of the family Acridiidae, akin to grasshoppers, highly destructive to vegetation.

lod (*MathSci*) The logarithm of the odds relating to an event.

lode (*CivEng*) An artificial dyke.

lode (*Geol, MinExt*) (1) A mineral deposit composed of a zone of veins. (2) Steeply inclined fissure of non-alluvial mineral enclosed by walls of country rock of different origin. See fig. at MINING.

loden (*Textiles*) A coarse-milled woollen fabric used for jackets and coats because of its weather-resisting properties.

lodestone (*Min*) Iron oxide. A form of magnetite exhibiting polarity, behaving, when freely suspended, as a magnet. Also *loadstone*.

Lodge–Cottrell precipitator (*ChemEng*) Plant for carrying out *electrostatic precipitation* of particles from gases.

lodging (*Agri*) Crop plants flattened by wind, rain, snow or, in the case of cereals, the weight of the seed head.

lodicules (*BioSci*) Small scales inserted below the stamens in the florets of most grasses; when the floret is mature the lodicules swell, forcing apart lemma and palea, and allowing stamens and stigmata to grow out.

loellingite (*Min*) See LÖLLINGITE.

loess (*Geol*) Homogeneous unstratified, usually calcareous, blanket deposit consisting mainly of quartz silt with a particle size 0·01–0·05 mm. It originates as wind-blown dust and vast accumulations cover large areas of China, E Europe and N and S America.

lofexidine (*Psych*) A drug used to alleviate withdrawal symptoms experienced by users of opiates.

Löffler boiler (*Eng*) A high-pressure boiler employing forced circulation, by pumping steam through small-diameter tubes. Part of the high-temperature steam is returned to the water drum, to produce the saturated steam supply to the pump.

Lofton–Merritt stain (*Paper*) A stain consisting of malachite green and basic fuchsin for distinguishing between unbleached sulphite and sulphate wood pulp fibres, under the microscope.

log (*MathSci*) Abbrev for LOGARITHM.

log (*Ships*) See NAUTICAL LOG.

logagraphia (*Psych*) Loss of the ability to express ideas in writing.

logarithm (*MathSci*) Of a number N to a given base b: the power to which the base must be raised to produce the number. It is written as $\log_b N$, or as $\log N$ if the base is implied by the context. There are two systems of logarithms in common use: (1) *Common* or *Briggs' logarithms*, with base 10. Any number, N, can be written in the form $N = 10^n \times M$ (where n is an integer and M is between 1 and 10) so that $\log N = n + \log M$. n is called the *characteristic* of the logarithm and $\log M$, which is obtained from tables, the *mantissa*. (2) *Natural* or *Napierian* or *hyperbolic logarithms* with base e ($= 2·718...$). To distinguish them, a logarithm to the base 10 is often written $\log x$ while a logarithm to the base e is written $\ln x$. However, in mathematical usage $\log x$ always means base e. The following conversions are useful: $\log_e x = \log_{10} x \times 2·302\,59$; $\log_{10} x = \log_e x \times 0·434\,29$.

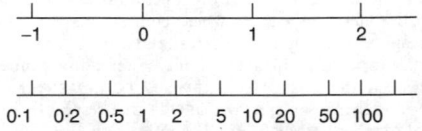

logarithm The upper scale gives the logarithm to the base 10 of the numbers below.

logarithmic amplifier (*Acous*) An amplifier with an output which is related logarithmically to the applied signal amplitude, as in decibel meters or recorders.

logarithmic array (*Radar, ICT*) Tapered END-FIRE ARRAY designed to operate over a wide range of frequency.

logarithmic capacitor (*ElecEng*) See LOGARITHMIC RESISTOR.

logarithmic damping ratio (*ElecEng*) In a plane wave, the logarithmic ratio of the amplitudes at two adjacent maxima or minima.

logarithmic decrement (*Phys*) Logarithm to base *e* of the ratio of the amplitude of diminishing successive oscillations.

logarithmic function (*MathSci*) The logarithmic function $\log x$ is defined by the equation

$$\log x = \int_1^x \frac{dt}{t}$$

It equals $\log_e x$. See E and EXPONENTIAL FUNCTION.

logarithmic horn (*Acous*) The horn of a loudspeaker which is of exponential form. Also a metal horn for microwave work.

logarithmic mean temperature difference (*ChemEng*) Universally used relationship in heat-exchange calculations. If temperature difference at one end of exchanger is T_1, and at other T_2, then the mean temperature at which heat exchange is calculated to occur is the log mean which is

$$\frac{T_1 - t_2}{\ln (T_1/t_2)}$$

and this is less than the arithmetic mean. In complex exchange conditions this may have to be corrected further to allow for special factors and is then usually said to be *weighted*.

logarithmic potentiometer (*ElecEng*) See LOGARITHMIC RESISTOR.

logarithmic resistor (*ElecEng*) A variable form of resistor for which the movement of a control is directly or inversely proportional to the fractional change of resistance. Such characteristics are used to counteract opposite characteristics in amplifiers etc. Similarly *logarithmic capacitor*, *logarithmic potentiometer*.

logarithmic spiral (*MathSci*) See EQUIANGULAR SPIRAL.

logarithmic strain (*Eng*) See STRAIN.

logatom (*Acous*) Artificial syllable without meaning of the form consonant–vowel–consonant, used in articulation testing.

log-dec (*Phys*) Abbrev for LOGARITHMIC DECREMENT.

log file (*ICT*) A FILE that holds details of all commands given to a computer over a period of time. Such a file is often used for auditing the use of a computer system or tracing the history of usage should a fault occur or in cases of suspected HACKING.

logger (*For*) One who works in the felling and extraction of timber. Also *bucheron* (Canada), *lumberjack* (US).

logger (*ICT*) An arrangement of electronic devices for obtaining an output indication which is proportional to the logarithm of the input amplitude or intensity. Required in modulation and noise meters.

loggia (*Arch*) A covered gallery or portico built into, or projecting from, the face of a building, and bounded by a colonnade on its open side.

logging wheels (*For*) A pair of wheels 2–3·5 m (7–12 ft) in diameter for transporting logs, which are slung below the axle. Also *katydid*.

logic (*ICT*) See LOGICAL OPERATION, MATHEMATICAL LOGIC.

logical design (*ICT*) The basic planning and synthesizing of a network of LOGIC ELEMENTS to carry out a particular function. See MACHINE ARCHITECTURE.

logical drive (*ICT*) A DISK DRIVE that is artificially created by means of software. Thus a large disk drive may be divided into a number of logical drives as an aid to their organization. It may or may not be mapped to a physical drive.

logical error (*ICT*) A mistake in the design of a program (eg a branch in the wrong statement).

logical operation (*ICT*) A non-arithmetic operation performed on binary variables, called Boolean variables, that can take the values TRUE or FALSE. The three basic operators are AND, NOT and OR. All other logical operators can be expressed in terms of these three; see NAND, NOR, XOR. All logical operations can be performed by a computer using 1 for TRUE and 0 for FALSE. See GATE.

logical operation (*MathSci*) See TRUTH FUNCTION.

logical shift (*ICT*) One where bits shifted from the end of the location are lost, and zeros are shifted in at the opposite end.

logic array (*ICT*) An integrated circuit consisting of an array of LOGIC GATES that are interconnected in a specified way. See PROGRAMMABLE LOGIC ARRAY, UNCOMMITTED LOGIC ARRAY.

logic bomb (*ICT*) Instructions inserted into a computer program such that a, usually harmful, operation will occur if preset conditions are fulfilled.

logic element (*ICT*) A GATE or combination of gates. Also *logic unit*.

logic programming (*ICT*) A style of programming based on reasoning in formal logic exemplified by resolution theorem proving which underlies PROLOG.

logic unit (*ICT*) See LOGIC ELEMENT.

log in (*ICT*) A method by which the remote terminal user enters a multi-access computer system. Also *log on*.

login script (*ICT*) A sequence of COMMANDS stored in a FILE that are executed each time a USER LOGS ON to the system. Each user may have a different script to accommodate their different required environments.

logistic equation (*BioSci*) An equation describing the typical increase of a population towards an asymptotic value:

$$\frac{dN}{dt} = rN\left(\frac{K - N}{K}\right)$$

where N is the population size, t is time, r is the innate capacity for increase and K is the maximum or asymptotic value of N.

logistic regression (*MathSci*) A particular form of REGRESSION in which the relationship is between the LOGIT of the probability of the occurrence of an event and the values of one or more possibly related variables.

logit (*MathSci*) The logarithm of the ratio of the probability that an event occurs to the probability that it does not occur.

log-normal (*MathSci*) Having a distribution such that the logarithms of its value have a NORMAL DISTRIBUTION.

LOGO (*ICT*) A programming language with list processing features, widely known for its TURTLE graphics.

log off (*ICT*) See LOG OUT.

log on (*ICT*) See LOG IN.

logotype (*Print*) A word, or several letters, cast as one piece of type; also, a block or line illustration of a trade mark.

log out (*ICT*) A method by which the remote terminal user leaves a multi-access computer system. Also *log off*.

log-periodic (*ICT*) A tapered END-FIRE ARRAY, that may consist of dipoles, folded dipoles or monopoles, fed in such a way that the maximum radiation is along the major axis of the array. At any given frequency, only one element and a few immediate neighbours are in operation; hence, the tapered form of the elements gives the antenna a wide frequency range.

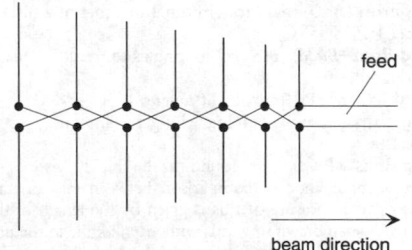

feed

beam direction

log-periodic dipole antenna

log washer (*MinExt*) A trough or tank set at a slope, in which ore is tumbled with water by means of one or more box girders with projecting arms set helically, so that as they rotate coarse lump is cleaned and delivered upslope while mud and sand overflows downslope.

logwood (*For*) The heartwood of *Haematoxylon campechianum*, having a distinctive sweet taste and a smell resembling that of violets. It yields a black dye, and also haematoxylin, a stain used in microscopy.

loktal base (*Electronics*) A valve base with a centre pin to lock the base securely in an appropriate socket. It has eight pins which extend directly through the glass envelope of the valve.

löllingite (*Min*) Arsenide of iron, $FeAs_2$, occurring as steel-grey crystals, prismatic in habit, belonging to the orthorhombic system. Also *loellingite*.

lomasome (*BioSci*) A compact mass of membranes in the cytoplasm adjacent to a plant cell wall, apparently invaginated from the plasmalemma, of obscure function.

lomentum (*BioSci*) A fruit, usually elongated, that develops constrictions as it matures, finally breaking across these into one-seeded portions. Adj *lomentose*.

London forces (*Chem*) Forces arising from the mutual perturbations of the electron clouds of two atoms or molecules, the forces varying as the inverse sixth power of the distance between the molecules. Also *van der Waals forces*.

London hammer (*Build*) A light type of cross-pane hammer. Also *Exeter hammer*.

London plane (*For*) See EUROPEAN PLANE.

London screwdriver (*Build*) One with flat blade and tapered end, usually favoured for large sizes.

London-shrunk (*Textiles*) A tTerm in the woollen and worsted trades to indicate that a fabric has been pre-shrunk. Wet fabric is allowed to dry naturally without tension.

lone pair (*Chem*) A pair of valence electrons unshared by another atom. Such lone pairs are responsible for the formation of co-ordination compounds.

long (*Glass*) Slow-setting.

long-and-short work (*Build*) A mode of laying quoins and of forming door and window jambs in rubble walling, the stones being alternately laid horizontally and set up on end; the latter stones are usually longer than the former.

long-bodied type (*Print*) Type cast on a larger body to avoid leading, eg 9-point on 11-point type will have the appearance of 9-point leaded 2-point font. Also *bastard font*.

long column (*CivEng*) A slender column which fails by bending rather than by crushing. It typically has a length 20–30 times its diameter.

long-day plant (*BioSci*) A plant that will flower better under conditions of long days and short nights, eg spinach, *Spinacia oleracea*. Cf SHORT-DAY PLANT. See DAY-NEUTRAL PLANT, PHOTOPERIODISM.

long descenders (*Print*) The length of the descenders (j, g, etc) is a feature of type design, long descenders being usually considered to confer elegance. See TYPEFACE. Certain typefaces are supplied with alternative long descenders to improve their appearance for bookwork, among them being Monotype Baskerville, Plantin, Times.

long dolly (*CivEng*) See FOLLOWER.

longeron (*Aero*) Main longitudinal member of a fuselage or nacelle.

long float (*Build*) A trowel so long as to need two plasterers to handle it.

longi- (*Genrl*) Prefix from Lt *longus*, long.

longicorn (*BioSci*) Having elongate antennae, as some beetles.

long inks (*Print*) Inks found to be highly viscous in an empirical method of comparison between inks of differing viscosity; a measure of this is given by the length of the ink thread when drawn upwards with an ink knife to the point at which it breaks. *Short inks* are those which are least viscous.

longipennate (*BioSci*) Having elongate wings or feathers.

longirostral (*BioSci*) Having a long beak or rostrum.

longitude (*Astron*) See LATITUDE AND LONGITUDE, CELESTIAL.

longitude (*Genrl*) See LATITUDE AND LONGITUDE, TERRESTRIAL.

longitude (*Surv*) The perpendicular distance of the mid-point of a survey line from the reference meridian.

longitudinal (*Aero*) A girder that runs fore and aft on the outside of a rigid airship frame. Longitudinals connect the outer rings of the transverse frames.

longitudinal axis (*Aero*) See AXIS.

longitudinal current (*ElecEng*) See CIRCULATING CURRENT.

longitudinal dune (*Geol*) A long, narrow sand-dune parallel to the direction of the prevailing wind.

longitudinal frame (*Ships*) A stiffening member of a ship's hull disposed longitudinally, as opposed to a transverse frame. It is supported at ends by either bulkheads or web frames, disposed transversely.

longitudinal heating (*ElecEng*) Dielectric heating in which electrodes apply a high-frequency electric field parallel to lamination. See GLUELINE.

longitudinal instability (*Aero*) The tendency of an aircraft's motion in the plane of symmetry to depart from a steady state, ie to pitch up or down, to rise or fall, or to vary in horizontal speed.

longitudinal joint (*Build*) A joint used to secure two pieces of timber together in the direction of their length. See SCARF.

longitudinal magnetization (*Acous*) The magnetization of a magnetic recording medium along an axis parallel to the direction of motion.

longitudinal magnification (*Phys*) The ratio of the length of the image to the length of the object in a lens system when the object is small and lies along the axis of the system.

longitudinal metacentre (*Ships*) The METACENTRE obtained by producing an angle of TRIM about a transverse axis by an external force. Cf TRANSVERSE METACENTRE.

longitudinal oscillation (*Aero*) A periodic variation of speed, height and angle of pitch. See PHUGOID OSCILLATION.

longitudinal stability (*Aero*) See STABILITY.

longitudinal study (*Psych, Med*) A research design that uses the same subjects over an extended period of time. Also *diachronic study*. See CROSS-SECTIONAL STUDY.

longitudinal valve (*BioSci*) In amphibians, a large, flap-like valve that traverses the CONUS ARTERIOSUS longitudinally and obliquely.

longitudinal wave (*Acous*) Propagating sound wave in which the motions of the relevant particles are in line with the direction of translation of energy.

long letters (*Print*) Letters with the long accent added (ā, ē, ō, etc).

long-oil (*Build*) Term applied to varnishes etc containing a high proportion, ie more than about 60%, of oil.

long-persistence screen (*Electronics*) Cathode-ray-tube screen coated with long afterglow phosphor (up to several seconds).

longprimer (*Print*) An old type-size, approximately 10-point.

long QT syndrome (*Med*) A genetic defect in a potassium channel that increases the risk of sudden death. It is characterized by a long QT interval on the ECG. Acquired long QT syndrome is usually caused by drug treatment or electrolyte disturbance.

long-range (*Aero*) Aircraft, ship or missile capable of covering great distances without refuelling.

long-range elasticity (*Eng*) A term used for elastomeric behaviour, often to distinguish it from the short-range Hookean elasticity shown by eg metals, ceramics and glasses.

long-range order (*Chem*) A SUBSTITUTIONAL SOLID SOLUTION in which an ordered arrangement of solute extends across large numbers of lattice units or the entire crystal.

long shoot (*BioSci*) A shoot of mostly woody plants with relatively long internodes that extends the canopy. Long shoots typically bear few or no flowers and in some species bear no foliage leaves but only scales (eg pines) or spines (eg *Berberis*). Cf SHORT SHOOT.

longshore current (*Geol*) A current which flows parallel to the shoreline.

longshore drift (*Geol*) The movement of material along the shore by a LONGSHORE CURRENT.

long shot (*ImageTech*) A scene photographed from a distance, showing the general setting of the action.

long-shunt compound winding (*ElecEng*) A field-winding arrangement for a compound-wound dc machine in which the shunt winding is connected across the external terminals, ie across the armature and the series winding.

long-sightedness (*Med*) See HYPERMETROPIA.

long superstructure (*Ships*) A superstructure which is sufficiently long to be included in calculations of the ship's main strength, and not simply as an excrescence.

long-tail pair (*Electronics*) A two-valve or two-transistor circuit with a resistor common to both cathode or emitter connections. The common resistor provides strong negative feedback, and the circuit is inherently stable. Antiphase outputs can be obtained from each anode or emitter, and if both inputs are energized, differential amplification can take place.

long-term exposure limit (*EnvSci*) Time-weighted average of exposure to a hazard mainly used in control of occupational exposure. Long-term in this context is often 8 h, the length of the standard working day.

long-term memory (*Psych*) According to the three-store model of memory, a memory system that keeps memories for long periods, has a very large capacity, and stores items in an organized and processed form. Cf SHORT-TERM MEMORY. See SENSORY STORE.

long-term potentiation (*BioSci*) Long-lasting (hours or days) increase in the strength of transmission at the neural synapse, that results from repetitive use. May be important as the cellular basis of memory. Abbrev *LTP*.

long tom (*MinExt*) Portable sluice in which rough concentrates made during treatment of alluvial sands or gravels are sometimes worked up to a better grade.

long ton (*Genrl*) A unit of mass, 2240 lb. See TON.

longwall coal-cutting machine (*MinExt*) A machine which severs coal in mechanized mining.

longwall working (*MinExt*) A method of mining bedded deposits, notably coal, in which the whole seam is removed, leaving no pillars. If advancing, work goes outwards from shaft and roads must be maintained through excavated areas. If retreating, ground behind workers can be allowed to cave if surface rights are not thus affected.

long waves (*ICT*) Low-frequency radio waves of wavelength >1000 m (frequency <300 kHz).

look-through (*Paper*) A term for the appearance of the paper by transmitted light.

lookup table (*ICT*) A table giving a set of values for a VARIABLE.

loom (*Textiles*) See WEAVING MACHINE.

loom efficiency (*Textiles*) The average number of PICKS per minute inserted by a loom expressed as a percentage of what is theoretically possible if the machine were running continuously.

looming (*EnvSci*) The vague enlarged appearance of objects seen through a mist or fog, particularly at sea.

looming (*Phys*) A particular form of mirage in which the images of objects below the horizon appear in a distorted form.

looming (*Textiles*) Drawing the threads of the warp through the eyes of the heald shaft and the reed, in the order arranged for the predetermined pattern.

looming response (*Psych*) The response of many animals, including humans, to a sudden increase in stimulus size by ducking or turning away from it.

loop (*Aero*) An aircraft manoeuvre consisting of a complete revolution about a lateral axis, with the normally upper surface of the machine on the inside of the path of the loop. See INVERTED LOOP.

loop (*Electronics*) Feedback control system.

loop (*ICT*) A sequence of instructions that is executed repeatedly until some specified condition is satisfied. See INFINITE LOOP, NESTED LOOP.

loop (*ImageTech*) (1) The length of slack film between two driven points, as between the intermittent mechanism and a sprocket in a motion picture camera or projector. (2) A length of film or tape joined end to end, for use as a continuous strip.

loop (*Phys*) (1) Closed graphical relationship, eg hysteresis loop. (2) Same as ANTINODE of displacement in standing waves.

loop actuating signal (*Electronics*) The combined input and feedback signals in a closed-loop system.

loop antenna (*ICT*) See FRAME ANTENNA.

loop cable (*ElecEng*) See TWIN CABLE.

loop coupling (*ElecEng*) Small loop connected between conductor and shield of a coaxial transmission line, for collecting energy from one of the series of resonators in a magnetron or a rhumbatron in a klystron valve.

loop dialling (*ICT*) See LOOP-DISCONNECT PULSING.

loop difference signal (*Electronics*) Output signal at point in feedback loop produced by input signal applied at the same point.

loop-disconnect pulsing (*ICT*) The normal subscriber dialling, whereby trains of impulses are set up by interrupting the loop current from the exchange. Also *loop dialling*.

loop diuretics (*Pharmacol*) A group of drugs which inhibit resorption from the loop of Henle in the renal tubule. They are powerful diuretics and include furosemide, bumetanide and ethacrynic acid.

loop feedback signal (*Electronics*) The part of the loop output signal fed back to the input to produce the LOOP ACTUATING SIGNAL.

loop gain (*Electronics*) Product gain of amplifier and feedback circuit; if greater than unity, the system is liable to sustain oscillations. See NYQUIST CRITERION.

loop galvanometer (*ElecEng*) A sensitive galvanometer in which the moving element is a U-shaped current-carrying loop of aluminium foil, the two sides of the loop being in magnetic fields of opposite direction.

looping-in (*ElecEng*) A term used in wiring work to denote a method of avoiding the use of a tee-joint by carrying the conductor to and from the point to be supplied.

looping mill (*Eng*) A rolling mill in which the product from one stand is fed into another in the opposite direction, usually above the first.

looping roller (*Print*) A roller on a web-fed press which moves to provide an intermittent flow to the web.

loop input signal (*Electronics*) An external signal applied to a feedback control loop in control systems.

loop output signal (*Electronics*) The extraction of the controlling signal from a feedback control.

loop pile (*Textiles*) Fabric made with loops protruding from the surface of a firm ground. See MOQUETTE.

loop-raised fabric (*Textiles*) A warp-knitted fabric in which some of the loops are plucked by wires in a subsequent raising process.

loopstick antenna (*ICT*) See FERRITE-ROD ANTENNA.

loop test (*ElecEng*) A method of test used for locating faults in electric cables. The faulty conductor is made to form part of a closed circuit or loop, an adjacent sound conductor usually forming another part. See ALLEN'S LOOP TEST, VARLEY LOOP TEST.

loop-type reactor (*NucEng*) An INDIRECT-CYCLE reactor in which the heat exchanger is outside the pressure vessel.

loop yarn (*Textiles*) A fancy yarn formed by wrapping an effect thread round a twisted core so that loops protrude from the yarn surface.

loose butt hinge (*Build*) A butt hinge in which one leaf may be lifted from the other, enabling eg a door to be easily removed.

loose centres (*Eng*) Heads similar to the tailstock of a lathe; used for supporting work on the table of a planing machine etc so that it may be rotated.

loose coupling (*ElecEng*) See WEAK COUPLING.

loose coupling (*Eng*) A shaft coupling capable of instant disconnection, as distinct from a FAST COUPLING.

loose eccentric (*Eng*) An eccentric used on small reversing steam engines. It rides freely on the shaft but is located by either of two stops on the shaft, which position and drive it for either ahead or reverse running.

loose gland (*Eng*) A ring used in making an EXPANSION JOINT between hot-water pipes. It slides on the spigot and compresses a rubber ring against the socket, to which it is bolted.

loose-leaf (*Print*) A binding system in which separate leaves are held together within a cover by means of a spring, ring, spiral, or other device, which by unlocking, a leaf may be removed or inserted at any point.

loose-leaf gold (*Build*) Real gold beaten into very thin leaves and supplied in small books with tissue between each leaf. Produces a better lustre than transfer leaf.

loose needle survey (*Surv*) A traverse with miner's dial, in which the magnetic bearing is read at each station.

loose piece (*Eng*) Part of a foundry pattern which has to be withdrawn from the mould after the main pattern, through the cavity formed by the latter.

loose pulley (*Eng*) A pulley mounted freely on a shaft; generally used in conjunction with a fast pulley to provide means of starting and stopping the shaft by shifting the driving belt from one to the other.

loparite (*Min*) A titanate of sodium, calcium and the rare earth elements; a variety of PEROVSKITE.

loph (*BioSci*) A crest connecting the cusps of a molar tooth.

lopho- (*Genrl*) Prefix from Gk *lophos* signifying crested or tufted.

lophobranchiate (*BioSci*) Having tuft-like, crest-like or lobe-like gills.

lophodont (*BioSci*) Of mammals, having cheek teeth with transverse ridges on the grinding surface.

lophophore (*BioSci*) A ciliated tentacle used for food gathering; found in some aquatic invertebrates.

LORAN (*Aero*) Abbrev for *long-range navigation*. An early but much developed hyperbolic navigation aid using on-board aircraft systems to translate the time difference between pulsed transmissions received from two or more ground stations to provide positional information.

loratadine (*Pharmacol*) A non-sedating antihistamine used to relieve symptoms of hay fever and urticaria. TN *Clarityn*.

lorazepam (*Pharmacol*) A short-acting *benzodiazepine* for short-term treatment of anxiety and insomnia.

lordoscoliosis (*Med*) A deformity of the spine in which LORDOSIS is associated with lateral curvature of the spine.

lordosis (*Med*) Deformity of the spine in which there is an increase of the forward convexity of the lower half of the spinal column.

lore (*BioSci*) The space between the beak and the eye in birds. Adj *loral*.

Lorentz contraction (*Phys*) See FITZGERALD–LORENTZ CONTRACTION.

Lorentz force (*Phys*) The force experienced by a point charge q moving with a velocity v in a field of magnetic induction B and in an electric field E. The force F is given by

$$F = q[(v \times B) + E]$$

Lorentz–Lorenz equation (*Chem*) The equation by which *molecular refraction* is defined:

$$[R] = \left(\frac{n^2 - 1}{n^2 + 2}\right)\frac{M}{\rho}$$

where $[R]$ is the molecular refraction, n is the refractive index, M is the molecular mass and ρ is the density.

Lorentz transformation (*Phys*) One of the relations between the co-ordinates of space and time of the same event as measured in two inertial frames of reference moving with a uniform velocity relative to one another.

They are derived from the postulates of the SPECIAL RELATIVITY theory.

Lorenz apparatus (*ElecEng*) A delicate apparatus for the absolute determination of the value of a resistance. A metal disk is rotated in an accurately known uniform field, and the emf produced between its centre and its periphery is balanced against the drop of potential caused by the field-producing current in the resistance to be measured.

Loschmidt number (*Chem*) (1) The number of molecules in unit volume of an ideal gas at stp, ie $2 \cdot 687 \times 10^{25}$ m^{-3}. (2) Also, esp on the Continent, name given to AVOGADRO'S NUMBER.

löss (*Geol*) See LOESS.

loss (*ICT*) Opposite of gain, the diminution of power, as in a transformer or line. Loss is realized as a standard in a resistive pad (attenuator), which introduces a known loss into a measuring circuit.

loss angle (*ElecEng*) The difference between 90° and the angle of lead of current over voltage in a capacitor.

Lossev effect (*Electronics*) The radiation due to the recombination of charge carriers injected into a p–i–n or p–n junction biased in the forward direction.

loss factor (*Acous*) Quantity to describe the damping of structure-borne sound in materials or structures.

loss factor (*Eng*) LOGARITHMIC DECREMENT divided by π, a measure of the energy dissipation in a material, esp polymeric. Equal to tan δ. See LOSS TANGENT.

loss factor (*ICT*) (1) The ratio of average power loss in a circuit or device to the power loss under peak loading. (2) The POWER FACTOR of a material multiplied by its DIELECTRIC CONSTANT. This varies with frequency and governs the amount of heat generated in a material. See LOSSY.

lossless line (*ICT*) See DISSIPATIONLESS LINE.

loss of charge method (*ElecEng*) A method of measuring very high resistances. The resistance is placed across the terminals of a charged capacitor and the rate at which the charge leaks away is observed.

loss of vend (*MinExt*) The difference in weight between raw (run-of-mine) coal and saleable product leaving washery.

loss tangent (*Eng*) The ratio of imaginary to real parts of dielectric constant. Also *tan δ, dissipation factor*. See DIELECTRIC LOSS, DIELECTRICS, FERROELECTRICS

lossy (*ElecEng, ICT*) Of a material or apparatus that dissipates energy, eg dielectric material or transmission line with a high attenuation. The attenuation loss will be in *decibels* (dB) while the rate of loss in the dielectric is proportional to its *loss factor*, ie to the product of POWER FACTOR and RELATIVE PERMITTIVITY.

lost flux (*ElecEng*) See LEAKAGE FLUX.

lost lead (*Print*) See LEAD OF THE WEB.

lost wax casting (*Eng*) See INVESTMENT CASTING.

Lotka's equations (*BioSci*) Mathematical relationships between: (1) the populations of two species living together and competing for food or space; and (2) the situation where one is the predator and the other the prey.

loud-hailer (*Acous*) See MEGAPHONE.

loudness (*Acous*) The intensity or volume of a sound as perceived subjectively by a human ear.

loudness contour (*Acous*) Line drawn on the audition diagram of the average ear which indicates the intensities of sounds that appear to the ear to be equally loud.

loudness level (*Acous*) Of a specified sound, the intensity of the *reftone* (1000 Hz) on the *phon* scale, which is adjusted to equal, in apparent loudness, the specified sound. The adjustment of equality is made either subjectively or objectively as in special sound-level meters.

loudspeaker (*Acous*) An electro-acoustic TRANSDUCER which accepts transmission currents and radiates (by horn or by diaphragm) corresponding sound waves. See ELECTRODYNAMIC LOUDSPEAKER, ELECTROMAGNETIC LOUDSPEAKER, ELECTROSTATIC LOUDSPEAKER, MAGNETOSTRICTION LOUDSPEAKER, PIEZOELECTRIC LOUDSPEAKER.

loudspeaker dividing network (*Acous*) See DIVIDING NETWORK.

loudspeaker microphone (*Acous*) A microphone and dynamic loudspeaker combined; useful for intercommunication systems.

loudspeaker response (*Acous*) The response measured under specified conditions over a frequency range, at a specified direction and distance.

Lou Gehrig's disease (*Med*) See AMYOTROPHIC LATERAL SCLEROSIS.

loughlinite (*Min*) A variety of sepiolite (see MEERSCHAUM) containing sodium.

louping ill (*Vet*) An encephalomyelitis affecting principally sheep and cattle, transmitted by ticks and caused by a virus, and characterized by acute fever and nervous symptoms. Occasionally may cause human infection. Vaccines available.

louvre (*Build*) A system of covering a space (eg window, air vent) with sloping slats of glass, metal, wood, etc, so as to allow ventilation while protecting against rain, wind, etc. The effect is also used architecturally, in indirect lighting, and purely decoratively in furniture design. Also *louver*.

love arrows (*Min*) See FLÈCHES D'AMOUR.

Lovibond comparator (*Chem*) A term applied to a range of instruments used in the estimation of concentrations of chemical compounds in solution by comparison of colour intensities with known standards.

Lovibond tintometer (*Chem*) An instrument for measuring the colours of liquids or solids by transmitted or reflected light and relating the colours to standards. The colour is matched against combinations of glass slides of three basic colours.

low (*EnvSci*) A region of low pressure, or a *depression*.

low-angle plane (*Build*) One with the iron set at 12° level uppermost, for use on end grain.

low-angle X-ray scattering (*Phys*) A method of structure determination for macromolecules, esp biomaterials and polymers. Where long-range order occurs, as in styrene–butadiene–styrene block copolymers, diffraction measurements allow calculation of domain spacing in three-dimensional array using the BRAGG EQUATION.

low aspect ratios (*Aero*) See ASPECT RATIO.

low-band (*ImageTech*) By comparison with HIGH-BAND, the original format specification.

low-carbon steel (*Eng*) Arbitrarily, steel containing from 0·04% to 0·25% of carbon. See panel on STEELS.

low-density polyethylene (*Chem*) Mp approx 115°C. Original polymer from ethylene made at high pressure by a free radical mechanism. Has high ratio of both short- and long-chain branches, so low DEGREE OF CRYSTALLINITY and hence low density (approx 920 kg m^{-3}). Used as insulant in cables etc and in packaging, agricultural and building sheet. Abbrev *LDPE*.

low Earth orbit (*Space, ICT*) An orbit much lower than the geostationary, typically for satellites forming part of a global mobile-telephone system. See panel on COMMUNICATIONS SATELLITE.

löweite (*Min*) A hydrated sulphate of sodium and magnesium, crystallizing in the trigonal system, occurring in salt deposits.

lower bound (*MathSci*) See BOUNDS OF A FUNCTION.

lower case (*Print*) The small letters as distinct from the capitals, the name deriving from the arrangement (now largely superseded) of the small letters in a lower case with the capitals in an *upper case*. Abbrev *lc*.

lower culmination (*Astron*) See CULMINATION.

lower deck (*Ships*) A deck below the weather deck. The term has no legal definition status, and is usually applied to a partial deck which acts simply as a platform and contributes nothing to main longitudinal strength.

lowering wedges (*CivEng*) See STRIKING WEDGES.

lower mean hemispherical candle-power (*Phys*) See MEAN HEMISPHERICAL CANDLE-POWER.

lower-pitch limit (*Acous*) The minimum frequency for a sinusoidal sound wave which produces a sensation of pitch.

lower quartile (*MathSci*) The argument of the cumulative distribution function corresponding to a probability of 0·25; (of a sample) the value below which occur a quarter of the observations in the ordered set of observations.

lower transit (*Astron*) See CULMINATION.

lowest common multiple (*MathSci*) Of a set of numbers, the smallest number which is exactly divisible by all numbers of the set. Abbrev *LCM*.

lowest-observed-adverse-effect level (*BioSci*) Lowest dose of a substance that causes an adverse effect on morphology, functional capacity, growth, development, or lifespan of a target organism under defined conditions of exposure. Abbrev *LOAEL*.

low fidelity (*Acous*) The opposite of HIGH FIDELITY.

low frequency (*ICT*) Vague term widely used to indicate audio frequencies as distinct from radio frequencies, but more correctly implying radio frequencies between 30 and 300 kHz.

low-frequency amplifier (*ICT*) An amplifier for audio-frequency signals.

low-hysteresis steel (*Eng*) Steel with from 2·5% to 4·0% silicon, a high PERMEABILITY and electric resistance, but low loss through HYSTERESIS.

low key (*ImageTech*) Describing a scene containing mostly dark tones with substantial areas of shadow, often used to create a dramatic or sinister mood.

low-level language (*ICT*) Machine-orientated programming language in which each program instruction corresponds to a single MACHINE-CODE INSTRUCTION. See ASSEMBLY LANGUAGE.

low-level modulation (*ICT*) See HIGH-LEVEL MODULATION.

low-level waste (*NucEng*) In general, those radioactive wastes which, because of their low activity, do not require shielding during normal handling or transport.

low-light TV (*Aero*) Sensor system, capable of operating at dawn and dusk, smaller and cheaper than radar.

low-melting-point alloys (*Eng*) See FUSIBLE ALLOYS, WOOD'S METAL.

low-noise block converter (*ImageTech*) A circuit which selects the wanted signal from a satellite transmission. See FEEDHORN.

low-pass filter (*ICT*) One that freely passes signals of all frequencies below a reference value known as the cut-off frequency, f_c.

low-power modulation (*ICT*) See HIGH-LEVEL MODULATION.

low-pressure compressor, stage, turbine (*Aero*) Stages of an AXIAL-FLOW TURBINE. Abbrev *LP compressor* etc.

low-pressure cylinder (*Eng*) The largest cylinder of a multiple-expansion steam engine (eg the third of a triple-expansion engine), in which the steam is finally expanded.

low-red heat (*Eng*) A temperature between 550 and 700°C at which an object so heated emits radiation which makes it appear dull red.

low-resolution graphics (*ICT*) A term generally applied to graphical display units where simple pictures can be built up by plotting relatively large blocks of colour or by using special graphics characters. Cf HIGH-RESOLUTION GRAPHICS.

lowry (*CivEng*) An open form of box-car (US).

low-stop filter (*ICT*) See HIGH-PASS FILTER.

low-temperature carbonization (*ChemEng*) Processing of bituminous coal at temperatures up to 760°C to produce smokeless fuel, gas and by-product oil.

low tension (*ICT*) Term loosely applied to the currents and voltages associated with the filament or heater circuits of a thermionic valve or tube.

low-tension detonator (*ElecEng*) The usual form of detonator, in which a charge is fired by heating a wire by an electric current.

low-tension ignition (*ElecEng*) Electric ignition of the charge in the cylinder of an internal-combustion engine by the interruption of a current-carrying circuit inside the cylinder, no special means being used to produce a high-voltage spark as in high-tension ignition.

low-tension magneto (*ElecEng*) A magneto for producing the current impulses necessary in a low-tension ignition system.

low-velocity scanning (*Electronics*) Scanning of target by electron beam under conditions such that the secondary emission ratio is less than one.

low voltage (*ElecEng*) Legally, any voltage not exceeding 250 volts dc or rms.

low-volt release (*ElecEng*) See UNDER-VOLTAGE RELEASE.

low-water alarm (*Eng*) An arrangement for indicating that the water level in a boiler is dangerously low. See LOW-WATER VALVE.

low-water valve (*Eng*) A boiler safety valve which is opened by a float in the water if the level of the latter falls dangerously low; generally fitted to stationary boilers such as the Lancashire.

low-wing monoplane (*Aero*) A monoplane wherein the main planes are mounted at or near the bottom of the fuselage.

lox (*Chem*) Abbrev for LIQUID OXYGEN.

LP compressor, stage, turbine (*Aero*) Abbrev for LOW-PRESSURE COMPRESSOR, STAGE, TURBINE.

LPE (*Electronics*) Abbrev for LIQUID-PHASE EPITAXY.

LPG (*Chem*) Abbrev for LIQUEFIED PETROLEUM GASES such as propane and butane, used for fuels.

LPS (*BioSci*) Abbrev for LIPOPOLYSACCHARIDE.

LPT (*ICT*) Abbrev for LINE PRINTER.

LPT port (*ICT*) A PORT used to connect the computer to many kinds of printer. Also *parallel port*.

L-rest (*Eng*) A lathe rest used in hand turning, shaped like an inverted L.

L-ring (*ElecEng*) A method of tuning a magnetron.

LRP (*Genrl*) Abbrev for *lead replacement petrol*.

LSB (*ICT*) Abbrev for LEAST SIGNIFICANT BIT.

l–s coupling (*Phys*) See RUSSELL–SAUNDERS COUPLING.

LSD (*Pharmacol*) See LYSERGIC ACID DIETHYLAMIDE.

L section (*ICT*) Section or half-section of a filter, having shunt and series arms.

L-shell (*Phys*) The ELECTRON SHELL in an atom corresponding to a principal quantum number of two. The shell can contain up to eight electrons. See panel on ATOMIC STRUCTURE.

LSI (*ICT*) Abbrev for LARGE-SCALE INTEGRATION.

LTC (*ImageTech*) Abbrev for *linear time code*. See TIME CODE.

LT, LTO (*Chem*) Abbrevs for *lithium tantalate*. See DIELECTRIC AND FERROELECTRIC MATERIALS.

LTP (*BioSci*) Abbrev for LONG-TERM POTENTIATION.

Lu (*Chem*) Symbol for LUTETIUM.

lubricants (*Eng*) Compounds (solid, plastic or liquid) entrained between two sliding surfaces. 'Wetting' lubricants adhere strongly to one or both such surfaces while leaving the intervening film fairly non-viscous. Solids include graphite, molybdenite, talc. Plastics include fatty acids and soaps, sulphur-treated bitumen and residues from petroleum distillation. Liquids include oils from animal, vegetable or mineral sources. See BOUNDARY LUBRICATION, FLUID LUBRICATION.

lubricity (*Aero*) A property of liquid fuels that also lubricate the feed pumps in the fuel system. Lubricity can vary markedly for different types of aviation jet fuel, and affects the wear and performance of components.

Lucas numbers (*MathSci*) The sequence of integers, starting with 2 and 1 (in that order) and that continues 3,4,7,11,18,29... each of which is the sum of the preceding two. Cf FIBONACCI NUMBERS.

Lucas theory (*Chem*) A theory that substituent groups attract or repel electrons from neighbouring groups and thus alter their reactivity, eg in benzene chlorine attracts electrons and deactivates the benzene ring, while hydroxyl and amine lend electrons and activate.

lucid dreaming (*Psych*) A dreaming state in which the dreamer feels awake and in conscious control over dream events.

luciferase (*BioSci*) An oxidizing enzyme that occurs in the luminous organs of certain animals (eg firefly) and acts on luciferin to produce light.

luciferin (*BioSci*) A protein-like substance that occurs in the luminous organs of certain animals and is oxidized by the action of LUCIFERASE.

Lucite (*Chem*) TN for POLYMETHYL METHACRYLATE (US).

Lüders bands (*Eng*) Blemishes which appear as bands across the surface of a low-carbon steel during the initial yielding stage of plastic deformation. The bands divide material which has yielded from that which has not and move across the surface while straining is continued, eventually disappearing when all the cross-section has yielded and the bulk material is undergoing plastic deformation. See YIELD POINT. Also *Lüders lines*.

Ludhamian (*Geol*) An early stage of the Pleistocene. See QUATERNARY.

Ludlow (*Geol*) The youngest series of the Silurian period. See PALAEOZOIC.

Ludlow (*Print*) A machine which produces slugs from handset matrices, used mainly for headings and display lines for newspapers etc.

Ludwig's angina (*Med*) A term applied to infection of the mouth spreading along the parapharangeal spaces to produce a hard swelling of the neck, with difficulty in swallowing and fever; usually caused by a β-haemolytic streptococcal infection.

lues (*Med*) A plague or pestilence. The term is now synonymous with SYPHILIS. Adj *luetic*.

luffer-boarding (*Build*) Sloping slats arranged as in a LOUVRE.

luffing jib crane (*Eng*) A common form of JIB CRANE, in which the jib is hinged at its lower end to the crane structure, so altering its radius of action. See DERRICKING JIB CRANE, LEVEL-LUFFING CRANE.

lug (*ElecEng*) On an accumulator plate, a projection to which the electrical connection is made. See TERMINAL LUG.

lug (*Eng*) See EAR.

lug sill (*Build*) A sill which is of greater length than the distance between the jambs of the opening, so that its ends have to be built into the wall.

lumbago (*Med*) A term applied to pain of the muscles and ligaments in the lumbar region or lower part of the back.

lumbar (*BioSci*) Pertaining to, or situated near, the lower or posterior part of the back; as the *lumbar vertebrae*.

lumbar puncture (*Med*) The process of inserting a needle into the lumbar subarachnoid space to obtain a specimen of cerebrospinal fluid.

lumber (*For*) The term employed in N America for sawn wood of all descriptions. See TIMBER.

lumberjack (*For*) US for LOGGER.

lumen (*BioSci*) The central cavity of a duct, sac, tubular organ or organelle.

lumen (*Phys*) Unit of LUMINOUS FLUX, being the amount of light emitted in unit solid angle by a small source of output 1 CANDELA. In other words, the lumen is the amount of light which falls on unit area when the surface area is at unit distance from a source of 1 candela. Symbol lm.

lumen-hour (*Phys*) Quantity of light emitted by a 1 lumen lamp operating for 1 hour.

lumenmeter (*Phys*) An integrating photometer in which the total luminous flux is integrated in a matt-white diffusing enclosure and measured through an opening.

luminaire (*ElecEng*) The British Standard term for any sort of electric-light fitting, whether for a tungsten or a fluorescent lamp.

luminaire (*ImageTech*) The eneral term for a studio light source with its mounting and controls.

luminance (*Phys*) The measure of brightness of a surface, eg candela per square metre of the surface radiating normally. Symbol *L*.

luminance channel (*ImageTech*) In a colour TV system, any circuit path intended to carry the LUMINANCE SIGNAL.

luminance flicker (*ImageTech*) Perceptible rapid periodic variations in brightness, more obvious at higher levels of picture brightness; the eye is much less sensitive to chrominance flicker.

luminance signal (*ImageTech*) Signal controlling the luminance of a colour TV picture, its point-by-point image brightness, as distinct from its CHROMINANCE; known as the *Y-signal*. Sometimes abbreviated to *luma* by analogy with *chroma*.

luminescence (*Phys*) The emission of light (other than from thermal energy causes) such as BIOLUMINESCENCE. Thermal luminescence in excess of that arising from temperature occurs in certain minerals. See FLUORESCENCE, PHOSPHORESCENCE.

luminescent centres (*Chem*) Activator atoms, excited by free electrons in a crystal lattice and giving rise to ELECTROLUMINESCENCE.

luminophore (*Chem*) (1) A substance which emits light at room temperature. (2) A group of atoms which can make a compound luminescent.

luminosity (*Astron*) The intrinsic or absolute amount of energy radiated per second from a celestial object. Its units in astronomy are usually absolute MAGNITUDE, rather than watts, largely for historical reasons.

luminosity (*Phys*) The visual perception of the brightness of an area. The density of luminous intensity in a particular direction.

luminosity coefficients (*Phys*) Multipliers of the TRISTIMULUS VALUES of a colour, so that the total represents the luminance of the colour according to its subjective assessment by the eye.

luminosity curve (*Phys*) A curve which gives the relative effectiveness of perception or of sensitivity of vision in terms of wavelength. See PHOTOPIC LUMINOSITY CURVE, SCOTOPIC LUMINOSITY CURVE.

luminosity factor (*Phys*) The ratio of the total luminous flux emitted by a light source at a given wavelength to the total energy emitted.

luminous efficiency (*Phys*) The ratio of luminous flux of lamp to total radiated energy flux; usually expressed in lumens per watt for electric lamps. Not to be confused with OVERALL LUMINOUS EFFICIENCY.

luminous flame (*Chem*) One containing glowing particles of solid carbon due to incomplete combustion.

luminous flux (*Phys*) The flux emitted within unit solid angle of 1 steradian by a point source of uniform intensity of 1 CANDELA. Unit of measurement is the LUMEN.

luminous flux density (*Phys*) The quantity of luminous flux, passing through a normal unit area, weighted according to an internationally accepted scale of differential visual sensitivity.

luminous gas flame (*Eng*) Thermal breakdown of the hydrocarbons into carbon and hydrogen arising from lack of primary combustion air, the carbon being heated to incandescence.

luminous intensity (*Phys*) See CANDELA.

luminous paints (*Build, Chem*) Paints which glow in the dark. Based on the salts, eg sulphide, silicate, phosphate, of eg zinc, cadmium, calcium, with traces of heavy metals, eg manganese. Phosphorescent paints glow for longer or shorter periods after exposure to light. Fluorescent paints glow continuously under the action of radioactive additives.

luminous sensitivity (*Phys*) In a photoconductive cell, the ratio of output current to incident luminous flux, at a constant electrode voltage.

Lummer–Brodhun photometer (*Phys*) A form of contrast photometer in which, by an arrangement of prisms, the surfaces illuminated respectively by the standard source and the source under test are next to one another, thus enabling an easy comparison to be made.

Lummer–Gehrcke interferometer (*Phys*) A very accurately worked, plane, parallel-sided glass plate, so arranged that light is internally reflected in a zigzag path through the plate. The rays which emerge into the air at each reflection are in a condition to produce INTERFERENCE FRINGES of a very high order. The instrument is used for studying the fine structure of spectral lines.

lump (*Textiles*) A piece of cloth taken straight from the loom; usually larger than normal.

lumped constant (*Electronics, Phys*) A single constant which is electrically equivalent to the total of that type of distributed constant existing in a coil or circuit. Also *lumped parameter.*

lumped impedance (*Electronics*) An impedance concentrated in a single component rather than being distributed throughout the length of a transmission line.

lumped loading (*ICT*) See LINE EQUALIZER.

lumped parameter (*Phys*) See LUMPED CONSTANT.

lump lime (*Build*) The CAUSTIC LIME produced by burning limestone in a kiln.

lumpy jaw (*Vet*) See ACTINOMYCOSIS.

lumpy skin disease (*Vet*) An acute virus disease of cattle in Africa, characterized by inflammatory nodules in the skin and generalized lymphadenitis.

lumpy withers (*Vet*) Small swellings on the withers of the horse caused by irritation of the subcutaneous bursae by the collar.

lumpy wool (*Vet*) See MYCOTIC DERMATITIS OF SHEEP.

Luna programme (*Space*) A highly successful evolutionary series of Soviet lunar missions carried out between 1959 and 1976. Included the first spacecraft to acquire pictures of the far side of the Moon.

lunar (*BioSci*) In tetrapods, the middle member of the three proximal CARPALS in the fore-limb, corresponding to part of the ASTRAGALUS in the hind-limb.

lunar bows (*EnvSci*) Bows of a similar nature to *rainbows* but produced by moonlight.

lunar maria (*Astron*) The Latin designation of the so-called 'seas' on the lunar surface, named before the modern telescope showed their dark areas to be dry plains filled with mafic volcanic rocks. Since 1959 spacecraft probes and landings (manned and unmanned) have provided much detailed information but their origin is still problematical. See MASCON, MOON.

lunar module (*Space*) Lander developed for the Apollo programme to carry astronauts from the main spacecraft in orbit of the Moon to the lunar surface.

lunar month (*Astron*) See SYNODIC MONTH.

Lunar Orbiter programme (*Space*) A series of NASA spacecraft (1966–7) used to survey the Moon at high resolution from a lunar orbit.

lunate (*Genrl*) Crescent-shaped; shaped like the new Moon. Also *lunulate.*

lunation (*Astron*) See SYNODIC MONTH.

lune (*MathSci*) The portion of the surface of a sphere intercepted by two great circles. The portion of a circular area that remains when a part is removed by an intersecting circle.

lunette (*Arch*) (1) A semicircular window or pediment over a doorway. (2) A small arched opening in the curved side of a vault.

lung (*BioSci*) The respiratory organ in air-breathing vertebrates. The lungs arise as a diverticulum from the ventral side of the pharynx; they consist of two vascular sacs filled with constantly renewed air.

lung book (*BioSci*) An organ of respiration in some Arachnida (scorpions, spiders), consisting of an air-filled cavity opening on the ventral surface of the body; it contains a large number of thin vascular lamellae, arranged like the leaves of a book. Also *book lung.*

Lunge nitrometer (*Chem*) Apparatus devised for the determination of oxides of nitrogen by absorption in a gas burette. May be used for other analytical processes which involve the measurement of a gas.

lung plague (*Vet*) See CONTAGIOUS BOVINE PLEUROPNEUMONIA.

lungworm disease (*Vet*) See HUSK.

lunitidal interval (*Genrl*) The time interval between the Moon's transit and the next high water at a given place.

lunula (*BioSci*) See LUNULE.

lunulate (*Genrl*) See LUNATE.

lunule (*BioSci*) A crescentic mark: specifically, the white crescent lying at the root of the nail. Also *lunula*. Adj *lunular*.

lupinosis (*Vet*) Poisoning of sheep, goats, cattle and horses, by plants of the genus *Lupinus*.

Lupus (Wolf) (*Astron*) A small southern constellation containing many bright stars.

lupus erythematosus (*Med*) A connective-tissue disease thought to be due to an auto-immune reaction. Occurs in two main forms: discoid, where the skin of the nose and cheeks is raised and reddened giving a butterfly appearance; systemic, where tissue throughout the body is involved, causing kidney disease, painful joints, anaemia and mental abnormality. See panel on AUTO-IMMUNITY.

lupus erythematosus cell (*BioSci*) A neutrophil leucocyte that contains in its cytoplasm homogeneous masses of phagocytosed nuclear material derived from other dead neutrophils. This occurs when antinuclear antibodies are present in the blood which opsonize the released nucleoproteins; their presence is diagnostic of systemic LUPUS ERYTHEMATOSUS. Abbrev *LE cell*.

lupus vulgaris (*Med*) Tuberculosis of the skin particularly affecting the face.

lustre (*Build*) The degree of sheen on a surface or coating.

lustre (*Min*) The degree of sheen of a mineral; depends upon the quality and amount of light that is reflected from its surface. The highest degree of lustre in opaque minerals is *splendent*, the comparable term for transparent minerals being *adamantine* (ie the lustre of diamond). *Metallic* and *vitreous* indicate less brilliant lustre, while *silky*, *pearly*, *resinous* and *dull* are self-explanatory terms covering other degrees of lustre.

lute (*Build*) A straightedge for levelling off clay in a brick mould by removing the excess.

lute (*Eng*) Fireclay mixture used to seal cracks in eg a crucible.

luteal (*BioSci*) Pertaining to, or resembling, the CORPUS LUTEUM.

lutein cells (*BioSci*) Yellowish-coloured cells occurring in the CORPUS LUTEUM of the ovary and containing fat-soluble substances.

luteinizing hormone (*Med*) A gonadotrophic mucoprotein hormone, secreted by the basophil cells of the anterior lobe of the pituitary gland, which causes growth of the corpus luteum of the ovary and also stimulates activity of the interstitial cells of the testis. Also *interstitial cell stimulating hormone*. Abbrev *LH*.

luteinoma (*Med*) A tumour occurring in the ovary, composed of cells resembling those of the CORPUS LUTEUM.

luteotrophic hormone (*Med*) See PROLACTIN.

Lutetian (*Geol*) A stage in the Eocene. See TERTIARY.

lutetium (*Chem*) Symbol Lu, at no 71, ram 174·97. A metallic element, a member of the rare earth group. It occurs in monazite and xenotime. Formerly called *cassiopeium*.

luthern (*Arch*) A vertical window set in a roof.

lutidines (*Chem*) Dimethyl-pyridines, which all occur in bone-oil and in coaltar; general formula $C_5(CH_3)_2H_3N$.

lutite (*Geol*) A consolidated rock composed of silt and/or clay, eg MUDSTONE, SHALE.

lux (*Phys*) Unit of illuminance or illumination in SI system, 1 lumen m^{-2}. Symbol lx.

luxation (*Med*) Dislocation.

Luxemburg effect (*ICT*) Cross-modulation of radio transmissions during ionospheric propagation. Caused by non-linearity in motion of the ions.

Lw (*Chem*) Symbol for LAWRENCIUM.

LWR (*NucEng*) Abbrev for LIGHT-WATER REACTOR.

lych gate (*Arch*) A roofed gateway entrance to a churchyard. Also *lich gate*.

lycopene (*BioSci*) Red carotenoid pigment found in tomatoes and other fruit; has antioxidant properties and is extracted and used as a dietary supplement.

Lycopsida (*BioSci*) The Lycopods, a class of Pteridophyta dating from the Early Devonian onwards. The sporophyte usually has roots, stems and leaves (MICROPHYLLS), is protostelic and has lateral sporangia. Includes the Proto-lepidodendrales, Lycopodiales, Lepidodendrales, Isoetales, Selaginellales.

Lycra (*Textiles*) TN for first SPANDEX fibre to be introduced into the market (US).

lyddite (*Chem*) See PICRIC ACID.

Lydian stone (*Min*) Black flinty jasper (*touchstone*), also other silicified fine-grained rocks. The name touchstone has reference to the use of lydite as a streak plate for gold: the colour left after rubbing the metal across it indicates to the experienced eye the amount of alloy. Also *lydite*.

lye (*Chem*) Strong solution of sodium or potassium hydroxide.

lying panel (*Build*) A door panel whose width is greater than its height.

lying press (*Print*) A small portable screw press in which books are held firmly during various operations. Also *laying press*.

Lyman series (*Phys*) One of the hydrogen series occurring in the extreme ultraviolet region of the spectrum. The series may be represented by the formula

$$v = N\left(\frac{1}{1^2} - \frac{1}{n^2}\right)$$

where $n = 2,3,4,\ldots$ (see BALMER SERIES), the series limit being at wavenumber $N = 109\,678$, which corresponds to wavelength 91·26 nm. The leading line, called *Lyman alpha*, has a wavelength of 121·57 nm, and is important in upper atmosphere research, as it is emitted strongly by the Sun.

Lyme disease (*Med*) An acute, relapsing inflammatory disease that is caused by the spirochaete *Borrelia bergdorfi*, transmitted to humans by the bite of ticks.

lymph (*BioSci*) A colourless circulating fluid occurring in the lymphatic vessels of vertebrates and closely resembling blood plasma in composition. Lymphocytes circulate within the LYMPHATIC SYSTEM. Adj *lymphatic*.

lymph-, lympho- (*Genrl*) Prefixes from Lt *lympha*, water.

lymphadenitis (*Med*) Inflammation of lymph glands.

lymphadenoid (*Med*) Resembling the structure of a lymphatic gland.

lymphadenopathy (*Med*) Swelling of the lymph nodes due to immune reactions. Persistent swelling may indicate non-resolving infection or the development of a lymph node tumour/lymphoma.

lymphangiectasis (*Med*) Dilatation and distension of lymphatic vessels, due usually to obstruction.

lymphangiology (*Med*) The anatomy of the lymphatics.

lymphangioma (*Med*) A nodular tumour consisting of lymphatic channels.

lymphangitis (*Med, Vet*) Inflammation of a lymphatic vessel or vessels. See EPIZOOTIC LYMPHANGITIS, ULCERA-TIVE LYMPHANGITIS.

lymphatic leukaemia (*Med*) See LEUKAEMIA.

lymphatic system (*BioSci*) In vertebrates, a system of vessels pervading the body, in which the lymph circulates; lymphatic vessels eventually drain into either the thoracic or right lymphatic duct and thence into the subclavian veins; LYMPH GLANDS and LYMPH HEARTS are found on its course.

lymph gland (*BioSci*) An aggregation of reticular connective tissue, crowded with lymphocytes, surrounded with a fibrous capsule, and provided with afferent and efferent lymph vessels.

lymph heart (*BioSci*) A contractile portion of a lymph vessel, which assists the circulation of the lymph and forces the lymph back into the veins.

lymph node (*BioSci*) Solid organ of the immune system that acts as a filter for antigens carried in the lymph from sites of tissue damage. Organized into T-lymphocyte-rich areas and B-lymphocyte follicles, it is within lymph nodes that lymphocytes are exposed to antigen and those that can respond are activated and proliferate.

lympho- (*Genrl*) See LYMPH-.

lymphoblast (*BioSci*) T- or B-lymphocyte that has been stimulated into proliferation by antigen exposure. Often referred to simply as a blast cell. Uncontrolled proliferation results in lymphoblastic leukaemia.

lymphocyte (*BioSci*) Spherical cell with a large round nucleus (often slightly indented) and very scanty cytoplasm. Diameter varies between seven and 12 μm. They are actively mobile. The term is essentially morphological and is used to refer to cells responsible for development of specific immunity, B-lymphocytes being associated with humoral and T-lymphocytes with cellular immunity. Lymphocytes are the predominant constituents of LYMPHOID TISSUES. A normal adult human has about 2×10^{12} lymphocytes. See panel on IMMUNE RESPONSE.

lymphocyte activation (*BioSci*) The change in morphology and behaviour of lymphocytes exposed to a mitogen or to an antigen to which they have been primed. The resulting cells are LYMPHOBLASTS.

lymphocyte function associated molecules (*BioSci*) Leucocyte cell-surface integrins important in adhesion; if defective (*leucocyte adhesion defect*) the consequences can be quite severe. An example is LFA-1 that binds ICAM-1.

lymphocytopenia (*Med*) Diminution, below normal, of the number of lymphocytes in the blood.

lymphocytosis (*Med*) Increase in the number of lymphocytes in the blood, as in certain infections.

lymphogenous (*BioSci*) Lymph-producing.

lymphogranuloma inguinale (*Med*) A virus venereal infection, characterized by enlargement of the glands in the groin; common in the tropics. Also *poradenitis venerea*.

lymphoid tissue (*BioSci*) Tissue that is particularly rich in lymphocytes (and accessory cells such as macrophages and reticular cells), particularly the lymph nodes, spleen, thymus, Peyer's patches, pharyngeal tonsils, adenoids, and (in birds) the Bursa of Fabricius.

lymphokine (*BioSci*) Generic name for proteins (other than antibodies or surface receptors) that are released by lymphocytes, stimulated by antigens or by other means, which act on other cells involved in the immune response. The term includes CYTOKINES, INTERFERONS, INTERLEUKINS.

lymphokine-activated killer cells (*BioSci*) Cytotoxic lymphocytes activated by CYTOKINES, particularly INTERLEUKIN 2, that acquire the ability to kill tumour but not normal cells. See KILLER CELLS.

lymphoma (*Med*) A disease of lymphoid tissue, sometimes involving bone marrow, in which the neoplastic cells originate from lymphocytes or mononuclear phagocytes. It includes Hodgkin's disease, reticulosarcoma, giant follicular lymphoma, lymphatic leukaemia and Burkitt's lymphoma. Lymphomas are clinically divided by histological appearance into two main groups, Hodgkin's and non-Hodgkin's lymphoma, with the former carrying the more favourable prognosis.

lymphosarcoma (*Med*) A term used to describe one of the lymphomas with an intermediate grade of malignancy.

lymphotoxin (*BioSci*) Actually several proinflammatory cytokines. Lymphotoxin-α is identical to TUMOUR NECROSIS FACTOR-$\beta(TNF\beta)$ but membrane-bound and other soluble forms also exist.

lymph sinuses (*BioSci*) Part of the LYMPHATIC SYSTEM in some lower vertebrates, consisting of spaces surrounding the blood vessels and communicating with the coelom by means of small apertures.

lymph vessels (*BioSci*) See LYMPHATIC SYSTEM.

Lynx (*Astron*) A faint northern constellation.

lyocytosis (*BioSci*) Histolysis by the action of enzymes secreted outside the tissue, as in insect metamorphosis.

lyolysis (*Chem*) The formation of an acid and a base from a salt by interaction with the solvent; ie the chemical reaction which opposes neutralization.

Lyon hypothesis (*BioSci*) Hypothesis, first advanced by Lyon, concerning the random inactivation of one of the two X chromosomes of the cells of female mammals. In consequence females are chimaeric for the products of the X chromosomes.

lyophilic colloid (*Chem*) A colloid (*lyophilic*, solvent-loving) which is readily dispersed in a suitable medium, and may be redispersed after coagulation.

lyophilization (*BioSci*) Freeze drying, removal of water by sublimation under vacuum.

lyophobic colloid (*Chem*) A colloid (*lyophobic*, solvent-hating) which is dispersed only with difficulty, yielding an unstable solution which cannot be reformed after coagulation.

lyosorption (*Chem*) The adsorption of a liquid on a solid surface, esp of solvent on suspended particles.

Lyot filter (*Astron*) See POLARIZING MONOCHROMATOR.

lyotropic material (*Phys*) A material for which the concentration determines the phase.

lyotropic series (*Chem*) Ions, radicals or salts arranged in order of magnitude of their influence on various colloidal, physiological and catalytic phenomena, an influence exerted by them as a result of the interaction of ions with the solvent. Cf HOFMEISTER SERIES.

lyra (*BioSci*) Any lyre-shaped structure, as the lyre pattern on a bone.

Lyra (Harp) (*Astron*) A small northern constellation which includes the star VEGA, the prototype of RR LYRAE VARIABLE stars and the RING NEBULA.

lyrate (*BioSci*) Shaped like a lyre. A pinnately lobed leaf with a large terminal and small lateral lobes.

Lyrids (*Astron*) One of two minor meteor showers: the April Lyrids show maximum activity on 21 April and were much more active in the past; the June Lyrids show maximum activity on 16 June.

lyriform organs (*BioSci*) Patches, consisting of well-innervated ridges of chitin, on the legs, palpi, chelicerae, and body of various ARACHNIDA; mechanoreceptive in function.

Lys (*Chem*) Symbol for LYSINE.

lyse (*Chem, BioSci*) To cause to undergo LYSIS.

lysergic acid diethylamide (*Med*) A compound which, when taken in minute quantities, produces hallucinations and thought processes outside the normal range. The results sometimes resemble schizophrenia. Also *lysergide*. Abbrev *LSD*.

Lysholm grid (*Radiol*) A type of grid interposed between the patient and film in diagnostic radiography in order to minimize the effect of scattered radiation.

Lysholm–Smith torque converter (*Eng*) A variable-ratio hydraulic gear of the Föttinger type, but in which multistage turbine blading gives high efficiency of transmission over a wide range: used in road and rail vehicles.

lysigenic (*BioSci*) Said of a space formed by the breakdown and dissolution of cells. Also, *lysigenous, lysogenous*. Cf REXIGENOUS, SCHIZOGENOUS.

lysine (*Chem*) *2,6-diaminohexanoic acid*. $H_2N(CH_2)_4CH (NH_2).OOH$. A 'basic' amino acid, as it contains two amino groups. The L- or S-isomer is a constituent of proteins. Symbol Lys, short form K.

lysis (*BioSci*) Decomposition or splitting of cells or molecules, eg hydrolysis. Destruction of cells or tissues by various pathological processes, eg autolysis, necrosis.

Lysithea (*Astron*) The tenth natural satellite of Jupiter, discovered in 1938. Distance from the planet 11 720 000 km; diameter 40 km.

lysogeny (*BioSci*) That part of the life cycle of a temperate phage in which it replicates in synchrony with its host. Cf LYTIC CYCLE.

Lysol (*Chem*) TN for a solution of cresols in soft soap. It is a well-known disinfectant.

lysosomal diseases (*Med*) Diseases, also known as *storage diseases*, in which a deficiency of a particular lysosomal enzyme leads to accumulation of the undigested substrate for that enzyme within cells. Not immediately fatal, but within a few years lead to serious neurological and skeletal disorders and eventually to death. Examples are the following diseases or syndromes: *Hurler, Hunter, San Fillipo, Gaucher's, Niemann–Pick, Pompe's, Tay–Sachs*.

lysosome (*BioSci*) A vesicular cytoplasmic organelle containing hydrolytic enzymes that degrade those cellular constituents which become incorporated into the vesicle.

lysozyme (*BioSci*) Enzyme present in egg-white, tears, nasal secretions, on the skin, and in monocytes and granulocytes. It lyses certain bacteria, chiefly Gram-positive cocci, by splitting the muramic acid-β(1-4)-N-acetylglucosamine linkage in their cell walls. The first *enzyme* to have its three-dimensional structure determined.

lyssa (*BioSci*) See LYTTA.

lytic cycle (*BioSci*) That part of the life cycle of a temperate bacteriophage in which it causes LYTIC INFECTION.

lytic infection (*BioSci*) Infection of a bacterium by a bacteriophage that replicates uncontrollably, destroying its host and eventually releasing many copies into the medium.

lytta (*BioSci*) In CARNIVORA, a rod of cartilage or fibrous tissue embedded in the mass of the tongue. Also *lyssa*.

M

M (*Chem*) General symbol for a metal or an electropositive radical. Abbrev for *mega-*, ie 10^6.

M (*ElecEng*) Symbol for MUTUAL INDUCTANCE.

M (*Genrl*) Symbol for MEGA-.

M (*Geol*) Symbol for mafic and related minerals.

M (*Chem*) Symbol for RELATIVE MOLECULAR MASS.

M (*Eng*) Symbol for BENDING MOMENT.

M (*Phys*) Symbol for: (1) absolute MAGNITUDE; (2) luminous emittance; (3) MAGNETIZATION per unit volume; (4) moment of force; (5) mutual inductance.

M_{CRIT} (*Aero*) See CRITICAL MACH NUMBER.

M_{ne} (*Aero*) Abbrev for the maximum permissible indicated MACH NUMBER: a safety limitation, the suffix means 'never exceed' because of strength or handling considerations. The symbol is used mainly in operational instructions.

M_{no} (*Aero*) Abbrev for *normal operating Mach number*, usually of a jet airliner, the term being used mainly in flight operations instructions for flight levels above 7600 m.

m (*Genrl*) Symbol for: (1) METRE; (2) MILLI-.

m (*Phys*) Symbol for: (1) ELECTROMAGNETIC MOMENT; (2) mass; (3) MOLALITY.

m- (*Chem*) Abbrev for: (1) META-; (2) MESO-.

μ (*Chem*) Symbol for CHEMICAL POTENTIAL.

μ (*Electronics*) Symbol for AMPLIFICATION FACTOR.

μ (*Genrl*) Symbol for: (1) MICRO-; (2) MICRON (obsolete, replaced by μm).

μ (*Phys*) Symbol for: (1) coefficient of FRICTION; (2) dipole moment; (3) MOBILITY; (4) PERMEABILITY; (5) REFRACTIVE INDEX.

μ- (*Chem*) Symbol signifying: (1) MESO-; (2) a bridging ligand.

M31 (*Astron*) See ANDROMEDA GALAXY.

M & E (*ImageTech*) Abbrev for MUSIC AND EFFECTS.

Ma (*Geol*) Symbol for a million years. Also *my*.

maar (*Geol*) The explosion vent of a volcano.

Maastrichtian (*Geol*) The highest stage in the Cretaceous. See MESOZOIC.

MAC (*ImageTech*) Abbrev for *multiplexed analogue components*. A 625-line TV system in which the COLOUR DIFFERENCE SIGNALS and LUMINANCE SIGNAL are compressed and sent in sequence, preceded by digital audio and data, thus obviating mutual interference. Variations in the number of audio channels and bandwidth are denoted by a prefix, eg D-MAC, D2-MAC. The system is designed for 4:3 and 16:9 wide-screen use. HD-MAC is a high-definition version.

macadamized road (*CivEng*) A road whose surface is formed with broken stones of fairly uniform size rolled into a 15–25 cm (6–10 in) layer, with gravel to fill the interstices. See TARMACADAM.

Macardle's disease (*Med*) An inborn error of metabolism leading to abnormal glycogen storage in the muscle and causing muscle pain and stiffness.

maceral (*Geol*) The organic material which comprises coal. It includes EXINITE, INERTINITE and VITRINITE.

maceration (*BioSci*) The process of soaking a specimen in a reagent in order to destroy some parts of it and to isolate other parts.

maceration (*FoodSci*) Grinding a raw material or mixture into a paste or slurry. A form of COMMINUTION.

Mach angle (*Aero*) In supersonic flow, the angle between the SHOCK WAVE and the airflow, or line of flight, of the body. The cosecant of this angle is equal to the Mach number.

Mach cone (*Aero*) In supersonic flow, the conical SHOCK WAVE formed by the nose of a body, whether stationary in a WIND TUNNEL or in free flight through the air.

machicolation (*Arch*) A projecting gallery on a fortification which had a pierced floor through which boiling oil was poured on the enemy.

machinability (*Eng*) The ease with which a metal or alloy can be machined.

machine (*Phys*) (1) A device for overcoming a resistance at one point by the application of a force at some other point. Typical simple machines are the inclined plane, the lever, the pulley and the screw. (2) Generally, any instrument for the conversion of motion. See MECHANICAL ADVANTAGE, VELOCITY RATIO.

machine architecture (*ICT*) See COMPUTER ARCHITECTURE.

machine code (*ICT*) Programming language in which each MACHINE-CODE INSTRUCTION can be recognized and executed by the computer without any intermediate translation.

machine-code instruction (*ICT*) A WORD that contains several codes: the operation code that identifies a particular, elementary operation, and others that are address codes for the data and the result. The instruction is linked to a series of hardware operations either directly or through MICROPROGRAMS. See ADDRESS CALCULATION, INSTRUCTION CYCLE, INSTRUCTION SET.

machine direction (*Paper*) The direction in the plane of a sheet or web of paper corresponding with the direction of travel of the paper machine. The direction at right angles to this is the *cross-direction*.

machine finished (*Paper*) Paper on which the requisite degree of surface finish has been attained by use of a CALENDER or calenders forming part of the paper-making machine.

machine glazed (*Paper*) Paper or board with a characteristic high finish on one side, produced by causing the damp web to adhere to the surface of a large-diameter, highly polished, steam-heated drying cylinder (MG or Yankee dryer). As water is evaporated the paper surface assumes the polish of the cylinder.

machine head (*Genrl*) The part of a stringed musical instrument containing the tuning pegs.

machine instruction (*ICT*) See MACHINE-CODE INSTRUCTION.

machine language (*ICT*) See MACHINE CODE.

machine moulding (*Eng*) The process of making moulds and cores by mechanical means, as eg by replacing hand-ramming by power-squeezing of the sand or by jolting on a vibrating table.

machine proof (*Print*) A proof taken on a printing machine as distinct from a proofing press, and implying a high standard in the result.

machine revise (*Print*) A proof taken on the printing machine and submitted for a final check before printing.

machine riveting (*Eng*) Clenching rivets by the use of mechanical or compressed-air hammers or hydraulic riveters. See HYDRAULIC RIVETER, PNEUMATIC RIVETER.

machine room (*Print*) That department of a printing establishment where the actual process of printing (by machine) is carried out.

machine tools (*Eng*) Any power-driven, non-portable machines designed primarily for shaping and sizing (metal) parts. See DRILLING MACHINE, KEY-WAY TOOL, LATHE, MILLING MACHINE, PLANING MACHINE, SHAPING MACHINE.

machine translation (*ICT*) The automatic production of text in one natural language from that in another.

machine-washable (*Textiles*) A textile article that may be washed in a domestic washing machine in aqueous detergent solutions at controlled temperatures suitable for the particular article. The conditions may be specified on a *care label* attached to the article.

machine wire (*Paper*) Originally a woven mesh of phosphor bronze for use on a FOURDRINIER or cylinder mould machine for forming the sheet of paper. Stainless steel is now used for machine wires and the term is retained for plastics cloths serving the same purpose.

machine word (*ICT*) See WORD.

machining allowance (*Eng*) The material provided beyond the finished contours on a casting, forging or roughly prepared component, which is subsequently removed in machining to size.

machmeter (*Aero*) A pilot's instrument for measuring FLIGHT MACH NUMBER.

Mach number (*Aero*) The ratio of the speed of a body, or of the flow of a fluid, to the speed of sound in the same medium. At *Mach* 1, speed is *sonic*; below *Mach* 1, it is *subsonic*; above *Mach* 1, it is *supersonic*, creating a *Mach* (or *shock*) *wave*. Hypersonic conditions in air are reached at *Mach* numbers exceeding 5. See SPEED OF SOUND.

MACHO (*Astron*) Abbrev for *massive astrophysical compact halo object*. Such objects, including large planets and brown dwarfs, are proposed as a possible form of the dark matter which is thought to make up more than 90% of the universe.

Mach principle (*Genrl*) The principle that scientific laws are descriptions of nature, are based on observation and alone can provide deductions which can be tested by experiment and/or observation.

mackerel sky (*EnvSci*) Cirrocumulus or altocumulus cloud arranged in regular patterns suggesting the markings on mackerel.

Mackie line (*ImageTech*) A boundary effect locally increasing the difference between areas of high and low image density, used to advantage in HIGH-DEFINITION developers.

mackle (*Print*) A defective printed sheet, having a blurred appearance due to incorrect impression. Adj *mackled*.

Maclaurin's series (*MathSci*) Modification of TAYLOR'S SERIES, putting $a = 0$.

macle (*Min*) Fr term for a TWINNED CRYSTAL (panel); in the diamond industry, more commonly used than *twin*, esp for twinned octahedra.

MacLeod's equation (*Chem*) The surface tension of a liquid is given as the difference between the densities of the liquid and its vapour raised to the fourth power, multiplied by a constant.

Macnaughten rules (*Med*) Rules formulated by the House of Lords where a defence of insanity must show the accused was labouring under such a defect of reason from disease of the mind, as not to know the nature or quality of the act he was doing; or if he did know it, he did not know that what he was doing was wrong.

Maco template (*Build*) Adjustable template consisting of a clamp which holds a stack of thin brass strips. When the strips are pressed edgewise against a moulding or curved surface, they conform with its shape and can be clamped in position to give an accurate replica.

Macpherson strut suspension (*Autos*) Widely used system for independent front suspension, in which a single lower WISHBONE locates the wheel in the fore and aft plane, and

springing and damping are accommodated by a combined coilspring/damper unit attached between wheel hub and chassis.

macr-, macro- (*Genrl*) Prefixes from Gk *makros*, large.

macro (*ICT*) (1) A series of keystrokes that are recorded in a FILE which may then be executed when required. An often-repeated phrase may be recorded as a macro in a WORD PROCESSING PROGRAM and entered into a document with a single keystroke. The file may also contain COMMANDS. In some APPLICATIONS, the macro may be changed in a TEXT EDITOR for additional sophistication. Also *script*. (2) See MACRO-INSTRUCTION.

macro- (*Genrl*) See MACR-.

macro assembler (*ICT*) An ASSEMBLER that offers the facility of expanding MACRO-INSTRUCTIONS.

macro-axis (*Crystal*) The long axis in orthorhombic and triclinic crystals.

macrocephaly (*Med*) Abnormal largeness of the head due to excess fluid, resulting in mental retardation. Also *macrocephalia*. Cf HYDROCEPHALUS.

macrocheilia (*Med*) Abnormal increase in the size of the lips.

macrocheiria (*Med*) Abnormally large hands.

macrocode (*ICT*) See MACRO-INSTRUCTION.

macrocycle (*Chem*) A large ring compound, usually heterocyclic, and often containing repeating units of the form $-CH_2CH_2X-$, where X is commonly O, S or NH. See CROWN ETHER.

macrocyte (*Med*) An abnormally large red cell in the blood.

macrocytosis (*Med*) The presence of many abnormally large red cells in the blood.

macrodactyly (*Med*) Congenital hypertrophy of a finger or fingers. Also *macrodactylia*.

macro-defect-free cement (*Build*) PORTLAND CEMENT of much enhanced tensile strength, made by the addition of approx 5% polyvinyl alcohol followed by shear-mixing and compressing to eliminate pores over 15 μm in size. Abbrev *MDF cement*. See panel on CEMENT AND CONCRETE.

macrogamete (*BioSci*) The larger of a pair of conjugating gametes, generally considered to be the female gamete.

macroglia (*BioSci*) A general term for neuroglial cells, including astroglia, EPENDYMA and OLIGODENDROGLIA.

macroglobulin (*BioSci*) Any globulin with a molecular weight above about 400 kDa but usually applied to IgM (900 kDa) or to a protease inhibitory molecule alpha-2-macroglobulin (820 kDa). Macroglobulinaemia refers to markedly raised levels of IgM in the blood, due to an abnormally proliferating clone of neoplastic plasma cells, as in WALDENSTROM'S MACROGLOBULINAEMIA.

macroglossia (*Med*) Abnormal enlargement of the tongue.

macrogols (*Pharmacol*) POLYETHYLENE GLYCOLS, used as ointment bases.

macrograph (*ImageTech*) A photograph reproducing an object much larger than its actual size.

macro-instruction (*ICT*) An instruction in a programming language that causes several other instructions in the same language to be carried out and that is automatically replaced by them in the program.

macro lens (*ImageTech*) A camera lens designed for extreme close-up photography, down to a subject distance of a few centimetres.

macrolide antibiotics (*Pharmacol*) A group of primarily bacteriostatic antibiotics that act by inhibiting bacterial protein synthesis. They are active against most aerobic and anaerobic Gram-positive cocci and Gram-negative anaerobes. Examples are *azithromycin, clindamycin, erythromycin* and *lincomycin*.

macromere (*BioSci*) In a segmenting ovum, one of the large cells that are formed in the lower or vegetal hemisphere.

macromolecular crystals (*Crystal*) Many proteins because of their regular three-dimensional structures will form crystals under suitable conditions which can then be analysed by X-RAY CRYSTALLOGRAPHY to determine their fine structure down to the level of 2×10^{-10} m. The interrelations of the amino acid side chains and the particular conformation of the active site of an enzyme can eg be analysed.

macromolecule (*Chem*) A term applied to a very large molecule such as haemoglobin (containing about 10 000 atoms) and sometimes to a polymer of high molecular mass.

macronucleus (*BioSci*) In CILIOPHORA, the larger of the two nuclei, composed of 'vegetative' or somatic chromatin, not germ-line chromatin. Cf MICRONUCLEUS.

macronutrient (*BioSci*) An element required in relatively large quantities by living organisms; for plants they are H, C, O, N, K, Ca, Mg, P, S. Cf MICRONUTRIENT. See ESSENTIAL ELEMENT.

macrophage (*BioSci*) A type of phagocytic cell of the mononuclear phagocyte system, derived from blood MONOCYTES that migrate into tissues and differentiate. They become 'activated' following the ingestion of degradable particles and by lymphokines such as interferon. Most macrophages are motile, although some remain in the same site (eg as Kupffer cells in liver sinusoids) for long periods; they can occasionally fuse together to form multinucleated giant cells. Macrophages are important accessory cells for presentation of antigens to T-lymphocytes.

macrophagous (*BioSci*) Feeding on relatively large particles of food. Cf MICROPHAGOUS.

macrophotography (*ImageTech*) A general term for photography at extremely close range, so that the image size is as large or larger than the actual subject.

macrophyll (*BioSci*) A large leaf. Also *megaphyll*.

macrophyric (*Geol*) A textural term for medium to fine-grained igneous rocks containing phenocrysts > 2 mm long. Cf MICROPHYRIC.

macro processing language (*ICT*) A language for processing MACRO-INSTRUCTIONS.

macroscopic (*Genrl*) Visible to the naked eye.

macroscopic state (*Chem*) One described in terms of the overall statistical behaviour of the discrete elements from which it is formed. Cf MICROSCOPIC STATE.

macrosection (*Eng*) A section of metal, ceramic, etc, mounted, cut, polished and etched as necessary to exhibit the MACROSTRUCTURE.

macrosmatic (*BioSci*) Having a highly developed sense of smell.

macrosome (*BioSci*) A large protoplasmic granule or globule.

macrosplanchnic (*BioSci*) Having a large body and short legs; as a tick.

macrospore (*BioSci*) See MEGASPORE.

macrosporophyll (*BioSci*) Leaf-like structure bearing megasporangia. Also *megasporophyll*.

macrostoma (*Med*) Abnormal width of the mouth due to a defect in development.

macrostructure (*Eng*) Specifically, the structure of a metal as seen by the naked eye or at low magnification on a ground or polished surface or on one which has been subsequently etched. Generally, the structure of any body visible to the naked eye. Cf MICROSTRUCTURE.

macrotous (*BioSci*) Having large ears.

macula (*BioSci*) A blotch or spot of colour; a small tubercle; a small shallow pit. Also *macule*.

macula (*Med*) A small discoloured spot on the skin, not raised above the surface; in particular a small yellow area seen on examination of the retina. Also *macule*. Adjs *macular, maculate*.

macula acustica (*BioSci*) Patches of sensory epithelium in the ear of dogfish and the middle ear of mammals. In the latter they are organs of static or tonic balance consisting of supporting cells and hair cells, the hairs being embedded in a gelatinous membrane containing crystals of calcium carbonate, ie otoliths.

macula lutea (*BioSci*) See YELLOW SPOT.

MAD (*Aero*) Abbrev for: (1) *Mutual assured destruction*, the strange situation during the Cold War in which nuclear

conflict would ensure destruction of both combatants, thus assuring (it was felt) stalemate or stand-off between the superpowers. (2) *Magnetic anomaly detector* (or *detection*), a sensitive magnetometer used to detect submarines or in surveying for minerals; the use of such an instrument.

Madagascar aquamarine (*Min*) A strongly dichroic variety of the blue beryl obtained, as gemstone material, from Madagascar.

Madagascar topaz (*Min*) See CITRINE.

mad cow disease (*Vet*) Colloq term for BOVINE SPONGIFORM ENCEPHALOPATHY. See panel on TRANSMISSIBLE SPONGIFORM ENCEPHALOPATHY

made end-papers (*Print*) A term for any of the several varieties of specially assembled end-papers, having cloth joints, doublures, etc, as used in hand binding.

made ground (*Build*) Ground formed by filling in natural or artificial pits with hard core or rubbish.

Madeira topaz (*Min*) A form of *Spanish topaz*. See CITRINE.

Madelung's constant (*Chem*) A constant representing the sum of the mutual potential coulombic attraction energy of all the ions in a lattice, in the equation for the lattice energy of an ionic crystal. Varies with lattice type.

MADGE (*Aero*) Abbrev for MICROWAVE AIRCRAFT DIGITAL GUIDANCE EQUIPMENT.

mad itch (*Vet*) See AUJESKY'S DISEASE.

madreporite (*BioSci*) In ECHINODERMATA, a calcareous plate with a grooved surface, perforated by numerous fine canaliculi and situated in an interambulacral position, through which water is passed to the axial sinus. Adj *madreporic*.

Madura foot (*Med*) A disease, endemic in India and occurring elsewhere, in which nodular, ulcerated swellings appear on the foot, due to infection with a fungus. Also *mycetoma*.

Mae West (*Aero*) Personal life jacket designed for aircrew, inflated by releasing compressed carbon dioxide.

mafic (*Min*) A mnemonic term for the ferromagnesian and other non-felsic minerals actually present in an igneous rock.

magamp (*Acous*) Abbrev for MAGNETIC AMPLIFIER.

magazine (*ImageTech*) A removable light-tight container for feeding film to and from a motion picture camera, printer or processing machine.

magazine projector (*ImageTech*) One capable of holding a number of slides, which are fed automatically for projection.

magazine reel stand (*Print*) A unit on a web-fed press which supports the reel in running position with one or more reels in the ready position.

Magellanic Clouds (*Astron*) Two dwarf galaxies, satellites of our own Galaxy, visible as cloudy patches in the night sky in the southern hemisphere. First recorded by F Magellan in 1519. They are around 180 000 light-years away and contain a few billion stars.

Magellanic Stream (*Astron*) A cool cloud of neutral hydrogen, containing around 10^9 solar masses, which surrounds the Magellanic Clouds and extends towards our galaxy, probably owing to the gravitational force exerted by our galaxy during a close encounter with the Magellanic Clouds around a billion years ago.

Magellan project (*Space*) A NASA space mission (1989) to map Venus at subkilometre resolution using side-looking radar from a Venus orbit.

Magendie's foramen (*BioSci*) In vertebrates, an aperture in the roof of the fourth ventricle of the brain, through which the cerebrospinal fluid communicates with the fluid in the spaces enclosed by the meningeal membranes.

magenta (*Chem*) See FUCHSIN.

maggot (*BioSci*) An acephalous, apodous, caterpillar-like larva such as that of certain Diptera.

maghemite (*Min*) An iron oxide with the crystal structure of magnetite but the composition of haematite. A cation-deficient spinel, produced by the oxidation of magnetite.

magic eye (*ICT*) A simple cathode-ray tube in which a metal fin is employed as a control electrode. The opening or closing of a fluorescent pattern is the indication of balance. Used as a detector for ac bridges and as a tuning indicator in radio receivers. Also *tuning indicator*.

magic number (*Phys*) The combined total number of protons and neutrons for which an atomic nucleus is especially stable, namely 2, 8, 20, 28, 50, 82 or 126.

magic square (*MathSci*) A square array of integers arranged in such a way that the sum of the numbers in each row, column and diagonal is equal.

magic-tee (*Radar*) A four-port waveguide junction, a combination of an E-plane and an H-plane junction, having the properties of a HYBRID JUNCTION.

magistral (*Eng*) Powdered roasted copper pyrite used in amalgamation of silver ores in a Mexican process.

magma (*Geol*) Molten rock, including dissolved water and other gases. It is formed by melting at depth and rises either to the surface, as lava, or to whatever level it can reach before crystallizing again, in which case it forms an igneous intrusion.

magmatic cycle (*Geol*) See IGNEOUS CYCLE.

magnesia (*Chem*) See MAGNESIUM OXIDE.

magnesia alba (*Chem*) Commercial basic magnesium carbonate.

magnesia alum (*Min*) See PICKERINGITE.

magnesia cement (*Build*) See SOREL'S CEMENT.

magnesia glass (*Glass*) Glass containing usually 3–4% magnesium oxide. Electric-lamp bulbs have been mainly made from this type of glass since fully automatic methods of production were adopted.

magnesia mixture (*Chem*) A mixture of magnesium chloride, ammonium chloride and ammonia solution used in chemical analysis for the estimation of phosphates.

magnesian spinel (*Min*) See SPINEL.

magnesite (*Eng*) Carbonate of magnesium, crystallizing in the trigonal system. Magnesite is a basic refractory used in open-hearth and other high-temperature furnaces; it is resistant to attack by basic slag. It is obtained from natural deposits (mostly magnesium carbonate, $MgCO_3$) which is calcined at high temperature to drive off moisture and carbon dioxide, before being used as a refractory.

magnesite flooring (*Build*) A composition of cement, magnesium compounds, sawdust and sand to form a hard continuous screed for floor surfaces. Also *jointless flooring*.

magnesium (*Chem*) Symbol Mg, at no 12, ram 24·312, mp 651°C, rel.d. 1·74, electrical resistivity $42 \times 10^{-8}\ \Omega$ m, bp 1120°C at 1 atm, specific latent heat of fusion 377 kJ kg^{-1}. A light metallic element in the second group of the periodic system. The sixth most abundant element of the Earth's crust, which contains 2·76% Mg, but the figure is much higher in the Earth as a whole. There is 0·13% in sea water. The metal is a brilliant white in colour, and magnesium ribbon burns in air, giving an intense white light. It is used as a deoxidizer for copper, brass and nickel alloys. Apart from its widespread use as a structural material in the aircraft and other industries, it is also widely used alloyed with 0·8% aluminium and 0·5% beryllium in the form of MAGNOX as a fuel cladding in gas-cooled reactors. New alloys with zirconium and thorium are used in aircraft construction.

magnesium carbonate (*Chem*) $MgCO_3$. See MAGNESITE.

magnesium orthodisilicate (*Chem*) Occurs in nature as SERPENTINE.

magnesium oxide (*Chem*) MgO. Obtained by igniting the metal in air. In the form of calcined MAGNESITE and DOLOMITE, it is used as a refractory material. See PERICLASE.

magnesium oxychloride cement (*Build*) See OXYCHLOR-IDE CEMENT.

magneson (*Chem*) 4-(4-nitrophenolazo)resorcinol. Used as a reagent for detection and determination of magnesium, with which it forms a characteristic blue colour in alkaline solution.

magnet (*Phys*) A piece of iron or other material which possesses the property of attracting or repelling other pieces of iron, and which also exerts a force on a current-carrying conductor placed in its vicinity.

magnetar (*Astron*) A dense neutron star with a magnetic field 10^{14} times stronger than that of the Earth.

magnet coil (*ElecEng*) See MAGNETIZING COIL.

magnet core (*Phys*) The iron core within the coil of an electromagnet.

Magnetherm (*EnvSci*) A method for producing magnesium by reduction of the oxide when heated in partial vacuum with calcium, aluminium and silicon oxides.

magnetic (*Phys*) Said of all phenomena depending on magnetism, and also of materials which are ferromagnetic or ferrimagnetic. See DIAMAGNETISM, PARAMAGNETISM and panel on FERROMAGNETICS AND FERRIMAGNETICS.

magnetic alloys (*Eng*) Generally, any alloy exhibiting FERROMAGNETISM, eg and most importantly SILICON IRON, but also iron–nickel alloys, which may contain small amounts of any of a number of other elements (eg copper, chromium, molybdenum, vanadium, etc), and iron–cobalt alloys.

magnetic amplifier (*Acous*) Amplifier in which the saturable properties of magnetic material are utilized to modulate an exciting ac, using an applied signal as *bias*; the signal output when rectified becomes a magnification of the input signal. Abbrev *magamp*.

magnetic annealing (*Eng*) Heat treatment of magnetic alloy in a magnetic field, used to increase its PERME-ABILITY.

magnetic anomaly (*Geol*) The value of the local magnetic field remaining after the subtraction of the dipole portion of the Earth's field. The local deviation measured in magnetic prospecting.

magnetic anomaly patterns (*Geol*) When new basaltic rock wells up along mid-oceanic ridges it spreads out symmetrically on either side away from the ridge. See PLATE TECTONICS (panel). As this new rock cools it assumes the direction of magnetism of the Earth's prevailing magnetic field, displaying remanent magnetism. However, the polarity of the Earth's field reverses from time to time and this is recorded in successive stripes of rock, which then show normal and reversed magnetization. The ages of these stripes and their distances away from the ridge allow the rate of sea-floor spreading to be calculated.

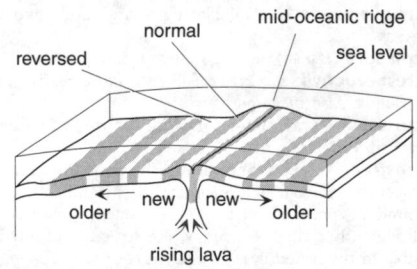

magnetic anomaly patterns Normal and reversed magnetization at a mid-oceanic ridge.

magnetic armature (*ElecEng*) Ferromagnetic element, the position of which is controlled by external magnetic fields.

magnetic axis (*Phys*) A line through the effective centres of the poles of a magnet.

magnetic balance (*ElecEng*) A form of fluxmeter in which the force required to prevent the movement of a current-carrying coil in a magnetic field is measured. Cf CURRENT BALANCE.

magnetic bearing (*Surv*) The horizontal angle between a survey line and magnetic north.

magnetic bias (*Acous*) A steady magnetic field added to the signal field in magnetic recording, to improve linearity of relationship between applied field and magnetic REMANENCE in recording medium.

magnetic blowout (*ElecEng*) A magnet coil fitted to circuit breakers etc which deflects any arc formed so as to lengthen it or force it against a cool surface, thereby extinguishing it. Also *blowout coil*.

magnetic bottle (*NucEng*) See MAGNETIC TRAP.

magnetic braking (*ElecEng*) A method of braking a moving system in which a brake is applied and released by an electromagnet.

magnetic bubble (*Phys*) See MAGNETIC BUBBLE MEMORY.

magnetic bubble memory (*ICT, Electronics*) Potentially cheap and reliable CHIP memory based on mixed ferrimagnetic GARNETS. Thin films of these materials can be prepared so that magnetization lies perpendicular to the plane of the film. Stable domains of reverse magnetization (bubbles) can be manipulated by a surface magnetic structure. The presence or absence of a magnetic bubble in a localized region of the film designates 0 or 1. Very compact, robust, high-capacity memories can be realized by this method but cheap hard drives made this commercially uninteresting.

magnetic card (*ICT*) One with a suitable surface that can be magnetized in selective areas so that data can be stored.

magnetic character reading (*Print*) Scanning and interpretation of characters printed with magnetic ink on documents, eg cheques; special lettering is necessary.

magnetic chuck (*Eng*) A chuck having a surface in which alternate steel elements, separated by non-magnetic material, are polarized by electro- or permanent magnets, so as to hold light, flat steel securely on the table of a grinding machine or other machine tool.

magnetic circuit (*ElecEng*) Complete path, perhaps divided, for magnetic flux, excited by a permanent magnet or electromagnet. The range of RELUCTANCE is not so great as resistance in conductors and insulators, so that leakage of the magnetic flux into adjacent non-magnetic material, esp air, is significant.

magnetic clutch (*Autos*) (1) A friction clutch without pressure springs in which a solenoid pulls the plates together. (2) A clutch in which ferromagnetic powder takes up the drive when drawn into an annular gap by a solenoid. Also *Smith's coupling*.

magnetic component (*ElecEng*) The magnetic field associated with an electromagnetic wave. The magnetic and electric fields are related through the intrinsic impedance of the medium,

$$E/H = \eta = \sqrt{\mu/\varepsilon}$$

magnetic confinement (*NucEng*) In fusion research, the use of shaped magnetic fields to confine a plasma.

magnetic confinement fusion system (*NucEng*) Fusion system in which plasma is confined, ie not allowed to come in contact with the walls of the chamber, by use of specially shaped magnetic fields. Cf INERTIAL FUSION SYSTEM.

magnetic controller (*ElecEng*) Unit in control system actuated by magnetics.

magnetic cooling (*Phys*) Cooling of a paramagnetic sample by conventional means whilst subjected to a magnetic field; if the sample is then thermally insulated and the field removed, the individual magnetic moments in the material are free to become disordered, but in doing so they take up heat from the sample, causing it to cool. Can achieve temperatures as low as 10^{-3} K.

magnetic core (*ICT*) See FERRITE-CORE MEMORY.

magnetic coupling (*ElecEng*) The magnetic flux linkage between one circuit and another.

magnetic creeping (*ElecEng*) A gradual increase in the intensity of magnetization of a piece of magnetic material, after a continued application of the magnetizing force.

magnetic cutter (*Acous*) A cutter used in recording in which the motions of the recording stylus are operated by magnetic fields.

magnetic damping (*ElecEng*) Slowing the movement of a conductor when it passes through a magnetic field by induced eddy currents, as eg the moving parts of instruments and electricity-integrating meters.

magnetic declination (*Surv*) The angular deviation of a magnetic compass, uninfluenced by local causes, from the true north and south. The declination varies at different points on the Earth's surface and at different times of the year. Also *magnetic deviation*.

magnetic deflection (*Electronics*) That of an electron beam in a cathode-ray tube, caused by a magnetic field established by current in coils where the beam emerges from the electron gun which forms the beam, the deflection being at right angles to the direction of the field.

magnetic deviation (*Surv*) See MAGNETIC DECLINATION.

magnetic difference of potential (*ElecEng*) A difference in the magnetic conditions at two points which gives rise to a magnetic flux between the points.

magnetic dip (*Geol*) See DIP.

magnetic dipole moment (*Phys*) A vector associated with a small current loop which gives rise to a magnetic field characteristic of a magnetic dipole. The magnitude is $\mu = iA$, where i is the current and A the area of the loop. An orbiting electron in an atom will have an *orbital* magnetic dipole moment and also have a *spin* magnetic dipole moment due to its intrinsic spin. Nuclei of atoms with non-zero spin also have magnetic dipole moments. See BOHR MAGNETON.

magnetic discontinuity (*ElecEng*) An air-gap, or a layer of non-magnetic material, in a magnetic circuit.

magnetic disk (*ICT*) Storage device consisting of a flat rotatable circular plate, coated on both surfaces with a magnetic material. Data are written to and read from a set of concentric circular tracks.

magnetic displacement (*ElecEng*) See MAGNETIC INDUCTION.

magnetic domain (*Eng*) Aggregated ferromagnetic group of atoms, usually well below 1 micrometre in diameter, forming part of a system of such groups. See DOMAIN and panel on FERROMAGNETICS AND FERRIMAGNETICS.

magnetic doublet radiator (*ElecEng*) A hypothetical radiator consisting of two equal and opposite varying magnetic poles whose distant field is equivalent to that of a small loop aerial.

magnetic drum (*ICT*) A device used for BACKING STORE that consists of a cylinder driven very uniformly at high speed, carrying a layer of magnetic material on which data are stored and read.

magnetic elongation (*ElecEng*) The slight increase in length of a wire of magnetic material when it is magnetized. See MAGNETOSTRICTION.

magnetic energy (*ElecEng*) The product of flux density and field strength for points on the demagnetization curve of a permanent magnetic material, measuring the energy established in the magnetic circuit. Normally required to be a maximum for the amount of magnetic material used. See MAGNETIC MATERIALS.

magnetic epoch (*Geol*) A long period of geological time in which the Earth's magnetic field was essentially of one polarity.

magnetic events (*Geol*) Geologically short periods during MAGNETIC EPOCHS when the magnetic field had a reversed polarity.

magnetic ferrites (*ElecEng*) See FERRITE.

magnetic field (*ElecEng*) Modification of space, so that forces appear on magnetic poles or magnets. Associated with electric currents and the motions of electrons in atoms.

magnetic field intensity (*ElecEng*) See MAGNETIC MATERIALS. Also *magnetic intensity, magnetizing field strength, magnetizing force*.

magnetic field strength (*ElecEng*) A measure of the strength of a magnetic field. Unit ampere per metre, symbol $A\,m^{-1}$. The CGS unit is the *oersted*: $1\,A\,m^{-1} = 4 \times 10^{-3}$ oersteds. Also *magnetic field intensity, magnetizing force*.

magnetic film (*ImageTech*) A full-width magnetic coating on a film base, slit and perforated to regular 35 mm or 16 mm motion picture standards.

magnetic flux (*ElecEng*) The surface integral of the product of the PERMEABILITY of the medium and the magnetic field intensity normal to the surface. The magnetic flux is conceived, for theoretical purposes, as starting from a positive fictitious north pole and ending on a fictitious south pole, without loss. When associated with electric currents, a complete circuit, the *magnetic circuit*, is envisaged, the quantity of magnetic flux being sustained by a magnetomotive force, mmf (coexistent with ampere-turns linked with the said circuit). Permanent magnetism is explained similarly in terms of molecular mmfs (associated with orbiting electrons) acting in the medium. Measured in *maxwells* (CGS) or *webers* (SI).

magnetic flux density (*ElecEng*) The basic magnetic quantity which accounts for the magnetization of a medium and the effects of any external magnetizing fields. See MAGNETIC MATERIALS.

magnetic focusing (*Electronics*) That of an electron beam by applied magnetic fields, eg in a cathode-ray tube or electron microscope.

magnetic forming (*Eng*) A fast, accurate production process for swaging, expanding, embossing, blanking, etc, in which a permanent or expendable coil is moved by electromagnetism to act on the workpiece.

magnetic gate (*ElecEng*) A gate circuit used in magnetic amplifiers.

magnetic head (*Acous*) Recording, reproducing or erasing head in magnetic recorder.

magnetic hysteresis (*ElecEng*) See HYSTERESIS LOOP.

magnetic hysteresis loop (*Phys*) See HYSTERESIS LOOP.

magnetic hysteresis loss (*ElecEng*) The energy expended in taking a piece of magnetic material through a complete cycle of magnetization. The magnitude of the loss per cycle is proportional to the area of the magnetic hysteresis loop. See MAGNETIC MATERIALS.

magnetic induction (*ElecEng*) Induced magnetization in magnetic material, by saturation, by coil excitation in a magnetic circuit, or by the simple method of stroking with another magnet. Also *magnetic displacement*.

magnetic ink (*Print*) Ink containing particles of ferromagnetic material, used for printing data so that magnetic character recognition is possible.

magnetic ink character recognition (*ICT*) Machine recognition of stylized characters printed in magnetic ink.

magnetic intensity (*ElecEng*) See MAGNETIC FIELD INTENSITY.

magnetic iron ore (*Min*) See MAGNETITE.

magnetic lag (*ElecEng*) The time required for the magnetic induction to adjust to a change in the applied magnetic field.

magnetic leakage (*ElecEng*) That part of the magnetic flux in a system which is useless for the purpose in hand and may be a nuisance in affecting nearby apparatus.

magnetic lens (*Phys*) The magnetic equivalent of an optical lens; a magnet or system of magnets is used to produce a field which acts on a beam of electrons in a similar way to a glass lens on light rays. Comprises current-carrying solenoids of suitable (usually quadrupole) design.

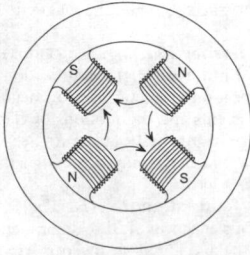

magnetic lens Quadrupole design.

magnetic levitation (*Phys*) A method of opposing the force of gravity using the mutual repulsion between two like magnetic poles.

magnetic link (*ElecEng*) A small piece of magnet steel placed in the immediate vicinity of a conductor carrying a heavy surge current, eg a transmission line tower carrying a lightning stroke current. The magnetization of the link affords a means of estimating the value of the current.

magnetic map (*Phys*) A map showing the distribution of the Earth's magnetic field.

magnetic materials (*Eng*) A magnetic material is characterized by the amount it is magnetized by an external magnetizing field. A magnetizing field (H) is associated with electric currents and in free space leads to a magnetic flux density (B). The relationship is linear with the constant μ_0 known as the permeability of free space which has the value of $4\pi \times 10^{-1}$: $dB = \mu_0\,dH$. Strictly B and H are vector quantities. When materials are present the consequence of their becoming magnetized by the field, H, can be incorporated in a relative permeability, μ_r, so that $B = \mu_r \mu_0 H$. Paramagnetic materials have μ_r slightly greater than unity and diamagnetic substances have μ_r slightly less than unity. The common magnetic materials do not have unique values of relative permeability and susceptibility but exhibit hysteresis. Their magnetization (M) depends not only on the magnetizing field (H) but also on the previous state of magnetization as shown by their magnetization, $M(H)$, curve. See fig. for examples. See panel on FERROMAGNETICS AND FERRIMAGNETICS. Fig. ▷

magnetic memory (*ICT*) See BACKING STORE.

magnetic mirror (*NucEng*) A device based on the principle that ions moving in a magnetic field towards a region of considerably higher magnetic field strength are reflected. This principle is used in MIRROR MACHINES. Mirrors may be simple, minimum β (more efficient) or *tandem* (solenoid plugged at both ends by minimum β-filters).

magnetic modulator (*ElecEng*) One using a magnetic circuit as the modulating element.

magnetic moment (*Phys*) A vector such that its product with the magnetic induction gives the torque on a magnet in a homogeneous magnetic field. See panel on FERROMAGNETICS AND FERRIMAGNETICS.

magnetic monopole (*Phys*) Single isolated magnetic pole. Some GRAND UNIFIED THEORIES (panel) of matter predict the existence of point-like defects in the structure of space–time which would behave like the isolated pole of a bar magnet. Monopoles would have to be quantized with the quantum g related to the quantum of electric charge e by $ge = h$ where h is PLANCK'S CONSTANT.

magnetic north (*Phys*) The direction in which the north pole of a pivoted magnet will point. It differs from the geographical north by an angle called the *magnetic declination*.

magnetic oxide of iron (*Min*) See MAGNETITE.

magnetic oxides (*Eng*) The iron oxides which are ferromagnetic and which, suitably fabricated from the powder form, provide efficient permanent magnets. See FERRITES.

magnetic particle clutch (*Eng*) A form of hydraulic coupling in which the fluid is a suspension of magnetizable particles, the viscosity of the fluid, and thus the degree of slip of the clutch, being variable by varying the intensity of magnetization.

magnetic particle inspection (*Eng*) A rapid, non-destructive test for fatigue cracks and other surface and subsurface defects in steel and other magnetic materials, in which the workpiece is magnetized so that local flux-leakage fields are formed at cracks or other discontinuities. The position of these fields is shown by dusting the workpiece with a magnetic powder which may be coloured red or black with the workpiece painted white to facilitate detection. Abbrev *MPI*.

magnetic pendulum (*ElecEng*) Suspended magnet executing torsional oscillations in any horizontal magnetic field.

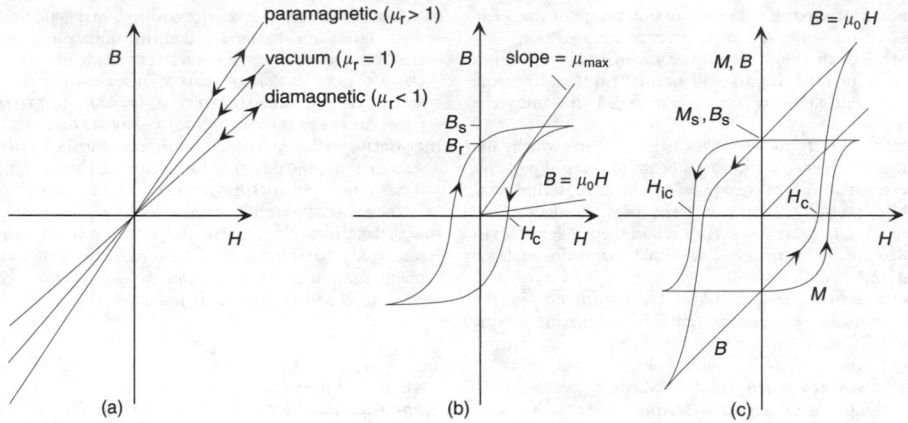

magnetic materials (a) non-magnets, (b) soft magnets and (c) hard magnets. B is the magnetic flux density, H the magnetizing field, μ the permeability and M the magnetization.

It forms the basis of the horizontal component magnetometer.

magnetic permeability (*ElecEng*) See PERMEABILITY.

magnetic polarization (*Chem*) The production of optical activity by placing an inactive substance in a magnetic field.

magnetic polarization (*Phys*) The vector J which is related to the magnetic field H and the magnetic induction B by

$$B = \mu_0 H + J$$

where μ_0 is the permeability of free space.

magnetic pole (*ElecEng*) A convenient conception, which cannot exist, deduced from the experimental indication of the direction of the magnetic field arising from a permanent magnet. If the latter is long in comparison with its cross-section and the ends are provided with soft-iron balls, the direction of the magnetic field, as indicated by iron filings, appears to radiate from the centres of such spheres, called *poles*. Experimentally, such poles appear as magnetic charges, from which are deduced the magnitude of magnetic poles, magnetic field strength, magnetic flux, magnetic potential and electrical units.

magnetic potential (*Phys*) A continuous mathematical function, the value of which at any point is equal to the potential energy (relative to infinity) of a theoretical unit north-seeking magnetic pole placed at that point.

magnetic potentiometer (*ElecEng*) A flexible solenoid (wound on a non-magnetic base) used with a ballistic galvanometer to explore the distribution of magnetic potential in a field etc.

magnetic pressure (*NucEng*) The pressure which a magnetic field is capable of exerting on a plasma.

magnetic printing (*ICT*) The transfer of recorded signal from one magnetic recording medium or element to another.

magnetic printing (*Print*) Printing ink containing magnetic elements and capable of being read both visually and by computer.

magnetic prospecting (*MinExt*) A form of GEOPHYSICAL PROSPECTING which measures either the distortion of the Earth's magnetic field by local accumulation of ferromagnetic materials, eg magnetite, pyrrhotite, chromite or ilmenite, or the magnetic susceptibility of rocks. Methods include the use of a simple DIP NEEDLE, instrumental MAGNETOMETER in field studies or airborne magnetometers (*aeromagnetic survey*).

magnetic pumping (*Phys*) The use of radio-frequency currents in coils over bulges in the tube of a stellarator or tokamak fusion reactor to modulate the steady axial field

and provide heat to the plasma. This process is most efficient when there is resonance between the radio-frequency signals and vibrations of the molecules of plasma. The process is then called *resonance heating*.

magnetic pyrites (*Min*) See PYRRHOTINE.

magnetic quantum numbers (*Phys*) Quantum numbers associated with the quantized orientation of the orbital and spin angular moments with respect to an applied magnetic field.

magnetic reaction (*ElecEng*) See ELECTROMAGNETIC REACTION.

magnetic recording (*Acous*) (1) The magnetic tape, from which the recorded signal may be reproduced. (2) The process of preparing a magnetic recording.

magnetic resonance imaging (*Radiol*) The use of NUCLEAR MAGNETIC RESONANCE of protons to produce proton density maps or images of the human body. Abbrev *MRI*.

magnetic reversal (*Geol*) A change of the Earth's magnetic field, to the opposite polarity. See MAGNETIC ANOMALY PATTERNS.

magnetic rigidity (*NucEng*) A measure of the momentum of a particle. It is given by the product of the magnetic intensity perpendicular to the path of the particle and the resultant radius of curvature of this path.

magnetic rotation (*Phys*) See FARADAY EFFECT.

magnetic saturation (*Phys*) The limiting value of the magnetic induction in a medium when its MAGNETIZATION is complete.

magnetic screen (*ElecEng*) A screen of a high-magnetic-permeability material such as MUMETAL, used to surround certain electrical and electronic components, in order to protect them from the effect of external magnetic fields. Also *magnetic shield*.

magnetic separator (*Phys*) A device for separating, by means of an electromagnet, any magnetic particles in a mixture from the remainder of the mixture, eg for separating iron filings from brass filings.

magnetic shell (*ElecEng*) A magnetized body of dimensions identical with a current-carrying coil, which may be considered instead of the latter when considering the forces acting in a system.

magnetic shield (*ElecEng*) See MAGNETIC SCREEN.

magnetic shift register (*ElecEng*) One in which the pattern of settings of a row of magnetic cores is shifted one step along the row by each fresh pulse.

magnetic shunt (*ElecEng*) A piece of magnetic material in parallel with a portion of a magnetic circuit, so arranged as to vary the amount of magnetic flux in that portion of the circuit.

magnetic slot-wedge (*ElecEng*) A slot-wedge of magnetic material which gives the same effect as a closed slot.

magnetic south (*Phys*) The direction in which the south pole of a pivoted magnet will point. It differs from the geographical south by an angle called the *magnetic declination*.

magnetic spectrometer (*NucEng*) One in which the distribution of energies among a beam of charged particles is investigated by means of magnetic focusing techniques.

magnetic stability (*Phys*) A term used to denote the power of permanent magnets to retain their magnetism in spite of the influence of eg external magnetic fields or vibration.

magnetic storm (*EnvSci*) Magnetic disturbance in the Earth, causing spurious currents in submarine cables; probably arises from variation in particle emission from the Sun, which affects the ionosphere.

magnetic surface wave (*Radar*) Magnetic wave propagated along the surface of a ferromagnetic garnet substrate. Used in microwave delay lines, filters, etc. Abbrev MSW.

magnetic susceptibility (*ElecEng*) The amount by which the relative permeability of a medium differs from unity, positive for a paramagnetic medium, but negative for a diamagnetic one. Equals intensity of magnetization divided by the applied field. See MAGNETIC MATERIALS.

magnetic suspension (*ElecEng*) The use of a magnet to assist in the support of eg a vertical shaft in a meter, thereby relieving the jewelled bearings of some of the weight.

magnetic tape (*ICT*) Flexible plastic tape, typically 6–50 mm wide, coated on one side with dispersed magnetic material, in which signals are registered for subsequent reproduction. Used for storing TV images, sound or computer data.

magnetic tape encoder (*ICT*) Input device that accepts data from a keyboard and writes it directly to magnetic tape.

magnetic tape reader (*ICT*) One that has a multiple head which transforms the pattern of registered magnetic signals into electrical pulse signals.

magnetic track (*ImageTech*) A magnetic stripe on cinematograph film for recording and reproducing sound.

magnetic transition temperature (*Phys*) See CURIE POINT.

magnetic transmission (*Autos*) See MAGNETIC CLUTCH.

magnetic trap (*NucEng*) Configuration of magnetic fields which will confine plasma for times long enough for it to react. Also *magnetic bottle*. See MAGNETIC CONFINEMENT.

magnetic tube (*Phys*) See TRAPPING REGION.

magnetic tuning (*ElecEng*) Control of a very-high-frequency oscillator by varying the magnetization of a rod of ferrite in the frequency-determining cavity by an externally applied steady field.

magnetic units (*Phys*) Units for electric and magnetic measurements in which μ_0, the permeability of free space, is taken as unity. Now replaced by SI units in which μ_0 is given the value $4\pi \times 10^{-7}$ H m^{-1}.

magnetic variables (*Astron*) Stars in which strong variable magnetic fields (up to 10^{-3} T) have been detected by the ZEEMAN EFFECT.

magnetic variation (*Genrl*) The angular difference between true north and magnetic north at a particular location.

magnetic variations (*EnvSci*) Both diurnal and annual variations of the magnetic elements (dip, declination, etc) occur, the former having by far the greater range. In the northern hemisphere, the declination moves to the west during the morning and then gradually back, the extreme range being nowhere more than 1°. The dip varies by a few minutes during the day. It is thought that these effects are caused by varying electric currents in the ionized upper atmosphere.

magnetic vector potential (*Phys*) See VECTOR POTENTIAL.

magnetic wire (*Acous*) A wire of magnetic material used in recording.

magnetism (*Phys*) The science covering magnetic fields and their effect on materials due to unbalanced spin of electrons in atoms. See ANTIFERROMAGNETISM, COULOMB'S LAW FOR MAGNETISM, DIAMAGNETISM, FERRIMAGNETISM, FERROMAGNETISM, PARAMAGNETISM and panel on FERROMAGNETICS AND FERRIMAGNETICS.

magnetite (*Min*) An oxide of iron, crystallizing in the cubic system. It has the power of being attracted by a magnet, but it has no power to attract particles of iron to itself, except in the form of lodestone. Also *magnetic iron ore*.

magnetization (*Phys*) Orientation from randomness of saturated DOMAINS in a ferromagnetic material. The magnetization per unit volume M is related to the field strength H and the magnetic induction B by

$$B = \mu_0(H + M) \text{ tesla}$$

where μ_0 is the permeability of free space. SI unit is A m^{-1}.

magnetization curve (*ElecEng*) See MAGNETIC MATERIALS.

magnetize (*Phys*) (1) To induce magnetization in ferromagnetic material by direct or impulsive current in a coil. (2) To apply alternating voltage to a transformer or choke having ferromagnetic material, generally laminated, in its core.

magnetizing coil (*ElecEng*) A current-carrying coil used to magnetize an electromagnet, such as the field coil of an electric generator or motor. Also *field coil, magnet coil*.

magnetizing current (*ElecEng*) (1) The current (direct or impulsive) in a coil for the magnetization of ferromagnetic material in its core. (2) The ac taken by the primary of a transformer, apart from a load current in the secondary.

magnetizing roast (*Eng*) Reduction of weakly magnetic iron ore or concentrate by heating in a reducing atmosphere to convert it to a more strongly ferromagnetic compound.

magneto (*ElecEng*) A small permanent-magnet electric generator capable of producing periodic high-voltage impulses; once used for providing the ignition of internal-combustion engines, and also for firing explosives etc.

magneto-acoustic effect (*Phys*) See BARKHAUSEN EFFECT.

magnetocaloric effect (*Phys*) The reversible heating and cooling of a medium when the magnetization is changed. Also *thermomagnetic effect*.

magnetochemistry (*Chem*) The study of the relation of magnetic properties to chemical structure. Particularly, the extent of paramagnetism in transition metal compounds may be related to the type of ligand bonded to the metal.

magnetocrystalline anisotropy (*Phys*) The tendency for magnetic moments (and therefore DOMAINS) to align preferentially with certain crystal axes. This is quantified in terms of a set of anisotropy constants (in J m^{-3}) for a given material.

magneto-electric (*Phys*) Of certain materials, eg chromium oxide, the property of becoming magnetized when placed in an electric field. Conversely, they are electrically polarized when placed in a magnetic field. May be used for measuring pulsed electric or magnetic fields.

magnetoencephalography (*Med*) Measurement of the weak magnetic signals produced by the brain. Abbrev MEG.

magnetogenerator (*ElecEng*) An electric generator in which the existing flux is obtained from permanent magnets.

magnetohydrodynamic generator (*ElecEng*) Abbrev MHD generator. See MAGNETOPLASMADYNAMIC GENERATOR.

magnetohydrodynamic instability (*NucEng*) See INSTABILITY.

magnetohydrodynamics (*Phys*) The study of the motions of an electrically conducting fluid in the presence of a magnetic field. The motion of the fluid gives rise to induced electric currents which interact with the magnetic field which in turn modifies the motion. The phenomenon has applications both to magnetic fields in space and to the possibility of generating electricity. If the free electrons in a plasma or high-velocity flame are subjected to a strong magnetic field, then the electrons will constitute a current flowing between two electrodes in a flame. Abbrev *MHD*. See panel on AERODYNAMICS.

magneto-ignition (*ElecEng*) An ignition system for internal-combustion engines in which the voltage necessary to produce the spark is generated by a magneto.

magneto-ionic (*Phys*) Said of components of an electromagnetic wave passing through an ionized region and divided into ordinary and extraordinary waves by the magnetic field of the Earth.

magnetometer (*Surv*) Any instrument for measurement either of the absolute value of a magnetic field intensity, or of one component of this, eg horizontal component magnetometer for Earth's magnetic field. See PROTON-PRECESSIONAL MAGNETOMETER.

magnetomotive force (*Phys*) Line integral of the magnetic field intensity round a closed path. Abbrev *mmf*.

magneton (*Phys*) See BOHR MAGNETON.

magneto-optical disk (*ImageTech*) A reusable VIDEO DISK, recorded by a laser operating in a magnetic field that produces a magnetic signal in the surface coating which, during playback, varies the axis of polarization of the laser beam as it is reflected, these differences being detected to reproduce the signal (see KERR EFFECT). Abbrev *MO disk*.

magneto-optical effect (*Phys*) See FARADAY EFFECT, KERR EFFECT.

magneto-optic rotation (*Chem*) See MAGNETIC POLARIZATION.

magnetophone (*Acous*) Any recording device involving the magnetization of a medium, eg magnetic tape.

magnetoplasmadynamic generator (*ElecEng*) A device which produces electrical energy from an electrically conducting gas (plasma) flowing through a transverse magnetic field. Abbrev *MPD generator*. Also *magnetohydrodynamic generator*.

magnetoresistance (*ElecEng*) The resistivity of a magnetic material in a magnetic field which depends on the direction of the current with reference to the field. If parallel to one another, the resistivity increases, but if mutually perpendicular, it decreases.

magnetoresistive (*Eng*) A material, a metal or semiconductor, that exhibits a change in electrical resistance when subjected to a magnetic field.

magnetosphere (*Astron*) The asymmetrically shaped volume round the Earth and other magnetic planets in which charged particles are subject to the planet's magnetic field rather than the Sun's. Its radius is least towards the Sun and greatest away from it.

magnetostatics (*Phys*) The study of steady-state magnetic fields.

magnetostriction (*Phys*) A phenomenon of elastic deformation of certain ferromagnetic materials, eg nickel, on the application of magnetizing forces. Used in ultrasonic transducers and, formerly, in memory in computers.

magnetostriction loudspeaker (*Acous*) Loudspeaker based on the magnetostriction effect, in which the length of materials varies according to an impressed magnetic field. The diaphragm is driven by a nickel or rare earth rod in which magnetostrictive changes are induced by an embracing coil.

magnetostriction microphone (*Acous*) Inverse of a MAGNETOSTRICTION LOUDSPEAKER.

magnetostriction transducer (*ElecEng*) Any device employing the property of magnetostriction to convert electrical to mechanical oscillations, eg by using a rod clamped at its centre and passing ac through a coil wound around the rod. Also *magnetostrictor*.

magnetostrictive filter (*ElecEng*) A filter network which utilizes magnetostrictive elements, bars or rods, with their energizing coils.

magnetostrictive oscillation (*Phys*) Oscillation based on the principle of the alteration of dimensions of a bar of magnetic material when the magnetic flux through it is changed. Nickel contracts with an increasing applied magnetic field but iron expands in weak fields and contracts in strong fields. The mechanical oscillatory system, eg a bar clamped at its centre, can be coupled magnetically to an amplifier to maintain the oscillation, which may be in the audio-frequency range.

magnetostrictive reaction (*Phys*) The inverse magnetostrictive effect, ie a change in magnetization under applied stresses.

magnetostrictor (*ElecEng*) See MAGNETOSTRICTION TRANSDUCER.

magnetotaxis (*BioSci*) A tactic response to a magnetic field; in magnetotactic bacteria the Earth's magnetic field is used as a guide to 'up' and 'down' in deep sediment.

magnet pole (*ElecEng*) See POLE PIECE.

magnetron (*Electronics*) (1) A two-electrode valve in which the flow of electrons from a large central cathode to a cylindrical anode is controlled by crossed electric and magnetic fields; the electrons gyrate in the axial magnetic field, their energy being collected in a series of slot resonators in the face of the anode. Magnetrons, used mainly as oscillators, can produce pulsed output power at microwave frequencies, with high peak power ratings. Used in microwave and radar transmitters and microwave cookers. See CAVITY MAGNETRON, CROSSED-FIELD TUBE. (2) A low-pressure gas discharge in which magnetic fields (usually from permanent magnets) near one or more electrodes aid the confinement of electrons. Such discharges are often used for high-rate sputtering.

magnetron amplifier (*Electronics*) A high-power microwave amplifier using a magnetron in such a configuration that it is not self-oscillatory.

magnetron critical field (*Electronics*) That which would just prevent an electron emitted from the cathode with zero energy from reaching the anode at a given value of anode voltage.

magnetron critical voltage (*Electronics*) The voltage which would just enable an electron emitted from the cathode with zero velocity to reach the anode at a given value of magnetic flux density.

magnetron modes (*Electronics*) Different frequencies of oscillation corresponding to different field configurations and selected by interconnecting cavities in various ways to control their phase differences.

magnet steel (*Eng*) A steel from which permanent magnets are made. It must have a high remanence and coercive force. Steels for this purpose may contain considerable percentages of cobalt (up to 35%) as well as nickel, aluminium, copper, etc. See PERMANENT MAGNET.

magnettor (*ElecEng*) A second-harmonic type of magnetic modulator which uses a saturable reactor to amplify dc, or low alternating-frequency, signals.

magnet yoke (*ElecEng*) Sometimes applied to the whole of the magnetic circuit of an electromagnet (or transformer etc), but strictly should refer only to the part which does not carry the windings.

magnification (*MathSci*) In a conformal mapping, given by $w = f(z)$, the magnification varies from point to point but is the same in all directions at a given point. The linear magnification at a point z_0 is given by $|f'(z_0)|$ and the surface magnification by $|f'(z_0)|^2$.

magnification (*Phys*) The ratio of the size of the image to that of the object in an optical system where the object is plane and lies perpendicular to the axis of the system. Also *transverse magnification*. See LONGITUDINAL MAGNIFICATION, MAGNIFYING POWER.

magnification factor (*Phys*) See Q.

magnifier (*Radiol*) Any thermionic amplifier, esp one used for the amplification of audio frequencies.

magnifying power (*Phys*) The ratio of the apparent size of the image of an object formed by an optical instrument to that of the object seen by the naked eye. For a microscope, it is necessary to assume that the object would be examined by the naked eye at the least distance of distinct vision, 25 cm. Unless otherwise stated, the *linear* magnification is assumed to be indicated. See LONGITUDINAL MAGNIFICATION.

magnitude (*Astron*) A measure of the apparent or absolute brightness of a celestial object, first used around 120 BC with the brightest stars referred to as first magnitude and the dimmest as sixth. This was put on a scientific basis by Pogson in 1854, who showed that equal magnitude steps were in logarithmic progression: a magnitude difference of one unit corresponds to a brightness ratio of 2·512; five magnitudes to 100 in actual brightnesses. The zero of the magnitude scale is essentially arbitrary. The *apparent magnitude* is the brightness measured at the Earth and it depends on distance and LUMINOSITY. More useful physically is the *absolute magnitude*, in which the observed apparent magnitude is corrected to that which it would have if placed at an (arbitrary) distance of 10 pc (32·6 light-years) from the Earth. The observed magnitude will depend on the technique used to measure the brightness: *visual magnitude* refers to observation by eye; *photographic magnitude* is measured from the blackening of photographic emulsion; *photoelectric magnitude* is determined photoelectrically; *bolometric magnitude* is the integrated brightness across the whole electromagnetic spectrum.

magnolia metal (*Eng*) A lead-base alloy, containing 78–84% lead; the remainder is mainly antimony, but small amounts of iron and tin are present. Used for bearings.

Magnoliidae (*BioSci*) A subclass or superorder of dicotyledons, comprising trees, shrubs and herbs characterized by a well-developed perianth, numerous centripetal stamens and an apocarpous gynoecium. It contains c.11 000 spp in 36 families, including the Magnoliaceae and the Ranunculaceae. It is more or less synonymous with the ranalian complex, generally regarded as the most primitive of extant angiosperms.

Magnoliophyta (*BioSci*) In some classifications, a division containing the angiosperms. It is subdivided into two classes: the Magnoliopsida (see DICOTYLEDONES) and the Liliopsida. See MONOCOTYLEDONES and panel on ANGIOSPERMS (FLOWERING PLANTS).

Magnoliopsida (*BioSci*) See DICOTYLEDONES.

magnon (*Phys*) Quantum of SPIN WAVE energy in magnetic material.

Magnox (*NucEng*) The name of a magnesium-containing alloy used in the first commercial power reactors built in the UK for cladding the natural uranium metallic fuel. Best known alloys are Magnox B and Magnox A 12. The latter is Mg with 0·8% Al and 0·01% Be. The reactors are called *Magnox reactors*.

Magnus effect (*Phys*) The force experienced by a spinning ball or cylinder in a fluid. The effect is responsible for the swerving of golf and tennis balls when hit with a slice.

mag-opt (*ImageTech*) International code name for a print of a motion picture film having both magnetic and photographic (*optical*) sound tracks.

Magslip (*ElecEng*) TN for a synchro system for remote control or indication.

mahlstick (*Build*) A slender stick padded at one end with cloth or leather; used as a support for the hand guiding the brush. Also *maulstick*.

mahogany (*For*) Hardwoods from the family *Meliaceae*. Also *Brazilian mahogany*, *Cuban mahogany*, *Uganda mahogany*. See AFRICAN MAHOGANY, AMERICAN MAHOGANY.

maiden (*Agri*) Female livestock animal of breeding age but not yet mated.

Maillard reaction (*FoodSci*) Non-enzymic browning caused by the reaction of the aldehydes or ketones present in reducing sugars with proteins and amino acids.

Maillifer screw (*Eng*) A type of multiple extrusion screw to aid the mixing of polymers with additives.

mailmerge (*ICT*) A facility within a WORD PROCESSOR whereby a FILE containing names and addresses may be combined with a standard letter to produce a set of 'personal' letters.

main airway (*MinExt*) In a colliery, one directly connected with the point of entry to the mine.

main and tail (*MinExt*) Single-track underground rope haulage by means of a rope to draw out the full wagons and the tail rope to draw back the empties.

main anode (*Electronics*) That carrying the load current in mercury-arc rectifiers which have an independent EXCITATION ANODE.

main beam (*Build*) In floor construction, a beam transmitting loads direct to the columns.

main circuit (*ElecEng*) See CURRENT CIRCUIT.

main contacts (*ElecEng*) The contacts of a switch which normally carry the current. Cf ARCING CONTACTS, which carry the current at the instant when the circuit is being interrupted.

main deck (*Ships*) Usually means the principal deck or the upper of two decks or the second deck from the top when there are more than two decks, excluding the decks covering deck houses or erections. Frequently used to name a deck in a passenger liner.

main distribution frame (*ICT*) A frame for rearranging the incoming lines to a telephone exchange into the numerical order required in the exchange. Abbrev *MDF*.

main exchange (*ICT*) One that has other exchanges, such as satellites, dependent on it for extension from them to other exchanges.

main field (*ElecEng*) The chief exciting field in an electric machine, as opposed to an auxiliary field, such as that produced by the compoles.

main float (*Aero*) The two single, or one central, float(s) which give buoyancy to a seaplane or amphibian.

mainframe (*ICT*) A computer with a variety of peripheral devices, a large amount of backing store and a fast central processing unit. The term came in with third-generation computers and is generally used in comparison with smaller or subordinate computers.

main line (*MinExt*) Pipeline connecting producing areas to a refinery. Also *trunk line*.

main memory (*ICT*) Computer storage that holds programs and data during execution. It is described in terms of LOCATIONS each of which has an ADDRESS. It is fast, read/write, RANDOM ACCESS storage largely built from CHIPS. Also *immediate access store* (IAS), *internal store*, *main store*, *primary store*. Also core store whether or not it uses ferrite core memory. Cf BACKING STORE. See ACCESS TIME, BUBBLE MEMORY, CACHE MEMORY, CHARGED-COUPLED DEVICE, MOS, PAGING, RAM, THIN-FILM MEMORY.

main plane (*Aero*) The principal supporting surface, or wing, of an aircraft or glider, which can be divided into centre, inner, outer and/or wing-tip sections.

main rope (*MinExt*) See MAIN AND TAIL.

main rotor (*Aero*) (1) The principal assembly or assemblies of rotating blades which provide lift to a rotorcraft. (2) The assembly of compressor(s) and turbine(s) forming the rotating parts of a gas turbine engine.

mains (*ElecEng*) The source of electrical power; normally the electricity supply system.

main sequence (*Astron*) The broad band in the Hertzsprung–Russell diagram on which the vast majority of stars (including the Sun) lie, running diagonally from high-temperature, high-luminosity stars to low-temperature, low-luminosity stars, in a smooth progression. A star spends most of its life on this main sequence, and throughout that time it converts hydrogen to helium. Once the hydrogen in the core is consumed, the star evolves away from the main sequence, becoming first a RED GIANT. See panel on HERTZSPRUNG–RUSSELL DIAGRAM.

mains frequency (*ElecEng*) Electricity ac supply frequency; 50 Hz in the UK, 60 Hz in the US, often 400 Hz in aircraft or ships. Also *power frequency*.

main store (*ICT*) See MAIN MEMORY.

mains transformer (*Electronics*) See POWER TRANSFORMER.

main switching unit (*ICT*) The highest level of switching in the PUBLIC SWITCHED TELEPHONE NETWORK, accepting traffic from REMOTE CONCENTRATOR UNITS and communicating only with other main switching units.

maintainability (*ICT*) A characteristic of a delivered system or program, often indicating the quality of design and coding deployed in its development – the economy (or otherwise) with which changes may be made, whether bug-fixes or extensions to functionality.

maintained tuning fork (*Acous*) Tuning fork associated with an electronic oscillator so that the latter supplies energy continuously to maintain the fork in steady oscillation. The frequency of the oscillation is substantially that of the free fork, and provides a method for establishing frequencies with great accuracy. Also *tuning fork oscillator*.

maintaining voltage (*ElecEng*) That which just maintains ionization and discharge.

main tanks (*Aero*) See FUEL TANKS.

main tie (*Build*) The lower tensional members of a roof truss, connecting the feet of the principal rafters.

main title (*ImageTech*) The section of a film or video programme giving the name of the production and the associated credits.

maisonette (*Arch, Build*) A two-storey self-contained dwelling incorporated in a building of three or more storeys.

maize (*FoodSci*) A cereal, native of N America but grown elsewhere, important as an animal feed, also milled to CORNFLOUR, a widely used thickening agent. Also *corn* (US), *sweetcorn*.

maize oil (*Chem*) See CORN OIL.

major axis (*MathSci*) Of an ellipse: see AXES.

major depression (*Psych*) A type of AFFECTIVE DISORDER characterized by major depressive episodes occurring without intervening manic episodes.

major histocompatibility complex (*BioSci*) The collection of genes coding for the major histocompatibility antigens. Abbrev *MHC*. See HLA SYSTEM and panel on IMMUNE RESPONSE.

majority carrier (*Electronics*) In a semiconductor, the electrons or HOLES, whichever carry most of the measured current. Cf MINORITY CARRIER.

majority emitter (*Electronics*) An electrode releasing majority carriers into a region of semiconductor.

majority voting system (*Aero*) In a redundant electrical or computer system a means whereby signals from all channels (>3) are continually monitored; any discrepancy in a single channel is recognized and 'voted out' of circuit so that the system can continue to function.

major lobe (*ICT*) That containing the direction of highest sensitivity or maximum radiation for any form of POLAR DIAGRAM.

major planet (*Astron*) A planet of the solar system, as opposed to an asteroid.

majuscule (*Print*) A capital letter as distinct from the small letter or MINISCULE.

make (*ICT*) The operation, partial or complete, of a telephone relay, when current is passed through its windings.

make-before-break contact (*ICT*) The group of contacts in a relay assembly so arranged that the one that moves makes contact with a front contact before it separates from a back contact, and so is never free, whether the relay is operated or not.

make-contact (*ICT*) A pair of contacts in a relay assembly that are brought together on operation of the relay and so close a circuit.

make even (*Print*) To arrange type so that the last word of a portion or 'take' of copy ends in a full line.

make-ready (*Print*) See MAKING-READY.

make-up (*Print*) The arrangement of type-matter and blocks into pages.

making-capacity (*ElecEng*) A term used in connection with switchgear rating to denote the ability of a switch to make a circuit under certain specified conditions.

making-current (*ElecEng*) A term used to denote the maximum peak of current which occurs at the instant of closing the switch.

making good (*Build*) The repair of defective work.

making paper (*Print*) See EXCESS FEED.

making-ready (*Print*) Operations involved in preparing a printing machine to run. Includes machine adjustments, setting ink levels, fitting and adjusting plates, or forme, adjusting impression, etc. Also *make-ready*.

making-up (*Textiles*) (1) Examination, packing and ticketing of fabrics before dispatch from the mills. (2) Factory manufacture of garments from the cut-out pieces of fabric.

makoré (*For*) W African hardwood (*Tieghemella*) whose heartwood is pinkish-red to red-brown, straight-grained, with a uniform, fine texture, and can have a MOIRÉ lustre.

Makrolon (*Plastics*) TN for POLYCARBONATE (Ger).

Maksutov telescope (*Astron*) Optical telescope in which the image-forming surfaces are spherical, and therefore easy to make. A deeply curved meniscus lens corrects for aberrations in the primary mirror. Design published by Maksutov in 1944.

malabsorption (*Med*) The failure of adequate intestinal absorption of one or more groups of nutrients: occurs in COELIAC DISEASE, SPRUE and chronic pancreatitis.

malachite (*Min*) Basic copper carbonate $CuCO_3Cu(OH)_2$, crystallizing in the monoclinic system. It is a common ore of copper, and occurs typically in green botryoidal masses in the oxidation zone of copper deposits.

malachite green (*Chem*) A triphenylmethane dyestuff of the rosaniline group. Obtained by the condensation of benzaldehyde with dimethylaminobenzene in the presence of $ZnCl_2$, HCl or H_2SO_4, and by oxidation of the resulting LEUCOBASE with PbO_2.

malacia (*Med*) Pathological softening of any organ or tissue.

malacology (*BioSci*) The study of molluscs.

malacophily (*BioSci*) Pollination by snails.

malacoplakia (*Med*) The occurrence of soft, rounded, pale plaques in the wall of the bladder, in chronic inflammation of the bladder.

Malacostraca (*BioSci*) The largest subclass within the Crustacea. Its organisms have the body clearly divided into head, thorax and abdomen, often with a carapace, and have 20 segments, eg crabs, lobsters, shrimps.

malacostracous (*BioSci*) Having a soft shell.

malaria (*Med*) Infection with *Plasmodium*, a genus of protozoa that live in red cells of the blood. The recurrent fever occurs in quartan, tertian or subtertian (malignant) forms according to the species of *Plasmodium*. Common in most tropical areas; transmitted by anopheline mosquitoes.

malathion (*Chem*) *S-[1,2-di(ethoxycarbonyl)ethyl]dimethyl phosphorothiolothionate*. Used as an insecticide.

mal de caderas (*Vet*) A chronic infectious disease of horses in S America; due to *Trypanosoma equinum*, and characterized by weakness of the hindquarters.

MALDI (*Chem*) Common abbrev for *matrix-assisted laser-desorption ionization*, a method for generating molecular ions in mass spectrometry.

MALDI-TOF mass spectroscopy (*Chem*) Abbrev for *MALDI time-of-flight mass spectrometry*.

mal du coit (*Vet*) See DOURINE.

male (*BioSci*) (1) Generally, an individual in which the gonads produce spermatozoa or some corresponding form of smaller and motile gamete. (2) In plants, a gametophyte that produces male but not female gametes or a sporophyte that produces microspores or pollen and, by derivation, individual seed plants or flowers that have functional stamens but not functional carpels. Cf FEMALE, HERMAPHRODITE.

male and female (*ElecEng, ICT*) Applied to a plug and its complementary socket respectively.

male and female (*Eng*) Trade terms applied to inner and outer members respectively of eg pipe fittings and thread pieces. See EXTERNAL SCREW-THREAD, INTERNAL SCREW-THREAD.

maleic acid (*Chem*) *cis*-Butenedioic acid. HOOCCHCHCOOH. Mp 130°C, with decomposition into its anhydride and water, readily soluble in water. It has the *cis*-configuration, whereas its isomer, FUMARIC ACID, has the *trans*-configuration. Important in polyester resins.

maleic anhydride (*Chem*) *cis*-Butenedioic anhydride. The anhydride of *maleic acid*. Important in establishing the structure of organic compounds containing conjugated double bonds. Also important industrially, mainly as an intermediate. Manufactured by passing a mixture of benzene vapour and air over heated vanadium pentoxide.

maleic hydrazide (*BioSci*) See GROWTH RETARDANT.

male pronucleus (*BioSci*) The nucleus of the spermatozoon.

male sterility (*BioSci*) A condition in which viable pollen is not formed. Used by plant breeders to ensure cross-pollination esp in the production of F1 HYBRID seed. See CYTOPLASMIC MALE STERILITY.

male thread (*Eng*) See EXTERNAL SCREW-THREAD.

male tool (*Eng*) Part of a tool with a core which mates with a female part to form a cavity.

malic acid (*Chem, FoodSci*) HOOCCH$_2$CH(OH)COOH. Mp 100°C. Hygroscopic needles, found in unripe fruit; it also occurs in wines. When attacked by certain ferments, butanoic, lactic and propanoic acids are produced. Heat causes the loss of a molecule of water, producing maleic and fumaric acids. It has been synthesized by various methods.

malignant (*Med*) Tending to go from bad to worse, esp cancerous (see TUMOUR).

malignant aphtha (*Vet*) See CONTAGIOUS PUSTULAR DERMATITIS.

malignant catarrhal fever (*Vet*) Uncommon disease of cattle caused by a herpes virus. Not apparently contagious between cattle, but sheep and deer may act as carriers. There is pyrexia, anorexia, depression, enlarged superficial lymph nodes, oculo-nasal discharges, keratitis and profuse salivation. Mortality 100%.

malignant oedema (*Vet*) An acute toxaemia of cattle, sheep, pigs and horses due to wound infection by *Clostridium septicum* or *Clostridium novyi* (*Cl. oedematiens*). Vaccines available.

malignant stomatitis (*Vet*) See CALF DIPHTHERIA.

malignite (*Geol*) An alkaline igneous rock, a *melanocratic* variety of nepheline syenite.

mall (*Build*) See BEETLE.

malleability (*Eng*) A property of metals and alloys which affects their alteration by hammering, rolling, extrusion. Temperature, as it affects crystallization, may alter resistance.

malleable cast-iron (*Eng*) A variety of cast-iron which is cast white, and then annealed at about 850°C to remove carbon (white-heart process) or to convert the cementite to rosettes of graphite (black-heart process). Distinguished from grey and white cast-iron by exhibiting some elongation and reduction in area in a tensile test. See PEARLITIC IRON.

malleable iron (*Eng*) Now usually means malleable cast-iron, but the term is sometimes applied to wrought-iron.

malleable nickel (*Eng*) Nickel obtained by remelting and deoxidizing electrolytic nickel and casting into ingot moulds. Can be rolled into sheet and used in equipment for handling food, for coinage, condensers and other purposes where resistance to corrosion, particularly by organic acids, is required.

mallein (*Vet*) A concentrated filtrate of broth cultures of *Actinobacillus mallei* (*Malleomyces mallei*) which have been killed by heat; used as an inoculum for the diagnosis of glanders in horses.

mallenders and sallenders (*Vet*) Psoriasis affecting the skin at the flexures of the carpus (knee) joint (mallenders) and tarsus (hock) joint (sallenders) in the horse.

malleolar (*BioSci*) Pertaining to, or situated near, the MALLEOLUS; in Ungulata, the reduced FIBULA.

malleolus (*BioSci*) A process of the lower end of the TIBIA or FIBULA.

mallet (*Build*) A wooden hammer, or one made of rawhide or rubber.

mallet-finger (*Med*) Permanent flexion of the end joint of a finger or thumb.

malleus (*BioSci*) In mammals, one of the ear ossicles; more generally, any hammer-shaped structure.

Mallophaga (*BioSci*) An order of the Psocopteroidea (Paraneoptera) comprising ectoparasites, usually of birds, with reduced eyes, flattened form, tarsal claws and biting mouthparts, eg biting lice.

Malm (*Geol*) The youngest epoch of the Jurassic period. See MESOZOIC.

malm (*Build*) An artificial imitation of natural marl made by mixing clay and chalk in a wash mill; the product is used as a clay for the manufacture of bricks. Also *washed clay*.

malm rubber (*Build*) A soft form of malm brick, capable of being cut or rubbed to special shapes.

malocclusion (*Med*) Imperfect positioning of the upper and lower teeth when the jaw is closed.

malonic acid (*Chem*) *Propanedioic acid*. HOOCCH$_2$COOH. Soluble in water, alcohol, ether; mp 132°C; it decomposes at a slightly higher temperature, giving ethanoic acid; it occurs in beetroot as its calcium salt and can be obtained from malic acid by oxidation with chromic acid. *Malonic ester*, or diethyl malonate, is a liquid of aromatic odour, bp 198°C, CH$_2$(COOC$_2$H$_5$)$_2$. The hydrogen of the methylene group is replaceable by sodium which in turn can be exchanged for an alkyl group. In this way, malonic ester is important for the synthesis of higher dibasic acids. Numerous derivatives of *malonylurea*, C$_4$H$_4$O$_3$N$_2$, barbituric acid, are used as hypnotics and anaesthetics.

Malpighian body (*BioSci*) In the vertebrate kidney, the expanded end of a uriniferous tubule surrounding a glomerulus of convoluted capillaries; in the vertebrate spleen, one of the globular or cylindrical masses of lymphoid tissue which envelops the smaller arteries. Also *Malpighian corpuscle*.

Malpighian cell (*BioSci*) A macrosclereid in the epidermis of the testa of a leguminous seed.

Malpighian corpuscle (*BioSci*) See MALPIGHIAN BODY.

Malpighian layer (*BioSci*) The innermost layer of the epidermis of most chordates, containing polygonal cells which continually proliferate, dividing by mitosis. Also *rete Malpighii*, STRATUM GERMINATIVUM, *basal layer*.

Malpighian tubes (*BioSci*) In insects, some Arachnida and myriapods, tubular glands of excretory function opening into the alimentary canal, near the junction of the midgut and hindgut.

malpresentation (*Med*) Abnormal posture of the fetus during birth.

MALT (*BioSci*) Abbrev for MUCOSAL ASSOCIATED LYMPHOID TISSUE.

malt (*FoodSci*) A product of the controlled germination (sprouting) of cereals to produce enzymes which convert the starch into fermentable sugar. The malting consists of steeping cereal grains in water at 10–12°C for up to 2 days, draining the grains and spreading to allow germination. After germinating for 4–5 days the grains, now malt, are kiln dried, then screened to remove the sprouted rootlets. Used in the brewing industry (mainly barley malt) and in other beverages. Malt flour is produced by milling kiln-dried malted barley or wheat and is used to colour and flavour bakery products.

Malta fever (*Med*) See UNDULANT FEVER.

maltese cross (*ImageTech*) Mechanism providing intermittent frame-by-frame movement in a motion picture film projector. See GENEVA MOVEMENT.

malthenes (*Chem*) Such constituents of asphaltic bitumen as are soluble in carbon disulphide and petroleum spirit. See ASPHALTENES, CARBENES.

maltobiose (*Chem*) See MALTOSE.

maltose (*Chem*) 4-O-(α-D-glucopyranosyl)–D-glucopyranose, *maltobiose*. $C_{12}H_{22}O_{11}$. A disaccharide, a white crystalline mass, dextrorotatory, formed by the action of diastase upon starch during the germination of cereals. It reduces Fehling's solution, and when hydrolysed is converted into D-glucose. In cereal chemistry, the *maltose figure* indicates the natural sugar content and diastatic activity of flour or meal. The unit of starch.

malt-sugar (*Chem*) See MALTOSE.

Malus's law (*Phys*) An expression giving the intensity I of the transmitted beam when a plane-polarized beam of light is incident on a polarizer: $I = I_0 \cos^2 \theta$, where I_0 is the intensity of incident beam and θ is the angle between the plane of vibrations of the beam and the plane of vibrations which are transmitted by the polarizer.

Malvaceae (*BioSci*) A family of c.1000 spp of dicotyledonous flowering plants (superorder Dilleniidae), comprising trees, shrubs and herbs that are cosmopolitan except in very cold regions. The flowers have five sepals (sometimes with an epicalyx), five free petals, many stamens with the filaments united at the base and five fused carpels. Includes cotton, okra and some ornamental plants, eg hollyhock.

malware (*ICT*) Generic term for all types of adverse software that may attack a computer or be downloaded to it without the user's knowledge, eg *adware, Trojan, virus,* etc.

mamilla (*BioSci*) A nipple.

mamillary body (*BioSci*) See CORPUS MAMILLARE.

mamma (*BioSci*) In female mammals, the milk gland; the breast. Adj *mammary*.

mamma (*EnvSci*) Clouds with rounded protuberances on their lower surfaces, like udders. They often occur below thunder clouds.

Mammalia (*BioSci*) A class of vertebrates. The skin is covered by hair (except in aquatic forms) and contains sweat and sebaceous glands and they are HOMOIOTHERMOUS. The young are born alive (except in the Prototheria) and are initially nourished by milk. Respiration is by lungs and a diaphragm is present; the circulation is double, and only the left systemic arch is present. Dentition is heterodont and there is a juvenile (deciduous) dentition preceding that of the adult. There is a double occipital condyle; the lower jaw articulates with the squamosal; the long bones and vertebrae have three centres of ossification. Generally there is an external ear and three auditory ossicles in the middle ear. The brain has large cerebral hemispheres, extensively expanded in hominins. See panel on VERTEBRATE EVOLUTION.

mammogenic (*Med*) Promoting growth of the duct and alveolar systems of the mammary gland, eg applied to hormones.

mammography (*Radiol*) The radiological examination of the breast.

mammoth (*BioSci*) Extinct Pleistocene ancestor of the elephant, with long tusks curling upwards and hairy skin which helped their survival along the borders of continental glaciers. They had a much wider distribution than present-day elephants, which are restricted to Africa and Asia.

mamu (*Phys*) Abbrev for MILLIMASS UNIT.

man (*BioSci*) The human race, all living races being included in the genus *Homo*, suborder Anthropoidea of the Primates. Distinguishing features include elaboration of the brain and behaviour, including communication by facial gestures and speech; the erect posture; the structure of the limbs (including the opposable thumb) and skull; the dentition with small canines; the long period of postnatal development associated with parental care. Political correctness demands replacement when possible by 'human' although in scientific usage man is synonymous with *Homo sapiens*. See HOMININ.

management information system (*ICT*) A computer system, usually consisting of a data warehouse and OLAP software, in which complex data are presented in a simplified and easy-to-understand format for senior managers. Often (and incorrectly) applied to any system performing day-to-day management tasks. Abbrev *MIS*.

Manchester encoding (*ICT*) A technique for sending DIGITAL DATA serially in which the data and timing (clock) pulses are combined.

Manchester yellow (*Chem*) See MARTIUS YELLOW.

Mandelbrot set (*MathSci*) The best-known FRACTAL set, defined in terms of the JULIA SET of $x^2 - c$, where c is a complex parameter. Fig. ▷

mandelic acid (*Chem*) *Phenyl-glycollic acid.* $C_6H_5CH(OH)COOH$. Mp 133°C. Glistening crystals, soluble in water. It occurs naturally in the form of its glycoside, AMYGDALIN, and can be synthesized by the hydrolysis of benzaldehyde cyanhydrin. Mandelic acid possesses an asymmetric carbon atom, and exists in a (+) and a (−) form and as the racemic compound.

mandible (*BioSci*) (1) In vertebrates, the lower jaw. (2) In Arthropoda, a masticatory appendage of the oral somite., (3) In Polychaeta and Cephalopoda, one of a pair of chitinous jaws lying within the buccal cavity. Adjs *mandibular, mandibulate*.

mandibular disease (*Vet*) A malformation, accompanied by bacterial infection, affecting the beak of young chicks; associated with feeding dry meal. Also *shovel beak*.

mandibular glands (*BioSci*) In some insects, glands opening near the articulation of the mandibles; in some lepidopterous larvae, they function as salivary glands, the true salivary glands secreting silk; also present in the hive bee and other adult hymenoptera.

mandrel (*Eng*) (1) Accurately turned rod over which metal is forged, drawn or shaped during working so as to create or preserve desired axial cavity. A tapered mandrel is also used for holding and locating a bored component so that external diameters can be machined true to the bore. Also *arbor* (US), *mandril*. (2) Internal metal part of pipe-forming die in polymer EXTRUSION. Held in position by spider screws.

mandrel (*Glass*) (1) A refractory tube used in the manufacture of glass tubing or rod. (2) A former used in making lamp-blown articles from tubing and in making precision-bore tubing.

manganates (*Chem*) See PERMANGANATES.

manganepidote (*Min*) See PIEDMONTITE.

manganese (*Chem*) Symbol Mn, at no 25, ram 54·94, valency 2,3,4,6,7, rel.d. at 20°C 7·39, mp 1220°C, bp 1900°C, electrical resistivity $5·0 \times 10^{-8}\ \Omega$ m, hardness in Mohs' scale, 6. A hard, brittle metallic element, in the seventh group of the periodic system, which exists in four polymorphic forms, α, β, γ and δ, the first two of which have complicated crystal structures. It is brilliant white in colour, with reddish tinge. The principal manganese ores are *hausmannite, psilomelane, pyrolusite*; pure manganese is obtained electrolytically. Manganese is mainly used in steel manufacture, as a deoxidizing and desulphurizing agent.

manganese alloys (*Eng*) Manganese is not used as the basis of alloys, but is a common constituent in those based on other metals. It is present in all steel and cast-iron, and in larger amount in special varieties of these, eg MANGANESE STEEL, SILICO-MANGANESE STEEL; also in many varieties of brass, in aluminium-bronze, and aluminium- and nickel-base alloys.

manganese bronze (*Eng*) Originally an alpha–beta brass containing about 1% of manganese; the term is now applied generally to HIGH-STRENGTH BRASS, with up to 4% manganese. See COPPER ALLOYS.

manganese dioxide (*Chem*) *Manganese (IV) oxide.* MnO_2. Black solid, insoluble in water; occurs in PYROLUSITE. Basic and (mainly) acidic. Forms manganites. Oxidizing agent; uses: depolarizer in LECLANCHÉ CELLS, decolorizing oxidant for green ferrous ion in glass, laboratory reagent.

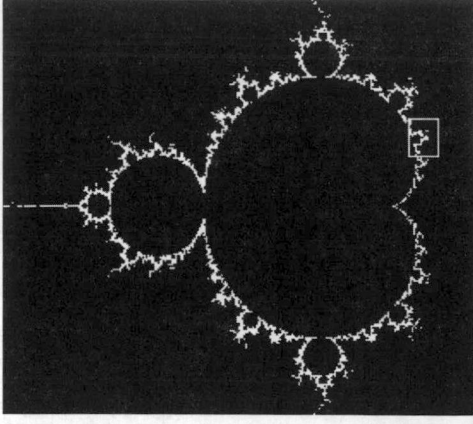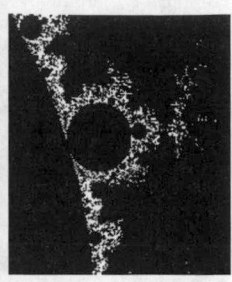

Mandelbrot set. The image on the right is an enlargement of the box on the left-hand image.

manganese epidote (*Min*) See PIEDMONTITE.

manganese garnet (*Min*) See SPESSARTITE.

manganese heptoxide (*Chem*) Manganese (*VII*) oxide. Mn_2O_7. A heavy, dark-coloured oil, prepared by adding concentrated sulphuric acid to potassium manganate (VII). Unstable, explosive, strong oxidizing agent. Acidic. Forms manganates (VII).

manganese nodules (*Geol*) Irregular small concretions with high concentrations of Mn together with Fe, Cu, Ni minerals. They are most common on the deep-ocean floors, notably the N Pacific where there is little sedimentation. See BIRNESSITE, TODOROKITE.

manganese spar (*Min*) See RHODOCHROSITE.

manganese steel (*Eng*) A term sometimes applied to any steel containing more manganese than is usually present in carbon steel (ie 0·3–0·8%), but generally to austenitic manganese steel (known as *Hadfield's steel*) which contains 11–14%. This steel is resistant to shock and wear. Used for railway crossings, rock-crusher parts, etc.

manganin (*Eng*) A copper-base alloy, containing 13–18% of manganese and 1·5–4% of nickel. The electrical resistivity is high (about $38 \times 10^{-8}\ \Omega\,m$), and its temperature coefficient low; it is therefore suitable for resistors. See COPPER ALLOYS.

manganite (*Min*) A grey or black hydrated oxide of manganese crystallizing in the monoclinic system (*pseudo-orthorhombic*). It is a minor ore of manganese, and also occurs in deep-sea MANGANESE NODULES.

mangano-manganic oxide (*Chem*) Manganese (*II,III*) oxide. Mn_3O_4. A *spinel*.

manganophyllite (*Min*) A phlogopite or biotite mica containing appreciable manganese.

manganosite (*Min*) An oxide of manganese (MnO) which crystallizes in the cubic system.

manganous oxide (*Chem*) Manganese (*II*) oxide. MnO. Basic. Forms manganese (II) salts. Occurs naturally as *manganosite*.

mange (*Vet*) Inflammation of the skin of animals due to infection by certain species of ascarids or mites. See CHORIOPTIC MANGE, DEMODECTIC MANGE, NOTOEDRIC MANGE, OTODECTIC MANGE, PSOROPTIC MANGE, SARCOPTIC MANGE.

mangerite (*Min*) An intermediate member of the charnockite rock series, the equivalent of hypersthene monzonite.

mangle (*Textiles*) A machine consisting of two or more rollers running in contact with each other. Wet fabric is passed through the nip between the rollers to have as much liquid as possible removed.

mania (*Med*) Mental abnormality where the mood is of extreme elation with speed of thought giving a flight of ideas. Patient has little or no insight and business or family affairs can be seriously damaged.

manic-depressive (*Psych*) See BIPOLAR DISORDER.

manifest dream content (*Psych*) See FREUD'S THEORY OF DREAMS.

manifold (*Eng*) A pipe or chamber with several openings. See EXHAUST MANIFOLD, INDUCTION MANIFOLD.

manifold (*MathSci*) A HAUSDORFF SPACE with a countable basis, which is a union of open sets each of which can be mapped homeomorphically to an open set in some fixed *n*-dimensional Euclidean space.

manifold bank (*Paper*) A thin bank paper, generally around $30\ g\,m^{-2}$.

manifold pressure (*Aero*) The absolute pressure in the induction manifold of a reciprocating aero-engine which, indicated by a cockpit gauge, is used together with rpm settings to control engine power output and fuel consumption. See SUPERCHARGE.

manifold pressure gauge (*Aero*) See BOOST GAUGE.

manila (*Textiles*) A rough bast fibre obtained from the leaves of *Musa textilis*, a banana-like plant grown in the Philippines; used in sacking, rope-making, etc.

manilla paper (*Paper*) Paper generally made from unbleached chemical wood pulp, with admixtures of waste paper and/or mechanical wood pulp and buff/brown in colour (unless dyed). Machine-glazed varieties are generally intended for conversion into envelopes. Solid or pasted varieties are generally used for files or folders.

manio (*For*) S American softwood (*Podocarpus chilensis*) similar and related to the PODO.

manipulator (*NucEng*) Remote handling device used eg with radioactive materials, the operator being protected by a lead-glass window while watching the process.

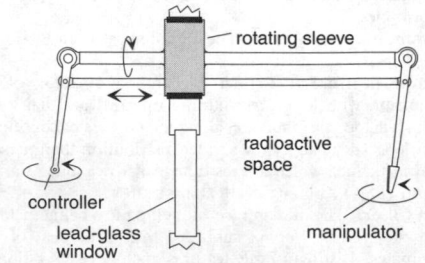

manipulator Limited movements drawn.

man lock (*CivEng*) An air lock enabling workers to pass into, and out of, spaces filled with compressed air.

man–machine interface (*ICT*) See USER INTERFACE.

manna (*BioSci*) See HONEY DEW.

mannans (*Chem*) The condensation polymers of mannose.

Mannerism (*Arch*) A recently coined term to describe the predominantly Italian architectural style which attempted to expand upon the classical ideals of the RENAISSANCE (1530–1600). The Mannerist phase is often referred to as *Proto-Baroque*, when the departure from the ancient Roman precedents, both in plan forms and decorative practices, became apparent.

Mannesmann process (*Eng*) A process for making seamless metal tubing from a solid bar of metal by the action of two eccentrically mounted rolls which simultaneously rotate the bar and force it over a mandrel. Also *Mannesmann piercing*.

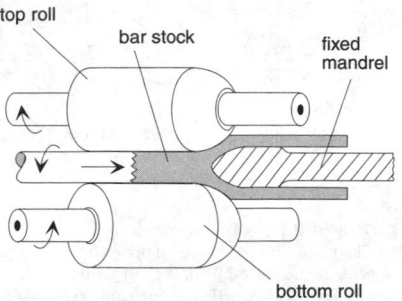

Mannesmann process The fixed mandrel protects the core of the bar from the elongation forced on its outer region causing the centre to separate.

Mannesmann tubes (*Eng*) Cylindrical piles made up of shorter lengths of steel tubes butt-welded together. The thickness of the individual pieces is varied to increase the bending strength. Used in large diameters (2 m) typically for single-pile mooring DOLPHINS.

Mannheim process (*Chem*) In manufacture of sulphuric acid, catalysis of SO_2, in two stages, first with iron (III) oxide and then with platinized asbestos.

mannitol (*Chem*) $HOCH_2(CHOH)_4CH_2OH$. Mp 166°C. A hexahydric alcohol, fine needles or rhombic prisms, soluble in water and in hot alcohol. It is found in many plants, and is an important osmotic diuretic used to treat cerebral oedema.

mannosans (*Chem*) See MANNANS.

mannose (*Chem*) An aldohexose. D-mannose can be obtained by the oxidation of MANNITOL, and is a stereoisomer of D-glucose.

manocryometer (*Chem*) A combination manometer and cryoscope for measuring the effect of pressure on melting point.

manoeuvre demand system (*Aero*) A FLY-BY-WIRE automatic control system in which the pilot's action determines a required manoeuvre, eg a pitch up of a certain value of *g*, the automatic control system then setting the control deflections to achieve the desired result while maintaining overall stability despite changes of speed.

manometer (*Phys*) An instrument used to measure the pressure of a gas or liquid. The simplest form consists of a U-tube containing a liquid (water, oil or mercury), one limb being connected to the enclosure whose pressure is to be measured, while the other limb is either open to the atmosphere or is closed or otherwise (eg by float) connected to a registering or recording mechanism. There are, however, numerous more sophisticated variations (eg balanced hollow ring, bell chamber, diaphragm and metallic bellows). See PRESSURE GAUGE.

manoscopy (*Chem*) The measurement of gas densities.

manostat (*Eng*) A device for keeping the pressure in an enclosure at a constant level.

manoxylic wood (*BioSci*) Secondary xylem in gymnosperms with very wide parenchymatous rays between the small groups of conducting cells, as in cycads. Cf PYCNOXYLIC WOOD.

man-riding (*MinExt*) Said of equipment on oil rigs which is used by personnel and is built to a higher standard of safety than material handling systems.

mansard roof (*Build*) A double-sloped pitched roof rising steeply from the eaves, and having a summit of flatter slope on both sides of the ridge. Also *curb roof*, *French roof*, *gambrel roof*.

mantel tree (*Build*) The lintel of a fireplace.

mantissa (*MathSci*) The positive, fractional part of a LOGARITHM.

mantle (*BioSci*) (1) In Urochordata, the true body wall lying below the test and enclosing the atrial cavity. (2) In Mollusca, Brachiopoda and Cirripedia, a soft fold of integument that encloses the trunk and which is responsible for the secretion of the shell or carapace.

mantle (*Geol*) That part of the Earth between CRUST and CORE, ranging from depths of approx 40 to 2900 km.

mantle cavity (*BioSci*) In Urochordata, the atrial cavity; in Mollusca, Brachiopoda and Cirripedia, the space enclosed between the mantle and the trunk.

mantled gneiss dome (*Geol*) A dome-like structure consisting of granite at the centre, surrounded by gneiss, and intruded into low-grade regionally metamorphosed rocks. Such structures are found in Precambrian shield areas.

Mantoux test (*BioSci*) A form of TUBERCULIN TEST commonly used in humans to indicate present or past infection with *Mycobacterium tuberculosis* (or previous immunization with BCG). A tuberculin preparation, usually PPD, is injected intracutaneously. A positive test indicates that delayed hypersensitivity is present.

manubrium (*BioSci*) (1) Any handle-like structure. (2) The anterior sternebra in mammals. (3) Part of the malleus of the ear in mammals. (4) The pendant oral portion of a medusa.

manufactured fibre (*Textiles*) A regenerated or synthetic fibre that has been manufactured, as distinct from a natural fibre. Manufactured fibres are obtained as continuous filaments which may be cut or broken into staple fibres.

manufacturing gauge (*Eng*) A gauge used by machine operators in the production of parts, as distinct from an inspection or master gauge.

manus (*BioSci*) The podium of the fore-limb in land vertebrates.

map (*BioSci*) Depending on the level of organization, it can refer to the ordering of genes in an eukaryotic chromosome or to the determination of the arrangement of DNA sequences in a gene or cluster of genes. Thus *gene mapping* involves the determination of the positions and relative distances of genes on chromosomes by means of their LINKAGE.

map (*MathSci*) See MAPPING.

map comparison unit (*Radar*) See CHART COMPARISON UNIT.

maple (*For*) Various species of the hardwood genus *Acer*, found in Europe, Asia and America. They have in common wood of a creamy-white to light-tan colour, generally straight-grained with a fine texture. Also *European maple*, *Japanese maple*, *rock maple*, *soft maple*.

maple syrup (*Chem*) Syrup made from sap tapped from the maple tree. The delicate flavour is probably due to the small protein content. Similarly *maple sugar*.

map matching (*Aero*) Navigation by autocorrelation of terrain with data stored in aircraft, missile or RPV, often in the form of film. See TERCOM.

map measurer (*Surv*) An instrument used to find the length of a route on a map. It consists of a small wheel which is made to roll over the route, in so doing actuating a needle which records the distance traversed. Also *opisometer*.

mapping (*MathSci*) In modern mathematical literature generally regarded as equivalent to FUNCTION esp between abstract spaces. Also *map*.

MAR (*Chem*) Abbrev for *microanalytical reagent*, a standard of purity which indicates that a reagent is suitable for use in *microanalysis*.

MAR (*ICT*) Abbrev for MEMORY ADDRESS REGISTER.

maraging (*Eng*) Heat treatment, 'martensite ageing', used to harden alloy steels (commonly those containing 18% nickel), involving precipitation of intermetallic compounds in a carbon-free martensite. These include nickel–iron martensites with high toughness and resistance to shock and saline corrosion.

marangoni convection (*Space*) The flow resulting from gradients in surface tension giving rise to the transfer of heat and mass; it is particularly relevant to MICROGRAVITY conditions when gravity-induced convection is absent.

Maranyl (*Plastics*) TN for nylon moulding material (UK).

marasmus (*Med*) Progressive wasting, esp in infants. Adj *marasmic*.

marble (*Geol*) The term strictly applies to a granular crystalline metamorphosed limestone, but in a loose sense it includes any calcareous or other rock of similar hardness that can be polished for decorative purposes.

marble bones (*Med, Vet*) See OSTEOPETROSIS, OSTEOPE-TROSIS GALLINARUM.

marblewood (*For*) A straight-grained, fine and evenly textured hardwood with figuring of dark black bands on a grey-brown background, from *Diaspyros kurzii*, a native of the Andaman and Nicobar Islands.

marbling (*Print*) The operation of decorating book-edges etc with a variegated 'marble' effect. Carried out in a trough containing gum (made from carragheen moss), on the surface of which pigments (containing ox-gall) are worked in intricate patterns.

Marburg disease (*Med*) Disease caused by infection by a filovirus. Marburg and EBOLA DISEASE are collectively known as *African haemorrhagic fever*. Mortality rate is high.

marcasite (*Min*) (1) A disulphide of iron which crystallizes in the orthorhombic system. It resembles iron pyrites, but has a rather lower density, is less stable, and is paler in colour when in a fresh condition. (2) In the gemstone trade, *marcasite* is pyrites, polished steel (widely used in ornamental jewellery in the form of small 'brilliants'), or even white metal.

marcella (*Textiles*) Cloth, often cotton, used for the fronts of dress shirts, and with a PIQUÉ structure and superimposed fancy woven design.

marcescent (*BioSci*) Withered but remaining attached to the plant.

marching modulus (*Eng*) A term describing a type of overcure of rubber where cross-linking continues at a slow rate after end of cure cycle.

marchioness (*Build*) A slate, 22 × 12 in (558 × 305 mm).

Marconi antenna (*ICT*) Original simple vertical wire, fed between base end and earth.

Marconi–Franklin beam array (*ICT*) A BROADSIDE ANTENNA consisting of a curtain of dipoles, mounted end to end and fed in phase; often with a second reflecting curtain. Gives a sharp beam with little upward radiation, for long-distance short-wave communication.

marcus (*Build*) A large hammer with an iron head.

mare (*Astron*) Any of various darkish level areas on the planets of the solar system and their moons. See LUNAR MARIA.

marekanite (*Geol*) A rhyolitic perlite broken down into more or less rounded pebbles, named from the type locality, Marekana River, in E Siberia.

Marek's disease (*Vet*) See FOWL PARALYSIS.

Marezzo marble (*Build*) An artificial marble made with Keene's cement.

Marfan syndrome (*Med*) A genetic disorder that affects the connective tissue in the body, usually leading to extreme tallness and cardiovascular and optical defects. Also *Marfan's syndrome*.

Marform process (*Eng*) A proprietary production process for forming and deep-drawing irregularly shaped sheet-

metal parts, in which a confined rubber pad is used on the movable platen of a press and a rigid punch on the fixed platen. The unformed portion of the blank is subjected to sufficient pressure during the progress of the operation to prevent wrinkling.

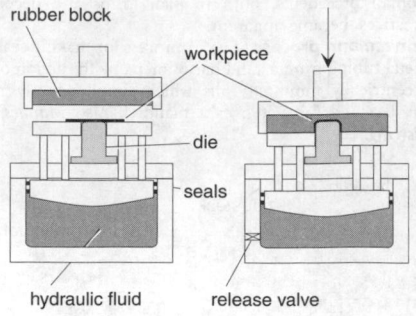

Marform process

margaric acid (*Chem*) *Heptadecanoic acid*. $CH_3(CH_2)_{15}COOH$. A long-chain saturated 'synthetic' fatty acid. It has an odd number of carbon atoms in the molecule in contrast with the even number of almost all the naturally occurring fatty acids.

margarine (*Chem, FoodSci*) A butter substitute made from various mixtures of animal and vegetable fats, suitably treated by heating and cooling, churning with milk, colouring, and adding concentrates of vitamins A and D.

margarite (*Geol*) An aggregate of minute sphere-like crystallites, arranged like beads, found as a texture in glassy igneous rocks.

margarite (*Min*) Hydrated silicate of calcium and aluminium, crystallizing in the monoclinic system, often as lamellae with a pearly lustre; one of the so-called BRITTLE MICAS.

margin (*Build*) The flat surface of stiles or rails in panelled framing.

margin (*ICT*) In an amplifier or system, the increase in the number of decibels in gain before oscillation occurs.

marginal (*BioSci*) A species or community occurring at the boundary between two distinct habitats.

marginal bars (*Build*) Glazing bars so arranged as to divide the glazed opening into a large central part bordered by narrow panes at the edges.

marginal meristem (*BioSci*) Meristem along the margin of a leaf primordium giving rise to the tissues of the blade.

marginal ore (*MinExt*) Ore which, at current market value of products from its excavation and processing, just repays the cost of its treatment.

marginal testing (*Electronics*) A form of test where the operation of a piece of equipment is tested with its operating conditions altered to decrease the safety margin against faults, eg an amplifier may be required to give a certain minimum gain with reduced supply voltage.

margin draft (*Build*) A smooth face round a joint in ashlar work.

margin lights (*Build*) Narrow panes of glass near the edges of a sash.

margin of safety line (*Ships*) A line drawn 3 in below the upper surface of the bulkhead deck at the side. A passenger ship is subdivided into watertight compartments in such a way that this line is not submerged if two adjacent watertight compartments are flooded.

margin trowel (*Build*) A box-shaped float for finishing internal angles.

marialite (*Min*) Silicate of aluminium and sodium with sodium chloride, crystallizing in the tetragonal system. It is one of the end-members in the isomorphous series of the scapolite group.

marihuana (*BioSci*) Also *marijuana*. See CANNABIS.

marinate (*FoodSci*) The steeping of meat or fish in an alcoholic, acetic or citric liquor (the *marinade*) often flavoured with herbs or other volatile compounds prior to cooking. Marinating provides flavour and tenderizes the flesh.

marine boiler (*Eng*) A cylindrical boiler, of large diameter and short length, provided with two or more furnaces in flue tubes leading to combustion chambers, surrounded by water, at the back. The gases pass through banks of fire-tubes to the smoke-box or uptakes at the boiler front. Also *Scotch boiler*. See DRY-BACK BOILER, YARROW BOILER.

marine chronometer (*Genrl*) A specially mounted chronometer for use on board ship in the determination of longitude.

marine coatings (*Build*) Paints or varnishes specially formulated to withstand exposure to salt water, sea air, marine growths, extreme changes in climatic conditions, etc.

marine compass (*Genrl*) See FLOATING COMPASS, GYRO-COMPASS.

marine engineering (*Ships*) That branch of mechanical engineering concerned with the design and production of propelling machinery and auxiliary equipment for use in ships.

marine engines (*Eng*) Engines used for ship propulsion.

marine erosion (*Geol*) The processes, both physical and chemical, which are responsible for the wearing away and destruction of coastlines.

marine glue (*Build*) A form of glue resisting the action of water.

marine park (*EnvSci*) An area of the sea bed that is set aside for wildlife conservation, as a marine equivalent of a nature reserve.

Mariner program (*Space*) A series of spacecraft launched by NASA to begin the exploration of the solar system (1962–75).

marine screw propeller (*Ships*) A boss, carrying two, three or four blades of helical form, which produces the thrust to drive a ship by giving momentum to the column of water which it displaces astern.

marine surveying (*Surv*) Hydrographical surveying undertaken in tidal waters.

Mariotte's law (*Phys*) See BOYLE'S LAW.

mark (*ICT*) Departure (positive or negative) from a neutral or no-signal state (space) in accordance with a code, using equal intervals of time. See MARK/SPACE RATIO.

mark and recapture (*BioSci*) A method of animal census, in which a marked sample of animals is returned to the population, allowed to mix with unmarked individuals, and then the proportion of the marked individuals in a second sample is used to derive an estimate of the population size. Also *Lincoln index*.

Markarian galaxy (*Astron*) A galaxy with excessive ultra-violet emission. See panel on GALAXY.

marker (*ICT*) A device used in a common control system that tests and selects incoming and outgoing circuits and causes switching apparatus to connect them together.

marker (*Radar*) A pip on a radar display for calibration of range and direction.

marker antenna (*Aero*) One giving a beam of radiation for marking air routes, often vertically for blind or instrument landing.

marker beacon (*Aero*) A radio beacon in aviation which radiates a signal to define an area above the beacon. See FAN BEACON, GLIDE-PATH BEACON, INNER BEACON, LOCALIZER BEACON, LOCATOR BEACON, MIDDLE BEACON, OUTER BEACON, RADAR BEACON, TRACK GUIDE, Z MARKER BEACON.

marker horizon (*Geol*) A layer of rock in a sedimentary sequence which, because of its distinctive appearance or fossil content, is easily recognized over large areas, facilitating correlation of strata.

marker light (*CivEng*) An indicating light on a signal post, to indicate the position or aspect of the main signal should its light have failed.

marker pulses (*ICT*) (1) Pulses superimposed on a cathode-ray-tube display for timing purposes. (2) Pulses used to synchronize transmitter and receiver, eg in time-division multiplex.

markers (*Aero*) See AIRPORT MARKERS.

markfieldite (*Geol*) A variety of diorite with a porphyritic structure and a granophyric groundmass.

Mark–Houwink equation (*Chem*) The basis for calculation of viscosity-average molecular mass of polymers, M, from intrinsic viscosity, η, measurements. The equation is:

$$\eta = KM^a$$

where K and a are experimentally determined constants.

marking blue (*Eng*) (1) A mixture of spirit and blue dye applied to metal objects before scratching dimensions onto work. (2) Used for bedding two surfaces together. See REDDLE.

marking gauge (*Build*) A tool for marking lines on the work parallel to one edge of it. It consists of a wooden bar having a projecting steel marking pin near one end and a sliding block adjustable for position along the bar.

marking knife (*Build*) A small steel tool having a chisel edge at one end and pointed at the other. It is used for setting out fine work.

marking-out (*Build, Surv*) Setting out boundaries and levels for a proposed piece of work.

marking-out (*Eng*) Setting out centre lines and other dimensional marks on material, as a guide for subsequent machining operations. See LAYING-OUT.

marking up (*Print*) (1) Adding typographic instructions to the copy or the layout as a guide to setting. (2) Adding codes to text for interpretation by a computer.

marking wave (*ICT*) In FREQUENCY-SHIFT KEYING a signal slightly different in frequency from the spacing signal that corresponds to the mark of the signal code. Also *keying wave*.

mark of reference (*Print*) A sign which directs the reader to a footnote. The commonest marks of reference are * asterisk, † dagger, ‡ double dagger, § section, ‖ parallel, ¶ paragraph, in order of use. SUPERIOR FIGURES are now more commonly used.

Markov chain (*MathSci*) A sequence of observations in which the probability of each member depends on the probability of the preceding event.

mark sense reader (*ICT*) Input device that reads special forms (or cards), usually by electrically sensing the marks made in predetermined positions. Also *mark sense device*.

mark/space ratio (*ICT*) The ratio of the time occupied by a MARK to that occupied by the space in between marks in a telecommunication channel or recording system using pulsed rather than continuous signals.

marl (*Build*) A brick earth which contains a high percentage of carbonate of lime; it is the best clay for making bricks without addition of other substances. Also *calcareous clay*.

marl (*Geol*) A general term for a very-fine-grained rock, either clay or loam, with a variable admixture of calcium carbonate.

Marlex (*Plastics*) TN for high-density polyethylene (UK).

marl yarn (*Textiles*) Folded yarn made from single yarns of two or more colours. May be continuous filament or woollen spun single yarn.

marmite disease (*Vet*) See GREASY PIG DISEASE.

marmoration (*Build*) A marble casing for a building.

marmoratum (*Build*) A cement containing pulverized marble and lime. Also *marmoretum*.

marmorization (*Geol*) The recrystallization of a limestone by heat to give a marble.

marquise (*Arch*) A projecting canopy over the entrance to a building.

married print (*ImageTech*) A positive print of a motion picture with both picture and sound track images correctly synchronized for projection.

marrow (*BioSci*) The vascular connective tissue that occupies the central cavities of the long bones in most vertebrates, and also the spaces in certain types of cancellated bone.

marrying (*ImageTech*) The synchronization and make-up of the separate picture and sound track negatives for printing together.

Mars (*Astron*) The fourth planet in the solar system, and distinctly red even to the unaided eye. It moves on an eccentric orbit at a mean distance of 1·52 AU and a period of 686·98 days. Its diameter is half and its mass 0·107 that of the Earth. Atmospheric pressure is 0·007 compared with the Earth, and the composition is 95% carbon dioxide with small quantities of nitrogen and other gases. Its remarkable surface features were revealed by the Mariner (1965, 1969, 1971) and Viking (1976) probes. There are numerous large extinct volcanoes, huge canyons and a great many impact craters. There are two permanent ice caps of frozen carbon dioxide. It has two tiny natural satellites, PHOBOS and DEIMOS. Also *red planet*. See appendix on Planets.

marsh (*EnvSci*) Vegetation which is seasonally waterlogged and non-peat forming.

Marshall valve gear (*Eng*) A radial gear of the *Hackworth* type, in which the straight guide is replaced by a curved slot to correct inequalities in steam distribution.

marsh gas (*Chem*) Methane.

Marsupialia (*BioSci*) The single order included in the mammalian group METATHERIA, and having the characteristics of the subclass, eg opossum, Tasmanian wolf, marsupial mole, kangaroo, koala bear.

marsupium (*BioSci*) A pouch-like structure occupied by the immature young of an animal during the later stages of development; as the abdominal pouch of metatherian mammals. Adj *marsupial*.

Martello Tower (*Arch*) Low round tower for guns, built as a coastal defence in the UK from 1793.

martempering (*Eng*) Austenitizing a steel prior to hardening and then quenching to just above the M_s temperature (see ISOTHERMAL TRANSFORMATION DIAGRAM) until the temperature has equalized throughout the section but not long enough for bainitic transformation to begin, and then quenching to form martensite. Applied to steel components which are likely to crack as a result of thermal gradients if quenched directly from the austenitizing temperature. See AUSTEMPERING.

martensite (*Eng*) Non-equilibrium microstructure formed in steel when the austenite phase is cooled too rapidly for carbon to diffuse out of solid solution to form Fe_3C and thus transform to pearlite. The entrapped carbon distorts the lattice and retards the shear transformation from FACE-CENTRED CUBIC to BODY-CENTRED CUBIC, causing the product of the transformation to be a tetragonal (BODY-CENTRED TETRAGONAL, BCT) lattice. This has a high dislocation density associated with it, and the BCT lattice lacks the five independent slip systems necessary for ductility. It becomes harder and less ductile the greater the carbon content. See MARTENSITIC TRANSFORMATION and panel on CRYSTAL LATTICE.

martensitic transformation (*Eng*) A type of rapid transition from one crystal structure to another by shear (ie displacement) rather than the more common, and much slower, diffusion, nucleation and growth. Many materials show this transformation – it forms the basis of SHAPE-MEMORY ALLOYS – but only in steels is the product much harder and less ductile than the untransformed material. Because of its importance, the name martensite usually refers to that in steel. See panel on STEELS.

Martin's cement (*Build*) A quick-setting hard plaster, made by soaking plaster of Paris in a solution of potassium carbonate.

martite (*Min*) A variety of HAEMATITE (Fe_2O_3) occurring in dodecahedral or octahedral crystals believed to be pseudomorphous after magnetite, and in part perhaps after pyrite.

martius yellow (*Chem*) Salts of dinitro-α-naphthol, used as a pigment and for colouring soap; poisonous. Also *Manchester yellow*.

Martonite (*Chem*) A tear gas, a mixture (4:1) of mono-bromopropanone and monochloropropanone.

marver (*Glass*) A flat cast-iron or stone (marble) block upon which glass is rolled during the hand method of working.

mascon (*Astron*) Abbrev for *mass concentration*. Mascons are regions of high gravity occurring within certain LUNAR MARIA. Their origin is still conjectural.

maser (*Phys*) Abbrev for *microwave amplification by stimulated emission of radiation*. A microwave oscillator that operates on the same principle as the LASER (panel). Maser oscillations produce coherent monochromatic radiation in a very narrow beam. Less noise is generated than in other kinds of microwave oscillators. Materials used are generally solid-state, but masers have been made using gases and liquids.

maser relaxation (*Phys*) The process by which excited molecules in the higher energy state revert spontaneously or under external stimulation to their equilibrium or ground state. Maser action arises when energy is released in this process to a stimulating microwave field which is thereby reinforced.

mash seam welding (*Eng*) An electrical resistance welding process in which the slightly overlapping edges of the workpiece are forged together during welding by broad-faced, flat electrodes.

mask (*BioSci*) A prehensile structure of the nymphs of certain dragonflies (Odonata), formed by the labium.

mask (*Electronics*) In semiconductor technology, a patterned metal film on a glass substrate containing selected open areas. During integrated circuit manufacture a series of different masks and lithographic processes is used to build up a complex circuit structure. See SEMICONDUCTOR DEVICE PROCESSING.

mask (*ImageTech*) Opaque material used to determine the size and shape of a photographic image in either the camera or print.

mask (*Print*) (1) Material which blocks off portions of an illustration in order to protect it. (2) Opaque material used to protect specified areas of a printing plate during exposure. (3) Material placed in position over a film assembly to prevent the marked areas from being exposed on the printing plate, the technique being used to create the separate plates in multicolour printing when only one film assembly carrying all the information required for the work is used.

maskelynite (*Min*) A glass which occurs in colourless isotropic grains in meteorites and has the composition of a plagioclase. It probably represents re-fused feldspar.

masking (*Acous*) Loss of sensitivity of the ear for specified sounds in the presence of other louder sounds.

masking (*ICT*) An operation that selects certain of the bits in a register for subsequent processing. A register of equal length holds a bit pattern called a mask with each bit set to '1' where a corresponding bit is to be selected and '0' otherwise.

masking (*ImageTech*) See COLOUR MASKING.

masking agent (*Pharmacol*) A chemical that conceals the presence of another substance, esp one taken by an athlete to avoid testing positive for an illegal drug.

masking frame (*ImageTech*) A device for holding printing paper in an enlarger, comprising a base plate to which is hinged a frame forming a border for the print, with movable metal strips for adjusting horizontal and vertical margins.

masking paper (*Paper*) A paper, generally semi-bleached and crêped, coated on one side with a self-adhesive compound. Usually sold in coil form and may be peeled from its support after use.

masking tape (*Paper*) See MASKING PAPER.

masochism (*Psych*) Sexual gratification through having pain inflicted on oneself.

masonry cement (*Build*) A cement incorporating inert material and an air entrainer to increase plasticity in mortar. Specially useful in minimizing possibility of damage in cold-weather work.

mason's joint (*Build*) Pointing finished with a projecting vee.

mason's mitre (*Build*) An effect similar to the MITRE but produced (particularly in stonework) by shaping the intersection out of the solid.

mason's putty (*Build*) A mixture of Portland cement, lime putty and stone dust, usually in the proportions 2:5:7, with water. Used for making fine joints, esp in ashlar work.

mason's scaffold (*Build*) A form of scaffold used in the erection of stone walls, when it is not convenient to leave holes for the support of one end of the putlogs; an inner set of standards and ledgers is used to provide this support.

mason's stop (*Build*) See MASON'S MITRE.

mass (*Phys*) The quantity of matter in a body. The *inertial mass* is the mass as a measure of resistance of a body to changes in its motion, and the *gravitational mass* is the mass as a measure of the attraction of one body to another. The general theory of relativity shows that inertial and gravitational mass are equivalent. By a suitable choice of units they can be made numerically equal for a given body, which has been confirmed to a high degree of precision by experiment. See MASS-ENERGY EQUATION, RELATIVISTIC MASS EQUATION, REST MASS.

mass absorption coefficient (*Phys*) See ABSORPTION COEFFICIENT.

mass action (*Chem*) See LAW OF MASS ACTION.

mass balance (*Aero*) A weight or mass attached ahead of the hinge line of an aircraft control surface to give static balance with no moment about the hinge, and to reduce to zero inertial coupling due to displacement of the control surface. Mass balancing is a precaution against control surface FLUTTER.

mass balance (*EnvSci*) The balance between inputs and outputs of a process.

mass concrete (*CivEng*) Concrete which is not reinforced. Also *bulk concrete*.

mass control (*Acous*) Said of mechanical systems, particularly those generating sound waves, when the mass of the system is so large that the compliance and resistance of the system are ineffective in controlling motion.

mass decrement (*Phys*) (1) The measured mass of an isotope less its mass number. (2) Sometimes the nuclear binding energy expressed in MASS UNITS. See MASS DEFECT and panel on BINDING ENERGY OF THE NUCLEUS.

mass defect (*Phys*) (1) The nuclear BINDING ENERGY expressed in ATOMIC MASS UNITS. (2) The measured mass of an isotope less its mass number. Also *mass decrement*; this is recommended by the BSI for (2) above. In the USA, it is usually used for (1) and *mass excess* for (2). See PACKING FRACTION and panel on BINDING ENERGY OF THE NUCLEUS.

mass effect (*Eng*) The tendency for hardened steel to decrease in hardness from the surface to the centre, as a result of the variation in cooling throughout the section. Becomes less marked as the rate of cooling required for hardening decreases, ie as the HARDENABILITY of the steel imparted by the content of alloying elements increases. See CONTINUOUS-COOLING TRANSFORMATION DIAGRAM.

mass-energy equation (*Phys*) The equation $E = mc^2$, where E is energy, m is mass and c is the speed of light. Confirmed deduction from Einstein's special theory of relativity that all energy has mass. If a body gains energy E, its inertia is increased by the amount of mass $m = E/c^2$. Derived from the assumption that all conservation laws must hold equally in all frames of reference and using the principle of conservation of momentum, of energy and of mass. See REST-MASS ENERGY.

masseter (*BioSci*) An elevator muscle of the lower jaw in higher vertebrates. Adj *masseteric*.

mass excess (*Phys*) See MASS DEFECT.

mass-flow hypothesis (*BioSci*) The hypothesis proposing that translocation in the PHLOEM results from a continuous flow of water and solutes, esp sugars, through the sieve tubes from source to sink, the flow being driven by a hydrostatic pressure difference caused by the osmotic movement of water following the loading of solutes into the sieve tubes at the source end and their removal (unloading) at the sink end.

mass-haul curve (*CivEng*) A curve used in the design of earthworks involving cuttings and embankments, the abscissae representing distance along the centre line, and the ordinates the excess of cutting over filling, ie the material requiring to be hauled to another position.

massicot (*Min*) Yellow lead oxide. A rare mineral of secondary origin, associated with galena.

mass law (*Acous*) The law describing the sound transmission through walls. The transmission coefficient is approximately proportional to the inverse of the mass per unit area and to the inverse of the frequency, ie the TRANSMISSION LOSS increases by approximately 6 dB when mass or frequency are doubled. There are many exceptions.

mass limitation (*NucEng*) The method of CRITICALITY CONTROL in a plant in which the total mass of fissile material is limited.

mass-luminosity law (*Astron*) A relationship between the mass and absolute magnitude of stars, the most massive stars being the brightest; applicable to all MAIN SEQUENCE stars but excludes eg white dwarfs.

mass manufacture (*Eng*) A method of producing discrete products in a continuous way using hand, automatic or robotic assembly methods.

mass number (*Phys*) Total number of PROTONS and NEUTRONS in a nucleus, each being taken as a unit of mass. Also *nuclear mass*, *nucleon number*.

mass ratio (*Aero, Space*) The ratio of the fully fuelled mass of a rocket-propelled vehicle or stage to that when all fuel has been consumed.

mass resistivity (*ElecEng*) The product of the volume resistivity and the density of a given material at a given temperature.

mass spectrograph (*Phys*) A vacuum system in which positive rays of various charged atoms are deflected through electric and magnetic fields so as to indicate, in order, the charge-to-mass ratios on a photographic plate, thus measuring the atomic masses of isotopes with precision. System used for the first separation for analysis and early use of the isotopes of uranium.

mass spectrometer (*Phys*) A MASS SPECTROGRAPH in which the charged particles are detected electrically instead of photographically.

mass spectrum (*Phys*) See SPECTRUM.

mass stopping power (*Phys*) See STOPPING POWER.

mass storage device (*ICT*) A term used to denote a large-capacity BACKING STORE, such as a large disk drive or magnetic drum.

mass transfer (*ChemEng*) The transport of molecules by convection or diffusion, as in the operations of extraction, distillation and absorption.

mass transfer coefficient (*ChemEng*) The molecular flux per unit driving force. Symbol k, K.

mass unit (*Phys*) See ATOMIC MASS UNIT.

mass wasting (*Geol*) Downhill gravity movement of rock material, eg *landslides*, *rockfall*.

mast (*BioSci*) Fruit of beech, oak and other forest trees.

mast (*Eng*) A slender vertical structure which is not self-supporting and requires to be held in position by guy-ropes. Cf PYLON.

mast cell (*BioSci*) A type of cell with basophil cytoplasmic granules that are similar to but smaller than those of basophil leucocytes in the blood. The granules contain mainly histamine, serotonin and heparin. Mast cells bind the Fc region of IgE antibodies; reaction with an antigen (or the action of anaphylatoxin) causes extrusion of the granule contents, and the release of various pharmacologically active substances. There are two populations of mast cells. One is normally present in connective tissues, in the neighbourhood of small blood vessels. The other, *mucosal mast cells*, is induced by IL-3 secreted by T-cells.

mastectomy (*Med*) Surgical removal of the breast.

master (*Acous*) In vinyl disk record manufacture, the metal (negative) disk obtained by plating the original lacquer surface on which a recording has been cut.

master (*Eng*) (1) The term applied to special tools, gauges, etc, used for checking the accuracy of others used in routine work. (2) The chief or key member of a system, eg the 'master cylinder' of a hydraulic brake mechanism.

master (*ImageTech*) (1) The original 16 MM reversal film exposed in the camera, after processing. (2) A positive print made for protection or duplication, not for normal projection. (3) The original videotape recorded in the camera videotape recorder, used for editing. (4) The finished edit master from which video copies can be produced. (5) In video disk production, a glass disk coated with a photoresist layer which is exposed by a laser and washed leaving pits in its surface, it is then electroplated with nickel and this is used as a negative stamper to press the disks.

master alloy (*Eng*) An alloy enriched in the required components which can be added to a melt to bring it to the correct composition.

masterbatch (*Plastics*) Concentrate of polymer and additives supplied direct to moulders for addition to hopper. Eliminates cost of intermediate mixing.

master clock (*ElecEng*) A clock emitting impulses at set time intervals and used to keep timekeeping equipment and other clocks in synchronization.

master clock (*ICT*) See CLOCK.

master connecting-rod (*Aero*) The main member of the master and articulated assembly of a radial aero-engine. It incorporates the crank-pin bearing, and the *articulated rods* of the other cylinders oscillate on it by means of wrist pins.

master controller (*ElecEng*) A controller from which it is possible to operate switches etc at remote positions.

master curve (*Eng*) See VISCOELASTICITY.

master cylinder (*Eng*) The container holding the fluid which operates the mechanism of hydraulic brakes or clutch.

master file (*ICT*) The principal source of information for a job. Cf TRANSACTION FILE.

master frequency meter (*ElecEng*) See INTEGRATING FREQUENCY METER.

master gain control (*Acous*) (1) In a broadcasting studio, the attenuator connected between the programme input and main amplifiers in order to regulate the gain as desired. (2) In a stereo amplifier, the control which adjusts simultaneously the gains of both channels to equalize them.

master gauge (*Eng*) A standard or reference gauge made to specially fine limits; used for checking the accuracy of inspection gauges.

master group (*ICT*) In a FREQUENCY-DIVISION MULTIPLEX system, an assembly of 10 SUPERGROUPS, formed by remodulation of supergroups onto that number of new carriers. (In some systems just five supergroups constitute a master group.) A master group will then contain 600 (or 300) voice channels.

master oscillator (*ICT*) One, often of low power, that establishes the frequency of transmission of a radio transmitter.

master station (*ICT*) (1) Transmitting station from which one or more satellite stations receive a programme for rebroadcasting. (2) In a radio navigation system depending on a chain of synchronized transmitters, eg LORAN, the main transmitter to which the signals from all the others are referred.

master switch (*ElecEng*) A switch for controlling the effect of a number of other switches or contactors; eg if the master switch is open, none of the other switches is operative.

master tap (*Eng*) A substandard screw-tap, sometimes used after the plug-tap when great accuracy is required. See SUBSTANDARD INSTRUMENT.

master telephone transmission reference system (*ICT*) High-quality transmission system provided with means for calibrating its transmitting, attenuating and receiving components in terms of absolute units. Used for comparison of quality between different systems. Located at CCITT laboratories in Geneva.

mastic (*Build*) A permanently waterproof but flexible seal used in building applications. Usually a synthetic polymer, eg silicone.

mastic (*Chem*) A pale-yellow resin from the bark of *Pistacia lentiscus*, used in the preparation of fine varnishes.

mastic asphalt (*Build*) A mixture of bitumen with stone chippings or sand, used for roofing, roads, paving and damp-proof courses.

mastication (*BioSci*) The act of reducing solid food to a fine state of subdivision or to a pulp; chewing.

mastication (*Eng*) A method used in rubber technology to increase processability of raw rubber stock, often on a calender mill. Natural rubber stock, eg, has too high a molecular mass for mixing, so must be milled to break down chains mechanically.

masticator (*Chem*) Apparatus consisting of two revolving and heated cylinders studded with teeth or knives; used for converting rubber into a homogeneous mass.

masticatory (*BioSci*) Pertaining to the trituration of food by the mandibles, teeth or gnathobases, prior to swallowing.

Mastigomycotina (*BioSci*) A subdivision or class containing those Eumycota or true fungi in which the spores and/or gametes are motile. Includes the Chytridiomycetes and, in some schemes, the Domycetes.

Mastigophora (*BioSci*) A class of Protozoa that possess one or more flagella in the principal phase. Such organisms may be amoeboid but usually have a pellicle or cuticle. They are often parasitic but rarely intracellular. They have no meganucleus. Reproduction is mostly by longitudinal binary fission; nutrition may be holophytic, saprophytic or holozoic. Also *Flagellata*.

mastitis (*Med, Vet*) Inflammation of the mammary gland.

mastodynia (*Med*) Pain in the breast.

mastoid (*BioSci*) Resembling a nipple; as a posterior process of the otic capsule in the mammalian skull.

mastoidectomy (*Med*) Excision of the (infected) air cells of the mastoid bone.

mastoiditis (*Med*) Inflammation of the air cells of the mastoid bone.

mat (*Print*) The commonly used contraction for MATRIX.

match-boarding (*Build*) See MATCHED BOARDS.

matched boards (*Build*) Boards specially cut at the edges to enable close joints to be made, either by tongue and groove or by rebated edge. Also *match-boarding, matching, match-lining*.

matched filter (*Electronics*) One used to maximize the signal-to-noise ratio in order to detect weak echoes or those where jamming is present. The filter has an impulse response which corresponds precisely to the spectrum of the transmitted pulse(s), which may be frequency modulated or coded. See CODING, PULSE COMPRESSION.

matched load (*ICT*) One with impedance equalling the characteristic impedance of a source, line or waveguide, so that there is no reflection of power, and the received power is the maximum possible. Also *matched termination*. See TERMINATION.

matching (*Build*) See MATCHED BOARDS.

matching (*ICT*) Adjusting a load impedance to match the source impedance with a transformer or network, so that maximum power is received, ie so that there is no reflection loss due to mismatch. The principle applies to many physical systems, eg non-reflecting optical surfaces, use of a horn in loudspeakers to match impedance of vibrator to that of air, matching load on electrical transmission line. Reactance in the load impedance can be neutralized or tuned out by an equal reactance of opposite sign.

matching (*Radar*) Said of the insertion of *matching sections* into radio-frequency transmission lines, with the aim of minimizing power reflections at a MISMATCH. A matching section may consist of specifically chosen lengths of waveguide, stripline or coaxial cable having a different impedance from the main system and connected in series so as to cause impedance transformation; alternatively, a *matching stub* may be connected in parallel with the transmission path.

matching stub (*ICT*) Short- or open-circuited stub line attached to main line to neutralize reactive component of load and so improve matching.

matching transformer (*ICT*) One expressly inserted into a communication circuit to avoid reflection losses when the load and source impedances differ.

match-lining (*Build*) See MATCHED BOARDS.

matchwood (*For*) Billets suitable for manufacturing into matches.

matelassé (*Textiles*) Woven fabric with a quilted effect made using two warps and two wefts, and used for formal dress wear.

material implication (*MathSci*) In logic, the relationship between two statements, p and q, when 'if p then q', defined as 'p cannot be true and q false'. See TRUTH FUNCTION.

materialization (*Phys*) The production of mass from energy according to the MASS–ENERGY EQUATION; a common example is by *pair production* (electron–positron) from gamma rays.

materials database (*Genrl*) A collection of materials information (physical and chemical properties of available grades) usually kept in computerized form. The largest polymer manufacturers, for instance, supply such databases for their own materials on request (eg *EPOS*, *CAMPUS*). Most materials selection systems have their own databases.

materials handling (*Eng*) The process, which includes mechanical handling, of transporting and positioning raw materials, semi-finished and finished products in connection with industrial operations, by conveyors, cranes, trucks, hopperfeeds, etc.

materials identification (*Eng*) Various analytical methods applied to the problem of identifying unknown materials. Some methods are of general applicability (eg energy dispersive analysis of X-rays (EDAX) combined with scanning electron microscopy, optical microscopy and thermal analysis), while others are material specific (eg gel permeation chromatography for polymers) or more useful when applied to a specific material (eg infrared and ultraviolet spectroscopy for polymers). In its widest sense, identification will include determining the elements and compounds present, their composition and phase, as well as such microstructural features as grain structure (for metals and ceramics) or spherulite structure (for polymers). See panel on TRACE ELEMENT ANALYSIS.

materials matching (*Eng*) The method of materials analysis where the unknown sample is identified by comparison with standard samples available to analyst. More comprehensive materials identification methods will be needed if standards are not available.

materials science (*Genrl*) The study of chemical and physical properties of elements, compounds, mixtures (blends and alloys) and minerals for understanding of atomic and/or molecular causes.

materials selection (*Eng*) Activity involving matching materials available to product specification, particularly in terms of mechanical properties (eg elastic modulus and strength), thermal properties (eg melting and glass transitions, thermal conductivity), chemical properties (eg corrosion resistance and environmental stress cracking) and electrical properties (eg electrical conductivity and tracking resistance). It also includes processing properties and ease of manufacture, as well as costing. Methods

include merit index analysis and those based on value judgements. Some methods are computerized and cover metals, ceramics and polymers (eg PERITUS), others, like PLASCAMS, only apply to polymers. Problems of application include neglect of difficult-to-quantify properties like appeal and marketability, the need to update data regularly, and the mismatch between standard material properties and actual properties of real products.

materials substitution (*Eng*) Replacement of one material in a product by another, such as wood by glass-reinforced plastic, or the latter by thermoplastic in small boats. Similarly, high-tensile steels for mild-steel sheet in car bodies. Such replacement often demands substantial design changes to allow for different material properties. Motives include cost savings, rise in productivity, better properties or safety, eg polyethylene terephthalate for glass in beverage bottles.

materials technology (*Eng*) The application of materials science to the development and practical use of conventional or new materials, esp for manufactured products (eg metal-matrix composites, high-temperature polymers, mixed-oxide superconductors). See panel on HIGH-PER-FORMANCE POLYMERS.

materials testing reactor (*NucEng*) See HIGH-FLUX REACTOR.

maternal effect (*BioSci*) A non-heritable influence of a mother on characters in her offspring, eg through her milk supply.

maternal immunity (*BioSci*) Passive immunity acquired by the newborn animal from its mother. In humans and other primates this is mainly by trans-placental transfer of IgG antibodies. In species that have thicker placentas, such as ungulates, antibodies are not transferred *in utero* but are acquired from the colostrum, and are absorbed intact from the gut during the first few days of life. The young of birds acquire maternal immunity from antibody in the egg yolk.

maternal inheritance (*BioSci*) Inheritance through the maternal cell line, eg through the oöcyte and eggs. Mitochondrial genes are maternally inherited.

mathematical induction (*MathSci*) The axiom of arithmetic that if a proposition is true of 1, and whenever it is true of any n it is also true of $n + 1$, then it is true of all positive integers.

mathematical logic (*MathSci*) The study of logic as a system of axioms and rules of inference. The attempt to reduce arithmetic and hence all of mathematics to logic has revealed a number of paradoxes, several of which have yet to be resolved. Also *symbolic logic*. Cf BOOLEAN ALGEBRA.

mathematical modelling (*Genrl*) The representation by mathematical expressions of a physical event as an aid to understanding the process.

mathematics (*Genrl*) The study of the logical consequences of sets of axioms. Pure mathematics, roughly speaking, comprises those branches studied for their own sake or their relation to other branches. The most important of these are algebra, analysis and topology. The term *applied mathematics* is usually restricted to applications in physics. Applications in other fields, eg economics, mainly statistical, are sometimes referred to as *applicable mathematics*.

maths co-processor (*ICT*) An additional PROCESSOR that performs FLOATING-POINT arithmetic and improves the speed of a system undertaking such arithmetic. Now often integral with the central processing unit.

mating (*Build*) Said of surfaces or pieces which come into contact or interlock.

mating type (*BioSci*) A group of individuals, within a species, that cannot breed among themselves but that are able to breed with individuals of other such groups. See INCOMPATIBILITY, MINUS STRAIN, PLUS STRAIN.

matric potential (*BioSci*) That component of the WATER POTENTIAL due to the interaction of the water with colloids and to capillary forces (surface tension). Often important component in soils and cell walls. Symbol ψ.

matrix (*BioSci*) More or less continuous matter in which something is embedded, eg the non-cellulosic substances of the cell wall in which the cellulose microfibrils lie in vascular plant cell walls, the intercellular ground substance of animal connective tissues.

matrix (*Build*) The lime or cement constituting the cementing material that binds together the aggregate in a mortar or concrete.

matrix (*Eng*) The component of a composite material in which the fibres or filler materials are embedded, and which variously transfers stresses to them, prevents fibres from buckling and protects their surfaces.

matrix (*ImageTech*) A network of electronic circuits to combine several signal sources in a specified mathematical arrangement, eg to transform colour co-ordinates in a colour TV system.

matrix (*MathSci*) An array of elements, eg real or complex numbers, arranged in a square or rectangular formation:

$$\begin{pmatrix} a & b & c \\ d & e & f \\ g & h & i \end{pmatrix}$$

matrix (*PowderTech*) The phase or phases which form the continuous skeleton of a powder body, thus forming the cells in which constituents imparting particular qualities may be held.

matrix (*Print*) The mould from which type is cast, produced from an impression with a punch; also the mould made from a relief surface in stereotyping and electrotyping. Usually contracted to *mat*.

matrixing (*ImageTech*) In colour TV, performing a colour co-ordinate transformation by computation with electrical or optical methods.

matrix metalloproteinases (*BioSci*) Proteolytic enzymes that degrade extracellular matrix. Include collagenases and elastases. Inhibitors are predicted to have benefits in arthritis and metastasis though this remains to be proven. Abbrev *MMPs*.

matroclinous (*BioSci*) Exhibiting the characteristics of the female parent more prominently than those of the male parent. Cf PATROCLINOUS.

matromorphic (*BioSci*) Resembling the mother.

matt (*Genrl*) Smooth but dull; tending to diffuse light; said eg of a surface painted or varnished so as to be dull or flat. Also *matte*.

matte (*ImageTech*) An opaque mask determining the image area exposed in a motion picture camera or printer. It may be a physical object in front of the camera lens or a strip of film with silhouette images of high density, used in special-effects work for image combination.

matte (*MinExt*) Fusion product consisting of mixed sulphides produced in the smelting of sulphide ores. In the smelting of copper, for instance, a slag containing the gangue oxides and a matte consisting of copper and iron sulphides are produced. The copper is subsequently obtained by blowing air through the matte, to oxidize the iron and sulphur.

matte box (*ImageTech*) An extension in front of the lens of a motion picture camera in which filters and mattes can be mounted.

matter (*Phys*) The substance of which the physical universe is composed. Matter is characterized by gravitational properties (on Earth by weight) and by its indestructibility under normal conditions.

matte shot (*ImageTech*) A special-effects shot involving the combination of images by the use of MATTES.

matt finish (*Build*) A surface or finish which has no sheen or gloss.

Matthiessen hypothesis (*ElecEng*) The hypothesis stating that the total electrical resistivity of a metal may be equated to the sum of the various resistivities due to the different sources of scattering of free electrons; this also applies to thermal resistivity. Also *Matthiessen rule*.

Matura diamonds (*Min*) Colourless (fired) zircons from Ceylon (Sri Lanka), which on account of their brilliancy are useful as gemstones. A misleading name.

maturation (*BioSci*) Final stages in the development of the germ cells; more generally, the process of becoming adult or fully developed.

maturation divisions (*BioSci*) The divisions by which the germ cells are produced from the primary spermatocyte or oöcyte, during which the number of chromosomes is reduced from the diploid to the haploid number. See MEIOSIS.

maturation of behaviour (*Psych*) Used in a variety of ways, although common to all is the idea of a genetically programmed behavioural repertoire that develops as the animal matures and that is, to a greater or lesser extent, relatively insensitive to environmental influence.

maturity (*Geol*) (1) A stage of the geomorphological cycle characterized by maximum relief and well-developed drainage. (2) The ultimate stage in the development of a sediment, characterized by stable minerals and rounded grains.

MATV (*ImageTech*) Abbrev for *master antenna television*, in which a number of individual receivers are fed from one communal antenna.

MAU (*ICT*) Abbrev for MEDIA ACCESS UNIT.

maul (*Build*) See BEETLE.

maulstick (*Build*) See MAHLSTICK.

Maunder diagram (*Astron*) See BUTTERFLY DIAGRAM.

mauveine (*Chem*) Perkin's mauve, the first synthetic dyestuff.

max gross (*Aero*) See WEIGHT.

maxilla (*BioSci*) In vertebrates, the upper jaw; a bone of the upper jaw; in Arthropoda, an appendage lying close behind the mouth and modified in connection with feeding. Pl *maxillae*. Adjs *maxillary, maxilliferous, maxilliform*.

maxillary (*BioSci*) (1) Pertaining to a maxilla or to the upper jaw. (2) A paired membrane bone of the vertebrate skull that forms the posterior part of the upper jaw.

maxillary glands (*BioSci*) See ANTENNAL GLANDS.

maxilliped (*BioSci*) In Arthropoda, esp Crustacea, an appendage behind the mouth, adapted to assist in transferring food to the mouth.

maxillofacial implant (*Med*) A type of implant composed of silicone or polyurethane rubber and polytetrafluorethylene for soft facial tissue (such as gum, skin, etc) or polymethyl methacrylate, vitallium, composite materials for harder tissue.

maximally flat (*ICT*) Said of amplifiers so designed that the circuit elements are transformed from filter sections incorporating stray admittances.

maximum and minimum thermometer (*EnvSci*) An instrument for recording the maximum and minimum temperatures of the air between two inspections, usually a period of 24 h. A type widely used is *Six's thermometer*.

maximum continuous rating (*Aero*) See POWER RATING.

maximum demand (*ElecEng*) The maximum load taken by an electrical installation during a given period. It may be expressed in kW, kV A or A.

maximum-demand indicator (*ElecEng*) An instrument for indicating the maximum demand which has occurred on a circuit within a given period.

maximum-demand tariff (*ElecEng*) A form of charging for electrical energy in which a fixed charge is made, depending on the consumer's maximum demand, together with a charge for each unit (kW h) consumed.

maximum equivalent conductance (*ElecEng*) The value of the equivalent conductance of an electrolytic solution at infinite dilution with its own solvent.

maximum flying speed (*Aero*) See FLYING SPEED.

maximum landing weight (*Aero*) See WEIGHT.

maximum permissible concentration (*Radiol*) The recommended upper limit for the dose which may be received during a specified period by a person exposed to ionizing radiation. Also *permissible dose*.

maximum permissible dose rate (*Radiol*) That dose rate which, if continued throughout the exposure time, would lead to the absorption of the maximum permissible dose. Similarly *maximum permissible flux*.

maximum permissible level (*Radiol*) A phrase used loosely to indicate maximum permissible concentration, dose or dose rate.

maximum point on a curve (*MathSci*) A peak on a curve. Formally, the curve $y = f(x)$ has a maximum at a if its derivative is 0 at a and $f(a) - f(b)$ is positive for all b in a neighbourhood of a.

maximum-reading accelerometer (*Aero*) See ACCELEROMETER.

maximum safe airspeed indicator (*Aero*) A pilot's *airspeed indicator* with an additional pointer showing the INDICATED AIRSPEED corresponding to the aircraft's LIMITING MACH NUMBER and also having a mark on the dial for the maximum permissible airspeed.

maximum sustainable yield (*EnvSci*) The maximum yield that can be obtained from a crop or a population without causing a progressive reduction. It is often grossly overestimated as eg in the case of fish stocks.

maximum take-off rating (*Aero*) See POWER RATING.

maximum take-off weight (*Aero*) See WEIGHT.

maximum tensile stress (*Eng*) See ULTIMATE TENSILE STRESS.

maximum tolerable concentration (*BioSci*) The highest concentration of a substance that does not cause the death of test organisms or species. Abbrev *MTC*.

maximum traction truck (*ElecEng*) A special form of bogie or truck often used on trams, and arranged so that the greater part of the weight comes on the driving wheels, thereby enabling the maximum tractive effort to be obtained.

maximum usable frequency (*ICT*) That which is effective for long-distance communication, as predicted from diurnal and seasonal ionospheric observation. Varies on an 11-year cycle. Abbrev *MUF*.

maximum value (*ElecEng*) See PEAK VALUE.

maximum weight (*Aero*) See WEIGHT.

max level speed (*Aero*) The maximum velocity of a power-driven aircraft at full power without assistance from gravity; the altitude should always be specified.

max weight (*Aero*) See WEIGHT.

maxwell (*ElecEng*) The CGS unit of magnetic flux, the MKSA (or SI) unit being the *weber*: 1 maxwell = 10^{-8} Wb.

Maxwell–Boltzmann distribution law (*Phys*) A law expressing the energy distribution among the molecules of a gas in thermal equilibrium.

Maxwell bridge (*ElecEng*) An early form of ac bridge which can be used for the measurement of both inductance and capacitance.

Maxwell experiment (*ImageTech*) The pioneer demonstration of three-colour additive synthesis, using three black-and-white negatives.

Maxwellian viewing system (*Phys*) In some light measurement instruments such as photometers, spectrophotometers and colorimeters, an arrangement in which the field of view is observed by placing the eye at the focus of a lens, instead of using an eyepiece.

Maxwell model (*Eng*) Conceptual model for a viscoelastic material consisting of a Hookean spring and a dashpot which contains a Newtonian fluid arranged in series. Inadequate for modelling real materials since responses are linear rather than non-linear, as in real systems. See VOIGT MODEL.

Maxwell primaries (*ImageTech*) The colours red, green and blue-violet, used in Maxwell's experiment.

Maxwell's circuital theorems (*Phys*) Generalized forms of FARADAY'S LAW OF INDUCTION and AMPÈRE'S LAW (modified to incorporate the concept of displacement current). Two of MAXWELL'S EQUATIONS are direct developments of the circuital theorems.

Maxwell's circulating current (*ElecEng*) A mesh or cyclic current inserted in closed loops in a complex network for analytical purposes.

Maxwell's demon (*Chem*) Imaginary creature who, by opening and shutting a tiny door between two volumes of gases, could in principle concentrate slower (ie colder) molecules in one and faster (ie hotter) molecules in the other, thus reversing the normal tendency towards increased disorder or entropy and breaking the second law of thermodynamics.

Maxwell's distribution law (*Phys*) The distribution of numbers of gas molecules which have given speeds, or kinetic energies in a gas of uniform temperature. The law can be deduced from the KINETIC THEORY OF GASES.

Maxwell's equations (*Phys*) Mathematical formulations of the laws of Gauss, Faraday and Ampère from which the theory of electromagnetic waves can be conveniently derived:

$$\text{div} B = 0; \quad \text{curl } H = \frac{\partial D}{\partial t} + j$$

$$\text{div} D = \rho; \quad \text{curl } E = \frac{\partial B}{\partial t}$$

Used to analyse the propagation of radio waves in free space, at all sorts of boundaries and in all guided-wave structures or transmission lines. Also *Maxwell's field equations*.

Maxwell's rule (*ElecEng*) A law stating that every part of an electric circuit is acted upon by a force tending to move it in such a direction as to enclose the maximum magnetic flux.

Maxwell's theorem (*Eng*) See RECIPROCAL THEOREM.

Maxwell's thermodynamic relations (*Phys*) Four mathematical identities relating the pressure, the volume, the entropy and the thermodynamic temperature for a system in equilibrium. They are expressed in the form of partial derivatives relating the quantities.

mayday (*ICT*) Verbal international radio-telephone distress call or signal, corresponding to SOS in telegraphy. Corruption of Fr *m'aidez*.

May's graticule (*PowderTech*) An eyepiece graticule used in microscopic methods of particle size analysis, having a rectangular grid for selecting particles, and a series of eight circles for sizing particles. The size of the circle diameters increases by $\sqrt{2}$ progression, and a series of parallel lines is superimposed on the rectangular grid.

maze (*NucEng*) See RADIATION TRAP.

maze (*Psych*) An apparatus consisting of a series of pathways in a more or less complicated configuration, beginning with a starting box, possibly including blind alleys, and ending in a goal box which generally contains a reward, this not being visible from the starting box. The simplest mazes are the T- and Y-mazes.

Mb See MEGABYTE.

MBE (*Electronics*) Abbrev for MOLECULAR BEAM EPITAXY.

MBR (*ICT*) Abbrev for MEMORY BUFFER REGISTER.

MBS (*Chem*) Abbrev for METHACRYLATE BUTADIENE STYRENE.

MCA (*ICT*) Abbrev for MICROCHANNEL ARCHITECTURE.

MCA (*Med*) Abbrev for UK *Medicines Control Agency*.

McBurney's point (*Med*) A point situated on a line joining the umbilicus to the bony prominence of the hip-bone at the upper end of the groin, and 38 mm from the latter; a point of maximum tenderness in appendicitis.

McCabe–Thiele diagram (*ChemEng*) A graphical method, based on vapour–liquid equilibrium properties, for establishing the theoretical number of separation stages in a continuous distillation process.

McColl protective system (*ElecEng*) A form of protective system used on electric power networks; it operates on the balanced principle embodying biased beam relays.

MCI (*ICT*) Abbrev for MEDIA CONTROL INTERFACE.

MCI (*Med*) Abbrev for *mild cognitive impairment*, a slight impairment of memory that may be a precursor of dementia.

McLeod gauge (*Chem*) Vacuum pressure gauge in which a sample of low-pressure gas is compressed in a known ratio until its pressure can be measured reliably. Used for calibrating direct-reading gauges.

MC, mc (*Genrl*) Abbrev for *metric carat*. See CARAT.

McNally tube (*Electronics*) A reflex klystron capable of being tuned electronically to a wide range of frequencies.

MCPA (*Chem*) *4-chloro-2-methyl-phenoxyacetic acid*. Used as a selective weedkiller. Also *methoxone, MCP*.

MCPB (*Chem*) *4-(4-chloro-2-methylphenoxy)butanoic acid*. Used as a weedkiller.

McQuaid–Ehn test (*Eng*) A method of showing grain size after heating a ferrous alloy to the austenitic temperature range. Grain sizes are classified from 1 (the finest) to 8.

Md (*Chem*) Symbol for MENDELEVIUM.

m-derived section, network, filter (*ICT*) A *T* or *pi* network section so designed that when two or more sections are joined to form a filter, their impedances match at all frequencies, although the individual sections may have different resonant frequencies.

MDF (*For*) Abbrev for MEDIUM-DENSITY FIBREBOARD.

MDF (*ICT*) Abbrev for MAIN DISTRIBUTION FRAME.

MDF cement (*Build*) Abbrev for MACRO-DEFECT-FREE CEMENT.

MDI (*Chem*) Abbrev for DIPHENYL-METHANE DIISOCYA-NATE.

M-discontinuity (*Geol*) Abbrev for MOHOROVIČIĆ DISCON-TINUITY.

M display (*Radar*) Modified form of *A display* in which range is determined by moving an adjustable pedestal signal along the baseline until it coincides with the target signal; range is read off the control which moves the pedestal.

MDPE (*Chem*) Abbrev for MEDIUM-DENSITY POLYETHY-LENE.

mdr-TB (*Med*) Abbrev for *multi-drug-resistant tuberculosis*.

ME (*Med*) Abbrev for *myalgic encephalomyelitis*. See CFS.

Me (*Chem*) Symbol for the methyl radical –CH₃.

meadow (*Agri*) Permanent grassland used for grazing and cutting for fodder.

mean (*MathSci*) Of a sample, the ARITHMETIC MEAN. Cf GEOMETRIC MEAN. See EXPECTATION. Cf MEDIAN, MODE.

mean calorie (*Phys*) See CALORIE.

mean chord (*Aero*) See STANDARD MEAN CHORD.

mean curvature (*MathSci*) See CURVATURE.

mean daily motion (*Astron*) The angle through which a celestial body would move in the course of 1 day if its motion in the orbit were uniform. It is obtained by dividing 360° by the period of revolution in days.

meander (*Geol*) Sharp sinuous curves in a stream particularly in the mature part of its course. The meanders are accentuated by continuing erosion on the convex side and deposition on the concave side of the stream course.

mean draught (*Ships*) Half of the sum of the forward and after draughts of a vessel; differs slightly from draught at half-length.

mean effective pressure (*Eng*) See BRAKE MEAN EFFEC-TIVE PRESSURE, INDICATED MEAN EFFECTIVE PRESSURE.

mean establishment (*Surv*) The average value of the lunitidal interval at a place.

mean free path (*Acous*) Average distance travelled by a sound wave in an enclosure between wall reflections; required for establishing a formula for reverberation calculations.

mean free path (*Phys*) The mean distance λ travelled by a particle between collisions, eg electrons in a solid or molecules in a gas. It is dependent on the molecular cross-section $\pi\sigma^2$ so that

$$\lambda = \frac{1}{\sqrt{2\pi}\, n\sigma^2}$$

where *n* is the number of molecules per unit volume. According to the kinetic theory of gases it is related to the viscosity η by $\lambda = k\eta/\rho u$ where ρ is the density, u is the mean molecular velocity and k is a constant between $\frac{1}{3}$ and $\frac{1}{2}$ depending on the approximation in the theory.

mean free time (*Phys*) Average time between collisions of electrons with impurity atoms in semiconductors; also of intermolecular collision of gas molecules.

mean hemispherical candle-power (*Phys*) The average value of the candle-power in all directions above or below a horizontal plane passing through the source; called the *upper* or *lower* mean hemispherical candle-power accord-ing to whether the candle-power is measured above or below the horizontal plane through the source.

mean horizontal candle-power (*Phys*) The average value of the candle-power of a light source in all directions in a horizontal plane through the source.

mean lethal dose (*Radiol*) The single dose of whole-body irradiation which will cause death, within a certain period, to 50% of those receiving it. Abbrev *MLD*.

mean life (*Phys*) (1) The average time during which an atom or other system exists in a particular form, eg for a thermal neutron it will be the average time interval between the instant at which it becomes thermal and the instant of its disappearance in the reactor by leakage or by absorption. Equal to $1\cdot443 \times$ half-life. Also *average life*. (2) The mean time between birth and death of eg a charge carrier in a semiconductor or a particle (such as an ion or a pion).

mean noon (*Astron*) The instant at which the mean Sun crosses the meridian at upper culmination at any place; unless otherwise specified, the meridian of Greenwich is generally meant.

mean normal curvature (*MathSci*) See CURVATURE.

mean one-way propagation time (*ICT*) The average time required for speech to travel between two nodes of a telephone network, including time taken for digital processing at each end. Between eg mobile units of the GLOBAL SYSTEM FOR MOBILE COMMUNICATION this can be as much as 180 ms.

mean place (*Astron*) The position of a star freed from the effects of precession, nutation and aberration, and of parallax, proper motion and orbital motion where appreciable. These corrections can be computed for any future date, and when applied to the mean place give the apparent place.

mean residence time (*Pharmacol*) The average time a drug molecule remains in the body or an organ after rapid intravenous injection. Abbrev *MRT*.

mean residence time (*Phys*) The mean period during which radioactive debris from nuclear weapons tests remains in the stratosphere.

mean sea level (*Surv*) In the UK the ordnance survey datum level, determined at Newlyn, Cornwall.

means, inequalities between (*MathSci*) $A \geqslant G \geqslant H$, where A is the arithmetic, G the geometric and H the harmonic mean of *n* positive numbers.

mean solar day (*Astron*) The interval, perfectly constant, between two successive transits of the mean Sun across the meridian.

mean solar time (*Astron*) Time as measured by the hour angle of mean Sun. When referred to the meridian of Greenwich it is called Greenwich Mean Time. Before 1925 this began at noon but, by international agreement, is now counted from midnight; it is thus the hour angle of mean Sun plus 12 h, and is identical to UNIVERSAL TIME.

mean-spherical candle-power (*Phys*) The average value of the candle-power of a light source taken in all directions.

mean-spherical response (*Acous*) The response of a microphone or loudspeaker taken over a complete sphere, the radius of which is large in comparison with the size of the apparatus. For a loudspeaker, this response (total response) determines the total output of sound power, and therefore, in conjunction with the acoustic properties of an enclosure, the average reverberation intensity in the enclosure. For a microphone, this response is substantially equal to the response for reverberant sound.

mean-square error (*MathSci*) The expected value of the square of the difference between an estimate of a parameter and its true value, taken with respect to the sampling distribution of the estimate.

mean stress (*Eng*) The mid-point of a range of stress. When it is zero, the upper and lower limits of the range have the same value but are in tension and compression respectively.

mean Sun (*Astron*) A fictitious reference point which has a constant rate of motion and is used in timekeeping in preference to the non-uniform motion of the real Sun. The mean Sun is imagined to follow a circular orbit along the celestial equator and is used to measure MEAN SOLAR TIME.

mean time between failures (*Aero, Eng*) A general measure of reliability used in specifications. Abbrev *MTBF*.

mean value theorem (*MathSci*) The theorem which states that an arc on a differentiable curve has at least one tangent parallel to its SECANT; ie for continuous $f(x)$ differentiable on (a, b) there exists a $\xi \in (a, b)$ such that

$$f'(\xi) = \frac{f(b) - f(a)}{b - a}$$

Cf ROLLE'S THEOREM.

mean-zonal candle-power (*Phys*) The average value of the candle-power of a light source taken in a given zone, the angular limits of the zone being stated.

measles (*Med*) An acute infectious fever caused by a virus; characterized by catarrh of the respiratory passages, conjunctivitis, KOPLIK'S SPOTS and a distinctive rash. Also *morbilli*.

measles of beef (*Vet*) Infection of beef by the cyst stage, 'Cysticercus bovis', of the human tapeworm *Taenia saginata*.

measles of pork (*Vet*) Infection of pork by the cyst stage, 'Cysticercus cellulosae', of the human tapeworm *Taenia solium*. Hence *measly pork*.

measure (*Print*) The width of the type-matter on a page or column, measured in ems of 12-point.

measured ore (*MinExt*) Proved quantity and assay grade of ore deposit as ascertained by competent measurement of exposures and an adequate sampling campaign.

measuring chain (*Build, Surv*) See CHAIN.

measuring frame (*Build*) A wooden box without top or bottom used as a measure for aggregates in mixing concrete.

measuring instrument (*ElecEng*) A device serving to indicate or record one or more of the electrical conditions in an electrical circuit. Literally, the term also includes but is not generally used for integrating meters.

measuring tape (*Build, Surv*) See TAPE.

measuring wheel (*Surv*) See PERAMBULATOR.

meat (*FoodSci*) Edible flesh of an animal, bird or fish, includes fat, rind, gristle and sinew, muscle and offal. Lean meat is muscle consisting of intracellular and extracellular proteins (18–22%), intramuscular fat and water (70–75%). Most of the fat in meat is extramuscular in that it is around the organs and just under the skin.

meatotomy (*Med*) Incision of the urinary meatus to widen it.

meatspace (*ICT*) Computing slang for the physical world, as opposed to cyberspace.

meatus (*BioSci*) A duct or channel, as the *external auditory meatus* leading from the external ear to the tympanum.

mechanical advantage (*Phys*) The ratio of the resistance (or load) to the applied force (or effort) in a MACHINE.

mechanical alloying (*Eng*) Producing an intimate dispersion of elements by milling together fine powders of the constituents in the desired proportions.

mechanical analogue (*Eng*) That which can be drawn between mechanical and electrical systems obeying corresponding equations, eg mechanical and electrical resonators.

mechanical bond (*CivEng*) The use of reinforcing bars with special ribs or after shaping, eg as bends or hooks, to supplement the bond between concrete and steel.

mechanical characteristic (*ElecEng*) See SPEED–TORQUE CHARACTERISTIC.

mechanical depolarization (*ElecEng*) Dissipation, by mechanical means, of the hydrogen bubbles causing polarization of an electrolytic cell.

mechanical deposits (*Geol*) Those of sediments which owe their accumulation to mechanical or physical processes.

mechanical efficiency (*Eng*) Of an engine, the ratio of the brake or useful power to the indicated power developed in the cylinders, ie the efficiency of the engine regarded as a machine. In other types of machines, the mechanical efficiency similarly accounts for the friction losses.

mechanical engineering (*Eng*) That branch of engineering concerned primarily with the design and production of all mechanical contrivances or machines, including prime movers, vehicles, machine tools and production machines.

mechanical equivalent of heat (*Phys*) Originally conceived as a conversion coefficient between mechanical work and heat (4·186 J = 1 cal) thereby denying the identity of the concepts. Now recognized simply as the SPECIFIC HEAT CAPACITY of water, $4·186 \text{ kJ kg}^{-1} \text{ K}^{-1}$.

mechanical equivalent of light (*Phys*) The ratio of the radiant flux, in watts, to the luminous flux, in lumens, at the wavelength for which the RELATIVE VISIBILITY FACTOR is a maximum. Its value is about $0·0015 \text{ W lm}^{-1}$.

mechanical filter (*ImageTech*) An arrangement of springs and masses interposed in the drive of a camera or projector, to smooth out variations in the required constant speed.

mechanical finishing (*Textiles*) A finish obtained by mechanical means such as raising or calendering as distinct from CHEMICAL FINISHING.

mechanical impedance (*Phys*) In a mechanical vibrating system, the ratio of the force and the velocity it produces. It consists of the *real* part, the mechanical resistance, which represents the power lost in the system, and the reactance which depends on the mass and COMPLIANCE of the system.

mechanically recovered meat (*FoodSci*) Meat stripped mechanically from bones remaining after butchery. Used as a binding material in some sausages and other meat products. Abbrev *MRM*.

mechanical overlay (*Print*) An OVERLAY prepared without handcutting of which there are several varieties.

mechanical paper (*Paper*) Paper composed substantially of MECHANICAL WOOD PULP.

mechanical plating (*Eng*) Forming a metallic layer on a surface by a hammering or tumbling process. See SHERADIZING.

mechanical refrigerator (*Eng*) Common domestic or industrial plant with a compressor for raising the pressure of the refrigerant, a condenser for removing the latent heat, a regulating valve for lowering its pressure and temperature by throttling, and an evaporator in which it absorbs heat at a low temperature.

mechanical resonance (*Phys*) In a mechanical vibrating system, the enhanced response to a driving force as the frequency of this force is increased through a resonant frequency at which the inertial reactance balances the reactance due to the compliance of the system.

mechanical scanning (*ImageTech*) Systems of SCANNING used in early forms of TV, both for the original scene and the displayed image; these included disks with a spiral of holes or lenses (NIPKOW DISK) and MIRROR DRUMS with a series of reflecting facets or prisms.

mechanical splicing (*ICT*) The joining of OPTICAL FIBRES by cleaving them to produce optically flat ends which are then clamped in alignment but not fused together. The required equipment is cheaper than that for FUSION SPLICING, but the resultant splice loss is generally higher.

mechanical stipple (*Print*) An aid to the making of line blocks and litho plates. The stipple is chosen from a large variety and applied to the drawing, to the negative made from it or to the block or plate during preparation. When used in line-colour work, a range of tones can be obtained. Also *mechanical tint*.

mechanical stoker (*Eng*) A device for stoking or firing a steam boiler. It receives fuel continuously by gravity, carries it progressively through the furnace, and deposits or discharges the ash. Also *automatic stoker*. See CHAIN GRATE STOKER, OVERFEED STOKER, UNDERFEED STOKER.

mechanical tint (*Print*) See MECHANICAL STIPPLE.

mechanical tissue (*BioSci*) Tissues such as COLLENCHYMA and SCLERENCHYMA whose primary function is the support of the plant body.

mechanical wood pulp (*Paper*) Pulp produced from wood entirely, or almost entirely, by mechanical means, eg by grinding logs (*groundwood*) or by passing wood chips through a refiner (*refiner mechanical pulp*).

mechanical working (*Eng*) Changing the shape of a metal or material by processes applying mechanical forces, eg rolling, forging, extrusion, spinning, pressing, etc.

mechanics (*Phys*) The study of the forces acting on bodies, whether moving (DYNAMICS) or at rest (STATICS).

mechanism (*Psych, Genrl*) (1) Philosophical viewpoint that considers all events and phenomena to be explicable within a mechanistic framework and thus is highly deterministic. (2) Theoretical process to explain phenomena. (3) An adaptive response, eg a defence mechanism. (4) A piece of machinery that does some specific task, eg clock, counting device.

mechano-chemical engine (*Chem*) A device designed to turn chemical energy directly into mechanical motion, as in muscle. Laboratory curiosities built from polyelectrolyte fibres can function in this way by cycling the PH of the solution enclosing them.

mechanomotive force (*Phys*) The rms value of an alternating mechanical force, in newtons, developed in a transducer.

mechanoreceptor (*BioSci*) A sense organ specialized to respond to mechanical stimuli, such as pressure or deformation.

Meckel's diverticulum (*Med*) A diverticular outgrowth from the lower end of the small intestine, as a result of the persistence in the adult of the vitelline or yolk-sac duct of the embryo.

meclozine HCl (*Pharmacol*) An *antihistamine* used for the treatment of motion sickness.

meconium (*BioSci*) (1) In certain insects, liquid expelled from the anus immediately after the emergence of the imago. (2) The first feces of a newborn animal.

medi-, medio- (*Genrl*) Prefixes from Lt *medius*, middle.

media (*ICT*) Collective name for materials (tape, disk, paper, cards, etc) used to hold data.

media access unit (*ICT*) A single device that combines the function of a transmitter and a receiver. Also *transceiver*.

media control interface (*ICT*) A standard control INTERFACE for MULTIMEDIA devices and FILES.

medial (*BioSci*) Situated near, or tending towards, the median axis.

median (*MathSci*) The central value in a set of observations ordered by value, dividing the ordered set into two equal parts; the argument of the cumulative distribution function of a random variable corresponding to a probability of one-half. Cf MEAN, MODE.

median effective dose (*BioSci*) The dose of a chemical or physical agent, such as radiation, that statistically is expected to produce a certain effect in test organisms under a defined set of conditions. Also ED_{50}.

median lethal dose (*BioSci*) The statistically calculated dose of a chemical or physical agent that is expected to kill 50% of organisms in a defined population under a particular set of conditions. Also LD_{50}.

mediastinitis (*Med*) Inflammation of the tissues of the mediastinum.

mediastinotomy (*Med*) The surgical exposure of the mediastinal region in the chest. Now used extensively in cardiac surgery.

mediastinum (*BioSci*) In higher vertebrates, the mesentery-like membrane that separates the pleural cavities of the two sides ventrally; in mammals, a mass of fibrous tissue representing an internal prolongation of the capsule of the testis.

media streaming (*ICT*) The delivery of media, such as video images or audio, over the Internet for viewing or listening in real time.

medical model (*Psych*) The conceptualization of mental disorders as a group of diseases analogous to physical diseases.

medio- (*Genrl*) See MEDI-.

Mediterranean fever (*Vet*) A disease of cattle and water buffalo in Mediterranean countries due to infection by the protozoon *Theileria annulata*; characterized by fever, ocunasal discharge, anaemia and diarrhoea, and transmitted by ticks.

medium (*BioSci*) Substance in which an organism or part exists naturally, eg *aqueous medium*, or has been placed experimentally. See CULTURE MEDIUM, MOUNTANT.

medium (*Build*) A liquid or a semi-liquid vehicle, such as water, oil, spirit, wax, which makes pigment and other components of paint workable. See VEHICLE.

medium (*Paper*) See PAPER SIZES.

medium-density fibreboard (*For*) A building board made by compressing finely ground wood chips in urea-formaldehyde adhesive. Available in sheets from 8 to more than 30 mm thick. Density from 700 to $800\,\mathrm{kg\,m^{-3}}$. Abbrev *MDF*.

medium-density polyethylene (*Chem*) Grade of polymer developed mainly for gas pipes, with enhanced resistance to ENVIRONMENTAL STRESS CRACKING. Intermediate density of approx $940\,\mathrm{kg\,m^{-3}}$ and T_m about 130°C. Abbrev *MDPE*.

medium Earth orbit (*Space*) See panel on COMMUNICATIONS SATELLITE. Abbrev *MEO*.

medium Edison screw-cap (*ElecEng*) An EDISON SCREW-CAP having a diameter of approximately 1 in and approximately seven threads per inch. Abbrev *MES*.

medium frequency (*ICT*) Between 300 kHz and 3 MHz. Also *hectometric waves*.

medium-scale integration (*ICT, Electronics*) A CHIP with not more than a few hundred logic GATES. Abbrev *MSI*.

medium screen (*ImageTech, Print*) The term for half-tones suitable for semi-smooth paper, usually either 100 or 120 lines per inch.

medium voltage (*ElecEng*) Legally, a voltage which is over 250 V and not greater than 650 V.

medium waves (*ICT*) Vague description of the range of wavelengths from about 200 to 1000 m; used generally for short- and medium-range broadcasting.

medroxyprogesterone (*Pharmacol*) A synthetic progestogen used for the treatment of menstrual disorders and, when given as a deep intramuscular injection, as a depot contraceptive.

medulla (*BioSci*) (1) The central portion of an organ or tissue, eg. the medulla of the mammalian kidney. Adj *medullary*. Cf CORTEX. (2) See PITH.

medulla (*Textiles*) Hollow cellular central portion of some animal fibres which is surrounded by the cortex.

medullablastoma (*Med*) A malignant and rapidly growing tumour occurring in the cerebellum.

medulla oblongata (*BioSci*) The hind-brain in vertebrates, excluding the cerebellum.

medullary bundle (*BioSci*) Vascular bundle in the pith of a stem.

medullary canal (*BioSci*) The cavity of the central nervous system in vertebrates; the central cavity of a shaft-bone.

medullary folds (*BioSci*) In a developing vertebrate, the lateral folds of a medullary plate, by the upgrowth and union of which the tubular central nervous system is formed.

medullary plate (*BioSci*) In a developing vertebrate, the dorsal plate-like area of ectoderm that will later give rise to the central nervous system.

medullary ray (*BioSci*) (1) The interfasicular region. (2) Ray stretching from pith to cortex and deriving from the interfasicular region. Also *primary ray*.

medullary sheath (*BioSci*) (1) In plants, the peripheral layers of cells of the PITH. The cells are usually small, sometimes thick-walled and sometimes more or less lignified. (2) See MYELIN SHEATH.

medullate, medullated (*BioSci*) Of plants, having a PITH; of nerve fibres, having a MYELIN SHEATH.

medullated protostele (*BioSci*) PROTOSTELE with a central core (medulla) of non-vascular tissue.

medusa (*BioSci*) In metagenetic Cnidaria, a free-swimming sexual individual.

Meehanite (*Eng*) TN for high-silicon cast-iron produced by INOCULATION.

meerschaum (*Min*) A hydrated silicate of magnesium. It is clay-like, and is shown microscopically to be a mixture of a fibrous mineral called parasepiolite and an amorphous mineral β-sepiolite. It is used for making pipes, and was formerly called in Morocco as a soap. Also *sepiolite*.

meeting post (*Build*) The vertical post at the outer side of a mitre gate of a lock, which is chamfered so as to fit against the corresponding edge of the outer leaf when the gates are shut. Also *mitre post*.

meeting rail (*Build*) The top rail of the lower sash, or the bottom rail of the upper sash, of a double-hung window.

meeting stile (*Build*) See SHUTTING STILE.

mega- (*Genrl*) Prefix denoting 1 million, or 10^6, eg a frequency of 1 *megahertz* is equal to 10^6 Hz; *megawatt* = 10^6 watts; *megavolt* = 10^6 volts. Symbol M.

megabit (*ICT*) Unit of 1 million or, sometimes, 2^{20} *bits*. Cf MEGABYTE.

megabit erlang per second (*ICT*) A measure of traffic in digital transmission systems such as ASYNCHRONOUS TRANSFER MODE, calculated by taking the demand per customer per service in ERLANGS, multiplying by the bit rate for each service and totalling over all services and all customers.

megabyte (*ICT*) Commonly 1 million but sometimes 2^{20} BYTES. Abbrev *MB*.

megacolon (*Med*) Abnormally large colon.

mega-electron-volt (*Phys*) See MEV.

megaflop (*ICT*) A unit of processor speed approximately equal to 10^6 floating-point operations per second.

megagamete (*BioSci*) See MACROGAMETE.

mega-gram-roentgen (*Phys*) See GRAM-ROENTGEN.

megahertz (*ICT*) The unit of frequency in which there are 1 million complete cycles of alternation per second. Used in preference to wavelength when the latter attribute of a wave or oscillation is very short. Abbrev *MHz*.

megakaryocyte (*BioSci*) A giant polyploid cell of bone marrow and other blood-forming organs, each of which gives rise to 3–4000 bloodplatelets. Also *myeloplax*.

megalecithal (*BioSci*) Said of eggs that contain a large quantity of yolk. Cf MICROLECITHAL.

megaloblast (*BioSci*) An embryonic cell that has a large spherical nucleus and haemoglobin in the cytoplasm, which will later give rise to ERYTHROBLASTs by mitotic division within the blood vessels.

megaloblastic anaemia (*Med*) A form of anaemia in which megaloblasts are found in the bone marrow, and often macrocytes in the peripheral blood. The usual cause is a deficiency in vitamin B_{12} or folic acid.

megamere (*BioSci*) See MACROMERE.

meganucleus (*BioSci*) See MACRONUCLEUS.

megaparsec (*Astron*) See PARSEC.

megaphanerophyte (*BioSci*) A tree over 30 m high.

megaphone (*Acous*) A horn to direct the voice. Can include microphone, amplifier and sound reproducer. Then called a *loud-hailer*.

megaphyll (*BioSci*) (1) A leaf, typically associated with a LEAF GAP and often relatively large and with a branching system of veins, assumed to have evolved from a branch system or directly from a telome truss, and supposed to be the sort of leaf possessed by seed plants and ferns. See TELOME THEORY. (2) Any large leaf. Adj *megaphyllous*. Cf MICROPHYLL.

megapixel (*ICT*) Abbrev for *a million pixels*.

megaripple (*Geol*) A sand wave with a wavelength of <1 m.

megascopic (*Genrl*) Visible to the naked eye

megasporangium (*BioSci*) A sporangium that contains megaspores. It corresponds to the ovule or nucellus of seed plant. See HETEROSPORY.

megaspore (*BioSci*) In a heterosporous species, a MEIOSPORE potentially developing into a female gametophyte. Also *macrospore*.

megasporophyll (*BioSci*) A leaf-like structure (or a structure thought to be homologous to a leaf) bearing megasporangia. Cf MICROSPOROPHYLL. See SPOROPHYLL.

megaton (*Phys*) Explosive force equivalent to 1 million tons of TNT, taken as 10^{12} calories. Used as a unit for classifying nuclear weapons.

megaureter (*Med*) Enormous dilation of a ureter with no demonstrable abnormality.

megavoltage therapy (*Radiol*) See SUPERVOLTAGE THERAPY.

megawatt days per tonne (*NucEng*) Unit for energy output from reactor fuel; a measure of burn-up. Expressed as $MW\,d\,t^{-1}$.

meglip (*Build*) A preparation of mastic varnish which has been diluted with linseed oil; used as a medium in the mixing of paints for fine work, eg GRAINING.

megohmmeter (*ElecEng*) Portable apparatus for indicating values of high resistance, containing a circuit excited by a battery, transistor converter, or dc generator. Not to be used for measuring insulation resistance of all types of capacitor.

Meibomian glands (*BioSci*) In mammals, sebaceous glands on the inner surface of the eyelids, between the tarsi and conjunctiva. Also *tarsal glands*.

meimechite (*Min*) An ultramafic volcanic rock containing phenocrysts of olivine.

meiomerous (*BioSci*) Having a small number of parts. N *meiomery*.

meionite (*Min*) Silicate of aluminium and calcium, together with calcium carbonate, which crystallizes in the tetragonal system. It is an end-member of the isomorphous series forming the scapolite group.

meiosis (*BioSci*) A process of cell division by which the chromosomes are reduced from the diploid to the haploid number. Adj *meiotic*. Fig. ▷

meiospore (*BioSci*) A spore, commonly one of four, containing a nucleus formed by meiosis and therefore haploid. The spores of bryophytes and vascular plants.

meiotic arrest (*BioSci*) The waiting at a particular stage of meiosis by the oöcyte until some stimulus is received, usually the entry of the sperm.

Meissner effect (*Phys*) The flux expulsion which occurs when a perfectly diamagnetic superconducting material is exposed to a weak magnetic field. See SUPERCONDUCTING LEVITATION and panel on SUPERCONDUCTORS.

meitnerium (*Chem*) An artificially manufactured radioactive chemical element, symbol Mt, at no 109, of the transactinide series. Also *unnilenium*.

MEK (*Chem*) Abbrev for METHYL ETHYL KETONE (butan-2-one).

Meker burner (*Chem*) A Bunsen burner using a wire mesh over its outlet as a stabilizer to support a wide, hot flame.

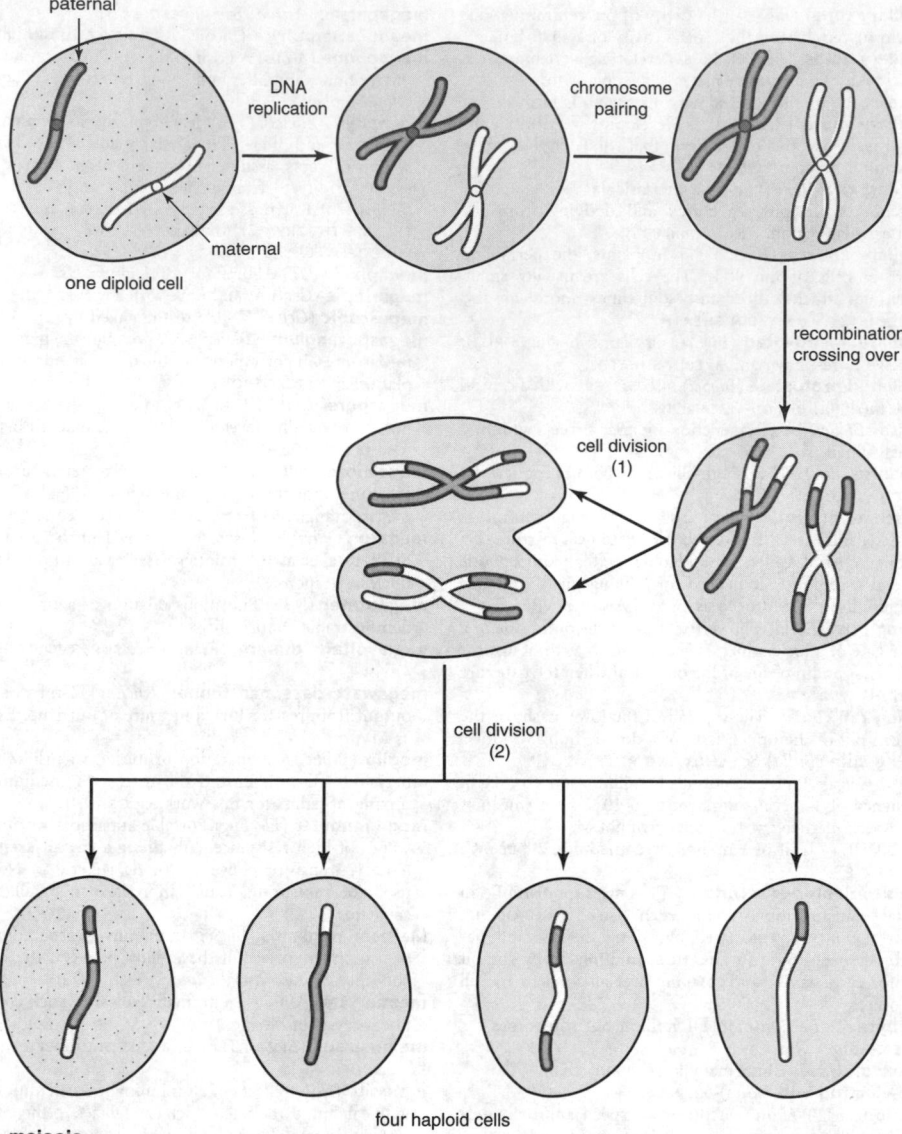

paternal

DNA replication

chromosome pairing

one diploid cell

maternal

recombination/ crossing over

cell division (1)

cell division (2)

four haploid cells

meiosis

mel (*Acous*) Unit of subjective pitch in sound, a pitch of 131 mels being associated with a simple tone of frequency 131 Hz.

melaconite (*Min*) Cupric oxide crystallizing in the monoclinic system. It is a black earthy material found as an oxidation product in copper veins, and represents a massive variety of TENORITE.

melaena (*Med*) The passage of black, pitch-like feces due to the admixture of altered blood, the result of haemorrhage in the alimentary tract.

melamine-formaldehyde methanal resin (*Plastics*) A synthetic resin derived from the reaction of melamine with methanal or its polymers. It is thermosetting and much used for moulding and laminating. Also *MF resin*.

melamine resins (*Plastics*) Synthetic resins derived from the reaction of melamine with formaldehyde or its polymers. See panel on THERMOSETS.

melan-, melano- (*Genrl*) Prefixes from Gk *melas*, gen *melanos*, black.

melange (*Geol*) A heterogeneous mixture of rock materials mappable as a rock unit. Mostly of tectonic origin but some are caused by large-scale slumping of sediments.

melange (*Textiles*) Yarns made from colour-printed top or slivers.

melanin (*Chem*) A dark-brown or black pigment occurring in hair and skin. Soluble only in alkali and formed by the oxidation of tyrosine.

melanism (*BioSci*) The abnormal situation, caused by overproduction of melanin, where a proportion of the individuals in an animal population are black or melanic (melanotic). See INDUSTRIAL MELANISM.

melanite (*Min*) A dark-brown or black variety of andradite garnet containing appreciable titanium.

melano- (*Genrl*) See MELAN-.

melanoblast (*BioSci*) A special connective tissue cell containing melanin.

melanocratic (*Geol*) A term applied to rocks which are abnormally rich in dark and heavy ferromagnesian minerals (to the extent of 60% or more). See LEUCOCRATIC.

melanocyte (*BioSci*) A cell found in mammalian skin that contains the pigment melanin within membrane-bound organelles (melanosomes).

melanodermia (*Med*) Also *melanoderma*. See LEUCODERMIA.

melanoglossia (*Med*) Black hairy tongue. An overgrowth of the papillae of the tongue, which are stained as the result either of bacterial action or of chemical action on certain food substances.

melanoma (*Med*) A tumour, of variable malignancy, arising from cells in the skin and retina that produce melanin. Becoming more common generally but esp in people of N European origin in sunny climates.

melanophore (*BioSci*) A chromatophore containing black pigment.

melanosis (*Med*) The abnormal deposit of the pigment melanin in the tissues of the body.

melanosporous (*BioSci*) Having black spores.

melanterite (*Min*) Hydrated ferrous sulphate which crystallizes in the monoclinic system. It usually results from the decomposition of iron pyrites or marcasite. Also *copperas*, *green vitriol*.

melanuria (*Med*) The presence in the urine of the pigment melanin.

Melastomaceae (*BioSci*) A family of c.3000 spp of dicotyledonous flowering plants (superorder Rosidae). They are mostly shrubs and small trees and are mostly tropical; they are important in S American flora but are of little economic value.

melatonin (*BioSci*) A hormone, *N-acetyl 5-methoxytryptamine*, secreted by the pineal gland. It is believed to play a role in establishment of circadian rhythms, and is said to help overcome the effects of jet lag.

melded fabric (*Textiles*) A fabric made from BICOMPONENT FIBRES in which the outer component melts at a lower temperature than the core. The cohesion is obtained by subjecting the pressed fibre mass to a temperature at which partial fusion occurs.

meld line (*Eng*) A feature on a moulded polymer surface where FLOW LINES have visibly met, without forming a WELD LINE.

melena (*Med*) See MELAENA.

melibiose (*Chem*) Naturally occurring di-saccharide based on glucose and α-galactose. (*D-galactosyl-α1 → 6-D-glucose*).

melilite (*Min*) Calcium magnesium aluminium silicate; crystallizing in the tetragonal system, and occurring in alkaline igneous and contact-metamorphosed rocks and slags.

melilitite (*Min*) An ultramafic volcanic rock composed essentially of melilite and pyroxene.

melilitolite (*Min*) An ultramafic plutonic rock composed essentially of melilite, pyroxene and olivine.

Melinex (*Plastics*) TN for polyethylene terephthalate (PET) in film form. Very strong film with extremely good transparency and electric properties. Similar to *Mylar*.

melioidosis (*Med*) An infectious disease, predominantly of tropical climates, caused by the bacterium *Burkholderia pseudomallei*; clinically and pathologically similar to glanders, a disease that is contracted by humans from infected domestic animals. Has been seen as a potential bioterror agent. Also *Whitmore's disease*.

melittin (*BioSci*) A peptide component of bee venom, responsible for the pain of the sting, that lyses cell membranes and activates phospholipase A_2.

melliphagus (*BioSci*) Honey-eating. Also *mellivorous*.

mellitic acid (*Chem*) Made by oxidizing carbon with strong nitric acid. Condensation with 1,3-ol:hydroxybenzene and aminophenols produces phthalein and rhodamine dyestuffs respectively.

mellivorous (*BioSci*) See MELLIPHAGUS.

meloxicam (*Pharmacol*) A non-steroidal anti-inflammatory drug used for short-term treatment of acute osteoarthritis and long-term treatment of rheumatoid arthritis.

meltback transistor (*Electronics*) A junction transistor in which the junction is formed by allowing the molten doped semiconductor to solidify.

melt down (*NucEng*) A type of nuclear reactor accident in which the fuel becomes excessively overheated so that it melts and collapses into or through the fabric of the reactor. Overheating may be caused by loss of *coolant* function or of MODERATOR.

melteigite (*Min*) A melanocratic intrusive rock with 10–30% nepheline, part of the *ijolite* series.

melt elasticity (*Eng*) See MELT MEMORY.

melt flow index (*Chem*) Measure of melt viscosity of polyolefins, esp polyethylene and polypropylene. Involves extrusion of polymer from standard orifice at temperature of 190°C (for polyethylenes). Melt flow index is the weight extruded in 10 min, and is inversely related to molecular mass (BS 2782). Abbrev *MFI*.

melt fracture (*Eng*) Break-up of extrudate at high shear rate. See EXTRUSION.

melting band (*EnvSci*) A bright horizontal band often observed in vertical cross-section (RHI) WEATHER RADAR displays. It is due to strong reflections from snowflakes which become covered with a film of water as they fall through the 0°C level and begin to melt.

melting point (*Chem*) The temperature at which a solid begins to liquefy. Pure crystalline materials, eutectics and some intermediate constituents melt at a constant temperature; alloys generally melt over a range. Glasses, whether inorganic, metallic or polymeric, have no well-defined transition from solid to liquid. Abbrev *mp*. Symbol T_m.

melting point test (*Build, CivEng*) A test for the determination of the melting point of a bitumen for use in building or road-making. Also *fusing point test*.

melting temperature (*Chem*) See MELTING POINT.

melting temperature of DNA (*BioSci*) The temperature at which half the molecules in a nucleic acid solution have undergone thermal DENATURATION.

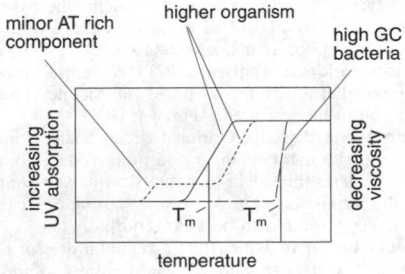

melting temperature of DNA The steeper the slope the more homogeneous the DNA.

melt memory (*Eng*) The tendency of a polymer to 'remember' its former state, esp during processing; eg DIE SWELL is caused by the extrudate reverting to its original state in the barrel prior to extrusion through the die. Also *melt elasticity*. See EXTRUSION, SHAPE MEMORY ALLOYS.

melton (*Textiles*) A strong, heavily milled and cropped fabric used for making overcoats. Made from pure wool or cotton warp and woollen weft and having a felt-like appearance. Cf KERSEY.

melt-spinning (*Textiles*) The formation of continuous filaments by extrusion of molten polymer. A subsequent drawing process is necessary to develop the desired mechanical properties, including strength, in the filaments.

melt transport (*Eng*) The analysis of the movement of molten or melting polymer along injection-moulding or extruding machines. Involves analysis of drag, pressure and leak flows in barrel, and depends on screw characteristics, channel depth, melt viscosity, pressure gradient and screw speed. See EXTRUSION.

melt viscosity (*Chem*) The viscosity of molten polymer of specific importance for injection moulding. Strongly dependent on polymer grade, temperature and shear rate.

member (*BioSci*) (1) Any part of a plant considered from the standpoint of morphology. (2) An organ of the animal body, esp an *appendage*.

member (*Build*) (1) A constituent part of a structural framework. (2) A division of a moulding.

membrana (*BioSci*) A thin layer or film of tissue; a membrane.

membrana tectoria (*BioSci*) A soft fibrillated membrane overlying Corti's organ.

membrana tympani (*BioSci*) A thin fibrous membrane forming the tympanum or ear drum.

membrane (*BioSci*) (1) A thin sheet-like structure, often fibrous, connecting other structures or covering or lining a part. (2) *Cell membrane, unit membrane*; a sheet (7–10 nm thick), composed characteristically of a bimolecular leaflet of lipid and protein, enclosing a cell, an organelle or a vacuole. See PIT MEMBRANE, TONOPLAST.

membrane attack complex (*BioSci*) Complex formed by the five terminal proteins (C5–C9) involved in the COMPLEMENT CASCADE which assembles on cell membranes and causes lysis of the cell.

membrane filter (*Chem*) Thin-layer filter made by fusing cellulose ester fibres or by β-bombardment of thin plastic sheets, so that they are perforated by tiny uniform channels. Also *molecular filter*.

membranella (*BioSci*) In Ciliophora, an undulating membrane formed of two or three rows of cilia.

membrane potential (*BioSci*) More correctly, *transmembrane potential difference*, the electrical potential difference across a plasma membrane.

membrane protein (*BioSci*) See INTEGRAL MEMBRANE PROTEIN.

meme (*Psych*) Practice, belief or other cultural feature passed on by non-genetic means.

memory (*Eng*) See MELT MEMORY.

memory (*ICT*) Part of a computer system where data and instructions are held. Also *storage, store*. See ACCESS TO STORE, BACKING STORE, MAIN MEMORY.

memory address register (*ICT*) A central processor register that holds the ADDRESS of the next memory location to be referenced. Abbrev *MAR*.

memory board (*ICT*) A printed circuit board containing INTEGRATED CIRCUITS for the purposes of data storage. In some systems this will be an EXPANSION BOARD connected to the MOTHERBOARD. Also *memory card*. See panel on PRINTED, HYBRID AND INTEGRATED CIRCUITS.

memory buffer register (*ICT*) A central processor register that acts as a BUFFER for all data transfers to and from MAIN MEMORY. Abbrev *MBR*.

memory capacity (*ICT*) The amount of information, usually expressed by the number of WORDS, that can be retained by a MEMORY. It may also be expressed in bytes. Also *data storage*. See ACCESS TO STORE.

memory card (*ICT*) See MEMORY BOARD.

memory effect (*Psych*) Error that arises from experimental subjects remembering previous testing and applying that knowledge to current testing.

memory leak (*ICT*) Colloq term for the situation in which programs fail to release all the memory taken to execute a procedure. Repeatedly calling such a procedure results in the steady accumulation of unusable memory and eventual system CRASH.

memory location (*ICT*) See STORE LOCATION.

memory manager (*ICT*) A MEMORY-RESIDENT PROGRAM that RELOCATES other SOFTWARE in the computer to optimize the use of available MEMORY.

memory map (*ICT*) A description often in the form of a diagram that shows the way in which storage is allocated, ie which locations are used for the APPLICATION PROGRAMS, OPERATING SYSTEM, VARIABLES, VIDEO DISPLAY and general working space. Fig. ▷

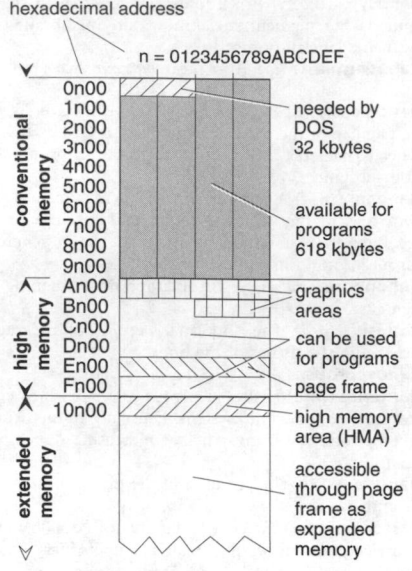

memory map A possible configuration of the PC standard, drawn two dimensionally in 64 kilobyte blocks. The HMA region is often used by parts of DOS.

memory mapping (*ICT*) A technique whereby a computer treats peripherals as part of main memory.

memory register (*ICT*) A REGISTER in the storage of a computer.

memory-resident program (*ICT*) A program stored in MEMORY that is available when another program is active, eg a program that enables a BACKGROUND printing job to be run concurrently with the main program. Also *terminate-and-stay-resident program, TSR program*.

memory span (*Psych*) The number of items a person can recall after just one presentation.

memory trace (*Psych*) The inferred change in the nervous system that persists after learning or experiencing something.

menarche (*Med*) The first menstruation.

mendelevium (*Chem*) Artificially produced element, symbol Md, at no 101, principal isotope ^{258}Md, half-life 54 days. Named after Mendeleyev, Russian chemist, associated with the PERIODIC TABLE.

Mendeleyev's law (*Phys*) The 'Law of Octaves'. If elements are listed according to increasing atomic weight, their properties vary but show a general similarity at each period of eight true rises.

Mendeleyev's table (*Chem*) See PERIODIC SYSTEM.

Mendelian character (*BioSci*) A character determined by a single gene and inherited according to MENDEL'S LAWS.

Mendelian genetics (*BioSci*) The genetics of characters determined by single genes with effects large enough to be easily recognizable.

Mendel's laws (*BioSci*) Laws dealing with the mechanism of inheritance. First law (law of segregation) states (in modern terminology) that the two alleles, one received from each parent, are segregated in gamete formation, so that each gamete receives one or the other with equal probability. This results in various characteristic ratios in the progeny depending on the parental genotypes and dominance. Second law (law of recombination) states that two characters determined by two unlinked genes are recombined at random in gamete formation, so that they segregate independently of each other, each according to the first law. This results in various dihybrid ratios.

mending (*Textiles*) Making good any imperfection in cloth caused during weaving by yarn breakages etc, or knots. Some faults are made good by hand sewing.

Ménière's disease (*Med*) A disorder characterized by attacks of dizziness, buzzing noises in the ears and progressive deafness; due to chronic disease of the labyrinth of the ear. Also *labyrinthine vertigo*.

menilite (*Min*) An alternative and more attractive name for LIVER OPAL; it is a grey or brown variety of that mineral.

meninges (*BioSci*) In vertebrates, envelopes of connective tissue surrounding the brain and spinal cord. Sing *meninx* (Gk membrane).

meningioma (*Med*) A tumour of the meninges of the brain and, more rarely, of the spinal cord.

meningism (*Med*) The presence of the symptoms of meningitis in conditions in which the meninges are neither diseased nor inflamed. Also *meningismus*.

meningitis (*Med*) Inflammation of the MENINGES characterized by severe headache, PHOTOPHOBIA and neck stiffness.

meningocele (*Med*) The hernial protrusion of the meninges through some defective part of the skull or the spinal column.

meningococcus (*Med*) A Gram-negative non-motile pyogenic coccus (*Neisseria meningitidis*) that is responsible for epidemic bacterial meningitis.

meningo-encephalitis (*Med*) Inflammation of the meninges and of the brain substance.

meningo-encephalocele (*Med*) A hernia of the meninges and brain through some defect in the skull.

meningomyelitis (*Med*) Inflammation of the meninges and of the spinal cord.

meningomyelocele (*Med*) Hernial protrusion of meninges and spinal cord through a defect in the spinal column.

meningovascular (*Med*) Pertaining to, or affecting, the meninges and the blood vessels (esp of the nervous system).

meniscus (*BioSci*) A small interarticular plate of fibrocartilage that prevents violent concussion between two bones; as the intervertebral disks of mammals.

meniscus (*Phys*) The departure from a flat surface where a liquid meets a solid, due to surface tension. The effect can be seen clearly through the wall of a glass tube. The surface of water in air rises up the wall of the tube, while that of mercury is depressed.

meniscus lens (*Phys*) A lens which is convex on one face and concave on the other, the radius of curvature of the concave face being greater than that of the convex face.

meniscus telescope (*Astron*) A compact instrument, developed by Maksutov in 1941, in which the spherical aberration of a concave spherical mirror is corrected by a meniscus lens. It differs from the Schmidt type in having a correcting plate with two spherical surfaces. It is much used in mirror lenses for miniature cameras.

menopause (*Med*) The natural cessation of menstruation in women.

menorrhagia (*Med*) Excessive loss of blood due to increased discharge during menstruation.

mensa (*BioSci*) The biting surface of a tooth.

Mensa (Table) (*Astron*) An inconspicuous southern constellation.

menstruation (*BioSci*) The periodical discharge from the uterus occurring in women from puberty to menopause. Also *xenomenia*.

mental age (*Psych*) A score devised by Binet to represent a child's test performance; it indicates the average age (50% of that age group) of those children who pass the same items as the person whose mental age is being computed, expressed in terms of years and months. See INTELLIGENCE QUOTIENT.

mental retardation (*Psych*) A condition characterized by a very low intellectual ability and a serious inability to cope with the environment.

mental set (*Psych*) The tendency to view new problems in the same fashion as old problems; also a predisposition to perceive or think in one way rather than another.

menthol (*Pharmacol*) A camphor compound: The $(-)$isomer is the chief constituent of peppermint oil. It is used as an antiseptic and local analgesic.

mentum (*BioSci*) In higher vertebrates, the chin; in some Gastropoda, a projection between the head and foot; in insects, the distal sclerite forming the basal portion of the labium. Adj *mental*.

menu (*ICT*) A line across the top of the screen listing the titles of the menus that the USER can choose from at that point in the APPLICATION PROGRAM. See also PULL-DOWN MENU, WIMP.

menu-bar (*ICT*) In a windowed environment, a bar showing options, often drop-down submenus, that can be selected by mouse click or keyboard.

MEO (*Space*) Abbrev for *medium Earth orbit*. See panel on COMMUNICATIONS SATELLITE.

mepacrine (*Pharmacol*) A yellow bitter powder, formerly widely used in the prophylaxis and treatment of malaria; replaced by CHLOROQUINE.

meprobamate (*Pharmacol*) Tranquillizing drug, used in motion sickness and alcoholism.

meralgia paraesthetica (*Med*) An affection of the nerve supplying the skin of the front and outer part of the thigh; characterized by pain, tingling and/or numbness.

meranti (*For*) A SE Asian hardwood, from the genus *Shorea*. Four varieties are recognized: dark-red, light-red, yellow and white.

merbromin (*Chem*) A green iridescent crystalline compound that forms a red solution in water, used as an antiseptic. TN *Mercurochrome*.

Mercalli scale (*Geol*) A scale of intensity used to measure earthquake shocks. The various observable movements are graded from 1 (very weak) to 12 (total destruction). See panel on EARTHQUAKE and appendix on Earthquake severity measurement scales.

mercaptans (*Chem*) Thio-alcohols. General formula, RSH. They form salts with sodium, potassium and mercury, and are formed by warming alkyl halides or sulphates with potassium hydrosulphide in concentrated alcoholic or aqueous solution. Ethyl mercaptan (ethanethiol), C_2H_5SH, is a liquid of nauseous odour; bp 36°C; it is readily oxidized to ethyl disulphide $(C_2H_5)_2S_2$, by exposure to the air. It is an intermediate for *sulphonal* and for rubber accelerators.

mercaptides (*Chem*) The salts of mercaptans.

Mercator's projection (*Geol*) A derivative of the simple *cylindrical projection*. The meridian scale is adjusted to coincide with the latitudinal scale at any given point, allowing true representation of shape but not of area, regions towards the poles being exaggerated in size. Long used by navigators, for straight lines drawn on it represent a constant bearing. A transverse form of Mercator is sometimes used in large-scale maps of limited E–W extent such as many of the UK Continental Shelf.

mercerization (*Textiles*) A process which greatly increases the lustre of cotton yarns and fabrics. It consists of treating the material under tension with aqueous sodium hydroxide (caustic soda) which causes swelling and results in increased strength and dye absorption properties. Invented by John Mercer in 1844.

merchant convertor (*Textiles*) A supplier of finished fabric who buys loom-state material and commissions its further processing.

merchant iron (*Eng*) A term (now obsolete) for bar iron made by repiling and rerolling puddled bar. All wrought-iron used to be treated in this way before being used for manufacture of chains, hooks, etc.

mercuric [mercury (II) chloride] (*Chem*) A soluble toxic substance $(HgCl_2)$ used as a pesticide. Prepared by sublimation from a mixture of mercury (II) sulphate and sodium chloride.

mercuric [mercury (II) iodide] (*Chem*) HgI_2. See NESSLER'S SOLUTION.

Mercurochrome (*Med*) TN for MERBROMIN.

mercurous [mercury (I) chloride] (*Chem*) See CALOMEL.

Mercury (*Astron*) The planet closest to the Sun, with a diameter 4878 km, orbital period 88 days and rotation period 59 days. The surface is heavily cratered and there is no true atmosphere, although traces of helium and argon from the SOLAR WIND or rocks are present. Its rel.d. of 5·5 is almost as large as the Earth's and may indicate an iron core. Daytime temperatures reach 450°C, falling to −180°C at night. There are no natural satellites. See appendix on Planets.

mercury (*Chem*) Chemical symbol Hg, at no 80, ram 200·59, rel.d. at 20°C 13·596, mp −38·9°C, bp 356·7°C, electrical resistivity $95·8 \times 10^{-8} \, \Omega$ m. A white metallic element which is liquid at atmospheric temperature. Metallic mercury occurs rarely but its principal ore is *cinnabar*, Hg S, which occurs in brilliant-red acicular crystals. A solvent for most metals, the products being called *amalgams*. Its chief uses are in the manufacture of batteries, drugs, chemicals, fulminate and vermilion. Used as metal in mercury-vapour lamps, arc rectifiers, power-control switches, and in many scientific and electrical instruments; ^{198}Hg is a mercury isotope made from gold in a reactor, for use in a quartz mercury-arc tube, light from which has an exceptionally sharp green line, because of even mass number of a single isotope. This considerably improves comparisons of end gauges in interferometers. Organic forms of mercury are particularly toxic (see MINAMATA DISEASE). Also *quicksilver*.

mercury arc (*Electronics*) An electrical discharge though ionized mercury vapour, giving off a brilliant blue-green light containing strong ultraviolet radiation. Used in hot- and cold-cathode tubes, for power switching and for illumination in fluorescent tubes, the ultraviolet radiation is used to excite visible radiation from a phosphor coating on the tube wall. See MERCURY-VAPOUR LAMP, MERCURY-VAPOUR TUBE.

mercury-arc converter (*ElecEng*) A *converter* making use of the properties of the MERCURY-ARC RECTIFIER.

mercury-arc rectifier (*ElecEng*) One in which rectification arises from the differential migration of electrons and heavy mercury ions in a plasma, formed by evaporation from a *hot spot* on a cathode pool of mercury. This vaporization has to be started by withdrawing an electrode from this pool, thus generating an arc. Used for ac/dc conversion for railways, trams, etc, before the introduction of silicon rectifiers. See THYRATRON.

mercury barometer (*EnvSci, Phys*) An instrument used for measuring the pressure of the atmosphere in terms of the height of a column of mercury which exerts an equal pressure. In its simplest form, it consists of a vertical glass tube about 80 cm long, closed at the top and having its lower open end immersed in mercury in a dish. The tube contains no air, the space above the mercury column being known as a *Torricellian vacuum*.

mercury cell (*Chem*) (1) Electrolytic cell with mercury cathode. (2) Dry cell employing mercury electrode, emf approx 1·3 V; eg MALLORY BATTERY, REUBEN–MALLORY CELL.

mercury delay line (*ICT*) One in which mercury is used as the medium for sound transmission, conversion from and to electrical energy being through suitable transducers at the ends of the mercury column.

mercury discharge lamp (*Electronics*) See MERCURY ARC.

mercury fulminate (*Chem*) $Hg(ONC)_2$. A crystalline solid prepared from ethanol, mercury and nitric acid; used as an initiator in detonators and percussion caps.

mercury intrusion method (*Chem*) Finding the distribution of sizes of capillary pores in a body by forcing in mercury, the radius being found from the pressure, and the percentage from the volume of mercury absorbed at each pressure.

mercury-pool cathode (*Electronics*) Cold cathode in valve where arc discharge releases electrons from surface of mercury pool, eg in MERCURY-ARC RECTIFIER.

Mercury program (*Space*) NASA project which resulted in the first US manned orbital flight by John Glenn on 20 February 1962.

mercury seal (*Chem*) A device which ensures that the place of entry of a stirrer into a piece of apparatus is gas-tight, while allowing the free rotation of the stirrer.

mercury switch (*ElecEng*) A switch in which the fixed contacts consist of mercury cups into which the moving contacts dip, or in which the mercury is contained in a tube which is made to tilt, thereby causing the mercury to bridge the contacts.

mercury-vapour cycle (*Eng*) The use of mercury in a closed loop for external combustion turbines. It has certain advantages over water, but is not widely used.

mercury-vapour lamp (*Electronics*) A quartz tube containing a mercury arc; used to provide ultraviolet rays for therapeutic and cosmetic treatment.

mercury-vapour pump (*ChemEng*) See GAEDE DIFFUSION PUMP.

mercury-vapour rectifier (*ElecEng*) See MERCURY-ARC RECTIFIER.

mercury-vapour tube (*Electronics*) Generally, any device in which an electric discharge takes place through mercury vapour. Specifically, a triode valve with mercury vapour, which is ionized by the passage of electrons, and reduces the space charge and the anode potential necessary to maintain a given current. The grid is effective in controlling the start of the discharge. See THYRATRON.

mereology (*Genrl*) The formal study of the relationship between the parts of a system and the whole.

merge (*ICT*) Combine two or more ordered files into a single ordered file.

mericlinal chimera (*BioSci*) A chimera in which one component does not completely surround the other; an incomplete PERICLINAL CHIMERA.

meridian (*Astron*) The great circle passing through the poles of the celestial sphere which cuts the observer's horizon in the north and south points, and also passes through the zenith. Also *meridian of longitude*.

meridian circle (*Astron*) A telescope mounted on a horizontal axis lying due east and west, so that the instrument itself moves in the meridian plane. It is used to determine the times at which stars cross the meridian, and is equipped with a graduated circle for deducing declinations. Also *transit circle*.

meridional (*BioSci*) Extending from pole to pole; as in *meridional furrow* in a segmenting egg.

merino wool (*Textiles*) Wool of fine quality from merino sheep or similar. The name is also used in the woollen trade for wool fibre recovered from fine woollen and worsted clothing rags.

Merioneth (*Geol*) The youngest epoch of the Cambrian period.

meristele (*BioSci*) A strand of vascular tissue, enclosed in a sheath of endodermis, forming part of a dictyostele.

meristem (*BioSci*) A group of actively dividing cells including INITIALS and their undifferentiated derivatives. See APICAL MERISTEM, INTERCALARY MERISTEM, LATERAL MERISTEM.

meristem culture (*BioSci*) The aseptic culture on a suitable medium of excised shoot apical meristems including one or a few leaf primordia. Used for MICROPROPAGATION and to obtain virus-free plants from virus-infected stocks because meristems are normally virus free. Also *shoot-tip culture*. See TISSUE CULTURE.

meristic (*BioSci*) Segmented; divided up into parts; pertaining to the number of parts, as in MERISTIC VARIATION. See MEROME.

meristic variation (*BioSci*) Variation in the number of organs or parts; as in the number of body somites of a metameric animal.

merit index (*Eng*) Criterion for materials selection, eg in terms of the best mechanical properties for least weight of material. Relevant for design of vehicles, aerospace products and devices fitted to or by hand (eg tools, building products). Products put in tension suggest a merit index of E/ρ, so for rope, a table of comparative indices can be made. The best materials with high merit indices can thus be filtered and further criteria applied (eg cost). See panel on HIGH-PERFORMANCE POLYMERS.

merlons (*Arch*) The projecting parts of a battlement.

mermaid's purse (*BioSci*) A popular name applied to the horny purse-like capsule in which the eggs of certain selachian fish (sharks, dogfish, skates, rays) are enclosed.

mero- (*Genrl*) Prefix from Gk *meros*, part.

meroblastic (*BioSci*) Said of a type of egg in which cleavage is restricted to a part of the surface, ie is incomplete, usually owing to the large amount of yolk. Cf HOLOBLASTIC.

merogamy (*BioSci*) Having gametes smaller than the vegetative cells and produced by a special division; union of such gametes.

merogenesis (*BioSci*) Segmentation, formation of parts.

merogony (*BioSci*) See SCHIZOGONY.

merome (*BioSci*) A body somite or segment of a metameric animal.

meromorphic function (*MathSci*) See ANALYTIC FUNCTION.

meroplankton (*BioSci*) Organisms which spend only part of their life history as plankton. Cf HOLOPLANKTON.

meros (*Arch*) The surfaces between the channels in a triglyph. Also *femur*.

merosmia (*Med*) Temporary deficiency of the sense of smell which may be organic in nature (eg sinusitis) or a result of hysteria.

merosthenic (*BioSci*) Having the hind-limbs exceptionally well developed, eg frogs, kangaroos.

mersawa (*For*) SE Asian timber from a variety of related hardwood species (*Anisoptera*). The heartwood is pink-tinged, yellowish-brown, with a lightly interlocked grain and a coarse but even texture.

Mersenne prime (*MathSci*) Any prime number which can be written as 2^P-1, where p is itself prime. There are 32 such primes known, but not all such numbers are prime. Cf FERMAT PRIME.

Merulius lacrymans (*Build*) See SERPULA LACRYMANS.

merycism (*Med*) The return, after a meal, of gastric contents to the mouth; they are then chewed and swallowed once more as in rumination in animals.

Merz–Hunter protective system (*ElecEng*) See SPLIT-CONDUCTOR PROTECTION.

Merz–Price protective system (*ElecEng*) A form of balanced protective system for electric power networks, in which the current entering a section of the network is balanced against that leaving it. If a fault occurs on the section this balance is upset, and a relay is caused to operate and trip circuit breakers to clear the faulty section from the network.

Merz slit (*Phys*) A variable-width bilateral slit for spectrographs. Characterized by the fixed position of the centre of the slit.

mes-, meso- (*Genrl*) Prefixes from Gk *mesos*, middle.

mesa (*Electronics*) A type of transistor in which one electrode is made very much smaller than the other, to control bulk resistance. Also, by selective etching, the base and emitter are raised above the region of the collector.

mesa (*Geol*) A flat-topped, steep-sided tableland. Small mesas are called *buttes*.

mesarch (*BioSci*) A strand of xylem having the PROTOXYLEM in the centre with METAXYLEM developing centripetally and centrifugally.

mesa transistor (*Electronics*) A transistor using the MESA method of construction; the semiconductor wafer is etched down in steps so that the base and emitter regions appear as plateaux above the collector. Connections are terminated at the edges of the material, instead of coming to the surface, as in the PLANAR PROCESS.

mesaxonic foot (*BioSci*) A foot in which the skeletal axis passes down the third digit, as in Perissodactyla. Cf PARAXONIC FOOT.

mescaline (*Pharmacol*) A hallucinogenic drug derived from the Mexican cactus, mescal.

mesectoderm (*BioSci*) Parenchymatous tissue formed from ectoderm cells which have migrated inwards.

mesencephalon (*BioSci*) The midbrain of vertebrates.

mesenchyma (*BioSci*) Parenchyma; embryonic mesodermal tissue of spongy appearance. Adj *mesenchymatous*.

mesenchyme (*BioSci*) Mesodermal tissue, comprising cells that migrate from ectoderm, or endoderm, or mesothelium into the blastocoel. Cf *mesothelium*.

mesenteric (*BioSci*) Pertaining to the mesenteron; pertaining to a mesentery.

mesenteric caeca (*BioSci*) Digestive diverticula of the mesenteron in many invertebrates (eg in Arachnida, Crustacea, Echinodermata, insects).

mesenteron (*BioSci*) See MIDGUT.

mesentery (*BioSci*) In Cnidaria, a vertical fold of the body wall projecting into the enteron. More generally, a fold of tissue supporting part of the viscera. Adjs *mesenterial, mesenteric*.

mesethmoid (*BioSci*) A median cartilage bone of the vertebrate skull, formed by ossification of the ethmoid plate. Also *internasal septum*.

mesh (*Build, CivEng*) Expanded metal or plastic (eg TENSAR) used as a reinforcement for concrete, asphalt, clay and many other building materials.

mesh (*ICT*) A complete electrical path (including capacitors) in the component branches of a complex network. Also *loop*.

mesh connection (*ElecEng*) A method of connecting the windings of an ac electric machine; the windings are connected in series so that they may be represented diagrammatically by a polygon. The DELTA CONNECTION is a particular example of this method.

mesh network (*ICT*) One formed from a number of impedances in series.

mesh voltage (*ElecEng*) The voltage between any two lines of a symmetrical polyphase system which are consecutive as regards phase sequence. Called *delta voltage* in a three-phase system, and *hexagon voltage* in a six-phase system.

mesiad (*BioSci*) Situated near, or tending towards, the median plane.

mesial, mesian (*BioSci*) In the median vertical or longitudinal plane.

mesic atom (*Phys*) Short-lived atom in which a negative MUON has displaced a normal electron.

mesitylene (*Chem*) 1,3,5-trimethyl-benzene. $C_6H_3(CH_3)_3$. Bp 164°C. A colourless liquid, occurring in coaltar.

mesityl oxide (*Chem*) $(CH_3)_2C=CHCOCH_3$. Bp 122°C. A colourless liquid of peppermint-like odour. Obtained from propanone by an ALDOL CONDENSATION type of reaction induced by ammonia.

mesmerism (*Psych*) An early name for *hypnosis*.

meso- (*Chem*) (1) Optically inactive by intramolecular compensation. Abbrev *m-*. (2) Substituted on a carbon atom situated between two hetero-atoms in a ring. Abbrev μ-. (3) Substituted on a carbon atom forming part of an intramolecular bridge. Abbrev μ-. From Gk *mesos*, middle.

meso- (*Genrl*) See MES-, MESO-.

mesobenthos (*BioSci*) Fauna and flora of the sea floor, at depths ranging from 100 to 500 fathoms (200–1000 m).

mesoblast (*BioSci*) The mesodermal or third germinal layer of an embryo, lying between the endoderm and ectoderm. Adj *mesoblastic*.

mesoblastic somites (*BioSci*) In developing metameric animals, segmentally arranged blocks of mesoderm, the forerunners of the somites.

mesocarp (*BioSci*) The middle, often fleshy, layer of the PERICARP.

mesocoele (*BioSci*) In vertebrates, the cavity of the midbrain; mid-ventricle; Sylvian aqueduct.

mesocolloid (*Chem*) A particle whose dimensions are 25–250 nm containing 100–1000 molecules.

mesoderm (*BioSci*) See MESOBLAST.

mesogaster (*BioSci*) In vertebrates, the portion of the dorsal mesentery which supports the stomach.

mesogastrium (*Med*) The region of the abdomen between the epigastrium and the hypogastrium. Pl *mesogastria*.

mesogloea (*BioSci*) In Cnidaria, a structureless layer of gelatinous material intervening between the ectoderm and the endoderm.

mesokaryote (*BioSci*) Having a nucleus with a nuclear envelope and with chromosomes, but having very little histone associated with the DNA of the chromosomes which remain condensed throughout interphase; the condition in the Dinophyceae. Also *dinokaryote*. Cf EUKARYOTE, PROKARYOTE.

mesokurtic (*MathSci*) Of a curve or distribution having the same concentration around the MEAN as a normal curve. Compare LEPTOKURTIC, PLATYKURTIC. See KURTOSIS.

mesolecithal (*BioSci*) A type of egg of medium size containing a moderate amount of yolk that is strongly concentrated in one hemisphere, ie the lower one in eggs floating in water. Found in frogs, Urodela, lungfishes, lower Actinopterygii and lampreys. Cf OLIGOLECITHAL, TELOLECITHAL.

mesolite (*Min*) A zeolite intermediate in composition between NATROLITE and SCOLECITE. Crystallizes in the monoclinic system, and occurs in amygdaloidal basalts and similar rocks.

Mesolithic (*Geol*) The middle division of the Stone Age. Cf NEOLITHIC, PALAEOLITHIC.

mesomerism (*Chem*) (1) See DESMOTROPISM. (2) See RESONANCE.

mesometrium (*BioSci*) The mesentery that supports the uterus and related structures.

mesomorph (*Psych*) One of Sheldon's somatotyping classifications: mesomorphs are hard, rectangular and well-muscled; delight in physical activity, love adventure and power, and are indifferent to people. See SOMATOTYPE THEORY.

mesomorphous (*Chem*) Existing in a state of aggregation midway between the true crystalline state and the completely irregular amorphous state. See LIQUID CRYSTALS.

meson (*Phys*) A hadron with a baryon number of 0. Mesons generally have masses intermediate between those of electrons and nucleons and can have negative, zero or positive charges. Mesons are BOSONS and may be created or annihilated freely. There are three groups of mesons: π-mesons (pions), K-mesons (kaons) and η-mesons.

mesonephric duct (*BioSci*) See WOLFFIAN DUCT.

mesonephros (*BioSci*) Part of the kidney of vertebrates, arising later in development than, and posterior to, the pronephros, and discharging into the WOLFFIAN DUCT; becomes the functional kidney in adult anamniotes. Adj *mesonephric*.

meson field (*Phys*) The field concerned with the interchange of protons and neutrons in the nucleus of an atom, mesons transferring the energy.

mesopause (*EnvSci*) The top of the mesosphere at about 80–85 km.

mesophilic bacteria (*BioSci*) Bacteria that grow best at temperatures of 20–45°C.

mesophyll (*BioSci*) The ground tissue of a leaf, located between the upper and lower epidermis and typically differentiated as chlorenchyma. See PALISADE, SPONGY LAYER.

mesophyte (*BioSci*) A plant adapted to habitats that are neither very wet nor dry. Cf HYDROPHYTE, XEROPHYTE.

mesorchium (*BioSci*) In vertebrates, the mesentery supporting the testis.

mesosilicate (*Min*) A silicate mineral whose atomic structure contains isolated groups of silicon–oxygen tetrahedra.

mesosphere (*EnvSci*) The region of the atmosphere lying between the STRATOPAUSE and MESOPAUSE (50–85 km), in which temperature generally decreases with height. See panels on STRATOSPHERE AND MESOSPHERE and TROPOSPHERE.

mesosternum (*BioSci*) In insects, the sternum of the mesothorax; in vertebrates, the middle part of the sternum, connected with the ribs; the gladiolus.

mesotarsal (*BioSci*) In insects, the tarsus of the second walking leg; in land vertebrates, the ankle joint or joint between the proximal and distal rows of tarsals.

mesothelioma (*Med*) A malignant tumour of the mesothelium, usually of the lung; frequently caused by exposure to asbestos fibres, particularly those of crocidolite.

mesothelium (*BioSci*) Mesodermal tissue comprising cells which form the wall of the coelom. Cf MESENCHYME.

mesotherapy (*Med*) The treatment of a specific area of the body by injecting a substance into the mesoderm.

mesothorax (*BioSci*) The second of the three somites composing the thorax in insects. Adj *mesothoracic*.

mesotrochal (*BioSci*) Having an equatorial band of cilia.

mesovarium (*BioSci*) In vertebrates, the mesentery supporting the ovary.

Mesozoic (*Geol*) The era embracing the Triassic, Jurassic and Cretaceous periods, characterized by ammonites and reptiles, together with brachiopods, lamellibranchs, gastropods and corals. The Triassic first period had an impoverished fauna and flora after the extinctions at the end of the Palaeozoic era. In the Jurassic there was a rich flora in the warm climate, when reptiles, notably dinosaurs, were dominant on land. In the Cretaceous, flowering plants spread and many large reptiles, ammonites, most belemnites and many brachiopod species became extinct and Chalk was the most important formation. See appendix on Geological time.

message board (*ICT*) A BULLETIN BOARD SYSTEM.

message box (*ICT*) A box that appears on the screen of a computer conveying information, eg about an error. Also *dialogue box*.

message pager (*ICT*) A pager that, in addition to simply signalling reception of its call sign, can also receive and display a short written message. See PAGING.

message switching (*ICT*) A method using sequentially organized switching mechanisms, of batching and storing messages for economical transmission on shared network lines.

messenger RNA (*BioSci*) The RNA whose sequence of nucleotide triplets determines the sequence of a polypeptide. Abbrev *mRNA*.

messenger wire (*ElecEng*) A strong suspension wire for holding aerial cables. Also *bearer cable*.

Messier catalogue (*Astron*) A listing of 108 galaxies, star clusters and nebulae drawn up by the French comet hunter Charles Messier in 1770. Objects are designated M1, M2, etc.

Messinian (*Geol*) A stage in the Miocene. See TERTIARY.

messuage (*Build*) A dwelling house and its adjacent land and buildings.

mestome (*BioSci*) Conducting tissue, with associated parenchyma, but without mechanical tissue.

mestome sheath (*BioSci*) An endodermis-like bundle sheath; the inner of two sheaths round the vascular bundle of some grasses, eg wheat. Also *mestom sheath*.

Met (*Chem*) Symbol for methionine.

meta- (*Chem*) Prefix indicating: (1) derived from an acid anhydride by combination with one molecule of water; (2) a polymer of; (3) a derivation of; (4) the 1,3 relationship of substituents on a benzene ring (abbrev *m*-), eg *m*-cresol.

meta- (*Genrl*) Prefix from Gk *meta*, after.

meta-aldehyde (*Chem*) See METALDEHYDE.

meta analysis (*MathSci*) The statistical procedure used to combine numerous and independent research results into one study. Each research study becomes one subject in the meta analysis.

metabolic burst (*BioSci*) When a phagocyte ingests a particle there is a burst of increased oxygen uptake, leading to the production of hydrogen peroxide, superoxide anion, singlet oxygen and the hydroxyl radical, which form the most important microbicidal mechanism of phagocytes (oxygen-dependent killing). Also *respiratory burst*. See MYELOPEROXIDASE–HALIDE SYSTEM.

metabolic cooperation (*BioSci*) The transfer between tissue cells in contact of low-molecular-weight metabolites such as nucleotides and amino acids. Transfer is via channels (*gap junctions*), and does not involve exchange with the extracellular medium.

metabolic half-life (*BioSci*) The time required for half of the quantity of a substance in the body to be metabolized.

metabolism (*BioSci*) The sum total of the chemical and physical changes constantly taking place in living matter. Adj *metabolic*.

metabolite (*BioSci*) A substance involved in metabolism, being either synthesized during metabolism or taken in from the environment.

metaboly (*BioSci*) The power possessed by some cells of altering their external form, eg as in Euglenida.

metabonomics (*BioSci*) The study of metabolic responses to external stimuli.

metaboric acid (*Chem*) See BORIC ACID.

metabotropic (*BioSci*) A category of neurotransmitter receptor that affects cell activity by means other than direct regulation of the cell's ion channels (cf IONOTRO-PIC).

metacarpal (*BioSci*) One of the bones composing the METACARPUS in vertebrates. Also *metacarpale*.

metacarpus (*BioSci*) In land vertebrates, the region of the fore-limb between the digits and the carpus.

metacentre (*Phys*) The point of intersection of: (1) a vertical line drawn through the centre of gravity of a body floating in equilibrium in a liquid; and (2) a second vertical line drawn through the centre of buoyancy (centre of gravity of the displaced liquid) when the body is displaced slightly from its equilibrium position. According to whether this point is above or below the centre of gravity of the body, the equilibrium is stable or unstable.

metacentric (*Ships*) The measure of stability of a vessel at small angles of heel, indicative of its behaviour when rolling.

metacentric height (*Phys*) The distance between the centre of gravity of a floating body and its METACENTRE.

metacercaria (*BioSci*) An infectious larva of a digenean TREMATODE parasite that encysts on vegetation or inside an intermediate host.

metachromatic (*BioSci*) Relating to *metachromatism*: of dyes, capable of staining cells so that they turn a different colour to that of the dye; of cells, taking a colour different to that of the dye with which they are stained.

metachronal rhythm (*BioSci*) The rhythmic beat of cilia on a cell surface in which the beat of adjacent cilia is slightly out of phase. Consequently it is seen as a series of waves passing over the ciliary surface.

metachrosis (*BioSci*) The ability, shown by some animals (as the chameleon), to change colour by expansion or contraction of chromatophores.

metacoele (*BioSci*) In Craniata, the cavity of the hind-brain; the fourth ventricle.

metacognition (*Psych*) Awareness and understanding of one's own thought processes.

metadata (*ICT*) Information about the characteristics of an item of data, usually collected in a data dictionary.

metadiscoidal placentation (*BioSci*) Having the placental villi at first scattered and then restricted to a disk as in primates.

metadyne (*ElecEng*) See AMPLIDYNE.

metafile (*ICT*) Graphical information capable of being transferred between systems or software.

metagenesis (*BioSci*) See ALTERNATION OF GENERATIONS.

metal (*Astron*) Loosely, any element heavier than helium.

metal (*Chem*) Any material whose FERMI SURFACE lies predominantly in its conduction band. More particularly an element which is held together by metallic bonds and shows characteristic properties, which include high reflectivities and electrical and thermal conductivities and relatively high density compared with non-metals. Most elements of the periodic table (see appendix on The periodic table) are metals. Further subdivided into alkali metals, alkaline earth metals, transition metals, actinides, lanthanides, etc. See panel on BONDING.

metal (*CivEng*) See ROAD METAL.

metal arc welding (*ElecEng*) A type of electric welding in which the electrodes are of metal, and melt during the welding process to form filler metal for the weld.

Metalastik (*Eng*) TN for type of INCLINED SHEAR MOUNT.

metal-clad switchgear (*ElecEng*) A type of switchgear in which each part is completely surrounded by an earthed metal casing. Cf METAL-ENCLOSED SWITCHGEAR.

metaldehyde (*Chem*) *Meta-aldehyde*. $(CH_3CHO)_4$. Long glistening needles which sublime at 115°C with partial decomposition into ethanal. Acetaldehyde is polymerized to metaldehyde by the action of acids at temperatures below 0°C. Sometimes used as a portable fuel, *meta-fuel*.

metal detector (*Eng*) An instrument, widely used in industrial production, for detecting the presence of embedded stray metal parts in food products. It is usually incorporated in a conveyor line and gives visible or audible warning or automatically stops the line. Also used for detecting buried metal.

metal electrode (*ElecEng*) A form of electrode used in METAL ARC WELDING.

metal-enclosed switchgear (*ElecEng*) A type of switchgear in which the whole equipment is enclosed in an earthed metal casing. Cf METAL-CLAD SWITCHGEAR.

metal-evaporated tape (*ImageTech*) Tape produced by evaporating metal in a vacuum and condensing it onto the BASE, giving a high recording density. Abbrev *ME tape*.

metal feeder (*Print*) A device which maintains a supply of metal in typesetting machines from an ingot fed at the required rate into the metal pot.

metal-film resistor (*ElecEng*) One formed by coating a high-temperature insulator with a very thin layer of metallic film, eg by vacuum deposition.

metal halide lamp (*ImageTech*) A compact-source enclosed mercury-arc lamp with metal halide additions, for ac operation. Abbrev *MH lamp*.

metal inert-gas welding (*Eng*) Arc welding with a metal electrode shielded by an inert gas such as argon or carbon dioxide. Abbrev *MIG*.

metal insulator (*ICT*) Waveguide or transmission line an odd number of quarter-wavelengths long, that has a high impedance. Used as a support or anchor without normal insulators at very high frequencies.

metal lathing (*Build*) EXPANDED METAL used to cover surfaces to provide a basis for plaster.

metallic bond (*Chem*) In metals, the valence electrons are not even approximately localized in discrete covalent bonds, but are delocalized and interact with an indefinite number of atomic nuclei. This gives rise both to the opacity and lustre of metals and to their electrical conductivity. See panel on BONDING.

metallic coated paper (*Paper*) Wrapping or decorative paper coated with powdered metal, eg tin, bronze, in a binder.

metallic conduction (*Electronics*) The transport of electrons which are freely moved by an electric field within a body of metal.

metallic glass (*Glass*) One of a range of metal alloys which can be produced in the form of a glass by very rapid cooling. They tend to have good corrosion resistance and the absence of grain boundaries on which to pin MAGNETIC DOMAINS means that they are well-suited for transformer cores.

metallic lens (*ICT*) One with slats or louvres that give varying retardation to a passing electromagnetic wave, so that it is controlled or focused. Can also be used for sound waves.

metallic lustre (*Min*) A degree of lustre exhibited by certain opaque minerals, comparable with that of polished steel.

metallic packing (*Eng*) A PACKING consisting of a number of rings of soft metal, or a helix of metallic yarn encircling the piston rod and pressed into contact therewith by a gland nut.

metallization (*Chem*) (1) The deposition of thin films of metal onto any surface for decorative or electrical purposes. Also the film itself. See SEMICONDUCTOR DEVICE PROCESSING. (2) The conversion of a substance, eg selenium, into a metallic form.

metallized yarn (*Textiles*) An effect yarn containing some metallic components which may be separate filaments or fibres made of metal or a thin metal strip (eg of anodized aluminium) which is protected by a transparent film (eg of cellulose acetate).

metallography (*Eng*) The study of metals and their alloys with the aid of various procedures, eg microscopy, X-ray diffraction, etc.

metalloid (*Chem*) An element having both metallic and non-metallic properties, eg arsenic.

metallo-organic chemical vapour deposition (*Electronics*) Metallic species introduced as gaseous organic complexes in epitaxial growth. See panel on SEMICONDUCTOR FABRICATION.

metallo-organic compounds (*Chem*) A rather broader category than ORGANO-METALLIC COMPOUNDS that includes metal complexes with organic ligands, even those without a direct M–C bond, as well as those in which the carbon atoms are linked directly with metal atoms.

metalloprotein (*BioSci*) A conjugated protein in which the prosthetic group is a metal.

metallurgical balance sheet (*MinExt*) A report in equation form of the products from treatment of a known tonnage of ore of specified assay value, yielding a known weight of concentrate and tailing, to which the head value can be attributed for economic and technical control.

metallurgical coke (*MinExt*) Coke of high strength to resist pressure and breakage, and of high purity. Used for smelting mineral ores.

metallurgy (*Eng*) The science and technology of metals and their alloys including methods of extraction and use.

metal matrix composite (*Aero, Eng*) A class of COMPOSITE MATERIALS that incorporates fibres (typically ceramics such as ALUMINA, SILICON CARBIDE) in a metallic matrix. They offer much better high-temperature performance than composites based on polymeric matrices. Abbrev MMC.

metal–oxide–semiconductor transistor (*Electronics*) An active semiconductor device in which a conducting channel is induced in the region between the electrodes by applying a voltage to an insulated electrode placed on the surface in this region. It is self-isolating by virtue of its construction, and so can be fabricated in a smaller area than a BIPOLAR TRANSISTOR.

metal–oxide–silicon (*Electronics*) An integrated-circuit technology based on device structures comprising conductors, silicon dioxide insulator and semiconducting silicon (principally field-effect devices). Abbrev MOS.

metal–oxide–silicon field-effect transistor (*Electronics*) Silicon field-effect transistor in which the gate electrode is insulated from the conducting channel. Transistor action involves INVERSION of the semiconductor beneath the gate. Abbrev MOSFET. Cf JUNCTION FIELD-EFFECT TRANSISTOR.

metal particle tape (*ImageTech*) Tape coated with pure metal particles rather than oxide.

metal pattern (*Eng*) A PATTERN made in cast-iron, brass or light alloy with the durability needed when a large number of castings are required, as in repetition work on moulding machines. See DOUBLE CONTRACTION.

metal powder spraying (*Eng*) See METAL SPRAYING.

metal rectifier (*ElecEng*) A rectifier utilizing a layer of oxide on a metal disk, eg copper oxide on a copper disk, many of which can be connected together in series or in parallel to give high-voltage or high-current rectifiers. Generally superseded by semiconductor rectifying devices. Also *dry-plate rectifier*.

metal rule (*Print*) See EM RULE, RULE.

metals (*CivEng*) The rails of a railway.

metal spinning (*Eng*) The shaping of thin sheet-metal disks into cup-shaped forms by the lateral pressure of a steel roller or a stick on the revolving disk, which is gradually pressed into contact with a former on the lathe faceplate.

metal spraying (*Eng*) A method of applying protective metal coatings or building up worn parts by spraying molten metal from a gun. The coating metal is supplied as wire or powder, melted by flame, and blown out of the gun as finely divided particles which form a mechanically adherent layer on the surface of the component. There is no diffusion or alloying.

metal trim (*Build*) Architraves and other finishings made out of pressed metal sheeting clipped or screwed in position around door or window openings.

metal valley (*Build*) A gutter, lined with lead, zinc or copper, between two roof slopes.

metamere (*BioSci*) See MEROME.

metameric match (*ImageTech*) A subjective match of two colours whose actual spectral composition differs.

metamerism (*BioSci*) Having a segmented body form. Adj *metameric*.

metamerism (*Chem*) Having two or more constitutional isomers.

metamerism (*Textiles*) A marked change in colour of material subjected to different lighting. Thus two such fabrics may match in daylight but appear different when examined in artificial light.

metamict (*Min*) Mineral which has been exposed to natural radioactivity so that its crystalline structure breaks down to a glassy amorphous state, eg zircon.

metamorphic facies (*Geol*) All those rocks that have reached chemical equilibrium under the same pressure–temperature range of METAMORPHISM.

metamorphism (*Geol*) The change in the mineralogical and structural characteristics of a rock as a consequence of heat and/or pressure.

metamorphosis (*BioSci*) Pronounced change of form and structure taking place within a comparatively short time, as the changes undergone by an animal in passing from the larval to the adult stage. Adj *metamorphic*.

metanephric duct (*BioSci*) Ureter of amniote vertebrates.

metanephridia (*BioSci*) Nephridia that open into the coelom, the open end (nephrostome) being ciliated. Cf PROTONEPHRIDIUM.

metanephros (*BioSci*) In amniote vertebrates, part of the kidney arising later in development than, and posterior to, the mesonephros; becomes the functional kidney, with a special METANEPHRIC DUCT. Adj *metanephric*.

metanilic acid (*Chem*) Meta-aminobenzene-sulphonic acid. $C_6H_4(NH_2)(SO_3H)$. An intermediate for dyestuffs.

ME tape (*ImageTech*) Abbrev for METAL EVAPORATED TAPE.

metaphase (*BioSci*) A stage in cell division at which the chromosomes are attached to the spindle but have not yet segregated. See fig. at MITOSIS and panel on CELL CYCLE.

metaphase plate (*BioSci*) The equatorial plane of the mitotic or meiotic spindle, approximately equidistant from the two poles, on which the chromosomes are arrayed.

metaphloem (*BioSci*) The later-formed primary PHLOEM, esp that which matures after the organ has ceased to elongate. Cf PROTOPHLOEM.

metaphosphoric acid (*Chem*) HPO_3. Formed as a viscous solid when phosphorus (V) oxide is left exposed to the air.

metaphysis (*Med*) The end of the shaft (diaphysis) of a long bone where it joins the epiphysis.

metaplasia (*BioSci*) Tissue transformation, as in the ossification of cartilage.

metaplasis (*BioSci*) The period of maturity in the life cycle of an individual.

metapodium (*BioSci*) (1) In vertebrates, the second podial region; metacarpus or metatarsus; palm or instep. (2) In insects, that portion of the abdomen posterior to the abdominal constriction in eg wasps. (3) In Gastropoda, the posterior part of the foot. Adj *metapodial*.

metapophysis (*BioSci*) In some mammals, a process of the vertebrae above the prezygapophysis which strengthens the articulation.

metapsychology (*Psych*) Theories and theorizing on psychological matters, such as the nature of the mind, which cannot be verified or falsified by experiment or reasoning.

metaraminol (*Pharmacol*) A *sympathomimetic* drug given to produce vasoconstriction and to increase blood pressure.

metarchon (*Chem*) An agent which, without being toxic, so changes the behaviour of a pest that its persistence is diminished, eg a confusing sex attractant.

metasitism (*BioSci*) Cannibalism.

metasoma (*BioSci*) In Arachnida, the posterior part of the abdomen, or hindermost tagma of the body, which is always devoid of appendages. Adj *metasomatic*.

metasomatism (*Geol*) The change in the bulk chemical composition of a rock by the introduction of liquid or gaseous material from elsewhere.

metastable state (*Chem*) A state which is apparently stable, often because of the slowness with which equilibrium is attained; said eg of a supersaturated solution.

metastasic (*Phys*) Said of electrons which move from one shell to another or are absorbed from a shell into the nucleus.

metastasis (*Med*) The transfer, by lymphatic channels or blood vessels, of diseased tissue (esp cells of malignant tumours) from one part of the body to another; the diseased area arising from such transfer. Adj *metastatic*, v *metastasize*.

metatarsal (*BioSci*) One of the bones composing the METATARSUS in vertebrates. Also *metatarsale*.

metatarsalgia (*Med*) A painful neuralgic condition of the foot, felt in the ball of the foot and often spreading thence up the leg.

metatarsus (*BioSci*) In insects, the first joint of the tarsus when it is markedly enlarged; in land vertebrates, the region of the hind-limb between the digits and the ankle.

Metatheria (*BioSci*) A subclass of viviparous mammals in which the newly born young are carried in an abdominal pouch which encloses the teats of the mammary glands. There is usually no allantoic placenta. The scrotal sac is in front of the penis, the angle of the lower jaw is inflexed, and the palate shows vacuities. The subclass contains only one order, the Marsupialia.

metathorax (*BioSci*) The third or most posterior of the three somites composing the thorax in insects. Adj *metathoracic*.

metaxenia (*BioSci*) Any effect that may be exerted by pollen on the tissues of the female organs.

metaxolone (*Pharmacol*) A skeletal muscle relaxant. Its mode of action is unclear.

metaxylem (*BioSci*) The later-formed primary XYLEM, esp that which matures after the organ has ceased to elongate; commonly with reticulately thickened or pitted walls. Cf PROTOXYLEM.

Metazoa (*BioSci*) A subkingdom of the animal kingdom, comprising multicellular animals that have two or more tissue layers, never possess choanocytes, usually have a nervous system and enteric cavity, and always show a high degree of co-ordination between the different cells composing the body. Cf PARAZOA, PROTOZOA.

metecdysis (*BioSci*) The period after a moult in Arthropoda when the new cuticle is hardening and the animal is returning to normal in its physiological condition.

metencephalon (*BioSci*) The anterior portion of the hind-brain in vertebrates, developing into the cerebellum, and,

in mammals, the pons (part of the MEDULLA OBLONGATA). Cf MYELENCEPHALON.

meteor (*Astron*) A small body which enters the Earth's atmosphere from interplanetary space and becomes incandescent by friction, flashing across the sky and generally ceasing to be visible before it falls to Earth. Also *shooting star*.

meteor crater (*Astron*) Circular unnatural crater of which the Arizona Meteor Crater is best known; believed to be caused by the impact of meteorites.

meteoric water (*Geol*) Groundwater of recent atmospheric origin.

meteorism (*Med, Vet*) Excessive accumulation of gas in the intestines. See TYMPANITES.

meteorite (*Astron, Min*) A mineral aggregate of cosmic origin (METEOR) which reaches the Earth from interplanetary space. See ACHONDRITE, AEROLITE, CHONDRITE, IRON METEORITE, PALLASITE, SIDERITE, STONY METEORITE.

meteoroid (*Astron*) A METEOR which is outside the Earth's atmosphere.

meteorological satellite (*EnvSci*) An artificial satellite orbiting the Earth either in a GEOSTATIONARY or SUN-SYNCHRONOUS ORBIT. Used for weather observation and forecasting, it carries TV cameras to transmit cloud formation changes, and *radiometers* to measure terrestrial and solar radiation.

meteorology (*Genrl*) The study of the Earth's atmosphere in relation to weather and climate.

meteor shower (*Astron*) A display of meteors in which the number seen per hour greatly exceeds the average. Occurs when the Earth crosses a METEOR STREAM in its orbit.

meteor stream (*Astron*) Streams of dust revolving about the Sun, whose intersection by the Earth causes meteor showers. Some night-time showers have orbits similar to those of known comets; daytime showers, detected by radio-echo methods, have smaller orbits, similar to those of minor planets.

meter (*ElecEng*) A general term for any electrical measuring instrument, but usually confined to integrating meters.

meter (*Phys*) US for METRE.

meter protection circuit (*ElecEng*) One designed to avoid transient overloads damaging a meter. It may employ a gas-discharge tube which breaks down at a dangerous voltage or a Zener diode or similar device.

metformin HCl (*Pharmacol*) An oral hypoglycaemic drug that lowers blood glucose levels and is used in the management of Type II diabetes.

methacrylate butadiene styrene (*Chem*) A COPOLYMER with tough mechanical properties. Abbrev *MBS*. See panel on RUBBER TOUGHENING.

methadone (*Pharmacol*) Methadone hydrochloride. An analgesic with similar but less marked properties, both good and bad, than morphine.

methaemoglobin (*Chem*) A compound of haemoglobin and oxygen, more stable than oxyhaemoglobin, obtained by the action of oxidizing agents on blood, eg nitrites and chlorates. US *methemoglobin*.

methaemoglobinaemia (*Med*) The condition in which haemoglobin, as modified by drugs or hereditary defects, forms methaemoglobin and is incapable of carrying oxygen. Patients are blue and have shortness of breath. US *methemoglobinemia*.

methaemoglobinuria (*Med*) The presence of methaemoglobin in the urine. US *methemoglobinuria*.

methanal (*Chem*) See FORMALDEHYDE.

methane (*Chem*) CH_4. Mp $-186°C$, bp $-164°C$. The simplest ALKANE, a gas, occurs naturally in oil wells and as marsh gas. *Fire damp* is a mixture of methane and air; coal-gas contains a large proportion of methane. It can be synthesized from its elements, and prepared by various methods, as by catalytic reduction of CO or CO_2, or by passing CO and H_2O over heated metal oxides, or by the action of water on aluminium carbide (methanide).

methanides (*Chem*) Carbides, such as aluminium and beryllium carbides, which give methane when decomposed by water.

methanoates (*Chem*) See FORMATES.

methanoic acid (*Chem*) See FORMIC ACID.

methanol (*Chem*) CH_3OH. Bp 66°C, rel.d. 0·8. A colourless liquid. It may be produced by the destructive distillation of wood; nowadays it is synthesized from CO and H_2 in the presence of catalysts. It is an important intermediate for numerous chemicals, and is used as a solvent and for denaturing ethanol. Poisonous when drunk, it damages the optic nerve to produce blindness and also produces severe metabolic acidosis. Also *methyl alcohol* or *wood alcohol*.

methene (*Chem*) See METHYLENE.

methicillin (*Pharmacol*) A semi-synthetic antibiotic drug used to treat infections by penicillin-resistant staphylococcal bacteria.

methine (*Chem*) The trivalent radical $CH\equiv$.

methine dyes (*Chem*) A group of dyestuffs important in the development of photography. Consist mainly of dyes containing two quinoline or benzthiazole groups joined by conjugated aliphatic chains.

methionine (*Chem*) 2-amino-4(methylsulphonyl) butanoic acid. $CH_3S(CH_2)_2CH(NH_2)COOH$. The L- or S- isomer is an essential amino acid for protein synthesis. Symbol Met, short form M.

method (*ICT*) In object-oriented programming an algorithm or function that forms part of a class and that defines the behaviour of an object of that class in response to a message.

method of moments (*MathSci*) A procedure of estimation of PARAMETER values in which sample MOMENTS based on data are equated to corresponding population moments; eg the sample mean and variance may be equated to the population mean and variance.

method study (*Eng*) A branch of work study, concerned with determining the best production methods and the corresponding equipment for making an article in the desired quantities at optimum quality and cost.

methotrexate (*Pharmacol*) A cytotoxic antimetabolite that inhibits dihydrofolate reductase and is used in treatment of some cancers and as an immunosuppressant in severe psoriasis and rheumatoid arthritis.

methoxone (*Chem*) See MCPA.

methoxychlor (*Chem*) 1,1,1-trichloro-2,2-di-(4-methoxyphenyl)ethane. Used as an insecticide. Also *dianysl-trichloroethane, DMDT, methoxy-DDT*.

methoxyl group (*Chem*) The monovalent radical $-OCH_3$. In certain compounds it can be estimated analytically by *Zeisel's method*.

Methylal (*Chem*) $(CH_3O)_2CH_2$. Bp 42°C. A colourless liquid; widely used as a solvent.

methyl alcohol (*Chem*) See METHANOL.

methylamines (*Chem*) Mono-, CH_3NH_2; di-, $(CH_3)_2NH$; and tri-, $(CH_2)_2N$, may be regarded as ammonia in which one or more of the hydrogen atoms is replaced by the methyl group. Ammoniacal liquids with fishy smell, occurring in herring brine. Used in manufacturing drugs, dyes, etc.

4-methyl-aminophenol (*Chem*) $NH_2C_6H_4OH$; a solid crystalline compound, soluble in acids and alkalis and readily oxidized in air, used as a photographic developer and widely marketed under the TN Rodinal.

methylated spirit (*Chem*) See DENATURED ALCOHOL.

methylation (*BioSci*) The addition of a methyl ($-CH_3$) group to a nucleic acid base, usually cytosine or adenine. In bacteria, this can protect the site against cleavage by the appropriate restriction enzyme. In eukaryotes, the transcription of certain genes is inhibited by DNA methylation, although the function is not fully understood.

methylation (*Chem*) The introduction of methyl ($-CH_3$) groups into organic compounds. Carried out by using methylating agents such as diazomethane or dimethyl sulphate.

methylbenzene (*Chem*) See TOLUENE.

methyl bromide (*Chem*) Bromomethane. CH_3Br. Bp 4°C, rel.d. (0°C) 1·732. Organic liquid widely used as a fire-extinguishing medium.

methyl cellulose (*Chem*) An ETHER prepared from CELLULOSE which gives highly viscous solutions in water. It is used in distempers and other water paints, foodstuffs, cosmetics, etc.

methyl cellulose (*Med*) A bulking agent used to increase fecal mass and relieve constipation.

methyl chloride (*Chem*) Chloromethane. CH_3Cl. Bp −23·7°C, rel.d. (0°C) 0·952. Used as local anaesthetic and refrigerant.

methylcyclohexanol (IV) (*Chem*) The methyl derivative of cyclohexanol, actually a mixture of the 2-, 3- and 4-isomers. Used widely as a solvent for fats, waxes, resins and in lacquers. Soaps prepared from it are used as detergents. Also *hexahydrocresol, sextol*.

methyldopa (*Pharmacol*) Centrally and peripherally active drug used to treat hypertension. Use becoming less common as newer drugs are introduced.

methylene (*Chem*) The group $-CH_2-$.

methylene blue (*BioSci*) Water-soluble dye that can be reduced to a colourless form and can be oxidized by atmospheric oxygen. Used as a stain in bacteriology and histology. Also *Basic Blue 9, Swiss blue, tetramethythionine chloride*.

methylene blue (*Chem*) A thiazine dyestuff prepared by oxidizing dimethylaniline in the presence of sodium thiosulphate and zinc chloride. It is fixed to cotton with the aid of tannin.

methylene chloride (*Chem*) Dichloromethane. CH_2Cl_2. Organic solvent widely used in paint strippers. Bp 41°C. By-product in the manufacture of trichloromethane.

methylene iodide (*Chem*) Di-iodomethane. CH_2I_2. High-rel.d. (3·32) organic liquid. Used in flotation processes for ore separation and in the density determination of minerals.

methyl ethyl ketone (*Chem*) Butan-2-one. $CH_3COC_2H_5$. Bp 81°C. A colourless liquid of ethereal odour, prepared by the oxidation of butan-2-ol; an important solvent. Its peroxide is widely used as a catalyst for curing polyester resins. Abbrev *MEK*.

methyl group (*Chem*) Monovalent radical CH_3-.

methyl iodide (*Chem*) Iodomethane. CH_3I. Bp 42·5°C. Prepared by adding iodine to methanol and red phosphorus. Useful reagent for the preparation of methyl ethers by reaction with primary alcohols in the presence of silver (I) oxide.

methyl mercury (*BioSci*) Toxic compound formed under anaerobic conditions in marine sediments from mercury. It enters the food chain via marine organisms that are eaten by fish, rendering them poisonous.

methyl methacrylate resins (*Plastics*) See ACRYLIC RESINS.

methylol group (*Chem*) The group $.CH_2OH$. Found in primary alcohols and glycols.

methyl orange (*Chem*) $(CH_2)_2NC_6H_4N=NC_6H_4SO_2O^- Na^+$. The sodium salt of helianthine. It is a chrysoidine dye, and is used as an indicator in volumetric analysis.

methylphenidate (*Pharmacol*) A dexamfetamine-like drug that stimulates the central nervous system and is used in ADD/ADHD. TN Ritalin.

methylprednisolone (*Pharmacol*) A corticosteroid used to treat inflammatory disorders.

methyl pyridines (*Chem*) See PICOLINES.

methyl rubber (*Chem*) The polymerization product of 2,3-dimethylbutadiene, $CH_2=C(CH_3)C(CH_3)=CH_2$, one of the first synthetic rubbers, much inferior to the natural product. It oxidizes easily, and can be vulcanized only by the addition of organic catalysts.

methyl salicylate (*Pharmacol*) The main constituent of oil of wintergreen, with a characteristic aromatic odour. Used as an ointment or liniment for treatment of rheumatism.

methyl sulphate (*Chem*) $(CH_3)_2SO_4$. Bp 188°C. A colourless syrupy oil, very poisonous, used for introducing the methyl group into phenols, alcohols and amines.

methyl violet (*Chem*) A triphenylmethane dyestuff consisting of a mixture of the hydrochlorides of tetra-, penta- and hexamethyl-pararosaniline.

methysergide (*Pharmacol*) An anti-serotinergic drug that is occasionally used to prevent migraine but that can cause adverse effects including retroperitoneal fibrosis.

Metis (*Astron*) A tiny natural satellite of Jupiter, discovered in 1979 by Voyager 2 mission. Distance from the planet 128 000 km; diameter 40 km.

metoecious (*BioSci*) See HETEROECIOUS.

metoestrus (*BioSci*) In mammals, the recuperation period after estrus.

metol (*ImageTech*) See 4-METHYL-AMINOPHENOL.

Metonic cycle (*Astron*) A period of 19 years, which is very nearly equal to 235 synodic months, this relationship having been introduced in Greece in 433 BC by the astronomer Meton; its effect is that after a full cycle the phases of the Moon recur on the same days of the year.

metope (*Arch*) A slab or tablet (generally of marble and ornamental) filling the space between the triglyphs in a Doric frieze.

metoprolol tartrate (*Pharmacol*) A cardioselective *beta blocker* used for the treatment and prevention of heart arrythmia and hypertension.

metoxenous (*BioSci*) See HETEROECIOUS.

metre (*Phys*) The SI fundamental unit of length, equal to 1·093 yards, currently defined as the length of the path travelled by light in a vacuum during a time interval of 1/299 792 458 of a second. Abbrev *m*. See SI UNITS.

metre bridge (*ElecEng*) A Wheatstone bridge in which 1 metre of resistance wire, usually straight, with a sliding contact is used to form two variable ratio arms in a Wheatstone bridge network.

metre-candle (*Phys*) See LUX.

metre–kilogram–second–ampere (*Phys*) See MKSA.

metric (*MathSci*) (1) A generalization of the familiar notion of distance to more general spaces. Let M be any set, and let d be a mapping from the set of ordered pairs (x,y) of members of M, into the real numbers. Then d is called a metric in M if it satisfies the following conditions. For all x,y,z, in M: (i) $d(x,x) = 0$; (ii) $d(x,y) \geqslant 0$; (iii) $d(x,y) = 0$ implies $x = y$; (iv) $d(x,y) = d(y,x)$; (v) $d(x,y) + d(y,z) \geqslant d(x, z)$. The value $d(x,y)$ is called the distance between x and y. See METRIC SPACE. (2) A differential expression of distance in a generalized vector space, ie $ds^2 = g_{\alpha\beta}dx^\alpha dx^\beta$. The coefficient $g_{\alpha\beta}$ forms a tensor known as the *fundamental metric tensor*.

metric quad demy (*Paper*) A size of paper for the production of demy quarto or demy octavo books.

metric quad royal (*Paper*) A size of paper for the production of royal quarto or royal octavo books. 102 × 127 cm.

metric screw-thread (*Eng*) A standard screw-thread in which the diameter and pitch are specified in millimetres with a 60° angle.

metric space (*MathSci*) A set M together with a metric d is called a metric space. See METRIC.

metric system (*Genrl*) A system of weights and measures based on the principle that each quantity should have one unit whose multiples and submultiples are all derived by multiplying or dividing by powers of ten. This simplifies conversion, and eliminates completely the complicated tables of weights and measures found in the traditional UK system. Originally introduced in France, it is the basis for the *Système International* (*SI*) now universally adopted. See SI UNITS.

metric ton (*Genrl*) A unit of weight equal to 1000 kg. Also *tonne*. See TON.

metric trait (*BioSci*) See QUANTITATIVE CHARACTER.

metritis (*Med*) Inflammation of the UTERUS.

metrology (*Genrl*) The science of measuring.

metronidazole (*Pharmacol*) An antibiotic active against certain bacteria and protozoa including *Trichomonas*, *Entamoeba* and *Giardia*. Often used to treat surgical and gynaecological sepsis.

metropolitan area network (*ICT*) A private data network covering an area as large as a town or city. See LOCAL AREA NETWORK, NETWORK, WIDE AREA NETWORK.

metrorrhagia (*Med*) Bleeding from the uterus between menstrual periods. Also *metrostaxis*.

metrostaxis (*Med*) See METRORRHAGIA.

-metry (*Chem*) A suffix denoting a method of analysis or measurement, eg acidimetry, iodimetry, nephelometry.

metyrapone (*Pharmacol*) Competitive inhibitor for 11β-hydroxylation in the synthesis of cortisol by the adrenal. Used to suppress cortisol output in some tumours of the adrenal gland and in the assessment of adrenal–pituitary interaction.

MeV (*Phys*) Abbrev for mega-electron-volts. Unit of particle energy; 10^6 ELECTRON-VOLTS.

mevalonic acid (*Chem*) *3,5-dihydroxy-3-methylpentanoic acid.* $CH_2OHCH_2C(OH)(CH_3)CH_2COOH$. A precursor in the biosynthesis of terpenes and steroids.

Mexican onyx (*Min*) A translucent, veined and particoloured aragonite found in Mexico and in the SW US.

mezzanine (*Build*) An intermediate floor constructed between two other floors in a building.

mezzotint (*Print*) An intaglio process in which printing is done from a copper plate, grained by rocking a semicircular toothed knife over the surface, the lighter tones being produced by scraping or burnishing away the grain to reduce the ink-holding capacity.

MF (*Aero*) Abbrev for *medium frequency*. Frequencies from 300 to 3000 kHz.

MF (*Paper*) Abbrev for MACHINE-FINISHED.

μF (*ElecEng*) Abbrev for *microfarad*.

MFI (*Plastics*) Abbrev for MELT FLOW INDEX.

MF keypad (*ICT*) Keypad, incorporated into a telephone and capable of producing multifrequency signals for operating electronic exchange equipment. Low-speed data (< 600 baud) can also be transmitted from an MF keypad.

MF resin (*Plastics*) Abbrev for MELAMINE-FORMALDEHYDE RESIN.

MF switching (*ICT*) Operation of electronic telephone exchange equipment by multifrequency tones, rather than by electrical impulses.

MG (*Build*) Abbrev for MAKE GOOD.

MG (*Paper*) Abbrev for MACHINE-GLAZED.

Mg (*Chem*) Symbol for MAGNESIUM.

mg (*Phys*) Abbrev for *milligram*, 10^{-3} g.

MG machine (*Paper*) A paper machine incorporating a Yankee or MG drying cylinder in the drying section to produce machine-glazed paper.

Mg point (*Glass*) See TRANSFORMATION POINTS.

MHC (*BioSci*) Abbrev for MAJOR HISTOCOMPATIBILITY COMPLEX.

MHC restriction (*BioSci*) The phenomenon whereby T-lymphocytes can only recognize foreign antigen on the surface of another cell, eg a virus-infected cell or an accessory cell, when it is associated with self-antigens of the major histocompatibility complex. Cytotoxic T-lymphocytes usually respond to foreign antigen from within eg virus-infected cells in association with Class I MHC antigens, whereas helper T-lymphocytes respond to foreign antigen presented by an accessory cell in association with Class II MHC antigens. See panel on IMMUNE RESPONSE.

M(H) curve (*Eng*) See MAGNETIC MATERIALS. Also *M–H*, *M/H curve*.

MHD (*Phys*) Abbrev for MAGNETOHYDRODYNAMICS.

MHD generator (*ElecEng*) Abbrev for *magnetohydrodynamic generator*.

M(H) loop (*Eng*) See HYSTERESIS LOOP, MAGNETIC MATERIALS. Also *M–H*, *M/H loop*.

mho (*ElecEng*) Name for the reciprocal of the ohm in the CGS system. See SIEMENS.

mianserin (*Pharmacol*) An antidepressant drug similar to the *tricyclic antidepressants* but with fewer antimuscarinic effects.

miarolitic structure (*Geol*) A structure found in an igneous rock, consisting of irregularly shaped cavities into which the constituent minerals may project as perfectly terminated crystals.

miaskite (*Geol*) A leucocratic biotite nepheline monzosyenite. See PLUTONIC ROCKS.

MIC (*ElecEng*) Abbrev for MINERAL-INSULATED CABLE.

mica (*Min*) A group of silicates which crystallize in the monoclinic system; they have similar chemical compositions and highly perfect basal cleavage. Mica is one of the best electrical insulators. See panel on SILICON, SILICA, SILICATES.

micaceous iron ore (*Min*) A variety of specular HAEMATITE (Fe_2O_3) which is foliated or which simulates mica in the flakiness of its habit.

micaceous sandstone (*Geol*) A sandstone containing conspicuous flakes of mica.

mica cone (*ElecEng*) See MICA V-RING.

micafolium (*ElecEng*) A composite insulating material consisting of a paper backing covered with mica flakes and varnish. Formerly much used for insulating wire and machine coils but widely replaced by polymers.

mica marks (*Eng*) Silvery streaks found on visible surfaces of injection mouldings. A defect caused by impurities, such as water, in thermoplastic granules.

micanite (*ElecEng*) Mica splittings bonded by varnish or shellac into a large sheet; mechanically weak at high temperatures.

mica-schist (*Geol*) Schist composed essentially of micas and quartz, the foliation being mainly due to the parallel disposition of the mica flakes. See SCHIST.

mica V-ring (*ElecEng*) A ring of V-shaped cross-section made of a mica compound and used to insulate a metal V-ring from the bars of the commutator which it supports.

micelle (*BioSci*) Crystalline region, inferred from X-ray diffraction data, within a microfibril of cellulose or similar structure.

micelle (*Chem*) (1) Colloidal aggregate of molecules formed in solution, esp soaps in water. Particles are often spherical, with hydrophobic chains in the centre surrounded by hydrophilic groups. Only formed above a certain limit, known as the critical micelle concentration or cmc. (2) Supposed form of crystallite found in partly crystalline polymers. See CRYSTALLIZATION OF POLYMERS, FRINGED MICELLE MODEL.

Michaelis–Menten equation (*BioSci*) An equation derived from a simple kinetic model of enzyme action that successfully accounts for the hyperbolic (adsorption–isotherm) relationship between substrate concentration S and reaction rate V: $V = V_{max} \times S/(S + K_m)$, where K_m is the Michaelis constant and V_{max} is maximum rate approached at very high substrate concentrations.

Michell bearing (*Eng*) A thrust or JOURNAL bearing in which pivoted pads support the thrust collar or journal in such a way that they tilt slightly under the wedging action of the lubricant induced between the surfaces by their relative motion. The fluid lubrication conditions thus produced result in a very low friction coefficient and power loss in the bearing.

Michelson interferometer (*Phys*) An early interferometer designed for the investigation of the fine structure of spectrum lines and for the evaluation of the standard metre in wavelengths of light. The principle of operation is similar to that of the FABRY–PÉROT INTERFEROMETER, ie using circular HAIDINGER FRINGES.

Michelson–Morley experiment (*Phys*) An attempt to detect and measure the relative velocity of the Earth and the ETHER by observation of the shift of interference fringes produced in an interferometer whose orientation could be changed by 90°. No shift was detected and so no relative motion measured. This result can be explained in terms of the principle of the constancy of the speed of light. See PRINCIPLE OF RELATIVITY.

MICR (*ICT*) See MAGNETIC INK CHARACTER RECOGNITION.

micrite (*Min*) The microcrystalline matrix of fine-grained limestones seen to have a semi-opaque appearance, when viewed microscopically, due to their small grain size. Also *micritic limestone*.

micro (*Geol*) Applied to names of rocks, it indicates the medium-grained form, eg *microdiorite*, *microsyenite*, *microtonalite*.

micro- (*Genrl*) Prefix from Gk *mikros*, small. When used of units it indicates the basic unit × 10^{-6}, eg 1 microampere (μA) = 10^{-6} amperes. Symbol μ.

micro-aerophile (*BioSci*) An organism that grows well at low-oxygen concentrations.

micro-ammeter (*ElecEng*) A most sensitive form of robust current-measuring instrument.

micro-analysis (*Chem*) A special technique of both qualitative and quantitative analysis, by means of which very small amounts of substances may be analysed.

micro-analytical reagent (*Chem*) See MAR.

microarray (*BioSci*) A small membrane or glass slide containing samples of biological material (eg oligonucleotides, proteins) arranged in a regular pattern, used as a tool in genetic analysis. See GENE ARRAY.

microbalance (*Chem*) Sensitive to 1 microgram, for use in micro-analysis; may be either beam or quartz fibre.

microbar (*Phys*) Unit of pressure equal to 10^{-6} bar = 10^{-1} N m^{-2} = 1 dyne cm^{-2}. Used in acoustics.

microbe (*BioSci*) An organism that can only be seen under the microscope. Often implies bacteria rather than all microscopic organisms.

microbeam analysis (*Chem*) Use of fine collimated beam of radiation for identification of materials, eg ENERGY DISPERSIVE ANALYSIS OF X-RAYS (EDAX), infrared analysis in optical microscopes.

microbicide (*BioSci*) A substance that kills microbes.

microbiological mining (*MinExt*) (1) Use of natural or GENETICALLY ENGINEERED strains of bacteria to enhance or induce acid leaching of metals from ores (*bacterial leaching*), either *in situ* within an ore deposit, or to promote leaching of metals from mine waste. (2) Use of bacteria to recover useful or toxic metals from natural drainage waters, mine drainage or waste water from tips (*bacterial recovery*). Also *biomining*.

microbiology (*Genrl*) The biology of microscopic or ultramicroscopic organisms, such as bacteria, viruses or fungi.

microbody (*BioSci*) Cytoplasmic organelle of eukaryotes up to about 1·5 μm diameter, bounded by a single membrane. Types include the GLYOXYSOME and PEROXISOME.

microbore (*Build*) Pump-assisted central heating system in which the hot-water pipes have diameters of 6, 8 or 10 mm.

microbot (*Eng*) A small mobile robot with dimensions in the mm range. Speculatively, sufficiently small robotic devices could be used eg within the intestinal tract. Also *microscopic robot*.

microburst (*Aero*) Dangerous vertical gust having a core $\approx$1·5 miles (2·5 km) in diameter in which downward velocities of 4000 ft m^{-1} (20 m s^{-1}) can occur down to low altitude.

microcanonical assembly (*Phys*) In statistical thermodynamics, an assembly which consists of a large number of systems each having the same energy.

microcarrier (*BioSci*) A small bead on which anchorage-dependent cells may be cultured. Large-scale cell culture is possible if a large number of beads are maintained in suspension by gentle agitation.

microcell (*ICT*) Very small mobile-telephone CELL, typically of radius around 500 m, as used in a PERSONAL COMMUNICATIONS NETWORK.

microcephaly (*Med*) Abnormally small size of the head. Also *microcephalia*.

microchannel architecture (*ICT*) A proprietary ARCHITECTURE developed by IBM for the PC. Not widely used. Cf EISA, ISA.

microchemistry (*Chem*) The preparation and analysis of very small samples, usually less than 10 milligrams but, in radioactive work, down to a few atoms.

microcircuit isolation (*ElecEng*) The electrical insulation of circuit elements from the electrically conducting silicon wafer. The two main techniques are DIODE ISOLATION and OXIDE ISOLATION.

microcircuits (*ElecEng*) Those with components formed in one unit of semiconductor crystal.

microclimate (*BioSci*) The climate in small places, eg very close to organisms or in specific habitats such as nests or on bare ground.

microcline (*Min*) A silicate of potassium and aluminium which crystallizes in the triclinic system. A feldspar, it resembles orthoclase, but is distinguished by its optical and other physical characters. See POTASSIUM FELDSPAR.

Micrococcaceae (*BioSci*) A family of bacteria belonging to the order *Eubacteriales* and comprising Gram-positive cocci. It includes free-living, parasitic, saprophytic and pathogenic species, eg *Staphylococcus aureus* (one cause of food poisoning).

microcode (*ICT*) Software equivalent of a MICRO-INSTRUCTION, written to extend the machine-code instruction of the computer without the addition of further hardwired logic. See fig. at MICROPROCESSOR.

microcomputer (*ICT*) Computer based on a MICROPROCESSOR. Originally this meant a cheap and relatively slow computer with a limited access store, a simple instruction set and only elementary backing store (eg cassette tapes, floppy disks) but see PERSONAL COMPUTER.

microcosmic salt (*Chem*) Sodium ammonium hydrogen phosphate. $NaNH_4HPO_4.4H_2O$. Present in urine.

microcrystalline texture (*Geol*) A term applied to a rock or groundmass in which the individual crystals can be seen only under the microscope.

microdensitometer (*ImageTech*) A densitometer capable of measuring the density variations of extremely small areas, down to the size of individual grains in a photographic image.

microdissection (*BioSci*) A technique for small-scale dissection, eg on living cells, using MICROMANIPULATORS and viewing the object through a microscope.

microelectrode (*BioSci*) An electrode with a tip less than 1 μm wide that can be used non-destructively to puncture a cell membrane, thus allowing intracellular recordings and measurements.

microelectronics (*Electronics*) The technology of constructing and utilizing complex electronic circuits and devices in extremely small packages by using integrated-circuit manufacturing techniques.

microencapsulation (*Paper*) A process whereby a substance in a state of extreme comminution is enclosed in sealing capsules from which the material is released by impact, solution, heat or other means. See NCR.

microengineering (*Electronics*) See MICROFABRICATION.

micro-environment (*BioSci*) The environment of small areas in contrast to large areas, with particular reference to the conditions experienced by individual organisms and their parts (eg leaves).

microfabrication (*Electronics*) Device or component manufacture (not necessarily electrical or electronic) on a small scale (typically of the order of micrometres), particularly using the techniques of SEMICONDUCTOR DEVICE PROCESSING. Also *microengineering*.

microfarad (*ElecEng*) A unit of capacitance equal to one-millionth of a farad; more convenient for use than the farad. Abbrev μF.

microfelisitic texture (*Geol*) The cryptocrystalline texture seen, under the microscope, in the groundmass of quartzfelsites and similar rocks; due to the devitrification of an originally glassy matrix.

microfibril (*BioSci*) Fine fibril, 5–30 nm wide, in a CELL WALL (panel); of cellulose in vascular plants and some algae but of other polysaccharide (eg xylans) in other algae. Cf MATRIX.

microfiche (*ICT*) Output medium consisting of large-capacity microfilm sheets that may be randomly accessed by a special optical reader/magnifier. Also used in non-computer applications and then usually A6 size with images on 60–98 pages reduced 20–25 times.

microfilament (*BioSci*) A filament of F actin, about 6 nm in diameter, found in the cytoplasm in long bundles, or as a meshwork, and involved in localized cell contractions and streaming. Cf MICROTUBULE. See STRESS FIBRES.

microfilaria (*BioSci*) The early larval stage of certain parasitic Nematoda.

microfilm (*ImageTech*) (1) A black-and-white film of extremely high resolution and fine grain, available in long rolls esp for MICROPHOTOGRAPHY. (2) A record produced by MICROPHOTOGRAPHY.

microfossils (*Geol*) Fossils or fossil fragments too small to be studied without using a microscope.

microgamete (*BioSci*) The smaller of a pair of conjugating gametes, generally considered to be the male gamete.

microgametocyte (*BioSci*) In Protozoa, a stage developing from a trophozoite and giving rise to male gametes.

microgap switch (*ElecEng*) A switch, used on low-power, low-voltage circuits, which relies for arc extinction on the lateral spread of the arc stream by mutual repulsion between contacts separated by thousandths of an inch.

microgeneration (*Eng*) The generation of energy on a small scale by individuals for their own use, eg using solar panels.

microglia (*BioSci*) A small type of neuroglia cell (occurring more frequently in grey matter than in white matter) having an irregular body and freely branching processes; can be phagocytic.

microglobulin (*BioSci*) Any small globulin. Used in respect of Bence Jones protein in urine or of β_2-microglobulin.

microgram (*Genrl*) Unit of mass equal to one-millionth of a gram (10^{-9} kg). Symbol μg. Also *microgramme*.

microgranite (*Geol*) A medium-grained, microcrystalline, acid igneous rock having the same mineral composition and texture as a granite.

micrographic texture (*Geol*) A distinctive rock texture in which the simultaneous crystallization of quartz and feldspar has led to the former occurring as apparently isolated fragments, resembling runic hieroglyphs, set in a continuous matrix of feldspar.

microgravity (*Space*) The condition of near-weightlessness induced by FREE FALL or unpowered space flight; it is characterized by the virtual absence of gravity-induced convection, hydrostatic pressure and sedimentation. The term also refers to the scientific discipline which is concerned with the evaluation of processes in a near-zero-g environment, particularly fluid physics, life sciences and materials science.

microgyria (*Med*) Abnormal smallness of the convolutions of the brain.

micro-incineration (*BioSci*) A technique for examining the distribution of minerals in slide preparations of tissue sections or cells. The organic material is vaporized by heat and the nature and position of the mineral ash determined by microscopic examination.

micro-instruction (*ICT*) Simple executable instructions wired or built into a computer. Cf MICROCODE.

microkeratome (*Med*) A tool used in eye surgery to make incisions of a predetermined depth in the cornea.

microkernel (*ICT*) The minimum possible collection of processor-specific operating-system functions. Other services like input/output, windows and communications are designed to have multiple PERSONALITIES that will allow them to run a variety of different software, eg Windows, UNIX, OS/2 and Mac. A feature of many operating systems under development.

microlecithal (*BioSci*) Said of eggs containing very little yolk. Cf MEGALECITHAL.

microlens CCD (*ImageTech*) A CCD ARRAY with a microscopic lens over each PIXEL to increase its light-gathering capacity. For specialized applications the microlens can be employed in place of a conventional lens.

microlight (*Aero*) Aircraft whose empty weight does not exceed 330 lb (150 kg). In the USA *ultralight* is used for weights up to 254 lb (115 kg).

microlite (*Geol*) A general term for minute crystals of tabular or prismatic habit found in microcrystalline rocks. These give a reaction with polarized light.

microlux (*Phys*) A unit for very weak illumination, equal to one-millionth of a LUX.

micromanipulator (*BioSci*) An instrument used to handle cells seen in a microscope, eg to remove a nucleus or inject RNA. The fine movements are controlled indirectly by pneumatic, mechanical or other means.

micromazia (*Med*) Failure of the female breast to develop after puberty.

micromere (*BioSci*) In a segmenting ovum, one of the small cells that are formed in the upper or animal hemisphere.

micromesh sieves (*PowderTech*) See ELECTRO-FORMED SIEVES.

micrometeorite (*Astron, Space*) An extremely small particle found in space, typically of mass less than 10^{-6} g, and diameter less than 10^{-4} m. It does not burn up in the Earth's atmosphere, but drifts down to the surface. Comets are probably abundant sources of new micrometeorites.

micrometer (*Astron*) An instrument which measures small angular separations in the telescope. It consists of three frameworks carrying spider-webs close to the image plane: one is fixed and the others are each adjustable by micrometer heads, by which the separation is read, with a graduated circle giving the angular relation of a double star. See MICROMETER GAUGE.

micrometer eyepiece (*BioSci*) See EYEPIECE GRATICULE.

micrometer gauge (*Eng*) A U-shaped length gauge in which the gap between the measuring faces is adjustable by an accurate screw whose end forms one face. The gap is read off a scale uncovered by a thimble carried by the screw, and by a circular scale which is engraved on the thimble. Commonly *micrometer*.

micrometer theodolite (*Surv*) A theodolite equipped with micrometers instead of the usual verniers for reading the horizontal and vertical circles.

micrometre (*Phys*) One-millionth of a metre. Symbol μm. Formerly *micron*.

micromicro- (*Genrl*) Prefix for one million millionth, or 10^{-12}: replaced in SI by PICO- (p).

micromodule (*ElecEng*) Sometimes said of circuits or components formed from the same crystal of material, eg germanium. An *integrated circuit*.

micron (*Genrl*) Obsolete but still popular measure of length equal to 10^{-6} m, symbol μ. Replaced in SI by MICRO-METRE, symbol μm.

micronized coal (*Eng*) Pulverized coal in which > 80% will pass through a 40 μm sieve.

micronucleus (*BioSci*) In Ciliophora, the smaller of the two nuclei which is involved with sexual reproduction. Cf MACRONUCLEUS.

micronutrient (*BioSci*) A *trace element* required in relatively small quantities by living organisms; for plants the micronutrients include Fe, B, Mn, Zn, Cu, Mo, Cl. Cf MACRONUTRIENT. See ESSENTIAL ELEMENT.

micropalaeontology (*Geol*) The study of MICROFOSSILS.

microperthite (*Min*) A feldspar which consists of intergrowths of potassium feldspar and albite in a microscopic scale.

microphage (*BioSci*) A small phagocytic cell in blood or lymph, chiefly the polymorphonuclear leucocytes (neutrophils). Adj *microphagocytic*.

microphagous (*BioSci*) Feeding on small particles of food. Cf MACROPHAGOUS.

microphanerophyte (*BioSci*) A PHANEROPHYTE, 2–8 m high.

microphone (*Acous*) An acousto-electric transducer, essential in all sound-reproducing systems. The fluctuating pressure in the sound wave is applied to a mechanical system, such as a ribbon or diaphragm, the motion of which generates an electromotive force, or modulates a current or voltage. See CARBON MICROPHONE, DIRECTIONAL MICROPHONE, ELECTROMAGNETIC MICROPHONE, HOT-WIRE MICROPHONE, LAPEL MICROPHONE, MOVING-COIL MICROPHONE, OLSON MICROPHONE, OMNIDIRECTIONAL MICROPHONE, PRESSURE MICROPHONE, PRESSURE-GRADIENT MICROPHONE.

microphone response (*Acous*) The response measured over the operating frequency range in a particular direction, or averaged over all directions. The characteristic response is usually given by the ratio of the open-circuit voltage generated by the microphone to the sound pressure ($N\,m^{-2}$) existing in the free progressive wave before introducing the microphone.

microphonic (*Electronics*) Said of a component which responds to acoustic vibrations and/or knocks.

microphonicity (*Acous*) See MICROPHONIC NOISE.

microphonic noise (*Acous*) Noise in the output of a valve related to mechanical vibration of the electrode system. Also *microphonicity*.

microphotography (*ImageTech*) Photography of normal sized objects, esp documents, plans and graphic materials, as greatly reduced images of small area which must be examined by magnification or enlarged projection.

microphyll (*BioSci*) (1) A small leaf, not associated with a leaf gap, assumed to have evolved from an enation and supposed to be the sort of leaf possessed by lycopods. See ENATION THEORY. (2) Any very small leaf, eg of heather or *Tamarix*. Adj *microphyllous*. Cf MEGAPHYLL.

Microphyllophyta (*BioSci*) A division of the plant kingdom, here treated as the class Lycopsida.

microphyric (*Geol*) A textural term descriptive of medium- to fine-grained igneous rocks containing phenocrysts <2 mm long. Cf MACROPHYRIC.

micropinocytosis (*BioSci*) Pinocytosis of small vesicles (around 100 nm in diameter) by a mechanism distinct from normal pinocytosis.

Micropodiformes (*BioSci*) An order of birds with a short humerus and long distal segments to the wings. Swifts and hummingbirds.

micropodous (*BioSci*) Having the foot, or feet, small or vestigial.

micropore filter (*BioSci*) A filter made of a meshwork of cellulose acetate or nitrate and with a defined pore size, dependent on the extent of cross-linking, eg Millipore filters. Small pore sizes that filter bacteria are used for sterilizing solutions for eg cell culture. Cf NUCLEOPORE FILTER.

microporosity (*Eng*) Minute cavities generally found in heavy engineering sections usually due to lack of efficient feeding during solidification or to release of dissolved gas.

microporous coatings (*Build*) Paints, stains or clear coatings which allow a surface to breathe. Primarily intended for use on timbers, these coatings do not trap moisture unlike conventional paint systems.

microporous materials (*Chem*) Polymers made permeable to gases or liquids by creation of very fine pores, used for filtration of liquids, battery separators or in clothing. Pores are made chemically with a blowing agent or physically, by microfibrillation, eg GORE-TEX.

microprism (*ImageTech*) Focusing device which breaks the image up into dots when it is not in focus.

microprocessor (*ICT, Electronics*) A computer CENTRAL PROCESSING UNIT (CPU) usually on one integrated-circuit chip, using very-large-scale integration (VLSI) technology. Microprocessors with over a million transistors are generally available as the main element of a personal computer or as part of an automatic control system. See REDUCED INSTRUCTION SET COMPUTER. Fig. ▷

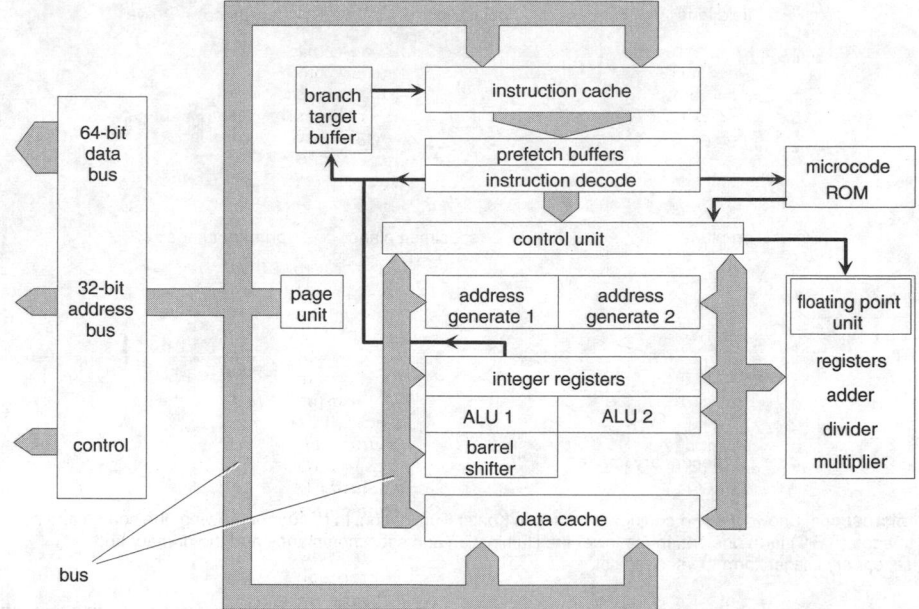

microprocessor Block drawing of a modern design with two integer pipelines and floating point unit. (Intel Pentium).

microprogram (*ICT*) A sequence of MICRO-INSTRUCTIONS used as part of the design of a processor that enables CONTROL logic to be established by SOFTWARE. Microprogrammed control is not quite as fast but much more flexible than control by HARDWARE.

micropropagation (*BioSci*) The use of small pieces of tissue, such as meristem grown in culture, to obtain large numbers of individuals. Needs sterile precautions and the proper nutrients.

micropsia (*Med*) The condition in which objects appear to the observer smaller than they actually are; may be due to retinal disease.

micropterous (*BioSci*) Having small or reduced fins or wings.

micropyle (*BioSci*) (1) A tiny opening in the integument at the apex of an ovule, through which the pollen tube usually enters. (2) The corresponding opening in the testa of the seed. Also *foramen*. (3) An aperture in the chorion of an insect egg through which a spermatozoon may gain admittance.

microradiography (*Radiol*) Exposure of small thin objects to SOFT X-RAYS, with registration on a fine-grain emulsion and subsequent enlargement up to 100 times. Also used to signify the optical reproduction of an image formed, eg by an electron microscope.

microreader (*ICT*) A device that produces an enlarged image of a microphotograph or microfilm.

microRNAs (*BioSci*) Small RNA molecules (≈ 20 nucleotides) that regulate gene expression in multicellular eukaryotes. Abbrev *miRNAs*. Also *small interfering RNAs*.

microsatellites (*BioSci*) Short sequences of di- or trinucleotide repeats of very variable length distributed widely throughout the genome. Analysis of microsatellite polymorphism is used in investigating genetic associations with disease. See MINISATELLITES.

microscope (*Phys*) An instrument used for obtaining magnified images of small objects. The *simple microscope* is a convex lens of short focal length, used to form a virtual image of an object placed just inside its principal focus. The *compound microscope* consists of two short-focus convex lenses, the objective and the eyepiece, mounted at opposite ends of a tube. For most optical microscopes, the magnifying power is roughly equal to $450/f_o f_e$, where f_o and f_e are the focal lengths of objective and eyepiece in centimetres. The *petrological* or *polarizing microscope* is widely used for the examination of thin sections of rocks and minerals. A polarizer (formerly a *Nicol prism*) polarizes the incident light below the stage, which may be rotated. A second polarizer, the *analyser*, can then be used to examine the light from the section to determine eg extinction angles, interference fringes. See PHASE-CONTRAST MICROSCOPY, ULTRAVIOLET MICROSCOPE and panel on ELECTRON MICROSCOPE. Fig. ▷

microscope count method (*PowderTech*) A technique for measuring the average particle diameter of a powder by microscope examination of a weighed drop of suspension.

microscopic (*Genrl*) Invisible or hardly visible without the aid of a microscope.

microscopic state (*Chem*) One in which the condition of each individual atom has been fully specified. Cf MACROSCOPIC STATE.

microscopic stress (*Eng*) One set up at a level of the grain size of a metal due to heat treatment etc.

Microscopium (Microscope) (*Astron*) A small faint southern constellation.

microscopy (*Genrl*) The study of phenomena using a microscope.

microsection (*Eng*) A section of metal, ceramic, etc, mounted, cut, polished and etched as necessary to exhibit the microstructure.

microsite (*ICT*) A self-contained set of pages within a larger website, usually dealing with a particular topic.

microsleep (*BioSci*) An episode lasting a few seconds during which external stimuli are not perceived; associated with narcolepsy or sleep deprivation.

microsmatic (*BioSci*) Having a poorly developed sense of smell.

Microsoft Internet Explorer (*ICT*) TN of a browser for accessing the WORLD WIDE WEB. See panel on INTERNET.

microsome (*BioSci*) The membranous pellet obtained by centrifugation of a cell homogenate after removal of the

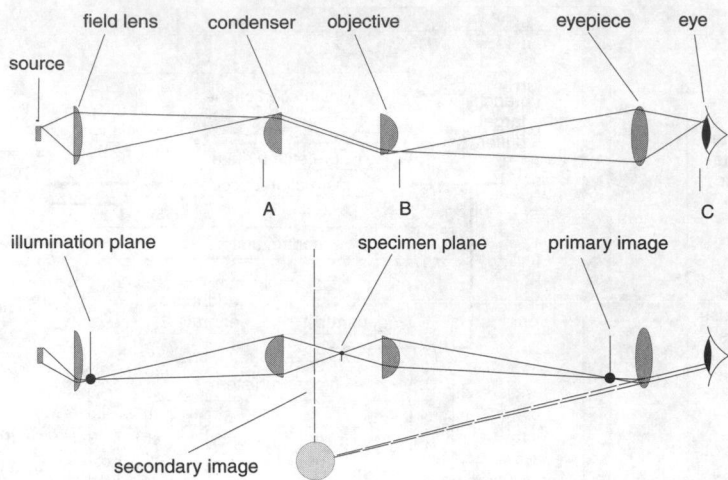

microscope Shows the two conjugate planes in Kohler illumination. In the upper drawing, the source and its images (A,B,C) form one set. In the lower the illumination and specimen planes and the primary and secondary images form the second set.

mitochondria and nuclei. It was originally believed to represent a cell organelle but is now known to consist of fragments of endoplasmic reticulum, plasma membrane, Golgi apparatus, etc.

microspacing (*ICT*) The ability to adjust the spacing between characters so that the text may be both left and right JUSTIFIED. This is necessary when PROPORTIONAL FONTS rather than MONO-SPACED fonts are being used.

microspecies (*BioSci*) See JORDANON SPECIES.

microspherulitic texture (*Geol*) A texture in which spherulites on a microscopic scale are distributed through the groundmass of an igneous rock.

microsplanchnic (*BioSci*) Having a small body and long legs, as a harvestman.

microsporangium (*BioSci*) The structure within which MICROSPORES are formed. In seed plants, the pollen sac.

microspore (*BioSci*) (1) In heterosporous species, a MEIOSPORE able to develop into a male gametophyte. (2) In seed plants, the pollen grain. (3) A small swarm spore or ANISOGAMETE of Sarcodina.

microsporocyte (*BioSci*) A cell that divides to give MICROSPORES, ie a microspore mother cell.

microsporophyll (*BioSci*) A leaf-like structure bearing microsporangia. Cf MEGASPOROPHYLL. See SPOROPHYLL.

microsporophyte (*BioSci*) A typically diploid cell in a MICROSPORANGIUM, that divides by meiosis to give four MICROSPORES.

microstoma (*Med*) An abnormally small mouth, due either to developmental defect or to contraction of scar tissue. Also *microstomia*.

microstrip (*ICT*) A microwave transmission line consisting basically of a dielectric sheet carrying a conducting strip on one side and an earthed conducting plane on the other. The strip with its image in the plane (see IMAGE CHARGE) forms a parallel-strip transmission line.

microstructure (*Eng*) Units of microscopic size (about 1 to 100 μm in diameter) which occur in materials. Such units include: spherulites, fibrils, lamellae in crystalline polymers (see CRYSTALLIZATION OF POLYMERS and panel on POLYMER SYNTHESIS); domains and rubber particles in amorphous polymers (see panel on RUBBER TOUGHENING); grain structure in metals (see panel on CREEP AND DEFORMATION) and ceramics; cell structure in natural materials.

microswitch (*ElecEng*) A switch operated by very small movements of a lever, the circuit being made or broken by spring-loaded contacts.

microtome (*BioSci*) An instrument for cutting thin sections of specimens, esp sections 1–10 μm thick for light microscopy. See ULTRAMICROTOME.

microtubule (*BioSci*) Tubular structure about 24 nm in diameter formed by the aggregation of TUBULIN dimers and small amounts of associated proteins in a helical array. They function as skeletal components within cells and may be arranged into complex structures such as centrioles, cilia and mitotic spindles. See fig. at CENTRIOLE.

microtubule-organizing centre (*BioSci*) Cytoplasmic site where MICROTUBULES are formed and organized; associated with or including centromeres, basal bodies, centrioles, etc. Abbrev *MTOC*.

microvillus (*BioSci*) A finger-like protrusion of the cell surface supported by a central bundle of actin microfilaments. The length and abundance vary characteristically between cells. When present in very high density, as on the apical surface of epithelial cells of the small intestine, microvilli form a *brush border*.

microwave background (*Astron*) A weak microwave background radiation which is detectable in every direction with almost identical intensity, discovered in 1963. The radiation spectrum has an equivalent black-body temperature of 2·7 K. It exhibits a slight asymmetry due to the motion of our Galaxy relative to this radiation, as well as COSMIC RIPPLES. The radiation is the relic of the early hot phase in the Big Bang universe. Also *cosmic background radiation*. See panel on COSMOLOGY.

microwave heating (*FoodSci, Phys*) Heating (induction or dielectric) of materials in which the current frequency is in the range $0·3 \times 10^{12}$ to 10^9 Hz. Extensively used in domestic microwave ovens; microwave energy is absorbed by water and other mobile polar molecules and can cause very rapid boiling throughout the product which becomes evenly cooked in a short time.

microwave radiation (*Phys*) Electromagnetic radiation corresponding to a wavelength range of approximately 0·3–30 cm (frequency 1–100 GHz); lies between infrared and radio waves. See panel on NON-IONIZING FIELDS AND RADIATION.

microwave resonance (*ElecEng*) One between microwave signals and atoms or molecules of medium.

microwave resonator (*ICT*) Effective tuned circuit for microwave signal. Usually a cavity resonator but tuned lines are also used.

microwaves (*ICT*) Those electromagnetic wavelengths between 1 mm and 30 cm, from 300 GHz to 1 GHz in frequency, thus bridging gap between normal radio waves and heat waves. See appendix on Electromagnetic spectrum.

microwave spectrometer (*Phys*) An instrument designed to separate a complex microwave signal into its various components and to measure the frequency of each; analogous to an OPTICAL SPECTROMETER.

microwave spectroscopy (*Phys*) The study of atomic and/ or molecular resonances in the microwave spectrum.

micturition (*BioSci*) In mammals, urinating.

Micum test (*MinExt*) A standard laboratory test for determining the mechanical strength of coke.

midband (*ICT*) Operating across a range of frequencies between those used by broadband and narrowband systems.

midbrain (*BioSci*) In vertebrates, that part of the brain derived from the second or middle brain vesicle of the embryo; the second or middle brain vesicle itself dorsally comprises the tectum, including optic lobes (CORPORA QUADRIGEMINA in mammals), and laterally the tegmentum and ventrally the crura cerebri (cerebral peduncles).

middle atmosphere (*EnvSci*) The stratosphere and mesosphere combined. See panels on STRATOSPHERE AND MESOSPHERE and TROPOSPHERE.

middle conductor (*ElecEng*) See NEUTRAL CONDUCTOR.

middle ear (*BioSci*) In Tetrapoda excluding Urodela, Gymnophiona and snakes, the cavity containing the auditory ossicles.

middle gear (*Eng*) See MID-GEAR.

middle lamella (*BioSci*) Layer of intercellular material, mostly pectic substances, developing from the cell plate and cementing together the primary walls of contiguous plant cells.

middle marker beacon (*Aero*) A marker beacon associated with the ILS, used to define the second predetermined point during a beam approach.

middle oils (*Chem*) Carbolic oils, obtained from coaltar distillation. Their boiling range is from about 210 to 240°C.

middle rail (*Build*) The rail next above the bottom rail in doors, framing and panelling.

middle shore (*Build*) An inclined shore placed between the bottom and top shores in a set of RAKING SHORES.

middle third (*CivEng*) The middle part of a solid structure, such as an arch or dam. It is equal in width, at any section, to one-third the width of the section, and is centrally disposed. The importance of the middle third is that, providing the line of resultant pressure lies wholly within it, no tensile forces come into play.

middleware (*ICT*) Software that allows two otherwise incompatible programs or networks to operate together.

middle wire (*ElecEng*) See NEUTRAL CONDUCTOR.

middlings (*MinExt*) In ore dressing, an intermediate product left after the removal of clean concentrates and rejected tailings. It consists typically of interlocked particles of desired mineral species and gangue, or of by-product minerals not responding to the treatment used. See RIDDLE.

mid-feather (*Build*) (1) See CROSS-TONGUE. (2) See PARTING SLIP.

mid-gear (*Eng*) The position of a steam-engine link motion or valve gear when the valve travel is minimal. Also *middle gear*.

midgut (*BioSci*) That part of the alimentary canal of an animal derived from the archenteron of the embryo.

MIDI (*ICT*) Abbrev for *musical instrument digital interface*. This is a standard protocol for communication between computers and musical instruments and vice versa.

MIDI file (*ICT*) A FILE containing all the data necessary to replay a tune or musical phrase on a MIDI device such as MUSIC SYNTHESIZER.

MIDI sequencer (*ICT*) A program that records and replays tunes and musical phrases stored as MIDI FILES.

mid-ocean ridge (*Geol*) A major, largely submarine, mountain range where two *plates* are being pulled apart

and new volcanic LITHOSPHERE is being created. See MAGNETIC ANOMALY PATTERNS and panel on PLATE TECTONICS.

mid-point protective system (*ElecEng*) A method of balanced protection used for protecting generators against faults between turns by balancing the voltage of one half of the winding against that of the other.

midrib (*BioSci*) (1) The main vein or nerve of a leaf. (2) Thickened region down the middle of a thallus.

midriff (*BioSci*) See DIAPHRAGM.

mid space (*Print*) A type space cast four to the em.

mid-wing monoplane (*Aero*) A monoplane wherein the main planes are located approximately midway between the top and bottom of the fuselage.

MIF (*BioSci*) Abbrev for MIGRATION INHIBITION FACTOR. Also *macrophage migration inhibitory factor*.

mifepristone (*Pharmacol*) A synthetic compound that inhibits the action of progesterone, used to induce abortion up to the 20th week of pregnancy.

MIG (*Eng*) Abbrev for METAL INERT GAS WELDING.

migmatite (*Geol*) A rock with both igneous and metamorphic characteristics; generally consisting of a host metamorphic rock injected by granitic material.

migraine (*Med*) Hemicrania or paroxysmal headache. A condition in which recurring headaches are often associated with vomiting and disturbances of vision.

migration (*BioSci*) Long-distance animal movement, often involving large populations and often seasonal.

migration (*Chem*) The movement of ions under the influence of an electric field and against the viscous resistance of the solvent. Measured by observation in thin tubes.

migration (*FoodSci*) The movement of molecules, eg flavourings, contaminants, between two foodstuffs in contact.

migration (*ICT*) (1) A change, either permanent or transitory but transparent to the user, of the physical mapping of operational service functions within a network. (2) A move by a network user from one type of service to another.

migration area (*NucEng*) One-sixth of the mean square distance covered by a neutron between creation and capture. Its square root is the *migration length*. See DIFFUSION AREA, SLOWING-DOWN AREA.

migration inhibition factor (*BioSci*) Lymphokine which acts on macrophages so as to increase their adhesiveness. When released by T-lymphocytes *in vivo* (eg in the peritoneal cavity) it causes macrophages to clump together. *In vitro* it inhibits their migration (eg out of a capillary tube); this can be used as a semi-quantitative means of detecting the existence of delayed-type hypersensitivity. Abbrev *MIF*.

migration length (*NucEng*) See MIGRATION AREA.

migratory cell (*BioSci*) See AMOEBOCYTE.

MII (*ImageTech*) TN for a sub-broadcast standard COMPONENT VIDEO using $\frac{1}{2}$ in metal tape in two sizes of cassette.

Mikulicz's disease (*Med*) A chronic inflammation of salivary and lacrimal glands; seen in, LEUKAEMIA, LYMPHOMA, SARCOIDOSIS and SJOGREN'S DISEASE.

mil (*Genrl*) A unit of length equal to 10^{-3} in, used in measurement of small thicknesses, ie thin sheets. Colloq *thou*.

MIL-1553 B (*Aero*) Standard requirement for airborne digital databus. Originally US but now international.

milanese fabric (*Textiles*) A warp-knitted fabric made with a warp containing twice as many threads as there are WALES in the fabric with resultant reinforcement of the fabric.

Milankovitch theory of climatic change (*EnvSci*) The theory that large oscillations in climate are related to changes in solar radiation received by the Earth as a result of (1) the variations in eccentricity of the Earth's orbit (periods of 10^5 and 4×10^5 years), (2) variations in the obliquity of tilt of the Earth's axis (period 4×10^4 years) and (3) the precession of the equinoxes (period 2×10^4 years). There is evidence for the two shorter cycles in data obtained from analysis of deep-sea bottom cores. See panel on CLIMATIC CHANGE.

milarite (*Min*) A hydrated silicate of aluminium, beryllium, calcium and potassium, crystallizing in the hexagonal system.

mild clay (*Build*) See LOAM.

mildew (*BioSci*) A plant disease in which fungal mycelium is visible on the surface of the host, eg downy mildew, powdery mildew. Also *mould*.

mild steel (*Eng*) Steel containing up to 0·25% by weight of carbon. See panel on STEELS.

mile (*Genrl*) A unit of length commonly used for distance measurement in the UK and the USA. A *statute mile* = 1760 yd = 1609·34 m. See GEOGRAPHICAL MILE, NAUTICAL MILE.

miliaria (*Med*) Inflammation of the sweat glands, accompanied by intense irritation of the skin. Also *prickly heat*.

miliary (*Med*) Like a millet seed; said of lesions which are small and multiple, eg *miliary tuberculosis*.

miliary tuberculosis (*Med*) A form of tuberculosis in which small tuberculous lesions are found in various organs of the body, esp in the meninges and in the lungs.

milium (*Med*) A whitish pimple formed on the skin, usually by a clogged sebaceous gland.

milk (*FoodSci*) A liquid produced by female mammals for the suckling of their young. In food technology, milk is normally taken to mean cow's milk which contains approx 88% water, the remainder being carbohydrates, fat and protein in order of magnitude. It is usually pasteurized to destroy pathogens. Long-life milk is ultra heat treated (UHT) and aseptically packed. Milk powder is produced by spray or roller drying. To improve solubility, dried milk powder can be agglomerated. See AGGLOMERATION, DRYING, PASTEURIZATION.

milk fever (*Vet*) A metabolic disease of unknown cause affecting the parturient cow, ewe, goat, sow and bitch; characterized by hypocalcaemia and muscular weakness and inco-ordination, tetany, loss of consciousness and death. The disease also occurs during pregnancy in the ewe and during lactation in the goat. Also *calving fever, lambing sickness* (ewe), *parturient eclampsia* (sow and bitch), *parturient fever* (cow), *parturient paresis*.

milk glands (*BioSci*) The mammary glands of a female mammal; in viviparous tsetse flies, special uterine glands by which the larva is nourished until it is ready to pupate.

milk let-down (*Agri*) The release of milk stored in the cow's udder stimulated by a calf suckling or the mechanical milking process.

milk-sugar (*Chem*) Lactobiose or LACTOSE.

milk teeth (*BioSci*) In diphyodont mammals, the first or deciduous dentition.

milk tetany (*Vet*) See CALF TETANY.

Milky Way (*Astron*) (1) The bright band across the sky which contains a high density of stars, corresponding to the cross-section of the disk structure of our Galaxy. (2) Loosely, our Galaxy as a whole.

mill (*Eng*) (1) A machine for grinding or crushing, eg *flour mill, paint mill*. (2) A factory fitted with machinery for manufacturing, eg *cotton mill, saw mill*.

mill (*MinExt*) In the UK, a crushing and grinding plant. In the USA, the whole equipment for comminuting and concentrating an ore.

millboards (*Paper*) Boards, usually very dense, manufactured on an intermittent board machine from a variety of furnishes.

milled (*Eng*) Having the edge grooved or fluted, as a coin or the head of an adjusting screw. See KNURLING TOOLS.

milled cloth (*Textiles*) Woven or knitted wool or woollen fabric in which the milling action causes felting. The cloth has a fibrous surface, the threads and structure being almost completely hidden. The process involves the

application of pressure and friction to the cloth, while it is wet, eg with soapy water.

milled lead (*Build*) Sheet lead formed from cast slabs by rolling.

millefiori glass (*Glass*) Glassware in which a large number of sections of glass rods of various colours form a pattern and are fused together or set in a clear glass matrix. From Italian, *thousand flowers*.

millennium bug (*ICT*) Deficiency in the BIOS of some computers and some older programs in which date codes had only the last two digits, so that the change of century in the year 2000 caused unpredictable errors.

Miller bridge (*ElecEng*) A bridge used to measure amplification factors of valves.

Miller circuit (*Electronics*) Amplifier in which negative feedback from output to input is regulated by a capacitor.

Miller effect (*Electronics*) The change in effective input impedance of an amplifier due to unwanted shunt voltage feedback, which makes the input impedance a function of the voltage gain. In particular the increase in effective input capacitance for a valve or transistor used as a voltage amplifier.

Miller indices (*Crystal*) Integers which determine the orientation of a crystal plane in relation to the three crystallographic axes. The reciprocals of the intercepts of the plane on the axes (in terms of lattice constants) are reduced to the smallest integers in ratio. Also *crystal indices*.

millerite (*Min*) Nickel sulphide, crystallizing in the trigonal system. It usually occurs in very slender crystals and often in delicately radiating groups. Also *capillary pyrite*.

Miller process (*MinExt*) Purification of bullion by removal of base metals as chlorides. Chlorine gas is bubbled through the molten metal.

millers' disease (*Vet*) See OSTEODYSTROPHIA FIBROSA.

millet-seed sandstone (*Geol*) A sandstone consisting essentially of small spheroidal grains of quartz; typical of deposits accumulated under desert conditions.

mill-fitting (*ElecEng*) See FACTORY FITTING.

mill-head (*MinExt*) A grade of ore accepted for processing after removal of waste rock and detritus. Also *assay grade*.

milli- (*Genrl*) Prefix from Lt *mille*, thousand. When attached to units, it denotes the basic unit $\times 10^{-3}$. Symbol m.

milliammeter (*Phys*) An ammeter calibrated and scaled in milliamperes; used for measuring currents up to about 1 ampere.

millibar (*EnvSci*) See BAR.

millicurie (*Phys*) One-thousandth of a CURIE.

millidarcy (*Geol*) See DARCY.

Millikan oil-drop experiment (*Phys*) First experiment to determine the value of the electronic charge. The vertical motion of very small charged oil drops in the electric field between two horizontal parallel plates was measured, and the charge deduced.

millilambert (*Phys*) A unit of brightness equal to 0·001 LAMBERTS; more convenient magnitude than the lambert.

millilitre (*Chem*) A unit of volume, one-thousandth of a litre, essentially equivalent to 1 cm^3 or 10^{-6} m^3.

millilux (*Phys*) Unit of illumination equal to one-thousandth of a LUX.

millimass unit (*Phys*) Unit equal to 0·001 ATOMIC MASS UNITS. Symbol mu.

millimetre (*Genrl*) The thousandth part of a metre. Abbrev *mm*. US *millimeter*.

millimetre pitch (*Eng*) See METRIC SCREW-THREAD.

millimicron (*ElecEng*) Obsolete term for nanometre, 10^{-9} m. Used in the measurement of wavelengths of eg light.

milling (*Eng*) A machine process in which metal is removed by a revolving multiple-tooth cutter, to produce flat or profiled surfaces, grooves and slots. See MILLING CUTTER, MILLING MACHINE.

milling (*FoodSci*) Grinding, shearing and crushing many raw materials into a form in which their usable and unusable

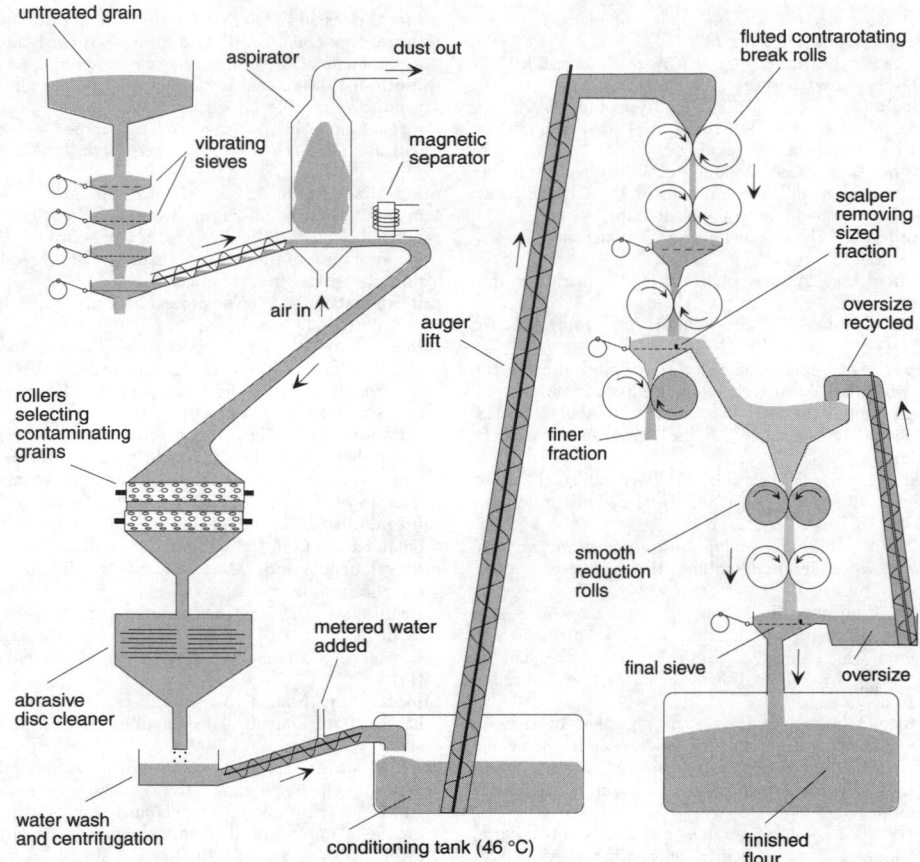

milling The processes of cleaning, conditioning and milling flour. Partly processed material is removed at various stages for both blending and further recycling but is not shown in the figure.

components may be separated, eg cereals for separating their bran or endosperm, deciduous fruits for juice extraction and seeds and nuts for extraction of their oils. Flour milling involves crushing the grains between fluted or serrated rollers; sieving to separate the bran from semolina and flour, and grinding and sieving to produce a finer flour. See EXTRACTION RATE, SEMOLINA.

milling (*MinExt*) Removing valueless material and harmful constituents from an ore, in order to render marketing more profitable. Also *comminution, dressing*.

milling (*Textiles*) A felting process carried out on already woven woollen fabrics to make them thicker and hairier, carried out in a milling machine by the agency of soap, alkali or acid (depending on the nature of the fabric and the dye), pressure and friction. Also *fulling*. See MILLED CLOTH.

milling crop (*Agri*) A grain crop grown for flour production.

milling cutter (*Eng*) A hardened steel disk or cylinder on which cutting teeth are formed by slots or grooves on the periphery and faces, or into which separate teeth are inserted; used in the milling machine for grooving, slotting, surfacing, etc. See END MILL, MILLING MACHINE.

milling grade (*MinExt*) The grade at which an ore is sufficiently rich to repay cost of processing.

milling machine (*Eng*) A machine tool in which a horizontal arbor or a vertical spindle carries a rotating multiple-tooth cutter, the work being supported and fed by an adjustable and power-driven horizontal table. See MILLING CUTTER.

Millington reverberation formula (*Acous*) A modified formula for calculating the period of reverberation time, taking into account the random disposition of the reflecting and acoustically absorbing surfaces in an enclosure. Not applicable if one partial surface has the absorption coefficient one (eg open window).

milling width (*MinExt*) The extent of the lode which will determine the tonnage sent daily from mine to mill.

million (*Genrl*) The number 1 000 000 (10^6).

million-electron-volt (*Phys*) See MEV.

million tons oil equivalent (*MinExt*) A standard basis for comparing between countries the annual production and consumption of different fuels (with different CALORIFIC VALUE) by relating each fuel to crude oil. Hydroelectric and nuclear power are also expressed in *oil equivalence* terms by calculating the amount of oil needed to produce an equivalent quantity of electrical energy: 1 mtoe is equivalent to about 1·5 million tons of coal. Abbrev *mtoe*.

millipede (*BioSci*) Any myriapod of the class Diplopoda, vegetarian cylindrical animals with many joints most of which bear two pairs of legs.

Millipore filter (*Chem*) TN for a type of MICROPORE FILTER.

milliradian (*Genrl*) Unit of 0·001 (10^{-3}) radians.

mill join (*Print*) An adhesion between the ends of two similar webs made at paper mill, the joints being sometimes marked for the printer's guidance.

Millon's reaction (*Chem*) A test for proteins, based on the formation of a pink or dark-red precipitate of coagulated proteins on heating with a solution of mercuric nitrate containing some nitrous acid.

mill race (*Eng*) Channel or FLUME by which water is led to a mill wheel or waterwheel.

mill scale (*Build, Eng*) A thin flaky layer of blue/black iron oxide found on new hot-rolled steel. Best removed by abrasive blasting before painting.

Millstone Grit (*Geol*) A long-established name for the middle coarse sandstone division of the Carboniferous, roughly corresponding to the Namurian.

mill tail (*Eng*) The channel conveying water away from a mill wheel.

mill wheel (*Eng*) A water wheel driving the machinery in a mill.

Milroy's disease (*Med*) See HEREDITARY ANGIONEUROTIC OEDEMA.

Mil-Spec (*Aero*) Abbrev for *military specification* issued in the US, which lays down basic requirements to be observed by design teams in the development of aircraft. Abbrev also *MS*.

milt (*BioSci*) The spleen; in fish, the testis or spermatozoa; to fertilize the eggs.

Mimas (*Astron*) The natural satellite orbiting closest to Saturn, discovered in 1789. Distance to the planet 186 000 km; diameter 390 km.

MIMD (*ICT*) Abbrev for *multiple instruction stream, multiple data stream*. Term describing the architecture of a PROCESSOR. See MULTIPROCESSOR.

mimetic diagram (*ElecEng*) In a control room of a large process plant or an electrical network, the animated diagram which indicates to the controller the state of operations, by coloured lights, recorders or indicating instruments.

mimetite (*Min*) A chloride–arsenate–phosphate of lead with As > P; cf PYROMORPHITE. It crystallizes in the hexagonal system, often in barrel-shaped forms, and is usually found in lead deposits which have undergone a secondary alteration.

mimicry (*BioSci*) The adoption by one species of the colour, habits, sounds or structure of another species. Adjs *mimetic, mimic*.

mimivirus (*BioSci*) A virus found in amoebae, much larger and having a more complex genetic structure than any other known virus.

Mimosa (*Astron*) A bright blue-white giant variable star in the constellation Crux. Distance approx 130 pc. Also *Beta Crucis*.

Minamata disease (*Med, EnvSci*) Poisoning by methylmercury in humans ingesting fish and shellfish contaminated by waste water (in 1956) from a chemical plant in Minamata, SW Kyushu, Japan. The disease is characterized by sensory and motor disturbances and fetal damage in pregnant women.

minaret (*Arch*) A lofty slender tower rising from a mosque or similar building and surrounded by a gallery near the top.

Mindel (*Geol*) The second major glacial stage of the Pleistocene epoch of the Alps. See GÜNZ, RISS, WÜRM.

mine (*MinExt*) (1) Subterranean excavation made in connection with exploitation of, or search for, minerals of economic interest. Terms *opencast*, *pit* and *quarry* are reserved for workings open to daylight. (2) Term (N England) for any coal seam irrespective of thickness or grade. See MINING.

mine detector (*Eng*) An electronic device for the detection of buried explosive mines (or buried metal) depending on the change in the electromagnetic coupling between coils in the search head.

mineral (*Min*) A naturally occurring substance of more or less definite chemical composition and physical properties. It has a characteristic atomic structure frequently expressed in the crystalline form or other properties.

mineral caoutchouc (*Min*) See ELATERITE.

mineral deposit (*Geol*) A naturally occurring body containing minerals of economic value.

mineral dressing (*MinExt*) See MINERAL PROCESSING.

mineral flax (*Build*) A fibrized form of asbestos once used in the manufacture of asbestos–cement sheeting.

mineral-insulated cable (*ElecEng*) One in which the conductor runs in an earthed copper sheath filled with magnesium oxide, which makes it fireproof and able to withstand excess loads. Also *copper-sheathed cable*. Abbrev MIC.

mineralization (*BioSci*) (1) The decomposition in soils of organic matter by micro-organisms with the release of the mineral elements (N, P, K, S, etc) as inorganic ions. (2) The incorporation of mineral into tissue, as with the calcification of cartilage to form bone.

mineralization (*Geol*) The process by which mineral(s) are introduced into a rock.

mineral nutrient (*BioSci*) An essential element, other than C, H or O, normally obtained as an inorganic ion taken up eg through the roots of a land plant, N, P, K, Ca, etc. See MACRONUTRIENT, MICRONUTRIENT.

mineralocorticoid (*BioSci*) A natural or synthetic corticosteroid that acts on the water and electrolyte balance of the body. Aldosterone is the most potent natural example.

mineralogy (*Genrl*) The scientific study of minerals.

mineral oils (*Chem*) Petroleum and other hydrocarbon oils obtained from mineral sources. Cf *vegetable oils*.

mineral processing (*MinExt*) Crushing, grinding, sizing, classification or separation of ore into waste and value by chemical, electrical, magnetic, gravity and physicochemical methods. First-stage extraction metallurgy. Also *beneficiation, mineral dressing, ore dressing, préparation mécanique* (*Fr*).

mineral vein (*MinExt*) A fissure or crack in a rock which has been subsequently lined or filled with minerals. See LODE.

mineral water (*FoodSci*) Water from a natural spring or deposit. Before exploiting a source hydrogeological details and a full chemical and microbiological analysis must be provided to ensure that the minimum standards for drinking water are met. In the UK specific regulations for natural mineral water specify that its composition cannot be altered in any way except by carbonation if this is declared on the label.

mineral wool (*Eng*) See ROCK WOOL.

miner's dip needle (*MinExt*) A portable form of DIP NEEDLE used for indicating the presence of magnetic ores.

miner's lamp (*MinExt*) A portable lamp specially designed to be of robust construction and adequate safety for use in mines.

Miner's rule (*Eng*) Empirical but widely used method of estimating the fatigue lifetime. See FATIGUE STRENGTH. Also *Miner's law*.

minette (*Geol*) (1) A lamprophyre composed essentially of biotite and orthoclase, occurring in dykes associated with major granitic intrusions. (2) Jurassic ironstones of Briey and Lorraine.

miniature camera (*ImageTech*) A term sometimes applied to any small camera, but normally used for those taking 35 mm film.

miniature Edison screw-cap (*ElecEng*) An Edison screw-cap for electric filament lamps, in which the screw-thread has a diameter of about 0·4 in and about 14 threads per inch.

miniature valve (*Electronics*) One in which all the dimensions are reduced to very small values, to keep down the interelectrode capacitances and the electron transit time, making it suitable for high-frequency, very-high-frequency and ultrahigh-frequency applications. Also *subminiature valve*.

minicomputer (*ICT*) Computer whose size and capabilities lie between those of a MAINFRAME and a PERSONAL COMPUTER. The term referred originally to a range of third-generation computers, cheaper and less well-equipped than contemporary mainframe machines. With changes in technology the term has become obsolete.

MiniDisc (*ICT*) TN for a small recordable compact disc.

minidish (*ICT*) A small satellite dish used to receive digital TV.

mini head drum (*ImageTech*) A smaller version of a standard head drum used in some domestic camcorders. The track length is maintained by increasing the TAPE WRAP and the WRITING SPEED by raising the rotational speed. Four heads are necessary because with two they would be in the wrong place every other revolution (head sequence is A1,B1,A2,B2).

minikin (*Print*) The smallest of the old type sizes, approximately 3-point.

minimal area (*BioSci*) A concept used in sampling vegetation: the minimum area which must be searched in order to find (nearly) all of the species. See SPECIES–AREA CURVE.

minimal tillage (*Agri*) A technique whereby soil disturbance is minimized during crop establishment to reduce the costs of inputs and preserve the natural soil structure.

minimum access program (*ICT*) Program routine involving minimum ACCESS TIME.

minimum blowing current (*ElecEng*) The minimum current which will cause melting of a fuse link under certain specified conditions.

minimum burner pressure valve (*Aero*) A device which maintains a safe minimum pressure at the burners of a gas turbine when it is idling.

minimum deviation (*Phys*) See ANGLE OF MINIMUM DEVIATION.

minimum discernible signal (*ICT*) The smallest input power to any unit that just produces a discernible change in output level.

minimum flying speed (*Aero*) The minimum speed at which an aircraft has sufficient lift to support itself in level flight in standard atmosphere. There is a close relationship with the weight, which affects the WING LOADING, and the term must be stated with the weight (and altitude if INTERNATIONAL STANDARD ATMOSPHERE, sea level is not implied) and true airspeed specified.

minimum ionization (*Phys*) The smallest possible value of the specific ionization that a charged particle can produce in passing through a given substance. It occurs for particles having velocities of around $0.95c$, where c is the speed of light.

minimum pause (*ICT*) The interval of lost time that is necessarily introduced into the operation of a dial to ensure that the selectors have time to complete their hunting.

minimum point on a curve (*MathSci*) A trough on a curve. Formally, the curve $y = f(x)$ has a minimum at a if its derivative is 0 at a and $f(a)-f(b)$ is negative for all b in a neighbourhood of a.

minimum sampling frequency (*ICT*) The lowest sampling rate that can provide an accurate reproduction of the signal in a pulse-code-modulation system. Equal to twice the maximum signal frequency.

minimum wavelength (*Phys*) The shortest wavelength emitted in an X-ray spectrum. It is determined by the maximum voltage applied to the X-ray tube. The emitted X-ray photon acquires all the energy of the electron accelerated by the voltage, V, so the minimum wavelength is given by

$$\lambda_m = \frac{hc}{eV} = \frac{1.2396 \times 10^{-6}}{V} \text{ m}$$

where h is Planck's constant, e is the electronic charge and c is the speed of light.

mining (*MinExt*) The process of extracting metallic or non-metallic mineral deposits, including gemstones, from below the Earth's surface. Mine is a subterranean excavation unlike quarry, pit and opencast which are reserved for workings open to daylight. Mining methods are devised for each deposit, so there are many variations but common terms are shown in the figure; some methods may be of considerable antiquity.

mining dial (*MinExt*) See DIAL.

mining engineering (*Genrl*) The branch of engineering chiefly concerned with the sinking and equipment of mine shafts and workings, and all operations incidental to the winning and preparation of minerals.

minion (*Print*) An old type size, approximately 7-point.

minisatellites (*BioSci*) A class of highly repetitive satellite DNA, comprising variable (typically 10–20) repeats of short (eg 64 bases) DNA sequences. The high level of polymorphism of such minisatellites makes them very useful in genomic mapping. Also *variable-number tandem repeats*. See also MICROSATELLITE.

miniscule (*Print*) A lower case, or small, letter.

minitrack (*ICT*) Phase-comparison angle-tracking system, used for tracking satellites.

Minkowski diagram (*Phys*) A space–time diagram used to represent the positions and times of events relative to an inertial reference frame.

Minkowski's inequality (*MathSci*)

$$\left(\sum_{r=1}^{n}|a_r + b_r|^\alpha\right)^{\frac{1}{\alpha}} \le \left(\sum_{r=1}^{n}|a_r|^\alpha\right)^{\frac{1}{\alpha}}\left(\sum_{r=1}^{n}|b_r|^\alpha\right)^{\frac{1}{\alpha}}$$

if $a \ge 1$.

minnesotaite (*Min*) The iron-bearing equivalent of talc. A major constituent of the siliceous iron ores of the Lake Superior region.

Minnesota Multiphasic Personality Inventory (*Psych*) A widely used self-report personality assessment device. The 2nd edition uses 567 items, each a statement about the person being tested that needs to be answered 'true'/'false'/'don't know' and can then be scored. The device has been empirically tested and is used to indicate various psychological problems. Abbrev *MMPI*.

minor (*MathSci*) A sub-determinant, ie a determinant obtained from another determinant by deleting a number

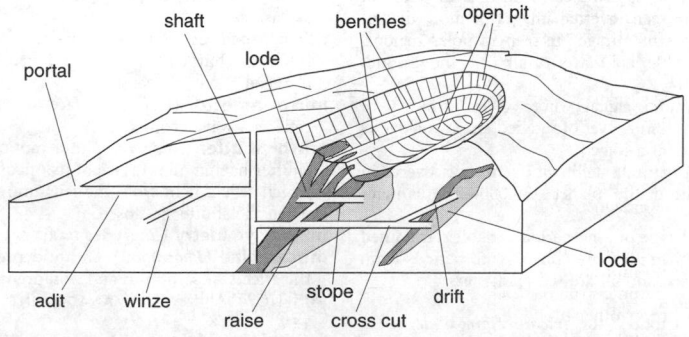

mining Schematic drawing of a section through a deep mine and an opencast pit.

of rows and columns. The minor of a particular element of a determinant is the determinant that remains when the row and column containing the element are deleted.

minor axis (*MathSci*) Of an ellipse: see AXES.

minor exchange (*ICT*) One directly connected to its group centre.

minor histocompatibility antigens (*BioSci*) Alloantigens that cause much slower graft rejection than that initiated by antigens of the MAJOR HISTOCOMPATIBILITY COMPLEX.

minor intrusions (*Geol*) Igneous intrusions of relatively small size, compared with PLUTONIC (MAJOR) INTRUSIONS. They comprise dykes, sills, veins and small laccoliths. The injection of the minor intrusions constitutes the dyke phase of a volcanic cycle.

minority carrier (*Electronics*) In a semiconductor, the electrons or HOLES which carry the lesser degree of measured current. See MAJORITY CARRIER.

minor planet (*Astron*) A term used generally in professional astronomy for ASTEROID.

minoxidil (*Pharmacol*) A vasodilatory drug used to treat severe hypertension and, externally, to stimulate the regrowth of hair in cases of male-pattern baldness.

minuend (*MathSci*) See SUBTRACTION.

minus colour (*ImageTech*) The complementary to a given colour, ie the remainder when that colour is subtracted from white light. Thus, *minus red* is blue-green, *cyan*. Hence, the equivalent term *subtractive colour*.

minus strain (*BioSci*) One of the two, arbitrarily designated, mating types of a heterothallic species. Cf PLUS STRAIN. See HETEROTHALLISM.

minute (*Genrl*) (1) A 60th part of an hour of time. (2) A 60th part of an angular degree. (3) A 60th part of the lower diameter of a column.

minverite (*Geol*) A basic intrusive rock, essentially a dolerite, containing a brown, soda-rich hornblende; named from the type locality, St Minver, Cornwall.

Miocene (*Geol*) An epoch of the Neogene subperiod of the Cenozoic era. See TERTIARY.

miosis (*Med*) Contraction of the pupil of the eye.

miospore (*Geol*) A spore or pollen grain arbitrarily defined in palaeopalynology as less than 200 micrometres (200 μm) in diameter.

MIPS (*ICT*) Abbrev for *millions of instructions per second*. The number of million instructions that can be executed by a computer in a second. Cf FLOPS.

Mir (*Space*) Advanced Soviet space station, developed from experience gained with SALYUT; launched in February 1986 and assembled in space. See panel on SPACE STATION.

Mira (*Astron*) A red-giant long-period variable star in the constellation Cetus, the prototype of the MIRA STARS. Distance approx 40 pc. Also *Omicron Ceti*.

mirabilite (*Min*) See GLAUBER SALT.

miracidium (*BioSci*) The ciliated first-stage larva of a trematode.

mirage (*EnvSci*) An effect caused by total reflection of light at the upper surface of shallow layers of hot air in contact with the ground, the appearance being that of pools of water in which are seen inverted images of more distant objects. Other types of mirage are seen in polar regions, where there is a dense, cold layer of air near the ground. See FATA MORGANA.

mirage (*ICT*) A radio signal whose transmission path includes reflection from a layer of rarefied air, in analogous manner to the optical mirage.

Miranda (*Astron*) A natural satellite of Uranus, discovered in 1948. Distance from the planet 130 000 km; diameter 470 km.

Mira star (*Astron*) A type of long-period variable star named after MIRA; more than 3000 are known, with periods from 2 months to 2 years, and all are red-giant stars.

mire (*EnvSci*) See BOG.

mired (*ImageTech*) Abbrev for *micro-reciprocal degree*, a measure of colour temperature being 1 million divided by the value in kelvins.

mirror (*Eng*) Very smooth region of tensile fracture surface in brittle materials, esp glass, associated with slow, but accelerating crack growth. Cf HACKLE, MIST. See FRACTOGRAPHY.

mirror (*Phys*) A highly polished reflecting surface capable of reflecting light rays without appreciable diffusion. The commonest forms are plane, spherical (convex and concave) and paraboloidal (usually concave). The materials used are glass silvered on the back or front, speculum metal or stainless steel.

mirror arc (*ImageTech*) A projector lamphouse system in which the source of light, a carbon or xenon arc, is at the focus of a parabolic mirror which concentrates the illumination on the gate.

mirror drum (*ImageTech*) A rotating cylinder or cone with a number of mirrors on its circumference to reflect a beam of light in MECHANICAL SCANNING; still of possible application in projection TV systems.

mirror finish (*Eng*) A very smooth, lustrous surface finish produced eg on stainless steels and other metals by electrolytic polishing or lapping, and on electroplated surfaces by mechanical polishing.

mirror galvanometer (*ElecEng*) A galvanometer having a mirror attached to the moving part, so that the deflection can be observed by directing a beam of light onto the mirror and observing the movement of the reflection of this over a suitable scale. Also *reflecting galvanometer*.

mirroring (*ICT*) The arrangement of processors or disks such that there is sufficient redundancy in a hardware configuration for a computer to continue to function in the event of component failure.

mirror lens (*ImageTech*) A compact TELEPHOTO system employing two concave reflecting surfaces to form the image; because a diaphragm cannot be included, exposure must be controlled by neutral density filters or by shutter time. Also *catoptric lens*.

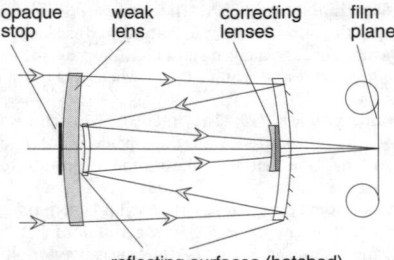

opaque stop   weak lens   correcting lenses   film plane

reflecting surfaces (hatched)

mirror lens Typical design with correcting lenses needed to avoid aspheric surfaces.

mirror machine (*NucEng*) Fusion machine using the MAGNETIC MIRROR principle to trap high-energy ions in a plasma.

mirror nuclides (*Phys*) Nuclides with the same number of nucleons, but with proton and neutron numbers interchanged.

mirror reflector (*ICT*) A surface or set of metal rods that reflects a wave geometrically.

mirror shutter (*ImageTech*) In a motion picture camera, a shutter having one surface of its opaque blades as mirrors, which reflect light from the lens into a viewfinder system when the shutter is closed.

mirror symmetry (*Phys*) See PARITY.

mirtazapine (*Pharmacol*) An antidepressant that enhances the effects of serotonin and noradrenaline in the brain.

MIS (*ICT*) Abbrev for MANAGEMENT INFORMATION SYSTEM.

miscarriage (*Med*) Expulsion of the fetus before the 28th week of pregnancy. Loosely, abortion.

miscibility (*Chem*) The property enabling two or more liquids to dissolve when brought together and thus form one phase.

miscibility gap (*Chem*) The region of composition and temperature in which two liquids form two layers or phases when brought together.

MISD (*ICT*) Abbrev for *multiple instruction stream, single data stream*. Term that describes the architecture of a PROCESSOR. See PIPELINING.

miser (*Build*) A large *auger* used for boring holes in the ground in wet situations.

misfire (*ElecEng*) The failure to establish, during an intended conducting period, a discharge or arc in a gas-discharge tube or a mercury-pool rectifier.

misfiring (*Autos*) The failure of the compressed charge to fire normally, generally due either to ignition failure or to an over-rich or weak mixture.

mismatch (*ICT*) Load impedance of incorrect value for maximum energy transfer. See MATCHING.

mismatch repair (*BioSci*) A DNA repair system that detects and replaces wrongly paired, ie mismatched, bases in newly replicated DNA.

misoprostol (*Pharmacol*) A synthetic prostaglandin analogue used chiefly in treating and preventing gastric ulcers, esp those caused by anti-inflammatory drugs.

mispickel (*Min*) See ARSENOPYRITE.

missense mutation (*BioSci*) A mutation that alters a triplet codon in DNA so that it specifies a different amino acid.

missile (*Aero, Space*) There are two basic types of missile, in the current sense, *guided* and *ballistic*. The former is controlled from its launch until it hits its target; the latter, always of long-range, surface-to-surface type, is controlled into a precision ballistic path so that its course cannot be deflected by countermeasures. See GUIDED MISSILE.

missile shield (*NucEng*) The structure placed over or round a reactor to prevent an explosively ejected fuel or control rod from piercing the containment building.

missing mass (*Astron*) See DARK MATTER.

mission (*Space*) The succession of events which must happen to achieve the objectives stipulated; it includes everything which must be done from conception to the delivery of the results (more loosely, the actual flight of the spacecraft).

mission adaptive wing (*Aero*) A wing whose section profile is automatically adjusted to suit different flight conditions, eg Mach number, lift and altitude.

mission control centre (*Space*) A room or building in which are assembled the means necessary to visualize and control a space system so that its mission objectives can be achieved.

mission specialist (*Space*) A member of the crew of the SPACE SHUTTLE whose responsibilities are concerned with mission aspects, such as the control of the orbiter's resources to a payload, the handling of payload equipment and the performance of the experiments in orbit.

Mississippian (*Geol*) A period of the Upper Palaeozoic era, lying between the Devonian and Pennsylvanian. It covers an approx time span from 360 to 330 million years and is the N American equivalent of the *Lower Carboniferous* in Europe. The corresponding system of rocks. See PALAEOZOIC.

missourite (*Min*) A melanocratic plutonic rock composed of clinopyroxene, olivine and leucite.

mist (*Chem*) A suspension, often colloidal, of a liquid in a gas.

mist (*Eng*) Cloudy-looking region of tensile fracture surface in brittle materials, esp glass, between MIRROR and HACKLE, associated with crack moving at up to its maximum velocity, but with insufficient energy to branch (or bifurcate). See FRACTOGRAPHY.

mist (*EnvSci*) A suspension of water droplets (radii less than $1\ \mu m$) reducing the visibility to not less than 1 km. See FOG.

mist coat (*Build*) A highly atomized thin coat of paint applied by spray, applied either before a 'full' coat or as a finishing on spraying cellulose.

mitochondrial diseases (*Med*) Illnesses, frequently neurological, that can be ascribed to defects in mitochondrial function.

mitochondrion (*BioSci*) A mobile cytoplasmic *organelle* of EUKARYOTES visible in the light microscope whose main function is the generation of ATP by AEROBIC RESPIRATION. Pl *mitochondria*. See panel on MITOCHONDRION.

mitogen (*BioSci*) Any agent that induces mitosis in cells. In immunology often used to refer to substances such as lectins or lipopolysaccharides that cause a wide range of T- or B-lymphocytes to undergo cell division.

mitosis (*BioSci*) The normal process of somatic cell division, in which each of the daughter cells is provided with a chromosome set identical to that of the parent cell. Adjective *mitotic*. Cf AMITOSIS, MEIOSIS. See panel on CELL CYCLE. Fig. ▷

mitospore (*BioSci*) A spore formed by mitosis and hence having the same number of chromosomes as the parent.

mitotic crossing-over (*BioSci*) The (rare) exchange of genetic material between homologous chromosomes during mitosis, resulting in genetic RECOMBINATION. See CROSSING-OVER.

mitotic death (*BioSci*) A phenomenon in which cells fatally damaged by ionizing radiation do not die until the next mitosis, at which point the radiation damage to the DNA becomes evident, particularly if there is chromosome breakage.

mitotic index (*BioSci*) The proportion in any tissue of dividing cells, usually expressed as per thousand cells.

mitotic spindle (*BioSci*) See SPINDLE and MITOSIS.

mitral (*BioSci*) Mitre-shaped, as the mitral valve, guarding the left atrioventricular aperture of the heart in higher vertebrates; of the olfactory bulb, composed of mitre-shaped cells. Also *mitriform*.

mitral (*Med*) Pertaining to, or affecting, the mitral valve or valves of the heart.

mitral stenosis (*Med*) Narrowing of the communication between the left atrium and the left ventricle of the heart, as a result of disease of the mitral valves.

mitral valve (*BioSci*) See BICUSPID VALVE.

mitre (*Build*) A joint between two pieces at an angle to one another, each jointing surface being cut at an angle.

mitre block (*Build*) A block of wood rebated along one edge and having saw cuts in the part above the rebate, with the kerfs inclined at 45° to the face of the rebate so as to guide the saw when cutting mouldings for a mitred joint.

mitre board (*Build*) See MITRE SHOOT.

mitre box (*Build*) An open-ended box having saw cuts in the sides at 45° to the length of the box; used like the mitre block but capable of taking deeper mouldings.

mitre cramp (*Build*) A clamp for holding together temporarily the two parts of a mitre joint. Also *mitre clamp*.

mitre-cut piston ring (*Eng*) A piston ring in which the ends are mitred at the joint, as distinct from stepped or square ends.

mitre dovetail (*Build*) See SECRET DOVETAIL.

mitred valley (*Build*) See CUT-AND-MITRED VALLEY.

mitre post (*Build*) See MEETING POST.

mitre saw (*Build*) See TENON SAW.

mitre-saw cut (*Build*) A device, such as a mitre block or box, for keeping the saw at the required angle to the work when cutting mouldings for a mitre joint. Also *mitre-sawing board*.

mitre shoot (*Build*) A block of wood rebated along one edge as a guide for a jointing plane, and having a pair of wood strips fixed to the top face of the part above the rebate, at 45° to the face of the rebate, so as to hold the mitre face of a moulding at a right angle to the plane while it is being shot. Also *mitre board*.

Mitochondrion

A mitochondrion is a mobile cytoplasmic ORGANELLE of EUKARYOTES visible in the light microscope, whose main function is the generation of ADENOSINE TRIPHOSPHATE (ATP) by AEROBIC RESPIRATION. The organelle consists of an outer membrane surrounding an inner membrane enclosing a protein-rich matrix. The outer membrane is freely permeable to METABO-LITES and of indeterminate function, the characteristic functions of the mitochondrion being associated with the inner membrane and the matrix proteins. The components necessary for oxidative phosphorylation, for the electron transfer chain (see below) and for the proton translocating ATPASE are located on highly convoluted infoldings of the inner membrane called *cristae*. Proton translocating ATPase molecules can be seen in the electron microscope as knob-like structures protruding from the surface of the cristae into the matrix. The enzymes which degrade the energy sources, glucose and fatty acids, by the TRICARBOXYLIC ACID CYCLE and BETA-OXIDATION, are distributed between the inner membrane and the matrix.

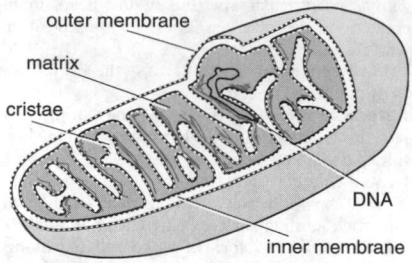

Fig. 1 **Mitochondrion**.

The biogenesis of mitochondria is complex. New mitochondria arise by the division of pre-existing mitochondria, while the synthesis of the mitochondrial proteins is under the co-ordinated activity of nuclear genes and mitochondrial DNA. The organelles contain all the components required for protein synthesis, RIBOSOMES, TRNA, etc. This limited autonomy of the mitochondria, together with features like the ribosomal nucleotide sequences and circular DNA, has led to the suggestion of an independent PROKARYOTE existence prior to their incorporation into the cytoplasm of present-day eukaryotes. Oxidative phosphorylation occurs across the cytoplasmic plasma membrane in modern prokaryotes. See the ENDOSYMBIOTIC HYPOTHESIS for an analogous hypothesis concerning CHLOROPLASTS.

Electron transfer chain

The energy derived from the oxidation of glucose by the tricarboxylic acid cycle is harnessed for energy-requiring processes in the cell in the form of ATP. As the amount of energy liberated from an oxidation step is proportional to the difference in reduction potentials of the oxidized and reduced components of the reaction, the energy from glucose can be released in suitably sized packets for ATP synthesis by the transfer of electrons from glucose to their ultimate acceptor, oxygen, along a series of components with appropriate spans of reduction potentials.

This series of components is known as the ELECTRON TRANSFER (or TRANSPORT) CHAIN. The transfer of electrons from $NADH^+$ to oxygen involves three large multiprotein complexes, composed of CYTOCHROMES, FLAVOPROTEINS and other metaloproteins, together with the smaller components of UBIQUINONE (also known as coenzyme Q) and cytochrome c which shuttle between the larger complexes.

Thus the first complex (NADH–ubiquinone reductase) accepts the electrons from $NADH^+$ and passes them on to ubiquinone; the next complex (ubiquinone–cytochrome c reductase) accepts these electrons from the oxidized ubiquinone and passes them to cytochrome c which is then reoxidized by the third large complex (cytochrome oxidase) on the last stage of their journey to the terminal oxygen. The stepwise increase of the standard reduction potentials of the three complexes ensures that the energy is released as three packets, each of sufficient magnitude to convert adenosine diphosphate (ADP) to ATP and thus account for the synthesis of the three molecules of ATP which is associated with the oxidation of each $NADH^+$. Succinate–ubiquinone reductase acts as a second point of entry by also being able to reduce ubiquinone, although, in this case, as the electrons enter the chain after the $NADH^+$–ubiquinone reductase complex, only two molecules of ATP are synthesized.

---

mitre sill (*Build*) The raised part of the bed of a canal lock against which the parts of the gates abut in closing. Also *clap-sill*, *lock-sill*.

mitre square (*Build*) A tool similar to the bevel, having the blade at 45° to the stock.

mitre wheels (*Eng*) See BEVEL GEAR.

mitriform (*BioSci*) See MITRAL.

Mitscherlich's law of isomorphism (*Chem*) A law stating that salts having similar crystalline forms have similar chemical constitutions.

mixed (*BioSci*) Said of nerve trunks containing motor and sensory fibres.

mixed bud (*BioSci*) A bud containing young foliage leaves and also the rudiments of flowers or of inflorescences.

mixed cell system (*ICT*) A cellular mobile-telephone system employing a wide variety of different CELL shapes and sizes. In eg urban environments, cells may need to be split into microcells to cope with high traffic densities, while long narrow cells may be required on motorways to minimize HANDOVER problems at high vehicle speeds.

Mitochondrion *(Cont.)*

The electron transfer chain is located in the inner membrane of the mitochondrion where it is so orientated that the passage of electrons is coupled to a flow of protons across the membrane, a process which leads to the actual synthesis of the ATP. See oxidative phosphorylation below.

The flow of electrons, and hence the oxidation of the food, requires a supply of ADP for conversion to ATP. This 'respiratory control' ensures that the energy source is consumed only when energy is required.

Oxidative phosphorylation
Glucose is oxidized to carbon dioxide and water with the concomitant synthesis of ATP by aerobic respiration. The energy for the synthesis of nearly all the ATP is derived from a series of oxidations which occur when the electrons liberated from glucose during respiration pass down the electron transfer chain to be accepted ultimately by oxygen, reducing it to water. A stoichiometric ratio of 3 moles of ATP per mole of oxygen consumed was established during the 1950s but the search for the metabolic intermediates was fruitless

until P Mitchell made the then revolutionary suggestion that there were no intermediates in the normally accepted sense and that the synthesis depended upon the integrity of the inner mitochondrial membrane and on the disposition of its components across it by a process which he termed CHEMIOSMOSIS.

During chemiosmosis the energy derived from the sequential oxidation of the components of the electron transfer chain is released in three packets which energize proton pumps driving protons across the inner membrane from the mitochondrial matrix. The resulting pH differential and membrane potential generate a 'proton motive force' across the inner mitochondrial membrane which causes a flow of protons back into the matrix through the proton translocating ATPase of the membrane where the ATP is synthesized. Respiratory poisons, such as dinitrophenol, which 'uncouple' the electron flow from ATP synthesis, exert their effect by themselves carrying protons through the membrane, thus short-circuiting the ATP-generating mechanism.

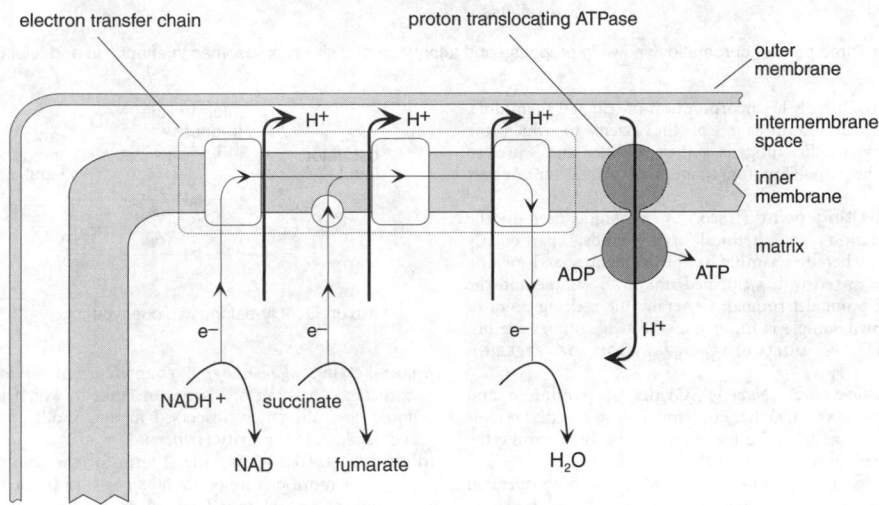

Fig. 2 **Oxidative phosphorylation** Showing the electron pathways through the compartments of the mitochondrion.

H^+ = hydrogen ions, e^- = electrons, ATP and ADP = adenine triphosphate and diphosphate, $NADH^+$ = oxidized nicotinamide adenine dinucleotide phosphate, NAD = nicotinamide adenine dinucleotide.

mixed coupling (*ICT*) Simultaneous inductive and capacitance coupling between two resonant circuits.

mixed crystal (*Crystal*) A crystal in which certain atoms of one element are replaced by those of another.

mixed-flow water turbine (*Eng*) An inward-flow reaction turbine in which the runner vanes are so curved as to be acted on by the water as it enters radially and leaves axially. Also *American water turbine*.

mixed grazing (*Agri*) Mixed types of grazing animals in the same area of pasture.

mixed highs (*ImageTech*) In some colour TV systems, high-frequency components of the picture, representing fine detail, are transmitted or recorded and reproduced as a mixed (MONOCHROME) signal to conserve the bandwidth necessary.

mixed inflorescence (*BioSci*) An inflorescence in which some of the branching is RACEMOSE and some is CYMOSE.

mixed lymphocyte culture (*BioSci*) A culture in which lymphocytes from two individuals are mixed for 3 to 5 days, at the end of which the number of dividing cells is

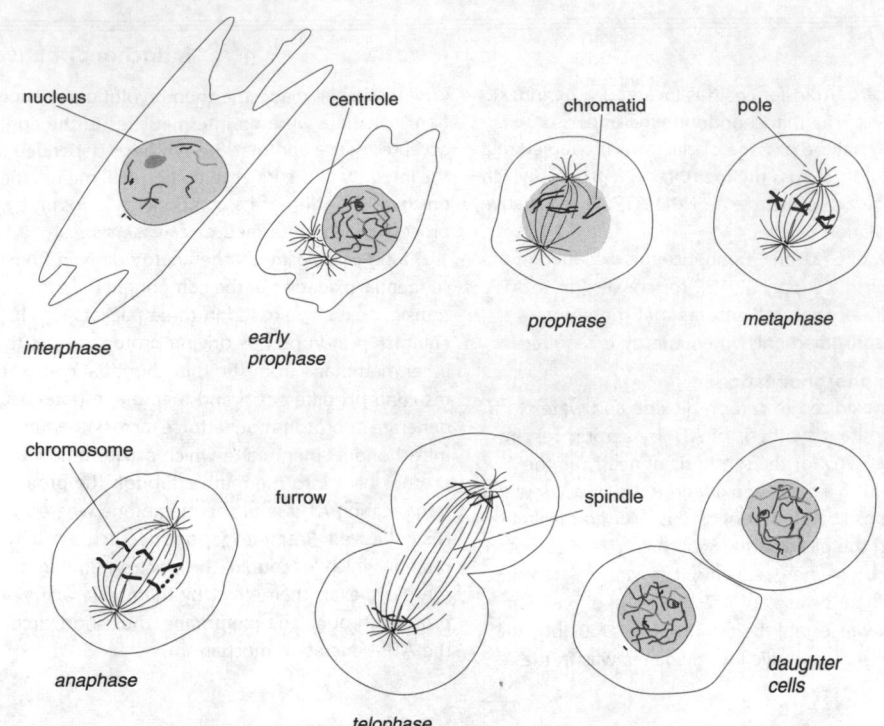

nucleus   centriole   chromatid   pole

interphase   early prophase   prophase   metaphase

chromosome   furrow   spindle

anaphase   daughter cells

telophase

mitosis Three pairs of chromatids shown in prophase and telophase and six chromosomes in anaphase and telophase.

measured, usually by incorporation of tritiated thymidine. This provides a measure of the extent to which the histocompatibility antigens of the two differ, and is used to assess the suitability for tissue transplantation. Abbrev *MLC*.

mixed melting point (*Chem*) A technique used in the identification of chemical compounds, particularly organic, whereby a sample of known identity and melting point is mixed with a purified unknown sample and the melting point determined. Generally the melting point of the known sample is lowered if the two samples are not identical. See MOLECULAR DEPRESSION OF FREEZING POINT.

mixed oxide fuel (*NucEng*) Oxides of plutonium and uranium mixed together and forming a breeder reactor fuel, neutrons from the fissile plutonium bombarding the fertile uranium. Abbrev *MOX*.

mixed-pressure turbine (*Eng*) A steam turbine operated from two or more sources of steam at different pressures, the low-pressure supply, from eg the exhaust of other engines, being admitted at the appropriate pressure stage.

mixed service (*ICT*) Service provided by a PRIVATE BRANCH EXCHANGE (PBX) to the main exchange for a number of extension lines only.

mixed-settler solvent extraction (*NucEng*) A method in which two immiscible liquids are agitated and then allowed to settle, both phases then being separated and mixed with fresh solvent and the process repeated to form a cascade for separating eg plutonium and uranium from fission products.

mixer (*Build, CivEng*) See CONCRETE MIXER.

mixer (*ICT*) (1) Collection of variable attenuators, hand-controlled, which allows the combination of several transmissions to be independently adjusted from zero to maximum. A group mixer or fader deals with several groups of transmissions. Hence mixer, the person who operates mixers. (2) Frequency conversion stage in a SUPERHET RECEIVER. See FREQUENCY CHANGER. Fig. ▷

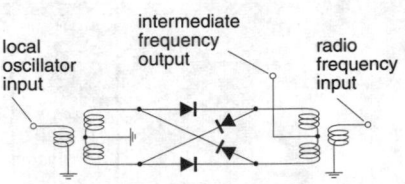

intermediate frequency output

local oscillator input

radio frequency input

mixer Doubly-balanced diode type.

mixer–settlers (*ChemEng*) A countercurrent *liquid–liquid extractor*, consisting of a series of tanks in which the two liquids are alternately dispersed in one another (*mixers*) and separated by gravity (*settlers*).

mixing (*ImageTech*) The general term for the combination of two or more picture or sound signals. In particular, the dubbing of several sound recordings into a single mixed track.

mixing (*Phys*) In gaseous isotope separation, the process of reducing the concentration gradient for the lighter isotope close to the diffusion barrier.

mixing (*Plastics*) For polymers, additives must usually be homogeneously distributed through the material for full effect. It is a problem for high-viscosity melts with low REYNOLDS NUMBER, where fluid flow is laminar. Special methods may be needed (eg CAVITY TRANSFER MIXING). Pigments like carbon black form clumps, which must be broken up for optimizing properties. Hence DISPERSIVE MIXING and DISTRIBUTIVE MIXING. See panel on TYRE TECHNOLOGY

mixing (*Textiles*) Mechanically blending staple fibres, esp wools or cottons of similar staple and colour, to obtain the most suitable material for spinning uniform yarns economically.

mixing efficiency (*NucEng*) A measure of the effectiveness of the mixing process in isotope separation.

mixing index (*NucEng*) A measure of quality of mixing process, the ratio of surface areas of particles after and before mixing, esp when mixing fillers into polymers.

mixing length (*EnvSci*) The average distance travelled by an eddy which is transporting heat, momentum or water vapour in the atmosphere.

mixotrophic (*BioSci*) Combining two or more fundamental methods of nutrition, as in certain Mastigophora which combine HOLOPHYTIC with SAPROPHYTIC nutrition or as in a partial parasite.

mixture (*Autos*) The combined flammable gas and air constituting the explosive charge.

mixture control (*Aero*) An auxiliary control fitted to a carburettor to allow variation of mixture strength with altitude. May be manually operated or automatic.

Mizar (*Astron*) The middle star in the handle of the Plough in the constellation Ursa Major. A multi-component star which forms an optical double with ALCOR. Distance 26 pc. Also *Zeta Ursae Majoris*.

mizzonite (*Min*) One of the series of minerals forming the scapolite group, consisting of a mixture of the meionite and marialite molecules. Mizzonite includes those minerals with 50–80% of the meionite end-member molecule. Found in metamorphosed limestones and in some altered basic igneous rocks.

MKSA (*Phys*) Abbrev for the *metre–kilogram–second–ampere* system of units, now superseded by SI UNITS.

MLD (*Radiol*) Abbrev for MEAN LETHAL DOSE.

M-lines (*Phys*) Characteristic X-ray frequencies from atoms due to the excitation of electrons from the M-shell. Only developed in atoms of high atomic number.

M-loop (*ImageTech*) Tape path employed in VHS videotape recorders, giving rather more than 180° wrap on the drum; from the M-shaped tape guide in the mechanism.

mm (*Genrl*) Abbrev for MILLIMETRE.

MMA (*Chem*) Abbrev for *monomethyl aniline, N-methyl-aminobenzene*. $C_6H_5NHCH_3$. Used occasionally as an ANTIKNOCK SUBSTANCE, as an alternative to lead tetraethyl.

MMC (*Eng*) Abbrev for METAL MATRIX COMPOSITE.

MMD (*Chem*) Abbrev for MOLECULAR MASS DISTRIBUTION.

mmf (*Phys*) Abbrev for MAGNETOMOTIVE FORCE.

$\mu\mu$F (*ElecEng*) Obsolete abbrev for *micromicrofarad*. Now replaced by PICOFARAD.

Mn (*Chem*) Symbol for MANGANESE.

mneme (*Psych*) A memory trace.

mnemonic (*ICT*) A memorizing aid or an abbreviation for an operation, particularly in ASSEMBLY LANGUAGE, eg JMP means 'jump', ADC means 'add with carry'.

mnemonics (*Psych*) Techniques or strategies to improve recall (eg rhyming).

Mo (*Chem*) Symbol for MOLYBDENUM.

mo (*Build*) Abbrev for *moulded*.

mobbing (*Psych*) A form of harassment directed at predators by potential prey.

mobile application part (*ICT*) The part of a mobile-telephone network dedicated to conveying operational messages, such as those containing the information that is exchanged between the HOME LOCATION REGISTER and VISITOR LOCATION REGISTERS.

mobile belt (*Geol*) A long zone of the Earth's crust associated with igneous activity and deformation. Traditionally associated with geosynclinal development. See panel on PLATE TECTONICS.

mobile element (*BioSci*) DNA sequence capable of excising itself from one chromosome and then reintegrating itself, or its copies, into different sites in the chromosome set.

mobile phase (*Chem*) See CHROMATOGRAPHY.

mobile services switching centre (*ICT*) Equipment that provides an interface between the fixed telephone network and the radio system of a mobile-telephone network. Not only acts as a fixed telephone exchange but is responsible for eg HANDOVER and location updating.

mobile station (*ICT*) In a mobile-telephone network, the user equipment carried in a car or on the person. Colloq *mobile*.

mobile station roaming number (*ICT*) In a mobile communications network, a temporary number assigned to a visiting mobile by a VISITOR LOCATION REGISTER. The HOME LOCATION REGISTER is advised of this number so that calls can be routed appropriately.

mobility (*Electronics*) (1) The freedom of particles to move, either in random motion or under the influence of fields or forces. (2) The average velocity of charge carriers per unit electric field in a given material. The mobility of electrons and holes differs greatly. In general mobility decreases with increasing temperature and with increasing defect and impurity densities.

Möbius function (*MathSci*) A function μ defined for positive integers as $\mu(1) = 1$, $\mu(n) = (-1)^r$ if n is the product of r distinct primes, and $\mu(n) = 0$ for all other positive integers.

Möbius strip (*MathSci*) The one-sided surface formed by joining together the two ends of a long rectangular strip, one end being twisted through 180° before the join is made.

Möbius transformation (*MathSci*) See BILINEAR TRANSFORMATION.

Mocha stone (*Min*) See MOSS AGATE.

mock leno (*Textiles*) Woven fabric with a cellular structure produced by allowing spaces to develop between groups of threads. The open structure makes it suited for resin INFILTRATION when making a COMPOSITE laminate by hand lay-up. Cf LENO FABRIC.

mock Moons (*EnvSci*) Lunar images similar to MOCK SUNS. Also *paraselenae*.

mock Suns (*EnvSci*) Images of the Sun, not usually very well defined, seen towards sunset at the same altitude as the Sun and 22° from it on each side. They are portions of the 22° ice *halo* formed by ice crystals which, for some reason, are arranged with their axes vertical. Also *parhelia*.

MOCVD (*Electronics*) Abbrev for METALLO-ORGANIC CHEMICAL VAPOUR DEPOSITION.

modacrylic fibre (*Textiles*) Fibres made from synthetic linear polymers containing 35–85% of acrylonitrile groups.

modafinil (*Pharmacol*) A stimulant drug that enhances wakefulness and vigilance, used in the treatment of narcolepsy.

modal fibre (*Textiles*) See POLYNOSIC FIBRES.

modal interval (*MathSci*) The class interval corresponding to the largest frequency in a tabulation of observations into class intervals.

modality (*BioSci*) A category of sensation, eg touch, smell or sight.

modal position (*ICT*) The position assumed by a telephone handset when its earpiece is in close contact with the ear of a person whose head has dimensions of modal value, itself derived from a representative sample.

modal value (*MathSci*) See MODE.

mode (*Geol*) The actual mineral composition of a rock expressed quantitatively in percentages. Cf NORM.

mode (*ICT*) In optical fibres, the manner in which light rays travel inside the fibre. There are a variety of paths because the light can be reflected internally at a variety of angles. See MONOMODE FIBRE, MULTIMODE FIBRE.

mode (*MathSci*) The most frequent value in a set of observations; the value of a random variable at which the corresponding probability density function is a maximum. Also *modal value*. Compare MEAN, MEDIAN.

mode (*Phys*) (1) One of several electromagnetic wave frequencies which a given oscillator may generate, or to which a given resonator may respond, eg magnetron modes, tuned line modes. In a waveguide, the mode gives the number of half-period field variations parallel to the transverse axes of the guide. Similarly, for a cavity resonator, the half-period variations parallel to all three axes must be specified. In all cases, different modes will be characterized by different field configurations. (2) Similarly, one of several frequencies of mechanical vibration which a body may execute or with which it may respond to a forcing signal. (3) A well-defined distribution of the radiation amplitude in a cavity which results in the corresponding distribution pattern in the laser output beam. In a multimodal system the beam will tend to diverge.

mode (*Space*) Situation or method of performing a specified task.

mode dispersion (*ICT*) In optical fibre communications, distortion or smearing of individual pulsed components of digital signals, caused by the different modes of propagation of the light inside the fibre arriving at the receiver at different times. Pulses need to be detected and regenerated before this leads to distortion and/or errors.

mode jump (*ICT*) Switch of a MAGNETRON or similar microwave generator from one mode to another. Also *mode shift*.

modelling (*Psych*) In behaviour therapy, the learning of a new behaviour by imitation of a model, usually overtly or covertly reinforcing the desired behaviour. Also *social learning*.

mode-locking (*Phys*) A technique for producing laser pulses of extremely short duration. Laser cavities have modes with frequency spacing $c/2L$ where c is the speed of light and L is the length of the cavity. Oscillations can occur in any mode as long as its frequency is within the natural line width of the laser transition. If a property of the cavity is modulated at a frequency of $c/2L$, then all the modes become coherently coupled. A train of extremely short pulses is emitted where the time duration is roughly the inverse of the line width. See panel on LASER.

modem (*ICT*) Abbrev for *modulator/demodulator*. Device that converts speech and analogue electrical impulses that can be transmitted as a frequency-modulated tone over telephone circuits.

mode number (*Electronics*) A number indicating the mode in which devices capable of operating with more than one field configuration are actually being used: eg in a cavity resonator, the mode numbers indicate the number of half-wavelengths in the field pattern parallel to the three axes; in a magnetron, the mode number gives the number of cycles through which the phase shifts in one circuit of the anode; and in a klystron, it gives the number of cycles of the field which occur while an electron is in the field-free drift space.

moder (*EnvSci*) A form of humus intermediate between MULL and MOR.

moderating ratio (*NucEng*) A figure of merit balancing a moderator's ability to capture neutrons and to slow them down.

moderation (*NucEng*) The slowing down of neutrons in a reactor to thermal energies. *Degradation* is the unintentional slowing of neutrons.

moderator (*NucEng*) Material such as water, heavy water, graphite used to slow down neutrons in a reactor. See LETHARGY OF NEUTRONS and panel on NUCLEAR REACTOR.

moderator control (*NucEng*) The control of a reactor by varying the position or quantity of the moderator.

modern face (*Print*) A style of type with contrasting thick and thin strokes, serifs at right angles, curves thickened, etc. See TYPE.

mode separation (*ICT*) The frequency difference between operation of a microwave tube in adjacent modes. See MODE JUMP.

mode shift (*ICT*) See MODE JUMP.

modification (*BioSci*) See RESTRICTION.

modification (*Eng*) Alteration of the structure or properties of a material by the addition of small quantities of another element, eg an addition of 0·05% sodium drastically changes the form of the aluminium–silicon eutectic structure and improves ductility.

modified atmosphere packing (*FoodSci*) Abbrev *MAP*. See CONTROLLED ATMOSPHERE PACKING.

modified NTSC (*ImageTech*) See NTSC.

modified refractive index (*EnvSci*) The sum of the refractive index of the atmosphere at a given height and the ratio of the height to the radius of the Earth.

modified starch (*FoodSci*) Physical, chemical or enzymic treatment to achieve specific properties, eg pre-gelatinization, a form of physical modification, enables starch to dissolve and gel in cold liquids; cross-linking to supplement hydrogen bonds improves stability to heat, agitation and acidity. See STARCH.

modifier (*BioSci*) A gene that influences the effect of another.

modifier (*ICT*) A code element used to alter the address of an operand. See ADDRESS CALCULATION.

modifier (*MinExt*) Modifying agent used in froth flotation to increase *either* the wettability *or* the water-repelling quality of one or more of the minerals being treated.

modiolus (*BioSci*) The conical central pillar of the cochlea.

MO disk (*ImageTech*) Abbrev for MAGNETO-OPTICAL DISK.

Modula (*ICT*) A development of the PASCAL PROGRAMMING LANGUAGE. Modula 2 programs may be constructed of separately COMPILED modules. Facilities are provided for PARALLEL PROCESSING.

modular (*Electronics*) A form of construction in which units, often with differing functions, are therefore quickly interchangeable.

modular design (*Arch*) A design on a grid of fixed dimensions, generally to facilitate the prefabricated manufacture of building components.

modular programming (*ICT*) An approach to programming in which separate logical tasks are programmed separately and joined later. See SEGMENTATION, STRUCTURED PROGRAMMING.

modular ratio (*CivEng*) The ratio between Young's modulus for steel and that for the concrete in any given case of reinforced concrete.

modulated amplifier (*ICT*) An amplifier stage during which modulation of the signal is carried out. Also *modulated stage*.

modulated carrier (*ICT*) A frequency that can be transmitted or received through space or a transmission circuit, with a superposed information signal, that, by itself, could not be effectively transmitted and received. Also *modulated wave*.

modulated continuous wave (*ICT*) Transmission in which a carrier is modulated by a tone and interrupted by keying. Abbrev *MCW*.

modulated stage (*ICT*) See MODULATED AMPLIFIER.

modulated wave (*ICT*) See MODULATED CARRIER.

modulating electrode (*Electronics*) That of a thermionic valve to which a voltage is applied to control the size of the beam current.

modulation (*Acous*) (1) Changing from one key to another in music. (2) The continual change from one fundamental frequency to another in speech.

modulation (*ICT*) The process of impressing information (code, speech, video, data, etc) onto a higher-frequency CARRIER. See AMPLITUDE MODULATION, DELTA MODULATION, DUAL MODULATION, FREQUENCY MODULATION, FREQUENCY-SHIFT TRANSMISSION, PHASE MODULATION, PULSE-AMPLITUDE MODULATION, PULSE-CODE MODULATION, PULSE-WIDTH MODULATION.

modulation capability (*ICT*) The maximum percentage modulation that can be used without exceeding a specified distortion level.

modulation condition (*ICT*) The condition of voltages and currents in an amplifier for a modulated signal when the carrier is steadily modulated to a stated degree, eg 100%.

modulation depth (*ICT*) A factor indicating the extent of amplitude modulation of a carrier. The difference divided by the sum of peak and trough values of a modulated wave. Often expressed as a percentage.

modulation distortion (*ICT*) Any distortion in the transmission of a signal introduced during the process of modulating that signal onto the carrier.

modulation frequency (*ICT*) One impressed upon a carrier wave in a modulator.

modulation index (*ICT*) In a FREQUENCY-MODULATION system and in the case of a simple sinusoidal modulating signal, the ratio of the frequency deviation to the frequency of the modulating signal.

modulation pattern (*ICT*) That on an oscilloscope when the amplitude-modulated wave is connected to the Y-deflection system and the modulation signal to the X-deflection plates. The result is a trapezoidal pattern that enables the modulation depth to be measured.

modulation rate (*ICT*) The reciprocal of the shortest time interval between successive significant instances of the modulating signal. If the interval is in seconds, the modulation rate is given in BAUD.

modulation transformer (*ICT*) One that applies the modulating signal to the carrier-wave amplifier in a transmitter.

modulator (*Acous*) Circuit in an electronic organ which changes the pitch of the notes.

modulator (*ICT*) Any circuit unit that modulates a radio carrier, at high level directly for transmission, or at low level for amplification of the modulated carrier before transmission.

module (*Arch*) The radius of the lower end of the shaft of a column.

module (*Build*) Unit of size used in the standardized planning of buildings and design of components.

module (*Eng*) Of a gear wheel, the pitch diameter divided by the number of teeth. The reciprocal of DIAMETRICAL PITCH.

module (*ICT*) (1) A section of CODE in the HIGH-LEVEL LANGUAGE MODULAR 2 capable of separate compilation. (2) A HARDWARE component capable of a discrete function.

module (*MathSci*) A *left module M* over a ring *R* is a set of elements with an additional operation defined under which they form a commutative group, and a multiplication defined on pairs $r \times m$, where r is in R and m in M, such that multiplication is associative and distributive over addition. A *right module* is defined analogously.

module (*Space*) A separate, and separable, compartment of a space vehicle.

modulus (*Eng*) Various moduli determine the deflection of a material under stress. Each is the ratio of stress to strain, which is a constant for a given material up to the elastic limit stress. See BULK MODULUS, ELASTICITY.

modulus (*Genrl*) Constant for conversion units between systems.

modulus (*MathSci*) (1) The absolute value of a number (say x), regardless of the sign; written $|x|$ or $\mathrm{mod}\,x$, eg $|3| = |-3| = 3$. (2) For a complex number $z = x + iy$, the value

$$|z| = +\sqrt{x^2 + y^2}$$

Also *absolute value*. See AMPLITUDE, COMPLEX NUMBER.

modulus of compression (*Phys*) See BULK MODULUS.

modulus of elasticity (*Phys*) See BULK MODULUS, COMPLEX MODULUS, ELASTIC CONSTANT, POISSON'S RATIO, SHEAR MODULUS, TENSILE MODULUS, YOUNG'S MODULUS.

modulus of rigidity (*Eng*) See ELASTICITY OF SHEAR.

modulus of rupture (*Eng*) A measure of the ultimate strength of the breaking load per unit area of a specimen, as determined from a torsion or a bending test.

Moebius process (*Eng*) An electrolytic process for parting gold–silver bullion. The electrolyte is silver nitrate. Bullion forms the anode, silver passes into solution and is deposited on the cathode. Gold remains on the anode.

moellon (*Build*) A rubble filling between the facing walls of a structure, sometimes laid in mortar.

Moerner's test (*Chem*) A test for the presence of tyrosine, based on the appearance of a green colour when the solution is heated with a mixture of formalin and sulphuric acid.

mofette (*Geol*) A volcanic opening through which emanations of carbon dioxide, nitrogen and oxygen pass. It marks the last phase of volcanic activity.

mogas (*Aero*) Abbrev for *motor gasoline*, 91–93 octane.

mohair (*Textiles*) The long fine hair from the angora goat, *Capra hircus aegagrus*.

Moho (*Geol*) Abbrev for MOHOROVIČIĆ DISCONTINUITY.

Mohorovičić discontinuity (*Geol*) The boundary, at which there is a marked change in seismic velocity, separating the Earth's crust above from the mantle below. Abbrev *Moho*. See panel on EARTH.

Mohr balance (*Chem*) A balance used to determine density by weighing a solid when suspended in air and in a liquid.

Mohr–Coulomb theory (*MinExt*) The resistance of rock to crushing is due to internal friction *plus* cohesion of bonding materials.

Mohr's salt (*Chem*) *Ammonium iron (II) sulphate*. $(NH_4)_2SO_4FeSO_4.6H_2O$.

Mohs' scale (*Min*) A scale of hardness based on series of minerals, each of which scratches the one beneath it on the scale. See panel on HARDNESS MEASUREMENTS.

moil (*Glass*) (1) Glass left on a *punty* or BLOWING IRON after the gather has been cut off or after a piece of ware has been blown and severed. (2) Glass originally in contact with the blowing mechanism or head, which becomes CULLET after the desired article is severed from it.

Moinian (*Geol*) This group, with various names such as 'Moine Schists', consists of Late Precambrian metamorphosed sediments, and forms much of the N Highlands of Scotland.

moiré effect (*ImageTech*) Pattern formed by interference or combination between two sets of regular divisions, for instance between two line screens in printing or between the TV RASTER and a striped object within the scene.

moiré fibre (*Textiles*) A ribbed fabric in which the yarns have been partially flattened by heat and pressure in calendering. This gives rise to the optical interference effect commonly known as *watered silk*.

moiré fringe (*Phys*) A set of dark fringes produced when two ruled gratings or uniform patterns are superimposed. The separation D of the moiré fringes is equal to d/θ where d is the line spacing of the grating and θ is the angle of intersection of the gratings. Moiré fringes can be used to measure the displacement of one ruled pattern with respect to the other to a high degree of precision.

moiré pattern (*ICT, Print*) A regular patterned effect formed by superimposing two or more sets of lines or dots of different pitch, or at certain angles; a defect to be avoided, esp in half-tone reproduction and in half-tone four-colour process work. In computer graphics it occurs when a half-tone image is scanned using half-tone format or when such an image is scaled (changed in size) in an APPLICATION PROGRAM after it is scanned.

moisture content (*Textiles*) See REGAIN.

moisture expansion (*Build, CivEng*) The increase in the volume of a material from absorption of moisture. Also *bulking*.

mol (*Chem*) See MOLE.

molality (*Chem*) The concentration of a solution expressed as the number of moles of dissolved substance per kilogram of solvent.

molal specific heat capacity (*Phys*) The SPECIFIC HEAT CAPACITY of 1 mole of an element or compound. Also *volumetric heat* (for gases).

molar absorbance (*Chem*) The ABSORBANCE of a solution with a concentration of $1\ \mathrm{mol\,dm^{-3}}$ measured in a cell of a thickness of 1 cm.

molar conductance (*Chem*) The conductance which a solution would have if measured in a cell large enough to contain 1 mole of solute between electrodes 1 cm apart.

molar conductivity (*Phys*) The ratio of the electrical conductivity of an electrolyte to its concentration in moles of per unit volume. Measured in S cm^2 mol.

molar gas constant (*Phys*) See GAS CONSTANT.

molar heat (*Chem*) See MOLAR HEAT CAPACITY.

molar heat capacity (*Chem*) The heat required to raise the temperature of a substance by 1 K. The symbol for that measured at constant volume is C_V and for that at constant pressure C_P.

molarity (*Chem*) The concentration of a solution expressed as the number of moles of dissolved substance per dm^3 of solution.

molars (*BioSci*) In mammals, the posterior grinding or cheek teeth that are not represented in the milk dentition.

molar surface energy (*Chem*) The surface energy of a sphere containing 1 mole of liquid; equal to $\gamma V^{2/3}$, where γ is the surface tension and V the molar volume. It is zero near the critical point and its temperature coefficient is often a colligative property. See EÖTVÖS EQUATION.

molar volume (*Chem*) The volume occupied by 1 mole of a substance under specified conditions. That of an IDEAL GAS at stp is $2 \cdot 2414 \times 10^{-2}$ m^3 mol^{-1}.

molasse (*Geol*) Sediments produced by erosion of mountain ranges following the final stages of an *orogeny*. Sandstones and other detrital rocks are the dominant products of this type of sedimentation. Cf FLYSCH.

molasses (*Chem*) Residual sugar syrups from which no crystalline sugar can be obtained by simple means. An important raw material of ethyl and other alcohols.

moldavite (*Min*) A type of TEKTITE.

Moldcool (*Eng*) TN for CAD package which helps the design of injection-moulding tools, using eg cooling analysis of mould cavity.

Moldflow (*Eng*) TN for computer analysis of polymer flow within runner, gate and cavity of injection-moulding tool.

molding (*Eng*) See MOULDING.

mole (*Chem*) The amount of substance that contains as many entities (atoms, molecules, ions, electrons, photons, etc) as there are atoms in 12 g of ^{12}C. It replaces in SI the older terms *gram-atom*, *gram-molecule*, etc, and for any chemical compound will correspond to a mass equal to the relative molecular mass in grams. Abbrev *mol*. See AVOGADRO'S NUMBER.

mole (*CivEng*) A breakwater or masonry pier.

mole (*Med*) (1) See NAEVUS. (2) A haemorrhagic mass formed in the Fallopian tube as a result of bleeding into the sac enclosing the embryo.

molecular (*Chem*) Pertaining to a molecule or molecules.

molecular association (*Chem*) The relatively loose binding together of the molecules of a liquid or vapour in groups of two or more.

molecular beam (*Chem*) Directed stream of un-ionized molecules issuing from a source and depending only on their thermal energy.

molecular beam epitaxy (*Electronics*) The deposition of molecular species on a crystalline substrate by effusing a molecular beam from an oven containing the pure material. See panel on SEMICONDUCTOR FABRICATION.

molecular biology (*BioSci*) The study of the structure and function of macromolecules in living cells. Notably successful in explaining the structure of proteins, and the role of DNA as the genetic material, but with the ultimate aim of explaining the biology of cells and organisms in molecular terms. It is not primarily concerned with metabolic pathways or with the chemistry of natural products.

molecular cloud (*Astron*) A very-large-scale dense cloud of interstellar gas containing a significant proportion of molecular material.

molecular diffusion (*NucEng*) The process used in gaseous diffusion plants to separate molecules of gas with slightly different molecular weights by forcing them through very small holes.

molecular distillation (*Chem*) The distillation, in a vacuum, of a labile material in which the temperature and time of heating are minimized by providing a condensation surface at such a distance from the evaporating surface that molecules can reach the condenser without intermolecular collision.

molecular electronics (*Electronics*) The technique of growing solid-state crystals so as to form transistors, diodes, resistors, in a single mass, for microelectronic devices.

molecular elevation of boiling point (*Chem*) The rise in the boiling point of a liquid which would be produced by the dissolution of 1 mol of a substance in 1 kg of the solvent, if the same laws held as in dilute solution.

molecular engineering (*Chem*) A term applied to chemical methods of manipulating molecules to achieve a specific effect, esp tailoring monomers and polymers with new and unusual properties. See panel on HIGH-PERFORMANCE POLYMERS.

molecular filter (*Chem*) See MEMBRANE FILTER.

molecular formula (*Chem*) A representation of the atomic composition of a molecule. When no structure is indicated, atoms are usually given in the order C, H, other elements alphabetically. Functional groups may be written separately, thus sulphanilamide may be written $C_6H_8N_2O_2S$ or $H_2NC_6H_4SO_2NH_2$. Cf STRUCTURAL FORMULA.

molecular genetics (*BioSci*) The study and manipulation of the molecular basis of heredity.

molecular heat (*Chem*) See MOLAR HEAT CAPACITY.

molecular mass distribution (*Chem*) Important structural determinant of polymer properties, characterized by specific mass averages. They include the number-average molecular mass ($\overline{M}_n$) and the weight-average molecular mass ($\overline{M}_w$) as well as the simple ratio between the two, the dispersion (D), which characterizes the breadth of the distribution:

$$\overline{M}_n = \frac{\sum_i n_i M_i}{\sum_i n_i}; \quad \overline{M}_w = \frac{\sum_i w_i M_i}{\sum_i w_i} = \frac{\sum_i n_i M_i^2}{\sum_i n_i M_i}; \quad D = \frac{\overline{M}_w}{\overline{M}_n}$$

where n_i is the number of molecules of molecular mass M_i. The weight of molecules of mass M_i is w_i and is the product ($n_i M_i$). Absolute values of M_n and M_w can be determined by osmometry and light scattering respectively, but most values are now computed from the molecular mass distribution itself, which is determined directly by GEL-PERMEATION CHROMATOGRAPHY. Abbrev *MMD*.

molecular models (*Crystal*) Three-dimensional models of the structures of many molecules including complex molecules like proteins and DNA have been used as an aid in both determining their structure and understanding their function. *Space filling models* use truncated spheres of diameters corresponding to the atomic radius which can be fitted together to give the proper bond angles. *Stick models* show only the positions of a special repeating feature in the structure such as the *alpha carbon* of a peptide. These positions can be represented by a marker on a rod (*stick*) which can be set to the appropriate three-dimensional co-ordinates. These mechanical models are becoming redundant as computer modelling has become more widely available.

molecular orbital (*Chem*) A wavefunction defining the energy of an electron in a molecule. Molecular orbitals are often constructed from linear combinations of the atomic orbitals of the constituent atoms.

molecular refraction (*Chem*) See LORENTZ–LORENZ EQUATION.

molecular rotation (*Chem*) One-hundredth of the product of the specific rotation and the relative molecular mass of an optically active compound.

molecular sieve (*Chem*) Framework compound, usually a synthetic zeolite, used to absorb or separate molecules. The molecules are trapped in 'cages', the sizes of which can be selected to suit solvent. See CLATHRATE.

molecular stopping power (*Phys*) The energy loss per molecule per unit area normal to the motion of the particle in travelling unit distance. It is approximately equal to the sum of the atomic stopping powers of the constituent atoms.

molecular streaming (*ChemEng*) See KNUDSEN FLOW.

molecular structure (*Chem*) The way in which atoms are linked together in a molecule.

molecular volume (*Chem*) The volume occupied by 1 mol of a substance in gaseous form at atp (approx $2 \cdot 2414 \times 10^{-2} \, \text{m}^3$).

molecular weight (*Chem*) More formally *relative molecular mass*. (1) The mass of a molecule of a substance referred to that of an atom of ^{12}C taken as 12·000. (2) The sum of the relative atomic masses of the constituent atoms of a molecule.

molecule (*Chem*) An atom or a finite group of atoms which is capable of independent existence and has properties characteristic of the substance of which it is the unit. Molecular substances are those which have discrete molecules, such as water, benzene or haemoglobin. Diamond, sodium chloride and zeolites are examples of non-molecular substances.

mole drain (*Agri*) A drain created in a stable clay soil by pulling a metal object, or *mole*, through it to leave a subsoil drain. These can remain effective for many years.

mole fraction (*Chem*) The fraction, of the total number of molecules in a phase, represented by a given component. Symbol x_i.

moler (*Build*) A diatomaceous earth used in the manufacture of lightweight temperature-resistant components.

moleskin (*Textiles*) A heavy fustian type of cotton fabric, with smooth face and twill back; traditionally used for working clothes.

Mollier diagram (*ChemEng*) A diagram, for any substance relating total heat and entropy, in which heat and work quantities are represented by line segments, not areas. Used to calculate the efficiency of steam engines or refrigeration cycles.

mollites ossium (*Med*) See OSTEOMALACIA.

Mollusca (*BioSci*) Unsegmented coelomate invertebrates with a head (usually well developed), a ventral muscular foot and a dorsal visceral hump. The skin over the visceral hump (the mantle) often secretes a largely calcareous shell and encloses a mantle cavity into which open the anus and kidneys, and in which are the CTENIDIA, originally used for gaseous exchange. There is usually a RADULA. Some have haemocyanin as a respiratory pigment, and the blood and nervous systems are well developed. The larva is often of the TROCHOPHORE type. Includes the ammonites, belemnites, bivalves, cephalopods and gastropods, all of which have made important contributions to the fossil record, eg chitons, slugs, snails, mussels, whelks, limpets, squids, cuttlefish and octopods.

molybdates (VI) (*Chem*) Salts of molybdic acid.

molybdenite (*Min*) Disulphide of molybdenum, crystallizing in the hexagonal system. It is the most common ore of molybdenum, and occurs in lustrous lead-grey crystals in small amounts in granites and associated rocks.

molybdenosis (*Vet*) Teart. A disease of cattle and sheep caused by an excess of molybdenum in the diet; characterized by chronic diarrhoea and emaciation. Occurs where high levels of molybdenum are present in the soil.

molybdenum (*Chem*) Symbol Mo, at no 42, ram 95·94, rel.d. at 20°C 10·2, hardness 147 (Brinell), mp 2625°C, bp 3200°C, resistivity approx 5×10^{-8} Ω m. A metallic element in the sixth group of the periodic system. Its main ore is *molybdenite*, MoS_2, a lead-grey mineral. It is also obtained from *wulfenite*, $PbMoO_4$. Its physical properties are similar to those of iron, its chemical properties to those of a non-metal. Used in the form of wire for filament supports, hooks, etc, in electric lamps and radio valves, for electrodes of mercury-vapour lamps, and for winding electric resistance furnaces. It is added to a number of types of alloy steels, certain types of Permalloy and Stellite. It seals well to Pyrex, spot-welds to iron and steel. Molybdenum is an essential trace element in plants and animals.

molybdenum blues (*Chem*) A variety of reducing agents (including *molybdenum* (III)) convert molybdates into colloidal molybdenum blues, the compositions of which approach $Mo_8O_{23}.xH_2O$.

molybdenum (VI) oxide (*Chem*) MoO_3. Behaves as an acid anhydride, forming molybdic acid. The essential starting point in the manufacture of molybdenum metal and most other compounds of molybdenum.

molybdite (*Min*) Strictly, molybdenum oxide. Much so-called molybdite is ferrimolybdite, a hydrated ferric molybdate which crystallizes in the orthorhombic system. It is commonly impure and occurs in small amounts as an oxidation product of molybdenite. Also *molybdic ochre*.

moment (*Aero*) See HINGE MOMENT, PITCHING MOMENT, ROLLING MOMENT, YAWING MOMENT.

moment (*ElecEng*) See ELECTRIC MOMENT.

moment (*MathSci*) The expected value of the nth power of a random variable, where n is an integer.

moment (*Phys*) (1) See COUPLE. (2) Of a force or vector about a point, the product of the force or vector and the perpendicular distance of the point from its line of action. In vector notation, R × F, where R is the position vector of the point, and F is the force or vector. (3) Of a force or vector about a line, the product of the component of the force or vector parallel to the line and its perpendicular distance from the line.

moment distribution (*CivEng*) A method of analysing the forces in a continuous structure by adjusting between the imposed loads, spans and sectional properties of the supporting members concerned.

moment of a magnet (*Phys*) See MAGNETIC MOMENT.

moment of force (*Phys*) See TORQUE.

moment of inertia (*Phys*) Of a body about an axis: the sum $\sum mr^2$ taken over all particles of the body where m is the mass of a particle and r its perpendicular distance from the specified axis. When expressed in the form Mk^2, where M is the total mass of the body ($M = \sum m$), k is called the *radius of gyration* about the specified axis. Also used erroneously for *second moment of area*.

moment to change trim one inch (*Ships*) The moment, taken about the *centre of flotation*, which will change the *trim* by 1 in. Expressed in foot-tons.

momentum (*Phys*) A dynamical quantity, conserved within a closed system. A body of mass M and whose centre of gravity G has a velocity v has a *linear momentum* of Mv. It has an *angular momentum* about a point O defined as the moment of the linear momentum about O. About G this reduces to $I\omega$ where I is the moment of inertia about G and ω the angular velocity of the body.

momentum wheel (*Space*) A flywheel, part of an attitude control system, which stores momentum by spinning; three wheels with their axes at right angles can serve to stabilize a satellite's attitude. Also *inertia wheel*.

mometasone furoate (*Pharmacol*) A potent topical steroid used for treatment of psoriasis, allergic dermatitis and allergic rhinitis.

mon-, mono- (*Chem*) Containing one atom, group, etc, eg monobasic.

mon-, mono- (*Genrl*) Prefixes from Gk *monos*, alone, single.

monacid (*Chem*) Containing one hydroxyl group, replaceable by an acid radical, with the formation of a salt.

monad (*BioSci*) (1) A flagellated unicellular organism. (2) A single pollen grain, not united with others. Cf TETRAD.

monadelphous (*BioSci*) Having all the stamens in the flower joined together by their filaments, eg many Leguminosae.

monandrous (*BioSci*) (1) Having one antheridium. (2) Having one stamen.

monarch (*BioSci*) Having a single strand of protoxylem in the stele.

monaural (*Acous*) Pertaining to use of one ear instead of two (cf BINAURAL). In sound recording, the (inaccurate) opposite of STEREOPHONIC. See MONOPHONIC.

monazite (*Min*) Monoclinic phosphate of the rare earth metals, $CePO_4$, containing cerium as the principal metallic constituent, and also some thorium. Monazite is exploited from beach sands, where it may be relatively abundant; one of the principal sources of rare earths and thorium.

monchiquite (*Geol*) An alkaline lamprophyre with phenocrysts of olivine, pyroxene and usually mica or amphibole, in a glassy groundmass.

Mönckeberg's sclerosis (*Med*) Degeneration of the middle coat of medium-sized arteries in old people, characterized by the deposit of calcium. Also *Mönckeberg's degeneration*.

Monday morning disease (*Vet*) See SPORADIC LYMPHANGITIS.

Mond gas (*Chem*) The gas produced by passing air and a large excess of steam over coal slack at about 650°C. See SEMIWATER GAS.

Mond process (*Eng*) A process used by Mond Nickel Co. in extracting nickel from a matte consisting of copper–nickel sulphides. The matte is roasted to obtain oxides, the copper is leached out with H_2SO_4, the nickel oxide is reduced to nickel with hydrogen, then the nickel is caused to combine with CO to form a carbonyl, which is decomposed by heating.

Monel metal (*Eng*) A nickel-base alloy containing nickel 68%, copper 29% and iron, manganese, silicon and carbon 3%. Has high strength (about $500\ MN\ m^{-2}$), good elongation (about 45%), and high resistance to corrosion. Used for condenser tubes, propellers, pump fittings, turbine blades, and for chemical and food-handling plant.

Monera (*BioSci*) The KINGDOM that contains all prokaryotic organisms (bacteria and cyanobacteria) in the Five Kingdom scheme, although in some schemes the Eubacteria remain in the Monera and the Archaea are elevated to kingdom status.

mongolism (*Med*) Deprecated term for DOWN'S SYNDROME.

mongrel (*BioSci*) The offspring of a cross between varieties or races of a species.

moniliasis (*Med, Vet*) Infection of the mucosa of the mouth and other parts of the digestive tract of birds and animals by yeast-like organisms of the genus *Monilia*. Also *candidiasis, thrush*.

monimostyly (*BioSci*) In vertebrates, the condition of having the QUADRATE immovably united to the SQUAMOSAL. Cf STREPTOSTYLY.

monitor (*ICT*) (1) A supervisory program that controls sequencing and TIME-SHARING procedures in a computer system. Also *executive program, supervisor program*. See OPERATING SYSTEM. (2) An arrangement for reproducing and checking any transmission without interfering with the regular transmission. (3) Casual term for a VIDEO DISPLAY UNIT.

monitor (*ImageTech*) A video display screen for critical picture presentation, not usually provided with radio-frequency reception circuits to act as a TV receiver.

monitor (*MinExt*) (1) In hydraulic mining, high-pressure jet of water used to break down loosely consolidated ground in opencast work. Also *giant*. (2) In ore treatment, a device which checks part of the process and sounds a signal, makes a record, or initiates compensating adjustment if the detail monitored requires it.

monitor (*Radiol*) Ionization chamber or other radiation detector arranged to give a continuous indication of intensity of radiation, as in radiation laboratories, radiation protection from fallout contamination, industrial operations or X-ray exposure.

monitoring (*Radiol*) Periodic or continuous determination of the amount of specified substances, eg toxic materials or radioactive contamination, present in a region or a person or of a flow rate, pressure, etc, in a (continuous) process, as safety measures.

monitoring loudspeaker (*Acous*) Loudspeaker of high quality and exactly matching others, for verifying quality of programme transmission before radiation or recording.

monitoring receiver (*ICT*) See CHECK RECEIVER.

monitoring station (*ICT*) National service for verifying frequencies of various transmitters within their allocated frequency bands; also for listening to international broadcasts and other traffic for the purposes of compiling bulletins and other traffic for the purposes of compiling bulletins and for security.

monitor position (*ICT*) The special position at any exchange at which operators deal with queries and complaints.

monk (*Print*) An area of printing with too much ink. See FRIAR.

monk bond (*Build*) A modification of the *Flemish bond*, each course consisting of two stretchers and one header alternately. Also *flying bond*.

monkey tail (*Build*) A vertical scroll at one end of a hand rail.

monkey-tail bolt (*Build*) A long-handled bolt for the top of a door, capable of operation from the floor.

mono (*Acous*) Abbrev for MONOPHONIC.

mono- (*Chem, Genrl*) See MON-.

monoamine oxidase (*BioSci*) An enzyme involved in the inactivation of catecholamine neurotransmitters. Inhibitors of monoamine oxidase are used as antidepressants.

monoamine-oxidase inhibitors (*Pharmacol*) A class of drugs used to treat very severe depressive illness. Little used because of their dangerous interactions with foods containing tyramine and with SYMPATHOMIMETICS.

monobasic (*Chem*) Containing one hydrogen atom replaceable by a metal with the formation of a salt.

monobath (*ImageTech*) A solution which develops and fixes an emulsion in one stage.

monobloc (*Autos*) The integral casting of all the cylinders of an engine, ie in the same cylinder block.

monocable (*CivEng*) The type of AERIAL ROPEWAY in which a single endless rope is used both to support and to move the loads.

monocardian (*BioSci*) Having a completely undivided heart.

monocarpellary (*BioSci*) Having, or consisting of, a single carpel.

monocarpic (*BioSci*) Dying at the end of its first fruiting season, as do annuals, biennials and some perennials. Cf POLYCARPIC. See HAPAXANTHIC.

Monoceros (Unicorn) (*Astron*) A faint constellation in the Milky Way.

monocerous (*BioSci*) Having a single horn.

monochasium (*BioSci*) An inflorescence in which the main stem ends in a flower and bears, below the flower, a lateral branch which itself ends in a flower. This branch may in turn bear further similar branches. Also *cymose inflorescence, monochasial cyme*. Cf DICHASIUM.

monochlamydeous (*BioSci*) Having a perianth of one whorl of members.

monochlamydeous chimera (*BioSci*) A periclinal chimera in which one component is present, at least at the shoot apex, as a single superficial layer of cells.

monochord (*Acous*) Primitive apparatus with a string and wooden resonator for demonstrating the properties of single stretched wires and the sounds generated thereby.

monochromatic (*Phys*) By extension from MONOCHROMATIC LIGHT, any form of oscillation or radiation characterized by a unique or very narrow band of frequency.

monochromatic filter (*ImageTech*) A filter which transmits light of a single wavelength, or, in practice, a very narrow band of wavelengths.

monochromatic light (*Phys*) Light containing radiation of a single wavelength only. No source emits truly monochromatic light, but a very narrow band of wavelengths can be obtained, eg the cadmium red spectral line, wavelength 643·8 nm with a HALF-WIDTH of 0·0013 nm. Light from some lasers has extremely narrow line width. Also *homogeneous light*.

monochromatic radiation (*Phys*) Electromagnetic radiation (originally visible) of one single-frequency component. By extension, a beam of particulate radiation comprising particles all of the same type and energy. *Homogeneous* or *monoenergic* is preferable in this sense.

monochromator (*Phys*) A device for converting heterogeneous radiation (electromagnetic or particulate) into a homogeneous beam by absorption, refraction or diffraction processes.

monochrome (*ImageTech*) Picture reproduction in one colour only, but generally used to describe a black-and-white image.

monochrome receiver (*ImageTech*) TV receiver which reproduces a black-and-white transmission, or a colour transmission in black and white. See COMPATIBLE COLOUR TV.

monochrome signal (*ImageTech*) That part of a TV signal controlling luminance only, not including the CHROMINANCE SIGNAL. In compatible systems, this signal provides the black-and-white picture.

monoclimax theory (*BioSci*) Idea, proposed by F E Clements in 1916, that all successional sequences in a region lead to a single climax vegetation of a type determined by the climate. Cf POLYCLIMAX THEORY.

monocline (*Geol*) A fold with one limb which dips steeply; the beds, however, soon approximate to horizontality on either side of this flexure.

monoclinic sulphur (*Chem*) See SULPHUR.

monoclinic system (*Crystal*) The style of crystal architecture in which the three crystal axes are of unequal lengths, having one of their intersections oblique and the other two at right angles. Also *oblique system*. See fig. at BRAVAIS LATTICES.

monoclinous (*BioSci*) Hermaphrodite flowers. Cf DICLINOUS.

monoclonal (*BioSci*) A description of a cell line, whether within the body or in culture, that has a single clonal origin. See MONOCLONAL ANTIBODY.

monoclonal antibody (*BioSci*) Antibody produced by a single clone of cells or a cell line derived from a single cell. Such antibodies are all identical and have unique amino acid sequences. Commonly used to refer to antibody secreted by a hybridoma cell line, but can also refer to the immunoglobulin produced *in vivo* by a B-cell clone such as a plasmacytoma if this has identifiable antibody properties. Abbrev mAb.

monocolpate (*BioSci*) A pollen grain having one elongated aperture in its wall (colpus).

monocoque (*Aero, Eng*) A structure in which all structural loads are carried by the skin. In a semimonocoque, loads are shared between skin and framework, which provides local reinforcement for openings, mountings, etc. See STRESSED SKIN CONSTRUCTION.

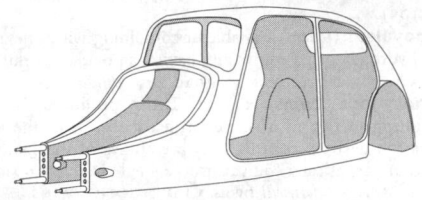

monocoque Citroën body shell of 1934.

monocoque (*Autos*) See CHASSIS.

Monocotyledones (*BioSci*) Monocotyledons, or monocots, the smaller of the two classes of angiosperm or flowering plants. These are mostly herbs of which the embryo has characteristically one cotyledon, the parts of the flower are in threes or sixes and the leaves often have the main veins parallel. Very few are woody, fewer still have secondary thickening. The class contains c.55 000 spp in 60 families, divided among four subclasses or superorders: Alismatidae, Arecidae, Commelinidae and Liliidae. Also *Liliopsida*. Cf

DICOTYLEDONES. See panel on ANGIOSPERMS (FLOWERING PLANTS).

monocotyledonous (*BioSci*) (1) Belonging to the MONOCOTYLEDONES. (2) An embryo or seed having one cotyledon.

monocule (*BioSci*) An animal possessing a single eye, as the water flea *Daphnia*. Adj *monocular*.

monoculture (*BioSci*) A culture, crop or plantation with only one species.

monocyclic (*BioSci*) Stamens or other floral parts in a single whorl.

monocyte (*BioSci*) A large motile phagocytic cell with an indented nucleus present in normal blood, where it is the blood representative of the MONONUCLEAR PHAGOCYTE SYSTEM. Monocytes are derived from promonocytes in the bone marrow. They remain in the blood for only a short time and then migrate into the tissues where they become macrophages.

monodactylous (*BioSci*) Having only a single digit.

monodisperse polymer (*Chem*) Polymer in which all chains are of equal length. Can be made by anionic polymerization.

monodisperse system (*Chem*) A colloidal dispersion having particles all of effectively the same size.

monodont (*BioSci*) Having a single persistent tooth, as the male narwhal.

monoecious (*BioSci*) (1) Flowering plants having separate male and female flowers on each individual plant. (2) Moss gametophytes, algae, etc, producing male and female gametes on the same individual. See HERMAPHRODITE.

monoenergic (*Phys*) See MONOCHROMATIC RADIATION.

monoestrous (*BioSci*) Exhibiting only one estrous cycle during the breeding season. Cf POLYESTROUS.

monofilament (*Plastics*) A single filament of indefinite length. Used for ropes, surgical sutures, etc, and, in finer gauge, for textiles.

monogenetic (*BioSci*) Multiplying by asexual reproduction; showing a direct life history; of parasites, having a single host.

monogenic function (*MathSci*) See ANALYTIC FUNCTION.

monogerm (*BioSci*) Varieties of sugar beet etc in which each small dry fruit, sown like a seed, produces a single seedling rather than the group of seedlings typically produced by such fruits.

monogony (*BioSci*) Asexual reproduction.

monohydric alcohols (*Chem*) Alcohols containing one hydroxyl group only.

monoid (*MathSci*) A semigroup with identity; equivalently, category with only one object.

monolayer (*Chem*) See MONOMOLECULAR LAYER.

monolayer culture (*BioSci*) A TISSUE CULTURE technique in which thin sheets of cells are grown, on glass or plastic, in a nutrient medium.

monolete (*BioSci*) A pollen grain or other spore with a single linear scar or aperture.

monolithic integrated circuit (*Electronics*) Electronic circuit formed by diffusion or ion implantation on a single crystal of semiconductor, usually silicon.

monomer (*Chem*) Small molecule with high chemical reactivity, capable of linking up with itself to produce polymers, or with similar molecules to make copolymers. See COPOLYMER and panel on POLYMERS.

monominerallic rocks (*Geol*) Rocks consisting essentially of one mineral, eg dunite and anorthosite.

monomode fibre (*ICT*) A STEPPED-INDEX optical fibre where the diameter of the inner core (of higher refractive index) is comparable with the wavelength of light; this results in there being only one mode of light propagation, so MODE DISPERSION is eliminated.

monomolecular layer (*Chem*) A film of a substance one molecule in thickness.

monomolecular reaction (*Chem*) A reaction in which only one species is involved in forming the activated complex of the reaction.

monomorphic (*BioSci*) Showing little change of form during its life history; structures with the same form or appearance. Cf POLYMORPHIC.

monomorphous (*Crystal*) Existing in only one crystalline form.

mononuclear phagocyte system (*BioSci*) A classification of phagocytic cells of which the typical mature form is the macrophage. They are all derived from bone marrow promonocytes. Their appearance and functional capacity differ according to the tissue in which they are situated. Macrophages in the alveoli of the lung, the peritoneal cavity, and moving in tissues and in the lymph are free to migrate. Others such as Kupffer cells, tissue histiocytes, osteoclasts and astroglia remain *in situ* for long periods of time.

monopack (*ImageTech*) See INTEGRAL TRIPACK.

monophagous (*BioSci*) Feeding on one kind of food only, eg Sporozoa, living always in the same cell, or phytophagous insects, with only one food plant. Also *monotrophic*.

monophasic (*BioSci*) Having an abbreviated life cycle, without a free active stage; said of certain trypanosomes. Cf DIPHASIC.

monophonic (*Acous*) Single-channel sound reproduction, recreating the acoustic source, as compared with *stereophonic* which uses two or more identical channels for auditory perspective. See MONAURAL.

Monophoto (*Print*) A FILMSETTING system developed from the Monotype hot-metal system, the caster being replaced by a photographic unit.

monophyletic (*BioSci*) Of a group of species that are descended from a single ancestral species. See POLYPHYLY.

monophyletic group (*BioSci*) (1) Generally, a group that is descended from a common ancestor. (2) In cladistics, a group comprising a common ancestor and all of its descendants. Cf PARAPHYLETIC GROUP, POLYPHYLETIC GROUP.

monophyodont (*BioSci*) Having only a single set of teeth, the permanent dentition. Cf POLYPHYODONT.

monopitch building (*Arch*) A simple building, often for farm use, where the roof slopes across the shorter dimension with access where the roof is highest.

monoplane (*Aero*) A heavier-than-air aircraft, either an aircraft or glider, having one main supporting surface.

monoplegia (*Med*) Paralysis of an arm or of a leg.

monoploid (*BioSci*) True HAPLOID.

monopodial (*BioSci*) Having a single pseudopod.

monopodial growth (*BioSci*) A pattern of growth in which a shoot continues to grow indefinitely and bears lateral shoots that behave similarly, eg in pines and other conifers. Also *indefinite growth, indeterminate growth*. Cf SYMPODIAL GROWTH. See RACEMOSE INFLORESCENCE.

monopod platform (*MinExt*) Drilling or production rig with one central leg, used in arctic conditions because of risk of ice damage to conventional designs.

monopole (*Acous*) Spherical radiator whose surface moves inwards and outwards with the same phase and amplitude everywhere. Any sound source which produces an equivalent sound field is also a monopole, eg any small source generating volume flow.

monopropellant (*Space*) Single propellant which produces propulsive energy as the result of a chemical reaction, usually induced by the presence of a catalyst. Cf BIPROPELLANT.

monopulse (*Radar*) A radar system with an antenna system with two or more overlapping lobes in its radiation pattern. From the transmission of a single pulse and analysis of error signals due to the target being off-axis in one or more lobes, detailed information about direction can be obtained. Used in many gun-control and missile guidance systems.

monorail (*CivEng*) A railway system in which carriages are suspended from, and run along, a single continuous elevated rail.

monorail (*ImageTech*) Said of technical cameras having the focusing movement of the bellows accommodated on a single rail, obviating the need for a baseboard.

monosaccharides (*Chem*) The simplest group of carbohydrates, classified into tetroses, pentoses, hexoses, etc, according to the number of carbon atoms, and into aldoses and ketoses depending on whether they contain a potential aldehyde or ketone group.

monosodium glutamate (*Chem, FoodSci*) Sodium hydrogen glutamate. $HOOCCH(NH_2)CH_2CH_2COONa$. A derivative of glutamic acid, used as condiment or flavouring. Found naturally at quite high levels in cheese, tomatoes and mushrooms and as a product of human metabolism. Once extracted from the seaweed *Laminaria japonica* but now obtained as a by-product of commercial fermentation. It is a white crystalline substance, soluble in water.

monosome (*BioSci*) The unpaired accessory or X chromosome. See SEX DETERMINATION.

monosomy (*BioSci*) A condition in which a particular chromosome is represented once only in an otherwise diploid complement; a sort of ANEUPLOIDY.

mono-spaced font (*ICT*) A FONT in which each character has the same width, eg Courier, a typewriter-like font. Cf PROPORTIONAL FONT.

monospermy (*BioSci*) Fertilization of an ovum by a single spermatozoon.

monosporous (*BioSci*) (1) Containing one spore. (2) Derived from one spore.

monostable (*Electronics*) Of a circuit or system, fully stable in one state only but metastable in another to which it can be driven for a fixed period by an input pulse.

monostichous (*BioSci*) Forming one row.

monosymmetric system (*Crystal*) See MONOCLINIC SYSTEM.

monotocous (*BioSci*) Producing a single offspring at a birth.

monotonic (*ICT*) Said of transient response when it increases continuously with time.

monotonic (*MathSci*) Of a function or sequence, either never increasing or never decreasing.

Monotremata (*BioSci*) The single order included in the mammalian group Prototheria and having the characteristics of the subclass, eg spiny ant-eater (*Tachyglossus*), duck-billed platypus (*Ornithorhynchus*).

monotrophic (*BioSci*) See MONOPHAGOUS.

monotropic (*Chem*) Existing in only one stable crystalline form, the other forms being unstable under all conditions.

Monotype (*Print*) See COMPOSING MACHINES.

monotypic (*BioSci*) A taxonomic group containing only one subordinate member, a family with a single genus or a genus with a single species.

monounsaturated (*Chem*) Containing only one double or triple bond per molecule (cf POLYUNSATURATED FATTY ACIDS).

monovalent (*Chem*) Capable of combining with one atom of hydrogen or its equivalent, having an oxidation number or co-ordination number of one. Also *univalent*.

monozygotic twins (*BioSci*) Twins produced by the splitting in two of a single fertilized egg, or of the early embryo derived from it. They are always of the same sex and are genetically equivalent to a single individual (ie are clones). Also *identical twins*. Cf DIZYGOTIC TWINS.

Monro's foramen (*BioSci*) A narrow canal connecting the first or second ventricle with the third ventricle in the brain of vertebrates.

monsoon (*EnvSci*) Originally winds prevailing in the Indian Ocean, which blow SW from April to October and NE from October to April; now generally winds which blow in opposite directions at different seasons of the year. Similar in origin to land and sea breezes, but on a much larger scale, in both space and time. Particularly well developed over S and E Asia, where the wet summer monsoon from the SW is the outstanding feature of the climate.

mons pubis (*Med*) A convex formation of subcutaneous fatty tissue over the pubic symphysis.

monster (*BioSci*) An abnormal form of a species.

montage (*ImageTech*) (1) A film sequence containing rapidly changing multiple images, often superimposed and dissolving together, to convey an integrated visual impression. (2) Composite photograph made from the juxtaposition of cut-up photographs arranged in a pattern.

montasite (*Min*) An asbestiform variety of the amphibole grunerite. It differs from AMOSITE in having less harsh and more silky fibres.

montebrasite (*Min*) A variety of AMBLYGONITE in which the amount of hydroxyl exceeds that of fluorine.

Monte Carlo method (*MathSci*) A procedure employed to obtain numerical solutions to mathematical problems by means of random sampling.

montelukast (*Pharmacol*) A leukotriene receptor antagonist used to treat asthma.

month (*Astron*) See ANOMALISTIC MONTH, SIDEREAL MONTH, SYNODIC MONTH, TROPICAL MONTH.

monticellite (*Min*) A silicate of calcium and magnesium which crystallizes in the orthorhombic system. It occurs in metamorphosed dolomitic limestones and, more rarely, in some ultrabasic igneous rocks.

montmorillonite (*Min*) A hydrated silicate of aluminium, one of the important clay minerals and the chief constituent of bentonite and fuller's earth. The montmorillonite groups of clay minerals are also collectively termed *smectites*.

monzodiorite (*Geol*) A plutonic rock intermediate between monzonite and diorite. Also *syenodiorite*. See PLUTONIC ROCKS.

monzogabbro (*Geol*) A plutonic rock intermediate between monzonite and gabbro. Also *syenogabbro*. See PLUTONIC ROCKS.

monzogranite (*Geol*) A variety of granite with roughly equal amounts of alkali feldspar and plagioclase. See ADAMELLITE, PLUTONIC ROCKS.

monzonite (*Min*) A coarse-grained igneous rock of intermediate composition, characterized by approximately equal amounts of orthoclase and plagioclase (near andesine in composition) together with coloured silicates in variety. Named from Monzoni in the Tyrol. Also *syenodiorite*. See PLUTONIC ROCKS.

monzonorite (*Min*) A norite containing some orthoclase.

Moon (*Astron*) The Earth's only natural satellite and closest neighbour in space. Its average distance is 382 000 km. Its diameter of 3476 km is a quarter that of Earth. The relative density is also lower, 3·3. Consequently the Moon has a mass only 0·012 that of the Earth. The Moon orbits the Earth in 27·32 days (relative to the stars), passing through the familiar cycle of *lunar phases*. The interval between successive new Moons is 29·53 days, the *lunar month*. The Moon's rotation is in resonance with its orbit, so the same face is always presented to the Earth apart from small effects of LIBRATION. Even to the unaided eye the Moon has a range of features, including darker LUNAR MARIA and lighter mountainous regions. The whole surface is heavily cratered; in parts the density is so great they overlap; larger ones have a central mountain. All are believed to have been created by impact with METEORITES. There is no water, atmosphere or magnetic field. The 386 kg of rock and dust retrieved in the Apollo missions have shown that the age of the Moon is around 4650 million years. There is a thick crust 30–100 km thick overlaying a solid mantle.

Mooney viscometer (*Eng*) A type of viscometer used widely in the rubber industry to obtain cure curves by measurement of torque needed to shear vulcanizing system.

moonstone (*Min*) A variety of alkali feldspar or sometimes plagioclase, which possesses a bluish pearly opalescence attributed to lamellar micro- or crypto-perthitic intergrowth. It is used as a gemstone.

moor (*EnvSci*) An area above the height of enclosed or improved land, typically used for sheep grazing. Vegetation composed of ericaceous plants and certain grasses, growing on acid soil.

Moore's law (*ICT*) The law formulated by Gordon E Moore (1929–), originally proposed in 1965, that states that the number of transistors that it is possible to fit onto a cheap silicon chip will grow constantly at an exponential rate, doubling every 12 (later revised to 24) months.

mooring tower (*Aero*) A permanent tower or mast for the mooring of airships. Provided with facilities for the transference of passengers and freight, and arrangements for replenishing ballast, gas and fuel.

moorland (*EnvSci*) See MOOR.

mopboard (*Build*) See SKIRTING BOARD.

mop-stick hand rail (*Build*) A timber hand rail having a circular section flattened on the under side.

moquette (*Textiles*) A heavy warp pile fabric used for upholstery. The pile may be cut or left as uncut loops.

mor (*EnvSci*) Crumbly layer of partially decomposed plant-like litter occurring under some coniferous trees and woody shrubs, in which the pH is acid and the soil fauna poor. See MODER, MULL.

moraine (*Geol*) Material laid down by moving ice. Moraines are found in all areas which have been glaciated, and are of several types. *Terminal moraines* are irregular ridges of material marking the farthest extent of the ice and representing debris pushed along in front of the ice. *Lateral moraines* are found along the sides of present-day and former glaciers. *Medial moraines* form by the combination of the two lateral moraines when two glaciers join. *Ground moraine* is an irregular sheet of TILL laid down beneath an ice sheet.

morbid (*Med*) Diseased: pertaining to, or of the nature of, disease.

morbidity (*Med*) The state of being diseased: the sick rate in a community.

morbilli (*Med*) See MEASLES.

mordant (*Build*) A preparation applied to surfaces to improve the adherence of paint or gold leaf.

mordant (*Chem, Textiles*) A compound, frequently a metallic salt or oxide, applied to a fabric to form a stable complex with a dyestuff. The product formed by the action of a dye on a mordant is called a *lake*. The complex has superior fastness on the fabric compared with the dyestuff itself.

mordanting (*ImageTech*) Adding to a silver image, or replacing a silver image, by a substance having the requisite affinity for a specified dye, which silver in itself has not.

mordenite (*Min*) A hydrated sodium, potassium, calcium, aluminium silicate. A zeolite crystallizing in the orthorhombic system and occurring in amygdales in igneous rocks and as a hydration product of volcanic glass.

more hug (*Print*) An increase in the wrap of the paper round the cylinder(s) of web-fed presses.

Morgagni's ventricle (*BioSci*) In the higher mammals, a paired pocket of the larynx, anterior to the vocal cords, and acting in the Anthropoidea as a resonator.

morganite (*Min*) A pink or rose-coloured variety of beryl, used as a gemstone.

morion (*Min*) A variety of smoky quartz which is almost black in colour.

morning star (*Astron*) Popularly, a planet, usually Venus and sometimes Mercury, seen in the eastern sky at or about sunrise; also, loosely, any planet which transits after midnight.

morph (*BioSci*) A specific form or shape of an organism, singled out for attention.

morph-, morpho-, -morph (*Genrl*) Prefixes and suffix from Gk *morphe*, form.

morphactins (*BioSci*) A group of synthetic compounds, based on fluorene-carboxylic acid, that have a variety of effects, mostly inhibitory, on plant growth and development.

morphallaxis (*BioSci*) A change of form during regeneration of parts, as the development of an antenna in certain Crustacea to replace an eye; gradual growth or development.

morphea (*Med*) US for MORPHOEA.

morpheme (*Psych*) The minimal linguistic unit that has meaning. Some are free-standing; others, eg prefixes, can only be used in combination.

morphine (*Pharmacol*) The principal alkaloid present in opium. Characterized by containing a phenanthrene nucleus in addition to a nitrogen ring. Extensively used as a hypnotic to obtain relief from pain.

morphing (*ImageTech*) A process in which the image of one object is smoothly changed into that of another by a computer, using the particular features and outlines of each as references by which to guide the process. From Gk *morphe*, form.

morpho- (*Genrl*) See MORPH-.

morphoea (*Med*) A disease in which thickened patches appear on the skin of the trunk. US *morphea*. Also *localized scleroderma*.

morphogenesis (*BioSci*) The origin and development of a part, organ or organism. Adj *morphogenetic*.

morphogenetic movements (*BioSci*) Movements of cells or of groups of cells in the course of development, eg the invagination of cells in gastrulation and the migration of neural crest cells.

morpholine (*Chem*) C_4H_9NO. Bp 128°C. A six-membered heterocyclic compound. It is a liquid, with strong basic properties. Miscible with water and used as a solvent in organic reactions.

morphology (*BioSci*) (1) The study of the structure and forms of organisms, as opposed to the study of their functions. (2) By extension, the nature of a member. Adj *morphological*.

morphology (*Chem, Eng*) A general term for the study of shapes of microstructural units in materials, such as spherulites in polymers and minerals, grain structure in metals and ceramics.

morphosis (*BioSci*) The development of structural characteristics; tissue formation. Adj *morphotic*.

morphotropic (*Crystal*) Refers to the effect on crystal structure of atomic substitutions.

Morquio–Brailsford disease (*Med*) A hereditary disease characterized by dwarfism, kyphosis and skeletal defects in the hip joint.

Morse code (*ICT*) A system used in signalling or telegraphy, which consists of various combinations of dots and dashes.

Morse equation (*Phys*) An equation which relates the potential energy of a diatomic molecule to the internuclear distance.

Morse key (*ICT*) A hand-operated device that opens and closes contacts which modulate currents with coded telegraph signals. See BUG KEY.

Morse taper (*Eng*) A system of matching tapered shanks and sockets used for holding tools in lathes, drills, etc. There are six sizes, with tapers approx 3°.

mortality (*EnvSci*) The death of individuals in a population. Cf NATALITY.

mortar (*Build, CivEng*) A pasty substance formed normally by the mixing of cement, sand and water, or cement, lime-sand and water in varying proportions. Used normally for the binding of brickwork or masonry. It is THIXOTROPIC and its working life can be extended by addition of soap solutions. Hardens on setting and forms the bond between the bricks or stones.

mortar (*Chem*) A bowl, made of porcelain, glass or agate, in which solids are ground up with a pestle.

mortar board (*Build*) See HAWK.

mortar mill (*Build*) A shallow pan in which ingredients are crushed mechanically by two rollers running on the ends of a horizontal bar rotating about a central axis and bearing on the base of the pan. Also *pan-mill mixer*.

mortar structure (*Geol*) A cataclastic structure resulting from dynamic metamorphism in which small grains produced by granulation occupy the interstices between larger grains.

mortice (*Build*) See MORTISE.

mortise (*Build*) A rectangular hole cut in one member of a framework to receive a corresponding projection on the mating member. Also *mortice*.

mortise bolt (*Build*) A bolt which is housed in a mortise in a door so as to be flush with its edge. Also *mortice bolt*.

mortise chisel (*Build*) A more robust type of chisel than the FIRMER, for use in cutting mortises. Also *framing chisel*, *heading chisel*, *mortice chisel*.

mortise gauge (*Build*) A tool similar to the MARKING GAUGE but having an additional marking pin, which is adjustable for position along the bar and allows parallel lines to be set out in marking tenons and mortises. Also *counter gauge*, *mortice gauge*.

mortise lock (*Build*) A lock sunk into a mortise in the edge of a door. Also *mortice lock*.

mortuary fat (*Med*) See ADIPOCERE.

morula (*BioSci*) A solid spherical mass of cells resulting from the cleavage of an ovum.

MOS (*Electronics*) Abbrev for METAL–OXIDE–SILICON.

mosaic (*BioSci*) (1) The disposition in three dimensions of the leaves of a shoot or a plant, resulting from phyllotaxis and leaf morphology, to maximize interception of light while minimizing mutual shading. (2) A patchy variation of the normal green colour of leaves; usually the result of infection by a virus. (3) Any multicellular organism displaying the effects of different alleles or genes in different parts of the body, but derived from a single embryo. This may arise from the mutation of an unstable gene, random inactivation of the X chromosome in female mammals (as in tortoiseshell cats) or the sorting out of dissimilar plastids. Cf CHIMERA.

mosaic (*Build*) Inlaid work on plaster or stone, formed with small cubes, tesserae, or irregular-shaped fragments of marble, glazed pottery or glass.

mosaic (*Chem*) The structure of crystalline solids due to dislocations, consisting of an irregular matrix or mosaic of otherwise perfect crystallites.

mosaic (*ImageTech*) (1) See PHOTO-MOSAIC. (2) Video effect in which the picture is represented by a number of small square or rectangular elements. Also *pixillation, tile*.

mosaic development (*BioSci*) Development when the eventual fate of cells of the developing embryo is determined, eg when the blastomeres are formed in the spirally segmented eggs of some invertebrates. A mosaic egg is thus one whose areas of future functional development are determined before the very earliest cleavages.

mosaic filter (*ImageTech*) A mosaic of microscopic colour filters (cyan, yellow, magenta and green) attached one to each PIXEL on a SOLID-STATE IMAGE SENSOR to provide colour video signals from a single sensor.

mosaic gold (*Chem*) Complex tin (IV) sulphide obtained by heating dry tin amalgam, ammonium chloride and sulphur in a retort. Sometimes used as a pigment.

mosaicism (*Med*) A condition descriptive of individuals who possess cell populations with mixed numbers of chromosomes, eg some cells have 45, others 47, making a mosaic of 45/47.

mosaic screen (*ImageTech*) A screen made up of a continuous pattern of very small dots or lines of the three primary colours, red, green and blue, used in early forms of additive colour photography, either as a separate layer or forming an integral part of the support for a black-and-white emulsion.

mosaic structure (*Eng*) Discontinuous structure of a compound or metal consisting of minute domains, each bounded by its discontinuity lattice at the interface with other domains of like composition.

Moscicki capacitor (*ElecEng*) A capacitor on the principle of the Leyden jar; sometimes used on transmission lines to act as a protective device against effects of high-frequency surges.

Moscovian (*Geol*) An epoch of the Pennsylvanian period.

Mortality

Major causes of mortality

Twenty conditions account for more than half of the 57 million deaths that occur annually throughout the world.

Almost one-quarter of deaths are due to heart disease and stroke, accounting for over 8 million deaths a year. Many of these deaths occur in relatively young people, and can be prevented. Circulatory diseases are an emerging major public health problem in most developing countries as Western lifestyles and associated risk factors of raised blood pressure, smoking, raised blood cholesterol, unhealthy diet, overweight and lack of physical exercise are adopted. Mortality rates in developed countries are currently declining, partly because of improved medical treatment but also because of improved prevention and health education. Cancer is responsible for more than 7 million deaths a year and more than twice this number of people worldwide will develop cancer each year. Lung cancer is the most common cancer death and is rising in some countries, especially in women. Breast cancer is the commonest cancer in women, but has been overtaken by lung cancer in some Western countries. Infectious diseases of all kinds account for one in five of all deaths and HIV/AIDS is now commoner than tuberculosis or malaria, accounting for 1 in 20 of all deaths worldwide. The numbers of people with diabetes are expected to increase more than twofold within the next 20 years, the majority of which will be non-insulin-dependent (Type II), typically a disease of adults. Diabetes is badly recorded as a cause of death in its own right and has many long-term complications including heart disease, kidney failure, blindness and peripheral vascular disease. Alzheimer's disease is expected to affect an increasing proportion of the population in the future and is also associated with increasing age. Road traffic accidents are a significant cause of a loss of useful life in developed countries, and can be prevented by non-medical interventions such as speed restrictions and traffic-calming measures.

		World mortality, 2002	
		Number	% of all deaths
	All causes	57 029 155	100.0
1	Tuberculosis	1 566 003	2.7
2	HIV/AIDS	2 777 175	4.9
3	Diarrhoeal diseases	1 797 972	3.2
4	Measles	610 818	1.1
5	Malaria	1 272 393	2.2
6	Respiratory infections	3 962 893	6.9
7	Maternal conditions	510 262	0.9
8	Oesophagus cancer	446 166	0.8
9	Trachea, bronchus, lung cancers	1 243 199	2.2
10	Breast cancer	477 196	0.8
11	Prostate cancer	269 292	0.5
12	Lymphomas, multiple myeloma	334 421	0.6
13	Diabetes mellitus	987 816	1.7
14	Alcohol use disorders	90 746	0.2
15	Alzheimer and other dementias	396 576	0.7
16	Hypertensive heart disease	911 397	1.6
17	Ischaemic heart disease	7 207 725	12.6
18	Cerebro-vascular disease	5 508 950	9.7
19	Road traffic accidents	1 191 796	2.1
20	Self-inflicted injuries	873 361	1.5
	Selected causes	32 436 157	56.9

Source: WHO Global Burden of Disease project

MoSCoW (*ICT*) A prioritized list of system requirements commonly used in DSDM. The requirements are ordered according to whether they M(ust), S(hould), C(ould) or W(on't) be delivered in the current increment.

Moseley's law (*Phys*) A law stating that for one of the series of characteristic lines in the X-ray spectra of atoms, the square root of the frequency of a line is directly proportional to the atomic number of the element plus a constant. This result stresses the importance of atomic number rather than atomic weight.

MOSFET (*Electronics*) Abbrev for METAL–OXIDE–SILICON FIELD-EFFECT TRANSISTOR.

mosquito (*BioSci*) The family Culicidae of the Diptera; have piercing proboscies and suck blood, often transmitting diseases (eg malaria, yellow fever, elephantiasis) while doing so; larvae and pupae aquatic, eg *Aedes*, *Anopheles*, *Culex*.

moss (*BioSci*) See BRYOPSIDA.

moss agate (*Min*) A variegated cryptocrystalline silica containing visible impurities, as manganese oxide, in moss-like or dendritic form. Also *Mocha stone*.

Mössbauer effect (*Phys*) The effect occurring when an atomic nucleus emits a gamma-ray photon. The nucleus must recoil to conserve linear momentum, and consequently there is a change of frequency of the radiation due to the movement of the source (the DOPPLER EFFECT). If the atom is firmly bound in a crystal lattice so that it may not recoil, the momentum is taken up by the whole lattice; an effect much used in the study of the structure of solids.

most economical range (*Aero*) The range obtainable when an aircraft is flown at the height, airspeed and engine conditions which give the lowest fuel consumption for the aircraft weight and the wind conditions prevailing.

most significant bit (*ICT*) The bit with the greatest place value in a word. Abbrev *MSB*.

mother (*Acous*) Metal positive which is made from the master in vinyl disk record manufacture.

motherboard (*ICT*) Printed circuit board that holds the principal components in a microcomputer or personal computer. It carries slots on its bus for the attachment of ADAPTER CARDS.

mother cell (*BioSci*) A cell that divides to give daughter cells; the term is applied particularly to cells that divide to give spores, pollen grains, gametes and blood corpuscles.

mother liquor (*Chem*) The solution remaining after a solute has been crystallized out.

mother of emerald (*Min*) A variety of *prase*, a leek-green quartz owing its colour to included fibres of actinolite; thought at one time to be the mother-rock of emerald.

mother of pearl (*BioSci*) The iridescent nacreous material from the shells of molluscs, used as a gemstone and for small ornamental objects. See BIVALVES.

mother of the chapel (*Print*) A woman elected by the associated employees of a printing department to represent them and to safeguard their interests.

mother set (*Print*) A set of printing plates (eg of a standard reference work) kept solely for the purpose of electrotyping or stereotyping further sets. Not for printing.

motile (*Psych*) In studies of imagery, somebody whose imagery naturally takes the form of feelings of action.

motional impedance (*ICT*) In an electromechanical transducer, eg a telephone receiver or relay, that part of the input electrical impedance due to the motion of the mechanism; the difference between the input electrical impedance when the mechanical system is allowed to oscillate and the same impedance when the mechanical system is stopped from moving, or blocked.

motion bars (*Eng*) See GUIDE BARS.

motion compensation (*MinExt*) Automatic machinery which maintains a constant downhole pressure on the drill bit bored from a floating platform.

motion parallax (*Psych*) Cues for detecting distance based on systematic movements in the visual field as the observer changes position.

motion picture camera (*ImageTech*) A camera for the exposure of cinematograph film, intermittently, one frame at a time, to record movement.

motion work (*Eng*) The set of gears within a clock that transfers drive from the ESCAPEMENT to the hands.

motivation (*Psych*) As a general term, mostly with reference to human behaviour, the desire to act in certain ways in order to achieve a goal; these may be transitory impulses or more persistent intentions.

motogen (*BioSci*) A general term for a molecule that stimulates cell motility.

motoneuron (*BioSci*) A motor neuron.

motor (*BioSci*) Pertaining to movement; as nerves that convey movement-initiating impulses to the muscles from the central nervous system.

motor (*Eng*) A machine used to transform power from some other form into mechanical motion.

motor areas (*BioSci*) Nerve centres of the brain concerned with the initiation and co-ordination of movement.

motor-boating (*ICT*) Very-low-frequency relaxation oscillation in an amplifier, arising from inadequate decoupling of common sources of current supply.

motor bogie (*ElecEng*) A bogie or truck on a railway locomotive or motor-coach which carries one or more electric motors.

motor cell (*BioSci*) One of a number of cells that together can expand or contract and so cause movement in a plant member.

motor end plates (*BioSci*) The special end organ in which a motor nerve terminates adjacent to a striated muscle.

motor generator (*ElecEng*) A motor connected to a supply of one voltage, frequency or number of phases, and driving a generator giving a different kind of output supply.

motor habits (*Psych*) Repetitive, non-functional patterns of motor behaviour that seem to occur in response to stress (eg nail biting, thumb sucking). See STEREOTYPING.

motorized fuel valve (*Eng*) See ADJUSTABLE-PORT PROPORTIONING VALVE.

motor meter (*ElecEng*) An integrating meter embodying a motor whose speed is proportional to the power flowing in the circuit to which it is connected, so that the number of revolutions made by the spindle is proportional to the energy consumed by the circuit. See INDUCTION METER.

motor-operated switch (circuit breaker) (*ElecEng*) A large switch or circuit breaker which is closed by means of an electric motor. Cf *solenoid-operated switch*.

motor starter (*ElecEng*) A device for operating the necessary circuits for starting and accelerating to full speed an electric motor, but not for controlling its speed when running. See CONTROLLER.

motor system (*BioSci*) Tissues and structures concerned in the movements of plant members, as in PULVINI.

motte (*Arch*) The steep mound, natural or constructed, on which medieval castles were often built within an open fortified space, or BAILEY; thus the term *motte and bailey*.

mottled iron (*Eng*) Cast-iron in which most of the carbon is combined with iron in the form of cementite (Fe_3C) but in which there is also a small amount of graphite. The fracture has a white crystalline appearance with clusters of dark spots, indicating the presence of graphite.

mottler (*Build*) Small short-haired brush used in graining to create the effect of mottling etc.

mottle yarn (*Textiles*) See MARL YARN.

mottramite (*Min*) Descloizite in which the zinc is almost entirely replaced by copper.

MOU (*Genrl*) Abbrev for MEMORANDUM OF UNDERSTANDING.

mould (*BioSci*) A fungus, esp one that produces a visible mycelium on the surface of its host or substrate. Mould is sometimes visible as powdery or thread-like deposits (hyphae), often coloured, on the surface of foods. It is readily destroyed at temperatures > 91°C. Most moulds spoil food but some help develop the flavour of cheeses. See MYCOTOXIN. US *mold*.

mould (*Build*) Zinc sheet or thin board cut to a given profile; used in running cornices etc. US *mold*.

mould (*Geol*) The impression of an original shape. Usually refers to fossils, but may be minerals or sedimentary structures. US *mold*. Cf CAST.

mould (*Print*) An impression of the type or blocks made in flong or plastic, and used for making stereotypes, electrotypes, or rubber or plastic duplicates. US *mold*.

mouldability index (*Plastics*) A term referring to the ease or difficulty of melt flow of a polymer. Thus polycarbonate has a high index and is thus difficult to mould, while high-impact polystyrene is easy to mould and has a low index. Temperature dependent.

mould board (*Agri*) See PLOUGHSHARE.

mould breathing (*Plastics*) Injection-moulding technique which allows moulding of products larger than the nominal locking force available, by opening the tool at the end of the cycle. See VERTICAL FLASH TOOL.

mould cycle (*Plastics*) The events during injection moulding from the initial injection of the material to the closing of the mould after removing the product whose cost depends on the length of this cycle.

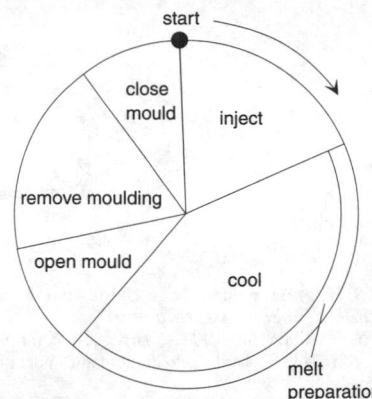

mould cycle

mould defects (*Plastics*) Moulding faults which are minimized by careful tool design and tool setting. They include sink, mica and burn marks, weld and flow lines, and voids.

moulded breadth (*Ships*) The breadth over the frames of a ship, ie heel of frame to heel of frame. It is the breadth termed *B* by Lloyd's Register, and is measured at the widest part of the hull.

moulded depth (*Ships*) The depth of a ship from the top of the keel to the top of the beam at side, measured at mid-length; referred to as Lloyd's *D*.

moulded dimensions (*Ships*) The dimensions used in the calculation and tabulation of the scantlings (sizes) of the various members of the structure. They are LENGTH BETWEEN PERPENDICULARS, MOULDED BREADTH and MOULDED DEPTH.

moulding (*Build*) A more or less ornamental band projecting from the surface of a wall etc.

moulding (*Eng*) Forming a molten material by pouring it into a preformed shape or mould and then allowing it to cool.

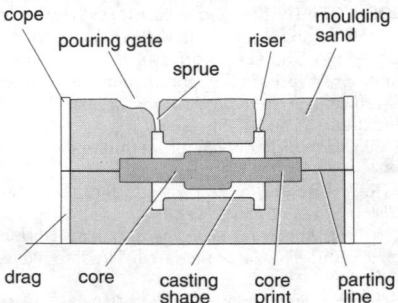

moulding Section through a moulding box for a simple cored casting.

moulding (*Plastics*) The moulding powder is weighed carefully into small containers or preformed in pill machines to pellets. The material is then placed in a steel die, heated to about 160°C, and subjected to pressure of 20–35 MN m^{-2}. Wood moulds are used to make articles from heated thermoplastic sheets. See BLOW MOULDING, COLD MOULDING, COMPRESSION MOULDING, INJECTION MOULDING, ROTATIONAL MOULDING, TRANSFER MOULDING.

moulding box (*Eng*) See FLASK.

moulding cutter (*Build*) A shaped tool usually mounted on a vertical spindle for cutting a moulding profile.

moulding machines (*Eng*) See MACHINE MOULDING.

moulding plane (*Build*) A plane with a shaped sole and blade for cutting mouldings. There are two types: *English* (held at an angle) and *Continental* (held upright).

moulding powder (*Plastics*) The finely ground mixture of binder, accelerator, colouring matter, filler and lubricant which is converted under pressure into the final moulding.

mouldings (*Build*) Strips of wood or plaster with a special cross-sectional profile for decorating surfaces etc.

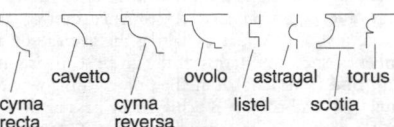

cavetto   ovolo / astragal / torus
cyma   cyma   listel scotia
recta   reversa

mouldings The classic mouldings in their roman form with circular curves.

moulding sands (*Eng*) Siliceous sands (containing clay or aluminium silicate as a binding agent) possessing naturally or by blending the qualities of fineness, plasticity, adhesiveness, strength, permeability and refractiveness. See DRY SAND, GREEN SAND, LOAM.

mould machine (*Paper*) See CYLINDER MOULD MACHINE.

mould oil (*Build*, *CivEng*) A substance applied to shuttering to prevent adherence of the concrete.

mould parting line (*Plastics*) A line on a moulding left by the junction of the male and female parts of the tool.

mould release agent (*Plastics*) Speciality material often applied in aerosol form to mould surfaces to aid release of polymer product at end of moulding cycle. Usually consists of silicone or fluoropolymer.

mould temperatures (*Plastics*) Temperatures at which injection-moulding tool parts are kept, often below ambient for crystalline polyolefins, but well above ambient for many engineering thermoplastics. Polycarbonate, eg, usually demands mould temperatures above about 80°C to minimize frozen-in strains.

moult (*Agri*) A change of feathers in naturally raised chickens each autumn, triggered by reducing light levels. It can also be triggered by feed stress, and causes hens to stop laying for up to 16 weeks. US *molt*.

moult (*BioSci*) (1) See ECDYSIS. (2) Periodic shedding of hair or feathers as seasons change.

mount (*ICT*) That part of a switching tube or cavity that enables it to be connected to a waveguide.

mount (*ImageTech*) Card or other backing surface to which a photograph is permanently attached for display.

mountain cork (*Min*) A variety of asbestos which consists of thick interlaced fibres. It is light and will float, and is of a white or grey colour.

mountain leather (*Min*) A variety of asbestos which consists of thin flexible sheets made of interlaced fibres.

mountain wood (*Min*) A compact fibrous variety of asbestos looking like dry wood.

mountants (*ImageTech*) Special adhesives for fixing prints on mounts, free from chemicals which might attack the silver or other image during the course of time.

mounted sprayer (*Agri*) Tractor-mounted or self-propelled spray equipment, usually of greater capacity than trailed equipment.

mouse (*ICT*) Input device, hemispherical in appearance. Moving the device around on a flat surface causes a cursor to move around the display screen. See WIMP.

mouse-mat (*ICT*) A small flat piece of fabric backed with foam rubber used as a surface on which to move a computer mouse. Also *mouse pad*.

mouse roller (*Print*) A small extra roller used to obtain better distribution of the ink on a machine.

mouth (*Acous*) See FLARE.

mouth (*Build*) The slot in the sole of a plane though which the cutting iron projects.

mouth parts (*BioSci*) In Arthropoda, the appendages associated with the mouth.

movable types (*Print*) Single types, as distinguished from Linotype slugs or from blocks.

move (*Glass*) A fixed number of articles to be made for a given rate of pay by a CHAIR.

movement (*Genrl*) The mechanism of a clock or watch, not including the case or dial.

movement area (*Aero*) That part of an airport reserved for the take-off, landing and movement of aircraft.

moving average (*MathSci*) A sequence derived from a given sequence of values by taking the average of a fixed number of successive terms of the given sequence in turn.

moving bed (*ChemEng*) A method of exposing reactants to a catalyst. Good contact is achieved by passing an upward flow of reactant gases through a bed of catalyst moving continuously downwards. Exhausted catalyst is removed from the bottom of the reactor for regeneration. See CATALYSIS.

moving-coil galvanometer (*ElecEng*) A galvanometer depending on the principle of the MOVING-COIL INSTRUMENT. Cf MOVING-MAGNET GALVANOMETER.

moving-coil instrument (*ElecEng*) An electrical measuring instrument depending for its action upon the force on a current-carrying coil in the field of a permanent magnet.

moving-coil loudspeaker (*Acous*) See ELECTRODYNAMIC LOUDSPEAKER.

moving-coil microphone (*Acous*) See ELECTRODYNAMIC MICROPHONE.

moving-coil regulator (*ElecEng*) A type of voltage regulator, for use on ac circuits, in which a short-circuited coil is made to move up and down the iron core of a specially arranged autotransformer.

moving-coil transformer (*ElecEng*) A type of transformer, occasionally used in constant-current systems, in which one coil is made to move relatively to the other for regulating purposes.

moving-coil voltmeter (*ElecEng*) One constructed like a galvanometer and used for dc measurements.

moving-conductor microphone (*Acous*) See RIBBON MICROPHONE.

moving-field therapy (*Radiol*) A form of crossfire technique in which there is relative movement between the beam of radiation and the patient, so that the entry portal of the beam is constantly changing. See CONVERGING-FIELD THERAPY.

moving form (*CivEng*) See CLIMBING FORM.

moving-iron (*ElecEng*) Describes a type of drive once used in microphones and loudspeakers, which depends on the motion of magnetic material which is part of a magnetic circuit.

moving-iron voltmeter (*ElecEng*) One used for ac measurements, depending on attraction of a soft-iron vane into the magnetic field due to current.

moving load (*CivEng*) A variable loading on a structure, consisting of the pedestrians or vehicles passing over it. Also *live load*.

moving-magnet galvanometer (*ElecEng*) A galvanometer depending for its action on the movement of a small permanent magnet in the magnetic field produced by the current to be measured. Cf MOVING-COIL GALVANOMETER.

moving-target indicator (*Radar*) A device for restricting the display of information to moving targets. Abbrev MTI.

moving trihedral (*MathSci*) Three mutually orthogonal unit vectors T, N and B along the tangent, principal normal

and binormal respectively at a point *P* on a space curve. The three planes perpendicular to the three vectors are called respectively the *normal*, *tangent* and *osculating* planes. The triad is determined by two limits, ie (1) the limiting position of a chord *PQ* as *Q* approaches *P*, which determines T, and (2) the limiting position of a plane *PQR* as *Q* and *R* approach *P*, which determines the osculating plane. Cf CURVATURE.

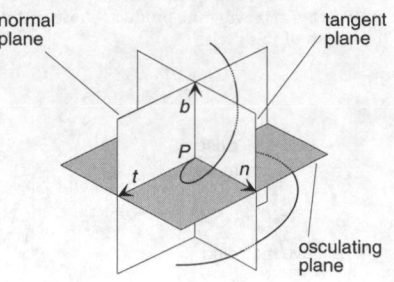

moving trihedral

MOX (*NucEng*) Abbrev for MIXED OXIDE FUEL.

mp (*Chem*) Abbrev for MELTING POINT.

MP3 (*ICT*) Abbrev for *MPEG-1 Layer 3*, a compressed file format that allows fast downloading of audio data from the Internet.

MP3 player (*ICT*) A device for storing and playing audio data that has been recorded using the MP3 format.

Mpc (*Astron*) Abbrev for *megaparsec*. See PARSEC.

MPD (*ElecEng*) Abbrev for *magneto-plasmadynamic*.

MPD (*Radiol*) Abbrev for MAXIMUM PERMISSIBLE DOSE.

MPEG (*ImageTech*) Abbrev for *Motion Picture Experts Group*. A body producing standards for compressed digital video signals. See FULL MOTION VIDEO, MP3.

mph (*Genrl*) Abbrev for *miles per hour*.

MP tape (*ImageTech*) Abbrev for METAL PARTICLE TAPE.

MQ (*ImageTech*) Abbrev for *metol-quinol* or *metol-hydroquinone* developers. See 4-METHYL-AMINOPHENOL.

MRAM (*ICT*) Abbrev for *magnetoresistive random access memory*.

MRI (*Radiol*) Abbrev for MAGNETIC RESONANCE IMAGING.

MRM (*FoodSci*) Abbrev for MECHANICALLY RECOVERED MEAT.

mRNA (*BioSci*) Abbrev for MESSENGER RNA.

MRSA (*Med*) Methicillin-resistant *Staphylococcus aureus*. MRSA infections are an increasing problem, particularly in hospitals, suggesting that bacteria are beginning to win the 'arms race' against antibiotics.

MS (*Aero*) See MIL-SPEC.

MSB (*ICT*) Abbrev for MOST SIGNIFICANT BIT.

MS-DOS (*ICT*) TN for an OPERATING SYSTEM written by Microsoft originally for the IBM PC.

M-shell (*Phys*) The ELECTRON SHELL in an atom corresponding to a principal quantum number of three. The shell can contain up to 18 electrons. See panel on ATOMIC STRUCTURE.

MSI (*ICT, Electronics*) Abbrev for MEDIUM-SCALE INTEGRATION.

MSW (*Radar*) Abbrev for MAGNETIC SURFACE WAVE.

Mt (*Chem*) Symbol for MEITNERIUM.

MTBE (*Chem*) Abbrev for *methyl tertiary-butyl ether*, used as a petrol additive in place of tetra-ethyl lead, to reduce toxic emissions.

MTBF (*Aero, Eng*) Abbrev for MEAN TIME BETWEEN FAILURES.

M-test of Weierstrass (*MathSci*) A test for uniform convergence. Both the series $\sum u_n(x)$ and the product $\prod (1+u_n(x))$ are uniformly convergent in the interval (a,b) if it is possible to find a convergent series of positive constants $\sum M_n$ such that $|u_n(x)| \leqslant M_n$, for all values of x in the interval (a,b).

MTF (*ImageTech*) Abbrev for *modulation transfer function*; a method of expressing the performance of a lens, photographic emulsion or complete image reproduction system by its degree of modulation (image contrast) for varying unit image size (frequency). This gives a better evaluation of subjective SHARPNESS than simple RESOLUTION values.

M-theory (*Astron*) See BRANES.

MTI (*Radar*) Abbrev for MOVING-TARGET INDICATOR.

MTOC (*BioSci*) Abbrev for MICROTUBULE-ORGANIZING CENTRE.

mtoe (*MinExt*) Abbrev for MILLION TONS OIL EQUIVALENT.

mucic acid (*Chem*) HOOC(CHOH)$_4$COOH. An acid obtained by the oxidation of galactose.

mucigen (*BioSci*) A substance occurring as granules or globules in chalice cells and later extruded as mucin.

mucilages (*Chem*) Complex organic compounds related to the polysaccharides, of vegetable origin, and having glue-like properties.

mucilaginous (*BioSci*) Pertaining to, containing or resembling mucilage or mucin.

mucinogen (*BioSci*) A substance producing or being the precursor of MUCINS, eg granules of mucous gland cells.

mucins (*Chem*) A group of glycoproteins occurring in mucus, saliva and other secretions. Widely distributed in nature, the polypeptide chains are densely glycosylated.

muck (*MinExt*) See DIRT.

mucocele (*Med*) A localized accumulation of mucous secretion in eg a hollow organ, the outlet of which is blocked.

mucomembranous colic (*Med*) A condition in which constipation is associated with abdominal pain and the passing in the stools of membranes or casts of mucus.

mucopolysaccharides (*Chem*) Heteropolysaccharides each containing a hexosamine in its characteristic repeating disaccharide unit.

mucoproteins (*Chem*) Conjugated proteins with carbohydrate side chains, which may include hexoses, hexosamines or glucuronic acids.

mucopurulent (*Med*) Consisting of mucus and pus.

mucosa (*BioSci*) See MUCOUS MEMBRANE.

mucosal associated lymphoid tissue (*BioSci*) Organized lymphoid tissue and lymphoid aggregates associated with mucosal surfaces; includes BRONCHUS ASSOCIATED LYMPHOID TISSUE and GUT ASSOCIATED LYMPHOID TISSUE. Abbrev *MALT*.

mucosal disease (*Vet*) Common togavirus infection of cattle, of which acute and subacute signs may be seen. Symptoms include fever, anorexia, diarrhoea, ulceration of the abdominal tract and oculo-nasal discharge. Mortality can be high. Also *bovine viral diarrhoea* (*BVD*).

mucosanguineous (*Med*) Consisting of mucus and blood.

mucous glands (*BioSci*) Glands secreting or producing mucus.

mucous membrane (*BioSci*) A tissue layer found lining various tubular cavities of the body (eg the gut, uterus, trachea, etc). It is composed of a layer of epithelium containing numerous unicellular mucous glands and an underlying layer of areolar and lymphoid tissue, separated by a basement membrane. Also *mucosa*.

muco-viscidosis (*Med*) See CYSTIC FIBROSIS.

mucro (*BioSci*) A short, sharp, terminal point.

mucronate (*BioSci*) Terminated by a short point (mucro).

mucus (*BioSci*) The viscous slimy fluid secreted by the mucous glands. Adjs *muciform, mucoid, mucous*.

mud (*Geol*) A fine-grained unconsolidated rock, of the clay grade, often with a high percentage of water present. It may consist of several minerals.

mud (*MinExt*) See DRILLING MUD.

mud acid (*MinExt*) Inorganic acids and chemicals used to promote flow in oil wells by acid treatment.

mud column (*MinExt*) The column of mud from the top to the bottom of a borehole.

mud drum (*Eng*) A vessel placed at the lowest part of a steam boiler: a similar plant to intercept and retain insoluble matter or sludge, as the lowest drum of a WATER-TUBE BOILER.

mud fever (*Vet*) See GREASE.

mud flow (*Geol*) Mass movement of fine-grained material in a highly fluid state. When abundant coarse material is also carried it is known as a *debris flow*.

mud hole (*Eng*) A hand hole in a mud drum, or in the bottom of a boiler, for the removal of scale and sludge.

mudline (*MinExt*) Separating line between clear overflow water and settled slurry in DEWATERING or thickening plant.

mud motor (*MinExt*) Hydraulic motor situated down-hole and actuated by the pressure of the DRILLING MUD forced down the borehole. Can be used for the main drilling, when the drill pipe does not rotate or to actuate subsidiary drills, reamers, etc. See DRAIN HOLES. See panel on DRILLING RIG.

mud pipe (*MinExt*) An outer casing which is run down through semi-liquid material found during drilling until it rests on firmer strata. It protects the casing proper which contains the DRILLING MUD.

mud pump (*MinExt*) One used in drilling of deep holes (eg oil wells) to force thixotropic mud to bottom of hole and flush out rock chips.

mudstone (*Geol*) An argillaceous sedimentary rock characterized by the absence of obvious stratification. Cf SHALE.

mud volcano (*Geol*) A conical hill formed by the accumulation of fine mud which is emitted together with various gases, from an orifice in the ground. Generally derived from a volcanic hot spring but sedimentary mud volcanoes can also be produced by earthquake activity, which generates the extrusion of liquid mud.

MUF (*ICT*) Abbrev for MAXIMUM USABLE FREQUENCY.

muffle furnace (*Eng*) A furnace in which heat is applied to the outside of a refractory chamber containing the charge.

muffler (*Autos*) US for SILENCER.

mugearite (*Geol*) A dark, finely crystalline, basic igneous rock which contains oligoclase, orthoclase, and usually olivine in greater amount than augite. Occurs typically at Mugeary in Skye.

mulch (*Agric*) In horticulture, matter placed on the soil surface in order eg to suppress the growth of weed seedlings, or to reduce temperature fluctuations at the soil surface; often organic, eg peat, shredded bark; sometimes inorganic, eg pebbles.

mule (*Textiles*) A machine, intermittent in action, which first drafts and twists the yarn on the outward run, then on the inward run winds it on a spindle. It can spin very fine counts but there are also woollen mules for heavy yarns and condenser mules for waste yarns. (Invented by Samuel Crompton, Bolton, c.1770.) Now only used for wool spinning.

mull (*Chem*) A liquid used to prepare specimens for infrared spectroscopy, by crushing and abrasion in a small pestle and MORTAR. See NUJOL.

mull (*EnvSci*) Loose crumbly layer of soil occurring in some deciduous woodlands, in which leaf litter is being broken down rapidly at near-neutral pH in the presence of a rich fauna including earthworms. Cf MODER, MOR.

Mullen burst test (*Paper*) The BURST TEST when performed on a Mullen instrument.

muller (*MinExt*) Pestle, dragstone, iron wearing plate or shoe used to crush and/or abrade rock entrained between it and a base plate over which it is moved.

Müllerian duct (*BioSci*) A duct that arises close beside the oviduct, or which, by the actual longitudinal division of the archinephric duct, in many female vertebrates becomes the oviduct.

Müllerian mimicry (*BioSci*) Resemblance in colour between two animals, both benefiting by the resemblance. Cf BATESIAN MIMICRY.

Müller's glass (*Min*) See HYALITE.

Müller's muscle (*BioSci*) The circular ciliary muscle of the vertebrate eye.

mullion (*Build*) A vertical member of a window frame separating adjacent panes. Also *monial, munnion*.

mullion structure (*Geol*) A linear structure found in severely folded sedimentary and metamorphic rocks in which the harder beds form elongated fluted columns. Mullions are parallel to fold axes.

mullite (*Min*) A silicate of aluminium, rather similar to sillimanite but with formula close to $Al_6Si_2O_{13}$. It occurs in contact-altered argillaceous rocks.

mulluscum contagiosum (*Med*) A contagious condition in which small, white, waxy nodules appear on the skin; believed to be due to infection with a virus.

multi-access (*ICT*) A system that allows several users to have apparently simultaneous access to a computer. Used mainly for INTERACTIVE computing on a TIME-SHARING basis.

multiarticulate (*BioSci*) Many-jointed.

multiaxial (*BioSci*) Having a main axis consisting of several more or less equal files of cells with or without subordinate branches. Cf UNIAXIAL.

multibreak switch (*ElecEng*) A switch in which the circuit is broken at two or more points in series on each pole or phase. Similarly *multibreak circuit breaker*.

multicasting (*ICT*) The transmission of a message in a LOCAL AREA NETWORK from one node to all others.

multicavity magnetron (*ICT*) One with many cavities cut in the inner face of the solid cylindrical anode, with the mouths of the cavities facing the cathode. Sizes of alternate cavities may differ to assist in mode separation. See MAGNETRON.

multicellular (*BioSci*) Consisting of a number of cells.

multicellular voltmeter (*ElecEng*) A form of electrostatic voltmeter in which a number of moving vanes mounted upon the spindle are drawn into the spaces between a corresponding number of fixed vanes.

multi-centred bonding (*Chem*) The bonding characteristic of ELECTRON-DEFICIENT compounds. In such compounds, 'bonds' can only be described that involve three or more atoms at one time.

multichannel (*ICT*) Any system that divides the frequency spectrum of a signal into a number of bands which are separately transmitted, with subsequent recombination.

multichip integrated circuit (*Electronics*) An electronic circuit comprising two or more semiconductor wafers which contain single elements (or simple circuits). These are interconnected to produce a more complex circuit and are encapsulated within a single pack. See panel on PRINTED, HYBRID AND INTEGRATED CIRCUITS.

multicipital (*BioSci*) Many-headed.

multicuspidate (*BioSci*) Teeth with many cusps.

multidimensional database (*ICT*) Often used in OLAP systems to bring together complex arrays of denormalized data in an intermediate repository for ease of analysis by users.

multi-electrode valve (*Electronics*) One comprising two or more complete electrode systems having independent electron streams in the same envelope. There is sometimes a common electrode, eg double-diode.

multienzyme complex (*BioSci*) A cluster of distinct enzymes that catalyse consecutive reactions of a metabolic pathway and that remain physically associated through purification procedures.

multi-exchange system (*ICT*) A group of local exchanges in an exchange area.

multifactorial (*BioSci*) Determined by many genes and non-genetic factors. See POLYGENIC.

multifee metering (*ICT*) Successive operations of a subscriber's meter to correspond with the multifee chargeable for the call.

multifrequency (*ICT*) Said of a signal, mostly used for multifrequency signalling or data transmission, consisting of several superimposed audio tones. According to CCITT specifications, 1 consists of 697 and 1209 Hz, 2 is 697 and 1336 Hz, 3 is 687 and 1477 Hz, 4 is 770 and 1209 Hz, etc.

multifrequency generator (*ICT*) The multifrequency inductor generator that is used for the multichannel voice-frequency telegraph system operated by teleprinter and transmitting over normal telephone lines.

Multigrade printing paper (*ImageTech*) TN for a VARI-ABLE-CONTRAST PRINTING PAPER.

multigravida (*Med*) A woman who has been pregnant more than once.

multigroup theory (*NucEng*) Theoretical reactor model in which presence of several energy groups among the neutrons is taken into account. See DELAYED NEUTRON GROUPS.

multi-impression (*Eng*) A term applied to injection-moulding tool possessing several identical cavities, so increasing productivity over the mould cycle time. Contrast family tools, where the cavities have different shapes.

multilayered structure (*BioSci*) Single broad band of microtubules forming a 'flagellar root' in the motile cells and gametes of the Charophyceae, bryophytes and vascular plants.

multilayer film (*ImageTech*) A photographic material comprising several distinct emulsion coatings and filter layers, esp in colour photography. See TRIPACK.

multilayer winding (*ElecEng*) A type of cylindrical winding, used chiefly for transformers, in which several layers of wire are wound one over the other with layers of insulation between.

multilive feed (*Eng*) A type of injection-moulding process where molten polymer (often with chopped fibres) can be fed at two or more channels into the mould recess. See INJECTION MOULDING.

multilobal (*Textiles*) A 'synthetic' fibre or filament which is extruded so that the cross-section has several rounded lobes. More precisely named according to the number of lobes, eg *trilobal*.

multilocular (*BioSci*) Having a number of compartments or loculi.

multimedia (*ICT*) The simultaneous availability on a communications network, computer system or electronic recording medium of audio, video (still and motion) and data transfer. Multimedia use of a LOCAL AREA NETWORK could involve eg users communicating by voice, showing each other documents and exchanging files, all within a common screen format.

multimode fibre (*ICT*) An optical fibre (usually a GRADED-INDEX one) with a core diameter sufficiently larger than the wavelength of light to allow propagation of light energy in a large number of different modes.

multinet growth (*BioSci*) Steady-state pattern postulated for the growth of the cell walls of some elongating plant cells in which the cellulose microfibrils, deposited more or less transversely on the cytoplasmic surface of the wall, become passively reorientated by the growth of the cell so as to lie more nearly longitudinally in the older, outer parts of the wall. See panel on CELL WALL.

multinucleate (*BioSci*) Having many nuclei. Also *polynucleate*.

multiparous (*BioSci*) (1) Having had two pregnancies resulting in live progeny. (2) Bearing many offspring at a birth.

multipath reception (*ICT*) Signals from the transmitter arriving by two or more paths: one direct, the other(s) by reflection from buildings or other obstructions. Can lead to distortion in radio and ghost images in TV.

multiple (*MathSci*) Any product of a given number with an integer. Formally, if a is an integer then ax is a multiple of x.

multiple-circuit winding (*ElecEng*) See LAP WINDING.

multiple decay (*Phys*) BRANCHING in a radioactive decay series. Also *multiple disintegration*.

multiple-disk clutch (*Eng*) A friction clutch similar in principle to the SINGLE-PLATE CLUTCH, but in which a smaller diameter is obtained by using a number of disks, alternately splined to the driving and driven members, loaded by springs, and usually run in oil. See FRICTION CLUTCH.

multiple duct (*ElecEng*) A cable duct having a number of tunnels for the reception of several cables.

multiple echo (*Acous, ICT*) The perception of a number of distinct repetitions of a signal, because of reflections with different delays of separate waves following various paths between the source and observer. Also *flutter echo*.

multiple-effect evaporation (*ChemEng*) A system of evaporation in which the hot vapour produced in one vessel, instead of being condensed by cooling water, is condensed in the calandria of another evaporator to produce further evaporation. To achieve this, the pressure in the second evaporator must be artificially lowered. Can, according to conditions, work as double, triple or quadruple effect.

multiple-expansion engine (*Eng*) An engine in which the expansion of the steam or other working fluid is divided into two or more stages, which are performed successively in cylinders of increasing size. See COMPOUND ENGINE, QUADRUPLE-EXPANSION ENGINE, TRIPLE-EXPANSION ENGINE.

multiple feeder (*ElecEng*) A feeder consisting of a number of cables connected in parallel: used where a single cable to carry the load would be prohibitively large.

multiple fission (*BioSci*) A method of multiplication found in Protozoa, in which the nucleus divides repeatedly without corresponding division of the cytoplasm, which subsequently divides into an equal number of parts leaving usually a residuum of cytoplasm. Cf BINARY FISSION.

multiple fruit (*BioSci*) A fruit formed from the maturing ovaries of a group of flowers, often with their receptacles and their floral parts, eg pineapple, fig. Also *aggregate fruit*.

multiple-hearth furnace (*Eng*) A type of roasting furnace consisting of a number of hearths (six or more). The charge enters on the top hearth and passes downwards from hearth to hearth, being RABBLED by rotating arms from centre to circumference and circumference to centre of alternate hearths.

multiple-hop transmission (*ICT*) Transmission that uses multiple reflection of the sky wave by the ground and ionosphere.

multiple intrusions (*Geol*) Minor intrusions formed by several successive injections of approximately the same magma.

multiple isomorphous replacement (*BioSci*) A method of solving the *phase problem* in the X-RAY CRYSTALLOGRAPHY of protein crystals. Such a crystal consists of an array of geometrically identical unit cells arranged in three dimensions in which each unit cell contains one or more identical asymmetric units, each containing one or a small number of protein molecules. For the method to work, the isomorphous crystals must have identical geometry and molecular structure, the only difference being the substitution of a heavy atom (platinum or mercury) at a small number of sites in each molecule. When diffraction patterns can be obtained from the unlabelled protein *and* two or more isomorphous derivatives, then the *phases* of the unlabelled crystal can be calculated and, together with the *amplitude* data, the molecular image obtained. Abbrev *MIR*.

multiple modulation (*ICT*) The use of a modulated wave for modulating a further independent carrier of much higher frequency.

multiple neuritis (*Med*) See POLYNEURITIS.

multiple personality disorder (*Psych*) An extreme form of dissociative reaction in which two or more complete personalities, each well developed and distinct, are found in one individual.

multiple point (*MathSci*) A point on the curve $f(x,y)$ at which the first non-zero derivative of f is of order n. Cf DOUBLE POINT.

multiple proportions (*Chem*) See LAW OF MULTIPLE PROPORTIONS.

multiple-retort underfeed stoker (*Eng*) A number of underfed inclined retorts arranged side by side with TUYÈRES between, resulting in a fuel bed the full width of the furnace walls. See SINGLE-RETORT UNDERFEED STOKER.

multiple sclerosis (*Med*) A chronic progressive disease in which patches of thickening occur throughout the central nervous system leading to progressive paralysis. Also *disseminated sclerosis*.

multiple screws (*Eng*) A design of extruder or injection-moulding screw where two or more steel helices are used on the rotor to aid mixing of polymer and additives.

multiple-spindle drilling machine (*Eng*) A drilling machine having two or more vertical spindles for simultaneous operation.

multiple star (*Astron*) A system in which three or more stars united by gravitational forces revolve about their common centre of gravity.

multiple-switch starter (*ElecEng*) A starter for an electric machine in which the steps of resistance are cut out, or other operations performed, by hand-operated switches. Similarly *multiple-switch controller*.

multiplet (*Phys*) (1) A group of optical spectrum lines showing fine structure with several components, ie triplet or more complex structures, owing to spin–orbit interactions in the atom. (2) See ISOTOPIC SPIN.

multiple-threaded screw (*Eng*) A screw of coarse pitch in which two or more threads are used to reduce the size of thread and maintain adequate core strength. Also *multistart thread*. See DIVIDED PITCH, MULTISTART WORM.

multiple-tool lathe (*Eng*) A heavy lathe having two large tool posts, one on either side of the work, each carrying several tools operating simultaneously on different parts of the work.

multiple-tuned antenna (*ICT*) A transmitting antenna system comprising an extensive horizontal 'roof' with a number of spaced vertical leads, each connected to earth through appropriate tuning circuits, and all tuned to the same frequency, the connection to the transmitter being made through one of them.

multiple-twin cable (*ICT*) A cable in which there are numbers of cores, each comprising two pairs twisted together.

multiple-unit control (*ElecEng*) The method of control by which a number of motors operating in parallel can be controlled from any one of a number of points; used on multiple-unit trains.

multiple-unit train (*ElecEng*) An electric or diesel train consisting of a number of motor coaches and trailers, all controlled from one driving position at the front or rear of the train.

multiple valve (*Electronics*) See MULTI-ELECTRODE VALVE.

multiplex (*ICT*) Use of one channel for several messages by TIME-DIVISION MULTIPLEX or FREQUENCY DIVISION.

multiplexer (*ICT*) An IO device that routes data from several sources to a common destination. Cf DEMULTIPLEXER.

multiplexer (*ImageTech*) An apparatus for transferring projected still and motion picture film to video, with the beam(s) directed by prisms to a video camera.

multiplex film (*Plastics*) Packaging film where different polymers are sandwiched together by co-extrusion, eg low-density polyethylene for strength and polyvinylidene chloride for barrier properties.

multiplex transmission (*ICT*) Transmission in which two or more signals modulate the carrier wave.

multiplicand (*MathSci*) See MULTIPLICATION.

multiplication (*Electronics*) See SECONDARY EMISSION.

multiplication (*MathSci*) The process of finding the *product* of two quantities which are called the *multiplicand* and the *multiplier*. Denoted by the multiplication sign ×, or by a dot, or by the mere juxtaposition of two quantities. Some branches of mathematics use special signs, eg the vector product of A and B is sometimes indicated by A ∧ B.

Numbers are multiplied in accordance with rules of arithmetic, but multiplication of other mathematical entities has to be defined specifically for the entities concerned.

multiplication constant (*Phys*) The ratio of the average number of neutrons produced by fission in one neutron lifetime to the total number of neutrons absorbed or leaking out in the same interval. Symbol *k*. Also *reproduction constant*.

multiplication factor (*NucEng*) See NEUTRON REPRODUCTION FACTOR.

multiplier (*Electronics*) See ELECTRON MULTIPLIER.

multiplier (*MathSci*) See MULTIPLICATION.

multiply-connected domain (*MathSci*) See CONNECTED DOMAIN.

multiplying camera (*ImageTech*) A camera for taking a number of small exposures on one negative, using a deflecting mirror or a number of lenses which can be traversed.

multiplying constant (*Surv*) A factor in the computation of distance by tacheometric methods (see TACHEOMETER). It is that constant value for the particular instrument by which the staff intercept must be multiplied in order to give the distance of the staff from the focus of the object glass. If the distance from the centre of the instrument is required, it is necessary to add (see ADDITIVE CONSTANT) the distance between the focus of the object glass and the centre of the instrument.

multipolar (*BioSci*) Said of nerve cells having many axons. Cf BIPOLAR, UNIPOLAR.

multipole generator (*ElecEng*) Generator designed to provide smoother dc output by increasing the number of poles.

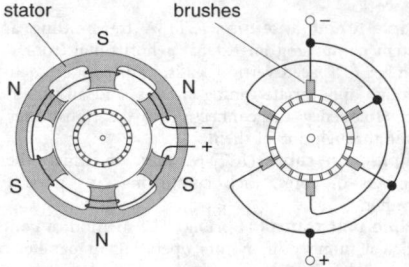

stator   brushes

multipole generator 6-pole DC machine showing pole polarities and commutator connections.

multipole moments (*Phys*) Measures (magnetic and electric) of the charge, current and magnet (via intrinsic spin) distributions in a given state. These static multipole moments determine the interaction of the system with weak external fields. There are also *transition* multipole moments which determine radiative transitions between two states.

multiposition (*ICT*) Said of a system in which output or final control can take on three or more preset values.

multipotent cell (*BioSci*) A progenitor cell that has the potential to give rise to a variety of cell types, the type selected resulting from environmental cues.

multiprocessor (*ICT*) A linked set of central processors that allows PARALLEL PROCESSING. See MIMD.

multiprogramming (*ICT*) The capacity of a central processor to handle more than one program at a time. See TIME SHARING.

multirow radial engine (*Aero*) A radial *aero-engine* with two or more rows of cylinders.

multisensor (*Aero*) The use of more than one sensor to obtain information.

multiseriate (*BioSci*) In several rows; vascular ray several cells wide.

multispeed supercharger (*Aero*) A gear-driven SUPERCHARGER in which a clutch system allows engagement of different ratios to suit changes in altitude.

multispindle automatic machine (*Eng*) A fully automatic, high-speed, special purpose lathe primarily for bar work, in which the work sequence is divided so that a portion of it is performed at each of the several spindle stations, one part being completed each time the tools are withdrawn and the spindles indexed.

multistage (*Electronics*) Said of a tube in which electrons are progressively accelerated by anode rings at increasing potentials, also of an amplifier with transistors or valves in series, or cascade.

multistage (*Space*) Said of a space vehicle having successive rocket-firing stages, each capable of being jettisoned after use.

multistage compressor (*Aero*) A gas turbine compressor with more than one stage; each row of blades in an AXIAL-FLOW COMPRESSOR is a stage, each IMPELLER is a stage in a CENTRIFUGAL-FLOW COMPRESSOR. In practice, all axial compressors are multistage, while almost all centrifugal compressors are single stage. Occasionally an initial axial stage is combined with a centrifugal delivery stage.

multistage supercharger (*Aero*) A supercharger with more than one impeller in series.

multistage turbine (*Aero*) A turbine with two or more disks joined and driving one shaft.

multistandards equipment (*ImageTech*) See MULTISYSTEMS EQUIPMENT.

multistart thread (*Eng*) See MULTIPLE-THREADED SCREW.

multistart worm (*Eng*) A worm in which two or more helical threads are used to obtain a larger pitch and hence a higher velocity ratio of the drive.

multisystems equipment (*ImageTech*) Video equipment which can handle incompatible TV systems, eg NTSC, PAL, SECAM. Also *multistandards equipment*. See 4·43 NTSC, NTSC PLAYBACK, STANDARDS CONVERTER.

multitasking (*ICT*) The ability of a computer to run more than one APPLICATION apparently simultaneously. This may take the form of a FOREGROUND TASK which will be the current application and a BACKGROUND TASK such as printing.

multithreading (*ICT*) A combination of microprocessor design and machine code that allows computer instructions to be carried out simultaneously and the results to be recombined in the right logical order.

multitone (*Acous*) A generator, thermionic or mechanical, which produces a spectrum of currents, ie with a large number of components, equally spaced in frequency.

multituberculate (*BioSci*) Said of tuberculate teeth with many cusps; having many small projections.

multi-turn current transformer (*ElecEng*) A current transformer in which there are several turns on the primary winding. Cf BAR-TYPE CURRENT TRANSFORMER.

multiuser system (*ICT*) A COMPUTER SYSTEM that supports many simultaneous users. This is done by sharing the PROCESSOR's time by various means such as TIME SHARING or by means of INTERRUPTS.

multivalent (*BioSci*) A group of three or more partly homologous chromosomes held together during the prophase of the first division of meiosis. Cf BIVALENT, UNIVALENT.

multivariate analysis (*MathSci*) Statistical analysis of several measurements of different characteristics on each observation.

multivibrator (*Electronics*) A relaxation oscillator circuit consisting of two active elements (transistors, valves, etc) connected so that when one is conducting, the other is not. An astable vibrator switches spontaneously between these two states, the frequency of switching depending on the time constants of the coupling elements or on the frequency of an external synchronizing voltage. A monostable vibrator can be triggered into an unstable state by an external signal, remaining in that state for a period determined by the time constants. The flip-flop or bistable multivibrator can remain indefinitely in either state, external triggering being required to change from one state to the other.

multivoltine (*BioSci*) Birds having more than one brood in a year. Cf BIVOLTINE, UNIVOLTINE.

multiwire antenna (*ICT*) An antenna consisting of a number of horizontal wires in parallel.

mu-meson (*Phys*) An elementary particle once thought to be a MESON but now known to be a LEPTON. See MUON.

Mumetal (*Eng*) TN of high-permeability, low-saturation magnetic alloy of about 80% nickel, requiring special heat treatment to achieve special low-loss properties. Useful in non-polarized transformer cores, ac instruments, small relays. Also for shielding devices from external field as in the CATHODE-RAY OSCILLOSCOPE.

mumps (*Med*) Epidemic or infectious *parotitis*. An acute infectious disease characterized by a painful swelling of the parotid gland; caused by a virus.

mundic (*Min*) See PYRITE.

munge (*ICT*) Computing slang for converting data so as to render them useless or incapable of automatic manipulation.

mungo (*Textiles*) A low grade of waste fibres recovered by pulling down old *hard*-woven wool rags, including tailors' cuttings and old felt. Cleaned and spun again as woollen yarns and used for weaving into lower-grade fabrics. Cf SHODDY.

Munsell colour system (*Print*) A system of colour notation devised by Albert Munsell which breaks colour into three attributes: (1) *hue* colour, red, blue, green, etc; (2) *value*, the lightness or darkness of a colour; (3) *chroma*, strength or saturation of a colour. Colour identification can be done rapidly with hand-held photometers.

Munsell scale (*Phys*) A scale of chromaticity values giving approximately equal magnitude changes in visual hue.

muntin (*Build*) The vertical framing piece separating the panels of a door. Also *munting*.

Muntz metal (*Eng*) Alpha–beta brass, 60% copper and 40% zinc. Stronger than alpha brass and used for castings and hot-worked (rolled, stamped or extruded) products. See COPPER ALLOYS.

muon (*Phys*) Fundamental particle with a rest mass equivalent to 106 MeV; it is one of the LEPTONS and has a negative charge and a half-life of about 2 μs. Decays to electron, neutrino and antineutrino. It participates only in WEAK INTERACTIONS. The *antimuon* has a positive charge and decays to positron, neutrino and antineutrino.

muramic acid (*Chem*) Monosaccharide derivative of glucose found in bacterial cell walls consisting of N-acetylglucosamine joined by an ethereal linkage to O-lactic acid.

muramyl dipeptide (*BioSci*) N-acetyl-muramyl-L-alanyl-D-isoglutamine. This is the simplest structural unit of bacterial peptidoglycans, which can mimic the adjuvant effects of mycobacteria in complete Freund's adjuvant. It causes release of IL-1 from macrophages, and is sometimes incorporated in a vaccine to increase its immunogenicity.

murein (*BioSci*) See PEPTIDOGLYCAN.

murexide (*Chem*) The dye ammonium purpurate, used as an indicator for calcium in COMPLEXOMETRIC TITRATION using EDTA.

Murex process (*MinExt*) Magnetic separation of desired mineral from pulp, in which the ore is mixed with oil and magnetite and then pulped in water. Magnetite clings selectively to the desired mineral.

Murgatroyd belt (*Glass*) A circumferential band in the side wall of a glass container, extending from the bottom upwards for about 20 mm. In this band, maximum stresses due to thermal shock occur.

muriatic acid (*Chem*) See HYDROCHLORIC ACID.

muricate (*BioSci*) Having a surface roughened by short, sharp points.

murmur (*Med*) An irregular sound which follows, accompanies or replaces the normal heart sounds and often indicates disease of the valves of the heart; similar sound heard over blood vessels.

Murphree efficiency (*ChemEng*) In the performance of a single plate in a FRACTIONATING COLUMN, the ratio of the actual enrichment of the vapour leaving the plate, to the enrichment which would have occurred if there had been sufficient time for the vapour to reach equilibrium with the liquid leaving the plate.

Murray red gum (*For*) See RED RIVER GUM.

Musca (Fly) (*Astron*) A small southern constellation.

muscarine (*Pharmacol*) A toxin (alkaloid) from the mushroom *Amanita muscaria* (Fly Agaric) that binds to (muscarinic) acetylcholine receptors.

muscarinic acetylcholine receptor (*BioSci*) A receptor for acetylcholine at the neural synapse that has its effect by the activation of G-PROTEINS (cf NICOTINIC ACETYLCHOLINE RECEPTOR). The binding of acetylcholine is inhibited by the alkaloid muscarine.

Musci (*BioSci*) The mosses. See BRYOPSIDA.

muscle (*BioSci*) A tissue whose cells can contract; a definitive mass of such tissue. See CARDIAC MUSCLE, STRIATED MUSCLE, UNSTRIATED MUSCLE, VOLUNTARY MUSCLE. Fig. ▷

muscle spindle (*BioSci*) A specialized bundle of muscle fibres innervated by sensory neurons. Stretching the muscle causes the neurons to fire; the muscle spindle thus functions as a stretch receptor.

muscology (*Genrl*) See BRYOLOGY.

muscovite (*Geol, Min*) The common or white mica; a hydrous silicate of aluminium and potassium, crystallizing in the monoclinic system. It occurs in many geological environments but deposits of economic importance are found in granitic pegmatites. Used as an insulator and a lubricant.

Muscovy glass (*Min*) Formerly a popular name for MUSCOVITE.

muscular dystrophy (*Med*) Hereditary diseases marked by hypertrophy or wasting of pelvic and limb muscles. Also *dystrophia myotonica*. See DUCHENNE MUSCULAR DYSTROPHY, FASCIOSCAPULOHUMERAL MUSCULAR DYSTROPHY.

musculature (*BioSci*) The disposition and arrangement of the muscles in the body of an animal.

musculocutaneous (*BioSci*) Pertaining to the muscles and the skin.

MUSE (*ImageTech*) Abbrev for *multiple sub-Nyquist sampling encoding*. A compression system used to reduce the BANDWIDTH of HI-VISION.

mush (*Aero*) The condition of flight at the stall when the aircraft tends to maintain ANGLE OF ATTACK while losing height rather than the sharp nose-down pitch which is more common.

mush (*ICT*) Noise and distortion in reception due to the interaction of waves from two or more radio transmitters.

mushroom (*BioSci*) Common name for the fruiting body of an AGARIC, esp of species of *Agaricus*.

mushroom bodies (*BioSci*) Paired nerve centres of the protocerebrum in insects, regarded as the principal association areas. Also *corpora pedunculata*.

mushroom construction (*CivEng*) Reinforced concrete construction composed only of columns and floor slabs, the columns having a spread at their head to counteract shear forces.

mushroom follower (*Eng*) A cam follower in the form of a mushroom, ie with a domed surface, as distinct from a roller-type follower.

mushroom valve (*Autos*) See POPPET VALVE.

mush winding (*Electronics*) A type of winding used for ac motors and generators in which the conductors are placed one by one into partially closed and lined slots, the end connections being subsequently insulated separately.

musical echo (*Acous*) Echo in which the repetition interval is so small that the impulses received appear to have the quality of a musical tone.

music and effects (*ImageTech*) A sound track containing recordings of both these but without speech or dialogue. Abbrev *M & E*.

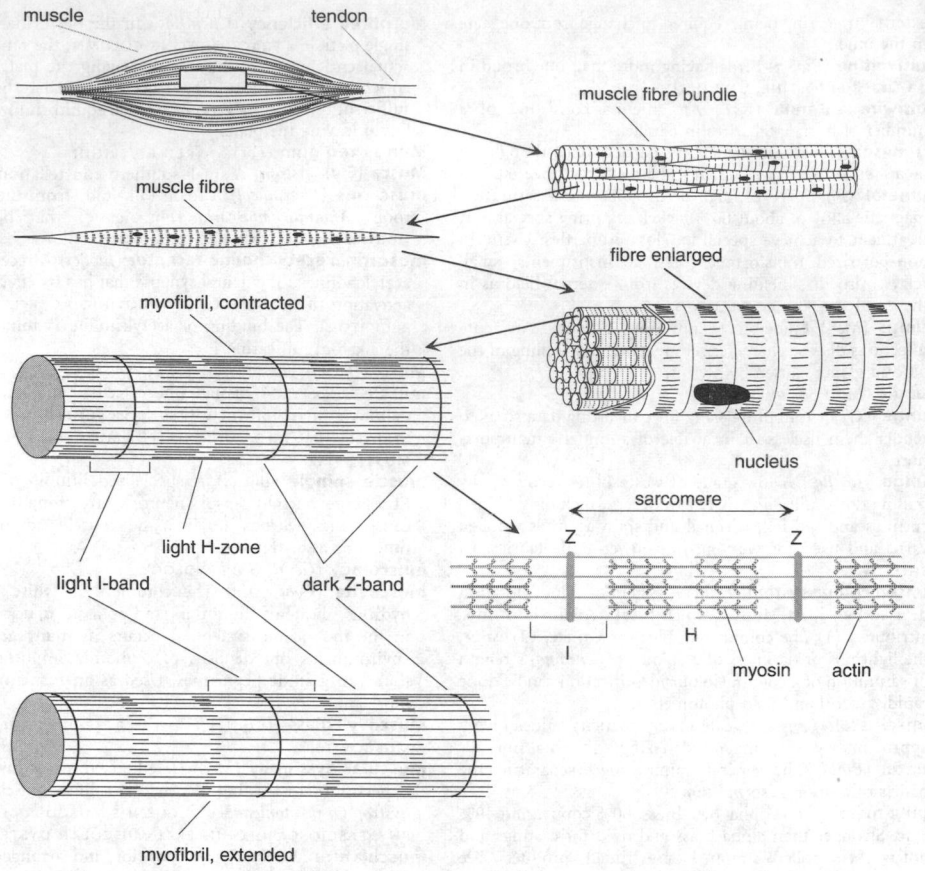

muscle **muscle** The relation between the muscle and the myofibril in which filaments of myosin slide along the actin.

music synthesizer (*ICT*) Output device that on receipt of digital signals generates sounds similar to musical notes. See MIDI.

muskeg (*EnvSci*) An unstable peat formation found in high latitudes, usually as deep and treacherous swamp, bog or marsh. Composed of partly decomposed sphagnum and other plant material, loosely compacted.

musk glands (*BioSci*) In some vertebrates, glands, associated with the genitalia, secreting scented musk.

muslin (*Textiles*) Lightweight, plain-woven cloths of open texture and soft finish, bleached, and dyed; used as dress fabrics. There are also unbleached butter, cheese, meat, etc, muslins for wrapping purposes.

mustard gas (*Chem*) Dichlorodiethyl sulphide. $(CH_2ClCH_2)_2S$. A poison gas manufactured from ethene and S_2Cl_2. See NITROGEN MUSTARD.

mustine (*Chem*) See NITROGEN MUSTARD.

mutagen (*BioSci*) A substance that causes MUTATION.

mutant (*BioSci*) An individual that displays the result of a *mutation*. A gene that has undergone MUTATION.

mutarotation (*Chem*) The change with time of the optical activity of a freshly prepared solution of an active substance. The sugars are the best-known class to exhibit the phenomenon. See ANOMERISM.

mutation (*BioSci*) A change, spontaneous or induced, that converts one allele into another (*point mutation*). More generally, any change of a gene or of chromosomal structure or number. A *somatic mutation* is one occurring in a somatic cell and not in the germ line. Silent mutations are ones that are not apparent because the nucleotide

change does not alter the amino acid specified by the codon and the protein product of the gene is unchanged.

mutation rate (*BioSci*) The frequency, per gamete, of mutations of a particular gene or a class of genes. Sometimes, esp in micro-organisms, the number per unit of time (esp generation time).

mute (*ImageTech*) A negative or positive film containing the picture record only, without any associated sound record.

muticate (*BioSci*) Without a point, mucro or awn. Unarmed. Also *muticous*.

muting (*Electronics*) Suppression of the output of a receiver or amplifier unless there is sufficient signal-to-noise ratio to comfortably ensure intelligibility.

muting circuit (*ICT*) Arrangement for attenuating amplifier output unless paralysed by a useful incoming carrier; used with automatic tuning devices. Also *muting switch*.

mutton (*Agri*) Meat from older sheep.

mutton (*Print*) See EM QUAD.

mutton cloth (*Textiles*) A plain circularly knitted cotton fabric often used in the unbleached state as a cleaning cloth.

mutton rule (*Print*) See EM RULE.

mutual capacitance (*Phys*) The capacitance calculated from the displacement current flowing between two conducting bodies when all adjacent conducting bodies are earthed.

mutual conductance (*Electronics*) The transconductance specifically applied to a thermionic valve. Differential change in a space or anode current divided by differential change of grid potential which causes it. Colloq termed *slope* of a valve, measuring the effectiveness of the valve as an amplifier in normal circuits. Expressed in milliamperes per volt and denoted by *gm*.

mutual coupling (*ElecEng*) See TRANSFORMER COUPLING.

mutual impedance (*ElecEng*) See TRANSFER IMPEDANCE.

mutual inductance (*ElecEng*) The generation of emf in one system of conductors by a variation of current in another system linked to the first by magnetic flux. The unit is the henry, when the rate of change of current of 1 ampere per second induces an emf of 1 volt.

mutual inductor (*ElecEng*) A component consisting of two coils designed to have a definite mutual inductance (fixed or variable) between them.

mutualism (*BioSci*) Any association between two organisms that is beneficial to both and injurious to neither. See SYMBIOSIS.

muzzle (*BioSci*) See RHINARIUM.

MWE, MWe (*NucEng*) Abbrev for *megawatt electric*; unit for electric power generated by a nuclear reactor.

my (*Geol*) A million years. Also *Ma*.

my-, myo- (*Genrl*) Prefixes from Gk *mys*, gen *myos*, muscle.

myalgia (*Med*) The sensation of pain in muscle.

myalgic encephalomyelitis (*Med*) See CFS.

myarian (*BioSci*) Based on musculature, as a system of classification; pertaining to the musculature.

myasthenia (*Med*) Muscular weakness.

myasthenia gravis (*Med*) A disease characterized by increasing muscular weakness on exercise, caused by faulty transmission at the neuromuscular junction. Auto-antibodies are present in the blood against antigens of the acetylcholine receptor on the post-synaptic membrane, and are the putative cause of the symptoms by interfering with or damaging this receptor. One form is associated with the presence of a thymoma, and thymectomy may reverse the condition.

myc-, myceto-, myco- (*Genrl*) Prefixes from Gk *mykes*, gen *myketos*, fungus.

Mycalex (*Phys*) TN for mica bonded with glass. It is hard, and can be drilled, sawn and polished; has a low power factor at high frequencies, and is a very good insulating material at all frequencies.

mycelium (*BioSci*) A mass of branching hyphae; the vegetative body (thallus) of most true fungi. Adj *mycelial*.

myceto- (*Genrl*) See MYC-.

mycetocytes (*BioSci*) Cells containing symbiotic microorganisms occurring in the MYCETOME.

mycetoma (*Med*) See MADURA FOOT.

mycetome (*BioSci*) A special organ in some species of insects, ticks and mites, inhabited by intracellular symbionts, eg bacteria, fungi, rickettsiae. Also *pseudovitellus*.

mycetophagous (*BioSci*) Fungus-eating.

myco- (*Genrl*) See MYC-.

mycobacteria (*BioSci*) Bacteria with unusual staining properties, unlike those of Gram-negative and Gram-positive organisms, being acid-fast. Many are intracellular parasites, causing serious diseases such as leprosy and tuberculosis.

mycobiont (*BioSci*) The fungal partner in a symbiosis, eg in a lichen.

mycology (*BioSci*) The study of fungi.

mycophenolate mofetil (*Pharmacol*) An immunosuppressant used together with ciclosporin and corticosteroids to prevent graft rejection.

mycophthorous (*BioSci*) A fungus parasitic on another fungus.

mycoplasma (*Med*) The smallest free-living organism known. *Mycoplasma pneumoniae* is an important cause of atypical pneumonia.

Mycoplasmatales (*BioSci*) A group of organisms associated with pleuropneumonia in cattle and contagious agalactia in sheep. They are very variable in form, consisting of cocci, rods, etc, and a cytoplasmic matrix; very similar to the L-FORMS of true bacteria. Morphologically similar, but non-pathogenic, organisms have also been found in animals and humans.

mycoplasmosis (*Vet*) Infection of animals and birds by organisms of the genus *Mycoplasma*, eg chronic respiratory disease of fowl, contagious agalactia of sheep and goats, contagious pleuropneumonia of cattle.

mycorrhiza (*BioSci*) A symbiotic association between a fungus and the roots or other structures of a plant. See ECTOTROPHIC MYCORRHIZA, ENDOTROPHIC MYCOR-RHIZA, VESICULAR–ARBUSCULAR MYCORRHIZA.

mycosis (*Med, Vet*) Any disease of humans or animals caused by a fungus.

mycosis fungoides (*Med*) A chronic, and usually fatal, disease in which multiple fungus-like tumours appear in the skin. Associated with lymphoma.

mycotic dermatitis of sheep (*Vet*) A dermatitis of sheep due to infection of the skin by the fungus *Dermatophilus dermatonomus*. Also *lumpy wool*.

mycotoxin (*BioSci; FoodSci*) A substance produced by a fungus and toxic to other organisms, esp to humans and animals, eg the AFLATOXINS. In foods mycotoxins tend to form only when conditions for mould growth are less than ideal.

mycotrophic plant (*BioSci*) A plant that lives in symbiosis with a fungus, ie has mycorrhizas.

mydriasis (*Med*) Extreme dilation of the pupil of the eye.

mydriatic (*Med*) Producing dilation of the pupil of the eye: any drug which does this.

mydriatic alkaloids (*Pharmacol*) Alkaloids which cause dilation of the pupil of the eye, eg atropine.

myel-, myelo- (*Genrl*) Prefixes from Gk *myelos*, marrow.

myelencephalon (*BioSci*) A region of the vertebrate hindbrain that forms part of the MEDULLA OBLONGATA. Cf METENCEPHALON.

myelin (*BioSci*) A white fatty substance that forms the MYELIN SHEATH of nerve fibres.

myelination (*BioSci*) The formation of a myelin sheath around an axon.

myelin sheath (*BioSci*) A layer of white fatty substance (myelin) that, in vertebrates, surrounds the axons of the central nervous system and acts as an insulator, thereby increasing the speed of neuronal impulses that move by saltatory conduction between NODES OF RANVIER. Produced by Schwann cells that wrap around the axon. Also *medullary sheath*.

myelitis (*Med*) Inflammation of the spinal cord.

myelo- (*Genrl*) See MYEL-.

myelocele (*Med*) A condition in which the spinal cord protrudes onto the surface of the body; due to a defect of the spinal vertebrae.

myelocoel (*BioSci*) The central canal of the spinal cord.

myelocyte (*BioSci*) A large amoeboid cell found in the marrow of the long bones of higher vertebrates. Bone marrow myelocytes give rise, by division, to myeloid cells which differentiate into LEUCOCYTES.

myelography (*Radiol*) The radiological examination of the space between the theca and the spinal cord, following the injection of air or other contrast media.

myeloid cell (*BioSci*) Cells that derive from stem cells in the bone marrow, and that mature to form the granular leucocytes (granulocytes) of blood. They differentiate into different forms (neutrophils, eosinophils or basophils) under the influence of COLONY STIMULATING FACTORS.

myeloma (*BioSci*) See PLASMACYTOMA.

myelomalacia (*Med*) Pathological softening of the spinal cord.

myelomatosis (*Med*) The occurrence of myelomata in several bones. See PLASMACYTOMA.

myelomeningocele (*Med*) Protrusion of the spinal cord on spinal membranes; due to a defect in the spinal column.

myelopathy (*Med*) Any disease affecting spinal cord or myeloid tissues.

myeloperoxidase–halide system (*BioSci*) A potent microbicidal mechanism in phagocytic cells that hydrolyses hydrogen peroxide (produced in the METABOLIC BURST) in the presence of halide ions.

myeloplast (*BioSci*) A leucocyte of bone marrow.

myeloplax (*BioSci*) See *megakaryocyte*.

myenteric (*BioSci*) Pertaining to the muscles of the gut; as a sympathetic nerve plexus controlling their movements.

myiasis (*Vet*) Parasitism of the tissues of animals by the larvae of flies of the suborder *Cyclorrhapha*.

myiophily (*BioSci*) See MYOPHILY.

Mylar (*Plastics*) TN for polyethylene terephthalate film (US).

mylonite (*Geol*) A banded, chert-like, cataclastic rock produced by the shearing and granulation of rocks associated with intense folding and thrusting.

mylonitization (*Geol*) The process by which rocks are granulated and pulverized and formed into mylonite.

myo- (*Genrl*) See MY-.

myoblast (*BioSci*) An embryonic muscle cell that will develop into a muscle fibre.

myocardial (*Med*) Pertaining to, or affecting, the myocardium.

myocardial infarction (*Med*) The infarction or death of heart muscle usually as a result of a CORONARY THROMBOSIS. Colloq *heart attack*.

myocarditis (*Med*) Inflammation of the muscle of the heart.

myocardium (*BioSci*) In vertebrates, the muscular wall of the heart.

myoclonia congenita (*Vet*) Six types of tremor recognized. Piglets show no tremor when asleep, slight tremor when lying down and marked tremor when standing. Aetiology varies in all six types. Also *congenital tremors, trembling*.

myoclonus (*Med*) (1) A condition in which there occur sudden shock-like contractions of muscles, often associated with epilepsy and progressive mental deterioration (*paramyoclonus multiplex*). (2) A sudden shock-like contraction of a muscle.

myocoel (*BioSci*) The coelomic space within a myotome.

myocomma (*BioSci*) A partition of connective tissue between two adjacent myomeres. Also *myoseptum*.

myocyte (*BioSci*) (1) A muscle cell. (2) A contractile cell.

myo-edema (*Med*) See MYOIDEMA.

myo-epithelial (*BioSci*) A term used to describe the epithelial cells of Cnidaria that are provided with tail-like contractile outgrowths at the base.

myofibril (*BioSci*) The contractile filament consisting of actin, myosin and associated proteins within muscle cells. See fig. at MUSCLE.

myogenic (*BioSci*) Said of contraction arising in a muscle independent of nervous stimuli. Cf NEUROGENIC.

myoglobin (*BioSci*) A haem protein related to haemoglobin but consisting of only one polypeptide chain. It is present in large amounts in the muscles of mammals such as whales and seals, where it acts as an oxygen store during diving. It was the first protein to have its three-dimensional structure worked out.

myohypertrophy (*Med*) Increase in the size of muscle fibres.

myoidema (*Med*) Mounding. A localized swelling of wasting muscle obtained when the muscle is lightly struck. Also *myo-edema*.

myolemma (*BioSci*) See SARCOLEMMA.

myology (*BioSci*) The study of muscles.

myoma (*BioSci*) A tumour composed of unstriped (leiomyoma) or striped (rhabdomyoma) muscle fibres.

myomectomy (*Med*) Surgical removal of a myoma, esp of a fibromyoma of the uterus.

myomere (*BioSci*) In metameric animals, the voluntary muscles of a single somite.

myometrium (*BioSci*) The muscular coat of the uterus.

myoneme (*BioSci*) In Protozoa, a contractile fibril of the ectoplasm.

myoneural (*BioSci*) Pertaining to muscle and nerve, as the junction of a muscle and a nerve.

myopathy (*Med*) Any one of a number of conditions in which there is progressive wasting of skeletal muscles from no known cause.

myophily (*BioSci*) Pollination by flies or other Diptera. Also *myiophily*.

myopia (*Med*) Short-sightedness. A condition of the eye in which, with the eye at rest, parallel rays of light come to a focus in front of the retina. Adj *myopic*.

myosarcoma (*Med*) A malignant tumour composed of muscle cells and sarcoma cells.

myoseptum (*BioSci*) See MYOCOMMA.

myosin (*BioSci*) A large protein with ATPase activity that assembles into the thick filaments that interdigitate in a regular hexagonal array with actin filaments (thin filaments) to form muscle fibrils. The relative movement of the two sets of filaments, driven by ATP-consuming conformational changes in the myosin, provides the molecular basis of the sliding filament mechanism of *muscular contraction*. Myosin was found originally in muscle but isoforms are found in a wide variety of cells, not always in filamentous form. See fig. at MUSCLE.

myosis (*Med*) See MIOSIS.

myositis (*Med*) Inflammation of striped muscle.

myositis ossificans progressiva (*Med*) A condition in which there is progressive ossification of the muscles of the body.

myostatin (*BioSci*) A blood protein that limits muscle growth.

myotasis (*Med*) Muscular tension. Adj *myotatic*.

myotome (*BioSci*) A muscle merome; one of the metameric series of muscle masses in a developing segmented animal.

myotonia atrophica (*Med*) A disease characterized by wasting of certain groups of muscles, difficulty in relaxing muscles after muscular effort, and general debility. See MUSCULAR DYSTROPHY.

myotonia congenita (*Med*) A rare and congenital malady characterized by extreme slowness in relaxation of muscles after voluntary effort. Also *Thomsen's disease*.

myriametric waves (*ICT*) Waves corresponding to VERY LOW FREQUENCY, longer than 10 km.

myriapod (*BioSci*) A general term denoting arthropods with many similar segments and comprising the classes Chilopoda and Myriapoda (centipedes and millipedes).

myringitis (*Med*) Inflammation of the drum of the ear.

myringoscope (*Med*) An instrument for viewing the drum of the ear.

myringotomy (*Med*) Incision of the drum of the ear.

myristic acid (*Chem*) *Tetradecanoic acid*. $CH_3(CH_2)_{12}COOH$. Mp 58°C. Crystalline solid. Found (as glycerides) in milk and various vegetable oils.

myristoylation (*BioSci*) Post-translational modification of proteins that are destined to be membrane-associated, by covalent addition of a myristoyl residue.

myrmecochory (*BioSci*) Distribution of seeds or other reproductive bodies by ants.

myrmecology (*Genrl*) The study of ants.

myrmecophagous (*BioSci*) Feeding on ants.

myrmecophily (*BioSci*) (1) Symbiosis with ants, eg *Acacia*. (2) Pollination by ants.

myrmekite (*Min*) An intergrowth of plagioclase and vermicular quartz.

Myrtaceae (*BioSci*) A family of c.3000 spp of dicotyledonous flowering plants (superorder Rosidae), comprising woody plants that are tropical and subtropical, esp Australian. It includes some sources of spices, eg cloves and allspice, and the genus *Eucalyptus*, trees important in the hardwood forests of Australia.

myrtle (*For*) A tree of the Leguminosae family giving a coarse-textured hardwood that is reddish-brown in colour.

mysophobia (*Psych*) Morbid fear of being contaminated.

myxamoeba (*BioSci*) The amoeboid stage of a slime mould. See MYXOMYCOTA.

myxo- (*Genrl*) Prefix from Gk *myxa*, mucus, slime.

Myxobacteriales (*BioSci*) A group of bacteria characterized by the absence of a rigid cell wall, gliding movements and the production of slime. Some genera produce characteristic fruiting bodies and spores in a fashion reminiscent of the much larger eukaryotic cellular slime moulds (MYXOMYCOTA).

myxoedema (*Med*) A condition due to deficiency of thyroid secretion; characterized by loss of hair, increased thickness and dryness of the skin, increase in weight, slowing of mental processes, and diminution of metabolism.

myxoma (*Med*) A term applied to a tumour composed of a clear jelly-like substance and star-shaped cells.

myxomatosis (*Vet*) A highly contagious and fatal viral disease of rabbits, characterized by tumour-like proliferation of myxomatous tissue esp beneath the skin of the head and body. Virus lethality has progressively decreased. Vaccine available.

Myxomycetes (*BioSci*) The true (or acellular) slime moulds. Class of slime moulds (MYXOMYCOTA) that alternate between haploid free-living phagocytic myxamoebae (that can interconvert into flagellated swarm cells) and a diploid multinucleate PLASMODIUM, formed by fusion of pairs of myxamoebae or swarm cells. The plasmodium eventually forms a fruiting body liberating haploid spores, eg *Physarum*.

Myxomycota (*BioSci*) The cellular slime moulds, wall-less heterotrophic organisms that are usually classified with the fungi. They are phagotrophic on bacteria or parasitic within plant cells. Dispersal is by means of zoospores and/or small, walled, wind-blown spores. They include Acrasiomycetes, Mysomycetes and Plasmodiophoromycetes. Also *Gymnomycota*.

Myxophyceae (*BioSci*) Old name for blue-green algae. See CYANOBACTERIA.

myxoviruses (*BioSci*) RNA containing viruses pathological to vertebrates, including the influenza virus and related species. Many have considerable genetic variability that results in continual changes in their antigenic status and consequent difficulty in producing an effective vaccine.

N

N (*Chem*) Symbol for NITROGEN.

N (*Eng*) Symbol for modulus of rigidity.

N (*Phys*) Symbol for NEWTON.

N (*Chem*) Symbol for: (1) AVOGADRO'S NUMBER; (2) number of molecules.

N (*ElecEng*) Symbol for number of turns.

N (*Phys*) Symbol for NEUTRON NUMBER.

N- (*Chem*) Symbol indicating substitution on the nitrogen atom.

N_A (*Phys*) Symbol for AVOGADRO'S NUMBER.

n (*Genrl*) Symbol for NANO-.

n (*Phys*) Symbol for NEUTRON.

n (*Chem*) Symbol for amount of substance.

n (*Electronics*) Number density expressed as particles per cubic metre (or per cubic centimetre), esp of negative carriers (electrons) in semiconductors. Cf P.

n- (*Chem*) Abbrev for *normal*, ie containing an unbranched carbon chain in the molecule.

ν (*Phys*) Symbol for: (1) FREQUENCY; (2) KINEMATIC VISCOSITY; (3) NEUTRINO; (4) POISSON'S RATIO.

Na (*Chem*) Symbol for SODIUM.

NA (*Eng*) Abbrev for NEUTRAL AXIS.

NA (*Phys*) Abbrev for NUMERICAL APERTURE.

NAA (*BioSci*) Abbrev for NAPHTHALENE ACETIC ACID.

NAA (*Chem*) Abbrev for *neutron activation analysis*. See panel on TRACE ELEMENT ANALYSIS.

nab (*Build*) The keeper part of a door lock.

nabla (*MathSci*) See DEL.

nacelle (*Aero*) A small streamlined body on an aircraft, distinct from the fuselage, housing engine(s), special equipment or crew.

nacre (*BioSci*) The iridescent calcareous substance, mostly calcium carbonate, composing the inner layer of a molluscan shell and formed by MANTLE cells. Mother of pearl. Also *nacreous layer*.

nacreous (*Min*) A term applied to the lustre of certain minerals, usually on crystal faces parallel to a good cleavage, the lustre resembling that of pearls.

nacreous clouds (*EnvSci*) Clouds composed of ice crystals in 'mother of pearl' formations, found at a height of 25–30 km. They may be WAVE CLOUDS.

nacreous layer (*BioSci*) See NACRE.

nacrite (*Min*) A species of clay mineral, identical in composition with kaolinite, from which it differs in certain optical characters and in atomic structure.

NAD (*BioSci*) See NICOTINAMIDE ADENINE DINUCLEOTIDE.

nadir (*Astron*) The pole of the horizon vertically below the observer; hence the point on the celestial sphere diametrically opposite the ZENITH.

NADP (*BioSci*) See NICOTINAMIDE ADENINE DINUCLEOTIDE PHOSPHATE.

naevus (*Med*) A birthmark or mole. (1) A pigmented tumour in the skin. (2) A patch or swelling in the skin composed of small dilated blood vessels.

nagana (*Vet*) A name for a group of animal diseases in Africa caused by trypanosomes. Also *fly disease, tsetse fly disease*.

nagging piece (*Build*) A horizontal timber between the studs in a partition.

nail (*BioSci*) In higher mammals, a horny plate of epidermal origin taking the place of a claw at the end of a digit.

nail punch (*Build*) A small steel rod tapering at one end, used to drive the head of a nail beneath the surface of the timber. Also *nail set*.

NaK (*NucEng*) Acronym for sodium (Na) and potassium (K) alloy used as coolant for liquid metal reactor. It is molten at room temperature and below.

naked (*BioSci*) Lacking a structure or organ.

naked-light mine (*MinExt*) Non-fiery mine, where safety lamps are not required.

name table (*ICT*) See SYMBOL TABLE.

Namurian (*Geol*) A stratigraphical series name in the Upper Carboniferous of Europe. Approximately corresponding to the Millstone Grit of England and Wales. See PALAEOZOIC.

NAND (*ICT*) A logical operator where (p NAND q) written ($p \cdot q$) takes the value FALSE if both p and q are TRUE, otherwise (p NAND q) is TRUE. See LOGICAL OPERATION.

NAND gate (*ICT*) A GATE with output signal 0 when all input signals are 1, otherwise output is 1. A NAND gate can be made from a single transistor and all other gates can be constructed from combinations of NAND gates. Also *NAND element*.

nanism (*Med*) The condition of being a dwarf (Lt *nanus*); dwarfism.

nano- (*Genrl*) Prefix for 10^{-9}, ie equivalent to millimicro or one thousand millionth. Symbol n.

nano-computer (*ICT*) A computer of microscopic size.

nanoelectronics (*ElecEng*) Electronics on a nanometer scale, below 100 nm in dimensions. Covers both molecular electronics and nanoscale devices.

nanoparticle (*Genrl*) Any particle with dimensions of around 5–40 nm. The properties of such particles may differ from that of bulk materials although there is less information on these effects than might be considered desirable.

nanophanerophyte (*BioSci*) A phanerophyte, with buds 25 cm to 2 m above soil level.

nanoplankton (*BioSci*) Plankton of microscopic size from 0·2 to 20 mm.

nanopore (*Genrl*) A microscopic pore or opening, specifically one between 10^{-7} m and 10^{-9} m.

nanotechnology (*Electronics*) The engineering of matter at a scale approaching that of individual atoms.

nanotube (*Chem*) A microscopic tubular form of carbon, structurally related to buckminsterfullerene; 'buckytube'. Can have diameters from ten to several hundred nm and lengths up to 100 μm and have remarkable stiffness and mechanical properties. Carbon nanotubes are used as probes for atomic force microscopy.

nap (*Textiles*) A fluffy surface on fabrics produced by the finishing process of raising. *Napping* is raising by means of a revolving cylinder covered with stiff wire brushes, or rollers covered by teasels.

napalm (*Chem*) A gel of inflammable hydrocarbon oils with soaps, used in warfare because of its cheapness, ease of application and ability to stick to the target while burning.

naphtha (*Chem*) A mixture of light hydrocarbons which may be of coaltar, petroleum or shale oil origin used as a solvent etc. *Coaltar naphtha* (boiling range approx 80–170°C) is characterized by the predominant aromatic nature of its hydrocarbons. Generally *petroleum naphtha* is a cut between gasoline and kerosine with a boiling range

120–180°C, but much wider naphtha cuts may be taken for special purposes, eg feedstock for high-temperature cracking for chemical manufacture. The hydrocarbons in petroleum naphthas are predominantly aliphatic.

naphthalene (*Chem*) $C_{10}H_8$. Mp 80°C, bp 218°C. Consists of two condensed benzene rings. Glistening plates, insoluble in water, slightly soluble in cold ethanol and ligroin, readily soluble in hot ethanol and ethoxyethane; sublimes easily and is volatile in steam. It occurs in the coaltar fraction boiling between 180 and 200°C. It forms an additive compound with picric acid (trinitrophenol). Naphthalene is more reactive than benzene, and substitution occurs in the first instance in the *alpha* position. It is an important raw material for numerous derivatives, many of which play a role in the manufacture of dyestuffs.

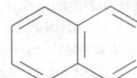

naphthalene

naphthalene acetic acid (*BioSci*) A synthetic plant growth regulator with AUXIN-like activity used esp in ROOTING COMPOUNDS.

naphthalene derivatives (*Chem*) Substitution products of naphthalene.

naphthalene di-isocyanate (*Chem*) A type of aromatic isocyanate monomer used to make polyurethanes. Abbrev *NDI*.

naphthenates (*Chem*) Metal salts of the naphthenic acids which occur naturally in crude petroleum. Mainly used as wood and textile fungicides, eg copper naphthenate, and also as paint driers.

naphthenes (*Chem*) *Cycloalkanes*, cyclic aliphatic hydrocarbons, like cyclohexane. Many occur in petroleum.

naphthionic acid (*Chem*) *1,4-naphthylamine-monosulphonic acid.* $H_2NC_{10}H_6SO_3H$. Obtained by the sulphonation of 1-naphthylamine. Intermediate for azo dyes.

naphthoic acids (*Chem*) *Naphthalene carboxylic acids.* $C_{10}H_7COOH$. There are two isomers, of which the 1-naphthoic acid crystallizes in fine needles, mp 160°C. On distillation with lime, they are decomposed into naphthalene and CO_2.

naphthols (*Chem*) $C_{10}H_7OH$. There is a 1-naphthol, mp 95°C, bp 282°C, and a 2-naphthol, mp 122°C, bp 288°C. Both are present in coaltar and can be prepared from the respective naphthalene sulphonic acids or by diazotizing the naphthylamines. They have a phenolic character: 1-naphthol is prepared on a large scale by heating α-naphthylamine under pressure with sulphuric acid; 2-naphthol is an antiseptic and can be used as a test for primary amines.

naphthylamines (*Chem*) $C_{10}H_7NH_2$. In these, 1-naphthylamine forms colourless prisms or needles, mp 50°C, bp 300°C, soluble in ethanol, and is of unpleasant odour, sublimes readily, and turns brown on exposure to the air; 2-naphthylamine, mp 112°C, bp 294°C, is odourless and forms colourless plates. Originally used as an anti-oxidant in rubber manufacture, but found to be a cause of bladder cancer.

Napierian logarithm (*MathSci*) See LOGARITHM.

Napier's analogies (*MathSci*) Formulae for solving SPHERICAL TRIANGLES.

Napier's bones (*MathSci*) A device for multiplying and dividing mechanically by means of rods, invented by John Napier (1550–1617).

Napier's compasses (*Eng*) A form of compasses having a needle point and a pencil holder pivoted at the end of one limb, and a needle point and a pen pivoted at the end of the other, both limbs being jointed for safe carrying.

napoleonite (*Geol*) A diorite containing spheroidal structures, about 2·5 cm in diameter, which consist of alternating shells essentially of hornblende and feldspars.

nappe (*Geol*) A major structure of mountain chains such as the Alps, consisting essentially of a great recumbent fold with both limbs lying approximately horizontally. It is produced by a combination of compressional earth movements and sliding under gravity, resulting in translation of the folded strata over considerable horizontal distances.

nappe (*MathSci*) A SHEET; one of the two sheets on either side of the vertex forming a cone.

nap roller (*Print*) A hand roller with a rough surface used for lithographic transfer work.

naproxen (*Pharmacol*) A non-steroidal anti-inflammatory drug.

naratriptan (*Pharmacol*) An agonist for a subclass of serotonin receptors (*5HT1 receptors*), used to treat acute migraine.

narcissism (*Psych*) An excessive preoccupation with oneself, something that is normal in the very young, but a form of personality disorder if it persists into adulthood.

narcolepsy (*Med*) Periodic attacks of an uncontrollable urge to sleep.

narcosis (*Med*) A state of unconsciousness produced by a drug; the production of a narcotic state.

narcotic (*Med*) Tending to induce sleep or unconsciousness; a drug which does this.

narcotine (*Chem*) $C_{22}H_{23}O_7N$. Mp 176°C. An alkaloid of the *isoquinoline* series; occurs in opium; forms colourless needles.

narcotize (*Med*) To subject to the influence of a narcotic.

nares (*BioSci*) Nostrils; nasal openings; as the internal or posterior nares to the pharynx, the external or anterior nares to the exterior. Adjs *narial, nariform.*

narra (*For*) See AMBOYNA.

narrow-band integrated services digital network (*ICT*) The original version of INTEGRATED SERVICES DIGITAL NETWORK offering two 64 Kbps bearer channels and a 16 Kbps data channel. Also *basic rate ISDN, ISDN 2*. See B-CHANNEL, BROADBAND INTEGRATED SERVICES DIGITAL NETWORK, D-CHANNEL.

narrow-base tower (*ElecEng*) A tower for overhead transmission lines having a base sufficiently small to be supported on a single foundation. Cf BROAD-BASE TOWER.

narrow-cut filter (*ImageTech*) A colour filter which transmits only a very limited part of the spectrum.

narrow gauge (*CivEng*) A railway gauge less than the standard 4 ft 8·5 in (1·435 m).

narrow gauge film (*ImageTech*) Motion picture film of 16 mm or 8 mm width.

NASA (*Space*) Abbrev for *National Aeronautics and Space Administration*, responsible for civil space activities in the USA, both research and development.

nasal (*BioSci*) Pertaining to the nose; a paired dorsal membrane bone covering the olfactory region of the vertebrate skull.

nasal sinusitis (*Med*) See SINUSITIS.

NASDA (*Space*) Abbrev for *National Space Development Agency*, the Japanese space agency mainly responsible for applications of space activities and launch systems.

Nasmyth focus (*Astron*) See panel on ASTRONOMICAL TELESCOPE.

nasolacrimal canal (*BioSci*) A passage through the skull of mammals, passing from the orbit to the nasal cavity, and through which the tear duct passes.

nasopalatine duct (*BioSci*) In some reptiles and mammals, a duct piercing the secondary palate and connecting the vomeronasal organs with the mouth.

nasopharyngeal duct (*BioSci*) The posterior part of the original vault of the vertebrate mouth that, in mammals, owing to the development of a secondary palate, carries air from the nasal cavity to the pharynx.

nasopharyngitis (*Med*) Inflammation of the nasopharynx.

nasosinusitis (*Med*) Inflammation of the air-containing bony cavities in communication with the nose.

nasoturbinal (*BioSci*) In vertebrates, a paired bone or cartilage of the nose that supports the folds of the olfactory mucous membrane.

nastic movement (*BioSci*) A plant movement in response to but not orientated by an external stimulus, eg the opening or closing of some flowers in response to increasing or decreasing temperature. See EPINASTY, HYPONASTY, PHOTONASTY, THERMONASTY. Also *nasty*. Cf TAXIS, TROPISM.

natal (*BioSci*) (1) Pertaining to birth. (2) Pertaining to the buttocks.

natality (*BioSci*) The inherent ability of a population to increase. *Maximum* or *absolute natality* is the theoretical maximum production of new individuals under ideal conditions. *Ecological* or *realized natality* is the population increase occurring under specific environmental conditions. Cf MORTALITY.

natatory (*BioSci*) Adapted for swimming. Also *natatorial*.

nates (*Med*) The buttocks.

National Bureau of Standards (*Genrl*) US federal department set up in 1901 to promulgate standards of weights and measures and generally investigate and establish data in all branches of physical and industrial sciences. Abbrev *NBS*.

National Nature Reserve (*EnvSci*) In the UK, extensive area set aside for nature conservation, originally under the auspices of the Nature Conservancy Council, an organization that has been fragmented into subsections whimsically renamed at frequent intervals (eg Natural England, formerly English Nature).

National Physical Laboratory (*Genrl*) UK authority for establishing basic units of eg mass, length, time, resistance, frequency, radioactivity. Founded by the Royal Society in 1900; now government controlled and engaged in a very wide range of research. Abbrev *NPL*.

native (*MinExt*) Said of naturally occurring metal, eg *native gold*, *native copper*.

native hydrocarbons (*Geol*) A series of compounds of hydrogen and carbon formed by the decomposition of plant and animal remains, including the several types of coal, mineral oil, petroleum, paraffin, the fossil resins, and the solid bitumens occurring in rocks. Many which have been allotted specific names are actually mixtures. By the loss of the more volatile constituents as natural gas, the liquid hydrocarbons are gradually converted into the solid bitumens, such as *ozocerite*. See ASPHALT, BITUMEN, COAL, MINERAL OILS.

nativism (*Psych*) In philosophy, the position that humans are born with some innate knowledge; in the study of perception, the view that some important abilities are innate.

natriuretic (*BioSci*) Of a substance or hormone, causing *natriuresis*, the elimination of extra sodium in the urine.

natrojarosite (*Min*) Hydrated sulphate of sodium and iron crystallizing in the trigonal system.

natrolite (*Min*) Hydrated silicate of sodium and aluminium crystallizing in the orthorhombic system. A sodium-zeolite. It usually occurs in slender or acicular crystals, and is found in cavities in basaltic rocks and as an alteration product of nepheline or plagioclase.

natron (*Min*) Hydrated sodium carbonate, occurring in soda-lake deposits.

natural (*FoodSci*) Usually applied to agricultural and horticultural materials obtained without any processing or chemical modification. Natural substances are retained in their raw state or extracted from it by physical separation, eg milling, pressing, aqueous extraction.

natural abundance (*Phys*) See ABUNDANCE.

natural ageing (*Eng*) Changes in physical or mechanical properties which take place in a material over a period of time at ambient temperature. It is usually a response to some previous heat treatment or operation which has left the material in a metastable condition. If such changes occur at raised temperatures, the processes are termed *artificial ageing*.

natural antibody (*BioSci*) Antibody present in the blood of normal individuals not known to have been immunized against the relevant antigen; eg in humans, antibodies against antigens of the ABO BLOOD GROUP SYSTEM or Forssman antibodies. They are generally induced by organisms in the gut owing to shared antigenic determinants.

natural background (*Phys*) In the detection of nuclear radiation, the radiation due to NATURAL RADIOACTIVITY and to cosmic rays, enhanced by contamination and fallout.

natural classification (*BioSci*) One based on many characters, and likely to have a predictive value.

natural draught (*Eng*) The draught or airflow through a furnace induced by a chimney and dependent on its height and the temperature difference between the ascending gases and the atmosphere.

natural evaporation (*EnvSci*) The evaporation that takes place at the surface of ponds, rivers, etc, which are exposed to the weather; it depends on solar radiation, strength of wind and relative humidity.

natural frequency (*ICT*) The frequency of free oscillations in a system. See NATURAL PERIOD.

natural frequency (*Phys*) See NORMAL MODES.

natural frequency of antenna (*ICT*) The lowest frequency at which an unloaded antenna system is resonant.

natural gas (*Geol*) Any gas found in the Earth's crust, including gases generated during volcanic activity (see PNEUMATOLYSIS, SOLFATARA). The term, however, is particularly applied to natural hydrocarbon gases which are associated with the production of petroleum. These gases are principally methane and ethane, sometimes with propane, butane, nitrogen, CO_2 and sulphur compounds (notably H_2S). The gas is found both above the petroleum and dissolved in it, but many very large gas fields are known which produce little or no petroleum. Natural gas has largely replaced TOWN GAS in many countries where it occurs abundantly (the US, Canada, Algeria, W Europe); since 1967 important finds in the North Sea have provided supplies for the UK gas-grid system, although the UK is now a net importer of gas.

natural glass (*Geol*) Magma of any composition is liable to occur in the glassy condition if cooled sufficiently rapidly. Acid (ie granitic) glass is commoner than basic (ie basaltic) glass; the former is represented among igneous rocks by pumice, obsidian and pitchstone; the latter by tachylite. Natural quartz glass occurs in masses lying on the surface of certain sandy deserts (eg the Libyan desert), while both clay rocks and sandstones are locally fused by basic intrusions. See BUCHITE, FULGURITE, TEKTITES.

natural immunity (*BioSci*) See INNATE IMMUNITY.

naturalistic observation (*Psych*) A research method where events are observed under normal or natural circumstances.

naturalized (*BioSci*) Introduced from another region but growing, reproducing and maintaining itself in competition with the native vegetation.

natural killer cell (*BioSci*) Abbrev *NK cell*. See KILLER CELLS.

natural language (*ICT*) Any naturally evolved human language.

natural load (*ElecEng*) A resistive load impedance, numerically equal to the characteristic impedance of the transmission line which it terminates, or the power which the line would transmit if it were so loaded.

natural magnet (*Phys*) See LODESTONE.

natural modes (*ICT*) See NORMAL MODES.

natural number (*MathSci*) A number used for counting objects; any element of the set [1,2,3 ...] or the set [0,1,2,3, ...]. The inclusion of zero is a matter of definition. Also *whole number*; see INTEGER.

natural period (*ICT*) The time of one cycle of oscillation arising from free oscillation, depending on inertia and elastance of a system. Reciprocal of natural frequency.

natural radioactivity (*Phys*) Radioactivity which occurs in nature. Such radioactivity indicates that the isotopes involved have a half-life comparable with the age of the Earth or result from the decay of such isotopes. Most such nuclides can be grouped in one of three RADIOACTIVE SERIES. It also accounts for 86% of all the radiation received by humans and originates from space, from natural radioactive elements, eg uranium-238, present in small amounts in rocks and soil, and from food, water and buildings. Uranium decays in a sequence of products (forming lead ultimately), one of which is radon: radioactive radon seeps through soil and may accumulate in poorly ventilated buildings. Radioactive elements naturally present in fossil fuels are released when the fuels are burned. See IONIZING RADIATION and panel on RADIATION.

natural resonance (*ICT*) The response of a system to a signal with period equal to its own natural period.

natural rubber (*Chem*) Polyisoprenes similar to that of *Hevea brasiliensis*, yielded by c.2000 plant spp (see BALATA, GUTTA PERCHA), although the rubber tree is the main and richest source of natural rubber. After tapping, the LATEX is concentrated from about 33% *dry rubber content* to 60% dry rubber content, and may then be coagulated direct to dry rubber or further processed in latex form. The main component is *cis*-polyisoprene of very high molecular mass $(5 \times 10^5–10^6)$ together with traces of resin, vegetable protein, minerals and plant sugars.

natural scale (*Acous*) The musical scale in which the frequencies of the notes within the octave are proportional to 24, 27, 30, 32, 36, 40, 45 and 48, and which can be realized in continuously variable-pitch instruments, such as the human voice and stringed instruments, but not in keyboard instruments, which use the TEMPERED SCALE. Also *just scale, just temperament*.

natural scale (*Surv*) A term applied to a section drawn with equal vertical and horizontal scales.

natural selection (*BioSci*) Darwin's theory, now generally accepted, that heritable characteristics that make an individual more reproductively successful will become increasingly common in successive generations because they make those individuals carrying the character better fitted to survival in the natural environment. Disadvantageous combinations of characters or deleterious mutations have a lower probability of being passed on and will gradually be eliminated. Natural selection pressures on randomly generated variants make species better adapted to their environment. Our modern understanding of genetics provides a clear basis for the evolution, in response to selection pressures, of successful genotypes. Artificial selection pressures for particular characteristics are the basis for conventional breeding programmes that have produced eg the diverse breeds of dogs.

natural slope (*CivEng*) The maximum angle at which soil in a cutting or bank will stand without slipping.

natural strain (*Eng*) See STRAIN.

natural uranium (*Chem*) That with its natural isotopic abundance, not depleted by the removal of ^{235}U.

natural uranium reactor (*NucEng*) A reactor in which natural, ie unenriched, uranium is the chief fissionable material.

natural wavelength of antenna (*ICT*) The free space wavelength corresponding to the natural frequency of an antenna.

nature identical (*FoodSci*) Substances which are chemically identical to natural substances.

nauplius (*BioSci*) The typical first larval form of Crustacea; egg-shaped, unsegmented, and having three pairs of appendages and a median eye; found in some members of every class of Crustacea, but often passed over, becoming an entirely embryonic stage.

Nauta mixer (*ChemEng*) A proprietary mixer comprising a stationary cone, with point vertically downwards, within which a screw arm rotates on its own axis parallel to the conical surface. The lower end of the screw is fixed in a universal joint at its lower end, and to an arm at its upper end, by means of which the whole screw, while rotating, is moved round the interior surface of the cone. Much used for mixing small quantities of one constituent with large quantities of another and mostly used on dry solids.

Nautical Almanac (*Astron*) An astronomical ephemeris published annually in advance, for navigators and astronomers. First published in 1767, it is now called the *Astronomical Ephemeris*. An abridged version, for the use of navigators, is given the original title *The Nautical Almanac*.

nautical log (*Ships*) A device for estimating the speed of a vessel. In the old-fashioned log, a line divided into equal spaces (knots) runs freely off a reel and is attached to a chip log, which is stationary in the water as the vessel travels. Time is measured by a log glass. The modern patent (or taffrail) log mechanically indicates the rate of travel by means of a submerged fly or rotator, whose revolutions are conveyed to a register on the rail of the vessel by a braided hemp line secured to the rotator.

nautical mile (*Genrl*) One-sixtieth of a degree of latitude, a distance varying with latitude. The *UK nautical mile* is 6080 ft (1853·18 m), differing slightly from the *international nautical mile*, 1852 m.

nautical twilight (*Astron*) The interval of time during which the Sun is between 6° and 12° below the horizon, morning and evening. See ASTRONOMICAL TWILIGHT, CIVIL TWILIGHT.

Nautiloidea (*BioSci*) A subclass of the Cephalopoda, having a wide central siphuncle and a planospiral chambered shell. They were abundant from the Early Cambrian to the Late Cretaceous, but are now represented by one genus, *Nautilus*, which lives in tropical seas. All chambers except the terminal living chamber contain gas which buoys up the heavy shell.

navaid (*Aero*) Abbrev for *navigational aid*.

naval brass (*Eng*) An alpha–beta brass (centred on 60% Cu/ 40% Zn) containing an addition of 1% tin to improve corrosion resistance. See COPPER ALLOYS, TOBIN BRONZE.

nave (*Arch*) The middle or main body of a basilica, rising above the *aisles*: the main part of a church, generally west of the crossing, including or excluding its aisles.

navel (*Textiles*) Part of the spinning head through which the yarn is withdrawn in ROTOR SPINNING.

navel (*BioSci*) In mammals, the point of attachment of the umbilical cord to the body of the fetus.

navel-ill (*Vet*) See JOINT-ILL.

navicular bone (*BioSci*) In mammals, one of the tarsal bones. Also *centrale* or *scaphoid*.

navicular disease (*Vet*) Chronic osteitis of the navicular bone in the foot of the horse.

navigable semicircle (*EnvSci*) The left-hand half of the storm field in the northern hemisphere, the right-hand half in the southern hemisphere, when looking along the path in the direction a *tropical revolving storm* is travelling. Cf DANGEROUS SEMICIRCLE.

navigation (*Build*) A name for a canalized river the flow of which is more or less under artificial control.

navigation (*Genrl*) The art of determining: (1) the position of a vessel or vehicle at a particular time; and (2) the course to be followed in order to arrive at a destination. Uses terrestrial and stellar observation, or radio and radar signals.

navigation (*Psych*) Complex forms of long-distance orientation by animals.

navigational system (*ICT*) Any system of obtaining bearings and/or ranges for navigational purposes by radio techniques.

navigation flame float (*Aero*) A pyrotechnic device, dropped from an aircraft, which burns with a flame while floating on the water. Used to determine the drift of the aircraft at night.

navigation lights (*Aero*) Aircraft navigation lights consist of red, green and white lamps located in the port wing tip, starboard wing tip and tail respectively.

navigation smoke float (*Aero*) A pyrotechnic device, dropped from an aircraft, which emits smoke while floating on the water. Used for ascertaining the direction of the wind or the drift of the aircraft.

navigation systems (*Genrl*) (1) Automatic systems using celestial navigation supplemented by radio or radar fixes from beacons, and which perform all calculations. Connecting these systems to flight control systems enables aircraft to fly themselves from airport to airport. (2) INERTIAL NAVIGATION SYSTEMS which do not rely on external beacons or transmissions, and are immune from interference. See GLOBAL POSITIONING SYSTEM, JOINT TACTICAL INFORMATION DISTRIBUTION SYSTEM, LORAN, OMEGA.

NAVSTAR (*Aero, Space*) Name for the GLOBAL POSITIONING SYSTEM using satellites spaced around the world.

Nb (*Chem*) Symbol for NIOBIUM.

N-bands (*BioSci*) See BANDING TECHNIQUES.

NBR (*Chem*) Abbrev for NITRILE–BUTADIENE RUBBER. Also *nitrile rubber*.

NBS (*Genrl*) Abbrev for NATIONAL BUREAU OF STANDARDS.

NBS smoke test (*Build*) Test method for grading burning polymers by the amount of smoke made, measured optically.

NBT (*BioSci*) Abbrev for NITROBLUE TETRAZOLIUM.

NCR (*Paper*) Abbrev for *No Carbon Required*. TN for stationery, esp for office machine use, in which simultaneous duplicate copies are obtained without the use of carbon paper. Microcapsules containing dyes on the verso of the 'top' copy are fractured by the impact of the writing medium (pen, typewriter, etc), the image being then transferred to the under copy or copies, the receiving surface of which has a coating of attapulgite with a starch or latex binder. See MICROENCAPSULATION.

Nd (*Chem*) Symbol for NEODYMIUM.

NDB (*Aero*) Abbrev for *non-directional beacon*. See BEACON.

NDI (*Chem*) Abbrev for NAPHTHALENE DIISOCYANATE.

N-display (*Radar*) A radar K-DISPLAY in which the target produces two breaks on the horizontal time base. Direction is proportional to the relative amplitude of the breaks, and range is indicated by a calibrated control which moves a pedestal signal to coincide with the breaks.

NDT (*Eng*) Abbrev for NON-DESTRUCTIVE TESTING.

Ne (*Chem*) Symbol for NEON.

neanic (*BioSci*) Said of the adolescent period in the life history of an individual.

neap tides (*Astron*) High tides occurring at the Moon's first or third quarter, when the Sun's tidal influence is working against the Moon's, so that the height of the tide is below the maximum in the approximate ratio 3:8.

Nearctic region (*BioSci*) One of the subrealms into which the HOLARCTIC REGION is divided; it includes N America and Greenland.

near-Earth object (*Astron*) A large celestial object, such as an asteroid, whose orbit may possibly bring it into collision with the Earth. Abbrev *NEO*.

near-end cross-talk (*ICT*) Cross-talk between two parallel circuits when both the listener and the speaker originating the inducing currents are at the same end of the parallelism. See FAR-END CROSS-TALK.

nearest-neighbour analysis (*BioSci*) A method for determining the frequency of pairs of adjacent bases in DNA. It has shown that there is a deficiency of the pair CG in most higher organisms.

near field (*Acous*) See FAR FIELD.

nearly free electron model (*Phys*) A model from which the band structure of simple metals can be calculated. The periodic part of the potential due to the crystal lattice is treated as a minor modification to the free electron gas model. See BAND THEORY OF SOLIDS.

near point (*Phys*) The nearest position to the eye at which an object can be seen distinctly. The object point conjugate to the retina when accommodation is exerted to its fullest extent.

neat cement (*Build, CivEng*) A cement mortar mixture made up with water only, without addition of sand.

neat size (*Build*) The net or exact size after preparation.

neat work (*Build*) The brickwork above the footings.

nebula (*Astron*) A term applied to any celestial object which appears as a hazy smudge of light in an optical telescope, such as extended clouds of gas or distant galaxies, its usage predating photographic astronomy. It is now more properly restricted to true clouds of INTERSTELLAR MEDIUM. See DARK NEBULA, EMISSION NEBULA.

nebula (*Med*) (1) A slight opacity in the cornea of the eye. (2) An oily preparation for use in an atomizer or nebulizer (eg a nasal spray).

nebular hypothesis (*Astron*) One of the earliest scientific theories of the origin of the solar system, stated by Laplace. It supposed a flattened mass of gas extending beyond Neptune's orbit to have cooled and shrunk, throwing off in the process successive rings which in time coalesced to form the several planets.

neck (*Arch*) The narrow moulding separating the capital of a column from the shaft.

neck (*BioSci*) (1) The upper tubular part of an archegonium and of a perithecium. (2) The lower part of the capsule of a moss, just above the junction with the seta. (3) More or less flexible link region between head and thorax.

neck (*Eng*) See NECKING.

neck (*Geol*) A plug of volcanic rock representing a former feeder channel of an extinct volcano.

neck (*Textiles*) In the process of extruding and drawing synthetic filaments a marked reduction in the cross-sectional area may occur. This part of the filament is called the neck and the occurrence is *necking*.

neck canal cell (*BioSci*) One of the cells in the central canal in the neck of an archegonium. Also *neck cell*.

Necker cube (*Psych*) Classic example of a reversible figure, one that can be seen in two different ways.

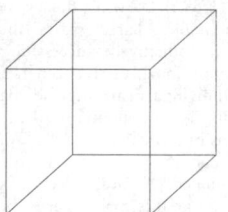

Necker cube

necking (*Eng*) The localized reduction in cross-section which occurs under uniaxial deformation of ductile materials, eg in a tensile test specimen shortly before failure. It reflects the reduction in rate of work hardening as degree of plastic strain increases, causing all the deformation to be confined to one region which progressively reduces in cross-section until fracture occurs.

necro- (*Genrl*) Prefix from Gk *nekros*, dead body.

necrobacillosis (*Vet*) Infection of animals by *Sphaerophorus necrophorus* (*Fusiformis necrophorus*). Also *bacillary necrosis*.

necrobiosis (*Med*) See NECROSIS.

necrophagous (*BioSci*) Feeding on the bodies of dead animals.

necrophorous (*BioSci*) Carrying away the bodies of dead animals; as certain beetles, which usually afterwards bury the bodies.

necropsy (*Med*) A postmortem examination of the body. Also *autopsy*.

necrosis (*BioSci*) Death of a cell (or of groups of cells) while still part of the living body. Also *necrobiosis*. Adj *necrotic*; v. *necrose*. Cf APOPTOSIS.

necrotic enteritis of swine (*Vet*) Infection of swine with *Campylobacter sputorum*, with various forms. Symptoms include weight loss, anorexia, diarrhoea, melaena and, in the case of the chronic form, poor weight gain. Also *porcine intestinal adenomatosis*.

necrotic stomatitis (*Vet*) See CALF DIPHTHERIA.

necrotizing fasciitis (*Med*) A bacterial infection that causes rapid decay of the fascia and soft tissue.

necrotroph (*BioSci*) An organism that feeds off dead cells and tissues. A necrotrophic parasite kills host cells and feeds on them once they are dead, eg the DAMPING-OFF fungi. Cf BIOTROPH, FACULTATIVE PARASITE.

nectar (*BioSci*) A sugary fluid exuded by plants, usually from some part of the flower but occasionally from somewhere else on the plant. It attracts insects that assist in pollination.

nectar (*FoodSci*) A class of fruit drink widely consumed on the European mainland. It consists of fruit juice with added sugar and possibly acidulant.

nectar guide (*BioSci*) See HONEY GUIDE.

nectarivorous (*BioSci*) Nectar-eating.

nectary (*BioSci*) A glandular organ or surface from which nectar is secreted.

necto- (*Genrl*) Prefix from Gk *nektos*, swimming.

necton (*BioSci*) See NEKTON.

nectopod (*BioSci*) An appendage adapted for swimming.

need (*Psych*) The specific physical or psychological conditions necessary for an individual's welfare and/or sense of well-being.

needle (*Acous*) See STYLUS.

needle (*BioSci*) A long, narrow, stiffly constructed leaf, characteristic of pines and similar plants.

needle (*Build*) A timber or steel beam used in the process of UNDERPINNING. It is laid horizontally at right angles to the wall (through which it passes) and is supported on both sides by DEAD SHORES, so as to take the load of the upper part of the walls.

needle (*ElecEng*) The moving magnet of a compass or galvanometer of the moving-magnet type. Sometimes also the moving element of an electrostatic voltmeter.

needle (*Textiles*) A simple pointed instrument with an eye for carrying the thread, used in sewing. In machine knitting, the device that forms and intermeshes the loops produced from the yarn supplied. Normally each needle forms one WALE of the fabric. There are many different kinds of needle used for different purposes. A bearded needle has a terminal hook that is flexed by pressing and released on removing the force. A latch needle has a terminal hook that is closed by a pivoting latch. The fabric loop overturns the latch and allows the loop to be cast off as another loop is being formed. A barbed needle in a needleloom causes the entangling of fibres in a method for making NON-WOVEN FABRICS.

needle beam (*CivEng*) A transverse floor beam supported across the chords of a bridge.

needle chatter (*Acous*) Noise arising from vibrations of the stylus of a record player being transferred to the tone arm and radiated.

needlecord (*Textiles*) A fine-ribbed CORDUROY used as a dresscloth.

needle lubricator (*Eng*) A crude form of lubricator consisting of an inverted stoppered flask attached to a bearing and containing a wire loosely fitting a hole in the stopper and touching the shaft.

needle paper (*Paper*) An acid-free, black paper for wrapping needles, pins, etc.

needle pickup (*Acous*) A pickup in which the sole moving part is the magnetic needle, which by its motion diverts magnetic flux and induces electromotive forces in coils on the magnetic circuit.

needle roller bearing (*Eng*) A ROLLER BEARING without cage, in which long rollers of small diameter are used, located endwise by a lip on the inner or outer race.

needles (*Print*) Removable points on web-fed presses, either fixed or retractable by cam action, used to control and convey the cut leading edge of the web or copy until it is severed, transferred, folded or stitched, as required; can be mounted on the folding transfer, cutting or collection cylinders.

needle scaffold (*Build*) A scaffold built up from cantilever or NEEDLE beams jutting out from an intermediate height in the building, avoiding erection from ground level.

needle scratch (*Acous*) Noise emanating from a record player, specifically due to irregularities in the contact surface of the groove. Also *surface noise*.

needle stone (*Min*) A popular term for clear quartz containing acicular inclusions, usually of rutile, but, in some specimens, of actinolite. Also *rutilated quartz*. The name has also been used for various acicular zeolites.

needle talk (*Acous*) Direct sound output from transducer of record player pickup.

needle traverse (*Surv*) A traverse in which the angles between successive lines, or the directions of the lines, are measured by means of a magnetic compass.

needle valve (*Eng*) A slender pointed rod working in a hole or circular seating; operated by automatic means, as in a carburettor float chamber, or by a screw, for the control of fluid.

needling (*Build*) The process of UNDERPINNING in which NEEDLES are used in the support of the upper part of the building.

needling (*Med*) Cutting the lens of the eye with a needle in the treatment of cataract. Also *discission*.

Néel temperature (*Phys*) The temperature at which the magnetic susceptibility of an antiferromagnetic material has a maximum value.

NEF (*Acous*) See NOISE EXPOSURE FORECAST VALUE.

negater (*ICT*) See NOT GATE.

negation (*MathSci*) The TRUTH FUNCTION operator on one variable that takes the value *true* when the given argument is *false*, and *false*, when it is *true*.

negative (*ImageTech*) An image, photographically or electronically produced, in which the tonal values of the original subject are reversed, bright areas being represented as dark, and vice versa. In a *colour negative*, in addition the hues of the original are represented by substantially complementary colours.

negative (*MathSci*) (1) Of a number, less than zero. (2) Of a statement, having negation as its principal operator.

negative (*Phys*) Designation to electric charge, introduced by Benjamin Franklin, now known to be exhibited by the electron, which, in moving, forms the normal electric current.

negative after-image (*Phys*) The image of complementary colour arising after visual fatigue from viewing a coloured object and then a white surface.

negative bias (*Electronics*) Static potential, negative with reference to earth, applied to electrode of valve or transistor, to obtain desired operating conditions.

negative binomial distribution (*MathSci*) The probability distribution of the total number of outcomes of a particular kind in a sequence of trials before the occurrence of exactly a prespecified number of outcomes of another kind, the probability of the outcome being constant at each trial and the different trials being statistically independent. When the prespecified number is one, the distribution is equivalent to a GEOMETRIC distribution.

negative booster (*ElecEng*) A series-wound booster used in connection with an earthed return power supply system, eg for a tramway. It is connected between two points on the earthed return path to reduce the potential between them and minimize the possibility of electrolysis due to leakage currents.

negative catalysis (*Chem*) The retardation of a chemical reaction by a substance which itself undergoes no permanent chemical change.

negative colour film (*ImageTech*) A multilayer photographic material intended to yield a COLOUR NEGATIVE image after exposure and processing.

negative conduction (*ICT*) The conductance (reciprocal of resistance) of negative resistance devices. See NEGATIVE RESISTANCE OSCILLATOR.

negative coupling (*ElecEng*) See POSITIVE COUPLING.

negative crystal (*Phys*) Birefringent material for which the velocity of the extraordinary ray is greater than that of the ordinary ray.

negative electricity (*Phys*) The phenomenon in a body when it gives rise to effects associated with excess of electrons. Cf POSITIVE ELECTRICITY.

negative electrode (*Electronics*) The anode of a primary cell (the electrode by which conventional current returns to the cell), but the cathode of a valve (connected to the negative side of the power supply).

negative feedback (*BioSci*) Situation in which the products of a process or reaction slow down or inhibit the reaction that produces them. Cf POSITIVE FEEDBACK.

negative feedback (*ICT*) The reduction of amplifier gain through feeding part of the output signal back to the input, in such a way that it is out of phase with the incoming signal. This gives more uniform performance, greater stability, reduced distortion and sometimes improved bandwidth.

negative feeder (*ElecEng*) The feeder connecting the negative terminal of a load to the negative bus-bars of the power supply. Also *return feeder*.

negative g (*Aero, Space*) (1) In a manoeuvring aircraft, any force acting opposite to the normal force of gravity. (2) The force exerted on the human body in a gravitational field or during acceleration so that the force of inertia acts in a foot-to-head direction, causing considerable blood pressure on the brain. *Negative g tolerance*, in practice, is the degree of tolerance $3g$ for 10–15 s. Also *minus g*.

negative glow (*Electronics*) In a medium-pressure gas-discharge tube, the glow between the cathode and Faraday dark space.

negative ion (*Electronics*) Radical, molecule or atom which has become negatively charged through the gain of one or more electrons. See ANION, ION.

negative-ion vacancy (*Electronics*) See HOLE.

negative mineral (*Phys*) A doubly refracting mineral in which the ordinary refractive index n_o is greater than the extraordinary refractive index n_e. Calcite is a negative mineral, for which the values of n_o and n_e are 1·66 and 1·48 respectively. See OPTIC SIGN.

negative modulation (*ICT*) That in which the carrier level is reduced as the level of the modulating signal is increased. Used in many TV systems, ie peak transmitter power corresponds to black, and the power decreases as brightness increases. Also *downward modulation*.

negative mutual inductance (*ElecEng*) See POSITIVE COUPLING.

negative phase sequence (*ElecEng*) A three-phase system in which the voltages and currents in each of the three phases reach their maximum values in the reverse order to conventional phase sequence, ie red, blue, yellow as opposed to red, yellow, blue. See PHASE SEQUENCE.

negative phase-sequence component (*ElecEng*) The symmetrical component of an unbalanced three-phase system of voltages or currents in which the phasesequence is in the opposite order to standard; ie it is in the order red, blue, yellow.

negative phase-sequence relay (*ElecEng*) A relay which operates when any negative phase-sequence components of current or voltage appear in the circuit to which it is connected.

negative plate (*ElecEng*) The plate of an accumulator or primary cell which is normally at the lower potential and to which the current from the circuit during discharge is said to return.

negative proton (*Phys*) See ANTIPROTON.

negative reaction (*BioSci*) A tactism or tropism in which the organism moves, or the member grows, from a region where the stimulus is stronger to one where it is weaker.

negative reinforcement (*Psych*) In conditioning situations, a stimulus, usually aversive, that increases the probability of escape or avoidance behaviour. Cf PUNISHMENT.

negative resistance (*Electronics*) A characteristic such that, when the current though a device increases, the voltage drop across it decreases. Most electrical gas discharges have this property, along with some valves and semiconductor devices, including the Gunn and tunnel diodes. See GUNN EFFECT, TUNNEL DIODE.

negative-resistance oscillator (*Electronics*) One in which a parallel-tuned resonant circuit or a cavity is connected to a negative-resistance device, the negative resistance compensating for the losses in the resonant circuit and allowing oscillation to become continuous. See GUNN EFFECT, TUNNEL DIODE.

negative scanning (*ImageTech*) Scanning a photographic negative, with reversal in the circuits, so that the reproduced image is the normal positive.

negative stagger (*Aero*) See STAGGER.

negative staining (*BioSci*) An important technique in electron microscopy, in which heavy metals which scatter electrons are deposited around the specimen. This is then seen in negative contrast.

negative transconductance (*Electronics*) Property of certain valves whereby an increase in positive potential on one electrode accompanies a decrease in current to another electrode.

negative video signal (*ImageTech*) A video signal in which increasing amplitude corresponds to decreasing light value in the transmitted picture. Black is taken as 100%, white about 30%, of the maximum amplitude in the signal.

neg-pos (*ImageTech*) (1) A process in which the film exposed in the camera is processed to form a negative image which is then used to make a separate positive by printing, in contrast to a REVERSAL PROCESS. (2) A PICTURE INVERSION facility found on some camcorders and film–video transfer devices which enables NEGATIVE film to be reproduced as positive images.

Negri bodies (*Vet*) Specific inclusion bodies found in the cytoplasm of nerve cells in the brain and spinal cord in animals affected with rabies.

neighbourhood (*MathSci*) Of a point: any open set containing that point.

Neisseriaceae (*BioSci*) A family of bacteria belonging to the order Eubacteriales. They are Gram-negative and characteristically occur as paired spheres. They are parasitic in mammals, eg *Neisseria gonorrhoeae* (gonorrhoea).

nekton (*BioSci*) Actively swimming aquatic organisms, as opposed to the passively drifting organisms or plankton. Also *necton*. Adj *nektonic*.

nematic (*Phys*) Said of a mesomorphous substance whose molecules or atoms are oriented in parallel lines; threadlike. Cf SMECTIC. See LIQUID CRYSTAL PHASES and panel on LIQUID CRYSTAL DISPLAYS.

nematoblast (*BioSci*) A cell that will develop a NEMATOCYST.

nematocyst (*BioSci*) An intracellular organelle, found in specialized cells in most Cnidaria (and a few Protozoa), consisting of a fluid-filled sac that is elongated at one end into a long, narrow, pointed hollow thread that normally lies inverted and coiled up within the sac. The sac can be everted and the contents injected into the prey when a small sensory projection (the cnidocil) is stimulated. Used for prey capture and defence. Also *cnidocyst*.

Nematoda (*BioSci*) A class of the phylum Aschelminthes, comprising unsegmented worms with an elongate rounded body pointed at both ends; marked by lateral lines and covered by a heavy cuticle composed of protein. They have a mouth and alimentary canal, a perivisceral cavity and a pseudocoele. Only longitudinal muscles are present. The nervous system consists of a circum-pharyngeal ring with a number of longitudinal cords. The sexes are separate and development is direct with larvae resembling the adults. Many species are of economic importance; mostly free-living in soil, although some are parasitic. Includes round worms, thread worms, eel worms. Various species can cause significant damage to crops, eg potatoes, and can be used as parasites to control pests, eg slugs and vine weevils. One nematode, *Caenorhabditis elegans*, is an important laboratory animal and was one of the first multicellular organisms to have its genome fully sequenced.

Nemertea (*BioSci*) A phylum of apparently non-metameric acoelomate worms with an elongate flattened body, a ciliated ectoderm and a dorsal eversible proboscis not connected with the alimentary canal. Most are marine, but some are fresh water or terrestrial. Ribbon worms. Also *Nemertini*.

NEMS (*Eng*) Abbrev for *nanoelectromechanical systems*.

NEO (*Astron*) Abbrev for *near-Earth object*.

neoblasts (*BioSci*) In many of the lower animals (Annelida, Ascidia, etc), large amoeboid cells widely distributed through the body which play an important part in REGENERATION.

neocerebellum (*Med*) Phylogenetically, the more recently developed part of the posterior lobe of the cerebellum, receiving predominantly afferent fibres via the pontocerebellar tract.

Neo-Classicism (*Arch*) An architectural style popular in Europe during the late 18th and early 19th centuries, when, in spite of the achievements of the BAROQUE phase, architects once more reverted to Greek and Roman ideals. It was a part of the ANTIQUARIAN movement, the final phase of the Renaissance. The term used more generally described any design which uses Classical ideals as a source of inspiration.

Neocomian (*Geol*) The oldest epoch of the Cretaceous period.

neo-Darwinism (*BioSci*) The modern version of Darwin's theory of evolution by NATURAL SELECTION, incorporating the discoveries of Mendelian and population genetics.

neodymium (*Chem*) Symbol Nd, at no 60, ram 144·24, rel.d. 6·956, mp 840°C. A metallic element, one of the RARE EARTH ELEMENTS. The metal is found in cerite, monazite and orthite. Neodymium glass is used for solid-state lasers and light amplifiers. Compounds of neodymium, iron and boron can be used to manufacture extremely high energy (BH_{max}) permanent magnets.

neo-Freudian (*Psych*) Any psychoanalytical school of thought or therapy based on modification of Freud's theories; such schools usually stress social rather than biological factors as important determinants of unconscious conflict.

Neogene (*Geol*) A subperiod of the Tertiary covering a time span from approx 25 to 2 million years ago. The corresponding system of rocks. See TERTIARY.

neohexane (*Chem*) 2,2-dimethylbutane. Bp 50°C. Antiknock fuel prepared by the addition of ethene to 2-methylpropane under pressure at increased temperatures.

Neolithic period (*Geol*) The later portion of the Stone Age, characterized by well-finished, polished stone implements, agriculture and domesticated farm animals. Cf PALAEOLITHIC PERIOD.

neologism (*Psych*) Verbal construction such as occurs in schizophrenia, manic depressive psychosis and some aphasias, in which the patient uses coined words, which may have meaning for him or ger but not for others, or else gives inappropriate meanings to ordinary words.

neomycin (*Pharmacol*) An antibiotic derived from *Streptomyces fradiae*; its sulphate is esp effective against external staphylococcal infections (skin, eyes, etc); also used internally.

neon (*Chem*) Symbol Ne, at no 10, ram 20·179, mp −248·67°C, bp −245·9°C. Light, gaseous, inert element, recovered from atmosphere. Historically important in that J J Thomson, through his parabolas for charge/mass of particles, found two isotopes in neon, the first non-radioactive isotopes to be recognized. Used in many types of lamp, particularly to start up sodium vapour discharge lamps. Pure neon was the first gas to be used for high-voltage display lighting, being bright orange in colour. Much used in cold-cathode tubes, reference tubes.

neonate (*Med*) A newly born infant.

neon induction lamp (*Phys*) A lamp consisting of a small tube containing neon at low pressure; luminescence is produced by the action of high-frequency currents in a few turns surrounding the tube.

neonychium (*BioSci*) A pad of soft tissue enclosing a claw of the fetus during the development of many mammals, to eliminate the risk of ripping the fetal membranes.

neopallium (*BioSci*) In mammals, that part of the cerebrum occupied with impressions from senses other than the sense of smell.

neoplasm (*Med*) A new formation of tissue in the body; a tumour. Adj *neoplastic*.

Neoprene (*Chem*) *Polychloroprene*. The first commercial synthetic rubber (US 1931). Chloroprene (3-chlorobut-1,2: 3,4-diene), the monomer, $CH_2=CClCH=CH_2$, is derived from acetylene and hydrochloric acid.

neossoptiles (*BioSci*) The down feathers found on a newly hatched bird.

neotenin (*BioSci*) In insects, the JUVENILE HORMONE that is produced by the corpora allata and that suppresses the development of adult characteristics at each moult except the last, when the corpora allata become inactive and metamorphosis occurs.

neoteny (*BioSci*) Retention of some juvenile characteristics by the sexually mature adult, eg some amphibians which have the appearance of tadpoles. Cf PAEDOGENESIS.

neotropical region (*BioSci*) One of the primary faunal regions into which the surface of the globe is divided. It comprises S America, the West Indian islands, and C America south of the Mexican plateau.

neovitalism (*BioSci*) The belief that a complete causal explanation of vital phenomena cannot be reached without invoking some extra-material concept.

Neozoic (*Geol*) The name (= *new life*) sometimes given to the Tertiary and Post-Tertiary rocks.

neper (*ICT*) Unit of attenuation. If current I_1 is attenuated to I_2 so that $I_2/I_1 = \exp(-N)$, then N is attenuation. In circuits matched in impedance, 1 neper = 8·686 dB. After John Napier (Lt *Nepero*), Scottish scientist, inventor of natural logarithms.

nepheline (*Min*) Silicate of sodium and aluminium, $NaAlSiO_4$, but generally with some potassium partially replacing sodium, which crystallizes in the hexagonal system. It is frequently present in igneous rocks with a high sodium content and a low percentage of silica, ie the undersaturated rocks. Also *nephelite*. See ELAEOLITE.

nepheline-syenite (*Geol*) A coarse-grained igneous rock of intermediate composition, undersaturated with regard to silica, and consisting essentially of nepheline, a varying content of alkali feldspar, with soda amphiboles and/or soda pyroxenes. Common hornblende, augite or mica are present in some varieties. See eg FOYAITE, LAURDALITE.

nephelinite (*Geol*) A fine-grained igneous rock normally occurring as lava flows, and resembling basalt in general appearance; it consists essentially of nepheline and pyroxene, but not of olivine or feldspar. The addition of the former gives *olivine-nephelinite*, and of the latter, *nepheline-tephrite*.

nephelite (*Min*) See NEPHELINE.

nephelometric analysis (*Chem*) A method of quantitative analysis in which the concentration or particle size of suspended matter in a liquid is determined by measurement of light absorption. Also *photoextinction method*, *turbidimetric analysis*.

nephograph (*ImageTech*) An instrument comprising electrically controlled cameras for photographing clouds etc, in order that their position in the sky may subsequently be determined.

nephology (*Genrl*) The study of clouds in meteorology.

nephoscope (*EnvSci*) An instrument for observing the direction of movement of a cloud and its angular velocity about the point on the Earth's surface vertically beneath it. If the cloud height is also known its linear speed may be calculated.

nephr-, nephro- (*Genrl*) Prefixes from Gk *nephros*, kidney.

nephrectomy (*Med*) Removal of a kidney.

nephric (*BioSci*) Pertaining to the kidney.

nephridiopore (*BioSci*) The external opening of a nephridium.

nephridium (*BioSci*) In invertebrates and lower Chordata, a segmental excretory organ consisting of an intercellular duct of ectodermal origin leading from the coelom to the exterior; more generally, an excretory tubule. Adj *nephridial*. Cf COELOMODUCT.

nephrite (*Min*) One of the minerals grouped under the name of *jade* ('New Zealand jade'); consists of compact and fine-grained tremolite or actinolite. It has been widely used for ornaments in the Americas and the East.

nephritis (*Med*) Inflammation of the substance of the kidney. See GLOMERULONEPHRITIS. Adj *nephritic*.

nephro- (*Genrl*) See NEPHR-.

nephrodinic (*BioSci*) Using the same duct for the discharge of both excretory and genital products.

nephrogenic tissue (*BioSci*) In the embryonic development of vertebrates, a relatively small intermediate region of the mesoderm lateral and ventral to the somites, from which derive the kidney tubules, their ducts, and the deeper tissues of the gonads. May be segmented, forming nephrotomes, or a continuous band of tissue.

nephrogonoduct (*BioSci*) Esp in invertebrates, a common duct for genital and excretory products.

nephrolithiasis (*Med*) The presence of stones in the kidney.

nephrolithotomy (*Med*) Removal of stones from the kidney through an incision in the kidney.

nephrologist (*Med*) A specialist in diseases of the kidney.

nephropathy (*Med*) Any disease of the kidneys.

nephropexy (*Med*) The fixation, by operative measures, of a kidney which is abnormally movable. Cf NEPHRORRHAPHY.

nephropore (*BioSci*) See NEPHRIDIOPORE.

nephroptosis (*Med*) Movable kidney; floating kidney. An abnormally mobile kidney, associated with general displacement downwards of other abdominal organs.

nephrorrhaphy (*Med*) The fixation, by suture, of a displaced kidney. Cf NEPHROPEXY.

nephros (*BioSci*) A kidney. Adj *nephric*.

nephrostome (*BioSci*) The ciliated funnel by which some types of nephridia open into the coelom.

nephrostomy (*Med*) The formation of an opening into the pelvis of the kidney for the drainage of urine.

nephrotic syndrome (*Med*) Increased permeability of kidney glomerulus basement membrane leading to albumen loss in urine (albuminuria), low plasma membrane and OEDEMA.

nephrotomy (*Med*) The making of an incision into the kidney.

nephrotoxin (*Med*) A poison or toxin which specifically affects the cells of the kidney.

nepionic (*BioSci*) Said of the embryonic period in the life history of an individual. From Gk *nepios*, infant.

neps (*Textiles*) A term applied in the cotton industry to small entanglements of fibres that cannot be unravelled; generally formed during the ginning process from dead or immature fibres.

Neptune (*Astron*) The eighth planet of the solar system, discovered in 1846 by J G Galle of the Berlin Observatory. It never gets brighter than magnitude 7·7, so it is never visible to the unaided eye. It orbits the Sun in 165 years, at a distance of 30·1 AU. The diameter is 49 492 km and the mass 17·2 times that of the Earth. There are seven natural satellites, including TRITON and NEREID. See appendix on Planets.

neptunean dyke (*Geol*) An intrusive sheet of sedimentary rock.

neptunism (*Geol*) The obsolete theory that all rocks including granite, basalt, etc, are deposited from water. Also *Wernerism*. Cf PLUTONISM.

neptunium (*Chem*) Element, at no 93, symbol Np; named after planet Neptune; produced artificially by nuclear reaction between uranium and neutrons. Has principal isotopes 237 and 239.

neptunium series (*Chem*) Series formed by the decay of artificial radioelements, the first member being plutonium-241 and the last bismuth-209, of which neptunium-237 is the longest lived (half-life $2 \cdot 2 \times 10^6$ years).

NEQ gate (*ICT*) See XOR GATE.

Nereid (*Astron*) The second substantial natural satellite of Neptune, discovered in 1949. Distance from the planet 5 515 000 km; diameter approx 340 km.

neritic zone (*Geol*) That portion of the sea floor lying between low-water mark and the edge of the continental shelf, at a depth of about 180 m. Sediments deposited here are of *neritic facies*, showing rapid alternations of the clay and sand grades; ripple marks etc indicate accumulation in shallow water.

Nernst bridge (*ElecEng*) An ac bridge for capacitance measurements at high frequencies.

Nernst effect (*Electronics*) A voltage which appears at opposite edges of a strip of metal that is conducting heat in the presence of a magnetic field which is perpendicular to the surface of the metal.

Nernst equation (*Chem*) A version of the GIBBS–HELMHOLTZ EQUATIONS relating the FREE ENERGY change in an electrochemical cell ΔG to the number of electrons transferred z, Faraday constant F and electrochemical potential E:

$$\triangle G = -zFE$$

Nernst heat theorem (*Chem*) As the absolute temperature of a homogeneous system approaches zero, so does the specific heat and the temperature coefficient of the free energy.

Nernst lamp (*Phys*) A lamp in which electric heating of a rod of zirconia in air gives rise to infrared rays for spectroscopy. The rod must be heated separately to start the lamp, as the material is insulating at room temperature.

Nernst's distribution law (*Chem*) When a single solute distributes itself between two immiscible solutes, then for each molecular species (ie dissociated, single or associated molecules) at a given temperature, there exists a constant ratio of distribution between the two solvents, ie $(C_1/C_2)_i = K_i$ for each molecular species, i.

Nernst theory (*Chem*) An explanation of the development of electrode potentials, based on the supposition that an equilibrium is established between the tendency of an electrode material to pass into solution and that of the ions to be deposited on the electrode.

nervation (*BioSci*) Also *nervature*. See VENATION.

nerve (*Arch*) Projecting rib on a vault surface. Also *nervure*.

nerve (*BioSci*) (1) In animals, a collection of axons leading to or from the central nervous system; also a nerve bundle or tract. Adjs *nervous*, *neural*. See NEURON. (2) In plants, a strand of conducting tissue and/or strengthening tissue in a leaf or leaf-like organ; a vein.

nerve (*Eng*) A term used to describe the defective state of vulcanized rubber where the degree of cross-linking varies randomly through the thickness of the product.

nerve block (*Med*) Production of insensibility of a part by injecting an anaesthetic into the nerve or nerves supplying it.

nerve centre (*BioSci*) An aggregation of nerve cells associated with a particular sense or function.

nerve ending (*BioSci*) The distal end of a nerve axon, normally a synapse.

nerve fibre (*BioSci*) An axon.

nerve gas (*Chem*) One which, by inhibiting the enzyme cholinesterase, rapidly and fatally acts on the nervous system. Many are derivatives of fluorophosphoric acid. Several have been made for military use, and some for insecticidal purposes.

nerve growth factor (*BioSci*) Multimeric polypeptide found in a variety of peripheral tissues that attracts developing neurites of the sensory and sympathetic systems to the tissues to form synapses. It then maintains the neurons. Abbrev *NGF*.

nerve impulse (*BioSci*) A regenerative electrical potential that travels along an AXON. See ACTION POTENTIAL.

nerve net (*BioSci*) The primitive type of nervous system found in Cnidaria, consisting of numerous multipolar neurons that form a net underlying and connecting the various cells of the body wall.

nerve plexus (*BioSci*) A network of interlacing nerve fibres.

nerve root (*BioSci*) The origin of a nerve in the central nervous system.

nerve trunk (*BioSci*) A bundle of nerve fibres united within a connective tissue coat.

nervous system (*BioSci*) The whole system of nerves, ganglia and nerve endings of the body of an animal, considered collectively.

nervure (*BioSci*) One of the chitinous struts that support and strengthen the wings of an insect.

nesosilicate (*Min*) A silicate mineral whose atomic structure contains isolated groups of silicon–oxygen tetrahedra. See SILICATES.

Nessler's solution (*Chem*) Used in the analysis of water for determination of free and combined ammonia. A solution of mercury (II) iodide in potassium iodide, made alkaline with sodium or potassium hydroxide. With a trace of ammonia it gives a yellow colour, but with larger amounts a brown precipitate.

nest (*BioSci*) An artefact built to provide temporary shelter as in some primates, or protection for the young and eggs in most birds, or for housing the colony in social insects.

nest (*Glass*) A cushion upon which glass is placed to be cut with a diamond.

nested loop (*ICT*) One contained within another LOOP.

nested PCR (*BioSci*) Polymerase chain reaction, in which two sets of primers (*nested primers*) are used sequentially. The initial PCR uses the 'outer' primer pairs, then a small aliquot is used as a template for a second round of PCR with the 'inner' primer pair.

nest epiphyte (*BioSci*) An EPIPHYTE in which the leaves and/or a tangle of stems and roots form a structure in which leaf litter collects, humifies and is used by the epiphyte to root into, as a source of mineral nutrients, eg the Bird's Nest Fern, *Asplenium nidus*.

net (*Textiles*) A firm open-mesh fabric made by weaving, knitting or knotting.

net assimilation rate (*BioSci*) Abbrev *NAR*. The net photosynthetic rate (ie total photosynthesis minus respiration for the plant) per unit leaf area. $NAR = (1/A)(dW/dt)$ where A is leaf area, W is dry weight and t is time. Also *unit leaf rate*.

Netlon (*Plastics*) TN for polymer net used in packaging, agriculture and netting. Made by extruding polymer melt through contra-rotating dies with matching slots in their mating faces.

netiquette (*ICT*) Standards for polite on-line behaviour.

NETNORTH (*ICT*) The Canadian constituent of the BITNET network.

net production (*EnvSci*) See PRODUCTION.

net pyrradiometer (*EnvSci*) An instrument for measuring the difference of the total radiations falling on both sides of a plane surface from the solid angle 2π respectively. Also *radiation balance meter*.

net register tonnage (*Ships*) Theoretically, the earning space of the ship, and the figure on which payment of harbour dues etc is based. It is the GROSS TONNAGE less DEDUCTED SPACES.

Netscape (*ICT*) TN of a BROWSER.

net tonnage (*Ships*) See NET REGISTER TONNAGE.

net transport (*NucEng*) The difference between the actual TRANSPORT in an isotope separation plant and that which would be obtained by the same plant with raw material of natural isotopic abundance.

net wing area (*Aero*) The GROSS WING AREA minus that part covered by the fuselage and any nacelles.

network (*ICT*) (1) Combination of electrical components, such as resistances, inductances, capacitances, etc; a network consisting only of these elements may be called passive, whereas one incorporating a source of energy, a valve, transistor, integrated circuit, etc, is active. (2) A number of interconnected communications facilities, eg telephones, telex machines, computer terminals, or a chain of transmitters interconnected so that they can provide the same programme material. Terms such as *ring* and *star* indicate the shape of the network.

network address (*ICT*) The unique designator of a specific, though not necessarily fixed, node in a network.

network analysis (*ElecEng*) The process of calculating theoretically the electrical properties of any network of passive and active components. Typical properties of interest include current flowing, voltages at different points and transfer functions.

network architecture (*ICT*) The communications equipment, PROTOCOLS and links that form a NETWORK and the methods by which the network is implemented.

network calculator (*ElecEng*) The combination of resistors, inductors, capacitors and generators used to simulate electrical characteristics of a power generation system, so that the effects of varying different operating conditions can be studied in computers. Also *network analyser*.

network computer (*ICT*) A simple computer that relies on a server, accessed over a network to provide application software and disk storage, functions normally found in a PC.

network drive (*ICT*) A DISK DRIVE that is shared between users on a NETWORK. The disk identity apparent to the users may be a LOGICAL DRIVE.

network layer (*ICT*) Level 3 of the OPEN SYSTEMS INTERCONNECTION (OSI) model that is responsible for establishing, maintaining and terminating the switched connection between systems.

network modifier (*Glass*) Metal ions such as Na^+, Ca^{2+}, K^+, B^{3+}, etc, in silicate glasses, which open up the network of silica tetrahedra and thus modify the properties of the glass.

network parameter control (*ICT*) Software in an ASYNCHRONOUS TRANSFER MODE network that protects QUALITY OF SERVICE by ensuring that users do not violate their contracts and by taking action to deal with any such illegal traffic.

network polymer (*Chem*) A polymer in which cross-linking has occurred.

network printer (*ICT*) A printer that is available and shared between the users of a NETWORK. See also LOCAL PRINTER.

network structure (*Eng*) The type of structure formed in alloys when one constituent exists in the form of a continuous network round the boundaries of the grains of the other. Even if the grains included in the cells are themselves duplex, they are regarded as individual grains.

network synthesis (*ElecEng*) The process of formulating a network with specific electrical requirements.

network theory (*BioSci*) (1) A theory postulated by N K Jerne in 1974 that the immune system is controlled by a network of interaction between antigen-binding sites (paratopes) that may be on immunoglobulin molecules or lymphocyte receptors. Each paratope is capable of binding an epitope on an external antigen and also an idiotope with a shape resembling the epitope present on another immunoglobulin molecule (the 'internal image'). An immunoglobulin that is an anti-idiotope will in turn be recognized by another molecule that is an anti-idiotope, and so on. (2) Mathematical analysis of network properties that are being applied, not always appropriately, to the complex interactions in eg intracellular signalling pathways.

network topology (*ICT*) The different layouts and configurations that may be used for building NETWORKS. See BUS TOPOLOGY, RING TOPOLOGY, STAR TOPOLOGY, TREE TOPOLOGY. Fig. ▷

bus   star   ring   tree

network topology Four common types.

network transfer function (*ICT*) Mathematical expression giving the ratio of the output of a network to its input. The natural logarithm of its value at any one frequency is the transfer constant, and gives the attenuation and phase shift for the signal propagated through the network.

Neumann function (*MathSci*) A BESSEL FUNCTION of the second kind.

Neumann lamellae (*Eng*) Straight, narrow bands appearing in the microstructure parallel to the crystallographic planes in the crystals of metals that have been deformed by sudden impact. They are actually crystallographic twins produced during rapid deformation processes, particularly in BODY-CENTRED CUBIC crystals, and are often observed in meteorites.

Neumann principle (*Crystal*) The physical properties of a crystal are never of lower symmetry than the symmetry of the external form of the crystal. Consequently the tensor properties of a cubic crystal, such as elasticity or conductivity, must have cubic symmetry, and the behaviour of the crystal will be isotropic.

neur-, neuro- (*Genrl*) Prefixes from Gk *neuron*, nerve.

neural (*BioSci*) See NERVE.

neural arch (*BioSci*) The skeletal structure arising dorsally from a vertebral centrum, formed by the neurapophyses and enclosing the spinal cord.

neural canal (*BioSci*) The space enclosed by the centrum and the neural arch of a vertebra, through which passes the spinal cord.

neural crest (*BioSci*) In a vertebrate embryo, a band of cells that lies parallel and close to the nerve cord and that will later give rise to the ganglia of the dorsal roots of the spinal nerves and a variety of other cell types.

neuralgia (*Med*) Paroxysmal intermittent pain along the course of a nerve. Adj *neuralgic*. See TIC DOULOUREUX.

neural network (*MathSci*) A digital simulation of the brain that creates connections between processing elements (the computer equivalent of neurons). The organization and weights of the connections determine the output, so neural networks can be 'trained' using a large database of prior examples in order to predict future events.

neural spine (*BioSci*) The median dorsal vertebral spine, formed by the fusion of the neurapophyses above the neural canal.

neural tube (*BioSci*) A tube formed dorsally in the embryonic development of vertebrates by the joining of the two upturned neural folds formed by the edges of the ectodermal neural plate, giving rise to the brain and spinal nerve cord.

neuraminic acid (*BioSci*) A 9-carbon amino sugar; the acylated derivative, sialic acid, occurs in glycoproteins and glycolipids and is responsible, via its carboxyl groups, for most of the negative charge of animal cell surfaces.

neuraminidase (*BioSci*) An enzyme produced by viruses of the myxovirus and paramyxovirus groups and by some bacteria which splits the glycosidic link between neuraminic acid or sialic acid and other sugars.

neurapophyses (*BioSci*) A pair of plates arising dorsally from the vertebral centrum, and meeting above the spinal cord to form the neural arch and spine. Sing *neurapophysis*.

neurasthenia (*Psych*) Psychiatric diagnostic category, of doubtful validity, in which there is mental and physical fatigue often accompanied by headaches, sleep disturbances and general malaise but no obvious cause.

neurectomy (*Med*) Excision of part of a nerve.

neurilemma (*BioSci*) See NEUROLEMMA.

neurine (*Chem*) *Trimethylvinylammonium hydroxide*. $CH_2\text{=}CHN(CH_3)_3OH$. Obtainable from brain substance and from putrid meat; related to *choline*, into which it can be transformed. It is a ptomaine base.

neurite (*BioSci*) A process growing out of a neuron that may be either a dendrite or an axon.

neuritis (*Med*) Inflammation of a nerve, eg *polyneuritis* (inflammation of many nerves), *optic neuritis* (inflammation of optic nerve).

neuro- (*Genrl*) See NEUR-.

neuroblastoma (*Med*) A malignant tumour composed of primitive nerve cells, arising in the adrenal gland or in connection with sympathetic nerve cells.

neuroblasts (*BioSci*) Cells of ectodermal origin that give rise to neurons.

neurocranium (*BioSci*) The brain case and sense capsules of a vertebrate skull.

neurocrine (*BioSci*) Neurosecretory; secretory property of nervous tissue.

neurocyte (*BioSci*) The body of a nerve cell. See NEURON.

neuroendocrinology (*Med*) The study of interactions between the *nervous system* and *endocrine organs*, particularly pituitary gland and hypothalamic region of brain.

neurofibrillary tangle (*BioSci*) Tangles of coarse neurofibrils within large neurons of the cerebral cortex are a characteristic pathological feature of the brain of patients with Alzheimer's disease. Whether this causes neuronal degeneration or is a secondary consequence remains contentious.

neurofibroma (*Med*) A tumour composed of fibrous tissue derived from the connective tissue sheath of a nerve. See MOLLUSCUM FIBROSUM.

neurofibromatosis (*Med*) Multiple tumours attached to peripheral nerves associated with pigmented skin patches.

neurogenesis (*BioSci*) The development and formation of nerves.

neurogenic (*BioSci*) The activity of a muscle or gland that is dependent on continued nervous stimuli. Cf MYOGENIC.

neuroglia (*BioSci*) The supporting tissue of the brain and spinal cord of vertebrates, composed of much branched fibrous cells that occur among the nerve cells and fibres. Also *glia*.

neurohaemal organs (*BioSci*) Organs that serve as a gateway for the escape of the products of neurosecretory cells from the neurons into the circulating blood, eg the CORPORA CARDIACA of insects.

neurohypophysis (*BioSci*) See PARS NERVOSA.

neurolemma (*BioSci*) The thin insulating MYELIN SHEATH, produced by Schwann cells, that surrounds nerve fibres in vertebrates.

neuroleptic drugs (*Pharmacol*) Antipsychotic drugs.

neurologist (*Med*) A specialist in diseases of the nervous system.

neurology (*Med*) The study of the nervous system.

neurolymphomatosis (*Vet*) See FOWL PARALYSIS.

neuromasts (*BioSci*) Sensory hair cells embedded in a gelatinous cupola found in the lateral line system of lower vertebrates and concerned with MECHANORECEPTION.

neuromodulation (*BioSci*) Alteration in the effectiveness of voltage- or ligand-gated ION CHANNELS by changing the characteristics of the current flow through them.

neuromuscular (*BioSci*) Pertaining to nerve and muscle, as a myoneural junction.

neuron (*BioSci*) A nerve cell and its processes. Also *neurone*. Fig. ▷

neuronitis (*Med*) Inflammation (or degeneration) of neurons.

neuropathology (*Med*) The study of pathology of diseases of the nervous system.

neuropathy (*Med*) Functional or structural disorders of the nervous system.

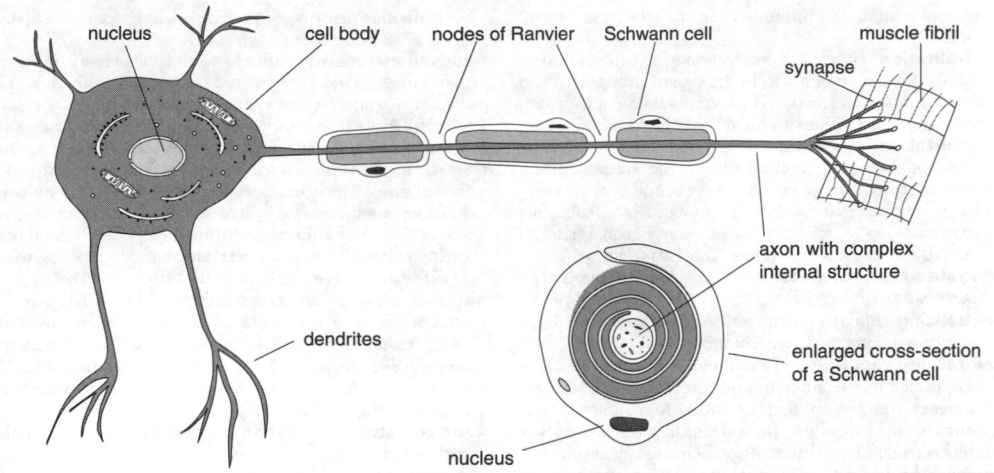

neuron Diagram of a typical motor neuron.

neuropeptide (*BioSci*) A peptide that either directly affects a nerve synapse (peptide neurotransmitter) or that has an indirect modulatory effect on the nervous system (peptide neuromodulator), eg BRADYKININ, GASTRIN, NEUROTENSIN, SECRETIN.

neurophysiology (*Genrl*) The physiology of the nervous system.

neuropil (*BioSci*) In vertebrates, a network of axons, dendrites and synapses within the central nervous system.

neuropile (*BioSci*) In arthropods, regions within the brain and the central portion of segmental ganglia consisting of dendrites and synapses.

neuropore (*BioSci*) The anterior opening by which the cavity of the central nervous system communicates with the exterior.

neurosecretory cell (*BioSci*) A special type of neuron in which the axon terminates against the wall of a blood vessel or sinus into which it secretes a hormone or other factor.

neurosis (*Psych*) A loose term for any mental disorder in which the individual experiences anxiety, or engages in behaviour to avoid experiencing it; there is no evidence of an organic component, and the individual remains in contact with reality.

neurosurgery (*Med*) Surgery on parts of the nervous system.

neurosyphilis (*Med*) Syphilitic infection of the nervous system.

neurotransmitter (*BioSci*) A substance that transmits a nerve impulse across a chemical SYNAPSE by binding to a LIGAND-GATED ION CHANNEL on the post-synaptic membrane, eg acetylcholine, gamma amino butyric acid (GABA), noradrenaline.

neurotrophin (*BioSci*) Any of a group of proteins that are structurally related to NERVE GROWTH FACTOR, and that are important in the survival or growth of certain classes of embryonic neurons.

neurotropic (*Med*) Having a special affinity for nerve cells, eg *neurotropic virus*.

neurula (*BioSci*) In the embryonic development of vertebrates, the stage after the gastrula in which the processes of organ formation begin, with the formation of the neural tube, mesodermal somites, notochord and archenteron.

neuter (*BioSci*) (1) Sexless. (2) Lacking functional sexual organs; a flower having neither functional stamens nor functional carpels; sterile. Also *neutral*.

neutral (*BioSci*) See NEUTER.

neutral (*ElecEng*) See NEUTRAL CONDUCTOR, NEUTRAL POINT.

neutral (*Glass*) Glass of high chemical durability.

neutral (*ImageTech*) Possessing no colour or hue; grey.

neutral (*Phys*) (1) Exhibiting no resultant charge or voltage. (2) Said of a return conductor of a balanced power supply, nearly at earth potential, if without a local earth connection.

neutralator (*ElecEng*) See EARTH REACTOR.

neutral autotransformer (*ElecEng*) See EARTH REACTOR.

neutral axis (*ElecEng*) A term used to denote the diametral plane in which the brushes of a commutator machine should be situated to give perfect commutation.

neutral axis (*Eng*) The line of zero stress in a beam subjected to bending; more accurately, a transverse section of the longitudinal plane which passes through the centre of area of the section. Also *neutral surface*.

neutral beam (*Phys*) A beam of high-energy atoms used to heat a plasma. As the atoms are neutral the beam is not affected by magnetic fields.

neutral compensator (*ElecEng*) See EARTH REACTOR.

neutral conductor (*ElecEng*) The middle wire of a dc THREE-WIRE SYSTEM or a distribution system, or the wire of a polyphase distribution system which is connected to the neutral point of the supply transformer or alternator. Also *middle conductor*, *middle wire*, *neutral*, *neutral wire*. Cf OUTER.

neutral current (*Phys*) Weak interaction in which electric charge is not transferred between particles; mediated by the Z BOSON. Cf CHARGED CURRENT.

neutral density filter (*ImageTech*) One which attenuates all colours uniformly so that the relative spectral distribution of the energy of the transmitted light has been unaltered.

neutral element (*MathSci*) See IDENTITY (2).

neutral equilibrium (*Phys*) The state of equilibrium of a body when a slight displacement does not alter its potential energy.

neutral flame (*Eng*) In welding, flame produced by a mixture at the torch of acetylene and oxygen in equal volumes.

neutral flux (*Eng*) A non-basic, non-acidic flux for modifying the fusibility of furnace slags, eg calcium fluoride.

neutral injection (*NucEng*) The additional heating of a plasma by injecting beams of accelerated atoms into it.

neutral inversion (*ElecEng*) A condition in which the phase to neutral voltages of a three-phase, star-connected system are unbalanced. Commonly due to FERROMAGNETIC resonance.

neutralization (*Chem*) The interaction of an acid and a base with the formation of a salt. In the case of strong acids and bases, the essential reaction is the combination of hydrogen ions with hydroxyl ions to form water molecules.

neutralization (*Electronics*) A method of counteracting oscillation-inducing feedback from the output of an amplifier to the input via the interelectrode capacitances of a valve. Reversed feedback is provided by a balancing capacitance to which is applied voltage equal and in antiphase to that on the anode. Also *balancing*.

neutralized series motor (*ElecEng*) See COMPENSATED SERIES MOTOR.

neutralizing (*Build*) Pretreatment of a surface to render it chemically neutral before decoration.

neutralizing capacitance (*Electronics*) An adjustable capacitor, placed in the anode to grid feedback path of a high-frequency valve amplifier, in order to counteract any tendency to instability from feedback via the valve's interelectrode capacitances. Also *balancing capacitance*.

neutralizing indicator (*ElecEng*) One indicating the degree of neutralization present in an amplifier.

neutralizing voltage (*ICT*) That fed back on the process of neutralization.

neutral mutation (*BioSci*) A mutation that has no selective advantage or disadvantage. Considerable controversy surrounds the question of whether such mutations can exist.

neutral point (*Aero, Space*) (1) See GRAVIPAUSE. (2) That cg (centre of gravity) position in an aircraft at which longitudinal stability is neutral. *Stick-fixed neutral point* is the cg position at which control column movement to trim a change in speed is zero. *Stick-free neutral point* is the cg position at which the stick force needed to trim a change in speed is zero.

neutral point (*ElecEng*) (1) The point at which the windings of a polyphase star-connected system of windings are connected together. (2) The mid-point of the neutral zone of a dc machine. Also *neutral*.

neutral point (*EnvSci*) A small region of the daylight sky from which scattered sunlight is unpolarized; such points were discovered by Arago, Babinet and Brewster.

neutral point (*Phys*) (1) A point in the field of a magnet where the Earth's magnetic field (usually the horizontal component) is exactly neutralized. (2) See GRAVIPAUSE.

neutral pump (*Phys*) A pump which transports only uncharged molecules or appropriately balanced pairs of ions so that there is no net transfer of charge.

neutral relay (*ICT*) US NON-POLARIZED RELAY.

neutral solution (*Chem*) An aqueous solution which is neither acidic nor alkaline. It therefore contains equal quantities of hydrogen and hydroxyl ions and has a PH value of 7.

neutral state (*Phys*) Said of ferromagnetic material when completely demagnetized. Also *virgin state*.

neutral surface (*Eng*) See NEUTRAL AXIS.

neutral wedge filter (*ImageTech*) A neutral grey filter, such as a wedge of grey glass cemented to a similar wedge of clear glass, which introduces a continuously variable attenuation of light, depending on the density or thickness of the grey glass introduced into the beam, without altering the relation between hues in the transmitted light.

neutral wire (*ElecEng*) See NEUTRAL CONDUCTOR.

neutral zone (*ElecEng*) That part of the commutator of a dc machine in which, when the machine is running normally, the voltage between adjacent commutator bars is approximately zero.

neutrino (*Phys*) A fundamental particle, a LEPTON, with zero charge and zero mass. A different type of neutrino is associated with each of the three charged leptons. Its existence was predicted by Pauli in 1931 to avoid beta decay infringing the laws of conservation of energy and angular momentum. As they have very weak interactions

with matter, neutrinos were not observed experimentally until 1956. See ANTINEUTRINO.

neutrino astronomy (*Astron*) A term applied to attempts to detect NEUTRINOS from the Sun, with the aim of discovering more exactly the conditions in the solar core.

neutron (*Phys*) Uncharged subatomic particle, mass approximately equal to that of the proton, which enters into the structure of atomic nuclei. Interacts with matter primarily by collisions. The spin quantum number is $+\frac{1}{2}$, rest mass is 1·008 665 amu, charge is zero and magnetic moment is $-1·9125$ nuclear Bohr magnetons. Although stable in nuclei, isolated neutrons decay by beta emission into protons, with a half-life of 11·6 minutes. See NEUTRON ENERGY.

neutron absorption cross-section (*Phys*) The cross-section for a nuclear reaction initiated by neutrons. For many materials this rises to a large value at particular neutron energies due to resonance effects, eg a thin sheet of cadmium forms an almost impenetrable barrier to thermal neutrons. Unit is the BARN.

neutron activation analysis (*Chem*) Irradiating a sample with thermal neutrons and studying the gamma rays emitted. Abbrev *NAA*. See panel on TRACE ELEMENT ANALYSIS.

neutron age (*NucEng*) See FERMI AGE.

neutron balance (*NucEng*) For a constant power level in a reactor, there must be a balance between the rate of production of both prompt and delayed neutrons, and their rate of loss due to both absorption and leakage from the reactor.

neutron current (*NucEng*) The net rate of flow of neutrons through a surface perpendicular to the direction in which they are migrating. *Neutron current* is a *vector* whereas *neutron flux* is a *scalar*.

neutron detection (*NucEng*) Observation of charged-particle recoils following collisions of neutrons with protons in a counter containing hydrogen gas or a compound of hydrogen; or of charged particles produced by interaction of neutrons with atomic nuclei, eg in a boron trifluoride *counter*. Absorption of neutrons by boron gives rise to α-particles which can be detected.

neutron diffraction (*Phys*) The coherent elastic scattering of neutrons by the atoms in a crystal. If the scattering is by the nucleus then the atomic structure of the crystal can be deduced from the measurements of the diffraction pattern. If the scattering is by atoms with electron configurations that have a magnetic moment, then details of the magnetic structure of the crystal can be determined. See NEUTRON MAGNETIC SCATTERING.

neutron diffusion (*NucEng*) The migration of neutrons from regions of high neutron density to those of low density, in a medium in which neutron capture is small compared with neutron scattering. See AGE THEORY, GROUP THEORY, TRANSPORT THEORY.

neutron economy (*NucEng*) Reducing the losses of neutrons in a reactor to a minimum.

neutron elastic scattering (*Phys*) Crystal diffraction of a beam of thermal neutrons with no loss of energy, ie no wavelength change. See NEUTRON INELASTIC SCATTERING.

neutron energy (*Phys*) (1) The binding energy of a neutron in a nucleus, usually several mega-electron-volts. (2) The energy of a free neutron which in a reactor will be classed in several groups: high-energy neutrons, energy >10 MeV; fast neutrons, energy 10 MeV to 20 keV; intermediate neutrons, energy 20 keV to 100 eV; epithermal neutrons, energy 100 eV to 0·025 eV; thermal or slow neutrons, energy approx 0·025 eV. See MULTIGROUP THEORY.

neutron excess (*Phys*) The difference between the neutron number and the proton number for a nuclide. Also *isotopic number*.

neutron flux (*Phys*) The number of neutrons passing through unit area in unit time, or the product of the number of neutrons per unit volume and their mean speed. In a nuclear reactor the flux is of the order of 10^{16} to $10^{18}\,\mathrm{m}^{-2}\,\mathrm{s}^{-1}$. See panel on NUCLEAR REACTORS.

neutron generation (*Phys*) The cycle of events between the emission of a neutron from a fissile element and the induction of further neutrons in the CHAIN REACTION. In a controlled fission reaction these events include neutron losses in the MODERATOR and from the reactor.

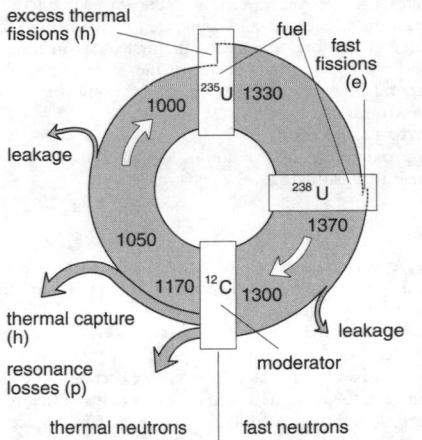

excess thermal fissions (h)

fuel

fast fissions (e)

^{235}U 1330

1000

leakage

^{238}U

1050

1370

1170 ^{12}C 1300

thermal capture (h)

leakage

resonance losses (p)

moderator

thermal neutrons

fast neutrons

neutron generation The fate of neutrons from their emission (top of the diagram).

neutron gun (*NucEng*) A block of moderating material with a channel through it used for producing a beam of fast neutrons.

neutron hardening (*Phys*) Increasing the average energy of a beam of neutrons by passing them through a medium which shows preferential absorption of slow neutrons.

neutron inelastic scattering (*Phys*) Crystal diffraction of a beam of thermal neutrons, whose energies are of the same order of magnitude as a quantum of lattice vibrational energy, a *phonon*, with exchanges of energy with the excited travelling waves. A detailed examination of the change in direction and energy of the scattered neutrons gives valuable information about the lattice dynamics of the crystal. Cf NEUTRON ELASTIC SCATTERING.

neutron leakage (*NucEng*) The escape of neutrons in a reactor from the core containing the fissile material; reduced by using a reflector.

neutron magnetic scattering (*Phys*) Magnetic scattering of neutrons due to interaction of the magnetic moment in a crystal (from both electron spin and electron orbital angular momenta) with the neutron spin magnetic moment. This is a powerful method for the study of magnetic structures.

neutron–nuclear scattering length (*Phys*) A measure of the ability of different nuclei to scatter neutrons; it is independent of the scattering angle. The scattering length is different for different isotopes of the same atom, so that, in addition to coherent scattering from a crystal containing more than one isotope, there will be incoherent scattering. Neutron–nuclear scattering is spin dependent, and this leads to further incoherent scattering from a nucleus with non-zero spin.

neutron number (*Phys*) The number of neutrons in a nucleus. Equal to the difference between the relative atomic mass (the total number of nucleons) and the atomic number (the number of protons).

neutron poison (*NucEng*) Any material, other than fissionable, which absorbs neutrons; used for the control of nuclear reactors. See panel on NUCLEAR REACTORS.

neutron radiography (*Radiol*) A beam of neutrons from a nuclear reactor traverse an object and impinge on an image-producing detector which reveals areas of absorption. Similar to X-ray radiography, but with the advantage that elements of low atomic number (eg H, Li, B) are strongly absorbed, unlike X-ray radiography in which low atomic numbers give low X-ray absorptions. Used as a

NON-DESTRUCTIVE TESTING technique and also for therapy but clinical trials do not demonstrate much, if any, therapeutic advantage.

neutron reproduction factor (*NucEng*) The factor, k, which is the ratio between the number of neutrons inducing fission in the fuel to the number required to maintain the chain reaction. If $k = 1$, the reactor is *critical* and operating at a steady state; if $k < 1$ it is *subcritical* and in a shutdown state; if $k > 1$ it is *supercritical* or *divergent*. Also *multiplication factor*.

neutron scatter plug (*NucEng*) A mechanical device in a fuel assembly designed to reflect those neutrons back into the core which would otherwise have escaped through the coolant pipe.

neutron shield (*NucEng*) Radiation shield erected to protect personnel from neutron irradiation. In contra-distinction to gamma-ray shields, it must be constructed of very light hydrogenous materials which will quickly moderate the neutrons which can be absorbed, for instance, by boron incorporated in the shield.

neutron source (*NucEng*) One giving a high neutron flux, eg for neutron activation analysis. Apart from reactors, these are chemical sources such as a radium–beryllium mixture emitting neutrons as a result of the (αv) reaction, and accelerator sources in which deuterium nuclei are usually accelerated to strike a tritium-impregnated titanium target, thus releasing neutrons by the (DT) reaction. The former are continuously active but the latter have the advantage of becoming inert as soon as the accelerating voltage is switched off.

neutron spectrometer (*NucEng*) An instrument for investigation of energy spectrum of neutrons. See CRYSTAL SPECTROMETER, TIME-OF-FLIGHT SPECTROMETER. Other techniques depend on the nuclear reaction of neutrons with ^{3}He (*helium-3 spectrometer*), ^{6}Li or hydrogen (*proton recoil spectrometer*).

neutron spectroscopy (*Phys*) Experimental determination of the intensity and change in energy (wavelength) of the neutrons scattered in a particular direction when a beam of monoenergetic neutrons are incident on a crystal. A powerful method of studying lattice vibrations and phonon energies.

neutron star (*Astron*) A small body of very high density (approx 10^{17} kg m^{-3}) resulting from a supernova explosion, following which the core of a massive star collapses under its own gravitational forces, the electrons and protons combining to form neutrons. See panel on PULSAR.

neutron therapy (*Med*) Use of neutrons for medical treatment.

neutron velocity selector (*NucEng*) (1) A device using a rotating CHOPPER or a rotating helix, or a pulsed accelerator, to provide a pulse of neutrons, the velocity being selected by time of flight (see TIME-OF-FLIGHT SPECTROMETER). (2) A device using neutron diffraction. See CRYSTAL SPECTROMETER.

neutron yield (*NucEng*) The average number of neutrons emitted at each fission; of the order of 2·5.

neutropenia (*Med*) Abnormal diminution in the number of neutrophil leucocytes in the blood.

neutrophil (*Med*) (1) Stainable by neutral dyes. (2) Casual abbreviation for NEUTROPHIL LEUCOCYTE.

neutrophil leucocyte (*BioSci*) The predominant type of circulating leucocyte. Neutrophils are POLYMORPHONUC-LEAR GRANULOCYTES and are involved in the body's natural defence against pathogens. Neutrophils move out of the blood stream into damaged tissue where they attack and phagocytose invading micro-organisms. The cytoplasm is readily stained by neutral dyes.

Nevadan orogeny (*Geol*) An OROGENY of Jurassic age affecting the western US.

névé (*EnvSci*) A more or less compacted snow ice occurring above the snowline; it consists of small rounded crystalline grains formed from snow crystals. Also *firn*.

Newall system (*Eng*) A commonly used system of limits and fits, using a hole basis.

Newcastle disease (*Vet*) An acute, highly contagious paramyxovirus disease of chickens, other domestic fowl and wild birds. Symptoms include loss of appetite, diarrhoea, and respiratory and nervous symptoms. Vaccines widely used. Also *fowl pest*, *pneumoencephalitis*, *pseudo-fowl plague*.

newel (*Build*) An upright post fixed at the foot of a stair or at a point of change of direction and used as a support for a balustrade. See fig. at STRING.

newel cap or drop (*Build*) Ornamental finish planted on the upper or lower end of a newel post.

newel joints (*Build*) The joints connecting the newel and the hand rail or string.

New Forest disease (*Vet*) See INFECTIOUS OPHTHALMIA.

Newlyn datum (*Surv*) See MEAN SEA LEVEL.

new Moon (*Astron*) The instant when the Sun and Moon have the same celestial longitude; the illuminated hemisphere of the Moon is then invisible.

New Red Sandstone (*Geol*) A name frequently applied to the combined Permian and Triassic systems, and particularly applicable in N Europe, where the palaeontological evidence is insufficient to allow of their separation. The term reflects the general resemblance between the rocks comprising these two systems and the Old Red Sandstone of Devonian age.

newsgroup (*ICT*) A group that exchanges views and information on-line via the Internet. See panel on INTERNET.

newsprint (*Paper*) Cheap printing paper intended for printing by letterpress or web offset for the production of newspapers, generally from 40 to 52 g m^{-2}. The furnish is generally predominantly a mechanical wood pulp but de-inked waste paper may also feature prominently.

New Style (*Astron*) A name given to the system of date reckoning established by the GREGORIAN CALENDAR. Abbrev *NS*.

newton (*ElecEng*) The unit of force in the SI system, being the force required to impart, to a mass of 1 kg, an acceleration of 1 m s^{-2}; 1 newton = 0.2248 lbf. Symbol *N*.

Newtonian flow (*Phys*) Fluid flow which obeys Newton's law of viscosity. See NEWTONIAN VISCOSITY and panel on AERODYNAMICS.

Newtonian fluid (*Phys*) Fluid flow which obeys Newton's law of viscosity. See NEWTONIAN VISCOSITY.

Newtonian mechanics (*Phys*) The theory of mechanics dealing with the relationships between force and motion for large objects, based on Newton's laws of motion. Also *classical mechanics*.

Newtonian telescope (*Astron*) A form of reflecting telescope designed by Newton, in which the object is viewed through an eyepiece in the side of the tube, the light reflected from the main mirror being deflected into it by a small plane mirror inclined at 45° to the axis of the telescope and situated just inside the principal focus. See panel on ASTRONOMICAL TELESCOPE.

Newtonian viscosity (*Phys*) Viscosity which is described by Newton's law of viscosity, which states that the shear stress is directly proportional to the shear rate, the proportionality constant (η) being known as the *coefficient of shear viscosity*. It is temperature dependent but independent of shear rate. Fluids like water are strictly Newtonian, but many fluids are non-Newtonian. See NON-NEWTONIAN FLOW.

Newton's disk (*Phys*) Motor-driven disk with sectors of primary colours, which appears white on fast rotation and with white illumination, demonstrating the synthesis of colour vision.

Newton's law of cooling (*Phys*) The rate of cooling of a hot body which is losing heat both by radiation and by natural convection is proportional to the difference in temperature between it and its surroundings. The law does not hold for large temperature excesses.

Newton's laws of motion (*Phys*) Three fundamental classical laws: (1) every body continues in a state of rest or uniform motion in a straight line unless acted upon by an external impressed force; (2) the rate of change of momentum is proportional to the impressed force and takes place in the direction of the force; (3) action and reaction are equal and opposite, ie when two bodies interact the force exerted by the first body on the second body is equal and opposite to the force exerted by the second body on the first. These laws were first stated by Newton in his *Principia* (1687).

Newton's method (*MathSci*) An iterative method of approximating the solution of an equation $f(x) = 0$. From an initial estimate x_0, the equation

$$x_{n+1} = x_n - \frac{f(x_n)}{f'(x_n)}$$

is repeatedly used to obtain the sequence $x_1, x_2, x_3, \ldots$ which will converge to the solution if the initial estimate is sufficiently accurate.

Newton's rings (*Phys*) Circular concentric interference fringes seen surrounding the point of contact of a convex lens and a plane surface. Interference occurs in the air film between the two surfaces. If r_n is the radius of the nth dark ring, R is the radius of curvature of the lens surface and λ the wavelength, then $r_n = \sqrt{(nR\lambda)}$. See CONTOUR FRINGES.

new variant Creutzfeldt–Jakob disease (*Med*) Abbrev *nvCJD*. See panel on TRANSMISSIBLE SPONGIFORM ENCEPHALOPATHY.

New Zealand greenstone (*Min*) Nephritic 'jade' of gemstone quality, from New Zealand.

next instruction register (*ICT*) See PROGRAM COUNTER.

N galaxy (*Astron*) A type of active galaxy with an extremely bright compact nucleus; may be a powerful radio source. The luminosity is between that of a quasar and a SEYFERT GALAXY. See panels on GALAXY and QUASAR.

NGC (*Astron*) Abbrev for the *New General Catalogue* of all nebulous objects known in 1888. Together with the supplementary Index Catalogue (IC) it lists 13 000 galaxies, clusters and nebulae.

NGF (*BioSci*) Abbrev for NERVE GROWTH FACTOR.

NGTE rigid rotor (*Aero*) A helicopter self-propelling rotor system evolved at the *National Gas Turbine Establishment* which uses the principle of the JET FLAP to obtain very high lift coefficients.

Ni (*Chem*) Symbol for NICKEL.

niacin (*Med*) Deficiency results in PELLAGRA. Also *vitamin B$_3$*; *nicotinic acid*.

niangon (*For*) W African hardwood (*Tarrietia*) with pale-pink to reddish-brown heartwood, with wavy, interlocked grain and coarse texture. Also *nyankom*.

nib (*Build*) The point of a crowbar.

nibble (*ICT*) A group of 4 BITS. Half of a BYTE.

nibbler (*CivEng*) Hydraulic pincers attached to an excavator and used for demolition of concrete slabs by combined bending and pincer action.

nibbling (*Eng*) A sheet metal process for cutting irregular and tightly curved paths in which a rapidly reciprocating cutter slots the sheet.

nibs (*Build*) Specks of solid matter in surface paint coatings, marring the finished appearance.

NICAM (*ImageTech*) Abbrev for *near instantaneous companded audio multiplex*. Digital stereo or bilingual audio which is compressed for transmission and expanded in the TV receiver.

niccolite (*Min*) Nickel arsenide, crystallizing in the hexagonal system. It is one of the chief ores of metallic nickel. Also *copper nickel*, *kupfernickel*.

niche (*EnvSci*) See ECOLOGICAL NICHE.

niched column (*Arch*) A column set back in a wall with a clear space between it and the wall.

Nicholson hydrometer (*Phys*) A hydrometer of the constant displacement type, used for determining the density of a solid.

Nichrome (*Eng*) TN for nickel–chromium alloy largely used for resistance heating elements because of its high resistivity ($\approx 110 \times 10^{-8}$ Ω m) and its ability to withstand high temperatures.

nicitating (*BioSci*) Winking; said of the third eyelid of the vertebrates, which by its movements keeps clean the surface of the eye.

nick (*BioSci*) (1) A cut between adjacent nucleotides in one strand of a duplex DNA molecule. (2) A particular combination of male and female parents giving desirable offspring, esp in the breeding of F1 HYBRIDS.

nick (*Print*) The groove(s) in the shank of a typefounder's type letter.

nickel (*Chem*) Symbol Ni, at no 28, ram 58·71, mp 1450°C, bp 3000°C, electrical resistivity (at 20°C) $10·9 \times 10^{-8}$ Ω m, rel.d. 8·9. Silver-white metallic element. Its ores are *nickeliferous pyrrhotite*, Fe_nS_{n+1} with Ni; *garnierite*, a nickeliferous serpentine, and *pentlandite*, (Fe, Ni)S. Primary nickel minerals are oxidized to many *nickel blooms*, green hydrated nickel salts. Used for structural parts of valves. It is magnetostrictive, showing a decrease in length in an applied magnetic field and, in the form of wire, was much used in computers for small stores, the data circulating and extracted when required. Used pure for electroplating, coinage, and in chemical and food-handling plants. See NICKEL ALLOYS.

nickel alloys (*Eng*) Nickel is the main constituent in Monel metal, permalloy and nickel–chromium alloys. It is also used in cupro-nickel, nickel–silver, various types of steel and cast-iron, brass, bronze and light alloys.

nickel antimony glance (*Min*) See ULLMANITE.

nickel arsenic glance (*Min*) See GERSDORFFITE.

nickel bloom (*Min*) See ANNABERGITE.

nickel–cadmium accumulator (*ElecEng*) A battery using nickel and cadmium compounds in potassium hydroxide electrolyte. Characterized by extremely low self-discharge rate; therefore used in emergency lighting systems etc.

nickel carbonyl (*Chem*) $Ni(CO)_4$. A volatile compound of nickel, bp 43°C, formed by passing carbon monoxide over the heated metal. The compound is decomposed into nickel and carbon monoxide by further heating. Formerly used on a large scale in industry for the production of nickel from its ores by the Mond process.

nickel–chromium steel (*Eng*) Steel containing nickel and chromium as alloying elements: 1·5–4% nickel and 0·5–2% chromium are added to impart HARDENABILITY which enables thick sections to be heat treated to levels of high tensile strength, hardness and toughness, eg highly stressed automobile and aero-engine parts, armour plate, etc. See panel on STEELS.

nickel electro (*Print*) An electro in which a thin layer of nickel is first deposited, the shell being completed by depositing copper. Superior to NICKEL-FACED ELECTRO.

nickel-faced electro (*Print*) An electro which has been given a facing of nickel to improve its quality. Cf NICKEL ELECTRO.

nickel–iron–alkaline accumulator (*ElecEng*) An ACCUMU-LATOR in which the positive plate consists of nickel hydroxide enclosed in perforated steel tubes, and the negative plate consists of iron or cadmium also enclosed in perforated steel tubes. The electrolyte is potassium hydrate, the emf 1·2 volts per cell. It is lighter than the equivalent lead accumulator. Also *nickel–iron–alkaline cell*. Cf *Edison accumulator*, *Ni–Fe accumulator*.

nickel silver (*Eng*) See COPPER ALLOYS, GERMAN SILVER.

nickel steel (*Eng*) Steel containing nickel as principal alloying element. Varying amounts, between 0·5% and 6·0%, are added to increase the strength in the normalized condition, to enable hardening to be performed in oil or air instead of water, or to increase the core strength of carburized parts. Larger amounts are added to steels for special purposes. See panel on STEELS.

nicker (*Build*) The side wing of a CENTRE-BIT, scribing the boundary of the hole to be cut.

nick translation (*BioSci*) An inexact phrase for a method of radioactively labelling a DNA molecule. The DNA is first nicked in one strand by a brief DNase treatment. This allows DNA polymerase I to both remove nucleotides from the exposed end and replace them by highly radioactive nucleotides.

Niclad (*Eng*) Composite laminated sheets made by rolling together sheets of nickel and mild steel, to obtain the corrosion resistance of nickel with the strength of steel.

Niclausse boiler (*Eng*) A French marine boiler which consists of a horizontal water and steam drum from which vertical double headers are suspended, carrying banks of FIELD TUBES slightly inclined downwards.

Nicol prism (*Phys*) A device for obtaining plane-polarized light, it consists of a crystal of Iceland spar which has been cut and cemented together in such a way that the ordinary ray is totally reflected out at the side of the crystal, while the extraordinary plane-polarized ray is freely transmitted. Largely superseded by POLAROID.

nicotinamide adenine dinucleotide (*BioSci*) A coenzyme that serves as an electron acceptor for many dehydro-genases. The reduced form (NADH) subsequently donates its electrons to the *electron transport chain*. Abbrev *NAD*.

nicotinamide adenine dinucleotide phosphate (*BioSci*) Phosphorylated derivative of NAD. It also serves as an electron carrier but the electrons are primarily used for *reductive biosynthesis*. Abbrev *NADP*.

nicotine (*Chem*) An alkaloid of the pyridine series, $C_{10}H_{14}N_2$. It occurs in tobacco leaves, is extremely poisonous and, in small quantities, is highly addictive. It is a colourless oil, of nauseous odour; bp 246°C at approx 101 kPa.

nicotine

nicotine replacement therapy (*Pharmacol*) A treatment to help people stop smoking, in which small quantities of nicotine are administered by patches, chewing gum, etc, until the craving for cigarettes is cured.

nicotinic acetylcholine receptor (*BioSci*) A protein receptor and ion channel in the post-synaptic membrane, for which acetylcholine is the ligand; the binding of acetylcholine is blocked by nicotine. Cf MUSCARINIC ACETYLCHOLINE RECEPTOR.

nicotinic acid (*Chem*) See vitamin B complex in panel on VITAMINS.

nidamental (*BioSci*) Said of glands that secrete material for the formation of an egg covering.

nidation (*BioSci*) In the estrous cycle of mammals, the process of renewal of the lining of the uterus between the menstrual periods.

nidged ashlar (*Build*) See NIGGED ASHLAR.

nidicolous (*BioSci*) Said of birds that remain in the parental nest for some time after hatching. Cf *nidifugous*.

nidification (*BioSci*) The process of building or making a nest. Also *nidulation*.

nidifugous (*BioSci*) Said of birds that leave the parental nest soon after hatching. Cf NIDICOLOUS.

nidus (*BioSci*) A nest; a small hollow resembling a nest; a nucleus.

niello (*Eng*) A method of decorating metal. Sunk designs are filled with an alloy of silver, lead and copper, with sulphur and borax as fluxes, and fired.

nielsbohrium (*Chem*) The name formerly suggested by Soviet scientists for the chemical element JOLIOTIUM. It was named after the Danish physicist Niels Bohr.

Niemann–Pick disease (*Med*) An autosomal recessive here-ditary disease in which large amounts of lipid are deposited in the reticulo-endothelial tissue, causing enlargement of the lymph glands, spleen and liver and leading to death by the age of 2.

Ni–Fe accumulator (*ElecEng*) TN of a NICKEL–IRON–ALKALINE ACCUMULATOR.

nifedipine (*Pharmacol*) A calcium channel blocker used for the prevention of angina and the treatment of hypertension.

nigericin (*BioSci*) An IONOPHORE, formerly used as an antibiotic.

niger morocco (*Print*) A goatskin tanned and dyed in Nigeria; limited range of colours, flexible, used for good-quality binding.

nigged ashlar (*Build*) A block of stone dressed with a pointed hammer. Also *nidged ashlar*.

night blindness (*Med*) Abnormal difficulty in seeing objects in the dark; due often to deficiency of vitamin A (retinol) in the diet. Also *nyctalopia*.

night bolt (*Build*) See NIGHT LATCH.

night latch (*Build*) A lock operated by key outside and knob inside, but fitted with a device to prevent operation from either side. Also *night bolt*.

night terror (*Psych*) A particularly harrowing variety of bad dream experienced by a child.

nigrescent (*BioSci*) Becoming blackish.

nigrite (*Min*) A pitch-like member of the asphaltite group.

nigrosines (*Chem*) Diphenylamine dyestuffs, used as black pigments, prepared by heating nitrobenzene or nitrophenol, aniline and phenylammonium chloride with iron filings.

Ni-hard (*Eng*) Cast-iron to which a ladle addition of about 4·5% nickel has been made, to render the alloy martensitic and abrasion resistant.

NII (*NucEng*) Abbrev for *Nuclear Installations Inspectorate*. A branch of the Health and Safety Executive in the UK responsible for the safety assessment and inspection of nuclear facilities.

nile (*NucEng*) A unit of 1 nile corresponds to a REACTIVITY of 0·01. In indicating reactivity changes, it is more usual to use the smaller unit, the millinile, equal to a change in reactivity of 10^{-5}.

nimbostratus (*EnvSci*) Grey cloud layer, often dark, the appearance of which is rendered diffuse by more or less continuously falling rain or snow, which in most cases reaches the ground. It is thick enough throughout to blot out the Sun. Low, ragged clouds frequently occur below the layer, with which they may or may not merge. Abbrev *Ns*.

NiMH (*Eng*) Abbrev for *nickel–metal hydride*, denoting a type of rechargeable battery.

Nimonic (*Eng*) TN for alloy series used in high-temperature work such as gas turbine blades with a basic composition of 80% nickel and 20% chromium and closely controlled amounts and combinations of other elements, eg titanium, aluminium, manganese, iron, carbon, cobalt, silicon, copper, boron, zirconium and molybdenum.

nine-point circle (*MathSci*) Of a triangle, the circle which passes through the mid-points of the three sides, the points of intersection of each side with the perpendicular to that side from the opposite vertex, and, if *c* is the point where the three perpendiculars intersect, the three points midway between *c* and the respective vertices.

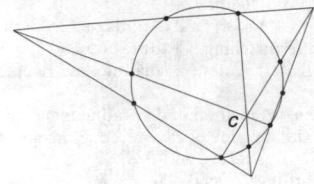

nine-point circle

ninhydrin (*Chem*) *Tri keto-hydrindene-hydrate*. Used as a reagent to detect proteins or amino acids, with which it forms a characteristic blue colour on heating.

niobite (*Min*) See COLUMBITE.

niobium (*Chem*) Symbol Nb, at no 41, ram 92·9064, mp 2500°C. Rare metallic element. Used in high-temperature engineering products (eg gas turbines and nuclear reactors) owing to the strength of its alloys at temperatures above 1200°C. Combined with tin (Nb_3Sn) it has a high degree of superconductivity. Occurs with tantalum in *columbite*, niobate and tantalite of iron and manganese.

nip (*Eng*) The manufacturing practice of giving different curvatures to the individual leaves of a laminated spring before assembling, so as to attain the most favourable distribution of working stresses.

nip (*Glass*) The gap between rollers in a sheet glass rolling machine.

nip (*MinExt*) See ANGLE OF NIP.

nip (*Paper*) The flat area at the point of contact between two horizontal parallel rolls.

nip (*Textiles*) The line of near contact between two rotating parallel rollers in a CALENDER through which material passes while being compressed. Produces dispersive mixing in polymers.

nip and tuck folder (*Print*) A type of folder on a rotary press in which a fold at right angles to the web is formed by a blade thrusting the web between folding jaws.

Nipkow disk (*ImageTech*) A disk with one or more sets of holes arranged in a spiral, the rotation of which provided a form of MECHANICAL SCANNING in some early TV systems, such as Baird's Televisor.

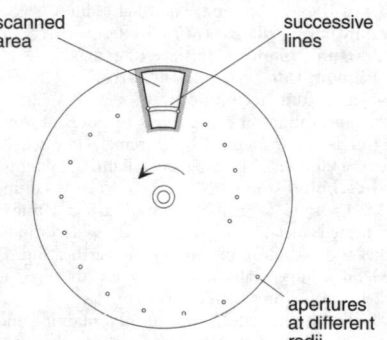

scanned area
successive lines
apertures at different radii

Nipkow disk Eighteen line scan.

nippers (*Build*) See STONE TONGS.

nipping (*Print*) See SMASHING.

nipple (*BioSci*) The mamma or protuberant part of the mammary gland in female mammals, bearing the openings of the milk-forming glands.

nipple (*Eng*) (1) A short length of externally threaded pipe for connecting two lengths of internally threaded pipe. (2) A small drilled bush, sometimes containing a non-return valve, screwed into a bearing for the supply of lubricant by a GREASE GUN. (3) A nut for securing (bicycle) wheel spokes.

nip rolls (*Paper*) A pair of horizontal, parallel steel rolls providing a nip through which a web of paper may be passed and so drawn through the machine to which they are fitted, eg a sheet cutter.

Ni-resist (*Eng*) A cast-iron consisting of graphite in a matrix of austenite. Contains carbon 3%, nickel 14%, copper 6%, chromium 2% and silicon 1·5%. Has a high resistance to growth, oxidation and corrosion.

Nissl bodies (*BioSci*) Aggregations of RIBOSOMES found within nerve cells.

nit (*Phys*) Unit of luminance equal to 1 CANDELA per square metre. Symbol nt.

niter (*Chem*) See NITRE.

nitometer (*Phys*) An instrument for measuring the brightness of small light sources.

Nitralloy (*Eng*) Steel specially developed for nitriding (which is less effective with ordinary steels). Contains carbon 0·2–0·3%, aluminium 0·9–1·5%, chromium 0·9–1·5% and molybdenum 0·15–0·25%.

nitramines (*Chem*) Amines in which an aminohydrogen has been replaced by the nitro group. They have the general formula $RNHNO_2$.

nitrate film (*ImageTech*) Film with a base of highly inflammable cellulose nitrate, celluloid; used for 35 mm motion pictures up to 1951 but now completely replaced by SAFETY FILM.

nitrate-reducing bacteria (*BioSci*) Facultative aerobes able to reduce nitrates to nitrites, nitrous oxide, or nitrogen under anaerobic conditions, eg *Micrococcus denitrificans*. This process is termed *denitrification*. A few bacteria use such reduction processes as hydrogen acceptor reactions and hence as a source of energy; in this case the end product is ammonia.

nitrates (*Chem*) Salts or esters of nitric (V) acid. Metal nitrates are soluble in water; decompose when heated. The nitrates of polyhydric alcohols and the alkyl radicals explode with violence. Uses: explosives, fertilizers, chemical intermediates.

nitrates (*Pharmacol*) Vasodilatory drugs used in treatment of angina and congestive heart failure. They cause dilatation of large veins and thus reduce the workload on the heart. An example is *nitroglycerin*.

nitration (*Chem*) (1) The introduction of nitro (NO_2) groups into organic compounds. Usually carried out by mixing concentrated nitric and sulphuric acids. (2) The final stage of nitrification in the soil.

nitrazepam (*Pharmacol*) Widely used non-barbiturate hypnotic, which can be addictive.

nitre (*Chem*) Potassium nitrate (V), also *saltpetre*. See CHILE NITRE, SODA NITRE.

nitric (V) acid (*Chem*) NHO_3. Bp 83°C, mp −41·59°C, rel.d. 1·5, miscible with water. A fuming unstable liquid. Old name *aqua fortis*. Prepared in small quantities by the action of conc sulphuric acid on sodium nitrates and on large scale by the oxidation of nitrogen or ammonia. An important intermediate of fertilizers, explosives, organic synthesis, metal extraction and sulphuric acid manufacture. See NITRATES, CHAMBER PROCESS.

nitric anhydride (*Chem*) Nitrogen (V) oxide. N_2O_5. Dissolves in water to give nitric acid.

nitric [nitrogen (II)] oxide (*Chem*) NO. Colourless gas. In contact with air it forms reddish-brown fumes of nitrogen dioxide.

nitric oxide (*BioSci*) A gas produced from L-arginine by the enzyme nitric oxide synthase. It acts as an intracellular and intercellular messenger in the vascular and nervous systems.

nitrides (*Chem*) Compounds of metals with nitrogen. Usually prepared by passing nitrogen or gaseous ammonia over the heated metal. Those of groups I and II metals are ionic compounds which react with water to release ammonia. Those of groups III to V exist as interstitial compounds, having great hardness and refractoriness, eg boron, titanium, iron.

nitriding (*Eng*) A process for producing hard surface on special types of steel by heating in gaseous ammonia. Components are finish-machined, hardened and tempered, and heated for about 4–10 h at 520°C. Case is about 0·50 mm deep and surface hardness is $1100H_V$. Also *nitrogen case-hardening*. See NITRALLOY.

nitrification (*BioSci*) The oxidation of ammonia to nitrite and nitrate by chemoautotrophic bacteria whose energy requirements come from these exergonic reactions.

nitrification (*Chem*) The treatment of a material with nitric acid.

nitrile rubber (*Eng*) Copolymer made from acrylonitrile and butadiene, composition depending on use. Widely used in solvent-resistant applications, eg petrol hose, brake hose.

nitriles (*Chem*) Alkyl cyanides of the general formula $RC≡N$. When hydrolysed they yield carboxylic acids or the corresponding ammonium salts; reduced, they yield amines.

nitrite (*FoodSci*) Curing agent for meats as the sodium or potassium salt, having a pronounced bacteriostatic effect and giving a characteristic flavour and red coloration. When cooked, the nitrite interacts with protein to form a preservative, particularly effective against spore-forming organisms, eg *Clostridium botulinum*. Because nitrites prevent oxygen transport in the blood, levels in cured products are strictly controlled in all countries. See CURING.

nitrites [nitrates (III)] (*Chem*) Salts or esters of nitrous acid, O=NOH.

nitroanilines (*Chem*) The 4- and 3-isomers are used as important intermediates in the preparation of azo dye-stuffs.

Nitrobacteriaceae (*BioSci*) A family of bacteria belonging to the order Pseudomonadales, important in nitrification processes in the soil and fresh water. They are autotrophic bacteria that derive energy from oxidation processes: *Nitrosomonas* from the oxidation of ammonia to nitrites and *Nitrobacter* from the oxidation of nitrites to nitrates. Also *nitrifying bacteria*.

nitrobenzene (*Chem*) $C_6H_5NO_2$. Mp 5°C, bp 211°C, rel.d. 1·2. A yellow liquid with an odour of bitter almonds, It is obtained by the action of a mixture of concentrated sulphuric acid and nitric acid on benzene. When reduced it yields aniline.

nitroblue tetrazolium (*BioSci*) A yellow dye that forms insoluble deep-blue granules of formazan when reduced. The nitroblue tetrazolium reduction test is often used to show whether phagocytic cells are functioning normally; formazan is formed during the METABOLIC BURST that follows phagocytosis. Abbrev *NBT*.

nitrocelluloses (*Chem*) The nitric acid esters of cellulose formed by the action of a mixture of nitric and sulphuric acids on cellulose. The cellulose can be nitrated to a varying extent ranging from two to six nitrate groups in the molecule. Nitrocelluloses with a low nitrogen content, up to the tetranitrate, are not explosive, and are used in lacquer and artificial silk and imitation leather book cloth manufacture. They dissolve in ether–alcohol mixtures, and in so-called lacquer solvents. A nitrocellulose with a high nitrogen content is gun-cotton, an explosive. Also *cellulose nitrates*. See PYROXYLINS.

nitro derivatives (*Chem*) Aliphatic or aromatic compounds containing the group –NO_2. The aliphatic nitro derivatives are colourless liquids which are not readily hydrolysed, but have acidic properties, eg the primary and secondary aliphatic nitro derivatives can form metallic compounds. Aromatic nitro derivatives are easily formed by the action of nitric acid on aromatic compounds. The nitro groups substitute in the nucleus and only exceptionally in the side chain.

nitrogen (*Chem*) Symbol N, at no 7, ram 14·0067, mp −209·86°C, bp −195°C. Gaseous element, forming a diatomic, colourless, odourless gas. Determination of its ram by Rayleigh and Ramsay led to the discovery of the *inert gases* in the atmosphere, eg argon etc. Approx 80% of the normal atmosphere is nitrogen, which is also widely spread in minerals, the sea, and in all living matter. Its abundance in the Earth's crust is only 19 ppm. It occurs in rare independent minerals, mainly nitrates in evaporite deposits; there is some in coal and petroleum and it is present in volcanic gases. Nitrogen occurs as a chemical component of minerals and all living matter. Gaseous nitrogen is used as an inert atmosphere in filament lamps, in sealed relays, in Van de Graaff generators, in high-voltage cables as an insulant and extensively in inorganic and organic chemistry. Liquid nitrogen is used as a coolant. See AMMONIA, NITRATES, NITRIC ACID, NITROGEN FIXATION.

nitrogenase (*BioSci*) The enzyme system that catalyses the reduction of gaseous nitrogen (dinitrogen, N_2) to ammonia in biological NITROGEN FIXATION. It is inactivated by oxygen.

nitrogen balance (*BioSci*) The state of equilibrium of the body in terms of intake and output of nitrogen; positive nitrogen balance indicates intake exceeds output, negative nitrogen balance denotes output exceeds intake.

nitrogen bases (*Chem*) See AMINES.

nitrogen case-hardening (*Eng*) See NITRIDING.

nitrogen chlorides (*Chem*) Three nitrogen chlorides, NH_2Cl, $NHCl_2$ and NCl_3, produced by the chlorination of ammonium ions. Unstable and explosive.

nitrogen cycle (*BioSci*) The sum total of the transformations undergone by nitrogen and nitrogenous compounds in nature in relation to living organisms.

nitrogen [dioxide] (IV) oxide (*Chem*) NO_2. Oxidizing agent; formed when one volume of oxygen is mixed with two volumes of nitric oxide. Below $17°C$ the formula is N_2O_4 and the molecule is colourless. Decomposed by water, forming a mixture of nitric acid and nitric oxide. See NITROGEN FIXATION.

nitrogen fixation (*BioSci*) The formation of chemical compounds from free atmospheric nitrogen (dinitrogen, N_2) and other elements, eg oxygen and hydrogen. Nitrogen is fixed (1) in the atmosphere into oxides, by lightning and UV radiation, estimated at 7×10^6 tonnes of nitrogen per annum, globally; (2) industrially to make nitrogenous fertilizers, eg nitrates, ammonia, ammonium salts and urea (50×10^6 tonnes per annum); and (3) biologically by free-living or symbiotic prokaryotic organisms into ammonia (150×10^6 tonnes per annum). Free-living nitrogen fixers include many cyanobacteria (notably those with heterocysts, eg *Anabaena*) and photosynthetic bacteria, and a variety of heterotrophic bacteria which can be aerobic (eg *Azotobacter*), facultatively aerobic (eg *Klebsiella*) or anaerobic (eg *Clostridium*). Some heterotrophic bacteria are esp associated with the RHIZOSPHERE of plant roots to the mutual benefit of bacteria and plant. Such symbiotic fixers include species of the bacterium *Rhizobium*, which inhabit root nodules of the pea and bean family (Leguminosae), of the actinomycete *Frankia* which inhabit root nodules in various plants, and of various cyanobacteria which form associations with a variety of other plants, including a few angiosperms and some lichens. At least two sorts of animal (termites and shipworms) are reported to harbour nitrogen-fixing symbiotic bacteria. Nitrogen is also fixed inadvertently by humans in automobile engines and other combusting devices to produce oxides of nitrogen, NOx, which are locally significant as aerial pollutants. Also *dinitrogen fixation*. See ACID RAIN, NITROGENASE.

nitrogen mustard (*Chem*) Any of various toxic compounds similar to MUSTARD GAS, but in which the sulphur is replaced by an amino nitrogen. Used in the treatment of various types of cancer, and as a military weapon.

nitrogen narcosis (*Med*) 'Rapture of the deep'. Intoxicating and anaesthetic effect of too much nitrogen in the brain, experienced by divers at considerable depths.

nitrogen oxides (*Chem, EnvSci*) Consisting mainly of NITRIC OXIDE, NITROGEN DIOXIDE and NITROUS OXIDE, they are produced by natural processes (lightning, bacterial NITRIFICATION and decomposition) and by burning fossil fuels. In coal-fired power stations, nitrogen oxides are derived mainly from the coal, but in oil-fired stations nitrogen oxides come mainly from the nitrogen in the air supporting combustion. As atmospheric pollutants they contribute to ACID RAIN. Control is achieved by changes in boiler and burner design, and by catalytic reduction to nitrogen using ammonia as a reductant. Abbrev *NOx*.

nitrogen pentoxide (*Chem*) See NITRIC ANHYDRIDE.

nitrogen [tetroxide] (IV) oxide (*Chem*) N_2O_4. See NITROGEN DIOXIDE.

nitroglycerine (*Chem*) $C_3H_5(ONO_2)_3$. Mp $11–12°C$. A colourless oil; insoluble in water; prepared by treating glycerine with a cold mixture of concentrated nitric and sulphuric acids. It solidifies on cooling, and exists in two physical crystalline modifications. In thin layers it burns without explosion, but explodes with tremendous force when heated quickly or struck. See DYNAMITE, EXPLOSIVE. Used medically in solution and tablets for angina.

Nitrolime (*Chem*) TN for an artificial fertilizer consisting of calcium cyanamide.

nitromethane (*Chem*) CH_3NO_2. Bp $99–101°C$. A liquid prepared from chloroacetic acid and sodium nitrate.

nitrophilous (*BioSci*) Plants growing characteristically in places where there is a good supply of fixed nitrogen.

nitroprussides (*Chem*) Formed by the action of nitric acid on either hexacyano ferrates (II) or (III). The nitroprusside ion is $[FeNO(CN)_5]^-$. Also *nitrosoferricyanides*.

nitrosamine (*FoodSci*) A compound produced by the reaction of nitrates and nitrites with primary and secondary amines under mildly acid conditions. There is evidence that such compounds could be carcinogenic.

nitroso compounds (*Chem*) Compounds containing the monovalent radical –NO.

nitroso-dyes (*Chem*) Dyestuffs resulting from reaction between phenols and nitrous acid.

nitrosoferricyanides (*Chem*) See NITROPRUSSIDES.

nitrotoluenes (*Chem*) $CH_3C_6H_4NO_2$. On nitration of toluene a mixture of 2- and 4-nitrotoluene is obtained with very little 3-nitrotoluene; 2-nitrotoluene, a liquid, has bp $218°C$; 4-nitrotoluene crystallizes in large crystals, mp $54°C$, bp $230°C$.

nitrous [nitric (III)] acid (*Chem*) The pale-blue unstable solution obtained by precipitating barium nitrite with dilute sulphuric acid is supposed to contain nitrous acid, HNO_2.

nitrous oxide (*Chem*) Laughing gas, N_2O. A colourless gas with a sweetish odour and taste, soluble in water, alcohol, ether and benzene. Nitrous oxide supports combustion better than air. The gas is manufactured by the decomposition of ammonium nitrate by heat. It is used for producing anaesthesia of short duration.

nitroxyl (*Chem*) The radical $–NO_2$ when attached to a halogen atom or a metal. Compounds containing the group are *nitroxyls*.

nitrozation (*BioSci*) The conversion of ammonia into nitrites by the action of soil bacteria (*Nitrosomonas*), being the second stage in the nitrification in the soil.

Nivarox (*Eng*) An alloy of iron and nickel with a small addition of beryllium, non-magnetic, rustless and of controllable elasticity, used for hairsprings.

NK cell (*BioSci*) Abbrev for *natural killer cell*. See KILLER CELLS.

NLM (*ICT*) Abbrev for *Netware loadable module*. A program stored on a FILE SERVER operating a LAN under Novell Netware. The module provides extra or advanced facilities to the network and its users.

nm (*Genrl*) Abbrev for *nanometre* = $10 \text{ Å} = 10^{-9}$ m.

NMDA (*BioSci*) Abbrev for N-METHYL-D-ASPARTIC ACID.

N-methyl-D-aspartic acid (*BioSci*) An agonist for a class of GLUTAMATE RECEPTOR (NMDA-receptor) found on some vertebrate nerve cells. Abbrev *NMDA*.

NMOS (*Electronics*) METAL–OXIDE–SILICON technology based on n-channel devices in a p-type substrate.

NMR (*Chem, Phys*) Abbrev for NUCLEAR MAGNETIC RESONANCE.

NMRI (*Med*) Abbrev for *nuclear magnetic resonance imaging*. Often shortened to *MRI*.

NMR spectroscopy (*Chem*) Abbrev for NUCLEAR MAGNETIC RESONANCE SPECTROSCOPY.

nN heterojunction (*Electronics*) A junction in a HETEROSTRUCTURE formed between a narrower band-gap material (n-type) and a wider gap material (N-type). Also *n–N heterojunction*.

nn junction (*Electronics*) A junction formed within a n-type semiconductor where the donor dopant concentration changes abruptly. Also *n–n junction*.

No (*Chem*) Symbol for NOBELIUM.

NOAA (*Space*) Abbrev for *National Oceanic and Atmospheric Administration*, a US body which manages and operates environmental satellites, and provides data to users worldwide.

nobelium (*Chem*) Artificially produced element. At no 102; symbol No. Principal isotope is 254.

noble gases (*Chem*) Elements helium, neon, argon, krypton, xenon and radon-222, much used (except the last) in gas-discharge tubes. (Radon-222 has short-lived radioactivity, half-life less than 4 days.) Their outer (valence) electron shells are complete, thus rendering them inert to all the usual chemical reactions, a property for which argon, the most abundant, finds increasing industrial use. The heavier ones, Rn, Xe, Kr, are known to form a few unstable compounds, eg XeF_4. Also *inert gases*, *rare gases*.

noble metals (*Eng*) Metals, such as gold, silver, platinum, etc, which have a relatively positive electrode potential, and which do not enter readily into chemical combination with non-metals. They have high resistance to corrosive attack by acids and other agents, and resist atmospheric oxidation. Cf BASE METAL.

nocardiasis (*Med*) Infection (usually of the lungs) with any one of a number of spore-forming fungi of the genus *Nocardia*.

nociceptive (*BioSci*) Descriptive of a sensory nerve ending that sends signals that cause pain in response to certain stimuli.

noctilucent (*BioSci*) Phosphorescent; light-producing.

noctilucent clouds (*EnvSci*) Thin but sometimes brilliant and beautifully coloured clouds of dust or ice particles at a height of 75–90 km. They are visible about midnight in latitudes greater than about 50° when they reflect light from the Sun below the horizon.

Noctovision (*ImageTech*) TN for a system of TV in which the light-sensitive elements respond to infrared light, and which can therefore be operated in darkness.

nocturia (*Med*) Passing excessive quantities of urine at night.

nodal gearing (*Eng*) The location of gear wheels, eg between a turbine and propeller shaft, at a nodal point of the shaft system with respect to torsional vibration.

nodal point (*ICT*) (1) The point in a high-frequency circuit where current is a maximum and voltage a minimum, or vice versa. (2) In electrical networks, a terminal common to two or more branches of a network or to a terminal of any branch. Also *node*.

nodal points of a lens (*Phys*) Two points on the principal axis of a lens or lens system such that an incident ray of light directed towards one of them emerges from the lens as if from the other, in a direction parallel to that of the incident ray. For a lens having the same medium on its two sides, the nodal points coincide with the principal points.

node (*Astron*) One of the two points at which the orbit of a celestial object intersects a reference plane, such as the ECLIPTIC or CELESTIAL EQUATOR. The path crossing from south to north is the *ascending node*; the *descending node* has the opposite sense.

node (*BioSci*) The position on a stem at which one or more leaves are attached. Cf INTERNODE.

node (*ElecEng*) (1) The point of minimum disturbance in a system of waves in tubes, plates or rods. The amplitude cannot become zero, otherwise no power could be transmitted beyond the point. (2) The point in an electrical network where two or more conductors are connected.

node (*Eng*) (1) A point, or more than one, of rest in vibrating body. (2) A junction point of two or more members in a structural frame.

node (*ICT*) (1) A data item within a TREE DATA STRUCTURE. (2) A point at which two or more communications lines meet. The term is usually applied to the switching device or computer that is situated at this point. See NODAL POINT.

node (*MathSci*) See DOUBLE POINT.

node (*Phys*) See ANTINODE.

nodes of Ranvier (*BioSci*) Constrictions of the neurolemma occurring at regular intervals along myelinated nerve fibres. See fig. at NEURON.

node voltage (*ICT*) That of a NODAL POINT in an electrical network.

nodose (*BioSci*) Bearing localized swellings or nodules. Also *nodular*.

nodular cast-iron (*Eng*) See DUCTILE CAST-IRON.

nodular structures (*Geol*) Spheroidal, ovoid or irregular bodies often encountered in both igneous and sedimentary rocks, and formed by segregation about centres. See CLAY IRONSTONE, DOGGERS, FLINT, SEPTARIA.

nodule (*BioSci*) Any small rounded structure on a plant, esp a swelling on a root inhabited by symbiotic, nitrogen-fixing bacteria or actinomycetes.

nodulizing (*MinExt*) Aggregation of finely divided material such as mineral concentrates, by aid of binder and perhaps kilning, into nodules sufficiently strong and heavy to facilitate subsequent use, such as charging into blast furnaces.

no-effect level (*BioSci*) The maximum dose of a substance that produces no detectable changes under defined conditions of exposure. Abbrev *NEL*.

no-fines concrete (*Build*, *CivEng*) Concrete without fine aggregate. Therefore open textured, with a comparatively low strength, but other advantages. It can be cast *in situ* using simple, even open-mesh shuttering, sets rapidly, has little capillary attraction and low moisture movement, and can be plastered and rendered readily owing to its texture.

nog (*Build*) A block of wood built into a wall to which joinery such as skirtings may be nailed.

noil (*Textiles*) (1) Short or broken fibres from silk after opening and combing. (2) Short fibre removed from wool during combing; often added to woollen blends.

noise (*Acous*) (1) Socially unwanted sounds. (2) Interference in a communication channel.

noise (*ICT*) Any undesired disturbance within a useful frequency band that tends to obscure its information content. Those caused by parallel services may be termed interference. See THERMAL NOISE, WHITE NOISE.

noise (*ImageTech*) Unwanted signals or background giving rise to visual disturbance in picture reproduction, often appearing as excessive graininess or irregular colour spots.

noise abatement climb procedure (*Aero*) Means of flying a civil aircraft from an airport so as to climb rapidly until the built-up area is reached and thereafter reducing power to just maintain a positive rate of climb until the area is overflown or 5000 ft is reached.

noise audiogram (*Acous*) Audiogram taken in the presence of a specified masking noise.

noise audiometer (*Acous*) An audiometer which measures the threshold of hearing of a deaf person's ear, the other ear or both ears, being subjected to a standardized noise in addition to the test sound.

noise background (*Acous*) In reproduction or recording, the total noise in the system with the signal absent.

noise bars (*ImageTech*) Horizontal bars of NOISE due to the video heads mistracking in slow and fast motion. See AUTOMATIC TRACKING.

noise control (*Acous*) The reduction of unwanted noise by various methods, eg absorption, isolation, antisound.

noise current (*ICT*) That part of a signal current conveying noise power.

noise diode (*ElecEng*) One operating as a NOISE GENERATOR, under temperature-limited conditions.

noise exposure forecast value (*Acous*) A method to evaluate the annoyance by fluctuating noises. Abbrev *NEF*.

noise factor (*Acous*) The ratio of noise in a linear amplification system to thermal noise over the same frequency band and at the same temperature. Also *noise figure*.

noise factor (*ICT*) The ratio, usually expressed in dB, of the noise power output per unit bandwidth at the output (of a communication system or of a receiver) to the portion of the noise power that is due to the input termination at an adopted standard temperature of 290 K.

noise figure (*Acous*) See NOISE FACTOR.

noise figure (*ICT*) A measure of quality for a communications element, indicating the degradation in signal-to-noise ratio that it introduces. It is given by the input signal-to-noise ratio divided by the output signal-to-noise ratio, or the difference between these quantities when working in DECIBELS. An ideal element has a ratio of unity, or 0 dB.

noise footprint (*Aero*) The contour beneath an aircraft of constant noise level, measured in decibels (dB) or derived units.

noise generator (*ElecEng*) A device, eg a diode, for producing a controlled noise signal for test purposes. Also *noise source*.

noiseless recording (*Acous*) The practice of making the sound track of a positive sound film as dense as possible, consistent with the accommodation of the modulation, in order to keep the photographic noise level as far below the recorded level as possible.

noise level (*Acous*) The LOUDNESS LEVEL of a noise signal.

noise limiter (*Acous*) A device for removing the high peaks in a transmission, thus reducing contribution of clicks to the noise level and eliminating acoustic shocks. Effected by biased diodes or other clipping circuits.

noise meter (*Acous*) See OBJECTIVE NOISE METER.

noise power (*Electronics*) That dissipated in a system by all noise signals present. See NYQUIST NOISE THEOREM.

noise ratio (*Acous*) See SIGNAL-TO-NOISE RATIO.

noise reduction (*Acous, ImageTech*) (1) A procedure in magnetic and photographic sound recording whereby the average density in the negative sound track is kept as low as possible, and hence in the positive print as high as possible, both consistent with linearity, so that dirt, random electrostatics and scratches produce less noise. (2) See NOISE CONTROL.

noise resistance (*Acous*) The resistance for which the thermal noise would equal the actual noise signal present, usually in a specific frequency band.

noise source (*Acous, ElecEng*) The object or system from which the noise originates. See NOISE GENERATOR.

noise suppressor (*Aero*) A turbojet propelling nozzle fitted with fluted members which induct air to slow and break up the jet efflux, thereby reducing the noise level.

noise suppressor (*ICT*) A circuit that suppresses noise between usable channels when these are passed through during tuning. An automatic gain control that cuts out weak signals and levels out loud signals. See MUTING.

noise temperature (*Electronics*) The temperature at which the thermal noise power of a passive system per unit bandwidth is equal to the noise at the actual terminals. The standard reference temperature for noise measurements is 290 K. At this temperature the available noise power per unit bandwidth (see NYQUIST NOISE THEOREM) is 4×10^{-21} watts.

noise transmission impairment (*Acous*) The transmission loss in decibels which would impair intelligibility of a telephone system to the same extent as the existing noise signal. Abbrev NTI.

noise voltage (*Acous*) A noise signal measured in rms volts.

no-load characteristic (*ElecEng*) See OPEN-CIRCUIT CHARACTERISTIC.

no-load current (*ElecEng*) The current taken by a transformer when it is energized but is giving no output, or by a motor when it is running but taking no mechanical load.

no-load loss (*ElecEng*) The losses occurring in a motor or transformer when it is operating but giving no output. Also *open-circuit loss*.

noma (*Med*) See CANCRUM ORIS.

nomadism (*BioSci*) The habit of some animals of roaming irregularly without regularly returning to a particular place. Cf MIGRATION.

Nomarski interference (*Phys*) A method of polarized optical microscopy of material surfaces producing a colour contour map, esp useful for semiconductor chip surfaces or others where it is important to observe slight changes in relief. Also used in imaging cells. See DIFFERENTIAL INTERFERENCE CONTRAST MICROSCOPE, POLARIZED LIGHT MICROSCOPY

nomeristic (*BioSci*) Of metameric animals, having a definite number of somites.

Nomex (*Plastics*) TN for aramid fibre based on meta-terephthalamide polymer, used in paper for a *stiff honeycomb* for aerospace applications and for fire-resistant clothing (US).

nominal (*MathSci*) A type of variable which can take unordered, qualitative values, which may be known as levels, eg hair colour.

nominal data (*MathSci*) Data that consist solely of non-numerical values that therefore cannot be compared arithmetically. These is so even when the values are actually numbers, such as telephone numbers. Compare INTERVAL SCALE, ORDINAL SCALE, RATIO SCALE.

nominal section (*ICT*) A network that is equivalent to a section of transmission line, based on the assumption of lumped constants.

nomogram (*MathSci*) Chart or diagram of scaled lines or curves for facilitating calculations. Those comprising three scales in which a line joining values on two determines a value on the third are frequently called *alignment charts*. Also *nomograph*.

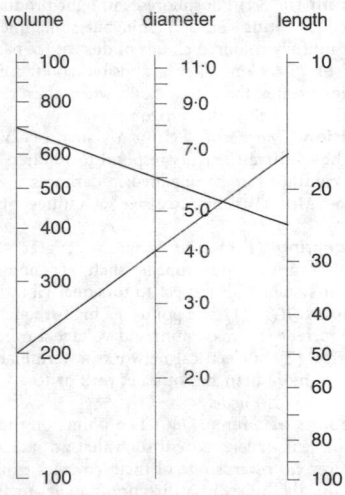

nomogram Relating the length, diameter and volume of a cylinder.

nomothetic (*Psych*) Adj describing any research, system or philosophy dealing with the formulation of universal laws. See IDIOGRAPHIC.

non-, nona- (*Chem*) Containing nine atoms, groups, etc.

nonagon (*MathSci*) A nine-sided polygon.

nonane (*Chem*) C_9H_{20}. Mp $-51°C$, bp $150°C$, rel.d. $0·72$. A paraffin hydrocarbon.

non-aqueous solvents (*Chem*) May be classed broadly into (1) water-like or levelling solvents, which are highly polar and form strong electrolytic solutions with most ionizable solutes, eg $NH_2(NH_4^+ NH_{2-})$, $SO_2(SO^{++} SO_{2--})$, $N_2O_4(NO^+ NO_3^-)$; (2) differentiating solvents, which bring out differences in the strength of electrolytes, eg weak amines or acids, ethers, halogenated hydrocarbons.

non-aqueous titration (*Chem*) Certain substances such as weak acids or bases, or compounds sparingly soluble in water, are better titrated in *non-aqueous solvents*. The strengths of weak acids are enhanced in PROTOPHILIC solvents, and of weak bases in PROTOGENIC solvents, making a sharp END-POINT possible. The use of AMPHI-PROTIC or *aprotic* solvents, of mixtures, makes available a wide choice of *levelling* or *differentiating properties*.

non-association cable (*ElecEng*) Cable which is not manufactured or designed in accordance with the standards of the Cable Makers' Association.

non-bearing wall (*Build*) A wall carrying no load apart from its own weight.

non-bleeder spray gun (*Build*) A spray gun which is fitted with an air valve which opens and closes with the action of the trigger.

non-caducous (*BioSci*) See INDECIDUATE.

non-coding DNA (*BioSci*) DNA that does not code for RNA or part of a polypeptide chain. The majority of DNA in eukaryotes is non-coding. Also *junk DNA*.

non-competitive inhibition (*BioSci*) Reversible inhibition of an enzyme by a compound that does not compete for the substrate-binding site but binds elsewhere.

non-conductor (*Electronics*) Under normal conditions, an electrical insulator in which there are very few free electrons.

non-conforming (*Eng*) Said of a product outside manufacturing limits but not necessarily defective.

non-convertible coatings (*Build*) Reversible paint films which can be softened with the application of the original solvent. The drying of these coatings involves physical rather than chemical change, eg spirit-based paints. Cf NON-REVERSIBLE COATINGS.

non-dairy (*FoodSci*) A term defining products which imitate the appearance, taste or function of dairy products without any raw materials of dairy origin.

non-declarative memory (*Psych*) A subsystem of long-term memory which consists of skills acquired through repetition and practice (eg riding a bicycle, playing the piano). Also *procedural memory*.

non-degenerate gas (*Electronics*) One formed of particles, the concentration of which is so low that the MAXWELL–BOLTZMANN DISTRIBUTION applies. Particular examples are the electrons applied to a conduction band by donor levels in an n-type semiconductor, and those resulting from the passage of electrons from the normal band to an impurity band of acceptor levels in a p-type semiconductor.

non-denominational number system (*MathSci*) Any system which is not DENOMINATIONAL. The best known is the Roman number system but a large variety have been devised for use in digital computers.

non-depolarizing muscle relaxant (*Pharmacol*) A group of drugs which block neuromuscular transmission by competing with acetylcholine at the receptor site. Used in anaesthesia to produce paralysis and muscle relaxation. Common examples are *alcuronium, gallomine* and *tubocurarine*.

non-destructive read-out (*ICT*) In a computer, stored data remain stored after they have been read. Abbrev *NDRO*.

non-destructive testing (*Eng*) Methods of inspecting materials and products without affecting their subsequent properties and performance. These include acoustic emission, eddy current testing, neutron and X-ray radiography, optical and ultrasound holography, magnetic particle inspection, penetrant flaw detection and ultrasonic imaging and testing. Abbrev *NDT*.

non-deterministic (*ICT*) Applies to a machine if it can take a number of possible new states, given its present state and its new input.

non-directional microphone (*Acous*) See OMNIDIRECTIONAL MICROPHONE.

non-disjunction (*BioSci*) Failure of one or more chromosomes of a set to move with the rest of its set towards the appropriate pole at anaphase.

non-dispersion-shifted fibre (*ICT*) A type of OPTICAL FIBRE having a stepped change in refractive index between core and cladding, so that the zero-dispersion wavelength is

at about $1 \cdot 3 \, \mu m$ rather than in the $1 \cdot 5 \, \mu m$ wavelength region at which efficient laser diodes operate.

non-dissipative network (*ICT*) One designed as if the inductances and capacitances are free from dissipation, and as if constructed with components of minimum loss.

non-equivalence gate (*ICT*) See XOR GATE.

non-essential organs (*BioSci*) The sepals and petals of flowers.

non-Euclidean geometry (*MathSci*) A system of geometry in which the axioms stated by Euclid are not satisfied, such as the geometry of the surface of a sphere on which there may be no 'lines' parallel to a given line through a given point.

non-ferrous alloy (*Eng*) Any alloy based mainly on metals other than iron, ie usually on copper, aluminium, lead, zinc, tin, nickel or magnesium.

non-flam film (*ImageTech*) See ACETATE FILM.

non-homologous pairing (*BioSci*) The pairing between regions of non-homologous chromosomes. In some cases short stretches of similar sequences, possibly repetitive, may be involved.

non-hydrostatic model (*EnvSci*) Numerical forecasting model in which the HYDROSTATIC APPROXIMATION is not used so that the effects of vertical accelerations can be accounted for.

Nonidet (*Chem*) TN for *nonionic detergent*, based on the condensation products of polyglycols with octyl or nonyl hydroxybenzenes.

non-inductive capacitor (*Electronics*) One constructed so that it has virtually no inductance, with staggered foil layers left with an entire layer of foil at each end for making connections. Currents then flow laterally rather than spirally on the foil layers.

non-inductive circuit (*ElecEng*) One in which effects arising from associated inductances are negligible.

non-inductive load (*ElecEng*) See NON-REACTIVE LOAD.

non-inductive resistor (*ElecEng*) A resistor having a negligible inductance, eg comprising a ceramic rod, or a special design of winding of resistance wire.

non-ionic detergents (*Chem*) Series of detergents in which the molecules do not ionize in aqueous solution, unlike soap and the sulphonated alkylates. Typical examples are the detergents based on condensation products of long-chain glycols and octyl or nonyl hydroxybenzenes.

non-ionizing fields (*Radiol*) The steady magnetic and electric fields produced by a steady electric current. See panel on NON-IONIZING FIELDS AND RADIATION.

non-ionizing radiation (*Phys, Radiol*) All radiation with a wavelength longer than about 100 nm in the far ultraviolet which has insufficient energy to produce ionization in matter. It includes all radio, microwave and infrared radiation as well as all the visible spectrum. Cf IONIZING RADIATION. See panels on NON-IONIZING FIELDS AND RADIATION and RADIATION.

non-isolated essential singularity (*MathSci*) A SINGULARITY that is a LIMIT POINT of a set of singularities of a function.

non-leakage probability (*Phys*) For neutrons in a reactor, the ratio of the actual multiplication constant to the infinite multiplication constant.

non-linear (*ElecEng*) The property of a component where applied voltage and current flowing are not directly proportional.

non-linear distortion (*ICT*) Distortion resulting from the situation where the output of a system or component does not bear the desired relation to the input. Amplitude, harmonic and intermodulation distortion are examples and results of non-linear distortion.

non-linear distortion factor (*ICT*) The square root of the ratio of the powers associated with alien tones to the powers associated with wanted tones in the output of a non-linear distorting device.

non-linear editing (*ImageTech*) A random access, disk-based video editing system in which compressed digital

Non-ionizing fields and radiation

A steady electric current will produce a steady magnetic field as well as a steady electric field. The strengths of such fields are measured in amperes per metre and volts per metre respectively, with their flux densities measured in TESLAS (T) and watts per square metre (W m^{-2}). The fields are not propagated and remain static around the source, declining linearly with distance from the source. Oscillating electric or magnetic fields are propagated as photons. Neither the static fields nor the long-wavelength photons from electromagnetic sources have sufficient energy to ionize an atom. The kind of effects found may well, therefore, be different from those produced by eg FREE RADICALS.

Typical natural static electrical fields are about 150 volts per metre but may reach thousands of volts per metre in a thunderstorm. Even higher electric fields are generated near TV sets or even by friction. The natural static magnetic field, that detected by a compass, is between 10 and 100 microteslas (μT), but small magnets found in the home may have a field 100 times higher, and the large superconducting magnets in magnetic resonance imaging machines will have flux densities of 2·5 teslas or higher.

Static electric fields do not penetrate the body but cause an electric charge to build up on the surface that can sometimes be felt by the movement of body hair. With very high voltages the surrounding air can become ionized and the surface voltage will discharge to earth as a spark and may be fatal. There is very little evidence that such charges can cause any internal effect; volunteers have stayed in 600 volts per metre charged compartments for several weeks without ill effects. Static magnetic fields, on the other hand, penetrate living tissue and any molecules which themselves behave as small magnets will become aligned in the field, this being the basis for the medical imaging procedures that exploit magnetic resonance.

All of the data from people exposed for short times in magnetic resonance imaging machines, and most of the data from animal experiments, show that magnetic fields up to several teslas fail to induce any behavioural difference, genetic or somatic chromosomal changes, or a wide range of other effects. However, vertigo and nausea have been reported in a few individuals exposed to fields of about 4 teslas near magnetic resonance machines. It has been suggested that average exposures should be restricted to 200 milli-teslas per day for those working near large magnets and to less than this figure for the general public.

Extremely-low-frequency radiation below 3 kHz
People under high-voltage ac power lines are subject to an electric field of about 10 kilovolts per metre and a magnetic field of 40 microteslas, but industrial welders

---

images and sound are stored on a database with instant access for viewing, logging and arranging and rearranging in any order.

non-linearity (*ElecEng, ICT*) The lack of proportionality between either applied voltage and resulting current, or between input and output, of an electrical network, component, amplifier or transmission line that causes distortion in a passing signal. See NON-LINEAR DISTORTION.

non-linear network (*ElecEng*) A network in which the electrical elements are not all linear with varying current, as rectifying semiconductor diodes.

non-linear resistance (*ElecEng*) A resistance in which the relation between potential difference and current is not proportional.

non-linear resistor (*ElecEng*) One which does not 'obey' Ohm's law, in that there is departure from proportionality between voltage and current. In semiconductor crystals, the ratio between forward and backward resistance may be 1:1000. Also *non-ohmic resistor*. See DIODE, TRANSISTOR.

non-magnetic steel (*ElecEng*) One containing approx 12% Mn, exhibiting no magnetic properties. Stainless steel is almost non-magnetic.

non-medullated (*BioSci*) See AMYELINATE.

non-Mendelian inheritance (*BioSci*) In eukaryotes, patterns of gene transmission not explicable in terms of segregation, independent assortment and linkage, the classical Mendelian genetics. May be due to cytoplasmic inheritance, gene conversion, meiotic drive, etc.

non-metal (*Chem*) An element which is generally of low density and a poor conductor of heat and electricity, being held together by strong covalent bonds, eg carbon as in diamond. They occur towards the upper right-hand corner

of the periodic table as drawn conventionally (see appendix on The periodic table). Together with metals, they form a vast range of ionic materials. See panel on BONDING.

non-metallic inclusions (*Eng*) See INCLUSION.

non-Newtonian flow (*Eng*) Fluid flow which deviates from Newton's law of viscosity, esp pseudo-plastic fluids, dilatant fluids and BINGHAM SOLIDS. See DILATENCY.

non-Newtonian liquids (*Phys*) See ANOMALOUS VISCOSITY.

non-ohmic resistor (*ElecEng*) See NON-LINEAR RESISTOR.

non-operable instruction (*ICT*) One whose only effect is to advance the PROGRAM COUNTER. Often written as 'continue'.

non-parametric test (*MathSci*) Any statistic that is designed for ORDINAL or NOMINAL data.

nonpareil (*Print*) An old type size, approximately 6-*point*, with which it is still synonymous.

nonpolar group (*Chem*) A group in which the electronic charge density is essentially uniform, and that cannot therefore interact with other groups by forming hydrogen bonds, or by strong dipole–dipole interactions. In an aqueous environment, nonpolar groups tend to cluster together, providing a major force for the folding of macromolecules and formation of membranes.

non-polarized relay (*ICT*) One in which there is no magnetic polarization. Operation depends on the square of the current in the windings, and is therefore independent of direction, as in telephone and ac relays.

non-quantized (*Phys*) See CLASSICAL.

non-reactive load (*ElecEng*) A load in which the current is in phase with the voltage across its terminals. Also *non-inductive load*.

Non-ionizing fields and radiation *(Cont.)*

and similar workers may receive over a thousand times as much, while in the home the magnetic field is probably 50 times less than the power-line figure. The electrical effects of these low-frequency fields is very like that for static fields, but electric fields will also occur within the body, typically of the order of a million times less than that outside. Low-frequency magnetic fields induce electric currents within the body which will depend on how the body is placed relative to the magnetic field and the radii of the conduction loops. Currents of 5 amperes per square metre could be induced by a magnetic field of 1 tesla.

The effect of an electric current passing through body fluids is to alter the electric potentials which exist across membranes and so change their properties which, for nerve cells, include the transmission of electrical impulses. One problem is that the currents are so small that they are less than the electrical noise generated by the thermal agitation of the atoms present (THERMAL NOISE), although it is possible to calculate that if the bandwidth of the current is very restricted, the effect could be detected above background noise for even very small currents.

A considerable amount of animal research has been done on whether low doses of electromagnetic radiation have any immediate effect on the fetus or longer-term genetic effect. Unfortunately they do not provide clear-cut results. Some defects in animals have been noted in some laboratories, but others have often failed to repeat their results. It has been suggested that high electric (20 kV m^{-1}) and magnetic (5 mT) fields which can be detected by the individual and may cause irritability should be avoided. It is thought that more subtle effects on nervous tissue will be avoided if external magnetic fields are kept below that needed to induce a secondary current of 10 mA m^{-2}. Further definitive research on the problems of genetic and developmental damage is clearly needed.

Microwaves

Above 100 kHz the predominantly electrical effects on living tissue induced by a varying magnetic field give way to the thermal excitation of the atoms and the consequent heating of the tissue. It therefore becomes important to measure the rate at which the energy is deposited for which a *specific energy absorption rate* (SAR) of watts per kilogram (W kg^{-1}) is used. An SAR of 1 W kg^{-1} is needed to raise body temperature by 1°C in an hour if cooling is neglected, with higher figures in a normal environment. An SAR of 0·4 W kg^{-1} h^{-1} has been adopted by several authorities for maximum exposure.

Most research indicates that effects other than those associated with temperature rise are unlikely with microwave radiation and that provided the overall temperature rise is kept below 1°C, there are no unfavourable biological effects.

See panel on RADIATION; appendices.

---

non-reactive power (*ElecEng*) The value of active power in an electrical system.

non-relativistic (*Phys*) Said of any procedure in which effects arising from relativity theory are absent or can be disregarded, eg properties of particles moving with speeds which are much smaller than that of light.

non-renewable resource (*EnvSci*) Naturally occurring resources such as fossil fuels which form over periods so long that they are available to humans only once (although in some cases they may be recycled) and the stock of them is limited.

non-resonant antenna (*ICT*) See APERIODIC ANTENNA.

non-return-flow wind tunnel (*Aero*) A straight-through wind tunnel in which the airflow is not recirculated.

non-reversible coatings (*Build*) Convertible coats, like oil-based paints, which, when dry, cannot be returned to the liquid state by the application of further coatings. Cf NON-CONVERTIBLE COATINGS.

non-selective herbicide (*Agri*) A formulation intended to kill all vegetation, eg when maintaining paths or clearing a site for construction.

nonsense mutation (*BioSci*) A base change that causes an amino-acid-specifying sequence to be changed into one in which there is premature termination of polypeptide chain synthesis.

nonsense syllable (*Psych*) A series of letters, usually consisting of two consonants with a vowel between them, that does not constitute a word; used in studies of learning and memory.

non-sequence (*Geol*) A break in the stratigraphical record, less important and less obvious than an unconformity, and deduced generally on palaeontological evidence.

non-singular matrix (*MathSci*) A square matrix the determinant of which is not equal to zero.

non-solid colour (*ICT*) See DITHER COLOUR.

non-specific immunity (*BioSci*) Mechanisms that protect the body against microbial invasion but that do not depend upon the mounting of a specific immune response. They include physical barriers to infection (skin, mucous membranes); enzyme inhibitors naturally present in the blood; activation by 'rough' variants of Gram-negative bacteria of complement via the alternative pathway; interferon; lysozyme; phagocytosis, etc. These mechanisms are important in protection against pathogenic and non-pathogenic organisms even if the adaptive immune system is fully functional.

non-spectral colour (*Phys*) A colour which is outside the range which contributes to white light, but which affects photocells.

non-specular reflection (*Phys*) Wave reflection of light or sound from rough surfaces, resulting in scattering of wave components, depending on relation between wavelength and dimensions of irregularities. Also *diffuse reflection*.

non-steroidal anti-inflammatory drugs (*Pharmacol*) A group of drugs which suppress inflammation and relieve pain probably by interfering with PROSTAGLANDIN synthesis. Common examples are *ibuprofen, indomethacin, phenylbutazone*. Abbrev *NSAID*.

non-stoichiometric compounds (*Chem*) Some solid compounds do not possess the exact compositions which are predicted from Daltonic or electronic considerations alone (eg iron (II) sulphide is FeS$_{1.1}$), a phenomenon which is associated with the so-called DEFECT STRUCTURE of *crystal* *lattices*. Often show semiconductivity, fluorescence and centres of colour. See panel on RUSTING.

non-sweating (*Vet*) See ANHIDROSIS.

non-symmetrical (*ElecEng*) See ASYMMETRICAL.

non-synchronous computer (*ICT*) See ASYNCHRONOUS COMPUTER.

non-synchronous motor (*ElecEng*) An ac motor which does not run at synchronous speed, eg an induction motor or an ac commutator motor. Also *asynchronous motor*.

non-tension joint (*ElecEng*) A joint in an overhead transmission line conductor which is designed to carry full-load current but not to withstand the full mechanical tension of the conductor.

non-theatrical (*ImageTech*) Describing the use of motion picture film outside the scope of the professional cinema entertainment industry.

nontronite (*Min*) A clay mineral in the montmorillonite group (smectites), containing appreciable ferric iron replacing aluminium.

non-verbal communication (*Psych*) A general term for any communication between individuals that does not use spoken language; includes 'body language' and facial expression. See PARALINGUISTICS.

non-viable (*BioSci*) Incapable of surviving.

non-volatile memory (*ICT*) One in a computer that holds data even if power has been disconnected. Most magnetic memories are non-volatile. Cf VOLATILE MEMORY.

non-woven fabric (*Textiles*) Cloth formed from a random arrangement or WEB of natural or synthetic fibres using adhesives, heat and pressure, needling techniques, etc, to confer adhesion. See FELT.

no-observed-adverse-effect level (*BioSci*) The greatest concentration or amount of a substance that causes no detectable adverse change in morphology, functional capacity, growth, development or lifespan of the target organism. Changes in detection methods render previous values obsolete. Abbrev *NOAEL*. Also *NOEL, no observed effect level*.

noon (*Astron*) The instant of the Sun's upper culmination at any place. See MEAN NOON, SIDEREAL DAY.

no-ops (*ICT*) See NON-OPERABLE INSTRUCTION.

nopaline (*BioSci*) One sort of OPINE.

NOR (*BioSci*) See NUCLEOLAR-ORGANIZING REGION.

NOR (*ICT*) A logical operator where (*p* NOR *q*) written $(\overline{p+q})$ takes the value TRUE when neither *p* nor *q* are TRUE. When either *p* or *q* are TRUE, it takes the value FALSE. See LOGICAL OPERATION.

noradrenaline (*Pharmacol*) A catecholamine neurohormone, the neurotransmitter at most synapses of the sympathetic nervous system (adrenergic neurons). It is also found in the adrenal glands where it is stored and released from chromaffin cells. It has greater affinity for alpha-adrenoreceptors than beta-adrenoreceptors and has similar properties to ADRENALINE but differs in structure only by not having the amino group substituted with a methyl group. See CATECHOLAMINES. US *norepinephrine*.

norbergite (*Min*) Magnesium silicate with magnesium fluoride or hydroxide. A member of the humite group, it crystallizes in the orthorhombic system and occurs in metamorphosed dolomitic limestones.

nordmarkite (*Geol*) An alkali quartz-bearing syenite, described originally from Nordmarken in Norway; consists essentially of microperthite, aegirine, soda amphibole and accessory quartz.

norepinephrine (*Chem*) See NORADRENALINE.

Norfolk latch (*Build*) A latch in which the fall bar is actuated within the limits of the keeper by a lifting lever passing through a slot in the door and operated by the pressure of the thumb at one end. Also *Canadian latch*.

NOR gate (*ICT*) A GATE with output 1 when neither of the inputs is 1. When either input signal is 1 the output is 0. All gates can, in theory, be constructed from combinations of NOR gates. Also *NOR element*.

norgine (*Chem*) See ALGINIC ACID.

norite (*Geol*) A coarse-grained igneous rock of basic composition consisting essentially of plagioclase (near labradorite in composition) and orthopyroxene. Other coloured minerals are usually present in varying amount, notably clinopyroxene.

norm (*Geol*) The theoretical composition of an igneous rock expressed in terms of standard mineral molecules, calculated from the chemical analysis as stated in terms of percentages of oxides. Cf *mode*.

norm (*MathSci*) Of a vector in a finite-dimensional real vector space: the square root of the scalar product of the vector with itself; the magnitude of the vector. A norm on a vector space is a consistent definition of the norms of its vectors, or equivalently of a scalar product.

norm (*Psych*) A set of shared expectations about how individuals should or do behave, or that is representative of a group or culture.

Norma (Level) (*Astron*) A small southern hemisphere constellation.

normal (*Chem*) Containing an unbranched chain of carbon atoms; eg *normal* propyl alcohol is $CH_3CH_3CH_2OH$, whereas the isomeric *iso*propyl alcohol, $(CH_2)_2\!=\!CHOH$, has a branched chain. In more modern nomenclature, this prefix is redundant.

normal (*MathSci*) Of a line or surface, a line drawn perpendicular to it.

normal axis (*Aero*) See AXIS.

normal calomel electrode (*Chem*) A calomel electrode containing molar potassium chloride solution.

normal curvature (*MathSci*) See CURVATURE.

normal distribution (*MathSci*) A distribution that is symmetrical around its mean, which is also the median and mode, usually represented by a bell-shaped curve. It is widely used in statistics to model the variation in a set of observations, as an approximation to other distributions, or as the asymptotic distribution of statistics from large samples. The normal distribution is indexed by two parameters, the mean and variance. Also *Gaussian distribution*. See STANDARD NORMAL DISTRIBUTION.

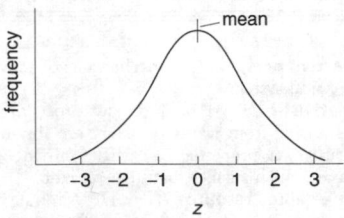

normal distribution

normal electrode potential (*Chem*) See STANDARD ELECTRODE POTENTIAL.

normal fault (*Geol*) A fracture in rocks along which relative displacement has taken place under tensional conditions, the *fault* hading to the downthrow side. Cf REVERSED FAULT.

normal flight (*Aero*) (1) All flying other than aerobatics including straight and level, climbing, gliding, turns and *sideslips* for the loss of height or to counteract drift. (2) A licensing category for certifying whether *airworthy*.

normal frequency (*Phys*) See NORMAL MODES.

normal functions (*MathSci*) See ORTHOGONAL FUNCTIONS.

normal induction (*ElecEng*) That represented by the curve of normal magnetization.

normality (*Chem*) An obsolescent concentration unit, abbreviated N. It is used mainly for acids or bases and for oxidizing or reducing agents. In the first case, it refers to the concentration of titratable H^+ or OH^- in a solution. Thus a solution of sulphuric acid with a concentration of $2\ mol^{-1}$ (2M) will be a 4N solution.

normalize (*ICT*) To represent a number in agreed FLOATING-POINT notation, usually that which provides maximum precision.

normalized data (*ICT*) In data modelling, data which conform to the requirements of one or more of the five normal forms of the relational data model. These prescribe the consistency of record types, elimination of redundancy, and the correct and complete identification of record keys in relation to unique facts within a record. In most physical implementations, such as those of a transaction processing system, this leads to a structure optimized for the rapid retrieval and update of uniquely keyed single records containing unique facts. In certain circumstances, such as the need to perform online ANALYTICAL PROCESSING in a DATA WAREHOUSE, these forms may be violated to provide a denormalized implementation with redundant data and ambiguous keys, enabling the easy retrieval of large sets of factual data.

normalizing (*Eng*) A heat treatment applied to steel. Involves heating above the critical range, followed by cooling in air, performed to refine the crystal structure and eliminate internal stress. See ANNEALING.

normally magnetized crust (*Geol*) See MAGNETIC ANOMALY PATTERNS.

normal magnetization (*ElecEng*) The locus of the tips of the magnetic hysteresis loops obtained by varying the limits of the range of alternating magnetization.

normal modes (*ICT*) In a linear system, the least number of independent component oscillations that may be regarded as constituting the free, or natural, oscillations of the system. Also *natural modes*.

normal modes (*Phys*) Simple vibrations of a system of a number of coupled oscillators in which every part has the same frequency. The associated frequencies are the *normal* or *natural frequencies*. If there are N degrees of freedom in the system then there are N normal modes. See COUPLED VIBRATIONS.

normal pressure (*Chem*) Standard pressure, $101 \cdot 325$ kN m^{-2} = 760 torr, to which experimental data on gases are referred.

normal radius of curvature (*MathSci*) See CURVATURE.

normal salts (*Chem*) Salts formed by the replacement by metals of all the replaceable hydrogen of the acid.

normal section of a surface (*MathSci*) A section of the surface made by a plane which contains a normal to the surface.

normal segregation (*Eng*) A type of segregation in which the concentration of low-melting-point constituents and inclusions tends to increase from the surface to the axial regions of cast metals. See INVERSE SEGREGATION.

normal solution (*Chem*) See NORMALITY.

normal state (*Phys*) See GROUND STATE.

normal subgroup (*MathSci*) A subgroup which contains all conjugates of its elements. Also *invariant subgroup, self-conjugate subgroup*.

normal temperature and pressure (*Chem*) The earlier term for STANDARD TEMPERATURE AND PRESSURE.

normative composition (*Geol*) The theoretical mineral chemical composition of a rock expressed in terms of the standard minerals of the NORM.

normed space (*MathSci*) A VECTOR SPACE which has a norm defined on it.

normoblast (*BioSci*) A stage in the development of an erythrocyte from an erythroblast when the nucleus has become reduced in size and the cytoplasm contains much haemoglobin.

norovirus (*Med*) A type of calicivirus that causes intestinal illness. Also *Norwalk virus*.

Northern blot (*BioSci*) See BLOTTING.

Northern Lights (*Astron*) See AURORA.

northing (*Surv*) A north latitude. Northerly displacement of a point with reference to observer's station.

north light roof (*Arch*) A pitched roof with unequal slopes, of which the steeper is glazed and arranged in such a way as to receive light from the north.

North Polar Sequence (*Astron*) A sequence of stars near the north celestial pole whose photographic magnitudes

have been accurately measured to calibrate the arbitrary zero point in the magnitude scale for the measurement of other stellar magnitudes. Photoelectric determinations of magnitude are now preferred.

north pole (*Phys*) See POLE.

Northrup furnace (*ElecEng*) See CORELESS INDUCTION FURNACE.

North Sea gas (*Geol*) See NATURAL GAS.

North Star (*Astron*) See POLARIS.

Northwestern blot (*BioSci*) See BLOTTING.

Norton's theorem (*ElecEng*) Any electrical source can be represented at its terminals by a constant-current generator in parallel with a shunt resistance. The value of the current generator is that which would flow with a short circuit at the generator terminals. The shunt resistance is that measured between the terminals with no current source. Often regarded as the dual of THÉVENIN'S THEOREM.

Norwegian quartz (*Build*) A white translucent quartz found in Norway, used for the facing of CLADDING slabs.

Noryl (*Plastics*) TN for blend of polystyrene and polyphenylene oxide (US). A rare example of compatible polymers, mixing at a molecular level.

nosean (*Min*) Silicate of sodium and aluminium with sodium sulphate, crystallizing in the cubic system. Occurs in extrusive igneous rocks which are rich in alkalis and deficient in silica, eg phonolite. Also *noselite*.

nose bit (*Build*) A type of SHELL BIT with a projecting tip for extracting the waste.

nosedive (*Aero*) See DIVE.

nose heaviness (*Aero*) The state in which the combination of the forces acting upon an aircraft in flight is such that it tends to pitch downwards by the nose.

noselite (*Min*) See NOSEAN.

nose ribs (*Aero*) Small intermediate ribs, usually from the front spar to the leading edge only, of planes and control surfaces. They maintain the correct wing contour under the exceptionally heavy air load at that part of the aerofoil.

nose suspension (*ElecEng*) A method of mounting a traction motor, by supporting one side of it on the axle and the other side on the framework of the truck.

nose-wheel landing gear (*Aero*) See TRICYCLE LANDING GEAR.

nosing (*Build*) A bead on the edge of a board, making it half-round.

nosocomial infections (*Med*) Hospital-acquired infections, the commonest at present being due to *Staphylococcus aureus* (particularly MRSA), *Escherichia coli*, *Klebsiella pneumoniae*, *Proteus mirabilis*, *Pseudomonas aeruginosa* and *Serratia marcescens*.

nosology (*Med*) Systematic classification of diseases; the branch of medical science which deals with this.

nosophobia (*Med*) Morbid fear of contracting disease.

nostrils (*BioSci*) The external nares.

NOT (*ICT*) A logical operator such that NOT p, written $\bar{p}$, takes the value FALSE if p is TRUE and vice versa. See LOGICAL OPERATION.

not (*Paper*) The unglazed surface of hand-made or mould-made drawing-papers. See HOT-PRESSED, ROUGH.

notch (*ElecEng*) A term often used to denote any of the various positions of a controller.

notch (*ImageTech*) A shallow cut on the edge of a motion picture negative to actuate the light change mechanism of a printer.

notch aerial (*Aero*) A radio aerial, usually for high frequency, formed by cutting a notch out of the aircraft's skin and covering it with a dielectric material to its original profile.

notch binding (*Print*) Unsewn binding style where a notch shape is cut out of the spine of the section, at intervals, during the folding operation allowing the adhesive to reach the centre pages of the section. See BURST BINDING.

notch board (*Build*) A notched board carrying the treads and risers of a staircase.

notch brittleness (*Eng*) Susceptibility to fracture, as disclosed by IZOD or CHARPY TEST, due to weakening effect of a stress-concentrating flow at the surface.

notched-bar test (*Eng*) A type of impact test using a notched specimen. See CHARPY TEST, IZOD TEST and panel on IMPACT TESTS.

notching (*CivEng*) The method of excavating cuttings for roads or railways in a series of steps worked at the same time.

notching (*Eng*) A presswork process similar to punching or piercing, in which material is removed from the edge of a strip or sheet, often as a means of location or registering.

notching, linking up (*Eng*) Movement of the gear lever of a locomotive or steam engine towards the centre of a notched quadrant, to decrease the valve travel and shorten the cut-off.

notch sensitivity (*Eng*) The extent to which the endurance of metals, as determined on smooth and polished specimens, is reduced by surface discontinuities, such as tool marks, notches and changes in section, which are common features of actual components. It tends to increase with the hardness and endurance limit.

notch toughness (*Eng*) Misleading use of toughness for the energy required to break standard specimens under standard conditions in eg IZOD or CHARPY TESTS. It is also used to mean the opposite of NOTCH BRITTLENESS.

note (*Acous*) An identifiable musical tone, whether pure or complex. See STRIKE NOTE, WOLF NOTE.

notebook (*ICT*) Hand-held computer, smaller than a LAPTOP COMPUTER and often with a very small keyboard. Useful for entering data to be transferred to a larger computer later.

NOT gate (*ICT*) The only single-input GATE. It gives an output signal when there is no input signal and vice versa, ie input 0, output 1 and input 1, output 0. Also *NOT element*.

notifiable disease (*Agri*) A term used for various highly infectious livestock diseases that, if suspected, must immediately be notified to a designated authority for confirmation, eg foot and mouth disease.

notochord (*BioSci*) In Chordata, a skeletal rod formed of turgid vacuolated cells. Adj *notochordal*.

notoedric mange (*Vet*) Mange on the face and ears of cats and rabbits, due to *Notoedres cati*.

notum (*BioSci*) The tergum of insects.

nova (*Astron*) Classically, any new star which suddenly becomes visible to the unaided eye. In modern astronomy, a star late in its evolutionary track which suddenly brightens by a factor of 10^4 or more. The rise in brightness is rapid (on a time-scale of days); after a few weeks it returns to the pre-nova magnitude. Thought to occur in close binary systems, in which one component is a WHITE DWARF: matter is transferred from the other, highly evolved, companion and triggers a new burst of nuclear reactions. See panel on REDSHIFT–DISTANCE RELATION.

novaculite (*Geol*) A fine-grained or cryptocrystalline rock composed of quartz or other forms of silica; a form of chert. Used as a whetstone.

novocaine (*Chem*) See PROCAINE.

novolak (*Chem*) A type of phenolic resin prepolymer made by reacting excess phenol with formaldehyde under acidic conditions. Also *two-stage resin*, since more formaldehyde must be added for final cross-linking. Both novolaks and resols are A-stage resins.

no-voltage release (*ElecEng*) A relay or similar device which causes the circuit to a motor or other equipment to be opened automatically if the supply voltage falls.

nowcasting (*EnvSci*) A system of rapid and very-short-range (1 to 2 h) forecasting of phenomena such as heavy rain and thunderstorms based, on real-time processing of simultaneous observations from a network of remote-sensing devices (including METEOROLOGICAL SATELLITES and WEATHER RADARS) combined with simple extrapolation techniques.

NOx (*Chem*) See NITROGEN OXIDES.

nox (*Phys*) Obsolete unit of scotopic illuminance: 1 nox = 10^{-3} lux.

noy (*Acous*) Unit of perceived noisiness by which equal-noisiness contours (eg for 10 noys, 20 noys) replace equal-loudness contours.

nozzle (*Aero*) See PROPELLING NOZZLE.

nozzle (*Eng*) (1) In impulse turbines, specially shaped passages for expanding the steam, thus creating kinetic energy of flow with minimum loss. (2) In oil engines, orifices, open or controlled by the injection valve, through which the fuel is sprayed into the cylinder. See CONVERGENT–DIVERGENT NOZZLE. (3) Generally, a convergent or convergent–divergent tube attached to the outlet of a pipe or a pressure chamber to convert efficiently the pressure of a fluid into velocity. (4) Narrow orifice at front of injection-moulding machine through which molten polymer passes to reach tool cavity. See INJECTION MOULDING.

nozzle (*ICT*) End of a waveguide that may be contracted in area.

nozzle guide vanes (*Aero*) In a gas turbine, a ring of radially positioned aerofoils which accelerate the gases from the combustion chamber and direct them onto the first rotating turbine stage.

Np (*Chem*) Symbol for NEPTUNIUM.

npin transistor (*Electronics*) Similar to npn transistor but incorporating a layer of intrinsic semiconductor between the base and the collector to extend the high-frequency range. Also *n–p–i–n transistor*.

NPK (*Agric*) Nitrogen, phosphorus and potassium as fertilizer.

NPL (*Genrl*) Abbrev for NATIONAL PHYSICAL LABORATORY.

NPL-type wind tunnel (*Aero*) The *closed-jet, return-flow* type is often called the original NPL type, and the *closed-jet, non-return flow* type the standard NPL type, as they were first used by the UK *National Physical Laboratory*.

npn transistor (*Electronics*) A bipolar transistor having a p-type base between an n-type emitter and an n-type collector. In operation, the emitter should be negative and the collector positive with respect to the base. Also *n–p–n transistor*.

NR (*Genrl*) Abbrev for NATURAL RUBBER.

NRME (*Build*) Abbrev for NOTCHED, RETURNED AND MITRED ENDS.

NRR (*Eng*) Abbrev for NON-REPEATABLE RUNOUT. See FLUID BEARING MOTOR

NSAID (*Pharmacol*) Abbrev for NON-STEROIDAL ANTI-INFLAMMATORY DRUGS.

N-shell (*Phys*) The ELECTRON SHELL in an atom corresponding to a principal quantum number of four. The shell contains up to 32 electrons and is the last shell to be filled completely by electrons in the naturally occurring elements. See panel on ATOMIC STRUCTURE.

nt (*Phys*) Symbol for NIT.

nτ (*Phys*) The Lawson criterion: the product of the plasma density n and the plasma confinement time τ for nuclear fusion processes. An $n\tau$ of 10^{20} m^{-3} s is required for fusion to produce useful energy.

N-terminal pair network (*ICT*) One having N-terminal pairs in which one terminal of each pair may coincide with a node.

NTI (*Acous*) Abbrev for NOISE TRANSMISSION IMPAIRMENT.

n-tier architecture (*ICT*) An approach to the design of a system that will typically distribute data storage, logical processing, information presentation and user interaction into distinctly encapsulated tiers of processing using software components hosted on different hardware platforms.

NTP (*Chem*) Abbrev for *normal temperature and pressure*. Previous term for STP, ie 0°C and 101·325 kPa.

N-truss (*Eng*) See WHIPPLE–MURPHY TRUSS.

NTS (*Build*) Abbrev for *not to scale*; on plans etc.

NTSB (*Aero*) Abbrev for NATIONAL TRANSPORTATION SAFETY BOARD, US.

NTSC (*ImageTech*) The US *National Television System Committee*, and the colour TV system so standardized: 4·43 NTSC (or *modified NTSC*) is the system with the COLOUR SUBCARRIER at the European frequency of 4·43 MHz instead of 3·58 MHz.

NTSC playback (*ImageTech*) A videotape recorder and video disk player facility which enable NTSC tapes/disks to be shown on a PAL TV by converting the colour. The 525 lines and 60 Hz FIELD FREQUENCY remain unchanged, but modern receivers can cope.

n-tuple (*MathSci*) An ORDERED SET with *n* members; an $n \times 1$ array; a ROW VECTOR or COLUMN VECTOR.

n-type conduction (*Electronics*) Conductivity associated with charge transport by electrons. Also *n-type conductivity*.

N-type semiconductor (*Electronics*) A relatively wide-band-gap n-type semiconductor.

n-type semiconductor (*Electronics*) One in which the electron conduction (negative) exceeds the hole conduction (absence of electrons), the DONOR impurity predominating.

nucellus (*BioSci*) Parenchymatous tissue in the ovule of a seed plant, more or less surrounded by the integuments and containing the embryo sac; equivalent to the MEGA-SPORANGIUM. See PERISPERM.

nuchal (*BioSci*) Pertaining to, or situated on, the back of the neck.

nuchal crest (*BioSci*) A transverse bony ridge forming across the posterior margin of the roof of the vertebrate skull for attachment of muscles and ligaments supporting the head.

nuchal flexure (*BioSci*) In developing vertebrates, the flexure of the brain occurring in the hinder part of the medulla oblongata, which bends in the same direction as the primary flexure.

nucivorous (*BioSci*) Nut-eating.

nuclear battery (*Phys*) A battery in which the electric current is produced from the energy of radioactive decay, either directly by collecting beta particles or indirectly, eg by using the heat liberated to operate a thermojunction. In general, nuclear batteries have very low outputs (often only microwatts) but long and trouble-free operating lives.

nuclear binding energy (*Phys*) The binding energy that holds together the constituent nucleons of the nucleus of an atom. It is the energy equivalence of the mass difference between the masses of the atom and the sum of the individual masses of its constituents. Units MeV or amu.

nuclear Bohr magneton (*Phys*) See BOHR MAGNETON.

nuclear breeder (*NucEng*) A nuclear reactor in which in each generation there is more fissionable material produced than is used up in fission.

nuclear budding (*BioSci*) The production of two daughter nuclei of unequal size by constriction of the parent nucleus.

nuclear charge (*Phys*) Positive charge arising in the atomic nucleus because of protons, equal in number to the atomic number.

nuclear chemistry (*Chem*) The study of reactions involving the transmutation of elements, either by spontaneous decay or by particle bombardment.

nuclear conversion ratio (*NucEng*) That of the fissile atoms produced to the fissile atoms consumed, in a breeder reactor.

nuclear cross-section (*NucEng*) See CROSS-SECTION.

nuclear disintegration (*Phys*) Fission, radioactive decay, internal conversion or isomeric transition. Also *nuclear reaction*.

nuclear emission (*Phys*) Emission of a gamma ray or particle from the nucleus of an atom as distinct from emission associated with orbital phenomena.

nuclear emulsion (*Phys*) Thick photographic coating in which the tracks of various fundamental particles are revealed by development as black traces.

nuclear energy (*Phys*) In principle, the binding energy of a system of particles forming an atomic nucleus. More usually, the energy released during nuclear reactions involving regrouping of such particles (eg fission or fusion processes).

The term *atomic energy* is deprecated as it implies rearrangement of atoms rather than of nuclear particles.

nuclear envelope (*BioSci*) See NUCLEAR MEMBRANE.

nuclear family (*BioSci*) In human genetics, a family providing data on the two parents and their children.

nuclear field (*Phys*) Postulated short-range field within a nucleus, which holds protons and neutrons together, possibly in shells.

nuclear fission (*Phys*) The spontaneous or induced disintegration of the nucleus of a heavy atom into two lighter atoms. The process involves a loss of mass which is converted into nuclear energy.

nuclear force (*Phys*) The force which keeps neutrons and protons together in a nucleus, differing in nature from electric and magnetic forces, gravitational forces being negligible. The force is of short range, is practically independent of charge, and arises from the exchange of *pions* between the nucleons. See EXCHANGE FORCES, MESON, SHORT-RANGE FORCE.

nuclear fragmentation (*BioSci*) The formation of two or more portions from a cell nucleus by direct break-up and not by mitosis.

nuclear fuel (*NucEng*) See FUEL, REACTOR CLASSIFICATION.

nuclear fusion (*Phys*) The process of forming atoms of new elements by the fusion of atoms of lighter ones. Usually the formation of helium by the fusion of hydrogen and its isotopes. The process involves a loss of mass which is converted into nuclear energy. The basis of FUSION REACTORS.

nuclear isomer (*Phys*) A nuclide existing in an excited metastable state. It has a finite half-life after which it returns to the ground state with the emission of a gamma quantum or by internal conversion. Metastable isomers are indicated by adding *m* to the mass number.

nuclear lamina (*BioSci*) A fibrous protein network lining the inner surface of the nuclear envelope. Proteins of the lamina are lamins A, B and C that have sequence homology to proteins of intermediate filaments. The extent to which this system also provides a scaffold within the nucleus is controversial.

nuclear localization signal (*BioSci*) In eukaryotes, peptide signal sequence that identifies a protein as being destined for the nucleus.

nuclear magnetic resonance (*Phys*) Certain atomic nuclei, eg hydrogen, fluorine-19, phosphorus-31, have a nuclear magnetic moment. When placed in a strong magnetic field, the nuclear moments can only take up certain discrete orientations, each orientation corresponding to a different energy state. Transitions between these energy levels can be induced by the application of radio-frequency radiation. This is known as nuclear magnetic resonance (abbrev *NMR*): eg the protons in water experience this resonance effect in a field of 0·3 T at a radiation frequency of 12·6 MHz. The resonant frequency depends on the magnetic field at the nucleus which in turn depends on the environment of the particular nucleus. NMR gives invaluable information on the structure of molecules. It has also been developed to provide a non-invasive clinical imaging of the human body, magnetic resonance imaging (MRI); multiple projections are combined to form images of sections through the body, providing a powerful diagnostic aid.

nuclear magnetic resonance spectroscopy (*Chem*) Routine analytical tool for detecting atomic nuclei with spin (^{1}H, ^{14}C, ^{15}N, ^{19}F, ^{31}P, etc) in molecules, by absorption at resonance. Nuclear magnetic field is modified by chemical environment (chemical shift), so protons etc in different parts of molecules can be differentiated, allowing structural characterization. It is esp valuable for polymers, enabling different stereoisomeric forms to be identified, and its sensitivity enables fingerprinting of unknown materials. Resolution depends on applied field strength, with many machines available now up to 800 MHz or more. Complementary to infrared and ultraviolet spectroscopy. Also *NMR spectroscopy*.

Nuclear reactors

Each atom of uranium-235 disintegrates to produce a pair of fission products, neutrons and energy. The elements of the fission pairs can differ considerably, the only condition being that their atomic numbers must add up to 92 and their atomic weights should add up to 235 less the mass of the neutrons emitted.

The energy produced is roughly equivalent to 200 MeV or $3 \cdot 2 \times 10^{-11}$ joules per disintegration, so that 1 gram of uranium-235 ($2 \cdot 56 \times 10^{21}$ atoms) would release $8 \cdot 19 \times 10^{10}$ joules, equivalent to burning nearly 14 barrels of crude oil.

The energy of 200 MeV is divided as follows:

affect nuclei, so neutrons of this energy are called *thermal* neutrons and reactors which mainly use thermal neutrons are THERMAL REACTORS. Other types of reactors, such as those used in *breeding*, the process of producing more fuel than was initially present, make use of high energy or fast neutrons; hence the terms *fast breeder* and *fast reactor*.

Moderators

A moderator needs to reduce the energy of neutrons without capturing many of them. In other words, neutrons should react and change direction on colliding with a nucleus, releasing energy, but should not induce fission. Such nuclei have low molecular weight. Carbon, beryllium, water or heavy water are good moderators and can be either interspersed with the fuel or placed around the fuel elements. As neutrons gradually lose energy they pass through

MeV	167	5	7	6	11	5
Form	Kinetic energy of products	β-particles	Prompt γ-particles	delayed-γ-particles	Neutrinos	Fast neutrons
Effect	Absorbed in fuel elements		Irradiate structure		Lost	To chain reaction

Prompt particles result from the neutrons emitted by the fission of the uranium-235. *Delayed* particles come from the neutrons emitted by the fission products of the primary disintegration. *To chain reaction* means that these neutrons are available for continuing the chain reaction.

Fast neutrons initially have an average energy, about 2 MeV, which is too high to be effective in inducing a further fission of uranium-235. For maximum effect their energy must be reduced to about 3 eV by passage through a MODERATOR (see diagram); 3 eV is in the range of the normal thermal processes which

energy levels at which they are likely to be captured by other elements, in particular the common isotope of uranium, present in the fuel.

Reactor poisons

Other elements, especially those present as fission products, also capture neutrons very effectively. Of these xenon-135 and samarium-149 capture neutrons about a million times more effectively than elements in moderators, so very little of these elements can shut down a reactor or POISON it. The gas xenon-135 readily fissions to form less damaging products in a reactor running at full power but can cause considerable

---

nuclear magneton (*Electronics*) See BOHR MAGNETON.

nuclear matrix (*BioSci*) The nuclear residue left after removal of chromatin, including pore complexes, lamina, nucleolar residues and ribonucleoprotein fibrils.

nuclear medicine (*Med*) The application of radionuclides in the diagnosis or treatment of disease.

nuclear membrane (*BioSci*) The double membrane, punctuated by NUCLEAR PORE COMPLEXES, which surrounds the interphase nucleus, the outer membrane being continuous with the membrane of the ENDOPLASMIC RETICULUM. Also *nuclear envelope*.

nuclear model (*Phys*) Theory or model giving an explanation of the properties of the atomic nucleus and its interactions with other particles. See COLLECTIVE MODEL, INDEPENDENT PARTICLE MODEL, LIQUID-DROP MODEL, OPTICAL MODEL, SHELL MODEL, UNIFIED MODEL.

nuclear number (*Phys*) See MASS NUMBER.

nuclear paramagnetic resonance (*Chem, Phys*) See NUCLEAR MAGNETIC RESONANCE.

nuclear photoeffect (*Phys*) See PHOTODISINTEGRATION.

nuclear physics (*Phys*) The science of forces and transformations within the nucleus of the atom.

nuclear pore complex (*BioSci*) Sites at which the two layers of the nuclear membrane are joined forming pores that connect the nucleoplasm and the cytoplasm, and which are surrounded on either side by symmetrical arrays of granules called *annuli*.

nuclear potential (*Phys*) The potential energy of a nuclear particle in the field of a nucleus as determined by the short-range forces acting. Plotted as a function of position it will normally represent some sort of POTENTIAL WELL.

nuclear power (*Phys*) Power generated by the release of energy in a nuclear reaction.

nuclear propulsion (*Space*) The use of the energy released by a nuclear reaction to provide propulsive thrust through heating the working fluid or providing electric power for an ion or similar propulsion system.

nuclear radius (*Phys*) The somewhat indefinite radius of a nucleus within which the density of nucleons (protons and neutrons) is experimentally found to be nearly constant. The radius in metres is $1 \cdot 2 \times 10^{-15}$ times the cube root of

Nuclear reactors (Cont.)

problems for a reactor starting up or running at lower power for a considerable time.

Control

Still other elements, such as boron and cadmium, have the ability to absorb neutrons with an efficiency between that of a nuclear poison and the light elements used as moderators. Therefore, the number of neutrons available to sustain the chain reaction can be controlled by mechanically withdrawing or inserting rods made of these substances. The NEUTRON FLUX will fall as the nuclear fuel is burnt until the amount of uranium-235 approaches half its initial value and the reactor has to be refuelled, but during the life of the fuel the neutron flux is maintained by slowly withdrawing the control rods. The same control elements are required to shut down the reactor and protect it against wear and tear in fuel elements and any other accidents.

Reactivity

Reactivity is the change in the number of neutrons or the neutron flux which follows a change in some factor like the position of a control rod. It is important in determining how easy a reactor will be to control. The flux of prompt neutrons, ie those coming from the primary disintegration, will change tenfold about a second after moving a fuel rod in a working reactor, and it would be difficult to achieve stability with such extreme sensitivity. Fortunately there is a small proportion (<1%) of delayed neutrons coming from the fission products whose flux alters more slowly with time. Fission reactors are therefore designed to operate with a prompt neutron flux just below that required to sustain a chain reaction, ie just below CRITICALITY, and to rely on the delayed neutrons to bring it up to this figure.

The reactor problem

These effects, and others like the proportion of neutrons which escape the core, contribute to the *reactor problem*, whose solution determines whether a design can sustain a chain reaction without risk of meltdown. They all depend not only on the fuel, the moderator and the coolant, but also on the positions and shapes of the components. Calculating the effects of various arrangements is no easy task. Much calculation and experiment were needed to determine the most suitable materials and their disposition in the early years of the nuclear age and so solve the reactor problem: ie in ensuring that just one neutron can survive to continue the chain, no more and no less.

See REACTOR CLASSIFICATION.

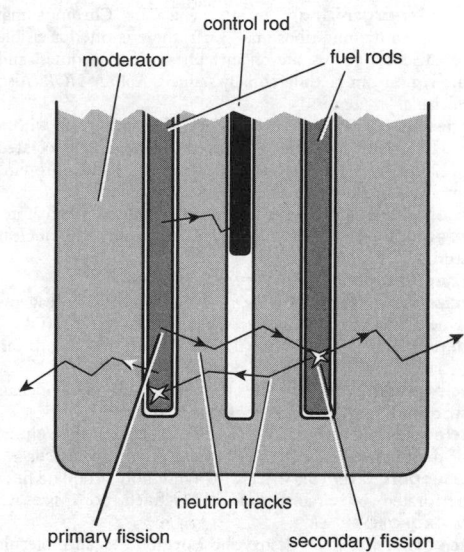

Schematic drawing of a fission reactor

---

the nuclear number (atomic mass). It is not a precise determinable quantity.

nuclear reaction (*Phys*) The interaction of a photon or particle with a target nucleus. An amount of energy, the Q-value, is released or absorbed, depending on whether the mass of the reaction products is less than or more than the mass of the reactants. Reactions fall into two broad categories: those in which the reaction proceeds via the formation of a *compound nucleus*; and the alternative mechanism by *direct interaction*. At higher energies *spallation* occurs in which the target nucleus splits up into a number of fragments.

nuclear reactor (*NucEng*) See panel on NUCLEAR REACTORS.

nuclear reactor oscillator (*NucEng*) A device producing variations in reactivity by oscillatory movement of sample to measure reactor properties, or neutron capture cross-section of sample.

nuclear sap (*BioSci*) See NUCLEOPLASM.

nuclear selection rules (*Phys*) Rules specifying the transitions of electrons or nucleons between different energy levels which may take place (*allowed transitions*).

The rules may be derived theoretically through wave mechanics but are not obeyed rigorously; so-called *forbidden transitions* do occur, but are highly improbable.

nuclear spindle (*BioSci*) See SPINDLE.

nuclear waste (*NucEng*) See WASTE.

nuclear weapon (*Phys*) A bomb or missile deriving its destructive force from the energy released by a nuclear reaction.

nuclear winter theory (*EnvSci*) The theory, based on model calculations, that nuclear war would be followed by a period of cold resulting from the attenuation of solar energy by dust and smoke in the atmosphere.

nuclease (*BioSci*) An enzyme that specifically cleaves the 'backbone' of nucleic acids. Nuclease specificity ranges from those like DNase and RNase that cut the phosphodiester bonds in any DNA or RNA, to RESTRICTION ENZYMES that only cut a particular four or six base pair sequence.

nucleating agent (*Chem*) A substance added to molten materials and solutions to accelerate the onset and increase the rate of crystallization, eg stearates in polypropylene.

nucleating agent (*EnvSci*) A substance used for seeding clouds to control rainfall and fog formation. See RAIN-MAKING.

nucleation (*Chem*) Initiation of processes, such as crystallization or fracture of materials. Hence homogeneous and heterogeneous nucleation of crystalline materials, depending on phase state of nuclei. Followed by growth processes, studied using eg the AVRAMI EQUATION.

nuclei (*Eng*) Centres or 'seeds' from which crystals begin to grow during solidification. In general, they are minute crystal fragments formed spontaneously in the melt, but frequently non-metallic inclusions act as nuclei. See CRYSTAL NUCLEI.

nucleic acid (*BioSci*) General term for natural polymers in which *bases* (purines or pyrimidines) are attached to a sugar phosphate backbone. Can be single- or double-stranded. Short molecules are called *oligonucleotides*. Also *polynucleotide*.

nucleocapsid (*BioSci*) The nucleic acid genome and coat (*capsid*) of a virus.

nucleogenesis (*Phys*) Theoretical process(es) by which nuclei could be created from possible fundamental dense plasma. See YLEM.

nucleolar-organizing region (*BioSci*) Chromosomal region containing ribosomal genes; there is often a visible constriction at this site on metaphase chromosomes, and the region can be differentially stained. Abbrev NOR. Also *nucleolar organizer*.

nucleolus (*BioSci*) Roughly spherical body occurring within a cell nucleus, consisting of ribosomal genes and associated polymerases, nascent RNA transcripts and proteins involved in ribosome assembly. Adj *nucleolar*.

nucleon (*Phys*) A proton or neutron in the nucleus of an atom.

nucleonics (*Phys*) The science and technology of nuclear studies.

nucleon number (*Phys*) See MASS NUMBER.

nucleophilic reagents (*Chem*) A term applied to reagents which react at positive centres, eg the hydroxyl ion (OH^-) in replacing Cl^- from a C–Cl link in which the C atom can be considered as being the positive centre.

nucleoplasm (*BioSci*) The protoplasm in the nucleus surrounding the chromatin. Cf CYTOPLASM.

nucleoplasmic ratio (*BioSci*) The ratio between the volume of the nucleus and of the cytoplasm in any given cell.

Nucleopore filter (*BioSci*) TN for a thin polycarbonate filter perforated by circular holes of defined pore size. Cf MICROPORE FILTER.

nucleoside (*BioSci*) A desoxyribose or ribose sugar molecule to which a purine or pyrimidine is covalently bound. See NUCLEOTIDE.

nucleosome (*BioSci*) A quaternary-level organizational unit of the eukaryotic chromosome consisting of a histone multimer around which is wrapped about 145 base pairs of DNA; each 'bead' is separated by less folded chromatin.

nucleosynthesis (*Astron*) The synthesis of elements other than hydrogen and helium by means of nuclear fusion reactions in stellar interiors and SUPERNOVA explosions. See panel on COSMOLOGY.

nucleotide (*BioSci*) A NUCLEOSIDE to which a phosphate group is attached at the $5'$ position on the sugar. The individual components of a NUCLEIC ACID.

nucleus (*Astron*) (1) The central core of a comet, around 1–10 km across, and consisting of icy substances and dust. (2) The central part of a galaxy or quasar, possibly the seat of unusually energetic activity within the galaxy. (3) The term used generally in astronomy to indicate any concentration of stars or gas in the central part of a NEBULA. See panels on GALAXY and QUASAR.

nucleus (*BioSci*) (1) A compartment within the interphase eukaryotic cell bounded by a double membrane (*nuclear envelope*) and containing the genomic DNA, with its associated functions of transcription and processing. (2) Any nut-shaped structure. (3) A nerve centre in the brain. (4) A collection of nerve cells on the course of a nerve or tract of nerve fibres.

nucleus (*Phys*) The central dense core of an atom, composed of protons (positively charged) and neutrons (no charge), and constituting practically all the mass. Its charge equals the atomic number; its diameter is in the range 10^{-15} to 10^{-14} m. With protons, equal to the atomic number, and neutrons to make up the atomic mass number, the positive charge of the protons is balanced by the same number of extra-nuclear electrons.

nuclide (*Phys*) An atomic nucleus as characterized by its atomic number, its mass number and nuclear energy state.

nude mouse (*BioSci*) A strain of almost hairless mouse that lacks a thymus and thus all or most of the T-lymphocyte population; incapable of rejecting allografts or xenografts.

nudicaudate (*BioSci*) Having the tail uncovered by fur or hair, as rats.

nuée ardente (*Geol*) A glowing cloud of gas and volcanic ash that moves rapidly downhill, as a density flow.

nugget (*Eng*) (1) A term for a welding bead. (2) A mass of gold or silver found free in nature.

Nujol (*Chem*) TN for a pure paraffin oil free of unsaturated compounds and used esp for making MULLS for infrared spectroscopy.

null cells (*BioSci*) Cells that lack characteristic surface antigenic markers, particularly lymphocytes that carry neither T- nor B- lymphocyte cell surface markers.

null hypothesis (*MathSci*) The default hypothesis that given data result from random variation. Compare ALTERNATIVE HYPOTHESIS.

null indicator (*ElecEng*) Any device for determining zero signal in a specified part of an electric circuit. Often used to determine balance conditions in bridge networks.

nullipara (*Med*) A woman who has never given birth to a child. Adj *nulliparous*.

nullisomic (*BioSci*) A cell or organism having one particular chromosome of the normal complement not represented at all in an otherwise diploid (or, more generally, euploid) complement. See ANEUPLOIDY, of which nullisomy is one sort.

null method (*ElecEng*) See ZERO METHOD.

null modem cable (*ICT*) (1) Originally a special cable connected to a modem and used for testing without connection to the telephone system, as it was wired to simulate the response from a telephone line; (2) more generally now a cable wired to connect two computers through their serial ports.

number (*MathSci*) According to context, see: COMPLEX NUMBER, INTEGER, NATURAL NUMBER, RATIONAL NUMBER, REAL NUMBER.

number-average molecular mass (*Chem*) See MOLECULAR MASS DISTRIBUTION.

number sizes (*Eng*) A series of drill sizes in which gauge numbers represent diameters, the higher numbers representing the smaller diameters, in irregular increments.

numbers, pyramid of (*EnvSci*) See PYRAMID OF NUMBERS.

numerable set (*MathSci*) See DENUMERABLE SET.

numerator (*MathSci*) See DIVISION.

numerical analysis (*MathSci*) The derivation of a particular numerical solution or set of solutions arising from the input of a single set of data into a mathematical model, as opposed to a general algebraic analytical solution.

numerical aperture (*Phys*) The product of the refractive index of the object space and the sine of the semi-aperture of the cone of rays entering the entrance pupil of the objective lens from the object point. The resolving power is proportional to the numerical aperture. Abbrev NA.

numerical control (*Eng*) The operation of machine tools from numerical data which have been recorded and stored on magnetic or punched tape or, esp formerly, on punched cards. See COMPUTER NUMERICAL CONTROL.

numerical forecast (*EnvSci*) A forecast of the future state of the atmosphere made by solving the equations of a numerical forecast model on a digital computer. The initial data are based upon observations, objective analysis

Numerical weather forecast

Abbrev *NWF*. A method of predicting the future state of the atmosphere and the associated weather, not by traditional methods whereby human forecasters study and analyse charts of meteorological observations and use their physical understanding and experience to forecast developments, but by mathematical calculation using a digital computer. For NWF it is necessary (1) to devise a conceptual model of the atmosphere and its mode of behaviour according to the laws of physics, a model which is inevitably simpler than the full complexity of nature, but not oversimple; (2) to write down the mathematical equations describing how the model works using the laws of motion, of conservation of energy, momentum and matter, together with the gas laws and those of thermodynamics including the phase changes of water; (3) to solve these equations given certain initial and boundary conditions by methods of numerical approximation that tend towards the 'true' solutions of the unapproximated equations.

Certain physical processes, such as the turbulent exchange of heat and moisture at the Earth's surface or the formation of clouds and the fallout of precipitation, are so complicated that they cannot be modelled in full detail but have to be parameterized, ie represented in a quasi-empirical and statistical way as functions of directly modelled values of wind, pressure, temperature and humidity. There is as much art and skill in NWF as there used to be in old-fashioned forecasting, but the relationship to fundamental physical processes is closer.

In modern operational NWF the forecasts produced are continually updated by the introduction of new observations in a type of rolling process. Some years ago a separate process of 'objective computer analysis' was performed on the new observations using the appropriate forecast as a first guess, and this objective analysis would have to be initialized, or subtly modified, so that it could be used as a set of initial conditions for the model equations without producing spurious 'shock waves' in the forecast. More recently, the absorption and initialization of new data have been carried out much more as a unified process.

The PRIMITIVE EQUATIONS now used for NWF describe, in theory, all scales of motion from sound waves to meteorological waves thousands of kilometres long. In the 1950s, however, when computers were less powerful, transformations of the equations were used which automatically filtered out unwanted non-meteorological solutions, albeit with some loss of detail and accuracy in the forecast.

At various standard intervals after the commencement of a forecast, the resulting data are obtained from the computer in whatever forms are operationally convenient for the large variety of uses to which they are put. For instance, data relevant to the operation of commercial aircraft are directly transmitted in numerical form to the airlines' own computers; for guidance to forecasters at official meteorological service outstations, information is supplied in a variety of graphical ways as well as numerically. Methods are also being developed whereby staff at remote locations can interrogate a central forecast data store in order to obtain data suited to their individual needs.

See panels on CLIMATIC CHANGE, STRATOSPHERE AND MESOSPHERE and TROPOSPHERE.

---

and the calculation from these of the conditions that should follow in accordance with the known laws of physics. Also *numerical weather prediction*. Abbrev *NWP*. See panel on NUMERICAL WEATHER FORECAST.

numerical forecast model (*EnvSci*) A set of differential equations, with suitable boundary conditions, for the production of a numerical forecast and usually describing an artificially simplified atmosphere. See panel on NUMERICAL WEATHER FORECAST.

numerical index of efficiency of screening (*MinExt*) See EFFICIENCY OF SCREENING, NUMERICAL INDEX OF.

numerical selector (*ICT*) A group selector that connects the caller to the called subscriber within the latter's own exchange.

numerical taxonomy (*BioSci*) A series of methods, based on the analysis of numerical data, for generating a classification (often in the form of dendrograms) of a group or groups of organisms. See OPERATIONAL TAXONOMIC UNITS.

numerical weather prediction (*EnvSci*) See NUMERICAL FORECAST.

numeric co-processor (*ICT*) Synonym for MATHS CO-PROCESSOR.

numeric keypad (*ICT*) Possessing only numeric keys, often provided in addition to a QWERTY KEYBOARD to speed numerical data entry.

nummulites (*Geol*) A group of extinct Foraminifera which were important rock-forming organisms in the Early Tertiary period. A nummulitic limestone is one which is composed mainly of their skeletal remains. See PROTOZOA.

nunatak (*Geol*) An isolated mountain peak which projects through an ice sheet.

nun's veiling (*Textiles*) Lightweight, plain-weave fabric made from worsted, silk or cotton yarns, usually dyed black, and used as the name suggests.

nuptial flight (*BioSci*) The flight of a virgin queen bee, during which she is followed by a number of males, copulation and fertilization taking place in mid-air.

nurse cells (*BioSci*) Follicle cells surrounding, or attached to, an ovum, with a nutritive and probably a morphogenetic function.

nurse crop (*Agri*) An additional crop sown with the main crop to aid establishment but not intended to be persistent, eg peas to aid establishment of red clover.

Nusselt number (*Phys*) The significant non-dimensional parameter in convective heat loss problems, defined by $Qd/k\Delta\theta$, where Q is rate of heat lost per unit area from a solid body, $\Delta\delta\theta$ is the temperature difference between the body and its surroundings, k is the thermal conductivity of the surrounding fluid and d is the significant linear dimension of the solid.

nut (*BioSci*) A hard, dry, indehiscent fruit formed from a syncarpous gynaecium, and usually containing one seed.

The term is used loosely for any fairly large to large, hard, dry, one-seeded fruit.

nut (*Eng*) A metal collar, screwed internally, to fit a bolt; usually hexagonal in shape, but sometimes square or round. See CASTLE NUT, LOCK NUT.

nut (*Print*) See EN QUAD.

nutating feed (*Radar*) That to a radar transmitter which produces an oscillation of the beam without change in the plane of polarization. The resulting radiation field is a *nutation field*.

nutation (*Astron*) An oscillation of the Earth's pole about the mean position. It has a period of around 19 years, and is superimposed on the precessional movement.

nutation (*BioSci*) Autonomic swaying movement of eg a growing shoot tip, esp circularly. See CIRCUMNUTATION.

nutation (*Phys*) The periodic variation of the inclination of the axis of a spinning top (or GYROSCOPE) to the vertical.

nutation field (*Radar*) See NUTATING FEED.

nutlet (*BioSci*) A small, one-seeded portion of a fruit that divides up as it matures. See SCHIZOCARP.

NU tone (*ICT*) Number unobtainable tone.

nutrient (*Med*) Conveying, serving as, or providing nourishment; nourishing food.

nutrient film technique (*BioSci*) A method of growing plants without soil, their roots in a gutter-like channel with a nutrient solution trickling over them; used commercially for growing eg tomatoes in greenhouses.

nutrient solution (*BioSci*) An artificially prepared solution containing some or all of the mineral substances used by a plant in its nutrition. Also *culture medium*.

nutrigenomics (*BioSci*) The study of the combined effect of diet and genetic make-up on health.

nutrition (*BioSci*) (1) The process of feeding and the subsequent digestion and assimilation of food material. Adj *nutritive*. (2) The study of the effects of food and drink in the diet on the growth and health of humans and animals.

nutritional encephalomalacia (*Vet*) A nervous disease of young chicks characterized by loss of balance, inability to stand and other nervous symptoms, associated with oedema, haemorrhages and degenerative changes in the brain; caused by vitamin E deficiency. Also *crazy chick disease*.

nutritional roup (*Vet*) An affection of the upper respiratory tract of the chicken, caused by vitamin A deficiency, and characterized by a discharge from the eyes and nostrils.

nut runner (*Eng*) A power tool or head fitted with a socket bit and adapted to drive and tighten a nut, often at controlled torque.

Nuvistor (*Electronics*) TN for a subminiature electron tube.

nvCJD (*Med*) Abbrev for *new variant Creutzfeldt—Jakob disease*. See panel on TRANSMISSIBLE SPONGIFORM ENCEPHALOPATHY.

NW (*Build*) Abbrev for *narrow widths*.

N-wave (*Acous*) See SUPERSONIC BOOM.

NWP (*EnvSci*) Abbrev for *numerical weather prediction*. See NUMERICAL FORECAST and panel on NUMERICAL WEATHER FORECAST.

nyankom (*For*) See NIANGON.

nybble (*ICT*) See NIBBLE.

nyctanthous (*BioSci*) Flowering at night.

nyctinastic movement (*BioSci*) A NASTIC MOVEMENT in which plant parts, esp flowers and leaves, take up different position by night and day. Also *nyctinasty, sleep movement*.

nyctipelagic (*BioSci*) Found in the surface waters of the sea at night only.

Nylander solution (*Chem*) An alkaline solution of bismuth subnitrate and Rochelle salt, giving a black colour on boiling with glucose; used to detect glucose in urine.

Nylatron (*Plastics*) TN for a range of filled nylon compounds reinforced for engineering purposes.

nylon (*Textiles*) Generic name for any long-chain synthetic polymeric amide which has recurring amide groups as an integral part of the main polymer chain, and which is capable of being formed into a filament in which the structural elements are orientated in the direction of the axis. See STEP POLYMERIZATION and panel on BIOLOGICAL ENGINEERING POLYMERS for structure of nylon 6.6.

nylon-11 (*Plastics*) TN Rilsan. Polyamide plastic with eleven carbon atoms between each amide group in backbone chain. Also *nylon-12*.

nylon plates (*Print*) See PHOTOPOLYMER PLATES.

nylon salt (*Chem*) See STEP POLYMERIZATION.

nylon wool separation (*BioSci*) Commonly used laboratory technique which separates T-lymphocytes from B-lymphocytes by passing the cell suspension through a small column packed with nylon wool; the non-adherent cells are mostly T-cells.

nymph (*BioSci*) In Acarina, the immature stage intervening between the period of acquisition of four pairs of legs and the attainment of full maturity; in Insecta, a young stage of Exopterygota intervening between the egg and the adult, and differing from the latter only in the rudimentary condition of the wings and genitalia.

Nyquist criterion (*ICT*) One governing the stability of an amplifier or system incorporating feedback. It demands that the NYQUIST DIAGRAM shall not enclose the point $X = -1$, $Y = 0$, where $\mu/\beta = X + jY$.

Nyquist diagram (*ICT*) For a feedback amplifier or system, a plot in rectangular co-ordinates of the factor μ/β for frequencies from zero (dc) to infinity, where μ is the gain without feedback and β is the fraction of the output voltage superimposed by the feedback loop onto the amplifier input.

Nyquist limit (*ICT*) The maximum rate of transmitting pulses through a system. If B is the effective bandwidth in Hz, then $2B$ is the maximum number of code elements per second (bauds) which can be received with certainty; $\frac{1}{2}B$ is known as the Nyquist interval. The Nyquist rate effects the minimum sampling rate allowable for accurate reconstruction of a signal in a pulse-coded system; if f is the maximum frequency occurring in the transmitted signal, then a minimum sampling rate of $2f$ is demanded.

Nyquist noise theorem (*Electronics*) One by which the thermal noise power P in a resistor at any frequency can be calculated from Boltzmann's and Planck's constants k and h and the temperature. At normal temperatures it reduces to $P = kT df$, where T is the temperature in K and df is the frequency interval.

Nyquist rate (*ICT*) See NYQUIST LIMIT.

Nyrim (*Eng*) TN for a *reaction injection-moulding* process using block polymers formed from caprolactam and polyethers, so a type of thermoplastic elastomer (Europe).

nystagmus (*Med*) An abnormal and involuntary movement of the eyeball seen as a flicking backwards and forwards when the eye is deviated. Adj *nystagmoid*.

nystatin (*Pharmacol*) An anti-fungal antibiotic, produced by a strain of *Streptomyces noursei*. Used in treatment of moniliasis and candidiasis and as a fungicide in tissue culture preparations.

NZB, NZW mice (*BioSci*) Inbred strains of mice (New Zealand Black/White) that develop spontaneous autoimmune diseases, including haemolytic anaemia, glomerulonephritis and, when interbred, a condition resembling systemic lupus erythematosus.

O (*Chem*) Symbol for OXYGEN.

O• (*Chem*) A symbol indicating that the radical is attached to the oxygen atom.

Ω (*MathSci*) Symbol for SOLID ANGLE.

Ω (*Phys*) Symbol for: (1) ANGULAR VELOCITY; (2) OHM.

−Ω (*Chem*) Symbol for the ultimate disintegration product of a radioactive series.

o, O, ∼ (*MathSci*) These symbols are defined as follows:

$$f(x) = \mathbf{o}\{\varphi(x)\} \quad \text{if } \frac{f(x)}{\varphi(x)} \to 0$$

$$f(x) = \mathbf{O}\{\varphi(x)\} \quad \text{if } |f(x)| < K\varphi(x)$$

$$f(x) \sim \varphi(x) \quad \text{if } \frac{f(x)}{\varphi(x)} \to 1$$

where the limiting value of x is either stated explicitly or implied: eg if $f(x) = x^2 + 1/x$ then $f(x) = O(1/x)$ if x is small, and $f(x) = O(x^2)$ if x is large.

o- (*Chem*) Abbrev for *ortho-*, ie containing a benzene nucleus substituted in the 1 and 2 positions.

ω (*Chem*) A symbol indicating: (1) substitution in the side chain of a benzene derivative; (2) substitution on the last carbon atom of a chain, farthest from a functional group; (3) SPECIFIC MAGNETIC ROTATION.

ω (*MathSci*) Symbol for SOLID ANGLE.

ω (*Phys*) Symbol for: (1) ANGULAR FREQUENCY; (2) ANGULAR VELOCITY; (3) DISPERSIVE POWER; (4) PULSATANCE.

oak (*For*) A strong, tough and dense hardwood, from European, Asian and American varieties of *Quercus*. The European oak is hard and very durable with a straight-grained, light-tan to biscuit-coloured heartwood. Also *American red oak*, *American white oak*, *Japanese oak*.

oakum (*Build*) Tarred, untwisted rope or hemp used for caulking joints.

O antigen (*BioSci*) The somatic antigen of Gram-negative bacteria, eg genera *Escherichia*, *Shigella*, *Salmonella*, etc. Expressed in the cell wall, antigens of this type have an outer side chain of repeating units of oligosaccharides that confers strain specificity. The polysaccharide is internally linked to lipid A, forming LIPOPOLYSACCHARIDE.

OB (*ImageTech*) Abbrev for *outside broadcast*, but used for any video production work on LOCATION, whether immediately transmitted or not.

ob- (*Genrl*) Prefix meaning *reversed*, *turned about*. Thus *obclavate* is reversed *clavate*, ie attached by the broad and not the narrow end.

obconic (*BioSci*) Cone-shaped but attached by the point. Also *obconical*.

obdiplostemonous (*BioSci*) Having stamens in two whorls, those in the outer opposite to the petals, eg *Geranium*. Cf DIPLOSTEMONOUS.

obduction (*Geol*) The process during collisions between TECTONIC PLATES whereby a piece of the subducted plate is broken off and pushed onto the overriding plate.

obeche (*For*) A low-density W African hardwood (*Triplochiton scleroxylon*), creamy-white to pale-yellow. Also *African whitewood*, *wawa*.

obelisk (*Arch*) A slender stone shaft, generally monolithic, square in section, and tapering towards the top, which is surmounted by a small pyramid.

Oberon (*Astron*) A natural satellite of Uranus, discovered in 1787. Distance from the planet 583 000 km; diameter 1520 km.

obesity (*Psych*) An excessive proportion of fat on the body; cultural norms vary. It is produced by a large range of factors including faulty metabolism and abnormal behaviour.

obex (*Med*) A triangular area of grey matter and thickened ependyma in the MEDULLA OBLONGATA.

object (*ICT*) (1) In the context of OBJECT LINKING AND EMBEDDING, information such as a drawing which may be LINKED or EMBEDDED into another document. (2) In the context of OBJECT-ORIENTED PROGRAMMING, a combination of DATA STRUCTURES, PROCEDURES and VARIABLES stored as an entity called a 'method': eg for a heating system certain objects could be defined; any subsequent manipulation of an object automatically invokes these and this is called 'passing a message to an object'. (3) An entity, eg a picture or a procedure in a software program, that can be individually manipulated.

object code (*ICT*) Translated version of a program that has been assembled or compiled. Cf SOURCE CODE.

object constancy (*Psych*) See OBJECT PERMANENCE.

objective (*Phys*) Usually the lens of an optical system nearest the object. Abbrev *OG*, objective glass for microscope work.

objective analysis (*EnvSci*) A method of processing the original observations by computer to give all values of the atmospheric variables needed to produce a *numerical forecast*, in contrast to a subjective analysis made by scrutinizing observations plotted on charts. The computer programs, after checking for transmission and other errors, produce interpolated grid-point values dynamically consistent with the equations of the forecast model. There are often options to allow manual incorporation of late observations and other data not easily processed automatically.

objective noise meter (*Acous*) Sound-level meter in which noise level to be measured operates a microphone, amplifier and detector, the last named indicating noise level on the phon scale. The apparatus is previously calibrated with known intensities of the *reference tone*, 1 kHz, suitable weighting networks and an integrating circuit being incorporated in the amplifier to simulate relevant properties of the ear in appreciating noise.

objective prism (*Astron, Phys*) A narrow-angle (approx 1°) prism placed in front of the objective lens or primary mirror of a telescope, which causes the image of each star to give a small spectrum. Those of many objects within a small field can thus be recorded simultaneously, either photographically or photoelectrically.

object language (*ICT*) A language into which a source program is translated by a compiler for subsequent repeated execution by a computer.

object linking and embedding (*ICT*) A method to share and transfer information between APPLICATION PROGRAMS, eg a drawing may be EMBEDDED into a text document.

object-oriented database (*ICT*) A database built to handle data in the form of OBJECTS.

object-oriented programming (*ICT*) An approach to programming using the concept of an OBJECT. The approach uses a HIERARCHICAL object structure that reflects the system to be modelled. Programming languages offering this approach include Smalltalk, C++, MODULA-2 and some versions of PASCAL. Abbrev *OOP*.

object permanence (*Psych*) Awareness that an object continues to exist even when attention is focused elsewhere: acquiring this is a key step in childhood development and occurs during the PREOPERATIONAL STAGE.

object program (*ICT*) See OBJECT CODE.

oblate (*Genrl*) Globose, but noticeably wider than long.

oblate ellipsoid (*MathSci*) An ellipsoid obtained by rotating an ellipse about its minor axis, ie like a squashed sphere. Cf PROLATE ELLIPSOID.

obligate (*BioSci*) Obliged to function in the way specified, eg an obligate anaerobe cannot grow (and may not survive) in the presence of free oxygen and an obligate parasite cannot live outside its host. Cf FACULTATIVE.

obligate parasite (*BioSci*) A parasite capable of living naturally only as a parasite. Cf FACULTATIVE PARASITE. A more useful distinction is between BIOTROPHS and NECROTROPHS.

obligate saprophyte (*BioSci*) An organism that lives on dead organic material and cannot attack a living host.

oblique (*MathSci*) Not perpendicular, not right-angled, such as co-ordinate axes that are not mutually perpendicular.

oblique aerial photograph (*Surv*) A photograph taken from the air, for purposes of aerial survey work, with the optical axis of the camera inclined from the vertical, generally at some predetermined angle.

oblique arch (*CivEng*) An arch whose axis is not normal to the face.

oblique circular cylinder (*MathSci*) See CYLINDER.

oblique fracture (*Med*) A fracture of a bone caused by an impact at an oblique angle.

oblique system (*Crystal*) See MONOCLINIC SYSTEM.

obliquity factor (*Phys*) See INCLINATION FACTOR.

obliquity of the ecliptic (*Astron*) The angle at which the CELESTIAL EQUATOR intersects the ECLIPTIC. At present this angle is slowly decreasing by 0·47 arcseconds per year, due to PRECESSION and NUTATION. Its value can be calculated for the year t from the formula $23°27'08.26'' - 0.4684(t-1900)$.

obliquus (*BioSci*) An asymmetrical or obliquely placed muscle.

oblongata (*BioSci*) See MEDULLA OBLONGATA.

OBM (*Surv*) Abbrev for ORDNANCE BENCH MARK.

obovate (*BioSci*) Having the general shape of the longitudinal section of an egg, not exceeding twice as long as broad, and with the greatest width slightly above the middle; hence, attached by the narrower end.

obovoid (*BioSci*) Solid, egg-shaped and attached by the narrower end.

obscuration (*Phys*) The fraction of incident radiation which is removed in passing through a body or a medium.

obscured glass (*Glass*) Glass which is so treated as to render it translucent but not transparent, eg by sandblasting or etching.

obsequent drainage (*Geol*) See DRAINAGE PATTERNS.

observational learning (*Psych*) Learning through the observation of another individual (model) which is accomplished without practice or direct experience.

observation well (*MinExt*) As oil extraction proceeds from a reservoir, special observation wells may be drilled, or allocated, for monitoring the changing fluid levels or conditions of pressure in the reservoir.

observatory (*Astron*) A building or station for making astronomical and physical observations.

obsession (*Psych*) The morbid persistence of an idea in the mind, against the wish of the obsessed person.

obsessive–compulsive disorder (*Psych*) Anxiety disorder in which there are persistent and recurrent thoughts (*obsessions*) and repetitive, ritualized behaviour patterns (*compulsions*) that give no pleasure, but relieve the tension that otherwise arises from the obsession.

obsidian (*Geol*) A volcanic glass of granitic composition, generally black with vitreous lustre and conchoidal fracture; occurs at Mt Hecla in Iceland, in the Lipari Isles, and in Yellowstone Park in the US. A green silica glass found in ploughed fields in Moravia is cut as a gemstone and sold under the name *obsidian*. True obsidian is used as a gemstone and is often termed *Iceland agate*.

obstetrician (*Med*) A medically qualified person who practises obstetrics.

obstetrics (*Med*) That branch of medical science which deals with the problems and management of pregnancy and labour.

obstruction lights (*Aero*) Lights fixed to all structures near airports which constitute a hazard to aircraft in flight.

obstruction markers (*Aero*) See AIRPORT MARKERS.

obstruent (*Med*) Obstructing; that which obstructs; an astringent drug.

obturator (*BioSci*) Any structure that closes off a cavity, eg all the structures that close the large oval foramen formed by the ischio-pubic fenestra; the foramen itself.

obtuse (*BioSci*) The blunt tip of a leaf etc with the sides forming an angle of more than 90°. Cf ACUTE.

obtuse angle (*MathSci*) An angle greater than 90° and less than 180°. Cf ACUTE ANGLE.

obvolvent (*BioSci*) Folded downwards and inwards, as the wings in some insects.

occam (*ICT*) A HIGH-LEVEL LANGUAGE used to program systems capable of PARALLEL PROCESSING.

occipital condyle (*BioSci*) (1) In Craniata, one or two projections from the skull that articulate with the first vertebra. (2) In insects, a projection from the posterior margin of the head which articulates with one of the lateral cervical sclerites.

occipitalia (*BioSci*) A set of cartilage bones forming the posterior part of the brain case in the vertebrate skull.

occipital lobe (*Med*) The posterior lobe in each cerebral hemisphere, dealing with the interpretation of vision. See panel on BRAIN STRUCTURE.

occiput (*BioSci*) The occipital region of the vertebrate skull forming the back of the head. Adj *occipital*.

occlusion (*BioSci*) Closure of a duct or aperture, eg the upper and lower teeth of a vertebrate.

occlusion (*Chem*) The retention of a gas or a liquid in a solid mass or on the surface of solid particles, esp the retention of gases by solid metals.

occlusion (*EnvSci*) The coming together of the WARM and COLD FRONTS in a depression so that the warm air is no longer in contact with the Earth's surface. If the *warm frontal zone* is cut off from contact with the surface it is a *cold occlusion*. If the *cold frontal zone* is cut off from contact with the surface it is a *warm occlusion*.

occlusor (*BioSci*) A muscle that by its contraction closes an operculum or other movable lid-like structure.

occultation (*Astron*) The hiding of one celestial body by another interposed between it and the observer, as the hiding of the stars and planets by the Moon, or the satellites of a planet by the planet itself.

occult deposition (*EnvSci*) Deposition of acid compounds and other pollutants on the surfaces of vegetation, buildings, etc, by direct contact with mist or cloud containing droplets of contaminated water. Acidity of the droplets can be high, but the volume deposited is much less than in WET DEPOSITION from rain.

occulting light (*Ships*) A navigation mark identified during darkness by a light extinguished for short periods in a distinctive pattern. See ALTERNATING LIGHT, FLASHING LIGHT.

occupational psychology (*Psych*) Another term for INDUSTRIAL/ORGANIZATIONAL PSYCHOLOGY. The scientific study of human behaviour in the workplace, including personnel selection and stress management.

ocean-floor spreading (*Geol*) See figs at MAGNETIC ANOMALY PATTERNS and panel on PLATE TECTONICS.

oceanic crust (*Geol*) That part of the Earth's crust which is normally characteristic of oceans. In descending vertical section, it consists of approximately 5 km of water, 1 km of sediments and 5 km of basaltic rocks.

oceanic ridge (*Geol*) See MID-OCEAN RIDGE.

oceanite (*Geol*) A type of basaltic igneous rock occurring typically in the oceanic islands as lava flows; characterized by a higher percentage of coloured silicates (olivine and pyroxene), and a lower percentage of alkalis, than in normal basalt.

oceanography (*Genrl*) The scientific study and description of the ocean.

ocellus (*BioSci*) A simple eye or eyespot in invertebrates; an eye-shaped spot or blotch of colour. Adj *ocellate*.

ochre (*Min*) (1) Naturally occurring red, yellow and brown iron oxides, or clays strongly coloured by iron oxides (limonite), formed by residual weathering and used as pigments. (2) Highly coloured alteration products from other metals, eg chrome ochre. See UMBER.

ochrea (*BioSci*) A sheath around the base of an internode formed from united stipules or leaf bases, eg dock (*Rumex*) and other Polygonaceae. Also *ocrea*.

ochre codon (*BioSci*) The STOP CODON UAA, one of three that cause termination of protein synthesis.

ochroleucous (*BioSci*) Yellowish-white.

ochrophore (*BioSci*) See XANTHOPHORE.

ochrosporous (*BioSci*) Having yellow or yellow-brown spores.

OCR (*ICT*) Abbrev for OPTICAL CHARACTER RECOGNITION.

ocrea (*BioSci*) See OCHREA.

oct-, octa-, octo- (*Genrl*) Prefixes from Gk *okta* denoting eight.

octa (*Aero, EnvSci*) See OKTA-.

octa- (*Chem*) Containing eight atoms, groups, etc.

octagon (*MathSci*) An eight-sided polygon.

octahedral system (*Crystal*) See CUBIC SYSTEM.

octahedrite (*Min*) A form of *anatase*, crystallizing in tetragonal bipyramids.

octahedrites (*Min*) A class of iron meteorites showing an octahedral internal structure.

octahedron (*Crystal*) A form of the cubic system which is bounded by eight similar faces, each being an equilateral triangle with plane angles of 60°. Pl *octahedra*.

octal notation (*ICT*) A system of representing numbers in base 8 with the digits 0,1,2,…,7.

octane (*Chem*) C_8H_{18} alkane hydrocarbon. There are 18 compounds of this formula. Straight chain or normal octane, a colourless liquid, bp 126°C, rel.d. 0·702 at 20°C, is found in petroleum.

octane number (*Autos*) The percentage, by volume, of *iso*-octane (2,2,4-trimethylpentane) in a mixture of *iso*-octane and *normal* heptane which has the same knocking characteristics as the motor fuel under test; it serves as an indication of the knock-rating of a motor fuel. See LEAD TETRAETHYL.

Octans (Octant) (*Astron*) An inconspicuous southern constellation.

octastyle (*Arch*) A building having a colonnade of eight columns in front.

octavalent (*Chem*) Capable of combining with eight atoms of hydrogen or their equivalent. Having an oxidation or co-ordination number of eight.

octave (*Acous*) The interval between any two frequencies having the ratio 2:1.

octave (*Chem*) See LAW OF OCTAVES.

octave analyser (*Acous*) A filter in which the upper cut-off frequency is twice the lower.

octave filter (*Acous*) A bank of filters for analysing the spectral energy content of complex sounds and noises, using adjacent octave bands of frequency over the whole audio range. One-half and one-third octave filters are also used.

octavo (*Print*) The eighth of a sheet, or a sheet folded three times to make eight leaves or 16 pages and written 8vo. See PAPER SIZES.

octet (*Chem*) See ELECTRON OCTET.

octet (*ICT*) A group of 8 BITS in a digital communication system. Unlike the bits of an 8 bit BYTE, these normally have no collective meaning and are handled separately.

octo- (*Genrl*) See OCT-.

octodecimo (*Print*) The 18th of a sheet or a sheet folded to make 18 leaves or 36 pages. Also *18mo*.

octopamine (*BioSci*) A catecholamine found in both vertebrates and invertebrates that may act as a hormone and NEUROTRANSMITTER, and as an AGONIST for adrenergic receptors.

octopine (*BioSci*) An OPINE.

octopod (*BioSci*) Having eight feet, arms or tentacles.

octyl alcohol (*Chem*) See CAPRYL ALCOHOL.

ocular (*BioSci*) Pertaining to the eye; capable of being perceived by the eyes.

ocular micrometer (*BioSci*) US term for an *eyepiece graticule*.

oculate (*BioSci*) Possessing eyes; having markings which resemble eyes.

oculist (*Med*) One skilled in the knowledge and treatment of refractive diseases of the eye.

oculomotor (*BioSci*) Pertaining to, or causing movements of, the eye; the third cranial nerve of vertebrates, innervating some of the muscles of the eyeball.

oculus (*Arch*) A round window.

OD (*Genrl*) Abbrev for ORDNANCE DATUM.

odd–even check (*ICT*) See PARITY CHECKING.

odd–even nuclei (*Phys*) Nuclei containing an odd number of protons and an even number of neutrons.

odd function (*MathSci*) f is an odd function if $f(-x) = -f(x)$. Cf EVEN FUNCTION.

odd legs (*Eng*) Callipers with two straight hinged legs, tapering distally from the hinge. One leg is curved distally and is placed against an edge. The other has a scribing point. Used mainly for locating centres from outside or inside edges and for other layout work.

odd number (*MathSci*) An integer that is not divisible by two; a number n that can be expressed as $n = 2m + 1$, where m is an integer.

odd–odd nuclei (*Phys*) Nuclei with an odd number of both protons and neutrons. Very few are stable.

odd parity (*ICT*) Binary representation in which the number of ones is odd. See PARITY CHECK.

odds (*MathSci*) The ratio of the probability that an event occurs to the probability that it does not occur.

odd sorts (*Print*) A general term for any unusual characters required for technical or foreign language setting.

odds ratio (*MathSci*) The ratio of two odds, used particularly in comparing and modelling conditional probabilities.

odograph (*Surv*) A recording PEDOMETER.

odometer (*Surv*) See PERAMBULATOR.

Odonata (*BioSci*) An order of primitive insects with two pairs of similar membranous, many-veined wings. They are generally diurnal and have large eyes. Both adults and larvae are predators. Dragonflies, damselflies.

odont-, odonto- (*Genrl*) Prefixes from Gk *odous, odontos*, denoting tooth.

odontalgia (*Med*) Toothache.

odontic (*Med*) Pertaining to the teeth.

odonto- (*Genrl*) See ODONT-.

odontoblast (*BioSci*) A dentine-forming cell, one of the columnar cells lining the pulp cavity of a tooth.

odontoclast (*BioSci*) A dentine-destroying cell, one of the large multinucleate cells that absorb the roots of the milk teeth in mammals.

odontogeny (*BioSci*) The origin and development of teeth.

odontograph (*Eng*) An approximate but practical guide for setting out the profiles of involute gear teeth, in which a pair of circular arcs is substituted for the true involute curve.

odontoid (*BioSci*) Tooth-like.

odontoid process (*BioSci*) A process of the anterior face of the centrum of the axis vertebra that forms a pivot on which the atlas vertebra can turn.

odontolite (*Min*) See BONE TURQUOISE.

odontology (*Med*) The study of the physiology, anatomy, pathology, etc, of the teeth.

odontoma (*Med*) Any of a variety of tumours that arise in connection with the teeth.

odontophore (*BioSci*) In molluscs, a feeding organ comprising the radula and radula sac, with muscles and cartilages.

odontostomatous (*BioSci*) Having jaws that bear teeth.

odorant (*Chem*) A strong smelling, volatile compound added to piped gas that is otherwise odourless to facilitate the detection of leaks, eg of natural gas. MERCAPTANS are often used.

odorimetry (*Chem*) Measurement of the intensity and permanency of odours. Also *olfactometry*.

odoriphore (*Chem*) A group of atoms which confer an odour on a compound.

ODR (*Eng*) Abbrev for OSCILLATING DISK RHEOMETER.

oedema (*Med*) A pathological accumulation of fluid in the tissue spaces and serous sacs of the body; sometimes the term is restricted to such accumulation in tissue spaces only: pulmonary oedema, fluid in the lung; sacral oedema, fluid at the base of the spine. US *edema*.

oedema disease (*Vet*) See BOWEL OEDEMA DISEASE. US *edema*.

oedematous (*Med*) Affected by OEDEMA.

Oedipus complex (*Psych*) In psychoanalytic theory, refers to unconscious conflicts, occurring during the PHALLIC STAGE of psychosexual development, that centre around the relations a male child forms with his parents. A fantasized form of sexual love for the opposite sex develops and a resentment of the same-sex parent. In girls these possessive feelings and the associated conflicts of guilt are called the *Electra complex*.

OEM (*ICT*) Abbrev for ORIGINAL EQUIPMENT MANUFACTURER.

oersted (*Phys*) CGS electromagnetic unit of magnetic field strength, such that 2π oersteds is a field at the centre of a circular coil 1 centimetre in radius carrying a current of 1 abampere (10 A). Now replaced by the SI unit $A\,m^{-1}$: $1\,A\,m^{-1} = 4\pi \times 10^{-3}$ oersteds.

oesophageal valve (*BioSci*) See CARDIAC VALVE.

oesophagectasis (*Med*) Pathological dilatation of the oesophagus. Also *oesophagectasia*. US *esophagectasis, esophagectasia*.

oesophagectomy (*Med*) Removal of the oesophagus or part of it.

oesophagitis (*Med*) Inflammation of the oesophagus.

oesophagoscope (*Med*) An instrument for viewing the interior of the oesophagus.

oesophagospasm (*Med*) See ACHALASIA OF THE CARDIA.

oesophagostomy (*Med*) Formation of an artificial opening into the oesophagus.

oesophagotomy (*Med*) Incision into the oesophagus.

oesophagus (*BioSci*) In vertebrates, the section of the alimentary canal leading from the pharynx to the stomach; usually lacking a serous coat and digestive glands; the corresponding portion of the alimentary canal in invertebrates. Adj *oesophageal*. US *esophagus*.

oestradiol, oestriol, oestrogen, oestrus (*Med*) See ESTRADIOL, ESTRIOL, ESTROGEN, ESTRUS.

oestriasis (*Vet*) Infection of the nasal cavity and sinuses of sheep and goats by larvae of the sheep nasal fly *Oestrus ovis*. A form of MYIASIS.

OFDM (*ICT*) Abbrev for *orthogonal frequency-division multiplexing*, a technique for transmitting large amounts of digital data over radio waves.

off-air (*ImageTech*) Received from a broadcast transmission.

off-flavour (*FoodSci*) An uncharacteristic taste often combined with an off-odour resulting from chemical or microbiological spoilage, incorrect processing, poor storage or poor-quality raw materials. See FOOD SPOILAGE, TAINT.

off gas (*NucEng*) Gas escaping from a reactor, diffusion plant or other installation.

off-gassing (*Space*) See OUTGASSING.

offhand grinding (*Eng*) Grinding in which the tool being sharpened is held in the hand and worked freely.

off-hook (*ICT*) The condition in which a telephone ringing circuit has been disconnected and the speech circuit connected by lifting the handset from the SWITCH-HOOK.

official (*BioSci*) Used in medicine. Also *officinal*.

off-label (*Pharmacol*) US term for a prescription drug used to treat a condition for which the Food and Drug Administration (FDA) has not approved it.

off-lap (*Geol*) The dispositional arrangement of a series of conformable strata laid down in the waters of a shrinking sea, or on the margins of a rising land mass, so that the successive strata cover smaller areas than their predecessors. Cf OVERLAP.

offlet (*Build*) See GRIP.

off line (*ICT*) Describes processing carried out by devices not under the control of the CENTRAL PROCESSOR.

off-line editing (*ImageTech*) Videotape editing involving an intermediate transfer to a low-cost system solely for preliminary selection and arrangement prior to the final CONFORMING of the original material.

off-peak load (*ElecEng*) The load on a generating station or power supply system taken at times other than the time of the system peak load, eg during the night.

offprints (*Print*) Separately printed copies of articles that have appeared in periodicals.

offset (*BioSci*) Short shoot arising from an axillary bud near the base of a shoot and producing a daughter plant at its apex, in eg the houseleek, *Sempervivum*. Also a bulbil or cormlet formed near base of parent bulb or corm.

offset (*Build*) A ledge formed at a place where part of a wall is set back from the face.

offset (*Surv*) The horizontal distance measured to a point from a main survey line, in a direction at right angles to a known point on the latter.

offset blanket (*Print*) The rubber covering on the cylinder on which the image is taken from the plate cylinder and from which it is offset to the stock being printed, being the distinguishing feature of an offset press. See OFFSET PRINTING, PRINTING.

offset printing (*Print*) A process in which the ink from a plate is received on a rubber-covered blanket cylinder from which it is transferred to the paper or other material, including sheet metal, the rubber surface making good contact even with rough surfaces. The standard method of lithographic printing, it is now being used to a lesser extent for relief printing, then called *letterset printing*.

offset rod (*Surv*) A wooden pole painted in bands of different colours of suitable length so that the pole may be used for the measurement of short distances.

offset scale (*Surv*) Short graduated scale used in plotting detail from field notes of *offsets*.

offset well (*MinExt*) A well drilled near to an existing well to explore or exploit a field further.

offstream (*ChemEng, MinExt*) A term for large-scale process plant that is not in production owing to maintenance, development work or other circumstances, eg oil wells, refineries or chemical plant.

off-the-film flash metering (*ImageTech*) A means of controlling an electronic flash exposure by measuring the light reflected off the film and focal-plane shutter blinds, which are patterned to approximate the reflectance of the film coating, and quenching the dedicated flash unit when complete.

OFHC (*Eng*) Abbrev for OXYGEN-FREE HIGH-CONDUCTIVITY COPPER.

OG (*Phys*) Abbrev for *objective glass*. See OBJECTIVE.

OG, ogee (*Arch*) See CYMA.

ogee wing (*Aero*) A wing of ogee plan form and very low aspect ratio which combines low *wave drag* in supersonic flight with high lift at high incidence through separation vortices at low speed.

ogival arch (*Arch*) A pointed arch of which each side consists of a reverse curve.

ohm (*Phys*) SI unit of electrical resistance, such that 1 ampere through it produces a potential difference across it of 1 volt. Symbol Ω.

ohm centimetre (*Phys*) CGS unit of RESISTIVITY.

ohmic contact (*ElecEng*) One in which the current flows with equal facility in both directions, with incremental junction voltage proportional to incremental current. Ohmic contacts in connections to semiconductors can usually be ensured by heavily doping the semiconductor so that electrons can tunnel through any barrier layers which may arise from the bulk and interface properties of the materials involved.

ohmic heating (*FoodSci*) Applied to food processing when a tubular vessel through which a liquid or semi-liquid food product passes forms part of an electrical circuit. The resistance of the electrolyte when current is passed causes heat to be generated sufficient to destroy micro-organisms and cook the product. Because there is no direct heating, there is no charring or deposition on the inside wall of the heating tube and it can cope with relatively large particles or chunks, eg meat in gravy. Used with aseptic packing for the production of ambient long-life products.

ohmic loss (*ElecEng*) Power dissipation in an electric circuit arising from circuit resistance and current flow, ie I^2R loss.

ohmic resistance (*ElecEng*) See DC RESISTANCE.

ohmic resistor (*ElecEng*) See LINEAR RESISTOR.

ohm metre (*Phys*) SI unit of RESISTIVITY. Symbol Ω m.

Ohm's law (*Phys*) A law stating that in metallic conductors, at constant temperature and zero magnetic field, the current I flowing through a component is proportional to the potential difference V between its ends, the constant of proportionality being the *conductance* of the component. So $I = V/R$ or $V = IR$, where R is the *resistance* of the component. The law is strictly applicable only to electrical components carrying dc and for practical purposes to those of negligible reactance carrying ac. Extended by analogy to any physical situation where a pressure difference causes a flow through an impedance, eg heat through walls, liquid through pipes.

Ohm's law of hearing (*Acous*) Law of psychoacoustics. A simple harmonic motion of the air is appreciated as a simple tone by the human ear. All other motions of the air are analysed into their harmonic components which the ear appreciates as such separately.

OHP (*ImageTech*) Abbrev for *overhead projector*. See PROJECTOR.

-oid (*Genrl*) Suffix after Gk *oides*, from *eidos*, form.

oil (*Genrl*) (1) Crude oil; see PETROLEUM. Source of fuels and chemical intermediates, and materials such as polymers or carbon black. (2) Viscous liquid of organic nature, having special properties, eg BITUMEN OF JUDEA, aromatic oil. See OILS.

oil absorption (*Chem*) A term usually applied to pigments. The amount of linseed oil a pigment will absorb to reach a given consistency as determined by certain standards.

oil-based mud (*MinExt*) Used when drilling very hot formations or through water-absorbing strata; faster but dirtier drilling.

oil-blast circuit breaker (*ElecEng*) A circuit breaker designed so that the vaporization of the oil at the arc causes a blast of fresh oil to cross and rapidly extinguish the arc.

oil-break (*ElecEng*) A term applied to switches, circuit breakers, fuses, etc, to indicate that the circuit is opened under oil.

oil circuit breaker (*ElecEng*) The usual type of circuit breaker used on high- and medium-power ac circuits; the contacts are immersed in oil for insulating and arc-rupturing purposes.

oilcloth (*Textiles*) A waterproof material obtained by coating a cotton fabric with oxidized linseed or other drying oil.

oil-control ring (*Autos*) See SCRAPER RING.

oil-cooled (*Eng*) Said of apparatus which is immersed in oil to facilitate cooling.

oil cooler (*Aero*) See FUEL-COOLED OIL COOLER.

oil cooler (*Autos*) A small air-cooled radiator sometimes used for cooling the lubricant after its return from the engine and before delivery to the oil tank.

oil-dilution system (*Aero*) In a reciprocating aero-engine, a device for diluting the lubricant with fuel as the engine is stopped so that there is less resistance when starting in cold weather.

oiled paper (*Paper*) Paper impregnated with non-drying oil, eg oiled wrapping, or with linseed oil, eg oiled manilla.

oil engine (*Autos*) See COMPRESSION–IGNITION ENGINE.

oil extension (*Eng*) Absorption of aromatic oil by rubber to aid processing. Similar effect to plasticization of rigid polymers like polyvinyl chloride. Most widely used for styrene butadiene rubber in manufacture of tyres. See panel on TYRE TECHNOLOGY.

oil-filled cable (*ElecEng*) Cable with a central duct, formed by an open spiral of steel tape into which oil is forced to eliminate gaseous voids and the consequent ionization. Used up to the highest voltages.

oilgas (*Chem*) A gas of high energy value, obtained by the destructive distillation of high-boiling mineral oils. It consists chiefly of methane, ethene, ethyne, benzene and higher homologues.

oil gland (*BioSci*) The preen gland or uropygial gland of birds, a cutaneous gland forming an oil secretion used in preening the feathers.

oil hardening (*Eng*) The hardening of steels of medium- and high-carbon content by heating to the austenitic condition, followed by quenching into a bath of oil, resulting in a cooling less sudden than is effected by water, and consequently a reduced risk of cracking.

oil-hardening steel (*Eng*) Any alloy steel which will harden when cooling in oil instead of in water. The limiting diameter or cross-section which will harden fully in this manner must also be stated, since this depends on the alloy composition and transformation characteristics of the steel.

oil-immersed (*ElecEng*) Electrical apparatus which is immersed in oil. See OIL-BREAK, OIL-COOLED, OIL-INSULATED.

oil-immersion objective (*BioSci*) A microscope objective that requires the addition of a thin film of oil, with the same refractive index as glass, between the front lens and the coverslip above the specimen. This increases the numerical aperture, and thus the resolving power, and is necessary to achieve the maximum magnification obtainable with a light microscope.

oiling (*Textiles*) Spraying wool, woollen blends, jute and some manufactured fibres with oil or oil/water emulsion during opening operations. It facilitates drawing, drafting and spinning, and reduces static electricity and waste.

oiling ring (*Eng*) A simple device used to feed oil to a JOURNAL bearing. It consists of a light metal ring, larger in diameter than the shaft, and riding loosely thereon, located at the mid-point of the brasses, the upper brass being slotted to receive it. The ring dips into an oil reservoir in the base of the housing and, as it rotates, feeds oil to the brasses.

Oklo natural fission reactor

One of the places where uranium has been mined is at Oklo, in what is now the Republic of Gabon in West Africa. In 1972 a sample of uranium was found to have a slightly lower abundance of uranium-235 than had been found anywhere else in the world. The normal value for this abundance is 0·720% to a high degree of accuracy, and the value found by French scientists at their Atomic Energy Commission was 0·717; subsequent samples had even less, down to 0·440. The only possible explanation for such a discrepancy is that uranium-235 fission had occurred.

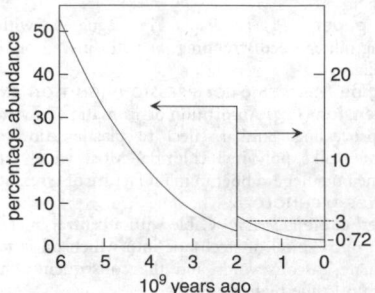

Isotope abundance of uranium-235.
Twice times magnified scale on right.

The half-life of uranium-235 is $0·7 \times 10^9$ years compared with $4·5 \times 10^9$ for uranium-238.

Therefore uranium-235 must have been much more abundant in the geological past. Indeed it is not unreasonable to suppose that the conditions at the formation of our galaxy did not favour one isotope against the other. Knowing the different half-lives of the isotopes, their relative abundance at earlier times can be calculated. This relation is illustrated in the diagram. It shows not only that 6×10^9 years ago, when our galaxy was formed, the relative abundance of uranium-235 was about 50%, but also that 2×10^9 years ago it was over 3%. Fission, moderated by natural water, would therefore have been possible.

Sceptics were confounded and fission confirmed by analysing the isotopic abundances of elements like ruthenium and neodymium in the deposit and comparing them with those in spent fission fuel and in nature. The Oklo abundances were much closer to those in reactor fission products and unlike those found elsewhere in nature.

Estimates of the size of the Oklo uranium deposit and the average amount of uranium-235 depletion indicate that perhaps 5000 kg of uranium-235 was involved and that fission could have occurred over a period of perhaps 1 million years. It would not have occurred rapidly because the water would have boiled and the reaction immediately halted until the rocks cooled enough to let more water seep back in. Nevertheless the energy produced was very large, equivalent in total to 100 million tons of TNT.

See panels on GALAXY and NUCLEAR REACTORS.

---

oil-insulated (*ElecEng*) Electrical apparatus which is insulated by immersion in oil.

oil length (*Build*) The ratio of oil to resin in a varnish.

oil-less circuit breaker (*ElecEng*) A circuit breaker not using oil either for quenching the arc or for insulation.

oil of bitter almonds (*Chem*) See BENZALDEHYDE.

oil of cloves (*Chem*) See CLOVE OIL.

oil of Judea (*Chem*) See BITUMEN OF JUDEA.

oil of turpentine (*Chem*) See TURPENTINE.

oil of vitriol (*Chem*) An old name for SULPHURIC ACID.

oil paints (*Build*) Paints in which the film former comprises drying oils, driers and resins, drying by oxidation.

oil pump (*Autos*) A small auxiliary pump, driven from an engine crankshaft, which forces oil from the sump or oil tank to the bearings; often of the gear type. See GEAR PUMP.

oil quench (*Eng*) Immersion, for purpose of tempering, of hot metal in oil.

oil-quenched fuse (*ElecEng*) A fuse having oil to quench the arc.

oils (*Chem*) A group of neutral lipids that are liquid at room temperature. (1) *Fixed (fatty) oils*, from animal, vegetable and marine sources, consist chiefly of glycerides and esters of fatty acids, and also include essential oils (volatile products, mainly hydrocarbons, with characteristic odours, derived from certain plants). (2) *Mineral oils* derive from petroleum, coal, shale, etc, and consist of hydrocarbons. See OIL.

oils (*ElecEng*) Different oils are used in transformers, cables and switchgear. Class A oils have a maximum of sludge of

0·1% and may be used in transformers above 80°C and in oil switches above 70°C. Class B oils have a maximum of 0·8% and may be used in transformers up to 75°C.

oil shale (*Geol*) An argillaceous sediment containing diffused KEROGEN in a state suitable for distillation into paraffin and other mineral oils by the application of heat. See SHALE OILS.

oil slick (*EnvSci*) A floating layer of oil on sea, lake, river or canal usually as a result of spillage.

oilstone (*Build*) A smooth stone, moistened with oil and used to impart a fine keen edge to a cutting tool.

oilstone slip (*Build*) See GOUGE SLIP.

oil string (*MinExt*) See PRODUCTION STRING.

oil sump (*Autos*) See SUMP.

oil switch (*ElecEng*) The usual type of switch used on high- and medium-power ac circuits; the contacts are immersed in oil for insulating and arc-rupturing purposes.

oil tanker (*Ships*) A ship for carrying oil in bulk. They are subdivided internally into a number of tanks by longitudinal and transverse bulkheads, with pumps for discharging.

oil varnishes (*Build*) Varnishes containing a drying oil, resin and driers. The oils used are linseed, China wood, soya bean, poppy-seed, cottonseed and castor oil.

OK (*Aero*) Abbrev for ODOURLESS KEROSENE.

Oklo natural fission reactor (*NucEng*) Uranium mine in Gabon, W Africa, where evidence of a one-time naturally occurring reactor was found. See panel on OKLO NATURAL FISSION REACTOR.

okta (*Aero, EnvSci*) One-eighth of the sky area used in specifying cloud cover for airfield weather condition reports. Also *octa*.

olanzapine (*Pharmacol*) An atypical antipsychotic drug used for the treatment of schizophrenia.

OLAP (*ICT*) Abbrev for *on-line analytical processing*, a form of data processing in which large volumes of data (typically stored in a data warehouse or multidimensional database) are retrieved on-line, summarized and presented in a form that allows the user to conduct interactive analysis.

Olbers' paradox (*Astron*) An apparent paradox expressed in 1826 by Heinrich Olbers: 'why is the sky dark at night?' In an infinitely large, unchanging universe populated uniformly with stars and galaxies, the sky would be uniformly bright, which is not the case. The paradox is now explained by postulating a finite expanding universe, and by the phenomenon of REDSHIFT.

old age (*Geol*) The final stage in the cycle of erosion in which base level is nearly attained and the landscape has little relief.

Old English (*Print*) A black-letter type in which early books were printed.

Old Face (*Print*) Classic form of roman type originating in the 15 and 16th centuries:

RQENbaegn

Oldham coupling (*Eng*) A coupling permitting misalignment of the shafts connected. It consists of a pair of flanges whose opposed faces carry diametrical slots, and between which a floating disk is supported through corresponding diametral tongues which are arranged at right angles.

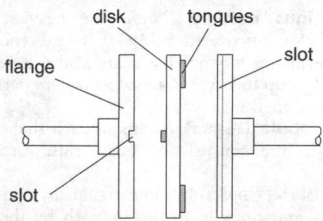

Oldham coupling

oldhamite (*Min*) Sulphide of calcium, usually found as cubic crystals in meteorites.

Old Red Sandstone (*Geol*) The continental facies of the DEVONIAN system in the UK, comprising perhaps 12 000 m of red, brown, or chocolate sandstones, red and green marls, cornstones, breccias, flags and conglomerates, yielding on certain horizons the remains of archaic fishes, eurypterids, plants and rare shelly fossils. Abbrev ORS.

Old Style (*Astron*) A name given to the system of date reckoning superseded by the adoption of the GREGORIAN CALENDAR.

Old Style (*Print*) Type based on the old style of roman letter used until the end of the 18th century. Century Old Style:

RQENbaegn

old style figures (*Print*) See HANGING FIGURES.

OLE (*ICT*) Abbrev for OBJECT LINKING AND EMBEDDING.

olecranon (*BioSci*) In land vertebrates, a process at the upper end of the ulna that forms the point of the elbow.

olefin (*Textiles*) Synthetic fibre or polyolefin fibre consisting of a long-chain polymer composed of at least 85% by weight of olefins, usually ethylene or propylene. Hence polyethylene and polypropylene are common.

olefins (*Chem*) See ALKENE, OLIGOMER. Several monomer units linked together.

oleic acid (*Chem*) $C_{18}H_{34}O_2$. Mp 14°C. A colourless liquid. It is an unsaturated acid of the formula $CH_3(CH_2)_7$ CH=CH$(CH_2)_7$COOH, with *cis* configuration about the double bond, and occurs as the glycerine ester in fatty oils. Oleic acid oxidizes readily on exposure to the air, turns yellow and becomes rancid.

olein (*Chem*) A glycerine ester of oleic acid.

oleo (*Aero*) Main structural member of the support assemblies of an aircraft's *landing gear*. Of telescopic construction and containing oil so that on landing the oil is passed under pressure through chambers at a controlled rate thereby absorbing the shock. Also *oleo leg*, *shock absorber*, *shock strut*.

oleo-pneumatic (*Aero*) Means of absorbing shock loads by a combination of air compression and oil pressure created by forcing the latter through an orifice.

oleoresinous (*Build*) Paints and varnishes the binder of which contains both oils and resins.

olestra (*Pharmacol*) A synthetic fat substitute that passes through the gastro-intestinal tract without being digested or absorbed.

oleum (*Chem*) A commercial name for *fuming sulphuric (disulphuric) acid*, $H_2S_2O_7$.

olfactometry (*Chem*) See ODORIMETRY.

olfactory (*BioSci*) Pertaining to the sense of smell; the first cranial nerve of vertebrates, running to the olfactory organ.

olfactory lobes (*BioSci*) Part of the fore-brain in vertebrates, concerned with the sense of smell, from which the olfactory nerves originate.

olig-, oligo- (*Genrl*) Prefixes from Gk *oligos*, a few, small.

oligaemia (*Med*) Diminution in the volume of the blood. Also *oligemia*.

oligo- (*Genrl*) See OLIG-.

Oligocene (*Geol*) The youngest epoch of the PALAEOGENE subperiod, the time succession being Palaeocene, Eocene and Oligocene. See TERTIARY.

Oligochaeta (*BioSci*) A class of Annelida with relatively few chaetae that are not situated on parapodia. There is a definite prostomium usually without appendages. They are always hermaphrodite and have only one or two pairs of male and female gonads in fixed segments of the anterior part of the body. Reproduction involves copulation and cross-fertilization; the eggs are laid in a cocoon and develop directly. Terrestrial and aquatic forms exist, eg earthworms.

oligoclase (*Min*) One of the plagioclase feldspars, consisting of the albite (Ab) and anorthite (An) molecules combined in the proportions of Ab_9An_1 to Ab_7An_3. It is found esp in the more acid igneous and metamorphic rocks.

oligodendrocyte (*BioSci*) See OLIGODENDROGLIA.

oligodendroglia (*BioSci*) Cells within the central nervous system that deposit the myelin sheath. Also *oligodendrocyte*.

oligodendroglioma (*Med*) A cerebral tumour derived from oligodendroglia.

oligolecithal (*BioSci*) A type of egg with little yolk, somewhat more concentrated in one hemisphere. Found in *Amphioxus* (a genus of Cephalochordata) and mammals. Cf MESOLECITHAL, TELOLECITHAL.

oligomenorrhoea (*Med*) Scantiness of the discharge which occurs during menstruation. US *oligomenorrhea*.

oligomer (*Chem*) Low-molecular-mass polymer formed from two or more monomer units linked together. Used as PREPOLYMERS in eg epoxy adhesives.

oligomerous (*BioSci*) Consisting of only few parts.

oligomycin (*BioSci*) A bacterial toxin that inhibits oxidative phosphorylation.

oligonucleotide (*BioSci*) A nucleic acid with few nucleotides. Cf POLYNUCLEOTIDE.

oligopeptides (*BioSci*) Short polymers of 10–20 amino acids. May be synthesized, but also occur naturally with often powerful biological effects.

oligopod (*BioSci*) (1) Having few legs or feet. (2) Said of a phase in the development of larval insects in which the thoracic limbs are large while the evanescent abdominal appendages of the polypod phase have disappeared.

oligosaccharides (*Chem*) Carbohydrates containing a small number (2–10) of monosaccharide units linked together, with elimination of water.

oligospermia (*Med*) Diminution, below the average, of the quantity of semen voided in an ejaculation.

oligotokous (*BioSci*) Bearing few offspring. Cf POLY-TOKOUS.

oligotroph (*BioSci*) An organism that can grow in a nutrient-poor environment.

oligotrophic (*EnvSci*) Said of a type of lake having steep and rocky shores and scanty littoral vegetation, and in which the HYPOLIMNION does not become depleted of oxygen in the summer. The hypolimnion is generally larger than the EPILIMNION, and primary productivity is low. Cf DYSTROPHIC, EUTROPHIC.

oligotrophication (*EnvSci*) A process in which the nutrients in an aquatic system become depleted, or the rate at which they are recycled is reduced; usually caused by acidification of the water, commonly resulting from pollution, esp from acid precipitation.

oligotrophophyte (*BioSci*) A plant growing in a soil poor in soluble mineral salts.

oligozoospermia (*Med*) Diminution of the number of spermatozoa in the semen.

oliguria (*Med*) Abnormally diminished secretion of urine.

oliphagous (*BioSci*) Feeding on few different kinds of food; as phytophagous insects that are limited to a few related food plants. Cf POLYPHAGOUS.

olistostrome (*Geol*) A sediment which consists of a jumbled mass of heterogeneous blocks of material, generated by gravity sliding.

olivary (*Med*) Of or relating to two masses of tissue (the olivary bodies or olivary nuclei) situated in the medulla oblongata. See panel on BRAIN STRUCTURE.

olive (*For*) A tree (*Olea europaea*) from the Mediterranean regions of Europe, important for its fruit which produces edible oil when compressed and is itself edible after processing with dilute NaOH to neutralize the bitter glucoside content. E African olive (*O. hochstetteri*) has a pale- to mid-brown, attractively marbled wood.

olivenite (*Min*) A hydrated arsenate of copper which crystallizes in the orthorhombic system. It is a rare green mineral of secondary origin found in copper deposits.

olive oil (*Chem*) A pale-yellow or greenish oil obtained from the fruit of *Olea europaea*; rel.d. 0·91–0·92, acid value 1·9–5, saponification value 185–196, iodine value 77–88.

Oliver filter (*MinExt*) Drum filter used in large-scale dewatering or filtration of mineral pulps, usually after THICKENING. Pulp is drawn by vacuum to filtering membrane as drum rotates slowly through a trough, and filtrate is drawn off while moist filter cake is scraped from down-running side of drum.

olivine (*Min*) (1) Orthosilicate of iron and magnesium, crystallizing in the orthorhombic system, which occurs widely in the basic and ultramafic igneous rocks, including olivine-gabbro, olivine-dolerite, olivine-basalt, peridotites, etc. See CHRYSOLITE. The clear-green variety is used as a gemstone under the name *peridot*. For *iron olivine*, see FAYALITE; for *magnesium olivine*, see FORSTERITE. (2) As a prefix for many rocks which contain olivine, eg *olivine-basalt, olivine-gabbro*.

Olsen ductility test (*Eng*) A test in which a standard-size ball is forced into a blank of sheet metal and the depth of cupping measured when the metal fractures.

Olson microphone (*Acous*) The original RIBBON MICRO-PHONE using a battery-excited magnet.

omasitis (*Vet*) Inflammation of the omasum.

omasum (*BioSci*) See PSALTERIUM.

ombrogenous (*BioSci*) (1) A mire or bog receiving water only by precipitation and hence extremely oligotrophic or poor in nutrients. Cf SOLIGENOUS. (2) Obtaining nutrients from rain.

ombrophile (*BioSci*) A plant adapted to rainy places. Also *ombrophyte*.

Omega (*Aero*) Long-range radio navigation aid of very low frequency covering the whole Earth from eight ground transmitters. It can be received down to sea level.

Omega (*MathSci*) See ORDINAL NUMBER.

Omega Centauri (*Astron*) A bright globular cluster in the constellation Centaurus which contains over 100 000 stars.

omega equation (*EnvSci*) A diagnostic equation for the vertical velocity in PRESSURE CO-ORDINATES dP/dt conventionally denoted by ω. It is obtained by eliminating the time derivatives from the thermodynamic and VORTICITY EQUATIONS, and applying the QUASI-GEOSTROPHIC APPROXIMATION. With the omission of some small terms the co-equation may be written

$$\nabla^2(\sigma\omega) - \frac{f(\zeta+f)}{g}\frac{\partial^2 w}{\partial p^2}$$

$$= \frac{1}{f}\nabla^2 J\left(\varphi, \frac{\partial\varphi}{\partial p}\right) - \frac{1}{g}J(\varphi, g+f)$$

where σ is a measure of atmospheric stability, f the CORIOLIS PARAMETER, ζ the vertical component of relative vorticity, φ the geopotential, g the acceleration of gravity, and J indicates an operator such that

$$J(u, v) \equiv \frac{\partial u}{\partial x}\frac{\partial v}{\partial y} - \frac{\partial u}{\partial y}\frac{\partial v}{\partial x}$$

omega fatty acid (*BioSci*) A fatty acid, found in unsaturated fat, belonging to one of two groups (omega-3 fatty acids and omega-6 fatty acids) regarded as being beneficial in combating heart disease, depression, cancer, etc.

omega-minus particle (*Phys*) The heaviest HYPERON (1672 MeV), discovered in 1964. Its existence produces strong evidence for the classification, developed from Lie algebra (group theory), of those elementary particles which interact strongly.

Omega Nebula (*Astron*) A famous arch-shaped emission nebula in the constellation Sagittarius. Also *Horseshoe Nebula*.

omental bursa (*BioSci*) In some mammals, a pouch formed ventrally and dorsally to the stomach by the mesentery supporting the stomach.

omentopexy (*Med*) The stitching of the omentum to the abdominal wall in the treatment of cirrhosis of the liver.

omentum (*BioSci*) In vertebrates, a portion of the serosa connecting two or more folds of the alimentary canal. Adj *omental*.

omeprazole (*Pharmacol*) A proton pump inhibitor used for the treatment of gastric and duodenal ulcers. TN *Losec*.

Omicron Ceti (*Astron*) See MIRA.

ommatidium (*BioSci*) One of the visual elements composing the compound eyes of ARTHROPODA.

ommatophore (*BioSci*) An eye stalk.

omnibus-bar (*ElecEng*) The original term from which BUS-BAR is derived.

omnidirectional (*Space*) Said of a simple aerial, mounted on a spacecraft, radiating energy equally in all directions.

omnidirectional antenna (*ICT*) One receiving or transmitting equally in all directions in a horizontal plane, although the radiation pattern may be directional in elevation.

omnidirectional microphone (*Acous*) A microphone whose response is essentially independent of the direction of sound incidence. Also *non-directional microphone*.

omnidirectional radio beacon (*Aero*) A very-high-frequency radio beacon radiating through 360° upon which an aircraft can obtain a bearing. Used for navigation by VOR with distance-measuring equipment. Abbrev ORB.

Omnimax (*ImageTech*) TN for a development of the IMAX motion picture film system, using a FISH-EYE lens to give wrap round images inside a tilted dome.

omnirange (*ICT*) A VHF radio navigation system in which transmitters on the ground allow a pilot to plot their exact position.

omnivore (*BioSci*) An animal which eats both plants and animals. Omnivorous fungi attack several hosts. Adj *omnivorous*.

omphacite (*Min*) An aluminous sodium-bearing pyroxene, occurring in eclogites as pale-green mineral grains.

omphalectomy (*Med*) Removal of the umbilicus.

omphalic (*BioSci*) Pertaining to the umbilicus.

omphalitis (*Med*) Inflammation of the umbilicus.

omphaloid (*BioSci*) Navel-shaped.

omphalophlebitis (*Vet*) Inflammation of the navel or umbilical cord. See JOINT ILL.

OMR (*ICT*) Abbrev for OPTICAL MARK READER.

once-through boiler (*Eng*) See FLASH BOILER.

once-through fuel cycle (*NucEng*) A reactor in which the fuel is not reprocessed when spent.

onchocerciasis (*Med*) Infestation of the skin and subcutaneous tissues with the nematode worm of the genus *Onchocerca volvulus* in W Africa. It causes skin tumours and eye diseases, including blindness.

oncogene (*BioSci*) A genetic locus, originally identified in RNA tumour viruses, that is capable of the TRANSFORMATION of the host cell. Oncogenes are implicated as the cause of certain cancers. See PROTO-ONCOGENE.

oncogenic (*Med*) Inducing, or tending to induce, the formation of tumours.

oncogenic virus (*BioSci*) Generally a virus able to cause cancer, but more specifically one carrying an ONCOGENE.

oncology (*Med*) That part of medical science dealing with new growths (tumours) of body tissue.

onco-mouse (*BioSci*) A transgenic mouse, genetically engineered to develop cancer, and produced as a tool for the development of anti-cancer drugs.

on-costs (*Build, Eng*) All items of expenditure that cannot be allocated to a definite job, ie all expenses other than the prime cost. Also *burden, establishment charges, loading, overhead expenses*.

Oncovirinae (*BioSci*) The subfamily of single-stranded RNA retroviruses (Retroviridae) that can cause tumours. They are budded from the host cell and are coated with host-derived membrane. No longer considered an appropriate classification since the genera are unrelated.

ondansetron (*Pharmacol*) A drug used to treat vomiting and nausea, esp that resulting from chemotherapy or radiotherapy.

Ondiri disease (*Vet*) See BOVINE INFECTIOUS PETECHIAL FEVER.

ondoscope (*Electronics*) Glow tube operated by strong microwave radiation fields, and used eg in tuning transmitters.

one-electron bond (*Chem*) A bond formed by the resonance of a single electron between two atoms or radicals of similar energy.

one-group theory (*NucEng*) Greatly simplified reactor model in which all neutrons are regarded as having the same energy. See MULTIGROUP THEORY.

one-hour rating (*ElecEng*) A form of rating commonly used for electrical machinery supplying an intermittent load, eg traction motors for suburban service. It indicates that the machine will deliver its specified rating for a period of 1 h without exceeding the specified temperature rises.

one light (*ImageTech*) Describes printing a complete reel of motion picture film at a single level of exposure, without variations from one scene to another.

one-particle model (*Phys*) A form of SHELL MODEL of the nucleus in which nuclear spin and magnetic moment are regarded as associated with one resident nucleon.

one-phase (*ElecEng*) See SINGLE-PHASE.

one-pipe system (*Build*) That in which both soil and waste are carried by a common pipe, fittings being protected with deep seal traps.

ONERA (*Aero*) Abbrev for *Office National d'Etudes et de Recherches Aerospatiales*, Fr.

one's complement (*ICT*) Formed from a binary number by changing each 1 bit to a 0 bit and vice versa. Thus the one's complement of 00 101 011 (= 43 base 10) is 11 010 100. Cf TWO'S COMPLEMENT.

one-shot circuit (*ICT*) See SINGLE CIRCUIT.

one-shot multivibrator (*Electronics*) See MONOSTABLE MULTIVIBRATOR.

one-to-one (*MathSci*) Of a map, mapping, function or transformation, *F*, from a set S to a set T, such that each element of one set is associated with a unique element of the other. If every element of S is mapped to a distinct element of T, then *F* is said to be a mapping from S into T and if every element of T is the image of at least one element of S, *F* is said to be a mapping from S onto T; a one-to-one mapping is both into and onto.

one-to-one transformer (*ElecEng*) A transformer having the same number of turns on the primary and secondary, used for circuit isolation.

one-way switch (*ElecEng*) A switch providing only one path for the current. Cf DOUBLE-THROW SWITCH (two-way switch).

on-hook (*ICT*) The condition in which a telephone speech circuit has been disconnected and the ringing circuit connected by placing the handset on the SWITCH-HOOK.

onion skin paper (*Paper*) A glazed translucent paper having an undulating surface caused by the special beating and drying of the paper.

on lap (*Geol*) An UNCONFORMITY above which beds are successively pinched out by younger beds. Largely synonymous with OVERLAP, and the reverse of OFF LAP.

on-line (*ICT*) Describes processing performed on equipment directly under the control of the CENTRAL PROCESSOR, while the user remains in communication with the computer.

on-line editing (*ImageTech*) Videotape editing using the original recordings on their own standard equipment for all stages of selection and arrangement as well as the final CONFORMING.

on-net to off-net calling (*ICT*) Calling from a node of a VIRTUAL PRIVATE NETWORK (VPN) to a public switched telephone network node that is not a VPN node, or from any other private network to any public node.

on-net to on-net calling (*ICT*) Calling from one node of a private network to another.

on-off control (*ICT*) A simple control system that is either on or off, with no intermediate positions. See CONTINUOUS CONTROL.

on-off keying (*ICT*) That in which the output from a source is alternately transmitted and suppressed to form signals.

Onsager equation (*Chem*) An equation based on the DEBYE–HÜCKEL THEORY, relating the EQUIVALENT CONDUCTANCE Λ_c at concentration c to that at zero concentration, of the form $\Lambda_c = \Lambda_0 - (A + B\Lambda_0)\sqrt{c}$, where A and B are constants depending only on the solvent and the temperature.

on-screen (*ICT*) As displayed on a TV or computer screen.

onstream (*ChemEng, MinExt*) A term describing the functional status of process plant in full production after the commissioning phase. Used usually of large-scale plant such as oil wells, refineries or continuous chemical plant.

onto (*MathSci*) See ONE-TO-ONE.

ontogeny (*BioSci*) The history of the development of an individual. Also *ontogenesis*. Cf PHYLOGENY. Adj *ontogenetic*.

on-top altitude clearance (*Aero*) AIR-TRAFFIC CONTROL clearance for VISUAL FLIGHT RULES flying above cloud, haze, smoke or fog.

onych-, onycho- (*Genrl*) Prefixes from Gk *onyx*, gen *onychos*, a nail or claw.

onychia (*Med*) Inflammation of the nail bed.

onycho- (*Genrl*) See ONYCH-.

onychocryptosis (*Med*) Ingrowing toenail.

onychogenic (*BioSci*) Nail-forming, nail-producing, occurring in the superficial cells of the nail bed.

onychogryphosis (*Med*) Thickening, twisting and overgrowth of the nails (usually of the toes) as a result of chronic infection and irritation. Also *onychogrypsis*.

onychomycosis (*Med*) A disease of the nails due to a fungus.

Onychophora (*BioSci*) A subphylum of rather annelid-like Arthropoda having trachea, a soft thin cuticle, and a body wall consisting of layers of circular and longitudinal muscles. The head is not marked off from the body, and consists of three segments, one preoral, bearing preantennae, and two postoral, bearing jaws and oral papillae respectively. There are a pair of simple vesicular eyes. All body segments are similar, each with a pair of parapodia-like limbs which end in claws and containing a pair of excretory tubules. Development is direct, eg the genus *Peripatus*.

onyx (*Min*) A cryptocrystalline variety of silica with layers of different colour, typically whitish layers alternating with brown or black bands. Used in cameos, the figure being carved in relief in the white band with the dark band as background.

onyx marble (*Min*) A banded form of calcite. *Oriental alabaster* is a beautifully banded form.

oö- (*Genrl*) Prefix from Gk *oon*, egg.

oöblastema (*BioSci*) A fertilized egg.

oöcium (*BioSci*) A brood pouch.

oöcyst (*BioSci*) In certain Protozoa, the cyst formed around two conjugating gametes; in Sporozoa, the passive phase into which the active vermiform phase changes in the host.

oöcyte (*BioSci*) An ovum prior to the formation of the first polar body; a female gametocyte.

oögamy (*BioSci*) (1) The union of gametes of dissimilar size, usually of a relatively large non-motile egg and a small active sperm. (2) In *Protozoa*, ANISOGAMY in which the female gamete is a hologamete. Cf ISOGAMY.

oögenesis (*BioSci*) The origin and development of ova.

oögonium (*BioSci*) In many algae and the Oomycetes, a unicellular female gametangium containing one or more eggs or oöspheres. In animals, an egg mother cell or oöcyte.

oölemma (*BioSci*) See VITELLINE MEMBRANE.

oölite (*Geol*) A sedimentary rock composed of oöliths. In most cases oöliths are composed of calcium carbonate, in which case the rock is an oölitic limestone, but they can also be made of chamosite or limonite, in which case the rock is an oölitic ironstone. Written by itself, the word can be assumed to refer to an oölitic limestone. Also *oolite*.

oölith (*Geol*) A more or less spherical concretion of calcium carbonate, chamosite, or dolomite, not exceeding 2 mm in diameter, usually showing a concentric-layered and/or a radiating fibrous structure. Also *oolith*.

oölitic (*Geol*) Pertaining to an OÖLITE.

oölogy (*Genrl*) The study of ova.

Oomycetes (*BioSci*) A group of fungus-like, non-photosynthetic organisms with a non-septate mycelium. They reproduce asexually by zoospores or dispersed sporangia that may germinate to give zoospores or a hypha, and sexually by oöspores. Heterokont flagellation and mitochondrial microvilli suggest affinity with the Heterokontophyta. They include some water 'fungi' and many plant parasites, eg *Pythium* (damping-off), the 'downy mildews' and the historically significant potato blight (*Phytophthora infestans*).

OOP (*ICT*) Abbrev for OBJECT-ORIENTED PROGRAMMING.

oöphorectomy (*Med*) Removal of an ovary.

oöphoritis (*Med*) Inflammation of an ovary.

oöphorosalpingectomy (*Med*) Removal of an ovary and a Fallopian tube.

oöphorostomy (*Med*) The formation of an opening into an ovarian cyst.

oöphorotomy (*Med*) Incision of an ovary.

Oort cloud (*Astron*) A huge orbiting reservoir of comets thought to exist at the edge of the solar system (at approximately 75 000–150 000 AU); believed to be the source of long-period comets.

oösperm (*BioSci*) See OÖBLASTEMA.

oöspore (*BioSci*) (1) A fertilized ovum. (2) In protozoa, an encysted zygote. (3) A thick-walled zygote with food reserves formed from fertilized oösphere in some algae and the Oomycetes.

oötheca (*BioSci*) An egg-case, as in the cockroach.

oötocoid (*BioSci*) Bringing forth the young in an immature condition and allowing them to complete their early development in a marsupium.

oötocous (*BioSci*) See OVIPAROUS.

ooze (*Geol*) (1) A fine-grained, soft, deep-sea deposit, composed of shells and fragments of foraminifera, diatoms and other organisms. (2) A soft mud.

opacity (*Build*) The HIDING POWER of a paint or other pigmented compound. Also *obliterating power*.

opacity (*ImageTech, Phys*) The reciprocal of the optical TRANSMISSION RATIO.

opal (*Min*) A cryptocrystalline or colloidal variety of silica with a varying amount of water. The transparent coloured varieties, exhibiting opalescence, are highly prized as gemstones.

opal agate (*Min*) A variety of opal, of different shade of colour and agate-like in structure.

opalescence (*Chem*) The milky, iridescent appearance of a solution, due to the reflection of light from very fine, suspended particles.

opalescence (*Min*) (1) In general, the milky, iridescent appearance of a mineral. (2) The play of colour exhibited by precious opal, due to interference at the surfaces of minutely thin films, the thicknesses of the latter being of the same order of magnitude as the wavelength of light.

opal glass (*Glass*) Glass which is opalescent or white; made by the addition of fluorides (eg fluorspar, cryolite) to the glass mixture.

opal suppressor (*BioSci*) A gene that codes for an altered transfer RNA so that the opal codon, UGA, is not treated as a STOP CODON and protein synthesis continues.

opaque (*Phys*) Totally absorbent of rays of a specified wavelength, eg wood is opaque to visible light but slightly transparent to infrared rays, and completely transparent to X-rays and waves for radio communication.

opcode (*ICT*) See CHARACTER CODE.

OPEC (*MinExt*) Abbrev for ORGANIZATION OF PETROLEUM EXPORTING COUNTRIES.

open account (*ICT*) One of a large number of accounts registered on magnetic tape, all of which are rerecorded with UPDATE at regular intervals, eg daily.

open aestivation (*BioSci*) Aestivation in which the leaves or perianth parts neither overlap nor meet by their edges.

open antenna (*ICT*) One that is open-ended and able to support a standing wave of current.

open architecture (*ICT*) A COMPUTER ARCHITECTURE that is compatible with HARDWARE and SOFTWARE from a range of manufacturers.

open back (*Print*) See HOLLOW BACK.

opencast (*MinExt*) Quarry. Open cut. Mineral deposit worked from surface and open to daylight. See MINING.

opencast mining (*Geol*) A form of quarrying used to extract coal and mineral deposits.

open circuit (*ElecEng*) A circuit providing infinite impedance. Cf SHORT CIRCUIT.

open-circuit characteristic (*ElecEng*) A term used for the curve obtained by plotting the emf generated by an electric generator on open circuit against the field current. Also *no-load characteristic*.

open-circuit grinding (*MinExt*) Size reduction of solids in which the material to be crushed is passed only once through the equipment, so that all the grinding has to be done in a single step. Generally less efficient than *closed-circuit grinding*.

open-circuit impedance (*Phys*) Input or driving impedance of line or network when the far end is free (open-circuited), not grounded or loaded.

open-circuit loss (*ElecEng*) See NO-LOAD LOSS.

open-circuit transition (*ElecEng*) A method used for changing the connections of traction motors from series to parallel in which the circuit is broken during reconnection. Cf BRIDGE TRANSITION.

open-circuit voltage (*ElecEng*) The voltage appearing across the terminals of an electric generator or transformer when it is delivering no load.

open cluster (*Astron*) A loose star cluster containing at most a few hundred stars; the stars of a cluster have a common motion through space, and are associated with dust and gas clouds, eg the HYADES, PLEIADES and PRAESEPE. Such clusters tend to be close to the galactic plane in our Galaxy. Also *galactic cluster*. See panel on HERTZSPRUNG–RUSSELL DIAGRAM.

open community (*BioSci*) A plant community that does not occupy the ground completely, so that bare spaces are visible.

open-diaphragm loudspeaker (*Acous*) The most common domestic type, in which a paper, plastic or doped-fabric cone diaphragm is driven by a circular coil at its apex. Mounted in a baffle or box, with or without ports or labyrinth.

open-end spinning (*Textiles*) The system in which the fibres being fed to the spinning head are highly drafted as near as practicable to the individual fibre state so creating a discreet break or open end in the fibre flow. The fibres are then caught and twisted onto the end of a rotating yarn. The main methods in use are ROTOR SPINNING and FRICTION SPINNING. Also *break spinning*.

open-field test (*Psych*) An experimental procedure in which an animal is released into an open area with no obstacles, and features of its behaviour, such as defecation, urination and locomotion, are observed, these sometimes being related to emotionality.

open file (*ICT*) In a program or in interactive computing, to make a file on backing store ready for writing to and reading from. Cf CLOSED FILE.

open floor (*Build*) A floor which is not covered by a ceiling, the joists being therefore on view.

open flow (*MinExt*) Running an oil well without any valves or constrictions at the casing head.

open-frame girder (*Eng*) A girder consisting of upper and lower booms connected at intervals by (usually) vertical members, and not braced by any diagonal members. Also *Vierendele girder, Vierendele truss*.

open fuse (*ElecEng*) A fuse mounted so that the fuse link is fully exposed, except for any external casing.

open-hearth furnace (*Eng*) A furnace, of reverberatory type, used in steel-making. See REVERBERATORY FURNACE.

open-hearth process (*Eng*) Now obsolete process for steel-making in a furnace having a dish-shaped hearth, eg Siemens–Martin process (1866). Cold pig iron and steel scrap are charged to the open hearth and melted by preheated gases with flame temperatures up to 1750°C produced by regeneration as they enter through a brick labyrinth, where heat is maintained by periodic reversal between these and exit gases. After melting (up to 1600°C) most impurities are in the covering slag. The remainder are removed by addition of iron ore, scale or limestone before tapping into ladles, where ferrosilicon, ferromanganese or aluminium may be added as deoxidants. See ACID PROCESS, BASIC PROCESS, STEEL-MAKING.

open inequality (*MathSci*) An inequality which defines an open set of points, eg $-1 < x < +1$.

opening (*Print*) The appearance of a pair of facing pages, given careful consideration in the best typography.

opening (*Textiles*) The separation of compressed fibres after leaving the bale-breaking machine. Large, high-speed spiked cylinders remove impurities into dust chambers and leave the fibre in a fluffy state for further treatment.

open interval (*MathSci*) An interval, such as $a < x < b$, the points of which form an open set. Cf CLOSED INTERVAL.

open-jet wind tunnel (*Aero*) A WIND TUNNEL in which the working section is not enclosed by a duct.

open loop (*ICT*) A signal path, system or amplifier without a feedback path; *open-loop gain* is that of an amplifier without its feedback loop.

open-loop recycling (*Eng*) Material reclamation from one system to another, eg collection of polyethylene terephthalate containers, granulation and melt spinning to fibre for low-grade filling. Also *secondary recovery*.

open mortise (*Build*) See SLOT MORTISE.

open newel stair (*Build*) A stair having successive flights rising in opposite directions, and arranged about a rectangular well-hole.

open pipe (*Acous*) A pipe which is partially or completely open at both ends, so that the wavelength of the fundamental resonance is approximately double the length of the air column.

open pore (*PowderTech*) A cavity within a particle of powder which communicates with the surface of the particle.

open reading frame (*BioSci*) A possible reading frame of DNA that is capable of being translated into protein, ie is not punctuated by stop codons. Abbrev *ORF*.

open reel (*ImageTech*) The use of magnetic tape on equipment with separate feed and take-up spools, in contrast to an enclosed CASSETTE.

open roof (*Build*) A roof which is not covered in by a ceiling, the trusses being exposed.

open sand (*Eng*) (1) A sand of good porosity or permeability, as distinct from a close sand. (2) The process of casting in an open mould when the finish of the top surface is immaterial.

open set (*MathSci*) A basic undefined concept of topology. Intuitively, it means a set which does not include any boundary points, eg the set of real numbers x such that $1 < x < 2$ does not include its boundaries 1 and 2. Note a set may be open, closed, both or neither.

open slating (*Build*) Slating in which gaps of 1–4 in (25–100 mm) are left between adjacent slates in any course.

open slot (*ElecEng*) A type of slot, used in the armatures of electric machines, in which the opening is the same width as the rest of the slot. Cf SEMICLOSED SLOT.

open source (*ICT*) A form of licensing by which software is made freely available for use and the source code can be extended or amended by third parties.

open string (*Build*) See CUT STRING.

open study (*Psych, Med*) A study or trial in which both experimenter and subject are aware of the purpose and what factors are being tested, eg whether drug or placebo is being taken. Unlike double-blind studies, the probability of bias is very high and the reliability of the data is doubtful.

open subroutine (*ICT*) One that is part of the main program and is copied into the program where required. Cf CLOSED SUBROUTINE.

open systems interconnection (*ICT*) A standard recommended by the ISO to facilitate communication between systems produced by different manufacturers. It is based on a model containing seven layers arranged hierarchically, each carrying out a different task, from physical interconnection to the specific application for which the communication link is being used. Each layer in particular equipment can communicate with the layers immediately above and below it, but communication between equipment can occur only at matching levels. Abbrev *OSI*. Fig. ▷

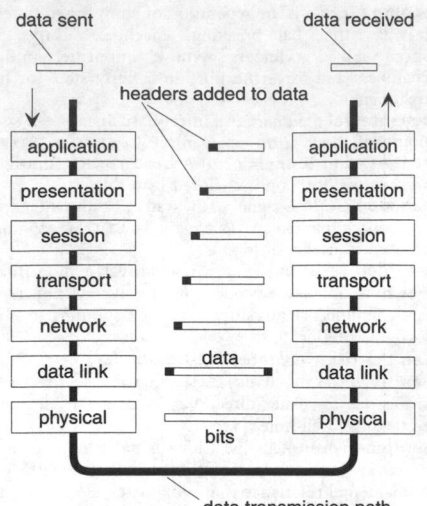

open systems interconnection Showing the seven layers; at each a header is added which is stripped out when it reaches the same layer after transmission.

open traverse (*Surv*) One which does not close on the point of origin, and is therefore not directly checked for accuracy or closing error.

open universe (*Astron*) A term used to describe the universe if it has a density less than a certain critical value. Under this condition space is hyperbolically curved and the universe has an infinite volume. Gravity would be too weak to ever halt the expansion of the universe.

open vascular bundle (*BioSci*) A bundle including cambium.

open well (*Arch*) A stair enclosing a vertical opening between the outer sides of the flights.

open-width washer (*Textiles*) A machine having a series of tanks each fitted with nip rollers through which continuous lengths of fabrics are passed while being held out to full width. The tanks contain the different solutions required for washing, eg soapy water, dilute acid or alkali, as well as clean water.

open window unit (*Acous*) See SABIN.

open-wire feeder (*ICT*) One supported from insulators on poles, forming a pole route between antenna and transmitter or receiver.

opepe (*For*) W African hardwood (*Nauclea diderrichii*) whose heartwood is orange-brown, with an interlocked grain, fairly coarse texture, and has a copper lustre.

operand (*ICT, MathSci*) An item or quantity on which an OPERATION is to be performed. In computing, its address. See ADDRESS FIELD.

operand field (*ICT*) See ADDRESS FIELD.

operant chain (*Psych*) A complex chain of behaviour built up by the operant technique of *shaping*.

operant behaviour (*Psych*) Behaviour that is spontaneous and voluntary, rather than an automatic, reflexive response to a stimulus (which is RESPONDENT BEHAVIOUR).

operant conditioning (*Psych*) A learning procedure in which a reinforcement follows a particular spontaneous response on a proportion of occasions so that an operant behaviour is brought under the control of a conditioning stimulus.

operant response (*Psych*) A response which acts on the environment to produce an event which affects the subsequent probability of that response.

operating characteristic (*ElecEng*) See LOAD CHARACTERISTIC.

operating diagram (*Eng*) A graph of volumetric melt flow in extruder barrel versus melt pressure. The intersection of screw and die operating curves defines the OPERATING POINT. See EXTRUSION.

operating duty (*ElecEng*) A term applied to a switch or circuit breaker to define the number and duration of operations used in specifying its performance.

operating factor (*ElecEng*) The ratio between the time a motor actually runs to the length of the duty cycle.

operating point (*Eng*) Optimum set of conditions, eg temperature, pressure, shear rate, for extrusion of polymers.

operating system (*ICT*) A collection of system programs that control the activities of the computer system such as JOB CONTROL, input/output and processing, eg DOS, UNIX, VM/CMS, VMS, etc. See MONITOR.

operation (*ICT*) The action of an operator or function that takes one or more pieces of data and produces a new piece of data, eg 'squared' is an arithmetic operation that produces 16 from 4. See LOGICAL OPERATION, RELATIONAL OPERATION.

operation (*MathSci*) Any procedure that generates an output given an input, such as the arithmetic operations addition and multiplication, and the logical operations negation and conjunction. See also ALGORITHM, FUNCTION.

operational amplifier (*Electronics*) High-gain, high-stability dc amplifier used with external feedback path. These were originally used to perform one specific mathematical operation in analogue computing but are now more widely used as the basic building blocks of electronic circuits.

operationalism (*Genrl*) The theory that defines scientific concepts by means of the operations used to prove or determine them. Also *operationism*.

operational research (*MathSci*) The branch of mathematics associated with construction of mathematical models in order to deduce the best possible course of action.

operational sphericity (*PowderTech*) A SHAPE FACTOR defined as the cube root of the ratio of the volume of a particle of powder to the volume of the circumscribing sphere.

operational taxonomic units (*BioSci*) The entities of any taxonomic rank, such as individuals, species, genera, etc, whose relationships are studied in NUMERICAL TAXONOMY. Abbrev *OTU*.

operation code (*ICT*) See MACHINE-CODE INSTRUCTION.

operator (*BioSci*) Sequence of DNA to which a REPRESSOR or ACTIVATOR can bind. Situated before the coding sequence of a gene and close to the PROMOTER.

operator (*ICT*) See OPERATION.

operator (*MathSci*) Mathematical symbol representing a specific OPERATION to be carried out on a particular operand, eg the differential operator $D \ (\equiv d/dx)$.

opercular apparatus (*BioSci*) In fish, the operculum, together with the BRANCHIOSTEGAL MEMBRANE and rays.

operculate (*BioSci*) (1) Possessing a lid. (2) Opening by means of a lid.

operculum (*BioSci*) (1) In plants, a lid or cover, composed of part of a cell wall or of from one to many cells, that opens to allow the escape of contents from some sort of container, esp the lid of an antheridium, a moss capsule, an ascus or other sporangium or the germ pore of a pollen grain. (2) In the eggs of some insects, a differentiated area of the chorion that lifts up when the larva emerges from the egg. (3) In the higher fish, a fold that articulates with the hyoid arch in front of the first gill slit, and extends backwards, covering the branchial clefts. A similar structure occurs in the larvae of amphibians. (4) In some tubiculous Polychaeta, an enlarged branch of a tentacle closing the mouth of the tube when the animal is retracted. (5) In spiders, a small plate partially covering the opening of a lung book. (6) In some Cirripedia, plates of the carapace that can be closed over the retracted thorax. (7) In some Gastropoda, a plate of chitinoid material, strengthened by calcareous deposits, that fits across the opening of the shell.

operon (*BioSci*) In bacteria a set of functionally related genes that share a common promoter and mRNA, thus securing co-ordinated transcription.

ophi- (*Genrl*) Prefix from Gk *ophis* signifying snake.

ophicalcite (*Geol*) See FORSTERITE-MARBLE.

ophiolite (*Geol*) A group of mafic and ultramafic igneous rocks ranging from extrusive spilites to intrusive gabbros, associated with deep-sea sediments. Ophiolites are commonly found in CONVERGENCE ZONES.

ophitic texture (*Geol*) A texture characteristic of dolerites in which relatively large pyroxene crystals completely enclose smaller, lath-shaped plagioclases. See POIKILITIC TEXTURE.

Ophiuchus (Serpent Bearer) (*Astron*) A large constellation on the celestial equator.

Ophiuroidea (*BioSci*) A class of Echinodermata with a dorsoventrally flattened star-shaped body and the arms sharply differentiated from the disk. Arms do not contain caecae of the alimentary canal and the tube feet lack ampullae and suckers, and lie on the lower surface, although not in grooves. There is no anus, the madreporite is aboral and there is a well-developed skeleton. Free-living; brittle-stars.

ophthalm-, ophthalmo- (*Genrl*) Prefixes from Gk *ophthalmos* denoting eye.

ophthalmectomy (*Med*) Excision of an eye.

ophthalmia (*Med*) Inflammation of various parts of the eye, esp of the conjunctiva.

ophthalmic (*BioSci*) Pertaining to or situated near the eye, as the *ophthalmic nerve*, which passes along the back of the orbit in lower vertebrates.

ophthalmo- (*Genrl*) See OPTHALM-.

ophthalmodynamometer (*Med*) An instrument used to measure the intraocular arterial pressure by means of external pressure on the eye.

ophthalmology (*Med*) The study of the eye and its diseases; a practitioner is an *ophthalmologist*.

ophthalmoplegia (*Med*) Paralysis of one or more muscles of the eye.

ophthalmoscope (*Med*) An instrument for inspecting the interior of the eye.

Opiliones (*BioSci*) A subclass of Arachnida with rounded bodies, the prosoma and opisthosoma broadly jointed and usually with long legs. They are mostly predacious; harvestmen.

opine (*BioSci*) Guanidoamino acids (either octopine or nopaline) synthesized and released by plant cells after infection with a Ti plasmid and used by *Agrobacterium tumefaciens* as carbon and nitrogen source. See CROWN GALL.

opioids (*Pharmacol*) A group of effective analgesics that includes the opiates such as morphine. Found in the opium poppy (*Papaver somniferum*) or are synthetic derivatives of the natural compounds. Tend to be addictive and are drugs of abuse. Weak opioids include *codeine* and *dihydrocodeine*; strong opioids include *morphine, diamorphine, fentanyl* and *phenazocine*.

opisthocoelous (*BioSci*) Concave posteriorly and convex anteriorly; said of vertebral centra.

opisthoglossal (*BioSci*) Having the tongue attached anteriorly, free posteriorly, as in frogs.

opisthomere (*BioSci*) A postoral somite.

opisthosoma (*BioSci*) In Chelicerata (spiders, scorpions, etc) the segments posterior to those bearing the legs; the abdomen. Cf PROSOMA.

opisthotonos (*Med*) Extreme arching backwards of the spine and the neck as a result of spasm of the muscles in these regions, eg in tetanus.

Oppenheimer–Phillips process (*Phys*) A form of STRIPPING in which a deuteron surrenders its neutron to a nucleus without entering it. Abbrev *OP process*.

OPP film (*Plastics*) Abbrev for ORIENTED POLYPROPYLENE FILM.

opportunistic infection (*Med*) An infection to which healthy people are resistant or from which they recover quickly, but that occurs in those whose immune systems have been compromised by illness. Such bacterial, fungal, protozoal and viral infections occur in patients with terminal cancer or with AIDS, and may be the immediate cause of death.

opportunistic species (*BioSci*) A species adapted to colonize temporary or local conditions.

opposed-cylinder engine (*Autos*) An engine with cylinders, or banks of cylinders, on opposite sides of the crankcase in the same plane, their connecting-rods working on a common crankshaft placed between the cylinders. Also *boxer*.

opposed-voltage protective system (*ElecEng*) A form of Merz–Price protective system in which the secondary voltages of current transformers, situated at each end of the circuit to be protected, are balanced against each other, so that there is normally no current on the pilots connecting them.

opposite (*BioSci*) Two organs, esp leaves, arising at the same level but on opposite sides of a stem. Cf ALTERNATE, DECUSSATE, WHORLED. See PHYLLOTAXIS.

opposition (*Astron*) The instant when the geocentric longitude of the Moon or of a planet differs from that of the Sun by 180°.

OP process (*Phys*) See OPPENHEIMER–PHILLIPS PROCESS.

opsin (*BioSci*) Apoprotein of visual pigment of the *rhodopsin* family.

opsonin (*BioSci*) Factors that are present in blood and other body fluids, and that bind to particles to increase their susceptibility to phagocytosis. They may be antibody, or products of complement activation (esp C3b), or some other substances such as fibronectin that bind to particles.

opsonization (*BioSci*) Coating of micro-organisms, particularly bacteria, with opsonin, thereby enhancing their uptake by phagocytic cells.

optic (*BioSci*) Pertaining to the sense of sight; the second cranial nerve of vertebrates.

optical activity (*Chem*) A property possessed by many substances whereby plane polarized light, in passing through them, suffers a rotation of its plane of polarization, the angle of rotation being proportional to the thickness of substance traversed by the light. In the case of molten or dissolved substances it is due to the possession of an asymmetric molecular structure, eg no mirror plane of symmetry in the molecule. See CHIRALITY, SPECIFIC ROTATION.

optical axial angle (*Min*) The angle between the two optical axes in biaxial minerals, usually denoted as 2V (when measured in the mineral) or 2E (in air).

optical axis (*Crystal*) Direction(s) in a doubly refracting crystal for which both the ordinary and the extraordinary rays are propagated with the same velocity. Only one exists in uniaxial crystals, two in biaxial.

optical axis (*Phys*) The line which passes through the centre of curvature of a lens' surface, so that the rays are neither reflected nor refracted. Also *principal axis*.

optical bench (*Phys*) A rigid bed along which optical components, mounted in suitable holders, may be moved. It is a device used in experimental work on linear optical systems.

optical black (*Phys*) The description of a body when it absorbs all radiation falling on its surface. No substance is completely black in this context.

optical bleaches (*Chem*) See FLUORESCENT WHITENING AGENTS.

optical brightener (*Textiles*) See BRIGHTENING AGENT.

optical centre of a lens (*Phys*) The point on the principal axis of a lens or lens system for which the incident direction of a ray passing through is parallel to the emergent direction.

optical character recognition (*ICT*) Machine recognition of characters by light-sensing methods. Abbrev *OCR*.

optical constants (*Phys*) The refractive index (*n*) and the absorption coefficient (*k*) of an absorbing medium. Together these determine the complex refractive index (*n*−*k*) of the medium.

optical crown (*Glass*) Glass of low dispersion made for optical purposes. See CROWN GLASS.

optical diffraction (*BioSci*) A technique using diffraction of visible light to obtain information about repeating patterns.

optical disk (*ICT*) A STORAGE MEDIUM consisting of marks etched on metal-coated plastic disk. The pattern of marks is read by a laser beam. The data content is set during manufacture but systems offering both read and write capability are available. See CD-ROM, DVD, PHOTO CD, WORM.

optical distance (*Phys*) The distance travelled by light (*d*) multiplied by the refractive index of the medium (*n*). Length of equivalent path in air (strictly vacuum). Also *optical path*.

optical double (*Astron*) A pair of stars which lie along almost the same line of sight, but which are too far apart to be physically related in a binary system.

optical electronic devices (*Phys*) Devices used to locate weakly radiating sources by the detection of their infrared emission. The radiation is collected by optical mirrors or lenses and concentrated on a sensitive infrared detector. Detailed maps of the Earth's surface, weather mapping and non-destructive testing of materials and components are some of the non-military applications.

optical fibres (*ICT*) Fibres made from high-purity, low-loss, low-dispersion glass and used as medium for telecommunications by transmission of high-frequency pulses of light. Basis of operation is the total internal reflection at the interface between the higher-refractive-index core and the lower-refractive-index sheath or cladding. Core is usually vitreous silica doped with germania (GeO_2). Used in FIBRE OPTICS. See MONOMODE FIBRE, MULTIMODE FIBRE.

optical flat (*Eng*) A flat glass disk having very accurately polished surfaces, used for testing by interference patterns the flatness of gauge anvils and other plane surfaces.

optical flat (*ImageTech*) A surface whose deviation from a true plane is small enough to be expressed in terms of wavelengths of light; for a plate the two surfaces must be parallel to within seconds of arc.

optical flint (*Glass*) Glass of high dispersion made for optical purposes. See FLINT GLASS.

optical glass (*Glass*) Glass made expressly for its optical qualities. The composition varies widely in both constituents and proportions, with very exacting requirements for freedom from streaks and bubbles. See CROWN GLASS, FLINT GLASS and panel on GLASSES AND GLASS-MAKING.

optical illusion (*Genrl*) Something that has an appearance which deceives the eye; a misunderstanding caused by such a deceptive appearance.

optical indicator (*Autos*) An engine indicator in which a ray of light is deflected successively by mirrors in directions at right angles, proportionately first to cylinder pressure, then to piston displacement, being finally focused on a ground-glass screen or photographic plate, on which it traces the INDICATOR DIAGRAM.

optical isomerism (*Chem*) The existence of isomeric compounds which differ in their CHIRALITY. Important for polymers. See STEREOREGULAR POLYMERS.

optical lever (*Phys*) A device for measuring the small relative displacement of two objects by means of angular displacement of a light beam.

optical mark reader (*ICT*) An INPUT DEVICE that uses an optical sensing method to read marks made in predetermined positions on paper forms or cards, eg pencil marks may be made against choices in a questionnaire and thus the responses may be automatically read into a COMPUTER SYSTEM.

optical maser (*Phys*) A MASER in which the stimulating frequency is visible or infrared radiation. See panel on LASER.

optical microscopy (*Phys*) A method of analysing the microstructure of materials using visible light source, in transmission (transparent materials like many polymers and glasses) or reflection (opaque materials like crystalline metals and ceramics). Specimen preparation method is important for accurate interpretation of structure, and includes ultramicrotomy for polymers and biomaterials, polishing and etching for metals. Special methods use eg polarized light or ultraviolet radiation.

optical model (*Phys*) A model of the nucleus which treats the target nucleus during nuclear reactions as a sphere that partly absorbs and partly transmits the incidental radiation, ie it has, in the optical analogy, a *complex* refractive index. The nucleus can thus be represented by a 'potential well' having real and imaginary components.

optical path (*Phys*) See OPTICAL DISTANCE.

optical printing (*ImageTech*) A method of printing motion picture film in which an image of the frame on one strip is formed on the other by means of a copy lens, thus allowing changes of size and position to be made.

optical pumping (*Phys*) A mechanism by which an external light source of suitable frequency stimulates the material to produce a POPULATION INVERSION for a particular energy transition in a laser. See panel on LASER.

optical pyrometer (*Phys*) An instrument which measures the temperatures of furnaces by estimating the colour of the radiation, or by matching it with that of a glowing filament.

optical range (*ICT*) That which a radio transmitter or beacon would have if it were radiating visible light.

optical rotary dispersion (*Chem*) The change of optical rotation with wavelength. Rotatory dispersion curves may be used to study the configuration of molecules. Abbrev *ORD*.

optical rotation (*Phys*) The rotation of the plane of polarization of a beam of light when passing through certain materials.

opticals (*ImageTech*) The general term for image effects made in a motion picture by laboratory printing operations, such as fades and dissolves.

optical sound (*ImageTech*) See PHOTOGRAPHIC SOUND.

optical spectrometer (*Phys*) A spectroscope for studying optical spectra. Fitted with a graduated circle it enables the angle of deviation of each spectrum line to be measured and so the wavelengths deduced.

optical spectrum (*Phys*) The visible radiation emitted from a source separated into its component frequencies.

optical square (*Surv*) A hand instrument for setting out right angles in the field. It works on the principle of the SEXTANT, the two reflecting surfaces being arranged in this case to yield lines of sight at a fixed angle of 90° apart.

optical stress analysis (*Eng*) See PHOTOELASTIC ANALYSIS.

optical track (*ImageTech*) A photographic sound track on motion picture film.

optical transfer function (*Phys*) A mathematical representation of the effect of a lens or other component in an optical system on the imaging of a point source.

optical tweezers (*BioSci*) A beam of light focused on a microscopic particle that exerts force, in the piconewton range, that is sufficient to move small organelles around under the microscope or to measure the forces that motor molecules are exerting. Also referred to as *laser tweezers* or an *optical trap*.

optical whites (*Chem*) See FLUORESCENT WHITENING AGENTS.

optic atrophy (*Med*) The condition of the optic disk (where the nerve fibres of the retina pass through the eyeball) resulting from degeneration of the optic nerve.

optic branch (*Phys*) A branch of the dispersion curve (frequency ω against wavenumber *q*) for crystal lattice vibrations for which ω is independent of *q* for small values of q. For a crystal containing n atoms per unit cell, the dispersion curve has $3n$ branches of which $3n-3$ are optic branches. The branches are characterized by different patterns of movement of the atoms. See ACOUSTIC BRANCH.

optic disc (*BioSci*) The blind spot of the eye where the optic nerve enters the retina.

optic lobes (*BioSci*) In vertebrates, part of the midbrain that is concerned with the sense of sight, and from which the optic nerves originate.

optic nerve (*BioSci*) The second cranial nerve that transmits information from the retina of the eye to the visual cortex of the brain.

optic neuritis (*Med*) See PAPILLITIS.

optics (*Phys*) The study of light. *Physical optics* deals with the nature of light and its wave properties; *geometrical optics* ignores the wave nature of light and treats problems of reflection and refraction from the ray aspect.

optic sign (*Min*) Anisotropic minerals are either optically positive or negative, indicated by $+$ or $-$ in technical descriptions. See NEGATIVE MINERAL, POSITIVE MINERAL.

optic tectum (*BioSci*) A region of the midbrain in which input from the optic nerve is processed. Retinally derived neurons of the optic nerve 'map' onto the optic tectum.

optimal damping (*Eng*) Adjustment of damping just short of CRITICAL, which allows a little overshoot. This attains the final reading of an indicating instrument most rapidly while showing that it is moving freely.

optimal proportions (*BioSci*) Describes the relative proportions of antibody and a soluble antigen that, when mixed together, produce the maximum degree of cross-linkage, such that all the antibody and all the antigen are included in the precipitate which forms. See LATTICE HYPOTHESIS.

optoelectronics (*Electronics*) The technology of the inter-conversion of electrical and optical signals. Cathode-ray-tube phosphors, indicators on hi-fi equipment, the lasers which are used to read compact disks and the light sensors in cameras are common examples exploiting this technology. The absorption and emission of light from materials is associated with electronic transitions between energy levels, differing in energy by an amount equivalent to that of the photons involved. Visible radiation has wavelengths in the range 400–800 nm but optoelectronic devices operate over a wider range from the infrared (3000 nm) to the near ultraviolet (200 nm), which correspond with photon energies from 0·4 to 6 eV.

optophone (*Electronics*) A photoelectric device for training the blind by converting printed words into sounds.

OR (*Aero*) Abbrev for: (1) OPERATIONAL REQUIREMENT; (2) OPERATIONAL RESEARCH.

OR (*ICT*) A logical operator such that (p OR q), written $p + q$, takes the value FALSE if both p and q are FALSE, otherwise (p OR q) takes the value TRUE. See LOGICAL OPERATION.

Oracle (*ICT*) (1) See TELETEXT. (2) Proprietary name for a widely used large-scale relational database management system.

oral (*BioSci*) Pertaining to the mouth.

oral characters (*Psych*) In psychoanalytic theory, refers to fixation at, or regression to, the oral stage of development; Freud considered that many oral habits reflected this (eg smoking).

oral contraception (*Med*) The use of synthetic hormones (estrogen and progestogen steroids in varying proportions), taken orally in pill form, to prevent conception by reacting on the natural LUTEINIZING and FOLLICLE-STIMULATING HORMONES and so inhibiting ovulation and/or fertilization. Colloq *the pill*.

oral stage (*Psych*) In psychoanalytic theory, the first stage of psychosexual development in which stimulation of the mouth and lips is the primary focus of bodily (libidinal) pleasure; occurs in the first year of life.

Orange book (*ImageTech*) The technical specifications that define the digital recordable CD (CD-R) standard. See PHOTO CD.

orange lead (*Chem*) Lead oxide. Pb_3O_4. Obtained by heating white lead (basic lead (II,II,V) carbonate) in air at approximately 450°C. Commercial varieties contain up to approximately 35% PbO_2.

orange peel (*Build*) A defect of paint surfaces applied by spray gun, in which more or less pronounced small depressions resemble an orange skin, attributable to incorrect thinning, wrong viscosity or too low a pressure.

ora serrata (*BioSci*) The edge of the retina.

ORB (*Aero*) Abbrev for OMNIDIRECTIONAL RADIO BEACON.

orbicular (*BioSci*) Flat, with a circular or almost circular outline.

orbiculares (*BioSci*) Muscles that surround an aperture; as the muscles that close the lips and eyelids in mammals.

orbicular structure (*Geol*) A structure exhibited by those plutonic igneous rocks which contain spherical orbs up to several centimetres in diameter, each showing a development of alternating concentric shells of different minerals, so deposited by rhythmic crystallization.

orbit (*Aero*) An aircraft circling a given point is said to orbit that point and AIR-TRAFFIC CONTROL instructions incorporate the term. See HOLDING PATTERN.

orbit (*Astron*) The path of a heavenly body (or eg an artificial satellite or spacecraft) moving about another under gravitational attraction. In unperturbed motion the orbit is a conic. See ELEMENTS OF AN ORBIT.

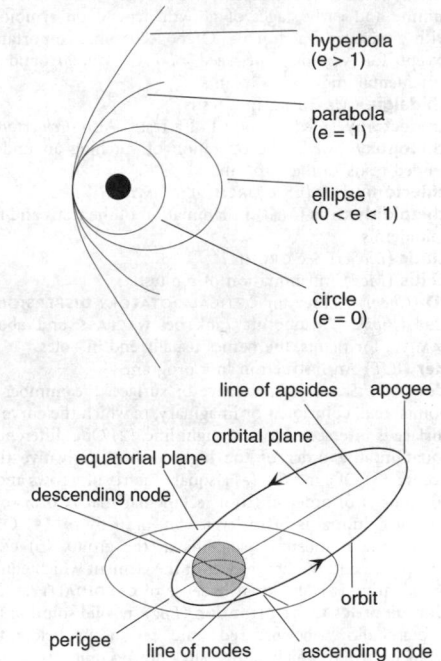

orbit The different types (above) and the geometry of an Earth satellite (below).

orbit (*BioSci*) (1) Generally, a space lodging an eye. (2) In vertebrates, the depression in the skull containing the eye. (3) In Arthropoda, the hollow that receives the eye or the base of the eye stalk. (4) In birds, the skin surrounding the eye.

orbital (*Phys*) The space-dependent part of the WAVE-FUNCTION for an electron in an atom. The properties of each electron in a many-electron atom may be reasonably described by its response to the potential due to the nucleus and to the other electrons. The wavefunction, which expresses the probability of finding the electron in a region, is specified by a set of four quantum numbers and defines the orbital of the electron. The state of the many-electron atom is given by defining the orbitals of all the electrons subject to the PAULI EXCLUSION PRINCIPLE.

orbital quantum number (*Electronics*) The spin of an orbital electron, indicating the angular momentum associated with the orbit. Zero for s states; one for p states; two for d states; and three for f states. See PRINCIPAL QUANTUM NUMBER.

orbital velocity (*Space*) See CIRCULAR VELOCITY.

orbit decay (*Space*) The change in orbit parameters of a space vehicle caused by air drag which becomes more rapid as the surface of a planet is approached owing to increasing atmospheric density, eventually resulting in entry or re-entry of the vehicle.

orbitosphenoid (*BioSci*) A paired cartilage bone of the vertebrate skull, forming the side wall of the brain case in the region of the presphenoid.

orbit shift coils (*Phys*) Coils placed on the magnetic pole faces of a betatron or synchrotron, so that by passing a current pulse through the coils the particles may be momentarily displaced from the stable orbit to strike a target.

orchic (*BioSci*) Pertaining to the testis. Also *orchitic.*

Orchidaceae (*BioSci*) The largest monocot family, the orchid family, comprising c.18 000 spp of flowering plants (superorder Liliidae). All are herbs but can be terrestrial or epiphytic and are cosmopolitan. Some are CAM plants. The flowers are zygomorphic, usually showy. The seeds are minute and early stages of growth depend on symbiosis with a mycorrhizal fungus. Of no economic importance except for vanillin (the seed pod of *Vanilla*) and as ornamental and florists' plants.

orchidalgia (*Med*) Pain in a testis.

orchidectomy (*Med*) Removal of a testis. Also *orchiectomy.*

orchidopexy (*Med*) The operation of stitching an undescended testis to the scrotum.

orchiectomy (*Med*) See ORCHIDECTOMY.

orchiepididymitis (*Med*) Inflammation of the testis and the epididymis.

orchitic (*BioSci*) See ORCHIC.

orchitis (*Med*) Inflammation of the testis.

ORD (*Chem*) Abbrev for OPTICAL ROTATORY DISPERSION.

order (*BioSci*) Taxonomic rank below CLASS and above FAMILY; for plants, the names usually end in –oles.

order (*ICT*) An instruction in a program.

order (*MathSci*) (1) Of a curve or surface: the number of points, real, coincident or imaginary, in which the curve or surface is intersected by a straight line. (2) Of a differential equation: the order of the highest-order derivative that occurs. (3) Of a matrix: (a) a square matrix of n rows and n columns is of order n; (b) a rectangular matrix of n rows and m columns is of order $n \times m$, or n by m. (4) Of a group: the number of elements in the group. (5) Of a group element: the first power of the element which equals the identity. (6) Also used in sense of CARDINALITY.

order–disorder transformation (*Eng*) In solid solution (eg a brass alloy) the ordered state can occur below the temperature at which lattice sites are occupied by single atomic species, and the disordered state can ensue above a critical temperature, with two or more elements in one primitive lattice.

ordered set (*MathSci*) A sequence of elements in which order is significant, so that {a,b,c} is not the same as {b,a,c}; a one-dimensional array. A *pair* is an ordered set of two elements. Also *n-tuple.*

ordered state (*Eng*) A solid-solution alloy existing as an ordered arrangement of atoms within the crystal lattice, eg $CuAu$ and Cu_3Au

order number (*Eng*) Of a torque impulse or a vibration, as the torsional oscillation of an engine crankshaft, the number of impulses or vibrations during one revolution of the shaft.

order of reaction (*Chem*) A classification of chemical reactions based on the index of the power to which concentration terms are raised in the expression for the instantaneous rate of the reaction, ie on the apparent number of molecules which interact. If $A + B \rightarrow$ products, then the rate of reaction is kA^aB^b; the order is a with respect to A and b with respect to B, and k is the *rate constant.*

ordinal (*MathSci*) A type of variable which can take ordered qualitative values, which may be known as levels, eg degrees of pain: none, mild, severe.

ordinal number (*MathSci*) In ordinary usage, first, second, third, etc, as opposed to one, two, three, etc, which are CARDINAL NUMBERS. Formally, ordinal numbers are properties of well-ordered sets such that two sets have the same ordinal if there is a one-to-one mapping from one set to the other which preserves their ordering. An infinite set can be well-ordered in different ways to give different ordinal numbers, while it has only one cardinal number. The first infinite ordinal, that of the infinite sequence of finite numbers 1,2,3,..., is called *omega.* It is followed by $\omega + 1, \omega + 2, ..., 2\omega, ..., \omega^2, ..., 3\omega^5 + \omega + 6, ..., \omega^\omega, \cdots$. All of these correspond to the first infinite cardinal ALEPH-0.

ordinal set (*MathSci*) A scale in which data are shown in order, but no sense can be given to their magnitude, such as a league table. Cf INTERVAL SCALE, RATIO SCALE.

ordinary differential equation (*MathSci*) See DIFFERENTIAL EQUATION.

ordinary ray (*Phys*) See DOUBLE REFRACTION.

ordinate (*MathSci*) See CARTESIAN CO-ORDINATES.

ordination (*BioSci*) A family of multivariate statistical techniques commonly used to plot ecological data sets collected from a large number of sites, on geometric axes so that similarity is represented by proximity.

Ordnance Bench Mark (*Surv*) A bench mark officially established with reference to the Ordnance Datum. Abbrev *OBM.*

Ordnance Datum (*Surv*) The level based on the mean sea level determined by tidal measurements at Newlyn, Cornwall, from which heights on UK maps are measured. Abbrev *OD.*

Ordnance Survey (*Surv*) Originally a triangulation of the UK made in 1791 by the military branch of the Board of Ordnance. Now a government survey which is concerned with land, buildings, roads, etc. Published as maps on which the National Grid is superimposed, giving the whole country a unified reference system. Abbrev *OS.*

Ordovician (*Geol*) The second-oldest period of the Palaeozoic era, covering an approx time span from 510 to 440 million years. Named after the *Ordovices,* an ancient tribe of the Welsh borders. Also, the corresponding system of rocks. See PALAEOZOIC.

ore (*Min*) A term applied to any metalliferous mineral from which the metal may be profitably extracted. It is extended to non-metals and also to minerals which are potentially valuable.

ore bin (*MinExt*) A storage system (usually of steel or concrete) which receives ore intermittently from the mine. A fine ore bin holds material crushed to centimetric size and keeps from 1 to 3 days' milling supply.

ore body (*MinExt*) Deposit, seam, bed, lode, reef, placer, lenticle, mass, stockwork, according to geological genesis.

ore dressing (*MinExt*) See MINERAL PROCESSING.

Oregon pine (*For*) See DOUGLAS FIR.

oreography (*Genrl*) See OROGRAPHY.

ore reserves (*MinExt*) Ore whose grade and tonnage has been established by drilling etc with reasonable assurance. See RESOURCES.

orexin (*BioSci*) A hormone that stimulates the appetite and is thought to be important for sleep regulation. Also *hypocretin.*

orf (*Med, Vet*) See CONTAGIOUS PUSTULAR DERMATITIS, which is transmissible to humans causing skin eruptions, usually on hands or forearms.

organ (*Acous*) A musical instrument, comprising ranks of pipes which radiate sound when blown by compressed air, the operation of the pipes being controlled by manuals or keyboards and by a set of pedals. Hence, any musical instrument producing synthetically (eg electronically) tones similar to those from pipes and operated from keyboards.

organ (*BioSci*) A part of the body of an animal or a plant adapted and specialized for the performance of a particular function.

organ culture (*BioSci*) A TISSUE CULTURE technique in which parts of organs can be kept alive and functional for a limited period for observation and experiment.

organdie (*Textiles*) Lightweight, cotton, plain-weave dress fabric, stiffened and made transparent by treatment with conc sulphuric acid. Cf MERCERIZATION.

organelle (*BioSci*) A defined structure within a cell, eg nucleus, mitochondrion, lysosome.

organ genus (*BioSci*) See FORM TAXON.

organic (*FoodSci*) Describes agricultural raw materials which have been grown without chemical fertilizers, pesticides or growth promoters. Organic foods have AGRICULTURAL INPUTS grown or processed from organic raw materials. For a food to be legally defined as organic in the EU, not less than 95% of the agricultural inputs must be from organic sources. If between 50% and 95% of the agricultural input is organic this can be declared, but only on the ingredient list.

organic brain disorders (*Psych*) See ORGANIC MENTAL DISORDERS.

organic chemistry (*Chem*) The study of the compounds of carbon. Owing to the ability of carbon atoms to combine together in long chains (catenate), these compounds are far more numerous than those of other elements. They are the basis of living matter.

organic disease (*Med*) See FUNCTIONAL DISEASE.

organic electrical conductor (*Phys*) An organic substance exhibiting electrical conductivity similar to that of metals. For instance, TTF–TCNQ (tetrathiofulvalene–tetracyano-quinodimethane) has a room temperature conductivity of $5 \times 10^4 \, S\,m^{-1}$ and its conductivity increases at low temperature, by a factor of at least 20 at 60 K. $(TMTSF)_2ClO_4$, where TMTSF is the tetramethyl-selenium derivative of TTF, is one of a series of organic conductors which show superconductivity below 1·2 K.

organic farming (*Agri*) A farming method that emphasizes environmental conservation and promotion of biodiversity to reduce habitats for pests and disease, without resort to synthetic pesticides. Soil fertility is maintained without synthetic fertilizers and is reliant on fallowing, nitrogen-fixing plants and organic composts and manures. This method is highly regulated with international certification and recognition.

organic fertilizer (*Agri*) A fertilizer derived from the decomposition or processing of plant or animal material, eg blood meal or fish emulsion.

organic mental disorders (*Med*) Behavioural or psycho-logical disorders that demonstrably arise from damage to brain tissue or to chemical imbalances in the nervous system.

organic phosphor (*Radiol*) Organic chemical used as solid or liquid scintillator in radiation detection.

organisms (*BioSci*) Animals, plants, fungi and micro-organisms.

Organization of Petroleum Exporting Countries (*MinExt*) An organization formed in 1960 with Iran, Iraq, Kuwait, Saudi Arabia and Venezuela as member states. It was joined by Qatar in 1961, Indonesia and Libya in 1962, United Arab Emirates in 1967, Algeria in 1969, Nigeria in 1971, and Ecuador and Gabon in 1973. It aims to co-ordinate the policies of these countries in relation to the production and distribution of oil, and to maximize their income from it. Abbrev *OPEC*.

organized (*BioSci*) Showing the characteristics of an organism; having the tissues and organs formed into a unified whole.

organo- (*Genrl*) Prefix from Lt *organum* denoting organ; (of a chemical compound) containing an organic radical.

organochlorines (*Chem*) Class of chemicals used in a wide range of applications; as insecticides (eg DDT), aerosol propellants, insulation fluids (PCBs), etc. Tend to be persistent in the environment and may accumulate in organisms at the upper end of the food chain with serious toxic effects.

organogenesis (*BioSci*) The process of the formation and development of organs. Also *organogeny*.

organography (*BioSci*) A descriptive study of the external form of plants, with relation to function.

organoleptic (*FoodSci*) Describing a food in terms of the senses, eg taste, aroma, texture (mouthfeel), appearance. Generally such tests are non-quantitative and are used mainly to compare raw materials or food products with given standards or bench marks.

organo-magnesium compounds (*Chem*) GRIGNARD REAGENTS, compounds of the type RMgX, where X is a halogen.

organo-metallic compounds (*Chem*) Wide range of compounds in which carbon atoms are linked directly with metal atoms, including the alkali metals. They are important reagents in polymerization reactions (see ZIEGLER–NATTA CATALYST) and many are formed by transition elements. The term is often used to include compounds without a direct metal–carbon bond although strictly these should probably be considered *metallo-organic compounds*. Examples include butyl lithium, C_4H_9Li; lead tetraethyl, $Pb(C_2H_5)_4$; zinc dimethyl, $Zn(CH_3)_2$; trimethyl gallium, $Ga(CH_3)_3$, the last used as a precursor in one of the growth processes for a GaAs (gallium arsenide) semiconductor. See EPITAXIAL GROWTH OF SEMICONDUCTORS.

organosilicone (*Chem*) Synthetic resin characterized by long thermal life and resistance to thermal ageing. See SILICONES.

organosol (*Chem, Plastics*) A coating composition based on a dispersed polyvinyl chloride resin mixed with a plasticizer and a diluent (see PLASTISOL), a colloidal solution in any organic liquid.

organza (*Textiles*) Plain-weave fabric made from untreated continuous filament silk, it is sheer and stiff (due to the silk gum); now also made from continuous 'manufactured' fibre yarns.

orgasm (*BioSci*) Culmination of sexual excitement. Adjs *orgasmic, orgastic*.

OR gate (*ICT*) GATE for which output is 0 only when all inputs are 0, otherwise output is 1. Also *OR element*.

oriel (*Arch*) A projecting window supported upon corbels or brackets.

Oriental alabaster (*Min*) See ONYX MARBLE.

Oriental almandine (*Min*) A name sometimes used for *corundum*, of gemstone quality, which is deep-red in colour, resembling true almandine (a garnet) in this, but no other, respect.

Oriental amethyst (*Min*) A misnomer for purple corundum or sapphire. Also *false amethyst*.

Oriental cat's eye (*Min*) See CYMOPHANE.

Oriental emerald (*Min*) A name sometimes used for *corundum*, of gemstone quality, resembling true emerald in colour.

Oriental region (*BioSci*) One of the primary faunal regions into which the land surface of the globe is divided. It includes the southern coast of Asia east of the Persian Gulf, the Indian subcontinent south of the Himalayas, southern China and Malaysia, and the islands of the Malay Archipelago north and west of Wallace's line.

Oriental ruby (*Min*) See RUBY.

Oriental sore (*Med*) A skin disease caused by the flagellate protozoan *Leishmania tropica*.

Oriental topaz (*Min*) A variety of *corundum*, resembling topaz in colour.

orientating reflex (*Psych*) First described by Pavlov, an animal's response to the sudden presentation of a novel stimulus. Includes turning the body and head so that the animal's attention can be focused on the source of stimulation (the 'what is it?' reaction).

orientation (*BioSci*) The position, or change of position, of a part or organ with relation to the whole; change of position of an organism under stimulus.

orientation (*Chem*) (1) The determination of the position of substituent atoms and groups in an organic molecule, esp in a benzene nucleus. (2) The ordering of molecules, particles or crystals so that they point in a definite direction.

orientation (*Eng*) The position of important sets of planes in a crystal in relation to any fixed system of planes. See PURE METAL CRYSTALS.

orientation (*Plastics*) (1) The way in which polymer chains are aligned in a moulded product. (2) The direction of fibres in COMPOSITE MATERIALS. Control of orientation is critical to optimize product strength and stiffness.

orientation (*Surv*) Fixing of line or plan in azimuth, with respect to true north.

orientation (*Textiles*) The direction of polymer molecules or crystallites in fibres relative to the fibre axis imposed either during growth (natural fibres) or while being extruded and drawn (manufactured and synthetic fibres).

orientation behaviour (*Psych*) The positioning of the body or of a behavioural sequence, with respect to some aspect of the external environment; it includes simple postural preferences as well as complex navigational behaviours.

oriented polypropylene film (*Plastics*) Film which has been biaxially stretched to improve its physical properties. Widely used for packaging because of its high gloss and clarity, high impact strength and low moisture permeability. Abbrev *OPP film*.

orifice gauge (*Eng*) A flow gauge consisting of a thin orifice plate clamped between pipe flanges, with pressure take-offs drilled into the adjacent pipes, or of a thick orifice plate similarly clamped but containing its own pressure take-offs.

orifice meter (*ChemEng*) A device for measuring the pressure drop across a constriction, or orifice, in a pipe to give the flow rate of a gas or liquid.

origin (*BioSci*) That end of a skeletal muscle attached to a portion of the skeleton which remains, or is held, rigid when the muscle contracts. Cf INSERTION.

origin (*MathSci*) A fixed point with respect to which the position of points, lines, etc, is located; the point of intersection of co-ordinates of axes.

original equipment manufacturer (*ICT*) A firm that makes basic computer hardware for other manufacturers to build into their products (eg microprocessors supplied to a washing machine manufacturer to use as control devices). Abbrev *OEM*.

orimulsion (*ChemEng*) An emulsion of bitumen, water and detergents, used as a fuel and originally obtained from Orinoco in Venezuela.

O-ring (*Eng*) A toroidal ring, usually of circular cross-section, made of neoprene or similar materials, used eg as an oil or air seal.

Orion (*Astron*) An equatorial constellation, a dominant sight in the winter sky of the northern hemisphere; includes the bright stars BETELGEUSE and RIGEL.

Orionids (*Astron*) A major meteor shower which shows maximum activity on 21 October with a rate of around 30 per hour.

Orion Nebula (*Astron*) A bright EMISSION NEBULA in the constellation Orion, which results from the illumination of a giant molecular cloud by the four hot stars which form the Trapezium. Lies in a region of active star formation. Distance approx 500 pc.

orlistat (*Pharmacol*) An orally administered drug that reduces the absorption of dietary fat by inhibiting the action of enzymes in the digestive system, used to treat obesity.

Orlon (*Chem*) TN for synthetic fibre based almost wholly on POLYACRYLONITRILE (PAN) with a small amount of a different monomer to serve as a dye receptor. Widely used in knitted fabrics, as imitation fur and in carpets.

orlop deck (*Ships*) The lowermost deck in a ship of several decks. It is simply a platform, and contributes nothing to main longitudinal strength; usually of small extent.

ormolu (*Eng*) An alloy of copper, zinc and sometimes tin, used (esp in the 18th century) for furniture mountings, decorated clocks, etc.

ornis (*BioSci*) A bird fauna. Adj *ornithic*.

ornithine (*Chem*) *2-6-diaminovaleric acid*. It is concerned in urea formation in animals (see ARGININE), and a derivative, *ornithuric acid*, is found in the excrement of birds.

ornithology (*BioSci*) The study of birds.

ornithophily (*BioSci*) Pollination by birds.

ornithopter (*Aero*) Any flying machine that derives its principal support in flight from the air reactions caused by flapping motions of the wings, this motion having been imparted to the wings from the source of power being carried.

ornithosis (*Med, Vet*) See PSITTACOSIS.

oro- (*Genrl*) Prefix from (1) Lt *os*, gen *oris*, mouth or (2) Gk *oros*, mountain.

oroanal (*BioSci*) Connecting, pertaining to, or serving as, mouth and anus.

orogenesis (*Geol*) See OROGENY.

orogenic belt (*Geol*) A region of the Earth's crust, usually elongated, which has been subjected to an *orogeny*. Recently formed orogenic belts correspond to mountain ranges, but older belts have often been eroded flat.

orogeny (*Geol*) The tectonic process whereby large areas are folded, faulted, metamorphosed and subject to igneous activity. Different periods of orogeny are given specific names, eg *Alpine, Caledonian, Laramide*.

orographic ascent (*EnvSci*) The upward displacement of air blowing over a mountain.

orographic rain (*EnvSci*) Rain caused by moisture-laden winds impinging on the rising slopes of hills and mountains. Precipitation is caused by the cooling of the moist air consequent upon its being forced upwards.

orography (*Genrl*) The description of mountains. Also *oreography*.

oroide (*Eng*) See FRENCH GOLD.

oronasal (*BioSci*) Pertaining to or connecting the mouth and the nose.

orphan (*Print*) The first line of a paragraph at the foot of a page. Should be avoided. See WIDOW.

orphan drug (*Pharmacol*) A drug for an uneconomically small patient population that is given special status and financial support to enable it to be produced.

orpiment (*Min*) Arsenic trisulphide, which crystallizes in the monoclinic system; commonly associated with realgar; golden-yellow in colour and used as a pigment.

orrery (*Astron*) A mechanical model of the solar system showing the relative motions of the planets by means of clockwork; much in vogue in the 18th century. Named after Charles Boyle, Earl of Orrery.

ORS (*Geol*) Abbrev for OLD RED SANDSTONE.

orthite (*Min*) See ALLANITE.

ortho- (*Chem*) (1) Derived from an acid anhydride by combination with the largest possible number of water molecules, eg orthophosphoric acid. (2) Consisting of diatomic molecules with parallel nuclear spins and an odd rotational quantum number, eg orthohydrogen. (3) The 1,2 relationship of substituents on a benzene molecule. Cf META (4), PARA (2).

ortho- (*Genrl*) Prefix from Gk *orthos*, straight.

orthocentre (*MathSci*) Of a triangle: the point of intersection of the perpendiculars from the vertices to their opposite sides.

orthoclase (*Min*) Silicate of potassium and aluminium, $KAISi_3O_8$, crystallizing in the monoclinic system; a feldspar, occurring as an essential constituent in granitic and syenitic rocks, and as an accessory in many other rock types. See MICROCLINE, SANIDINE and panel on TWINNED CRYSTALS.

orthodiagraph (*Radiol*) An X-ray apparatus for recording exactly the size and form of organs and structures inside the body.

orthodromic (*BioSci*) Characteristic of nerve fibres that conduct, or are able to conduct, impulses in the normal direction as opposed to the *antidromic* direction.

orthoferrosilite (*Min*) The ferrous iron end-member of the orthopyroxene group of silicates.

orthognathous (*BioSci*) With the long axis of the head at right angles to that of the body, and the mouth directed downwards. Cf PROGNATHOUS.

orthogneiss (*Geol*) A term applied to gneissose rocks which have been derived from rocks of igneous origin. Cf PARAGNEISS.

orthogonal cutting (*Eng*) A cutting process used to analyse the forces occurring in machining, in which a straight-edged cutting tool moves relatively to the workpiece in a direction perpendicular to its cutting edge.

orthogonal functions (*MathSci*) A set of functions $f_r(x)$ such that, over the range considered, $\int f_s(x)f_s(x)dx = 0$, except when $r = s$. If the integral equals unity when $r = s$ the functions are also said to be *normal* or *orthonormal*. If also the equation $\int F(x)f_r(x)dx = 0$ for all r implies that $F(x)$ is identically zero, then the set is said to be *complete*.

orthogonal matrix (*MathSci*) A MATRIX that is equal to the inverse of its transpose. Any two rows, or any two columns, will be orthogonal vectors.

orthogonal vectors (*MathSci*) Two vectors whose scalar product is zero.

orthograde transport (*BioSci*) Axonal transport from the cell body of the neuron towards the synaptic terminal. The opposite of RETROGRADE TRANSPORT and dependent on a different mechano-chemical protein interacting with microtubules.

orthograph (*Arch*) A view showing an elevation of a building or of part of a building.

orthographic projection (*Eng*) A method of representing solid objects in two dimensions by viewing on three mutually perpendicular plane surfaces, using parallel rays or projectors perpendicular to the surfaces; commonly used as the basis of engineering drawing.

orthohydrogen (*Chem*) Hydrogen molecule in which the two nuclear spins are parallel, forming a triplet state.

orthokeratology (*Med*) A technique for improving the vision of people affected by myopia and astigmatism, in which the cornea is temporarily reshaped by a special contact lens that is then removed and reused as necessary.

orthokinesis (*BioSci*) Kinesis in which the speed or frequency of movement is increased (*positive orthokinesis*) or decreased.

orthologous genes (*BioSci*) Genes related by common phylogenetic descent, in contrast to paralogous genes.

orthomorphic (*Genrl*) (of a map). Conformal, representing small areas in their true shape.

Orthomyxoviridae (*BioSci*) Single-stranded RNA viruses. The RNA acts as mRNA in the host cell. Mature progeny leave cells by budding out of the host-cell plasma membrane and are thus enveloped. Influenza viruses are the major members of this group.

orthonormal (*MathSci*) See ORTHOGONAL FUNCTIONS.

orthopaedics (*Med*) That branch of surgery which deals with deformities arising from injury or disease of bones or of joints. Adj *orthopaedic*. US *orthopedics*.

orthophosphoric acid (*Chem*) H_3PO_4. Formed when phosphorus pentoxide is dissolved in water and the solution is boiled. The highest hydrated stable form of phosphoric acid.

orthophyric (*Geol*) A textural term applied to medium- and fine-grained syenitic rocks consisting of closely packed orthoclase crystals of stouter build than in the typical

trachytic texture. The term actually implies the presence of porphyritic orthoclase crystals.

orthopnoea (*Med*) DYSPNOEA so severe that the patient is unable to lie down; a symptom of heart failure. Adj *orthopnoeic*. US *orthopnea*.

Orthopoxviridae (*BioSci*) A genus of double-stranded DNA viruses that preferentially infect epithelial cells. Includes variola (smallpox) and vaccinia.

Orthoptera (*BioSci*) An order of the Insecta, comprising large insects with biting mouthparts and posterior legs often with enlarged femora for jumping. The fore-wings are toughened (tegmina) and overlap when folded. Such insects have unjointed cerci and a well-developed ovipositor, and possess a variety of stridulatory organs. Grass-hoppers, locusts, crickets, cockroaches.

orthoptic circle (*MathSci*) For a conic, the locus of a point from which tangents to the conic are perpendicular to each other. Except for the parabola, this locus is a circle concentric with the conic. For the parabola, the locus is its directrix. Also *director circle*.

orthoptic treatment (*Med*) The non-operative treatment of squint by specially devised stereoscopic exercises.

orthopyroxene (*Min*) A group of pyroxene minerals crystallizing in the orthorhombic system, eg enstatite, hypersthene.

orthoquartzite (*Geol*) A pure quartz sandstone.

orthorhombic system (*Crystal*) The style of crystal architecture which is characterized by three crystal axes, at right angles to each other and all of different lengths. It includes such minerals as olivine, topaz and barytes. See fig. at BRAVAIS LATTICES.

orthosilicic acid (*Chem*) $Si(OH)_4$.

orthostatic (*Med*) Associated with or caused by the erect posture, eg *orthostatic albuminuria*.

orthostyle (*Arch*) A colonnade formed of columns arranged in a straight line.

orthotropism (*BioSci*) A TROPISM in which a plant part becomes aligned directly towards (positive-) or away from (negative-) the source of the orientating stimulus, eg most seedling shoots which are negatively *orthogravitropic* and positively *orthophototropic*. Cf DIATROPISM, PLAGIO-TROPISM.

orthotropous (*BioSci*) An ovule that is straight and on a straight stalk, so that the micropyle points away from the stalk. Also *atropous*.

OS (*Build*) Abbrev for *one side*.

OS (*Surv*) Abbrev for ORDNANCE SURVEY.

Os (*Chem*) Symbol for OSMIUM.

os (*BioSci*) (1) An opening, as the *os uteri* (Lt *os*, gen *oris*, mouth). Pl *ora*. (2) A bone, as the *os coccygis* (Lt *os*, gen *ossis*, bone). Pl *ossa*.

os (*Geol*) See ESKER.

OS/2 (*ICT*) An OPERATING SYSTEM developed for PERSONAL COMPUTERS by IBM as a successor to PC-DOS.

OS & W (*Build*) Abbrev for *oak, sunk and weathered*.

osazones (*Chem*) The diphenylhydrazones of monosaccharides, obtained by the action of two molecules of phenylhydrazine on one molecule of the monosaccharide. They are sparingly soluble in water, can be purified by recrystallization, and serve to identify the respective monosaccharides.

oscillating capacitor (*ElecEng*) See VIBRATING CAPACITOR.

oscillating die press (*Eng*) That in which the stock moves continuously and the high-speed punch and die set moves with it for a sequence of presses before returning to its starting position for the next cycle.

oscillating disk rheometer (*Eng*) A device for measuring flow rate, such as the Monsanto rheometer, which measures torque needed to shear rubber, typically during a CURE CYCLE. Abbrev ODR.

oscillating neutral (*ElecEng*) A phenomenon occurring in three-phase, unearthed, star-connected systems, due to third harmonic voltages, which results in a distorted phase voltage waveform.

oscillating sequence (*MathSci*) Also *oscillating series*. See DIVERGENT sequence, divergent series.

oscillation (*Aero*) See LONGITUDINAL OSCILLATION, PHUGOID OSCILLATION.

oscillation (*Eng*) See HUNTING.

oscillation (*ICT*) Sustained and very stable periodic alteration of current in a tuned circuit or other resonant structure or cavity. Maintained by the supply of synchronous pulses of energy lost through dissipation and output.

oscillation (*Phys*) Any motion that repeats itself, eg a particle in periodic motion moving back and forth over the same path. If the oscillations are not precisely repeated owing to frictional forces which dissipate the energy of motion, the oscillations are said to be *damped*. See CENTRE OF OSCILLATION.

oscillation frequency (*ICT*) That determined by the balance between the inertia reactance and the elastic reactance of a system, eg open- or short-circuited transmission line, cavity, resonant circuit, quartz crystal. If C = capacitance and L = self-inductance of circuit, then frequency $f = 1/[2\pi\sqrt{(LC)}]$. In a mechanical oscillating system $f = 2\pi\sqrt{(M/S)}$, where M = mass, and S = restoring force per unit displacement.

oscillator (*ICT*) A source of ac of any frequency that is sustained in a circuit by a valve or transistor using positive-feedback principle or by a negative-resistance device. There are two types: (1) stable-type, in which frequency is determined by a line or a tuned (LC) circuit, the waveform being substantially sinusoidal; and (2) relaxation-type, in which frequency is determined by resistors and capacitors, the waveform having considerable content of harmonics. Also applied to mechanical systems, velocities being equivalent to currents.

oscillator crystal (*ICT*) A piezoelectric crystal used in an oscillator to control the frequency of oscillation.

oscillator drift (*ICT*) See FREQUENCY DRIFT.

oscillatory discharge (*ICT*) That of a capacitor through an inductor when the resistance of circuit is sufficiently low and current persists after the capacitor has completely discharged, so that it charges again in the reverse direction. This process is repeated until all initial energy is dissipated in resistance, including radiation.

oscillatory scanning (*ImageTech*) That in which the scanning spot moves repeatedly to and fro across the image, so that successive lines are scanned in opposite directions.

oscillatory zoning (*Min*) The compositional variation within a crystal which consists of alternating layers rich in the two end-members of an isomorphous solid-solution series.

oscillogram (*Electronics*) A record of a waveform obtained from any oscillograph. Usually a photograph of cathode-ray-tube display.

oscillograph (*Electronics*) An oscilloscope with a photographic recording system to register waveforms displayed.

oscilloscope (*Electronics*) (1) Equipment incorporating a cathode-ray tube, time-base generators, triggers, etc, for the display of a wide range of waveforms by electron beam. (2) Formerly, mechanical or optical equipment with a corresponding function, eg Duddell oscilloscope.

osculating circle (*MathSci*) See CURVATURE.

osculating orbit (*Astron*) The instantaneous ellipse whose elements represent the actual position and velocity of a comet or planet at a given instant (*epoch of osculation*).

osculating plane (*MathSci*) See MOVING TRIHEDRAL.

osculating sphere (*MathSci*) At a point P on a space curve, a sphere that has four-point contact with the curve at P, ie the limiting sphere through four neighbouring points P, Q, R and S, on the curve, as Q, R and S tend to P. Its centre and radius are called respectively the *centre* and *radius* of *spherical curvature*.

osculation (*MathSci*) See POINT OF OSCULATION.

osculum (*BioSci*) In Porifera, an exhalant aperture by which water escapes from the canal system. Adjs *oscular*, *osculiferous*.

oseltamivir (*Pharmacol*) An antiviral drug used to treat certain types of influenza. TN *Tamiflu*.

Osgood–Schlatter disease (*Med*) Osteochondritis of the tibial tubercle.

O-shell (*Phys*) The ELECTRON SHELL in an atom corresponding to a principal quantum number of five. In naturally occurring elements the shell is never completely filled but has most electrons, 21, for the element uranium. See panel on ATOMIC STRUCTURE.

OSI (*ICT*) Abbrev for OPEN SYSTEMS INTERCONNECTION.

Osler's nodes (*Med*) Painful papules on the digits in cases of bacterial endocarditis.

osmeterium (*BioSci*) In the larvae of certain Papilionidae (Lepidoptera), a bifurcate sac, exhaling a disagreeable odour, which can be protruded through a slit-like aperture in the first thoracic segment.

osmic acid (*Chem*) An erroneous name for OSMIUM TETROXIDE.

osmiophilic (*Chem*) Having an affinity for, staining readily with osmic acid, eg certain components and organelles of cells.

osmiridium (*Eng*) A very hard, white, naturally occurring alloy of osmium (17–48%) and iridium (49%) containing smaller amounts of platinum, ruthenium and rhodium. Used for the tips of pen-nibs.

osmium (*Chem*) Symbol Os, at no 76, ram 190·2, mp 2700°C. A metallic element, a member of the platinum group. Osmium is the densest element, rel.d. (at 20°C) 22·48. Like platinum it is a powerful catalyst for gas reactions, and is soluble in aqua regia, but unlike platinum when heated in air it gives an oxide, volatile OsO_4. Alloyed with iridium it forms an extremely hard material, eg OSMIRIDIUM.

osmium tetroxide (*Chem*) *Osmium* (VIII) *oxide*, OsO_4, yellow crystals which give off an ill-smelling, poisonous vapour. Its aqueous solution is used as a histological stain for fat, a catalyst in organic reactions, a fixative in electron microscopy and a stain for double-bonded polymers in blends and copolymers.

osmole (*BioSci*) The amount of a solute that when dissolved in water gives a solution of the same osmotic pressure as that expected from 1 mole of an ideal non-ionized solute. The total osmotic concentration or *osmolarity* of complex solutions is usually estimated by measuring the vapour pressure or the freezing-point depression of the solution. Ordinary sea water is approximately 1200 milliosmolar (1200 osmol m^{-3}), mammalian isotonic saline is about 290 milliosmolar. *Osmolality* (abbrev *Osm*) expresses equivalent figures per kilogram of solvent.

osmometer (*Chem*) An apparatus for the measurement of osmotic pressures.

osmometry (*Chem*) A method for determining molecular mass of molecules, esp soluble polymers. Measurement of osmotic pressure across a semipermeable membrane allows calculation of the number-average molecular mass, M_n. It is an absolute method for molecular mass, M, but relatively slow compared with GEL-PERMEATION CHROMATOGRAPHY (GPC) which also gives the whole distribution. See MOLECULAR MASS DISTRIBUTION.

osmoreceptors (*BioSci*) Cells specialized to react to osmotic changes in their environment, eg cells that react to osmotic changes in the blood or tissue fluid and that are involved in the regulation of secretion of anti-diuretic hormone by the neurohypophysis.

osmoregulation (*BioSci*) The process by which animals regulate the amount of water in their bodies, and the concentration of various solutes and ions in their body fluids.

osmosis (*Chem*) Diffusion of a solvent through a semipermeable membrane into a more concentrated solution, tending to equalize the concentrations on both sides of the membrane.

osmotic coefficient (*Chem*) The quotient of the van't Hoff factor and the number of ions produced by the dissociation of one molecule of the electrolyte.

osmotic potential (*BioSci*) π_p. That component of the WATER POTENTIAL due to the presence of solutes; equal to minus the osmotic pressure. Also *solute potential*, π_s.

osmotic pressure (*Chem*) The pressure which must be applied to a solution, separated by a SEMIPERMEABLE MEMBRANE from pure solvent, to prevent the passage of solvent through the membrane. For substances which do not dissociate, it is related to the concentration (*c*) of the solution and the absolute temperature (*T*) by the relationship $\pi = cRT$, where *R* is the gas constant. Symbol π.

osmotic shock (*BioSci*) A method of lysing cells or organelles by placing them in hypo-osmotic fluid so that solvent enters the membrane-bound structure by osmosis and causes swelling and rupture of the membrane.

osmotrophy (*BioSci*) Nutrition based on the uptake of soluble materials.

os penis (*BioSci*) A bone developed in the middle line of the penis in some mammals, such as bats, whales, some rodents, carnivores and primates.

osphradium (*BioSci*) A sense organ of certain aquatic Mollusca, consisting usually of a patch of columnar ciliated epithelium and concerned in the assessment of suspended silt in the water entering the mantle chamber. Adj *osphradial*.

ossa (*BioSci*) See OS (2).

osseous (*BioSci*) Bony; resembling bone.

ossicle (*BioSci*) A small bone; in Echinodermata, one of the skeletal plates; in Crustacea, one of the calcified toothed plates of the gastric mill.

ossification (*BioSci*) The formation of bone; transformation of cartilage or mesenchymatous tissue into bone. V *ossify*.

oste-, osteo- (*Genrl*) Prefixes from Gk *osteon*, bone.

ostearthritis (*Med*) See OSTEOARTHRITIS.

Osteichthyes (*BioSci*) A class of fishes characterized by possession of a bony skeleton and a swim bladder. It is by far the largest class, with representatives in marine, estuarine and fresh-water habitats from the tropics to polar latitudes. See panel on VERTEBRATE EVOLUTION.

osteitis (*Med*) Inflammation of a bone.

osteitis deformans (*Med*) See PAGET'S DISEASE OF BONE.

osteitis fibrosa (*Med*) A condition in which there may be (1) a single cyst in a bone or (2) cysts in many bones; the latter (generalized osteitis fibrosa; von Rechlinghausen's disease of bone) is due to loss of calcium salts from the bone, and is associated with a tumour of the parathyroid glands.

osteo- (*Genrl*) See OSTE-.

osteoarthritis (*Med*) A group of conditions in which the cartilage of joints is gradually worn away, and the bone adjacent to it remodelled. Also *ostearthritis*.

osteoarthropathy (*Med*) Strictly, any disease affecting both bones and joints. Specifically, symmetrical enlargement of the bones of the hands and the feet with thickening of the fingers and toes associated esp with chronic diseases of the lungs or of the heart.

osteoblast (*BioSci*) A bone-forming cell.

osteochondritis (*Med*) Inflammation of both bone and cartilage. See PERTHE'S DISEASE.

osteochondritis dessicans (*Med*) A rare disease characterized by avascular necrosis of bone and cartilage which eventually resolves spontaneously.

osteochondroma (*Med*) A tumour composed of bony and of cartilaginous elements.

osteochondrosis (*Med*) A disease in which abnormal growth of cartilage or bone leads to degeneration of the cartilage, usually in the joints. Pl *osteochondroses*.

osteoclasis (*Med*) The absorption and destruction of bone tissue by osteoclasts.

osteoclast (*BioSci*) A bone-destroying cell, esp one that breaks down any preceding matrix, chondrified or calcified during bone formation.

osteocranium (*BioSci*) The bony brain-case that replaces the chondrocranium in higher vertebrates.

osteocyte (*BioSci*) A bone cell derived from an osteoblast.

osteodermis (*BioSci*) An ossified or partially ossified dermis: membrane bones formed by ossification of the dermis. Adj *osteodermal*.

osteodystrophia fibrosa (*Vet*) A disease of the skeletal system in animals in which excessive amounts of calcium and phosphorus are withdrawn from the bones and replaced by fibrous tissue. Caused by an excess of phosphorus or a deficiency of calcium in the diet in the horse (*bran disease, millers' disease, big head of horses*), calcium deficiency in the pig (*snuffles*), and as a complication of chronic nephritis in the dog (*rubber jaw*). Also *osteofibrosis*.

osteofibrosis (*Vet*) See OSTEODYSTROPHIA FIBROSA.

osteogenesis (*BioSci*) See OSSIFICATION.

osteogenesis imperfecta (*Med*) A condition in which a child is born with abnormally brittle bones, multiple fractures occurring. Also *fragilitas ossium*.

osteoid (*Med*) Resembling bone.

osteology (*BioSci*) The study of bones.

osteoma (*Med*) A tumour composed of bone.

osteomalacia (*Med, Vet*) A condition in which the bones soften as a result of absorption of calcium salts from them, usually due to a deficiency of vitamin D in the diet. Asian immigrants and the elderly are particularly at risk. If it occurs in childhood while the bones are still forming it causes rickets. In cattle one form is associated with a deficiency of phosphorus resulting in both calcium and phosphorus being withdrawn from the bone. Also *boglame, stiff sickness*.

osteomyelitis (*Med*) Inflammation of the bone marrow and of the bone.

osteopathy (*Med*) A method of healing, based on the hypothesis that abnormalities in the human framework (bones, muscles, ligaments, etc) ultimately cause damage by interfering with the blood and nerve supply to the body, thereby allowing other factors in ill-health to exert their influence unduly. These abnormalities are often the direct single cause of much suffering and they can be removed by skilled manual adjustment.

osteopetrosis (*Med*) Congenital osteosclerotic anaemia; marble bones. A rare condition in which the bones become solid as a result of obliteration of the bone marrow by bone, associated with enlargement of the liver and of the spleen, and with anaemia. Also *Albers–Schönberg disease*.

osteopetrosis gallinarum (*Vet*) A chronic virus infection of chickens in which there is an excessive simulation of bone formation, leading to thickening and deformation of the bones. Also *marble bones, thick leg disease*.

osteophagia (*Vet*) An appetite for bones and dead animals, exhibited by herbivorous animals suffering from a deficiency of phosphorus and calcium salts in the diet.

osteophyte (*Med*) A bony excrescence or outgrowth from the margin of osteoarthritic joints or from diseased bone.

osteoporosis (*Med*) Decrease in bone density and mass, often occurring in old age.

osteosarcoma (*Med*) A malignant tumour derived from osteoblasts, composed of bone and sarcoma cells.

osteosclereid (*BioSci*) Sclereid having a columnar middle and enlarged ends like a stylized thigh bone.

osteosclerosis (*Med*) Abnormal thickening of bone.

osteotomy (*Med*) Cutting of a bone.

ostiolate (*BioSci*) Having an opening.

ostiole (*BioSci*) A pore, esp one by which spores or gametes escape.

ostium (*BioSci*) (1) In general, a mouth-like aperture. (2) In Porifera, an inhalant opening on the surface. (3) In Arthropoda, an aperture in the wall of the heart by which blood enters the heart from the pericardial cavity. (4) In mammals, the internal aperture of a Fallopian tube. Adj *ostiate*.

ostracod (*BioSci*) Small arthropods (0·4–1·5 mm long) that belong to the subclass Ostracoda and that have a bivalve shell. They range from the Lower Cambrian to the present day and are used for zoning in the Jurassic. The name should not be confused with OSTRACODERMS (fossil fishes). Also *ostracode*.

Ostracoda (*BioSci*) A subclass of the Crustacea, with or without compound eyes; having a bivalve shell with adductor muscle; cephalic appendages well-developed and complex; not more than two pairs of trunk limbs, often parthenogenetic, eg *Cypris*.

ostracode (*BioSci*) See OSTRACOD.

ostracoderms (*BioSci*) Fossil agnathan fishes. See panel on VERTEBRATE EVOLUTION.

Ostwald colour atlas (*ImageTech*) A system of colour relations arranged according to hue, luminosity and saturation.

Ostwald's dilution law (*Chem*) The application of the law of mass action to the ionization of a weak electrolyte, yielding the expression

$$\frac{\alpha^2}{(1-\alpha)V} = K$$

where α is the degree of ionization, V the DILUTION (2), and K the ionization constant, for the case in which two ions are formed.

Ostwald's theory of indicators (*Chem*) The assumption that all INDICATORS (1), are either weak acids or weak bases, in which the colour of the ionized form differs markedly from that of the undissociated form.

Ostwald viscometer (*Chem*) A type of capillary viscometer used for measuring viscosity of dilute polymer solutions, and hence, by a series of steps, viscosity-average molecular mass.

ot-, oto- (*Genrl*) Prefixes from Gk *ous*, gen *otos*, ear.

otalgia (*Med*) Earache.

OTC drugs (*Pharmacol*) 'Over the counter' drugs for which a doctor's prescription is not required.

otic (*Med*) Pertaining to the ear or to the auditory capsule; one of the cartilage bones of the auditory capsule.

otitis (*Med*) Inflammation of the ear. Otitis externa, a term for various inflammatory conditions of the external ear. Otitis media, inflammation of the middle ear. Otitis interna, inflammation of the inner ear.

oto- (*Genrl*) See OT-.

otocyst (*BioSci*) (1) In many aquatic invertebrates, a sac lined by sensory hairlets, filled with fluid, and containing a calcareous or siliceous concretion (otolith) for sensing equilibrium. (2) In vertebrates, part of the internal ear that is similarly constructed (the UTRICLE).

otodectic mange (*Vet*) Mange affecting the external ear canal, particularly of dogs and cats, caused by *Otodectes cyanotis*.

otolith (*BioSci*) The calcareous concretion that occurs in an otocyst.

otology (*Med*) The branch of medicine dealing with the ear and its diseases. A practitioner is an *otologist*.

otorhinolaryngology (*Med*) The branch of medicine dealing with diseases of the ear, nose and throat.

otorrhoea (*Med*) A discharge, esp of pus, from the ear. US *otorrhea*.

otosclerosis (*Med*) The formation of spongy bone in the capsule of the labyrinth of the ear, associated with progressive deafness.

otoscope (*Med*) An instrument for viewing the external canal of the ear and the eardrum.

otter boards (*Ships*) Oblong boards, bound with iron, attached to the sides of a trawl net eccentrically to the towing warps; they keep the mouth of the net open. Often abbrev *otter*. As used in mine-sweeping, the otter is a heavy steel frame with horizontal vanes.

Otto cycle (*Autos*) The working cycle of a four-stroke engine: suction, compression, explosion at constant volume, expansion and exhaust, occupying two revolutions of the crankshaft.

otto de rose (*Chem*) See ATTAR OF ROSES.

ottoman (*Textiles*) A woven cloth with a flat, prominent rib in the weft direction; originally with a silk warp and a worsted weft. Used in tailoring.

ottrelite (*Min*) A manganese-bearing chloritoid mineral occurring in schists, a product of the metamorphism of certain argillaceous sedimentary rocks.

OTU (*BioSci*) Abbrev for OPERATIONAL TAXONOMIC UNITS.

ouabain (*BioSci*) A plant alkaloid that binds specifically to and inhibits SODIUM–POTASSIUM ATPASE. Also *strophanthin G*.

Ouchterlony test (*BioSci*) A precipitin test in which antigen and antibody are allowed to diffuse towards one another in a gel medium.

Oudin test (*BioSci*) A precipitin test in which antigen diffuses into antibody incorporated in a gel medium.

ounce (*Genrl*) The twelfth part of the (legally obsolete) pound troy, $\frac{1}{16}$ of a pound avoirdupois.

out-and-in bond (*Build*) The mode of laying ashlar quoins, so that they will be headers and stretchers alternately.

out-and-out (*Print*) Spacing out pages to their exact finished size with no allowance for trim between them.

outband (*Build*) A jamb stone laid as a stretcher and recessed to take a frame.

out-box (*ICT*) A file for storing electronic mail that has been or is to be sent to another computer.

outbreeding (*BioSci*) Sexual reproduction between unrelated individuals, thus increasing heterozygosity. Cf INBREEDING. See ALLOGAMY.

outcrop (*Geol*) An occurrence of a rock at the surface of the ground.

outcross (*BioSci*) A cross to a strain with a different genotype.

outer (*Print*) An imposition in *sheetwork* which contains the first and last pages of the section, as distinct from the *inner* imposition printed on the reverse of the outer, and always containing the second and second-last pages.

outer conductor (*ElecEng*) See EXTERNAL CONDUCTOR.

outer dead centre (*Eng*) The position of the crank of a reciprocating engine or pump when the piston is at the end of its outstroke, ie when the piston is nearest the crankshaft. Also *bottom dead centre*.

outer marker beacon (*Aero*) A marker beacon, associated with the ILS or with the STANDARD BEAM APPROACH SYSTEM, which defines the first predetermined point during a beam approach.

outer planet (*Astron*) Any of the planets in the solar system whose orbits lie outside the asteroid belt.

outer section (*Aero*) See MAIN PLANE.

outer string (*Build*) The STRING farthest from the wall.

outfall (*Build*) The discharge point of a sewer.

outfall sewer (*Build*) The main sewer carrying away sewage material from a town.

outgassing (*Chem*) Removal of occluded, absorbed or dissolved gas from a solid or liquid. For metals and alloys, done by heating *in vacuo*.

outgassing (*Electronics*) Removal of maximum amount of residual gas in a valve envelope by baking the whole valve before sealing.

outgassing (*Geol*) The release of juvenile gases from molten rocks, leading to the development of the Earth's atmosphere and oceans.

outgassing (*Space*) Spontaneous liberation of gas from a material in a space environment. To avoid contamination the material is left for some time before nearby instruments are used. Also *off-gassing*.

out-gate (*Eng*) See RISER.

outgoing feeder (*ElecEng*) A feeder along which power is supplied from a substation or generating station.

outgroup (*Psych*) See INGROUP/OUTGROUP.

outlier (*Geol*) A remnant of a younger rock which is surrounded by older strata.

outlier (*MathSci*) With reference to a particular statistical model, an observation with a probability so low that it may be disregarded for some purposes.

outline font (*ICT*) A FONT built into a POSTSCRIPT-compatible printer.

outline letters (*Print*) Display types in which the outline only of the letter is shown; sometimes issued as part of a type family, but may also be a specifically designed type style on its own.

outliner (*ICT*) A facility available in many WORD PROCESSORS and DESKTOP PUBLISHING programs allowing the structure of a document including page formats, chapters, headings and TYPEFACES to be defined before the main body of the text is written.

out of balance (*Eng*) Said of a rotating machine element which is imperfectly balanced, or of a mechanism or machine which contains such an element.

out-of-band signalling (*ICT*) Signalling in a telephone system by using either an additional speech channel or signals within the user channel but just outside the normal speech band of 300 Hz–3·4 kHz; a typical frequency is 3·825 kHz.

out of phase (*ElecEng, ICT*) See PHASE.

out of wind (*Build*) A term applied to a flat surface; a plane surface which is not twisted, timber free from warp or twist.

output (*ICT*) (1) Information leaving a device, data resulting from processing. (2) To give out information, to print or transfer to auxiliary storage the data resulting from processing. (3) Audio, electric or mechanical signal delivered by instrument or system to a load.

output capacitance (*Electronics*) (1) The capacitive component of the output impedance or transducer, amplifier, or other circuit or device. (2) The anode–cathode capacitance of a thermionic valve.

output characteristic (*Electronics*) See TRANSISTOR CHARACTERISTICS.

output coefficient (*ElecEng*) See SPECIFIC TORQUE COEFFICIENT.

output device (*ICT*) Peripheral that translates signals from the computer into a human-readable form or into a form suitable for reprocessing by the computer at a later stage.

output gap (*Electronics*) An interaction gap through which an output signal can be withdrawn from an electron beam.

output impedance (*ElecEng*) That presented by the device to the load and which determines REGULATION (voltage drop) of source when current is taken. In a linear source, the backward impedance when the emf is reduced to zero. Also *source impedance*. See THÉVENIN'S THEOREM.

output meter (*ElecEng*) That which measures output voltage of an oscillator, amplifier, etc. Calibrated in VOLTS, or POWER LEVEL in dB in relation to ZERO POWER LEVEL (1 milliwatt) when circuit is properly terminated. See VOLUME UNIT.

output noise (*Genrl*) See THERMAL NOISE.

output regulation (*ElecEng*) Of a power supply, the variation of voltage with load current.

output transformer (*ElecEng*) One which couples the last stage in an amplifier to the load, eg a loudspeaker or line.

output valve (*Electronics*) One designed for delivering power to a load, eg line or loudspeaker, voltage gain not being relevant. Final stage of any multivalve amplifier.

output winding (*ElecEng*) That from which power is withdrawn in a transformer, transductor or magnetic amplifier.

outrigger (*Build*) (1) A projecting beam carrying a suspended scaffold. (2) Timbers built across a gable end to hold a rafter for a projecting verge.

outset (*Print*) A section placed on the outside of the main section, and sometimes called a 'wrap round'.

outside crank (*Eng*) An overhung or single-web crank attached to a crankshaft outside the main bearings.

outside cylinders (*Eng*) The steam cylinders carried outside the frame of a locomotive, working onto crank pins in the driving wheels.

outside gouge (*Build*) A firmer gouge having the bevel ground upon the convex side of the cutting edge.

outside lap (*Eng*) The amount by which the slide valve of a steam engine overlaps the edge of the steam ports when in mid-position. Also *steam lap*.

outside lining (*Build*) The external member of a cased frame.

outside loop (*Aero*) See INVERTED LOOP.

out-takes (*ImageTech*) TAKES of scenes photographed and printed for a production but not used in the finally edited version.

out-to-out (*Build*) A term applied to an overall measurement across a piece of framing.

outturn sheet (*Paper*) A representative sample sheet of a particular batch of paper.

outwash fan (*Geol*) A sheet of gravel and sand, lying beyond the margins of a sheet of till, deposited by meltwaters from an ice sheet or glacier.

ova (*BioSci*) Pl of OVUM.

ovalbumin (*Chem*) Same as EGG ALBUMEN.

oval pistons (*Autos*) (1) Pistons, originally round, worn oval through friction at the thrust faces. (2) Pistons purposely turned slightly oval, to compensate for the unequal diametral expansion.

ovals of Cassini (*MathSci*) Curves defined by the bipolar equation $rr' = k$. Each consists of either two ovals, one surrounding each reference point, or a single oval surrounding both reference points, according as $k < c$ or $k > c$ respectively, where c is the distance between the two reference points. If $k = c$, it reduces to the LEMNISCATE OF BERNOULLI.

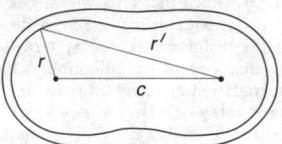

ovals of Cassini Drawn for two values of k.

oval window (*BioSci*) See FENESTRA OVALIS.

ovarian (*Med*) Pertaining to or connected with the ovary.

ovarian follicle (*BioSci*) In mammals the group of cells around the primary oöcyte proliferate and produce a surrounding non-cellular layer. A space opens up in the follicle cells and the whole structure is then the *Graafian follicle*.

ovariectomy (*Med*) Surgical removal of one or both ovaries.

ovariole (*BioSci*) In insects, one of the egg tubes of which the ovary is composed.

ovary (*BioSci*) (1) A female gonad; a reproductive gland producing ova. (2) The hollow structure, the basal part of a carpel or of a syncarpous gynoecium, that contains the ovules. (3) Also *pistil*. Adj *ovarian*.

ovate (*BioSci*) Egg-shaped with the broadest part nearer to the point of attachment.

oven-dry paper (*Paper*) See BONE-DRY PAPER.

oven-type furnace (*Eng*) Industrial heat-treatment furnace fired under the hearth, the live gases flowing directly into the heating chamber through live-gas flues disposed along each side of the hearth. Also *semimuffle-type furnace*.

overall echo loudness rating (*ICT*) A measure of physical echo suppression in a mobile-telephone system, the sum of two components, both expressed in DECIBELS: the weighted terminal coupling loss of the near-end terminal, representing the isolation between earpiece and microphone of the handset; and the overall loudness rating of the far-end terminal.

overall efficiency (*ElecEng*) The ratio of useful output to total input power.

overall luminous efficiency (*Phys*) The ratio of luminous flux of a lamp to total energy input. Not to be confused with LUMINOUS EFFICIENCY.

overall merit (*ICT*) A system of rating a radio channel (esp short wave) on a scale 0–5, derived from signal strength, fading, interference, modulation, depth, distortion.

overblowing (*Eng*) Continuing to force oxygen through molten steel in the BESSEMER PROCESS after the carbon has been removed, resulting in oxidizing of iron.

overburden (*Build*) The encallow or overlying stratum of soil. Generally applied to brickfields or to an area of gravel for quarrying.

overburden (*MinExt*) Earth or rock overlying the valuable deposit. In smelting, a furnace is *overburdened* when the ratio of ore to flux or fuel is too high.

overcasting (*Print*) The method of sewing used to make separate leaves into sections for binding. Also *whipping*, *whipstitching*.

overcloak (*Build*) That part of the overlapping edge of a sheet lead etc which extends over the ROLL to the flat surface beyond.

overclock (*ICT*) To modify a computer component so that it operates at a higher clock speed than was intended by the original manufacturer.

overcompensate (*Psych*) To go too far in trying to correct a fault that one (believes one) suffers from.

overcompounded generator (*ElecEng*) A compound-wound dc generator in which the series winding is so designed that the voltage rises as the load increases.

overcompounded motor (*ElecEng*) A compound-wound dc motor in which the series winding is so designed that the speed rises with an increase in load.

overcure (*Eng*) Rubber technology term for effects occurring towards end of vulcanization, which may involve chain degradation or increased cross-linking.

overcurrent relay (*ElecEng*) A relay which operates as soon as the current exceeds a certain predetermined value. Also *overload relay*.

overcurrent release (*ElecEng*) A device for tripping an electric circuit when the current exceeds a predetermined value. Also *overload release*.

overdispersion (*BioSci*) A non-random distribution of organisms, in which the organisms are aggregated (also contagious or heterogeneous distribution). The pattern is described empirically by a negative binomial frequency distribution, and the variance is greater than the mean. It is a common pattern for distribution of parasites within their host population. Cf RANDOM DISTRIBUTION, UNDER-DISPERSION.

overdispersion (*MathSci*) The increased variability in a set of data above that which may be expected under a particular model.

overdoor (*Build*) (1) An ornamental doorhead. (2) A pediment.

overdrive (*Autos*) A method of reducing engine rpm in relation to road speed, using a separate epicyclic gear unit. Now largely supplanted by an additional high-ratio gear within the main gearbox.

overdub (*ICT*) To add (new sound) to a recording. The new sound added to a recording.

overexposure (*ImageTech*) Excess of exposure of any sensitive surface, above that required for the proper gradation of light and shade.

overfeed fabric (*Textiles*) A warp-knitted fabric made with excess yarn being fed on one warp to form loops and underflaps that appear as a pile.

overfeed stoker (*Eng*) A MECHANICAL STOKER consisting of a hopper from which the fuel is continuously fed onto the bars of an inclined stepped grate, mechanically oscillated or rocked to cause the burning fuel to descend towards an ash table.

overflow (*ICT*) Occurs when arithmetic operations produce results which are too large to store.

overflow flag (*ICT*) A single bit that is set to 1 when OVERFLOW occurs during an arithmetic operation.

overfold (*Geol*) A fold with both limbs dipping in the same direction, but more steeply inclined than the other. Cf ISOCLINAL FOLD.

overfold (*Print*) A lip or overhang formed by the leading edge of the section or copy when the fold is out of centre. Over-adjustment of the folding mechanism will result in underfold.

overgassing (*Eng*) A condition occurring in gas-heated furnace or appliance when the burners are calibrated for, and operated at, a higher gas rate than that actually required.

overgrainer (*Build*) A long-haired brush used for graining in marbling. The standard overgrainer is hog hair. Special overgrainers such as the pencil or fantail types are available for imitating certain types of grain.

overgraining (*Build*) A coat of graining colour applied over grained work so as to produce shades across the work.

overgrowth (*Crystal*) See CRYSTALLINE OVERGROWTH.

overhand stopes (*MinExt*) Stopes in which severed ore from an inclined seam or lode gravitates downwards to tramming level. Also *overhead stopes*. Cf BACK STOPES, UNDERHAND STOPES.

overhang (*Aero*) (1) In multiplanes, the distance by which the tip of one of the planes projects beyond the tip of another. (2) In a wing structure, the distance from the outermost supporting point to the extremity of the wing tip.

overhang (*ElecEng*) See ARMATURE END CONNECTIONS.

overhaul period (*Aero*) See TIME BETWEEN OVERHAULS.

overhead camshaft (*Autos*) One running across the cylinder heads of an engine, operating the valves directly or through rockers. See CAMSHAFT, TIMING GEAR.

overhead-contact system (*ElecEng*) An electric traction system in which the current is collected from a contact wire suspended above the track by means of current collectors mounted on the roof of the vehicle. The term may also refer to the actual contact wire and its supporting structure.

overhead crossing (*ElecEng*) A device used on an OVER-HEAD-CONTACT SYSTEM of electric traction to allow the crossing of two contact wires and the passage of a current collector along either wire.

overhead expenses (*Build*) See ON-COSTS.

overhead projector (*ImageTech*) See PROJECTOR. Abbrev OHP.

overhead transmission line (*ElecEng*) A transmission line in which the conductors are supported above the earth.

overhead travelling-crane (*Eng*) A workshop crane consisting of a girder along which a wheeled crab can be traversed. The girder is mounted on wheels running on rails fixed along the length of the shop, near the roof. Also *shop traveller*.

overhead valves (*Autos*) In a vertical petrol or oil engine, inlet and exhaust valves working in the surface of the head opposite the piston, in either a vertical or inclined position.

overheated (*Eng*) Said of metal which has been heated in preparation for hot-working, or during a heat-treating operation, to a temperature at which rapid grain growth occurs and large grains are produced. The structure and properties can be restored by treatment, and in this respect it differs from burning (see BURNT METAL).

overlap (*ElecEng*) The period (often expressed in electrical degrees) which is required before commutation can be completed successfully in an electrical converter. Occurs because of source impedance in the ac supply to the converter.

overlap (*Geol*) The relationship between conformable strata laid down during an extension of the basin of sedimentation (eg on the margins of a slowly sinking land mass), so that each successive stratum extends beyond the boundaries of the one lying immediately beneath. Cf OFF-LAP and OVERSTEP.

overlapping covers (*Print*) The term is used only for pamphlet binding, where the cover may overlap as distinct from being CUT FLUSH.

overlapping genes (*BioSci*) Some small DNA viruses exploit the degeneracy of the genetic code by making different proteins from overlapping sequences of DNA. Achieved by displacing the READING FRAME by one or two bases.

overlap span (*ElecEng*) See SECTION GAP.

overlap test (*ElecEng*) A test used for locating a fault in a cable; the resistance between the cable and earth is measured, first with the far end of the cable earthed, and then with it free.

overlay (*ICT*) A section of computer code that is loaded into an area of memory that was previously allocated to another section of the same executing program. See SEGMENTATION.

overlay (*ImageTech*) The combination of two video images using information from one to KEY it into the other.

overlay (*Print*) (1) To adjust the impression surface of a letterpress machine in order to increase the pressure on dark tones and decrease it on light. (2) Translucent paper or transparent film covering original artwork to protect it from damage, or to enable instructions to the camera operator, plate-maker or printer to be shown. The overlay sheet may also indicate how the artwork should be broken down for multicolour printing.

overlay network (*ICT*) A communications network installed to increase the capacity of an existing network but having few or no connections with it.

overlearning (*Psych*) A learning procedure where training or practice on what is being learned continues beyond the point where learning can be said to be adequate (learning to criterion). Overlearning often results in improvements in efficiency and in changes in the organization of performance (eg from conscious to *automatic* control).

overload (*ElecEng*) Exceeding the level at which operation can continue satisfactorily for an indefinite period. This may lead to distortion or overheating and consequent damage although temporary overloads are often permissible. See OVERLOAD CAPACITY.

overload capacity (*ElecEng*) Excess capacity of a generator over that of its RATING, generally for a specified time.

overload protective system (*ElecEng*) A system of protecting an electric power network by means of overcurrent relays. To provide discrimination, the relays have time lags, graded so that the relays more remote from the supply point have shorter lags.

overload relay (*ElecEng*) See OVERCURRENT RELAY.

overload release (*ElecEng*) See OVERCURRENT RELEASE.

overlocking (*Textiles*) Using an overseaming machine to join two pieces of fabric using a double-chain stitch.

overmodulation (*ICT*) Attempted modulation to depth exceeding 100%, ie to such a degree that amplitude falls to zero for an appreciable fraction of the modulating cycle, with marked distortion.

overpoled copper (*Eng*) See POLING.

overpotential (*Chem*) The extra voltage which must be applied to an electrode to initiate the electrode reaction in an electrochemical cell, over and above the equilibrium electrode potential. Symbol η. See TAFEL PLOT.

overproof (*Chem*) See PROOF.

overreach (*Vet*) An error of gait in the horse, in which the toe of the hind-foot strikes the heel of the fore-foot.

overrigid (*Eng*) See REDUNDANT.

overrun (*FoodSci*) A measure of the density of an aerated product such as ice cream or whipped cream. Using a fixed-volume container, the percentage overrun is given by $100(w_u-w_a)/w_a$, where w_u and w_a are the weights of the unaerated and aerated products respectively: ie when a product has been whipped to half its original density, the overrun will be 100% (in the calculation, volume can be discounted because it is constant throughout).

overrun (*Print*) (1) To carry words from the end of one line of type to the beginning of the next, and so on until the matter fits. Insertions or deletions frequently necessitate overrunning. (2) Running an extra quantity of printed paper in excess of the order. The extra copies can be used for setting up for further printings or in the finishing department.

overs (*Print*) Extra sheets allowed to a job to provide for ordinary SPOILAGE during printing and in subsequent operations.

oversailing courses (*Build*) Brick or stone courses projecting from a wall for the sake of appearance only, as distinct from corbels, which are normally load-carrying.

oversaturated (*Geol*) Refers to an igneous rock in which excess silica crystallizes as a separate silica mineral or as a glass. See UNDERSATURATED.

overscanning (*ImageTech*) The deflection of an electron beam beyond the phosphor in a TV reproducer or a cathode-ray oscilloscope.

overshoot (*ICT*) For a step change in signal amplitude undershoot and overshoot are the maximum transient signal excursions outside the range from the initial to the final mean amplitude levels.

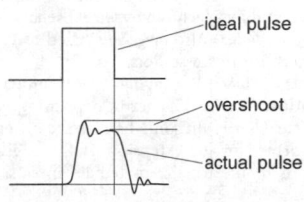

overshoot

overshoot (*ImageTech*) A transient excess change at the start or end of a pronounced variation of signal; in video pictures it can appear as fringes at the boundaries of large changes of image brightness.

overshot duct (*Print*) An ink duct using a blade mounted above the roller which rotates in a trough of ink.

overshot ink fountain (*Print*) See OVERSHOT DUCT.

overshot tool (*MinExt*) See FISHING TOOL.

overshot wheel (*Eng*) A water wheel in which the discharge flume or head race is at the top, the water flowing tangentially into the bucket near the top of the wheel.

oversite concrete (*Build*) A concrete layer covering a building site within the external walls, serving to keep out ground air and moisture, and also providing a foundation for the floor.

overspeed protection (*ElecEng*) Protection, usually by means of a centrifugally operated device, against excessive speed of an electric machine; used on inverted rotary converters and some dc motors.

overspun wire (*Acous*) A wire from a musical instrument, eg for a low note in a piano, round which a loading wire is tightly spun to lower its fundamental pitch.

oversteer (*Autos*) The tendency of a vehicle to exaggerate the degree of turn applied to the steering wheel. Cf UNDERSTEER.

overstep (*Geol*) The structural relationship between an unconformable stratum and the outcrops of the underlying rocks, across which the former transgresses. Cf OVERLAP.

overstrain (*Eng*) The result of stressing an elastic material beyond its YIELD POINT; a new and higher yield point results, but the elastic limit is reduced.

overstrike (*Print*) The act of printing one character on top of another.

overthrust (*Geol*) A fault of low HADE along which one slice or block of rock has been pushed bodily over another, during intense compressional earth movements. The horizontal displacement along the THRUST PLANE may amount to several kilometres.

overtone (*Acous*) In a complex tone, any of the components above the fundamental frequency.

overtone crystal (*Electronics*) A piezoelectric crystal operating at a higher frequency than the fundamental for any given mode of vibration.

overtype mode (*ICT*) A mode of operation in an EDITOR whereby characters typed in by the user at the CURSOR overwrite existing text. Cf INSERT MODE.

overvoltage (*Electronics*) A voltage higher than normal or predetermined limiting value, esp when likely to cause damage to or destroy electronic components or circuits.

overvoltage protective device (*ElecEng*) A device giving protection to electrical apparatus against the possibility of damage caused by a voltage above normal.

overvoltage release (*ElecEng*) A device arranged to trip an electrical circuit when the voltage exceeds a predetermined value.

overwrite mode (*ICT*) A method of STORING DATA whereby an updated version of a FILE overwrites the original.

overwriting (*ICT*) Erasing of a data item or file from memory or backing store by writing another in its place.

ovi-, ovo- (*Genrl*) Prefixes from Lt *ovum* denoting egg.

oviduct (*BioSci*) The tube that leads from the ovary to the exterior and by which the ova are discharged. Adj *oviducal*.

oviferous (*BioSci*) Also *ovigerous*. Used to carry eggs, as the *ovigerous legs* of sea spiders.

oviparous (*BioSci*) Egg-laying. Cf VIVIPAROUS.

oviposition (*BioSci*) The act of depositing eggs.

ovipositor (*BioSci*) In some fish (as the bitterling), a flexible tube formed by the extension of the edges of the genital aperture in the female; in female insects, the egg-laying organ.

ovisac (*BioSci*) A brood pouch; an egg receptacle.

ovo- (*Genrl*) See OVI-.

ovo-lacto (*Genrl*) A type of vegetarian whose diet excludes meat but permits eggs and milk products.

ovolo (*Arch*) A quarter-round convex moulding. See fig at MOULDINGS.

ovotestis (*BioSci*) A genital gland that produces both ova and spermatozoa, as in some Gastropoda.

ovoviviparous (*BioSci*) Producing eggs that hatch out within the uterus of the mother.

ovulation (*BioSci*) The formation of ova; in mammals, the release of the ovum from the ovary.

ovule (*BioSci*) The structure in a seed plant, consisting of embryo sac, nucellus and integuments, that after fertilization develops into the seed.

ovule culture (*BioSci*) The culture on a suitable medium of excised ovules, eg in an attempt, by adding pollen, to obtain *in vitro* fertilization and viable seed in crosses normally frustrated by the failure of the pollen to grow on the stigma or through the style.

ovum (*BioSci*) A non-motile female gamete. An egg or egg cell.

Owen bridge (*ElecEng*) An ac bridge of the four-arm (or Wheatstone) type, used for the measurement of inductance.

own ends (*Print*) See SELF END-PAPERS.

ox (*Agri*) A castrated male bovine, typically kept for draught use.

oxalates (*Chem*) Salts and esters of oxalic acid. Also *ethandioates*.

oxalic acid (*Chem*) HOOCCOOH. Mp 101°C, mp (anhydrous) 190°C. A dibasic acid which crystallizes with two molecules of water, it sublimes readily, occurs in many plants, and is obtainable by the oxidation of many organic substances. Strong oxidizing agents convert it to CO_2.

oxaluria (*Med*) The presence of crystals of oxalates in the urine.

oxalyl (*Chem*) The bivalent acid radical O=ĊĊ=O.

oxamic acid (*Chem*) $H_3NCOCOOH$. Mp 210°C (with decomposition). A crystalline powder.

oxamide (*Chem*) $H_2N.CO.CO.NH_2$, the normal amide of oxalic acid, a crystalline powder which sublimes when heated. Also *ethandiamide*.

oxbow lake (*Geol*) A meander loop which has been cut off.

ox-eye (*Arch*) An oval-shaped dormer window.

Oxford hollow (*Print*) A hollow back, in which the case is kept separated from the spine of the sections by a flattened tube which is glued to each, giving extra strength to the binding.

Oxfordian (*Geol*) A stage in the Upper Jurassic. See MESOZOIC.

Oxford shirting (*Textiles*) A plain-weave, warp-striped cotton shirting in which two ends weave as one.

oxidant (*Aero*) The oxygen-bearing component in a bipropellant rocket, usually liquid oxygen, high-test hydrogen peroxide, or nitric acid.

oxidase (*BioSci*) One of a group of enzymes occurring in plant and animal cells and promoting oxidation.

oxidates (*Geol*) Those sedimentary rocks and weathering products whose composition and geochemical behaviour are mainly determined by the oxidation process. This category includes mainly those minerals and rocks containing iron and manganese.

oxidation (*Chem*) The addition of oxygen to a compound. More generally, any reaction involving the loss of electrons from an atom. It is always accompanied by reduction.

oxidation (*Print*) Surface corrosion in the form of light spots which occur on lithographic plates which have been improperly protected by gumming up. The spots accept ink and usually print with a characteristic clustered formation.

oxidation number (*Chem*) For simple atoms or ions, the charge in units of the charge on the electron. For more complex groups, a formal oxidation number is often applied to specific atoms, particularly the central atom of a co-ordination compound. Thus, assuming that the ligands are chloride ions (o.n. = −1), the o.n. of copper in the complex ion $[CuCl_4]^{2-}$ may be deduced to be +2.

oxidation of polymers (*Chem*) The effect of atmospheric oxygen on polymers, particularly at elevated temperature and during exposure to ultraviolet light. Starts at weak points in chain (eg tertiary carbon atoms) or chain defects (eg head-to-head junctions), with formation of carbonyl groups. If further activated, they break the backbone chain and the molecular mass falls.

oxidation potential (*Chem*) A measure of the electron concentration of a system in internal equilibrium. Symbol Eh. Also *redox potential*.

oxidation–reduction indicators (*Chem*) Substances which exist in oxidized and reduced forms having different colours, used to give approximate values of oxidation–reduction potentials.

oxidation–reduction potential (*Chem*) See STANDARD ELECTRODE POTENTIAL.

oxidative decarboxylation (*BioSci*) A decarboxylation that is coupled with the oxidation of a substrate. The four oxidative decarboxylations of the TCA CYCLE generate the electrons whose passage along the electron transport chain produces ATP by OXIDATIVE PHOSPHORYLATION.

oxidative phosphorylation (*BioSci*) The process by which glucose is oxidized to carbon dioxide and water with the concomitant synthesis of ATP by aerobic respiration. See panel on MITOCHONDRION.

oxide-coated cathode (*Electronics*) One coated with oxides of the alkali and alkaline earth metals, to produce thermionic emission at relatively low temperatures.

oxide deposition (*Electronics*) In semiconductor processing, the deposition of silicon dioxide as a passivation layer, eg by CHEMICAL VAPOUR DEPOSITION or as a SPIN ON GLASS.

oxide-film arrester (*ElecEng*) A lightning arrester using lead peroxide (a good conductor) which changes to red lead (a good insulator) when heated by the passing of a current.

oxide fuel (*NucEng*) Fuel made from the oxides of fissile elements, which can operate at much higher temperatures than the metal.

oxide growth (*Electronics*) In silicon semiconductor processing the growth of silicon dioxide by oxidation of substrate silicon at high temperatures ($\approx 1000°C$) in an oxygen ('dry') or steam ('wet') environment.

oxide isolation (*Electronics*) The isolation of the circuit elements in a microelectronic circuit by forming a layer of silicon oxide around each element.

oxides (*Chem*) Compounds of oxygen with another element. Oxides are formed by the combination of oxygen with most other elements, particularly at elevated temperatures, with the exception of the noble gases and some of the noble metals. Many are used in ceramics, eg MgO, and glasses, eg SiO_2, B_2O_3.

oxidizer (*Chem*) A substance that removes electrons from another and produces heat; in the case of a rocket, a gas.

oxidizing agent (*Chem*) A substance which is capable of bringing about the chemical change known as oxidation in another substance. It is itself reduced.

oxidizing flame (*Chem*) Outer cone of a non-luminous gas flame, which contains an excess of air over fuel.

oxidizing roast (*Eng*) Heating of sulphide ores or concentrates to burn off part or all of the sulphur as dioxide.

oxidoreductase (*BioSci*) An oxidase that uses molecular oxygen as the electron acceptor.

oximes (*Chem*) Compounds obtained by the action of hydroxylamine on aldehydes or ketones, containing the bivalent oximino group=NOH attached to the carbon. The oximes of aldehydes are termed *aldoximes* and those obtained from ketones *ketoximes*. Collectively, hydroximino-alkanes.

oxine (*Chem*) 8-hydroxy-quinoline is used as a reagent in analysis of metals. When the H in the –OH group is substituted by a metal, insoluble compounds result, but their solubility varies according to temperature, concentration and other conditions, thus making it possible to use the differences for analysis.

oxo (*Chem*) The radical O = in organic compounds, eg ketones (R_2C=O).

oxonium salts (*Chem*) Derivatives of a hypothetical oxonium hydroxide, H_3OOH, a base with a trivalent oxygen atom. Such substances are readily produced from several heterocyclic compounds containing oxygen, eg dimethylpyrone, which forms salts with hydrochloric acid etc by direct addition to the oxygen atom.

oxyacetylene cutting (*Eng*) Cutting material with a flame resulting from the combustion of oxygen and acetylene. If the material is a metal which oxidizes readily and the temperature is high enough, the jet of oxygen alone can continue the cutting action. See FLAME CUTTING.

oxyacetylene welding (*Eng*) Welding with a flame resulting from the combustion of oxygen and acetylene.

oxybutinin (*Pharmacol*) An antimuscarinic drug used to treat urinary incontinence by acting on the musculature of the bladder wall.

oxycelluloses (*Chem*) Products formed by the action of oxidizing agents on cellulose. They dissolve in dilute alkaline solution and have strong reducing properties. When boiled with hydrochloric acid they yield furfuraldehyde quantitatively. This reaction serves for the analytical estimation of oxycelluloses.

oxychloride cement (*Build*) A strong, extremely hard-setting cement used in making composition floors, magnesite flooring. A chemical combination of magnesium oxide and chloride.

oxycodone (*Pharmacol*) An opioid analgesic derived from thebaine.

oxydactylous (*BioSci*) Having narrow-pointed digits.

oxygen (*Chem*) A non-metallic element, symbol O, at no 8, ram 15·9994, valency 2. Molecular oxygen is a diatomic, colourless, odourless gas which supports combustion and is essential for the respiration of most forms of life. Mp $-218·4°C$, bp $-183°C$, density $1·429\,04\,g\,dm^{-3}$ at stp, formula O_2. An unstable form is ozone, O_3. Oxygen is the most abundant element on Earth, forming 21% by volume of the atmosphere, 89% by weight of the water, and nearly 50% by weight of the rocks of the Earth's crust. It is manufactured from liquid air, and is used in gas welding (usually with acetylene), in steel manufacture, in medical practice and in anaesthesia; liquid oxygen is much used in rocket fuels. Ultraviolet radiation forms ozone, O_3, from atmospheric oxygen. See OZONE LAYER.

oxygen cracks (*Eng*) Surface effect of oxidation of polymers, usually forming a randomly oriented pattern. In rubbers, forms a crazy-paving pattern readily distinguishable from OZONE CRACKS.

oxygen demand (*EnvSci*) The oxygen required to satisfy the biological and chemical demands (*BOD* and *COD*) of contaminated water. The biological demand is from microbial flora involved in digesting organic constituents and the chemical demand is for the oxidation of compounds that are not biologically oxidized. Usually given as parts per million of oxygen taken up by a sample over a defined period in the dark.

oxygen dissociation curve (*BioSci*) The sigmoid curve describing the saturation of haemoglobin with oxygen when related to the partial pressure of oxygen.

oxygen electrode (*BioSci*) A sensitive method to detect oxygen consumption; involves a PTFE (Teflon) membrane.

oxygen index (*Chem*) Test for flammability of polymers, defined as the fraction of oxygen in a nitrogen mixture needed to maintain a candle-like flame on a burning specimen (held in a standard way). Polymers with an oxygen index (abbrev *OI*) greater than the normal fraction of oxygen in air are held to be flammable, those with an OI less than 0·21, inflammable. But see FLAMMABILITY. Also *limiting oxygen index* (LOI).

oxygen-isotope determinations (*Geol*) A method of using the ^{16}O and ^{18}O isotope ratio measurements from cores taken from Greenland and Antarctic ice sheets or from marine fossils in cores from the sea bottom. The results may be used to estimate (1) the temperature at which the original snow fell before turning to ice, (2) the sea temperature at the time of deposition of marine fossils and (3) the global ice volume, thus giving a chronology of the ice ages during the Pleistocene. See panel on RADIOMETRIC DATING.

oxygen lancing (*Eng*) A process like OXYACETYLENE CUTTING, used principally for cutting heavy sections of steel or cast-iron, in which oxygen is fed to the cutting zone through a length of steel tubing which is consumed as the cutting action proceeds. The cutting zone or the end of the tube has to be preheated to commence cutting.

oxygen scavenger (*Eng*) An additive to boiler feed-water which removes traces of oxygen and helps to prevent corrosion, eg hydrazine or tannin.

oxyhaemoglobin (*Chem*) The product obtained by the action of oxygen upon haemoglobin. The oxygen is readily given up when oxygen tension is low in surrounding medium.

oxyhornblende (*Min*) See BASALTIC HORNBLENDE.

oxyhydrogen welding (*Eng*) A method of welding in which the heat is produced by the combustion of oxygen and hydrogen.

oxyntic (*BioSci*) Acid-secreting.

oxyphile (*Geol*) Descriptive of elements which have an affinity for oxygen, and therefore occur in the oxide and silicate minerals of rocks rather than in the sulphide minerals or as native elements. Also *lithophile*.

oxyphobic (*BioSci*) Unable to withstand soil acidity.

oxyproline (*Chem*) *4-hydroxypyrrolidine-2-carboxylic acid*. Obtained by cleavage of gelatine.

oxytetracycline (*Pharmacol*) A broad-spectrum antibiotic used in a wide variety of infections.

oxytocin (*Med*) An octapeptide hormone secreted by the posterior lobe of the pituitary body (neurohypophysis) which stimulates the uterine muscle to contract, and causes milk ejection from lactating mammary glands.

ozocerite (*Min*) A mineral paraffin wax, of dark-yellow, brown or black colour. Also *ozokerite*.

ozoena (*Vet*) Catarrh of the frontal and maxillary sinuses of the horse.

ozokerite (*Min*) See OZOCERITE.

ozone (*Chem*) O_3. Produced by the action of ultraviolet radiation or electrical corona discharge on oxygen or air. It is a powerful oxidizing agent which absorbs harmful short-wave ultraviolet radiation in the atmosphere which would otherwise harm life on Earth.

ozone cracks (*Eng*) Deep, sharp cracks oriented perpendicular to strain axis, commonly found on external surfaces of rubber products, esp tyres made from natural rubber and butadiene polymers. Caused by chain scission at double bonds in polymers owing to atmospheric attack by ozone (O_3), esp found in air polluted from sunlight photolysis, and near electrical equipment. Inhibited by anti-ozonants. Cf OXYGEN CRACKS.

ozone depletion (*EnvSci*) Damage to the OZONE LAYER caused by compounds such as CHLOROFLUOROCARBONS (CFCs).

ozone hole (*EnvSci*) Hole in the OZONE LAYER, which allows ultraviolet rays into the Earth's atmosphere. See panel on ATMOSPHERIC POLLUTION.

ozone layer (*EnvSci*) That region of the stratosphere, between about 20 and 40 km above the Earth's surface, where ozone makes up a greater proportion of the air than at any other height. Although this proportion is only a few parts per million, it nevertheless exerts a vital influence by absorbing much of the ultraviolet radiation in sunlight and preventing it from reaching the Earth's surface where it has considerable biological effect; in addition, it changes the thermal structure of the middle atmosphere owing to the heating caused by such absorption and is also a contributor to the greenhouse effect. Ozone is formed, mainly in the tropical stratosphere, by complex photochemical reactions and is distributed over the whole world by large-scale eddy diffusion. It is removed predominantly by catalytic chain reactions involving molecules such as H, OH, NO, NO_2, Cl and Br which initiate thousands of ozone-destroying cycles before they are themselves removed; the supply of such active molecules has been much increased since 1950 by human-made pollutants including chlorofluorocarbons and this is affecting the subtle dynamic equilibrium of the ozone layer. See panels on ATMOSPHERIC POLLUTION, STRATOSPHERE AND MESOSPHERE and TROPOSPHERE.

ozonides (*Chem*) Explosive organic compounds formed by the addition of an ozone molecule to a double bond.

ozonizer (*Chem*) An apparatus in which oxygen is converted into ozone by being subjected to an electric brush discharge. Ozone was formerly thought to be beneficial in air-conditioning, but is now known to be toxic.

P

P (*Chem*) Symbol for PHOSPHORUS.

P (*Genrl*) Symbol for: (1) PETA-; (2) POISE.

P (*Phys*) Symbol for: (1) ELECTRIC POLARIZATION; (2) power; (3) pressure.

[P] (*Chem*) Symbol for parachor.

Π (*Chem*) Symbol for pressure, esp OSMOTIC PRESSURE.

Π (*ElecEng*) Symbol for PELTIER COEFFICIENT.

∏ (*Genrl*) Symbol for product.

Ψ (*Phys*) Symbol for: (1) ELECTRIC FLUX; (2) MAGNETIC FIELD STRENGTH.

P$_{fr}$ (*BioSci*) The physiologically active form of the plant pigment PHYTOCHROME, having a peak of absorption at $\lambda = 730$ nm ('far red') that converts it to P$_r$ (to which it also changes slowly in the dark). See P$_R$.

P$_i$ (*BioSci*) In biochemical equations inorganic phosphate, eg $H_2PO_4{}^-$.

P$_r$ (*BioSci*) The form of the plant pigment PHYTOCHROME having a peak of absorption at $\lambda \approx 660$ nm which converts it to P$_{fr}$.

p (*BioSci*) (1) Symbol for *pressure potential*. (2) See CHROMOSOME MAPPING.

p (*Electronics*) Symbol for the number density of positive carriers in semiconductors expressed as holes per cubic metre (or per cubic centimetre). Cf N.

p (*Genrl*) Symbol for: (1) PICO-; (2) PROTON.

p (*Phys*) Symbol for: (1) ELECTRIC DIPOLE MOMENT; (2) IMPULSE; (3) MOMENTUM; (4) pressure.

p- (*Chem*) Symbol for PARA-.

φ (*Chem*) Symbol for: (1) the phenyl radical C_6H_5; (2) *amphi-*, ie containing a condensed double aromatic nucleus substituted in the 2 and 6 positions.

φ (*Phys*) Symbol for: (1) heat flow rate; (2) LUMINOUS FLUX; (3) MAGNETIC FLUX; (4) PHASE DIFFERENCE; (5) WORK FUNCTION.

π (*Genrl*) The ratio of the circumference of a circle to its diameter. $\pi = 3{\cdot}141\,592\,653$ to ten places.

ψ (*Chem*) Symbol for PSEUDO-.

P2P (*ICT*) Abbrev for *peer-to-peer*, a system in which the workload of a network is evenly distributed among the workstations.

P450 (*BioSci*) A family of cytochromes responsible for detoxification of compounds in the liver. Extensive polymorphism in the *CYP* genes for these enzymes, of which there are approximately 50 in humans, accounts for much of the individual variation in drug sensitivity.

P680 (*BioSci*) A (chlorophyll a)–protein complex (absorption peak at $\lambda = 684$ nm) that acts as the light trap in PHOTOSYSTEM II.

P700 (*BioSci*) A (chlorophyll a)–protein complex (absorption peak at $\lambda = 700$ nm) that acts as the light trap in PHOTOSYSTEM I.

p53 (*BioSci*) A gene activated when DNA is damaged and, in response to other factors in the cell, either facilitates repair or causes the cell to die. A defective p53 gene is often found in cancer cells.

PA (*Acous*) Abbrev for PUBLIC-ADDRESS SYSTEM.

PA (*Chem*) Abbrev for POLYAMIDE, often followed by numbers, eg PA 6,6, which represent the number of carbon atoms in repeat unit(s).

Pa (*Chem*) Symbol for PROTACTINIUM.

Pa (*Genrl*) Abbrev for PASCAL.

PAA (*Chem*) Abbrev for POLYACRYLIC ACID.

PABX (*ICT*) Abbrev for PRIVATE AUTOMATIC BRANCH EXCHANGE.

pacemaker (*Med*) Electronic device implanted in to chest wall with conducting wire to the heart to regulate abnormal heart rhythms by means of electrical pulses. Used when the sino-atrial or atrio-ventricular nodes, which usually regulate heart rhythm, are deficient or diseased. Powered by lightweight long-life cells, eg lithium battery.

pacemaker region (*Med*) A region of the body which determines the activity of other parts of the body; eg the sino-atrial node from which originate the impulses causing the heart beat.

Pachuca tank (*MinExt*) Large vertically set cylindrical vessel used in chemical treatment of ores, in which pulp is reacted with suitable solvents for long periods, while the contents are agitated by compressed air. Also *Brown tank*.

pachyderma (*Med*) See PACHYDERMIA.

pachydermatocele (*Med*) A soft flabby tumour, composed of fibrous and nervous tissue, which hangs over the face or the ears.

pachydermatous (*BioSci*) Thick-skinned.

pachydermia (*Med*) Abnormal thickness of the skin. Also *pachyderma*.

pachymeningitis (*Med*) Inflammation of the DURA MATER.

pachyphyllous (*BioSci*) Having thick leaves.

pachytene (*BioSci*) The third stage of meiotic prophase, intervening between zygotene and diplotene, in which condensation of chromosomes commences. Also *bouquet stage*. See fig. at MEIOSIS.

Pacinian corpuscles (*BioSci*) In vertebrates, skin receptors in which the nerve ending is surrounded by many concentric layers of connective tissue. Sensitive to pressure.

pack (*MinExt*) Waste rock, mill tailings, etc, used to support excavated stopes. Also *fill*.

package (*Textiles*) Trade term for a cop, cheese, cone, etc, of yarn, indicating that it is in convenient form for transport, or further processing.

packaged (*ElecEng*) (1) Description of electrical components of circuits which are sealed from ambient conditions. Many forms are possible, including plastic enclosures for discrete transistors, ceramic and plastic enclosures for integrated circuits, resin potting for printed circuit boards and hermetically sealed metal boxes for complete systems. (2) Said of equipment complete for use, made up from a series of subassemblies.

packaged (*NucEng*) Said of a reactor of limited power which can be packaged and erected easily on a remote site.

package dyeing (*Textiles*) A method of dyeing yarn already on a package.

package program (*ICT*) A program designed for a particular application, eg a spreadsheet program.

packaging (*Genrl*) Material used to surround and thus protect products from their environment. Paper and board now giving way to thermoplastics, esp polyolefins, SAN, for their design flexibility, as in blister packs and shrink-wrap film. Foamed polymers, eg foam polystyrene, provide additional protection with crush-proof properties and good thermal and electrical insulation. Some materials present recycling problems, esp polyvinyl chloride, which can form HCl gas when burnt. Plasticizer migration into food can be a problem with plasticized polyvinyl chloride film.

packed column (*ChemEng*) A column used for distillation or absorption, which consists of a shell filled either with random material such as coke or other broken inert material, or with one of the proprietary rings or similar, usually ceramic or stainless steel (*Lessing, Pall* or *Raschig rings*).

packed file (*ICT*) A file compressed to save space on a disk or in transmission. See ZIP.

packed-hole assembly (*MinExt*) See BOTTOM-HOLE ASSEMBLY.

packer (*MinExt*) Expandable plugs sent down the borehole to seal off a section, often prior to making a more permanent seal with cement. Can be used outside the casing, between casing and drill tube or within the drill tube.

packet (*ICT*) Self-contained message or component of a message, comprising address, control and data signals, that can be transferred as an entity within a communications network.

packet sniffer (*ICT*) See SNIFFER.

packet switching (*ICT*) In data communications, a method in which messages are assembled into one or more PACKETS, including address and control codes, that can be sent independently through a NETWORK, collected and then reassembled into the original information at the destination. Individual packets do not necessarily travel by the same route and only occupy a *channel* during transmission, unlike conventional switching when a channel is established and remains open for the duration of the whole transmission, whether data are flowing or not.

packet switching system (*ICT*) A method of sending data between computers connected to a WIDE AREA NETWORK. Each PACKET is appropriately routed by the switching device.

pack-hardening (*Eng*) CASE-HARDENING, using a solid carburizing medium which is packed around the low-carbon-steel objects and the whole then heated for a time and at a temperature to allow carbon to diffuse into the surface to the desired depth. After the carburization process, the objects must be heat treated to develop hardness at the surface.

packing (*Build*) The operation of filling in a double or hollow wall.

packing (*Eng*) (1) Material inserted in STUFFING BOXES to make engine and pump rods pressure tight; traditionally hemp, but now often special compositions, O- and other shaped rings, or of metal in METALLIC PACKING. (2) Spacers between clamped surfaces.

packing (*ICT*) See DATA COMPACTION.

packing density (*ICT*) A measure of the quantity of data that can be held per unit length on a storage medium, eg bits per cm of magnetic tape.

packing fraction (*Phys*) The fraction

$$\frac{(M - A)}{A}$$

where M is the mass of a nuclide expressed in ATOMIC MASS UNITS (which differs slightly from a whole number) and A is the mass number. The MASS DECREMENT is $(M-A)$. See panel on BINDING ENERGY OF THE NUCLEUS.

packing pressure (*Eng*) Extra pressure applied towards end of injection-moulding cycle to ensure complete filling of cavity.

packing ratio of DNA (*BioSci*) The ratio of the calculated length of a helical DNA molecule to its length after organization into more compact form, in chromosomes, as NUCLEOSOMES together with higher-order coiling or folding. The packing ratio of metaphase chromosome DNA is about 10 000:1.

pad (*ICT*) (1) Small, preset, adjustable capacitor, to regulate the exact frequency of oscillation of an oscillator, or a tuned circuit in an amplifier or filter. Also *padder, trimmer*. (2) Fixed attenuator inserted in a transmission line or waveguide.

pad (*Textiles*) A padding mangle used to apply chemicals (including dyestuff) in solution to open-width fabrics in which the excess is mechanically expressed.

padauk (*For*) Three varieties of hardwood (*Pterocarpus*): African-, Andaman-, and Burma-padauk. Their heartwood has attractive shades of red to brown, with straight to interlocked grain and a moderately coarse texture.

paddle (*ICT*) A JOYSTICK whose operating lever only moves in one dimension.

paddle hole (*Build*) The opening in a lock gate through which water flows from the high-level pond to the lock chamber, or from the lock chamber to the low-level pond.

paddle mixer (*Plastics*) A type of machine used for mixing paints and polyvinyl chloride plastisols.

paddle plane (*Aero*) See CYCLOGYRO.

paddles (*Print*) Rotating curved or shaped fingers of the fly which transfer copies from the folder to the delivery belt of web-fed rotary presses.

paddle-wheel fan (*Eng*) See CENTRIFUGAL FAN.

paddle-wheel hopper (*Eng*) A hopper feed for small articles, in which a continuously rotating paddle pushes the articles along an inclined track towards the delivery chute.

paddy (*FoodSci*) Unmilled rice with the husk on. Also *paddy rice* or *rough rice*.

pad oxide (*Electronics*) In silicon semiconductor processing, a thin oxide growth used to protect active areas during early stages of fabrication.

pad roller (*ImageTech*) See HOLD-DOWN ROLLER.

padsaw (*Build*) See KEYHOLE SAW.

pad stone (*Build*) A stone TEMPLATE.

paed-, paedo-, paid-, paido-, ped-, pedo- (*Genrl*) Prefixes from Gk *pais, paidos*, denoting child, boy.

paediatrician (*Med*) Physician who specializes in the study of childhood and the diseases of children. Also *paediatrist*. US *pediatrician*.

paediatric (*Med*) Pertaining to paediatrics, the branch of medicine concerned with the study of childhood and the diseases of children. US *pediatric*.

paediatrist (*Med*) See PAEDIATRICIAN.

paedo- (*Genrl*) See PAED-.

paedogenesis (*BioSci*) Sexual reproduction by larval or immature forms.

paedomorphosis (*BioSci*) See NEOTENY.

paedophilia (*Psych*) Committing sexual offences against children; sexual gratification through sexual activity with children.

PAF (*BioSci*) Abbrev for PLATELET ACTIVATING FACTOR.

PAGE (*BioSci*) See POLYACRYLAMIDE GEL ELECTROPHORESIS.

page (*ICT*) See PAGING.

page cut-off (*Print*) A control on the ink duct of sheet-fed rotary presses which allows ink feed to be cut off from one page.

page description language (*ICT, Print*) The computer language that interprets a stream of instructions, usually in plain ASCII, into a description of where each letter, drawing and half-tone is to be placed on the printed page. It can accept either a bit-mapped page of text and graphics or information in VECTOR GRAPHICS form in which each element, such as part of a letter or picture element, is described as a vector with its beginning, intermediate and end positions. The latter form is more powerful because it can be made independent of the printer and because a letter from eg a standard font can be easily transformed into any required size. This can therefore provide output both for medium-resolution laser printers and high-resolution typesetters in DESKTOP PUBLISHING. Most widely used is POSTSCRIPT. Abbrev *PDL*.

Page effect (*ElecEng*) A click heard when a bar of iron is magnetized or demagnetized. See MAGNETOSTRICTION.

page impression (*ICT*) A request to a web server, from a client browser or a request broker, for a web page, that may be made up of many files.

page proofs (*Print*) Proofs taken when the type is made up into pages and before imposition.

Paget's disease of bone (*Med*) A chronic disease characterized by progressive enlargement and softening of bones, esp of the skull and of the lower limbs. Also *osteitis deformans*.

Paget's disease of the nipple (*Med*) A condition in which chronic eczema of the nipple is associated with underlying cancer of the breast.

pagination (*Print*) The allotting of numbers (folios) to the pages of a book, roman numerals being traditionally but not invariably used for the preliminary matter and arabic numerals thereafter, beginning again with the numeral '1' on the first text page, but sometimes continuing the sequence where the preliminary folios left off.

paging (*ICT*) (1) A one-way radio communication service that, by means of a network of transmitters, broadcasts the identity code or call sign of a small pocket receiver, or pager. The pager monitors the broadcast channel continuously but responds by 'bleeping' or vibrating only when it recognizes its own call sign. See MESSAGE PAGER. (2) Technique in which MAIN MEMORY is divided into segments called pages. Large user programs may cover several pages, possibly too many to fit into the available store. The operating system transfers pages between main memory and backing store to ensure that the correct page is in main store at any stage during the execution of the program. See SEGMENTATION.

paging (*Print*) Applying numbers to the pages of an account book or to sets of stationery with a hand-numbering tool or a pedal-operated paging machine.

pagodite (*Min*) This is like ordinary massive *pinite* in its amorphous compact texture and other physical characters, but contains more silica. The Chinese carve the soft stone into miniature pagodas and images. Also *agalmatolite*.

pahoehoe (*Geol*) Lava with a glassy, smooth, ropy surface.

paid-, paido- (*Genrl*) See PAED-.

paint (*Build*) A suspension of solids in a liquid which, when applied to a surface, dries to a more or less opaque, adhering solid film. The solids are colour- or opacity-imparting, finely ground pigments together with EXTENDERS; the liquid consists of suitable oils, solvents, resins, aqueous colloidal solutions or dispersions, etc, together with other lesser ingredients. See OIL PAINTS.

paint remover (*Build*) A liquid applied to paint to facilitate removal. Can be solvent- or caustic-based. Also *paint stripper*.

pair (*Electronics*) Two electrons forming a non-polar valency bond between atoms.

pair (*MathSci*) An ORDERED SET of two elements.

paired-associate learning (*Psych*) A procedure in which a list of pairs is presented in which one item serves as stimulus and the other as response; the subject must learn to respond with the second item when the first item of a pair is presented.

paired cable (*ElecEng*) One in which multiple conductors are arranged as twisted pairs but not quadded.

pairing (*BioSci*) The process by which *homologous chromosomes* are brought together during meiosis preparatory to being distributed one to each gamete.

pairing (*ImageTech*) Deviation from exactness of interlacing of horizontal lines in a reproduced image, reducing vertical definition. Also *twinning*.

pairing energy (*Phys*) A component of the binding energy of the nucleus that represents an increase in this energy where the number of neutrons and the number of protons are even. See EVEN–EVEN NUCLEI.

pair production (*Electronics*) Creation of a POSITRON and an ELECTRON when a PHOTON passes into the electric field of an atom. See COMPTON EFFECT, PHOTOELECTRIC EFFECT.

paisanite (*Geol*) A sodic microgranite containing *riebeckite* as the principal coloured mineral.

PAL (*BioSci*) Abbrev for PHENYLALANINE AMMONIA-LYASE.

PAL (*ImageTech*) Abbrev for *phase alternation line*, the colour TV coding system generally used for European broadcasting, in which one of the two chrominance signals is reversed in phase for each alternate scanning line.

palae-, palaeo- (*Genrl*) Prefixes from Gk *palaios*, ancient. US *pale-, paleo-*.

Palaearctic region (*BioSci*) One of the subrealms into which the Holarctic region is divided; it includes Europe and N Asia, together with Africa north of the Sahara.

palaeo- (*Genrl*) See PALAE-.

palaeobotany (*Geol*) The study of fossil plants. See PALAEONTOLOGY.

Palaeocene (*Geol*) The oldest epoch of the TERTIARY period.

palaeocerebellum (*Med*) Phylogenetically, the older part of the cerebellum, comprising such parts as pyramis, UVULA and paraflocculus.

palaeoclimatology (*Geol*) The study of climatic conditions in the geological record, using evidence from fossils, sediments and their structures, geophysics and geochemistry.

palaeocurrent (*Geol*) An ancient current whose direction can often be worked out by examination of sedimentary structures (eg *cross bedding, ripple marks, sole structures*).

palaeoecology (*Geol*) The study of fossil organisms in terms of their mode of life, their interrelationships, their environment, their manner of death and their eventual burial.

palaeoencephalon (*Med*) Phylogenetically, the ancient brain; the brain excluding the cerebral cortex and appendages.

Palaeogene (*Geol*) A period lying above the Cretaceous and below the *Neogene*, containing the EOCENE, OLIGOCENE and PALAEOCENE epochs. It covers a time span from approx 65 to 25 million years. The corresponding system of rocks. See TERTIARY.

palaeogenetic (*Genrl*) Originating in the past.

palaeogeography (*Geol*) The study of the relative positions of land and water at particular periods in the geological past.

Palaeolithic period (*Geol*) The oldest stone age, characterized by successive 'cultures' of stone implements, made by extinct peoples. Cf NEOLITHIC PERIOD.

palaeomagnetism (*Geol*) The study of the Earth's ancient magnetism. *Remanent magnetization* in both igneous and sedimentary rocks provides a means of determining former magnetic poles. See MAGNETIC ANOMALY PATTERNS.

palaeontology (*Geol*) The study of fossil animals and plants, including their morphology, evolution and mode of life.

palaeopathology (*Med*) The study of disease of previous eras from examination of bodily remains or evidence from ancient writings.

Palaeozoic (*Geol*) A major era of geological time comprising the Cambrian, Ordovician, Silurian, Devonian, Carboniferous and Permian periods. See appendix on Geological time.

palaeozoology (*Geol*) The study of fossil animals. See PALAEONTOLOGY.

palagonite (*Geol*) A hydrous altered basaltic glass. It occurs as infillings in rocks, and is a soft-brown or greenish-black cryptocrystalline substance. Named from Palagonia, Sicily.

palama (*BioSci*) The webbing of the feet in birds of aquatic habit.

palate (*BioSci*) In vertebrates, the roof of the mouth; in insects, the epipharynx. Adjs *palatal, palatine*.

palatine (*BioSci*) Pertaining to the palate; a paired membrane bone of the vertebrate skull that forms part of the roof of the mouth.

palatoplegia (*Med*) Paralysis of the palate.

pale-, paleo- (*Genrl*) US for PALAE-, PALAEO-.

palea (*BioSci*) The usually thin and membranous, upper or inner of the two bracts (lemma and palea) that enclose a grass floret. Sometimes, synonymous with glume. Also *pale, palet, valvule*.

paleo- (*Genrl*) US for *palaeo-*. See PALAE-.

palindromic sequence (*BioSci*) A nucleic acid sequence that is identical to its complementary strand when each is read in the correct direction (eg. TGGCCA). Palindromic sequences are often the recognition sites for *restriction enzymes*.

palingenesis (*Geol*) The production of new magma by the complete or partial melting of previously existing rocks. See GRANITIZATION.

palingenesis (*BioSci*) The reproduction of truly ancestral characters during ontogeny. Adj *palingenetic*.

palisade (*BioSci*) CHLORENCHYMA in which the cells are elongated at right angles to the surface of the organ. Palisade mesophyll is characteristically present towards the upper surface of dorsiventral leaves of mesophytic dicotyledons. Palisade layers also occur in the outer cortex of many photosynthetic stems. See SPONGY LAYER.

Palladian (*Arch*) An architectural style named after Andrea Palladio, (1508–80), the most influential architect of the RENAISSANCE. The style is characterized by the manner in which certain CLASSICAL motifs such as PORTICOS, arches and columns are grouped together.

Palladian window (*Arch*) A type of window comprising main window with arched head and, on either side, a long narrow window with square head.

palladinized asbestos (*Chem*) Asbestos fibres saturated with a solution of a palladium compound, which is subsequently decomposed to give finely divided palladium dispersed throughout the asbestos.

palladium (*Chem*) A metallic element, symbol Pd, at no 46, ram 106·4. The metal is white, mp 1549°C, bp 2500°C, rel.d. 11·4. It is extracted from copper–nickel ores. Used as a catalyst in hydrogenation. Native palladium is mostly in grains and is frequently alloyed with platinum and iridium. See GOLD.

pallaesthesia (*Med*) Insensibility of bone to vibratory stimuli.

Pallas (*Astron*) The second asteroid to be discovered (1802) and the third largest, with a diameter of 540 km.

pallasite (*Min*) A group name for stony meteorites which contain fractured or rounded crystals of olivine in a network of nickel–iron.

pallescent (*BioSci*) Becoming lighter in colour with age.

pallet (*Acous*) The flap of wood, faced with felt or leather, which is raised to permit the flow of air to eg wind-chests in the mechanism of an organ.

pallet (*Build*) A thin strip of wood built into the mortar joint of a wall, to which joinery may be nailed.

pallet (*Eng, Print*) A platform on which sheets of paper (or other goods) can be stacked before printing and between printings, and designed for transporting by fork-lift truck. Also *stillage*.

pallet brick (*Build*) A purpose-made brick with a groove in one edge to receive a fixing strip or PALLET.

pallet truck (*Genrl*) See FORK-LIFT TRUCK.

palli-, pallio- (*Genrl*) Prefixes from Lt *pallium*, mantle.

palliative (*Med*) Affording temporary relief from pain or discomfort; a medicine which does this.

pallio- (*Genrl*) See PALLI-.

pallium (*BioSci*) (1) The mantle in Brachiopoda or Mollusca, a fold of integument that secretes the shell. (2) In the vertebrate brain, that part of the wall of the cerebral hemispheres excluding the corpus striatum and rhinencephalon. Adjs *pallial, palliate*.

Palmae (*BioSci*) The palm family, c.2800 spp of monocotyledonous flowering plants (superorder Arecidae). Trees, mostly tropical, typically with a single trunk (which does not undergo secondary thickening) bearing a crown of pinnate or palmate leaves. The fruit is a one-seeded berry or a drupe. The most important economic products

are coconuts, palm oil, dates, copra and fibres. Also *Arecaceae*.

palmar (*BioSci*) Pertaining to the palm of the hand.

palmate (*BioSci*) (1) In plants, having four or more equal divisions, lobes or veins radiating from a common point rather in the manner of the fingers of a hand; as in a palmately compound, lobed or veined leaf respectively. (2) Having webbed feet.

palmatine (*Chem*) An alkaloid of the *isoquinoline* group, obtained from calumba root, *Jatrorrhiza palmata*, generally isolated in the form of the sparingly soluble iodide, $C_{21}H_{22}O_4NI.2H_2O$, which crystallizes in yellow needles, mp 240°C.

palmelloid form (*BioSci*) A condition in algae in which non-motile cells divide within a mucilaginous matrix to give large gelatinous masses containing many cell generations.

palmisect (*BioSci*) A leaf blade etc cut almost to the centre in a palmate fashion.

palmitic acid (*Chem*) $C_{15}H_{31}COOH$. Mp 63°C, bp 269°C. A normal fatty acid. It occurs as glycerides in vegetable oils and fats.

palmitins (*Chem*) The glycerine esters of palmitic acid.

palm-kernel oil (*Chem*) A yellowish oil from the nuts of *Elaeis guineensis*, mp 26–30°C, rel.d. 0·95, saponification value 247, iodine value 13·5, acid value 8·4.

palm oil (*Chem*) A reddish-yellow fatty mass from the fruit of *Elaeis guineensis*, mp 27–43°C, rel.d. 0·90–0·95, saponification value 196–205, iodine value 51–57, acid value 24–200.

palmtop (*ICT*) A portable personal computing device providing common office automation functions such as email, diary and word processing, small enough to be held in the palm of the hand.

palp (*BioSci*) See PALPUS. Adj *palpal*.

palpation (*Med*) Physical examination by touch.

palpebra (*Med*) An eyelid.

palpebral fissure (*Med*) The space between the upper and lower eyelids.

palpitation (*Med*) Subjective awareness of rapid or irregular heart beat.

PALplus (*ImageTech*) A wide-screen development of PAL. The picture is shown in LETTERBOX form on 4:3 aspect ratio receivers and expanded to full height on 16:9 receivers, with a 'helper' signal used to restore the vertical resolution to normal. The system also reduces picture defects such as CROSS-COLOUR NOISE.

palpus (*BioSci*) In Crustacea and insects, a jointed sensory appendage associated with the mouthparts; in Polychaeta, sensory appendage of prostomium. Also *palp*.

palygorskite (*Min*) A group of clay minerals, hydrated magnesium aluminium silicates, in appearance resembling cardboard or paper, having a fibrous structure. Also *attapulgite*.

palynology (*BioSci*) The study of fossil pollen and spores. Because pollen is very resistant to degradation it often provides important clues to the vegetation in ancient deposits. See POLLEN ANALYSIS.

PAN (*Chem*) Abbrev for: (1) PEROXYACETYL NITRATE; (2) POLYACRYLONITRILE.

pan (*Build*) A panel of brickwork, or lath and plaster, in half-timbered work.

pan (*EnvSci*) (1) A compact layer of soil particles, lying some distance beneath the surface, cemented together by organic material, or by iron and other compounds, and relatively impermeable to water. (2) A depression in the surface of a salt marsh, in which salt water stands for lengthy periods.

pan (*ImageTech*) (1) Movement of a camera in a horizontal plane about a vertical axis (from panoramic). (2) Abbrev for PANCHROMATIC.

panache (*CivEng*) See PENDENTIVE.

panama (*Textiles*) Plain-weave, worsted fabric used for tropical suiting and weighing about 200 g m^{-2}; may also be made from non-woollen fibres.

pan-and-scan (*ImageTech*) The continuous selection of a limited area from the picture of a WIDE-SCREEN motion picture film for presentation within a narrower format, such as that of TV.

pan-and-tilt head (*ImageTech*) A camera mounting allowing rotation about a vertical axis (*panning*) and a horizontal one (*tilting*), esp in cinematography and TV.

pancake coil (*ICT*) Inductor in which the conductor is wound spirally.

pancaking (*Aero*) The alighting of an aircraft at a relatively steep angle, with low forward speed.

pancarditis (*Med*) Concurrent inflammation of the three main structures of the heart: the pericardium, the myocardium and the endocardium.

panchromatic (*ImageTech*) A photographic emulsion substantially sensitive to the whole of the visible spectrum. Abbrev *pan*.

pancreas (*BioSci*) A moderately compact structure in vertebrates, mostly consisting of exocrine glandular tissue with one or more ducts opening to the small intestine, but also with scattered islets (ISLETS OF LANGERHANS) of endocrine tissue. The former produces enzymes involved in digestion, while the latter secrete the hormones insulin and glucagon. See PANCREATIC ACINAR CELLS.

pancreatectomy (*Med*) Surgical removal of the pancreas.

pancreatic acinar cells (*BioSci*) Cells of the pancreas that secrete digestive enzymes; the archetypal secretory cell upon which much of the early work on the sequence of events in the secretory process was done.

pancreatin (*Med*) Pancreatic enzyme preparation used to compensate for reduced or absent exocrine secretion of the pancreas in children with cystic fibrosis and in adults following pancreatectomy or chronic pancreatitis.

pancreatitis (*Med*) Inflammation of the pancreas.

pancreatolith (*Med*) A calculus in the pancreas.

pancreatotomy (*Med*) Incision of the pancreas.

pancreozymin (*BioSci*) A polypeptide hormone, secreted by the intestinal wall, that stimulates the pancreas to secrete digestive enzymes.

pandemic (*Med*) An epidemic, occurring over a wide area such as a country or a continent.

pandiculation (*Med*) The combined action of stretching the body and the limbs and yawning.

pandurate (*BioSci*) Shaped like the body of a fiddle. Also *panduriform*.

pane (*Build*) The end of a hammer head opposite to hammering face; made to various shapes for particular operations such as riveting etc. Also *pean, peen, pein*. See BALL-PANE HAMMER, CROSS-PANE HAMMER, STRAIGHT-PANE HAMMER.

panel (*ElecEng*) A sheet of metal, plastic or other material upon which instruments, switches, relays, etc, are mounted. Also *switchboard panel*.

panel (*Print*) A piece of material (leather, paper) fixed to the spine of a book, containing the title and author's name.

panel absorber (*Acous*) Light panels mounted at some distance in front of a rigid wall in order to absorb sound waves incident on them.

panel back (*Print*) A style of FINISHING the spine of a book by decorated panels between raised bands.

panel beating (*Eng*) A sheet-metal working craft, now mainly used in automobile body repairing, in which complex shapes are produced by stretching and gathering the sheet locally, with subsequent finishing by PLANISHING and WHEELING.

panel heating (*Build*) A system of heating a building in which heating units or coils of pipes are concealed in special panels, or built in wall or ceiling plaster. Also *concealed heating*.

panelled framing (*Build*) Doors and frames formed of MUNTINS, RAILS and STILES framed together with mortise-and-tenon joints, and having panels fitted into the spaces.

panel mounting (*ElecEng*) The normal method of accommodating a collection of non-portable apparatus. Each piece or unit is constructed separately on its standard panel, which is mounted with others on a standard vertical rack, the different panels being provided with terminal blocks so that the units can be wired together after assembly.

panel pins (*Build*) Light, narrow-headed nails of small diameter used chiefly for fixing plywood or hardboard.

panels (*Eng*) In a truss or open-web girder, the framed units of which the truss is composed; the divisions separated by the vertical members.

panel saw (*Build*) A hand saw used for panelling, having seven teeth to the inch.

panel switch (*ElecEng*) See FLUSH SWITCH.

PAN fibres (*Textiles*) Abbrev for *polyacrylonitrile fibres*. See ACRYLIC FIBRES.

Pangaea (*Geol*) A hypothetical supercontinent that existed in the geological past and consisted of all the present continents before they split up.

pangamic (*BioSci*) Of indiscriminate mating.

panhead rivet (*Eng*) One with a head shaped like a truncated cone.

panhysterectomy (*Med*) Complete surgical removal of the uterus.

panic attack (*Psych*) An attack of intense terror and anxiety, usually lasting several minutes, though possibly continuing for hours; apprehension often persists for long periods after the panic attack.

panic bolt (*Build*) A special form of door bolt which is released by pressure at the middle of the bolt, often used on fire-escape doors.

panic disorder (*Psych*) A PANIC ATTACK occurring in the absence of any phobic stimulus.

panicle (*BioSci*) Strictly, a branched raceme with each branch bearing a raceme of flowers, eg oat. Loosely, any branched inflorescence of some degree of complexity. Adj *paniculate*.

panidiomorphic (*Geol*) A term applied to igneous rocks with well-developed crystals.

pan-mill mixer (*Build*) See MORTAR MILL.

panmixis (*BioSci*) RANDOM MATING within a population, esp a model system. Adj *panmictic*. Also *panmixia*.

panniculitis (*Med*) Inflammation of the subcutaneous fat in any part of the body.

panniculus carnosus (*BioSci*) In some mammals, an extensive system of dermal musculature covering the trunk and part of the limbs, by means of which the animal can shake itself.

panning (*BioSci*) Commonly used laboratory technique for isolating cell subsets from mixed populations. Antibodies specific for a particular subset are coated onto plastic dishes, the mixed population is added and incubated. The desired cells adhere to the antibody and the rest of the cells can be washed away.

panning (*MinExt*) Use by a prospector or plant worker of gold pan, batea, plaque or dulong to concentrate heavier minerals in a crushed sample by washing away the lighter ones.

panning head (*ImageTech*) Head or platform for a camera, permitting panoramic motion.

pannose (*BioSci*) Felted.

pannus (*Med*) (1) The appearance of a curtain of blood vessels round the margin of the cornea in eg TRACHOMA or PHLYCTENULAR KERATITIS. (2) Inflammatory granulation tissue that invades the joint in rheumatoid arthritis.

panophthalmitis (*Med*) Inflammation of all the structures of the eye. Also *panophthalmia*.

panoramic attenuator (*Acous*) Rerecording device by which a one-channel recording is reproduced and distributed to, say, three tracks, generally magnetic, to give an illusion, when these are reproduced, of stereophonic sound reproduction. Used with wide cinema screens.

panoramic camera (*ImageTech*) A camera intended to take very-wide-angle views, generally by rotation about an axis and by exposing a roll of film through a vertical slit.

Paper and paper-making

Paper is a mat of cellulose fibres held together by hydrogen bonds. Any cellulose-based, natural material can be used for the fibres, and hemp ropes, cotton rags and ESPARTO grass have all been widely used. Now, over 90% of paper is made from wood pulp, containing fibres 0·5 to 3 mm long.

A suspension of about 2% fibres in water is subjected to a high shear rate, known as BEATING or FIBRILLATION, partially to break down their cell-wall structures. See STRUCTURE OF WOOD (panel). A combination of hydrolysis (swelling the fibres and breaking the cellulose–cellulose hydrogen bonds) and the shearing action splits the cell walls into fibrils, which splay out from the parent fibres. The beaten suspension is then spread either over a fine-mesh 'conveyor belt' (the WEB or FELT) in continuous paper-making or onto a rotating, perforated cylinder to allow drainage which can be vacuum-assisted. The fibrous mat is turned into paper by the consolidating force provided by the surface tension of the draining water. It has to be sufficiently high to ensure that intersecting fibres and fibrils are close enough to form hydrogen bonds. This is aided by the enhanced plasticity of the swollen fibres. About 1–2% of all the hydrogen bonds in the cellulose are used in the bonding of paper.

The properties of the resulting paper are determined by the fibrillation stage and can be extensively modified by the use of additives and coatings. Density, elastic modulus and strength (particularly when wet), as well as opacity, colour, gloss, liquid absorbance and printability, can all be changed. Typically, papers show the following range of properties: density, 200–1400 kg m^{-3}; tensile modulus, 1–5 GN m^{-2}; tensile strength, 30–80 MN m^{-2}; breaking strain, 4–12%.

---

panoramic receiver (*ICT*) One in which tuning sweeps over wide ranges, with synchronized display on a cathode-ray tube of output signals.

panradiometer (*Phys*) An instrument for measuring radiant heat irrespective of the wavelength.

pansexual (*Psych*) Including all or many different forms of sexuality.

pansinusitis (*Med*) Inflammation of all the air-containing sinuses which communicate with the nasal cavity.

pantellerite (*Min*) A peralkaline leucocratic rhyolite.

panting (*Ships*) Pulsating movement of the shell plating of the ship developing in the bows and stern while the ship is under way.

pantobase (*Aero*) The fitment of a landplane with HYDROSKIS, enabling it to taxi, take off from and alight on water or snow.

pantograph (*ElecEng*) A sliding type of current collector for use on traction systems employing an overhead contact wire. The contact strip of the collector is mounted on a hinged diamond-shaped structure, so that it can move vertically to follow variations in the contact wire height.

pantograph (*Eng*) A mechanism by which a point is constrained to copy, to any required scale, the path traced by another point. It is based on the geometry of a parallelogram; used in engraving machines etc.

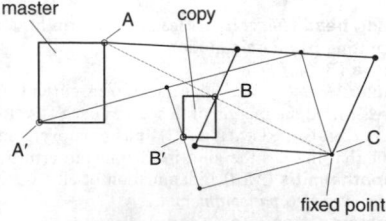

pantograph The scale is determined by the ratio of BC to AC. Two positions are drawn.

Pantone Matching System (*Print*) TN for a colour matching system extensively used by graphics designers to specify colours to be used in printing. A book of colour swatches is used, each having a number and an indication of the ink mix required to achieve the desired hue, thus acting as an aid to the printer. Abbrev *PMS*.

pantophagous (*BioSci*) Omnivorous.

pantothenic acid (*BioSci*) See vitamin B complex in panel on VITAMINS.

papain (*Chem*) A protein-digesting enzyme present in the juice from the fruits and leaves of the papaya tree (*Carica papaya*); commercially produced as a meat tenderizer.

Papanicolaou's stain (*BioSci*) A complex stain for detecting malignant cells in cervical smears, hence the *Pap test*.

papaverine (*Chem*) An alkaloid occurring in opium, colourless prisms, mp 147°C, optically inactive. It is 3,4-dimethoxybenzyl-4',5'-dimethoxy-*iso*quinoline.

papaya (*For*) A tree, *Carica papaya*, native to S America but common in the tropics. The trunk, leaves and fruit yield papain, a protease used commercially as a meat-softener.

paper (*Genrl*) Consists of continuous webs of suitable vegetable fibres, freed from non-cellulose constituents and deposited from an aqueous suspension. See CELLULOSE and panel on PAPER AND PAPER-MAKING.

paper capacitor (*ElecEng*) One which has thin paper as the dielectric separating aluminium-foil electrodes, all wound together and waxed.

paper chromatography (*Chem*) A type of CHROMATO-GRAPHY using a sheet of special grade filter paper as the adsorbent. Advantages: microgram quantities; bands can be formed in two dimensions and cut with scissors.

paper foils (*Print*) BLOCKING FOILS with a glassine substrate. These are the cheapest foils and the most widely used.

paper-making (*Paper*) See panel on PAPER AND PAPER-MAKING.

paper negatives (*ImageTech*) Negatives made with emulsions on paper supports, instead of the more usual film or glass. They have the advantages of lightness, cheapness, ease of retouching, and the possibility of large dimensions.

paper sizes (*Paper*) Now increasingly based on ISO SIZES; previous UK sizes, shown in inches below, for esp book printing were based on broadsides (ie unfolded sheets) of *crown* (15 × 20), *demy* ($17\frac{1}{2}$ × $22\frac{1}{2}$), *medium* (18 × 23) and *royal* (20 × 25), which were then divided into quarter (*quarto*, or *4to*) and eighth (*octavo* or *8vo*) sizes. See ALBERT, ELEPHANT, FOOLSCAP, QUAD.

papier mâché (*Paper*) Substance made from paper mashed in water and built up in layers with glue, either over a preform or into a mould. Used for furniture in the 18th century, imitation interior wood-, stone- and plasterwork in the 19th century, and aircraft fuel tanks in World War II.

papilionaceous (*BioSci*) Having some likeness to a butterfly, esp the flowers of the Papilionaceae, including the pea.

papilla (*BioSci*) (1) A small nipple-shaped conical projection of soft tissue, eg on the skin or lining of the alimentary canal. (2) A conical mass of soft tissue or pulp projecting into the base of a developing feather or tooth. Adjs *papillary, papillate.*

papillae foliatae (*BioSci*) In some mammals, two small oval areas at the back of the tongue, marked by a series of alternating transverse ridges and richly provided with taste buds.

papilledema (*Med*) See PAPILLOEDEMA.

papillitis (*Med*) Inflammation of the disk or head of the optic nerve within the globe of the eye. Also *optic neuritis.*

papilloedema (*Med*) Swelling and congestion of the disk or head of the optic nerve within the globe of the eye, as a result of increase of pressure within the skull or of severe hypertension. US *papilledema.* Also *choked disk.*

papilloma (*Med*) A tumour (usually innocent) resulting from the new growth of the cells of the skin or of the mucous membrane.

papovavirus (*Med*) A group of small DNA viruses which have the capacity to produce tumours in animals and humans.

pappus (*BioSci*) Modified calyx in the Compositae, consisting of a ring of feathery hairs or the like around the top of the fruit, as in the dandelion, where it aids in wind dispersal.

Pappus's theorem (*MathSci*) Either of the theorems that: (1) The surface area and the volume respectively swept out by revolving a plane area about a non-intersecting coplanar axis is the product of the distance moved by its centroid and either its perimeter or its area. Frequently attributed to Guldin who rediscovered it in the 17th century. (2) If *ABC* and *A'B'C'* are two straight lines in a plane, then the three points of intersection of lines *AB'* and *A'B*, *AC'* and *A'C*, *BC'* and *B'C* lie on a straight line. This is in fact *Pascal's theorem* applied to the 'hexagon' *AB'CA'BC'* inscribed in the degenerate conic consisting of two straight lines.

Pap test (*Med*) A procedure in which cells are scraped from the cervix for histopathological examination involving Papanicolaou's stain. It is used to detect cancer and pre-cancerous changes and can also show non-cancerous conditions, such as infection or inflammation. Also *Pap smear.*

papulae (*BioSci*) The dermal gills of Echinodermata, small finger-shaped, thin-walled respiratory projections of the body wall.

papule (*Med*) A small, circumscribed, solid elevation above the skin, as in chicken pox. Adj *papular.*

papulopustular (*Med*) Of papules and pustules.

papyrus (*Paper*) An early form of writing material prepared from the water reed (*Cyperus papyrus* or *Papyrus antiquorum*) of the same name and used for records up to the 9th century AD.

PAR (*Aero*) Abbrev for PRECISION-APPROACH RADAR.

PAR (*BioSci*) Abbrev for PHOTOSYNTHETICALLY ACTIVE RADIATION.

para- (*Genrl*) Prefix from Gk *para*, beside.

para- (*Chem*) (1) Containing a benzene nucleus substituted in the 1,4 positions. Abbrev *p-*. (2) Consisting of diatomic molecules with antiparallel nuclear spins and an even rotational quantum number.

para(4)-aminosalicylic acid (*Pharmacol*) Used medically, usually in the form of the sodium salt, in the treatment of tuberculosis. Usually administered in conjunction with ISONIAZID. Abbrev PAS.

parabiosis (*BioSci*) The surgical linkage of two organisms so that their circulatory systems interconnect.

parabola (*MathSci*) The intersection of a cone with a plane which is parallel to the side of the cone. See fig. at CONIC.

parabolic flight (*Space*) The flying of a special parabolic trajectory by a suitably fitted aeroplane to reproduce the conditions of FREE FALL over a period of minutes.

parabolic microphone (*Acous*) A microphone with a parabolic reflector to give enhanced directivity for high audio frequencies and hence greater range for wanted sounds amid ambient noise.

parabolic mirror (*Phys, ICT*) A mirror shaped as a paraboloid of revolution. Theoretically produces a perfectly parallel beam of radiation if a source is placed at the focus (or vice versa). Such mirrors are used in reflecting telescopes and car headlamps.

parabolic nozzle (*Eng*) A nozzle of parabolic section with a high coefficient of discharge, placed in a pipe to measure the flow of a gas. The pressure drop is measured by a manometer.

parabolic point on a surface (*MathSci*) One at which one of the principal curvatures, and therefore also the Gaussian curvature, is zero. Cf ELLIPTICAL, HYPERBOLIC, UMBILICAL POINTS ON A SURFACE.

parabolic reflector (*ICT*) See DISH, PARABOLIC MIRROR.

parabolic spiral (*MathSci*) A spiral with polar equation $r^2 = a^2\omega$. Also *Fermat's spiral.*

parabolic velocity (*Astron*) The velocity which a body at a given point would require to describe a parabola about the centre of attraction; smaller values give an ellipse, larger values a hyperbola. Also ESCAPE VELOCITY, since it is the upper limit of velocity in a closed curve.

paraboloid (*MathSci*) See QUADRIC.

paraboloid of revolution (*MathSci*) A solid figure formed by the revolution of a parabola about its axis. Cf QUADRIC.

parabrake (*Aero*) See BRAKE PARACHUTE.

paracentesis (*Med*) The puncture of body cavities with a hollow needle, for the removal of inflammatory or other fluids. Also *tapping.* See AMNIOCENTESIS.

paracetamol (*Pharmacol*) Antipyretic and analgesic drug.

parachor (*Chem*) A quantity which may be regarded as the molecular volume of a substance when its surface tension is unity; in most cases it is practically independent of temperature. Its value is given by the expression

$$\frac{M\gamma^{\frac{1}{4}}}{\rho_L - \rho_V}$$

where M is the molecular mass, γ is the surface tension, ρ_L and ρ_V are the densities of the liquid and vapour respectively.

parachute (*Aero*) An umbrella-shaped fabric device of high drag (1) to retard the descent of a falling body, or (2) to reduce the speed of an aircraft or item jettisoned therefrom. Commonly made of silk or nylon, sometimes of cotton or rayon where personnel are not concerned. See ANTI-SPIN PARACHUTE, AUTOMATIC PARACHUTE, BRAKE PARACHUTE, RIBBON PARACHUTE, RING SLOT PARACHUTE, PILOT CHUTE. See PARASHEET.

parachute flare (*Aero*) A pyrotechnic flare, attached to a parachute released from an aircraft to illuminate a region.

paracrine (*BioSci*) A form of signalling in which the target cell is close to the signal-releasing cell. Neurotransmitters and neurohormones are usually considered to fall into this category.

paracytic (*BioSci*) A stomatal complex having one or more subsidiary cells parallel to each guard cell, as in many monocotyledons.

paradigm shift (*Genrl*) A radical change in one's assumptions or way of thinking.

paradoxical sleep (*Psych*) Stage of sleep when dreaming is assumed to occur. It is characterized by rapid eye movements (REM), loss of muscle tone, and an EEG very similar to the waking state (hence paradoxical). Also *REM sleep.*

paraeiopod (*BioSci*) In Crustacea, a walking leg.

paraesthesia (*Med*) An abnormal sensation, such as tingling, tickling and formication.

paraffins (*Chem*) A term for the whole series of saturated aliphatic hydrocarbons of the general formula C_nH_{2n+2}. They are indifferent to oxidizing agents, and not reactive, hence the name *paraffin* (Lt *parum affinis*, little allied). Also *alkane hydrocarbons*.

paraffin wax (*Chem*) Higher homologues of alkanes, wax-like substances obtained as a residue from the distillation of petroleum; mp 45–65°C, rel.d. 0·9, resistivity 10^{13} to 10^{17} ohm metres, permittivity 2–2·3.

Parafil (*Eng*) TN for parallel-laid cable constructed from high-performance fibre (eg aramid) enclosed by thermoplastic protective sheath, and fitted with special end connectors.

paraformaldehyde (*Chem*) $(HCHO)_2$. A self-addition product of formaldehyde (methanal) obtained by the evaporation of an aqueous solution of methanal. It is a white crystalline mass, soluble in water. Also *paramethanal*.

paraganglia (*BioSci*) In higher vertebrates, small glandular bodies, occurring in the posterior part of the abdomen, that show a chromaphil reaction and are believed to secrete adrenaline.

paraganglioma (*Med*) See PHAEOCHROMACYTOMA.

paraglider (*Aero*) (1) Inflatable hypersonic re-entry kite of highly swept-back wing shape with rounded leading edge, proposed for Gemini spacecraft, but not used. (2) Modified parachute-like 'wing' that can be used to fly using thermals, usually launched from a hillside. Also *parapent*.

paragnathous (*BioSci*) Having jaws of equal length; as birds which have upper and lower beak of equal length.

paragneiss (*Geol*) A term given to gneissose rocks which have been derived from detrital sedimentary rocks. Cf ORTHOGNEISS.

paragonimiasis (*Med*) Invasion of the lungs by the lung fluke *Paragonimus westermanii*; the condition is endemic in the Far East, and infection results from eating fresh-water crustaceans.

paragonite (*Min*) A hydrated sodium aluminium silicate. It is a sodium mica, has a yellowish or greenish colour, and is usually associated with metamorphic rocks. Differs from MUSCOVITE chiefly in containing sodium rather than potassium.

paragraphia (*Psych*) Faulty spelling, misplacement of letters and words, and use of wrong words in writing, as a result of a lesion in the brain.

parahydrogen (*Chem*) A hydrogen molecule in which the two nuclear spins are antiparallel, forming a singlet state.

parainfluenza virus (*BioSci*) Species of the Paramyxoviridae. There are four types. Type 1 is also known as Sendai virus; types 2–4 cause mild respiratory infections in humans.

parakeratosis (*Med*) Faulty formation of the horny layer of the skin, with scaling of the skin.

paralalia (*Psych*) A form of speech disturbance, particularly that in which a different sound or syllable is produced from the one which is intended.

paraldehyde (*Chem*) $(CH_3CHO)_3$. A self-addition product of ethanal, obtained by the action of concentrated sulphuric acid upon ethanal. It is a colourless liquid, bp 124°C, and can be converted again into ethanal by distillation with dilute sulphuric acid. A common hypnotic. Used as a chemical intermediate in the synthesis of *pentaerythritol*.

paralexia (*Psych*) A defect in the power of seeing and interpreting written language, with meaningless transposition of words and syllables, due to a lesion in the brain.

paralimnion (*EnvSci*) The zone of a lake floor between the water's edge or shoreline, and the lakeward margin of rooted vegetation.

paralinguistics (*Psych*) The study of those aspects of communication that are not linguistic, eg tone of voice, pause, emphasis, etc; how things are said, rather than what is said.

parallactic angle (*Astron*) The angle in the astronomical triangle formed at the heavenly body by the intersection of the arcs drawn to the zenith and to the celestial pole. See ASTRONOMICAL TRIANGLE.

parallax (*Astron*) The apparent displacement in the position of any celestial object caused by a change in the position of the observer. Specifically, the change due to the motion of the Earth through space. The observer's position changes with the daily rotation of the Earth (DIURNAL PARALLAX), the yearly orbit (ANNUAL PARALLAX) and motion through space generally (*secular parallax*). The term 'parallax' is often loosely used by astronomers to be synonymous with 'distance' because the annual parallax is inversely proportional to distance. See SPECTROSCOPIC PARALLAX.

parallax (*Phys*) Generally, the apparent change in the position of an object seen against a more distant background when the viewpoint is changed. Absence of parallax is often used to adjust two objects, or two images, at equal distances from the observer.

parallax stereogram (*ImageTech*) The use of a line screen in front of a positive transparency of alternate strips of two views of an object, made by exposing an emulsion in a similar arrangement with a large lens and two apertures representing the two eyes.

parallel (*ElecEng*) Two circuits are said to be *in parallel* when they are connected so that any current flowing divides between the two. Two machines, transformers or batteries are said to be *in parallel* when the terminals of the same polarity are connected together.

parallel (*MathSci*) Of a family of curves, remaining a constant distance apart; in particular two straight lines are said to be parallel if they do not meet when extended indefinitely. In projective geometry parallel lines are said to meet at POINTS AT INFINITY.

parallel (*Surv*) A *parallel of latitude* is an imaginary line around the Earth's surface connecting points of equal terrestrial latitude.

parallel arithmetic unit (*ICT*) One in which the digits are operated on concurrently.

parallel-axis theorem (*Phys*) A theorem expressing the moment of inertia I of a body about any axis, in terms of its moment of inertia, I_a, about a parallel axis through G, its centre of gravity. Thus $I = I_a + Md^2$, where d is the perpendicular distance between the axes and M the mass of the body.

parallel body (*Ships*) That portion of a ship's form wherein the fullest transverse shape is maintained constant.

parallel circuit (*ElecEng*) (1) Electric or magnetic circuit in which current or flux divides into two or more paths before joining to complete the circuit. (2) See VOLTAGE CIRCUIT.

parallel data transmission (*ICT*) The BITS making up a character are sent simultaneously along separate data lines. Cf SERIAL DATA TRANSMISSION.

parallel descent (*BioSci*) The appearance of similar characteristics in groups of animals or plants that are not directly related in evolutionary descent. Also *parallel evolution, parallelism*.

parallelepiped (*MathSci*) A solid figure bounded by six parallelograms, opposite pairs being identical and parallel.

parallel-feeder protection (*ElecEng*) A type of balanced protective equipment relying for its action on the fact that the current in two parallel feeders will normally be equal, this balance being upset if a fault occurs on one of the feeders.

parallel folding (*Print*) When two or more successive folds in a sheet are in the same direction they are described as *parallel*. Cf RIGHT-ANGLED FOLDING.

parallelism (*BioSci*) See PARALLEL DESCENT.

parallel motion (*Eng*) (1) Mechanism comprised of a parallel arrangement of links, constructed so that motion of one point can induce similar motion at another. The relative motion can be larger or smaller. See PANTOGRAPH. (2) Mechanism constraining a straightedge to move parallel with the two edges of a drawing board which are at right angles.

parallelodromous (*BioSci*) Having parallel veins.

parallelogram (*MathSci*) A quadrilateral with two pairs of opposite sides parallel.

parallelogram of forces (*Phys*) The PARALLELOGRAM RULE FOR ADDITION OF VECTORS when the vectors are forces.

parallelogram rule for addition of vectors (*MathSci*) The representation of the sum of two vectors as the diagonal through *O* of the parallelogram of which the two vectors, when represented by lines through *O*, are adjacent sides.

parallel-plate capacitor (*Phys*) A device for storage of electric charge, consisting of two plane-parallel conducting electrodes of area A m^2, separated by a dielectric of absolute PERMITTIVITY ε F m^{-1}, thickness d m. Its capacitance is given by $C = \varepsilon A/d$ farads; charge stored when potential difference between plates is V volts is CV coulombs.

parallel-plate chamber (*NucEng*) Ionization chamber with plane-parallel electrodes.

parallel-plate waveguide (*ICT*) One formed by two parallel conducting or dielectric planes. Often realized in atmospheric propagation under suitable meteorological conditions. See SURFACE DUCT.

parallel port (*ICT*) A connection on a computer (or port) through which PARALLEL DATA TRANSMISSION takes place. A parallel printer would be attached to this connector. See LPT PORT, SERIAL PORT.

parallel printer (*ICT*) A printer that receives its data 1 BYTE or one WORD at a time. See PARALLEL DATA TRANSMISSION, SERIAL PRINTER.

parallel processing (*ICT*) The simultaneous performance of two or more tasks on a computer system. It may involve executing several instructions and/or processing several distinct data items. See ARRAY PROCESSOR, MIMD, MISD, MULTIPROCESSOR, PIPELINING, SIMD, THIRD-GENERATION COMPUTER.

parallel resonance (*ElecEng*) See SHUNT RESONANCE.

Parallel Roads (*Geol*) The strandlines of a glacial lake which occupied Glen Roy (Invernesshire, Scotland) during the Pleistocene period, when the lower part of the valley was blocked by ice.

parallel ruler (*Eng*) A drawing instrument consisting of two straightedges so linked together by connecting pieces that their edges are always parallel, although the distance between them may be varied.

parallel slot (*ElecEng*) The most usual shape of slot for the armature windings of electric machines, the slot having parallel sides. Cf *taper slots*.

parallel-T network (*ICT*) See TWIN-T NETWORK.

parallel-wire resonator (*ElecEng*) See LECHER WIRES.

paralogous genes (*BioSci*) Genes that result from duplication of existing genes and then divergence of function. Cf ORTHOLOGOUS GENES.

paralyser (*Chem*) See CATALYTIC POISON.

paralysis (*Med*) The loss in any part of the body of the power of movement, or of the capacity to respond to sensory stimuli. See DIPLEGIA, HEMIPLEGIA, MONOPLEGIA, PARAPLEGIA, QUADRAPLEGIA.

paralysis agitans (*Med*) See PARKINSON'S DISEASE.

paralysis time (*NucEng*) The time for which a radiation detector is rendered inoperative by an electronic switch in the control circuit. See DEAD TIME.

paralytic ileus (*Med*) A condition in which, from various causes, there is extensive paralysis of the intestines, leading to persistent vomiting, pain being absent.

paramagnetism (*Phys*) The phenomenon in materials in which the susceptibility is positive and whose permeability is slightly greater than unity. An applied magnetic field tends to align the magnetic moments of the atoms or molecules and the material acquires magnetization in the direction of the field; it disappears when the field is removed. Used to obtain very low temperatures by adiabatic demagnetization. See FERROMAGNETISM.

paramere (*BioSci*) Half of a bilaterally symmetrical structure; one of the inner pair of gonapophyses in a male insect.

parameter (*Crystal*) The *parameters* of a plane consist of a series of numbers which express the relative intercepts of that plane upon the crystallographic axes. Given in terms of the established unit lengths of those axes.

parameter (*ICT*) (1) A derived constant of a transmission circuit or network, which is more convenient for expressing performance or for use in calculations. (2) Name or value made available to a subprogram (eg a subroutine or procedure) from a calling program or vice versa.

parameter (*MathSci*) (1) *Mathematics.* Generally, a variable in terms of which it is convenient to express other interrelated variables which may then be regarded as being dependent upon the parameter. More specifically, the quantitative design elements which relate to a specific system. (2) *Statistics.* The population value of a particular characteristic which describes the distribution of a random variable.

paramethanal (*Chem*) See PARAFORMALDEHYDE.

parametric amplifier (*ICT*) One that uses a device, commonly a VARACTOR DIODE, whose reactance can be varied periodically by an external pump signal, that is usually at a much higher frequency than that to be amplified; energy can then be transferred from the pump to the signal frequency, without altering the latter. Abbrev *paramp*.

parametric diode (*Electronics*) One whose series capacitance can be varied by the biasing voltage. Usually a VARACTOR diode. See PARAMETRIC AMPLIFIER.

parametric equations (*MathSci*) Of a curve or surface: equations which express the co-ordinates of points on the curve or surface explicitly in terms of other variables (PARAMETERS), which are, for these purposes, regarded as independent variables.

parametric resonance (*ICT*) The condition of a parametric amplifier when energy transfer from the pump circuit through any resonant circuits to the signal is a maximum.

parametritis (*Med*) Inflammation of the pelvic cellular connective tissue in the region of the uterus, eg in the puerperium. Also *pelvic cellulitis*.

parametrium (*Med*) The subperitoneal connective tissue surrounding the uterus, esp that in the region of the cervix.

paramnesia (*Psych*) False memory as eg déjà vu.

paramorph (*BioSci*) A general term for any taxonomic variant within a species, usually used when more accurate definition is not possible.

paramorph (*Min*) A mineral species which has changed its molecular structure without any change of chemical constitution, eg aragonite altered to calcite. Cf PSEUDO-MORPH.

paramp (*ICT*) Same as PARAMETRIC AMPLIFIER.

paramphistomiasis (*Med*) Invasion of the human intestine by trematode parasites of the family *Paramphistomidae*.

paramudras (*Geol*) Flint nodules of exceptionally large size and doubtful significance occurring in the Chalk exposed on the east coast of England.

paramylon (*BioSci*) Reserve polysaccharide allied to starch, but not giving a blue coloration with iodine. A linear $\beta 1 \rightarrow$ 3 glucan present as highly refractive solid bodies in the cytoplasm of the Euglenophyceae, Xanthophyceae and Haptophyceae. Also *paramylum*.

paramyosin (*BioSci*) A protein found in the thick filaments of invertebrate muscle, and in high concentration in the catch muscle that operates the shells of bivalve molluscs such as mussels.

Paramyxoviridae (*BioSci*) Enveloped viruses with a single negative strand of RNA. The main members are Newcastle disease virus, measles virus, and the parainfluenza viruses.

parana pine (*For*) S American softwood (*Araucaria angustifolia*) that is not a true pine (*Pinus*). Its heartwood is soft, non-durable, pale-brown, with straightish grain and uniform texture.

paranephric (*BioSci*) Situated beside the kidney.

paranephros (*BioSci*) See SUPRARENAL BODY.

paranoia (*Med*) A delusion of grievance beyond all bounds of reality. Occurs in a variety of mental diseases including SCHIZOPHRENIA.

paranoid disorder (*Psych*) A personality disorder characterized by extreme suspiciousness in all situations, and with almost all people; delusions of persecution or grandeur may occur, but without the serious disorganization associated with schizophrenia.

paranoid schizophrenic (*Psych*) One of Bleuler's four subtypes of schizophrenia; characterized by delusions of persecution or grandness, with hallucinations and a loss of contact with reality.

paraphase amplifier (*Electronics*) Push–pull stage incorporating PARAPHASE COUPLING. Also *see-saw amplifier*.

paraphase coupling (*Electronics*) A push–pull stage, or series of stages, in which reversed phase is obtained by taking a fraction of the output voltage of the first valve of the amplifier and applying it to a similar balancing first valve, which operates succeeding stages in normal push–pull, the number of transfers being minimized thereby.

paraphasia (*Psych*) A defect of speech in which words are misplaced and wrong words substituted for right ones; due to a lesion in the brain.

paraphilia (*Psych*) Any sexual pattern in which arousal is caused by something other than what is considered a normal sexual object or activity.

paraphimosis (*Med*) Persistent retraction of the inner lining of the prepuce behind the glans penis in a case of phimosis.

paraphonia (*Med*) Alteration of the voice as a result of disease.

paraphrenia (*Med*) A form of SCHIZOPHRENIA with wild, improbable delusions of persecution.

paraphyletic group, paraphyly (*BioSci*) In CLADISTICS, a group that includes a common ancestor and some, but not all, of its descendants. Cf MONOPHYLETIC GROUP, POLY-PHYLETIC GROUP.

paraphysis (*BioSci*) (1) In plants, a sterile filament borne among the reproductive structures of many algae, fungi and bryophytes. (2) In vertebrates, a thin-walled sac developed as an outgrowth from the non-nervous roof of the telencephalon, represented in mammals by the pineal organ. Pl *paraphyses*. Adj *paraphysate*.

parapineal organ (*BioSci*) See PARIETAL ORGAN.

paraplegia (*Med*) Paralysis of the lower part of the body and of the legs.

parapodium (*BioSci*) (1) In Mollusca, a lateral expansion of the foot. (2) In Polychaeta, a paired fleshy projection of the body wall of each somite used in locomotion. Adj *parapodial*.

parapophyses (*BioSci*) A pair of ventrolateral processes of a vertebra arising from the sides of the centrum.

paraprotein (*BioSci*) Immunoglobulin derived from an abnormally proliferating clone of neoplastic plasma cells. The immunoglobulin and the cells making it will all have the same Ig class, subclass and light-chain determinants.

parapsid (*BioSci*) In the skull of reptiles, the condition when there is one temporal vacuity, this being high behind the eye, usually with the post-frontal and supratemporal meeting below. Found in Mesosaurs and Ichthyosaurs. See TEMPORAL VACUITIES for other types.

parapsychology (*Psych*) The study of alleged phenomena, the *paranormal*, that are beyond the scope of ordinary psychology, eg ESP, psychokineses, etc.

paraquat (*Chem*) 1,1'-*dimethyl*-4,4'-*dipyridylium salts*, used as a weedkiller. When taken orally causes severe and often irreversible damage to lungs, liver and kidneys.

paraquinones (*Chem*) Quinones in which the two quinone oxygen atoms are in *para* position.

pararosaniline (*Chem*) A triamino-triphenyl-methanol, obtained by oxidation of a mixture of 4-toluidine (1 mol) and aminobenzene (2 mol). As oxidizing agents, arsenic acid or nitrobenzene are used. Acids effect the elimination of water and the formation of a dyestuff with a quinonoid structure, eg the hydrochloride of pararosaniline or parafuchsine.

pararosolic acid (*Chem*) See AURINE.

paraselenae (*EnvSci*) See MOCK MOONS.

parasexual cycle (*BioSci*) A genetic system in some fungi that allows limited RECOMBINATION as a result of the doubling of the chromosomes in a nucleus, followed by crossing over and the gradual return to the haploid state by progressive chromosome loss or haploidization.

parasheet (*Aero*) A simplified form of parachute for dropping supplies, made from one or more pieces of fabric with parallel warp in the form of a polygon, to the apices of which the rigging lines are attached.

parasitaemia (*BioSci*) Infection of a host by a parasite or the level of infection by the parasite, depending upon context. US *parasitemia*.

parasite (*BioSci*) See PARASITISM.

parasitic antenna (*ICT*) Unfed dipole element which acts as a DIRECTOR or REFLECTOR. See YAGI ANTENNA.

parasitic bronchitis (*Vet*) See HUSK.

parasitic capture (*Phys*) Neutron capture in a reactor not followed by fission.

parasitic castration (*BioSci*) Castration brought about by the presence of a parasite, as in the case of a crab parasitized by *Sacculina*.

parasitic loss (*ElecEng*) Power loss in equipment which is not associated with its principal use, eg eddy currents induced in the different parts of an electric machine.

parasitic male (*BioSci*) A dwarf male in which all but the sexual organs are reduced, and that is entirely dependent on the female for nourishment, as in some deep-sea angler fish (Ceratioids).

parasitic oscillation (*Electronics*) Unwanted oscillation of an amplifier, or oscillation of an oscillator at some frequency other than that of the main resonant circuit. Generally of high frequency, it may occur during a portion of each cycle of the main oscillation. Also *spurious oscillation*.

parasitic stopper (*Electronics*) Components which attenuate a feedback path which would otherwise maintain unwanted oscillations.

parasitism (*BioSci*) A close internal or external partnership between two organisms which is detrimental to one partner (the *host*) and beneficial to the other partner (the *parasite*); the latter often obtains its nourishment at the expense of the nutritive fluids of the host. The term usually refers to such a feeding relationship but other forms exist. See SOCIAL PARASITISM.

parasitoid (*BioSci*) An animal that is parasitic in one stage of the life history and subsequently free-living in the adult stage, as the parasitic Hymenoptera.

parasitology (*Genrl*) The study of parasites and their habits (usually confined to animal parasites).

parasphenoid (*BioSci*) In some of the lower vertebrates, a membrane bone of the skull, which forms part of the cranial floor.

parastas (*Arch*) See PILASTER.

parasymbiosis (*BioSci*) The condition when two organisms grow together but neither assist nor harm one another.

parasympathetic nervous system (*BioSci*) In vertebrates, a subdivision of the autonomic nervous system. The action of these nerves tends to slow down activity in the glands and smooth muscles which they supply, but promotes digestion, and acts antagonistically to that of the SYMPA-THETIC NERVOUS SYSTEM. Parasympathetic nerves are cholinergic. Also *craniosacral system*.

paratenic host (*BioSci*) An animal in which a parasite, accidentally ingested, undergoes no further maturation but remains alive and potentially infectious to the definitive host, should the paratenic host itself be ingested. Also *transport host*.

parathion (*Chem*) 2,2-*diethyl*-2-p-*nitrophenyl-thiophosphate*. Used as an insecticide.

parathyroid (*BioSci*) An endocrine gland of vertebrates found near the thyroid or embedded in it. Two pairs are usually present, deriving from the third and fourth pairs of gill pouches. Probably only secretes PARATHYROID HORMONE.

parathyroid hormone (*BioSci*) A hormone, secreted by the parathyroid, that regulates the calcium balance in the blood plasma and hence the deposition of bone.

paratonic movement (*BioSci*) A plant movement in response to an external stimulus, eg taxis, tropism. See AUTONOMIC MOVEMENT.

paratope (*BioSci*) Same as *antigen binding site*. See NETWORK THEORY.

paratuberculosis (*Vet*) See JOHNE'S DISEASE.

paratyphoid (*Med*) Enteric fever due to infection by *Salmonella* spp other than *S. typhi*; similar to, but milder than, typhoid fever.

paraxial (*ImageTech*) The path of a ray which is parallel to the axis of an optical system.

paraxial focus (*Phys*) The point at which a narrow pencil of rays along the axis of an optical system comes to a focus.

paraxonic foot (*BioSci*) A foot in which the skeletal axis passes between the third and fourth digits, as in the even-toed 'ungulates', Artiodactyla. Cf MESAXONIC FOOT.

Parazoa (*BioSci*) An animal subkingdom comprising multicellular organisms with relatively unspecialized cells that are less interdependent than in the Metazoa. Contains the single phylum PORIFERA. Cf METAZOA, PROTOZOA.

par-boiled rice (*FoodSci*) Rice steeped in hot water to loosen the husk and then dried prior to milling which allows a high proportion of the vitamins and minerals in the husk to be dissolved in the water and transferred to the endosperm. See RICE.

Par C (*Build*) Abbrev for *Parian cement*.

parcel (*For*) Any quantity of standing, felled or converted timber forming a unit or item for purposes of trade.

parcel plating (*ElecEng*) The electrodeposition of a metal over a selected area of an article, the remainder being covered with a non-conductor to prevent deposition.

parchmentizing (*Paper*) The process of passing paper through sulphuric acid, or zinc chloride, which causes the fibres to swell and the paper to become translucent, dense and greaseproof, possessing high wet-strength properties.

parencephalon (*BioSci*) A cerebral hemisphere.

parenchyma (*BioSci*) Tissue composed of mature, vacuolated, but relatively unspecialized cells. Typically blunt-ended, somewhat elongated cells with thin or evenly thickened cell walls, they are not adapted for water transport but are sometimes photosynthetic, as chlorenchyma, or able to act as a store. See SPONGY LAYER.

parent (*Phys*) A radioactive particle which undergoes decay to give a new *daughter* product.

parent directory (*ICT*) The DIRECTORY immediately above the current directory.

parenteral (*Med*) Said of the administration of therapeutic agents by any way other than through the alimentary tract.

parent exchange (*ICT*) Of an automatic exchange system, the manual exchange that handles the assistance (emergency) traffic.

parentheses (*Print*) Marks of punctuation ( ) used to enclose a definition, explanation, reference, etc, or interpolations and remarks made by the writer of the text. Sing *parenthesis*.

parent metal (*Eng*) A term used in welding to denote the metal of the parts to be welded.

parent peak (*Phys*) The component of a mass spectrum coming from the undissociated molecule.

parent program (*ICT*) A master copy of a program from which subsequent copies are made. These copies are called child programs.

paresis (*Med*) Slight or incomplete paralysis. See GENERAL PARESIS (general paralysis of the insane, GPI).

paresis juvenilis (*Med*) General paresis occurring in a child or young person.

paresthesia (*Med*) See PARAESTHESIA.

parfocal (*BioSci*) Microscope objectives mounted in such a way that changing objectives does not cause the specimen to go out of focus.

pargasite (*Min*) A monoclinic amphibole of the hornblende group, particularly rich in magnesium, sodium, calcium and aluminium. It occurs chiefly in metamorphic rocks.

pargeting (*Build*) Covering the interior of a flue with mortar made of cow dung or, in more recent practice, covering the undersurfaces of slates with plaster. Also *pergeting*.

parge-work (*Build*) An ancient form of external plastering using cow dung. Also *parging*.

parging (*Build*) See PARGE-WORK.

parhelia (*EnvSci*) See MOCK SUNS. Sing *parhelion*.

Parian cement (*Build*) A hard plaster made from an intimate mixture of gypsum and borax which has been calcined and then ground to powder.

parietal (*BioSci*) (1) Pertaining to, or forming part of, the wall of a structure. (2) In botanical usage, peripheral. (3) A paired dorsal membrane bone of the vertebrate skull, situated between the auditory capsules.

parietal foramen (*BioSci*) Small rounded aperture in the middle of the united parietals of the skull. Site of the pineal eye.

parietal organ (*BioSci*) The anterior diverticula of the pineal apparatus. When present may be developed as an eye-like organ, the pineal eye.

parietal placentation (*BioSci*) Placentas that develop along the fused margins of the carpels of a unilocular ovary, eg violet.

parietes (*BioSci*) The walls of an organ or a cavity. Sing *paries*.

paring chisel (*Build*) A long chisel with a thinner blade than a firmer tool, used for finishing off work by hand. It is not intended to be struck with a mallet.

paring gouge (*Build*) See SCRIBING GOUGE.

paripinnate (*BioSci*) A pinnate leaf, having no terminal leaflet; even pinnate.

Paris green (*Chem*) *Copper ethanoato-arsenate*, a bright blue-green pigment once used in anti-fouling compositions but now mainly in wood preservatives and as an agricultural chemical. Also *emerald green*, *Schweinfurt green*.

parison (*Glass*) Short length of moulded or extruded glass or polymer which is the precursor for subsequent BLOW MOULDING.

parison variator (*Glass*) A device fitted to extrusion blow moulding machine which automatically controls wall thickness of PARISON to give even wall in final product.

parity (*MathSci*) The property of being ODD or EVEN, eg the parity of four is even.

parity (*Med*) The condition or fact of having borne children.

parity (*Phys*) The principle related to space-reflection symmetry (*mirror symmetry*). The conservation of parity principle states that no fundamental difference can be made between right and left, or in other words the laws of physics are identical in a right- or in a left-handed system of co-ordinates. The law is obeyed for all phenomena described by classical physics but was recently shown to be violated by the weak interactions between elementary particles.

parity bit (*ICT*) Extra binary digit appended to a binary word to produce even or odd parity for a subsequent PARITY CHECK.

parity check (*ICT*) A test for corruption applied to binary data to confirm EVEN PARITY or ODD PARITY.

parity checking (*ICT*) In data communications, an odd–even check used when the total number of zeros or ones is always made odd or even by adding the appropriate extra digit; a check may then be made at the receiver for even or odd parity.

park (*ICT*) To move the read/write HEADS of a HARD-DISK DRIVE to a position away from areas of the disk where information is stored. Now usually done automatically but needed before equipment is moved to reduce the risk of accidental damage to a vital part of the disk's surface.

Parkesine (*Plastics*) TN for the first thermoplastic resin (1865), based on cellulose nitrate, camphor, vegetable oil and other ingredients.

Parke's process (*Eng*) A process used for desilverizing lead. It depends on the fact that when zinc is added to molten lead it combines with any gold or silver present to form compounds that have a very slight solubility in lead. The bullion-rich zinc is then skimmed off.

parking orbit (*Space*) The waiting orbit of a spacecraft between two phases of a mission.

Parkinson's disease (*Med*) 'Shaking palsy'. A progressive disease due to degeneration of dopaminergic nerve cells in the basal ganglia of the brain; characterized by rigidity of muscles (the body being fixed in a posture of slight flexion), mask-like expression of the face, and a coarse tremor, esp of the hands. Similar effects (*Parkinsonism*) can be the result of drug or other intoxication. Also *paralysis agitans*. See RETROPULSION.

parliament hinge (*Build*) See H-HINGE.

paronychia (*Med*) A felon or whitlow. Purulent inflammation of the tissues in the immediate region of the finger nail.

parosmia (*Med*) Abnormality of the sense of smell.

parotid gland (*BioSci*) In some Anura, an aggregation of poison-producing skin glands on the neck; in mammals, a salivary gland situated at the angle of the lower jaw.

parotitis (*Med*) Inflammation of the parotid gland. See MUMPS.

paroxetine (*Pharmacol*) An antidepressant drug, of the *SSRI* class.

parpoint work (*Build*) Stone-wall construction in which the squared stones are laid as stretchers, with occasional courses of headers.

parquet (*Build*) A floor covering of hardwood blocks glued and pinned to the main flooring.

parrot coal (*Min*) See BOGHEAD COAL.

parrot disease (*Med, Vet*) See PSITTACOSIS.

pars (*BioSci*) A part of an organ. Pl *partes*.

pars anterior (*BioSci*) See PARS DISTALIS.

pars distalis (*BioSci*) The anterior part of the adenohypophysis of the pituitary.

parsec (*Astron*) The unit of length used for distances beyond the solar system. It is the distance at which the ASTRONOMICAL UNIT subtends 1 arcsecond, and is therefore 206 265 AU, $3 \cdot 086 \times 10^{13}$ km, $3 \cdot 26$ light-years. From *parallax* *second*. Symbol pc. Kiloparsec (kpc) and megaparsec (Mpc), for 1000 pc and 1 000 000 pc respectively, are widely used in galactic and extragalactic contexts.

parsing (*ICT*) Breaking down of high-level programming language statements into their component parts during the translation process.

pars intermedia (*BioSci*) In higher vertebrates, part of the posterior lobe of the pituitary body, which is derived from the hypophysis at first but tends to become spread over the surface of the PARS NERVOSA as development proceeds.

pars nervosa (*BioSci*) In higher vertebrates, part of the posterior lobe of the pituitary body, developed from the infundibulum.

Parson's steam turbine (*Eng*) A REACTION TURBINE in which rings of moving blades of increasing size are arranged along the periphery of a drum of increasing diameter. Fixed blades in the casing alternate with the blade rings. Steam expands gradually through the blading, from inlet pressure at the smallest section to condenser pressure at the end.

parthen-, partheno- (*Genrl*) Prefixes from Gk *parthenos*, virgin.

parthenocarpy (*BioSci*) The production of fruit without seeds either spontaneously or by artificial induction by AUXINS.

parthenogenesis (*BioSci*) The development of a new individual from a single, unfertilized gamete, often an egg. Adj *parthenogenetic*.

parthenospore (*BioSci*) A spore formed without previous sexual fusion. See AZYGOSPORE.

partial (*Acous*) Any one of the single-frequency components of a complex tone; in most musical complex tones, the partials have harmonic frequencies with respect to a fundamental.

partial agonist (*BioSci*) An agonist for a receptor population that does not produce a maximal response even if all the receptors are occupied.

partial capacitance (*Phys*) The capacitance between all pairs of conductors (including earth) in a circuit; the effective capacitance under specified conditions may be calculated from this network. See CAPACITANCE COEFFICIENTS.

partial common (*ICT*) A trunk that is common to some of the groups in a grading unit.

partial differential coefficient (*MathSci*) If $\psi = f(x,y,z,\ldots)$, then the partial differential coefficient of ψ with respect to x is the limit

$$\lim_{h \to 0} \frac{f(x+h,y,z,\ldots) - f(x,y,z,\ldots)}{h}$$

if it exists: written $\partial\psi/\partial x$, $f_x(x,y,z,\ldots)$, or $f'_x = (x,y,z,\ldots)$.

partial earth (*ElecEng*) An earth fault having an appreciable resistance.

partially oriented yarn (*Textiles*) A synthetic polymer in continuous filament form that already has a substantial degree of molecular orientation but which requires further orientation. This may be done by drawing during a subsequent process such as *texturing*. Abbrev POY.

partial melting (*Geol*) The process by which a rock, subjected to high temperature and pressure, is partly melted and the liquid removed, to solidify to a rock of different composition from the parent.

partial order (*MathSci*) A binary relation (usually written $\leqslant$) defined on a set S which satisfies the following properties for all *a,b,c* in S: (1) $a \leqslant a$ (reflexive); (2) $a \leqslant b$ and $b \leqslant c$ then $a \leqslant c$ (transitive); (3) $a \leqslant b$ and $b \leqslant a$ then $a = b$ (antisymmetric).

partial parasite (*BioSci*) (1) A plant capable of photosynthesis but dependent on another plant for water and mineral nutrients, eg mistletoe. (2) A plant capable of living independently but able to become parasitic in suitable circumstances, ie a *facultative parasite*.

partial pressures (*Chem*) The pressure exerted by each component in a gas mixture. See DALTON'S LAW OF PARTIAL PRESSURES.

partial pressure suit (*Aero*) A laced airtight overall for aircrew members in very-high-flying aircraft. It has inflatable cells to provide the wearer with an atmosphere and external body pressure in the event of cabin pressure failure. Essential for survival above 50 000 ft (15 000 m).

partial pyritic smelting (*Eng*) Blast furnace smelting of copper ores in which some of the heat is provided by oxidation of iron sulphide and some by combustion of coke. See PYRITIC SMELTING.

partial reinforcement (*Psych*) Conditions in which a response is reinforced only some of the time; such responses are more resistant to extinction than responses acquired through CONTINUOUS REINFORCEMENT. Also *intermittent reinforcement*.

partial roasting (*MinExt*) Roasting carried out to eliminate some but not all of the sulphur in an ore or sulphidic concentrate.

partial teleconference (*ICT*) A TELECONFERENCE in which, normally because of BANDWIDTH restrictions, some participants are unable to have a video link, and contribute only by speech.

partial umbel (*BioSci*) One of the smaller group of flowers which together make up a compound umbel.

partial veil (*BioSci*) In some basidiomycete fruiting bodies, eg mushrooms, a membrane joining the edge of the cap to the stalk, rupturing to leave an annulus. Cf UNIVERSAL VEIL.

particle (*MinExt*) A single piece of solid material, usually defined (when small) by its mesh, or size passing through a specified size of sieve.

particle (*Phys*) (1) A useful concept of a small body which has a finite mass but negligible dimensions so that it has no MOMENT OF INERTIA about its CENTRE OF MASS. (2) A volume of air or fluid which has dimensions very small compared with the wavelength of a propagated sound wave, but large compared with molecular dimensions.

particle accelerator (*Electronics*) See ACCELERATOR.

particle exchange (*Phys*) The interaction between fundamental particles by the exchange of another fundamental particle. See GAUGE BOSON.

particle-induced X-ray emission (*Phys*) X-rays emitted and studied when a sample has been bombarded with protons or heavy ions. Abbrev *PIXE*. See panel on TRACE ELEMENT ANALYSIS.

particle mean size (*PowderTech*) The dimension of a hypothetical particle such that, if a powder were wholly composed of such particles, such a powder would have the same value as the actual powder in respect of some stated property. See PARTICLE SIZE.

particle physics (*Phys*) The study of the properties of fundamental particles and of fundamental interactions. Also *high-energy physics*.

particle porosity (*PowderTech*) The ratio of the volume of open pores to the total volume of the particle. See POROSITY.

particle scattering (*Phys*) See SCATTERING.

particle size (*Geol*) The general dimensions of grains in a rock, esp a sediment. Many definitions have been used. One of the more common is the WENTWORTH SCALE (see table). Field definitions are a little less precise. 'If the grains can be distinguished then it is at least silt grade; if it doesn't feel gritty on the teeth, then it is clay.' Also *grain size*.

Dimension	Name
greater than 256 mm	boulder
64 – 256 mm	cobble
4 – 64 mm	pebble
2 – 4 mm	gravel
$\frac{1}{16}$ – 2 mm	sand
$\frac{1}{256}$ – $\frac{1}{16}$ mm	silt
less than $\frac{1}{256}$ mm	clay

particle size (*PowderTech*) The magnitude of some physical dimension of a particle of powder. When the particle is a sphere, it is possible to define the size uniquely by a unit of length. For a non-spherical particle, it is not possible to define a size without specifying the method of measurement. See PARTICLE MEAN SIZE.

particle size analysis (*PowderTech*) The study of methods for determining the physical structure of powdered materials.

particle velocity (*Acous*) In a progressive or standing sound wave, the alternating velocity of the particles of the medium, taken as either the maximum or rms velocity.

particular average (*Ships*) Loss or damage to marine property whose cost is borne by the owner (unless insured against it).

particular integral (*MathSci*) Generally, a solution of a differential equation formed by assigning values to the arbitrary constants in the complete primitive. Non-singular solution of a differential equation containing no arbitrary constants.

particulates (*MinExt*) Microscopic airborne material such as sand and volcanic ash but also industrial dust from power stations and other manufacturing processes.

parting bead (*Build*) A bead fixed to the cased frame of a double-hung window to separate the inner and outer sashes.

parting of bullion (*Eng*) See INQUARTATION.

parting sand (*Eng*) Dry sand sprinkled on the parting face of a mould to prevent adhesion of the two surfaces at the joint when the cope is rammed.

parting slip (*Build*) A thin lath of wood or zinc which keeps the sash weights apart within the cased frame of a double-hung window. Also *mid-feather*, *wagtail*.

partite (*BioSci*) Split almost to the base.

partition (*ICT*) (1) To subdivide the available storage space on a HARD-DISK DRIVE. Each section may be used to hold programs and data for different OPERATING SYSTEMS or to provide additional LOGICAL DRIVES. (2) To subdivide the WORKING STORE such that several APPLICATION PROGRAMS may be stored simultaneously and each be executed for a short period of time in a system using a MULTI-PROGRAMMING OPERATING SYSTEM.

partition (*MathSci*) (1) The division of a set into a collection of disjoint subsets so that their union is the whole set; membership of the same subset is then an EQUIVALENCE RELATION. (2) A decomposition of a positive number n into the sum of positive integers $n = a_1 + a_2 + \ldots + a_r$, where $a_1 \geqslant a_2 \geqslant \ldots \geqslant a_r$; eg the partitions of 4 are 4, $3 + 1$, $2 + 2$, $2 + 1 + 1$ and $1 + 1 + 1 + 1$.

partition chromatography (*Chem*) See CHROMATOGRAPHY.

partition coefficient (*Chem*) The ratio of the equilibrium concentrations of a substance dissolved in two immiscible solvents. If no chemical interaction occurs, it is independent of the actual values of the concentrations. Also *distribution coefficient*.

partition noise (*Electronics*) That arising when electrons are abstracted from a stream by a number of successive electrodes, as in a travelling-wave tube.

partition plate (*Build*) The upper horizontal member of a wooden partition, capping the studding and providing support for joists etc. Also *head piece*.

partridge disease (*Vet*) A popular term for infection of the gut of partridges by the nematode worm *Trichostrongylus tenuis*.

parts rationalization (*Eng*) Design method which seeks to minimize the number of parts in a specific product, thereby saving assembly costs and material, eg in replacing many metal parts with a single injection moulding.

parturient (*Med*) Of or pertaining to parturition.

parturient fever (*Vet*) Also *parturient eclampsia*, *parturient paresis*. See MILK FEVER.

parturition (*BioSci*) In viviparous animals, the act of bringing forth young.

paruresis (*Med*) Difficulty in urinating in the presence of other people.

parvifoliate (*BioSci*) Having leaves that are small in relation to the size of the stem.

parvovirus (*Vet*) See CANINE PARVOVIRUS INFECTION, PORCINE PARVOVIRUS INFECTION.

Parylene (*Plastics*) TN for POLYXYLYLENE film.

PAS (*BioSci*) (1) Abbrev for *periodic acid-Schiff reagent*. (2) See PARA(4)-AMINOSALICYLIC ACID.

PASCAL (*ICT*) Programming language designed to encourage structured programming.

pascal (*Genrl*) The SI derived unit of pressure or stress, equal to 1 newton per square metre. Symbol Pa.

Pascal's theorem (*MathSci*) The theorem that the intersections of pairs of opposite sides of a hexagon inscribed in a conic are collinear. Cf BRIANCHON'S THEOREM, PAPPUS'S THEOREM (2).

Pascal's triangle (*MathSci*) An easily remembered summary of the BINOMIAL COEFFICIENTS in the expansion of $(1 + x^n)$ for $n = 0,1,2,3 \ldots$. Starting with a single 1 in the top line, each digit is the sum of the two above and on either side of it. The triangle, up to $n = 5$, is as follows:

$$
\begin{array}{ccccccccc}
& & & & 1 & & & & \\
& & & 1 & & 1 & & & \\
& & 1 & & 2 & & 1 & & \\
& 1 & & 3 & & 3 & & 1 & \\
1 & & 4 & & 6 & & 4 & & 1 \\
\end{array}
$$
$$ 1 \quad 5 \quad 10 \quad 10 \quad 5 \quad 1 $$

Paschen–Back effect (*Phys*) The splitting of spectrum lines into a number of components by very strong magnetic fields. The fields are strong enough to decouple the spin and orbital angular momenta and the lines are split into three components. See ZEEMAN EFFECT.

Paschen circle (*Phys*) A type of mounting in a concave-grating spectrograph which employs only a small part of the Rowland circle, the slit, grating and plate being fixed in position.

Paschen series (*Phys*) A series of lines like the BALMER SERIES, in the infrared spectrum of hydrogen. Their wavenumbers are given by the same expression as that for the BRACKETT SERIES, but with $n_1 = 3$.

Paschen's law (*Phys*) The observation that the breakdown voltage, at constant temperature, is a function only of the product of the gas pressure and the distance between parallel-plate electrodes.

Pasiphae (*Astron*) The eighth natural satellite of Jupiter, discovered in 1908. Distance from the planet 23 500 000 km; diameter 50 km.

passage beds (*Geol*) The general name given to strata laid down during a period of transition from one set of geographical conditions to another; eg the Downtonian stage consists of strata intermediate in character (and in position) between the marine Silurian rocks below and the continental Old Red Sandstone above.

passage cell (*BioSci*) An endodermal cell in a root, usually opposite the protoxylem, that retains unthickened cell walls in an endodermis in which most of the cells have developed secondary walls.

pass band (*ICT*) The frequency range within which a filter allows signals to pass with minimum attenuation.

passenger leucocytes (*BioSci*) Donor-derived leucocytes carried in an organ to be transplanted. These cells can give rise to GRAFT VERSUS HOST reactions.

Passeriformes (*BioSci*) An order of birds containing those that perch, containing about half the known bird species. They are mostly small, live near the ground and have four toes arranged to allow gripping of the perch. The young are helpless at hatching. Rooks, finches, sparrows, tits, warblers, robins, wrens, swallows and many others.

passings (*Build*) The overlap of one sheet of lead past another in FLASHINGS etc.

passivation (*CivEng*) Protection of steel reinforcement against carbonates and chlorides by the concrete cover which, if broken, allows ingress of water, destruction of the passivity and rapid deterioration. A main cause of failure of tower blocks built in the 1960s.

passivation (*Eng*) (1) To mask the normal electropotential of a metal. A treatment to give greater resistance to corrosion in which the protection is afforded by surface coatings of films of oxides, phosphates, etc. See panel on RUSTING. (2) For non-metallic materials, eg oils, the introduction of substances which have a stabilizing action in preventing chemical reaction under service conditions. Cf ACTIVATION.

passivation layer (*Eng*) A thin coating for the purpose of passivating a surface In semiconductor processing, esp important in protecting high-field surface regions (eg where p–n junctions cut the semiconductor surface) from the environment.

passive (*Electronics*) Said of transducers or filter sections without an effective source emf.

passive–aggressive behaviour (*Psych*) Indirectly expressed resistance to the demands of others, eg forgetting appointments, losing important objects.

passive cutaneous anaphylaxis (*BioSci*) A test *in vivo* to reveal the presence of mast-cell-sensitizing antibody. Antibody is injected into the skin and antigen is injected intravenously shortly thereafter, together with a blue dye that binds to serum albumin. Where the antigen reacts with cell-fixed antibody, vascular permeability is increased, and the dye leaks out to give a blue spot in proportion to the amount of antibody attached.

passive electrode (*ElecEng*) In electrical precipitation equipment, the earthed electrode on which the particles are deposited. Also *collecting electrode*.

passive homing guidance (*Aero*) A missile guidance system which homes on to radiation (eg infrared) from the target.

passive immunization (*BioSci*) Use of antibody from an immune individual to provide temporary immunity in a non-immune individual, eg with diphtheria antitoxin or tetanus antitoxin.

passive intermodulation production (*ICT*) A problem sometimes experienced when high-powered transmitters are operated near metal structures. These reradiate the signal but, particularly where corrosion is present, may do so non-linearly, allowing signal components at different frequencies to modulate each other and produce new, possibly illegal, frequencies not present in the original.

passive margin (*Geol*) A continental margin characterized by thick, relatively undeformed sediments, deposited at the trailing edge of a *lithospheric plate*. See ACTIVE MARGIN.

passive network (*Electronics*) Consists of components in which there is no emf or other source of energy.

passive optical network (*ICT*) Equipment that uses purely optical means to operate on OPTICAL FIBRE signals without converting or amplifying them. See PASSIVE SPLITTER.

passive permeability (*BioSci*) The flux of solutes across a cell membrane by simple diffusion at a rate proportional to the difference in concentration of the solute across the membrane. Also *passive transport*. Cf ACTIVE TRANSPORT (panel).

passive radar (*Radar*) That using microwaves or infrared radiation emitted from source, and hence not revealing the presence or position of the detecting system. Military use.

passive satellite (*Space*) See ACTIVE SATELLITE.

passive splitter (*ICT*) (1) Any device that is able to distribute signal power between more than one channel but does not incorporate an amplifier. (2) A PASSIVE OPTICAL NETWORK used to distribute several channels from a single incoming fibre to several outgoing fibres.

passive transport (*BioSci*) See PASSIVE PERMEABILITY.

passivity (*MinExt*) The lack of response of metal or mineral surface to chemical attack such as would take place with a clean, newly exposed surface. It is due to various causes, including insoluble film produced by ageing, oxidation or contamination; rundown of surface energy at discontinuity lattices; adsorbed layers. Phenomenon prevents use of cyanide process to dissolve large particles of gold, but is sometimes used to aid froth flotation by rendering specific minerals passive to collector agents.

pass-over offset (*Build*) The local bend which enables one pipe to pass over another otherwise in the same plane.

pass sheet (*Print*) The sheet which has been signed by overseer and proofreader and serves as a standard, particularly for colour and register.

password (*ICT*) A sequence of characters that must be keyed in before a user can gain access to a computer system.

pasta (*FoodSci*) An extruded paste made with water and durum wheat flour. Commercial pasta is dried to reduce the moisture content to below 2·5% to ensure long life. Eggs, spinach, herbs and tomato powders can be added to give other colour variants.

paste (*Chem*) Thick dispersion of powder in a fluid. Applied eg to fluid dispersion of polyvinyl chloride in a plasticizer used for a variety of dipping processes such as glove-making, leather cloth and soft toys.

paste (*Glass*) (1) Glass used for imitation gemstones. (2) The combined ingredients of PORCELAIN. (3) See STRASS.

paste (*ICT*) See CUT AND PASTE.

paste (*Print*) (1) Operating an automatic paster to replace the old reel with a new. (2) Securing two webs together during the run of the press.

pasteboard (*Paper*) A laminated product formed by pasting two or more webs or sheets of paper together. The middle layer(s) may if required be of cheaper material than the facings.

pasted filament (*ElecEng*) An electric-lamp filament prepared by squirting through a die a paste formed of powdered metal, usually tungsten, together with a binding material, the latter being subsequently removed by heat treatment.

pasted plate (*ElecEng*) A plate in which the active material of the plate is applied in the form of a paste; used for lead–acid accumulators.

paste mould (*Glass*) A mould for blowing light-walled hollow-ware. As a good finish is needed, the moulds are coated with adherent carbon ('paste'), which is wetted before each blowing operation.

paster tab (*Print*) A strip of gummed paper with a weakened section or other device used to secure the prepared end of the new reel prior to joining it to the old reel on an AUTOMATIC REEL CHANGE.

pasteurellosis (*Vet*) The group name for diseases caused by organisms of the genus *Pasteurella*. *Pasteurella multocida* causes an acute septicaemic and pneumonic disease in several species of animals. Vaccines widely used. See FOWL CHOLERA (fowl), HAEMORRHAGIC SEPTICAEMIA (cattle and sheep), RABBIT SEPTICAEMIA (rabbit), SWINE PLAGUE (pig).

Pasteur filter (*Chem*) See CERAMIC FILTER.

pasteurization (*FoodSci*) Inactivating enzymes and reducing the population of micro-organisms by heating to a minimum temperature between 65 and 100°C for a given period. Effective against vegetative cells but not spores. See FLASH PASTEURIZATION.

pasteurization unit (*FoodSci*) The equivalent heat required to cause a tenfold reduction in the total count of organisms. Abbrev *PU*.

Pastonian (*Geol*) A temperate stage of the Pleistocene. See QUATERNARY.

pasture (*Agri*) Grazed grassland that may be permanent or a LEY.

patagium (*BioSci*) (1) A lobe-like structure at the side of the prothorax in some Lepidoptera. (2) In bats and some other flying mammals, a stretch of webbing between the fore-limb and the hind-limb. (3) In birds, a membranous expansion of the wing. Adj *patagial*.

patand (*Build*) A sill resting on the ground as a support for a post. Also *patin*.

patch (*ICT*) Small fragment of code provided by a software supplier to enable users to modify or correct their own copy of software without requiring a complete replacement.

patch (*Med*) A piece of adhesive material impregnated with a medicinal drug and put on the skin so that the drug can be gradually absorbed.

patch bay (*ICT*) A section of equipment that includes all connectors which terminate units of equipment; inter-connection can be altered and/or test equipment connected.

patch board (*ICT*) One on which simple programs can be set up and modified by making electrical connections between logic elements.

patch clamp (*BioSci*) A type of VOLTAGE CLAMP in which a patch electrode of tip diameter approximately 5 μm (cf MICROELECTRODE) is held against a cell's plasma membrane to form an electrically tight seal, enabling current flowing through individual ION CHANNELS to be measured.

patch in (*ICT*) Temporary insertion of spare apparatus in a circuit by patch cords, usually in a patch bay or field. Similarly removal of defective apparatus is termed *patch out*.

patching up (*Print*) (1) In letterpress, as part of MAKING READY, applying patches of thin paper which will be included in the packing to improve the impression where required. (2) In lithography, the arranging of positives or negatives, in required position on a suitable carrier, preliminary to making a plate.

patch map (*ICT*) In a MIDI system part of a CHANNEL MAP to translate instrument volume settings and sounds for a channel.

patch out (*ICT*) See PATCH IN.

patch panels (*ICT*) Panels of plugs used for making connections between incoming and outgoing lines.

patch test (*Med*) A test for allergy, consisting of the application to the skin of small pads soaked with the allergy-producing substance; if supersensitivity exists inflammation develops at the places where the substance was applied.

patella (*BioSci*) In higher vertebrates, a sesamoid bone of the knee joint or elbow joint.

patent (*BioSci*) Said of leaves and branches that spread out widely from the stem.

patent (*Genrl*) A document which gives an inventor monopoly rights over the manufacture or marketing of a new and non-obvious device, process, material or chemical for (usually) 20 years. It is divided into two parts, the specification and the claim. The claim part represents the kernel of the invention and its wording is critical to the validity of the invention.

patent glazing (*Build*) The name for devices for securing together glass sheets without using putty in sashes, eg roof coverings etc, the connection being made with special metal sections.

patenting (*Eng*) A process applied in the manufacture of steel wire to obtain high tensile strength. Consists of austenizing the wire at 900°C, cooling quickly to 550–600°C and allowing to transform to a very fine pearlitic microstructure. The wire may then be cold drawn to achieve tensile strength levels in excess of 1600 MN m^{-2} without becoming brittle.

patent log (*Ships*) See NAUTICAL LOG.

patera (*Arch*) A circular ornament in relief on friezes. Pl *paterae*.

paternoster (*Eng*) A lift for goods or passengers which consists of a series of floored but doorless compartments moving slowly and continuously up and down on an endless chain. Not considered compatible with current safety regulations.

path (*ICT*) (1) The route to the location of a file in a directory structure. (2) Channel through which signals can be sent, esp forward, through and feedback paths of servo systems.

path attenuation (*ICT*) The fall-off in amplitude of a radio wave with distance from the transmitter.

path environment variable (*ICT*) A VARIABLE used by the OPERATING SYSTEM to locate files. This is usually set before the APPLICATION PROGRAMS are executed. See ENVIRONMENT VARIABLE.

pathetic muscle (*BioSci*) The superior oblique muscle of the vertebrate eye.

pathetic nerve (*BioSci*) The fourth cranial nerve of vertebrates, running to the superior oblique muscle. Also *trochlear nerve*.

pathfinder elements (*Min*) Chemical elements such as Ag, As and Sb are trace elements that are enriched in almost all types of gold deposits and have higher average crustal abundances than gold. Consequently, identification of anomalous concentrations of these is a valuable tool in gold exploration. Other pathfinder elements associated with some gold deposits are B, Ba, Bi, C, Cd, Cu, Hg, Pb, Te, Ti and Zn.

path layer (*ICT*) In a generalized telecommunications network, that part which represents possible routes through the system rather than physical media or types of service.

pathogen (*BioSci, Med*) Any disease-producing micro-organism or substance.

pathogenesis (*Med*) The development or production of a disease process. Adj *pathogenic*.

pathognomonic (*Med*) Specially indicating a particular disease.

pathological (*Med*) Concerning pathology: morbid, diseased.

pathology (*Genrl*) The study of the causes and nature of disease and the resulting changes in the affected organism. See also PHYTOPATHOLOGY.

patin (*Build*) See PATAND.

patina (*Chem*) The thin, often multicoloured, film of atmospheric corrosion products formed on the surface of a metal or mineral.

Patra Test Bench (*Print*) See PIRA TEST BENCH.

patristic similarity (*BioSci*) Similarity due to common ancestry.

patroclinous (*BioSci*) Exhibiting the characteristics of the male parent more prominently than those of the female parent. Cf MATROCLINOUS.

patronite (*Min*) Ore of vanadium, perhaps VS_4.

patten (*Arch*) The base of a column or pillar.

patter (*Build*) A kind of float, made of thick wood, used to consolidate and level cement surfaces.

pattern (*Electronics*) The luminous trace on the screen of a CRO as traced out by the electron beam.

pattern (*Eng*) A wood, metal or plaster copy, in one piece or in sections, of an object to be made by casting. Made slightly larger than the finished casting to allow for contraction of casting while cooling; and suitably tapered to facilitate withdrawal from the mould. See DOUBLE CONTRACTION, METAL PATTERN, PLATE MOULDING.

pattern (*ICT*) The pattern of the radiation field from an aerial system as shown by a polar diagram of field strength and bearing.

patterning (*Electronics*) See SEMICONDUCTOR DEVICE PROCESSING.

pattern-maker's hammer (*Build*) A light type of cross-pane hammer with a long handle.

pattern-maker's rule (*Eng*) One with graduations lengthened so as to compensate for the cooling contraction which must be allowed for in the cast object.

pattern recognition (*ICT*) The automatic recognition of patterns, using a specialized input device linked to a processor. See CHARACTER RECOGNITION, SPEECH RECOGNITION.

pattern staining (*Build*) A defect most apparent on ceiling areas where light and dark patches follow the layout of the ceiling joists. It involves the transmission of heat and dirt through the surface and is difficult to eradicate but is reduced by improved thermal insulation.

Pattinson process (*Eng*) Obsolescent process used for the separation of small quantities of silver from lead by partially solidifying a molten bath of the two metals and separating the remaining liquid. This process is repeated several times and the silver is concentrated in the liquid. Cf PARKE'S PROCESS.

patulin (*FoodSci*) A MYCOTOXIN found in apples and due to mould growth usually arising from damaged or rotted fruit in storage. It is a carcinogen resistant to low pH and tolerant of high temperature. The maximum levels are regulated by law, and in the UK not more than 1 part per billion is tolerated.

Paul–Bunnell test (*Med*) Test used in the diagnosis of infectious mononucleosis.

Pauli exclusion principle (*Phys*) Fundamental law of quantum mechanics, explaining the electronic structure of atoms and also the general nature of the periodic table. See panel on ATOMIC STRUCTURE.

paunch (*BioSci*) See RUMEN.

pavement epithelium (*BioSci*) A variety of epithelium consisting of layers of flattened cells.

pavement light (*Build*) A panel formed of glass blocks framed in concrete, iron or steel, built into a pavement surface over an opening to the basement of a building, into which it admits light.

pavilion (*Arch*) An ornamental, detached structure which has a roof but is usually not entirely enclosed by walls. Used on sports fields, or as a place for entertainments.

pavings (*Build*) Very hard purpose-made bricks, usually of the dark-blue Staffordshire variety, having a surface chequered by grooves to make it less slippery.

pavior (*Build*) (1) A specially hard brick used in the construction of pavement surfaces. (2) A worker who lays bricks or setts to form pavement surfaces.

Pavlovian (*Psych*) Relating to the work of the Russian physiologist Ivan Pavlov (1849–1936) on conditioned reflexes, used more generally of reactions, responses, etc, that are automatic or unthinking.

Pavo (Peacock) (*Astron*) A small southern constellation containing several bright stars.

pawl (*Eng*) A pivoted catch, usually spring-controlled, engaging with a ratchet wheel or rack to prevent reverse motion, or to convert its own reciprocating motion into an intermittent rotary or linear motion.

PAX (*ICT*) Abbrev for PRIVATE AUTOMATIC EXCHANGE.

pax (*Aero*) Airline passengers.

Paxolin (*Plastics*) TN for a paper-reinforced laminated plastic usually of the phenolic class: used in the manufacture of sheets, tubes, cylinders and laminated mouldings and for low-grade printed circuit boards.

pay (*MinExt*) Pay dirt, ore, rock, streak, etc. Any mineral deposit which will repay efficient exploitation.

pay-as-you-go (*ICT*) A term used of mobile phones, denoting that the user, the outright purchaser of the phone, pays only for calls made using credit bought in advance; (of other services, eg the Internet) denoting that charges are made only when the service is used.

payload (*Aero, Space*) (1) That part of an aircraft's load from which revenue is obtained. (2) That part of a military aircraft which is devoted to offensive or defensive actions (bullets, bombs, chaff, etc). (3) That part of a spacecraft additional to that used for structure and for maintaining its essential functions; usually this implies the instruments and supporting hardware for performing experiments. (4) That part of a launch system which is the 'useful' mass placed on the desired trajectory or orbit.

payload integration (*Space*) The process of bringing together individual experiments, their support equipment and SOFTWARE into a payload entity, in which all interfaces are compatible and whose operation has been fully checked out.

payload specialist (*Space*) Highly qualified scientist member of the crew of the Space Shuttle, whose sole responsibility is the operation of the experiments of a payload. He or she is not necessarily a professional astronaut.

payload type (*ICT*) A field in the header of an ASYNCHRONOUS TRANSFER MODE CELL that indicates the general type of data in the cell. Its uses include identification of maintenance cells and of the last cell of a multicell message.

pay string (*MinExt*) The pipe through which oil or gas passes from the *pay zone* to the well head.

pay-TV (*ImageTech*) General term for a specialized form of TV distribution, such as by cable or by coded signals, for which a charge is made.

Pb (*Chem*) Symbol for LEAD.

PBT (*Chem*) Abbrev for POLYBUTYLENE TEREPHTHALATE.

PBX (*ICT*) Abbrev for PRIVATE BRANCH EXCHANGE.

PC (*Chem*) Abbrev for POLYCARBONATE.

PC (*ICT*) Abbrev for PERSONAL COMPUTER.

pc (*Astron*) Abbrev for PARSEC.

PCB (*Chem*) Abbrev for POLYCHLORINATED BIPHENYLS.

PCB (*ICT*) Abbrev for *printed circuit board*. See panel on PRINTED, HYBRID AND INTEGRATED CIRCUITS.

PC-DOS (*ICT*) A variant of MS-DOS marketed by IBM for its personal computers.

PCE (*Eng*) See PYROMETRIC CONE EQUIVALENT.

PCI (*ICT*) Abbrev for PERIPHERAL COMPONENT INTERCONNECT.

PCL (*ICT*) Abbrev for PRINTER CONTROL LANGUAGE.

PCM (*Aero, ICT*) Abbrev for PULSE-CODE MODULATION.

PCM audio (*ImageTech*) Abbrev for PULSE-CODE MODULATION AUDIO.

PCMCIA (*ICT*) Abbrev for *Personal Computer Memory Card International Association*.

PCNB (*Chem*) See QUINTOZENE.

PCO cycle (*BioSci*) Abbrev for *photorespiratory carbon oxidation cycle*. See PHOTORESPIRATION.

PCR (*BioSci*) Abbrev for POLYMERASE CHAIN REACTION.

PCR cycle (*BioSci*) Abbrev for *photosynthetic carbon reduction cycle*. See panel on CALVIN CYCLE.

Pd (*Chem*) Symbol for PALLADIUM.

pd (*Phys*) Abbrev for POTENTIAL DIFFERENCE.

PDA (*ICT*) Abbrev for PERSONAL DIGITAL ASSISTANT.

PDC (*ImageTech*) Abbrev for PROGRAM DELIVERY CONTROL.

PDF (*ICT*) Abbrev for PORTABLE DOCUMENT FORMAT, a format that makes documents accessible using a browser.

P-display (*Radar*) That of a PPI unit. Map display produced by intensity modulation of a rotating radial sweep.

PDL (*ICT*) Abbrev for PAGE DESCRIPTION LANGUAGE.

PDN (*ICT*) Abbrev for PUBLIC DATA NETWORK.

PDS (*Eng*) Abbrev for PRODUCT DESIGN SPECIFICATION.

PE (*Chem*) Abbrev for POLYETHYLENE.

pE (*Chem*) Negative logarithm of effective electron concentration in a redox system. Its use is analogous to that of pH in acid–base reactions. See RH.

peacock ore (*Min*) A name given to BORNITE or sometimes CHALCOPYRITE, because they rapidly become iridescent from tarnish.

peak (*Phys*) The instantaneous value of the local maximum of a varying quantity. It is $\sqrt{2}$ times the rms value for a sinusoid.

peak cell rate (*ICT*) A measure, used to characterize ASYNCHRONOUS TRANSFER MODE traffic for planning purposes. It indicates the maximum rate at which CELLS can be offered to the network.

peak dose (*Radiol*) Maximum absorbed radiation dose at any point in an irradiated body, usually at a small depth below the surface, due to secondary radiation effects.

peak envelope power (*ICT*) See PEAK POWER.

peak factor (*ElecEng*) The ratio of peak value of any ac or voltage to rms value. Also *crest factor*.

peak forward voltage (*ElecEng*) The maximum instantaneous voltage in the forward flow direction of anode current as measured between the anode and cathode of a diode or thyristor.

peaking (*ICT*) Inclusion of series or shunt resonant units in eg TV circuits to maintain response up to a maximum frequency.

peaking network (*ICT*) An interstage coupling circuit that gives a peak response at the upper end of the frequency range that is handled. This is achieved with a resonant circuit and minimizes the fall-off in the frequency response produced by stray capacitances. Not to be confused with *peaking circuit*.

peaking transformer (*ICT*) One in which the core is highly saturated by current in the primary, thus providing a peaky emf in the secondary as the flux in the core suddenly changes over.

peak inverse voltage (*ElecEng*) See INVERSE VOLTAGE.

peak joint (*Build*) The joint between the members of a roof truss at its ridge, nowadays often steel plates with projecting teeth hydraulically pressed into the timber.

peak limiter (*ICT*) A circuit for avoiding overload of a system by reducing gain when the peak input signal reaches a certain value.

peak load (*ElecEng*) The maximum instantaneous rate of power consumption in the load circuit. In a power-supply system, the peak load corresponds to the maximum power production of the generator(s).

peak molecular mass (*Chem*) Symbol M_p. See MOLECULAR MASS DISTRIBUTION.

peak power (*ICT*) Average radio-frequency power at maximum modulation, ie of envelope of transmission.

peak programme meter (*ICT*) One bridged across a transmission circuit to indicate the changes in volume of the ultimate reproduction of sound, averaging peaks over 1 ms.

peak sideband power (*ICT*) The average sideband power of a transmitter over one radio-frequency cycle at the highest peak of the modulation envelope.

peak-to-peak amplitude (*Phys*) See DOUBLE AMPLITUDE.

peak value (*Phys*) The maximum positive or negative value of an alternating quantity. Also *amplitude*, *crest value*, *maximum value*.

peak voltmeter (*ElecEng*) Measures the peak value of an alternating voltage, eg by biasing a diode so that it just conducts. Also *crest voltmeter*.

peak white (*ImageTech*) The level in a TV signal corresponding to white.

pear (*For*) Hardwood (*Pyrus communis*) from Europe and W Asia. Its heartwood is pinkish-brown, straight-grained and has a very fine and even texture.

pearl (*BioSci*) An abnormal concretion of nacre formed inside a mollusc shell round a foreign body such as a sand particle or a parasite. See GEM.

pearl (*Print*) An old type size, approximately 5-point.

pearlite (*Eng*) A microconstituent of steel and cast-iron comprising an intimate mechanical mixture of ferrite and cementite (iron carbide), so called because in etched sections it appears iridescent and resembles mother of pearl. It is produced at the eutectoid by the simultaneous formation of ferrite and cementite from austenite, and normally consists of alternate plates or lamellae of these two constituents (see, however, GRANULAR PEARLITE). A carbon steel containing 0·8% of carbon consists entirely of pearlite when cooled under equilibrium conditions. See EUTECTOID STEEL, HYPEREUTECTOID STEEL, HYPOEUTECTOID STEEL.

pearlitic iron (*Eng*) A grey cast-iron consisting of graphite in a matrix which is predominantly pearlite, and consequently stronger than one with a ferritic matrix.

pearl spar (*Min*) See DOLOMITE.

pearl white (*Chem*) See BISMUTH (III) CHLORIDE.

pear oil (*Chem*) See AMYL ACETATE.

Pearson product-moment correlation (*MathSci*) A correlation statistic used primarily for two sets of data on a ratio or interval scale. The most commonly used correlational technique.

peas (*MinExt*) See COAL SIZES.

peat (*Geol*) Layers of dead vegetation, in varying degrees of alteration, resulting from the accumulation of the remains of marsh vegetation in swampy hollows in cold and temperate regions. Geologically, peat may be regarded as the youngest member of the series of coals of different rank, including brown coal, lignite and bituminous coal, which link peat with anthracite. Peat is very widely used as a fuel, after being air-dried, in districts where other fuels are scarce and in some areas, eg in Russia and Ireland, it is used to fire power stations. It is low in ash, but contains a high percentage of moisture and is bulky; specific energy content about 16 MJ kg^{-1} or 7000 Btu lb^{-1}.

peau de soie (*Textiles*) Fine-silk, satin-weave fabric, sometimes reversible, with a ribbed or grained appearance; from Fr *skin of silk*.

pebble (*Min*) A small rounded fragment of rock between 4 and 64 mm diameter. Adj *pebbly*, 'containing scattered pebbles'. See PARTICLE SIZE, WENTWORTH SCALE.

pebble-bed reactor (*NucEng*) One with a cylindrical core into which spherical fuel pellets are introduced at the top and extracted at the base.

pebble-dashing (*Build*) A rough finish given to a wall by coating it with rendering, onto which, while it is still soft, small stones are thrown. Cf ROUGH CASTING.

pebble mill (*MinExt*) BALL MILL in which selected pebbles or large pieces of ore are used as grinding media.

peck (*Genrl*) Obsolete term for a measure of capacity for dry goods, 2 gallons, or a quarter of a bushel.

pecking order (*Psych*) The classic example of social behaviour in farmyard hens in which animals within a group form some consistent dominance hierarchy most apparent in their aggressive interactions, which hen pecks and which is pecked; the term is used metaphorically of similar dominance hierarchies.

peckings (*Build*) Under-burnt, badly-shaped bricks, used only for temporary work or for the inside of walls.

pecten (*BioSci*) (1) Any comb-like structure. (2) In some vertebrates (reptiles and birds), a vascular process of the inner surface of the retina. (3) In Scorpionidea, tactile sensory organs under the mesosoma.

pectinate (*BioSci*) Comb-like.

pectineal (*BioSci*) (1) Generally, comb-like. (2) A process of the pubis in birds. (3) A ridge on the femur to which is attached the pectineus muscle, one of the protractors of the hind-limb.

pectines (*BioSci*) Comb-like chitinous structures of mechanoreceptive function attached to the ventral surface of the second somite of the mesosoma in Scorpionidea.

pectins (*Chem*) Calcium–magnesium salts of polygalacturonic acid, partially joined to methanol residues by ether linkage. They occur in the cell walls, esp in the *middle lamellae* and *primary walls* of vascular plants. They are soluble in water and can be precipitated from aqueous solutions by excess alcohol. Acid solutions gel with 65–70% of sucrose, the basis of their use in jam-making.

pectization (*Chem*) The formation of a jelly.

pectolite (*Min*) A silicate of calcium and sodium, with a variable amount of water, which crystallizes in the triclinic system. It occurs in aggregations like the zeolites in the cavities of basic eruptive rocks, and as a primary mineral in some alkaline igneous rocks.

pectorales (*BioSci*) In vertebrates, muscles connecting the upper part of the fore-limb with the ventral part of the pectoral girdle. Sing *pectoralis*.

pectoral fins (*BioSci*) In fish, the anterior pair of fins.

pectoral girdle (*BioSci*) In vertebrates, the skeletal framework with which the anterior pair of locomotor appendages articulate.

pectoriloquy (*Med*) Conduction, to the chest wall, of the sound of words spoken or whispered by the patient and clearly heard through the stethoscope; indicative of consolidation or cavitation of the lung.

peculiar (*Print*) A term describing any unusual type character, such as certain accents.

peculiar velocity (*Astron*) The velocity of a star when the general motion of it and its neighbours round the galactic centre is subtracted.

ped- (*Genrl*) See PAED-.

pedal curve (*MathSci*) Of a given curve with respect to a given point: the locus of the point of intersection of a tangent to the curve and the perpendicular to the tangent from the given point.

pedate leaf (*BioSci*) A leaf with three divisions of which the two laterals are forked once or twice.

pedes (*BioSci*) See PES.

pedesis (*Phys*) See BROWNIAN MOVEMENT.

pedestal (*ImageTech*) The difference in signal between the BLACK LEVEL of the picture and the BLANKING level. Also lift.

pediatric, pediatrician (*Med*) See PAEDIATRIC, PAEDIATRICIAN.

pedicel (*BioSci*) (1) A stalk bearing one flower. (2) The stalk of a sedentary organism, or free organ, as eg the optic pedicel in some Crustacea. (3) The second joint of the antennae in insects. Adj *pedicellate*. Cf PEDUNCLE.

pedicellaria (*BioSci*) In Echinodermata, a small pincer-like calcareous structure on the body surface with two or three jaws provided with special muscles and capable of executing snapping movements; it may be stalked or sessile.

pedicle (*BioSci*) (1) Generally any pillar-like process supporting an organ. (2) In the vertebrae of the frog, a pillar-like process springing from the centrum and extending vertically upwards to join the flat, nearly horizontal lamina, which forms the roof of the neural canal. Intervertebral foramina, for the passage of the spinal nerves, occur between successive pedicles.

pediculosis (*Med*) Infestation of the body with lice.

pediment (*Arch*) A triangular or segmental part surmounting the portico in the front of a building.

pediment (*Geol*) A broad and relatively flat rock surface abutting a mountain range, in an arid environment. It may be covered by a veneer of alluvium.

pedion (*Crystal*) A crystal form consisting of a single plane; well shown by some crystals of tourmaline which may be terminated by a pedion, with or without pyramid faces.

pedipalp (*BioSci*) In Chelicerata, the appendage on which the gnathobase functions as a jaw. Borne by the first postoral somite, it may be a tactile organ or a chelate weapon.

pedo- (*Genrl*) See PAED-.

pedology (*Genrl*) The scientific study of soils.

pedometer (*Surv*) An instrument for counting the number of paces (and hence the approximate distance) walked by its wearer.

peduncle (*BioSci*) (1) A stalk bearing several flowers. (2) In Brachiopoda and Cirripedia, the stalk by which the body of the animal is attached to the substratum. (3) In some Arthropoda, the narrow portion joining the thorax and abdomen or the prosoma and opisthosoma. (4) In vertebrates, a tract of white fibres in the brain. Adj *pedunculate*. Cf PEDICEL.

PEEK (*Chem*) Abbrev for POLYETHER ETHER KETONES.

peek (*ICT*) To read the contents of a single specified location of MEMORY. Cf POKE.

peeling (*Eng*) A layer of a coating becoming detached because of poor adhesion.

peel test (*Eng*) Mechanical test for adhesive bonding between two surfaces in which the separation force is applied in such a way as to cause progressive failure of the adhesive bond along a linear interface rather than simultaneously over the whole area.

peening (*Eng*) Using hammer blows or shot blasting to cold-work the surface layers of a metal object.

peer-to-peer system (*ICT*) A method of sharing information between computers whereby each acts as a SERVER for the others as well as a CLIENT. Each device has equal status. In such a system there is no need to have a central FILE SERVER. Abbrev *P2P*.

PEG (*BioSci*) Abbrev for POLYETHYLENE GLYCOL.

peg-and-cup dowels (*Eng*) Metal pegs and sleeves inserted in adjoining parts of a split pattern to hold them in register while ramming the mould.

Pegasus (Winged Horse) (*Astron*) A conspicuous northern constellation.

peggies (*Build*) Slates 10–14 in (254–356 mm) in length.

pegmatite (*Geol*) A term originally applied to granitic rocks characterized by intergrowths of feldspar and quartz, as in graphic granite; now applied to igneous rocks of any composition but of particularly coarse grain, occurring as offshoots from, or veins in, larger intrusive rock bodies, representing a flux-rich residuum of the original magma.

pel (*ICT*, *ImageTech*) An earlier, now less common, word for a PIXEL.

pelagic (*BioSci*) Living in the middle depths and surface waters of the sea.

pelagic deposits (*Geol*) A term applied to any accumulation of sediments under deep-water conditions.

pelargonic acid (*Chem*) *Nonan-1-oic acid*. $CH_3(CH_2)_7COOH$. Mp 12·5°C. Oxidation product of oleic acid. Occurs naturally in fusel oil from beet and potatoes.

Péléan eruption (*Geol*) A type of eruption characterized by lateral explosions generating NUÉES ARDENTES.

Pelecaniformes (*BioSci*) A varied order of birds, mainly fish-eating and colonial nesters, with four-toed webbed feet, bodies adapted for diving, and long beaks with wide gapes and sometimes with a pouch. Pelicans, cormorants, gannets.

Pelecypoda (*BioSci*) See BIVALVIA.

Pelé's hair (*Min*) Long threads of volcanic glass which result from jets of lava being blown aside by the wind in the volcano of Kilauea, Hawaii.

pele tower (*Arch*) A term used in N England and Scotland to describe a small tower which was used to provide a retreat in the event of a sudden, unexpected attack.

pelitic gneiss (*Geol*) A gneissose rock derived from the metamorphism of argillaceous sediments.

pelitic schist (*Geol*) A schist of sedimentary origin, formed by the dynamothermal metamorphism of argillaceous sediments such as clay and shale.

pellagra (*Med*) A chronic disease due to a dietary deficiency of the nicotinic acid component of the vitamin B complex, associated with protein deficiency. It is characterized by gastro-intestinal disturbances, a symmetrical erythema of the skin, mental depression and paralysis. Seen in those whose diet consists predominantly of maize where the nicotinic acid is in bound form and there is a lack of the tryptophan precursor of nicotinic acid. Also *maidismus*.

pellet (*Build*) A moulding with a line of spherical protuberances.

pellet (*NucEng*) Nuclear fusion fuel contained in concentric spheres of glass, plastic and other materials. They are hit by a burst of laser energy to produce fusion.

pellet (*Plastics*) A compressed mass of moulding material of prescribed form and weight.

pelleted seed (*BioSci*) Seed coated with a layer of inert material, esp to make smaller, angular seeds into larger, rounded bodies that can be drilled more precisely, sometimes also to incorporate pesticides etc.

pelletization (*MinExt*) The treatment of finely divided ore, concentrate or coal to form aggregates some 1–1·5 cm in diameter, for furnace feed, transport, storage or use (eg as coal briquettes). Powder, with suitable additives, is rolled into aggregates (green balls) in pelletizing drum and then hardened in a furnace by specialized baking methods.

pellicle (*BioSci*) (1) Layer of interlocking, helically wound, proteinaceous strips just below the plasmalemma of Euglenophyceae. (2) A thin cuticular investment, as in some Protozoa. Adj *pelliculate*.

pellicle (*ImageTech*) An extremely thin transparent film which can be used as a semi-reflecting surface without producing double reflections. Also *pellicule*.

pellicle (*Phys*) A strippable photographic emulsion used to form a STACK in nuclear research emulsion techniques.

pellicule (*ImageTech*) See PELLICLE.

Pellin–Broca prism (*Phys*) A four-sided prism, which can be imagined to be formed by placing together two 30° prisms and one 45° prism. It is used in wavelength spectrometers where the wavelength of the light received in the telescope is altered by rotation of the prism.

pelma (*BioSci*) See PLANTA.

pelorus (*Ships*) A pivoted dial, marked in compass points or degrees, which is on a stand with a 'lubber line' and fitted with sight vanes. It was set manually to match the compass card and once used to take bearings.

peltate (*BioSci*) A flattened rounded plant organ attached to its stalk at about the middle of its lower surface, eg the leaf of the nasturtium (*Tropaeolum maius*).

Peltier coefficient (*ElecEng*) Energy absorbed or given out per second, due to the PELTIER EFFECT, when unit current is passed through a junction of two dissimilar metals. Symbol Π.

Peltier effect (*ElecEng*) A phenomenon whereby heat is liberated or absorbed at a junction when current passes from one metal to another.

Pelton wheel (*Eng*) An impulse water turbine in which specially shaped buckets attached to the periphery of a wheel are struck by a jet of water, the nozzle being either deflected or valve-controlled by a governor.

pelvic fins (*BioSci*) In fish, the posterior pair of fins.

pelvic girdle (*BioSci*) In vertebrates, the skeletal framework with which the posterior pair of locomotor appendages articulate.

pelvimetry (*Med*) Estimation, by the use of X-rays but more usually ultrasound, of the size and shape of the female pelvis.

pelvis (*BioSci*) The pelvic girdle or posterior limb girdle of vertebrates, a skeletal frame with which the hind-limbs or fins articulate; in mammals, a cavity, just inside the hilum of the kidney, into which the uriniferous tubules discharge and which is drained by the ureter. Adj *pelvic*.

pemphigus (*Med*) Inflammatory condition of the skin characterized by the eruption of crops of blisters, the mucous membranes at times also being involved.

pen (*BioSci*) In Cephalopoda, the shell or cuttle bone.

pen (*ICT*) See LIGHT PEN.

pencatite (*Geol*) A crystalline limestone which contains brucite and calcite in approximately equal molecular proportions.

pencil (*MathSci*) Of lines: a number of lines passing through a fixed point (*vertex*). Of planes: a number of planes that pass through a common line. See also RAY.

pencil (*Phys*) A narrow beam of light, having a small angle of convergence or divergence.

pencil hardness (*Eng*) See PENCIL HARDNESS TEST.

pencil hardness test (*Eng*) Simple test for hardness by scratching surface with pencils of varying hardness (8H to 4B). See panel on HARDNESS MEASUREMENTS.

pencil overgrainer (*Build*) An OVERGRAINING brush consisting of a row of pencil brushes in a setting and bound with metal. Best quality are made from sable hair with hog hair as cheaper substitute. Used to add equidistant lines in graining work

pencil stone (*Min*) The compact variety of pyrophyllite, used for slate-pencils. The term *pencil ore* has been used for the broken splinters of radiating massive haematite, as they give a red streak.

pendentive (*Arch*) A spherical triangle formed by a dome springing from a square base.

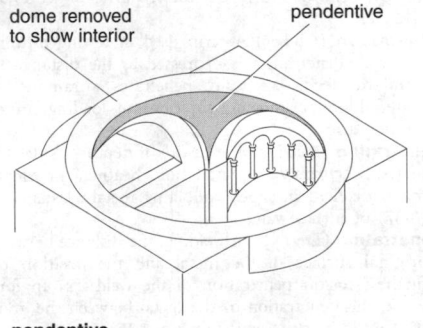

dome removed to show interior   pendentive

pendentive

pendentive dome (*Arch*) A dome covering a square area to which it is linked at the corners by pendentives.

pendulous placentation (*BioSci*) See APICAL PLACENTATION.

pendulum (*Phys*) The simple pendulum consists of a small, heavy bob suspended from a fixed point by a thread of negligible weight. Such a pendulum, when swinging freely with small amplitude, has a periodic time given by

$$T = 2\pi\sqrt{l/g}$$

where l is the length of the thread and g is the acceleration due to gravity. May be used as the time-controlling element of a pendulum clock. The theoretical length of a pendulum, in mm, is then given by $L = 993 \cdot 6t^2$, where t is the time of swing in seconds and $993 \cdot 6$ mm is the length of a pendulum beating seconds in London. For household clocks, the pendulum beats $0 \cdot 5$ s or less. For long-case clocks and regulators, a seconds pendulum is used; for tower clocks, it may be up to 2 s. See CENTRE OF OSCILLATION, COMPOUND PENDULUM.

pendulum damper (*Aero*) A short heavy pendulum, in the form of pivoted balance weights, attached to the crank of a radial aero-engine in order to neutralize the fundamental torque impulses and so eliminate the associated critical speed.

pendulum governor (*Eng*) An engine governor, the many forms of which involve the principle of the conical pendulum. Heavy balls swing outwards under centrifugal force, so lifting a weighted sleeve and progressively closing the engine throttle valve. See PORTER GOVERNOR, WATT GOVERNOR.

pendulum impact test (*Eng*) A method of loading an impact test by a dropping weight with impact speeds which range from 1 to 6 m s^{-1}. See panel on IMPACT TESTS.

penecontemporaneous (*Geol*) Describes any process occurring in a rock very soon after its formation.

peneplain (*Geol*) A gently rolling lowland, produced after long-continued denudation.

penetrance (*BioSci*) The proportion of individuals of a specified genotype in which a particular gene exhibits its effect.

penetrant (*Chem*) (1) Substance which increases the penetration of a liquid into porous material or between contiguous surfaces, eg alkyl-aryl sulphonate. (2) A WETTING AGENT.

penetrant flaw detection (*Eng*) The use of a dye, frequently fluorescent, which will penetrate any minute crack or flaw in a component. After immersion the component is wiped dry and any subsequent seepage from fissures is detected, eg by irradiation at the exciting wavelength or by drawing the liquid out into a white absorbent coating applied after drying off. Also *penetrant inspection*.

penetrant inspection (*Eng*) See PENETRANT FLAW INSPEC-TION.

penetrating shower (*Phys*) Cosmic-ray shower containing MESONS and/or other penetrating particles. See CASCADE SHOWER.

penetration (*CivEng*) A term used in testing bituminous material. Penetration is expressed as the distance that a standard needle vertically penetrates a sample of the material under known conditions of loading, time and temperature.

penetration (*ElecEng*) A measure of depth of SKIN EFFECT of eddy currents in induction heating, or depth of magnetic field in superconducting metals. Usually to $1/e$ ($0 \cdot 37$) of surface value.

penetration (*Eng*) (1) In welding, the distance between the original surface of the metal and the position of the furthest fusion penetration of the weld. (2) In foundry work, the penetration of the metal between the grains of the moulding sand which may mar the surface.

penetration depth (*ICT*) That thickness of hollow conductor of the same external dimension which, if the current were uniformly distributed throughout the cross-section, would have the same effective resistance as the solid conductor. See SKIN DEPTH.

penetration depth (*Phys*) (1) Depth to which an electromagnetic field is able to penetrate a conductor. See SKIN DEPTH. (2) Depth to which a magnetic field is able to penetrate a superconducting material.

penetration factor (*Phys*) The probability of an incident particle passing through the nuclear potential barrier.

penetration theory (*ChemEng*) The theory that mass transfer across an interphase into a stirred liquid takes place by diffusive penetration of the solute into the liquid surface, which is continually being renewed; hence the rate of mass transfer is proportional to the square root of the *diffusion coefficient*.

penetration twins (*Min*) See INTERPENETRATION TWINS. Cf JUXTAPOSITION TWINS.

penetrometer (*Agri*) A calibrated device for measuring resistance to penetration of tissues such as fruit flesh, used as a ripening indicator.

penetrometer (*Radiol*) A device for the measurement of the penetrating power of radiation by comparison of the transmission through various absorbers.

penicillamine (*Pharmacol*) A drug used for the treatment of rheumatoid arthritis with effects similar to gold (sodium aurothiomalate) although the mode of action is unclear. Also a chelating agent used to treat copper accumulation in Wilson's disease.

penicillins (*Pharmacol*) A group of antibiotics which are bactericidal and act by interfering in bacterial cell-wall synthesis.

penis (*BioSci*) The male copulatory organ in most higher vertebrates; a form of male copulatory organ in various invertebrates, eg Platyhelminthes, Gastropoda. Adj *penial* or *penile*.

penis envy (*Psych*) The Freudian concept of a woman's subconscious wish for male characteristics.

Penman–Monteith equation (*EnvSci*) Describes the dependence of evapotranspiration or transpiration rates on climatological variables and surface properties of the vegetation.

pennae (*BioSci*) See PLUMAE.

pennate (*Genrl*) Generally, winged.

pennine (*Min*) A silicate of magnesium and aluminium with chemically combined water. It crystallizes in the monoclinic system and is a member of the chlorite group. Also *penninite*.

Pennsylvanian (*Geol*) The period of the Upper Palaeozoic era lying between the Mississippian and Permian. It covers an approx time span from 330 to 290 million years and is the N American equivalent of the Upper Carboniferous in Europe. The corresponding system of rocks. See PALAEO-ZOIC.

pennyweight (*Genrl*) Troy weight of $\frac{1}{20}$ ounce or $1 \cdot 5552$ g, a unit widely used in valuation of gold ores and sale of bullion. Symbol dwt.

pen ruling (*Print*) The ruling of paper on a pen-ruling machine as distinct from DISK RULING.

Pensky–Martens test (*Chem*) Standard test for determining the flash and fire points of oils. Based on closed or open cups depending on the nature of the oil under test.

penstock (*Eng*) The valve-controlled water conduit between the intake and the turbine in a hydroelectric or similar plant.

pent-, penta- (*Chem*) Containing five atoms, groups, etc.

pent-, penta- (*Genrl*) Prefixes from Gk *pente*, five. Used in the construction of compound terms, eg *pentactinal*, five-rayed.

pentachlorophenol (*Chem*) C_6Cl_5OH. Mp 189°C. Widely used fungicidal and bactericidal compound, used particularly for timber protection.

pentad (*EnvSci*) A period of 5 days; being an exact fraction of a normal year, it is useful for meteorological records.

pentadactyl (*BioSci*) Having five digits.

pentadactyl limb (*BioSci*) The characteristic free appendage of Tetrapoda with five digits.

pentaerythritol (*Chem*) $C(CH_2OH)_4$. Mp 260°C. Condensation derivative of ethanal and methanal. Used in the production of surface finishes.

pentaerythritol tetranitrate (*Chem*) $C(CH_2ONO_2)_4$. A detonating explosive, abbrev *PETN*.

pentagon (*MathSci*) A five-sided polygon.

pentagonal dodecahedron (*Crystal*) A form of the cubic system comprising twelve identical pentagonal faces.

pentahydric alcohols (*Chem*) Alcohols containing five hydroxyl groups, eg arabitol, $HOCH_2(CHOH)_3$ CH_2OH, xylitol (stereoisomeric) and rhamnitol, $HOCH_2(CHOH)_4CH_3$.

pentamerous (*BioSci*) Having five members in a whorl.

pentamethylene (*Chem*) See CYCLOPENTANE.

pentamethylene-diamine (*Chem*) See CADAVERINE.

pentamethylene glycol (*Chem*) *Pentane-1,5-diol.* $CH_2OH(CH_2)_3CH_2OH$. Bp 239°C, rel.d. 0·994. Organic solvent used in syntheses.

pentanes (*Chem*) C_5H_{12}. Bp 36°C, rel.d. 0·63. Low-boiling paraffin hydrocarbons.

pentaprism (*ImageTech*) Five-sided prism which corrects lateral inversion, used on reflex cameras to allow eye-level viewing.

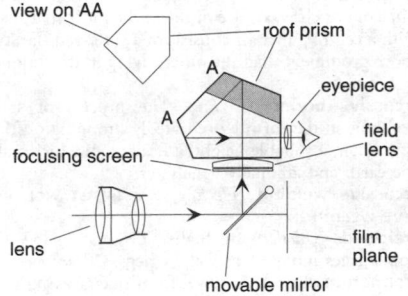

pentaprism As used on a single-lens reflex camera.

pentarch (*BioSci*) A stele having five strands of protoxylem as in roots of many dicotyledons.

pentavalent (*Chem*) Capable of combining with five atoms of hydrogen or their equivalent. Having an oxidation number or co-ordination number of five.

pentazocine (*Pharmacol*) A powerful synthetic drug with opiate action.

penthouse (*Arch*) An individual dwelling situated on the roof of a building but forming an integral part of the building.

penthouse (*MinExt*) Protective covering for workers at bottom of shaft. Also *pentice*.

Pentium (*ICT*) The fifth generation of Intel microprocessors designed for personal computers.

pentlandite (*Min*) A sulphide of iron and nickel which crystallizes in the cubic system. It commonly occurs intergrown with pyrrhotite.

pentode valve (*Electronics*) A five-electrode thermionic tube, comprising an emitting cathode, control grid, a screen (or auxiliary grid) maintained at a positive potential with respect to that of the cathode, a suppressor grid maintained at about cathode potential, and an anode. It has characteristics similar to those of a screened-grid valve, except that secondary emission effects are suppressed.

Penton (*Plastics*) TN for a chlorinated polyether widely used as a coating material for vessels etc where very good chemical resistance is required. Relatively expensive but cheaper than the fluorinated plastics. Film form called *Pentaphane*. See POLYTETRAFLUOROETHYLENE.

pentosans (*Chem*) $(C_5H_8O_4)_x$. Polysaccharides, comprising arabinans and xylans.

pentose phosphate pathway (*BioSci*) An alternative metabolic route to the Embden–Meyerhof pathway for breakdown of glucose. It is important as a source of pentoses, eg for nucleic acid biosynthesis. In plants, part of the pathway functions in the formation of hexoses from CO_2 in photosynthesis.

pentoses (*Chem*) A group of monosaccharides containing five carbon atoms in the molecule, and having the formula $HOCH_2(CHOH)_3CHO$ and $CH_3(CHOH)_4CHO$. Important pentoses are l-arabinose, l-xylose, rhamnose and fucose. Pentoses cannot be fermented. They are characterized by the fact that they yield furfuraldehyde or its homologues on boiling with dilute acids. A qualitative test for pentoses is the occurrence of a bright-red colour when they are boiled with HCl and phloroglucinol.

pentose shunt (*BioSci*) A series of metabolic reactions that converts glucose-6-phosphate into ribose-5-phosphate with concomitant generation of NADPH.

Pentothal (*Chem*) TN for THIOPENTONE.

pentyl- (*Chem*) Of or relating to an alkyl radical derived from pentane.

penumbra (*Astron*) See UMBRA.

PEO (*Chem*) Abbrev for POLYETHYLENE OXIDE.

PEP carboxylase (*BioSci*) Abbrev for *phosphoenolpyruvate carboxylase*. The enzyme that catalyses the reaction of phosphoenolpyruvate and CO_2 to give oxaloacetic acid, esp as the first step in both the HATCH–SLACK pathway and CRASSULACEAN ACID METABOLISM in both of which the oxaloacetic acid is then reduced to malic acid.

peppermint (*FoodSci*) A pungent natural herb whose flavour has a pronounced cooling effect and composed of menthol, menthone and certain terpenes. The optical isomer l-menthol has the strongest smell and cooling effect.

Pepper's ghost (*Phys*) An illusion used to introduce a 'ghost' into a stage play. A plane sheet of glass is placed vertically at an angle of 45° to the line of vision of the audience. Thus actors in the wings can be superimposed by reflection on actors on stage.

pepsin (*BioSci*) Peptidase of the gastric juice that is able to function optimally under the acidic conditions of the stomach.

peptic ulcer (*Med*) An ulcer of the stomach or of the duodenum.

peptidase (*BioSci*) An enzyme that hydrolyses the peptide bonds of a protein. They hydrolyse different sites according to the amino acids adjacent to the peptide bond under attack, eg CHYMOTRYPSIN, TRYPSIN. The modern term for all proteases.

peptide (*BioSci, Chem*) A sequence of amino acids held together by peptide bonds. With rare exceptions peptides are unbranched chains joined by peptide bonds between their α-amino and α-carboxyl groups. Peptides can vary in length from dipeptides with two amino acids to polypeptides with several hundred. See OLIGOPEPTIDES, POLYAMIDES, POLYPEPTIDES.

peptide bond (*BioSci*) The bond formed by the condensation of the amino group and carboxyl group of a pair of amino acids. Peptides are constructed from a linear array of amino acids joined together by a series of peptide bonds.

peptide map (*BioSci*) A distinctive two-dimensional pattern of peptide fragments produced on a gel after partial hydrolysis of a particular protein.

peptide nucleic acid (*BioSci*) A synthetic nucleic acid mimic, in which the sugar–phosphate backbone is replaced by a peptide-like polyamide. Abbrev *PNA*.

peptidoglycan (*BioSci*) Cross-linked complex of polysaccharide (chains of N-acetyl glucosamine and N-acetyl muramic acid) and peptides found in the cell wall of bacteria, and in particularly high concentration in Gram-negative bacteria. Also *murein*.

peptization (*Chem*) The production of a colloidal solution of a substance, esp the formation of a sol from a gel. Also *deflocculation*.

per- (*Chem*) (1) A prefix which properly should be restricted to compounds which are closely related to hydrogen peroxide, and thus contain two oxygen atoms linked together, eg *percarbonates, perchromates, persulphates*. (2) A prefix which is loosely used to denote that the central atom of a compound is in a higher state of oxidation than the usual, eg *perchlorates*, *permanganates*.

per-acid (*Chem*) A true per-acid is either formed by the action of hydrogen peroxide on a normal acid, or yields hydrogen peroxide by the action of dilute acids. Organic per-acids are assumed to contain the group –COOOH, eg perbenzoic acid, C_6H_5COOOH.

peralkaline (*Min*) An igneous rock in which the molecular proportion of $Na_2O + K_2O$ exceeds that of Al_2O_3. This usually produces alkaline pyroxenes or amphiboles in the rock.

perambulator (*Surv*) An instrument for distance measurement consisting of a large wheel (often 6 ft or 2 m in circumference) attached at its axis to a long handle, so that it may be wheeled along the distance to be measured. A recording mechanism records the number of revolutions of the wheel and is calibrated in order to give distance traversed directly. See AMBULATOR, ODOMETER.

perborate (*Chem*) A salt consisting of an oxidized borate, used as a bleaching agent.

percentage articulation (*ICT*) That of elementary speech-sounds received correctly when logatoms are called over a telephone circuit in the standard manner.

percentage differential relay (*ElecEng*) A differential relay which operates at a current which, instead of being fixed, is a fixed percentage of the current in the operating coils.

percentage modulation (*ICT*) See DEPTH OF MODULATION.

percentage registration (*ElecEng*) The registration of an integrating meter expressed as a percentage of the true value.

percentage tachometer (*Aero*) An instrument indicating the rpm of a turbojet engine as a percentage, 100% corresponding to a preset optimum engine speed. Provides better readability and greater accuracy and also enables various types of engine to be operated on the same basis of comparison.

percentile (*MathSci*) Any of the 99 values of a variable which divide a distribution ordered by magnitude into 100 equal parts. Thus the 37th (per)centile is the value below which 37% of the population lie. The 25th centile is the first *quartile*; the 50th centile is the MEDIAN; and the 80th centile is the 8th *decile*. Also *centile*.

perception (*Psych*) The individual's apprehension of the world and of the body through the action of various sensory systems.

perceptual constancy (*Psych*) The ability to realize that objects are unchanged despite changes noticed by the senses: eg that buildings remain a constant size even though the retinal image enlarges as they are approached.

perceptual defence (*Psych*) The tendency to identify anxiety-provoking stimuli less readily than neutral stimuli, esp if presented for only brief periods.

perceptual learning (*Psych*) Learning about aspects of the stimulus in situations where there is no external reinforcement for doing so; it is usually demonstrated by an increased ability on a subsequent discrimination task.

perching (*Textiles*) Inspection of cloth after weaving, and at various stages of finishing, for possible defects; the cloth is drawn over rollers some distance apart and, while hanging, near vertically, from the front roller, is examined in a good light.

perchlorates (*Chem*) *Chlorates (VII)*. Salts of perchloric acid, $HClO_4$. They are all soluble in water, though those of potassium and rubidium only slightly. The alkali perchlorates are isomorphous with the corresponding permanganates (manganates (VII)).

perchloric (chloric (VII)) acid (*Chem*) $HClO_4$. Mp $-112°C$, bp $110°C$. A colourless fuming liquid. A powerful oxidizing agent; harmful to the skin; monobasic, and forms perchlorates (VII).

perchloroethene (*Chem*) *Tetrachloroethene.* CCl_2CCl_2. Bp $121°C$, rel.d. (at $20°C$) $1·623$. Organic solvent used for dry-cleaning. Also *perchloroethylene*.

perchromates (*Chem*) See PEROXYCHROMATES.

Perciformes (*BioSci*) The largest order of bony fish with about 7000 spp; an advanced and successful group inhabiting marine and fresh-water habitats. Basses, perches, tuna.

percolating filter (*Build*) A bed of filtering material, such as broken stone or slag, through which liquid sewage passes and used in the final or oxidizing stage in sewage treatment. Also *continuous filter*. See CONTACT BED.

Percoll (*BioSci*) TN for an inert colloid of silica coated with polyvinyl pyrrolidone, used in density gradient centrifugation for the separation of cells, subcellular particles and viruses.

percurrent (*BioSci*) Extending throughout the entire length.

percussion (*Med*) The act of striking with one finger lightly and sharply against another finger placed on the surface of the body, so as to determine, by the sound produced, the physical state of the part beneath.

percussion drilling (*MinExt*) A system in which a string of tools falls freely on rock being penetrated. Also, pneumatic drilling in which hammer blows are struck on drill shank. See CABLE TOOL DRILLING.

percussion figure (*Min*) A figure produced on the basal pinacoid or cleavage face of mica when it is sharply tapped with a centre punch. It consists of a six-rayed star, two rays more prominent than the others, lying in the unique plane of symmetry.

percussive boring (*CivEng*) The process of sinking a borehole in the earth by repeatedly dropping on the same spot, from a suitable height, a heavy tool which pulverizes the earth and gradually penetrates.

percussive welding (*ElecEng*) See RESISTANCE PERCUSSIVE-WELDING.

pereiopods (*BioSci*) In higher Crustacea, the thoracic appendages modified as walking legs. Cf *pleopod*.

perennation (*BioSci*) Survival from one growing season to the next with, usually, a period of reduced activity between.

perennial (*BioSci*) A plant that lives for more than two years. Most flower in most years after the first or second but some are HAPAXANTHIC.

perfect (*BioSci*) (1) A flower having functional stamens and carpels. (2) A fungus reproducing sexually.

perfect (*Print*) To print the second side of a sheet of paper and so complete it.

perfect binding (*Print*) Same as ADHESIVE BINDING, UNSEWN BINDING. Named after the first machine designed for the purpose.

perfect combustion (*Eng*) That of which the products contain neither unburnt gas nor excess air.

perfect crystal (*Crystal*) A single crystal in which the arrangement of the atoms is uniform throughout.

perfect dielectric (*Phys*) A dielectric in which all the energy required to establish an electric field is returned when the field is removed. In practice, only a vacuum conforms, other dielectrics dissipating heat to varying extent.

perfect frame (*Eng*) A frame which has just sufficient members to keep it stable in equilibrium under the intended load.

perfect gas (*Chem*) See IDEAL GAS.

perfect number (*MathSci*) An integer which is equal to the sum of all its factors including unity. The first four are 6, 28, 496 and 8128, after which they become enormous. Cf ABUNDANT NUMBER, DEFICIENT NUMBER, FRIENDLY NUMBERS.

perfector (*Print*) A type of machine which prints both sides of the paper in one pass through the machine.

perfluoroalkoxy polymers (*Chem*) Melt processable fluoropolymer, a copolymer of polytetrafluorine ethylene and perfluoro (propyl vinyl ether). High T_m of approx $300°C$ with good mechanical properties. Used for high-performance insulation for cable and wire, and chemically resistant valves.

perfoliate (*BioSci*) The basal part of the lamina of a leaf, encircling the stem completely, so that the stem appears to pass through the leaf.

perforate (*BioSci*) (1) Having holes. (2) Containing small rounded transparent dots that give the appearance of holes. (3) Having apertures, said esp of shells. (4) Of gastropod shells, having a hollow columella.

perforated brick (*Build*) A clay brick manufactured with vertical perforations to reduce weight, improve insulating properties and reduce capillary attraction through a wall.

perforating (*MinExt*) After a well has reached the producing zone the base is sealed with cement and the sides need to be pierced to allow ingress. Often done by a special gun with radial bores from which charges fire projectiles through the casing. Cf PIT.

perforating press (*Eng*) A power press for removing small areas of metal by punching.

perforation (*BioSci*) A hole in the common wall between two successive elements of a vessel, resulting from the dissolution of the wall. Cf PIT.

perforation plate (*BioSci*) The part of the common wall between two consecutive elements of a xylem vessel that has one or more perforations.

perforations (*ImageTech*) Holes of precise dimensions and spacing along one or both edges of a strip of motion picture film for transport and registration.

perforin (*BioSci*) A protein, present in K and NK cells, that resembles the C9 component of complement. It forms ring-like tubular structures which become inserted into the cell membranes of target cells and cause leakage of their contents.

performance index (*Genrl*) Alternative term for MERIT INDEX, applied to both materials and products. Also *performance indicator*.

performance test (*Psych*) Mental tests consisting primarily of motor or perceptual items and not requiring verbal ability. Cf VERBAL TEST.

perfusion (*BioSci*) The passage of a fluid through a compartment; in physiology, often used for the technique of passing fluid through the vessels of a specific tissue or organ.

pergeting (*Build*) See PARGETING.

Perhydrol (*Chem*) TN for a 30% solution of hydrogen peroxide.

peri- (*Chem*) A term for the 1,8 positions in naphthalene derivatives.

peri- (*Genrl*) Prefix from Gk *peri*, around.

perianal (*Med*) The region around the anus.

perianth (*BioSci*) The set of sterile structures that typically surrounds the stamens and carpels of a flower. It may be differentiated into an outer, often green and protective, calyx of sepals and an inner, often coloured, corolla of petals. See PERIANTH SEGMENT.

perianth segment (*BioSci*) A member of the perianth, esp if there is no differentiation into sepals and petals. Also *tepal*.

periapsis (*Astron*) The closest point of approach of an orbiting body (eg planet, comet or spacecraft) to the primary body. Cf APOAPSIS.

periarticular (*Med*) Said of the tissues immediately around a joint.

periastron (*Astron*) The point in an orbit about a star in which the body describing the orbit is nearest to the star; applied to the relative orbit of a double star.

periblast (*BioSci*) In meroblastic eggs, the margin of the blastoderm merging with the surrounding yolk. See PERIPLASM.

periblastic (*BioSci*) Of cleavage, superficial.

periblem (*BioSci*) A HISTOGEN, the precursor of the cortex.

peribranchial (*BioSci*) Surrounding a gill or gills.

pericardectomy (*Med*) Surgical removal of the pericardium.

pericardiomediastinitis (*Med*) Inflammation both of the pericardium and of the MEDIASTINUM.

pericardiotomy (*Med*) Incision of the pericardium.

pericarditis (*Med*) Inflammation of the pericardium.

pericardium (*BioSci*) The space surrounding the heart; the membrane enveloping the heart. Adj *pericardial*.

pericarp (*BioSci*) Fruit wall derived from the ovary wall.

pericellular (*BioSci*) Surrounding a cell.

pericentriolar region (*BioSci*) A rather amorphous region of electron-dense material surrounding the CENTRIOLE in animal cells: the major MICROTUBULE-ORGANIZING CENTRE of the cell.

perichaetium (*BioSci*) (1) A cup-like sheath surrounding the archegonia in some liverworts. (2) The group of involucral leaves around the archegonia of a moss.

perichondritis (*Med*) Inflammation of the perichondrium, esp of the perichondrium of the cartilages of the larynx.

perichondrium (*BioSci*) The envelope of areolar connective tissue surrounding cartilage.

perichordal (*BioSci*) Encircling or ensheathing the notochord.

periclase (*Min*) Native magnesia. Oxide of magnesium, which crystallizes in the cubic system. It is commonly found in metamorphosed magnesian limestones, but readily hydrates to the much commoner brucite.

periclinal (*BioSci*) Parallel to the nearest surface. Cf ANTICLINAL.

periclinal chimera (*BioSci*) A plant or shoot in which tissues of one genetic constitution form a complete layer throughout, the remaining tissues being of different genetic constitution. See CHIMERA, TUNICA-CORPUS CONCEPT.

pericline (*Min*) A variety of ALBITE which usually occurs as elongated crystals. The name is also used for a type of twinning in feldspars (the *pericline law*).

pericolitis (*Med*) Inflammation of the peritoneum covering the colon.

pericranium (*BioSci*) The fibrous tissue layer which surrounds the bony or cartilaginous cranium in vertebrates.

pericycle (*BioSci*) A layer, one or more cells thick, of ground tissue at the periphery of the stele next to the endodermis; present in roots, rare in stems.

pericycloid (*MathSci*) See ROULETTE.

periderm (*BioSci*) Secondary protective tissue often replacing epidermis in longer-lived stems and roots and consisting of cork, cork cambium and phelloderm. See also PERISARC.

peridesmium (*BioSci*) The coat of connective tissue that ensheathes a ligament.

perididymis (*BioSci*) The fibrous coat that encapsulates the testis in higher vertebrates.

peridinin (*BioSci*) Major carotenoid accessory pigment in the Dinophyceae.

peridium (*BioSci*) A general term for the outer wall of the fruit body of a fungus, when the wall is organized as a distinct layer or envelope surrounding the spore-bearing organs partially or completely.

peridot (*Min*) See OLIVINE.

peridotite (*Geol*) A coarse-grained ultramafic igneous rock consisting essentially of olivine, with other mafic minerals such as hypersthene, augite, biotite and hornblende, but free from plagioclase. See DUNITE, KIMBERLITE.

perigee (*Astron, Space*) The point nearest to the Earth on the APSE LINE of a central orbit having the Earth as a focus.

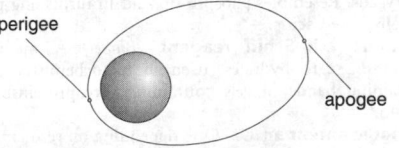

perigee					apogee

perigee   A satellite orbit round the Earth.

perigee motor (*Space*) See APOGEE MOTOR.

perigon (*MathSci*) See ROUND ANGLE.

perigynium (*BioSci*) See PERICHAETIUM.

perigynous (*BioSci*) A flower having the perianth and stamens inserted on a flat or cup-shaped structure which arises below and which is not fused to the ovary. Also the perianth and stamens so inserted, as in the blackberry. Cf EPIGYNOUS, HYPOGYNOUS, SUPERIOR.

perihelion (*Astron, Space*) The point in the solar orbit of a planet, comet or spacecraft at which it is nearest to the Sun. Pl *perihelia*.

perikinetic (*Phys*) Pertaining to BROWNIAN MOVEMENT.

perilune (*Astron*) The point in the lunar orbit of an object when it is closest to the Moon's surface.

perilymph (*BioSci*) The fluid that fills the space between the membranous labyrinth and the bony labyrinth of the internal ear in vertebrates. Adj *perilymphatic*. Cf ENDO-LYMPH.

perimedullary zone (*BioSci*) See MEDULLARY SHEATH.

perimeter (*Med*) An instrument, in the form of an arc, for measuring a person's field of vision.

perimeter track (*Aero*) A TAXI TRACK round the edge of an airport.

perimetritis (*Med*) Inflammation of the PERIMETRIUM covering the uterus.

perimetrium (*Med*) The peritoneum covering the uterus.

perimysium (*BioSci*) The connective tissue that binds muscle fibres into bundles and muscles.

perinaeum (*BioSci*) See PERINEUM.

perinatal (*Med*) Said of the period from the seventh month of pregnancy to the first week of life.

perineal glands (*BioSci*) In some mammals, a pair of small glands beside the anus that secrete a substance with a characteristic odour.

perineoplasty (*Med*) Repair of the perineum by plastic surgery.

perineorrhaphy (*Med*) Stitching of the perineum torn during childbirth.

perinephric (*Med*) Said of the tissues round the kidney, eg *perinephric abscess*.

perinephritis (*Med*) Inflammation of the tissues round the kidney.

perineum (*BioSci*) The tissue wall between the rectum and the urinogenital ducts in mammals. Also *perinaeum*. Adjs *perineal, perineal*.

perineurium (*BioSci*) The coat of connective tissue that ensheathes a tract of nerve fibres.

period (*Chem*) The elements between an alkali metal and the rare gas of next highest atomic number, inclusive, consisting of 2, 8, 18 or 32 elements.

period (*Geol*) A major unit of geological time, eg *Silurian*. See panel on GEOLOGICAL COLUMN.

period (*MathSci*) See PERIODIC FUNCTION.

period (*NucEng*) In a reactor, time in which the neutron flux changes by a factor of *e*.

period (*Phys*) Time taken for one complete cycle of an alternating quantity. Reciprocal of frequency.

periodates (*Chem*) Iodates (VII). Formed by the oxidation of iodates (V). Periodates can then form heteropolybasic compounds of the types $M_5[I(WO_4)_6]$ and $M_5[I(MoO_4)_6]$.

periodic acid (*Chem*) Iodic (VII) acid. H_5IO_6. A weaker acid and a stronger oxidizing agent than iodic acid. May be regarded as orthoperiodic acid. Exists in deliquescent crystals. Resembles phosphoric acid in furnishing partially dehydrated acids.

periodic acid-Schiff reagent (*BioSci*) A method for staining carbohydrates, used in histochemistry and for staining bands in gels containing glycoproteins. Abbrev PAS.

periodic antenna (*ICT*) One depending on resonance in its elements, thereby presenting a variation in input impedance as the operating frequency is varied.

periodic current (*ICT*) An oscillating current whose values recur at equal intervals.

periodic function (*MathSci*) A function f such that for all x, $f(x) = f(x + \omega)$ where the *period* ω is the smallest value for which this is true, eg $\sin x$ has a period of 2π and $\tan x$ a period of π. A function of a complex variable may have two or more independent complex periods (ie none being a real

rational multiple of any other), in which case it is referred to as *doubly* or *multiply periodic*.

periodicity (*BioSci*) Rhythmic activity.

periodicity (*Chem*) The location of an element in the periodic table.

periodicity (*Phys*) See FREQUENCY.

periodic law (*Chem*) See PERIODIC SYSTEM.

periodic ophthalmia (*Vet*) Recurrent IRIDOCYCLITIS in the horse; believed to be a form of LEPTOSPIROSIS. Also *Moon blindness*.

periodic precipitation (*Chem*) See LIESEGANG RINGS.

periodic rating (*ElecEng*) Rating of a component for continuous use with a specified periodically varying load.

periodic respiration (*Med*) Any waxing or waning of the pattern of breathing. In Cheyne–Stokes respiration periods of HYPERVENTILATION alternate with periods of APNOEA and indicate disease of the brain or severely impaired circulation to the brain.

periodic reverse (*Eng*) Changing the polarity in some electrolytic procedures to eg clean the electrodes.

periodic system (*Chem*) The classification of chemical elements into periods (corresponding to the filling of successive electron shells) and groups (corresponding to the number of valence electrons). Original classification by relative atomic mass (Mendeleyev, 1869). Formerly *periodic law*.

periodic table (*Chem*) The most common arrangement of the PERIODIC SYSTEM. See panel on BONDING and appendix on The periodic table.

period–luminosity law (*Astron*) A relationship between period and absolute magnitude, discovered by Henrietta Leavitt to hold for all CEPHEID VARIABLE stars; it enables the distance of any observable Cepheid to be found from observation of its light curve and apparent magnitude, this indirectly deduced distance being called the *Cepheid parallax*.

period meter (*NucEng*) An instrument for measurement of reactor period.

period of decay (*Phys*) See HALF-LIFE.

period of revolution (*Astron*) The mean value, derived from observations, of one complete revolution of a planet or comet about the Sun, or of a satellite about a planet.

periodontitis (*Med*) Inflammation of the membrane investing that part of the tooth seated in the jaw.

period range (*NucEng*) See START-UP PROCEDURE.

peri-oöphoritis (*Med*) Inflammation of the peritoneum investing the ovary and of the cortex of the ovary.

perioral (*BioSci*) Surrounding the mouth, as the *perioral membrane* of Ciliophora that surrounds the cytopharynx.

periosteum (*BioSci*) The covering of areolar connective tissue on bone.

periostitis (*Med*) Inflammation of the periosteum.

periostracum (*BioSci*) The horny outer layer of a molluscan shell.

periotic (*BioSci*) In higher vertebrates, a bone enclosing the inner ear and formed by the fusion of the otic bones. Also *petrosal*.

peripheral (*Genrl*) Situated or produced around the edge.

peripheral (*ICT*) The term used to describe any unit in a computer system such as BACKING STORE, INPUT DEVICE, OUTPUT DEVICE that is connected to the central processor.

peripheral component interconnect (*ICT*) A fast bus used in most modern personal computers. Abbrev *PCI*.

peripheral lymphoid tissue (*BioSci*) Secondary lymphoid tissue, not necessarily located peripherally.

peripheral vascular resistance (*Med*) An expression of the state of contraction of the arterioles which governs the overall resistance to blood flow. Measured as the drop in pressure in the arterial bed divided by the blood flow.

periplasm (*BioSci*) (1) The space between the plasma membrane and outer membrane of Gram-negative bacteria. It contains proteins secreted by the cell and a rigid peptide–oligosaccharide complex, the peptidoglycan. (2) A bounding layer of protoplasm surrounding an egg just beneath the vitelline membrane, as in insects.

periplasmic space (*BioSci*) A space between the cell wall and the plasmalemma.

periproct (*BioSci*) The area surrounding the anus.

perisarc (*BioSci*) In some Hydrozoa, the chitinous layer covering the polyps etc. Cf COENOSARC.

periscope (*Phys*) An optical instrument comprising an arrangement of reflecting surfaces whose purpose is to enable an observer to view along an axis deflected or displaced with respect to the axis of the observer's eye; useful for viewing radioactive sources.

perisperm (*BioSci*) Storage tissue, derived from nucellus (and hence wholly maternal), present in some seeds in which the endosperm does not develop, eg many Caryophyllaceae.

perissodactyl (*BioSci*) Having an odd number of digits. Cf ARTIODACTYL.

Perissodactyla (*BioSci*) An order of mammals containing the 'odd-toed' hooved animals, ie those with a mesaxonic foot with the skeletal axis passing down the third digit. Horses, tapirs, rhinoceros and extinct forms.

peristaltic (*BioSci*) Compressive; contracting in successive circles; said of waves of contraction passing from mouth to anus along the alimentary canal. Cf ANTI-PERISTALTIC, SYSTALTIC. N *peristalsis*.

peristaltic pump (*Eng*) One in which the flow is produced by peristaltic action, eg by rollers passing in succession over a length of flexible tube. Very useful in medicine because only the tube needs to be sterile.

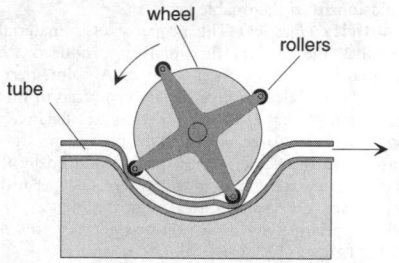

peristaltic pump Rotating the wheel causes the rollers to move. Tube clamps not shown.

peristerite (*Min*) A whitish variety of albite, or oligoclase, which is beautifully iridescent.

peristome (*BioSci*) (1) The margin of the aperture of a gastropod shell. (2) In some Ciliophora, a specialized food-collecting, frequently funnel-shaped, structure surrounding the cell mouth. (3) More generally, the area surrounding the mouth. Adj *peristomial*.

peristomium (*BioSci*) In Annelida, the somite in which the mouth is situated: in some forms (as *Nereis*), two somites have been fused to form the apparent peristomium.

peristyle (*Arch*) A colonnade encircling a building.

perisystole (*BioSci*) The period between diastole and systole in cardiac contraction.

peritectic (*Chem*) Physical reaction appearing in phase diagrams in which liquid reacts with a solid already separated to form a new solid phase. See fig. at PHASE DIAGRAM.

peritectoid (*Chem*) Similar reaction to PERITECTIC, except that a liquid phase is not involved; ie two solid phases reacting together to form a new solid phase.

perithecium (*BioSci*) A more or less flask-shaped ASCOCARP with a pore or ostiole at the top through which the asci are discharged.

peritoneal cavity (*BioSci*) In vertebrates, that part of the coelom containing the viscera; the abdominal body cavity.

peritoneal dialysis (*Med*) The passage of fluids through the peritoneal cavity to lower the blood urea in certain cases of renal failure.

peritoneum (*BioSci*) In vertebrates, a serous membrane that lines the peritoneal cavity and extends over the mesenteries and viscera. Adj *peritoneal*.

peritonitis (*Med*) Inflammation of the peritoneum.

peritrichous (*BioSci*) Bacteria having flagella distributed all over the surface of the cell.

peritrochoid (*MathSci*) See ROULETTE.

peritrophic (*BioSci*) Surrounding the gut; as the *peritrophic membrane* of insects, a membranous tube lining the stomach and partially separated from the stomach epithelium by the peritrophic space.

PERITUS (*Eng*) TN for a computer-aided materials selection procedure (Gk for *expert*).

perivascular sheath (*BioSci*) A sheath of connective tissue around a blood or lymph vessel.

perivitelline (*BioSci*) Surrounding an egg yolk.

perivitelline membrane (*BioSci*) See VITELLINE MEMBRANE.

Perkin's mauve (*Chem*) See MAUVEINE.

Perkin's synthesis (*Chem*) The synthesis of unsaturated aromatic acids by the action of aromatic aldehydes upon the sodium salts of fatty acids in the presence of a condensing agent, eg acetic (ethanoic) anhydride.

perknite (*Geol*) A family of coarse-grained ultramafic igneous rocks which consist essentially of pyroxenes and amphiboles, but contain no feldspar.

PERL (*ICT*) Abbrev for *practical extraction & report language* (also *Perl*), a high-level computer programming language derived from the C programming language. Favoured by many World Wide Web developers.

perlite (*Geol*) An acid and glassy igneous rock which exhibits perlitic structure.

perlitic structure (*Geol*) A structure found in glassy igneous rocks, which consists of systems of spheroidal concentric cracks produced during cooling.

Perlon (*Chem*) TN for polycaprolactam, nylon 6, a synthetic fibre.

permafrost (*Geol*) The permanently frozen soil in arctic and subarctic regions.

permalink (*ICT*) A hyperlink designed to provide a permanent connection to a web page.

permalloy (*Eng*) An alloy with high PERMEABILITY (name from *permeability alloy*) and low HYSTERESIS LOSS. Originally 78·5% Ni–21·5% Fe, but now includes compositions containing additions of Cu, Mo, Cr, Co, Mn, etc.

permanent dentition (*BioSci*) In mammals, the second set of teeth, which replaces the milk dentition.

permanent hardness (*Chem*) Of water, the hardness which remains after prolonged boiling is *permanent hardness*; due to the presence of calcium and magnesium chlorides or sulphates.

permanent implant (*Radiol*) An implant with radioactive material of short half-life, eg radon, arranged so that the prescribed dose is delivered by the time that the radioactive material has decayed, thus rendering removal of the sources unnecessary.

permanent load (*Eng*) The dead loading on a structure, consisting of the weight of the structure itself and the fixed loading carried by it, as distinct from any other loads.

permanent magnet (*ElecEng*) Ferromagnetic body which retains an appreciable magnetization after excitation has ceased. Cobalt steel, sintered and ceramic materials and various ferritic alloys are used on a large scale in loudspeakers, relays, small motors, magnetrons, etc. See MAGNETIC MATERIALS.

permanent mould (*Eng*) A metal mould (other than an ingot mould) used for the production of castings, eg in diecasting.

permanent pasture (*Agri*) Established grassland with few other plant types; naturally established woody types are occasional.

permanent set (*Eng*) (1) An extension remaining after load has been removed from a test piece, when the ELASTIC LIMIT has been exceeded. (2) Permanent deflection of any structure after being subjected to a load, which causes the elastic limit to be exceeded.

permanent store (*ICT*) See NON-VOLATILE MEMORY.

permanent threshold shift (*Acous*) Permanent hearing loss after exposure to high sound levels. Permanent hearing loss is what results from TEMPORARY THRESHOLD SHIFT. Abbrev *PTS*.

permanent virtual circuit (*ICT*) A FRAME RELAY service in which the logical channel between two terminals is not assigned dynamically at the time of connection, as it would be with a switched virtual circuit, but is fixed at the time of initial subscription to the service.

permanent way (*CivEng*) The ballast, sleepers and rails forming the finished track for a railway, as distinct from a temporary way.

permanent wilting point (*BioSci*) The water content of a soil at which a plant will wilt and not recover without additional water, even if shaded or left overnight. For most crop plants this corresponds to a soil MATRIC POTENTIAL of about -15 bar ($-1\cdot5$ MPa). Cf HYDRAULIC CAPACITY.

permanganates (*Chem*) *Manganates (VII)*. Oxidizing agents, the best known being *potassium permanganate* (manganate (VII)).

permanganic acid (*Chem*) Manganic (VII) acid. $HMnO_4$. Powerful oxidizing agent. Decomposes in the presence of organic matter.

permeability (*ElecEng*) The property of a material which describes the magnetization developed in that material when excited by an mmf source. *Absolute permeability* is the ratio of flux density produced to the magnetic field strength producing it: $\mu = B/H$ henrys per metre. *Relative permeability* is the ratio of magnetic flux density produced in a material to the value in free space produced by the same magnetic field strength: $\mu_r = \mu/\mu_0$, where $\mu_0 =$ permeability of free space, ie $4\pi \times 10^{-7}$ henrys per metre. Hence $B = \mu_0\mu_r H$ tesla.

permeability (*Eng*) The ability of the sand grains to allow passage of gases in foundry work.

permeability (*Geol*) The ability of a rock to transmit fluids, esp water, oil and gas.

permeability (*Phys*) The rate of diffusion of gas or liquid under a pressure gradient through a porous material. Expressed, for thin material, as the rate per unit area and, for thicker material, per unit area of unit thickness.

permeability bridge (*ElecEng*) A device for measuring magnetic properties of a sample of magnetic material, fluxes in different branches of a divided magnetic circuit being balanced against each other. See EWING PERMEABILITY BRIDGE, HOLDEN PERMEABILITY BRIDGE.

permeability coefficient (*PowderTech*) The volume of incompressible fluid per second which will flow through a unit cube of a porous mass or a packed powder across which unit pressure difference is maintained. Using cgs units, a flow rate of 1 $cm^3\ s^{-1}$ through a 1 cm cube under 1 dyne cm^{-2} pressure gives a permeability coefficient of 1 darcy. Using SI units, a flow rate of 1 $m^3\ s^{-1}$ through a 1 m cube under $1\ N\ m^{-2}$ pressure gives a permeability coefficient of 1 kilodarcy (1000 darcys). Abbrev *p.c.*

permeability equations (*PowderTech*) Mathematical equations derived to describe the flow of fluids through packed powder beds and porous media in general.

permeability surface area (*PowderTech*) The surface area of a particle calculated from the permeability of a powder bed under stated conditions.

permeability tuning (*ICT*) Adjusting a tuned circuit by varying the inductance of a coil by altering position on axis of a sintered iron core.

permeameter (*Phys*) An instrument for measuring static magnetic properties of ferromagnetic sample, in terms of magnetizing force and consequent magnetic flux.

permeameter (*PowderTech*) An instrument for measuring the specific surface area of a powder by measuring the resistance offered to a flowing fluid by a packed powder bed.

permeameter cell (*PowderTech*) That portion of a permeameter in which the plug or bed of the powder under test is assembled. Also *permeability cell*.

permeance (*ElecEng*) The reciprocal of the RELUCTANCE of a magnetic circuit.

permeation (*Phys*) The flow of a fluid through a porous material. If J_x is the flux as defined by the first of FICK'S LAWS OF DIFFUSION, then

$$J_x = \frac{P_n(p_1 - p_2)^{\frac{1}{2}}}{l}$$

(Richardson's law) where P_n is the permeability, l is the thickness of material and $(p_1 - p_2)$ is the pressure difference across the material.

Permian (*Geol*) The youngest geological period of the Palaeozoic era, lying between the Carboniferous and the Triassic. It covers a time span of approx 290 to 250 million years and is named after the type area of Perm in Russia. The corresponding system of rocks. See PALAEOZOIC.

permissible dose (*Radiol*) See MAXIMUM PERMISSIBLE DOSE.

permissive temperature (*BioSci*) The temperature at which a TEMPERATURE-SENSITIVE MUTANT organism will grow. Cf RESTRICTIVE TEMPERATURE.

permissive waste (*Build*) Dilapidations in a building which are the result of neglect on the part of a tenant.

permitted explosives (*MinExt*) Those which may be used in mines, under specified conditions, where a danger of explosion from flammable gas exists.

permittivity (*ElecEng*) The property of a material which describes the electric flux density produced when the material is excited by an emf source. Absolute permittivity is the ratio of electric flux density produced to the electric field strength: $\varepsilon = D/E$ farads per metre. Relative permittivity is the ratio of electric flux density produced in a material to the value in free space produced by the same electric field strength: $\varepsilon_r = \varepsilon/\varepsilon_0$ where $\varepsilon_0 =$ permittivity of free space, ie $8\cdot854 \times 10^{-12}\ F\ m^{-1}$. Hence $D = \varepsilon_0\varepsilon_r E$. Relative permittivity is also termed *dielectric constant*. See DIELECTRIC.

permonosulphuric acid (*Chem*) H_2SO_5, a powerful oxidizing reagent; prepared by anodic oxidation of concentrated sulphuric acid.

Permo-Trias (*Geol*) The Permian and Triassic systems considered together, as is commonly done in areas such as the UK where the rocks are of similar facies.

permutation (*MathSci*) The different ways of selecting a number of items from a given set when the order of selection matters. There are $n(n-1)(n-2)\ldots 2\cdot1 = n!$ ways of ordering n items, all different, and $n(n-1)(n-2)\ldots(n-r+1) = n!/(n-r)!$ ways of selecting r of the n items. The *degree* of the permutation is the number of elements in the set to be permuted. A permutation is *odd* or *even* depending on whether the number of exchanges of position required to obtain it from $|1,2,3,\ldots,n|$ is odd or even; eg the permutation

$$|1, 2, 3| \rightarrow |3, 1, 2|$$

has degree 3 and is odd.

Permutit (*Chem*) TN for natural or synthetic zeolite used as ion exchanger to substitute sodium for calcium or magnesium ions in hard water, to obtain softer water, the spent (ie calcium and magnesium) zeolite being regenerated (ie to sodium zeolite) periodically by treatment with concentrated sodium chloride solution.

pernicious anaemia (*Med*) A disease characterized by atrophic gastritis with achlorhydria and lack of gastric INTRINSIC FACTOR, which leads to failure of absorption of dietary vitamin B_{12} and consequent megaloblastic anaemia and perhaps peripheral neuritis. Gastric secretions and the blood of affected persons contain auto-antibodies against intrinsic factor and against a microsomal antigen present in gastric parietal cells. Also *Addison's anaemia*.

peroral (*BioSci*) Through the mouth.

perosis (*Vet*) A disease of chickens, turkeys and other birds characterized by swelling and deformity of the hock joint leading to dislocation of the gastrocnemius tendon; caused by a dietary deficiency of manganese or choline. Also *slipped tendon*.

perovskite (*Min*) Calcium titanate, with rare earths, which crystallizes in the monoclinic system but is very close to being cubic. An accessory mineral in melilite-basalt, and in contact-metamorphosed impure limestones. The crystal form of artificial ceramics which are superconductors above 80 K. Important crystal structure of some ferro-electrics, such as those in the barium titanate family. See panel on SUPERCONDUCTORS.

peroxidases (*BioSci*) Enzymes that catalyse the reduction of hydrogen peroxide. In cells, may be localized in PEROXI-SOMES, and in some phagocytic cells the MYELOPEROX-IDASE–HALIDE SYSTEM is responsible for killing ingested organisms. Peroxidases are also used in ENZYME-LINKED IMMUNOSORBENT ASSAY to generate a coloured reaction product, and lactoperoxidase is used to catalyse the binding of RADIOIODINE to proteins and cells.

peroxides (*Chem*) Strictly limited to compounds containing the ion O_2^{2-} or the organic function —O–O–, those containing the ion HO_2a or the group HO–O– being hydroperoxides. The term is loosely used for oxides of high-valent metals, eg lead peroxide is PbO_2 but is more properly called lead (IV) oxide.

peroxide value (*FoodSci*) A measure of the level of oxidation of a fat or oil containing polyunsaturated fatty acids, the cause of RANCIDITY. Determined by treating an acetified solution of fat in chloroform with potassium iodide and then titrating the iodine liberated by peroxide against a standard thiosulphate solution. Abbrev *PV*.

peroxisome (*BioSci*) Small membrane-bound cytoplasmic organelle containing oxidizing enzymes and catalase; common in phagocytic cells and in leaf cells where they contain some of the enzymes of the glycollate pathway.

peroxyacetyl nitrate (*Chem*) Found in smog and formed by the action of sunlight on the primary emission products, hydrocarbons and nitrogen oxides. Damaging to living systems at very low concentrations.

peroxychromates (*Chem*) Formed by the action of hydrogen peroxide on chromates (VI), and containing a peroxide group –O–O–. In alkaline solution it forms M_3CrO_8, red salts. In acid, $MCrO_6$, blue salts.

perpends (*Build*) The vertical joints in brickwork, usually kept in line in alternate courses.

perpetual motion (*Phys*) (1) The continual operation of a machine which creates its own energy, thereby violating the first law of thermodynamics. (2) The operation of any device which converts heat completely into work, thereby contravening the second law of thermodynamics. See THERMODYNAMICS.

perrhenates (*Chem*) Rhenates (VII). See RHENIUM OXIDES.

perrhenic acid (*Chem*) Rhenic (VII) acid. See RHENIUM OXIDES.

perron (*Build*) An external staircase to a building, from ground level to the first floor.

per-salts (*Chem*) Salts corresponding to PER-ACIDS.

Perseids (*Astron*) A major meteor shower visible for up to 2 weeks before and after peaking on 12 August each year, the date on which the Earth crosses the orbit. The maximum hourly rate is around 70 meteors. Associated with a comet seen in 1862.

Perseus (*Astron*) A northern hemisphere constellation which lies in the Milky Way.

perseveration (*Psych*) Meaningless repetition of an action or utterance.

persistence (*Electronics*) Continued visibility or detection of luminous radiation from a gas-discharge tube or luminous screen when the exciting agency has been removed. Long-persistence cathode-ray tubes are used for retaining a display which has been momentarily excited. Also *afterglow*.

persistence (*ICT*) In object-oriented design the capture and storage of the transitory state of an object in a permanent medium such as a relational database, from which it can be subsequently retrieved and reconstructed.

persistence characteristic (*Electronics*) A graph showing decay with time of the luminous emission of a PHOSPHOR, after excitation is cut off.

persistence of vision (*Phys*) Brief retention of image on retina of eye after optical excitation has ended. An essential factor in cinematography and TV.

persistent (*BioSci*) Continuing to grow or develop after the normal period for the cessation of growth or development, as teeth (cf DECIDUOUS); said also of structures which remain present in the adult although they normally disappear in the young.

persistron (*Electronics*) Electroluminescent photoconducting device, which gives a steady display on being impulsed.

person (*BioSci*) In colonial forms, an individual organism.

personal communications network (*ICT*) (1) Any network that provides a PERSONAL COMMUNICATIONS SERVICE. (2) One of a number of mobile-telephone networks established in the UK from 1993 onwards, operating at 1·8 GHz and intended to provide competition for the two established 940 MHz systems.

personal communications service (*ICT*) A system designed to provide communication to a person rather than a place, by means of services such as CALL FORWARDING or automatic connection to a mobile-telephone network. See UNIVERSAL PERSONAL TELECOM-MUNICATIONS.

personal computer (*ICT*) A low-cost computer, often standalone but nowadays increasingly part of a network in an office etc, with a very wide range of available software and now with processing power, storage and memory greater than minicomputers of a few years ago. Commonly applied to both computers conforming to the IBM-designed architecture and the Apple Macintosh. Abbrev *PC*. Cf WORKSTATION.

Personal Computer Memory Card International Association (*ICT*) An organization that has set standards for small cards that can carry storage, modem, etc, functions esp in laptop computers. Abbrev *PCMCIA*.

personal digital assistant (*ICT*) A very small computer in which information like addresses and telephone numbers can be entered, often by a stylus. Abbrev *PDA*.

personal dosimeter (*Radiol*) A small ionization chamber which loses charge during irradiation, the state of charge being immediately visible unlike the film badge which is processed later. Also *pocket chamber*. See fig. at QUARTZ FIBRE ELECTROSCOPE.

personal identification device (*ICT*) A device, often a card, that is inserted into a terminal to establish the holder as an authorized user of the system. Sometimes used with a PIN. Abbrev *PID*.

personal identification number (*ICT*) A number allocated to a user of an on-line computer system or any card-reading device. The number is unique and must be input to the system by the user as a means of establishing the user's identity. Abbrev *PIN*.

personal information manager (*ICT*) A software application that organizes personal information, eg addresses. Abbrev *PIM*.

personality (*ICT*) An operating system has multiple personalities if it can run software written for a number of operating systems. They can be dominant or subsidiary. The ability of OS/2 to run Windows is an example. See MICROKERNEL.

personality (*Psych*) The integrated organization of all the psychological, intellectual, emotional and physical characteristics of an individual which determines the unique adjustment he or she makes to the world.

personality disorders (*Psych*) An inflexible and well-established behaviour pattern that is maladaptive for the individual in terms of social or occupational functioning; usually recognizable by adolescence.

personal services environment (*ICT*) The parts of a network within which the requirements unique to an individual, such as the need for call screening or billing of calls to a special account, are recognized.

personal space (*Psych*) The space immediately surrounding one's body that an individual considers private; it varies with cultural and other factors.

personal telecommunications number (*ICT*) A telephone number assigned to an individual rather than a terminal, with the intention that calls in a UNIVERSAL PERSONAL TELECOMMUNICATIONS system should reach that individual wherever he or she may be.

personal video recorder (*ICT*) A device used to make digital recordings of TV broadcasts on a hard disk. Abbrev PVR.

personate (*BioSci*) A COROLLA that is two-lipped with the throat almost closed by the lower lip, eg *Antirrhinum*.

person-centred therapy (*Psych*) A therapeutic technique, based on humanistic theory, that is non-directive and empathic.

personnel monitoring (*Radiol*) Monitoring for radioactive contamination of any part of a person, their breath or excretions, or any part of their clothing.

perspective (*Acous*) See ACOUSTIC PERSPECTIVE.

Perspex (*Plastics*) TN for thermoplastic POLYMETHYL METHACRYLATE. Also *Lucite*.

Perthe's disease (*Med*) A deformed condition of the epiphysis of the head of the femur in young children, associated with a painful limp. Also *osteochondritis deformans juvenilis*.

perthite (*Min*) The general name for megascopic intergrowths of potassium and sodium feldspars, both components having been miscible to form a homogeneous compound at high temperatures, but the one having been thrown out of solution at a lower temperature, thus appearing as inclusions in the other. Perthite may also be formed by the replacement of potassium feldspar by sodium feldspar. See MICROPERTHITE.

perthosite (*Geol*) A type of soda-syenite consisting to a very large extent of perthitic feldspars, occurring at Ben Loyal and Loch Ailsh in Scotland.

perturbation (*Astron, Space*) Any disturbance to a planned trajectory or orbit, caused by the effects of eg drag, gravitation or solar pressure.

perturbation theory (*Phys*) A method of approximate solution of certain problems in quantum mechanics. Often different interactions have different orders of magnitude; smaller effects can be accounted for by considering perturbations on the eigenstates of the dominant interaction in the system when treated alone. Also applicable to the solution of the behaviour of any complex system.

pertusate (*BioSci*) Perforated; pierced by slits.

pertussis (*Med*) An acute infectious disease, due to infection with the bacillus *BORDETELLA PERTUSSIS*; characterized by catarrh of the respiratory tract, also by periodic, recurring spasms of the larynx, which end in the prolonged crowing inspiration known as the 'whoop'. Also *whooping cough*.

pertussis toxin (*BioSci*) A bacterial *AB* toxin crucial to the pathogenicity of *Bordetella pertussis*, the causative agent of whooping cough.

pervaporation (*EnvSci*) A separation process in which liquid permeates through a membrane into gaseous phase on the other side, selectively leaving some dissolved substances behind. It is used as a cheaper alternative to distillation.

perveance (*Electronics*) The constant G in the Child–Langmuir equation, $I = GV^{3/2}$, which governs the current in a space-charge-limited thermionic diode; relevant to the optimization of electron-gun design in electron-beam tubes and accelerators.

pes (*BioSci*) The podium of the hind-limb in land vertebrates. Pl *pedes*.

pes arcuatus (*Med*) A condition of the foot in which the balls of the toes approximate to the heel, so that the foot is shortened and the instep abnormally high. Also *pes cavus* or *claw foot*.

pes planus (*Med*) A condition of the foot in which the longitudinal arch is lost, so that the foot is flattened and turned outwards. Also *pes valgus* or *flat foot*.

pessary (*Med*) A medicated appliance for insertion into the vagina to treat vaginal disease.

pesticide (*Agri*) An agrochemical that controls pests, pathogens or weeds. Pesticides include herbicides, fungicides, insecticides, nematocides, acaricides and bactericides.

pestle (*Chem*) A club-shaped instrument, of glass or porcelain, for grinding and pounding solids in a mortar.

PET (*Chem*) Abbrev for POLYETHYLENE TEREPHTHALATE.

peta- (*Genrl*) Prefix (probably from Gk *penta*) denoting one thousand million million, 10^{15}. Symbol P.

petal (*BioSci*) A member of the COROLLA, often brightly coloured and conspicuous.

petalite (*Min*) A silicate of lithium and aluminium which crystallizes in the monoclinic system. It typically occurs in granite pegmatites.

petalody (*BioSci*) The transformation of stamens or carpels into petals. See DOUBLE.

petaloid (*BioSci*) Looking like a petal; petal-shaped, as the dorsal parts of the AMBULACRA in certain Echinoidea.

pet cock (*Eng*) A small plug cock for draining condensed steam from steam-engine cylinders, or for testing the water level in a boiler.

petechia (*Med*) A small red spot due to minute haemorrhage into the skin.

Petersburg standard (*For*) See STANDARD.

Petersen coil (*ElecEng*) A reactor placed between the neutral point of an electric-power system and earth. The value of the reactance is such that, when an earth fault occurs, the current through the reactor exactly balances the capacitance current flowing through the fault, so that any tendency to arcing is suppressed. Also *arc-suppression coil*.

pethidine (*Pharmacol*) A synthetic analgesic and hypnotic, having action similar to that of morphine.

petiolate (*BioSci*) A leaf having a stalk or petiole. Cf SESSILE.

petiole (*BioSci*) The stalk of a leaf.

petiolule (*BioSci*) The stalk of a leaflet of a compound leaf.

petit mal (*Med*) A form of epileptic attack in which convulsions are absent and certain transient phenomena, eg brief loss of consciousness, occur.

PETN (*Chem*) Abbrev for PENTAERYTHRITOL TETRANITRATE.

petrifaction (*Geol*) The process by which organic remains are changed in composition by molecular replacement but with the original structure largely unchanged.

petrified (*Geol*) A term describing organic remains which have undergone the process of PETRIFACTION. *Petrified wood* is wood which has had its structure replaced by eg calcium carbonate or silica. Many of the original minute structures are preserved.

petro- (*Genrl*) Prefix from Gk *petra* signifying rock.

petrochemicals (*Chem*) Those derived from crude oil or natural gas. They include light hydrocarbons such as butene, ethene and propene, obtained by fractional distillation or catalytic cracking.

petrographic province (*Geol*) A region characterized by a group of genetically related rocks, eg the Andes. See COMAGMATIC ASSEMBLAGE.

petrography (*Geol*) Systematic description of rocks, based on observations in the field, on hand specimens, and on thin microscopic sections. Cf PETROLOGY.

petrol (*Autos*) UK term for a light hydrocarbon liquid fuel for spark-ignition engines. Modern motor fuels are blends of several products of petroleum, such as straight-run distillate, thermal and catalytically cracked gasoline alkylates, isomerates, benzene (benzole) and, rarely, alcohol. It boils in the range of 30–200°C. An important measurement point is the engine test or octane rating of the fuel. US *gas, gasoline.*

petrol–electric generating set (*ElecEng*) A small generating plant using a petrol engine as the prime mover.

petrol engine (*Autos*) A reciprocating engine, working on the Otto four-stroke or the two-stroke cycle, in which the air charge is carburetted by a petrol spray from a carburettor, or alternatively by FUEL INJECTION. In four-stroke engines, inlet and exhaust valves control the entry of charge and the exit of exhaust gases; in two-stroke engines, the piston is usually made to act as both inlet and exhaust valve. Ignition of the combustible mixture is effected by sparking plug, operated either by coil and battery or by magneto.

petroleum (*Chem*) Naturally occurring green- to black-coloured mixtures of crude hydrocarbon oils, found as earth seepages or obtained by boring. Petroleum is widespread in the Earth's crust, notably in the US, the former USSR and the Middle East. In addition to hydrocarbons of every chemical type and boiling range, petroleum often contains compounds of sulphur, vanadium, etc. Commercial petroleum products are obtained from crude petroleum by distillation, cracking, chemical treatment, etc.

petroleum coke (*MinExt*) Nearly pure carbon formed during the refining of crude oil by high-temperature carbonization of the heavy residues.

petroleum jelly (*Chem*) A mixture of petroleum hydrocarbons; used for making emollients, for impregnating the paper covering of electric cables, and as a lubricant. Also *petrolatum.* See LIQUID PARAFFIN.

petroleum reservoirs (*Geol, MinExt*) Natural subsurface porous and permeable rocks to which oil or gas has migrated and accumulated under an adequate trap formed by an overlying or up-dip impermeable cap. There are various kinds of structural traps, some of considerable complexity. See ANTICLINAL TRAP, FAULT TRAP, PIERCEMENT SALT DOME, STRATIGRAPHIC TRAP.

petrol injection (*Autos*) See FUEL INJECTION.

petrology (*Geol*) That study of rocks which includes consideration of their mode of origin, present conditions, chemical and mineral composition, their alteration and decay.

petrol pump (*Autos*) A small pump of the diaphragm type, operated either mechanically from the camshaft, or electrically. It draws petrol from the tank and delivers it to the carburettor or fuel injection system. US *gas pump.*

petrosal (*BioSci*) See PERIOTIC.

petrous (*BioSci*) Stony, hard (as a portion of the temporal bone in higher vertebrates); situated in the region of the petrous portion of the temporal bone.

petticoat (*ElecEng*) One of the umbrella-shaped shields commonly provided on pin-type insulators in order to increase the length of the leakage path which will remain dry under rain conditions.

petzite (*Min*) A telluride of silver and gold. It is steel-grey to black and often shows tarnish.

Petzval curvature (*Phys*) The curvature of the image surface of a lens system in which spherical aberration, coma and astigmatism have been corrected.

pewter (*Eng*) An alloy containing 80–90% tin and 10–20% lead. In modern pewter, antimony may replace tin, and 1–3% copper is usual.

Peyer's patches (*BioSci*) Nodules of lymphoid tissue in the submucosa of the small intestine, important for the development of immunity (or of immunological unresponsiveness) to antigens present in the gut.

PF (*Build*) Abbrev for *plain face.*

pF (*ElecEng*) Abbrev for PICOFARAD.

Pfannkuch protection (*ElecEng*) A protective system for use with the cables of an electric-power system; some of the strands of the cable are lightly insulated from the others and have an emf applied between them which causes a current to flow and operate relays if the insulation is destroyed by a fault.

Pferdestärke (*Eng*) See CHEVAL-VAPEUR.

PFI & R (*Build*) Abbrev for *part fill in and ram.*

PF resins (*Plastics*) Abbrev for *phenol formaldehyde resins.* See PHENOLIC RESINS and panel on THERMOSETS.

Pfund series (*Phys*) A series of lines in the far-infrared spectrum of hydrogen. Their wavenumbers are given by the same expression as that for the BRACKETT SERIES but with $n_1 = 5$.

PGC (*Chem*) Abbrev for PYROLYSIS GAS CHROMATOGRAPHY.

PGD (*Med*) Abbrev for *pre-implantation genetic diagnosis*, a technique in which genetic testing of an *in vitro* fertilized embryo is carried out before the decision to implant the embryo is taken. An increasing range of diagnostic tests is becoming available as the genetic basis for various disorders becomes clear.

***p*-group** (*MathSci*) A group for which every element's order is a power of *p*, where *p* is a prime number.

pH (*Chem*) A logarithmic index for the hydrogen ion concentration in an aqueous solution. Used as a measure of acidity of a solution; given by $pH = \log_{10}(1/[H^+])$, where $[H^+]$ is the hydrogen-ion concentration or activity. A pH below 7 indicates acidity, and one above 7 alkalinity, at 25°C.

PHA (*BioSci*) Abbrev for PHYTOHAEMAGGLUTININ.

phacoidal structure (*Geol*) A rock structure in which mineral or rock fragments of lens-like form are included. The term is applicable to igneous rocks containing softened and drawn-out inclusions; also to metamorphic rocks such as crush-breccias and crush-conglomerates; and to certain gneisses. (Gk *phakos*, lentil.)

phacolith (*Geol*) A minor intrusion of igneous rock occupying the crest of an anticlinal fold. Its form is due to the folding, hence it is not the cause of the uparching of the roof. Cf LACCOLITH.

phacomalacia (*Med*) Pathological softening of the lens of the eye.

Phaedra complex (*Psych*) The difficult relationship which can arise between a new step-parent and the (usually teenage) son or daughter of the original marriage.

phaeic (*BioSci*) Dusky. N *phaeism.* Also *phaeochrous.*

phaeo-, pheo- (*Genrl*) Prefixes from Gk *phaios*, dusky.

phaeochromacytoma (*Med*) A tumour composed of chromaffin cells, esp those of the adrenal medulla; by secreting noradrenaline may cause high blood pressure. Also *paraganglioma.*

phaeochrous (*BioSci*) See PHAEIC.

Phaeophyceae (*BioSci*) A class of eukaryotic algae (brown algae) in the division Heterokontophyta. There are branching filamentous and parenchymatous types most of which are marine, littoral and sublittoral, the brown seaweeds. Can be isogamous, anisogamous or oögamous, may show alternation of generations (isomorphic or heteromorphic), or be almost diplontic. They are the source of alginic acid. Includes kelps and wracks.

phag-, phago-, -phage, -phagy (*Genrl*) Prefixes and suffixes from Gk *phagein*, to eat.

phage (*BioSci*) See BACTERIOPHAGE.

phagedaena (*Med*) Rapidly spreading and destructive ulceration. Also *phagedena.*

phago- (*Genrl*) See PHAG-.

phagocyte (*BioSci*) A stationary or motile cell (eg AMOEBOCYTE, MACROPHAGE) which is capable of engulfing particulate material, eg bacteria, often by extending PSEUDOPODS.

phagocytosis (*BioSci*) The ingestion of cells or particles that first attach to and then become surrounded by the cell membrane. This is invaginated to form an intracellular vesicle (phagosome), towards which lysosomes move and fuse to release their content of hydrolytic enzymes into the vesicle, which becomes a phagolysosome. Cells of the mononuclear phagocyte system and polymorphonuclear leucocytes are the main phagocytic cells in mammals.

phagotrophy (*BioSci*) Heterotrophic nutrition in which cells ingest solid food particles.

-phago (*Genrl*) See PHAG-.

phalanges (*BioSci*) In vertebrates, the bones supporting the segments of the digits; fiddle-shaped rings composing the reticular lamina of the organ of Corti. Sing *phalanx*.

Phalangida (*BioSci*) See OPILIONES.

phalanx (*BioSci*) See PHALANGES.

phalaris staggers (*Vet*) A nervous disease of sheep in Australia caused by eating the grass *Phalaris tuberosa*; believed to be caused by a toxic substance in the grass when eaten by cobalt-deficient sheep.

phallic stage (*Psych*) In psychoanalytic theory, the third stage of psychosexual development in which the child is preoccupied with his or her genitals; from the third to fifth or sixth year of life. See ELECTRA COMPLEX, OEDIPUS COMPLEX.

phallus (*BioSci*) The penis of mammals; the primordium of the penis or clitoris of mammals. Adj *phallic*.

phanero- (*Genrl*) Prefix from Gk *phaneros*, visible.

phanerocrystalline (*Min*) Said of an igneous rock in which the crystals of all the essential minerals can be discerned by the naked eye.

phanerophyte (*BioSci*) Woody plant with perennating buds more than 25 cm above the soil surface. See RAUNKIAER SYSTEM.

Phanerozoic (*Geol*) The span of *obvious life*. More precisely the unit of geological time that comprises the Palaeozoic, Mesozoic and Cenozoic eras. See appendix on Geological time.

phantasy (*Psych*) See FANTASY.

phantom antenna (*ICT*) See ARTIFICIAL ANTENNA.

phantom limb (*Psych*) Sensations that appear to arise from an amputated limb that is still neurologically represented in the brain.

phantom material (*Radiol*) That producing absorption and back-scatter of radiation very similar to human tissue, and hence used in models to study appropriate doses, radiation scattering, etc. A phantom is a reproduction of (part of) the body in this material. Also *tissue equivalent material*.

phantom ring (*Phys*) The coloured ring of light which appears to surround an observer's shadow when the shadow falls on an extended bank of fog or cloud.

pharate (*BioSci*) In insects, refers to a phase of development when the old cuticle of one stage is separate from the hypodermis of the next stage, but has not yet been ruptured and cast off. The 'pupa' of many insects actually represents a pharate adult and the 'prepupa' a pharate pupa.

pharmacodynamics (*Pharmacol*) The biochemical and physiological processes determining the effects of drugs on organisms. See PHARMACOKINETICS.

pharmacogenetics (*Pharmacol*) The study of genetic causes of individual variations in drug response; the term is often used interchangeably with PHARMACOGENOMICS.

pharmacogenomics (*Pharmacol*) Genome-wide analysis of the genetic determinants of drug efficacy and toxicity rather than individual genetic differences (polymorphisms). Often used interchangeably with PHARMACOGENETICS.

pharmacokinetics (*Pharmacol*) The pharmacokinetics of a drug relate to the rate and extent of uptake (absorption), transformation as a result of metabolism, the distribution of the drug and its metabolites in the tissues, the metabolic breakdown of the drug and the elimination of the drug or its metabolites from the body (excretion). Commonly abbreviated as *ADME*.

pharmacolite (*Min*) A hydrated arsenate of calcium which crystallizes in the monoclinic system. It is a product of the late alteration of mineral deposits which carry arsenopyrite and the arsenical ores of cobalt and silver.

pharmacology (*Genrl*) The scientific study of the action of chemical substances on living systems. The subject field is used in this dictionary for drugs and pharmacologically active substances used in medicine.

pharmacosiderite (*Min*) Hydrated arsenate of iron. It crystallizes in the cubic system, and is a product of the alteration of arsenical ores.

pharming (*BioSci*) The commercial production of substances from transgenic plants or animals for medical (pharmaceutical) use.

pharming (*ICT*) The covert redirection of computer users from legitimate websites to counterfeit sites in order to gain confidential information about them. By analogy with PHISHING.

pharyngismus (*Med*) Spasm of the muscles of the pharynx.

pharyngitis (*Med*) Inflammation of the pharynx.

pharyngoplasty (*Med*) The surgical alteration, or reconstruction, of the pharynx in congenital and acquired disease.

pharyngoplegia (*Med*) Paralysis of the muscles of the pharynx.

pharyngotomy (*Med*) Incision into the pharynx.

pharynx (*BioSci*) In vertebrates, that portion of the alimentary canal between the mouth and the oesophagus which serves both eating and respiration; in invertebrates, the corresponding portion of the alimentary canal lying immediately posterior to the buccal cavity, usually having a highly muscular wall. Adj *pharyngeal*.

phase (*Astron*) Of the Moon, the name given to the changing shape of the visible illuminated surface of the Moon due to the varying relative positions of the Earth, Sun and Moon during the synodic month. Starting from new Moon, the phase increases through crescent, first quarter, gibbous, to full Moon, and then decreases through gibbous, third quarter, waning to new Moon again. The inferior planets show the same phases, but in the reverse order; the superior planets can show a gibbous phase, but not a crescent.

phase (*Chem*) (1) The sum of all those portions of a material system which are identical in chemical composition and physical state, and are separated from the rest of the system by a distinct interface called the *phase boundary*. (2) The particular state of a substance as a solid, liquid or gas.

phase (*ElecEng, ICT*) The fraction of cycle of a periodic waveform (usually sinusoidal) that has been completed at a specific reference time, eg at the start of a cycle of a second waveform of the same frequency. Expressed as an angle, one cycle corresponds to 2π radians or 360°. The terms *in phase*, *in quadrature* and *antiphase* correspond to phase angles between two signals of 0° (or 360°), 90° and 180° respectively. See POLYPHASE, SINGLE PHASE, TWO PHASE, THREE PHASE.

phase advancer (*ElecEng*) A component connected in the secondary circuit of an induction motor to improve its POWER FACTOR.

phase–amplitude modulation (*ICT*) An efficient modulation scheme widely used for data transmission over telephone channels, in which two carriers of the same frequency but differing in phase by 90° are independently modulated by AMPLITUDE SHIFT KEYING. The result is a signal constellation consisting typically of 16 points in phase–amplitude space. Also *quadrature amplitude modulation*.

phase angle (*ElecEng, ICT*) See PHASE. Also given by δ, where $\tan \delta = $ (reactance/resistance) for an ac circuit or an acoustic system.

phase change (*Eng*) See PHASE TRANSITION.

phase compensation (*ICT*) See LINE EQUALIZER.

phase constant (*ICT*) Imaginary part of propagation constant of a line, or of the transfer constant of a filter section. It is expressed in radians per unit length or section. See IMAGE PHASE CONSTANT.

phase-contrast microscopy (*BioSci*) A simple non-quantitative form of interference microscopy of great utility in visualizing live cells. Small differences in optical path length due to differences in refractive index and thickness of structures are visualized as differences in light intensity.

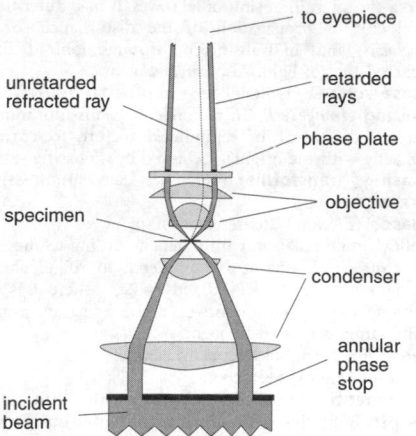

phase contrast Using annular illumination.

phase converter (*ElecEng*) On applying a single-phase voltage to one phase of a rotating three-phase induction motor, it will be found that three-phase voltages, which are approximately balanced, may be derived from the three terminals of the machine.

phase corrector (*ICT*) A circuit for correcting phase distortion.

phased array (*Radar*) An antenna consisting of an array of identical radiators (waveguides, horns, slots, dipoles, etc) with electronic means of altering the phase of power fed to each of them. This allows the shape and direction of the radiation pattern to be altered without mechanical movement and with sufficient rapidity to be made on a pulse-to-pulse basis.

phase defect (*ElecEng, ICT*) The phase difference between the actual current in a capacitor and that which would flow in an equivalent ideal (loss-free) capacitor.

phase delay (*ICT*) The delay, in radians or seconds, for transmission of wave of a single frequency through whole or part of a communication system. Also *phase retardation.*

phase delay distortion (*ICT*) That of a signal transmitted over a long line, so that difference in time of arrival of components of a complex wave is significant.

phase deviation (*ICT*) In phase modulation, maximum difference between the phase angle of the modulated wave and the angle of the non-modulated carrier.

phase diagram (*Eng*) A mapping to show the stable states in which a system of chemical element(s) or molecular components can exist under particular physical conditions, usually of temperature and pressure but also of concentration of component(s) where appropriate. Otherwise known as a *constitution diagram, equilibrium diagram* or *thermal equilibrium diagram,* since it represents equilibrium states of a system. The diagrams are essential for understanding the behaviour of crystalline materials (metals, ceramics) and the effect of thermal treatments such as involved in melting and control of mechanical properties. A typical diagram for a single-component system is illustrated in (a) in the figure. Because this is a unary (one-component) system, the two physical variables temperature and pressure can be represented by the axes of a two-dimensional diagram and show the conditions under which each of the phases solid, liquid and vapour exist. In the areas, which are called phase fields, any of the three physical states may exist under conditions of temperatures and pressures within the natural limits of the phase field boundaries. Two of the phases may co-exist under conditions represented by the lines on the diagram. At one unique combination of pressure and temperature all three phases co-exist (the triple point). Binary diagrams are more complex than those for unary systems and new types of relationship are possible when three phases co-exist. In particular three phases of fixed compositions can exist in equilibrium at one unique temperature (the invariant).

phase difference (*ElecEng*) The phase angle by which one periodic waveform lags or leads another of the same frequency.

phase discriminator (*Electronics*) A circuit preceding the demodulator in a phase-modulation receiver. It converts the carrier to an amplitude-modulated form.

phase distortion (*ICT*) Found in the waveform of transmitted signal on account of non-linear relation of wavelength constant of a line, or image phase constant of amplifier or network, with frequency.

phase equalization (*ICT*) See LINE EQUALIZER.

phase equalizer (*ElecEng*) See INTERPHASE TRANSFORMER.

phase focusing (*Electronics*) An effect used in electron bunching whereby electroncs lagging in phase tend to gain energy from the field and those leading tend to surrender it. A similar effect occurs in synchrotron charged-particle accelerators.

phase inversion (*ICT*) Any arrangement for obtaining an additional voltage of reverse phase for driving both sides of a push–pull amplifier stage.

phase lag (*ElecEng, ICT*) If a periodic disturbance A completes a cycle when a second disturbance B has completed $\varphi/2\pi$ of a cycle, B is said to *lead A* by a phase angle φ. Conversely, A *lags* on B by φ or *leads B* by $(2\pi-\varphi)$.

phase lead (*ElecEng, ICT*) See PHASE LAG.

phase-locked loop (*ICT*) Control loop incorporating a voltage-controlled oscillator and phase-sensitive detector in order to lock a signal to a stable reference frequency. Used in frequency synthesizers and synchronous detectors.

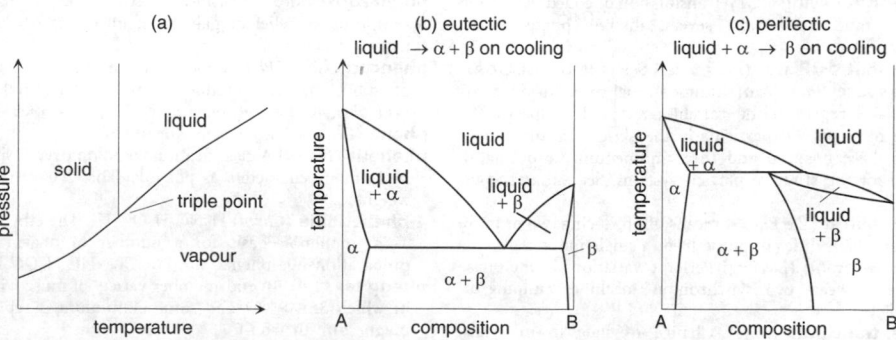

phase diagram (a) Single component system, (b) and (c) binary systems showing two types of 3-phase equilibria.

phase margin (*ICT*) Variation in phase of unity gain from that which would give instability. Readily measured from a Nyquist diagram.

phase meter (*ICT*) One for measuring the phase difference between two signals of the same frequency.

phase modifier (*ElecEng*) A term for a synchronous capacitor which is used for varying the power factor of the current in a transmission line for voltage regulation.

phase modulation (*ICT*) Periodic variation in phase of a high-frequency current or voltage in accordance with a lower impressed modulating frequency. Occurs as an unwanted by-product of amplitude modulation, but can be independently produced.

phase plate (*Phys*) A quarter-wave retarding plate used in PHASE-CONTRAST MICROSCOPY.

phase resonance (*ICT*) That in which the induced oscillation differs in phase by $\pi/2$ from the forcing disturbance. Also *velocity resonance*.

phase retardation (*ICT*) See PHASE DELAY.

phase reversal (*Chem*) An interchange of the components of an emulsion; eg under certain conditions an emulsion of an oil in water may become an emulsion of water in the oil.

phase rule (*Chem*) A generalization of great value in the study of equilibria between phases. In any system, $P + F = C + 2$, where P is the number of phases, F the number of DEGREES OF FREEDOM (1) and C the number of COMPONENTS.

phase-sensitive detector (*Electronics*) One in which the output is proportional to the phase difference between the incoming signal or carrier and a reference signal.

phase separation (*BioSci*) A purification method involving the separation of partially miscible solvents that contain different concentrations of the compound of interest that has partitioned differentially.

phase sequence (*ElecEng*) The order in which the phase voltages of a polyphase system reach their maximum values. If the phases of a three-phase system are given the standard colourings of red, yellow, blue, this phase sequence is said to be a *positive phase sequence*. See NEGATIVE SEQUENCE, ZERO SEQUENCE.

phase shifter (*ICT*) Waveguide or coaxial line component that produces any selected phase delay in signal transmitted.

phase-shifter control (*Electronics*) One that changes the phase angle (measured on the waveform of the incoming voltage) at which the applied voltage causes conduction through a silicon-controlled rectifier or thyristor, ignitron, thyratron or any other controlled power-switching device.

phase-shifting circuit (*ICT*) One in which the relative phases of components of a waveform or signal can be continuously adjusted.

phase-shifting transformer (*ElecEng*) One for which the secondary voltage can be varied continuously in phase. Usually with rotatable secondary winding and polyphase primaries. Also *phasing transformer*.

phase-shift keying (*ICT*) Transmission of coded data using phase modulation with several discrete phase angles. Abbrev *PSK*.

phase-shift oscillator (*Electronics*) See LINE OSCILLATOR.

phase space (*Phys*) A multidimensional space in which co-ordinates represent the variables required to specify the state of a system; particularly the six-dimensional space with three position and three momentum co-ordinates. Used for the study of particle systems. See MICROSCOPIC STATE.

phase splitter (*ElecEng*) A means of producing two or more waves which differ in phase from a single input wave.

phase swinging (*ElecEng*) Periodic variations in the phase angle between two synchronous machines running in parallel.

phase transition (*Phys*) Change of state in materials, whether first order (eg melting, boiling), second order (eg CURIE POINT for ferromagnetic materials, GLASS TRANSITION TEMPERATURE) or higher order.

phase velocity (*Phys*) The velocity of propagation of any one phase state, such as a point of zero instantaneous field, in a steady train of sinusoidal waves. It may differ from the velocity of propagation of the disturbance, or *group velocity*, and, in transmission through ionized air, may exceed that of light. Also *wave velocity*.

phase voltage (*ElecEng*) See VOLTAGE TO NEUTRAL.

phasing (*ImageTech*) In facsimile transmission and reception, adjustment of reproduced picture to correspond exactly with the original, achieved by a phasing signal.

phasing transformer (*ElecEng*) See PHASE-SHIFTING TRANSFORMER.

phasor (*ElecEng*) Representation of ac or voltage which allows manipulation with others according to the rules of vectors. Each phasor is considered to rotate about the origin at ω rads per second ($= 2\pi f$, where f is the ac frequency) and all phasors on a given diagram must have the same associated frequency.

Phe (*Chem*) Symbol for phenylalanine.

phellem (*BioSci*) See CORK.

phelloderm (*BioSci*) Parenchymatous tissue formed centripetally by the cork cambium (phellogen) as part of the periderm.

phellogen (*BioSci*) See CORK CAMBIUM.

phen-, pheno- (*Genrl*) Prefixes from Gk *phainein* denoting showing, visible; related to benzene (see PHENE).

phenakite (*Min*) A silicate of beryllium, crystallizing in the rhombohedral system. It is commonly found as a product of pneumatolysis. Sometimes cut as a gemstone, having brilliance of lustre but lacking fire. The name (Gk 'the deceiver') refers to the frequency with which it has been confused with quartz.

phenanthrene (*Chem*) $C_{14}H_{10}$, white, glistening plates, mp 99°C, bp 340°C; its solutions show a blue fluorescence. It occurs in coaltar, and can be synthesized by various reactions, the most important of which is the synthesis by Pschorr, based upon the condensation of 2-nitrobenzaldehyde with sodium phenylacetate yielding 2-phenyl-2-nitrocinnamic acid, which, on reduction, diazotization and subsequent elimination of N_2 and H_2O, yields phenanthrene-10-carboxylic acid. This acid yields phenanthrene on distillation. The 9-10-phenanthrene bridge is readily attacked by reagents, yielding diphenyl derivatives.

phenanthrene

phenates (*Chem*) Phenolates, phenoxides. Salts formed by phenols, eg $C_6H_5ON^-$, sodium phenate (phenolate, phenoxide).

phenazopyridine (*Pharmacol*) An analgesic used for symptomatic relief of pain from infections of the lower urinary tract.

phencyclidine (*Pharmacol*) An anaesthetic and drug of abuse that can produce marked behavioural effects: *1-(1-phenylcyclohexyl) piperidine, PCP, Angel dust*.

phene (*Chem*) An old name for BENZENE.

phenetic (*BioSci*) A classification based on overall similarity in as many characters as possible. The characters can be weighted.

4-phenetidine (*Chem*) $H_2NC_6H_4OC_2H_5$. The ethyl ether of 4-aminophenol, basis for a number of pharmaceutical preparations, eg phenacetin, $H_3CCONHC_6H_4OC_2H_5$.

phengite (*Min*) An end-member variety of muscovite mica, in which Si:A > 13:1 and some aluminium is replaced by magnesium or iron.

Phenidone (*ImageTech*) TN for *1-phenyl-3-pyrazolidinone*, a developing agent with characteristics similar to *metol*.

pheno- (*Genrl*) See PHEN-.

phenobarbital (*Pharmacol*) Barbiturate used mostly as an anticonvulsant in the treatment of epilepsy. Formerly *phenobarbitone*.

phenocopy (*BioSci*) An environmentally caused copy of a genetic abnormality. It is non-heritable.

phenocrysts (*Geol*) Large (megascopic) crystals, usually of perfect crystalline shape, found in a fine-grained matrix in igneous rocks. See PORPHYRITIC TEXTURE.

phenogram (*BioSci*) A branching diagram or dendrogram, reflecting the overall similarities of groups of organisms.

phenol (*Chem*) *Carbolic acid*. C_6H_5OH. Mp 43°C, bp 183°C. Colourless hygroscopic needles, chief constituent of the coaltar fraction boiling between 170 and 230°C, soluble in sodium and potassium hydroxide, forming Na and K phenates or phenolates. It forms with bromine 2,4,6-tribromophenol. Sodium phenolate reacts with CO_2 under certain conditions, yielding sodium salicylate. It is a strong disinfectant. See PHENOLS.

phenol A (*Chem*) See BISPHENOL A.

phenolic acids (*Chem*) A group of aromatic acids containing one or more hydroxyl groups attached to the benzene nucleus. The 2-hydroxy acids are volatile in steam, soluble in cold chloroform, and give a violet or blue coloration with iron (III) chloride. The 3-hydroxy acids are the most stable acids. Important phenolic acids are *gallic acid, salicylic acid, tannin*.

phenolic resins (*Plastics*) A large group of plastics, made from a phenol (phenol, 3-cresol, 4-cresol, catechol, resorcinol or quinol) and an aldehyde (methanal, ethanal, benzaldehyde or furfuraldehyde). An acid catalyst produces a soluble and fusible resin (modified resin) used in varnishes and lacquers, while an alkaline catalyst results in the formation, after moulding, of an insoluble and infusible resin. Phenolics may be moulded, laminated or cast. See panel on THERMOSETS.

phenolics (*BioSci*) A large and diverse group of plant secondary metabolites. The phenol group includes FLAVONOIDS, LIGNIN and TANNINS.

phenology (*BioSci*) The study of plant development in relation to the seasons.

phenoloxidase (*BioSci*) Any of a group of enzymes found in many plants and invertebrate animals. In arthropods, phenoloxidase acts on tyrosine to produce the black pigment, melanin, which is formed during wound-healing and internal defence.

phenolphthalein (*Chem*) A triphenylmethane derivative, obtained by the condensation of phthalic anhydride and phenol by the action of concentrated sulphuric acid. It forms colourless crystals which dissolve with a red colour in alkalis. It is used as an indicator in volumetric analysis. The colourless substance has the lactone formula, whereas its coloured salts have a quinonoid structure.

phenol red (*Chem*) *Phenolsulphonphthalein*. Used as an indicator, with pH range of 6·8–8·4, over which it changes from yellow to red. Often added to cell culture medium which tends to become acid (yellow) as it is exhausted; also used for testing the functioning of the kidney.

phenols (*Chem*) *Hydroxybenzenes*. A group of aromatic compounds having the hydroxyl group directly attached to the benzene nucleus. They give the reactions of alcohols, forming esters, ethers, thiocompounds, but also have feeble acidic properties and form salts or phenolates by the action of NaOH or KOH. Phenols are divided into mono-, di-, tri-, tetra- and polyhydric phenols. Phenols are more reactive than the benzene hydrocarbons.

phenomenology (*Psych*) In philosophy, the study of the psychic awareness that accompanies experience and that is the source of all meaning for the individual. In psychiatry, it refers to the description and classification of an individual's mental activity, including subjective experience and perceptions, mental performance (eg memory) and the somatic accompaniments of mental events (eg heart rate).

phenothiazines (*Pharmacol*) A group of antipsychotic drugs, thought to act by blocking dopaminergic transmission in the brain.

phenotype (*BioSci*) The observable characteristics of an organism as determined by the interaction of its GENOTYPE and its environment. Adj *phenotypic*.

phenotypic variance (*BioSci*) A measure of the amount of variation among individuals in the observed values of a quantitative character.

phenoxyacetic acids (*BioSci*) Synthetic compounds, many with auxin-like activity, including the important selective weedkillers 2,4-D, 2,4,5-T and MCPA.

phenoxy resins (*Plastics*) A range of polymers now replaced by polycarbonates.

phentermine (*Pharmacol*) An appetite-suppressing drug, a *sympathomimetic* amine, used in treating obesity.

phenylacetic acid (*Chem*) $C_6H_5CH_2COOH$. Mp 76°C. Colourless crystals. Also *phenylethanoic acid*.

phenylalanine (*Chem*) *2-amino-3-phenylpropanoic acid*. $C_6H_5CH_2CH(NH_2)COOH$. The L- or S-isomer is a constituent of proteins. Symbol Phe, short form F.

phenylalanine ammonia-lyase (*BioSci*) Enzyme catalysing the elimination of ammonia from L-phenylalanine in the synthesis of many SECONDARY METABOLITES, eg phenolic acids, flavonoids, alkaloids, lignins, tannins, etc. Abbrev PAL.

phenylamine (*Chem*) *Aminobenzene*. See ANILINE.

phenylbenzene (*Chem*) See DIPHENYL.

phenylbutazone (*Pharmacol*) Analgesic and antipyretic drug used in the treatment of rheumatic conditions.

phenyl cyanide (*Chem*) See BENZONITRILE.

phenylenediamines (*Chem*) $C_6H_4(NH_2)_2$. Obtained by the reduction of the dinitro, the nitroamino or the aminoazo compounds. The diamines are crystalline substances with strong basic properties. There are three isomers.

phenylethanoic acid (*Chem*) See PHENYLACETIC ACID.

phenylethanol (*Chem*) $C_6H_5CH_2CH_2OH$. An aromatic alcohol of pleasant odour, bp 220°C, a constituent of rose oil.

phenylethanone (*Chem*) See ACETOPHENONE.

phenylglycine (*Chem*) White, crystalline solid; mp 127°C. Used as an intermediate in the manufacture of indigo dyes.

phenyl group (*Chem*) The aromatic group C_6H_5-.

phenylhydrazine (*Chem*) $C_6H_5NHNH_2$. Mp 23°C. A colourless crystalline mass. It is easily oxidized and is a strong reducing agent. It forms salts with acids. Phenylhydrazine reacts readily with aldehydes and ketones, forming phenylhydrazones, which are crystalline substances and serve to identify the respective aldehydes and ketones.

phenylhydrazones (*Chem*) The reaction products of PHENYLHYDRAZINE with aldehydes and ketones, formed by the elimination of two hydrogen atoms of the amino group and the oxygen atom of the aldehyde or ketone group as water.

phenylketonuria (*Med*) A rare, autosomal recessive, hereditary condition causing mental retardation in which the amino acid phenylalanine cannot be converted to tyrosine and accumulates in the blood, to be ultimately excreted in the urine as phenylpyruvic and other acids. Of particular interest because it can be detected soon after birth and mental retardation prevented by dietary restriction of phenylalanine. Abbrev *PKV*. Also *phenylalaninaemia*.

phenyl methyl ether (*Chem*) See ANISOLE.

phenylmethylsulphonyl fluoride (*BioSci*) A substance widely used as a broad spectrum peptidase inhibitor. Abbrev *PMSF*.

phenytoin (*Pharmacol*) A drug widely used in the treatment of epilepsy. Of particular value in prevention of tonic and atonic seizures, it can also be used to control abnormal heart rhythms.

pheo- (*Genrl*) See PHAEO-.

pheromone (*BioSci*) A chemical substance, secreted by an animal, that influences the behaviour of other animals (cf

HORMONE) of the same species, eg the QUEEN SUBSTANCE of honey bees. See SEMIOCHEMICAL.

phi coefficient (*Psych, MathSci*) An index of correlation used when both variables are binary; eg true/false, yes/no or on/off.

phi grade scale (*Geol*) A logarithmic scale used for the mechanical analysis of sediments. It is expressed as the negative logarithm to base 2 from +10 for $\frac{1}{1024}$ mm to −5 for 32 mm diameter of grain on the Wentworth scale. Symbol φ. Also *Phi scale*.

Philadelphia chromosome (*Med*) A distinctive, small chromosome found in patients suffering from chronic myelocytic leukaemia, corresponding to chromosome 22 after a reciprocal TRANSLOCATION with chromosome 9.

phillipsite (*Min*) A fibrous zeolite; hydrated silicate of potassium, calcium and aluminium, usually grouped in the orthorhombic system.

Phillips screw (*Eng*) A screw with a cross-shaped slot, requiring a special screwdriver. Also *cross-headed screw*.

phimosis (*Med*) Narrowness of the prepuce (foreskin) so that it cannot be drawn back over the glans penis.

phi phenomenon (*Psych*) The apparent motion when two or more stationary light sources are flashed in succession and the light appears to move from one location to the other.

phishing (*ICT*) The practice of sending unsolicited emails in an attempt to acquire an Internet user's sensitive information, eg bank account details, by appearing to be a bona fide business person.

phleb-, phlebo- (*Genrl*) Prefixes from Gk *phleps*, gen *phlebos*, vein.

phlebectomy (*Med*) Surgical removal of a vein or part of a vein.

phlebitis (*Med*) Inflammation of the coats of a vein.

phlebo- (*Genrl*) See PHLEB-.

phlebography (*Radiol*) The radiological examination of veins following the injection of a CONTRAST MEDIUM. When applied to the peripheral venous system, called *venography*.

phlebolith (*Med*) A concretion in a vein due to calcification of a thrombus.

phlebosclerosis (*Med*) Thickening of a vein due chiefly to a pathological increase in the connective tissue of the middle coat.

phlebotomus fever (*Med*) See SANDFLY FEVER.

phlebotomy (*Med*) The cutting or needling of a vein for the purpose of letting blood.

phlegmasia alba dolens (*Med*) A term formerly used for painful swelling of the leg, the skin of which is shiny and white, occurring in women after childbirth; due to thrombosis of veins.

phlegmon (*Med*) Purulent inflammation, with necrosis of tissue. Adj *phlegmonous*.

phloem (*BioSci*) Tissue with the major function of transporting metabolites, esp sugars, from sources to sinks. See MASS FLOW HYPOTHESIS. The tissue consists of SIEVE ELEMENTS, and companion cells and/or parenchyma cells, often with fibres or sclereids.

phloem ray (*BioSci*) That part of a vascular ray which traverses the secondary phloem.

phlogisticated air (*Chem*) An old name for nitrogen.

phlogiston (*Chem*) A substance erroneously thought (mainly during the 18th century) to be lost to the atmosphere during combustion.

phlogopite (*Min*) Hydrous silicate of potassium, magnesium, iron and aluminium, crystallizing in the monoclinic system. It is a magnesium mica, and is usually found in metamorphosed limestones or in ultrabasic igneous rocks. Not so good an electrical insulator as MUSCOVITE at low temperatures, but keeps its water of composition until 950°C. Also *amber mica*.

phloroglucinol (*BioSci*) A stain for lignin when acidified.

phlycten (*Med*) A small round, grey or yellow nodule, occurring on the conjunctiva where it covers the sclera and cornea of the eye (Gk *phlyctaina*, a bleb). Also *phlyctenula* or *phlyctenule*.

phlyctenular conjunctivitis (*Med*) An inflammation of the conjunctiva covering the sclera and cornea of the eye and giving rise to phlyctens. Also *eczematous conjunctivitis*.

phlyctenule (*Med*) See PHLYCTEN.

pH meter (*Chem, Phys*) Specialized millivolt meter which measures potential difference between reference electrodes in terms of pH value of the solution in which they are immersed.

phobia (*Psych*) An excessive, irrational and persistent anxiety leading to avoidance of a specific situation or object which presents no apparent threat.

phobic disorder (*Psych*) A disorder characterized by intense anxiety to some external object or situation, and avoidance of the phobic stimulus.

Phobos (*Astron*) One of the two natural satellites of Mars, discovered in 1877. Distance from the planet 9380 km; diameter 27 km.

phocomelia (*Med*) Seal-like limb. A congenital malformation in which the proximal portion of the limb is absent. *Phocomelus*, one so affected.

Phoebe (*Astron*) The ninth natural satellite of Saturn, discovered in 1898. Distance from the planet 12 950 000 km; diameter 220 km.

Phoenix (*Astron*) A southern hemisphere constellation.

Pholidota (*BioSci*) An order of old-world mammals that have imbricating horny scales, interspersed with hairs, over the head, body and tail. They have a long snout, no teeth, and a long thin tongue. The hind-feet are plantigrade, and the fore-feet have long curved claws. They are nocturnal animals, feeding on ants or termites. Pangolins.

phon (*Acous*) Unit of the objective loudness on sound-level scale which is used for deciding the apparent loudness of an unknown sound or noise, when a measure of loudness is required. This is effected either by subjective comparison by the ear, or by objective comparison with a microphone amplifier and a weighting network. The reference sound pressure level at 1 kHz is 0·0002 microbar = 20 μN m^{-2}.

phonation (*BioSci*) Sound-production.

phoneme (*ICT*) The smallest unit of speech used in linguistic analysis, eg the sound represented by s. English speakers have a repertoire of about 40 phonemes and utter about 10 phonemes per second. The minimal unit of speech that is recognized by a fluent speaker of the language in question. Not necessarily represented by a single letter in a written alphabetic language.

phonetics (*Acous*) The study of speech and vocal acoustics. Used to describe the system of symbols which uniquely represent the spoken word of any language in writing, enabling the reader to pronounce words accurately in spite of spelling irregularities.

phonmeter (*Acous*) Apparatus for the estimation of loudness level of a sound on the phon scale by subjective comparison. Also *phonometer*.

phonochemistry (*Chem*) The study of the effect of sound and ultrasonic waves on chemical reactions.

phono connector (*ICT*) See PHONO PLUG.

phonograph (*Acous*) See GRAMOPHONE.

phonolite (*Geol*) A fine-grained igneous rock of intermediate composition, consisting essentially of nepheline, subordinate alkali feldspar (sanidine) and sodium-rich coloured silicates. Termed also *clink stone*, because it rings under the hammer when struck.

phonometer (*Acous*) See PHONMETER.

phonon (*Phys*) A quantum of lattice vibrational energy in a crystal. The thermal vibrations of the atoms can be described in terms of *normal modes* of oscillation, each mode specifying a correlated displacement of the atoms. The energy of a mode is quantized and can only exchange energy with other modes in units of hv, where v is the mode frequency and h is PLANCK'S CONSTANT. Lattice vibrations can be described in terms of waves in the lattice and the phonon is the particle of the field of the mechanical energy of a crystal (cf PHOTON as a particle of the electromagnetic field). Phonons have been identified by neutron inelastic scattering experiments.

phonon dispersion curve (*Phys*) A curve showing frequency as a function of wavevector for the modes of lattice vibrations in a crystal. Determined by neutron spectroscopy, their interpretation provides a powerful method of testing various models of interatomic forces in crystals.

phonon drag (*Phys*) In the PELTIER EFFECT due to electron–PHONON interactions, the phonons gain momentum from the electrons and the electron current 'drags' the phonons with it.

phonon–phonon scattering (*Phys*) The interaction of one mode of lattice vibration with another. One of the processes determining the thermal conductivity of a solid.

phono plug (*ICT*) Coaxial connector commonly used in non-professional audio equipment. Also *phono connector*.

phorbol esters (*BioSci*) Polycyclic compounds isolated from croton oil, the commonest being phorbol myristoyl acetate (PMA). Potent *tumour promoters*, they are diacyl glycerol analogues and irreversibly activate protein kinase C.

-phore (*Chem*) A suffix which denotes a group of atoms responsible for the corresponding property, eg *chromophore*.

-phore (*Genrl*) Suffix from Gk *pherein*, carry.

phoresy (*BioSci*) Transport or dispersal achieved by clinging to another animal, eg certain mites which achieve dispersal by attaching themselves to various insects.

Phoronidea (*BioSci*) A small phylum of hermaphrodite unsegmented coelomate animals of tubicolous habit. They have a U-shaped gut, a dorsal anus, and a LOPHOPHORE in the form of a double horizontal spiral. Marine forms occur in the sand and mud of the sea bottom.

phosgene (*Chem*) $COCl_2$. A very poisonous, colourless, heavy gas with a nauseating, choking smell; bp 8°C. It is manufactured by passing carbon monoxide and chlorine over a charcoal catalyst. It is used for the manufacture of intermediates in the dyestuff industry, and as a *war gas*. Also *carbonyl chloride*.

phosgenite (*Min*) A chlorocarbonate of lead, crystallizing in the tetragonal system. It is found in association with cerussite.

phosphatase (*BioSci*) An enzyme that dephosphorylates its substrate by hydrolysis of the phosphate ester bond. Acid and ALKALINE PHOSPHATASES have different pH optima; the former are found within LYSOSOMES.

phosphatase test (*FoodSci*) Phosphatase is present in raw milk and its absence is used to check whether pasteurization has been carried out properly. An immediate test is necessary because the enzyme can reactivate.

phosphate (*Agri*) A phosphorus-containing fertilizer that stimulates root and shoot growth. A component of compound fertilizers.

phosphate coatings (*Build*) Coatings of iron or zinc phosphate on steel made by treating with an acid which may be simply dilute phosphoric acid, as in the original Coslett process, or may contain a variety of metal phosphates and other additives. They provide a key for subsequent painting or may be filled with oil and dyed, as a decorative and protective film.

phosphate fixation (*BioSci*) The reaction of orthophosphate ions with Ca, Al or Fe ions to give low-solubility hydroxyphosphates which renders much of the phosphate unavailable to plants. Occurs in most mineralized or artificially fertilized soils.

phosphates (V) (*Chem*) Salts of phosphoric acid. There are three series of orthophosphates: MH_2PO_4, M_2HPO_4 and M_3PO_4; the first yield acid, the second are practically neutral, and the third alkaline, aqueous solutions. Metaphosphates, MPO_3, and pyrophosphates, $M_4P_2O_7$, are also known. All phosphates give a yellow precipitate on heating with ammonium molybdate in nitric acid.

phosphatic deposits (*Geol*) Beds containing calcium phosphate which are formed esp in areas of low rainfall, and which may be exploited as sources of phosphate. See PHOSPHATIC NODULES.

phosphatic nodules (*Geol*) Rounded masses containing calcium phosphate, which are formed on the sea floor.

phosphatides (*Chem*) Fat-like substances containing phosphoric acid. In some, glycero-phosphoric acid is combined with two molecules of fatty acid and one molecule of a hydroxy base which may be choline (in *lecithins*), aminoethanol (in *kephalins*) or serine. In others, the base sphingosine is combined with fatty acid, phosphoric acid and choline. Also *phospholipins*.

phosphatidylcholine (*BioSci*, *FoodSci*) A phosphatide in which the choline forms the organic base. Important component of biological membranes, and a widely used natural emulsifier obtained from egg yolk and soya beans. Previously termed *lecithin*.

phosphatidylinositol (*BioSci*) A minor phospholipid in eukaryotes that has a very important role in the response of cells to external ligands. The derivatives phosphatidyl inositol-phosphate (PIP) and phosphatidyl inositol-bisphosphate (PIP$_2$) are formed and broken down in cell membranes, and the trisphosphate inositol moiety (InsP$_3$) is important as a SECOND MESSENGER within the cytoplasm.

phosphating (*Eng*) The treatment of a metal surface with hot phosphoric acid before painting in order to inhibit corrosion.

phosphaturia (*Med*) The presence of an excess of phosphates in the urine.

phosphene (*Med*) Area of luminosity in the visual field. It is caused by pressure on the eyeball.

phosphides (*Chem*) Binary compounds of metals with phosphorus, eg calcium phosphide Ca_3P_2.

phosphine (*Chem*) *Phosphorus (III) hydride*. PH_3. A colourless, evil-smelling gas which usually burns spontaneously in air to form phosphorus (V) oxide. It has reducing properties and precipitates phosphides from solutions of many metallic salts.

phosphines (*Chem*) Derivatives of PH_3, obtained by the substitution of hydrogen for alkyl radicals; classified according to the extent of substitution into *primary*, *secondary* and *tertiary* phosphines. They correspond closely to the amines, except that they are easily oxidized even in the air, they are only feebly basic, and the P atom has a tendency to pass from the tri- to the quinquevalent state.

phosphites (*Chem*) *Phosphates (III)*. Salts of phosphorous acid.

phosphocreatine (*BioSci*) A compound present in high concentration (about 20 mM) in striated muscle, where it serves as an energy reserve. Also *creatine phosphate*.

phosphodiesterase inhibitors (*Pharmacol*) Compounds that inhibit the breakdown of the important intracellular second messenger cyclic AMP, although strictly speaking the cAMP-phosphodiesterase is only one of a number of such enzymes. Allowing cAMP to accumulate potentiates the action of the sympathetic nervous system. Most commonly encountered examples are the methylxanthines such as *theophylline*.

phosphoenolpyruvate (*BioSci*) The phosphate ester of the enol form of pyruvic acid, $CH_2{=}COPO_3H_2COOH$. An important metabolite in glycolysis and the substrate for PEP CARBOXYLASE.

phospholipase (*BioSci*) An enzyme that degrades a phospholipid. There are three main types, phospholipase A, C and D: A removes a hydrocarbon chain from the phospholipid, C cleaves the glycerol–phosphate ester bond and D the phosphate–base linkage.

phospholipid (*BioSci*) Any of a group of phosphate-containing lipids, the major structural lipids of most cellular membranes, eg phosphatidyl phospholipids, sphingomyelins.

phospholipid bilayer (*BioSci*) A lamellar organization of phospholipids that are packed as a bilayer with hydrophobic acyl tails inwardly directed and polar head groups on the outside surfaces. It is this bilayer that forms the basis of membranes in cells.

phosphonium salts (*Chem*) These salts are formed when phosphine is brought into contact with hydrogen chloride, hydrogen bromide or hydrogen iodide. Formed in a similar way to ammonium compounds.

phosphoproteins (*BioSci*) Proteins that have been enzymically phosphorylated so that they contain phosphate groups. An important functional modifier.

phosphor (*Chem, Electronics*) Generic name for materials which exhibit luminescence, esp the components of fluorescent coatings on cathode-ray tubes, such as zinc-doped zinc oxide.

phosphor bronze (*Eng*) A term sometimes applied to alpha (low-tin) bronze deoxidized with phosphorus, but generally it means a bronze containing 10–14% of tin and 0.1–0.3% of phosphorus, with or without additions of lead or nickel. Used, in cast condition, where resistance to corrosion and wear is required, eg gears, bearings, boiler fittings, parts exposed to sea water, etc. See COPPER ALLOYS.

phosphorescence (*BioSci*) Luminescence; production of light, usually (in animals) with little production of heat, as in glow-worms. Adj *phosphorescent*.

phosphorescence (*Chem*) Greenish glow observed during slow oxidation of white phosphorus in air. Also used more generally for other forms of luminescence.

phosphorescence (*Phys*) Luminescence which persists for more than 0.1 ns after excitation. See FLUORESCENCE, PERSISTENCE.

phosphoric acid (*Chem*) H_3PO_4. See ORTHOPHOSPHORIC ACID.

phosphorite (*Min*) The fibrous concretionary variety of APATITE. Also *rock-phosphate*.

phosphorized copper (*Eng*) Copper deoxidized with phosphorus. Contains a small amount (about 0.02%) of residual phosphorus, which lowers the conductivity. See COPPER ALLOYS.

phosphorofluoric acid (*Chem*) See HEXAFLUOROPHOSPHORIC ACID.

phosphorous (*Chem*) Relating to trivalent phosphorus, eg *phosphorus (III) oxide*, P_4O_6, which with cold water forms *phosphorous (III) acid*.

phosphorous (phosphorus (III)) acid (*Chem*) H_3PO_3. Formed by the action of cold water on phosphorous oxide; decomposes on heating; forms phosphites; reducing agent.

phosphorus (*Chem*) Symbol P, at no 15, ram 30.9738, valencies 3,5. A non-metallic element in the fifth group of the periodic system. White phosphorus is a waxy, poisonous, spontaneously flammable solid, mp 44°C, bp 282°C, rel.d. 1.8–2.3. Red phosphorus is non-poisonous and ignites in air only when heated above 300°C, mp 500–600°C. Its density and presumable structure vary with the method of preparation. Black has rel.d. 2.7, and is a metallic substance obtained at high temperature and pressure. Phosphorus occurs widely and abundantly in minerals (as phosphates) and in all living matter. There are numerous independent phosphate minerals in pegmatites and ores, and among their alteration products, but mainly the phosphate is *apatite*, a widespread accessory mineral of igneous rocks, which also occurs in sedimentary phosphate deposits, guano and bones. Manufactured by heating calcium phosphate with sand and carbon in an electric furnace. It is used mainly in the manufacture of phosphoric acid for phosphate fertilizers and plasticizers and in phosphating steel as a protection against corrosion; also used in matches and organic synthesis.

Phosphorus (*Astron*) An ancient name for Venus when it appears as a 'morning star'.

phosphorus oxychloride (*Chem*) $POCl_3$. Liquid; fumes in air; slowly hydrolysed by water, forming phosphoric and hydrochloric acids. It is formed when compounds containing a hydroxyl group are treated with phosphorus (V) chloride. Also *phosphoryl chloride*.

phosphorus pentahalides (*Chem*) Phosphorus (V) halides. These, *phosphorus (V) bromide* (PBr_5, *phosphorus (V) chloride* (PCl_5) and *phosphorus (V) fluoride* (PF_5), are formed by the action of the dry halogen on the trihalide. The properties of the (V) halides are similar. They transform hydroxyl compounds into the corresponding halides.

phosphorus pentoxide (*Chem*) Phosphorus (V) oxide. P_4O_{10}. Powerfully hygroscopic white solid obtained by burning phosphorus in air. Chiefly used to manufacture orthophosphoric acid by reaction with water; also as a drying agent.

phosphorylase (*BioSci*) An enzyme that catalyses the cleavage of a bond by the addition of orthophosphate (cf HYDROLYSIS). Thus glycogen phosphorylase liberates the terminal glucose of glycogen as glucose-1-phosphate.

phosphoryl bromide (*Chem*) $POBr_3$. Formed in a similar manner to PHOSPHORUS OXYCHLORIDE.

phosphoryl chloride (*Chem*) See PHOSPHORUS OXY-CHLORIDE.

phosphoryl fluoride (*Chem*) POF_3. May be made by the action of hydrofluoric acid on phosphorus (V) oxide; similar in properties to the other phosphoryl compounds.

phot (*Phys*) CGS unit of illumination: 1 phot = 1 lm cm^{-2} = 10^4 lm m^{-2}.

photic zone (*EnvSci*) The zone of the sea where light penetration is sufficient for photosynthesis, corresponding to the limnetic zone of fresh-water habitats. Also *euphotic zone*. Cf APHOTIC ZONE.

photoaffinity labelling (*BioSci*) A technique for covalently attaching a label or marker onto a molecule such as a protein. The label, which is often fluorescent or radioactive, contains a group that forms a covalent link to an adjacent molecule when illuminated (usually with ultra-violet light).

photobiology (*BioSci*) The study of light as it affects living organisms.

photobleaching (*BioSci*) Light-induced change in a chromophore, resulting in the loss of its absorption of light of a particular wavelength.

photocatalysis (*Chem*) The acceleration or retardation of rate of chemical reaction by light.

photocathode (*Electronics*) An electrode from which electrons are emitted on the incidence of radiation in a photocell. It is SEMITRANSPARENT when there is photo-emission on one side, arising from radiation on the other, as in the signal plate of a TV camera tube.

photo CD (*ImageTech*) A write-once disk which records photographic images in compressed digital form (together with audio, graphics and text). Its dye coating changes in reflectance where heated by the recording laser, varying the intensity of the reflected laser light when played (like the pits in a conventional CD). See ORANGE BOOK, TABLE OF CONTENTS.

photocell (*Electronics*) See PHOTOELECTRIC CELL.

photocell sensitivity (*Electronics*) The ratio of output current to level of illumination. Expressed in milliampere/lumen or microampere/lumen.

photochemical cell (*Electronics*) Photocell comprising two similar electrodes, eg of silver, in an electrolyte; illumination of one electrode results in a voltage between them. Also *Becquerel cell*, *photoelectrolytic cell*.

photochemical effect (*Chem*) Chemical effects of radiation, eg a light-catalysed reaction.

photochemical equivalence (*Chem*) See EINSTEIN LAW OF PHOTOCHEMICAL EQUIVALENCE.

photochemical induction (*Chem*) The lapse of an appreciable time between the absorption of light by a system and the occurrence of the resulting chemical reaction.

photochemical smog (*EnvSci*) Air pollution caused by light-driven reactions between nitrogen oxides and oxygen, leading to the formation of ozone that in turn reacts with hydrocarbons and other pollutants to produce a visible haze. This type of smog tends to form late in the day and can be damaging to plants and irritating to eyes. Large cities located in natural basins (eg Los Angeles) are particularly prone to this problem, which is distinct from normal smog.

photochemistry (*Chem*) The study of the chemical effects of radiation, chiefly visible and ultraviolet, and of the direct production of radiation by chemical change.

photochromic (*Glass*) A term describing a material whose TRANSMITTANCE varies with the intensity of light incident upon it. Also *photosensitive*.

photochromics (*Phys*) Light-sensitive materials that can be used in optical memory devices. Colour centres in the material form the basis of the processes of reading and writing the data.

photocomposition (*Print*) See PHOTOTYPESETTING.

photoconductive camera tube (*ImageTech*) One in which the optical image is focused onto a surface, the electrical resistance of which is dependent on illumination. See CHARGE-COUPLED DEVICE.

photoconductive cell (*Electronics*) Photoelectric cell using photoconductive element (often cadmium sulphide) between electrodes.

photoconductivity (*Electronics*) Property possessed by certain materials, such as selenium, of varying their electrical conductivity under the influence of light.

photocopying (*ImageTech*) Copying with photographic methods, such as making multiple prints from a negative. Now more generally used for XEROGRAPHY.

photocurrent (*Electronics*) That released from the sensitized cathode of a photocell on the incidence of light, the electrons which form the current being attracted to the anode which is maintained positive with respect to the cathode. The true photocurrent may be augmented, in a gas-filled cell, through collision ionization.

photodegradation (*Textiles*) Chemical degradation of a material such as a fibre caused by the absorption of light, esp ultraviolet radiation. Usually leads to loss of strength of the material as the polymer breaks down.

photodiode (*Electronics*) A semiconductor diode fitted with a small lens which can focus light on the p–n junction. Certain types of junction exhibit marked variation of reverse current with illumination; this allows them to be used as compact and rugged light sensors.

photodisintegration (*Phys*) The ejection of a neutron, proton or other particles from an atomic nucleus following the absorption of a PHOTON. A gamma-ray photon with energy >2·23 MeV can cause a deuteron nucleus to emit both a neutron and a proton.

photodissociation (*Chem*) Dissociation produced by the absorption of radiant energy.

photoelastic analysis (*Eng*) A method of examining transparent polymer models of structures etc to isolate stress concentrations and other weak zones. Magnitude of the STRESS CONCENTRATION FACTOR can be found knowing the nominal applied stress and the stress-optical coefficient. Mapping isochromatic zones in transparent moulded polymers placed between crossed, circularly polarizing filters (eg POLAROID sheets) gives an idea of FROZEN-IN STRAIN due to chain orientation. It can be used for NON-DESTRUCTIVE TESTING analysis of eg injection-moulded products subjected to severe stress in service. Analysis in white light gives coloured fringes, useful in finding fringe order.

photoelasticity (*Phys*) A phenomenon whereby strain in certain materials causes the material to display BIREFRINGENCE. Coloured fringes are observed when the transmitted light is viewed through crossed polarizers.

photoelectric absorption (*Phys*) The part of the absorption of a beam of radiation associated with the emission of photoelectrons. In general, it increases with increasing atomic number but decreases with increasing quantum energy when COMPTON ABSORPTION becomes more significant.

photoelectric cell (*Electronics*) See PHOTOEMISSIVE CELL.

photoelectric constant (*Phys*) The ratio of PLANCK'S CONSTANT to the electronic charge. This quantity is readily measured by experiments on photoelectric emission and forms one of the principal methods by which the value of Planck's constant may be determined.

photoelectric effect (*Electronics*) Any phenomenon resulting from the absorption of photon energy by electrons, leading to their release from a surface, when the photon energy exceeds the WORK FUNCTION (see PHOTOEMISSION), or otherwise allowing conduction when the incident energy exceeds an atomic binding energy. See PHOTOCONDUCTIVITY, PHOTOVOLTAIC CELL. Emission of X-rays on the impact of high-energy electrons on a surface is an inverse photoelectric effect.

photoelectricity (*Electronics*) Emission of electrons from the surface of certain materials by quanta exceeding a certain energy. See PHOTOELECTRIC WORK FUNCTION.

photoelectric multiplier (*Electronics*) See PHOTOMULTIPLIER.

photoelectric photometer (*Phys*) Photometer in which the light from the source under test is measured by the current from a photoelectric cell.

photoelectric photometry (*Astron*) The determination of stellar magnitude and colour index by means of a photoelectric device used at the focus of a large telescope.

photoelectric threshold (*Electronics*) The limiting frequency for which the quantum energy is just sufficient to produce photoelectric emission. Given by equating the quantum energy to the work function of the cathode.

photoelectric tube (*Electronics*) Any transducer which has an electrical output (current, voltage) corresponding to incident light.

photoelectric work function (*Electronics*) The energy required to release a photoelectron from a cathode. It should correlate closely with the THERMIONIC WORK FUNCTION.

photoelectric yield (*Electronics*) The proportion of incident quanta on a photocathode which liberate electrons.

photoelectroluminescence (*Electronics*) The enhancement of luminescence from a fluorescent screen during excitation by ultraviolet light or X-rays when an electric field is applied, ie the field enhancement of luminescence.

photoelectrolytic cell (*Electronics*) See PHOTOCHEMICAL CELL.

photoelectromagnetic effect (*Electronics*) See PHOTOMAGNETOELECTRIC EFFECT.

photoelectron (*Electronics*) One released from a surface by a photon, with or without kinetic energy.

photoelectron spectroscopy (*Chem, Phys*) The study of the energy of *photoelectrons* emitted from a material during irradiation by visible or ultraviolet light or X-rays, in order to analyse the properties of surfaces, interfaces and bulk materials. It can also be used to deduce binding energies for deep core levels in atoms with a high degree of precision. Also *electron spectroscopy, photoionization spectroscopy*. See IONIZATION POTENTIAL.

photoemission (*Electronics*) The emission of electrons from surface of a body (usually an electropositive metal) by incidence of light.

photoemissive camera tube (*ImageTech*) A tube operating on the photoemissive principle, the image falling on its photocathode, causing this to emit electrons in proportion to the intensity of the light in the picture elements.

photoemissive cell (*Electronics*) A category of PHOTOELECTRIC TUBES in which light falling on a sensitized cathode causes current to flow to an anode. The current may be enhanced (at the expense of linearity) by using a gas-filled envelope and operating above the ionization potential of the gas. PHOTOMULTIPLIER tubes are photoemissive.

photo-engraving (*Print*) See PROCESS-ENGRAVING.

photofission (*Phys*) Nuclear fission induced by gamma rays.

Photoflood lamp (*ImageTech*) A tungsten lamp run at excess voltage, with correspondingly reduced life, to raise its colour temperature.

photogenic (*BioSci*) Emitting light, light-producing, eg *photogenic bacteria*.

photoglow tube (*Electronics*) A gas-discharge tube in which conduction is enhanced by radiation falling on the cathode.

photogrammetry (*ImageTech*) The use of photographic records for precise measurement of distances or dimensions, eg aerial photography for surveying.

photographic borehole survey (*MinExt, Surv*) A check on orientation and angle of long borehole by insertion of a special camera which photographs a magnetic needle and a clinometer at known distance down.

photographic efficiency (*Phys*) Of a light source, the fraction of the light energy in the emitted spectrum which is usefully registered on a photographic emulsion.

photographic-emulsion technique (*Phys*) The study of the tracks of ionizing particles as recorded by their passage through a nuclear emulsion.

photographic photometry (*Phys*) The measurement of intensity of radiation by comparing a photographic image of the source with that of a standard source.

photographic recording (*Acous*) The registering of a modulated track on photographic film, so that it can be scanned by a constant beam of light, fluctuations of which, after being converted into corresponding electric currents by a photocell, can be amplified and reproduced as sound by loudspeakers.

photographic sound (*ImageTech*) A system of recording and reproducing sound on motion picture film in the form of a photographic sound track of varying width of modulation; the preferred term for *optical sound*.

photographic surveying (*Surv*) A method of surveying employing the principles of INTERSECTION by means of a special instrument called a phototheodolite, with which a series of photographs is taken of the points whose positions are required, each point appearing in at least two different photographs. Also *phototopography*.

photographic zenith tube (*Astron*) An instrument for the exact determination of time; it consists of a fixed vertical telescope which photographs stars as they cross the zenith; instrumental and observational errors are thus eliminated, the instrument being entirely automatic. Abbrev *PZT*.

photogravure (*Print*) An intaglio printing process using copper cylinders or plates etched through *carbon tissue*. The Helio-Klischograph is used to scan coloured separations into depressions cut in the cylinder, thus dispensing with process camera separation and carbon tissue.

photohalide (*Chem*) A halogen salt which is sensitive to light, eg silver bromide.

photoionization (*Phys*) The production of ions by light or other electromagnetic radiation of sufficient energy to remove an electron from an atom.

photoionization spectroscopy (*Chem*) See PHOTOELECTRON SPECTROSCOPY.

photolithography (*Electronics*) The process of pattern transfer using optics, used in semiconductor fabrication. See LITHOGRAPHY and panel on SEMICONDUCTOR FABRICATION.

photolysis (*Chem*) The decomposition or dissociation of a molecule as the result of the absorption of light.

photolysis of water (*BioSci*) The notional splitting of water molecules into oxygen, electrons and protons in the light reactions of photosynthesis. The term emphasizes the fact that the released oxygen comes from water not carbon dioxide.

photomagnetoelectric effect (*Electronics*) The generation of electric current by absorption of light on surface of semiconductor placed parallel to magnetic field. It is due to transverse forces acting on electrons and holes diffusing into the semiconductor from the surface. Also *photoelectromagnetic effect*.

Photomaton (*ImageTech*) Automatic machine which takes a number of photographs in succession and delivers a strip or sheet of finished prints in a few minutes. Prints are produced by automatic reversal of the image on a celluloid and paper base.

photomechanical (*Print*) Any process by which a printing surface is prepared mechanically with the aid of photography.

photomeson (*Phys*) A MESON, usually a pion, resulting from the interaction of a photon with a nucleus.

photometer (*Phys*) An instrument for comparing the luminous intensities of two sources of light. Most photometers employ the principle that, if equal illumination is produced on similar surfaces illuminated normally by two light sources, the ratio of their intensities equals the square of the ratio of their distances from the surfaces.

photometer bench (*Phys*) A bench upon which is mounted the apparatus for carrying out photometric tests by comparison with a standard lamp. The apparatus consists of a mounting for the standard lamp, the photometer itself, a mounting for the lamp under test, and equipment for moving any or all of these and determining their position.

photometer head (*Phys*) The part of a photometric system which contains the device for comparing the luminous intensities of two light sources.

photometric integrator (*Phys*) The part of an integrating photometer which actually sums up the light flux, eg the globe of the ULBRICHT SPHERE PHOTOMETER.

photometric surface (*Phys*) A surface used for photometric comparisons.

photometry (*Astron*) The accurate quantitative measurement of the amount of electromagnetic energy (usually visible light) received from a celestial object. Techniques are visual, photographic and photoelectric.

photometry (*Chem*) Volumetric analysis in which the endpoint of a reaction is determined from colour changes detected by photoelectric means.

photometry (*Phys*) The measurement of the luminous intensities of light sources and of luminous flux and illumination. See PHOTOMETER.

photomicrography (*ImageTech*) Photography through a microscope. Not to be confused with MICROPHOTOGRAPHY.

photomorphogenesis (*BioSci*) The control of plant morphogenesis by the duration and nature of the light. Cf ETIOLATION, PHOTOPERIODISM, PHYTOCHROME.

photo-mosaic (*ImageTech*) A light-responsive surface made up of minute photoemissive particles deposited on an insulating support, used in some early TV camera tubes.

photomultiplier (*Electronics*) A photocell with series of dynodes used to amplify emission current by electron multiplication, used in eg scintillation detectors.

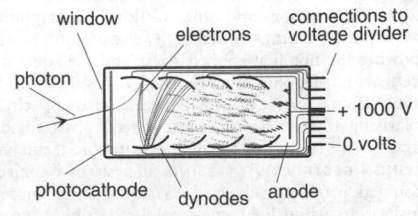

photomultiplier End window type.

photon (*Phys*) A quantum of visible light or other electromagnetic radiation of energy $E = h\nu$ where h is PLANCK'S CONSTANT and ν is the frequency. The photon has zero rest mass, but carries momentum $h\nu/c$, where c is the speed of light. The introduction of this 'particle' is necessary to explain the COMPTON EFFECT, the

PHOTOELECTRIC EFFECT, atomic line spectra, and other properties of electromagnetic radiation. See GAUGE BOSON.

photonastic movement (*BioSci*) NASTIC MOVEMENT resulting from change in illumination. Also *photonasty*.

photonegative (*Phys*) Said of a material for which electrical conductivity decreases with increasing illumination. Cf PHOTOPOSITIVE.

photoneutron (*Phys*) A neutron resulting from the interaction of a photon with a nucleus.

photonics (*ICT*) A concept for computing and data transmission using light, ie photons in place of electrons.

photon noise (*Electronics*) That occurring in photocells as a result of the fluctuations in the rate of arrival of light quanta at the photocathode.

photopeak (*Radiol*) The energy of the predominant photons released during the decay of a radionuclide.

photoperiodicity (*BioSci*) The controlling effects of the length of day on such phenomena as the flowering of plants, the reproductive cycles of mammals, migration and diapause of insects, and seasonal changes in the feathers of birds and the hair of mammals.

photoperiodism (*BioSci*) The response by an organism to day-length. See DAY-NEUTRAL PLANT, LONG-DAY PLANT, PHYTOCHROME, SHORT-DAY PLANT.

photophilous (*BioSci*) Light-seeking, light-loving; said of plants that inhabit sunny places.

photophobia (*Med*) Intolerance of the eye to light with spasm of the eyelids.

photophore (*BioSci*) A luminous organ of fish.

photophoresis (*Chem*) The migration of suspended particles under the influence of light.

photophosphorylation (*BioSci*) The production in photosynthesis of ATP from ADP and inorganic phosphate using energy from light. In cyclic photophosphorylation only PHOTOSYSTEM I is involved and NADP is not reduced; in non-cyclic photophosphorylation PHOTOSYSTEM I and II are both involved and NADP is reduced.

photophthalmia (*Med*) Burning pain in the eyes, lacrimation, photophobia, and swelling and spasm of the eyelids as a result of exposure to an intensely bright light (eg a naked arc light or laser); due to the action of ultraviolet rays. Also *electric-light ophthalmia*.

photophygous (*BioSci*) Shunning strong light.

photopic luminosity curve (*Phys*) A curve giving the relative brightness of the radiations in an equal-energy spectrum when seen under ordinary intensity levels. See SCOTOPIC LUMINOSITY CURVE.

photopic vision (*Phys*) Vision based on CONES, and therefore sensitive to colour. Possible only with adequate ambient illumination. Cf SCOTOPIC VISION.

photopigment (*BioSci*) See PHOTOSYNTHETIC PIGMENT.

photopolymerization (*Chem*) Polymerization initiated by visible light or ultraviolet photons, which react with monomer molecules to give free radicals. See CHAIN POLYMERIZATION.

photopolymer plates (*Print*) Relief printing plates made of a synthetic material, on a flexible or rigid base as required. After printing down with ultraviolet light through a negative, during which the exposed parts are polymerized, the plate is completed by washing away the unexposed areas with a caustic solution.

photopositive (*Phys*) Said of a material for which conductivity increases with increasing illumination. Cf PHOTONEGATIVE.

photoproton (*Phys*) A proton resulting from the interaction of a photon with a nucleus.

photopsy (*Med*) The appearance of flashes of light in front of the eyes, due to irritability of the retina. Also *photopsia*.

photoreceptor (*BioSci*) (1) A sensory cell specialized for the reception of light, ie rods and cones of the vertebrate eye. (2) A molecule (eg *phytochrome*) that absorbs stimulating light in phototaxis etc.

photoresist (*Electronics*) Photosensitive material used in photolithography. See fig. at PHOTOLITHOGRAPHY. See RESIST, SEMICONDUCTOR DEVICE PROCESSING.

photoresist process (*Electronics*) The process removing selectively the oxidized surface of a silicon slice semiconductor. The photoresist material is an organic substance polymerizing on exposure to ultraviolet light and in that form resisting attack by acids and solvents. See fig. at PHOTOLITHOGRAPHY.

photorespiration (*BioSci*) Light-stimulated respiration (O_2 uptake and CO_2 production) resulting from the production of phosphoglycollic acid by the oxygenase reaction of RIBULOSE 1,5-BISPHOSPHATE CARBOXYLASE OXYGENASE and its subsequent metabolism, in the so-called photorespiratory carbon oxidation cycle, to phosphoglyceric acid and CO_2.

photosedimentation (*PowderTech*) A method of particle size analysis in which concentration changes within a suspension are determined by measuring the attenuation of a beam of light passing through it.

photosensitive (*Glass*) See PHOTOCHROMIC.

photosensitive (*ImageTech, Phys*) The property of being sensitive to action of visible or invisible radiation.

photosensitizing dye (*BioSci*) INTERCALATING DYE that binds to DNA and makes it susceptible to breakage when exposed to ultraviolet light.

photosphere (*Astron*) The visible surface of the Sun on which sunspots and other physical markings appear; it is the limit of the distance that we can see into the Sun. See panel on SUN AS A STAR.

Photostat (*ImageTech*) TN for photographic apparatus (also for any print made by it) designed for rapidly copying flat original on sensitized paper, giving a negative image.

photosynthate (*BioSci*) The substances produced in photosynthesis, esp sugars.

photosynthesis (*BioSci*) The use of energy from light to drive chemical reactions, most notably the reduction of carbon dioxide to carbohydrates coupled with the oxidation either of water to free oxygen or of hydrogen sulphide to sulphur. See C3 PLANTS, C4 PLANTS, CAM PLANTS, DARK REACTIONS, LIGHT REACTIONS, PHOTOPHOSPHORYLATION, PHOTORESPIRATION, PHOTOSYSTEMS I AND II, PHOTOSYNTHETIC PIGMENT, Z SCHEME. Adj *photosynthetic*. See panel on CALVIN CYCLE.

photosynthetically active radiation (*BioSci*) Light with a wavelength $\lambda = 400$–700 nm, active in photosynthesis. Abbrev *PAR*.

photosynthetic bacteria (*BioSci*) Bacteria that can carry out photosynthesis. Light is absorbed by bacteriochlorophyll and carotenoids.

photosynthetic carbon reduction cycle (*BioSci*) Abbrev *PCR cycle*. See panel on CALVIN CYCLE.

photosynthetic pigment (*BioSci*) A pigment involved in the absorption of light in photosynthesis. In plants and algae, chlorophyll a and accessory pigments such as carotenoids and phycobilins. Also *photopigment*.

photosynthetic quotient (*BioSci*) The ratio of carbon dioxide absorbed to oxygen released in a photosynthesizing structure or organism. Abbrev *PQ*. Also *assimilatory quotient*.

photosystem I (*BioSci*) Reaction centre in photosynthesis comprising chlorophyll a, as P700, and other pigments and molecules, in which energy absorbed as light is used to transfer electrons from a weak oxidant to a strong reductant. The latter either reduces FERREDOXIN and ultimately NADP or is used in cyclic photophosphorylation. Abbrev *PS I*.

photosystem II (*BioSci*) Reaction centre in photosynthesis comprising chlorophyll a, as P680, and other pigments and molecules, in which energy absorbed as light is used to transfer electrons from a strong oxidant to a weak reductant, the former oxidizing water to oxygen. Abbrev *PS II*.

phototaxis (*BioSci*) Locomotory response or reaction of an organism or cell to the stimulus of light. Adj *phototactic*.

phototheodolite (*Surv*) A photographic camera of fixed known focal length, with horizontal and vertical lines engraved on a glass plate at the image plane, on which the film is registered. It is mounted on a tripod and fitted with levelling screws, a graduated horizontal circle and a telescope. See PHOTOGRAPHIC SURVEYING.

phototopography (*Surv*) See PHOTOGRAPHIC SURVEYING.

phototransduction (*BioSci*) The transformation of light energy by photoreceptors (eg rods and cones in the retina of the eye) into a change in electrical potential.

phototransistor (*Electronics*) A three-electrode photosensitive semiconductor device. The emitter junction forms a photodiode and the signal current induced by the incident light is amplified by transistor action.

phototrophic (*BioSci*) Said of organisms obtaining their energy from sunlight.

phototropism (*BioSci*) A TROPISM in which a plant part becomes orientated with respect to the direction of light.

phototropy (*Chem*) (1) The property possessed by some substances of changing colour according to the wavelength of the incident light. (2) The reversible loss of colour in a dyestuff when illuminated at a definite wavelength.

phototube (*Electronics*) See PHOTOELECTRIC TUBE.

phototypesetting (*Print*) The technique by which text and certain illustrations can be set onto bromide paper or film.

photovaristor (*Electronics*) Material, eg cadmium sulphide or lead telluride, in which the varistor effect, ie non-linearity of current–voltage relation, is dependent on illumination.

photovoltaic cell (*Electronics*) A class of photoelectric cell which acts as a source of emf and so does not need a separate battery.

photovoltaic effect (*Electronics*) The production of an emf across the junction between dissimilar materials when it is exposed to light or ultraviolet radiation.

phragma (*BioSci*) (1) A septum or partition. (2) An apodeme of the endothorax formed by the infolding of a portion of the tergal region of a somite. (3) An endotergite.

phragmoplast (*BioSci*) A complex of interdigitating microtubules aligned more or less parallel to the earlier spindle microtubules, developing at the end of mitosis and spreading like an expanding doughnut to the side walls while the cell plate forms in the centre. Characteristic of vascular plants, bryophytes and some algae.

phreak (*ICT*) A person who hacks into a telephone system in order to make free calls.

phreatic gases (*Geol*) Those gases of atmospheric or oceanic origin which are generated by contact with ascending magma.

phrenicectomy (*Med*) The cutting of the phrenic nerve in order to paralyse the diaphragm on one side; done in the treatment of lung disease. Also *phrenicotomy*.

phrenology (*Psych*) Historically, the notion that a person's skull formations reflect mental abilities and personality characteristics.

pH stat (*BioSci*) Metabolic reactions that collectively maintain constant pH in the cytoplasm.

phthalates (*Chem*) A family of organic compounds used as plasticizers in PVC films. They may be persistent and bioaccumulate in fatty tissue although their toxicity is disputed.

phthaleins (*Chem*) Triphenylmethane derivatives obtained by the action of phenols upon phthalic (benzene 1,2-dicarboxylic) anhydride.

phthalic acid (*Chem*) *Benzene-O-dicarboxylic acid*. $C_6H_4(COOH)_2$. Mp 213°C. Colourless prisms or plates, soluble in water, alcohol, ether. When heated above the melting point it yields its anhydride. It is prepared by catalytic air oxidation via the anhydride.

phthalic anhydride (*Chem*) *Phthalic (benzene 1,2-dicarboxylic) anhydride*. Long prisms which can be sublimed, mp 128°C, bp 284°C. Used in the production of plasticizers, paints and polyester resins. Prepared by air oxidation of naphthalene in the presence of a catalyst such as vanadium (V) oxide.

phthalic anhydride

phthalic glyceride resins (*Plastics*) See ALKYD RESINS.

phthalimide (*Chem*) *Benzene 1,2-dicarboximide*. Colourless crystals, mp 238°C, obtainable by passing ammonia over heated phthalic anhydride. The imide hydrogen is replaceable by Na or K.

phthalocyanines (*Chem*) A group of green and blue organic colouring matters, used mainly as pigments and formed by condensation polymerization of four molecules of phthalonitrile with one atom of a metal, such as copper phthalocyanine. Widely used in almost all pigmented products. *Green*: Consists of halogenated copper phthalocyanine, and available in a number of shades from deep blue-green to a bright green with only a slight blue tone.

phthalonitrile (*Chem*) *1,2-dicyanobenzene*. Crystalline solid; mp 140°C. PHTHALOCYANINE colours are made by heating it with metallic compounds.

phthiriasis (*Med*) Infestation with lice. See PEDICULOSIS.

phthisis (*Med*) Former term for: (1) wasting of the body; (2) pulmonary tuberculosis. Adj *phthisical*.

phugoid oscillation (*Aero*) A longitudinal fluctuation in speed of long periodicity, ie a velocity modulation, in the motion of an aircraft, accompanied by rising and falling of the nose.

pH value (*Chem*) See PH.

phyc-, phyco- (*Genrl*) Prefixes from Gk *phykos*, seaweed, and, by extension, algae.

phycobilin (*BioSci*) One of a number of red (phycoerythrin) or blue-green (phycocyanin) ACCESSORY PIGMENTS, found in Rhodophyceae, Cryptophyceae and cyanobacteria.

phycobiont (*BioSci*) The algal partner in a lichen or other symbiotic association.

phycocyanin (*BioSci*) Blue phycobilin found in some algae, and esp in cyanobacteria.

phycoerythrin (*BioSci*) Red phycobilins found in some algae, esp red algae (Rhodophyta).

phycology (*BioSci*) The study of the algae.

Phycomycetes (*BioSci*) A probably unnatural group of fungi and fungus-like organisms, comprising the Mastigomycotina (including the Oomycetes) and the Zygomycotina.

phyletic classification (*BioSci*) A scheme of classification based on presumed evolutionary descent.

phyllid (*BioSci*) The leaf of a moss or liverwort.

phyllite (*Min*) A name which has been used in different senses: (1) for the pseudohexagonal platy minerals (*phyllosilicates*) including mica, chlorite and talc (by some Fr authors); (2) for argillaceous rocks in a condition of metamorphism between slate and mica-schist (by most English authors). Phyllite in the latter (usual) sense is characterized by a silky lustre due to the minute flakes of white mica which, however, are individually too small to be seen with the naked eye.

phyllobranchia (*BioSci*) A gill composed of numerous thin plate-like lamellae.

phylloclade (*BioSci*) A flattened leaf-like stem, functioning as a photosynthetic organ. Also *cladode*.

phyllode (*BioSci*) A flat, more or less expanded, leaf-like petiole functioning as a photosynthetic organ, usually in the absence of a leaf blade.

phyllody (*BioSci*) The abnormal, often pathological, development of leaves in place of the normal parts of a flower.

phylloplane (*BioSci*) The leaf surface, esp as a habitat for micro-organisms. Also *phyllosphere*.

phyllopodium (*BioSci*) The thin leaf-like swimming foot characteristic of some crustaceans. Cf STENOPODIUM.

phyllosilicates (*Min*) Those silicate minerals having an atomic structure in which SiO_4 groups are linked to each other to form continuous sheets, eg *talc*. See SILICATES.

phyllosphere (*BioSci*) See PHYLLOPLANE.

phyllotaxis (*BioSci*) The arrangement of leaves on a stem. Also *phyllotaxy*.

phylo- (*Genrl*) Prefix from Gk *phylon*, race.

phylogeny (*BioSci*) The evolutionary development or history of groups of organisms. Also *phylogenesis*. Adj *phylogenetic*. Cf ONTOGENY.

phylum (*BioSci*) A category or group of related forms constituting one of the major subdivisions of the animal kingdom. A *division* in plants.

physiatrics (*Med*) (N America) Physiotherapy.

physical chemistry (*Chem*) The study of the dependence of physical properties on chemical composition, and of the physical changes accompanying chemical reactions.

physical connection (*ICT*) A transmission means between two or more users of a NETWORK that usually consists of electrical conductors along which signals are transmitted.

physical containment (*BioSci*) The construction of laboratory or workstation so as to prevent the contamination of the worker or the environment by harmful organisms.

physical cross-link (*Phys*) A type of cross-link which involves van der Waals' or ionic bonds in copolymers rather than chemical covalent bonds.

physical drive (*ICT*) A DISK DRIVE that physically exists rather than a LOGICAL DRIVE that is artificially created by means of software. A physical drive may be associated with or mapped to several logical drives.

physical electronics (*Electronics*) The branch of electronics which is concerned with the physical details of the behaviour of electrons in vacuum, in gases and in conductors, semiconductors and insulators.

physical geology (*Geol*) The processes involved in the inorganic evolution of the Earth and esp its morphology.

physical layer (*ICT*) Level 1 of the OPEN SYSTEMS INTERCONNECTION (OSI) model, responsible for the electrical or mechanical interface to the communications medium.

physical metallurgy (*Eng*) The study of the properties of metals and their alloys and the effect of composition, heat treatment, environment and other factors upon them.

physical optics (*Phys*) A branch of the study of light dealing with phenomena such as diffraction and interference, which are best considered from the standpoint of the wave theory of light. See GEOMETRICAL OPTICS.

physical vapour deposition (*Electronics*) A thin-film deposition process involving physical mechanisms such as evaporation or SPUTTERING, as distinct from one involving chemical reactions. Abbrev *PVD*. Cf CHEMICAL VAPOUR DEPOSITION.

physics (*Genrl*) The study of matter and energy with the aim of describing phenomena in terms of fundamental laws.

physio- (*Genrl*) Prefix from Gk *physis*, nature.

physiography (*Geol*) The science of the surface of the Earth and the interrelations of air, water and land.

physiological (*Genrl*) Relating to the functions of plant or animal as a living organism.

physiological anatomy (*BioSci*) The study of the relation between structure and function.

physiological drought (*BioSci*) A condition in which a plant is unable to take in water because of low temperature, or because the water available to it holds substances in solution which hinder absorption by the plant.

physiological psychology (*Psych*) The study of anatomy and physiology in relation to psychological phenomena.

physiological race (*BioSci*) A group of individuals within the morphological limits of a species but differing from other members of the species in habits (as host, larval food,

etc), eg the several races of a parasitic fungus each confined to a different host. Also *biological form, forma specialis*.

physiology (*Genrl*) The sciences of processes of life in animals and plants.

physoclistous (*BioSci*) Of fish, having no pneumatic duct connecting the air bladder with the alimentary canal. Cf PHYSOSTOMOUS.

physostigmine (*Pharmacol*) A reversible *acetylcholine esterase inhibitor* derived from the Calabar bean, used as a drug in treating glaucoma and for treating anticholinergic syndrome. Also *eserine*.

physostomous (*BioSci*) Of fish, having a pneumatic duct. Cf PHYSOCLISTOUS.

-phyte (*Genrl*) See PHYTO-.

phytic acid (*BioSci*) Inositol hexaphosphate, present in eg seeds as its calcium magnesium salt, phytin, perhaps as a phosphorus-storage compound.

phyto-, -phyte (*Genrl*) Prefix and suffix from Gk *phyton*, plant.

phytoalexin (*BioSci*) A substance produced by a plant in response to some stimulus. Often phenolic or terpenoid, they inhibit the growth of some micro-organisms esp pathogenic fungi.

phytochemistry (*BioSci*) The study of the chemical constituents and esp the SECONDARY METABOLITES of plants.

phytochrome (*BioSci*) A blue-green, phycobilin-like pigment reported from seed plants and a wide variety of other photosynthetic eukaryotes, acting as the light receptor molecule in a number of morphogenetic processes including photoperiodism, reversal of etiolation and the germination of some seeds and spores. See P_{FR}, P_R.

phytoferritin (*BioSci*) An iron–protein complex, similar to the ferritin of animals, apparently a store of iron.

phytohaemagglutinins (*BioSci*) Lectins extracted from the beans of *Phaseolus vulgaris*. They bind to N-acetyl-beta-D-galactosamine residues and can agglutinate certain erythrocytes which bear these on their surface. The purified lectins are potent mitogens of T-lymphocytes. Abbrev *PHA*.

phytohormone (*BioSci*) A plant hormone.

phytology (*Genrl*) See BOTANY.

phytonutrient (*BioSci*) Any of various organic substances derived from plants that are believed to have health-giving properties.

phytopathology (*BioSci*) The study of plant diseases, esp of plants in relation to their parasites. Also *plant pathology*.

phytophagous (*BioSci*) Plant-feeding. Also *phytophilous*.

phytoplankton (*BioSci*) The photosynthetic members of the plankton.

phytoplankton blooms (*BioSci*) Very high densities of plankton which arise quickly, and persist for short times, usually at regular times of the year.

phytosanitary certificate (*BioSci*) A certificate of health for plants or parts of plants for export.

phytosociology (*BioSci*) The study of the association of plant species.

phytotoxic substance (*BioSci*) A substance toxic to plants. Also sometimes PHYTOTOXIN.

phytotoxin (*BioSci*) A plant substance toxic to animals or other organisms.

phytotron (*BioSci*) A large and elaborate GROWTH ROOM for plants, or a collection of such.

pi (*MathSci*) See π.

Piacenzian (*Geol*) The highest stage of the Neogene (Pliocene). See TERTIARY.

pia mater (*BioSci*) In vertebrates, the innermost of the three membranes surrounding the brain and spinal cord; a thin vascular layer.

pian (*Med*) See YAWS.

piano nobile (*Arch*) The principal floor of an Italian palace, occurring at first-floor level and containing the main living areas.

Pianotron (*Acous*) Piano in which the normal vibration of the strings is used to modulate the potential applied to electrostatic screw pickups, with amplification and loudspeakers.

pi-attenuator (*ICT*) An attenuator network of resistors arranged as in the pi-network. Also *π-attenuator*.

piazza (*Arch*) (1) An enclosed court in a building. (2) A colonnade or arcade.

PIB (*Plastics*) Abbrev for POLYISOBUTYLENE, or *polybutene*.

pica (*Med*) Unnatural craving for unusual food.

pica (*Print*) An old type size, approximately 12-point. See EM, POINT.

pi characters (*Print*) (1) Special characters or symbols not usually included in a type FONT, including special LIGATURES, accented letters, mathematical symbols, etc. Also *sorts*. (2) OUTSIDE SORTS on line-composing machines which cannot be accommodated in a magazine but which are stored in small cases beside the machine and inserted in the line by hand. Also *side sorts*. (3) Outside sorts on phototypesetting machines which are positioned in a separate pi character store.

Piciformes (*BioSci*) An order of birds, containing climbing, insectivorous and wood-boring species. The beak is hard and powerful, the tongue long and protrusible, and the feet zygodactylous, eg woodpeckers, toucans.

pick (*Build*) A double-headed tool, pointed at both ends, having the handle fastened into the middle of the head (as in a hammer), used for rough digging.

pick (*Textiles*) One traverse of the WEFT-carrying device through the WARP array to lay a weft thread in weaving a cloth. A single weft thread in a fabric. Picks per inch (ppi) or per centimetre (ppcm) are used to denote the total number of weft threads per unit length of cloth.

pick-and-pick (*Textiles*) A woven fabric with alternate picks of yarns of two different colours or kinds.

pick-and-place (*Eng*) Robotic device for automatic removal of injection mouldings etc from tool cavity at end of cycle.

pick-axe (*Build*) A tool similar to a pick but having one end edged so that it may cut.

picker (*Textiles*) Buffalo hide, rubber, or plastic component that strikes the shuttle to propel weft across loom.

pickeringite (*Min*) Magnesia alum. Hydrated sulphate of aluminium and magnesium, crystallizing in the monoclinic system. It usually occurs in fibrous masses, and is formed by the weathering of pyrite-bearing schists.

picket (*Surv*) A short-ranging rod about 2 m (6 ft) long.

picking (*Build*) Chiselling or picking small indentations on a wall surface as a key for the finish. Also *stugging, wasting*.

picking (*Print*) Removal of part of the paper surface owing to faults in the paper or in printing.

picking (*Textiles*) (1) Removing extraneous matter (outstanding hairs, wrongly coloured fibres, slubs, etc) from the face of fabrics. (2) Propelling the shuttle carrying the weft through the warp shed in weaving.

picking belt (*MinExt*) Sorting belt, on which run-of-mine ore is displayed so that pickers can remove waste rock, debris or a special mineral constituent, which is not to be sent to the mill for treatment.

pickling (*ChemEng*) The removal of mill scale, lime scale or salt-water deposits, eg in a ship's engine or evaporators, by circulating a suitable acid containing inhibitors. Usually up to ten times diluted sulphuric acid. Inhibitors include glue and acidic long-chain molecular compounds such as HBF_4.

pickling (*FoodSci*) The preservation of fruit, vegetables, also fish, meat and eggs, in acid and salt liquor (brine). Vinegar (or acetic acid) lowers pH and salt reduces water activity to prevent spoilage and growth of food poisoning organisms, mustard and spices being often added for additional flavour.

Pick's disease (*Med*) A rare cause of pre-senile dementia. Now more commonly termed *frontotemporal dementia*

pickup device (*ImageTech*) An imaging device in a video camera. See CAMERA TUBE, SOLID-STATE IMAGE SENSOR.

pickup head (*Acous*) (1) Mechanical–electrical transducer, often piezo, actuated by a sapphire or diamond stylus which rests on the sides of the groove on a vinyl disk

record, and by tracking this groove generates a corresponding voltage (or, for stereo reproduction, two voltages) for driving an audio amplifier. Also *pickup*. (2) Device to measure structure-borne sound, usually employing the piezoelectric, electrodynamic or electrostatic principle. See ACCELEROMETER.

pickup needle (*Acous*) Loosely applied to any needle in a record player pickup.

pickup reaction (*Phys*) Nuclear reaction in which an incident particle collects a nucleon from a target atom and proceeds with it.

pickup tube (*ImageTech*) US for CAMERA TUBE.

pickup well (*Autos*) A small petrol reservoir arranged between the metering jet and the spraying tube in some carburettors; it provides a temporarily enriched mixture during acceleration.

pico- (*Genrl*) SI prefix for one million millionth, or 10^{-12}. Formerly *micromicro-*. Symbol p.

picocell (*ICT*) A very small area of coverage in a wireless communication or data system, typically a single room within a building.

picofarad (*ElecEng*) Unit equal to 10^{-12} farads. Also *micromicrofarad*. Abbrev p*F*.

picolines (*Chem*) Methyl pyridines, $CH_3C_5H_4N$. The three isomers are 2-picoline, bp 129°C; 3-picoline, bp 142–143°C; 4-picoline, bp 144–145°C.

Picornaviridae (*BioSci*) Viruses, with a single positive strand of RNA and an icosahedral capsid. There are two main classes: enteroviruses such as poliovirus, and rhinoviruses such as common cold virus, foot-and-mouth disease virus and hepatitis A.

picosecond (*Genrl*) One million millionth part of a second, 10^{-12} s. Abbrev ps.

picotite (*Min*) A dark-coloured spinel containing iron, magnesium, aluminium; a chromium-bearing hercynite.

picramic acid (*Chem*) Red crystalline solid; mp 168°C, obtained by reduction of PICRIC ACID (trinitrophenol). Used in manufacture of azo dyes.

picric acid (*Chem*) *2,4,6-trinitrophenol*. $C_6H_2(NO_2)_3.OH$. Mp 122°C. Yellow plates or prisms, made by nitrating phenol; slightly soluble in water. It is a strong acid and dyes wool and silk yellow. Used for the preparation of explosives; lyddite or melenite is compressed or fused picric acid.

picrite (*Geol*) An ultramafic coarse-grained igneous rock, consisting essentially of olivine and other ferromagnesian minerals, together with a small amount of plagioclase. Also used for volcanic rocks.

Pictor (Easel) (*Astron*) A small inconspicuous southern constellation.

picture (*ImageTech*) In TV usage, the picture is the complete image, or FRAME, made up from two interlaced fields; hence *picture frequency* is the same as FRAME FREQUENCY.

picture element (*ICT*) See PIXEL.

picture inversion (*ImageTech*) Conversion of negative to positive image (or vice versa) when carried out electronically. In facsimile transmission, it will correspond to the reversal of the black-and-white shades of the recorded copy.

picture monitor (*ImageTech*) Cathode-ray tube for exhibiting TV picture or related waveform for purposes of control.

picture noise (*Radar*) See GRASS.

picture signal (*ImageTech*) That portion of a TV signal which carries information relative to the picture itself, as distinct from synchronizing portions. See PEDESTAL.

picture slip (*ImageTech*) Horizontal or vertical displacement or distortion of the received picture due to loss of synchronization.

Picturesque (*Arch*) A design concept which originated in the late 18th century and which idealized classical landscape paintings by such artists as Claude and Poussin. It mainly influenced landscape gardening where buildings, and frequently sham ruins or follies, became part of a controlled but informal landscape. On a smaller scale, domestic architecture became more asymmetrical, while still retaining GOTHIC or Italianate forms. See FOLLY.

picture/sync ratio (*ImageTech*) The ratio of the total amplitude of the TV waveform assigned to picture information to that which is assigned to the synchronizing pulses and flyback times.

picture tube (*ImageTech*) Cathode-ray tube specifically designed for the reproduction of TV images.

PID (*ICT*) Abbrev for PERSONAL IDENTIFICATION DEVICE.

Pidgeon process (*Eng*) Reduction of magnesium oxide in the presence of ferrosilicon in the making of magnesium.

pie (*Print*) To upset type-matter accidentally.

piece (*Textiles*) Fabric of agreed length. See BOLT.

piece dyeing (*Textiles*) Dyeing lengths of fabric.

piece goods (*Textiles*) Fabric sold by length.

piedmont glacier (*Geol*) A glacier of the 'expanded foot' type; one which, after being restricted within a valley, spreads out on reaching the flat ground into which the latter opens.

piedmont gravels (*Geol*) Accumulations of coarse breccia, gravel and pebbles brought down from high ground by mountain torrents and spread out on the flat ground where the velocity of the water is checked. Literally, mountain foot gravels, typical of the outer zone of arid areas of inland drainage such as the Lop Nor Basin in Chinese Turkestan.

piedmontite (*Min*) A hydrated silicate of calcium, aluminium, manganese and iron, crystallizing in the monoclinic system; a member of the zoisite group. Also *manganepidote, piemontite*.

piedroit (*Build*) A pier projecting from a wall but having neither cap nor base. Cf PILASTER.

piend (*Build*) Hip rafter. See ARRIS.

piend check (*Build*) The rebate cut along a lower corner of a stone step to enable it to sit upon the step below.

piend rafter (*Build*) The rafter at the junction where two roof slopes meet forming an external angle. An *angle rafter*.

Pieper system (*ElecEng*) See AUTOMIXTE SYSTEM.

pier (*Build*) (1) The part of a wall between doors and windows. (2) See BUTTRESS.

pierced (*Print*) Said of a block with an internal portion removed to accommodate type.

piercement salt dome (*Geol*) A petroleum reservoir in which the plastic salt has squeezed up and bent or ruptured the overlying beds, to produce potential reservoirs. These occur on the flanks of the salt dome or above it and in small associated fault traps. Also *diapiric salt dome*.

Pierce oscillator (*Electronics*) Original crystal oscillator in which positive feedback from anode to grid in a triode valve is controlled by piezoelectric mechanical resonance of a suitably cut quartz crystal.

piercing (*Eng*) See MANNESMANN PROCESS.

pieze (*Phys*) Unit of pressure in the metre–tonne–second system, equivalent to 10^3 N m^{-2} or 1 kN m^{-2}.

piezo- (*Genrl*) Prefix from Gk *piezein*, to press.

piezochemistry (*Chem*) The study of the effect of high pressures in chemical reactions.

piezoelectric crystal (*Electronics*) One showing piezoelectric properties which may be shaped and used as a resonant circuit element or transducer (eg a microphone, pickup, loudspeaker, depth-finder, etc).

piezoelectric effect (*Phys*) Electric polarization arising in some ANISOTROPIC (ie not possessing a centre of symmetry) crystals (BARIUM TITANATE, QUARTZ, ROCHELLE SALT) when subject to mechanical strain. An applied electric field will likewise produce mechanical deformation.

piezoelectric loudspeaker (*Acous*) A PIEZOELECTRIC CRYSTAL for generating sound waves (direct piezoeffect).

piezoelectric microphone (*Acous*) A PIEZOELECTRIC CRYSTAL which produces electric potentials in response to incident sound waves (inverse piezoeffect).

piezoelectric pickup (*Acous*) Piezoelectric transducer, producing an emf due to mechanical drive arising from

vibration. Widely used for vinyl disk reproduction and vibration measurements.

piezoelectric resonator (*ICT*) A crystal used as a standard of frequency, controlling an electronic oscillator.

piezoid (*ElecEng*) Blank of piezo crystal, adjusted to a required resonance, with or without relevant electrodes.

piezomagnetism (*Phys*) An effect analogous to the PIEZO-ELECTRIC EFFECT in which mechanical strain in certain materials produces a magnetic field and vice versa.

piezoresistivity (*Phys*) Strain-dependent resistivity exhibited by single-crystal semiconductors. Mechanical strain modifies carrier mobility in the crystal; the basis of semiconductor strain gauges.

PIF (*ICT*) Abbrev for PROGRAM INFORMATION FILE.

pig (*Eng*) A handleable mass of metal (eg cast-iron, copper or lead) cast in a simple shape for transportation or storage, and subsequently remelted for purification, alloying, casting into final shapes, or into ingots for rolling.

pig (*Glass*) An iron block laid against the pot mouth as a support for the BLOWING-IRON.

pig (*MinExt*) See GO-DEVIL.

pig bed (*Eng*) A series of moulds for iron PIGS, made in a bed of sand. Connected to each other and to the tap hole of the blast furnace by channels, along which the molten metal runs. The first subdivisions of the liquid run into larger sections called sows, and the pigs are the smaller sections which connect to these, resembling a row of piglets sucking the mother's teats when viewed from above the pig bed. Now replaced by pig casting machines using moulds on a moving belt.

pigeon fancier's disease (*Med*) Interstitial lung disease caused by an allergic reaction to pigeon antigens common among pigeon keepers.

pigeon-holed (*Build*) Said of a wall built with regular gaps in it, eg HONEYCOMB WALL.

pigeonite (*Min*) One of the monoclinic pyroxenes, intermediate in composition between clinoenstatite and diopside. It is poor in calcium, has a small optic axial angle and occurs in quickly chilled lavas and minor intrusions.

pigeon's milk (*BioSci*) In pigeons and doves, a white slimy secretion of the epithelium of the crop in both sexes. It contains protein and fat, and is produced during the breeding season under the influence of PROLACTIN, being regurgitated to feed the young.

pig iron (*Eng*) The crude iron produced in the blast furnace and cast into pigs which are used for making steel, cast-iron or wrought-iron. Principal impurities are carbon, silicon, manganese, sulphur and phosphorus. Composition varies according to the ores used, the smelting practice and the intended usage.

pigment (*BioSci*) Substances that impart colour to the tissues or cells of animals and plants.

pigment (*Build, Chem*) Insoluble, natural or artificial, black, white or coloured materials reduced to powder form which, when dispersed in a suitable medium, are able to impart colour and/or opacity. Ideally, pigments should maintain their colour under the most unfavourable conditions; in practice, may tend to fade or discolour under the influence of acids, alkalis, other chemicals, sunlight, heat and other conditions. May also show BLEEDING. They are used in many industries, eg in paint, printing ink, plastics, rubber, paper, textiles, etc. The two main types are the natural earth pigments which tend to be drab and limited in colour range and synthetic pigments with a comprehensive range of colours.

pigmentary colours (*BioSci*) Colours produced by the presence of drops or granules of pigment in the integument, as in most fish. Cf STRUCTURAL COLOURS.

pigment cell (*BioSci*) See CHROMATOPHORE.

pigment foils (*Print*) BLOCKING FOILS, which leave a black, white or coloured matt transfer.

pigtail (*ElecEng*) The short length of flexible conductor connecting the brush of an electric machine to the brush-holder.

pilaster (*Build*) A square pier projecting from a wall, having both a cap and a base. Cf PIEDROIT.

pilaster strip (*Build*) A pilaster without a cap.

Pilat process (*Chem*) A method of separating the fractions of asphalt oils without distillation, but dissolving out the asphalt with propane and saturating the residual oil with methane under pressure, with the result that the lubricating oil fractions separate out in the order of decreasing viscosity.

pile (*CivEng*) A column or sheeting which is sunk into the ground to support vertical loading or to resist lateral pressures.

pile (*ElecEng*) Abbrev for *thermopile*, a close packing of thermocouples in series, so that alternate junctions are exposed for receiving radiant heat, thus adding together the emfs due to pairs of junctions.

pile (*Eng*) A number of wrought-iron bars arranged in an orderly pile which is to be heated to a welding heat and rolled into a single bar. Obsolete.

pile (*Med*) See HAEMORRHOID.

pile (*NucEng*) Original name for a reactor made from the pile of graphite blocks which formed the moderator of the original nuclear reactor which first went critical on 2 December 1942 in Chicago, Illinois.

pile (*Phys*) A stack of glass plates used to plane-polarize light. If the light is incident at the Brewster angle, the transmitted light is partially plane-polarized normal to the plane of the reflected polarized light; with the stack of plates the transmitted light becomes increasingly plane-polarized.

pile (*Textiles*) A covering on the surface of a fabric, formed by threads that stand out from it. Pile in loop form is termed *loop pile* or *terry*; if the loops are cut, it is termed *cut pile*. The latter is produced by weaving two cloths face to face, and cutting them apart. Some carpets and MOQUETTES are made this way.

pileate (*BioSci*) Crested.

pile delivery (*Print*) On a printing press, a delivery arrangement which can accommodate a large pile of sheets, the delivery board lowering automatically to floor level as the sheets are delivered.

pile-driver (*CivEng*) A framed construction erected above the spot where a pile is to be driven into the ground. It is provided with a heavy weight which runs in upright guides and is so arranged that it may fall by gravity onto the head of the pile and drive it in. Now superseded by more efficient plant. Also *pile frame*.

pile hoop (*CivEng*) An iron or steel band fitted around the head of a timber pile to prevent brooming.

pile shoe (*CivEng*) The iron or steel point fitted to the foot of a pile to give it strength to pierce the earth and so assist driving.

pileus (*BioSci*) The cap of the fruiting body of some fungi, eg mushrooms or toadstools (Basidiomycotina).

pileus (*EnvSci*) See CAP.

piliferous layer (*BioSci*) The epidermis, bearing root hairs, of a young root.

piling (*Print*) A build-up of ink on the plate, blanket or printing rollers during printing.

pill (*Textiles*) Small unsightly balls of fibre that accumulate during wear on the surface of certain items of apparel.

pillar (*ElecEng*) A structure of pillar form for containing switch or protective gear. Also *switchgear pillar*.

pillar (*MinExt*) Column of unserved ore left as roof support in stope.

pillar-and-stall (*MinExt*) See BORD-AND-PILLAR.

pillar drill (*Eng*) A drilling machine in which the spindle and table are supported by brackets on a pillar along which they may slide.

Pilling–Bedworth ratio (*Eng*) Volume ratio of metal oxide to that of its parent metal which provides a guide to the volume changes accompanying oxidation.

pillow distortion (*ImageTech, Phys*) See PINCUSHION DISTORTION.

pillow lava (*Geol*) A lava flow exhibiting pillow structure, generally formed in a subaqueous environment.

pillow structure (*Geol*) A term applied to lavas consisting of ellipsoidal and pillow-like masses which have cooled under subaqueous conditions. The spaces between the pillows consist, in different cases, of chert, limestones or volcanic ash.

pilomotor response (*Med*) A phenomenon commonly known as goose-flesh or goose-bumps, in which hairs on the skin stand erect and there is pimpling of the skin.

pilose (*BioSci*) Hairy, with long, soft hairs.

pilot (*ElecEng*) In power systems, a conductor used for auxiliary purposes, not for the transmission of energy. Also *pilot wire*.

pilot (*Textiles*) Heavyweight, woollen cloth that has been highly milled and raised, used in naval jackets and overcoats.

pilotaxitic texture (*Geol*) The term applied to the groundmass of certain holocrystalline igneous rocks in which there is a felt-like interweaving of feldspar microlites. Cf HYALOPILITIC TEXTURE.

pilot balloon (*EnvSci*) A small rubber balloon, filled with hydrogen, used for determining the direction and speed of air currents at high altitudes; the balloon is observed by means of a theodolite after being released from the ground.

pilot carrier (*ICT*) In a suppressed carrier system (as in single-sideband working) a small portion of original carrier wave transmitted to provide a reference frequency with which local oscillator at the receiving end may be synchronized.

pilot chute (*Aero*) A small parachute which extracts the main canopy from its pack.

pilot controller (*ElecEng*) See MASTER CONTROLLER.

piloted head (*Eng*) Gas burner head or nozzle having a bypass by which low-pressure feeder flames are produced around the main flame, to secure positive retention when the velocity of the combustible mixture exceeds the FLAME SPEED.

pilot electrode (*Electronics*) Additional electrode or spark gap which triggers and makes certain main discharge, eg in a discharge spark gap, for creating ions, or in a TR switch, or a mercury-arc rectifier. Also *ignitor, keep-alive electrode, starter, trigger electrode*.

pilot engine (*CivEng*) A separate locomotive preceding a train as a precaution against accidents to the latter.

pilot gauge (*Eng*) A plug gauge which has a circumferential groove near its free end to facilitate entry into the hole to be gauged.

pilotherm (*Phys*) A thermostat in which the temperature control is brought about by the deflection of a bimetallic strip, thereby switching on and off the electric heating current.

pilotis (*Arch*) Reinforced concrete columns carrying a building, leaving an open area at ground level.

pilot lamp (*ElecEng*) One giving visual indication of the closing of a circuit.

pilot nail (*Build*) A temporary nail used in fixing shuttering. May have two heads. Also *stitch nail*.

pilot pin (*Eng*) A rounded or tapered pin, projecting from a tool or machine which enters a hole in a mating part before engagement to align it. It is used in press tools to position the strip material.

pilot pins (*ImageTech*) Pins forming part of the mechanism of a motion picture camera, printer or projector which engage with the perforations to ensure accurate location and steadiness of the film at the period of exposure.

pilot plant (*ChemEng*) Smaller version of projected industrial plant, used to gain experience and data for the design and operation of the final plant.

pilot tone (*ICT*) Any constant, single-frequency signal injected into a channel to aid the setting up of correct frequency or amplitude in the receiving equipment. Examples are (1) carrier frequency reinserted at low level in SINGLE-SIDEBAND radio to facilitate demodulation; (2)

19 kHz tone added to the signal modulating an FM stereo transmitter to allow recovery of the suppressed 38 kHz stereo subcarrier.

pilot valve (*Eng*) A small balanced valve, operated by a governor or by hand, which controls a supply of oil under pressure to the piston of a SERVOMOTOR or relay connected to a large control valve, which it is desired to operate. Also *relay valve*.

pilot voltmeter (*ElecEng*) A voltmeter used in a power station or substation to indicate the voltage at the remote end of a feeder to which it is connected by means of a PILOT wire.

pilot wave (*ICT*) A carrier oscillatory current or voltage that is amplified in a high-efficiency amplifier independently of the side frequencies, that are added subsequently.

pilot wire (*ElecEng*) See PILOT.

pilot wire (*ICT*) In a multicore transmission cable, a wire that is solely concerned with detecting deterioration of the main insulation of the cable.

pilot-wire regulator (*ICT*) An automatic device in transmission circuits, eg for compensating changes in transmission arising from temperature variations.

PIM (*ICT*) Abbrev for PERSONAL INFORMATION MANAGER.

pimelic acid (*Chem*) $HOOC(CH_2)_5COOH$. Mp 105°C. A saturated dibasic acid of the oxalic acid series, crystals.

pi meson (*Phys*) See MESON. Also π-*meson*.

pi mode (*Electronics*) Operation of a multicavity magnetron, whereby voltages on adjacent segments of the anode differ by π radians (half-cycle). Also π-*mode*.

pimpling (*NucEng*) Small swellings on the surface of fuel during burn-up.

PIN (*ICT*) Abbrev for PERSONAL IDENTIFICATION NUMBER.

pin (*BioSci*) The long-styled form of such heterostyled flowers as the primrose, *Primula vulgaris*, with the pinhead-like stigma conspicuous at the top of the corolla tube. Cf THRUM. See HETEROSTYLY.

pin (*Build*) (1) A small wooden peg or nail. (2) The male part of a dovetail joint.

pin (*Eng*) A cylinder or tube, often heat-treated, used to connect members in a structure or machine, usually when freedom of angular movement at the joint is required.

pin (*NucEng*) Very slender fuel cans used eg in fast reactors or, in a group, to form certain types of fuel element, eg in water-cooled reactors. Also *fuel pin*. See FUEL ELEMENT.

pinacocytes (*BioSci*) The flattened epithelial cells forming the outer part of the dermal layer in sponges.

pinacoid (*Crystal*) An open crystal form which includes two precisely parallel faces.

pinacol (*Chem*) *Tetramethyl-ethan 1,2-diol*. $Me_2C(OH)-C(OH)Me_2$. Crystallizes with $6(H_2O)$. The anhydrous substance, mp 38°C, bp 172°C, is obtained by the reduction and condensation of acetone by the action of metallic sodium. Pinacol forms *pinacolone* by the elimination of water and intramolecular transformation in the presence of dilute acids. It is the simplest member of a series of tetra alkyl glycols known as *pinacols*. Also *pinacone*.

pinacolone (*Chem*) *3,3-dimethylbutan-2-one*. $CH_3COC(CH_3)_3$. Bp 106°C, rel.d. 0·800 at 16°C. A colourless liquid produced by the action of dilute sulphuric acid upon *pinacol*.

pincers (*BioSci*) Claws adapted for grasping, as CHELICERAE.

pinch (*Electronics*) An airtight glass seal through which pass the electrode connections in a thermionic valve or electron tube.

pinchbeck alloy (*Eng*) Red brass, 6–12% zinc. See COPPER ALLOYS.

pinch effect (*NucEng*) In a plasma carrying a large current, the constriction arising from the interaction of the current with its own magnetic field, just as two wires each carrying a current in the same direction experience an attractive force. The principle is used in fusion machines to confine the plasma.

pinch-off (*Electronics*) Cut-off of the channel current by the gate signal in a field-effect transistor.

pinch-off voltage (*Electronics*) In a FIELD-EFFECT TRANSISTOR (FET), the voltage at which the current flow between source and drain ceases because the CHANNEL between these electrodes is completely depleted. For enhancement-mode FETs, using an n-type channel, the pinch-off voltage is positive; for depletion-mode FETs with a p-type channel, it is negative.

pincushion distortion (*ImageTech, Phys*) Curvilinear distortion of an optical or electronic image in which horizontal and vertical straight lines appear concave, bowed inwards. Also *negative distortion*, *pillow distortion*. Cf BARREL DISTORTION.

PIN diode (*Electronics*) A semiconductor diode with a layer of intrinsic semiconductor material between the p-type and n-type material which would normally constitute the junction. Applications include high-frequency and microwave switching or attenuation, voltage-controlled resistors, photodiodes. Also *p–i–n diode*.

pine (*For*) Softwood trees of the genus *Pinus*. See PITCH PINE, SCOTS PINE.

pine (*Vet*) A disease of sheep and cattle caused by low levels of cobalt in the soil leading to a deficiency in the diet; characterized by debility, emaciation and anaemia. Also *bush sickness, enzootic marasmus, vinquish*.

pineal apparatus (*BioSci*) In some vertebrates, two median outgrowths from the roof of the diencephalon, one (originally the left) giving rise to the PARIETAL ORGAN and the other (originally the right) to the PINEAL ORGAN.

pineal body (gland) (*BioSci*) See PINEAL ORGAN.

pinealectomy (*Med*) Surgical removal of the pineal gland.

pineal eye (*BioSci*) An anatomically imprecise term referring to eye-like structures formed by one or other of the two outgrowths of the PINEAL APPARATUS. In Cyclostomata it derives from the PINEAL ORGAN, but in lizards it derives from the PARIETAL ORGAN.

pinealoma (*Med*) A tumour of the pineal gland.

pineal organ (*BioSci*) One of the outgrowths of the pineal apparatus. In Cyclostomata it forms an eye-like functional photosensitive structure involved in a diurnal rhythm of colour change, and may be sensitive to light in some fish. It persists in higher vertebrates, and may function as an endocrine gland.

pin efficiency (*ICT*) In a SWITCH realized as a printed circuit card or integrated circuit, the ratio of the number of signal pins to the total number of pins including those needed for control.

pinene (*Chem*) There are four terpenes known as pinenes. α-Pinene, $C_{10}H_{16}$, is the chief constituent of turpentine, eucalyptus, juniper oil, etc; bp 155–156°C. It forms a hydrochloride, $C_{10}H_{17}Cl$, a white crystalline mass of camphor-like odour, mp 131°C. As it contains a double bond it forms a dibromide which can be converted into a glycol.

pine oil (*MinExt*) Commercial frothing agent widely used in flotation of ores. Distillate of wood, varying somewhat in chemical composition according to timber used and scale of heating.

pi-network (*ICT*) A section of circuit with one series arm preceded and followed by shunt arms to the return leg of circuit. Also π-*network*.

ping (*Acous*) Brief pulse of medium-frequency sound, reproduced from the subaqueous reflection of asdic ultrasonic signals. Its length in space will be equal to the product of ping duration time and the speed of sound.

ping (*ICT*) A network tool used to test whether a particular host is reachable across an IP network. An 'echo request' packet (*ping?*) is sent to the target host and an 'echo response' indicates connectivity.

pingo (*Geol*) A raised area in permafrost due to the local expansion of an ice mass.

pinguecula (*Med*) A yellow, triangular patch on the conjunctiva covering the sclera.

pin hinge (*Build*) A form of butt hinge which has a removable pin connecting the two leaves.

pinhole porosity (*Eng*) Small rounded voids within the body of a casting caused by evolution of gas during solidification. Also *pinholing*.

pinholes (*ImageTech*) Small clear spots in a processed film caused by air bells or dust particles preventing the action of the developer.

pinholes (*Paper*) Small holes through the paper or coating caused by defects in manufacture.

pinholing (*Build*) A painting or varnishing defect in which the surface becomes pitted with small holes.

pinholing (*Eng*) See PINHOLE POROSITY.

pin insulator (*ElecEng*) An insulator which is supported from the cross-arm by a pin.

pinion (*Eng*) The smaller of a pair of high-ratio toothed spur wheels. In clocks it normally has less than twelve teeth or leaves.

pinite (*Min*) Hydrated silicate of aluminium and potassium which is usually amorphous. It is an alteration product of cordierite, spodumene, feldspar, etc, close to muscovite in composition. See PAGODITE.

pin joint (*Eng*) A joint between members in a structural framework in which moments are not transmitted from one member to another.

pink-eye (*Med*) Acute mucopurulent conjunctivitis (the inflammation of the conjunctiva making it red) due to infection with various bacteria.

pink-eye (*Vet*) See EQUINE INFLUENZA, INFECTIOUS OPHTHALMIA.

pinking (*Autos*) See KNOCKING.

pink-noise generator (*ElecEng*) A random-noise generator providing a frequency spectrum with higher amplitudes at the low-frequency end of the audible spectrum. See WHITE NOISE.

pin mark (*Print*) A mark near the top of the type shank, made in casting type founders' type.

pinna (*BioSci*) (1) A leaflet that is part of a pinnate leaf. (2) In fish, a fin. (3) In mammals, the outer ear. (4) In birds, a feather or wing.

pinnate (*BioSci*) (1) Generally, feather-like; bearing lateral processes. (2) Having four or more regular divisions, lobes, veins, etc, arranged in two rows along a common midrib, rachis or stalk, rather in the manner of the barbs of a feather, as in a pinnate leaf. Many ferns have leaves that are two, three or more pinnate.

pinnatifid (*BioSci*) A leaf blade etc cut about halfway into the lobes in a PINNATE fashion.

pinnatiped (*BioSci*) Having the digits of the feet united by flesh or membrane. Also *pinniped*. Cf FISSIPED.

pinning (*Electronics*) The restriction of the passage of a DISLOCATION or a DOMAIN WALL by an impurity or other crystalline defect.

pinning-in (*Build*) The operation of inserting small splinters of stone in the joints of coarse masonry.

pinniped (*BioSci*) See PINNATIPED.

pinnule (*BioSci*) One of the lobes or divisions of a leaf that is two or more PINNATE.

pinocytosis (*BioSci*) 'Cell drinking'. Uptake by cells of vesicles containing fluid from the environment. There are two forms. *Micropinocytosis* involves small droplets or microvesicles being pinched off from the cell membrane and interiorized, carrying with them any materials selectively adsorbed at the cell surface. This process is common to many cells. Uptake of ligand-receptor complexes into clathrin-coated vesicles is an important mechanism whereby ligands can be carried into the interior. *Macropinocytosis*, or the ingestion of large vesicles or vacuoles, similar to PHAGOCYTOSIS, is carried out by cells of the mononuclear phagocyte system.

pinpoint (*Aero*) An aircraft's ground position as fixed by direct observation. Cf FIX.

pin-punch (*Eng*) A cylindrical or cone-shaped tool for removing pins connecting members in a structure or machine.

pint (*Genrl*) A unit of capacity or volume in the imperial system. Equal to $\frac{1}{8}$ imperial gallon or 0·5682 dm³. In the US equal to 0·473 dm³ (liquid) or 0·551 dm³ (dry). Contains 20 fluid ounces (US 16 fl oz).

pinta (*Med*) A contagious skin disease, characterized by patches of coloured pigmentation; probably due to infection with a spirochaete; occurs in tropical America. Also *caraate, mal de los pintos*.

pintle (*Eng*) (1) The pin of a hinge. (2) The king pin of a wagon. (3) An iron bolt on which a chassis turns. (4) One of the metal braces on which a rudder swings, supported by a 'dumb-pintle' at its heel. (5) The plunger or needle of an oil-engine injection valve, opened by oil pressure on an annular face, and closed by a spring.

pin-twisting (*Textiles*) The formation of FALSE-TWIST by passing continuous filament yarn through a rotating tube and around a hard-wearing ceramic pin.

pioglitazone (*Pharmacol*) A thiazolidinedione glucose sensitizer used for the treatment of Type II diabetes.

pion (*Phys*) See MESON.

pioneer crop (*Agri*) The first crop raised on derelict land to begin remediation towards sustained crop production.

pioneer species (*BioSci*) Species that are early colonizers of new environment. Generally used of the flora rather than fauna.

PIP (*ImageTech*) Abbrev for *picture-in-picture*. A small picture inserted into the TV or video picture to view an alternative channel.

pip (*Radar*) Significant deflection or intensification of the spot on a cathode-ray tube giving a display for identification or calibration. Also *blip*.

pipe (*Acous*) Simple form of acoustic resonator. See PIPE RESONANCE. Examples include organ pipes and the RIJKE TUBE.

pipe (*Eng*) Tapering cavity formed in the upper parts of an ingot or casting due to the specific volume contraction when the liquid solidifies. May extend as far as halfway down depending on the degree of contraction, the rate of solidification and whether a feeder head is used.

pipe (*Geol*) A vertical conduit into the crust through which volcanic materials have passed. It may be filled with volcanic breccia and is often mineralized. See CHIMNEY.

pipe (*ICT*) The direction of the result of one program or command into another.

pipeclay (*Geol*) A white clay, nearly pure and free from iron, used in the pottery industry.

pipe coupling (*Build*) A short collar with female threads at both ends for joining screwed lengths of pipe.

pipe factor (*MinExt*) Compensating factor used when samples are taken from casings which go into or through running sands or gravels, so that the amount of material raised from the section traversed by the drill does not correspond with the volume enclosed by the pipe.

pipe fitting (*Eng*) Any piece of the wide variety of pipe connecting pieces used to make turns, junctions and reductions in piping systems.

pipelining (*ICT*) A form of PARALLEL PROCESSING in which the processing of a number of instructions is overlapped, enabling streams of instructions to be decoded and executed concurrently.

pipe moulding (*Eng*) The production of cast-iron pipes either by moulding in green sand, using split patterns, or by the process of CENTRIFUGAL CASTING.

piperazine (*Chem*) *Diethene-diamine*, a cyclic compound forming colourless crystals, mp 104°C, bp 145°C. It is a strong base and has the property of forming salts with uric acid which are relatively easily soluble in water; it is therefore used in medicine.

pipe resonance (*Acous*) Acoustic resonance of a pipe when the length, allowing for an end-correction, is an integer number of half-wavelengths when it is open at both ends, and odd multiples of a quarter-wavelength when it is closed at one end.

pipe resonator (*Acous*) An acoustic resonator, in the form of a pipe, which may be open at one or both ends, as in organ pipes, or entirely closed. Resonance arises from the stationary waves set up by a plane-progressive wave being reflected at the ends, open or closed.

piperidine (*Chem*) $C_5H_{11}N$, a heterocyclic reduction product of pyridine. It is a colourless liquid, of peculiar odour, bp 106°C. It is soluble in water and ethanol, has relatively strong basic properties, and the imino hydrogen is replaceable by alkyl or acyl radials.

piperine (*Chem*) An optically inactive alkaloid occurring in pepper, $C_5H_{10}NC_{12}H_9O_3$, piperyl-piperidine, which crystallizes in prisms, mp 129°C.

piperitone (*Chem*) A terpene ketone. The laevorotatory isomer is found in eucalyptus oils, whilst the dextro isomer occurs in Japanese peppermint oil. Colourless oil with peppermint odour.

pipe roller (*Print*) See IDLER.

piperonal (*Chem*) *Methylene-protocatechuic aldehyde.* $CH_2O_2{=}C_6H_3CHO$. A phenolic aldehyde of very pleasant odour, used as a perfume under the name of *heliotropin*.

pipe sample (*MinExt*) One obtained by driving open-ended pipe into a heap of material and withdrawing the core it collects.

pipe stopper (*Build*) An expanding form of drain plug for closing the outlet of drainpipes which are to be tested.

pipette (*Chem*) A dispenser of small measured quantities of liquid. Originally of glass with suitable graduation(s), now largely superseded by the *automatic pipette*, in which a spring-loaded plunger sucks the known volume into a disposable tip. Depressing the plunger then dispenses the liquid. There are automatic multiple versions available for many routine purposes.

pipette method (*PowderTech*) A sedimentation technique in which concentration changes within a sedimenting *suspension* are measured by pipetting samples out of the suspension and determining the concentration of the samples.

pipe wrench (*Build*) A device which turns a pipe or rod about its axis. Also *cylinder wrench*.

pip–pip tone (*ICT*) A voice-frequency signal comprising two pulses of tone, indicating that a particular switching stage has been reached.

piqué (*Textiles*) Fabric with pronounced cord effects. In woven fabrics the cords run in the weft direction but in warp-knitted fabrics the cords are in the other direction.

Pira test bench (*Print*) A small compact laboratory designed by the Printing Division of the Research Association for the Paper, Board, Printing and Packaging Industries, equipped to test paper, ink and printing surfaces before printing and to diagnose troubles encountered during and after printing.

pirn (*Textiles*) Small wooden bobbin which fits the loom shuttle and carries the weft thread.

piroplasmosis (*Vet*) See BABESIOSIS.

Pisces (Fishes) (*Astron*) A large faint northern constellation, lying between Aquarius and Aries.

Piscis Austrinus (Southern Fish) (*Astron*) A small southern hemisphere constellation, which includes the star FOMALHAUT.

piscivorous (*BioSci*) Fish-eating.

PISE (*Electronics*) Abbrev for *plasma and ion surface engineering.* See CHEMICAL VAPOUR DEPOSITION, DRY ETCHING, PHYSICAL VAPOUR DEPOSITION, SURFACE ENGINEERING.

pi-section filter (*ICT*) Unbalanced filter in which one series reactance is preceded and followed by shunt reactances. Also *π-section filter.*

pisé de terre (*Build*) A kind of cob wall used sometimes in cottage construction, the cob usually being moulded between forms.

pisiform (*BioSci*) Pea-shaped; as one of the carpal bones of humans.

pisolite (*Geol*) A type of limestone built of rounded bodies (*pisoliths*) similar to oöliths, but of less regular form and 2 mm or more in diameter.

pisolitic (*Geol*) A term descriptive of the structure of certain sedimentary rocks containing pisoliths (see PISOLITE above). Calcite-limestones, dolomitic limestones, laterites, iron ores and bauxites may be pisolitic.

pistacite (*Min*) See EPIDOTE.

pistil (*BioSci*) Ovary, style and stigma; either of a single carpel in an apocarpous flower, or of the whole gynoecium in a syncarpous flower. Adj *pistillate*.

piston (*Acous*) A small disk-like element which vibrates with the frequency of the sound which it emits. Pistons can be driven externally by electrical forces or internally by self-excitation (musical instruments).

piston (*ElecEng*) Closely fitting sliding short circuit in a waveguide.

piston (*Eng*) A cylindrical metal piece which reciprocates in a cylinder, either under fluid pressure, as in engines, or to displace or compress a fluid, as in pumps and compressors. Leakage is prevented by spring rings, rubber packing, hat leather, etc.

piston attenuator (*ElecEng*) An attenuator employing a waveguide system beyond cut-off. The attenuation in dB is linearly proportional to length, but the minimum attenuation is high.

pistonphone (*Acous*) A device in which a rigid piston is vibrated. Used for calibrating microphones.

piston pin (*Autos*) See GUDGEON PIN.

piston ring (*Eng*) A ring, of rectangular section, fitted in a circumferential groove in a piston, and springing outwards against the cylinder wall to prevent leakage. It is cut through at one point to increase its springiness and allow fitting. See JUNK RING, MITRE-CUT PISTON RING, SCRAPER RING.

piston rod (*Eng*) The rod connecting the piston of a reciprocating engine with the crosshead.

piston slap (*Autos*) The light knock caused by a worn or loose piston slapping against the cylinder wall when the connecting-rod thrust is reversed.

piston valve (*Eng*) A commonly used steam-engine slide valve in which the sealing or sliding surfaces of the valve are formed by two short pistons attached to the valve rod, working over cylindrical port faces in the steam chest.

pit (*BioSci*) A localized, thin area of a cell wall, typically where the primary wall is not covered by a secondary wall. Plasmodesmata may be present. A pit is usually one-half of a pit-pair and the term is sometimes loosely used to mean a pit-pair. Cf PERFORATION. See BORDERED PIT, SIMPLE PIT.

pit (*MinExt*) (1) A place where minerals are dug. (2) The shaft of a mine. The *pit eye* is the bottom, whence daylight is visible; the *pit frame* is the superstructure carrying poppet head and sheaves. The *pit head* is the surface landing stage.

pit-brow (*MinExt*) See BROW.

pit cavity (*BioSci*) The space within a pit from pit membrane to cell lumen.

pitch (*Acous*) The subjective property of a simple or complex tone which enables the ear to allocate its position on a frequency scale. If the fundamental of a complex tone is absent the pitch of this fundamental is still recognized because of subjective difference tones amongst the partials. See CONCERT PITCH.

pitch (*Aero*) (1) The distance forward in a straight line travelled by a propeller in one revolution at zero slip; often used as a colloq term though wrongly applied to the blade incidence. (2) Angular displacement along the lateral axis. See PITCHING. (3) Spacing between evenly spaced items, eg rivets.

pitch (*Build*) Of a plane, the angle between the blade and the sole, normally 45°, but in moulding planes up to 55° and in some smoothing planes 50° (York pitch). In planes with reversed bevel it is less, eg block planes approx 20° and low-angle planes approx 12°.

pitch (*Chem*) A dark-coloured, fusible, more or less solid material containing bituminous or resinous substances, insoluble in water, soluble in several organic solvents. Usually obtained as the distillation residue of tars.

pitch (*ElecEng*) A term used in connection with electric machines to denote the distance measured along the armature periphery between various parts. See POLE PITCH, SLOT PITCH, WINDING PITCH.

pitch (*ImageTech*) (1) The distance between successive perforations on the edges of film. (2) The track separation on a video disk or recorded videotape. (3) The separation between holes in a shadowmask. See SHADOWMASK TUBE.

pitch (*MinExt*) Orientation of a linear element, eg mine tunnel, mineral lineation, *slickensides*, on an inclined surface, whereby the angle of pitch is measured between the inclined element and the horizontal. See FOLDING.

pitch (*NucEng*) The distance between centres of adjacent fuel channels in a reactor.

pitch (*Paper*) Residual resinous material that may be present in unbleached sulphite or mechanical pulps. A form of CONTRARY that can block the machine wires or felts of paper-making machines.

pitch (*Ships*) See PITCHING PERIOD.

pitchblende (*Min*) The massive variety of uraninite. Radium was first discovered in this mineral. This and helium are due to the disintegration of uranium.

pitchboard (*Build*) A triangular board used as a template for setting out stairs, the sides of which correspond to the RISE, the GOING and the PITCH.

pitch circle (*Eng*) (1) In a toothed wheel, an imaginary circle along which the tooth pitch is measured, and with respect to which tooth proportions are given. For two wheels in mesh, the pitch circles roll in contact. See fig. at GEAR WHEEL. (2) Circle drawn, eg on a flange, around whose circumference hole centres are positioned for drilling.

pitch control (*Aero*) The *collective* and *cyclic pitch* (ie blade incidence) *controls* of a helicopter's main rotor(s).

pitch cylinder (*Eng*) Cylinder coaxial with a screw-thread and which intersects the flanks of the thread symmetrically.

pitch diameter (*Eng*) The diameter of the pitch circle of a gear wheel. See fig. at GEAR WHEEL.

pitch edge (*Print*) The leading edge of the paper as it enters the printing, or some other, machine.

pitched work (*Build*) Stone facing work for the slopes of jetties, breakwaters, etc, executed by pitching the stones into place with some regularity.

pitcher (*BioSci*) An urn-shaped or vase-shaped modification of a leaf, or part of a leaf, developed by certain plants; it serves as a means of trapping insects and other small animals which are killed and digested.

pitcher (*Build*) Thick-edged mason's chisel for rough-dressing stone.

pitch face (*Build*) A stone surface left with a rough finish produced by the hammer.

pitching (*Aero, Ships*) The angular motion of a ship or aircraft in a vertical plane about a lateral axis.

pitching (*CivEng*) (1) The foundation layer of well-rammed and consolidated broken stone upon which a road surfacing is built. Layer now referred to as sub-base and is laid by purpose-built machine. (2) Picking up a PILE and placing through temporary supports prior to driving.

pitching moment (*Aero*) The component of the couple about the lateral axis, acting on an aircraft in flight.

pitching period (*Ships*) The movement of a ship in waves about a transverse horizontal axis is known as 'pitching' as the bow goes down and 'scending' as the bow rises. The time taken for the complete movement, down and up, is the pitching period.

pitching-piece (*Build*) See APRON PIECE.

pitching tool (*Build*) A chisel with a very blunt edge, used to knock off superfluous stone.

pitch line (*Eng*) The line along which the pitch of a rack is marked out, corresponding to the pitch circle of a gear wheel.

pitch of propeller (*Ships*) The theoretical distance the propeller would advance through a solid in one revolution.

pitch pine (*For*) Wood from the trees *Pinus rigida* and *P. palustris* (American pitch pine). The softwood timber is very resinous, moderately durable and yellow-orange to reddish-brown in colour.

pitch setting (*Aero*) The blade angle of adjustable- or variable-pitch PROPELLERS.

pitchstone (*Geol*) A volcanic glass which has a pitch-like (resinous) lustre and contains crystallites and microlites. It is usually of acid to subacid composition, contains a notable amount of water (4% or more), and is usually intrusive.

pith (*BioSci*) Ground tissue in the centre of a shoot or root. Also *pith medulla*.

pith-ball electroscope (*ElecEng*) Primitive detector of charges, whereby two pith-balls, suspended by silk threads, attract or repel each other, according to inverse-square law.

pithed (*BioSci*) Having the central nervous system (spinal cord and brain) destroyed.

pith ray (*BioSci*) See MEDULLARY RAY.

PI-3-kinases (*BioSci*) Abbrev for *phosphatidylinositol-3-kinases*, lipid kinases that phosphorylate phosphatidylinositol phosphate on the 3 position. Such kinases are key signalling enzymes, acting downstream of many receptors. Also *PI kinases*.

pit membrane (*BioSci*) That part of the middle lamella and primary wall that lies across the distal end of a pit cavity.

pitmen (*MinExt*) Men employed in shaft inspection and repair.

pit moulding (*Eng*) A process by which extremely large castings are moulded in a pit, with brick-lined sides and cinder-lined bottom carrying connecting vent pipes to floor level.

Pitot-static tube (*Aero*) Tube inserted nearly parallel to flow stream. It has two orifices, one facing flow and hence receiving total pressure, and the other registering the static pressure at the side. The pressure difference between the two orifices registers dynamic air pressure ($\frac{1}{2}\rho v^2$, where ρ is the air density and v the velocity). This is displayed on the airspeed indicator.

Pitot tube (*Aero*) See PITOT-STATIC TUBE.

pitted (*BioSci*) Tracheids and vessel elements of METAXYLEM and SECONDARY XYLEM that have a secondary wall interrupted by pits.

pitting (*Eng*) (1) Corrosion of metal surfaces, eg boiler plates, due to local chemical action. (2) A form of failure of gear teeth, due to imperfect lubrication under heavy tooth pressure.

pitting factor (*Eng*) In assessment of metal corrosion, the depth of penetration of the deepest pit divided by average loss of thickness as calculated from loss of weight.

pituitary gland (*BioSci*) The major endocrine gland of vertebrates, formed by the fusion of a downgrowth from the floor of the diencephalon (the infundibulum) and an upgrowth of ectoderm from the roof of the mouth (hypophyseal pouch). The former becomes the neurohypophysis, comprising the neural lobe and infundibulum, and the latter becomes the adenohypophysis, comprising the pars intermedia. The pars intermedia and the neural lobe together form the posterior lobe. The adenohypophysis produces several hormones affecting growth, adrenal cortex activity, thyroid activity, reproduction (gonadotrophic hormones) and melanophore cells. The neurohypophysis secretes two hormones, OXYTOCIN and VASOPRESSIN.

pityriasis (*Med*) A term common to various skin diseases in which fine and loose (branny) scales appear.

pivot bridge (*Eng*) A form of swing bridge in which the vertical pivot is located at the middle of the length of the bridge.

pivot factor (*ElecEng*) In an electrical indicating instrument, the (full-scale torque)/(mass of movement), a measure of freedom from error due to friction in the bearings.

pivot irrigation (*Agri*) A continuous length of horizontal water pipe mounted at intervals on motorized trucks, with one end fixed at the centre of the field. The system moves in a circle, spraying water as it rotates.

pivot jaw (*ElecEng*) A fixed jaw to which the blade of a switch is pivoted.

pivot joint (*BioSci*) An articulation permitting rotary movements only.

PIXE (*Chem*) Abbrev for *particle-induced X-ray emission*. See panel on TRACE ELEMENT ANALYSIS.

pixel (*ICT, ImageTech*) From picture element. (1) The smallest element with controllable colour and brightness in a video display or in COMPUTER GRAPHICS. (2) An individual light-sensitive element in a SOLID-STATE IMAGE SENSOR. Also *pel*.

pixillation (*ImageTech*) Video effect in which the whole picture is broken down into a comparatively small number of square elements; by association with PIXEL.

pK (*Chem*) Value which is the logarithm to the base 10 of the reciprocal of the dissociation constant of a weak electrolyte. A high pK value indicates the substance has a small value for the dissociation constant and is very weak electrolyte.

PKU (*Med*) See PHENYLKETONURIA.

PL/1 (*ICT*) Abbrev for *programming language 1*. Used for both scientific and business computing.

PLA (*ICT*) See PROGRAMMABLE LOGIC ARRAY.

placebo (*Pharmacol*) A pharmacologically inactive substance which is administered as a drug either in the treatment of psychological illness or in the course of drug trials.

place bricks (*Build*) See GRIZZLE BRICKS.

placenta (*BioSci*) (1) The part of the ovary of a flowering plant where the ovules form and remain attached while they mature as seeds. (2) Any mass of tissue to which sporangia or spores are attached. (3) In Eutherian mammals, a flattened structure formed by the intimate union of the allantois and chorion with the uterine wall of the mother; it serves for the respiration and nutrition of the growing young. Adjs *placental, placentate, placentiferous, placentigerous*.

Placentalia (*BioSci*) See EUTHERIA.

placentation (*BioSci*) In plants, the arrangement of the placentas in an ovary, and of the ovules on the placenta. In animals, the method of union of the fetal and maternal tissues in the placenta.

placenta vera (*BioSci*) A deciduate placenta in which both maternal and fetal parts are thrown off at birth. Cf SEMIPLACENTA.

placer deposits (*Geol*) Superficial deposits, chiefly of fluviatile origin, rich in heavy ore minerals such as cassiterite, native gold, platinum, which have become concentrated in the course of time by long-continued disintegration and removal from the neighbourhood of the lighter associated minerals. Also *placers*. See AURIFEROUS DEPOSIT.

place–value system (*MathSci*) See POSITIONAL NOTATION.

placode (*BioSci*) Any plate-like structure; in vertebrate embryos, an ectodermal thickening giving rise to an organ primordium.

Placodermi (*BioSci*) A class of Gnathostomata comprising the earliest jawed vertebrates, known from fossils from the Silurian to Permian. All had a heavy defensive armour of bony plates and hyoid gill slits with no spiracle. The hyoid arch did not support the jaw. Most possessed paired fins.

placoid (*BioSci*) Plate-shaped, as the scales and teeth of Selachii.

plage (*Astron*) A dark or bright area on calcium or hydrogen spectroheliograms, identified as an area of cool gas or heated gas respectively on the Sun's surface; *flocculi* refers to small patches only.

plagio- (*Genrl*) Prefix from Gk *plagios*, slanting, oblique.

plagioclase feldspars (*Min*) An isomorphous series of triclinic silicate minerals which consist of ALBITE and ANORTHITE combined in all proportions. They are essential constituents of the majority of igneous rocks. See ANDESINE, ANORTHITE, BYTOWNITE, LABRADORITE, OLIGOCLASE.

plagiotropism (*BioSci*) A TROPISM in which a plant part becomes aligned at an angle to the source of the orientating stimulus. Cf DIATROPISM, ORTHOTROPISM.

plague (*Med*) A disease of rodents due to infection with the *Yersinia pestis* bacterium, transmitted to humans by rat fleas. Large infestations of fleas invariably precede epidemics. In humans the disease is characterized by enlargement of lymphatic glands (bubonic plague), severe prostration, a tendency to septicaemia and occasional involvement of the lungs.

plain conduit (*ElecEng*) See PLAIN STEEL CONDUIT.

plain coupler (*ElecEng*) A short length of tubing serving to connect the end of two adjacent pieces of plain steel conduit in line with each other in an electrical installation. Also *sleeve*.

plain fabric (*Textiles*) (1) In a knitted plain fabric all the loops are of the same type and are linked together in the same manner. (2) In a woven plain fabric a weft thread passes *over* and *under* each succeeding warp thread whereas the next weft thread passes *under* and *over* the same warp threads. This is the simplest of all weaves. See panel on FIBRE ASSEMBLIES.

plain flap (*Aero*) A wing flap in which the whole trailing edge (apart from the ailerons) is lowered so as to increase the camber. Also occasionally *camber flap*.

plain old telephone service (*ICT*) Colloq term for the type of service available before technical developments such as INTELLIGENT NETWORKS and OPTICAL FIBRE were available.

plain steel conduit (*ElecEng*) Conduit consisting of light-gauge steel tubing not having the ends screwed; used for containing the conductors in electrical installation. Also *plain conduit*. Cf SCREWED STEEL CONDUIT.

plain tile (*Build*) The ordinary flat tile, with two nibs for hanging from the battens.

plaiting (*Textiles*) See CUTTLING.

planapochromat (*BioSci*) A microscope objective that is corrected for spherical aberration and chromatic aberration at three wavelengths.

planar diode (*Electronics*) One produced using the PLANAR PROCESS, giving a diode with high forward conductance, low reverse leakage, fast recovery time and low junction capacitance.

planar graph (*MathSci*) A graph which can be drawn in the plane in such a way that edges intersect only at vertices. They satisfy EULER'S FORMULA.

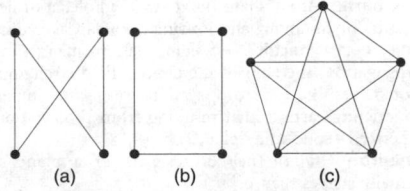

(a)   (b)   (c)

planar graph (a) and (b) are planar graphs; (c) is not.

planar process (*Electronics*) A silicon transistor and integrated-circuit manufacturing technique in which an oxide layer, fractions of a micron thick, is grown on a silicon substrate. A series of etching, ion implantation, diffusion and deposition steps is then undertaken to produce the active regions and junctions inside the substrate. See SEMICONDUCTOR DEVICE PROCESSING.

planar transistor (*Electronics*) A transistor made using the PLANAR PROCESS.

planceer (*Build*) A soffit, esp the under surface of the corona in a cornice. Also *plancier*.

planceer piece (*Build*) A horizontal timber to which the soffit boards of an overhanging eave are fastened.

plancier (*Build*) See PLANCEER.

Planckian colour (*Phys*) The colour or wavelength–intensity distribution of the light emitted by a black body at a given temperature.

Planckian locus (*Phys*) The line of a CHROMATICITY DIAGRAM joining points with co-ordinates corresponding to black-body radiators.

Planck length (*Phys*) A length scale thought to be of importance in quantum gravity, which may represent the shortest possible distance between points; equals $\sqrt{Gh/2\pi c}$, where G is the gravitational constant, h is Planck's constant and c is the speed of light; value 1.62×10^{-35} m. The corresponding PLANCK MASS is 2.1×10^{-8} kg.

Planck mass (*Phys*) See PLANCK LENGTH.

Planck's constant (*Phys*) The fundamental constant which is the basis of PLANCK'S LAW. It has the dimensions of energy × time, ie action. The present accepted value is 6.626×10^{-34} J s.

Planck's law (*Phys*) The basis of quantum theory, that the energy of electromagnetic waves is confined in indivisible packets or quanta, each of which has to be radiated or absorbed as a whole, the magnitude being proportional to frequency. If E is the value of the quantum expressed in energy units and v is the frequency of the radiation, then $E = hv$, where h is known as *Planck's constant* and has dimensions of energy × time, ie action, and is 6.626×10^{-34} J s. See PHOTON and panel on QUANTUM THEORY.

Planck's radiation law (*Phys*) An expression for the distribution of energy in the spectrum of a black-body radiator:

$$E_v \, dv = \frac{8\pi h v^3}{c^3 (e^{hv/kT} - 1)} dv$$

where E_v is the energy density radiated at a temperature T within the narrow frequency range from v to $v + dv$; h is PLANCK'S CONSTANT, c the speed of light, e the base of the natural logarithms and k BOLTZMANN'S CONSTANT.

plançon (*For*) A log of hardwood timber roughly sawn or hewn to an octagonal shape, with a minimum of 10 in (250 mm) between opposite faces.

plane (*Build*) A woodworking tool used for the purpose of smoothing surfaces, reducing the size of wood and, in specialized forms, for grooving, rebating and other purposes. See JAPANESE PLANES.

plane (*For*) See EUROPEAN PLANE.

plane (*MathSci*) A flat surface; one whose radii of curvature are infinite at all points.

plane baffle (*Acous*) Plane board, with a hole, at or near the centre, for mounting and loading a loudspeaker unit.

plane earth factor (*Phys*) In electromagnetic wave propagation, the ratio of the electric field strength which would result from propagation over an imperfectly conducting earth to that resulting from propagation over a perfectly conducting plane.

plane-iron (*Build*) The cutting part of a plane, which actually shapes the work.

plane of collimation (*Surv*) The imaginary surface swept out by the LINE OF COLLIMATION of a levelling instrument, when its telescope is rotated about its vertical axis.

plane of polarization (*Phys*) In the reflection of light, the plane containing the incident and reflected rays and the normal to the reflecting surface. The magnetic vector of plane-polarized light lies in this plane. The electric vector lies in the *plane of vibration* which is that containing the plane-polarized reflected ray and the normal to the plane of polarization. The description of plane-polarized light in terms of the plane of vibration is to be preferred as this specifies the plane of the electric vector. Fig. ▷

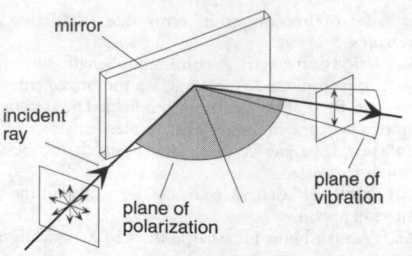

plane of polarization At a mirror.

plane of saturation (*CivEng*) The natural level of the GROUND WATER. See WATER TABLE.

plane of symmetry (*Crystal*) In a crystal, an imaginary plane on opposite sides of which faces, edges or solid angles are found in similar positions. One-half of the crystal is hence a mirror image of the other.

plane of symmetry (*MathSci*) See SYMMETRY.

plane polarization (*Phys*) The production of a transverse wave such that the vibrations are confined to one direction. For electromagnetic waves the direction of the electric vector of a plane-polarized wave is the *plane of vibration*; the magnetic vector lies in a plane at right angles to this. Light may be plane-polarized by polarizing filters, esp POLAROID sheet, and by BREWSTER ANGLE reflection. Polarization of radio waves and microwaves occurs as a result of the way these waves are transmitted from aerials.

planer (*Build*) A machine for smoothing or reducing the thickness of wood by means of a high-speed rotating cutter.

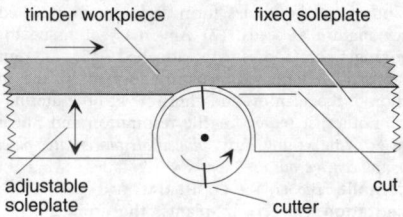

planer

planer (*Print*) A flat piece of wood or rubber which is placed on a forme of type and tapped with a mallet to level the surface.

planer tools (*Eng*) Planing machine cutting tools, similar to those used for turning, clamped vertically in a block pivoted in the CLAPPER BOX on the head.

plane stock (*Build*) The body of a plane holding the plane-iron in position.

plane strain (*Eng*) Stress state in which the strains are biaxial, so that the associated stresses are triaxial. Important esp in the interior of cracked materials under stress where through-the-thickness strains are constrained. See FRACTURE TOUGHNESS, GRIFFITH EQUATION.

plane stress (*Eng*) Stress state in which the stresses are biaxial, so that the associated strain field is triaxial. Important esp at the surfaces of cracked materials under stress where stresses normal to the material surface cannot exist. See FRACTURE TOUGHNESS, GRIFFITH EQUATION.

plane surveying (*Surv*) Surveying which makes no correction for curvature of Earth's surface.

planet (*Astron*) The name given in antiquity to the seven heavenly bodies, including the Sun and Moon, which were thought to travel among the fixed stars. The term is now restricted to those bodies, including the Earth, which revolve in elliptical orbits about the Sun; in the order of distance they are: Mercury, Venus, Earth, Mars, Jupiter,

Saturn, Uranus, Neptune and Pluto. The two planets Mercury and Venus, which revolve within the Earth's orbit, are designated *inferior planets*, the planets Mars to Pluto are *superior planets*. Planets reflect the Sun's light and do not generate light and heat. According to the International Astronomical Union (in 2006) a planet is an object that orbits the Sun, forms itself into a sphere, and has enough gravitational pull to clear its path of space debris; on this basis Pluto no longer qualifies as a planet.

plane table (*Surv*) A drawing-board mounted on a tripod so that the board can be levelled and also rotated about a vertical axis and clamped in position. An alidade completes the essential parts of a plane table. It is set up at ends of a suitable baseline where required survey points can be seen from these ends. By intersecting sights a rough plan can then be produced.

planetarium (*Astron*) A building in which an optical device displays the apparent motions of the heavenly bodies on the interior of a dome which forms the ceiling of the auditorium.

planetary electron (*Phys*) See BOHR MODEL.

planetary gear (*Eng*) Any gear wheel whose axis describes a circular path round that of another wheel, eg the bevel wheels carried by the crown wheel of DIFFERENTIAL GEAR.

planetary nebula (*Astron*) A shell of glowing gas surrounding an evolved star, from which it is ejected. There is no connection with planets: the name derives from the visual similarity at the telescope between the disk of such a nebula and that of a planet. They represent late stages in the evolution of stars 1–4 times as massive than the Sun.

plane-tile (*Build*) See CROWN-TILE.

planetoid (*Astron*) See ASTEROID.

planetology (*Geol*) The study of the composition, origin and distribution of matter in the planets of the solar system.

planet X (*Astron*) A hypothetical planet once thought to orbit beyond Pluto to explain apparent perturbations in the orbits of the planets, but now believed not to exist.

plane wave (*Phys*) A wave for which equiphase surfaces are planes.

planigraphy (*Radiol*) See EMISSION TOMOGRAPHY.

planimeter (*Eng*) Integrating instrument for measuring mechanically the area of a plane figure, eg an indicator diagram. A tracing point on an arm is moved round the closed curve, whose area is then given to scale by the revolutions of a small wheel supporting the arm.

planing bottom (*Aero*) The part of the under-surface of a flying-boat hull which provides hydrodynamic lift.

planing machine (*Eng*) A machine for producing large flat surfaces. It consists of a gear-driven reciprocating work table sliding on a heavy bed, the stationary tool being carried above it by a saddle, which can be traversed across a horizontal rail carried by uprights. See CLAPPER BOX.

planisher (*Eng*) (1) Hammer or tool for planishing. (2) A ROLLING MILL.

planishing (*Eng*) Giving a finish to metal surfaces by hammering.

plankton (*BioSci*) Animals and plants floating in the waters of seas, rivers, ponds and lakes, as distinct from animals which are attached to, or crawl upon, the bottom; esp minute organisms and forms, possessing weak locomotor powers or unable to swim actively. See NEKTON.

planning grid (*Arch*) Squared grid scaled in MODULES used in designing for modular construction.

plano-convex (*ImageTech, Phys*) Said of a lens with one surface flat and the other curved.

planogamete (*BioSci*) A motile or wandering gamete; a zoogamete.

planographic process (*Print*) The process in which the printing image is on a level with the plate, which is specially treated to accept ink on the printing image and reject ink on the non-printing areas. See COLLOTYPE, LITHOGRAPHY.

planospore (*BioSci*) See ZOOSPORE.

planozygote (*BioSci*) A motile zygote.

plan-position indicator (*Radar*) The screen of a cathode-ray tube with an intensity-modulated and persistent radial display, which rotates in synchronism with a highly directional antenna. The surrounding terrain is thus painted with relevant reflecting objects, such as ships, aircraft and physical features. Abbrev *PPI*. See AZIMUTH-STABILIZED PPI.

plant (*BioSci*) A photosynthetic organism or one related to it. It will always include the seed plants, almost always the pteridophytes and bryophytes, usually the algae and the fungi.

plant (*Eng*) (1) The machines, tools and other appliances required to carry on a mechanical or constructional business; the term sometimes includes also the building and the site and, in the case of a railway, the rolling stock. (2) The permanent appliances needed for the equipment of an institution.

planta (*BioSci*) The sole of the foot in land vertebrates; the flat apex of a proleg in insects. Adj *plantar*.

Plantae (*BioSci*) The plant kingdom.

plantar fasciitis (*Med*) Inflammation of a band of tissue (the *plantar fascia*) that supports the arch of the foot.

plantation (*Build*) A slate size, 330×280 mm, 13×11 in.

plant cell culture (*BioSci*) The culture of explants, ie cells, tissues or organs from plants, under aseptic conditions in or on a sterile growth medium typically containing sugar(s), as an energy and carbon source, mineral salts and growth substances, sometimes solidified with eg agar. Cells in an explant will grow and divide indefinitely in an appropriate medium. Depending on the conditions, cell growth and division may be more or less disorganized, forming individual cells or clumps of mostly parenchyma cells with little differentiation (callus cultures on solid media and suspension cultures in liquid media). Alternatively, organized growth may persist or develop: isolated shoots or roots may grow more or less indefinitely and, given appropriate levels of growth substances, organized meristems may develop in callus cultures to give roots, shoots or small plants (plantlets). Sometimes plantlets arise through a developmental process resembling that of normal embryos and are then termed embryoids. Techniques exist for the 'weaning' of plantlets and their continued growth in ordinary soil.

planted moulding (*Build*) A wooden moulding of the required section and secured to the surface to be decorated.

plant genetic manipulation (*BioSci*) The use of various techniques, except in most usages those of conventional breeding, to produce plants containing foreign DNA as part of their genomes. See PLANT CELL CULTURE and panel on GENETIC MANIPULATION.

plant growth substances (*BioSci*) Substances that, at low concentration, influence plant growth and differentiation. Formerly referred to as *plant hormones* or *phytohormones*. The major classes are abscisic acid, auxin, cytokinin, ethylene and gibberellin.

plantigrade (*BioSci*) Walking on the soles of the feet, as humans. Cf DIGITIGRADE, UNGULIGRADE.

planting (*Build*) Placing a moulding etc on a surface.

plant load factor (*ElecEng*) The ratio of the total number of kW h supplied by a generator or generating station to the total number of kW h which would have been supplied if the generator or generating station had been operated continuously at its maximum continuous rating.

plant pathology (*BioSci*) See PHYTOPATHOLOGY.

planula (*BioSci*) A larval form of some invertebrates, esp Cnidaria; it consists of an outer layer of ciliated ectoderm and an inner mass of endoderm cells.

plaque (*Med*) A layer of amorphous material adhering to the surfaces of teeth.

plaque (*MinExt*) White-enamelled saucer-shaped disk used in spot checking of products made during ore treatment. It has taken the place of the old vanning shovel. A sample is gently manipulated on it with added water, to separate the light from the heavy constituents.

plaque assay (*BioSci*) (1) An assay for a virus in which a dilute solution of the virus is applied to a culture dish containing a layer of the host cells. After incubation the 'plaques', areas in which cells have been killed (or transformed), can be recognized, and the number of infective virus particles in the original suspension estimated. (2) An assay for cells producing antibody against erythrocytes or against antigen that has been bound to the erythrocytes. The cell is surrounded by a clear plaque of haemolysis.

PLASCAMS (*Eng*) TN for a computer-aided polymer selection routine using input value judgements (on a scale of 1–10) of properties needed in the product being designed.

plasm (*BioSci*) Protoplasm, esp in compound terms, as *germ plasm.*

-plasm (*Genrl*) See PLASMA-.

plasma (*Electronics*) The positive column in a gas discharge.

plasma (*Med*) The watery fluid containing salts, protein and other organic compounds, in which the cells of the blood are suspended. When blood coagulates it loses certain constituents (eg fibrinogen and salts) and becomes serum.

plasma (*Min*) A bright-green translucent variety of cryptocrystalline silica (*chalcedony*). It is used as a semiprecious gem.

plasma (*Phys*) Very high-temperature ionized gaseous discharge in which there is no resultant charge, the number of positive and negative ions being equal, in addition to unionized molecules or atoms.

plasma-, plasmo-, -plasm (*Genrl*) Prefixes and suffix from Gk *plasma*, gen *plasmatos*, anything moulded.

plasma-arc cutting (*Eng*) Cutting metal at temperatures approaching 3500°C by means of a gas stream heated by a tungsten arc to such a high temperature that it becomes ionized and acts as a conductor of electricity. Heat is not obtained by a chemical reaction; the process can therefore be used to cut any metal.

plasma beta (*NucEng*) See BETA VALUE.

plasma cell (*BioSci*) Name given to the end stage of differentiation of B-lymphocytes into cells wholly devoted to synthesis and secretion of immunoglobulins. They have a very highly developed endoplasmic reticulum and a prominent Golgi apparatus, and are prominent in sites of intensive antibody synthesis. Plasma cell tumours occur and are termed *plasmacytomas* or *myelomas.*

plasmacytoma (*Med*) A tumour of plasma cells. Plasmacytomas are often preferentially localized in the bone marrow, where they produce typical erosion of the local bone (hence the alternative term *myeloma*). They continue to secrete an immunoglobulin product, although this may sometimes have sections of the normal amino acid sequence missing. Also *plasmoma.*

plasma deposition (*Electronics*) A coating process involving exposure to an ionized gas containing precursor material in the form of particles, radicals, ions or atoms.

plasma-enhanced processing (*Electronics*) Material processing such as CHEMICAL VAPOUR DEPOSITION assisted by the passage of electric current through the process gases.

plasma etching (*Electronics*) Etching by means of physical and/or chemical reactions at a surface exposed to a PLASMA. See DRY ETCHING.

plasma heating (*NucEng*) In fusion research, plasmas may be heated by ohmic heating, compression by magnetic fields, injection of high-energy neutral atoms, and by CYCLOTRON RESONANCE HEATING.

plasmalemma (*BioSci*) See PLASMA MEMBRANE.

plasmalogen (*BioSci*) Phosphatide in which a hydrocarbon chain is bound to a glycerol carbon by an unsaturated ether bond rather than an ester link.

plasma membrane (*BioSci*) The bounding membrane of cells that controls the entry of molecules and the interaction of cells with their environment. Like most cell membranes it consists of a lipid bilayer traversed by proteins. *Plasmalemma* is the commoner term in botany.

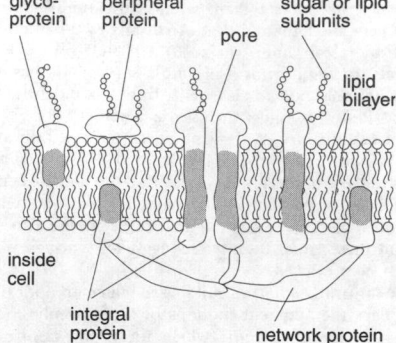

Plasma membrane Schematic diagram with lipophilic domains of proteins shaded.

plasma screen (*ICT*) Same as GAS PLASMA DISPLAY.

plasma spraying (*Eng*) Coating method in which material to be deposited is fed as a powder into a carrier gas flowing through an arc discharge, wherein it is melted. The method is useful for rapid deposition of a wide range of materials.

plasma temperature (*NucEng*) Temperature expressed in degrees K (thermodynamic temperature) or electron-volts (kinetic temperature): 1 keV = 10 000 K.

plasma torch (*Phys*) A torch in which solids, liquids or gases are forced through an arc within a water-cooled tube, with consequent ionization; de-ionization on impact results in very high temperatures. Used for cutting and depositing carbides.

plasmid (*BioSci*) A genetic element containing nucleic acid and able to replicate independently of its host's chromosome. Often carries genes determining ANTIBIOTIC RESISTANCE. Much used in recombinant DNA procedures.

plasmin (*Med*) A substance in blood capable of destroying fibrin as it is formed.

plasminogen (*Med*) The precursor of plasmin in the blood.

plasmo- (*Genrl*) See PLASMA-.

plasmocyte (*BioSci*) See LEUCOCYTE.

plasmodesma (*BioSci*) A fine tube of protoplasm that connects the protoplasts of two adjacent cells through the intervening wall. See PRIMARY PIT-FIELD, SYMPLAST. Pl *plasmodesmata.*

Plasmodium (*BioSci*) The genus of parasitic Protozoa that causes malaria.

plasmodium (*BioSci*) A multinucleate (syncytial) mass of naked (wall-less) protoplasm, which moves in an amoeboid fashion and constitutes the thallus as in the Myxomycetes. Cf PSEUDOPLASMODIUM.

plasmogamy (*BioSci*) Fusion of cytoplasm as distinct from fusion of nucleoplasm. Occurs when protoplasts or gametes fuse. In most organisms the latter is followed more or less immediately by karyogamy (fusion of nuclei); in some fungi it may result in a HETEROKARYON. See DIKARYOPHASE.

plasmoid (*Phys*) Any individual section of a plasma with a characteristic shape.

plasmolysis (*BioSci*) The process in which the protoplast of a plant cell shrinks away from the wall following water loss due to exposure to a solution of higher osmotic pressure, the wall being permeable to the solute but the plasmalemma not. Cf CYTORRHYSIS.

plasmoma (*Med*) See PLASMACYTOMA.

plaster (*Build*) A general name for plastic substances which are used for coating wall surfaces, and which set hard after application. See ACOUSTIC PLASTER.

plasterboard (*Build*) A building board made of plaster with paper facings, used as a base for plaster or providing a finish of its own.

plasterer's putty (*Build*) A preparation similar to FINE STUFF made by dissolving pure lime in water and passing it through a fine sieve.

plaster mould casting (*Eng*) Casting of small, precision parts of non-ferrous alloys in plaster moulds which are destroyed when the casting is removed.

plaster of Paris (*Chem*) Partly dehydrated gypsum, $2CaSO_4.H_2O$ (hemihydrate). When mixed with water, it evolves heat and quickly solidifies, expanding slightly; used for making casts.

plastic (*BioSci*) A genotype having a phenotype that varies markedly with conditions.

plastic (*Genrl*) A generic name for an organic material, usually a synthetic (see panel on POLYMERS) or semi-synthetic (casein and cellulose derivatives) polymer but also certain natural substances (shellac, bitumen, but excluding natural rubber), which under heat and pressure become plastic, and can then be shaped or cast in moulds, extruded as eg a rod or tube, or used in the formation of eg laminated products, paints, lacquers or glues. Plastics are THERMOPLASTIC or THERMOSETTING. See panel on THERMOSETS.

plastic bronze (*Eng*) Bronze containing a high proportion of lead; used for bearings. Composition: 72–84% copper, 5–10% tin and 8–20% lead plus zinc, nickel and phosphorus.

plastic clay (*Build*) See FOUL CLAY.

plastic deformation (*Eng*) Permanent change in the shape of a piece of material resulting from eg slip processes within the constituent crystals or by irreversible displacement of polymer chains, brought about by the application of mechanical force exceeding the elastic limit.

plastic deformation (*Geol*) The permanent deformation of a rock or mineral following the application of stress.

Plasticine (*Eng*) TN for heavily filled polymer dough used for modelling etc.

plasticity (*Eng*) A property of certain materials by which the deformation due to a stress is largely retained after removal of the stress.

plasticity (*Psych*) In a general sense, adaptability or a flexible cognitive or behavioural style; more specifically the ability of a different area of the brain to take over functions normally carried out elsewhere in order to compensate for damage.

plasticization (*Plastics*) The effect produced by addition of a PLASTICIZER to a polymer; generally lowers T_g, increases elongation to break but lowers modulus of material.

plasticized PVC (*Plastics*) A POLYVINYL CHLORIDE (panel) and PLASTICIZER blend widely used as a substitute for leather in eg upholstery. Also used in stationery products, for conveyor belting, toys, cable covering, hose, etc.

plasticizer cracking (*Plastics*) Loss of integrity of thermoplastics, esp in presence of plasticized polyvinyl chloride. Bleeding of small molecules onto plastic surface causes cracking or crazing.

plasticizers (*Chem*) High-boiling liquids used as ingredients in lacquers and certain plastics, eg polyvinyl chloride; they do not evaporate but preserve the flexibility and adhesive power of the cellulose lacquer films or the flexibility of plastic sheet and film. Well-known plasticizers are dioctylphthalate (DOP), diallanylphthalate (DAP), triphenyl phosphate, tricresyl phosphate, high-boiling glycol esters, etc. DOP and DAP are commonly used in polyvinyl chloride for flexible film etc. Long-chain plasticizers have been developed where migration is a problem, eg blood and serum bags for medical use. Phosphate plasticizers are used where flame resistance is needed, eg coal-mine conveyor belting.

plastic moulding (*Eng*) A process for manufacturing articles of plastic materials, using either injection-moulding machines for the rapid production of small articles or hydraulic presses for compression moulding of relatively large articles.

plastic paint (*Build*) Paint in which the covering agent is a plastic.

plastic rail-bond (*ElecEng*) A rail-bond made by inserting plastic non-conducting material between the rail itself and the fish-plate.

plastic sulphur (*Chem*) Allotropic form of elemental sulphur, created by polymerization of monomeric sulphur molecules (an eight-membered ring compound). It exhibits both ceiling and floor temperatures.

plastic surgery (*Med*) That branch of surgery which deals with the repair and restoration of damaged or lost parts of the body.

plastid (*BioSci*) One of a class of cytoplasmic organelles in plants and eukaryotic algae, comprising the chloroplasts and related organelles, surrounded by two membranes and containing DNA, eg *amyloplast, chloroplast, chromoplast, etioplast, leucoplast*.

plastisol (*Plastics*) Paint-like mixture of plasticizer and powdered polymer, usually polyvinyl chloride, used in a variety of processes to make flexible products. ROTOCASTING is used to make footballs, toys, etc, while KNIFE or REVERSE ROLL COATING produces conveyor belting.

plastochron (*BioSci*) The interval of time between the appearance of successive leaf primordia at the shoot apex or between other similar successive events. Also *plastochrone*.

plastocyanin (*BioSci*) A copper-containing protein involved in the transfer of electrons from PHOTOSYSTEM II to PHOTOSYSTEM I in photosynthesis.

plastoquinone (*BioSci*) Terpenoid involved in the transfer of electrons from PHOTOSYSTEM II to PHOTOSYSTEM I in photosynthesis.

plastron (*BioSci*) (1) The ventral part of the bony exoskeleton in Chelonia; any similar structure. (2) In some aquatic insects (eg Coleoptera and Hemiptera) a thin air film over certain parts of the body held by minute hydrofuge hairs. This serves as a physical gill rather than as a store of air. Adj *plastral*.

plastron (*Med*) The sternum and the costal cartilages.

platband (*Build*) (1) An IMPOST. (2) A flat moulding which projects from the wall surface by less than its breadth. (3) A door or window lintel.

plate (*BioSci*) See PLAX.

plate (*Build*) Usually *wall plate*, the top horizontal timber of a wall, supporting parts of the structure. Also *platt*.

plate (*ChemEng*) In an industrial fractionating column, a tray over which the liquid flows, with nozzles or bubble caps so that the vapour can bubble though the liquid. See MURPHREE EFFICIENCY, STILL.

plate (*Electronics*) (1) Each of the two extended conducting electrodes which, with a dielectric between, constitutes a capacitor. (2) US for ANODE in valves.

plate (*Eng*) A large flat body of steel, thicker than sheet, which is produced by the working of ingots, billets or slabs, in a rolling mill.

plate (*Geol*) The rigid structures of the lithosphere, of about continental size, consisting of the crust and the upper mantle, floating on the viscous lower mantle. See panel on PLATE TECTONICS.

plate (*ImageTech*) General term for the picture, which may be a still transparency or a motion picture print, used as the background in composite photography.

plate (*Print*) (1) Any original and duplicate for letter-press printing; also plates for lithographing, and the several kinds of intaglio plate. (2) An illustration, esp one that is printed separately from the text which it illustrates.

plate amalgamation (*MinExt*) Trapping of metallic gold on an inclined plate made of copper or an alloy, which has been coated with a pasty film of mercury. Method largely superseded by use of STRAKE.

plateau (*NucEng*) See GEIGER REGION.

plateau-basalts (*Geol*) Basic lavas of basaltic composition resulting from fissure eruptions and occurring as thin, widespread flows, forming extensive plateaux (eg the Deccan in India). See DYKES, SILLS.

plateau eruptions (*Geol*) Volcanic eruptions by which extensive lava flows are spread in successive sheets over a wide area and eventually build a plateau; as in Idaho. See FISSURE ERUPTION.

plateau gravel (*Geol*) Deposits of sandy gravel occurring on hilltops and plateaux at heights above those normally occupied by river-terrace gravels. Originally deposited as continuous sheets, plateau gravel has been raised by earth movements to its present level and deeply dissected. Of Pliocene or Early Pleistocene age in the main.

plateau length (*NucEng*) The voltage range which corresponds to the plateau of a Geiger counter tube.

plateau slope (*NucEng*) The ratio of the percentage change in count rate for a constant source, to the change of operating voltage. Measured for a median voltage corresponding to the centre of the Geiger plateau. Often expressed as percentage change in count rate for 100 volt change in potential.

plate cam (*Eng*) A flat, open cam for sliding movement used eg in automatic lathes.

plate clutch (*Eng*) See DISK CLUTCH.

plate columns (*ChemEng*) Distillation or absorption columns which contain plates of various types, spaced at regular intervals with nothing between the plates.

plate count (*BioSci*) The number of bacterial or fungal colonies growing on an agar plate that contains appropriate nutrients etc. It is used to estimate contamination levels, but is limited because not all micro-organisms will grow readily under such conditions.

plate cylinder (*Print*) The cylinder on a printing press to which the printing plate is attached.

plate dissipation (*Electronics*) See ANODE DISSIPATION.

plate exchanger (*ChemEng*) Heat exchanger comprising either a series of alternating flat and ribbed plates forming the flow channel, or flat plates which are hollow with internal flow passages immersed in or forming part of the wall of the vessels.

plate frame (*ElecEng*) The nickel-plated framework for supporting the perforated steel tubes of the electrode of a nickel–iron accumulator.

plate gauge (*Eng*) A limit gauge or single external gauge formed by cutting slots of the required gauge width in a steel plate, the surfaces of which are hardened. See LIMIT GAUGE.

plate girder (*Eng*) A built-up steel girder consisting of a single web plate along each edge of which is welded a plate to form the flanges. In older construction a pair of angles, top and bottom of the web, joined the flange plates.

plate glass (*Glass*) Glass of superior quality, originally cast on an iron bed and rolled into sheet form, and afterwards ground and polished. Modern methods have largely superseded this, save in special circumstances. Also *polished plate*. See FLOAT GLASS.

plate group (*ElecEng*) The complete unit, consisting of an accumulator plate or plates, terminal bar and terminal lug, forming the electrode of an accumulator cell. Also *plate section*.

platelet (*BioSci*) An anucleate discoid cell, approximately 3 μm in diameter, found in large numbers in blood. Such cells are important for blood coagulation and for haemostasis.

platelet activating factor (*BioSci*) A lipid released by various cells, including basophil leucocytes and monocytes, in the presence of antigen. Induces platelet aggregation and degranulation. Activity is transient since platelet activating factor is inactivated by phospholipase A which is also released. Abbrev *PAF*.

platelet-derived growth factor (*BioSci*) The major mitogen in serum for growth in culture of cells of connective tissue origin, believed to play a role in wound healing *in vivo*. Abbrev *PDGF*.

plate link chain (*Eng*) A chain comprising pairs of flat links connected by pins.

plate-lug (*ElecEng*) A projection on an accumulator plate used for connecting it to a terminal bar.

plate modulation (*Electronics*) See ANODE MODULATION.

plate moulding (*Eng*) A method of mounting the halves of a split pattern on opposite sides of a wood or metal plate, placed between the cope and drag, thus eliminating the making of the joint faces.

platen (*Eng*) (1) The work table of a machine tool, usually slotted for clamping bolts. (2) In plural, the steel plates to which fixed and moving parts of a tool are fitted in injection moulding. See INJECTION MOULDING.

plate out (*NucEng*) The coating (or plating) of exposed surfaces in a reactor with material held in suspension in the coolant. See TRAMP URANIUM.

plate proof (*Print*) A proof taken from a plate as distinct from one taken direct from type.

plate rectifier (*ElecEng*) One of large area for large output currents, eg for electrolytic bath supply or electric traction.

plate section (*ElecEng*) See PLATE GROUP.

plate support (*ElecEng*) A support from which the plates of an accumulator are suspended or on which they rest.

plate tectonics (*Geol*) The interpretation of the Earth's structures and processes in terms of the movements of large plates of lithosphere acting as rigid slabs floating on a viscous mantle. See panel on PLATE TECTONICS.

platform (*ICT*) The hardware and operating system of a computer, such as Apple's Macintosh platform.

platform (*Space*) A term sometimes applied to a spacecraft used as a base for experiments in space research, usually unmanned.

platform gantry (*Build*) A gantry formed to support a platform on which is erected a scaffold used for the handling of materials.

platforming (*ChemEng*) Catalytic process for reforming low-grade into high-grade petroleum components using platinum; hence the name. See CATALYTIC REFORMING.

platform tree (*MinExt*) *Christmas tree* with all the necessary valves for controlling the flow of oil from a producing platform.

platinammines (*Chem*) Compounds of platinum and ammonia of the form $Pt(NH_3X)_4$.

platinates (IV) (*Chem*) See PLATINIC HYDROXIDE.

platinectomy (*Med*) The operative removal of the stapedial footplate in the middle ear.

plating sequence (*Print*) The order in which printing plates are secured to the press.

plating up (*Print*) The securing of plates to the printing cylinders.

platinic hydroxide (*Chem*) *Platinum (IV) hydroxide*. $Pt(OH)_4$. Dissolves in acids to form platinic salts and in bases a series of salts called *platinates* (IV). A type of compound formed by the other members of the platinum group of metals.

platinic oxide (*Chem*) *Platinum (IV) oxide*. PtO_2. Dark-grey powder formed when platinic (platinum (IV)) hydroxide is heated. Also *platinum dioxide*.

platinite (*Eng*) Alloy containing iron 54–58% and nickel 42–46%, with a trace of carbon. Has the same coefficient of expansion as platinum, and is used to replace it in some light bulbs and measuring standards.

platinized asbestos (*Chem*) Asbestos permeated with finely divided platinum. Used as a catalyst.

platinoid (*Eng*) Alloy containing copper 62%, zinc 22%, nickel 15%. Has high electrical resistance and is used for resistances and thermocouples.

platinous hydroxide (*Chem*) *Platinum (II) hydroxide*. $Pt(OH)_2$. Soluble in the haloid acids (hydrochloric acid etc), forming platinous (platinum (II)) salts.

platinous oxide (*Chem*) PtO. Formed when platinous hydroxide is gently heated.

platinum (*Chem*) A metallic element, symbol Pt, at no 78, ram 195·09, rel.d. (at 20°C) 21·45, electrical resistivity (at 20°C) $9·97 \times 10^{-8}$ Ω m, mp 1773·5°C, bp 3910°C, Brinell hardness 47. Platinum is the most important of a group of six closely related rare metals, the others being osmium, iridium, palladium, rhodium and ruthenium. It is heavy, soft and ductile, immune to attack by most chemical reagents and to oxidation at high temperatures. Used for making jewellery, special scientific apparatus, electrical contacts for high temperatures and for electrodes subjected to possible chemical attack. Also used as a basic metal for resistance thermometry over a wide temperature range. Native platinum is usually alloyed with iron, iridium, rhodium, palladium or osmium, and crystallizes in the cubic system. Its abundance in the Earth's crust is only 0·01 ppm, occurring as the metallic element and in the arsenide *sperrylite*, $(PtAs_2)$, sulphides and a few other binary compounds.

platinum black (*Chem*) Platinum precipitated from a solution of the (IV) chloride by reducing agents. A velvety-black powder. Uses: catalyst and gas absorber, eg in the HYDROGEN ELECTRODE.

platinum dioxide (*Chem*) *Platinum (IV) oxide.* See PLATINIC OXIDE.

platinum metals (*Chem*) A block of six transition metals with similar physical and chemical properties. Specifically ruthenium, osmium, rhodium, iridium, palladium and platinum.

platinum tetrachloride (*Chem*) *Platinum (IV) chloride*, $PtCl_4$. Formed by dissolving platinum in aqua regia. Similar chlorides are formed with the other platinum metals.

platinum thermometer (*ElecEng*) See RESISTANCE THERMOMETER.

PLATO (*ICT*) See COMPUTER-ASSISTED INSTRUCTION.

platonic solid (*MathSci*) The five REGULAR CONVEX SOLIDS described by Plato.

platter (*ICT*) One of the circular plates in a hard-disk drive. See fig. at HARD DISK.

platter (*ImageTech*) Large horizontal turntable supporting a very long reel of film for projection.

P lattice (*Crystal*) Abbrev for *primitive crystal lattice*. See fig. at UNIT CELL.

platting (*Build*) The top course of a brick clamp.

platycephalic (*Med*) Having a flattened or broad head, with a breadth–height index of <70. Also *platycephalus*.

platydactyl (*BioSci*) Having the tips of the digits flattened.

Platyhelminthes (*BioSci*) A phylum of bilaterally symmetrical, triploblastic Metazoa; usually dorsoventrally flattened. The space between the gut and the integument is occluded by parenchyma; the excretory system consists of ramified canals containing flame cells. There is no anus, coelom or haemocoel. Genitalia are usually complex and hermaphrodite. Flat worms.

platykurtic (*MathSci*) Of a curve or distribution, concentrated around the MEAN to a lesser extent than a normal curve. Compare LEPTOKURTIC, MESOKURTIC. See KURTOSIS.

platysma (*BioSci*) A broad sheet of dermal musculature in the neck region of mammals.

platyspermic (*BioSci*) Having seeds that are flattened in transverse section, as in the Cordaitales and the conifers. Cf RADIOSPERMIC.

plax (*BioSci*) A flat plate-like structure, as a lamella or scale.

play (*Eng*) Limited movement between mating parts of a mechanism, due either initially to the type of fit and dimensional allowance specified or subsequently to wear.

play (*Psych*) Activity which occurs largely in the young of warm-blooded mammals and can involve almost any behaviour; tends to occur in isolation from their normal function, is carried out voluntarily rather than in conditions of necessity, and often merges behaviours from different functional systems. Exaggerated movements often occur and may be preceded or accompanied by a signal that the activity is a playful one (eg play fights).

playa lake (*Geol*) Shallow lake formed in a flat arid or semi-arid region in the wet season and drying out in the summer.

playback (*Acous*) A recording technique in which an existing partial recording of a piece of music (eg the beats)

is fed into an earphone or loudspeaker while other instruments, orchestra groups or soloists are recorded simultaneously.

playback equalizer (*Acous*) A resistance–capacitance network introduced into an interstage coupling so that all frequencies are reproduced with equal intensity in the recording of music.

play therapy (*Psych*) A form of child psychotherapy which uses play as the means of communication to reveal unconscious conflicts, and which encourages children to vent their feelings through symbolic play.

pleasure principle (*Psych*) In Freudian theory, the motive to seek immediate pleasure and gratification without regard to consequence. See ID.

pleated-diaphragm loudspeaker (*Acous*) A loudspeaker in which the radiating element is a pleated diaphragm, the pleats being radial and the rim clamped. It is driven by a pin at the centre.

pleated sheet (*Chem*) A type of secondary structure formed in proteins by lateral hydrogen bonds, esp beta-pleated sheet in silk. See BIOMATERIALS. Fig. p918 ▷

Plectomycetes (*BioSci*) A class of fungi in the Ascomycotina in which the fruiting body (ascocarp) is a cleistothecium. It includes *Eurotium* and allied genera which are the perfect stages of many species of *Aspergillus* and *Penicillium*. See DEUTEROMYCOTINA.

plectostele (*BioSci*) A type of protostele in which the xylem and phloem are arranged in alternating plates across the stele.

plei-, pleio-, pleo-, plio- (*Genrl*) Prefixes from Gk *pleion*, more.

Pleiades (*Astron*) The open cluster in the constellation Taurus, of which the seven principal stars, forming a well-known group visible to the naked eye, each have a separate name. Also *Seven Sisters*.

pleio- (*Genrl*) See PLEI-.

pleiomerous (*BioSci*) Having a large number of parts.

pleiomorphic (*BioSci*) Having more than one shape. Also *pleomorphic, pliomorphic.*

pleiotropy (*BioSci*) The condition when a molecule, esp a gene, a hormone or cyclic AMP, affects more than one character. Adj *pleiotropic*.

Pleistocene (*Geol*) The epoch of geological time following the Tertiary, and covering a time span of approx the last 2 million years. It was during this period that ice covered a large part of the northern hemisphere; hence it has been called the *Great Ice Age*.

Pleistogene (*Geol*) See QUATERNARY.

plenum chamber (*Aero*) A sealed chamber pressurized from an air intake. Centrifugal flow turbojets having double-entry impellers (see DOUBLE-ENTRY COMPRESSOR) have to be mounted in plenum chambers to ensure even air pressure on both impeller faces.

plenum system (*Build*) An air-conditioning system in which the air propelled into the building is maintained at a higher pressure than the atmosphere. The conditioned air is usually admitted to rooms from 2·5–3 m above floor level, while the vitiated air is extracted at floor level on the same side of the room.

pleo- (*Genrl*) See PLEI-.

pleocholia (*Med*) Excessive formation of bile pigment.

pleochroic haloes (*Min*) Dark-coloured zones around small inclusions of radioactive minerals which are found in certain crystals, notably biotite. The colour and pleochroism of the zones are stronger than those of the surrounding mineral, and result from radioactive emanations during the conversion of uranium or thorium into lead.

pleochroism (*Min*) The property of a mineral by which it exhibits different colours in different crystallographic directions on account of the selective absorption of transmitted light.

pleochromatic (*BioSci*) Presenting different colours according to changes in the environment or with different physiological conditions.

Plate tectonics

The concept of CONTINENTAL DRIFT or the movement of vast land masses was first proposed by A L Wegener in 1910, but lack of an obvious mechanism meant his theory was ignored for many years. Plate tectonics is now the accepted explanation in which huge plates of the LITHOSPHERE are considered to act as relatively rigid slabs floating on the relatively plastic ASTHENO-SPHERE. Major structures and processes in the crust are associated with the movement of these plates, including MID-OCEANIC RIDGES, mountain building, oceanic trenches, major TEAR FAULTS, earthquake zones and volcanic belts.

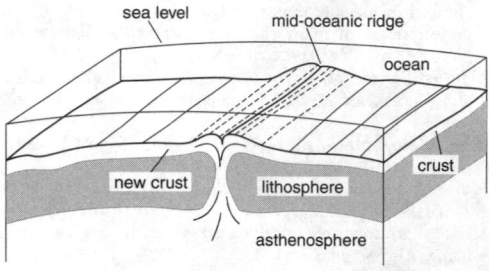

Fig. 2 **Sea-floor spreading at a divergent plate boundary**.

The major plates of the Earth and some of these associated features are shown in Fig. 1 (opposite). The plates move away from the mid-oceanic ridges at *divergent plate boundaries*. The space between the separating plates is filled by volcanic rock, welling up from below to form new crust (new sea floor), a process called SEA-FLOOR SPREADING (Fig. 2). While the sea-floor crust has been moving away from the

mid-oceanic ridges, the polarity of the Earth's magnetic field has reversed many times, resulting in MAGNETIC ANOMALY PATTERNS in the new crust. There are also many transform faults across the axes of spreading.

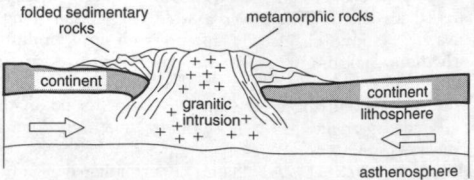

Fig. 3 **Mountain range at the convergence of two continental plates**.

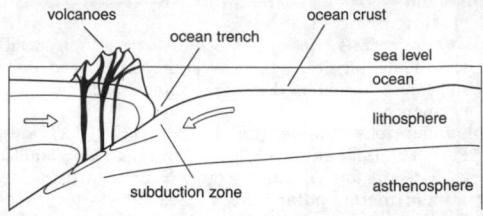

Fig. 4 **Collision of oceanic plates**.

At the leading edge, as the plate moves, there is a *convergent plate boundary*. The collision of the plates results in many earthquakes, and in continental regions may cause the development of mountain chains, eg the Himalayas between the Indian and Eurasian plates (Fig. 3). When plates bearing oceanic crust collide, one may plunge beneath the other to produce a SUB-DUCTION ZONE. Island arcs with volcanoes and deep oceanic trenches may then develop (Fig. 4).

If the plates slide past each other, they may form *shear boundaries*, eg along the San Andreas fault in California, where the Pacific plate moves against the North American plate.

See panels on EARTH and EARTHQUAKE.

---

pleocytosis (*Med*) An increase in the number of white blood cells esp in the cerebrospinal fluid.

pleomorphic (*BioSci*) See PLEIOMORPHIC.

pleonaste (*Min*) Oxide of magnesium, iron and aluminium, with Mg:Fe from three to one, crystallizing in the cubic system. It is a member of the spinel group and may be dark-green, brown or black in colour. Also *ceylonite*.

pleopod (*BioSci*) In Arthropoda (esp Crustacea), an abdominal appendage adapted for swimming.

plerome (*BioSci*) A HISTOGEN, the precursor of the stele.

plesiochronous digital hierarchy (*ICT*) Digital network system in which low-data-rate channels are multiplexed to form a smaller number of higher-rate channels. Since the systems from which the lower-rate channels originate are not truly synchronous, this demands the addition of padding bits to accommodate the resulting time differences, making extraction of an individual channel impossible without wastefully demultiplexing all channels. Cf SYNCHRONOUS DIGITAL HIERARCHY.

plethora (*Med*) Fullness of the blood vessels; plethoric appearance, suffused, reddened face. Pulmonary plethora is an increase in the blood vessels as shown on a chest X-ray.

plethysmograph (*Med*) Apparatus for measuring variations in the size of bodily parts and in the flow of blood through them. See ELECTROARTERIOGRAPH.

pleur-, pleuro- (*Genrl*) Prefixes from Gk *pleura*, side.

pleura (*BioSci*) The serous membrane lining the pulmonary cavity in mammals and birds.

pleurapophysis (*BioSci*) A lateral vertebral process; usually applied to the true ribs.

pleurisy (*Med*) Inflammation of the pleura, which may be either dry or accompanied by effusion of fluid into the pleural cavity (pleural effusion).

pleuro- (*Genrl*) See PLEUR-.

pleurocarp (*BioSci*) A moss bearing the archegonia, and therefore the capsule and its stalk, on a short side branch, not at the tip of a main stem or branch. Cf ACROCARP.

pleurodont (*BioSci*) Having the teeth fastened to the side of the bone which bears them, as in some lizards.

pleurodynia (*Med*) Pleural pain with chest muscle involvement, often thought to be viral in origin.

pleurogenous (*BioSci*) Borne on a lateral position.

pleurogenous (*Med*) Having origin in the pleura, eg *pleurogenous cirrhosis* of the lung. Also *pleurogenic*.

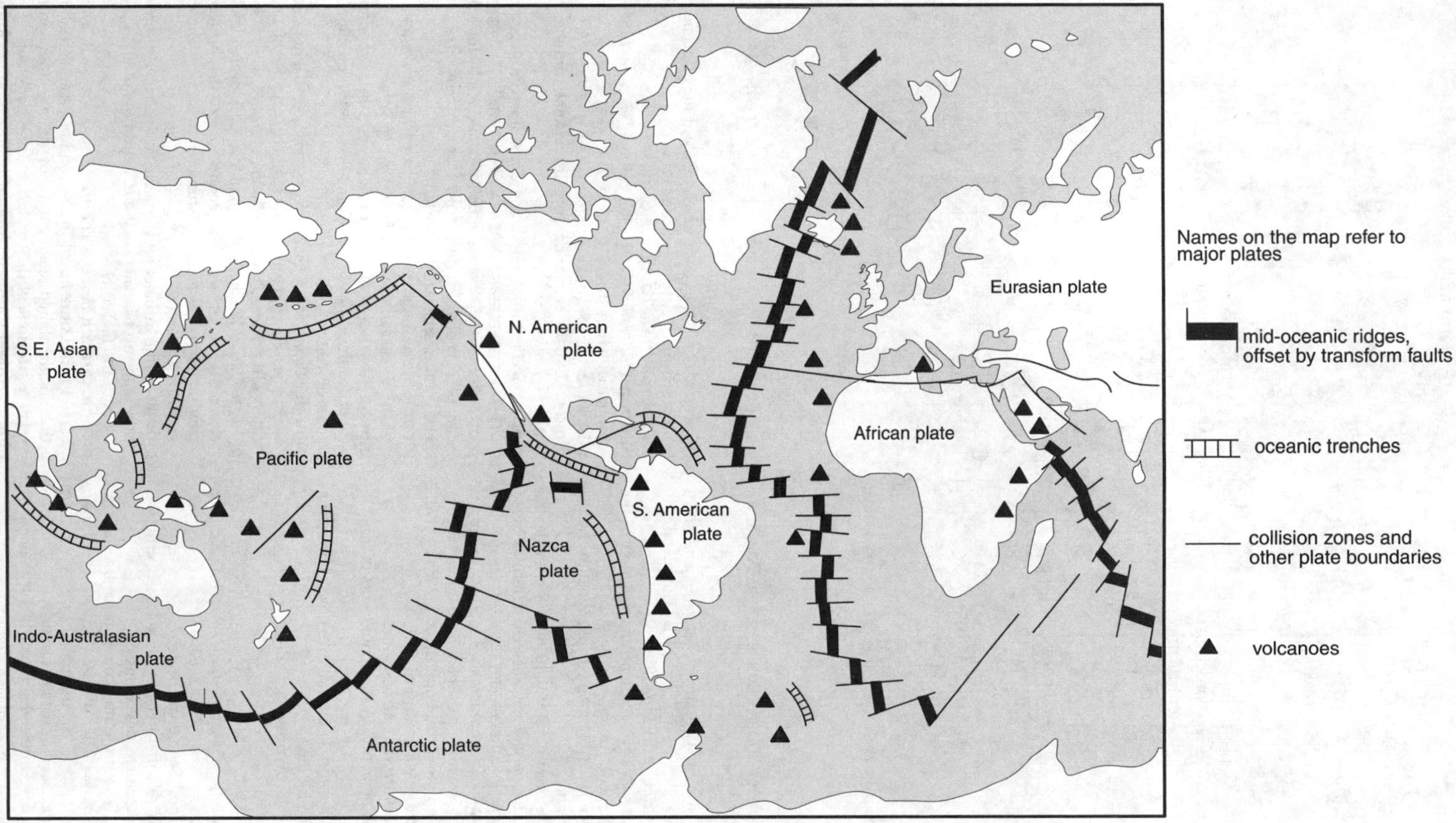

Names on the map refer to major plates

▆▆ mid-oceanic ridges, offset by transform faults

▥ oceanic trenches

— collision zones and other plate boundaries

▲ volcanoes

Eurasian plate

African plate

N. American plate

S. American plate

Nazca plate

Antarctic plate

Pacific plate

S.E. Asian plate

Indo-Australasian plate

Fig. 1 **Plate tectonics.**

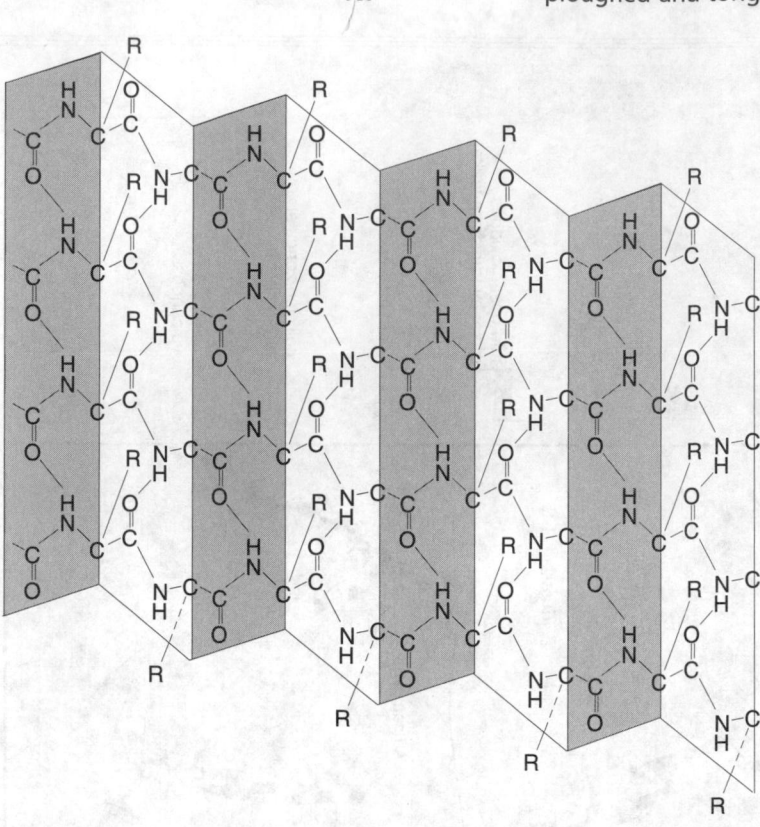

pleated sheet Parallel conformation. Only a few of the side chains (R) situated behind the sheet are shown.

pleuron (*BioSci*) In some Crustacea, a lateral expansion of the tergite; more generally, in Arthropoda. The lateral wall of a somite. Pl *pleura*. Adj *pleural*.

pleuropericarditis (*Med*) Concurrent inflammation of the pleura and of the pericardium.

pleuropneumonia (*Med*) Combined inflammation of the pleura and of the lung.

pleuropneumonia (*Vet*) A contagious disease of cattle due to infection by *Mycoplasma mycoides*; characterized by an exudative fibrinous pneumonia and pleurisy.

pleuropneumonia-like organisms (*BioSci*) A group of organisms, closely resembling the pleuropneumonia organisms, isolated from the throat and vagina. See MYCOPLAS-MATALES. Abbrev *PPLO*.

p-level (*Phys*) See PRINCIPAL SERIES.

plexitis (*Med*) Inflammation of the components of a nerve plexus.

plexus (*BioSci*) A network; a mass of interwoven fibres, as a *nerve plexus*. Adj *plexiform*.

plica (*BioSci*) A fold of tissue; a fold-like structure. Adjs *plicate, pliciform*.

plicate (*BioSci*) Folded.

Pliensbachian (*Geol*) A stage in the Lower Jurassic. See MESOZOIC.

plimsoll mark (*Ships*) See LOAD LINES.

Plinian eruption (*Geol*) A type of volcanic eruption characterized by repeated explosions.

plinth (*Build*) (1) The projecting course or courses at the base of a building. (2) The base of a pedestal, bookcase, wardrobe, etc.

plinth block (*Build*) See ARCHITRAVE BLOCK.

plinth course (*Build*) A projecting course laid at the base of a wall.

plio- (*Genrl*) See PLEI-.

Pliocene (*Geol*) The epoch which followed the Miocene and preceded the Pleistocene. See TERTIARY.

pliomorphic (*BioSci*) See PLEIOMORPHIC.

pliotron (*Electronics*) Large hot-cathode vacuum tube for industrial use, with one or more grids.

plissé (*Textiles*) See SEERSUCKER.

ploidy (*BioSci*) Pertaining to chromosome number, eg *diploid, haploid, polyploid*.

plotter (*ICT*) Output device that draws lines on paper. See DIGITAL PLOTTER, INCREMENTAL PLOTTER.

plotter font (*ICT*) A series of points connected by lines that can be scaled to produce FONTS of different sizes on a PLOTTER. Similar to VECTOR FONT.

Plough (*Astron*) A group of bright stars which make up part of the constellation Ursa Major. US *Big Dipper*.

plough (*Agri*) A trailed or draught-animal-drawn implement for inverting the soil as the first step in preparation for crop cultivation. US *plow*. Fig. ▷

plough (*Build*) (1) A form of grooving or shaping plane which has an adjustable FENCE and is capable of being fitted with various irons. (2) To cut a groove. US *plow*.

plough (*ElecEng*) The device under a tram which makes electrical contact with the conductor in the protected conduit placed between the running rails. US *plow*. See PLOUGH CARRIER.

plough (*Print*) A hand tool for cutting the edges of books, now only occasionally used for hand-bound books. US *plow*.

plough carrier (*ElecEng*) The frame under a tram, which carries the PLOUGH used in the conduit system. It allows the plough to slide laterally to follow any variations in the relative positions of the conduit and the track rails.

ploughed-and-tongued joint (*Build*) A joint formed between the square butting edges of two boards, each

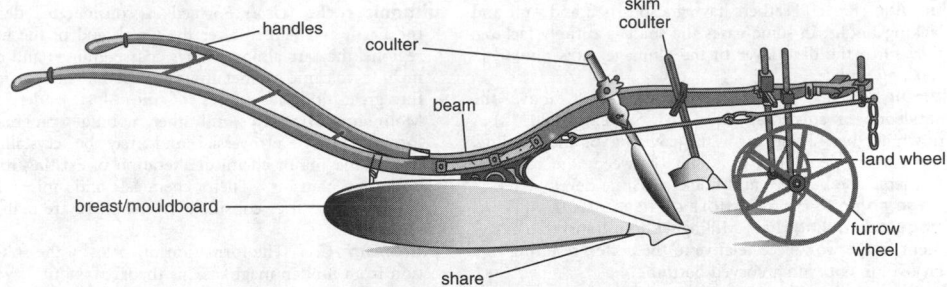

An early example of a horse-drawn plough.

having a plough groove into which a common tongue is inserted.

ploughshare (*Agri*) The horizontal blade of a plough, set into a mould board, that cuts into the soil when ploughing. It follows a vertically mounted blade or COULTER.

plow (*Genrl*) US for *plough*.

plug (*Build*) A wooden or plastic piece driven into a hole cut into brick or blockwork, finished off flush, to provide a fixing for joinery or fittings. In Scotland a *dook*.

plug (*Geol*) A vertical cylinder of solidified magma or pyroclastic material which represents the feeder pipe of a former volcano.

plug (*NucEng*) A piece of absorbing material used to close the aperture of a channel through a reactor core or other source of ionizing radiation.

plug-and-play (*ICT*) Peripherals that, when added to a computer, present an identification signal that allows the operating system to configure the system properly.

plug centre bit (*Build*) A form of centre bit in which the projecting central point is replaced by a plug of metal, adapting the bit for use in holes already drilled.

plug cock (*Build, Eng*) A simple valve in which the plug, usually conical, is drilled across its axis, allowing flow only when the drilled hole is in line with inlet and outlet. Also *plug tap*.

plug compatible (*ICT*) Interchangeable devices that use the same sockets, eg a card in a computer. They can have different functions but have complete hardware inter-changeability.

plug gauge (*Eng*) A gauge, made in the form of a plug for testing the diameter of a hole; in a plug limit gauge two plugs are provided, a 'go' and a 'not go'. See LIMIT GAUGE.

plugging (*Build*) The operation of making a hole in a wall or partition, and driving in a wall PLUG.

plugging (*ElecEng*) Braking an electric motor by rearranging the connections so that it tends to run in the reverse direction.

plug-in (*Electronics*) A single component, or subassembly, which has plug-in terminals so that all connections can be made simultaneously by pushing the unit into a socket; usually employed where rapid interchangeability of oper-ating ranges or functions is desired.

plug-in unit (*ICT*) Any panel or component that can be inserted and interchanged in a computing system, esp a cross-connection panel, that can be set up independently.

plug tap (*Eng*) (1) The final parallel tap required to finish an internal thread in a blind hole. Also *bottoming tap*, *third tap*. (2) A PLUG COCK.

plug welding (*Eng*) A method of welding in which apertures in one part allow penetration through to the other part to be joined, the apertures to be filled by welding in the process.

plumae (*BioSci*) Feathers having a stiff shaft and a firm web, and usually possessing small hooks; they appear on the surface of the plumage and determine the contours of the body in addition to forming the RECTRICES and REMIGES. Adjs *plumate*, *plumigerous*, *plumose*, *plumous*.

plumb (*Build, CivEng*) Vertical, from the use of a PLUMB LINE.

plumb- (*Chem*) From Lt *plumbum*, lead; eg plumbic chloride (PbCl$_4$), plumbous chloride (PbCl$_2$).

plumbago (*Min*) See GRAPHITE.

plumbates (II) (*Chem*) See LEAD (II) HYDROXIDE.

plumb-bob (*Surv*) A small weight or 'bob' hanging at the end of a cord, used to centre survey instrument over signal mark. Also *plummet*.

plumber's solder (*Build*) A lead–tin alloy of varying ratio from 1:1 to 3:1 for different classes of work, the melting points being always lower than that of lead itself.

Plumbicon (*ImageTech*) TN for an improved camera tube of the photoconductive VIDICON type employing lead oxide layers. Also *Leddicon*.

plumbing (*Build*) (1) The craft of working lead for structural purposes, or for the installation of domestic water-supply systems, sanitary fittings, etc. (2) The operation of arranging vertically.

plumbing (*ICT*) Colloq term for waveguides and their jointing in establishing microwave systems. Also for piped vacuum systems.

plumbing unit (*Build*) A prefabricated assembly of pipes and fittings, generally of storey height, used in multistorey flats.

plumbism (*Med*) Lead poisoning.

plumbites (*Chem*) See LEAD (II) HYDROXIDE.

plumb level (*Surv*) A LEVEL TUBE for showing plumb direction, with a small bubble at right angles to the main one. When the latter is held vertical the small bubble is central.

plumb line (*Surv*) A cord with a PLUMB-BOB attached to the end.

plumb rule (*Build*) A narrow board formerly used for determining verticals; it has at one end a point of suspension for a plumb-bob, which is free to swing in an egg-shaped hole at the other end of the board.

plume (*BioSci*) Any feather-like structure; a light, hairy or feathery appendage on a fruit or seed serving in wind dispersal.

plume (*EnvSci*) (1) Snow blown over the ridge of a mountain. (2) The stream of gases emerging from a chimney that retains its identity for some time before finally dispersing.

plume (*Geol*) Ascending partly molten material from the mantle believed to be responsible for intraplate volcanism.

plummer block (*Eng*) A JOURNAL bearing for line shafting etc consisting of a box-form casting holding the ball races, roller races or bearing brasses, split horizontally to take up wear (brasses) or to facilitate assembly (races).

Plummer–Vinson syndrome (*Med*) The association of iron-deficiency anaemia and difficulty in swallowing due to the development of webs of tissue at the top of the oesophagus. Also *Patterson–Kelly syndrome*.

plummet (*Surv*) See PLUMB-BOB.

plumose (*BioSci*) Hairy; feathered.

plumulae (*BioSci*) Feathers having a soft shaft and vane and lacking hooks; in some cases the shaft is entirely lacking; they form the deep layer of the plumage. Adjs *plumulaceous, plumulate.*

plumule (*BioSci*) (1) An embryonic shoot above the cotyledon or cotyledons in a seed. See EPICOTYL. (2) A down feather. These form the covering of the nestling, sometimes persisting in adult between the contour feathers. The barbules and hooks are little developed.

plunge angle (*Surv*) See ANGLE OF DEPRESSION.

plunge-cut milling (*Eng*) Milling without transverse movement of the workpiece relative to the cutter, resulting in a groove or slot with a curved bottom.

plunge grinding (*Eng*) See PROFILE GRINDING.

plunger (*ElecEng*) A device, in practice an annular disk, for altering the length of a short-circuited coaxial line. It needs no electrical contact for very high frequencies because a small clearance gives sufficient capacitance for an effective short circuit.

plunger (*Eng*) The ram or solid piston of a force pump.

plunge router (*Build*) The commonest form of hand-held ROUTER in which two pillars attach the motor and cutter to a plate through which the cutter projects. The plate is spring-loaded and adjustable so that the cutter can be exposed for different depths of cut when the router is depressed.

plunge saw (*Build*) A circular saw mounted on a counter-balanced arm hinged in the vertical plane so that the saw is brought down on top of the work but cannot move along the arm. A circular table with a fence allows the work to be adjusted at any angle to the saw. A similar device with a grinding wheel or slitting saw is used for metal. Cf RADIAL ARM SAW.

plunging fold (*Geol*) One whose axis is not horizontal. The angle between the axis and the horizontal is called the *plunge*. See FOLDING.

plunging shot (*Surv*) Downward theodolite sight. To *plunge* is to transit the instrument.

pluriglandular (*Med*) Pertaining to, affected by, or affecting, several (ductless) glands.

plurilocular (*BioSci*) A sporangium or ovary divided into several compartments by septa. Cf UNILOCULAR.

pluripotent stem cell (*BioSci*) Cells in a stem cell line capable of differentiating into several, but not all, final differentiated types. The exact boundaries between multi-potency, pluripotency and TOTIPOTENCY are rather ill-defined.

plus (+) (*BioSci*) Symbol used before the generic or specific epithet to denote, respectively, an intergeneric or inter-specific GRAFT CHIMERA.

plush (*Textiles*) (1) Woven fabric with cut pile, longer and less dense than velvet, on one side. Warp pile is generally made by weaving two cloths together, with a pile warp common to both, which is afterwards cut. (2) One type of weft-knitted fabric has the long loops on the back of the cloth, which is sometimes called knitted TERRY. In warp-knitted plush fabrics the loops may be cut or left uncut.

plus strain (*BioSci*) One of two arbitrarily designated, mating types of a heterothallic species. Also *(+) strain*. See HETEROTHALLISM.

Pluto (*Astron*) Ninth planet of the solar system, discovered by Clyde Tombaugh on 18 February 1930 as a result of a systematic search. The orbit is very elliptical and inclined at 17° to the ecliptic. For a proportion of its 248 yr orbit it is actually closer to the Sun than Neptune. Its diameter is around 2300 km (less than our Moon) with a relative density similar to water. Pluto has one moon, CHARON. Possibly the two objects are escaped satellites of Neptune. In 2006 Pluto's status as a planet was challenged by the International Astronomical Union and strictly it should now be referred to as asteroid number 134340, a 'dwarf planet'. See appendix on Planets.

plutonic intrusions (*Geol*) A term applied to large intrusions which have cooled at great depth beneath the surface of the Earth. Also *major intrusions*.

plutonic rocks (*Geol*) Formed at considerable depth in the Earth and named after the Greek god of the infernal regions, they are almost always coarse-grained and do not include volcanic or metamorphic rocks. They do include the great granitic masses of the Alps, Andes, Rocky Mountains, Himalayas and other mountain ranges. Their formation is controversial and may be crystallization from a magma or chemical alteration of existing rocks or both mechanisms. Their chemical and mineralogical composition varies considerably and there are many rock names.

plutonism (*Geol*) The formation of rocks by the solidification from molten magma. This theory was put forward in the 18th century. Cf NEPTUNISM.

plutonites (*Geol*) See PLUTONIC ROCKS.

plutonium (*Chem*) An element. Symbol Pu, at no 94, product of radioactive decay of *neptunium*. It has seven isotopes: the fissile isotope plutonium-239, produced from uranium-238 by neutron absorption in a reactor, is the most important for the production of nuclear power and in weapons.

plutonium reactor (*NucEng*) One in which plutonium-239 is used as the fuel.

pluviometry (*EnvSci*) The study of precipitation, including its nature, distribution and techniques of measurement.

plywood (*Build*) A board consisting of a number of thin layers of wood glued together so that the grain of each layer is at right angles to that of its neighbour.

PM (*Build*) Abbrev for *purpose-made*.

PM (*Eng*) Abbrev for POWDER METALLURGY.

PM₁₀ (*EnvSci*) Particulate material with a size of around $10\ \mu\mathrm{m}$.

Pm (*Chem*) Symbol for PROMETHIUM.

PMBX (*ICT*) Abbrev for PRIVATE MANUAL BRANCH EXCHANGE.

PMC (*Build*) Abbrev for *plaster-moulded cornice*.

PMDD (*Med*) See PREMENSTRUAL DYSPHORIC DISORDER.

PMMA (*Chem*) Abbrev for POLYMETHYLMETHACRYLATE.

PMOS (*Electronics*) Metal–oxide–silicon technology based on p-channel devices in an n-type substrate.

PMS (*Print*) See PANTONE MATCHING SYSTEM.

PMSF (*BioSci*) See PHENYLMETHYLSULPHONYL FLUORIDE.

PMX (*ICT*) Abbrev for PRIVATE MANUAL EXCHANGE.

PND (*Med*) Abbrev for POSTNATAL DEPRESSION.

pn boundary (*Electronics*) The surface on the transition region between p-type and n-type semiconductor material, at which the donor and acceptor concentrations are equal. Also *p–n boundary*.

pneum-, pneumo-, pneumat-, pneumato- (*Genrl*) Prefixes from Gk, *pneuma*, gen *pneumatos*, breath.

pneumathode (*BioSci*) A more or less open outlet of the ventilating system of a plant, usually consisting of loosely packed cells on the surface of the plant through which exchange of gases between the air and the interior of the plant is facilitated.

pneumatic (*Eng*) Operated by, or relying on, air pressure or the force of compressed air.

pneumatic (*BioSci*) Containing air, as in physostomous fish, the *pneumatic duct* leading from the gullet to the air bladder, and in birds, those bones which contain air cavities.

pneumatically operated (*ElecEng*) A term describing a switch or circuit breaker which is closed by a piston operated by compressed air.

pneumatic brake (*Eng*) A continuous braking system, used on some railway trains, in which air pressure is applied to brake cylinders throughout the train. See AIR BRAKE (1), CONTINUOUS BRAKE, ELECTROPNEUMATIC BRAKE.

pneumatic conveyer (*Eng*) A system by which loose material is conveyed through tubes by air in motion, the air velocity being created by the expansion of compressed air through nozzles.

pneumatic drill (*Eng*) A hard-rock drill in which compressed air is arranged to reciprocate a loose piston which hammers the shank of the bit or an intermediate piece, or in which the bit is clamped to a piston rod.

pneumatic flotation cell (*MinExt*) One in which low-pressure air is blown in and diffused upwards through the cell.

pneumatic lighting (*MinExt*) Use of small turbomotor driven by compressed air to drive a small dynamo connected to an electric lamp, thus avoiding wiring extensions underground where a compressed-air service exists.

pneumatic loudspeaker (*Acous*) A loudspeaker in which a jet of high-pressure air is modulated by a transducer. Also *stentorphone*.

pneumatic pick (*Eng*) A road contractor's tool in which, by mechanism similar to that of a PNEUMATIC DRILL, a straight pick is hammered rapidly by a reciprocating piston driven by compressed air.

pneumatic riveter (*Eng*) A high-speed riveting machine similar in arrangement to a HYDRAULIC RIVETER but in which a rapidly reciprocating piston driven by compressed air delivers 1000–2000 blows per minute.

pneumatic tools (*Eng*) Hand tools, such as riveters, scaling and chipping hammers, and drills, driven by compressed air. See PNEUMATIC DRILL, PNEUMATIC PICK, PNEUMATIC RIVETER.

pneumatic trough (*Chem*) A vessel used for the collection of gases over water.

pneumatic tube conveyor (*Eng*) A system in which small objects enclosed in suitable containers are transported along tubes, the container acting as a moving piston which is impelled either by means of pressure or by vacuum.

pneumato- (*Genrl*) See PNEUM-.

pneumatocele (*Med*) (1) A hernial protrusion of lung through some defect in the chest wall. (2) Any air-containing swelling.

pneumatocyst (*BioSci*) (1) Any air cavity used as a float. (2) In fish, the AIR BLADDER. (3) The cavity of a PNEUMATO-PHORE.

pneumatolysis (*Geol*) The alteration of rocks by the concentrated volatile constituents of a magma, effected after the consolidation of the main body of magma. See GREISENIZATION, KAOLINIZATION, TOURMALINIZATION.

pneumatophore (*BioSci*) A specialized root of swamp plants and mangroves which grows vertically upwards into the air from the root system in the mud and which, containing much aerenchyma, apparently facilitates gas exchange for the submerged roots. Also *breathing root*.

pneumaturia (*Med*) The passing of urine containing gas or air.

pneumectomy (*Med*) See PNEUMONECTOMY.

pneumo- (*Genrl*) See PNEUM-.

pneumococcal polysaccharide (*BioSci*) Type-specific polysaccharide present in the capsules of *Streptococcus pneumoniae*. The capsules inhibit phagocytosis and thus contribute to virulence.

pneumococcus (*BioSci*) A Gram-positive diplococcus, a causative agent of pneumonia, though it may occur normally in throat and mouth secretions.

pneumoconiosis (*Med*) A disease of the lungs due to inhalation of dust in excessive quantities, usually occupational, the chief being silicosis, asbestosis and (coal miner's) anthrocosis.

pneumoencephalitis (*Vet*) See NEWCASTLE DISEASE.

pneumohaemopericardium (*Med*) The presence of air and blood in the pericardial sac.

pneumohaemothorax (*Med*) The presence of air and blood in the pleural cavity.

pneumohydropericardium (*Med*) The presence of air and a clear effusion in the pericardial sac.

pneumolith (*Med*) A concretion in the lung, formed usually as a result of calcification of a chronic tuberculous focus.

pneumon-, pneumono- (*Genrl*) Prefixes from Gk *pneuma*, gen *pneumonos*, lung.

pneumonectomy (*Med*) Removal of lung tissue.

pneumonia (*Med*) A term generally applied to any inflammatory condition of the lung accompanied by consolidation of the lung tissue; more esp lobar pneumonia, in which the consolidation affects one or more lobes of the lung. *Pneumonitis* is lung inflammation that is not necessarily infectious in origin.

pneumono- (*Genrl*) See PNEUMON-.

pneumonomycosis (*Med*) A term applied to disease of the lung caused by any one of a number of various fungi.

pneumo-oil switch (*ElecEng*) A switch or circuit breaker in which the operation is carried out partly by pneumatic means, and partly by hydraulic means using oil as the medium. Also *oil breaker*.

pneumopericardium (*Med*) (1) The presence of air or gas in the peritoneal cavity. (2) The injection of air into the peritoneal cavity for radiographic purposes.

pneumopyopericardium (*Med*) The presence of air and pus in the pericardial cavity.

pneumopyothorax (*Med*) The presence of air and pus in the pleural cavity.

pneumostome (*BioSci*) (1) In Arachnida, the opening to the exterior of the lung books. (2) In Pulmonata, the opening to the exterior of the lung formed by the mantle cavity.

pneumothorax (*Med*) (1) The presence of air or gas in the pleural cavity. (2) The therapeutic injection of air or gas into the pleural cavity for the purpose of collapsing a diseased lung (artificial pneumothorax).

pnip transistor (*Electronics*) Similar to p–n–p transistor with a layer of an intrinsic semiconductor between the base and the collector to extend the high-frequency range. Also *p–n–i–p transistor*.

pn junction (*Electronics*) The boundary between p- and n-type semiconductors, having marked rectifying characteristics; used in diodes, photocells, transistors, etc. Also *p–n junction*.

pnp transistor (*Electronics*) A junction transistor having an n-type base region between a p-type emitter and a p-type collector. For normal operation, the emitter should be held positive and the collector negative with respect to the base. Also *p–n–p transistor*.

Po (*Chem*) Symbol for POLONIUM.

Poaceae (*BioSci*) See GRAMINEAE.

poaching (*Agri*) The breakdown of structure in saturated soils that are repeatedly compacted by eg livestock trampling. This leads to waterlogging.

PO box (*ElecEng*) Abbrev for POST OFFICE BOX.

PO bridge (*ElecEng*) Abbrev for POST OFFICE BRIDGE.

pock (*Med*) A pustule; any small elevation of the skin, containing pus, occurring in an eruptive disease (formerly seen in smallpox).

Pockels' effect (*ElecEng*) The ELECTRO-OPTICAL EFFECT in a piezoelectric material.

pocket (*Build*) The hole in a pulley stile through which the counterpoise weights are passed into the box of a sash and frame.

pocket chamber (*Radiol*) See PERSONAL DOSIMETER.

pocket chisel (*Build*) See SASH POCKET CHISEL.

pockmark (*Geol*) Concave cone-shaped depressions that occur in profusion in unconsolidated fine-grained sea-bed sediments off Canada and in the North Sea. They are typically 15–45 m across and 5–10 m deep, and were first recognized in 1970. They may be due to the escape of biogenic gas and their presence may be a potential hazard for engineering structures.

pod (*Aero*) See ENGINE POD.

pod (*BioSci*) See LEGUME.

pod-, podo-, -pod (*Genrl*) Prefixes and suffix from Gk *pous*, gen *podos*, foot.

podagra (*Med*) See GOUT.

podal (*BioSci*) Associated with or descriptive of the foot.

podauger (*Build*) An auger having a straight groove cut in its length to hold the chips.

podcast (*ICT*) A set of audio files downloaded from a website into a device such as an MP3 player, to be listened to at the convenience of the user.

podcatching (*ICT*) The act of downloading podcasts onto a digital audio player.

podex (*BioSci*) The anal region. Adj *podical*.

Podicipitiformes (*BioSci*) An order of birds, containing compact-bodied species of cosmopolitan distribution. They are almost completely aquatic, and build floating nests. The toes are lobate and the feet are placed far back. Grebes.

podite (*BioSci*) A walking leg of Crustacea.

podium (*Arch*) A continuous low wall under a row of columns.

podium (*BioSci*) (1) In land vertebrates, the third or distal region of the limb; manus or pes; hand or foot. (2) Any foot-like structure, as the locomotor processes or tube feet of Echinodermata. Pl *podia*. Adj *podial*.

podo (*For*) Varieties of S and E African softwood (*Podocarpus*) whose heartwood is soft, non-durable, light yellowish-brown, straight-grained and has a uniform texture with no defined growth rings.

podo- (*Genrl*) See POD-.

podomere (*BioSci*) In Arthropoda, a limb segment.

podsol (*EnvSci*) A common soil type developed on siliceous mineral soil in areas of very high rainfall and low evaporation. The soil has an ash-coloured layer below the surface, from which minerals and clay particles have been washed, and an orange-brown deeper layer where some of these have accumulated. Sometimes there is an impermeable layer of oxidized iron. The overall reaction is acid and the soil usually supports only a calcifuge flora. Also *podzol*.

poecilitic (*Geol*) See *poikilitic*.

Poetsch process (*MinExt*) Freezing of waterlogged strata by circulation of refrigerated brine through the boreholes surrounding the section through which a shaft or tunnel is to be driven.

Poggendorff compensation method (*Chem*) A method of measuring an unknown emf by finding the point at which it just opposes the steady fall of potential along a wire.

pogo effect (*Space*) Unstable, longitudinal oscillations induced in launch system, mainly due to fuel SLOSHING and engine vibration.

Pogonophora (*BioSci*) A phylum of coelomate animals possibly related to the Hemichordata, without an alimentary canal or bony tissue, and having a simple nervous system. There is a muscular heart, and the blood contains haemoglobin. The extremely long bodies are divided into three sections, the posterior one serving to anchor the animals in the chitinous tubes in which they live, and the anterior part bearing well-developed tentacles which probably serve to gather food, digest it, and absorb the products of digestion.

poikil-, poikilo- (*Genrl*) Prefixes from Gk *poikilos*, many-coloured.

poikilitic texture (*Geol*) Texture in igneous rocks in which small crystals of one mineral are irregularly scattered in larger crystals of another, eg small olivines embedded in larger pyroxenes, as in some peridotites. Also *poecilitic texture*.

poikilo- (*Genrl*) See POIKIL-.

poikiloblastic (*Geol*) A textural term applicable to metamorphic rocks in which small crystals of one mineral are embedded in large crystals of another. The texture is comparable with the POIKILITIC TEXTURE of igneous rocks.

poikilocyte (*Med*) A malformed red blood cell. The presence of such cells in the blood (*poikilocytosis*) is seen in eg severe anaemia.

poikilohydric (*BioSci*) Lacking structures or mechanisms to regulate water loss and, hence, having water content determined rapidly by the water potential of the environment, as in algae, lichens, bryophytes, submerged vascular plants. Cf HOMOIOHYDRIC.

poikilosmotic (*BioSci*) Of an aquatic animal, being in osmotic equilibrium with its environment, the

concentration of its body fluids changing if the environment becomes more dilute or more concentrated. Marine invertebrates are frequently poikilosmotic. Cf HOMOIOSMOTIC.

poikilothermal (*BioSci*) See COLD-BLOODED.

point (*Agri*) The tip of a cultivator blade that is continually eroded as it is taken through the soil. It is replaceable on modern machinery.

point (*ElecEng*) In electric-wiring installations, a termination of the wiring for attachment to a light-fitting socket outlet or other current-using device.

point (*MathSci*) See DOUBLE POINT, MULTIPLE POINT, SINGULAR POINT ON A CURVE.

point (*Print*) The unit of measurement for type and materials. The 12-point em (the foundation of the system) measures 0·166 in (approx $\frac{1}{6}$ in) (4·21 mm); 1 point measures 0·0138 in (0·351 mm); 72 points or 6 ems measure 0·996 in (25·3 mm). The old type sizes such as nonpareil, brevier, pica (approx 6-point, 8-point, 12-point) have largely been discarded. Symbol *pt*.

point-and-click (*ICT*) Of or relating to a computer interface in which the user moves a cursor on a screen by manipulating a mouse and clicks on a mouse button to select or activate a program etc.

point biserial correlation (*MathSci*) A correlational technique used when one variable is numeric and the other is binary, eg age and sex, or income and whether or not self-employed.

point block (*Arch*) A high block of housing with a central core of lifts, stairs and services.

point-contact diode (*Electronics*) A semiconductor which uses a slender wire filament, touching a small piece of semiconductor material, to provide rectifying action. Junction capacitance is kept to a minimum and high-frequency operation is possible. Sometimes used as a detector in very-high-frequency and microwave devices.

point counter tube (*Electronics*) A radiation-counter tube, using gas ionization amplification, having its central electrode in the form of a point or a small sphere.

point defect (*Crystal*) See DEFECT.

pointed arch (*Build*) An arch which rises on each side from the springing to a central apex.

pointed ashlar (*Build*) A block of stone with face-markings made with a pointed tool.

pointer (*Build*) A tool used for raking out old mortar from brickwork joints prior to pointing.

pointer (*ICT*) (1) An ICON on the screen that moves as a pointing device such as a MOUSE is moved. This icon may be in the shape of an arrow but may change depending on the current operation being carried out by the USER. See WIMP. (2) A REGISTER or MEMORY LOCATION whose contents point to some other location, eg in a LINKED LIST pointers will be used to link the elements of the list together. In the context of data storage on a DISK, a pointer may be used to specify the storage location of a particular record by specifying its SECTOR and TRACK.

Pointers (*Astron*) Popular name for the two stars Alpha and Beta Ursae Majoris, lying in the Great Bear; they are roughly in line with the Pole Star and so help to identify it.

point estimation (*MathSci*) Estimating a specific value of a population statistic based on sample statistics.

point gamma (*ImageTech*) The contrast gamma for a specified level of brightness. In TV reception the instantaneous slope of the curve connecting log(input voltage) and log(intensity of light output).

pointing (*Build, CivEng*) The process of raking out the exposed jointing of brickwork and refilling with, normally, cement mortar.

point mutation (*BioSci*) A mutation that causes the replacement of a single base pair in DNA with another pair. Depending upon the change, this may or may not alter the amino acid that is specified by an *open reading frame*.

point of inflexion (*MathSci*) A point at which a curve changes from convex to concave. For the curve $y = f(x)$, the point where $x = a$ is a point of inflexion if $f''(a) = 0$, and the first non-zero higher order derivative at $x = a$ is of odd order.

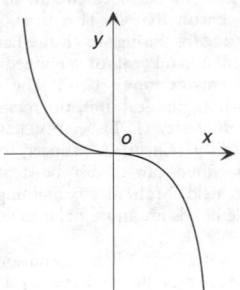

point of inflexion on a curve

point of lay (*Agri*) The time from which pullets are capable of egg-laying. Abbrev *POL*.

point of osculation (*MathSci*) A multiple point on a curve through which two branches, having a common tangent at the point, pass.

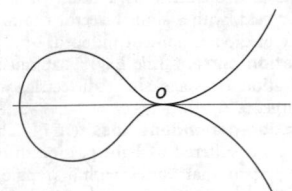

point of osculation

point-of-sale terminal (*ICT*) Input device that records and displays details about goods as they are sold, and transmits the information to a computer for stock control. Also *POS terminal*. See BAR CODE READER, EFTPOS, KIMBALL TAG.

point quadrat (*BioSci*) A device for measuring the canopy cover by lowering a thin pointed shaft into the vegetation many times.

points (*CivEng*) Movable tapered blades or tongues of metal for setting alternative routes of running rails. Each such blade is pivoted at the *heel*, its *toe* being locked against the stock rail, *facing points* if the train approaches the toe, *trailing points* if the train approaches the heel.

points at infinity (*MathSci*) Imaginary points added to projective geometry in order to be able to generalize that all pairs of lines intersect. See PARALLEL LINES.

point size (*Print*) See POINT.

point-to-multipoint (*ICT*) A system that uses radio to extend a public telephone network to subscribers in sparsely populated areas, typically covering up to 1024 subscribers within a radius of 500 km.

poise (*Phys*) CGS unit of viscosity. 1 poise is when a tangential force of 1 dyne per unit area maintains unit velocity gradient between two parallel planes. Equal to 1 dyne cm^{-2} s, or in SI units 0·1 N m^{-2} s. Named after the physicist Poiseuille. Symbol P. See COEFFICIENT OF VISCOSITY, VISCOSITY.

Poiseuille's formula (*Phys*) An expression for the volume of liquid per second Q, which flows through a capillary tube of length L and radius R, under a pressure P, the viscosity of the liquid being η:

$$Q = \frac{\pi P R^4}{8 L \eta}$$

Strictly, it is limited to laminar flow only. It is widely applied to non-Newtonian liquids which give an *apparent viscosity*.

poison (*Med*) Any substance or matter which, introduced into the body in any way, is capable of destroying or seriously impairing life. Poisons include products of decomposition or of bacterial organisms, and in some cases the organisms themselves; also very numerous chemical substances forming the residue of industrial, agricultural and other processes (see POLLUTION). Poisons are generally classified as irritants (eg cantharides, arsenic) and corrosives (eg strong mineral acids, caustic alkalis); systemics and narcotics (eg prussic acid, opium, barbiturates, henbane); narcotic-irritants (eg nux vomica, hemlock). Among gases, carbon monoxide, hydrogen sulphide, sulphur dioxide, ammonium sulphide and numerous others (eg fumes of leaded petrol) are of significance in industry and daily life. See WAR GASES.

poison (*NucEng, Phys*) Any contaminating material which, because of high-absorption cross-section, degrades intended performance, eg fission in a nuclear reactor, radiation from a phosphor. Also nuclear poison. See CATALYTIC POISON and panel on NUCLEAR REACTORS.

poisoning (*ChemEng, MinExt*) Loading of resin sites in ion exchange with ions of unwanted species which therefore prevent the capture of those required by the process. In liquid–liquid ion exchange, fouling of the organic solvent in similar manner. In chemical reactions, the absorption on a catalyst surface of any substance that diminishes the catalytic reaction rate.

Poisson distribution (*MathSci*) A probability distribution often applied to the number of occurrences of particular (esp rare) events in a particular time period.

Poisson's equation (*MathSci*) The equation $\nabla^2 V = -4\pi\rho$, where ∇ is the vector operator del, V is the electric or gravitational potential function, and ρ is the charge or mass density respectively. Cf LAPLACE'S EQUATION.

Poisson's ratio (*Phys*) One of the four elastic constants of an isotropic material, symbol v. It is defined as the ratio of the lateral contraction per unit breadth, w, to the longitudinal extension per unit length, l, when a piece of the material is stretched, ie

$$v = -\frac{\Delta w/w}{\Delta l/l}$$

Its value can range from -1 (no change in proportions) to $+\frac{1}{2}$ (no change in volume), but for most substances its value lies between 0·2 and 0·4. The relationships between Poisson's ratio, YOUNG'S MODULUS E, the SHEAR MODULUS G and the BULK MODULUS K are given by

$$v = \frac{E}{2G} - 1 = \frac{3K - 2G}{2(2K + G)}$$

poke (*ICT*) To store data in a single specified MEMORY LOCATION. Cf PEEK.

pokeweed mitogen (*BioSci*) Mitogenic lectins obtained from *Phytolacca americana*. There are five different substances of which one is active for both T- and B-lymphocytes, but the others act only on T-lymphocytes.

polar axis (*Astron*) (1) The diameter of a sphere which passes through the poles. (2) In an equatorial telescope, the axis, parallel to the Earth's axis, about which the whole instrument revolves in order to keep a celestial object in the field.

polar axis (*Crystal*) A crystal or symmetry axis to which no two- or fourfold axes are normal; thus the arrangements of faces at the two ends of such an axis may be dissimilar. The principal axis of tourmaline is a polar axis of threefold symmetry, the top of the crystal being terminated by pyramid faces, the bottom end by a single plane in some cases.

polar body (*BioSci*) A small cell produced during the maturation divisions (*meiosis*) of the ovum containing one daughter set of chromosomes. The polar body takes no further part in gametogenesis.

polar bond (*Chem*) Covalent bond between elements of differing electronegativity.

polar control (*Aero*) See TWIST AND STEER.

polar co-ordinates (*MathSci*) A method of locating a point, *P*, in space in terms of the orientation and length of the line joining it to the origin, *O*, defined in terms of rectangular cartesian co-ordinates as follows. In the plane: (r,θ), where *r* is the distance, and θ is the angle from the *x*-axis to *OP*. In three dimensions, *spherical polar coordinates* of *P* are (r,θ,φ), where *r* is the distance *OP*, θ is the angle between the *z*-axis and *OP* (ie the *colatitude*), and φ is the angle between the *x*-axis and the projection *OP* on the *xy* plane (ie the *longitude*). The *cylindrical polar coordinates* of *P* in three dimensions are (r,θ,z), where *r* is the projection of *OP* on the *xy* plane, θ is the angle between the *x*-axis and that projection, *z* is the height above the *xy* plane. It should be noted that these systems use the symbols in different senses. Other notations are also sometimes used.

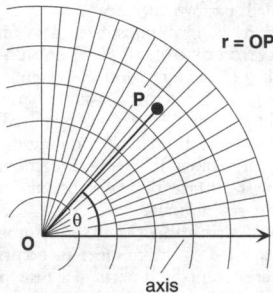

polar co-ordinates Two dimensions.

polar crystal (*Crystal*) A crystal, such as sodium chloride, with ionic bonding between atoms.

polar curve (*Phys*) A curve drawn in polar co-ordinates showing the field distribution around a radiator or receiver of radiated energy, to show its directional efficacy: eg a contour of equal field strength around a transmitting antenna; or, in a receiving antenna, a contour path of a mobile transmitter producing a constant signal at the receiver. Similar diagrams are prepared for sound fields round electro-acoustical transducers, for sensitivity curves around eg photocells or scintillation counters, and for all other energy detectors or radiators with directional sensitivity. Also *polar diagram*.

polarimeter (*Chem*) An instrument in which the optical activity of a liquid is determined by inserting Nicol prisms in the path of a ray of light before and after traversing the liquid.

polarimetry (*Chem*) The measurement of optical activity, esp in the analysis of sugar solutions.

Polaris (*Astron*) The brightest star in the constellation Ursa Minor which currently lies (by chance) within 1° of the north celestial pole. Its altitude is approximately equal to the latitude of the observer. Owing to the effects of PRECESSION, the north pole is drifting away from this particular star. Also *North Star*, *Pole Star*.

polariscope (*Phys*) An instrument for studying the effect of a medium on polarized light. Interference patterns enable elastic strains in doubly refracting materials to be analysed. It may consist of a polarizer and an analyser, with facilities for placing transparent specimens between them. The analyser is usually a Nicol prism. The polarizer may be also a Nicol prism or a PILE of plates. Modern polariscopes use light-polarizing films, eg POLAROID film, instead. See PHOTOELASTICITY.

polarity (*BioSci*) The existence of a definite axis.

polarity (*Chem*) A permanent property of a molecule which has an unsymmetrical electron distribution. All heteronuclear diatomic molecules are polar. See DIPOLE MOMENT.

polarity (*Phys*) (1) Distinction between positive and negative electric charges. (2) General term for the

difference between two points in a system which differ in one respect, eg potentials of terminals of a cell or electrolytic capacitor, windings of a transformer, video signal, legs of a balanced circuit, ac phase.

polarity chron (*Geol*) The time span of a POLARITY CHRONOZONE during which the Earth's magnetic polarity was predominantly of, or remained of, one polarity.

polarity chronozone (*Geol*) The fundamental polarity chronostratigraphical unit; the rocks of a polarity chron.

polarity diversity (*ICT*) See DIVERSITY RECEPTION.

polarizability (*Chem*) A property of a molecule that its electron cloud can readily be deformed by an external electron field. Molecules containing heavy atoms and/or multiple bonds are more polarizable than those which do not.

polarization (*Chem*) The separation of the positive and negative charges of a molecule by an external agent.

polarization (*ICT*) Orientation of the electric field of a horizontally propagating electromagnetic wave. The polarization of radio transmissions is normally horizontal, vertical or circular (in which two spatially orthogonal components with a phase difference of 90° are radiated simultaneously).

polarization (*Phys*) (1) Non-random orientation of electric and magnetic fields in an electromagnetic wave. (2) Change in a dielectric as a result of sustaining a steady electric field, with a similar vector character; measured by density of dipole moment induced.

polarization current (*ElecEng*) That causing or caused by polarization (soakage) in a dielectric, with possible late discharge.

polarization-dependent loss (*ICT*) That part of the attenuation suffered by light passing through an OPTICAL FIBRE system that varies with its angle of polarization, normally originating in couplers and multiplexers rather than the fibre itself.

polarization error (*ICT*) Error in determining the direction of arrival of radio waves by a direction-finder when the desired wave is accompanied by downward components that are out of phase. Generally occurs at night.

polarization mode dispersion (*ICT*) Differential GROUP DELAYS between the principle states of polarization of an OPTICAL FIBRE, causing spreading of received pulses.

polarized beam (*Phys*) (1) In electromagnetic radiation, a beam in which the vibrations are partially or completely suppressed in certain directions. (2) In CORPUSCULAR RADIATION, a beam in which the individual particles have non-zero spin and in which the distribution of the values of the spin component varies with the direction in which they are measured.

polarized capacitor (*ElecEng*) Electrolytic capacitor designed for operation only with fixed polarity. The dielectric film is formed only near one electrode and thus the impedance is not the same for both directions of current flow.

polarized light microscopy (*Min*) A method of examining thin, transparent materials using polarized light. Esp useful for mineral sections, biomaterials and polymers, all of which may show birefringence. Can give data for identification of material, possibly improve contrast (eg spherulite structure in polymers) and aid microstructure determination. One variant of method used for surfaces and thin transparent biological material is NOMARSKI INTERFERENCE.

polarized plug (*ElecEng*) One which can be inserted into a socket in only one position.

polarized relay (*ElecEng*) See RELAY.

polarizer (*Phys*) Sheet of birefringent polymer (esp POLAROID) used as a light filter to produce either plane- or circularly polarized light. Combined with analyser for eg photoelastic analysis. See NICOL PRISM, PILE.

polarizibility (*Phys*) The degree to which a medium may be electrically (or magnetically) polarized. Electric polarization of a material arises from the displacement of dipoles, ions and electrons within the constituent matter. As a result it is a function of frequency, varying strongly when any component undergoes resonance.

polarizing angle (*Phys*) See BREWSTER ANGLE.

polarizing filter (*ImageTech*) A filter allowing the passage of light which is polarized in one direction only; used in photography for the control of surface reflections and for darkening blue skies and also for image separation in some systems of stereoscopy. Also *polar filter*.

polarizing microscope (*BioSci*) A microscope that contains a polarizer (eg a Nicol quartz prism) and an analyser, and that can thus be used to examine birefringent objects.

polarizing monochromator (*Astron*) A filter consisting of a succession of quartz crystals and calcite or Polaroid sheets; the light passing through is restricted to a narrow band, useful in observing the solar chromosphere. Also *Lyot filter*.

polar line and plane (*MathSci*) Of a point (the POLE) with respect to a conic or quadric respectively: the locus of harmonic conjugates of the pole with respect to the pairs of points intersected on the conic or quadric by lines through it. When the pole is outside the conic or quadric, the polar is the line or plane determined by the points of contact of tangents from the pole to the conic or quadric.

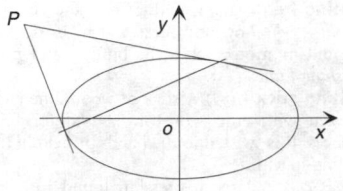

polar line and plane

polar molecule (*Phys*) A molecule with unbalanced electric charges, usually valency electrons, resulting in a dipole moment and orientation.

polar nuclei (*BioSci*) The two (typically haploid) nuclei that migrate from the poles to the centre of the embryo sac of an angiosperm and there fuse with the second male gamete to form the, typically triploid, primary ENDOSPERM nucleus.

polarograph (*Chem*) An instrument which records the current–voltage characteristic for a polarized electrode. It is used in chemical analysis.

Polaroid (*ImageTech*) TN for a range of photographic and optical products, including a transparent light-polarizing plastic sheet (*Polarscreen*) and methods of INSTANT PHOTOGRAPHY in black and white and colour.

Polaroid camera (*ImageTech*) TN for a camera for INSTANT PHOTOGRAPHY using a colour DIFFUSION-TRANSFER REVERSAL process.

Polaroid sheet (*Plastics*) TN for thermoplastic sheet containing optically active compound which acts as filter for visible light, giving either plane- or circularly polarized light on transmission. Birefringent material which has replaced Nicol prisms in most analytical applications, eg photoelasticity, polarized light microscopy. Two sheets (polarizer and analyser) are used to examine birefringent materials, such as polymer products, sandwiched between them.

polaron (*Electronics*) An electron in substance, trapped in potential well produced by polarization charges on surrounding molecules; analogous to *exciton* in semiconductor.

polar platform (*Space*) An unpressurized platform in SUN-SYNCHRONOUS ORBIT, used particularly for remote sensing, which is capable of in-orbit servicing and payload exchange.

polar reciprocation (*MathSci*) Transformation in which lines are replaced by points and points by lines, usually by replacing a line by its pole and a point by its polar with respect to a conic. Also *dualizing*.

polar response curve (*Acous*) The curve which indicates the distribution of the radiated energy from a sound reproducer for a specified frequency. Also the relative response curve of a microphone for various angles of incidence of a sound wave for a given frequency. Generally plotted on a radial decibel scale.

polar second moment of area (*Eng*) The polar second moment of area is a measure of resistance to twisting of a section. The polar second moment of area J about an axis zz, perpendicular to the plane of the section, is found from

$$J_{zz} = \int Az^2\, dA$$

where dA is an element of the total area and z the distance from zz. Cf SECOND MOMENT OF AREA.

Polar Sequence (*Astron*) See NORTH POLAR SEQUENCE.

polar wandering (*Geol*) Movement of the magnetic poles and of the poles of the Earth's rotation through geological time. Can be partly explained by plate tectonics. See panel on PLATE TECTONICS.

polder (*CivEng*) A piece of low-lying land reclaimed from the water or artificially protected therefrom.

pole (*Astron*) See CELESTIAL POLE.

pole (*BioSci*) (1) Generally, a point, apex or an opposite point (as aboral pole). (2) Specifically, in mitosis and meiosis, one end of the spindle, where the spindle microtubules come together.

pole (*ElecEng*) (1) That part of an electric machine which carries one of the exciting windings. (2) A term sometimes used to denote one of the terminals of a dc generator, battery or electric circuit, eg *negative pole, positive pole*. (3) A wooden, steel or concrete column for supporting the conductors of an overhead transmission or telephone line.

pole (*Electronics*) That part of the anode between adjacent cavities in a multiple cavity magnetron.

pole (*Genrl*) Generally, the axis or pivot on which anything turns; one of the ends of the axis of a sphere, esp of the Earth.

pole (*MathSci*) (1) Of a function: if a function $f(z)$ is expanded in LAURENT'S EXPANSION, in an annular domain surrounding a singularity $z = a$,

$$f(z) = \sum_{n=1}^{\infty} b_n (z-a)^n + \sum_{n=1}^{\infty} c_n (z-a)^{-n}$$

The second term is called the *principal part* of $f(z)$ at $z = a$. If the principal part has a finite number of terms, m say, then $f(z)$ is said to have a pole of order m at $z = a$. If the principal part is an infinite series, $f(z)$ is said to have an *isolated essential singularity* at $z = a$. (2) Of a line or plane with respect to a conic or quadric: see POLAR LINE AND PLANE. (3) Of a circle (or part thereof) on a sphere: the ends of a diameter of the sphere normal to the circle.

pole (*Phys*) The part of a magnet, usually near the end, towards which the lines of magnetic flux apparently converge or from which they diverge, the former being called a *south pole* and the latter a *north pole*.

pole arc (*ElecEng*) The length of the pole face of an electric machine measured circumferentially around the armature surface.

pole barn (*Agri*) A barn based on a pattern of poles set in the ground that are used to support rafters and roofing, plank walls and other framing.

pole bevel (*ElecEng*) A portion of the pole face of an electric machine, near the pole tip, which is made to slope away from the armature surface instead of being concentric with it, the object being to obtain a more satisfactory shape of flux wave.

pole-changing control (*ElecEng*) A method of obtaining two or more speeds from an induction motor by altering stator–winding connections to give a different number of poles.

pole core (*ElecEng*) See POLE SHANK.

pole end-plate (*ElecEng*) A thick plate placed at each end of the laminations of a laminated pole.

pole face (*ElecEng*) That surface of the pole piece of an electric machine which faces the armature.

pole face loss (*ElecEng*) Losses which occur in the iron of the pole face of an electric machine due to the periodic flux variations caused by the armature teeth.

pole-finding paper (*ElecEng*) Paper prepared with a chemical solution, which, when placed across two poles of an electric circuit, causes a red mark to be made where it touches positive.

pole horn (*ElecEng*) The portion of the pole shoe of an electric machine which projects circumferentially beyond the pole shank.

pole piece (*ElecEng*) Specially shaped magnetic material forming an extension to a magnet, eg the salient poles of a generator, motor or relay, for controlling flux.

pole pitch (*ElecEng*) The distance between the centre lines of two adjacent poles on an electric machine; it is measured circumferentially around the surface of the armature of the machine.

pole plate (*Build*) A horizontal member supporting the feet of the common rafters and carried upon the tie beams of the trusses.

pole shading (*ElecEng*) See SHADED POLE.

pole shank (*ElecEng*) The part of a pole piece around which the exciting winding is placed. Cf POLE SHOE.

pole shim (*Phys*) See SHIM.

pole shoe (*ElecEng*) That portion of the pole piece of an electric machine which faces the armature; it is frequently detachable from the pole shank.

Pole Star (*Astron*) See POLARIS.

pole strength (*Phys*) The force exerted by a particular magnetic pole upon a *unit pole* situated at unit distance from it. See COULOMB'S LAW FOR MAGNETISM.

pole tip (*ElecEng*) The edge of the pole face of an electric machine which runs parallel to the axis of the machine. Hence *leading pole tip, trailing pole tip.*

pole tip (*ImageTech*) See HEAD TIP.

pole top switch (*ElecEng*) A switch which may be mounted at the top of a transmission line pole; arranged for hand operation by some mechanical device.

poling (*Eng*) (1) In the fire refining of copper, the impurities are eliminated by oxidation and the residual oxygen may in turn be removed by the reducing gases produced when green logs (poles) are burned in the molten metal. If the final oxygen content is too high the metal is said to be *underpoled,* if too low, *overpoled*; when the desired balance of small residual oxygen content is achieved it is termed *tough pitch.* Poling with green tree trunks is now superseded by injection of reducing gas mixtures which has similar effect. (2) The application of an electric field to a ferroelectric material in order to establish a remanent polarization. *Depoling* is the removal of polarization.

poling boards (*CivEng*) Rough vertical planks used to support the sides of narrow trenches after excavation; placed in pairs on opposite sides of the trench at intervals along its length, each pair being wedged apart by wooden struts.

poling boards (*MinExt*) Fore-poling boards are used in tunnelling through loose (running) rock, and are driven horizontally ahead to support roof.

polioencephalitis (*Med*) Inflammation of the grey matter of the brain.

polioencephalomyelitis (*Med*) Inflammation of the grey matter both of the brain and of the spinal cord.

poliomyelitis (*Med*) Inflammation of the grey matter of the spinal cord. It is due to infection, chiefly of the motor cells of the spinal cord, with a virus; characterized by fever and by variable paralysis and wasting of muscles. Also *acute anterior poliomyelitis, infantile paralysis.*

polished face (*CivEng*) A fine surface finish produced on granite, or other natural stone, by rubbing down a sawn face with iron sand, fine grit and polishing powder in turn.

polished foil (*Print*) Blocking foil, which leaves a glossy transfer. These are available from eggshell to full gloss and will mark surfaces ranging from cheap paper to polypropylene.

polished plate (*Build*) A superseded sheet glass over $\frac{3}{16}$ in thick, used for shop windows.

polished rod (*MinExt*) In an oil pumping station the actuating rod which passes through the stuffing box at the top of the well; usually attached at the top to the cable going round the HORSEHEAD; at the lower end joined to the PONY RODS which are connected to the displacement pump.

polished specimen (*MinExt*) Characteristic hand specimen of ore, metal, alloy, compacted powder, etc, one face of which is ground plane and mirror smooth by abrasive powder and/or polishing laps, or electrolytic methods.

polishing (*Eng*) Sample preparation method for rigid materials when examined by reflection optical microscopy. Sample is abraded with succession of finer-grade emery papers, then diamond polished (to 0·25 μm) to give highly reflective surface. Art in method is to remove traces of all previous scratches. Usually followed by etching to expose microstructure (grains etc).

polishing head (*Eng*) A headstock and spindle carrying a *polishing wheel* or mop rotated at high speed by belt drive or built-in motor; used for buffing and polishing. Also *polishing lathe.*

polishing stick (*Eng*) A stick of wood, one end of which is charged with emery or rouge, used for finishing small surfaces. It is twisted in the hands or held in the chuck of a drilling machine.

polje (*Geol*) A large depression found in some limestone areas, due in part to subsidence following underground solution.

poll-adze (*Build*) An adze having a blunt head or poll opposite to the cutting edge.

pollard (*For*) A tree that has been topped to obtain a head of shoots, usually above the height to which browsing animals can reach.

pollards (*Agri*) A mixture of fine bran and some endosperm tissue produced as a by-product of grain milling.

polled (*Agri*) Animals without horns, either naturally or from being dehorned.

pollen (*BioSci*) The microspores of seed plants. The individual pollen grains may contain an immature or mature male gametophyte. See ACRITARCHS.

pollen analysis (*BioSci*) A method of investigating the past occurrence and abundance of plant species by a study of pollen grains and other spores preserved in peat and sedimentary deposits. Also *palynology.* See SPOROPOLLE-NIN.

pollen chamber (*BioSci*) A cavity in the micropylar end of the nucellus of the ovule of some gymnosperms in which pollen grains lodge as a result of pollination and where they develop further and germinate.

pollen count (*EnvSci*) A graded assessment of the level of plant pollen in the atmosphere; of importance to sufferers from HAY FEVER.

pollen flower (*BioSci*) A flower that produces no nectar but has abundant pollen which attracts pollinating insects.

pollen mother cell (*BioSci*) A MICROSPOROCYTE of a seed plant. Such cells are typically diploid and are found in the pollen sac; they divide by meiosis to form a TETRAD of pollen grains.

pollen sac (*BioSci*) A chamber in which pollen grains (microspores) are formed in seed plants, eg such chambers in the ANTHER.

pollen tube (*BioSci*) A tubular outgrowth of the intine, produced on the germination of a pollen grain. In angiosperms it grows to the embryo sac and there delivers the male gamete(s), ie siphonogamy.

poll-evil (*Vet*) Inflammation of the bursa of the ligamentum nuchae of the horse.

pollex (*BioSci*) The innermost digit of the anterior limb in Tetrapoda.

pollination (*BioSci*) The transfer of pollen from the pollen sac to the micropyle in gymnosperms (see POLLINATION DROP) or to the stigma in angiosperms, usually by means of wind, insects, birds, bats or water.

pollination drop (*BioSci*) A drop of sugary fluid that is secreted into the micropyle at the time of pollination in many gymnosperms, and which traps pollen grains that may then float up to the nucellus or be drawn there by the reabsorption of the drop.

polling (*ICT*) Automatic, sequential testing of each potential source of input to a computer, usually a terminal, to find its operational status. It may be ready to transmit data, in use or not in use. See MULTIACCESS.

pollinium (*BioSci*) A mass of pollen grains that is held together by a sticky secretion or retained within the pollen sac wall. It is transported, usually by an insect, in the pollination of eg orchids.

pollucite (*Min*) A rare hydrated alumino-silicate of caesium, occurring as clear colourless or white crystals with cubic symmetry. It occurs in granite pegmatites.

pollution (*EnvSci*) Modification of the environment by release of noxious materials, rendering it harmful or unpleasant to life.

Pollux (*Astron*) A bright orange giant star in the constellation Gemini. Distance 10·7 pc. Also *Beta Geminorum*.

poloidal field (*NucEng*) The magnetic field generated by an electric current flowing in a ring. Cf TOKAMAK, TOROIDAL FIELD.

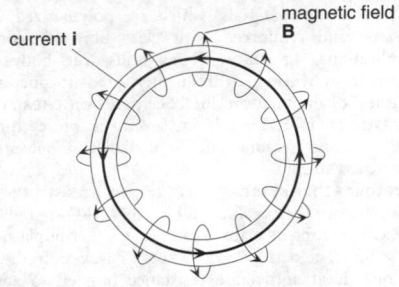

poloidal field

polonium (*Chem*) Radioactive element, symbol Po, at no 84. Important as an α-ray source relatively free from γ-emission.

poly- (*Chem*) (1) Containing several atoms, groups, etc. (2) A prefix denoting a polymer, eg polyethene.

poly- (*Genrl*) Prefix from Gk, *polys*, many.

polyacetals (*Plastics*) See POLYOXYMETHYLENE.

polyacetylene (*Plastics*) Electrically conductive polymer possessing conjugated double bonds along its backbone chain. Material still at developmental stage owing to poor mechanical properties, processing problems, cost, etc.

polyacrylamide gel (*BioSci*) Transparent cross-linked acrylamide gels. See POLYACRYLAMIDE GEL ELECTROPHORESIS.

polyacrylamide gel electrophoresis (*BioSci*) A technique used for separating nucleic acids or proteins on the basis of charge, shape and size. The highly cross-linked polymer of acrylamide and bis-acrylamide forms a gel matrix through which macromolecules move under the influence of an electric field. Proteins are frequently separated in the presence of the detergent, *sodium dodecylsulphate*, which forms polyanionic complexes with the protein whose mobility is a simple function of size. Abbrev *PAGE*.

polyacrylate (*Chem*) A polymer of an ester of acrylic acid or its chemical derivatives.

polyacrylic acid (*Chem*) Polyelectrolyte widely used as additive in range of different materials (eg building products) to increase viscosity. Unlike most synthetic polymers, it is water-soluble owing to charged carboxylic groups on its backbone chain.

polyacrylimide (*Chem*) A derivative of methacrylic acid and methacrylonitrile copolymer, where treatment with urea creates cyclic aliphatic units in the backbone chain (with imide functional units). Such partly cyclized material offers enhanced heat resistance (to 160°C), and is used mainly in rigid foam products, eg composite helicopter blade cores. Urea also makes the product foam by gas formation, helping to press other components to shape in such uses.

polyacrylonitrile (*Chem*) Fibre-forming polymer (under the name 'acrylic') which is also the starting material for carbon fibre. Abbrev *PAN*.

polyadelphous (*BioSci*) A flower having the stamens joined by their filaments into several separate bundles.

polyalkane (*Chem*) A hydrocarbon polymer essentially of long-chain molecules with only saturated carbon atoms in the main chain. More commonly called POLYOLEFINS.

polyamide (*Textiles*) Natural or synthetic fibres composed of polymers having the same amide group (–CO–NH–) repeated along the chain. Examples of the natural fibres are silk, wool and hair. For synthetic polyamides see NYLON. Polyamides made with *aromatic* groups attached to the amide links are assigned to a different class, the ARAMID FIBRES. See panel on HIGH-PERFORMANCE POLYMERS.

polyandrous (*BioSci*) A flower having a large, indefinite number of stamens.

polyandry (*BioSci*) The practice of a female animal mating with more than one male.

polyanion (*BioSci*) Any macromolecule carrying many negative charges. The most common in cell biological systems is nucleic acid.

polyarch (*BioSci*) A stele having many protoxylem strands, as in many roots, esp of monocots.

polyarteritis nodosa (*Med*) Disease characterized by inflammation of the arteries in many parts of the body causing fever, malaise and often leading to renal failure and hypertension. Thought to be an auto-immune process.

polyarthritis (*Med*) Inflammation affecting several joints at the same time.

polybasic acids (*Chem*) Acids with two or more replaceable hydrogen atoms in the molecule.

polybasite (*Min*) Sulphide of silver and antimony, often with some copper, crystallizing in the monoclinic system.

polybenzimidazoles (*Plastics*) Family of heat-resistant polymers with aromatic heterocyclic structures in the main chain. T_g generally above 400°C. See panel on THERMOSETS.

polybutadienes (*Plastics*) See panel on POLYMERS.

polybutene (*Plastics*) Isotactic, crystalline polyolefin made from 2-methyl propylene. Of very high molecular mass (up to 3 million), it shows high creep resistance and excellent electrical and moisture-resisting properties. Mainly used for extruded small-bore hot- and cold-water pipes.

polybutylene terephthalate (*Plastics*) Step polymer made from 1,2-butanediol and terephthalic acid and used as a thermoplastic material. Abbrev *PBT*.

polycarbonate (*Plastics*) Amorphous transparent and rigid polymer, made by step-growth mechanism from carbonate functional group (–O–CO–O–). Normally tough but notch sensitive, so care needed in design. Rigid backbone chain with bis-phenol A aromatic group in repeat unit, so high T_g of 145°C. Widely injection moulded for battery cases, containers and housings for consumer products. TNs *Lexan, Makrolon*. Abbrev *PC*. See CR39.

polycarpellary (*BioSci*) Consisting of many carpels.

polycarpic (*BioSci*) Potentially able to fruit many times, as most perennial plants are. Cf MONOCARPIC. See HAPAXANTHIC.

polycarpous (*BioSci*) See APOCARPOUS.

polycation (*BioSci*) Any macromolecule with many positively charged groups. At physiological pH the most commonly used in cell biology is poly-L-lysine; this is often used to coat surfaces, thereby increasing the adhesion of cells (that have net negative surface charge).

Polychaeta (*BioSci*) A class of marine Annelida that have locomotor appendages (parapodia) bearing numerous setae. There is usually a distinct head. The perivisceral cavity is subdivided by septa; the sexes are generally separate, with numerous gonads; and development after external fertilization involves metamorphosis, with a free-swimming trochosphere larva, eg marine bristle-worms.

polychasium (*BioSci*) A CYMOSE INFLORESCENCE in which the branches arise in sets of three or more at each node.

polychlorinated biphenyls (*Chem*) Chlorinated hydrocarbons of high stability and good dielectric properties which are used as a coolant in transformers etc. Highly persistent in the environment if not disposed of correctly, having been found worldwide and causing the death of many animals.

polychloroprene (*Chem*) Polymer made from chlorine-substituted butadiene, with repeat unit structure same as polyisoprene (Cl replaces methyl group). Important rubber for its resistance to weathering, so used in bridge bearings, rubber dinghies, etc.

polychromasia (*Med*) The diffuse bluish staining of the young immature red blood corpuscles with eosin and methylene blue. Also *polychromatophilia*.

polychromatic (*Genrl*) Many-coloured.

polychromy (*Arch*) Any form of decoration employing several colours, but esp applicable to Late Gothic Revival (1850–60), when colour was achieved, not only by applying paint, but by using contrasting bands of brickwork externally, and tiles and mosaics on floors and walls. This method of integrating colour has no parallels in early GOTHIC architecture and has become the hallmark of the Victorian era.

polyclimax theory (*BioSci*) Idea proposed by A G Tansley in 1939, that not one but several climax vegetation types are possible in one climatic region because the environment is influenced by local factors such as soil and the activity of animals. See CLIMAX.

polyclonal activators (*BioSci*) General term for substances that activate many clones of lymphocytes as opposed to an antigen which only activates clones which have receptors which recognize it. See LECTIN, MITOGEN.

polyclonal antibody (*BioSci*) Used to describe whole SERUM raised against a particular antigen to distinguish it from a MONOCLONAL ANTIBODY. Such serum contains many different antibodies against different EPITOPES of the antigen.

polycormic (*BioSci*) A woody plant having several strong vertical trunks.

polycotyledonous (*BioSci*) Having more than two cotyledons, eg pine.

polycrystalline materials (*Eng*) Common state of crystalline metals and ceramics, formed by mass of interlocking single crystals. Their crystallographic axes are usually random, so giving isotropic properties. If preferential orientation of the axes occurs, properties will be ANISOTROPIC (eg rolled metal sheet).

polycyclic (*BioSci*) Vascular tissues of a stem having two or more concentric DICTYOSTELES, SOLENOSTELES or rings of vascular bundles.

polycyclic (*Chem*) Containing more than one ring of atoms in the molecule.

polycystic (*Med*) Containing many cysts, eg a *polycystic kidney*.

polycythaemia (*Med*) Increase in the haemoglobin content of the blood, because of either a reduction in plasma volume or an increase in red cell numbers. US *polycythemia*. Also *erythraemia*.

polydactylism (*Med*) Having more than the normal number of digits. Also *polydactyly*. Adj *polydactylous*.

polydentate (*Chem*) Of a LIGAND, attaching via more than one atom, resulting in *chelate* formation.

polydipsia (*Med*) Excessive thirst.

polyelectrolyte (*Chem*) Polymer in which some or all of the repeat units possess ionic groups. Includes many biomaterials (eg proteins and polypeptides) as well as synthetics (eg polyacrylic acid).

polyembryony (*BioSci*) The development of more than one embryo in one ovule, seed or fertilized ovum.

polyester (*ImageTech, Textiles*) Polyester fibres and films are linear polymers which have the ester group (–CO–O–) repeated along the chain. The commonest polyester fibre is made by the polymerization of ethylene diol and terephthalic acid. Used as a film BASE having greater mechanical strength and stability than cellulose acetate.

polyester amide (*Plastics*) A polymer in which the structural units are linked by ester and amide (and/or thio-amide) groupings.

polyester rubber (*Plastics*) Thermoplastic elastomer made from polytetramethylene ether glycol and polybutylene terephthalate by block copolymerization. By varying block size and polyester composition, a wide range of hardnesses can be made. It is used in moulded products for its high abrasion resistance (eg in skimobile caterpillar tracks). Not to be confused with acrylic rubbers.

polyesters (*Plastics*) Polymers having ester functional group (–CO–O–) in repeat unit. Includes thermoplastics polyethylene terephthalate, polybutylene terephthalate, certain liquid crystal polymers, polyester rubbers, and thermosets acrylic rubbers, polyester resins, etc.

polyester thermoset resins (*Plastics*) A range of polymers formed by the step polymerization of polyhydric alcohols and polycarboxylic acids or anhydrides. Maleic and fumaric acids and ethylene and propylene alcohol are the usual starting materials, which are polymerized with eg styrene and reinforced with glass fibre for composite applications, eg canoes, boat hulls, car body panels. Phthalic anhydride is frequently used to increase the stiffness of the final product. See panel on THERMOSETS.

polyestrous (*BioSci*) Exhibiting several estrous cycles during the breeding season. UK, but obsolete, *polyestrus*. Cf MONOESTROUS.

polyether ether ketones (*Plastics*) Heat-resistant polymers with benzene rings linked by ether (–O–) and ketone (–CO–) groups in main chain. Tough thermoplastics with $T_g \approx 144°C$ and $T_m \approx 335°C$ for Victrex polyether ether ketone. Heat and flame resistance has led to aerospace applications.

polyether imides (*Plastics*) Type of melt-processable polyimide with ether links in main chain. TN Ultem.

polyethers (*Plastics*) Long-chain glycols made from alkene oxides such as ethylene oxide. Used as intermediates in the production of POLYURETHANES, antistatic agents and emulsifying agents. See PTMEG.

polyethylene (*Plastics*) A flexible waxy translucent thermoplastic, formed by the polymerization of ethene (ethylene), that is a good insulator, easily moulded and blown, and resistant to acids. It is used in the form of film or sheeting to package food products, clothing, etc, and to make pipes, moulded products, and electrical insulators. In the UK it is also known by the TN Polythene. Types of polyethylenes include: HIGH-DENSITY POLYETHYLENE, linear low-density polyethylene, LOW-DENSITY POLYETHYLENE, medium-density polyethylene, SCLAIR, ULTRAHIGH MOLECULAR MASS POLYETHYLENE; copolymers include: EPR, ETHYLENE VINYLACETATE COPOLYMER, , polypropylene–copolyethylene; fibres include: Dyneema, Spectra. Also *polyethyene*.

polyethylene glycol (*BioSci*) A hydrophilic hydrocarbon polymer that interacts with the membranes of cells in culture, and promotes intercellular fusion and production of viable hybrids. Used particularly in the formation of hybridomas for MONOCLONAL ANTIBODY production. Abbrev *PEG*.

polyethylene oxide (*Chem*) Water-soluble polymer with repeat unit –CH$_2$–CH$_2$–O–. Widely used in cosmetics, toothpaste, etc, as an inert matrix. Produces laminar flow in turbulent water streams at very low concentrations, so used by firefighters for increasing output from hose etc. Abbrev *PEO*.

polyethylene terephthalate (*Plastics*) The chemical name for the polyester forming the basis of Terylene and Melinex. Abbrev *PET*.

polyformaldehydes (*Plastics*) See POLYOXYMETHYLENE

polygamous (*BioSci*) (1) In plants, having staminate, pistillate and hermaphrodite flowers on the same plant and on distinct individual plants. (2) In animals, mating with more than one of the opposite sex during the same breeding season. N *polygamy*.

polygenes (*BioSci*) The genes that control *quantitative characters* and whose individual effects are too small to be detected.

polygenic (*BioSci*) Of a character whose genetic component is determined by many genes with individually small effects.

polygon (*MathSci*) A many-sided closed plane figure with straight sides.

polygonal roof (*Build*) A roof which in plan forms a figure bounded by more than four straight lines.

polygonal rubble (*Build*) A form of ragwork in which the rubble wall is built up of stones having polygonal faces.

polygoneutic (*BioSci*) Having several broods in a year.

polygon of forces (*Phys*) A polygon whose sides are parallel to and proportional to forces acting at a point, the directions of the forces being cyclic around the polygon. The polygon is closed if the forces are in equilibrium, otherwise the closing side of the polygon is parallel and proportional to the equilibrant of the forces.

polygraph (*Psych*) A recording device which can pick up physiological changes in the form of electrical impulses and record them. See GALVANIC SKIN RESPONSE. Sometimes referred to, perhaps optimistically, as a *lie detector*.

polygynous (*BioSci*) Said of a male animal that mates with more than one female.

polyhalite (*Min*) $K_2MgCa_2(SO_4)_4.2H_2O$. A hydrated sulphate of potassium, magnesium and calcium. Found in salt deposits.

polyhedron (*MathSci*) A three-dimensional solid with polygonal faces. Cf *regular convex solids*.

polyhydantoin (*Plastics*) Heat-resistant polymer having mixture of aromatic and aliphatic rings in its main chain. Used as wire enamel and insulating film for eg electric motors. Can also be used as solder-resistant base for flexible circuitry.

polyhydric (*Chem*) Containing a number of hydroxyl groups in the molecule, eg *polyhydric alcohols*, alcohols with three, four or more hydroxyl groups.

polyimides (*Chem*) Polymers with the functional –CO–NR–CO– group, and therefore akin to polyamides. Commercial materials are aromatic, so are heat resistant to over 300°C. Film used in electronic circuitry (eg flexible circuits), bulk in rod form. See panel on THERMOSETS.

polyisoprene (*Chem*) Main polymer in natural rubber, and also available synthetically. Also *cis-polyisoprene, isoprene rubber*. See panel on ELASTOMERS.

polylysine (*BioSci*) A polycationic polymer with multiple positive charges, used in cell biology and microscopy to mediate adhesion of living or fixed cells to culture surfaces or glass.

polymastia (*Med*) The presence of supernumerary breasts. Also *polymastism*.

polymer (*Plastics*) A material built up from a series of smaller units (*monomers* or *protomers*). See panel on POLYMERS.

polymerase (*BioSci*) Enzymes producing a polynucleotide sequence, complementary to a pre-existing *template* polynucleotide. DNA polymerase requires a PRIMER from which to start polymerization whereas RNA polymerase does not.

polymerase chain reaction (*BioSci*) A method of amplifying DNA sequences a few hundred bases long by over a

million times without using methods of genetic manipulation that need biological vectors. It requires two oligonucleotide PRIMERS that flank the sequence but bind to opposite strands of the DNA to be amplified. The reaction is a cycle in which the template DNA, and later the newly synthesized polynucleotide, is first denatured at high temperature (95°C), then annealed to the primers at a lower temperature (about 50°C) followed by the polymerase reaction at an intermediate temperature (about 70°C). The cycle is repeated some 25 times. Usually involves the use of a thermostable polymerase like that isolated from the bacterium *Thermus aquaticus* (Taq polymerase) that does not need to be replaces after each cycle. Different DNA POLYMERASES can be used to give greater fidelity when this is important. Results suggest that the method can amplify a single target sequence in a million cells; more recent modifications of the method allow quantitation of the starting content of the specific sequence being amplified. See panels on DNA AND THE GENETIC CODE and GENETIC MANIPULATION.

polymer crystallization (*Plastics*) The phenomenon where crystallizable polymers form crystallites, spherulites, etc, at a characteristic temperature not identical with polymer melting point, T_m. See panel on POLYMERS.

polymer engineering (*Plastics*) Application of polymer science to practical problems involving properties and use of polymeric materials in demanding environments.

polymeric glass (*Glass*) See panel on GLASSES AND GLASS-MAKING.

polymer identification (*Chem*) Variety of methods used to determine components in polymeric material. Complete analysis specifies repeat unit(s) and molecular mass distribution (MMD) of chains plus identity of fillers and other additives. Methods include IR, UV and NUCLEAR MAGNETIC RESONANCE SPECTROSCOPY for soluble thermoplastics and PYROLYSIS GAS CHROMATOGRAPHY and swelling tests for thermosets. Controlled degradation to identifiable soluble products at ambient temperature may also be useful.

polymerization (*Chem*) The combination of several molecules to form a more complex molecule, usually by a step- or chain-growth polymerization mechanism. See panels on POLYMERS and POLYMER SYNTHESIS.

polymer melting (*Chem*) The phenomenon of crystalline polymers melting at a characteristic temperature, dependent on rate of heating, impurities present, polymer type and molecular mass. Not to be confused with polymer crystallization temperature. See VISCOELASTICITY and panel on POLYMERS.

polymerous (*BioSci*) Having many members in a whorl, eg many petals.

polymers (*Chem*) Long-chain molecules built up by multiple repetition of groups of atoms known as repeat units. See STEREOREGULAR POLYMERS and panel on POLYMERS.

polymer selection (*Eng*) See CAMPUS, CAPS, EPOS, MATERIALS SELECTION, PLASCAMS.

polymer synthesis (*Chem, Plastics*) See panel on POLYMER SYNTHESIS.

polymethyl methacrylate (*Plastics*) The chemical name for Perspex, Lucite, etc. Rigid glassy thermoplastic used for glazing, optical devices, baths, aircraft windows, etc. Usually of very high molecular mass (approx 1 million) in sheet form. Used as a positive resist for electron-beam lithography in semiconductor processing. Abbrev *PMMA*.

poly(4-methylpent-1-ene) (*Plastics*) Isotactic, transparent aliphatic polyolefin prone to brittle cracking.

polymorph (*Aero*) The term applied by Barnes Wallis to his supersonic aircraft designs incorporating *variable sweep* wings.

polymorphic function (*ICT*) One that can handle a number of DATA TYPES.

polymorphic transformation (*Eng*) A change in a pure metal from one form to another, eg the change from γ- to α-iron.

Polymers

Polymer structure

Polymers can be of different kinds described by the sequence of repeat units along the chains. Copolymers are an important way of modifying the physical properties of the parent HOMOPOLYMER (see Fig. 1). Thus styrene–acrylonitrile (SAN) possesses a GLASS TRANSITION temperature (T_g) above the boiling point of water and so will remain rigid when used eg for holding hot coffee. Polystyrene possesses a T_g of about 97°C, so will collapse when exposed to boiling water at 100°C. Polystyrene is a brittle material at ambient temperatures, but copolymerization of styrene with polybutadiene gives high-impact polystyrene, a tougher material which can be used for stressed applications. An even tougher terpolymer, ABS plastic, is created by copolymerizing styrene and acrylonitrile with polybutadiene.

By increasing the proportion of butadiene to styrene units, elastomers will be produced. At about 75% by weight of polybutadiene, styrene–butadiene rubber is the most common synthetic rubber and widely used in car tyre treads. It is a RANDOM COPOLYMER where the repeat units are dispersed irregularly along the linear chains (Fig. 2).

The material is vulcanized in the conventional way to produce a network polymer. Alternatively, a different kind of material not requiring vulcanization can be produced by anionic polymerization where a much more regular BLOCK COPOLYMER such as styrene–butadiene–styrene is produced (Fig. 3).

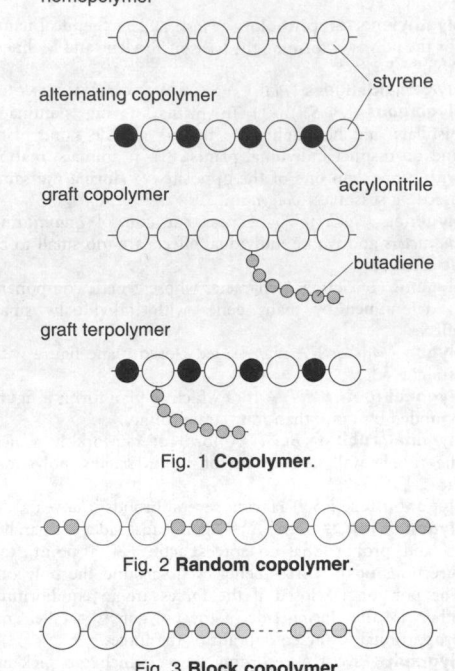

Fig. 1 **Copolymer**.

Fig. 2 **Random copolymer**.

Fig. 3 **Block copolymer**.

Crystallization of polymers

Polymer materials are not intrinsically crystallizable: chain irregularity found in random copolymer units or *atacticity* prevents the necessary close approach of chains. But when stereoregular chains are found, as in isotactic polystyrene or polypropylene, then crystallization can occur, resulting in an improvement in thermal and mechanical properties that is usually offset by a loss in optical properties, higher density and greater contraction after moulding. The CRYSTAL

---

polymorphism (*BioSci*) (1) The presence in a population of two or more alleles of a particular gene. It may be the result of selective advantage of the heterozygotes or of the rarer forms under some conditions. (2) The occurrence of different structural forms at different stages of the life cycle of the individual. Adjs *polymorphic*, *polymorphous*.

polymorphism (*ICT*) The inclusion of POLYMORPHIC FUNCTIONS in programs.

polymorphism (*Min*) The property possessed by certain chemical compounds of crystallizing in several forms which are structurally distinct; thus TiO_2 (titanium dioxide) occurs as the mineral species *anatase*, *brookite* and *rutile*.

polymorphonuclear leucocyte (*BioSci*) A cell of the myeloid series that, in its mature form, has a multilobed nucleus and cytoplasm containing granules. Such cells are present in inflammatory exudates and in the blood where, unlike erythrocytes, they have only a short residence time. The term encompasses neutrophil, eosinophil and basophil leucocytes.

polyneuritis (*Med*) A widespread affection of many peripheral nerves with flaccid paralysis of muscles and/or loss of skin sensibility, due to infection or poisoning with various agents, such as lead, alcohol, arsenic, diphtheria toxin, etc. Also *multiple neuritis*.

polynomial (*MathSci*) An algebraic expression of the form $a_0x^n + a_1x^{n-1} + \dots + a_n$, where $a_0, a_1, \dots, a_n$ are members of a field.

polynorbornene (*Plastics*) Speciality rubber with aliphatic ring system in main chain. When plasticized, gives very soft elastomer for rollers etc.

polynosic fibres (*Plastics*) Modified rayon fibres with improved properties, such as good dimensional stability on washing, arising from better uniformity of fibres.

polynuclear aromatic hydrocarbons (*EnvSci*) The collective name for hydrocarbons with multiple rings, such as anthracene, naphthalene, chrysene, often found as pollutants downwind of fires in which fossil fuel is undergoing incomplete combustion. Most are potential carcinogens. Abbrev *PAHs* or *PNAs*.

polynucleotide (*BioSci*) A long nucleic acid chain, usually of only deoxyribonucleic acid (DNA) or ribonucleic acid (RNA). Longer than an oligonucleotide.

polyolefin (*Textiles*) A fibre of film made from a linear polymer obtained from an olefin, esp ethylene (giving polyethylene) or propylene (giving polypropylene).

polyolefins (*Plastics*) Polymers based on olefin monomers, such as polyethylenes, polypropylene and poly(4-methylpent-1-ene). Also *polyalkanes*.

melting point (T_m) is always higher than the *glass transition temperature* (T_g), and correlation of the two transitions suggests that T_g (kelvin) $\sim (2/3)T_m$ (kelvin), although there is a wide divergence.

At a molecular level, polymers crystallize to form chain-folded LAMELLAE, which themselves twist and branch to form FIBRILS, then a wheatsheaf, which in turn grows into a SPHERULITE (see Fig. 4).

Impingement of spherulites finally occurs, although amorphous material between the lamellae and fibrils prevents 100% crystallization. Under certain conditions, eg high pressure, extended chain crystals can be formed in crystallizable polymers.

See panels on GLASSES AND GLASS-MAKING, HIGH-PERFORMANCE POLYMERS and POLYMER SYNTHESIS.

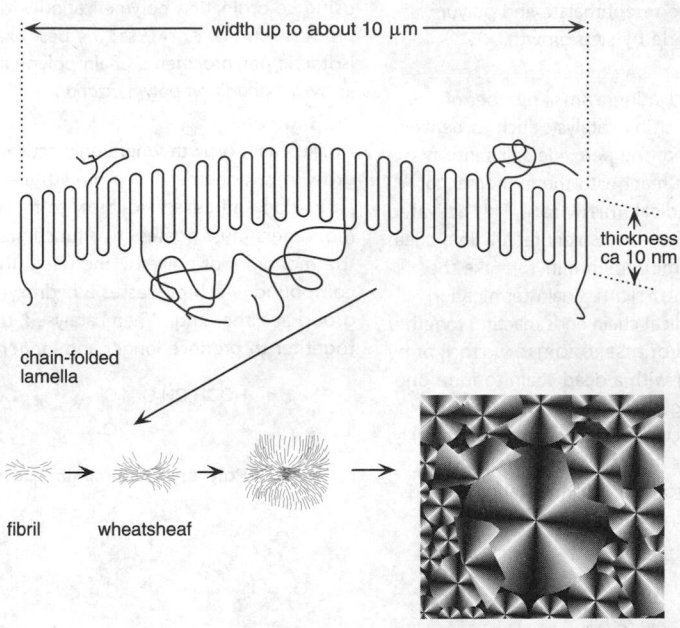

width up to about 10 μm

thickness ca 10 nm

chain-folded lamella

fibril   wheatsheaf

interpenetrating spherulites

Fig. 4 **Stages in the crystallization of polymers**.

polyoma (*BioSci*) A small DNA virus that can induce a wide variety of cancers in mice, hamsters and other species. Similar to SV40.

polyoses (*Chem*) Polysaccharides.

polyoxymethylene (*Plastics*) Moulding polymer widely used in engineering products (eg gear wheels) for its strength and stiffness. Highly crystalline with $T_m \approx 165°C$. Homopolymer is easily depolymerized, so the commercial polymer is copolymerized with ethylene oxide for extra stability or the chain ends may be capped with a stable group. Also *acetal resin, polyformaldehyde, polyoxymethylene resin*. Abbrev POM. TNs *Kematal, Delrin*.

polyp (*BioSci*) An individual of a colonial animal.

polyp (*Med*) A smooth, soft, pedunculated tumour growing from mucous membrane. Also *polypus*.

polyparabanic acids (*Plastics*) Heat-resistant polymer similar in structure to polyhydantoin, and used in similar applications.

polypeptide (*BioSci*) A linear condensation of amino acids that, alone or associated with others, forms a *protein* molecule.

polypeptide antibiotics (*Pharmacol*) Bactericidal antibiotics (bacitracin, colistin, Polymyxin B) with activity against Gram-negative aerobic bacilli including *Pseudomonas aeruginosa*. Polymyxin B and colistin are not active against *Proteus* spp and have no activity against Gram-positive organisms. Both act by disrupting the bacterial cell membrane.

polypetalous (*BioSci*) Having a corolla of separate petals, not fused with each other. Cf GAMOPETALOUS.

polyphagous (*BioSci*) Feeding on many different kinds of food, as Sporozoa which exist in several different cells during one life cycle, or phytophagous insects with many food plants.

polyphase (*ElecEng*) Said of ac power supply circuits (usually three) carrying currents of equal frequency with uniformly spaced phase differences. Normally using common return conductor.

polyphase motor (*ElecEng*) An electric motor designed to operate from a polyphase supply.

polyphenol (*Chem*) A compound with two or more phenolic hydroxyl groups.

polyphenylene oxide (*Plastics*) Aromatic polyether, poly-(2,6-dimethyl-p-phenylene ether). Its rigid backbone chain gives a high $T_g \approx 208°C$, but is mainly used blended with polystyrene (see NORYL). Abbrev *PPO*.

Polymer synthesis

Polymer chains of high molecular mass can be made by either chain growth or step growth. Each method creates quite different kinds of polymer with characteristic properties. Chain-growth polymers include polyethylene, vinyl polymers and polybutadiene, while nylon-66, polyethylene terephthalate and polyurethane rubber are made by step growth.

Chain growth

This is initiated by activating a small number of MONOMER molecules using catalysts such as benzoyl peroxide. When heated, the peroxide splits into two FREE RADICALS which react with the monomer to create another free radical (INITIATION). This activated monomer molecule then reacts very rapidly in a chain reaction with further monomer units to make the polymer chain (PROPAGATION). Chain termination occurs by two free radical chain ends reacting together (RECOMBINATION and/or DISPROPORTIONATION) or by a free radical reacting with a dead chain to form one chain and a branching chain (CHAIN BRANCHING). Polymer of high molecular mass is formed very quickly, and specific chemicals (TERMINATORS) are used commercially to control such polymerizations. This is shown schematically in Fig. 1.

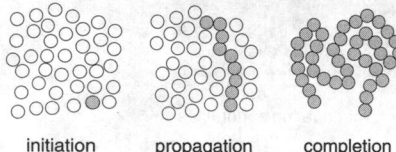

initiation　　　propagation　　　completion

Fig. 1 Chain polymerization.

Extra control of the very rapid propagation step is achieved by conducting the reaction as an emulsion or as a suspension in a water matrix. This also helps to remove the large amounts of heat produced, as all polymerization reactions are exothermic.

Control of polymer chain configuration with free radical ends is not easy, and it tends to produce atactic polymers (eg vinyls), branched polymers (eg low-density polyethylene) or random copolymers (with two

or more monomers). The proportion of the different repeat units in a copolymer is determined by the copolymer equation; the distribution of molecular mass is usually very broad. Altering the ionic charge of the catalyst enables a greater control over the polymer structure. Anionic polymerization allows block copolymers to be produced easily since the negatively charged chain end cannot branch and will only react with specific compounds. Monodisperse polymers can also be easily produced. Control of tacticity is achieved using co-ordination polymerization with a solid ZIEGLER–NATTA CATALYST; the best example is isotactic polypropylene. Chain polymerization is also known as *addition polymerization*.

Step growth

Unlike chain growth where only activated chain ends grow into monomers, step growth uses reactions like amidation and ESTERIFICATION to form linear chains. Monomers must possess two functional groups. Thus the monomer of polyethylene terephthalate is the compound which possesses a hydroxyl and carboxyl group at either end. When catalysed, they will react together to produce longer and longer chains:

$$HOCH_2CH_2O-\overset{\displaystyle |}{\underset{\displaystyle O}{C}}-\langle\bigcirc\rangle-CO_2H$$

Fig. 2 Polyethylene terephthalate monomer.

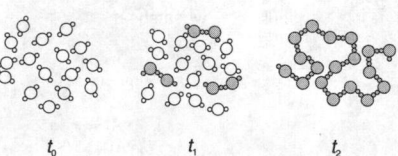

t_0　　　　　t_1　　　　　t_2

Fig. 3 Step polymerization Polyethylene terephthalate at successive times.

The reaction is usually carried out in the melt without solvent and the water produced must be removed from the system to achieve high molecular masses and prevent hydrolysis, which causes depolymerization. Compared with chain growth, reaction rates are relatively low and high-mass polymer is not produced until the very final stages.

See panels on HIGH-PERFORMANCE POLYMERS and POLYMERS.

---

polyphenylenes (*Plastics*) Heat-resistant, intractable polymers consisting of benzene rings linked together in chains. Processability improved by inserting ether, ketone or sulphide links between rings, eg PPO, PPS, PEEK, etc.

polyphenylene sulphide (*Plastics*) Heat-resistant aromatic polymer with sulphide functional linking group (–S–) in main chain. Claimed heat stability to 500°C, also fire resistant (oxygen index to 0·53) but rather brittle. Used in under-the-bonnet moulded parts. Abbrev *PPS*.

polyphosphates (*FoodSci*) Substances which accelerate the water binding properties of sodium chloride when added

to meat or fish products and prevent discoloration. The most important are pyrophosphates and tripolyphosphates which when added, eg, to bacon give a less salty product that still retains most of its moisture during cooking.

polyphosphates (*Plastics*) Glassy materials based on polymers of the phosphate group, $(PO_4)^{3-}$, susceptible to hydrolysis but widely used in protective coatings, eg in car steel bodies. See PHOSPHATING.

polyphyletic group (*BioSci*) (1) A group of species whose members derive from two or more independent ancestral lines. (2) In CLADISTICS, a group that does not include the most recent ancestor of its members. Also *polyphyly*. Cf MONOPHYLETIC GROUP, PARAPHYLETIC GROUP.

polyphyllous (*BioSci*) Having a perianth of separate members, not fused with each other. Cf GAMOPHYLLOUS.

polyphyly (*BioSci*) See POLYPHYLETIC GROUP.

polyphyodont (*BioSci*) Having more than two successive dentitions. Cf DIPHYODONT, MONOPHYODONT.

polyplexer (*Radar*) A device acting as duplexer and lobe switcher.

polyploid (*BioSci*) Having more than twice the normal haploid number of chromosomes. The condition is known as POLYPLOIDY. *Artificial polyploidy*, which can be induced by the use of chemicals (notably COLCHICINE), is of economic importance in producing hybrids with desired characteristics.

polyposis (*Med*) The development of many polyps in an organ such as the intestine.

polypropylene (*Plastics*) Polymers based on propylene (propene) monomer. The main parent material is isotactic and highly crystalline with repeat unit $-CH(CH_3)-CH_2-$. The T_m is high (approx 170°C) and the T_g is about -5°C, which makes it susceptible to cracking in a sharp frost. Copolymers with ethylene lower the T_g and improve toughness. Density of approx 905 kg m^{-3} (lowest of all solid polymers). Used for small boat hulls, rope and string (fibrillated fibre), milk crates, battery cases, etc. Atactic polypropylene is an important paper additive. Abbrev *PP*. See panel on POLYMERS.

polyprotodont (*BioSci*) Having numerous pairs of small subequal incisor teeth. Cf DIPROTODONT.

polypus (*Med*) See POLYP.

polypyromellitimide (*Plastics*) Specific type of polyimide used as heat-resistant film. TN Kapton.

polyribosome (*BioSci*) See POLYSOME.

polyrod antenna (*ICT*) One comprising a number of tapered dielectric rods emerging from a waveguide.

polysaccharides (*Chem*) Polyoses; a group of complex carbohydrates such as starch, cellulose, etc. They may be regarded as derived from x monosaccharide molecules by the elimination of $x-1$ molecules of water. Polysaccharides can be hydrolysed step by step, ultimately yielding monosaccharides.

polysepalous (*BioSci*) Having a calyx of separate sepals, not fused with each other. Cf GAMOSEPALOUS.

polysilicates (*Chem*) Minerals and materials based on polymers, sheets and networks of the silicate group $(SiO_4)^{4-}$. Widely exploited in ceramics, refractories and building materials. See panel on SILICON, SILICA, SILICATES.

polysilicon (*Electronics*) Polycrystalline silicon, deposited eg by CHEMICAL VAPOUR DEPOSITION; used in semiconductor technology as a refractory gate material for METAL–OXIDE–SILICON devices.

polysiloxanes (*Chem*) Oligomers and polymers based on the repeat unit and prepared by the hydrolysis of chlorosilanes R_3SiCl, R_2SiCl_2, ethers R_3SiOR, $R_2Si(OR)_2$, or mixtures. Frequently used as hydrophobic liquids on brickwork, cement, etc. See SILICONE RUBBERS.

polysome (*BioSci*) An assembly of ribosomes held together by their association with a molecule of MESSENGER RNA. Also *polyribosome*.

polysomy (*BioSci*) A chromosome complement in which some of the chromosomes are present in more than the normal diploid number, eg a *trisomic*.

polyspermy (*BioSci*) Penetration of an ovum by several sperm.

polyspondyly (*BioSci*) The condition of having more than two vertebral centra corresponding to a single myotome. Adjs *polyspondylic, polyspondylous*.

polystely (*BioSci*) The condition in which the vascular tissue in a stem exists as two or more separate steles interconnecting only at intervals such as where the stem branches, eg some *Selaginella* spp.

polystichous (*BioSci*) Arranged in several rows.

polystyrenes (*Plastics*) See COPOLYMER, STYRENE RESINS.

polysulphide rubbers (*Eng*) Polymers having linked sulphur atoms (eg $-S-S-$) in their backbone chain, and made by step mechanism using chain extension of thiol end groups. Vulcanized form used for lining aircraft fuel tanks, owing to fluid resistance. Extensively used in prepolymer form as viscous fluid, which when mixed with cross-linking agent, can be applied to joints in buildings etc as sealant.

polysulphones (*Eng*) Engineering polymers with aromatic backbone chains and T_g from 190 to 230°C. Step polymers based on the sulphonyl functional group $(-SO_2-)$, and non-crystalline, so transparent. Mainly used in heat-resisting products, eg microwave oven parts.

polyteny (*BioSci*) Special case of POLYPLOIDY in which chromatids remain very closely paired after duplication, through all or part of their length. In some cases homologues are also closely paired, eg in SALIVARY GLANDS of dipterans.

polytetrafluoroethylene (*Plastics*) Polymer, with the repeat unit $-CF_2-CF_2-$. The principal fluoropolymer, noted for its very marked chemical inertness and heat resistance. It is difficult to shape by normal moulding methods and is often sintered. Widely used for a variety of engineering and chemical purposes, and as a coating for 'non-stick' kitchen equipment. Abbrev *PTFE*. TNs *Fluon, Teflon*. Cf PENTON.

polytetramethylene ether glycol (*Plastics*) A low-molecular-mass prepolymer ($M_r = 2000$–3000), used in polyurethanes and polyester rubber. Made by polymerization of tetrahydrofuran. Abbrev *PTMEG*.

polythene (*Plastics*) See POLYETHYLENE.

polytokous (*BioSci*) Giving birth to many young; prolific; fecund. N *polytoky*.

polytrifluorochloroethylene (*Plastics*) A plastic material with many of the good properties of polytetrafluoroethylene (PTFE), but somewhat easier to fabricate as it can be injection moulded. Abbrev *PTFCE*.

polytrophic (*BioSci*) (1) Generally, obtaining food from several sources. (2) In insects, said of ovarioles in which nutritive cells alternate with the oöcytes. Cf ACROTROPHIC.

polyunsaturated fatty acids (*FoodSci*) Fatty acids containing two or more double bonds in their carbon chain. Among vegetable oils, the most important are linoleic acid and linolenic acid which have two and three double bonds in their C_{18} chains respectively. Fish oils have up to 26 carbon atoms and six double bonds. These fatty acids are less likely to increase serum cholesterol, raised levels of which are associated with cardiovascular diseases.

polyurethanes (*Plastics*) Versatile group of polymers forming tough material, either plastic or elastomeric. Step polymer containing the urethane group $(-NH-CO-O-)$ in its backbone chain. Formed by reaction of diols, dihydroxy polyesters, or hydroxyl-terminated polyethers (eg polytetramethylene ether glycol) with diisocyanates (eg tolylene-2,4-di-isocyanate). Cross-linking can be achieved with POLYOLS to give thermosets. Used as a casting polymer for shaped products, for foam rubber and as an extremely tough lining for eg ball mills. Also widely used for large car-body parts when made by REACTION INJECTION MOULDING.

polyuria (*Med*) Excessive secretion of urine.

polyvalent (*Chem*) Having a valency greater than two.

polyvinyl acetal (*Chem*) (1) Group comprising polyvinyl acetal, polyvinyl formal, polyvinyl butyral and other compounds of similar structure. (2) A thermoplastic material derived from a polyvinyl ester in which some or all of the ester groups have been replaced by hydroxyl groups and some or all of these hydroxyl groups replaced by acetal groups. Generally a colourless solid of high tensile strength.

polyvinyl acetate (*Plastics*) *Polymerized vinyl acetate*, used chiefly in adhesive and coating compositions. Abbrev *PVA*. See VINYL POLYMERS.

polyvinyl alcohol (*Plastics*) Water-soluble polymer made by hydrolysis of polyvinyl acetate, and modified chemically for fibre (vinal) and polyvinyl butyral.

polyvinyl butyral (*Plastics*) A polyvinyl resin made by reacting butanal with a hydrolyzed polyvinyl ester such as the acetate. Used chiefly as the interlayer in safety-glass windscreens to halt cracks and retain broken fragments.

polyvinyl carbazole (*Plastics*) Photoconductive polymer used in XEROGRAPHY. Made from polyvinyl alcohol and having large side-chain based on N-containing heterocyclic ring (carbazole group).

polyvinyl chloride (*Chem, Plastics*) PVC. Best known and most widely used of the vinyl plastics. See VINYL POLYMERS and panel on POLYVINYL CHLORIDE.

polyvinyl chloride fibres (*Textiles*) Fibres of the linear polymer derived from vinyl chloride, ie the repeat units are $-CH_2-CHCl-$. See panel on POLYVINYL CHLORIDE.

polyvinyl formal (*Plastics*) A polyvinyl resin made by reacting formaldehyde with a hydrolysed polyvinyl ester. Used in lacquers and other coatings, varnishes and some moulding materials.

polyvinylidene chloride (*Plastics*) Based on vinylidene chloride (1,1-dichloroethylene). Usually used as copolymer with small amounts of vinyl chloride or acrylonitrile. Basis of fibres, films for packaging, coatings for PET beer and cider bottles and chemically resistant piping. Highly crystalline and gas-impermeable polymer.

polyvinylidene chloride fibres (*Textiles*) Fibres from the polymer derived from vinylidene chloride, ie the repeat units are $-CH_2-CCl_2-$.

polyvinylidene fluoride (*Plastics*) Highly crystalline polymer with a repeat unit $-CH_2-CF_2-$. *Melt-processable* fluoropolymer of high toughness and piezoelectric properties in the drawn, oriented state. Used for large-area flexible transducers, esp in liquids (eg hydrophones, medical monitors).

polyvinyl polymers (*Plastics*) See VINYL POLYMERS.

polyvinyl pyrrolidone (*Chem*) Water-soluble polymer having a pyrrolidone cyclic unit as a pendant side group in the repeat unit. Widely used in cosmetic products, and as a copolymer with methyl methacrylate etc in cross-linked soft contact lenses. The polyvinyl pyrrolidone part helps the lens swell in eye-fluid.

polyxylylene (*Plastics*) Polymer made by polymerizing paraxylene in film form. Heat stable, used as dielectric film.

Polyzoa (*BioSci*) See BRYOZOA, ECTOPROCTA.

POM (*Pharmacol*) Abbrev for *prescription-only medicine*, one that can only be obtained with a prescription from an authorized medical practitioner.

POM (*Plastics*) Abbrev for POLYOXYMETHYLENE.

pomace (*FoodSci*) Skin and cell material remaining after the extraction of juice from apples and pears which contains pectin extracted commercially.

pome (*BioSci*) False fruit produced by members of the subfamily Pomoideae of the Rosaceae. The true fruit forms the core, containing the seeds and is surrounded by a greatly expanded, fleshy receptacle, eg apple.

pommel (*Build*) (1) An ornament in the shape of a ball, eg a ball finial. (2) See PUNNING.

Ponceaux (*Chem*) A group of dyestuffs prepared by the interaction of various diazo-salts with naphthol-sulphonic acids.

Poncelet wheel (*Eng*) An undershot water wheel with curved vanes; of higher efficiency than the flat-vane type. See UNDERSHOT WHEEL.

pond (*Build*) A reach or level stretch of water between canal locks. Also *pound*.

pondermotive force (*MinExt*) In high-tension separation, the electrostatic force exerted on a particle as it passes through the field of a corona-type separator. In magnetic separation, the flux field intensity together with density of particle determines its deflection from a straight path.

pongee (*Textiles*) Lightweight (75 g m^{-2}), plain-weave cloth of wild silk, lighter and more regular in appearance than SHANTUNG.

pons (*BioSci*) A bridge-like or connecting structure; a junction. Pl *pontes*. Adj *pontal*.

pons Varolii (*BioSci*) In mammals, a mass of transversely coursing fibres joining the cerebellar hemispheres.

pontal flexure (*BioSci*) The flexure of the brain occurring in the same plane as the cerebellum; it bends in the reverse direction to the primary and nuchal flexures and tends to counteract them.

pontie (*Glass*) Also *pontil*. See PUNTY.

pontoon (*CivEng*) A flat-bottomed floating vessel for the support or transport of plant, materials or workers.

pontoon bridge (*CivEng*) A bridge carried on PONTOONS.

pony girder (*Eng*) A secondary girder carried across side-by-side cantilevers.

pony motor (*ElecEng*) An auxiliary motor used to bring synchronous machinery up to speed before synchronizing.

pony rods (*MinExt*) In an oil well the rods which connect the POLISHED ROD to the pump. Also *drill rods*.

Ponzo illusion (*Psych*) Visual illusion in which both bars are actually the same length.

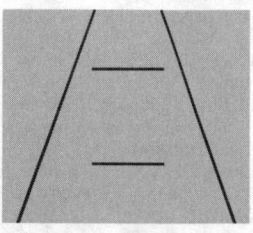

Ponzo illusion

pool cathode (*Electronics*) An emissive cathode which consists of a liquid conductor, eg mercury.

pool reactor (*NucEng*) See SWIMMING-POOL REACTOR.

pool-type reactor (*NucEng*) A dual-cycle reactor in which all the components of the primary cooling circuit (heat exchangers, pumps, etc) are contained within the primary pressure vessel, the situation in a pressurized-water reactor. Cf LOOP-TYPE REACTOR.

poor lime (*Build*) Lime in which the proportion of impurities insoluble in acids is in excess of 15%.

POP (*EnvSci*) Abbrev for *persistent organic pollutant*.

POP (*ImageTech*) (1) Abbrev for *picture outside picture*, small pictures that can be displayed beside a 4:3 picture on a 16:9 receiver to view alternative channels. (2) Abbrev for PRINTING-OUT PAPER.

POP-2 (*ICT*) Programming language used for LIST processing.

poplar (*For*) A common European hardwood tree (*Populus*) whose heartwood is creamy-white to grey and straight-grained, with a fine, even texture. Also *aspen, robusta*.

poplin (*Textiles*) A plain-woven fabric, with fine lines or cords running across the cloth (due to the *ends* per inch greatly exceeding the *picks* per inch); often of cotton and mercerized. Used for shirtings, pyjamas and dresses.

poppet head (*MinExt*) Headframe of hoisting shaft. The poppet is the bearing in which the winding pulley is set.

poppets (*Ships*) Temporary structures erected beneath a ship's hull to transfer the weight to the sliding ways, before and during launching.

poppet valve (*Autos*) The mushroom- or tulip-shaped valve, made of heat-resisting steel, commonly used for inlet and exhaust valves. It consists of a circular head with a conical face which registers with a corresponding seating round the port, and a guided stem by which it is lifted from its seating by the rocker or tappet. Also *mushroom valve*. See STELLITED VALVES, VALVE INSERTS.

popping (*Build*) A defect in plasterwork made from improperly slaked lime.

popping (*FoodSci*) See EXPLOSIVE DECOMPRESSION.

popping (*ICT*) The deletion of a data item from a STACK.

Polyvinyl chloride

Polyvinyl chloride (PVC) is one of the most versatile thermoplastics because of the way that its properties can be modified by additives. It is a cheap commodity polymer as it is made from chlorine (Cl_2) and ethylene (C_2H_4), both important chemical intermediates produced from common salt, NaCl, and from petrochemicals such as naphtha from crude oil. Another more efficient route for ethylene production, used in the USA, the UK and Norway, is from the ethane content of liquefied petroleum gases by CATALYTIC CRACKING. The PVC is then made by addition, dehydrochlorination and POLYMERIZATION.

Until about 40 years ago, PVC materials were of low strength due to thermal degradation during processing. Developments in polymerization (FREE RADICAL,

emulsion and suspension) and in additives such as the STABILIZERS have improved PVC beyond recognition. The applications of uPVC (u = unplasticized) include water pipes by extrusion, connections and supports for guttering, waste water, etc, made by injection moulding. High-impact PVC is rubber-toughened by graft copolymerization with methacrylate butadiene–styrene (MBS) or ACRYLATE RUBBER, and is widely used for door and window frames. Plasticized PVC is formed by compounding PVC powder with ESTERS like di-octyl phthalate (DOP) and di-alphanyl phthalate (DAP): they are absorbed and held within the PVC matrix by van der Waals bonding. Chain flexibility is increased because the glass transition temperature (T_g) is lowered from 80°C to less than ambient. Such materials are found as flexible film in stationery products (files, folders, etc), imitation leather for car upholstery and serum bags etc for medical applications. One of the most widespread applications is for bottling detergents etc, where the material competes with POLYOLEFINS.

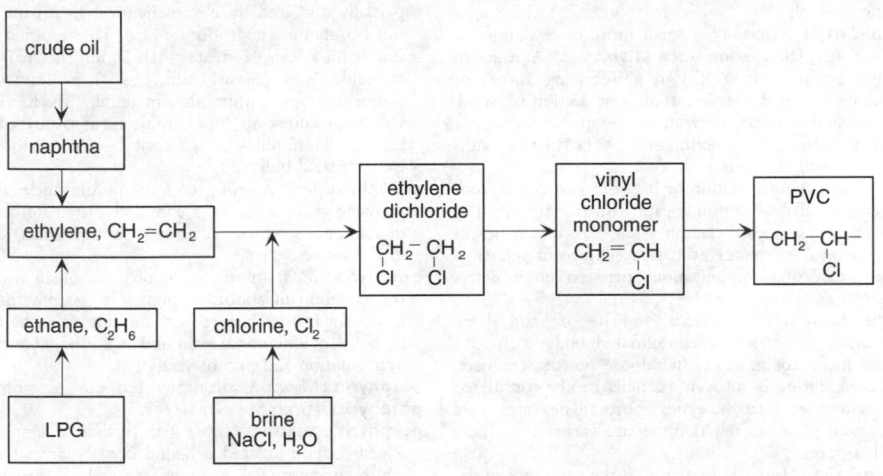

Polyvinyl chloride Synthetic pathway.

---

popping-back (*Autos*) An explosion through the inlet pipe and carburettor of a petrol engine, due to a weak, slow-burning mixture.

pop rivet (*Eng*) A proprietary design of hollow rivet, requiring work access from one side only for placing and clinching. Similar to CHOBERT RIVET.

population (*BioSci*) Any specified reproducing group of individuals.

population diffusion coefficient (*BioSci*) A coefficient that describes the tendency of a population of motile cells or organisms to diffuse through the environment. Its use presupposes movement in a random walk.

population inversion (*Phys*) The process employed in the operation of MASERS and lasers. According to the BOLTZMANN PRINCIPLE, the number of particles with a higher energy will be less than the number with a lower energy for a system in equilibrium at a given temperature. In a non-equilibrium situation it is possible to invert this so that the number of particles in the higher energy level is

greater than the number in the lower level. See OPTICAL PUMPING and panel on LASER.

population mean (*MathSci*) The true mean of the entire population, often estimated using the sample mean. Abbreviated with the lower case Greek letter μ.

population types (*Astron*) The two broad types of stellar population. Population I includes hot blue stars such as those in the Sun's neighbourhood; they are found in the arms of spiral galaxies, and share in the galactic rotation. Population II stars are found in the central regions of galaxies and in globular clusters, where dust and gas are absent; they are red stars, having high velocities and do not share in the regular rotation of the system.

pop-up menu (*ICT*) A MENU that appears wherever the user has placed the CURSOR. It may be activated by pressing a combination of keys or a special HOT KEY.

pop-up program (*ICT*) MEMORY-RESIDENT SOFTWARE that is activated by pressing a combination of keys or a

HOT KEY, or when an event takes place such as a message being received from an external source eg ELECTRONIC MAIL.

pop valve (*Eng*) A boiler safety valve in which the head of the wing valve is so shaped as to cause the steam to accelerate the rate of lift when a small lift occurs, giving rapid pressure release.

POR (*Plastics*) Abbrev for PROPYLENE OXIDE RUBBER.

porcelain (*Glass*) Originally objects made of mother-of-pearl from the eponymous shell, but in the 16th century applied by the Portuguese to the ceramic whiteware developed in China. Now refers to any impermeable and translucent ware made from earthy raw materials, in particular KAOLIN (or China clay). 'Hard' porcelain is fired together with its glaze at up to 1400°C giving a much harder surface than 'soft' porcelain which is fired unglazed at about 1250°, and then glazed and fired at about 1100°C. See BONE CHINA. Cf EARTHENWARE, FAÏENCE, STONEWARE, TERRACOTTA.

porcelain clay (*Geol*) See CHINA CLAY.

porcelain insulator (*ElecEng*) An insulator for supporting high-voltage electric conductors; made of a hard-quality porcelain.

porch (*ImageTech*) An interval in the TV signal at BLANKING level. See BACK PORCH, FRONT PORCH.

porcine parvovirus infection (*Vet*) Infection which affects 75% of UK pig farms. Disease manifests itself with infertility, early embryonic death, abortion and still births. Vaccine available.

pore (*BioSci*) (1) Generally, a small more or less rounded aperture. (2) The aperture of a STOMA. (3) A rounded aperture (rather than a slit) in a dehiscing anther or capsule. (4) A vessel seen in a transverse section of wood. (5) A more or less circular germinal aperture in the wall of a pollen grain. Also *porus*. See APERTURATE. Adjs *poriferous, poriform, porous*.

pore (*MinExt*) A cavity within or between particles in rock or aggregate. If these communicate with free surface they are open, if not, closed. Number and size of cavities in given volume determines POROSITY, degree of interconnection (ease of communication) between pores determines PERMEABILITY.

pore distribution (*PowderTech*) (1) The size and shape distribution of pores in a consolidated body such as a sintered metal compact. (2) In a loose powder compact, the space confined by adjacent particles can be considered as a system of interconnecting pores. The term *pore distribution* is used to describe the size variation in these so-called pores.

porencephalia (*Med*) The presence, in the substance of the brain, of cysts or cavities containing colourless fluid; due to a defect in development. Also *porencephaly*.

poricidal (*BioSci*) Opening or dehiscing by pores.

Porifera (*BioSci*) A phylum of sessile, aquatic, filter feeding animals. They lack sense organ systems but possess tissues, eg epithelial cells (pinacocytes) and flagellated cells (choanocytes). They have skeletons of silica or collagen. Sponges.

porins (*BioSci*) Transmembrane matrix proteins, found in the outer membranes of Gram-positive bacteria, that associate as trimers to form channels through which small hydrophilic molecules can pass. Similar porins are found in outer mitochondrial membranes.

porogamy (*BioSci*) The entry of the pollen tube into the ovule, in an angiosperm, by way of the micropyle (the commoner route). Cf CHALAZOGAMY.

poromeric (*Chem*) Permeable to water vapour; said of a polyurethane-base synthetic leather formerly used in the manufacture of shoe 'uppers'. See GORE-TEX.

porometer (*BioSci*) An instrument for investigating the opening of STOMATA by measuring the rate of flow of air (or other gas) through the leaf (viscous flow porometry) or the rate of diffusion of water vapour through the leaf (diffusion porometry).

porosity (*Build*) The percentage of space in a material which can absorb water.

porosity (*Eng*) (1) The fraction of voids in a material. If the porosity of a material is p, then its density ρ is related to p by

$$p = \frac{\rho_0 - \rho}{\rho_0}$$

where ρ_0 is the pore-free density. (2) In castings, unsoundness caused by shrinkage during cooling, or BLOWHOLES. (3) In compaction of powders, the percentage of voids in a given volume under specified packing conditions. Also *powder porosity*.

porosity (*Geol*) Of rocks, the ratio, usually expressed as a percentage, of the volume of the void space to the total volume of the rock.

porous bearing (*Eng*) A bearing produced by powder metallurgy which, after sintering and sizing, is impregnated with oil by a vacuum treatment. Bearing porosity can be readily controlled and may be as high as 40% of the volume.

porous dehiscence (*BioSci*) The liberation of pollen from anthers, and of seeds from fruits through pores in the wall of the containing structure.

porous pot (*ElecEng*) An unglazed earthenware pot serving as a diaphragm in a two-fluid cell.

porous silicon (*Electronics*) Bulk silicon which has been partially dissolved by electrochemical treatment in solutions containing hydrofluoric acid. The resulting porosity can be on a scale of greater than 50 nm (macro), between 50 and 2 nm (meso) and less then 2 nm (micro). Nanostructures within the material, which are made without recourse to lithography, form QUANTUM WIRES having high luminescent efficiency which do not occur in the untreated bulk.

porphin (*Chem*) A group of four pyrrole nuclei linked by methene groups, having a complete system of conjugated double bonds, which accounts for the (reddish) colour of its derivatives.

porphyria (*Med*) Inborn error of metabolism resulting in the excretion of abnormal pigments (porphyrins) in the urine which turn dark red on standing. It is characterized by bouts of abdominal pain and vomiting, abnormal skin pigmentation and photosensitivity.

porphyrin (*Chem*) A substituted porphin free from metal.

porphyrite (*Geol*) See PORPHYRY.

porphyritic texture (*Geol*) The texture of igneous rocks which contain isolated euhedral crystals larger than those which constitute the groundmass in which they are set.

porphyroblastic (*Geol*) A textural term applicable to metamorphic rocks containing conspicuous crystals in a finer groundmass, the former being analogous with the phenocrysts in a normal igneous rock, but having developed in the solid.

porphyry (*Geol*) A general term used rather loosely for igneous rocks which contain relatively large isolated crystals set in a fine-grained groundmass, eg *granite porphyry*.

porpoising (*Aero*) Oscillating symmetrical movements of a seaplane, flying boat or amphibian, when planing; pitching instability on the water, as distinct from instability under airborne conditions. Also for landplanes during take-off and landing due to undercarriage forces.

porrect (*BioSci*) Directed outwards and forwards.

port (*Eng*) An opening, generally valve-controlled, by which a fluid enters or leaves the cylinder of an engine, pump, etc. See PISTON VALVE, POPPET VALVE, SLEEVE VALVE, SLIDE VALVE.

port (*Genrl*) Place of access to a system, used for introduction or removal of energy or material, eg *glovebox port*.

port (*ICT*) Point at which signals from peripheral equipment enter the computer.

porta (*BioSci*) Any gate-like structure. Adj *portal*.

portability (*ICT*) The ability of a program to be used on computers with different operating systems. See ACROBAT.

portable (*ICT*) Applied to SOFTWARE that can be easily implemented on different types of computers.

portable colour duct (*Print*) An interchangeable ink duct which can be attached to the inking system of a printing unit and used to provide an additional colour.

portable document format (*ICT*) A file format, produced by a program such as ACROBAT, that preserves the text and graphics format on the page in a way that makes it readable by computers with different operating systems and over the Internet. Abbrev *PDF*.

portable electrometer (*ElecEng*) A portable form of the absolute attracted disk type of electrometer.

portable engine (*Eng*) A steam or internal-combustion engine carried on road wheels but not self-propelled thereby.

portable ink pump (*Print*) A pump-operated PORTABLE COLOUR DUCT.

portable instrument (*ElecEng*) An electrical measuring instrument specially designed for carrying about for testing purposes. Cf SUBSTANDARD INSTRUMENT, SWITCHBOARD INSTRUMENT.

portable planer (*Build*) A tool with flat sole plate in front of which cutters revolve. A second sole plate is adjustable relative to the first and controls the depth of cut.

portable substation (*ElecEng*) A substation comprising the converting or transforming plant and the necessary switch and protective gear, mounted on a railway truck or other vehicle in order that it can be quickly moved to any site for dealing with special loads or other emergency conditions.

portal (*Build*) (1) A structural frame consisting essentially of two uprights connected at the top by a fixing. (2) An arch spanning a doorway or gateway.

portal (*ICT*) A website, often incorporating a search engine, that provides access to a range of other sites, usually with a common relationship of content or subject, on the World Wide Web.

portal system (*BioSci*) A vein that breaks up at both ends into sinusoids or capillaries; as (in vertebrates) the hepatic portal system, in which the hepatic portal vein collects from the capillaries of the alimentary canal and passes the blood into the sinusoids and capillaries of the liver, and the RENAL PORTAL SYSTEM. Also *portal circulation*.

Porter governor (*Eng*) A pendulum-type governor in which, usually, the ends of two arms are pivoted to the spindle and sleeve, and carry heavy balls at their pivoted joints. The sleeve carries an additional weight. See PENDULUM GOVERNOR.

portico (*Arch*) A covered colonnade at one side of a building (usually the entrance side).

Portland cement (*Build, CivEng*) The usual binder for concrete, named for its resemblance, when set, to Portland stone. Invented by Joseph Aspdin in 1824. See panel on CEMENT AND CONCRETE.

Portlandian (*Geol*) The youngest stage of the Jurassic. See MESOZOIC.

portlandite (*Min*) Calcium hydroxide, $Ca(OH)_2$, occurring rarely in nature but also in Portland cement, hence the name.

porus (*BioSci*) Same as PORE sense (5).

poset (*MathSci*) A non-empty set on which a PARTIAL ORDER is defined.

position (*ICT*) In radio navigation, a set of co-ordinates used to specify location and elevation.

positional cloning (*BioSci*) Identification of a gene based on its location in the genome. Typically, this will result from linkage analysis based on a mutation in the target gene, followed by a chromosome walk from the nearest known sequence.

positional notation (*MathSci*) A system of representation of a number by a sequence of digits in which each digit represents a multiple of successive powers of the BASE. It is

used in the conventional Arabic numerals. Thus in the number 37, the 3 represents ten times as much as it does in the number 53; the number 7246 represents $(7 \times 10^4) + (2 \times 10^3) + (4 \times 10^2) + (6 \times 10^1)$. The *binary system* uses base 2, the *octal system* base 8, and the *decimal system* base 10. Also called *denominational number system*, *place-value notation*. Cf ROMAN NUMERALS.

position angle (*Astron*) A measure of the orientation of one point on the celestial sphere with respect to another. The position angle of any line with reference to a given point is the inclination of the line to the hour circle passing through the point; it is measured from 0° to 360° from the north point round through east.

position effect (*BioSci*) Differential effect on transcription and repression of the same gene in different chromosomal locations.

position error (*Aero*) That part of the difference between the *equivalent* and *indicated airspeeds* due to the location of the PRESSURE HEAD or STATIC VENT. Position error is not a constant factor, but varies with airspeed due to the variations in the airflow around an aircraft at different ANGLES OF ATTACK.

position-finding (*ICT*) Determination of location of transmitting station (eg an aircraft) by taking a number of bearings by direction finders that receive a signal from the transmitter.

positive (*ImageTech*) An image in which tones are reproduced in the same relation as in the original subject; where colour is reproduced, the hues are substantially those of the original.

positive (*MathSci*) A term describing a number greater than zero.

positive (*Phys*) Said of a point in a circuit which is higher in electric potential than earth; the designation of charge, introduced by Benjamin Franklin, now known to be opposite to that on the electron.

positive after-image (*Phys*) The continued image perceived after visual fatigue in the retina, when the object is replaced by a dark surface or the eye is closed.

positive amplitude modulation (*ImageTech*) That in which the amplitude of a video signal increases with an increase in luminance of the picture elements.

positive coarse pitch (*Aero*) An extreme blade angle which is reached and locked after engine failure to reduce the drag of a non-feathering propeller.

positive column (*Electronics*) Luminous plasma region in gas discharge, adjacent to positive electrode.

positive coupling (*ElecEng*) When two coils are inductively coupled so that the magnetic flux associated with the mutual inductance is in the same direction as that associated with the self-inductance in each coil, the mutual inductance and the coupling are both termed *positive*. If these fluxes oppose the mutual inductance and coupling they are known as *negative*.

positive distortion (*Phys*) See BARREL DISTORTION.

positive electricity (*Phys*) The phenomenon in a body when it gives rise to effects associated with deficiency of electrons, eg positive electricity appears on glass rubbed with silk. Cf NEGATIVE ELECTRICITY.

positive electrode (*Electronics*) (1) That connected to a positive supply line. (2) The anode in a voltameter. (3) The cathode in a primary cell.

positive electron (*Phys*) See POSITRON.

positive emulsion (*ImageTech*) An emulsion intended for printing positive images from a negative, generally characterized by fine grain and relatively high contrast rather than photographic speed.

positive feedback (*BioSci*) The products or consequences of a reaction or process increase the rate at which the reaction that produces them occurs; without suitable controls, positive feedback may lead to an exponential increase. Cf NEGATIVE FEEDBACK.

positive feedback (*ICT*) FEEDBACK such that the signal fed back to the input of the system or amplifier is in phase with the input signal; tends to increase gain at the expense of stability, high levels resulting in oscillation.

positive feeder (*ElecEng*) A feeder connected to the positive terminal of a dc supply.

positive film stock (*ImageTech*) Unexposed film specifically intended for making prints from motion picture negatives.

positive g (*Space*) The force exerted on the human body in a gravitational field or during acceleration so that the force of inertia acts in a head-to-foot direction. See NEGATIVE G.

positive ion (*Phys*) An atom (or a group of atoms which are molecularly bound) which has lost one or more electrons, eg the alpha particle, which is a helium atom less its two electrons. In an electrolyte, the positive ions (*cations*), produced by dissolving ionic solids in a polar liquid like water, have an independent existence and are attracted to the cathode. Negative ions likewise are those which have gained one or more electrons.

positive magnetostriction (*Phys*) See JOULE MAGNETOSTRICTION.

positive mineral (*Min*) A mineral in which the ordinary ray velocity is greater than that of the extraordinary ray, ie ω is less than ε. Quartz is a positive mineral for which $\omega = 1.544$ and $\varepsilon = 1.553$. See OPTIC SIGN.

positive ore (*MinExt*) Ore blocked out on four sides in panels sufficiently small to warrant assumption that the exposed mineral continues right through the block as a calculable tonnage.

positive phase sequence (*ElecEng*) See PHASE SEQUENCE.

positive reaction (*BioSci*) A tactism or tropism in which the plant moves, or the plant member grows, from a region where the stimulus is weaker to one where it is stronger.

positive reinforcement (*Psych*) A situation in which a response is followed by a positive event or stimulus (the positive reinforcer or reward) which increases the likelihood that the response will be repeated.

positive taxis (*BioSci*) See TAXIS.

positive video signal (*ImageTech*) One in which increasing amplitude corresponds to increasing light value in the transmitted image. White is 100%, and the black level makes about 30% of the maximum amplitude of signal.

positron (*Phys*) The antiparticle of the ELECTRON, of the same mass and opposite charge (ie charge $+e$). Produced in the decay of radioisotopes and PAIR PRODUCTION by X-rays of energy much greater than 1 MeV.

positronium (*Phys*) The combination of a positron and an electron forming a hydrogen-atom-like system, the positron taking the place of the proton. The system has a very short lifetime (less than 10^{-7} s) and disappears with the emission of gamma-ray photons.

posology (*Genrl*) The science of quantity or dosage.

possible ore (*MinExt*) Ore probably existing, as indicated by the apparent extension of proved deposits, but not yet entered and sampled.

POST (*ICT*) Abbrev for POWER-ON SELF-TEST.

post (*Glass*) The formed GATHER prior to drawing by the hand process.

post (*Paper*) A term used for a pile of sheets or alternate felts and sheets in hand-made mills.

post- (*Genrl*) Prefix from Lt *post*, after.

postal (*Paper*) A non-preferred size of printing board 572×725 mm (22×28 in).

Postal, Telephone and Telegraph Administration (*ICT*) A general term used to denote a supplier of telecommunications services. Abbrev *PTTA*.

post and pan (*Arch*) Half-timbering formed with brickwork or lath and plaster panels. Also *post and pane*, *post and petrail*.

post-capillary venules (*BioSci*) Small vessels through which blood flows after leaving the capillaries and before reaching the veins. The main route by which leucocytes migrate into inflammatory sites is between the endothelial cells of post-capillary venules. Specialized venules with high, rather than flattened, endothelial linings are present in the thymus-dependent area of lymph nodes. Through these, lymphocytes recirculate from blood to lymph.

postcardinal (*BioSci*) Posterior to the heart; as the *postcardinal sinus* of Selachii.

postcaval vein (*BioSci*) In higher vertebrates, the posterior vena cava conveying blood from the hind parts of the body and viscera to the heart. Called *inferior vena cava* in humans.

postclimax (*BioSci*) A relict of a former climax held under edaphic control in an area where the climate is no longer favourable for its development.

post-consumer recycling (*Genrl*) Reclamation of different materials in products after being discarded by consumer. Products such as lead–acid batteries, cars, paper and textile goods have well-established recycling routes but newer products containing large quantities of thermoplastics tend to be difficult to recycle easily. Incineration is one route where part of the feedstock energy of the polymer is recouped usefully (eg to raise steam). Also *secondary recycling*. See REUSABILITY.

postdeflection acceleration (*Electronics*) In a cathode-ray tube, acceleration of the beam electrons after deflection, so reducing the power required for this. Abbrev *PDA*.

post-dipping lameness (*Vet*) Fever and lameness in sheep due to cellulitis of the limbs caused by the bacterium *Erysipelothrix insidiosa* (*E. rhusiopathiae*); occurs after sheep have been dipped in baths contaminated by the organism.

post-emergent treatment (*Agri*) Any treatment made after a crop plant or its associated weeds have become visible.

poster (*Print*) A large sign printed on paper, for advertisement or propaganda purposes. Posters are usually printed by lithography, offset or direct, very large ones being executed in convenient sections. In the UK, the one-sheet poster measures 30×20 in (double crown) and larger sizes are two-sheet, four-sheet, etc.

posterior (*BioSci*) (1) In plants, the side nearer the axis of a bud, flower or other lateral structure. (2) In a bilaterally symmetrical animal, further away from the head region; behind. Cf ANTERIOR.

posterior distribution (*MathSci*) The representation as a probability distribution of the POSTERIOR PROBABILITIES of a PARAMETER.

posterior probability (*MathSci*) The probability of an event calculated after (posterior to) observation of the outcomes of a test.

posterization (*ImageTech*) Video effect in which the picture is reproduced in a limited number of flat tones and colours, without detailed gradation.

POS terminal (*ICT*) Abbrev for POINT-OF-SALE TERMINAL.

postern (*Arch*) A private door or gate, generally at the back or side of a building.

post-fertilization stages (*BioSci*) The developmental processes that go on between the union of the gametic nuclei in the embryo sac and the maturity of the seed.

post-forming sheet (*Plastics*) A grade of laminated sheet, of the thermosetting type, suitable for drawing and forming to shape when heated.

postganglionic (*Med*) Refers to the axon of a ganglionic cell of the autonomic nervous system which innervates the effector organ. Cf PREGANGLIONIC.

postharvest (*Agri*) The period between maturity of the crop and the time of its final consumption.

post head (*ElecEng*) A post or pillar at which cables supplying a third-rail traction system may be terminated and connection made to the conductor rail.

postheating (*Eng*) Annealing or tempering a WELDMENT to remove strain or prevent local hardening.

posthitis (*Med*) Inflammation of the prepuce.

post-hypnotic suggestion (*Psych*) Suggestions made to an individual during a hypnotic trance, which determine a behavioural or experiential response occurring after the individual has returned to ordinary consciousness.

postical (*BioSci*) Relating to or belonging to the back or lower part of a leaf or stem.

posting (*NucEng*) The transfer of highly radioactive material around a plant in such a way as to never expose personnel to dangerous radiation.

post insulator (*ElecEng*) A porcelain insulator built in the form of a post; used for supporting bus-bars etc in a high-voltage outdoor substation.

postmortem (*ICT*) A program designed to locate and diagnose a fault in an executing computer program.

postnatal depression (*Med*) The feelings of acute depression which some mothers experience in the period after giving birth. Abbrev *PND*.

Post Office box (*ElecEng*) A Wheatstone bridge in which the resistances making up the arms are contained in a box and varied by means of plugs. Also, *Post Office bridge*. Abbrevs *PO box*, *PO bridge*.

post pallet (*Build*) A PALLET with corner posts, used when a normal pallet and load cannot support another above it.

postpartum (*Med*) Occurring after childbirth; eg *postpartum* haemorrhage.

post-parturient haemoglobinuria (*Vet*) Uncommon disease of milk cows. Symptoms first appear within a month of calving and include dullness, anorexia, reduced milk yield, haemoglobulinuria, jaundice and recumbency. Caused by fall in blood phosphorus levels.

post-production (*ImageTech*) General term for the operations of editing, sound mixing, etc, which take place after a motion picture or video production has been shot.

post-scoring (*ImageTech*) The arrangement of a musical accompaniment or other sound effects after a motion picture production has been photographed.

PostScript (*ICT*) TN for a proprietary PAGE DESCRIPTION LANGUAGE commonly used when documents are transferred from a desktop publishing system to a printer.

postsynaptic cell (*BioSci*) In a chemical synapse, the cell that receives a signal (binds neurotransmitter released from the presynaptic cell) and responds with depolarization.

postsynaptic potential (*BioSci*) In a synapse, a change in the resting potential of a postsynaptic cell following stimulation of the presynaptic cell. Binding of the neurotransmitter signal causes ion channels to open in the postsynaptic cell, and each channel opening causes a small depolarization (a *miniature end-plate potential*); these sum to produce an excitatory postsynaptic potential.

post-synchronization (*ImageTech*) Adding sound to previously photographed motion picture material.

post-tensioned concrete (*CivEng*) A method whereby concrete beams or other structural members are compressed, after casting and reaching the requisite compressive strength, to enable them to act in the same way as PRESTRESSED CONCRETE. Achieved by the use of high tensile wires or rods threaded through preformed ducts in the member(s).

Post-Tertiary (*Geol*) Name assigned to geological events which occurred after the close of the Tertiary era, ie during Pleistocene and Recent times.

post-translational modification (*BioSci*) Changes that are made to proteins after peptide bond formation (*translation*) has occurred and thus only indirectly coded by DNA. Examples include glycosylation, acylation, limited proteolysis, phosphorylation, isoprenylation.

post-traumatic stress disorder (*Psych*) An anxiety disorder that follows the experience of traumatic events such as natural disasters, war, rape, etc. The syndrome includes dreams re-enacting the event, recurrent thoughts and images, a sense of psychological numbness and disengagement with normality. Also *post-traumatic stress syndrome*. Abbrev *PTSD*, *PTSS*.

post-trematic (*BioSci*) Posterior to an aperture; as, in Selachii, that branch of the ninth cranial nerve which passes posterior to the first gill cleft. Cf PRETREMATIC.

postulate (*MathSci*) An archaic term for an *axiom*.

postventitious (*BioSci*) Delayed in development.

postzygapophysis (*BioSci*) A facet or process on the posterior face of the neurapophysis of a vertebra, for articulation with the vertebra next behind. Cf PREZYGA-POPHYSIS.

pot (*ElecEng*) Abbrev for POTENTIOMETER.

pot (*Glass*) A vessel of fireclay holding from a few kg to about 2 tonnes, according to the type of manufacture; used to contain the glass during melting in the pot furnace. Such pots may be open or closed (ie provided with a hood to prevent furnace gases from acting on the glass).

potamous (*BioSci*) Living in rivers and streams.

pot annealing (*Eng*) See CLOSE ANNEALING.

potash (*Agri*) A fertilizer based on potassium carbonate and other potassium salts used to enhance fruit and flower production. A component of compound fertilizers.

potassium (*Chem*) Symbol K, at no 19, ram $39 \cdot 102$, mp $63°C$, bp $762°C$, rel.d. $0 \cdot 87$. A very reactive alkali metal, soft and silvery white. In the form of the element, it has little practical use, although its salts are used extensively. In combination with other elements it is found widely in nature. It shows slight natural radioactivity due to ^{40}K (half-life $1 \cdot 30 \times 10^9$ years). May be used as coolant in liquid-metal reactors, usually as an alloy with sodium.

potassium alum (*Min*) Synonym for ALUM.

potassium antimonyl tartrate (*Chem*) See TARTAR EMETIC.

potassium–argon dating (*Geol*) A method of determining the age in years of geological material, based on the known decay rate of ^{40}K to ^{40}Ar. See panel on RADIOMETRIC DATING.

potassium bichromate (*Chem*) See POTASSIUM DICHROMATE.

potassium bromate (*FoodSci*) An oxidizing agent used as an improver in flour which has no action on dry flour, but reproduces after dough-making the beneficial changes in the flour protein due to ageing, eg increasing the elasticity and reducing the fermentation period and extensibility of gluten.

potassium bromide (*Chem*) KBr. Used in photography and formerly in medicine as a sedative.

potassium carbonate (*Chem*) K_2CO_3. *Potash*. White deliquescent powder, soluble in water. Basic. Manufactured by extraction from Stassfurt and other potassium salt beds, or from the electrolysis of potassium chloride.

potassium channel (*BioSci*) An ION CHANNEL across a cell membrane that is selective for the passage of potassium ions.

potassium chlorate (V) (*Chem*) $KClO_3$. Detonates with heat; used in the manufacture of matches, fireworks and explosives, and in the laboratory as a source of oxygen.

potassium chloride (*Chem*) KCl. Occurs extensively in nature. With sodium chloride, it is extracted on a commercial scale from the waters of the Dead Sea.

potassium cyanide (*Chem*) KCN. White, deliquescent solid, smelling of bitter almonds. Extremely poisonous. In the fused condition, is a powerful reducing agent. Used in chemical analysis and in metallurgy, eg in the cyanide process for extracting gold. Also in HT salts, fumigants, pharmacy, photography. On the industrial scale, its use has largely been superseded by that of the cheaper sodium cyanide.

potassium dichromate (VI) (*Chem*) $K_2Cr_2O_7$. Used in analytical chemistry. Mixed with sulphuric acid, it is used as a cleanser of laboratory vessels, particularly after contamination with organic matter.

potassium feldspar (*Min*) Silicate of aluminium and potassium, $KAlSi_3O_8$, occurring in two principal crystalline forms: *orthoclase* (monoclinic) and *microcline* (triclinic). Both are widely distributed in acid and intermediate rocks, esp in granites and syenites and the fine-grained equivalents. See ADULARIA, FELDSPAR, SANIDINE.

potassium ferricyanide (ferrocyanide) (*Chem*) *Potassium hexacyano ferrate (III), (II).* $K_3Fe(CN)_6$, $K_4Fe(CN)_6$. Used in chemical analysis. Gives characteristic colour reactions.

Also used in dyeing, etching, blue print paper and as a fertilizer. With other iron salts it gives *Prussian blue*.

potassium hexachloroplatinate (*Chem*) K_2PtCl_6. Results from the reaction of chloroplatinic acid and potassium chloride.

potassium hydride (*Chem*) KH. A saline hydride with the NaCl structure. Formed when potassium is heated in hydrogen.

potassium hydrogen fluoride (*Chem*) KHF_2. Formed when potassium fluoride is dissolved in hydrofluoric acid and the solution evaporated. Contains the ion $[F—H—F]^-$.

potassium iodate (V) (*Chem*) KIO_3. Potassium salt of iodic acid.

potassium iodide (*Chem*) KI. Used in chemical analysis, medicine, photography.

potassium mica (*Min*) See MUSCOVITE, SERICITE.

potassium nitrate (*Chem*) KNO_3. Salt of potassium and nitric acid. Strong oxidizing agent. Used in pyrotechnics, explosives, HT salts, glass manufacture and as a fertilizer. Also known as *nitre, saltpetre*.

potassium oxalate (*Chem*) *Potassium ethandioate*. The normal salt, $K_2C_2O_4.H_2O$, is soluble in water. The acid salt, KHC_2O_4, is less soluble and occurs in many plants. A compound of these two, potassium quadroxalate, $K_3HC_4O_8.2H_2O$, is used for bleaching and removing iron stains.

potassium permanganate (*Chem*) *Potassium manganate (VII)*. $KMnO_4$. Dark purple-brown crystals and solution, with characteristic sweet–bitter taste. Strong oxidizing agent. Used in analytical chemistry and as a disinfectant.

potassium propionate (*Chem*) *Potassium propanoate*. CH_3CH_2COOK. Compound used as a fungicide in edible products such as baked goods.

potassium sorbate (*Chem*) $CH_3(CH=CH)_2COOK$. Anti-fungal agent in edible products used in place of sorbic acid when high water solubility is needed.

potassium-sparing diuretics (*Pharmacol*) A class of mild diuretics that act on the kidney to promote excretion of water without loss of potassium ions. Eg *amiloride*.

potato blight (*BioSci*) A destructive disease of the potato caused by either of the parasitic fungi *Alternaria solani* (early blight) or *Phytophthora infestans* (late blight).

potency (*BioSci*) In toxicology, a term for the relative toxicity of an agent as compared with a given or implied standard or reference agent.

potential (*BioSci*) See LATENT.

potential (*Phys*) Scalar magnitude, negative integration of the electric (or magnetic) field intensity over a distance. Hence all potentials are relative, there being no absolute zero potential, other than a convention, eg earth, or at infinite distance from a charge.

potential attenuator (*ElecEng*) Same as *potentiometer (1)*, as contrasted with a normal attenuator which adjusts power.

potential barrier (*Phys*) The maximum in the curve covering two regions of potential energy, eg at the surface of a metal, where there are no external nuclei to balance the effect of those just inside the surface. Passage of a charged particle across the boundary should be prevented unless it has energy greater than that corresponding to the barrier. Wave-mechanical considerations, however, indicate that there is a definite probability for a particle with less energy to pass through the barrier, a process known as *tunnelling*. Also *potential hill*. Cf POTENTIAL WELL.

potential coefficients (*Phys*) Parts of total potential of a conductor produced by charges on other conductors, treated individually.

potential-determining ions (*MinExt*) Ions which leave the surface of a solid immersed in an aqueous liquid before saturation point (equilibrium) has been reached.

potential difference (*Phys*) (1) The difference in potential between two points in a circuit when maintained by an emf or by a current flowing through a resistance. In an electrical

field it is the work done or received per unit positive charge moving between the two points. Abbrev *pd*. SI unit is the VOLT. (2) The line integral of magnetic field intensity between two points by any path.

potential divider (*ElecEng*) See VOLTAGE DIVIDER.

potential drop (*Phys*) The difference of potential along a circuit because of current flow through the finite resistance of the circuit.

potential energy (*Phys*) Universal concept of energy stored by virtue of position in a field, without any observable change, eg after a mass has been raised against the pull of gravity. A body of mass m at a height h above the ground possesses potential energy mgh, since this is the amount of work it would do in falling to the ground. In electricity, potential energy is stored in an electric charge when it is taken to a place of higher potential through any route. A body in a state of tension or compression (eg a coiled spring) also possesses potential energy.

potential evapotranspiration (*EnvSci*) The theoretical maximum amount of water vapour conveyed to the atmosphere by the combined processes of evaporation and transpiration from a surface covered by green vegetation with no lack of available water in the soil.

potential fuse (*ElecEng*) A fuse used to protect the voltage circuit of a measuring instrument or similar device.

potential galvanometer (*ElecEng*) A galvanometer having a resistance sufficiently high to enable it to be used as a voltmeter.

potential gradient (*ElecEng*) Potential difference per unit length along a conductor or through a dielectric; equal to slope or curve relating potential and distance. Also *electromotive intensity*.

potential hill (*Phys*) See POTENTIAL BARRIER.

potential indicator (*ElecEng*) An instrument which shows whether a conductor is alive. Also *charge indicator*.

potential instability (*EnvSci*) The condition of a layer of the atmosphere which is in a state of STATIC STABILITY but in which instability would appear if it were lifted bodily until it became saturated. US *convective instability*.

potential temperature (*EnvSci, Phys*) The temperature which a specimen of gas would have if it were brought to standard pressure adiabatically. The potential temperature is given by the expression:

$$\theta = T\left(\frac{P_0}{P}\right)^{\frac{\gamma-1}{\gamma}}$$

where T is the absolute temperature, P the pressure of the gas, P_0 the standard pressure, and γ the ratio of the specific heat capacities (= 1·40 for air).

potential theory (*MathSci*) That branch of applied mathematics which studies the properties of a potential function without reference to the particular subject (eg hydrodynamics, electricity and magnetism, gravitational attraction) in which the function is defined.

potential transformer (*ElecEng*) An undesirable synonym for VOLTAGE TRANSFORMER.

potential trough (*Phys*) The region of an energy diagram between two neighbouring hills, eg arising from the inner electron shells of an atom.

potential vorticity (*EnvSci*) The vorticity which a column of air between two isentropic surfaces would have if it were brought by an adiabatic process to an arbitrary standard latitude and then stretched or shrunk to an arbitrary standard thickness. It is a conservative air mass property for adiabatic processes. If the original values of the vertical component of vorticity, the CORIOLIS PARAMETER, and the *thickness* are ζ_0, f_0 and h_0, while those after the standardization are ζ_s, f_s and h_s, then

$$\frac{(\zeta_0 + f_0)}{h_0} = \frac{(\zeta_s + f_s)}{h_s}$$

See ERTEL POTENTIAL VORTICITY.

potential well (*Phys*) (1) Localized region in which the potential energy of a particle is appreciably lower than that outside. Such a well forms a trap for an incident particle which then becomes *bound*. Quantum mechanics shows that the energy of such a particle is quantized. Applied particularly to the variation of potential energy of a nucleon with distance from the nucleus. (2) Similarly, the region between atoms created by the balance between attractive and repulsive interatomic forces. See CHEMICAL BOND.

potentiation (*BioSci*) The phenomenon whereby a substance or physical agent at a level that has no effect, or no adverse effect, enhances the effect, or harm, done by another substance or physical agent. A specialized example is the augmented quantal release of transmitter at the nerve synapse after repeated stimulation; it is long-lasting compared with *facilitation* (minutes to hours). See LONG-TERM POTENTIATION.

potentiometer (*ElecEng*) (1) Precision-measuring instrument in which an unknown potential difference of emf is balanced against an adjusted potential provided by a current from a steady source. (2) Three-terminal voltage divider, often shortened to *pot*. The resistance change with shaft rotation of slider position may follow various laws, eg linear, logarithmic, cosine, etc, potentiometers. A form of *rheostat*.

potentiometer braking (*ElecEng*) A braking method used for series motors; the series field and a rheostat are connected in series across the supply, and the armature is connected across the field and a variable proportion of the rheostat.

potentiometer card generator (*ElecEng*) Potentiometer wound on a card whose shape determines the rate of change of the function.

potentiometer function generator (*ElecEng*) One in which functional values of voltage are applied to points on a potentiometer, which becomes an interpolator.

potentiometer-type field rheostat (*ElecEng*) A field rheostat which is connected across the supply, the field winding being connected between one pole of the supply and a variable tapping on the rheostat. See REVERSIBLE POTENTIOMETER-TYPE FIELD RHEOSTAT.

potentiometric titration (*Chem*) A titration in which the end-point is indicated by a change in potential at an electrode immersed in the solution. This change in potential occurs as the solution changes from having excess substance to be determined to having excess titrant.

potette (*Glass*) A hood shaped like a pot, but with no bottom, which is placed in a tank furnace so that it reaches below the glass level. It protects a worker gathering glass on a pipe or iron from furnace gases; also, the glass here is somewhat cooler than that in the main part of the furnace, where melting is taking place. Also *boot*, *hood*.

pot furnaces (*Glass*) Furnaces in which are set a number of *pots*. They may be: (1) direct-fired from below; (2) gas-fired from below through a central opening in the circular siege, using the recuperative principle; (3) fired through ports in the siege or in the walls, the waste gases escaping through similar openings. In the last-named process, which holds generally for non-circular furnaces, the regenerative principle may be used.

Potier construction (*ElecEng*) A graphical construction for determining the reactance, armature reaction, and regulation of a synchronous generator from the open-circuit, short-circuit and zero-power-factor characteristics.

Potier reactance (*ElecEng*) The reactance of a synchronous machine as determined by the POTIER CONSTRUCTION.

pot magnet (*ElecEng*) One embracing a coil or similar space, excited by current in the coil or a permanent magnet in the central core. Main use is with a circular gap at an end of the core for a moving coil. Miniature split-sintered pot magnets are also used to contain high-frequency coils.

potometer (*BioSci*) An apparatus to measure the rate of uptake of water by a plant or a detached shoot etc and often, thus, indirectly to estimate transpiration.

pot still (*Chem*) A still consisting of a boiling vessel with condenser attached. The use of a fractionating column is optional. Used for batch distillation.

potstone (*Min*) A massive variety of STEATITE, more or less impure.

Potter–Bucky grid (*Radiol*) A type of lead grid designed to avoid exposure of film to scattered X-radiation in diagnostic radiography. Mechanical oscillation of the grid eliminates reproduction of the grid pattern on the radiograph.

potters' clay (*Geol*) See BALL CLAY.

Pott's disease (*Med*) Tuberculous infection of the spinal column. Also *spinal caries*.

Pott's fracture (*Med*) Fracture dislocation of the ankle joint, the lower parts of the tibia and the fibula being broken.

potty-putty (*Eng*) TN for a silicone polymer. Also *bouncing putty*. See VISCOELASTICITY.

pouch (*BioSci*) Any sac-like or pouch-like structure; as the abdominal brood pouch of marsupials.

poughite (*Min*) A hydrated sulphate and tellurite of iron, crystallizing in the orthorhombic system.

poultice corrosion (*Eng*) That which occurs in pockets or on ledges particularly in cars subject to salt spray and dirt.

pound (*Build*) See POND.

pound (*Genrl*) The unit of mass in the old UK system of units established by the Weights and Measures Act (1856), and until 1963 defined as the mass of the Imperial Standard Pound, a platinum cylinder kept at the Board of Trade. In 1963 it was redefined as 0·453 592 37 kg. The US pound is defined as 0·453 592 427 7 kg.

poundal (*Phys*) Unit of force in the foot–pound–second system. The force that produces an acceleration of 1 ft s^{-2} on a mass of 1 pound. Symbol pdl: 32·2 pdl = 1 lbf (lb wt); 1 pdl = 0·138 255 N.

pound-calorie (*Genrl*) See CENTIGRADE HEAT UNIT.

pounding (*Ships*) The heavy falling of the fore end of the ship into the sea when it has been lifted clear of the water by wave action. Also striking the ground under the ship due to wave action.

Poupart's ligament (*Med*) The inguinal ligament.

Pourbaix diagram (*Eng*) Used in the prediction of metallic corrosion. See panel on RUSTING.

pouring basin (*Eng*) Part of the passage system for bringing molten metal to the mould cavity in metal casting. It is provided on large moulds, next to the sprue hole top, to simplify pouring and to prevent slag from entering the mould. See fig. at MOULDING.

pour point (*Chem*, *MinExt*) The lowest temperature at which a petroleum-based oil, chilled under test conditions, will flow.

powder (*Eng*, *PowderTech*) Discrete particles of dry material in the range 0·1–1000 μm.

powder core (*ElecEng*) The core of powdered magnetic material with an electrically insulating binding material to minimize the effects of eddy currents which are used for high-frequency transformers and inductors with low loss in power.

powder density (*Plastics*) The mass in grams of 1 cm^3 of loose moulding powder.

powderless etching (*Print*) The etching of line blocks in one stage without recourse to DRAGON'S BLOOD between a series of etches, using specially designed etching baths with several features including close temperature regulation, planetary movement of the plate, and controlled application of the etching fluid. A measured quantity of special inhibitor is added to the etchant and this forms a protecting film over the metal to control the progress of the etch, by protecting the sides of the lines from undercutting, and stopping the etch when a suitable depth is reached. Originally introduced for line work only, on magnesium alloy, it has been adapted for half-tone and is particularly suitable for combined work, using micrograin zinc or copper.

powder metallurgy (*Eng*) The working of metals and certain carbides in powder form by pressing and sintering. Used to produce self-lubricated bearings, tungsten filaments and shaped cutting inserts (carbide). See CEMENTED CARBIDES.

powder method (*Min*) See POWDER PHOTOGRAPHY.

powder photography (*Min*) A method of identification of minerals or crystals, in which the powdered preparation, mounted vertically in a special camera in which it rotates, is subjected to a suitably modified beam of X-rays. The pattern, characterized by a set of concentric rings produced by rays diffracted at the Bragg angle relative to the incident beam, is diffracted onto a surrounding strip of film to give positive identification. Also *powder method*.

powder porosity (*PowderTech*) The ratio of the volume of voids plus the volume of open pores to the total volume of the powder. Cf POROSITY.

powder-post beetle attack (*For*) Attack on sapwood of timber for its starch content by Bostrychidae in the tropics and Lyctidae in temperate regions.

powder technology (*Genrl*) The technology covering the production and handling of powders, particle size analysis and the properties of powder aggregates.

powdery mildew (*BioSci*) One of several plant diseases of, eg cereals and apples, caused by biotrophic fungi of the order Erysiphales.

power (*MathSci*) (1) An EXPONENT or INDEX, denoted by a small numeral placed above and to the right of a numerical quantity, which indicates the number of times that quantity is multiplied by itself. (2) The complement of the probability of an ERROR OF THE SECOND KIND, the probability of the correct rejection of the status quo, working or null HYPOTHESIS in favour of a specified alternative.

power (*Phys*) Rate of doing work. Measured in JOULES per second and expressed in watts ($1\,W = 1\,J\,s^{-1} = 10^7\,ergs^{-1}$). The foot–pound-second unit of power is the *horsepower*, which is a rate of working equal to $550\,ft\,lbf\,s^{-1}$; 1 horsepower is equivalent to 745·7 W.

power amplification (*ElecEng*) Increasing the power of a given signal to a level sufficient to drive output devices or circuits, the ratio of input to output level being expressed in decibels. See APPLIED POWER.

power amplifier (*ElecEng*) The stage designed to deliver the required power output with a specified degree of non-linear distortion, gain not being considered. In some cases, the voltage gain is fractional (or, in dB, negative). Also *power unit*.

power-assisted controls (*Aero*) Primary flying controls wherein the pilot is aided by electric motors or double-acting hydraulic jacks.

power-assisted steering (*Autos*) In this system, a hydraulic ram connects to the steering linkage and assists the steering effort. The ram is powered by a hydraulic pump driven off the engine and is controlled by a valve which responds to movements of the steering wheel.

power breeder (*NucEng*) A nuclear reactor which is designed to produce both useful power and fuel.

power budget (*ICT*) A calculation carried out in the planning of a communications link that, taking into account such factors as amplifier noise, non-linearity and ageing of components, indicates how much power is required to achieve a given measure of performance over the lifetime of the link.

power circuit (*ElecEng*) That portion of the wiring of an electrical installation which is used to supply apparatus other than fixed lighting, usually through socket outlets.

power coefficient (*NucEng*) The change of reactivity of a reactor with increase in power. In a heterogeneous reactor, owing to temperature differences of the fuel, moderator and coolant, it differs from the isothermal temperature coefficient of reactivity which assumes temperatures throughout the core change by the same amount.

power component (*ElecEng*) See ACTIVE COMPONENT.

power control rod (*NucEng*) Movable rod used to control the power level of a nuclear reactor. Usually a neutron-absorbing rod containing cadmium or boron steel, but may be a fuel rod or part of the moderator. See panel on NUCLEAR REACTORS.

power controls (*Aero*) A primary flying control system where movement of the surfaces is done entirely by a power system, commonly hydraulic, but sometimes electrohydraulic or electric. The system is always at least duplicated (power source, supply lines and operating rams) and both circuits are usually running continuously. Reversion to manual control was common in early systems, but the trend is towards a multiplication of reserve circuits and supply sources. See POWER-ASSISTED CONTROLS, Q-FEEL.

power demodulator (*ElecEng*) See POWER DETECTOR.

power density (*NucEng*) Energy released per second per unit volume of a reactor core.

power detector (*ElecEng*) A rectifier which accepts a relatively large carrier voltage and thereby achieves low non-linear distortion in the demodulation process. Also *power demodulator*.

power drag line (*Eng*) An excavator or tracked crane fitted with a bucket which is filled by dragging it through the material with a drag rope passing through a fairlead mounted on the car body and tipped by a rope passing over the jib-head sheave. Larger machines are called walking draglines or walkers.

powered supports (*MinExt*) Pit props held to the roof of a coal seam by hydraulic pressure. In fully mechanized collieries they form part of a mechanically operated system.

power efficiency (*Eng*) The ratio of power delivered by a transducer (optical, mechanical, acoustical, electrical, etc) to the power supplied; usually quoted as a percentage.

power factor (*ElecEng*) The ratio of the total power (in watts) dissipated in an electric circuit to the total equivalent volt-amperes applied to that circuit. In single and balanced three-phase systems, it is equal to $\cos\varphi$, where φ is the phase angle between the applied voltage and the applied current in a single-phase circuit, or between the phase voltage and phase current in a balanced three-phase circuit. In normal dielectrics, it is exactly equal to $G(G^2+\omega C)^{2-0.5}$, where C = capacitance, G = shunt conductance, $\omega = 2\pi \times$ frequency, and thus nearly equals $G\omega^{-1}C^{-1}$. Abbrev *pf*.

power-factor indicator (*ElecEng*) An instrument which reads the power factor of a circuit directly. Also *power-factor meter*.

power feed (*Eng*) The power-operated feed motion of tables, heads, etc, of machine tools, used when the duration of feed, loads and type of cut justify it.

power frequency (*ElecEng*) The frequency of the ac supply mains (UK 50 Hz, US 60 Hz). Also *mains frequency*.

power gain (*ICT*) The ratio of the power delivered to the load, by an amplifier, transducer or system compared with the power absorbed at the input. Usually expressed in dB.

power hammer (*Eng*) Any type of hammer which is operated, either continuously or intermittently, by power, eg by directly coupling the hammer to a steam or pneumatic cylinder.

power-law index (*Eng*) Exponent of shear rate, symbol n, for modelling flow behaviour of non-Newtonian polymer melts etc. Power-law equation is

$$\tau = k(\dot\gamma)^n$$

where k is the viscosity coefficient and $\dot\gamma$ is the shear rate or consistency index. For Newtonian fluids, $n = 1$ and k is the shear viscosity (η). Dilatant behaviour occurs if $n > 1$, pseudo-plasticity if $n < 1$. Also *flow behaviour index*. See NEWTONIAN VISCOSITY.

power level (*ICT*) See TRANSMISSION LEVEL.

power-level diagram (*ICT*) Diagram indicating how maximum power levels vary at different points of a transmission channel, thereby indicating how various losses are neutralized by appropriate amplifier gains.

power-level indicator (*ICT*) See LEVEL INDICATOR.

power line (*ElecEng*) (1) US for *mains*. (2) See BUS-LINE.

power loading (*Aero*) The gross weight of a propeller-driven aircraft divided by the take-off power of its engine(s). For jet aircraft, *thrust loading*.

power loss (*ElecEng*) (1) The ratio of the power absorbed by a transducer to that delivered to the load. (2) The energy dissipated in a passive network or system.

power of lens (*ImageTech*) The relative focusing power of a lens, measured in dioptres, which is the reciprocal of the focal length in metres.

power-on self-test (*ICT*) Checks carried out automatically by the BIOS software when the computer is first switched on. These procedures verify that essential hardware devices such as memory, display and keyboard are connected and operating correctly. Errors are reported to allow a fault diagnosis to be carried out if necessary. Abbrev *POST*.

power output (*Eng*) The net useful power delivered by a *prime mover* for external use.

power-pack (*ElecEng*) Power-supply unit for an amplifier, eg in a radio or TV receiver which supplies the necessary steady voltages

power ramping (*NucEng*) Fairly rapid increase in power of a reactor after a prolonged period at some lower level.

power range (*NucEng*) See START-UP PROCEDURE.

power rating (*Aero*) The power, authorized by current regulations, of an aero-engine under specified conditions, eg maximum take-off rating, combat rating, maximum continuous rating, weak-mixture cruising rating, etc. The conditions are specified by rpm and, for piston engines, MANIFOLD PRESSURE and torque (in large engines), for turboprops JET-PIPE TEMPERATURE and torque, for turbojets exhaust GAS TEMPERATURE, for rocket motors COMBUSTION-CHAMBER pressure.

power reactor (*NucEng*) One designed to produce useful power.

power relay (*ElecEng*) One which operates at a specified power level.

power series (*MathSci*) A series of the form $\sum a_n x^n$.

power set (*MathSci*) The set $P(S)$ of all subsets of a given set S. If S has n elements then $P(S)$ has 2^n elements. $P(S)$ always has more members than S, even when S is infinite.

power shovel (*Eng*) An excavator equipped with a short stiff jib to which is fitted the dipper arm carrying the digging bucket. Modern machines have a crawl action allowing the effective length of the dipper arm to be varied to assist penetration of hard material and allow better control of the shape of the cut. Also *crawl shovel, forward shovel, front-end shovel*.

power supply (*Phys*) (1) Arrangement for delivering available power from a source, eg public mains, in a form suitable for circuit components such as valves or transistors; generally involves a transformer, rectifier, smoothing filter, circuit breaker or other protection and frequently incorporating electronic regulation. In a full-wave supply, use is made of a full-wave rectifier and filter. (2) US for MAINS.

power take-off (*Agri*) Coupling to allow machinery to be powered by the engine of a tractor, or other farm vehicle. Abbrev *PTO*.

power transformer (*Electronics*) One with its primary connected to the ac mains supply, and one or more secondary windings which supply the lower- or higher-voltage supplies required for the various devices and subassemblies within a complete piece of electronic equipment. Also *mains transformer*.

power transistor (*ElecEng*) One capable of being used at power rating of greater than about 10 W, and generally requiring some means of cooling.

power unit (*Aero*) An engine (or assembly of engines) complete with any extension shafts, reduction gears or propellers.

power unit (*ElecEng*) See POWER AMPLIFIER.

powerwash (*Build*) A machine used to direct a high pressure jet of water, perhaps containing added chemicals, at a surface to clean dirt and debris. Used to clean masonry and as a preparation for painting.

pox (*Med*) Pl of *pock*; hence popular names for diseases characterized by pustules, eg *chickenpox, smallpox*; specifically (vulgar), syphilis.

pox viruses (*BioSci*) A group of fairly large viruses, brick-shaped in shadow-cast electron micrographs, usually characterized by the formation of cytoplasmic inclusion bodies in the cells they invade; usually cause skin lesions, eg smallpox virus.

POY (*Textiles*) Abbrev for PARTIALLY ORIENTED YARN.

Poynting–Robertson effect (*Astron*) An effect whereby small particles of dust in the solar system slowly fall into the Sun. Solar radiation causes them to lose angular momentum, and as a result they drift closer to the Sun. For particles smaller than 1 micron, RADIATION PRESSURE is great enough to counteract the effect and indeed blow the finest dust out of the solar system altogether.

Poynting's theorem (*ICT*) That the rate of flow of energy through a surface is equal to the surface integral of the Poynting vector formed by the components of field lying in the plane of the surface. Used for calculating the power radiated from antennas, or through waveguide systems.

Poynting vector (*ElecEng*) One whose flux, through a surface, represents the instantaneous electromagnetic power transmitted through the surface. Equal to vector cross-product of the electric and magnetic fields at any point. In electromagnetic wave propagation, where these fields have complex amplitudes, half the vector product of the electric field intensity and the complex conjugate of the magnetic field intensity is termed the *complex Poynting vector*, and the real part of this gives the time average of the power flux.

Pozidrive (*Genrl*) TN of a type of screwdriver with a cross-shaped tip or the screws that it drives.

pozzuolana (*CivEng, Geol*) A volcanic dust, first discovered at Pozzuoli in Italy, which has the effect, when mixed with mortar, of enabling the latter to harden either in air or under water. Also *pozzolana*.

PP (*Plastics*) Abbrev for POLYPROPYLENE.

p–p (*ImageTech*) Abbrev for *peak to peak*. The magnitude of an alternating voltage measured between negative and positive peaks.

PPC (*Build*) Abbrev for *plain plaster cornice*.

pP heterojunction (*Electronics*) A junction in a HETEROSTRUCTURE formed between a narrower band-gap material (p-type) and a wider gap material (P-type). Also *p–P heterojunction*.

PPI (*Pharmacol*) Abbrev for PROTON PUMP INHIBITOR.

PPI (*Radar*) Abbrev for PLAN-POSITION INDICATOR.

ppi (*Textiles*) Abbrev for *picks per inch*. See PICK.

pp junction (*Electronics*) One between p-type crystals having different electrical properties. Also *p–p junction*.

ppm (*Chem*) Abbrev for *parts per million*.

ppm (*ICT*) Abbrev for *pages per minute*. This term is used to specify the printing speed of a printer.

PPO (*Plastics*) Abbrev for POLYPHENYLENE OXIDE.

PPQ bar (*BioSci*) Abbrev for PTERYGOPALATOQUADRATE BAR.

P-protein (*BioSci*) Phloem-protein. Present in phloem cells, esp in sieve elements.

PPS (*Plastics*) Abbrev for POLYPHENYLENE SULPHIDE.

ppt (*Chem*) Abbrev for *precipitate*. Also *ppte*.

Pr (*Chem*) Symbol for: (1) PRASEODYMIUM; (2) the propyl radical C_3H_7-.

PRA (*NucEng*) Abbrev for PROBABILISTIC RISK ASSESSMENT.

praben (*Chem*) Any of several chemicals, used as preservatives, that are esters of para-hydroxybenzoic acid $C_7H_6O_3$.

practical units (*Phys*) Obsolete system of electrical units, whereby the ohm, ampere and volt were defined by physical magnitudes. Replaced now by SI units, in which they are defined in terms of arbitrarily fixed units of length, mass, time and electric current.

Prader–Willi syndrome (*Med*) A rare genetic condition characterized mainly by excessive appetite, leading to obesity, and also by mental handicap, growth and behavioural problems.

prae- (*Genrl*) See PRE-.

praecoces (*BioSci*) Birds which when hatched have a complete covering of down and are able at once to follow the mother on land or into water to seek their own food. Cf ALTRICES.

Praesepe (*Astron*) A well-known open star cluster in the constellation Cancer. Contains over 100 stars and can almost be resolved by the naked eye.

Prandtl number (*ChemEng*) A dimensionless group much used in heat exchange calculations. It is given by (specific heat capacity at constant pressure × dynamic viscosity)/ (thermal conductivity).

prase (*Min*) A translucent and dull leek-green variety of CHALCEDONY.

praseodymium (*Chem*) Symbol Pr, at.no. 59, ram 140·9077, rel.d. 6·48, mp 940°C. A metallic element, a member of the rare earth group. It closely resembles neodymium and occurs in the same minerals.

Pratt truss (*Eng*) See WHIPPLE–MURPHY TRUSS.

Prausnitz–Kustner reaction (*BioSci*) An obsolete skin reaction for the detection and measurement of human reaginic (IgE) antibodies. Now replaced by *in vitro* assays.

PRBS (*ICT*) Abbrev for PSEUDO-RANDOM BINARY SEQUENCE.

pre- (*Genrl*) Prefix from Lt *prae-*, in front of, before.

pre-adaptation (*BioSci*) Change of structure preceding appropriate change of habit.

pre-amplifier (*Acous*) Amplifier with at least a stage of valve or transistor gain following a high-impedance source from which the level is too low for line transmission and clearance above noise level.

prebiotic (*BioSci*) The very different conditions existing on Earth before the appearance of life, which provided an environment in which the first living organisms could evolve from non-living molecules.

preboarding (*Textiles*) See BOARDING.

Precambrian (*Geol*) That period of geological time before the beginning of the PHANEROZOIC. It represents about 90% of all geological time.

precast (*CivEng*) Said of concrete blocks etc which are cast separately before they are fixed in position.

precast stone (*CivEng*) See RECONSTRUCTED STONE.

precautionary principle (*Genrl*) The increasingly prevalent idea that, if the consequences of an action are unknown but might have the potential to cause major or irreversible negative effects, then that action should be avoided. Opponents would argue that this is a recipe for inaction and that it stifles progress.

precaval vein (*BioSci*) The anterior vena cava conveying blood from the head and neck to the right auricle. Called *superior vena cava* in humans.

precession (*Eng*) An effect exhibited by a rotating body, such as a gyroscope, when a torque is applied to it in such a way as to tend to change the direction of its axis of rotation. If the speed of rotation and the magnitude of the applied torque are constant, the axis slowly generates a cone as its precessional motion.

precession (*Phys*) A regular cyclic motion of a dynamical system in which, with suitably chosen co-ordinates, all except one remain constant, eg the regular motion of the inclined axis of a top around the vertical.

precession of the equinoxes (*Astron*) The westward motion of the equinoxes caused mainly by the attraction of the Sun and Moon on the equatorial bulge of the Earth. This luni-solar precession together with the smaller planetary precession combine to give the general precession amounting to 50·27″ per annum. The equinoxes thus make one complete revolution of the ecliptic in 25 800

years, and the Earth's pole turns in a small circle of radius 23°27′ about the pole of the ecliptic, thus changing the co-ordinates of the stars.

prechordal (*BioSci*) Anterior to the notochord or to the spinal cord.

precious stones (*Min*) See GEM.

precipitable water (*EnvSci*) The total mass of water in a vertical atmospheric column of unit area, or its height if condensed in liquid form.

precipitation (*BioSci*) The formation of a visible aggregate when two solutions are mixed. See PRECIPITIN TEST.

precipitation (*Build*) The process of assisting the settlement of suspended matters in sewage by the addition of chemicals to the sewage before admission to the sedimentation tanks.

precipitation (*Chem*) The formation of an insoluble solid by a reaction which occurs in solution. It is widely used for the separation and identification of substances in chemical processes and analyses. N and v *precipitate*.

precipitation (*EnvSci*) Moisture falling on the Earth's surface from clouds; it may be in the form of rain, hail or snow.

precipitation hardening (*Eng*) In a metal or alloy, precipitation of one phase in the lattice of another of different ionic diameter. This keys the structure against 'creep' or slip under stress. Produced by precipitation during cooling of a supersaturated solution from temperatures T_1 to T_2 in the figure, and of concentration m. See AGEING, TEMPER-HARDENING.

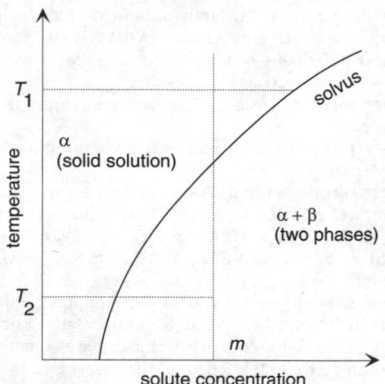

precipitation hardening

precipitator (*MinExt*) (1) A device, usually in flue stacks, to remove solid particles from the effluent gas. Can be electrostatic, mechanical and/or chemical in action. (2) A device for purifying boiler feed-water by adding chemicals to precipitate dissolved material.

precipitin (*BioSci*) An antibody that is analogous to AGGLUTININ, but whose action is characterized by clouding and precipitation.

precipitin test (*BioSci*) A serological test in which the reaction between soluble antigen and antibody results in the formation of a visible macromolecular precipitate. The test depends on the ratio of antigen and antibody being optimal.

precise levelling (*Surv*) Particularly accurate levelling in which the allowable discrepancy between two determinations of the level difference between two bench marks M km apart is very low, of the order of $0.003\sqrt{M}$ or less.

precision-approach radar (*Aero*) A primary radar system which shows the exact position of an aircraft during its approach for landing. Abbrev *PAR*.

precision drill (*Agri*) A seed drill that places individual seeds into the seed bed.

precision grinding (*Eng*) Grinding to a high finish and accurate dimensions on an appropriate machine. Cf OFFHAND GRINDING.

preclimax (*BioSci*) A seral stage which just precedes the CLIMAX.

precocial animals (*BioSci*) Precocious; used particularly of birds that hatch with a downy covering and can forage immediately.

pre-combustion chamber (*Autos*) A small chamber formed in the cylinder head of some compression–ignition engines into which the oil fuel is injected at the end of the compression stroke. The high pressure caused by the partial combustion of the fuel expels the rich mixture through a neck or perforated throat plate into the engine cylinder, where combustion is completed. Derived from the oil engine of Akroyd Stuart (1892). Also *antechamber*.

preconscious (*Psych*) Memories, ideas, etc, that are currently absent from the conscious mind but that can be readily recalled.

precoracoid (*BioSci*) An anterior ventral bone of the pectoral girdle in amphibians and reptiles, corresponding to the epicoracoid of Monotremes.

precordial (*Med*) Situated or occurring in front of the heart.

precursor (*Phys*) A nuclide which precedes another in a radioactive decay chain.

predation (*BioSci*) A form of species interaction in which an individual of one species of animal (the *predator*) directly attacks, kills and eats one of another species (the *prey*). The predator is usually larger than its prey. Cf PARASITE.

predator (*BioSci*) See PREDATION.

predicate (*MathSci*) The term in a statement of the form 'A is B' that ascribes a property to the other term. See PREDICATE CALCULUS.

predicate calculus (*MathSci*) A branch of MATHEMATICAL LOGIC that deals with the properties and relations of propositions that are analysed into subject and predicate. In 'A is B', A is the subject, B is the *predicate*. In 'All As are Bs', 'No As are Bs' and 'Some As are Bs', 'all', 'some' and 'none' are *quantifiers*.

predictive autofocus control (*ImageTech*) A through-the-lens automatic focusing system in a SINGLE-LENS REFLEX camera, which tracks a moving subject and predicts where it will be at the completion of the shutter-release sequence.

pre-distorting network (*ICT*) A network that anticipates subsequent frequency distortion in the transmission path, as along a line, so that the line distortion has not entirely to be compensated at the receiving end.

pre-distortion (*ICT*) The principle of altering the response of a circuit to compensate, fully or partially, anticipated distortion; the aim is to make transmission as high as practicable above the anticipated noise level.

prednisolone (*Pharmacol*) Synthetic steroid closely related to CORTISONE and used to suppress inflammatory and allergic disorders.

prednisone (*Pharmacol*) A prodrug of PREDNISOLONE, used for the treatment of inflammatory and allergic disorders.

Preece's formula (*ElecEng*) A formula stating that the fusing current of a wire is proportional to the rated current to the power of $\frac{3}{2}$.

pre-eclampsia (*Med*) The occurrence in pregnancy of oedema, high blood pressure and albuminuria. If untreated may progress to ECLAMPSIA.

pre-emergence treatment (*Agri*) Any treatment made after a crop is drilled but before the crop plant, or associated weeds, have become visible.

pre-emphasis (*ICT*) The process of increasing the strength of some frequency components of a signal to assist these components to override noise or other distortion in the system; mainly to ensure good high-frequency reproduction in frequency-modulation sound broadcasting and in recording. The original levels are restored by DE-EMPHASIS.

preen gland (*BioSci*) See OIL GLAND.

preening (*BioSci*) A form of grooming behaviour performed by birds as part of feather maintenance.

prefabricated building (*Build*) A building for which walls, roofs and floors are constructed off the site. See SYSTEM BUILDING.

prefading (*Acous*) Listening to programme material and adjusting its level before it is faded up for transmission or recording.

preferendum (*BioSci*) The part of the range of a species in which it functions most successfully.

preferential mating (*BioSci*) See SEXUAL SELECTION.

preferential routing (*ICT*) A service sometimes offered to users of a VIRTUAL PRIVATE NETWORK whereby they, rather than other customers, are offered any available high-quality digital terrestrial links.

preferred orientation (*Eng*) During slip, metal crystals change their orientation; when a sufficient amount of deformation has been performed, the random orientation of the original crystals is converted into an arrangement in which a certain direction in all the crystals is parallel to the direction of deformation. Produces a texture in a rolled or wrought product.

preferred values (*Eng*) Where a range of components (eg screws, resistors) is made for use in manufactured articles, and they are identical except for one variable (eg diameter, resistance), their values usually follow a geometric progression, rounded at each stage to a convenient 'preferred' value, giving a roughly constant proportional increment of variable from one component to the next. For instance, resistors, accurate to $\pm 10\%$, are valued 10, 12, 15, 18, 22, ..., 68, 82, 100 ohms, each value being approx 20% higher than its predecessor.

prefloration (*BioSci*) See AESTIVATION.

prefoliation (*BioSci*) See VERNATION.

pre-frontal lobotomy (*Med*) See LEUCOTOMY.

preganglionic (*BioSci*) Refers to the axon of a neuron of the autonomic nervous system which originates in the central nervous system and terminates at a synapse with a ganglion. Cf POSTGANGLIONIC.

pre-gelatinized starch (*FoodSci*) See MODIFIED STARCH.

pregnancy (*Med*) Gestation. The state of being with child.

pregnancy test (*Med*) Any physiological test used to diagnose pregnancy at an early stage and with a very high degree of certainty.

pregnancy toxaemia (*Vet*) An energy deficiency of sheep brought on by increased demand of the fetus at the end of pregnancy. Assumed to be due to adreno-cortical problems and/or failure of liver metabolism. Ketone bodies accumulate in the tissues and are excreted in the urine and breath. Symptoms include depression, anorexia, nervous signs and, later on, recumbency. Also *ketosis, twin lamb disease*.

pregnanediol (*Chem*) Steroid found in the urine during pregnancy and at one stage of the menstrual cycle. It is excreted partly free and partly combined with glycuronic acid.

pregnant solution (*MinExt*) In the CYANIDING process for recovering gold, the gold-bearing solvent prior to precipitation and recovery. Also *pregs*.

pregs (*MinExt*) See PREGNANT SOLUTION.

prehallux (*BioSci*) In amphibians and mammals, a rudimentary additional digit of the hind-limb.

preheating time (*Electronics*) The minimum time for heating a cathode before other voltages are applied, thus ensuring full emission. Automatic delays are often incorporated in large amplifiers and radio transmitters.

prehensile (*BioSci*) Adapted for grasping.

prehnite (*Min*) Pale-green and usually fibrous hydrated silicate of calcium and aluminium, crystallizing in the orthorhombic system. It occurs in altered igneous rocks.

pre-ignition (*Autos*) The ignition of the charge in a petrol-engine cylinder before normal ignition by the spark; caused by overheated plug points, the presence of incandescent carbon, etc.

pre-impregnation (*Plastics*) Reinforcing tapes etc that have been impregnated with resin and partially cured before assembly in a structure which is then finally cured in a mould. Abbrev *prepreg*.

prelacteal (*BioSci*) In mammals, said of teeth developed prior to the formation of the milk dentition.

preliminary matter (*Print*) The pages of a book preceding the actual text. The order should be bastard title (half-title), frontispiece or 'advertisement', title page, copyright page, dedication, errata, preface, contents, list of illustrations, introduction. Frequently abbrev to *prelims*.

pre-loaded (*ICT*) Denotes software, operating system and applications already installed on a personal computer at the time of purchase.

premature ejaculation (*Psych*) Inability of the male to postpone ejaculation long enough to satisfy the female.

premaxillary (*BioSci*) A paired membrane bone of the vertebrate skull which forms the anterior part of the upper jaw; anterior to the maxilla. Also *premaxilla*.

premeiotic mitosis (*BioSci*) The nuclear division immediately preceding the organization of nuclei which will divide by meiosis.

premenstrual dysphoric disorder (*Med*) A severe form of premenstrual tension. Abbrev PMDD.

premiere (*ImageTech*) The first formal public presentation of a completed motion picture or video production.

premix (*ImageTech*) The combination of a number of sound-track components in preparation for the main mixing operation, to reduce the total number of channels to be handled.

premolars (*BioSci*) In mammals, the anterior grinding or cheek teeth, which are represented in the milk dentition.

premorse (*BioSci*) Looking as if the end had been bitten off.

prenylation (*BioSci*) Post-translational addition of prenyl groups to a protein. Farnesyl, geranyl or geranyl–geranyl groups may be added. The consequence is usually to promote membrane association. Also *isoprenylation*.

pre-operational thinking (*Psych*) In Piaget's theory, the period (pre-operational stage) from about 2 to 6 years when children's thinking is characterized by an ability to represent objects and events internally, but an inability to manipulate these representations in a logical way, thus accounting for the errors children make when asked to perform certain tasks.

preoperculum (*BioSci*) In fish, an anterior membrane bone forming part of the gill cover.

préparation mécanique (*MinExt*) See MINERAL PROCESSING.

prepollex (*BioSci*) In some vertebrates (amphibians, mammals) a rudimentary extra digit of the fore-limb.

prepolymer (*Plastics*) Low-molecular-mass polymer, often a viscous liquid, used as an intermediate material prior to final shaping (eg polytetramethylene ether glycol).

prepotent stimulus (*Psych*) A stimulus that takes precedence over other stimuli presented at the same time.

prepreg (*Aero*) Fibrous composite material consisting of unidirectional fibres embedded in matrix of resin prepared in the form of sheet or strip ready for forming, by combining several plies arranged in different directions into the final product.

prepreg (*Plastics*) Abbrev for PRE-IMPREGNATION.

pre-press proof (*Print*) A proof made before the printing plate is made, eg from the film elements of the job. Proofs are made to check quality of the film image, esp of the colour separation. It can also simulate the final printed product, eg as a guide to printer and customer.

pre-prophase band (*BioSci*) A band of microtubules 2–3 μm wide which forms and moves to the plasmalemma shortly before prophase begins, predicting the position where the new cell wall will join the old.

prepubic (*BioSci*) Pertaining to the anterior part of the pubis; in front of the pubis, as bony processes in some marsupials and rodents.

prepuce (*BioSci*) In mammals, the loose flap of skin which protects the glans penis. Adj *preputial*.

pre-roll (*ImageTech*) Increasing the accuracy of edits by having the EDIT CONTROLLER automatically rewind the tapes in the EDIT RECORDER and SOURCE PLAYER for several seconds and then switch both into play for their CONTROL TRACK pulses to be compared and synchronized before the edit point, when the edit machine is switched to record. See BACKSPACING.

presbyopia (*Med*) Long-sightedness and impairment of vision due to loss of accommodation of the eye in advancing years.

prescoring (*ImageTech*) In motion picture production, the use of previously recorded sound, such as a musical sequence, for cueing the action of dancers or singers.

preselection (*ICT*) A type of automatic switching used in telephone exchanges.

preselector gearbox (*Autos*) A gearbox, generally epicyclic, in which the gear ratio is selected, before it is actually required, by a small lever, being afterwards engaged by pressure on a pedal.

presensitized plate (*Print*) A lithographic plate with a light-sensitive coating applied by the manufacturer and supplied ready for exposure.

presentation (*Med*) The relation which the long axis of the fetus bears to that of the mother; various presentations are defined in terms of the presenting part, as *breech presentation, shoulder presentation*, etc.

presentation layer (*ICT*) Level 6 of the OPEN SYSTEMS INTERCONNECTION (OSI) model that provides the means to represent the format of the information exchange.

preservation (*FoodSci*) The treatment of food to prevent the growth of pathogens and spoilage organisms, ensuring the food remains wholesome and palatable for the expected shelf-life. Preservation techniques include one or any combination of: reduction of water activity (drying, concentration); thermal processing (pasteurization, sterilization); changing the chemical composition (pickling, curing, fermenting); physical methods (controlled atmosphere, freezing, chilling); exclusion of contaminants (barrier packaging, aseptic packaging).

preservation index (*FoodSci*) An empirical quantity used for pickles and sauces as a guide to evaluate their microbiological stability. Expressed as $PI = 100A/(100-S)$ where A is the total acetic acid and S is the total solids. In general a product with a $PI > 3.6$ should not be prone to microbial spoilage.

preservative (*FoodSci*) A food additive which has an anti-microbial effect, eg sulphur dioxide, sorbic acid, benzoic acid.

preset (*Electronics*) (1) To establish an initial value or condition, generally by setting one or more controls in advance. (2) A control, eg a variable resistance or capacitor, not as readily accessible or easily altered as the main controls, which can be adjusted to obtain initial values or operating conditions.

preset guidance (*Aero*) The guidance of controlled missiles by a mechanism which is set before launching and is subsequently unalterable.

pre-shrunk (*Textiles*) See COMPRESSIVE SHRINKAGE.

press (*Eng*) A machine used for applying pressure to a workpiece, via a tool, usually for the purpose of carrying out cold-working operations like cutting, bending, drawing and squeezing.

press (*Print*) (1) Any hand-operated machine used for proofing or printing small runs. (2) A general term for the printing stage of a job; exemplified in such phrases as 'going to press', 'in the press'. (3) A general term for the printing and publishing industry, particularly newspapers.

press-and-blow machines (*Glass*) Machines in which the PARISON is formed by the pressing action of a plunger forced into a mass of plastic glass dropped in a parison mould; the parison is then blown to the shape of the finished ware in another mould.

pressboard (*ElecEng*) Compressed paper in thick sheets. See ELEPHANTIDE PRESSBOARD.

press brake (*Eng*) A power press akin to a guillotine, used mainly for producing long bends, as in corrugating or seaming, but also for embossing, trimming and punching.

pressed amber (*Min*) See AMBROID.

pressed brick (*Build*) A high-quality brick moulded under pressure, as a result of which it has sharp arrises and a smooth face, making it esp suitable for exposed surface work.

presser (*Textiles*) The device that closes the beards on machines having bearded needles.

presser-foot (*Textiles*) On weft-knitting machines, the flexible thin-metal device that controls the position of the loops formed during knitting.

press fit (*Eng*) A class of fit for mating parts, tighter than a sliding fit and used when the parts do not normally have to move relative to each other.

press forging (*Eng*) A forging process using a vertical press to apply a slow squeezing action as distinct from the sudden impact employed in drop forging. Press forging tends to work the material uniformly throughout the section whereas deformation in drop forging tends to concentrate near the surface.

pressing (*Acous*) Vinyl disk record formed by pressure, with or without heat; the negative of the recording on a stamper is transferred to a large number of pressings for distribution.

pressing (*Textiles*) The application of pressure, often accompanied by heat and/or steam, to smooth fabrics or garments. Also the same process used to introduce chosen creases or pleats. In garment manufacture flat-bed presses are used.

pressing boards (*Print*) Glazed boards used for removing the impression from printed sheets.

press proof (*Print*) The last proof checked over before going to press.

press rolls (*Paper*) The heavy horizontal rolls situated in the press section of the paper- or board-making machine in various configurations to consolidate the web and remover water at the nips between them. The rolls may be constructed of granite, rubber-covered metal or with a perforated metal shell over a fixed internal suction box.

press tool (*Eng*) A tool for cutting, usually comprising at least a punch and a die, or for forming or assembling, used in a manual or in a power press.

pressure (*Genrl*) A measure of uniform stress defined as the force per unit area. See ATMOSPHERIC PRESSURE, STRESS.

pressure altitude (*Aero*) Apparent altitude of the local ambient pressure related to the International Standard Atmosphere.

pressure angle (*Eng*) In toothed gearing, the angle between the common tangent to the pitch circles of two teeth at their point of contact and the common normal at that point. Two angles are in common use: $14 \cdot 5°$ and $20°$.

pressure bag moulding (*Eng, Plastics*) See BAG MOULDING.

pressure bomb (*BioSci*) A thick-walled metal vessel used in investigations of plant water relations, eg to apply pressure by compressed air to an excised leaf placed within the vessel with its petiole emerging. Water is forced from the leaf when the air pressure equals the WATER POTENTIAL of the leaf cells.

pressure broadening (*Phys*) The broadening of spectral lines due to increase in pressure and thus due to the effect of neighbouring atoms on the radiating atom.

pressure cabin (*Aero*) An airtight cabin which is maintained at greater than atmospheric pressure for the comfort and safety of the occupants. Above 20 000 ft (6000 m) a differential of $6\frac{1}{2}$ lbf in^{-2} (45 kN m^{-2}) is usual, and above 40 000 ft (12 000 m) one of $8\frac{1}{4}$ lbf in^{-2} (57 kN m^{-2}). Pressurization can be either by a shaft-driven *cabin blower*, or by air bled from the compressor of turbine main engines.

pressure cable (*ElecEng*) A paper-insulated power cable operated under a hydrostatic pressure greater than atmospheric pressure by means of gas (usually nitrogen) contained in an outer steel pipe or, in more modern forms, an outer reinforced plastic sheath; this minimizes ionization.

pressure capsule (*Eng*) See SYLPHON BELLOWS.

pressure circuit (*ElecEng*) See VOLTAGE CIRCUIT.

pressure co-ordinates (*EnvSci*) A system of co-ordinates used in NUMERICAL FORECASTING in which the vertical ordinate is pressure, *p*. In this system, the vertical velocity *w* is replaced by the total derivative, following the motion, of the pressure, ie *dp/dt*.

pressure diecasting (*Eng*) A process by means of which precision castings of various alloys are made by squirting liquid metal under pressure into a metal die. See DIECASTING.

pressure drag (*Aero*) The summation of all aerodynamic forces normal to the surface, resolved parallel to free stream direction; sum of FORM DRAG and INDUCED DRAG.

pressure forging (*Eng*) See DROP FORGING.

pressure gauge (*Eng*) (1) A flattened tube bent to a curve, which tends to straighten under internal pressure, thus indicating, by the movement of an indicator over a circular scale, the fluid pressure applied to it. Also *Bourdon gauge*. (2) A liquid manometer.

pressure gradient (*EnvSci*) (1) The rate of change of the atmospheric pressure horizontally in a certain direction on the Earth's surface as shown by isobars on a weather chart. (2) The rate of change of pressure with distance over the ground, normal to the isobars. The force acting on the air is the *pressure-gradient force*.

pressure-gradient microphone (*Acous*) A microphone which offers so little obstruction to the passage of a sound wave that the diaphragm, in practice a ribbon, is acted on by the difference in the excess pressures on the two sides, and therefore tends to move with the particle velocity in the wave. Also *velocity microphone*.

pressure head (*Aero*) A combination of a STATIC PRESSURE and a PITOT TUBE which is connected to opposite sides of a differential pressure gauge, for giving a visual reading corresponding to the speed of an airflow. Also *Pitot-static tube*.

pressure helmet (*Aero, Space*) A flying helmet for the crew of high-altitude aircraft or spacecraft for use with a *pressure* or *partial pressure suit*. Usually of plastic, with a transparent face-piece, which may be in the form of a visor, the helmet incorporates headphones, microphone and oxygen supply, and there is usually a feeding trap near the mouth.

pressure in bubbles (*Phys*) A spherical bubble of radius *r*, formed in a liquid for which the surface tension is *T*, contains air (or some other gas or vapour) at a pressure which exceeds that in the liquid in its immediate vicinity by

$$\frac{2T}{r}$$

The excess pressure within a soap bubble in air is

$$\frac{4T}{r}$$

since the soap film has two surfaces.

pressure jet (*Aero*) A type of small jet-propulsion unit fitted to the tips of helicopter rotor blades, in which small size (to give low drag for AUTOROTATION) is of greater importance than the losses due to ejecting the efflux at pressures as high as 2 or 3 atmospheres.

pressure leaching (*MinExt*) Chemical extraction of values from ore pulp in autoclaves, perhaps followed by precipitation as metal or refined salt.

pressure microphone (*Acous*) A microphone in which the electrical signal is proportional to the fluctuating component of the pressure in front of the microphone as opposed to the PRESSURE-GRADIENT MICROPHONE.

pressure of atmosphere (*Phys*) See ATMOSPHERIC PRESSURE, BAROMETRIC PRESSURE.

pressure pad (*ImageTech*) The device which keeps the film in a gate so that it remains exactly in focus.

pressure-pattern flying (*Aero*) The use of barometric pressure altitude to obtain the most favourable winds for long-distance, high-altitude aerial navigation.

pressure probe (*BioSci*) A device for measuring the TURGOR PRESSURE within a plant cell by inserting into the cell a fine, fluid-filled capillary connected to a pressure transducer.

pressure ratio (*Aero*) The absolute air pressure, prior to combustion, in a GAS TURBINE, RAMJET or PULSE-JET, divided by the ambient pressure: analogous to the *compression ratio* of a reciprocating engine.

pressure roller (*Eng*) A roller sometimes used in centreless grinding of short, heavy workpieces to assist rotation of the workpiece.

pressure suit (*Aero*) An airtight fabric suit, similar to that of a diver, for very-high-altitude and space flight. It differs from the *partial pressure suit* in being loose-fitting, with bellows or other form of pressure-tight joint, to permit limited movement by the wearer.

pressure suppression (*NucEng*) A method of reducing the pressure of a coolant liquid after an accident by passing the steam generated into cold-water ponds and so condensing it.

pressure-tube reactor (*NucEng*) A reactor in which the fuel elements are contained in a large number of separate tubes through which the coolant water flows, rather than in a single pressure vessel, eg the Canadian CANDU and UK SGHWR reactors.

pressure-type capacitor (*ElecEng*) One with a dielectric of inert gas under high pressure, used for high-voltage work as it is self-healing after breakdown.

pressure unit (*Acous*) A metal or plastic dome forming the diaphragm of a small moving-coil loudspeaker unit situated in the throat of a horn, for use at intense acoustic pressures.

pressure vessel (*Eng*) A container for fluids stored under pressure above or below atmospheric. Classified as fired or unfired. Usually cylindrical with dished ends for ease of construction and support, or spherical for very high pressures.

pressure vessel (*NucEng*) Reactor-containment vessel, usually made of thick steel or prestressed concrete, capable of withstanding high pressure and used in gas-cooled and light-water reactors. See REACTOR VESSEL.

pressure waistcoat (*Aero*) A double-skinned garment, covering the thorax and abdomen, through which oxygen is passed under pressure on its way to the wearer's lungs to aid breathing at great heights, ie above 40 000 ft (12 000 m).

pressure welding (*Eng*) Welding parts which are pressed tightly together, as in FORGE WELDING or in various electrical resistance welding processes or in COLD WELDING.

pressurized (*Aero*) Fitted with a device that maintains nearly normal atmospheric pressure, eg in an aircraft.

pressurized-water reactor (*NucEng*) A reactor using water cooling at a pressure such that its boiling point is above the highest temperature reached. Abbrev *PWR*.

presswork (*Print*) Printing of the job, quality of the result depending on care and attention given; also *machining*, but not in the case of hand press printing.

Prestel (*ICT*) See VIDEOTEXT.

presternum (*BioSci*) (1) the anterior part of the sternum. (2) The reduced sternum of whalebone whales. (3) In Anura, an anterior element of the sternum, of paired origin and doubtful homologies.

prestressed concrete (*CivEng*) Concrete beams and other structural members, in which the whole member is placed in compression by the tension in strained steel or other tendons within it. This enables the member to withstand higher bending and tensile loads than without prestressing. Developed by the Fr engineer Freysinnet, in 1927. The two variants are pre-tensioned and post-tensioned. Cf REINFORCED CONCRETE. See panel on CEMENT AND CONCRETE.

presystolic (*Med*) Pertaining to, or occurring just before, the beginning of the systole of the heart, eg *presystolic murmur.*

pre-tensioned concrete (*CivEng*) A form of PRESTRESSED CONCRETE in which the reinforcing cables are tensioned before the concrete sets. See panel on CEMENT AND CONCRETE.

pre-tensioner (*Autos*) A device (usually used in conjunction with the traditional locking system) that causes a seat belt to tighten in the event of a sudden forward movement by the wearer.

pre-TR cell (*Radar*) A gas-filled radio-frequency switching valve which protects the TR TUBE in a radar receiver from excessive power. Also acts as a block to receiver frequencies other than the fundamental.

pretrematic (*BioSci*) Anterior to an aperture, as (in Selachii) that branch of the ninth cranial nerve which passes anterior to the first gill cleft. Cf POST-TREMATIC.

preview (*ImageTech*) (1) A special showing of a motion picture or video production to a limited audience before general public presentation. (2) In TV, viewing the picture immediately before transmission or recording.

prezygapophysis (*BioSci*) A facet or process on the anterior face of the neurapophysis of a vertebra, for articulation with the vertebra next in front. Cf POSTZYGAPOPHYSIS.

PRF (*ICT*) See PULSE REPETITION FREQUENCY.

Priabonian (*Geol*) A stage of the Eocene. See TERTIARY.

priapism (*Med*) Abnormally, and often painfully, persistent erection of the penis unaccompanied by sexual desire; may be due to local disease or to hypercoaguability of the blood.

Priapulida (*BioSci*) A phylum of coelomate, superficially segmented worm-like animals, living in mud. They have a straight gut with an anterior mouth and a posterior anus. The nervous system is not separated from the epidermis and the urinogenital system is simple, with solenocytes.

Price's guard wire (*ElecEng*) A conductor placed around the edge of a piece of insulating material under test; it is arranged to be at the same potential as the surface of the material to prevent a leakage current from the surface to earth.

pricking-up (*Build*) The operation of scoring the surface of the first coat of plaster to provide a key for the next: the whole operation of laying and scoring such a coat.

prickle (*BioSci*) A hard, sharp-pointed outgrowth of the epidermis, a multicellular trichome which is not vascularized, the 'thorn' of the rose.

prickly heat (*Med*) See MILIARIA.

Pridoli (*Geol*) The youngest epoch of the Silurian period.

prill (*Eng*) (1) To make granular or crystalline solids fluid, eg for extraction from slag. (2) To turn into pellet form by melting and letting the drops solidify in falling. (3) Bullion bead produced by cupellation of lead button during fire assay of gold or silver.

primacord fuse (*MinExt*) A fuse based on PENTAERYTHRITOL TETRANITRATE, with detonating effect. Speed of detonation, 7000 m s^{-1}.

primacy effect (*Psych*) (1) In impression formation, the fact that attributes noted early on carry a greater weight than attributes noted at a later time. (2) In memory, the tendency for the first items on a list to be remembered better than other items on the list.

primaquine (*Pharmacol*) Anti-malarial drug derived from quinoline; used prophylactically and therapeutically.

primaries (*BioSci*) In birds, the remiges attached to the manus.

primary (*BioSci*) Original, first formed, as *primary meristem, primary body cavity*; principal, most important, as *primary axis, primary feathers.*

primary (*Chem*) A substance which is obtained directly, by extraction and purification, from natural raw material; eg benzene, phenol, anthracene are coaltar *primaries*. Cf INTERMEDIATE.

primary (*ElecEng*) A primary winding. See SECONDARY WINDING.

primary acids (*Chem*) Acids in which the carboxyl group is attached to the end carbon atom of a chain, ie to the $-CH_2-$ group.

primary additive colours (*Phys*) A set containing the minimum number of spectral colours (red, green, blue) which can be adjusted in intensity and, when mixed, visually make a match with a given colour. This match with real colour cannot be perfect, since one primary may have to be negative in intensity. A typical set of primary additive spectral colours is, in nanometres: red (640), green (537), blue (464). For colour TV and photography, original colour has to be separated into such arbitrary components.

primary alcohols (*Chem*) Alcohols containing the group $-CH_2OH$. On oxidation, they form aldehydes and then acids containing the same number of carbon atoms as the alcohol.

primary amines (*Chem*) Amines containing the amino group $-NH_2$. Primary amines are converted into the corresponding alcohol by the action of nitrous acid, nitrogen being eliminated.

primary body (*BioSci*) That part of the plant body formed directly from cells cut off from the apical meristems.

primary body cavity (*BioSci*) The blastocoel or segmentation cavity formed during cleavage, or that part of it which is not subsequently obliterated by mesenchyme.

primary bonds (*Chem*) A term usually applied to the strong covalent, ionic or metallic bonds in materials. See panel on BONDING.

primary bow (*EnvSci*) See RAINBOW.

primary carbon atom (*Chem*) Carbon atom linked to one carbon atom and three hydrogen atoms.

primary cell (*ElecEng*) Voltaic cell in which the electrical energy comes from a chemical reaction between the constituents. As this reaction is irreversible the cell cannot be recharged electrically.

primary cell culture (*BioSci*) Of animal cells, the cells taken from a tissue source and their progeny grown in culture before subdivision and transfer to a subculture.

primary cell wall (*BioSci*) See PRIMARY WALL.

primary circuit (*Build*) Pipe circuit in which water circulates between a boiler (or other heater) and a hot-water storage vessel.

primary circuit (*NucEng*) The circulation of the PRIMARY COOLANT through a reactor core.

primary coil (*ElecEng*) A coil in which a current produces the magnetic flux necessary for the operation of a machine or apparatus.

primary colours (*Genrl*) The colours red, yellow and blue. See PRIMARY ADDITIVE COLOURS, PRIMARY SUBTRACTIVE COLOURS.

primary constants (*ElecEng*) Those of capacitance, inductance, resistance and leakance of a conductor to earth or to a return conductor, per unit length of line.

primary constriction (*BioSci*) The region at which two chromatids are joined in metaphase chromosomes, and to which spindle microtubules attach. It appears narrower and more condensed than the arms.

primary coolant (*NucEng*) The fluid circulated through the reactor to remove heat. In *direct-cycle* reactors such as the boiling-water reactor this drives the turbines directly, but in other types the heat is passed via a heat exchanger to a SECONDARY COOLANT. Also *reactor coolant*.

primary crushing (*MinExt*) Reduction of run-of-mine ore as severed to somewhere below 6 in (15 cm) diameter lump, performed in jaw or gyratory breakers.

primary current (*Electronics*) That formed by PRIMARY ELECTRONS, as contrasted with secondary electrons, which reduce or even reverse it.

primary dispersion (*MinExt*) In GEOCHEMICAL PROSPECTING, the diffusion of metals or other elements through the bedrock surrounding an ore body. Also *halo*.

primary electrons (*Electronics*) (1) Those incident on a surface whereby secondary electrons are released. See PRIMARY IONIZATION. (2) Those released from atoms by internal forces and not by external radiation as with secondary electrons.

primary emission (*Electronics*) Electron emission arising from the irradiation (including thermal heating) or the application of a strong electric field to a surface.

primary energy (*Genrl*) Refers to energy such as crude oil, natural gas and coal, which is used in its natural form to provide *secondary energy* like electricity.

primary flexure (*BioSci*) The flexure of the midbrain by which, in vertebrates, the fore-brain and its derivatives are bent at a right angle to the axis of the rest of the brain.

primary flow (*Electronics*) That of CARRIERS when they determine the main properties of a device.

primary gneissic banding (*Geol*) Exhibited by certain igneous rocks of heterogeneous composition, possibly due to the admixture of two magmas only partly miscible, injection of magma along bedding or foliation planes in the country rocks or selective mobilization of a rock under metamorphosis.

primary growth (*BioSci*) The growth that results from division and expansion of cells produced at apical meristems. Cf SECONDARY GROWTH.

primary immune response (*BioSci*) The response made by an animal to an antigen on the first occasion that it encounters it. Characteristically low levels of antibody are produced after several days and these gradually decline. However, the immune system has been 'primed' so that a secondary response can be evoked on subsequent challenge with the same antigen. Responses of CELL-MEDIATED IMMUNITY follow a similar pattern. See panel on IMMUNE RESPONSE.

primary ionization (*Phys*) (1) In collision theory, the ionization produced by the primary particles, in contrast to total ionization, which includes the *secondary ionization* produced by delta rays. (2) In counter tubes, the total ionization produced by incident radiation without gas amplification.

primary luminous standard (*Phys*) A standard of luminous intensity which is reproducible from a given specification.

primary meristem (*BioSci*) Any of the three meristematic tissues derived in a pattern appropriate to the organ from the apical meristem: the GROUND MERISTEM, PROCAMBIUM and PROTODERM. Cf SECONDARY MERISTEM.

primary metal (*Eng*) Ingot metal produced from newly smelted ore, with no addition of recirculated scrap.

primary nitro-compounds (*Chem*) Nitro-compounds containing the group $-CH_2NO_2$.

primary node (*BioSci*) The node at which the cotyledons are inserted.

primary phloem (*BioSci*) Phloem tissue that differentiates from the procambium during the primary growth of a vascular plant, consisting of protophloem and metaphloem. Cf SECONDARY PHLOEM.

primary pit field (*BioSci*) A thin area in the primary wall of a plant cell, often penetrated by *plasmodesmata*, within which one or more pits may develop if a secondary wall is formed.

primary process thinking (*Psych*) In psychoanalytic theory, the mode of thinking characteristic of unconscious mental activity; it is governed by the pleasure principle (*id*) rather than the reality principle (*ego*). Cf SECONDARY PROCESS THINKING.

primary production (*BioSci*) See PRODUCTION.

primary production (*MinExt*) Production which occurs when oil and gas flow naturally to the well bore without assistance. Also *primary recovery*. Cf SECONDARY PRODUCTION, TERTIARY PRODUCTION.

primary radar (*Radar*) One in which the incident power from the transmitter is reflected from the target to form the return signal or *echo*. Cf SECONDARY RADAR.

primary radiation (*Phys*) Radiation which is incident on the absorber, or which continues unaltered in photon energy and direction after passing through the absorber. Also *direct radiation*.

primary ray (*BioSci*) See MEDULLARY RAY.

primary reinforcer (*Psych*) A stimulus that increases the probability of preceding responses even if the stimulus has never been experienced before.

primary separation plant (*NucEng*) That part of a fuel reprocessing plant where the uranium and plutonium are separated from each other and from other fission products coming after the HEAD END, but before final purification.

primary sere (*BioSci*) A SERE starting from a new, bare surface not previously occupied by plants. Land newly exposed on a rising coast or by a retreating glacier.

primary service area (*ICT*) That within which reception from a broadcast transmitter gives acceptable reproduction of sound and/or vision; for medium- and short-wave broadcasting interference and fading are the usual limiting factors. In very-high-frequency and ultra-high-frequency, NOISE and MULTIPATH RECEPTION are more likely to intrude.

primary solid solution (*Eng*) A constituent of alloys that is formed when atoms of an element B are incorporated in the crystals of a metal A. In most cases solution involves the substitution of B atoms for some A atoms in the crystal structure of A (substitutional solid solution), but in a few instances the B atoms are situated in the interstices between the A atoms (interstitial solid solution).

primary standard (*Genrl*) A standard agreed upon as representing some unit (eg length, mass, emf) and carefully preserved at a national laboratory. Cf SECONDARY STANDARD.

primary store (*ICT*) See MAIN MEMORY.

primary stress (*Eng*) An axial or direct tensile or compressive stress, as distinct from a bending stress resulting from deflection.

primary structure (*Aero*) All components of an aircraft structure, the failure of which would seriously endanger safety, eg wing or tailplane spars, main fuselage frames, engine bearers, portions of the skin which are highly stressed.

primary structure (*Chem*) Covalent-bonded atoms and groups in biopolymers, determining properties of the substances. In natural homopolymers or copolymers, specified by repeat units and sequencing; most complex in proteins, where amino acid sequence is usually unique. See BIOMATERIALS.

primary subtractive colours (*Phys*) A set containing the minimum number of spectral colours (cyan, magenta, yellow), which, when subtracted in the right intensity from a given white, result in a match with a given colour. These are complementary to the PRIMARY ADDITIVE COLOURS.

primary succession (*BioSci*) A succession beginning on an area not previously occupied by a community, eg a newly exposed rock or sand surface. Cf SECONDARY SUCCESSION.

primary tissue (*BioSci*) Tissue formed from cells derived from primary meristems.

primary tumour (*BioSci*) The mass of tumour cells at the original site of the neoplastic event and from which secondary tumours may arise by metastasis.

primary voltage (*ElecEng*) That which is applied to the input side of a transformer.

primary wall (*BioSci*) The earlier-formed part of the CELL WALL (panel) characteristically laid down while the cell is expanding and typically richer in pectins than the SECONDARY WALL.

primary wave (*Geol*) See panel on EARTHQUAKE.

primary xylem (*BioSci*) Xylem tissue that differentiates from the procambium during primary growth in a vascular plant, consisting of PROTOXYLEM and METAXYLEM. Cf SECONDARY XYLEM.

Primates (*BioSci*) An order of mammals. The dentition is complete, but unspecialized, the brain, esp the neopallium, is large and complex. Characteristically have pentadactyl limbs and well-developed eyes directed forwards. Basically arboreal, although this is not true for all species. The uterus is a single chamber and few young are produced, with parental care lasting a long time after birth; eg lemurs, tarsiers, monkeys, apes and humans.

prima vera (*For*) C American hardwood tree (*Tabebuia donnell-smithii*) whose heartwood is yellowish-rose streaked with orange, red and brown, and straight- to wavy-grained with a moderately coarse texture.

prime contract (*Aero*) That for the whole aircraft or weapon system from design and manufacture to test and supply, including the management of subcontractors for completion to time and cost.

primed (*BioSci*) The state in which a cell or organism, as a result of prior exposure, will mount a more substantial response, not always of exactly the same type, upon second exposure to a signal or substance. For instance, prior contact with an antigen will prime an animal to produce a secondary response on subsequent challenge with the same antigen. See PRIMING.

prime mover (*Eng*) An engine or other device by which a natural source of energy is converted into mechanical power. See GAS ENGINE, INTERNAL-COMBUSTION ENGINE, OIL ENGINES, PETROL ENGINE, STEAM ENGINE, STEAM TURBINE, WATER TURBINE.

prime number (*MathSci*) An integer that has precisely two divisors, ie that is divisible only by itself and one. It is known that there are infinitely many prime numbers, but there is no algorithm for generating them. The first prime numbers are 2, 3, 5, 7, 11, 13, 17, 19, 23, 29, 31, 37, ... Every natural number greater than one may be resolved uniquely into a product of prime numbers, eg $8316 = 2^2 \times 3^3 \times 7 \times 11$.

prime number theorem (*MathSci*) A theorem stating that the number of prime numbers less than or equal to x is asymptotic to $x/\log x$.

primer (*BioSci*) Nucleotide sequence with a free 3′–OH group, needed to initiate synthesis by DNA polymerase. A short primer sequence base-paired to a specific site on a longer DNA strand can initiate polymerization from that site. See POLYMERASE CHAIN REACTION.

primer (*Build*) The first coat of paint on a bare surface, formulated to provide a suitable surface for the next coats, and allowing for absorption in wood, the tendency to corrode in metals and the high alkalinity of concrete.

primer (*MinExt*) Cartridge in which detonator is placed in order to initiate explosion of string of high-explosive charges in borehole.

primes (*Eng*) Metal sheet and plate of the highest quality and free from visible imperfections.

primigravida (*Med*) A woman who is pregnant for the first time.

priming (*BioSci*) (1) A treatment that does not itself elicit a response from a system but induces an increased capacity to respond to a different stimulus. (2) The process of initiating polymerization of nucleic acid from the 3′ end of a DNA strand.

priming (*Eng*) (1) The delivery by a boiler of steam containing water in suspension, due to violent ebullition or frothing. (2) The operation of filling a pump intake with fluid to expel the air. (3) The operation of injecting petrol into an engine cylinder to assist starting. (4) Application of a first protective coating to eg clean and prepared steelwork to prevent subsequent corrosion before final painting.

priming pump (*Aero*) A manual or electric fuel pump which supplies the engine during starting where an injection carburettor or fuel injection pump is fitted.

priming valve (*Eng*) A valve fitted on the suction side of a pump to assist in priming.

primipara (*Med*) A woman who gives, or has given, birth to a child for the first time.

primitive (*BioSci*) Original, first-formed, of early origin, the ancestral condition, eg *primitive streak*.

primitive crystal lattice (*Crystal*) See UNIT CELL.

primitive equation model (*EnvSci*) A NUMERICAL FORECAST model that uses the PRIMITIVE EQUATIONS, not the FILTERED EQUATIONS.

primitive equations (*EnvSci*) The fundamental equations of motion of a fluid modified only by the use of the HYDROSTATIC APPROXIMATION and the neglect of viscosity. The primitive equations comprise three prognostic equations (the x and y components of the momentum equation and the thermodynamic equation of energy) and three diagnostic equations (the continuity equation, the HYDROSTATIC APPROXIMATION and the equation of state). These equations form a closed set in the dependent variables of velocity, pressure, density and temperature. The solutions include gravity waves but not vertically propagating sound waves. See panel on NUMERICAL WEATHER FORECAST.

primitive streak (*BioSci*) In developing birds and reptiles, a thickening of the upper layer of the blastoderm along the axis of the future embryo; represents the fused lateral lips of the blastopore.

primordial (*Phys*) A term used to describe radionuclides thought to have been present near the time of the origin of the universe. See COSMOGONIC.

primordial germ cells (*BioSci*) In the early embryo, cells which will later give rise to the germ cells within the gonads.

primordium (*BioSci*) An organ, cell or other structure in the earliest stage of development or differentiation. Also *anlage*. Adj *primordial*.

Prince Rupert's drops (*Glass*) Entertaining laboratory demonstration. Solid drops of silicate glass with long thin tails are formed by quenching gobs of the molten glass by dropping into water. The quenching puts the surface of the drop into compression and the interior into tension (see *thermally toughened glass* under SAFETY GLASS). The drops shatter into a powder when their tails are broken off since the rupture of the compressive layer 'releases' the internal tensile stresses.

principal (*Build*) See ROOF TRUSS.

principal axis (*ICT*) The direction of maximum sensitivity or response for a transducer or antenna.

principal axis (*Phys*) (1) At any given point O: one of the three mutually perpendicular axes $Oxyz$ of a body such that the three products of inertia about the co-ordinate planes are all zero. The principal axes at the centre of gravity are the axes of symmetry if any exist. (2) See OPTICAL AXIS.

principal direction (*MathSci*) See CURVATURE (3).

principal normal (*MathSci*) See MOVING TRIHEDRAL.

principal part (*MathSci*) Of a function $f(z)$ at a singularity. See POLE.

principal planes of a lens (*Phys*) Conjugate planes, perpendicular to the principal axis of a lens system, for which the transverse magnification is unity. Also *unit planes*.

principal points of a lens (*Phys*) Two points on the principal axis of a lens or lens system where the principal planes intersect the axis. If the object distance l is measured from one of the principal points and the image distance l' from the other, then the simple lens formula

$$\frac{1}{f} = \frac{1}{l'} - \frac{1}{l}$$

can be used to give f, the equivalent focal length (cartesian CONVENTION OF SIGNS). Also *Gaussian points*. See NODAL POINTS, PRINCIPAL PLANES.

principal quantum number (*Phys*) A number n which characterizes solutions of the SCHRÖDINGER EQUATION. When QUANTUM MECHANICS is applied to any particle moving in a central potential, eg the electron in a hydrogen atom, the ANGULAR MOMENTUM and the z component of the angular momentum are quantized with quantum numbers l and m_l. For a particular l and m_l the solutions of the Schrödinger equation are well-behaved only if the energy is also quantized; these solutions are denoted by n. In many-electron atoms each electron has an orbital specified by n, l, m_l and the spin quantum number m_s; the assignment of the orbitals depends on the PAULI EXCLUSION PRINCIPLE. In general the energy depends on both n and l. The value of n denotes the ELECTRON SHELL and that of l the subshell that the electron occupies.

principal radius of curvature (*MathSci*) See CURVATURE (3).

principal rafter (*Build*) A rafter forming part of the roof truss proper and supporting the purlins.

principal ray (*Phys*) From an object point lying off the axis, the ray passing through the centre of the entrance pupil of the system.

principal series (*Phys*) A series of optical spectrum lines observed in the spectra of alkali metals. Has led to energy levels for which the orbital quantum number is unity to be designated *p-levels*.

principal stress (*Eng*) The component of a stress which acts at right angles to a surface, occurring at a point at which the shearing stress is zero.

principal values (*MathSci*) A conventionally chosen single period of the inverse of a periodic function such as the trigonometric functions, usually those with smallest absolute value. Thus, since $\sin n\pi = 0$ for all integral n, $\arcsin(0)$ is many-valued, but the principal value, sometimes written with initial capital as $\text{Arcsin}(0)$, is zero.

principle of equivalence (*Phys*) A statement which forms a basic principle in GENERAL RELATIVITY: observers have no means of distinguishing whether their laboratories are in uniform gravitational fields or accelerated frames of reference.

principle of least action (*Phys*) A principle stating that the actual motion of a conservative dynamical system between two points takes place in such a way that the action has a minimum value with reference to all other paths between the points which correspond to the same energy.

principle of least constraint (*Eng*) A principle stating that the motions of any number of interconnected masses under the action of forces deviate as little as possible from the motions of the same masses if disconnected and under the action of the same forces. The motions are such that the constraints are a minimum, the constraint being the sum of the products of each mass and the square of its deviation from the position it would occupy if free.

principle of least time (*Phys*) See FERMAT'S PRINCIPLE OF LEAST TIME.

principle of reinforcement (*Psych*) Skinner's term for the LAW OF EFFECT.

principle of relativity (*Phys*) A universal law of nature which states that the laws of mechanics are not affected by a uniform rectilinear motion of the system of co-ordinates to which they are referred. Einstein's relativity theory is based on this principle, and on the postulate that the observed value of the speed of light is constant and is independent of the motion of the observer. See SPECIAL RELATIVITY.

principle of superposition (*Phys*) A principle that the resultant disturbance at a given place and time caused by a number of waves traversing the same space is the vector sum of the disturbances which would have been produced by the individual waves separately. The principle is the basis for the explanation of interference and diffraction effects.

principle of the equipartition of energy (*Chem*) A principle stating that the total energy of a molecule in the normal state is divided up equally between its different capacities for holding energy, or DEGREES OF FREEDOM.

print (*Eng*) See CORE PRINTS.

print (*ImageTech*) The image, usually a positive one, obtained by exposing a photographic material through another image, such as a negative.

printed circuit (*Electronics*) An electronic subassembly consisting of an insulating board or card with copper conductors laminated on them. See panel on PRINTED, HYBRID AND INTEGRATED CIRCUITS.

printed circuit board (*ICT*) The board carrying a PRINTED CIRCUIT. See panel on PRINTED, HYBRID AND INTEGRATED CIRCUITS.

printer (*ICT*) Output device producing characters or graphic symbols on paper. See DAISY-WHEEL PRINTER, DOT-MATRIX PRINTER, DYE SUBLIMATION PRINTER, GOLF-BALL PRINTER, IMPACT PRINTER, INKJET PRINTER, LASER PRINTER, LINE PRINTER, THERMAL PRINTER, XEROGRAPHIC.

printer (*ImageTech*) A machine for the exposure of photographic paper or film to produce prints, either by contact or by optical means. Exposure may be made one picture at a time or, in the case of motion pictures, with the film moving continuously.

printer control language (*ICT*) A standard set of commands to control printers developed by Hewlett Packard, often used as an alternative to POSTSCRIPT. It provides access to many features such as GRAPHICS and FONTS.

printer fonts (*ICT*) FONTS that are stored within the printer or SOFT FONTS that are sent before printing starts.

printer point (*ImageTech*) See LIGHT CHANGE POINTS.

print hammer (*ICT*) A component causing the contact between the character, ribbon and paper in an IMPACT PRINTER.

printing (*ImageTech*) The operation of making a still or motion picture PRINT by the exposure of photographic paper or film.

printing (*Print*) Any process of producing copies of designs or lettering by transferring ink to paper (or other material) from a printing surface. There are traditionally three main classes according to the method of application of the ink to the printing surface: (1) Relief, or LETTERPRESS, printing surfaces have the ink-carrying parts in relief, so that rollers deposit ink on these parts only, as in printer's type. (2) PLANOGRAPHIC printing surfaces are prepared so that parts accept the ink from the rollers, although there is no difference in level; the ink-accepting parts may be greasy, the remainder being moist and ink-rejecting, as in LITHOGRAPHY. (3) INTAGLIO printing surfaces have the ink-carrying portions hollowed out; the whole surface is covered with ink and then cleaned off, leaving the hollows filled with ink, which is lifted out when the paper is pressed into contact, as in photogravure. All classes can be adapted for use with a cylindrical printing surface which can be printed at high speed by continuous rotation against another cylinder, with the paper to be printed running between them. Additionally a number of newer methods have been developed, typically for shorter runs in the office, eg laser printing, dye-transfer printing, bubblejet printing, of which the last two can be used for colour printing to a varying standard.

printing diameter (*Print*) The correct diameters of printing cylinder and impression surface.

printing down (*Print*) A stage in the making of printing surfaces, for any of the main processes, in which the surface, after being made light-sensitive, is exposed to suitable light through a negative (or, in some cases, a positive).

printing ink (*Print*) A mixture of carbon black, or other pigments, in a vehicle of mineral oil, linseed oil, etc. Inks are formulated to dry by penetration, evaporation, oxidation, or by a combination of these, and also can be *cold-set*, *heat-set* or *moisture-set*.

printing-out paper (*ImageTech*) A once popular photographic paper which produced an image on printing which needed only fixing, no development. Abbrev POP.

printout (*ICT*) Printed output from the computer.

printout mask (*Print*) Opaque mask used to cover the image areas of a plate during a second exposure to remove unwanted work from the printing plate. Also *burnout mask*.

print queue (*ICT*) A method of storing requests for printing in order until the printer is available. In most systems, once the job or file has been placed in the queue, the user's computer will be able to continue to execute programs without having to wait until the printer has completed the printing.

print server (*ICT*) A computer and/or software that provides access to a NETWORK PRINTER for users. Printing requests will be placed in sequence in a PRINT QUEUE.

print-through (*Acous, ICT*) In magnetic tape recording, the transfer of a recording from one layer to another when the tape is spooled or reeled giving rise to a form of distortion; also *transfer*.

prion protein (*BioSci*) A protein of unknown function that normally undergoes rapid turnover in the brain. Defective prion protein may be the infectious agent responsible for SPONGIFORM ENCEPHALOPATHIES. See panel on TRANSMISSIBLE SPONGIFORM ENCEPHALOPATHY.

prior distribution (*MathSci*) The representation as a probability distribution of the PRIOR PROBABILITIES of a PARAMETER.

prior probability (*MathSci*) The probability of an event calculated independently of (prior to) observation of the outcomes of a test.

prisere (*BioSci*) Same as PRIMARY SERE.

prism (*Crystal*) A hollow (open) crystal form consisting of three or more faces parallel to a crystal axis.

prism (*MathSci*) A solid of which the ends are similar, equal and parallel polygons, and of which the sides are parallelograms.

prism (*Phys*) Transparent solid (usually glass) whose faces are triangles, used in a number of optical instruments. Equilateral prisms are used at minimum deviation in spectroscopes for forming spectra, and 90° prisms are used for totally reflecting a ray through a right angle in binoculars, periscopes and rangefinders.

prismatic (*Genrl*) Prism-shaped; composed of prisms.

prismatic astrolabe (*Surv*) An instrument for observing stars at an altitude of 60° (in some instruments, 45°) at different azimuths around the horizon, these observations being used for the computation of latitude and local time.

prismatic binoculars (*Phys*) Binocular telescopes in which the tubes, instead of being straight, are effectively shortened by using total reflecting prisms to 'fold' the light paths. The prisms at the same time produce an erect image.

prismatic coefficient (*Ships*) The ratio between the immersed volume of the vessel and the volume of an enclosing prism with a constant transverse section identical with the maximum immersed cross-section area of the vessel.

prismatic compass (*Surv*) A hand-held form of surveyor's compass in which the eye vane carries a prism reflecting a view of a graduated ring, attached to and moving round with the compass needle.

prismatic layer (*BioSci*) In the shell of Mollusca, a layer consisting of calcite or aragonite lying between the periostracum and the nacreous layer. In the shell of Brachiopoda, the inner layer of the shell, composed mainly of calcareous, but partly of organic, material.

prismatic spectrum (*Phys*) A spectrum formed by refraction in a prism, as contrasted with a *grating spectrum* formed by diffraction.

prismatic sulphur (*Chem*) See SULPHUR.

prismatic system (*Crystal*) See ORTHORHOMBIC SYSTEM.

prism light (*Build*) A pavement light in which glass prisms internally reflect light.

prismoid (*MathSci*) A body which has plane-parallel polygonal ends and is bounded by plane sides.

prismoidal formula (*CivEng*) A formula used in the calculation of earthwork quantities. It states that the

Printed, hybrid and integrated circuits

A circuit board manufactured in the late 1980s would have as a main subassembly a printed circuit board (PCB) and many of the components mounted on it would themselves be complex hybrid and integrated circuits. The three types – printed, hybrid and integrated – are distinguished by the type of the substrate material and the associated processing methods.

PCBs are built from copper-clad polymeric boards on which conducting tracks are defined by PHOTO-LITHOGRAPHY. Components (resistors, capacitors, transistors as well as the hybrid and integrated subassemblies) are fixed by soldering, either directly onto the surface or else after passing leads through locating holes from one side of the board to the other. Multilayer boards can be made with eg buried earth planes or power lines, and complicated assemblies use both sides of a board, allowing dense packing. Track and contact spacings less than 1 mm can be achieved. Most PCBs are now assembled by robot placement of components, with soldering taking place in a single pass through a soldering machine.

Hybrid circuits use ceramic substrates on which some components are placed, as with PCBs, but the conducting tracks and other components are formed in situ; hence the technology is a hybrid of the printed and integrated circuit approaches. Tracks, resistors and insulating layers are built onto the substrate first by screen printing with special inks, which are subsequently fired to change them to conducting metal or insulating glassy enamel; hence the use of a refractory substrate. Other components are then soldered in place as for PCBs. Hybrids are robust in hostile chemical, thermal and mechanical environments, and are especially useful at microwave frequencies.

Integrated circuits (ICs), or chips, may contain millions of interconnected components within a few square millimetres, capable of carrying out complex electronic operations. A silicon IC is built, eg onto a wafer of single-crystal silicon, by a series of deposition, patterning and etching steps, performed simultaneously on hundreds of devices; the wafer is divided into individual chips only after all processing is complete. Each IC is then fitted with external connection leads and packaged for incorporation into one of the larger subassemblies. See SEMICONDUCTOR DEVICE PROCESSING.

---

volume of any prismoid is equal to one-sixth its length multiplied by the sum of the two end-areas plus four times the mid-area.

prism square (*Surv*) A form of OPTICAL SQUARE in which the fixed angle of 90° between the lines of sight is obtained by reflection from the surfaces of a suitably shaped prism.

privacy (*ICT*) Recognition of the private nature of certain data. In consideration of privacy, safeguards are usually built into systems that hold confidential data to prevent unauthorized access. Now often also safeguarded by legal constraints. See DATA PROTECTION ACT.

privacy system (*ICT*) See INVERTER, SCRAMBLER.

private automatic branch exchange (*ICT*) A small automatic exchange on a subscriber's premises, for internal telephone connections, with extensions over the public telephone system through lines to the local exchange. Abbrev *PABX*.

private automatic exchange (*ICT*) An automatic exchange on private premises; not connectable with the public telephone system. Abbrev *PAX*.

private branch exchange (*ICT*) An automatic or manual exchange on a subscriber's premises that is used for internal connections, with extension through the local exchange to the public telephone system. Abbrev *PBX*.

private exchange (*ICT*) An exchange in a private establishment that is not connected in any way with the public telephone service. Abbrev *PX*.

private key (*ICT*) In cryptography, one of a pair of keys required to write and read a secure message. The private key is known only to the recipient of the message and is used to unlock its contents.

private mobile radio (*ICT*) The use by an organization of a radio network independent of the PUBLIC SWITCHED TELEPHONE NETWORK for mobile communication between its agents.

pro- (*Genrl*) Prefix from Gk and Lt *pro*, before in time or place, used in the sense of 'earlier', 'more primitive' or 'placed before'.

Pro (*Chem*) Symbol for PROLINE.

proactive interference (*Psych*) Interference with memorizing due to prior learning.

probabilistic risk assessment (*NucEng*) An almost complete analysis of all aspects of a plant's operation in regard to the risk of accident. Abbrev *PRA*. See panel on RISK ASSESSMENT.

probability density (*Phys*) A measure of the probability of finding an electron at a certain point. Quantum mechanics suggests that electrons must not be regarded as being located at a defined point in space, but as forming a cloud of charge surrounding the nucleus, the cloud density at a given location indicating the probability of finding an electron there. See UNCERTAINTY PRINCIPLE.

probability density function (*MathSci*) The first derivative of the cumulative distribution function of a continuous random variable, often identified as the probability that a random variable takes a value in an infinitesimal interval divided by the length of the interval.

probability function (*MathSci*) For a random variable which can only take specific discrete values, possibly infinite in number, the expression, as a function, of the probabilities that the variable can take each of these values.

probable ore (*MinExt*) Inferred ore, partly exposed and sampled but not fully *blocked-out* in panels. See BLOCKING-OUT.

proban (*Chem*) Flameproof finish for textile fabrics based on a phosphorus compound, tetrakis (hydroxymethyl) phosphonium chloride.

proband (*BioSci*) In human genetics, the affected individual who brings a family to the notice of the investigator.

probang (*Vet*) A flexible tube which may be passed into the oesophagus and stomach of animals; used for relieving obstructions and administering fluids.

probe (*Acous*) See SOUND PROBE.

probe (*BioSci*) A radioactive or otherwise labelled single-stranded sequence of nucleic acid that hybridizes to another single-stranded nucleic acid, usually separated into discrete spots on a nitrocellulose filter. The label then detects the presence of a sequence complementary to the probe sequence. See BLOTTING.

probe (*Electronics*) (1) Electrode of small dimensions compared with the gas volume, placed in gas-discharge tube to determine the space potential. (2) Magnetic or conducting device to extract power from a waveguide. (3) Coil or semiconductor sensing element associated with a fluxmeter.

probe (*Med*) A surgical instrument with a blunt end, used for exploring wounds, sinuses and cavities.

probe (*Phys*) Portable radiation-detector unit, cable-connected to counting or monitoring equipment.

probe (*Space*) A space vehicle, esp one unmanned, sent to explore near and outer space, to collect and transmit data back to Earth. If a planet or its environment is explored, the vehicle is sometimes referred to as a *planetary probe*.

probenecid (*Pharmacol*) A drug that promotes excretion of uric acid, used in treating gout.

problem-solving behaviour (*Psych*) Diverse strategies used by animals and humans to overcome difficulties in attaining a desired goal; it implies a higher order of intelligent behaviour than simple TRIAL AND ERROR strategies.

Proboscidea (*BioSci*) An order of mammals having a long prehensile proboscis with the nostrils at the tip (trunk), large lophodont molars, and a pair of incisors of the upper jaw enormously developed as tusks. Surviving members are large forest-living herbivorores in Africa and India. Includes elephants and the extinct mammoths.

proboscis (*BioSci*) (1) Generally, an anterior trunk-like process. (2) In Turbellaria and Polychaeta, the protrusible pharynx. (3) In Nemertinea, a long protrusible muscular organ lying above the mouth. (4) In some insects, the suctorial mouthparts. (5) In Hemichorda, a hollow club-shaped or shield-shaped structure in front of the mouth. (6) In Proboscidea, the long flexible prehensile nose.

procaine (*Pharmacol*) Procaine butyrate, borate and hydrochloride, used as local anaesthetics. Also *novocaine*.

procambium (*BioSci*) The cells of PRIMARY MERISTEM which are typically longer than broad, which differentiates into primary xylem and primary phloem and, in some cases, cambium. Also *provascular tissue*.

procartilage (*BioSci*) An early stage in the formation of cartilage in which the cells are still angular in form and undergoing constant division; embryonic cartilage.

procaryote (*BioSci*) See PROKARYOTE.

procedure (*ICT*) See SUBROUTINE.

Procellariiformes (*BioSci*) An order of birds containing wandering ocean species, often very large, with long narrow wings. They lay one white egg, often in a burrow. Petrels, shearwaters, albatrosses.

process (*BioSci*) An extension or projection.

process annealing (*Eng*) Heating steel sheet or wire between cold-working operations to slightly below the critical temperature and then cooling slowly. The process is similar to tempering but will not give as much softness and ductility as full annealing. See ANNEALING.

process camera (*ImageTech*) A large copying camera specifically designed for use in photomechanical reproduction processes in colour and black and white, with particular emphasis on stability and uniformity of image.

process chart (*Eng*) A chart in which a sequence of events is portrayed diagrammatically by means of conventional symbols.

process control (*Electronics*) In a complicated industrial or chemical process, control of various sections of the plant by electronic, hydraulic or pneumatic means, taking rates of flow, accelerations of flow, changes of law, temperatures and pressures into account automatically. Now generally computer-based.

process energy (*Genrl*) Energy needed to manufacture specific materials to shape and assemble them into finished products. The term usually extends to natural resources in the Earth's crust. See ECOBALANCE.

process-engraving (*Print*) A relief printing plate made with the aid of the process camera followed by etching; or by means of an ELECTRONIC ENGRAVING machine.

process factor (*Phys*) See SEPARATION FACTOR.

processing (*ImageTech*) The sequence of chemical reactions, washing and drying involved in treating an exposed photographic material, paper or film, to produce a permanent visible image which can be safely handled in further operations.

processing aid (*FoodSci*) Material which assists in a process but should not be significantly present in the finished product, eg filtering aids. See FININGS, KIESELGUHR.

processing routes (*MinExt*) A term used in considering alternative methods of treating a specified ore or concentrate. Main 'routes' are physical, chemical and pyrometallurgical.

process metallurgy (*Eng*) The science and technology of extracting metals from their ores and purifying them.

processor (*ICT*) A device that can perform logical and arithmetic operations. See CENTRAL PROCESSOR.

process value (*FoodSci*) The value F_0 which equals $D_T(\log A - \log B)$ where A is the spore number before process, B is the spore number after process, and D_T is the time in minutes to kill 90% of organisms at temperature T, taken as $250°F$ ($\approx 120°C$). It is the equivalent time in minutes at which a can should be processed at $250°F$ to achieve a given reduction in the population of spore-forming organisms. In sterilizing canned foods, an F_0 value of 3 would reduce the chance of survival of *Clostridium botulinum* spores to less than 1 in 10^{12}. Often higher values ($F_0 = 10$–20) are used for greater safety.

prochlorite (*Min*) See RIPIDOLITE.

Prochlorophyceae (*BioSci*) Prokaryotic algae with the pigmentation of green algae (chlorophyll a and b, and no phycobilins) rather than that of cyanobacteria. Two or three species have so far been identified. They are of interest as possibly representing the group from which green plant chloroplasts may have evolved. See ENDOSYMBIOTIC HYPOTHESIS.

procidentia (*Med*) A falling down or prolapse, esp severe prolapse of the uterus.

procoelous (*BioSci*) Concave anteriorly and convex posteriorly; said of vertebral centra.

proct-, procto- (*Genrl*) Prefixes from Gk *proktos*, anus.

proctal (*BioSci*) Anal.

proctalgia (*Med*) Neuralgic pain in the rectum.

proctectomy (*Med*) Surgical removal of the rectum.

proctitis (*Med*) Inflammation of the rectum.

procto- (*Genrl*) See PROCT-.

proctoclysis (*Med*) The slow injection of large amounts of fluid into the rectum.

proctodaeum (*BioSci*) That part of the alimentary canal which arises in the embryo as a posterior invagination of ectoderm. Adj *proctodaeal*. Cf MIDGUT, STOMODAEUM.

proctodynia (*Med*) Pain in or around the anus.

proctologist (*Med*) A surgeon specializing in diseases of ANUS and RECTUM.

proctoscope (*Med*) An endoscope used for inspecting the mucous membrane of the rectum.

proctosigmoiditis (*Med*) Inflammation of the rectum and of the sigmoid flexure of the colon.

proctotomy (*Med*) Surgical incision of the anus or rectum for the relief of stricture.

procumbent (*BioSci*) Lying loosely on the ground surface.

procurement (*Aero*) Organizational procedure for obtaining equipment, supplies, services and personnel.

procuticle (*BioSci*) In the cuticle of insects, a multilaminar layer initially present below the epicuticle and pierced by pore canals running perpendicularly to it. In soft transparent areas this undergoes no apparent change after formation, but in other cases the outer part becomes hard, dark sclerotized exocuticle, the inner unchanged part then being called the endocuticle. See SCLEROTIN, TANNING.

Procyon (*Astron*) A bright yellow-white visual binary star in the constellation Canis Minor, which is also a spectroscopic binary and optical double star. Also *Alpha Canis Minoris*.

prod mark (*Geol*) See BOUNCE MARK.

prodromal (*Med*) Premonitory of disease.

pro-drug (*Pharmacol*) A compound that is inactive in its original form but is converted by the metabolic processes of the body into an active drug.

producer (*BioSci*) In an ecosystem, one of the autotrophic organisms, largely green plants, which are able to manufacture complex organic substances from simple inorganic compounds. Cf CONSUMER, DECOMPOSER.

product (*MathSci*) See MULTIPLICATION.

product design (*Eng*) All those methods used to plan, manufacture and evaluate industrial and consumer articles, includes both industrial and engineering design activities as well as selling, costing, legal constraints (eg product liability), etc.

product design specification (*Eng*) Detailed outline of product function and the environment in which it will perform, with design geometry and materials of parts needed to fulfil that function. Abbrev *PDS*.

production (*BioSci*) Biomass, or heat of combustion of the biomass, expressed on an area basis (units: $g\,m^{-2}$ or $MJ\,m^{-2}$). Primary production is production by green plants. Secondary production is biomass produced by heterotrophic organisms. The rate of production is called the *productivity* (units: $g\,m^{-2}\,yr^{-1}$). Gross primary productivity is the rate of community photosynthesis, whereas net primary productivity is the rate of community photosynthesis minus the rate of community respiration.

production choke (*MinExt*) Aperture at the well head which limits the flow of oil from a well to the most economical or best allowable rate.

production platform (*MinExt*) Offshore platform from which the flow of oil from many wells is controlled and stored before onward transmission to the refinery. See panel on DRILLING RIG.

production reactor (*NucEng*) A reactor designed for large-scale production of transmutation elements such as plutonium. The reactors are characterized by the short residence time of the fuel pins which are therefore reprocessed before there has been time for a large build-up of fission products. They are generally military reactors.

production string (*MinExt*) The smallest casing in an oil well which reaches from the producing zone to the well head, up which the oil passes.

productivity (*BioSci*) See PRODUCTION.

product liability (*Genrl*) Risks associated with manufactured products when sold in the open market; specifically, risks to personal safety caused by faulty design, manufacturing defects, poor materials selection and/or marketing warnings. See STRICT LIABILITY.

product life cycle (*Genrl*) The life of an industrial or consumer product between factory gate and scrapping. Real life cycles of a specific product may vary enormously, but specification life cycles are based on a notional formula based on market research, engineering design and materials performance.

product rule (*MathSci*) The rule used in differentiating the product of two differentiable functions f and g: $(fg)' = f'g + g'f$.

products (*Chem*) The elements and/or compounds formed in a chemical reaction. They are written on the right-hand side of a chemical equation.

products of inertia (*Phys*) Of a body about two planes, the sum $\sum mxy$ taken over all particles of the body where m is the mass of a particle and x and y the perpendicular distances of the particle from the specified planes.

products pipeline (*MinExt*) That which runs from a refinery to distributors and may carry many different products separated by BATCHING SPHERES.

pro-ecdysis (*BioSci*) The period of preparation for a moult in Arthropoda during which the new cuticle is laid down and the old one ultimately detached from it.

pro-embryo (*BioSci*) The structure formed by the first few cell divisions of the zygote of seed plants, before differentiation into suspensor and the embryo proper.

pro-estrus (*BioSci*) In mammals, the coming on of heat in the estrus cycle. Also *pro-oestrus*.

profile (*Build*) A temporary guide set out at corners, normally with small timbers, to act as a guide for the foundations of a building.

profile (*Surv*) A longitudinal section, usually along the centre line of a proposed work such as a railway.

profile drag (*Aero*) The two-dimensional drag of a body, excluding that due to lift; the sum of the surface friction and form drag.

profile grinding (*Eng*) The grinding of cylindrical work without traversing the wheel whose periphery is profiled to the form required and extends over the full length of the work. Also PLUNGE GRINDING.

profiling (*Eng*) Producing the profile of a die or other workpiece, (1) with a modified milling machine, incorporating a tracing mechanism, or (2) with a grinding machine using a wheel dressed to correspond to the required profile.

proflavine (*Pharmacol*) Deep orange-coloured crystalline powder, used in dilute solution as an antiseptic.

progeria (*Med*) Premature old age. Occurring in children, the condition is characterized by dwarfism, falling out of hair, wrinkling of the skin and senile appearance.

progestational (*Med*) Used of the luteal phase of the estrous or menstrual cycle during which the endometrium is prepared for nidation; of the proliferative reaction of the endometrium towards the fertilized ovum or an irritant foreign body mimicking the ovum; of the hormones which bring about these effects.

progesterone (*BioSci*) A hormone produced in the corpus luteum as an antagonist of estrogens. It promotes proliferation of uterine mucosa and the implantation of the blastocyst; it prevents further follicular development.

proglottis (*BioSci*) One of the reproductive segments forming the body in Cestoda; produced by strobilation from the back of the scolex. Pl *proglottides*.

prognathous (*BioSci*) Having protruding jaws; having the mouth parts directed forwards. Cf ORTHOGNATHOUS.

prognosis (*Med*) A forecast of the probable course of an illness.

prognostic chart (*EnvSci*) A chart of the METEOROLOGICAL ELEMENTS which are expected to exist in the near future. A forecast weather chart.

progradation (*Geol*) Extension of the shoreline seawards by wave or current action.

program (*ICT*) A complete, structured sequence of PROGRAM STATEMENTS that direct a computer to implement an algorithm. Cf SUBROUTINE.

program counter (*ICT*) A register which contains the address of the next machine-code instruction to be expected. Also *instruction address register (IAR)*, *next instruction register*, *sequence control register*.

program exposure modes (*ImageTech*) A selection of automatic exposure settings biased to cope with particular subjects or conditions.

program flowchart (*ICT*) A flowchart used to describe the sequence of operations within a computer program, and may form part of the DOCUMENTATION of a finished program for maintenance.

program generator (*ICT*) Software that assists users to write their own programs, by expanding simple statements into program code. See FOURTH-GENERATION LANGUAGE.

program information file (*ICT*) In the context of WINDOWS, a file that provides information about how Windows should run a non-Windows application; eg specify a directory for starting the software or whether the application will be presented on a full screen or within a WINDOW. Abbrev *PIF*.

programmable logic array (*ICT*) One that allows the designer to choose the interconnections during manufacture.

programmable read-only memory (*ICT*) A type of ROM into which the program may be written after manufacture, by a customer, but that is fixed from that time on. Abbrev *PROM*. Cf EPROM.

programmed cell death (*BioSci*) The death of a cell in a multicellular organism that appears to be deliberately programmed as part of eg a morphogenetic process such as the development of the stalk in the fruiting body of cellular slime moulds. It may occur by APOPTOSIS (with which it is not synonymous) or by necrosis.

programme delivery control (*ImageTech*) Coded teletext signals which broadcasters use for timing programmes and inserting commercials using the AUTOMATED LIBRARY SYSTEM (ALS). This has been adapted for consumer timer setting as STARTEXT. Abbrev *PDC*.

programmed instruction (*Psych*) The process of learning from a systematic presentation of data constructed so that each step leads to the next. The learner has no need of recourse to material other than the programme and progresses by answering questions at each stage which are necessary for the understanding of the rest. Such programmes can be presented on a *teaching machine*. Also *programmed learning*.

programme level (*ICT*) The level, related to a datum power level of 1 milliwatt in 600 ohms, as indicated by a volume-unit (VU) meter, as defined for this purpose. Measured in decibels.

programmer (*ICT*) A person responsible for writing computer programs. See APPLICATION PROGRAMMER, SYSTEMS PROGRAMMER.

programming (*ICT*) Working out a detailed sequence of steps in a programming language with the aim of producing a program.

programming language (*ICT*) Artificial language devised to enable people to instruct machines. See HIGH-LEVEL LANGUAGE, LOW-LEVEL LANGUAGE.

program relocation (*ICT*) The moving of a program from one area of memory to another. This will be done by the operating system to make best use of available memory, eg during compilation or within a MULTIPROGRAMMING environment when the main store is PARTITIONED. See RELOCATABLE PROGRAM.

program software (*ImageTech*) See SOFTWARE.

program statement (*ICT*) A basic unit of a program; an instruction in a programming language that is translated by the compiler or interpreter into several MACHINE-CODE INSTRUCTIONS.

progressive heating (*ElecEng*) Same as SCANNING HEATING.

progressive interlace (*ImageTech*) Scanning of the TV image first with one field containing all the odd-numbered lines and then with a second field of all the even-numbered lines; this is the normal pattern for INTERLACED SCANNING.

progressive metamorphism (*Geol*) The progressive changes in mineral composition and texture observed in rocks within the aureole of contact metamorphism round igneous intrusions; also in rocks which have experienced regional metamorphism of varying degrees of intensity. The particular degree of metamorphism in the latter case is indicated by the 'metamorphic grade' of the rock.

progressive press tool (*Eng*) A *press* tool which performs several operations simultaneously but at successive stations.

progressive proofs (*Print*) In colour printing, a set of proofs supplied to the printer as a guide to colour and registration, each colour being shown both separately and imposed on the preceding ones.

progressive scan CCD (*ImageTech*) A solid-state image-capture device in which the signals are collected in a single pass from, say, the top to the bottom of the image area, unlike the interlaced scans intended for TV pictures.

proinsulin (*BioSci*) An inactive precursor form of insulin that is converted by enzymes into active insulin.

projected area (*Eng*) An area of moulding at right angles to injection direction. See INJECTION MOULDING.

projected diameter (*PowderTech*) The diameter of a circle which has the same area as the projected profile of the particle.

projectile (*Phys*) A body projected by force.

projection (*MathSci*) (1) Of a figure in three-dimensional space onto a plane: the points in the plane from which perpendiculars to the plane pass through points in the original object. (2) A transformation that can be produced by a succession of steps in which one plane is projected on to another plane by joining each point on the first plane to a fixed point outside the planes and then projecting it to the point where the joining line meets the second plane. A projection of *spaces* in higher dimensions is defined similarly.

projection (*Psych*) In psychoanalytic theory, a DEFENCE MECHANISM, in which we ascribe to others various feelings, thoughts and motives, esp of an undesirable sort, which really belong to the self, in order to ward off the anxiety associated with them.

projection distance (*ImageTech*) The distance between the projector and the screen.

projection lamp (*ImageTech*) The source of illumination for projecting the image of motion picture film, transparency sides, etc, onto a screen. See DIGITAL MICROMIRROR DEVICE, LCD PROJECTOR, LIGHT VALVE PROJECTOR, LIQUID CRYSTAL LIGHT VALVE.

projection lantern (*Phys*) See PROJECTOR.

projection lens (*ImageTech*) A lens designed to form an image from motion picture film, transparency or video display on a screen with a substantial degree of enlargement.

projection room (*ImageTech*) Enclosure in a cinema from which the projectors are operated. Also *booth, box*.

projection TV (*ImageTech*) Optical presentation of a TV picture on a separate open screen, in contrast to the direct viewing of a cathode-ray-tube image; generally used for showing a large picture to an audience.

projection welding (*Eng*) A process similar to spot welding but allowing a number of welds to be made simultaneously. One of the workpieces has projections at all points or lines to be welded, the other is flat. Both are held under pressure between suitable electrodes, the current flowing from one to the other at the contact points of the projections.

projective properties (*MathSci*) Of a figure, properties unaltered by projection, eg the class or order of a curve.

projective technique (*Psych*) A method of psychological testing, used in personality assessment, in which relatively unstructured stimuli are presented (eg an inkblot or a picture) and elicit subjective responses from the subject; these are presumed to involve the *projection* of the subject's personality onto the test material. Also *projective test*.

projective transformation (*MathSci*) A transformation projecting one figure into another.

projector (*Phys, ImageTech*) Apparatus for the presentation of an image on a screen, usually with magnification, from photographic transparencies such as slides, from motion picture film or from electronically generated video sources.

prokaryon (*BioSci*) The nuclear material, unbounded by membrane, of prokaryotes. See EUKARYOTE, PROKARYOTE.

prokaryote (*BioSci*) A major division of living organisms that have no defined nucleus. Their genetic material is usually a circular duplex of DNA. They have no endoplasmic reticulum. Bacteria are the prime example but also included are the Cyanobacteria (formerly blue-green algae), the Actinomycetes and the Mycoplasmata. Cf EUKARYOTE. Also *procaryote*. Adj *prokaryotic*.

prolactin (*Med*) Lactogenic hormone; a mammotrophic protein hormone, secreted by cells of the anterior lobe of the pituitary gland, which stimulates the developed mammary gland to secrete milk; has a similar effect on the pigeon crop gland; also called *luteotrophin* in respect of its other, gonadotrophic, properties.

prolactinoma (*Med*) A pituitary tumour secreting PROLACTIN.

prolamellar body (*BioSci*) Three-dimensional lattice composed of tubules, formed within an ETIOPLAST and rapidly converting into THYLAKOIDS on illumination.

prolan (*BioSci*) Former general name for gonadotrophic hormones of mammals found in various tissues and body fluids during pregnancy; prolan A is equivalent to FOLLICLE-STIMULATING HORMONE and prolan B to LUTEINIZING HORMONE.

prolapse (*Med*) The falling out of place or sinking of an organ or part of the body.

prolate cycloid (*MathSci*) See ROULETTE.

prolate ellipsoid (*MathSci*) An ellipsoid obtained by rotating an ellipse about its major axis, ie like a rugby ball. Cf OBLATE ELLIPSOID.

proleg (*BioSci*) One of several pairs of fleshy conical retractile projections borne by the abdomen in larvae of most Lepidoptera, sawflies and scorpion-flies; used in locomotion.

proliferation (*BioSci*) Growth or extension by the multiplication of cells. Adjs *proliferative, proliferous*.

proliferation (*NucEng*) In nuclear policy, the spread of nuclear weapons capability to countries not possessing such weapons.

prolificacy (*Agri*) The breeding performance of livestock, typically expressed as the number of live births per year.

prolification (*BioSci*) Development of buds in the axils of sepals and petals.

proline (*Chem*) *Pyrrolidine-2-carboxylic acid*. The L- or S-isomer is a constituent of proteins, particularly COLLAGEN. Strictly speaking proline is not an amino acid but an *imino* acid, but this nicety is generally neglected. Symbol Pro, short form P.

PROLOG (*ICT*) Abbrev for *programming in logic*. Programming language based on mathematical logic.

Pro Logic (*ImageTech*) TN for an analogue SURROUND SOUND system for consumer use, providing front, centre and rear channels. See AC-3.

PROM (*ICT*) Abbrev for PROGRAMMABLE READ-ONLY MEMORY.

promenade deck (*Ships*) On a passenger ship, the upper deck on which passengers walk.

promenade tile (*Build*) See QUARRY TILE.

promeristem (*BioSci*) The initial cells and their immediate derivatives in an apical meristem.

prometaphase (*BioSci*) The stage in mitosis and meiosis, between prophase and metaphase.

promethazine (*Pharmacol*) An antihistamine drug with sedative properties. It is used for treating hay fever, urticaria, as an antiemetic, and to relieve insomnia.

promethium (*Chem*) A radioactive member of the RARE EARTH ELEMENTS, symbol Pm, having no known stable isotopes in nature. Its most stable isotope, ^{145}Pm, has a half-life of over 20 years.

prominence (*Astron*) A streamer of glowing gas visible in the outer layers of the solar atmosphere. Several types are seen in the upper CHROMOSPHERE and lower CORONA. All consist of regions of higher density and lower temperature than the surrounding gas, which is why they can be seen. Although they are best seen at the rim of the Sun during an eclipse, they are frequently detectable above the PHOTOSPHERE by using a SPECTROHELIOGRAPH.

promontory (*BioSci*) A projecting structure; a small ridge or eminence.

promoter (*BioSci*) (1) A DNA region that lies in front of the coding sequence of a gene; it binds RNA polymerase and therefore signals the start of the gene. See TUMOUR PROMOTER.

promoter (*Chem*) A substance which increases the activity of a catalyst.

promoter (*MinExt*) See COLLECTOR AGENT.

prompt (*ICT*) Character(s) displayed on a video display unit in an interactive computer system to indicate that the user is expected to respond.

prompt critical (*NucEng*) The condition in which a reactor could become critical solely with PROMPT NEUTRONS. In this situation the reactor is difficult to control because the reactor PERIOD is much reduced.

prompt gamma (*Phys*) The gamma radiation emitted at the time of fission of a nucleus.

prompt neutrons (*NucEng*) Nucleons released in a primary nuclear fission process with practically zero delay.

pronation (*BioSci*) In some higher vertebrates, movement of the hand and forearm by which the palm of the hand is turned downwards and the radius and ulna brought into a crossed position. Cf SUPINATION. Adj *pronate*.

pronator (*BioSci*) A muscle effecting pronation.

pronephros (*BioSci*) In Craniata, the anterior portion of the kidney, functional in the embryo but functionless and often absent in the adult. Also *archinephros, fore-kidney, head-kidney*. Adj *pronephric*. Cf MESONEPHROS, METANEPHROS.

pronotum (*BioSci*) The notum of the prothorax in insects. Adj *pronotal*.

pronucleus (*BioSci*) The nucleus of a germ cell after the maturation divisions.

pro-oestrus (*BioSci*) See PRO-ESTRUS.

proof (*Chem*) In alcoholometry, a designation (*proof-spirit*) for spirituous liquid containing 49·28% of alcohol by weight, 57·10% by volume, with rel.d. of 0·920 at 15·6°C. Proof spirit is taken as the standard strength of alcoholic liquids for fiscal purposes. A spirituous liquid which is x% *overproof* contains as much alcohol in 100 vol as in 100+x vol of proof spirit; x% *underproof* signifies the opposite condition.

proof (*Eng*) The impression, often in soft metal, of a die for inspection purposes.

proof (*MathSci*) A step-by-step verification of a mathematical statement.

proof (*Print*) The impression taken from a printing surface for checking and correction only, not as representative of the finished appearance and quality of the work.

proof by contradiction (*MathSci*) See REDUCTIO AD ABSURDUM PROOF.

proof corrections (*Print*) Additions, deletions or amendments to a proof, made clearly in ink, in the margin, in accordance with agreed standards.

proofing press (*Print*) A machine constructed to print proofs, or to check quality of printing plates. Models are available for all three main printing processes and range from simple hand-operated versions to power-driven versions capable of giving impression and register results comparable with a production machine.

proof load (*Aero*) The load which a structure must be able to withstand, while remaining serviceable.

proof load (*Eng*) The load applied to a structure or component sufficiently in excess of the design load as to prove its capability of sustaining the latter. See PROOF TEST.

proof plane (*ElecEng*) A piece of conducting material, mounted on an insulating handle, which may be used for receiving or removing charges in electrostatic experiments.

proof stress (*Eng*) The stress required to produce a certain amount of permanent set in metals which do not exhibit a sudden yield point. Usually it is the stress producing a strain of 0·1, 0·2 or 0·5%. See STRENGTH MEASURES.

proof test (*Eng*) A mechanical test carried out on a manufactured component to ensure that it is capable of meeting and exceeding foreseeable service requirements, eg gun barrels, pressure vessels, chains and lifting gear. The proof test usually stresses the component to double the SAFE WORKING LOAD to be used in the application. Not to be confused with PROOF STRESS.

pro-otic (*BioSci*) An anterior bone of the auditory capsule of the vertebrate skull.

prop (*Build*) A post, usually relatively short and made of timber, used as a strut.

prop (*MinExt*) Sturdy supporting post set across a lode or seam underground to hold up roof after excavation. See POWERED SUPPORTS.

propagation (*BioSci*) The reproduction of a plant by asexual or sexual means, esp in horticulture.

propagation (*ICT*) The mode of transit of a radio signal in a particular environment, eg by reflection from ionized atmospheric layers for long-distance traffic or by multiple reflections from buildings in an urban mobile-telephone network.

propagation (*Phys*) Transmission of energy in the form of waves (eg acoustic, electromagnetic or water) in the direction normal to a wavefront, which is generally spherical, or part of a sphere, or plane.

propagation (*Plastics*) Repeated addition of monomer molecules to an activated polymer chain end, which may be anionic, cationic or free radical in nature. See CHAIN POLYMERIZATION and panel on POLYMER SYNTHESIS.

propagation constant (*Phys*) A measure of diminution in magnitude and retardation in phase experienced by a current of specified frequency in passing along unit length of a transmission line or waveguide, or through one section of a periodic lattice structure. Given by the natural logarithm of the ratio of output to input current or of acoustic particle velocity.

propagation loss (*Phys*) The transmission loss for radiated energy traversing a given path. Equal to the sum of the *spreading loss* (due to increase of the area of the wavefront) and the *attenuation loss* (due to absorption and scattering).

propagule (*BioSci*) Any structure, sexual or asexual and independent from the parent, that serves as a means of reproduction. Also *disseminule*.

propane (*Chem*) C$_3$H$_8$. Bp −45°C. An alkane hydrocarbon, a colourless gas at atmospheric pressure and temperature, found in crude petroleum. In liquid form it constitutes *liquefied petroleum gas* (LPG), a clean-burning fuel, used by some as a petrol substitute. Cracking yields propylene.

propanoic acid (*Chem*) See PROPIONIC ACID.

propanol (*Chem*) Propyl alcohol. C$_3$H$_7$OH. A monohydric aliphatic alcohol, existing in two isomers: (1) propan-1-ol (*n*-propyl alcohol), CH$_3$CH$_2$CH$_2$OH, bp 97°C, rel.d. 0·804, obtained from fusel oil, miscible with water in all proportions; (2) propanol-2-ol, (*iso*-propyl alcohol), CH$_3$CH(OH)CH$_3$, bp 81°C, rel.d. 0·789, which can be prepared by the reduction of acetone with sodium amalgam, or by the hydrolysis of propylene sulphate obtained by the absorption of propylene in sulphuric acid. It also results as a by-product in the synthesis of methanol by the Fischer–Tropsch process.

propanone (*Chem*) See ACETONE.

propantheline bromide (*Pharmacol*) Anticholinergic drug which, by blocking parasympathetic innervation, reduces secretion and mobility of the stomach and intestine.

Propathene (*Plastics*) TN for POLYPROPYLENE (UK).

propellant (*Space*) Comprehensive name for the combustibles for a chemical rocket motor. For liquid propellants it comprises the *fuel* (hydrocarbons, such as kerosine and hydrazine) and the *oxidant* (such as liquid oxygen and fluorine). With solid propellants, combustible materials (such as perchlorate and aluminium powder) are prepared prior to firing *in situ*.

propeller (*Aero*) An assembly of radially disposed blades of aerofoil shape which by reason of rotation in air and blade angle of attack produces thrust or power. Basic classifications are: (1) a *left-hand* propeller which rotates

counterclockwise when viewed from the rear; (2) a *right-hand propeller* which rotates clockwise when viewed from the rear; (3) a *pusher propeller* which is mounted at the rear of an engine; (4) a *tractor propeller* mounted at the front; (5) a *fixed-pitch propeller*, blades mounted at a fixed angle; (6) an *adjustable-pitch propeller*, blades alterable only when stationary; (7) a *variable-pitch propeller*, blades movable by mechanical means during rotation to optimize the PITCH for different speeds and engine revolutions (abbrev *VP propeller*). Types of VP propeller: *constant-speed propeller* which maintains a preselected constant rpm; *controllable-pitch propeller*, with which a desired blade angle may be selected, usually *fine pitch* (or high rpm), *manifold pressure* or *torque*, as appropriate; a *feathering propeller* has an extension of the blade angle to reduce drag when an engine is stopped in flight; a *reverse-pitch propeller* has an additional control which allows the blades to turn to a negative angle to aid stopping after landing (also *braking propeller*); a *swivelling propeller* has a rotatable axis of rotation so that its thrust direction can be used for control on *airships* or for transition in VTOL aircraft. See BLADE ANGLE, FEATHERING PUMP, HELICAL TIP SPEED, PITCH.

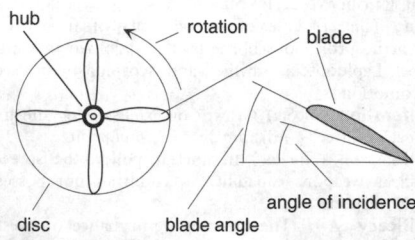

propeller Cross-section of one blade at right.

propeller (*Ships*) See MARINE SCREW PROPELLER.

propeller brake (*Aero*) A shaft brake to stop, or prevent windmilling of, a turboprop, principally to avoid inconvenience on the ground.

propeller efficiency (*Aero*) The ratio of the actual thrust power of a propeller to the torque power supplied by the engine shaft; 80–85% is a typical value.

propeller fan (*Eng*) A fan consisting of an impeller or rotor carrying several blades set at an angle to the axis, working in a cylindrical casing sometimes provided with fixed blades; usually driven by a direct-coupled motor.

propeller governor (*Aero*) A mechanical means of controlling propeller speed.

propeller hub (*Aero*) The detachable fitting by which a propeller is attached to the power-driven shaft, usually also containing the variable-pitch control gear.

propeller post (*Ships*) See STERN FRAME.

propellers (*Print*) Individually mounted rollers, draw rollers or bosses, with peripheries of rubber, synthetic material or knurled metal, usually operating in conjunction with a driven roller to assist in controlling the reels on web-fed presses.

propeller shaft (*Autos*) The driving shaft which conveys the engine power from the gearbox to the rear axle of a motor vehicle. It is connected through universal joints to permit displacement of the rear axle on the springs.

propeller singing (*Acous*) Phenomenon occurring in ship propellers at certain rates of revolution. A strong, almost pure tone is radiated from the propeller into the water.

propeller solidity (*Aero*) The proportion of the disk occupied by blades, measured at 70% radius as standard.

propeller turbine engine (*Aero*) See TURBOPROP.

propeller-type water turbine (*Eng*) A water turbine having a runner similar to a four-bladed ship's propeller. It gives a high specific speed under low heads, thus reducing the size of a direct-coupled generator. See KAPLAN WATER TURBINE.

propelling nozzle (*Aero*) The constricting nozzle at the outlet of a turbojet EXHAUST CONE or *jet pipe* which reduces the gases to slightly more than ambient atmospheric pressure and accelerates them to raise their kinetic energy, thereby increasing the thrust.

propene (*Chem*) C_3H_6, $CH_2=CHCH_3$, *propylene*, an alkene hydrocarbon, a gas, bp $-48°C$. An important by-product of oil refining used for the manufacture of many organic chemicals including propanone, glycerine and polypropene.

propenol (*Chem*) See ALLYL ALCOHOL.

properdin (*BioSci*) A component of the alternative pathway of complement activation that complexes with C3b and stabilizes the alternative pathway C3 convertase.

proper fraction (*MathSci*) See DIVISION.

proper motion (*Astron*) A component of a star's own motion in space which is at right angles to the line of sight, so that it constitutes a real change in the position of the star relative to its neighbouring stars as viewed from Earth.

proper subset (*MathSci*) A SUBSET that is not identical with the given set so that there is at least one member of the set that is not a member of the subset.

proper time (*Phys*) In special relativity, time as measured by an observer's own stationary clock, as distinct from time as visible on some other observer's clock, where the two observers are moving relative to one another.

prophage (*BioSci*) A phage genome that replicates in synchrony with its host. May be integrated into the host genome.

prophase (*BioSci*) The first stage of mitosis or meiosis, during which chromosomes condense and become recognizably discrete. See diagram at MITOSIS and panel on CELL CYCLE.

prophylactic (*Med*) Tending to prevent or to protect against disease, esp infectious disease; any agent which does this.

prophylaxis (*Med*) The preventive treatment of disease.

prophyll (*BioSci*) The first leaf in most monocots or either of the first two leaves in most dicots, on a shoot.

propiolic acid (*Chem*) $CH≡CCOOH$. Mp $6°C$, bp $144°C$. Acetylene-carboxylic acid; silky crystals, soluble in water and ethanol; it forms an explosive silver salt. Also *propargylic acid*.

propionic acid (*Chem*) CH_3CH_2COOH. Mp $-36°C$, bp $141°C$. A monobasic fatty acid, a colourless liquid. A constituent of pyroligneous acid; it is formed in certain fermentations. A more up-to-date process involves the use of the 'oxo' process with ethylene as the starting material. Also *propanoic acid*.

propionyl group (*Chem*) The monovalent radical CH_3CH_2CO-.

proplastid (*BioSci*) Small undifferentiated plastid.

proportional counter (*NucEng*) One which uses the *proportional region* in a tube characteristic, where the gas amplification in the tube exceeds unity but the output pulse remains proportional to initial ionization.

proportional limit (*Eng*) The point on a stress–strain curve at which the strain ceases to be proportional to the stress. Also *limit of proportionality*.

proportional region (*NucEng*) The range of operating voltage for a counter tube in which the gas amplification is greater than unity and the output pulse is proportional to the energy released by the initial event.

propositional calculus (*MathSci*) The branch of mathematical logic dealing with propositions and statements related by operators such as AND,, OR, NOT and 'if…then', without regard to the internal structure of the propositions. See BOOLEAN ALGEBRA. Cf PREDICATE CALCULUS.

proppants (*MinExt*) Materials like sand or special formulations used to keep open fissures in an oil-bearing sediment. See FRACKING.

propranolol (*Pharmacol*) A non-selective β-adrenoreceptor antagonist used in the treatment of hypertension, angina

and thyrotoxicosis. It slows the heart rate and decreases the output of the heart, but causes bronchoconstriction.

proprietary name (*Pharmacol*) Trade name of a drug; eg Zantac is the proprietary name for ranitidine. With a few exceptions drugs are described in the dictionary only under their generic names.

proprioceptor (*BioSci*) A sensory nerve ending receptive to internal stimuli, particularly signalling the relative positions of body parts. Also *interoceptor*. Adj *proprioceptive*.

prop root (*BioSci*) Adventitious root arising on a stem above soil level, growing into the soil and serving as additional support for the stem, as in maize or some mangroves. Also *stilt root*.

proptosis (*Med*) Displacement forwards or protrusion of a part of the body, esp of the eye.

propulsion reactor (*NucEng*) Nuclear reactor designed to supply energy for the propulsion of a vehicle, at present invariably a ship.

propulsive duct (*Aero*) Generic term for the simplest form of reaction–propulsion aero-engine having no compressor/turbine rotor. Thrust is generated by initial compression due to forward motion, the form of the duct converting kinetic energy into pressure, the addition and combustion of fuel, and subsequent ejection of the hot gases at high velocity. See PULSE-JET, RAMJET. Also *Athodyd* (*Aero Thermodynamic Duct*)

propulsive efficiency (*Aero*) (1) The propulsive horsepower divided by the torque horsepower. (2) In a turbojet, the net thrust divided by the gross thrust.

propulsive lift (*Aero*) The means of providing a force on an aircraft in the lift direction, ie at right angles to the longitudinal airflow over the aircraft, by engine power. Makes VTOL possible by VECTORED THRUST using engine exhaust gases, as in the *Harrier*, which also manoeuvres in flight by this means.

propyl alcohol (*Chem*) See PROPANOL.

propylene (*Chem*) $CH_2=CHCH_3$. Bp $-48°C$. An alkene hydrocarbon, a gas. An important by-product of oil refining used for the manufacture of many organic chemicals including propanone, glycerine and polypropylene. Also *propene*.

propylene dichloride (*Chem*) See 1,2-DICHLOROPROPANE.

propylene glycol (*Chem*) *Propan-1,2-diol*. $CH_3CH(OH)$-CH_2OH. Bp $188°C$. Used as a solvent, humectant and plasticizer, and as a chemical intermediate.

propylene oxide rubber (*Plastics*) A speciality rubber based on the repeat unit $-OCH(CH_3)CH_2-$. Made with allyl glycidyl ether as co-monomer. Similar to EPICHLOR-HYDRIN RUBBER. Abbrev *POR*.

propyne (*Chem*) *Allylene*. $CH_3C≡CH$. Methyl acetylene.

proscapula (*BioSci*) See CLAVICLE.

proscenium (*Build*) The stage frame in a theatre, fitted with curtains and a fireproof safety curtain to cut off the stage from the auditorium.

proscolex (*BioSci*) See CYSTICERCUS.

prosector (*Med*) One who dissects dead bodies for anatomical demonstration and teaching.

prosencephalon (*BioSci*) In Craniata, the part of the forebrain that gives rise to the cerebral hemispheres and the olfactory lobes.

prosocoele (*BioSci*) In Craniata, the cavity of the fore-brain or first brain vesicle in the embryo; fore-ventricle.

prosody (*Psych*) Aspects of language such as stress, pitch, intonation, pauses, etc.

prosoma (*BioSci*) (1) In Arachnida, the region of the body comprising all the segments in front of the segment bearing the genital pore. (2) In Acarina, both the mouth- and foot-bearing segments.

prosopagnosia (*Psych*) Inability to recognize faces, usually associated with lesions in the inferior temporal lobe. Adj *prosopagnosic*.

prospect (*Geol, MinExt*) An area which shows sufficient promise of mineral wealth to warrant exploration. Methods of search include aerial survey, magnetometry, geophysical and geochemical tests, seismic probe, electroresistivity measurement, pitting, trenching and drilling.

prostacyclin (*BioSci*) An unstable prostaglandin released by mast cells and endothelium, a potent inhibitor of platelet aggregation; it also causes vasodilation and increased vascular permeability.

prostaglandins (*BioSci*) A group of complex fatty acids, found in most human tissue, that act as local tissue hormones. They regulate blood supply, acid secretion of the stomach, vascular permeability, platelet aggregation and temperature.

prostanoids (*BioSci*) A collective term for prostaglandins, prostacyclins and thromboxanes: a slightly narrower category than EICOSANOIDS.

prostate (*BioSci*) (1) In eutherian mammals (except Edentata and Cetacea), including humans, a gland associated with the male urogenital canal. (2) In Cephalopoda, a gland of the male genital system associated with the formation of spermatophores.

prostatectomy (*Med*) Removal of the prostate gland.

prostatism (*Med*) Difficulty in urinating encountered in elderly men, usually indicative of abnormal enlargement of the prostate gland (benign prostatic hyperplasia).

prostatitis (*Med*) Inflammation of the prostate gland.

prostatorrhoea (*Med*) Chronic gleety or mucous discharge from the prostate. US *prostatorrhea*.

prosthesis (*Med*) Engineered device replacing or restoring tissue or organs but not normally exposed to body fluids such as blood etc, eg artificial limbs, hearing aids, etc. Adj *prosthetic*.

prosthetic group (*BioSci*) Non-proteinaceous entity essential for an enzyme's activity. It is functionally equivalent to a *co-enzyme* and differs only in being tightly bound to its protein.

prosthetics (*Med*) That branch of medicine concerned with prosthesis.

prostomium (*BioSci*) In annelid worms, that part of the head region anterior to the mouth.

protactinium (*Chem*) Symbol Pa, at no 91, half-life of 2×10^4 years. A radioactive element. One radioactive isotope is ^{233}Pa which is an intermediate in the preparation of the fissionable ^{233}U from thorium.

protamines (*BioSci*) Family of short, basic proteins that are bound to sperm DNA in place of histones.

protandry (*BioSci*) The maturation of the male organs before the female organs are receptive. A form of DICHOGAMY. Cf PROTOGYNY.

protanopic (*Med*) Colour blind to red.

protease (*BioSci*) Deprecated term for a PEPTIDASE.

protease inhibitors (*Pharmacol*) Any inhibitor of an enzyme that breaks down proteins (a peptidase), but commonly used as shorthand for drugs, eg *indinavir*, that inhibit the action of the protease involved in producing mature virus particles and that are used in combination therapy for AIDS.

protected cultivation (*Agri*) Crop production under glass or polythene structures ranging from large houses and tunnels to small cloches.

protected-type (*ElecEng*) Said of electric machinery or other apparatus in which any internal rotating or live parts are protected against accidental mechanical contact without impeding ventilation. See SCREEN-PROTECTED MOTOR.

protected zone (*ICT*) In computer security, a layer within an organization's network that is behind the *de-militarized zone* and is hence protected from incoming access from an external network.

protection cap (*ElecEng*) See FENDER.

protective coating (*Chem*) A layer of a relatively inert substance, on the surface of another, which diminishes chemical attack of the latter, eg Al_2O_3 on metallic Al. Also includes paints and many polymers.

protective colloid (*Chem*) A lyophilic colloid which is adsorbed on the dispersed phase of a lyophobic colloid,

thus decreasing its tendency to coagulate, eg albumen in mayonnaise.

protective furnace atmosphere (*Eng*) Inert gas produced from the products of combustion of gas and air in predetermined proportions, the atmosphere being first cooled, cleaned, dehydrated and desulphurized before delivery to heating chambers of furnaces operated for bright annealing of non-ferrous metals and bright treatment of steels. See FURNACE ATMOSPHERE.

protective gap (*ElecEng*) A spark gap between an electric circuit and earth or across a piece of equipment, which breaks down at a voltage below that which might cause damage.

protective gear (*ElecEng*) The apparatus associated with a protective system; eg relays, instrument transformers, pilots, etc.

protective layer (*BioSci*) In an ABSCISSION ZONE, a layer of cells lying immediately proximal to the abscission layer and protecting the underlying tissues from desiccation and invasion by parasites after abscission has occurred.

protective system (*ElecEng*) An arrangement of apparatus designed to isolate a piece of electrical apparatus should a fault occur on it.

protector (*Electronics*) A tube in which glow discharge from a cold cathode prevents high voltage across a circuit.

protein (*BioSci, FoodSci*) Any of thousands of different organic compounds that play a central role in the structure and functioning of all living cells. See panel on PROTEIN.

protein A (*BioSci*) A protein in the cell walls or extracts made from certain strains of *Staphylococcus aureus* which binds to the Fc fragment of IgG from a variety of species. This property has made protein A a useful reagent for isolating IgG and for detecting it in complexes. Biologically it has an anti-phagocytic effect.

protein G (*BioSci*) Protein derived from streptococci with strong affinity for the Fc domain of immunoglobulin molecules. Protein G is very similar in its binding properties to PROTEIN A, but binds to a broader range of immunoglobulin subclasses.

protein kinase (*BioSci*) Any of a group of enzymes that catalyse the transfer of phosphate from adenosine triphosphate to the hydroxyl side chains of a protein, resulting in a change in protein function. Two major classes are recognized: those that phosphorylate on serine or threonine residues and those that phosphorylate tyrosine residues (see PROTEIN TYROSINE KINASES). Protein kinase A (*PKA*) is cyclic AMP-dependent protein kinase; protein kinase C (*PKC*) is actually a family of protein serine/threonine kinases, activated by phospholipids, that play an important part in intracellular signalling.

protein quality (*FoodSci*) Protein quality classified by different criteria, one of the simplest being BIOLOGICAL VALUE on a scale 0–1 where zero is given to proteins without ESSENTIAL AMINO ACIDS and 1 to proteins which have amino acids close to human requirements (eg egg, human milk), other proteins having values in between. In general animal proteins have high values while vegetable proteins have values below 0·5, with the exception of soya.

protein structure (*BioSci*) The three-dimensional structure of a protein resulting from the sequence of amino acids in the polypeptide chain, the binding of non-protein moieties (eg the haem in haemoglobin), and the association with other protein subunits. Determined primarily by X-RAY CRYSTALLOGRAPHY of protein crystals, down to a resolution of 0·2 nm, although protein below approx 20 kDa can be analysed with NMR methods.

proteoclastic (*BioSci*) See PROTEOLYTIC.

proteoglycan (*BioSci*) A protein glycosylated with a variety of polysaccharide side chains (GLYCOSAMINOGLYCANS) to form a high-molecular-weight complex, present in connective tissues such as bone and cartilage, and on cell surfaces.

proteolysis (*BioSci*) The degradation of proteins into peptides and amino acids by cleavage of their peptide bonds.

Protein

Protein molecules consist of one or a small number of POLYPEPTIDE chains each of which is a linear polymer of up to several hundred amino acids linked through their amino and carboxylate groups by PEPTIDE BONDS. The amino acid side chains can have a positive or negative charge, a short aliphatic chain or an aromatic residue. Because the 20 amino acids can be arranged in nearly any sequence, the potential diversity of structure and function is enormous. See appendices.

The properties of each polypeptide depend on the amino acid sequence (its PRIMARY STRUCTURE) which itself determines the correct folding of the chain in three dimensions in its *native conformation* and thus give its specific biological activity. See ALPHA HELIX, BETA-PLEATED SHEET, SECONDARY STRUCTURE,

TERTIARY STRUCTURE. Because the three-dimensional structure is largely dependent on weak forces, it is usually readily disrupted by extremes of pH or heat with a resulting loss of biological activity (see DENATURATION). Covalent cross-links, particularly disulphide bonds, when present, provide a more stable component of higher orders of structure but, in most proteins, are not themselves sufficient to stabilize the native conformation.

If the polypeptide sequence contains many hydrophilic amino acids, the resulting proteins are water soluble, eg most enzymes, but polypeptides may also be assembled into extensive polymeric complexes in structures such as the CILIA, CYTOSKELETON, mitotic spindles, etc. If hydrophobic amino acids predominate, the protein is water-insoluble and fibrous, eg hair, silk and COLLAGEN. These fibrous proteins, in contrast to those which are water-soluble, usually have extremely stable secondary and tertiary structures, and are less prone to denature.

See panel on ENZYME.

---

proteolytic (*BioSci*) Said of enzymes that catalyse the breakdown of proteins into simpler substances, eg trypsin. Also *proteoclastic*.

proteome (*BioSci*) All the proteins encoded by the genome of an organism. Though all are coded for, not all are expressed in every cell and their differential expression, temporally and spatially, is key to understanding how cells and organisms work.

proteomics (*BioSci*) The study of PROTEOMES, by analogy with genomics.

proter-, protero- (*Genrl*) Prefixes from Gk *proteros*, before, former.

protero- (*Genrl*) See PROTER-.

proterokont (*BioSci*) A bacterial flagellum, not homologous with the flagella of higher organisms.

Proterozoic (*Geol*) A division of the Precambrian comprising the less ancient rocks of that system, and lying above the Archaean. See PRECAMBRIAN.

prothallus (*BioSci*) (1) The gametophyte of the pteridophytes, growing photosynthetically at the soil surface, when it may resemble a thallose liverwort, or heterotrophically in association with a fungus underground, and bearing, when mature, archegonia and/or antheridia. The embryo and young sporophyte, developing from a fertilized egg in the archegonium, are at first dependent on the prothallus. (2) Sometimes the gametophyte of gymnosperms.

prothorax (*BioSci*) The first or most anterior of the three thoracic somites in insects. Cf MESOTHORAX, METATHORAX.

Protista (*BioSci*) A paraphyletic group in some classifications of mostly unicellular organisms. Usually includes the Protozoa, Euglenophyceae, Crytophyceae, Dinophyceae and slime moulds, sometimes the flagellate members of the Chlorophyta and Heterokontophyta, and in older usages the bacteria and cyanobacteria.

protium (*Chem*) Lightest isotope of hydrogen, of mass unity (^1H), most prevalent naturally. The other isotopes are deuterium (^2H) and tritium (^3H).

proto- (*Genrl*) Prefix form Gk *protos*, first.

protocercal (*BioSci*) See DIPHYCERCAL.

Protochordata (*BioSci*) A division of Chordata comprising the subphyla Hemichordata, Urochordata and Cephalochordata, that do not have a cranium, vertebral column, or specialized anterior sense organs. Cf VERTEBRATA. See panel on VERTEBRATE EVOLUTION.

protocol (*ICT*) Agreed set of operational procedures to enable data to be transferred between systems.

protoco-operation (*BioSci*) See CO-OPERATION.

protoderm (*BioSci*) PRIMARY MERISTEM that gives rise to the epidermis and which may arise from independent INITIALS in the apical meristem.

protogenic (*Chem*) Capable of supplying a hydrogen ion (proton).

protogyny (*BioSci*) The maturation of the female organs before the male organs liberate their contents. See DICHOGAMY. Cf PROTANDRY.

protomorphic (*BioSci*) Primordial; primitive.

proton (*Phys*) The nucleus of the hydrogen atom; of positive charge and atomic mass number of unity. With neutrons, protons form the nucleus of all atoms, the number of protons being equal to the atomic number. It is the lightest BARYON (mass 1·007 276 amu) and the most stable. Beams of high-energy protons produced in particle accelerators are used to study elementary particles.

protonema (*BioSci*) The juvenile stage of the gametophyte of mosses and liverworts.

protonephridial system (*BioSci*) The excretory system of Platyhelminthes, consisting of flame cells and ducts.

protonephridium (*BioSci*) A larval nephridium, usually of the flame cell type.

protonic solvent (*Chem*) One that yields a proton, H^+, as the cation in SELF-DISSOCIATION.

proton motive force (*BioSci*) The electrochemical gradient that is derived from a membrane potential together with a proton gradient across the membrane. Such gradients operate across the inner mitochondrial membrane and the THYLAKOID membrane. Essential for the generation of ATP during oxidative phosphorylation and photosynthesis.

proton number (*Phys*) The number of protons in an atomic nucleus, equal to the number of electrons in the neutral atom. Also *atomic number*.

proton-precessional magnetometer (*Phys*) Precision magnetometer based on the measurement of the LARMOR FREQUENCY of protons in a sample of water. See NUCLEAR MAGNETIC RESONANCE, PROTON RESONANCE.

proton–proton chain (*Phys*) A series of thermonuclear reactions which is initiated by a reaction between two protons. It is thought that the proton–proton cycle is more important than the CARBON CYCLE in stars which are relatively cool.

proton pump inhibitors (*Pharmacol*) Drugs that inhibit the secretion of gastric acid by acting on the cellular proton ATPase, used to treat gastric and duodenal ulcers. An example is *omeprazole*. Used in combination with an H2-receptor antagonist and an antibiotic in the eradication of *Helicobacter pylori*, a cause of peptic ulceration. Abbrev PPI.

proton resonance (*Phys*) A special case of NUCLEAR MAGNETIC RESONANCE. Since the nuclear magnetic moment of protons is now well known, that of other nuclei is found by comparing their resonant frequency with that of the proton.

proton synchrotron (*Phys*) A SYNCHROTRON in which the accelerated particles are protons. It is capable of producing particle energies of more than 200 GeV.

proton-translocating ATPase (*BioSci*) Primary electrogenic ACTIVE TRANSPORT (panel) system, powered by the hydrolysis of ATP, pumping protons out of a plant cell across the PLASMALEMMA or into the vacuole across the TONOPLAST. The resulting pH and electrical potential gradients drive a number of secondary active transport processes coupled to the return movements of the protons. Also H^+-*ATPase*. See panel on MITOCHONDRION.

proto-oncogene (*BioSci*) A gene that is required for normal function of the organism, but which when altered can become an ONCOGENE.

protophilic (*Chem*) Able to combine with a hydrogen ion.

protophloem (*BioSci*) The first-formed PRIMARY PHLOEM; characteristically maturing while the organ is elongating.

protoplanet (*Astron*) A planet in the early stages of formation.

protoplasm (*BioSci*) The living material within a cell divided into discrete structures, eg mitochondria, ribosomes, nuclei, chromosomes and nucleoli in EUKARYOTES and chromosome and ribosomes in PROKARYOTES. Adj *protoplasmic*.

protoplast (*BioSci*) (1) The living part of a plant cell including the nucleus, cytoplasm and organelles, all bounded by the plasmalemma, but excluding any cell wall. (2) The above structure isolated from its cell wall usually by treatment of a tissue with wall-degrading enzymes or mechanically. See PROTOPLAST CULTURE, PROTOPLAST FUSION.

protoplast culture (*BioSci*) The aseptic culture on suitable media of protoplasts isolated from plant tissue, esp for the regeneration of plants from somatic hybrids and hybrids resulting from the fusion of protoplasts from different plants.

protoplast fusion (*BioSci*) The artificial production of SOMATIC HYBRIDS and CYBRIDS by the fusion of PROTOPLASTS from different plants. Used in attempts to transfer genetic information between sexually incompatible species. See panel on GENETIC MANIPULATION.

protoporphyrin (*Chem*) *1,3,5,8-tetramethyl-2,4-divinyl-6,7-dipropanoic acid-porphin*. It combines with iron (II) to give reduced haematin, the prosthetic group of haemoglobin.

protopsis (*Med*) Protrusion of the eye.

protostar (*Astron*) A star in the early stages of formation which has resulted from the gravitational fragmentation of a cloud of gas and dust, but which has not yet collapsed sufficiently for nuclear reactions to begin.

protostele (*BioSci*) A STELE without leaf gaps or, sometimes, with a solid core of xylem.

protostome (*BioSci*) An invertebrate phylum, comprising organisms in which the mouth forms from the embryonic blastopore. Major protostome phyla are Annelida, Mollusca and Arthropoda. See DEUTEROSTOME.

Prototheria (*BioSci*) A subclass of primitive mammals that probably left the main stock in the Mesozoic and are found only in Australasia. The adults have no teeth, the cervical vertebrae bear ribs, the limbs are held laterally, the shoulder girdle has precoracoids and an inter-clavicle, and large yolky eggs are laid. The young are fed after hatching on milk produced by specialized sweat glands, whose ducts do not unite to open on nipples. There is only one order, the Monotremata. Duck-billed platypus, spiny anteater.

prototroph (*BioSci*) A WILD-TYPE organism able to grow in its unsupplemented medium. Cf AUXOTROPH.

prototropic change (*Chem*) A term applied in the form of isomerism known as TAUTOMERISM when the 'movement' of a proton is involved.

prototropy (*Chem*) The reversible conversion of tautomeric forms by migration of hydrogen as a proton.

prototype (*BioSci*) An ancestral form; an original type or specimen.

prototype (*Eng*) Generally, the first or original type or model from which anything is developed.

prototype filter (*ICT*) Basic type that has the specified nominal cut-off frequencies, but that must be developed into derived forms to obtain further desirable characteristics, such as constancy of image impedance with frequency.

protoxylem (*BioSci*) The first-formed primary xylem; typically maturing while the organ elongates, having narrow tracheary elements with annular or helical thickening and parenchyma only, and becoming stretched and crushed as the organ elongates.

Protozoa (*BioSci*) A phylum of unicellular or acellular animals. Their nutrition is holophytic, holozoic or saprophytic and they reproduce by fission or conjugation. Their locomotion is by means of cilia, flagella or pseudopodia. They may be free-living or parasitic. Representative fossil skeletons are shown in the figure. The singular individual is a *protozoon*.

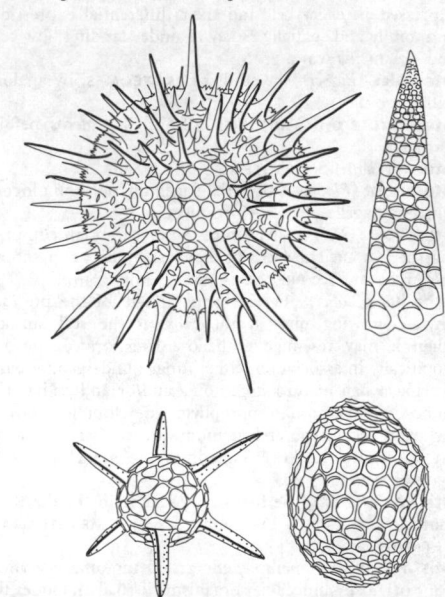

Protozoa Typical radiolarian skeletons.

protractor (*BioSci*) A muscle which by its contraction draws a limb or a part of the body forwards or away from the body. Cf RETRACTOR.

protractor (*Eng*) An instrument used by the draughtsperson for measuring or setting out angles on paper etc.

proud (*Build, Eng*) A part or portion of a part projecting above another or above its surroundings, 'standing proud'.

proud flesh (*Med*) The popular name for granulation tissue formed during wound healing.

proustite (*Min*) Sulphide of silver and arsenic which crystallizes in the trigonal system. It is commonly associated with other silver-bearing minerals. Also *light-red silver ore, ruby silver ore.* Cf PYRARGYRITE.

provascular tissue (*BioSci*) See PROCAMBIUM.

proved reserves (*MinExt*) Tonnages of economically valuable ore which have been tested adequately by being blocked out into panels and sampled at close intervals.

proventriculus (*BioSci*) (1) In birds, the anterior thin-walled part of the stomach, containing the gastric glands. (2) In Oligochaeta and insects, the gizzard, a muscular thick-walled chamber of the gut posterior to the crop. (3) In Crustacea, the stomach or gastric mill.

provitamin (*Chem*) A vitamin precursor, eg β-carotene gives vitamin A.

proxemics (*Psych*) The study of the spatial features of human social interaction, eg PERSONAL SPACE, territoriality, crowding.

Proxima Centauri (*Astron*) The nearest star to the Sun; a faint companion in the star system *Alpha Centauri* in the constellation Centaurus, its distance being 4·3 light-years.

proximal (*BioSci*) Pertaining to or situated at the inner end, nearest to the point of attachment. Cf DISTAL.

proximity effect (*Phys*) Increase in effective high-frequency resistance of a conductor when it is brought into proximity with other conductors, owing to eddy currents induced in the latter. It is esp prominent in the adjacent turns of an inductance coil.

proximity fuse (*Radar*) Miniature radar carried in guided missiles, shells or bombs so that they explode within a preset distance of the target.

proxy (server) (*ICT*) An intermediate network service via which client devices make outgoing requests to other Internet servers which the proxy then accesses on their behalf.

Prozac (*Pharmacol*) TN for fluoxetine hydrochloride, an SSRI drug.

PrPc (*Med*) Abbrev for the protease sensitive and soluble form of prions. See panel on TRANSMISSIBLE SPONGIFORM ENCEPHALOPATHY.

PrPsc (*Med*) Abbrev for the protease insensitive and insoluble form of prions. See panel on TRANSMISSIBLE SPONGIFORM ENCEPHALOPATHY.

pruinose (*BioSci*) Having a bloom on the surface, esp a whitish bloom like hoar frost.

pruniform (*BioSci*) Shaped like a plum.

prunt (*Glass*) A disk of glass bearing a monogram or badge, eg of the owner or vintner, and fused onto bottles or drinking vessels.

prurigo (*Med*) A term common to various skin diseases the chief characteristic of which is papular eruption and intense itching. Adj *pruriginous.*

pruritus (*Med*) Severe persistent itching, a characteristic of numerous skin diseases.

Prussian blue (*Chem*) Ferric ferrocyanide, a colour pigment. See POTASSIUM FERRICYANIDE.

prussic acid (*Chem*) A solution of HYDROGEN CYANIDE in water.

PS (*Plastics*) Abbrev for POLYSTYRENE.

PSA (*Med*) Abbrev for *prostate-specific antigen*, a protein whose presence in large quantities in blood is an indication of possible prostate cancer.

psalterium (*BioSci*) In ruminant mammals, the third division of the stomach. Also *manyplies, omasum.*

psammitic gneiss (*Geol*) A gneissose rock which has been produced by the metamorphism of arenaceous sediments.

psammitic schists (*Geol*) Schists formed from arenaceous sedimentary rocks. Cf PELITIC GNEISS, PELITIC SCHIST.

psammophyte (*BioSci*) A plant adapted to growing on sand or sandy soils.

pseud-, pseudo- (*Genrl*) Prefixes from Gk *pseudes*, false.

pseudautostyly (*BioSci*) A type of jaw suspension in which the upper jaw is fused with the ethmoidal, orbital and otic regions of the cranium. Cf AUTOSTYLY. Adj *pseudautostylic.*

pseudo- (*Chem*) A prefix which is sometimes used to indicate a tautomeric, isomeric, or closely related compound. Symbol, ψ.

pseudo- (*Genrl*) Prefix from the Gk *pseudes* false, lying or fraudulent. Used of something pretending to be other than it is, resembling something else (eg pseudopod).

pseudoacid (*Chem*) A substance which can exist in two tautomeric forms, one of which functions as an acid, eg nitromethane: $CH_3NO_2 \rightleftarrows H^+ + CH_2 = NO_2$.

pseudoalums (*Chem*) A name sometimes given to double sulphates of the alum type, where there is a bivalent element in place of the univalent element of ordinary alums.

pseudo-aposematic (*BioSci*) Warning or aposematic coloration borne by animals that are not dangerous or distasteful, but show BATESIAN MIMICRY of animals which are.

pseudobase (*Chem*) A substance which can exist in two tautomeric forms, one of which functions as a base. Cf PSEUDOACID.

pseudobrachium (*BioSci*) In some fish, an appendage used for propulsion along a substratum or on dry land; formed by modification of the pectoral fin.

pseudobulb (*BioSci*) Swollen, solid, above-ground stem of some orchids, acting as a storage organ.

pseudocarp (*BioSci*) See FALSE FRUIT.

pseudocode (*ICT*) Instructions written in symbolic language that must be translated into an acceptable program language or direct into machine language before they can be executed.

pseudocoele (*BioSci*) (1) In higher vertebrates, a space enclosed by the inner walls of the closely opposed cerebral hemispheres; the fifth ventricle. (2) A body formed from a persisting blastocoel. Also *pseudocoelom.* Cf COELOM.

pseudocopulation (*BioSci*) Attempts of a male insect to mate with a flower that resembles a female of its species, as in the pollination mechanisms of many orchids.

pseudocowpox (*Vet*) A common ZOONOSIS caused by infection with pseudocowpox virus. Lesions appear on the teats of cows over 2 years old. Vesicles, pustules and papules are present, followed by thick scabs which drop off after about a month. Also *milker's nodule.*

pseudocubic (*Min*) See PSEUDOSYMMETRY.

pseudocyesis (*BioSci*) In some mammals, uterine changes following estrus and resembling those characteristic of pregnancy. Also *pseudopregnancy.*

pseudodementia (*Psych*) Mental conditions in which there are symptoms which suggest dementia, but which are caused by other factors (eg drug use or depression).

pseudodont (*BioSci*) Having horny pads or ridges in place of true teeth, as monotremes.

pseudo-elastic method (*Eng*) See QUASI-ELASTIC METHOD.

pseudo fowl plague (*Vet*) See NEWCASTLE DISEASE.

pseudogamy (*BioSci*) A form of APOMIXIS in which, although fertilization does not occur, the stimulus of pollination is necessary for seed production.

pseudogene (*BioSci*) A copy of a gene that is defective and therefore not transcribed.

pseudogout (*Med*) An acute arthritis often affecting a single joint, often knee or hip. The clinical course is like gout but the disease is due to deposits of calcium pyrophosphate crystals around the joint.

pseudoheart (*BioSci*) (1) In Oligochaeta, one of a number of paired contractile anterior vessels by which blood is pumped from the dorsal to the ventral vessel. (2) In Echinodermata, the axial organ.

pseudohexagonal (*Min*) See PSEUDOSYMMETRY.

pseudoleucite (*Min*) An aggregate showing the crystal shape of LEUCITE but consisting mainly of potassium feldspar and nepheline.

pseudomalachite (*Min*) Phosphate and hydroxide of copper which resembles malachite and crystallizes in the monoclinic system.

pseudomerism (*Chem*) A type of tautomerism in which only one form is known, although derivatives exist of both forms, eg HCN (known) and HNC (unknown) giving rise to RCN (cyanides) and RNC (isocyanides).

pseudometamerism (*BioSci*) The condition of repetition of parts, found in some Cestoda, which bears a superficial resemblance to metamerism.

pseudo-momentum (*Phys*) See CRYSTAL MOMENTUM.

Pseudomonadaceae (*BioSci*) A family of Gram-negative rod-shaped bacteria belonging to the order Pseudomonadales. They occur in water and soil, and also include animal and plant pathogens, eg *Pseudomonas aeruginosa* (blue pus), *Xanthomonas hyacinthi* (yellow rot of hyacinth). *Acetobacter* species are used in production of vinegar.

Pseudomonadales (*BioSci*) One of the two main orders of true bacteria, distinguished by the polar flagella of motile forms. They are Gram-negative and may be spiral, spherical or rod-shaped cells. See EUBACTERIALES.

pseudomorph (*Min*) A mineral whose external form is not the one usually assumed by its particular species, the original mineral having been replaced by another substance or substances.

pseudoparenchyma (*BioSci*) A false tissue made of interwoven fungal hyphae.

pseudoperianth (*BioSci*) Cylindrical sheath growing up around the archegonium and young sporophyte of some liverworts.

pseudo-plasticity (*Phys*) A type of NON-NEWTONIAN FLOW, where viscosity falls with increasing shear rate. Most molten polymers are pseudo-plastic.

pseudopod (*BioSci*) (1) A broad finger-like protrusion of the cell surface that may be used in amoeboid cells for locomotion. Also *pseudopodium*. (2) A foot-like process of the body wall, characteristic of some insect larvae.

pseudo-potential method (*Phys*) A method of calculating the band structure of metals. See BAND THEORY OF SOLIDS.

pseudopregnancy (*Psych*) A psychosomatic condition marked by many of the symptoms of pregnancy. Also *pseudocyesis*.

pseudopterygium (*Med*) The adherence of a tip of a fold of oedematous conjunctiva to a corneal ulcer, thus simulating a pterygium.

pseudorabies (*Vet*) See AUJESKY'S DISEASE.

pseudo-random binary sequence (*ICT*) A fixed-length binary sequence that satisfies many of the tests for a true random sequence. Normally generated using a shift register with logical feedback. An n-stage register provides a sequence $p < 2^n - 1$ bits. The maximum length sequence is $(2^n - 1)$.

pseudo-random bit sequence (*ICT*) A series of binary digits, usually generated by means of a shift register having several feedback loops to intermediate bits, having all the characteristics of a truly random sequence except that its pattern repeats over a period of typically 2^{15} bits. Such sequences are often used for testing digital links. Also *pseudo-random binary sequence*.

pseudo-random number generator (*ICT*) Software that generates sequences of numbers that satisfy statistical tests for a RANDOM NUMBER sequence, but can be repeated by starting again at the same point.

pseudoring (*MathSci*) See RING.

Pseudoscorpionidea (*BioSci*) An order of Arachnida, resembling Scorpionidea, but with no tail. The pedipalps are large, chelate, and contain poison glands. Small carnivorous forms are found under stones, leaves, bark and moss; they are occasionally found in houses. Also *Chelonethida, false scorpions*.

pseudosolution (*Chem*) A colloidal solution or suspension.

pseudosymmetry (*Min*) A term applied to minerals whose symmetry elements place them on the borderline between two crystal systems, eg a mineral with the c-axis very nearly equal to the b- and a-axes might, on casual inspection, appear cubic, though actually tetragonal. It would be described as possessing *pseudocubic symmetry*. The phenomenon is due to slight displacement of the atoms from the positions which they would occupy in the class of higher symmetry. Also applied when the pseudosymmetry is due to twinning.

pseudotachylite (*Geol*) Flinty crush-rock, resulting from the vitrification of rock fragments produced during faulting under conditions involving the development of considerable heat by friction, as in the Glencoe CAULDRON SUBSIDENCE.

pseudotetragonal (*Min*) See PSEUDOSYMMETRY.

pseudovilli (*BioSci*) Projections from the surface of the trophoblast in some mammals, as distinct from the true villi, which are definite outgrowths.

pseudovitellus (*BioSci*) See MYCETOME.

P-shell (*Phys*) The ELECTRON SHELL in an atom corresponding to a principal quantum number of six. In naturally occurring elements it is never completely filled and is the outermost shell for most stable heavy elements. See panel on ATOMIC STRUCTURE.

psi (*Genrl*) Abbrev for *pounds per square inch*.

ψ/J particle (*Phys*) See J/ψ PARTICLE.

psilomelane (*Min*) A massive hydrated oxide of manganese which contains varying amounts of barium, potassium and sodium. It is a secondary mineral formed by alteration of manganese carbonates and silicates, and is used as an ore of manganese.

ψp (*BioSci*) Symbol for PRESSURE POTENTIAL.

psi particle (*Phys*) See J/ψ PARTICLE.

Psittaciformes (*BioSci*) An order of birds containing one family. They are mainly vegetarian, with powerful hooked beaks; the feet are typically zygodactylous. The birds are often vividly coloured, and capable of mimicry. Parrots, cockatoos.

psittacosis (*Med, Vet*) An acute or chronic contagious disease of wild and domestic birds which is transmittable to other animals and humans, and is caused by *Chlamydia psittaci*. Results in respiratory and systemic infections including, in humans, a disease resembling pneumonia. Also *ornithosis, parrot disease*.

PSK (*ICT*) Abbrev for PHASE-SHIFT KEYING.

psophometer (*ICT*) Noise-measuring instrument, incorporating weighting circuits, that relates its readings to the effects of noise as perceived by the human ear, rather than straightforward signal-level comparisons.

psophometric voltage (*ICT*) That which measures the noise in communication circuits arising from interference of any kind.

psoriasis (*Med*) A chronic disease of the skin in which red scaly papules and patches appear, esp on the outer aspects of the limbs.

psoroptic mange (*Vet*) Mange of animals due to mites of the genus *Psoroptes*. Sheep scab is caused by *Psoroptes communis ovis*.

PSS (*ICT*) Abbrev for PACKET SWITCHING SYSTEM.

PST (*Eng*) Abbrev for *lead scandium tantalate*. See DIELECTRIC AND FERROELECTRIC MATERIALS.

p-state (*Electronics*) That of an orbital electron when the orbit has angular momentum of one Bohr unit.

PSTN (*ICT*) Abbrev for PUBLIC SWITCHED TELEPHONE NETWORK.

PSu (*Plastics*) Abbrev for POLYSULPHONES.

psych-, psycho- (*Genrl*) Prefixes from Gk *psychē*, soul, mind.

psychiatrist (*Psych*) A medically qualified doctor specializing in mental disorders.

psychiatry (*Psych*) The study, diagnosis, treatment and prevention of mental disorders.

psychism (*BioSci*) The doctrine, difficult to sustain, that living matter possesses attributes not recognized in non-living matter.

psycho- (*Genrl*) See PSYCH-.

psychoanalysis (*Psych*) (1) A theory of personality developed by Freud in which the ideas of unconscious conflict and PSYCHOSEXUAL DEVELOPMENT are central. (2) A method of therapy based on Freud's theory of personality which attempts to help the individual gain insight into his or her unconscious conflicts using a variety of psychoanalytic techniques. See FREE ASSOCIATION, FREUD'S THEORY OF DREAMS, TRANSFERENCE.

psychodynamic (*Psych*) Psychological systems and theories emphasizing change and development or those that make motivation and drive more central. Also used to refer to the psychoanalytic approach of Freud, the psychodynamic approach.

psychogalvanic reflex (*BioSci*) The decrease in the electrical resistance of the skin under the stimulation of various emotional states.

psychogenic (*Med*) Having a mental origin.

psychogenic disorders (*Psych*) Disorders whose origins are psychological rather than organic.

psychokinesis (*Psych*) The alleged ability of some people to alter physical reality in the absence of any known mechanism for accomplishing it (eg bending metal objects without touching them).

psychology (*Psych*) A broad field of study with many subspecialities. In general, the study of emotion, cognition and behaviour, and their interactions. But see separate entries for CLINICAL, COGNITIVE, COMPARATIVE, DEVELOPMENTAL, DYNAMIC, EGO, EXISTENTIAL, EXPERIMENTAL, FORENSIC, HEALTH, HUMANISTIC, INDUSTRIAL/ORGANIZATIONAL and PHYSIOLOGICAL specialities of psychology, and PSYCHOPATHOLOGY, SOCIAL PSYCHOLOGY; see also ETHOLOGY.

psychometrics (*Psych*) The application of mathematical and statistical concepts to psychological data, particularly in the areas of mental testing and experimental data.

psychopath (*Psych*) A medical–legal term referring to a behaviour disorder characterized by repetitive, antisocial behaviour with emotional indifference and without guilt, where the individual does not learn from experience or punishment. It is a category not recognized in Scottish law. Also *antisocial personality*.

psychopathology (*Psych*) The study of psychological disorders.

psychopharmacology (*Psych*) The study and use of drugs that influence behaviour, emotions, perception and thought, by acting on the central nervous system.

psychophily (*BioSci*) Pollination by butterflies.

psychophysics (*Psych*) The branch of psychology that studies the relationship between characteristics of the physical stimulus and the psychological experience they produce.

psychophysiological disorders (*Psych*) Physical disorders which are thought to be due to emotional factors but which involve genuine organic changes (eg high blood pressure). Formerly *psychosomatic disorders*.

psychophysiology (*Psych*) The measurement of physiological processes such as heart rate and blood pressure in relation to various mental and emotional states. See PSYCHOPHYSIOLOGICAL DISORDERS.

psychoprophylaxis (*Med*) A method of psychological and physical training for childbirth, aimed at making labour painless.

psychosexual development (*Psych*) (1) In psychoanalytic theory, a progressive series of stages in which the source of bodily pleasure changes during development, defined by the zone of the body through which this pleasure is derived. See ANAL STAGE, GENITAL STAGE, ORAL STAGE, PHALLIC STAGE. (2) The term *psychosexual* also refers to mental aspects of sexual phenomena.

psychosexual disorders (*Psych*) Seriously impaired sexual performance of various kinds, or unusual methods of sexual arousal.

psychosis (*Psych*) A very general term used to describe mental illnesses which result in a severe loss of mental and emotional function, in contrast to NEUROSIS, where the individual remains competent to cope with reality.

psychosomatic (*Psych*) See PSYCHOPHYSIOLOGICAL DISORDERS.

psychosomatic medicine (*Med*) A branch of medicine that stresses the relationship of bodily and mental happenings, and combines physical and psychological techniques of investigation. Particular attention is paid to the possibility of physical disease (ie duodenal ulcer, asthma) being induced by mental states.

psychosurgery (*Med*) Surgery to remove or destroy brain tissue, intended to relieve the symptoms of either mental or physical disease.

psychotherapist (*Psych*) An individual, usually a physician, who practises PSYCHOTHERAPY.

psychotherapy (*Psych*) The treatment of mental and emotional disturbance by psychological means, often in an extended series of therapist–client sessions; refers to several different forms of treatment and techniques. The term is usually restricted to treatments supervised or conducted by trained psychologists or psychiatrists.

psychotomimetic (*Pharmacol*) Said of drugs which cause bizarre psychic effects in humans as well as marked behavioural changes in animals, eg esters of N-methyl-3-hydroxypiperidine.

psychrometer (*EnvSci*) See WET AND DRY BULB HYGROMETER.

psychrophilic (*BioSci*) Growing best at a relatively low temperature, esp (of a micro-organism) having a temperature optimum below 20°C. Cf MESOPHILIC, THERMOPHILIC.

Pt (*Chem*) Symbol for PLATINUM.

pt (*Print*) Symbol for POINT.

PTCR (*Electronics*) Abbrev for *positive temperature coefficient resistor*. See SILICON RESISTOR.

Ptd.A. (*Build*) Abbrev for *pointed arch*.

pter-, ptero- (*Genrl*) Prefixes from Gk *pteron* denoting feather or wing.

Pteridophyta (*BioSci*) (1) A division of the plant kingdom containing all the vascular plants that do not bear seeds, ie the ferns, clubmosses, horsetails, etc. There is an alternation of generations of, typically, a smaller, more or less thalloid, independent gametophyte and larger, longer-lived sporophyte usually with roots, stems and leaves. They are usually divided into eight classes: Rhyniopsida, Psilotopsida, Zosterophyllopsida, Lycopsida, Trimerophytopsida, Sphenopsida, Filicopsida and Progymnospermopsida. (2) Sometimes, confusingly, the ferns alone.

Pteridospermopsida (*BioSci*) The seed ferns and allies. Class of extinct gymnosperms, mostly extant from the Carboniferous to the Jurassic.

ptero- (*Genrl*) See PTER-.

pteropod ooze (*Geol*) A calcareous deep-sea deposit which contains the remains of a large number of gastropods which swam by means of wing-like extensions of the foot.

pterygial (*BioSci*) (1) Generally, pertaining to a fin or wing. (2) In fish, an element of the fin skeleton.

pterygium (*Med*) The encroachment on to the cornea from the side of a thickened, vascular, wing-shaped area of the conjunctiva.

pterygium (*BioSci*) In vertebrates, a limb.

pterygoid (*BioSci*) A paired cartilage bone of the vertebrate skull, formed by the ossification of the front part of the PTERYGOPALATOQUADRATE BAR; a membrane bone which replaces the original pterygoid in some vertebrates; more generally, wing-shaped.

pterygopalatoquadrate bar (*BioSci*) In fish with a cartilaginous skeleton, the rod of cartilage forming the upper jaw. Also *PPQ bar*.

pterylosis (*BioSci*) In birds, the arrangement of the feathers in distinct feather tracts or pterylae, whose form and arrangement are important in classification.

PTFCE (*Plastics*) See POLYTRIFLUOROCHLOROETHYLENE.

PTFE (*Plastics*) Abbrev for POLYTETRAFLUOROETHYLENE.

ptilinum (*BioSci*) In certain Diptera (Cyclorrhapha), an expansible membranous cephalic sac by which the anterior end of the puparium is thrust off at emergence.

PTMEG (*Plastics*) Abbrev for POLYTETRAMETHYLENE ETHER-GLYCOL.

Ptolemaic system (*Astron*) The final form of Gk planetary theory as described in Claudius Ptolemy's treatise. In this the Earth was the centre of the universe, the planets, including the Sun and Moon, being supposed to revolve round it in motions compounded of eccentric circles and epicycles; the fixed stars were supposed to be attached to an outer sphere concentric with the Earth.

Ptolemy's theorem (*MathSci*) The theorem that the product of the diagonals of a cyclic quadrilateral is equal to the sum of the products of the opposite sides.

ptomaines (*Chem*) Poisonous amino compounds produced by the decomposition of proteins, esp in dead animal matter. The ptomaines include substances such as putrescine, cadaverine, choline, muscarine, neurine. Few of the ptomaines are known to be poisonous by mouth, food poisoning (botulism) being caused by specific bacteria, eg *Clostridium botulinum*.

ptosis (*Med*) (1) Paralytic dropping of the upper eyelid. (2) Downward displacement of any bodily organ.

P-trap (*Build*) A trap used in sanitary pipes with the inlet vertical and the outlet inclined slightly below the horizontal.

PTS (*Acous*) Abbrev for PERMANENT THRESHOLD SHIFT.

PTTA (*ICT*) Abbrev for a POSTAL, TELEPHONE AND TELEGRAPH ADMINISTRATION.

ptyalin (*BioSci*) Common name for the enzyme, α-amylase, found in saliva.

ptyalism (*Med*) Excessive secretion of saliva.

p-type conduction (*Electronics*) That associated with charge transport by holes. Also *p-type conductivity*.

P-type semiconductor (*Electronics*) Relatively wide-band-gap P-TYPE SEMICONDUCTOR.

p-type semiconductor (*Electronics*) That arising in a semiconductor containing ACCEPTOR impurities, with conduction by (positive) holes.

ptyxis (*BioSci*) The manner in which an individual unexpanded leaf, sepal or petal is folded, rolled or coiled in the bud. Also *vernation*.

PU (*FoodSci*) Abbrev for PASTEURIZATION UNIT.

Pu (*Chem*) Symbol for PLUTONIUM.

puberty (*Med*) Sexual maturity.

puberulent (*BioSci*) Minutely pubescent.

pubescence (*BioSci*) A covering of fine hairs or down. Adj *pubescent*.

pubiotomy (*Med*) The operation of cutting the pubic bone to one side of the midline, so as to facilitate childbirth in difficult labour.

pubis (*BioSci*) In Craniata, an element of the pelvic girdle (contraction of *os pubis*). Adj *pubic*.

public-address system (*Acous*) Sound-reproducing system for large-space and outdoor use, usually with high-powered horn radiators, or columns of open-diaphragm units which concentrate radiation horizontally.

public data network (*ICT*) A communications system intended for carrying digital data that is open for anyone who wishes to subscribe to it.

public key (*ICT*) In cryptography, one of a pair of keys required to write and read a secure message. The public key may be widely known and is used to lock the contents of a message. See CRYPTOSYSTEM.

public land mobile network (*ICT*) A mobile-telephone network based on terrestrial BASE STATIONS and accessible to the public via the PUBLIC SWITCHED TELEPHONE NETWORK.

public switched telephone network (*ICT*) The traditional telephone network based on fixed links and SWITCHES

through which any two terminals can be connected to each other.

publishers' binding (*Print*) See EDITION BINDING.

pucella (*Glass*) An implement for opening out the top of a wine glass in the off-hand process.

puddingstone (*Geol*) A popular term for *conglomerate*. Hertfordshire Puddingstone, consisting of rounded flint pebbles set in a siliceous sandy matrix, is a good example.

puddle (*CivEng*) See CLAY PUDDLE.

puddle clay (*CivEng*) A plastic material produced by mixing clay thoroughly with about one-fifth of its weight of water. It is used in engineering construction to prevent the passage of water, eg for cores for earthen reservoir dams. Also *clay puddle*.

puddle mill (*CivEng*) A mixer for combining clay and water to form PUDDLE CLAY.

puddling (*MinExt*) Concentration of diamond from 'blue ground' clays (weathered kimberlite) by forming an aqueous slurry in a mechanized stirring pan in which the heavier fraction is retained from periodic retrieval, while the lighter fraction overflows.

puerilism (*Psych*) Reversion to a child-like state of mind.

puerperal (*Med*) Pertaining to or ensuing upon childbirth, eg *puerperal fever*, a condition caused by infection of the genital tract in the course of childbirth.

puerperium (*Med*) Strictly, the period between the onset of labour and the return to normal of the generative organs: usually the first 5 or 6 weeks after the completion of labour.

puff ball (*BioSci*) Fruiting body of some fungi, esp of the order Lycopodiales in the Gasteromycetes.

puffs (*BioSci*) Visibly decondensed bands of polytene chromosomes (see POLYTENY) in which active transcription of RNA is occurring.

pug (*Build*) To prevent leakage by packing cracks with clay; the material so used.

pug (*MinExt*) In metalliferous mining, the parting of soft clay which sometimes occurs between the walls of a vein and the country rock.

pugging (*Build*) A special mixture carried on boards between the floor joists, once widely used to insulate the room against sounds and smells from below. In Scotland *deafening*.

pug mill (*CivEng*) Mixing machine for wet materials, as used in the making of mortar.

pulaskite (*Geol*) A light-coloured alkali syenite consisting largely of alkali feldspar with subordinate ferromagnesian minerals and often a small amount of nepheline.

Pulfrich refractometer (*Phys*) Instrument for measuring the refractive index of oils and fats.

pull (*Print*) An impression taken for checking.

pull (*Textiles*) A sample of fibres pulled by hand from a large quantity of raw material (eg a bale) and used for analysis; particularly for determining fibre length.

pull-down (*ImageTech*) The movement of film from one frame to the next in a motion picture camera or projector mechanism.

pull-down menu (*ICT*) A MENU that can be 'pulled down' below its name on the menu bar after selecting it. A variation 'drops down' on being selected. See WIMP.

pulled coil (*ElecEng*) An armature coil wound with parallel sides on a suitable former and then pulled out to the correct coil span.

pullet (*Agri*) A hen that has begun to lay but has yet to moult.

pullet disease (*Vet*) See AVIAN MONOCYTOSIS.

pulley (*Eng*) A wheel on a shaft, sometimes having a crowned or cambered rim, for carrying an endless belt, or grooved for carrying a rope or chain. A 'fast pulley' is one that is keyed to the shaft and revolves with it; a 'loose pulley' is not attached to the shaft. The term 'pulley' is also applied to a small grooved wheel over which a sash cord etc runs and similar wheels taking a VEE BELT.

pulley mortise (*Build*) A form of joint between the end of a ceiling joist, which is tenoned, and the binding joist, which is mortised, so as to let in the ceiling joist in a position such that the lower faces of both are in the same plane.

pulley stile (*Build*) One of the upright sides of the frame of a double-hung window, to which is secured the pulley over which the sash cord passes.

pulling (*ICT*) Variation in frequency of an oscillator when the load on it changes.

pulling (*Textiles*) (1) The conversion of rags and short lengths of yarn into fibres for reuse. (2) The removal of the wool fibres from skins taken from dead sheep. See FELLMONGERING, SHEARING, SKIN WOOL.

pulling by crystal (*Crystal*) Growing both metallic and non-metallic crystals by slowly withdrawing the crystal from a molten surface.

pulling figure (*ICT*) The stability of an oscillator, measured by the maximum frequency change when the phase angle of the complex reflection coefficient at the load varies through 360° and its modulus is constant and equal to 0·2.

pulling focus (*ImageTech*) Alteration of focus during a shot so that the principal subject is kept in focus despite varying distances.

pulling tools (*MinExt*) The procedure whereby the drill string and bits are removed from the bore and stacked in the derrick before reuse. See ROUND TRIP.

pull-off (*ElecEng*) A fitting used in connection with the overhead contact wire of an electric traction system for retaining the contact wire in the correct position above the track on curves.

pullorum disease (*Vet*) A disease of young chicks caused by *Salmonella pullorum*; affected chicks appear drowsy and huddled, may develop diarrhoea and often die. Also *bacillary white diarrhoea* (abbrev BWD).

pull-out (*Aero*) The transition from a dive or spin to substantially normal flight.

pull-out distance (*Aero*) A naval term for the distance travelled by the hook of an aircraft while arresting on the deck of a carrier. Also *run-out distance*.

pull-out torque (*ElecEng*) (1) The value of the torque at which a synchronous motor falls out of synchronism. (2) The maximum torque of an induction motor.

pull-over mill (*Eng*) A rolling mill using a single pair of rolls. The metal, after passing through the rolls, is pulled back over the top roll to be fed through the mill a second time.

pull switch (*ElecEng*) See CEILING SWITCH.

pulmonary (*BioSci*) (1) In land vertebrates, pertaining to the lungs. (2) In pulmonate Mollusca, pertaining to the respiratory cavity.

pulmonary adenomatosis (*Vet*) A slowly developing respiratory disease of sheep caused by a retrovirus in conjunction with a herpes virus. Fluid collects in the lungs, and if the hind-legs are raised fluid pours out of the nostrils.

pulmonary osteoarthropathy (*Med*) Increased curvature of the nail bed with painful distal finger joints and radiological evidence of new bone growth. May indicate the presence of a tumour of the lung.

pulmonary valvotomy (*Med*) The surgical restoration of pulmonary valve function, when the valve will not open properly.

Pulmonata (*BioSci*) An order of Gastropoda. The members are hermaphrodite, exhibit torsion, and have a shell but no operculum. The mantle cavity forms a lung with no ctenidium but a vascular roof and a small aperture (the pneumostome); eg land and fresh-water snails, land slugs.

pulmonate (*BioSci*) Possessing lungs or lung-books; air-breathing.

pulmonectomy (*Med*) See PNEUMONECTOMY.

pulmones (*BioSci*) Lungs. Sing *pulmo*. Adj. *pulmonary*.

pulp (*BioSci*) A mass of soft spongy tissue situated in the interior of an organ, as *dental pulp, spleen pulp*.

pulp (*For*) Material formed after mechanical or chemical treatment of timber, used feedstock for paper etc.

pulp (*MinExt*) Finely ground ore freely suspended in water at a consistency which permits flow, pumping or settlement in quiet conditions.

pulp (*Paper*) Generic term for any fibrous raw material, often cellulose-based, after preparation in a pulp mill by grinding, refining, digesting with chemicals and/or bleaching. Pulp may be a suspension in water or in the form of fluff or dried or semi-dried sheets.

pulp (*Textiles*) Cellulose fibres obtained either from cotton linters or from wood by chemical purification.

pulp board (*Paper*) A printing board of the same FURNISH throughout.

pulping (*For*) Any process, mechanical or chemical, that disintegrates the fibres in wood. Also *defibration*.

pulp wash (*FoodSci*) Liquor collected by pressing the re-wetted pulp left after citrus fruit has been pressed to extract the juice. It is then concentrated for use in citrus fruit DRINKS. However, current legislation does not allow the addition of pulp wash.

pulpy kidney disease (*Vet*) An acute and fatal toxaemia of lambs due to enteric infection by the bacterium *Clostridium perfringens* (*Cl. welchii*), type D; the kidneys of affected lambs frequently show a characteristic degenerative change after death. Vaccines available.

pulsar (*Astron*) A source of cosmic radio emission in the form of rapid regular radio pulses, with period of approx 30 ms to 4 s. See panel on PULSAR.

pulsatance (*Phys*) See ANGULAR FREQUENCY.

pulsating current (*ElecEng*) One taking the form of a succession of isolated pulses, and usually unidirectional, which changes in magnitude in a regularly recurring manner.

pulsating star (*Astron*) A variable star which periodically contracts and expands with associated changes in brightness.

pulsator (*MinExt*) Mineral jig of Harz type. See HARZ JIG.

pulse (*ElecEng, ICT*) One *step* followed by a *reverse step*, after a finite interval. A unidirectional flow of current of non-repeated waveform, ie consisting of a transient and a zero-frequency component greater than zero. Measured by peak value, duration or integration of magnitude over time. Obsoletely, *impulse*.

pulse (*FoodSci*) Edible ripe, dry seeds of legumes. Useful source of dietary fibre, protein and polyunsaturated fatty acids. May contain toxic substances which are readily inactivated by boiling.

pulse (*Med*) See PULSE WAVE.

pulse amplifier (*ICT*) One with a very wide frequency response that can amplify pulses without distortion of the very short rise time of the leading edge.

pulse amplitude (*ICT*) That of the crest relative to the quiescent signal level. Sometimes a mean taken over the pulse duration.

pulse-amplitude modulation (*ICT*) That which is impressed upon a pulse carrier as variations of amplitude. They may be either unidirectional or bidirectional according to the system employed.

pulse bandwidth (*ICT*) The frequency band occupied by Fourier components of the pulse that have appreciable amplitude and that make an appreciable contribution to the actual pulse shape.

pulse-chase (*BioSci*) An experimental protocol used to determine cellular pathways, such as precursor–product relationships. A sample (organism, cell or cellular organelle), is exposed for a relatively brief time to a (radioactively) labelled molecule, the *pulse*. It is then replaced in an excess of the unlabelled molecule, the chase (*cold chase*). The sample is then examined at various later times to determine the fate of label incorporated during the pulse.

pulse code (*ICT*) Coding of information by pulses, either in amplitude, length, or absence or presence in a given time interval.

pulse-code modulation (*ICT*) Pulse modulation in which the magnitude of the signal is sampled and each sample

Pulsar

Pulsating star. Pulsars were discovered in 1967 by A Hewish and S J Bell. A pulsar is a radio source characterized by extremely regular bursts of radio waves. The periods of the known pulsars vary from a couple of milliseconds up to 4 seconds. The extreme regularity of the burst can only be explained if pulsars are very massive but small (a few km in diameter) rapidly rotating objects. Such objects are magnetized neutron stars, which emit strong radio waves in a small cone. The radio beam is analogous to a light beam from a lighthouse, and a pulse is observed when the cone sweeps across the radio telescope.

The periods of pulsars gradually become longer as their rotation slows. The rate of decline indicates that a pulsar will remain detectable for a few million years.

The distribution of pulsars is concentrated towards the plane of the Milky Way, and this indicates that they are the relics of relatively young stars. The nearest is about 100 pc from the Sun, and the Galaxy probably contains about 10^5 pulsars altogether, some 1% of which are catalogued. A pulsar is formed as an end-point in the evolution of a massive star. Although the details are uncertain, the general picture is that stars of more than a few solar masses end spectacularly as SUPERNOVA explosions, in which a huge remnant of the outer layers is flung into space and the exhausted nuclear core implodes to form a NEUTRON STAR. Conservation of angular momentum and magnetic flux during the collapse accounts for the rapid spin of the neutron star. One of the fastest pulsars is directly associated with the CRAB NEBULA, and is the central engine responsible for its continuing X-ray and radio emission.

Precise measurements of the pulse arrival times and the positions of pulsars have important applications in ASTROMETRY. The accuracy of these measurements is so great that all data have to be reduced to values at the centre of mass or *barycentre* of the solar system, with full relativistic correction for the motion of the Earth. It is partly for this reason that DYNAMICAL TIME (see TIME and appendices) replaced EPHEMERIS TIME in 1984: a clock on Earth gains 1·6 ms per year relative to one at the barycentre. Accurate timing measurements enable the relationship between the ECLIPTIC and EQUATORIAL co-ordinate systems to be more closely defined. PROPER MOTIONS can be measured for some pulsars, and they indicate space velocities of 170 km s^{-1}. The radio signals from pulsars act as important probes of the magnetic field in our galaxy (20 nT in the solar neighbourhood) as well as electron densities.

A few pulsars are members of BINARY systems, in which the pulsar is in orbit around a companion star. These pulsars have periods of just a few milliseconds. It appears that transfer of matter from the companion to the pulsar has, via angular momentum transfer, led to a speeding up of the pulsar. In one case the surface of the accelerated pulsar is now travelling at one-tenth the velocity of light. Timing observations of binary pulsars are a highly sensitive test of the GENERAL THEORY OF RELATIVITY, and they may yet provide evidence for the radiation of GRAVITY WAVES.

Historically, the discovery of pulsars was of immense value in showing that highly collapsed stars, which had been predicted in the 1930s, really did exist. Once neutron stars were firmly established, theorists speculated about stellar mass black holes, which probably exist in some X-ray binaries. High-energy astrophysics gained substantial momentum as a discipline as theorists sought to apply the physics of pulsars to a better understanding of energy sources in quasars and active galactic nuclei.

See panels on BLACK HOLE, COSMOLOGY, GALAXY, QUASAR and SUN AS A STAR.

---

approximated to a nearest reference level (called quantization). Each sample is then represented by binary code, the succession of coded samples becoming the transmitted signal. The transmission of such binary information is highly resistant to noise and interference because detection is only of the presence or absence of pulses. Abbrev *PCM*.

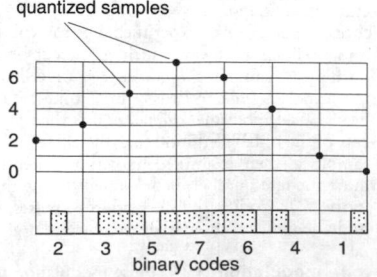

pulse-code modulation Three bit codes.

pulse code modulation audio (*ImageTech*) Digital audio used in the 8 MM/HI8 format, recorded on extensions of the video tracks so that it can be replaced; or in VHS/S-VHS where it is sandwiched between the FM audio and the video in DEPTH MULTIPLEX RECORDING. Abbrev *PCM audio*.

pulse compression (*Radar*) Techniques which permit high-range resolution (for which short pulses are necessary) while transmitting relatively long pulses in order to increase transmitter power and thereby enhance detection capability. Commonly, the pulse is frequency- or phase-modulated; modulation information is fed to the pulse-compression circuits in the receiver so that a matched-filter effect allows the receiver to respond only to echoes bearing the same modulation. See CODING.

pulsed columns (*ChemEng*) Columns used for LIQUID-LIQUID EXTRACTION in which there is a series of horizontal plates containing many small holes. Liquids flow continuously through the column but a reciprocating pump is attached to cause pulsations in the flow, which leads to high-velocity mixing as the liquids are pulsed through the perforated plates. Can also be applied to packed or even simple empty columns.

pulsed Doppler radar (*Radar*) One in which the Doppler shift of the signals received from a moving target is used to measure its velocity. The pulsed Doppler technique, but not a CW RADAR using Doppler measurement, can also give range and position information.

pulse decay time (*ICT*) Time for decay between the arbitrary limits of 0·90 and 0·10 of the maximum amplitude. Also *pulse fall time*.

pulse delay circuit (*ICT*) One through which the propagation of a signal takes a known time.

pulsed-field gel electrophoresis (*BioSci*) A variant of AGAROSE GEL ELECTROPHORESIS that allows fractionation of very large DNA fragments (up to 2 million base pairs) by applying the electric field in pulses from different angles.

pulse dialling (*ICT*) Telephone dialling system in which digits are transmitted as electrical pulses, the number of pulses corresponding to the value of each digit.

pulsed inertial device (*NucEng*) Fusion system based on INERTIAL CONFINEMENT.

pulse discriminator (*ICT*) Any circuit capable of discriminating between pulses varying in some specific respect, eg duration, amplitude or interval.

pulsed-radar system (*Radar*) One transmitting short pulses at regular intervals and displaying the reflected signals from the target on a screen. See CW RADAR.

pulse droop (*ICT*) The exponential decay of amplitude that is often experienced with nominally rectangular pulses of appreciable duration.

pulse duration (*ICT*) (1) Time interval for which the amplitude exceeds a specified proportion (usually $1/\sqrt{2}$) of its maximum value. Also *pulse length, pulse width*. (2) The duration of a rectangular pulse of the same maximum amplitude and carrying the same total energy.

pulse duty factor (*ICT*) The ratio of pulse duration to spacing.

pulse fall time (*ICT*) See PULSE DECAY TIME.

pulse-forming line (*Radar*) An artificial line which generates short high-voltage pulses for radar.

pulse-frequency modulation (*ICT*) Pulse modulation in which the repetition rate of transmitted pulses is varied in accordance with the level of the modulating signal.

pulse generator (*ICT*) One supplying single or multiple pulses, usually adjustable for PULSE REPETITION FREQUENCY amplitude and width. May be self-contained or require sine wave input signal.

pulse-height analyser (*Electronics*) (1) A single or multi-channel pulse-height selector followed by equipment to count or record the pulses received in each channel. The multichannel units are known as 'kicksorters'. (2) One which analyses statistically the magnitudes of pulses in a signal.

pulse-height discriminator (*ICT*) A circuit that produces an output pulse only when it receives an input whose amplitude exceeds a certain value or lies within a certain range of values. Also *amplitude discriminator*.

pulse-height selector (*Electronics*) Circuit which accepts pulses with amplitudes between two adjacent levels and rejects all others. An output pulse of constant amplitude and profile is produced for each such pulse accepted. The interval between the two reference amplitudes is termed the window or channel width and the lower level the threshold.

pulse interleaving (*ICT*) Adding independent pulse trains on the basis of time-division multiplex along a common path. Also *pulse interlacing*.

pulse-interval modulation (*ICT*) See PULSE-FREQUENCY MODULATION.

pulse ionization chamber (*NucEng*) Ionization chamber for the detection of individual particles by their primary ionization. Must be followed by a very-high-gain stable amplifier.

pulse-jet (*Aero*) A propulsive duct with automatic air intake valves, or a frequency-tuned jet pipe, so that pressure builds up between 'firings', thus achieving thrust at reasonable economy at moderate airspeeds, eg 200–400 mph (350–650 km h^{-1}).

pulse labelling (*BioSci*) See PULSE-CHASE. The technique of adding a pulse of radioactive material to a cell and then studying the subsequent metabolic stages.

pulse modulation (*ICT*) MODULATION in which pulses of electrical energy, rather than a signal of constant amplitude and frequency, constitute the information carrier. Most commonly found in PULSE-CODE MODULATION.

pulse-position modulation (*ICT*) PULSE MODULATION in which the modulating signal is used to alter the timing of individual pulses within a pulse train. Also *pulse-phase modulation*.

pulse regeneration (*ICT*) Correction of pulse to its original shape after phase or amplitude distortion. Also *pulse restoration*.

pulse repeater (*ICT*) See REGENERATOR.

pulse repetition frequency (*Radar*) The average number of pulses in unit time. Also *pulse repetition rate*.

pulse reshaping (*Electronics*) See PULSE SHAPING.

pulse restoration (*ICT*) See PULSE REGENERATION.

pulse rise time (*Electronics*) Time required for amplitude to rise from 0·10 to 0·90 of its maximum value.

pulse shaping (*Electronics*) Adjustment of a pulse to square-wave or other form by electronic means.

pulse spectrum (*ICT*) The distribution, as a function of frequency, of the magnitudes of the Fourier components of a pulse.

pulse spike (*ICT*) A subsidiary pulse superimposed upon a main pulse.

pulse stretcher (*ICT*) An electronic unit used to increase the time duration of a pulse.

pulse transformer (*ICT*) One designed to accept the very wide range of frequencies required to transmit pulse signals without serious distortion.

pulse wave (*Med*) A wave of increased pressure travelling along the arterial system and detected as the pulse. Generated by the discharge of blood from the heart into the aorta during ventricular systole, the contraction of the heart.

pulse-width modulation (*ICT*) PULSE MODULATION in which the duration of pulses is varied in accordance with the amplitude of the modulating signal.

pulsometer pump (*Eng*) A steam pump in which an automatic ball valve, the only moving part, admits steam alternately to a pair of chambers, so forcing out water which has been sucked in by condensation of the steam after the previous stroke.

pulsus alternans (*Med*) Situation when the arterial pulse has a strong beat or wave followed by a weak one. It indicates heart failure.

pulsus paradoxus (*Med*) When there is a marked decrease in the arterial pulse during inspiration. Often found in diseases affecting the PERICARDIUM.

pulverized fuel (*MinExt*) Finely ground solid fuels which can be fed by air blast into the combustion chamber of large furnaces.

pulvinated (*Build*) Said of a frieze which presents a bulging face.

pulvinule (*BioSci*) The small PULVINUS of a leaflet.

pulvinus (*BioSci*) (1) The swollen base of a petiole or pinna, containing motor tissue responsible for sleep movements etc as in many Leguminosae. (2) A thickened region of grasses at a node of the stem capable, by growth, of re-erecting a lodged CULM.

pumice (*Geol*) An acid vesicular glass, formed from the froth on the surface of some particularly gaseous lavas. The sharp edges of the disrupted gas vesicles enable pumice to be used as an abrasive. It floats on water.

pummel (*Build*) See PUNNING.

pump (*BioSci*) Molecular mechanism in a membrane that brings about the active transport (electrogenic or neutral) of a solute, eg a proton-translocating ATPase. Cf CARRIER.

pump (*Eng*) A machine driven by some prime mover, and used for raising fluids from a lower to a higher level, or for imparting energy to fluids.

pump (*ICT*) (1) In a PARAMETRIC AMPLIFIER, the external signal source, with a higher frequency than the signal frequency, that causes the parametric device to vary its reactance in a periodic manner. (2) The laser used to provide the optical power that an ERBIUM-DOPED FIBRE AMPLIFIER converts to optical signal power.

pumped storage (*ElecEng*) Pumping of water to high-level storage during off-peak periods of electric power generation, for return to source via turbines at periods of high demand.

pumped tube (*Electronics*) Transmitting valve, X-ray tube or other electronic device which is continuously evacuated during operation.

pumpellyite (*Min*) A complex greenish hydrated silicate of calcium, magnesium, iron and aluminium crystallizing in the monoclinic system. Found in low-grade metamorphic rocks and in amygdales in some basaltic rocks.

pump frequency (*Electronics*) The frequency of an oscillator used in a MASER or PARAMETRIC AMPLIFIER to provide the stored energy released by the input signal.

pumping speed (*Phys*) The rate at which a pump removes gas in creating a near-vacuum; measured in dm^3 or ft^3 per minute, stp, against a specified pressure.

pump jack (*MinExt*) Motor-operated well-head pump in which the reciprocating motion of the HORSEHEAD is transmitted to a displacement pump downhole. See POLISHED ROD.

pump-line (*ElecEng*) A cable extending throughout the length of an electric train for the control of auxiliary apparatus such as air compressors.

pump rod (*MinExt*) Small-diameter rods screwed together and used to connect a downhole pump to the surface. Also *pony rod*.

punch (*Eng*) (1) A tool for making holes by shearing out a piece of material corresponding in outline to the shape of the punch; a machine incorporating such a tool. (2) A hand tool, struck with a hammer, for marking a surface, or for displacing metal as in riveting. See CENTRE PUNCH, HOLLOW PUNCH, PIN PUNCH.

punch (*Print*) The first stage in the making of type, the letter being cut, nowadays usually by machine, in mild steel which, after hardening, is struck into copper or one of its alloys, to make a MATRIX from which the type is cast. Nearly obsolete.

punch and pin register (*Print*) A method of obtaining quick and accurate positioning of film elements onto carrier flats and subsequently, in plate-making and on the press, by pre-punching all the film elements and printing plates of the job. Pins fitted through the punched holes of the film and plates are used as register points when making a series of plates in process colour work. The press clamp bar has pins which locate in the holes punched on the plates, thus making for quick and accurate plate changing.

punched card (*ICT*) Once widely used standard-size, machine-readable card printed with columns of numbers up to 80 through which a pattern of holes was punched to represent binary-coded data and instructions. The inventor of the punched-card tabulating machine was Herman Hollerith who used it on the 1890 US census. Obsolete.

punched screens (*MinExt*) Suitably perforated robust plates which act as industrial screens.

punching (*ElecEng*) See LAMINATION.

punching machine (*Eng*) A machine for punching holes in plates, the punch being driven either mechanically by a crank and reciprocating block, or hydraulically.

punch-through (*Electronics*) Collector–emitter voltage breakdown in transistors.

punctate (*BioSci*) With translucent or coloured dots, or shallow pits.

punctate basophilia (*Med*) See BASOPHILIA.

punctuated equilibrium (*BioSci*) A concept of the process of evolution in which the fossil record is interpreted as long periods of stasis interrupted by relatively short periods of rapid change and speciation.

punctum (*BioSci*) A minute aperture; a dot or spot in marking. Adj *punctate*.

punctured (*Build*) A term applied to a variety of rusticated work distinguished by holes picked in the faces of the stones, either in lines or irregularly.

puncture test (*Paper*) A test measuring the energy required to force a standard pyramidal puncture head operated by the fall of a loaded sector-shaped pendulum through a sample of container board.

pungent (*BioSci*) (1) Ending in a point stiff and sharp enough to prick. (2) Acrid to taste.

punishment (*Psych*) In conditioning situations, the weakening of a response which is followed by an aversive or noxious stimulus, or by the withdrawal of a pleasant one.

punning (*Build, CivEng*) (1) The operation of ramming or consolidating the surface of hard-core, concrete, earth, etc, with repeated blows from a heavy-headed tool, called a punner. (2) The operation of consolidating concrete by means of constant driving in of a metal rod to distribute the mortar and aggregate evenly within the mix. Now usually done by mechanical vibrator.

punt (*Glass*) The bottom of a container.

punty (*Glass*) A short iron rod, at one end of which is either a button of hot glass or a suitably shaped piece of metal, which is applied hot to the end of a partially formed glass article in order that (1) it may be cracked off the blowpipe and manipulated on the punty, or (2) in the case of tube drawing, the mass of glass may be drawn out between punty and blowpipe. Also *pontie, pontil, puntee*.

pupa (*BioSci*) An inactive stage in the life history of an insect during which it does not feed and reorganization is taking place to transform the larval body into that of the adult imago. Adj *pupal*.

puparium (*BioSci*) The hardened and separated last larval skin which is retained to form a covering for the pupa in some Diptera.

pupil (*BioSci*) The central opening of the iris of the eye. Adj *pupillary*.

pupilometer (*Phys*) An instrument for measuring the size and shape of the pupil of the eye and its position with respect to the iris.

pupiparous (*BioSci*) Giving birth to offspring which have already reached the pupa stage, as some two-winged flies, eg *Glossina*, the tsetse fly.

Puppis (Ship's Stern) (*Astron*) A southern constellation which lies partly in the Milky Way.

purchase (*Eng*) (1) Mechanical advantage or leverage. (2) Mechanical appliance for gaining mechanical advantage.

pure clay (*Build*) See FOUL CLAY.

pure colour (*Phys*) Colour with CIE CO-ORDINATES lying on the SPECTRUM LOCUS or on the PURPLE BOUNDARY.

pure culture (*BioSci*) See AXENIC CULTURE.

pure line (*BioSci*) A group of individuals with their ancestors and descendants, usually the product of continued *inbreeding*, that breed true among themselves and which are, therefore, presumably homozygous at most loci. Also *inbred line*.

pure-metal crystals (*Eng*) The crystals of which a solid pure metal is composed. Each crystal in a given metal has a similar structure consisting of the same atoms arranged in the same way, though one crystal differs from another in orientation.

pure tone (*ICT*) One having no harmonics.

Purex (*NucEng*) TN for tri-*n*-butyl phosphate, which, diluted in kerosene, is used for extracting uranium and plutonium from other fission products in spent fuel. Cf BUTEX.

Purex process (*NucEng*) The process, now almost universally used in the reprocessing of uranium and plutonium, having superseded the older *Butex process*.

purgative (*Med*) An evacuant, eg cascara etc, used to treat functional constipation.

purge (*Aero*) To clean and flush out liquids (usually propellants) from tanks to prevent build-up of explosive mixture; dry nitrogen or helium is used.

purging (*Eng*) The operation at beginning of batch moulding, where previous polymer and accumulated debris are removed using a special polymer purging agent, often the polymer for the next run.

purines (*Chem*) A group of cyclic di-ureides, of which the most important are ADENINE and GUANINE which are part of the nucleotide chains of DNA and RNA. Caffeine, hypoxanthine and theophylline are also purines.

Purkinje cell (*BioSci*) A class of neuron in the cerebellum; the only neurons that convey signals away from the cerebellum.

Purkinje effect (*Phys*) An effect in which the maximum sensitivity of the human eye shifts towards the blue end of spectrum at very low illumination levels.

Purkinje fibres (*BioSci*) Specialized muscle cells in the heart that are involved in regulating the beat.

purl fabrics (*Textiles*) Knitted fabrics in which the reverse side stitches are brought to the surface for effect; used extensively for pullovers etc.

purlin (*Build*) A member laid horizontally on the principal rafters or between walls and supporting the common rafters.

puromycin (*BioSci*) An antibiotic that acts as an aminoacyl tRNA analogue and causes premature termination of protein synthesis.

purple boundary (*Phys*) Straight line joining the ends of the SPECTRUM LOCUS on a CHROMATICITY DIAGRAM. The co-ordinates of all real colours fall within the loop formed by these two lines.

purple copper ore (*Min*) Bornite.

purpleheart (*For*) Varieties of C American hardwood tree (*Peltogyne*) whose heartwood is a deep purple-violet and straight- to wavy-grained, with a moderate to fine texture. Also *amaranth*, *violetwood*.

purple of Cassius (*Chem*) Produced by adding a mixture of tin (IV) and tin (II) chlorides to a very dilute solution of gold (III) chloride; hydrated tin (IV) oxide is precipitated and the gold (III) chloride reduced to metal. The red-to-violet colour is due to the precipitation of finely divided gold on the tin (IV) hydroxide. Used in the making of ruby glass.

purpose-made brick (*Build*) A brick which has been specially moulded to shape suiting it for use in a particular position, eg an arch brick shaped like the voussoir of an arch.

purposive behaviour (*Psych*) Behaviour that is carried out with the design of achieving a desired end; it may be conscious or unconscious in its nature.

purpura (*Med*) The condition in which small spontaneous haemorrhages appear beneath the skin and the mucous membranes, forming purple patches; these may occur as a result of depleted or defective platelets or capillary damage. Adj *purpuric*.

purpura haemorrhagica (*Vet*) An acute, non-contagious disease occurring mainly in horses, characterized by mild fever, subcutaneous oedema and haemorrhages in the mucous membranes. In the horse it usually follows an infectious disease, such as strangles or equine influenza, and is thought to be allergic in origin.

purpuric acid (*Chem*) Barbituryl iminoalloxan. The ammonium salt is *murexide* (see MUREXIDE test).

purulent (*Med*) Forming or consisting of pus; resembling or accompanied by the formation of pus; of the nature of pus.

pus (*Med*) The yellowish fluid formed by suppuration, consisting of serum, pus cells (white blood cells), bacteria and the debris of tissue destruction. Also *matter*.

push-broom sensor (*Space*) A term applied to a detecting instrument which employs a line of detectors in juxtaposition for recording a line of a scene without recourse to mechanical scanning.

pushed punt (*Glass*) A concave bottom to a container. Also *push-up*.

push-fit (*Eng*) Loose-tolerance mechanical joint between two parts (eg Lego bricks).

pushing (*ICT*) The insertion of a data item onto a STACK.

push processing (*ImageTech*) See FORCED DEVELOPMENT.

push–pull (*Acous*) A term applied to sound tracks which carry sound recordings in antiphase. They are *class-A* when each carries the whole waveform, and *class-B* when each carries half the waveform, both halves being united optically or in a push–pull photocell.

push–pull amplifier (*Electronics*) A balanced amplifier using two valves or transistors working in phase opposition, each device conducting alternate halves of the input signal. Used for the reduction of harmonic distortion in power-output stages.

push–push amplifier (*Electronics*) One which uses two similar transistors or valves operating in phase opposition but connected to a common load. By this means, even-order harmonics are emphasized.

push rod (*Autos*) A rod through which the tappet of an overhead-valve engine operates the rocker arm, when the camshaft is located in the crankcase.

push technology (*ICT*) Unrequested messages and other information sent to a client, in contrast to normal Internet transactions where the client pulls information from a server. Could also include response to a general request, eg for a daily weather forecast. Also *webcasting*. See panel on INTERNET.

push-up (*Glass*) See PUSHED PUNT.

pustular stomatitis (*Vet*) See HORSE POX.

pustule (*BioSci*) A blister-like spot, on a leaf, stem, fruit, etc, from which erupts a fruiting structure of a fungus.

pustule (*Med*) A small elevation of the skin containing pus. Adjs *pustular*, *pustulous*.

putamen (*BioSci*) (1) In birds, the shell membrane of the egg. (2) In higher vertebrates, the lateral part of the lentiform nucleus of the cerebrum.

putlog (*Build*) A transverse bearer which in a bricklayer's scaffold is fixed at one end to the LEDGER and at the other end is wedged into a hole left by the bricklayer in the wall; used to support scaffold boards.

putrefaction (*Chem*) The chemical breaking down or decomposition of plants and animals after death. This is caused by the action of anaerobic bacteria and results in the production of obnoxious or offensive substances.

putrescine (*Chem*) 1,4-diaminobutane. $H_2N(CH_2)_4NH_2$. Mp 27°C. Crystals formed during the putrefaction of flesh.

putty (*Build*) A mixture of whiting and linseed oil, sometimes including white lead, forming a plastic substance for sealing panes of glass into frames. See GLAZIER'S PUTTY, PLASTERER'S PUTTY.

putty and plaster (*Build*) See GAUGED STUFF.

putty powder (*Build*) Tin oxide, used for polishing glass.

puy (*Geol*) A small volcanic cone, esp in the Auvergne, France.

puzzle box (*Psych*) A box in which an animal is confined, and from which it can escape only by performing a particular series of manipulations which it must discover by trial and error or problem-solving behaviour.

puzzolano (*Build*, *CivEng*) See POZZUOLANA.

PV (*FoodSci*) Abbrev for PEROXIDE VALUE.

PVA (*Plastics*) Abbrev for POLYVINYL ACETATE, POLYVINYL ALCOHOL. But see PVAC and PVAL.

PVAC (*Plastics*) Abbrev for POLYVINYL ACETATE.

PVAL (*Plastics*) Abbrev for POLYVINYL ALCOHOL.

_p_-value (*MathSci*) The probability of observing an outcome as extreme as, or more extreme than, that actually arising from a particular experiment or sample when a particular hypothesis is true. A low *p*-value is taken to indicate evidence against the particular hypothesis.

PVB (*Plastics*) Abbrev for POLYVINYL BUTYRAL.

PVC (*Plastics*) Abbrev for POLYVINYL CHLORIDE

PVD (*Electronics*) Abbrev for PHYSICAL VAPOUR DEPOSITION.

PVDC (*Plastics*) Abbrev for POLYVINYLIDENECHLORIDE.

PVDF (*Plastics*) Abbrev for POLYVINYLIDENEFLUORIDE. See DIELECTRIC AND FERROELECTRIC MATERIALS.

PVK (*Plastics*) Abbrev for POLYVINYL CARBAZOLE.

PVP (*Plastics*) Abbrev for POLYVINYL PYRROLIDONE.

PVR (*ICT*) Abbrev for PERSONAL VIDEO RECORDER.

P-wave (*Geol*) See panel on EARTHQUAKE.

pwn (*ICT*) To beat or dominate an opponent on-line (originally Internet gaming, now more generally used in chatrooms and blogs). Origins obscure, but thought to be a mistyping of 'own'.

PWR (*NucEng*) Abbrev for PRESSURIZED WATER REACTOR.

PX (*ICT*) Abbrev for PRIVATE EXCHANGE.

Py (*Chem*) Symbol for the PYRIDINE nucleus.

py-, pyo- (*Genrl*) Prefixes from Gk *pyon*, pus.

pyaemia (*Med*) The condition in which infection of the blood with bacteria, from a septic focus, is associated with the development of abscesses in different parts of the body. Also *pyemia*. Adj *pyaemic*.

pycastyle (*Build*) See PYCNOSTYLE.

pycn-, pycno-, pykn-, pykno- (*Genrl*) Prefixes from Gk *pyknos*, compact, dense.

pycnidiospore (*BioSci*) A spore formed within a pycnidium.

pycnidium (*BioSci*) A roundish fructification formed by many species of fungi, containing fertile hyphae and pycnidiospores but no asci. Appears to have no sexual function.

pycno- (*Genrl*) See PYCN-.

Pycnogonida (*BioSci*) An order of the Chelicerata. Marine animals with long legs that contain diverticulae of the digestive system and with reduced opisthosoma. Sea spiders.

pycnometer (*Chem*, *CivEng*) (1) A small, graduated glass vessel, of accurately defined volume, used for determining the relative density of liquids. (2) Similar device for measuring the relative density of (soil) particles less than about 5 mm in diameter. Also *density bottle*.

pycnosis (*BioSci*) The shrinkage of the stainable material of a nucleus into a deeply staining knot, usually a feature of cell degeneration.

pycnostyle (*Build*) A colonnade in which the space between the columns is equal to $1\frac{1}{2}$ times the lower diameter of the columns. Also *pycastyle*.

pycnoxylic wood (*BioSci*) Secondary xylem in gymnosperms composed mainly of tracheids, with relatively narrow rays, as in eg conifers. Cf MANOXYLIC WOOD.

pyelitis (*Med*) Inflammation of the pelvis of the kidney. (Gk *pyelos*, trough.)

pyelocystitis (*Med*) Inflammation of the pelvis of the kidney and the bladder.

pyelography (*Radiol*) Radiography of the pelvis or the kidney and ureter, after these have been filled with a CONTRAST MEDIUM which may have been given intravenously (intravenous pyelogram, IVP, IVU) or retrogradely into the ureter (retrograde pyelogram). See UROGRAPHY.

pyelolithotomy (*Med*) The operation for removal of a stone from the pelvis of the kidney.

pyelonephritis (*Med*) Inflammation of the renal pelvis, the part of the kidney which collects urine from the renal tubules before it is passed into the bladder. Pyelonephritis is a common cause of renal failure.

pyemia (*Med*) See PYAEMIA.

pygal (*BioSci*) Pertaining to the posterior dorsal extremity of an animal; in Chelonia, a posterior median plate of the carapace. From Gk *pyge*, rump.

pygostyle (*BioSci*) In birds, a bone at the end of the vertebral column formed by the fusion of some of the caudal vertebrae.

pyinkado (*For*) Durable timber from *Xylia dolabriformis*, found in India and Myanmar (Burma).

pykn-, pykno- (*Genrl*) See PYCN-.

pyknolepsy (*Med*) A form of epilepsy in which there are sudden attacks of momentary loss of consciousness; eventually the attacks disappear.

pyknometer (*Chem*) See PYCNOMETER.

pylephlebitis (*Med*) Inflammation of the portal vein (the vein formed by veins running from the spleen, stomach and intestines, and entering the liver) with or without thrombosis; in suppurative pylephlebitis abscesses form in the liver.

pylon (*ElecEng*) See TOWER.

pylon (*Eng*) A slender vertical structure which is self-supporting. Also *tower*. Cf MAST.

pylorectomy (*Med*) Excision of the pylorus.

pyloric stenosis (*Med*) Narrowing of the lower opening of the stomach. May result from scar tissue associated with peptic ulcers. Patient is unable to empty stomach except by vomiting.

pyloroplasty (*Med*) An operation for widening the lumen of the pylorus when this has been pathologically narrowed.

pylorospasm (*Med*) Spasm of the circular muscle of the pyloric part of the stomach.

pylorus (*BioSci*) In vertebrates, the point at which the stomach passes into the intestine. Adj *pyloric*.

pyo- (*Genrl*) See PY-.

pyocolpos (*Med*) A collection of pus in the vagina, the result of infection of a HAEMATOCOLPOS which has been inadequately treated.

pyogenic (*Med*) Having the power to produce pus.

pyometra (*Med*) A collection of pus in the cavity of the uterus.

pyonephrosis (*Med*) Accumulation of pus in the pelvis of the kidney.

pyopneumothorax (*Med*) The presence of pus and air or gas in the pleural cavity.

pyorrhoea (*Med*) Literally, a flow of pus. The term now used as a synonym for pyorrhoea alveolaris, a purulent inflammation of the periosteum round a tooth. US *pyorrhea*.

pyosalpingitis (*Med*) Purulent inflammation of a Fallopian tube.

pyosalpinx (*Med*) Accumulation of pus in a Fallopian tube.

pyralspite (*Min*) A group name for the *pyrope*, *almandine* and *spessartine* garnets.

pyramid (*BioSci*) A conical structure, eg part of the MEDULLA OBLONGATA in vertebrates. Adj *pyramidal*.

pyramid (*Crystal*) A crystal form with three or more inclined faces which cut all three axes of a crystal. See BIPYRAMID.

pyramid (*MathSci*) A polyhedron with one polygonal face (the *base*), each side of which is joined to a common point (the *vertex*) by a triangular face.

pyramidal disease (*Vet*) Exostosis affecting the extensor (pyramidal) process of the third phalanx of the foot of the horse.

pyramidal horn (*Acous*) Horn with linear flare-out in both planes. Cf SECTORAL HORN.

pyramidal system (*Crystal*) See TETRAGONAL SYSTEM.

pyramidal tract (*BioSci*) In the brain of mammals, a large bundle of motor axons carrying voluntary impulses from particular areas of the cerebral cortex.

pyramid of numbers (*BioSci*) The relative decrease in numbers at each stage in a food chain, characteristic of animal communities.

pyranometer (*EnvSci*, *Phys*) An instrument for measuring either the diffuse or the total global solar radiation by its heating action on two blackened metallic strips of different thickness which thereby assume different temperatures. Also *solarimeter*.

pyrargyrite (*Min*) Sulphide of silver and antimony which crystallizes in the trigonal system. It is commonly

associated with other silver-bearing minerals; cf PROUS-
TITE. Also *dark-red silver ore*.

pyrazinamide (*Pharmacol*) Bactericidal drug used in
treatment of tuberculosis.

pyrazines (*Chem*) Six-membered heterocyclic rings containing two nitrogen atoms in the 1,4 positions.

pyrazole (*Chem*) $C_5H_5N_3$. Mp 70°C, bp 185°C. Long
needles. It is a weak secondary base. Fuming sulphuric
acid forms a sulphonic acid. Pyrazole and its derivatives
can be halogenated, nitrated, diazotized, and generally
treated in a similar way to benzene or pyridine.

pyrazoles (*Chem*) Heterocyclic compounds containing a
five-membered ring consisting of three carbon and two
nitrogen atoms. Pyrazole derivatives are formed by the
condensation of hydrazines with compounds containing
two –CO groups, or a –CO and a –COOH group, in the
beta position, or which contain a –CO or –COOH group
attached to a doubly linked carbon atom. Pyrazoles have a
similar chemical character to benzene and pyridine.

pyrene (*Chem*) A tetracyclic hydrocarbon obtained from the
coaltar fraction boiling above 360°C, forming colourless,
monoclinic crystals, mp 148°C; soluble in ether, slightly
soluble in ethanol and insoluble in water. Has carcinogenic
properties.

pyrene

pyrenocarp (*BioSci*) See PERITHECIUM.

pyrenoid (*BioSci*) A small, dense, rounded, refractile,
proteinaceous body within or associated with the chloroplast, in some members of at least most classes of
eukaryotic algae. It is often surrounded by the appropriate
storage carbohydrate.

Pyrenomycetes (*BioSci*) A class of fungi in the Ascomycotina in which the fruiting body (ascocarp) is usually a
PERITHECIUM. It includes the POWDERY MILDEWS, *Claviceps* (ergot) and *Neurospora* (used in genetic research), etc.

pyrethrins (*Chem*) Active constituents of pyrethrum flowers
used as standard contact insecticide in fly-sprays etc;
remarkable for the very rapid paralysis ('knock-down'
effect) produced on flies, mosquitoes, etc. Pyrethrum root
is the source of a similar substance used as a sialagogue.
Chemically modified pyrethrins, which have greater
persistence and other desirable properties, are now
available.

pyretic (*Med*) Pertaining to fever.

Pyrex (*Glass*) TN for a borosilicate glass used for domestic
ovenware and laboratory glassware. Its low thermal
expansion coefficient makes it much less susceptible
to thermal shock than conventional soda–lime–silica
glass.

pyrexia (*Med*) Fever. An increase above normal of the
temperature of the body. Adj *pyrexial*.

pyrgeometer (*EnvSci, Phys*) An instrument consisting of a
number of blackened and polished surfaces, designed to
measure the loss of heat by radiation from the Earth's
surface. The surfaces exhibit differential cooling depending
on the radiation loss.

pyrheliometer (*EnvSci*) An instrument for measuring direct
solar radiation, excluding the diffuse and reflected
components.

pyribole (*Min*) A group name for pyroxene and amphibole.

pyridazines (*Chem*) Six-membered heterocyclic rings containing two nitrogen atoms in the 1,2 positions.

pyridine (*Chem*) A heterocyclic compound containing a
ring of five carbon atoms and one nitrogen atom, having
the formula C_5H_5N. It occurs in the coaltar fraction with a
boiling range between 80 and 170°C; a colourless liquid of

pungent, characteristic odour, bp 114°C; a very stable
compound and resists oxidation strongly. Fig. ▷

pyridine

pyridine alkaloids (*BioSci*) A group of ALKALOIDS based on
the pyridine ring, including coniine from hemlock.

pyridoxal (*BioSci*) See vitamin B complex in panel on
VITAMINS.

pyriform (*BioSci*) Pear-shaped, as the *pyriform organ* of a
Cyphonautes larva.

pyrimidine (*Chem*) A six-membered heterocyclic compound containing two nitrogen atoms in the 1,3 positions.
CYTOSINE, THYMINE and URACIL are the important
pyrimidine bases found in DNA and RNA.

pyrimidine

pyrite (*Min*) Sulphide of iron (FeS_2) crystallizing in the
cubic system. It is brassy yellow and is the commonest
sulphide mineral of widespread occurrence. Pyrite(s) is
sometimes used to include COPPER PYRITES, MAGNETIC
PYRITES, etc. Also *fool's gold, iron pyrites, mundic*.

pyritic smelting (*Eng*) Blast furnace smelting of sulphide
copper ores, in which heat is partly supplied by oxidation
of iron sulphide.

pyro- (*Chem*) A prefix used to denote an acid (and the
corresponding salts) which is obtained by heating a normal
acid, and thus contains relatively less water, eg *pyrosul-
phuric acid*, $H_2S_2O_7$.

pyro- (*Genrl*) Prefix from Gk *pyr* signifying fire, heat or
fever.

pyroborates (*Chem*) Generally known as *borates*.

pyroboric acid (*Chem*) See BORIC ACID.

Pyrochlor (*Chem*) TN for non-flammable transformer oil.
Mixture of 60% hexachlorodiphenyl and 40% trichlor-
obenzene.

pyrochlore (*Min*) A complex niobate of sodium, calcium
and other bases, with iron, uranium, zirconium, titanium,
thorium and fluorine; crystallizes in the cubic system. It is
found in nepheline-syenites and in alkaline pegmatites.

pyroclast (*Geol*) Crystal, glass or rock fragment generated
by a disruptive volcanic eruption and not subjective to any
secondary redeposition processes.

pyroclastic rocks (*Geol*) A name given to fragmental
deposits of volcanic origin.

pyroclimax (*EnvSci*) See FIRE CLIMAX.

pyrocondensation (*Chem*) A molecular condensation
caused by heating to a high temperature, eg the formation
of biuret from urea.

pyroelectric effects (*MinExt*) In high-tension (electrostatic) separation, the electrical charging of particles by
heating.

pyroelectricity (*Min*) Polarization developed in some
hemihedral crystals by an inequality of temperature.

pyrogallol (*Chem*) *1,2,3-trihydroxy-benzene*. $C_6H_3(OH)_3$.
Mp 132°C. White plates. It sublimes without decomposi-
tion, is soluble in water, and is a strong reducing agent.
Used as an absorbing agent for oxygen in gas analysis. Also
Pyrogallic acid.

pyrogenic (*Chem*) Resulting from the application of a high
temperature.

pyrogens (*Med*) Bacterial polysaccharides which produce
feverish reactions. An endogenous pyrogen is INTERLEUKIN-1.

pyroligneous acid (*Chem*) An aqueous distillate obtained
by the destructive distillation of wood, which contains
ethanoic acid, methanol, acetone and other products.

pyrolusite (*Min*) Manganese dioxide crystallizing in the tetragonal system. It typically occurs massive and as a pseudomorph after manganite, and is used as an ore of manganese, as an oxidizer and as a decolorizer.

pyrolysis (*Chem*) The decomposition of a substance by heat.

pyrolysis gas chromatography (*Chem*) A method for analysing the products of thermal decomposition of substances using mass spectroscopy. Useful for intractable polymers (eg thermosets) which cannot be easily analysed using ordinary methods (eg infrared, ultraviolet and nuclear magnetic resonance spectroscopy). Abbrev *PGC*.

pyrolytic mining (*MinExt*) See UNDERGROUND GASIFICATION.

pyromeride (*Geol*) An anglicized Fr term for nodular rhyolite. It is a quartz-felsite or devitrified rhyolite containing spherulites up to several centimetres in diameter which impart a nodular appearance to the rock.

pyrometallurgy (*Eng*) The treatment of ores, concentrates or metals when dry and at high temperatures. Techniques include smelting, refining, roasting, distilling, alloying and heat treatment.

pyrometer (*Phys*) An instrument for measuring high temperatures. See DISAPPEARING-FILAMENT PYROMETER, OPTICAL PYROMETER, RADIATION PYROMETER, SEGER CONES.

pyrometric cone equivalent (*Eng*) A measure of the softening or melting point of a refractory, carried out by means of comparing with standard SEGER CONES placed alongside in a furnace. Abbrev *PCE*.

pyrometric cones (*Eng*) See SEGER CONES.

pyromorphite (*Min*) Phosphate and chloride of lead, crystallizing in the hexagonal system. It is a mineral of secondary origin, frequently found in lead deposits; a minor ore of lead.

pyrones (*Chem*) Six-membered heterocyclic compounds containing a ring of five carbon atoms and one oxygen atom, one of the former being oxidized to a CO group. According to the position of the CO and the O in the molecule, there are α-pyrones (1,2) and γ-pyrones (1,4).

pyroninophilic cells (*BioSci*) Cells stained with methyl green pyronin stain that have bright-red cytoplasm. This indicates the presence of large amounts of RNA, and implies very active protein synthesis. It is characteristic of plasma cells.

pyrope (*Min*) The fiery-red garnet, magnesium aluminium silicate crystallizing in the cubic system. It is often perfectly transparent and then prized as a gem, being ruby-red in colour. It occurs in some ultrabasic rocks and in eclogites.

pyrophilous (*BioSci*) Growing on ground that has been recently burnt over.

pyrophoric (*Eng*) Materials and mixtures liable to spontaneous combustion, esp finely divided metals. See PYROPHORIC METALS.

pyrophoric metals (*NucEng*) Those liable to spontaneous combustion under conditions which may arise in a nuclear reactor. The nuclear fuels U, Th and Pu are all pyrophoric.

pyrophosphoric acid (*Chem*) HPO_3, obtained by the loss, through heating, of one H_2O molecule from orthophosphoric acid, H_3PO_4.

pyrophyllite (*Min*) A soft hydrated aluminium silicate crystallizing in the monoclinic system. It occurs in metamorphic rocks; often resembles talc.

pyrostibnite (*Min*) See KERMESITE.

pyrotechnics (*Genrl*) The science and art of making fireworks.

pyrotechny (*Chem*) The study and manufacture of fireworks.

Pyrotenax (*ElecEng*) TN of a type of MINERAL-INSULATED CABLE for low voltages. It is very tough, non-flammable and heat-resisting.

pyroxene group (*Min*) A most important group of rock-forming ferromagnesian silicates which, although falling into different systems (orthorhombic, monoclinic), are closely related in form, composition and structure. They are silicates of calcium, magnesium and iron, sometimes with manganese, titanium, sodium or lithium. See AEGIRINE, AUGITE, DIALLAGE, DIOPSIDE, ENSTATITE, HYPERSTHENE, ORTHOPYROXENE and panel on SILICON, SILICA, SILICATES.

pyroxenite (*Geol*) A coarse-grained, holocrystalline igneous rock, consisting chiefly of pyroxenes. It may contain biotite, hornblende or olivine as accessories. See HYPERSTHENITE.

pyroxilins (*Chem*) Nitrocelluloses with a low nitrogen content, containing from two to four nitrate groups in the molecule. Used in an ethanol-ethoxyethane solution to form collodion. Pyroxilin is a synonym for GUNCOTTON.

pyrrhotine (*Min*) Iron sulphide, approx Fe_7S_8, with variable amount of sulphur. Hexagonal. Nickel sulphide may be associated with it, as at Sudbury, Ontario, a major source of the world's nickel. Also *magnetic pyrites, pyrrhotite*.

pyrrole (*Chem*) A heterocyclic compound having a ring of four carbon atoms and one nitrogen. A colourless liquid of chloroform-like odour, bp 131°C, rel.d. 0·984. Pyrrole is a secondary base, and is found in coaltar and in bone oil. Numerous natural colouring materials are derivatives of pyrrole, eg chlorophyll and haemoglobin.

pyrrole

pyrrolidine (*Chem*) The final reduction product of pyrrole, a colourless, strongly alkaline base, bp 86°C.

pyrroline (*Chem*) A reduction product of pyrrole obtained by treating it with zinc and glacial acetic acid. It is a colourless liquid, bp 91°C, and is a strong secondary base.

Pyruma (*Build*) TN for a fireclay cement used in forming heat-resistant joints.

Pythagoras's theorem (*MathSci*) The theorem that the square on the hypotenuse of a right-angled triangle is equal in area to the sum of the squares on the other two sides.

Python (*ICT*) A high-level programming language.

pyuria (*Med*) The presence of pus in the urine.

pyxidium (*BioSci*) A capsule dehiscing by means of a transverse circular split, the top coming off like a lid. Also *pyxis*.

Pyxis (Mariner's Compass) (*Astron*) A small and insignificant southern constellation, lying partly in the Milky Way.

pyx, trial of the (*Eng*) Periodic official testing of sterling coinage.

Q

Q (*Build*) Symbol for quantity of water discharged, usually in $m^3 s^{-1}$.

Q (*ChemEng*) Symbol for THROUGHPUT.

Q (*Phys*) (1) Symbol for CHARGE. (2) Symbol of merit for an energy-storing device, resonant system or tuned circuit. Parameter of a tuned circuit such that $Q = \omega L/R$, or $1/\omega CR$, where L is the inductance, C the capacitance and R the resistance, considered to be concentrated in either inductor or capacitor. Q is the ratio of shunt voltage to injected emf at the resonant frequency $\omega/2\pi$. $Q = f_r/(f_1-f_2)$, where f_r is the resonant frequency and (f_1-f_2) is the bandwidth at the half-power points. For a single component forming part of a resonant system, it equals 2π times the ratio of the peak energy to the energy dissipated per cycle. For a dielectric it is given by the ratio of displacement to conduction current. Also *magnification factor, Q-factor, quality factor, storage factor*.

Q_{10} (*BioSci*) See TEMPERATURE COEFFICIENT.

q (*Chem*) Symbol for the quantity of heat which enters a system.

QA (*Aero*) Abbrev for QUALITY ASSURANCE.

QAM (*ICT*) Abbrev for QUADRATURE AMPLITUDE MODULATION.

Q-band (*ICT*) Frequency band mostly in radar, 36–46 GHz; now superseded by Ka-band. See K-BAND.

Q-banding (*BioSci*) See BANDING TECHNIQUES and panel on CHROMOSOME.

QCD (*Phys*) Abbrev for QUANTUM CHROMODYNAMICS.

Q-code (*Aero*) Telecommunications code using three-letter groups: QAA–QNZ for aeronautics; QOA–QQZ for maritime uses; QRA–QUZ for all services. Examples: QAH = 'What is your height above...?'; QAM = 'What is the latest met. report?'; QBA = 'What is the horizontal visibility at...?'.

QED (*Phys*) Abbrev for QUANTUM ELECTRODYNAMICS.

Q-factor (*Phys*) See Q.

q-feel (*Aero*) A term given (because of the use of $q = \frac{1}{2}\rho v^2$, ie DYNAMIC PRESSURE) to a device which applies an artificial force on the control column of a power-controlled aircraft proportional to the aerodynamic loads on the control surfaces, thereby simulating the natural 'feel' of the aircraft throughout its speed range.

Q fever (*Med, Vet*) An infection by the rickettsial microorganism *Coxiella burnetti* that produces a flu-like illness and may cause pneumonia. In rare and serious cases endocarditis occurs. In sheep and goats abortion is the only symptom.

Q-gas (*Chem*) One based on helium (98·2% He, 1·8% butane), widely used in gas-flow counting.

Qiana (*Textiles*) TN for a speciality nylon with silken properties when spun into fibre (US).

QIC (*ICT*) Abbrev for QUARTER-INCH CARTRIDGE.

Q-meter (*Phys*) Laboratory instrument which measures the Q-factor of a component.

QPP amplifier (*ICT*) Abbrev for QUIESCENT PUSH–PULL AMPLIFIER.

QPSK (*ICT*) Abbrev for QUATERNARY PHASE-SHIFT KEYING.

QS (*Build*) Abbrev for (1) QUICK SWEEP; (2) QUANTITY SURVEYOR.

Q-shell (*Phys*) The ELECTRON SHELL in an atom corresponding to a principal quantum number of seven. It is the outermost shell for heavy radioactive elements. See panel on ATOMIC STRUCTURE.

Q-signal (*ICT*) The first of three-letter code for standard messages in international telegraphy. See Q-CODE.

Q-signal (*ImageTech*) In the NTSC colour system, that corresponding to the narrow-band axis of the chrominance signal.

QSO (*Astron*) See panel on QUASAR.

Q-sort technique (*Psych*) A method of personality assessment in which the subject sorts a series of words or phrases into those characteristic/uncharacteristic of themselves.

Q-switching (*Phys*) A means of producing high instantaneous power from a laser. The cavity resonator has its reflectivity or 'Q' controlled externally. Q is made small while the population inversion is built up to its peak value. The reflectivity is then increased and the resultant high Q produces an intense burst of energy which almost completely empties the high energy states in a time of about 10^{-8} s. Switching is by a KERR CELL shutter or by rotating one of the mirrors.

quad (*ElecEng*) Either four insulated conductors twisted together (*star-quad*) or two twisted pairs (*twin-quad*). Normally a single structural unit of a multiconductor cable.

quad (*Paper*) Prefix to denote a paper size which is four times the area of that of the basic size (broadside), ie both dimensions of the basic size are doubled.

quad (*Print*) See QUADRAT.

quadr-, quadri- (*Genrl*) Prefixes from Lt *quattuor*, four.

quadrant (*BioSci*) A section of a segmenting ovum originating from one of the four primary blastomeres.

quadrant (*Eng*) A slotted segmental guide through which an adjusting lever (eg a reversing lever) works. It is provided with means for locating the lever in a number of angular positions. See LINK MOTION.

quadrant (*MathSci*) (1) Of a circle, a quarter of the circle bounded by two perpendicular radii. (2) Of the cartesian plane, the area bounded on two sides by the positive or negative segments of the axes; the first quadrant is the one in which both variables (x and y) are positive, and the quadrants are numbered 1 to 4 anticlockwise.

quadrant (*Surv*) An angle-measuring instrument of the sextant type, but embracing an angle of 90° or a little more.

quadrantal deviation (*Ships*) Those parts of the DEVIATION which vary as the sine and cosine of twice the COMPASS COURSE, thus changing their sign quadrantally with change in direction of the ship's head.

quadrantal point (*Astron*) One of the four points of the compass which in moving from north correspond to the headings NE(45°), SE(135°), SW(225°) and NW(315°). Cf CARDINAL POINT.

quadrant dividers (*Build*) A form of dividers in which one limb moves over an arc fixed rigidly to the second limb and may be secured to it by tightening a binding screw.

quadrant electrometer (*ElecEng*) See DOLEZALEK QUADRANT ELECTROMETER.

Quadrantids (*Astron*) A major meteor shower which shows maximum activity on 3 January with a rate of around 110 per hour.

quadraphonics (*Acous*) See QUADROPHONICS.

quadrat (*BioSci*) A small area (say $0.1–10 \ m^2$) of vegetation marked out for ecological study; a device of laths or strings to mark out such an area.

quadrat (*Print*) A piece of metal less than type height, for spacing. Also *quad*.

quadrate (*BioSci*) (1) Generally, square to squarish in cross-section or in face view. (2) A paired cartilage bone of the vertebrate skull formed by ossification of the posterior part of the PPQ BAR, or the corresponding cartilage element prior to its ossification; except in mammals, it forms part of the jaw articulation.

quadratic equation (*MathSci*) An algebraic equation of the second degree, ie $ax^2 + bx + c = 0$, whose solution is

$$x = \frac{-b \pm \sqrt{b^2 - 4ac}}{2a}$$

quadratic system (*Crystal*) See TETRAGONAL SYSTEM.

quadrature (*Astron*) The position of the Moon or a superior planet in elongation 90° or 270°, ie when the lines drawn from the Earth to the Sun and the body in question are at right angles.

quadrature (*ElecEng, ICT*) See PHASE.

quadrature (*ImageTech*) The relation between two waves of the same frequency but one-quarter of a cycle (90°) out of phase, as in TV colour difference signals.

quadrature amplitude modulation (*ICT*) Modulation system involving phase and amplitude modulation of a carrier, used in microwave and satellite communication links. Because it always allows high-power amplifier stages to operate close to their peak power output, more efficient use may be made of Earth and satellite amplifiers. Abbrev QAM.

quadrature component (*ElecEng*) See REACTIVE COMPONENT.

quadrature reactance (*ElecEng*) A term used in the two-reaction theory of synchronous machines to denote the ratio which the synchronous reactance drop produced by the quadrature component of the armature current bears to the actual value of the quadrature component.

quadrature transformer (*ElecEng*) A transformer designed so that secondary emf is 90° displaced from primary emf

quadratus (*BioSci*) A muscle of rectangular appearance, eg *quadratus femoris*.

quadri- (*Genrl*) See QUADR-.

quadric (*MathSci*) The three-dimensional surface represented by a general second-degree equation in three variables. By a suitable choice of co-ordinates such an equation can be reduced to one of the following standard equations:

$$(1) \quad \pm\frac{x^2}{a^2} \pm \frac{y^2}{b^2} \pm \frac{z^2}{c^2} = 1$$

an ellipsoid (three pluses), a hyperboloid of one sheet (one minus), a hyperboloid of two sheets (two minuses), an imaginary (virtual) quadric (three minuses).

$$(2) \quad ax^2 + by^2 = 2cz$$

an elliptic paraboloid (*a* and *b* of same sign), a hyperbolic paraboloid (*a* and *b* of opposite sign).

$$(3) \quad ax^2 + by^2 + cz^2 = 0$$

a cone.

$$(4) \quad ax^2 + 2hxy + by^2 = 1$$

a cylinder. Plane sections of quadrics are conics. For a cylinder, sections parallel to the plane $z = 0$ determine its type, which is elliptic, parabolic or hyperbolic. For an ellipsoid, *a*, *b* and *c* are the lengths of its principal semi-axes.

quadriceps (*BioSci*) A muscle having four insertions, as one of the thigh muscles of primates.

quadrilateral (*MathSci*) A four-sided polygon.

quadrilateral speed–time curve (*CivEng*) A simplified form of speed–time curve used in making preliminary calculations regarding energy consumption and average speed of railway trains. The acceleration and coasting

portions of the curve are sloping straight lines and the braking portion is neglected, so that the curve becomes a quadrilateral. Cf TRAPEZOIDAL SPEED–TIME CURVE.

quadriplegia (*Med*) Paralysis of both arms and both legs.

quadripole (*ICT*) A network with two input and two output terminals. A balanced wave-filter section.

quadrivalent (*BioSci*) A group of four at least partly homologous chromosomes held together by chiasmata during the prophase of the first division of meiosis, commonly found during meiosis in tetraploids. See fig. at MEIOSIS.

quadrivalent (*Electronics*) A term describing an atom with four electrons in its valency shell.

quadrophonics (*Acous*) A system of sound transmission using a minimum of four speakers fed by four, or sometimes three, separate channels. Also *quadraphonics*.

quadrumanous (*BioSci*) Of vertebrates, having all four podia constructed like hands, as in apes and monkeys.

quadruped (*BioSci*) Of vertebrates, having all four podia constructed like feet, as cattle.

quadruple-expansion engine (*Eng*) A steam engine in which the steam is expanded successively in four cylinders of increasing size, all working on the same crankshaft. Also *multiple-expansion engine*. Cf TRIPLE-EXPANSION ENGINE.

quadruple point (*Chem*) A point on a concentration–pressure–temperature diagram at which a two-component system can exist in four phases.

quadruplex (*ImageTech*) Videotape recording and reproduction system using four rotating heads to produce transverse tracks on 2-inch-wide magnetic tape.

quadruplex system (*ICT*) A system of Morse telegraphy arranged for simultaneous independent transmission of two messages in each direction over a single circuit.

quadrupole (*Phys*) A collection of charges such that the potential at a point distance *r* from their centre of mass may be expressed by an infinite series of terms in inverse powers of *r*. The inverse third-power term is the *quadrupole potential*.

quadrupole moment (*Phys*) The moment derived from the series expansion (see QUADRUPOLE) of charges multiplied by space co-ordinates. The sum of the quadratic terms is the *quadrupole moment*, which is possessed by most metals.

quadrupoles (*Acous*) Radiator producing a sound field of two adjacent dipoles in antiphase. The eddies in a subsonic jet of gas are quadrupoles.

qualia (*Psych*) The qualitative and subjective experience of something such as the smell of lavender, the taste of banana or the pain of a burn. Whether they necessarily correlate with a distinct sensory input can be debated, but presumably in the absence of colour receptors the qualia associated with colour cannot be perceived or, possibly, understood.

qualification test (*Space*) An evaluation of a flight article or its equivalent to verify that it functions correctly under the specified conditions of space flight; normally the test conditions are more severe than those expected.

qualitative analysis (*Chem*) Identification of the constituents of a sample without regard to their relative amounts. It often refers to elemental analysis, but may also refer to the detection of acid–base or redox properties in a sample. See QUANTITATIVE ANALYSIS.

quality (*Acous*) (1) In sound reproduction, the degree to which a sample of reproduced sound resembles a sample of the original sound. The general description of freedom from various types of acoustic distortion in sound-reproducing systems. See HIGH FIDELITY. (2) The timbre or quality of a note which depends upon the number and magnitude of harmonics of the fundamental.

quality (*Eng*) The condition of a saturated vapour, particularly steam, expressed as the ratio per cent of the vaporized portion to the total weight of liquid and vapour.

quality (*Radiol*) In radiography, an indication of the approximate penetrating power. Higher voltages produce higher-quality X-rays of shorter wavelength and greater penetration. The term dates from before the nature of X-rays was completely understood.

quality assurance (*Genrl*) A systematic procedure or set of procedures designed to ensure that a product or procedure matches specific standards and is therefore fit for purpose. Quality standards may be externally specified, even legally binding, or may simply be criteria abouteg the maximum and minimum permitted dimensions of a machined component.

quality control (*Eng*) A form of inspection involving sampling of parts in a mathematical manner to determine whether or not the entire production run is acceptable, a specified number of defective parts being permissible.

quality factor (*Phys*) See Q.

quality factor (*Radiol*) A measure of relative biological effectiveness. See panel on RADIATION.

quality of service (*ICT*) The overall performance level of a network connection as perceived by its users, covering such aspects as speech quality, digital error rate and delay jitter.

quality systems (*Genrl*) Ways of managing materials, components and products so as to ensure high-quality control of manufactured products at all levels, as specified ineg BS 5750.

quantal hypothesis (*Psych*) The hyphothesis that continuous increases in a physical stimulus bring about stepwise, quantal changes in sensation.

quantifier (*MathSci*) See PREDICATE CALCULUS.

quantile (*MathSci*) The argument of the cumulative distribution function corresponding to a specified probability; (of a sample) the value below which occur a specified proportion of the observations in the ordered set of observations.

QuantiMet (*Eng*) TN for machine which analyses material surfaces for microstructural variables such as grain size, diameter, orientation, etc. Based originally on the optical microscope, extended to electron optical examination. Uses computer techniques to perform statistical analyses based on stereological methods.

quantitative analysis (*Chem*) Identification of the relative amounts of substances making up a sample. It usually refers to elemental analysis, but may refer to any constituent of the sample. In addition to chemical methods, virtually every physical property can be a basis for some analytical method, and spectroscopic and electrochemical techniques are particularly often employed.

quantitative character (*BioSci*) A character displaying CONTINUOUS VARIATION. Cf UNIT CHARACTER.

quantitative genetics (*BioSci*) The genetics of QUANTITATIVE CHARACTERS. Also *biometrical genetics*.

quantitative structure–activity relationship (*BioSci*) Quantitative analysis of the relationship between molecular structure and biological activity of compounds using regression analysis and using as parameters physicochemical constants, indicator variables, or theoretically calculated values. Abbrev QSAR.

quantity of electricity (*ElecEng*) The product of flow of electricity (current) and time during which it flows. The term may also refer to a charge of electricity. See COULOMB.

quantity of light (*Phys*) The product of luminous flux and time during which it is maintained; usually stated in lumen-hours.

quantity of radiation (*Radiol*) The product of intensity and time of X-ray radiation. Not measured by energy, but by energy density and a coefficient depending on ability to cause ionization.

quantity surveyor (*Build, CivEng*) One who measures up from drawings and prepares a bill (or schedule) of quantities showing the content of each item. This is then used by contractors for estimating. The quantity surveyor also periodically measures and assesses the value of the work done.

quantization (*ICT*) In PULSE-CODE MODULATION, the division of the amplitude range of a continuously variable signal, eg speech or video, into discrete levels for the purposes of sampling and coding. See DIGITIZE.

quantization (*Phys*) In quantum theory, the division of energy of a system into discrete units (*quanta*), so that continuous infinitesimal changes are excluded.

quantization distortion (*ICT*) The distortion that arises in the mapping of a continuous signal onto a number of discrete levels so that it may be coded for digital transmission. Cf QUANTIZATION NOISE.

quantization distortion unit (*ICT*) A measure of QUANTIZATION DISTORTION equivalent to the transition from analogue to 64 Kbps A-law digital code and back again.

quantization noise (*ICT*) Noise introduced into a circuit using PULSE-CODE MODULATION because there are too few levels of quantization to describe the waveform accurately.

quantometer (*Eng*) An instrument showing by spectrographical analysis the percentages of the various metals present in a metallic sample.

quantum (*Phys*) (1) General term for the indivisible unit of any form of physical energy; in particular the *photon*, the discrete amount of electromagnetic radiation energy, its magnitude being hv where v is the frequency and h is PLANCK'S CONSTANT. See GRAVITON, MAGNON, PHONON, ROTON. (2) An interval on a measuring scale, fractions of which are considered insignificant.

quantum chromodynamics (*Phys*) The theory of strong interactions between elementary particles including the interaction that binds protons and neutrons to form a nucleus. It assumes that strongly interacting particles are made of QUARKS and that GLUONS bind the quarks together. Abbrev QCD.

quantum dot (*Chem*) A small particle of semiconductor material, typically a few nanometres in diameter; the small size implies that quantum mechanical effects can be more significant than in the (macroscopic) bulk material.

quantum efficiency (*Phys*) The number of electrons released in a photocell per photon of incident radiation of specified wavelength.

quantum electrodynamics (*Phys*) A relativistic quantum theory of electromagnetic interactions. It provides a description of the interaction of electrons, muons and photons and hence the underlying theory of all electromagnetic phenomena. Abbrev QED.

quantum electronics (*Phys*) The study of the amplification or generation of microwave power in solid crystals, governed by quantum mechanical laws.

quantum field theory (*Phys*) The overall theory of fundamental particles and their interactions. Each type of particle is represented by appropriate *operators* which obey certain commutation laws. Particles are the quanta of fields in the same way as photons are the quanta of the electromagnetic field. So GLUON fields and INTERMEDIATE VECTOR BOSON fields can be related to strong and weak interactions. Quantum field theory accounts for the LAMB SHIFT.

quantum gravity (*Phys*) The theory that would unify gravitational physics with modern QUANTUM FIELD THEORY.

quantum Hall effect (*Phys*) An effect in which Hall resistivity changes by steps so that it is a fraction of h/e^2 where h is PLANCK'S CONSTANT and e is the electronic charge. Observed in two-dimensional semiconductors (eg METAL–OXIDE–SILICON) at high magnetic fields and ultralow temperatures. See HALL EFFECT.

quantum mechanics (*Phys*) A generally accepted theory replacing classical mechanics for microscopic phenomena. Quantum mechanics also gives results consistent with classical mechanics for macroscopic phenomena. Two equivalent formalisms have been developed: matrix mechanics (developed by W Heisenberg) and WAVE MECHANICS (developed by E Schrödinger). The theory accounts for a very wide range of physical phenomena. See CORRESPONDENCE PRINCIPLE, STATISTICAL MECHANICS.

Quantum theory

Quantum theory is based on the basic hypothesis that there are small fundamental particles, which cannot be subdivided and which form the building blocks of everything else. These particles make up matter, light and other forms of energy. Modern physics uses quantum theory to describe the behaviour of atoms and subatomic particles.

History

While modern quantum theory was developed during the first half of the 20th century, the debate is very old as to whether the world is made up of fundamental particles which cannot be divided, or of a smooth continuum which can be arbitrarily and indefinitely divided. In Western natural philosophy, the earliest theories postulated a uniform basic material which Anaximander (6th century BC) called the *infinite* (and which in the 19th century had a revival as the *ether* to support ELECTROMAGNETIC WAVES). On the other side, Leukippos and Demokritos (5th century BC) proposed that everything was made of small identical and undividable (ie atomic) particles moving in an empty space (or vacuum). During the 19th century opinion was divided as to which was the right answer. Even some chemists who used the periodic table saw it as a useful tool but remote from the truth. The first evidence that matter might consist of fundamental particles came in 1897 with the discovery of the ELECTRON, which carries a unit charge. That light also comes in quanta (PHOTONS) was not realized fully until 1905. A body in thermal equilibrium emits a characteristic spectrum of radiation. If the body absorbs all incident radiation, all emitted radiation is purely due to its temperature; such radiation is BLACK-BODY RADIATION and the energy emitted at a given frequency is entirely determined by this frequency and the temperature of the black body, as proven by Kirchhoff. Classical physics and the then modern thermodynamics could give a description which worked either for low-frequency radiation (RAYLEIGH–JEANS LAW) or for high-frequency radiation (WIEN'S LAWS). In 1900 Planck guessed an equation, which worked for all wavelengths (PLANCK'S CONSTANT). To provide a theoretical derivation he was forced to assume that the total energy is made up of indistinguishable energy elements. When in 1905 Einstein examined the photoelectric effect (the release of electrons from certain metals or semiconductors by incident light) he found that classical electromagnetic theory did not agree with the experimental results. Einstein

proposed a theory in which Planck's black-body radiation and Maxwell's equations were combined. He noted that Planck's description of the exchange between the black body and radiation 'presupposes implicitly that energy can be absorbed and emitted by the individual resonator only in quanta of magnitude $h\nu$, ie that the energy of a mechanical structure capable of oscillations, as well as the energy of radiation, can be transferred only in such quanta, in contradiction to the laws of mechanics and electrodynamics'.

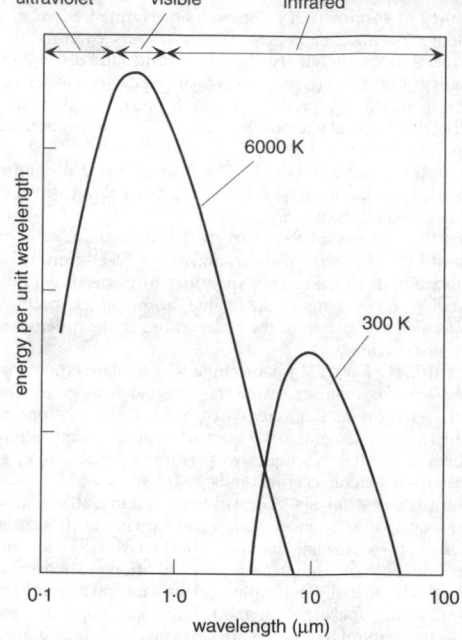

Black-body radiation

Modern development

The quantization of light allowed Bohr to construct a model of the hydrogen atom as a nucleus surrounded by an electron. The electron can only occupy discrete (quantized) energy levels, and jumps between levels result in the emission of a light quantum (photon) with exactly the frequency of the energy difference. Sommerfeld then generalized the theory to heavier atoms.

The quantum theory of light is in contrast to the very successful classical optics, which assumes a wave-like nature of light resulting in, among other things, refraction of light and interference patterns. Einstein wrote in 1924, 'There are therefore now two theories of light, both indispensable, and—as one must admit today despite 20 years of tremendous effort on the part of theoretical physicists—without any logical connection.' De Broglie noticed the duality of the light and the role of light quanta in the construction of a stable atom and showed using relativity theory that electrons also have a wave–particle duality: a wavelength or frequency is associated with the corpuscular electron.

Inspired by de Broglie, Schrödinger derived a wave mechanical description, the SCHRÖDINGER EQUATION

$$\nabla^2 \psi(x,y,z) + \frac{8\pi^2 m}{h^2}[E - U(x,y,z)\psi(x,y,z) = 0$$

which has waves as the solutions for ψ. The solutions, however, are only valid for a discrete (quantized) set of energy values E, so-called EIGENVALUES. The ensemble of eigenvalues forms a spectrum which in the case of the hydrogen atom is identical to Bohr's spectrum. Thus the quantum nature of matter and energy is an automatic consequence of the wave nature of the quantity ψ.

While the Bohr–Sommerfeld atomic model seemed plausible, it was impossible to give a theoretical proof of its validity. To overcome this, Heisenberg developed a theory independently of Schrödinger using only observable quantities and thus avoiding uncertain concepts such as the electron orbits. His result was a matrix algebra in terms of *observables* such as momentum p and position q. The theory was called matrix mechanics or QUANTUM MECHANICS. One striking feature of Heisenberg's matrix mechanics was that the order of multiplying the quantities, or measuring the observables, affected the result. In classical mechanics the result of multiplying p and q does not depend on the order of the operation and $pq - qp = 0$. In quantum mechanics, however,

$$pq - qp = \frac{h}{2\pi i}$$

because both p and q are matrices rather than scalar numbers, as in classical mechanics.

Dirac recognized the analogy between classical Hamiltonian mechanics and matrix mechanics, and formulated a complete and general form of quantum mechanics based on two types of numbers, c-numbers (classical numbers) and q-numbers (quantum numbers) (see DIRAC EQUATION). It was later shown that these three seemingly different types of quantum theories are all mathematically equivalent. For Heisenberg p and q are matrices, for Schrödinger q is a number and p a differential operator, and for Dirac they are special numbers obeying a special algebra.

An unresolved question was what the ψ meant, until Born provided an accepted interpretation of

ψ as a probability distribution: ie $|\psi(x, y, z)|^2 d\tau$ is the probability of finding the electron in a volume element $d\tau$ at the co-ordinates x, y, z. Quantum mechanics, which was developed as a deterministic mechanical theory, turned out to be a theory of probabilities introducing chance as a fundamental property into physics. This was difficult for many physicists to accept, among them Einstein, who remarked, 'God does not play dice.' The wave–particle duality of light and matter is intrinsically linked to the element of chance in quantum mechanics. Heisenberg showed that it was impossible to devise an experiment which demonstrated both aspects of the duality simultaneously; it either demonstrated the light or matter as a wave or as a particle. This also extends to the measurement of the observables; it is possible to measure either the position accurately or the momentum. Whenever one of the observables is measured to a given accuracy, the accuracy to which the other can be measured is limited by the Heisenberg UNCERTAINTY PRINCIPLE,

$$\Delta p \Delta q \cong \frac{h}{2\pi}$$

for the pair of momentum and position, or for the pair of energy and time

$$\Delta E \Delta t \cong h$$

The correspondence principle postulates that as a quantum mechanical system is made larger and larger, the relative uncertainty becomes smaller and smaller so that for large systems quantum theory and classical theory, which contains no element of uncertainty, agree with each other.

The quantum theories of Heisenberg, Schrödinger and Dirac initially excluded relativistic effects, though Dirac succeeded in incorporating relativistic effects into quantum theory by introducing a new variable, SPIN. Modern relativistic quantum theory of particles interacting with light or other electromagnetic radiation is QUANTUM ELECTRODYNAMICS. All modern physics describing events at an atomic or subatomic level is based on quantum theories. The theory describing the interactions of quarks and hadrons through the force of STRONG INTERACTION is QUANTUM CHROMODYNAMICS.

While quantum theories were extremely successful in fundamental physics, they have recently also inspired possible practical applications. Current research is investigating the development of quantum computers.

See GENERAL RELATIVITY and appendices.

quantum number (*Phys*) One of a set of numbers describing possible quantum states of a system, eg nuclear spin. See PRINCIPAL QUANTUM NUMBER, SPIN and panel on ATOMIC STRUCTURE.

quantum statistics (*Phys*) Statistics of the distribution of particles of a specified type in relation to their energies, the latter being quantized. See BOSE–EINSTEIN STATISTICS, FERMI–DIRAC STATISTICS.

quantum theory (*Phys*) The theory developed from PLANCK'S LAW to account for black-body radiation, the PHOTOELECTRIC EFFECT and the COMPTON EFFECT and to form the BOHR MODEL of the atom and its modification by Sommerfeld. See panel on QUANTUM THEORY.

quantum tunnelling (*Phys*) See POTENTIAL BARRIER, TUNNEL EFFECT.

quantum voltage (*Phys*) The voltage through which an electron must be accelerated to acquire the energy corresponding to a particular quantum.

quantum wire (*Electronics*) A nanostructure proportioned like a wire so that electron behaviour is strongly constrained by quantum mechanical effects in two dimensions.

quantum yield (*Phys*) The ratio of the number of photon-induced reactions occurring to the total number of incident photons.

quaquaversal fold (*Geol*) A dome-like structure of folded sedimentary rocks which dip uniformly outwards from a central point. See DOME.

quarantine (*Med*) Isolation or restrictions placed on the movements of individuals associated with a case of a communicable disease; place or period of detention of travellers coming from infected or suspected countries, or of animals on importation.

quark (*Phys*) A type of fundamental particle that forms the constituents of HADRONS. There are currently believed to be six types (or FLAVOURS) of quarks (and their antiquarks): UP, DOWN, CHARM, STRANGE, TOP, BOTTOM. In quark theory, the baryon is composed of three quarks of different COLOUR, an antibaryon is composed of three antiquarks, and a meson is composed of a quark and an antiquark. No quark has been observed in isolation. See appendix on Subatomic particles.

quarl (*Eng*) See BURNER FIRING BLOCK.

quarrel (*Build*) The diamond-shaped pane of glass used in FRET-WORK.

quarries (*Build*) Same as QUARRY TILES.

quarry (*MinExt*) (1) An open working or pit for granite, building stone, slate or other rock. (2) An underground working in a coal mine for stone to fill the goaf. Distinction between quarry and mine somewhat blurred in law, but usage implies surface workings.

quarry-faced (*Build*) A term applied to a building stone whose face is hammer-dressed before leaving the quarry. See ROCK FACE.

quarry-pitched (*Build*) A term applied to stones which are roughly squared before leaving the quarry.

quarry stone bond (*Build*) A term applied to the arrangement of stones in rubble masonry.

quarry tile (*Build*) The common unglazed, machine-made paving tile not less than $\frac{3}{4}$ in (20 mm) in thickness. Also *promenade tile*.

quart (*Genrl*) One-quarter of a gallon, or 2 pints (UK 1·14 l, US 0·946 l in liquid measure, 1·1 l in dry measure).

quartan (*Med*) In which a febrile paroxysm recurs every fourth day (ie at an interval of 72 h). *Quartan malaria* is associated with infection with the parasite *Plasmodium malariae*.

quartation (*Eng*) See INQUARTATION.

quarter (*Astron*) The phase of the Moon at quadrature. The first quarter occurs when the longitude of the Moon exceeds that of the Sun by 90°, the last quarter when the excess is 270°. The two other quarters are the new Moon and full Moon.

quarter (*Genrl*) (1) The fourth part of a hundredweight, equivalent to 28 (or in the US 25) pounds avoirdupois. (2) A unit equal to 8 bushels.

quarter bend (*Build*) Union connecting two pipes at 90°.

quarter bond (*Build*) The ordinary brickwork bond obtained by using a $2\frac{1}{4}$ in (57 mm) closer.

quarter-bound (*Print*) A term applied to a book having its back and part of its sides covered in one material and the rest of its sides in another.

quarter-chord point (*Aero*) The point on the CHORD LINE at one-quarter of the chord length behind the leading edge. SWEEPBACK is usually quoted by the angle between the line of the quarter-chord points and the normal to the aircraft fore-and-aft centre line.

quarter evil (*Vet*) See BLACKLEG.

quarter girth (*For*) The girth of a log or tree divided by four. A measure commonly used in countries where volumes are reckoned in HOPPUS FEET. Abbrev *qg*.

quarter ill (*Vet*) See BLACKLEG.

quarter-inch cartridge (*ICT*) A standard for computer tapes. Abbrev *QIC*.

quartering (*For*) A piece of timber of square section 2–6 in (50–150 mm) side.

quartering (*MinExt*) A method of obtaining a representative sample for analysis or test of an aggregate with occasional shovelfuls, of which a heap or cone is formed. This is flattened out and two opposite quarter parts are rejected. Another cone is formed from the remainder which is again quartered, the process being repeated until a sample of the required size is left.

quarter lines (*Ships*) The aggregation of waterlines, buttocklines, sections and diagonals indicative of a ship's form, drawn on a scale of $\frac{1}{4}$ in = 1 ft. See FAIRING.

quarter page folder (*Print*) A supplementary device to give a third fold in line with the run of the paper on web-fed presses.

quarter-phase systems (*ElecEng*) See TWO-PHASE systems.

quarters (*Build, CivEng*) See FLANKS.

quarter sawing (*For*) A mode of converting timber, adopted when it is desired that the growth rings shall all be at no less than 45° to the cut faces. Also *rift sawing*.

quarter-space landing (*Build*) A landing extending across only half the width of a staircase.

quarter turn (*Build*) A WREATH subtending an angle of 90°.

quarter-wave antenna (*ICT*) One whose overall length is approximately a quarter of free space wavelength corresponding to frequency of operation. Under these conditions it is oscillating in its first natural mode, and is half a dipole.

quarter-wave bar (*ICT*) See QUARTER-WAVE LINE.

quarter-wavelength stub (*ICT*) Resonating two-wire or coaxial line, approximately one quarter-wavelength long, of high impedance at resonance. Used in antennas, as insulating support for another line, and as a coupling element.

quarter-wave line (*ICT*) Quarter-wavelength section of transmission line designed to operate as a matching device between lines of different impedance levels.

quarter-wave plate (*Phys*) A plate of quartz, cut parallel to the optical axis, of such thickness that a retardation of a quarter of a period is produced between ordinary and extraordinary rays travelling normally through the plate. By using a quarter-wave plate, with its axis at 45° to the axes of a polarizer, circularly polarized light is obtained.

quartet (*BioSci*) A set of four related cells in a segmenting ovum. Also *quartette*.

quartic equation (*MathSci*) An algebraic equation of the fourth degree, ie $ax^4 + bx^3 + cx^2 + dx + e = 0$. Its resolution into a pair of quadratic equations, and hence its solution, depends upon the solution of a subsidiary cubic equation.

quartile (*MathSci*) The argument of the cumulative distribution function corresponding to a probability of either $\frac{1}{4}$ (first or *lower quartile*) or $\frac{3}{4}$ (third or *upper quartile*); (of a sample) the value below which occurs a quarter (first or *lower quartile*) or three-quarters (third or *upper quartile*) of the observations in the ordered set of observations. A measure of the variability of a distribution or data set.

quarto (*Print*) The quarter of a sheet, or a sheet folded twice to make four leaves or eight pages; written 4to. See PAPER SIZES.

quartz (*Min*) Crystalline silica, SiO_2, occurring either in prisms capped by rhombohedra (low-temperature quartz, stable up to 573°C) or in hexagonal bipyramidal crystals (high-temperature quartz, stable above 573°C). Widely distributed in rocks of all kinds; igneous, metamorphic and sedimentary; usually colourless and transparent (rock crystal), but often coloured by minute quantities of impurities as in citrine, cairngorm, etc; also finely crystalline in the several forms of chalcedony, jasper, etc. See CRISTOBALITE, TRIDYMITE, TWINNING.

quartz crystal (*ICT*) A disk or rod cut in the appropriate directions from a specimen of piezoelectric quartz, and accurately ground so that its natural resonance shall occur at a particular frequency.

quartz-diorite (*Geol*) A coarse-grained holocrystalline igneous rock of intermediate composition, composed of quartz, plagioclase feldspar, hornblende and biotite, and thus intermediate in mineral composition between typical diorite and granite. Also *tonalite*.

quartz-dolerite (*Geol*) A variety of dolerite which contains interstitial quartz usually intergrown graphically with feldspar, forming patches of micropegmatite. A dyke-rock of worldwide distribution, well represented by the Whin Sill rock in N England.

quartz-fibre balance (*Chem*) A very sensitive spring balance, the spring being a quartz fibre.

quartz-fibre electroscope (*Radiol*) A personal radiation monitor whose state can be viewed at any time. The fibre is charged periodically and discharged by radiation.

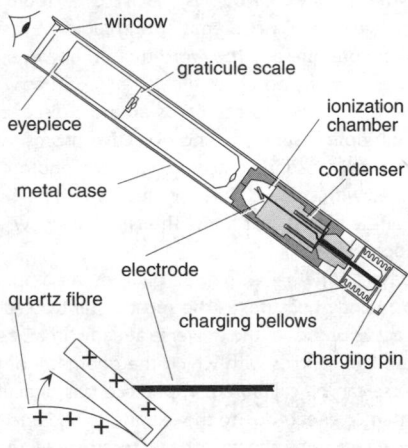

quartz-fibre electroscope Ions discharge fibre so that it can spring back to its rest position.

quartz glass (*Glass*) See VITREOUS SILICA. Also *fused silica*, *silica glass*.

quartz–iodine lamp (*ElecEng*) Compact high-intensity light source, consisting of a bulb with a tungsten filament, filled with an inert gas containing iodine (sometimes bromine) vapour. The bulb is of quartz, glass being unable to withstand the high operating temperature (600°C). Used for car-lamps, cine projectors, etc. Also *quartz–halogen lamp*, *tungsten–halogen lamp*.

quartzite (*Geol*) The characteristic product of the metamorphism of a siliceous sandstone or grit. The term is also used to denote sandstones and grits which have been cemented by silica.

quartz-keratophyre (*Geol*) A type of soda-trachyte carrying accessory quartz.

quartz lamp (*Radiol*) One which contains a mercury arc under pressure, a powerful source of ultraviolet radiation.

quartz oscillator (*ICT*) One whose oscillation frequency is controlled by a piezoelectric quartz crystal.

quartz porphyrite (*Geol*) A porphyrite carrying quartz as an accessory constituent; the representative in the medium grain-size group of the fine-grained dacite.

quartz porphyry (*Geol*) A medium-grained igneous rock of granitic composition occurring normally as minor intrusions, and carrying prominent phenocrysts of quartz.

quartz resonator (*ICT*) A standard of frequency comparison making use of the sharply resonant properties of a piezoelectric quartz crystal.

quartz topaz (*Min*) See CITRINE.

quartz wedge (*Min*) A thin wedge of quartz which provides a means of superposing any required thickness of quartz on a mineral section being viewed under a polarizing microscope, the wedge being cut parallel to the optical axis of a prism of quartz crystal. It enables the sign of the birefringence of biaxial minerals to be determined from their interference figure in convergent light.

quartz wind (*Acous*) A form of ACOUSTIC STREAMING near ultrasonic transducers operated at high amplitudes.

quasar (*Astron*) A distant, compact, object far beyond our Galaxy which looks star-like on a photograph but has a redshift characteristic of an extremely remote object. See panels on BLACK HOLE, QUASAR and REDSHIFT–DISTANCE RELATION.

quasi- (*Genrl*) Prefix meaning partial, to some extent, almost, approximate.

quasi-biennial oscillation (*EnvSci*) Alternation of easterly and westerly wind regimes in the equatorial stratosphere with an interval between successive corresponding maxima of from 24 to 30 months. A new regime starts above 30 km and propagates downwards at about 1 km per month. Abbreviation *QBO*. See panel on STRATOSPHERE AND MESOSPHERE.

quasi-bistable circuit (*ElecEng*) An astable circuit which is triggered at a high rate as compared with its natural frequency.

quasi-duplex (*ICT*) A circuit that operates apparently duplex, but actually functions in one direction only at a time, eg a long-distance telephone or a radio link that is automatically switched by speech.

quasi-elastic method (*Eng*) A method of stress analysis for non-linear and/or time-dependent materials, esp polymers, in which elastic moduli in the elastic equations are replaced by the values of the corresponding secant modulus or creep modulus, at the required levels of strain or time, respectively.

quasi-equivalence (*BioSci*) A term used to refer to the way in which subunits pack into a quasi-crystalline array as eg in viral coat assembly. There is usually some strain in the packing.

quasi-Fermi levels (*Electronics*) Energy levels in a semiconductor from which the number of electrons or holes available for conduction under non-equilibrium conditions, esp when light is falling on the semiconductor, can be calculated in the same way as from the true FERMI LEVEL which applies under equilibrium conditions.

quasi-geostrophic approximation (*EnvSci*) An approximation to the dynamical equations governing atmospheric flow, esp the VORTICITY EQUATION, whereby the horizontal wind is replaced by the GEOSTROPHIC WIND in the term representing the vorticity, but not in the term representing the DIVERGENCE.

quasi-longitudinal wave (*Acous*) Special type of wave occurring in plates and bars. The particle motion is mainly longitudinal and has a small transverse component caused by lateral contraction.

quasi-optical waves (*Phys*) Invisible electromagnetic waves with similar wavelength and laws of propagation to visible light.

Quasar

Quasi-stellar radio source. When radio astronomers made catalogues in the 1950s, the positional accuracy of their sources was not high enough to enable optical astronomers to find visible counterparts. By 1960 improvements in radio interferometers had shown that a handful of them seemed to be unusual stars in our galaxy. They had novel emission line spectra seemingly associated with highly excited states of rare elements. In 1963, lunar occultation was used to pinpoint source 3C 273 with unprecedented accuracy, and this too matched a radio source. Its optical spectrum had a familiar pattern of lines, the BALMER SERIES of hydrogen; what was unusual was the REDSHIFT of $z = 0.16$, then thought of as a high value. Once the possibility of such high redshifts was indicated, other spectra could be explained: source 3C 48 has a redshift of 0·37, and its emission lines are due to neon and oxygen. Over 20 quasars have been found with redshifts exceeding 4·0, corresponding to distances of about 4000 Mpc.

The optical spectrum of a typical quasar is non-thermal. Quasars have intense broad emission lines characteristic of highly ionized gas with a temperature of 10 000 K, the line widths corresponding to velocities of 10 000 km s^{-1} or so. In addition there are narrow-width lines associated with forbidden transitions. Higher redshift quasars have complex systems of absorption lines also. These have multiple redshifts, associated with matter expelled from the quasars perhaps, or existing in intervening galaxies along the line of sight.

The active core of a quasar is only about 1 parsec across, as shown in high-resolution maps made with interferometers. Some objects are highly variable on time-scales of a few days, which would indicate an active region only the size of our solar system, but with the energy output hundreds of times larger than a normal galaxy. In objects where it is possible to measure the motion of individual components within the core, the separation sometimes changes so fast that the apparent velocity of separation is up to ten times the speed of light. This *superluminal motion* is partly illusory: matter travelling at close to the velocity of light along a direction close to the line of sight will have an apparent transverse velocity greater than light. However, when motion is analysed completely relativistically, with proper allowance for the redshift of the object, the paradox is resolved.

The optical and radio continuum spectra are typical of SYNCHROTRON RADIATION. The power-law spectrum indicates that most emission is coming from relativistic electrons travelling through magnetic fields of about 10 nT. An important goal of theoretical research is to find mechanisms for accelerating the electrons up to the requisite energy, and to replenish the supply over millions of years. The rather simple physics of the synchrotron process shows that the energy requirement in relativistic electrons is about 10^{53} J. It appears unlikely that conventional nuclear reactions in stars can account for this energy. It is believed that a quasar consists of a supermassive black hole in the centre of a young galaxy, which sucks in passing stars and gas clouds. An ACCRETION DISC is formed around the black hole which is accelerated and heated by friction, and emits the characteristic radiation.

Other objects, which have the same optical properties but not necessarily strong radio emission, are known as *quasi-stellar objects* (QSOs). Only about 10% of the known QSOs also emit strong radio waves and are therefore quasars. The word quasar, however, is often used more loosely to include all QSOs as well as quasi-stellar radio sources. QSOs are now discovered through optical surveys using objective prisms. Automatic plate measuring machines can scan photographs obtained with SCHMIDT TELESCOPES to find new QSOs very efficiently, so that a few thousand are now catalogued.

Quasars, together with radio galaxies, are important in cosmology because, as the most distant objects, they act as probes of the universe at early times, as well as standard candles with which the geometry of the universe can be calibrated. In practice it is difficult to disentangle effects due to the expansion and ageing of the universe from effects intrinsic to an evolving population of radio sources.

See panels on BLACK HOLE, COSMOLOGY, GALAXY, PULSAR and REDSHIFT–DISTANCE RELATION.

---

quasi-stationary front (*EnvSci*) A FRONT which is moving slowly and irregularly so that it cannot be described as either a COLD FRONT or a WARM FRONT.

quasi-stellar object (*Astron*) See panel on QUASAR.

quasi-stellar radio source (*Astron*) See panel on QUASAR.

quasi S-VHS playback (*ImageTech*) The facility for playing S-VHS tapes in a VHS video recorder – with VHS resolution. Also *SQPB*.

Quaternary (*Geol*) The geological period which succeeded the Tertiary. It includes the Pleistocene and Holocene epochs and covers a time span of approx the last 2 million years.

quaternary (*Chem*) Consisting of four components etc; also, connected to four non-hydrogen atoms.

quaternary ammonium bases (*Chem*) Bases derived from the hypothetical ammonium hydroxide NH_4OH, in which the four hydrogen atoms attached to the nitrogen are replaced by alkyl radicals, eg $(C_2H_5)_4NOH$, tetraethyl-ammonium hydroxide.

quaternary diagram (*Eng*) Phase diagram of four-component system.

quaternary phase-shift keying (*ICT*) Used in microwave links and satellite communications to double the channel capacity of conventional binary phase-shift keying without changing the bandwidth. The phase of the carrier can be set by modulation to any one of four positions.

quaternary structure (*BioSci*) The fourth-order level of structural organization of proteins. Tertiary structure defines the shape of single protein molecules; quaternary structure the way in which dimers or multimers are arranged.

qubit (*ICT*) The basic unit of information in quantum computing, consisting of an individual atom or subatomic particle considered as forming a binary system by representing its spin state as 0 or 1.

Queckensted's sign (*Med*) Increase in the pressure of the cerebrospinal fluid when the jugular veins in the neck are compressed; if this manoeuvre causes no rise of pressure in the spinal fluid, an obstruction is present at a higher level.

queen (*BioSci*) In social insects, a sexually reproducing female.

queen (*Build*) A slate 36 × 24 in (914 × 610 mm).

queen bee substance (*BioSci*) See QUEEN SUBSTANCE.

queen bolt (*Build*) A long iron or steel bolt serving in place of a timber queen post.

queen closer (*Build*) A half-brick made by cutting the brick lengthwise.

queen post (*Build*) The two spaced vertical ties required for roofs of more than about 30 ft (approx 10 m) span, where the central support of the tie-beam by the king post is insufficient. See fig. at ROOF TRUSS.

queen post roof (*Build*) A timber roof having two queen posts but no king post. See fig. at ROOF TRUSS.

Queensland blue gum (*For*) See RED RIVER GUM.

Queensland walnut (*For*) Not a true walnut (*Juglans*), but an Australian hardwood tree (*Endiandra palmerstonii*) whose heartwood is light- to dark-brown streaked with pink, grey, green and black, with an irregular, interlocked grain.

queen substance (*BioSci*) A pheromone produced by queen honeybees (*Apis mellifera*; Hymenoptera) consisting of 9-ketodecanoic acid. Its effects include the suppression of egg laying and queen cell construction by workers.

quench (*ElecEng*) Resistor or resistor–capacitor shunting a contact, to reduce high-frequency sparking when a current is broken in an inductive circuit.

quenched cullet (*Glass*) Cullet made by running molten glass into water. Also *dragaded cullet, dragladled cullet.*

quencher (*Phys*) Material introduced into a luminescent substance to reduce the duration of phosphorescence.

quench frequency (*ICT*) The lower-frequency signal used to quench intermittently a high-frequency oscillator, eg in a super-regenerative receiver.

quenching (*Eng*) The process of cooling by plunging a heated object into a fluid, with the purpose of retaining the material in a metastable state. Quenching into water gives a more rapid cooling rate than into oil. The term also applies to cooling in salt and molten-metal baths or by means of an air blast. Applied to steels heated above their upper critical temperature in order to harden them prior to tempering and to other alloys for solution treatment prior to PRECIPITATION HARDENING. See panel on STEELS.

quenching (*ICT*) Suppression of oscillation, particularly periodically, as in a super-regenerative receiver.

quenching (*NucEng*) The process of inhibiting continuous discharge, by choice of gas and/or external valve circuit, so that discharge can occur again on the incidence of a further photon or particle in a counting tube. Essential in a GEIGER–MÜLLER COUNTER.

quenching media (*Eng*) See HARDENING MEDIA.

quenching oscillator (*ICT*) One with a frequency slightly above the audible limit, and that generates the voltage necessary to quench the high-frequency oscillations in a super-regenerative receiver.

quench oil (*ChemEng*) Oil injected into the product stream leaving a cracking or reforming heater. It lowers the temperature of the stream and thus stops (quenches) any further, undesired, chemical reaction.

quench time (*NucEng*) That required to quench the discharge of a Geiger tube. DEAD TIME for internal quenching, PARALYSIS TIME for electronic quenching, although dead time is often used synonymously for the other two terms.

quercertin (*BioSci*) A yellow pigment released from quercertin glycoside by intrinsic enzymes or from local micro-organisms in pickled onions and naturally occurring in other food plants. It is not toxic, although it may be mutagenic. Inhibits F-type ATPases.

query language (*ICT*) A method of retrieving information INTERACTIVELY from a database without having to write a complex program. Simple COMMANDS such as FIND postcode = 'SO9 2QU' are used. See STRUCTURED QUERY LANGUAGE.

queue (*ICT*) A LIST for which insertions are made at one end and deletions at the other. The arrangement is called FIFO. Cf STACK.

queuing (*ICT*) The situation that arises in a digital network or other system when data arrive at a device faster than the device can process them, in which data are allowed to accumulate in a BUFFER until a reduction in data rate allows them to be dealt with. See TIME-SHARING SYSTEM.

quick-break switch (*ElecEng*) A switch having a spring or other device to produce a quick break, independently of the operator.

quicking (*ElecEng*) Electrodeposition of mercury on a surface before regular plating.

quicklime (*Chem*) See CAUSTIC LIME, LIME.

quick make-and-break switch (*ElecEng*) See SNAP SWITCH.

quick return mechanism (*Eng*) A reciprocating motion, for operating the tool of a shaping machine etc, in which the return is made more rapidly than the cutting stroke, so as to reduce the 'idling' time.

quicksand (*CivEng*) Loose sand mixed with such a high proportion of water that its bearing pressure is very low. Also *running sand.*

quick-setting inks (*Print*) A general term for inks formulated to set quickly, allowing handling of the stock soon after printing.

quick-setting level (*Surv*) See FIXED-NEEDLE SURVEYING.

quicksilver (*Chem*) See MERCURY.

quick sweep (*Build*) A term applied to circular work in which the radius is small.

quidding (*Vet*) A condition in the horse in which food is expelled from the mouth after being chewed; usually due to disease of the mouth.

quiescent (*Electronics*) A general term for a system waiting to be operated, as a valve ready to amplify or a gas-discharge tube to fire. Also *preset.*

quiescent carrier transmission (*ICT*) One for which the carrier is suppressed in the absence of modulation.

quiescent centre (*BioSci*) A region within the apical meristem of many roots, in which the cells either do not divide, or divide very much more slowly than the cells around it.

quiescent current (*ICT*) Current in an active device in the absence of a driving or modulating signal. Also *standing current.*

quiescent operating point (*ICT*) The steady-state operating conditions of a valve or transistor in its working circuit but in the absence of any input signal.

quiescent period (*ICT*) That between pulses in a pulse transmission.

quiescent push–pull amplifier (*ICT*) Thermionic valve or transistor amplifier, in which one side alone passes current for one phase, the other side passing current for the other phase. Abbrev *QPP amplifier.*

quiescent tank (*Build*) A form of sedimentation tank in which sewage is allowed to rest for a certain time without flow taking place.

quiet automatic volume control (*ICT*) The application of this is known as 'quieting'. Also DELAYED AUTOMATIC GAIN CONTROL.

quieting sensitivity (*ICT*) The minimum input signal required by a frequency-modulation radio receiver to give a specified signal-to-noise ratio at the output.

quill (*BioSci*) See CALAMUS.

quill (*ElecEng*) A form of drive used for electric locomotives in which the armature of the driving motor is mounted on a quill surrounding the driving axle, but connected to it only by a flexible connection. This enables a small amount of relative motion to take place between the motor and the driving axle. Also *quill drive*.

quill (*Eng*) A hollow non-rotating shaft in which another shaft rotates under power, for providing axial movement as in a drilling machine spindle. Also *quill drive*.

quill drive (*ElecEng*) See QUILL.

quill feathers (*BioSci*) In birds, the RECTRICES and REMIGES.

quinacrine (*BioSci*) A fluorescent dye that intercalates into DNA helices. See BANDING TECHNIQUES.

quinaldine (*Chem*) 2-methylquinoline. $C_{10}H_9N$. Bp 246°C. A colourless refractive liquid, which occurs to the extent of 25% in quinoline obtained from coaltar.

quinapril (*Pharmacol*) An *angiotensin-converting enzyme inhibitor* used for the treatment of hypertension.

Quincke's method (*Phys*) A method for determining the magnetic susceptibility of a substance in solution by measuring the force acting on it in terms of the change of height of the free surface of the solution when placed in a suitable magnetic field.

quincuncial aestivation (*BioSci*) A common type of IMBRICATE aestivation of a five-membered calyx or corolla in which two members overlap their neighbours by both edges, two are overlapped on both edges and one overlaps one neighbour and is overlapped by the other, as in eg calyx of roses, corolla of Caryophyllaceae.

quinhydrone (*Chem*) $C_6H_4O_2+C_6H_4(OH)_2$. An additive compound of one molecule of 1,4-quinone and one molecule of 1,4-dihydroxybenzene. It crystallizes in green prisms with a metallic lustre.

quinhydrone electrode (*Chem*) A system consisting of a clean, polished, gold or platinum electrode dipping into a solution containing a little quinhydrone, for determining pH values, making use of the pH dependence of the redox properties of the system.

quinine (*Chem*) $C_{20}H_{24}O_2N_2 \cdot 3H_2O$. Mp 177°C. An alkaloid of the quinoline group, present in Cinchona bark. It is a diacid base of very bitter taste and alkaline reaction. It crystallizes in prisms or silky needles; the hydrochloride and sulphate are used as a febrifuge but have been largely superseded as a remedy for malaria, although they are still used in the treatment of leg cramps.

quinizarine (*Chem*) A synonym for 1,4-dihydroxy-anthra-quinone.

quinol (*Chem, ImageTech*) See HYDROQUINONE.

quinoline (*Chem*) A heterocyclic compound consisting of a benzene ring condensed with a pyridine ring. It is a colourless, oily liquid, mp −19·5°C, bp 240°C, rel.d. 1·08, of characteristic odour, insoluble in water, soluble in most organic solvents. It is found in coaltar, in bone oil and in the products of the destructive distillation of many alkaloids. It can be synthesized by heating a mixture of aniline, glycerine and nitrobenzene with concentrated sulphuric acid. See SKRAUP'S SYNTHESIS.

quinoline

quinolone antibiotics (*Pharmacol*) The quinolones and fluoroquinolones are bactericidal and inhibit the activity of DNA *gyrase*. The older quinolones, nalidixic acid and cinoxacin, are active only against Enterobacteriaceae but the newer fluoroquinolones (eg *ciprofloxacin*) have a broader spectrum of activity.

quinones (*Chem*) Compounds derived from benzene and its homologues by the replacement of two atoms of hydrogen with two atoms of oxygen, and characterized by their yellow colour and by being readily reduced to dihydric phenols. According to their configuration they are divided into 1,2-quinones and 1,4-quinones.

quinonoid formula (*Chem*) A formula based upon the diketone configuration of 1,4-quinone (BENZOQUINONE), involving the rearrangement of the double bonds in a benzene nucleus; adopted to explain the formation of dyestuffs, eg coloured salts of compounds of the triphenyl-methane series.

quinoxalines (*Chem*) A group of heterocyclic compounds consisting of a benzene ring condensed with a diazine ring: They can be obtained by the condensation of 1,2-diamines with 1,2-diketones.

quinqu-, quinque- (*Genrl*) Prefixes from Lt *quinque*, five.

quinsy (*Med*) Acute inflammation of the tonsil with the formation of pus around it. Also *acute suppurative tonsillitis, peritonsillar abscess*.

quint- (*Genrl*) Prefix from Lt *quintus*, fifth.

quintal (*Phys*) Unit of mass in the metric system, equal to 100 kg. Symbol q.

quintessence (*Astron*) Classically, the fifth element, after earth, air, fire and water – the ephemeral substance that prevented the planets from falling to the centre of the celestial sphere. Recently, the term has been used in scalar field models for dark energy and can be considered as a dynamical field that is repulsed by gravity and induces a vacuum energy, exerting a negative outward pressure that is thought to cause the acceleration of the expansion of the universe.

quintic equation (*MathSci*) An algebraic equation of the fifth degree. Unlike equations of lower degree, its general solution (and that of equations of higher degree) cannot be expressed in terms of a finite number of root extractions.

quintozene (*Chem*) *Pentachloronitrobenzene*. Used as a fungicide. Also *PCNB*.

quintuple point (*Chem*) A point on a concentration–pressure–temperature diagram at which a three-component system can exist in five phases.

quire (*Paper*) Paper quantity; 25 sheets, or $\frac{1}{20}$ of a ream.

quire spacing (*Print*) On a rotary printing press, as the product is delivered it is separated into quires or batches by the KICKER which delivers a KICK COPY at the required interval.

quirewise (*Print*) Sections which after printing are folded and insetted one in the other. This method allows the booklet to be stitched instead of stabbed.

quirk (*Build*) The narrow groove alongside a bead sunk flush with a surface.

quirk-bead (*Build*) See BEAD-AND-QUIRK.

quirk float (*Build*) A plasterer's trowel specially shaped for finishing mouldings.

quirk moulding (*Build*) A moulding having a small groove in it.

quirk-router (*Build*) A form of plane for shaping quirks.

quitclaim (*MinExt*) A deed of relinquishment of a claim or portion of mining ground.

quittor (*Vet*) A chronic suppuration of the lateral cartilage and its surrounding tissues within a horse's foot.

quoin (*Build*) An exterior angle of a building, esp one formed of large squared cornerstones projecting beyond the general faces of the meeting wall surfaces. Also *angle stone*.

quoin (*Print*) A wooden wedge or a metal device used to lock up formes.

quoin header (*Build*) A brick laid at the external angle of a building to be a header in the wall proper and a stretcher in the return wall.

quoin post (*Build*) See HEEL POST.

quotation marks (*Print*) If double quotes (" ") are used to indicate a quotation, the single quotes (' ') are used for a quotation within the passage quoted; the reverse procedure is becoming more popular. In handset matter the opening quotes are inverted commas, the closing quotes being apostrophes.

quotations (*Print*) Metal spaces of varying widths, 1, 2, 3 or 4 em, used for filling blanks in pages or formes.

quotient (*Genrl*) A measure of the extent or significance of something.

quotient (*MathSci*) See DIVISION.

quotient group (*MathSci*) The group G/N whose elements are the COSETS of N in G, where N is a normal subgroup of G. The product of two cosets is defined to be equal to the coset which contains the product of an element of the first coset and an element of the second. Also *factor group*.

quotient rule (*MathSci*) The rule used in differentiating the quotient of two differentiable functions f and g:

$$\left(\frac{f}{g}\right)' = \frac{f'g - g'f}{g^2}$$

Q-value (*NucEng*) (1) Quantity of energy released in a given nuclear reaction. Normally expressed in MeV, but occasionally in atomic mass units. (2) Ratio of thermonuclear power output to power needed to maintain the plasma.

QWERTY keyboard (*ICT*) One laid out in the standard typewriter pattern.

R

R (*Build*) Abbrev for *render*.

R (*Chem*) A general symbol for an organic hydrocarbon radical, esp an alkyl radical.

R (*radiol*) Symbol for ROENTGEN unit in X-ray dosage.

R (*Phys*) Symbol for RANKINE SCALE of temperature.

R (*Chem, Phys*) Symbol for: (1) the GAS CONSTANT; (2) the RYDBERG CONSTANT.

R- (*Chem*) Prefix denoting right-handed. See CAHN–INGOLD–PRELOG SYSTEM for absolute configuration.

[R] (*Chem*) With subscript, a symbol for molecular refraction.

R$_F$ (*Chem*) Term used in chromatography. The ratio of the distance moved by a particular solute to that moved by the solvent front.

R$_H$ (*Chem*) See RH VALUE.

r (*BioSci*) The instantaneous population growth rate defined as

$$r = \frac{1}{N}\frac{dN}{dt}$$

where N is the population size and t is time. For any organism there is a maximum, r, achieved in ideal conditions, called the innate capacity for increase r_{max}.

r (*Chem*) With subscript, a symbol for *specific refraction*.

r- (*Chem*) Abbrev for RACEMIC.

ρ (*Phys, Genrl*) Symbol for: (1) DENSITY; (2) RESISTIVITY; (3) RESONANCE ESCAPE PROBABILITY.

ρ- (*Chem*) Symbol for *pros-*, ie containing a condensed double aromatic nucleus substituted in the 2,3 positions.

r$_h$ (*Chem*) See RH VALUE.

RA (*Astron*) Abbrev for RIGHT ASCENSION.

Ra (*Chem*) Symbol for RADIUM.

ra- (*Chem*) Symbol for *radio-*, ie a radioactive isotope of an element, eg *ra*-Na, *radio-sodium*.

Raabe's convergence test (*MathSci*) If, for the series of positive terms $\sum u_n$, the product

$$n\left(\frac{u_n}{u_n+1}-1\right) \to \sigma$$

then $\sum u_n$ is convergent if $\sigma > 1$ and divergent if $\sigma < 1$. This is a special case of KUMMER'S CONVERGENCE TEST.

rabbet (*Build*) A corruption of REBATE.

rabbeted lock (*Build*) A lock which is fitted into a recess cut in the edge of a door.

rabbit (*MinExt*) Apparatus for checking the internal diameter of pipes etc to ensure that tools can pass. Also *pig*. See GO-DEVIL.

rabbit (*NucEng*) See SHUTTLE.

rabbit septicaemia (*Vet*) An acute septicaemic and pneumonic disease of rabbits caused by the bacterium *Pasteurella multocida*; a mild form of the disease, in which the infection is confined to the upper respiratory tract, is called *snuffles*.

rabble (*MinExt*) Mechanized rake used to loosen sluice bed or move ore or concentrate through a kiln or furnace.

rabies (*Med*) An acute disease of dogs, wolves and other carnivores, due to infection with a virus, and communicable to humans by the bite of the infected animals. In humans the disease is characterized by intense restlessness, mental excitement, muscular spasms (esp of the mouth and throat), convulsions and paralysis. Also *hydrophobia*.

Rabl configuration (*BioSci*) A spatial arrangement of interphase chromosomes with centromeres clustering at one side of the nucleus and telomeres at the other.

race (*BioSci*) A population, within a species, that is genetically distinct in some way, often geographically separate; a breed of domesticated animals. See PHYSIOLOGICAL RACE.

race (*Build*) A channel conveying water to or away from a hydraulically operated machine.

race (*Eng*) The inner or outer steel rings of a BALL BEARING or ROLLER BEARING.

race (*Geol, Build*) Fragments of limestone sometimes found in certain brick earths of a hard marly character.

racemates (*Chem*) See RACEMIC ISOMERS.

raceme (*BioSci*) A simple (unbranched) RACEMOSE INFLORESCENCE in which the flowers are visibly stalked. Cf SPIKE.

racemic acid (*Chem*) See TARTARIC ACID.

racemic isomers (*Chem*) Optically inactive mixtures consisting of equal quantities of enantiomorphous stereoisomers, R- and S-forms. Chemical synthesis of compounds with asymmetric molecules from optically inactive starting materials gives racemic mixtures. These can be resolved into the optically active components by various methods, eg by coupling with an optically active substance, such as an alkaloid, and subsequent fractional crystallization, by the action of lower plant organisms eg bacteria, moulds, yeasts, etc, which attack only one of the isomers, leaving the other one intact. Also *racemates*.

racemization (*Chem*) The transformation of an optically active substance into racemic inactive form, either by an isomerization through a symmetrical intermediate or by a reaction by which a new substance is formed via a similar intermediate or transition state.

racemose (*BioSci*) Shaped like a bunch of grapes; said esp of glands.

racemose inflorescence (*BioSci*) One in which the main axis (and, in a compound raceme, each of its main branches) does not end in a flower but continues to grow bearing flowers in ACROPETAL succession on its lateral branches, eg raceme, spike, panicle. Also *indefinite* or *indeterminate inflorescence*. Cf CYMOSE INFLORESCENCE, MIXED INFLORESCENCE.

race-track (*NucEng*) A discharge tube or ion-beam chamber where particles are constrained to an oval path.

rachi-, rachio- (*Genrl*) Prefixes from Gk *rhachis*, spine. Also *rhachi-, rhachio-*.

rachilla (*BioSci*) (1) The axis of the spikelet of a grass, on which are borne the glumes and the florets. (2) A secondary (or tertiary etc) axis of a pinnately compound leaf.

rachio- (*Genrl*) See RACHI-.

rachiodont (*BioSci*) Having some of the anterior thoracic vertebrae with the hypapophysis enlarged, forwardly directed, and capped with enamel to act as an egg-breaking tooth, as certain egg-eating snakes.

rachis (*BioSci*) (1) The main axis of an inflorescence or of a pinnately compound leaf etc. (2) The shaft of a feather. (3) The vertebral column. Also *rhachis*. Adj *rachidial*.

rachitic (*Med*) Affected with or pertaining to rickets.

R-acid (*Chem*) *2-naphthol-3,6-disulphonic acid*. Used in preparation of azo-dyes for wool.

rack (*Eng*) Component on which a linear array of gear-form teeth are cut.

rack (*MinExt*) See RAGGING FRAME.

rack-and-pinion (*Eng*) A method of transforming rotary into linear motion, or vice versa; accomplished by a pinion or small gear wheel which engages a straight, toothed RACK. See PITCH LINE.

rack-and-pinion steering gear (*Autos*) One in which a pinion carried by the steering column engages with a rack attached to the divided track rod.

racked (*Build*) Said of temporary timbering which is braced so as to stiffen it against deformation.

racking (*MinExt*) The operation of separating ore by washing on an inclined plane.

racking (*Ships*) Distortion of the ship's transverse shape.

racking back (*Build*) The procedure adopted when the full length of a wall is not built at once, the unfinished end being stepped or 'racked back' at an angle, so that on finishing there is not a vertical line of junction which might cause cracking due to uneven settlement.

rack mounting (*ICT*) The use of standard racks, of varying height but otherwise of standard dimensions, for mounting panels carrying electronic equipment or assemblies with a uniform scheme of wiring; such mounting gives both accessibility and compactness.

rack railway (*CivEng*) A device for overcoming adhesion difficulties, as met on mountain railways. A PINION on the locomotive axle engages with a RACK laid parallel with, usually in the centre of, the normal rails, which continue to carry the weight. Several forms exist. Also *mountain railway*.

rack saw (*Build*) A saw having wide teeth.

rad (*MathSci*) Abbrev for RADIAN.

rad (*Radiol*) Former unit of radiation dose which is absorbed, equal to $0.01 \, \mathrm{J\,kg^{-1}}$ of the absorbing (often tissue) medium. See GRAY.

radappertization (*FoodSci*) Treatment of food with radiation at levels between 10 and 50 kGy, which effectively sterilizes the food but causes some physical and chemical changes making it inappropriate except for special diets, eg for immunocompromised patients or astronauts.

radar (*Genrl*) Abbrev for *radio detection and ranging*. In general, a system using pulsed radio waves, in which reflected (*primary radar*) or regenerated (*secondary radar*) pulses lead to measurement of distance and direction of a target.

radar absorbing material (*Aero*) That attached to or built into an aircraft skin which responds to radar waves by attenuating their return echo, thus reducing the radar signal.

radar astronomy (*Astron*) The use of pulses of radio waves to detect the distances and map the surface morphology of objects in the solar system. It has been applied with great success to the mapping of Venus.

radar beacon (*Radar*) A fixed radio transmitter whose radiations enable a craft to determine its own direction or position relative to the beacon by means of its own radar equipment. See TRANSPONDER.

radar indicator (*Radar*) Display on a cathode-ray tube of a radar system output, either as a radial line or a co-ordinate system for range and direction. The echo signal gives a brightening of the luminous spot, which remains for some seconds because of afterglow. Also *radar screen*. See A-, B-, C-DISPLAY, etc, PLAN-POSITION INDICATOR.

radar mapping (*Aero*) Cartography using radar data, esp from sideways-looking radar (*SLAR*). May also be used as a navigation system.

radar performance figure (*Radar*) The ratio of peak transmitter power to minimum signal detectable by the receiver.

radar range (*Radar*) Usually given as that at which a specified object can be detected with 50% reliability.

radar range equation (*Radar*) A mathematical expression, for PRIMARY RADAR, which relates transmitter power, antenna gain, wavelength, effective area of the target, receiver sensitivity and RADAR RANGE.

Radarsat (*Space*) A Canadian remote sensing satellite; designed to gather information on crop conditions, ice distribution and geological data, unaffected by clouds.

radar scan (*Radar*) (1) The circular, rectangular or other motion of a radar as it searches for a target. (2) The physical movement of a radar antenna or of the radial line on a PPI display.

radar screen (*Radar*) See RADAR INDICATOR.

radial (*MathSci*) Radiating from a common centre; pertaining to a RADIUS.

radial arm saw (*Build*) A versatile machine in which a circular saw is mounted below an arm on which it slides. The arm can be rotated radially around a column fixed behind the work table so that cuts can be made at any angle across the table. The saw will also pivot to cut compound angles and for RIPPING.

radial commutator (*ElecEng*) A commutator for a dc machine in which the commutator bars are arranged radially from the axis to form a disk instead of a cylinder.

radial drill (*Eng*) A large drilling machine in which the drilling head is capable of radial adjustment along a rigid horizontal arm carried by a pillar.

radial ducts (*ElecEng*) In an electric machine, ventilating ducts which run radially from the shaft.

radial engine (*Aero*) An aircraft or other engine having the cylinders arranged radially at equal angular intervals round the crankshaft. See DOUBLE-ROW RADIAL ENGINE, MASTER CONNECTING-ROD.

radial feeder (*ElecEng*) See INDEPENDENT FEEDER.

radial longitudinal section (*BioSci*) A section cut longitudinally along a diameter of a more or less cylindrical organ. Abbrev RLS. Cf TANGENTIAL LONGITUDINAL SECTION.

radial-ply (*Autos*) Term applied to tyres which have a semi-rigid breaker strip under the tread and relatively little stiffening in the walls. Cf CROSS-PLY. See panel on TYRE TECHNOLOGY.

radial recording (*Acous*) See LATERAL RECORDING.

radial runout (*Eng*) Variation in the plane normal to its axis of a rotating part. Its eccentricity rather than its wobble. Cf AXIAL RUNOUT.

radial symmetry (*BioSci*) The condition in which an organ or the whole of an organism can be divided into two similar halves by any one of several planes which include the centre line. Cf BILATERAL SYMMETRY.

radial system (*ElecEng*) A distribution system in which the cables radiate out from a generating or supply station. If a fault occurs, all consumers beyond the fault are cut off.

radial valve gear (*Eng*) A steam-engine valve gear in which the slide valve is given independent component motions proportional to the sine and cosine of the crank angle respectively.

radial velocity (*Astron*) See LINE-OF-SIGHT VELOCITY.

radian (*MathSci*) The SI unit of plane angular measure, defined as the angle subtended at the centre of a circle by an arc equal in length to the radius. Abbrev *rad*. Since 2π radians = 360°, 1 radian = 57.295 779 513 1°, and 1 degree = 0.017 453 292 5... radian.

radiance (*Phys*) Of a surface, the luminous flux radiated per unit area.

radian frequency (*Phys*) See ANGULAR FREQUENCY.

radiant (*Astron*) The point on the celestial sphere from which a series of parallel tracks in space, such as those followed by the individual meteors in a shower, appear to originate.

radiant flux (*Phys*) The time rate of flow of radiant electromagnetic energy.

radiant-flux density (*Phys*) A measure of the radiant power per unit area that flows across a surface. Also *irradiance*.

radiant heat (*Phys*) Heat transmitted through space by INFRARED RADIATION.

RAD/DSDM Methodology

Methods of rapid application development (RAD) evolved in the 1990s in response to the perceived failure of formal or structured methods to control the completion time and cost of systems development projects. Massive increases in processor power and storage capacity had allowed the design and development of more and more complex systems and function-rich applications software. The result was larger, slower and more expensive projects unable to keep pace with either technological innovation or changes in the customer's needs, but highly adept at generating documentation. A different approach was required. One solution was RAD, and dynamic systems development methodology (DSDM), in which projects were broken down into subprojects on which small multidisciplinary teams of technologists and customers worked collaboratively for short, intensive periods to deliver small increments of system functionality quickly and in a ruthlessly prioritized order. Unfairly characterized as 'licensed hacking', this method in fact had a sophisticated approach to quality management, based upon reversible change, pervasive rather than phased testing, and fitness for purpose as the sole criteria for acceptance of a product. More justified criticisms were, perhaps, that any system involving complex algorithms was considered unsuitable for development; that it worked best on small systems with a highly demonstrable user interface; that it shifted time and money devoted to finishing systems out of the development project and into the domain of systems maintenance. A typical DSDM project will follow a standard life cycle, moving from feasibility to implementation in no more than six months:

(1) A feasibility phase, in which the problem domain is defined and an outline of possible technical solutions proposed.
(2) A business study, in which processes and data are defined, requirements prioritized and technical architecture agreed.
(3) One or more functional model iterations, in which a working model of the system is produced.
(4) One or more design and build iterations in which elements of system functionality are built by means of iterated prototyping in a strictly limited period known as a timebox. Each phase aims to build incrementally upon the functionality already delivered until the list of prioritized requirements has been exhausted or an externally imposed completion date has been reached.
(5) An implementation phase in which the system goes 'live' and users are trained.

This method of systems development was highly successful in achieving acceleration of delivery, flexibility of response to changing requirements and improved quality of product – where a careful preliminary evaluation process indicated that the project was likely to be suited to this approach. However, it remains unproven as a means of successfully delivering large-scale, highly complex systems. Techniques such as Agile Development and eXtreme Programming have subsequently extended the practice of collaborative working and iterative prototyping beyond the project management framework proposed by DSDM.

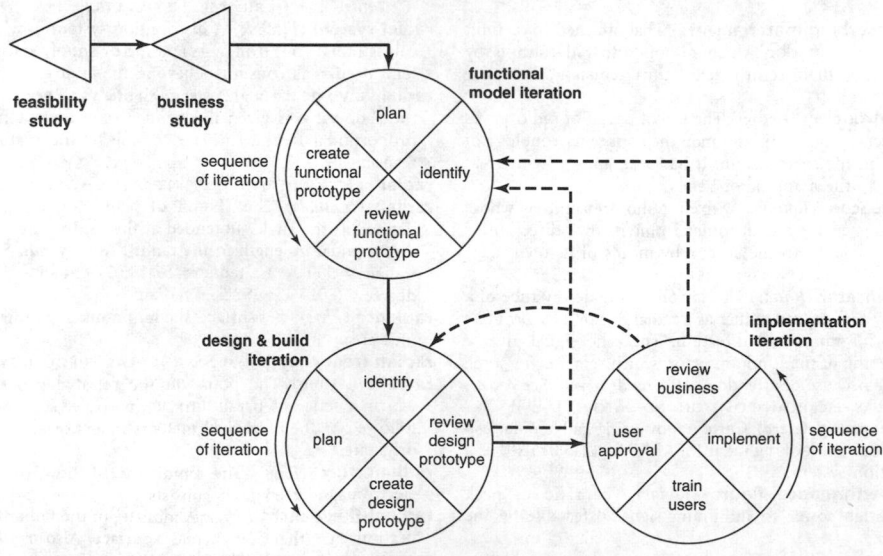

RAD/DSDM project methodology – generalised view

radiant intensity (*Phys*) The energy emitted per second per unit solid angle about a given direction.

radiant point (*MathSci*) See RAY.

radiant-tube furnace (*Eng*) Modified form of muffle furnace, the heating chamber having a series of steel alloy tubes in which the fuel is burned, thus excluding products of combustion from the heating chamber.

radiant-type boiler (*Eng*) A water-tube boiler having one or more drums and a circulation system consisting of vertically or horizontally inclined banks of tubes, heating surfaces of bare or protected water tubes forming the walls of the combustion chamber; firing is generally by pulverized fuel.

radiate (*BioSci*) A capitulum, having RAY FLORETS.

radiated power (*ICT*) The actual power level of the radio signals transmitted by an antenna. See EFFECTIVE RADIATED POWER.

radiating brick (*Build*) See COMPASS BRICK.

radiating circuit (*ICT*) Any circuit capable of sending out power in the form of electromagnetic waves into space; esp the antenna circuit of a radio transmitter.

radiating surface (*Phys*) The effective area of a radiator available for the transmission of heat by radiation.

radiation (*Phys*) The energy disseminated from a source which falls off as the inverse square of the distance from the source in the absence of absorption. See PLANCK'S RADIATION LAW, STEFAN–BOLTZMANN LAW, WIEN'S LAWS and panel on RADIATION.

radiation (*Surv*) A method of plane-table surveying in which a point is located on the board by marking its direction with the alidade and measuring off its distance to scale from the instrument station. See PLANE TABLE.

radiation area (*NucEng*) Area to which access is controlled because of a local radiation hazard.

radiation balance meter (*EnvSci*) See NET PYRRADIO-METER.

radiation burn (*Med*) A burn caused by overexposure to radiant energy.

radiation chemistry (*Chem*) Study of radiation-induced chemical effects (eg decomposition, polymerization etc). Cf RADIOCHEMISTRY.

radiation counter (*NucEng*) One used in nuclear physics to detect individual particles or photons.

radiation cross-linking (*Chem*) Cross-linking reaction induced by gamma rays etc. Used on polyethylene to make shrink-on sleeves for cable connections etc.

radiation danger zone (*Radiol*) A zone within which the MAXIMUM PERMISSIBLE CONCENTRATION or MAXIMUM PERMISSIBLE DOSE RATE is exceeded.

radiation diagram (*Phys*) See RADIATION PATTERN.

radiation dose (*Radiol*) The amount of radiation absorbed by a substance.

radiation efficiency (*ICT*) The ratio of actual power radiated by an antenna to that provided by the drive. Also *antenna efficiency*.

radiation field (*Phys*) See FIELD.

radiation flux density (*Phys*) Rate of flow of radiated energy through unit area of surface normal to the beam (for particles this is frequently expressed in number rather than energy). Also *radiation intensity*.

radiation hazard (*Radiol*) The danger to health arising from exposure to ionizing radiation, either due to external irradiation or to radiation from radioactive materials within the body.

radiation impedance (*Phys*) Impedance per unit area. Measured, eg by the complex ratio of the sound pressure to the velocity at the surface of a vibrating body which is generating sound waves, or by the corresponding electromagnetic quantities.

radiation intensity (*Phys*) See RADIATION FLUX DENSITY.

radiation length (*Phys*) The path length in which relativistic charged particles lose e^{-1} of their energy by radiative collisions. See BREMSSTRAHLUNG.

radiation loss (*ICT*) That power radiated from a non-shielded radio-frequency transmission line.

radiation pattern (*Phys*) Polar or cartesian representation of radiation in space from any source and, in reverse, effectiveness of reception. Also *radiation diagram*.

radiation pattern (*ICT*) See ANTENNA FIELD.

radiation polymerization (*Chem*) Chain reaction initiated by gamma rays etc, which react with monomer to give free radicals. See CHAIN POLYMERIZATION.

radiation potential (*Phys*) See IONIZATION POTENTIAL.

radiation pressure (*Phys*) Minute pressure exerted on a surface normal to the direction of propagation of an incident wave. It is due to the rate of transfer of momentum by the wave. For electromagnetic waves incident on a perfect reflector this pressure is equal to the energy density in the medium. In quantum physics the radiation pressure can be described as due to the transfer of the momentum of the PHOTONS as they strike the surface. Radiation pressures are very small, eg $10^{-5}\,\mathrm{N\,m^{-2}}$ for sunlight at the Earth's surface. For sound waves in a fluid the pressure gives rise to *streaming*, ie a flow of the fluid medium.

radiation prospecting (*MinExt*) A form of GEOPHYSICAL PROSPECTING which utilizes the radioactivity of uranium-, thorium- or radium-bearing minerals to identify potentially economic concentrations. Gamma-ray detectors (GEIGER–MÜLLER TUBES and SCINTILLATION COUNTERS) may be hand-held or airborne.

radiation pyrometer (*Phys*) A device for ascertaining the temperature of a distant source of heat, such as a furnace, by allowing radiation from the source to face, or be focused on, a thermojunction connected to a sensitive galvanometer. The deflection of the latter gives, after suitable calibration, the required temperature. For temperatures of approx 500–1500°C.

radiation resistance (*ICT*) That part of impedance of an antenna system related to power radiated; the power radiated divided by the square of the current at a specified point, eg at the junction with the feeder.

radiation sickness (*Med*) Illness, characterized by nausea, vomiting and loss of appetite, after excessive exposure to radiation either from radiation therapy or accidentally. If the exposure has been great it will cause bone marrow suppression with loss of blood cells, leading to anaemia, inability to overcome infection and internal bleeding.

radiation source (*NucEng*) Any device producing radiation of any kind, eg a lamp, an X-ray machine, a star, a nuclear reactor. See SEALED SOURCE.

radiation therapy (*Med*) The use of any form of radiation, eg electromagnetic, electron or neutron beam, or ultrasonic, for treating disease.

radiation trap (*NucEng*) (1) Beam trap for absorbing intense radiation beam with a minimum of scatter. (2) Maze or labyrinth formed by entry corridor with several right-angled bends, used for approach to multicurie radiation sources on some accelerating machines.

radiative collision (*Phys*) A particle collision in which kinetic energy is converted into electromagnetic radiation. See BREMSSTRAHLUNG.

radiative equilibrium (*Astron*) The normal state of matter inside stars in which the temperature in every part generates a gas pressure which exactly balances the pressure due to the self-gravity of the star – there is no convection in this idealized situation. The panel on SUN AS A STAR shows the radiative zone.

radiator (*Autos*) A device for dissipating heat created in a water-cooled engine. It consists of thin-walled tubes, or narrow passages of honeycomb form, through which the water is conducted, and across which an airstream is induced either by the motion of the vehicle or by a fan.

Radiation

The energy disseminated from a source which falls off as the inverse square of the distance from the source in the absence of absorption. This definition includes acoustic waves but it mainly refers to energy in the ELECTROMAGNETIC SPECTRUM (see appendices), which requires no supporting medium, and to the particles emitted during RADIOACTIVE DECAY.

The electromagnetic spectrum extends from the longest radio waves of wavelengths up to 10^5 m through infrared radiation, the visible spectrum and ultraviolet waves down to X-rays and gamma rays with a wavelength of 10^{-11} m. Non-ionizing radiations which extend from radio to ultraviolet waves are generally thought to be less damaging to biological systems, although short-wave ultraviolet, because it is strongly absorbed by nucleic acids and causes the formation of thymidine DIMERS, is highly mutagenic if it reaches the chromosomes. Ionizing radiation is much more hazardous because it can penetrate living tissue and release considerable energy when it collides with biological molecules.

Ionizing radiation
This form is frequently confused in the public's mind with all forms of radiation, but only refers to radiation that produces ionization in matter. Examples are alpha particles, beta particles, gamma rays, X-rays and cosmic rays; examples of non-ionizing radiation are light, infrared and radio-frequency radiation. Ionization is the process by which a neutral atom or molecule acquires or loses an electric charge to produce an electrically charged ion.

The *alpha particle* (α-particle) consists of two protons plus two neutrons emitted by a *radionuclide*, an unstable isotope of an element which undergoes natural radioactive decay; the *beta particle* (β-particle) is an electron emitted by the nucleus of a radionuclide; *gamma rays* are electromagnetic energy without mass or charge emitted by a radionuclide and *cosmic rays* are highly energized radiation from outer space. The time taken for a radionuclide to lose half its activity is its *half-life*, symbol $\tau_{\frac{1}{2}}$ (often $T_{\frac{1}{2}}$).

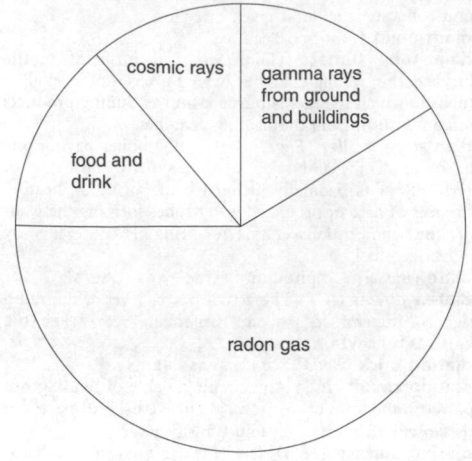

The proportion of natural radiation from different sources

Of the ionizing radiation received by everyone continuously, on average about 87% is *natural radiation*. The remaining 13% is *artificial radiation*. Twelve of these percentage points are from medical X-rays, nuclear medicine investigations and radiotherapy. The remaining 1% is composed of nuclear fallout (0·4%), exposure to radiation from work (0·2%), nuclear discharges (0·1%) and miscellaneous sources (0·4%). Fifty-nine per cent of the natural ionizing radiation received on average by the UK population is RADON GAS from the ground. Radon is a heavy radioactive element formed by the disintegration of radium and occurs especially in areas of granitic rocks. Out of doors it disperses but may accumulate in buildings. Uranium, thorium and potassium-40 are the principal sources of natural radioactivity in rocks. Gamma rays, amounting to 16% of natural radiation, are emitted by radioactive materials in the Earth and by some natural building materials taken from the Earth. Cosmic rays amount to about 11%, and 14% comes from food and drink, where the source of radiation is largely an isotope of potassium (potassium-40 or ^{40}K).

There is still considerable controversy about how hazardous such small excess doses are and it relates to whether the dose–response curve measured in experimental animals at high doses can be linearly extrapolated to zero or whether there is some

---

radiator (*ICT*) Any part of the antenna or transmission line that radiates electromagnetic waves, either directly into space or in conjunction with other elements or structures that concentrate the radiation into specific directions or spaced beams. A radiator may be a length of wire, one quarter- or one half-wavelength long or a multiple thereof, or a loop, spiral or helix, or the open end of a waveguide or a slot in a conducting surface.

radiator (*Phys*) In radioactivity, the origin of alpha, beta and/or gamma rays. Also *source*.

radiator flaps (*Aero*) See GILLS.

radical (*BioSci*) (1) Pertaining to the root. (2) Leaves, flowers, etc, arising at soil level from a root stock, rhizome or the base of a stem, as in rosette plants. Cf CAULINE.

radical (*Chem*) (1) A molecule or atom which possesses an odd number of electrons, eg the methyl radical CH_3–. It is often very short-lived, reacting rapidly with other radicals or other molecules. See FREE RADICAL. (2) A group of atoms which passes unchanged through a series of reactions, but is normally incapable of separate existence. This is now more usually called a *group*.

radical axis (*MathSci*) The straight line from each point of which the tangents to two circles are equal. Fig. ▷

Radiation (Cont.)

threshold below which radiation becomes dispropor-tionately ineffective. The question is important because, despite the small size of any expected effect, it could be spread across a very large population. *Repair enzymes* which locate and repair damaged DNA strands using the undamaged strand as a template are widespread and relevant to this discussion because they would be expected to be most effective when dealing with small amounts of damage. The main problem, however, is the difficulty of measuring the effects of a small excess of radiation over background in experimental animals. It has been estimated to require the observation of perhaps a million mice throughout their natural lives in order to give a significant result. Other approaches have been to study disease incidence in people who live in situations of higher-than-average background radiation due eg to granite in buildings or at high altitudes where the cosmic radiation is much increased, but even in these situations it is difficult to detect significant effects.

The four special radiation units defined here progressively relate the initial radioactive disintegra-tion to its final biological effect on a given tissue.

Activity and exposure

The SI unit is the *becquerel* (symbol Bq) named after A Henri Becquerel who first discovered radioactivity in 1896 in a uranium salt. It is simply one disintegration per second. This is an exceedingly small amount of radioactivity and substances of *megabecquerel* (10^6 Bq) activity will be routinely handled in a biological laboratory and *gigabecquerel* (10^9 Bq) and even *ter-abecquerel* (10^{12} Bq) activity would be common in the nuclear industry. For instance, 1 gram of plutonium-238 has an activity of over half a terabecquerel (600 000 million alpha particles emitted per second). The older unit was the *curie* (symbol Ci), based on the number of disintegrations per second from 1 gram of radium and equal to 3.7×10^{10} becquerels. Marie Curie discovered radium in 1898. Exposure is now defined in terms of coulombs per kilogram of dry air, with the coulomb being the charge transported by 1 ampere flowing for 1 second. The older unit is the *roentgen* (or *röntgen*), after W K Röntgen, the discoverer of X-rays

in 1895. It is defined as the radiation which causes 2.58×10^{-4} coulombs of electric charge in 1 kilogram of air. Exposure therefore takes into account the effec-tiveness with which a given kind of radiation will ionize a standard material, like air.

Absorbed dose and dose equivalent

Although exposure can be related to substances like water or air with similar properties to living tissue, it does not define the actual dose received by the tissue. The special SI unit for absorbed dose is the GRAY (symbol Gy), named after L H Gray, the British radiation biologist, and equal to the absorption of 1 joule of energy in 1 kilogram of tissue. The older unit was the RAD, exactly 100 times smaller than the gray. A distinction may be made between the total absorbed dose and the dose from the secondary charged ionizing particles liberated by an incident uncharged ionizing particle. This is called the *kerma* (K) dose and has the same unit as the absorbed dose.

We now come to *dose equivalent*, a quantity used in radiological protection. This is the absorbed dose multiplied by a *quality factor*, depending on the kind of radiation, and a *weighting factor*, depending on the sensitivity of the tissue irradiated. The unit is the *sievert* (symbol Sv), named after R Sievert, a Swedish radio-biologist who was once chairman of the International Committee for Radiological Protection. Its units are, like the gray, joules per kilogram. All electrons, positrons and X-rays have a quality factor of 1, but it can range up to 20 for radiation which transfers 100 or more keV per micrometre of tissue (*high LET radiation*). Thermal neutrons also have a quality factor of 1, but for fast neutrons and alpha particles it is now 20. The weighting factor is 1 for the whole body but less for individual tissues. The older unit was the REM, an abbreviation for ROENTGEN EQUIVALENT MAN and with the absorbed dose measured in *rads*.

It follows that for whole-body irradiation by elec-trons, positrons or X-rays, the sievert and the gray are equal. Related units which are frequently used include person (or man) grays, the average dose multiplied by the population at risk.

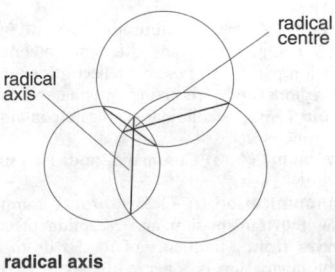

radical axis

radicand (*MathSci*) A quantity written under the root sign in a mathematical expression indicating the root to be calculated. In $\sqrt{234}$, 234 is the radicand.

radicidation (*FoodSci*) The treatment of food with radiation at levels between 2.5 and 10 kGy, which is sufficient to kill spoilage organisms and non-spore-forming pathogens, with only minor damage to the foodstuff.

radicivorous (*BioSci*) Root-eating.

radicle (*BioSci*) The primary root of an embryo, normally the first organ to emerge when a seed germinates.

radiculectomy (*Med*) The operation of cutting the roots of spinal nerves.

radiculitis (*Med*) Inflammation of the root of a spinal nerve.

radio (*ICT*) Generic term applied to methods of signalling through space, without connecting wires, by means of electromagnetic waves generated by high-frequency ac.

radio- (*Chem*) A prefix denoting an artificially prepared radioactive isotope of an element. Abbrev *ra-*.

radioactivation analysis (*Phys*) The study of artificially created radioactive nuclei, produced by bombarding a material with neutrons in a nuclear reactor, to give information about the isotopes present in the material.

radioactive atom (*Phys*) An atom which decays into another species by emission of an alpha or beta ray (or by electron capture). Activity may be natural or induced.

radioactive chain (*Phys*) See RADIOACTIVE SERIES.

radioactive dating (*Geol*) See panel on RADIOMETRIC DATING.

radioactive decay (*Phys*) See DISINTEGRATION CONSTANT, HALF-LIFE, RADIOACTIVE ATOM.

radioactive decay by heavy-ion emission (*Phys*) Radioactive decay by the emission of nuclei heavier than the alpha particle. The probability of this occurring is very small but carbon-14 rather than alpha decay has been observed from radium-223.

radioactive equilibrium (*Phys*) Eventual stability of products of radioactivity if contained, ie rate of formation (quantitative) equals rate of decay. Particularly important between radium and radon.

radioactive isotope (*Phys*) Naturally occurring or artificially produced isotope exhibiting radioactivity; used as a source for medical or industrial purposes. Also *radioisotope*.

radioactive series (*Phys*) One of the three series to which most naturally occurring radioactive isotopes belong; they show how they are related through radiation and decay. Each series involves the emission of an alpha particle, which decreases the mass number by four, and beta and gamma decay which do not change the mass number. The natural series have members having mass number $4n$ (thorium series), $4n+2$ (uranium–radium series) and $4n+3$ (actinium series), where n is an appropriate integer. Members of the $4n+1$ (plutonium series) can be produced artificially. Also *radioactive chain*.

radioactive standard (*Phys*) A radiation source for calibrating radiation measurement equipment. The source has usually a long half-life and during its decay the number and type of radioactive atoms at a given reference time is known.

radioactive tracer (*Phys*) Small quantity of radioactive preparation added to corresponding non-active material to *label* or *tag* it so that its movements can be followed by tracing the activity. (The chemical behaviour of radioactive elements and their non-active isotopes is identical.)

radioactive waste (*NucEng*) See WASTE.

radioactivity (*Phys*) Spontaneous disintegration of certain natural heavy elements (eg radium, actinium, uranium, thorium) accompanied by the emission of alpha rays, which are positively charged helium nuclei; beta rays, which are fast electrons; and gamma rays, which are short-wavelength electromagnetic waves. The ultimate end-product of radioactive disintegration is an isotope of lead. See ARTIFICIAL RADIOACTIVITY, INDUCED RADIOACTIVITY.

radio-allergosorbent test (*BioSci*) A method for measuring extremely small amounts of IgE antibody specific for various allergens. Blood serum is reacted with allergen-coated particles which are then washed to remove non-reacting proteins. Radiolabelled anti-human IgE is then added and this binds to the IgE antibody, bound to the particles via the allergen. Provided that the amount of allergen supplied and the anti-IgE are present in excess, the radioactivity on the particles after washing is proportional to the amount of allergen-specific antibody in the serum sample. Abbrev *RAST*.

radio altimeter (*Aero*) A device for determining height, particularly of aircraft in flight, by electronic means, generally by detecting the delay in reception of reflected signals, or change in frequency. Also *radar altimeter*.

radio approach aids (*Aero*) Those which assist landing in bad visibility, notably ILS and MLS. Also *radio* or *electronic landing aids*.

radio astronomy (*Astron*) The exploration of the universe by detecting radio emission from a variety of celestial objects. The frequency spans a vast range from 10 MHz to 300 GHz. A variety of antennas are used, from single dishes to elaborate networks of telescopes forming intercontinental radio INTERFEROMETERS. The principle sources of radio emission are: the Sun, Jupiter, interstellar hydrogen, emission nebulae, pulsars, supernova remnants, radio galaxies, quasars, and the cosmic background radiation of the universe itself.

radio beacon (*ICT*) Stationary radio transmitter that transmits steady beams of radiation along certain directions for guidance of ships or aircraft, or one that transmits from an omnidirectional antenna and is used for the taking of bearings, using an identifying code. Also *aerophare*.

radio beam (*ICT*) Concentration of electromagnetic radiation within narrow angular limits, such as is emitted from a highly directional antenna.

radio bearing (*ICT*) Direction of arrival of a radio signal, as indicated by a loop, goniometer, interferometer or any directional receiving system as used for navigational purposes.

radiobiology (*BioSci*) The branch of science involving study of effect of radiation and radioactive materials on living matter.

radio broadcasting (*ICT*) The transmission by means of radio waves, of a programme of sound or picture for general reception. The separation of the frequency channels is decided by international agreement.

radiocaesium (*Chem*) Namely ^{137}Cs, a radioactive isotope recovered from the waste of nuclear reactors in nuclear power plants. Useful for mass radiation and sterilization of foodstuffs. Also for high-intensity X-ray radiation of surface tumours in place of much more expensive radium. Half-life 37 years. Also *radiocesium*.

radiocarbon (*Chem*) Namely ^{14}C, a weakly radioactive isotope undergoing beta decay with a half-life of 5770 years. It is present in the atmosphere in roughly constant amount, as it is produced from ^{14}N by cosmic rays. It is used in some tracer studies. It can also be used to date the time of death of once-living material (and hence the likely time of manufacture of an artefact). This is because living material has the same ratio of ^{14}C to ^{12}C as the atmosphere. After death, however, the ^{14}C decays and is not replaced.

radiocarbon dating (*Geol*) A method of determining the age in years of fossil organic material or water bicarbonate, based on the known decay rate of ^{14}C to ^{14}N. See panel on RADIOMETRIC DATING.

radiochemical purity (*Chem*) The proportion of a given radioactive compound in the stated chemical form. Cf RADIOISOTOPIC PURITY.

radiochemistry (*Chem*) The study of science and techniques of producing and using radioactive isotopes or their compounds to study chemical compounds. Cf RADIATION CHEMISTRY.

radio circuit (*ICT*) Communication system including a radio link, comprising a transmitter and antenna, the radio transmission path, with possible reflections or scatter from ionized regions, and a receiving antenna and receiver.

radiocolloid (*Phys*) A colloidal aggregate containing radioactive atoms.

radio command (*Aero*) Command guidance using a radio link.

radio communication (*ICT*) Any form of communication involving the transmission and reception of electromagnetic waves, from a frequency of 10 kHz up to more than 10 GHz. Information is conveyed by MODULATION of the information it is desired to impart onto a CARRIER. The information may be letters represented by code (eg Morse), speech, telemetry, pictures (either facsimile or TV), digital signals or computer data. In broadcasting, radio communication is a one-way process serving many listeners or viewers, or it may be two-way as in telecommunication systems. In the latter, communication may be between two mobile users in different vehicles or from a mobile vehicle and a fixed station, from one microwave tower to another in terrestrial communication (see RADIO LINK) or from one Earth station to another via a communications satellite. See panel on COMMUNICATIONS SATELLITE.

Radiocommunications Agency (*ICT*) An executive agency of the Department of Trade and Industry, established in 1990 to regulate all UK civil radio matters other than telecommunications policy, broadcasting policy and the radio equipment market.

radio compass (*ICT*) Originally a rotating loop, later rendered more sensitive by a goniometer system and by display on a cathode-ray tube. Any device, depending on radio, that gives a bearing. See ADCOCK ANTENNA.

radio control (*Aero*) Control of vehicle trajectory by commands transmitted over a radio link.

radio countermeasures (*Aero*) Those ELECTRONIC COUNTERMEASURES concerned with telecommunications.

radio direction-finding (*ICT*) Passive reception of direction-finding signals from radio beacons or navigational transmitters, as distinct from active radar. Abbrev *RDF*.

radioelement (*Phys*) An element exhibiting natural radioactivity.

radio fix (*Aero, Space*) The position of a space vehicle, aircraft or other vehicle obtained using a radio navigation aid by traditional crossing of position lines; the same applied to a fixed radio emitter.

radio frequency (*ICT*) One suitable for radio transmission, above 10^4 Hz and below 3×10^{12} Hz approx. Abbrev *RF*. Also *radio spectrum*.

radio-frequency heating (*ElecEng*) See DIELECTRIC HEATING, INDUCTION HEATING.

radio-frequency identification device (*ICT*) A small device encoding data that can be remotely interrogated and that can be attached to or incorporated into a product, animal or person. RFID tags contain the necessary circuitry to receive and respond to radio-frequency queries from an RFID transceiver. Because passive tags require no internal power source they can be made extremely small. Abbrev *RFID*.

radio-frequency plasma (*NucEng*) PLASMA produced by either capacitively or inductively coupled radio-frequency fields.

radio-frequency spectrometer (*NucEng*) A type of MASS SPECTROMETER used in the study of ions in plasmas.

radio-frequency spectroscopy (*Phys*) See ELECTRON SPIN RESONANCE, NUCLEAR MAGNETIC RESONANCE.

radio galaxy (*Astron*) A galaxy (about one in a million) which is an intense source of cosmic radio waves, caused by synchrotron emission of relativistic electrons.

radiogenic (*Phys*) Said of stable or radioactive products arising from radioactive disintegration.

radiogoniometer (*ICT*) Rotating coil within crossed field established by crossed loops for direction-finding.

radiography (*Radiol*) The process of image production using X-rays.

radio heating (*Phys*) See HIGH-FREQUENCY HEATING.

radio horizon (*ICT*) In the propagation of electromagnetic waves over the Earth, the line that includes the part of the Earth's surface which is reached by direct rays.

radioimmunoassay (*BioSci*) A very sensitive method that uses an antibody to bind the analyte of interest; the concentration in the sample, which does not need to be purified, is estimated by the extent to which the sample competes for binding added radioactively labelled antigen of known (low) concentration.

radio interferometer (*Astron*) Radio telescope which records the result of interference of radio waves from astronomical objects; uses separate antennas (close together or thousands of kilometres apart) to make high-resolution measurements of angular distances.

radio in the loop (*ICT*) The use of radio for some part of the connection between a CENTRAL OFFICE and the CUSTOMER PREMISES EQUIPMENT.

radio-iodine (*Chem*) Radioactive isotopes ^{125}I, ^{131}I and ^{132}I are useful in diagnosis and treatment of thyroid gland disorders; ^{125}I and ^{131}I are used in organ function and blood volume studies.

radioisotope (*Phys*) See RADIOACTIVE ISOTOPE.

radioisotope thermoelectric generator (*NucEng*) Thermoelectric generator powered by heat from a radioactive source and suitable for long periods of maintenance-free operation on remote sites, eg for lighting marine navigational buoys or powering spacecraft. Also *Ripple* (UK) or *SNAP* (US) generators. See SPACE REACTORS and panel on NUCLEAR REACTORS.

radioisotopic purity (*Chem*) The proportion of the activity of a given compound which is due to material in the stated chemical form. Cf RADIOCHEMICAL PURITY.

radiolabelling (*BioSci*) Incorporating a radioisotope in a cell or molecule to identify and follow its subsequent fate.

Radiolaria (*BioSci*) An order of marine planktonic Sarcordina, the members of which have numerous fine radial pseudopodia which do not anastomose; there is usually a skeleton of siliceous spicules.

radiolarian chert (*Geol*) A cryptocrystalline siliceous rock in part composed of the remains of Radiolaria. Most described examples seem to be of shallow-water origin, such as that which reaches a thickness of 3000 m in New South Wales and contains 20 million Radiolaria per cubic centimetre. Also *radiolarite*. See PROTOZOA.

radiolarian ooze (*Geol*) A variety of non-calcareous deep-sea ooze, deposited at such depth that the minute calcareous skeletons of such organisms as Foraminifera pass into solution, causing a preponderance of the less soluble siliceous skeletons of Radiolaria. Confined to the Indian and Pacific Oceans, and passes laterally into red clay. See PROTOZOA.

radiolarite (*Geol*) See RADIOLARIAN CHERT.

radio link (*ICT*) Self-contained two-way communication that forms part of a more extensive broadcasting or telecommunications network; broadcasters may use transportable radio links (usually microwave) to feed sound and vision into their permanent networks, while fixed installations on tall towers and masts, using highly directional antennas and operating on a line-of-sight basis, provide high-capacity channels as part of most countries' telecommunication networks. International channels, via satellites, may also be described as radio links.

radio local area network (*ICT*) A LOCAL AREA NETWORK that uses radio connections to its terminals in order to avoid cabling and allow flexibility of terminal location.

radiolocation (*Radar*) Former term for RADAR.

radiology (*Radiol*) The science and application of X-rays, gamma rays and other penetrating ionizing or non-ionizing radiations.

radioluminescence (*Phys*) Luminous radiation arising from rays from radioactive elements, particularly in mineral form.

radiolysis (*Phys*) Chemical decomposition of materials induced by ionizing radiation.

radiometer (*Acous, Phys*) An instrument devised for the detection and measurement of electromagnetic radiant energy and acoustic energy, eg thermopile, bolometer, microradiometer. See CROOKES RADIOMETER, RAYLEIGH DISK.

radiometric age (*Geol*) The radiometrically determined age of a fossil, mineral, rock or event, generally given in years. See panel on RADIOMETRIC DATING.

radiometric dating (*Geol*) The method of obtaining a geological age by measuring the relative abundance of radioactive parent and daughter isotopes in geological materials. See panel on RADIOMETRIC DATING.

radiometric surveying (*Surv*) See AERIAL RADIOMETRIC SURVEYING.

radiometry (*Phys*) The measurement of radiated electromagnetic energy, esp infrared.

radio microphone (*ICT*) One with a miniature radio transmitter, allowing freedom from a cable for the speaker. Its transmissions are picked up by a receiver nearby and fed to a public address or broadcasting system.

radiomimetic (*Pharmacol*) Said of drugs which imitate the physiological action of X-rays, notably in suppressing new cell growth, particularly those used in treating cancer.

radionuclide (*Phys*) Any nuclide (isotope of an element) which is unstable and undergoes natural radioactive decay.

radionuclide imaging (*Radiol*) The use of radionuclide substances to image the normal or abnormal physiology or anatomy of the body. Technetium-99 is an important radionuclide used for diagnostic imaging in medicine.

radiopaque (*Radiol*) Opaque to radiation (esp X-rays).

radiophotoluminescence (*Phys*) Luminescence revealed by exposing to light a material which has been previously irradiated.

radio range (*ICT*) (1) Specific system of radio homing for aircraft, in which crossed loops are separately modulated with complementary signals, which coalesce on reception when the aircraft is on course. (2) For transmissions that are not affected by ionospheric or tropospheric phenomena, the optical or line-of-sight path between two points.

radio receiver (*ICT*) Any device that converts radio waves into sound or other intelligible signals. Most common types are SUPERHET and TUNED RADIO FREQUENCY. The former shows superior sensitivity and immunity to noise and interference for most applications. More complex forms of receiver may be used for radar and for satellite communications, though the most sophisticated receivers can usually be placed into one of the two categories quoted.

radio relay station (*ICT*) An intermediate station receiving a signal from the primary transmitter and reradiating it to its destination.

radioresistant (*Radiol*) Able to withstand considerable radiation doses without injury as certain bacteria, eg *Micrococcus radiodurans*.

radiosensitive (*Radiol*) Quickly injured or changed by irradiation. The gonads, the blood-forming organs, and the cornea of the eye are the most radiosensitive organs in humans.

radiosonde (*EnvSci*) An instrument for measuring temperature, pressure and humidity at successive levels in the atmosphere, which is carried upwards on a balloon and transmits the measurements by radio. The balloon also carries a radar target so that upper winds may be derived from ground measurements.

radio sonobuoy (*Aero*) Sonar device immersed or dropped into water; can be active (emitting) or passive, directional or non-directional, providing a radio read-out, usually on command.

radio spectrum (*ICT*) See RADIO FREQUENCY.

radiospermic (*BioSci*) Having seeds that are rounded in cross-section, as in the Cycadopsida. Cf PLATYSPERMIC.

radio telegraph (*ICT*) Using a radio channel for telegraph purposes, eg by interrupted carrier, change of frequency, or modulation with interrupted audio tone.

radio telephony (*ICT*) Use of a radio channel for transmission of speech. Methods include simple modulation, suppressed carrier and one sideband, inverted sidebands and carrier, scrambling before modulation, one in a group modulation or pulse modulation.

radio telescope (*Astron*) An instrument for the collection, detection and analysis of radio waves from any cosmic source. See panel on ASTRONOMICAL TELESCOPE.

radiotherapy (*Radiol*) Theory and practice of medical treatment of disease, particularly any of the forms of cancer, with large doses of X-rays or other ionizing radiations.

radiothermoluminescence (*Phys*) Luminescence released upon heating a substance previously exposed to radiation. Now used for personal dosimetry.

radiotoxicity (*Radiol*) The amount of damage done to an organ or tissue by the ingestion of radioactive substance.

radiotropospheric duct (*EnvSci*) Stratum in which, because of a negative gradient of refractive modulus, there is an abnormal concentration of radiated energy.

radio wave (*Phys*) An electromagnetic wave of radio frequency (approx 3 kHz to 300 GHz).

radium (*Chem*) A radioactive metallic element, one of the alkaline earth metals. Symbol Ra, at no 88, ram 226, mp 700°C. Because of its chemical similarity to calcium, a large fraction of any radium ingested is deposited in bone, but large-scale surveys have shown that for the normal population only about 850 millibecquerels is to be found in the skeleton with a smaller quantity in the soft tissues. Radium is still used in some forms of cancer therapy in a procedure called brachytherapy, where the small container (or radium needle) can be placed close to the tumour. There is a considerable dose to other organs and danger to medical staff who have to position the needle unless the needle can be remotely loaded. Radium has six isotopes and is white and resembles barium in its chemical properties. It occurs in bröggerite, cleveite, carnotite, pitchblende and in certain mineral springs. Pitchblende and carnotites are the chief sources of supply.

radium cell (*Radiol*) A sealed container in the shape of a thin-walled tube (usually metal) normally loaded into larger containers, eg RADIUM NEEDLE.

radium emanation (*Chem*) See RADON.

radium needle (*Radiol*) A container in the form of a needle, usually platinum–iridium or gold alloy, designed primarily for insertion into tissue. Little used now.

radium therapy (*Radiol*) Radiotherapy by the use of radiations from radium, now almost entirely superseded by X-rays.

radius (*BioSci*) (1) In land vertebrates, the pre-axial bone of the antebrachium. (2) In insects, one of the veins of the wing. (3) In Echinodermata and Cnidaria, one of the primary axes of symmetry. Adj *radial*.

radius (*Eng*) A rod attached to the die or block of a WALSCHAERT'S VALVE GEAR of a steam locomotive for transmitting its motion to the end of the COMBINATION LEVER pivoted to the valve rod.

radius (*MathSci*) A straight line joining the centre of a conic (particularly a circle) to the curve; the length of such a line.

radius brick (*Build*) See COMPASS BRICK.

radius gauge (*Eng*) A sheet-metal strip with accurately formed external and internal radii of specified size, used as a profile gauge to determine size of an unknown radius by comparison. Usually one of a set of gauges of graded sizes.

radius of action (*Aero*) Half the range in still air of a military aircraft, taking safety and operational requirements into account; the total range is out and home again.

radius of atom (*Chem*) See ATOMIC RADII.

radius of convergence (*MathSci*) See CIRCLE OF CONVERGENCE.

radius of curvature (*MathSci*) See CURVATURE.

radius of gyration (*Phys*) See MOMENT OF INERTIA.

radius of inversion (*MathSci*) See INVERSION.

radius of spherical curvature (*MathSci*) See OSCULATING SPHERE.

radius of torsion (*MathSci*) See CURVATURE.

radius rod (*Build*) A rod pivoted at one end for marking out a circle or arc. Also *gig stick*.

radius vector (*Astron*) The line joining the focus of an orbit to the body which moves about it, such as the line from the Sun to any of the planets or comets.

Radiometric dating

It is fairly easy to measure the amount of an unstable ISOTOPE relative to the amount of a PARENT or DAUGHTER PRODUCT or of a stable isotope of the same element. Because we know the time needed for half the isotope to disappear or decay (the HALF-LIFE), this ratio can be used to calculate when a FOSSIL or artefact, like pottery, or a geological formation was formed, provided certain assumptions are made. The best known of these assumptions is that used in RADIOCARBON DATING, namely that the ratio of radioactive carbon-14 to stable carbon-12 has remained constant in the atmosphere over a long period and that once carbon-14 has been incorporated into a fossil or artefact it can no longer exchange with the carbon-14 in the atmosphere. Carbon-14 is formed in the atmosphere by the bombardment of nitrogen-14 by cosmic-ray neutrons.

Radiocarbon dating
The half-life of carbon-14 is 5730 years and, if the assumptions about the stability of the Earth's atmosphere and of carbon-14 are correct, then a measurement of the amount of carbon-14 still present should give the age of the fossil or artefact. The problem is that only 1 atom of carbon-14 is present for every 10^{12} atoms of carbon-12 and so the methods of measurement must be very sensitive. It also means that times much longer than 50 000 years (10 half-lives) are difficult to study.

The original method of measuring carbon-14 was to count the BETA PARTICLES which result from its decay. Thus, 1 gram of natural carbon from a present-day sample, such as a newly dead bone, will contain about 10^{10} atoms of carbon-14 which will produce 15 disintegrations per minute. If the bone had been 10 000 years old (2 half-lives) this number would fall to 4 and it would take 2 days' counting to give a result

significant at the 1% level. This method is therefore time-consuming and not very accurate, particularly for older and smaller samples. More recently methods have been developed using PARTICLE ACCELERATORS, chiefly tandem VAN DE GRAAFF GENERATORS and CYCLOTRONS which measure the ratio of carbon-14 to carbon-12 atoms directly. This is much more sensitive, extending the method to 100 000 years ago and milligram quantities. The main difficulties are in removing contaminating ions of the same mass number such as nitrogen-14 and CH_2, and quite complicated procedures are necessary to achieve acceptable degrees of purity.

Radiocarbon dating covers the comparatively recent past; other methods are more useful in determining much longer times such as that when the measured atoms were incorporated into a mineral, which may have happened as the Earth cooled. Their problem is that, while we know the present-day ratios, we do not know how many daughter nuclei were present to start with. This is overcome by using a second stable isotope of the same element as the daughter product and assuming that, originally when the MINERAL crystallized, the ratio of the two isotopes was the same. This method has been very fully investigated by comparing the two ratios: (1) the amount of the daughter, strontium-87 to the parent, rubidium-87 and (2) the ratio of strontium-87 to the stable isotope, strontium-86. For different minerals over a range of compositions, plotting one ratio against the other gives a straight line, indicating an age of 4.5×10^9 years.

There are a number of similar methods which may be more suitable for other minerals and situations. Some of these are summarized in the following table:

Production by COSMIC RAYS (panel) occurs in the atmosphere and is assumed to have stayed constant over long periods. PRIMORDIAL production occurred at the time of the formation of our galaxy although the entrapment of the atoms concerned would have occurred when the minerals cooled. See FISSION-TRACK DATING.

Procedure	Isotope	Half-life	Production	Principal use
Beryllium	^{10}Be	1·6 My	Cosmic rays	Age of deep marine deposits
Carbon	^{14}C	5730 y	Cosmic rays	Age of fossils, human artefacts
Chlorine	^{36}Cl	0·3 My	Cosmic rays	Water migration in glaciers etc
Fission track	^{238}U	4500 My	Primordial	Age since firing of glassy substances
Potassium/Argon	^{40}K	$1·3 \times 10^9$ y	Primordial	Age since minerals formed
Rubidium/Strontium	^{87}Rb	$4·8 \times 10^{10}$ y	Primordial	Age since minerals formed; age of Earth
Thermal luminescence	Any			Time since last heated

radix (*BioSci*) The root or point of origin of a structure, as the *radix aortae*.

radix (*MathSci*) See BASE.

radix point (*MathSci*) The dot separating the whole from the fractional part of a number expressed in POSITIONAL NOTATION, such as the decimal point when the base is 10.

radome (*Radar*) Housing for radar equipment, transparent to the signals, eg a plastic shell on aircraft or a balloon on the ground.

radon (*Chem*) Symbol Rn, at no 86, ram 222, half-life of ^{222}Rn 3·82 days, bp−65°C, mp−150°C. A zero-valent, radioactive element, the heaviest of the noble gases. It is formed by the disintegration of radium and over half the natural ionizing radiation received by the UK population comes from radon, which can accumulate in buildings in areas where radium occurs in rocks. It has seven isotopes, including actinon (at no 219, half-life 4 s, from actinium) and thoron (at no 220, half-life 54 s, from thorium). Also *radium emanation* (obsolete). See IONIZING RADIATION.

radon seeds (*Radiol*) Short lengths of gold capillary tubing containing radon used in treatment of malignant and non-malignant neoplasms.

radula (*BioSci*) In Mollusca, a mechanism for rasping consisting of a strip of epithelium bearing numerous rows of horny or chitinous teeth. Adjs *radular, radulate, raduliform*.

radurization (*FoodSci*) The treatment of food with radiation at levels between 1 and 2.6 kGy; this inhibits sprouting, delays ripening, kills insect pests and extends shelf-life. Mimics pasteurization, and is generally used on foods with a high moisture content.

RAE (*Genrl*) Abbrev for (1) ROYAL AIRCRAFT ESTABLISHMENT, Farnborough, UK. (2) Research Assessment Exercise.

RAeS (*Aero*) Abbrev for ROYAL AERONAUTICAL SOCIETY, UK.

raffinate (*Chem, MinExt*) In an extraction process in which a solvent is passed through a solid mixture of a desired product and an undesired material, the *raffinate* is the solution of desired product. The *extract* is the undesired material. In solvent refining practice in the oil industry, raffinate is that portion of the oil being treated that remains undissolved, not being removed by the selective solvent.

raffinate (*MinExt*) Liquid layer in solvent extraction system from which required solute has been extracted, eg in the chemical extraction of uranium from its ores, the liquid left after the uranium has been extracted by contact with an immiscible solvent.

raffinose (*Chem*) *Melitriose*. $C_{18}H_{32}O_{16}\cdot5H_2O$. A non-reducing trisaccharide found in sugarbeet, in molasses, in cotton-seed cake, etc. On hydrolysis it gives D-glucose, D-fructose and D-galactose.

raft foundation (*Build*) A layer of concrete, usually reinforced, extending under the whole area of a building and projecting outside the line of its walls, to provide a foundation where the ground is too soft for the loading.

rag-bolt (*Eng*) A foundation bolt for eg concrete with a long tapered head of increasing size towards its end, and having jagged points on its surface to prevent withdrawal.

rag content papers (*Paper*) Papers the fibrous FURNISH of which contains 25% or more by weight of rag (cotton etc) fibres.

ragging (*MinExt*) Rough concentration or washing, for a low ratio of concentration. In mineral processing, grooves cut on surface of roll to improve grip on feed. In jigging, the bed of heavy mineral or metal shot maintained on the jig screen.

ragging frame (*MinExt*) Tilting table, which may be worked automatically, on which finely ground ore is treated by sluicing. Also *rack, reck*.

raggle (*Build*) A narrow groove cut into a masonry or brickwork to receive the edge of FLASHINGS which are to be fixed to it. Also *raglet*.

rag stone (*Build*) A general term for coarse-grained sandstone, often with a calcareous cement, eg Kentish rag.

rag-work (*Build*) A term applied to wall construction in which undressed flat stones of about the thickness of a brick are built up into a wall the outer faces of which are left rough.

RAID architecture (*ICT*) Abbrev for *redundant array of inexpensive disks*. A method of using a number of HARD DISKS to store data in such a way that should a disk fail, the data can be recovered from the remainder. There are five levels of sophistication: level 1 is simple duplication of data, whereas in level 5 systems any one disk may be removed and the data recovered while the system continues to operate normally. This technology is often used in file servers and most computer systems where a high degree of reliability or FAULT TOLERANCE is required.

rail (*Build*) (1) A horizontal member in framing or panelling. (2) The upper member in a balustrade.

rail (*CivEng*) A steel bar, usually of special section, laid across sleepers to provide a track for the passage of rolling stock with flanged wheels. Now always flat-bottomed and specified by weight per unit length. See CONTINUOUS WELDED RAIL.

rail bender (*Eng*) A short stiff steel girder with claws at the ends and a central boss carrying a heavy screw. Also *jim crow*.

rail bond (*ElecEng*) An electrical connection between two adjacent lengths of track or conductor rail on a railway.

rail chair (*CivEng*) See CHAIR.

rail gauge (*Eng*) See BROAD GAUGE, GAUGE, NARROW GAUGE, STANDARD GAUGE.

rail guard (*CivEng*) See CHECK RAIL.

rail post (*Build*) A newel post.

railroad disease (*Vet*) See TRANSIT TETANY.

railway curve (*Eng*) A drawing instrument similar to a French curve but cut at the edge to an arc of large radius. Used for drawing arcs when these are too large for beam compasses. Sets are supplied to cover a wide range of radii.

rain (*EnvSci*) The result of condensation of excess water vapour when moist air is cooled below its dew point. Rain falls when droplets increase in size until they form drops whose weight is equivalent to the frictional air resistance. The greater proportion of raindrops have a diameter of 0·2 cm or less; in torrential rain a small proportion may reach 0·4 cm. Rain effects important geological work by assisting in the mechanical disintegration of rocks; also chemically, in bringing about solution of carbonates etc and, through the agency of running water, in redistributing the products of erosion and disintegration.

rain band (*EnvSci*) An absorption band in the solar spectrum on the red side of the D lines, produced by water vapour in the Earth's atmosphere.

rainbow (*EnvSci, Phys*) A rainbow is formed by sunlight which is refracted and internally reflected by raindrops, the concentration of light in the bow corresponding to the position of minimum deviation of the light. The angular radius of the primary bow is 42°, this being equal to 360° minus the angle of minimum deviation for a spherical drop. The colours, ranging from red outside to violet inside, are due to dispersion in the water. See SECONDARY BOW.

rainbow quartz (*Min*) See IRIS.

rain chamber (*MinExt*) Washing tower or other space where rising dust and fumes are brought into contact with descending sprays of water.

rainforest (*EnvSci*) The natural forest of the humid tropics. Developing where rainfall is heavy (> 2500 mm yr^{-1}) and characterized by a great richness of species, very tall trees (> 30 m), lianes and epiphytes.

rain gauge (*EnvSci*) An instrument for measuring the amount of rainfall over a given period, usually 24 h. The usual form consists of a sharp-trimmed funnel, 5 in (12·5 cm) in diameter, leading into a narrow-necked graduated collecting vessel. The *Dines tilting siphon* is of the continuously recording type, noting both the time and the amount of rainfall.

rain gun (*Agri*) A single-nozzle sprinkler that spreads crop irrigation water in a repeating, controlled arc.

rain-making (*EnvSci*) Artificial stimulation of precipitation by scattering solid carbon dioxide on supercooled clouds, or by silver iodide nucleation.

rain prints (*Geol*) More or less circular, vertical or slanting pits occurring on the bedding planes of certain strata; believed to be the impressions of heavy raindrops falling on silt or clay, hard enough to retain the impression before being covered by a further layer of sediment.

rain shadow (*EnvSci*) A dry area, often a desert, on the sides of mountains away from the sea, due to the deposition of most of the moisture from the winds blowing off the ocean on the slopes facing the ocean. The higher the mountains, the greater the effect.

rain stage (*EnvSci*) That part of the condensation process taking place at temperatures above 0°C so that water vapour condenses to water liquid.

rain-wash (*Geol*) The creep of soil and superficial rocks under the influence of gravity and the lubricating action of rain.

rainwater pipe (*Build*) See DOWNPIPE.

raised band (*Print*) A band which shows on the back of a book when bound. This indicates that the book has been sewn FLEXIBLE.

raised beach (*Geol*) Beach deposits which are found above the present high-water mark; due to the relative uplift of the land or to a falling sea level. See EUSTATIC MOVEMENTS.

raised bog (*EnvSci*) A type of *Sphagnum* bog, originating from a valley bog or a fen by the upward growth of the vegetation and the failure of the dead plant material to decompose. The consequent raised bog is convex.

raised panel (*Build*) A panel whose surface stands PROUD of the general surface of the framing members.

raiser (*Build*) See RISER.

raising (*Textiles*) Textile process in which wire-covered rollers or teazles on cylinders revolve over the cloth surface to produce a nap.

raising-plate (*Build*) A horizontal timber resting on part of a structure and supporting a superstructure. Also *reason-piece*. See POLE PLATE, WALL PLATE (2).

rake (*Arch*) In a theatre, the upward slope, from the horizontal, of both the stage and the auditorium.

rake (*Build*) A long-handled tool with projecting teeth at one end, used for mixing plaster.

rake (*Eng*) Angular relief, eg SIDE RAKE, TOP RAKE, given to the faces of cutting tools to obtain the most efficient cutting angle. The face of a cutting tool with negative rake has no angular relief but, on the contrary, meets the workpiece with the cutting edge trailing, the tool being stronger in shear, impact and abrasion than one having positive rake.

rake (*MinExt*) (1) A forked tool for loading coal underground. (2) An irregular vein of ironstone. (3) Train or *journey* of mineral trucks. (4) Another name for PITCH.

rake (*Ships*) Refers to a part of a ship not perpendicular to the datum line.

rake classifier (*MinExt*) Inclined tank into which ore pulp is fed continuously, the slow settling portion overflowing and the coarser material gravitating down, to be gathered and raked up to a top discharge.

raker (*Build*) See RAKING SHORE.

raker set (*Build*) A pattern of saw teeth in which one straight tooth alternates with two teeth set in opposite directions.

raking bond (*Build*) A form of bond sometimes used for very thick walls, or for strengthening the bond in footings carrying heavy loads. The courses are built diagonally across the wall, successive courses crossing one another in respect of rake; triangular BATS are added to enable square facework to be completed. Also *diagonal bond*. See HERRINGBONE BOND. Fig. ▷

raking bond

raking cornice (*Arch*) A cornice decorating the slant sides of a pediment.

raking flashing (*Build*) That used at the sides of, eg a wall or chimney projecting from a sloping roof.

raking-out (*Build*) The operation of preparing mortar joints in brickwork for pointing.

raking pile (*CivEng*) A pile which is driven in at an angle to the vertical. Also *batter pile*.

raking shore (*Build*) An inclined baulk of timber, one end of which rests upon a sole plate (or sleeper) on the ground while the other presses against the wall to which temporary support is to be given.

râle (*Med*) A bubbling or crackling sound produced in a diseased lung by the passage of air over or through secretions in it. Also *crackles*, *crepitations*.

raloxifene (*Pharmacol*) A drug that acts selectively on estrogen receptors in bones, heart and arteries but is not estrogenic in uterus and breast. Used to protect against osteoporosis.

RALS (*Aero*) Abbrev for REMOTE AUGMENTED LIFT SYSTEM.

RAM (*Aero*) Abbrev for RADAR ABSORBING MATERIAL.

RAM (*ICT*) Abbrev for RANDOM ACCESS MEMORY.

ram (*Agri*) An entire male sheep that has reached sexual maturity. Also *tup*.

ram (*Chem*) Abbrev for RELATIVE ATOMIC MASS. Also *RAM*.

ram (*CivEng*) (1) The head of a pile-driver. (2) To consolidate the surface of loose material by punning.

ram (*Eng*) (1) The reciprocating head of a press. (2) The reciprocating slide of a shaping machine on which the cutting tool is mounted. (3) The falling weight of a pile-driver.

ram-air turbine (*Aero*) A small turbine motivated by ram (ie free stream) air; used (1) to drive fuel pumps, hydraulic pumps or electrical generators in guided weapons because of the absence of shaft drives in rockets and ramjets, (2) as an emergency power source for driving hydraulic pumps or electrical generators for high-speed aircraft, particularly those with power controls. Abbrev *RAT*.

Raman optical activity (*Chem*) The differential scattering of circularly polarized light by optically active (chiral) substances as observed by Raman spectroscopy. Abbrev *ROA*.

Raman scattering (*Phys*) The scattering of light by molecules in which there is a change of frequency due to the molecules gaining or losing energy as a result of transitions between vibrational or rotational energy levels. The scattering of light by the optic modes of vibration in a crystal, ie photon–phonon scattering. See SCATTERING.

Raman spectroscopy (*Chem*) A method making use of RAMAN SCATTERING for chemical analysis. Like INFRARED SPECTROSCOPY, it investigates molecular vibrations and rotations.

RAM cache (*ICT*) See CACHE MEMORY.

RAM disk (*ICT*) See RAM DRIVE.

RAM drive (*ICT*) A portion of main memory that is used as if it were a hard-disk drive, so increasing the speed of operation of software that makes extensive use of the hard disk. Also *RAM disk*, *silicon disk*, *virtual drive*.

ramentum (*BioSci*) Thin, chaffy, brownish scale, esp on the stem, petiole or leaf of a fern.

ramet (*BioSci*) Any physically and physiologically independent individual plant, whether grown from a sexually

produced seed or derived by vegetative reproduction. Cf
GENET. See CLONE.

rami communicantes (*Med*) The preganglionic myelinated
nerve fibres of sympathetic nervous system connecting
spinal nerves and sympathetic chain of ganglia.

ramie (*Textiles*) A BAST fibre from the stems of *Boehmeria
nivea*; can be bleached to give rather brittle white fibres
that may be used in garments blended with other fibres.
Formerly used for the manufacture of gas mantles.

ramiform (*BioSci*) Branching.

ramin (*For*) A pale-coloured hardwood from trees native to
swamp rainforests of the island of Borneo.

ram intake (*Aero*) A forward-facing engine (or accessory)
air intake which taps the kinetic energy in the airflow and
converts it into pressure energy by diffusion; in supersonic
flight very high pressure ratios can be obtained.

ramipril (*Pharmacol*) An *ACE inhibitor* used to treat
hypertension.

ramisection (*Med*) The operation of cutting the sympathetic nerves between the spinal cord and the sympathetic
ganglia. Also *ramisectomy*.

ramjet (*Aero*) The simplest PROPULSIVE DUCT deriving its
thrust by the addition and combustion of fuel with air
compressed solely as a result of forward speed. In subsonic
flight kinetic energy is converted into pressure by
DIFFUSER, or widening duct, which also slows it sufficiently
to permit combustion to be maintained; at about Mach 1
the shock wave generated by the air-intake lip improves the
compression when it decelerates the air to subsonic velocity
prior to diffusion. At high MACH NUMBERS, 1·5 and
upward, two shock waves are required for the dual purpose
of raising the pressure and slowing the air for combustion
to pressure ratios of 6:1 are attainable at Mach 2, 36:1 at
Mach 3. In supersonic flight the jet efflux of a ramjet has to
be accelerated to high velocity by a VENTURI or CON-
VERGENT–DIVERGENT NOZZLE.

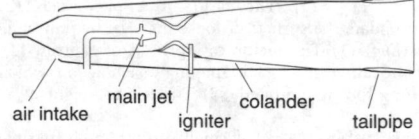

air intake   main jet   colander   tailpipe
     igniter

ramjet Cross-section.

rammer (*Build*, *CivEng*) A punner. See PUNNING.

rammer (*Eng*) Hand tool of various forms for packing the
sand of a mould evenly round the pattern. Ramming
consolidates the moulding sand before the pattern is
removed from the moulding box.

Rammstedt's operation (*Med*) Incision of the pylorus
down to the mucous membrane, done in the treatment of
congenital hypertrophy of the pylorus.

ramp (*Aero*) (1) Parking area for an aircraft at an airport. (2)
Inner wall of supersonic intake creating shock waves and
improving pressure recovery. Frequently movable as on
Concorde. (3) Inclined launcher for missile or RPV.

ramp (*Build*) A sudden rise in a hand rail when it is concave
upwards. Cf KNEE.

ramp (*CivEng*) An inclined surface, often in place of steps.

rampant arch (*CivEng*) An arch whose abutments are not in
the same horizontal line.

rampant centre (*Build*) A centre for a RAMPANT ARCH.

ramp voltage (*ElecEng*) Steadily rising voltage, as in a
sawtooth waveform.

Ramsar site (*EnvSci*) An area of wetland considered to be of
international importance and worthy of conservation
according to the criteria established at a convention in
the Iranian city of Ramsar in 1971.

Ramsauer effect (*Phys*) Sharp decrease to zero of scattering
cross-section of atoms of inert gases, for electrons of energy
below a certain critical value.

Ramsay and Young's rule (*Chem*) The ratio of the boiling
points of two liquids of similar chemical character is
approximately constant, independent of the pressure at
which they are measured.

Ramsden circle (*Phys*) The EXIT PUPIL of a telescope, found
as a ring of light at the eyepiece of a telescope focused on a
diffuse source at infinity. The magnification is the
OBJECTIVE diameter divided by that of this circle.

Ramsden eyepiece (*Phys*, *Surv*) An eyepiece often used in
an optical instrument in which crosswire measurements are
to be made. It consists of two similar plano-convex lenses
separated by a distance equal to two-thirds the focal length
of each, and having their convex faces towards each other.
The focal plane is just outside the system.

ramus (*BioSci*) (1) In general, any branch-like structure, a
ramification. (2) The barb of a feather. (3) In vertebrates,
one lateral half of the lower jaw, the mandible. (4) In
Rotifera, part of the trophi.

ramwing (*Aero*) Special kind of aircraft designed to fly very
low over water thereby gaining advantage in lift–drag ratio
by capturing the ground effect. May one day rival the low-
cost ocean freight ship.

ranalian complex (*BioSci*) A group of families, including
the Ranunculaceae, containing what are thought to be the
most primitive extant flowering plants. See MAGNOLIIDAE.

rance (*Build*) A shore.

rancidity (*FoodSci*) FOOD SPOILAGE caused by either
hydrolysis or oxidation of fats during storage, particularly
at higher temperatures, characterized by the off-flavours
and odours which develop. Hydrolytic rancidity needs
moisture and an enzyme, yeast or mould which will
hydrolyse the triglycerides and liberate free fatty acid.
Oxidative rancidity can usually be inhibited by natural or
added anti-oxidants in the fat or oil. Rancidity is also
inhibited below 5°C.

R & M (*Aero*) Abbrev for *reliability and maintainability*, USA.

random (*Build*) Said of rubble masonry in which the stones
are of irregular shape and the work is not coursed.

random access (*ICT*) A process in which records are
directly accessed in any order by their known address. Also
direct access.

random access memory (*ICT*, *Electronics*) A computer
memory that can be read from and written to by the
programmer. It is usually made on a CHIP. Each memory
location can be identified by an address. Unlike data on
magnetic tape or disks, data in random access memory are
volatile. The time required for reading out data is
independent of their location in the memory. Abbrev
RAM.

random assignment (*Psych*, *MathSci*) The assignment of
subjects to experimental groups on the basis of chance.

random coil (*BioSci*) A section of a polypeptide chain that is
not folded into any specific secondary structure.

random coil (*Chem*) Conformation of single polymer chain
in non-crystalline state and at equilibrium, a sphere of
diameter determined by its molecular mass. Present in
rubbers at rest, entangled with adjacent chains, but as
isolated coils in dilute polymer solutions. See panel on
ELASTOMERS.

random coincidence (*NucEng*) Simultaneous operation of
two or more coincidence counters as a result of their
discharge by separate incident particles arriving together
(instead of common discharge by a single particle as is
normally assumed in interpreting the readings).

random copolymer (*Chem*) A polymer where the repeat
units are dispersed irregularly along the linear chains. See
panel on POLYMERS.

random distribution (*BioSci*) A spatial distribution pattern
of organisms which is described empirically by the Poisson
probability curve, and where variance equals the mean. See
NORMAL DISTRIBUTION. Cf OVERDISPERSION.

random mating (*BioSci*) The phenomenon whereby any
individual in a population has an equal chance of mating
with any other of the opposite sex.

random noise (*Acous*) Noise due to the aggregate of a large number of elementary disturbances with random occurrence in time. Also *stochastic noise*.

random number (*ICT*) One from a sequence without any detectable bias or pattern. In a computer, random numbers are generally produced by an algorithm which is a pseudorandom number generator.

random number sequence (*MathSci*) A sequence of numbers that are not generated by any algorithm and exhibit no regular pattern, eg the numbers drawn in a lottery. No completely satisfactory definition has yet been proposed.

random searching (*BioSci*) A process of completely unorganized 'search' by which some ecologists suggest that some animal populations find food, mates and suitable places to live.

random sequence welding (*Eng*) Adding welding beads at random along a seam to minimize distortion.

random-tooled ashlar (*Build*) A block of stone finished with groovings irregularly cut.

random variable (*MathSci*) The mathematical representation of a variate associated with a stochastic phenomenon.

random walk (*BioSci*) A description of the path followed by a cell or particle when there is no bias in movement and the direction of movement at any instant is not influenced by the direction of travel in the preceding period. Although the behaviour of moving cells in a uniform environment can be described as a random walk in the long term, this is not true in the short term because of persistence.

random winding (*ElecEng*) See MUSH WINDING.

Raney nickel (*Chem*) A nickel catalyst, used for hydrogenation, produced by the action of alkali on a nickel–aluminium alloy.

range (*Aero*) The maximum horizontal distance covered by a projectile.

range (*ICT*) The maximum distance of radio transmitter at which effective reception is possible (not normally constant).

range (*MathSci*) (1) See CODOMAIN. (2) The difference between the largest and smallest values in a set of observations.

range (*Phys*) (1) Length of track along which ionization is produced by a nuclear particle. (2) Distance of effective operation of a force.

range (*Surv*) To fix points, either by eye or with the aid of an instrument, to be in the same straight line.

rangefinder (*ImageTech*) A device for measuring the distance of a remote object, esp to assist setting focus of a camera lens. It may be directly coupled to the lens mount for manual or automatic operation.

range-height indicator (*Radar*) A display used in conjunction with a PLAN-POSITION INDICATOR for airport control. It displays a vertical plane on which the elevation and bearing of the target may be seen.

range of stress (*Eng*) The range between the upper and lower limits of a cycle of stress in a fatigue test. The midpoint is the *mean stress*.

range tracking (*Radar*) The process of continuously monitoring the delay between the transmission of a pulse and reception of an echo. Tracking requires that the time elapsed between pulse and echo be measured, that the echo be identified as the target rather than random noise, and that the range–time history of the target be maintained.

ranging figures (*Print*) See LINING FIGURES.

ranging rod (*Surv*) A wooden pole used to mark the stations conspicuously, or to assist *ranging*.

ranine (*BioSci*) Pertaining to, or situated on, the under surface of the tongue.

ranitidine (*Pharmacol*) See HISTAMINE H2-RECEPTOR ANTAGONISTS.

rank (*MathSci*) (1) *Mathematics*. Of a matrix: a matrix is of rank r if it contains at least one determinant of order r ($\neq 0$) and all higher-order determinants are zero. (2) *Statistics*. The number in serial order corresponding to a given data value when all values are placed in ascending order of magnitude; to place in ascending order of magnitude.

Rankine cycle (*Eng*) A composite steam plant cycle used as a standard of efficiency, comprising introduction of water by a pump to boiler pressure, evaporation, adiabatic expansion to condenser pressure, and condensation to initial point.

Rankine efficiency (*Eng*) The efficiency of an ideal engine working on the RANKINE CYCLE under given conditions of steam pressure and temperature.

Rankine scale (*Phys*) Absolute scale of temperature, based on degrees Fahrenheit. See FAHRENHEIT SCALE.

Rankine's formula (*CivEng*) An empirical formula giving the collapsing load for a given column. It states that

$$P = \frac{f_c A}{1 + a(l/k)^2}$$

where P = the collapsing load, f_c = safe compressive stress for very short lengths of the material, A = area of cross-section, l = the length of the pin-jointed column, k = the least radius of gyration of the section, a is a constant for the material equal to $f_c/\pi^2 E$ where E is Young's modulus for the material.

rankinite (*Min*) A monoclinic calcium disilicate, $Ca_3Si_2O_7$, found in highly metamorphosed siliceous limestones.

rank of coal (*Geol*) A classification related to the percentage of carbon in dry mineral-free coal. The original vegetation has been modified by heat, pressure and chemical change after burial. Rank increases from *peat*, through *lignite* to *bituminous coal* and finally *anthracite*.

rank of selectors (*ICT*) The whole set of selectors concerned with a specified stage in setting up a call through an exchange.

rank-ordered array (*MathSci*) A table consisting of data in order of highest to lowest or lowest to highest where each value is given a numbered rank depicting its distance from the highest or lowest score.

rank test (*MathSci*) A statistical procedure carried out on the ranks rather than the values of the observations.

ranula (*Med*) A cystic tumour formed on the lower surface of the tongue or on the bottom of the mouth, caused by blocking or dilatation of mucous gland.

Ranunculaceae (*BioSci*) The buttercup family, containing c.1800 spp of dicotyledonous flowering plants (superorder Magnoliidae), most of which are herbs in northern temperate areas. The floral parts are free, the stamens numerous and hypogynous. Many are poisonous. They are of little economic importance other than as ornamentals. *Anemone*, *delphinium*, *helleborus*, etc.

Ranvier's nodes (*BioSci*) See NODES OF RANVIER.

Raoult's law (*Chem*) A law stating that the vapour pressure of an ideal solution at any temperature is the sum of the vapour pressure of each component, P_i, multiplied by the mole fraction of that component, x_i: $P = \sum P_i x_i$.

Rapakivi granite (*Geol*) A type described from a locality in Finland, characterized by the occurrence of rounded pink crystals of orthoclase surrounded by a mantle of whitish sodic plagioclase. A widely used textural term. Often used as an ornamental stone for facing buildings.

raphe (*BioSci*) (1) A ridge on an ovule or seed representing that part of the stalk that is fused to the ovule. (2) A broad junction, as between the halves of the vertebrate brain. Also *rhaphe*.

raphide (*BioSci*) A needle-shaped CRYSTAL, usually calcium oxalate, usually one of a bundle in the vacuole of a cell, esp in a leaf.

rapid eye movement (*Med*) Occurs at certain stages during dreaming sleep which it defines. It is believed to be necessary for brain repair. Abbrev *REM*.

rapping (*Eng*) The process of loosening a pattern in a mould to facilitate its withdrawal. A spike or lifting screw is inserted in the pattern and tapped smartly in every direction.

RAPRA (*Genrl*) Abbrev for RUBBER AND PLASTICS RESEARCH ASSOCIATION.

raptatory (*BioSci*) Adapted for snatching or robbing, as birds of prey. Also *raptorial*.

raptor (*BioSci*) General term for diurnal birds of prey. See ACCIPITRIFORMES, FALCONIFORMES.

rare earth elements (*Chem*) A group of metallic elements possessing closely similar chemical properties. They are mainly trivalent, but otherwise similar to the alkaline earth elements. The group consists of the lanthanide elements 57–71, plus scandium (21) and yttrium (39). Extracted from MONAZITE, and separated by repeated fractional crystallization, liquid extraction or ion exchange.

rare earths (*Chem*) The oxides (M_2O_3) of the rare earth elements.

rarefaction (*Med*) Abnormal decrease in the density of bone as a result of absorption from it of calcium salts, as in infection of bone.

rarefaction (*Phys*) The decrease in density along a low-pressure wavefront of a sound wave.

rare gases (*Chem*) See NOBLE GASES.

Raschel (*Textiles*) A one- or two-bar warp-knitting machine, fitted with latch needles.

Raschig rings (*ChemEng*) PACKED COLUMN fillers which are hollow open cylinders of diameter equal to length.

rash (*Med*) Any skin eruption.

Rasmussen report (*NucEng*) The influential US report on reactor safety issued in the mid-1970s. Also *Reactor Safety Study*. See panel on RISK ASSESSMENT.

rasorial (*BioSci*) Adapted for scratching.

rasp (*Build*) Coarse type of file for wood with teeth in the form of raised points.

Raspall test (*Paper*) A test to detect the presence of rosin in paper. A drop of concentrated solution of cane-sugar is placed on the paper, blotted and a drop of concentrated sulphuric acid applied to the same spot. A red coloration confirms rosin is present.

RAST (*BioSci*) Abbrev for RADIO-ALLERGOSORBENT TEST.

raster (*ICT, ImageTech*) In the display on a TV or RASTER GRAPHICS screen, the grid pattern of vertical and horizontal divisions outlining all the small elements of which the picture is composed. See RESOLUTION.

raster burn (*ImageTech*) Deterioration of the scanned area of the screen of a TV picture or camera tube as a result of use.

raster graphics (*ICT*) COMPUTER GRAPHICS based on TV technology in which the screen display is produced by an electron beam scanning one RASTER line at a time to cover the screen from top to bottom 30 times per second. The image is produced by modifying the intensity of the electron beam to each PIXEL from a map of the pixels in the computer memory. See BIT-MAPPED DISPLAY.

raster image processing (*Print*) A device used in printers to interpret the instructions from front-end systems, resulting in the final made-up pages.

RAT (*Aero*) Abbrev for RAM-AIR TURBINE.

ratchet brace (*Eng*) A drilling brace in which the drill spindle is rotated intermittently by a ratchet wheel engaged by a pawl on a hand lever; used in confined spaces, repair work. Also *ratchet tool*.

ratchet mechanism (*Eng*) A mechanism comprising a RATCHET WHEEL and a PAWL with which it engages. It is used to convert reciprocating motion into intermittent rotary or linear motion in one direction only.

ratchet screwdriver (*Build*) One with a ratchet mechanism to make it operate in one direction only.

ratchetting (*NucEng*) Intermittent movement of fuel elements arising from thermal cycling and differential expansion effects.

ratchet wheel (*Eng*) A wheel with inclined teeth used in a RATCHET MECHANISM.

rate constant (*Chem*) The speed of a chemical reaction, in moles of change per cubic metre per second, when the active masses of all the reactants are unity. If Rate = $k[A]^x[B]^y$, for

$A + B \rightarrow$ products, and x, y are partial orders, then k is the rate constant. See ORDER OF REACTION.

rated altitude (*Aero*) The height measured in the INTERNATIONAL STANDARD ATMOSPHERE, at which a piston aero-engine delivers its maximum power. Cf POWER RATING.

rated blowing-current (*ElecEng*) The current at which a fuse-link is specified by the maker to melt and break the circuit.

rated breaking-capacity (*ElecEng*) The rms current, or the kVA at the rated voltage, which a circuit breaker is specified by the maker to interrupt without damage.

rated capacity (*Eng*) General term for the output of equipment which can continue indefinitely in conformity with a criterion, eg heating, distortion of signals or of waveform. See CONTINUOUS RATING, INTERMITTENT RATING, PERIODIC RATING.

rate-determining step (*ChemEng*) Where a process consists of a series of consecutive steps, the overall rate of the process is largely determined by the step with the slowest rate, so that efforts to speed up the process must chiefly be directed to this step.

rated impedance (*ElecEng*) Particularly applied to a loudspeaker, in which impedance rises with frequency, with an added sharp rise at frequency of base resonance. That resistance, equal in magnitude to the modulus of the minimum impedance above this resonant frequency, which replaces the loudspeaker when measuring the power applied to the loudspeaker during testing.

rated making-capacity (*ElecEng*) The maximum asymmetrical current which a circuit breaker can make at the rated voltage.

rate fixing (*Eng*) The determination of the time allocation for carrying out a specific task of work, usually as a basis for remuneration.

rate gyro (*Aero, Space*) A single degree-of-freedom gyro which measures an angular rate by precession of the gyro against a spring restraint, ie a gyroscope with a single gimbal which produces a couple proportional to the rate of rotation.

ratemeter (*NucEng*) See COUNT RATEMETER.

rate of climb (*Aero*) Generally rate of ascent from Earth. In performance testing, the vertical component of the air path of an ascending aircraft, corrected for standard atmosphere. See INTERNATIONAL STANDARD ATMOSPHERE.

rate of climb indicator (*Aero*) See VERTICAL SPEED INDICATOR.

rate of crystallization (*Chem*) Different materials crystallize at different rates when cooled from melt, most markedly in polymers. High-density polyethylene shows very rapid rate but inflexible chains such as polyethylene terephthalate crystallize very slowly. See CRYSTALLIZATION OF POLYMERS.

Rathke's pouch (*BioSci*) In developing vertebrates, the diverticulum formed from the dorsal aspect of the buccal cavity ectoderm which gives rise to the ADENOHYPOPHYSIS. Also *craniobuccal pouch*.

rating (*ElecEng*) Specified limit to operating conditions, eg current rating etc.

ratio (*MathSci*) The ratio $a:b$ is equivalent to the quotient a/b.

ratio arms (*ElecEng*) Two adjacent arms of a Wheatstone bridge, the resistances in which can be made to have one of several fixed ratios.

ratio detector (*ICT*) Detector circuit used for frequency-modulated carriers.

ratio error (*ElecEng*) A departure of the ratio between the primary and secondary voltages or currents of a voltage or current transformer from the rated value.

ratio-imaging fluorescence microscope (*BioSci*) A technique for measuring intracellular pH or calcium levels using a fluorescent probe, by which two different excitation wavelengths are used and the emitted light levels at each wavelength compared.

rational emotive therapy (*Psych*) A directive form of psychotherapy based on Albert Ellis's theory that cognitions control our emotions and behaviours; therefore, changing the way we think about things will affect the way we feel and behave.

rational horizon (*Surv*) See TRUE HORIZON.

rationalization (*Psych*) In psychoanalytic theory, a mechanism of defence by means of which unacceptable thoughts or actions are given acceptable reasons which justify it, and also hide its true motivation.

rationalized units (*Phys*) Systems of electrical units for which the factor 4π is introduced in Coulomb's laws so that it is absent from more widely used relationships. In HEAVISIDE–LORENTZ UNITS (rationalized Gaussian units) this is done directly, thus modifying values for the unit charge and unit pole. In MKSA units it is done indirectly by modifying the values of the permittivity and permeability of free space.

rational number (*MathSci*) A number which can be expressed as the ratio of two integers, eg 3/4.

ratio of compression (*Eng*) See COMPRESSION RATIO.

ratio of slenderness (*Build*) The ratio between the length or height of a pillar and its smallest radius. See EFFECTIVE COLUMN LENGTH.

ratio of specific heat capacities (*Phys*) The ratio of specific heat capacity of a gas at constant pressure to that at constant volume. It has a constant value of about 1·67 for monatomic gases, 1·4 for diatomic gases and approaches unity for polyatomic gases. This ratio, denoted by γ, enters into the ADIABATIC EQUATION.

ratio scale (*MathSci*) Any scale of measurement possessing equal-magnitude intervals and a fixed zero. Compare INTERVAL SCALE, ORDINAL SCALE.

ratio schedule of reinforcement (*Psych*) A program of reinforcement in which a certain number of responses are necessary in order to produce the reward. On FIXED RATIO SCHEDULES the number of responses is always the same; on VARIABLE RATIO SCHEDULES the number of responses varies from trial to trial.

ratite (*BioSci*) A running bird such as an ostrich.

RATOG (*Aero*) Abbrev for *rocket-assisted take-off gear*. See TAKE-OFF ROCKET.

rat-race (*ICT*) See HYBRID JUNCTION.

rat-tail file (*Build*) Small round file, used for enlarging holes etc.

rattle (*BioSci*) The series of horny rings representing the modified tail-tip scale in rattlesnakes (Colubridae).

rattle (*Paper*) The crackling noise when paper is handled. It indicates the degree to which the fibre has been hydrated in the process of beating, and can be augmented by the addition of starch and other additives. Cf SCROOP.

rat-trap bond (*Build*) A form of bond in which a 9 in wall is built up of bricks on edge, so arranged as to enclose a 9×3 in cavity.

Rauber's cells (*BioSci*) In mammals, cells of the trophoblast situated immediately over the embryonic plate.

Raunkiaer system (*BioSci*) A classification of the vegetative or life forms of plants according to the positions of the perennating (resting) buds and the protection they receive during an unfavourable season of cold or drought.

rauwolfia serpentina (*Pharmacol*) One of a number of compounds obtained from the dried roots of trees of the *Rauwolfia* genus, including *R.canescens* and *R.vomitoria*. Formerly used for the treatment of hypertension.

raw data (*Genrl*) The initial data gathered that have not yet been graphed, organized or analysed.

rawhide hammer (*Eng*) A hammer the head of which consists of a close roll of hide projecting from a short steel tube; used by fitters to avoid damaging a finished surface.

raw material (*Genrl*) Starting point for manufacture of useful materials. Raw materials for polymers include oil, natural gas and liquid petroleum gas; for cement, coal, limestone and clay; for steel, iron ore and coking coal, oil or natural gas; for other metals, metal ore and reducing agent; for glass, silica sand and other metal oxides; for ceramics, metal oxides; for semiconductors, silicon plus dopants (eg arsenic or gallium). See INTERMEDIATE.

raw stock (*ImageTech*) General term for motion picture film before exposure and processing.

ray (*BioSci*) (1) In plants, a panel of tissue, usually mostly parenchyma, one to several cells wide and a few to many cells high, produced by ray initials in the cambium and extending radially into the secondary xylem (*xylem ray*) and secondary phloem (*phloem ray*), and with the functions of radial transport and storage. See MEDULLARY RAY, RAY TRACHEID. (2) In animals, a skeletal element supporting a fin or a sector of a radially symmetrical animal.

ray (*Geol*) A linear landform on the surface of the Moon radiating outwards from a crater. Probably caused by ejecta from volcanic activity or the impact of a meteorite.

ray (*MathSci*) A straight line, esp one of a PENCIL of such lines, through a given point (the *radiant point*).

ray (*Phys*) General term for the geometrical path of the radiation of wave energy, always in a direction normal to the wavefront, but with possible reflection, refraction, diffraction, divergence, convergence and diffusion. By extension, also the geometrical path of a beam of particles in an evacuated chamber. This may be curved in electric or magnetic field. See PARTICLE.

ray floret (*BioSci*) Usually one of the outer ring florets of the capitulum of the Compositae, regardless of its morphology. Cf DISK FLORET.

ray initial (*BioSci*) A more or less isodiametric cell, one of a group of such in the vascular CAMBIUM, each giving rise to one of the radial files of cells making up a ray. Cf FUSIFORM INITIAL.

Rayleigh criterion (*Phys*) Criterion for the resolution of interference fringes, spectral lines and images. The limit of resolution occurs when the maximum of intensity of one fringe or line falls over the first minimum of an adjacent fringe or line. For a telescope with a circular aperture of diameter D this criterion gives the smallest angular separation of the two images of point objects as $1·22\lambda/D$, where λ is the wavelength of the light.

Rayleigh disk (*Acous*) A small, light, circular disk (in water a lead disk is used) pivoted about a vertical diameter, hung by a fine thread of glass or quartz. If placed at an angle to a progressive sound wave, the disk experiences a torque which depends on the square of the velocity of the volume element in the medium. It provides a useful method of measurement of the absolute value of the velocity, calculated from the measured torque using a formula due to König. Historically used for calibrating microphones.

Rayleigh distillation (*Chem*) A simple distillation in which the composition of the residue changes continuously during the course of the distillation.

Rayleigh fading (*ICT*) Rapid variation of radio signal strength in a multipath mobile environment due to the presence of many randomly phased sinusoidal components arriving at the receiver antenna from all directions in the horizontal plane. Under these conditions, the resultant signal has the statistical characteristics of the Rayleigh distribution.

Rayleigh–Jeans law (*Phys*) An expression for the distribution of energy in the spectrum of a black-body radiator:

$$E_v \, dv = \frac{8\pi v^2 \, kT}{c^3} \, dv$$

where E_v is the energy density radiated at a temperature T within a narrow range of frequencies from v to $v + dv$, k is BOLTZMANN'S CONSTANT and c the speed of light. The formula holds only for low frequencies. See PLANCK'S RADIATION LAW.

Rayleigh limit (*Phys*) One-quarter of a wavelength, the maximum difference in optical paths between rays from an object point to the corresponding image point for perfect definition in a lens system.

Rayleigh refractometer (*Phys*) An instrument for measuring the refractive index of a gas by an optical interference method. Each of two interfering light beams passes through a tube which may contain air or gas or be evacuated. By observing the shift in the interference fringes when one of the tubes is evacuated and the other contains gas, the refractive index of the gas may be calculated using the expression

$$n = 1 + \frac{s\lambda}{l}$$

where *l* is the length of the tube, λ the wavelength used, and *s* the number of fringes shifted.

Rayleigh scattering (*Phys*) See SCATTERING.

Rayleigh's criterion (*Acous*) Criterion to predict the occurrence of thermo-acoustic feedback.

Rayleigh wave (*Acous*) Non-dispersive surface wave on a solid body with a free surface. Important for earthquakes.

Raynaud's disease (*Med*) Constriction of blood supply to digits producing a white finger or toe, most often in response to cold. *Reynaud's phenomenon* applies when the condition indicates an underlying disease; more common is the *disease* where there is no identifiable cause.

rayon (*Textiles*) Manufactured fibre produced by viscose process, a regenerated cellulose. Viscose solution is spun into an acid bath where the fibre coagulates to produce a characteristic three-lobed cross-section fibre. Finds application in tyre cord, textiles, etc.

ray theory (*Acous*) A model in acoustics based on the assumption that sound propagates in the form of rays perpendicular to the wavefronts.

ray tracheid (*BioSci*) Tracheids shorter than the ordinary axial tracheids. They are found at the top and bottom margins of the ray with their long axes in the radial direction. They occur in the wood of some conifers, eg pines.

Rb (*Chem*) Symbol for RUBIDIUM.

R-banding (*BioSci*) See BANDING TECHNIQUES and panel on CHROMOSOME.

RBMK reactor (*NucEng*) Graphite-moderated, boiling-water-cooled, pressure tube reactor unique to the former USSR (the type involved in the Chernobyl accident).

RC (*Build*) Abbrev for REINFORCED CONCRETE, *rough cutting*.

RC coupling (*Electronics*) Abbrev for RESISTANCE–CAPACITANCE COUPLING.

RCS (*Space*) Abbrev for REACTION CONTROL SYSTEM.

RCTC (*ImageTech*) Abbrev for *rewritable consumer time code*. See TIME CODE.

RDA (*FoodSci*) Abbrev for RECOMMENDED DAILY ALLOWANCE.

RDF (*ICT*) Abbrev for RADIO DIRECTION-FINDING.

R-display (*Radar*) An expanded A-DISPLAY in which an echo can be magnified by an expanded sweep for close examination.

rDNA (*BioSci*) Abbrev for RECOMBINANT DNA, RIBOSOMAL DNA.

RDT & E (*Aero*) Abbrev for *research, development, test and evaluation*, US.

Re (*Chem*) Symbol for RHENIUM.

reach (*Genrl*) A clear uninterrupted stretch of water.

reactance (*Phys*) The imaginary part of the impedance. Reactances are characterized by the storage of energy rather than by its dissipation as in resistance.

reactance chart (*ElecEng*) A chart of logarithmic scales so arranged that it is possible to read directly the reactance of a given inductor or capacitance at any frequency.

reactance coupling (*ElecEng*) Coupling between two circuits by a reactance common to both, eg a capacitor or inductor.

reactance drop (*ElecEng*) The decrease in the available voltage at the terminals of a circuit caused by the reactance voltage set up within that circuit.

reactance modulation (*ElecEng*) Use of a variable reactance, eg capacitor or inductor, or a reactance valve, to effect frequency modulation.

reactance relay (*ElecEng*) An impedance relay which operates as soon as reactances of the circuit to which it is connected fall below a predetermined value.

reactance rise (*ElecEng*) The increase in the available voltage at the terminals of a circuit caused by the reactance voltage set up within that circuit.

reactance voltage (*ElecEng*) The voltage produced by current flowing through the reactance of a circuit; equal to the product of the current (amps) and the reactance (ohms).

reactants (*Chem*) The elements and/or compounds which interact to form PRODUCTS and are written on the left-hand side of a chemical equation.

reaction (*BioSci*) Any change in behaviour of an organism in response to a stimulus.

reaction (*Chem*) (1) See CHEMICAL REACTION. (2) The acidity or alkalinity of a solution.

reaction (*Phys*) The equal and opposite force arising when a force is applied to a material system; in particular the force exerted by the supports or bearings on a loaded mechanical system.

reaction chain (*Chem*) See CHAIN REACTION.

reaction chamber (*Aero*) The chamber, usually cylindrical but sometimes spherical, in which the reaction or combustion of a rocket's fuel and oxidant occur.

reaction control system (*Space*) A set of small thrusters, suitably placed on a spacecraft to control its attitude in pitch, roll and yaw. Abbrev RCS.

reaction formation (*Psych*) In psychoanalytic theory, a defence mechanism by which an unacceptable impulse is mastered by establishing behaviour of an opposing tendency; eg hate is converted to oversolicitous love.

reaction injection moulding (*Eng*) Injection-moulding process where polymerization occurs in mould cavity. Since viscosities are much lower, locking forces are much less, so less substantial machinery is used. Used for polyurethanes and nylons, eg car bumpers and exterior trim. Abbrev *RIM* or *RRIM* when reinforcing chopped fibres are added to polymer. See INJECTION MOULDING.

reaction isochore (*Chem*) See VAN'T HOFF'S REACTION ISOCHORE.

reaction isotherm (*Chem*) See VAN'T HOFF'S REACTION ISOTHERM.

reaction kinetics (*Chem*) The study of the speed of chemical reactions or the rate of transformation of reactants to products. See ORDER OF REACTION.

reaction order (*Chem*) See ORDER OF REACTION.

reaction pair (*Geol*) Two minerals of different composition which exhibit the reaction relationship (see REACTION PRINCIPLE). Thus forsterite at high temperature is converted into enstatite at a lower temperature, by a change in the atomic structure involving the addition of silica from the magma containing it. Forsterite and enstatite form a *reaction pair*.

reaction principle (*Geol*) The conversion of one mineral species stable at high temperature into a different one at lower temperatures, by reaction between the crystal phase and the liquid magma containing it. The change may be continuous over a wide temperature range (*continuous reaction*), or may occur at a fixed temperature only (*discontinuous reaction*).

reaction products (*Chem*) The substances formed in a chemical reaction. Those on the right-hand side of a reaction as written.

reaction propulsion (*Aero, Space*) The scientifically correct expression for all forms of jet and rocket propulsion; they act by ejection of a high-velocity mass of gas, from which the vehicle reacts with an equal and opposite momentum, according to Newton's third law of motion. See panel on ROCKET.

reaction rate (*Phys*) The rate of fission in a nuclear reactor.

reaction rim (*Geol*) The peripheral zone of mineral aggregates formed round a mineral or rock fragment by reaction with magma during consolidation of the latter. Thus quartz caught up by basaltic magma is partially resorbed, at the same time being surrounded by a reaction rim of granular pyroxene.

reaction series (*Geol*) One in which the minerals of igneous rocks are arranged in the order of temperature at which they crystallize from magmas.

reaction sintering (*Eng*) See ACTIVATED SINTERING.

reaction time (*Psych*) The interval between the presentation of a signal and the subject's response to it.

reaction turbine (*Eng*) A turbine in which the fluid expands progressively in passing alternate rows of fixed and moving blades, the kinetic energy continuously developed being absorbed by the latter.

reaction wood (*BioSci*) Abnormal types of wood tissue grown in response to bending loads due to eg steady winds or to non-vertical growth, producing as it matures tensile or compressive forces that tend to maintain a growing branch at its appropriate orientation (in spite of increasing mass) or to correct misorientation. Reaction wood makes unsatisfactory timber and pulp. In SOFTWOODS, compression wood is formed, in HARDWOODS, tension wood.

reactivation (*Electronics*) When a thoriated tungsten filament loses its emission, the raising of temperature without anode voltage to bring fresh thorium to the surface.

reactive (*Chem*) Readily susceptible to chemical change.

reactive anode (*Ships*) See SACRIFICIAL ANODE.

reactive component of current (voltage) (*ElecEng*) Preferred term for component of vectors representing ac (voltage) which is in quadrature (90°) with the voltage (current) vector. Also *idle component, inactive component, quadrature component, wattless component.*

reactive depression (*Psych*) A type of depression clearly linked to environmental events (eg after a death).

reactive dye (*Textiles*) A dye that is fixed by reacting chemically with the fibre molecules. The best-known examples are used for dyeing cellulosic fibres (ie cotton and viscose).

reactive factor (*ElecEng*) The ratio of reactive volt-amperes to total supply volt-amperes.

reactive ion etching (*Electronics*) DRY ETCHING process which proceeds by the combined action of chemical species and ion bombardment.

reactive iron (*ElecEng*) Iron inserted in the leakage-flux paths of a transformer to increase its leakage reactance.

reactive load (*ElecEng*) A load in which current lags behind or leads on the voltage applied to its terminals.

reactive oxygen species (*BioSci*) Oxygen-containing radicals or reactive ions such as superoxide, singlet oxygen and hydroxyl radicals. Such species are the product of the respiratory burst in phagocytes and are responsible for bacterial killing as well as incidental damage to surrounding tissue.

reactive power (*ElecEng*) The reactive volt-amperes, ie the product of a voltage of a circuit and the reactive component of the current.

reactive schizophrenia (*Psych*) Those cases of schizophrenia in which onset is sudden and linked to some precipitating event in the environment.

reactive voltage (*ElecEng*) That component of the phasor representing voltage of an ac circuit which is in quadrature (at 90°) with the current.

reactive volt-ampere hour (*ElecEng*) A unit used in measuring the product of reactive volt-amperes in a circuit and the time during which they have been passing.

reactive volt-ampere-hour meter (*ElecEng*) An integrating meter which measures and records the total number of reactive volt-ampere-hours which have passed in the circuit to which it is connected.

reactive volt-amperes (*ElecEng*) The product of the reactive voltage and the amperes in the circuit, or the reactive current (amperes) and voltage of the circuit; measure of the wattless power in the circuit. Abbrev *V Ar* (volt-amperes-reactive).

reactivity (*NucEng*) The departure of the multiplication constant of a reactor from unity, measured in different ways, ie CENT, DOLLAR, INHOUR, NILE, or simply *per cent reactivity*. See panel on NUCLEAR REACTORS.

reactivity worth (*NucEng*) Change of reactivity of a reactor caused by the addition or removal of a material or piece of equipment. The control rods of a reactor could be *worth* 10% of the total reactivity. Also *worth*. See panel on NUCLEAR REACTORS.

reactor (*ElecEng*) Circuit component which stores energy, ie a capacitor or an inductor.

reactor classification (*NucEng*) Assembly of nuclear fuel and (usually) moderator in steady operation of which a self-sustaining chain reaction occurs. The neutron flux (and therefore power) is regulated by neutron absorbing rods (eg of boron steel, cadmium) or otherwise (eg movement of fuel, moderator or reflector; or by use of neutron poisons or spectral shift). Used for neutron irradiation and for releasing nuclear energy for electricity production. Reactors are classified according to their fuel, moderator and coolant (or less frequently according to their size, power output, or function). Several hundred types of reactors have been tested or suggested based on the following alternatives: fuel: uranium-235; plutonium-239; uranium-233; moderator: light water; heavy water; graphite; beryllium; organic liquid (or none in fast reactors); coolant: gas; light water; heavy water; organic liquid; liquid metal. Fuels are classified as natural, slightly enriched or highly enriched according to the extent to which the proportion of fissile material has been increased beyond its normal isotopic abundance. See panel on NUCLEAR REACTORS.

reactor noise (*NucEng*) Random statistical variations of neutron flux in a reactor.

reactor oscillator (*NucEng*) A device which produces periodic variations of reactivity by mechanical oscillation of neutron-absorbing sample in reactor core. Used to measure reactor properties, or nuclear cross-section of sample.

reactor problem (*NucEng*) Determining the optimum composition and disposition of the components of a nuclear reactor. See panel on NUCLEAR REACTORS.

reactor product decay (*NucEng*) The slow loss of radioactivity and toxicity of the fission products of a reactor after it is shut down.

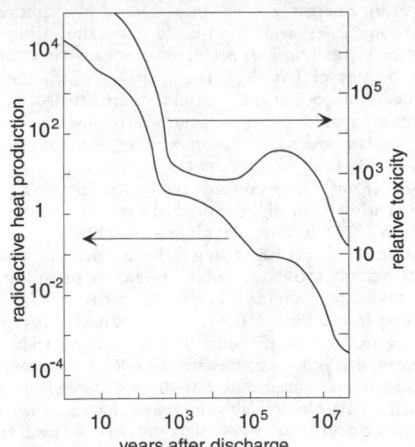

reactor product decay

reactor safety report (*NucEng*) See RASMUSSEN REPORT and panel on RISK ASSESSMENT. Fig. ▷

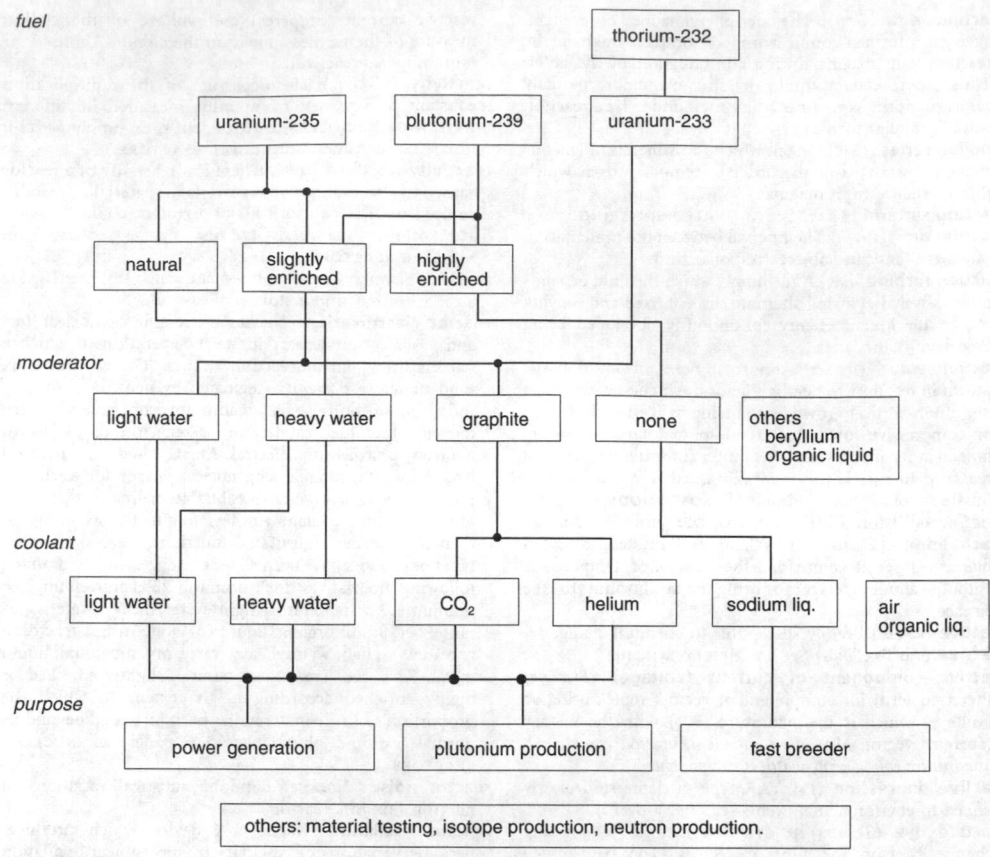

reactor

reactor simulator (*NucEng*) Analogue digital computer which simulates variations in reactor neutron flux produced by changes in any operating parameter. Used for training and for investigating reactor effects.

reactors in space (*NucEng*) See SPACE REACTORS.

reactor trip (*NucEng*) Rapid reduction of reactor power to zero by emergency insertion of control mechanisms and (in some cases) removal of liquid moderator. Also *scram*.

reactor types (*NucEng*) See REACTOR CLASSIFICATION.

reactor vessel (*NucEng*) The vessel in which the core, moderator, coolant and control rods are situated. The fuel rods are spaced so that the neutrons are slowed sufficiently to initiate secondary fission without encountering uranium-238 in the fuel. See PRESSURE VESSEL.

reader (*Print*) One who reads and corrects printers' proofs, comparing them with the original copy. Also *proofreader*.

read in (*ICT*) To insert data into a computer.

reading (*ICT*) (1) Registering and transferring to IMMEDIATE ACCESS STORE the data on eg BAR CODE or BACKING STORE. (2) See CHARACTER RECOGNITION.

reading frame (*BioSci*) The sets of triplet nucleotides that are being read by the ribosome to specify amino acids in the protein that is being synthesized. An mRNA molecule can be read in three different reading frames, depending on the starting base. Usually alternative reading frames contain many STOP CODONS and are not used except in some small viruses. See OPEN READING FRAMES, OVERLAPPING GENES.

reading microscope (*Phys*) Also *reading telescope*. See CATHETOMETER.

readme file (*ICT*) A text file supplied with computer software that contains information about the software, such as advice on installation and bugs.

read-only file (*ICT*) A file whose ATTRIBUTE has been set such that the file may be read, not modified or deleted.

read-only memory (*ICT, Electronics*) A random access memory in which the data pattern is fixed during manufacture and cannot be changed subsequently. Abbrev *ROM*. See EPROM.

read-out pulse (*ICT*) Pulse applied to binary cells to extract the bit of information stored.

ready-mixed concrete (*Build*) A method whereby concrete constituents are carried in a special vehicle and mixed with water to the right consistency while the vehicle is travelling. Used for medium quantities or where sites are too congested for concrete-mixing plant.

reagent (*Chem*) A substance or solution used to produce a characteristic reaction in chemical analysis.

reagent feeder (*MinExt*) Appliance which dispenses chemicals in continuously moving flow of ore or pulp at a controlled rate.

reagin (*BioSci*) Antibody that fixes to tissue cells of the same species so that, in the presence of antigen, histamine and other vasoactive agents are released. The term was used to describe such antibodies in humans before IgE was identified, and is occasionally still used. Also *reaginic antibody*.

real absorption coefficient (*Phys*) See TRUE ABSORPTION COEFFICIENT.

real axis (*MathSci*) See ARGAND DIAGRAM.

realgar (*Min*) A bright-red monosulphide of arsenic; monoclinic. Occurs associated with orpiment.

real image (*Phys*) See IMAGE.

reality principle (*Psych*) In psychoanalytic theory, the mental activity that leads to instinctual gratification by accommodating to the demands of the real world; it is acquired during development. Cf PLEASURE PRINCIPLE.

real memory (*ICT*) The MAIN MEMORY in a computer system that also uses VIRTUAL MEMORY.

real modulus (*Eng*) Also *storage modulus*. See COMPLEX MODULUS.

real number (*ICT*) Any number with a fractional (or decimal) part. Cf INTEGER.

real number (*MathSci*) Any rational or irrational number. Real numbers can be defined as DEDEKIND CUTS on the rationals.

real part (*MathSci*) See COMPLEX NUMBER.

real-time analyser (*Acous*) Analyser which calculates the spectrum (FOURIER ANALYSIS) of a signal so fast that no input data are excluded from the analysis. Uses *fast Fourier transform* (FFT) (a particularly fast numerical method for calculating Fourier transforms).

real-time clock (*ICT*) Electronic unit which keeps track of the date and the time of day in a special register that may be accessed by the programmer.

real-time encoding (*ImageTech*) Compressing digital video at the same rate as the incoming signal.

real-time PCR (*BioSci*) A method in which the rate of accumulation of PCR products is measured in real time using a fluorescent marker. The signal increases in direct proportion to the amount of PCR product; this allows, from the kinetics, an estimation of the original concentration of the target.

real-time system (*ICT*) Computing system that is designed to receive data, process them and respond within a time frame set by outside events, eg air-traffic control, automatic banking.

ream (*Glass*) A non-homogeneous layer in flat glass.

ream (*Paper*) Twenty quires; now usually 500 sheets; was 480.

reamer (*Eng*) A hand- or machine-operated tool for finishing drilled holes. It consists of a cylindrical or conical head on which cutting edges are formed by longitudinal or spiral flutes, or in which separate teeth are inserted.

rear projection (*ImageTech*) See BACK PROJECTION.

Réaumur scale (*Phys*) A temperature scale ranging from 0 to 80°R (freezing point and boiling point of pure water at normal pressure).

rebate (*Build*) A recess cut into the corner of a piece of timber. Also *rabbet*.

rebate plane (*Build*) A plane specially adapted for cutting a groove in the corner of a board.

rebecca-eureka (*Radar*) Radar system on aircraft carrying low-power interrogator transmitters (rebecca), working with fixed beacon responders (eureka), sending coded signals when triggered by interrogator pulses.

reboot (*ICT*) To restart a computer. See BOOT, WARM BOOT.

rebore (*Autos*) The treatment of a worn cylinder by boring it out and replacing the piston by a (necessarily) larger one.

rebound resilience (*Eng*) The ratio of rebound to drop height of a rubber ball, but temperature and rate sensitive owing to viscoelasticity. Rubbers of low hysteresis (eg polybutadiene 'superballs') are more resilient than high-hysteresis rubbers like butyl at ambient temperature.

rebreather (*Genrl*) An aqualung that recycles air exhaled by the diver.

recalescence (*Eng*) The release of heat in ferromagnetic material as it cools through a temperature at which a change in crystal structure occurs, normally associated with change in magnetic properties.

recall (*Psych*) A method of measuring retention in which some material must be produced from memory. Cf RECOGNITION.

recapitulation theory (*BioSci*) States that stages in the evolution of the species are reproduced during the developmental stages of the individual, ie ontogeny tends to recapitulate phylogeny. Superficially apparent in some instances. Also *biogenetic law, Haeckel's law*.

receiver (*ICT*) Final unit in transmission system where received information is stored, recorded or converted into the necessary form.

receiver response (*Acous*) The response of a telephone receiver operating into a real or artificial ear; expressed as a ratio of the square of the sound pressure in the specified cavity to the electrical power applied to the receiver.

recency effect (*Psych*) (1) In recall, the tendency to recall items from the end of the list more readily than those in the middle. (2) In impression formation, the fact that attributes noticed later carry a greater weight than those observed earlier. See PRIMACY EFFECT.

Recent (*Geol*) Holocene. See QUATERNARY.

receptacle (*BioSci*) (1) Generally, a structure on which reproductive organs are borne. (2) The swollen tip of a thallus with conceptacles in brown algae. (3) Area bearing archegonia or antheridia in liverworts, or sporangia in ferns. (4) The end of a stalk on which is borne either the parts of a single flower or the involucre and florets of a head or capitulum.

receptaculum (*BioSci*) (1) A receptacle; a sac or cavity used for storage. (2) A sac in which ova are stored, as in some Oligochaeta.

receptaculum seminis (*BioSci*) A sac in which spermatozoa are stored, as in many invertebrates; a spermotheca.

reception wall (*Build*) See RETENTION WALL.

receptive (*BioSci*) Capable of being effectively pollinated or fertilized.

receptor (*BioSci*) (1) At a molecular level, usually a membrane-bound or membrane-enclosed molecule that binds to or responds to something more mobile (the ligand), with high specificity, eg cell surface receptors for hormones or growth factors. The binding of ligand to a receptor on a cell membrane is often followed by transduction of a signal across the membrane and a response on the part of the cell. Some receptors, eg steroid hormone receptors, are cytoplasmic proteins that move into the nucleus following ligand binding. (2) Physiologically, an element of the nervous system specially adapted for the reception of stimuli, eg a sense organ or a sensory nerve ending.

receptor downregulation (*BioSci*) A phenomenon observed in many cells: following stimulation with a ligand the number of receptors for that ligand on the cell surface diminishes because internalization exceeds replenishment.

receptor-mediated endocytosis (*BioSci*) The internalization of ligands bound to certain receptors on the cell surface, which become clustered into COATED PITS and enter the cell via COATED VESICLES and ENDOSOMES.

receptor tyrosine kinase (*BioSci*) A class of membrane receptor that phosphorylates tyrosine residues on proteins. Many play significant roles in development or control of cell division.

recess (*BioSci*) A small cleft or depression, as the *optic recess*.

recessed arch (*Arch*) See COMPOUND ARCH.

recessed pointing (*Build*) A method of pointing designed to prevent peeling off; the mortar at all joints is pressed back about $\frac{1}{4}$ in (6 mm) from the face of the wall.

recessive (*BioSci*) Describes a gene (allele) that shows its effect only in individuals that received it from both parents, ie in homozygotes. Also describes a character due to a recessive gene. Cf DOMINANT.

rechip (*ICT*) To change the electronic identity of (a stolen mobile phone).

reciprocal (*MathSci*) The INVERSE of a number or quantity under multiplication; the reciprocal of a is the number $1/a$, ie a^{-1}.

reciprocal cross (*BioSci*) A cross made both ways with respect to sex, ie A♂ × B♀, and B♂ × A♀. Consistent differences between the offspring of such crosses suggests CYTOPLASMIC INHERITANCE.

reciprocal diagram (*Eng*) See FORCE DIAGRAM.

reciprocal hybrids (*BioSci*) A pair of hybrids obtained by crossing the same two species, in which the male parent of one belongs to the same species as the female parent of the other, eg mule and hinny.

reciprocal lattice (*Phys*) For a direct crystal lattice defined in terms of three vectors $\mathbf{a}_1$, $\mathbf{a}_2$, $\mathbf{a}_3$, the lattice whose vectors are $\mathbf{b}_1$, $\mathbf{b}_2$, $\mathbf{b}_3$, defined by $\mathbf{a}_i \cdot \mathbf{b}_i = \gamma$ and $\mathbf{a}_i \cdot \mathbf{b}_j = 0$. It follows that

$$\mathbf{b}_1 = \frac{\gamma}{V}(\mathbf{a}_1 \times \mathbf{a}_3) \text{ etc}$$

where V is the volume of the unit cell of the direct lattice. In crystallographic work γ is chosen as one, but in solid-state physics as 2π. The reciprocal lattice is extensively used to discuss diffraction and scattering effects by crystals and in the band theory of solids.

reciprocal networks (*ElecEng*) Those the product of whose impedances remains constant at all frequencies; thus an inductance is reciprocal to a capacitance.

reciprocal polar triangles (*MathSci*) See CONJUGATE TRIANGLES.

reciprocal proportions (*Chem*) See LAW OF EQUIVALENT PROPORTIONS.

reciprocal theorem (*Eng*) When a linear elastic body subject to a set of forces F_1 has a second set of forces F_2 applied to it, the work done by the displacements of F_1 in response to F_2 is equal to the work that would have been done by the displacements of F_2 in response to the application of F_1. This also applies to the case when moments as well as forces are applied to a body. Although often called *Maxwell's reciprocal theorem*, it was first stated in its general form by E Betti in 1872.

reciprocal translocation (*BioSci*) Mutual exchange of non-homologous portions between two chromosomes.

reciprocating compressor (*Eng*) Any compressor which employs a piston working in a cylinder, the piston causing the periodic compression of the working fluid.

reciprocating engine (*Eng*) An engine which employs a piston working in a cylinder, the piston being caused to oscillate by the periodic pressure of the working fluid.

reciprocating pump (*MinExt*) One which uses the displacing action of a plunger, piston or diaphragm to move water in a pulsated stream. Also *pulsometer pump*, using steam or compressed air in a valved system for similar pumping.

reciprocation (*ICT*) The operation of finding a reciprocal network to a given network. Used in electric wave filters.

reciprocation (*MathSci*) See POLAR RECIPROCATION.

reciprocity (*Eng*) The principle enunciated in the RECIPROCAL THEOREM.

reciprocity calibration (*ElecEng*) Absolute calibration of microphone by use of reversible microphone–loudspeaker. See RECIPROCITY THEOREM.

reciprocity constant (*ElecEng*) See RECIPROCITY THEOREM.

reciprocity failure (*ImageTech*) Photographic exposure is determined by the product of light intensity and time and within limits an increase of one may be compensated by a proportionate decrease of the other (*reciprocity rule*). But for extremely short or long exposure times or at very low or high light intensities this relation fails to apply, esp with colour materials.

reciprocity principle (*Phys*) The principle that the interchange of radiation source and detector will not change the level of radiation at the latter, whatever the shielding arrangement between them.

reciprocity theorem (*Acous*) A theorem of acoustics which says that under certain conditions the sound source and the receiver can be swapped, thereby the output of the receiver being unchanged.

reciprocity theorem (*ElecEng*) The interchange of emf at any one point in a network and the current produced at any other point results in the same current for the same emf. In an electrical network comprising two-way passive linear impedances, the so-called transfer impedance is given by the ratio of the emf introduced in a branch of the network to the current measured in any other branch. By the reciprocity theorem this ratio is equal in phase and magnitude to that observed if positions of current and emf

are interchanged. In its application to the calibration of transducers the reciprocity theorem concerns the quotient of the value of the ratio of open-circuit voltage at output terminals of the transducer (when used as a sound receiver) to the value of the free-field sound pressure (referred to some arbitrarily selected point of reference near the transducer) divided by the value of the sound pressure at a distance d from the point of reference to the current flowing at the input terminals of the transducer (used as a sound transmitter). The value of this quotient, termed the *reciprocity constant*, is independent of the constructional nature of the transducer.

recirculating-ball thread (*Eng*) The helical analogue of the radial ball bearing, in which a helical ball track is ground on the shaft as a thread. The nut has a complementary helical ball track at each end and the balls recirculate in a space provided outside the rolling balls. Cf LINEAR BALL BEARING.

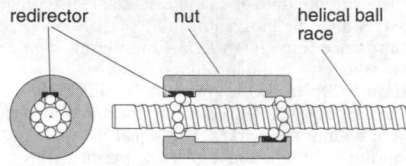

recirculating-ball thread The redirector lifts a ball from one turn of the race to the next.

recirculating ball-type steering (*Autos*) Steering gear resembling the screw and nut type, but using a half-nut with an eccentric ball race which is operated by a spiral cam.

recirculating heating system (*Eng*) Heating industrial ovens and low-temperature furnaces with the atmosphere of the working chamber under constant recirculation throughout the complete system.

reck (*MinExt*) See RAGGING FRAME.

recognition (*Psych*) A method of measuring retention in which a stimulus has to be identified as having occurred before.

recoil atom (*Phys*) An atom which experiences a sudden change of direction or reversal, after the emission from it of a particle or radiation. Also *recoil nucleus*.

recoil nucleus (*Phys*) See RECOIL ATOM.

recoil particles (*Phys*) Those arising through collision or ejection, eg Compton recoil electrons.

recombinant (*BioSci*) (1) An organism containing a combination of alleles differing from either of its parents. (2) Recombinant DNA contains sequences from different sources, made usually as the result of laboratory procedures *in vitro*. (3) An individual, gamete or chromosome resulting from *recombination*.

recombinant inbred strains (*BioSci*) Inbred strains, mostly of mice, that have been made by crossing two different inbred parental strains until they are homozygous at most loci. Because genes of the parental strains have become reassorted these RI strains provide a means of assessing the functions of gene products in a different genomic background. Abbrev *RI strains*.

recombinant protein (*BioSci*) A protein produced by expression, in a culture system, of a gene that has been isolated and possibly modified. The method has been used to produce human insulin on a large scale. One problem can be getting the appropriate post-translational modification if eg mammalian protein is produced in bacteria, although various alternative approaches, including transgenic animals and plants, and the use of eukaryotic cell expression systems, have been developed.

recombinase activating gene (*BioSci*) The essential gene for the recombination of the genetic elements involved in the production of functional lymphocyte antigen receptors. Abbrev *rag*.

recombination (*BioSci*) Reassortment of genes or characters in combinations different from those in the parents, in the case of *linked* genes by *crossing over*.

recombination (*Chem*) Chain termination reaction where two active free-radical chain ends combine to form dead polymer. See CHAIN POLYMERIZATION.

recombination (*Electronics*) Neutralization of free electron and hole in semiconductor, thus eliminating two current carriers. The energy released in this process must appear as a light photon or less probably as several phonons. See EXCITON, PHONON, PHOTON.

recombination (*Phys*) Neutralization of ions in gas by combination of charges or transfer of electrons. Important for ions arising from the passage of high-energy particles.

recombination coefficient (*Electronics*) The ratio of the rate of recombination per unit volume to the product of the densities of positive and negative current carriers.

recommended daily allowance (*FoodSci*) The daily amount of vitamin or mineral which should be consumed to maintain good health. In nutritional labelling of foods, the amount of vitamin or mineral is normally declared as a percentage of this allowance. Abbrev *RDA*.

reconcentration (*MinExt*) Additional treatment of a mineral product to raise its grade or to separate out one constituent.

reconditioned carrier (*Electronics*) Isolation of a pilot carrier for reinsertion of a carrier adequate for demodulation.

reconfigure (*ICT*) To set up in a new or different way, in particular of hardware or software.

reconstructed stone (*CivEng*) Artificial stone made of concrete blocks faced to resemble natural stone. Also *precast stone*.

reconstruction (*Psych*) The tendency to fill in the gaps in memory and then believe this rewriting of history. See FALSE MEMORY SYNDROME, RECOVERED MEMORY.

record (*Acous*) See RECORDING.

record (*ICT*) A collection of related items of data, treated as a unit.

recorder (*Acous*) (1) A machine for registering a sound magnetically, photographically or on plastic. (2) Musical instrument basically consisting of a *whistle* coupled with a tube-shaped resonator. Its resonance frequency can be varied by opening or closing holes along the wall of the tube.

recorder (*Eng*) An instrument which measures and records the quantity measured usually by a pen actuated by a motor in a servo system. Recording is on a chart moved at a speed to give a time-scale.

recording (*Acous*) (1) The process of making a record of a received signal. (2) A disk, tape or film on which a sound record is stored. (Disk recordings are commonly known simply as *records*.)

recording altimeter (*Aero*) A barographic type of instrument which traces height against time.

recording amplifier (*Acous*) An amplifier preceding the recording heads of any type of recorder.

recording head (*Acous*) The transducer (magnetic, electric, mechanical or electro-optical) used to record sound on tape, disk or film.

recording stylus (*Acous*) The instrument that cuts the groove in an original disk recording.

recovered memory (*Psych*) The memory of repressed childhood experiences, esp of being sexually abused, (apparently) recovered by psychoanalysis.

recovery (*Eng*) (1) First stage in the annealing process of cold-worked metals, in which some restoration of original properties (eg hardness, ductility, resistivity) is achieved by a reduction in the density of dislocations through their thermally stimulated mutual annihilation at temperatures around $0.4T_m$. Cf GRAIN GROWTH, RECRYSTALLIZATION. (2) The extent of the return of a stressed material to its original state after distortion under load. See panel on CREEP AND DEFORMATION.

recovery (*ICT*) The process of returning to normal after an error, which may include ensuring that the data have not been corrupted.

recovery (*MinExt*) The percentage of schedule tonnage actually mined.

recovery pegs (*Surv*) Special reference pegs established in known survey relation to the working setting-out pegs, so that the location of these can be re-established if disturbed.

recovery position (*Med*) In the medical treatment of unconscious or semiconscious casualties, a position of the body on its side with the face tilted slightly upwards.

recovery rate (*Radiol*) That at which recovery occurs after radiation injury. It may proceed at different rates for different tissues.

recovery time (*Electronics*) The time required for the control electrode of a gas tube to regain control.

recovery time (*Phys*) (1) For a Geiger tube, the period between the end of the dead time and the restoration of full normal sensitivity. (2) For a counting system, the minimum time interval between two events recorded separately.

recovery time (*Radar*) That required by a TRANSMIT–RECEIVE TUBE in a radar system to operate (usually measured to the point where receiving sensitivity is 6 dB below maximum).

recovery voltage (*ElecEng*) The normal frequency or dc voltage which appears across the contacts of a switch, circuit breaker or fuse after it has interrupted the circuit.

recreational drug (*Pharmacol*) A drug taken for non-medical reasons, eg mood enhancement, but often with addictive effects.

recrystallization (*Chem*) The process of reforming crystals, usually by dissolving them, concentrating the solution, and thus permitting the crystals to reform. Frequently performed in the process of purification of a substance.

recrystallization (*Eng*) A stage in the annealing process of cold-worked metals above about 0.4–$0.5T_m$, in which deformed crystals are replaced by a new generation of crystals, which begin to grow at certain points in the deformed metal and eventually absorb the deformed crystals. The new crystals have more equal axes and contain far fewer dislocations than the deformed ones. See ANNEALING. Cf GRAIN GROWTH, RECOVERY.

recrystallization annealing (*Eng*) See RECRYSTALLIZATION.

recrystallization temperature (*Eng*) That marking a change in crystal form at which new crystals nucleate and grow to consume the existing structure. The range of temperature through which strain-hardening disappears, approximately $0.6T_m$.

rectal gills (*BioSci*) In the larvae of some Odonata, tracheal gills in the form of an elaborate system of folds in the wall of the rectum, used in respiration.

rectangle (*MathSci*) A parallelogram whose angles are right angles.

rectangular axes (*MathSci*) Co-ordinate axes which are mutually at right angles.

rectangular loop hysteresis (*ElecEng*) Colloq expression for hysteresis curve of ferromagnetic or ferroelectric materials suitable for use in bistable or switching circuits. Characterized by very steep slope followed by unusually sharp onset of saturation.

rectangular notch (*CivEng*) A notch plate having a rectangular notch cut in it; used for the measurement of large discharges over weirs.

rectangular pulse (*ICT*) Idealized pulse with infinitely short rise and fall times and constant amplitude. Pulse amplitudes and durations are often specified in terms of those of the nearest equivalent pulse carrying the same energy (subtending the same area under the curve).

rectangular scan (*Electronics*) Any scanning system producing a rectangular field.

recti- (*Genrl*) Prefix from Lt *rectus*, straight.

rectification (*Chem*) Purification of a liquid by distillation.

rectification (*ElecEng*) The conversion of ac into dc using some form of rectifier, usually a solid-state one.

rectification efficiency (*ElecEng*) The ratio of dc output power to ac input. Often expressed as a percentage.

rectified airspeed (*Aero*) See CALIBRATED AIRSPEED.

rectified spirit (*Chem*) Distilled ethanol containing only 4·43% water. This is a constant-boiling (azeotropic) mixture with a boiling point of 78°C.

rectifier (*ElecEng*) Component for converting ac into dc.

rectifier instrument (*ElecEng*) An instrument for measuring ac which first rectifies the current and then measures it as dc.

rectifier leakage current (*ElecEng*) That passing through a rectifier under reverse bias conditions, due to finite reverse conduction.

rectifier ripple factor (*ElecEng*) The amount of ac voltage in the output of a rectifier after dc rectification. It is measured, in per cent, by the ratio of the rms value of the ac component to the algebraic average of the total voltage across the load. See RIPPLE.

rectifier stack (*ElecEng*) A pile of rectifying elements (usually semiconductor) series-connected for higher-voltage operation.

rectifier voltmeter (*ElecEng*) One in which applied voltage is rectified in a bridge circuit before measurement with a dc meter.

rectifying detector (*ElecEng*) A detector of electromagnetic waves which depends for its action on rectification of high-frequency currents, as opposed to one using thermal, electrolytic breakdown or similar effects.

rectifying valve (*Electronics*) Any thermionic valve in which direct use is made of unilateral or asymmetrical conductivity effects, as opposed to one used primarily for amplification, eg a diode used as a rectifier or a triode used as a detector.

rectilinear (*MathSci*) Of a plane figure: bounded by straight lines.

rectilinear lens (*ImageTech*) A lens which provides images with no distortion, as far as parallel lines are concerned. Not otherwise well-corrected.

rectilinear scan (*ImageTech*) A raster in which a rectangular area is scanned by a series of parallel lines.

rectirostral (*BioSci*) Having a straight beak.

recto (*Print*) A right-hand page of a book, bearing an odd page number. Cf VERSO.

recto- (*Med*) Prefix from Lt *rectus*, straight, but in medicine often used in terms pertaining to the rectum.

rectocele (*Med*) A prolapse or protrusion of the lower part of the vaginal wall, carrying with it the anterior wall of the rectum.

rectrices (*BioSci*) In birds, the stiff tail feathers used in steering. Sing *rectrix*. Adj *rectricial*.

rectum (*BioSci*) The posterior terminal portion of the alimentary canal leading to the anus. Adj *rectal*. From Lt *rectum intestinum*, straight intestine.

rectus (*BioSci*) A name used for various muscles that are of equal width or depth throughout their length, eg the *rectus abdominis* in vertebrates.

recumbent fold (*Geol*) An overturned fold with a more or less horizontal axial plane.

recuperative air heater (*Eng*) An air heater in which heat is transmitted from hot gases to the air through conducting walls, the flows of gas being continuous and unidirectional.

recuperator (*Eng*) An arrangement of flues which enables the hot gases leaving a furnace to be utilized in heating the incoming air (and sometimes gas). Outgoing hot gases and incoming cold gases pass in opposite directions through parallel flues and heat is transferred through the dividing walls.

recurrence (*Phys*) See REGGE TRAJECTORY.

recurrent (*BioSci*) Returning towards point of origin.

recurrent nova (*Astron*) A type of nova which has shown more than one outburst of light, as T Coronae Borealis in 1866, 1898 and 1933; they show smaller ranges of brightness than most novae.

recurrent vision (*Phys*) The perception of repeated images of brightly illuminated objects when the source of illumination is suddenly removed.

recurring decimal (*MathSci*) A decimal fraction in which, after a certain point, a number or set of numbers is repeated indefinitely. The figures which recur are indicated by dots placed above them, eg $0·\dot{3} \equiv 0·333\,333...$, $0·5\dot{2}4\dot{3} \equiv 0·5243\,243\,243...$. Also *repeater, repeating decimal*.

recursion (*MathSci*) A process by which a term in a sequence may be computed from one or more of the preceding terms.

recursive subprogram (*ICT*) One that includes among its program statements a call to the subprogram itself.

recurvirostral (*BioSci*) Having the beak bent upwards.

recycling (*Genrl*) See CLOSED-LOOP RECYCLING, IN-HOUSE RECYCLING, POST-CONSUMER RECYCLING.

recycling (*Phys*) Repetition of fixed series of operations, eg biodegradation of organic material followed by regrowth; applied also to isotope separation.

recycling ratio (*Plastics*) In injection moulding, the ratio of waste thermoplastic (eg cold runners) to total shot weight; must be kept constant for quality control and low (eg 0·10) for critical, stressed products owing to polymer chain degradation during each moulding cycle.

red algae (*BioSci*) See RHODOPHYCEAE.

red beds (*Geol*) Red sedimentary deposits, mainly sandstones, siltstones and shales, coloured by iron oxides (haematite) and resulting from the arid continental conditions of their formation. In W Europe found particularly in the Old Red Sandstone (Devonian) and New Red Sandstone (Permo-Trias).

red blood corpuscle (*BioSci*) See ERYTHROCYTE.

red body (*BioSci*) See RED GLAND.

red brass (*Eng*) Copper–zinc alloy containing 15% zinc; used for plumbing pipe, hardware, condenser tubes, etc. Red casting brass may contain up to 5% of lead and/or tin in place of the zinc.

red clay (*Geol*) A widespread deep-sea deposit; essentially a soft, plastic clay consisting dominantly of insoluble substances which have settled from the surface waters; these substances are of volcanic, partly of cosmic origin, and include nodules of manganese and phosphorus, crystals of zeolites and rare organic remains such as shark's teeth.

red corpuscle (*BioSci*) See ERYTHROCYTE.

Red Data Book (*BioSci*) A catalogue of rare and endangered species prepared by the International Union for Conservation of Nature and Natural Resources, started in 1966, and covering the whole world.

red deal (*For*) A light-yellow SOFTWOOD obtained from the SCOTS FIR.

red diesel (*Genrl*) Diesel fuel that is intended for use by agricultural vehicles and boats and is therefore subject to a reduced rate of excise duty and is chemically marked and dyed red to identify it.

reddle (*Eng*) A mixture of red lead and oil wiped over one of two surfaces to be bedded together to indicate high spots to be removed by scraping.

reddle (*Min*) A red and earthy mixture of haematite, often with a certain admixture of clay.

red dwarf (*Astron*) A small faint red star, often one of the lowest-luminosity main-sequence stars.

red-eye (*ImageTech*) Red pupils caused by on-camera flash light reflected from the blood vessels in the subject's retinae. See RED-EYE REDUCTION.

red-eye reduction (*ImageTech*) Using a low-power preflash or separate light to partially close the iris of the eye to reduce RED-EYE.

red giant (*Astron*) A cool, giant luminous star with its hydrogen exhausted by nuclear reaction to helium. See panel on HERTZSPRUNG–RUSSELL DIAGRAM.

red gland (*BioSci*) In some fish, a network of small blood vessels found in the wall of the air bladder, responsible for secretion or absorption of gas. Also *red body, rete mirabile*.

red gum (*For*) See RED RIVER GUM.

red hardness (*Eng*) The ability to retain hardness at temperatures at which a material becomes red hot. Applies to tool materials like high-speed steel and cemented carbides which become heated in use by friction. Plain carbon and low-alloy steels would soften under such conditions whereas those which do not are said to possess red hardness.

redhead (*ImageTech*) TN for a small 800 W quartz lighting unit.

red heat (*Eng*) As judged visually, a temperature between 500 and 1000°C.

redia (*BioSci*) The secondary larval stage of Trematoda, possessing a pair of locomotor papillae and a rudimentary pharynx and intestine, and capable of paedogenetic reproduction.

redintegration (*Psych*) Restoration to wholeness; the recurrence of a complete mental state when any single element of it recurs, eg when a piece of music reminds one of an occasion on which it was played.

redirected behaviour (*Psych*) Behaviour directed at inappropriate or irrelevant objects, often as a result of CONFLICT or FRUSTRATION.

redistilled zinc (*Eng*) Zinc from which impurities have been removed by selective distillation. The process takes advantage of the different boiling points of zinc (907°C) and the impurities lead (1750°C) and cadmium (767°C). A purity of 99.9% zinc is obtainable.

red lead (*Build*) An anticorrosive priming paint used on ferrous metals. Other coatings are now generally used in preference because of toxicity problems.

red lead (*Chem*) Dilead (II) lead (IV) oxide, Pb_3O_4. Formed by heating lead (II) oxide in air at approximately 450°C. It occurs as red and yellow crystalline scales. Commercial varieties contain up to approximately 35% PbO_2. Widely used as an anticorrosive pigment in iron and steel primers.

Redler conveyor (*Eng*) An enclosed conveyor in which an endless chain is continuously dragged along the bottom of a casing and brings along with it material fed into the trough. For conveying coal, cement, flue dust, etc.

red light (*BioSci*) For plant responses mediated by phytochrome, light of wavelength around 630 nm.

Red list (*EnvSci*) A UK list of substances whose discharge to water should be minimized. It includes a number of heavy metals, and insecticides such as dieldrin and DDT.

red mud (*Eng*) In the BAYER PROCESS, iron-rich residue from digestion of bauxite with caustic soda.

red muscles (*BioSci*) In vertebrates, muscles rich in sarcoplasm and myoglobin and red in colour.

redox (*Chem*) Abbrev for *reduction–oxidation*. See OXIDATION POTENTIAL.

red oxide of copper (*Min*) See CUPRITE.

red oxide of zinc (*Min*) See ZINCITE.

red pine (*For*) See DOUGLAS FIR.

red planet (*Astron*) See MARS.

Red River gum (*For*) A strong, durable, red-brown hardwood from Australia, with a close and even texture, but resinous with frequent gum pockets. Also (*Queensland*) *blue gum*, (*Murray*) *red gum*.

redruthite (*Min*) A name frequently applied to the mineral CHALCOCITE because of its occurrence, among other Cornish localities, at Redruth.

redshift (*Astron*) The displacement of features in the spectra of astronomical objects, particularly galaxies and quasars, towards the longer wavelengths, generally interpreted as a result of the DOPPLER EFFECT due to their recessional velocities. The expansion of the universe means that all but the nearest galaxies have redshifted spectra; indeed this very feature led to the discovery of the expanding universe in the 1920s. The relationship between redshift and distance is given in the HUBBLE RELATION and panels on GALAXY, QUASAR and REDSHIFT–DISTANCE RELATION.

redshift–distance relation (*Astron*) Described by E Hubble from the galaxy spectra obtained in 1921–5 by V M Slipher, which showed that most other galaxies have spectra shifted towards the longer (red) wavelengths, indicating that they are travelling away from Earth, and that the velocities are large. See HUBBLE RELATION, REDSHIFT and panel on REDSHIFT–DISTANCE RELATION.

red-short (*Eng*) Lacking ductility or malleability when red hot. See HOT-SHORT.

red silver ore (*Min*) For *dark-red silver ore*, see PYRARGYRITE; for *light-red silver ore*, see PROUSTITE.

red snow (*BioSci*) Lying snow coloured by the growth near the surface of algae, esp *Chlamydomonas nivalis*, containing haematochrome.

red spot (*Astron*) See JUPITER.

red tide (*BioSci*) Water containing dinophytes or other organisms in sufficient numbers to colour it red. Often referred to as a *bloom*. Some dinophytes contain sufficient toxin to make shellfish feeding on them fatally poisonous to humans.

reduced instruction set computer (*ICT*) One using a central processor with a very small INSTRUCTION SET which allows faster processing and simpler design but at the cost of breaking complex instructions into their components and executing each. Abbrev *RISC*.

reduced level (*Surv*) The elevation of a point above or below a specified datum. US *elevation*.

reduced mass (*Phys*) The quantity $mM/(m + M)$ used in the study of the relative motion of two particles, masses m and M, about their common centre of gravity. This is used in place of the smaller mass; movement of the larger is then ignored.

reduced viscosity (*Plastics*) Specific viscosity of a dilute polymer solution divided by concentration. Also *viscosity number*.

reducer (*Build*) See REDUCING SOCKET.

reducer (*ImageTech*) A solution which acts on the silver image and dissolves it away by chemical or abrasive action, thus reducing contrast and/or density.

reducing agent (*Chem*) A substance which is capable of bringing about the chemical change known as *reduction* in another substance, itself being *oxidized*.

reducing atmosphere (*Eng*) One deficient in free oxygen, and perhaps containing such reactive gases as hydrogen and/or carbon monoxide; used in a reducing furnace to lower the oxygen content of mineral.

reducing flame (*Chem*) One containing excess fuel over oxygen, hence capable of acting as a *reducing agent*.

reducing socket (*Build*) Used to connect pipes of different sizes, being threaded appropriately for screwed pipes or having different internal diameter for CAPILLARY FITTINGS. Also *reducer, reducing pipe-joint*.

reducing surface (*Phys*) A prepared surface, used in photometry, which reflects only a certain predetermined proportion of the luminous flux falling on it.

reductio ad absurdum (*MathSci*) A method of proof that proceeds by showing that the negation of the desired proposition leads to a contradiction, eg to prove that $\sqrt{2}$ is irrational, one proves that the assumption that $\sqrt{2}$ can be expressed as a/b leads to a contradiction.

reduction (*Chem*) Any process in which an electron is added to an atom or an ion. Four common types of reduction are removal of oxygen from a molecule, the liberation of a metal from its compounds, and diminution of positive valency of an atom or ion. Always occurs accompanied by oxidation.

reduction (*MinExt*) The extraction of gold from ore. The *reduction officer* is the official in charge of mill, extraction plant or reduction works (S Africa).

reduction division (*BioSci*) See MEIOSIS.

reduction factor (*Phys*) The ratio of the mean spherical luminous intensity of a light source to its mean horizontal luminous intensity.

reduction of levels (*Surv*) The process of computing reduced levels from staff readings booked when levelling.

Redshift–distance relation

A variety of methods have been used to determine distances to galaxies. For nearby galaxies the regular variations of the CEPHEID VARIABLE stars give absolute magnitudes, which yield a distance when compared with observed magnitudes. Other candidate objects where brightness comparisons give a distance indication are the brightnesses of O-type stars, NOVAE, SUPERNOVAE and clouds of ionized hydrogen. These methods work to a distance of about 65 million light-years from Earth.

The spectrum of a GALAXY (panel) tells us its velocity relative to Earth. When V M Slipher obtained the first galaxy spectra in 1921–5 there were two surprising results: most galaxies had spectra which were shifted towards the red, indicating that they were going away from us; and the velocities were large, up to 2000 km s^{-1} in the original sample. The amount of shifting to the red, or size of the *redshift*, is directly proportional to velocity due to the Doppler effect, unless the velocity is a fair fraction of the speed of light. So, if the wavelength of the BALMER SERIES of hydrogen is found to be, say, 1% longer for some galaxy, we speak of a redshift of $z = 0.01$, corresponding to a velocity of 3000 km s^{-1}.

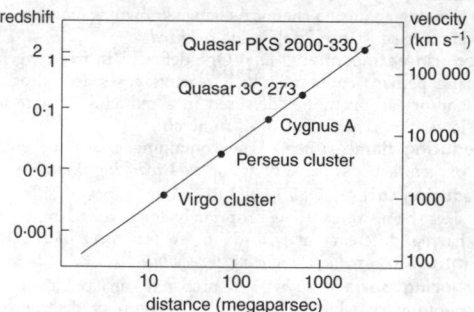

Hubble diagram

E Hubble at Mount Wilson built on this discovery of redshifts. He obtained velocities from the redshifts and distances for faint and distant galaxies. In 1929 he found a simple linear relationship between the distance (d) to a galaxy and the velocity (V): $V = Hd$. This relation came to be known as the Hubble law, now more properly called the HUBBLE RELATION. A graph showing the correlation between velocity (or redshift) and distance (or absolute luminosity) is termed the HUBBLE DIAGRAM (see diagram). The scale factor H was historically termed the *Hubble constant*. However, it varies with time as the universe evolves, and is therefore now called the HUBBLE PARAMETER. Determining this parameter is a major goal of extragalactic astronomy: its value is $50-100 \text{ km s}^{-1}$ per megaparsec of distance. The reciprocal of this quantity has the dimension of time and is known as the *Hubble time*. As a rough indication of the age of the universe, its value is 10–20 billion years.

Within the framework of the BIG BANG (panel) universe it is possible for the future of the universe to be either infinite expansion or ultimate collapse back to a singularity. In principle this can be decided by determining how the Hubble diagram curves at extreme distances, and this is one part of the mission of the Hubble Space Telescope. Recently, astronomers found hints from studies of supernovae that the expansion rate of the universe may be increasing with time. If so, the universe is likely to expand for ever.

Although the Hubble parameter has an uncertainty of perhaps a factor of two, this has not prevented astronomers from turning the Hubble relation round and using it to estimate distances to remote galaxies and QUASARS (panel). If the redshift z is determined, the distance is obtained directly from $d = cz/H$, where c is the velocity of light. More than 20 quasars are known with redshifts of about 4.0 or larger. These are the most distant observable objects.

See panels on COSMOLOGY, GALAXY, PULSAR and QUASAR.

---

reduction roasting (*Eng*) Use of heat in a controlled atmosphere to lower oxygen content of a mineral, eg iron (II) to iron (III) oxide by reaction with CO. Also *reducing roast*.

reduction to soundings (*Ships*) Correction of depths observed in tidal waters for comparison with charted depths referred to chart datum.

redundancy (*ICT*) The presence of components that improve the reliability of a (computer) system. Redundancy may involve (1) multiple copies of critical hardware, *hardware redundancy*; (2) alternative programs for critical operations, *software redundancy*; (c) error-correcting codes, *information redundancy*; (d) repeating critical operations several times, *time redundancy*.

redundant (*Eng*) A term applied to a structural framework having more members than it requires to be perfect. Also *overrigid*.

Redux bonding (*Aero*) TN for a method of joining primary sheet-metal aircraft structures with a two-component adhesive under controlled heat and pressure. It is widely used for making HONEYCOMB STRUCTURE sandwiches, for doubling sheet metal and for attaching STRINGERS or skin stiffeners.

red variable (*Astron*) See MIRA STAR.

redwater (*Vet*) (1) A disease of cattle caused by infection of the erythrocytes by the protozoon *Babesia bovis*, which is transmitted by the tick *Ixodes ricinus*; characterized by fever, diarrhoea, anaemia and haemoglobinuria (*moor ill*).

A similar tick-borne disease, called *Texas* or *tick fever*, is caused by *Babesia bigemina*. Also *babesiosis*, *blackwater*, *piroplasmosis*. (2) A disease of cattle, occurring in America, caused by *Clostridium haemolyticum*. Also *haemorrhagic disease*, *infectious icterohaemoglobinuria*.

redwood (*For*) (1) See SEQUOIA. (2) Name given to RED DEAL in N England.

Redwood second (*Eng*) Unit used in viscometry. See REDWOOD VISCOMETER.

Redwood viscometer (*Eng*) One of several designs of standard VISCOMETER in which viscosity is determined in terms of number of seconds required for the efflux of a certain quantity of liquid through an orifice under specified conditions.

red zinc ore (*Min*) See ZINCITE.

reed (*Acous*) Vibrating tongue of wood or metal, for generating air vibrations in musical instruments. Cane wood reeds are used for tongue action, as in the clarinet and saxophone, and a metal reed in organ reed pipes.

reed (*BioSci*) See ABOMASUM.

reed (*Textiles*) An arrangement of flattened steel wires fixed in a frame on a weaving machine to separate warp threads and fix their spacing, guide the weft insertion device, and beat up the weft into the fell of the cloth.

reed loudspeaker (*Acous*) Small loudspeaker with a driving mechanism in which the essential element is a magnetic reed, which is drawn into the gap between pole pieces on a permanent magnet by the currents in the driving coils.

reed pipe (*Acous*) Organ pipe in which pitch of the note is determined by vibration of a reed, the associated pipe reinforcing the generated note by resonance.

reed relay (*ElecEng*) One whose contacts are in the form of short straight springs (reeds of magnetic material which overlap slightly). Operation is by energizing a coil close to, or coaxial with, the reeds, which are then magnetized in such a sense that they are drawn together and the circuit is made.

reeds (*Build*) Moulding in form of several side-by-side beads sunk below the general surface.

Reed–Solomon code (*ICT*) A system of digital coding that treats symbols (groups of bits) rather than individual bits as the units to which error detection and correction are applied. It is highly effective for burst errors, in which a large number of errors may affect only a few neighbouring symbols. For this reason it is used on CDs, where scratches introduce localized error bursts.

reef (*Geol*) See CORAL REEF.

reef (*MinExt*) Originally an Australian term for a LODE. Now used for a gold-bearing tabular deposit or flattish lode.

reef knolls (*Geol*) Large masses of limestone formed by reef-building organisms; found typically in the Craven district of Yorkshire where they have weathered out as rounded hills above the lower ground on the shales. These are of Carboniferous age.

reef picking (*MinExt*) On the Rand, removal of gold-bearing blanket ore from barren waste rock, a reversal of the more usual hand sorting.

reel (*ImageTech*) (1) A flanged spool for holding film or tape. (2) A roll of film, esp one comprising a complete programme or a specific part of a programme.

reel barge (*MinExt*) A vessel carrying a very-large-diameter reel on which long lengths of oil pipe are wound and paid out during laying.

reel bogie (*Print*) A special truck (*reel truck*) used for transporting a reel of paper to a rotary press.

re-entrant (*Eng*) Angle or shape in moulding tool which would normally prevent product removal at end of cycle. If small, can often be jumped off, but if large, must be circumvented by using a side core or an alternative process, eg BLOW MOULDING. See UNDERCUT.

re-entrant angle (*MathSci, Surv*) Of a closed figure, eg a polygon, an angle which points inward, being greater than 180° as viewed from the interior. Cf SALIENT ANGLE (2).

re-entrant horn (*Acous*) A horn for coupling a sound-reproducing diaphragm with the outer air. To conserve space, the horn divides at a distance from the throat and, after convolutions, unites before expanding to the flare.

re-entrant polygon (*MathSci*) A polygon in which one or more internal angles is greater than 180°. Such an angle is called a *re-entrant angle*.

re-entrant program (*ICT*) One where a single copy of the program may be shared between several users at the same time.

re-entrant winding (*ElecEng*) A term used in connection with armature windings for dc machines; a *singly* (or *doubly*) *re-entrant winding* is one containing one (or two) independent closed circuits. The majority of windings are singly re-entrant.

re-entry (*MinExt*) Finding and connecting to a capped well on the sea bottom.

re-entry (*Space*) The period of return to Earth (or any other planet) when a spacecraft passes through the atmosphere to land on the surface. See panel on ROCKET.

re-entry corridor (*Space*) A narrow corridor which is available to a spacecraft returning to Earth (or any other body with an atmosphere) so that it can make a safe entry. Also *entry corridor*. See panel on ROCKET.

re-entry thermal protection (*Space*) The shielding of a body from the intense heat generated during atmospheric deceleration when its high kinetic energy is transferred to the atmosphere as heat. Stagnation temperatures of many thousands of degrees may be generated. Thermal protection is achieved by the following, alone or in combination: (1) an ablation shield which vaporizes and carries the heat away, a material of large latent heat (eg glass resin) being used; (2) a heat sink of high thermal capacity (eg copper); (3) a good thermal insulator (eg fibreglass); (4) radiative cooling with a high thermal emissivity surface.

refection (*BioSci*) In rabbits, hares and probably other herbivores, the habit of eating freshly passed feces. Also *autocoprophagy*.

reference address (*ICT*) One that is used as a locating point for a group of related addresses. See ADDRESS CALCULATION.

reference climatological station (*EnvSci*) A meteorological station where a homogeneous series of observations over weather elements over a period of at least 30 years have been, or are expected to be, made under approved conditions.

reference diode (*Electronics*) See ZENER DIODE.

reference electrode (*Chem*) An electrode used as a standard relative to which a varying potential is measured. See SATURATED CALOMEL ELECTRODE.

reference equivalent (*ICT*) The number of decibels by which a given piece of telephonic apparatus differs from the standardized piece of apparatus in the master transmission reference system.

reference frame (*Phys*) A set of position and time co-ordinates used to describe a physical experiment. Of special importance is the INERTIAL REFERENCE FRAME, in which no external forces act, so that an object remains at rest or moves at constant speed in a straight line.

reference level (*ICT*) A specified level of power, voltage or current in a circuit or system, to which all other levels of those quantities are referred, usually as a ratio expressed in dB; 1 mW is frequently taken as a reference signal power level, and is described as 0 dBm (zero decibels with respect to 1 mW). If voltages or currents are taken as reference levels, then it is customary and desirable to state the impedance level at which the measurement is to take place.

reference mark (*Print*) See MARK OF REFERENCE.

reference mark (*Surv*) A distant point from which angular distances to other marks may be taken at a station. Also *reference object*.

reference noise (*ICT*) Circuit noise level corresponding to that produced by 10^{-12} W of power at a frequency of 1 kHz.

reference power (*ICT*) See REFERENCE LEVEL.

reference system (*ICT*) See MASTER TELEPHONE TRANSMISSION REFERENCE SYSTEM.

reference voltage (*ElecEng*) (1) Closely controlled dc signal obtained from stable reference, eg a ZENER DIODE or STANDARD CELL. Often used for calibration purposes. (2) An ac voltage used to give both amplitude and phase reference, used eg in power system protection circuits.

reference volume (*ElecEng*) That transmission voltage which gives zero recording level on the standard volume unit meter.

refiner (*Paper*) A machine comprising a disk (or disks) or a cone, which rotates at high speed in a close-fitting casing or shell. Both the rotor and stator are fitted with metal bars and the material to be treated passes through the narrow gap between the two. Refiners of suitable types are used in the manufacture of mechanical wood pulp from chips or for treating paper stuff to effect a beating action.

refiner mechanical wood pulp (*Paper*) Mechanical wood pulp produced by subjecting chips of wood to the action of a refiner.

refinery (*Eng*) (1) Plant for refining crude oil to produce fuels and lubricants. (2) Plant where impure metals or mixtures are treated by electrolysis, distillation, liquation, pyrometallurgy, chemical or other methods to produce metals of a higher purity or specified composition.

refining (*Glass*) See FINING.

refining of metals (*Eng*) Operations performed after crude metals have been extracted from their ores, to obtain them at a higher purity.

reflectance (*Phys*) See REFLECTION FACTOR.

reflected (*BioSci*) Said of a structure, esp a membrane, that is folded back on itself.

reflected ray (*ICT*) See REFLECTED WAVE.

reflected wave (*ICT*) (1) One propagated back along a waveguide or transmission line system as a result of a mismatching at the termination. (2) See IONOSPHERIC WAVE.

reflected wave (*Phys*) A wave turned back from a discontinuity in a continuous medium.

reflecting galvanometer (*ElecEng*) See MIRROR GALVANOMETER.

reflecting level (*Surv*) An instrument, used for levelling, which employs the principle that a ray of light which strikes a reflecting plane at right angles is reflected back in the same direction. In its practical forms, it usually consists of a hanging mirror which takes up a position in the vertical plane, and has an unsilvered part through which a distant staff may be seen and also a reference horizontal line upon it. When the eye is in such a position that the image of the pupil is bisected by the horizontal line, the line of sight to staff is horizontal.

reflecting telescope (*Astron*) A telescope using a mirror to bring light rays to a focus; since the late 18th century, the world's largest telescopes have been reflectors. Several configurations are used under different circumstances: CASSEGRAIN, COUDÉ, GREGORIAN, MAKSUTOV, NEWTONIAN and SCHMIDT telescopes.

reflection (*ICT*) Reduction of power from the maximum possible, because a load is not matched to the source and part of the energy transmitted is returned to the source. Also, reduction in power transmitted by a wave filter due to iterative impedance becoming highly reactive outside the pass bands. In all instances, loss of power (the *reflection* or *return loss*) is measured in dB below maximum, ie when properly matched. See MISMATCH.

reflection (*NucEng*) The return of neutrons to a reactor core after a change of direction experienced in the shield surrounding the core.

reflection (*Phys*) The change of direction which occurs when an electromagnetic wave or sound wave strikes a surface and is thrown back.

reflection coefficient (*Acous*) The complex ratio of reflected pressure to incident pressure when a plane sound wave is incident on a discontinuity. The complex ratio includes changes in amplitude and in phase during the reflection.

reflection coefficient (*ICT*) The ratio of electric voltage or field amplitude for reflected wave to that for incident wave. Given by $(Z_2 - Z_1)/(Z_2 + Z_1)$, where Z_1 and Z_2 are the impedances of the medium (or line) and the load respectively. For acoustic reflection Z is the acoustic impedance.

reflection density (*ImageTech*) A measure of the light absorption of a surface, the logarithm of the reciprocal of the REFLECTION FACTOR.

reflection factor (*ICT*) (1) The ratio of current delivered to load to that which would be delivered to a perfectly matched load. (2) See REFLECTION COEFFICIENT.

reflection factor (*Phys*) The ratio of the luminous flux reflected from a surface to that falling upon it. Also *coefficient of reflection, reflectance*.

reflection laws (*Phys*) Two laws concerning wave propagation: (1) incident beam, reflected beam and normal to surface are coplanar; (2) the beams make equal angles with the normal.

reflection layer (*Phys*) One of the layers of very-low-pressure atmosphere partly ionized by particles from the Sun. Each layer progressively refracts electromagnetic waves, so that (below a definite frequency) they are effectively reflected downwards. See E-LAYER, F-LAYER.

reflection loss (*ICT*) See REFLECTION.

reflection nebula (*Astron*) A cloud of gas and dust in space which shines through the reflection of light from surrounding stars rather than through generating its own light.

reflection point (*ICT*) The point at which there is a discontinuity in a transmission line, and at which partial reflection of a transmitted electric wave occurs.

reflective mulch (*Agri*) A reflective covering of polythene or other material placed on the ground between rows of orchard trees to reflect light back into the canopy to promote fruit growth. It may also be used to repel insect pests that are sensitive to increased levels of ultraviolet light.

reflectivity (*Phys*) The proportion of incident energy returned by a surface of discontinuity.

reflectometer (*ICT*) A directional coupler connected to a transmission line or waveguide to extract a small proportion of the reflected power returning from a mismatched load; used for standing-wave measurements. See TIME-DOMAIN REFLECTOMETER.

reflectometer (*Phys*) An instrument measuring the ratio of the energy of a reflected wave to that of the incident wave in any physical system.

reflector (*Electronics*) The electrode in a reflex klystron connected to negative potential and used to reverse the direction of an electron beam. US *repeller*.

reflector (*ICT*) Part of an antenna array which reflects energy that would otherwise be radiated in a direction opposite to that intended.

reflector (*NucEng*) A layer of material (eg graphite) designed to scatter or reflect neutrons back into a reactor without absorbing many.

reflector (*Phys*) A device consisting of a bright metal surface shaped so that it reflects light falling on it in a desired direction.

reflector sight (*Aero*) Mirror gunsight which projects the aiming reticule and computed correction information for speed and deflection onto a transparent glass screen. Used first in World War II. Modern sights (head-up display) incorporate elaborate radar information and firing instructions which allow an attack and breakaway under completely blind conditions.

reflex (*Psych*) A simple, automatic, involuntary and stereotyped response to some stimulus (eg an eye-blink).

reflex action (*BioSci*) An automatic or involuntary response to stimulus.

reflex angle (*MathSci*) An angle greater than 180°. Compare: ACUTE, OBTUSE.

reflex arc (*BioSci*) The simplest functional unit of the nervous system, consisting of an afferent sensory neuron conveying nerve impulses from a receptor to the CNS, generally the spinal cord. The impulses are passed, either directly or via an internuncial or association neuron, to a motor neuron, which conveys them to a peripheral effector such as a muscle.

reflex camera (*ImageTech*) One which incorporates a negative-sized glass screen onto which the image is reflected for composing and focusing, either through a camera lens (*single-lens reflex*) or through a separate lens of the same focal length (*twin-lens reflex*). In the single-lens type the reflecting mirror is retracted when the shutter is released.

reflexed (*BioSci*) Bent back abruptly.

reflexive (*MathSci*) A relation, R, in a set S is reflexive if it applies from x to x for all elements x in S, ie $R(x,x)$ for all x in S.

reflex klystron (*Electronics*) A single-cavity klystron in which the electron beam is reflected back through the cavity resonator (see RHUMBATRON) by using a negative reflector. The beam, arriving back at the cavity in antiphase, can return energy in such a way as to cause oscillation; used as a microwave oscillator.

reflex projection (*ImageTech*) A method of composite photography by FRONT PROJECTION.

reflux (*Chem*) Boiling a liquid in a flask, with a condenser attached so that the vapour condenses and flows back into the flask, thus providing a means of keeping the liquid at its bp without loss by evaporation.

reflux oesophagitis (*Med*) Inflammation of the lower end of the OESOPHAGUS caused by regurgitation of stomach contents. US *reflux esophagitis*.

reflux ratio (*ChemEng*) In distillation design, the ratio of liquid reflux to vapour at any point in a column. In distillation plant operation, the reflux returned per unit quantity of condensate removed as product.

reflux valve (*CivEng*) A non-return valve used in pipelines at rising gradients to prevent water which is ascending the gradient from flowing back in the event of a burst lower down.

reforming process (*ChemEng*) A group of proprietary processes in which low-grade or low-molecular-weight hydrocarbons are catalytically reformed to higher-grade or higher-molecular-weight materials, eg PLATFORMING.

refracting telescope (*Astron*) A telescope using lenses to bring light rays to a focus, first applied to astronomy by Galileo. In its modern form, using lenses corrected for chromatic aberration, this telescope is still used by visual observers and amateurs.

refraction (*Phys*) A phenomenon which occurs when a wave crosses a boundary between two media in which its phase velocity differs. This leads to a change in the direction of propagation of the wavefront in accordance with SNELL'S LAW.

refractive index (*Phys*) For a transparent medium, the ratio of the phase velocity of electromagnetic waves in free space to that in the medium. It is given by the square root of the product of the complex relative permittivity and complex relative permeability. Symbol n. In anisotropic media, which include most minerals, there are two refractive indices (n_o and n_e) in uniaxial minerals and three (α, β, γ) in biaxial minerals. See GEM, REFRACTION, SNELL'S LAW.

refractive modulus (*EnvSci*) One million times the excess of MODIFIED REFRACTIVE INDEX above unity in M units.

refractivity (*Chem*) Specific REFRACTION.

refractometer (*Phys*) An instrument for measuring refractive indices. Refractometers used for liquids, such as the PULFRICH REFRACTOMETER, usually measure the critical angle at the surface between a liquid and a prism of known refractive index. See RAYLEIGH REFRACTOMETER. Fig. ▷

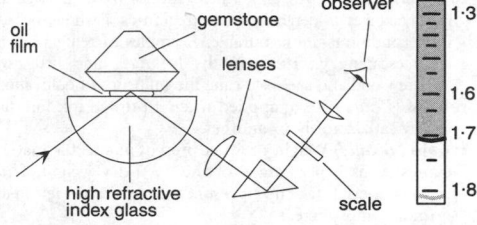

refractometer Observer's scale on right.

refractometric solids (*FoodSci*) The refractive index of a sugar solution measured using a refractometer and indicated on a scale which has been calibrated to directly show % soluble solids, degrees Brix or degrees Beaume.

refractor (*Phys*) A device by which the direction of a beam of light is changed by causing it to pass through the boundary between two transparent materials of different relative index.

refractories (*Eng*) Materials used in lining furnaces etc. They must resist high temperatures, changes of temperature, the action of molten metals and slags, and hot gases carrying solid particles. China clay, ball clay and fireclay are all highly refractory, the best qualities fusing at above 1700°C. Other materials are silica, magnesite, dolomite, alumina and chromite. See SILICA.

refractory alloy (*Eng*) An alloy which is: (1) difficult to work at high temperatures; (2) heat-resistant or having a very high melting temperature.

refractory cement (*Build*) A form of cement capable of withstanding very high temperatures.

refractory clay (*Eng*) See REFRACTORIES.

refractory concrete (*Build, CivEng*) Concrete able to withstand very high temperatures such as in the lining of chimneys. Constituents are normally high-alumina cement and crushed fire brick but other specially prepared aggregates are also used.

refractory metals (*Eng*) A term applied to transition group elements in the periodic table which have high melting points. They include chromium, titanium, platinum, tantalum, tungsten and zirconium.

refractory ore (*MinExt*) (1) Gold ore non-responsive to amalgamation process. (2) Ore of mineral or rock used in fabrication of REFRACTORIES, ie materials used in lining furnaces etc, eg CHROMITE, KYANITE.

refractory period (*BioSci*) For an organism or an excitable tissue, the unresponsive period immediately following a previous response.

refrangibility (*Phys*) Ability to be refracted.

refresh (*ICT*) To update a device with data, eg refresh a screen.

refrigerant (*Eng*) Substances suitable for use as working fluids in a two-phase refrigeration cycle. Ammonia and Freon-22 are most important in industrial refrigeration, Freon-11 and -12 in commercial and domestic use where non-toxic refrigerants are needed. Other examples include carbon dioxide and sulphur dioxide.

refrigeration cycle (*Eng*) Any thermodynamic cycle which takes heat at a low temperature and rejects it at a higher. The cycle must receive power from an external source.

refrigerator (*Eng*) A machine or plant by which mechanical or heat energy is used to produce and maintain a low temperature.

refringent (*Phys*) Refractive.

Refsum's disease (*Med*) A rare recessive genetic disorder in which there is an inability to metabolize a specific fatty acid, causing neuropathy, deafness, ATAXIA and CARDIO-MYOPATHY.

refuelling machine (*NucEng*) See CHARGE–DISCHARGE MACHINE.

refugium (*BioSci*) An area where species have survived the great changes undergone by the region as a whole, because local conditions are favourable. Examples of refugia are the areas escaping glaciation in the Ice Ages, and hedgerows (where woodland species escape the influence of cultivation).

refusal (*Print*) A term applied when a printed ink film fails to key satisfactorily to another.

regain (*Textiles*) Weight of water present in a textile material expressed as a percentage of the oven-dry weight. Dried textile materials take up or *regain* moisture when left in any normal atmosphere.

regatta (*Textiles*) Twill fabric, usually cotton, containing alternate stripes of white and colour, used in eg nurses' uniforms.

regelation (*Phys*) The process by which ice melts when subjected to pressure and freezes again when pressure is removed. Regelation operates when forming a snowball by pressure, in the flow of glaciers, and in the slow passage through a block of ice by a weighted loop of wire.

Regency (*Arch*) The last phase of English NEO-CLASSICAL movement which occurred during the regency of George, Prince of Wales (1810–20).

regenerated cellulose (*Chem*) Chemical dissolution of normally insoluble natural cellulose and reclamation from solution to produce fibre etc; techniques include cuprammonium method (now no longer used), hydrolysis of cellulose acetate and viscose process giving rayon.

regeneration (*BioSci*) Regrowth of tissues or organs, such as amphibian limbs, after injury; the formation of new plants from cultured tissues. See TISSUE CULTURE.

regeneration (*Electronics*) Same as POSITIVE FEEDBACK, but particularly applied to a super-regenerative receiving circuit, which oscillates periodically through self-quenching.

regeneration (*ICT*) Replacement or reforming of stored data, eg in a computer register or VOLATILE MEMORY.

regeneration (*MinExt*) (1) Reconstitution of liquid used in chemical treatment of ores before returning it to head of attacking process (eg in cyanide process). (2) Freshening of 'poisoned' ion-exchange resins.

regeneration (*NucEng*) Reprocessing of nuclear fuel by removal of fission products.

regenerative air heater (*Eng*) An air heater in which heat-transmitting surfaces of metallic plates, wire mesh or bricks are exposed alternately to the heat-surrendering gases and to the air.

regenerative braking (*ElecEng*) A method of braking for electric motors in which the motors are operated as generators by momentum of the equipment being braked, returning the energy to the supply.

regenerative detector (*ElecEng*) One in which the high-frequency components in the output are fed back to the input, thus increasing gain and selectivity.

regenerative furnace (*Eng*) A furnace in which the hot gases pass through chambers containing firebrick structures, to which the sensible heat is given up. The direction of gas flow is reversed periodically, and cold incoming gas is preheated in the chambers.

regenerative receiver (*ICT*) One with positive feedback for the carrier, enhancing efficiency of amplification and demodulation.

regenerator (*Eng*) Labyrinth which transfers heat of exit gases to air entering furnace, or feed-water to boiler.

regenerator (*ICT*) Circuits, used in electrical and/or optical communication systems, using PULSE-CODE MODULATION and placed at intervals along the transmission path. They detect incoming and retransmit stronger and more sharply defined output pulses. The pattern of pulses is unaltered, retaining the meaning of the transmitted information.

Regge trajectory (*Phys*) A graph relating spin angular momentum and energy for a nuclear particle. Possible quantized values of spin correspond to large discrete energy increments on the graph. This enables recurrences of nuclear particles to be predicted, the extra energy corresponding to the greater rest mass expected to be

associated with such particles. A *recurrence* is a particle identical in all respects, except energy (or mass) and spin momentum, with a known particle, and is regarded as being a higher-energy equivalent of the normal particle.

region (*MathSci*) See DOMAIN (1).

regional metamorphism (*Geol*) All those changes in mineral composition and texture of rocks due to compressional and shearing stresses, and to rise in temperature occasioned by intense earth movements over a widespread area. The characteristic products are the crystalline schists and gneisses.

regional roaming (*ICT*) A type of mobile-telephone ROAMING designed to deal with the situation where a number of networks are deployed in the same country, but with complementary coverage, in which a mobile registered with one network will switch to another when entering a region that its 'home' network does not cover.

region of limited proportionality (*Phys*) The range of operating voltages for a counter tube in which the gas amplification depends on the number of ions produced in the initial ionizing events as well as on the voltage. For larger initial events the counter saturates.

register (*Build*) (1) A metal damper to close a chimney. (2) A grilled aperture to allow the passage of hot or cold air.

register (*ICT*) (1) A location in the CENTRAL PROCESSOR that is used for special purposes only and is sometimes protected, eg ACCUMULATOR, CONTROL REGISTER, INDEX REGISTER. (2) Mechanical, electrical or electronic device that stores and displays one item of data.

register (*ImageTech, Print*) Exact correspondence of superimposed work, eg when the separate colours in colour photography are printed or projected together to reproduce the original picture, or when semiconductor processing masks are aligned with features defined in previous steps.

register tonnage (*Ships*) See NET REGISTER TONNAGE.

registered breadth (*Ships*) The breadth measured of the shell plating at widest part.

registered depth (*Ships*) The depth measured from top of CEILING to top of deck beam at midlength at the centre line of the vessel. Deck to which it is measured is usually stated.

registered dimensions (*Ships*) Dimensions appearing on the Certificate of Registry. Their main purpose is to identify the ship and they are also called *identification dimensions*. They are REGISTERED LENGTH, REGISTERED BREADTH, REGISTERED DEPTH.

registered length (*Ships*) The length from the fore side of the stem at the top to after side of stern post or, in a vessel without a stern post, to the centre of the rudder stock.

register length (*ICT*) The number of BITS that can be stored in a computer REGISTER.

register lock-up (*Print*) Mechanism allowing fine positioning of plates on the cylinders of web-fed presses.

register marks (*Print*) Fine lines, cross marks or similar, added to artwork to provide reference points and thus aid fitting and positioning of images during film assembly, plate-making and printing.

register pin (*ImageTech*) See PILOT PIN.

register rollers (*Print*) Adjustable rollers that provide a means of varying the web length between one unit of a web-fed press and another.

register sets (*Print*) A combination of *mixed forme base* and HONEYCOMB BASE, each supplied in a variety of accurately sized units, to be assembled with type to the size required for a particular plate, for which it is used to provide both a mount and a means of attaining register.

register sheet (*Print*) The sheet used in obtaining correct register or position.

reglet (*Arch*) (1) A flat narrow rectangular moulding. (2) A FACETTE.

reglette (*Surv*) The short graduated scale attached at each end of the special measuring tape or wire used in baseline measurement.

Regnault's hygrometer (*EnvSci*) A type of hygrometer in which the silvered bottom of a vessel contains ethoxyethane, through which air is bubbled to cool it, its temperature being indicated by a thermometer.

regolith (*Astron*) The layer of fine powdery material on the Moon produced by the repeated impact of METEORITES.

regolith (*Geol*) The mantle of rock material that overlies bedrock.

regrating (*Build*) Operation of redressing the faces of old hewn stonework.

regression (*BioSci*) A tendency to return from an extreme to an average condition, as when a tall parent gives rise to plants of average stature.

regression (*Geol*) The retreat of the sea from the land (stratigraphical usage).

regression (*MathSci*) A model of the relationship between the expected value of a random variable and the values of one or more possibly related variables.

regression (*Psych*) In psychoanalytic theory, a defence mechanism that involves a reversion to an earlier and less threatening mode of functioning, usually reverting to child-like behaviour.

regression analysis (*MathSci*) Analysis of the correlation between a dependent variable and one or more independent variables to predict or estimate a value of the former if the others are known.

regression testing (*ICT*) The phase of a testing cycle in which results are compared against a previous control version of the system or software, enabling the impact of changes to be understood.

regression therapy (*Psych*) A therapeutic technique that aims to return a patient to the emotional state of a child in order to identify the causes of psychological problems.

regular (*BioSci*) A radially symmetrical ACTINOMORPHIC flower.

regular convex solids (*MathSci*) Solids having congruent all faces bounded by plane surfaces and all corners. They are (1) *regular tetrahedron*, four equilateral triangular faces, (2) *regular hexahedron* or *cube*, six equal squares as faces, (3) *regular dodecahedron*, twelve regular pentagons as faces, (4) *regular octahedron*, eight equilateral triangles as faces, (5) *regular icosahedron*, 20 equilateral triangles as faces. Also *Platonic solids*.

regular-coursed (*Build*) Said of rubble walling built up in courses of the same height.

regular function (*MathSci*) See ANALYTIC FUNCTION.

regular polygon (*MathSci*) A POLYGON all of whose sides are equal and all of whose angles are equal, eg an equilateral triangle.

regular reflection factor (*Phys*) The ratio of the luminous flux regularly reflected from a surface to the total flux falling on the surface.

regular transmission (*Phys*) Transmission of light through a surface in such a way that the beam of light, after transmission, appears to proceed from the light source.

regular transmission factor (*Phys*) The ratio of the luminous flux regularly transmitted through a surface to the total luminous flux falling on the surface.

regulating rod (*NucEng*) Fine CONTROL ROD of reactor.

regulation (*Electronics*) (1) Fractional change in voltage level when a load is connected across a power supply, due to internal resistance. (2) Difference between minimum and maximum voltage drops across a reference diode over its range of operating currents. (3) The process of controlling a physical quantity (speed, temperature, position voltage, etc), by a control system or network employing negative feedback.

regulator (*Electronics*) A device or circuit which maintains a desired quantity (eg voltage, current, frequency or a mechanical property) at a predetermined level, usually by comparison with a reference source.

regulator cells (*ElecEng*) Cells at the end of a battery of accumulator cells which can be switched in and out of circuit in order to adjust the voltage of the battery as a whole. Also *end cell*.

regulator gene (*BioSci*) A gene whose product controls the rate at which the product of another gene is synthesized.

regulatory body (*Genrl*) Organizations set up by governments to oversee the proper use of method or technology. Particularly important in the nuclear industry, where such bodies have power to license use, construction and disposal.

regulatory lymphocyte (*BioSci*) Lymphocytes, of the T-cell lineage, that regulate the activities of other cells of the immune system. Also *regulatory T-cells*, T_{reg}.

Regulus (*Astron*) A bright blue-white multiple-component star in the constellation Leo. Distance 26 pc. Also *Alpha Leonis*.

regulus (*MathSci*) One of the sets of lines forming a RULED SURFACE.

regulus of antimony (*Eng*) Commercially pure metallic antimony. Also, the impure metallic mixture which is produced during smelting. 'Regulus' was the alchemical name for antimony, which readily combines with gold.

regurgitation (*BioSci*) (1) The bringing back into the mouth of undigested or partially digested food. (2) The flowing of blood in reverse direction to the circulation in the heart as a result of valvular disease, eg aortic regurgitation.

reheat (*Aero*) The injection of fuel into the jet pipe of a turbojet for the purpose of obtaining supplementary thrust by combustion with the unburnt air in the turbine efflux. Reheat is the UK, and original, term, but is gradually being superseded by the US term *afterburning*, with *afterburner* for the device itself.

reheating (*Eng*) The process of passing steam, which has been partially expanded in a steam turbine, back to a superheater before subjecting it to further expansion. Reheating is also used sometimes repeatedly, in heat-treating processes, like annealing, and in pneumatic systems for operating power tools. Also *resuperheating*.

reheating furnace (*Eng*) The furnace in which metal ingots, billets, blooms, etc, are heated to temperature required for hot-working.

Reichert–Meissl number (*Chem*) A standard used in butter analysis. A Reichert–Meissl number of n means that the soluble volatile fatty acids liberated from 5 g of butter fat under specified conditions require n cm^3 of 0·05M barium hydroxide solution for their neutralization.

Reimer–Tiemann reaction (*Chem*) The synthesis of phenolic aldehydes by heating a phenol with trichloromethane in the presence of conc KOH. The intermediate dichloro derivative is hydrolysed to an aldehyde. The CH=O group takes up the 2 or 4 position with respect to the hydroxyl group.

re-imposition (*Print*) (1) Transferring the page from one chase to another, the latter being perhaps a machine chase or a foundry chase. (2) Altering position of pages in a forme to suit a different size of paper or the requirements of printing and binding equipment.

reinforced concrete (*CivEng*) Concrete which is strong in compression, reinforced with steel which is strong in tension, designed so as to take advantage of both materials.

reinforced plastics (*Plastics*) General term for plastic composite materials, whether thermosetting or thermoplastic, in which the basic plastic has been reinforced by incorporating a fibrous material, eg paper, cloth, aramid, carbon or glass fibre, usually leading to enhanced stiffness, sometimes higher toughness, but reduced impact strength.

reinforcement (*Acous*) Sound reproduction, using loudspeakers in different positions, in which the received enhanced level appears to come from the actual source as required, eg in theatres. See HAAS EFFECT.

reinforcement (*Psych*) Situations when a response is predictably followed by an event, the *reinforcer*, and the event can be shown to increase or alter the future probability of the response. Reinforcement can be positive when the reinforcer is a reward for desirable behaviour, or negative when the desired behaviour is produced to avoid an unpleasant stimulus. See also PARTIAL REINFORCEMENT.

reinforcement schedule (*Psych*) Various schedules are used experimentally by behavioural psychologists. These include *interval schedules* where reinforcement is given for the first response made after a certain fixed or variable period of time has passed, *ratio schedules* where the number of responses required to elicit a reinforcing stimulus can be fixed or variable between tests, protocols where reinforcement is continuous and every response receives reinforcement, or is intermittent or partial.

Reinluft process (*EnvSci*) A process for removing oxides of sulphur and nitrogen from flue gases using catalysed oxidation at high temperature and adsorption onto carbon.

reinsertion (*ImageTech*) See DC RESTORATION.

Reiss microphone (*Acous*) Carbon transmitter in which a large quantity of carbon granules between a cloth or mica diaphragm and a solid backing, such as block of marble, are subjected to the applied sound wave. Characterized by high damping of the applied vibrational forces, and freedom from carbon noise by virtue of packing amongst the granules.

reiterated sequences (*BioSci*) See REPEATED DNA.

reiteration (*Surv*) A method of checking angular measurements made with a theodolite (and of securing greater accuracy) by repeating the observations after reversing face (turning the sighting telescope through 180°). Cf REPETITION.

Reiter's syndrome (*Med*) The association of ARTHRO-PATHY, CONJUNCTIVITIS and URETHRITIS thought to be due to a mycoplasma infection and, in the majority, of venereal origin.

rejection (*Med*) The process by which the body rejects tissue transplanted into it.

rejection band (*ICT*) See STOP BAND.

rejector circuit (*ICT*) Parallel combination of inductance and capacitance, tuned to the frequency of an unwanted signal, to which it offers a high impedance when placed in series with a signal channel.

rejuvenation (*Geol*) A term applied to the action of a river system which, following the uplift of the area drained by it, can resume down-cutting in the manner of a younger stream.

rejuvenescence (*BioSci*) Renewal of growth from old or injured parts.

relapsing fever (*Med*) A term applied to a number of diseases which are transmitted by lice or ticks and which are due to infection by various spirochaetes; characterized by recurrent attacks of fever and by enlargement of the liver and spleen. Also *spirochaetosis*.

relation (*MathSci*) A PREDICATE of two arguments; the set of all ordered pairs (x,y) with some property. Where x and y are members of sets S and T respectively, any subset p of $S \times T$ can be regarded as a *relation from S to T*.

relational database (*ICT*) Database using a relational DATA MODEL. The data are stored in the form of several two-dimensional tables or flat files. The tables embody different ideas about the data but contain overlapping information.

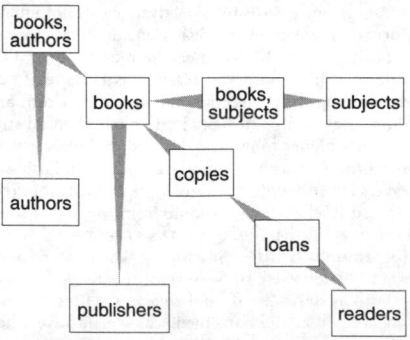

relational database Ten related databases for a loan library. The shaded triangles show how a selection in one database points to alternatives in another.

relational operator (*ICT*) Symbol used to express a relationship to be tested, eg >, <, =. See OPERATOR.

relative abundance (*BioSci*) A rough measure of population density, relative eg to time (as the number of birds seen per hour) or percentage of sample plots occupied by a species of plant.

relative abundance (*Phys*) See ABUNDANCE.

relative address (*ICT*) An address in memory determined by an offset from a previously specified address. Cf ABSOLUTE ADDRESS.

relative atomic mass (*Chem*) Mass of atoms of an element formerly in *atomic weight units* but now more correctly given on the *unified scale* where 1 u is $1 \cdot 660 \times 10^{-27}$ kg, where u is the ATOMIC MASS UNIT. For natural elements with more than one isotope, it is the average for the mixture of isotopes. Abbrevs *ram*, RAM. See panel on ATOMIC WEIGHT.

relative bearing (*Ships*) Angle between direction of ship's head and of an object.

relative biological effectiveness (*Radiol*) The inverse ratio of the absorbed dose of ionizing radiation to the absorbed dose of 200 kV X-rays which would produce an equivalent biological damage. Definition dates from early radiological measurements. Quality factors are rounded-off values used in the definition of the sievert. Abbreviation *rbe*. See panel on RADIATION.

relative density (*Phys*) The ratio of the mass of a given volume of a substance to the mass of an equal volume of water at a temperature of 4°C. Originally *specific gravity*.

relative efficiency (*Eng*) In an internal-combustion engine, the ratio of actual indicated thermal efficiency to that of some ideal cycle, such as AIR STANDARD CYCLE, at the same compression ratio.

relative excess risk (*BioSci*) A measure that is based on the component of risk due to the hazard or compound under investigation, removing the risk due to background exposure experienced by all in the population. Abbrev *RER*.

relative growth rate (*BioSci*) Abbrev *R*. Mathematical expression of growth given by

$$R = \frac{1}{W} \frac{dW}{dt}$$

where W is weight and t time.

relative humidity (*EnvSci*) The ratio of vapour pressure in a sample of moist air to the SATURATION VAPOUR PRESSURE WITH RESPECT TO WATER at the same temperature.

relatively prime (*MathSci*) Of two integers i and j having no common factors other than one. Also *coprime*.

relative molecular mass (*Chem*) Preferred term for MOLECULAR WEIGHT. Symbol M_r.

relative permeability (*ElecEng*) See PERMEABILITY.

relative permittivity (*ElecEng*) See PERMITTIVITY.

relative plateau slope (*NucEng*) See PLATEAU SLOPE.

relative risk (*BioSci*) (1) The ratio of the risk of disease or death among those exposed to the risk among the unexposed. Also *risk ratio*. (2) The ratio of the cumulative incidence rate in those exposed to the cumulative incidence rate in the unexposed population. Also *rate ratio*.

relative stopping power (*Phys*) See STOPPING POWER.

relative viscosity (*Phys*) The ratio of time for solution to fall a standard distance in a capillary viscometer (eg Ostwald or Ubbelohde) to time for solvent to fall the same distance in the same device. Also *viscosity ratio*.

relative visibility factor (*Phys*) The ratio of apparent brightness of a monochromatic source to that of a source of wavelength 550 nm having the same energy.

relativistic (*Phys*) Said of any deviation from classical physics and mechanics based on relativity theory.

relativistic mass equation (*Phys*) Equation giving the mass of a *relativistic* particle which has been accelerated up to a velocity (v) comparable with the speed of light (c):

$$m = \frac{m_0}{\sqrt{1 - v^2/c^2}}$$

where m_0 is the rest mass (mass at low velocities). The increase in mass has to be considered in cyclotron, betatron and linear accelerator design.

relativistic particle (*Phys*) Particle having a velocity comparable with that of light.

relativistic quantum mechanics (*Phys*) Quantum mechanics consistent with special relativity, and thus applicable to fast-moving systems; originally developed by UK physicist P Dirac in 1928. It predicts the observed fine detail of atomic spectra, and is essential in particle physics.

relativity (*Phys*) See GENERAL RELATIVITY, SPECIAL RELATIVITY and panel on COSMOLOGY.

relaxation (*Eng*) Exponential return of a system to equilibrium after sudden disturbance. The time constant of an exponential function is the RELAXATION TIME.

relaxation method (*CivEng*) A method of solving structural equations by making an initial estimate of the solution and then systematically reducing the errors in estimation.

relaxation oscillator (*ICT*) One that generates relaxation oscillations. Characterized by peaky or rectangular waveforms and the possibility of being pulled into step (locked) by an independent source of impulses of nearly the same frequency.

relaxation spectrum (*Chem*) A curve describing relaxation times of polymers as a function of temperature and strain. Can be derived from the master curve. See VISCOELASTICITY.

relaxation time (*BioSci*) In excitable tissues, the period during which activation subsides after cessation of a stimulus.

relaxation time (*Chem*) The time constant of *relaxation*, eg from the higher to the lower energy level in nuclear magnetic resonance.

relaxation time (*Eng*) The time for a perturbed system to return to equilibrium. More specifically the time taken for all or part of a polymer chain to respond to stress or strain stimulus. An important polymer property which determines the shape of the master curve. Temperature-sensitive since thermal motion of chains affects chain flexibility. Symbol τ. See VISCOELASTICITY.

relaxation time (*Phys*) A time-scale which characterizes exponential decay. For instance, the drift velocity $\langle v_x \rangle$ of electrons in a conductor will decay to zero after the field is removed according to

$$\langle v_x \rangle = \langle v_x \rangle_0 e^{-t/\tau}$$

where $\langle v_x \rangle_0$ is the velocity at time $t = 0$, and τ is the relaxation time for the decay.

relay (*ElecEng*) A device whereby a variation in the current in one electric circuit controls the current in a second circuit. *Allström*: a sensitive form using a light beam and photocell. *Control*: one operated by permitting the next step in a control circuit. *Differential*: one operating on the difference between eg two currents. *Frequency*: one operating with a selected change in the supply frequency. *Polarized*: one in which the movement of the armature depends on the current direction on the armature control circuit.

relay cell (*Med*) An internuncial neurone or interneurone of the central nervous system, particularly one forming a link between afferent and efferent neurons of a reflex arc.

relay spring (*ElecEng*) The spring which keeps a relay in the unoperated state.

relay valve (*Eng*) See PILOT VALVE.

rel.d. (*Phys*) Abbrev for RELATIVE DENSITY.

relearning (*Psych*) A method of measuring retention on a learning/memory task; the material is relearned again some time after the original learning. The difference between the original learning and relearning time is a measure of the original learning.

release (*ICT*) In automatic telephony, the release of apparatus that has been seized for making a connection. In manual telephony, the positive disengagement of apparatus on cessation of a conversation.

release (*ImageTech*) The trigger arrangement for releasing the shutter and effecting exposure in a camera.

release (*Print*) A term used in hot press stamping to describe the tendency of foil substrate to adhere to surface being marked. A foil with 'good release' is suitable for use with high-speed automatic machines. Release properties depend also on correct choice of temperature and pressure.

release mesh (*MinExt*) That at, and below, which screen size mineral is released from a closed crushing or grinding circuit and passed to next stage of treatment.

release paper (*Paper*) Papers treated so that an adhesive surface will become easily detached from them without rupture of the paper surface. Used as protective backing of self-adhesive materials.

release print (*ImageTech*) A print of a cinematograph film for public use in cinemas.

releaser (*Psych*) A term originating in classical ethology; it refers to aspects of stimulus (the sign stimulus) which are esp effective in releasing a specific response in all individuals of a species. It also implies that both the relevant stimulus features and the response to them have become mutually adapted through evolution. Also *social releaser*.

Relenza (*Pharmacol*) TN for a drug used in treating influenza; acts to inhibit the neuraminidase that allows the virus to spread in the subject's system. Proprietary name for *zanamivir*.

relevé (*BioSci*) A list of the plant species at any site with visual estimation of canopy cover.

reliability (*ElecEng*) The probability that equipment or a component will continue to function when required. Expressed as average percentage of failure per 1000 h of availability. Also *fault rate*.

relic (*MinExt*) Block of ore temporarily or permanently left close to a drive through solid rock, forming a wall between this and the stoped portion of the deposit. Cf REMNANT.

relict (*BioSci*) A species that has survived while other related ones have become extinct. In some cases they are remnant populations that once had a much wider range, eg some alpine plants restricted in the UK to a few high mountain areas. In other cases they have survived relatively unchanged from an older period while other members of the group have become extinct (eg horseshoe crabs).

relief block (*Print*) A letterpress printing block (eg line, half-tone) which can be used with printing type.

relief map (*Surv*) One with contour lines, shading, or colouring to indicate changes in surface configuration of the area mapped.

relief process (*ImageTech*) A printing process in which a photographic image processed to form a gelatine relief (matrix) is used to transfer dye to another layer on a separate support. See DYE TRANSFER, IMBIBITION.

relief process (*Print*) See PRINTING.

relief well (*MinExt*) One drilled into a reservoir to reduce the pressure in a burning or blown-out well or to inject water to flood it. A *killer well*.

relieving (*Eng*) (1) Interrupting a bearing surface, such as a machine tool slide or plain JOURNAL bearing, so as to improve alignment or lubrication conditions. (2) In cutting tools, removing material adjacent to a cutting edge so as to improve the flow of chips or cooling and lubrication conditions.

relieving gear (*Ships*) Any gear attached to the rudder, directly or indirectly, in such a way as to reduce stresses on the steering gear due to wave action on the rudder.

relight (*Aero*) A term used for igniting an aircraft gas turbine in flight after shut down.

relocatable program (*ICT*) One that runs regardless of where it is loaded into memory.

reluctance (*ElecEng*) The ratio of magnetomotive force, *mmf*, applied to a magnetic circuit or component to the flux in that circuit. It is the magnetic dual of RESISTANCE and the reciprocal of PERMEANCE.

reluctance pickup (*ElecEng*) A transducer for detecting vibrations or reproducing records. The signal causes a change in the reluctance of a magnetic circuit, which induces an emf linked to it.

reluctivity (*ElecEng*) The reciprocal of *permeability*.

REM (*Med*) Abbrev for RAPID EYE MOVEMENT.

rem (*Radiol*) Abbrev for ROENTGEN EQUIVALENT MAN.

remainder (*MathSci*) See DIVISION.

remanence (*Phys*) Magnetization remaining after an exciting magnetic field has been removed from ferromagnetic materials. See HYSTERESIS LOOP, RESIDUAL MAGNETIZATION.

remanent flux density (*Phys*) See RESIDUAL FLUX DENSITY.

remanent magnetization (*Geol*) That magnetization formed in a rock at the time of its formation, by the Earth's magnetic field or during some subsequent event. See MAGNETIC ANOMALY PATTERNS.

remiges (*BioSci*) In birds, the large contour feathers of the wing. Sing *remex*.

remiped (*BioSci*) Having the feet adapted for paddling, as many aquatic birds.

remission (*Med*) An abatement (often temporary) of the severity of a disease; the period of such abatement.

remittent (*Med*) Of a fever, characterized by remissions in which the temperature falls, but not to normal; as in malaria.

remnant (*MinExt*) Block of ore or stope pillar left *well clear* of the underground travelling ways on completion of stoping. Cf RELIC.

remodulation (*ICT*) Transferring modulation from one carrier to another, as in the frequency changer in a supersonic heterodyne receiver.

remote access (*ICT*) Access through a terminal physically separated from the computer.

remote augmented lift system (*Aero*) Engine designed for STOVL aircraft employing nozzles, remote from the engine, powered by compressed air ducted from the compressor and heated by fuel burnt in the nozzle in a manner similar to an *afterburner*.

remote concentrator unit (*ICT*) Equipment associated with a local exchange or CENTRAL OFFICE that switches its many individual customer lines onto the much smaller number of fixed channels connecting it with a MAIN SWITCHING UNIT.

remote control (*ElecEng*) Control of distant equipment by electrical or other signals, in response to information about the equipment's state.

remote cut-off (*Electronics*) A characteristic whereby a large negative bias is needed for complete cut-off of conduction output in a valve or other amplifying device.

remote handling equipment (*NucEng*) Apparatus developed to enable an operator to manipulate highly radioactive materials from behind a suitable shield, or from a safe distance.

remote mass-balance weight (*Aero*) A MASS-BALANCE weight which, usually because of limitations of space, is mounted away from the control surface, to which it is connected by a mechanical linkage.

remote sounding (*EnvSci*) See EARTH OBSERVATION.

remould (*Autos*) Used tyre which has had a new tread vulcanized to the casing, including a coating of rubber on the walls. Cf RETREAD.

removable isolated singularity (*MathSci*) If a function $f(z)$, having a singularity at $z = a$, has no negative powers of $(z-z_0)$ in LAURENT'S EXPANSION, then the singularity may be removed by defining $f(a) = a_0$, where a_0 is the first coefficient of $(z-z_0)$ in the expansion.

removes (*Print*) Quotations etc set in smaller type than the main text. The difference in size is usually 2 points. Thus a book the text of which is set in 12-point or pica should have its quotations set in 10-point.

REM sleep (*Psych*) See PARADOXICAL SLEEP.

Renaissance (*Arch*) The cultural movement, inspired by classical ideals, which spread from Tuscany throughout Europe in the 15th century, and which had a major effect on architectural style. The movement passed through various stages in its development, these stages being known generally as Early Renaissance, High Renaissance, Baroque and Antiquarian; however, each European country produced regional variations, and in the British Isles the phases relating to the above are known as Elizabethan, Jacobean, Stuart and Georgian respectively, the Georgian lasting until about 1830.

renal (*BioSci*) Pertaining to kidneys.

renal colic (*Med*) Severe loin pain usually due to obstruction of the ureter by a stone.

renal portal system (*BioSci*) In some lower vertebrates, the part of the venous system which transfers blood from the posterior part of the body to the kidneys.

renaturation (*BioSci*) The converse of DENATURATION. Complementary strands of DNA or DNA and RNA will reform duplex molecules. The kinetics of the process depends on the number of copies and the concentration of the molecules. Usually achieved by heating single strands at about 20°C below T_m. The basis of molecular PROBES.

render and set (*Build*) Two-coat plasterwork on walls.

rendered (*Build*) A term applied to laths which are split rather than sawn. The fibres are not severed so maximum strength is retained, ie no short grain. Also *split* or *riven lath*.

render, float and set (*Build*) Three-coat plasterwork on walls.

rendering (*Build*) The operation of covering brick and stonework with a coat of coarse stuff; the coating itself.

rendezvous (*Space*) The meeting and bringing together of two spacecraft in orbit at a planned place and time; also the type of mission where a spacecraft encounters a target body with zero relative velocity at a preplanned time and place.

rendzina (*EnvSci*) Shallow, dark, intrazonal soil, rich in calcium carbonate, developed on limestone, esp chalk.

renewable resource (*EnvSci*) A resource that either is not depleted by its use (eg solar radiation, the motion of wind, rivers, waves or tides) or is replenished at a rate comparable with the rate at which it is consumed (eg agricultural produce).

reni- (*Genrl*) Prefix meaning relating to the kidneys.

reniform (*BioSci*) Kidney-shaped.

renin (*Med*) Peptidase liberated from the glomerular apparatus in the kidney into the blood stream when blood pressure is reduced. This reacts with angiotensinogen to produce ANGIOTENSIN I.

rennet (*Chem, FoodSci*) A commercial preparation used in the production of cheese and junket (clotted milk); obtained from the mucous membrane of the stomach of calves, and containing the peptidase chymosin (*rennin*).

renormalization (*Phys*) A mathematical procedure in quantum field theories for avoiding infinite results in calculations by a careful redefinition of basic quantities such as mass and charge. The requirement of renormalization, as displayed eg by quantum electrodynamics, is regarded as prerequisite for a useful theory.

Reoviridae (*BioSci*) Viruses with a segmented double-stranded RNA genome: only one strand of the RNA acts as template for mRNA. Included in this group are several pathogenic viruses including orbivirus, a tick-borne virus that causes Colorado tick fever.

rep (*Textiles*) A plain-weave fabric with weft ribs formed by using coarse and fine yarns in alternate order in both warp and weft. Also *repp*.

repeated DNA (*BioSci*) A sequence that occurs more than once in the haploid genome. Such sequences are often short and occur many times.

repeater (*ICT*) A device used to regenerate, amplify and retransmit signals. By this means the transmission distance of a link may be increased. Commonly used with ETHERNET systems. Repeaters function at the physical layer, layer 1 of the OPEN SYSTEMS INTERCONNECTION (OSI) model.

repeater (*MathSci*) See RECURRING DECIMAL.

repeating back (*ImageTech*) Sliding back for a camera, in which images for colour separation can be taken successively and side by side.

repeating decimal (*MathSci*) See RECURRING DECIMAL.

repeat unit (*Chem*) See panel on POLYMERS.

repeller (*Electronics*) US for REFLECTOR.

reperfusion injury (*Med*) Damage that occurs to tissue when blood flow is restored after a period of ischaemia. It can exacerbate the effects of stroke or cardiac ischaemia and can follow extended periods of arterial occlusion during surgical operations. Also *ischaemia–reperfusion injury*.

repertoire (*Psych*) The complete set of behaviours or an organism or a species.

repetition (*Surv*) A method of checking angular measurements made with a theodolite by repeating the observation after unclamping the lower plate and sighting on the back station so that the vernier reading is unaltered, and then sighting forwards to get a new reading on the vernier, which should be double the previous reading. Cf REITERATION.

repetition compulsion (*Psych*) In psychoanalytic theory, the factor in mental life which compels early patterns of behaviour to be repeated, irrespective of pleasure/displeasure thereby experienced by the individual.

repetition rate (*ICT*) The rate at which recurrent signals, usually pulses, are repeated.

replaceable hydrogen (*Chem*) Those hydrogen atoms in the molecule of an acid which can be replaced by atoms of a metal on neutralization with a base.

replacement (*Geol*) The process by which one type of rock occupies the space previously occupied by another rock; also applies to minerals.

replica plating (*BioSci*) Typically, transferring the *pattern* of bacterial colonies on an agar plate by impressing velvet, stretched over a holder, onto the agar and then placing the velvet in turn onto a number of further sterile plates. If the latter contain, say, different antibiotics, a sensitive strain can be selected by noting colonies which fail to grow and then picking them from the master plate.

replication (*BioSci*) Duplication of genetic material, usually prior to cell division.

replication fork (*BioSci*) The fork where duplex DNA becomes split into two double strands as replication moves from the *origin of replication*.

replicon (*BioSci*) A part of a DNA molecule that is replicated from a single origin. In prokaryotes there is usually one origin per genome, but in eukaryotes there are many spaced along the chromosome.

replum (*BioSci*) A partition across an ovary or fruit formed by ingrowth from the placentas, not by the walls of the carpels. Also *false septum*. Cf SEPTUM.

report call (*ICT*) A call made to ascertain whether a desired subscriber is available for connection.

reporter gene (*BioSci*) A gene that encodes an easily assayed product that is coupled to the upstream sequence of another gene and transfected into cells. The reporter gene can then be used to see which factors activate response elements in the gene of interest.

report generator (*ICT*) Software that gives the non-specialist user the capability of producing reports from one or more files, through easily constructed statements.

repp (*Textiles*) See REP.

representative sample (*MinExt*) Sample cut from the bulk of material or ore deposit in such a way as to make it reasonably representative of the whole body.

repression (*Psych*) In psychoanalytic theory, the process by which an unacceptable thought, impulse or memory is rendered unconscious.

repressor (*BioSci*) A protein that binds to an OPERATOR site and prevents transcription of the associated gene.

reprocessing (*NucEng*) See FUEL REPROCESSING.

reproducer (*Acous*) (1) Complete sound reproduction system. (2) Loudspeaker.

reproducibility (*Genrl*) The precision with which a measured value can be repeated in a process or component.

reproduction (*Acous*) Replay of recorded sounds.

reproduction (*BioSci*) The generation of new individuals in the perpetuation of the species. Adj *reproductive*.

reproduction constant (*Phys*) See MULTIPLICATION CONSTANT.

reproduction proof (*Print*) Print taken of a letterpress forme, on ART PAPER or BARYTA PAPER, from which a plate is to be made photomechanically.

reproductive behaviour (*Psych*) The varied activities that lead to production and rearing of offspring. Includes AGONISTIC BEHAVIOUR, courtship and other mate interactions, as well as maternal behaviour.

reproductive cloning (*BioSci*) Cloning, by nuclear transfer, of an organism that is then allowed to develop fully. In the case of mammals this means that the blastocyst must be transferred into a surrogate mother, successfully implant and come to term. In most countries this is illegal for humans, although the technique has been used in sheep and a few other species. Strictly speaking, the embryo is chimeric, since only nuclear genes have been transferred and mitochondrial genes are derived from the egg into which the nucleus was transplanted.

Reptilia (*BioSci*) A class of Craniata. They are pentadactyl and have shelled amniote eggs. The vertebrae are gastrocentrous, the kidney metanephric and the skin completely covered by epidermal scales or, sometimes, by bony plates. They are poikilothermous, breathe by lungs, and retain aortic arches. Known as fossils from Late Carboniferous, they were dominant and various in the Mesozoic (dinosaurs) but became less numerous in the Cretaceous. Living forms include lizards, snakes, turtles, tortoises, crocodiles and alligators. See panel on VERTEBRATE EVOLUTION.

repugnatorial glands (*BioSci*) In Arthropoda, glands, usually abdominal in position, that secrete a pungent or corrosive substance used in self-defence.

repulsion (*BioSci*) Situation in which two specified *non-allelic* genes are on different but homologous chromosomes, having come from different parents. Cf COUPLING.

repulsion–induction motor (*ElecEng*) Single-phase induction motor having, in addition to the squirrel-cage winding on the rotor, a commutator winding with its brushes short-circuited, so that the motor starts as a repulsion motor with a high starting torque and runs with the characteristics of an induction motor.

repulsion motor (*ElecEng*) A type of single-phase commutator motor in which power is supplied to the stator winding with the armature winding short-circuited through the brushes.

repulsion-start induction motor (*ElecEng*) A repulsion motor having a centrifugal device which short-circuits all the commutator bars when the motor reaches a certain speed, so that it runs as a single-phase induction motor, and starts as a repulsion motor with a high starting torque.

requirement (*Space*) The demand placed on any element of a space system (such as spacecraft, payload, subsystem, communications network or ground organization) which has to be satisfied by that element.

reradiation (*ICT*) Radiation from resonating elements, such as masts, antennas, telephone lines, giving errors in bearings or misplaced TV images.

rere arch (*Arch*) A flat soffit arch laid over splayed jambs.

rerecording (*Acous*) Recording acoustic waveforms immediately upon reproduction from the same or any other type of recording medium as that in use.

re-reeler (*Print*) An auxiliary unit to rewind the reel on web-fed presses for subsequent operations.

reroute on busy (*ICT*) An INTELLIGENT NETWORK service whereby the system automatically tries one or more prearranged alternative numbers if the first number dialled is busy.

rerun (*ICT*) Repeat of part of a program for a computer. Many programs incorporate a rerun for every few minutes' running time. In the event of error only the part subsequent to the previous rerun point is repeated. See FAULT TOLERANCE.

réseau (*Astron*) A reference grid of fine fiducial lines or points used in image analysis and measurement.

réseau (*ImageTech*) A mosaic formed by rulings of fine coloured lines in red, green and blue used as a screen in some early additive processes of colour photography.

resection (*Med*) Cutting off a part of a bodily organ, esp the ends of bones and other structures forming a joint.

resection (*Surv*) Positional fix of a point which is not going to be occupied by sighting it from two or more known stations.

reserpine (*Pharmacol*) An alkaloid obtained from *Rauwolfia serpentina*, formerly used for the treatment of hypertension.

reserve buoyancy (*Aero*) Potential buoyancy of a seaplane or amphibian which is in excess of that required for normal floating. The downward force required for complete immersion.

reserve buoyancy (*Ships*) Watertight volume above the load water line.

reserved word (*ICT*) Identifier used by the system and therefore not available to a user.

reserve factor (*Aero*) The ratio of actual strength of an aircraft structure to estimated minimum strength for a specified load condition.

reserves (*MinExt*) Block of ore proved by development to warrant extraction. This is normally done by driving levels and winzes or raises so as to expose its four sides in a rectangular panel which is sufficiently sampled and tested. Cf RESOURCES.

reservoir (*MinExt*) See PETROLEUM RESERVOIRS.

reservoir (*NucEng*) Any volume in an isotope separation plant which is for the purpose of storing material or to ensure smooth operation.

reservoir pressure (*MinExt*) The pressure in a natural oil or gas reservoir which causes flow to the borehole. Maintenance of a proper pressure in the reservoir by not allowing too high an extraction rate is important in ensuring good recovery.

reset (*ElecEng*) General term for returning any circuit or equipment to its initial state, ready for restarting.

reset (*ICT*) To return a computer to its initial state (eg by restoring all registers to known values). See BOOT.

reset circuit (*ElecEng*) One which, when operated, performs a RESET.

reshaping (*ElecEng*) Restoring a distorted waveform to its intended shape.

residence time (*Genrl*) The term given by dividing the volume of any reservoir or pool or processing unit by the rate of flow through it. The expression gives the average time spent in the pool by the substance or organism under consideration.

resident program (*ICT*) See MEMORY-RESIDENT PROGRAM.

residual (*MathSci*) The difference between an observed value and that predicted by a statistical model.

residual activity (*NucEng*) In a nuclear reactor the remaining activity after the reactor is shut down following a period of operation.

residual deposits (*Geol*) Accumulation of rock waste resulting from weathering *in situ*. They cover the whole range of grain size, from residual boulder beds to residual clays.

residual errors (*Phys*) Errors which remain in an observation despite all attempts to eliminate them.

residual field (*Phys*) Magnetic field remaining in a magnetic circuit after removal of the magnetizing force.

residual flux density (*Phys*) The magnetic flux density of a bulk material sample when a magnetizing field is restored to zero after saturation has been achieved. Also *remanent flux density*.

residual gas (*Electronics*) Small amount of gas which inevitably remains in a 'vacuum' tube after pumping. If present to excess, it causes erratic operation of the tube, which is said to be 'soft'.

residual induction (*Phys*) See RESIDUAL FLUX DENSITY.

residually excited linear predictive coding (*ICT*) A type of LINEAR PREDICTIVE CODING in which the spectral coefficients derived from speech are used to inverse-filter it, giving a residual signal of which the low-frequency components are transmitted. The high-frequency components are reconstructed at the receiver and the reconstituted residual signal used to excite a synthesizer controlled by the spectral coefficients to recreate the original speech.

residual magnetization (*Phys*) Magnetization remaining in a ferromagnetic material after the exciting magnetic field has been removed. See REMANENCE.

residual resistance (*Phys*) Electrical resistance persisting at temperatures near zero on the absolute scale, arising from crystal irregularities and impurities, and in alloys.

residual stress (*Eng*) Stresses remaining within a material after processing. For equilibrium, residual tensile stresses must be balanced by compressive ones and vice versa. Also *internal stress*.

residual volume (*Med*) Volume of air left in the lungs after the strongest possible forced expiration, usually of the order of 1500 cm^3 in humans.

residue (*FoodSci*) Substance, eg pesticide, fertilizer, hormone, plastic monomer, present in a food product as a result of CARRYOVER or MIGRATION usually in parts per million, billion or trillion. Statutory acceptable limits for residues are in force in most countries.

residue (*MathSci*) (1) The nth *power residues* of p are the remainders when r_n, for $r = 1,2,3,\ldots$, are divided by p. Thus the quadratic residues of 13 are 1,3,4,9,10 and 12. (2) Of an analytic function $f(z)$ at an isolated singularity,

$$\frac{1}{2\pi c} \int_c f(z)\,dz$$

where c is a simple closed curve around the singularity, and the singularity is deleted from the region contained therein.

resilience (*Eng*) Stored energy of a strained material, or the work done per unit volume of an elastic material by a bending moment, force, torque or shear force, in producing strain.

resilience (*EnvSci*) The capacity of ecosystems and populations to return to a previous state after they have been disturbed.

resilin (*BioSci*) An elastomeric protein occurring in insects. Cf ELASTIN.

resin (*Build, Chem*) A constituent of varnishes or paints, classified as (1) natural resin, eg copal, rosin, amber; (2) synthetic resin, eg alkyd, phenolic, polyurethane. Natural resins are chiefly obtained as the exudation from plants or trees and have a glassy appearance with a slightly yellow-brown colouring. They consist of highly polymerized acids and neutral substances mixed with terpene derivatives. Synthetic resins are specially formulated to impart particular qualities to coatings such as hardness, flexibility, chemical resistance, etc.

resin (*Plastics*) The term resin was widely but loosely used to describe any synthetic plastics material. Now, more precisely, it is applied to a thermosetting polymer prior to curing, eg epoxy resin, polyester resin. See panel on THERMOSETS.

resinates (*Chem*) Calcium, magnesium, aluminium, iron, nickel, cobalt, zinc, tin, manganese and lead salts or ROSIN, obtained by fusion of rosin with metal oxides.

resin-bonded plywood (*Build*) Plywood in which the wood veneers are held together with synthetic resin, glues, or glue-impregnated paper and finally formed with pressure and heat.

resin duct (*BioSci*) A duct of schizogenous origin lined with resin-secreting cells, as in the leaves and sometimes wood of conifers. Also *resin canal*.

resin-in-pulp (*MinExt*) Ion-exchange method of continuously treating ores by acid leaching. Baskets containing ION-EXCHANGE RESINS are jigged through tanks containing finely ground ore pulp as it flows from vessel to vessel. Abbrev *RIP*.

resinous substances (*Chem*) Term applied to (1) true resins; (2) substances resembling true resins in their physical properties.

resin poisons (*MinExt*) In ion-exchange processes, substances which reduce efficiency of resin loading by masking activated resin sites.

resin soaps (*Chem*) See SOAPS.

resist (*Electronics*) Polymer material used in LITHOGRAPHY. Positive resists become more soluble in a developer after exposure to radiation (eg ultraviolet) whereas negative resists become less so.

resist (*ImageTech*) A coating of chemically neutral substance placed over a surface when the latter has to be protected at some stage in processing, as in etching or selective dyeing.

resist (*Print*) When making printing surfaces, the protection over the lines and dots of the image which preserve it from the action of etching fluid.

resistance (*Phys*) In electric and acoustic fields, the real part of the impedance characterized by the dissipation of energy as opposed to its storage. Electrical resistance may vary with temperature, polarity, field illumination or the purity of materials. SI unit is the OHM, symbol Ω. Reciprocal of CONDUCTANCE. See IMPEDANCE, OHM'S LAW, REACTANCE.

resistance (*Psych*) The opposition encountered during psychoanalytic treatment to the process of making unconscious memories and impulses conscious.

resistance box (*ElecEng*) One containing carefully constructed and adjusted resistors, which can be introduced into a circuit by switches or keys. At higher frequencies there are disturbing inductive and capacitive effects which complicate measurements using resistance boxes, but which are mitigated by suitable design. The boxes are then described as *non-reactive*.

resistance butt-seam welding (*ElecEng*) Resistance welding process in which coaxial roller electrodes conduct current to the edges of the seam to be joined, the mechanical pressure being applied independently. Extensively used in tube-making from strip.

resistance butt-welding (*ElecEng*) A resistance welding process in which the two parts to be joined are butted together. See RESISTANCE BUTT-SEAM WELDING, RESISTANCE FLASH-WELDING, RESISTANCE UPSET-BUTT WELDING.

resistance–capacitance coupling (*ElecEng*) That in which signal voltages developed across a load resistance are passed to the subsequent stage through a dc blocking capacitor.

resistance–capacitance oscillator (*ElecEng*) One producing a sine waveform of frequency determined by phase shift in a resistance–capacitance section of an artificial line.

resistance coupling (*ElecEng*) See DIRECT COUPLING.

resistance drop (*ElecEng*) Voltage drop produced by a current flowing through the resistance of a circuit; equal to the product of current and effective resistance.

resistance flash-welding (*ElecEng*) A resistance welding process in which an arc is struck and maintained between the parts until the correct temperature is attained, after which the current is cut off and the parts are forced together by mechanical pressure. Also *flash-butt welding*.

resistance grid (*ElecEng*) A resistance unit generally used for heavy currents. Made up of a cast-iron grid designed so that current enters one end and passes through all the sections in series, before leaving at the other end.

resistance lap-welding (*ElecEng*) A resistance welding process in which the two parts to be joined overlap one another. See RESISTANCE SEAM-WELDING, RESISTANCE SPOT-WELDING, RESISTANCE STITCH-WELDING.

resistance noise (*Electronics*) See THERMAL NOISE.

resistance percussive-welding (*ElecEng*) A resistance welding process in which a heavy electric current is discharged momentarily across the electrodes, and a momentary mechanical force is applied simultaneously.

resistance projection-welding (*ElecEng*) A variant of resistance spot-welding in which current is concentrated at the desired points by projections on one of the parts.

resistance pyrometer (*ElecEng*) See RESISTANCE THERMO-METER.

resistance seam-welding (*ElecEng*) A resistance welding process in which the welding electrodes consist of two rollers having mechanical pressure between them through which the work passes while the current flows continuously or intermittently, producing a line of overlapping welds.

resistance spot-welding (*ElecEng*) A resistance welding process in which the electrodes consist of two points and cause welding in one spot.

resistance stitch-welding (*ElecEng*) A form of resistance spot-welding consisting of a series of overlapped spot welds to form a seam weld.

resistance strain gauge (*ElecEng*) Foil, wire or thin-film resistor which has a value which varies with mechanical strain. Normally used in a bridge circuit. See STRAIN GAUGE.

resistance thermometer (*ElecEng*) One using resistance changes for temperature measurement. Resistance element may be platinum wire for extreme precision or semiconductor (*thermistor*) for high sensitivity. Also *resistance pyrometer* for higher temperature.

resistance upset-butt welding (*ElecEng*) A resistance welding process in which mechanical pressure is first applied to the joint and then current is passed until welding temperature is reached and the weld is 'upset'. Also *slow-butt welding*.

resistance welding (*ElecEng*) Pressure welding, in which the heat to cause fusion of the metals is produced by the welding current flowing through the contact resistance between the two surfaces to be welded, these being held together under mechanical pressure. See RESISTANCE BUTT-WELDING, RESISTANCE FLASH-WELDING, RESISTANCE PERCUSSIVE-WELDING, RESISTANCE SEAM-WELDING, RESISTANCE SPOT-WELDING.

resistant (*BioSci*) Not readily attacked by a parasite, disease or drug. See ANTIBIOTIC RESISTANCE.

resistant variety (*Agri*) A crop plant selected and multiplied at a commercially important level because of inheritable resistance to a specific disease.

resistive component (*ElecEng*) That part of the impedance of an electrical system which leads to the absorption and dissipation of energy as heat.

resistive load (*ElecEng*) Terminating impedance which is non-reactive, so that the load current is in phase with the source emf. Reactive loads are made entirely resistive by adding inductors or capacitors in series or shunt (tuning).

resistivity (*ElecEng*) Intrinsic property of a conductor, which gives the resistance in terms of its dimensions. If R is the resistance in ohms, of a wire l m long, of uniform cross-section a m^2, then $R = \rho l / a$, where the resistivity ρ is in ohm metres (*not* ohm m^{-3}). Also, erroneously, *specific resistance*.

resistor (*ElecEng*) Electric component designed to introduce known resistance into a circuit and to dissipate accompanying loss of power. Types are wirewound, composition, metal film, etc.

resit (*Chem*) C-stage phenolic resin made from a resol.

resitol (*Chem*) B-stage phenolic resin made from a resol.

resol (*Chem*) A type of phenolic resin prepolymer made by reacting phenol with excess formaldehyde under basic conditions. Also called *one-stage resin*, since no other agents need be added for final cross-linking. Both resols and novolaks are A-stage resins.

resolution (*Chem*) The separation of an optically inactive mixture or compound into its optically active components.

resolution (*ICT*) (1) A rule of inference that is used for automatic theorem proving. Also *resolution principle*. (2) The number of elements per unit length available for display, scanning or printing by the particular device; eg 1200 dots per inch is considered to be excellent resolution. See GRAPHICS RESOLUTION.

resolution (*ImageTech*) The ability of an imaging system to differentiate between closely spaced objects. With film the resolving power is usually expressed as the maximum number of light and dark line-pairs per millimetre which can be observed. With video the resolution can be expressed in terms of BANDWIDTH, lines or PIXELS.

resolution (*Med*) Retrogression of the phenomena of inflammation; the subsidence of inflammation. V *resolve*.

resolution of forces (*Phys*) The process of substituting two forces in different directions for a single force, the latter being equal to the resultant of the two components. If these are at right angles to each other, the one which makes an angle θ with the original force P is equal to $P\cos\theta$, the other being $P\sin\theta$.

resolution time (*Phys*) The minimum time interval between two events recorded separately by a detector of ionizing radiation, or the maximum time between two events recorded as coinciding.

resolution-time correction (*Phys*) Correction applied to observed counting rate for random events as measured by a detector of ionizing radiation, which allows for those events not recorded because of finite resolution time.

resolvant equation (*MathSci*) An equation which is used in the solution of a higher-order equation.

resolving power (*Phys*) The ability of an instrument to produce separate images or distinguish two similar quantities; eg in an astronomical telescope to measure the angular separation of two images which are close together, or in a mass spectrograph to separate particles of different masses. See RESOLUTION.

resolving power of the eye (*Phys*) The angle subtended by a small object which can just be determined visually.

resolving time (*Phys*) See RESOLUTION TIME.

resonance (*Aero*) See GROUND RESONANCE.

resonance (*Chem*) A description of a molecule whose structure cannot be represented by a single LEWIS STRUCTURE but only as a mixture of two or more of them. More appropriately called *mesomerism*, it is used in the sense that a mule may be said to be a resonance of a horse and a donkey.

resonance (*Phys*) (1) The vibrations of maximum velocity amplitude resulting when a vibrating system (mechanical or acoustical) is set into forced vibrations by a periodic driving force with the applied frequency at or near the natural frequency of the system; corresponds to minimum mechanical or acoustical impedance. The *sharpness of resonance* is measured by the ratio of the dissipation to the inertia of the system which also determines the rate of decay of the vibrating system when it is impulsed. See DECAY FACTOR, QUALITY FACTOR. (2) A very unstable meson or HYPERON state frequently created during nuclear reactions. These decay through the strong interaction, with a half-life of the order of 10^{-23} s. Consequently such particles are undetectable and their formation as an intermediate step in the reaction can only be inferred from indirect measurements.

resonance bridge (*ElecEng*) One for which balance depends upon adjustment for resonance.

resonance curve (*Phys*) A curve showing variation of current in a resonant circuit in series with an emf as the ratio of the resonance frequency to the frequency of the generator is varied through unity.

resonance escape probability (*NucEng*) In a reactor the probability of a fission neutron slowing down without being captured in ^{238}U resonances. Symbol ρ.

resonance form (*Chem*) See LEWIS STRUCTURE.

resonance heating (*Phys*) See MAGNETIC PUMPING.

resonance lamp (*ElecEng*) One which depends on the absorption and reradiation of a prominent line from a mercury arc, excited in mercury vapour.

resonance level (*Phys*) An excited level of the compound system which is capable of being formed in a collision between two systems, such as between a nucleon and a nucleus.

resonance potential (*Electronics*) See EXCITATION POTENTIAL.

resonance radiation (*Phys*) Emission of radiation from gas or vapour when excited by photons of higher frequency.

resonance scattering (*Phys*) See SCATTERING.

resonance step-up (*ElecEng*) The ratio of the voltage appearing across a parallel tuned circuit to the emf acting in the circuit (usually induced in the coil) when the circuit is resonant at the applied frequency. See Q.

resonance test (*Aero*) A test in which an aircraft, while suspended by cables or supported on inflated bags, is excited by forced oscillations over a range of frequencies, so as to establish the natural frequencies and modes of oscillation of the structure.

resonant cavity (*ICT*) One in which resonant effects result from the possibility of a modal pattern of electric and magnetic fields, as in magnetrons, klystrons, waveguides. Also applies in acoustics.

resonant circuit (*Electronics*) One consisting of an inductor and a capacitor in series or parallel. The series circuit has an impedance which falls to a very low value at the resonant frequency; that of the parallel circuit rises to a very high value.

resonant frequency (*ICT*) That at which reactance of a series resonant circuit, or susceptances of a parallel resonant circuit, balance out; numerically equal to $1/2\pi\sqrt{(LC)}$ Hz, where L is inductance in henrys and C capacitance in farads.

resonant gap (*Radar*) The interior volume of the resonant structure of a TRANSMIT–RECEIVE TUBE in which the electric field is concentrated.

resonant line (*ICT*) Parallel wire or coaxial transmission line open- or short-circuited at the ends and an integral number of quarter-wavelengths long. Used in some radio-frequency oscillators.

resonant mode (*Electronics*) Field configuration in a tuned cavity. In general, resonance occurs at several related frequencies corresponding to different configurations.

resonator (*Electronics*) Any device exhibiting a sharply defined electric, mechanical or acoustic resonance effect, eg a stub, piezoelectric crystal or Helmholtz resonator.

resonator grid (*ICT*) Electrode traversed by an electron beam and that provides a coupling to a resonator.

resorcinol (*Chem*) *1,3-dihydroxy-benzene*. $C_6H_4(OH)_2$. Mp 111°C, bp 276°C. A dihydric phenol, colourless crystal. With formaldehyde used in the preparation of cold-setting adhesives. Also *resorcin*.

resorption (*Geol*) The partial or complete solution of a mineral or rock fragment by a magma, as a result of changes in temperature, pressure or composition of the latter.

resources (*MinExt*) Mineral deposits estimated to be present worldwide from geological data and which are reasonably assured of being recoverable with present technology and at economic prices. Cf RESERVES.

respiration (*BioSci*) A term applied to events that occur at the level of the whole organism or its constituent cells. (1) At organism level it relates to the exchange of oxygen and carbon dioxide between the organism and its environment, 'breathing'. Small organisms can exchange gases across their body surface but larger animals require a richly vascularized respiratory surface, such as gills or lungs, and mechanisms for the movement of water or air over the respiratory surface. (2) At the cellular level respiration consists of the metabolic processes that degrade foodstuffs with the synthesis of ATP. It is of two major types: *aerobic*, which requires oxygen as a terminal electron receptor; and *anaerobic*, where some other terminal acceptor is used. Aerobic respiration of glucose consists of its total oxidative degradation by GLYCOLYSIS and the TCA CYCLE, to carbon dioxide and water with the generation of 36 molecules of ATP per glucose molecule, made up of two molecules by SUBSTRATE LEVEL PHOSPHORYLATION and 34 molecules by oxidative phosphorylation. See FERMENTATION and panel on MITOCHONDRION.

respiratory burst (*BioSci*) See METABOLIC BURST.

respiratory centre (*BioSci*) In vertebrates, a nerve centre of the hind-brain that regulates the respiratory movements.

respiratory failure (*Med*) Occurs when the respiratory system is no longer able to maintain normal tensions of oxygen or carbon dioxide in the body.

respiratory movement (*BioSci*) The muscular movements associated with the supply of air or water to the respiratory organs.

respiratory organs (*BioSci*) The specialized structures such as lungs and gills that enable oxygen to be transferred to the body fluids.

respiratory pigment (*BioSci*) In the blood of many animals, a coloured protein with a metal-containing prosthetic group, the complex having a high affinity for oxygen. Used for oxygen transport, eg haemoglobin, haemocyanin.

respiratory quotient (*BioSci*) The ratio of moles CO_2 evolved to moles O_2 absorbed in respiration; unity when the substrate is carbohydrate, lower when protein or fat. Abbrev *RQ*.

respiratory substrate (*BioSci*) Any chemical compound broken down during respiration to release the chemical energy stored in its bonds.

respiratory system (*BioSci*) See RESPIRATORY ORGANS.

respiratory valve (*BioSci*) In some fish, eg trout, a pair of transverse membranous folds, one attached to the floor, the other to the roof of the mouth, which prevent water from escaping through the mouth during expiration.

respond (*Build*) (1) A pilaster which forms a pair with another. (2) A REVEAL.

respondant (*Psych*) In classical conditioning, a response that is elicited by a known stimulus (eg a knee jerk).

responder (*ICT*) That part of a transponder which replies automatically to the correct interrogation signal.

response (*ICT*) That of a transmission system at any particular frequency is given by the ratio of the output to input level. If these levels are defined on a logarithmic scale, eg in *dB*, the response of the complete system is the sum of responses of the separate parts. See POLAR CURVE.

response (*Psych*) The effect of stimulation; it may, as in muscular and glandular responses, be easily observable and measurable, but it may also be an inferred response which is not immediately apparent in behaviour.

response curve (*ICT*) That which exhibits the trend of the response of a communication system or a part thereof, for the range of frequency over which the system or part is intended to operate. Usually plotted in *dB* often against a logarithmic frequency scale.

response latency (*Psych*) The time elapsing between the onset of a stimulus and the beginning of an animal's response to it.

responser (*Radar*) Receiver of secondary radar signal from a TRANSPONDER.

response time (*Electronics*) Time constant of change in output of an electronic circuit after a sudden change in input, or of indication given by any instrument after change in signal level.

response time (*ICT*) In INTERACTIVE COMPUTING the time it takes for a system to respond to an input from the user.

rest (*Eng*) (1) The tool-attaching device or tool rest. (2) Metal bar mounted in the tool post to support a hand-held tool for light cutting or forming, eg thread chasing or SPINNING.

rest bend (*Build*) A 90° bend off a horizontal drainpipe, fitted with a flat seating for connection to a vertical. Also *duckfoot bend*.

restenosis (*Med*) Re-occlusion of the coronary arteries after angioplasty or after replacement with blood vessels from elsewhere. Probably due to excessive proliferation of vascular smooth muscle that inappropriately thickens the intima and narrows the lumen.

restiform (*BioSci*) Rope-like.

resting nucleus (*BioSci*) A nucleus of a cell that is not undergoing active growth and division.

resting potential (*BioSci*) The electrical potential across the membrane of a cell, when an action potential is not occurring; almost all animal cells are negative inside, with potentials ranging from -20 to -100 mV.

resting spore (*BioSci*) A thick-walled spore able to endure drought and other unfavourable conditions, and normally remaining quiescent for some time before it germinates.

restitution nucleus (*BioSci*) A single nucleus formed following failure of the chromosomes to separate properly at anaphase and hence containing, say, twice the expected chromosome number.

rest mass (*Phys*) The mass of a particle measured by an observer at rest relative to it. In nuclear reactions, the total rest mass of the particle involved need not be conserved. Cf MASS. See RELATIVISTIC MASS EQUATION.

rest-mass energy (*Phys*) c^2 times the REST MASS of the particle, where *c* is the speed of light.

restorative (*Med*) Capable of restoring to health, consciousness or good condition; any remedy which does this.

restore (*ICT*) The return of a variable address or word or cycle index to its initial value.

restoring moment (*Aero*) A moment which, after any rotational displacement, tends to restore an aircraft to its normal attitude.

restrainer (*ImageTech*) An ingredient of a developer which slows the development of unexposed silver halide, thus reducing tendency to fog.

restriction (*BioSci*) The ability of some bacteria to restrict their susceptibility to lysis by phage or other genetic elements by cleaving the invading DNA with restriction enzymes. Their own DNA is made immune by methylating the susceptible sites (*modification*).

restriction endonuclease (*BioSci*) A class of endonucleases able to cleave DNA at a specific nucleotide sequence. In some bacteria they form part of a defensive mechanism against infection by bacteriophage (*restriction*), in which the bacterium will methylate specific sites in its own DNA (called *modification*) to prevent cleavage by its intrinsic restriction enzymes that are still able to destroy foreign unmethylated DNA. Different enzymes, obtained from a wide range of organisms, have different specificities, often recognizing four or six base pairs. Because of this specificity, restriction enzymes will cleave a sample of DNA into defined polynucleotide fragments which can then be separated according to their length. The pattern of fragments will depend on both the source and the enzyme used. This procedure is sufficiently sensitive to be able to detect a difference of one base pair in certain circumstances. The enzymes are a prerequisite for the procedures of genetic manipulation. Where the cleavage sites are not directly opposite each other, as is often the case, the cleaved DNA has *sticky ends* which facilitate the joining of one piece of DNA to another cut by the same enzyme, as in the insertion of a eukaryote DNA into that of a bacterial VECTOR. Also *restriction enzyme*. See panel on GENETIC MANIPULATION.

restriction fragment (*BioSci*) Because of their sequence specificity, RESTRICTION ENDONUCLEASES will cleave DNA into defined polynucleotide fragments, which can be separated on an agarose gel and show as bands.

restriction fragment length polymorphism (*BioSci*) A restriction fragment identified by blotting whose length is variable in the population. Those which map close to sites of genetic diseases are a useful aid to antenatal diagnosis in families at risk. Abbrev *RFLP*.

restriction map (*BioSci*) A map of a DNA sequence showing the position of sites recognized and cut by various restriction endonucleases.

restrictive temperature (*BioSci*) The temperature at which a TEMPERATURE-SENSITIVE MUTANT organism will not grow. Cf PERMISSIVE TEMPERATURE. See panel on CELL CYCLE.

restriking voltage (*ElecEng*) The high-frequency transient voltage which appears across the contacts of a switch, circuit breaker or fuse immediately after it has interrupted a circuit, and which is superimposed on the recovery voltage.

resultant (*Phys*) For two forces, the single force obtained by the parallelogram of forces which is equivalent to the two forces.

resuperheating (*Eng*) See REHEATING.

resupinate (*BioSci*) Inverted, eg the flowers of orchids in which, because of a 180° twist in the stalk, what appears to be the lower petal is, in fact, morphologically the upper petal.

resurgent gas (*Geol*) Superheated steam and other volatiles which play an active role in volcanic action, and which are derived from the water included in sedimentary rocks at the time of accumulation.

resuscitation (*Med*) The restoration of circulation and respiration after these functions have ceased.

resveratrol (*Pharmacol*) A fungicidal phenol with antioxidant properties, found in grape skins. It is thought to protect against heart disease.

retaining mesh (*MinExt*) In sizing ore before further treatment, the screen aperture above which size the material is arrested.

retaining wall (*CivEng*) A wall built to support earth at a higher level on one side than on the other. Also *revetment*.

retake (*ImageTech*) Re-photography of a scene.

retardation (*Psych*) A slowing of mental or intellectual progress.

retardation coil (*ElecEng*) Inductor for separating dc from ac particularly from a rectifier or supply with ripple.

retardation test (*ElecEng*) A method of determining the iron, friction and windage losses of electrical machinery by determining the rate at which it slows down after being run up to speed and then disconnected from the supply.

retarded fields (*ElecEng*) Electric or magnetic fields derived from retarded potentials.

retarded hemihydrate plaster (*Build*) A plaster based on calcium sulphate hemihydrate (PLASTER OF PARIS) but with the addition of retarder in varying quantities. Particularly useful as an undercoat on plasterboard, fibreboard or metal lath, because of low expansion.

retarded potentials (*ElecEng*) Electric and magnetic potentials that are delayed as they move from point to point in a system because of the finite speed of propagation of waves.

retarder (*Build*) A substance which delays or prevents the setting of cement.

retarder (*Chem*) A 'negative' catalyst which is added to a reaction system to slow down the reaction rate.

retarder (*CivEng*) (1) An arrangement of braking surfaces placed alongside, and parallel with, the running rails in a shunting yard; operated from a signal box by electric, pneumatic, hydraulic or mechanical means. Also *wagon retarder*. (2) A substance applied to the inside of concrete FORM to delay surface hardening.

retarding field (*Electronics*) Electric field such as between a positively charged grid and a lower potential outer grid in a valve, so that electrons entering this region lose energy to the field.

retarding-field oscillator (*ICT*) One that depends on the electron transit time of a positive grid oscillator valve. See BARKHAUSEN–KURZ OSCILLATION.

rete (*BioSci*) A net-like structure. Pl *retia*.

rete Malpighii (*BioSci*) See MALPIGHIAN LAYER.

rete mirabile (*BioSci*) See RED GLAND, BODY.

retention (*Med*) The abnormal keeping back in the body of matter (eg urine) normally evacuated.

retention wall (*Build*) A thin wall built alongside an external wall of a building leaving a $\frac{1}{2}$–1 in (12–25 mm) cavity between, which is later filled with waterproofing material to form a vertical damp-proof course.

retentivity (*Phys*) Capacity to hold a response after the stimulus is removed. In ferromagnetic materials, equivalent to the RESIDUAL FLUX DENSITY.

reticul-, reticulo- (*Genrl*) Prefixes from Lt *reticulum*, net.

reticular (*BioSci*) Resembling a net; of or pertaining to the reticuloendothelial system.

reticular activating system (*Psych*) Part of the brain stem once considered to be an activating centre and thus involved in arousal, attention, sleep, and control of reflexes; the reticular formation in the core of the brain stem is certainly involved with all these activities.

reticular tissue (*BioSci*) A form of connective tissue in which the intercellular matrix is replaced by lymph; it derives its name from the network of collagenous fibres which it shows. Also *retiform tissue*.

reticulated (*Build*) A term applied to a variety of rusticated work distinguished by irregularly shaped sinkings separated by narrow margins of the regular width.

reticulated foam (*Eng*) A FOAM in which the cell structure is delineated by rod-like struts and ties of material rather than by cell walls.

reticulate thickening (*BioSci*) Secondary wall deposition in the form of an irregular network in tracheids or vessel elements of METAXYLEM.

reticulation (*Print*) The mottled appearance occurring when a wet ink is applied to another ink film which has dried to a smooth non-porous surface. This causes the top film to be drawn into minute beads.

reticule (*Surv*) A cell carrying cross-hairs and fitting into the diaphragm of a surveying telescope. Also *graticule*.

reticulitis (*Vet*) Inflammation of the *reticulum*.

reticulo- (*Genrl*) See RETICUL-.

reticulocytosis (*Med*) An increase in the number of reticulocytes or immature red cells in the blood, seen in HAEMOLYTIC ANAEMIAS or after treatment has commenced in other anaemias.

reticuloendothelial system (*BioSci*) A term formerly used to describe the system of cells that have the ability to take up certain dyes and particles (such as carbon in the form of India ink) when injected into the living animal. It has been replaced by MONONUCLEAR PHAGOCYTE SYSTEM.

reticulum (*BioSci*) In ruminant mammals, the second division of the stomach, or *honeycomb bag*; any net-like structure. Adj *reticular*.

Reticulum (Net) (*Astron*) A small southern constellation.

retiform tissue (*BioSci*) See RETICULAR TISSUE.

retina (*BioSci*) The light-sensitive layer of the eye in all animals. The human retina contains two types of sensitive element: ROD and CONE. Adj *retinal*.

retinal disparity (*Psych*) The difference between the two images sent to the brain by each eye, an important cue in distance perception; the closer the object, the greater the differences in the images.

retinal fatigue (*Med*) Retention of images after removal of excitation, due to chemical changes in the retina.

retinal illumination (*Phys*) The luminous flux received by unit area of the retina. It equals KLS/l^2, where K is a transmission factor of the eye, L is the luminance of a uniformly diffusing surface, S is the pupil area and l is the distance of retina from the second nodal point.

retinene (*BioSci*) Vitamin A_1 aldehyde, a component of rhodopsin.

retinite (*Min*) A large group of resins, characterized by the absence of succinic acid.

retinitis (*Med*) Inflammation of the retina.

retinitis pigmentosa (*Med*) A hereditable disease in which chronic and progressive degeneration of the choroid occurs in both eyes, with progressive loss of vision.

retinoblastoma (*Med*) A tumour of the retina composed of small round cells, arising from embryonic retinal cells; it is locally destructive and forms metastases.

retinochoroiditis (*Med*) Inflammation of the retina and choroid.

retinoic acid (*BioSci*) An organic compound thought to be a morphogen in chick limb-bud development; the aldehyde (retinal) is involved in photoreception and some derivatives are used in treating acne. Also *Vitamin A*.

retinol (*BioSci, FoodSci*) See vitamin A in panel on VITAMINS.

retinoscopy (*Med*) A method of estimating the refractive index of the eye by reflecting light onto it from a mirror and observing the movements of the shadow across the pupil. Also *shadow test, skiascopy*.

retinulae (*BioSci*) In Arthropoda, the visual cells of the compound eye, forming the base of each ommatidium.

retort (*FoodSci*) (1) A pressure vessel in which canned, bottled or other aseptically packed product can be thermally processed, using steam to destroy enzymes and kill micro-organisms; and then cooled with water to achieve a vacuum in the pack. (2) To carry out a thermal process using a pressure vessel.

retrace (*ImageTech, Radar*) See FLYBACK.

retractable landing gear (*Aero*) An alighting gear which can be withdrawn completely or nearly so from its operative position to reduce drag.

retractable radiator (*Aero*) A liquid cooler for an aero-engine, capable of being withdrawn out of the airstream, for reducing drag and controlling the temperature of the cooling liquid.

retractile (*BioSci*) Capable of being withdrawn, as the claws of most Felidae.

retraction lock (*Aero*) A device preventing inadvertent retraction of the landing gear while an aircraft is on the ground. Also *ground safety lock*.

retractor (*BioSci*) A muscle which, by its contraction, draws a limb or a part of the body towards the body. Cf PROTRACTOR.

retransfer (*Print*) See TRANSFER.

retread (*Autos*) Tyre carcass to which new, extruded tread is bonded and revulcanized. More frequently used with large truck tyres, where the cost of carcass is higher than the car tyre. Cf REMOULD.

retreating systems (*MinExt*) Systems in which the removal of ore or coal is commenced from the boundary of the deposit, which is then worked towards the entry through undisturbed rock.

retrices (*BioSci*) In birds, the stiff tail feathers used in steering. Sing *retrix*. Adj *retricial*.

retrieval (*Psych*) Memory. The process of searching for and bringing stored information into consciousness.

retrieval cue (*Psych*) Environmental or internal stimuli that help the retrieval of an experience.

retro- (*Genrl*) Prefix from Lt *retro*, backwards, behind.

retroactive interference (*Psych*) Interference in memory as a result of later learning.

retrobulbar neuritis (*Med*) Inflammation of that part of the optic nerve behind the eyeball.

retrocaecal (*Med*) Behind the caecum. Also *retrocecal*.

retrocerebral glands (*BioSci*) In insects, a collective name for a number of endocrine glands in the head, behind the brain, concerned with postembryonic development and metamorphosis. See CORPORA ALLATA, CORPORA CARDI-ACA.

retroflexed (*Med*) Said of the uterus when its body is bent back on the cervix. Cf RETROVERSION. N *retroflexion*.

retrofocus lens (*ImageTech*) See INVERTED TELEPHOTO LENS.

retrograde amnesia (*Psych*) A type of amnesia that often occurs after a head injury, or from electrical shock; there is loss of memory for events leading up to the injury, although the period of time that is lost to memory varies with the conditions of injury.

retrograde axonal transport (*BioSci*) The transport of vesicles along microtubules from the synaptic region of an axon towards the cell body.

retrograde metamorphism (*Geol*) See RETROGRESSIVE METAMORPHISM.

retrograde motion (*Astron*) (1) Motion of a comet (or satellite) whose orbit is inclined more than 90° to the ecliptic (or to the planet's equatorial plane). (2) Apparent motion of a planet from east to west among the stars,

caused by a combination of its true motion with that of the Earth.

retrograde vernier (*Surv*) A vernier in which *n* divisions on the vernier plate correspond to $(n+1)$ divisions on the main scale.

retrogressive metamorphism (*Geol*) A term descriptive of those changes which are involved in the conversion of a rock of high metamorphic grade to one of lower grade, through the advent of metamorphic processes less intense than those which determined the original mineral content and texture of the rock. Also *retrograde metamorphism*.

retrolental fibroplasia (*Med*) Damage to retina in the newborn caused by excessive oxygen concentrations.

retroperitoneal (*Med*) Situated or occurring behind the peritoneum.

retropharangeal (*Med*) Situated or occurring in the tissues behind the pharynx.

retropulsion (*Med*) The running backwards of a patient with paralysis agitans or PARKINSONISM; the patient's centre of gravity is displaced backwards, the rigidity of this posture making it difficult for the patient to recover his or her balance.

retrorocket (*Space*) A small rocket motor used for reducing the velocity of a space vehicle in landing, or in any manoeuvre calling for a thrust in the direction opposite to the motion.

retrotransposon (*BioSci*) A transposon that replicates by producing an RNA transcript which is then translated back into a DNA copy, and is inserted at a different site on the chromosome or in a different chromosome in the same cell.

retroversion (*Med*) The abnormal displacement backwards of the uterus, with or without RETROFLEXION.

retrovirus (*BioSci*) A virus of higher organisms whose genome is RNA, but which can insert a DNA copy of its genome into the host's chromosome. Important because they include the ONCOGENIC VIRUSES, and because they can be used as VECTORS for the introduction of DNA sequences into eukaryotic cells.

retting (*Textiles*) Soaking flax straw in ponds, canals, tanks, etc, for bacteria to soften the woody tissue to enable the fibres to be separated by scutching (beating).

Rett's syndrome (*Med*) A severe progressive genetic neurological disorder that mainly affects baby girls, causing dyspraxia and impaired learning and communication.

return (*ICT*) To transfer control (exit) from a subprogram back to a calling program. A subprogram may have more than one return instruction.

return (*Radar*) Refers to radar reflections, eg land (or ground) return, sea return.

return airway (*MinExt*) One leading foul air away from the mine workings to the upcast shaft. Also *return aircourse*.

return bead (*Build*) A double-quirk bead formed on the exterior angle of a timber.

return crank (*Eng*) A short crank which replaces an eccentric in the WALSCHAERT'S VALVE GEAR on outside cylinder locomotives.

return feed (*ElecEng*) See NEGATIVE FEEDER.

return-flow system (*Aero*) A gas turbine combustion system in which the air is turned through 180° so that it emerges in the opposite direction to that in which it entered. Also *reverse-flow system*.

return-flow wind tunnel (*Aero*) One in which the air is circulated round a closed loop to preserve its momentum and so reduce the power requirement.

returning charge (*MinExt*) In custom smelting, that imposed by the smelter per unit of mineral treated. It may be modified by penalties or premiums if the ore or concentrate varies from a specified composition.

return line flyback (*Electronics*) Faint trace formed on the screen of a cathode-ray tube by the beam during the flyback period. Usually suppressed. Also *return trace*.

return loss (*ICT*) See REFLECTION.

returns (*MinExt*) Oil-rig term for the material carried back by the returning drilling mud; provides essential information about conditions downhole.

return trace (*Electronics*) See RETURN LINE FLYBACK.

return wall (*CivEng*) A short length of wall built perpendicularly to one end of a longer wall.

retuse (*BioSci*) Having a slight notch at a more or less obtuse apex.

Reuben–Mallory cell (*Chem*) Small robust primary cell, of a very level discharge characteristic, having a zinc anode and a (red) mercuric oxide cathode, which also depolarizes. Made in minute sizes for hearing aids, internal radio transmitters, watches.

Reuleaux valve diagram (*Eng*) See VALVE DIAGRAM.

reusability (*Eng*) The potential for a product to be used again following normal use; thus car and truck tyres with good carcasses can be retreaded at the end of their useful first life.

reusability (*Space*) The property of space hardware which may be used more than once. A system may be partially or fully reusable but, in either case, it implies recovery and refurbishment before reuse. Cf EXPENDABLE LAUNCH VEHICLE.

revalé (*Build*) Said of a cornice, moulding, etc, finished when the work is in position.

reveal (*Build*) The depth of wall revealed, beyond the frame, in the sides of a door or window opening. Also *respond*, and (Scottish) *ingo* or *ingoing*.

revehent (*BioSci*) Carrying back.

reverberation bridge (*Acous*) A method of measuring the reverberation time in an enclosure; the rate of decay of the sound intensity is balanced against the adjusted and known decay of the discharge of a capacitance through a resistance.

reverberation chamber (*Acous*) A room with a long reverberation time and diffuse sound field. The long reverberation time is achieved by highly reflective walls. Also *echo chamber*.

reverberation response (*Acous*) The response of a microphone for reverberant sound, ie for the simultaneous arrival of sound waves of random phase, magnitude and direction. Substantially equal to the mean spherical response at each frequency of interest.

reverberation response curve (*Acous*) The response curve of a microphone to reverberant sound waves. Plotted with the response in decibels as ordinate, on a logarithmic frequency base.

reverberation time (*Acous*) The time in seconds required for the decay of the average sound intensity in a closed room over an amplitude range of 1 million, or 60 dB. A space with many *scatterers* also creates reverberation, eg the ocean.

reverberatory furnace (*Eng*) A furnace in which the charge is melted on a shallow hearth by flame passing above the charge and heating a low roof. Firing may be with coal, pulverized coal, oil or gas. Much of the heating is done by radiation from the roof. Has many applications.

reversal colour film (*ImageTech*) Film in which the negative image in the colour layers is reversed in processing to give a positive transparency.

reversal design (*Psych*) Experimental protocol that includes the removal of treatment to determine if the subject reverts to baseline.

reversal of control (*Aero*) Reversal of a control moment (or couple) which occurs when displacement of the control surface results in such high forces that distortion of the main structure counteracts the effect of the surface. This overloading is a function of airspeed, since control forces increase proportionally to the square of the velocity, and *reversal speed* is the lowest EAS at which reversal occurs.

reversal of spectrum lines (*Phys*) The appearance of a spectral line as a broad, diffuse bright line with a narrow dark line down the centre. The effect is caused by cool vapour surrounding a hot source such as an electric arc, which produces a narrow absorption line on the short range of continuous spectrum given by the same vapour, at a high temperature, at the centre of the arc. Only certain lines are thus affected.

reversal process (*ImageTech*) A method of processing a photographic emulsion to produce a positive image on the original film which has been exposed in the camera, without making a print; the areas initially unexposed form the final picture. A similar process can also be used for making a copy negative from a processed original negative.

reversed drainage (*Geol*) See DRAINAGE PATTERNS.

reversed fault (*Geol*) A type of FAULT in which compression has forced the strata on the side towards which the fracture is inclined to override the strata on the downthrow side. Cf NORMAL FAULT.

reversed field pinch (*NucEng*) A toroidal magnetic trap in which the toroidal field changes sign in the outer region of a plasma discharge. Found in one type of experimental fusion reactor. Abbrev *RFP*.

reversed magnetization (*Geol*) See MAGNETIC ANOMALY PATTERNS.

reverse engineering (*Eng*) Disassembly of a finished product for analysis of materials, design and manufacture. Usually performed on a competitor's product, so that innovations can be incorporated into one's own. Also used with computer software.

reverse genetics (*BioSci*) The process of removing a gene from an organism, altering it in a known way and reinserting it. The organism is then tested for any altered function.

reverse osmosis (*Chem*) A method for purifying water by forcing it through a suitable semipermeable membrane, using hydrostatic pressure to counter the osmotic gradient.

reverse Polish notation (*ICT*) A form of postfix notation (ie the operator follows the operands). This allows a STACK to be used for evaluation. Cf INFIX. Abbrev *RPN*.

reverse proxy (*ICT*) A form of proxy server that controls incoming requests from the Internet to a group of web servers, generally in order to enable protection from hostile access.

reverse roll coating (*Textiles*) A process for coating fabrics or textiles with flexible, elastomeric layer, often plasticized polyvinyl chloride. Involves passing fabric between NIP of two CALENDER rolls, onto one of which polyvinyl chloride plastisol is fed. Gives an even surface, which is cured by heating in oven or by hot air blast, when plasticizer diffuses into polyvinyl chloride particles to form a homogeneous coating.

reverse transcriptase (*BioSci*) An enzyme, found in retroviruses, that catalyses the formation of double-stranded DNA from an RNA template. It is used as a tool in genetic engineering.

reverse transcription (*BioSci*) The process whereby double-stranded DNA is synthesized from an RNA template using the enzyme REVERSE TRANSCRIPTASE.

reverse video (*ICT*) A method of highlighting text to give emphasis by reversing the foreground and background colours. If the usual display is white on a black background, reverse video would be black on white.

reversibility (*Psych*) The ability mentally to reverse operations and recall the original state, thereby enabling recognition that the qualities of an object remain the same despite changes in appearance.

reversible (*Phys*) Said of a process whose effects can be reversed so as to bring a system to its original thermodynamic state.

reversible cell (*ElecEng*) See ACCUMULATOR.

reversible coatings (*Build*) Any coating which, when dry, resoftens on the application of a second coat or its own thinner or solvent.

reversible colloid (*Chem*) See LYOPHILIC COLLOID.

reversible potentiometer-type field rheostat (*ElecEng*) A rheostat for controlling the field current of an electric machine which can also reverse the current.

reversible reaction (*Chem*) A chemical reaction which can occur in both directions, and which is therefore incomplete, a mixture of reactants and reaction products being obtained, unless the equilibrium is disturbed by removing one of the products as rapidly as it is formed. Examples of reversible reactions are the formation of an ester and water from an alcohol and an acid, the dissociation of vapours, eg ammonium chloride, and the ionic dissociation of electrolytes. All reactions are in principle reversible, but often limited by kinetic effects. Also *equilibrium reaction*.

reversible saturation-adiabatic process (*EnvSci*) An idealized, alternating condensation–evaporation process occurring in the atmosphere, assuming that none of the condensation products are removed by precipitation.

reversible transducer (*ICT*) One for which the loss is independent of the direction of transmission.

reversible unit (*Print*) A printing unit on a web-fed press which can print in either direction of rotation, having reversible drives.

reversing commutator (*ElecEng*) Any form of reversing switch, particularly the type in which brushes bear on conducting segments let into a insulated cylinder, rotated to open or close the circuits.

reversing face (*Surv*) The process of transiting a theodolite telescope, thereby changing its position from face left to face right, or vice versa.

reversing field (*ElecEng*) In a commutator machine, a field of opposite polarity to that in which an armature coil had previously been moving; designed to produce a reversed emf to assist commutation. The field may be produced by a compole, or by shifting the brushes from the neutral axis.

reversing gear (*Eng*) (1) In machine tools, used to reverse the rotation of one spindle relative to another. On a lathe it is needed to change rotation of lead screw relative to spindle, in order to change from left-handed to right-handed screw cutting. (2) Steam engine; see JOY'S VALVE-GEAR, LINK MOTION, WALSCHAERT'S VALVE GEAR.

reversing layer (*Astron*) The lower part of the Sun's chromosphere where the absorption lines of the solar system are formed by 'reversal' from bright emission lines to dark absorption lines.

reversing mill (*Eng*) A type of rolling mill in which the stock being rolled passes backwards and forwards between the same pair of rolls, which are reversed between passes. See CONTINUOUS MILL, PULL-OVER MILL, THREE-HIGH MILL.

reversing switch (*ElecEng*) A switch used for reversing the connections in an electric circuit.

reversion (*BioSci*) The process by which a mutant phenotype is restored to normal by another mutation of the same gene, ie is back-mutation. But sometimes used in the sense of suppression.

reversion (*Eng*) Rubber technology term for drop in modulus at end of cure cycle due to chain degradation or breakdown of cross-links. One type of OVERCURE.

revertive control system (*ICT*) A register-controlled system of automatic switching in which a selector, when set in motion by a positioning signal from a register, transmits progress signals back to the register to enable it to determine when the selector has reached the desired position.

revetment (*CivEng*) A retaining wall.

review room (*ImageTech*) A small cinematograph theatre at a studio, production centre or processing laboratory for the presentation of a film for detailed examination.

revise (*Print*) A second or third proof supplied in order that corrections made on the preceding proof may be checked over; to prepare and submit such a proof.

revolute (*BioSci*) Margin or apex of a leaf, rolled outwards or downwards (ie towards abaxial surface). Cf INVOLUTE. See VERNATION.

revolution (*Astron*) The term generally reserved for orbital motion, as of the Earth about the Sun, as distinct from ROTATION about an axis.

revolving centre (*Eng*) A CENTRE which is mounted on rolling bearings and revolves with the workpiece so as to obviate relative motion and thus friction between centre and workpiece.

rewritable (*ICT*) Data that are capable of being recorded on the same medium from which they have been read.

rewrite (*ICT*) To return data to memory when they have been erased during reading or modified.

rexigenous (*BioSci*) A space in a tissue formed by the rupture of cells. Also *rhexigenous*. Cf LYSIGENOUS, SCHIZOGENOUS.

Reynold's number (*ChemEng, BioSci*) The dimensionless number defined as $R_e = (\rho v d)/\mu$ where ρ = density of a fluid with viscosity μ, travelling at velocity v in a pipe of diameter d. Below $R_e = 2000$, flow is *laminar*. Above $R_e = 4000$, flow is *turbulent*. Between these values, $2000 < R_e < 4000$, flow is in transition between *laminar* and *turbulent*. The Reynold's number also relates the elements of inertial and viscous drag that hinder movement through fluid medium. For cells and microscopic organisms the Reynold's number is very small; viscous drag is dominant, and inertial resistance can be neglected.

RF (*ICT*) Abbrev for RADIO FREQUENCY.

RF converter (*ImageTech*) Abbrev for *radio-frequency converter*. A circuit which modulates video and audio signals onto a RADIO-FREQUENCY carrier for feeding into the aerial input of a TV.

RFID (*ICT*) Abbrev for RADIO-FREQUENCY IDENTIFICATION DEVICE.

RFLP (*BioSci*) Abbrev for RESTRICTION FRAGMENT LENGTH POLYMORPHISM.

RFP (*NucEng*) Abbrev for REVERSED FIELD PINCH.

RFS (*Build*) Abbrev, for RENDER, FLOAT AND SET.

RFTP (*ICT*) Abbrev for *radio-frequency transmission protocol*, a standard for wireless communication between computers in a limited geographical space.

RF welding (*Eng*) A method of joining similar thermoplastics using radio-frequency radiation to heat surfaces of parts.

Rf (*Chem*) Symbol for RUTHERFORDIUM.

Rg (*Chem*) Chemical symbol for ROENTGENIUM.

RGB signals (*ImageTech*) The three signals which correspond directly to the colour primaries, red, green and blue, as distinct from the COLOUR DIFFERENCE SIGNALS which are used for transmission.

Rh (*Med*) See RHESUS FACTOR.

rH (*Chem*) See RH VALUE.

rhabdom (*BioSci*) In the compound eyes of Arthropoda, the structure containing the visual pigment and concerned with phototransduction.

rhabdomeres (*BioSci*) One of the constituent portions of the rhabdom, secreted by a single visual cell.

rhabdomyoma (*Med*) A tumour composed of striated muscle cells.

rhabdomyosarcoma (*Med*) A malignant rhabdomyoma.

Rhabdoviridae (*BioSci*) Viruses with a single negative strand RNA genome and an associated virus-specific RNA polymerase. The capsid is bullet-shaped and enveloped by a membrane that is formed when the virus buds out of the plasma membrane of infected cells. Includes rabies virus, vesicular stomatitis virus and a number of plant viruses.

rhachi-, rhachio- (*Genrl*) See RACHI-.

rhachis (*BioSci*) See RACHIS.

Rhaetian (*Geol*) Uppermost stage of the Triassic. *Rhaetic* is an old lithostratigraphic term. See MESOZOIC.

rhamnose (*Chem*) *6-deoxy mannose*. $CH_3(CHOH)_4CHO$. A methylpentose obtained from several glucosides; crystallizes with one H_2O, mp $93°C$. On distillation with sulphuric acid it yields 5-methyl-furfural.

rhamphotheca (*BioSci*) In birds, the horny coverings ensheathing the upper and lower jaws.

rhaphe (*BioSci*) See RAPHE.

Rhea (*Astron*) The fifth natural satellite of Saturn, discovered in 1672. Distance from the planet 527 000 km; diameter 1530 km. It is the second-largest moon in Saturn's system.

Rheiformes (*BioSci*) An order of birds, containing two species of large running bird found on the S American pampas, occupying an ecological niche approximately similar to that of the emu. They have three toes. Rheas.

rhenic (VII) acid (*Chem*) See RHENIUM OXIDES.

Rhenish bricks (*Build*) Very light bricks made of calcareous material bound together with dolomitic lime. Also *floating bricks.*

rhenium (*Chem*) Symbol Re, at no 75, ram 186·2, rel.d. 21, mp 3000°C. Valencies, 2,3,4,6,7. A metallic element in the subgroup manganese, technetium, rhenium. A very rare element, occurring in molybdenum ores. A small percentage increases the electrical resistance of tungsten. Used in high-temperature thermocouples.

rhenium oxides (*Chem*) Re_2O_7, ReO_3, ReO_2 and Re_2O_3. The volatile (VII) oxide Re_2O_7 is formed when the metal or its compounds are heated in air. Rhenium (VI) oxide is the anhydride of rhenic acid, H_2ReO_4. The (VII) oxide dissolves in water to form perrhenic acid, which forms metallic perrhenates (rhenates (VII)). Cf PERMANGANATES.

rheo- (*Genrl*) Prefix from Gk *rheos*, current, flow.

rheobase (*Med*) The minimal electrical stimulus (volts) that will produce a physiological response and below which no stimulus will excite however long it is maintained; it is half the strength of that for CHRONAXIE.

rheolaveur (*MinExt*) Coal-cleaning plant where raw coal is sluiced through a series of troughs, the high-ash fraction gravitating down to a separate discharge.

rheology (*Phys*) The scientific study of fluid flow; the critical study of elasticity, viscosity and plasticity, which is of particular importance for polymer processing and solid/fluid mixtures.

rheomorphism (*Geol*) The flow of rocks resulting from severe deformation, esp applied to those undergoing partial melting as a result of heating to a high temperature.

rheopectic fluids (*Phys*) See THIXOTROPY.

rheoreceptors (*BioSci*) Receptors, found in fish and certain amphibians, that respond to stimulus of water current. Some fish also respond to electric currents through the lateral line system, but these should properly be termed galvanoreceptors.

rheostat (*ElecEng*) Electric component in which resistance is introduced into a circuit, operated by hand or mechanically.

rheotaxis (*BioSci*) A taxis in response to the direction of flow of a fluid.

rhesus blood group system (*Med*) A human blood group system, so called because the antigen involved is also present on rhesus monkey red blood cells and was first detected when these were used to immunize rabbits. The rhesus blood group system is genetically complex and there are several alleles. The most important is that known as the D-antigen. Antibodies against rhesus antigens do not occur naturally in the blood but may be produced after transfusion into a rhesus(D)-negative person of rhesus(D)-positive blood or in a rhesus(D)-negative mother who bears a rhesus(D)-positive child. In the former case a subsequent transfusion of positive blood may cause a TRANSFUSION REACTION, and in the latter give rise to ERYTHROBLASTOSIS FETALIS in the child.

rhesus factor (*Med*) Blood group ANTIGENS possessed by 85% of the population (rhesus-positive). Of importance in blood transfusion during pregnancy. A rhesus-positive baby born to a rhesus-negative mother with rhesus antibodies may develop HAEMOLYTIC DISEASE OF THE NEWBORN.

rhesus monkey (*BioSci*) One of the species *Macaca mulatta*, of the macaque monkeys, native to SE Asia. Robust and intelligent, they have been widely used in medical research. See RHESUS FACTOR.

rheumatic fever (*Med*) An acute inflammatory disease, involving the heart and the joints, that generally follows a few weeks after an infection by *Streptococcus pyogenes* of Lancefield Group A. The characteristic lesions are degeneration and necrosis of fibrous tissue and nodules of necrotic fibrous tissue surrounded by macrophages, lymphocytes and plasma cells. These are probably some form of hypersensitivity reaction to Group A streptococci which share antigenic determinants with the sarcolemma of heart muscle.

rheumatism (*Med*) A general term for a wide range of diseases characterized by painful inflammation and degeneration particularly of joints and muscles.

rheumatoid arthritis (*Med*) Chronic inflammatory polyarthritis, which may be accompanied by systemic disturbances such as fever, anaemia and enlargement of lymph nodes. The synovia of joints are infiltrated with granulomata containing plasma cells, lymphocytes, macrophages and germinal centres, causing inflammation and swelling of particularly the small joints of the extremities. RHEUMATOID FACTOR is present in the blood as well as locally. Rheumatoid arthritis is generally considered an auto-immune disease, but the initiating cause or causes are not known.

rheumatoid factor (*BioSci*) Antibody reactive with determinants present on the heavy-chain constant region of immunoglobulins of many species, including that in which the antibody is made. It causes the inflammation characteristic of rheumatoid arthritis.

rhexigenous (*BioSci*) See REXIGENOUS.

rhexis (*Med*) Rupture of a bodily structure, esp of a blood vessel.

RHI (*EnvSci*) Abbrev for *range-height indicator* radar system, used in weather forecasting, that automatically scans various elevation angles while spinning around 360° of azimuth.

rhin-, rhino- (*Genrl*) Prefixes from Gk *rhis*, gen *rhinos*, nose.

rhinal (*BioSci*) Pertaining to the nose.

rhinarium (*BioSci*) In mammals, the moist skin around the nostrils, also known as the muzzle, which is lacking in anthropoids.

rhinencephalon (*BioSci*) The olfactory lobes of the brain in vertebrates.

rhinitis (*Med*) Inflammation of nasal mucous membrane. For allergic rhinitis see HAY FEVER.

rhino- (*Genrl*) See RHIN-.

rhinocoele (*BioSci*) The cavity of the rhinencephalon; olfactory ventricle of the craniate brain.

rhinopharyngitis (*Med*) Inflammation of the nose and the pharynx.

rhinophyma (*Med*) Overgrowth of the subcutaneous tissue and the skin of the nose as a result of enlargement of the sebaceous glands which may develop in acne rosacea.

rhinoplasty (*Med*) The repair of a deformed, diseased or wounded nose by plastic surgery.

rhinorrhoea (*Med*) Discharge of mucus from the nose. US *rhinorrhea.*

rhinoscope (*Med*) A speculum for viewing the interior of the nose.

rhinotomy (*Med*) Incision into the nose.

rhinovirus (*BioSci*) A genus of the Picornaviridae family comprising viruses that infect the upper respiratory tract, including common cold virus and foot-and-mouth disease virus.

rhizo- (*Genrl*) Prefix from Gk *rhiza*, root, root-like.

Rhizobaceae (*BioSci*) A family of bacteria belonging to the order Eubacteriales (Bergey classification). It includes the symbiont *Rhizobium* which is important in nitrogen fixation by leguminous plants.

rhizodermis (*BioSci*) The outermost layer of cells of a root in its primary state.

rhizoid (*BioSci*) An outgrowth from an alga, fungus, bryophyte or pteridophyte gametophyte, attaching to or growing into the substrate and serving in anchorage and, possibly, absorption.

rhizome (*BioSci*) A stem, usually underground, often horizontal, typically non-green and root-like in appearance but bearing scale leaves and/or foliage leaves, eg nettle, many *Iris* spp. Cf STOLON.

rhizomorph (*BioSci*) A strand, like a length of thin string, composed of densely packed hyphae, by means of which some fungi spread, eg boot-lace or honey fungus, *Armillaria mellea*, dry-rot fungus *Serpula (Merulius) lacrymans*.

rhizophagous (*BioSci*) Root-eating.

Rhizopoda (*BioSci*) See SARCODINA.

rhizopodium (*BioSci*) Long, very fine, sometimes branched, cytoplasmic process from an algal cell, esp in the Chrysophyceae.

rhizosphere (*BioSci*) The zone of soil in the immediate vicinity of an active root, influenced by the uptake and output of substances by the root and characterized by a microbial flora different from the bulk soil.

rhod-, rhodo- (*Genrl*) Prefixes from Gk *rhodon*, rose.

rhodamine (*BioSci*) A FLUOROCHROME commonly conjugated with antibodies for use in INDIRECT IMMUNO-FLUORESCENCE.

rhodamines (*Chem*) Dyestuffs of the triphenylmethane group, closely related to fluorescein. They are obtained by the condensation of phthalic anhydride with *N*-alkylated *m*-aminophenols in the presence of sulphuric acid.

rhodanizing (*Eng*) The process of electroplating with rhodium, esp on silver, to prevent tarnishing.

rhodeose (*Chem*) $C_6H_{12}O_5$. A methylpentose sugar, an isomer of rhamnose.

rhodium (*Chem*) A metallic element of the platinum group. Symbol Rh, at no 45, ram 102·9055, rel.d. (at 20°C) 12·1, mp approx 2000°C; electrical resistivity approx $5·1 × 10^{-8}$ Ω m. A noble silvery white metal, it resembles platinum and is alloyed with the latter to form the positive wire of the platinum–rhodium–platinum thermocouple. Used for plating silver and silverplate to prevent tarnishing, in catalysts and in alloys for high-temperature thermocouples.

rhodo- (*Genrl*) See RHOD-.

rhodochrosite (*Min*) Manganese carbonate which crystallizes in the trigonal system, occurring as rose-pink rhombohedral crystals. It is a minor ore of manganese. Also *manganese spar* or *dialogite*.

rhodonite (*Min*) Manganese silicate, generally with some iron and calcium, crystallizes in the triclinic system. It is rose-coloured, and is sometimes used as an ornamental stone.

rhodophane (*BioSci*) A coloured oily substance, globules of which are found in the CONES of birds and in parts of the retina in some other forms.

Rhodophyceae (*BioSci*) The class of eukaryotic red algae containing chloroplasts with chlorophyll a and phycobilins, and single thylakoids. They do not have flagellate stages. Their reserve carbohydrate is floridian starch ($\alpha1 \rightarrow 4$ glucan) in the cytoplasm. Their life cycles are often complex, eg triphasic alternation of generations; sexual reproduction is oogamous. They may be unicellular, filamentous or parenchymatous. Most are marine, littoral and sublittoral (the red seaweeds); some live in fresh water or soil and some are parasitic on other red algae. They are a source of agar and carragheen; a few are eaten, eg *Porphyra*, laver-bread.

rhodopsin (*BioSci*) The light-sensitive protein present in the eye. Its light sensitivity is due to the prosthetic group of 11-cis-retinol.

rhombic antenna (*ICT*) Directional short-wave antenna comprising an equilateral parallelogram of conductors, each several quarter-wavelengths long, usually arranged in a horizontal plane.

rhombic dodecahedron (*Crystal*) A crystal form of the cubic system, consisting of twelve exactly similar faces, each of which is a regular rhombus. Does not occur in the orthorhombic system, despite its name.

rhombic system (*Crystal*) See ORTHORHOMBIC SYSTEM.

rhombohedral class (*Crystal*) A class of the trigonal system, a characteristic form being the RHOMBOHEDRON, which is exhibited by crystals of quartz, calcite, dolomite, etc. See fig. at BRAVAIS LATTICES.

rhombohedron (*Crystal*) A crystal form of the trigonal system, bounded by six similar faces, each a rhombus or parallelogram.

rhomb-porphyry (*Geol*) A medium-grained rock of intermediate composition, usually occurring in dykes and other minor intrusions; characterized by numerous phenocrysts of anorthoclase which are rhomb-shaped in cross-section, set in a finer-grained groundmass. Related to laurvikite among the coarse-grained and to kenyte among the fine-grained rocks.

rhomb-spar (*Min*) An old-fashioned synonym for DOLOMITE.

rhombus (*MathSci*) A quadrilateral with all its sides equal. Loosely, a diamond shape.

Rhometal (*Eng*) An alloy of permalloy type. Contains 64% iron and 36% nickel. Used in high-frequency electric circuits.

rhometer (*Eng*) One for the measurement of impurity content of molten metals by means of the variation in electrical conductivity.

rhonchus (*Med*) A harsh, prolonged sound, heard on auscultation, produced by air passing over narrowings in the bronchial tubes; found in ASTHMA and BRONCHITIS.

rhumbatron (*Electronics*) A type of cavity resonator used in eg a klystron. It acts as a tuned circuit comprising a parallel-disk capacitor surrounded by a single-turn toroidal inductance, and is used to velocity-modulate an electron beam passing through holes in the capacitor disks. See BUNCHER.

rH value (*Chem*) The logarithm, to base 10, of the reciprocal of hydrogen pressure which would produce the same electrode potential as that of a given oxidation–reduction system, in a solution of the same PH value. The greater the oxidizing power of a system, the greater the rH value. Also R_H, r_h.

rhyncho- (*Genrl*) Prefix from Gk *rhynchos*, beak, snout, proboscis.

Rhynchocephalia (*BioSci*) An order of the Lepidosauria with two temporal vacuities and a large parietal foramen which, in the one living form, *Sphenodon punctatum*, contains a non-functional median eye. They are known as fossils from the Middle Trias; *Sphenodon*, or Tuatara, survives in coastal islands off New Zealand.

rhynchodont (*BioSci*) Having a toothed beak.

rhynchophorous (*BioSci*) Having a beak.

Rhynie Chert (*Geol*) A silicified peaty bed containing well-preserved plant remains as well as spiders, scorpions and insects; discovered at Rhynie, Aberdeenshire, in the Middle Old Red Sandstone (ie Devonian Age).

rhyolite (*Geol*) General name for fine-grained igneous rocks having a similar chemical composition to granite, commonly occurring as lava flows, although occasionally as minor intrusions, and generally containing small phenocrysts of quartz and alkali-feldspar set in a glassy or cryptocrystalline groundmass. Sometimes *liparite*. See OBSIDIAN, PITCHSTONE, PUMICE, VOLCANIC ROCKS.

rhythmic crystallization (*Geol*) A phenomenon exhibited by rocks of widely different composition but characterized by development of orbicular structure.

rhythmic sedimentation (*Geol*) A more or less consistently repeated sequence of two or more rock units which can be recognized as forming a pattern, eg BOUMA CYCLE, CYCLOTHEM, VARVED CLAYS.

rhytidome (*BioSci*) Dead, outer bark, consisting of layers of periderm with some cortex and/or secondary phloem. Also *scale bark, scaly bark*.

ria (*Geol*) A normal valley drowned by a rise of sea level relative to the land. Cf FIORDS, in the production of which glacial action plays an essential part. A good example of a ria type of coastline is SW Ireland, the rias being long synclinal valleys lying between anticlinal ridges.

rib (*Aero*) A fore-and-aft structural member of an *aerofoil* which has the primary purpose of maintaining the correct contour of the covering, but is usually a stress-bearing component of the main structure. Ribs are usually set either parallel with the longitudinal axis or at right angles to the front spar. Cf NOSE RIBS.

rib (*BioSci*) (1) Any small ridge or rib-like structure. (2) In vertebrates, an element of the skeleton in the form of a curved rod connected at one end with a vertebra; it serves to support the body walls enclosing the viscera.

rib (*Build, CivEng*) (1) A curved member of a centre or ribbed arch. (2) A moulding projecting for purposes of ornamentation from a ceiling or vault surface. (3) The vertical portion of a T-beam.

rib (*Textiles*) A prominent line running along or across woven or knitted fabrics, and forming a cord effect.

rib and panel (*Build*) A term applied to a vault formed of separate ribs and panels, the latter being supported on the former.

ribbed arch (*CivEng*) An arch composed of many side-by-side ribs spanning the distance between the springings.

ribbed flutings (*Build*) Flutings separated by a flat or slightly convex FACETTE.

ribbon (*ImageTech*) A loop of fine metal filament in a LIGHT VALVE, the opening in which is varied by the modulating currents.

ribbon (*Textiles*) A decorative closely woven narrow fabric nearly always made from lustrous continuous filament yarns.

ribbon microphone (*Acous*) A special type of electrodynamic microphone for measuring sound velocities. A very thin aluminium ribbon in a magnetic field acts as a diaphragm. Incident sound waves make the ribbon vibrate and an electrical voltage is induced proportional to the wave velocity. See PRESSURE-GRADIENT MICROPHONE.

ribbon parachute (*Aero*) A parachute in which the canopy is made from light webbing with spaces between, instead of conjoined fabric gores, so as to give greater strength against ripping for deployment at high speed. Commonly used for BRAKE PARACHUTES.

ribbon strip (*Build*) A horizontal timber attached to vertical timbers as a support for joists. Also *girt strip, ledger board.*

rib mesh (*Build*) See EXPANDED METAL.

riboflavin (*BioSci*) See vitamin B complex in panel on VITAMINS.

ribonuclease (*BioSci*) A widely distributed type of enzyme that cleaves RNA. Some act as endonucleases, others as exonucleases; they generally recognize their targets by tertiary structure rather than sequence.

ribonucleic acid (*BioSci*) See RNA.

ribonucleoprotein (*BioSci*) A term describing complexes of RNA and protein that are involved in a wide range of cellular processes. These include ribosomes, the signal recognition particle, and a complex involved in termination of transcription.

ribose (*Chem*) $C_5H_{10}O_5$. A pentose, a stereoisomer of arabinose. D-ribose occurs in ribose nucleic acid (RNA).

ribosomal DNA (*BioSci*) Genes specifying the several kinds of ribosomal RNA molecules. Abbrev *rDNA.*

ribosomal RNA (*BioSci*) See RIBOSOME. Abbrev *rRNA.*

ribosome (*BioSci*) A complex bead-like structure consisting of three subunits of RNA and protein that can associate with mRNA and is the site of synthesis in the cytoplasm of polypeptides encoded by the mRNA. See fig. at ENDO-PLASMIC RETICULUM.

ribotype (*BioSci*) The RNA complement of a cell, by analogy with the phenotype or genotype.

ribozyme (*BioSci*) An RNA with catalytic capacity, and thus a non-protein enzyme.

ribulose (*BioSci*) A five-carbon ketose sugar. See RIBULOSE BISPHOSPHATE.

ribulose bisphosphate (*BioSci*) The 1,5-bisphosphate ester of the sugar ribulose, which is a substrate for RIBULOSE 1,5-BISPHOSPHATE CARBOXYLASE OXYGENASE. Abbrev *RuBP.* Also *ribulose diphosphate, RuDP.*

ribulose 1,5-bisphosphate carboxylase oxygenase (*BioSci*) The enzyme that catalyses both the reaction of ribulose bisphosphate (RuBP, five carbons) and CO_2 to give two molecules of phosphoglyceric acid (PGA, three carbons) as the first step of the CALVIN CYCLE (panel) in photosynthesis, and also the apparently wasteful reaction of RuBP with oxygen to give one molecule each of PGA and phosphoglycollic acid (two carbons). The former, carboxylation, reaction results in the net fixation of carbon (1C per RuBP); the phosphoglycollate produced in the latter, oxygenation, reaction is mostly metabolized to PGA and CO_2 ($\frac{1}{2}$ molecule each) resulting in a net loss of CO_2 ($\frac{1}{2}$ C per RuBP) (photorespiration). Oxygen and CO_2 compete, and oxygenation becomes more important at higher temperatures. Under temperate conditions in an ordinary (C3) plant the ratio of carboxylation to oxygenation might be 4:1, representing the loss of 30% of the potential fixation. RUBISCO in higher plants is made of eight large (55 kDa) subunits, coded by CHLOROPLAST DNA and made in the chloroplast, and eight small (15 kDa) subunits, coded by nuclear DNA and made in the cytoplasm. There is one reaction site. RUBISCO has a low reaction rate; in a typical (C3) leaf about half the soluble protein is RUBISCO, making it the most abundant protein on earth. Alternative names include *RuBPc/o, RuBP carboxylase, carboxydismutase* and *fraction I protein.* Abbrev *RUBISCO.*

Riccati equation (*MathSci*) A differential equation of the form

$$\frac{dy}{dx} = py^2 + qy + r$$

where *p, q, r* are functions of *x* alone.

rice (*FoodSci*) A cereal from plants of the genus *Oryza* grown in wet areas where the paddy fields are often deliberately waterlogged for much of the growing period. Long-grain rice is grown in Indonesia, India and the USA and S America, while short-grain rice is produced in some Mediterranean countries and Japan. Consumed mainly as cooked grains after husking and polishing (milling). Brown rice has been husked but not polished. See PARBOILED RICE.

rice paper (*FoodSci, Paper*) Edible paper from the pith of a small tree *Tetrapanax papyrifer*, a native of Taiwan. A substitute is made from rice flour and gum.

Richardson–Dushman equation (*Electronics*) The original Richardson formula, as modified by Dushman, for the emission of electrons from a heated surface, current density being

$$I = AT^2 \exp\left(-\varphi/kT\right)$$

where *T* is the absolute temperature, *A* is a material constant, *k* is Boltzmann's constant, with *φ* the work function of the surface.

Richardson effect (*Electronics*) See EDISON EFFECT.

Richardson number (*EnvSci*) A non-dimensional number R_i arising in the study of shearing flow in the atmosphere. If *g* is the acceleration of gravity, *β* is a measure of vertical stability, commonly

$$\frac{1}{\theta}\frac{\partial\theta}{\partial z}$$

where *θ* is the *potential temperature*, and

$$\frac{\partial u}{\partial z}$$

is a characteristic vertical wind shear, then

$$R_i = \frac{g\beta}{(\partial u/\partial z)^2}$$

Turbulence is likely to be suppressed if $R_i > 1$. See panel on ATMOSPHERIC BOUNDARY LAYER.

rich lime (*Build*) See FAT LIME.

rich mixture (*Autos*) A combustible mixture in which the fuel is in excess of that physically correct for the air.

richness (*BioSci*) The number of species in a defined area.

richterite (*Min*) Hydrated metasilicate of sodium, calcium and magnesium, occurring as monoclinic crystals in alkaline igneous rocks and in thermally metamorphosed limestones and skarns. A member of the amphibole group.

Richter scale (*Geol*) See panel on EARTHQUAKE and appendix on Earthquake severity measurement scales.

rich text format (*ICT*) A MARKUP LANGUAGE that can describe text layout precisely and which is portable between different operating systems. Abbrev *RTF*.

ricin (*BioSci*) A highly toxic lectin from seeds of the castor bean, *Ricinus communis*. This *AB toxin* slowly but progressively inactivates ribosomes, thus blocking protein synthesis and leading eventually to death. The binding subunit has specificity for N-acetyl galactosamine.

ricinoleic acid (*Chem*) $CH_3(CH_2)_5CHOHCH_2CH=CH(CH_2)_7COOH$. An oily liquid which, in glyceride form, is the chief constituent of castor oil.

rickets (*Med*) A nutritional disease of childhood characterized by defective ossification and softening of bones: due to deficiency of vitamin D and failure to absorb and utilize calcium salts. Also *rachitis*.

rickettsiae (*Med*) Small, Gram-negative bacteria, obligate intracellular parasites, found in the tissues of lice, ticks, mites and fleas. They can cause disease, eg scrub typhus, in humans and other animals.

rictus (*BioSci*) Of birds, the mouth aperture Adj *rictal*.

riddle (*MinExt*) Strong coarse sieve used to size gravel, furnace clinker, etc. Large pieces removed by hand in riddling are called *knockings*, remaining on-screen material *middlings* and through-passing particles *fells*, *undersize* or *smalls*.

Rideal–Walker test (*Chem*) A test for germicidal power of a disinfectant, carbolic acid (phenol) being taken as the standard. A series of dilutions of disinfectant is tested with a typhoid broth culture, samples being taken at short intervals and subjected to incubation.

rider (*Chem*) A small piece of platinum wire used on a chemical balance as a final adjustment. Now mostly obsolete, except on museum pieces.

rider (*MinExt*) (1) A *horse*, ie mass of country rock occurring in a mineral deposit. (2) A thin seam of coal above a thick one. (3) A guide for a *bowk*, in sinking.

rider rollers (*Print*) Steel rollers in the inking system of a printing machine which secure the inkers (and distributors), act as a necessary link between their tacky surfaces, and contribute to efficient distribution and inking.

rider shore (*Build*) An inclined baulk of timber used in a system of raking shores for a high building. It abuts at its lower end against a length of timber laid along the back of the outer RAKING SHORE, instead of against the sole plate on the ground. Also *riding shore*.

ridge (*Build, CivEng*) The summit line of a roof; the line on which the rafters meet.

ridge (*EnvSci*) An outward V-shaped extension of the isobars from a centre of high pressure.

ridge-board (*Build*) A horizontal timber at the upper ends of the common rafters, which are nailed to it.

ridge capping (*Build*) The covering applied over a ridge to protect the intersection of the sloping roof surfaces.

ridge course (*Build*) The last (ie the top) course of slates or tiles on a roof, cut to length as required.

ridge-pole (*Build, CivEng*) A timber member laid horizontally along the ridge of a roof. Also *ridge-piece*.

ridge roll (*Build*) A ridge of rounded section, over which a zinc or lead flashing is formed and secured as a covering for the top ends of the ridge courses to seal the top of the roof.

ridge roof (*Build*) A pitched roof whose sloping surfaces meet to form an apex or ridge.

ridge stop (*Build*) A piece of sheet lead shaped over the junction between a roof ridge and a wall; used in cases where the one runs into the other and a watertight joint has to be made.

ridge tile (*Build*) A purpose-made tile specially shaped for use as a covering over the ridge of a roof.

ridging (*Build*) The operation of covering the ridge of a roof with specially shaped ridge tiles or other material.

riding lamps (*Aero*) Lamps displayed at night by a float plane or flying boat when moored or at anchor. Colours and positions as in the maritime code.

riding shore (*Build*) See RIDER SHORE.

riebeckite (*Min*) A dark-blue hydrated silicate of sodium and iron found in alkaline igneous rocks as monoclinic prismatic crystals. A member of the amphibole group: the blue asbestos *crocidolite* is a variety occurring in metamorphosed ironstones.

Riedel's disease (*Med*) A rare chronic inflammation of the thyroid gland, which becomes enlarged and hard as the result of excessive formation of dense fibrous tissue. Also *chronic thyroiditis*.

Riedel's lobe (*Med*) An anomalous downward prolongation of the right lobe of the liver.

Rieke diagram (*ICT*) Polar form of load–impedance diagram representing the components of the complex reflection coefficient of the oscillator load in a microwave oscillator.

Riemann surface (*MathSci*) An extension of the complex plane whereby a multivalued function can be regarded as a one-valued function. The surface has to be designed for the particular function concerned, eg for the two-valued function $w = \pm\sqrt{z}$, the complex plane is extended by superimposing a similar plane above it, cutting both planes from the origin along the real axis to infinity, and joining the opposite edges of the two cuts cross-wise. The result is that regaining a starting point by tracing a circle around the origin now takes two revolutions (argument increase of 4π) because both the original and the additional planes have to be covered. Thus effectively the complex plane has been doubled and the two-valued function $w = \pm\sqrt{z}$ can be laid out as a one-valued function of position. That portion of a multivalued function which lies on a single sheet of a Riemann surface is called a *branch of a function*, and the connecting point of the several sheets of the surface a *branch point*.

Riemann zeta function (*MathSci*) The function $\zeta(z)$ of a complex variable defined by

$$\zeta(z) = \sum_{n=1}^{\infty} \frac{1}{n^z}$$

rifampicin (*Pharmacol*) A semi-synthetic antibiotic used in treating tuberculosis (in combination with streptomycin and isoniazid), leprosy and meningitis. Its mode of action is to inhibit bacterial DNA-dependent RNA polymerase, leading to suppression of RNA synthesis.

riffler (*Eng*) A file bent so as to be capable of operating in a shallow depression.

riffler (*MinExt*) A device for dividing a stream of crushed material, eg coal, into truly representative samples.

riffler (*Paper*) See SAND TRAP.

rift and grain (*Geol*) The two directions, approximately at right angles to one another, along which granite and other massive igneous rocks can be split, rift being the easier of the two.

rift valley (*Geol*) See GRABEN.

Rift Valley fever (*Med, Vet*) An infectious disease of cattle and sheep in Africa, characterized by high fever and hepatitis, and caused by a virus; probably transmitted by mosquitoes. Humans are also susceptible.

rig (*MinExt*) A well-boring plant, eg for oil.

rig (*NucEng*) An experimental set-up designed to investigate a particular property, eg the effect of neutrons on cladding. Often arranged, in the nuclear industry, in association with a reactor or other nuclear plant.

riga last (*For*) A unit of timber measure containing 80 ft³ (2·265 m³) of squared or 65 ft³ (1·84 m³) of round timber.

Rigel (*Astron*) A prominent bright blue-white supergiant variable star in the constellation Orion, the seventh brightest in the sky. Distance approx 250 pc. Also *Beta Orionis*.

rigging (*Aero*) (1) The operation of adjusting and aligning the various components, notably flight and engine control, of an aircraft. (2) In airships and balloons, the system of wires by which the weight to be lifted is distributed over the envelope or gas-bag.

rigging angle of incidence (*Aero*) See ANGLE OF INCIDENCE.

rigging diagram (*Aero*) The drawing giving the manufacturer's instructions as to the positioning and aligning of the components and control systems of an aircraft.

rigging line (*Aero*) See SHROUD LINE.

rigging position (*Aero*) The position in which an aircraft is set up in order to effect the adjustment and alignment of the various parts, ie with the lateral axis and an arbitrarily chosen longitudinal datum line horizontal.

rigging screw (*Eng*) See SCREW SHACKLE.

right angle (*MathSci*) One-quarter of a complete rotation: 90° or π/2 radius.

right-angled folding (*Print*) Each fold is at right angles to the preceding one. Cf PARALLEL FOLDING.

right ascension (*Astron*) One of the two co-ordinates, used with DECLINATION for specifying position on the CELESTIAL SPHERE in the equatorial co-ordinate system. It is the angular distance measured eastwards along the CELESTIAL EQUATOR from the vernal EQUINOX to the intersection of the hour circle passing through the body (ie the celestial equivalent of longitude). Its units are hours, minutes and seconds, and 1 hour of right ascension is 15°; the Earth's daily rotation takes the celestial sphere through 1 hour of right ascension in 1 hour of SIDEREAL TIME. Abbrev *RA*.

right circular cone (*MathSci*) See CONE (2) .

right circular cylinder (*MathSci*) See CYLINDER (2).

right-click (*ICT*) To press and release the right-hand button on a computer mouse, providing a result distinct from left-clicking the same item.

right-handed engine (*Aero*) An aero-engine in which the propeller shaft rotates clockwise with the engine between the observer and the propeller.

right-hand rule (*Phys*) (1) See AMPÈRE'S RULE. (2) For a moving wire in a magnetic field, the rule that if the thumb, first finger and second finger of the right hand are extended at right angles to each other, with the thumb pointing in the direction of motion of the wire and the first finger representing the direction of the magnetic field, the second finger points in the direction of induced current flow in the wire.

right helicoid (*MathSci*) The surface generated by a straight line moving so that it always intersects a helix and cuts its axis perpendicularly. A right circular helicoid is like a spiral staircase.

righting reflex (*Psych*) A reflexive response to falling which ensures that the animal lands upright.

right-reading (*Print*) When making plates the various printing processes have their particular requirements from the process camera or filmsetting system, eg right-reading negative for direct letterpress, or *wrong-reading* positive for deep-etch offset; the direction is read from the emulsion side.

rigid arch (*CivEng*) A continuous arch without hinges or joints, the arch being rigidly fixed at the abutments.

rigid expanded polyurethane (*Plastics*) Thermosetting FOAM used in thermal insulation and in providing light structural reinforcement. Reaction products of di-isocyanates with polyesters or polyethers in the presence of water. See POLYURETHANE.

rigidity (*Phys*) See ELASTICITY OF SHEAR.

rigidity modulus (*Eng*) Obsolete term for SHEAR MODULUS.

rigid polyvinyl chloride (*Plastics*) Polyvinyl chloride in its unplasticized form, widely used for applications such as chemical and building pipework. Also *rigid PVC*. See panel on POLYVINYL CHLORIDE.

rigid support (*ElecEng*) A support for an overhead transmission line designed to withstand, without appreciable bending, a longitudinal load as well as transverse and vertical loads.

Rigil Kent (*Astron*) See ALPHA CENTAURI.

rigor (*BioSci*) A state of rigidity and irresponsiveness into which some animals pass on being subjected to a sudden shock as a defensive mechanism; 'shamming dead'.

rigor (*Med*) A sudden chill of the body, accompanied by a fit of shivering that heralds the onset of fever.

rigor mortis (*Med*) Stiffening of the body following death.

Rijke tube (*Acous*) An open-ended vertical tube with a gauze stretched across inside the lower half of the tube. When the gauze is heated, a loud sound of the tube's resonance frequency is produced.

rille (*Astron*) A winding valley, with a U-shaped cross-section, found in the LUNAR MARIA.

rill stoping (*MinExt*) Overhand or upward stoping, in which ore is detached from above the miner so as to form an inverted stoped pyramid spreading from a winze at its apex, through which broken ore is withdrawn.

Rilsan (*Plastics*) TN for nylon-11, nylon-12, etc.

RIM (*Eng*) Abbrev for REACTION INJECTION MOULDING.

rima (*BioSci*) A narrow cleft. Adjs *rimate, rimose, rimiform*.

rima glottidis (*Med*) The gap in the larynx between vocal cords in front and arytenoid cartilages of the larynx behind.

rim lock (*Build*) A lock distinct from a MORTISE LOCK in that its metal case is screwed to the face of the door.

rimmed steel (*Eng*) Low-carbon steel that has not been completely deoxidized before casting. Gas is evolved during solidification and, after initial solidification of a sound rim, small bubbles form within the body of the casting and the consequent volume increase counteracts solidification shrinkage. Rimmed ingots contain no PIPE and the yield of useful metal is thereby increased. Impurities and inclusions are concentrated in the interior parts of the ingot. The internal voids close up during subsequent rolling or forging, but the rim of relatively pure and inclusion-free metal is always present and enhances the surface quality. Also *rimming*. See KILLED STEEL.

Rinco process (*Print*) Reproduction proofs taken with white ink on black paper and photographed to yield positives suitable for printing down, particularly for photogravure.

rinderpest (*Vet*) An acute, highly contagious and often fatal myxovirus infection of cattle, sheep and goats; characterized by fever and ulceration of the mucous membranes, esp of the alimentary tract, causing severe diarrhoea and discharges from the mouth, nose and eyes. Notifiable in the UK, and vaccines used elsewhere. Also *cattle plague*.

ring (*BioSci*) See ANNULUS.

ring (*ICT*) See NETWORK, RING TOPOLOGY.

ring (*MathSci*) A set, S, with two operations called *addition* and *multiplication* that satisfy the following conditions: (1) for all elements x, y in S, $x + y$ and xy belong to S; (2) the set S is an Abelian group with respect to the addition operation; (3) the multiplication is (a) associative, (bi) distributive with respect to the addition operation: eg the positive and negative multiples of three with zero, ie the set $\{\dots -9, -6, -3, 0, 3, 6, 9, \dots\}$, with the usual addition and multiplication of numbers. A ring which is such that all its non-zero elements form an Abelian group with respect to the multiplication operation is called a *field*: eg the rational numbers with the usual addition and multiplication of numbers. Some writers require that a ring should have an identity for multiplication, and call the rest *pseudorings*.

ring armature (*ElecEng*) An electric-machine armature having a ring winding.

ringbone (*Vet*) An exostosis on the phalangeal bones of the horse's foot. *High ringbone* involves the first interphalangeal joint and *low ringbone* involves the second interphalangeal joint. *False ringbone* is an exostosis affecting the first or second phalanx but not involving a joint.

ringbound (*Vet*) See RINGWOMB.

ring complex (*Geol*) See RING DYKE.

ring counter (*ICT*) A number of counting circuits in complete series, for sequence operating in counting impulses.

ring course (*Build*) The course farthest from the intrados of an arch.

ring culture (*BioSci*) A system for growing eg tomatoes in greenhouses, using bottomless containers filled with fresh compost (through which the mineral elements are supplied) resting on a bed of sand or other inert material (through which water is supplied). The system makes it easier to avoid the infection of the roots by fungi that is common in greenhouse crops grown in the soil.

Ring drier (*ChemEng*) A proprietary device for drying solids in which the wet solid is entrained in a hot airstream which travels in a circular duct itself arranged in a large-diameter circle.

ring dyke (*Geol*) An almost vertical intrusion of igneous rock which rose along a more or less cylindrical fault which had an approximately circular outcrop. In some Tertiary instances several successive ring dykes, separated by 'screens' of country rock and approximately concentric, form ring complexes. See DYKES.

Ringelmann smoke chart (*Genrl*) One of a series of six charts, numbered 0 to 5 and shaded from white to black, indicating a shade of grey against which density of smoke may be gauged.

ringer equivalent number (*ICT*) A rating for telephone systems indicating the current drawn when the ringing voltage is applied by the local exchange to indicate an incoming call. A standard electromechanical bell has a ringer equivalent of unity.

Ringer solution (*BioSci*) A physiological salt solution isotonic for mammalian cells.

ring fire (*ElecEng*) Thin streaks of fire appearing round the commutator of an electric machine. It is due to small particles of copper or carbon which have become embedded in the insulation between the commutator and raised to incandescence by the current, and indicates that the commutator needs cleaning.

ring gauge (*Eng*) A hardened steel ring having an internal diameter of specified size within very small limits of error; used to check the diameter of finished cylindrical work.

ring gland (*BioSci*) See WEISMANN'S RING.

ringing (*BioSci*) The removal of the outer tissues from a strip encircling or partly encircling a stem or trunk, eg experimentally to interrupt phloem transport while leaving xylem transport more or less undisturbed, the phloem being peripheral to the xylem in most stems. Used horticulturally to encourage flowering and fruiting in over-vigorous fruit trees or done by rabbits or deer and, if extensive, often fatal.

ringing (*ICT*) Extended oscillation in a tuned circuit, at its natural frequency, continuing after an applied voltage or current has been shut off, dying away according to its DECAY CONSTANT, but running into the next oscillation.

ringing (*ImageTech*) Light and dark bands at the edge of image areas of large brightness difference, due to unwanted oscillation in the system.

ring latch (*Build*) A latch in which the fall bar is operated by a handle in the shape of a ring, pivoted at the top so that it always falls into the vertical position.

ring main (*ChemEng*) Closed loop of piping through which chemicals in solution, or such finely divided materials as powdered coal, are circulated in suspension past suitable draw-off points.

ring main (*ElecEng*) A domestic ac wiring system in which a number of outlet sockets are connected in parallel to a ring circuit which starts and finishes at a mains supply point. All plugs used in the power outlets are fitted with individual fuses.

ring modulator (*ElecEng*) Four rectifying elements in complete series, which act as a switch, being fed with appropriate currents at the corners.

Ring Nebula (*Astron*) A famous PLANETARY NEBULA in the constellation Lyra. Confusingly, not a ring nebula.

ring nebula (*Astron*) A nebulous arc or ring which is ionized by high-energy radiation from a WOLF–RAYET STAR.

ring-opening polymerization (*Chem*) The polymerization of a cyclic monomer unit. See STEP POLYMERIZATION.

ring oscillator (*ElecEng*) One in which a number of active components feed each other in a circle and in which the frequency is determined by a ring cut from a quartz crystal, suspended at its nodes to minimize damping; used eg as a standard time-keeper (quartz crystal clock) at 10^5 Hz.

ring-porous (*BioSci*) Wood that has much larger and/or more vessels in the early wood in each annual ring than in the late wood, so that the early wood may appear in cross-sections of stems as a ring of small holes. Characteristic of some mainly north-temperate deciduous trees, eg chestnut, elm and deciduous oaks. Cf DIFFUSE-POROUS.

rings and brushes (*Crystal*) Name applied to the patterns produced when convergent or divergent plane-polarized light, after passing through a doubly refracting crystal cut perpendicular to the optical axis, is examined by an analyser. See INTERFERENCE FIGURE.

ring shift (*ICT*) See END-AROUND SHIFT.

ring size (*MinExt*) A description of rock too large for handling by screening, in accordance with the diameter of a ring which can be slipped over it.

ring slot parachute (*Aero*) A parachute the canopy of which has slots all round its circumference, to give it greater stability.

ring spanner (*Eng*) One in the form of a notched ring in which diametrically opposite notches fit over the nut.

ring spinning (*Textiles*) The most widely used spinning method. Fibres in the form of drafted roving are twisted together to form yarn by a tiny traveller rotating in a ring around a collection package on a spindle which is driven at a speed greater than that of the traveller. Many of these spinning heads are mounted together on a spinning frame.

ring-spot (*BioSci*) An area, eg on a leaf, surrounded by a ring (or concentric rings) of chlorotic, necrotic or abnormally dark green tissue; a characteristic symptom in some virus diseases.

ring stress (*MinExt*) That adjacent to the walls of an unsupported underground excavation.

ringtone (*ICT*) A characteristic sound or tune made by a mobile phone when a call or message is received.

ring topology (*ICT*) A layout for a NETWORK that takes the form of a closed loop with devices attached to the ring. See fig. at NETWORK TOPOLOGY. See TOKEN RING.

ring twisting (*Textiles*) Forming a folded yarn by twisting together two or more single yarns by using a *ring-and-traveller machine*. See RING SPINNING.

ring winding (*ElecEng*) (1) A helical winding arranged on a ring of iron or other material. Also *toroidal winding*. (2) A form of armature winding in which the armature core is a hollow cylinder with each turn of the winding threaded through the centre.

ringwomb (*Vet*) Incomplete dilatation of the cervix at parturition in the ewe. Also *ringbound*.

ringworm (*Med*) A contagious disease characterized by formation of ring-shaped patches on the skin; due to infection with moulds, esp the three genera *Microsporon*, *Trichophyton* and *Epidermophyton*. Also *tinea*.

rINNs (*Pharmacol*) Abbrev for *recommended international non-proprietary names* for medicinal substances. British Approved Names (BANs) have been harmonized with rINNs.

rio rosewood (*For*) See BRAZILIAN ROSEWOOD.

RIP (*MinExt*) Abbrev for RESIN-IN-PULP.

RIP (*Print*) Abbrev for *rest in proportion*. An instruction to reproduce all remaining artwork in a batch in proportion to that marked.

rip (*Build*) To saw timber along the direction of the grain.

rip-bit (*MinExt*) See JACKBIT.

ripcord (*Aero*) (1) A cable used for opening the pack of a personal parachute. (2) An emergency release for gas in an aerostat envelope.

ripening (*ImageTech*) A stage in the manufacture of a photographic emulsion in which the silver halide grains reach their optimum size. Also *digestion*.

Riphean (*Geol*) The later part of the Proterozoic. See PRECAMBRIAN.

ripidolite (*Min*) A species of the chlorite group of minerals, crystallizing in the monoclinic system. It is essentially a hydrated silicate of magnesium and aluminium with iron. Also *prochlorite*.

ripper (*Build*) Slater's tool with a cranked handle and a long flat blade ending in an arrow-shaped head. Used for removing slates by cutting the fixing nails.

ripping saw (*Build*) See RIP-SAW.

ripple (*ElecEng*) The ac component in the output of a rectifier delivering dc. May be reduced by a series choke and shunting (smoothing) capacitor, or ZENER DIODE. Measured as a percentage of the steady (average) current.

ripple (*Phys*) Small wave on the surface of a liquid for which the controlling force is not gravity, as for large waves, but surface tension. The velocity of ripples diminishes with increasing wavelength, to a minimum value which for water is 23 cm s^{-1} for a wavelength of 1·7 cm.

ripple control (*ElecEng*) A method of controlling street lighting or other equipment from some central point by means of a high-frequency signal superimposed on the current-carrying conductors of an electric power system. The information contained in the ripple signal is decoded and the appropriate load is disconnected by operation of a local circuit breaker.

ripple filter (*ElecEng*) A low-pass filter which is designed to reduce the ripple current but at the same time permits the free passage of dc, eg from a rectifier. Also *smoothing circuit*.

ripple finish (*Build*) See WRINKLE FINISH.

ripple frequency (*ElecEng*) The frequency of the ripple current in rectifiers etc. Usually double the supply frequency in a full-wave rectifier.

Ripple generator (*NucEng*) See RADIOISOTOPE THERMO-ELECTRIC GENERATOR.

ripple marks (*Geol*) Undulating ridges and furrows found on the bedding planes of certain sedimentary rocks, due to the action of waves or currents of air or water on the sediments before they were consolidated. Such ripple and rill marks can be seen in the process of formation today on most sandy beaches, on sand dunes and in deserts.

ripple tank (*Phys*) Water-lined tank which uses the property of 'shallow' surface waves (velocity proportional to depth) to demonstrate the refraction and focusing of waves.

ripple trays (*ChemEng*) Distillation column trays (or plates) which consist of thin metal formed into a series of parallel channels and perforated to provide space for rising vapour.

rip-saw (*Build*) A saw designed for cutting timber along the grain. Also *ripping saw*.

RISC (*ICT*) Abbrev for REDUCED INSTRUCTION SET COMPUTER.

rise (*Build, CivEng*) (1) The vertical distance from the centre of span of an arch in the line of the springings to the centre of the intrados. Also *versed sine*. (2) The vertical height from end supports to ridge of a roof. (3) The height of a step in a staircase.

rise and fall system (*Surv*) A system of reduction of levels in which the staff reading at each successive point after the first is compared with that preceding it, and the difference of level entered as a *rise* or a *fall*. See COLLIMATION SYSTEM.

rise and run (*Build*) A term applied to the amount of any given slope quoted as a given *rise* (vertical distance) in a given *run* (horizontal distance).

risen moulding (*Build*) A moulding decorating a panel and projecting beyond the general surface of the surrounding framing.

rise of floor line (*Ships*) A tangent to the curve of the BILGE to meet the extremity of the flat of the keel.

riser (*Build*) The vertical part of a step. See fig. at STRING.

riser (*Eng*) In a mould, a passage up which the metal flows after filling the mould cavity. It allows dirt to escape, indicates that the mould is full, and supplies metal to compensate for contraction on solidification. Also *out-gate*. See fig. at MOULDING.

rise time (*ICT*) The time for a pulse signal in an amplifier or filter to rise to from 10% to 90% of the maximum amplitude. Also *build-up time*.

rising and falling saw (*Build*) A circular saw whose spindle can be raised or lowered relative to the work table to vary the depth of the saw cut.

rising arch (*CivEng*) An arch whose springing line is not horizontal.

rising butt hinge (*Build*) A butt hinge with a loose leaf which, when opened, rises on the centre pin due to helical bearing surfaces on the two leaves. This enables a door, on opening, to rise above a carpet and to close automatically. See fig. at COUNTER-FLAP HINGE.

rising front (*ImageTech*) A sliding panel for carrying the lens in a field or technical camera; used to diminish foreground and avoid distortion of perspective when photographing high buildings, trees, etc.

rising main (*Build*) Mains water supply where it enters premises.

rising shaft (*MinExt*) A shaft which is excavated from below upward. Cf SINKING SHAFT.

risk (*BioSci*) (1) The probability of adverse effects caused under specified circumstances to an organism, a population or an ecological system. (2) The expected frequency of occurrence of a harmful event arising from exposure to a substance, agent or process.

risk assessment (*BioSci*) Identification and quantification of the potential risk resulting from exposure to an agent, substance, process, etc. It can only be a prediction.

risk assessment (*NucEng*) A description of the safety of a plant in terms of the frequency and consequence of any possible accident. See panel on RISK ASSESSMENT.

risperidone (*Pharmacol*) An atypical antipsychotic used in the treatment of acute and chronic psychoses, including schizophrenia.

Riss (*Geol*) A glacial stage in the Pleistocene epoch of the Alps. See QUATERNARY.

RI strains (*BioSci*) Abbrev for RECOMBINANT INBRED STRAINS.

risus sardonicus (*Med*) Wrinkling of the forehead and retraction of the angles of the mouth due to a spasm of the facial muscles (as in tetanus), giving the appearance of a grin.

Ritalin (*Pharmacol*) TN for METHYLPHENIDATE.

Ritchie wedge (*Phys*) A photometer head in the form of a wedge with two white diffusing surfaces set 90° apart.

ritonavir (*Pharmacol*) A *protease inhibitor* used as a drug in the treatment of HIV.

Rittinger's law (*ChemEng*) A law stating that the energy required in a crushing operation is directly proportional to the area of fresh surface produced: $E = k_r(1/d_2 - 1/d_1)$, where E is the energy used in crushing, k_r is a constant, depending on the characteristics of the material and on the type and method of operation of the crusher, and d_1 and d_2 are the average initial and final linear dimensions of the material crushed.

ritualization (*Psych*) The evolutionary process by which a behaviour pattern is modified to enhance its communication value, usually through exaggeration or repetition of some of its elements.

rituximab (*Pharmacol*) A therapeutic antibody used to treat arthritis and non-Hodgkin's lymphoma.

Risk assessment

A description of the safety of any kind of plant, eg chemical or nuclear, in terms of the frequency and consequence of any possible accident. The safety of the first nuclear power plants was assessed by identifying design-based accidents and determining whether the engineered safeguards could prevent them. The consequences of the maximum credible accident were also calculated. This DETERMINISTIC approach has failed on a number of occasions because essentially trivial malfunctions and operator error have resulted in quite unforeseen accidents.

Designers and nuclear safety agencies have therefore increasingly turned to a much more complete analysis of all aspects of a plant's operation called *probabilistic risk assessment* (PRA), which can be split naturally into four stages: (1) The identification of accidents which might occur. This will include everything from a weld failure in the main coolant circuit to a fuse blowing in the control board. (2) Estimating the frequency with which a possible accident occurs, ie its probability per unit time. (3) Determining the quantified consequences of each accident to the plant itself, its operators, the surrounding population and the environment. (4) Using the risk information, ie the frequencies and consequences, as a guide in making decisions about eg whether the plant is acceptably safe, how it could most cost-effectively be made safer, or the nature of the weak points of the design.

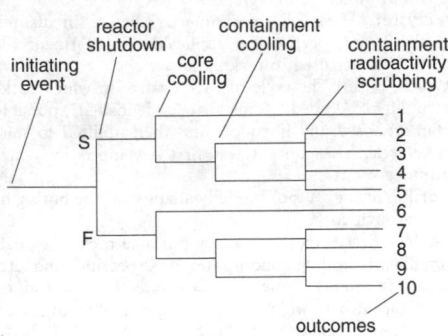

Risk assessment Success (S) or failure (F) indicated by an upward or downward line at each stage.

These simple concepts are applied to nuclear reactors at three levels of analysis:

Damage to the reactor core. This involves identifying the initiating events and then, assuming the success or failure of each emergency system, constructing an event tree (see diagram) which will have different end-points depending on which system failed or succeeded. The probability of each event is estimated from the history of previous incidents and the events may be very detailed, down to the level of eg individual pumps or particular operator errors.

Effect on the containment barrier. Again an event tree has to be made for the possible accident sequences and will involve eg fuel cladding, the reactor vessel and coolant, as well as the containment building. It will result in end-points which can be divided into different categories depending on the proportions and amounts of the radioactivity released. This analysis will involve knowledge of the physics and chemistry of the fission products and of their movements. It should give information about improving the design of the containment barriers.

Dispersion of radioactivity and its effects. This requires the modelling of how a cloud of radioactivity will move under different conditions, its fallout onto land and sea, and its subsequent ingestion by humans or incorporation into food chains. It will require the analysis of risk to society in an area perhaps extending over hundreds of kilometres. From the results, the number of people who may be affected can be calculated, and predictions made about the area of land to be evacuated and the cost of decontamination.

Probabilistic risk assessment is a very ambitious endeavour because it seeks to include every possible accident sequence. There are inevitably areas where there is insufficient information, which has led to criticism of the whole concept of this kind of analysis. This has, in turn, prompted a more detailed investigation of the gaps in knowledge that have been revealed, and attempts to quantify the degree of uncertainty which may, in places, be very large.

See panel on NUCLEAR REACTORS.

---

rivastigmine (*Pharmacol*) A drug used to treat Alzheimer's disease.

river capture (*Geol*) Natural diversion of a stream or river by a neighbouring stream with greater power of erosion that erodes headworks and taps the flow of the stream into its own channel. See DRAINAGE PATTERNS.

rivers (*Print*) In widely spaced text matter the spaces in successive lines can form channels of space running down the page, a tendency which is reduced by close spacing. Also *streets*.

rivers, geological work of (*Geol*) This involves *corrasion* (wearing away) of their banks and beds, *corrosion* (solvent and chemical action of river water), *hydraulic action* and *attrition* of the transported material.

river terrace (*Geol*) A nearly flat surface along the side (or sides) of a valley marking the position of a former flood plain when the river was at a higher level. The terrace is usually built up of gravel, sand or alluvium, or may be a *bench* cut into solid rock, and it can occur on either side of a valley where the meandering river has not removed all of the earlier deposits.

rivet (*Eng*) A headed shank for making a permanent joint between two or more pieces. It is inserted in a hole which is made through the pieces, and 'closed' by forming a head on the projecting part of the shank by hammering or other means. The head may be rounded flat, pan-shaped or countersunk. See BLIND RIVET.

riveted joint (*Eng*) A joint between plates, sheets or strips of materials (usually metallic) secured by rivets. See BUTT JOINT, LAP JOINT.

riveting machine (*Eng*) A machine for closing, clinching or setting rivets. See HYDRAULIC RIVETER, PNEUMATIC RIVETER.

RJ45 connector (*ICT*) An electrical connector frequently used to terminate unshielded twisted pair cables for both voice and data transmission. The 'RJ' is derived from 'remote jack'.

RLL (*ICT*) Abbrev for *run-length limited*.

rmm (*Chem*) Abbrev for RELATIVE MOLECULAR MASS.

rms power (*Phys*) Abbrev for *root-mean-square power*. The effective mean power level of an alternating electric supply.

rms value (*Phys*) See ROOT-MEAN-SQUARE VALUE.

Rn (*Chem*) Symbol for RADON.

RNA (*BioSci*) Abbrev for *ribonucleic acid.*, a single-stranded macromolecule similar in structure to DNA but in which the base group thymine is replaced by uracil. Like DNA, RNA can hold genetic information (as in viruses) but is also the primary agent for transferring information from DNA to the protein-synthesizing machinery of cells. See MRNA.

road bed (*CivEng*) Foundation carrying the sleepers, rails, chairs, points and crossings, etc, of a railway track.

road line paint (*Build*) A quick-drying bitumen-resistant coating used for marking road lines.

road metal (*CivEng*) Broken stone for forming the surfaces of macadamized roads. Also *metal, metalling*.

road studs (*CivEng*) Rubber or metal pads, the former generally incorporating small reflectors (cat's eyes), built into the surface of a road to define traffic lanes at night or in fog.

roak (*MinExt*) A SEAM in mining. Also *roke*.

roaming (*ICT*) The ability to make and receive calls on a mobile-telephone network other than the one with which the mobile is registered and using the same telephone number. In the GLOBAL SYSTEM FOR MOBILE COMMUNICATION, international roaming is possible.

roaring (*Vet*) A respiratory affection of horses in which an abnormal sound is produced in the larynx during inspiration; caused by paralysis of the laryngeal muscles or by inflammatory conditions affecting the larynx. *Whistling* is a similar condition in which a sound of higher pitch is produced.

roasting (*Eng*) The operation of heating sulphide ores in air to convert to oxide. Sometimes the sulphur-bearing gases produced are used to make sulphuric acid. See CHLORIDIZING ROASTING, SULPHATING ROASTING, SWEET ROASTING.

roasting (*FoodSci*) Dry cooking food over an open fire or in an enclosed oven.

roasting furnace (*Eng*) A furnace in which finely ground ores and concentrates are roasted to eliminate sulphur. Part or all of the necessary heat may be provided by the burning sulphur. The essential feature is free access of air to the charge. This is done by having a shallow bed which is continually RABBLED. Many types have been devised; multiple-hearth is most widely used.

Robertsonian translocation (*BioSci*) A balanced translocation in which the breakpoints in the translocated chromosomes cannot be identified.

robinia (*For*) N American hardwood (*Robinia pseudoacacia*) whose heartwood is golden brown, straight-grained, with a coarse texture.

Robinson bridge (*ElecEng*) An ac bridge used for the measurement and control of frequency. Also *Robinson–Wien* bridge.

robot (*ICT*) A computer-controlled machine that is able to sense, grip and move objects.

robotics (*ICT, Eng*) The study of the design and use of robots, particularly for their use in manufacture and related processes. Distinguished from automatic handling devices (eg pick-and-place units) by their flexibility for multiprogramming. Thus they can be used for part handling, assembly, painting, packing, etc.

robustness (*ICT*) The ability of a system to cope with errors during execution.

Roche limit (*Astron*) The lowest orbit at which a satellite can withstand the tides raised within it by the primary body.

Rochelle salt (*Chem*) A crystal of sodium potassium tartrate, having strong piezoelectric properties, but high damping. Used as a *biomorph* in microphones, loudspeakers and pickups. Its disadvantage is the limited range of piezoelectric property (-18 to $23°C$) and high temperature coefficient. See PIEZOELECTRIC EFFECT, PIEZOELECTRICITY.

roche moutonnée (*Geol*) A mound of bare rock which is usually smoothed on the upstream side and roughened by plucking on the downstream side, as a result of a moving ice sheet.

Rochon prism (*Phys*) A device for producing linearly polarized light. Made of two wedge-shaped prisms of quartz; the first has its optical axis parallel to an incident beam and the second has its optical axis perpendicular to it. The ordinary ray is undeviated in this configuration.

rock (*Geol*) Any mineral matter making up the Earth. As used by geologists, the term also includes unconsolidated material such as sand, mud, clay and peat, in addition to the harder materials described as rock in conventional usage.

rock and roll (*ImageTech*) Film transport control system, used esp in sound mixing and dubbing, in which a number of interlocked paths can be simultaneously stopped at any point and run forward or back without losing the precise synchronization between them.

rock burst (*MinExt*) Sudden failure of stope pillars, walls or other rock buttresses adjacent to underground works, with explosively violent disintegration.

rock crystal (*Min*) Colourless quartz whether in distinct crystals or not; particularly applicable to quartz of the quality formerly used in making lenses.

rock cycle (*Geol*) The cycle of rock change in which rocks are uplifted, eroded, transported, deposited, possibly metamorphosed and intruded, and then uplifted to start a new cycle. The concept was first developed by James Hutton.

rock drill (*CivEng*) A tool specially adapted to the boring of holes through rock.

rocker (*MinExt*) A short and easily portable rocking trough or cradle for washing concentrates, gold-bearing sand, etc.

rocker arms (*Autos*) Pivoted levers operated by push rods or overhead camshaft, which carry the tappets of an overhead valve system. Also VALVE ROCKERS.

rocker gear (*ElecEng*) The hand wheel or other device for moving a brush-rocker.

rocket (*Aero, Space*) (1) A missile projected by a rocket system. (2) A system, or a vehicle, powered by reaction or rocket propulsion. See panel on ROCKET.

rocket engine (*Space*) A device for providing thrust produced by the expulsion of hot gases from a combustion chamber. Oxygen for burning the fuel would normally be carried in the vehicle, not collected from the atmosphere. See REACTION PROPULSION, ROCKET PROPULSION and panel on ROCKET.

rocket equation (*Space*) The relationship between burnout velocity (v_b) and the gas exhaust velocity (v_e) for a rocket:

$$v_b = v_e \ln \frac{M}{M_b}$$

where M and M_b are the masses of the rocket at launch and burnout respectively. See panel on ROCKET.

rocket propulsion (*Aero, Space*) Reaction propulsion using internally stored oxygen for combustion of the fuel. Propellants may be single compounds, eg hydrogen peroxide, or separately stored fuel, eg liquid hydrogen, and oxidizer, eg liquid oxygen. Used primarily where there is insufficient atmospheric oxygen, eg above 70 000 ft (20 km), or where the lightness and compactness of the motor offset the high propellant weight, eg for assisted take-off and missiles. See panel on ROCKET.

rocket tester (*Build*) A rocket giving off dense smoke; used to test a drain for leaks.

rock face (*Build*) The form of face given to a building-stone which has been QUARRY-FACED.

rock flour (*Geol*) A term used for finely comminuted rock material found at the base of glaciers and ice sheets. It is mud-like and is composed largely of unweathered mineral particles.

rock-forming minerals (*Geol*) The minerals which occur as dominant constituents of igneous rocks, including quartz, feldspars, feldspathoids, micas, amphiboles, pyroxenes and olivine.

rock head (*MinExt*) See STONE HEAD.

rock milk (*Min*) A very soft white variety of calcium carbonate resembling cotton which breaks easily in the fingers; it is sometimes deposited in caverns or about sources holding lime in solution. Also *rock meal*.

rock phosphate (*Min*) See PHOSPHORITE.

rock roses (*Min*) See DESERT ROSE.

rock salt (*Min*) See HALITE.

rock wall failure (*MinExt*) Collapse due to one or more of the following: rock fall, simple dropping; rock flow, slope failure; plane shear, failure along weakness plane; rotational shear, stress where soil has dropped leaving a rounded cavity.

Rockwell hardness test (*Eng*) A method of determining the hardness of metals by indenting them with a hard steel ball or a diamond cone, first applying a light load and then increasing to a specified higher load, and measuring the additional depth of penetration. See panel on HARDNESS MEASUREMENTS.

rock wool (*Eng*) Fibrous insulating material made by blowing steam through molten slag. Also *mineral wool, slag wool*.

Rocky Mountain fever (*Med*) A disease, more or less limited to the US, characterized by fever, headache, muscular pains, enlargement of the spleen, and a macular eruption on the skin; the disease is spread by a tick and is associated with infection by *Rickettsia rickettsi*. Also *blue disease, black fever, spotted fever*.

Rococo (*Arch*) The name given to an ornate form of decoration which developed from BAROQUE and which was particularly popular in Paris in the 18th century, being the style of the interiors, furniture and art of Louis XV's reign. Basic motifs were drawn from free-flowing plant and shell forms, often without organic coherence, and often rendered in stucco or carved wood with gold paint.

rod (*BioSci*) One of the colour-insensitive, light-perceptive elements of the vertebrate retina. Rods respond to lower illumination levels than CONES.

rod (*Genrl*) (1) A pole or perch ($5\frac{1}{2}$ yd or $16\frac{1}{2}$ ft). (2) A square pole ($272\frac{1}{4}$ ft^2). (3) Of brickwork, 272 ft^2 of standard thickness of $1\frac{1}{2}$ bricks or 306 ft^3.

rod (*NucEng*) A rod-shaped reactor fuel element or control absorber, or sample, intended for irradiation in reactor.

rodding (*Build*) (1) Clearing a stoppage in a pipe by inserting a rod to break down the obstruction and remove it. (2) Descaling encrusted pipework with scrapers attached to jointed rods.

rodding eye (*Build*) The removable cover on a small opening in a drainpipe which enables an obstruction to be removed by drain rods.

Rodentia (*BioSci*) An order of generally small mammals with never more than a single pair of chisel-shaped upper incisors that have open roots. The lower incisors can move like scissors as there is no anterior symphysis between the mandibles. Canines are never present but there is a wide diastema between the gnawing incisors and the grinding cheek teeth, which vary in numbers and frequently have persistently open roots. The glenoid cavities are elongated anteroposteriorly, with the lower jaw being moved forwards for gnawing and backwards for grinding, and the jaw muscles are greatly enlarged. Members are herbivorous with a large caecum. They are almost universally distributed, and are of considerable economic and medical importance as pests of stored food and carriers of plague fleas. They are adaptive with a wide radiation: terrestrial, amphibian, burrowing, arboreal, gliding and saltatorial. Agoutis, beavers, gophers, guinea pigs, hamsters, mice, porcupines, rats, squirrels, voles.

rodent ulcer (*Med*) A slow-growing ulcerating cancer of the skin which usually affects the upper part of the face; arises from the basal cells of the skin and is of low malignancy. Also *basal cell carcinoma*.

rodman (*Surv*) A staffman (US).

Roe chlorine number (*Paper*) A test method to determine the bleachability of a sample of pulp by measuring the amount of chlorine gas it will absorb under the specified conditions of test.

roentgen (*Radiol*) Unit of X-ray or gamma dose, for which the resulting ionization liberates a charge of each sign of $2\cdot58 \times 10^{-4}$ C per kilogram of air. Symbol R.

roentgen equivalent man (*Radiol*) Former unit of biological dose given by the product of the absorbed dose in R and the relative biological efficiency of the radiation. Abbrev *rem*. Now replaced by EFFECTIVE DOSE EQUIVALENT, unit the SIEVERT (Sv): 1 rem = 0·01 Sv.

roentgenium (*Chem*) An artificially produced radioactive transuranic element (symbol Rg; at no 111), formerly called *unununium*.

roentgenology (*Radiol*) US for RADIOLOGY.

roentgen ray (*Phys*) See X-RAY.

Rogallo wing (*Aero*) Delta-shaped wing formed by three spars which meet at the apex and are covered with fabric. In flight the fabric becomes convex due to reduced air pressure over the top surface. Can be folded compactly and, although originally intended for re-entering spacecraft as a *paraglider*, has become very popular for MICROLIGHT aircraft.

roger (*Aero*) Radio code for 'I have received and understood all of your last transmission'.

rogue (*BioSci*) (1) A plant that is not true to type. (2) To remove such plants from a crop, esp one grown for the production of seed.

rogue value (*ICT*) A special value, outside the generally expected range, that is used to terminate a list of data items; eg in a list of positive values, -1 would be a suitable rogue value. Also *terminator*.

Rohypnol (*Pharmacol*) TN for a powerful sedative drug (*flunitrazepam*, a benzodiazepine) used for short-term treatment of insomnia. Also used in so-called 'date-rape'.

roke (*MinExt*) See ROAK.

role (*Psych*) A pattern of behaviour, and the expectation of it, associated with individuals who hold a particular position in a society or social group.

roll (*Aero*) Aerobatic manoeuvre consisting of a complete revolution about the longitudinal axis. In a *slow roll* the centre line of the aircraft follows closely along a horizontal straight line; an *upward roll* is similar, but considerable height is gained; a *hesitation roll* is one where the pilot brings the aircraft momentarily to rest in its rolling motion. A vertical upward or downward roll is usually called an *aileron turn* because these are the only control surfaces involved. A *flick roll* is an entirely different, very rapid and violent manoeuvre in which the aircraft makes its revolution along a helical path; high structural stresses are imposed and many countries ban this aerobatic. A *half-roll* is lateral rotation through 180°.

Rocket

A system or vehicle propelled into and in space which is based on the rocket or *reaction* principle. A rocket carries its own source of energy in the form of propellants which are made to react and produce a stream of gas (or other) particles which are ejected with a high exhaust velocity. In accordance with Newton's third law of motion, a force is produced on the rocket system which is in the opposite direction to that of the ejected gases. This force is termed the thrust (*T*) and can be deduced from the rate of momentum change. It is related to the exhaust velocity (*V*$_e$) by the equation

$$T = mV_e (T \text{ in newtons}) \text{ or } T = mV_e/g_o \ (T \text{ in kgf})$$

where *m* is the mass of propellant expended per second and g_o is the acceleration due to gravity at the Earth's surface. It follows that a high exhaust velocity is desirable. The ratio *T/m*, thrust divided by propellant consumption, is simply V_e using MKS units. If *T* is measured in kgf, then $T/m = V_e/g_o$, referred to as the specific impulse (I_{sp}) which has the units of seconds. Specific impulse can then be defined as the thrust in kgf obtained from 1 kg of propellant burnt in 1 second. Again the total impulse is *Tt*, where *t* is the burning time, and $I_{sp} = Tt/mt$, so specific impulse can be regarded as the total impulse per unit mass of propellant consumed.

Rocket engines

Rocket engines are essentially rather simple machines which provide thrust from the reaction of propellants in a combustion chamber to propel a vehicle to which it is attached. The skill resides in the optimum design of the propellant injectors, combustion chamber, ignition system and nozzle, and of the oxidizer-to-fuel ratio, all of which are essential for efficient operation.

The reaction normally takes place between a fuel and an oxidizer. Two forms of liquid propellant are available. A *monopropellant*, eg H_2O_2, relies on the chemical reaction or decomposition of a single liquid, usually in the presence of a catalyst. A *bipropellant*, eg liquid O_2 and liquid H_2, consists of two liquids, the fuel and the oxidizer, which are stored separately. Solid propellants can also be used and are then contained within the combustion chamber, making the system inherently simple. Ignition is usually effected by a spark, after which the heat produced maintains the burning process. Sometimes the fuel and oxidizer are designed to ignite spontaneously on mixing; such propellants are called HYPERGOLIC.

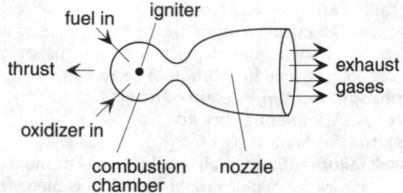

Fig. 1 **Rocket engine** Liquid fuel type.

In a liquid rocket engine, such as that in Fig. 1, the propellants are introduced into a combustion chamber where they are burnt, with the products of combustion being ejected through the nozzle at high velocity. Since a rocket engine works most efficiently in a vacuum, it is ideally suited for propulsion in space.

Re-entry

Ballistic missiles and some spacecraft must return to Earth or sometimes another planet and therefore pass

---

roll (*Build*) A joint between the edges of two lead sheets on the flat, the edges being overlapped over a 50 mm (2 in) diameter wood roll fastened to the surface to be covered.

roll (*Print*) A tool used to impress designs on leather book cases. Usually brass, set in a long handle, which is necessary because of the pressure required in its operation.

roll (*Ships*) The phenomenon of a ship's behaviour in waves, wherein it changes its angle of HEEL. See ROLLING PERIOD.

roll-capped (*Build*) Said of ridge tiles which are finished with a roll or cylindrical projection along the apex.

roll damper (*Aero*) See DAMPER.

rolled gold (*Eng*) Composite sheet made by soldering or welding a sheet of gold onto both sides of a thicker sheet of silver, or other suitable metal, and rolling the whole down to the thickness required.

rolled laminated tube (*Plastics*) Tube produced by winding, under heat, pressure and tension, a synthetic-resin-impregnated or coated fabric or paper onto a former.

rolled-steel joist (*Eng*) H-section steel beams with flanges having an 8° taper and a maximum size about 254 × 203 mm. Now supplanted by UNIVERSAL BEAMS, UNIVERSAL BEARING PILES, UNIVERSAL COLUMNS. Abbrev RSJ. See BEAM.

rolled-steel sections (*Eng*) Steel bars rolled into I- or T-shaped channel, angle, cruciform or similar cross-sections for different applications in structural work, each section being made in graded standardized sizes. See BEAM.

roller (*Agri*) A trailed implement, now usually with clusters of metal rings, used to firm the surface of cultivated soil.

roller (*ImageTech*) Flanged drum providing a guided path over which film passes, eg through a processing machine.

roller bearing (*Eng*) A shaft bearing consisting of inner and outer steel races between which a number of parallel or tapered steel rollers are located by a cage; suits heavier loads than the BALL BEARING.

roller bit (*MinExt*) Drilling bit with three or more conical rollers carrying teeth. The rollers have axes inclined to that of the drill pipe and rotate individually with it. Also *Tricone bit*. See panel on DRILLING RIG.

roller chain (*Eng*) A driving or transmission chain in which the links consist of rollers and side plates, the rollers being mounted on pins which connect the side plates. See DRIVING CHAIN.

roller coating (*Build*) An industrial process for coating strips of metal with paint etc on a coating machine. A cylindrical roller picks up the paint from a trough, and

Rocket *(Cont.)*

through an atmosphere. During re-entry (sometimes called *entry*) the spacecraft decelerates and is subject to intense heating generated by atmospheric friction. The magnitude of the problem is indicated by the deceleration which may be greater than 10 g and the stagnation temperature which may reach many thousands of degrees. Furthermore, a sheath of ionized air around the spacecraft can black out radio communication for several minutes.

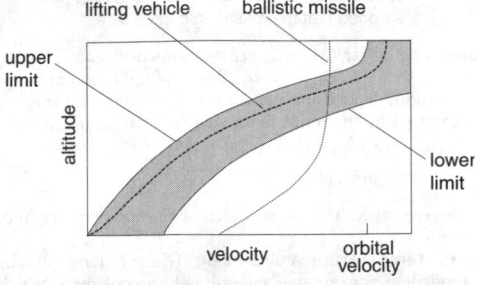

Fig. 2 **Re-entry corridor** Showing upper and lower limits.

For a safe passage through the atmosphere a manned spacecraft must keep within a narrow RE-ENTRY CORRIDOR, which widens as the velocity is reduced and is also dependent on atmospheric density. It can be as low as about 10 km, giving rise to stringent guidance requirements. The lower side of the corridor or *undershoot boundary* is fixed by the excessive heating experienced below it, and the upper side or *overshoot boundary* marks that region above which the atmospheric density is so low that the spacecraft cannot be slowed down and skips back into space.

Re-entering spacecraft must be designed to withstand the hazardous re-entry conditions, and so that their trajectory and angle of entry are optimized. Re-entry may be *direct*, as for ballistic missiles, or *lifting* when a spacecraft's lifting capabilities are used to reduce the high heating rates but, thereby, increase the total heat transferred. In both cases, large values of the drag parameter, $m/C_D A$ (where m is the mass, C_D the drag coefficient and A the drag reference area), imply large stagnation heating rates that are inversely proportional to the square root of the nose radius. In addition, lifting re-entry provides lower g-loads and more manoeuvrability, and is preferable for manned flight. The re-entry vehicle can be shielded from the intense heat by the following techniques, alone or in combination: (1) An ablation shield which vaporizes and carries the heat away. A material of large latent heat (eg glass resin) is used. (2) A heat sink of high thermal capacity (eg copper). (3) A good thermal insulator (eg fibreglass). (4) Radiative cooling using a high thermal emissivity surface.

As an example, the US Space Shuttle Orbiter re-enters in about 55 minutes over a distance of 8000 km. Its velocity is reduced from $7 \cdot 8$ km s^{-1} to $0 \cdot 1$ km s^{-1} and a maximum deceleration of only 3 g is encountered. The vehicle's angle of attack is gradually reduced and its aerodynamic surfaces utilized. A CROSS-RANGE of 1600 km on each side of its nominal re-entry track is possible. Temperatures of 1600°C at the stagnation points and 300°C over leeward surfaces are encountered. To protect the aluminium structure, reinforced carbon–carbon inserts are used at the leading edges and SILICA tiles over the main body.

See panels on ATMOSPHERIC BOUNDARY LAYER and SPACE STATION.

---

distributes it over one or more rollers, from the last of which it is picked up by the strip.

roller conveyor *(Eng)* A CONVEYOR comprising a series of closely spaced rollers set so that their crests can support articles to be conveyed. They may be power-driven, or freely revolving. Tapered ones are used to form bends in the conveyor line. See GRAVITY ROLLER CONVEYOR.

roller mill *(Eng)* In its simplest form, consists of two rolls of suitable material, mounted with their axes horizontal, running in opposite directions; used for crushing and mixing operations.

Rolle's theorem *(MathSci)* The theorem which states that if a real-valued function $f(x)$ is continuous on the interval $[a,b]$ and differentiable on (a,b), and if $f(a) = f(b)$, then $f'(\xi) = 0$ for some $\xi \in (a,b)$. See MEAN VALUE THEOREM.

roll feed *(Eng)* A mechanism incorporating nipping rollers for feeding strip material to power presses. The rollers turn intermittently by a set distance and are timed in relation to the press so as to advance the required length of strip, correctly timed, for each press stroke.

roll film *(ImageTech)* Film, wound on a spool, with opaque paper fitting closely to the end-cheeks of the spool, excluding light sufficiently to allow loading and unloading in daylight.

roll forging *(Eng)* A hot forging process for reducing or tapering short lengths of bar stock in the manufacture of axles, chisels, knife blades, tapered tubes, etc. Parts are placed by tongues into profiled grooves in pairs of forging rollers.

roll forming *(Eng)* A cold forming process in which a series of mating rolls progressively forms strip metal into tube or section, usually of complex shape, at high rolling speeds. Often associated with a continuous resistance welding machine.

rolling *(Aero)* The angular motion of an aircraft tending to set up a rotation about a longitudinal axis. One complete revolution is called a ROLL.

rolling *(Ships)* See ROLLING PERIOD.

rolling bearing *(Eng)* A ball bearing, roller bearing, needle bearing or other type of bearing in which elements roll between load-bearing surfaces.

rolling-circle replication *(BioSci)* The method by which an un-nicked circle of double-stranded DNA is replicated esp in bacteria. See panel on BACTERIA.

rolling instability *(Aero)* See LATERAL INSTABILITY.

rolling lift bridge *(Eng)* A type of BASCULE BRIDGE in which the bascule has, at the shore end, a surface of segmental profile rolling on a flat bearing.

rolling load (*Eng*) See MOVING LOAD.

rolling mills (*Eng*) Sets of rolls used in rolling metals into numerous intermediate and final shapes, eg blooms, billets, slabs, rails, bars, rods, sections, plates, sheets and strip. The roll in contact with the material is the work roll and others mounted behind to prevent it bowing are referred to as back-up rolls. For foils and very thin sheet, clusters of back-up rolls may be required on account of the high pressures necessary.

rolling moment (*Aero*) The component of the couple about the longitudinal axis acting on an aircraft in flight.

rolling period (*Ships*) The movement of a ship in waves about a longitudinal horizontal axis is known as *rolling*. The rolling period is the time taken for a complete roll from the upright position first to one side then to the other and back to upright again. Cf PITCHING PERIOD.

rolling resistance (*Eng*) A component of the force that arises when a wheel rolls on a surface; distortion of the contact surfaces due to the normal force between them destroys the ideal line contact and opposes the motion.

rolling stock (*CivEng*) A general and collective term for all coaches, trucks, etc, which run along a railway track.

roll-off frequency (*ICT*) The frequency where the response of an amplifier or filter is 3 dB below maximum.

roll-on, roll-off ferry (*Ships*) Vehicular ferry with wide doors at bow and stern to facilitate the loading of vehicles, characterized by a large open deck near the water line.

rollover (*ICT*) An explanatory message or image that is displayed when the cursor rests over a certain point on a computer screen. Also *mouseover*.

rolls (*MinExt*) Crushing rolls are pairs of horizontally mounted cylinders, faced with manganese steel, between which ore is crushed as they rotate inward.

ROM (*ICT*) Abbrev for READ-ONLY MEMORY.

roman (*Print*) Ordinary upright type, as distinct from ITALIC or sloping.

Roman mosaic (*Build*) See TESSELLATED PAVEMENT.

Roman numerals (*MathSci*) A system of representing integers in which each symbol has an absolute value, the number being represented by the sum of those values modified by the convention of subtracting smaller values from succeeding larger values. Thus, since I, V, X and C represent 1, 5, 10 and 100 respectively, CCCXXVII = 327. Compare POSITIONAL NOTATION.

Romberg's sign (*Med*) Sign present when a patient, standing with feet close together and eyes closed, sways more than when his or her eyes are open; it indicates disease of the sensory tracts in the spinal cord.

roméite (*Min*) Naturally occurring hydrated antimonite of calcium, sometimes with manganese and iron; crystallizes in the cubic system, often as brownish octahedra.

rone (*Build*) In Scotland, an EAVES GUTTER. Also *rhone*.

röntgen (*Radiol*) See ROENTGEN.

roof antenna (*ICT*) See FLAT-TOP ANTENNA.

roof boards (*Build*) Boards fixed to the rafters to provide a fixing and undercovering for the felting, slates, tiles, etc. Also *sarking* in Scotland.

roof bolting (*MinExt*) A method of roof support in which steel bolts are inserted in drill holes so as to pin supporting steel beams under the roof of a stope.

roof guard (*Build*) A device fitted to a roof to prevent snow from sliding off it. Also *snow boards*.

roofing slate (*Geol*) A term widely applied to rocks of fine grain in which regional metamorphism has developed good slaty cleavage.

roof pendant (*Geol*) A mass of country rock projecting downwards, below the general level of the roof, into an intrusive rock body.

roof truss (*Build*) The structural framework built to support the roof covering for a building. Fig. ▷

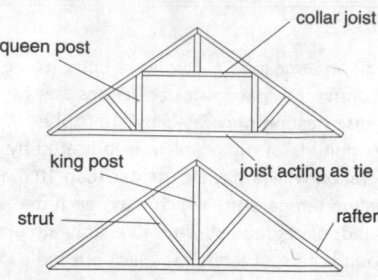

roof truss Timbers are butted and held by studded plates pressed into each side.

room-and-pillar (*MinExt*) See BORD-AND-PILLAR.

room index (*ElecEng*) The coefficient of utilization of lamps in a room depends on the shape of the room. This shape is expressed by the room index, k. For rectangular rooms

$$k = \frac{0 \cdot 9w + 0 \cdot 1l}{h}$$

where w = width of room, l = length of room, h = height of room.

room temperature vulcanizing (*Chem*) Term usually applied to silicone and polysulphide prepolymers, which when mixed with cross-linking agent on site can be applied to form final shape. Abbrev *RTV*.

root (*BioSci*) (1) The typically descending axis of a plant and other axes that are anatomically similar and/or clearly homologous. In most vascular plants, roots may be recognized by their endogenous origin, lack of leaves and possession of a root cap. Roots typically function in anchorage and in the absorption of water and mineral salts in the soil. See AERIAL ROOT, ROOT TUBER. Cf STEM. (2) See FLAGELLAR ROOT.

root (*MathSci*) (1) A factor of a quantity that when multiplied by itself a specified number of times produces that quantity again, eg CUBE ROOT, SQUARE ROOT. (2) A value of a variable that satisfies a given equation.

root cap (*BioSci*) A hollow cone of cells protecting the apical meristem of a growing root, which is renewed from within as it wears. Also *calyptra*.

root diameter (*Eng*) (1) Of a gear wheel, the diameter at the bottom of the tooth spaces. (2) Of a thread, the diameter at the bottom surfaces which join the adjacent sides or flanks of a thread.

root directory (*ICT*) The top-level DIRECTORY. Below this, subdirectories may be defined to help categorize and store files systematically by the user.

rooter (*Electronics*) A circuit designed to give an output amplitude proportional to the square root of the input amplitude. Used in compressors for reducing dynamic range in sound reproduction, and for gamma correction in video, to compensate for TV camera-tube characteristics.

root hair (*BioSci*) A tubular outgrowth from a cell in the epidermis of a young root, possibly important in the uptake of the more slowly diffusing mineral ions, eg phosphate.

rooting compound (*BioSci*) A substance containing AUXINS in which a cutting is dipped to promote root growth.

rootkit (*ICT*) Software embedded within an operating system that performs operations without informing the user.

root locus (*ElecEng*) The locus for the roots of the closed-loop response for a system, derived by plotting the poles and zeros of the open-loop response in the complex plane. Used in the study of system stability.

root mean square (*MathSci*) Square root of sum of squares of individual observations divided by total number of observations. Abbrev *rms*.

root-mean-square power (*Phys*) See RMS POWER.

root-mean-square value (*Phys*) The measure of any alternating waveform, the square root of the mean of the squares of continuous ordinates (eg voltage or current) through one complete cycle. For a simple sinusoid, it is $1/\sqrt{2}$ times the peak value. Used because the energy associated with any wave depends on its intensity, ie upon the square of the amplitude. Abbrev *rms value*. Also *effective value*.

root node (*ICT*) The NODE at the top level of a TREE.

root nodule (*BioSci*) A globular structure formed on the roots of some plants, notably legumes and alder, by symbiotic association between the plant and a nitrogen-fixing micro-organism (*Rhizobium* in the case of legumes and *Frankia* in the case of alder and a variety of other plants).

root of the joint (*Eng*) In welding, the place where the original components were closest together.

root of the weld (*Eng*) The place where the weld comes closest to the ROOT OF THE JOINT.

root planing (*Med*) A treatment for gum disease in which the parts of teeth below the gum are cleaned and smoothed.

root pressure (*BioSci*) The positive pressure that may develop in the xylem when water uptake by osmosis follows ion uptake in the root, and transpiration is low. It may result in guttation or bleeding.

Roots blower (*Eng*) A form of ROTARY PUMP.

rootstock (*BioSci*) (1) A rhizome, esp a short, erect one. (2) A stock for grafting, esp one from a clone selected for desirable effects on the scion, eg dwarfing, early fruiting.

root test (*MathSci*) A test for absolute convergence. The series:

$$\sum_{i=1}^{\infty} |a_i|$$

converges if there is a positive $r<1$, and an N such that

$$\sqrt[n]{|a_n|} < r$$

for $n>N$. The series diverges if no such r exists.

root tuber (*BioSci*) A swollen adventitious root acting as a storage organ, eg dahlia. Cf TAPROOT, TUBER.

rope (*Radar*) See WINDOW.

rope brake (*Eng*) An absorption dynamometer consisting of a rope encircling a brake drum or flywheel, one end of the rope being loaded by weights and the other supported by a spring balance. The effective torque absorbed is obtained by multiplying the drum radius by the difference of the tensions.

rope processing (*Textiles*) Wet processing, including scouring, of fabric moving in loose rope form through large porcelain rings (*pot-eyes*) placed to reduce friction on the fabric. Nevertheless the fabrics sometimes suffer because of the development of rope marks (eg creases) running in the warp direction when this system is used.

ropewick applicator (*Agri*) A herbicide-saturated rope, charged by capillary action from a reservoir, that is trailed across the foliage of the weeds.

ropiness (*Build*) A defect in applied paint, usually attributed to a lack of skill by the person applying the coating or to the use of excessively heavy or thick paint. Appears as 'tram lines' or visible brushmarks.

ropy lava (*Geol*) Ses PAHOEHOE.

ro-ro (*Ships*) Abbrev for ROLL-ON, ROLL-OFF FERRY.

Rorschach inkblot test (*Psych*) A PROJECTIVE TEST which requires subjects to look at inkblots and report on what they see; the answers are used to interpret fantasy life, personality, intelligence, and also as an aid to psychiatric diagnosis.

rosacea (*Med*) A chronic inflammatory skin disease, usually beginning in middle age and characterized by redness and papules esp in the centre of the face. Cause unknown.

Rosaceae (*BioSci*) The rose family, comprising c.3300 spp of dicotyledonous flowering plants (superorder Rosidae). Varieties include trees, shrubs and herbs; they are wide-spread, esp in northerly temperate regions. The flowers are polypetalous, and perigynous or epigynous. The family includes many important tree and bush fruits of temperate regions, eg apple, pear, plum, cherry, almond, raspberry, strawberry, and also many ornamentals including the rose.

rosaniline (*Chem*) *Triamino-diphenyl-tolyl-hydroxy-methane*. A base of the fuchsin dyes. It is obtained by oxidation of an equimolecular mixture of aniline, 2-toluidine and 4-toluidine.

roscoelite (*Min*) This mineral is essentially *muscovite* in which vanadium has partly replaced the aluminium. Its colour is clove-brown to greenish-brown.

ROSE (*ICT*) Abbrev for *research open systems in Europe*. The principal development project for information exchange within the ESPRIT community; a basis for network research.

rose (*Arch*) A decorative circular escutcheon through which eg the spindle of a door handle or the flexible cord of a pendant *luminaire* may pass. See CEILING ROSE, a more practical extension of the latter use.

Rose crucible (*Chem*) A crucible the lid of which is fitted with an inlet tube. It is used for igniting substances in a current of gas.

Rosenmüller organ (*BioSci*) See EPIDIDYMIS.

roseola (*Med*) A rose-coloured rash following a fever, occurring in infancy and presumed due to a virus.

rose opal (*Min*) A variety of opaque common opal having a fine red colour.

rose quartz (*Min*) Quartz of a pretty rose-pink colour, due probably to titanium in minute quantity. The colour is apt to be destroyed by exposure to strong sunlight. See BOHEMIAN GEMSTONE.

rose topaz (*Min*) The yellow-brown variety of topaz changed to rose-pink by heating. These crystals often contain inclusions of liquid carbon dioxide.

rosette (*BioSci*) (1) Generally, any rosette-shaped structure. (2) In some Oligochaeta, a large ciliated funnel by which the contents of the vesiculae seminales pass to the exterior. (3) In some Crinoidea, a thin calcareous plate (rosette plate, rosette ossicle) formed by the coalescence of the basal plates.

Rosette Nebula (*Astron*) A prominent EMISSION NEBULA in the constellation Monoceros. Distance approx 1400 pc.

rosette plant (*BioSci*) A plant in which the leaves radiate out at about soil level and which has a more or less leafless flowering stem. See HEMICRYPTOPHYTE.

rosetting (*BioSci*) A process by which cells in suspension interact by adhesion with other immune system cells or with red blood cells. The resultant cell aggregates can be detected microscopically and are referred to as rosettes. A method of separating cell subsets from mixed populations by centrifugation, as rosetted cells sediment faster than their counterparts.

Rose–Waaler test (*Med*) A test for the presence of rheumatoid factor in blood.

rose window (*Arch*) A circular window with radial bars. Also *Catherine wheel, marigold window.*

rosewood (*For*) See BRAZILIAN ROSEWOOD.

Rosidae (*BioSci*) A subclass or superorder of dicotyledons that includes trees, shrubs and herbs. The flowers are mostly polypetalous; the stamens (if numerous) develop centripetally, rarely with parietal placentation. There are c.60 000 spp in 108 families including Leguminosae, Rosaceae, Crassulaceae, Myrtaceae, Melastomaceae, Euphorbiaceae and Umbelliferae.

rosiglitazone (*Pharmacol*) A thiazolidinedione, a glucose sensitizer used in the treatment of Type II diabetes.

rosin (*Chem*) *Colophony*. The residue from the distillation of turpentine. The colour varies from colourless to yellow, red, brown and black. Rel.d. 1·08, mp 100–140°C. Wood-rosin is obtained by the extraction of long-leaf pine wood;

chief sources, the US and France. Used as a soldering flux, in varnish, soap and size manufacture, and (in the form of resinates) as a drier in paint. See RESINATES.

rosinates (*Chem*) See RESINATES.

Rosiwal intercept method (*Geol*) Particle size analysis technique based on measurement of the intercepts made by a line drawn through a selection of particle images on a photomicrograph or in the field of view of a projection microscope.

rosolic acid (*Chem*) Quinonoid triarylcarbinol anhydride. An acidic dyestuff of the triphenylmethane series, obtained by oxidizing a mixture of phenol and 4-cresol with arsenic (V) acid and sulphuric acid. Green glistening crystals, insoluble in water, dissolving in alkalis with a red colour.

Rossby number (*EnvSci*) A non-dimensional number Ro defined as the ratio, for a particular class of motions in a rotating fluid such as the Earth's atmosphere, of inertial forces to CORIOLIS FORCES. $Ro = U(fL)^{-1}$ where U is a characteristic velocity, L a characteristic length and f the CORIOLIS PARAMETER. If Ro is large, then the effect of the Earth's rotation may be neglected.

Rossby wave (*EnvSci*) A wave in the general atmospheric circulation, in one of the principal zones of westerly winds, characterized by large wavelength (approx 6000 km), significant amplitude (approx 3000 km) and slow movement. First described by C G Rossby.

Rossi–Forell scale (*Geol*) A scale of apparent intensity of earthquake movements, now replaced by the MERCALLI SCALE.

rostellum (*BioSci*) A small beak-like outgrowth, esp one from the column of the flower of some orchids.

rostrum (*BioSci*) (1) In birds, the beak; a beak-shaped process. (2) In Cirripedia, a ventral plate of the carapace. (3) In some Crustacea, a median anterior projection of the carapace. (4) The anterior end of the rostro-caudal axis. Adjs rostral, rostrate.

rostrum camera (*ImageTech*) A moveable camera mounted on a vertical stand above a horizontal table also incorporating movements, used for reproducing artwork, graphics, animation drawings, etc.

rot (*BioSci*) The disintegration of tissue resulting from the activity of invading fungi or bacteria.

rotachute (*Aero*) A 'parachute', usually for stores or the recovery of missiles, in which the normal retarding canopy is replaced by rotor blades, which are freely revolving and act like a ROTOR.

rotameter (*Phys*) An indicating and measuring device for the rate of flow of gases and liquids.

rotaplane (*Aero*) A heavier-than-air aircraft which derives its lift or support from the aerodynamical reaction of freely rotating rotors. See ROTORCRAFT.

rotary amplifier (*ElecEng*) A rotary generator, the output of which is field-controlled by another generator or amplifier.

rotary combustion engine (*Autos*) See WANKEL ENGINE.

rotary converter (*ElecEng*) The combination of electric motor and dynamo used to change the form of electrical energy, eg dc to ac or one dc voltage to another.

rotary disk contactor (*ChemEng*) See ROTATING DISK CONTACTOR.

rotary drier (*MinExt*) Tubular furnace sloped gently from feed to discharge end, through which moist material is tumbled as it rotates slowly, while rising hot gases remove moisture.

rotary drill (*MinExt*) The drill downhole connected by the drilling pipe to the rotary table in the oil derrick or drilling platform. It may consist of several drills, reamers and stabilizing COLLARS, themselves joined by special tool joints. It is usually lubricated and cleared by DRILLING MUD forced down the drilling tube and out through nozzles in the drill head. The oil and debris pass up through the annular space between the drill tube and casing. The most important component fixed to the rotary table is the KELLY. See panel on DRILLING RIG.

rotary engine (*Aero*) An early type of aero-engine in which the crankcase and radially disposed cylinders revolved

round a fixed crankshaft; not to be confused with the modern radial engine.

rotary erase head (*ImageTech*) See FLYING ERASE HEAD.

rotary field (*ElecEng*) See ROTATING FIELD.

rotary indexing machine (*Eng*) A transfer machine, used for cutting, assembling or other production work, in which workpieces or tools are carried on a circular turntable which rotates intermittently.

rotary lip-seal (*Eng*) Special type of rubber seal for rotating shafts, eg crankshaft seal in a car engine.

rotary machine (*Print*) A machine in which the printing surface is a cylinder or a plate attached to a cylinder. Used in all the major printing processes and can be sheet- or web-fed.

rotary pump (*Eng*) A pump, similar in principle to a GEAR PUMP, in which two specially shaped members rotate in contact within a common housing, no valves being required; suited to large deliveries at low pressure.

rotary regenerative heater (*Eng*) An air heater consisting of a slowly revolving rotor made up of concentric rings of corrugated and flat plates, which pass alternately and continuously through the hot gases and the air drawn across opposite halves of the rotor.

rotary shutter (*ImageTech*) The rotating vanes which cut off the light from the screen while the frames are being moved and located in the picture gate of a projector.

rotary strainer (*Paper*) Cleaning device comprising a drum with slits or perforations on its surface, slowly rotating in a shallow vat to which the uncleaned stock is introduced. In passing through the orifices, the stock is purged of contraries that are too large.

rotary switch (*ElecEng*) A switch operated by a rotating handle capable of rotation in one direction only.

rotary table (*MinExt*) Heavy circular component mounted just above the derrick floor which carries the KELLY BUSHING. The table is rotated by the draw-works machinery and thus rotates the KELLY which slides in the bushing. The table thus turns the drill string. See fig. at KELLY. See panel on DRILLING RIG.

rotary valve (*Autos*) A combined inlet and exhaust valve in the form of a ported cylinder rotating on cylindrical faces in the cylinder head, usually parallel with the crankshaft.

rotate (*BioSci*) A corolla that is wheel-shaped, with the petals or lobes spreading out at right angles to the axis of the flower.

rotating amplifier (*ElecEng*) A form of dc generator in which the electrical power output can be accurately and rapidly controlled by a small electrical signal applied to the control field of the machine. Mainly used in industrial closed-loop systems.

rotating anode (*Electronics*) A high-power X-ray tube in which the anode rotates continuously to bring a fresh area of its surface into the electron beam; this allows higher output without melting the target.

rotating camera (*ImageTech*) A photographic camera used to record panoramic scenes. Short-rotation models (up to 150°) employ a rotating lens, with full-rotation models the camera rotates (up to 360°). Both use a scanning vertical slit to expose the film.

rotating core (*Eng*) Special core used to create internal screw-thread in injection-moulded product.

rotating crystal method (*Crystal*) A widely used method of X-ray analysis of the atomic structure of crystals. A small crystal, less than 1 mm in maximum size, is rotated about an axis at right angles to a narrow incident beam of X-rays. The diffraction of the beam by the crystal is recorded photographically or with a detector.

rotating disk contactor (*ChemEng*) A liquid–liquid extraction device which consists of a column with annular stators fitted to it and a central shaft carrying rotors of diameter nearly equal to the holes in the stators. Liquids flow countercurrent through the column and under the effect of high speed of rotation of the shaft and rotors improved contact is obtained.

rotating field (*ElecEng*) One in which the magnitude is constant at a point but whose direction is rotating about a point in a fixed reference system.

rotating-field magnet (*ElecEng*) The rotating portion of an electric machine, usually a synchronous motor or generator, in which the field poles rotate and the armature is stationary.

rotating joint (*ICT*) Short length of cylindrical waveguide, constructed so that one end can rotate relative to the other; used to couple two waveguide systems of rectangular cross-section.

rotation (*Agri*) See CROP ROTATION.

rotation (*Astron*) The term generally confined to the turning of a body about an axis passing through itself, eg rotation of the Earth about its polar axis in one sidereal day.

rotational field (*ElecEng*) A field in which the *circulation* is, in some parts, not always zero.

rotational grazing (*Agri*) Moving grazing livestock between pastures to provide regular periods of recovery for individual areas.

rotational isomerism (*Chem*) A type of conformational isomerism, where steric hindrance between adjacent side groups in repeat units favours particular conformers, eg transin polyethylene giving zigzag conformation.

rotational moulding (*Plastics*) The process by which an object is shaped by rotating the mould to which the POWDER or PLASTISOL material (eg polyvinyl chloride) has been added, in the heating oven and during cooling; used for some hollow articles. Also *rotocasting, rotomoulding*.

rotation axes of symmetry (*Crystal*) Symmetrically placed lines, rotation about which causes every atom in a crystal structure, as revealed by X-ray analysis, to occupy identical positions a given number (2, 3, 4, 6) of times. Cf SCREW AXES.

rotation of a vector (*MathSci*) See CURL.

rotation of the plane of polarization (*Chem, Phys*) A property possessed by optically active substances. See OPTICAL ACTIVITY.

rotation shift (*ICT*) See END-AROUND SHIFT.

rotation speed (*Aero*) The speed during take-off at which the nose wheel of an aircraft is raised from the ground prior to LIFT-OFF.

rotator (*Phys*) A device for rotating the plane of a wave in a waveguide.

rotatory dispersion (*Chem, Phys*) Variation of rotation of the plane of polarized light with wavelength for an optically active substance.

rotatory evaporator (*Chem*) A device for facilitating the evaporation of a liquid, generally under reduced pressure, by continuously rotating the flask in which it is contained.

rotatory power (*Chem, Phys*) See OPTICAL ROTATION.

rotavator (*Agri*) See CULTIVATOR.

rotavirus infection (*Vet*) A common cause of calf scour, but vaccine is available.

rotenone (*Chem*) See DERRIS.

Rotifera (*BioSci*) A class of small, unsegmented, pseudocoelomate animals, phylum Aschelminthes. A distinctive anterior ciliary apparatus is used for locomotion and food gathering. Aquatic. Also *wheel animalcules*.

Rotliegendes (*Geol*) The lower series of the Permian. See PALAEOZOIC.

rotocasting (*Plastics*) See ROTATIONAL MOULDING.

Rotocure (*Plastics*) TN for continuous vulcanization process where product of constant cross-section such as conveyor belting or flooring is made by reeling around a heated vulcanization drum. Care is needed to match speed of drum and cross-linking kinetics.

rotogravure (*Print*) Photogravure printing on a ROTARY MACHINE.

rotomoulding (*Plastics*) See ROTATIONAL MOULDING.

roton (*Phys*) The quantum of rotational energy analogous to the PHONON.

rotor (*Aero*) A system of revolving aerofoils producing lift, acting on a plane at right angles to the driving shaft.

rotor (*Autos*) The revolving arm of a distributor.

rotor (*BioSci*) A muscle which by its contraction turns a limb or a part of the body on its axis.

rotor (*ElecEng*) See ARMATURE.

rotor (*EnvSci*) A large closed eddy which may form under LEE WAVES of large amplitude; often associated with severe turbulence.

rotor cloud (*EnvSci*) A whirling quasi-stationary cloud that forms in the upper part of a rotor to the lee of a range of hills. The *helm bar* near Cross Fell in Cumbria is a well-known example.

rotor core (*ElecEng*) That portion of the magnetic circuit of an electric machine which lies in the rotor.

rotorcraft (*Aero*) Any aerodyne which derives its lift from a rotor, or rotors.

rotor head (*Aero*) The structure at the top of the rotor pylon, including the hub member to which the blades of a rotorcraft are attached.

rotor hinge (*Aero*) A hinge for the blades of a rotorcraft. See DRAG HINGE, FEATHERING HINGE, FLAPPING HINGE.

rotor hub (*Aero*) The rotating portion of the rotor head of a rotorcraft to which the rotor blades are attached.

rotor spinning (*Textiles*) A widely used OPEN-END SPINNING system. A stream of fibres from a roving enters a rapidly rotating cell or rotor. The fibres are temporarily held on the circumference by centrifugal force. The yarn is drawn out of the rotor while being twisted through a NAVEL into the doffing tube and is collected on a suitable package.

rotor starter (*ElecEng*) A motor starter used for slip-ring induction motors; it cuts out resistance previously inserted in the rotor circuit.

rotor-tip jets (*Aero*) Propulsive jets in the tips of a rotorcraft's blades that are used to obtain a drive with minimum torque reaction; they may be PRESSURE JETS, PULSE JETS, RAMJETS, combustion units fed with air and fuel from the fuselage, or small ROCKET units.

rotoscope (*ImageTech*) TN for a device providing frame-by-frame projection of a film to form the background of animation drawings or for the analysis of movement.

rottenstone (*Geol*) A material used commercially for polishing metals; formed by the weathering of impure siliceous limestones, the calcareous material being removed in solution by percolating waters.

rotula (*BioSci*) In higher vertebrates, the kneecap.

rotunda (*Arch*) A building or room which is circular in plan and is covered by a dome.

rot v (*MathSci*) See CURL.

rouge (*Chem*) Hydrated oxide of iron in a finely divided state; used as a polish for metals.

rough (*Paper*) The unglazed surface of drawing papers specially induced by a coarse felt. See HOT-PRESSED, NOT.

rough ashlar (*Build*) A block of freestone as taken from the quarry.

rough brackets (*Build*) Pieces of wood nailed to the sides of the CARRIAGE, to provide intermediate support for treads of a wooden stair.

rough-cast (*Build*) A rough finish given to a wall by coating it with a plaster containing gravel or small stones.

rough coat (*Build*) The first coat of plaster applied to a wall surface.

rough colony (*BioSci*) A bacterial colony produced by mutation from a smooth colony. The morphological change is frequently accompanied by physiological changes, eg altered virulence.

rough endoplasmic reticulum (*BioSci*) The cisternal form of ENDOPLASMIC RETICULUM, bearing ribosomes on the cytoplasmic surface. It is the site of synthesis of protein for export from the cell.

rough grounds (*Build*) Unplaned strips of wood used as *grounds* when the attached joinery will entirely cover them.

roughing (*MinExt*) Production of an impure concentrate as an early stage in ore processing, thus reducing bulk for more thorough treatment.

roughing-in (*Build*) The first coat of three-coat plaster work.

roughing tool (*Eng*) A lathe or planer tool, generally having a round-nosed or obtuse-angled cutting edge, used for roughing cuts.

rough proof (*Print*) A print of the work in hand, much below the finished quality to be expected, submitted for checking purposes only.

rough-string (*Build*) Part of a staircase. See CARRIAGE.

rough trimmed (*Print*) A design feature in many well-produced books where the tail edges are merely cleaned up by cutting the farthest projecting leaves, the pages having been deliberately positioned to ensure a variation in the tail margin. See CUT EDGES, TRIMMED SIZE.

roulette (*MathSci*) The locus of any point, or envelope of any line, moving with a first curve which rolls without slipping on a second curve, referred to respectively as *point* or *line roulette*. The locus of a point on a circle rolling on a straight line is called a *curtate cycloid*, a *cycloid* or a *prolate cycloid* respectively depending on whether the point is inside, on the circumference or outside the rolling circle. The locus of a point on a circle rolling on another circle is called an *epicycloid* or *hypocycloid* respectively depending on whether the rolling circle is outside or inside the fixed circle, if the point is on the circumference, or an *epitrochoid* or *hypotrochoid* respectively if the point is not on the circumference of the rolling circle. Epicycloids and epitrochoids in which the rolling circle completely encloses the fixed circle are sometimes called *pericycloids* and *peritrochoids* respectively. Curtate and prolate cycloids are sometimes called *trochoids*.

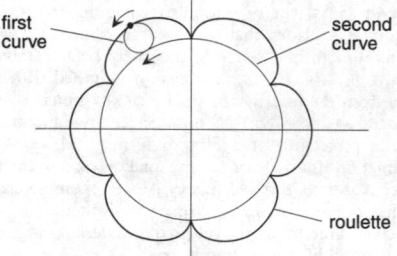

roulette Both first and second curves are circular in this figure.

round (*Build*) A rung of a ladder.

round angle (*MathSci*) A complete rotation, an angle of $360°$ or 2π radians. Also a *perigon*.

round dance (*BioSci*) Circular movements of a worker bee on returning from a foraging trip which communicates information that a food source is less than 150 m from the hive. See WAGGLE DANCE.

round heart disease (*Vet*) A disease of chickens characterized by enlargement and muscular degeneration of the heart and sudden death; the cause is unknown.

rounding (*ICT*) Approximating a number by its nearest equivalent, using a set number of significant figures. Also *rounding-off*. Cf TRUNCATING.

rounding (*Print*) The process of giving the back of a book a convex shape before casing; usually performed along with BACKING.

rounding error (*ICT*) Error introduced by rounding.

rounding off (*MathSci*) The reduction of the number of digits in a number by omitting the least significant; eg the number π, whose 'length' is infinite, can be rounded off to 3·14, 3·1416, 3·141 59, etc. The rounding off is always to the nearest number of the required 'length', so 3·141 is *rounded down* to 3·14, but 3·141 59 is *rounded up* to 3·1416. If the number to be rounded off is exactly halfway, eg 3·1415, it may be either rounded up or rounded down, but always rounding in the same direction may introduce

systematic errors, so that it may be better to alternate rounding up and down.

round key (*Eng*) A circular bar or pin, sometimes threaded, fitted in a hole drilled half in the shaft and half in the boss, parallel to the shaft axis; used for light work to avoid fitting. See KEY.

round of beam (*Ships*) See CAMBER.

rounds (*Build*) The general name for planes having a concave sole and cutting iron, used for forming rounded surfaces. Cf HOLLOWS.

round step (*Build*) A step finished with a semicircular end.

round trip (*MinExt*) Removing and dismantling the drill string to replace the drill bit and then reassembling and placing it back *downhole*. Also *trip*.

roundworm (*BioSci*) A name applied to a number of parasitic nematodes, esp those of the genus *Ascaris*, including the large intestinal roundworm of humans (*A. lumbricoides*).

roup (*Vet*) A term applied to symptoms of oculonasal discharge and swelling of the face and wattles of fowl, which occur in infections of the upper respiratory tract such as infectious coryza, fowl pox and infectious laryngotracheitis. Also *infectious coryza*.

Rous' sarcoma (*Vet*) A tumour, occurring in fowls, which is transmitted by an oncogenic RNA virus. The first demonstration that viruses can cause some cancers.

Rousseau diagram (*ElecEng*) A diagram by the use of which the total output (in lumens) of a light source can be obtained, if the polar curve of the lamp about the vertical axis is known.

Roussin's salts (*Chem*) Formed when sodium trisulphide is added to a solution of iron (II) chloride saturated at $-2°C$ with nitrogen (II) oxide, and converted by sodium sulphide into so-called *Roussin's red salt*; by treatment with dilute acids this is converted into *Roussin's black salt*.

rout (*Build*) To cut out wood from the bottom of a groove or sinking with a router plane.

router (*Build*) (1) Power-operated machine, fixed or hand-held, which holds and rotates various types and shapes of cutter for shaping or recessing. Similar to a spindle moulder but the spindle points downwards. See PLUNGE ROUTER. (2) A plane adapted to work on circular sashes; operated in the manner of a spokeshave. (3) The side wing of a CENTRE-BIT, which removes the material in forming the hole.

router (*ICT*) A computer on a network that directs data to their destination.

router plane (*Build*) A plane having a central projecting cutting iron, adapted to smoothing the bottom of a groove or sinking.

Routh's rule (*Phys*) A rule summarizing the values of the radius of gyration (k) of a rectangular lamina, an elliptical lamina and an ellipsoid about a principal axis through its centre of gravity. According to the rule, k^2 equals the sum of the squares of the other semi-axes divided by three, four or five respectively.

routine (*ICT*) See SUBROUTINE.

routiner (*ICT*) Apparatus that tests, as a routine, all machine-switching apparatus in an exchange, so that faults may be rapidly detected and rectified, and contacts kept clean.

routing (*ICT*) The function of selecting the path for transmission of data across a NETWORK.

routing machine (*Print*) A machine having a revolving tool which removes unwanted metal from printing plates.

roux (*FoodSci*) Dough made by heating flour and fat while continuously mixing and used to thicken sauces or gravies. Choux pastry is made by baking a roux thickened with liquid whole egg.

roving (*Textiles*) A continuous strand of fibres sufficiently drawn to a diameter suitable for drafting and spinning into yarn, eg by RING SPINNING frames.

Rowland circle (*Phys*) A circle having the radius of curvature of a concave diffraction grating as diameter. It has the property that, if the slit is placed anywhere on the circumference of the circle, the spectra of various orders are formed in exact focus also round the circumference of the circle. This fact is used in designing mountings for the concave grating.

rowlock (*Arch*) A term applied to a course of bricks laid on edge.

rowlock-back (*Arch*) A term applied to a wall whose external face is formed of bricks laid flat in the ordinary manner, while the back is formed of bricks laid on edge.

row vector (*MathSci*) A matrix consisting of a single row. Compare COLUMN VECTOR.

royal (*Paper*) A former imperial paper size retained under the metric system for printing boards, 52×64 cm (20×25 in). Note also the following sizes recommended for book publishing papers. Metric small quad royal 96×127 cm, metric quad royal 102×127 cm. See PAPER SIZES.

royals (*MinExt*) In the CYANIDING process for recovering gold, a term for the sludge, rich in gold, formed when the PREGNANT SOLUTION is precipitated on zinc dust or onto resin or carbon.

rpm (*Genrl*) Abbrev for *revolutions per minute*.

RPN (*ICT*) Abbrev for REVERSE POLISH NOTATION.

RPV (*Aero*) Abbrev for *recoverable pilotless aircraft* (vehicle).

RRIM (*Eng*) Abbrev for *reinforced reaction injection moulding*. See REACTION INJECTION MOULDING.

RR Lyrae variable (*Astron*) A type of variable star with a period of less than 1 day; common in globular clusters and used, like CEPHEID VARIABLES, to measure galactic distances.

RSA encryption (*ICT*) An encryption algorithm based on a public key system used to secure authentication requests between networks.

RSF (*Build*) Abbrev for *rough sunk face*.

RSJ (*Build*) Abbrev for ROLLED STEEL JOIST.

RSS (*ICT*) Abbrev for *rich site summary* (or, popularly, *really successful syndication*), a system that allows computer users to view the content of many web pages in a single screen.

RSS feed (*ICT*) Abbrev for *really simple syndication*, a file format enabling the syndication of web content, esp blog postings, to those who have subscribed to receive the material.

r-strategist (*BioSci*) Organism that assigns much of its resources to reproduction; usually opportunistic and colonizing species (weeds) with high fecundity and low competitive ability. Cf K-STRATEGIST.

RTF (*ICT*) Abbrev for RICH TEXT FORMAT.

RTV (*Chem*) Abbrev for ROOM TEMPERATURE VULCANIZING.

Ru (*Chem*) Symbol for RUTHENIUM.

Rubarth's disease (*Vet*) See INFECTIOUS CANINE HEPATITIS.

rubber (*Chem*) Generic term for any elastomer, but also used specifically for natural rubber (caoutchouc) whose main source is the tree, *Hevea brasiliensis*, originally a native of C and S America, but since the end of the 19th century widely grown in plantations in SE Asia. Commercial rubber consists of caoutchouc, a polymerization product of isoprene, of resin-like substances, nitrogenous substances, inorganic matter and carbohydrates. The caoutchouc portion is soluble in CS_2, CCl_4, trichloromethane or benzene, forming a viscous colloidal solution. When heated, rubber softens at 160°C and melts at about 220°C. Rubber easily absorbs a large quantity of sulphur either by heating or in the cold by contacting with S_2Cl_2 etc. This process is called VULCANIZATION. Carbon black, in a fine state of division, is used as a reinforcing filler; other substances, produced by the condensation of aldehydes with amines, retard the oxidation of vulcanized rubber. The uses of rubber are innumerable. See CRÊPE RUBBER, SYNTHETIC RUBBER and panel on RUBBER TOUGHENING. For *foamed rubber*, see EXPANDED PLASTICS.

Rubber and Plastics Research Association (*Genrl*) A leading research and development organization for the polymer industry.

rubber blanket (*Print*) See OFFSET PRINTING.

rubber forming (*Eng*) Pressing a rubber pad over a die to form a sheet placed between them.

rubber hardness (*Eng*) Various empirical measures of shear modulus of rubbers, by resistance of a rubber to an impressed indentor needle. *International Rubber Hardness* and *Shore A* scales are roughly equal and proportional to shear modulus which in turn is proportional to degree of cross-linking. See panel on HARDNESS MEASUREMENTS.

rubber jaw (*Vet*) See OSTEODYSTROPHIA FIBROSA.

rubber plates (*Print*) Extensively used for all grades of work. Flexible, they can be affixed easily to rotary cylinders; have good inking qualities; are durable; and require precision grinding to exact thickness and KISS IMPRESSION. Hand-cut rubber plates are extensively used for bold designs, particularly on packages.

rubber reclaim (*Eng*) Rubber produced by chemical action (eg alkali digestion) on waste products such as worn tyres. Of limited use as recycled material for new tyre compounds.

rubbers (*Build*) See CUTTERS.

rubber seal (*Eng*) A common form of a device for preventing leakage of fluid from one part of a container to another, whether moving or static. Care is needed in design, since poor tolerances can cause extrusion of the seal under pressure. Material is also critical to performance, depending on the nature of fluids, so oil-resisting rubbers like nitrile, chloroprene rubber and Viton are favoured for engine and fuel seals. Often an O-ring, but many other shapes exist for special uses.

rubber springs (*Eng*) Devices designed to isolate a structure from its environment, ranging from bridge bearings and earthquake mounts to antivibration mountings for engines and torsion bushes.

rubber stippler (*Build*) A tool with rubber prongs used in broken colour work and texture painting. The stippler is used to manipulate the material on the surface.

rubber toughening (*Eng, Plastics*) Many homopolymers are brittle, glassy solids at ambient temperatures but can be toughened by copolymerization with elastomeric polymers. See panel on RUBBER TOUGHENING.

rubbing stone (*Build*) An abrasive stone with which the bricklayer rubs smooth the bricks which have been cut to a special shape. See GAUGED ARCH.

rubble (*Build*) Rough uncut stones, of no particular size or shape, used for rough work and filling in between facing walls etc.

rubeanic acid (*Chem*) *Dithio-oxamide*. An orange crystalline powder, sparingly soluble in water but soluble in alcohol; used as a reagent to detect small amounts of copper, with which it forms a black precipitate.

rubefacient (*Med*) Producing reddening of the skin; any agent which does this, a counter-irritant.

rubella (*Med*) A mild acute viral infectious disease mainly affecting children, characterized by slight fever, enlargement of glands in the neck and at the back of the head, and a pink papular macular rash. Infection in pregnancy can cause severe fetal abnormalities. Sometimes resembles mild measles, hence its common name, *German measles*.

rubellite (*Min*) The pink or red variety of tourmaline, sometimes used as a semiprecious gemstone.

Rubiaceae (*BioSci*) A family of c.7000 spp of dicotyledonous flowering plants (superorder Asteridae). They are cosmopolitan; most tropical species are trees and shrubs, while all the (fewer) temperate species are herbs. The leaves are opposite and have stipules, the ovary is inferior and usually of two carpels. Includes coffee and *Cinchona*, the source of quinine.

rubicelle (*Min*) A yellow or orange-red variety of spinel; an aluminate of magnesium.

rubidium (*Chem*) Symbol Rb, at no 37, ram 85·47, mp 38·5°C, bp 690°C, rel.d. 1·532. A metallic element in the first group of the periodic system, one of the alkali metals. The element is widely distributed in nature, but occurs

Rubber toughening

Many homopolymers like polystyrene, polymethyl methacrylate and polyvinyl chloride are brittle, glassy solids at ambient temperatures, and of restricted use for stressed, structural application. Such materials can be toughened by copolymerization with elastomeric polymers, such as polybutadiene and polyacrylate. The latter's polymer chains are highly flexible and non-crystalline (being atactic) at ambient temperatures, but, like most polymers, they are incompatible with the host matrix and phase separate to form spherical domains up to about 2 μm in diameter. Polymerization is effected by dispersing a stable emulsion in the MONOMER followed by catalytic initiation in a chain-growth reaction. Since the rubber chains are covalently linked to the matrix chains, the material can be reprocessed (ie it is thermoplastic) with reformation of the domain structure.

When a bulk sample is stressed in tension, each particle behaves as a stress concentration zone and two CRAZES form at right angles to the applied strain (see diagram).

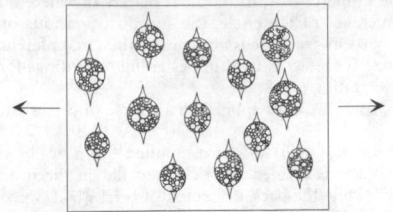

Rubber toughening Tension applied to elastomeric particles embedded in a glossy plastic.

Each craze absorbs energy and if particles are fine enough and well distributed through the matrix, a very large amount of energy will be absorbed, ie the material is toughened. When the craze size approaches the wavelength of light (approx 0.4 μm), they become visible owing to light scattering, an effect known as *stress* or *strain whitening*. The method of rubber toughening has been applied to polystyrene in high-impact polystyrene and acrylonitrile–butadiene–styrene polymers, to polymethyl methacrylate in impact-resistant acrylic (eg Plexidur sheet) using polybutyl acrylate and to polyvinyl chloride in high-impact PVC using methacrylate–butadiene–styrene terpolymer (MBS) or polyacrylate elastomers. It has also been applied to epoxy resins and nylon 6,6 in eg Zytel.

See panels on HIGH-PERFORMANCE POLYMERS and TYRE TECHNOLOGY.

---

only in small amounts; the chief source is carnallite. The metal is slightly radioactive.

rubidium–strontium dating (*Geol*) A method of determining the age in years of geological material, based on the known decay rate of ^{87}Rb to ^{87}Sr. See panel on RADIOMETRIC DATING.

RUBISCO (*BioSci*) Abbrev for RIBULOSE 1,5-BISPHOSPHATE CARBOXYLASE OXYGENASE.

RuBP carboxylase (*BioSci*) Abbrev for RIBULOSE 1,5-BISPHOSPHATE CARBOXYLASE OXYGENASE. Also *RuBPC-ase*.

rubric (*Print*) A heading or passage printed in red, the main text being in black. Also the marginal headings of minutes etc although printed in black.

ruby (*Min, Phys*) The blood-red variety of the mineral corundum, the oxide of aluminium (Al_2O_3), which crystallizes in the trigonal system. Also *true ruby* (to distinguish it from the various types of FALSE RUBY) and *Oriental ruby*, though the adj *Oriental* is quite unnecessary, since it merely stresses the fact that rubies come from the East (Myanmar (Burma), Thailand, Sri Lanka, Afghanistan). See BALAS RUBY, RUBY SPINEL. Used as an optically pumped, solid-state laser medium in which the Cr^{3+} ions are the active component.

ruby (*Print*) An old type size approximately $5\frac{1}{2}$-point.

ruby silver ore (*Min*) See PROUSTITE, PYRARGYRITE.

ruby spinel (*Min*) That variety of magnesian spinel, $MgAl_2O_4$, which has the colour, but none of the other attributes, of true ruby. Also *almandine spinel*. Spinel ruby is a deceptive misnomer.

ruche (*Textiles*) A narrow woven or knitted fabric made with extended weft threads which may be bunched together for extra effect. The material is used as decorative trim round the edges of upholstery or applied to dresses.

rudaceous (*Geol*) Used of a sedimentary rock which is coarser in grain size than sand.

rudder (*Aero*) A movable surface in a vertical plane for control of an aircraft in angles of yaw (ie movement in a horizontal plane about a vertical axis). Usually located at the rear end of the body and controlled by the pilot through a system of rods and/or cables or electric signals in FLY-BY-WIRE.

rudder (*Ships*) Broad, flat device, varying in form, hinged vertically to, or behind, the stern post of a vessel; the rudder serves to change the vessel's course when it is moved from a position in line with the keel.

rudder bar (*Aero*) A mechanism consisting of differential foot-operated levers by which the pilot actuates the rudder of a glider or an aircraft, or controls the pitch of a helicopter tail rotor through mechanical, hydraulic or electrical relaying devices. Also *rudder pedals*.

rudder post (*Ships*) See STERN FRAME.

rudenture (*Arch*) A cylindrical moulding carved in imitation of a rope.

ruderal (*BioSci*) A plant that grows usually on rubbish heaps or waste places.

rudiment (*BioSci*) The earliest recognizable stage of a member or organ.

Rudistes (*Geol*) A group of heavily built lamellibranchs of coral-like form which are characteristic of the Cretaceous rocks formed in the southern ocean (the Tethys) of the period; Rudistids also occur in the Cretaceous Trinity Series of Texas and Mexico.

RUDP (*BioSci*) Abbrev for *ribulose diphosphate*. See RIBULOSE 1,5-BISPHOSPHATE.

Ruffini's organs (*BioSci*) In vertebrates, a type of cutaneous sensory nerve ending concerned with the perception of heat.

rufous (*BioSci*) Red-brown.

rugose (*BioSci*) Wrinkled. Dim *rugulose*.

rule (*Build*) See FLOATING RULE.

rule (*Print*) Type-high brass or metal strip of various thicknesses and designs; a dash or score (see EM RULE, EN RULE).

rule-based system (*ICT*) A software system that represents knowledge by a set of simple conditional sentences. See EXPERT SYSTEM, INFERENCE ENGINE.

rule border (*Print*) A frame of rules fitted around an advertisement or other displayed matter.

ruled surface (*MathSci*) A surface generated by motion of a straight line with one degree of freedom, eg a cone.

rule of mixtures (*Chem*) A method of determining properties of a mixture (eg a composite material) by summing for all constituents their value of the corresponding property multiplied by the volume fraction present (ie a volume-weighted average of the properties of the components). Applies generally to the density of any composite, and under certain conditions, to such properties as elastic modulus, tensile strength, thermal and electrical conductivity, and dielectric constant. Does not, in general, apply to TOUGHNESS.

ruling (*Print*) The operation of making lines on writing, account-book and ledger paper, etc; the paper is conveyed on an endless belt and makes contact with suitably adjusted disks or pens.

ruling gradient (*CivEng*) The maximum gradient permissible for any given section of road or railway. Also *limiting gradient*.

rumble (*Acous*) Low-frequency noise produced in disk recording when turntable is not dynamically balanced.

rumen (*BioSci*) The first division of the stomach in ruminants and Cetacea, being an expansion of the lower end of the oesophagus used for storage of food; the paunch.

rumenotomy (*Vet*) The operation of cutting into the rumen.

rumination (*BioSci*) The regurgitation of food that has already been swallowed, and its further mastication before reswallowing. Adj and n *ruminant*.

run (*Build*) The part of a pipe which is in the same direction as that to which it is connected.

run (*ICT*) See EXECUTE.

run (*Print*) The number of copies to be printed.

run (*Surv*) In a level tube, the movement of a bubble with change of inclination.

run around (*Print*) Type set, alongside an illustration, to less than the full measure.

runaway electron (*Electronics*) One under an applied electric field in an ionized gas which acquires energy from the field at a greater rate than it loses through particle collision.

runaway star (*Astron*) A young star of spectral type O or B which has unusually high velocity, possibly following the supernova explosion of its companion in a close binary system.

runite (*Min*) See GRAPHIC GRANITE.

run-length coding (*ICT*) A method of data compression used for digital FAX and similar black and white images.

run-length limited (*ICT*) The coding strategy used in magnetic and optical recording to eliminate the error caused by long runs of zeros and ones that cannot be counted accurately. Abbrev RLL.

runner (*BioSci*) A stem growing more or less flat on the ground, with long internodes, rooting at the nodes and/or the tip and there producing new plantlet(s) from axillary or terminal bud(s), as in strawberry.

runner (*Eng*) The channel through which molten metal or plastic flows to a mould or cavity for shaping. Also applied to the solid material left in the channel at the end of the moulding cycle. Plastic runners are often cold, needing the solid product to be recycled after granulation. Hot runners are now increasingly used in INJECTION MOULDING, so

reducing or eliminating the waste problem. Also *runner-gate*. See INGATE.

runners (*Print*) Marginal figures for reference purposes indicating the number of each line in a poem or play.

runner stick (*Eng*) See GATE STICK.

running (*Build*) The operation of forming a plaster moulding, cornice, etc, *in situ*, or on a bench. Running a HORSED mould along the material while it is still plastic. See HORSING UP.

running bond (*Build*) The same as STRETCHING BOND.

running fit (*Eng*) A fit for rotating or sliding parts with sufficient clearance to support an oil film. National standards exist for these and other fits.

running heads (*Print*) The headings at the top of the page, the usual arrangement being title of book on left-hand page and title of chapter on right-hand page.

running on (*Print*) The actual printing of an edition after the MAKE-READY operations have been completed.

running rule (*Build*) A wood strip fixed temporarily to serve the same purpose as a RUNNING SCREED.

running screed (*Build*) A band of plaster laid on the surface of a wall as a guide to the movement of a horsed mould in the process of running a moulding. See HORSING UP.

running shoe (*Build*) The zinc part of a horsed mould, giving protection to the wood and facilitating running. See HORSING UP.

running tapes (*Print*) Tapes travelling at the press speed for the purpose of leading or conveying the web of paper.

running to seed (*BioSci*) See BOLTING.

running trap (*Build*) Tubular trap used in sanitary pipes having the inlet and outlet in horizontal alignment.

runoff (*Agri*) (1) Water from precipitation or irrigation that is not absorbed into the soil but runs on the surface away from the cultivated area. It can contaminate surface and groundwater sources. (2) Excess liquid that is shed from foliage surfaces after spraying.

runoff (*CivEng, Geol*) The resultant discharge of a river from a catchment area; surface water as distinct from that rising from deep-seated springs.

run-of-mine coal (*MinExt*) Coal raised from the mine before screening or other treatment.

run on (*Print*) An indication that a new paragraph is not to be made. Marked in copy and proof by a line running from the end of one piece of matter to the beginning of the next. US *run in*.

run-on chapters (*Print*) Chapters in a book which do not commence on a new page but after a few lines of space.

run out (*Eng*) Errors in concentricity of moving parts introduced by its bearings. Cf NON-REPEATABLE RUN OUT.

run-out (*ImageTech*) The end of a print, ie the length of film between the last effective frame and the end.

runt disease (*BioSci*) A disease that develops after injection of allogeneic lymphocytes into immunologically immature experimental animals. It is characterized by loss of weight, failure to thrive, diarrhoea, splenomegaly and often death. This is an example of a GRAFT-VERSUS-HOST REACTION.

run-through ruling (*Print*) A term which indicates that the ruling continues from edge to edge, horizontally or vertically, without interruption. See STOPPED HEADING.

run time (*ICT*) (1) The length of time between the beginning and completion of the execution of a program. (2) The time during which a program is being executed.

run-time error (*ICT*) See EXECUTION ERROR.

run-time system (*ICT*) The complete set of instructions that must be in main memory to enable a user's program to be executed.

runway threshold (*Aero*) The usable limit of a runway; in practice it is usually the current downwind end which is intended.

runway visual markers (*Aero*) See AIRPORT MARKERS.

runway visual range (*Aero*) In bad weather, the horizontal distance at which black-and-white markers of standard size are visible, the figure being transmitted to pilots approaching by AIR-TRAFFIC CONTROL. Abbrev RVR.

Rusting

Rusting is the name given to the atmospheric corrosion of iron and steel. Unlike the other transition metals in their group, they corrode readily in the presence of oxygen and water, and this is enhanced by the presence of ions such as sulphides and chlorides common in industrial and/or marine atmospheres. The economic consequences of corrosion, in terms of prevention, repair and replacement, are enormous (well over £1 billion per annum in the UK alone), and a major part of this is due to rusting.

Rust is the mixture of oxides of iron which forms on iron and steel surfaces in oxygenated aqueous environments. These oxides are:

green rust, hydrated ferrous oxide, $FeO.OH$, GOETHITE or LEPIDOCROCITE;

brown rust, hydrated ferric oxide, Fe_2O_3 with $Fe(OH)_3$, HAEMATITE;

black rust, mixed ferrous and ferric oxides, Fe_3O_4 (= $FeO.Fe_2O_3$), MAGNETITE.

Above a relative humidity threshold of about 60% (less in the presence of, for instance, hygroscopic dust), there is sufficient water around for some droplets to condense on the metal surface. These provide the electrolyte for electrochemical corrosion to start with

$Fe \rightarrow Fe^{2+} + 2e^-$ at the anode, and
$\frac{1}{2} O_2 + H_2O + 2e^- \rightarrow 2OH^-$ at the cathode.

This then proceeds by a number of reactions to yield a mixture of nominally $FeO.OH$, Fe_2O_3 and Fe_3O_4 which readily interconvert by migration of ions. Each of these oxides tends to be non-stoichiometric owing to the close structural relationship between them. Their lack of STOICHIOMETRY leads to defect structures which provide diffusion paths through their lattices. In Fe_2O_3 and Fe_3O_4, vacancy defects appear in the oxide anion sublattice, so that oxide ions can diffuse through the film towards the metal–oxide interface. At the same time, and also owing to the non-stoichiometry, electrons can diffuse in the other direction (all three oxides are semiconductors), thus completing an electrochemical circuit.

The net result is that further oxidation takes place at the interface between the metal and the oxide which does not slow down as the oxide layer thickens and the diffusion paths lengthen because the volume approximately doubles during the reaction. This expansion against the existing iron oxide layer leads to SPALLING, which exposes fresh metal to further direct attack.

In dry, unpolluted atmospheres, a thin ($\approx 4\,nm$), protective oxide film can form on iron surfaces. This is anhydrous FeO, or wüstite, which has a slightly non-stoichiometric composition (ie there are not an equal number of Fe^{2+} and O^{2-} ions).

Pourbaix diagram

The plot of electrode potential against pH value which defines regions over which various ions and products are stable. It is used in electrochemistry and the prediction of metallic corrosion behaviour as shown in the diagram for an iron–water–air system at 25°C.

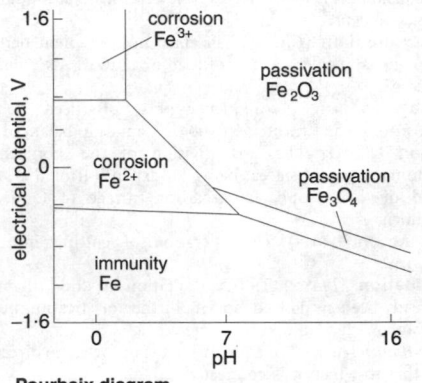

Pourbaix diagram

This exhibits three different types of zone:

(1) Corrosion zones where metal ions are stable and therefore corrosion occurs.

(2) Passivation zones where a solid film of oxide or hydroxide forms on the surface. If this sticks and is impermeable to air and water, corrosion will not continue unless the film breaks or comes off. Such zones are normally not stable.

(3) Zones of immunity are where the metal is stable with no tendency to ionize. Corrosion does not and cannot take place in such conditions.

Rupelian (*Geol*) A stage in the Oligocene. See TERTIARY.
Rupert's drops (*Glass*) See PRINCE RUPERT'S DROPS.
rupicolous (*BioSci*) Living or growing on or among rocks.
rupture (*Med*) (1) Forcible breaking or tearing of a bodily organ or structure. (2) To break or to burst (said of a blood vessel or viscus). (3) HERNIA.
rupturing capacity (*ElecEng*) See BREAKING CAPACITY.
rushes (*ImageTech*) The first positive prints made from motion picture negative immediately after processing.

Russell–Saunders coupling (*Phys*) Extreme form of coupling between orbital electrons of atoms. The angular and spin momenta of the electrons combine and the combined momenta then interact. Also *l–s coupling*.
Russell's test (*ElecEng*) A method of determining the insulation resistance of a three-wire dc distribution network. The value is obtained by calculation from readings of an electrostatic voltmeter connected between

the neutral wire and earth, both with and without a known resistance in parallel.

russet (*BioSci*) A brownish, roughened, corky layer or patch on the surface of a fruit (or other organ) as a varietal characteristic or as the result of disease or of injury from insects or spraying.

rust (*BioSci*) One of a number of plant diseases, some economically very important, caused by biotrophic fungi of the order Uredinales and often recognizable by the rounded or elongated pustules of rust-coloured spores on stems or leaves, eg black rust of cereals caused by *Puccinia graminis*.

rust (*Eng*) The product of oxidation of iron or its alloys, due either to atmospheric attack or to electrolytic effect of cell action round impurities. See panel on RUSTING.

rusticated ashlar (*Build*) Ashlar work in which the face stands out from the joints, at which the arrises are bevelled. The face may be finished rough or smooth or tooled in various ways.

rustic joint (*Build*) A sunken joint between adjacent building-stones.

rustics (*Build*) Bricks having a rough-textured surface, often multicoloured. Also TEXTURE BRICKS.

rusting (*Chem, Eng*) The atmospheric corrosion of iron and steel. See panel on RUSTING.

rust joint (*Build*) A watertight joint between adjoining lengths of guttering or pipes.

rusty gold (*MinExt*) Native gold which has become surface-filmed by adherent staining substances and is non-amalgamable and non-treatable by cyanide process in consequence.

rut (*BioSci*) (1) The noise made by certain animals, such as deer, when sexually excited. (2) Estrus (Also *oestrus*). (3) To be sexually excited, ie to be in the estrous period. (4) To copulate.

ruthenium (*Chem*) Symbol Ru, at no 44, ram 101·07, mp 2400°C, rel.d. 12·26. A metallic element. The metal is silvery-white, hard and brittle. It occurs with the platinum metals in osmiridium, and is used in certain platinum alloys.

ruthenium red (*BioSci*) An electron-dense stain used in electron microscopy for identifying GLYCOSAMINOGLY-CANS on the outer surfaces of cells.

Rutherford atom (*Phys*) Earliest modern concept of the atomic structure, in which all the positive charge and nearly all the mass of the atom is in the nucleus. Electrons, equal in number to the atomic number, occupy the rest of the volume and make the atom electrically neutral.

rutherfordium (*Chem*) An artificially manufactured chemical element of the transactinide series (symbol Rf, at no 106), with a half-life of less than a second, discovered in 1974, but not officially named until 1994, after confirmation of its discovery. Also *unnilhexium*.

Rutherford scattering (*Phys*) See SCATTERING.

rutilant (*BioSci*) Brightly coloured in red, orange or yellow.

rutilated quartz (*Min*) See NEEDLE STONE.

rutile (*Min*) Titanium dioxide which crystallizes as reddish-brown prismatic crystals in the tetragonal system. It is found in igneous and metamorphic rocks, and in sediments derived from these, also in quartz (see FLÈCHES D'AMOUR), and it is a source of titanium. See panel on TWINNED CRYSTALS.

RWP (*Build*) Abbrev for *rainwater pipe*.

Ryazanian (*Geol*) The oldest stage of the Cretaceous. See MESOZOIC.

rybat (*Build*) An INBAND or OUTBAND.

Rydberg constant (*Phys*) The constant R appearing in the Rydberg formula, which relates the frequencies of atomic spectrum lines in a given series. It was first deduced from spectroscopic data but has since been shown to be a universal constant:

$$R = \frac{2\pi^2 e^4}{ch^3} M_r$$

where M_r is the reduced mass of the electrons, e is the electronic charge, c is the speed of light and h is Planck's constant.

Rydberg formula (*Phys*) A formula, similar to that of Balmer, for expressing the wavenumbers (v) of the lines in a spectral series:

$$v = R \left[\frac{1}{(n+a)^2} - \frac{1}{(m+b)^2} \right]$$

where n and m are integers and $m > n$, a and b are constants for a particular series, and R is the RYDBERG CONSTANT.

R–Y signal (*ImageTech*) A component of colour TV chrominance signal. Combined with luminance (Y) signal it gives primary red component.

S

S (*BioSci*) Abbrev for SVEDBERG UNIT, referring to the sedimentation coefficient of proteins analysed in an ultracentrifuge.
S (*Chem*) Symbol for: (1) black, in names of dyestuffs; (2) SULPHUR.
S (*Phys*) Symbol for: (1) POYNTING VECTOR; (2) SIEMENS.
S (*Phys*) Symbol for: (1) area; (2) ENTROPY.
S- (*Chem*) Prefix denoting left-handed. See CAHN–INGOLD–PRELOG SYSTEM for absolute configuration.
∑ (*Genrl*) Symbol for *sum of*.
s (*Genrl*) Symbol for SECOND (time).
s (*Chem, Phys*) Symbol for: (1) distance along a path; (2) SOLUBILITY; (3) SPECIFIC ENTROPY.
s- (*Chem*) Abbrev for: (1) symmetrically substituted (also *sym-*); (2) *secondary*, ie substituted on a carbon atom which is linked to two other carbon atoms; (3) *syn-*, ie containing the corresponding radicals on the same side of the plane of the double bond between a carbon and a nitrogen atom or between two nitrogen atoms.
σ (*Chem*) Symbol for the diameter of a molecule.
σ (*Phys*) Symbol for: (1) CONDUCTIVITY; (2) normal stress; (3) nuclear CROSS-SECTION; (4) Stefan–Boltzmann constant (see STEFAN–BOLTZMANN LAW); (5) SURFACE CHARGE DENSITY; (6) SURFACE TENSION; (7) WAVENUMBER.
sabin (*Acous*) Obsolete unit of acoustic absorption; equal to the absorption, considered complete, offered by 1 ft^2 of open window to low-frequency reverberant sound waves in an enclosure. Also *open window unit*.
Sabine reverberation formula (*Acous*) Earliest formula (named after investigator) for connecting the reverberation of an enclosure, T seconds, with the volume, V in cubic metres, and the total acoustic absorption in the enclosure, $\sum aS$, where a is the absorption coefficient of a surface of S square metres. The formula is $T = 0.16V/\sum aS$.
sabkha (*Geol*) A flat salt-encrusted coastal plain, common in Arabia.
sable (*Build*) (1) Hair obtained from a small animal of the weasel family and used as the filling for signwriters' brushes. (2) Heraldic term for black.
sabulose (*BioSci*) Growing in sandy places. Also *sabuline*.
sac (*BioSci*) Any sort of bag-like structure or pouch.
saccadic eye movements (*Psych*) The rapid, ballistic movements of the eyes used in scanning a scene; these involuntary eye movements occur about every quarter of a second even when the eyes are fixated on an object.
saccate (*BioSci*) Bag-like or pouch-like.
saccharides (*Chem*) CARBOHYDRATES, which according to their complexity are usually divided into *mono-*, *di-*, *tri-* and *polysaccharides*.
saccharimeter (*Chem*) A hydrometer which is used to determine the concentration of sugar in a solution.
saccharimetry (*Chem*) The estimation of the percentage of sugar present in solutions of unknown strength, esp by measurements of optical activity.
saccharin (*Chem, FoodSci*) 2-sulphobenzimide. A white crystalline powder, 300 times as sweet as sugar, not very soluble in water. The imido-hydrogen is replaceable by Na, forming a salt which is readily soluble in water. It is used in medicine in cases where sugar is harmful, eg in diabetes, and as a non-nutritive sweetener. Fig. ▷

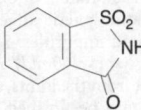

saccharin

saccharobiose (*Chem*) Cane-sugar or sucrose.
saccharoidal textures (*Geol*) Granular textures which resemble sugar; found esp in limestones and marbles.
Saccharomyces cerevisiae (*BioSci*) The YEAST used widely in bread and alcohol manufacture. It can also be used as an eukaryotic host for growing and expressing DNA sequences.
saccule (*BioSci*) (1) A small sac. (2) The lower chamber of the auditory vesicle in vertebrates. Also *sacculus*. Adj *sacculate*.
sacculiform (*BioSci*) Shaped like a little bag.
sacculus (*BioSci*) See SACCULE.
saccus (*BioSci*) A large, hollow, pouch-like projection of the outer part of the wall of a pollen grain.
sacking (*Textiles*) Coarse fabrics of jute, flax or polyolefin used for making sacks.
sacralgia (*Med*) Pain in the sacral region.
sacralization (*Med*) A developmental anomaly in which one or both transverse processes of the fifth lumbar vertebra become abnormally large and strong, appearing to form part of the sacrum.
sacral ribs (*BioSci*) Bony processes uniting the sacral vertebrae to the pelvis, distinct in reptiles but fused to the transverse processes in other Tetrapoda.
sacral vertebrae (*BioSci*) In higher Craniata, those vertebrae which articulate with the ilia of the pelvis via sacral ribs, there being one in the frog and two in the lizard, coming between the lumbar vertebrae and the caudal vertebrae (if any). In birds and mammals they are fused with other vertebrae to form the SACRUM.
sacrificial anode (*Ships*) A metal electrode used to prevent the corrosion of a structure. The anode dissolves and requires regular replacement. Commonly made of magnesium alloy containing 6% aluminium and 3% zinc. The zinc coating on galvanized steel acts as a sacrificial anode. See CATHODIC PROTECTION.
sacrificial protection (*Eng*) The prevention of ELECTROLYTIC CORROSION in a component by providing another electrochemically more active metal close by and electrically connected to it. See SACRIFICIAL ANODE.
sacrificial-tape welding (*Eng*) Joining method for similar thermoplastics in safety-critical products, like small boats and some types of battery cases. Involves placing a braid of metal and thermoplastic fibre in joint, passing an electric current through the braid while the two parts are loaded.
sacroiliac joint (*BioSci*) In some Craniata, the almost immovable joints between the SACRUM and the two ilia of the pelvis. The articular surfaces of the bones are partly covered with cartilage and partly roughened for the attachment of the sacroiliac ligament.
sacrum (*BioSci*) (1) In the skeleton of some Craniata, part of the vertebral column which articulates immovably with the ilium of the pelvis at the SACROILIAC JOINT. It is composed of several fused vertebrae, including the SACRAL

VERTEBRAE. (2) In birds it consists of one thoracic vertebra, five or six lumbar vertebrae, the two sacral vertebrae and the anterior five caudal vertebrae. It is sometimes called the *synsacrum*. (3) In mammals it comprises varying numbers of vertebrae in different orders (eg four in the rabbit and five in humans), the first one or two being regarded as sacral, and the others as caudal. The sacral vertebrae have low spines and expanded ventral surfaces for the attachment of muscles.

SAD (*Psych*) Abbrev for SEASONAL AFFECTIVE DISORDER.

saddle (*CivEng*) A block surmounting one of the towers of a suspension bridge, providing bearing or fixing for the suspension cables.

saddle (*Eng*) The part of a lathe which slides on the bed, between headstock and tailstock.

saddle-back board (*Build*) A narrow board, chamfered along each of the upper edges, which is fixed on the floor across the threshold of a doorway so that the gap beneath the door will be small when the latter is shut and large enough when it opens to accommodate a carpet.

saddle-back coping (*Build*) A coping stone whose upper surface slopes away on both sides from the middle.

saddle bar (*Build*) A metal bar fixed across a window to support glazing held in lead cames.

saddle coils (*Electronics*) Rectangularly formed coils which are bent around the neck of a cathode-ray tube; used for magnetic deflection of the beam.

saddle key (*Eng*) A key sunk in a key way in the boss, but having a concave face which bears on the surface of the shaft, which it grips by friction only. See KEY.

saddle point (*NucEng*) A point on the plot of potential energy against distortion for nucleus at which fission will occur, instead of return to equilibrium.

saddle scaffold (*Build*) A scaffold erected over a roof from standards on both sides of the building; used for repair work on eg a chimney at the middle of the roof.

saddle-stitching (*Print*) A method of wire-stitching in which the book is placed astride a saddle-shaped support and stitched through the back.

saddle stone (*Build*) See APEX STONE.

sadism (*Psych*) Sexual gratification through the infliction of pain on others; pleasure in cruel behaviour.

sado-masochism (*Psych*) The pairing of a SADIST and a MASOCHIST to satisfy their mutually complementary sexual needs.

SAE (*Aero*) Abbrev for *Society of Automotive Engineers* (US). Gives name to a widely used viscosity scale for classifying motor oils.

safe (*Print*) The condition of a press when locked. See LOCK.

safe area (*ImageTech*) The area of a film frame which, when transmitted on TV, is reasonably certain to be produced on a domestic receiver.

safe edge (*Eng*) The edge of a file on which no teeth are cut.

safeguard (*CivEng*) See CHECK RAIL.

safe light (*ImageTech*) Lighting fixture in a photographic dark-room whose intensity and colour of visible illumination permits the safe handling of unprocessed materials.

safe load (*CivEng*) The maximum working load which a member or structure is designed to carry. See FACTOR OF SAFETY.

safety arch (*Build*) See DISCHARGING ARCH.

safety barrier (*Aero*) A net which is erected on the forward part of the deck of an aircraft carrier to stop any aircraft which misses the *arrester gear*. A cable and/or nylon net which can be quickly raised to prevent an aircraft from overrunning the end of a runway. The barrier is held by friction brakes or weights so that it imposes a 1 or 2*g* deceleration on the aircraft.

safety cage (*MinExt*) A cage fitted with a 'safety catch' to prevent it from falling if the hoisting rope breaks.

safety coupling (*Eng*) A friction coupling adjusted to slip at a predetermined torque, to protect the rest of the system from overload.

safety-critical (*Eng*) A part or component, the loss or damage of which will endanger or destroy the parent product. See FAILSAFE.

safety cut-out (*ElecEng*) An overload protective device in an electric circuit.

safety factor (*Eng*) The provision of extra margin in stress calculations etc to allow for errors and uncertainties. Thus a safety factor of two allows for twice the allowable stress calculated for the product specification.

safety factor (*NucEng*) Of a fusion system, the ASPECT RATIO multiplied by the ratio of toroidal to poloidal field. This requires to be greater than unity for magnetohydrodynamic stability. Also *q*.

safety film (*ImageTech*) See ACETATE FILM.

safety fuse (*ElecEng*) A protective fuse in part of an electric circuit.

safety glass (*Glass*) (1) *Laminated glass*, formed of a sandwich of a thin (0·4–0·750 mm) layer of polymeric material, usually polyvinyl butyral, between glass sheets. The functions of the layer are to hold the glass fragments in place following fracture and to act as a barrier to penetration. (2) So-called *toughened glass*, formed by putting the surfaces into compression, either by chilling them with air jets from close to the softening point (*thermal toughening*), or by exchanging sodium ions in the surfaces for larger ions (*chemical toughening*). Stronger than untreated glass, with blunter fracture fragments, but TOUGHNESS is unaffected. (3) Glass incorporating wire mesh to hold fracture fragments in place, but weaker than plain glass. See WIRED GLASS.

safety height (*Aero*) The height below which it is unsafe to fly on instruments because of high ground.

safety lamp (*MinExt*) Oil-burning miners' lamp which will not immediately ignite firedamp or gas in a coal mine, eg a DAVY LAMP. Also used for detecting gas.

safety lintel (*Arch*) A lintel doing the work of a relieving arch, and serving to protect another more decorative lintel used for architectural reasons.

safety plug (*Eng*) See FUSIBLE PLUG.

safety rail (*CivEng*) See CHECK RAIL.

safety rods (*NucEng*) Rods of neutron-absorbing material capable of rapid insertion into a reactor core to shut it down, in case of emergency.

safety speed (*Aero*) The lowest speed above stalling at which the pilot can maintain full control about all three axes. It is particularly applicable to multi-engined aircraft, where it is taken to be the minimum speed at which control can be maintained after complete failure of the engine most critical to directional control.

safety switch (*ElecEng*) See EMERGENCY SWITCH.

safety valve (*Eng*) A valve, spring or dead-weight loaded, fitted to a boiler or other pressure vessel, to allow fluid to escape to the atmosphere when the pressure exceeds the maximum safe value.

safe working load (*Eng*) Maximum load permitted in service for items used for lifting, ie chains, slings, hoists, etc, usually to 20% of their minimum breaking load. It is a statutory requirement that such items are subjected to PROOF TEST, normally to 40% of the minimum breaking load before being allowed to enter service. All such items must be subjected to regular inspections for wear, corrosion and mechanical deterioration. Abbrev *SWL*.

safranines (*Chem*) A group of azine dyestuffs. They are 2,8-diamino derivatives and have also a phenyl or a substituted phenyl group attached to the nitrogen in position 10.

sag correction (*Surv*) A correction applied to the observed length of a baseline, to correct for the sag of the measuring tape.

SAGE (*Aero*) Abbrev for *semi-automatic ground environment*. Air defence system whereby information is received from radar and other sources and is processed at a central station to give an evaluation of a situation.

saggar (*Eng*) A clay box in which pottery is packed for baking.

sagging (*Glass*) Forming glass by reheating until it conforms with the mould or form on which it rests.

sagging (*Ships*) Occurs when the ends of the ship are supported on wave crests while the middle is in a trough or when the middle is more heavily loaded than the ends. If the ship actually bends it is said to be *sagged*. Cf HOGGING.

Sagitta (Arrow) (*Astron*) A small constellation which lies in the Milky Way.

sagittal (*BioSci*) Elongate in the median vertical longitudinal plane of an animal (thus on the imaginary line separating left from right), as the *sagittal suture* between the parietals, the *sagittal crest* of the skull; used also of SECTIONS (1) .

sagittal field (*Phys*) The image surface formed by the sagittal foci of a series of object points lying in a plane at right angles to the axis.

sagittal focus (*Phys*) The focus of an object point lying off the axis of an optical system in which the image is drawn out by the astigmatism of the system into a line radial to the optical axis.

Sagittarius A (*Astron*) The brightest member of a group of radio sources at the centre of our Galaxy, also a powerful infrared emitter. Possible sources of power include accretion onto a black hole or intense star formation.

Sagittarius (Archer) (*Astron*) A southern constellation, lying between Scorpius and Capricornus, containing bright second- and third-magnitude stars. It lies in the direction of the centre of our Galaxy.

sagittate (*BioSci*) Shaped like an arrowhead with the barbs pointing backwards.

sago (*FoodSci*) Starch from the stem of the SAGO palm, a native of the Moluccas.

Sahelian drought (*EnvSci*) The pattern of drought in the northern regions of W Africa, where several years of below-average rainfall often occur in succession.

sahlite (*Min*) A mineral of the clino-pyroxene group, intermediate in composition between diopside and heden-bergite. Also *salite*.

sailcloth (*Textiles*) (1) Woven fabrics designed for use in sailing ships and yachts. Originally closely woven cotton or linen canvases were used but these have been largely displaced by nylon for spinnakers, and polyester or aramid for foresails and mainsails. The weaves are carefully designed to give just the right air porosity under a wide range of wind velocities. (2) A ribbed cotton fabric with a structure between those of poplin and repp, used in dresswear.

sailing courses (*Build*) See OVERSAILING COURSES.

sail-over (*Build*) To project over. See OVERSAILING COURSES.

sailplane (*Aero*) A glider designed for sustained motorless flight by the use of air currents. The most advanced methods of streamlining and very high *aspect ratio* are used to reduce *drag* to the barest minimum.

Saint Anthony's fire (*Med*) An old name applied to ERYSIPELAS and to ERGOTISM.

Saint David's (*Geol*) The middle epoch of the Cambrian period.

Saint Elmo's fire (*Phys*) A visible electric discharge from isolated points above the ground, eg ship's mast and aircraft, associated with thunderstorms.

Saint Vitus's dance (*Med*) See CHOREA.

Sakmarian (*Geol*) The lowest stage of the Permian system in E Europe and former USSR.

sal-ammoniac (*Min*) Chloride of ammonia, which crystallizes in the cubic system. It is found as a white encrustation around volcanoes, as at Etna and Vesuvius.

salbutamol (*Pharmacol*) A *sympathomimetic* drug that stimulates beta-adrenoreceptors in the airway. It acts as a bronchodilator for the relief of acute asthma attacks and symptoms of chronic bronchitis and emphysema.

salic minerals (*Geol*) Those minerals of the *norm* which are rich in silicon and aluminium, including quartz, feldspars and feldspathoids.

salicylic acid (*Chem, Pharmacol*) An antiseptic and an important intermediate for a number of derivatives, eg aspirin.

salience (*Pharmacol*) Any aspect of a stimulus that particularly stands out, sometimes as a result of emotional or cognitive factors rather than its physical properties.

salient (*Surv*) A jutting-out piece of land.

salient angle (*MathSci, Surv*) An angle in a closed figure which points outwards, being less than 180°. Cf RE-ENTRANT ANGLE.

Salientia (*BioSci*) A superorder of Amphibia. The adults are four-legged, the hind-limbs being esp well-developed, and short-bodied, with no tail. Toads and frogs. Also *Anura*, *Batrachia*.

salient junction (*Build*) See EXTERNAL ANGLE.

salient pole (*ElecEng*) A type of field pole protruding beyond the periphery of the circular yoke in the case of a stator field system, or the circular core in the case of a rotor field system.

salient-pole generator (*ElecEng*) An ac generator whose rotor field system is of the salient-pole type, eg in slow-speed water-turbine-driven generators.

salina (*Geol*) Also *saline lakes*. See SALT LAKES.

salinometer (*Phys*) A HYDROMETER for measuring the density of sea water, the stem being scaled in arbitrary units; used by engineers for estimating the amount of dissolved solids in feed-water.

salite (*Min*) See SAHLITE.

saliva (*BioSci*) The watery secretion produced by the salivary glands, whose function is to lubricate the passage of food and, sometimes, to carry out part of its digestion. In insects saliva may contain amylase, invertase, protease and lipase, according to the usual diet, and in some blood-sucking insects it contains anticoagulants. In mammals it contains water, mucin and, in humans and some herbivores, the amylase ptyalin, which catalyses the breakdown of starch to maltose.

salivary gland chromosome (*BioSci*) A polytene chromosome (see POLYTENY) found in the salivary glands of larval dipterans, eg *Drosophila melanogaster*. It is conspicuously banded, whether stained or not, and used for gene mapping and other studies of chromosome organization.

salivary glands (*BioSci*) Glands present in many land animals, the ducts of which open into or near the mouth.

sallenders (*Vet*) See MALLENDERS AND SALLENDERS.

sally (*Build*) A re-entrant angle cut into the end of a timber, so as to allow it to rest over the arris of a cross-timber.

salmeterol (*Pharmacol*) A drug that stimulates beta-2 adrenenoreceptors, esp in bronchial smooth muscle. It is used by inhalation for the treatment of asthma and is longer-acting than salbutamol.

Salmonella (*BioSci*) A group of Gram-negative, carbohy-drate-fermenting, non-sporing bacilli. These organisms are all pathogenic to animals and include *S. typhi* and *S. paratyphi*. They are associated with food poisoning in humans. They are usually found in poultry, eggs, milk and meat products, and occasionally exist on 'low-risk products' such as chocolate; they are generally destroyed by heating foods to above 70°C throughout.

salmonellosis (*Med, Vet*) A form of food poisoning due to infection with *Salmonella* spp. It is most often associated with *S. dublin* or *S. typhimurium*. All warm-blooded animals can be affected; it is characterized by vomiting, diarrhoea and abdominal pain. It is an important zoonosis and therefore a notifiable veterinary disease in the UK. A vaccine is available.

Salmoniformes (*BioSci*) An order of fresh-water and ANADROMOUS Osteichthyes of great commercial and sporting importance. Salmon, trout, char, pike.

salping- (*Genrl*) Prefix from Gk *salpinx*, gen *salpingos*, trumpet, referring esp to the Fallopian tubes. See SALPINX.

salpingectomy (*Med*) Removal of a Fallopian tube.

salpingitis (*Med*) Inflammation of a Fallopian tube.

salpingo-oöphorectomy (*Med*) Removal of a Fallopian tube and of the ovary on the same side.

salpingo-oöphoritis (*Med*) Inflammation of both the Fallopian tube and the ovary.

salpingorrhaphy (*Med*) The suturing of a Fallopian tube to the ovary on the same side, after a part of the latter has been removed.

salpingostomy (*Med*) The operative formation of an opening into a Fallopian tube whose natural opening has been closed by disease.

salpingotomy (*Med*) Incision into a Fallopian tube.

salpinx (*BioSci*) (1) Generally, a trumpet-shaped structure. (2) A structure, adapted for the reception of pollen, at the distal end of the nucellus of ovules of many seed ferns. Also *lagenostome*. (3) The Eustachian tube. (4) The Fallopian tube. Adj *salpingian*.

SALR (*EnvSci*) Abbrev for SATURATED ADIABATIC LAPSE RATE.

salsuginous (*BioSci*) Growing on a salt marsh.

SALT (*BioSci*) Abbrev for SKIN ASSOCIATED LYMPHOID TISSUE.

salt (*Chem*) A compound which results from the replacement of one or more hydrogen atoms of an acid by metal atoms or electropositive radicals. Salts are generally crystalline at ordinary temperatures, and form positive and negative ions on dissolution in water, eg chlorides, nitrates, carbonates, sulphates, silicates and phosphates. For *common* or *rock salt*, see HALITE.

salt (*FoodSci*) See SODIUM CHLORIDE.

saltation (*BioSci*) A sudden heritable variation in a species. The term is now applied more often to large morphological changes which occur during evolution over a time period shorter than that required by similar changes earlier or later. The phenomenon is said to be difficult to explain by NATURAL SELECTION.

saltatorial (*BioSci*) Used in, or adapted for, jumping, as the third pair of legs in grasshoppers. Also *saltatory*.

saltatory conduction (*BioSci*) The process of nervous conduction along a myelinated nerve axon where the impulse jumps from one NODE OF RANVIER to the next.

salt bath (*Eng*) A bath of molten salts used for heat treatment, ie for hardening, tempering or solution treatments. Salt baths give rapid, uniform heating and protect against oxidation. Different salts are used for different temperatures, eg for tempering of steels, sodium and potassium nitrate are used. For hardening of steels, sodium cyanide, and sodium, potassium, barium and calcium chlorides. An electric salt-bath furnace is a conductor-type electric furnace in which the salt is melted by the passage of the current.

salt dome (*Geol, MinExt*) A diapiric salt plug which has arched up, or broken through, the sediments into which it has been intruded. Such rock salt deposits have become plastic under pressure and the plugs or domes so formed are impervious to oil. Differences in gravity between the dome and the intruded rocks can be measured at the surface and may indicate the presence of an oil reservoir. See PETROLEUM RESERVOIRS.

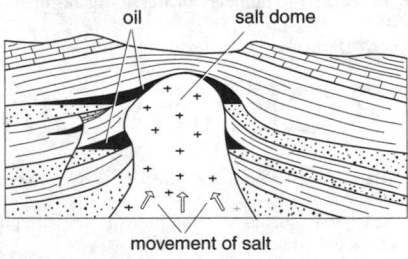

salt dome Shows associated oil traps but no gas.

salt gland (*BioSci*) (1) A structure at the leaf surface which actively secretes sodium chloride in many salt marsh and mangrove species. (2) A gland located just above the eye in marine birds such as the albatross, and in various other terrestrial birds and reptiles; responsible for excretion of excess salt from the diet.

saltigrade (*BioSci*) Progressing by jumps, as grasshoppers.

salting (*MinExt*) Fraudulent enrichment of ore samples, made to increase apparent value of a mine. Originally, to sprinkle salt in dry mines to allay dust.

salting-out (*Chem*) The removal of an organic compound from an aqueous solution by the addition of salt.

salt lakes (*Geol*) Enclosed bodies of water, eg lake, lagoon, marsh, spring, etc, in areas of inland drainage, whose concentration of salts in solution is much higher than in ordinary river water. Also *salina*, *saline lakes*. See SODA LAKES.

salt marsh (*BioSci*) A marsh characterized by saline soil, most often in estuaries and subject to marine inundation. See HALOSERE.

saltpetre (*Min*) See POTASSIUM NITRATE.

sal volatile (*Chem*) Ammonium carbonate, the main constituent of smelling salts.

Salyut (*Space*) The first-generation Soviet space station, capable of docking with the Soyuz crew ferry and Progress resupply vehicle. The first station was flown in 1971, the last (Salyut 7) over 1982–6. See panel on SPACE STATION.

samara (*BioSci*) A dry, indehiscent fruit of which part of the wall forms a flattened wing, eg ash key.

samariform (*BioSci*) Winged, like an ash key.

samarium (*Chem*) Symbol Sm, at no 62, ram 150·35, mp 1350°C, bp 1600°C, hard and brittle, rel.d. 7·7. A metallic element. Found in allanite, cerite, gadolinite and samarskite. It is feebly, naturally radioactive, but the stable isotope samarium-149, produced by the decay of promethium-149 in fission reactors, has a very high neutron capture cross-section of $5·3 \times 10^4$ barns and is therefore a reactor poison, but one which does not have a large effect on reactor kinetics.

sampled data tracking (*Radar*) That used with high-speed electronic beam switching, using PHASED ARRAYS, allowing data to be gathered on each of several tracks virtually simultaneously.

sample space (*Genrl*) The set of all possible outcomes of an experiment.

sampling (*BioSci*) The survey of a small but representative part of a population or stand of vegetation, with the intention of obtaining an estimate of some characteristics of the whole, eg age distribution or species present.

sampling (*ICT*) The process of measuring at regular intervals the level of a varying (analogue) waveform, in order to convert it to digital form or to achieve TIME-DIVISION MULTIPLEX. To allow the reconstruction of the waveform, the sampling rate must exceed twice the highest frequency component of the sampled waveform. See NYQUIST LIMIT, RATE.

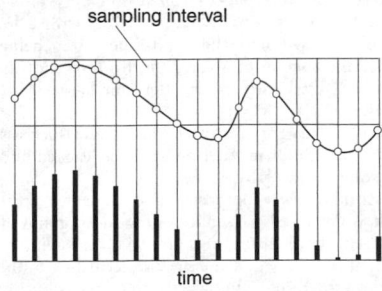

sampling

sampling (*Phys*) The selection of an irregular signal over stated fractions of time or amplitude (pulse height).

sampling distribution (*MathSci*) The probability distribution describing the variation of a statistic in repeated sampling, or hypothetical repetitions of the same experiment.

sampling error (*MathSci*) Variation due to a sample from a population necessarily giving only incomplete information about the population.

SAN (*Plastics*) Abbrev for STYRENE ACRYLONITRILE.

sand (*Eng*) See MOULDING SANDS.

sand (*Geol*) A term popularly applied to loose, unconsolidated accumulations of detrital sediment, consisting essentially of rounded grains of quartz. In the mechanical analysis of soil, sand, according to international classification, has a size between $\frac{1}{16}$ and 2 mm. See SILT. In coral sand the term implies a grade of sediment the individual particles of which are fragments of coral, not quartz. See PARTICLE SIZE, WENTWORTH SCALE.

sandalwood (*For*) A scented wood from several trees of the family *Santalaceae*, found in the E Indies. Important for its essential oil and used for joss-sticks and small ornaments. One species is found in Australia.

sand blasting (*Eng*) A method of cleaning metal or stone surfaces by sand, steel shot or grit blown from a nozzle at high velocity; also used for forming a key on the surface of various materials requiring a finish, such as enamel. See GRIT BLASTING.

sand casting (*Eng*) The formation of shapes by pouring molten metal into a cavity shaped in sand in a MOULDING FLASK.

sand colic (*Vet*) Colic caused by the collection of sand in the intestines.

sandcrack (*Vet*) A fissure of the horse's hoof.

sand dunes (*Geol*) Rounded or crescentic mounds of loose sand which have been piled up by wind action on sea coasts or in deserts. See BARCHAN.

sand fill (*MinExt*) Underground support of worked-out stopes by return of ore tailings from mill, usually by hydraulic flow.

sand-fly fever (*Med*) An acute disease caused by infection with a virus conveyed by the bite of a sand-fly *Phlebotomus papatasii*; characterized by a three-day fever, pains in the joints and the back, diarrhoea and a slow pulse. Also *phlebotomus fever*.

sand lime bricks (*Build*) Bricks made by mixing suitable sand with approx 6% of hydrated lime and water, moulding under high pressure and then curing in steam at high pressure.

Sandmeyer's reaction (*Chem*) The replacement of the diazonium group, $-N_2^+$, in a diazonium compound by chlorine, bromine, or the cyanogen radical, which is effected by heating a solution of the diazonium compound with eg a conc solution of cuprous chloride in hydrochloric acid. In this case the diazonium group is replaced by Cl, with evolution of gaseous N_2.

sandpaper (*Build*) Stout paper or cloth with a thin coating of fine sand glued onto one side, for use as an abrading material. Cf EMERY PAPER, GLASSPAPER.

sand-pump dredger (*CivEng*) A vessel with a long pipe reaching down into the sand, the latter being raised under the suction of a centrifugal pump and discharged into the vessel itself or an attendant barge. Also *suction dredger*.

sands (*MinExt*) Particles of crushed ore of such a size that they settle readily in water and may be leached by allowing the solution to percolate. See SLIMES.

sandstone (*Geol*) Compacted and cemented sedimentary rock, which consists essentially of rounded grains of quartz, between the diameters of 0·06 and 2 mm, with a variable content of 'heavy mineral' grains. According to the nature of the cementing materials the varieties *calcareous sandstone*, *ferruginous sandstone*, *siliceous sandstone* may be distinguished; *glauconitic sandstone*, *micaceous sandstone*, etc, are so termed from the presence in quantity of the mineral named.

sand trap (*Paper*) An inclined trough across which bars are set at intervals. During the passage of the pulp to strainers, any heavy particles such as sand sink to the bottom, and are retained by the bars. Also a *riffler*.

sand volcano (*Geol*) A structure formed by sand flowing upwards through an overlying bed of sediment and spilling out onto the surface; not related to any type of volcanic activity or volcanic rock.

sandwich (*Phys*) Photographic nuclear research emulsion forming a series of thin layers with intervening layers in which some event is to be studied.

sandwich beam (*Build*) See FLITCH BEAM.

sandwich beam (*Eng*) Composite structural beam comprising outer 'skin' layers which are typically a fibre-reinforced COMPOSITE MATERIAL (eg CARBON or GLASS-FIBRE-REINFORCED PLASTIC) encasing a lightweight 'core' which may be a FOAM or HONEYCOMB material. Its chief advantage is a high ratio of bending stiffness to weight because the main load-bearing elements (the skins) are situated some distance away from the NEUTRAL AXIS. Cf STRUCTURAL FOAM.

sandwich compounds (*Chem*) Compounds in which a metal atom is 'sandwiched' between two rings, eg dibenzene chromium. Ferrocene (dicyclopentadienyl iron) is another example.

sandwich construction (*Aero*) Structural material, mainly used for skin or flooring, possessing exceptionally good stiffness for weight characteristics. It consists of two approximately parallel thin skins with a thick core having different mechanical properties, so that the tensile and compressive stresses develop in the skin, and the core both stabilizes these surfaces and gives great strength in bending; core materials range from balsa wood through metal-foil honeycomb (light-alloy or steel) to corrugated sheet.

sandwich irradiation (*Radiol*) The irradiation of tissues from opposite sides.

sandwich moulding (*Plastics*) Injection-moulding process by which surface of product formed is solid but the interior is foamed. It normally involves two separate injection units.

sandwich technique (*BioSci*) (1) A technique for the detection of antibody or antibody-producing cells in histological preparations. A first layer of antigen is applied and allowed to react with the antibody in the section. After washing, this is followed by a second layer of fluorochrome-labelled antibody specific for the antigen. The antigen is 'sandwiched' between the two layers of antibody. (2) Any technique in which multiple 'layers' of reagent such as antibody are used.

sanidine (*Min*) A form of potassium feldspar similar in chemical composition to orthoclase, but physically different, formed under different conditions and occurring in different rock types. It is the high-temperature form of orthoclase, to which it inverts below 900°C. Occurs in lavas and dyke rocks.

sanitation (*FoodSci*) The cleaning of plant, equipment and the production environment with any combination of steam, water and chemical sterilants to remove soiling and to reduce the numbers of micro-organisms to a safe level.

sans serif (*Print*) A typeface without serifs, eg

RQENbaegn

Santonian (*Geol*) A stage of the Upper Cretaceous. See MESOZOIC.

sap (*BioSci*) An aqueous solution present in xylem, phloem, cell or vacuole, released on wounding.

sapele (*For*) An African hardwood from *Entandrophragma cylindricum*, with a mahogany-like, silky grain.

saphir d'eau (*Min*) Fr 'water sapphire'. A misnomer for an intense-blue variety of the mineral cordierite, occurring in water-worn masses in the river gravels of Ceylon (Sri Lanka); used as a gemstone.

saponification (*Build*) The action of alkali on oil paint whereby the paint is softened causing a defect.

saponification (*Chem*) The hydrolysis of esters into acids and alcohols by the action of alkalis or acids, or by boiling with water, or by the action of superheated steam. It is the reverse process to ESTERIFICATION if acids are used, but when alkalis are used then soaps result, hence the term.

saponification number (*Chem*) The number of milligrams of potassium hydroxide required to saponify 1 g of a fat or oil.

saponification value (*FoodSci*) A measure of the average chain length of a fatty acid. The amount of potassium hydroxyde in milligrams that will saponify 1 g of fat. High values (normally over 200) are obtained with fats and hydrogenated oils, while oils tend to have values below 195.

saponins (*Chem*) Steroid vegetable glycosides that act as emulsifiers of oils. They dissolve the red corpuscles, irritate the eyes and organs of taste and are toxic to lower animals, eg *digitonin*, found in *Digitalis purpurea*.

saponite (*Min*) Hydrated aluminosilicate of magnesium. A clay mineral of the smectite (montmorillonite) group, occurring as white soapy masses in serpentinite. Also *bowlingite*.

sapphire (*Min*) The fine blue transparent variety of crystalline corundum (predominantly Al_2O_3), of gemstone quality; obtained chiefly from Sri Lanka, Kashmir, Thailand, Cambodia and Australia. Also the single-crystal alumina used as a substrate for SILICON ON INSULATOR technology.

sapphire needle (*Acous*) A vinyl disk record reproducing stylus ground from natural sapphire; by virtue of its hardness in comparison with that of the record surface, it is relatively hard-wearing although inferior in this respect to the diamond stylus.

sapphirine (*Min*) A silicate of magnesium and aluminium with lesser iron, crystallizing in the monoclinic system. It occurs as blue grains in metamorphosed, aluminous, silica-poor rocks.

sapr-, sapro- (*Genrl*) Prefixes from Gk *sapros*, rotten, rancid.

saprobe (*BioSci*) An organism such as a bacterium or fungus which obtains its nourishment osmotrophically from dead organic matter. Cf SAPROPHYTE.

saprobic classification (*EnvSci*) A system in which aquatic organisms are arranged hierarchically according to their tolerance for decaying organic matter and which can be used as a measure of the quality of water. Groups include *oligosaprobic* (unpolluted water), *beta-mesosaprobic* (mildly polluted water), *alpha-mesosaprobic* (water where the decomposition of organic matter is partly aerobic and partly anaerobic) and *polysaprobic* (grossly polluted water where decomposition is anaerobic).

saprogenous (*BioSci*) Growing on decaying matter.

sapropel (*Geol*) Slimy sediment laid down in stagnant water, largely consisting of decomposed algal material. A source material for oil and natural gas.

sapropelite (*Geol*) A term applied to coals derived from algal materials. Cf HUMITE.

saprophilous (*BioSci*) Saprogenous.

saprophyte (*BioSci*) An organism living heterotrophically and osmotrophically on dead organic matter. Adj *saprophytic*. Cf SAPROBE.

saprotrophy (*BioSci*) Heterotrophic nutrition based on non-living (dead) organic matter.

sapwood (*BioSci*) The outer, lighter-coloured, younger part of the wood of a tree, surrounding the heartwood, and used for conduction of nutrients and storage of carbohydrates. Usually less than $\frac{1}{3}$ of the total radius, it is enclosed by the CAMBIUM. Also *alburnum*. Cf HEARTWOOD. See panel on WOOD, STRUCTURE.

saquinavir (*Pharmacol*) A *protease inhibitor* used as a drug in the treatment of HIV.

SAR (*Radar*) Abbrev for SYNTHETIC APERTURE RADAR.

SARAH (*Radar*) Abbrev for *search and rescue homing*. A system for facilitating rescue when aircraft go down at sea, consisting of a small beacon transmitter which sends a coded pulse to rescue craft. Also used to guide support vessels to spacecraft after a sea landing.

Saran (*Plastics*) US name for synthetic fibre or film based on a copolymer of vinylidene chloride (at least 80% by weight) and vinyl chloride – a *chlorofibre*.

sarc-, sarco- (*Genrl*) Prefixes from Gk *sarx*, gen *sarkos*, flesh.

sarcodic (*BioSci*) Pertaining to or resembling flesh. Also *sarcodous, sarcoid*.

Sarcodina (*BioSci*) A class of Protozoa with pseudopodia, containing both irregular, amoeboid forms and others possessing regular calcareous or siliceous tests, eg Radiolaria, Foraminifera.

sarcodous (*BioSci*) Also *sarcoid*. See SARCODIC.

sarcoidosis (*Med*) A disease, cause unknown, characterized by granulomatous lesions; affects the lungs particularly, and other organs, including liver, skin and brain.

sarcolemma (*BioSci*) The plasma membrane of a striated muscle fibre.

sarcoma (*Med*) A malignant tumour of connective tissue origin (eg of fibrous tissue, bone, cartilage); the tumour invades adjacent tissue and organs, and metastases are formed via the blood stream. Pl *sarcomata* or *sarcomas*. Cf CARCINOMA. Adj *sarcomatous*.

sarcomatosis (*Med*) The presence of many sarcomata in the body.

sarcomere (*BioSci*) The basic contractile unit of the MYOFIBRIL. See fig. at MUSCLE.

sarcophagous (*BioSci*) Feeding on flesh.

sarcoplasmic reticulum (*BioSci*) A network of tubules, associated with muscle fibrils, that acts as the source of calcium ions that stimulate contraction.

sarcoptic mange (*Vet*) Mange of animals due to mites of the genus *Sarcoptes*.

sarcosine (*Chem*) *Monomethyl-glycine*. H_3CHNCH_2COOH. It is obtained by the decomposition of creatine or caffeine. Crystals, mp 212°C, readily soluble in water. It may be synthesized from chloroacetic ester and amino-methane, or by hydrolysis with barium hydroxide of methylaminoethanonitrile.

sarcosporidiosis (*Vet*) Infection of the muscles of pigs, sheep, horses, cattle, goats and birds by the parasitic Sarcosporidia.

Sardinian (*Textiles*) A heavy, woollen twill overcoating fabric whose face is raised to a dense NAP and then rubbed into PILLS.

sardonyx (*Min*) A form of chalcedony in which the alternating bands are reddish-brown and white. Cf ONYX.

Sargent diagram (*NucEng*) Log–log plot of radioactive decay constant against maximum β-ray energy, for various β-emitters. Most of the points relating to natural heavy radioisotopes lie on one or other of two straight lines.

sarin (*Chem*) An organic compound, developed as a lethal nerve gas, that inhibits acetylcholine esterase. *O-isopropyl methylphosphonofluoridate*. Also GB.

sarking (*Build*) See ROOF BOARDS.

sarking felt (*Build*) A bituminous underlining placed beneath slates or tiles and above the SARKING.

saros cycle (*Astron*) A cycle of 18 years 11 days, which is equal to 223 synodic months, 19 eclipse years and 239 anomalistic months. After this period the centres of the Sun and Moon return to their same relative positions and the same pattern of eclipses is repeated; known to the ancient Babylonians, it was used to predict eclipses.

SARS (*Med*) Abbrev for *severe acute respiratory syndrome*. A rapidly progressive viral respiratory disease caused by a coronavirus that first appeared in China, Vietnam, Hong Kong, Singapore and Canada in 2003. Spread is by person to person, but airborne spread is a possibility. The illness, which usually affects adults, is associated with a case fatality rate of 11%. The initial outbreak ended in the autumn of 2003 and only sporadic cases have been reported since.

sarsen (*Geol*) Irregular masses of hard sandstones which are found in the Reading and Bagshot Beds of the Tertiary system in S England. They often persist as residual masses after the softer sands have been denuded away.

sartorius (*BioSci*) A thigh muscle of Tetrapoda that by its contraction causes the leg to bend inwards.

sash (*Build*) A framing for window panes.

sash and frame (*Build*) A cased frame in which counter-weighted sashes slide vertically.

sash bar (*Build*) A *transom* or a *mullion*.

sash centres (*Build*) The points about which a pivoted sash is moved.

sash cramp (*Build*) A contrivance for holding parts of a frame in place during construction. It usually consists of a steel bar along which slide two brackets between which the work is fixed, one of the brackets being pegged into a hole in the bar while the other is adjustable for position by means of a screw. In the US the bar is generally tubular. Also *clamp*.

sash door (*Build*) A door which has its upper part glazed.

sash fastener (*Build*) A fastening device secured to the meeting rails of the sashes of a double-hung window, serving to fix both sashes in the shut position. Also *sash lock*.

sash fillister (*Build*) A special plane for cutting grooves in stuff for sash bars.

sash lock (*Build*) See SASH FASTENER.

sash mortise chisel (*Build*) One somewhat lighter than a mortise chisel, used for mortising softwood.

sash pocket chisel (*Build*) A strong-bladed chisel with a narrow edge, used for cutting the pocket in the pulley stile of a sash and frame.

sash rail (*Build*) See TRANSOM.

sash saw (*Build*) A saw similar to the tenon saw but slightly smaller and finer; used for making window sashes.

sash stuff (*Build*) The timber prepared for use in the making of sashes.

sash weights (*Build*) Weights which are used as counterpoises in balancing the sashes of windows.

sassafras (*For*) A N American hardwood tree (*Sassafras officinale*), with a light- to dark-brown, straight-grained, coarse-textured heartwood. Its bark yields sassafras oil, used in perfumery and cosmetics, and its root an extract used for flavouring.

sassolite (*Min*) The mineral boric acid, H_3BO_3.

sateen (*Textiles*) A smooth-surfaced fabric produced by a weft-faced, sateen weave, often treated to make it lustrous. Cf SATIN.

satellite (*Astron*) (1) Any small body orbiting under gravitational forces in a closed path around a much more massive body, eg a planet orbiting the Sun. (2) A manufactured device launched into orbit around the Earth, the Moon or other planets. Satellites serve a variety of functions: relaying globally telephone and TV signals; covert intelligence gathering of all kinds; remote sensing of the Earth and its environment; weather forecasting; and as platforms for astronomical telescopes.

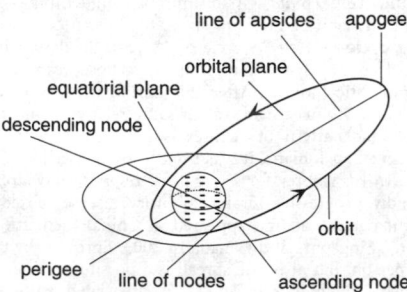

satellite The geometry of an Earth satellite.

satellite (*BioSci*) (1) The part of a chromosome distal to the SECONDARY CONSTRICTION. (2) See SATELLITE DNA, SIMPLE SEQUENCE DNA.

satellite (*ICT*) See panel on COMMUNICATIONS SATELLITE.

satellite computer (*ICT*) One used to relieve a central processing device of relatively simple but time-consuming operations, such as compiling, editing and controlling input and output devices. See DISTRIBUTED COMPUTING.

satellite DNA (*BioSci*) DNA from eukaryotic chromosomes that separates from the remainder of the genome on BOUYANT DENSITY centrifugation. It contains closely related repeated sequences with a base composition different from the rest of the DNA.

satellite ephemeris (*ICT*) A table predicting the variation of position with time for an Earth satellite, particularly a LOW EARTH ORBIT satellite.

satellite exchange (*ICT*) A small automatic telephone exchange that is dependent on a main automatic exchange for completion of its calls to subscribers other than those connected to it.

satellite station (*ICT*) (1) One that rebroadcasts a transmission received directly, but on another wavelength. (2) See EARTH STATION.

satellite station (*Surv*) In triangulation by theodolite, one resected and thus fixed for reference purposes, but not occupied.

Saticon (*ImageTech*) TN for an improved camera tube of the Vidicon type, having a photoconductive layer of doped selenium.

satin (*Textiles*) A lustrous, smooth-surfaced fabric produced by a warp-faced, satin weave. Cf SATEEN.

satin spar (*Min*) Name given to fine fibrous varieties of calcite, aragonite and gypsum, the gypsum variety being distinguished from the others by its softness (it can be scratched by a fingernail).

satin walnut (*For*) Misleading UK name for the HEART-WOOD of AMERICAN RED GUM.

satinwood (*For*) See EAST INDIAN SATINWOOD.

Sativex (*Med*) TN for a selective cannabinoid developed to relieve spasticity in patients with multiple sclerosis, neuropathic pain and side effects of chemotherapy.

saturable reactor (*ElecEng*) An inductor in which the core is saturable by turns carrying dc, which controls the inductance. Used for modulation, control of lighting and developed in the magnetic amplifier.

saturated (*Min*) Igneous rocks that lack silica minerals and feldspathoids so that in the normal mode they are neither *oversaturated* nor *undersaturated*.

saturated adiabatic (*EnvSci*) A curve on an AEROLOGICAL DIAGRAM representing the temperature changes of a parcel of saturated air subjected to an adiabatic process, the state of saturation being maintained.

saturated adiabatic lapse rate (*EnvSci*) The temperature lapse rate of air which is undergoing a *reversible natural adiabatic process* as shown by the *saturated adiabatic* lines on an AEROLOGICAL DIAGRAM. Abbrev *SALR*.

saturated calomel electrode (*Chem*) A calomel electrode containing saturated potassium chloride solution.

saturated compounds (*Chem*) Compounds which do not contain any free valencies and to which no hydrogen atoms or their equivalent can be added, ie which contain neither a double nor a triple bond.

saturated fatty acids (*BioSci*) A term referring, in eukaryotic membranes, to stearic, palmitic and myristic acids, which have linear aliphatic chains with no double bonds. Prokaryotes have numerous branched-chain saturated fatty acids.

saturated humidity mixing ratio (*EnvSci*) The *humidity mixing ratio* of air which is saturated at a specified temperature and pressure. Saturation may be defined with reference either to liquid water or (below 0°C) to ice.

saturated solution (*Chem*) A solution which can exist in equilibrium with excess of the dissolved substance as a second phase.

saturated steam (*Eng*) Steam at the same temperature as the water from which it was formed, as distinct from steam subsequently heated. See DRY STEAM.

saturated vapour (*Phys*) A vapour which is sufficiently concentrated to exist in equilibrium with the liquid form of the same substance.

saturation (*Electronics, Radar*) (1) The condition in which a further increase in one controlling variable produces no further increase in the resultant effect. (2) The condition occurring when a transistor is driven so hard (by a large base current) that the base–collector junction, reverse-biased in normal operation, becomes forward-biased. Recovery from this condition takes a long time and can impede high-speed switching operation. (3) In a thermionic valve, the condition when the anode current approaches total electron emission current available from the cathode. (4) In an amplifier, saturation occurs when the output power into the load approaches the limit of power available from the output stages. Similarly in a radar when the input signal drives the receiver to a point when its output can increase no further; characterized in such cases by a severe increase in non-linear distortion.

saturation (*ImageTech, Phys*) The degree to which a colour departs from white and approaches the pure colour of the spectral line; dull or pale colours are said to have low saturation, vivid colours high saturation. Cf DESATURATION.

saturation (*NucEng*) The condition where a field applied across the ionization chamber is sufficient to collect all ions produced by incident radiation.

saturation (*Phys*) For magnetic materials, the application of sufficient magnetizing field to achieve the maximum (saturation) magnetization. Likewise with analogous phenomena.

saturation activity (*NucEng*) The maximum level of artificial radioactivity, induced in a given sample, by a specific level of irradiation when the rate of formation equals the rate of decay.

saturation coefficient (*Build*) The ratio between the natural capacity of a material (such as a building-stone) to absorb moisture and its porosity.

saturation current (*ElecEng*) The steady current in a winding of an iron-cored transformer which causes the inductance of the winding to be seriously reduced.

saturation current (*Electronics*) See SATURATION.

saturation curve (*Phys*) The characteristic curve relating magnetic flux density to the strength of the magnetic field.

saturation factor (*ElecEng*) The ratio of the increase of field excitation to the increase of generated voltage which it produces.

saturation limit (*ElecEng*) The maximum flux density economically attainable.

saturation of the air (*EnvSci*) The air is said to be saturated when the relative humidity is 100%.

saturation scale (*Phys*) Minimum visual steps of saturation in scale of spectrum colours, varying with wavelength.

saturation vapour pressure (*EnvSci*) For water, the maximum WATER VAPOUR PRESSURE which can occur when the water vapour is in contact with a free water surface at a particular temperature. The water vapour pressure existing when effective evaporation ceases.

saturation voltage (*ElecEng*) The voltage applied to a device to operate under saturation conditions.

Saturn (*Astron*) Sixth planet of the solar system in order from the Sun, second largest at 752 times the volume of the Earth. Encircled by an extensive ring system. Density is 0·70, a low value because the planet is predominantly gaseous, consisting mainly of hydrogen and helium. The atmosphere is rich in methane and ethane. The rotation period varies from 10·233 hours at the equator to nearly 11 hours at the poles; this rapid rotation makes Saturn the most oblate of all planets, the equatorial radius (60 000 km) being 11 per cent more than the polar radius. There are numerous rings with diameters from 67 000 to 480 000 km, and 18 satellites. See appendix on Planets.

Saturn Nebula (*Astron*) A bright planetary nebula in the constellation Aquarius which resembles the planet Saturn and its rings.

Saturn rocket (*Space*) One of a successful series of NASA rocket launchers, first launched in October 1961, developed from military rockets; used to launch the Apollo spacecraft and to deploy the Skylab space station in Earth orbit.

sauconite (*Min*) One of the montmorillonite group of clay minerals in which zinc has replaced magnesium. See SMECTITES.

saussurite (*Geol, Min*) A fine-grained mixture of zoisite and other minerals resulting from the more or less complete alteration of feldspar. Sometimes simulates JADE.

savanna (*EnvSci*) (1) Extensive area in which grasses are an important part of the vegetation. (2) Common vegetation type in dry parts of Africa, consisting of trees and grasses, usually burnt every year.

save (*ICT*) To store a file, program or document on BACKING STORE. This makes a copy of the current file or program held in WORKING STORE onto backing store. With many systems, the previously held version of the file on backing store is retained as well. In eg MS-DOS, the old version of the file has its file extension changed to .BAK.

saveall (*Paper*) Any plant designed to recover fine fibres, loading and similar suspended matter from surplus backwater.

SAW (*Radar, ICT*) Abbrev for SURFACE ACOUSTIC WAVE DEVICE.

SAW delay line (*ICT*) One in which the delay is determined by the length of a piece of piezoelectric material along which a radio-frequency signal is launched into a surface acoustic mode by appropriate electrodes.

SAW filter (*ICT*) Filter depending on the propagation of surface acoustic waves through a resonant structure mounted on a piezoelectric material. Advantages include highly predictable and stable performance and very sharp cut-off characteristics; insertion loss is higher than for conventional filters, but amplification can compensate for this.

saw gumming (*Eng*) A term for the process of cleaning up burrs and machining marks in the gullets of circular saws.

sawtooth oscillator (*ICT*) See RELAXATION OSCILLATOR, TIME-BASE GENERATOR.

sawtooth roof (*Build, Eng*) A roof formed by multiple spans of short trusses giving a sawtooth appearance in elevation.

sawtooth wave (*Electronics*) One generated by a TIME-BASE GENERATOR for scanning in a cathode-ray tube, for uniform sweep and high-speed return. See FLYBACK.

sax (*Build*) An axe used for shaping slates; it has a pointed peen for piercing the nail-holes. Also *saixe, slate axe*.

saxicole (*BioSci*) Growing on or among rocks. Also *saxicolous*.

saxonite (*Geol*) A coarse-grained, ultrabasic rock, consisting essentially of olivine and orthopyroxene, usually hypersthene. A hypersthene-peridotite.

saxony (*Textiles*) A high-quality woollen cloth made from merino wool.

Sb (*Chem*) Symbol for ANTIMONY.

SBA (*Aero*) Abbrev for STANDARD BEAM APPROACH.

SBAC (*Aero*) Abbrev for *Society of British Aerospace Companies*.

S-band (*ICT*) Loose definition, due to international disagreement, of microwave band in the 2–3 GHz region.

SBC (*ElecEng*) Abbrev for SMALL BAYONET CAP.

SBR (*Plastics*) Abbrev for STYRENE–BUTADIENE RUBBER. See panel on ELASTOMERS.

Sc (*Chem*) Symbol for SCANDIUM.

scab (*BioSci*) A discrete localized superficial lesion characterized by roughening, abnormal thickening and esp cork formation; a disease in which such scabs are formed, eg potato and apple scab.

scab (*Vet*) See MANGE.

scabbling (*Build*) The operation of rough-dressing a stone face with an axe, prior to smoothing it.

scabbling hammer (*Build*) The pointed hammer used in the rough-dressing of a nigged ashlar.

scabellum (*BioSci*) In Diptera, the dilated basal portion of a haltere.

scabies (*Med, Vet*) A contagious skin disease caused by the acarine parasite *Sarcoptes scabiei*, the female of which burrows in the horny layer of the skin. See also MANGE.

scabies (*Vet*) See MANGE.

scabrid (*BioSci*) Rough to the touch; scaly.

scabrous (*BioSci*) Having a surface roughened by small wart-like upgrowths. Dim *scaberulous*.

scaffold (*BioSci*) The protein core of histone-depleted metaphase chromosomes left after nuclease treatment.

scaffold (*Build*) A temporary erection of timber or steel-work, used in the construction, alteration or demolition of a building. Also *scaffolding*.

scaffold/radial loop model (*BioSci*) A model of metaphase chromosome structure that postulates a non-histone protein core to which the linear DNA molecule has an ordered series of attachment points every 30–90 kbp, with the intervening DNA forming a loop packed by super-coiling or folding.

scalable font (*ICT*) A FONT that may be adjusted in size as required rather than one that is available in one fixed POINT SIZE. Also *scalable typeface*.

scalar (*MathSci*) A real number or element of a field; a quantity with a magnitude but not a direction. The term *scalar* is used in vector geometry and in vector and matrix algebra to contrast numbers with vectors and matrices. See VECTOR SPACE.

scalar matrix (*MathSci*) A matrix that is a SCALAR multiple of a UNIT MATRIX. A number (or scalar), λ, may be represented by the scalar matrix λI, where I is the unit matrix, which has λs in the principal diagonal and zeros elsewhere.

scalar product (*MathSci*) Of two vectors, a scalar (ie real number) equal to the product of the magnitudes of the two vectors and the cosine of the angle between them. Denoted by a dot, eg a.b; cf VECTOR PRODUCT.

scale (*Acous*) (1) The ratio of length to diameter of organ pipe, a factor which, among others, determines the timbre of a note. (2) A set of (usually) twelve notes forming an octave or frequency range of two to one.

scale (*BioSci*) (1) A thin, flat, semi-transparent plant member, usually of small size, and green only when very young, if then. See BUD SCALE, SCALE LEAF. (2) A small exoskeletal outgrowth of tegumentary origin, of chitin, bone or some horny material, usually flat and plate-like.

scale (*Eng*) Numerical factor relating measured quantity to indication of instrument.

scale (*ImageTech*) (1) General term for the range of brightness or density over which satisfactory tonal reproduction can be expected. (2) The range of exposure steps available in a printer or enlarger.

scale (*MathSci*) The ratio between the linear dimensions of a representation or model and those of the object repre-sented, as on a map or technical drawing.

scale bark (*BioSci*) (1) Bark that becomes detached in irregular patches. (2) See RHYTIDOME.

scale leaf (*BioSci*) Leaf, often thin, more or less flattened against the stem or other leaves, and photosynthetic, protective (bud scales) or holding reserves (bulb scales).

scalene triangle (*MathSci*) A triangle in which no two sides are equal.

scale-of-ten (*ICT*) Ring, or other, system of counting elements that divides counts by ten.

scale-of-two (*Electronics*) Any bistable circuit which can divide counts by two, when operated by pulses. See FLIP-FLOP.

scaler (*Electronics*) An instrument, incorporating one or more scaling circuits, used to register a count.

scaling (*Acous*) Adjustment of the notes of a musical instrument to a specified scale, eg natural or tempered scale.

scaling circuit (*Electronics*) One which divides counts of pulses by an integer, so that they are more readily

indicated, to a required degree of accuracy. If scalers of ten are used in cascade, indications of counts in decimal numbers are possible.

scaling hammer (*Eng*) See BOILERMAKER'S HAMMER.

scaly bark (*BioSci*) See SCALE BARK.

scaly leg (*Vet*) A form of mange affecting the feet and legs of fowl, due to the mite *Cnemidocoptes mutans* burrowing in the skin.

scan (*Radar*) Systematic variation of a radar beam direction for search or angle tracking. See A-DISPLAY, B-DISPLAY, etc.

scandent (*BioSci*) Climbing.

scandium (*Chem*) Symbol Sc, at no 21, ram 44·956. A metallic element, classed with the rare earth metals. It has been found in cerite, orthite, thortveitite, wolframite and euxenite; discovered in the last named. Scandium is the least basic of the rare earth metals.

scanner (*ICT*) Generally any device that may be passed over a document (or the document passed over it) to transfer the information into the computer, eg a BAR CODE scanner. More specifically the term is used to describe a device used to scan a document to transfer a graphic image into a DESKTOP PUBLISHING SYSTEM. See FLATBED SCANNER.

scanner (*ImageTech*) (1) The DRUM in a videotape recorder. (2) An outside broadcast van used as the control room.

scanner (*Radar*) Mechanical arrangement for covering a solid angle in space, for the transmission or reception of signals, usually by parallel lines or *scans*.

scanning (*ImageTech*) The systematic coverage of an area by a spot moving in a series of progressive lines, esp by the use of an electron beam in a TV camera or display tube.

scanning (*Radar*) Coverage of a prescribed area by a directional radar antenna or sonar beam.

scanning coils (*ImageTech*) Coils mounted in cathode-ray tube, and carrying suitable currents for deflecting electron beam, so as to sweep the picture area.

scanning electron microscope (*Phys*) A form of electron microscope in which a very fine beam of electrons at 5–100 kV is made to scan a chosen area of specimen as a *raster* of parallel contiguous lines. Abbrev *SEM*. See panel on ELECTRON MICROSCOPE.

scanning frequency (*ImageTech*) The number of times per second that an area is completely scanned.

scanning heating (*Phys*) INDUCTION HEATING where the workpiece is moved continuously through the heating region, as in ZONE REFINING of germanium. Also *progressive heating*.

scanning linearity (*ImageTech*) Uniformity of scanning speed for a cathode-ray oscilloscope or TV receiver. This is necessary to avoid waveform or picture distortion.

scanning loss (*Radar*) That which arises from relative motion of a scanning beam across a target, as compared with zero relative motion.

scanning slit (*ImageTech*) In a system for photographic sound on film, the narrow illuminated aperture across which the track passes for recording and reproduction.

scanning speech (*Med*) A disturbance of speech in which the utterance is slow and halting, the words being broken into syllables; a sign of a lesion in the nervous system, as in disseminated sclerosis.

scanning speed (*ImageTech*) That of a scanning spot across the screen of a cathode-ray tube. Usually accurately specified for a cathode-ray oscilloscope to facilitate time measurements.

scanning transmission electron microscope (*Phys*) One which uses field emission from a very fine tungsten point as the source of electrons. See panel on ELECTRON MICROSCOPE.

scanning tunnelling electron microscope (*Phys*) A microscope which uses a probe with an 'atomic micro-tip' floated, using SUPERCONDUCTING LEVITATION, over the surface being scanned. See panel on ELECTRON MICROSCOPE.

scan plates (*Print*) A general term for plates made by ELECTRONIC ENGRAVING.

scansorial (*BioSci*) Adapted for climbing trees.

scantling (*Build*) Stones more than 2 m (6 ft) long.

scantling (*For*) A piece of timber 50–100 mm (2–4 in) thick and 50–110 mm wide.

scape (*BioSci*) (1) The flowering stem, nearly or quite leafless, arising from a rosette of leaves and bearing a flower, several flowers or a crowded inflorescence; eg the dandelion. (2) The basal joint of the antenna in insects. Adj *scapigenous*.

scaph-, scapho- (*Genrl*) Prefixes from Gk *skaphe*, boat.

scaphoid (*BioSci*) See NAVICULAR BONE.

Scaphopoda (*BioSci*) A class of bilaterally symmetrical Mollusca with a tubular shell open at both ends, a reduced foot used for burrowing, a head with many prehensile processes, a radula, and separate cerebral and pleural ganglia. There are no ctenidia and the circulatory system is rudimentary. The larva is a trochosphere.

scapolite (*Min*) A group of minerals forming an isomorphous series, varying from meionite, a silicate of aluminium and calcium with calcium carbonate, to marialite, a silicate of aluminium and sodium with sodium chloride. Common scapolite is intermediate in composition between these two end-member minerals. See DIPYRE, MIZZONITE. The scapolites crystallize in the tetragonal system and are associated with altered lime-rich igneous and metamorphic rocks. A transparent honey-yellow variety is cut as a gemstone.

scappling (*Build*) Also SCABBLING.

scapula (*BioSci*) In vertebrates, the dorsal portion of the pectoral girdle, the shoulder blade; any structure resembling the shoulder blade. Adj *scapular*.

scapulars (*BioSci*) In birds, small feathers attached to the HUMERUS, and lying along the side of the back.

scarcement (*Build*) A ledge formed at a place where part of a wall is set back from the general face of the wall.

scarf (*Build*) A joint between timbers placed end to end, notched and lapped, and secured together with bolts or straps.

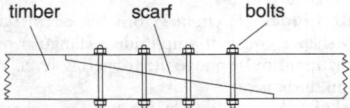

timber   scarf   bolts

scarf One of several forms.

scarfed joint (*ElecEng*) A cable joint in which the conductor ends are tapered off so that, after soldering, there is no appreciable increase in conductor diameter at the joint.

scarfing (*Eng*) (1) Tapering the ends of materials for a lap joint, so that the thickness at the joint is substantially the same as that on either side of it. (2) Preparing metal edges for forge welding.

scarification (*BioSci*) Any treatment, eg with sulphuric acid or by mechanical abrasion, that makes the coat of a seed more permeable to water and promotes imbibition and germination.

scarifier (*CivEng*) A spiked mechanical picking appliance for breaking up road surfaces as a preliminary to remetalling.

scarlatina (*Med*) See SCARLET FEVER.

scarlet fever (*Med*) An acute infectious fever due to infection of the throat with a haemolytic streptococcus; characterized by sore throat, headache, raised temperature and a punctate erythema of the skin, which subsequently peels. Also *scarlatina*.

scarp face (*Geol*) See ESCARPMENT.

SCART (*ImageTech*) Abbrev for *Syndicat des Constructeurs d'Appareils Radio Recepteurs et Televiseurs*. Defined a 21-pin connector for stereo audio, video, and Y/C or RGB SIGNALS. Also *euro-connector*.

Scatchard plot (*BioSci*) A method for analysing data for freely reversible ligand/receptor binding interactions. The graphical plot is: [bound ligand]/[free ligand] against [bound ligand] where the square brackets indicate concentration; the slope gives the negative reciprocal of the binding affinity, the intercept on the *x*-axis the number of receptors.

scatter diagram (*MathSci*) A graph plotting two discrete variables, with axes at right angles to one another.

scatterer (*Acous*) An object which causes scattering, eg curved hard boards or fish bladders in the sea. Used in REVERBERATION CHAMBERS.

scattering (*Phys*) General term for irregular reflection or dispersal of waves or particles, eg in acoustic waves in an enclosure leading to diffuse reverberant sound, COMPTON EFFECT on electrons, light in passing through material, electrons, protons and neutrons in solids, radio waves by ionization. See BACK SCATTER, FORWARD SCATTER. Particle scattering is termed *elastic* when no energy is surrendered during scattering process, otherwise it is *inelastic*. If due to electrostatic forces it is termed COULOMB SCATTERING; if short-range nuclear forces are involved it becomes *anomalous*. Long-wave electromagnetic wave scattering is *classical* or *Thomson*, while for higher frequencies *resonance* or *potential* scattering occurs according to whether the incident photon does or does not penetrate the scattering nucleus. Coulomb scattering of alpha particles is *Rutherford scattering*. For light, scattering by fine dust or suspensions of particles is *Rayleigh scattering*, while that in which the photon energy has changed slightly, due to interaction with vibrational or rotational energy of the molecules, is RAMAN SCATTERING. *Shadow scattering* results from interference between scattered and incident waves of the same frequency. See ACOUSTIC SCATTERING, ATOMIC SCATTERING.

scattering amplitude (*Phys*) The ratio of the amplitude of a scattered wave at unit distance from the scattering nucleus to that of the incident wave.

scattering cross-section (*Phys*) Effective impenetrable cross-section of scattering nucleus for incident particles of low energy. The radius of this cross-section is the *scattering length*.

scattering mean free path (*Phys*) The average distance travelled by a particle between successive scattering interactions. It depends upon the medium traversed and the type and energy of the particle.

scatterometer (*EnvSci*) An instrument carried in a METEOROLOGICAL SATELLITE for measuring the light scattered from the surface of the sea, thus yielding information on the height and movement of waves which can be used to derive estimates of the surface wind.

scavenge pump (*Autos*) An oil-suction pump used to return used oil to the oil tank from the crankcase of an engine using the dry sump system of lubrication.

scavenging (*Eng*) Addition made to molten metal to counteract an undesired substance.

scavenging (*MinExt*) Final stage of froth flotation in which a low-grade concentrate or middling is removed.

scavenging stroke (*Eng*) See EXHAUST STROKE.

scc (*ElecEng*) Abbrev for the obsolete SINGLE COTTON-COVERED WIRE.

scenario (*ICT*) The detailed description of a set of real-world processes, from which test cases may be derived, that can be used to evaluate whether a system or program is fit for purpose.

scenario (*ImageTech*) See SCRIPT.

scending (*Ships*) See PITCHING PERIOD.

scene (*ImageTech*) The component of a motion picture or video production which is intended to be recorded as uninterrupted action.

scene-slating attachment (*ImageTech*) A device attached to a motion picture camera for identifying individual takes on the film. Used in place of CLAPPER BOARD.

scent-marking (*BioSci*) A form of communication between individuals of a species involving the deposition of glandular secretion onto ground or some other surface; the scent dissipates at a slow rate to form a relatively long-lived signal.

Schafer's method (*Med*) A method of artificial respiration in which the patient lies prone, the head supported on one forearm, and the operator, his or her knees on either side of the patient's hips, exerts pressure with each hand over the lower ribs at the back at intervals of from 3 to 5 seconds. Named after the physiologist E Sharpey-Schafer. Cf KISS OF LIFE.

scheduling (*ICT*) A method by which central processing unit time is allotted in a MULTI-ACCESS system. The scheduling algorithm may include a system of priorities.

Scheele's green (*Chem*) Copper (II) hydrogen arsenite ($CuHAsO_3$). Poisonous. Formerly used in wallpapers.

scheelite (*Min*) An ore of tungsten. It occurs in association with granites and pegmatites, has the composition calcium tungstate and crystallizes in the tetragonal system.

schema (*Psych*) A mental pattern or body of knowledge that provides a framework within which to place newly acquired knowledge. Pl *schemata*.

scheme arch (*CivEng*) See SKENE ARCH.

Schering bridge (*ElecEng*) An ac bridge of the four-arm (or Wheatstone) type usually used to measure capacitance and power factor of small capacitors.

Schick test (*Med*) A test to assess the degree of susceptibility or immunity of individuals to diphtheria.

Schiff's bases (*Chem*) (1) A term for benzylidene anilines, eg $C_6H_5CH=NC_6H_5$. (2) More generally, any IMINE.

Schiff's reagent (*Chem*) A reagent, consisting of a solution of fuchsine decolorized by sulphurous acid, for testing the presence of aldehydes. Their presence is shown by a red-violet colour.

schillerization (*Geol*) A play of colour (in some cases resembling iridescence due to tarnish) produced by the diffraction of light in the surface layers of certain minerals.

schillerspar (*Min*) See BASTITE.

Schimmelbusch's disease (*Med*) A condition characterized by the formation of cysts in the breast, and by hyperplasia of the epithelium of the glandular tissue of the breast.

schist (*Geol*) A metamorphic rock which has a tendency to split on account of the presence of folia of flaky and elongated minerals, such as mica, talc and chlorite; formed from original sedimentary or igneous rocks by the action of regional metamorphism.

schistosity (*Geol*) The tendency in certain rocks to split easily along weak planes produced by regional metamorphism and due to the abundance of mica or other cleavable minerals lying with their cleavage planes parallel.

schistosomiasis (*Med*) Infestation by parasitic worms of the *Schistosoma* genus, transmitted via water by a complex cycle involving snails as intermediate hosts; common in many parts of the tropics. The chief varieties affect mainly the bladder (*S. haematobium*, see BILHARZIASIS) and rectum (*S. manśonii*).

schizo- (*Genrl*) Prefix from Gk *schizein*, to cleave.

schizocarp (*BioSci*) A fruit that is derived from a syncarpous ovary and that, when mature, becomes divided into separate, one-seeded, indehiscent parts (mericarps).

schizocoel (*BioSci*) A coelom produced within the mass of mesoderm by splitting or cleavage. Cf ENTEROCOEL. Adj *schizocoelic*.

schizogamy (*BioSci*) In Polychaeta, a method of reproduction in which a sexual form is produced by fission or germination from a sexless form.

schizogenesis (*BioSci*) Reproduction by fission.

schizogenous (*BioSci*) A space in a tissue, formed by the separation of cells by splitting their common wall along the middle lamella. Cf LYSIGENOUS, REXIGENOUS, SCHIZOLYSIGENOUS.

schizogony (*BioSci*) In Protozoa, vegetative reproduction by fission.

schizoid (*Med*) A term used to describe withdrawn and introspective individuals with dissociation between emotions and intellect. Normal but tending to SCHIZOPHRENIA.

schizolysigenous (*BioSci*) A space in a tissue, formed by both dissolution of cells and separation at the middle lamella. Cf LYSIGENOUS, REXIGENOUS, SCHIZOGENOUS.

schizont (*BioSci*) In Protozoa, a mature feeding individual about to reproduce by SCHIZOGONY.

schizophrenia (*Psych, Med*) A group of psychoses marked by severe distortion and disorder of thought, perception, motivation and mood; delusion and hallucination are common, as are bizarre behaviours and social withdrawal. The term was invented by Eugene Bleuler and was previously known as *dementia praecox*.

Schizosaccharomyces pombe (*BioSci*) A species of fission yeast commonly used for studies on cell cycle control because there is a distinct G2 phase to the cycle. It is only distantly related to the budding yeast *Saccharomyces cerevisiae*.

schlieren photography (*Aero, Phys*) Technique by which the flow of air or other gas may be photographed, the change of refractive index with density being made apparent under a special type of illumination. The method is used in studying the behaviour of models in transonic and supersonic wind tunnels.

Schmidt lines (*NucEng*) Two parallel lines in plot of nuclear magnetic moment against nuclear spin, as a result of the spin of an odd unpaired proton or neutron. Experimental magnetic moments for the majority of such nuclides lie between these lines. Also known as *Schmidt limits*.

Schmidt optical system (*Phys*) An optical system, for telescopes and for projection work, which uses a spherical mirror instead of a parabolic mirror. The resulting spherical aberration is corrected by using a moulded transparent plastic plate in front of the mirror to give a CATADIOPTRIC system.

Schmidt telescope (*Astron*) A wide-field optical telescope which uses a combination of spherical mirror, a Schmidt correction plate and a curved reflecting plate at the focus of the mirror to minimize spherical aberration and coma. Used in extensive sky surveys. See panel on ASTRONOMICAL TELESCOPE.

Schmitt trigger (*Electronics*) Bistable circuit giving accurately shaped constant-amplitude rectangular pulse output for any input pulse above the triggering level. Widely used as a pulse shaper.

schnorkel (*Ships*) A retractable tube or tubes containing pipes for discharging gases from, or for taking air into, a submerged submarine or other underwater vessel; a tube for bringing air to a submerged swimmer. Also *snorkel*, *snort*.

school (*BioSci*) See SCHOOLING.

schooling (*BioSci*) In fish, refers to groups of individuals who maintain a constant distance and orientation from their neighbours, and who all swim at a constant pace; the primary function of schooling is as an anti-predator strategy.

school phobia (*Psych*) Severe reluctance to attend school.

schorlomite (*Min*) A black variety of andradite garnet richer in titania (5–20% TiO_2) than melanite.

schorl-rock (*Geol*) A rock composed essentially of aggregates of black tourmaline (schorl) and quartz. A Cornish term for the end product of tourmalinization. See TOURMALINE.

Schottky defect (*Chem*) Deviation from the ideal crystal lattice by removal of some of the molecules to the surface, leaving within the lattice equivalent numbers of randomly spaced anion and cation vacancies.

Schottky diode (*Electronics*) A semiconductor diode formed by contact between a semiconductor layer and a metal coating. Hot carriers (electrons in n-type material and holes in p-type material) are emitted from the Schottky barrier of the semiconductor and move into the metal coating. Majority carriers predominate, so that injection or storage of minority carriers is not present to limit switching speeds.

Schottky effect (*Electronics*) (1) The removal of electrons from the surface of a semiconductor when a localized electric field is present at the surface; electrons can then surmount a semiconductor potential barrier and enter regions where allowed energy levels are available. (2) The increase in cathode current in a valve, beyond that available by normal thermionic emission, on account of the effective lowering of the work function of the cathode by the localized electric field.

Schottky noise (*Electronics*) Strictly, noise in the anode current of a thermionic valve due to random variations in the surface condition of the cathode. Frequently extended to include SHOT NOISE.

Schottky TTL (*Electronics*) A term describing logic switching and memory circuits where each transistor involved in the switching operation has a Schottky diode connected across it to reduce the number of charge carriers in the base when the transistor is on. This means that the transistors can be turned off more rapidly and allows the circuits to function more rapidly.

Schrage motor (*ElecEng*) A variable-speed induction motor employing a commutator winding on the rotor from which an emf is collected and injected into the stator winding.

Schrödinger equation (*Phys*) Fundamental equation of WAVE MECHANICS. Solutions of this equation are *wavefunctions* for which the square of the amplitude expresses the probability density for a particle or a set of particles. If the system is isolated then a time-independent form of the equation is applicable. Solutions for this version for bound particles show that the energy for the system must be quantized. See panel on QUANTUM THEORY.

Schrödinger's cat (*Phys*) A thought experiment introduced by E Schrödinger in 1935. A box contains a radioactive source, a bottle of poison and a live cat. The equipment is so arranged that the detection of radioactive decay breaks the bottle and poisons the cat. If the experiment lasts just long enough for a 50% chance of decay, quantum mechanical analogy suggests the cat is neither alive nor dead until the box is opened. This paradox highlights the difficulties of interpretation in quantum mechanics.

Schuler pendulum (*Eng*) A theoretically ideal pendulum which will not be affected by the Earth's rotation, having a length equal to the Earth's radius and hence a period of 84 min. A stable platform servomechanism can be constructed to simulate pendular motion of the above period, and is then said to be 'Schuler-tuned'. Conversely a gyroscopically stabilized platform, constrained to move parallel to the Earth's surface in its motion over the Earth, will possess a period of 84 min and will be conditionally stable. Damping to reduce the maximum error may be introduced by rate feedback.

Schwann cell (*BioSci*) A neuroglial cell that deposits the MYELIN SHEATH along myelinated AXONS. See fig. at NEURON.

schwannoma (*Med*) A tumour growing from the sheath of a nerve (neurofibroma) and containing cells resembling those of the neurolemma.

Schwarzchild radius (*Astron*) The critical radius at which an object becomes a black hole if collapsed or compressed indefinitely; at this radius the escape velocity is equal to the speed of light.

Schwarz's inequality (*MathSci*)

$$\left(\int f(x)\, g(x)\, dx \right)^2 \le \left(\int [f(x)]^2\, dx \right) \left(\int [g(x)]^2\, dx \right)$$

unless $f(x) = kg(x)$, where k is independent of x. This is the integral analogue of CAUCHY'S INEQUALITY.

Schweitzer's reagent (*Chem*) A reagent for cellulose. It consists of a 0·3% solution of precipitated copper (II) hydroxide in a 20% ammonium hydroxide solution. This mixture is a solvent for cellulose, which can be re-precipitated by the addition to the solution of mineral acids.

sciatic (*BioSci*) Situated in, or pertaining to, the ischial or hip region.

sciatica (*Med*) Inflammation or irritation of the fibrous elements of the sciatic nerve, resulting in pain and tenderness along the course of the nerve in the buttock and the back of the leg; less strictly, any pain along the course of the sciatic nerve.

SCID (*BioSci*) Abbrev for SEVERE COMBINED IMMUNODEFICIENCY SYNDROME.

science (*Genrl*) The ordered arrangement of ascertained knowledge, including the methods by which such knowledge is extended and the criteria by which its truth is tested. The older term *natural philosophy* implied the contemplation of natural processes per se, but modern science includes such study and control of nature as is, or might be, useful to humankind. *Theoretical science* derives hypotheses and theories, and deduces critical tests whereby uncoordinated observations and properly ascertained facts may be brought into the body of non-theoretical science.

scientific alexandrite (*Min*) Synthetic corundum coloured with vanadium oxide and resembling true alexandrite in some of its optical characters.

scientific emerald (*Min*) Beryl glass coloured with chromic oxide, resembling true emerald in colour.

scintillation (*Astron*) The twinkling of stars, a phenomenon due to the deflection, by the strata of the Earth's atmosphere, of the light rays from what are virtually point sources.

scintillation (*Phys*) Minute light flash caused when alpha, beta or gamma rays strike certain PHOSPHORS, known as *scintillators*. The latter are classed as liquid, inorganic, organic or plastic according to their chemical composition.

scintillation camera (*Radiol*) An imaging device which may have either a single sodium iodide crystal or multiple crystals, which is capable of detecting and recording the spatial distribution of an internally administered radio-nuclide. Also *gamma camera*.

scintillation counter (*NucEng*) A counter consisting of a PHOSPHOR or *scintillator*, eg NaI(Tl), which, when radiation falls on it, emits light which is detected and amplified by a photomultiplier, the height of the pulses from which are proportional to the energy of the event. These pulses are further amplified and passed to a single- or multichannel PULSE-HEIGHT ANALYSER, to measure the energy and intensity of the radiation. Also *scintillation spectrometer*.

scintillation proximity assay (*BioSci*) An assay system in which a specific receptor, eg antibody, is bound to a bead that will emit light when stimulated by nearby β-emission from a radioisotopically labelled ligand (antigen).

scintillation spectrometer (*NucEng*) See SCINTILLATION COUNTER.

scintling (*Build*) Placing half-dry raw bricks diagonally and a little distance apart, so as to admit air between them.

scion (*BioSci*) A piece of a plant; in horticulture usually a young, often dormant, shoot that is inserted into the STOCK when a graft is made.

sciophyte (*BioSci*) A plant that is adapted to living in shady places.

scirrhous carcinoma (*Med*) A hard cancer, in which there is an abundance of connective tissue and few cells.

scirrhous cord (*Vet*) See BOTRYOMYCOSIS.

scissors truss (*Build*) A type of truss used for a pitched roof, consisting of two principal rafters braced by two other members, each of which connects the foot of a rafter to an intermediate point in the length of the other rafter.

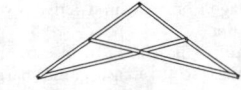

scissors truss

Sclair (*Plastics*) TN for polyethylene which cross-links during ROTATIONAL MOULDING using free radical catalyst. This gives lower creep rate for uses like kayak canoes.

scler-, sclero- (*Genrl*) Prefixes from Gk *skleros*, hard.

sclera (*BioSci*) The tough fibrous outer coat of the vertebrate eye. Adj *sclerotic*.

sclere (*BioSci*) A skeletal structure; a sponge spicule.

sclereid (*BioSci*) A short SCLERENCHYMA CELL, star-shaped, rod-shaped or rounded. Such cells are often found in tissues of other cell types but also exist in some seed coats as a tissue of sclereids.

sclereide (*BioSci*) (1) A general term for a cell with a thick, lignified wall, ie any sclerenchymatous cell. (2) A thick-walled cell mixed with the photosynthetic cells of a leaf, giving them mechanical support. (3) A STONE CELL.

sclerencephalia (*Med*) Hardening of the brain.

sclerenchyma (*BioSci*) (1) Tissue composed of sclerenchyma cells, or the collective term for such cells. (2) Hard skeletal tissue, as of corals.

sclerenchyma cell (*BioSci*) Cells with thick, usually lignified, walls, often dead when mature, and usually having a supporting function in the plant; either a SCLEREID or a FIBRE. Cf COLLENCHYMA.

sclerified (*BioSci*) See SCLEROTIZATION.

sclerite (*BioSci*) A hard skeletal plate or spicule.

scleritis (*Med*) Inflammation of the sclera of the eye. See EPISCLERITIS.

sclero- (*Genrl*) See SCLER-.

sclerodactylia (*Med*) Sclerodermia of the hands, the skin being drawn tightly over the fingers.

scleroderma (*Med*) A condition of hardness and rigidity of the skin as a result of overgrowth of fibrous tissue in the dermis and subcutaneous tissue, the fat of which is replaced by the fibrous tissue. Also *sclerodermia*.

scleroma (*Med*) A condition in which hard nodules of granulomatous tissue appear in the nose or occasionally in the trachea.

sclerometer (*Eng*) An instrument used for measuring the hardness of minerals or metals by impressing a polished surface with a diamond point.

scleronychia (*Med*) Thickening and dryness of the nails.

sclerophyll (*BioSci*) A leaf with well-developed sclerenchyma and hence tough and fibrous or leathery. Such leaves are usually evergreen and are typical of trees and shrubs in places with rather warm and dry, esp Mediterranean, climates, eg olive and *Eucalyptus*.

scleroproteins (*BioSci*) Insoluble proteins forming the skeletal parts of tissues, eg keratin from hoofs, nails, hair, etc, chondrin and elastin from ligaments.

scleroscope hardness test (*Eng*) The determination of the hardness of metals by measuring the rebound of a diamond-tipped hammer dropped from a given height. See SHORE SCLEROSCOPE and panel on HARDNESS MEASUREMENTS.

sclerosed (*Med*) See SCLERIFIED.

sclerosis (*BioSci*) The hardening of cell walls or of tissues by thickening and deposition of extracellular material. In plants this can involve lignification.

sclerosis (*Med*) An induration or hardening, as of the arteries. See MULTIPLE SCLEROSIS.

sclerotic (*BioSci*) The sclera of the eye; pertaining to the sclera.

sclerotic cell (*BioSci*) See SCLEREID.

sclerotin (*BioSci*) In the cuticle of insects and some other arthropods, a protein that has become strengthened and dark through cross-linkage by the action of quinones. See TANNING.

sclerotium (*BioSci*) A hard mass of fungal hyphae, often black on the outside, crust-like to globular, and serving as a resting stage from which an active mycelium or spores are formed later.

sclerotization (*BioSci*) (1) In insects, the process by which most of the PROCUTICLE becomes hardened and darkened to form tough, rigid, discrete sclerites of EXOCUTICLE. See TANNING. (2) In plants, hardening of tissue, eg by secondary wall formation and/or lignification. Also *sclerified*, *sclerosed*.

sclerotomy (*Med*) Operative incision of the sclera.

scobicular (*BioSci*) Looking like sawdust. Also *scobiform*.

scoinson arch (*Build*) See SQUINCH.

scolecite (*Min*) A member of the zeolite group of minerals; a hydrated silicate of calcium and aluminium, occurring usually in fibrous or acicular groups of crystals.

scolex (*BioSci*) The terminal organ of attachment of a tapeworm (Cestode). Pl *scoleces*, wrongly *scolices*. Adjs *scolecid*, *scoleciform*.

scoliosis (*Med*) Abnormal curvature of the spine laterally.

scolophore (*BioSci*) A subcuticular spindle-shaped nerve ending in insects, sensitive to mechanical vibrations. Also *scolopidium*.

scolopidium (*BioSci*) A campaniform sensillum acting as a strain sensor. Arrays of scolopidia form the chordotonal organ of some insects.

scombrotoxin (*FoodSci*) Causative agent of *scombroid poisoning* that occurs following ingestion of foods containing high levels of histamine and possibly other vasoactive amines and compounds. Histamine and other amines are formed by the growth of certain bacteria and the subsequent action of their decarboxylase enzymes on histidine and other amino acids. Spoilage of fishery products is a common cause, but the production processes for Swiss cheese can also give high histamine levels. Also called *histamine poisoning*.

sconchion (*Build*) An inside quoin, as laid in SPLAYED JAMBS. Also *scontion*.

scoop (*Med*) A spoon-like instrument for clearing out cavities.

scoop (*NucEng*) The device used for removing gas from a centrifuge in an enrichment plant.

scopa (*BioSci*) The pollen brush of bees, consisting of short stiff spines on the posterior metatarsus.

scope (*Electronics*) Colloq term for OSCILLOSCOPE.

'scope (*ImageTech*) Popular term for anamorphic systems of WIDE-SCREEN cinematography, by derivation from CINE-MASCOPE.

scopolamine (*Pharmacol*) See HYOSCINE.

scopophilia (*Psych*) Deriving sexual pleasure from visual sources.

scorbutic (*Med*) Pertaining to scorbutus (scurvy).

scorch (*BioSci*) Necrosis, like that of leaf margins looking as if seared by heat, caused by infection, mineral deficiency, weather conditions, etc.

scorch (*Eng*) A term used in rubber technology for premature vulcanization, before final shape of product has been achieved.

score (*Print*) See EM RULE, EN RULE.

scoria (*Geol*) A cavernous mass of volcanic rock which simulates a clinker.

scorification (*Eng*) Assay in which impurities are slagged while bullion metals dissolve in molten lead, from which they are later separated by CUPELLATION.

scorifier (*Eng*) A crucible of bone ash or fireclay used in assaying and in the metallurgical treatment of precious metals. See SCORIFICATION.

scorodite (*Min*) An orthorhombic hydrated arsenate of iron and aluminium.

Scorpionidea (*BioSci*) An order of Arachnida with the posterior part of the body (opisthosoma) being divided into a distinct mesosoma and metasoma, and consisting of twelve segments and a telson. The chelicerae and pedipalps are chelate and there are four pairs of walking legs. The mesosomatic segments carry the genital operculum, the pectines and four pairs of lung books, the metasoma form a flexible tail with a terminal sting. Such creatures are viviparous and terrestrial. Known as fossils from the Silurian. Scorpions.

Scorpius (Scorpion) (*Astron*) Southern constellation lying between Libra and Sagittarius. Includes Scorpius X-1, the first X-ray star discovered, and ANTARES.

scotch (*Build*) See SCUTCH.

Scotch block (*CivEng*) Attachment to running rails, to prevent the passage of rolling stock.

Scotch boiler (*Eng*) See MARINE BOILER.

Scotch bond (*Build*) A bond in which a course of headers alternates with three courses of stretchers. Also *English garden-wall bond*.

Scotch crank (*Eng*) A form of crank, used on a DIRECT-ACTING PUMP, in which a square block, pivoted on the overhung crank pin, works in a slotted cross-head carried by the common piston rod and ram. Also *Scotch yoke*.

Scotch yoke (*ElecEng*) A triangular framework sometimes used for coupling one traction motor to two driving wheels of a locomotive.

scotia (*Arch*) See CAVETTO and fig. at MOULDINGS.

scoto- (*BioSci*) Prefix from Gk *skotos*, darkness; dark (in the sense of not illuminated).

scotoma (*Med*) (1) A blind or partially blind area in the visual field, the result of disease of, or damage to, the retina or optic nerve or visual cortex. (2) The appearance of a black spot in front of the eye, as in choroiditis. Pl *scotomata*.

scotomization (*Psych*) A defence mechanism where the individual fails to perceive consciously parts of the environment or of him- or herself. Term derived from SCOTOMA, a blind spot in the visual field.

scotophilia (*Psych*) Preference for darkness, typical of nocturnal animals.

scotophor (*Electronics*) Material which darkens under electron bombardment, used for screen of cathode-ray tube in storage oscilloscopes. Recovers upon heating. Usually potassium chloride.

scotopic luminosity curve (*Phys*) The curve giving relative brightness of the radiations in an equal-energy spectrum when seen at a very low intensity level. See PHOTOPIC LUMINOSITY CURVE.

scotopic vision (*Phys*) Vision which occurs at low illumination levels through the medium of the retinal RODS. Cf PHOTOPIC VISION.

Scots fir (*For*) Synonym for SCOTS PINE.

Scots pine (*For*) Softwood (*Pinus sylvestris*) native to many parts of Europe and the best-known UK species for producing commercial timbers. The heartwood is a pale reddish-brown and resinous, with well-defined growth rings. Also *Baltic redwood, red deal, Scots fir, Scots pine, yellow deal*.

Scott connection (*ElecEng*) A method of connecting two single-phase transformers so as to convert a three-phase three-wire ac supply to a two-phase three-wire supply, and vice versa.

Scottish topaz (*Min*) A term applied in the gemstone trade to yellow transparent quartzes, resembling Brazilian topaz in colour, used for ornamental purposes. Not a true topaz. See CAIRNGORM, CITRINE.

scourer (*Eng*) A flour-milling machine in which the wheat, for cleaning purposes, is subjected to the action of revolving beaters in a ventilated casing.

scouring (*Agri*) A condition in which animals produce copius watery feces, often causing soiling of the perianal region. May be due to diet (lush new grass) or infection.

scouring (*Genrl*) Said of the eroding action of water flowing at high velocity.

scouring (*Textiles*) Processes for removing from textile materials fats, oils, waxes and other impurities which may be natural (eg wool grease) or which may have been added to aid processing (eg SIZE). For cotton and flax, boiling aqueous sodium hydroxide solutions are used but other fibres require milder treatments eg with aqueous sodium carbonate or neutral solutions of detergent (for wool).

SCR (*Electronics*) Abbrev for SILICON-CONTROLLED RECTI-FIER.

scram (*NucEng*) General term for emergency shutdown of a plant, esp of a reactor when the safety rods are automatically and rapidly inserted to stop the fission process.

scrambler (*ICT*) Multiple modulating and demodulating system that interchanges and/or inverts bands of speech, so that speech in transmission cannot be intelligible, the reverse process restoring normal speech at receiving end.

scram rod (*NucEng*) An emergency safety rod used in a reactor.

scrap (*Eng*) (1) Materials reclaimed from products at the end of their specification life cycle or product life cycle. Scrap metal, glass and ceramic, paper, textile and some polymers are routinely reclaimed from used products. The extent of recycling depends on raw material or intermediate prices as well as the degree of mixing of different materials (and hence the costs of separation). See TRAMP ELEMENT. (2) Defective products unfit for sale.

scraper (*Build*) A thin flat steel blade with a square straightedge on which a burr is raised; used to pare wood from a surface which is being finally dressed.

scraper (*CivEng*) An earth-moving machine either towed or self-propelled, essentially a box on wheels which scrapes off earth to fill the box, moves and deposits it at the discharge point.

scraper board (*Print*) A coated cardboard, the coating being either white on black or black on white; it can be readily scraped away to expose the underlayer. Used by commercial artists to obtain the effect of wood engraving, the result being made into a printing plate photomechanically.

scraper plane (*Build*) Tool shaped like a SPOKESHAVE but with a thin blade set at a slight angle from the perpendicular. Used for cleaning up hardwood and removing tears left after ordinary planing.

scraper ring (*Autos*) A ring usually fitted on the skirt of a petrol- or oil-engine piston, to prevent excessive oil consumption. It may have a bevelled upper edge or a slotted groove, the oil being scraped off the cylinder wall and led back to the sump through holes in the piston wall. Also *oil control ring*.

scrapie (*Vet*) Progressive, fatal nervous disease of sheep and goats, recognized for the last 300 years. The cause is not fully understood but is probably a proteinaceous infectious particle or prion. Characterized by trembling, pruritus, wool loss, staggering and loss of condition. Incubation period can be several years. See panel on TRANSMISSIBLE SPONGIFORM ENCEPHALOPATHY.

scratch (*Build*) A tool with an upright shaped blade fixed in a wooden body; used for working small mouldings.

scratch (*ICT*) Release of storage area for subsequent reuse.

scratch-coat (*Build*) The first of three coats applied in plastering. It consists of coarse stuff.

scratcher (*Build*) A tool used to make scratch marks in a cement surface to provide a grip for a subsequent coat.

scratch file (*ICT*) Temporary storage area, usually held in backing store, for use by a program during execution. Also *work file*.

scratch pad (*ICT*) (1) Section of immediate access store reserved for temporary information for immediate use. Used to transfer data between application programs. Also *working store*. (2) A facility on a mobile phone that enables the user to record information in the phone's memory during a call .

scratch tape (*ICT*) Magnetic tape used for a scratch file.

scray (*Textiles*) A trough, probably on wheels, in which fabric (wet or dry) is collected and taken to the next process.

scree (*Geol*) The accumulation of rock debris strewn on a hillside or at a mountain foot, resulting from mechanical weathering of rocks.

screeching (*Aero*) A cacophonous form of unstable combustion that can occur with rockets, and occasionally in turbine engines, causing very rapid damage due to resonance stresses on the jet pipe or nozzle. Also *howl*.

screed (*Build*) (1) A band of plaster laid on the surface of a wall as a guide to the thickness of a coat of plaster to be applied subsequently. (2) A layer of concrete or mortar used to provide a finish to a concrete floor or the gradient on a floor or flat roof.

screed (*CivEng*) A horizontal beam which is moved on screed rails to give a level or shaped finish to a concrete slab, commonly fitted with vibrators.

screed-coat (*Build*) A coat laid level with the screeds.

screen (*BioSci*) To investigate a large number of organisms for the presence of a particular property as in screening for a mutation for *antibiotic resistance*.

screen (*Build, CivEng*) A large sieve used for grading fine or coarse aggregates.

screen (*ElecEng*) Electrode interposed between two other electrodes, to reduce the electrostatic capacitance between them. See ELECTROSTATIC SHIELD, FARADAY CAGE.

screen (*ImageTech, Print*) (1) The surface on which a picture is presented by projection of a photographic or electronic image. (2) The meshwork of lines at right angles, ruled on glass, used to convert the subject of a half-tone illustration into dots for photomechanical reproduction.

screen (*MinExt*) Perforated or woven cloths (metal, fibre, rods, bars), used to size ore or products as part of treatment required to regulate concentration. Include *grizzlys*, *trommels*, and mechanically and electrically vibrated screens. Also *screening*.

screen burning (*Electronics*) Gradual falling off in luminosity, sometimes accompanied by discoloration, in the fluorescent screen of a cathode-ray tube, particularly if operated under adverse conditions.

screen capture (*ICT*) See SCREEN DUMP.

screen dump (*ICT*) To print out or save as a file a representation of current contents of the screen. Also *screen capture*. See DUMP.

screened-grid valve (*Electronics*) Four-electrode valve, with cathode, control grid, screen and anode. Used as a high-frequency amplifier, where the screen, of unvarying potential, prevents positive feedback, and so greatly enhances stability. See TETRODE.

screened horn balance (*Aero*) A HORN BALANCE which is screened from the airflow by the fixed surface in front of it.

screen editing (*ICT*) Changing stored data or programs by altering text displayed on a video display unit, using a CURSOR to indicate the position.

screened wiring (*ElecEng*) Insulated conductors enclosed in a sheath of woven or other form of wire, connected to earth and used to prevent induction and for mechanical protection. Formerly *steel tube*.

screen font (*ICT*) The FONT that is displayed on the computer screen which approximates to the printer font, ie the one that will appear on the printer document.

screening (*ElecEng*) Use of a screen, in the form of a metal or gauze can, normally earthed, so that electrostatic effects inside are not evident outside, or vice versa.

screening (*Med*) The performance of various tests on a large population of apparently healthy people in order to identify the presence of certain diseases or disorders.

screening (*MinExt*) See SCREEN.

screening (*Phys*) An effect in which the nucleus of an atom is screened by its surrounding electrons, resulting in a reduction in the electric field surrounding the nucleus.

screening constant (*Phys*) A number which, when subtracted from the atomic number (Z) of an atom, gives the *effective* atomic number so that X-ray spectra may be described by 'hydrogen-like' formulae. It arises from the nuclear charge ($+Ze$) being screened by the inner electron shells.

screening protector (*ElecEng*) See LINE CHOKING COIL.

screenings (*Build, CivEng*) The residue from a sieving operation.

screen pack (*Eng*) Gauze fitted behind polymer extrusion die to filter out impurities. See EXTRUSION.

screen printing (*Textiles*) A process in which the coloured printing pastes are forced by a squeegee through selected meshes (left open) of screens onto the face of the fabric being printed. The screens, which may be flat or cylindrical, are made of a woven fabric or of metallic wire mesh and one is required for each colour.

screen process printing (*Print*) See SILK-SCREEN PRINTING.

screen-protected motor (*ElecEng*) A protected type of electric motor in which the openings for ventilation are covered with wire-mesh screens.

screen saver (*ICT*) A program which, after a period of user inactivity, temporarily replaces the screen image with, usually, a moving pattern to prevent the screen elements from burning out.

screw (*Eng*) (1) The common fastening with a threaded shank. (2) In plastics technology a steel cylinder with external flight(s) shaped in a helical form, so that when rotated in cylinder of injection-moulding or extrusion machine, the screw will transport molten polymer through the nozzle. Sometimes known as an *Archimedes screw*, it is also widely used for transporting powder and other materials or fluid (eg water).

screw-and-nut steering gear (*Autos*) One in which a square-threaded screw formed on the lower end of the steering column engages with a nut provided with trunnions, which work in blocks sliding in a short slotted arm, connecting with the remainder of the steering system.

screw-auger (*Build*) An auger having a helical groove cut in its surface so as to carry away the chips from the cutting edge.

screw axes (*Min*) Axes of symmetry about which the atoms in a mineral are symmetrically disposed. Rotation about eg a fourfold screw axis will carry an atom 1 into the positions successively occupied by similar atoms 2, 3 and 4, after rotations of 90°, 180°, 270° and 360°. Cf ROTATION AXES OF SYMMETRY.

screw box and tap (*Build*) A device for making wooden screws on furniture legs, wooden cramps, etc.

screw characteristics (*Eng*) Geometric properties of injection-moulding or extrusion screws, which may vary depending on polymer type etc. Includes compression ratio, L/D ratio, flight angle and multiple screws. See EXTRUSION.

screw chases (*Print*) Chases used in newspaper work. They are tightened by screws, which obviate the use of separate quoins.

screw chasing (*Eng*) See CHASER.

screw conveyor (*Eng*) See WORM CONVEYOR.

screw-cutting lathe (*Eng*) A metal turning lathe provided with a lead screw driven by CHANGE WHEELS, for traversing the pointed tool used in screw cutting.

screw dislocation (*Crystal*) A defect within a crystal in which the BURGERS VECTOR is parallel to the line of the dislocation. See DISLOCATION.

screwdriver (*Eng*) A tool with shank terminating in a blade of size and shape to fit the slot in screws. See PHILLIPS SCREWDRIVER, RATCHET SCREWDRIVER.

screwed steel conduit (*ElecEng*) Light steel tubing, having screwed ends for connecting up in lengths by means of sockets, in which electrical installation wiring is run. Cf PLAIN STEEL CONDUIT.

screwing die (*Eng*) An internally threaded hardened steel block, sometimes split in halves, on which cutting edges are formed by longitudinal slots. Held in a stock, lathe or screwing machine for cutting external threads.

screwing machine (*Eng*) A form of lathe adapted for the continuous production of screws or screwed pieces, by means of DIES.

screw jack (*Eng*) See JACK.

screw micrometer (*Eng*) See MICROMETER GAUGE.

screwnail (*Eng*) A nail in whose surface shallow helical depressions are formed, so that as it is driven in place with blows from a hammer it turns like a screw. Also *drivenail*.

screw pile (*CivEng*) A pile having a wide projecting helix or screw at the foot, formerly used on alluvial ground.

screw plate (*Eng*) A hardened steel plate in which a number of screwing dies of different sizes are formed for repairing or sizing screws.

screw plug (*Build*) A drain plug consisting of a rubber ring held between two steel disks which, on being screwed together, force the ring out to close the drainpipe in which the plug is placed.

screw press (*Eng*) See FLY PRESS.

screw propeller (*Eng*) See MARINE SCREW PROPELLER, PROPELLER.

screw shackle (*Eng*) A long nut screwed internally with a right-hand thread at one end and a left-hand thread at the other, serving to connect the ends of two rods which are to be joined together, and providing a means of adjusting the total length. Also *bottle screw, rigging screw, turnbuckle*. See COUPLING.

screw-thread (*Eng*) A helical ridge of approximately triangular (or V), square or rounded section, formed on a cylindrical core, the pitch and core diameter being standardized under various systems. See BRITISH ASSOCIATION SCREW-THREAD, BRITISH STANDARD WHITWORTH SCREW-THREAD, METRIC SCREW-THREAD, SELLERS SCREW-THREAD, UNIFIED SCREW-THREAD, etc.

scribbler (*Textiles*) A machine for carding wool.

scriber (*Eng*) A pointed steel tool used for making an incised mark on timber or metal, to guide a subsequent cutting operation.

scribing block (*Eng*) A tool for gauging the height of some point on a piece of work, above a surface plate or machine table. It consists of a base supporting a pivoted column, to which a scriber is slidably clamped. Also SURFACE GAUGE.

scribing gouge (*Build*) One sharpened with the bevel on the inside or concave face of the cutting edge, used for work where an upright cut is necessary.

scrieve board (*Ships*) A formation of portable portions of flat wooden boards whereon are scrieved (or scribed) the ship's transverse frame sections and lines indicative of shell seams, decks, stringers, etc. Scrieve boards are used for setting the soft iron, to which the frames etc are turned.

scrim (*Build, Textiles*) An open woven fabric used for reinforcement in bookbinding and upholstery, and for the base cloths of certain non-woven fabrics. Sometimes used in building to cover the joints between fibre- or plasterboards before applying a finish, although paper tape is now generally used.

script (*ICT*) A list of commands in a high-level programming language that can be executed by a computer.

script (*ImageTech*) The written outline of a film or video production detailing the settings, action and dialogue for each scene throughout (*scenario*). If camera directions are included it becomes the *shooting script*.

script (*Print*) A style of type which imitates handwriting.

scripting language (*ICT*) A high-level programming language that uses an INTERPRETER to execute programs.

scrobiculate (*BioSci*) Having the surface dotted all over with small rounded depressions; pitted.

scrobiculus (*BioSci*) A small pit or rounded depression.

scrofula (*Med*) Caseating tuberculosis of the lymphatic glands. Adj SCROFULOUS.

scrofuloderma (*Med*) Tuberculous infection of the skin from the bursting of a deep-seated tuberculous abscess; a subcutaneous tuberculous abscess. Also *scrofulodermia*.

scroll (*ImageTech*) Video transition effect in which one picture is displaced vertically by another.

scroll bar (*ICT*) A bar that appears at the edges of a WINDOW usually with a SCROLL BOX within it and arrows displayed at either end. By means of clicking on the arrows or dragging the SCROLL BOX, other parts of the window area can be brought into view.

scroll box (*ICT*) A small box in a SCROLL BAR that is used to adjust the position of information currently displayed within the WINDOW.

scroll chuck (*Eng*) A self-centring chuck for holding round work, having jaws slotted to engage with a raised spiral or scroll on a plate which is rotated by a key, so as to advance the jaws while maintaining their concentricity. Esp THREE-JAW CHUCK.

scrolling (*ICT*) In a VIDEO DISPLAY UNIT displaying text, the action of moving lines up the screen losing old lines from above as new lines appear below.

scroop (*Textiles*) The crunching noise and the corresponding HANDLE that is obtained when certain fabrics, particularly of silk, are crushed in the hand. The noise is rather like that obtained when dry snow is squeezed into balls. Cf RATTLE.

Scrophulariaceae (*BioSci*) A family of c.3000 spp of dicotyledonous flowering plants (superorder Asteridae). Such plants are cosmopolitan and are almost all herbs. The flowers are gamopetalous, the ovary superior and the fruit usually a capsule. Examples include the foxglove from which the important cardiac drug, digitalis, was once derived, and some ornamental plants, eg *Antirrhinum*.

scrotum (*BioSci*) In mammals, a muscular sac forming part of the ventral body wall into which the testes descend. Adj *scrotal*.

scrubber (*ChemEng*) (1) Part of a process plant used to purify gas streams. Contaminated gas is passed upwards through a column, and mixes intimately with a counter-current of liquid solvent, reactant solution or slurry which is passing downwards. This removes the impurities by absorption. Desulphurization of flue gases to remove sulphur dioxide is carried out in this way before the scrubbed gases are discharged into the atmosphere. (2) A kind of *dampener* that uses baffles to change the velocity of a flowing gas stream by altering its direction and flow area.

scrub typhus (*Med*) See TYPHUS.

SCSI (*ICT*) Abbrev for *small computer systems interface*. A standard defined for connecting PERIPHERAL DEVICES such as CD-ROM DRIVES to COMPUTERS.

scuffing (*Eng*) (1) A sign of inadequate lubrication in which ADHESIVE WEAR produces scratches and tears in a mating surface. (2) Surface marks on a moulding running in the direction of mould opening, due to tool wear or damage. A type of MOULD DEFECT.

scull (*Glass*) The glass remaining in a ladle after most of the molten glass has been poured out.

Sculptor (*Astron*) A small faint southern constellation.

scum (*Build*) A surface formation of lime crystals appearing on new cement work.

scumble (*Build*) A light-coloured, low-opacity paint allowing a darker background colour to GRIN THROUGH deliberately. Used in GRAINING.

scumming (*Print*) Defective printing due to ink building up in the non-image area during lithographic printing.

S-curve (*Surv*) See REVERSE CURVE.

scurvy (*Med*) A nutritional disease due to deficiency in the diet of vitamin C (ascorbic acid), characterized by anaemia, apathy, sponginess of the gums, ulceration of the mouth and haemorrhages into the skin. Lime juice was described by James Lind (1716–94) as a cure. Also *scorbutus*.

scutch (*Build*) The bricklayer's cutting tool for dressing bricks to special shapes. Also *scotch*.

scutching (*Textiles*) (1) The process of mechanically separating and cleaning closely packed cotton fibres before forming a yarn. (2) The process of separating flax fibres from the woody part of deseeded or retted flax straw.

scute (*BioSci*) An exoskeletal scale or plate. Adj *scutate*.

scutellum (*BioSci*) More or less shield-shaped structure, possibly a modified cotyledon, attached to the side of the embryo in a grass grain, and which at germination secretes hydrolytic enzymes into, and absorbs sugars etc from, the adjacent endosperm.

Scutum (Shield) (*Astron*) A small southern constellation in the Milky Way.

scybalum (*Med*) A round, hard and dry fecal mass in the intestine.

Scyphomedusae (*BioSci*) See SCYPHOZOA.

Scyphozoa (*BioSci*) A class of Cnidaria in which the polyp stage is inconspicuous and may be completely absent. Where present called a scyphistoma which gives rise to the ephyra larvae. A velum and nerve ring are generally absent, the gonads are endodermal and there is no skeleton. Jellyfish.

Scythian (*Geol*) The oldest epoch of the Triassic period.

SDD (*ICT*) Abbrev for SUBSCRIBER DIRECT DIALLING.

SDI (*ImageTech*) Abbrev for SERIAL DIGITAL INTERFACE.

SDI (*Space*) Abbrev for *Strategic Defense Initiative*, a US military programme commonly referred to as Star Wars. Intended to provide a defensive shield for the destruction of hostile ballistic missiles; discontinued in 1993.

SDS (*BioSci*) See SODIUM DODECYL SULPHATE.

SDS gel electrophoresis (*BioSci*) POLYACRYLAMIDE GEL ELECTROPHORESIS in the presence of the denaturant SDS that binds to protein and confers negative charge in proportion to the molecular weight.

SDSL (*ICT*) Abbrev for *symmetric digital subscriber line*, a telephone connection that carries ingoing and outgoing messages at the same speed. Cf ADSL.

SE (*Build*) Abbrev for *stopped end*.

Se (*Chem*) Symbol for SELENIUM.

sea anchor (*Ships*) A float to which a ship may be attached by a hawser to ride out a gale. Used in small boats in the form of a canvas DROGUE.

sea breeze (*EnvSci*) See LAND AND SEA BREEZES.

sea cell (*Phys*) Primary electrolyte cell which functions as a source of electric power when immersed in sea water. A battery of such cells is possible, even though cells are partially short-circuited by the sea water. Fitted eg to life-belts, so that an indicating light is produced automatically for use at night.

sea clutter (*Radar*) CLUTTER generated by rough sea surfaces; potential source of difficulty with sea-skimming guided missiles.

sea-floor spreading (*Geol*) The process by which new oceanic crust is generated at oceanic ridges by the convective upwelling of magma. The plates on either side of the *divergent junction* move very gradually apart. See MAGNETIC ANOMALY PATTERNS and panel on PLATE TECTONICS.

seal (*Build*) The water contained in a trap, which prevents the flow of air or gases from one side to the other. Also *water seal*.

seal (*Electronics*) In a vacuum tube, the point at which the tube is closed after pumping, and the act of closing off.

seal (*Print*) A small printing plate usually used to indicate the edition of a newspaper. See SEAL CYLINDER, SEAL UNIT.

seal cylinder (*Print*) The auxiliary cylinder to which the *seal* is attached.

sealed cover (*Build*) An airtight cast-iron or precast concrete cover fitting into a frame to cover a manhole.

sealed face production line (*NucEng*) A method for handling and fabricating radioactive material in which the remotely controlled equipment is sited behind a continuous impervious face with suitable inspection and other ports, as in the figure below. Materials are brought to the site in sealed containers and automatically transferred to the production line and, after fabrication, moved to sealed containers for any further operations. Alternative to separate GLOVE BOXES. Fig. ▷

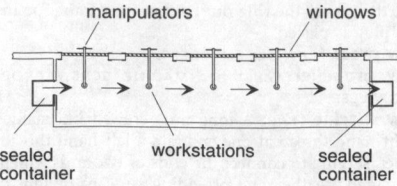

sealed face production line Arrows show the movement of the work.

sealed pressure balance (*Aero*) An AERODYNAMIC BALANCE, used mainly on ailerons, consisting of a continuous projection forwards of the hinge line within a cavity formed by close-fitting SHROUDS projecting rearwards from the main surface, the gap between the balance and main surface being sealed to prevent communication of pressure between lower and upper surfaces. Sometimes called a *Westland–Irving balance* after its inventor and the company which developed it.

sealed source (*Radiol*) A radioactive source for eg medical or calibration purposes, in a radiation leakproof container. It is opened by remote control only in its shielded site of operation.

sea level (*Surv*) The datum line from which heights are measured in surveying; the term is also used more loosely to denote the mean surface level of the oceans or the surface of the geoid.

sea-level pressure (*EnvSci*) Atmospheric pressure at mean sea level deduced from the pressure measured at the level of the observing station by taking into account the theoretical effect of a column between the two levels. Use of sea-level pressure on synoptic charts reveals the true meteorological patterns which would otherwise be totally obscured by the effect of altitude on the observations.

sea-level static thrust (*Aero*) See STATIC THRUST.

sea lily (*BioSci*) See CRINOIDEA.

sealing box (*ElecEng*) A box in which the end of a paper-insulated cable is hermetically sealed.

sealing-in (*ElecEng*) The making of an airtight joint between the filament wires and the glass envelope of an incandescent lamp or vacuum tube.

sealing-off (*ElecEng*) The final sealing of the exit to the evacuating pumps of an incandescent lamp or vacuum tube.

seal plug (*NucEng*) A removable plug at the end of the coolant tubes in a CANDU-type reactor which allows access for the refuelling machine.

seal unit (*Print*) The small auxiliary printing unit used for printing the *seal*, usually in a second colour on front page.

seam (*Build*) See WELT.

seam (*Eng*) (1) A surface defect in worked metal, the result of a blowhole being closed but not welded; it remains as a fine crack. (2) A ridge in casting, the effect of enclosure of impurity or scale in working.

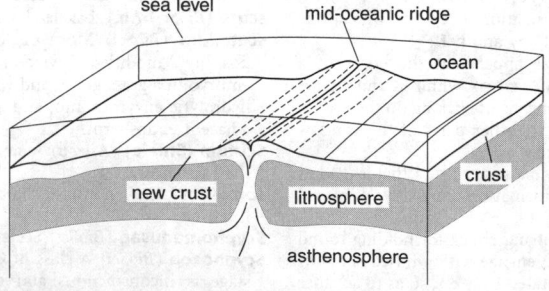

sea-floor spreading Volcanic rock welling up at ridge.

seam (*MinExt*) (1) A tabular, generally flat deposit of coal or mineral; a stratum or bed. (2) A joint or fissure in a coal bed.

sea marker (*Aero*) Any device dropped from an aircraft onto water to make an observable patch from which the drift of the aircraft may be determined. Usually filled with a fluorescent substance for use during the day, and with a flame-producing device for night use.

seamer (*FoodSci*) Machine used in the canning of foods to form the DOUBLE SEAM between the can body and end after filling.

seaming machine (*Eng*) A press for forming and closing longitudinal and circumferential interlocking joints used in the manufacture of containers from sheet metal.

seamless tube (*Eng*) Tube other than that made by bending over and welding the edges of flat strip. May be made by extrusion (non-ferrous metals), by piercing a hole through a billet and then rolling down over a mandrel to form a tube of the required dimensions, or by the MANNESMANN PROCESS.

seamount (*Geol*) An elevation from the ocean floor which may be flat topped (a *guyot*) or peaked (a *sea peak*).

seam roll (*Build*) See HOLLOW ROLL.

seam welding (*ElecEng*) (1) See RESISTANCE SEAM-WELDING. (2) Uniting sheet plastic by heat arising from dielectric loss, the electric field being applied by electrodes carrying a high-frequency displacement current. Also *high-frequency welding*, *jig welding*.

seaplane (*Aero*) An aircraft fitted with means for taking off from and alighting on water. See FLOAT SEAPLANE, FLYING BOAT, HYDROSKIS.

seaplane tank (*Aero*) A long, narrow water tank with a powered carriage carrying equipment by which the water performance of a seaplane model can be observed and precisely measured.

search coil (*ElecEng*) See EXPLORING COIL.

search engine (*ICT*) Information-retrieval software that can search the Internet for words or phrases entered by the user. It has access to a database of previous searches to save time. See panel on INTERNET.

search image (*BioSci*) In behavioural ecology, a predator's pre-conception of what its prey looks like and where it is found.

search image (*Psych*) The perceptual phenomena of an increased accuracy of discrimination for certain objects in the environment, eg a predator's improved ability to see camouflaged prey against its background.

search radar (*Radar*) Radar designed to cover a large volume of space and to give a rapid indication of any target which enters it.

search strategy (*ICT*) A systematic procedure for retrieving a predefined item from among data. See BINARY SEARCH, DATA RETRIEVAL, INFORMATION RETRIEVAL, SEQUENTIAL ACCESS, SQL.

search, vote and lock (*ICT*) A technique for wide area coverage in PRIVATE MOBILE RADIO, in which a number of base stations radiate the same message, each on a different frequency, and the mobile unit scans the available frequencies to find the base station giving the best received signal before locking on to that frequency.

seasonal affective disorder (*Psych*) Mood disorder characterized by depression, lethargy and sleep disturbances usually occurring during winter months. Treatment with high-intensity light may reduce the severity of the disorder. Abbrev *SAD*.

season cracking (*Eng*) A form of STRESS CORROSION cracking first experienced in the late 19th century in brass cartridge cases during the subtropical monsoon season, caused by a combination of high residual stress from the forming process and a mildly corrosive environment containing traces of ammonia. The cracking can be overcome by a simple stress-relieving anneal immediately after manufacture.

seasoning (*For*) The process in which the moisture content of timber is brought down to an amount suitable for the purpose for which the timber is to be used. See KILN DRYING.

seat earth (*Geol*) A fossil soil which underlies a coal seam. Also *seggar*.

seating (*Eng*) A surface for the support of another piece, eg the end of a girder, or a masonry block.

seaweed (*BioSci*) Macroscopic marine alga. Most seaweeds belong to the Phaeophyceae or Rhodophyceae, some to the Chlorophyceae.

sebaceous (*BioSci*) Producing or containing fatty material, as the *sebaceous glands* of the scalp in humans.

sebaceous cyst (*Med*) A cyst formed as a result of blockage of the duct of a sebaceous gland, often present on the face, scalp or neck.

sebacic acid (*Chem*) $HOOC–(CH_2)_6–COOH$, a dibasic acid. Mp 129°C. Obtained by heating castor oil with sodium hydroxide. A white crystalline solid. Its esters are used in the production of resins and plasticizers.

sebiferous (*BioSci*) Conveying fatty material.

sebiparous (*BioSci*) Sebaceous.

seborrhea (*Med*) Adj *seborrheic*. US for SEBORRHOEA, SEBORRHOEIC.

seborrhoea (*Med*) Overactivity of the sebaceous gland, resulting in an abnormally greasy skin. US *seborrhea*.

seborrhoeic dermatitis (*Med*) An inflammatory disease of the skin characterized by the presence of reddish patches covered with greasy scales; esp of the scalp, causing 'scurfy head'. US *seborrheic dermatitis*.

sebum (*BioSci*) The fatty secretion, produced by the sebaceous glands, which protects and lubricates hair and skin.

sec (*MathSci*) See TRIGONOMETRICAL FUNCTIONS.

SECAM (*ImageTech*) The colour TV system adopted in France and E Europe, from *SEquential Couleur A Memoire*.

secant (*MathSci*) (1) See TRIGONOMETRICAL FUNCTIONS. (2) A line joining any two points on a curve.

secant modulus (*Eng*) ELASTIC (or quasi-elastic) MODULUS derived from a non-linear STRESS–STRAIN CURVE by taking the ratio of the stress to the strain at a particular point on the curve, which must be specified in terms of the level of stress or strain, eg '0·5% strain secant modulus'. The creep moduli of polymers are almost invariably secant moduli. Cf TANGENT MODULUS.

secodont (*BioSci*) Having teeth adapted for cutting.

second (*Genrl*) (1) 1/60 of a minute of time, or 1/86 400 of the mean solar day; once defined as the fraction 1/31 556 925·9747 of the tropical year for the epoch 1900 January 0 at 12 h ET. Since 1965 defined, in terms of the resonance vibration of the caesium-133 atom, as the interval occupied by 9 192 631 770 cycles. This was adopted in 1967 as the SI unit of time interval. Symbol s. (2) Unit of angular measure, equal to 1/60 of a minute of arc; indicated by the symbol ". Also *arcsecond*. (3) In duodecimal notation $\frac{1}{12}$ of an inch; indicated by '''. (4) Unit for expressing flow times in capillary viscometers (Redwood, Saybolt Universal, or Engler), eg an Engler second is that viscosity which allows 200 cm³ of fluid through an Engler viscometer in 1 second.

secondary (*BioSci*) (1) Arising later; of subsidiary importance. (2) In insects, the hind-wing. (3) In birds, a quill feather attached to the forearm, and also called the *cubital*.

secondary alcohols (*Chem*) Alcohols containing the group –CH(OH)–. When oxidized they yield ketones (alkanones).

secondary amines (*Chem*) Amines containing the imino (–NH) group. They yield nitrosoamines with nitrous acid (nitric (III) acid).

secondary battery (*ElecEng*) A number of secondary cells connected to give a larger voltage or a larger current than a single cell.

secondary beam (*Build*) In floor construction, a beam carried by MAIN BEAMS and transmitting loads to them.

secondary body cavity (*BioSci*) See COELOM.

secondary bonds (*Chem*) A term usually applied to the weak van der Waals and hydrogen bonds in materials. See panel on BONDING.

secondary bow (*EnvSci*) A *rainbow* having an angular radius of 52°, the red being inside and the blue outside, usually fainter than the primary bow. It is produced in a manner similar to the primary bow except that two internal reflections occur in the raindrops.

secondary carbon atom (*Chem*) Carbon atom linked to two other carbon atoms and two hydrogen atoms with stability between the more stable primary and less stable tertiary atoms.

secondary cell (*ElecEng*) See ACCUMULATOR.

secondary cell wall (*BioSci*) See SECONDARY WALL.

secondary circuit (*Build*) Pipe circuit in which water circulates to and from a hot-water storage vessel.

secondary coil (*ElecEng*) A coil which links the flux produced by a current flowing in another coil (*primary coil*).

secondary colours (*Genrl*) Colours produced by mixing primary colours.

secondary constants (*ElecEng*) Those for a transmission line which are derived from the *primary constants*. They are the *characteristic impedance* (*impedance level*), as of an infinite line, and the *propagation constant* (*attenuation and phase delay constant*).

secondary constriction (*BioSci*) A non-centromeric constriction of the chromosomes, often at the site of the NUCLEOLAR ORGANIZING REGION.

secondary coolant (*NucEng*) In a reactor, a separate stream of coolant which is converted to steam by the PRIMARY COOLANT in a heat exchanger (steam generator) to power the turbine.

secondary depression (*EnvSci*) A DEPRESSION embedded in the circulation of a larger primary depression.

secondary dispersion (*MinExt*) In GEOCHEMICAL PROSPECTING, the dispersion of elements or minerals from an ore body or HALO by physical agents such as stream, river or groundwater flow, glacial ice, wind or wave action.

secondary electrode (*ElecEng*) See BIPOLAR ELECTRODE.

secondary electrons (*Electronics*) Those given off during the process of SECONDARY EMISSION.

secondary emission (*Electronics*) The emission of electrons from a surface (usually conducting) by the bombardment of the surface by electrons from another source; the number may greatly exceed that of the primaries, depending on the velocity of the latter and the nature of the surface.

secondary energy (*Genrl*) Energy, such as electricity, provided through the conversion of a raw source (eg crude oil, natural gas or coal).

secondary enrichment (*Geol*) The addition of minerals to, or the change in the composition of the original minerals in, an ore body, either by precipitation from downward-percolating waters or by upward-moving gases and solutions. The net result of the changes is an increase in the amount of metal present in the ore at the level of secondary enrichment.

secondary gneissic banding (*Geol*) A prominent mineral banding exhibited by coarse-grained crystalline rocks which have been subjected to intense regional metamorphism, involving rock-flowage. Often it is difficult to distinguish from PRIMARY GNEISSIC BANDING.

secondary growth (*BioSci*) See SECONDARY THICKENING (1).

secondary hardness (*Eng*) Further increase in hardness produced in tempering high-speed steel after quenching.

secondary immune response (*BioSci*) The response of the body to an antigen with which it has already been primed (see PRIMARY IMMUNE RESPONSE). There is a very rapid production of large amounts of antibody over a few days, followed by a slow exponential fall. The response of CELL-MEDIATED IMMUNITY follows a similar pattern. See panel on IMMUNE RESPONSE.

secondary ion mass spectrometry (*Phys*) A surface analytical technique in which sputtering by primary ions leads to the ejection of some surface species as secondary ions. Subsequent mass analysis is used to identify the secondary ions. Abbrev SIMS.

secondary leakage (*ElecEng*) The magnetic leakage associated with the secondary winding of a transformer.

secondary memory (*ICT*) See BACKING STORE.

secondary meristem (*BioSci*) (1) Meristem producing secondary tissues. (2) Meristem derived by dedifferentiation from differentiated cells, eg the interfasicular cambium, cork cambium.

secondary metabolites (*BioSci*) Applied to those compounds which do not function directly in biochemical activities like photosynthesis, respiration and protein synthesis which support growth. They include alkaloids, terpenoids, flavonoids which may function in defence against insects, fungi and herbivores, in ALLELOPATHY or as attractants to pollinators or fructivores.

secondary metal (*Eng*) Pure metal or alloy retrieved from engineering scrap and residues refined and worked up for return to manufacturing industry. Distinguished from primary metal produced directly from ore and appearing in metallic form for the first time.

secondary mineral (*Geol, MinExt*) (1) One formed after the formation of the rock enclosing it. (2) That of minor interest in an ore body undergoing exploitation.

secondary nitro-compounds (*Chem*) Nitro-compounds containing the $-CH(NO_2)$ group.

secondary phloem (*BioSci*) Phloem formed by the activity of a cambium.

secondary pollutant (*EnvSci*) A polluting compound or RADICAL not emitted in that form, but produced by the chemical reaction of other pollutants already existing in the environment, eg polluting ozone is formed from nitrogen oxides and hydrocarbons emitted from car engines.

secondary-process thinking (*Psych*) In psychoanalytic theory, logic-bound thinking, governed by the REALITY PRINCIPLE (*ego*); it involves the ego-functions of remembering, reasoning and evaluation that mediate between instinctual needs and adaptation to the external world.

secondary production (*MinExt*) That in which means like pumping gas or water into an oil reservoir are needed to assist the flow to the well bore. Also *secondary recovery*. Cf PRIMARY PRODUCTION, TERTIARY PRODUCTION.

secondary radar (*Radar*) Radar which involves transmission of a second signal when the incident signal triggers a TRANSPONDER beacon.

secondary radiation (*Phys*) Radiation produced by interaction of primary radiation and an absorption medium.

secondary recycling (*Genrl*) See POST-CONSUMER RECYCLING.

secondary reinforcement (*Psych*) An initially neutral stimulus that acquires reinforcing properties through pairing with another stimulus that is already reinforcing.

secondary service area (*ICT*) That surrounding a radio or TV broadcasting station where satisfactory reception is not guaranteed, and is only possible given good transmission conditions or a favourable location.

secondary sexual characters (*BioSci*) Features that distinguish between the sexes other than the reproductive organs.

secondary shutdown system (*NucEng*) In the unlikely event of failure of control rods to enter and shut down reactor, a system such as, in a gas-cooled reactor, the insertion of nitrogen which absorbs neutrons much more strongly than does carbon dioxide. See EMERGENCY SHUTDOWN SYSTEM.

secondary spectrum (*Phys*) The residual longitudinal chromatic aberration in a lens corrected to bring two wavelengths to the same focus.

secondary standard (*Genrl*) A copy of a PRIMARY STANDARD for general use in a standardizing laboratory.

secondary stress (*Eng*) A bending stress, resulting from deflection, as distinct from a direct tensile or compressive stress. Obsolete.

secondary structure (*BioSci*) The first level of three-dimensional folding of the backbone of a polymer. Thus a polypeptide chain can be folded into an α-helix or β-pleated sheet and the nucleotide chains of DNA into a double helix.

secondary substances (*BioSci*) Plant biochemicals which are involved in no known biosynthetic pathways, but which are often detected in high concentrations in leaves and other organs, and so are presumed to be chemical defences.

secondary succession (*BioSci*) A succession proceeding in an area from which a previous community has been removed, eg a ploughed field. Cf PRIMARY SUCCESSION.

secondary surveillance radar (*Radar*) See ATCRBS. Abbrev SSR.

secondary thickening (*BioSci*) (1) The increase in girth of a stem or root that results from the activity of a cambium after elongation has ceased. (2) Confusingly, the formation by a cell of a secondary wall.

secondary transitions (*Plastics*) Transitions in polymers secondary to T_g or T_m, such as those involving increase in side chain flexibility. Often detected in accurate damping experiments.

secondary voltage (*ElecEng*) The voltage at the terminals of the secondary winding of a transformer.

secondary wall (*BioSci*) A later-formed part of the plant cell wall (sometimes the major part) laid down on the cytoplasmic side of the primary wall after cell expansion has ceased, typically richer than the primary wall in cellulose. See panel on CELL WALL.

secondary wave (*Geol*) See panel on EARTHQUAKE.

secondary wave (*ICT*) A wave deriving from the main or desired wave forming a communication link but arising when this wave is partially reflected, refracted or scattered.

secondary winding (*ElecEng*) A winding which links the flux produced by a current flowing in another winding, ie the primary winding.

secondary xylem (*BioSci*) Xylem formed by the activity of a cambium. Wood.

second-channel interference (*ICT*) That due to a signal coinciding with the IMAGE FREQUENCY of a SUPERHET radio receiver.

second detector (*ICT*) In superhet receivers, a detector that demodulates the received signal after passing through the intermediate-frequency amplifier.

second development (*ImageTech*) The second development in a reversal process, after the first image has been removed by bleaching and the remaining silver halide rendered developable.

second-generation computer (*ICT*) One produced around 1955–64 when the valve was being replaced by the TRANSISTOR which consumed less power and was much more reliable. Other important changes included the handling of input/output without involving the central processor, and the development of HIGH-LEVEL LANGUAGES and FLOATING-POINT NOTATION. Machines included IBM 7090 and 7094, Atlas. See COMPUTER GENERATIONS.

second isomorphism theorem (*MathSci*) Theorem which states that if H is a normal subgroup of G and K is a normal subgroup of H, then the quotient group G/H is isomorphic to $(G/K)/(H/K)$.

second messenger (*BioSci*) The intracellular signalling substance produced in response to the binding of hormones or growth factors to receptors on the cell surface. The effect may be to amplify the signal and also allows for integration of signals from different hormones. Major examples are cyclic AMP, inositol phosphate and diacyl glycerol.

second moment of area (*Eng*) The second moment of area is a measure of resistance to bending of a loaded section. A plank stood on edge is less easily bent when loaded than one which has the wide dimension horizontal. This is because the second moment of area on edge is very much greater than that when the plank is laid flat. The second moment of area, I, about an axis xx in the same plane is defined as

$$I_{xx} = \int_A y^2 \, dA$$

where dA is an element of the total area and y the perpendicular distance from xx. I_{xx} can also be obtained from the second moment of area about a parallel axis through the centroid of the area, I_{gg}, by the equation $I_{xx} = I_{gg} + Ah_2$, where A is the total area and h the perpendicular distance between the two axes. Cf POLAR SECOND MOMENT OF AREA.

second-operation work (*Eng*) Machining work carried out after cutting from bar or other material and for which a second clamping operation is necessary.

seconds (*Build*) Bricks similar to CUTTERS but of a slightly uneven colour.

second tap (*Eng*) A tap used, after a taper tap, to carry the full thread diameter further down the hole, or to give the finished size of thread in a through hole.

second ventricle (*BioSci*) In vertebrates, the cavity of the right lobe of the cerebrum.

secrecy system (*ICT*) Privacy system. See INVERTER, SCRAMBLER.

secretagogue (*Med*) A substance, eg a hormone, that stimulates secretion.

secret dovetail (*Build*) An angle joint between two timbers in which neither shows end grain, the visible external parts being mitred, while the dovetails are kept back from both faces. Also *dovetail mitre, mitre dovetail*.

secretin (*BioSci*) A polypeptide hormone secreted by the intestinal wall which stimulates the pancreas to secrete bicarbonate ions.

secretion (*BioSci*) The release of synthesized product or ions from cells. Release may be of membrane-bounded vesicles (merocrine secretion) or of vesicle content following fusion of the vesicle with the plasma membrane (apocrine secretion). In holocrine secretion in eg sebaceous glands, whole cells are released.

secretor (*BioSci*) A person who secretes ABO blood group substances into mucous secretions such as gastric juice, saliva and ovarian cyst fluid. Over 80% of humans are secretors. The status is genetically determined.

secretory (*BioSci*) Secretion-forming.

secretory duct (*BioSci*) A duct containing material secreted by its lining of epithelial cells, eg resin duct.

secretory piece (*BioSci*) A large polypeptide attached to dimers of the secreted form of IgA. It has strong affinity for mucus, thus prolonging retention on mucous surfaces, and it may inhibit destruction of IgA by peptidases in the gut.

section (*BioSci*) (1) A thin slice of biological or mineral material sufficiently transparent to be studied with the light or electron microscope or its virtual equivalent in magnetic resonance imaging etc. (2) A taxonomic group, esp a subdivision of a genus, but sometimes of a higher rank.

section (*ICT*) Unit of a ladder network, derived through design techniques to give specified transmission performance with respect to frequency.

section (*Print*) A reference mark (§) directing the reader's attention to a footnote.

section (*Surv*) (1) The representation to scale of the variations in level of the ground surface along any particular line. (2) Drawing which shows a plane through a solid object, succession of geological strata, mine, etc.

section gap (*ElecEng*) An arrangement for dividing the overhead contact wire of an electric traction system into sections, both electrically and mechanically, without interfering with the smooth passage of the current collector. Also *overlap span*.

section modulus (*Eng*) The ratio of SECOND MOMENT OF AREA to distance of the farthest stressed element from the NEUTRAL AXIS. It is an important property of structural members, for calculating bending stresses. Symbol Z. Also, loosely, *elastic modulus*. Units cm³.

section mould (*Build*) A templet with a profile corresponding to the required shape and used for marking the ends of the timber and for checking progress.

section switch (*ElecEng*) A switch whose function is to connect or disconnect two sections of an electric circuit, generally two bus-bar sections.

sector (*ICT*) Smallest addressable portion of the track on a magnetic tape, disk or drumstore. See fig. at HARD DISK.

sector (*MathSci*) A plane figure enclosed by two radii of a circle (or of an ellipse) and the arc cut off by them.

sectoral chimera (*BioSci*) A chimera in which one component forms a longitudinal strip of tissue down a shoot, or leaf. The strip includes more than one layer of the tunica and/or corpus but does not include the apical meristem.

sectoral horn (*Acous*) Waveguide horn with two surfaces parallel to side of guide and the other two flared out, classified according to whether the flaring is in the plane of the electric or magnetic field.

sector disk (*Phys*) Rotating disk with angular sector removed, interposed in path of beam of radiation. Used to produce known attenuation, or to chop or modulate intensity of transmitted beam.

sector display (*Radar*) A form of radar display used with continuously rotating antennas, *not* with sector scanning. (So termed because the display uses a long-persistence cathode-ray tube excited only when the antenna is directed into a sector from which a reflected signal is received.) Cf SECTOR SCAN.

sectorial (*BioSci*) Adapted for cutting.

sector regulator (*CivEng*) A form of DRUM WEIR. It consists of a hollow reinforced concrete sector of a cylinder placed transversely across the direction of flow and capable of rotation about a horizontal axis on the downstream side, with accommodation for the sector in a special pit in the bed of the stream.

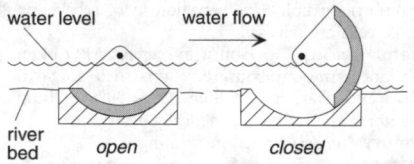

sector regulator

sector scan (*Radar*) Scan in which the antenna moves through only a limited sector. (Not to be confused with a SECTOR DISPLAY.)

sectroid (*Build*) The curved surface between adjacent groins on a vault surface.

secular acceleration (*Astron*) A non-periodic term in the mathematical expression for the Moon's motion by which the mean motion increases approx $11''$ per century; caused by perturbations and by tidal friction in shallow seas.

secular changes (*Geol*) Changes which are extremely slow and take many centuries to accomplish; they may apply to climate, levels of land and sea, or, as in geomagnetism, to long-period changes in the magnetic fields at any place.

secular equilibrium (*Phys*) Radioactive equilibrium where parent element has such long life that activities remain effectively constant for long periods.

secund (*BioSci*) Having the lateral members, leaves or flowers, all turned to one side.

security (*ICT*) Establishment and application of safeguards to protect data, software and computer hardware from accidental or malicious modification, destruction or disclosure. See COMPUTER MISUSE ACT, DATA PROTECTION ACT, PRIVACY.

security paper (*Paper*) Anti-counterfeit paper intended to discover the falsification of documents and to reveal attempts at so doing. Achieved by special, intricate watermarking, the addition of selected chemicals to the stock or incorporation in the furnish of synthetic or animal fibres, threads, planchettes, etc.

sedentary (*BioSci*) Said of an animal that remains attached to a substratum.

sedimentary rocks (*Geol*) All those rocks which result from the wastage of pre-existing rocks. They include the fragmental rocks deposited as sheets of sediment on the floors of seas, lakes and rivers and on land; also deposits formed of the hard parts of organisms, and salts deposited from solution, in some cases by organic activity. Igneous and metamorphic rocks are excluded.

sedimentary structure (*Geol*) Any physical structure in a sedimentary rock that was formed at the time of its deposition, eg CROSS-BEDDING, SOLE MARK.

sedimentation (*Chem*) A method of analysis of suspensions, by measuring the rate of settling of the particles under gravity or centrifugal force and calculating a particle parameter from the measured settling velocities. It is also used for measuring the molecular mass of soluble polymers. The same method can be applied to biological macromolecules using the ULTRACENTRIFUGE. It gives an absolute measure of $\overline{M}_z$, a higher molecular mass than $\overline{M}_w$. See MOLECULAR MASS DISTRIBUTION.

sedimentation balance (*PowderTech*) A device widely used in particle size analysis by sedimentation techniques. The accumulation of particles at the bottom of a suspension vessel is automatically recorded, the particle size distribution being derived from the rate of accumulation of the particles at various times.

sedimentation coefficient (*BioSci*) The ratio of the velocity of sedimentation of a molecule to the centrifugal force required to produce this sedimentation. It is a constant for a particular species of molecule, and the value is given in Svedberg units (S) that are non-additive.

sedimentation potential (*Chem*) The converse of ELECTROPHORESIS; a difference of potential which occurs when particles suspended in a liquid migrate under the influence of mechanical forces, eg gravity.

sedimentation tank (*Build*) A tank into which sewage from the detritus pit is passed so that suspended matters may sink to the bottom, from which they can be removed.

sedimentation techniques (*PowderTech*) Groups of methods of particle size analysis in which concentration changes within a suspension are measured, the changes being due to differential rates of settling of various sizes of particle.

sedimentation test (*Med*) The measurement of the rate of sinking of red blood cells (erythrocytes) in drawn blood placed in a tube; the rate is increased in disease and in pregnancy. Also *erythrocyte sedimentation rate*, abbrev *ESR*.

Seebeck effect (*ElecEng*) Phenomenon by which a (thermoelectric) emf is set up in a circuit in which there are junctions between different bodies, metals or alloys, the junctions being at different temperatures. See THERMOELECTRIC EFFECT.

seed (*BioSci*) The matured ovule of a seed plant containing usually one embryo, with, in some species, endosperm or perisperm, surrounded by the seed coat or testa.

seed (*Glass*) Small bubbles.

seed bank (*BioSci*) (1) The total seed content of the soil. (2) A repository for seeds to try to prevent loss of rare species or particular varieties.

seed bed (*Agri*) An area of soil that has been cultivated and prepared to the point of readiness for sowing.

seed crystal (*Chem*) A crystal introduced into a supersaturated solution or a supercooled liquid in order to initiate crystallization.

seed drill (*Agri*) An implement used to plant seeds with precision in terms of depth and spacing.

seed leaf (*BioSci*) Also COTYLEDON.

seed plant (*BioSci*) A member of those plant groups that reproduce by seeds, ie the gymnosperms and the angiosperms; the division Spermatophyta.

seedy toe (*Vet*) An affection of the horse's hoof in which the wall of the hoof becomes separated from the subcorneal tissue, forming a space which becomes filled with abnormal, crumbly, horn; causes lameness when severe.

seeing (*Astron*) A term used by telescopic observers to describe the quality of observing conditions as influenced by turbulence in the Earth's atmosphere.

seep hose (*Agri*) A hose that distributes irrigation water from the source to the growing crop, allowing gradual seepage of water into the soil along its length.

seerloup (*Textiles*) A plain-weave, lightweight cloth similar to gingham, containing some coloured, coarse warp threads which form loops on the cloth surface.

seersucker (*Textiles*) A woven fabric with stripes or checks of puckered material alternating with flat sections. The effect is called *plissé*, Fr for *pleated*.

see-saw amplifier (*Electronics*) See PARAPHASE AMPLIFIER.

Seewer governor (*ElecEng*) A hydraulic turbine governor for controlling the speed of high-pressure PELTON WHEELS; a needle valve varies the divergence of the conical pressure jet issuing from the nozzle.

SEG (*ImageTech*) Abbrev for SPECIAL-EFFECTS GENERATOR.

seg (*Textiles*) Dress or suiting dyed fabrics of simple twill weave, often made from wool but also from other fibres or wool-containing blends.

Seger cones (*Eng*) Small cones of clay and oxide mixtures, calibrated within defined temperature ranges at which the cones soften and bend over. Used in furnaces to indicate, within fairly close limits, the temperature reached at the position where the cones are placed. Also *fusion cones*, *pyrometric cones*.

segment (*BioSci*) (1) One of the joints of an articulate appendage. (2) One of the divisions of the body in a metameric animal. (3) A cell or group of cells produced by cleavage of an ovum. Adj *segmental*.

segment (*ElecEng*) One of many elements, insulated from one another, which collectively form a commutator.

segment (*MathSci*) A plane figure enclosed by the *chord* of a circle (or of an ellipse) and the arc cut off by it. The segment of a sphere or of an ellipsoid is the portion cut by a plane.

segmental (*BioSci*) In metameric animals, repeated in each somite, as *segmental arteries*, *segmental papillae*.

segmental arch (*CivEng*) An arch having the shape of a circular arc struck from a point below the springings.

segmental core disk (*ElecEng*) An armature core disk made up in segments; used when a disk in a single piece would be so large as to be unwieldy.

segmental interchange (*BioSci*) The exchange of portions between two chromosomes which are not homologous.

segmentation (*BioSci*) Meristic repetition of organs or of parts of the body; the early divisions of a fertilized ovum, leading to the formation of a blastula or analogous stage.

segmentation (*ICT*) The process of dividing the program into sections (segments or modules) that can be independently executed or changed. See MODULAR PROGRAMMING, OVERLAY.

segmentation cavity (*BioSci*) See BLASTOCOEL.

Segrè chart (*Phys*) A chart on which all known nuclides are represented by plotting the number of protons vertically against the number of neutrons horizontally. Stable nuclides lie close to a line which rises from the origin at 45° and gradually flattens at high atomic masses. Nuclides below this line tend to be beta emitters whilst those above tend to decay by positron emission or electron capture. Data for half-life, cross-section and disintegration energy are frequently added.

segregation (*BioSci*) The separation of the two alleles in a heterozygote when gametes are formed, each carrying one or the other; and, consequently, the appearance of more than one genotype in the progeny of a heterozygote.

segregation (*Eng*) Non-uniform distribution of impurities, inclusions and alloying constituents in metals. Arises from the process of freezing, and usually persists throughout subsequent heating and working operations. See INVERSE SEGREGATION, NORMAL SEGREGATION.

seiche (*EnvSci*) An apparent tide in a lake (originally observed on Lake Geneva) due to the pendulous motion of the water when excited by wind, earth tremors or atmospheric oscillations.

Seidlitz powder (*Chem*) Effervescent powder. A mixture of sodium hydrogen carbonate with tartaric acid, acid sodium tartrate, or some similar acid or salt. Purgative.

seif dune (*Geol*) A longitudinal sand dune developed parallel to the dominant wind direction.

seism-, seismo- (*Genrl*) Prefixes from Gk *seismos*, shaking.

seismic prospecting (*MinExt*) Determining underground structure by seismic methods which are used esp in petroleum exploration. In seismic reflection surveying, explosives or other energy sources produce sudden pulses of short duration that are reflected and detected by small detectors or geophones. The signals from each geophone are amplified, fed into sophisticated data processing equipment and arranged to produce a seismic reflection record, a method often known as reflection shooting. If the waves from the energy source reach a bed of rock through which the sound waves move faster they are then transmitted along it and the method is refraction shooting. Fig. ▷

seismo- (*Genrl*) See SEISM-.

seismology (*Geol*) The study of earthquakes, particularly their shock waves. Studies of the velocity and refraction of seismic waves enable the deeper structure of the Earth to be investigated. See panel on EARTHQUAKE.

seismonasty (*BioSci*) A NASTIC MOVEMENT in response to a shock, esp mechanical shock, as in eg *Mimosa pudica*.

seizing signal (*ICT*) One sent from the outgoing end of a circuit at the start of a call and having the primary function of preparing the apparatus at the incoming end of the circuit for the reception of subsequent signals.

seizure (*Eng*) The locking or partial welding together of sliding metallic surfaces normally lubricated, eg a JOURNAL or bearing. Also *seizing-up*.

selcal (*Aero*) An automatic signalling system used to notify the pilot that the aircraft is receiving a call. It makes constant monitoring of the receiving equipment unnecessary. A contraction of *selective calling*.

selectance (*ICT*) The ratio of sensitivities of a receiver to two specified channels.

selectin (*BioSci*) Any of a group of cell adhesion molecules that bid to carbohydrates via a lectin-like domain. Some are involved in lymphocyte homing.

selection (*BioSci*) See NATURAL SELECTION.

selection bias (*Psych*) Errors in the selection and placement of subjects into groups that results in differences between groups which could affect the results of an experiment.

selection rule (*Phys*) A restriction on the transitions between quantum states of atoms, molecules or nuclei. The rules are derived theoretically by quantum mechanics. See NUCLEAR SELECTION RULES.

selective absorption (*Phys*) Absorption of light, limited to certain definite wavelengths, which produces so-called absorption lines or bands in the spectrum of an incandescent source, seen through the absorbing medium. See ATOMIC ABSORPTION SPECTROSCOPY, FRAUNHOFER LINES, KIRCHHOFF'S LAWS.

selective assembly (*Eng*) The assembly of mating parts selected by trial for their accuracy so as to obtain the required precision of fit.

selective cell discard (*ICT*) A method of congestion control in an ASYNCHRONOUS TRANSFER MODE network, in which a congested node discards any CELLS explicitly identified as belonging to a connection that is failing to comply with its contract, as well as those in which the CELL LOSS PRIORITY bit is set.

selective dump (*ICT*) A DUMP of full contents of a specific part of a computer memory.

selective emission (*ElecEng*) The property of an incandescent body whereby it emits radiation, predominantly of one frequency.

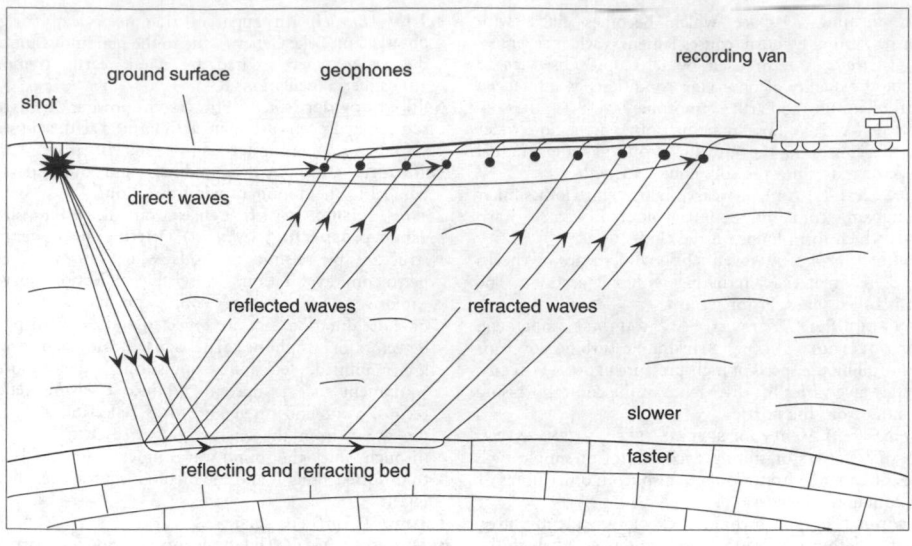

seismic prospecting

selective fading (*ICT*) That affecting some parts of a broadband signal more than others, eg sound and not vision in TV reception.

selective freezing (*Eng, Crystal*) A process involved in solidification, as a result of which the first crystals formed differ markedly in composition from the melt. Thus, in a EUTECTIC SYSTEM (except the eutectic alloy), crystals based on one component are formed from a melt containing two or more, and these early crystals may be separated from the remainder of the melt if desired.

selective mating (*BioSci*) See PREFERENTIAL MATING.

selective network (*ICT*) One for which the loss and/or phase shift are a function of frequency.

selective pesticide (*Agri*) A pesticide that has limited, lethal impact on flora and fauna other than the target populations. Such pesticides are effective owing to differential toxicity, or careful timing or placement of the pesticide.

selective protection (*ElecEng*) A term applied to methods of protecting power transmission networks in which an automatic disconnection of the faulty section occurs without disturbance of the remainder of the network.

selective resonance (*ICT*) Resonance that occurs at one or more discrete frequencies, instead of extending over a band of frequencies as in some forms of filter.

selectivity (*ICT*) The ability of a receiver to distinguish by tuning between specified wanted and unwanted signals. Measured by frequency difference for the half-power points of the pass band of the receiver. Often aided by directive reception.

selector (*ICT*) A unit device in older telephone exchanges, operated either by the dialled impulses originated by the subscriber, or by impulses arising within the exchange.

selector forks (*Autos*) In a gearbox, forked members whose prongs engage with grooves cut in bosses which they move along a splined shaft for changing gear. They are operated through rods by the gear lever.

selector valve (*Aero*) A valve used to direct the flow of the hydraulic fluid or compressed air in a system into the desired actuating current.

selenite (*Chem*) Selenate (IV). The salt of the hypothetical selenous acid $HSeO_2$, eg sodium selenite, $NaSeO_2$.

selenite (*Min*) The colourless and transparent variety of GYPSUM which occurs as distinct monoclinic crystals, esp in clay rocks.

selenium (*Chem*) Symbol Se, at no 34; ram 78·96; valencies 2,4,6. A non-metallic element. A number of allotropic forms are known. *Red selenium* is monoclinic; mp 180°C; rel.d. 4·45. *Grey (metallic) selenium*, formed when the other varieties are heated at 200°C, is a conductor of electricity when illuminated; mp 220°C; bp 688°C; rel.d. 4·80; electrical resistivity 12×10^{-8} ohm metres. Selenium is widely distributed in small quantities, usually as selenides of heavy metals. It is obtained from the flue dusts of processes in which sulphide ores are used, and from the anode slimes in copper refining. It is used as a decolorizer for glass, in red glass and enamels, and in photoelectric cells and rectifiers. Selenium is similar to sulphur in chemical properties, but resembles tellurium more closely still.

selenium cell (*Electronics*) Early photoconductive cell which depends on the change in electrical resistance of selenium when illuminated.

selenium halides (*Chem*) Selenium has a greater affinity for the halogens than sulphur. Selenium (VI) fluoride, SeF_6, is the only (VI) halide; (IV) halides are known. No compounds with iodine.

selenodont (*BioSci*) Having cheek teeth with crescentic ridges on the grinding surface.

selenography (*Astron*) The study of the physical geography of the Moon.

selenology (*Astron*) The scientific study of the Moon.

selenophone (*Acous*) Original system of photographically recording sound on paper, the track being reproduced by scanning with a focused slit, the modulated reflected light being received into a photocell.

self (*BioSci*) Self-fertilize or self-pollinate. Cf CROSS.

self-absorption (*Phys*) See SELF-SHIELDING.

self-actualization (*Psych*) The realization of one's potential through self-understanding and self-awareness; for some theorists a major motivating force in psychological development.

self-aligned gate (*Electronics*) Descriptive of a design procedure and processing step in semiconductor technology in which the gate electrode of a MOSFET is fashioned early on and subsequently used as the mask for ion implantation of source and drain regions. This ensures accurate registration and minimizes stray capacitances or faults associated with misaligned gates.

self-aligning ball bearing (*Eng*) A BALL BEARING in which the two rows of balls roll between an inner race and a spherical surface in the outer race, thus allowing considerable shaft deviation from the normal.

self-annealing (*Eng*) A term applied to metals such as lead, tin and zinc, which recrystallize at ambient temperature and consequently exhibit little STRAIN-HARDENING when cold-worked.

self-assembly (*BioSci*) The process of forming structures from subunits (*protomers*) without any external source of information about the structure to be formed such as a priming structure or template.

self-baking electrode (*ElecEng*) An arc-furnace electrode in the form of a hollow tube, into which a paste-like electrode material is continuously fed as it becomes hard-baked and burns away in the furnace.

self-balance protection (*ElecEng*) A method of protecting transformers and ac generators from internal faults, based on the fact that the instantaneous sum of the phase currents in a symmetrical three-phase system is always zero.

self-capacitance (*Phys*) See CAPACITANCE.

self-centring chuck (*Eng*) A lathe chuck for cylindrical work in which the jaws are always maintained concentric by a scroll in a *scroll chuck*, or sometimes by radial screws driven by a ring gear operated by a key. Also *universal chuck*.

self-centring lathing (*Build*) Expanded metal with raised ribs, greatly stiffening the sheet and enabling it to be used with the minimum of framing. Also *stiffened expanded metal*.

self-cleansing (*Build*) A term for that velocity of flow of sewage material which prevents deposition of solids.

self-clocking (*ICT*) See panel on RUN-LENGTH ENCODING.

self-compatible (*BioSci*) An individual plant or a clone, capable of self-fertilization. See HOMOTHALLISM.

self-concept (*Psych*) The subjective perception of oneself.

self-conjugate directions (*MathSci*) See CONJUGATE DIRECTIONS.

self-conjugate triangle (*MathSci*) See SELF-POLAR TRIANGLE.

self-consistent field (*Phys*) The approximate wavefunction of a system of many electrons found by an iterative method. Assuming the electrons occupy levels similar to that of hydrogen, the electrostatic field in which the electrons exist is guessed, and then a new set of energy levels and a new field is calculated. The process is repeated until the system is self-consistent.

self-cure (*BioSci*) In animals infested with intestinal nematodes, the expulsion of the majority of the worms about 10 days after the initial establishment of the infection, probably due to an immune response by the host.

self-discharge (*ElecEng*) (1) Loss of capacity of PRIMARY CELL or ACCUMULATOR as a result of internal leakage. (2) Loss of charge from capacitor due to the finite insulation resistance between its plates.

self-dissociation (*Chem*) The weak tendency of water and some other liquids (eg liquid ammonia) with strongly polar molecules (see ASSOCIATED LIQUID) to break up into their component ions, such as H^+ (proton) and HO^- (hydroxyl), the former of which usually attaches itself to a complete water molecule H_2O, forming a hydronium ion, H_2OH^+.

self-documenting program (*ICT*) One that informs the user how to use the program as it runs.

self end-papers (*Print*) Instead of having separate end-papers, the first two and last two leaves of the book are left blank, the first and the last leaves being pasted down on the cover. Also *own ends*.

self-excitation (*ElecEng*) A form of machine excitation in which the supply to the field system is obtained either from the machine itself or from an auxiliary machine which is coupled to it.

self-faced (*Build*) A term applied to stone, eg flagstone, which splits along natural cleavage planes leaving faces which do not have to be dressed.

self-fertilization (*BioSci*) The fertilization of an egg by a male gamete from the same individual or the same clone

(genet). Cf CROSS-FERTILIZATION, SELF-POLLINATION. See AUTOGAMY.

self-fluxing ore (*Eng*) Mineral charged to smelter which contains its own slag-forming constituents.

self-hardening steel (*Eng*) See AIR-HARDENING STEEL.

self-heterodyne (*ICT*) See AUTODYNE.

self-incompatible (*BioSci*) An individual plant or a clone, incapable of self-fertilization. See HETEROTHALLISM.

self-induced emf (*ElecEng*) That induced in an electric circuit as a result of a change in the current flowing in it.

self-inductance (*Phys*) A phenomenon in which, if the current in a circuit changes, the magnetic flux linked with the circuit changes and induces an emf in such a direction as to oppose the change causing it (Lenz's law). Unit is the HENRY. See MUTUAL INDUCTANCE.

self-inductance coefficient (*ElecEng*) See INDUCTANCE COEFFICIENT.

selfing (*BioSci*) Self-fertilization, self-pollination.

selfish DNA (*BioSci*) A class of DNA sequence thought to have been selected during evolution only by its ability to spread and duplicate itself in the genome of higher organisms, with minimal damage to the 'host'.

self-levelling level (*Surv*) An instrument which levels automatically by means of a pendulum-operated system of prisms.

self-lubricating bearing (*Eng*) Plain, sintered-metal bearing made by powder metallurgy techniques, having a porous structure impregnated with lubricant or polymer, eg PTFE. No external lubricant need be supplied during normal service lifetime. Used extensively in small electric motors in eg domestic appliances. Also applied to all-plastic (eg acetal) bearings.

self-phase modulation (*ICT*) Change of signal phase with amplitude due to non-linearity in an OPTICAL FIBRE. The effect is to introduce chirp (shift of spectral maximum with time) into the transmitted data. This, when acted on by the CHROMATIC DISPERSION of the fibre, can result in severe pulse distortion.

self-polar triangle (*MathSci*) A triangle whose sides are the polars of the opposite vertices with respect to a conic. Also *self-conjugate triangle*. Cf CONJUGATE TRIANGLES.

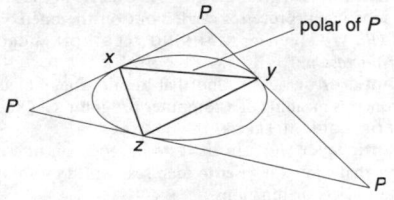

self-polar triangle The triangle *xyz*.

self-pollination (*BioSci*) The transfer of pollen to a stigma from an anther of the same flower or individual or clone (genet).

self-quenching (*Phys*) Said of counter tubes which do not depend on an external circuit for quenching, the residual gas providing sufficient resetting for the next operation of detecting a further photon or particle.

self-rectifying (*Radiol*) Said of an X-ray tube when an alternating voltage is applied directly between target and cathode.

self-regulating (*NucEng*) Said of a system when departures from the required operating level tend to be self-correcting, and in particular of a nuclear reactor where changes of power level produce a compensating change of reactivity, eg through negative temperature coefficient of reactivity. Also *load following*.

self-scattering (*Phys*) The scattering of radioactive radiation by the body of the material which is emitting the radiation.

self-serving bias (*Psych*) The tendency to believe that success derives from personal virtue and failure from external malignance.

self-shielding (*ICT*) A coaxial line is self-shielding in that the return transmission current is in the inside surface of the outer conductor, while the interfering currents, if of sufficiently high frequency, are on the outside surface of the outer conductor.

self-shielding (*NucEng, Phys*) Displaying the absorption of ionizing radiation by a material from which the radiation is emitted.

self-skinning foams (*Eng*) The term applied to polymer foams which collapse to form a solid surface when they collide with the mould walls, eg in REACTION INJECTION MOULDING. Usually achieved by using volatile liquid blowing agent.

self-starting rotary converter (*ElecEng*) A synchronous converter designed to start up from the ac supply as an induction motor, thus requiring no separate starting motor.

self-sterile (*BioSci*) Not capable of producing viable offspring by self-fertilization.

self-sterility (*BioSci*) In a hermaphrodite animal or plant, the condition in which self-fertilization is impossible or ineffective.

self-synchronizing (*ElecEng*) A term applied to a synchronous machine that can be switched onto the ac supply without being in exact synchronism with it.

self-tapping screw (*Eng*) A screw made of hard metal which cuts its own thread when driven into softer materials.

self-thinning curve (*BioSci*) A curve describing the survival of individuals in a crowded population with time.

self-tolerance (*BioSci*) A process by which the individual's immune system is prevented from reacting against its own antigenic determinants.

Sellers screw-thread (*Eng*) The US standard thread, having a profile angle of 60°C, and a flat crest formed by cutting off $\frac{1}{8}$ of the thread height. Abbrev *USS*.

Sellotape (*Plastics*) TN for transparent polymer film with adhesive backing.

selvedge (*Textiles*) The strong edge at both sides of a woven cloth. It sometimes bears a woven trademark or name, but its function is to give strength to the fabric in the loom and in subsequent processes carried out on the open width.

SEM (*Phys*) Abbrev for SCANNING ELECTRON MICROSCOPE (or *microscopy*).

semantic error (*ICT*) One that results in ambiguous or erroneous meaning of a computer program. Cf EXECUTION ERROR, LOGICAL ERROR.

semantic layer (*ICT*) In OLAP technology, an intermediate layer that allows the user to map real-world meanings onto data objects or functions.

semantic memory (*Psych*) The part of DECLARATIVE MEMORY that is concerned with meanings, general principles, theories, etc, and is not tied to specific episodes.

semantics (*ICT*) The meaning attached to words or symbols.

semantic web (*ICT*) The idea that universal meaning can be given to documents or pages on the Internet, rendering them consistently understandable to machine processing.

semantide (*BioSci*) A molecule carrying information, as in a gene or messenger RNA.

sematic (*BioSci*) Warning; signalling; serving for warning or recognition, as *sematic colours*.

semeiology (*Med*) The branch of medical science which is concerned with the symptoms of disease. From Gk *semeion*, sign.

semeiotic (*Med*) Pertaining to, or relating to, the symptoms and signs of disease.

semen (*BioSci*) The fluid formed by the male reproductive organs in which the spermatozoa are suspended. Adj *seminal*.

semi- (*Genrl*) Prefix from Lt *semi*, half.

semi-automatic (*ElecEng*) Said of an electric control in which manual initiation is followed by further operations which proceed automatically.

semi-automatic exposure control (*ImageTech*) A photo-electric device coupled to the lens of a camera which operates indicators, eg pointers, which can be aligned by moving the iris ring, thus obtaining the correct aperture setting for the available light.

semi-autonomous (*BioSci*) Systems or processes that are not wholly independent of other systems or processes.

semicarbazide (*Chem*) $H_2NCONHNH_2$. Mp 96°C A base forming salts, eg hydrochloride; may be prepared from potassium cyanate and hydrazine hydrate. It reacts with aldehydes and ketones, forming SEMICARBAZONES.

semicarbazones (*Chem*) The reaction products of aldehydes or ketones with semicarbazide. The two amino hydrogen atoms of the semicarbazide react with the carbonyl oxygen of the aldehydes or ketones, forming water, and the two molecular groups then combine to form the semicarbazone.

semichemical pulp (*Paper*) Pulp produced from the raw material by a combination of chemical and mechanical means. The relatively light digestion is insufficient to resolve all the non-cellulose matter or to permit dispersion of the fibres without some mechanical treatment.

semicircular canals (*BioSci*) Structures forming part of the labyrinth of the inner ear of most vertebrates, there usually being three at right angles to each other, two being vertical and one horizontal. Movements of the head cause movement of the endolymph in the canals, which moves the gelatinous cupula attached to the sensory hairs of a neuromast sense organ in the swelling (called *ampullae*) at the base of the canals, initiating nerve impulses which travel to the brain via the eighth cranial nerve. They thus serve as organs of dynamic equilibrium.

semicircular deviation (*Ships*) Those components of the DEVIATION which vary as the sine and cosine of the COMPASS COURSE. That is, deviations which have the same sign in one semicircle of courses and the opposite sign in the other semicircle.

semiclosed slot (*ElecEng*) A slot in an electric machine whose width narrows sharply at the top so that conductors must be inserted from the ends.

semiconductor (*Electronics*) An element or compound having higher resistivity than a conductor, but lower resistivity than an insulator. Semiconductor materials are the basis of diodes, transistors, thyristors, photodiodes and all integrated circuits. See panel on INTRINSIC AND EXTRINSIC SILICON.

semiconductor detector (*Electronics*) See SEMICONDUCTOR RADIATION DETECTOR.

semiconductor device processing (*Electronics*) Methods by which various layers of material are grown and deposited onto selected areas of the surface of a semiconductor wafer. See panel on SEMICONDUCTOR FABRICATION.

semiconductor diode (*Electronics*) Two-electrode, point contact or junction, semiconducting device with asymmetrical conductivity.

semiconductor diode laser (*Phys*) A laser in which the lasing medium is p- or n-type semiconductor diode, eg gallium arsenide. Capable of continuous output of a few milliwatts at wavelengths in the range 700–900 nm. See OPTOELECTRONICS.

semiconductor fabrication (*Electronics*) See panel on SEMICONDUCTOR FABRICATION (p 1076).

semiconductor junction (*Electronics*) One between DONOR and ACCEPTOR impurity semiconducting regions in a continuous crystal; produced by one of several techniques, eg alloying, diffusing, doping, drifting, fusing, growing, etc.

semiconductor radiation detector (*Electronics*) Semiconductor diodes, eg silicon junction, are sensitive under reverse voltage conditions to ionization in the junction depletion layer, and can be used as radiation counters or monitors.

semiconductor trap (*Electronics*) Lattice defects in a semiconductor crystal that produce potential wells in which electrons or holes can be captured.

semiconservative replication (*BioSci*) The system of replication of DNA found in all cells in which each daughter cell receives one old strand of DNA and one strand newly synthesized at the preceding S phase.

semidiameter (*Astron*) Half the angular diameter of a celestial body.

semi-elliptic spring (*Autos*) A carriage spring, so called because when a pair is used, one inverted and attached by its ends to the other, the arrangement resembles an ellipse.

semi-enclosed (*ElecEng*) Said of electric motors in which ventilation is provided but access to live parts necessitates opening the case.

semigroup (*MathSci*) A set S together with an operation * in S such that (1) the set S is closed with respect to *, ie for all elements x, y in S, $x*y$ is an element of S; (2) the operation * in S is associative, ie for all elements x, y, z in S, $(x*y)*z = x*(y*z)$. For instance, the set of natural numbers with either addition or multiplication.

semi-immersed liquid-quenched fuse (*ElecEng*) A liquid-quenched fuse in which the fuse-link is above the liquid before operation but drawn down into it during or after fusion.

semi-interquartile range (*MathSci*) A measure of the variability of a distribution or data set defined as half the difference between the first and third QUARTILES.

semimajor axis (*Astron*) Half of the longest distance across an ellipse.

semimetallic foils (*Print*) Coloured BLOCKING FOILS which give a metallic effect when viewed from certain angles. Used for stamping the reverse side of moulded acrylate badges.

semimonocoque construction (*Aero*) See MONOCOQUE and STRESSED-SKIN CONSTRUCTION.

semimuffle-type furnace (*Eng*) See OVEN-TYPE FURNACE.

seminal (*BioSci*) Pertaining to the seed. See SEMINAL ROOTS.

seminal receptacle (*BioSci*) See VESICULA SEMINALIS.

seminal roots (*BioSci*) Adventitious roots produced at the base of the stem of young seedlings of eg cereals and grasses.

seminiferous (*BioSci*) Semen-producing or semen-carrying.

seminoma (*Med*) A malignant tumour of the testis arising from the germinal cells.

semiochemical (*BioSci*) A chemical substance produced by an animal and used in communication. See PHEROMONE, ALLOMONE.

semiotics (*BioSci*) The study of communication.

semi-oviparous (*BioSci*) Giving birth to imperfectly developed young, as the marsupials.

semipalmate (*BioSci*) Having the toes partially webbed.

semipermeable membrane (*Chem*) A membrane which permits the passage of solvent but is impermeable to dissolved substances.

semiplacenta (*BioSci*) An indeciduate placenta in which only the fetal part is thrown off at birth.

semirigid airship (*Aero*) See AIRSHIP.

semisteel (*Eng*) Cast-iron with low carbon content incorporating a proportion of melted-down scrap steel.

semisubmersible (*MinExt*) A seagoing vessel with adjustable buoyancy. A drilling rig of this type is towed to its location floating high in the water for passage, but once on site is flooded and partially submerged to provide a stable platform no longer riding on the waves. Emergency vessels of this type may have firefighting facilities for attending well or platform accidents in offshore oil fields.

semitone (*Acous*) The difference of pitch between two sounds with a frequency ratio equal to the twelfth root of two on the even-tempered scale.

semitransparent mirror (*ImageTech*) A mirror which partially reflects and partially transmits light rays without appreciable diffusion.

semitransparent photocathode (*Electronics*) One where the electrons are released from the opposite side to the incident radiation.

semitubular rivet (*Eng*) A rivet which has been drilled hollow partway up the shank to provide a thin shank wall for easy setting.

semiwater gas (*Chem*) A mixture of carbon monoxide, carbon dioxide, hydrogen and nitrogen obtained by passing a mixture of air and steam continuously through incandescent coke. Its calorific value is low, about 45 MJ m^{-3}.

semiworsted spun (*Textiles*) Yarn made from carded sliver or roving.

semolina (*FoodSci*) The coarse granules of endosperm mixed with fine bran particles which are a by-product from the initial stages of flour milling, the basic raw material in the production of PASTA.

sempervirent (*BioSci*) Evergreen.

sender (*ICT*) US for TRANSMITTER.

Senegal gum (*Chem*) See GUM ARABIC.

senescent (*BioSci*) Said of that period in the life history of an individual when its powers are declining prior to death.

senile-degenerative disorders (*Psych*) Deterioration of intellectual, emotional and motor functioning with advancing age.

senile dementia (*Med*) Progressive DEMENTIA in the elderly with loss of memory for recent events and a decline in all mental abilities, resulting from severe organic deterioration of the brain.

senile plaque (*Med*) A characteristic feature of the brains of patients with Alzheimer's disease and aged monkeys, consisting of a core of amyloid fibrils surrounded by dystrophic neurites.

senility (*BioSci*) Condition of exhaustion or degeneration due to old age.

Senonian (*Geol*) The youngest epoch of the Cretaceous period. See MESOZOIC.

sensation curves (*ImageTech*) Curves which give the relative response of the eye to different colours having the same intensity.

sensation unit (*Acous*) Original name of the DECIBEL; so called because it was erroneously thought that the subjective loudness scale of the ear is approximately logarithmic.

sense (*MathSci*) Either of the orientations of a directed line; thus the line (or vector) AB and the line BA have the same direction but opposite sense.

sense of absolute pitch (*Acous*) Ability to recognize a single tone without using a reference tone.

sense of relative pitch (*Acous*) Ability to recognize a given interval.

senses (*Psych*) Any number of responses to stimulation through the specialized sense organs (ie eyes, ears, etc). The responses of these organs translate into neural impulses, often referred to as *sensation*.

sense strand (*BioSci*) That strand of a double-stranded DNA molecule which is transcribed into messenger or other RNA.

sensible heat (*Phys*) Heat which effects a change in the body which is detectable by the senses; ie it causes the temperature of the body to change. Measured by the product of the specific heat capacity, the mass of the body and the change of temperature.

sensible horizon (*Surv*) See VISIBLE HORIZON.

sensiferous (*BioSci*) Sensitive. Also *sensigerous*.

sensillum (*BioSci*) In insects, small sense organs of varied function on the integument typically comprising a cuticular and/or hypodermal structure. Pl *sensilla*.

sensing (*Radar*) Removal of 180° ambiguity in bearing as given by simple vertical loop antenna, by adding signal from open aerial.

sensitive (*BioSci*) Capable of receiving stimuli.

sensitive drill (*Eng*) A small drilling machine in which the drill is fed into the work by a hand lever attached directly

Semiconductor fabrication

Semiconductor devices are manufactured by a sequence of processes during which various layers of material are grown and deposited, and dopants are introduced in controlled quantities into selected areas of the surface of a semiconductor wafer. Operations are carried out on wafers of several centimetres' diameter at a scale as small as 1 micrometre. The sequence of processes is specific to the devices being fabricated. All semiconductor processing takes place in extremely clean, dust-free environments because of the small scale of the engineering, and uses high-purity chemicals and gases because of the sensitivity of the materials to impurities (below the parts per billion level).

BOULES (rods) of electronics-grade semiconductor material are produced by the FLOAT ZONE and CZOCHRALSKI PROCESSES. Wafers are then cut and polished to produce the high-quality substrates on which devices are built. Often the next stage involves growing an epitaxial layer of the active semiconductor on the substrate which then determines the crystalline orientation of the deposit, while allowing the purity and composition of the active layer to be independently controlled. Epitaxial layers can be grown (with simultaneous doping) in several ways.

Epitaxy

In *liquid-phase epitaxy* (LPE), layers are grown from a molten source on an appropriate substrate, eg the controlled cooling of a solution of arsenic in gallium on a gallium arsenide (GaAs) substrate deposits a layer of gallium arsenide.

Vapour phase epitaxy (VPE) introduces material for the growing layer via a gas or vapour, eg CHEMICAL VAPOUR DEPOSITION (CVD) from a mixture of silicon tetrachloride and hydrogen onto a silicon substrate at 1200°C deposits silicon. Layers grown too cool or too quickly are polycrystalline.

In *metallo-organic chemical vapour deposition* (MOCVD), metallic species are introduced as gaseous organic complexes, eg trimethyl indium vapour with hydrogen and phosphine (PH_3), when passed over a heated substrate, allows epitaxial growth of indium phosphide (InP).

In *molecular beam epitaxy* (MBE), each species is supplied as a molecular beam effusing from an oven containing the pure material and is particularly suited to the growth of compound semiconductors. Deposition is carried out in an ultrahigh vacuum, ensuring the highest levels of purity and simultaneous analysis of the growing material.

Photolithography

To produce local N-TYPE regions in a P-TYPE layer it is necessary to introduce donor dopant atoms into restricted regions. In silicon technology this is done by first growing silicon dioxide over the entire surface. Next, PHOTOLITHOGRAPHY is used to open apertures through the oxide and down to the underlying p-type silicon, according to the desired pattern, which is stored as a MASK. Shown in Fig. 1 is how one such aperture can be opened using either a negative or a

---

to the drilling spindle, the operator being thus given sensitive control of the rate of drilling.

sensitive period (*Psych*) Periods of time during development when an individual is particularly sensitive to environmental and social experiences and which affect learning in a variety of ways in a wide range of species.

sensitive time (*Phys*) The period for which conditions of supersaturation in a cloud chamber or bubble chamber are suitable for the formation of tracks.

sensitive tint plate (*Min*) A thin, optically orientated plate of a crystal, usually gypsum, used to measure the optical properties of minerals and other crystalline substances with a polarizing microscope.

sensitive volume (*BioSci, Phys*) The portion of an ionization chamber or counter tube across which the electric field is sufficiently intense for incident radiation to be detected. The portion of living cells believed to be susceptible to ionization damage. See TARGET THEORY.

sensitivity (*MathSci*) The proportion of true positives that are detected by a system designed to discriminate between two categories, known conventionally as positive and negative. Compare SPECIFICITY.

sensitivity (*Phys*) General term for ratio of response (in time and/or magnitude) to a driving force or stimulus, eg galvanometer response to a current, or the minimum signal required by a ratio of output level to illumination in a camera tube or photocell.

sensitivity guide (*Print*) A strip of photographic film where the emulsion is in a series of graduated continuous tones each of a known density. Each density step is numbered and the strip when printed down can be used as a guide to control exposure.

sensitization (*BioSci*) (1) Administration of an antigen to provoke an immune response so that, on later challenge, a more vigorous secondary response will ensue. This involves the recruitment of PRIMED cells. (2) Coating of cells with antibody, eg for use in complement fixation tests.

sensitization (*Chem*) The process by which a sol of a lyophilic colloid becomes lyophobic in character with the result that it may readily be coagulated by electrolytes.

sensitized cheque paper (*Paper*) Paper for cheques and similar financial documents, containing chemicals intended to reveal attempts at falsification. Cf SECURITY PAPER.

sensitized paper (*Paper*) Paper that has been coated to render it suitable for a reprographic process, generally light-activated, eg dyeline, blueprint, photographic. Also paper made from stock to which selected chemicals have been added, or coated, so that a colour reaction is produced on exposure to other reagents.

sensitizer (*Chem*) A substance, other than the catalyst, whose presence facilitates the start of a catalytic reaction.

sensitizer (*ImageTech*) Chemical, usually dye, used to increase the sensitivity of photographic emulsions, generally or to specific colours.

sensitometer (*ImageTech*) Instrument providing a controlled series of graduated exposures on a sensitized material for the examination of its reproduction characteristics.

sensitometry (*ImageTech*) The study of the effect of light on sensitized materials and their response to subsequent positive photoresist. The role of the photoresist is to protect the silicon dioxide from exposure in those areas where access to the silicon is not required, the oxide being etched away from the unprotected areas in a subsequent step. The dopant is then introduced either by diffusing dopant atoms in from the surface or else by injecting ionized dopant atoms at high energy (ion implantation).

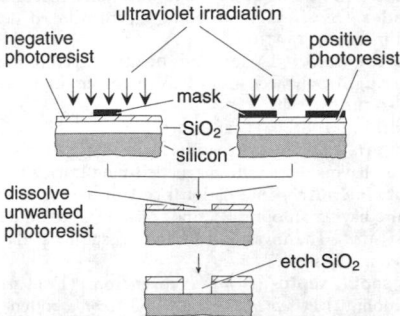

Fig. 1 **Photolithography** Opening a window in a silicon dioxide wafer.

Diffusion processes enable dopant to penetrate several micrometres below the surface but require prolonged high-temperature ($\approx 1000°C$) exposure and have to be carefully planned to anticipate the effects of a series of diffusion cycles in complex device structures. Ion implantation penetrates only the first micrometre below the surface but allows a degree of control over

Semiconductor fabrication *(Cont.)*

the dopant profile and does not require prolonged high temperatures.

The fabrication of one device may require several cycles of oxidation, masking and doping, and finishes with the opening up of contact apertures to the underlying semiconductor, the deposition of a metal interconnection layer and its subsequent patterning by further photolithography. Shown in Fig. 2 is a schematic section through the surface of a wafer where *complementary metal–oxide–silicon* (CMOS) devices have been fabricated. The changes in level of the silicon surface are a consequence of the various oxide growth cycles during which silicon from the wafer is consumed.

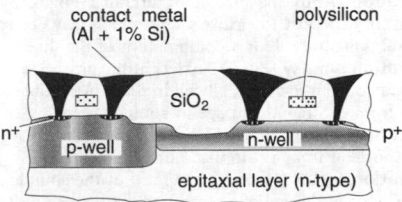

Fig. 2 **Semiconductor device processing** Cross-section through a wafer of a CMOS device.

One wafer may contain several hundred separate integrated circuits, processed simultaneously. After testing and identifying defective circuits, the wafer is cut up (*diced*) into chips, the faulty ones discarded and the remainder then individually packaged, using fine wire to connect from the chip to the more robust legs of the finished device.

---

processing, generally by the measurement of the densities so produced.

sensor (*Eng*) General name for detecting device used to locate (or detect) the presence of matter (or energy, eg sound, light, radio or radar waves).

sensor (*ICT*) A device that is able to detect a change in a physical quantity (eg light level) or an event (eg paper in printer) and produce an electrical signal suitable for a computer. This signal may be in analogue or digital form and may be passed to an interface that is itself connected to the computer.

sensorimotor development (*Psych*) The development of co-ordination between perception and action (eg hand–eye co-ordination).

sensorimotor intelligence stage (*Psych*) According to J Piaget, the first period of intellectual development (0–2 yr) in which the infant's interactions with the environment consist of motor responses to classes of sensory stimuli (eg looking, grasping, etc). During this period the infant progresses from simple reflex actions to complex ways of playing with and manipulating objects, which leads eventually to internal representations of the world.

sensorium (*BioSci*) The seat of sensation; the nervous system. Adj *sensorial*.

sensory (*BioSci*) Directly connected with the sensorium; pertaining to, or serving, the senses.

sensory adaptation (*Psych*) The reduced ability to sense a stimulus after prolonged or repeated exposure.

sensory deprivation (*Psych*) Experimental work, mostly with humans, in which total sensory input is reduced

beyond normal conditions through the use of special chambers or devices (eg translucent goggles).

sensory memory (*Psych*) The brief storage of information brought in through the senses; typically only lasts for 1 to 2 seconds. Also *sensory information store*. See ICONIC and ECHOIC MEMORY.

sensory neuron (*BioSci*) (1) A neuron that receives input from sensory cells. (2) Sensory cells such as cutaneous mechanoreceptors and muscle receptors.

sensory store (*Psych*) The portion of the memory system that maintains representations of sensory information for very brief intervals; divided into ECHOIC and ICONIC memory. See LONG-TERM MEMORY, SHORT-TERM MEMORY.

sentinel (*ICT*) A symbol used to indicate the end of a specific block of information in a data processing system. Also *flag*, *marker*, *tag*.

sentinel pile (*Med*) An oedematous mass of rolled-up anal mucous situated at the margin of the anus at the lower end of an anal fissure.

sentinel species (*EnvSci*) A species that is particularly sensitive to environmental stressors and thus can be used as a marker for environmental quality.

sepal (*BioSci*) A member of the CALYX; typically a green, more or less leaf-like structure, several enclosing the rest of the flower in the bud, but sometimes petaloid.

separated lift (*Aero*) Lift generated by wings of very low aspect ratio (usually of *delta* or *gothic* plan form) at high angles of incidence (approx 20°) through separated vortices causing large suction forces. The JET FLAP is also a separated lift system.

separate excitation (*ElecEng*) A form of machine excitation in which the supply to the field system is obtained from a separate dc source.

separates (*Print*) Same as OFFPRINTS.

separate system (*Build*) A system of sewerage in which two sewers are provided, one for the sewage proper and the other for the rainwater. Cf COMBINED SYSTEM.

separating calorimeter (*Eng*) A device for mechanically separating and measuring the water associated with very wet steam; used in conjunction with the THROTTLING CALORIMETER in determining dryness fractions.

separating drum (*Eng*) An auxiliary steam-collecting drum attached by tubes to the upper drum of some water-tube boilers to avoid priming or foaming.

separating funnel (*Chem*) A funnel with a tap at the bottom and a stoppered top, in which two immiscible liquids can be dispersed by shaking, then separated by settling and drawing off the lower layer; used in LIQUID–LIQUID EXTRACTION.

separation (*Aero*) The spacing of aircraft arranged by AIR-TRAFFIC CONTROL to ensure safety, which may be vertical, lateral, longitudinal, or a combination of the three.

separation anxiety (*Psych*) Anxiety at the prospect of being separated from someone; all children are susceptible to this fear from 6–8 months of age to about 2 years.

separation energy (*Phys*) The energy required to separate one nucleon from a complete nucleus.

separation factor (*NucEng*) The ratio of the abundance of an isotope at the end of a separation system or unit to that at the start of the process. It is usually only slightly greater than unity. US *enrichment factor*.

separation filters (*ImageTech*) The three filters used in separation methods of colour photography and printing.

separation layer (*BioSci*) See ABSCISSION LAYER.

separation point (*Phys*) The point at which streamline flow (LAMINAR FLOW or TURBULENT FLOW) separates from the surface of a body.

separation potential (*NucEng*) A dimensionless function used in the definition of the SEPARATIVE WORK of an uranium enrichment plant. It is given by

$$V(N) = (2N - 1) \ln \left[\frac{N}{1 - N} \frac{1 - N_0}{N_0} \right]$$
$$+ \frac{(1 - 2N_0)(N - N_0)}{N_0(1 - N_0)}$$

where N and N_0 are the concentrations of the product and initial materials respectively.

separative efficiency (*NucEng*) In a single stage, the ratio of actual concentration to the change in the theoretical value. See panel on URANIUM ISOTOPE ENRICHMENT.

separative element (*NucEng*) One unit of a cascade forming a complete isotope separation plant. See panel on URANIUM ISOTOPE ENRICHMENT.

separative power (*NucEng*) The quantity of material a SEPARATIVE ELEMENT is capable of enriching. It is given by the expression

$$\frac{\theta}{1 - \theta} L \frac{(\alpha - 1)^2}{2} \text{ mol s}^{-1}$$

where α is the SEPARATION FACTOR and θ, the CUT. L is the number of mol s^{-1} of material.

separative work (*NucEng*) Measures the amount of separation an enrichment plant can achieve and defined as $PV(N_P) + WV(N_W) - FV(N_F)$, where P, W, F are the masses of product, waste and feed (ie initial) materials, $V(N_P)$, $V(N_W)$ and $V(N_F)$ are the *value functions* of isotope concentrations N_P, N_W, N_F respectively. Measured in *separative work units* (kg SWU or tonnes SWU). It is not the weight of enriched material drawn from the plant. See panel on URANIUM ISOTOPE ENRICHMENT.

separator (*ChemEng*) An item of process plant for separating one substance from a mixture with another, the two substances usually being in two different PHASES, eg oil from water, oil from gas, gas from oil, or ash from flue gases.

separator (*ElecEng*) A thin sheet of porous polyvinyl chloride separating the plates of a secondary cell.

separator (*Eng*) A trap in a pipe containing a gas with condensed vapours. Removes eg water to produce dry steam.

separator (*ICT*) A FLAG used to separate items of data.

separator (*MinExt*) Concentrating machine, used to separate constituent minerals of mixed ore from one another.

separator (*Phys*) An electromagnet used to select iron and steel from mixed scrap.

sepdumag (*ImageTech*) International code for motion picture with two magnetic sound tracks on a separate film.

Sephadex (*BioSci*) TN for beads of cross-linked dextran used in GEL FILTRATION.

Sepharose (*BioSci*) TN for beads of agarose gel from which the charged polysaccharides have been removed, used in GEL FILTRATION.

sepiolite (*Min*) See MEERSCHAUM.

sepmag (*ImageTech*) International code name for a motion picture having a separate magnetic sound track.

sepopt (*ImageTech*) International code name for a motion picture having an optical sound track on a separate film.

sepsis (*Med*) The invasion of bodily tissue by pathogenic bacteria. Adj *septic*.

sept-, septi-, septo- (*Genrl*) Prefixes from: (1) Lt *septum*, partition; (2) Lt *septem*, seven; (3) Gk *septos*, rotten.

septaria (*Min*) Concretionary nodules containing irregular cracks which have been filled with calcite or other minerals. Also *septarian nodules*.

septate (*BioSci*) Divided into cells, compartments or chambers by walls or partitions.

septate fibre (*BioSci*) A plant fibre cell in which the lumen is divided into several compartments by transverse septa.

septavalent (*Chem*) See HEPTAVALENT.

septechlorites (*Min*) A group of sheet silicates closely related chemically to the chlorites, and structurally to the serpentines and kandites. Includes CHAMOSITE and GREENALITE.

septi- (*Genrl*) See SEPT-.

septicaemia (*Med*) The invasion of the blood stream by bacteria and their multiplication therein; associated with high fever, chills, and petechial haemorrhages into the skin. Adj *septicaemic*. US *septicemia*.

septicidal (*BioSci*) A dehiscent fruit, opening by breaking into its component carpels leaving the placental axis standing, as in *Hypericum*.

septic shock (*Med*) A condition of clinical shock caused by ENDOTOXIN in the blood.

septic tank (*Build*) A tank in which sewage is left for a time when a scum forms on the surface and the sewage below is partly purified by the action of the anaerobic bacteria present.

septifragal (*BioSci*) A dehiscent fruit, opening by the breaking away of the outer wall leaving the septa standing.

septo- (*Genrl*) See SEPT-.

septum (*BioSci*) A partition separating two cavities, eg a cell wall or multicellular structure acting as a partition as in a fungal hypha, between cells or between adjacent chambers in an ovary. Adj *septal*.

septum (*ICT*) Dividing partition in a waveguide.

septum transversum (*BioSci*) See DIAPHRAGM.

sequence (*BioSci*) The linear order of bases in a nucleic acid or of amino acids in a protein.

sequence (*ElecEng*) The order in which the several phases of a polyphase ac supply undergo their cyclic variation of voltage.

sequence (*ImageTech*) The unit of the scenario, involving one general idea or happening and a number of scenes, each of which may include a number of shots.

sequence (*MathSci*) An ordered set of numbers, usually derived according to a rule, each member being determined either directly or from the preceding terms.

sequence control register (*ICT*) See PROGRAM COUNTER.
sequencer (*ICT*) See MIDI SEQUENCER.
sequence register (*ICT*) See PROGRAM COUNTER.
sequence valve (*Aero*) A type of automatic selector valve in a hydraulic or pneumatic system, much used in aircraft, whereby the action of one component is dependent upon that of another.
sequencing (*BioSci*) Biochemical procedure for determining the sequence of a nucleic acid or protein. DNA sequencing typically involves cloning the DNA of interest, to produce sufficient material, and usually application of the Sanger (DIDEOXY SEQUENCING) method. Protein sequencing involves controlled fragmentation (proteolysis followed by Edman degradation) of the protein and often then by mass spectrometric analysis of the fragments. The sequences of the fragments can be 'assembled' to give the full sequence.
sequential access (*ICT*) The process of storing or retrieving data items by first reading through all previous items to locate the one required. Also *serial access*. Cf BINARY SEARCH.
sequential colour systems (*ImageTech*) Colour TV systems in which colour information for each channel is transmitted sequentially. Systems may be *field-*, *line-* or *dot-sequential*.
sequential memory (*ICT*) See SERIAL ACCESS MEMORY.
sequential operation (*ICT*) One in which all instructions are carried out sequentially.
sequential scanning (*ImageTech*) A system in which all the lines of the picture are scanned in strict sequence from top to bottom, not interlaced.
sequential transmission (*ImageTech*) A technique of transmitting pictures so that the picture elements are selected at regular times and are then delivered to the communication channel in the correct sequence.
sequestered iron (*BioSci*) See SEQUESTRENE.
sequestering agent (*Chem*) One which removes an ion or renders it ineffective, by forming a complex with the ion. See COMPLEXONES.
sequestrant (*FoodSci*) Substance which can bind metal ions, thus preventing them from catalysing deteriorative reactions in foodstuffs.
sequestrectomy (*Med*) The surgical removal of a SEQUESTRUM.
sequestrene (*BioSci*) Preparations of chelated mineral elements, esp iron and some trace elements, used horticulturally to correct such mineral deficiencies as LIME-INDUCED CHLOROSIS by application to leaves (foliar feeding) or to the soil. Also *sequestered iron, sequestrol*.
sequestrum (*Med*) A piece of bone, dead as a result of infection and separated off from healthy bone.
sequoia (*For*) A fast-growing, N American softwood tree (*Sequoia sempervirens*), with a spongy bark and a dull reddish-brown, straight-grained, medium- to fine-textured heartwood. It is closely related to the giant redwood (*S. gigantea*), the world's largest known tree. Also *redwood*.
Ser (*Chem*) Symbol for SERINE.
sere (*BioSci*) Particular example of plant communities which succeed each other. Hydroseres originate in water, xeroseres occur in dry places, and lithoseres develop on rock surfaces. Adj *seral*. See PRIMARY SERE.
serein (*EnvSci*) The rare phenomenon of rainfall out of an apparently clear sky.
serge (*Textiles*) Dress or suiting dyed fabrics of simple twill weave, often made from wool but also from other fibres or wool-containing blends.
serial access (*ICT*) See SEQUENTIAL ACCESS.
serial access memory (*ICT*) Computer memory where storage locations can be accessed only in predetermined sequences, eg MAGNETIC BUBBLE MEMORY, MAGNETIC TAPE. Cf RAM.
serial arithmetic unit (*ICT*) One in which the digits of a number are operated on sequentially.

serial computer (*ICT*) A computer that operates successively on each BIT of a WORD. Only the very earliest machines were totally serial. Cf PARALLEL PROCESSOR.
serial digital interface (*ImageTech*) A standard connection for component digital equipment. See COMPONENT VIDEO.
serial interface (*ICT*) An INTERFACE to which a serial device may be connected. The interface will contain circuitry to transform the serial data, 1 BIT arriving at a time, into parallel data, 1 BYTE or WORD arriving at a time, for further processing.
serial learning (*Psych*) Refers to a learning or memory recall task in which the subject is required to repeat a list of items in the same order as they were presented.
serial port (*ICT*) A connection on a computer into which a serial device such as a printer or MODEM may be plugged in. Also *COM port*. See SERIAL INTERFACE.
serial-position effect (*Psych*) The observation that in verbal learning, items at the beginning and end of a list are recalled better than those in the middle of the list.
serial printer (*ICT*) A printer that receives its data 1 BIT at a time. Cf PARALLEL DATA TRANSMISSION, PARALLEL PRINTER.
serial radiography (*Radiol*) A technique for making a number of radiographs of the same subject in succession.
serial recall (*Psych*) See SERIAL LEARNING.
serial store (*ICT*) See SERIAL ACCESS MEMORY.
sericite (*Min*) A fine-grained white potassium mica, like muscovite in chemical composition and general characters but occurring as a secondary mineral, often as a decomposition product of orthoclase.
series (*Geol*) A time-stratigraphic unit intermediate between SYSTEM and STAGE, and corresponding to an EPOCH of geological time.
series (*MathSci*) The sum of the terms of a *sequence*.
series (*Phys*) Said of electrical components when a common current flows through them.
series arm (*ICT*) Part of a filter that is in series with one leg of the transmission line.
series capacitor (*ElecEng*) A capacitor connected in series with a transmission line or distribution circuit to compensate for the inductive reactance drop and thereby improve the regulation.
series characteristic (*ElecEng*) The characteristic graph relating terminal voltage and load current in the case of a series-wound dc machine.
series-characteristic motor (*ElecEng*) An electric motor having a speed torque characteristic similar to that of a dc series motor, ie one in which the speed falls with an increase of torque. Also *inverse-speed motor*.
series field (*ElecEng*) Two variable-vane capacitors, usually with air dielectric, with the moving vanes on the same rotating shaft; used in high-frequency circuits with the two capacitances in series, to obviate taking the current through a rubbing contact or through a pigtail, the latter being inductive.
series motor (*ElecEng*) An electric motor whose main excitation is derived from a field winding in series with the armature.

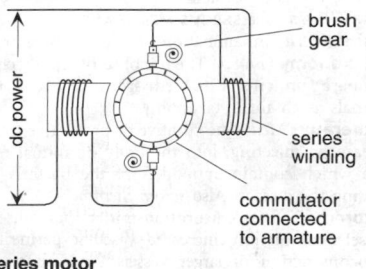

series motor

series–parallel controller (*ElecEng*) A method of controlling the speed and tractive effort of an electric tractor having one or more pairs of series motors, whereby the motors can be connected either in series or in parallel.

series–parallel network (*ElecEng*) One in which the electrical components are composed of branches which are successively connected in series and/or in parallel.

series resonance (*ElecEng*) The condition of a tuned circuit when it offers minimum impedance to an ac voltage supply connected in series with it (due to the circuit reactances neutralizing each other). The term *tunance* is sometimes used in place of resonance if this condition is attained by adjustment of a component value and not of frequency.

series stabilization (*ICT*) A technique of stabilization using amplifier feedback in which the feedback and amplifier circuits are in series at each end of the amplifier.

series system (*ElecEng*) (1) Circuit comprising electrical component as connected in series. (2) The constant-current system of dc distribution developed by Thury, in which generators and motors are all connected in series to form a single dc circuit. Also *Thury system*.

series transformer (*ElecEng*) A power transformer operating under constant-current instead of constant-voltage conditions. See CURRENT TRANSFORMER.

series winding (*ElecEng*) A field winding connected in series with the armature of the motor.

serif (*Print*) The short strokes of a letter, at the extremities of the main strokes and hair lines.

RQENbaegn

serine (*Chem*) 3-hydroxy-2-aminopropanoic acid. HOCH$_2$CH(NH$_2$)COOH. A polar amino acid. The L- or S-isomer is a constituent of protein. Symbol Ser, short form S.

serine peptidase (*BioSci*) Any of a group of enzymes that hydrolyse proteins, and that have serine and histidine at the active site, eg trypsin, subtilisin. Also *serine proteinase*.

sero- (*Genrl*) Prefix denoting serum.

serogroup (*BioSci*) Any group of bacteria or other micro-organisms (that may comprise several serotypes) that have a certain antigen in common.

serological determinants (*BioSci*) Antigenic determinants on cells that are recognized by and accessible to antibodies, as opposed to determinants which are not recognized by antibodies but only by T-lymphocytes.

serological typing (*BioSci*) A technique used for the identification of pathogenic organisms, eg bacteria, particularly strains within a species, when morphological differentiation is difficult or impossible. It is based on antibody–antigen reactions, specific proteins of the organism acting as antigens.

serology (*Med*) The study of sera. See SERUM.

serophyte (*Med*) Any micro-organism which will grow in the presence of fresh serum exuding into a wound, eg STAPHYLOCOCCUS, STREPTOCOCCUS.

seropurulent (*Med*) Said of a discharge or effusion which is both serous and purulent.

serosa (*BioSci*) See SEROUS MEMBRANE.

serositis (*Med*) Inflammation of a serous membrane.

serotaxonomy (*BioSci*) The use of serological techniques to compare proteins extracted from different plants or animals as an aid in taxonomy.

serotherapy (*Med*) The curative or preventive treatment of disease by injecting, into the body of animal or human, sera which contain antibodies to the bacteria or toxins causing the disease. Also *serum therapy*.

serotonin (*BioSci*) A neurotransmitter that causes smooth muscle contraction, increased vascular permeability and vasoconstriction of larger vessels. An important neurotransmitter in the central nervous system where serotinergic neurons are apparently involved in mood etc. (Hence the pharmacological effects of selective serotinin reuptake

inhibitors (SSRIs).) It is present in platelets, from which it is released on activation, and also present in the mast cells of some species. Also *5-hydroxytryptamine*, abbrev 5-HT.

serous (*BioSci*) Watery; pertaining to, producing or containing a watery fluid or serum.

serous membrane (*BioSci*) One of the delicate membranes of connective tissue which line the internal cavities of the body in Craniata; the chorion. Also *serosa*.

Serpens (Serpent) (*Astron*) A constellation in the northern hemisphere; bisected by Ophiuchus into two distinct sections: Caput (head) and Cauda (body).

serpentine (*Min*) Hydrated magnesium silicate which crystallizes in the monoclinic system. The three chief polymorphic forms are ANTIGORITE, CHRYSOTILE and LIZARDITE. The serpentine minerals occur mainly in altered ultrabasic rocks, where they are derived from olivine or from enstatite. Usually dark-green, streaked and blotched with red iron oxide, with talc, etc. The translucent varieties are used for ornamental purposes; those with a fibrous habit form one type of asbestos.

serpentine-jade (*Min*) A variety of serpentine, resembling bowenite, occurring in China.

serpentinization (*Geol*) A type of metamorphism effected by water, which results in the replacement of the original mafic silicates in peridotites by the mineral serpentine and secondary fibrous amphibole.

Serpukhovian (*Geol*) The youngest epoch of the Mississippian period.

Serpula lacrymans (*Build*) Formerly *Merulius lacrymans*, the fungus which commonly causes DRY ROT in timber. Small droplets of water often form on the mycelium, hence the name.

serrate (*BioSci*) Leaf margin toothed like a saw.

serrated roller (*Print*) A knurled or serrated roller on web-fed presses which gives a pull to the paper and minimizes ink pickup.

Serravallian (*Geol*) A stage of the Miocene. See TERTIARY.

Serret–Frenet formulae (*MathSci*) See FRENET'S FORMULAE.

serrulate (*BioSci*) Minutely serrate.

serum (*BioSci*) (1) The watery fluid that separates from blood or lymph in coagulation. (2) Blood serum containing antibodies, taken from an animal that has been inoculated with bacteria or their toxins, used to immunize people or animals. Serum obtained from deliberately immunized individuals may be referred to as antiserum. Adj *serous*.

serum albumin (*BioSci*) A globular protein obtained from blood and body fluids, having a transport and osmoregulatory function. A crystalline, water-soluble substance, not precipitated by NaCl.

serum hepatitis (*Med*) See INFECTIOUS HEPATITIS.

serum sickness (*Med*) A hypersensitivity reaction to the injection of foreign antigens in large quantity, esp those contained in antisera used for passive immunization. This is an example of a type 3 hypersensitivity reaction.

serum therapy (*Med*) See SEROTHERAPY.

server (*ICT*) (1) In NETWORKS, a NODE that permits other nodes to access its resources. (2) See FILE SERVER.

server application (*ICT*) (1) An application program running on a server to which the client machine makes access. See CLIENT–SERVER SYSTEM. (2) In WINDOWS, an application which creates OBJECTS which may then be EMBEDDED or linked to other documents.

service access code (*ICT*) The string of DUAL-TONE MULTIFREQUENCY digits dialled by a VIRTUAL PRIVATE NETWORK user in order to gain access as such rather than as an ordinary public network user.

service area (*ICT*) That surrounding a broadcasting station where the signal strength is above a stated minimum and not subject to fading.

service band (*ICT*) That allocated in the frequency spectrum and specified for a definite class of radio service, for which there may be a number of channels.

service capacity (*ElecEng*) The power output of an electric motor, as specified on the maker's nameplate.

service ceiling (*Aero*) The height at which the rate of climb of an aircraft has fallen to a certain agreed amount (in UK practice, originally, 100 ft min^{-1}, 30 m min^{-1}, but for jet aircraft 500 ft min^{-1}, 150 m min^{-1}).

service control point (*ICT*) An ADVANCED INTELLIGENT NETWORK node containing the customer database and software used to control the services offered by the network. It is typically a modified SWITCH or a commercial computer.

service mains (*ElecEng*) Cables of small conductor cross-section which lead the current from a distributor to the consumer's premises.

service management system (*ICT*) The element of an ADVANCED INTELLIGENT NETWORK that holds the master copy of the network databases, maintains the SERVICE CONTROL POINT databases and collects statistics and measurements.

service reservoir (*Build*) A small reservoir supplying a given district, and capable of storing the water which is filtered during the hours of small demand for use when the requirements become greater. Also *clear water reservoir, distribution reservoir.*

service switching point (*ICT*) A special exchange that provides the user access to an ADVANCED INTELLIGENT NETWORK by recognizing service access codes and sending appropriate requests to the SERVICE CONTROL POINT.

service tanks (*Aero*) See FUEL TANKS.

service tee (*Build*) Having a female thread on the branch and one end of the RUN and a male thread on the other.

serving (*ElecEng*) (1) A layer of jute, tape or yarn, impregnated with bitumen or similar substance, once used to prevent the steel-wire armouring biting into the lead sheath of a cable. Also *bedding.* (2) The process of covering a cable with some form of mechanically strong insulating and binding tape.

servlet (*ICT*) A Java program residing on a web server receiving and responding to messages from a number of clients.

servo-amplidyne system (*ElecEng*) One in which an AMPLIDYNE, together with a control amplifier, is used in order to amplify mechanical power.

servo amplifier (*ElecEng*) One designed to form the part of a servomechanism from which output energy can be drawn.

servo brakes (*Autos*) Power-assisted brakes worked by a hydraulic servo, mechanically from the transmission or in 'vacuum brakes' by differential air pressure.

servocontrol (*Aero*) A reinforcing mechanism for the pilot's effort. It may consist of SERVO TABS.

servo link (*ElecEng*) A mechanical power amplifier which permits low-strength signals to operate control mechanisms that require fairly large powers.

servomotor (*ElecEng*) A motor (electric, hydraulic, etc) for use in an automatic control system for eg the operation of a large valve by a governor of small power. See PILOT VALVE.

servo tab (*Aero*) A control surface TAB moved directly by the pilot, the moment from which operates the main surface, the latter having no direct connection with the pilot.

sesamoid (*BioSci*) A small rounded ossification forming part of a tendon usually at, or near, a joint, as the PATELLA.

sesqui (*Chem*) Containing two kinds of atom, radical, etc, in the proportion of 2:3. Means $1\frac{1}{2}$.

sesquiterpenes (*Chem*) A group of terpene derivatives of the empirical formula $C_{15}H_{24}$, with three isoprene units.

sessile (*BioSci*) (1) Having no stalk. (2) Fixed and stationary.

session (*ICT*) When two pieces of hardware, software or other components in a network or two users are connected together for the purpose of exchanging information they are 'in session'. 'Current session' means until the connection is broken.

session layer (*ICT*) Level 5 of the OPEN SYSTEMS INTERCONNECTION (OSI) model which is concerned with establishing communications SESSIONS between systems.

set (*Build*) (1) The alternating lateral deflection of the teeth of a saw so that the kerf is wider than the sawblade. (2) See NAIL PUNCH.

set (*CivEng*) Of a driven pile, the penetration under a blow of the pile hammer, generally expressed as the number of blows for 25 mm movement.

set (*Eng*) (1) Percentage residual deformation left in a material, esp viscoelastic polymers, after deformation for a given time (often after period of recovery following removal of load). Also *compression set, permanent set.* (2) A smith's tool similar to a short, stiff cold chisel; used for cutting bars etc without heating. Also *cold sett, sett.*

set (*Genrl*) The direction of a current of water.

set (*ICT*) (1) To give a value to a VARIABLE or PARAMETER. (2) To store the value of one in a REGISTER or FLAG.

set (*MathSci*) Any collection of distinct entities ('elements') treated as a mathematical object in its own right. The properties of sets are described by *set theory*, using the operations of UNION, INTERSECTION and COMPLEMENTATION. See SUBSET, UNIVERSAL SET, VENN DIAGRAM.

set (*MinExt*) A frame of timber used in a shaft or tunnel.

set (*Print*) (1) The width of a type character. (2) To COMPOSE type-matter.

set (*Psych*) See MENTAL SET.

set (*Textiles*) See SETT.

seta (*BioSci*) (1) Generally, a small bristle-like structure. (2) The stalk supporting the capsule of the sporophytes of mosses (Bryopsida). (3) A chaeta. Adjs *setaceous, setiferous, setiform, setigerous, setose, setulose.*

set-aside land (*Agri*) Areas of arable land taken out of production as part of the European Union arable area payments scheme, designed to prevent overproduction of cereals.

set flush (*Print*) A typographic instruction to set all lines without indentation.

SETI (*Space*) Abbrev for the *search for extraterrestrial intelligence*, ie investigating the possibility of intelligent life in the universe other than on Earth.

set-off (*Build*) See OFFSET.

set-off (*Print*) Smudging of ink from one sheet to reverse of another before ink has dried. Obviated by interleaving (see SLIP SHEETS) or by using ANTI-SET-OFF SPRAY.

set point (*ElecEng*) See CONTROL POINT.

set screw (*Eng*) A screw, usually threaded along the entire shank length, which is used to prevent relative motion by exerting pressure with its point.

set solid (*Print*) A typographic instruction to use no leading between lines.

set stocking (*Agri*) Grazing management where livestock are on one area of pasture for a prolonged period.

sett (*Build*) A small rectangular block of stone 6 in deep by 3–4 in wide, and from 6 to 9 in long; formerly used for surfacing roads where traffic was heavy. The best setts were of either Scottish or Welsh granite.

sett (*Eng*) See COLD SETT.

sett (*Textiles*) The number of threads per in or per cm in the weft and/or warp of woven fabrics. In a square sett fabric the two values are equal and the yarns in both directions are of the same count. In an unbalanced sett fabric these values are significantly different. Also *set.*

setting (*Build*) The hardening of a lime, cement, mortar or concrete mixture, or a plaster.

setting (*Eng*) Operation in INJECTION MOULDING for optimizing production from new tools. It involves balancing mould temperatures, pressures, machine parameters so that mould defects are minimized.

setting (*Textiles*) Treatments usually by heating and cooling in dry or steamy atmospheres that confer stability on textile materials. Heat setting is particularly important for many fabrics.

setting coat (*Build*) The finishing coat of plaster; a thin layer, about $\frac{1}{8}$ in (3 mm) thick, of fine stuff. Also *skimming coat*.

setting point (*Chem*) The temperature at which a melted wax, when allowed to cool under definite specified conditions, first shows the minimum rate of temperature change.

setting rule (*Print*) See COMPOSING RULE.

setting stick (*Print*) See COMPOSING STICK.

settlement (*Build, CivEng*) The subsidence of a wall, structure, etc.

settling (*PowderTech*) Classification effected by the rate of fall in a fluid which may have a horizontal component of velocity.

settling tank (*Build*) See SEDIMENTATION TANK.

set-top box (*ICT*) A device that allows a conventional TV set to receive a digital signal.

sett paving (*CivEng*) Pavement constructed with SETTS on a suitable foundation. A causeway in Scotland.

set-up (*ImageTech*) The ratio between black-and-white reference levels measured from blanking level for facsimile transmission.

set-up (*Surv*) Location of theodolite above a station point.

set-up instrument (*ElecEng*) Also *set-up-scale instrument*, *set-up-zero instrument*. See SUPPRESSED-ZERO INSTRUMENT.

set-work (*Build*) Two-coat plasterwork on lath.

Seven Sisters (*Astron*) See PLEIADES.

severe combined immunodeficiency syndrome (*Med*) The most severe form of congenital immunological deficiency state, in which both T- and B-lymphocytes are absent, resulting in a lack of both antibody-based and cytotoxic immune responses. Also known as *Swiss-type hypogammaglobulinaemia*, first described in that country. Death in early life from infection occurs if the condition is untreated, but it may be cured by bone marrow transplantation. Abbrev *SCID*.

severy (*Arch*) See CIVERY.

sewage (*Build*) Liquid contained in a sewer.

sewage farm (*Build*) A place where sewage is treated for use as manure or a farm on which such manure is used. See *land treatment*.

sewage gas (*Build*) A self-generated combustible gas collected from the digesting tanks of sewage sludge. General composition 66% CH_4 and 33% CO_2, with energy density in the region of 25 MJ m^{-3}. The gas has a very slow rate of flame propagation. Also *sludge gas*.

sewerage (*Build*) The network of sewers serving a community.

sewing (*Print*) The operation of joining the gathered sections of a book by sewing.

sex (*BioSci*) (1) The sum total of the characteristics, structural and functional, which distinguish male and female organisms, esp with regard to the part played in reproduction. (2) As a verb to determine sex. Adj *sexual*.

sex-, sexi- (*Genrl*) Prefixes from Lt *sex*, six.

sexavalent (*Chem*) See HEXAVALENT.

sex cells (*BioSci*) See GAMETES.

sex chromosome (*BioSci*) In many organisms (including vertebrates) sex is determined by the possession of a particular combination of chromosomes. In some cases, presence or absence of one special chromosome, known as the *accessory* or *X chromosome*, is the determining factor, eg in the insect *Pyrrhocoris apterus* males and females have 13 and 14 chromosomes respectively. In many species there are two sex chromosomes, and the sex of the individual depends on whether it has two identical chromosomes, the *homogametic sex*, or one of each of the two types, the *heterogametic sex*. Where the female is homogametic, as in mammals, the two chromosomes are designated XX, and the male's chromosomes are known as XY. Where the male is homogametic, as in birds, the male's chromosomes are called WW and the female's chromosomes are known as ZW. See panel on CHROMOSOME.

sex determination (*BioSci*) The mechanisms by which the sex of an individual is determined. These may be genetic or environmental. See SEX CHROMOSOME, TEMPERATURE-DEPENDENT SEX DETERMINATION.

sex gland (*BioSci*) See GONAD.

sexi- (*Genrl*) See SEX-.

sex-limited character (*BioSci*) A character developed only by individuals belonging to a particular sex.

sex-linked (*BioSci*) Of genes, characters or diseases, located on a SEX CHROMOSOME, eg in mammals on the X chromosome. A sex-linked character is associated with sex in transmission; it appears in one sex in one generation and appears, or is transmitted by, the other sex in the next generation (*criss-cross inheritance*). The X-linked diseases characteristically only affect the heterogametic sex (males) in which only a single copy of the altered allele is present, but are transmitted by females who are heterozygous for the trait.

sex mosaic (*BioSci*) An individual showing characteristics of both sexes; an intersex, gynandromorph.

sex reversal (*BioSci*) The gradual change of the sexual characters of an individual, during its lifetime, from male to female or vice versa. Also *sex transformation*.

sex roles (*Psych*) A set of attitudes, behaviours, perceptions and feelings which are commonly held to be associated with either being male or being female.

Sextans (Sextant) (*Astron*) A very faint equatorial constellation between Leo and Hydra.

sextant (*Surv*) A reflecting instrument in the form of a quadrant, for measuring angles up to about 120°. It consists essentially of two mirrors: a fixed *horizon glass*, half-silvered and half-plain glass, and a movable *index glass*, to which is attached an arm moving over a scale graduated to read degrees directly. The index glass reflects an image of one signal, or body, into the silvered part of the horizon glass, and this image is brought into coincidence with the other signal or body as seen through the plain part of the same glass. The sextant is used chiefly for measuring the altitude of the Sun at sea, the reflected image of the Sun being made to touch the visible horizon. See ASTROLABE.

sextodecimo (*Print*) The 16th of a sheet or a sheet folded four times to make 16 leaves or 32 pages. Also 16mo.

sex transformation (*BioSci*) See SEX REVERSAL.

sexual behaviour (*Psych*) All behaviour leading to the fertilization of eggs by sperm.

sexual coloration (*BioSci*) Characteristic colour difference between the sexes, esp marked at the breeding season. See EPIGAMIC.

sexual dimorphism (*BioSci*) Marked differences between the males and females of a species, esp differences in superficial characters, such as colour, shape, size, etc.

sexual dysfunction (*Psych*) Inhibition of arousal or weakness of the psychophysiological aspects of the sexual response in the absence of any organic disorder.

sexual organs (*BioSci*) The gonads and their accessory structures; reproductive system.

sexual orientation (*Psych*) A feeling of attractedness or arousal preferentially associated with a particular gender.

sexual reproduction (*BioSci*) The union of gametes or of gametic nuclei, preceding the formation of a new individual.

sexual selection (*BioSci*) Selection occurring as a result of mate selection.

Seyfert galaxy (*Astron*) A member of a small class of galaxies with brilliant nuclei and inconspicuous spiral arms. The intensely bright nuclei possess many of the properties of quasars. They are strong emitters in the infrared, and are also detectable as radio and X-ray sources. Carl Seyfert discussed this morphological type in 1943. See panels on GALAXY and QUASAR.

Sezary syndrome (*Med*) A disease syndrome characterized by general redness and thickening of the skin. The skin is infiltrated with lymphocytes with an unusual hairy appearance, and large numbers of similar lymphocytes (Sezary cells) are present in the blood. They have been shown to be T-lymphocytes, but what causes them to move into the skin is not known.

SF (*Eng*) Abbrev for STRUCTURAL FOAM.

sferics (*EnvSci*) Lightning flashes or other natural electrical impulses esp in relation to the determination of their location by simultaneous radio direction-finding using a number of aerials. The word is derived from atmo*spherics*.

SGHWR (*NucEng*) Abbrev for STEAM-GENERATING HEAVY-WATER REACTOR.

S-glass (*Glass*) A glass-fibre composition of high tensile strength and high modulus (percentage by weight: SiO_2 65%, Al_2O_3 25%, MgO 10%), developed for reinforcing composite materials in aerospace applications.

SGML (*Print, ICT*) Abbrev for STANDARD GENERALIZED MARKUP LANGUAGE, a form of coding of electronic data for printed applications.

sgraffito (*Build*) A surface decoration in which two finishing coats of contrasting colours are applied, one over the other. Before the upper one has set, parts are removed according to some design exposing the coat below. Also *graffito*.

shackle (*Eng*) U-shaped machine element for connecting two links of a chain or two ends of a wire rope.

shackle insulator (*ElecEng*) A porcelain insulator whose ends are secured to metal shackles.

shade (*Surv*) A disk of coloured glass used in telescope or theodolite when making Sun observations.

shade (*Textiles*) (1) The colour of a material, usually one that has been dyed. (2) To modify the colour of a fabric being dyed to bring it nearer to that required.

shaded pole (*ElecEng*) A pole having a short-circuited ring around one section, thus altering the phase of the flux over that section. Sometimes used to make small single-phase motors self-starting.

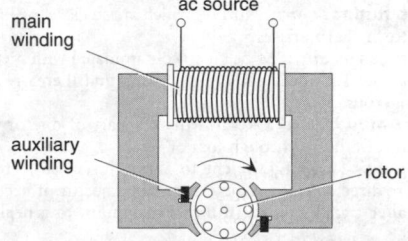

shaded pole The auxiliary winding is a copper ring retarding the flux at the edge of the pole.

shade plant (*BioSci*) A plant adapted to living at low light intensities. Cf SUN PLANT.

shading (*ImageTech*) Unwanted variation of brightness within the picture area, not forming part of the image; where this originates in the camera, correction signals can be inserted.

shadow (*ImageTech*) Ineffectiveness of reception because of an obstacle, eg due to the topography of the terrain, between the transmitter and the receiver.

shadow (*Phys*) Of an obstacle cast by a point source of light, the geometrical projection of the obstacle, except for small-scale diffraction effects at the edge. See UMBRA.

shadow casting (*PowderTech*) A method, carried out in high-vacuum equipment, of determining the thickness of a particle of powder or other structure on a microscope slide. A beam of vaporized metal (usually gold or chromium) is directed towards the specimen slide at an oblique angle. The thickness of the particles can be measured from the angle of approach of the beam and the dimensions of the shadow cast by the particles.

shadow fringe test (*Phys*) A technique for examining the optical quality of glass. The shadows formed on

transmission of a beam of light limited laterally are examined, since inequalities in the refractive index appear as fringes in the shadow.

shadowing technique (*BioSci*) A technique of shadow casting used in electron microscopy, in which a very thin non-granular film of a metal, eg chromium, gold, uranium, is deposited obliquely onto the surface of the specimen prior to examination. This gives a three-dimensional effect and improves the clarity of the surface contours.

shadow loss (*ICT*) Attenuation of radio signals due to the presence of obstacles. In a mobile network, the varying shadow loss has a log-normal distribution (ie Gaussian when expressed in DECIBELS) made up of a global component applying to all signals plus a component for each separate path.

shadow-mark (*Paper*) A defect of paper showing in the *look-through* as a faint reproduction of the holes of the suction couch or press rolls.

shadow mask (*ICT*) A perforated metal sheet situated behind the phosphor screen in some colour TV tubes.

shadow mask tube (*ImageTech*) A type of directly viewed three-gun cathode-ray tube for colour TV display, in which beams from three electron guns converge on holes in a shadow mask placed behind a tricolour phosphor-dot screen.

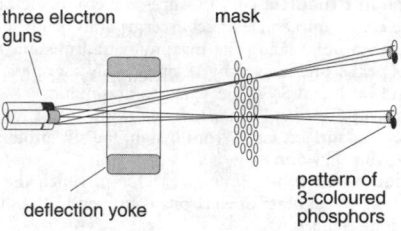

shadow mask tube In practice the mask lies directly in front of the phosphors with one pattern for each of the 200 000 holes in the mask.

shadow photography (*ImageTech*) High-speed technique using an electric spark or similar light source to photograph the shadow of a fast-moving object such as a projectile. Exposures of 10^{-6} s have been attained.

shadow scattering (*Phys*) See SCATTERING.

shadow stripes (*Textiles*) Shadow stripes in which stripes are produced by using warp or weft yarns of different directions of twist, ie *S*- or *Z-twist*. The shadow effect is due to the differing reflectivity of the different twists. See TWIST DIRECTION and panel on FIBRE ASSEMBLIES.

shadow zone (*Acous*) See ZONE OF SILENCE.

shadow zone (*Geol*) See panel on EARTHQUAKE.

shaft (*Arch*) The principal portion of a column between the capital and the base.

shaft (*BioSci*) (1) The part of a hair distal to the root. (2) The straight cylindrical part of a long limb bone. (3) The rachis, or distal solid part of the scapus of a feather.

shaft (*CivEng, MinExt*) A passage, usually vertical, leading from ground level into an underground excavation, for purposes of ventilation, access, etc. See fig. at MINING.

shaft furnace (*Eng*) One in which ore and fuel are charged into the top and gravitate vertically, reacting as they proceed to bottom discharge.

shaft governor (*Eng*) A compact type of spring-loaded governor used for controlling the speed of small oil engines etc. It is arranged to rotate about the crankshaft axis, and is sometimes housed in the flywheel. See SPRING-LOADED GOVERNOR.

shafting (*Eng*) See LINE SHAFTING.

shaft pillar (*MinExt*) Solid block of coal or ore left unworked round the bottom of a shaft or pit for support.

shaft station (*MinExt*) A room excavated underground adjacent to shaft, to accommodate special equipment such as pumps, crushing machine, truck tipples, ore sorting equipment and surge storage bins.

shaft turbine (*Aero*) Any gas turbine aero-engine wherein the major part of the energy in the combustion gases is extracted by a turbine and delivered, through appropriate gearing, by a shaft. See AERO-ENGINE, FREE TURBINE.

shake (*For*) A partial or complete separation between adjacent layers of fibres in wood. See CUP SHAKE.

shaken baby syndrome (*Med*) A collection of symptoms, including brain damage and paralysis, that can occur when an infant is shaken violently by an adult.

shaking grate (*Eng*) A grate for a hand-fired boiler furnace in which the pivotally supported fire-bars can be rocked by hand levers in order to break up clinker.

shaking table (*MinExt*) See WILFLEY TABLE.

shale (*Geol*) A consolidated clay rock which possesses closely spaced well-defined laminae. Cf MUDSTONE. See OIL SHALE.

shale oils (*MinExt*) Oils obtained by the pyrolysis of oil shale at approx 550°C and characterized by a large proportion of unsaturated hydrocarbons, eg alkenes and di-alkenes.

shank (*Build*) (1) Shaft of column, pillar, etc. (2) Shaft of tool, connecting head and handle.

shank (*Eng*) A ladle for molten metal.

shank (*Print*) See BODY.

Shannon's theorem (*ICT*) Concerns the ultimate capacity of a communication channel in terms of its bandwidth and signal-to-noise ratio. The maximum transmission rate in bits per second is given by $W\log_2(1 + SN)$, where W is the bandwidth and SN is the signal-to-noise ratio.

shantung (*Textiles*) Plain-weave silk cloth, with a randomly irregular surface; made from tussah, the silk produced by the wild silkworm.

shaped-beam tube (*Electronics*) One in which the cross-section of the beam of electrons is formed to the shape of various characters.

shaped-conductor cable (*ElecEng*) A three-phase cable in which the conducting cores are specially shaped so as to give the best utilization of the total available cross-section of the cable.

shape factor (*Eng*) The ratio of single loaded to total unloaded surface area of rubber in laminated bearing; the higher it is, the greater the resistance to downward thrust from the structure.

shape factor (*PowderTech*) A parameter descriptive of the shape of a particle of a powder or the ratio of two average particle sizes determined by techniques in which the shapes of the particles of a powder influence the measured parameter in different ways.

shape memory alloys (*Eng*) Alloys which undergo mechanical twinning, or a martensitic or similar reversible solid-state transformation which involves a dimensional change, occurring usually over a narrow temperature range. This enables a shape produced in one state to be recovered if the temperature is altered back again despite the shape having changed in the interim. Useful for thermostats and on–off switches and for clamping devices such as pipe couplings where a loose sleeve can be caused to contract and clamp tightly by heating or cooling to induce the shape recall dimensions.

shaper tools (*Eng*) Cutting tools similar to those used on planing machines, and similarly supported in a CLAPPER BOX.

shaping (*Psych*) The training of a response by successively reinforcing responses that are increasingly similar to the target behaviour, until that behaviour is reached. Also *conditioning by successive approximations*.

shaping (*Textiles*) A process used to change the width of knitted fabrics or garments by changing the number of stitches in the course or wale directions. Includes *fully fashioned*. In stitch-shaped garments the dimensions are changed by altering the stitch length and/or structure.

shaping machine (*Eng*) A machine tool for producing small flat surfaces, slots, etc. It consists of a reciprocating

ram carrying the tool horizontally in guide ways, and driven by a QUICK RETURN MECHANISM. Either the tool or the table is capable of traverse.

shaping network (*ICT*) One that determines or restores the shape of a pulse, esp in radar and computing.

shard (*Geol*) A fragment of volcanic glass, often with curved edges. Glass shards are important constituents of some pyroclastic rocks.

shared-channel broadcasting (*ICT*) See COMMON-FREQUENCY BROADCASTING.

shared memory (*ICT*) Fast memory to which more than one processor has access.

shared protection ring (*ICT*) A network arrangement that provides continuity of service in the event of a link failure. All network nodes are connected in a ring, with the capacity of each link divided between working and protection functions. In the event of failure, switches operate on both sides of the failed link to route traffic from the working capacity through the spare protection capacity.

shareware (*ICT*) Software distributed on the expectation that a regular user will later pay for it.

sharkskin (*Eng*) Defective surface on an extrudate, shown by lines at right angles to polymer extrusion direction. See EXTRUSION.

sharkskin (*Textiles*) Woven or warp-knitted fabric with a characteristically firm construction and stiff handle, often with a dull appearance, used as a dress and suiting material.

sharp (*Build, CivEng*) Said of sand the grains of which are angular, not rounded.

sharp coat (*Build*) A thin oil paint which contains much pigment and little medium but with a high proportion of solvent and used as primer or sealer.

sharpening image (*Print*) Printing image losing its printing area from the edges by eg half-tone dots becoming smaller. One cause is attack by an overactive fountain solution.

sharp flutings (*Arch*) Flutings which are so close together as to form sharp arrises.

sharp gas (*MinExt*) Mine air so contaminated with methane as to burn inside the Davy-type lamp and therefore to be dangerous.

sharp mouth (*Vet*) Overgrowth of a part of one or more teeth of a horse through loss or wear.

sharpness (*ICT*) Equivalent to SELECTIVITY, but referring more directly to the change in circuit adjustment necessary to alter signal strength from its maximum to a negligible value.

sharpness (*ImageTech*) The subjective impression of the amount of detailed information provided in a picture image; it is affected by the type of subject and viewing conditions as well as the reproduction characteristics of the system. See DEFINITION, MTF, RESOLUTION.

sharpness of resonance (*Phys*) The rapidity with which resonance phenomena arise and then disappear as the frequency of excitation of a constant driving force is varied through the resonant frequency.

sharp paint (*Build*) Oil paint drying rapidly to give a flat surface.

sharp series (*Phys*) Series of optical spectrum lines observed in the spectra of alkali metals. Has led to energy levels for which the orbital quantum number is zero being designated *s-levels*.

shavehook (*Build*) A tool used to remove paint from moulded areas during burning off or when using liquid paint removers. Available with three shapes of head and pulled across the surface: (1) pear shape; (2) triangular shape; (3) combination shape made to suit various contours.

shaving (*Acous*) Machining the surface of a master vinyl disk recording to give fresh surface for further use.

shear (*BioSci*) To cut the long, stiff DNA duplex by hydrodynamic means.

shear (*Eng, Phys*) A type of deformation in which parallel planes in a body remain parallel but are relatively displaced in a direction parallel to themselves with a tendency for adjacent planes to slide over each other. A rectangle, if subjected to a shearing force parallel to one side, becomes a parallelogram. See ELASTICITY OF SHEAR, STRAIN, TORSION.

shear (*Textiles*) See SHEARING.

shearer loader (*MinExt*) A machine which cuts coal from the seam and loads it in the same operation to a conveyor belt working parallel to the face. In fully mechanized mining the assembly, together with roof support props, is moved hydraulically and can be remotely controlled.

shear force (*Eng*) A force which tends to cause sliding of adjacent layers, relative to each other, in a material.

shear heating (*Phys*) See ADIABATIC HEATING.

sheariness (*Build*) A paint defect similar to FLASHING, possibly caused by failure to keep a wet edge during painting. Most apparent when viewed across the sheen, most prone on flat or semi-glass finishes.

shearing (*NucEng*) See CHOPPING.

shearing (*Textiles*) (1) Cutting the wool from a living sheep (cf FELLMONGERING, PULLING, SKIN WOOL). (2) Trimming a pile on a fabric to a uniform height. Also *cropping*.

shear-legs (*Eng*) A large lifting device used in shipyards etc, resembling a crane in which a pair of inclined struts take the place of a jib. Also *sheer-legs, sheers*.

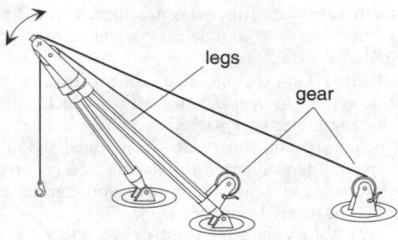

shear-legs Simple hand-operated design.

shear lip (*Eng*) Small, often linear zone on fracture surface where brittle crack meets edge of specimen, and yielding occurs. See BRITTLE FRACTURE.

shear modulus (*Eng*) One of the four basic elastic constants of an isotropic, linear elastic material. Usual symbol G, sometimes μ. Defined by $G = \tau/\gamma$, where τ is the shear stress and γ is the shear strain. Related to the other constants by

$$G = \frac{E}{2(1+\nu)} = \frac{3EK}{9K-E}$$

where E is YOUNG'S MODULUS, K is the BULK MODULUS, and ν is POISSON'S RATIO. Also *rigidity modulus*.

shear mouth (*Vet*) An increase in the obliquity of the wearing surfaces of the molar teeth of horses.

shear pin (*Eng*) A pin used as a safety device to connect elements in a power transmission system, which is strong enough to transmit permissible loads but will fail by shearing when these loads are exceeded.

shear ram (*MinExt*) Hydraulically operated sliding jaws designed to cut off flow near the BLOWOUT PREVENTER. It compresses the pipe and cuts it. The well is then *shut in*, but the gear above can be removed.

shear rate (*Eng*) The rate at which the velocity of a fluid under shear changes through its thickness (γ). Units are reciprocal seconds. Basic parameter in Newtonian viscosity. See EXTRUSION, INJECTION MOULDING.

shears (*Eng*) See WAYS.

shear strain (*Eng*) The angular displacement of adjacent parallel planes in a body subject to shear. In a body of length l twisted through an angle θ when loaded in torsion, the shear strain γ at a distance r from the torsional axis is $\gamma = r\theta/l$.

shear stress (*Eng*) Shearing force divided by the area over which it is acting. Shear stress occurs across the section of a beam loaded transversely and also across shaft sections subject to torque. The magnitude of the stress varies across the section depending on its shape.

shear wave (*Acous*) TRANSVERSE WAVE without compression of the medium.

shear wave (*Geol*) See panel on EARTHQUAKE.

shear zones (*Geol*) Bands in metamorphic rocks consisting of crushed and brecciated material and many parallel fractures. See STRAIN-SLIP CLEAVAGE.

sheath (*BioSci*) An enclosing or protective structure, eg elytron of some insects, leaf sheath. Can be a tissue layer that surrounds other tissues, eg bundle sheath.

sheath (*ElecEng*) The covering on a cable.

sheath (*Electronics*) Excess of positive or negative ions in a plasma, giving a shielding or SPACE-CHARGE effect.

sheath (*NucEng*) The can protecting a nuclear fuel element.

sheath-circuit eddies (*ElecEng*) The paths of currents in the sheaths of separate cables which flow only when the sheaths are bonded. See SHEATH EDDIES.

sheath current (*ElecEng*) The eddy current flowing in the metallic sheath of an ac cable.

sheath eddies (*ElecEng*) Currents which are induced in the sheath of a single cable, and which flow even when the sheaths are isolated. Cf SHEATH-CIRCUIT EDDIES.

sheath effects (*ElecEng*) The phenomena associated with the metallic sheaths of cables carrying ac.

sheathing (*Build*) Close boarding nailed to the framework of a building to form the walls or the roof.

sheathing paper (*Build*) A flexible waterproof lining material made from bitumen reinforced with fibre, and faced with stout kraft paper.

sheath voltage (*Electronics*) Electrostatic potential across a space-charge SHEATH region. In low-pressure plasma processing (SPUTTERING, ETCHING, etc) use is often made of the fact that positive ions are accelerated by the sheath voltage onto adjacent surfaces.

sheave (*Eng*) Grooved pulley for use with vee-belts, ropes or round belts.

shed (*Phys*) Minute unit of nuclear cross-section, 10^{-52} m^2 or 10^{-24} barn.

shed (*Textiles*) Opening created by dividing the warp threads during weaving so that the shuttle or other device can take the weft through to be beaten up into cloth.

sheen (*Build*) The degree or lustre, shine or reflection on a surface or finish.

sheep ked (*Vet*) A blood-sucking, wingless fly, *Melophagus ovinus*, which lives on the wool and skin of sheep.

sheep pox (*Vet*) A highly contagious disease of sheep caused by a virus and characterized by a papulo-vesicular eruption of the skin and mucous membranes of the respiratory and alimentary tracts.

sheep scab (*Vet*) See PSOROPTIC MANGE.

sheer-legs (*Eng*) Also *sheers*. See SHEAR-LEGS.

sheerstrake (*Ships*) The top strake or line of plating below, but extending a little above, the freeboard deck.

sheet (*Aero*) The general term for aircraft structural material under 0·25 in (6 mm) thick; above that it is usually called plate.

sheet (*MathSci*) One of the separate continuous surfaces of a CONIC.

sheet (*Print*) A term applied to any one piece of printing paper printed or plain.

sheet anchor (*Ships*) A third bower anchor carried abaft the starboard bower for use in an emergency. Formerly *waist anchor*.

sheet-fed (*Print*) A term applied to a rotary machine indicating that it prints separate sheets and not a reel or web.

sheet furnace (*Eng*) One in which metal sheet is heated before further size reduction in a rolling mill.

sheet glass (*Glass*) A form of FLAT GLASS, now largely superseded by FLOAT GLASS, used largely for glazing purposes; produced by drawing a continuous film of glass upwards from a molten bath and, after a suitable time interval for cooling, cutting up the product into sheets. It is not of such good quality, or so flat, as PLATE GLASS which is ground and polished. See panel on GLASSES AND GLASS-MAKING.

sheeting (*CivEng*) Horizontal boards used to support the sides of narrow trenches during excavation in very loose soils, each pair of boards on opposite sides of the trench being wedged apart with struts.

sheeting (*Textiles*) A medium-weight, closely woven, plain- or twill-weave fabric used for bedding.

sheet lead (*Build*) Lead in a form in which it is commonly used in building construction. Trade practice refers to it in terms of the mass of unit area, eg 7-lb sheet lead, meaning $7\,\text{lb}\,\text{ft}^{-2}$.

sheet lightning (*EnvSci*) Diffuse illumination of clouds by distant lightning of which the actual path of the discharge is not seen.

sheet materials (*Chem*) Materials in which structure is formed of very strong covalently bonded sheets of atoms, held together by relatively weak bonds. Thus mica and talc minerals contain sheets of silicon–oxygen, held together electrostatically, so are easily cleaved. Graphite has a similar structure of carbon–carbon sheets.

sheet pavement (*CivEng*) A road surfacing formed of continuous material such as concrete. See CONTRACTION JOINTS.

sheet piling (*CivEng*) A continuous retaining wall made from sheet steel piles so shaped that adjacent members interlock as in the drawing.

sheet severer (*Print*) An automatic device for cutting the web to prevent wrap-around.

sheet wander (*Print*) Undesirable lateral movement of the running web of paper.

sheet-work (*Print*) Work in which two formes are used, one for each side of the paper, to give one complete copy of the job or section per sheet. Cf WORK-AND-TURN.

shelf-back (*Print*) See SPINE.

shelf-life (*Build*) The length of time which a paint will remain usable when stored.

shelf-life (*FoodSci*) The time during which any changes taking place during food storage have no adverse effects on its safety or its ORGANOLEPTIC or visual characteristics. It depends on the nature of the product, the method of its processing, its packaging and its storage conditions.

shelf stable (*FoodSci*) A product which remains palatable and free of spoilage organisms during storage or display at ambient temperature for a finite time after its packaging has been opened or removed.

shelfware (*ICT*) A system or set of programs developed for reasons other than commercial marketability, and that are expected to achieve few or no sales or usage.

shell (*BioSci*) A hard outer case or exoskeleton of inorganic material, chitin, lime, silica, etc.

shell (*ICT*) The availability of OPERATING-SYSTEM commands within an APPLICATION PROGRAM. Thus the user may be able to transfer temporarily to the operating system without completely leaving the current application. Term originated in the UNIX operating system.

shell (*Phys*) See ELECTRON SHELL.

shellac (*Build*) The refined form of lac, the secretion of the insect *Lacifer lacca*, parasitic on trees in SE Asia. Consists of a mixture of a red dye, wax and natural polymer etc. The polymer fraction is a low-molecular-mass polyester with numerous hydroxyl (–OH) and carboxylic acid (–COOH) side and end groups on the polymer chains. Thermal polymerization at 150–160°C gives a hard, brittle solid used for electrical insulation when reinforced with mica particles. Also used for FRENCH POLISHING.

shell bit (*Build*) A bit shaped like a narrow gouge, used for screw holes. Also *gouge bit*.

Shelldyne (*Aero*) TN for synthetic fuel having a higher than normal density for expendable turbojets.

shell gland (*BioSci*) In some invertebrates, a glandular organ which secretes the materials for the formation of the shell. Also *shell sac*.

shell ligament (*BioSci*) The dorsal ligament joining the valves of the shell in bivalve Mollusca.

shell model (*Phys*) A model of the nucleus of the atom with a nucleon moving independently in the common field representing the effect of the other nucleons. This leads to the nucleons being arranged in shells as for the ELECTRON SHELLS in atomic structure.

shell program (*ICT*) A skeleton of a program that can be developed for particular applications by the user.

shell reamer (*Eng*) A reamer in the form of a hollow cylinder, end-mounted on an arbor. This construction is used for economy in relatively large reamers as reamers are made of expensive materials.

shell sac (*BioSci*) See SHELL GLAND.

shell shock (*Psych*) See POST-TRAUMATIC STRESS DISORDER.

shell star (*Astron*) One of a number of stars of spectral type O, B or A, which is surrounded by a shell of luminous gas giving bright emission lines.

shell-type transformer (*ElecEng*) A transformer in which the magnetic circuit surrounds the windings more or less completely.

shelterbelt (*For*) Planting of trees in narrow strips to give wind shelter for crops, animals or other planting.

shelter deck (*Ships*) A term correctly interchangeable with AWNING DECK. Sometimes used informally for a deck above a WEATHER DECK.

shepherd satellites (*Astron*) Small moons whose gravitational fields serve to confine narrow rings around some of the outer planets.

sheradizing (*Eng*) A zinc–zinc oxide corrosion-resistant coating applied to iron and steel articles which also acts as a paint key. Objects are packed in a finely divided mixture of zinc powder and zinc oxide. Then heated at 350–450°C for a period depending on the coating thickness required.

sheridanite (*Min*) A mineral in the chlorite group poor in iron and relatively low in silica.

SHF (*ICT*) Abbrev for SUPER-HIGH FREQUENCY.

shide (*Build*) See SHINGLE.

shield (*CivEng*) Hollow steel cylinder adapted for use in driving a tunnel through loose or water-bearing ground, protecting workers and machinery at the face. It is driven forward as excavation proceeds by means of hydraulic jacks bearing on preceding lines of tunnel.

shield (*ElecEng, NucEng*) Screen used to protect persons or equipment from electric or magnetic fields, X-rays, heat, neutrons, etc. In a nuclear reactor the shield surrounds it to prevent the escape of neutrons and radiation into a protected area. See BIOLOGICAL SHIELD, FARADAY CAGE, MAGNETIC SHIELD, NEUTRON SHIELD, THERMAL SHIELD.

shield (*Electronics*) A metallic housing or screen made of earthed mesh or thin sheet, placed around a component or circuit to suppress or isolate electromagnetic fields which might otherwise cause interference.

shield (*Geol*) A large stable area of the Earth's crust consisting of Precambrian rocks. Effectively synonymous with CRATON, eg Canadian shield.

shielded box (*NucEng*) GLOVE BOX protected by lead walls and lead-glass windows, and with facilities for manipulation of contents by remote handling equipment.

shielded line (*ElecEng*) A line or circuit which is specially shielded from external electric or magnetic induction by shields of highly conducting or magnetic material.

shielded line (*ICT*) See SELF-SHIELDING.

shielded metal arc welding (*Eng*) The use of covered consumable electrodes to provide protection for the weld, as the covering vaporizes.

shielded nuclide (*Phys*) A nuclide, found among fission fragments, but known not to have been formed as a result of beta decay. It is therefore assumed to have been a direct fission product.

shielded pair (*ICT*) Balanced pair of transmission lines within a screen, to mitigate interference from outside.

shielded twisted pair cable (*ICT*) A data cable consisting of pairs of wires twisted together surrounded by a metal screening sheath to reduce interference. Cf UNSHIELDED TWISTED PAIR CABLE.

shielding (*ElecEng*) (1) Prevention of interfering currents in a circuit, due to external electric fields. Any complete metallic shield earthed at one point is adequate. (2) Use of high-permeability material, eg Mumetal, for shielding devices susceptible to a magnetic field, eg cathode-ray beam; the field, direct or alternating, is shunted away from spaces where it would cause interference.

shielding (*NucEng, Radiol*) Protective use of low-atomic-number materials to thermalize strong neutron beams, or concrete, lead or other heavy materials to shield against gamma radiation when it might be harmful to operator or measuring system. Concrete is commonly used for large areas and lead for smaller.

shielding pond (*NucEng*) See SHIELD POND.

shielding windows (*NucEng*) Dense glass blocks or liquid-filled tanks used as windows for inspecting the interior of shielded boxes.

shield pond (*NucEng*) Deep tank of water used to shield operators from highly radioactive materials stored and manipulated at the bottom. Also *shielding pond*.

shift (*Build*) See BREAKING JOINT.

shift (*Electronics*) Movement of a pattern on a CRT phosphor, by imposition of steady voltages, eg X-shift, Y-shift.

shift (*ICT*) (1) In computing, an operation that moves the bits held in a memory location to the left or right as specified. There are three different types of shift. See ARITHMETIC SHIFT, END-AROUND SHIFT, LOGICAL SHIFT. (2) In telecommunications, the double use of code, using one code for changing over, as in Telex, teleprinter, teletype, analogous with typewriter keyboard. In teleprinters, one shift is capital letters, the other figures and special signs (case shift).

shift (*Phys*) (1) Change in wavelength of spectrum line due eg to DOPPLER EFFECT, RAMAN SCATTERING or ZEEMAN EFFECT. (2) Change in value of energy level (*level shift*) for quantum mechanical system arising from interaction or perturbation.

shift factor (*Eng*) A term described by the WILLIAMS–LANDEL–FERRY EQUATION, which is a measure of the distance a stress relaxation curve needs to be moved in order to create a master curve. Symbol *a*. See VISCOELASTICITY.

shifting dullness (*Med*) Impaired resonance on PERCUSSION of the abdominal flanks which shifts when the patient rolls on his or her side and indicates free fluid in the peritoneal cavity, ASCITES.

shifting of brushes (*ElecEng*) The displacement of the brushes of a commutator motor from the neutral position.

shikimic acid (*BioSci*) *3,4,5-trihydroxy-1-cyclohexene-1-carboxylic acid*. An important cyclic intermediate in the synthesis, in plants, of the aromatic amino acids and other aromatic compounds from non-aromatic precursors.

shim (*Eng*) A thin strip of material, used singly or in multiples, to take up space between clamped parts.

shim (*Phys*) A packing piece consisting of a thin sheet of magnetic material for placing behind a pole piece in a magnetic circuit to adjust an air-gap.

shim (*Print*) A sheet of metal or plastic on which flexible plates can be mounted and then secured to the plate cylinder, or to the bed. Sometimes called *draw sheet*, thus risking confusion with the top sheet of the impression cylinder.

shimamushi fever (*Med*) An acute febrile disease associated with infection by rickettsiae, transmitted by the bite of a larval mite; it is characterized by fever, enlargement of lymphatic glands in the neck, axilla and groin, conjunctivitis, and a dark-red macular rash. Also *flood fever, Japanese river fever, tsutsugamushi fever*.

shimming (*Phys*) Adjustment of magnetic field with soft iron shims or, by extension, small compensating coils.

shimmy (*Aero*) The violent oscillation of a castoring wheel (in practice the nose or tail wheel of an aircraft) about its castor axis, which occurs when the coefficient of friction between the surface and the tyre exceeds a critical value. It is usually suppressed by a friction, spring or hydraulic device called a *shimmy damper* (see DAMPER) or by a twin-tread tyre.

shimmy (*Autos*) See WHEEL WOBBLE.

shim rod (*NucEng*) Coarse control rod of reactor. It is usually positioned so that the reactor will be just critical when the rod is near the centre of its travel path. It is designed to move slowly unless it is also used as a SAFETY ROD, when a magnetic clutch allows it to drop rapidly into the core. See panel on NUCLEAR REACTORS.

shinbone (*BioSci*) The *tibia*.

shiner (*Build*) A thin flat stone laid on edge in a rubble wall, the width of the stone being equal to the depth of at least two courses of the other stones.

shiner (*Print*) See LIGHT TABLE.

shiners (*Paper*) Particles, usually from mica in the china clay, or undissolved alum, showing on the surface of the paper.

shingle (*Build*) A thin, flat, rectangular piece of wood laid in the manner of a slate or tile, as a roof covering or for the sides of buildings. Normally of tapercut red cedar. Also *shide*.

shingle (*Geol*) Loose detritus, generally of coarser grade than gravel though finer than boulder beds, occurring typically on the higher parts of beaches on rocky coasts.

shingles (*Med*) See HERPES ZOSTER.

shin splints (*Med*) Inflammation of the muscles around the shinbone caused by strenuous exercise, esp running on hard surfaces.

shipboard aircraft (*Aero*) Any aircraft designed or adapted for operating from an aircraft carrier; special modifications are strengthened landing gear, strong points for catapulting, arrester hook and, if large, folding wings.

ship caisson (*Build*) A floating caisson shaped like a ship, and capable of being floated into position across the entrance to a basin, lock or graving dock, and then sunk into grooves in the sides and bottom of the entrance.

shiplap (*Build*) A term applied to parallel boards having a rebate cut in each edge, the two rebates being on opposite faces. They are esp adapted for use as sheathing.

shippers (*Build*) Bricks which are sound and hard-burned, but not of good shape.

shipping fever (*Vet*) See EQUINE INFLUENZA, HAEMORRHAGIC SEPTICAEMIA.

shipping pneumonia (*Vet*) See HAEMORRHAGIC SEPTICAEMIA.

shish-kebabs (*Plastics*) Polymer microstructure where lamellae form at right angles (ie epitaxially) to an oriented fibril, strung out along it at regular intervals like the eponymous product of the Middle East.

shiva (*Phys*) Powerful laser capable of producing pulses of up to 15 kJ of energy, for use in nuclear fusion experiments.

shivering (*Vet*) A disease of horses, of unknown cause; characterized by involuntary spasmodic contractions of the muscles of one or both hind-limbs and tail.

shives (*Paper*) Undigested particles of wood showing as pale-yellow to brown splinters in the wood pulp or paper.

shives (*Textiles*) (1) Vegetable matter found in wool fleece. (2) Small particles of woody tissue removed from flax fibres in SCUTCHING and HACKLING.

shm (*Phys*) Abbrev for SIMPLE HARMONIC MOTION.

shoad (*MinExt*) Water-worn fragments of vein minerals found on the surface away from the outcrop. Also *float-ore, shode*.

shoal (*Genrl*) A submerged sand bank.

shock (*Med*) Acute peripheral circulatory failure due to diminution in the volume of circulating blood and usually characterized by a low blood pressure, a weak thready pulse and diminished urine output following severe haemorrhage, sepsis and fluid loss. Shock may also occur with a preserved circulating blood volume when the heart pump ls defective, eg in myocardial infarction and pulmonary embolism.

shock (*Psych*) An unexpected and intense experience which compels a total reorientation to life. See TRAUMA.

shock absorber (*Aero*) See OLEO.

shock absorber (*Autos*) See DAMPER.

shock heating (*Phys*) Heating, esp of a plasma, by the passage of a shock wave.

shockproof switch (*ElecEng*) A switch having all its external metallic parts covered, or protected by insulating material, in order to guard against the possibility of electric shock. Also *all-insulated switch*, *Home Office switch*.

shock tube (*Space*) A device which generates high-speed flows of air over short periods of time by the passage of a shock wave down a tube; used to simulate re-entry conditions in a shock tunnel.

shock wave (*Aero, Phys*) A surface of discontinuity in which the airflow changes abruptly from SUBSONIC to SUPER-SONIC, ie from viscous to compressible fluid conditions, thus causing an abrupt rise in pressure and temperature. When a shock wave is caused by the passage of a supersonic body the airflow will decelerate to subsonic conditions through another shock wave. A supersonic body normally sets up a conical shock wave with its nose, the angle ($= \mathrm{cosec}\,(1/M)$, where M is the Mach number) becoming increasingly acute the higher the speed, together with subsidiary shock waves from projections on the body and a decelerating shock wave from its tail. The nose and tail shock waves, either attached or travelling on after the passage of an aircraft, are the source of the pressure waves causing SONIC BOOMS. Abbrev *shock*.

shock wave (*Phys*) A wave of high amplitude, in which the group velocity is higher than the phase velocity, leading to a steep wavefront. Occurs eg when an explosive is detonated. The speed at which the chemical reaction travels through the material is higher than the speed of sound in the material, hence a shock wave occurs. Strong shocks cause luminosity in gases, and so are useful for spectroscopic work. Also *blast wave*.

shoddy (*Textiles*) (1) Short, fibrous material obtained from old, cleaned, loosely woven or knitted wool cloth after treatment, in a rag-tearing machine. Sprayed with oil, it is reused in blends for cheap suits and coatings. (2) The short dirty fibres which drop from woollen cards.

shode (*MinExt*) See SHOAD.

shoe (*Build*) The short bent part at the foot of a DOWNPIPE, directing the water away from the wall.

shoe (*MinExt*) The replaceable steel wearing part of the head of a stamp or muller of a grinding pan.

Shone ejector (*Build*) A type of EJECTOR using compressed air to raise sewage from a lower level.

shonkinite (*Geol*) A coarse-grained, feldspar-rich syenite, consisting largely of pyroxenes and some olivine. Named after the laccolith, Shonkin Sag, Montana.

shoo flies (*Print*) A feature of the TWO-REVOLUTION press; they direct the gripper edge of the printed sheet clear of the opened grippers and the strippers with which they work in conjunction.

shoot (*BioSci*) A stem with all its branches and appendages developed from a bud.

shooting (*Build*) The operation of truing with a *jointing plane* the edges of timbers which are to be accurately fitted together.

shooting (*ImageTech*) The original photography or recording of a film or video production.

shooting board (*Build*) A prepared board used to steady a piece of timber whilst shooting the edges. It has a stop against which the piece of timber abuts endwise, and a guide surface against which the jointing plane runs.

shooting plane (*Build*) See JOINTING PLANE.

shooting star (*Astron*) See METEOR.

shooting stick (*Print*) A short tapered length of hard material used with a mallet to lock up and unlock formes by tapping the wooden quoins.

shoot-tip culture (*BioSci*) Same as MERISTEM CULTURE.

shop priming (*Build*) A term used to describe the process of applying a first coat of paint to an item before delivery to site. Must be done with correct primer to have any beneficial effect.

shop rivet (*Eng*) A rivet which is put in when the work is being erected on the floor of the assembly shop prior to delivery to the site but may be replaced later.

shoran (*Radar*) Abbrev for *short-range navigation*. A precision position-fixing system using a pulse transmitter in an aircraft or other vehicle and TRANSPONDERS at two known fixed points.

Shore hardness (*Eng*) Two fundamentally different ways of assessing the hardness of materials. One, *Shore A*, for softer materials such as elastomers and plastics, measures the depth of penetration of an indenter under specified conditions; the other (see SHORE SCLEROSCOPE) is a rebound test for harder materials such as metals and rocks. See panel on HARDNESS MEASUREMENTS.

Shore scleroscope (*Eng*) An instrument for determining a hardness value for materials by measuring the rebound of a diamond-tipped hammer dropped from a given height. See SHORE HARDNESS.

shoring (*Build*) The method of temporarily supporting by shores, ie props of timber or other material in compression, the sides of excavations and esp unsafe buildings. See RAKING SHORE.

short (*Glass*) A term used for a glass that is fast-setting, ie has a short freezing range.

short (*MinExt*) Brittle.

short circuit (*ElecEng*) The reduction of potential difference between two points in a circuit to zero by connection of a conductor of zero impedance, in which no power is dissipated. If unintended, damage may occur unless the circuit is opened quickly elsewhere.

short-circuit calculator (*ElecEng*) An assembly of variable impedances or resistances which can be connected to represent in miniature the circuits of a power system. If a low voltage is applied and a short circuit put on the system, the currents which flow represent to scale the short-circuit currents which would flow in the actual system under similar conditions. Cf NETWORK CALCULATOR.

short-circuit characteristic (*ElecEng*) The characteristic graph relating emf or excitation to load current in the case of a machine operating under short-circuit conditions.

short-circuit impedance (*ElecEng*) Input impedance of a network when the output is short-circuited or grounded.

short-circuiting device (*ElecEng*) A switching device on the rotor of a slip-ring induction motor; operated by a mechanical clutch, which short-circuits the rotor windings when the motor has gained speed.

short-circuit protector (*ElecEng*) A device for preventing damage from excessive currents caused by a short circuit. See OVERCURRENT RELEASE, OVERLOAD PROTECTIVE SYSTEM.

short-circuit ratio (*ElecEng*) The ratio of the field ampere turns in a synchronous generator at normal voltage and no load to the field ampere turns on short circuit with full-load stator current flowing. The value is important in evaluating and comparing the regulation and stability of machines.

short-circuit test (*ElecEng*) A low-voltage test carried out on an electrical machine with its output terminals short-circuited and full-load current flowing.

short-circuit voltage (*ElecEng*) The emf necessary to cause full-load current to flow under short-circuit conditions.

short column (*Eng*) A column the diameter of which is so large that bending under load may be neglected, and in which failure would occur by crushing; commonly assumed as a column of height less than 20 diameters.

short-cord winding (*ElecEng*) An armature winding employing coils whose span is less than the pole pitch.

short-day plant (*BioSci*) A plant that naturally flowers or shows other morphogenetic change only, or better, as the days shorten. It requires the stimulus of dark period(s) longer than some critical length.

short descenders (*Print*) The length of the descenders (in g, j, etc) is a feature of type design; for bookwork LONG DESCENDERS are usually considered more elegant, but many successful display types have very short descenders. See TYPEFACE.

shortening (*FoodSci*) Fat or oil added to dough that retards the development of gluten to give a 'short' or 'non-chewy' texture. Vegetable shortenings are usually hydrogenated with benefit to the manufacturing process.

shortening capacitor (*ICT*) One inserted in series with an antenna to reduce its natural wavelength.

short inks (*Print*) See LONG INKS.

short-oil (*Build*) A term applied to varnishes etc with a low proportion, ie less than about 40%, of oil content. Cf LONG-OIL.

short-period comet (*Astron*) A comet with a period of less than 150 years, the orbit of which lies entirely within the solar system, eg HALLEY'S COMET. They have been captured into the solar system through gravitational interaction with Jupiter or Saturn.

short-period stability (*Aero*) The rapid-incidence adjustment of a stable vehicle to a disturbance.

short-range force (*Phys*) A non-coulomb force which acts between nucleons and is responsible for the stability of the nucleus.

short-range order (*Crystal*) SUBSTITUTIONAL SOLID SOLUTION in which the strictly ordered arrangement exists over only a few lattice units within the crystal.

short shoot (*BioSci*) In many, esp woody, plants, a side shoot with very short internodes, on which are borne most (eg larch) or all (eg pine) of the foliage leaves and most or all of the fruit (eg most apples). See SPUR. Cf LONG SHOOTS.

short shot (*Eng*) A defect in INJECTION MOULDING where molten polymer fails to reach end of tool cavity.

short-sightedness (*Med*) See MYOPIA.

short take-off and vertical landing (*Aero*) A class of V/STOL aircraft, usually with supersonic capability, whose downward-vectored, reheated, engine efflux is too energetic to permit regular vertical take-off operations from conventional surfaces. Abbrev STOVL.

short tandem repeat (*BioSci*) Short sequences of DNA, normally of length 2–5 base pairs, repeated in tandem fashion. Polymorphisms in these sequences are useful markers in DNA analysis for forensics cases and paternity testing. Abbrev STR.

short-term memory (*Psych*) According to the three-store model of memory, a memory system that keeps memories for short periods, has a limited storage capacity and stores items in a relatively unprocessed form. See LONG-TERM MEMORY, SENSORY STORE.

short-time breakdown voltage (*ElecEng*) The voltage required to break down a cable in a short time (minutes).

short-time rating (*ElecEng*) The output which an electric machine can deliver for a specified short period ($\frac{1}{2}$ h or 1 h) without exceeding a specified safe temperature.

short ton (*Genrl*) A unit of mass, 2000 lb. See TON.

short wave (*ICT*) General designation of radio transmission with wavelengths between about 10 and 200 m.

short-wave therapy (*Med*) Treatment with radio-frequency energy of about 30 MHz. Therapeutic results are due mainly to the heat produced in body tissues. See DIATHERMY.

shoshonite (*Min*) A potassic variety of basaltic trachyandesite.

shot (*ImageTech*) The unit element of action recorded in motion picture or video production; for each shot there may be several TAKES with the same camera and lighting set-up but repeating the action for improved performance.

shot blasting (*Eng*) Blasting with metal shot to remove scale and other deposits from castings etc.

shot drilling (*MinExt*) Boring deep holes by means of hard steel shot fed down rotating hollow cylinder.

shot effect (*Textiles*) The variable colour seen when certain fabrics are viewed at different angles. The effect is obtained by having the warp threads of one colour and the weft threads of a contrasting colour.

shot firer (*MinExt*) A miner who tests for gas and then fires explosive charges in colliery.

shot hole (*CivEng*) A hole bored in rock for the reception of a blasting charge.

shot noise (*Electronics*) Broadband noise associated with the quantized nature of electric charge.

shot peening (*Eng*) Bombarding with rounded shot locally to harden surface layers. Results in significant improvement in fatigue resistance.

shot weight (*Eng*) The mass of polymer required for each moulding cycle. Similarly, *shot volume*. See INJECTION MOULDING.

shoulder (*Electronics*) The part of a characteristic curve where a distinct levelling off is perceptible; eg in an amplifier, where the output power is plotted against input power, a shoulder is seen where the output power approaches the maximum power which can be delivered by the output stages.

shoulder (*Print*) The space from the foot of the BEVEL to the edge of the type body.

shouldered arch (*Arch*) A lintel supported over a door opening upon corbels.

shoulder girdle (*BioSci*) See PECTORAL GIRDLE.

shoulder heads (*Print*) Subheadings set flush to the left.

shoulder nipple (*Build*) A nipple threaded at each end only. Also *barrel nipple*.

shoulder notes (*Print*) Notes which are printed, only one per page, in the outer margin level with the first line of text.

shoulder plane (*Build*) Metal type of rebate plane with reversed cutting bevel; used for fine trimming, esp of wide shoulders.

shoulders (*Build*) The abutting surfaces left on each side of a tenon; they abut against the cheeks of the mortise.

shovel beak (*Vet*) See MANDIBULAR DISEASE.

shovelware (*ICT*) Data, esp originating from traditional media, published in electronic form without appropriate adaptation to the new format.

shower (*Phys*) The result of the impact of a high-energy cosmic ray and photons in the upper atmosphere; a very large number of ionizing particles and photons may be produced, directed downwards towards the Earth in a narrow cone.

showerproofing (*Textiles*) A light proofing given to fabrics by treating them with metallic salts, insoluble soap, or silicone-based preparations. The thermal and ventilating properties and the general appearance are not much affected by these treatments. The term has no precise meaning and heavy rain should be expected to penetrate coats made from such material.

shower unit (*Phys*) The mean path length for a reduction of 50% of the energy of cosmic rays as they pass through matter.

show-through (*Print*) The appearance of print through another sheet placed on top of the printed sheet; paper should be sufficiently opaque to prevent this. Cf STRIKE-THROUGH.

shread head (*Build*) See JERKIN HEAD.

shrinkage (*CivEng*) The difference in space occupied by material before excavation and after settlement in an embankment. Also the contraction of concrete after placing.

shrinkage (*Eng*) Dimensional misfit between cool plastic product and tool cavity used to make it. Useful for ease of removal at end of moulding cycle, but needs careful control for final performance of product. Closely related to orientation and polymer crystallization.

shrinkage (*Textiles*) The reduction in dimensions of a fibre, yarn or fabric that takes place in processing or in wear. Particularly effective shrinking treatments are wetting including any laundering cycle.

shrinkage allowance (*Eng*) The difference in diameter, when both are cold, of two parts to be united by shrinking. See SHRINKING-ON.

shrinkage porosity (*Eng*) Cavities produced within a solidified mass of eg ingot or casting, due to specific volume contraction on solidification. See PIPE. Usually appears as ragged edged cavities associated with the last regions to solidify in contrast to the smooth rounded voids resulting from GAS POROSITY.

shrinkage rule (*Eng*) See PATTERN-MAKER'S RULE.

shrinking-on (*Eng*) The process of fastening together two parts by heating the outer member so that it expands sufficiently to pass over the inner and on cooling grips it tightly, eg in the attachment of steel tyres to locomotive wheels.

shrink-resistant finish (*Textiles*) A chemical or mechanical treatment applied, particularly to a fabric, to increase its dimensional stability in use. See COMPRESSIVE SHRINKAGE.

shrink-ring commutator (*ElecEng*) A high-speed commutator in which the segments are held together by a steel ring shrunk on over a layer of insulation.

shrink-wrap (*Plastics*) Thermoplastic film used to cover products, then heat-treated so that film shrinks onto product to provide a seal against the atmosphere.

shroud (*Aero*) (1) Rearward extension of the skin of a fixed aerofoil surface to cover the whole or part of the leading edge of a movable surface, eg flap, elevator, hinged to it. (2) See JET PIPE SHROUD.

shroud (*Eng*) (1) Circular webs used to stiffen the sides of gear teeth. See FULL SHROUD, HALF SHROUD. (2) An outer or peripheral strip used to strengthen turbine blading. (3) A semicircular deflecting wall formed at one side of an inlet port in some internal combustion engines to promote air swirl in the cylinder. Also *shrouding*.

shroud (*ICT*) The extension of metal parts in valves, and other electrical devices subject to high voltages, so that parts of the insulating dielectric are not excessively stressed.

shroud (*Space*) Streamlined covering, part of a launch system, to protect the payload during launch and to reduce aerodynamic drag; it is ejected when a sufficiently high altitude is reached. Also *fairing*.

shrouded balance (*Aero*) An AERODYNAMIC BALANCE in which the area of the hinge line moves within a space formed by shrouds projecting aft from the upper and lower fixed surfaces.

shroud line (*Aero*) Any one of the cords attaching a parachute's load to the canopy. Also *rigging line*.

shroud tube (*NucEng*) A tube which lies between the position of the control rod in the core and the access for the control-rod actuator, guiding and restraining the control rod when it is partly or fully withdrawn from the core.

shuffs (*Build*) See CHUFFS.

shunt (*CivEng*) To divert a train from one track to another, esp to allow another train to pass along the principal track.

shunt (*Med*) A short circuit usually between blood vessels allowing an abnormal circulation of blood from a high-pressure blood vessel or heart chamber to a lower-pressure blood vessel or chamber.

shunt (*Phys*) The addition of a component to divert current in a known way, eg from a galvanometer, to reduce temporarily its effective sensitivity.

shunt characteristic (*ElecEng*) The characteristic graph relating terminal voltage and load current for a shunt-wound dc machine.

shunt circuit (*ElecEng*) (1) Electric or magnetic circuit in which current or flux divides into two or more paths before rejoining to complete the circuit. Also *parallel circuit*. (2) See VOLTAGE CIRCUIT.

shunt-excited antenna (*ICT*) An antenna consisting of a vertical radiator (frequently the mast itself) directly earthed at the base, and connected to the transmitter through a lead attached to it a short way above ground.

shunt field (*ElecEng*) The main field winding of a motor when shunt connected.

shunt-field relay (*ICT*) One with two coils on opposite sides of a closed magnetic circuit, so that a bridging magnetic circuit takes no flux while the currents in the two coils magnetize the circuit when one current is reversed.

shunt-field rheostat (*ElecEng*) A rheostat for insertion in the shunt field of a dc shunt machine; used to vary the speed of a shunt motor or the voltage of a shunt generator.

shunt motor (*ElecEng*) One whose main excitation is derived from a shunt-field winding.

shunt resonance (*ElecEng*) The condition of a parallel tuned circuit connected across an ac voltage supply when maximum impedance is offered to the supply, and the circulating loop current is also a maximum. The term *tunance* is sometimes used in place of resonance if this condition is attained by adjustment of a component value and not of frequency. Also *parallel resonance*.

shunt trip (*ElecEng*) A solenoid-type device, connected in shunt across either the main or an auxiliary supply, which trips a circuit breaker.

shunt voltage regulation (*ElecEng*) That performed by control of a variable impedance in parallel with the output.

shunt winding (*ElecEng*) A field-winding connected in shunt across the armature circuit of a motor.

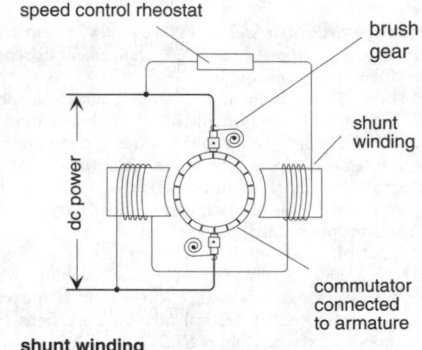

shunt winding

shutdown (*NucEng*) The reduction of power level in a nuclear reactor to the lowest possible value by maintaining the core in a subcritical condition.

shutdown amplifier (*NucEng*) See TRIP AMPLIFIER.

shutdown heating (*NucEng*) Heat coming from the continued decay of the fission products formed in a reactor. See DECAY HEAT.

shutdown power (*NucEng*) See DECAY HEAT, SHUTDOWN HEATING.

shuting (*Build*) See EAVES GUTTER.

shutter (*ImageTech*) A device in a camera for exposing the sensitive material to the image formed by the lens for a known period at the required instant.

shuttering (*CivEng*) See FORM, FORMWORK.

shutting stile (*Build*) The stile of a door further from the hinges. Also *meeting stile*.

shuttle (*NucEng*) A container for samples to be inserted in, and withdrawn from, nuclear reactors, where they are made radioactive by irradiation with neutrons. Also *rabbit*.

shuttle (*Textiles*) (1) Loom accessory which carries the weft (in the form of a cop or pirn) across a loom, through the upper and lower warp threads. Often made of hardwood (eg persimmon), it is boat-shaped, with a metal tip at each end. One end has a porcelain eye through which the weft passes. Some modern WEAVING MACHINES do not have a shuttle. (2) In a sewing machine, a sliding or rotating device that carries the lower thread to form a lock-stitch.

shuttle armature (*ElecEng*) A simple form of armature with only two slots used on small dc machines. The armature is a single coil connected to a two-part commutator. Also *H-armature*.

shuttle box (*Textiles*) Box-like extension at each side of a loom, from which the shuttle is thrown to and fro when loom is working.

shuttle guard (*Textiles*) Robustly constructed metal guard fixed to loom to retain or/and keep down a shuttle which by accident flies out of loom.

shuttleless weaving (*Textiles*) See WEAVING MACHINE.

shuttle vector (*BioSci*) A cloning VECTOR that can replicate in cells of more than one type of organism, and can thus be used to transfer genes from one organism to another, eg yeast and the bacterium *E. coli*.

Shwartzman reaction (*BioSci*) The phenomenon whereby the intravenous administration of endotoxin, followed by a second intravenous dose 24 hours later, causes renal tubular necrosis and adrenal haemorrhages. If the first dose is given into the skin a second skin dose results in destruction of venules and haemorrhagic necrosis at the skin site. Although apparently a hypersensitivity reaction, there is no identifiable immunological basis. Also *Sanarelli–Shwartzman phenomenon*.

SI (*Genrl*) Abbrev for *Système International* (d'Unités). See SI UNITS.

Si (*Chem*) Symbol for SILICON.

sial (*Geol*) The discontinuous shell of granitic composition which forms the foundation of the continental masses and which is in turn underlain by the *sima*. So called because it is essentially composed of *si*liceous and *al*uminous minerals.

sialagogue (*Pharmacol*) Stimulating the flow of saliva, from Gk *sialon*; any medicine which does this. Also *sialogogue*.

sialic acid (*BioSci*) See NEURAMINIC ACID.

sialogogue (*Pharmacol*) See SIALAGOGUE.

sialolith (*Med*) A calculus in a salivary gland.

sialons (*Eng*) Hard, tough ceramic based on β-silicon nitride, Si_6N_8, in which a proportion of the Si is replaced by Al and a proportion of the N by O, to give a range of compositions, $Si_{(6-z)}Al_zO_2N_{(8-z)}$. Some also contain yttrium to form yttrium aluminium garnet (YAG) which further improves high-temperature properties. Used for high-performance cutting tools, bearings and parts for the combustion zones of internal-combustion engines.

siblings (*BioSci*) Brothers and/or sisters. *Full siblings* have both parents in common, *half siblings* have one parent in common. Abbrev *sib*.

sickle cell anaemia (*Med*) A HAEMOLYTIC ANAEMIA due to an inherited abnormality in the haemoglobin molecule which causes the red blood cells to adopt a sickle-shaped deformity. The polymorphism persists because people who are heterozygotic are more resistant to malaria.

side-and-face cutter (*Eng*) A milling cutter with plain or staggered teeth on the sides as well as on the periphery, widely used for cutting slots.

sideband (*ICT*) Those bands on both sides of the CARRIER FREQUENCY that contain additional frequencies, themselves constituting the information to be conveyed, introduced by the process of MODULATION. In amplitude modulation the sideband frequencies are equal to the carrier frequency plus or minus the modulating signal frequency. In pulse and frequency modulation it becomes more complex, but the essential information is contained in just one of the sidebands, without the carrier. See SINGLE-SIDEBAND SYSTEM, SUPPRESSED-CARRIER SYSTEM.

side bones (*Vet*) Ossification of the lateral cartilages of a horse's foot.

side chains (*Chem*) Alkyl groups, or long chains, which replace hydrogen in ring compounds.

side draught (*Autos*) Said of a carburettor in which the mixture is drawn in at right angles to the force of gravity.

side drift (*CivEng*) An ADIT.

side keelson (*Ships*) See KEELSON.

sidelobe (*Electronics, ICT*) A pronounced region of the radiation pattern of an acoustic radiator or radio or radar antenna, in a direction other than that of the main beam. Fig. ▷

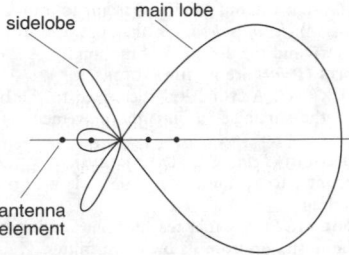

sidelobe

side pond (*Build*) A storage space at the side of a canal lock chamber, the two being interconnected by a sluice, so that the normal loss of water occurring in the process of passing a vessel through the lock may be reduced.

side rail (*CivEng*) See CHECK RAIL.

sidereal day (*Astron*) The interval of time between successive passages of the vernal EQUINOX across the same MERIDIAN. It is 23 h 56 min 4·091 s of mean solar time. In order to prevent the *sidereal day* changing in the middle of the night, when observations are taking place, the sidereal day begins at sidereal noon, when the vernal equinox crosses the local meridian. The time for the Earth to rotate once relative to the distant stars is longer than the sidereal day by about 0·008 s, due to the PRECESSION OF THE EQUINOXES.

sidereal month (*Astron*) The interval (27·321 66 sidereal days) for the Moon to complete one orbit of the Earth relative to the distant stars.

sidereal period (*Astron*) The interval between two successive positions of a celestial body in the same point with reference to the stars; applied to the Moon and planets to indicate their complete revolution with respect to the stellar background considered to be fixed in position.

sidereal time (*Astron*) Time measured by considering the movement of the Earth relative to the distant stars (rather than the Sun, which is the basis of civil time). The sidereal time at any instant is the same as the RIGHT ASCENSION of objects exactly on the MERIDIAN.

sidereal year (*Astron*) The interval between two successive passages of the Sun in its apparent annual motion through the same point relative to the fixed stars; it amounts to 365·25636 days, slightly longer than the tropical year, owing to the annual PRECESSION OF THE EQUINOXES.

side-rebate plane (*Build*) A rebate plane with its cutting edge on the side, not sole, of the tool.

siderite (*Min*) (1) Ferrous carbonate, crystallizing in the trigonal system and occurring in sedimentary iron ores and in mineralized veins. Also *chalybite*. (2) A name for iron meteorites as a class.

sideropenia (*Med*) Deficiency of iron.

siderophile (*Geol*) Descriptive of elements which have an affinity for iron, and whose geochemical distribution is influenced by this property.

siderophore (*BioSci*) Any of a group of naturally occurring compounds that chelate ferric ions.

siderophyllite (*Min*) The iron and aluminium-rich end-member of the biotite micas.

siderosis (*Med*) (1) PNEUMOCONIOSIS due to the inhalation of metallic particles by workers in tin, copper, lead and iron mines, and by steel grinders. (2) Excessive deposit of iron in the body tissues.

siderostat (*Astron*) An instrument designed on the same principle as the COELOSTAT to reflect a portion of the sky in a fixed direction; applied specially to a form of telescope called the *polar siderostat*, in which the observer looks down the polar axis onto a mirror.

sideslip (*Aero*) The component of the motion of an aircraft in the plane of its lateral axis; generally a piloting or stability error, but also used intentionally to obtain a steep glide descent without gaining speed or to improve weapon aiming. *Angle of sideslip* is that between the plane of symmetry and the direction of motion.

side sorts (*Print*) See PI CHARACTERS.

side stick (*Vet*) A cylindrical stick fixed to the head-collar and to the surcingle to limit the movement of a horse's head.

side stitching (*Print*) See FLAT STITCHING.

side thrust (*Acous*) Radial force on pickup arm caused by stylus drag.

side tone (*ICT*) A signal reaching the receiver of a radio telephone station from its own transmitter.

side tool (*Eng*) A cutting tool in which the cutting face is at the side, and which is fed laterally along the work.

side valve (*Autos*) Said of an engine having the valves situated at the side of the cylinder block with their posts on the same side of the combustion chamber as the piston and their tappets operated directly by the camshaft.

sidewall (*Eng*) The tyre wall. See panel on TYRE TECHNOLOGY.

sideways RAM (*ICT*) Additional RAM that is used to store application software like SIDEWAYS ROM except that the software will be loaded from backing store before it is available for use. Sideways RAM may also be used as a RAM DRIVE.

sideways ROM (*ICT*) A method of providing APPLICATION SOFTWARE, stored on a number of ROMS. The software is made available to the user as required by means of BANK SWITCHING.

siding (*CivEng*) A short length of side line onto which a train from the main line may be shunted to allow the passage of another on the main line. Colloq *hole*.

SIDS (*Med*) Abbrev for SUDDEN INFANT DEATH SYNDROME.

siege (*Build*) See BANKER.

siege (*Glass*) The floor of a tank furnace or pot furnace.

Siegennian (*Geol*) A stage in the Lower Devonian. See PALAEOZOIC.

siemens (*Phys*) SI unit of electrical conductance; reciprocal of OHM. Symbol S.

Siemens dynamometer (*ElecEng*) A dynamometer-type instrument arranged for measuring current or power.

Siemen's furnace (*Eng*) A gas-heated REVERBERATORY FURNACE.

Siemens–Halske process (*MinExt*) Chemical extraction of sulphidic copper from its ores or concentrates with sulphuric acid and ferrous sulphate.

Siemens–Martin process (*Eng*) See OPEN-HEARTH PROCESS.

Siemen's ozone tube (*Chem*) Apparatus used in the preparation of ozone by the corona discharge of electricity.

sieve analysis (*Build, CivEng*) A simple method of assessing the suitability of sand or other aggregate for concrete by passing a representative sample through progressively finer sieves and measuring the quantity passed on each occasion.

sieve analysis (*PowderTech*) The measurement of the size distribution of a powder by using a series of sieves of decreasing mesh aperture.

sieve area (*BioSci*) See SIEVE ELEMENT.

sieve element (*BioSci*) The cell type in which translocation occurs in the phloem of vascular plants, characterized by the presence of sieve areas in the walls and the disappearance of the TONOPLAST at maturity. The sieve area is a small area of the common wall between two adjoining sieve elements; it develops from a primary pit field and is perforated by many pores (less than 1–10 μm diameter) through which the protoplasts connect and the translocating solutes move. In angiosperms, the sieve elements are sieve tube members (or elements) organized into sieve tubes, each consisting of a number of such members joined end to end and connected by sieve plates (highly differentiated sieve areas, typically with relatively large pores) in their common end walls. A sieve tube member has one or more COMPANION CELLS and its nucleus disappears as the cell matures. Fig. ▷

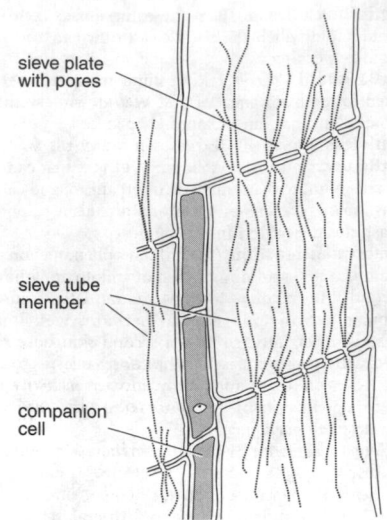

sieve element

sieve mesh number (*PowderTech*) The number of apertures occurring in the surface of a sieve per linear inch. Unless the size of the wire used in weaving the mesh is specified, the mesh number does not uniquely specify the aperture size.

sieve plate (*BioSci*) Part of the wall of a sieve tube member, bearing one or more highly differentiated sieve areas typically with relatively large pores. See SIEVE ELEMENT.

sieve plate (*ChemEng*) Distillation column plate (or tray) with large number of small holes through which the vapour rises, and fitted with weirs and downcomers to retain liquid through which the vapour rises.

sievert (*Radiol*) The SI unit for dose equivalent, measured in grays times a quality factor for the type of radiation and a weighting factor for the tissue irradiated. Numerically equivalent to gray for electrons and X-rays irradiating the whole body. Symbol Sv. See EFFECTIVE DOSE EQUIVALENT and panel on RADIATION.

sieve tube (*BioSci*) A series of sieve tube members, connected end to end and with sieve plates in the common walls between adjacent members. See SIEVE ELEMENT.

sig. fig. (*MathSci*) Abbrev for SIGNIFICANT FIGURES.

sight (*Phys*) Sensation produced when light impinges on the photosensitive cells of the eye.

sight (*Surv*) Bearing from instrument at known point on a distant signal or topographical feature. See BACK OBSERVATION, FORE SIGHT, INTERMEDIATE SIGHT.

sight-feed lubricator (*Eng*) A small glass tube through which oil dropping from a reservoir can be seen, or which is filled with water so that oil from the pump rises in visible drops on its way to the oil pipe.

sight lines (*ImageTech*) The extreme angles from which the screen can be seen in a cinema.

sight rail (*Surv*) An above-ground horizontal wooden rail fixed to two upright posts, one on each side of a trench excavation for a sewer, drain, etc. Used with others to establish a reference line from which the sewer etc may be laid at the required gradient.

sight rule (*Surv*) See ALIDADE.

sigma co-ordinates (*EnvSci*) A co-ordinate system used in numerical forecasting in which the vertical ordinate σ is pressure p divided by surface pressure p_s, ie $\sigma = p/p_s$.

sigma particle (*Phys*) HYPERON triplet, rest mass equivalent to 1190 MeV, hypercharge 0, isotopic spin 1.

sigma phase (*Eng*) A microconstituent causing embrittlement sometimes found in stainless and heat-resistant steels. Based on FeCr but can take other elements into solid solution, it tends to form after long periods at temperatures in the range 450–1000°C.

SIGMET message (*EnvSci*) A warning issued by an aviation meteorological watch and forecast office of the occurrence or expected occurrence of one or more meteorological hazards to aircraft including thunderstorms, severe CLEAR AIR TURBULENCE, marked LEE WAVES and severe icing.

sigmoid curve (*Radiol*) An S-shaped curve which is often obtained in dose–effect curves in radiobiological studies.

sigmoidectomy (*Med*) Excision of part of the sigmoid flexure of the colon.

sigmoid flexure (*BioSci*) An S-bend.

sigmoidoscope (*Med*) An endoscope for viewing the mucous membrane of the rectum and pelvic colon.

sigmoidostomy (*Med*) The surgical formation of an opening (artificial anus) in the sigmoid flexure of the colon.

sign (*MathSci*) Of a number or expression: either of the properties POSITIVE (+) and NEGATIVE (−).

sign (*Med*) Any objective evidence of disease or bodily disorder, as opposed to a SYMPTOM, which is a subjective complaint of a patient.

signal (*ICT*) General term referring to a conveyor of information, eg an audio waveform, a video waveform, series of pulses in a computer. Colloq, the message itself. In radio, the signal modulates a carrier, and is recovered during reception by demodulation.

signal (*Surv*) A device, such as a ranging rod, heliostat, etc, used to mark a survey station.

signal code (*ICT*) In voice-frequency signalling, the plan for representing each of the required signalling functions as a voice-frequency signal.

signal component (*ICT*) That part of a signal which continues uniform in character throughout its duration. In a multicomponent signal with spaces between current pulses a space may be regarded as a signal component.

signal distortion (*Phys*) Modification of the information content of a signal, sometimes irreversibly; eg the suppression or introduction of HARMONICS.

signal element (*ICT*) The portion of a signal occupying the smallest interval of the signal code.

signal frequency shift (*ICT*) The bandwidth between white and black signal levels in frequency-modulation facsimile transmission systems.

signal generator (*ICT*) Oscillator designed to provide known voltages, typically from 1 to less than 1 μV, over a range of frequencies. Used for testing or ascertaining performance of radio-related equipment. It may be amplitude-, frequency- or pulse-modulated.

signal level (*ICT*) The difference between the level of a signal at a point in a transmission system and the level of the arbitrarily specified reference signal.

signalling system (*ICT*) The means by which the messages needed to set up and control calls, as distinct from the calls themselves, are conveyed across a network. Signalling may take place over the user links (as with the signals sent by a telephone keypad) or over dedicated data channels. See COMMON CHANNEL SIGNALLING.

signal output current (*Electronics*) The absolute difference between the output current and the dark current of a phototube or a camera tube.

signal peptides (*BioSci*) Short N-terminal peptide sequences of newly synthesized membrane proteins which direct the protein towards the appropriate membrane, facilitate its transfer across it and are usually deleted during the subsequent maturation of the protein. Also *leader peptides*.

signal recognition particle (*BioSci*) A complex of proteins and RNA, found in the cytoplasm of cells, that assists the translocation of proteins across the membrane of the endoplasmic reticulum.

signal–response coupling (*BioSci*) See SIGNAL TRANSDUCTION.

signal shaping (*ICT*) Use of specially designed electrical network to correct distortion produced during transmission or propagation of signals.

signal-to-cross-talk ratio (*ICT*) In line telephony, the ratio of the test level in the disturbed circuit to the level of the cross-talk at the same point which is caused by the disturbing circuit operating at the test level.

signal-to-noise ratio (*Acous*) The ratio of a wanted signal to that of unwanted random (eg thermal) noise, usually expressed in decibels. Abbrevs *SNR*, *S/N ratio*.

signal transduction (*BioSci*) The cascade of processes by which a signal at the outer surface of a cell (eg hormone, neurotransmitter) interacts with a cell surface receptor and causes a response to take place within the cell, typically by stimulating an increase in a SECOND MESSENGER such as calcium ions or cyclic AMP.

signal windings (*ElecEng*) US term for CONTROL TURNS (or *windings*) of a saturable reactor.

sign and magnitude (*ICT*) A method of representing numbers in a binary word by coding the sign of the number in a SIGN BIT and the magnitude of the number in the remaining bits. Also *sign and modulus*.

signature (*Print*) See SECTION.

signature mark (*Print*) A number or letter of the alphabet placed at the tail on the first page of a section as a guide to the binder in GATHERING.

sign bit (*ICT*) Single bit, used to indicate the sign of a number, usually 0 for positive, 1 for negative.

signed minor (*MathSci*) See COFACTOR.

significance (*MathSci*) A threshold value of probability at or below which the results of a statistical investigation are held to justify rejecting a particular hypothesis.

significant figures (*MathSci*) Of a number: those digits which make a contribution to its value, from the leftmost non-zero digit to the rightmost non-zero digit; eg in the number 00·1230, the first two zeros are insignificant and the digits 1,2,3 are significant. The last zero may be significant, indicating that the number is accurate to four places of decimals, but this convention is not universal. Also *significant digits*. See also ROUNDING OFF.

sign stimulus (*Psych*) Part of a complex stimulus configuration which is relevant to a particular response and evokes the strongest response (eg the red breast of the robin).

Sikes hydrometer (*Chem*) A hydrometer used for determining the strengths of mixtures of alcohol and water.

silage (*Agri*) High-moisture-content FORAGE, commonly from grasses, fermented in storage for subsequent feeding to ruminant livestock during the winter. Also *ensilage*, the process of making such food.

silal (*Eng*) High-silicon (6%) cast-iron suitable for chemical plant and heat-resistant applications.

silanes (*Chem*) A term given to the silicon hydrides: silane, SiH_4, disilane $H_3Si–SiH_3$, trisilane, $H_3Si(SiH_2)SiH_4$, etc. Cf METHANE. Silane and chlorosilanes are used in chemical vapour deposition processes for silicon and in silicon dioxide deposition in semiconductor technology.

Silastic (*Plastics*) TN for a range of silicone rubbers. Noted for very good heat resistance and a wide temperature range of application. Excellent chemical resistance and electrical properties.

sildenafil (*Pharmacol*) Any of several compounds, esp sildenafil citrate (TN Viagra), that increase blood flow to the penis, used in treating male impotence.

silencer (*Autos*) An expansion chamber fitted to the exhaust pipe of an internal-combustion engine to dampen the noise of combustion. US *muffler*.

silent mutation (*BioSci*) Mutations that have no effect on phenotype because they do not affect the activity of the protein product of the gene, usually because of codon ambiguity.

silent period (*ICT*) Stated period within each hour during which all marine transmissions must close down and listen on the international distress frequency of 500 kHz.

silesia (*Textiles*) A smooth-faced, cotton lining fabric for garments, originally a plain weave but now a twill.

Silesian (*Geol*) The Upper Carboniferous of W Europe. See PALAEOZOIC.

silex (*MinExt*) Silica brick used to line grinding mills when contamination by abraded steel must be avoided.

silica (*Chem*) Dioxide ((IV) oxide) of silicon, SiO_2, occurring in many crystalline forms and as an essential constituent of the silicate groups of minerals. See panel on SILICON, SILICA, SILICATES.

silica gel (*Chem*) Hard amorphous granular form of hydrated silica, chemically inert but very hygroscopic. Used for absorbing water and vapours of solvents in eg desiccators, electronic equipment. When saturated, it may be regenerated by heat.

silica glass (*Glass*) See VITREOUS SILICA.

silica glass (*Min*) Fused quartz, occurring in shapeless masses on the surface of the Libyan desert, in Moravia, in parts of Australia and elsewhere; believed to be of meteoritic origin. See TEKTITES.

silica poisoning (*MinExt*) Loading of resins used in ion-exchange process with silica, thus reducing the efficiency of reaction with desired ions.

silicates (*Min*) The largest group of minerals, of widely different and, in some cases, extremely complex composition, but all composed of silicon, oxygen and one or more metals, with or without hydrogen. See panel on SILICON, SILICA, SILICATES.

siliceous deposits (*Geol*) Those sediments, incrustations or deposits which contain a large percentage of silica in one or more of its modes of occurrence. They may be chemically or mechanically formed, or may consist of the siliceous skeletons of organisms such as diatoms and Radiolaria. See SILICIFICATION.

siliceous sinter (*Geol*) Cellular quartz or translucent to opaque opal, found as incrustations or fibrous growths and deposited from thermal waters containing silica or silicates in solution.

silicic acid (*Chem*) An acid formed when alkaline silicates are treated with acids. Amorphous, gelatinous mass. Dissociates readily into water and silica.

silicides (*Chem*) Compounds formed by the combination of silicon with other elements, chiefly metals.

silicification (*Geol*) The process by which silica is introduced as a cement into rocks after their deposition, or as an infiltration or replacement of organic tissues or of other minerals such as calcite. See NOVACULITE.

silicole (*BioSci*) A plant that grows on soils which are rich in silica and are usually acid.

silico-manganese steel (*Eng*) See MANGANESE ALLOYS.

silicon (*Chem*) A non-metallic element. Symbol Si, at no 14, ram 28·086, valency 4. See panels on INTRINSIC AND EXTRINSIC SILICON, SEMICONDUCTOR FABRICATION and SILICON, SILICA, SILICATES.

silicon brass (*Eng*) A corrosion-resistant alloy based on copper and zinc but with a small addition of silicon. See COPPER ALLOYS.

silicon bronze (*Eng*) A non-corroding alloy based on copper and tin.

silicon carbide (*Chem*) SiC. Formed by fusing a mixture of carbon and sand or silica in an electric-arc furnace (see ACHESON FURNACE). Used as an abrasive and refractory. Also a wide-band-gap semiconductor with potential for high-temperature applications. See CARBORUNDUM.

silicon-controlled rectifier (*Electronics*) A three-terminal semiconductor switching device consisting of a sandwich of p–n–p–n-type materials. It is normally open-circuit, but application of an appropriate control signal to the GATE allows it to conduct, in one direction only, like a conventional rectifier. It continues to conduct even with the gate signal removed, until it is reverse-biased by the voltage it is intended to switch. Used in voltage control of power circuits. Abbrev SCR.

silicon copper (*Eng*) An alloy (20–30% Si), a 'getter' used to remove oxygen from molten copper alloys.

silicon detector (*ICT*) Stable silicon crystal diode for demodulation.

silicon dioxide (*Chem*) The oxide of silicon (SiO_2). In semiconductor technology, silicon dioxide grown on silicon wafers is a key material in device manufacturing processes and also is vital as a dielectric within device structures. See SEMICONDUCTOR DEVICE PROCESSING and panel on SEMICONDUCTOR FABRICATION. Also *silica*.

silicon disk (*ICT*) See RAM DRIVE.

silicone resins (*Build*) A group of resins with particular properties which benefit coatings, eg resistance to heat, acids, alkalis, oils, salts and the ability to repel water, making them useful for masonry water repellants. Silicones are also used in polishes.

silicone rubbers (*Chem*) *Polysiloxanes*. They have a main backbone chain which is inorganic and the repeat unit is $-Si(RR')O-$. Different types are created by varying the substituent organic groups (R and R') on the silicon atom. The commonest type has two methyl groups ($-CH_3$) in this position (MQ systems). Copolymers where different substituents occur (V = vinyl, P = phenyl, F = fluorine-containing groups) modify properties, eg PVMQ is a low-temperature-resistant rubber, cross-linked via vinyl side groups. Generally used for gaskets where resistance to temperature extremes is needed, despite low strength. Room temperature vulcanizing materials are also used for water-resistant seals, eg aquarium seals, building seals.

silicones (*Chem*) Open-chain and cyclic organosilicon compounds containing $-SiR_2O-$ groups. The simpler substances are oligomeric oils of very low melting point, the viscosity of which changes little with temperature, used as lubricants, shock-absorber fluids, constituents of polishes, etc. Polymers, stable to heat and cold, and chemically inert, are exceptionally good electrical insulators which are highly hydrophobic and used in electric motors etc. Also found in gaskets and a wide variety of special applications. See SILICONE RUBBERS.

silicon hydrides (*Chem*) See SILANES.

silicon iron (*Eng*) Iron or low-carbon steel to which 0·75–4·0% silicon has been added. Has low magnetic hysteresis, and is resistant to mild acids. Used for sheets for transformer cores. Typical composition: silicon 4%, manganese under 0·1%, phosphorus 0·02%, sulphur 0·02%, carbon 0·05%.

silicon nitride (*Eng*) Hard engineering ceramic used for cutting tools and in wear-resistant and high-temperature applications. Composition Si_3N_4 but exists in two crystalline forms, α and β; formed from powders by HOT ISOSTATIC PRESSING. See SIALONS.

silicon on insulator (*Electronics*) A semiconductor technology in which silicon devices are defined in a thin layer of silicon on an insulating substrate; devices are isolated by etching through to the substrate between active regions. Abbrev SOI. See SILICON ON SAPPHIRE.

silicon on sapphire (*Electronics*) A SILICON ON INSULATOR technology in which silicon is grown on a single-crystal alumina substrate, with which there is a close enough match of lattice constant to allow epitaxial growth. Abbrev SOS. See EPITAXIAL GROWTH OF SEMICONDUCTORS.

silicon rectifier (*Electronics*) A semiconductor diode rectifier usually based on p–n junction in silicon crystal.

silicon resistor (*Electronics*) A resistor of special silicon material which has a fairly constant positive temperature coefficient, making it suitable as a temperature-sensing element.

silicon tetrachloride (*Chem*) *Tetrachlorosilane*. $SiCl_4$. Formed by the action of chlorine on a mixture of silica and carbon, or silicon. Liquid.

silicon tetrafluoride (*Chem*) *Tetrafluorosilane*. SiF_4. A gaseous compound formed by the action of hydrofluoric acid on silica. Readily hydrolyses into silica and hydrofluoric acid.

silicosis (*Med*) PNEUMOCONIOSIS due to the inhalation of particles of silica by masons and miners who work in the presence of silica.

Silicon, silica, silicates

Silicon is the non-metallic element, symbol Si, obtained from the dioxide of silica, SiO_2. The element, silicon, is the second commonest in the Earth's crust, after oxygen, amounting to 27%, but it is never found in the elemental form. It has at no 14, ram 28·086, valence 4, three stable isotopes, rel.d 2·33, mp 1410°C, bp 2355°C. It has a blue-grey metallic appearance and is produced by reduction of quartz sand, SiO_2, with carbon. It is used as an alloying agent for steels and other metals, and in semiconductors.

Silica, or silicon (IV) oxide, SiO_2, forms the basis of silicates which, in combination with other elements, chiefly aluminium, make up the bulk of the Earth's crust. It occurs in crystalline form as quartz, cristobalite, tridymite; as cryptocrystalline chalcedony; as amorphous opal; and as an essential constituent of the silicate groups of minerals, over a thousand of which are known. It is the main ingredient of the silicate glasses (see panel on GLASSES AND GLASSMAKING) and, in mineral form, is important in PORTLAND CEMENT (see panel on CEMENT AND CONCRETE), as well as in most ceramic and refractory products. Refractory materials containing a high proportion of silica (over 90%) are known as *acid refractories*, eg GANNISTER. They are used as a lining for metallurgical and other furnaces to resist the high temperatures and attack by ACID SLAGS.

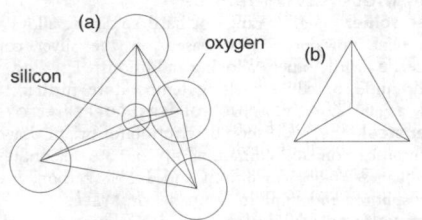

Fig. 1 **Silica** (a) Tetrahedral structure and (b) conventional representation.

The basis of these materials is the SiO_4^{4-} tetrahedron (Fig. 1(a)) in which four oxygens surround a silicon to share most of the silicon's four outer electrons, the bonding being about 65% covalent and 35% ionic. The smaller silicon is located in the interstice between the four, much larger, oxygens. Charge neutrality is maintained by each oxygen atom being shared by two neighbouring tetrahedra. The tetrahedra can combine together in many different ways, producing chains and extended two- and three-dimensional networks leading to a range of crystalline forms (eg Fig. 2). The low mobility in the melt of these tetrahedral structures means that it is often relatively easy to form glasses when cooling, since there is insufficient time for the necessary rearrangements for crystal formation to take place.

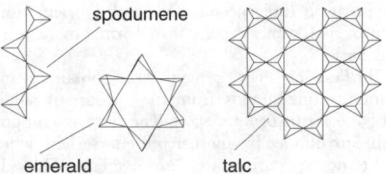

Fig. 2 **Silicates** Networks of silicate tetrahedra.

Silica is important in biological structures, especially in some plants. In DIATOMS the cell walls are siliceous as are the skeletons of RADIOLARIA. There are huge quantities of siliceous minerals used, largely for structural purposes, and very substantial amounts for other purposes, eg in the chemical industry, manufacture of glass, abrasives, fillers, insulation, etc.

Gemstones
Coloured varieties of quartz, some used as gemstones, include amethyst (purple), cairngorm (brownish-yellow), citrine (yellow), rose quartz (rose-pink), milky quartz (translucent white) and smoky quartz (grey-brown), as well as rock crystal (clear). Opal is hydrated silica much prized in its precious varieties as a gemstone.

Geology
Among the principal rock groups, the OLIVINES, which are common constituents of basic and ultrabasic rocks, have SiO_4 tetrahedra linked by DIVALENT atoms (mainly Mg, Fe^{2+}) in sixfold coordination. In PYROXENES, which occur in all types of igneous rock, two of the four corners of each tetrahedron are linked to the next tetrahedron, forming continuous chains with the composition $(SiO_3)_n$, which are linked laterally by CATIONS. AMPHIBOLES have two chains of silica tetrahedra, linked together, with the composition $(Si_4O_{11})_n$. The double chains are joined by cations and hydroxyl ions. MICAS have a sheet structure of tetrahedra linked together in a hexagonal pattern, with cations linking sheets above and below, and additional (OH) ions occurring in the structure. Sheet structures also form the CHLORITES, clay minerals, eg kaolinite, illite, montmorillonite (smectite) and vermiculite groups. The total number of structures within the silicates is very large.

See panel on INTRINSIC AND EXTRINSIC SILICON.

siliqua (*BioSci*) A capsule with the general characters of a SILICLE, but at least four times as long as it is broad. Also *silique*.

silk (*BioSci*) A fluid substance secreted by various Arthropoda. It is composed mainly of fibroin, together with sericin and other substances, and hardens on exposure to air in the form of a thread. Used for spinning cocoons, webs, egg cases, etc.

silk (*Min*) A sheen resembling that of silk, exhibited by some corundums, including ruby, and due to minute tubular cavities, or to rutile needles, in parallel orientation. The colour of such stones is paler than normal by reason of the inclusions.

silk (*Textiles*) (1) The protein fibre obtained in long continuous fine strands from the cocoon of silkworms, esp of the moth *Bombyx mori*. The fibre is composed of fibroin surrounded by another protein, sericin, which is a gum removed during wet processing. Wild silk (eg TUSSAH) is a similar fibre from the cocoon of other moths. (2) An imitation (*artificial silk*) made by forcing a viscous solution of modified cellulose through small holes.

silk-screen printing (*Print*) A mesh-stencil process in which ink is squeezed through the open parts of the mesh onto the surface to be printed, which may be of any material, smooth or rough, flat or curved. The stencil may be prepared by hand, or by photography, or by a mixture of both. Screens of metal and fibres other than silk are also used, and SCREEN PROCESS PRINTING has become the preferred name.

sill (*Build*) (1) The horizontal timber, stone, etc, at the foot of an opening, as for a door (also *threshold*), window, embrasure, etc. (2) The top level of a weir, or the lowest level of a notch.

sill (*Geol*) Minor intrusion of igneous rock concordantly injected between and more or less parallel to the bedding planes of the country rocks. Sills occasionally break across the bedding, eg as the Great Whin Sill which may be traced across much of N England. See fig. at DYKES.

sillénite (*Min*) Cubic bismuth trioxide. Cf BISMITE.

sillimanite (*Min*) An orthorhombic aluminium silicate. The high-temperature polymorph of Al_2SiO_5, occurring in high-grade metamorphosed argillaceous rocks. *Fibrolite* is a fibrous variety, used as a gemstone. *Kyanite* and *andalusite* are other aluminium silicates.

silo (*Aero*) Underground chamber housing a guided missile which is ready to be fired.

silo (*Agri*) A free-standing, cylindrical storage structure for grains or fodder. Less commonly, a pit dug for similar purposes.

siloxane (*Chem*) Any of various polymers containing silicon and oxygen with a wide range of industrial applications.

Silsbee rule (*ElecEng*) A wire of radius r cannot carry a superconducting current greater than $1/2rH_c$, where H_c is the critical field. The self-magnetic field would then destroy superconductivity.

silt (*Geol*) Material of an earthy character intermediate in grain size between sand and clay. See PARTICLE SIZE, WENTWORTH SCALE.

silt box (*Build*) A removable iron box placed at the bottom of a gulley; serves to accumulate the deposited silt for periodic removal.

Siluminite (*ElecEng*) TN designating materials composed principally of asbestos or mica for electrical insulating materials.

Silurian (*Geol*) The youngest period of the Lower Palaeozoic, covering a time span between approx 440 and 408 million years ago. Named after the *Silures*, an ancient Welsh tribe. The corresponding system of rocks. See PALAEOZOIC.

Silurian (*Paper*) A mottled effect in special papers produced by adding deeply dyed fibres to the main furnish. Under certain circumstances this can be a paper defect.

Siluriformes (*BioSci*) An order of mainly fresh-water, bottom-living Osteichthyes that have barbels used in detecting food and usually do not have scales. Catfish.

silver (*Chem*) A pure white metallic element, symbol Ag, at no 47, ram 107·868, rel.d. (at 20°C) 10·5, mp 960°C, bp 1955°C, casting temp 1030–1090°C, Brinell hardness 37, electrical resistivity approx $1·62 \times 10^{-8}$ Ω m. The metal is not oxidized in air. Occurs massive, or assumes arborescent or filiform shapes. The principal minerals are native silver, *argentite* (AgS) and complex sulphides, but most production is from argentiferous lead, copper, zinc, tin and gold ores. The best electrical conductor and the main constituent of photographic emulsions. Native silver often has variable admixture of other metals: gold, copper or sometimes platinum. Used for ornaments, mirrors, cutlery, jewellery, etc, and for certain components in food and chemical industries where cheaper metals fail to withstand corrosion.

silver amalgam (*Min*) *Arquerite*, a solid solution of mercury and silver, which crystallizes in the cubic system. Of rare occurrence, it is found scattered in mercury or silver deposits.

silver glance (*Min*) See ARGENTITE.

silver grain (*For*) The light-greyish, shining flecking seen in oak timber, caused by vascular rays exposed in preparing the timber when it is cut radially through the centre of a log. Also seen in beech.

silver (I) halides (*Chem*) Silver fluoride, AgF; silver iodide, AgI; silver chloride, AgCl; and silver bromide, AgBr. The last three are sensitive to light and are of basic importance in photography.

silvering (*Glass*) Traditional process of forming a reflective film, eg for mirrors, on a clean surface of glass by pouring onto it an ammoniacal silver solution, mixed with Rochelle salt or with a nitric acid/cane sugar/alcohol mixture. The silver film so formed is then washed, backed with varnish and painted. Alternative nowadays is to use sputtering.

silver lead ore (*Min*) Galena containing silver. When 1% or more of silver is present it becomes a valuable ore of silver. Also *argentiferous galena*.

silver leaf (*Build*) Metallic silver prepared in leaves in loose or transfer form; similar use as gold leaf. Tarnishes rapidly and best protected with a clear lacquer.

silverlock bond (*Build*) See RAT-TRAP BOND.

silver oxide (*Chem*) See ARGENTIC [SILVER (II)] OXIDE, ARGENTOUS [SILVER (I)] OXIDE.

silver solder (*Eng*) A range of hard soldering alloys with excellent flow properties, based on the silver copper eutectic with other alloying additions, eg zinc, tin, cadmium, to reduce melting temperature further to as low as 600°C. See BRAZING SOLDERS, COPPER ALLOYS.

silver steel (*Eng*) Workshop description for bright-drawn or ground carbon steel containing up to 0·3% silicon, 0·45% manganese, 0·5% chromium and 1·25% carbon; low in phosphorus and sulphur. See panel on STEELS.

silver voltameter (*ElecEng*) An electrolytic cell used for determining accurately the average value of a current from the quantity of silver deposited from a silver nitrate solution.

silviculture (*For*) The planting and care of forests.

sim (*ICT*) A contraction of *simulation*. Hence *sim game*, a computer game that attempts to replicate the sort of events encountered in the real world.

sima (*Geol*) The lower layer of the Earth's crust with the composition of a basic or ultrabasic igneous rock. Such rocks contain silicon and magnesium as their principal constituents, hence the name.

SIM card (*ICT*) Abbrev for *subscriber identification module*, a removable electronic card inside a mobile phone that stores information about the subscriber.

SIMD (*ICT*) Abbrev for *single instruction stream, multiple data stream*. Term that describes the architecture of a PROCESSOR. See ARRAY PROCESSOR, ASSOCIATIVE MEMORY.

simian virus 40 (*BioSci*) See SV40.

similarity coefficient (*BioSci*) Index of similarity between two stands of vegetation, based on their species composition.

similar polygons (*MathSci*) Polygons whose corresponding angles are equal and in which the sides about the corresponding angles are in direct proportion.

SIMM (*ICT*) Abbrev for *single in-line memory module*. A small printed circuit board carrying a number of MEMORY CHIPS that plugs into a suitable socket in a motherboard or adapter card. Using nine 4 Mb, 16 Mb or larger chips on each SIMM, a very large amount of memory can be put in a small space.

Simmonds' disease (*Med*) A rare disease due to destruction of the pituitary gland; characterized by cachexia, atrophy of the skin and the bones, premature senility, loss of hair and loss of sexual function. Also *hypophyseal cachexia*.

simple (*BioSci*) Consisting of one piece or component; unbranched; not COMPOUND.

simple curve (*Surv*) A curve composed of a single arc connecting two straights.

simple eye (*BioSci*) See OCELLUS.

simple fruit (*BioSci*) A fruit formed from one pistil.

simple graph (*MathSci*) A graph in which no edge connects a vertex to itself and no two vertices are joined by more than one edge.

simple group (*MathSci*) A group which has no normal subgroups except for the identity and the group itself.

simple harmonic motion (*Phys*) The motion of a particle (or system) for which the force on the particle is proportional to its distance from a fixed point and is directed towards the fixed point. The particle executes an oscillatory motion about the point. The motion satisfies the equation $(d^2x/dt^2) = -\omega^2 x$ where x is the displacement of the particle and ω is a constant for the motion. The majority of small-amplitude oscillatory motions are simple harmonic, eg the oscillations of a mass suspended by a spring, the swing of a pendulum, the vibrations of a violin string, the oscillations of atoms or molecules in a solid, or the oscillations of air as a sound wave passes. When such a motion takes place in a resistive medium, eg air, the oscillations die away with time; the motion is then said to be *damped*. Abbrev *shm*.

simple leaf (*BioSci*) A leaf in which the lamina consists of one piece, which, if lobed, is not cut into separate parts reaching down to the midrib.

simple network management protocol (*ICT*) A protocol used for managing NETWORKS, esp TCP/IP.

simple pit (*BioSci*) A pit of which the cavity does not become markedly narrower towards the cell lumen. Cf BORDERED PIT.

simple press tool (*Eng*) A PRESS TOOL which performs only one operation at each stroke of the press, as distinct from a COMPOUND PRESS TOOL and a PROGRESSIVE PRESS TOOL.

simple sequence DNA (*BioSci*) A block of a DNA sequence that consists of many repeats of a short, unit sequence. The repeats are not necessarily identical.

simple steam engine (*Eng*) An engine with one or more cylinders in which the steam expands from the initial pressure to the exhaust pressure in a single stage. Cf COMPOUND STEAM ENGINE.

simplex (*ICT*) Transmission of data in only one direction.

simplex winding (*ElecEng*) An armature winding through which there is only one electrical path per pole.

simply connected domain (*MathSci*) See CONNECTED DOMAIN.

Simpson's rule (*MathSci*) The theorem that the area under the curve $y = f(x)$ from $x = x_0$ to $x = x_2$ is approximately $\frac{1}{3}[f(x_0) + 4f(x_1) + f(x_2)][x_2 - x_0]$, where x_0, x_1 and x_2 are equally spaced.

SIMS (*Phys*) Abbrev for SECONDARY ION MASS SPECTRO-METRY.

Sims' speculum (*Med*) A speculum, shaped like a duck's bill, for viewing the lining of the vagina and the cervix uteri.

simulated line (*ICT*) See ARTIFICIAL LINE.

simulation (*BioSci*) Mimicry; assumption of the external characters of another species in order to facilitate the capture of prey or escape from enemies. V *simulate*.

simulation (*ICT, Electronics*) The representation of physical systems and phenomena by computers, models and other equipment.

simulation by computer (*Psych*) The investigation of thought processes by the use of computers programmed to imitate them.

simulcasting (*ICT*) Inelegant abbrev for SIMULTANEOUS BROADCASTING.

simultaneity (*Phys*) A basic consequence of SPECIAL RELATIVITY. Two events that are simultaneous according to one observer may occur at different times according to another observer in another reference frame moving relative to the first.

simultaneous broadcasting (*ICT*) (1) Transmission of one programme from two or more transmitters. (2) Simultaneous TV and radio (usually stereophonic) broadcasts from live concerts etc. Latter also *simulcasting*.

simvastatin (*Pharmacol*) A STATIN used to lower the amount of fatty substances, eg cholesterol, in the blood.

sin (*MathSci*) See TRIGONOMETRICAL FUNCTIONS.

Sindanyo (*ElecEng*) TN designating materials, composed principally of asbestos, for the mounting of switchgear of all types for electrical insulation work generally, and for arc shields, barriers, furnace linings and other purposes.

sine (*MathSci*) See TRIGONOMETRICAL FUNCTIONS.

sine bar (*Eng*) A hardened steel bar carrying two rollers of standard diameter accurately spaced to some standard distance; used with SLIP GAUGES to set out angles to close limits.

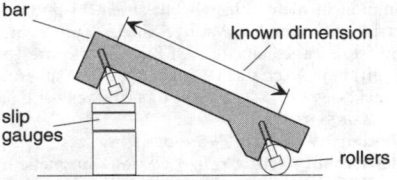

sine bar Bar and rollers are precision ground.

sine condition (*Phys*) A condition which must be satisfied by a lens if it is to form an image free from aberrations (other than chromatic). It may be stated as $n_1 l_1 \sin\alpha_1 = n_2 l_2 \sin\alpha_2$, where n_1 and n_2 are the refractive indices of the media on the object and image sides of the lens respectively, l_1 and l_2 are the linear dimensions of the object and image, and α_1 and α_2 are the angles made with the principal axis by the conjugate portions of a ray passing between object and image.

sine galvanometer (*ElecEng*) A galvanometer in which the coil and scale are rotated to keep the needle at zero. The current is then proportional to the sine of the angle of rotation. The arrangement can be made more sensitive than the TANGENT GALVANOMETER.

Sinemurian (*Geol*) A stage in the Lower Jurassic. See MESOZOIC.

sine potentiometer (*ElecEng*) Voltage divider in which the output of an applied direct voltage is proportional to the sine of the angular displacement of a shaft.

sine wave (*Phys*) The waveform of a single frequency, indefinitely repeated in time, the only waveform whose integral and differential have the same waveform as itself. Its displacement can be expressed as the sine (or cosine) of a linear function of time or distance, or both. In practice there must be a transient at the start and finish of such a wave.

singing (*ICT*) Oscillation in a telephone system caused by feedback across a source of gain because of mismatch in the circuit.

singing tube (*Acous*) (1) A tube with a flame inside which under certain conditions excites the tube resonance. (2) See RIJKE TUBE.

single-acting cylinder (*Eng*) Fluid-powered cylinder in which the piston is displaced in one direction by the fluid and returned mechanically, usually by a spring.

single-acting engine (*Eng*) A reciprocating engine in which the working fluid acts on one side of the piston only, as in most internal-combustion engines.

single-blind (*Genrl*) Denoting a comparative experiment or trial in which the identities of the control group are known to the experimenters but not the subjects.

single-bond (*Chem*) A covalent bond involving the sharing of one pair of electrons.

single bridging (*Build*) Bridging in which a pair of diagonal braces are used to connect adjacent floor joists at their middle points.

single-catenary suspension (*ElecEng*) A catenary suspension system in which the conductor wire is hung from a single catenary or bearer wire.

single-cell protein (*BioSci*) Protein-rich material from cultured algae, fungi (including yeasts) or bacteria, used (potentially) for food or as animal feed. Abbrev *SCP*.

single-channel per carrier (*ICT*) Used in satellite communications for a system where each carrier is dedicated to one telephone call or data transmission for its duration. Cf TIME-DIVISION MULTIPLE ACCESS.

single-channel pulse-height analyser (*ICT*) See PULSE-HEIGHT ANALYSER.

single-core cable (*ElecEng*) A cable having only one conductor.

single crystal (*Crystal*) A crystal formed by propagation of unit cell in three dimensions, usually from a single nucleus. Common in many minerals but absent in polycrystalline metals and ceramics. Massive single crystals of silicon grown for semiconductors. Of microscopic size (typically $0.1\ \mu m$) in most crystalline polymers, where spherulites are normal. See CRYSTALLIZATION OF POLYMERS and panel on POLYMER SYNTHESIS.

single-density disk (*ICT*) See DENSITY.

single-domain particle (*Phys*) Part of a magnetic material of sufficiently small size that it is energetically unfavourable for it to be further subdivided into DOMAINS.

single-electrode system (*ElecEng*) The electrode of an electrolytic cell and the electrolyte with which it is in contact. Also *half-cell, half-element*.

single-ended (*ElecEng*) Unit or system designed for use with an unbalanced signal, having one input and one output terminal permanently earthed.

single-entry compressor (*Aero*) A centrifugal compressor which has vanes on one face only.

single Flemish bond (*Build*) A form of bond combining English bond for the body of the wall with Flemish bond for the facework.

single floor (*Build*) A floor in which the bridging joists span the distance from wall to wall without intermediate support.

single-hung window (*Build*) A window having top and bottom sashes, of which only one (usually the bottom sash) is balanced by sash cord and weights so as to be capable of vertical movement.

single jersey (*Textiles*) Weft-knitted fabrics of a simple plain nature although patterned effects may be included eg with a JACQUARD.

single laths (*Build*) Wood laths 1 in (25 mm) by $\frac{1}{8} - \frac{3}{16}$ in (3–5 mm) thick in section.

single-lens reflex (*ImageTech*) A camera in which the image is viewed and focused by way of a movable mirror behind the lens and a PENTAPRISM in front of the eyepiece. The mirror is rotated out of the way just before the FOCAL PLANE SHUTTER opens. Easily used with different focal length lenses and with automatic exposure and focusing devices. Abbrev *SLR*.

single phase (*ElecEng*) The electrical power transmission system used for domestic ac supply having two conductors, one of which may be the earth or at earth potential and between which there is a sinusoidally alternating potential difference.

single-phase induction regulator (*ElecEng*) An induction regulator for use on a single-phase circuit; the voltages in primary and secondary are always in phase. Cf THREE-PHASE INDUCTION REGULATOR.

single-plate clutch (*Eng*) A FRICTION CLUTCH in which the disk-shaped or annular driven member, fabric-faced, is pressed against a similar face on the driving member by springs, being withdrawn against them through a thrust collar; used in automobiles.

single pole (*ElecEng*) Of a switch or relay contact which makes, breaks or changes over connection on one pole only, ac or dc, of a circuit.

single quotes (*Print*) See QUOTATION MARKS.

single-rate prepayment meter (*ElecEng*) A prepayment meter in which the circuit is broken and the supply cut off after a predetermined number of units have been consumed.

single-retort underfeed stoker (*Eng*) An UNDERFEED STOKER consisting of a retort along the bottom of which coal is fed by a steam-driven ram or a screw conveyor, air being supplied through TUYÈRES round the upper edge of the retort and into the sealed ash pit below.

single-revolution (*Print*) A design of letterpress machine in which the cylinder rotates continuously while the bed reciprocates, and prints one impression during each revolution. Cf STOP-CYLINDER, TWO-REVOLUTION.

single-row ball bearing (*Eng*) BALL BEARING comprising a set of balls arranged in a single plane, as distinct from a double-row bearing.

singles (*Build*) Roofing slates about 12 × 18 in (305 × 457 mm).

singles (*MinExt*) See COAL SIZES.

single-sideband suppressed carrier (*ICT*) See VESTIGIAL SIDEBAND.

single-sideband system (*ICT*) A form of amplitude modulation in which one sideband and the carrier are eliminated, either by filtering after modulation or, more commonly, by using balanced modulators and phase-shift circuits. On reception, the carrier has to be reinserted before conventional demodulation can occur. Such transmissions make better use of transmitter power, occupy less bandwidth and therefore improve the signal-to-noise ratio. Abbrev *SSB*. See DOUBLE-SIDEBAND SYSTEM, SUPPRESSED-CARRIER SYSTEM.

single-spindle automatic (*Eng*) AUTOMATIC SCREW MACHINE or lathe with one tool turret and one chucking spindle, producing one part per operating cycle.

single-stack system (*Build*) Where soil and waste discharges are piped together within certain design limitations and without vertical ventilating pipework other than a main vertical stack.

single-stage-to-orbit (*Space*) Space system which can launch a payload into orbit without staging, ie using one engine which can provide the necessary thrust throughout all the flight regimes of the complete ascent.

singlet (*Chem*) A state in which there are no unpaired electrons.

singlet oxygen (*BioSci*) An uncharged but unstable species of oxygen produced in the METABOLIC BURST in phagocytic cells. Symbol $^1O_2$.

single-turn coil (*ElecEng*) An armature coil consisting of a single turn of copper bar.

single-turn transformer (*ElecEng*) A current transformer in which the primary winding takes the form of a single straight conductor of heavy cross-section, to which the cable or bus-bar is connected.

single-valued (*MathSci*) Of a function having only one value for each set of values of the independent variables.

single-wave rectification (*ElecEng*) See HALF-WAVE RECTIFICATION.

single-wire circuit (*ElecEng*) One with a single live wire, including coaxial or concentric, with sheath, earth or frame return.

single-wire feeder (*ICT*) One for an antenna, similar to an ordinary downlead, but connected to the antenna in such a manner that it is terminated in its characteristic impedance, so that no standing waves are formed on it.

single yarn (*Textiles*) The thread obtained from one unit of a spinning machine.

singularity (*Astron*) A point in space–time at which matter is compressed to an infinitely great density.

singularity (*MathSci*) Of a function: a point where the function ceases to be ANALYTIC.

singular matrix (*MathSci*) A matrix which has no inverse, or equivalently one whose determinant is zero.

singular point on a curve (*MathSci*) A point on the curve $f(x,y) = 0$ at which

$$\frac{\partial f}{\partial x} = \frac{\partial f}{\partial y} = 0$$

One at which there is either no real tangent or two or more tangents. Cf DOUBLE POINT.

singular solution (*MathSci*) Of a differential equation: a solution which cannot be derived as a particular integral from a complete primitive.

sinistral fault (*Geol*) A tear fault in which the rocks on one side of the fault appear to have moved to the left when viewed across the fault. Cf DEXTRAL FAULT.

sinistrorse (*BioSci*) Helical, twisted or coiled in the sense of left-hand screw-thread or of an S-helix. Cf DEXTRORSE.

sink (*BioSci*) A region within a plant (or a cell) where a demand exists for particular metabolites, eg growing shoots, roots and developing tubers (sinks for photosynthate) or mitochondria (sinks for oxygen).

sink (*ICT*) Unstable operating region on RIEKE DIAGRAM.

sink (*Print*) A depression in the printing surface of a plate.

sinker (*Textiles*) The mechanism in a knitting machine that pushes a length of thread over the spring needles to form a new course of loops.

sinker bar (*MinExt*) A heavy bar attached to the cable above the drilling tools used in percussive drilling.

sink–float process (*MinExt*) See HEAVY MEDIA SEPARATION.

sinking (*Build*) A recess cut below the general surface of the work.

sinking (*CivEng*) The operation of excavating for a shaft, pit or well.

sinking shaft (*MinExt*) Shaft excavated from above downwards. Cf RISING SHAFT.

sink mark (*Eng*) A depression on outer surface of moulding. Often an inevitable defect created at thick sections.

sino-auricular (*Med*) Applied to structures located in the right atrium near the opening of the venae cavae, corresponding to the sinus venosus, eg sinu-atrial (or sino-auricular) node. Also *sinu-auricular*.

Sinope (*Astron*) The ninth natural satellite of Jupiter, discovered in 1914. Distance from the planet 23 700 000 km; diameter 40 km.

sinter (*Chem*) To coalesce into a single mass under the influence of heat, without actually liquefying.

sinter (*Min*) A concretionary deposit of opaline silica which is porous, incrusting, or stalactitic in habit; found near geysers, as at Yellowstone National Park (US). Also *geyserite*.

sintered carbides (*Eng*) See CARBIDE TOOLS, SINTERING.

sintered crucible (*Chem*) A crucible with a permeable sintered base used as a combination filter and crucible.

sintering (*ElecEng*) The process of consolidating the filament of an electric lamp by passing a relatively high current through it when in a vacuum.

sintering (*Eng*) The coalescing or fusing together of small particles to form larger masses. Used for ceramics, in powder metallurgy, and for ores and concentrates. High temperatures are frequently required because sintering occurs by solid-state diffusion, either along the external surfaces or through the crystalline interior. In the latter, crystal defects such as grain boundaries are the preferred

paths. Sometimes impurities in the batch or deliberate additions (fluxes) can speed up the sintering processes or allow lower temperatures to be used. Also used as a shaping process for intractable polymers, especially polytetrafluoroethylene as in hip-joint implant sockets. See HOT ISOSTATIC PRESSING, LIQUID PHASE SINTERING and panel on CERAMICS PROCESSING.

sinuate (*BioSci*) A leaf margin with rounded teeth and notches; wavy.

sinu-auricular (*Med*) See SINO-AURICULAR.

sinuitis (*Med*) See SINUSITIS.

sinus (*BioSci*) A cavity or depression of irregular shape.

sinus arrhythmia (*Med*) A normal speeding up and slowing down of heart rate due to alterations in tone in the vagus nerve.

sinusitis (*Med*) Inflammation of any one of the air-containing cavities of the skull which communicate with the nose. The condition is called ethmoid, frontal, maxillary or sphenoid sinusitis, according to the site affected. It may be acute or chronic, purulent or non-purulent. Also *nasal sinusitis*.

sinusoid (*BioSci*) In vertebrates, a sinus-like blood space connected usually with the venous system and lying between the cells of the surrounding tissue or organ.

sinusoidal (*ElecEng*) An alternating quantity is said to be *sinusoidal* when its trace, plotted to a linear time base, is a sine wave.

sinusoidal current (*ElecEng*) One which varies sinusoidally with time, having a frequency, amplitude and phase. It flows in each direction alternately for equal periods.

sinusoidal spirals (*MathSci*) Curves with polar equation $r^n = a^n \cos n\theta$. Not true spirals since the tracing point does not recede continuously. For certain values of n the curves have specific names, eg $n = -2$ (hyperbola), $n = -1$ (straight line), $n = -\frac{1}{2}$ (parabola), $n = \frac{1}{2}$ (cardioid) and $n = 2$ (lemniscate).

sinusoidal wave (*Phys*) See HARMONIC WAVE.

sinus venosus (*BioSci*) In a vertebrate embryo, the most posterior chamber of the developing heart; in lower vertebrates, the tubular chamber into which this develops and which receives blood from the veins or sinuses, and passes it into the auricle.

SIP (*ICT*) Abbrev for *session initiation protocol*, a widely used protocol for VOICE OVER INTERNET TRANSMISSION

sipho-, siphono- (*Genrl*) Prefixes from Gk *siphōn*, gen *siphōnos*, tube.

siphon (*BioSci*) A tubular organ serving for the intake or output of fluid, as the pallial siphons of many bivalve Mollusca. Adj *siphonate*.

siphon (*CivEng*) (1) A pipe system comprising a rising leg and a falling leg, typically in the shape of an inverted 'U': pressure at the inlet to the rising leg is atmospheric pressure; as liquid rises in the rising leg the pressure falls below atmospheric pressure and reaches a minimum at or near the apex, recovering to atmospheric pressure in the down leg. The lift of a siphon can equal or exceed the equivalent of atmospheric pressure expressed in *head* of the fluid which enters the siphon; this results from entrained gases emerging throughout the rising leg. (2) A pipe or aqueduct crossing a valley and rising again to somewhat less than its inlet level, so as to have the necessary hydraulic gradient. More correctly *inverted siphon*.

siphonaceous (*BioSci*) See SIPHONEOUS.

Siphonaptera (*BioSci*) An order of wingless insects ectoparasitic on warm-blooded animals. Such insects are laterally compressed and have mouthparts for piercing and sucking. Their coxae are large and their tarsi five-jointed with prominent claws. They undergo complete metamorphosis. The larvae are legless, the pupae have free wings and legs and are enclosed in a cocoon. Adults may be vectors for diseases, eg myxomatosis, plague. Fleas. Also *Aphaniptera*.

siphoneous (*BioSci*) Having large, tubular, multinucleate cells without cross walls; coenocytic. Also *siphonaceous*.

siphono- (*Genrl*) See SIPHO-.

siphonogamy (*BioSci*) A reproductive process in which non-motile male nuclei are carried to the egg cell through a pollen tube, as in conifers and angiosperms. Cf ZOOIDOGAMY.

siphonostele (*BioSci*) A stele with a more or less continuous ring of vascular tissue surrounding a pith, ie a medullated protostele or a solenostele.

siphon spillway (*CivEng*) A siphon connecting the upstream and the downstream sides of a reservoir dam, thus enabling flood waters to pass, as in the case of a BYE-CHANNEL.

siphuncle (*BioSci*) In Nautiloidea, a narrow vascular tube extending from the visceral region of the body through all the chambers of the shell to its apex. Adj *siphunculate*.

Siporex (*Build*) TN for a material of light density and high insulation value manufactured with sand, cement and a catalyst and cured under high-pressure steam conditions.

Sipunculida (*BioSci*) A phylum of marine worm-like animals that lack segmentation and that have an anterior introvert with a mouth surrounded by tentacles. The gut is U-shaped and opens dorsally. These animals are mostly detritus-feeding burrowing forms.

siren (*Acous*) A powerful source of noise of a more or less pure tone; the noise is usually generated by the periodic escape of compressed air through a rotary shutter.

Sirenia (*BioSci*) An order of large aquatic mammals of herbivorous habit. The fore-limbs are fin-like and the hind-limbs lacking; there is a horizontally flattened tail fin. The skin is thick with little hair and underlying blubber. There are two pectoral mammae, no external ears, and the neck is very short. Manatees, dugongs.

Sirius (*Astron*) The brightest star in the sky and the sixth nearest. It lies in the constellation Canis Major, and its faint companion was the first star to be recognized as a white dwarf. Also *Dog Star*.

sisal (*Textiles*) The BAST fibre from the sisal plant (*Agave sisalana*). Grown in Mexico and used to make string and cord but being displaced for many uses by polyethylene or polypropylene.

SISD (*ICT*) Abbrev for *single instruction stream, single data stream*. Term which describes the architecture of a PROCESSOR.

sister cell (*BioSci*) One of the two cells formed by the division of a pre-existing cell.

sister-chromatid exchange (*BioSci*) Reciprocal exchange of DNA between the chromatids of a single chromosome. See fig. at MEIOSIS.

sister nucleus (*BioSci*) One of the two nuclei formed by the division of a pre-existing nucleus.

SIT (*Eng*) Abbrev for SPONTANEOUS IGNITION TEMPERATURE.

site (*ICT*) (1) A place where there are interconnected computers belonging to an organization. (2) Computer connected to the Internet, often providing World Wide Web information for an organization. See panel on INTERNET.

site error (*ICT*) In radio direction-finding, that due to distortions in the electromagnetic field caused by obstructions in the vicinity of the navigational antenna system.

site licence (*NucEng*) Permission to operate a nuclear site given by the relevant authority and without which work cannot proceed.

Site of Special Scientific Interest (*EnvSci*) In the UK, an area deemed originally by the Nature Conservancy Council to be important for nature conservation, but not large enough to become a National Nature Reserve. Abbrev *SSSI*.

site rivet (*Eng*) See FIELD RIVET.

site-specific mutagenesis (*BioSci*) The possibility of altering a DNA sequence at a defined position by, eg synthesizing an alternative sequence and reinserting into its host chromosome. A technique for REVERSE GENETICS.

sitfast (*Vet*) A small hard lump on the skin of a horse's back, due to necrosis of the skin caused by pressure of the saddle or harness.

sitka spruce (*For*) A N American softwood tree (*Picea sitchensis*), with a light pinkish-brown, straight-grained, medium-textured heartwood.

sitosterol (*Chem*) The β form $C_{29}H_{50}O$, a sterol derivative, found in corn oil, which closely resembles cholesterol. Also occurs in other forms: α_1-, α_2-, α_3- and γ-sitosterol.

situs inversus (*Med*) A congenital anomaly with the heart and abdominal viscera situated in the right side of the chest.

SI units (*Genrl*) A system of coherent metric units (*Système International d'Unités*) proposed for international acceptance in 1960. It was developed from the Giorgi/MKSA system (see MKSA) by the addition of the KELVIN and CANDELA, and later the MOLE, as base units. There are numerous *derived units* (eg JOULE, NEWTON) and scales of decimal multiples and submultiples, all with agreed symbols. See appendices on Units of measurement and SI derived units.

six-phase (*ElecEng*) A term applied to circuits or systems of supply making use of six alternating voltage phases, vectorially displaced from each other by $\pi/3$ radians.

Six's thermometer (*EnvSci*) A form of MAXIMUM AND MINIMUM THERMOMETER consisting of a bulb containing alcohol joined to a capillary stem bent twice through 180°C. A long thread of mercury is in contact with the alcohol in the stem, and this mercury moves as the alcohol in the bulb expands and contracts. Each end of the mercury thread pushes a small steel index in front of it, one of which registers the maximum temperature and the other the minimum.

six–twelve potential (*Chem*) See LENNARD-JONES POTENTIAL.

size (*Build*) Material added to surfaces to equalize porosity prior to hanging a wall covering. Can be either glue or cellulose size.

size (*MathSci*) With reference to a TEST, the probability of an error of the first kind.

size (*Paper, Textiles*) Film-forming substances (eg starch), sometimes containing lubricants, that are applied to warp threads to protect them during weaving, to paper to reduce its rate of water absorption and to glass fibres to protect their surfaces and reduce the incidence of strength-impairing flaws. Also *sizing*.

size distribution (*PowderTech*) Proportions of each size of particles in a powder or colloidal system. Expressed either as cumulative or frequency distribution.

size-exclusion chromatography (*BioSci*) Chromatographic separation of particles according to size by a solid phase of material of restricted pore size which functions as a molecular sieve. Such materials include dextran, cross-linked agarose and, for HPLC, coated silica.

size factor compound (*Chem*) An intermetallic compound or phase in which the composition is determined by the best fit of the different sizes of atom within a crystal lattice.

size fraction (*PowderTech*) A portion of a powder composed of particles between two given size limits, expressed in terms of the weight, volume, surface area or number of particles.

size-grading (*Chem*) The process of determining the frequency distribution of particles of different sizes in a material.

sizing (*Paper, Textiles*) See SIZE.

Sjogren's disease (*Med*) Chronic inflammatory disease of salivary and lacrimal glands often accompanied by inflammation of the conjunctiva. Blood of persons with this condition often contains antinuclear antibodies and rheumatoid factor. It is associated with RHEUMATOID ARTHRITIS and SYSTEMIC LUPUS ERYTHEMATOSUS.

skarn (*Geol*) A rock containing calcium silicate minerals produced by metasomatic alteration of limestone close to the contact of an igneous intrusion. Skarns are sometimes enriched in ore minerals such as magnetite or scheelite. US *tactite*.

skate (*CivEng*) See RETARDER.

skeletal muscle (*BioSci*) See VOLUNTARY MUSCLE.

skeleton (*BioSci*) The rigid or elastic, internal or external, framework, usually of inorganic material, which gives support and protection to the soft tissues of the body and provides a basis of attachment for the muscles, forming a system of jointed levers which they can move. Adjs *skeletal*, *skeletogenous*.

skeleton (*Surv*) The network of survey lines providing a figure from which the shape and salient features of the survey may be determined.

skeleton steps (*Build*) Steps in a stair, of a construction such that there are no risers but only treads fixed at suitable positions above one another between side supporting pieces.

skeleton-type switchboard (*ElecEng*) A switchboard consisting of a metal framework upon which the switches and other apparatus are mounted. Also *frame-type switchboard*.

skelp (*Eng*) Mild steel strip from which tubes are made by drawing through a bell at welding temperature, to produce lap-welded or butt-welded tubes.

skene arch (*CivEng*) An arch having the shape of a circular arc subtending less than 180°. Also *scheme arch*.

skew (*MathSci*) (1) Of a pair of lines, not lying in the same plane; neither intersecting nor parallel. (2) Of a distribution, not symmetrical, having significantly more values distant from the mean on one side than the other. See also KURTOSIS.

skew arch (*Arch, CivEng*) An arch which has its axis or line of direction oblique to its face.

skewback (*CivEng*) The courses of stones from which an arch springs (on the top of a pier) where the upper and lower beds are oblique to each other.

skewbacks (*Glass*) Specially shaped blocks, supporting a furnace crown or other arch. Also *springers*.

skew bevel gear (*Eng*) See HYPOID BEVEL GEAR.

skew butt (*Build*) See SKEW CORBEL.

skew coil (*ElecEng*) An asymmetrical coil inserted in the armature winding of an alternator having an odd number of pole pairs.

skew corbel (*Arch*) The projecting masonry or brickwork supporting the foot of a gable coping.

skewed pole (*ElecEng*) A field pole whose cross-section is a parallelogram instead of the usual rectangle.

skewed slot (*ElecEng*) A slot whose diameter is not parallel to the axis of rotation.

skew fillet (*Build*) See TILTING FILLET.

skew flashing (*Build*) Flashing fixed down a gable wall.

skew lines (*MathSci*) Two lines in three-dimensional space which do not lie in the same plane.

skew nailing (*Build*) The operation of driving nails in obliquely.

skewness (*MathSci*) The degree of asymmetry about the central value of a distribution. See KURTOSIS.

skew rebate plane (*Build*) A rebate plane with its cutting edge arranged obliquely across the sole.

skew table (*Build*) A stone which is bonded in with a gable wall, as a support for the foot of the coping.

skew wall (*Acous*) In a studio, a wall which does not form a face of a parallelepiped; the walls are so arranged that continuous reflections between opposite walls are obviated.

skiagram (*Radiol*) Obsolete term for a radiograph, the film produced after exposure to X-rays. Also *skiagraph*.

skiatron (*Electronics*) A cathode-ray tube having a DARK-TRACE SCREEN.

skids (*Build*) Small pieces of timber packed under a surface to bring it to the plane.

ski jump ramp (*Aero*) Curved ramp fitted at the forward flight deck of an aircraft carrier to give improved take-off performance to vectored-thrust V/STOL aircraft. The end slope is typically 8–15° and allows (1) shorter take-off at given weight, (2) take-off at higher weight and (3) operations in higher wave states.

skillet (*Eng*) Mould for casting bullion.

skimming coat (*Build*) See SETTING COAT.

skin (*Aero*) The outer surface other than fabric of an aircraft structure; the outer surface in a SANDWICH CONSTRUCTION.

skin (*BioSci*) The protective tissue layers of the body wall of an animal, external to the musculature. Formed of collagen fibres and elastin. Animal skin is tanned to make leather. See EPIDERMIS.

skin (*Eng*) The hard surface layer found on iron castings due to the rapid cooling effect of the mould, or on steel plates, strip and sheet due to rolling, or on other materials or products due to the surface-hardening effect of the finishing process.

skin (*ICT*) The combination of CSS and HTML that determines the look and feel of a website, document or page and that can be easily substituted, thus transforming appearance without altering functionality or data.

skin associated lymphoid tissue (*BioSci*) The cells of the immune system that are normally resident in the dermis and epidermis of the skin. Abbrev SALT.

skin depth (*ICT*) Owing to the SKIN EFFECT the current density in a conductor, induced by a high-frequency electromagnetic field, falls off rapidly below the surface. Skin depth is that at which the current density has decreased by 1 NEPER compared with the current density at the surface.

skin depth (*Phys*) The distance into a conductor of conductivity σ and absolute permeability μ, at which the amplitude of an electromagnetic wave (frequency $\omega/2\pi$) falls to $1/e = 0.3679$ of its value at the surface, given by

$$\delta = \sqrt{2/\sigma\omega\mu}$$

In copper at 50 Hz, δ is about 10^{-2} m. See SKIN EFFECT.

skin dose (*Radiol*) Absorbed or exposure radiation dose received by or at the skin of a person exposed to sources of ionization. Cf TISSUE DOSE.

skin effect (*ElecEng*) That phenomenon by which high-frequency currents tend to be confined to the thin layer (skin) of conductors. It follows, from

$$R = \frac{\rho l}{A}$$

(see RESISTIVITY), that the resistance R of a given conductor will be higher at high frequencies, because of the effective reduction in cross-sectional area A for the current to flow. See DEPTH OF PENETRATION, LITZ WIRE, SKIN DEPTH.

skin friction (*Aero*) See SURFACE-FRICTION DRAG.

skin graft (*Med*) The transplantation of a piece of skin from one part of the body to another where there has been an injury, especially a burn.

skinner (*ElecEng*) The length of insulated wire between the point of connection to a solder tag and the cable form from which it emerges.

Skinner box (*Psych*) A chamber designed for the study of OPERANT CONDITIONING; it is provided with mechanisms for an animal to operate and an automatic device for presenting rewards according to schedules of reinforcement preset by the experimenter.

skin-sensitizing antibody (*BioSci*) Antibody capable of attachment to skin cells so that, on subsequent combination with an antigen, an immediate-type hypersensitivity reaction (type 1) occurs. Mainly IgE.

skin test (*BioSci*) Any test in which substances are injected into or applied to the skin in order to observe the host's response to them. Used extensively in the study of hypersensitivity and immunity, eg tuberculin test, Schick test.

skintled (*Build*) Said of brickwork in which the bricks are laid irregularly, so as to leave an uneven surface on the wall; also of a similar effect produced by protruding mortar squeezed from the joints.

skin tuberculosis of cattle (*Vet*) A benign lymphangitis producing nodules in the skin of cattle, caused by bacteria which resemble in some respects the causative organism of true tuberculosis.

skin wool (*Textiles*) Wool removed from the fleeces of slaughtered sheep by chemical or biochemical processes. Cf FELLMONGERING, PULLING, SHEARING.

skip (*ICT*) See HOP.

skip distance (*ICT*) A region of no-signal between the limit of reception of the direct (*ground*) wave and the first downcoming reflection from an ionized layer; prominent with short waves.

skip printing (*ImageTech*) Printing only selected frames of a motion picture film, omitting those in between; usually at regular intervals, such as every third frame, to produce an apparent increase in the speed of movement.

skip slitter (*Print*) A cam-operated slitter giving intermittent cuts, whereby both broadsheet and tabloid sections can be incorporated in a single copy by means of cylinder collection.

skip-tooth saw (*Build*) A saw from which alternate teeth are cut away.

skirt (*ChemEng*) A cylindrical support on large vessels and fractionating columns welded to the main part of the cylindrical shell and enclosing the bottom of the vessel or column.

skirt (*ICT*) Lower side portions of a resonance curve, which should be symmetrical.

skirting board (*Build*) A board covering the plaster wall where it meets the floor. Also *baseboard*, *mopboard*, *washboard*.

skirting board (*MinExt*) At delivery onto belt conveyor, boards which direct falling rock towards the centre of the belt, or prevent spillover.

skittle pot (*Glass*) A small pot, in shape resembling a skittle, which can be set in a furnace in some small corner to melt a special glass, eg a colour. Some small firms use a furnace holding only four or six of these, fired by coke.

skiver (*Print*) A split sheepskin used in bookbinding.

skot (*Phys*) Obsolete unit of scotopic luminance. 1 skot = $3{\cdot}1931 \times 10^{-4}$ cd m^{-2}.

skotograph (*ImageTech*) A developable image produced in a photographic emulsion by radiation from organic tissue in the dark.

Skraup's synthesis (*Chem*) The synthesis of quinoline by heating aminobenzene with glycerine and sulphuric acid, nitrobenzene or arsenic (V) acid acting as oxidizing agent.

skull (*BioSci*) In vertebrates, the brain case and sense capsules, together with the jaws and the branchial arches.

skutterudite (*Min*) Grey or whitish arsenide of cobalt, which crystallizes in the cubic system. Often contains appreciable nickel and iron substituting for cobalt.

Skylab (*Space*) A manned space station placed in orbit in May 1973. It re-entered the Earth's atmosphere in July 1979. See panel on SPACE STATION.

skylight filter (*ImageTech*) A slightly pink filter that absorbs ultraviolet light.

Skype (*ICT*) TN for a network that allows telephone calls to be made using VOIP technology.

sky ray (*ICT*) Also *sky wave*. See IONOSPHERIC WAVE, SPACE WAVE.

skyrmion (*Phys*) In non-linear field theory, a SOLITON with spin and statistics different from those of the underlying fields.

SL (*Build*) Abbrev for *short lengths*.

slab (*Build*) An outer piece of a log cut away in the process of slabbing.

slab (*CivEng*) Normally a reinforced concrete floor supported, at intervals, on beams and/or columns.

slab (*Eng*) Metal, twice as wide as thick, and thus intermediate between an ingot and a plate in a rolling mill.

slab and beam floor (*CivEng*) A reinforced concrete floor in which the beams are designed to be homogeneous with the slab and to provide continuity and reduce the amount of concrete and steel.

slabbing (*Build*) The operation of squaring a log.

slab coil (*ElecEng*) A coil in the form of a flat spiral; the term is normally applied to inductance coils.

slab resolver (*Electronics*) Four contacts on a square, giving $(\pm R \cos \theta)$ and $(\pm R \sin \theta)$, concentric with a slab potentiometer fed with $\pm R$ at its ends, θ being angular displacement.

slab serif (*Print*) A type style with strokes of uniform thickness and with straight serifs of the same thickness as the strokes. Also *Egyptian*.

slab tail (*Aero*) A one-piece horizontal tail surface, pivoted and power operated so as to serve as a stabilizing tailplane, elevator and, through a lower gearing, trimming tab.

SLAC (*Phys*) Abbrev for *Stanford Linear Acceleration Center*, US. A linear accelerator which at the centre gives electron beams of energy about 50 GeV.

slack (*MinExt*) Small coal dirt, as in *slack heap*, a tip or dump.

slack sheet (*Print*) Insufficient tension of the web between units of the press.

slack-water navigation (*CivEng*) River or canal navigation rendered possible by the construction of dams across the stream at intervals, dividing it into separate reaches, communication being maintained by the use of locks. Also *still-water navigation*.

slade (*Build*) An inclined pathway.

slag (*Eng*) The top layer of the two-layer melt formed during smelting and refining operations. In smelting it contains the GANGUE minerals and the flux; in refining, the oxidized impurities. See fig. at BLAST FURNACE.

slag cement (*CivEng*) An artificial cement made by granulating slag from blast furnaces by chilling it in water and then grinding it with lime, to which it imparts hydraulic properties.

slag hole (*Eng*) In a smelting furnace, aperture above that through which molten metal is withdrawn, used to top off accumulated slag. See fig. at BLAST FURNACE.

slag wool (*Eng*) Fibrous material very similar to fibreglass but made from slag produced by an iron blast furnace as distinct from silicate glass.

slaked lime (*Build*, *Chem*) See CAUSTIC LIME.

slaking (*Build*) The process of combining quicklime with water.

slamming stile (*Build*) The upright member of a door case against which the door shuts and into which the bolt of a rim lock engages.

slant azimuth (*ImageTech*) In magnetic recording, head AZIMUTH settings other than 90°. See AZIMUTH RECORDING.

slant-azimuth recording (*ImageTech*) See AZIMUTH RECORDING.

slant range (*Surv*) The distance along sloping sight between two points as measured by tacheometer or tellurometer.

slant rig (*MinExt*) One with special facilities for drilling at an angle from the vertical. The head-string components are hauled up rails set at the angle.

slap dash (*Build*) A rough finish given to a wall by coating it with a plaster containing gravel or small stones.

slat (*Aero*) An auxiliary aerofoil which constitutes the forward portion of a slotted aerofoil, the space between it and the main portion of the structure forming the slot.

slat (*Build*) A thin, flat strip of wood.

slat conveyor (*Eng*) See APRON CONVEYOR.

slate (*Geol*) A fine-grained metamorphic rock with good fissility along cleavage planes. Spotted slate is an argillaceous rock altered by low- or moderate-grade metamorphism to produce porphyroblasts which impart a speckled appearance to the rock.

slate axe (*Build*) See SAX.

slate boarding (*Build*) Close boarding laid as an underlining to, and a support for, roofing slates. Termed *sarking* in Scotland.

slate cramp (*Build*) A piece of slate about $7 \times 2\frac{1}{4} \times 1$ in cut to a narrow waist at the middle, and fitted flush into mortises in adjacent stones to bind them together.

slate hanging (*Build*) Similar to WEATHER TILING, slates being used instead of tiles.

slating and tiling battens (*Build*) Any pieces of square-sawn converted timber between $\frac{1}{2}$ and $1\frac{1}{4}$ in thick and from 1 to $3\frac{1}{2}$ in wide; commonly used as a basis for slating and tiling where SLATE BOARDING is not used.

slatted lens (*ICT*) One with shaped metal slats, parallel to E or H vector in wave from waveguide. Also used for low-frequency acoustic waves. Also *egg-box lens*.

slaty cleavage (*Geol*) The property of splitting easily along regular, closely spaced planes of fissility, produced by pressure in fine-grained rocks.

slave VTR (*ImageTech*) A videotape recorder used for duplicating SOFTWARE in real time. See HIGH-SPEED VIDEOTAPE DUPLICATOR.

sleber number (*Paper*) A measure of the bleachability of a pulp determined by mixing 5 g of a sample in water with 50 ml of bleach solution and allowing to stand for 1 hour at 20°C. Thereafter the amount of available chlorine consumed is measured by titrating the filtrate.

sledge-hammer (*Build*) A heavy double-faced or straight-pane hammer, weighing up to 100 lb (45 kg), swung by both hands.

sledger (*CivEng*) A machine for the first stage of crushing of rock in quarrying. Also *scalper*.

sleekers (*Eng*) Moulders' tools with a face of various shapes for smoothing over small irregularities in the sand of the mould. Also *smoothers*. See CORNER TOOL.

sleep (*Psych*) A state characterized by prolonged periods of immobility, often with an associated and species-typical posture, and an increased reluctance to respond to stimulation; most animals sleep during a particular period of the day and for a species-characteristic duration.

sleep apnoea syndrome (*Med*) See APNOEA.

sleeper (*Build*) A horizontal timber supporting a vertical shore or post, and distributing the load over the ground. Also *sole plate*.

sleeper (*CivEng*) A timber, steel or prestressed concrete beam passing transversely beneath the railway track for support and gauge maintenance. Also *cross-sill*, *cross-tie*, *sleeping-car*. US *tie*.

sleeper plate (*Build*) A wall plate resting upon a sleeper wall.

sleeper wall (*Build*) A low wall built under the ground storey of buildings having no basement, as a support for the floor joists. When in brick, the wall is built honeycombed to leave spaces for ventilation, and when in stone, small piers at intervals provide the support required.

sleepiness (*Build*) A defect similar to loss of gloss.

sleeping sickness (*Med*) See TRYPANOSOMIASIS.

sleep mode (*ICT*) After a period of inactivity, the ability of a computer to save energy by restricting power to, usually, a monitor, the power being restored when activity resumes.

sleep movement (*BioSci*) See NYCTINASTIC MOVEMENT.

sleet (*EnvSci*) A mixture of rain and snow, or partially melted snow. US *ice pellets*.

sleeve (*ElecEng*) See PLAIN COUPLER.

sleeve (*Eng*) A tubular piece, usually one machined externally and internally.

sleeve antenna (*ICT*) A single vertical half-wave rectifier, whose lower half is a metal sleeve through which the concentric feed line runs. The upper radiating portion, one quarter-wavelength long, connects to the centre of the line.

sleeve dipole (*ICT*) Same as SKIRT dipole.

sleeve joint (*ElecEng*) A conductor joint formed by a sleeve fitting over the conductor ends. It is either pinned or soldered to the conductors.

sleeve piece (*Build*) A brass or copper pipe joint used between a lead pipe and one of another material. Also *thimble*.

sleeving (*ElecEng*) Tubular flexible insulation for threading over bare conductors.

slenderness ratio (*Build*) See RATIO OF SLENDERNESS.

s-level (*Phys*) See SHARP SERIES.

slewing rollers (*Print*) Rollers whose angle can be adjusted to alter the run of the web to maintain sidelay register, manual control being superseded by electronic equipment.

slew rate (*Electronics*) The maximum rate of change of output voltage for an amplifier when a voltage step is applied at the input. Normally measured in $V\mu s^{-1}$.

sley (*Textiles*) (1) Heavy metal or wooden beam carrying the shuttle race board, reed and shuttle boxes. As it moves backwards the shuttle is picked from one box to the other carrying the weft. When the sley comes forward the reed beats up the weft to form cloth. (2) The part of a lace machine, between the beams and the thread guides, which functions in keeping the threads properly arranged.

slice (*Paper*) The part of the flow box, head box or breast box where the stock is discharged over the apron to the machine wire. It generally takes the form of a hinged lip with numerous control points to ensure the right flow of stock and even level across the machine.

slice (*Radiol*) The cross-sectional portion of the body which is scanned for the production of images in eg COMPUTER-AIDED TOMOGRAPHY and MAGNETIC RESONANCE IMAGING.

slicing (*MinExt*) Removal of a layer from a massive ore body. In top slicing this is horizontal, a mat of timber separating it from the overburden. Side and bottom slicing are also practised.

slick bit (*Build*) A type of interchangeable scraping bit attached to a shank for use in electric drills.

slickensides (*Geol*) Smooth, grooved, polished surfaces produced by friction on fault planes and joint faces of rocks which have been involved in faulting.

slicker (*Eng*) A small implement used by a moulder for smoothing the surface of a mould.

slide (*ImageTech*) A still-picture transparency in a mount of standard size, such as 5×5 cm, intended for viewing by projection.

slide (*MinExt*) A crack or plane along which movement has taken place; the clay filling of such a crack; a fault.

slide bars (*Eng*) See GUIDE BARS.

slide resistance (*ElecEng*) A rheostat whose ohmic value is adjusted by sliding a contact over the resistance wire.

slide rest (*Eng*) A slotted table carrying the tool post of a lathe. It is mounted on the saddle or carriage, and is capable of longitudinal and cross traverse.

slide rule (*Genrl*) A device for the mechanical performance of arithmetic processes, such as multiplication or division. It consists of one rule sliding within another, so that their adjacent similar logarithmic scales permit of the addition and subtraction, corresponding to the multiplication and division, of the numbers engraved thereon.

slide-valve (*Acous*) The slide, containing holes, which is drawn across the supply of air to a rank of organ pipes, to stop the pipes speaking when a key is depressed. Operated by draw-stops through trackers and stickers.

slide-valve (*Eng*) A steam-engine inlet and exhaust valve shaped like a rectangular lid. It is reciprocated inside the steam chest, over a face in which steam ports are cut, so as alternately to admit steam to the cylinder and connect the ports to exhaust through the valve cavity. See D SLIDE-VALVE, PISTON VALVE.

slide-valve lead (*Eng*) The amount by which the steam port of a steam engine is already uncovered by the valve when the piston is at the beginning of its working stroke.

slide wire (*ElecEng*) The wire along which a contact is moved in a potential divider. Similarly a concentric tube with a slot and a probe contact becomes a coaxial slotted line for measuring standing-wave voltages at high frequencies.

sliding bevel (*Build*) See BEVEL.

sliding caisson (*Eng*) A floating body used to open or shut the entrance to a dock or basin, and capable of being drawn for the former purpose into a recess at right angles to the channel.

sliding contact (*ElecEng*) Tangential movement between contacting metal surfaces, to remove film and establish conduction contact. Wear of contact is proportional to total use.

sliding filament model (*BioSci*) A generally accepted model for the way in which contraction occurs in the sarcomere of striated muscle, by the sliding of the thick myosin filaments relative to the thin actin filaments.

sliding growth (*BioSci*) The sliding of the wall of one cell past that of the next, as has been postulated to occur during the growth of fibres.

sliding-mesh gearbox (*Autos*) One in which the ratio is changed by sliding one pair of wheels out of engagement and sliding another pair in.

sliding sash (*Build*) A sash that moves horizontally on runners, as distinct from a *balanced sash* sliding vertically.

sliding ways (*Ships*) The portion of a ship's launching ways which move with the ship on launching.

slime mould (*BioSci*) See MYXOMYCOTA.

slime plug (*BioSci*) Accumulation of phloem protein on a sieve area.

slimes (*MinExt*) Particles of crushed ore which are of such a size that they settle very slowly in water and through a bed which water does not readily percolate. Such particles must be leached by agitation. By convention these particles are regarded as less than $\frac{1}{400}$ in (0·0635 mm) in diameter (mesh number 200). *Primary slimes* are naturally weathered ore, or associated clays. *Secondary slimes* are produced during comminution. See ANODE SLIME.

slip (*Build*) A long narrow piece of wood the thickness of a mortar joint, built into brickwork and to which joinery may be nailed.

slip (*ElecEng*) The fraction by which the rotor speed of an induction motor is less than the rotational speed of the stator field.

slip (*Eng*) The process involved in the plastic deformation of metal crystals in which the change in shape is produced by parts of the crystals sliding with respect to each other along certain crystallographic planes. The force required to cause block slip in perfect crystals would be very high and in practice it is the gliding of DISLOCATIONS along the slip planes which causes the movement to occur at the much lower levels of stress found in practice.

slip (*ImageTech*) Vertical shift of TV image because of imperfect field synchronization. Similarly for line, but the shift is horizontal.

slip (*Ships*) (1) The difference between the distance travelled by the ship through the water and the distance the propeller would have moved in a solid. Expressed as a percentage of the latter. (2) A sloping masonry or concrete surface for the support of a vessel in process of being built or repaired.

slip angle (*Autos*) Slight angle of the tyres to the actual track of the wheels when cornering. Some slip angle is required for the tyres to generate a cornering force.

slip bands (*Eng*) Steps or terraces produced on the polished surface of metal crystals as a result of the parts moving with respect to each other during SLIP.

slip casting (*Eng*) A method of making plaster and ceramic components by extracting water into the mould by capillary action. See panel on CERAMICS PROCESSING.

slip correction (*PowderTech*) Correction for KNUDSEN FLOW in permeability equations.

slip dock (*Ships*) A dock from which the water can be discharged, and which is equipped with a SLIP (2).

slip feather (*Build*) A wooden tongue for a PLOUGHED-AND-TONGUED JOINT, distinguished as a CROSS-TONGUE, FEATHER TONGUE or STRAIGHT TONGUE according to grain direction.

slip flow (*Aero*) The molecular shearing which replaces normal gas-flow conditions at HYPERSONIC velocities above *Mach 10* where the mean free path is of the same order of dimensions as the body. See MACH NUMBER.

slip form (*CivEng*) A *form* that can be moved slowly as work progresses.

slip gauge (*Eng*) Accurately ground and lapped rectangular block or plate used singly or in combination with others, the distance between end faces forming a gauging length.

slip joint (*MinExt*) Special COUPLING used on floating platforms, which is splined or keyed to allow relative vertical movement in the drilling pipe while transmitting rotary torque.

slip meter (*ElecEng*) A device for measuring the SLIP of an induction motor.

slip mortise (*Build*) A CHASE MORTISE or a SLOT MORTISE.

slipped bank multiple (*ICT*) A bank of outgoing trunks so connected that they are tested in the same order by all switches but starting from different points, unlike the STRAIGHT BANK MULTIPLE.

slipped tendon (*Vet*) See PEROSIS.

slipper (*CivEng*) See RETARDER.

slipper brake (*ElecEng*) An electromechanical brake acting directly on the rails of a tramway.

slipper piston (*Autos*) A light piston having the lower part or skirt cut away between the thrust *faces*, to save weight and reduce friction.

slipper satin (*Textiles*) A heavy, smooth, high-quality satin made from continuous filament yarns (silk or manufactured) used particularly for wedding dresses and evening shoes.

slipper tank (*Aero*) An AUXILIARY TANK mounted externally, close up under wing or fuselage.

slip planes (*Crystal*) The particular set or sets of crystallographic planes along which slip takes place in metal and other crystals. These are usually the most widely spaced set or sets of planes in the crystals concerned. See GLIDING PLANES.

slip proof (*Print*) A proof taken from a galley of type-matter before it is made up into pages.

slip regulator (*ElecEng*) A regulating resistance connected in series with the rotor of a slip-ring type of induction motor to alter the slip, and thus vary the speed of the machine.

slip-ring rotor (*ElecEng*) The rotor of a slip-ring induction motor; it has a two- or three-phase winding brought out to slip rings. Cf CAGE ROTOR.

slip rings (*ElecEng*) The rings mounted on, and insulated from, the rotor shaft of an ac machine, which form the means of leading the current into or away from the rotor winding.

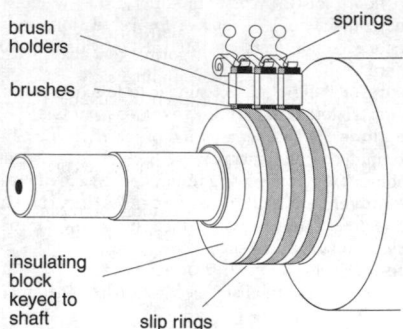

brush holders

springs

brushes

insulating block keyed to shaft

slip rings

slip rings As for a 3-phase generator.

slips (*Print*) Name given to ends of tapes or cords after sewing and before covering.

slip sheets (*Print*) Sheets of paper, oiled manilla being very suitable, placed between sheets as they are printed to avoid SET-OFF; largely superseded by ANTI-SET-OFF SPRAY. Also name given to sheets placed in a letterpress machine packing.

slip sill (*Build*) A sill of length equal to the distance between the jambs of the opening, so that it can be placed in position after the shell of the building has been completed.

slip stone (*Build*) See GOUGE SLIP.

slipstream (*Aero*) The helical airflow from a propeller, faster than the aircraft.

slip tank (*Aero*) Same as DROP TANK.

slipway (*Ships*) See SLIP (2).

slit (*ImageTech*) See SCANNING SLIT.

slitless spectroscope (*Phys*) See OBJECTIVE PRISM.

slitter marks (*Print*) Two marks in the centre gutter of a printed sheet as a guide to slitting either on the printing machine or on the guillotine.

slitters (*Print*) Circular rotating knives used (1) to separate printed sheets on sheet-fed printing of folding machines, (2) to divide the printed web on reel-fed rotaries.

sliver (*Textiles*) Continuous untwisted strand of fibres as formed by carding or combing.

slope (*Build*) See SPLAY BRICK.

slope (*Electronics*) See MUTUAL CONDUCTANCE.

slope (*MathSci*) A measure of the inclination of a line with respect to a fixed line, usually horizontal. On a graph, the tangent of the angle between a line and the x-axis. Also *gradient*.

slope correction (*Surv*) A correction applied to the observed length of a baseline to correct for differences of level between the ends of the measuring tape.

slope deflection method (*CivEng*) A method of structural analysis normally used where only the bending moment at every point is evaluated in terms of the loads applied. The reactions are then determined and the analysis can be completed by the application of normal statics.

sloped roman (*Print*) Fonts in which the letters are not cursive but have the same form as the roman. Sometimes use as an alternative to italic.

slope staking (*Surv*) The locating and pegging of points at which proposed earth slopes in cutting or bank will meet the original ground surface.

sloshing (*Space*) Bulk motion of liquid propellants in their tanks when subject to accelerations, particularly at rocket launches. This must be accounted for in the structural design of a rocket vehicle.

slot (*ElecEng*) An axially cut trench, cut out of the periphery of the stator or rotor of an electric machine, into which the current-carrying conductors forming the winding are embedded.

slot (*ICT*) See EXPANSION SLOT.

slot antenna (*ICT*) Radiating element formed by metal surrounding a slot.

slot binding (*Print*) See NOTCH BINDING.

slot-fed dipole (*ElecEng*) One normal to a coaxial line, and coupled to it by adjacent longitudinal slots.

slot leakage (*ElecEng*) In an electric machine, the leakage flux that passes across the slots.

slot link (*Eng*) See LINK MOTION, RADIAL VALVE GEAR.

slot mortise (*Build*) A mortise made in the end of a member.

slot permeance (*ElecEng*) The total permeance of the several parallel portions of the slot-leakage flux path.

slot pitch (*ElecEng*) The distance between successive slots around an armature.

slot ripple (*ElecEng*) The harmonic ripple in the emf wave of an electric machine. It arises from the regular variation in the permeability of the magnetic path between stator and rotor, caused by the repeated change in air-gap length as the rotor and stator slots pass one another and the intervening teeth.

slotted aerofoil (*Aero*) Any aerofoil having an air passage (or slot) directing the air from the lower to the upper surface in a rearward direction. Slots may be permanently open, closable, automatic or manually operated.

slotted core (*ElecEng*) The usual type of armature core, in which slots are provided for the windings. Cf SMOOTH CORE.

slotted flap (*Aero*) A trailing-edge *flap* which opens a slot between itself and the main aerofoil as it is lowered or extended.

slotted line (*ICT*) Rigid coaxial line, with slot access for contact with central conductor. Used for impedance and voltage standing-wave ratio measurements at wavelengths comparable with length of slotted line.

slotting machine (*Eng*) A machine tool resembling a SHAPING MACHINE but in which the ram has a vertical motion and is balanced by a counterweight, the tool cutting on the down stroke, towards the table.

slotting tools (*Eng*) Cutting tools used for keyway cutting etc in a SLOTTING MACHINE; they are of narrow edge and deep, stiff section, with top and side clearance but little rake.

slough (*BioSci*) The cast-off outer skin of a snake.

slough (*Med*) A mass of dead, soft, bodily tissue in a wound or infected area; to form dead tissue (said of the soft parts of the body); to come away as a slough.

slow (*Acous*) Measuring mode of a SOUND-LEVEL METER with a time constant of 1 s.

slow-acting relay (*ICT*) A relay designed to operate at an appreciable time after the application of voltage. A copper sleeve or slug is placed over the core, or a short-circuited winding is used.

slow associated control channel (*ICT*) A sub-band signalling channel time-multiplexed onto the traffic channel of a digital mobile-telephone system and used by the mobile to report received signal strength and quality to its base station.

slow-break switch (*ElecEng*) A knife switch with a single rigid blade forming the moving part of each pole.

slow-butt welding (*ElecEng*) See RESISTANCE UPSET-BUTT WELDING.

slow fever (*Vet*) See BOVINE ACETONAEMIA.

slowing-down area (*NucEng*) In reactor physics calculations, one-sixth of the mean square distance travelled by neutrons from their source to reach thermal energy. See FERMI AGE.

slowing-down density (*NucEng*) In reactor theory, the rate at which neutrons slow down past a given energy, per unit volume.

slowing-down length (*NucEng*) The square root of the slowing-down area.

slowing-down power (*NucEng*) The increase in LETHARGY OF NEUTRONS per unit distance travelled in medium.

slow motion (*ImageTech*) Achieved by running film through a MOTION PICTURE CAMERA at a faster than normal speed but projecting at normal speed, or by slowing down the projection rate. With video, expensive equipment is required to achieve slow-motion recording, so it is more usual to slow down the videotape during playback. See AUTOMATIC TRACKING, NOISE BARS.

slow muscle (*BioSci*) Striated muscle used for long-term activity (eg postural support) and that depends on oxidative metabolism. It has many mitochondria and abundant myoglobin.

slow neutron (*Phys*) See NEUTRON.

slow-reacting substance A (*BioSci*) A pharmacologically active material comprising LEUKOTRIENES C, D and E, released by MAST cells and other cells in the course of immediate hypersensitivity reactions. It causes contraction of smooth muscle, esp bronchial muscle, and increased vascular permeability. The term is that applied when it was first discovered. Abbrev *SRS-A*.

slow-running cut-out (*Aero*) See FUEL CUT-OFF.

slow-scan TV (*ImageTech*) A TV system in which the sequential field scanning rate is much slower than in broadcasting standards, thus allowing picture transmission by telephone line although with imperfect reproduction of movement.

slow shutter (*ImageTech*) A video camera facility in which FRAMES from the SOLID-STATE IMAGE SENSOR are collected in a digital store before recording to increase the exposure time. See FAST SHUTTER.

slow virus (*Med*) A transmissible agent which causes disease after very long incubation periods often of many years. Also *lentivirus*.

slow-wave sleep (*Psych*) Sleep characterized by the presence of high-amplitude, slow-wave changes in potential, as measured on the EEG. Also *NREM sleep, quiet sleep*.

slow-wave structure (*ElecEng, ICT*) A circuit or transmission path where the PHASE VELOCITY is much less than the speed of light: eg in a helix, the velocity reduction is in the ratio of the pitch of the helix to the circumference. Used to match an electromagnetic wave to the slower electron beam in eg a TRAVELLING-WAVE TUBE.

SLR (*ImageTech*) Abbrev for SINGLE-LENS REFLEX.

slub (*Textiles*) A fault in any drafted, twisted yarn, which appears as a thicker part, with little twist. Slubs can also be deliberately made in yarn at regular or random intervals for ornamentation and special weave effects. They occur naturally in TUSSAH silk.

slubbings (*Textiles*) Relatively thick, open but coherent strands of fibres; converted into yarn by subsequent spinning processes.

sludge (*Build*) A slime produced by the precipitation of solid matters from liquid sewage in sedimentation tanks. See SEWAGE GAS.

sludger (*CivEng*) A long cylindrical tube, fitted with a valve at the bottom and open at the top, used for raising the mud which accumulates in the bottom of a boring during the sinking process. Also *sand pump, shell pump*.

sludging (*Build*) (1) Free-running mud. (2) The process of filling the crevices left in the dried clay of an embankment formed by the method of FLOOD FLANKING.

slug (*Glass*) Any non-fibrous glass in a glass-fibre product.

slug (*ICT*) Thick copper band, comparable with a portion of a winding on a telephone-type relay which, through induced eddy currents, retards the operation and fall-off of the relay.

slug (*NucEng*) Unit of fuel in nuclear reactor, either rod or slab of fissile material encased in a hermetic can of Al, Be, Magnox, Zr or stainless steel. Also *cartridge*. See FUEL ROD.

slug (*Phys*) Unit of mass in the gravitational system of units. A force of 1 lbf (pound-force) acting on a mass of 1 slug gives it an acceleration of $1 \, \mathrm{ft \, s^{-2}}$. See FUNDAMENTAL DYNAMICAL UNITS.

slug tuning (*ICT*) Alteration of inductance in radio-frequency tuning circuits, by inserting a magnetic core or a copper disk or cylinder.

sluice (*Build*) A water channel equipped with means of controlling the flow, enabling a sudden rush of water to be used at harbours or canal locks for the purpose of cleaning out silt, mud, etc, obstructing navigation.

sluice (*MinExt*) A long trough for washing gold-bearing sand, clay or gravel. Also *launder, sluice box*.

sluice gate (*Build*) A barrier plate free to slide vertically across a water channel to control the flow.

sluicing (*Build*) The process of deepening a navigation channel by discharging water from a reservoir through a sluice.

slump (*Geol*) Downslope gravity movement of unconsolidated sediments, esp in a subaqueous environment.

slumping (*NucEng*) The movement of molten fuel; not necessarily as the result of an accident but most dramatically seen after the Chernobyl accident in the Ukraine.

slump test (*CivEng*) A test for the consistency of concrete, made with a metal mould in the form of a frustum of a cone with the following internal dimensions: bottom diameter 200 mm (8 in), top diameter 100 mm (4 in), height 300 mm (12 in). This is filled with the concrete, deposited and punned in layers 100 mm (4 in) thick, and then the mould is removed and the height of the specimen measured when it has finished subsiding.

slur (*Print*) A printing fault in which the image lacks sharpness, caused by drag or movement of the paper, plate or forme, blanket or image carrier or combination thereof.

slurry (*Agri*) A thick suspension of fine solids in a liquid. Farm slurry typically refers to a suspension of FYM in water.

slurry (*MinExt*) A thin paste produced by mixing some materials, esp Portland cement, with water, sufficiently fluid to flow viscously. Used eg to repair (*fettle*) slag-eroded brickwork in smelting furnace etc.

slurry reactor (*NucEng*) One in which fuel or BLANKET material exists as a slurry carried by the coolant fluid.

slushed-up (*Build*) A term applied to brickwork the joints of which are filled with mortar.

slushing compound (*Eng*) A rust-inhibiting liquid composition consisting of mineral oil and anticorrosive additives, such as barium petroleum sulphonates.

slush moulding (*Plastics*) A method based on (1) injecting metal into a die in the pasty stage between LIQUIDUS and SOLIDUS and (2) using certain plastics, particularly polyvinyl chloride, in PLASTISOL form. This is placed in a hollow heated mould which is rotated until the paste forms the solid replica of the mould configuration. Used eg for dolls' heads.

slush pulp (*Paper*) Pulp which is pumped direct from the pulp mill to the paper mill for use without passing through the pulp drying stage.

Sm (*Chem*) Symbol for SAMARIUM.

small bayonet cap (*ElecEng*) A bayonet cap of about 16 mm (0·75 in) diameter; used for small lamps, eg automobile head and side lamps.

small-bore (*Build*) A term applied to pump-assisted hot-water central heating systems with 0·5 in or 15 mm copper or stainless-steel pipes.

small capital (*Print*) A letter having the form of a capital but the height of a lower-case letter, eg c; indicated in manuscript or proof by two lines under the letter. See EVEN SMALL CAPS.

small circle (*MathSci*) A section of a sphere by a plane not passing through its centre.

small Edison screw-cap (*ElecEng*) An Edison screw-cap; having a screw-thread of about 12·5 mm (0·5 in) diameter and about 3·5 threads per cm.

small nuclear RNA (*BioSci*) A discrete set of RNA molecules, found in ribonucleoprotein particles (SnRNPs), which are responsible for processing HNRNA to give MRNA.

small office/home office (*ICT*) A large market for computer hardware and software. Abbrev *SOHO*.

small offset (*Print*) A term applied to sheet-fed offset lithographic machines with a sheet size below that of about 375 × 500 mm.

small pica (*Print*) An old type size, approximately 11-point.

smallpox (*Med*) An acute, highly infectious viral disease characterized by fever, severe headache, pain in the loins, and a rash which is successively macular, papular, vesicular and pustular, affecting chiefly the peripheral parts of the body. Until recent times, one of the major killing diseases of humans but has now been eradicated. Also *variola*.

smallpox vaccination (*Med*) A method of producing active immunity against smallpox (*variola*), discovered by E Jenner in Gloucestershire, UK, in 1796. It is no longer practised now smallpox has been eliminated on a world-wide scale, but is of great historical significance. Vaccinia (cowpox) virus was used; this shares antigens with variola and thus conferred protective immunity.

smalls (*MinExt*) See RIDDLE.

small-scale integration (*ICT*) A CHIP with about 1–10 logic GATES. Abbrev *SSI*.

small-signal parameters (*Electronics*) See TRANSISTOR PARAMETERS.

smallwares (*Textiles*) General name given to tapes, ribbons and other narrow fabrics made on special narrow looms or braiding machines.

smaltite (*Min*) Cobalt arsenide, crystallizing in the cubic system and usually associated with *chloanthite*, nickel arsenide.

smaragdite (*Min*) A fibrous green amphibole, pseudomorphous after pyroxene in such rocks as eclogite.

smart (*Aero*) Originally applied to guided and self-homing bombs for attacking point targets, now used for any device showing 'artificial intelligence' capability.

smart card (*ICT*) A card similar to a credit card but containing a microcontroller, memory and communications interface. Contacts on the card receive power and transfer data when it is inserted in a reader. Telecommunications applications include subscriber identification, billing, authentication and service profiling. See SUBSCRIBER IDENTITY MODULE.

smart materials (*Eng*) Metals, ceramics or polymers which respond to an external stimulus in a specific, controlled way. They include PHOTOCHROMIC lenses which darken in bright sunlight, plastic film containing LIQUID CRYSTALS which change colour with changing ambient temperature, PIEZOELECTRIC CRYSTALS which generate a voltage when stressed. Widely used in consumer products etc for controlling function by sensing changes in the environment. Also *intelligent materials*.

smashing (*Print*) The pressing of a book in a machine after sewing, thereby crushing and expelling air. Also *crushing*, *nipping*.

SMC (*Aero*) Abbrev for STANDARD MEAN CHORD.

smear head (*Eng*) A type of solid head of extruder screw which aids mixing of polymers and additives. But see CAVITY TRANSFER MIXING.

smear metal (*Eng*) Particles and serrated edges produced by metal cutting and welded or combined into an amorphous metal substance by the cutting heat generated.

smear test (*Med*) Diagnostic test for cancer by laboratory examination of the cells of the blood and other fluids. Most commonly applied to cervical smears for the detection of precancerous states; see PAP TEST.

smear test (*NucEng*) A method of estimating the loose, ie easily removed, radioactive contamination upon a surface. Made by wiping the surface and monitoring the swab.

smectic (*Phys*) Said of a mesomorphous substance whose atoms or molecules are oriented in parallel planes. Cf NEMATIC.

smectites (*Min*) A group of clay minerals including montmorillonite, beidellite, nontronite, saponite, sauconite and hectorite. They are 'swelling' clay minerals and can take up water or organic liquids between their layers, and they show cation exchange properties.

SMEDI (*Vet*) Abbrev for *swine mummification embryonic death and infertility*. An enterovirus disease of pigs.

smegma (*Med*) A thick greasy secretion of the sebaceous glands of the glans penis.

smell (*Med*) The sensation produced by stimulation of the mucous membrane of the olfactory organs.

smelting (*Eng*) Fusion of an ore or concentrate with suitable fluxes, to produce a melt consisting of two layers, a top slag of flux and GANGUE minerals with molten impure metal below.

SMIL (*ICT*) Abbrev for *synchronized multimedia integration language*, a recommended means of describing the timing and characteristics of a multimedia presentation using XML.

Smith chart (*ElecEng*) Polar chart with circles for constant resistance and reactance, lines for phasor angles, and standing-wave ratio circles. Used for impedance calculations, esp with data from slotted lines for very high frequencies and from waveguides. See VOLTAGE STANDING-WAVE RATIO.

Smith's coupling (*Autos*) See MAGNETIC CLUTCH (2).

smithsonite (*Min*) Carbonate of zinc, crystallizing in the trigonal system. It occurs in veins and beds and in calcareous rocks, and is commonly associated with hemimorphite. The honeycombed variety is known as *dry bone ore*. Formerly (in UK) *calamine*.

smog (*EnvSci*) General name for fog that is contaminated by pollutants such as smoke, unburnt hydrocarbons, etc. Smog is often trapped near the ground by a temperature inversion. It should be distinguished from *photochemical smog*.

smoke (*PowderTech*) Visible cloud of airborne particles derived from combustion, or from chemical reaction; the particles are mainly smaller than 10 μm.

smoke control area (*Genrl*) An area, statutorily defined, in which emission of smoke from chimneys is prohibited.

smoke point (*Chem*) In the testing of kerosine, the maximum flame height (in millimetres) at which a kerosine will burn without smoking, under prescribed conditions.

smoker (*Geol*) See BLACK SMOKER, CHIMNEY, WHITE SMOKER.

smoke test (*Build*) A test for new or suspect drainpipes in which dense smoke is put into the plugged pipes and its escape searched for.

smoking (*FoodSci*) Preserving meat or fish by hanging them in a smoke-filled chamber and due to the antimicrobial and drying effects of the chemicals in smoke. Smoking also confers a characteristic flavour.

smoky quartz (*Min*) Dark greyish-brown transparent quartz, used as a gemstone. See CAIRNGORM.

smooth ashlar (*Build*) A block of stone dressed ready for use, usually for the stone facing of walls.

smooth colony (*BioSci*) A bacterial colony of a typical regular, glistening appearance; sometimes developed from a rough colony by mutation and frequently differing from it physiologically, eg altered sensitivity to bacteriophage.

smooth-core rotor (*ElecEng*) A rotor carrying a field winding embedded in tunnels, used in high-speed steam-turbine-driven generators.

smooth endoplasmic reticulum (*BioSci*) A tubular form of endoplasmic reticulum without ribosomes but important in lipid synthesis, well developed in gland cells that secrete terpenoids, flavonoids, etc.

smoother (*ElecEng*) A combination of capacitors and inductors for removing the ripple from rectified power supplies.

smoothers (*Eng*) See SLEEKERS.

smoothing choke (*ElecEng*) An inductor in a filter circuit which attenuates ripple in a given dc supply.

smoothing circuit (*ElecEng*) See RIPPLE FILTER.

smoothing plane (*Build*) A bench plane about 8 in (20 cm) in length used to give a smooth even finish to timber.

smooth mouth (*Vet*) Smooth and polished grinding surface of the molar teeth of horses.

smooth muscle (*BioSci*) See UNSTRIATED MUSCLE.

SMPTE (*ImageTech*) Abbrev for *Society of Motion Picture and Television Engineers*. The US body responsible for numerous standards and recommended practices.

SMR (*Eng*) Abbrev for STANDARD MALAYSIAN RUBBER.

SMR (*Med*) Abbrev for STANDARDIZED MORTALITY RATIO.

smudge (*Build*) A lampblack and glue size mixture which is painted over lead surfaces so that solder will not adhere.

smut (*BioSci*) One of a number of plant diseases, some economically important, caused by biotrophic fungi of the order USTILAGINALES and characterized by the production of masses of usually black spores within the host. See BUNT.

smut (*MinExt*) (1) Bad soft coal containing earthy matter. (2) Worthless outcrop material of a coal seam.

Sn (*Chem*) Symbol for TIN (Lt *stannum*).

sn (*MathSci*) See ELLIPTIC FUNCTIONS.

SNA (*ICT*) Abbrev for SYSTEMS NETWORK ARCHITECTURE.

snakewood (*For*) (1) Strong, tough and very durable hardwood from a C and tropical S American tree (*Piratinera*). The heartwood is red-brown with darker markings resembling a snake skin. Also *leopard wood*, *letterwood*, *tortoiseshell wood*. (2) The *Strychnus nux vomica*, a tree species yielding strychnine.

snaking (*Aero*) An uncontrolled oscillation in yaw, usually at high speed, of approximately constant, but small, amplitude ($\approx 1°$).

snap (*ElecEng*) Sudden action in magnetic amplifiers with excessive positive feedback, arising from hysteresis in the core.

snap (*Eng*) (1) A form of punch with a hemispherically recessed end, used to form rivet heads. (2) A limit gauge of plate or calliper type.

snap-fit (*Eng*) A joint where closure is achieved over a small lip, usually in plastics products; reversible, allowing disassembly.

snapped header (*Build*) A half-length brick, sometimes used in Flemish bond.

snapshot (*ImageTech*) A photograph taken with a simple camera using a short exposure time, generally without any special preparation of subject or lighting.

snap switch (*ElecEng*) A switch which makes and breaks the circuit with a quick snap; it comprises blades whose rate of motion is controlled by a spring. Also *quick make-and-break switch*.

snap the line (*Build*) To pluck a well-chalked string, held taut in position, against work to mark a straight line.

snare (*Med*) A wire loop for removing soft tumours such as nasal polypi.

snarl (*Textiles*) Small extra thickness of yarn where the twist has run back on itself; caused by excessive twist and inadequate tension on the yarn.

S/N curve (*Eng*) Abbrev for *stress–number of cycles to failure curve*. See panel on FATIGUE.

sneck (*Build*) The lifting lever which passes through a slot in a door and actuates the fall bar. See NORFOLK LATCH.

snecked (*Build*) Said of rubble walls in which the stones are roughly squared but of irregular size and uncoursed.

sneezewood (*For*) A S African tree, *Ptaeroxylon utile*, that gives a hard and strong timber. It has a smell akin to pepper.

Snell's law (*Phys*) A law relating angles of incidence and refraction of a ray of light at the boundary between media of different refractive index:

$$n_1 \sin \theta_1 = n_2 \sin \theta_2$$

where n_1 and n_2 are the refractive indices on each side of the surface, and θ_1 and θ_2 are the corresponding angles. The two rays and the normal at the point of incidence on the boundary lie in the same plane.

sniffer (*ICT*) A software tool used to identify, capture and decode packets of data being transmitted over a network.

snippet (*ICT*) A short piece of code, usually written in a scripting language by a third party and made available over the Internet, that can be embedded in a web page or application, eg to provide access to a service or website of the third party.

SNMP (*ICT*) Abbrev for SIMPLE NETWORK MANAGEMENT PROTOCOL.

Snoek effect (*Chem*) When an INTERSTITIAL SOLID SOLUTION is subjected to small elastic strains, solute atoms tend to move to interstices which align with the strain direction in order to minimize the total strain energy in the lattice. If the direction of strain is altered the atoms jump to adjacent sites in order to realign with the new strain axis. This effect can be used to study diffusion rates at ambient temperature by measuring the increase in internal friction associated with repeated realignments caused by eg a torsional pendulum with a period of roughly 1 second.

snow (*EnvSci*) Precipitation in the form of small ice crystals, which may fall singly or in flakes, ie tangled masses of snow crystals. The crystals are formed in the cloud from water vapour.

snow (*ICT*) The effect of electrical noise on display of intensity-modulation signals. The effect is random and resembles falling snow and may be seen, eg on a TV screen in the absence of a signal, or when the signal is weak.

snow boards (*Build*) Horizontal boards, about 20 cm high, fixed over roof gutters to prevent snow from sliding off in mass. The snow, on melting, drops into the gutter through gaps left between the boards. Also *gutter boards, snow guards*.

snow guards (*Build*) See SNOW BOARDS.

snow load (*Build*) The unit loading assumed in the design of a roof to allow for the probable maximum amount of snow lying upon it.

snow pack (*EnvSci*) An accumulation of packed snow that lingers throughout the winter.

snow stage (*EnvSci*) That part of the condensation process taking place at temperatures below 0°C so that water vapour condenses directly to ice.

SNP (*BioSci*) Abbrev for *single nucleotide polymorphism*.

SNR (*Acous*) Also *S/N ratio*. Abbrevs for SIGNAL-TO-NOISE RATIO.

snubbing (*MinExt*) In high-pressure oil wells the process by which drill pipe is removed through the BLOWOUT PREVENTER stack while retaining pressure at all times.

snuffles (*Med*) Symptoms resulting from nasal discharge and obstruction of the nose in infants.

snuffles (*Vet*) A term popularly applied to several diseases of animals in which the nasal cavities are affected. See ATROPHIC RHINITIS (pig), OSTEODYSTROPHIA FIBROSA (pig), RABBIT SEPTICAEMIA (rabbit).

soakaway (*Build*) A pit excavated to receive surface water and allow it to seep away.

soakers (*Build*) Small pieces of sheet lead or zinc bonded in for watertightness with the slates or tiles of a roof at joint with walls or at valleys and hips.

soaking (*Eng*) A phase of a heating operation during which metal or glass is maintained at the requisite temperature until uniformly heated, and/or until any required phase transformation has occurred.

soaps (*Build*) Bricks of size $9 \times 2\frac{1}{4} \times 2\frac{1}{4}$ in, which are often pierced for use as air bricks.

soaps (*Chem*) (1) The alkaline salts of fatty acids, chiefly palmitic, stearic or oleic acids. *Soft soaps* contain the potassium salts, whereas the sodium salts are *hard soaps*. *Metallic soaps*, water-insoluble compounds of fatty acids with bases of copper, aluminium, lithium, calcium, etc, are used as waterproofing agents and as the bases of many greases. (2) The alkali salts of resins, so-called *resin soaps*.

soapstone (*Min*) See STEATITE.

soaring (*Aero*) The art of sustained motorless flight by the use of thermal upcurrents and other favourable air streams.

social facilitation (*Psych*) An increase in the performance of a behaviour as a consequence of its performance by other individuals nearby (eg yawning in primates) or, in some cases, simply as a result of having an audience.

socialization (*Psych*) The process whereby the child learns the norms of its society and acquires the attitudes and behaviour that conform to them.

social-learning theory (*Psych*) A theory of behavioural development that emphasizes the importance of observation and mimicry of the behaviour of others.

social organization (*Psych*) The totality of all social relationships among members of a particular group.

social parasitism (*Psych*) A parasitic relationship between members of one species or with individuals of another species which involves exploiting aspects of the host's social behaviour, eg their nesting or hunting behaviour. A common example is the female cuckoo which lays its egg in another bird's nest.

social perception (*Psych*) Several related areas of study: (1) how individuals perceive others, eg the judgment of emotions in others, studies of impression formation (or person perception); (2) the psychological processes that underlie these activities in the perceiver; (3) the study of how social factors influence perception (the effects of group pressure on the perception of an event).

social phobia (*Psych*) The fear of performing certain actions when exposed to the scrutiny of others.

social psychology (*Psych*) The branch of psychology concerned with interpersonal relations and the interactions with and between groups, social institutions and society as a whole.

social symbiosis (*Psych*) A relationship between members of different species from which one or both derive some advantage; the nature of the benefit for each species can be quite different, eg one partner may gain protection from predators while the other gains a nesting site, or food supply.

societal risk (*Radiol*) The chance of radiation damage to large numbers of people, the result of bomb testing or nuclear accidents.

society (*BioSci*) A minor plant community within an association, dominated by a species which is not a general dominant of the association, eg an alder-dominated community on wet ground within an oak wood.

socket (*Build*) (1) A pipe end enlarged to pass over a same-sized pipe to make a joint. US *bell* or *hub*. (2) Pipe fitting for joining two pipes in line.

socket (*MinExt*) A portion of drill hole left undisturbed after blasting, and liable to contain unexploded charge.

socket chisel (*Build*) A robust type of chisel used for mortising; it has a hollow tapering end to the steel shank into which the handle fits.

socket head screw (*Eng*) A screw which has a hexagonal recess in its head, so that it can be turned by means of a hexagonal bar formed into a key. Also *cap screw*.

socket spanner (*Eng*) A head with a recess having six or more flats to fit a hexagonal nut in various positions; often used with alternative handles which fit into a square hole at the end opposite to the hexagonal recess as part of a socket set.

socle (*Arch*) A plain projecting block or plinth at the base of a pedestal, wall or pier.

soda-ash (*MinExt*) Impure (commercial) sodium carbonate. Widely used in pH control of FLOTATION PROCESS.

soda lakes (*Geol*) Salt lakes the water of which has a high content of sodium salts (chiefly chloride, sulphate and acid carbonate). These salts also occur as an efflorescence around the lakes.

soda–lime–silica glass (*Glass*) The commonest type of glass, used (with small variations in composition) for windows, containers and electric-lamp bulbs. Contains silica, soda and lime in approx proportions (percentage by weight): SiO_2 70–74%, Na_2O 12–15%, CaO 8–12%, plus MgO 0–4%, Al_2O_3 1–2%. CROWN GLASS is of this type, but in OPTICAL CROWN barium oxide is now normally substituted for lime.

sodalite (*Min*) A cubic feldspathoid mineral, essentially silicate of sodium and aluminium with sodium chloride, occurring in certain alkali-rich syenitic rocks.

sodamide (*Chem*) $Na^+ NH_2^-$; an ionic compound formed when ammonia gas is passed over hot sodium.

soda nitre (*Min*) Sodium nitrate, crystallizing in the trigonal system. It is found in great quantities in N Chile, where beds of it are exposed at the surface and are known as *caliche*. Also *Chile saltpetre, nitratine*.

sodar (*Acous*) Acoustic method for finding atmospheric layers. In principle like sonar, used in meteorology.

soda recovery (*Paper*) A process for recovering a substantial amount of inorganic soda from the spent liquor of an alkaline digestion by concentrating it, burning off the organic matter and treating the resultant soda ash with caustic.

sodium (*Chem*) A metallic element, one of the alkali metals. Symbol Na, at no 11, ram 22·9898, mp 97·5°C, bp 883°C, rel.d. 0·978, valency 1. Sodium does not occur in nature in the free state and is a soft silvery-white metal, which reacts violently with water. Sodium is the seventh commonest element in the Earth's crust, with an abundance of 2·27%, and 1·06% in sea water. Its principal mode of occurrence is in complex silicates, esp in igneous and metamorphic rocks. It is also present as various salts in evaporite deposits which are commercial sources of sodium. It has eight isotopes. In nuclear engineering it has been mainly used as a coolant in fast-breeder reactors, where its high thermal conductivity, low neutron capture cross-section and high boiling point make it almost the only coolant capable of removing the heat from the tightly packed cores of these reactors.

sodium aluminate (*Chem*) See ALUMINATE.

sodium benzoate (*FoodSci*) A food preservative which is soluble in water and effective against yeasts and some bacteria. It has optimum functionality at a pH between 2·5 and 4·0. Statutory maximum levels are enforced.

sodium channel (*BioSci*) A protein in the nerve cell membrane that is the VOLTAGE-GATED ION CHANNEL responsible for electrical excitability of neurons; its action can be blocked by neurotoxins such as TETRODOTOXIN. Also *sodium gate*.

sodium chloride (*Chem*) NaCl; a white crystalline salt, soluble in water, obtained from underground deposits of the mineral HALITE, and from sea water. Used for seasoning and preserving food, and as a deicing and water-softening agent, and an intermediate in the manufacture of chlorine, sodium hydroxide, hydrochloric acid, etc. Also *common salt*.

sodium citrates (*FoodSci*) Compounds which are used in the food industry as ACIDITY REGULATORS. They are readily soluble in water and are used as buffering agents and sequestrants in many foods including soft drinks and preserves; in dairy products they have useful emulsification properties.

sodium-cooled reactor (*NucEng*) One in which liquid sodium is used as the primary coolant, as in the FAST REACTOR.

sodium cromoglycate (*Pharmacol*) A drug that is of value in the prophylaxis of allergic asthma. It appears to act by inhibiting release of bronchoconstricting agents from the mast cells in the airways.

sodium cyanide (*Chem*) NaCN. See POTASSIUM CYANIDE.

sodium cyclamate (*Chem*) The sodium salt of N-cyclohexylsulphamic acid (see CYCLAMATES). A white, crystalline powder readily soluble in water. Used as an artificial sweetening material, eg in soft drinks and diabetic sweetening tablets, but banned in some countries because of possible health hazards in long-term use.

sodium dodecyl sulphate (SDS) (*Chem*) Anionic detergent ($Cl_2H_{25}SO_4Na$) widely used for denaturing and solubilizing proteins. SDS residues attach to the protein and confer charge in rough proportion to size, which is important for electrophoretic separation methods. A common ingredient of shampoos and other personal care products. Also *sodium lauryl sulphate* (SLS). Abbrev SDS. See POLYACRYLAMIDE GEL ELECTROPHORESIS.

sodium gate (*BioSci*) See SODIUM CHANNEL.

sodium hydroxide (*Chem*) NaOH, a deliquescent substance, with a soapy feel, whose solution in water is strongly alkaline. It is manufactured by treating quicklime with hot sodium carbonate solution; and its main industrial use is the manufacture of soap. Also *caustic soda*. See CASTNER–KELLNER PROCESS.

sodium iodide scintillation crystal (*Radiol*) A high-density photon absorber that converts the energy of a photon from radioactivity to a light photon.

sodium laureth sulphate (*Chem*) Sodium lauryl ether sulphate; Also *sodium polyethoxyethyl dodecylsulphate*, similar to sodium lauryl sulphate. See SODIUM DODECYL SULPHATE.

sodium lauryl sulphate (*Chem*) Abbrev SLS. See SODIUM DODECYL SULPHATE.

sodium nitrate (*Min*) See SODA NITRE.

sodium–potassium ATPase (*BioSci*) The adenosine-triphosphatase that actively transports sodium and potassium ions across cell membranes to establish a sodium gradient. Abbrev $Na^+ K^+–ATPase$.

sodium thiosulphate (*Chem, ImageTech*) See HYPO.

sodium vapour lamp (*Phys*) An electric lamp of the gaseous-discharge type whose electrodes operate in an atmosphere of sodium vapour.

sodoku (*Med*) A disease due to infection with the microorganism *Spirillum minus*, conveyed by the bite of a rat; it is characterized by inflammation of the skin around the bite, relapsing fever, swelling of the lymphatic glands, and a red, patchy rash. Also *rat-bite fever*.

soffit (*Build, CivEng*) (1) A term often used for INTRADOS, but more particularly applied to that part of the intrados in the immediate vicinity of the keystone. See fig. at ARCH. (2) The under surface of a stair or floor or of the head of an opening such as a door or window opening.

soft (*Electronics*) Said of valves and tubes when there is appreciable gas pressure within the envelope, such as gas-discharge tubes and photocells. Particularly said of valves when gas is released from the envelope or electrodes. See HARD, OUTGASSING.

soft (*Glass*) Having a relatively low softening point.

soft acids and bases (*Chem*) See HARD ACIDS AND BASES.

soft agar (*BioSci*) Semi-solid AGAR that is used to increase the viscosity of media for the suspension culture of certain animal cells. Ability to grow in suspension correlates with transformation and malignant potential

soft boot (*ICT*) See WARM BOOT.

soft commissure (*BioSci*) In mammals, the point at which the thickened sides of the DIENCEPHALON touch one another across the constructed third ventricle.

softener (*Build*) A brush used in graining and marbling for blending colour, or reducing the harsh appearance of the work. Hog hair softeners are used in oils and badger hair softeners are used in water colour.

softening (*Eng*) (1) The end result of ANNEALING or tempering, ie a reduction in hardness and strength. (2) A process, for removing arsenic, antimony and tin from lead, after drossing.

softening (*Textiles*) The application of a chemical (the softener) that softens the handle and increases the fluffiness of fabrics.

softening point (*Glass*) One of the reference temperatures in glass-making. See panel on GLASSES AND GLASS-MAKING

softening-point test (*Build*) See MELTING-POINT TEST.

soft-focus lens (*ImageTech*) A lens which images a point source with a slight halo, giving a softness of outline to the whole picture sometimes preferred for portraiture.

soft font (*ICT*) See DOWNLOADABLE FONT.

soft iron (*Eng*) Iron low in carbon with low hysteresis and unable to retain magnetism; useful as solenoid cores etc.

soft-iron armature (*ElecEng*) The attracted part of an electromagnet retaining little residual magnetism.

soft-iron instrument (*ElecEng*) An undesirable synonym for *moving-iron instrument*.

softness (*Eng*) Tendency to deform easily. It is indicated in a TENSILE TEST by low ultimate tensile stress and large reduction in cross-section before fracture. Usually the elongation is also high. In a notched-bar test, specimens bend instead of fracturing, and the energy absorbed is relatively small. See BRITTLENESS, TOUGHNESS.

soft palate (*BioSci*) In mammals, the posterior part of the roof of the buccal cavity which is composed of soft tissues only.

soft radiation (*Radiol*) General term used to describe radiation whose penetrating power is very limited, eg low-energy X-rays.

soft return (*ICT*) A line feed that is inserted by a word processor or text editor to JUSTIFY the text. Its position may change when the text is edited or the FONT SIZE changed. See HARD RETURN, WORD-WRAP.

soft-rock geology (*Geol*) An informal term for the geology of sedimentary rocks.

soft rot (*BioSci*) Rot in which tissues (usually parenchyma) soften because of the destruction of middle lamellae and cell walls, as in many bacterial and fungal diseases of stored fruits and vegetables.

soft sectored disk (*ICT*) One that is formatted into sectors with information written on the disk by a program.

soft segment (*Chem*) A term applied to the elastomeric part of block copolymers, esp polyurethanes and block polyesters, usually made of oligomeric polyglycols or polyesters.

soft-shelled egg (*Vet*) A bird's egg in which lime salts are deficient or absent.

soft soaps (*Chem*) See SOAPS.

soft solder (*Eng*) Alloys of lead and tin used in soldering. Tin content varies from 63% to 31%. The remainder is mainly lead, but some types contain about 2% antimony and others contain cadmium. The best-known types are *plumber's solder* and *tinman's solder*. Owing to the toxicity of lead in plumbing systems, food cans, etc, solders based on tin–copper and tin–silver are preferred.

soft space (*ICT*) A space character that is inserted to JUSTIFY the text. It may be moved or removed entirely if the text is edited or if the FONT SIZE is changed.

software (*ICT, ImageTech*) (1) A general term for all types of program and their associated documentation. (2) Pre-recorded tape or disk carrying motion picture and video productions. Also *program software, videogram*. Cf HARD-WARE. See APPLICATIONS SOFTWARE, SYSTEMS SOFT-WARE.

software house (*ICT*) A commercial organization that specializes in the preparation of APPLICATIONS SOFTWARE or SYSTEMS SOFTWARE.

software package (*ICT*) Fully documented program, or set of programs, designed to perform a particular task.

soft water (*Chem*) Generally, water free from calcium and magnesium salts (cf HARD WATER), naturally or artificially. See WATER SOFTENING.

softwood (*BioSci*) The wood from a conifer and coniferous trees generally. Cf HARDWOOD.

SOG (*Electronics*) Abbrev for SPIN ON GLASS.

SOHO (*ICT*) Abbrev for SMALL OFFICE/HOME OFFICE.

SOI (*Electronics*) Abbrev for SILICON ON INSULATOR.

soil (*Geol*) The material, normally unconsolidated and directly below the ground surface, composed of rock material, weathered to a greater or lesser extent, including organic matter, and able to support plant life. There is a wide range of compositions and textures. In civil engineering soil is taken to include a broader definition of soft unconsolidated materials, eg some geological clays and sands. See SEAT EARTH, SOIL MECHANICS.

soil-acting herbicide (*BioSci*) A plant poison absorbed from the soil by the roots; eg sodium chlorate.

soil flora (*BioSci*) Fungi, bacteria and algae living in the soil.

soil improver (*Agri*) Organic material worked into soil to improve aeration and water-holding capacity. It may also contain nutrients.

soil injection (*Agri*) Mechanical placement of materials beneath the soil surface with a minimum of mixing or stirring.

soil mechanics (*CivEng*) The process of determining the properties of any soil, eg water content, bulk, density, permeability, shear strength, etc.

soil persistence (*Agri*) The length of time during which the toxic effects of a pesticide can be measured in soil samples.

soil pipe (*Build*) A vertical cast-iron or plastic pipe conveying waste matter from water closets etc to the drains. Abbrev SP.

soil release agent (*Textiles*) A chemical added to fabrics to make it easier to remove stains by washing.

soil sampler (*CivEng*) A hollow circular tool, with a sharp edge, for extracting specimens of soil for examination or analysis. Also *soil borer, soil pencil*.

soil sampling (*MinExt*) Taking samples of soil or over-burden as part of a GEOCHEMICAL PROSPECTING exercise to identify anomalous concentrations of metals, or the presence of TRACERS.

soil solarization (*Agri*) A technique to elevate soil temperatures above the lethal level for pests and pathogens. This is achieved by covering moistened soil with plastic film for extended periods in summer sunshine.

soil sterilization (*Agri*) The application of steam or chemical fumigants to the soil to eradicate pests and disease. In some instances the soil may not support plant growth until after a recovery period.

soil structure (*BioSci*) The nature of a soil in terms of the manner in which the particles are aggregated into larger bodies or peds.

soil texture (*BioSci*) The nature of a soil in terms of the proportions of mineral particles of various sizes (sand, silt and clay).

soil water deficit (*Agri*) The volume of water required to restore soil to its *field capacity*.

sol (*Chem*) A colloidal solution, ie a suspension of solid particles of colloidal dimensions in a liquid.

Solanaceae (*BioSci*) A family of c.2500 spp of dicotyledonous flowering plants (superorder Asteridae). It comprises cosmopolitan plants: mostly herbs, but some are shrubs and small trees. The petals are fused, the ovary is superior and the fruit is usually a berry although less often a capsule. It includes potatoes, tomatoes, aubergines, sweet peppers, chillies, cayenne pepper, tobacco. Many species are poisonous and often contain tropane alkaloids of medicinal importance, eg belladonna or deadly nightshade (atropine) and henbane (hyocyamine).

solanine (*FoodSci*) A glycoalkaloid. A natural toxin which develops in green or sprouting potatoes.

solar (*BioSci*) Having branches or filaments radially arranged.

solar antapex (*Astron*) The point on the celestial sphere diametrically opposite to the SOLAR APEX.

solar apex (*Astron*) The point on the celestial sphere towards which the solar system as a whole is moving at the rate of 20 km s^{-1}. It is located in the constellation Hercules at equatorial co-ordinates RA 18 h and declination $+30°$ approx.

solar array (*Space*) A bank of solar cells, mounted on a panel structure, and extended from a satellite, which converts solar energy into electrical energy using photovoltaic conversion. Also *solar battery, solar cell array, solar paddle*.

solar attachment (*Surv*) See SHADE.

solar cell (*Electronics*) Photoelectric cell using silicon, which collects photons from the Sun's radiation and converts the radiant energy into electrical energy with reasonable efficiency. Used in satellites and for remote locations lacking power supplies, eg for radiotelephony in the desert.

solar constant (*Astron, Phys*) The total electromagnetic energy radiated by the Sun at all wavelengths per unit time through a given area, normal to the solar beam, at the mean distance of the Earth and after correction for loss by absorption in the Earth's atmosphere. Its value is $1\cdot37 \text{ kW m}^{-2}$. It is not, in fact, truly constant, and variations of the order of $0\cdot1\%$ are detectable.

solar corona (*Astron*) See CORONA.

solar day (*Astron*) See APPARENT SOLAR DAY, MEAN SOLAR DAY.

solar eclipse (*Astron*) See ECLIPSE.

solar energy utilization (*Phys*) Exploitation of solar energy by conversion: (1) to thermal energy using a working fluid; (2) to electrical energy using a photovoltaic cell; (3) to mechanical energy using the radiation pressure (using solar sails in space); or (4) to chemical energy by photosynthesis.

solar flare (*Astron*) See FLARE.

solar flocculi (*Astron*) See PLAGE.

solar gain (*Build*) Heat gain in a building due entirely to the Sun.

solar granulation (*Astron*) See GRANULATION.

solarimeter (*Phys*) See PYRANOMETER.

solarization (*Glass*) Changes of the light transmission properties of a glass or other transparent material as a result of exposure to sunlight or other radiation in or near the visible spectrum.

solarization (*ImageTech*) The reversal of an image because of excessive exposure to light, so that an intended negative appears to be a positive.

solar paddle (*Space*) See SOLAR ARRAY.

solar panel (*Electronics*) Arrays of solar cells fitted to spacecraft and satellites in order to gather solar energy for conversion into electrical power for the equipment on board. In some satellites these may be in the form of extensible 'paddles' which are stowed away during launching. Non-electrical devices transferring solar energy to water are widely used domestically.

solar parallax (*Astron*) The mean value of the angle subtended at the Sun by the Earth's equatorial radius. Equal to $8\cdot794\,148''$.

solar plexus (*BioSci*) In higher mammals, a ganglionic centre of the autonomic nervous system. It is situated in the anterior dorsal part of the abdominal cavity; nerves radiate from it in all directions.

solar prominence (*Astron*) See PROMINENCE.

solar radio noise (*Astron*) Radio emission from the atmosphere of the Sun, which is investigated by many techniques in radio astronomy. Sudden bursts are associated with solar FLARES in particular. Sunspots, and the activity in their vicinity, are also strong sources of radio emission.

solar rotation (*Astron*) The non-uniform rotation of the Sun, which takes place in the same direction as the orbital motion of the planets. The period is 24·65 days at the Sun's equator, but increases to around 34 days near the poles. Because of the Earth's motion, the equatorial solar rotation has a synodic period of 27 days, and this interval is apparent in the recurrence of eg magnetic storms and aurorae.

solar sailing (*Space*) A means of movement in space using the pressure of the SOLAR WIND on a large surface or solar sail, suitably deployed.

solar system (*Astron*) The term designating the Sun and the attendant bodies moving about it under gravitational attraction; comprises eight major planets (excluding Pluto), and a vast number of asteroids, comets and meteors.

solar time (*Astron*) Time measured by considering the rotation of the Earth relative to the Sun. Mean solar time is established by reference to the mean Sun, whereas apparent solar time is the time shown by a sundial; the difference between the two can amount to 16·4 minutes.

solar wind (*Astron, Space*) A continuous plasma stream of protons and electrons, emitted by the Sun, that moves at very high velocities $(250–800 \text{ km s}^{-1})$ and pervades all interplanetary space.

solation (*Chem*) The liquefaction of a gel.

solder (*Eng*) A general term for alloys, frequently of eutectic composition, used for joining metals together. The principal types are soft solder (lead–tin alloys) and brazing solders (alloys of copper, silver and zinc). Cf HOT-MELT ADHESIVE.

solder-covered wire (*ElecEng*) Copper wire coated with solder instead of tin, to facilitate connections between components.

soldered dot (*Build*) A method for fixing sheet lead to woodwork, which has had small depressions worked in it. The lead is fixed by splayed screws within the depression which is then soldered over.

soldering (*Eng*) Hot joining of metals by adhesion using, as a thin film between the parts to be joined, a metallic bonding alloy having a relatively low melting point. A FLUX is usually necessary to aid wetting of the surfaces to be joined.

solder paint (*Eng*) A mixture of powdered solder and flux, applied by brush to surfaces to be joined by SOLDERING to form the bonding film.

soldier (*BioSci*) In some social insects, a form with esp large head and mandibles adapted for defending the community, for fighting, and for crushing hard food particles.

soldier (*Build*) A term applied to a course of bricks laid so that they are all standing on end.

sole (*Build*) The lower surface of the body of a plane.

sole (*Eng*) (1) The bed plate of a marine engine; secured through bearers to the hull of the ship. Also *sole plate*. (2) A timber base for supporting the feet of RAKING SHORES. (3) The plate supporting a leg of a process vessel.

sole mark (*Geol*) A physical structure found on the underside of a bed of sandstone, or siltstone, that is the *mould* of the top surface of the bed on which it lies. The mould may represent a sedimentary structure (eg a groove or a ripple mark) or the remains of a TRACE FOSSIL. Also *bottom structure*.

Solenhofen stone (*Geol*) An exceedingly fine-grained and even-bedded limestone, thinly stratified, of Upper Jurassic age, occurring in SE Bavaria; formerly widely used in lithography.

solenocyte (*BioSci*) In invertebrates and lower Chordata, an excretory organ consisting of a hollow cell with branched processes. Bunches of beating cilia in the lumen maintain a current of liquid.

solenoid (*ElecEng*) Current-carrying coil, of one or more layers. Usually a spiral of closely wound insulating wire, in the form of a cylinder, not necessarily circular. Generally used in conjunction with an iron core, which is pulled into the cylinder by the magnetic field set up when current is passed through the coil.

solenoidal field (*ElecEng*) One in which divergence is zero, and the vector is constant over any section of a tube of force.

solenoidal magnetization (*Phys*) The distribution of the magnetization on a piece of magnetic material when the poles are at the ends. Cf LAMELLAR MAGNETIZATION. Also *circuital magnetization*.

solenoid brake (*ElecEng*) An electromechanical brake in which the brake toggle is operated by the plunger of a solenoid.

solenoid model (*BioSci*) The organization of nucleosomes and spacer regions into a *solenoid coil*, containing 6–7 nucleosomes per turn, with a diameter of 30 nm.

solenoid-operated switch (*ElecEng*) A switch in which the closing force is provided by a solenoid. Cf MOTOR-OPERATED SWITCH, PNEUMATICALLY OPERATED SWITCH.

solenoid relay (*ElecEng*) A relay in which the contacts are closed by the action of a solenoid-operated plunger.

solenostele (*BioSci*) Stele with a central pith and consisting of an annulus of xylem completely enclosed by an endodermis, and with phloem between the endodermis and the xylem throughout.

sole piece (*Build*) The plate to which the feet of the shores, in a system of RAKING SHORES, are secured, and which forms an abutment for them at their lower ends.

sole piece (*Ships*) See STERN FRAME.

sole plate (*Build*) See SLEEPER.

sole plate (*Eng*) See SOLE.

solfatara (*Geol*) A volcanic orifice which is in a dormant or decadent stage and from which gases (esp sulphur dioxide) and volatile substances are emitted.

sol-gel process (*Glass*) Chemical route to glass formation, using reactions in solution to produce a gel precursor, which yields the glass on drying out, thus avoiding the melting stage. See panel on GELS.

solid (*Chem*) A state of matter in which the constituent molecules or ions possess no translational motion, but can only vibrate about fixed mean positions. A solid has a definite shape and offers resistance to a deforming force.

solid (*MathSci*) A figure having three dimensions.

solid angle (*MathSci*) A cone with a given area as its base and a given point as its vertex. See STERADIAN.

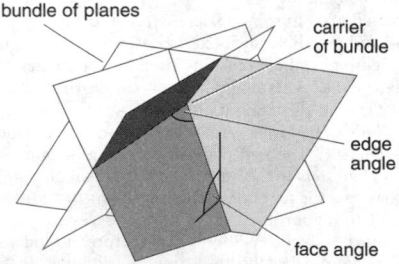

solid angle The three shaded planes include one of the eight solid angles defined by a bundle of three planes.

solid colour (*ICT*) A colour displayed on the SCREEN when all the PIXELS are the same colour. Cf DITHER COLOUR.

solid diffusion (*Eng*) The movement of atoms through a solid material, as when carbon diffuses into or out of steel during carburizing or decarburizing respectively, or when dopants are diffused into semiconductors. See FICK'S LAWS OF DIFFUSION.

solid floor (*Build*) A floor laid on a concrete subfloor. Cf SUSPENDED FLOOR.

solid head (*Autos*) A cylinder or cylinder block cast in one piece, as distinct from one with a detachable head.

solidification range (*Chem*) The range of temperature in which solidification occurs in alloys and silicate melts etc, other than those which freeze at constant temperature. It extends from a point on the liquidus to one on the solidus.

solidity (*Aero*) A measure of the effective area of a propeller, usually measured as the ratio of the sum of the blade chords to the circumference at a standard radius. See CHORD LINE.

solid matter (*Print*) Text-matter set without space between the lines; or, in mechanical typesetting, type-matter cast on its own body size rather than on a larger one.

solid newel (*Build*) The centre post of a winding stair, as distinct from a HOLLOW NEWEL.

solid panel (*Build*) A panel whose surface is in line with the faces of the stiles.

solid pole (*ElecEng*) A field pole of an electric machine which is not built up from laminations.

solid propellant (*Space*) Rocket propellant in solid state, usually in caked, plastic-like, form, comprising a fuel-burning compound of fuel and oxidizer.

solid solubility (*Eng*) The extent to which one metal is capable of forming solid solutions with another.

solid solution (*Chem*, *Phys*) An arrangement of atoms or molecules of different species within the same crystal lattice. See INTERSTITIAL SOLID SOLUTION, SUBSTITUTIONAL SOLID SOLUTION.

solid state (*Electronics*) Pertaining to a circuit, device or system, which depends on some combination of electrical, magnetic or optical phenomena within a material which is usually a crystalline semiconductor. Loosely applied to all active devices or circuits which do not rely on valves or tubes.

solid-state capacitor (*Electronics*) See VARACTOR.

solid-state detector (*Phys*) A detector of ionizing radiation which uses an energy-sensitive solid-state device. See LITHIUM-DRIFTED SILICON DETECTOR.

solid-state image sensor (*ImageTech*) A video camera chip using an array of photodiodes (pixels) to build up an electrical charge proportional to the image light. Also *solid-state pickup*. Colloq *chip*. See CCD ARRAY, CHARGE-COUPLED DEVICE, FIT CCD, FT CCD, HYPER HAD, IT CCD.

solid-state maser (*ICT*) One, commonly made from ruby, kept at a few degrees above absolute zero to ensure low-noise operation and placed in an intense magnetic field. A high-frequency pump signal raises electrons to a higher energy level than normal; the lower-frequency signal, ie that to be amplified, causes high-energy electrons to revert to their lower levels and, in so doing, absorb photons, the low-frequency signal becoming amplified in the process.

solid-state physics (*Phys*) The branch of physics which covers all properties of solid materials, including electrical conduction in crystals of semiconductors and metals, superconductivity and photoconductivity.

solid-state pickup (*ImageTech*) See SOLID-STATE IMAGE SENSOR.

solid-state polymerization (*Chem*) Polymerization carried out in a crystalline monomer.

solid-state welding (*Eng*) Welding which does not involve melting, brazing or soldering, but may involve pressure.

solid-type cable (*ElecEng*) See STRAIGHT-TYPE CABLE.

solidus (*Chem*) The line on a phase diagram representing temperatures above which mixtures begin to melt and below which mixtures are completely solid under equilibrium conditions. See PHASE DIAGRAM.

solidus (*Print*) The diagonal or oblique stroke (/).

solifluction (*Geol*) Soil creep on sloping ground, characteristic of, though not restricted to, regions subjected to periods of alternating freezing and thawing. Also *solifluxion*.

soligenous (*EnvSci*) A mire or fen receiving water that has passed through a mineral soil and hence oligotrophic to relatively eutrophic and fairly rich in nutrients depending on the nature of the soil. Cf OMBROGENOUS.

solitaria phase (*BioSci*) One of the two main phases of the locust (Orthoptera), which occurs when nymphs are reared in isolation. They adjust their colour to match their background and lack the higher activity and gregarious tendencies of the GREGARIA PHASE.

soliton (*Phys*) A solitary wave; a quantum which corresponds to a solitary wave in its transmission.

soliton propagation (*ICT*) A phenomenon observed in certain non-linear systems whereby energy is propagated by solitary waves called solitons rather than by a continuous wavetrain. The effect can be used for efficient pulse transmission in OPTICAL FIBRE systems.

sollar (*Build*) A loft which is open to sunlight.

solochrome black (*Chem*) Indicator for the complexometric titration of both calcium and magnesium ions in hard water (red colour) with EDTA. Often used in parallel with MUREXIDE, which responds to calcium only. The uncomplexed indicator is blue.

solstice (*Astron*) (1) One of the two instants in the year when the Sun reaches its greatest excursion north (*summer solstice*) or south (*winter solstice*) of the equator. (2) One of the two points on the ECLIPTIC midway between the EQUINOXES.

solubility (*Chem*) The extent to which one substance will dissolve in another. Usually expressed as the mass or the quantity of a substance which will dissolve in $1 \, dm^3$ of water.

solubility curve (*Chem*) The curve showing the variation of the solubility of a substance with temperature.

solubility of polymers (*Chem, Plastics*) The extent to which polymers pass into solution. Unlike small-molecule substances or materials, solubilization may be very slow owing to the time needed for large-chain molecules to diffuse into the fluid. The tendency of a polymer to be soluble in a fluid can be assessed by matching their SOLUBILITY PARAMETERS. Good solubility occurs if they match exactly, poor solubility if they differ greatly, although hydrogen bonding and degree of crystallinity can affect the outcome. Uncrosslinked rubbers and amorphous thermoplastics tend to be soluble in organic solvents rather than water, although slight absorption can occur, esp if the polymer is hydrogen-bonded (eg nylons). Polyelectrolytes tend to behave in the opposite way. Polyolefins are usually difficult to dissolve in any solvent at ambient temperature, high temperatures (160°C) being needed.

solubility parameter (*Chem*) The square root of cohesive energy density, symbol δ. Used to assess polymer solubility, by matching δ (solvent) and δ (polymer). See SOLUBILITY OF POLYMERS.

solubility product (*Chem*) The equilibrium constant defining the solubility of an ionic substance in water. It is equal, in a saturated solution, to the product of the ACTIVITIES of the ions each raised to the power of the number of ions of that type in the formula.

soluble complex (*BioSci*) Applied to antigen–antibody complexes in soluble form rather than in precipitates. Occur *in vivo* or *in vitro* when there is an excess of antigen over antibody so that a large lattice is not formed.

soluble oil (*Eng*) Cutting fluid consisting of oil and an emulsifier to which water is added.

soluble starch (*Chem*) A product of the hydrolysis of starch obtained by treating starch with dilute acids, or by boiling with glycerine, or by the action of diastase.

solute (*Chem*) A substance which is dissolved in another.

solute potential (*BioSci*) Same as OSMOTIC POTENTIAL.

solution (*Chem*) Homogeneous mixture of two or more components in a single phase. Often refers specifically to a solution in water (an aqueous solution).

solution heat treatment (*Eng*) Heating suitable alloys (eg DURALUMIN) in order to take the hardening constituent into solution. This is followed by quenching, to retain the solid solution, and the alloy is then age hardened at atmospheric or elevated temperature. See PRECIPITATION HARDENING.

solution mining (*MinExt*) Winning of soluble salts by use of percolating liquor introduced through shafts, drives and/or bores. Resulting saturated solution is pumped to surface for further treatment.

solution polymerization (*Chem*) Polymerization conducted in an inert solvent where components are homogeneously dissolved. Contrast with emulsion and suspension polymerizations. See CHAIN POLYMERIZATION.

solutizer process (*ChemEng*) Process for removing mercaptans from petroleum fractions by two-stage treatment with caustic soda and sodium cresylate solution.

solvation (*Chem*) The association or combination of molecules of solvent with solute ions or molecules.

Solvay's ammonia soda process (*Chem*) A process based on the fact that when a conc solution of sodium chloride is saturated with ammonia, and carbon dioxide is passed through, sodium hydrogen carbonate is precipitated and ammonium chloride remains in solution. Used for the manufacture of sodium carbonate from chloride.

solvent (*Build*) In painting a liquid capable of dissolving the BINDER and added to make it work more freely.

solvent (*Chem*) That component of a solution which is present in excess, or whose physical state is the same as that of the solution.

solvent bonding (*Textiles*) A process in which an organic liquid is used to soften fibres so that they adhere to each other and form a non-woven fabric. Cf SOLVENT WELDING.

solvent degradation (*NucEng*) The solvents used in fuel reprocessing plants have limited lifetimes because of radiation from the spent fuel.

solvent extraction (*MinExt*) In chemical extraction of values from ores or concentrates, selective transfer of desired metal salt from aqueous liquor into an immiscible organic liquid after intimate stirring together followed by phase separation.

solvent naphtha (*Chem*) Middle- and high-boiling benzene hydrocarbons chiefly consisting of toluene and xylene, obtained from the fractionation of light tar oils after the benzene fractions have been distilled off.

solvent processing (*Textiles*) Scouring, dyeing and finishing processes carried out in organic liquids rather than in aqueous solutions.

solvent welding (*Eng, Plastics*) Use of a good solvent or mixture of solvents to create a joint between similar or identical thermoplastics. Formation of a good bond may take some time, since solvent must diffuse away through the solid material before a solid joint is formed. A similar polymer is often added to the solvent to give a solvent cement to aid bond formation.

solvolysis (*Chem*) See LYOLYSIS.

solvus (*Chem*) See PHASE DIAGRAM.

soma (*BioSci*) The body of an animal, as distinct from the germ cells. Cf GERMEN. Pl *somata*. Adj *somatic*.

somaclonal variation (*BioSci*) Variability commonly found among plants that have been regenerated from TISSUE CULTURES.

somatic (*BioSci*) Of cells of the body, as distinct from the *germ line*.

somatic cell (*BioSci*) One of the non-reproductive cells of the parent body, as distinct from the reproductive or germ cells.

somatic cell hybrid (*BioSci*) A cell formed by the fusion of cells from the same or different species, in which there is also nuclear fusion.

somatic doubling (*BioSci*) A doubling of the number of chromosomes in the nuclei of somatic cells.

somatic effects of radiation (*Radiol*) The biological effects which do not affect the gonads and are therefore not transmitted to the offspring. Somatic effects are much more common than the genetic effects of radiation.

somatic hybridization (*BioSci*) The production of hybrid cells by the fusion of non-gametic nuclei, either naturally in the PARASEXUAL CYCLE of some fungi or artificially, such as by PROTOPLAST FUSION.

somatic mutation (*BioSci*) Mutations that occur in the genes of cells of the body other than the germ cells and that give rise to sperm or ova. They are therefore not inheritable. Such mutations occur rarely in most cells, but the genes controlling synthesis of the variable region of immunoglobulin molecules exhibit a greatly increased rate of mutation.

somatic pairing (*BioSci*) A closely paired arrangement of CHROMATIDS in polytene chromosomes (see POLYTENY).

somatoblast (*BioSci*) In development, a cell that will give rise to somatic cells.

somatoform disorders (*Psych*) Conditions in which psychological conflicts take on a somatic or physical form; includes the hypochondrias and conversion disorders.

somatogenic (*BioSci*) Arising as the result of external stimuli; developing from somatic cells, as opposed to germ cells.

somatomedin (*BioSci*) A peptide GROWTH HORMONE, produced in the liver and released in response to somatotropin, which stimulates growth of bone and muscle.

somatopleure (*BioSci*) The outer body wall of coelomate animals; the outer layer of the mesoblast which contributes to the outer body wall. Cf SPLANCHNOPLEURE. Adj *somatopleural*.

somatostatin (*BioSci*) A peptide of the hypothalamus that controls the secretion of *somatotropin* by the pituitary.

somatotropin (*BioSci*) A peptide GROWTH HORMONE, secreted by the anterior pituitary, that causes the liver to produce *somatomedins*, which stimulate the growth of bones and muscle.

somatotropism (*BioSci*) Directed growth movements in plants so that the members come to be placed in a definite position in relation to the substratum.

somatotype theory (*Psych*) A theory proposed by Sheldon who suggested that bodily characteristics reflect clusters of personality traits; the theory is no longer held in any serious regard, although the terms ECTOMORPH, ENDO-MORPH and MESOMORPH still occasionally appear.

somite (*BioSci*) One of the divisions or segments of the body in a metameric animal; a mesoblastic segment in a developing embryo.

sommaite (*Geol*) An alkaline igneous rock, similar to essexite but with leucite in place of nepheline.

Sommerfeld atom (*Phys*) Atomic model developed from the BOHR MODEL, but allowing for elliptical orbits with radial, azimuthal, magnetic and spin quantum numbers. Modern theories modify this by regarding the electrons as forming a cloud, the density of which is described in terms of their wavefunction. See panel on ATOMIC STRUCTURE.

somnambulism (*Psych*) (1) The fact or habit of walking in the sleep. (2) A hysterical state of automatism in which the patient performs acts of which he or she is unaware at the time or after emerging from the state.

sonar (*Acous*) Abbrev for *sound navigation and ranging*. See ASDIC.

sonde (*EnvSci*) Small telemetering system in satellite, rocket or balloon.

sone (*Acous*) Unit of loudness equal to a tone of 1 kHz at a level of 40 dB above the threshold of the listener.

son file (*ICT*) See GRANDFATHER FILE.

sonic boom (*Aero*) Noise phenomenon due to the shock waves projected outwards and backwards through the atmosphere from leading and trailing edges of an aircraft travelling at SUPERSONIC SPEED. The waves are disconti-nuities of atmospheric pressure and are heard as a characteristic double report which may be of sufficient intensity to cause damage to buildings etc.

sonic fatigue (*Acous*) The deterioration (cracks) and failure which are caused in materials (eg metal, concrete) by strong stress fluctuations of sound waves of high intensity. Sometimes happens in mechanical systems if the eigen-frequency is excited. Important for spacecraft and aircraft.

sonic line (*Phys*) The locus of field points in two-dimensional flow where the medium attains the speed of sound under local conditions.

sonics (*Phys*) General term for the study of mechanical vibrations in matter.

sono- (*Genrl*) Prefix from Lt *sonus*, sonic.

sonobuoy (*ICT*) Equipment dropped and floated on the sea to pick up underwater noise and transmit a bearing of it to aircraft; three such bearings enable the aircraft to 'fix' the source of the noise, eg from submarines.

sonogram (*Acous*) Three-dimensional representation of a sound signal, the three co-ordinates being frequency, time and intensity. Intensity is often represented by shading.

sorbic acid (*Chem*) Hexa-2,4-dienoic acid. $CH_3(CH=CH)_2COOH$. Solid, mp 133–135°C. Used as a preservative for pharmaceuticals, cosmetics, etc. Antifungal for edible products.

sorbite (*Eng*) Product formed by tempering of MARTENSITE to produce rounded particles of CEMENTITE in a ferrite matrix, thus distinguishing it from pearlite which forms directly from austenite by EUTECTOID decomposition.

sorbitol (*Chem*) A hexahydric alcohol, isomeric with mannitol.

sorbose (*Chem*) A ketohexose, an isomer of fructose.

sordes (*Med*) Foul, dark-brown crusts which collect on the lips and the teeth in prolonged fever (eg in typhoid fever).

sore-heels (*Vet*) See GREASE.

Sorel's cement (*Build*) Calcined magnesite (MgO) mixed with a solution of magnesium chloride of a concentration of about 20° Baumé. It sets within a few hours to a hard mass. The basis of artificial flooring cements.

Sorensen's formol titration (*Chem*) See FORMOL TITRA-TION.

sore-shins (*Vet*) An inflammation of the periosteum of the large metacarpus (shin bone), or occasionally of the metatarsus, of young horses.

Soret effect (*Phys*) See THERMAL DIFFUSION.

soroban (*MathSci*) Japanese abacus.

sorocarp (*BioSci*) A fruiting body formed by some cellular slime moulds; includes both the stalk and the terminal spore mass.

sorosilicate (*Min*) A silicate mineral whose atomic structure contains paired silicon–oxygen tetrahedra (Si_2O_7 groups), eg *melilite*. See SILICATES.

sorption (*Chem*) A general term for the processes of absorption, adsorption, chemisorption and persorption.

sorting (*Build*) The process adopted when a roof is to be covered by slates of different sizes; the largest slates are nailed at the eaves and the smallest at the ridge.

sorting (*MathSci*) The process of selecting the elements of a set with a common attribute from a random collection.

sorts (*Print*) Particular type letters as distinct from complete fonts. A case is 'out of sorts' when one or more of the boxes is empty. 'Outside sorts' are required for mathematical and foreign language setting.

sorus (*BioSci*) (1) Generally, a cluster of reproductive structures on a thallus. (2) In ferns, a cluster of sporangia on a leaf, often covered by an INDUSIUM which acts as a protective structure.

SOS (*Electronics*) Abbrev for SILICON ON SAPPHIRE.

sough (*CivEng*) A drain at the foot of a slope, eg an embankment, to receive and carry away surface waters from it.

sound (*Acous*) The periodic mechanical vibrations and waves in gases, liquids and solid elastic media. Specifically the sensation felt when the eardrum is acted upon by air vibrations within a limited frequency range, ie 20 Hz to 20 kHz. Sound of a frequency below 20 Hz is called infrasound and above 20 kHz ultrasound.

sound (*Med*) A solid rod used for exploring hollow viscera (eg the bladder, the uterus) or for dilating stenosed passages.

sound absorption factor (*Acous*) See ACOUSTIC ABSORPTION FACTOR.

sound analyser (*Acous*) A device which measures each frequency, amplitude or phase of sound. See OCTAVE ANALYSER, OCTAVE FILTER, REAL-TIME ANALYSER.

sound articulation (*Acous*) The percentage of all elementary speech sounds received correctly, when LOGATOMS are called over a circuit or in an auditorium, in a standard manner. See INTELLIGIBILITY.

sound barrier (*Acous*) The divide between subsonic and supersonic speeds.

sound board (*ICT*) An EXPANSION CARD that enhances a computer's ability to produce good-quality sound. Also *sound card*.

sound boarding (*Build*) Boards fitted in between the joists of a floor, to carry the PUGGING or deafening which is to insulate the room from sound and smell from the room below. *Deafening boarding* in Scotland. Obsolete.

sound bridge (*Acous*) (1) Rigid connection piece between the two plates of a DOUBLE WALL, increasing the sound transmission considerably. (2) Structures such as gaps, pipes and chimneys which increase the sound transmission in buildings.

sound camera (*ImageTech*) A continuous motion camera in which photographic film is exposed when transferring from magnetic recording to optical sound track.

sound card (*ICT*) See SOUND BOARD.

sound channel (*ImageTech*) The carrier frequency with its associated sidebands which are involved in the transmission of the sound in TV.

sound energy density (*Acous*) See ENERGY DENSITY OF SOUND.

sound field (*Acous*) The volume filled with a medium in which sound waves propagate.

sound gate (*ImageTech*) The position on a motion picture projector where the film passes the SCANNING SLIT in photographic sound reproduction.

sound head (*ImageTech*) That unit in a projector which reproduces the sound track on the edge of the film.

sounding (*Surv*) The depth of an underwater point below some chosen reference datum. Cf REDUCED LEVEL.

sounding balloon (*EnvSci*) A small free balloon carrying a meteorograph, used for obtaining records of temperature, pressure and humidity in the upper atmosphere.

sounding line (*Surv*) A stout cord, divided into fathoms and feet and weighted at one extremity with a lead weight; used in finding soundings.

sounding rocket (*Space*) Unmanned rocket-powered vehicle used for research purposes (studies of eg atmosphere, astronomy, microgravity) which does not go into Earth orbit, but follows a suborbital trajectory.

sound insulation (*Build*) The property, possessed in varying degrees by different materials, of blocking the transmission of sound. High density offers greatest resistance.

sound intensity (*Acous*) The flux of sound power through unit area normal to the direction of propagation. If p is the acoustic pressure and v the velocity of the medium particles in the direction of propagation, the intensity is the time average of the product pv.

sound intensity level (*Acous*) At any audio frequency, the intensity of a sound, expressed in decibels above an arbitrary level, 10^{12} W m^{-2}, which is equivalent, in air, to a pressure of 20 μPa. Also SOUND PRESSURE LEVEL.

sound interval (*Acous*) The interval between two sounds is the ratio of their fundamental frequencies.

sound level (*Acous*) Loose term for SOUND INTENSITY LEVEL and SOUND PRESSURE LEVEL.

sound-level meter (*Acous*) Microphone–amplifier–indicator assembly which indicates total intensity in decibels above an arbitrary zero. Also includes suitable weighting networks and time averaging. See FAST and SLOW.

sound locator (*Acous*) Apparatus for determining the direction of arrival of sound waves, particularly the noise from aircraft and submarines.

sound–picture spoilation (*ImageTech*) The leakage of the sound signal of a TV transmission into the vision circuit of a receiver, giving rise to alternate dark and light horizontal bands in the screen picture.

sound pressure (*Acous*) Fluctuating mean component of the pressure in the medium containing a sound wave, as opposed to the constant component, eg atmospheric or hydrostatic pressure. Also *acoustic pressure*.

sound pressure level See SOUND INTENSITY LEVEL.

sound probe (*Acous*) Usually a very small microphone to minimize the disturbance of the sound field which is being measured. It is often equipped with a fine tube which is inserted in the field.

sound ranging (*Acous*) The determination of the locality of a source of sound, eg from guns, by simultaneously recording through spaced microphones and making deductions from the differences of times of arrival.

sound recording (*Acous*) The practice of registering sound so that it can be reproduced at some subsequent time. See RECORDER, RECORDING.

sound reduction index (*Acous*) See TRANSMISSION LOSS.

sound reflection (*Acous*) The return of sound waves from discontinuities (eg surface, tube end). See REFLECTION COEFFICIENT.

sound-reflection factor (*Acous*) The percentage of energy reflected from a discontinuity; proportional to the square of the REFLECTION COEFFICIENT.

sound reinforcement system (*Acous*) A system used to increase the uniform sound intensity in large halls by employing public-address systems. Also by the control of reverberation time using an electronic method, involving microphones, resonators and loudspeakers.

sound-reproducing system (*Acous*) A system comprising sound recording, transducers and equipment for sound reproduction.

soundseeing (*ICT*) Common content of a PODCAST comprising ambient sounds designed to give the listener an audio tour of a remote location.

sound spectrograph (*Acous*) An electronic instrument which makes SONOGRAMS.

sound speed (*Phys*) See SPEED OF SOUND.

sound stage (*ImageTech*) The main floor of a motion picture studio on which sets are built and the artists perform during shooting.

sound track (*ImageTech*) The track on magnetic tape or ciné film on which sound signals have been or can be recorded.

sound velocity (*Phys*) See SPEED OF SOUND.

source (*ElecEng*) An active pair of terminals, which can deliver power to a load.

source (*Electronics*) The electrode from which majority carriers flow into the CHANNEL of a field-effect transistor. Comparable with the emitter in a conventional bipolar transistor.

source (*Phys*) See RADIATOR.

source code (*ICT*) A program as written using a programming language; it must be assembled, compiled or interpreted before it can be executed. Also *source program*. Cf OBJECT PROGRAM.

source impedance (*ElecEng*) See OUTPUT IMPEDANCE.

source language (*ICT*) The language in which the program is first written.

source player (*ImageTech*) The machine that plays the master videotape in an EDIT SUITE, feeding the EDIT RECORDER.

source program (*ICT*) See SOURCE CODE.

source range (*NucEng*) See START-UP PROCEDURE.

source resistance (*ElecEng*) See INTERNAL RESISTANCE.

source strength (*Radiol*) Activity of radioactive source expressed in disintegrations per second.

sour crude (*MinExt*) Crude oil containing significant amounts of sulphur compounds, eg HYDROGEN SULPHIDE and MERCAPTANS, giving it an unpleasant odour. Hydrogen sulphide is normally removed from the crude oil before shipment. Cf SWEET CRUDE.

sour gas (*MinExt*) Natural gas containing gaseous impurities such as hydrogen sulphide (H_2S), hydrogen cyanide (HCN) or carbon dioxide (CO_2). Sour gas will normally be treated to remove these impurities before use. See SCRUBBER.

souring (*Textiles*) The treatment of yarn or cloth with dilute acid, frequently after an alkali process to ensure that no alkali remains.

sous vide (*FoodSci*) Technology whereby food is vacuum packed prior to pasteurization. Foods are prepared with strict temperature control and hygiene. The safety of sous vide products greatly depends on maintaining temperature below 3°C during transportation and storage to prevent the growth of any surviving spores of *Clostridium botulinum*.

South African jade (*Min*) See TRANSVAAL JADE.

Southern blot (*BioSci*) See BLOTTING.

Southern Cross (*Astron*) See CRUX.

Southern Lights (*Astron*) See AURORA.

southern oscillation (*EnvSci*) A slow fluctuating exchange of air between the eastern tropical Pacific on the one hand, and the Indian Ocean and Indonesia on the other, with a corresponding negative correlation between annual mean pressure values over the two areas. The interval between corresponding points in successive cycles varies from 1 to 5 years, and the oscillation is linked to variations in sea-surface temperature and the pattern of rainfall.

southing (*Surv*) Measured difference southwards from a reference latitude.

south pole (*Phys*) See POLE.

sovite (*Min*) A coarse-grained calcite carbonatite, commonly containing biotite and apatite.

sövite (*Chem*) Laboratory apparatus for the continuous extraction of a solid substance with a solvent, consisting of a distillation flask, a reflux condenser, and a cylindrical vessel fitted between them to which a syphon system is attached.

sow (*Agri*) A female pig after the first farrowing.

soya flour (*FoodSci*) A rich source of vegetable protein (over 50%) in which all the essential amino acids are present and produced by milling de-fatted soybeans. It has good water binding properties and is used to improve the keeping and eating qualities of bakery products. SOYA PROTEIN ISOLATE has a protein content of around 90% and is used as a binder and emulsifier in sausages and similar products and in making vegetarian meat analogues.

Soyuz (*Space*) A one- to three-cosmonaut vehicle used by the former USSR for ferrying crews to and from its orbital stations, first launched in 1967. See panel on SPACE STATION.

SP (*Build*) Abbrev for SOIL PIPE.

sp (*BioSci*) Abbrevs for SPECIES (sing).

space (*Print*) A keystroke less than type height and thinner than a quadrat; used to separate words and justify lines of type.

space (*Genrl*) The regions, sometimes with near-vacuum conditions, surrounding all bodies in the universe. See panel on SPACE.

space (*ICT*) The period of time in transmission during which a Morse key is open, ie not in contact.

space charge (*Electronics*) As distinct from surface charge, the term applied to the local net charge when it is distributed through a finite volume, such as in the depletion region in a semiconductor device structure (p–n junction etc). Gradients in electrostatic fields are associated with regions of space charge.

space-charge limitation (*Electronics*) The condition in a thermionic valve when the electron current leaving a

cathode is limited by balance between attractive electric forces from other electrodes and repulsion within space charge. See CHILD–LANGMUIR EQUATION.

spacecraft (*Space*) Vehicle containing all the necessary subsystems to support a payload for the performance of a particular space mission. It may be manned or unmanned. See panel on SPACE.

space current (*Electronics*) See THERMIONIC CURRENT.

spaced antennas (*ICT*) Those used with diversity systems, or to enhance directivity.

spaced-loop direction-finder (*ICT*) One including two loops spaced sufficiently in terms of the wavelength to enhance their normal directivity, as exhibited by the polar diagram of response.

spaced slating (*Build*) Slating laid with gaps between adjacent slates in any course.

space dyeing (*Textiles*) Production of irregularly multi-coloured yarns by applying various colours at intervals along a single yarn or pad of yarns often by a printing process.

space environment (*Space*) The extraterrestrial conditions existing in a particular region between Earth and distant bodies. Specifically, it involves the phenomena of near-vacua, fields, particles and related effects.

space factor (*ElecEng*) The ratio of the active cross-sectional area of an insulated conductor to the total area occupied by it.

space frame (*Aero*) The collection of connected struts, rods or wires which are stressed when a whole assembly is stressed, eg trestle bridge, half-timbered cottage. Cf MONOCOQUE.

space group (*Crystal*) The classification of crystal lattice structures into groups with corresponding symmetry elements.

space junk (*Space*) Debris in space consisting of burned-out rocket stages and other artificial objects; most fragments are only a few centimetres across but present hazards to functional satellites and must be tracked by radar.

Spacelab (*Space*) Reusable orbital research laboratory which extends the Space Shuttle's capabilities and is carried in its cargo bay.

space lattice (*Crystal*) Three-dimensional regular arrangement of atoms characteristic of a particular crystal structure. There are 14 such simple symmetrical arrangements, known as BRAVAIS LATTICES. See SYMMETRY CLASS and panel on CRYSTAL LATTICE.

space parallax (*Acous*) The difference in bearing between a moving object, such as a machine in flight, and the direction of arrival of the sound waves emitted by it. This arises from the comparable velocity of flight with that of the propagation of sound waves.

space parasite (*BioSci*) A plant that inhabits intercellular spaces in another plant, obtaining shelter but possibly taking nothing else.

space programmes (*Space*) Peaceful and military projects in space; major contributors are the former USSR, the USA, the European Space Agency (ESA), Japan and China. Unmanned programmes have produced communications and remote sensing satellites, as well as interplanetary surveys. Manned activities have resulted in Moon landings and extended stays in orbiting space stations. Emerging nations including Brazil, the Arab States, Pakistan, India and Indonesia have their own space activities. See panel on SPACE STATION.

space qualified (*Space*) Said of space systems, subsystems and components which meet the specifications relevant to their use in space.

space reactor (*Space*) Energy source of a nuclear-powered space vehicles. Previous designs have used radiation from enriched plutonium to provide heat which is then converted to electricity; small conventional fission reactors are also used.

space-reflection symmetry (*Phys*) See PARITY.

space research (*Space*) Investigation of the space environment and its effects, and its use as a vantage point for

Space

That region of near vacuum surrounding all bodies in the universe. In practice, *near-Earth space* is the extra-atmospheric region just surrounding our planet, INTERPLANETARY SPACE between planets, interstellar space between stars, and *intergalactic space* between galaxies. Although normally regarded as void, space is not entirely empty. Interplanetary space is permeated by COSMIC RAYS (panel) and ELECTROMAGNETIC RADIATION, as well as force and MAGNETIC FIELDS, and contains electrons, protons, neutral hydrogen and small particles of dust. Heavier atoms and molecules such as sodium and formaldehyde have been detected in interstellar space. The density of this matter is extremely low and it is less for each region as one progresses outwards from near-Earth space (the density for interplanetary space is about 100 particles per cm^3 and for interstellar space ten times less). Also, it varies considerably with the local conditions. The vacuum of space is much greater than that obtainable on Earth.

Two areas of human space activity can be recognized: exploration and exploitation. In the former, the physical nature of the universe and its contents are investigated, and the properties of space and its constituents defined. Space exploration is performed with the aid of satellites and probes, suitably instrumented for measurements *in situ* or the observation of more distant objects. The exploitation of space involves the use of space and space-related phenomena in our everyday lives. This includes using manufactured objects like weather, observation and COMMUNICATIONS SATELLITES (panel). Later, the exploitation of space could well lead, among other things, to the harnessing of solar energy and obtaining material from extraterrestrial bodies.

Space flight can be achieved using unmanned or manned spacecraft. To date, unmanned vehicles have been put into various low-Earth orbits (LEOs), or GEOSYNCHRONOUS orbits (GEOs) and deep-space elliptical orbits on near-Earth space missions. They have been used to explore the planets of our solar system and interplanetary space itself. Manned exploration has been confined to LEO and the Moon, although one can envisage future missions involving manned interplanetary flights and a human presence in GEO.

Spacecraft

Spacecraft are commonplace in the exploration and exploitation of space. They are vehicles containing all the necessary subsystems to support a payload for the performance of a space mission. By their very nature,

they must be self-supporting at all times and mission events are controlled by on-board software programs or ground intervention. The subsystems provide for the electrical power for the operation of the individual subsystems themselves and for the payload; communications to and from the ground; thermal control of the spacecraft; data handling of input data and commands; propulsion for course changes and reaction control thrusters; attitude control and guidance; and control for executing the necessary trajectory and pointing manoeuvres to satisfy mission requirements. A spacecraft may be manned or unmanned. In the former, additional subsystems, such as life support and specific crew-related items, are also needed.

Microelectronics, computers, new high-strength materials and similar highly sophisticated devices have allowed the production of relatively small, lightweight and efficient spacecraft, in the design of which safety and reliability must be paramount. The overall mass must also be controlled during the design process so as to be consistent with the available launch system and mission objectives.

The spacecraft requires some support from the ground for mission control and data dissemination, and this is provided by a dedicated ground segment.

Space shuttle

A sophisticated form of spacecraft which is a ground-to-orbit and return transportation system capable of lifting payloads into orbit and used for easy access to and departure from a permanent station or platform in space. In the latter role it can be employed for construction, crew transfer and logistic support. The US Space Shuttle is the only design now in service, as the Soviet Energiya-launched shuttle, BURAN, has been abandoned.

The manned part of the Space Shuttle (carrying up to seven persons) is the Orbiter and is launched by rocket propulsion using its own engines which burn propellants (liquid oxygen and hydrogen) stored in a large external tank together with two solid-fuel boosters. The latter drop away after 2 minutes and the tank is then jettisoned just before orbit is reached. The orbital manoeuvring system (OMS), burning a HYPER-GOLIC mixture of monomethyl hydrazine and nitrogen tetroxide, is used to insert the Orbiter into orbit, where the desired orientation is achieved by small rocket engines using the same hypergolic mixture. On completion of the mission, a retrorocket firing causes the reusable Orbiter to re-enter the atmosphere, and it lands horizontally like a glider. The Space Shuttle can also be used for in-orbit experiments, for the capture and repair of satellites in orbit, and the return of payloads to Earth. It has been used extensively for the resupply of the Russian space station, MIR, and it will be used for the building and resupply of the INTERNATIONAL SPACE STATION.

Space station

Several manned modules and/or unmanned platforms, launched separately but joined together in orbit to form a base which permits a permanent presence in space for exploration and exploitation of the environment, and as a staging post and refurbishment centre for other space activities. The Russian MIR space station was operational between 1986 and 2001. Its successor was the INTERNATIONAL SPACE STATION, the first two modules of which were launched and joined together in orbit in 1998. Other modules soon followed and the first crew arrived in 2000. Accidents with the US Space Shuttle have delayed completion. SALYUT (USSR, first use 1971) and SKYLAB (US, first use 1973) are examples of previous long-duration manned orbital stations.

The Soviet Mir had a crew of between three and six people and was developed from the earlier Salyut stations. It consisted (see Fig. 1) of a central living module of length 13·15 m and maximum diameter of 4·2 m with access to six docking ports, of which five were in the multiple docking module, and which allowed specialized modules such as the Kvant astrophysics laboratory to be added. The station had

further modules added and eventually was larger than the figure suggests. Mir operated between 300 and 400 km in an orbit inclined at 51°.

The International Space Station (see Fig. 2) consists of four manned modules attached to a single truss structure and interconnected by resource nodes. Solar panels attached to the 150 m truss can generate 75 kW of electrical power, although more use of solar dynamic generators will be made as the technology becomes available. The module diameters are 4 m with lengths up to 12 m. By 2006 the habitable volume amounted to 15 000 cubic feet (4250 m^3). Up to eight people can live in the habitation module and the three laboratories are equipped for space-related experiments, particularly in microgravity science. The Japanese Experimental Module (JEM) has its own logistics module and the Columbus Attached Laboratory, based on Spacelab experience, is part of ESA's Columbus programme. The latter also includes a man-tended free-flying laboratory and a polar platform. The Columbus Polar Platform is unmanned and will work in conjunction with a similar US platform, complementing the space station. The USA will ensure logistic supplies for the space station, via the Space Shuttle and its logistic modules. The space station will operate in a circular orbit at an altitude of 450 km and an inclination of 28·5°.

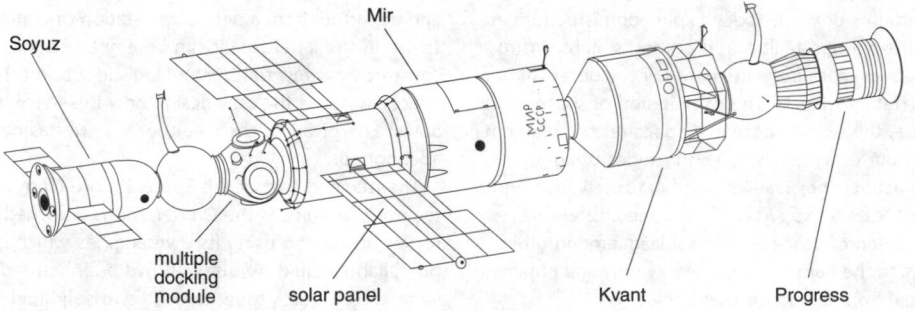

Fig. 1 **Mir** The space station developed from Salyut and launched by the USSR. Kvant was an astrophysics laboratory.

viewing the Earth and deep space. Generally, research which is performed with the aid of a space system.

Space Shuttle (*Space*) Manned ground-to-orbit and return transportation system first launched in 1981 but due to be phased out after various accidents. See panel on SPACE.

space station (*Space*) Several manned modules and/or unmanned platforms, launched separately and then joined to form a base for a permanent presence in space. See panel on SPACE STATION.

space suit (*Space*) Specially designed suit which allows an astronaut to operate in a space environment. The design includes the provision of a pressurized oxygen supply, and provides for temperature control and the purification of exhaled gases. Also *pressure suit*.

space switching (*ICT*) Routing of calls from one network node to another by selecting and connecting them to a physical channel permanently linking those nodes. Cf TIME SWITCHING.

space system (*Space*) The total assembly of space-related hardware and software; it can refer to the launch system plus payload or a spacecraft plus payload and includes the related ground segment.

space–time (*Phys*) Normal three-dimensional space plus the dimension of time, modified by gravity in GENERAL RELATIVITY.

space velocity (*Astron*) The rate and direction of a star's motion in space of three dimensions, as deduced from its observable components: the radial velocity in the line of sight deduced using a spectroscope, and the velocity perpendicular to the line of sight found by observing proper motions.

space wave (*ICT*) A wave from an antenna that is not a ground wave, but which travels rectilinearly in space, apart from reflection (negative refraction and bending) when it enters an ionized region; also sky wave (or ray).

Space station (*Cont.*)

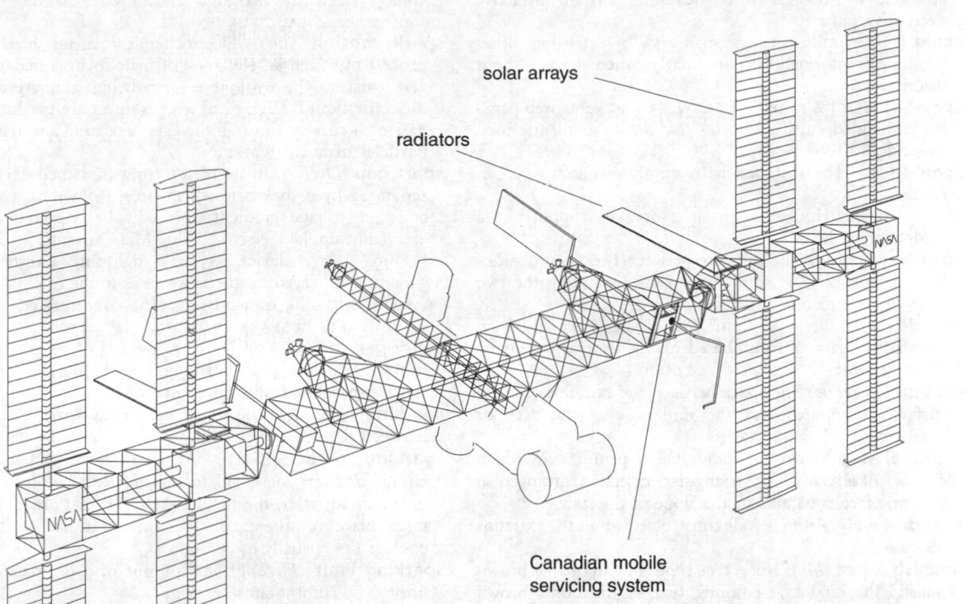

Fig. 2 **Space station** The International Space Station in an early configuration which has been much modified. Habitation and laboratory modules attach below the main boom at the position shown in the top drawing.

spacing material (*Print*) Lower than type height; there are spaces for each size of type and a wide variety made to 12-pt sizes for making up pages and spacing them apart. See FURNITURE, LEADS, QUADRAT, QUOTATIONS.

spacing wave (*ICT*) Emitted wave corresponding to spacing impulses in a code, eg Morse code. Also *back wave*.

spadiceous (*BioSci*) Shaped like a palm branch. Also *spadiciform, spadicose*.

spadix (*BioSci*) A spike, the axis of which is fleshy. The characteristic inflorescence of the arum family; commonly associated with a SPATHE.

spall (*Build*) A fragment detached by weather action, by internal movement or by the process of chiselling. Also *galet*.

spallation (*Phys*) Any nuclear reaction when several particles result from a collision, eg cosmic rays with atoms of the atmosphere, or a chain reaction in a nuclear reactor or weapon.

spallation neutron source (*Phys*) A powerful pulsed neutron source for research. Protons, accelerated in a synchrotron, are focused onto a target of uranium-238; 25–30 neutrons per proton are released, moderated and collimated into beams.

spalling (*Build, CivEng*) The flaking-off of material from a surface due to eg impact (as in flint knapping), expansion of subsurface material (see panel on RUSTING), ingress of water and subsequent freezing (common in stonework), etc.

spalted (*For*) Split; splintered; (of wood) having been attacked by fungus, often resulting in an attractive decorative pattern.

spam (*ICT*) Junk electronic mail and, by extension, other unwanted information particularly when sent without discrimination.

spambot (*ICT*) A variety of CRAWLER that scans web pages for embedded email addresses that are subsequently used to send and receive spam.

span (*Aero*) The distance between the wing tips of an aircraft.

span (*CivEng*) Horizontal distance between supports of a bridge, arch, etc.

span (*ElecEng*) (1) The distance between two transmission-line towers. (2) The number of slots separating the two sides of an armature coil (also *throw*).

spandex (*Textiles*) A term for elastic fibre used in foundation garments etc, based on thermoplastic polyurethane rubber.

spandrel (*CivEng*) The space between the extrados or outer surface of an arch and the underside of the deck or entablature. See fig. at ARCH.

spandrel step (*Arch*) An individual step in a stair, which consists of a solid block, triangular in section, arranged so that one face is parallel to the slope of the stair.

spandrel wall (*Arch*) A wall constructed upon the extrados of an arch.

Spanish topaz (*Min*) Not a true topaz but an orange-brown quartz, the colour resembling that of the honey-brown Brazilian topaz. It is often amethyst which has been heat-treated. See CITRINE.

span loading (*Aero*) The gross weight of an aircraft or glider divided by the square of the span.

spanning tree (*MathSci*) A graph with the same set of vertices and a subset of the edges of a particular graph G.

span pole (*ElecEng*) The pole to which the span wires are attached.

span saw (*Build*) See FRAME SAW.

span wire (*ElecEng*) One of several wires by which the trolley wire of a tramway or trolleybus system is suspended from street poles or buildings.

spar (*Aero*) A main spanwise member of an aerofoil or control surface. The term can be applied either to individual beam(s) designed to resist bending, or to the box structure of spanwise vertical webs, transverse ribs and skin which form a torsion box.

spar (*For*) A round timber more than 6 in (150 mm) in diameter in the middle.

spar (*Min*) Transparent to translucent crystalline mineral with vitreous lustre and clean cleavage planes, eg *fluorspar*. See ICELAND SPAR.

Sparagmite (*Geol*) A comprehensive term which includes the Late Precambrian rocks of Scandinavia. These, like the Torridonian Sandstone of N Scotland, consist of conglomerates and red feldspathic grits and arkoses.

spar frame (*Aero*) A specially strong transverse fuselage, or hull, frame to which a wing spar is attached.

spark (*ElecEng*) The breakdown of insulation between two conductors, such that the field is sufficient to cause ionization and rapid discharge. See ARC, FIELD DIS-CHARGE, LIGHTNING.

spark absorber (*ElecEng*) A resistance and/or capacitor placed across a break in an electric circuit to damp any possible oscillatory circuit which would tend to maintain an arc or spark when a current is interrupted.

spark chamber (*NucEng*) Radiation detector for rendering visible the tracks of ionizing particles by the sparks formed following the ionization. Consists of a stack of parallel metal plates with the electric field between them raised nearly to the breakdown point.

spark coil (*ElecEng*) Induction or Ruhmkorff coil used as the source of high voltage in a spark transmitter. Obsolete for radio, but used in motor cars.

sparker (*MinExt*) Marine seismic method employing a high-voltage electrical discharge under water as the energy source. See SEISMIC SURVEYING.

spark erosion (*Eng*) Electrochemical metal machining process in which an electrode of male form is maintained very close to the workpiece, both being submerged in a dielectric liquid. High local temperatures are produced by passing a current through the gap, detaching and repelling particles from the workpiece.

spark gap (*ElecEng*) In simplest form, two shaped electrodes separated by a dielectric which breaks down at a quasi-constant voltage gradient; this may be triggered by an externally applied electric field. Main uses: (1) voltage-limiting safety device as in a lightning arrester; (2) generator of electromagnetic waves as in the original spark system of radio transmission; (3) concentrated energy deposition as in SPARK EROSION.

spark-gap modulation (*ICT*) A method of pulse modulation in which a pulse-forming line discharges across a spark gap in the transmitter circuit.

sparking (*ElecEng*) The occurrence of a spark discharge between the brushes and the surface of a commutator.

sparking contact (*ElecEng*) An auxiliary contact used on circuit breakers; designed to make circuit before, and to break circuit after, the main contact, so that any sparking takes place on the auxiliary contact. It has removable contact tips, usually of carbon.

sparking limit (*ElecEng*) The output of a dc machine as limited by commutator sparking.

sparking plug (*Autos*) A plug screwed into the cylinder head of a petrol engine for ignition purposes, a spark gap being provided between an insulated central electrode and one or more earthed points.

sparking potential (*ElecEng*) Potential difference between the ends of an insulator, sufficient to cause a spark discharge through or over the insulator. Also *sparkover potential*.

sparkless commutation (*ElecEng*) A term applied to methods of current commutation in which the reactance voltage is neutralized before actual commutation occurs, so that the formation of a commutation spark or arc is avoided.

spark machining (*Eng*) See SPARK EROSION.

spark photography (*ImageTech*) High-speed technique employing a high-intensity electric spark for illumination. Used in ballistics and in *schlieren photography*.

spark resistance (*ElecEng*) That between electrodes after a discharge has commenced; if excessive and in an oscillatory circuit, it causes loss of power and a high decrement.

spark spectrum (*Phys*) A spectrum produced by means of an electric spark. The high temperature reached will generate the spectrum lines of multiply ionized atoms as well as of uncharged and singly ionized ones (as distinct from ARC SPECTRUM). Also evaporation of metal from the electrodes leads to additional lines not associated with the gas through which the discharge takes place.

sparteine (*Chem*) $C_{15}H_{26}N_2$, an alkaloid of the quinuclidine group, obtained from the branches of the common broom, *Cytisus scoparius*; a colourless oil, bp 188°C, sparingly soluble in water, soluble in ethanol, trichloromethane or ethoxyethane. It resembles coniine in its physiological action.

spasm (*BioSci*) An involuntary contraction of muscle fibres. Adj *spasmodic*.

spasmodic torticollis (*Med*) A nervous disorder in which the muscles of either side, or both sides, of the neck are in a state of continuous or of intermittent spasm.

spasmus nutans (*Med*) Rhythmic nodding of the head seen in babies in the first year of life. Also *nodding spasm*.

spastic (*Med*) Of the nature of spasm (sudden involuntary contraction) of muscle: characterized or affected by muscular spasm: rigid, or in a state of continuous spasm. May be used to describe *spasm of colon*, *spasm of ureter* or *spastic paralysis*.

spastic paralysis (*Med*) Paralysis of the voluntary movements caused by prenatal brain damage; characterized by spasm of muscles.

spathe (*BioSci*) A large, sometimes coloured or showy bract that subtends and may enclose a SPADIX.

spathic iron (*Min*) See SIDERITE (1).

spathulate (*BioSci*) See SPATULATE.

spatial filtering (*Phys*) The removal of part of the optical diffraction pattern of an object by opaque masks before an image is reconstructed; the high-frequency components in the image formation may in this way be removed or enhanced. Removal rounds sharp edges in the image; enhancement sharpens edges.

spatial summation (*Psych*) The additive effect of stimuli spread over space (eg the brightness of a beam of light has a lower absolute threshold for larger diameters of beam).

spatter finish (*Build*) A decorative finish produced by spraying thick flecks of paints onto a surface using a low 'atomizing' pressure. Can also be produced by flicking colour from a brush.

spatula (*BioSci*) Any spoon-shaped structure.

spatulate (*BioSci*) Shaped like a spoon or paddle. Also *spathulate*.

spavin (*Vet*) Chronic arthritis of the hock joint of a horse.

spawn (*BioSci*) (1) The mycelium of a fungus, esp the mycelial preparations used to propagate the cultivated mushroom. (2) To deposit eggs or discharge spermatozoa. (3) A collection of eggs, such as that deposited by many fish.

spay (*Vet*) To remove or destroy the ovaries.

SPC exchange (*ICT*) See STORED-PROGRAM CONTROL.

speaking pair (*ICT*) When wires are grouped for trunking through automatic switching, the pair carrying the speech currents is termed the speaking pair, as contrasted with the guard wire, private wire or meter wire.

Spearman's rho (*MathSci*) A statistic that measures the correlation between two sets of discrete data, such as heights and weights of the same individuals.

spear pyrites (*Min*) Twin crystals of marcasite which show re-entrant angles, in form somewhat like the head of a spear. Cf COCKSCOMB PYRITE.

special character (*ICT*) Any character recognized by a particular computing system which is not ALPHANUMERIC, eg CONTROL CHARACTER.

special effects (*ImageTech*) General term for scenes in motion picture and video productions where the original image recorded by the camera is substantially modified by subsequent technical operations, for instance combining pictures from several sources as though they formed a single shot.

special-effects generator (*ImageTech*) A video device which produces SPECIAL EFFECTS electronically.

specialist (*BioSci*) Organism with a restricted food source, living in a restricted habitat, often displaying specific behaviour or structural adaptations. Cf GENERALIST.

speciality materials (*Eng*) Commercial materials made in very low quantities compared with general purpose materials, generally of high price but with specific effect or properties not shared with others.

special linear group (*MathSci*) The group containing all the $n \times n$ matrices of determinant 1 with entries from the field F. Usually denoted by $SL_n(F)$. Cf GENERAL LINEAR GROUP.

special relativity (*Phys*) A system of mechanics applicable at high velocities (approaching the speed of light) in the absence of gravitation; a generalization of Newtonian mechanics, formulated by A Einstein (1905). Its fundamental postulates are that the speed of light c is the same for all observers, no matter how they are moving; that the laws of physics are the same in all inertial frames; and that all such frames are equivalent. On this basis, no object may have a velocity in excess of the speed of light, and two events which appear simultaneous to one observer need not be so for another. The theory predicts the contraction of length (along the direction of motion) and TIME DILATION for bodies moving at high speed relative to an observer, as well as changes in mass; the predictions have been verified in studies of high-energy particles.

special rules zone (*Aero*) A three-dimensional space, under AIR-TRAFFIC CONTROL, wherein aircraft must obey special instructions.

special theory of relativity (*Phys*) See SPECIAL RELATIVITY.

special unitary groups (*Phys*) A scheme which predicts that as far as STRONG INTERACTIONS are concerned, elementary particles can be grouped into multiplets, the particles in each multiplet being considered as different states of the same particle. This unitary group SU(3) has been successful in correlating the range of particles and in predicting the existence of hitherto undiscovered particles, notably the particle Ω^-. The *isospin* unitary group SU(2) is a subgroup of SU(3). Other groups are being explored in connection with the explanation of strong interactions in terms of QUARKS.

speciation (*BioSci*) The formation of new biological species including formation of polyploids. It is usually considered to require isolation of a subpopulation set of the ancestral species, either geographically, by occupying a different niche (*adaptive radiation*), or through acquisition of behavioural changes that restrict mating, so that distinct genetic variations accumulate and prevent further interbreeding. See SYMPATRIC SPECIATION.

species (*BioSci*) A group of individuals that (1) actually or potentially interbreed with each other but not with other such groups, (2) show continuous morphological variation within the group but which is distinct from other such groups. Taxonomically, species are grouped into genera and divided into subspecies and varieties or, horticulturally, into cultivars. In the system of BINOMIAL NOMENCLATURE of plants and animals, the second name (ie the name by which the species is distinguished from other species of the same genus) is termed the *specific epithet* or *specific name*. The latter, however, correctly refers to the full name, eg *Lilium candidum* (Madonna Lily), where *candidum* is the *specific epithet*.

species–area curve (*BioSci*) A curve relating the number of species found (*y*-axis) to the area over which the observer searched (*x*-axis). The shape of the curve provides information about DIVERSITY and RICHNESS.

specific (*BioSci*) Of a parasite, restricted to a particular host. N *specificity*.

specific (*Genrl*) A term often used to indicate that the property described relates to unit mass of the substance involved, eg specific entropy is entropy per kilogram.

specific (*Med*) A treatment or medicine effective against a particular disease.

specific activity (*Radiol*) See ACTIVITY.

specification (*Eng*) (1) A detailed description, including dimensions and other quantities of the function, construction, materials quality of a manufactured article. Also applied to an engineering project. (2) A description by an applicant for a patent of the operation and purpose of the invention.

specific characters (*BioSci*) The constant characteristics by which a species is distinguished.

specific charge (*Phys*) The ratio of the electric charge to mass of an elementary particle. For slow-moving electrons $e/m = 1·759 \times 10^{11}\,\text{C}\,\text{kg}^{-1}$. This value decreases with increasing velocity because of the relativistic increase in mass.

specific damping (*ElecEng*) The attenuation constant per km of a cable.

specific depression (*Phys*) See DEPRESSION OF FREEZING POINT.

specific dielectric strength (*ElecEng*) The DIELECTRIC STRENGTH of an insulating material, expressed in V mm^{-1}.

specific dynamic action (*BioSci*) The special calorigenic property of foodstuffs, and particularly of proteins, of raising the metabolic rate after ingestion by an amount in excess of their calorific value. It may be expressed as the ratio of the calories in excess of the basal to urinary nitrogen in excess of the basal.

specific electric loading (*ElecEng*) The electric loading, in ampere-conductors, of the armature of a machine per cm of circumference.

specific excess power (*Aero*) Thrust power available to an aircraft in excess of that required to fly at a particular constant height and speed, thus being usable for climbing, accelerating or turning.

specific fuel consumption (*Autos*) The rate at which fuel is used by an engine per unit energy or power delivered; generally expressed in pounds/BHP-hour or kg MJ^{-1}.

specific gravity (*Phys*) See RELATIVE DENSITY.

specific gravity bottle (*Phys*) See DENSITY BOTTLE.

specific heat capacity (*Phys*) The quantity of heat which unit mass of a substance requires to raise its temperature by 1 degree. Abbrev *shc*. This definition is true for any system of units, including SI, but whereas in all earlier systems a unit of heat was defined by putting the shc of water equal to unity, SI employs a single unit, the joule, for all forms of energy including heat, which makes the shc of water $4 \cdot 1868$ kJ kg^{-1} K^{-1}. Gases have two values of specific heat capacity: c_p, the shc when the gas is heated and allowed to expand against a constant pressure, and c_v, the shc when the gas is heated while enclosed within a constant volume. See MECHANICAL EQUIVALENT OF HEAT.

specific humidity (*EnvSci*) The ratio of the mass of water vapour in a sample of moist air to the total mass of the air.

specific impulse (*Aero*) The thrust of a rocket motor divided by the rate of propellant (fuel and oxidant) consumption, or the total impulse (thrust × time) divided by the total propellant mass. It is a basic performance measure of a rocket motor and equals the effective jet velocity divided by g. Symbol I_{sp}.

specific inductive capacity (*ElecEng*) See PERMITTIVITY.

specific ionization (*Phys*) The number of ion pairs formed by ionizing particle per centimetre of path. Also the *total specific ionization* to avoid confusion with the *primary specific ionization* which is defined as the number of ion clusters produced for unit length of track.

specificity (*MathSci*) The proportion of true negatives that are detected by a system designed to discriminate between two categories, known conventionally as positive and negative. Compare SENSITIVITY.

specific latent heat (*Phys*) The heat which is required to change the state of unit mass of a substance from solid to liquid, or from liquid to gas, without change of temperature. Most substances have a latent heat of fusion and a latent heat of vaporization. The specific latent heat is the difference in enthalpies of the substance in its two states. Unit J kg^{-1}.

specific magnetic loading (*ElecEng*) The average flux density (ie the total magnetic loading divided by the peripheral area) in the armature of a machine.

specific modulus (*Eng*) A merit index, usually the ratio of the Young's modulus of a material to its density. Also *specific stiffness, stiffness-to-weight ratio*.

specific name (*BioSci*) See SPECIES.

specific output (*ElecEng*) The ratio of the electrical output of a machine to its weight, its volume or some other function of its dimensions. Cf SPECIFIC TORQUE COEFFICIENT.

specific permeability (*ElecEng*) Also RELATIVE PERMEABILITY. See PERMEABILITY.

specific power (*NucEng*) US for FUEL RATING.

specific reaction rate (*Chem*) See RATE CONSTANT.

specific refraction (*Chem*) The molecular refraction of a compound, defined by the Lorentz–Lorenz equation, divided by the molecular weight. Symbol *r*.

specific resistance (*ElecEng*) See RESISTIVITY.

specific rotation (*Chem, Phys*) The angle through which the plane of polarization of a ray of sodium D light would be rotated by a column of liquid 1 dm in length, containing 1 g of an optically active substance per cm^3.

specific strength (*Eng*) A merit index, usually the ratio of the tensile strength of a material to its density. Also *strength-to-weight ratio*.

specific surface (*PowderTech*) The surface area of the particles in a unit mass of the powder determined under stated conditions. Also sometimes used to describe the surface area of the particles in unit volume of the powder.

specific temperature rise (*ElecEng*) The temperature rise of an electric machine per unit of radiating surface.

specific torque coefficient (*ElecEng*) A coefficient used in the design of electric machines, giving a figure representing the torque per unit of volume enclosed by the air-gap periphery. Also *Esson coefficient, output coefficient*.

specific viscosity (*Phys*) Relative viscosity minus unity.

specific volume (*Phys*) The reciprocal of the DENSITY of a material.

speckle interferometry (*Astron*) A technique using the principle of interference of light which enables very small angles, such as the diameters of stars, to be measured directly.

spectacle crown (*Glass*) White crown glass of refractive index $1 \cdot 523$, used for ophthalmic purposes.

spectacle flint (*Glass*) A glass of high refractive index which is fused to SPECTACLE CROWN in the manufacture of bifocal spectacle lenses. See FLINT GLASS.

spectinomycin (*Pharmacol*) A bacteriostatic antibiotic that binds to the bacterial ribosome, thus inhibiting protein synthesis. It is used in the treatment of gonococcal infections.

Spectra (*Eng, Plastics*) TN for gel-spun polyethylene fibre (US). See panel on HIGH-PERFORMANCE POLYMERS.

spectral characteristic (*Phys*) A graph of photocell sensitivity, as related to wavelength of radiation.

spectral colour (*Phys*) Colour with degrees of saturation between no hue and a pure spectral colour on the rim of the chromaticity diagram.

spectral distribution curve (*Phys*) A curve showing the relationship between radiant energy and wavelength for a light source.

spectral efficiency (*ICT*) A term used to compare radio systems by relating their traffic capacity to the amount of electromagnetic spectrum that they occupy.

spectral line (*Phys*) A component consisting of a very narrow band of frequencies isolated in a spectrum. These are due to similar quanta produced by corresponding electron transitions in atoms. The lines are broadened into bands when the equivalent process takes place in molecules.

spectral sensitivity (*ImageTech*) The comparative response of an emulsion to exposures of light of different wavelengths but constant intensity.

spectral series (*Phys*) A group of related spectrum lines produced by electron transitions from different initial energy levels to the same final one. The recognition and measurement of series has been of great importance in atomic and quantum theories.

spectral-shift-controlled reactor (*NucEng*) One in which loss of reactivity which would occur on burn-up is compensated by *softening* the neutron spectrum, eg by varying the heavy water/light water ratio in the reactor coolant, or by change of coolant temperature. Abbrev *SSCR*.

spectral transmission (*ImageTech*) The relative transmission, opacity or density of a filter in respect of light of different wavelengths.

spectral type (*Astron*) The Harvard classification of a star according to its spectrum; the graded types are represented by the letters (W) O B A F G K M S (R N), forming a sequence (called the *main sequence*) of descending temperature, the O-type stars being hot, white and gaseous, while the cooler M-type show molecular band spectra. See panel on SUN AS A STAR.

spectre of the Brocken (*EnvSci*) See BROCKENSPECTRE.

spectrin (*BioSci*) A family of closely related cytoskeletal proteins, which consist of an α–β heterodimer of two high-molecular-weight polypeptides. The eponymous member of the spectrin family constitutes a major component of the erythrocyte membrane. Other members are found in brain and intestinal epithelium.

spectrograph (*Phys*) Usually a spectroscope designed for use over a wide range of frequencies (well beyond visible spectrum) and recording the spectrum photographically. The MASS SPECTROGRAPH separates particles of different specific charge in a manner analogous to the separation of spectrum lines in an optical spectrum.

spectroheliogram (*Astron*) The recorded result of an exposure on the Sun by the SPECTROHELIOGRAPH.

spectroheliograph (*Astron*) An instrument for photo-graphing the Sun in monochromatic light. It consists essentially of a direct-vision spectroscope, with a second slit instead of an eyepiece, which can be set so that only light of a desired wavelength passes through it onto a photographic plate.

spectrohelioscope (*Astron*) An instrument in principle the same as the SPECTROHELIOGRAPH, but adapted for visual use by the employment of a rapidly oscillating slit which, by the persistency of vision, enables an image of the whole solar disk to be viewed in light of one wavelength; it also detects the velocities of moving gases in the solar atmosphere by an adjustment called the *line-shifter*.

spectrometer (*Phys*) An instrument used for measurement of wavelength or energy distribution in a heterogeneous beam of radiation.

spectrophotometer (*Phys*) An instrument for measuring photometric intensity of each colour or wavelength present in an optical spectrum.

spectroradiometer (*Phys*) A spectrometer for measurements in the infrared.

spectroscope (*Phys*) General term for instrument (eg a spectrograph or spectrometer) used in spectroscopy. The basic features are a slit and collimator for producing a parallel beam of radiation, a prism or grating for dispersing different wavelengths through differing angles of deviation, and a telescope, camera or counter tube for observing the dispersed radiation.

spectroscopic binary (*Astron*) A binary whose components are too close to be resolved visually, but are detected by the mutual shift of their spectral lines owing to their varying velocity in the line of sight.

spectroscopic parallax (*Astron*) The indirect method of deducing the distances of stars too far away to have detectable annual parallaxes; it involves the inferring of their absolute magnitudes from spectroscopic evidence which then, combined with the observed apparent magnitudes, gives their distances.

spectroscopy (*Phys*) The practical side of the study of spectra, including the excitation of the spectrum, its visual or photographic observation, and the precise determination of wavelengths.

spectrum (*Phys*) The arrangement of components of a complex colour or sound in order of frequency or energy, thereby showing distribution of energy or stimulus among the components. A mass spectrum is one showing the distribution in mass, or in mass-to-charge ratio of ionized atoms or molecules. The mass spectrum of an element will show the relative abundances of the isotopes of the element.

spectrum analyser (*ICT*) (1) Swept-frequency receiver, linked to a cathode-ray tube or chart recorder, that can be used to display a number of signals, with their relative spacing and amplitudes, over a wide frequency band. (2) Pulse-height analyser for a single signal with complex SIDEBANDS, eg an amplitude- or frequency-modulated signal or a pulsed signal such as a radar transmission.

spectrum colours (*Phys*) The continuous range of merging colours which white light is shown to be composed of

when split into different wavelengths, traditionally red, orange, yellow, green, blue, indigo, violet.

spectrum line (*Phys*) Isolated component of a spectrum formed by radiation of almost uniform frequency; due to photons of fixed energy radiated as the result of a definite electron transition in an atom of a particular element.

spectrum locus (*Phys*) Curved line on the CIE chromaticity diagram representing the monochromatic hues.

specular density (*ImageTech*) The photographic density in an image measured with parallel light, as contrasted with diffuse density, when the total light passed is measured, including that dispersed.

specular iron (*Min*) A crystalline rhombohedral variety of haematite which possesses a splendent metallic lustre often showing iridescence.

specular reflectance (*Phys*) The quotient of reflected to incident luminous flux for a polished surface.

specular reflection (*Phys*) General conception of wave motion in which the wavefront is diverted from a polished surface, so that the angle of the incident wave to the normal at the point of reflection is the same as that of the reflected wave. Applicable to heat, light, radio and acoustic waves. See REFLECTION LAWS.

specular transmittance (*Phys*) See TRANSMITTANCE.

speculum (*Eng*) Alloy of one part tin to two of copper, providing wide spectral reflection from a GRATING ruled on its highly polished surface.

speculum (*Med*) A hollow or curved instrument for viewing a passage or cavity of the body, eg vaginal speculum, nasal speculum.

speech (*Acous*) The fundamental method of communicating thoughts, which consists of regulating the pitch and intensity of voiced sounds, and the intensity of unvoiced sounds, by the larynx, and in modifying the spectral content of these elementary sounds by posturing the cavities of the mouth (assisted by the nasal cavities), which form double or triple Helmholtz resonators.

speech clipping (*ICT*) Removal of high peaks in speech, to get higher loading of transmitters with some loss in intelligibility.

speech coding (*ICT*) Conversion of speech to digital code not simply by sampling its time waveform but by parameterizing those features, eg fundamental frequency and distribution of spectral energy, necessary for its recognition, with the object of reducing the bandwidth required for transmission or storage. See LINEAR PRE-DICTIVE CODING.

speech frequency (*ICT*) See VOICE FREQUENCY.

speech inverter (*ICT*) See INVERTER.

speech recognition (*ICT*) The process of analysing a spoken word and comparing it with those known to the computer system. See PATTERN RECOGNITION.

speech recognizer (*ICT*) Equipment intended to recognize spoken input, eg in an automatic telephone enquiry system. Current devices can respond to moderate voca-bularies of isolated words or connected speech with restricted vocabulary. In both cases, training to an individual speaker improves performance.

speech scrambler (*ICT*) See SCRAMBLER.

speech-sounds (*Acous*) The distinctive elements in speech, such as vowels and consonants. About 30 English speech-sounds are sufficient for recognition in telephonic work, but phoneticians recognize about 60. See LOGATOM.

speech synthesizer (*ICT*) Output device that generates sound similar to human speech on receipt of digital signals.

speed (*ImageTech*) (1) The sensitivity of a photographic material as rated by a standard method. The current international system (*ISO*), gives ratings on an arithmetic scale which closely match those of US (*ASA*) and UK (*BS*) practice. The Russian *GOST* values are numerically equivalent. The logarithmic *DIN*-scale standard in Ger-many is also popular in Europe: here doubling the speed is indicated by an increase of 3°. For conversion, 400 ASA = 27 DIN. (2) A measure of the light transmitting power of a lens, usually stated as its F-NUMBER or T-NUMBER.

speed (*Phys*) (1) The rate of change of distance with time of a body moving in a straight line or in a continuous curve (cf VELOCITY, a vector expressing both magnitude and direction). Units of speed include metres per second (m s^{-1}), feet per second (ft s^{-1}), miles per hour (mph), kilometres per hour (km h^{-1}), knots. (2) Angular speed, the rate of rotation, expressed in eg revolutions per minute or radians per second.

speed-adjusting rheostat (*ElecEng*) A rheostat arranged in the field or armature circuit of an electric motor for varying the motor speed.

speed bulges (*Aero*) Streamlined bulges on the fuselage, or nacelles near the trailing edge of the wing, which meet the requirements of AREA RULE for a smooth area distribution where it is impractical to give the fuselage a wasp-waist to reduce transonic WAVE DRAG.

speed control (*ElecEng*) The method by which the speed of an electric motor may be varied.

speed dialling (*ICT*) A facility of some telephones that enables numbers to be stored in the telephone's memory and dialled automatically by pressing a single button or a short combination of buttons.

speed–distance curve (*Eng*) The curve showing the relation between the speed of a moving object or vehicle and the distance it has travelled.

speed–frequency (*ElecEng*) The product of rotor speed and the number of pole pairs in an induction motor.

speed governing (*ElecEng*) The method of keeping the speed of a prime mover independent of the load which it is driving.

speed gun (*Genrl*) A hand-held radar or laser device used by police at the side of a road to measure and record the speed of passing vehicles.

speed indicator (*Eng*) See SPEEDOMETER.

speed of light (*Phys*) Constant and universal value of the speed of light in a vacuum, *defined* (1983) to be exactly $2 \cdot 997\ 924\ 58 \times 10^8$ m s^{-1}. This enables the SI fundamental unit of length, the metre, to be defined in terms of this value.

speed of rotation (*Phys*) For a rotating body, the number of rotations about the axis of rotation divided by the time (see SPEED). Units are revolutions per second, per minute or per hour, or radians per second, per minute or per hour. The axis of rotation may have a translatory speed of its own. See MOMENT OF INERTIA.

speed of sound (*Phys*) Speed at which sound waves travel. In dry air at stp it is 331·4 m s^{-1} (750 mph); in fresh water 1410 m s^{-1}; and in sea water 1540 m s^{-1}. The above values are used for sonar ranging but do not apply to explosive shock waves. They must be corrected for variations of eg temperature and humidity.

speedometer (*Autos*) A TACHOMETER fitted to the gearbox or propeller shaft of a road vehicle, so graduated as to indicate the speed in mph, km h^{-1}, or both. It may be centrifugal, magnetic, air-vane, chronometric or electrical.

speed–time curve (*Eng*) A curve of vehicle speed plotted against running time. The area beneath it represents the distance covered between any two given instants.

speed–torque characteristic (*ElecEng*) The curve showing the relation between the speed of a motor and the torque developed. Also *mechanical characteristic*.

speedy-cut (*Vet*) Injury of the fore-leg of a horse near the knee, made by the shoe of the opposite foot.

speise (*MinExt*) Metallic arsenides and antimonides produced in the smelting of cobalt and lead ores. Also *speiss*.

speleology (*Gen*) The study and exploration of caves. Also *spelaeology*.

speleothems (*Geol*) Secondary calcium carbonate encrustations deposited in caves by running water.

spelling checker (*ICT*) A facility found in a WORD PROCESSOR or DESKTOP PUBLISHING PROGRAM to check the spelling within a document. Each word is checked against a dictionary. Most systems allow the creation of supplementary or user-defined dictionaries to include words, acronyms and proper nouns particular to the user. Also *spell checker*.

spelter (*Eng*) Zinc of about 97% purity, containing lead and other impurities.

spent fuel (*NucEng*) Reactor fuel element which must be replaced due to (1) burn-up or depletion, (2) poisoning by fission fragments, (3) swelling and/or bursting. The fissile material is not exhausted and so-called spent fuel is normally subsequently reprocessed.

spent hen (*Agri*) A hen no longer commercially efficient as an egg producer, usually culled and sold for processing.

sperm (*BioSci*) See SPERMATOZOON.

sperm-, sperma-, spermi-, spermo-, spermato- (*Genrl*) Prefixes from Gk *sperma*, gen *spermatos*, seed.

spermaceti (*Chem*) A glistening white wax from the head of the sperm whale, consisting mainly of cetyl palmitate, $C_{15}H_{31}COOC_{16}H_{33}$; mp 41–52°C, saponification number 120–135, iodine number 0. Used in the manufacture of cosmetics and ointments.

spermaduct (*BioSci*) See SPERMIDUCT.

spermagonium (*BioSci*) See SPERMOGONIUM.

spermary (*BioSci*) See TESTIS.

spermatheca (*BioSci*) A sac or cavity used for the reception and storage of spermatozoa in many invertebrates; the RECEPTACULUM SEMINIS.

spermatic (*BioSci*) Pertaining to spermatozoa or testis.

spermatid (*BioSci*) A cell formed by division of a secondary spermatocyte, and developing into a spermatozoon without further division.

spermato- (*Genrl*) See SPERM-.

spermatoblast (*BioSci*) A spermatid.

spermatocele (*Med*) A cyst of the epididymis or of the tubules of the testis as a result of blocking of the ducts of the epididymis; contains a clear fluid and spermatozoa.

spermatocide (*Pharmacol*) Any agent (esp chemical) which kills spermatozoa. Adj *spermatocidal*. Also *spermicide*.

spermatocyte (*BioSci*) A stage in the development of the male germ cells, arising by growth from a spermatogonium or by division from another spermatocyte, and giving rise to the spermatids.

spermatogenesis (*BioSci*) Sperm formation; the maturation divisions of the male germ cells by which spermatozoa are produced from spermatogonia.

spermatogonium (*BioSci*) A sperm mother cell; a primordial male germ cell. Adj *spermatogonial*.

spermatophore (*BioSci*) A packet of spermatozoa enclosed within a capsule.

Spermatophyta (*BioSci*) A division containing the seed plants, ie the gymnosperms and the angiosperms. Also *Magnoliophyta*.

spermatorrhoea (*Med*) Involuntary, frequent discharge of seminal fluid in the absence of sexual excitement or intercourse. US *spermatorrhea*.

spermatozoid (*BioSci*) A motile, flagellated, male gamete, as found in most algae, some fungi and in bryophytes, pteridophytes, cycads and *Ginkgo*. Also *antherozoid*.

spermatozoon (*BioSci*) The male gamete, typically consisting of a head containing the nucleus, a middle piece containing mitochondria, and a tail whose structure is similar to that of a flagellum. Pl *spermatozoa*. Abbrev *sperm*.

spermaturia (*Med*) The presence of spermatozoa in the urine.

sperm cell (*BioSci*) A male gamete, motile or not.

spermi- (*Genrl*) See SPERM-.

spermiducal glands (*BioSci*) In many vertebrates, glands opening into or near the spermiducts.

spermiduct (*BioSci*) A duct by which sperms are carried from the testis to the external genital opening; the vas deferens. Also *spermaduct*. Adj *spermiducal*.

spermo- (*Genrl*) See SPERM-.

spermogonium (*BioSci*) A flask-shaped structure in which spermatia are formed as in some ascomycetes, rusts and perhaps some lichens.

sperrylite (*Min*) Platinum diarsenide, crystallizing in the cubic system; has a brilliant metallic lustre and is tin-white in colour.

spessartite (*Min*) A lamprophyre composed of hornblende and plagioclase feldspar, with other mafic minerals and subordinate alkali feldspar.

spew pip (*Eng*) Small knob extruded from holes in tyre mould during final step in tyre manufacture. Removed before sale. See panel on TYRE TECHNOLOGY.

SPF/DB (*Aero*) Abbrev for SUPER PLASTIC FORMING/DIFFUSION BONDING.

SPG (*ImageTech*) Abbrev for SYNC PULSE GENERATOR.

sp. gr. (*Chem*) Abbrev for SPECIFIC GRAVITY.

sphagnicolous (*BioSci*) Living in peat moss.

Sphagnum (*BioSci*) The bog mosses. A genus of mosses (Bryopsida) of c.350 spp with upright, branching gametophytes, the leaves of which are very absorbent because of a regular pattern of dead cells with holes through their walls. *Sphagnum* spp dominate many bogs. Sphagnum peat is acidic and lacking in mineral nutrients. Cf SEDGE peat.

sphalerite (*Min*) Zinc sulphide which crystallizes in the cubic system as black or brown crystals with resinous to adamantine lustre. The commonest zinc mineral and ore, deposits with fluorite, galena, etc. Also *blende, zinc blende*.

S phase (*BioSci*) The period in the cell cycle during which the nuclear DNA content doubles. See panel on CELL CYCLE.

sphen-, spheno- (*Genrl*) Prefixes from Gk *sphen*, wedge.

sphene (*Min*) Calcium titanium silicate, with varying amounts of iron, manganese and the rare earths, it crystallizes in the monoclinic system as lozenge-shaped black or brown crystals and occurs as an accessory mineral in many igneous and metamorphic rocks. Also *titanite*.

Spenisciformes (*BioSci*) An order of birds in which flight feathers are lacking and the wings are stiff and used as paddles in swimming. The feet are webbed; the bones are solid and there are no air sacs. These flightless marine birds, with their streamlined bodies, are powerful swimmers and divers. They are confined to the southern hemisphere. Penguins.

spheno- (*Genrl*) See SPHEN-.

sphenoid (*Crystal*) A wedge-shaped crystal-form consisting of four triangular faces. The tetragonal and orthorhombic analogue of the cubic tetrahedron.

sphenoidal (*BioSci*) Wedge-shaped.

sphenoiditis (*Med*) Inflammation of the air-containing sinus in the sphenoid bone.

Sphenopsida (*BioSci*) The horsetails and allies; a class of Pteridophyta dating from the Devonian onwards. The sporophytes have roots, stems and whorled leaves (microphylls). The sporangia are borne usually reflexed on sporangiophores arranged in whorls, often on terminal cones. The species are mostly homosporous. The spermatozoids are multiflagellate. The class includes the Sphenophyllales, CALAMITALES and EQUISETALES.

spherical aberration (*Phys*) Loss of definition of images formed by optical systems and arising from the geometry of a spherical surface. Parabolic mirrors are used in astronomical telescopes to avoid this defect. Combinations of lenses can be used to reduce this effect. See SCHMIDT OPTICAL SYSTEM.

spherical astronomy (*Astron*) The branch of astronomy concerned with the position of heavenly bodies regarded as points on the observer's celestial sphere. It comprises all diurnal and seasonal phenomena and the precise assignment of co-ordinates to the heavenly bodies. See ASTROMETRY.

spherical candle-power (*Phys*) The illumination on a sphere of unit radius having the source of light at its centre.

spherical curvature (*MathSci*) See OSCULATING SPHERE.

spherical excess (*Surv*) The amount by which the sum of the three angles of a spherical triangle exceeds 180°. It is equal to the area of the triangle divided by the square of the sphere's radius.

spherical polar co-ordinates (*MathSci*) See POLAR CO-ORDINATES.

spherical radiator (*ICT*) Same as ISOTROPIC RADIATOR.

spherical roller bearing (*Eng*) A roller bearing having two rows of barrel-shaped rollers of opposite inclination, working in a spherical outer race, thus providing a measure of self-alignment.

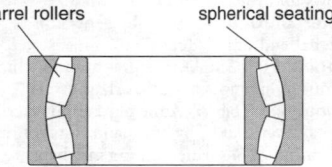

barrel rollers   spherical seating

spherical roller bearing Cage not drawn.

spherical triangle (*MathSci*) A triangle formed by three GREAT CIRCLES on the surface of a sphere.

sphericity (*PowderTech*) The ratio of the surface area of a particle to the surface area of the sphere having the same volume as the particle.

spherics (*ICT*) See ATMOSPHERICS.

spherocytosis (*Med*) An autosomal dominant hereditary disease in which the red cells of the blood are smaller than normal, biconvex instead of biconcave, and abnormally fragile. Leads to frequent bouts of mild haemolysis and jaundice with splenomegaly. Also *acholuric jaundice*.

spheroid (*MathSci*) The surface obtained by rotating an ellipse about one of its principal axes. It is therefore an ellipsoid in which two of its three principal axes are equal. If the axis of rotation is the major axis of the ellipse it is called a *prolate spheroid*, otherwise it is an *oblate spheroid*.

spheroidal graphite cast-iron (*Eng*) See SPHERULITIC GRAPHITE CAST-IRON.

spheroidal jointing (*Geol*) Spheroidal cracks found in both igneous and sedimentary rocks. Some are due to cooling and resultant contraction in the igneous rock body; others are due to a shell-like type of weathering.

spheroidal state (*Phys*) The state adopted by water when dropped upon a clean, horizontal, red-hot metal plate; it gathers into spheroidal drops which roll about, rather like mercury drops, without boiling. The latter is prevented by a cushion of steam on which the drop rides, the rapid evaporation of the drop showing that its temperature is near the boiling point.

spheroidal structure (*Geol*) A structure exhibited by certain igneous rocks, which appear to consist of large rounded masses, surrounded by concentric shells of the same material. Presumably a cooling phenomenon, comparable with perlitic structure, but on a much bigger scale, and exhibited by crystalline, not glassy, rocks.

spheroidizing (*Eng*) A process of producing, by heat treatment, a structure in which the cementite in steel or cast-iron is in a spheroidal distribution, giving improved ductility and machinability.

spherometer (*Phys*) An instrument for measuring the curvature of a lens surface.

spherosome (*BioSci*) A small (<1.0 µm), spherical, refractile body rich in lipid, in the cytoplasm of plant cells. Probably bounded by a half unit-membrane. Cf LIPID BODY.

spherulite (*Chem, Plastics*) Common morphology of crystalline polymers. See CRYSTALLIZATION OF POLYMERS and panels on POLYMERS and POLYMER SYNTHESIS.

spherulite (*Geol*) A crystalline spherical body built of exceedingly thin fibres radiating outwards from a centre and terminating on the surface of the sphere, which may vary in diameter in different cases from a fraction of a millimetre to that of a large apple.

spherulitic graphite cast-iron (*Eng*) Cast-iron in which the graphite formed during solidification is induced to form as isolated spheroidal particles, as distinct from interconnected flakes. This greatly improves the FRACTURE TOUGHNESS and imparts a degree of ductility. Also *spheroidal graphite cast-iron*. See DUCTILE CAST-IRON.

spherulitic texture (*Geol*) A type of rock fabric consisting of spherulites, which may be closely packed or embedded in an originally glassy groundmass. Commonly exhibited by rhyolitic rocks.

sphincter (*BioSci*) A muscle that by its contraction closes or narrows an orifice. Cf DILATOR.

sphincter of Oddi (*Med*) Fibres of muscle surrounding duodenal end of bile duct.

sphingolipid (*BioSci*) A structural lipid of which the parent structure is sphingosine rather than glycerol.

sphingomyelin (*BioSci*) A membrane phospholipid derived from SPHINGOSINE by the addition of a long-chain hydrocarbon, phosphate and organic base.

sphingosine (*BioSci*) A hydrophobic amino alcohol that is a component of the phospholipids known as *sphingomyelins*.

sphygmogram (*Med*) A tracing of the movements of the pulse made by a sphygmograph.

sphygmograph (*Med*) An instrument for recording the movements of the arterial pulse by means of tracings.

sphygmomanometer (*Med*) An instrument for measuring the arterial blood pressure, an inflatable bag being applied to the arm and pressure increased to occlude the artery. On deflation the systolic blood pressure is that at which sounds are heard by auscultation over the artery as blood flow recommences. The diastolic blood pressure is that at which sounds due to flow turbulence disappear.

sphygmus (*BioSci*) The pulse; the beat of the heart and the corresponding beat of the arteries.

Spica (*Astron*) A bright blue-white eclipsing binary star in the constellation Virgo, the 15th-brightest star in the sky. Distance 65 pc. Also *Alpha Virginis*.

spica (*Med*) A figure-of-eight bandage with turns that cross one another.

spicate (*BioSci*) Bearing or pertaining to spikes; spike-like.

spice (*FoodSci*) Strongly aromatic berries, seeds, barks of various tropical plants which are milled or ground to a fine powder to provide the characteristic flavour of many speciality foods, and to flavour pickles, oils, etc. Broadly classified as hot (eg peppers, ginger, mustard) and aromatic (eg cinnamon, cumin, dill, nutmeg). Concentrated spice extracts are produced by solvent extraction and distillation to separate aroma and flavour compounds and natural oleoresins.

spicule (*BioSci*) (1) Generally, a small pointed process. (2) One of the small calcareous or siliceous bodies, forming the skeleton in many Porifera and Cnidaria. Adjs *spicular, spiculate, spiculiferous, spiculiform*.

spiculum (*BioSci*) (1) Any spicule-like structure. (2) In snails, the dart.

spider (*Acous*) See INSIDE SPIDER.

spider (*ElecEng*) The centre part of an armature core, upon which the core stampings are built up.

spider (*Eng*) See CATHEAD.

spider (*ICT*) A computer program that automatically scans pages on the Internet, extracting and caching important data for subsequent retrieval by a SEARCH ENGINE.

spider (*ImageTech*) (1) A three-armed base for tripod legs. (2) A multiple distribution box for lighting cables in a studio.

spider lines (*Eng*) Linear defects on plastic pipe analogous to weld lines in injection moulding. See EXTRUSION.

spiders (*Eng*) Special fixtures which hold mandrel in extruder barrel for manufacture of polymer pipe or tubing. See EXTRUSION.

spider wrench (*Build*) A BOX SPANNER having heads of different sizes at the ends of radial arms. Also *spider spanner*.

spiegeleisen (*Eng*) Pig iron containing 15–30% manganese and 4–5% carbon. Added to steel as a deoxidizing agent and to raise the manganese content of the steel. Also *spiegel*.

spigot (*Eng*) Short raised step on the face of a component usually for location with a mating recess.

spike (*BioSci*) A racemose inflorescence with sessile, often crowded, flowers on an elongated axis, eg *Plantago*.

spike (*Build*) A stout nail more than 4 in (10 cm) long.

spike (*ElecEng*) A short pulse of voltage or current.

spike (*ICT*) A burst of electrical interference often carried through the mains electricity supply that may corrupt data or damage computer systems. For this reason much equipment is fitted with a 'spike suppressor' in the mains supply or connected to a specially filtered mains supply. See PULSE SPIKE.

spike (*Phys*) A zone surrounding the track of charged particles in which atoms have been displaced (or heating has occurred, producing a *thermal spike*).

spike (*Radar*) Initial rise in excess of the main pulse in transmission.

spikelet (*BioSci*) (1) Generally, a small or secondary spike. (2) In grasses, the basic unit of the inflorescence usually consisting of a short axis (rachilla) bearing two bracts (glumes) and one or more florets.

spilite (*Geol*) A fine-grained igneous rock of basaltic composition, generally highly vesicular and containing the sodium feldspar, albite. The pyroxenes or amphiboles are usually altered. These rocks are frequently developed as submarine lava flows and exhibit pillow structure.

spill burner (*Aero*) A gas turbine burner wherein a portion of the fuel is recirculated instead of being injected into the combustion chamber.

spill door (*Aero*) Auxiliary door mounted in an engine nacelle which opens to spill excess air provided by the intake but not needed by the engine. Designed to minimize drag; often spring-loaded.

spillway (*CivEng*) See BYE-CHANNEL.

spillway dam (*CivEng*) A reservoir dam over which flood water is allowed to flow to a downstream escape channel situated at the foot of the dam.

SPIM (*ICT*) Abbrev for *spam instant messaging*, unsolicited or junk messages transmitted to recipients through the medium of instant messaging.

spin (*Aero*) A continuous, but not necessarily even, spiral descent with the mean ANGLE OF INCIDENCE to the relative airflow above the STALLING angle. In a *flat spin* the mean angle of incidence is nearer the horizontal than the vertical, while in an *inverted spin* the aircraft is actually upside down.

spin (*Phys*) The intrinsic angular momentum of an electron, nucleus or elementary particle. Spin is quantized in integral multiples of half of *Dirac's constant*, ie $h/2$, where h is PLANCK'S CONSTANT h divided by 2π. Those particles with spin of odd multiples are *fermions* (eg electrons, proton, neutron) and those with even multiples are *bosons* (eg photon, phonon). It is the quantized electron spin angular momentum combined with the orbital angular momentum that gives rise to the fine structure in atomic line spectra.

spina (*BioSci*) A small sharp-pointed process.

spina bifida (*Med*) Developmental malformation of the bony spinal canal, spinal segments failing to meet. Often associated with defect at brain base preventing circulation of cerebrospinal fluid and causing brain damage and HYDROCEPHALUS. There may also be leg paralysis, deformity of the feet, or other abnormalities (see SYRINGOMYELOCELE). Also *rachischisis*.

spinacene (*Chem*) See SQUALENE.

spinal (*BioSci*) Pertaining to the vertebral column or to the spinal cord.

spinal anaesthesia (*Med*) A form of regional anaesthesia. It is produced by injecting a solution of a suitable drug within the spinal theca (intrathecally), causing temporary paralysis of the nerves with which it comes into contact.

spinal canal (*BioSci*) The tubular cavity of the vertebral column which houses the spinal cord.

spinal cord (*BioSci*) In Craniata, that part of the dorsal tubular nerve cord posterior to the brain.

spinal reflex (*BioSci*) A reflex situated in the spinal cord in which higher nerve centres play no part.

spin avoidance system (*Aero*) One designed to detect the onset of a spin (primarily high angles of attack) and warn the pilot or make an angular correction of pitch.

spin bath (*ChemEng*) The acid bath with various additives into which the spinning solution is injected to form the thread in viscose rayon production.

spin chute (*Aero*) See ANTI-SPIN PARACHUTE.

spindle (*BioSci*) (1) In MITOSIS and MEIOSIS, a spindle-shaped structure (ie widest in the middle and tapering towards the poles), containing longitudinally running MICROTUBULES, formed within the nucleus or the cytoplasm at the end of prophase, with centrioles, if present, usually at the poles. It forms a structural framework for the movements of chromosomes and chromatids. (2) A special sensory receptor in muscle, *muscle spindle*.

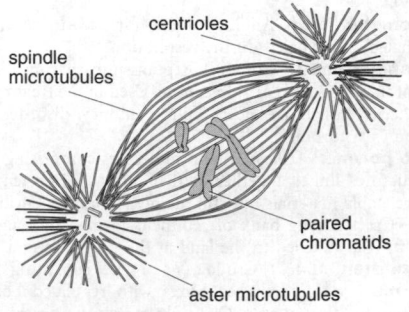

centrioles

spindle microtubules

paired chromatids

aster microtubules

spindle

spindle (*Eng*) (1) The tubular member revolving with the headstock of a lathe, through which bar material may be introduced into the chuck or collet. (2) Generally, a machine element acting as a revolving axis or a pin on which another element revolves.

spindle fibre (*BioSci*) One of the microtubules of the mitotic or meiotic spindle.

spindleless reel stand (*Print*) One supporting the reel on free-running cones at each side.

spindle moulder (*Build*) A machine with a revolving spindle to which cutters of various shapes can be fixed. The spindle projects up through a hole in the machine table and the work is fed onto the cutter. Cf ROUTER.

spin-draw (*Textiles*) A process for producing synthetic fibres in which drawing takes place as extrusion proceeds.

spin-draw texturing (*Textiles*) In the manufacture of synthetic fibres, the combination on one machine of the extrusion, drawing and *texturing* processes.

spine (*BioSci*) A stiff, sharp-pointed process relatively small in proportion to the organism, tissue or cell from which it projects. Adjs *spinate, spiniform, spinose, spinous*.

spine (*Print*) The back of a book, ie the edge where the gathered sections are sewn together. The spine faces outwards when the book is placed on a shelf, hence the name often used, *shelf-back*.

spinel (*Min*) A group of closely related oxide minerals crystallizing in the cubic system, usually in octahedra. Chemically, spinels are aluminates, chromates or ferrates of magnesium, iron, zinc, etc, and are distinguished as *iron spinel* (hercynite), *zinc spinel* (gahnite), *chrome spinel* (picotite) and *magnesian spinel*. See BALAS RUBY, CHROMITE, FRANKLINITE, MAGNETITE, RUBY SPINEL, SYNTHETIC SPINEL.

spinel ruby (*Min*) See RUBY SPINEL.

spin-labelling (*BioSci*) The technique of introducing a molecular group with an unpaired electron into a

substance, so that the labelled marker can be detected using ELECTRON SPIN RESONANCE spectroscopy.

spinner (*Aero*) A streamlined fairing covering the hub of a propeller and rotating with it.

spinner (*Agri*) Trailed machinery that broadcasts granular fertilizer using the momentum from a large, spinning disk contained in an open hopper.

spinneret (*BioSci*) In spiders, one of the spinning organs, consisting of a mobile projection bearing at the tip a large number of minute pores by which the silk issues.

spinneret (*Textiles*) A disk with fine holes through which the molten polymer or a solution of it is forced to form the continuous filaments of a synthetic fibre in the spinning process. Also *spinnerette*.

spinner-gate (*Eng*) An ingate incorporating a small whirl-chamber into and from which the metal flows tangentially, so releasing any dirt, which rises to the top of the chamber or up a riser. Also *whirl-gate*. See INGATE.

spinning (*Eng*) Forming a metal disk into a hollow shape by rotating it in a lathe over a former or chuck, by pressing a spinning tool against it.

spinning (*Textiles*) (1) The drafting and twisting together of staple fibres to form a yarn. (2) The extrusion of continuous filaments of synthetic fibres (or of silk) through spinnerets.

spinning glands (*BioSci*) The silk-producing glands of Arthropoda.

spinning jenny (*Textiles*) First spinning machine employing vertical spindles, invented by James Hargreaves, of Blackburn, in Lancashire, in 1767.

spinning tunnel (*Aero*) See VERTICAL WIND TUNNEL.

spinode (*MathSci*) See DOUBLE POINT.

spin on glass (*Electronics*) An organic silicate used to deposit planarizing layers over semiconductor devices during fabrication; it is subsequently transformed to silicon dioxide by heating. Abbrev *SOG*.

spin–orbit coupling (*Electronics*) The interaction between the intrinsic and angular momentum of particles, esp electrons.

spinous (*BioSci*) Covered with spines. Also *spinose*.

spinous process (*BioSci*) A process of (1) the proximal end of the tibia, (2) the sphenoid bone; the neural process or spine of a vertebra.

spin polarization (*Phys*) For a beam of particles, the preferential orientation of the particle spin.

spin quantum number (*Phys*) The contribution to the total angular momentum of the electron of that due to the rotation of the electron about its own axis.

spin-stabilized (*Space*) Said of a spacecraft whose attitude is stabilized by causing it to spin about a rotationally symmetric axis.

spin-stabilized satellite (*ICT*) An earlier generation of communications satellites like Intelsat IV, in which the cylindrical body of the satellite was rotated rapidly to achieve positional stability. See DE-SPUN ANTENNA.

spinule (*BioSci*) A very small spine or prickle.

spin wave (*Phys*) Coherent deviations of the spins of magnetic ions in ferromagnetic, ferrimagnetic and anti-ferromagnetic materials from their preferred directions of orientation; these are propagated in space and time like waves. See MAGNON.

spin welding (*Eng, Plastics*) A method of joining similar thermoplastics by rotating parts against one another so that frictional heat causes fusion bond.

spiny vesicle (*BioSci*) Same as COATED VESICLE.

spiracle (*BioSci*) (1) In insects and some Arachnida, one of the external openings of the tracheal system. (2) In fish, the first visceral cleft, opening from the pharynx to the exterior between the mandibular and the hyoid arches. (3) In amphibian larvae, the external respiratory aperture. (4) In Cetacea, the external nasal opening. Adjs *spiracular, spiraculate, spiraculiform*.

spiral (*MathSci*) A plane curve traced by a point which winds about a fixed pole from which it continually recedes.

See ARCHIMEDES' SPIRAL, CORNU'S SPIRAL, EQUIANGULAR SPIRAL, HYPERBOLIC SPIRAL, PARABOLIC SPRIAL, SINUSOIDAL SPIRAL.

spiral binding (*Print*) A speciality binding in which holes are punched near spine of book and a spiral of wire or plastic is fed through them.

spiral cleavage (*BioSci*) A type of segmentation of the ovum occurring in many Turbellaria, most Mollusca and all Annelida. The early micromeres rotate with respect to the macromeres, so that the micromeres lie opposite to the furrows between the macromeres; the direction of rotation (viewed from above) is normally clockwise (dexiotropic) but in 'reversed cleavage' it is anticlockwise (laeotropic). Also *alternating cleavage*.

spiral flow test (*Plastics*) The measurement of polymer melt flow in a specially designed tool with a narrow, spiral cavity. The distance along the spiral the melt flows at a specific temperature gives some idea of molecular mass for moulding purposes. See MELT FLOW INDEX.

spiral fracture (*Med*) A fracture of a bone caused by a sudden violent twisting movement.

spiral galaxy (*Astron*) The second-commonest morphological galaxy type, characterized by a large nuclear bulge of stars, surrounded by a pair of conspicuous spiral arms. These arms contain gas, dust and newly formed star clusters. Our own Galaxy is a spiral galaxy. See panel on GALAXY.

spiral gear (*Eng*) A toothed gear for connecting two shafts whose axes are at any angle and do not intersect. The teeth are of spiral form (ie parts of a multiple-threaded screw) and engage as in a WORM GEAR.

spiral instability (*Aero*) That form of lateral instability which causes an aircraft to develop a combination of sideslipping and banking, the latter being increasingly too great for the turn. This causes the machine to follow a spiral path.

spiral ratchet screwdriver (*Build*) A RATCHET SCREWDRIVER, which can be made to turn in either direction by downward pressure on a sleeve which applies torque to helical grooves in the shank.

spiral reel (*ImageTech*) In a light-tight developing tank, a reel on which film has been wound in a spaced spiral so that, on rotation, the developer has free access to the emulsion.

spiral roller (*Print*) (1) An ink roller with spiral channels or grooves to improve ink supply and distribution. (2) An idling roller with a spiral groove to smooth out unevenness in the web of paper.

spiral spring (*Eng*) A spring formed by coiling a steel ribbon into an elongated spiral or a helix of increasing diameter. When compressed completely it forms a true spiral.

spiral stairs (*Build*) Circular stairs of small diameter and usually open. More correctly helical stair.

spiral time base (*Electronics*) Arrangement for causing the fluorescent spot to rotate in a spiral path at a constant angular velocity, to obtain a much longer baseline than is possible with linear deflection. Used for detailed delineation of events relatively widely spaced in time, with or without a memory through long-glow or photography.

spiral valve (*BioSci*) In lampreys, Elasmobranchii and some lung fish, part of the intestinal canal, which is provided with an internal spiral fold to increase its absorptive surface.

spire (*Arch*) A slender tower tapering to a point.

spirillum (*BioSci*) A bacterium that is a variation of the rod form, ie a curved or corkscrew-shaped organism. Such bacteria vary from 0·2 to 50 μm in length. One or more flagella may be present. Also the name of a particular genus, which includes *Spirillum minus* (rat-bite fever).

spirit (*Chem*) An aqueous solution of ethanol, esp one obtained by distillation.

spirit duplicating (*Print*) The matter to be duplicated is typed, written or drawn on the underside of a smooth-surfaced paper with a special carbon paper in contact with

the smooth surface, several colours of carbon being available, and multicoloured arrangements easily made. The image transferred from the carbon paper is sufficient for about 250 copies, being moistened by a volatile fluid between each impression on the duplicating machine.

spirit level (*Surv*) See LEVEL TUBE.

spirit stain (*Build*) A stain for wood; colouring matter dissolved in methanol.

spirit varnish (*Build*) A varnish made by dissolving certain alcohol-soluble gums or resins like shellac in industrial alcohol.

spirochaetes (*BioSci*) Filamentous flexible bacteria showing helical spirals; without true flagella. They are divided into two families: the Spirochaetaceae and the Treponemataceae. There are both saprophytic and parasitic members. Pathogenic species include the causative agents of syphilis (*Treponema pallidum*), spirochaetosis (relapsing fever) in humans (*Borrelia recurrentis*), infectious jaundice (LEPTOSPIROSIS, *Leptospira icterohaemorrhagiae*) and yaws (*Treponema pertenue*).

spirochaetosis (*Med*) See LEPTOSPIROSIS, RELAPSING FEVER.

spirometer (*Med*) An instrument for measuring the air inhaled and exhaled during respiration.

spironolactone (*Pharmacol*) A potassium-sparing diuretic that inhibits action of aldosterone. Used in the treatment of oedema, congestive heart failure, kidney disorders and Conn's syndrome.

spiro polymers (*Chem*) Chain molecules comprising cyclic structures linked together at their apices. High-temperature stable materials but still mainly laboratory curiosities.

spit (*Geol*) A long bank of sediment formed by longshore drift. It is attached to the land at the upstream end.

spitzkasten (*MinExt*) Crude CLASSIFIERS, consisting of one or more pyramid-shaped boxes with regulated holes in down-pointing apexes. Ore pulp streaming across either settles (*coarse particles*) to bottom discharge or overflows (*fines*). In the *spitzlutten*, efficiency is improved by adding upflow of low-pressure hydraulic water.

splanchnic (*BioSci*) Visceral.

splanchno- (*Genrl*) Prefix from Gk *splanchnon*, inward parts.

splanchnocoel (*BioSci*) In vertebrates, the larger posterior portion of the coelom which encloses the viscera, as opposed to the pericardium.

splanchnomegaly (*Med*) Enlargement of bodily organs.

splanchnopleure (*BioSci*) The wall of the alimentary canal in coelomate animals; the inner layer of the mesoblast which contributes to the wall of the alimentary canal. Cf SOMATOPLEURE. Adj *splanchnopleural*.

splash baffle (*Electronics*) See ARC BAFFLE.

splash marks (*Eng*) Defective marks on surface of injection-moulded products caused by contamination of melt by volatiles, eg water. Recognized by silvery streaks radiating from gate. Also *mica marks, silvering*.

splashproof fitting (*ElecEng*) See WEATHERPROOF FITTING.

splat (*Build*) A cover strip for joints between adjacent sheets of building board.

splay brick (*Build*) A purpose-made brick bevelled off on one side. Also *cant brick, slope*.

splayed coping (*Build*) See FEATHER-EDGED COPING.

splayed grounds (*Build*) GROUNDS with splayed or rebated edges, providing a key for holding the plaster to the wall in cases where the grounds also serve as SCREEDS.

splayed jambs (*Build*) Internal jambs or sides of a door or window opening which slope away from the opening to admit more light or to increase the width.

splayed skirting (*Build*) A skirting board having its top edge chamfered.

splaying arch (*CivEng*) An arch in which the opening at one end is less than that at the other end, so that the arch is funnel shaped. Also *fluing arch*.

splay marks (*Eng*) Marks on surface of moulded products, consisting of fine grooves concentric to gate, and caused by low pressurization towards end of cycle.

spleen (*BioSci*) The largest lymphoid organ in the body, located in the abdomen. All lymphocytes enter via the blood stream. The 'white pulp' of the spleen is lymphoid tissue; 'red pulp' contains myleoid cells, red blood cells and platelets.

splenectomy (*Med*) Removal of the spleen.

splenomegaly (*Med*) Abnormal enlargement of the spleen.

splenopexy (*Med*) Fixation, by suture, of the spleen to the abdominal wall.

splenotomy (*Med*) Incision of the spleen.

splice (*ImageTech*) A join in motion picture film, made either by cementing the base or by applying transparent adhesive tape. Also a join in magnetic tape made by adhesive tape.

spliced joint (*ElecEng*) A cable joint in which the conductor strands are spliced, in the manner of a rope.

splice loss (*ICT*) Signal attenuation in an OPTICAL FIBRE due to misalignment or other defects at a joint. With current methods, the loss per joint can be kept below 0·1 dB.

splicer (*ImageTech*) A device for joining motion picture film or magnetic tape.

splicing (*BioSci*) Specifically refers to the excision of the INTRONS from an mRNA and the rejoining of EXONS to form the mature message. Loosely used for joining nucleic acid molecules as in *gene splicing*.

splicing (*Textiles*) Joining two pieces of yarn or rope together by intertwining the constituent strands manually or by machine so that the joint is nearly the same diameter as the original and certainly smaller than a knot.

splines (*Eng*) A number of relatively narrow keys formed integral with a shaft, somewhat resembling long gear teeth; produced by milling longitudinal grooves in the shaft, external splines; similarly, the grooved ways formed in a hole into which the splined shaft is to fit, internal splines. Used, instead of keys, for maximum strength.

splints (*Vet*) Exostoses on the small metacarpal or metatarsal bones of the horse.

split-anode magnetron (*Electronics*) Early type with split anode, to give a push–pull output from electrons from the filament cathode, when gyrating in a (nearly) coaxial magnetic field.

split-beam cathode-ray tube (*Electronics*) A tube containing only one electron gun, but with the beam subdivided so that two traces are obtained on the screen.

split bearing (*Eng*) A shaft bearing in which the housing is split, the bearing bush or brasses being clamped between the two parts.

split boards (*Print*) In hand binding two layers of board between which the tapes are securely held.

split-brain (*Psych, Med*) A condition in which the corpus callosum and some other fibre bundles are surgically cut so that the two cerebral hemispheres are isolated; used to treat patients with severe epilepsy who do not respond to drug treatment. Also *hemispherectomy*.

split compressor (*Aero*) An AXIAL-FLOW TURBINE compressor in which front and rear sections are mounted on separate concentric shafts (being powered by separate turbines) as a means of increasing the pressure ratio without incurring difficulties with *surge*. Also *two-spool compressor*.

split-conductor cable (*ElecEng*) A cable in which each conductor is divided into two sections lightly insulated from each other and connected in parallel at the ends. Used with special schemes of protection.

split-conductor protection (*ElecEng*) A current-balance system of feeder protection avoiding the use of pilot wires by splitting each phase conductor into two parallel sections lightly insulated from each other.

split course (*Build*) A course of bricks which have been cut lengthwise, so that the depth of the course is less than that of a brick.

split crankcase (*Autos*) An engine crankcase split horizontally at about the centre line of the crankshaft. Cf BARREL-TYPE CRANKCASE.

split duct (*Print*) A process in which the forme or plate can be printed in two or more colours by separating each from the other across the inking system.

split fitting (*ElecEng*) A bend, elbow or tee used in electrical installation work, which is split longitudinally so that it can be placed in position after the wires are in the conduit.

split flap (*Aero*) A trailing-edge flap in which only the lower surface of the aerofoil is lowered.

split-flow reactor (*NucEng*) One in which the coolant enters at the central section and flows outwards at both ends (or vice versa).

split fractions (*Print*) Fonts of figures and spaces which can be assembled to form any required fraction, the horizontal rule being cast, sometimes on the numerator, sometimes on the denominator; a font of 10-pt split fractions would be cast on a 5-pt body.

split-image rangefinder (*ImageTech*) Optical device generally used in conjunction with the camera lens to assist reflex focusing. See fig. at BIPRISM.

split pastern (*Vet*) Fracture of the first phalanx of the foot of the horse.

split-phase (*ElecEng*) A term denoting a circuit arrangement for changing a single-phase to a two-phase supply.

split pin (*Eng*) One formed of wire with hemispherical section, bent round till the flat faces meet. See COTTER PIN.

split-pole converter (*ElecEng*) A synchronous converter in which the flux distribution under the poles can be varied by means of auxiliary windings on the individual pole limits.

split pulley (*Eng*) A belt pulley split diametrically, the halves being bolted together on the shaft; used when a solid pulley cannot be fitted.

splits (*Build*) Bricks of the same length and breadth as ordinary bricks but of smaller thickness.

splits (*Textiles*) Fabrics, usually woven on wide machines, that have spaces in the warp direction so that by cutting two or more narrower fabrics are obtained.

split screen (*ImageTech*) A shot in which two or more images separately recorded appear in the same picture, often with the boundary between them made invisible.

splitting limits (*MinExt*) Divergence between assay of ore, concentrate or metal made by vendor and purchaser, inside which mutual adjustment will be made without going to arbitration.

splitting ratio (*NucEng*) See CUT.

splog (*ICT*) Computing slang for an automatically generated and maintained weblog (BLOG), usually offering products for sale.

spodogram (*BioSci*) A preparation of the ash of a plant, esp from a section, used in investigating structure by light or electron microscopy.

spodumene (*Min*) A silicate of aluminium and lithium which crystallizes in the monoclinic system; a pyroxene. It usually occurs in granite-pegmatites, often in very large crystals. The rare emerald-green variety *hiddenite* and the clear lilac-coloured variety *kunzite* are used as gems. See fig. at SILICATES.

spoil (*CivEng*) The excess of cutting over filling on any given construction. Also *waste*.

spoilage (*Print*) (1) Accidental spoilage, the cost of reprinting being budgeted for in the costing system. (2) Ordinary spoilage, allowed for on each job. Also *spoils*.

spoil bank (*CivEng*) An earthwork bank formed by depositing spoil.

spoiler (*Aero*) A device for changing the airflow round an aerofoil to reduce, or destroy, the lift. There are three principal types: (1) a small fixed spanwise ridge on the wing-root leading edge along the line of the STAGNATION POINT which improves lateral stability at the stall by ensuring that it starts at the root; (2) controllable devices at, or near, both wing tips which, by destroying lift on the side raised, impart a rolling moment to the aircraft; (3) small-chord spanwise flaps on top of the wing of a sailplane which can be raised to destroy a large part of the lift so as to make landing of the lightly loaded aircraft more positive. Similar devices are often now fitted to jet aircraft so that by destroying lift at touchdown better braking can be achieved. See THRUST-.

spoils (*Print*) See SPOILAGE.

spokeshave (*Build*) A form of double-handled plane which is used in shaping convex or concave surfaces.

spondyl (*BioSci*) A VERTEBRA. Adj *spondylous*.

spondylitis (*Med*) Inflammation of the vertebrae.

spondylitis deformans (*Med*) A condition in which the ligaments of the spinal column become ossified and the vertebrae fused together, so that the spine is bent, rigid and immobile.

spondylolisthesis (*Med*) A forward displacement of the fifth lumbar vertebra (carrying the vertebral column with it) on the sacrum.

sponge (*BioSci*) See PORIFERA.

sponge (*ElecEng*) Loose, fluffy cathode deposit in electrolysis.

sponge (*Eng*) Porous metal formed by chemical reduction or decomposition process without fusion.

sponge beds (*Geol*) Deposits, either calcareous or siliceous, which contain a large proportion of the remains of spicular organisms belonging to the phylum Porifera.

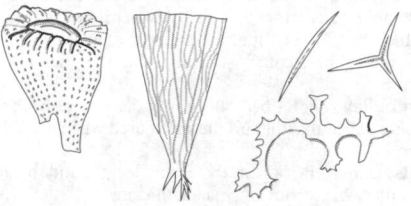

sponge beds Skeletons (left) and spicules (right, much enlarged).

sponge stipple (*Build*) A method of producing an even but irregular pattern by dabbing colour on using natural sponge.

spongiform encephalopathy (*BioSci*) Any of a group of fatal diseases that have a long incubation period and produce spongiform degeneration of the grey matter of the cortex. The causative agents are now thought to be PRION PROTEIN particles. Examples are BOVINE SPONGIFORM ENCEPHALITIS (BSE) in cattle, CREUTZFELDT–JAKOB DISEASE in humans, KURU, SCRAPIE in sheep, . See panel on TRANSMISSIBLE SPONGIFORM ENCEPHALOPATHY.

spongin (*BioSci*) A horny skeletal substance, occurring usually in the form of fibres, in various groups of Porifera (sponges).

spongioblastoma (*Med*) A soft, rapidly growing, malignant tumour occurring in the brain (or in the spinal cord) and derived from cells of the supporting structure of the brain.

spongioblasts (*BioSci*) Columnar cells of the neural canal giving rise to neuroglia cells.

spongy layer (*BioSci*) Chlorenchyma in which the cells are irregularly lobed leaving very large continuous intercellular spaces. Spongy MESOPHYLL (or spongy PARENCHYMA) is usually present towards the lower surface of dorsiventral leaves of mesomorphic dicotyledons. See PALISADE.

spongy platinum (*Chem*) The spongy mass resulting from the calcination of ammonium chloroplatinate (IV).

sponson (*Aero*) A short, wing-like projection from a flying-boat hull to give lateral stability on the water.

spontaneous behaviour (*Psych*) Behaviour occurring in the apparent absence of any stimuli.

spontaneous combustion (*Build*) The ignition of a substance or material without direct application of flame. Can occur with rags soaked in flammable liquids.

spontaneous emission (*Phys*) A process involving the emission of energy in an atomic system without external stimulation. Spontaneous emission is a strictly quantum effect. Cf LASER (panel), STIMULATED EMISSION.

spontaneous fission (*Phys*) Nuclear fission occurring without absorption of energy. The probability of this increases with increasing values of the fission parameter, Z^2A^{-1} (where Z is the atomic number and A is the relative atomic mass) for the fissile nucleus. *Induced* fission is that caused by the impact of nuclear particles.

spontaneous generation (*BioSci*) See ABIOGENESIS.

spontaneous ignition temperature (*Eng*) The temperature at which a liquid or gaseous fuel will ignite in the presence of air or oxygen, measured, for liquid fuels, by allowing a drop to fall into a heated pot. Abbrev *SIT*.

spontaneous process (*Chem*) A process which occurs of its own accord, such as dissolution of a soluble salt or the melting of ice when warmed. The FREE ENERGY for such processes is zero or negative in sign.

spontaneous recovery (*Psych*) The return of an extinguished response (see EXTINCTION) following a rest period in which neither the conditioned nor the unconditioned stimulus are presented, when an animal is returned to the original conditioning situation.

spontaneous remission (*Psych*) Recovery without treatment.

spontaneous symmetry breaking (*Phys*) In physical systems, the consequence of symmetry reducing in an unpredictable way as a system changes to one of lower energy. For instance, when a ferromagnetic material such as iron is cooled to below its Curie temperature, its atoms align in a certain direction even though the equations describing the process do not distinguish this one direction from any other.

spoofing (*ICT*) The practice of breaching system security by successfully masquerading as an authorized user of that system.

spool (*ElecEng*) See BOBBIN.

spool (*ImageTech*) A flanged core on which film or magnetic tape is wound for storage and transport; the term *reel* in this sense is deprecated.

spooling (*ICT*) Temporary storage of input or output data on magnetic disk or tape, as a means of compensating for slow operating speeds of peripheral devices or when queuing different output streams to one device.

sporadic-E (*ICT*) A layer of intense ionization that appears unpredictably in the E-LAYER; can lead to the propagation of very-high-frequency signals over anomalously great distances.

sporadic lymphangitis (*Vet*) An acute disease of horses characterized by fever and lameness due to lymphangitis affecting the limbs. The disease is often recurrent and occurs esp in working horses that have been rested for a day or two on full working rations; the cause is not known. Also *Monday morning disease*, *weed*.

sporadic simple group (*MathSci*) One of the 26 finite simple groups which do not belong to any of the known classes of finite simple groups.

sporangium (*BioSci*) A hollow, walled, structure in which spores are produced.

spore (*BioSci*) Any of a wide variety of reproductive bodies, usually unicellular; often a resting stage to avoid adverse conditions, or a means of dissemination. Bacterial spores can be extremely long-lived and resistant. See APLANOSPORE, CONIDIUM, ENDOSPORE, EXOSPORE, HYPNOSPORE, ZOOSPORE, ZYGOSPORE.

spore mother cell (*BioSci*) A cell that gives rise to a spore, esp one which divides by meiosis to give four cells which develop into spores. Also *sporocyte*.

spore print (*BioSci*) The marks obtained by placing eg the cap of a mushroom or toadstool, gills downward, on a piece of paper and allowing the spores to fall onto the paper.

spores (*Geol*) See ACRITARCH.

spori-, sporo- (*Genrl*) Prefixes from Gk *sporos*, seed.

sporocarp (*BioSci*) A hard multicellular structure enclosing sporangia in some fungi and some heterosporous ferns.

sporocyst (*BioSci*) The tough resistant envelope secreted by, and surrounding, a protozoan SPORE.

sporocyte (*BioSci*) Same as SPORE MOTHER CELL.

sporogenesis (*BioSci*) Spore formation.

sporogenous (*BioSci*) Producing or bearing spores.

sporogenous layer (*BioSci*) Same as HYMENIUM.

sporogonium (*BioSci*) Same as SPOROPHYTE in bryophytes.

sporogony (*BioSci*) In Protozoa, propagation, usually involving sexual processes and always ending in the formation of spores.

sporont (*BioSci*) A stage in the life history of some Protozoa which, as a gametocyte, gives rise to gametes, which in turn, after a process of syngamy, may give rise to spores.

sporophore (*BioSci*) Any structure that bears spores.

sporophyll (*BioSci*) A leaf that bears SPORANGIA.

sporophyte (*BioSci*) The typically diploid generation of the life cycle of a plant showing alternation of generations. It produces, by meiosis, the spores that germinate to give the GAMETOPHYTE.

sporopollenin (*BioSci*) Polymerized carotenoids, a constituent of the cell walls of pollen grains (the exine) and of many spores. It is exceedingly resistant to decay. See POLLEN ANALYSIS.

sporotrichosis (*Med*) An infection of the skin (and rarely of muscles and bones) with the fungi of the genus *Sporotrichum*, causing granulomatous lesions.

Sporozoa (*BioSci*) A class of parasitic Protozoa, members of which are usually at some stage intracellular. In the principal phase they have no external organs of locomotion or are amoeboid; they lack a meganucleus, and form large numbers of spores after syngamy, which constitute the infectious stage.

sporozoite (*BioSci*) In Protozoa, an infectious stage developed within a spore.

sport (*BioSci*) See BUD SPORT.

sporulation (*BioSci*) The production of spores.

spot (*Electronics*) A point on a phosphor which becomes visible through impact of electrons in a beam. See ION BURN.

spot beam (*ICT*) In a communications satellite system, radiation from a satellite antenna designed to illuminate only a small region of the Earth's surface. As well as conserving power, this allows the system to employ FREQUENCY REUSE.

spot beam (*Space*) A concentration of radio waves by an antenna–reflector so that a particular area is highly illuminated with radiation, resulting in the concentration of power over a small area and a high signal strength.

spot board (*Build*) The square wooden board on which the plasterer works up the coarse or fine stuff prior to applying it to the walls.

spot face (*Eng*) Machined surface produced in the vicinity of a hole in a casting or other rough part, to permit a bolt head, washer or adjacent part to seat evenly.

spot level (*Surv*) The reduced level of a point (usually on ground surface), not necessarily lying along a traverse or survey line.

spot light (*ImageTech*) A focusable studio lamp, capable of producing a narrow beam.

spot meter (*ImageTech*) A photometer or exposure meter which takes its reading from a small area, eg subtending 2° at the point of observation.

spot price (*MinExt*) The price agreed for an immediate shipment or parcel. For crude oil and products in Europe, the price is set in the Rotterdam market and differs from the price negotiated for a long-term contract or supply, or from one set by an organization such as OPEC.

spot priming (*Build*) A term for applying the first coat of paint to small areas, used eg for painting steelwork when only localized areas need treatment.

spot speed (*ICT*) In facsimile recording, the speed of the recording or scanning spot within the allotted time. In TV, the product of the number of spots in a scanning line multiplied by the number of scanning lines per second.

spotted gum (*For*) *Eucalyptus maculata*, from New South Wales and Queensland, growing up to 45 m in height.

spotted slate (*Geol*) See SLATE.

spotting (*Eng*) (1) The operation of turning a short length of a bar or forging to form a JOURNAL, by which the work is to be supported by the jaws of a steady rest. See STEADY. (2) A method of finishing plates or other flat surfaces with a regular pattern of circular patches.

spotting drill (*Eng*) A flat drill having a point so shaped as to centre and face the end of a bar at one operation. See CENTRE DRILL.

spotting out (*Print*) When a photographic emulsion is broken by pinhole blemishes etc it can be rendered opaque by the application of high-opacity paint, always to the non-emulsion side of the film.

spot treatment (*Agri*) The application of spray to a restricted area, or to an individual plant or animal.

spot welding (*ElecEng*) A process of welding in which metal sheets or wires are pressed together between two electrodes and a pulse of heavy current passed. Produces a nugget of fused metal holding the sheets together. Usually made in seams or rows which cannot be too close as the welding current would then shunt through neighbouring welds and the metal not reach its fusion temperature. Often required to resist only light and temporary stresses, eg many steel meshes used for reinforcement are spot welded purely to resist stresses arising from handling and not for resisting applied structural loads. Also *resistance spot-welding*.

spot wobble (*ImageTech*) A vertical oscillatory movement given to the scanning spot to make the space between the scanned lines less obvious.

spp (*BioSci*) Abbrev for SPECIES (pl).

sprag (*MinExt*) (1) Timber prop; short piece of wood, used to prevent the wheels of a train from revolving. (2) Slanting prop used to support coal face. Also *gib*.

sprag clutch (*Eng*) One incorporating balls and wedging surfaces, which allows power to be transmitted from one rotating shaft to another in one direction only.

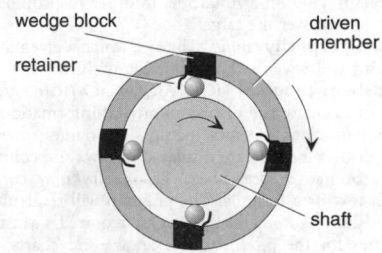

sprag clutch

sprain (*Med*) A wrenching of a joint with tearing or stretching of its ligaments, damage to the synovial membrane, effusion into the joint, occasionally rupture of muscles or tendons attached to the joint, but without dislocation.

spray (*Genrl*) Dispersion of liquid under pressure as droplets suspended in air.

spray bonded (*Textiles*) A method of treating a WEB of fibres with a spray of adhesive so that a non-woven fabric is obtained.

spray column (*MinExt*) A tower packed with coarse material, eg coke, or set with grids or trays down which a liquid trickles countercurrent to rising gas so as to facilitate interaction. See GLOVER TOWER.

spray discharge (*MinExt*) In high-tension separation the corona radiated from the electric-discharge wire (18 000 volts or more) onto a passing stream of finely divided mineral particles. In classification, a spray-shaped discharge from the hydrocyclone.

spray drift (*Agri*) Spray particles that land outside the intended area of application.

spray drying (*ChemEng*) Rapid drying of a solution or suspension by spraying into a flow of hot gas, the resultant powder being separated by a cyclone. Used to prepare powdered milk, detergent, fertilizer, etc.

sprayer (*Agri*) A machine that delivers liquids as a continuous emission of fine droplets over a required area.

spray gate (*Eng*) An ingate consisting of a number of small separate gates, fed from the runner; used for shallow castings where there is insufficient depth for a single large gate. See INGATE.

spray gun (*Build, CivEng*) A tool used in spray painting to 'atomize' paint or other coating, like cement, and facilitate its application in a controlled pattern.

spraying (*Eng*) The process of coating the surface of an article by projecting onto it a spray of molten metal or other substance. The adhesion is mechanical and does not involve welding or diffusion. Metal spraying is often used for building up worn parts of old machinery. See PLASMA SPRAYING.

spray recovery (*Agri*) The amount of spray deposited on a crop after runoff and drift, used as a measure of efficiency.

spray spectrum (*Genrl*) The range of droplet diameters present in a spray emission.

spray tower (*ChemEng*) A plant for purifying gases, which pass up through a tower into which a suitable liquid is sprayed from the top.

spread (*BioSci*) The establishment of a species in a new area.

spread (*Print*) (1) An illustration or diagram occupying each side of the fold. (2) A story or feature run across the two centre pages of magazine or newspaper. (3) Refers to a technique in photography or plate-making whereby an image is slightly enlarged to ensure image overlap when printed.

spreader (*Agri*) A machine used to disperse dry, granulated or powdered materials.

spreader (*ICT*) Wooden or metal spar for keeping the wires of a multiwire antenna spaced apart.

spread factor (*Phys*) See DISTRIBUTION COEFFICIENTS.

spreading agent (*BioSci*) A substance added to a solution for eg spraying on a fungicide, in order to promote even distribution over the target.

spreading capacity (*Build*) The area which a given quantity of paint will cover without thinning unduly.

spreadsheet program (*ICT*) An APPLICATION PROGRAM used to calculate and display numerical information, often of a financial or statistical nature. The name comes from the fact that the data are displayed in rows and columns as an accountant's spreadsheet. Each entry may contain a value, text or a formula. The program will recalculate the spreadsheet automatically or as required. Most have facilities for the production of graphs and charts; this is called 'presentation graphics'.

spread spectrum (*ICT*) Modulation technique used for security and to increase immunity from noise and interference. The normally narrow-band information signal is spread over a much wider range of frequencies in a pseudo-random (*noise-like*) manner; the receiver is adapted to correlate these deviations in order to retrieve the original signal.

sprig (*Build*) A small nail with little or no head.

sprig (*Eng*) A small nail pushed into a weak edge of sand in a mould to reinforce it during pouring. See BRAD.

spring (*Eng*) A device capable of deflecting so as to store energy, used to absorb shock or as a source of power or to maintain pressure between contacting surfaces or to measure force. See CARRIAGE SPRING, CONTACT SPRING, HELICAL SPRING, RELAY SPRING, SPIRAL SPRING.

spring-back (*Eng*) Elastic recovery of a workpiece after completion of a bending operation.

spring-back (*Print*) A style of stationery binding in which the book opens quite flat to facilitate writing close to the back of the margin.

spring balance (*Chem*) A balance in which the weight of the sample is balanced by the extension of a spring.

spring bows (*Eng*) Small compasses whose two limbs are not hinged together but are connected by a bow of spring steel, the distance apart of the marking points being adjusted by means of a screw. Also *bow compasses.*

spring constant (*Eng*) The ratio of load to deflection of a spring, measured eg in $N\,mm^{-1}$. Also *spring rate.* See STIFFNESS.

spring control (*ElecEng*) Controlling the movement of an indicating instrument with a spring.

spring cramp (*Build*) One in the shape of a broken ring made of tensile steel, used for light work.

springer (*CivEng*) The lowest VOUSSOIR on each side of an arch. Also *springing.* See fig. at ARCH.

springing (*Arch*) The line where the intrados of an arch meets the abutment or pier. See fig. at ARCH.

springing line (*Arch*) The line joining the springings on both sides of an arch.

spring-loaded governor (*Eng*) An engine governor consisting of rotary masses which move outwards under centrifugal force and are controlled by a spring. See HARTNELL GOVERNOR.

spring-loaded idler (*Print*) See JOCKEY ROLLER.

spring needle (*Textiles*) One with hook or beard which is flexed by the pressers on a knitting machine. Used to produce close and even texture in knitted fabrics. Also *bearded needle.*

spring pawl (*Eng*) See PAWL.

spring points (*CivEng*) Points on a railway which are normally held closed by springs in determining the route when they are facing points, but can be passed through as trailing points. See POINTS.

spring rate (*Eng*) See SPRING CONSTANT.

spring safety valve (*Eng*) See SAFETY VALVE.

spring tab (*Aero*) A *balance tab* connected so that its angular movement is geared to the compression or extension of a spring incorporated in the main control circuit. Its primary purpose is to reduce the effort required by the pilot to overcome the air loads on the main control surface resulting from high airspeeds. Cf SERVO TAB, TRIMMING TAB.

spring tide (*Astron*) See TIDE.

spring wood (*BioSci*) Same as EARLY WOOD.

sprinkler (*Build*) A pipe system installed in a building with spray nozzles sealed with a fusible alloy which melt and release water in the event of a fire.

sprite (*ImageTech*) An individual graphic in a computer/video game.

sprocket (*Build*) A small wedge-shaped piece of wood nailed to the top of the lower end of a common rafter to reduce the slope of the roof near the eaves.

sprocket (*Eng*) See SPROCKET WHEEL.

sprocket (*ImageTech*) A cylinder with regularly spaced teeth which engage the perforation holes of motion picture film to provide its movement through various forms of equipment.

sprocket holes (*ImageTech*) Deprecated term for the PERFORATIONS in motion picture film.

sprocket noise (*ImageTech*) A 96 Hz hum occurring in optical sound reproduction if the film is misplaced so that the perforations modulate the exciter lamp beam at the scanning slit.

sprocket wheel (*Eng*) A toothed wheel used for chain drives, as on the pedal shaft and rear hub of a bicycle.

sprout depressant (*Agri*) A formulation to prevent or slow the emergence of potato sprouts from tubers.

spruce (*For*) Trees of the genus *Picea*, giving a valuable structural softwood. See CANADIAN SPRUCE, SITKA SPRUCE, WHITEWOOD.

sprue (*Eng*) See GATE.

sprue (*Med*) A disease affecting the gastro-intestinal tract and causing malabsorption of vitamins and nutrients. Characterized by loss of energy, loss of weight, anaemia, inflammation of the tongue and the mouth, and by the frequent passage of pale, bulky, acid, frothy stools, there being inability to absorb adequately fat, glucose and calcium. The term is usually applied to the disease in the tropics where it appears to have an infectious origin.

Non-tropical sprue or coeliac sprue is a synonym for COELIAC DISEASE.

sprue (*Plastics*) Waste plastic formed during injection-moulding processes, being the material setting in the main inlet passages of the mould.

SPS (*Phys*) Abbrev for *super proton synchrotron*. A particle accelerator at CERN, Geneva, with an output beam of 500 GeV protons for fixed-target experiments; modified to perform 270 GeV antiproton colliding-beam experiments. Used to discover (1983) W^+, W^- and Z GAUGE BOSONS that mediate weak interactions.

spud (*MinExt*) To begin actual well-drilling operations; 'spudding in'.

spudding bit (*MinExt*) Large drill bit for making the initial *top hole* which takes the ANCHOR STRING or TOP CASING. Very deep, high-pressure wells may require over 1000 ft of top hole into which the casing is cemented.

spun-dyed (*Textiles*) Manufactured fibres formed from substances to which the colouring matter has been added before the filaments are formed.

spun silk (*Textiles*) Yarn made from silk waste which is spun in a manner very similar to the woollen systems.

spun yarn (*Textiles*) A yarn made from staple fibres twisted together.

spur (*BioSci*) (1) A projection, usually containing nectar, that arises from the base of a petal, sepal or gamopetalous corolla, etc, eg *Aquilegia, Linaria*. (2) A short shoot, esp one of the condensed lateral fruiting shoots of many fruit trees. See also CALCAR.

spur (*Build*) A strut.

spur (*Geol*) A hilly projection extending from the flanks of a valley.

spur gear (*Eng*) Gearwheel with straight teeth machined parallel to its axis.

spur gearing (*Eng*) A system of gears with straight teeth connecting two parallel shafts.

spuriae (*BioSci*) In birds, the feathers of the bastard wing.

spurious coincidences (*NucEng*) Those recorded by a coincidence counting system, when a single particle has not passed through both or all the counters in the system. They usually result from the almost simultaneous discharge of two counters by different particles.

spurious counts (*NucEng*) Those arising in counter tubes from voltage leakages in the counter and defects in external quenching circuits.

spurious oscillation (*ICT*) See PARASITIC OSCILLATION.

spurious pregnancy (*Med*) See PSEUDOCYESIS.

spurious pulse (*NucEng*) One arising from self-discharge of particle counter leading to erroneous signals.

spurious radiation (*ICT*) Undesired transmission, eg harmonics of carrier or modulation, outside specified band, causing interference with reception of other transmissions.

spurious response ratio (*ICT*) The ratio of field strengths of signals producing spurious and required response in telecommunication receiving equipment, eg of image frequency or intermediate frequency relative to required signal frequency.

Sputnik (*Space*) A series of former USSR artificial satellites; Sputnik 1, launched in October 1957, was the first ever human-made object to orbit about the Earth.

sputtering (*Electronics*) The non-thermal removal of atoms from a surface under energetic ion bombardment. Sputtering yields are strongly dependent on the ion and its energy but rise steadily from thresholds at a few eV to several atoms per ion in the keV range. The ejected atoms deposit on any adjacent surfaces. In plasma sputtering the ions come directly from a low-pressure gas discharge. Sputtering is used for thin-film deposition (eg metallization layers in semiconductor technology), for surface (discharge) cleaning and for surface analysis (eg SECONDARY ION MASS SPECTROMETRY). Different types of plasma source can be used, eg dc sputtering, RF sputtering, magnetron sputtering. Sputtering is particularly useful for making films from refractory materials.

sputum (*Med*) Matter composed of secretions from the nose, throat, bronchi or lungs, which is spat out.

spycam (*ICT*) A camera set up for hidden surveillance purposes, esp a webcam relaying pictures to a computer.

spyware (*ICT*) Computer software, usually loaded without the user's knowledge or consent, that gathers information about the user and transmits it to another computer user.

SQ (*Build*) Abbrev for SQUINT QUOIN.

SQL (*ICT*) Abbrev for STRUCTURED QUERY LANGUAGE.

SQPB (*ImageTech*) See QUASI S-VHS PLAYBACK.

squalene (*Chem*) A symmetrical triterpene, formerly known as *spinacene*, originally detected in shark oil but also occurring in mammalian and plant tissue.

squall (*EnvSci*) A temporary sharp increase in the wind speed, lasting for some minutes.

squama (*BioSci*) A scale or a scale-like structure.

Squamata (*BioSci*) An order of diapsid reptiles in which the skull has lost either one or both temporal vacuities, the quadrate is movably articulated with the skull, and there is no inferior temporal arch. Snakes and lizards.

squamiform (*BioSci*) Scale-like.

squamosal (*BioSci*) A bone that forms the principal component of the cheek region in the skull of higher vertebrates. Articulates with the quadrate and pterygoid bones.

squamous-cell carcinoma (*Med*) A carcinoma that develops from the squamous layer of the epithelium.

squamous epithelium (*BioSci*) Epithelium consisting of one or more layers of flattened scale-like cells; pavement epithelium.

squamule (*BioSci*) A small scale. Adj *squamulose*.

square (*For*) A piece of square-section timber of side up to 6 in (150 mm).

square (*MathSci*) (1) A regular four-sided plane figure, a rectangle whose adjacent sides are equal. (2) The product of a quantity with itself: $a^2 = a \times a$.

square drilling (*Eng*) Rotating and oscillating a specially shaped drill so that it follows a square-shaped guide bush to produce a square hole. See SPARK EROSION.

squared rubble (*Build*) Walling in which the stones are roughly squared to rectangular faces but are of irregular size.

square folding (*Print*) See RIGHT-ANGLED FOLDING.

square law (*Electronics*) Said of any device, such as a rectifier or (de)modulator, in which the output is proportional to the square of the input amplitude.

square law (*Phys*) The law of inverse squares expressing the relation between the amount of radiation falling upon unit area of a surface and the distance of the surface from the source.

square-law capacitor (*ElecEng*) Variable-vane capacitor, used for tuning, in which capacitance is proportional to the square of the scale reading, so that wavelength of circuit which it tunes becomes directly proportional to it.

square-law demodulator (*ElecEng*) See SQUARE-LAW DETECTOR.

square-law detector (*ICT*) A DEMODULATOR in which the output is proportional to the square of the amplitude-modulated input voltage.

square-law rectifier (*ElecEng*) One in which the rectified output current is proportional to the square of the applied alternating voltage.

square rebate plane (*Build*) A rebate plane with its cutting edge square across the sole.

square root (*MathSci*) The square root of a number is that quantity which when multiplied by itself gives the number, written $\sqrt{x}$.

squares (*Print*) The protrusion of the case of a book beyond the leaves.

square staff (*Build*) An ANGLE STAFF of square-section material, as distinct from an ANGLE BEAD.

square step (*Build*) An individual stone step in a stair which consists of a solid block, rectangular in section, either lapping over the back edge of the step below or rebated to fit over it.

square thread (*Eng*) A screw-thread of substantially square profile.

square wave (*Electronics*) Pulse wave with very rapid (theoretically zero) rise and fall times; and pulse duration equal to half-period of repetition. Mark/space ratio of unity.

squaring shears (*Eng*) Manual or power-operated press used for shearing sheets of steel.

squaring the circle (*MathSci*) The insoluble ancient problem of constructing, with straightedge and compasses, a square with area equal to that of a given circle.

squaring-up (*Build, CivEng*) A process following TAKING-OFF in drawing up a bill of quantities, superficial areas of items being calculated by multiplying the relevant dimensions entered on the dimensions paper.

squarrose (*BioSci*) Of leaves, hairs, scales, etc, sticking out more or less at right angles to the stem or other structure.

squash (*BioSci*) Spreading of tissue or chromosomes on a microscope slide by application of pressure.

squeegee (*ImageTech*) Rubber or plastic roller or blade, or a jet of compressed air, directed at the surface of a film during processing to wipe away surplus liquid.

squeeze (*Print*) The amount of impression between plate and impression cylinders.

squeezed print (*ImageTech*) Colloq term for a motion picture print in which the image is anamorphically compressed horizontally, as in CINEMASCOPE.

squeezer (*Eng*) A moulding machine, operated by hand, compressed air, hydraulic power or magnetic means, in which the sand is squeezed or compressed into the box and round the pattern by a ram.

squeeze track (*ImageTech*) In VARIABLE-AREA TRACK recording, reducing the image width to a minimum during unmodulated passages.

squegging (*ICT*) Colloq engineering term for an unwanted condition in which an electronic oscillator operates in short bursts rather than continuously, normally owing to operation of devices in their extreme non-linear region or to inadequate power supply design.

squelch (*ICT*) A form of automatic gain control where the gain of the receiver is reduced in response to certain characteristics of the input; eg in order to suppress background noise at very low signal levels.

SQUID (*Electronics*) Abbrev for SUPERCONDUCTING QUANTUM INTERFERENCE DEVICE.

squid (*Aero*) A dynamically stable condition of a fully deployed parachute canopy which will not fully distend.

squinch (*Arch*) A small arch running diagonally across the corner of a square tower or room, to support a side of an octagonal tower or spire above. Also a *scoinson arch*.

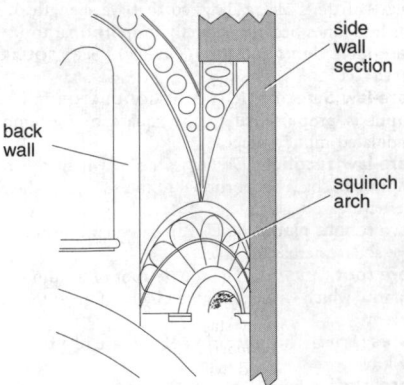

side wall section

back wall

squinch arch

squinch Spans the diagonal between back and side walls.

squint (*Build*) A specially shaped brick for use as a SQUINT QUOIN.

squint (*ICT*) The difference between the geometrical axis of an aerial array and the axis of the radiation pattern.

squint (*Med*) See STRABISMUS.

squint quoin (*Build*) A quoin enclosing an angle which is not a right angle. Abbrev *SQ*.

squirrel-cage motor (*ElecEng*) An induction motor whose rotor winding consists of a number of copper bars distributed in slots round the periphery, with the ends welded to two heavy copper end rings, the whole forming a rigid cage embedded in the rotor. See fig. at CAGE WINDING.

squirrel-cage rotor (*ElecEng*) The rotor of a SQUIRREL-CAGE MOTOR.

squirrel-cage winding (*ElecEng*) The winding of a SQUIRREL-CAGE MOTOR. See CAGE WINDING.

Sr (*Chem*) Symbol for STRONTIUM.

sr (*MathSci*) Symbol for STERADIAN.

SRAM (*ICT*) (1) Abbrev for *static random access memory*, random access memory which does not need to be continually refreshed and is not lost when power is lost. Also *non-volatile memory*. (2) Abbrev for SIDEWAYS RAM.

S–R theory (*Psych*) Abbrev for *stimulus and response theory*. The stimulus–response theory of learning which holds that the basic components of learning are S–R bonds, stimuli and responses which become forged together as learning proceeds.

SSADM (*ICT*) Abbrev for *structured systems analysis design method*. A formally structured method for SYSTEMS ANALYSIS AND DESIGN.

ss and sc lathe (*Eng*) A sliding, surfacing and screw-cutting lathe, ie one suitable for working on the periphery and on the end faces of workpieces, and capable of cutting a screw-thread using a single-point tool.

SSB (*ICT*) Abbrev for *single sideband modulation*, an efficient refinement of AMPLITUDE MODULATION.

SSI (*ICT*) Abbrev for SMALL-SCALE INTEGRATION.

SSL (*ICT*) Abbrev for *secure sockets layer*, a proprietary standard for secure communication between computers in a network.

ssp, sspp (*BioSci*) Abbrevs for SUBSPECIES (sing and pl).

SSRIs (*Pharmacol*) Abbrev for *selective serotonin re-uptake inhibitors*. SSRIs are antidepressant drugs that inhibit re-uptake of serotonin released in the brain, thereby prolonging its action as a neurotransmitter. They are less sedative than the tricyclic antidepressants. Examples are *citalopram*, *fluoxetine* (TN Prozac), *paroxetine*, *sertraline* and the related drug *venlafaxine*.

SSSI (*EnvSci*) Abbrev for *site of special scientific interest*.

SST (*Aero*) Abbrev for *supersonic transport aircraft*.

s state (*Electronics*) State of zero orbital angular momentum.

ST (*Build*) Abbrev for *surface trench*.

stabbing (*Build*) Making a brickwork surface rough to provide a key for plasterwork.

stabbing (*MinExt*) The process of locating one tube above and in line with another so that they can be screwed together.

stabbing (*Print*) A special kind of flat stitching required for very thick books, pairs of wires being necessary (driven from the front page and from the back), clenching not being possible; often (wrongly) used as a synonym for flat stitching.

stabilate (*BioSci*) A population, usually, of a micro-organism preserved in a viable condition on a unique occasion by eg freezing.

stabilator (*Aero*) See ALL-MOVING TAIL.

stabilizer (*FoodSci*) An ADDITIVE included in a formulation to prevent separation of incompatible components.

stability (*Aero*) The quality whereby any deviation from steady motion tends to decrease. A given type of steady motion is stable if an aircraft will return to that state of motion after a disturbance, without intervention by the pilot. An aircraft has three axes about which its stability is defined with three associated degrees of freedom (angular, normal displacement and change of velocity). See ARTIFICIAL STABILITY, CONTROL.

stability (*BioSci*) The ability of an ecosystem to resist change.

stability (*ElecEng*) The property of any electrical circuit or system (eg electricity transmission network or closed-loop controller) whereby changes, usually sudden, in operating conditions (eg electrical load or speed of machine) can be coped with by the electrical circuit or system, without loss of control within the designed range.

stability derivatives (*Aero*) Quantitative expressions for the variation of forces and moments on an aircraft due to disturbances from steady motion.

stability limits (*Aero*) (1) The extreme angles of incidence (maximum or minimum) to which a taxiing, taking-off or landing seaplane can be trimmed without PORPOISING. (2) Now refers to the range of centre of gravity positions between which an aircraft can fly with acceptable safety.

stability test (*ElecEng*) A test in which the cable is subjected to its working voltage (or a higher voltage) while it is alternately heated and allowed to cool. The power factor is measured during each heating and cooling period. If the power factor increases steadily during the test, the cable is said to be unstable.

stabilization (*Electronics*) Maintenance of a quantity (voltage, current, frequency, gain, etc) against variations induced by supply voltage fluctuations, changing load conditions, temperature and ageing. See REGULATION.

stabilized feedback amplifier (*ICT*) One in which amplification is stabilized against changes in supply voltages etc by the application of negative feedback.

stabilized glass (*Glass*) (1) Glass heat-treated so that it is in an equilibrium state corresponding to some particular temperature. (2) Glass heat-treated so as to suffer no permanent change of dimensions or properties over a particular range of temperature. (3) Glass resistant to darkening by high-energy, short-wave radiation.

stabilized yarn (*Textiles*) Yarn that has been subjected to processes (eg heating and cooling under controlled tension) that reduce its ability to change in length or to twist and snarl.

stabilizer (*Aero*) In the US, *horizontal stabilizer* is *tail plane*, and *vertical stabilizer* is *fin*. See fig. at CONTROL. See AUTOMATIC STABILIZER.

stabilizer (*Chem*) (1) Important additive for polyvinyl chloride, which was formerly very unstable owing to chain defects etc. There is a variety of chemicals which are thought to react with such defects, so inhibiting their action. Includes lead sulphate, carbonate; cadmium–barium salts; organo-tin compounds; and, lately, thiol systems. (2) A negative catalyst (cf RETARDER). (3) A substance which makes a solution stable.

stabilizer (*FoodSci*) An ADDITIVE included in a formulation to prevent separation of incompatible components.

stabilizer (*ImageTech*) A solution used to render the processed image as permanent as possible by neutralizing chemical residue.

stabilizer tube (*Electronics*) Gas-discharge tube, the voltage across which is much more stable than a voltage applied to it in series with a resistor.

stabilizing choke (*ElecEng*) A reactive choke coil inserted in series with an electric-discharge lamp to compensate for its negative resistance characteristic.

stable (*Chem*) A term describing systems not exhibiting sudden changes, esp chemical or physical transformations.

stable (*Phys*) Used to indicate the incapability of following a stated mode of spontaneous change; eg *beta stable* means incapable of ordinary beta disintegration but capable of eg isomeric transition or alpha disintegration.

stable equilibrium (*Phys*) The state of equilibrium of a body when any slight displacement increases its potential energy. A body in stable equilibrium will return to its original position after a slight displacement.

stable oscillation (*ICT*) One for which amplitude and/or frequency will remain constant indefinitely. A statically stable system may be dynamically unstable and follow a divergent oscillation when subjected to a disturbance. In a dynamically stable system any induced oscillation will be convergent, ie of decreasing amplitude.

stable platform (*ICT*) A structure that can be controlled in position with great precision, eg by gyroscopes, and that forms a base for other information to be measured and transmitted by telemetry, eg from satellites.

stable pneumonia (*Vet*) See EQUINE INFLUENZA.

stachyose (*Chem*) $C_{24}H_{42}O_{21} \cdot 4H_2O$, a tetrasaccharide, found in the roots of *Stachys tuberifera* and of several Labiatae; mp (anhydrous) 170°C.

stack (*ICT*) A LIST for which all insertions and deletions are made at one end only. The arrangement is called last-in, first-out. Abbrev *LIFO*. Cf QUEUE.

stack (*Phys*) A pile of photographic plates exposed to radiation together, and used to study tracks of ionizing particles.

stacked array (*ICT*) An antenna array in which the radiators are stacked one above the other and connected in phase so as to give the antenna directional properties.

stacking fault (*Phys, Crystal*) A fault which occurs in close-packed crystal systems appearing as a mismatch of one of the close-packed planes as the crystal grows or is deformed. It amounts to a half DISLOCATION. Stacking faults are significant in the deformation and creep of metals with close-packed crystal structures, since they affect the ease with which a dislocation may cross from one slip plane to another. See DEFORMATION MAP and panel on CREEP AND DEFORMATION.

stack pointer (*ICT*) A POINTER used to point to the ADDRESS of the data item at the top of a STACK. In the case of the stack used by the PROCESSOR, this pointer will be held in a special REGISTER rather than in WORKING STORE for high-speed access.

stadia hairs (*Surv*) The two additional horizontal lines (*stadia lines*), one on each side of the central line, fitted to the diaphragm of a telescope to be used in TACHEOMETRY.

stadia rod (*Surv*) A special form of levelling staff bearing bold graduations suitable for the long sights usual in stadia tacheometry.

stadium (*BioSci*) An interval in the life history of arthropods between two consecutive ECDYSES.

staff (*Build*) See ANGLE STAFF.

staff (*Med*) A grooved rod introduced into the urethra as a guide for cutting a stricture.

staff (*Surv*) See LEVELLING STAFF.

staff angle (*Build*) See ANGLE STAFF.

staff bead (*Build*) See ANGLE BEAD.

stage (*Build, CivEng*) A ledge or working platform associated with scaffolding.

stage (*Geol*) A chronostratigraphic succession of rocks which were deposited during an *age* of geological time. A subdivision of a geological series.

stage (*NucEng*) Unit of cascade in isotope separation plant, consisting of a single separative element, or group of these elements, operating in parallel on material of same concentration.

stage (*Space*) A section or part of a launch system which fires for a certain time only and then is separated from the main system; when more than one stage is used, the technique is termed STAGING.

stage efficiency (*ElecEng*) The ratio of ac output power to dc input power for any stage of an electronic amplifier.

stage micrometer (*BioSci*) A device for measuring the magnification achieved with a given microscope or for calibrating an eyepiece graticule. It usually consists of a small accurate scale mounted on a microscope slide.

stage separation factor (*NucEng*) See SEPARATION FACTOR.

stage theory (*Psych*) Any theory of development that proposes that an individual must pass sequentially, in a set order, through various stages of development. Piaget's theory of cognitive development is a classic example.

stagger (*Aero*) The horizontal distance between the leading edges of the wings of a multiplane as projected vertically. If the upper plane is ahead of the lower, stagger is positive; if behind, it is negative.

staggered (*Chem*) In CONFORMATIONAL ANALYSIS, this represents a conformation in which the substituents of one atom in a bond are situated as far as possible from those of the other.

staggered tuning (*ICT*) An attempt to get a wide-band response by a number of tuned circuits, having slightly different frequencies of resonance.

staggering (*ElecEng*) A term signifying the displacement of the brushes of a commutator motor from the neutral zone.

stagger-tuned amplifier (*ICT*) One with couplings tuned to different frequencies to give a band-pass response.

stag-headed (*BioSci*) Of a tree having the upper branches dead with regrowth of the crown from new branches.

staging (*Build*) A robust scaffold of timbers on metal supports, braced together and capable of handling substantial loads.

staging (*Space*) The principle of increasing the velocity achieved by a launch system and its payload by using more than one propulsive stage. Tandem (nose-to-tail) or parallel (side-by-side) staging may be employed, each stage being jettisoned after the fuel has been expended, thus increasing the mass ratio and therefore the efficiency of the whole system.

stagnation point (*Aero*) The point at or near the nose of a body in motion in a fluid where the flow divides and where fluid pressure is at a maximum, and the fluid is at rest. Theoretically there is another stagnation point near the trailing edge.

stagnation temperature (*Aero*) The temperature which would be reached if a flowing fluid were brought to rest adiabatically, which is almost applicable in supersonic flight for the leading edges and air intakes, where the air in the boundary layer of a body is drastically and rapidly decelerated. Also *total temperature*.

stagnicolous (*BioSci*) Living in stagnant water.

stainer (*Build*) A pigment added to paint when a final colour is required which is different from that of the base used. Common pigments are: (1) *earth pigments*, ochres, umbers, siennas, Venetian red, red oxide and malachite; (2) *synthetic pigments*, chromes, Monastral blue, ultramarine, Prussian blue and lakes.

staining (*BioSci, Chem*) A general method of increasing contrast in thin polymer and biomaterial films (sections) for microscopical examination by selective absorption of dyestuff or, in electron microscopy, of electron-dense heavy metals.

staining power (*Chem*) The degree of intensity of colour which a coloured pigment will impart when mixed with a standard white pigment under standardized conditions.

stainless steel (*Eng*) Corrosion-resistant steels of a wide variety of compositions, but always containing a high percentage of chromium (12–25%) since the stainless property derives from chromium oxide film on the surface. Exhibits passivity and therefore highly resistant to corrosive attack by organic acids, weak mineral acids, atmospheric oxidation, etc. Used for cutlery and domestic appliances, furnace parts, chemical plant equipment, stills, valves, turbine blades, ball bearings, etc. Widely used material in nuclear reactors because of its good anti-corrosion properties. Despite its high thermal neutron cross-section, an alloy containing iron (54·5%), chromium (20%), nickel (25%) with 0·5% niobium is the fuel cladding in gas-cooled reactors because it can operate at high temperatures (>800°C). See panel on STEELS.

staking (*Eng*) Fastening operation similar to rivet setting, in which a projection on one part is upset by means of a punch so as to fit tightly against a mating feature in another part.

stalactite (*Geol*) A concretionary deposit of calcium carbonate which is formed by percolating solutions and hangs icicle-like from the roofs of limestone caverns.

stalagmite (*Geol*) A concretionary deposit of calcium carbonate, precipitated from dripping solutions on the floors and walls of limestone caverns. Stalagmites are often complementary to stalactites, and may grow so that they eventually join with them.

stalagmometer (*Chem, Phys*) Apparatus which measures the surface tension of liquid in terms of mass of a drip leaving a specified orifice.

stale seed bed (*Agri*) A seed bed left undisturbed after initial preparation to allow weed germination. Latterly the weeds are destroyed by herbicide application and the crop sown with minimal soil disturbance.

staling (*BioSci*) The accumulation with time of metabolites in a culture medium which results in the slowing down of growth.

stall (*Aero*) The progressive breakdown of the lift-producing airflow over an aerofoil, which occurs near the angle of maximum lift.

stall (*Eng*) Of an engine, stopping due to the too sudden application of a load or brake.

stall (*MinExt*) The working compartment or room in the BORD-AND-PILLAR method of working coal; a coal miner's working place.

stalling speed (*Aero*) The airspeed of an aircraft at which the wing airflow breaks down.

stalling torque (*ElecEng*) The overload torque which is sufficient to slow down to zero the speed of an electric motor operating under load.

stallion (*Agri*) An entire, reproductively capable, male equine.

stall-warning indicator (*Aero*) A device fitted to those aircraft which do not give any positive warning of the approach of the stall by BUFFETING. Usually operated by the change of pressure and movement of the STAGNATION POINT near the stall, the warning may be audible, visual or by a *stick-shaker* (or forward-pushing electric motor). See STICK PUSHER.

stamen (*BioSci*) The microsporophore (microsporophyll), or male reproductive organ, of flowering plants, ie the structure, within the flower, that bears the pollen. See ANDROECIUM, ANTHER, FILAMENT.

staminal (*BioSci*) Pertaining to a stamen; derived from a stamen.

staminate (*BioSci*) Of flowers, male.

staminode (*BioSci*) An imperfectly developed, vestigial, anther-less or petaloid stamen.

stamp (*MinExt*) (1) To crush. (2) A freely falling weight, attached to a long rod and lifted by means of a cam; once widely used for crushing ores.

stamping (*ElecEng*) See LAMINATION.

stamping press (*Print*) See BLOCKING.

stanchion (*CivEng*) A pillar, usually of steel, for the support of a superstructure.

stand (*BioSci*) Any living assemblage of land plants.

standard (*For*) Volume measure for timber, particularly softwoods, equal to 165 ft^3. Also *Petersburg standard*. In the USA, 1 standard = $16\frac{2}{3}$ ft^3. See BOARD FOOT, PETERSBURG STANDARD.

standard (*Genrl*) (1) Established unit of measurement, or reference instrument or component, suitable for use in calibration of other instruments. Basic standards are those possessed or laid down by national laboratories or institutes (eg NPL, BSI). (2) A detailed description, including dimensions and other quantities, of the function, construction, materials, and quality of a manufactured article or an engineering project. (3) Any specification for a material, process or test method developed by an official body (eg BS, DIN, ASTM) or by private treaty.

standard atmosphere (*Aero*) See INTERNATIONAL STANDARD ATMOSPHERE.

standard atmosphere (*EnvSci*) Hypothetical atmosphere approximating to the average state of the real atmosphere in which pressure and temperature are defined at all heights. Internationally agreed standard atmospheres are used as bases for assessing the performance of altimeters, aircraft, etc.

standard atmosphere (*Phys*) Unit of pressure, defined as 101 325 N m^{-2}, equivalent to that exerted by a column of mercury 760 mm high at 0°C. Symbol atm.

standard atmosphere (*Textiles*) In order to get reproducible and consistent results during testing, it is necessary to test textile materials at standard conditions. These are 20°C (27°C in the tropics) and 65% relative humidity. It is necessary to condition the materials for some time before they are tested to ensure they have reached equilibrium.

standard beam approach (*Aero*) A system of radio navigation which provides an aircraft with lateral guidance and marker-beacon indications at specific points during its approach. Abbrev *SBA*.

standard calomel electrode (*Chem*) A half-element consisting of mercury, a paste of mercury and calomel (mercury (I) chloride), and a saturated solution of potassium chloride saturated with calomel; used as a standard potential difference in emf measurements.

standard cell (*Phys*) See WESTON STANDARD CADMIUM CELL.

standard chamber (*NucEng*) Ionization chamber used for calibration of radioactive sources, or of absolute values of exposure doses.

standard deviation (*MathSci*) The square root of the ratio of the sum of the squared deviations from the *mean* of a set of observations to either the sample size (the *population standard deviation*) or the sample size minus one (the *sample standard deviation*); the square root of the variance.

standard electrode potential (*Chem*) The potential of a chemical element dipping into a solution of its ions at unit activity, referred to that of hydrogen under a pressure of 1 atmosphere as zero.

standard error (*MathSci*) The STANDARD DEVIATION of an estimate of a population statistic. If the mean is 10 and the standard error of the mean is 2 then the true score is likely to fall between 8 and 12 or 10 $\pm$ 2. Also *standard error of the mean*, Abbrev *sem*.

standard filter (*ImageTech*) A filter which, when placed in front of a specified source, eg a tungsten lamp, gives a standard white light of black-body temperature 4800 K.

standard frequency (*ElecEng*) Namely 50 or 60 Hz (cycles/ second), the standards of POWER FREQUENCY in most countries of the world.

standard function (*ICT*) A subprogram provided by a compiler or other translator that carries out a task such as the computation of a mathematical function (eg log, square root).

standard gauge (*CivEng*) The railway gauge employed in most countries of the world, of width 4 ft 8$\frac{1}{2}$ in (1·435 m).

standard generalized markup language (*Print, ICT*) A specification for controlling a document format that uses tags to control fonts, line spacing, the appearance of a page, etc. Tag specifications are held in a dictionary, peculiar to the document, which can be transmitted with it. Abbrev *SGML*.

standard heat of formation (*Phys*) ENTHALPY change when a compound is formed from its constituent elements under standard conditions (temperature 298·1 K, pressure 1 atmosphere or 101·325 kN m^{-2}). By convention, the enthalpies of formation of the elements are defined as zero. Standard free energy of formation is defined in a similar way.

standard illuminant (*Phys*) A type of illumination source used for accurate colour measurements. The CIE specifies various alternative standards: (1) Illuminant A has the profile of a black-body radiator with a colour temperature of 2854 K and represents an incandescent lamp; (2) series C are produced by filtering Illuminant A and are rather unsatisfactory; (3) the D series of illuminants are based upon natural daylight, (colour temperature 4810 K) but

are difficult to produce artificially; (4) illuminant series F represent various fluorescent sources. Methods for producing these are no longer published.

standardized mortality ratio (*Med*) The ratio of observed to expected deaths in a subpopulation. Abbrev *SMR*.

standard knot (*For*) A knot which is 1$\frac{1}{2}$ in (40 mm) or less in diameter.

Standard Malaysian Rubber (*Eng*) A strict specification for various grades of NATURAL RUBBER. Abbrev *SMR*.

standard mean chord (*Aero*) The average chord, ie gross wing area divided by the span.

standard measurement (*Build*) The method recommended by the Chartered Surveyors' Institution for measurement of building works.

standard normal distribution (*MathSci*) The NORMAL DISTRIBUTION with mean 0 and variance 1, which is extensively tabulated.

standard oxidation–reduction potential (*Chem*) The potential established at an inert electrode dipping into a solution containing equimolecular amounts of an ion or molecule in two states of oxidation.

standard page (*Print*) The largest size of page on any particular rotary press, there being no standard size.

standard propagation (*ICT*) With standard refraction, the propagation of radio waves over a perfectly smooth earth with uniform electrical characteristics.

standard radio atmosphere (*ICT*) A radio atmosphere in which the index of refraction decreases by 39 $\times$ 10^{-6} km^{-1} above the Earth's surface.

standard refraction (*ICT*) Refraction arising in a STANDARD RADIO ATMOSPHERE.

standards converter (*ImageTech*) Equipment to convert signals from one colour TV system to another, eg from NTSC to PAL.

standard score (*Psych, MathSci*) Any derived score based upon transforming the data using the standard deviation. Standard scores can then be compared with one another. Also *z-score*. See STANINES as an example.

standard signal generator (*ICT*) A precision oscillator whose output is calibrated as regards frequency and amplitude, and sometimes depth of modulation; used for testing radio equipment, receivers, etc.

standard solenoid (*ElecEng*) A laboratory standard of inductance consisting of an air-cored solenoid with a secondary coil located at its centre. The dimensions are such that a value of the mutual inductance between the windings can be calculated from them.

standard solution (*Chem*) A solution whose concentration is accurately known. Such solutions are used in VOLUMETRIC ANALYSIS.

standard specification (*Eng*) A specification incorporating widely accepted quantities or features, themselves standards, to ensure interchangeability, quality and reliability for least cost. Usually drawn up by national standards institutions.

standard state (*Phys*) The condition of elements and compounds under standardized conditions of temperature (298·1 K) and external pressure (1 atmosphere, 101·325 kN m^{-2}).

standard temperature and pressure (*Genrl*) A temperature of 0°C and a pressure of 101 325 N m^{-2}. Abbrevs *stp*, *STP*. See STANDARD ATMOSPHERE.

standard time (*Astron*) The civil time in any of the time zones established by international agreement. These are about 15° of longitude wide, equal to 1 hour. Within a zone all civil clocks are set the same standard time or rather local SOLAR TIME. Zones usually differ by a whole hour, but there are a few cases of half-hour zones (eg S Australia).

Standard Wire Gauge (*Eng*) A standard series of sizes denoted by arbitrary numbers, widely used to describe the diameter of wire. Abbrev *SWG*.

stand-by losses (*ElecEng*) That part of the power expended in a generating station to maintain plant in instant readiness to take a sudden load.

standing crop (*EnvSci*) The total dry mass of organisms present in an area, obtained by harvesting sample plots, drying and weighing the harvested biomass and expressing the standing crop in units such as g m^{-2}.

standing current (*ICT*) See QUIESCENT CURRENT.

standing hay (*Agri*) Grass crop that has matured and dried out whilst remaining uncut. May be subsequently cut and stored or livestock may be given access directly.

standing-off dose (*Radiol*) Absorbed dose after which occupationally exposed radiation workers must be temporarily or permanently transferred to duties not involving further exposure. Doses are normally averaged over 13-week periods and standing-off would then continue for the remainder of the corresponding period.

standing panel (*Build*) A door panel whose height is greater than its width.

standing wave (*ICT*) One in which, for any component of the electromagnetic field, the ratio of its instantaneous amplitude at one point to that at any other point does not vary with time. In transmission lines, waveguides, etc, the result of reflections from a load that is not perfectly matched to the transmission line or source. Earlier term *stationary wave*.

standing wave (*Phys*) A wave resulting from interference between waves travelling in opposite directions (eg on a guitar string, where waves run along the string, reflected from end to end); the ratio of the instantaneous value of displacement at one point to that at another does not change. Also *stationary wave*.

standing-wave indicator (*ICT*) Any device that may be attached to or inserted in a transmission line or waveguide to indicate the presence of standing waves and assist correct matching. May be a simple neon lamp on a slide, a sensitive detector inserted into the line or waveguide in a SLOTTED LINE arrangement, or a REFLECTOMETER.

standing-wave meter (*ICT*) One designed to measure voltage standing-wave ratio in transmission lines and waveguides.

standing-wave ratio (*ICT*) Where standing and progressive waves are superimposed, the SWR is the ratio of the amplitudes at nodes and antinodes. For a transmission line or waveguide, it is equal to

$$\frac{1-r}{1+r}$$

where r is the coefficient of reflection at the termination. It may alternatively be defined by the reciprocal of this value as shown by its value being numerically greater than unity.

standing ways (*Ships*) The portion of a ship's launching ways which are fixed to the ground. The sliding ways move on these ways and are positioned by an upstanding rib integral with the fixed ways.

stand-off bomb (*Aero*) A small, fast, powered, unmanned aircraft or rocket containing a nuclear warhead, released from a bomber to fly many hundreds of miles to the target. It is automatically piloted and navigated, usually by a DOPPLER NAVIGATOR and/or INERTIAL NAVIGATION system. Propulsion can be by ramjet, rocket, turbojet or in combination.

stand pipe (*Eng*) (1) An open vertical pipe connected to a pipeline, to ensure that the pressure head at that point cannot exceed the length of the stand pipe. (2) Pipe connected to water main for attaching a hose.

stand pipe (*NucEng*) The connection between the charge face and the interior of a reactor vessel, giving access to the fuel channels, eg for refuelling.

stand sheet (*Build*) A window having no frame. Also *dead light, fast sheet, fixed sash*.

standstill (*ElecEng*) A term pertaining to the electrical behaviour of a machine when it is at rest.

standstill torque (*ElecEng*) The load torque which would bring an electric motor to a standstill.

Stanford–Binet scale (*Psych*) See BINET INTELLIGENCE SCALE.

Stanford Linear Accelerator Center (*Phys*) See SLAC.

stannane (*Chem*) Tin hydride. SnH_4.

stannates (*Chem*) See STANNITES (II).

stannic (IV) acid (*Chem*) Acids of two types, formed by action of alkalis on solutions of tin (IV) chloride and by the action of nitric acid on the metal; called respectively α-stannic (IV) acid and metastannic acid or β-stannic (IV) acid.

stannic oxide (*Chem*) Tin (IV) oxide. SnO_2. Formed (1) by combustion of tin, (2) when stannic acids are calcined. Forms alkali stannates (IV) when fused with alkali carbonates. See CASSITERITE.

stannite (*Min*) Sulphide of copper, iron and tin, which crystallizes in the tetragonal system and usually occurs in tin-bearing veins. Also *bell-metal ore, tin pyrites*.

stannites (II) (*Chem*) Stannates (II). Salts of stannous acid. Formed when stannous hydroxide is dissolved in alkaline solutions.

stannous hydroxide (*Chem*) Tin (II) hydroxide. $Sn(OH)_2$. Precipitated when sodium hydroxide is added to a solution of tin (II) chloride. When heated in carbon dioxide, forms black tin (IV) oxide, SnO, which, heated in air, forms tin (IV) oxide, SnO_2.

stanozolol (*Pharmacol*) An anabolic steroid that is illegally used as a performance-enhancing drug by some athletes.

Stanton number (*ChemEng*) A dimensionless parameter. $St = h/(C_p\rho v)$ where h is the heat transfer coefficient, C_p the specific heat capacity at constant pressure, ρ the density of a fluid travelling and v the fluid velocity.

stapedectomy (*Med*) Excision of the stapes.

stapes (*BioSci*) (1) In amphibians, a small nodule of cartilage in connection with the FENESTRA OVALIS of the ear. (2) In mammals, the stirrup-shaped innermost auditory ossicle. Adj *stapedial*.

Staphylococcus (*BioSci*) A Gram-positive coccus of which the individuals tend to form irregular clusters. The commonest types, associated with various acute inflammatory and suppurative conditions, including mastitis in animals, are S.aureus (golden yellow colonies) and S. albus (white colonies).

staphyloma (*Med*) Local bulging of the weakened sclera of the eye (as in GLAUCOMA or MYOPIA); bulging of a corneal scar in which the iris of the eye has become fixed.

staphylorrhaphy (*Med*) The operation of closing a cleft in the soft palate.

staple fibre (*Textiles*) Natural or artificial fibres that are comparatively short in length, eg 10–500 mm.

star (*Astron*) A sphere of matter held together entirely by its own gravitational field and generating energy by means of NUCLEAR FUSION reactions in its deep interior. The pressure of the mass of material overlying the core is sufficient to cause nuclear reactions, the principle one of which is the transmutation of hydrogen into helium. During this process about 0·5% of the mass is converted into electromagnetic radiation. The minimum mass needed to make a star is probably 1/20th the mass of the Sun, the maximum about 70 times as great. See MAIN SEQUENCE and panels on HERTZSPRUNG–RUSSELL DIAGRAM and SUN AS A STAR.

star (*ICT*) See NETWORK, STAR TOPOLOGY.

star (*ImageTech*) Radiating tracks in photographic emulsion, arising from particle disintegration on collision, and dispersal of energy to other particles.

starburst galaxy (*Astron*) A galaxy, usually spiral, in which an unusually large burst of star formation is taking place over an extended region; this results in a high infrared luminosity.

starch (*Chem*) Amylum. $(C_6H_{10}O_5)_x$. Polysaccharide found in all assimilating (green) plants. A white hygroscopic powder which can be hydrolysed to dextrin and finally to D-glucose. Diastase converts starch into maltose. Starch does not reduce Fehling's reagent and does not react with phenylhydrazine. It forms a blue compound with iodine.

star chart (*Astron*) A systematic and accurately made map of the heavens in which the star positions are generally plotted according to equatorial co-ordinates.

starch grain (*BioSci*) A rounded or irregular mass of starch. These are found within chloroplasts in green algae, and within chloroplasts, amyloplasts or other plastids in vascular plants and bryophytes.

starch gum (*Chem*) See DEXTRIN.

starch plant (*BioSci*) A plant in which carbohydrate is stored as starch. Cf SUGAR PLANT.

starch sheath (*BioSci*) (1) A one-layered cylinder of cells lying on the inner boundary of the cortex of a young stem, with prominent starch grains in the cells. It is homologous with an endodermis. (2) A layer of starch grains around a PYRENOID in an algal cell.

star cluster (*Astron*) See GLOBULAR CLUSTER, OPEN CLUSTER.

star connection (*ElecEng*) A method of connecting a three-phase load or source such that one terminal of each phase is connected to a common point, the neutral point. Currents flowing in the lines are equal to the corresponding phase currents. Line voltages are $\sqrt{3}$ times the corresponding phase voltage. Also *Y-connection*. See DELTA CONNECTION, STAR POINT, VOLTAGE TO NEUTRAL.

star–delta starter (*ElecEng*) A starting switch for an induction motor which, in one position, connects the stator windings in star for starting and, in the other position, reconnects the windings in delta when the motor has gained speed.

Stark effect (*Phys*) Splitting of atomic energy levels, and of corresponding emission spectrum lines, by placing source in region of strong electric field. Cf ZEEMAN EFFECT.

Stark–Einstein equation (*Phys*) An equation giving the energy absorbed per mole for a photochemical reaction:

$$E = Nh\nu$$

where N is AVOGADRO'S NUMBER, h is PLANCK'S CONSTANT and ν *is the frequency of the absorbed light*.

stark rubber (*Eng*) Natural rubber which has crystallized over several years by storage at low temperature. It is very stiff due to a high DEGREE OF CRYSTALLINITY, and is difficult to process.

starlite (*Min*) A name suggested (from a fancied resemblance to starlight) for the blue zircons which are heat-treated and used as gemstones.

star magnitude (*Astron*) See MAGNITUDE.

star–mesh transformation (*ElecEng*) A technique for simplifying a network, in which any number of branches meeting at a point are replaced by an equivalent mesh, reducing the number of connections.

star network (*ElecEng*) One with many branches connected at a point; a T- or Y-network has three branches.

star observation (*Surv*) Use of theodolite to locate or orient a ground station by sighting on a star.

star point (*ElecEng*) The common junction of the several phases of a star-connected three-phase system.

star polymer (*Chem*) Macromolecule created by multifunctional chain extending agent reacting with growing and active chain ends. Usually formed by anionic polymerization.

star-quad (*ElecEng*) See QUAD.

star quartz (*Min*) The prefix 'star' has reference to the narrow-rayed star of light exhibited by varieties of quartz, ruby and sapphire. The star is seen to best advantage when they are cut *en cabochon*. It is caused by reflections from exceedingly fine inclusions lying in certain planes. Also *star ruby, star sapphire*. See ASTERISM.

starred signature (*Print*) When a section is to be insetted in another it is given the same signature mark followed by an asterisk (*) or the figure 2.

star-streaming (*Astron*) A phenomenon, discovered from analysis of observed stellar motions (after removing the effects of the observer's own motions), by which the stars are found to have two preferential directions of motion, one towards the point RA 90°, dec 15°S, and the other towards RA 285°, dec 64°S; the first stream contains about 60% of the observed stars. The effect is due to the rotation of the Galaxy.

star target (*Print*) A circular symbol with numerous radii, tapering to the centre but not meeting, leaving a small, clear inner circle, used particularly on lithographic plates as a guide to maintenance of colour and quality of printing. Can be used to detect image loss or gain, slur, etc.

start bits (*ICT*) In asynchronous data transfer, a sequence of bits that signal the sending of a character. The end of a character or transmission is signified by a sequence of STOP BITS. See ASYNCHRONOUS TRANSFER MODE.

start codon (*BioSci*) A triplet sequence of nucleotides (usually AUG) in mRNA that signals the initiation of translation and hence the first amino acid in a polypeptide chain.

starter (*ElecEng*) A device for starting an electric motor and accelerating it to normal speed. Also *motor starter*.

starter (*Electronics*) See PILOT ELECTRODE.

starter gap (*Electronics*) The conducting path between the pilot electrode and the electrode to which the starting voltage is applied in a glow-discharge tube.

starter voltage (*Electronics*) That applied to the pilot electrode of a cold-cathode discharge tube or mercury-arc rectifier; cf STARTING VOLTAGE.

startext (*ImageTech*) A coded TELETEXT signal identifying and accompanying each broadcast TV programme to start and stop accurately video timer recordings when a startext decoder is employed. From *start text*. See PROGRAMME DELIVERY CONTROL.

starting current (*ElecEng*) The current drawn by a motor from the mains when starting up.

starting resistance (*ElecEng*) A fixed resistance connected in series with the main circuit of a motor when starting up. This added resistance serves to limit excessive currents during start-up.

starting sheet (*Eng*) A sheet of pure metal used as the initial cathode on which the metal being refined is deposited during electrolytic refining.

starting torque (*ElecEng*) The torque developed by a motor at starting.

starting-up time (*NucEng*) See START-UP TIME.

starting voltage (*Electronics*) The voltage which initiates current passing in a gas-discharge tube after non-conduction; much greater than that required to maintain conduction. Also *ionizing voltage, striking voltage*.

starting voltage (*Phys*) See THRESHOLD VOLTAGE.

starting winding (*ElecEng*) An auxiliary winding on the armature of a single-phase motor (enabling it to start up as a two-phase machine) or of a synchronous converter.

startle colours (*BioSci*) Bright colours on the body or wings of animals which often resemble vertebrate eyes and which are normally concealed. They are exposed on being disturbed and are anti-predator devices.

star topology (*ICT*) A NETWORK layout where each TERMINAL or NODE is connected individually back to a central computer, file server or hub. See fig. at NETWORK TOPOLOGY.

start-up disk (*ICT*) A DISK on which the INITIALIZATION FILES, OPERATING SYSTEM and other BOOT files are stored which some computers need to start operating.

start-up procedure (*NucEng*) That followed when bringing a nuclear reactor into operation. It involves four successive stages: (1) *source range*, where a neutron source is introduced to generate the required neutron flux (this may not be necessary in a reactor which has already been operating); (2) *counter range*, where reactor is just critical but counters are required to monitor neutron-flux changes; (3) *period range*, where changes in reactivity are monitored on period meter; (4) *power range*, where reactor is operating within its designed power ratings.

start-up time (*NucEng*) The time required by an instrument or system (eg nuclear reactor, chemical plant, etc) to reach equilibrium operating conditions. Also *starting-up time*.

star voltage (*ElecEng*) See VOLTAGE TO NEUTRAL.

Star Wars (*Space*) See SDI.

star wheel (*Eng*) (1) Toothed wheel moved one tooth per revolution of an adjacent shaft by a pin attached to that shaft. Cf GENEVA MOVEMENT. (2) A continuously or intermittently rotating disk with scalloped circumference, used to guide bottles or other containers onto a conveyor at correct intervals.

stasis (*BioSci*) Cessation of growth.

stasis (*Med*) (1) Complete stoppage of the circulation of blood through the capillaries and smallest blood vessels in a part. (2) Arrest of the contents of the bowel at any point from obstruction or weakness of the bowel wall.

stassfurtite (*Min*) A massive variety of boracite which sometimes has a subcolumnar structure and resembles a fine-grained white marble or granular limestone. From Stassfurt, Germany. See BORACITE.

stat- (*Phys*) Prefix to name of unit, indicating derivation in obsolete electrostatic system of units, eg *statampere*, *statohm*.

state (*Phys*) The energy level of a particle as specified by the appropriate quantum numbers.

state-dependent learning (*Psych*) Learning in which the recall depends on the degree of similarity between the physiological state of the individual at the time of training and at the time of testing.

state-dependent memory (*Psych*) The theory that information learned, eg in a happy frame of mind, is more easily remembered when in the same emotional state.

state function (*Phys*) A quantity in thermodynamics which has a unique value for each state of a system. Internal energy, ENTHALPY and ENTROPY are examples. The value associated with a given state is independent of the process used to bring about that state.

statement number (*ICT*) One used to label a specific statement in a program so that the user and the machine can refer back to it subsequently. Such numbers do not indicate the order in which instructions must be carried out, or the number of the address at which the instruction will be stored. Also *instruction number, line number*.

statenchyma (*BioSci*) A tissue consisting of cells containing STATOLITHS.

state of matter (*Chem*) Traditionally all matter is in one of three states: solid (fixed volume and shape), liquid (fixed volume, shape that of container), gaseous (filling the containing vessel).

static (*ElecEng*) Non-movable or non-rotating; eg a transformer or rectifier is a static converter.

static (*ICT*) Said of all electrical disturbances to a radio system that arise through electrostatic induction, particularly from lightning flashes.

statical stability (*Ships*) The ability of a ship to return to its initial position when forcibly inclined.

static balancer (*ElecEng*) See AC BALANCER.

static capacitor (*ElecEng*) A capacitor, described as static to distinguish it from *rotating*, often used in electric supply systems for power factor correction.

static characteristic (*ElecEng*) A curve, or set of curves, which describes the relation between specified voltages and currents of electrodes under unvarying conditions, as compared with DYNAMIC CHARACTERISTIC, which implies operations under time-varying conditions.

static convergence (*ImageTech*) The adjustment to a colour TV whereby the three coloured RASTERS are made to coincide in the centre of the screen.

static discharge wick (*Aero*) Wicks, usually of cotton impregnated with metallic silver, or of nichrome wire, fitted at the trailing edges of an airplane's flight control surfaces, by which static electricity is discharged into the atmosphere.

static electricity (*Phys*) See DYNAMIC ELECTRICITY.

static electrification (*ElecEng*) The tendency of insulators to accumulate an electric charge, esp when rubbed by another material. High voltages can be generated, leading to a spark discharge which may be hazardous in the presence of flammable gases. See TRIBOELECTRIC SERIES.

static fatigue (*Eng*) The phenomenon of a material failing at a smaller load than that required to cause short-term failure, after a period of constant loading by the smaller load; the load necessary to produce static fatigue decreases with increasing time under load. In brittle materials, static fatigue is due to the slow growth of subcritical cracks to a length at which they will propagate catastrophically; in ductile and/or viscoelastic materials it is due to the progress of plastic or viscoelastic deformation (ie creep) to the point where catastrophic yielding can occur. Also *creep rupture*, esp in relation to polymeric materials. Assisted by ENVIRONMENTAL STRESS CRACKING.

static friction (*Eng*) See FRICTION.

static impedance (*ElecEng*) The electrical impedance of a machine or transducer when it is stopped from moving. In loudspeakers, it has the same meaning as BLOCKED IMPEDANCE.

static instability (*EnvSci*) Atmospheric state such that a parcel of air moved from its initial level experiences a hydrostatic force tending to remove it further from this level. Also *hydrostatic instability*.

static inverter (*Aero*) A non-rotating device for converting dc to ac supply, usually of high voltage, for radio and instrument services.

static jet thrust (*Aero*) See STATIC THRUST.

static line (*Aero*) A cable joining a parachute pack to the aircraft, so that when the wearer jumps, the parachute is automatically deployed.

static machine (*ElecEng*) See ELECTROSTATIC GENERATOR.

static marks (*ImageTech*) Marks caused by electrostatic discharge near the surface of undeveloped photographic film which become visible after processing.

static memory (*ICT*) Memory that needs no *refreshing* once information is stored. Cf DYNAMIC MEMORY.

static pressure (*Aero*) The pressure at any point on a body moving freely with a fluid in motion; in practice, the pressure normal to the surface of a body moving through a fluid.

static-pressure tube (*Aero*) A tube with openings placed so that when the air is moving past it the pressure inside is that of still air. See PITOT-STATIC TUBE, STATIC VENT.

statics (*Phys*) The branch of applied mathematics which studies the way in which forces combine with each other usually so as to produce equilibrium. Until the early part of the 20th century the term also embraced the study of gravitational attractions, but this is now normally regarded as a separate subject.

static stability (*Aero*) Positive static stability in an aircraft means that if it is disturbed from a trimmed speed, the disturbance will be reduced.

static stability (*ElecEng*) The stability of a transmission system with reference to gradual changes in load demand.

static thrust (*Aero*) The net thrust (kN or lbst) of a jet engine at *International Standard Atmosphere* sea level and without translational motion.

static vent (*Aero*) An opening, usually in the fuselage, found by experiment, where there is minimum POSITION ERROR and which is used instead of the STATIC-PRESSURE TUBE.

statins (*Pharmacol*) A group of lipid-lowering drugs that act by inhibiting HMG Co-A reductase, a key enzyme in cholesterol biosynthesis. Examples are *atorvastatin* (TN Lipitor), *pravastatin, rosuvastatin, simvastatin*.

station (*ElecEng*) In general, a generating station. Specifically, a key point on an electricity supply system.

station (*Eng*) The location, on a transfer machine or on an assembly line, at which a workpiece or an assembly is halted for the placing of a component or for the execution of a machining or fastening operation.

station (*ICT*) The location of radio transmitters and/or receivers with antennas, for sending or receiving radio signals.

station (*Surv*) (1) A point at an apex of a triangle in a skeleton, or otherwise situated in a line of the skeleton. (2) A point whose reduced level is to be found.

stationary orbit (*Space*) The circular orbit of a satellite which holds it above a fixed point on its parent body's equator. If the parent body is the Earth, the term *geostationary* is used and the orbit lies about 35 784 km above the equator. A special case of SYNCHRONOUS ORBIT.

stationary phase (*Chem*) See CHROMATOGRAPHY.

stationary point (*MathSci*) A point at which the derivative of a function is zero. Includes maximum and minimum (*turning points*).

stationary point on a curve (*MathSci*) A point at which the tangent is parallel to the *x*-axis. For the curve $y = f(x)$, the point where $x = a$ is a stationary point if $f'(a) = 0$. All turning points are stationary points but not all stationary points are turning points.

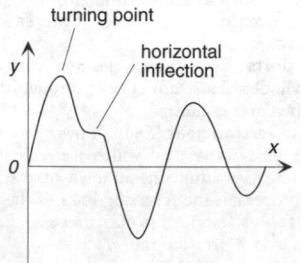

stationary point on a curve

stationary points (*Astron*) Those points in the apparent path of a planet where its direct motion in right ascension changes to retrograde motion or vice versa.

stationary wave (*ICT*) Earlier alternative to STANDING WAVE.

stationary wave (*Phys*) See STANDING WAVE.

station keeping (*Space*) The manoeuvres necessary to adjust a GEOSTATIONARY satellite's orbit so that its position in space is correct, its ground coverage does not vary and data transmission is optimized; the manoeuvres are usually effected by small jets or rockets.

station pointer (*Surv*) An instrument for obtaining a mechanical solution of the THREE-POINT PROBLEM. It consists of a full-circle protractor with one fixed radial arm and two movable radial arms, which can be set to the correct mutual directions of the three points.

statistic (*MathSci*) A numerical quantity calculated from a set of observations.

statistical chain (*Chem, Eng*) A model of a polymer chain used as the basis for the KINETIC THEORY OF ELASTICITY. The links in the chain are freely rotating, unlike a real chain, where rotational isomers occur owing to steric hindrance. The chains are free to pass through one another, while real chains experience the EXCLUDED VOLUME effect. Nevertheless, the model accounts for many elastomeric properties. See panel on ELASTOMERS.

statistical diameters (*PowderTech*) Statistical average of a specified parameter on microscopic methods of particle size analysis.

statistical energy analysis (*Acous*) A method to calculate the density of vibratory energy in coupled structures. Used in structure-borne sound problems.

statistical error (*Phys*) The error which inevitably accompanies measurement of frequencies of random events. As a result of statistical fluctuations, the average count N of eg radioactive decays measured by a radiation detector has a statistical error of $\sqrt{N}$.

statistical mechanics (*Phys*) Theoretical predictions of the behaviour of a macroscopic system by applying statistical laws to the behaviour of component particles. *Quantum mechanics* is an extension of classical statistical mechanics introducing the concepts of the quantum theory, esp the PAULI EXCLUSION PRINCIPLE. *Wave mechanics* is a further extension based on the SCHRÖDINGER EQUATION, and the concept of particle waves.

statistical time-division multiplexing (*ICT*) A method whereby a multiplexer apportions time on a dynamic basis only to those channels which are active.

statistical weight (*Phys*) For an energy level of a system which has a number of quantized states, the number of states having that given energy. See DEGENERACY.

statistics (*Genrl*) The classification, tabulation and study of numerical facts.

statocyst (*BioSci*) (1) An organ or cell for the perception of the position of the body in space, consisting usually of a sac lined by sensory cells and containing a free hard body or bodies (*statoliths*), either introduced or secreted; in plants these may be starch granules or other inclusions within a single cell. (2) An otocyst.

statolith (*BioSci*) A starch grain, or other solid body in a *statocyst* cell, that moves in response to gravity and appears to function as a gravity sensor.

stator (*Aero*) The row of fixed, radially disposed aerofoils which forms an essential part of the dynamics of an axial compressor or axial turbine.

stator (*ElecEng*) Stationary part of machine, esp a dynamo or motor.

stator blade (*Aero*) A small fixed aerofoil, usually of thin, highly cambered section, and of approximately parallel chord, mounted in the outer case of an axial compressor or turbine. See EXHAUST STATOR BLADES.

stator core (*ElecEng*) The assembly of laminations forming the magnetic circuit of the stator of an ac machine.

stator–rotor starter (*ElecEng*) A combined stator-circuit switch and rotor-circuit regulating resistance for use with slip-ring induction motors.

stator winding (*ElecEng*) That part of the electrical winding of a machine accommodated in the stator.

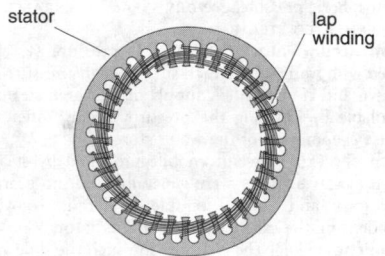

stator winding Multipole as in a three-phase induction motor.

status epilepticus (*Med*) A succession of severe epileptic convulsions with no recovery of consciousness between each convulsion.

statute mile (*Genrl*) Linear measure equal to 1760 yd or 5280 ft (1·61 km).

staurolite (*Min*) Silicate of aluminium, iron and magnesium, with chemically combined water, sometimes occurring as brown cruciform twins, and crystallizing in the orthorhombic system. It is typically found in medium-grade regionally metamorphosed argillaceous sediments.

stay (*Eng*) See BRACE.

stay tap (*Eng*) A long tap for threading the holes for stays connecting adjacent plates in boilers, thus ensuring that the two holes are threaded in correct pitch relation.

stay tubes (*Eng*) Boiler fire tubes acting as stays to the flat surfaces which they join; sometimes threaded and nutted to the plates for extra strength.

stay wire (*ElecEng*) One of several steel cables by which a transmission-line pole is secured to the ground.

STD (*ICT*) Abbrev for SUBSCRIBER TRUNK DIALLING.

St David's (*Geol*) See SAINT DAVID'S.

STDM (*ICT*) Abbrev for STATISTICAL TIME-DIVISION MULTI-PLEXING.

Steadicam (*ImageTech*) TN for a camera support designed to allow shake-free hand-held shooting, consisting of a body harness and counterbalancing weights.

steady (*Eng*) (1) A device for supporting long, heavy or slender work in turning. The fixed type clamps to the lathe bed and carries three radial jaws, adjustable to bear on the rough-turned work. The travelling type fixed to the moving carriage usually has two jaws supporting the work close to the cutting tool. (2) In milling, a device used to couple the machine table to the overarm support to give added rigidity and reduce vibration.

steady flow (*Phys*) See VISCOUS FLOW.

steady pin (*Eng*) A pin which permits mechanical parts to be fitted together accurately with one fixing screw. Cf DOWEL.

steady state (*ICT*) Said of any oscillation system that continues unchanged indefinitely.

steady state (*Phys*) A state in dynamic equilibrium, with entropy at its maximum.

steady-state theory (*Astron*) A model of the universe in which matter is continuously created to fill the voids formed as the universe expands and the galaxies move away from each other; the universe is then uniform in space and time. The alternative Big Bang theory is the currently accepted model. See panel on COSMOLOGY.

stealth (*Aero*) The technology of reducing the observable characteristics of military aircraft and missiles. Means include reducing size, noise, IR emissions from engines and from hot surfaces, radar reflections from intakes and between surfaces.

steam (*Phys*) Water in the vapour state; formed when specific latent heat of vaporization is supplied to water at boiling point. The specific latent heat varies with the pressure of formation, being approximately 2257 kJ kg^{-1} at atmospheric pressure. See DRY STEAM, SATURATED STEAM, SUPERHEATED STEAM.

steam accumulator (*Eng*) A large pressure vessel, partly filled with water, into which surplus high-pressure steam is blown and condensed. A supply of saturated steam is thus available by lowering the pressure at the outlet valve to cause evaporation of the water stored.

steam car (*Eng*) An automobile propelled by steam. Oil-fired FLASH BOILERS were generally used, no gearbox was necessary and control simple. Water is recovered by condensing the exhaust steam in a radiator.

steam chest (*Eng*) The chamber in which the slide valve of a steam engine works, and to which the steam pipe is connected.

steam coal (*MinExt*) Two varieties classified originally by the National Coal Board are *dry steam coal* (also *semianthracite*) and *coking steam coal* (rank 202, 203, 204).

steam distillation (*Chem*) The distillation of a substance by bubbling steam through the heated liquid. It is a useful method of separation for substances which are practically insoluble in water. The rapidity with which a substance distils in steam depends on its vapour pressure and on its vapour density.

steam dome (*Eng*) See DOME.

steam driers (*NucEng*) The last drops of water in a flow of steam can be removed by making the steam change direction abruptly, when the momentum of the water carries it onto a surface which leads it back into the boiling water. Used in boiling-water reactors.

steam economizer (*Eng*) See ECONOMIZER.

steam–electric generating set (*ElecEng*) Main parts of a STEAM GENERATING STATION.

steam engine (*Eng*) An external combustion engine whose working fluid is steam.

steam gauge (*Eng*) A gauge for indicating or recording steam pressure in a boiler or other part of a steam system.

steam generating station (*ElecEng*) A generating station in which the prime movers driving the electric generators are operated by steam, eg steam turbines or reciprocating steam engines.

steam generator (*Eng*) A steam boiler.

steaming up (*Vet*) The practice of increasing the nutritional plane of dairy cattle a few weeks before calving.

steam injector (*Eng*) See INJECTOR.

steam jacket (*Eng*) A jacket formed round a steam-engine cylinder; supplied with live steam to prevent excessive condensation of the working steam in the cylinder.

steam lap (*Eng*) See OUTSIDE LAP.

steam locomotive (*Eng*) A self-propelled steam engine and boiler integrally mounted on a frame which is fitted with wheels driven by the engine. The term is usually restricted to locomotives used to haul passenger or goods traffic on a railway, but various kinds of road locomotives were once common, such as the traction engine.

steam nozzle (*Eng*) See CONVERGENT–DIVERGENT NOZZLE, NOZZLE.

steam ports (*Eng*) Passages leading from the valve face to the cylinder of a steam engine; through them the steam is supplied and exhausted.

steam reversing gear (*Eng*) A power reversing gear, used in steam locomotives, by which movement of the driver's reversing lever admits steam to an auxiliary cylinder, whose piston operates the reversing links of the valve gear.

steam tables (*Eng*) A list of figures giving the properties of steam over a pressure range.

steam trap (*Eng*) A device into which condensed steam from steam pipes etc is allowed to drain, and which automatically ejects it without permitting the escape of steam.

steam turbine (*Eng*) A machine in which steam is made to do work by expanding so as to create kinetic energy, which is then partly absorbed by causing the steam to act on moving blades attached to a disk or drum. See BACK-PRESSURE TURBINE, DISK-AND-DRUM TURBINE, EXTRACTION TURBINE, IMPULSE TURBINE, MIXED-PRESSURE TURBINE, REACTION TURBINE.

stearic acid (*Chem*) $C_{18}H_{36}O_3$. A monobasic fatty acid; mp 69°C, bp 287°C; obtained from mutton suet, or by reducing oleic acid. It occurs free in a few plants, as glycerides in many fats and oils, and as esters with the higher alcohols in certain waxes.

stearin (*Chem*) A term for the glyceryl ester of stearic acid. The name is also applied to a mixture of stearic acid and palmitic acid.

steatite (*Min*) A coarse, massive or granular variety of talc, greasy to the touch. On account of its softness it is readily carved into ornamental objects. Also *soapstone*.

steatorrhoea (*Med*) The presence of an excess of fat in the stools, due either to failure of absorption or to deficiency of the fat-splitting enzymes in the digestive juices, as a result of disease of the pancreas. US *steatorrhea*.

steckling (*Agri*) An alternative term used for a cutting in vegetative propagation, commonly used to describe sugar beet seedlings.

steel (*Eng*) An important group of engineering materials based on the iron–carbon system, which may contain up to 2% carbon. See ALLOY STEEL, STAINLESS STEEL and panel on STEELS.

steel-cored aluminium (*ElecEng*) An electrical conductor consisting of a layer or layers of low-resistance aluminium wire surrounding a core of galvanized steel strands of high tensile strength.

steel-cored copper conductor (*ElecEng*) A conductor made in the same way as STEEL-CORED ALUMINIUM, except that the steel core is covered by a layer of insulating tape, to prevent corrosion of the surrounding copper.

steel-making (*Eng*) The process of making steel from pig iron, with or without admixture with steel scrap. Includes

Steels

A very important versatile group of engineering materials based on the iron carbon system, which may contain up to 2% carbon although usually below 1% and accompanied by other elements in small amounts. When carbon is the principal alloying component they are referred to as *plain carbon steels*, but, even so, these normally contain up to 0·8% manganese and 0·3% silicon together with 0·05% maximum of sulphur and phosphorus as impurities. *Low-alloy steels* contain, in addition to carbon, up to 5% in total of manganese, nickel, chromium, vanadium and molybdenum, at least two and often three or four of these elements being present in combination. Larger quantities of some of these elements can be added to impart specific properties in amounts significantly above 5%; such materials are referred to as *high-alloy steels*.

Carbon forms an INTERSTITIAL SOLID SOLUTION with iron, and dramatic differences in properties can be achieved by controlling the crystal structure by heat treatment. This is because iron can occur in two forms in the same phase: the BODY-CENTRED CUBIC (see panel on CLOSE PACKING OF ATOMS) alpha form, stable up to 803°C, with an equilibrium solubility of carbon of only 0·03%; and the FACE-CENTRED CUBIC gamma form with a maximum of 2%. With more than 2% carbon, a new phase appears during solidification and such materials are the family of cast irons, the useful range of which extends to 4·3% carbon. Steels may be cast or wrought, although some in the latter group are even nowadays mistakenly described as 'cast steel', a description resulting from the long obsolete method of making high-quality steels by the crucible process.

Steels are conveniently divided into three broad categories, depending on their carbon content which determines their potential response to heat treatment:

Mild steels contain up to 0·25% carbon and essentially can only be strengthened by WORK HARDENING, although some modern compositions develop higher yield and tensile strengths by PRECIPITATION HARDENING.

Medium carbon steels, 0·3–0·7% carbon, are heat treated to produce a wide range of engineering properties. Heat treatment usually consists of two stages. In the first, the steel is hardened at about 860°C when the steel is wholly austenitic and then submerged in oil or water to achieve a martensitic condition which is hard and extremely brittle. In the second, the steel is reheated to a lower temperature which allows some diffusion of carbon and relief of the lattice strain, a process called tempering or drawing. Mechanical properties change during this process as indicated schematically in the diagram, and the tempering temperature is chosen to produce the desired combination for that particular carbon content.

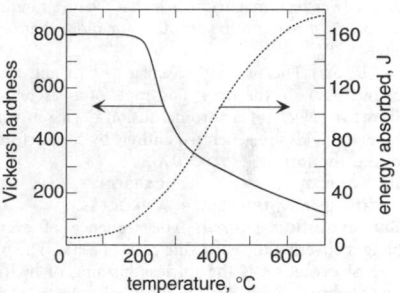

Steel Effect of tempering temperature on hardness and energy absorbed (toughness).

High-carbon steels respond to heat treatment like medium carbon steels but they also contain free CEMENTITE (Fe_3C) which tends to reduce their ductility and shock resistance. However, the cementite present in their microstructures makes them particularly suitable for retaining the sharp edge required for metal forming, machining and cutting tools.

the BESSEMER, CRUCIBLE, DUPLEX, ELECTRIC-ARC, HIGH-FREQUENCY INDUCTION and OPEN-HEARTH process.

steel-tank rectifier (*ElecEng*) A mercury-arc rectifier in which the arc chamber is of steel. Cf GLASS-BULB RECTIFIER.

steel tape (*Surv*) See BAND CHAIN.

steel tower (*ElecEng*) The framed steel structure carrying a high-voltage transmission line. Also *pylon*.

steeping (*Textiles*) To leave a yarn or fabric in a liquid usually without agitation; to soak the fabric, squeeze and leave wet.

steer (*Agri*) See BULLOCK.

steerable antenna (*ICT*) Any antenna in which the direction of its greatest radiation or sensitivity can be altered or steered; this may be done by electronic means (see ACTIVE ARRAY, PHASED ARRAY) or by mechanically rotating or elevating the antenna array or dish.

steering (*ICT*) Alteration by mechanical or electrical means of the direction of maximum sensitivity of a directional antenna, eg radar, radio telescope or Earth station.

steering arm (*Autos*) An arm rigidly attached to a stub axle, to which it transmits angular movement from the motion of the steering rod, and attached to it by a ball joint.

steering box (*Autos*) The housing which encloses the steering gear and provides an oil bath for the working surfaces. It is rigidly attached to a side member of the chassis frame. It contains gears which transmit the action of the steering column to the steering rod. In Europe, steering-box systems have been largely supplanted by RACK- AND-PINION gear for cars, although they are still standard on commercial vehicles.

steering gear (*Autos*) The two geared members attached to the steering column and the drop-arm spindle respectively. They transmit motion from the steering wheel to the stub axles through the drop arm, steering rod or drag link, steering arms and track rod. See CAM-TYPE STEERING GEAR, RACK-AND-PINION STEERING GEAR, SCREW-AND-NUT STEERING GEAR, WORM-AND-WHEEL STEERING GEAR.

steering rod (*Autos*) See DRAG LINK.

Stefan–Boltzmann law (*Phys*) The law stating that the total radiated energy from a black body per unit area per unit time is proportional to the fourth power of its absolute temperature, ie $E = \sigma T^4$ where σ is equal to $5 \cdot 6696 \times 10^{-8}$ W m^{-2} K^{-4} (the Stefan–Boltzmann constant).

steganography (*ICT*) A method of digital watermarking in which data, esp in graphics files, are so concealed that they can only be read by a special program.

Steiner's tricusp (*MathSci*) A hypocycloid in which the radius of the rolling circle is one-third or two-thirds that of the fixed circle. It has three cusps. Also *deltoid*.

steining (*CivEng*) The process of lining a well with bricks, stone, timber or metal, so as to prevent the sides from caving in. Also *steaning*.

Steinmann trinity (*Geol*) An association of cherts, spilites and serpentines, characteristic of a former ocean-floor environment.

Steinmetz coefficient (*Phys*) The constant of proportionality in the STEINMETZ LAW. Also *hysteresis coefficient*.

Steinmetz law (*Phys*) An empirical law stating that the energy loss per unit volume in a ferromagnetic material during each cycle of a hysteresis loop is proportional to $B_m{}^n$, where B_m is the maximum value of the magnetic flux density attained and n is $1 \cdot 6$ for many materials but some newer ferromagnetic alloys have values ranging from $1 \cdot 5$ to $2 \cdot 5$; n may not be constant for more than a limited range of B_m.

stele (*BioSci*) The primary vascular system and associated ground tissue of the stems and roots of a vascular plant.

stellarator (*NucEng*) A toroidal fusion device in which the magnetic fields are generated entirely by conductors placed around the torus. See TOKAMAK.

stellar energy (*Astron*) See CARBON CYCLE, PROTON–PROTON CHAIN, TRIPLE ALPHA PROCESS.

stellar evolution (*Astron*) The sequence of events and changes covering the entire life cycle of a star. The principal stage of evolution is the nuclear burning of hydrogen to form helium, with a consequential release of energy. Eventually the hydrogen in the core is exhausted, and the star becomes a RED GIANT. In the final stages of evolution there are several paths: the formation of a WHITE DWARF, or, for very massive stars, the build-up to a supernova explosion with possible core collapse to form a NEUTRON STAR or black hole. See MAIN SEQUENCE and panels on BLACK HOLE and HERTZSPRUNG–RUSSELL DIAGRAM.

stellar interferometer (*Astron, Phys*) A device, developed by A A Michelson, by means of which, when fitted to a telescope, it is possible to measure the angular diameters of certain giant stars (all of which are below the limit of resolution of even the largest telescopes) by observations of interference fringes at the focus of the telescope.

stellar magnitude (*Astron*) See MAGNITUDE.

stellar population (*Astron*) See POPULATION TYPES.

stellar wind (*Astron*) Radial outflow of material from the atmosphere of a very hot star, analogous to the SOLAR WIND.

stellate (*BioSci*) Radiating from a centre, like a star.

stellate hair (*BioSci*) A hair which has several radiating branches.

Stellite (*Eng*) TN for a series of alloys with cobalt, chromium, tungsten and molybdenum in various proportions. The range is chromium 10–40%, cobalt 35–80%, tungsten 0–25% and molybdenum 0–10%. Very hard. Used for cutting tools and for protecting surfaces subjected to heavy wear.

St Elmo's fire (*Phys*) See SAINT ELMO'S FIRE.

Stelvetite (*Plastics*) TN for PVC-coated steel sheet used for a variety of purposes including cabinets, wall sections, domestic equipment.

STEM (*BioSci, Phys*) Abbrev for SCANNING–TRANSMISSION ELECTRON MICROSCOPE (or *microscopy*).

stem (*BioSci*) The above-ground axis of a plant, and other axes (above or below ground) that are anatomically similar and/or clearly homologous. In most vascular plants, stems may be recognized by their bearing foliage leaves and/or scale leaves.

stem (*Print*) See BODY.

stem-and-leaf diagram (*MathSci*) The representation of data as a table in which the values falling in each class interval are explicitly listed, eg the square numbers in the interval [0,100] can be displayed as:

Stem	Leaf
1–25	1,4,9,16,25
26–50	36,49
51–75	64
76–100	81

stem-and-leaf plot (*MathSci*) An arrangement of a set of values into rows with common leading digits (stems), entries (leaves) in each row being the remaining digit of each value, truncated if necessary.

stem bar (*Ships*) The extreme forward end of a ship. The hull proper is secured thereto. Sometimes *stem post*.

stem cell (*BioSci*) An undifferentiated cell that can proliferate indefinitely, or that can differentiate into a specialized cell such as the erythrocyte. The stem cell's name usually ends in –*blast*, as in *erythroblast*. See EMBRYONIC STEM CELL.

stem correction (*Phys*) Correction to a temperature reading required if only the bulb of a thermometer is immersed, and the stem may be at a different temperature from the bulb.

stempipe (*MinExt*) See DRILLING PIPE.

stem post (*Ships*) See STEM BAR.

stem succulent (*BioSci*) A plant with a succulent, photosynthetic stem and with the leaves small or sometimes represented by spines. Many are CAM PLANTS, eg cacti. Cf LEAF SUCCULENT.

stench trap (*Build*) See AIR TRAP.

stencil (*Print*) (1) A plate of fine metal, plastic, waxed paper, etc, used for design and lettering, perforated in the required pattern so that ink, paint or other colouring substance may be passed through according to that pattern. (2) A term applied to the prepared mesh in SCREEN PROCESS PRINTING.

steno- (*Genrl*) Prefix from Gk *stenos*, narrow.

stenode (*ICT*) Supersonic heterodyne receiver in which there is very sharp tuning in intermediate-frequency circuits, using piezoelectric quartz crystals, with frequency correction of audio signal after demodulation.

stenohaline (*BioSci*) Capable of existence within a narrow range of salinity only. Cf EURYHALINE.

stenophyllous (*BioSci*) Having narrow leaves.

stenopodium (*BioSci*) The typical biramous (forked) limb of Crustacea, having slender EXOPODITE and ENDOPODITE. Cf PHYLLOPODIUM.

stenosis (*Med*) Narrowing or constriction of any duct, orifice or tubular passage as a result of disease. Adj *stenosed*.

stent (*Med*) A device used as a temporary splint inside a bodily vessel to keep it open; such a device used to support or immobilize a body part onto which skin has been grafted.

stenter (*Textiles*) A machine that holds a fabric by its selvedges while keeping it taut in open width and transporting it through a long heated chamber. The machine may hold the fabric by pins or clips. The machine is used to dry fabrics, to heat-set them or to fix chemicals (eg resins) on them. Also *tenter*.

stentorphone (*Acous*) See PNEUMATIC LOUDSPEAKER.

step (*Aero*) The step discontinuity in the bottom of a flying-boat hull, to facilitate take-off from the water surface by allowing the forebody to plane and the afterbody to be clear of the forebody wake. Typical values range from 4% to 12% of the maximum beam. Low values are likely to induce PORPOISING.

step (*ElecEng*) Synchronous machines are said to keep *in step* when they remain in synchronism with each other.

step (*ICT*) See TRANSIENT.

step-and-repeat (*Print*) A method of contacting and repeating a single image in predetermined position onto film or printing plate to create multiple images on the printed paper, eg in label work or postage stamps.

step-by-step method (*ElecEng*) A method of determining the hysteresis curve of a magnetic material, in which the field strength is increased and reversed in steps.

step-down transformer (*ElecEng*) One in which energy is transferred from a high to a low voltage. It has more primary than secondary turns. Cf STEP-UP TRANSFORMER.

step-faults (*Geol*) A series of tensional or normal faults which have a parallel arrangement, throw in the same direction, and hence progressively 'step down' a particular bed.

step function (*ICT*) A function that is zero for all time preceding a certain instant, and has a constant infinite value thereafter.

step function (*MathSci*) A function that makes an instantaneous change in value from one constant value to another. Its FOURIER ANALYSIS shows an infinite number of harmonics present.

Stephanian (*Geol*) The uppermost stage of the Carboniferous system, corresponding in the UK to the beds above the Coal Measures. See PALAEOZOIC.

stephanite (*Min*) A sulphide of silver and antimony which crystallizes in the orthorhombic system. It is usually associated with other silver-bearing minerals. Also *brittle silver ore*.

Stephenson's link motion (*Eng*) See LINK MOTION.

step irons (*Build*) See FOOT IRONS.

step-out well (*MinExt*) See APPRAISAL WELL.

stepped (*Print*) When type is to be printed close to an irregularly shaped block, the mount must be cut in steps to allow this.

stepped flashing (*Build*) Used where a brick chimney projects from a sloping roof, the lead being cut in steps so that the horizontal edges of the 'steps' may be secured into RAGLETS cut in the joints of the brickwork. Also *skeleton flashing*.

stepped-index fibre (*ICT*) An optical fibre in which the transition from the high refractive index of the information-bearing core to the lower refractive index of the cladding is abrupt. Also *step-index fibre* or, in the USA, *step-index fiber*. Cf GRADED-INDEX FIBRE. See MONOMODE FIBRE, MULTIMODE FIBRE.

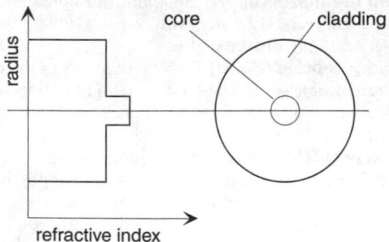

stepped-index fibre

stepping (*CivEng*) Laying foundations in horizontal steps on sloping ground. See BENCHED FOUNDATION.

stepping (*Surv*) The process of chaining over sloping ground by making the measurement in horizontal lengths with the chain always held horizontally, one end in the air.

stepping motor (*ElecEng*) A motor whose rotor moves through a fixed angle in response to a pulse from a controlling element. Widely used for the accurate positioning of machinery by eg a computer.

step polymerization (*Chem, Plastics*) The use of reactions like amidation and esterification to achieve stepwise polymerization of plastics. Also *step-growth*. See panels on POLYMERS and POLYMER SYNTHESIS.

step printer (*ImageTech*) That which prints one frame at a time.

step-up instrument (*ElecEng*) See SUPPRESSED-ZERO INSTRUMENT.

step-up transformer (*ElecEng*) One in which energy is transferred from a low to a high voltage. It has more secondary than primary turns. Cf STEP-DOWN TRANSFORMER.

step wedge (*ImageTech*) A series of graded exposures given to a photographic material and the resultant image produced after processing, for SENSITOMETRIC examination.

steradian (*MathSci*) The SI unit of solid angular measure. It is defined as the solid angle subtended at the centre of a sphere by an area on its surface numerically equal to the square of the radius. Symbol *sr*.

Sterba antenna (*ICT*) A stacked BROADSIDE array with a reflecting curtain, which may be PARASITIC or fed in the same way as the main radiating curtain. It can be uni- or bidirectional; used for short-wave communication.

stercolith (*Med*) A hard fecal concretion, impregnated with calcium salts, in the intestine. Also *stercorolith*. (Lt *stercus*, gen *stercoris*, dung.)

stercoraceous (*Med*) Consisting of or pertaining to feces. Adj *stercoral*.

stere (*For*) A stacked cubic metre of timber, which equals 35·3 stacked cubic feet.

stereo (*Acous*) Abbrev for *stereophonic*.

stereo- (*Genrl*) Prefix from Gk *stereos* meaning solid, hard, three-dimensional.

stereoblock polymer (*Plastics*) Homopolymer possessing regular sequences of different stereoisomers.

stereocamera (*ImageTech*) One equipped with two matched lenses or a beam-splitting device for taking photos in stereo pairs. Also *binocular camera, stereoscopic camera*.

stereochemistry (*Chem*) The branch of chemistry dealing with arrangement in space of the atoms within a molecule. See STEREOREGULAR POLYMERS.

stereognosis (*Med*) The ability to recognize similarities and differences in the size, weight, form and texture of objects brought into contact with the surface of the body.

stereogram (*ImageTech*) General term for photographs intended to be viewed in an apparatus to give a three-dimensional appearance. Also *stereograph*. See PARALLAX STEREOGRAM.

stereoisomerism (*Chem*) The existence of different substances whose molecules possess an identical connectivity but different arrangements of their atoms in space. Also *alloisomerism*. See GEOMETRICAL ISOMERISM, OPTICAL ISOMERISM.

stereokinesis (*BioSci*) Movement of an organism in response to contact stimuli.

stereology (*Eng*) Quantitative method of relating two-dimensional analysis of structure of materials (eg as seen in an optical microscope) to their structure as it exists in three dimensions. Based on methods of mathematical statistics and often used in the QUANTIMET analysis.

stereome (*BioSci*) A general term for the mechanical tissue of the plant.

stereome cylinder (*BioSci*) A cylinder of strengthening tissue lying in a stem, usually just outside the phloem.

stereomicrophone system (*Acous*) Dual microphone with eg interlocking figure-of-eight polar diagrams; used to provide signals for both channels of a stereophonic sound-reproduction system.

stereophonic recording (*Acous*) (1) Use of adjacent tracks on magnetic tape, with multiple recording and reproducing channels. (2) Use of a spiral cut on disks, two channels being represented by stylus motions at right angles, each at 45° to the surface; reproduction is by one stylus operating a double transducer.

stereophony (*Acous*) A method of sound reproduction which attempts to give the listener an effect of auditory perspective similar to that created by the original sound. Methods used require two loudspeakers and three microphones or a pair of microphones placed close together. Adj *stereophonic*. Also *auditory perspective, localization*.

stereoregular polymers (*Chem, Plastics*) In vinyl polymers like polypropylene every other carbon atom along the main chain is asymmetric. Since such chiral atoms can exist in L- or D-forms (the one the mirror image or enantiomer of the other), three different types of polypropylene can exist as shown in the figure. Iso- and syndiotactic polymers are produced by solid-state catalysts (eg ZIEGLER–NATTA CATALYSTS), the chain regularity created by stereospecific sites on the catalyst surface. Such polymers are usually highly crystalline unlike atactic polymers which are totally non-crystalline and amorphous. Polyethylene, isoprene rubber and cis-polybutadiene can also be prepared using similar catalysts to give stereospecific polymers with enhanced properties.

stereoregular polymers Showing effect of the asymmetric carbon atoms which alternate along the main chain.

stereoscope (*Phys*) A device for producing an apparently binocular (three-dimensional) image by presenting differing plane images to the two eyes.

stereoscopic camera (*ImageTech*) See STEREOCAMERA.

stereoscopy (*Phys*) Sensation of depth obtainable with binocular vision due to small differences in parallax producing slightly differing images on the two retinas.

stereospondyly (*BioSci*) The condition of having the parts of the vertebrae fused to form one solid piece. Adj *stereospondylous*. Cf TEMNOSPONDYLY.

stereotaxis (*BioSci*) The response or reaction of an organism to the stimulus of contact with a solid body, as the tendency of some animals to insert themselves into holes or crannies, or to attach themselves to solid objects. Adj *stereotactic*.

stereotaxis (*Med*) A procedure, often using X-rays, carried out at a precise localization in a tissue, eg *stereotactic surgery*, where a precise area of brain is identified for the surgeon to transect.

stereotype (*Print*) A DUPLICATE PLATE made from the original surface (type and/or blocks) or from an existing plate; for metal plates, the mould is made in flong; for rubber or plastic plates, a thermosetting plastic sheet is used.

stereotype (*Psych*) An oversimplified and very generalized belief about groups of people which is applied to individuals identified as members of the group.

stereotype alloys (*Eng*) Lead-based alloys with 5–10% tin and 10–15% antimony, once widely used in printing processes.

stereotyped behaviour (*Psych*) Behaviour patterns which are performed on different occasions with very little variation in their component parts, typical of many animal displays. Animals under stress, eg in close confinement, may develop very fixed and idiosyncratic behaviours, eg rigid pacing actions.

stereotypy (*Psych*) The repetition of senseless movements, actions or words.

steric hindrance (*Chem*) The retarding influence by virtue of the size of neighbouring groups on reactions in organic molecules. In polymers, interference between adjacent atoms or side groups in repeat units or short-chain lengths, giving rise to different energy states. See CHAIN FLEXIBILITY.

sterile (*BioSci*) (1) Unable to breed. (2) Free from living organisms, esp culture media, foodstuffs, surfaces, medical supplies, etc, which are free from micro-organisms that could cause spoilage or infection.

sterile flower (*BioSci*) (1) A flower with neither functional carpels nor functional stamens. (2) Sometimes, a male flower.

sterile line (*ElecEng*) One that has no direct connection with adjoining circuits and is therefore unaffected by them.

sterilization (*BioSci*) (1) The loss of sexual reproductive function. (2) The removal of unwanted organisms by heat, radiation, chemicals or by filtration.

Sterling board (*For*) TN for processed wood board and used for flooring, panelling, etc.

sterling silver (*Eng*) A silver alloy with not more than 7·5% base metal. Legal requirement is that articles described as silver shall have minimum 92·5% silver content.

S-terminal (*ImageTech*) A four-pin connector for separated Y/C signals.

sternal (*BioSci*) See STERNUM.

sternebrae (*BioSci*) In mammals, a median ventral series of bones which alternate with the ribs.

stern frame (*Ships*) In a twin-screw ship a heavy bar attached to the keel and to a strong transverse plate at the top, with gudgeons to carry the rudder. In a single-screw ship it surrounds the propeller aperture and is almost rectangular. The forward vertical part is the *propeller post* and the after part is the *rudder post* (sometimes called the *stern post*), the lower horizontal part is the *sole piece* and the upper part is the *arch piece*.

Stern–Gerlach experiment (*Phys*) Atomic beam experiment which provided fundamental proof of the quantum theory prediction that the magnetic moment of atoms can only be orientated in certain fixed directions relative to an external magnetic field.

Sterno (*Genrl*) TN for a form of flammable hydrocarbon jelly used as cooking fuel.

stern post (*Ships*) See STERN FRAME.

sternum (*BioSci*) (1) The ventral part of a somite in arthropods. (2) The breast bone of vertebrates, forming part of the pectoral girdle, to which, in higher forms, are attached the ventral ends of the ribs. Adj *sternal*.

sternutation (*Med*) The act of sneezing; a sneeze.

steroid hormones (*BioSci*) Lipophilic hormones that have a four-membered ring system with various substitutions. See panel on STEROID HORMONES.

steroid receptors (*BioSci*) Discrete domains in the cell that are responsible for DNA binding, steroid binding and gene activation and repression. See panel on STEROID HORMONES.

steroid regulated gene (*BioSci*) Genes whose expression is modulated by steroid hormones to affect the physiology of target tissues. See panel on STEROID HORMONES.

steroids (*Chem*) Compounds containing the perhydrocyclopentenophenanthrene nucleus. They include the sterols, bile acids, sex hormones, adrenocortical hormones, cardiac glycosides, sapogenins and some alkaloids.

steroid therapy (*BioSci*) Immunosuppressive therapy whereby steroids are used to downregulate uncontrolled immune responses such as in autoimmunity or chronic inflammatory conditions. Steroid therapy can be given either systemically or locally to an affected area of the body.

sterols (*Chem*) A group of steroid alcohols obtained originally from the non-saponifiable portions of the lipid extracts of tissues. The best known is CHOLESTEROL.

Steroid hormones

A class of lipophilic hormones which are synthesized from CHOLESTEROL and consist of a four-membered ring system with various substitutions. In mammals there are three major divisions: the adrenal steroids, ALDOS-TERONE and CORTISOL; the sex steroids, progesterone, ESTROGEN and TESTOSTERONE, which are synthesized by the gonads; and vitamin D_3 which is converted into its active form in the liver and kidney. The *adrenal steroids* influence body homeostasis, and control GLY-COGEN and mineral metabolism as well as mediating the stress response. They also affect the immune and nervous systems. The *sex steroids* determine the control and development of the embryonic reproductive system, control reproduction and reproductive behaviour in the adult and the development of secondary sexual characteristics. Vitamin D plays an important role in the regulation of calcium and phosphorus homeostasis and is necessary for normal bone development.

Steroid hormones travel to their target tissues in the blood and, since they are poorly soluble in water, are transported by high-specificity carrier proteins such as corticosteroid-binding globulin, sex-hormone-binding globulin and progesterone-binding protein. They also associate with low affinity with α1-acid glycoprotein and serum albumin. Steroids enter cells by diffusing across the membrane since they are readily lipid soluble. They influence the physiology of particular target tissues by regulating the expression of genes called steroid-regulated genes.

Steroid-regulated genes
The activity of steroid responsive genes may be regulated either by the direct effect on the rate of transcription initiation by RNA polymerase II or by other less-well-characterized mechanisms, such as the selective stabilization of certain mRNAs. The rate of transcription of particular steroid-regulated genes may be increased or depressed by steroid. In both cases the steroid–receptor complex recognizes specific DNA sequences upstream or within the gene – these are known as *steroid response elements*.

Steroid response elements have the properties of inducible transcriptional ENHANCERS: they can act independently of their orientation in relation to the TRANSCRIPTION site and they can also act at varying distances upstream or downstream of the mRNA start

site. Further, they can stimulate transcription from heterologous PROMOTERS. Like other enhancers, steroid response elements interact with sequence-specific DNA-binding proteins (STEROID RECEPTORS). In this case the activity of the enhancer binding factor is induced by interaction with a LIGAND, the steroid. Comparison of the DNA sequences of various steroid response elements has revealed that they contain one or more copies of a short imperfect inverted repeat sequence and that as little as 15 base pairs containing one inverted repeat can act as a steroid response element. Three different classes of steroid hormone – the androgens, glucocorticoids and progestins – are known to act through similar 15 bp DNA sequences. The response elements for estrogen and ecdysone (an insect steroid hormone) are related but distinct.

Upon binding to a response element, the steroid–receptor complex is thought to stimulate the rate of transcription by making protein–protein contacts with other transcription factors or possibly RNA polymerase II (see diagram). Genes which are repressed by steroid hormones contain response elements similar to those stimulated by steroid. In some instances the negative response elements overlap binding sites for essential transcription factors and binding of a receptor may then interfere with binding or functioning of such factors, resulting in a decrease in gene activity.

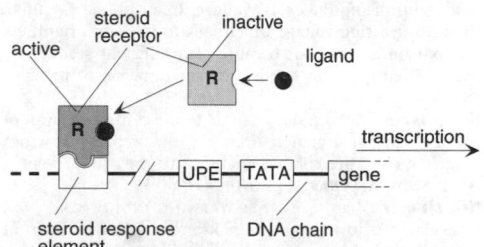

Steroid regulated gene The ligand binds to an inactive steroid receptor which is then able to bind to a steroid response element. This increases DNA transcription involving various upstream promoter elements (UPE and TATA box).

Upon interacting with ligand (●), a steroid receptor (R) becomes active and can then bind to a steroid response element (SRE), which may be close to or far upstream of the promoter. This binding leads to an increase in the rate of transcription, which may involve protein–protein interactions between receptor and transcription factors which recognize upstream promoter elements (UPEs), the TATA box binding factor or RNA polymerase II.

---

stet (*Print*) A former reader's mark in the margin of a proof indicating that the correction marked is to be ignored, replaced by a circled tick; in the text a dashed line is drawn under the original wording.

stethoscope (*Med*) An instrument for the study of sounds generated inside the human body.

Stevenson screen (*EnvSci*) A form of housing for meteorological instruments consisting of a wooden cupboard having a double roof and louvred walls, these serving to protect the instruments from sunlight and wind while permitting free ventilation. The base of the screen should be about 1 m above the ground.

sthene (*Genrl*) Unit of force in the metre–tonne–second system, equivalent to 10^3 N.

STI (*Med*) Abbrev for *sexually transmitted infection.*

stib- (*Chem*) Root word denoting *antimony*, from the Lt *stibium*. As in *stibine*, *stibnite*.

stibic (*Chem*) Referring to the (V) and (III) oxidation state of antimony respectively. Now normally *antimonous* and *antimonic*. Also *stibious*.

stibine (*Chem*) SbH$_3$. Antimony (III) hydride. A poisonous gas. Less stable than arsine.

stibnite (*Min*) Antimony sulphide, which crystallizes in grey metallic prisms in the orthorhombic system. It is sometimes auriferous and also argentiferous. It is widely distributed but not in large quantity, and is the chief source of antimony. Formerly *antimony glance*. Also *antimonite*.

stibious (*Chem*) See STIBIC.

stichtite (*Min*) A lilac or pink trigonal hydrated and hydrous carbonate of magnesium and chromium.

stick-and-rag work (*Build*) Plasterwork formed of canvas stretched across wooden frame and coated with a thin layer of gypsum plaster.

stick force (*Aero*) The force exerted on the CONTROL COLUMN by the pilot when applying aileron or elevator control.

stick-force recorder (*Aero*) A device attached to the control column of an aircraft by which the pilot's effort is measured and transmitted to a recording instrument.

sticking (*Build*) Shaping a STUCK MOULDING.

sticking probability (*Phys*) The probability of an incident particle, which reaches the surface of a nucleus, being absorbed and forming a compound nucleus.

sticking voltage (*Electronics*) The potential in an electron-beam tube above which electrons collected at screen cannot all be dispersed, leading to negative charge accumulating and neutralizing the excess voltage. In a cathode-ray tube, that accelerating voltage which fails to increase brightness of spot on a phosphor because of insufficient secondary-electron emission or conduction for dispersal of incident electrons.

stick pusher (*Aero*) A device fitted to the control column of some high-performance aircraft with swept-back wings which moves the column sharply forwards to prevent a stall. See STALL-WARNING INDICATOR.

stick shaker (*Aero*) See STALL-WARNING INDICATOR.

stick–slip motion (*Eng*) The motion of sliding surfaces of various materials, in which the force to start them moving is greater than the force to keep them moving.

sticky (*ICT*) Used for a website that tends to attract visitors and retain their attention.

sticky end (*BioSci*) The staggered cut made in DNA by some RESTRICTION ENZYMES. One chain is longer by one or two bases than the other. This single-stranded end is able to base-pair with a complementary end on another molecule cut by the same enzyme, causing the two molecules to stick together. Also *cohesive end*.

stiction (*Phys*) Abbrev for *static friction*. See FRICTION.

stiffened expanded metal (*Build*) See SELF-CENTRING LATHING.

stiffener (*Aero*) A member attached to a sheet for the purpose of restraining movement normal to the surface. Usually of thin-drawn or extruded light alloy, L-, Z- or U-section, attached by riveting, metal bonding or spot welding. See INTEGRAL STIFFENERS, STRINGER.

stiffener (*Eng*) A steel angle or bar riveted or welded across the web of a girder to stiffen it.

stiff lamb disease (*Vet*) (1) See WHITE MUSCLE DISEASE. (2) Arthritis of lambs due to infection by *Erysipelothrix insidiosa* (*E. rhusiopathiae*).

stiff neck (*Med*) See TORTICOLLIS.

stiffness (*Eng*) The ability to resist bending, or any type of elastic or viscoelastic deformation. Formally, it is the reciprocal of COMPLIANCE and equals the ratio of force to deflection.

stiffness control (*Acous*) In a mechanically vibrating system, the condition in which the motion is mainly determined by the stiffness of the retaining springs and negligibly by the resistance and mass of the system.

stiffness criterion (*Aero*) The relationship between the stiffness, strength and other structural properties which will prevent *flutter* or dangerous aero-elastic effects.

stiffness-to-weight ratio (*Eng*) See MERIT INDEX.

stiff ship (*Ships*) A ship with a large transverse metacentric height and short rolling period. Cf TENDER SHIP.

stiff sickness (*Vet*) See OSTEOMALACIA.

stifle (*Vet*) The femorotibial joint of animals.

stigma (*BioSci*) (1) Generally, a spot or mark of distinctive colour, as on the wings of many butterflies. (2) The part of the carpel of a flowering plant that is adapted for the reception and germination of the pollen. (3) In Protozoa, an eye spot. (3) In Arthropoda, one of the external apertures of the tracheal system. (5) In Urochordata, a gill slit. Pl *stigmata*.

stigmata (*Med*) Physical characteristics of a disease process or syndrome.

stigmator (*ImageTech*) A secondary cylindrical lens orientated at right angles to the primary to remove astigmatic errors in electronic image-forming systems, eg ELECTRON MICROSCOPES (panel).

stilb (*Phys*) Unit of luminance for a surface, equal to 1 cd cm^{-2} or 10^4 cd m^{-2}.

stilbene (*Chem*) *S-di-phenylethene*. C$_6$H$_5$CH=CHC$_6$H$_5$. Mp 125°C, bp 306°C. Can occur as two isomers, the *trans* solid form and the *cis* liquid form.

stilbestrol (*Pharmacol*) A drug used as an ESTROGEN. Also *stilboestrol*.

stilbite (*Min*) A zeolite; silicate of sodium, calcium and aluminium with chemically combined water; crystallizes in the monoclinic system, the crystals frequently being grouped in sheaf-like aggregates. Found both in igneous rock cavities and in fissures in metamorphic rocks. Also *desmine*.

stile (*Build*) An upright member in framing or panelling. Often incorrectly spelt *style*.

Stiles–Crawford effect (*Phys*) The effect that light entering the eye near to the margin of the pupil is less effective in producing a sensation of brightness than the same amount of light entering through the centre of the pupil.

still (*Chem*) Apparatus for the distillation of liquids, consisting of a reboiler, a fractionating column and arrangement for reflux.

stillage (*Print*) See PALLET.

still air range (*Aero*) The theoretical ultimate range of an aircraft without wind and with allowances only for take-off, climb to cruising altitude, descent and alighting.

Still's disease (*Med*) Acute polyarthritis of children (resembling rheumatoid arthritis), with fever and enlargement of the spleen and of the lymphatic glands.

still video back (*ImageTech*) An alternative back for a photographic camera, which enables it to operate as a STILL VIDEO CAMERA.

still video camera (*ImageTech*) A camera in which a still picture is recorded electronically on a magnetic floppy disk or card for immediate reproduction on a domestic TV receiver or monitor without any photographic processing.

still video recorder (*ImageTech*) Any device, tape or disk, which can record still video images such as a converted digital audio tape recorder.

stilpnomelane (*Min*) Monoclinic hydrated iron magnesium potassium aluminium silicate resembling biotite. It occurs in metamorphosed sediments and in iron ores.

stilted arch (*Arch*) An arch rising from points below its centre, and having the form of a circular arch above its centre.

stilt root (*BioSci*) See PROP ROOT.

stimulated Brillouin scattering (*ICT*) A non-linear effect in OPTICAL FIBRES in which, above a certain threshold, nearly all the light emitted by the transmitter is reflected back towards it. The effect is stronger for narrower spectral line widths; modulated systems are normally unaffected below powers of about 17 dBm.

stimulated emission (*Phys*) A process by which an incident photon of frequency v stimulates an atom to make a transition from energy E_2 to energy E_1 where

$$v = (E_2 - E_1)h$$

h being PLANCK'S CONSTANT. The atom is left in the lower energy state as *two* photons of the same frequency emerge, the incident one and the emitted one. An essential process in the operation of a laser. See panel on LASER.

stimulus (*Psych*) An aspect of an environment internal or external to the individual which produces some response, although this is not always an immediate response nor an easily observable one. Pl *stimuli.*

stimulus control (*Psych*) A phrase that refers to the relationship between a given stimulus and a given response. This relationship can be demonstrated by eliminating all other stimuli associated with that response and all other responses associated with that stimulus.

stimulus discrimination (*Psych*) The ability to differentiate between stimuli and thus respond to one and not the other.

stimulus generalization (*Psych*) The principle that when a subject has been conditioned to make a response to a stimulus, other similar stimuli will tend to evoke the same response, although to a lesser degree; the greater the similarity to the original stimulus, the greater this tendency will be.

stimulus–secretion coupling (*BioSci*) By analogy with *excitation–contraction coupling*, the events that link receipt of a stimulus with the release of materials from membrane-bounded vesicles. An important example is the release of neurotransmitter substances from the presynaptic terminal of a nerve.

stimulus threshold (*Psych*) The value of a quantified stimulus which elicits a particular response at a definite intensity. See ABSOLUTE THRESHOLD, DIFFERENCE THRESHOLD.

sting (*BioSci*) A sharp-pointed organ by means of which poison can be injected into an enemy or a victim, as the poisonous fin-spines of some fishes or the ovipositor of a worker wasp. See URTICARIA.

stinging hair (*BioSci*) An epidermal hair capable of injecting an irritating fluid into the skin of an animal when its tip is broken by contact, as in nettles.

stink damp (*MinExt*) Underground ventilation tainted by sulphuretted hydrogen.

stinkwood (*For*) A strong, hard, but non-durable timber from the lauraceous S African hardwood tree *Ocotea bullata* (also *Cape olive, Cape walnut*), known for its beautiful and distinctive figuring. It has an unpleasant smell when green.

stipe (*BioSci*) A stalk, esp: (1) of the fruiting body of a fungus; (2) the part connecting holdfast and lamina of a large algal thallus.

stipe (*Geol*) One of the branches of a fossil GRAPTOLITE.

stipes (*BioSci*) A stalk-like structure; an eyestalk. Pl *stipites.* Adjs *stipiform, stipitate.*

stipple (*Print*) See MECHANICAL STIPPLE.

stippling (*Build*) The operation of breaking up the smoothness of a paint, distemper, plaster or cement surface by dabbing it repeatedly with a hair or rubber *stippler.*

stipular trace (*BioSci*) The vascular tissue running into a stipule.

stipule (*BioSci*) In many dicotyledons, one of a pair of appendages that start development as outgrowths of the flank of a leaf primordium. Stipules often serve to protect the leaves in the bud and máture as leaf-like photosynthetic structures or as spines, scales, etc.

Stirling engine (*Eng*) An 'external' combustion reciprocating engine, patented by a Scottish clergyman, Robert Stirling, in 1827. It consists essentially of a cylinder in which two pistons (a working piston and a displacer) operate. When the air (or a suitable gas) in the cylinder is heated it expands, driving the working piston. The second piston transfers the air to a cold region for cooling; it is then recompressed by the working piston and transferred by the displacer to the hot region to start the cycle again. Such an engine is very much quieter and cleaner than a petrol or diesel engine. A modern version has been developed in the Netherlands in which helium under pressure is used as the working medium.

Stirling's approximation (*MathSci*)

$$n! = \sqrt{2\pi}\, n^{(n+\frac{1}{2})} \exp\left(-n + \frac{1}{12n} - \frac{1}{360n^3} + \cdots\right)$$

stirrup (*CivEng*) A vertical steel rod which loops together the top and bottom reinforcing bars of a reinforced concrete beam and helps to resist the shear.

stishovite (*Min*) A high-density form of silica. Synthesized at $1\cdot6 \times 10^{10}$ N m^{-2} and 1200°C, and also found occurring naturally in Meteor Crater, Arizona, in shock-loaded sandstone.

stitch (*Textiles*) (1) In sewing: fastening with thread carried through the fabric by a needle. A variety of stitches (eg buttonhole, chain, lock) are used for different purposes. (2) In knitting: a single loop.

stitch-bonded fibre (*Textiles*) A WEB of fibres stitched together to form a non-woven fabric.

stitch density (*Textiles*) The number of stitches per unit area in a knitted fabric.

stitch finish (*Textiles*) A finish, usually a lubricant, added to yarn or fabric to aid the penetration of the needle and thread in sewing.

stitching (*Print*) Joining the sections of an insetted book (see INSET (2)) along the back by means of thread or wire.

stitch welding (*ElecEng*) SEAM WELDING, using small mechanically operated electrodes, similar to a sewing machine.

STM-1 (*ICT*) Abbrev for SYNCHRONOUS TRANSPORT MODULE 1.

stoa (*Arch*) A covered colonnade or portico.

stochastic (*MathSci*) Developing in accordance with a probabilistic model; random.

stochastic noise (*Acous*) See RANDOM NOISE.

stock (*BioSci*) (1) Usually a rooted stem into which a SCION is placed in grafting. (2) The perennial part of a herbaceous perennial. (3) A strain maintained for breeding or propagation.

stock (*Build*) The principal part of a tool, eg the body of a plane, in which the cutting iron is held, or the stouter arm of a bevel, in which the blade is fastened.

stock (*ImageTech*) See RAW STOCK.

stock (*Print*) The general term for the material being printed: paper, board, foil, etc.

stock board (*Build*) A bottom made to fit the mould used in the handmoulding of bricks.

stock brush (*Build*) A brush used to moisten surfaces with water, prior to plastering, so that the surface will not absorb moisture from the plaster.

stock chest (*Paper*) A vessel, usually of upright cylindrical form, generally tile-lined, to contain stock from which the paper machine draws its supply. Fitted with a propeller or agitator to ensure a uniform suspension and even mixing of additives such as filler, size, colour, etc, which are usually added at this point.

Stockholm syndrome (*Psych*) A mental condition sometimes experienced by hostages and kidnap victims in which positive feelings develop towards their captors esp after their release. These feelings are difficult to reconcile with normal moral standards.

stocking density (*Agri*) The number of animals held per unit area of land at a given time.

stocking rate (*Agri*) The number of animals that can be raised in a given area over a specified time.

stockless anchor (*Ships*) A form of anchor in which there is no crosspiece on the shank and the arms are pivoted so that both of them can engage at the same time; the shank can be drawn into the hawsepipe of the ship.

stock pile (*MinExt*) (1) Temporarily stored tonnage of ore, middlings, concentrates or saleable products. (2) A country's holdings of strategic minerals.

stock rail (*CivEng*) The outer fixed rail against which the POINT works at a turn-out.

stocks (*Build*) Bricks which are fairly sound and hard-burned but are more uneven in colour than SHIPPERS; the bricks most used for ordinary building purposes.

stocks (*Ships*) The massive timbers supporting a ship in course of construction.

stockwork (*Geol*) An irregular mass of interlacing veins of ore; good examples occur among the tin ores of Cornwall and in the Erzgebirge. Ger *Stockwerk*.

stoichiometry (*Chem*) The determination of exact proportion of elements to make pure chemical compounds. Non-stoichiometric compouds or salts have non-integral numbers of atoms in their formulae. See VACANCIES.

stokes (*Phys*) The CGS unit of kinematic viscosity (10^{-4} m^2 s^{-1}). Symbol St.

Stokes–Adams syndrome (*Med*) Sudden loss of consciousness, with or without convulsions, in heart block.

Stokes' law (*Phys*) (1) Expression for the resisting force F offered by a fluid of dynamic viscosity η to a sphere of radius r, moving through it at steady velocity v:

$$F = 6\pi\eta\, rv$$

Hence it can be shown that the TERMINAL VELOCITY of a sphere of density ρ, falling under gravitational acceleration g through fluid of density ρ_0, is given by

$$v = \frac{2gr^2}{9\eta}(\rho - \rho_0)$$

Applies only for viscous flow with REYNOLDS NUMBER less than 0·2. (2) Incident radiation is at a higher frequency and shorter wavelength than the reradiation emitted by an absorber of that incident radiation.

Stokes layer (*Acous*) Very thin boundary layer along an interface between a fluid and a solid in which the velocity and temperature fluctuations in a sound wave are reduced because of friction and thermal conductivity respectively. Important for sound absorption. Also *AC-boundary layer*.

Stokes' line (*Phys*) A line in a spectrum satisfying STOKES' LAW, ie a line seen in the Raman spectrum on the long-wavelength side of the Rayleigh line when monochromatic light is scattered.

Stoke's radius (*BioSci*) The apparent radius of a molecule, sedimenting under centrifugal force, calculated from Stokes' law; a feature of the tertiary structure and thus informative.

Stokes' theorem (*MathSci*) The theorem that the surface integral of the curl of a vector function equals the line integral of that function around a closed curve bounding the surface, ie

$$\int_L \mathbf{E}\, dl = \int_S \operatorname{curl} \mathbf{E}\, ds$$

STOL (*Aero*) Abbrev for *short take-off and landing*, a term applied to aircraft with high-lift devices and/or deflected engine thrust enabling them to operate from small airstrips, 1000 ft (300 m) or less being the criterion.

stolon (*BioSci*) (1) See RUNNER. (2) An arching stem that forms a new rooted plant at the tip, eg blackberry. (3) A slender horizontally growing underground stem that forms a new plant at the end. (4) A tubular outgrowth in hydroid colonies of Cnidaria and Entoprocta from which new individuals or colonies may arise. Adj *stolonate*.

stoma (*BioSci*) Any small aperture. Used specifically for a pore in the epidermis of a leaf or stem etc of a vascular plant, of variable aperture, surrounded and controlled by two GUARD CELLS and providing regulated gas exchange between the tissues and the atmosphere. Also *stomate*. Pl *stomata*, *stomates*.

-stoma (*Genrl*) Suffix from Gk *stoma*, mouth, applied esp in zoological nomenclature. Pl *-stomata*.

stomach (*BioSci*) In vertebrates, the sac-like portion of the alimentary canal between the oesophagus and the intestine. The term is loosely applied in invertebrates to any sac-like expansion of the gut behind the oesophagus. Adj *stomachic*.

stomach insecticide (*Chem*) One acting on ingestion and applicable only to insects which eat as distinct from sucking insects which draw food in liquid form from host plant or animal; may be used on foliage against leaf-eating insects, or as poison-bait ingredient against locusts etc. Lead arsenate, DDT, Gammexane.

stomatal (*Genrl*) Adj from *stoma*, an aperture. Also *stomate*, *stomatiferous*, *stomatose*, *stomatous*.

stomatal complex (*BioSci*) A stoma with its guard cells and any subsidiary cells.

stomate (*BioSci*) See STOMA, STOMATAL.

stomatiferous (*Genrl*) See STOMATAL.

stomatitis (*Med*) Inflammation of mucous membrane of the mouth.

stomatitis (*Vet*) See HORSE POX.

stomatogastric (*BioSci*) Pertaining to the mouth and stomach; said esp of that portion of the autonomic nervous system which controls the anterior part of the alimentary canal.

stomatose (*Genrl*) Also *stomatous*. See STOMATAL.

stomium (*BioSci*) A part of the wall of a fern sporangium composed of thin-walled cells where splitting begins during dehiscence.

stomodaeum (*BioSci*) That part of the alimentary canal which arises in the embryo as an anterior invagination of ectoderm. Cf MIDGUT, PROCTODAEUM. Adj *stomodaeal*.

-stomy (*Genrl*) Suffix from Gk *stoma*, mouth, referring esp to the formation of an opening by surgery.

stone (*Genrl*) Unit of mass equal to 14 lb (0·4536 kg).

stone (*Print*) The smooth, milled cast-iron surface on which formes are locked up.

stone cell (*BioSci*) A more or less isodiametric sclereid, eg in the fruit of the pear. Also *brachysclereid*.

stone head (*MinExt*) (1) First solid rock met while sinking a shaft or drill hole, also *rock head*. (2) A heading or tunnel in stone.

Stoner–Wohlfarth model (*Phys*) A model describing the magnetization curves of an aggregation of single-domain particles.

stone saw (*Build*) (1) A smooth-faced blade which in use is fed with an abrasive such as sand, carborundum or diamond powder, as it cuts its way through stone. (2) Diamond-tipped circular saw.

stone tongs (*Build*) An accessory used in hoisting blocks of stone. It resembles a large pair of scissors with the points curved inwards. These clip into the sides of the block, while chains connect the loops of the tongs to the hoisting ring. Also *nippers*.

stoneware (*Build*) A material used for some sanitary fittings etc; made from plastic clays of the Lias formation, with a small amount of sharp sand etc, added to reduce shrinkage.

stoneware (*Glass*) A form of ceramic ware which is dense, impermeable and hard enough to resist scratching with a steel point, but differs from PORCELAIN in being more opaque and having a smaller proportion of glassy phase. Cf BONE CHINA, EARTHENWARE, FAIENCE, TERRACOTTA.

stoneworts (*BioSci*) See CHARALES.

stony meteorites (*Geol*) Those meteorites which consist essentially of rock-forming silicates. See ACHONDRITE, AEROLITES, CHONDRITE.

stool (*BioSci*) A tree or shrub cut back to ground level and allowed to produce a number of new shoots, as in COPPICE management, as a method of managing fruit bushes (cf LEG) or to provide shoots for making cuttings etc.

stool (*Med*) The FECES from one bowel movement.

stoop (*Build*) A low platform outside the entrance door of a house. Also *stoep*.

stop (*Build*) (1) A projecting piece set in the top of a bench at one end and adjustable for height. It is used to steady work which is being planed. (2) An ornamental termination to a stuck moulding. (3) See DOOR STOP.

stop (*ImageTech, Phys*) (1) Circular opening which sets the effective aperture of a lens, eg the iris of a camera or the rim of the objective in a telescope. Also *diaphragm*. (2) An F-NUMBER, esp as marked on the iris scale of a camera lens.

stop band (*ICT*) The frequency band in which a filter highly attenuates signals; in this band, its impedance is highly reactive, causing incoming signals to be reflected.

stop-bath (*ImageTech*) See ACID STOP.

stop bit (*ICT*) In asynchronous data transfer, one of a sequence of bits which signal the end of a character or transmission. See ASYNCHRONOUS TRANSFER MODE.

stop-cock (*Build*) A short pipe opened or stopped by turning a key or handle.

stop codon (*BioSci*) Specific triplet sequences, which in mRNA do not code for an amino acid but cause protein synthesis to stop. They are UAA (ochre), UAG (amber) and UGA (opal). Also *chain terminator*. See AMBER MUTATION and panel on DNA AND THE GENETIC CODE.

stop-cylinder (*Print*) A letterpress printing machine in which the cylinder makes one revolution and prints a sheet as the bed travels in the printing direction, and then remains stationary while the bed returns to its starting position. Cf SINGLE-REVOLUTION, TWO-REVOLUTION.

stop down (*ImageTech*) To reduce the working aperture of a lens in order to increase depth of field and/or exposure time.

stope (*MinExt*) (1) To excavate ore from a reef, vein or lode. (2) Space formed during extraction of ore underground. Types are flat, open, overhand, rill, shrinkage and underhand, with variations to suit shape, geology and size of deposit. See fig. at MINING.

stop-frame animation (*ImageTech*) Shooting one or two frames of film or video for each change of the artwork or model.

stoping (*Geol*) A mining term applied by R A Daly to a process in the emplacement of some igneous rock bodies, by which blocks of the overlying country rock are wedged off and sink into the advancing magma.

stop-motion (*ImageTech*) Of or relating to *stop-motion animation*, a technique in which filming is repeatedly stopped to allow very slight changes of position in the subjects being filmed, creating the illusion of movement when the film is run.

stop moulding (*Build*) A STUCK MOULDING terminating in a stop.

stopped end (*Build*) A square end to a wall.

stopped heading (*Print*) In paper ruling, the heading for account books and other work may require the ruling to stop at more than one stage both vertically and horizontally.

stopped mortise (*Build*) See BLIND MORTISE.

stopped pipe (*Acous*) See CLOSED PIPE.

stopper (*Electronics*) A simple circuit element (resistance, or resistance and capacitance in combination) to obviate parasitic oscillations.

stopping (*Build*) Plastic material used to fill holes and cracks in timber, eg before painting.

stopping equivalent (*Phys*) The thickness of a standard substance which would produce the same energy loss as the absorber under consideration. The standard substance is usually air at stp but can be eg Al, Pb, H_2O. See AIR EQUIVALENT.

stopping motions (*Textiles*) Electrical or mechanical devices employed on many textile machines when a fault develops in raw material feeding arrangements (openers, scutchers, spinning-frames, etc) or a yarn breaks in winding, warping or weaving.

stopping-off (*ElecEng*) Coating a conducting surface with a resist to prevent electrodeposition.

stopping-out (*ImageTech*) See BLOCKING-OUT.

stopping-out (*Print*) (1) Painting out with a protecting varnish the darker tones of a half-tone block, in stages during the etching, to obtain the best rendering of the subject. (2) Painting out unwanted background or detail on a negative before printing down. (3) Painting out non-image areas of the mesh in screen printing.

stopping potential (*Electronics*) Reverse difference of potential required to bring electrons to rest against their initial velocity from either thermal or photoelectric emission.

stopping power (*Phys*) Energy loss resulting from a particle traversing a material. The *linear stoppage power* S_L is the energy loss per unit distance and is given by $S_L = -dE/dx$, where x is path distance and E is the kinetic energy of the particle. The *mass stopping power* S_M is the energy lost per unit surface density traversed and is given by $S_M = S_L/\rho$, where ρ is the density of the substance. If A is taken as the ram of an element and n the number of atoms per unit volume, then the *atomic stopping power* S_A of the element is defined as the energy loss per atom per unit area normal to the motion of the particle, and is given by $S_A = S_L/n = S_M A/N$, where N is AVOGADRO'S NUMBER. The *relative stopping power* is the ratio of the stopping power of a given substance to that of a standard substance, eg air or aluminium.

stop press (*Print*) See FUDGE.

stop valve (*Eng*) The main steam valve fitted to a boiler to control the steam supply and to allow isolation of the boiler from the main steam pipe.

stop watch (*Genrl*) A watch, usually having seconds and minutes hands only, which is started and stopped by pressure of the winding knob. The normal type reads to $\frac{1}{5}$ s, special types read to $\frac{1}{50}$ s. See CHRONOGRAPH.

storage (*ICT*) General term covering all units of computer equipment used to store data (and programs). Also *memory*.

storage (*NucEng*) The keeping of radioactive waste material in a facility, either made specially or naturally occurring, with the intention of treating it further or of *disposing* of it, normally elsewhere. Storage implies further treatment, disposal does not.

storage battery (*ElecEng*) See ACCUMULATOR.

storage capacity (*ICT*) The maximum number of *bits* that can be stored, located and recovered in the main memory of a computer.

storage disorder (*Med*) Any disease in which a metabolic defect results in the abnormal accumulation of a substance in the body (eg fat, carbohydrate, protein or iron).

storage element (*ICT*) One unit in a memory, capable of retaining 1 BIT of information.

storage factor (*Phys*) See Q.

storage heater (*ElecEng*) A heater with large thermal capacity, used to store heat during off-peak periods and release it over a longer period. A fan may be used to increase the rate of heat output.

storage modulus (*Eng*) See COMPLEX MODULUS.

storage oscilloscope (*Electronics*) One in which a trace is retained indefinitely or until deliberately wiped off, incorporating a STORAGE TUBE.

storage tube (*Electronics*) That which stores charges deposited on a plate or screen in a cathode-ray tube, a subsequent scanning by the electron beam detecting, reinforcing or abolishing the charge.

store (*ICT*) See MEMORY.

store (*ImageTech*) A device in which signal information can be accumulated for subsequent retrieval, eg two interlaced fields may be stored and released in the sequential order as a complete frame.

store-and-forward (*ICT*) A service allowing messages or data to be transmitted immediately even when a complete path to the receiving terminal is not available at the time of transmission. Data are held at an intermediate node until final delivery can be effected.

store cattle (*Agri*) Animals raised to achieve full skeletal growth but with muscle development somewhat below full potential. They are usually sold on to finishers to gain muscle mass to commercial levels.

stored program (*ICT*) In computer design, the fundamental idea that a program can be stored in the same way as data. See FIRST-GENERATION COMPUTER.

stored-program control (*ICT*) Said of automatic, mostly fully electronic, telephone exchanges in which switching functions are controlled by stored logic and traffic-handling capacity is optimized in accord with current conditions.

store lambs (*Agri*) Lambs not sold for slaughter by the autumn of their first year but kept for sale or finishing in the following spring.

store location (*ICT*) Basic unit within a MAIN MEMORY, capable of holding a single BYTE or WORD. Also *cell*.

storey rod (*Build*) A pole on which is marked the level of the courses and which gives a guide to the ultimate level.

storied (*BioSci*) A term describing a vascular cambium and the secondary xylem or wood derived from it with the cells arranged in horizontal tiers (ie with the end walls more or less aligned). Also *stratified*.

storied cork (*BioSci*) A protective layer of suberized cells that develops around the stems of woody monocotyledons (eg palms) in which the cells occur in radial files, each file of several cells being derived from a single precursor. Also *protective cork*.

storm (*ICT*) See IONOSPHERIC STORM.

storm-centre (*EnvSci*) The position of lowest pressure in a cyclonic storm.

storm window (*Build*) (1) A window arranged with double sashes enclosing air, which acts as a sound and heat insulator. (2) A small upright window set in a sloping roof surface so as not to project beyond it. Cf DORMER.

stover (*Agri*) Dried stalks of field crops that are left after harvest and seed removal and are used as animal feed.

stoving (*Build*) Industrial process for quickly drying specially formulated paints by radiant or convected heat above 180°F (82°C). For lower temperatures, see FORCE DRYING.

STOVL (*Aero*) See SHORT TAKE-OFF AND VERTICAL LANDING.

stowage factor (*Ships*) The space required to contain unit weight of a commodity allowing for all packing, dunnage and unavoidable lost space between units. Usually, measured in cubic feet per ton or cubic metres per tonne.

stp (*Chem, Phys*) Abbrev for *standard temperature and pressure*: a temperature of 0°C and a pressure of 101 325 N m^{-2}. Also *STP*.

STP cable (*ICT*) Abbrev for SHIELDED TWISTED PAIR CABLE.

STR (*BioSci*) Abbrev for SHORT TANDEM REPEAT.

strabismus (*Med*) Squint. A condition in which the visual axes of the eyes assume an abnormal position relative to each other.

strabotomy (*Med*) The surgical operation of curing STRABISMUS by dividing one or more muscles of the eye.

straddle milling (*Eng*) The use of two or more side-cutting milling-cutters on one arbor so as to machine eg both side faces of a workpiece at one operation.

straddle scaffold (*Build*) See SADDLE SCAFFOLD.

straggling (*Phys*) Variation of range or energy of particles in a beam passed through absorbing material, arising from random nature of interactions experienced. Additional straggling may arise from instrumental effects such as noise, source thickness and gain instability.

straight angle (*MathSci*) An angle of 180° or π radians.

straight arch (*Build*) See FLAT ARCH.

straight-bar machine (*Textiles*) A knitting machine, with bearded needles on a movable bar, used to make plain or rib-knitted shaped articles.

straight eight (*Autos*) An eight-cylinder in-line engine, as distinct from an eight-cylinder V-type engine.

straighteners (*Aero*) See HONEYCOMB.

straight-flute drill (*Eng*) A conical pointed drill having backed-off cutting edges, formed by cutting straight longitudinal flutes in the shank; more rigid than a twist drill and often used for soft metals.

straight joint (*Build*) A continuity of vertical joints in brickwork.

straight-line capacitor (*ICT*) Variable capacitor whose value varies linearly with scale reading.

straight-line frequency capacitor (*ICT*) Variable capacitor whose value is inversely proportional to the square of the scale reading, so that the frequency of the circuit that it tunes is directly proportional thereto.

straight-line wavelength capacitor (*ICT*) Variable capacitor whose value is proportional to square of scale reading, so that it can tune a circuit with a linear relationship between scale reading and wavelength.

straight-pane hammer (*Eng*) A fitter's hammer the head of which has a flat striking face at one end and a blunt chisel-like edge, parallel with the shaft, at the other.

straight receiver (*ICT*) See TUNED RADIO-FREQUENCY RECEIVER.

straight run (*Print*) Running a web-fed press without cylinder collection and sometimes without turner bars to give the maximum number of copies to the cylinder revolution.

straight tongue (*Build*) A wooden tongue for a PLOUGHED-AND-TONGUED JOINT cut so that the grain is parallel to the grooves.

straight-type cable (*ElecEng*) A cable which has oil-impregnated paper as the dielectric. Used up to 66 kV in the form of single-core H-type cable. Also *solid-type cable*.

strain (*BioSci*) A variant group within a species, often breeding true and maintained in culture or cultivation, with more or less distinct morphological, physiological or cultural characteristics. The term is not used in formal taxonomy.

strain (*Eng, Phys*) When a material is distorted by forces acting on it, it is said to be in a state of strain, or strained. Strain is the ratio

$$\frac{\text{change in dimension}}{\text{original dimension}}$$

and thus has no units. The main types of strain are DIRECT (tensile or compressive) STRAIN:

$$\frac{\text{elongation } or \text{ contraction}}{\text{original length}}$$

SHEAR STRAIN:

$$\frac{\text{deflection in direction of shear force}}{\text{distance between shear forces}}$$

VOLUMETRIC (or bulk) STRAIN:

$$\frac{\text{change of volume}}{\text{original volume}}$$

These definitions, known as *engineering strains*, are strictly only applicable to infinitesimal strains, and cannot be added at large deformations, where *true* (logarithmic or natural) *strain* is more convenient. For direct strains, true strain is

$$\log_e(1 + \text{engineering strain})$$

which is additive. For the large elastic deformations of elastomers, the *extension* (or elongation) *ratio*, or

$$\frac{\text{deformed length}}{\text{original length}}$$

is used.

strain age embrittlement (*Eng*) Loss of ductility and fracture toughness resulting from strain ageing processes in cold-worked metals.

strain ageing (*Eng*) An increase in metal strength and hardness that proceeds with time, after cold-working. It takes place slowly at ambient temperature and is accelerated by heating. It is most pronounced in iron and steel, but also occurs in other metals. May result in STRAIN AGE EMBRITTLEMENT.

strain birefringence (*BioSci*) See BIREFRINGENT.

strain crystallization (*Chem, Eng*) Crystallization of elastomers when strained to high extension ratios ($\lambda > 5$). The crystallites are aligned along the strain axis and increase the breaking strength of the rubber. On retraction, the crystallites melt so the effect is reversible. The heat given out in stretching is partly due to the JOULE–THOMSON EFFECT, and partly to the heat of crystallization. The hot-stretching of fibres exploits strain crystallization.

strain disk (*Glass*) A glass disk of calibrated birefringence, used as a comparative measure of the degree of annealing of glass.

strain energy (*Eng*) The energy stored in an elastically deformed body equal to the work done on the body in deforming it (ie the area under the elastic portion of the force–deflection curve).

strain energy release rate (*Eng*) The elastic strain energy required for unit area of crack propagation. At its critical value, it equals the TOUGHNESS, G_c, of the material. Also CRACK DRIVING FORCE.

strainer (*Paper*) Any piece of equipment intended to clean pulp, stuff or stock by passing it through metal plates containing perforations or slits of appropriate size so that larger particles are held back and removed.

strain gauge (*Eng*) Metal or semiconductor filament on a backing sheet by which it can be attached to a body to be subjected to strain, so that the filament is correspondingly strained. The strain alters the electrical properties of the filament which is the basis of measurement.

strain-hardening (*Eng*) Increase in resistance to deformation (ie in hardness) produced by deformation. See COLD-WORKING, WORK-HARDENING.

straining sill (*Build*) A piece of scantling lying on the tie-beam of a timber roof and butting against the feet of the queen posts, or between the feet of the queens and princesses to keep them apart.

strain insulator (*ElecEng*) An insulator inserted in the span wire of an overhead contact-wire system.

strain point (*Glass*) One of the reference temperatures in glass-making. See panel on GLASSES AND GLASS-MAKING.

strain–slip cleavage (*Geol*) A cleavage in which the cleavage planes are parallel shear planes; between each pair the rocks are puckered into small sigmoidal folds.

strain viewer (*Phys*) Eyepiece or projection unit of a polariscope.

strain whitening (*Eng, Plastics*) Yielding effect noticed in tough polymers. Also stress whitening. See panel on RUBBER TOUGHENING.

strait work (*MinExt*) (1) Narrow headings in coal. (2) A method of working coal by driving parallel headings and then removing the coal between them.

strake (*MinExt*) Gently sloped, flat table used for catching grains of heavy waterborne mineral. See BLANKET STRAKE, TYE.

strake (*Ships*) A row of plates positioned end to end.

strand (*ElecEng*) One of several wires which together constitute a stranded conductor.

stranded cable (*ElecEng*) One whose core (or cores) consists of stranded conductor.

stranded caisson (*CivEng*) A watertight box, having a solid floor, which is floated over the site where a bridge pier is to be constructed. Construction goes on in the dry on the floor of the box, which sinks finally to a previously levelled bed under water, the sides of the box being kept always above water. Also *American caisson*.

stranded conductor (*ElecEng*) One woven from individual wires or strands, like a rope.

stranding effect (*ElecEng*) An increase (20–30%) of the stress at the surface of the conductor, caused by stranding. A usual increase is 25%. Stranding effect is overcome by sector-shaped conductors or by lead sheathing the conductor.

strand plant (*BioSci*) A seashore plant growing just above the normal upper limit of the tide.

strange (*Phys*) One of the six FLAVOURS of QUARKS with a mass of 150 MeV and a charge of $-e/3$. Also the property of the strange quark (strangeness), which is 1 for the strange and -1 for the antistrange quark, and zero for all other particles.

stranger anxiety (*Psych*) A common fear infants have of unfamiliar people, onset usually the end of the first year, until the child is 2 years old or so; related to separation anxiety in that the two tend to co-occur.

strangler (*Autos*) US for CHOKE (2).

strangles (*Vet*) A contagious disease of horses, due to infection by *Streptococcus equi*, characterized by rhinitis and suppurative adenitis. Uncommonly, purpura haemorrhagica, guttural pouch empyema or laryngeal hemiplegia may follow.

strangury (*Med*) Slow and painful urination.

strap (*Build*) A metal plate or band securing timbers together at a joint.

S-trap (*Build*) A trap used in sanitary pipes in which the outlet leg is parallel with the inlet leg.

strapdown (*Aero*) Any device mounted to an aircraft so that its attitude changes with that of the aircraft, eg navigation systems like FIBRE-OPTICS GYROS. In contrast, ordinary gyroscopes maintain a constant attitude.

strap hinge (*Build*) A hinge having one long leaf for securing to a heavy door or gate. Also *joint hinge*.

strapping (*Build*) A general term for battens fixed to the internal faces of walls as a support for laths and plaster or for plasterboard.

strapping (*ICT*) Alternate connection of segments in a magnetron, to stabilize phases and mode of resonance in the cavities.

strapping wires (*ElecEng*) Parallel single-wire connections between a pair of two-way electric-light switches for dual control of a lighting point.

strass (*Glass*) A very dense glass of high refractive power; used largely in making artificial jewellery. Also *paste*.

Strategic Defense Initiative (*Aero*) A military programme commonly referred to as *Star Wars*. Strategic Defense Initiative was intended to provide a defensive shield based on satellite, laser and high-energy particle technology for the destruction of incoming hostile ballistic missiles while they were still in the atmosphere. Now abandoned. Abbrev *SDI*.

strategic minerals (*Genrl*) Minerals considered essential for the security of a nation but not available in sufficient quantity from domestic sources in time of war.

strategy, *K* and *r* (*BioSci*) See K-STRATEGIST, R-STRATEGIST.

stratification (*BioSci*) (1) Banding seen in thick cell walls, due to presence of wall layers differing in water content, chemical composition and physical structure. (2) Grouping of vegetation into two or more fairly well-defined layers of different height, as trees, shrubs and ground vegetation in a wood. (3) Vertical structure or layering within a terrestrial or aquatic environment. (4) Method of breaking dormancy period of seeds by storage in moist sand, often at around 4°C.

stratification (*Geol*) The layering in sedimentary rocks due to chemical, physical or biological changes in the sediment. Also *bedding*. See LAMINATION.

stratification (*MathSci*) The division of a population to be sampled into subsets, within each of which a sample of observations will be taken.

stratified (*BioSci*) Multilayered. Also storied. See STRATIFIED EPITHELIUM.

stratified charge combustion (*Autos*) A technique in gasoline engines for burning mixtures too weak to be ignited by normal spark ignition. A pocket of enriched mixture is provided close to the sparking plug which ignites normally and fires the remainder.

stratified epithelium (*BioSci*) A type of epithelium consisting of several layers of cells, the outer ones flattened and horny, the inner ones polygonal and protoplasmic.

stratiform (*BioSci*) Arranged in layers. Also *stratose*.

stratigraphical break (*Geol*) The geological record is incomplete, the succession of strata being broken by unconformities and non-consequences, these representing longer or shorter periods of time during which no sediment was deposited or erosion predominated.

stratigraphical level (*Geol*) See HORIZON.

stratigraphic column (*Geol*) See panel on GEOLOGICAL COLUMN.

stratigraphic facies (*Geol*) See FACIES, STRATIGRAPHIC.

stratigraphic trap (*Geol*) Petroleum reservoir in which lenticular bands of porous sandstone pass laterally and vertically into impervious clay or shale. More difficult to locate than structural traps.

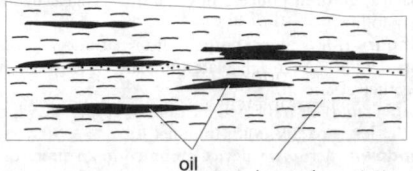

oil

impervious strata with sand lenses

stratigraphic trap Gas not shown.

stratigraphy (*Geol*) The definition and description of the stratified rocks of the Earth's crust, their relationships and structure, their arrangement into chronological groups, their lithology and the conditions of their formation, and their fossil contents. The subject does not exclude igneous and metamorphic rocks where these are part of the succession.

stratocumulus (*EnvSci*) Grey and/or whitish patch, sheet or layer of cloud which almost always has dark parts, composed of tessellations, rounded masses, rolls, etc, which are non-fibrous (except for VIRGA) and which may or may not be merged; most of the irregularly arranged small elements have an apparent width of more than 5°. Abbrev *Sc*. See panel on ATMOSPHERIC BOUNDARY LAYER.

stratopause (*EnvSci*) The top of the stratosphere, at about 50–55 km above the surface of the Earth. See panel on STRATOSPHERE AND MESOSPHERE.

stratose (*BioSci*) See STRATIFORM.

stratosphere (*EnvSci*) The region of the atmosphere between the TROPOPAUSE and the STRATOPAUSE, in which temperature generally increases with height. See panels on STRATOSPHERE AND MESOSPHERE and TROPOSPHERE.

stratotype (*Geol*) The type representative of a named stratigraphic unit or of a stratigraphic boundary.

stratum (*BioSci*) A layer of cells; a tissue layer.

stratum (*Geol*) A single bed of rock bounded above and below by divisional planes of STRATIFICATION. A stratum differs from a lamination only in thickness. Pl *strata*. Adj *stratified*.

stratum contours (*Geol*) Contours drawn on the surface of a bed of rock. The position of the outcrop of the bed can be predicted from the intersection of these contours with the surface of the ground.

stratum corneum (*BioSci*) The outer layers of the skin in vertebrates where the cells are flattened and often reduced to dead keratinized shells that will later be shed as squames.

stratum germinativum (*BioSci*) The lowest layer of the skin in vertebrates consisting of a single layer of cells that

proliferate, one of the daughter cells moving outwards, the other retaining proliferative capacity. Basal cell carcinomas derive from this layer. Also *basal layer, Malpighian layer*.

stratum granulosum (*BioSci*) The layers of cells lying immediately above the basal layer in vertebrate epidermis in which the cells are progressively more keratinized as they progress outwards towards the stratum corneum.

stratum lucidum (*BioSci*) A thin, clear layer of dead keratinocytes lying below the *stratum corneum* of thick skin.

stratum Malpighii (*BioSci*) See STRATUM GERMINATIVUM.

stratus (*EnvSci*) Generally grey cloud layer with a fairly uniform base, which may give drizzle, ice prisms or snow grains. When the Sun is visible through the cloud, its outline is clearly discernible. Stratus does not produce halo phenomena, except possibly at very low temperatures, and sometimes it appears in the form of ragged patches. Abbrev *St*.

straw (*Agri*) Ripened, harvested plant stems after the seeds or grains have been removed.

strawberry footrot (*Vet*) A proliferative dermatitis of sheep affecting particularly the lower parts of the legs; caused by the fungus *Dermatophilus pedis*.

stray capacitance (*Electronics*) Any occurring within a circuit other than that intentionally inserted by capacitors, eg capacitance of connecting wires, giving rise to PARASITIC OSCILLATION.

stray field (*ElecEng*) A magnetic field set up in the neighbourhood of electric machines or current-carrying conductors, which serves no useful purpose and which may interfere with the operation of measuring instruments etc.

stray flux (*ElecEng*) The leakage flux in an ac machine or transformer.

stray induction (*ElecEng*) The equivalent induction of the leakage flux effective in producing a reactive voltage drop.

stray losses (*ElecEng*) The stray load losses of an electric machine, due to STRAY FIELDS and harmonic flux pulsations in the iron circuit and EDDY CURRENTS in the windings.

stray radiation (*Phys*) Direct and secondary radiation from irradiated objects which is not serving a useful purpose.

stray resonance (*ElecEng*) That arising from unwanted inductance and capacitance, eg in leads between conductors, in leads inside canned capacitors, between turns of inductors.

strays (*ICT*) US for ATMOSPHERICS.

streak (*BioSci*) An elongated chlorotic or necrotic spot as a symptom of virus infection. Also *stripe*.

streak (*Min*) The colour of the powder obtained by scratching a mineral with a knife or file or by rubbing the mineral on paper or an unglazed porcelain surface (*streak plate*). For some minerals, this differs from the body colour.

streaking (*ImageTech*) A defect of the TV image showing a trail at the image boundary instead of a sharp transition. Also *smearing*.

stream anchor (*Ships*) An anchor of lighter weight than a bow anchor, used at the stern.

stream factor (*ChemEng*) That proportion of the time, during which a complete plant is in operation, that any individual item of the plant is working.

stream feeder (*Print*) An automatic feeder in which a sheet is separated and lifted at its rear edge and moved forward to the feed board so that its front edge is placed beneath the previous sheet. In this way a continuous stream of paper moves forward to the front and side lays at a moderate speed relative to the printing speed of the press.

streaming (*BioSci*) Flowing of cytoplasm, either unidirectionally (eg in a growing fungal hypha) or in a circulation (eg cyclosis).

streaming effect (*NucEng*) See CHANNELLING EFFECT.

streaming potential (*Chem*) The difference of electrical potential induced between the two ends of a capillary by forcing a liquid through it.

Stratosphere and mesosphere

Above the TROPOSPHERE is the *stratosphere*, a region where, in contrast, temperature increases with height and it is therefore one of great static stability. The stratosphere accounts for most of the atmospheric mass not in the troposphere, and is bounded at about 50 km by the STRATOPAUSE. Above this lies the *mesosphere* which together with the stratosphere is known as the *middle atmosphere*. This is a region where complex photochemical reactions, including the formation and destruction of ozone, take place under the influence of solar radiation. Here chlorine and chlorofluorocarbons (CFCs) and other human-made pollutants exert a widespread and deleterious effect by changing the rate of these reactions. Recently, therefore, there has been a greatly increased interest in the middle atmosphere.

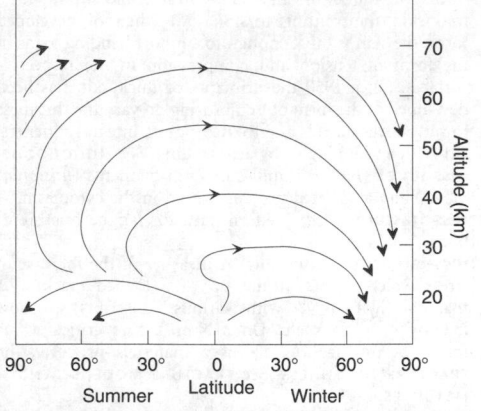

Schematic diagram of stratospheric circulation

The chief energy inputs to the middle atmosphere are (1) the difference between the absorption of short-wave radiation (particularly by ozone) and the emission of long-wave radiation (chiefly by carbon dioxide and ozone); (2) air movements in the troposphere that propagate energy upwards through the tropopause, most significant being the very long, quasi-stationary planetary waves occurring in the winter hemisphere, and the gravity waves including atmospheric tides and mountain lee-waves.

The main features (see diagram) of the general stratospheric circulation are the *polar night jet*, a strong cyclonic vortex surrounding the winter pole with a maximum of wind speed in middle latitudes, a moderate anticyclonic vortex centred on the summer pole, covering much of the summer hemisphere and with a relatively weak circulation at the equinoxes. In addition, there is a semi-annual oscillation of the zonal wind field which is especially marked at the equator, the QUASI-BIENNIAL OSCILLATION (QBO) which is a strong oscillation of the zonal wind in the tropics with an irregular period of about 26 months and, in winter, in the northern hemisphere SUDDEN WARMING in which the polar vortex occasionally, but not every year, breaks down completely with mean zonal winds becoming easterly and temperatures rising by up to 50 K in a few days.

In mid-winter, sudden warmings are usually followed by a re-establishment of the normal cyclonic vortex, but in the late winter they mark the beginning of the summer circulation regime. In the southern hemisphere, the polar winter vortex is more nearly circular than in the northern, has stronger winds, is colder, is not subject in the same way to violent distortions, such as reversible sudden warmings, and lasts until well after the equinox. These differences are due to the widely different distributions of land, sea and mountains in the two hemispheres.

See panels on ATMOSPHERIC BOUNDARY LAYER, ATMOSPHERIC POLLUTION, EARTH and TROPOSPHERE.

---

streamline (*Phys*) A line in a fluid such that the tangent at any point follows the direction of the velocity of the fluid particle at the point, at a given instant. When the streamlines follow closely the contours of a solid object in a moving fluid, the object is said to be of streamline form.

streamline burner (*Eng*) See FANTAIL BURNER.

streamline flow (*Phys*) Path taken by fluid molecules or minute suspended particles. Usually qualified as LAMINAR FLOW or TURBULENT FLOW. See VISCOUS FLOW.

streamline motion (*Phys*) The steady motion of a fluid in LAMINAR FLOW past a body with neither abrupt changes in direction nor close curves.

streamlines (*EnvSci*) A set of lines on a chart showing the direction of the horizontal wind at some particular level.

streamline wire (*Aero*) High-tensile steel wire of elliptical, not true streamline, cross-section, used to reduce the drag of external bracing wires. Fitted principally to biplanes and some early types of monoplane.

stream order (*Geol*) Classification of streams according to their hierarchical position in the drainage network. In the most widely used system, first-order streams (1) are the outermost tributaries, which join together to form a second-order stream (2). Only streams of the same order, eg 2, 2, can combine to produce a stream of higher order. Streams 2, 2, would produce a stream of order 3 below their confluence, but streams 2, 3, would produce a stream that was still third order. See DRAINAGE PATTERNS.

stream sampling (*MinExt*) Sampling of stream or river water to identify anomalous concentrations of dissolved metals; in gravels to find chemical or mineral concentrations, eg *tracers*. See GEOCHEMICAL PROSPECTING.

stream tin (*Min*) Cassiterite occurring as derived grains in sands and gravels in the beds of rivers.

street ell (*Build*) An elbow with a male thread at one end.

streets (*Print*) See RIVERS.

strength measures (*Eng*) There are a large number of quantities used to characterize a material's resistance to failure. Few have as well-defined properties as eg Young's modulus or electrical resistivity; many depend on details of how they are measured, specimen geometry and, esp for brittle materials, on the flaw size distribution in the surface. One group of these quantities is based around the tensile test, schematic results from which are sketched in the figure SCHEMATIC TENSILE STRESS–STRAIN CURVES. See ELASTIC LIMIT, FRACTURE STRESS, TENSILE STRENGTH, YIELD STRESS and panel on FATIGUE.

strengths of acids (*Chem*) The extent to which an acid dissociates in a given solvent, usually water. The strengths of acids may be related to the structures of the dissociated and undissociated forms.

strength-to-weight ratio (*Eng*) See MERIT INDEX.

strepto- (*Genrl*) Prefix from Gk *streptos* denoting bent, flexible, twisted quality.

streptococcus (*BioSci*) A Gram-positive coccus of which the individuals tend to be grouped in chains. Many forms possess species-specific capsular polysaccharides by which they can be divided into groups, which include the causative agents of scarlet fever, erysipelas and one form of mastitis. Other streptococci include *Diplococcus pneumoniae* (one cause of pneumonia). Some types occur normally in the mouth, throat and intestine.

streptokinase (*Med*) Enzyme which activates plasminogen to form plasmin which degrades fibrin and breaks up thrombi. This fibrinolytic action makes it more potent than anticoagulants. Used in life-threatening pulmonary embolism and in early treatment of myocardial infarction.

streptomycin (*Pharmacol*) A water-soluble *aminoglycoside antibiotic* derived from the bacterium *Streptomyces griseus*. It is used mainly in the treatment of tuberculosis, usually in conjunction with other drugs to prevent the development of resistance.

streptostyly (*BioSci*) In vertebrates, the condition of having the QUADRATE movably articulated with the SQUAMOSAL. Cf MONIMOSTYLY.

stress (*Eng, Phys*) (1) The force per unit area acting on a material and tending to change its dimensions, ie cause a STRAIN. The stress in the material is the ratio of applied force to the area of material resisting the force, ie

$$\frac{force}{area}$$

True stress is evaluated in terms of the actual area, *engineering stress* in terms of the original area before the force was applied. The two main types of stress are *direct* or *normal* (ie tensile or compressive) stress (usual symbol σ) and *shear* stress (usual symbol τ). The SI UNITS are $N\,m^{-2}$ (= Pa) and its multiples, $kN\,m^{-2}$, $MN\,m^{-2}$ (= $N\,mm^{-2}$), $GN\,m^{-2}$, but $lbf\,in^{-2}$, $tonf\,in^{-2}$ and bar were frequently used. (2) To stress may also mean 'to calculate stresses'; thus a structure has been stressed when the distribution and levels of stress in it have been determined by calculation.

stress (*Psych*) (1) Excessive and aversive environmental factors that produce physiological responses in the individual. (2) The psychological tension produced by environmental stressors.

stress concentration (*Eng*) An abrupt, local increase of otherwise uniform stress created by some geometric configuration at the surface or within the body of a component. It may vary in size from a fine score mark or pit, through inclusions or voids within the microstructure, to a major change in section of a large engineering structure. The sharper the change, the greater the stress concentration, and the more brittle the material, the greater its damaging effect; a fine score by a glass cutter guides the fracture of a large sheet of glass, whereas the same mark on a steel plate would have virtually no effect. The figure shows diagrammatically the stress-concentrating effect of a notch at the surface of a section subject to uniform tensile stress. The tip of the notch is where stress-induced failures are most likely to start. See panel on FATIGUE. Fig. ▷

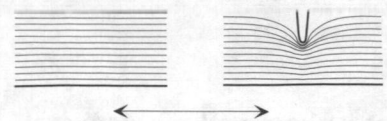

stress concentration A uniform bar (left) and a notched bar (right) both subject to a uniform tensile strain showing concentration of lines of force.

stress concentration factor (*Eng*) The factor by which the stress is increased at a stress-concentrating feature in a loaded body. Abbrev *scf*.

stress corrosion (*Eng*) Postulated enhanced rate of chemical attack on a material when it is subjected to an applied stress. Advanced as a mechanism for ENVIRONMENTAL STRESS CRACKING and STATIC FATIGUE by which crack growth is accelerated owing to the increased reaction rate in the highly stressed neighbourhood of a crack tip.

stress diagram (*Eng*) See FORCE DIAGRAM.

stressed-skin construction (*Aero*) The general term for aircraft structures in which the skin (usually light alloy, formerly plywood, occasionally plastic such as fibreglass, and in supersonic aircraft, titanium alloy or steel) carries a large proportion of the loads. In the more elementary forms, the framework may take bending and shear, with a thin skin transmitting torsion, but when of developed form, the skin is thick enough to support bending loads in the form of tension and compression in the respective surfaces. Since 1950 the principle of thick skin has been developed in the form of 'sculpturing' to vary the thickness to suit the local loads and to incorporate integral stiffeners, either by machining or by acid etching. See MONOCOQUE.

stress fibres (*BioSci*) Bundles of actin filaments (*microfilaments*) and associated proteins found in the cytoplasm.

stress fracture (*Med*) Medical term for fatigue fracture of bone.

stress-intensity factor (*Eng*) A measure of the increase in stress which occurs at the tip of a loaded crack in a material. Symbol K, with various subscripts to show loading conditions etc. Units $N\,m^{-\frac{3}{2}}$. The critical value for crack propagation, K_C, is a materials property, the FRACTURE TOUGHNESS. See FRACTURE MODE, STRENGTH MEASURES.

stress marks (*ImageTech*) Marks on photographic materials caused by mechanical pressure or friction on the emulsion surface before processing, usually heavier in density than the normal image.

stress–number curve (*Eng*) A curve obtained in fatigue tests by subjecting a series of specimens of a given material to different ranges of stress and plotting the range of stress (*S*) against the number of cycles required to produce failure (*N*). Abbrev *S/N* curve. See panel on FATIGUE.

stress-optical coefficient (*Eng*) Material constant, symbol *C*, which relates birefringence Δv to applied stress, σ: $\Delta n = C\sigma$. Used for determining degree of frozen-in strain in polymer products and in photoelastic analysis. Highest values of *C* are possessed by some polyurethanes, polycarbonate, etc, so most useful for PHOTOELASTICITY.

stress relaxation (*Eng*) The decrease in stress with time in a viscoelastic material held at constant strain (or deformation). See VISCOELASTICITY.

stress-relaxation modulus (*Eng*) Modulus obtained from a stress-relaxation experiment, where sample is held at constant strain and the decaying stress recorded. Given by the equation

$$E_r(t) = \sigma(t)\,\varepsilon^{-1}$$

Commonly used for viscoelastic materials such as polymers, where moduli can be used in stress calculations provided time-scale (*t*) is taken into account.

stress relief (*Eng*) See STRESS RELIEF ANNEALING.

stress relief annealing (*Eng*) Heating materials to a temperature ($0.3T_m$ for metals) below that liable to alter the crystalline structure and with the object of reducing or eliminating any harmful residual stresses arising from other processes. Also *stress relief*. See ANNEALING.

stress relieving (*ChemEng*) A process used in the manufacture of welded pressure vessels for severe duties, particularly involving strong caustic solutions or very thick walls, in which the completed vessel is raised to a high temperature, usually red hot, held for a time and then allowed to cool in still air. The exact temperatures and times are determined by CODES OF CONSTRUCTION.

stress–strain curve (*Eng*) A curve similar to a load–extension curve, except that the load is divided by the original cross-sectional area of the test piece and expressed in units of stress, while the extension is divided by the length over which it is measured and expressed as a ratio. Depending on the material, the curve can reflect elastic or viscoelastic behaviour, brittle fracture, high elasticity, plastic flow, work-hardening and residual strain (permanent set). See STRENGTH MEASURES.

stress zone (*MinExt*) The depth of rock surrounding an underground excavation, eg a stope, which is now bearing the transferred stress originally supported by the removed ore.

stretch (*ImageTech*) The introduction of additional frames at regular intervals in a motion picture record, usually by selected repetition, to extend the action or slow down its presentation. An example is printing silent films shot at 16 pictures per second for projection at 24 pps.

stretch breaking (*Textiles*) See CONVERTING.

stretched diaphragm (*Acous*) A diaphragm in a microphone or loudspeaker which has its rigidity increased by radial stretching, frequently by screwing onto it a rim near its edge. Resonance then becomes marked, and the tension is adjusted so that the major resonant frequency nears the upper limit of the desired transmission frequency band. Trend of the response curve is then adjusted by altering the damping.

stretcher (*Build*) A brick laid parallel to the course. See fig. at ENGLISH BOND.

stretcher strains (*Eng*) See LÜDERS LINES.

stretch fabric or yarn (*Textiles*) A fabric or yarn that stretches easily and recovers to its original dimensions. It may contain elastic material to increase its extensibility and recovery.

stretch forming (*Eng*) A process for forming large sheets of thin metal into symmetrical shapes by gripping the sheet edges in horizontally sliding stretcher jaws and moving a forming punch, without a die, vertically between them against the sheet.

stretching bond (*Build*) The form of bond, used largely for building internal partition walls $4\frac{1}{2}$ in (114 mm) thick, in which every brick is laid as a stretcher, each vertical joint lying between the centres of the stretchers above and below, so that angle closers are not required. Also *stretcher bond*. See CHIMNEY BOND.

stria (*Genrl*) A faint ridge or furrow; a streak; a linear mark.

striae (*Geol*) Parallel lines or grooves occurring on glaciated pavements, roches moutonnées, etc; produced by rock material frozen into the base of a moving ice sheet; also seen on slickensided rock surfaces along which movement has taken place during faulting.

striae (*Min*) Parallel lines occurring on the faces of some crystals; caused by oscillation between two crystal forms. The striated cubes of pyrite are good examples.

striae atrophicae (*Med*) Greyish-white bands of atrophied skin in areas where the skin has been unduly stretched, as in pregnancy (striae gravidarum).

stria medullaris (*BioSci*) See HABENULA.

striated muscle (*BioSci*) Contractile tissue in which the SARCOMERES are aligned, eg skeletal and cardiac muscle. Cf UNSTRIATED MUSCLE.

striation (*Electronics*) A phenomenon in low-pressure gas discharge, which forms luminous bands (*striae*) across the line between electrodes.

striation (*Genrl*) The appearance of faint ridges or furrows; linear markings.

strickle board (*Eng*) A board profiled along one edge to the required shape of the surface of a loam mould or core; used to sweep or strike the loam to the correct section. See LOAM.

strict inequality (*MathSci*) An INEQUALITY that excludes the possibility of the terms being equal, such as $a > b$, as contrasted with the *weak inequality*, $a \geq b$, which permits equality.

strict liability (*Genrl*) Direct responsibility to user or consumer by manufacturer for defective product. See PRODUCT LIABILITY.

stricture (*Med*) Any abnormal narrowing of a duct or passage in the body, esp the narrowing of the urethra due to gonorrhoeal inflammation. See STENOSIS.

striding level (*Surv*) Sensitive spirit level which can be placed astride a theodolite by resting the V-shaped ends of its legs on the trunnion axis, enabling the latter to be accurately levelled.

stridor (*Med*) A harsh vibrating noise produced by any obstruction in the respiratory tubes, eg in diphtheria of the larynx.

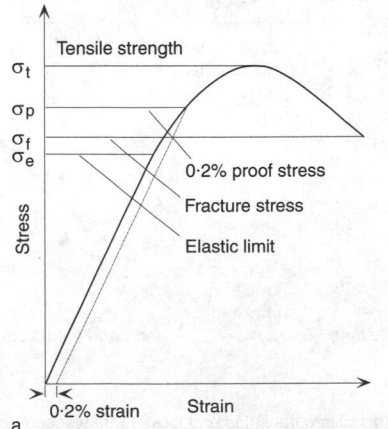
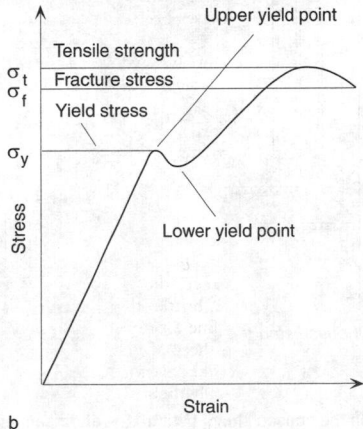

Schematic tensile stress-strain curves. Without a yield point in (a) and with one in (b).

stridulating organs (*BioSci*) The parts of the body concerned in sound production by STRIDULATION.

stridulation (*BioSci*) Sound production by friction of one part of the body against another, as in some insects.

Strigiformes (*BioSci*) An order of birds containing nocturnal birds of prey with hawk-like beaks and claws. The plumage is adapted for silent flight. The eyes are large and immovable and the retina contains mainly rods; such birds probably hunt mainly by sound. Owls.

strigose (*BioSci*) With stiff, appressed hairs or bristles.

strike (*Geol*) The horizontal direction which is at right angles to the dip of a rock. See DRAINAGE PATTERNS.

strike (*Print*) To drive a hardened steel punch into a brass or copper bar, so producing a matrix from which types are cast. The term is also used to describe the impression itself.

strike (*Vet*) See BLOWFLY MYIASIS.

strike fault (*Geol*) A fault aligned parallel to the strike of the strata which it cuts. Cf DIP FAULT.

strike lines (*Geol*) See STRATUM CONTOURS.

strike note (*Acous*) The note, largely subjective, which is initially prominent when a bell is struck. It rapidly attenuates, leaving the hum note and some overtones.

striker (*Build*) A long-handled paint brush with the head set at an angle used on bridge, roof or pipe painting. Also *spout brush*.

strike–slip fault (*Geol*) A fault whose movement is parallel to its strike.

strike-through (*Print*) Ink percolating through from the other side of a paper, due to lack of opacity or sizing; often seen in newsprint. Cf SHOW-THROUGH.

striking (*Build, CivEng*) The operation of removing temporary supports or shuttering from a structure.

striking (*Eng*) (1) In electroplating, the flash deposition of a very thin layer of one metal to facilitate subsequent plating with another. (2) See STRIKING-UP.

striking plate (*Build*) A metal plate screwed to the jamb of a door case in such a position that when the door is being shut the bolt of the lock strikes against, and rubs along, the plate, finally engaging in a hole in the latter.

striking potential (*Electronics*) That sufficiently large to break down a gap and cause an arc, or start discharge in a cold-cathode tube.

striking-up (*Eng*) The process of generating a loam mould surface by means of a STRICKLE BOARD. Also *striking*.

striking voltage (*Electronics*) See STARTING VOLTAGE.

striking wedges (*CivEng*) A pair of wedge-shaped blocks of hardwood packed beneath each end of a centre and placed in contact, with their thin ends pointing in opposite directions, so that, by moving them relatively, the centre may be gradually lowered on completion of the work. Also *casing wedges, lowering wedges*.

string (*Acous*) Cylindrical body whose length is much greater than the diameter, commonly made out of gut or wire. A stretched string can be excited to vibrate with the fundamental frequency depending on parameters such as the length and tension. Used in stringed instruments (eg violin, piano) where it is coupled with a resonator.

string (*Astron*) See COSMIC STRING.

string (*Build*) A sloping wooden joist supporting the steps in wooden stairs. See CUT STRING.

string (*ElecEng*) The series of insulator units combining to form a suspension insulator.

string (*ICT*) Series of letters and/or numbers in order but not meaningful to the computer.

string (*ImageTech*) The fine metallic strip in a LIGHT VALVE.

string (*MinExt*) The succession of tubes and other drilling and well-top equipment joined together make a *string*. Also *drill string*. See panel on DRILLING RIG.

string chart (*ElecEng*) A diagram from which the relation between the sag of an overhead line and the temperature may be rapidly obtained.

string course (*Build*) See BELT.

string efficiency (*ElecEng*) The ratio of the flashover voltage of a suspension-insulator string to the product of the flashover voltage of each unit and the number of units forming the string.

string electrometer (*ElecEng*) An electrometer consisting of two metal plates oppositely charged between which a conducting fibre is displaced from a middle position in proportion to the voltage between the plates.

stringer (*Aero*) A light auxiliary member parallel with the main structural members of a wing, fuselage, float or hull, mainly for bracing the transverse frames and stabilizing the skin material. See STIFFENER.

stringer (*Build*) A long horizontal member in a structural framework.

stringer (*NucEng*) A group of reactor fuel elements strung together for insertion into one channel of the core.

string galvanometer (*ElecEng*) See EINTHOVEN GALVANOMETER.

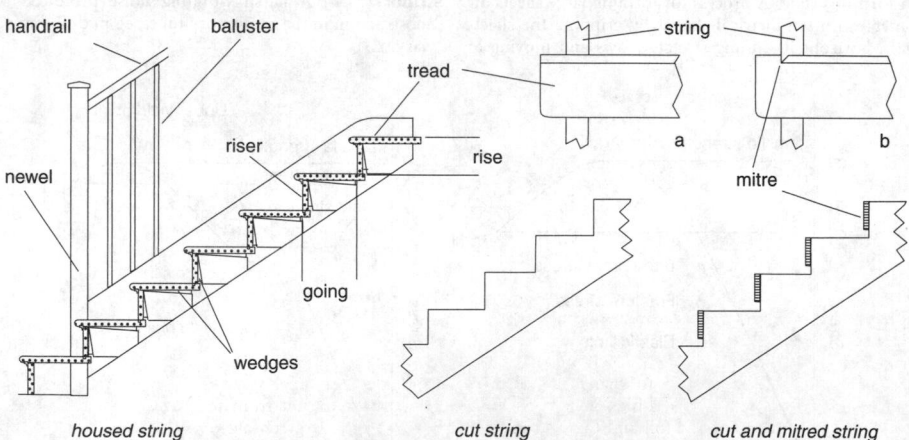

string In the housed string, both treads and risers are mortised into the string and locked with wedges. In the cut string, treads and risers overlap the string (inset a) while in the cut and mitred string, the treads overlap but the risers are mitred to the string (inset b).

string galvanometer (*ImageTech*) A vibrator used in variable-area SOUND RECORDING , the mirror of which deflects a light beam across the recording slit.

stringhalt (*Vet*) A disease of horses, characterized by involuntary sudden and excessive flexion of one hind-limb or both.

string theory (*Astron*, *Phys*) A theory in fundamental physics that attempts to construct a model of elementary particles from one-dimensional entities rather than the zero-dimensional 'points' of conventional particle physics. See COSMIC STRING, SUPERSTRING THEORY.

strip cropping (*Agri*) Growing crops in narrow strips that follow the contours of the land to minimize erosion.

stripe (*BioSci*) See STREAK.

stripe (*ImageTech*) Magnetic coating applied as one or more narrow bands to motion picture film for sound recording and reproduction.

striped muscle (*BioSci*) See STRIATED MUSCLE.

stripe filter (*ImageTech*) A filter with fine stripes of red, green and blue attached to the face of a pickup device to provide a colour video signal from a single pickup.

strip grazing (*Agri*) Restricting the access of livestock to strips of pasture within a larger area, by means of temporary, often electric, fencing.

stripline (*ICT*) Waveguide formed of strips of copper on dielectrics, formed by etching a printed circuit.

strip mining (*MinExt*) Form of opencast work, in which the overburden is removed (stripped) after which the valuable ore is excavated, the work usually being done in a series of benches, steps or terraces.

strippable coatings (*Build*) Coatings for the inside of spray booths etc, which can be removed mechanically, bringing away all overspray. These are formulated from well-plasticized ethyl cellulose or similar systems.

stripped atom (*Phys*) Ionized atom from which at least one electron has been removed.

stripper (*NucEng*) The section of an isotope-separation plant which strips the selected isotope from the waste stream.

strippers (*Print*) A feature of delivery arrangement of some printing machines, including the two-revolution, by which the printed sheet is led away after being released from the grippers.

stripping (*Chem*) Removal of an electrodeposit by any means, ie by chemical agent or by reversed electrodeposition.

stripping (*ImageTech*) The process of removing the negative emulsion film from its glass support for transfer to another glass or other support (as in process-block making).

stripping (*MinExt*) Removal of barren overburden in opencast work.

stripping (*Phys*) A phenomenon observed in deuteron (or heavier nuclei) bombardment in which only a portion of the incident particle merges with the target nucleus, the remainder proceeding with most of its original momentum practically unchanged in direction.

stripping (*Print*) (1) The procedure of assembling negative and positive film elements on a carrier film to make a *flat* for use in plate-making. (2) Where metal inking rollers of a press will not accept ink.

stripping (*Textiles*) Removing a dye or finish from a fabric, usually before reprocessing to achieve the desired effect.

stripping foil (*Phys*) The foil (of an element whose outer electrons are loosely bound) through which ions pass in a tandem van de Graaff accelerator to change the sign of their charge. If the foil is placed at a high-voltage position they will be first attracted to the foil and then accelerated away from it.

strip-wound armature (*ElecEng*) An armature whose winding consists of conductors in the form of copper strip.

strobe (*Electronics*) (1) General term for detailed examination of a designated phase or epoch of a recurring waveform or phenomenon. (2) Enlargement or intensification of a part of a waveform as exhibited on a cathode-ray

tube. (3) Process of viewing mechanical vibrations with a STROBOSCOPE; colloq term for the stroboscope itself.

strobe effect (*ImageTech*) Jerkiness of moving objects in a video picture caused by using the camera's FAST SHUTTER. Normal exposure equals the FIELD frequency (50 Hz European), but a fast shutter only exposes at the end of that time and so leaves gaps during which the objects have moved.

strobe lighting (*Aero*) An anticollision lighting system based on the principle of a capacitor-discharge flash tube. A capacitor is charged to a very high voltage which is then discharged in a controlled sequence as a high-intensity flash of light, usually blue-white, through xenon-filled tubes located at wing tips and tail of an aircraft.

strobe lighting (*ImageTech*) (1) Electronic flash for still photography of moving objects. (2) In cinematography, synchronization of the flash repetition with the camera shutter at the frame rate provides a very short exposure period so that fast-moving objects are sharply imaged instead of blurred.

strobe marker (*Electronics*) A pulse much shorter than a repeated waveform, for examination of a display.

strobila (*BioSci*) In Scyphozoa, a scyphistoma in process of production of medusoids by transverse fission; in Cestoda, a chain of proglottides. Also *strobile*. Adjs *strobilaceous*, *strobilate*, *strobiliferous*, *strobiloid*.

strobilate (*BioSci*) Bearing or pertaining to a strobilus or cone.

strobile (*Genrl*) See STROBILA, STROBILUS.

strobilization (*BioSci*) (1) Generally, the production of strobilae. (2) In Scyphozoa, transverse fission of a scyphistoma to form medusoids. (3) In Cestoda, production of proglottides by budding from the back of the scolex. (4) In some Polychaeta, reproduction by gemmation.

strobilus (*BioSci*) (1) The cone-like reproductive structure of most gymnosperms and some pteridophytes, consisting of a well-defined group of packed sporophores or sporophylls bearing sporangia and arranged around a central axis. (2) An angiosperm inflorescence of similar appearance to (1). Also *cone*.

stroboscope (*Electronics*) A flashing lamp, of precisely variable periodicity, which can be synchronized with the frequency of rotating machinery or other periodic phenomena, so that, when viewed by the light of the stroboscope, they appear to be stationary.

stroke (*Eng*) The travel or excursion of a piston, press ram or other (reciprocating) part of a machine.

stroke (*Med*) An apoplectic seizure or a sudden attack of paralysis. It is now usually called a CEREBRO-VASCULAR ACCIDENT.

stroked (*Build*) A term applied to the face of an ashlar which has been so tooled as to present a regular series of small flutings.

stroma (*BioSci*) (1) The matrix of the chloroplast, in which the dark reactions of photosynthesis take place. Cf GRANUM. (2) A mass of fungal tissue formed from intertwined adherent hyphae (plectenchyma), eg the major part of a mushroom. Cf SCLEROTIUM. (3) A supporting framework, as the connective tissue framework of the ovary or testis in mammals. Pl *stromata*. Adjs *stromate*, *stromatic*, *stromatiform*, *stromatoid*, *stromatous*, *stromoid*.

stroma lamellae (*BioSci*) Thylakoids that cross the stroma of a chloroplast, interconnecting the grana.

stromatolites (*BioSci*) Rounded, multilayered structures up to approx 1 m across, found in rocks back to at least 2800 million years ago. What are, apparently, the present-day equivalents result from the growth, under special conditions, of blue-green algae. The matted algal strands trap sediment and also precipitate lime.

stromatoporoid limestone (*Geol*) A calcareous sediment, rich in remains of the reef builder *Stromatopora*, important from the Cambrian to Cretaceous.

Strombolian eruption (*Geol*) A type of volcanic eruption characterized by frequent small explosions as trapped gases break through overlying viscous lava.

strong clay (*Build*) See FOUL CLAY.

strong electrolyte (*Chem*) An ELECTROLYTE which is completely ionized even in fairly concentrated solutions.

strong force (*Phys*) See STRONG INTERACTION.

strong interaction (*Phys*) One of the four fundamental forces. An interaction between particles involving QUARKS (and so BARYONS and MESONS) completed in 10^{-23} s. It is the strong interaction that binds protons and neutrons together in the nuclei of atoms. The strong interaction is mediated by *gluons* which couple to the colour of quarks. See panel on GRAND UNIFIED THEORIES.

strongyloidiasis (*Med*) Infestation of humans with the nematode worm *Strongyloides stercoralis*, the worm living in the intestines and causing diarrhoea; common in the tropics. Also *strongyloidosis*.

strontium (*Chem*) Metallic element, symbol Sr, at no 38, ram 87·62, rel.d. 2·54, mp 800°C, bp 1300°C. Silvery-white in colour, it is found naturally in *celestine* and in *strontianite*; it also occurs in mineral springs. Similar chemical qualities to calcium. Compounds give crimson colour to flame and are used in fireworks. It has 13 isotopes: the radioactive isotope, ^{90}Sr, is produced in the fission of uranium and has a long life, hence its presence in 'fallout' after a nuclear explosion; the RADIOGENIC isotope ^{87}Sr is produced by radioactive decay of ^{87}Rb, and is used (as the ratio ^{87}Sr/^{86}Sr) in RADIOMETRIC DATING (panel).

strontium titanate (*Min*) A colourless isotropic, highly refringent, artificial gemstone that is a simulant for DIAMOND.

strontium unit (*Phys*) Unit used to measure the concentration of radioactive strontium-90 in calcium; 1 $SU = 10^{-12}$ Ci g^{-1}.

Stroop effect (*Psych*) The difficulty, due to interference effects, observed when eg subjects are asked to state the colour of ink in which a colour word is written (eg if the word blue is written in red ink).

strophic movement (*BioSci*) See STROPHISM.

strophiole (*BioSci*) See CARUNCLE.

strophism (*BioSci*) A growth movement in which an organ or its stalk twists in response to a directional stimulus, eg the twisting of leaf bases and petioles on many horizontal branches in response to light and/or gravity resulting in the horizontal orientation of the leaf blades. Also *strophic movement*.

Strouhal number (*Acous*) A factor in the equation for the frequency of an AEOLIAN TONE. The equation relates the frequency f with the speed v and the thickness d of the obstacle by $f = kv/d$, where k is the Strouhal number.

Strowger exchange (*ICT*) The old type of telephone exchange that used entirely electromechanical switching; after A B Strowger, their inventor.

struck (*Build*) (1) Taken away, dismantled as for scaffolding or shuttering. (2) Said of the recessed joints of mortar on an outside wall.

struck (*Vet*) (1) A form of enterotoxaemia of sheep caused by *Clostridium perfringens* (*Cl.welchii*), type C. (2) Said of sheep affected with BLOWFLY MYIASIS.

struck core (*Eng*) A loam core formed by revolving the built-up core, loam-covered, against a STRICKLE BOARD.

struck-joint pointing (*Build*) See WEATHERED POINTING.

structural (*BioSci*) Of changes, aberrations, etc, in the number or arrangement of chromosomes. See STRUCTURAL GENE.

structural colours (*BioSci*) Colour effects produced by some structural modification of the surface of the integument, as the iridescent colours of some beetles. Cf PIGMENTARY COLOURS.

structural damping (*Aero*) See DAMPING.

structural foam (*Eng*) Products whose interior is foamed to decrease overall weight, and hence improve bending stiffness-to-weight ratio. Abbrev *SF*.

structural foam moulding (*Eng*) Injection-moulding process where a blowing agent like AZODICARBONAMIDE is added to polymer so that a foamed interior is formed in the final product.

structural formula (*Chem*) A representation of the chemical structure of a substance which shows not only its composition but also its CONNECTIVITY and the order of the bonds connecting the atoms.

structural gene (*BioSci*) The stretch of DNA specifying the amino acid sequence of a polypeptide, as distinct from the interspersed and associated DNA, some of which is concerned with control of gene expression.

structuralism (*Psych*) (1) Outmoded school of experimental psychology that sought to understand mental experience by introspection. (2) Theory of cognitive development devised by Piaget. (3) Various sociological and anthropological approaches that focus on societal and organizational structures and the way they affect members of those groups.

structural timber (*Build*) (1) Canadian name for CARCASSING TIMBER. (2) Any timber acting as a support.

structure (*Aero, Space*) (1) In aircraft, the rigid construction which is designed to house the crew, payload, fuel and subsystems, and made to withstand the forces of aerodynamics, inertia, propulsion and landing, together with the reactions arising from the release of armament. (2) Framework or ensemble of rigid elements which is designed to withstand a variety of mechanical and thermal influences (eg thrust forces, bending moments, aerodynamic heating effects) during launch and flight of a spacecraft, and to provide protective support for its subsystems and payload.

structure (*Chem*) Assembly of atoms and molecules in a material. See PRIMARY STRUCTURE. Also for larger assemblies, see MACROSTRUCTURE, MICROSTRUCTURE, SECONDARY STRUCTURE, TERTIARY STRUCTURE.

structure-borne sound (*Acous*) Sound in solid bodies, as opposed to sound in gases (eg airborne sound) and sound in liquids (eg water sound).

structured programming (*ICT*) Orderly approach to programming that emphasizes breaking large and complex tasks into successively smaller sections. Also *top-down programming*. See JACKSON STRUCTURED PROGRAMMING.

structured query language (*ICT*) A standardized method for obtaining information from databases. Abbrev *SQL*.

structure–property relation (*Chem*) Any clear dependence of the physical or chemical properties of a material on atomic, molecular structure or microstructure, thus allowing control of mechanical properties of metals and ceramics by grain size (see HALL–PETCH EQUATION) or the strength of polymers by the control of molecular mass. See MOLECULAR MASS DISTRIBUTION.

struma (*Med*) (1) See SCROFULA. (2) See GOITRE.

strut (*CivEng*) A timber or adjustable metal member used for bracing purposes during excavation or construction works.

strut (*Eng*) Any light structural member or long column which sustains an axial compressive load. Failure occurs by bending before the material reaches its ultimate compressive stress. See COLUMN.

Struthioniformes (*BioSci*) An order of birds retaining only two toes and whose feathers lack an aftershaft. They extend their rudimentary wings when running. Known from the Pliocene onwards. Ostriches.

strutting (*Build*) The process of using props to give temporary support between two surfaces.

struvite (*Min*) Magnesium ammonium phosphate hexahydrate, crystallizing in the orthorhombic system. Found in guano and dung, and common in human calculi.

strychnine (*Chem*) $C_{21}H_{22}N_2O_2$. Mp 265°C. A monoacidic alkaloid base, very poisonous, causing tetanic spasms. It occurs in the seeds of *Strychnos ignatii* and *Strychnos nux vomica*, in *Upas tieuté* and in *Lignum colubrinum*, Strychnine is almost insoluble in water, but is readily soluble in chloroform and benzene.

strychnine bases (*Chem*) A group of alkaloids obtained from *Strychnos nux vomica*. They include BRUCINE and STRYCHNINE.

stub (*Build*) A small projection on the under surface at the top edge of a tile, enabling it to be hung on a batten.

stub (*ICT*) An auxiliary section of a waveguide or transmission line connected at some angle with the main section. See COAXIAL STUB, QUARTER-WAVELENGTH STUB.

stub antenna (*ICT*) A quarter-wavelength rod or wire.

stub axle (*Autos*) A short dead axle. If carrying a steered wheel it is capable of limited angular movement about a swivel pin carried by the end of the axle beam.

stubble (*Agri*) The basal portions of crop plants that remain in the soil after the upper parts have been harvested.

stubble mulch (*Agri*) Crop stubble or residues left *in situ* as soil cover during the growth of a succeeding crop, as a means of conserving water, adding nutrient and reducing soil erosion. Also *trash cover*.

stub plane (*Aero*) A short length of wing projecting from the fuselage, or hull, of some types of aircraft to which the main planes are attached.

stub tenon (*Build*) A very short tenon for fitting into a blind mortise. Also *joggle*.

stub-tooth gear (*Eng*) A gear tooth of smaller height and of more robust form than that normally employed; used in the manufacture of automobile gears.

stub tuning (*ElecEng*) Use of shunt stubs connected to short-circuited section of line or waveguide in order to produce matched conditions. In the single-stub tuner the stub susceptance is $-jb$ and it is connected to the main transmission line at a point where the transformed load admittance is $y = 1 + jb$ (normalized).

stuc (*Build*) Plasterwork resembling stone.

stucco (*Build*) A smooth-surfaced plaster or cement rendering applied to external walls. Also *stuke*.

stuck moulding (*Build*) A moulding shaped out of solid timber.

stud (*Build*) The vertical members in a timber partition framework.

stud (*Eng*) A shank, or headless bolt, generally screwed from both ends and plain in the middle. It is permanently screwed into one piece, to which another is then secured by a nut.

Student's *t*-test (*MathSci*) See *T*-DISTRIBUTION.

stud partition (*Build*) A wooden partition based on rough vertical timbers.

stuff (*Build*) (1) See COARSE STUFF, FINE STUFF. (2) Timber sawn or manufactured from logs.

stuff (*Paper*) Paper stock or pulp ready for making into paper.

stuffer box (*Textiles*) A box into which continuous filament yarn is packed closely so resulting in the yarn becoming crimped. The yarn may then be heated in order to set the crimp and produce a TEXTURED YARN.

stuffing-box (*Eng*) A cylindrical recess provided in eg a cylinder cover, at the point at which the piston rod emerges; it is filled with packing which is compressed by a GLAND to make a pressure-tight joint.

stugging (*Build*) See PICKING.

stuke (*Build*) See STUCCO.

stupor (*Med*) A state of mental and physical inertia: inhibition of instinctive activity and indifference to social environment.

sturdy (*Vet*) See COENURIASIS.

Sturm's theorem (*MathSci*) A theorem by which the number of real roots of an algebraic equation which lie in any given interval can be determined. It utilizes sign changes in the partial remainders that occur in calculating the HCF of $f(x)$ and $f'(x)$, where $f(x) = 0$ is the given equation.

Stuttgart disease (*Vet*) The name formerly given to a disease of the dog characterized by apathy, stomatitis and gastro-enteritis, but which probably was, in most cases, the uremic stage of nephritis caused by *Leptospira canicola*. Vaccines widely used. Also *canine typhus*.

Stüve diagram (*EnvSci*) An AEROLOGICAL DIAGRAM with rectangular axes temperature and $P^{(\gamma-1)/\gamma}$, where P is pressure and γ is the ratio of the specific heats of a perfect gas.

St Vitus's dance (*Med*) See CHOREA.

S-twist (*Textiles*) See TWIST DIRECTION and panel on FIBRE ASSEMBLIES.

stye (*Med*) Staphylococcal infection of a sebaceous gland of the eyelid. Also *hordeoleum*, *sty*.

style (*BioSci*) The part of the carpel between the ovary and the stigma, often relatively long and thin.

style (*Build*) See STILE.

style of the house (*Print*) The customary style of spelling, punctuation, capitalization, etc, used in a printing establishment. It is followed in the absence of contrary instructions. Also *house style*.

style sheet (*ICT*) See TEMPLATE FILE.

stylet (*BioSci*) A small pointed bristle-like process.

styliform (*BioSci*) Bristle-shaped.

stylo- (*Genrl*) Prefix from Gk *stylos*, pillar.

stylobate (*Arch*) A continuous pedestal supporting a row of columns.

stylolite (*Geol*) An irregular suture-like boundary found in some limestones. In three dimensions it has a tooth-and-socket arrangement and appears to have been formed by pressure solution after deposition.

stylopodium (*BioSci*) (1) The swollen base of a style, as in some Umbelliferae. (2) The proximal segment of a typical pentadactyl limb; brachium or femur; upper arm or thigh.

stylus (*Acous*) A needle for cutting or replaying a disk recording. With the introduction of lightweight pickups, the more durable sapphire- or diamond-tipped reproducer styli replaced the earlier steel or chrome-plated needles.

stypsis (*Med*) The application or use of styptics.

styptic (*Med*) Astringent; tending to stop bleeding by coagulation.

styrene (*Chem*) Phenylethene, $C_6H_5CH{=}CH_2$. Intermediate for PS, ABS, HIPS, etc. It is a colourless aromatic liquid, bp $145°C$, widely used in polyester thermoset resins. See COPOLYMER.

styrene acrylonitrile (*Chem*) A copolymer. Abbrev *SAN*. See COPOLYMER.

styrene–butadiene rubber (*Eng, Plastics*) The main synthetic rubber used in tyre treads. Formerly GRS. See panels on ELASTOMERS and TYRE TECHNOLOGY.

styrene–butadiene–styrene (*Plastics*) Abbrev *SBS*. See COPOLYMER and panel on POLYMERS.

styrene joint (*ElecEng*) A cable joint filled with hot liquid styrene, which polymerizes on cooling into a very hard solid and prevents displacement of the cores.

styrene resins (*Plastics*) Polymers and copolymers based on styrene monomer. The parent homopolymer is a transparent, brittle and rigid material with a T_g of nearly $100°C$. The backbone chain is atactic so the solid is amorphous. Copolymers like ABS, high-impact polystyrene and SAN are now more important for their improved properties. See COPOLYMER and panel on POLYMERS.

styrol resins (*Chem*) Obsolescent name for STYRENE RESINS.

sub- (*Genrl*) Prefix from Lt *sub*, under, used in the following senses: (1) deviating slightly from, eg *subtypical*, not quite typical; (2) below, eg *subvertebral*, below the vertebral column; (3) somewhat, eg *subspatulate*, somewhat spatulate; (4) almost, eg *subthoracic*, almost thoracic in position.

subacute (*Med*) Said of a disease whose symptoms are less pronounced than those of the acute form; between acute and chronic.

subacute combined degeneration of the cord (*Med*) A condition in which there is degeneration of motor and sensory nerve tracts in the spinal cord, giving rise to paraplegia and loss of sensibility of the skin, the disease being associated with vitamin B_{12} deficiency. Also *anaemic spinal disease*.

subadditive function (*MathSci*) See ADDITIVE FUNCTION.

subaqueous loudspeaker (*Acous*) Sound source for producing sound in water.

subaqueous microphone (*Acous*) See HYDROPHONE.

subarachnoid haemorrhage (*Med*) Haemorrhage into the space between the arachnoid and the pia mater, esp as a result of rupture of an aneurysm of one of the arteries.

subassembly (*Eng*) Assembly which can be handled and stored as a unit and subsequently incorporated in a more complex assembly.

subatomic (*Chem*) Said of particles or processes at less than atomic level, eg radioactivity, production of X-rays, nuclear shells, etc.

subaudio frequency (*Acous*) A frequency below those usefully reproduced through a sound-reproducing system or part of such system.

subcarrier (*ICT*) One frequency that is modulated over a narrow range by a measured quantity, and then used to modulate (with others) a carrier that will be finally demodulated on reception.

subchelate (*BioSci*) In Arthropoda, having the distal joint of an appendage modified so that it will bend back and oppose the penultimate joint, like the blade and handle of a penknife, to form a prehensile weapon. Cf CHELATE.

subchord (*Surv*) The chord length from a tangent point on a railway or highway curve to the adjacent chainage peg around the curve when this is less than the full chord distance employed in setting out the chainage pegs.

subcircuit (*ElecEng*) One of several lighting circuits supplied from a common branch distribution fuse-board.

subclavian (*BioSci*) Passing beneath or situated under the clavicle, as the *subclavian artery*.

subclimax (*BioSci*) Vegetation held more or less permanently at some stage of a succession before the CLIMAX.

subconscious (*Psych*) Distant from the focus of attention but capable of being consciously recalled; (of memories, motives, intentions, thoughts, etc) of which the individual is only dimly aware but which exert an influence on his or her behaviour.

subcooled water (*EnvSci*) See SUPERCOOLED WATER.

subcortical (*BioSci*) Below the cortex or cortical layer; as certain cavities in sponges.

subcritical (*Phys*) Said of an assembly of fissile material, for which the multiplication factor is less than unity.

subculture (*BioSci*) A culture of a micro-organism, tissue or organ prepared from a pre-existing culture.

subcutaneous (*BioSci*) Situated just below the skin.

subdirectory (*ICT*) A directory that is contained within another.

subdorsal (*BioSci*) Situated just below the dorsal surface.

subduction zone (*Geol*) The area where a plate moves under an overriding plate. Associated with regions of high seismic activity. See panel on PLATE TECTONICS.

subduction zone

subdural (*Med*) Situated beneath the dura mater, eg *subdural abscess, subdural haemorrhage*.

suberin (*BioSci*) A mixture of fatty substances, esp of cross-linked polyesters of long-chain ($> C_{20}$) ω-hydroxy and ω-dicarboxylic aliphatic acids, in some plant cell walls, esp cork.

suberin lamella (*BioSci*) A layer of wall material impregnated with SUBERIN.

suberization (*BioSci*) The deposition of SUBERIN on or in a cell wall.

subfactorial (*MathSci*) Operation giving the number of different ways of arranging *n* objects so that no object occupies its original position. Cf FACTORIAL *N*.

subfloor (*Build*) See COUNTER-FLOOR.

subgenital (*BioSci*) Below the genital organs, as the subgenital pouches of the jellyfish, *Aurelia*.

subgenual organ (*BioSci*) In many insects, chordotonal organs in the fibia adapted for perceiving vibrations of the substrate.

subglacial drainage (*Geol*) The system of streams beneath a glacier or ice sheet; formed chiefly of melt-waters. Cf ENGLACIAL streams.

subgroup (*MathSci*) A subset of the members of a given GROUP that themselves form a group under the same operation; eg the integers with addition constitute a subgroup of the rational numbers with addition.

subharmonic (*Acous*) Having a frequency which is a fraction of a fundamental. Subharmonics appear in some forms of non-linear distortion.

subhedral (*Geol*) See HYPIDIOMORPHIC.

subimago (*BioSci*) In Ephemeroptera (mayflies) the stage in the life history emerging from the last aquatic nymph. It has wings and moults to give the true imago. This ECDYSIS is unique among insects, involving the casting of a delicate pellicle from the whole body, including the wings. Adj *subimaginal*.

subinvolution (*Med*) Partial or complete failure of the uterus to return to the normal state after childbirth.

subitize (*Psych*) To perceive, or be capable of perceiving, the number of items in a group at a glance without actually counting.

subjective noise meter (*Acous*) A noise meter for assessing noise levels on the phon scale, the loudness of the noise level being measured by ear with the adjusted reference tone, 1000 Hz. See OBJECTIVE NOISE METER.

subjective reality (*Psych*) Perception of reality that is made by an individual and that may differ from those made by others.

subject matching (*Psych*) Selecting a subject group to match the characteristics of the population being studied in terms of age, sex or some other characteristic.

sublevel caving (*MinExt*) A method of mining massive ore deposit in which ore is drawn down to a delivery road under-running the deposit, and overburden is allowed to cave in, the process being repeated.

sublevel stoping (*MinExt*) A method in which ore is blasted in stopes and drawn down to a sublevel in the footwall through ore passes.

sublimate (*Chem*) The product of sublimation.

sublimation (*Chem*) The vaporization of a solid (esp when followed by the reverse change) without the intermediate formation of a liquid.

sublimation (*Psych*) In psychoanalytic theory, a defence mechanism in which forbidden impulses are gratified in socially acceptable ways, eg intellectual curiosity as a sublimation of childhood voyeurism.

sublimed white lead (*Chem*) Basic lead (II) sulphate fume.

subliminal perception (*Psych*) The phenomenon whereby a stimulus presented below the threshold of conscious awareness may influence behaviour. Also *subception*.

sublingua (*BioSci*) In marsupials and lemurs, a fleshy fold beneath the tongue.

sublittoral plant (*BioSci*) A plant which grows near the sea, but not on the shore.

sublittoral zone (*BioSci*) In a lake, the lake bottom below the PARALIMNION, extending from the lakeward limit of rooted vegetation to the upper limit of the HYPOLIMNION.

subluxation (*Med*) Partial, incomplete dislocation of a joint.

submarginal ore (*MinExt*) Developed ore which, at current market price of extracted values, cannot be profitably treated.

submarine cable (*ICT*) Long-distance cable laid along the sea bed. Coaxial in form, with SUBMARINE REPEATERS at intervals to amplify signals. In shallow water, with a danger from anchors or trawling, the cables may be armoured or even buried in the sea bed. In deep water, lightweight cables without armouring but with a central core of high-tensile steel are used to prevent stretch during laying. Some may contain OPTICAL FIBRE channels.

submarine canyon (*Geol*) A trench on the continental shelf. It sometimes has tributaries.

submarine fan (*Geol*) A fan of terrigeneous material formed at the foot of SUBMARINE CANYONS and large rivers, often as TURBIDITE deposits.

submarine repeater (*ICT*) A REPEATER built into a watertight and pressure-resistant housing; long-life components are used to ensure high reliability, owing to the high cost of raising cables and repeaters for repair.

submaxillary (*BioSci*) Situated beneath the lower jaw.

submerged arc welding (*ElecEng*) Automatic *arc-welding* process, using a single, bare electrode which passes through a blanket of granular, fusible flux laid along the seam to be welded, so that the entire welding action takes place beneath this blanket. This eliminates spatter losses and protects the joint from oxidation.

submerged heating (*ElecEng*) Induction heating of workpiece submerged in quenching liquid.

submicron (*Phys*) A particle of diameter less than a MICRON.

submillimetric waves (*ICT*) Microwaves at a frequency exceeding 300 GHz, at which the wavelength is 1 mm.

subminiature valve (*Electronics*) See MINIATURE VALVE.

subnormal (*MathSci*) Of a curve: the projection onto the y-axis of that part of a normal to the curve lying between its point of normality with the curve and its intersection with the y-axis.

suboutcrop (*MinExt*) See BLIND APEX.

subpress (*Eng*) Unit comprising a carrier plate for the punch and another for the die of a press tool, connected by pillars sliding in bushes so that the alignment of punch and die is maintained accurately during working of the press tool. Also *die set*.

subprogram (*ICT*) See SUBROUTINE.

subroutine (*ICT*) A set of PROGRAM STATEMENTS performing a specific task, but which requires to be initiated by a calling program. Also *function, procedure, routine, subprogram*. See CLOSED SUBROUTINE, OPEN SUBROUTINE.

subscriber identity module (*ICT*) A device, usually a SMART CARD, used to personalize a telephone in a GLOBAL SYSTEM FOR MOBILE COMMUNICATION network. The telephone itself is not registered on the network, but once the module is inserted calls can be billed to the appropriate user.

subscriber network interface (*ICT*) Equipment linking the SWITCHED MULTIMEGABIT DATA SERVICE (SMDS) to customer premises equipment. Its function is to translate user data to and from SMDS format.

subscriber's line (*ICT*) The line connecting the subscriber's main telephone instrument to the exchange, as contrasted with an extension line from this instrument.

subscriber's station (*ICT*) See SUBSTATION.

subscripted variable (*ICT*) Reference to an individual element of an ARRAY (eg CASS(20) or LAR(3,5))

subsequent drainage (*Geol*) See DRAINAGE PATTERNS.

subsequent pickup (*ICT*) The pickup occurring at the end of a pulse train due to the change in line-relay holding conditions.

subset (*BioSci*) A term used to classify functionally or structurally different populations of cells within a single cell type. Used esp of T-lymphocytes (helper, suppressor, cytotoxic).

subset (*MathSci*) A set, all of whose members are members of a given set; eg the positive even integers constitute a subset of the set of positive integers. According to this definition each set is a subset of itself, and the empty set is a subset of every set. Subsets other than these are referred to as *proper subsets*.

subsidence (*Build, CivEng*) (1) The sinking or caving-in of the ground. (2) The settling of a structure etc to a lower level.

subsidiary cell (*BioSci*) An epidermal cell associated with the guard cells of a stoma and morphologically different from the other epidermal cells. Also *accessory cell*.

subsoil (*Geol*) Residual deposits lying between the soil above and the bedrock below, the three grading into one another.

subsoil drain (*CivEng*) A drain laid just below ground level to carry off waters from saturated ground. Formerly earthenware, now corrugated and perforated plastic pipe is laid at the bottom of a trench and covered with broken stones.

subsoiling (*Agri*) Deep ploughing to break up soil compaction at significant depth.

subsonic (*Acous*) Said of an object or flow which moves with a speed less than that of sound. See panel on AERODYNAMICS.

subsonic speed (*Aero*) Any speed of an aircraft where the airflow round it is everywhere below Mach 1. See MACH NUMBER.

subspecies (*BioSci*) A taxonomic subdivision of a species, with some morphological differences from the other subspecies and often with a different geographical distribution or ecology.

substance (*Chem*) A kind of matter, with characteristic properties, and generally with a definite composition independent of its origin.

substance (*Glass*) The thickness of flat glass; in the case of sheet glass, formerly expressed as mass per unit area. Largely superseded by millimetres.

substance (*Paper*) Thickness. The preferred term is GRAMMAGE or BASIS WEIGHT.

substance dependence (*Psych*) Current terminology for what used to be called *addiction*: a pattern of substance use in which the individual develops tolerance and exhibits withdrawal symptoms if the substance is withheld.

substance P (*BioSci*) A peptide neurotransmitter, present in nerve cells and intestinal tissue, that increases capillary permeability, and induces vasodilation and the contraction of intestinal smooth muscle.

substandard (*ImageTech*) Deprecated term for NARROW GAUGE FILM, meaning widths smaller than the *standard* 35 mm.

substandard instrument (*ElecEng*) A laboratory instrument whose accuracy is very great and which has been calibrated against an international standard of measurement.

substantia (*BioSci*) Substance; matter.

substantia nigra (*BioSci*) A region of darkly pigmented dopaminergic neurons in the ventral midbrain, thought to control movement and damaged in Parkinsonism.

substantive dyes (*Chem*) Dyestuffs which can dye cotton and other fibres direct without the aid of a mordant. Many are derived from benzidine and its derivatives.

substantive variation (*BioSci*) Variation in the constitution of an organ or organism, as opposed to variation in the number of parts. Also *qualitative variation*.

substantivity (*Textiles*) The attractive force that enables a textile material to remove dyestuff from a dyebath.

substation (*ElecEng*) A switching, transforming or converting station intermediate between the generating station and the low-tension distribution network.

substation (*Surv*) Apex of a subsidiary triangle in a skeleton.

substellar point (*Astron*) The point on the Earth's surface, regarded as spherical, where it is cut by a line from the centre of the Earth to a given star; hence the point where the star would be vertically overhead, the point whose latitude is equal to the star's declination. Applied also to the Sun and Moon as *subsolar point* and *sublunar point* respectively.

substitution (*Chem*) The replacement of one group, esp hydrogen, by another, eg halogen, alkyl, hydroxyl, etc.

substitutional solid solution (*Chem*) A type of solution in which atoms of solute(s) replace those of the solvent in the same crystal lattice, eg nickel and copper or the alpha solid solutions of aluminium, tin and/or zinc in copper.

substrate (*BioSci*) (1) A reactant in a reaction catalysed by an enzyme. (2) The surface or medium on or in which an organism lives and from which it may derive nourishment. See SUBSTRATUM.

substrate (*Electronics*) The surface onto which thin or thick films are deposited. Any object on which layers are to be deposited or fashioned, such as the wafer of a semiconductor onto and into which electronic circuitry is to be integrated.

substrate (*ImageTech*) A layer applied to the film base before coating the emulsion to ensure complete adhesion.

substrate level phosphorylation (*BioSci*) The conversion of ADP to ATP which is brought about by the concomitant hydrolysis of some other HIGH-ENERGY PHOSPHATE COMPOUND.

substratum (*BioSci*) The surface on which a cell or organism lives, to which it is attached and over which it may be able to move. Used as an alternative to SUBSTRATE in this sense, it removes ambiguity.

subsynchronous (*ElecEng*) Below SYNCHRONISM.

subsystem (*Space*) Constituent part of a space vehicle system which performs a particular function; eg *electrical power subsystem*, *data handling subsystem*. It is the sum of the coherent subsystem performances which provides a certain system capability.

subtangent (*MathSci*) Of a curve, the projection onto the *x*-axis of that part of a tangent lying between its point of tangency with the curve and its intersection with the *x*-axis.

subtangent

subtectal (*BioSci*) Lying beneath the roof, as of the skull; in some fish, a cranial bone.

subtend (*BioSci*) To be situated immediately below, eg as a leaf is situated immediately below the bud in its axil.

subtend (*MathSci*) Of a figure: to define an angle at some point. Thus a segment of a straight line or arc subtends the angle at a given point formed by the lines joining the ends of the segment to that point. The chord joining the points at the end of an arc is also said to subtend the arc.

subtense bar (*Surv*) A horizontal bar, bearing two targets fixed at a known distance apart, used as the distant base in one system of tacheometry.

subthreshold (*Psych*) Of a stimulus intensity that is insufficient to produce a specified response; subliminal.

subtitle (*ImageTech*) Wording superimposed on the lower part of a film or TV picture, often to give a translation of the accompanying sound track dialogue or to provide an outline of a spoken commentary for the benefit of the hard-of-hearing.

subtraction (*MathSci*) The INVERSE OPERATION to addition. Denoted by the minus sign — so that $a - b = c$ when $a = b + c$. Thus in the statement $m - s = d$, the terms m, s and d are referred to as the *minuend*, *subtrahend* and *difference* respectively.

subtractive-coloured light (*Phys*) The monochromatic illumination obtained from a polychromatic light source by the aid of an appropriate absorption screen.

subtractive printer (*ImageTech*) A photographic or motion picture printer or an enlarger in which the intensity and colour of the exposing light is altered by the use of filters, in contrast to an ADDITIVE PRINTER.

subtractive process (*ImageTech*) A colour process in which the red, green and blue components of the original subject are reproduced as three superimposed images in the complementary (*subtractive*) colours of cyan, magenta and yellow respectively; this is the basis of all modern colour photography and cinematography as well as colour printing on paper.

subtrahend (*MathSci*) See SUBTRACTION.

subtransient reactance (*ElecEng*) The reactance of the armature winding of a synchronous machine corresponding to the leakage flux which occurs in the initial stage of a short circuit. This flux is smaller than that corresponding to the transient reactance due to eddy currents set up in the rotor during the first one or two half-cycles of a short circuit.

subulate (*BioSci*) Awl-shaped, tapering from base to apex.

succession (*BioSci*) The sequence of communities (ie a sere) which replace one another in a given area, until a relatively stable community (ie the climax) is reached, which is in equilibrium with local conditions.

succinamic acid (*Chem*) Monoamidobutandioic (*succinic*) acid. $H_2NCOCH_2CH_2COOH$.

succinic acid (*Chem*) Butandioic acid. $HOOCCH_2CH_2$-$COOH$. Dibasic acid; mp 185°C, bp 235°C, with partial decomposition into its anhydride. It occurs in the juice of sugar cane, in the castor-oil plant and in various animal tissues, where it plays an important part in metabolism (see TCA CYCLE). Succinic anhydride is used in the manufacture of alkyd thermoset resins.

succinite (*Min*) A variety of AMBER, separated mineralogically because it yields succinic acid.

succinyl (*Chem*) The bivalent acid residue $-COCH_2CH_2CO-$.

succise (*BioSci*) Ending below abruptly, as if cut off.

succulent (*BioSci*) (1) Juicy, having a high water content. (2) A plant with succulent stems or leaves; most are xerophytes and CAM PLANTS, eg cacti or halophytes. Salicornia.

succus entericus (*BioSci*) A collective name for the enzymes secreted by glandular cells in the walls of the duodenum.

succussion (*Med*) The act of shaking a patient to detect the presence of fluid in a pleural cavity already containing air (pneumothorax).

sucker (*BioSci*) (1) An upward-growing shoot arising from the base of a stem or adventitiously from a root. (2) A suctorial organ adapted for adhesion or imbibition, as one of the muscular sucking disks on the tentacles of Cephalopoda. (3) The suctorial mouth of animals like the leech and the lamprey. (4) A newly born whale. (5) One of a large number of fishes having a suctorial mouth or other suctorial structure, as the remora (*Echeneis*), members of the genus *Lepadogaster*, etc.

sucking booster (*ElecEng*) A booster whose function is to overcome voltage drop occurring in a feeder.

suckler herd (*Agri*) A breeding herd of cows with calves suckled by the mother rather than the milk being taken for human consumption.

sucralose (*Chem*) An artificial sweetener made from sucrose.

sucrase (*BioSci*, *FoodSci*) See INVERTASE.

sucrol (*Chem*) See DULCIN.

sucrose (*Chem*, *FoodSci*) Saccharobiose. $C_{12}H_{22}O_{11}$. Mp 160°C. A disaccharide carbohydrate; it crystallizes in large monoclinic crystals, is optically active and occurs in beet, sugar cane and many other plants. Hydrolyses to glucose and fructose. Colloq *cane-sugar*, *sugar*.

sucrose gradient (*BioSci*) A density gradient used in centrifugation to separate molecules on the basis of their sedimentation velocity.

suction box (*Paper*) A shallow compartment extending across a paper-making machine connected to a vacuum pump for the removal of water from the web or felt. Generally positioned under the forming web on the wire table or in a perforated roll, eg couch or press.

suction couch roll (*Paper*) A *couch roll* consisting of a perforated metal roll within which is a stationary suction box to remove water from the web on the machine wire.

suction dredger (*CivEng*) See SAND-PUMP DREDGER.

suction feed gun (*Build*) A spray gun with a paint cup fitted beneath the gun. The stream of compressed air which passes over the cup creates a vacuum causing atmospheric pressure to force the paint up to the gun where it is atomized. Used for small quantities of thin materials.

suction pressure (*BioSci*) An obsolete term equivalent to minus the water potential of a cell.

suction roll (*Paper*) Roll consisting of a perforated metal sleeve rotating about a fixed internal suction box.

suction valve (*Eng*) See FOOT VALVE (1).

suctorial (*BioSci*) Drawing in; imbibing; tending to adhere by producing a vacuum; pertaining to a sucker.

suctorial mouthparts (*BioSci*) Tubular mouthparts adapted for the imbibition of fluid nourishment; found in some insects and many ectoparasites.

sudamina (*Med*) Whitish vesicles on the skin, due to retention of sweat in the sweat glands. Sing *sudamen*. Adj *sudaminal*.

sudden infant death syndrome (*Med*) Death of an apparently healthy infant for which there seems no obvious cause. Often associated with sleeping on chest but the precise cause still not understood. Also *cot death*, SIDS.

sudden oak death (*BioSci*) A destructive disease of oak trees caused by the fungus *Phytophthora ramorum*.

sudden warming (*EnvSci*) A rapid rise in temperature of the polar stratosphere of up to 50 K in a few days occurring in winter or early spring. It is associated with a breakdown of the winter polar stratospheric vortex and may be either temporary or, in spring, permanent. See panel on STRATOSPHERE AND MESOSPHERE.

sudor (*Med*) Sweat or perspiration.

sudoriferous (*BioSci*) Sweat-producing; sweat-carrying. Also *sudoriparous*.

sudorific (*Med*) Connected with the secretion of sweat: stimulating the secretion of sweat: a drug which does this.

sudoriparous (*BioSci*) See SUDORIFEROUS.

suffructescent (*BioSci*) Somewhat woody; diminutively shrubby; woody at base with herbaceous stems, eg alpine willows. Also *suffruticose*.

sugar (*Chem, FoodSci*) (1) A water-soluble, crystalline mono- or oligosaccharide. (2) The common term for *sucrose*, or *cane-sugar*, $C_{12}H_{22}O_{11}$. See REFINED SUGAR.

sugar beet (*FoodSci*) A biennial root crop from which sucrose is extracted. Approx 130 g of sugar can be extracted from 1 kg of beet. Widely grown throughout Europe.

sugar beet pulp (*Agri*) The residues of sugar beet after sugar extraction. Can be used as high-energy fodder for cattle and sheep, but unsuitable for horses because they swell during digestion.

sugar cane (*FoodSci*) A perennial grass and the principal source of sugar grown in tropical and subtropical countries. RAW CANE-SUGAR is crystallized from the juice from the crushed canes and can be further refined to produce white sugar.

sugar charcoal (*Chem*) Highly pure form of charcoal derived from sucrose.

sugar plant (*BioSci*) A plant in which carbohydrate is stored as sugar. Cf STARCH PLANT.

sugar soap (*Build*) An alkaline cleansing or stripping preparation for paint surfaces.

Suhl effect (*Electronics*) The reduction in lifetime of holes injected into an n-type semiconducting filament, by deflecting them to the surface using a powerful transverse magnetic field. Reverse of HALL EFFECT.

sulcus (*BioSci*) (1) Generally, a groove or furrow, as one of the grooves on the surface of the cerebrum in mammals.

(2) In Dinoflagellata, a longitudinal groove in which a flagellum lies. (3) In Anthozoa, the 'ventral' siphonoglyph.

sulfasalazine (*Pharmacol*) A combination of aminosalicylic acid and the sulfonamide sulfapyridene. It is used to treat inflammatory bowel disease and some cases of rheumatoid arthritis. Also *sulphasalazine*.

sulfate, sulfur (*Chem*) Alternative spellings for sulphate and sulphur, standard in the USA. With a few exceptions, in the case of British approved names for drugs, the UK form (sulphate etc) is used throughout.

sulfonamides (*Pharmacol*) Synthetic bacteriostatic antibiotics with a wide spectrum of activity against most Gram-positive and many Gram-negative organisms. Sulfonamides inhibit multiplication of bacteria by acting as competitive inhibitors of p-aminobenzoic acid in the folic acid metabolism cycle. Formerly *sulphonamides*.

sulfotep (*Chem*) Bis-*OO*-diethylphosphorothionic anhydride, used as an insecticide. Also *dithio*, *dithioTEPP*, *thiotep*.

sulfur (*Chem*) US for SULPHUR.

sullage (*CivEng*) The mud and silt deposited by flowing waters.

sulphamic acid (*Chem*) The monoamide of sulphuric acid: $HO-SO_2-NH_2$. Prepared by the ammonolysis of sulphuric acid; commercially, by the reaction of urea (carbamide) with fuming sulphuric acid.

sulphamide (*Chem*) The diamide of sulphuric acid: $SO_2(NH_2)_2$. Prepared by the ammonolysis of sulphuryl chloride.

sulphane (*Chem*) A compound of the formula HS_xH, analogous to an alkane. See SULPHUR HYDRIDES.

sulphanilic acid (*Chem*) 4-*aminobenzene-sulphonic acid*. $H_2NC_6H_4SO_3H$. Crystallizes with $2H_2O$; sparingly soluble in water.

sulphate of ammonia (*Chem*) $(NH_4)_2SO_4$. Commercially the most important of the ammonium salts, particularly for use as fertilizer. Produced partly as a by-product of gas works, coke ovens, etc, but now largely by direct synthesis.

sulphate of iron (*Min*) See MELANTERITE.

sulphate of lead (*Min*) See ANGLESITE.

sulphate of lime (*Min*) See ANHYDRITE, GYPSUM.

sulphate of strontium (*Min*) See CELESTINE.

sulphate-resisting cement (*CivEng*) Cement manufactured for use in concrete to resist normal concentrations of sulphates as in most flues; also for underwater work.

sulphates (*Chem*) Salts of sulphuric acid. Produced when the acid acts on certain metals, metallic oxides, hydroxides and carbonates. The acid is dibasic, forming two salts: normal and acid sulphates.

sulphating roasting (*Eng*) Roasting of minerals carried out under conditions designed to convert part of the contained sulphur (sulphide mineral) to sulphate.

sulphation (*ElecEng*) The formation of the insoluble white sulphate of lead ($PbSO_4$) in the plates of a lead–acid type of secondary cell, a process which diminishes the efficiency and capacity of the cell.

sulphides (*Chem*) Salts of hydrogen sulphide. Many sulphides are formed by direct combination of sulphur with the metal.

sulphide toning (*ImageTech*) A process of toning photographic prints in which the silver is converted to silver sulphide via silver bromide.

sulphide zone (*MinExt*) Primary (unaltered) zone of sulphide-mineral lode, underlying leached (superficial) zone and that of secondary enrichment in which there has been redeposition of values oxidized from leached zone by penetrating water.

sulphinic acids (*Chem*) Acids containing the monovalent sulphinic acid group –SOOH.

sulphite process (*Paper*) One of the acid pulp digestion processes using a bisulphite liquor with some free sulphur dioxide.

sulphites (*Chem*) *Sulphates (IV)*. Salts of sulphurous (sulphuric (IV)) acid. The acid forms two series of salts, acid sulphites or bisulphites and normal sulphites.

sulphite wood pulp (*Paper*) Chemical wood pulp in which digestion is carried out by the sulphite process.

sulphocyanides (*Chem*) See THIOCYANATES.

sulphonation (*Chem*) The reversible process of forming sulphonic acids by the action of conc sulphuric acid on aliphatic or aromatic compounds.

sulphones (*Chem*) Compounds having the formula $R-SO_2-R'$. The sulphur is hexavalent.

sulphonic acids (*Chem*) Acids containing the monovalent sulphonic acid group $-SO_2OH$.

sulphonylureas (*Pharmacol*) A group of drugs that act by augmenting insulin secretion. Used in the treatment of Type II diabetes.

sulphosol (*Chem*) A colloidal solution in conc sulphuric acid.

sulphoxides (*Chem*) Compounds having the formula $R-SO-R'$. The sulphur is in the (IV) oxidation state.

sulphoxylic acid (*Chem*) The hypothetical oxy acid of sulphur, $S(OH)_3$.

sulphur (*Chem*) A non-metallic element occurring in many allotropic forms. Symbol S, at no 16, ram 32·06, valencies 2, 4, 6. Rhombic (β-)sulphur is a lemon-yellow powder; mp 112·8°C, rel.d. 2·07. Prismatic (monoclinic) (β-) sulphur has a deeper colour than the rhombic form; mp 119°C, rel.d. 1·96, bp 444·6°C. Chemically, sulphur resembles oxygen, and can replace the latter in many compounds, organic and inorganic. It is abundantly and widely distributed in nature with an abundance in the Earth's crust of 340 ppm, and 900 ppm in sea water. It occurs as the native element in volcanic regions in fumaroles and hot springs, and in sediments, esp the cap rocks of salt domes. Hydrogen sulphide (H_2S) and sulphur oxides (SO_2, SO_3) also occur in fumarolic or volcanic gases. H_2S is recovered from natural gas. Many ore bodies are of sulphides of metals (galena, PbS, sphalerite, ZnS, etc). In the evaporite deposits there are many sulphate minerals. Sulphur is used in the manufacture of sulphuric acid and carbon disulphide; in the preparation of gunpowder, matches, fireworks and dyes; as a fungicide, and in medicine; and for vulcanizing rubber. Vapour bath at boiling point is used as fixed point in platinum resistance thermometry. US *sulfur*.

sulphur bacteria (*BioSci*) Bacteria that live in situations where oxygen is scarce or absent, and that act upon compounds containing sulphur, liberating the element. They occur in two families of true bacteria, the *Thiorhodaceae* (purple sulphur bacteria) and the *Thiobacteriaceae* (colourless sulphur bacteria).

sulphur cement (*Build*) A cement made of sulphur and pitch mixed in equal parts; used to fix cast-iron work as in the first Tay Bridge.

sulphur dioxide (*Chem*) Sulphur (IV) oxide. SO_2. A colourless gas formed when sulphur burns in air. Dissolves in water to give sulphurous (sulphuric (IV)) acid. Also *sulphurous anhydride*. See SULPHUR OXIDES.

sulphuretted hydrogen (*Chem*) See HYDROGEN SULPHIDE.

sulphur hydrides (*Chem*) Four well-defined hydrides: H_2S, H_2S_2, H_2S_3 and H_2S_5. Also *sulphanes*.

sulphuric acid (*Chem*) A strong dibasic acid, H_2SO_4. The conc acid is a colourless oily liquid; rel.d. 1·85, bp 338°C; it dissolves in water with the evolution of heat, and is very corrosive, largely owing to its dehydrating action. It is manufactured from sulphur dioxide, obtained by burning either pyrites or sulphur, by the CONTACT PROCESS or the CHAMBER PROCESS. It is an important heavy chemical, used extensively in the dyestuffs and explosives industries; as a drying agent in chemical process; in the manufacture of other acids, eg HCl, HF, phosphoric acid; in fertilizers, pickling liquors and leaching solutions; in petroleum treatment and the manufacture of alkyl sulphonate detergents; in rayon production, etc. The salts of sulphuric acid are called *sulphates* (VI).

sulphuric anhydride (*Chem*) See SULPHUR TRIOXIDE.

sulphurous acid (*Chem*) An aqueous solution of sulphur dioxide, which contains the hypothetical compound H_2SO_3. The corresponding salts, sulphites, are well known.

sulphurous anhydride (*Chem*) See SULPHUR DIOXIDE.

sulphur oxides (*Chem*) A series of oxides: SO, S_2O_3, SO_3, S_2O_7 and SO_4.

sulphur trioxide (*Chem*) Sulphur (VI) oxide. SO_3. Dissolves in water to give sulphuric acid.

sumatriptan (*Pharmacol*) An antagonist of vascular $5HT_1$ receptors, used for the treatment of migraine.

summation (*Med*) The production of an effect by repetition of a causal factor which would be insufficient in a single application, as *summation of contractions*, the production of a state of tetanic contraction by a series of stimuli.

summation check (*ICT*) Figure added to a WORD, indicating a summation of the digits so that accuracy of processing can be verified. Totals arrived at by two methods for verifying processing of data.

summation instrument (*ElecEng*) An instrument for indicating, integrating or recording the sum total of the energy, power or current in two or more circuits.

summation of losses (*ElecEng*) The process of adding together the individual losses, after allowing for any corrections, in order to obtain the guaranteed efficiency of an electric machine.

summation panel (*ElecEng*) A switchboard panel on which are mounted the instruments for measuring and recording the total output of a number of generators.

summation tone (*Acous*) See COMBINATION TONE.

summer annual (*BioSci*) A plant that completes its life cycle over a few weeks in the summer, surviving the winter as seed. Cf WINTER ANNUAL.

summer draught (*Ships*) The DRAUGHT when loaded to the SUMMER-LOAD WATERLINE.

summer egg (*BioSci*) In many fresh-water animals, a thin-shelled, rapidly developing egg laid during the warm season. Cf WINTER EGG.

summer-load waterline (*Ships*) The waterline to which a ship may be loaded in summer. It is indicated in the freeboard markings.

summer mastitis (*Vet*) Mastitis of non-lactating cows due to infection by *Corynebacterium pyogenes*, or occasionally other organisms; possibly transmitted by flies.

summer solstice (*Astron*) See SOLSTICE.

summer wood (*BioSci*) See LATE WOOD.

summit canal (*Build*) A canal crossing a summit; one, therefore, to which water must be supplied.

sump (*Autos*) The lower part of the crankcase of an automobile engine, which usually acts as an oil reservoir.

sump (*CivEng*) A small hole dug usually at the lowest part of an excavation to provide a place into which water can drain and from which it can be pumped at intervals to keep the working part of the excavation dry. See CATCH PIT.

sump (*MinExt*) The prolongation of a shaft or pit, to provide for the collection of water in a mine. The pump sump is that from which casual water or ore pulp is delivered to the mine or mill pumps.

Sumpner test (*ElecEng*) A back-to-back load test, or regenerative test, on two similar transformers.

Sumpner wattmeter (*ElecEng*) An iron-cored type of dynamometer wattmeter for use on ac circuits.

Sun (*Astron*) The central object of our solar system and the nearest star to Earth, at an average distance of 149 600 000 km. The source of energy is nuclear reactions in the central core where the temperature is around 15 000 000 K and the relative density 155. The core extends to a quarter of the solar radius and includes half the mass. Our Sun is nearly 5 billion years old, and is about halfway through its expected life cycle. Every second it annihilates 5 billion tonnes of matter, to release 3×10^{26} W of energy. See AURORA, FACULAE, SUNSPOT and panels on HERTZSPRUNG–RUSSELL DIAGRAM and SUN AS A STAR.

sunburner (*Glass*) An excessive local thickness of material in a mouth-blown glass article.

Sun as a star

The Sun is 150 million km from Earth, so close that it can be studied in far greater detail than any other star (see solar constants below). The layer we see is the PHOTOSPHERE, the lowest layer of the solar atmosphere, with a temperature of about 5800 K. High-resolution observations show individual convection cells, formed through the phenomenon of GRANULATION. The upper photosphere oscillates with a period of 5 minutes and these *solar oscillations* give information about the deep interior. In the spectrum of the photosphere there are hundreds of thousands of FRAUNHOFER LINES, caused by the absorption of the inner radiation by the cooler regions of the upper solar atmosphere. Their careful study provides information about the abundances of chemical elements, and shows that the Sun consists of nearly 75% hydrogen, 25% helium and about 1% heavier elements.

Above the photosphere lies the *chromosphere*, which is about 10 000 km thick and is visible as a pinkish glow during a total solar eclipse. Much of its matter is arranged in spiky cylinders called *spicules*. The temperature is about 15 000 K, high enough to excite emission lines, seen as the FLASH SPECTRUM, during eclipse. The outermost layer is the CORONA where temperatures reach from 0·5 to 2 million K. At optical wavelengths plumes and streamers of matter extend far away from the Sun and X-ray telescopes have shown cool and dark zones in the corona, termed CORONAL HOLES.

The solar interior reaches 15 million K at the core (see diagram), high enough for the proton–proton reaction in which protons fuse to form helium-4 nuclei, the source of solar energy, releasing enormous quantities of neutrinos. A major problem in solar astronomy is that there are far fewer neutrinos than any plausible model of solar evolution can explain. Experiments at an underground detector in Japan suggest we may not see the expected number of neutrinos because they can flip into different, undetectable types as they travel.

Because of the presence of free electrons the interior is opaque to photons; one consequence of this is the long diffusion time, about a million years, for the energy released today to work its way to the photosphere. The present age of the Sun, 5 billion years, means that about half of the available protons have already been processed in nuclear fusion reactions.

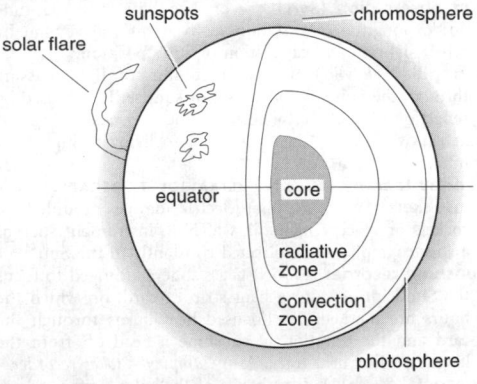

Sun Cutaway showing interior.

Much activity is superimposed on the basic structure of the Sun of which SUNSPOTS are the most obvious signs. Sunspots are cool dark regions (4500 K) in the photosphere which can last a few hours or several months. The activity is seen as the number of sunspots rises and falls in a cycle of approximately 11 years, and this affects the Earth's ionosphere and therefore telecommunications. SOLAR FLARES are bright hydrogen arcs which emit ultraviolet radiation and X-rays. They are usually associated with sunspots and may be accompanied by magnetic storms on the Earth and by the AURORA BOREALIS. The SOLAR WIND is a steady flow of matter from the corona and is responsible for keeping the tails of comets always pointing away from the Sun.

The Sun is located about 8 kiloparsecs from the centre of our galaxy and is about 10 parsecs above the galactic plane. Its orbital velocity of 250 km s^{-1} implies that it travels round our galaxy in about 220 million years, and it rotates about its own axis in just over 27 days. The table gives the main solar constants.

See panels on BLACK HOLE, COSMIC RAYS, COSMOLOGY and EARTH.

Radius	$6·96 \times 10^5$ km	Mass	$1·989 \times 10^{30}$ kg
Mean density	$1·409 \times 10^{32}$ kg m^{-3}	Surface gravity	$2·74 \times 10^2$ ms^{-1}
Escape velocity	$617·7$ km s^{-1}	Apparent magnitude	$-26·74$
Absolute magnitude	$+4·83$	Luminosity	$3·83 \times 10^{26}$ W
Effective temperature	5770 K	Central temperature	$1·5 \times 10^7$ K
Spectral class	G2V		

sun cracks (*Geol*) Polygonal cracks, usually in a fine-grained sedimentary rock, indicative of desiccation at the time of formation. Common in beds laid down under arid conditions.

sun gear (*Eng*) See EPICYCLIC GEAR.

sunk face (*Build*) A term applied to a stone in whose face a panel is sunk by cutting into the solid material. Abbrev *SF*.

sunk fence (*Build*) See HA-HA.

sunk key (*Eng*) A key which is sunk into key ways in both shaft and hub. See KEY.

sun observation (*Surv*) Use of theodolite to fix latitude and/or longitude of a station, or to orient a survey line by direct sighting and calculation of Sun's position.

sun pillar (*EnvSci*) A vertical column of light passing through the Sun, seen at sunset or sunrise. It is caused by reflection of sunlight by horizontal ice crystals.

sun plant (*BioSci*) A plant adapted to living at high light intensities. Cf SHADE PLANT.

sun-ray therapy (*Med*) See ULTRAVIOLET THERAPY.

sunseeker (*Space*) A photoelectric device mounted in rockets or space vehicles, in which an instrument such as a spectrograph can be directed constantly to the Sun.

sunshine recorder (*EnvSci*) Glass sphere arranged to focus the Sun's image onto a bent strip of card, on which the hours are marked. The focused heat burns through the card and the duration of sunshine is read off from the length of the burnt track. Also *Campbell–Stokes recorder*.

sunspot (*Astron*) A disturbance of the solar surface which appears as a relatively dark centre (*umbra*), surrounded by a less dark area (*penumbra*); spots occur generally in groups, are relatively short-lived, and with few exceptions are found in regions between 30°N and 30°S latitude. Their frequency shows a marked period of about 11 yr (the *sunspot cycle*). They have intense magnetic fields and are sometimes associated with magnetic storms on the Earth. See panel on SUN AS A STAR.

sunstone (*Min*) See AVENTURINE FELDSPAR.

sunstroke (*Med*) A condition produced by exposure to high atmospheric temperature and characterized by a rapid rise of bodily temperature, convulsions and coma. Also *heat hyperpyrexia, insolation*.

Sun-synchronous orbit (*Space*) A near-polar orbit in a plane inclined at a small angle to the Earth's axis which precesses slowly in space so that the phase of the orbit is constant relative to the Sun. This permits a satellite to pass over a point on the Earth at practically the same local time each day regardless of the longitude, at a height of about 1000 km, crossing the equator about 15 times a day in each direction.

sun wheel (*Eng*) A gearwheel round which one or more planet wheels or planetary pinions rotate in mesh.

super-, supra- (*Genrl*) Prefixes from Lt *super*, over, above.

Super-8 (*ImageTech*) A system of narrow-gauge cinematography on 8 mm film, mainly for amateur use, having a larger frame area than the original *Regular 8*.

Super-16 (*ImageTech*) A 16 mm motion picture film with a greater frame width than normal 16 mm. The negative is used for enlargement to 35 mm for wide-screen prints. It is also used as an originating medium for TV productions shot on location. See TELECINE.

superadditive function (*MathSci*) See ADDITIVE FUNCTION.

superaerodynamics (*Aero*) Aerodynamics at very low air densities occurring above 100 000 ft, ie for spacecraft on ascending and re-entry trajectories. The mean free path of the molecules is long compared with vehicle length. The physics is described by *free-molecule-flow* and Newtonian aerodynamics. *Magnetohydrodynamic* features are likely to be significant.

superalloy (*Eng*) Alloy capable of service at high temperatures, usually above 1000°C, eg for turbine blades and components. See HEAT-RESISTING ALLOY.

superantigens (*BioSci*) A group of antigens that activate multiple clones of T-lymphocytes by directly cross-linking

the T-cell antigen receptor and major histocompatibility complex class II molecules on an ANTIGEN PRESENTING CELL.

superaudio frequency (*Acous*) A frequency above those usefully transmitted through an audio-frequency reproducing system, or part of such.

supercalendered paper (*Paper*) Paper to which the surface finish has been applied by passing it through a supercalender. The gloss is usually greater than that attained by machine calenders. See CALENDERED PAPER.

supercharger (*Autos*) A compressor, commonly of the Rootes rotary vane or centrifugal type, used to supply air or combustible mixture to an internal-combustion engine at a pressure greater than atmospheric; driven either directly by the engine or by an exhaust gas turbine. See TURBOCHARGER.

supercharging (*Aero, Eng*) (1) In aero-engines, maintenance of ground-level pressure in the inlet pipe up to the rated altitude by means of a centrifugal or other blower. Necessary for flying at heights at which the air pressure is low and normal aspiration would be insufficient. (2) In other internal-combustion engines, the term is used synonymously with boosting. See BOOST.

superciliary (*BioSci*) Pertaining to, or situated near, the eyebrows; above the orbit.

supercirculation (*Aero*) A form of boundary layer control, creating and controlling high lift in which air (usually derived from a jet-engine compressor) is blown supersonically over the leading edge of a plain flap so that it carries the main airflow downwards below the actual surface as an invisible extension. May also involve blowing over the leading edge. See JET FLAP.

supercluster (*Astron*) A group of galaxy clusters, physically bound loosely by their mutual gravitational attraction. Typical scale 100 Mpc.

supercoiling (*BioSci*) In circular DNA or closed loops of DNA, the phenomenon whereby twisting of the DNA about its own axis changes the number of turns of the double helix; if the twist is anticlockwise the supercoiling is negative, if clockwise (the same as the DNA helix) it is positive. DNA that shows no supercoiling is said to be relaxed. The DNA of eukaryotes largely exists as supercoils associated with protein in the nucleosome.

supercomputer (*ICT*) A high-performance MAINFRAME typically used for the solution of numerical problems in the sciences.

superconducting amplifier (*Electronics*) One using superconductivity to give noise-free amplification.

superconducting gyroscope (*Electronics*) Frictionless gyroscope supported in vacuum through magnetic field produced by currents in superconductor.

superconducting levitation (*Phys*) An effect in which if a small permanent magnet is placed on a surface which is then made superconducting by lowering its temperature below its critical temperature, the magnet will rise and float above the surface. The superconductor as a perfect diamagnetic excludes the flux associated with the magnet and so provides a large enough repulsive force to balance the weight of the magnet. See MEISSNER EFFECT, SUPERCONDUCTIVITY and panel on ELECTRON MICROSCOPE.

superconducting magnet (*Electronics*) Very powerful electromagnet, made possible by the exceptionally large current-carrying capacity of superconductors.

superconducting memory (*Electronics*) One made up of thin-film devices which can be made to change from their low-temperature superconducting state to a normal resistive state by the brief application of a magnetic field, generated by current pulse on a control wire. The entire memory is maintained at the superconducting temperature and dissipates power only during the read or write operation; large, high-density memories can be built using these techniques.

superconducting quantum interference device (*Electronics*) One of a family of devices capable of measuring extremely small currents, voltages and magnetic fields. Based on two quantum effects in superconductors: (1) FLUX QUANTIZATION; and (2) the JOSEPHSON EFFECT. They can detect changes in magnetic flux densities of ≈ 1 nT (10^{-9} T). Applications include the study of fields generated by the action of the human brain and in magneto-biological research. Abbrev SQUID.

superconductivity (*Phys*) The property of some pure metals and metallic alloys of having negligible resistance to the flow of an electric current at very low temperatures. See MEISSNER EFFECT and panel on SUPERCONDUCTORS.

superconductor (*Phys*) A substance which exhibits super-conductivity. See panel on SUPERCONDUCTORS.

supercooled (*Chem*) Cooled below normal boiling or freezing point without change of phase.

supercooled water (*EnvSci*) Water which continues to exist as a liquid at temperatures below 0°C. Also *subcooled water*.

supercritical (*Phys*) Said of an assembly of fissile material for which the multiplication factor is greater than unity.

superego (*Psych*) In psychoanalytical theory, the part of the personality that has incorporated moral and ethical values and is responsible for self-imposed standards of behaviour; comprises the conscience or internal parent and the ego ideal.

superelevation (*Surv*) The amount by which the outer rail of a railway curve is elevated above the inner rail to counteract the effect of the centrifugal force of the moving train. Superelevation is also applied in the construction of highway curves. See CANT.

super exchange energy (*Phys*) The EXCHANGE ENERGY in ferrimagnetic materials which is analogous to that in ferromagnetic materials. It is associated with the coupling of neighbouring dipole moments of ions through an interaction with the electron spins on intermediate oxygen ions.

superfetation (*Med*) Fertilization of an ovum in a woman already pregnant, some time after fertilization of the first ovum. Also *superfoetation*.

superficial deposits (*Geol*) See DRIFT.

superficial radiation therapy (*Radiol*) X-ray therapy (usually by soft radiation produced at less than 140 kVp) of the skin or of any surface of the body made accessible.

superfinishing (*Eng*) An abrasive process, resembling honing and lapping, for removing SMEAR METAL and scratches and ridges produced by grinding and machining operations from bearing surfaces.

superfluid (*Phys*) Condensed degenerate gas in which a significant proportion of the atoms are in their lowest permitted energy state. In practice this occurs only for liquid helium II.

superfluidity (*Phys*) The state of a SUPERFLUID. At temperatures below 2·19 K, the lambda point, this behaviour occurs for the helium-4 isotope which shows a striking change in its liquid properties; in particular its viscosity effectively vanishes. At temperatures above the lambda point, helium-4 is known as helium I, and below, helium II. Superfluid helium II behaves as if it consists of two parts, a superfluid and a normal fluid. The lower the temperature, the greater the fraction of superfluid in the mixture. The superfluid part consists of helium atoms in their lowest quantum state.

supergene enrichment (*MinExt*) Part of a mineral vein, lode or massive deposit where material removed from the *leached zone* is re-precipitated. See SECONDARY ENRICHMENT.

supergiant star (*Astron*) Star of late type and abnormally high luminosity, such as BETELGEUSE and ANTARES; they are of enormous size and low density.

superglue (*Plastics*) Popular name for cyanoacrylate adhesive, supplied as fluid monomer or prepolymer, which polymerizes when in contact with surfaces.

supergrain (*Agri*) BREWER'S GRAINS from the whisky industry that contain the yeasts from fermentation in addition to the non-starchy grain residues.

supergravity (*Astron, Phys*) A particular version of a SUPERSYMMETRY theory which postulates GRAVITONS and *gravitinos* as carriers of the gravitational force.

supergrid (*ElecEng*) The national electric power network at voltages of 275 kV and 400 kV. See GRID.

superheat (*Aero*) The increase (positive) or decrease of the temperature of the gas in a gas-bag as compared with the temperature of the surrounding air. Similarly, *super-pressure*.

superheated steam (*Eng*) Steam heated at constant pressure out of contact with the water from which it was formed, ie at a higher temperature than that of saturation.

superheterodyne receiver (*Phys*) See SUPERHET RECEIVER.

superhet receiver (*ICT*) Abbrev for *supersonic heterodyne receiver*. One in which the frequency of the incoming signal is reduced in a mixer or frequency changer, by HETERODYNING with another frequency from the LOCAL OSCILLATOR. The lower-frequency output from the mixer, the intermediate frequency or IF, is taken through one or more steps of a selective amplifier, before conventional demodulation. There may also be stages of signal-frequency amplification before the mixer, and two stages of frequency conversion and two separate IF amplifiers. Advantages include better gain and selectivity; disadvantages are the extra components and complication, and the possibility of receiving IMAGE RESPONSES.

super-high frequency (*ICT*) A radio frequency in the range 3000 to 30 000 MHz. Abbrev SHF.

superimpose (*ImageTech*) Adding one image on top of another, so that both are visible.

superimposed drainage (*Geol*) A river system unrelated to the geological structure of the area, as it was established on a surface since removed. Cf CONSEQUENT DRAINAGE.

superimposition (*Print*) Coloured blocking foils are frequently used as a base for overstamping by gold, silver, or other coloured foil. This 'superimposition' is of particular importance to bookbinders and display-card printers.

superior (*BioSci*) An ovary in a flower that is HYPOGYNOUS or PERIGYNOUS.

superior (*Genrl*) Placed above something else; higher, upper (as the *superior* rectus muscle of the eyeball).

superior (*Print*) A term used to describe small figures or letters printed above the general level of the line. They are used instead of MARKS OF REFERENCE, and in mathematical work etc; thus x^2, e^x, 10^6.

superiority complex (*Psych*) Overvaluation of one's worth, often affected to cover a sense of inferiority.

superior planet (*Astron*) See PLANET.

superior vena cava (*BioSci*) See PRECAVAL VEIN.

SuperJANET (*ICT*) A greatly enhanced version of JANET (Joint Academic Network), operating over OPTICAL FIBRE links and having sufficient bandwidth to allow the exchange of still and moving images as well as text.

superlattice (*Chem*) An ordering of solute atoms in a SUBSTITUTIONAL SOLID SOLUTION which results in a regular pattern of solute atoms superimposed upon the solvent lattice structure, like Cu_3Au and Ni_3Al (face-centred cubic) and CuZn (body-centred cubic). The ordering may extend over a few lattice units (short-range) or over the whole crystal (long-range) when it may significantly affect mechanical properties.

superluminal motion (*Astron*) Motion which is apparently faster than the speed of light and due to geometrical effects. Can occur in some components of astronomical radio sources.

supermassive black hole (*Astron*) See panel on BLACK HOLE.

supermassive star (*Astron*) Hypothetical stars of around 1000 solar masses or more; the existence of such objects has not been proven.

supernatant liquid (*Chem*) The clear liquid above a precipitate which has just settled out.

Superconductors

Superconductors are a class of materials which lose their resistance to the flow of dc when cooled below a characteristic transition temperature, the critical temperature, T_c. It is a property exhibited by tens of metallic elements and thousands of compounds. Superconductors usually behave as poor metals at room temperature with dc resistivities typically around 10^{-3} Ω m. This resistivity becomes immeasurably small on transition to the superconducting state (ie less than 10^{-25} Ω m). Superconductivity is a quantum mechanical effect resulting from a cooperative interaction between conduction electrons close to the FERMI SURFACE. It is usually interpreted within the framework proposed by Bardeen, Cooper and Schrieffer (the BCS THEORY) which has been applied successfully to most known superconducting materials.

In a normal metal the lattice vibrations (PHONONS) scatter the conduction electrons and hence contribute significantly to its electrical resistance. In a superconductor, however, the same lattice vibrations bind electrons together in COOPER PAIRS. These electron pairs are then able to move unimpeded through the lattice when an electric field is applied. If induced in a sample of toroidal geometry, therefore, a dc supercurrent will flow indefinitely, ie without any dissipation of energy.

A second manifestation of superconductivity is that bulk superconductors behave as perfect diamagnets when exposed to a weak magnetic field. A state of zero magnetic field is maintained in the interior of the superconductor by eddy currents induced in a thin layer at the surface, known as the PENETRATION DEPTH, approx 100 nm thick. This flux expulsion is known as the MEISSNER EFFECT and it is the hallmark of superconductivity.

A superconductor will revert to its normal conducting state when the applied magnetic field exceeds a critical value. At this point the sum of the stored magnetic energy and the energy of the free superconducting state exceeds that of the normal state. Superconductors are classified as either type I or type II according to their behaviour in an applied magnetic field. Type I materials lose their properties abruptly, whereas type II superconductors undergo a gradual transition as the magnetic field is increased. The diamagnetic response of type I and II superconductors to an applied magnetic field is shown in the diagrams.

Superconductivity is also destroyed by current densities in excess of a critical value. The normal state is separated from the superconducting state by an ENERGY GAP. Properties such as CRITICAL FIELD, critical CURRENT DENSITY and energy gap increase sharply as the temperature of the material is reduced below T_c before levelling off as absolute zero is approached. For most practical applications superconductors have to be stabilized at a temperature less than $(2/3)T_c$ to yield close to optimum properties.

High-temperature superconductors

The discovery in 1986 of high-temperature superconducting ceramics which exhibit T_c values up to 125 K is significant in that liquid nitrogen can be used as a coolant (boiling point approx 77 K) rather than the more expensive and difficult to contain liquid helium (boiling point approx 4 K). Yttrium barium copper oxide ($YBa_2Cu_3O_{7-x}$, $T_c \approx 92$ K) and bismuth calcium strontium copper oxide ($Bi_2Ca_2Sr_2Cu_3O_z$, $T_c \approx 105$ K), in particular, offer great potential for a variety of commercial ceramic and thin-film applications. It is unlikely that these materials will replace liquid-helium-cooled Nb_3Sn ($T_c \approx 18$ K) in superwire applications in the short term, however, owing to their refractory nature.

Superconductors find application in zero-resistance devices such as high-field electromagnets, Josephson devices such as SQUIDS (SUPERCONDUCTING QUANTUM INTERFERENCE DEVICES) which are used to detect minute magnetic fields ($< 10^{-10}$ T) and magnetic devices such as NMR-based medical imaging systems.

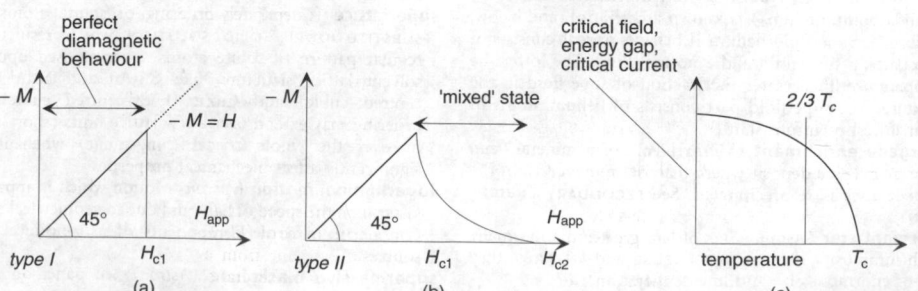

Superconductivity (a) Type I superconductor and (b) type II superconductor behaviour in a weak magnetic field; (c) the change in properties of a type I superconductor below the critical temperature (T_c) and as the temperature approaches absolute zero.

supernormal stimulus (*Psych*) A stimulus that surpasses a natural stimulus in its ability to evoke a response.

supernova (*Astron*) Novae of absolute magnitude −14 to −16; several have been recorded in our own Galaxy, and many hundreds in other spiral galaxies. The violent outburst results from the gravitational collapse of a massive star, the outer layers being explosively ejected as a gas shell; the core may be left to collapse forming a NEUTRON STAR or black hole. See panels on BLACK HOLE and PULSAR.

supernumary chromosomes (*BioSci*) Also B-CHROMO-SOMES.

superovulation (*BioSci*) Hormone-induced excess OVULATION. See INSEMINATION.

superoxide anion (*BioSci*) O_2^-. An oxygen molecule that carries an extra unpaired electron, and is therefore a free radical. One of the reactive oxygen species generated in neutrophil leucocytes and mononuclear phagocytes when activated eg by ingestion of particles or immune complexes.

superoxide dismutase (*BioSci*) Any of a range of metalloenzymes that catalyse the formation of hydrogen peroxide and oxygen from superoxide, and thus protect against superoxide-induced damage.

superoxides (*Chem*) Compounds of the alkali and alkaline earth metals containing the O_2^- group, eg $K^+(O_2^-)$. Differ from peroxides in yielding oxygen as well as hydrogen peroxide on hydrolysis.

superparamagnetism (*Phys*) The apparent paramagnetic behaviour exhibited by small single-domain particles when their thermal energy is comparable with the magnetization energy penalty exacted by their departure from an easy axis.

superphosphate (*Chem*) Superphosphate of lime, an agricultural fertilizer; a mixture of calcium sulphate and dihydrogen calcium phosphate; made by treating bone ash or basic slag (calcium phosphate) with sulphuric acid.

superplastic forming/diffusion bonding (*Aero*) Means of manufacturing by joining parts of structures together at high temperature and pressure. Abbrev *SPF/DB*.

superplasticity (*Eng*) The extremely high ductility shown by certain alloys when deformed at low strain rates at moderately high temperatures. Zinc–aluminium alloys near their eutectoid composition behave this way around 275°C. Normally ductile metals work-harden as they deform and neck down locally which rapidly leads to fracture. Superplastic materials do not neck down and can undergo considerable plastic deformation provided the rate is kept below a critical level, thus resembling thermoplastic materials. Similar forming processes are possible but at lower production rates because of the slow deformation rate needed.

superposed circuit (*ICT*) An additional channel obtained from one or more existing circuits, normally provided for other channels, in such a way that all the channels can be used simultaneously without mutual interference.

superposition (*Phys*) The state of being in coincidence; something placed vertically over or on something else, eg two interfering waveforms.

superposition, law of (*Geol*) Strata which overlie other strata are always younger, except in strongly folded areas.

superposition theorem (*ElecEng*) That any voltage/current pattern in a linear network is additive to any other voltage/current pattern.

superpressure (*Aero*) See SUPERHEAT.

super proton synchrotron (*Phys*) See SPS.

super-refraction (*EnvSci*) Refraction greater than standard refraction.

super-regeneration (*ICT*) Regeneration, or feedback, leading to oscillation that is broken up or quenched at a frequency above the upper limit of audibility by a separate oscillator circuit suitably connected to the main amplifying circuit. Amplifiers using this phenomenon can achieve extremely high gain and sensitivity with the minimum of circuit components.

super-regenerative receiver (*ICT*) One with sufficient positive feedback to result in a quenched supersonic oscillation (squegging), with consequent increase in sensitivity but also increase in distortion of demodulated signals.

super-releasers (*Psych*) Stimuli with very marked effects, eliciting behaviour in a very effective manner. Also *supernormal stimulus*.

supersaturation (*Chem*) Solution containing solute in excess of equilibrium. Condensation can take place on nuclei, particularly ions, eg those produced by high-speed charged particles, exhibiting a track of minute but visible water drops, as in a WILSON CHAMBER.

superscalar (*ICT*) Of a processor able to perform more than one instruction simultaneously. See MULTITHREADING.

supersonic (*Phys*) Faster than the speed of sound in that medium. Erroneously used for ultrasonic. See MACH NUMBER, ULTRASONIC and panel on AERODYNAMICS.

supersonic boom (*Acous*) Shock wave produced by an object moving supersonically. At a large distance from the object the time history of the pressure has the shape of an N and is therefore called *N-wave*.

supersonic speed (*Aero*) Applies to aircraft when its speed exceeds that of local sound. Applies to airflow anywhere when local speed exceeds that of sound.

supersonic wind tunnel (*Aero*) A wind tunnel in which the stream velocity in the working section exceeds the local speed of sound.

super stall (*Aero*) This phenomenon appeared with the adoption of high tailplanes for swept-wing jet aircraft. When the disturbed airflow from a stalled wing renders the tail controls inoperative, the aircraft will remain in a stable, substantially level attitude, while descending very rapidly. Recovery is by releasing the tail parachute to raise the tail clear of the wing wake so that the elevators again become operative.

superstitious behaviour in animals (*Psych*) Behaviour that is produced by the joint action of REINFORCEMENT and accident; certain acts which happen to coincide with reinforcement will tend to increase; these are often of a bizarre and fixed nature.

superstring theory (*Astron, Phys*) A version of STRING THEORY that incorporates ideas of supersymmetry in which all classes of elementary particles are placed on an equal footing. The astronomical context is that these classes of theory may have applied to matter in the very early universe. See COSMIC STRING.

superstructure (*CivEng*) The part of a structure carried upon any main supporting level.

superstructure (*Ships*) All structures other than masts that project above the upper deck of a ship.

supersulphated cement (*CivEng*) Cement manufactured using a high proportion of granulated blast-furnace slag.

supersymmetry (*Astron, Phys*) A theory which attempts to link all four fundamental forces, and postulates that each force emerged separately during the expansion of the very early universe.

supertraits (*Psych*) Eysenck's term for two distinct categories of personality traits, *introversion–extroversion* and *neuroticism*, each a continuum. Individuals differ in their relative positions on each scale.

supertwist (*ICT*) A particular type of liquid crystal display used on (primarily) portable computers.

supervisor (*ICT*) See MONITOR.

supervisor program (*ICT*) See MONITOR.

supervisory control (*ElecEng*) A method of remote control of electrical plant from a distant centre in which back-indication of the several control operations is given to the control centre.

supervoltage therapy (*Radiol*) Application of voltage, over a million volts, to X-ray tubes or accelerators in therapy. Also *megavoltage therapy*.

supination (*BioSci*) In some higher vertebrates, movement of the hand and forearm by which the palm of the hand is turned upwards and the radius and ulna are brought parallel to one another. Cf PRONATION. Adj *supinate*.

supinator (*BioSci*) A muscle effecting supination.

supplemental (*BioSci*) Additional; extra; supernumerary, as (in some Foraminifera) *supplemental skeleton*, a deposit of calcium carbonate outside the primary shell. Also *supplementary*.

supplemental chords (*MathSci*) Of a circle, ellipse or hyperbola: the chords joining any point on the curve to the ends of any diameter. Diameters parallel to supplemental chords are conjugate (perpendicular for a circle).

supplementary (*BioSci*) See SUPPLEMENTAL.

supplementary angles (*MathSci*) Two angles whose sum is 180°. Cf COMPLEMENTARY ANGLES.

supplementary lens (*ImageTech*) A lens placed in front of the main lens of a camera to alter its effective focal length without changing the distance from the film plane. Also *afocal attachment*.

supply frequency (*ElecEng*) The electrical frequency, in Hz, of an ac supply.

supply meter (*ElecEng*) An instrument for measuring the total quantity of electrical energy supplied to a consumer during a certain period.

supply point (*ElecEng*) A point on an electric power system from which electrical energy may be drawn.

supply station (*ElecEng*) See GENERATING STATION.

supply terminals (*ElecEng*) Those at which connection may be made to a supply point.

supply voltage (*ElecEng*) The voltage across a pair of supply terminals.

suppository (*Med*) A conical or cylindrical plug of a medicated mass for insertion into the rectum, vagina or urethra.

suppressed-carrier system (*ICT*) One in which the carrier wave is not radiated but is supplied by an oscillator at the receiving end. See DOUBLE-SIDEBAND SYSTEM, SINGLE-SIDEBAND SYSTEM.

suppressed-zero instrument (*ElecEng*) An indicating or graphic instrument in which the zero position or first scale reading is off-scale, ie beyond the range of travel of the pointer. Also *inferred-zero instrument, set-up instrument, set-up-scale instrument, set-up-zero instrument, step-up instrument*.

suppression (*BioSci*) (1) The absence of some organ or structure normally present. (2) The process by which a mutant phenotype is restored to normal by a mutation at another locus. Cf REVERSION.

suppression (*ICT*) Elimination of specified data or digits, eg initial zeros.

suppression (*Med*) The stoppage of discharge, as by obstruction of a duct.

suppression (*Psych*) In psychoanalytical theory, the conscious exclusion of unacceptable thoughts, ideas or impulses; a defence mechanism. In contrast, repression is considered an unconscious mechanism.

suppressor (*ElecEng*) A component, such as a capacitor or resistor, or both, which damps high-frequency oscillation liable to arise on breaking a current at a contact, causing radio interference.

suppressor cell (*BioSci*) A lymphocyte capable of suppressing antibody production or a specific cell-mediated response made by other cells. Suppression may be antigen-specific or non-specific. Also *T-suppressor cell*.

suppressor grid (*Electronics*) That between anode and screen in pentode valves, to repel secondary electrons back to the anode.

suppressor-grid modulation (*Electronics*) Insertion of the signal voltage into the (suppressor-)grid circuit of a valve which is amplifying the carrier. Also *grid modulation*.

suppressor mutation (*BioSci*) A base change that suppresses the effect of mutations elsewhere. Thus a base change at the anticodon site of a tRNA can suppress lethal mutations which would otherwise result in chain termination, or the insertion of an unacceptable peptide into a protein.

suppuration (*Med*) The softening and liquefaction of inflamed tissue, with the production of pus. Adj *suppurative*.

supra- (*Genrl*) See SUPER-.

supradorsal (*BioSci*) On the back; above the dorsal surface; a dorsal intercalary element of the vertebral column.

supra-occipital (*BioSci*) A median dorsal cartilage bone of the vertebrate skull forming the roof of the brain case posteriorly.

suprarenal (*BioSci*) Situated above the kidneys.

suprarenal body (*BioSci*) In higher vertebrates, one of the endocrine glands lying close to the kidney and releasing into the blood secretions having important effects on the metabolism of the body. Also *adrenal gland, suprarenal gland*. See ADRENAL CORTEX, ADRENAL MEDULLA.

supremum (*MathSci*) Least upper bound. See BOUNDS OF A FRACTION.

surcharge (*CivEng*) A term applied to the earth supported by a retaining wall at a level above the top of the wall.

surd (*MathSci*) An irrational root or the sum of such roots, eg $\sqrt{2}$ or $\sqrt{3} + 3\sqrt{5}$.

surface absorption coefficient (*Phys*) See ABSORPTION COEFFICIENT.

surface acoustic wave (*Radar, ICT*) Acoustic wave, that may have frequencies corresponding to the microwave bands, travelling along the optically polished surfaces of a piezoelectric substrate, at a velocity about 10^{-5} that of light. Used in microwave components and amplifiers. Abbrev *SAW*.

surface acoustic wave device (*Radar, ICT*) A device that manipulates electronic signals by coupling them with (Rayleigh) surface acoustic waves on a piezoelectric crystal; used in high-frequency filter circuits and delay lines. Surface acoustic waves travel at a speed close to that of sound waves in the bulk material. See DIELECTRIC AND FERROELECTRIC MATERIALS.

surface active agent (*Chem*) A substance which has the effect of reducing the surface tension of water and other liquids or solids, eg a detergent or soap. Also *surfactant*.

surface activity (*Chem*) The influence of certain substances on the surface tension of liquids.

surface barrier (*Electronics*) Potential barrier across surface of semiconductor junction due to diffusion of charge carriers.

surface boundary layer (*EnvSci*) The atmospheric layer, extending to a height of about 100 m, in which the motion is controlled predominantly by the presence of the Earth's surface. It forms the lowest part of the FRICTION LAYER.

surface charge (*ElecEng*) See BOUND CHARGE.

surface charge density (*Phys*) The amount of electric charge per unit surface area.

surface chemistry (*Chem*) That of the interface or interphase between two systems in which the substrate of one or both has become ionically unbalanced.

surface combustion (*Eng*) Bringing a combustible mixture of gas and air into contact with a suitable refractory material so as to produce flameless or nearly flameless combustion, the surface of the refractory material being maintained in a state of incandescence.

surface compressibility (*Chem*) The compressibility of the layer of adsorbed molecules in response to surface tension forces.

surface concentration excess (*Chem*) The excess concentration (may be negative) of a solute per unit area in the surface layer of a solution.

surface condenser (*Eng*) A steam condenser for maintaining a vacuum at the exhaust pipe of a steam engine or turbine. It consists of a chamber in which cooling water is circulated through tubes, and which is evacuated by an air pump. See CONDENSER, CONDENSER TUBES.

surface conductivity (*Phys*) See SURFACE RESISTIVITY.

surface duct (*ICT*) Atmospheric propagation duct for which the Earth's surface forms the lower boundary. Cf WAVEGUIDE.

surface energy (*Phys*) The free potential energy of a surface, equal to the SURFACE TENSION multiplied by the surface area.

surface engineering (*Eng*) The treatment of surfaces to confer on them properties distinct from those of the bulk; eg surface heat treatments, carburizing and nitriding, laser annealing, thin-film coating and the techniques associated with semiconductor fabrication.

surface-friction drag (*Aero*) That part of the drag represented by the components of the pressures at points on the surface of an aerofoil, resolved tangential to the surface. Also *skin friction*.

surface gauge (*Eng*) See SCRIBING BLOCK.

surface-grinding machine (*Eng*) A grinding machine for finishing flat surfaces. It consists of a high-speed abrasive wheel, mounted above a reciprocating or rotating work table on which flat work is held, often by a MAGNETIC CHUCK.

surface hardening (*Eng*) Processes for locally increasing the hardness of surfaces of finished components to improve wear resistance or other performance-related properties, eg fatigue resistance. They include those in which elements are diffused into the surface layers, eg CARBURIZING, FLAME and INDUCTION HARDENING, NITRIDING.

surface irradiation (*Radiol*) Irradiation of a part of the body by applying a mould or applicator loaded with radioactive material to the surface of the body.

surface leakage (*ElecEng*) That along the surface of a non-conducting material or device. May vary widely with contamination, humidity, etc. It sets a practical limit to the value of high resistors for use with electrometers etc.

surface lifetime (*Electronics*) The lifetime of current carriers in the surface layer of a semiconductor (where recombination occurs most readily). Cf VOLUME LIFETIME.

surface loading (*Aero*) The average force per unit area, normal to the surface, on an aerofoil under specified aerodynamic conditions.

surface measure (*For*) A method of measuring timber in quantity, by the area of one face, irrespective of thickness. Cf BOARD MEASURE.

surface noise (*Acous*) (1) See NEEDLE SCRATCH. (2) Underwater noise produced by waves on the sea surface.

surface of operation (*Build*) A plane surface for use as a reference from which the rest of the work can be set out.

surface oil resistance time (*Paper*) Abbrev SORT. An indication of the printing ink hold-out properties of a paper by measuring the resistance to penetration by a drop of liquid paraffin spread by a roller over a sample supported on an inclined plane, under the specified conditions of test. Results are expressed in seconds.

surface pipe (*MinExt*) See ANCHOR STRING.

surface plasmon resonance (*BioSci*) Alteration in light reflectance as a result of the binding of molecules to a surface from which total internal reflection is occurring. This effect is used in the proprietary Biacore instrument, which detects the binding of ligand to surface-immobilized receptor or antibody.

surface plate (*Eng*) A rigid cast-iron plate whose surface is accurately scraped flat; used to test the flatness of other surfaces or to provide a truly plane datum surface in marking off work for machining.

surface plates (*Print*) A general name for litho plates which are not DEEP ETCH or BIMETALLIC.

surface potential (*BioSci*) The electrostatic potential due to charged groups and adsorbed ions at a surface. It is usually measured as the zeta potential at the Helmholtz slipping plane outside the surface.

surface pressure (*Chem*) The two-dimensional analogue of gas pressure. Defined as the difference between the surface tension of a pure liquid and that of a surface active solution, it represents the tendency of the adsorbed surfactant molecules to spread over the clean liquid surface. See GIBB'S ADSORPTION THEOREM.

surface recombination velocity (*Electronics*) Electron–hole recombination on surface of semiconductor occurs more readily than in the interior, hence the carriers in the interior drift towards the surface with a mean speed termed the surface recombination velocity. It is defined as the ratio of the normal component of the impurity current to the volume charge density near the surface.

surface resistivity (*Phys*) Resistivity between opposite sides of a unit square inscribed on the surface. Its reciprocal is *surface conductivity*.

surface sterilization (*Phys*) Sterilization using low-energy radiation (eg ultraviolet) which penetrates thin surface layers only.

surface strength (*Paper*) The resistance of a paper to an adhesive force acting normally to the surface.

surface tension (*Phys*) A property possessed by liquid surfaces whereby they appear to be covered by a thin elastic membrane in a state of tension, the surface tension being measured by the force acting normally across unit length in the surface. The phenomenon is due to unbalanced molecular cohesive forces near the surface. Units of measurement are $N\,m^{-1}$. See CAPILLARITY, LIQUID-DROP MODEL, PRESSURE IN BUBBLES.

surface wave (*Phys*) (1) A wave propagated along the surface of a liquid or solid. For deep-water waves (the wavelength less than the water depth), the phase velocity depends on gravitational forces, surface tension and wavelength. For shallow-water waves (the wavelength greater than the depth), the phase velocity depends only on the depth and is independent of the wavelength. See RIPPLE TANK, TSUNAMI, WAVE. (2) A component of an electromagnetic wave radiated from a relatively low antenna, which depends on the nature of the surface. See GROUND WAVE.

surface wind (*EnvSci*) The wind at a standard height of 10 m (33 ft) above ground. Differs from the GEOSTROPHIC WIND and the GRADIENT WIND because of friction with the Earth's surface.

surface wiring (*ElecEng*) A wiring installation in which the insulated conductors are attached to the surfaces of a building, either enclosed in conduit or secured by cleats.

surfactant (*Chem*) Abbrev for *surface active agent*. A compound that reduces the surface tension of its solvent, eg a detergent or soap dissolved in water.

surfactant flooding (*MinExt*) Recovery enhancement process in oil wells in which surface-tension-reducing compounds are forced into the surrounding strata and release oil held there.

surfer (*ICT*) A person who browses the INTERNET without a specific destination site in mind.

surge (*Aero*) Unstable airflow condition in the compressor of a gas turbine due to a sudden increase (or decrease) in mass airflow without a compensating change in pressure ratio.

surge (*ElecEng*) A large but momentary increase in the voltage of an electric circuit.

surge absorber (*ElecEng*) A circuit device which diverts, and may partly dissipate, the energy of a surge, thus preventing possible damage to apparatus or machines connected to a transmission line. Also *surge modifier*.

surge arrester (*ElecEng*) See LIGHTNING ARRESTER.

surge bin (*MinExt*) Hopper (dry material) or reservoir with means of agitation (ore pulps), used to minimize irregularities in process delivery and flow. Also *surge tank*.

surge-crest ammeter (*ElecEng*) An instrument for recording a surge on a transmission line by measurement of the residual magnetism in a piece of magnetic material which has been magnetized by the surge current.

surge generator (*ElecEng*) See IMPULSE GENERATOR.

surge impedance (*Phys*) See CHARACTERISTIC IMPEDANCE.

surge modifier (*ElecEng*) See SURGE ABSORBER.

surge point (*Autos*) Of a centrifugal supercharger, the value of the mass airflow at which, during throttling of the delivery, surging occurs. See SURGING (1).

surge tank (*MinExt*) See SURGE BIN.

surge tank (*NucEng*) One used to absorb irregularities in flow like a suitably pressurized tank connected to eg the secondary steam-generating circuit of a reactor. It is able to compensate for changes in flow rate through the pumps by accepting or supplying condensate to the system.

surgical spirit (*Chem*) Ethanol, to which is added small amounts of oil of wintergreen and castor oil; used chiefly for sterilizing the skin in surgical operations.

surging (*Autos*) (1) In centrifugal superchargers, an abrupt decrease or severe fluctuation of the delivery pressure as the weight of air delivered is reduced. See SURGE POINT. (2) In valve springs, the coincidence of some harmonic of the cam lift curve with the spring's natural frequency of vibration, leading to irregular action and failure.

Surlyn (*Chem*) TN for ionomer resin (US).

surmounted (*Arch*) A term applied to a vault springing from points below its centre and having the form of a circular arc above its centre.

surra (*Vet*) A form of trypanosomiasis affecting horses, dogs, cattle, elephants and camels, occurring in Asia and Sudan, caused by *Trypanosoma evansi*; symptoms include emaciation and subcutaneous oedema, usually fatal. Transmitted by biting flies.

surround sound (*ImageTech*) An audio system used in feature films, video SOFTWARE and broadcast TV programmes, to provide front stereo, central dialogue, and ambient and directional effects, which can move spatially. See AC-3, PRO LOGIC.

surveillance radar (*Radar*) A plan-position indicator radar showing the position of aircraft within an air-traffic control area or zone.

surveillance TV (*ImageTech*) Use of closed-circuit TV for prevention and detection of crime, including video recording for evidence; in unattended areas it may employ SLOW SCAN TV and be activated by unexpected movement.

surveying (*Surv*) Measurement of the relative positions of points on the surface of the Earth and/or in space, to enable natural and artificial features to be depicted in their true horizontal and vertical relationship by drawing them to scale on paper.

Surveyor (*Space*) A series of NASA robotic soft-lander missions to the Moon (1966–8) in preparation for the manned Apollo landings.

survival curve (*Radiol*) One showing the percentage of organisms surviving at different times after they have been subjected to large radiation dose. Less often, one showing percentage of survivals at given time against size of dose.

survivorship curve (*BioSci*) The number or percentage of an original population surviving, plotted against time, indicating the mortality rate at different ages.

susceptance (*Phys*) Imaginary part of ADMITTANCE, equals

$$-\frac{X}{R^2 + X^2}$$

in a circuit of impedance $R + jX$.

susceptibility (*ElecEng*) See ELECTRIC SUSCEPTIBILITY, MAGNETIC SUSCEPTIBILITY.

susceptibility curves (*ElecEng*) Curves of susceptibility plotted to a base of magnetic field strength.

susceptor (*FoodSci*) A foil laminated to paper, often a part of the packaging, placed in contact with food in a microwave oven. The microwave energy results in arcing and the resulting radiant energy browns the surface of the food. Often patterned to give a browning pattern on the food.

susceptor phase advancer (*ElecEng*) A phase advancer which injects into the secondary circuit of an induction motor an emf which is a function of the open-circuit secondary emf Cf EXPEDOR PHASE ADVANCER.

suspect terrane (*Geol*) See TERRANE.

suspended scaffold (*Build*) A form of scaffold used in the construction, repair, cleaning, etc, of buildings in which

light working platforms are slung from higher points in the building.

suspended span (*CivEng*) The middle length of a bridge span connecting, and carried upon, the cantilever arms, when the span is built after the arms. See panel on BRIDGES AND MATERIALS.

suspension (*Autos*) A system, primarily of springs (leaf or coil) and dampers (usually hydraulic), designed to support the body of a vehicle and to protect it, and hence its occupants, from road shocks. See HYDROLASTIC.

suspension (*Chem*) A system in which denser particles, which are at least microscopically visible, are distributed throughout a less dense liquid or gas, settling being hindered either by the viscosity of the fluid or by the impacts of its molecules on the particles.

suspension bridge (*CivEng*) A bridge in which the deck is suspended from massive cables supported by towers near the sides of the waterway and anchored at each end of the bridge. For large-span bridges, experience has shown that the possibility of harmonic action must be taken into account and that wind tunnel experiments are needed to determine the necessary degree of lateral stiffness. See panel on BRIDGES AND MATERIALS.

suspension cable anchor (*CivEng*) The anchorage, which may have various forms, of the cables of a suspension bridge.

suspension culture (*BioSci*) A method of culturing large quantities of cells which are kept in vessels continuously stirred and aerated. Sterile nutrient media can be added and spent media removed. Used eg for producing cell products like interferon or antibodies.

suspension insulator (*ElecEng*) A freely hanging insulator made of units connected in series, by which an overhead line is suspended from the arm of a transmission-line tower; it cannot withstand any other force than a tension. Also *chain insulator*.

suspension polymerization (*Chem*) See CHAIN POLYMERIZATION.

suspensoid (*Chem*) See LYOPHOBIC COLLOID.

suspensor (*BioSci*) A file or files of cells that develop from the proembryo of a seed plant and anchor the embryo in the embryo sac and push it into the endosperm.

suspensorium (*BioSci*) In vertebrates, the apparatus by which the jaws are attached to the cranium.

suspensory (*BioSci*) Pertaining to the SUSPENSORIUM; serving for support or suspension.

Sussex garden-wall bond (*Build*) The form of GARDEN-WALL BOND in which one header and three stretchers are laid in each course.

sustainable (*EnvSci*) Involving the long-term use of resources that do not damage the environment.

sustainable agriculture (*Agri*) Resource-conservative farming that minimizes environmental damage and delivers productivity and economic viability at a level that does not endanger future production.

sustained oscillations (*ICT*) Externally maintained oscillations of a system at or very near its natural resonant frequency. Cf FORCED OSCILLATIONS, FREE OSCILLATIONS.

suture (*BioSci*) (1) The line at the junction of fused parts. (2) A line of weakness along which splitting may occur, as in a dehiscent fruit. (3) A line of junction of two structures, as the line of junction of adjacent chambers of a nautiloid shell (see fig. at AMMONOIDS). (4) A synarthrosis or immovable articulation between bones, as between the bones of the cranium. (5) Junctions of exoskeletal cuticular plates in insects. Adj *sutural*.

suture (*Med*) Surgical stitch, or group or row of such stitches.

sutured (*Geol*) A textural term descriptive of the sinuous interlocking grain boundaries of rocks which have undergone extensive recrystallization, eg quartzites.

Sv (*Radiol*) Abbrev for SIEVERT.

SV 40 (*BioSci*) A small virus normally infecting monkey cells (simian virus 40), either causing a lytic infection or being integrated into the host chromosome. It has a circular chromosome which has been fully sequenced, and vectors derived from it are used to transfer inserted DNA into mammalian cells.

SVC (*ImageTech*) Abbrev for STILL VIDEO CAMERA.

SVD (*Vet*) Abbrev for SWINE VESICULAR DISEASE.

Svedberg unit (*BioSci*) See SEDIMENTATION COEFFICIENT.

SVGA (*ICT*) Abbrev for *super video graphics adapter*. A higher-resolution version of the VGA standard, typically 800 by 600 pixels or higher. See VGA.

S-VHS (*ImageTech*) Abbrev for *super VHS*. TN for a high-band version of the VHS format, using a Y/C signal. Both consumer and industrial equipment is available. See VHS-C, W-VHS.

swab (*Med*) (1) Any small mass of cotton wool or gauze used for mopping up blood or discharges, or for applying antiseptics to the body, or for cleansing surfaces (eg the lips, the mouth). (2) A specimen of a secretion taken on a swab for bacteriological examination.

swage (*Eng*) Smith's tool (dolly) used in shaping metal.

swaging (*Eng*) Reducing of cross-section of metal rod or tube by forcing it through a tapered aperture between two grooved dies.

swallowtail (*Build*) See DOVETAIL.

swamp fever (*Vet*) See INFECTIOUS ANAEMIA OF HORSES.

Swan cube (*Phys*) The prism system used in the LUMMER–BRODHUN PHOTOMETER. Consists of two 45° prisms placed with their hypotenuse faces together, but with one face ground so that the prisms are touching only over the central part of the faces.

swan-neck (*Build*) The bend formed in a hand rail when a knee and a ramp are joined together without any intermediate straight length.

swan-neck chisel (*Build*) Chisel curved for lock mortising. Also *swan-neck lock mortise chisel*.

swan-neck insulator (*ElecEng*) A pin-type insulator with a bent pin, arranged so as to bring the insulator into approximately the same horizontal plane as that of the support.

swap file (*ICT*) A FILE that is used to store temporarily a program or data while the MAIN MEMORY is used to execute another program or store other data. When this is complete, the original program or data will be transferred from backing store into main memory. Commonly used in MULTIPROGRAMMING or MULTITASKING environments.

sward (*Agri*) The surface layer of grassland containing the leaves and stems of grasses and small herbaceous plants.

swarf (*Eng*) (1) The cuttings from a machining operation. (2) US for the mixture of abrasive, bond and metal particles formed during grinding.

swarm (*BioSci*) A large number of small animals in movement together; esp a number of bees emigrating from one colony to establish another under the guidance of a queen.

swarm cell (*BioSci*) (1) A flagellated naked cell in Myxomycetes, interconvertible with myxamoeba, capable of encysting and of acting as an isogamete. (2) A flagellated reproductive cell, esp a zoospore. Also *swarmer*.

swarmer (*BioSci*) See SWARM CELL.

swash letters (*Print*) Ornamental italic letters with tails and flourishes, commonly used only at the beginning or end of a word.

$\mathcal{A\,G\,J\,K\,M\,N\,P\,Q}$

swash plate (*Eng*) A circular plate mounted obliquely on a shaft; sometimes used in conjunction with working cylinders mounted axially parallel with the shaft, as a substitute for an engine or pump crank mechanism.

swatch (*Textiles*) A collection of small fabric samples.

swath (*Agri*) (1) The target area covered by one pass of the spraying equipment. (2) The cut area behind a mower or harvester.

S-wave (*Geol*) Abbrev for *secondary wave*. See EARTHQUAKE.

swayback (*Vet*) A nervous disease of newborn and young lambs characterized by degenerative changes in the cerebrum of the brain, causing inability or difficulty in standing or walking. Associated with a low copper content of tissues in the lamb and its ewe and preventable by administering copper to the pregnant ewe. Also *enzootic ataxia*.

swaybrace (*Eng*) See BRACE.

sway rod (*Build*) A member inserted in a structural framework to resist wind forces. Also *wind bracing*.

sweat cooling (*Aero*) Cooling of a component by the evaporation of a fluid through a porous surface layer; used for high-performance gas turbine blades or hypersonic vehicles.

sweated joint (*Eng*) See SWEATING.

sweating (*Build*) A term applied to a surface showing traces of moisture due either to formation of condensate or to water exuding through a porous surface.

sweating (*Eng*) The operation of soldering pieces together by 'tinning' the surfaces and heating them while pressed into contact. See SOFT SOLDER.

Swedish iron (*Eng*) Wrought-iron of high purity, with some 300 MN m^{-2} ultimate tensile strength, and 33% elongation before fracture.

Swedish standards (*Build*) A range of photographic standards for comparison with actual metal surfaces being prepared for painting. The photographic plates show varying degrees of corrosion from SA1 to SA3 which is white metal generally obtained by abrasive blasting. UK equivalent standard is BS 4232. Standard referred to on painting specifications where specific preparation is required.

sweep (*Aero*) The angle, in plan, between the normal to the plane of symmetry and a specified spanwise line on an aerofoil. Most commonly, the quarter-chord line is used, but leading and trailing edges are sometimes stipulated. Sweep increases longitudinal stability by extending the centre of pressure and delays compressibility drag by reducing the chordwise component of the airflow. *Sweep-back*, the more usual, is the aft displacement of the wings and *forward sweep* the opposite.

sweep (*ImageTech*) The movement of the electron beam across the surface of the cathode-ray tube.

sweepback (*Aero*) Aircraft wings making an acute angle with the fuselage, the wing tips being towards the tail.

sweep circuit (*Electronics*) That which supplies deflecting voltage to one pair of plates or coils of a cathode-ray tube, the other pair being connected to the source of current or voltage under examination. See LINEAR SCAN.

sweeper (*Electronics*) Frequency-swept oscillator, particularly at microwave frequencies.

sweeps (*Eng*) Dust and debris in jeweller's workshops, gold refineries, bullion assay offices, etc, collected and treated periodically to recover valuable contents.

sweep-saw (*Build*) A thin-bladed saw which is held taut in a special frame and used for making curved cuts. Also *turning-saw*.

sweet (*Glass*) Glass that is easily workable.

sweet clover disease (*Vet*) A fatal, haemorrhagic disease of cattle and other animals caused by feeding sweet clover (genus *Melilotus*), which has been damaged during hay-making or ensiling, causing the formation of a toxic substance, dicoumarin, which interferes with blood coagulation.

sweet crude oil (*MinExt*) One containing no hydrogen sulphide, H_2S. Cf SOUR CRUDE.

sweetening (*ImageTech*) Electronic enhancement of a video image by edge sharpening and noise reduction.

sweetening (*MinExt*) A refining process that improves the smell of some light products without necessarily reducing their sulphur content; thus MERCAPTANS may be oxidized to less offensive disulphides. Sodium plumbite, hypochlorite or an oxidizing copper salt are used.

sweet gas (*MinExt*) Natural gas that is free of malodorous sulphur compounds and therefore smells 'sweet'. Most natural gas is contaminated by such compounds, eg hydrogen sulphide (H_2S) and MERCAPTANS in addition to carbon dioxide, and needs SWEETENING treatment before it is ready for use.

sweet roasting (*Eng*) Ignition of metal sulphides to remove bulk of sulphur and arsenic as gaseous fume, leaving the mineral as its oxide.

swell (*Acous*) (1) Mechanism for altering the volume of sound in an organ by opening or closing shutters. Operated by the *swell pedal*. (2) In an electronic organ, volume control operated by potentiometer.

swell (*MinExt*) Volumetric increase due to crushing, which creates more void space in a given weight of rock.

swelled head (*Vet*) See BIG HEAD DISEASE OF SHEEP.

swelled rules (*Print*) Decorative rules in a variety of sizes with a central swelling, occasionally tooled. Also *Bodoni rules*.

swelling (*Arch*) See ENTASIS.

swelling (*NucEng*) Change of volume of fuel rods which may occur during irradiation.

swelling (*Plastics*) A process where polymers or biomaterials absorb fluids and expand. Negligible in rigid thermosets, it is widespread in thermoplastics prior to solubilization (depending on the SOLUBILITY PARAMETER of the solvent). In cross-linked rubbers, the chains cannot pass into solution, but the degree of swelling is a guide to the type of rubber and the degree of cross-linking.

swell pedal (*Acous*) The foot-operated organ lever for regulating the loudness of stops drawn on the swell manual. See GRAND SWELL.

SWG (*Eng*) Abbrev for STANDARD WIRE GAUGE.

swim bladder (*BioSci*) See AIR BLADDER.

swimmerets (*BioSci*) In some Crustacea, paired biramous abdominal appendages used in part for swimming.

swimming-pool reactor (*NucEng*) One in which the fuel elements are immersed in a deep pool of water which acts as coolant, moderator and shield. Also *pool reactor*.

swine erysipelas (*Vet*) A disease of pigs caused by infection by *Erysipelothrix insidiosa* (*E. rhusiopathiae*). The acute form of the disease is a septicaemia characterized by fever, urticaria-like patches on the skin and sometimes lameness due to arthritis; in the chronic form endocarditis occurs. Vaccines and antisera widely used. Also *diamond-skin disease*.

swine fever (*Vet*) A highly contagious disease of pigs, caused by a togavirus and sometimes complicated by secondary bacterial infection; symptoms may include fever, diarrhoea, pneumonia and nervous symptoms. Notifiable in the UK with a slaughter policy for infected herds, but vaccines are available. Also *hog cholera*.

swine influenza (*Vet*) An acute viral infection of pigs characterized by fever, coughing and respiratory distress. Enzootic pneumonia due to *Mycoplasma hyopneumoniae* and pleuropneumonia due to *Haemophilus pleuropneumoniae* give similar symptoms.

swine paratyphoid (*Vet*) A disease of pigs caused by *Salmonella cholerae suis*, and characterized by septicaemia in the acute form and necrotic enteritis in the chronic form. Vaccination widely used.

swine plague (*Vet*) Considered to be the respiratory symptoms of SWINE FEVER rather than a separate entity.

swine pox (*Vet*) A disease of pigs characterized by the formation of papules, vesicles and pustules on the skin, caused by a virus related to that of VACCINIA. A similar disease, in which papules and scabs develop, is caused by an unrelated virus. Also *variola*.

swine vesicular disease (*Vet*) Caused by an enterovirus related to human *coxsackie B5*. Symptoms include pyrexia, vesicle formation on coronary band of heels and pastern area, marked lameness and sometimes sloughing of hooves. Similar in appearance to FOOT AND MOUTH DISEASE. Notifiable in the UK.

swing (*Aero*) The involuntary deviation from a straight course of an aircraft while taxiing, taking-off or alighting.

swing (*Eng*) Lathe dimension, equal to the radial size of the largest workpiece that can be rotated or swung in it.

swing (*ICT*) Extreme excursion from positive peak to negative peak in an alternating voltage or current waveform.

swing arm (*Autos*) The part of a motorcycle chassis to which the rear wheel is attached.

swing back (*ImageTech*) The back of a camera which can tilt upwards or sideways, or both, so that distortion of objects (such as the vertical lines of buildings) may be minimized, or objects at different distances may be brought into focus.

swing front (*ImageTech*) The provision for tilting the front of a camera, with the lens, so that distortion of an object due to its receding along the axis can be minimized.

swinging choke (*ElecEng*) Iron-cored inductor with saturable core, used in smoothing circuits where decreasing impedance with increasing current improves regulation.

swinging grippers (*Print*) Found on sheet-fed rotaries and on high-speed single- and two-revolution machines, the sheets being transferred at speed from the feedboard to grippers on the continuously rotating impression cylinder.

swinging post (*Build*) See HINGING POST.

swing-wing (*Aero*) See VARIABLE SWEEP.

swirl chamber (*Autos*) A type of diesel-engine combustion chamber, separated from the cylinder by a short vent and shaped so as to set up an eddy in the air drawn in through the intake valve.

swirl sprayers (*Aero*) Fuel injectors in a gas turbine which impart a swirling motion to the fuel.

swirl vanes (*Aero*) Vanes which impart a swirling motion to the air entering the flame tube of a gas turbine combustion chamber.

swirl vane separator (*Eng*) A STEAM DRIER in which the central steam flow rotates propeller-like blades, causing the impinging water droplets to be flung onto an outer surface and lead down to the boiling water again.

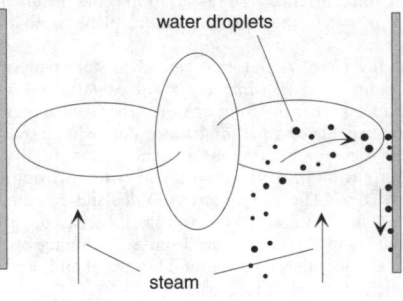

water droplets

steam

swirl vane separator

Swiss lapis (*Min*) An imitation of lapis lazuli, obtained by staining pale-coloured jasper or ironstone with blue pigment. Also *German lapis*.

Swiss screw-thread (*Eng*) A metric thread having a profile angle of 47°30′. The Thury screws are numbered exactly as BA screws, which are generally identical with the exception of 22, 23 and 25 (Thury).

Swiss-type automatic (*Eng*) A single-spindle automatic lathe in which several single-point tools are arranged radially around the stock, used primarily for small precision parts for watches and instruments.

switch (*CivEng*) US for *turnout*. A device for moving a small section of a railway track so that rolling stock may pass from one line of track to another.

switch (*ElecEng*) A device for opening and closing an electric circuit.

switch (*Electronics*) Electronic circuit for switching between two independent inputs, eg by valves, tubes or transistors.

switch (*ICT*) Equipment used within telephone exchanges and elsewhere to set up routes from one network node to another, by SPACE SWITCHING, TIME SWITCHING or a combination of the two. See CROSSPOINT SWITCH.

switch (*Phys*) Reversal by saturation of the residual flux in a ferrite core.

switch base (*ElecEng*) The insulating base on which a switch is mounted.

switch blades (*CivEng*) See POINTS.

switchboard (*ElecEng*) An assembly of switch panels.

switchboard instrument (*ElecEng*) An electric measuring instrument arranged for mounting on a switchboard.

switchboard model (*Chem, Plastics*) Suggested morphology of bulk crystallized polymers, with lamellar single crystals connected to one another by tie molecules, which form the amorphous regions in most crystalline polymers. An alternative model involves chain-folded crystals. See CRYSTALLIZATION OF POLYMERS and panel on POLYMER SYNTHESIS.

switchboard panel (*ElecEng*) See PANEL.

switch-box (*ElecEng*) An enclosure housing one or more switches operated by means of an external handle.

switch-desk (*ElecEng*) A control desk on which a number of miniature switches are mounted, each of which serves to initiate some control operation.

switched multimegabit data service (*ICT*) A digital data transmission standard defined by Bellcore in the USA to provide a high-speed public connectionless packet switching service capable of extending a LOCAL AREA NETWORK across a wider area. Access to the system is defined at 1·544 Mbps and 44·736 Mbps.

switched virtual private network (*ICT*) A form of VIRTUAL NETWORK in which transport paths do not exist permanently but are assigned on demand on a per-call basis and thus committed only for the duration of a call. Cf DEDICATED VIRTUAL PRIVATE NETWORK.

switch-fuse (*ElecEng*) A knife switch carrying a fuse in each blade.

switchgear (*ElecEng*) The generic name for that class of electrical apparatus whose sole function is to open and close electric circuits.

switchgear pillar (*ElecEng*) See PILLAR.

switch-hook (*ICT*) Normally a combined switch and handset rest forming part of a telephone, but in general any switch that is closed to signal to the local exchange that service is required and opened to indicate that a call is terminated.

switching (*ICT*) The provision of point-to-point connections between constantly changing sources of information and their intended recipients.

switching constant (*ElecEng*) The ratio of SWITCHING TIME and MAGNETIC FIELD INTENSITY.

switching time (*ElecEng*) The time required for complete reversal of flux in a ferroelectric or ferromagnetic core.

switch panel (*ElecEng*) An insulating panel on which a switch is mounted.

switch plant (*BioSci*) A plant with small, scale-like or fugacious leaves and long, thin, photosynthetic stems, eg *Cytisus, Ephedra*.

switch plate (*ElecEng*) A plate for covering one or more flush switches. Also *flush plate*.

switch region (*BioSci*) The sequence of amino acids at the junction of the variable and constant regions of immunoglobulin light or heavy chains. The sequence in this region is coded for by the D and J exons of the immunoglobulin gene. It determines which class of constant region is joined to the variable region.

switch-starter (*ElecEng*) A combination of knife switch and starting regulator, in which the circuit is closed and the resistance progressively cut out in one continuous movement.

swivelling propeller (*Aero*) See PROPELLER.

swivel pin (*Autos*) See KING PIN.

SWL (*Eng*) Abbrev for SAFE WORKING LOAD.

sycamore (*For*) Timber from the largest hardwood tree (*Acer pseudoplatanus*) native to Europe and W Asia. It is creamy-white and lustrous, straight-grained but often curly, with a fine, even texture (the curly grained wood is used for the backs of violins, hence known as *fiddleback*). In N America, 'sycamore' is the American plane or buttonwood (*Platanus occidentalis*). See EUROPEAN PLANE.

sycosis barbae (*Med*) Inflammation of the hair follicles of the beard region, due to infection with STAPHYLOCOCCUS.

Sydenham's chorea (*Med*) See CHOREA.

syenite (*Geol*) A coarse-grained igneous rock of intermediate composition, composed essentially of alkali feldspar to the extent of at least two-thirds of the total, with a variable content of mafic minerals, of which common hornblende is characteristic. See PLUTONIC ROCKS.

syenite-porphyry (*Geol*) An igneous rock of syenitic composition and medium grain size, commonly occurring in minor intrusions; it consists of phenocrysts of feldspar and/or coloured silicates set in a microcrystalline groundmass.

syenodiorite (*Geol*) See MONZONITE.

syenogranite (*Geol*) A variety of granite composed of alkali feldspar with subordinate plagioclase, quartz and biotite or hornblende. See PLUTONIC ROCKS.

syllable articulation (*ICT*) See ARTICULATION.

sylphon bellows (*Eng*) A thin-walled cylindrical metal bellows consisting of a number of elements arranged concertina-fashion, responding to external or internal fluid pressure; used in pressure-governing systems.

sylvanite (*Min*) Telluride of gold and silver, which crystallizes in the monoclinic system and is usually associated with igneous rocks and, in veins, with native gold. It is an ore of gold. Also *yellow tellurium*.

Sylvian aqueduct (*BioSci*) In vertebrates, the cavity of the mesencephalon.

Sylvian fissure (*BioSci*) In mammals, a deep lateral fissure of the cerebrum.

sylvine (*Min*) Potassium chloride, which crystallizes in the cubic system. It occurs in bedded salt deposits (*evaporites*), and as a sublimation product near volcanoes; it is a source of potassium compounds, used as fertilizers. Also *sylvite*. See SYLVINITE.

sylvinite (*Min*) A general name for mixtures of the two salts sylvine and halite, the latter predominating, occurring at Stassfurt, Germany, and elsewhere. Also used as a commercial name for SYLVINE.

sylvite (*Min*) See SYLVINE.

sym- (*Chem, Genrl*) See SYN-.

symbiosis (*BioSci*) An intimate partnership between two organisms (*symbionts*), in which the mutual advantages normally outweigh the disadvantages. See MUTUALISM.

symbiosis (*Psych*) See SOCIAL SYMBIOSIS.

symbiotic star (*Astron*) A close binary system involving accretion from a giant star onto its companion (usually a white dwarf); displays irregular variability and spectrum shows lines characteristic of both stars superimposed.

symblepharon (*Med*) Adhesion of the eyelid to the globe of the eye.

symbol (*Chem*) See CHEMICAL SYMBOL.

symbol (*Psych*) Words, objects, events which represent or refer to something else, established by convention. In psychoanalytic theory, a symbol's referent is unconscious, and the meaning is hidden from the individual's unconscious awareness (eg in dreams).

symbolic address (*ICT*) In an ASSEMBLY LANGUAGE program the ADDRESS of a memory location, identified by means of a symbol rather than an ABSOLUTE ADDRESS. See ADDRESS CALCULATION.

symbolic logic (*MathSci*) See MATHEMATICAL LOGIC.

symbolic method (*ElecEng*) A method of ac circuit analysis using complex numbers to represent the circuit voltages and currents. *In-phase* or *real* components are plotted in the direction of the *x*-axis and *quadrative* or *unreal* components are plotted in the direction of the *y*-axis. The operator j (equals $\sqrt{-1}$ and signifying vector rotation by 90° in an anticlockwise direction) is used to identify quadrative components. See COMPLEX NUMBERS.

symbolic programming (*ICT*) See ASSEMBLY LANGUAGE.

symbol table (*ICT*) A table maintained by a compiler or assembler relating names to machine addresses. Also *name table*.

symmetrical (*BioSci*) See ACTINOMORPHIC.

symmetrical components (*ElecEng*) A method of calculating voltages and currents in an unbalanced three-phase network in which quantities are represented by combinations of symmetrical positive, negative and zero phase sequence components.

symmetrical deflection (*Electronics*) The application of a voltage to a pair of deflection plates in a CRO such that they vary symmetrically above and below an average value, which is equal to the final anode potential of the tube. This procedure minimizes the possibility of trapezoidal distortion of the screen image.

symmetrical flutter (*Aero*) See FLUTTER.

symmetrical grading (*ICT*) Grading in which all groups of selectors are equally favoured in seeking outlets.

symmetrical network (*ElecEng*) One which can be divided into two mirror half-sections.

symmetrical short circuit (*ElecEng*) An ac short circuit in which each phase carries the same current.

symmetrical winding (*ElecEng*) A term applied to an armature winding which fulfils certain conditions of electrical symmetry.

symmetric dyadic (*MathSci*) See CONJUGATE DYADICS.

symmetric matrix (*MathSci*) A square matrix which is equal to its transpose, or, equivalently, one whose elements are symmetric around the main diagonal.

symmetric relation (*MathSci*) A relation is symmetric if, when it applies from x to y, then it also applies from y to x: ie $R(x,y)$ if and only if $R(y,x)$ for all x and y; eg 'is parallel to', 'is the complement of'. 'Is greater than' is not a symmetric relation.

symmetry (*BioSci*) (1) The method of arrangement of the constituent parts of the animal body. (2) In higher animals, the disposition of such organs as show bilateral or radial symmetry.

symmetry (*Crystal*) The quality possessed by crystalline substances by virtue of which they exhibit a repetitive arrangement of similar faces. This is a result of their peculiar internal atomic structure, and the feature is used as a basis of crystal classification.

symmetry (*MathSci*) (1) A geometric configuration is said to be symmetrical about a point, line or plane if any line through the point or any line perpendicular to the line or plane cuts the configuration in pairs of points equally spaced from the point, line or plane, which are then referred to respectively as the *centre of symmetry*, the *axis of symmetry* and the *plane of symmetry*. (2) A function of several variables is symmetric if it is unchanged when any two of the variables are interchanged.

symmetry breaking (*Phys*) See SPONTANEOUS SYMMETRY BREAKING.

symmetry class (*Crystal*) Crystal lattice structures can show 32 combinations of symmetry elements, each combination forming a possible symmetry class.

sympathectomy (*Med*) Excision or cutting of a part of a sympathetic nerve. Also *sympatheticectomy*.

sympathetic nervous system (*BioSci*) In some invertebrates (crustaceans and insects), a part of the nervous system supplying the alimentary canal, heart and reproductive organs and spiracles. In vertebrates, a subdivision of the autonomic nervous system. The action of these nerves tends to increase activity, speed the heart and circulation, and slow digestive processes. Also *thoracicolumbar system*. Cf PARASYMPATHETIC NERVOUS SYSTEM.

sympathetic ophthalmia (*Med*) The phenomenon in which perforating injury to one eye may be followed by inflammatory disease in the sound eye, characterized by lymphocyte infiltration and granuloma formation, esp in the uveal tract. Probably an auto-immune reaction.

sympathetic reaction (*Chem*) See INDUCED REACTION.

sympathomimetic (*Med*) Mimicking the sympathetic nervous system.

sympathomimetic drugs (*Pharmacol*) A class of drugs that mimic the stimulation of the sympathetic nervous system to produce TACHYCARDIA and increase output from the heart, eg *isoprenaline*. Some members can increase heart output without tachycardia (eg *dobutamine*). Others mimic the β_2-sympathetic stimulation to produce bronchodilatation and vasodilatation (eg *salbutamol*).

sympatric (*BioSci*) Two species or populations having a common or overlapping geographical distribution. Cf ALLOPATRIC.

sympatric speciation (*BioSci*) A controversial theory proposing that new species may arise without isolation. See SPECIATION.

sympetalous (*BioSci*) Gamopetalous. See GAMOPETALY.

symphyseotomy (*Med*) The operation of cutting through the pubic joint to facilitate the birth of a child. Also *symphysiotomy*.

symphysis (*BioSci*) (1) Union of bones in the middle line of the body, by fusion, ligament or cartilage, as the *mandibular symphysis*, the *pubic symphysis*. (2) Growing together or coalescence of parts, as acrodont teeth with the jaw. (3) The point of junction of two structures: chiasma; commissure. Adj *symphysial*.

symplast (*BioSci*) The continuum of protoplasts, linked by plasmodesmata and bounded by the plasmalemma.

symplastic (*BioSci*) Pertaining to the SYMPLAST. See SYMPLASTIC GROWTH.

symplastic growth (*BioSci*) A type of growth of a tissue, in which touching walls grow equally so that cell contacts persist. Cf INTRUSIVE GROWTH, SLIDING GROWTH. Also *co-ordinated growth*.

sympodial growth (*BioSci*) A pattern of growth in which, after a period of extension, a shoot ceases to grow and one or more of the lateral buds next to the apical bud grow out and repeat the pattern. Cf MONOPODIAL GROWTH. See CYMOSE INFLORESCENCE. Also *definite growth, determinate growth*.

sympodium (*BioSci*) A branch system that shows SYMPODIAL GROWTH.

symport (*BioSci*) A mechanism of transport across a membrane in which two different molecules move in the same direction. Often, one molecule can move up an electrochemical gradient because the movement of the other molecule is more favourable.

symptom (*Med*) Evidence of disease or disorder as experienced by the patient (eg pain, weakness, dizziness); any abnormal sensation or emotional expression or thought accompanying disease or disorder of the body or the mind; less accurately, any objective evidence of disease or bodily disorder. Cf SIGN.

symptomatology (*Med*) The study of symptoms; a discourse or treatise on symptoms; the branch of medical science concerning symptoms of disease.

syn-, sym- (*Chem*) See *s-*

syn-, sym- (*Genrl*) Prefixes from Gk *syn*, with, generally signifying fusion or combination.

synaesthesia (*Psych*) A condition in which one sensory input stimulates two senses so that eg sounds are heard but also vividly and repeatedly perceived as being coloured. Various forms of cross-modality have been described; they often go unrecognized because they are considered to be normal by the individual in question.

synaldoximes (*Chem*) The stereoisomeric forms of aldoximes in which the H and OH groups are on the same side of the plane of the double bond.

synandrium (*BioSci*) A group of united anthers or microsporangia.

synandrous (*BioSci*) Having the stamens united to one another.

synangium (*BioSci*) A number of sporangia fused into a single structure.

synapse (*BioSci*) The region where one nerve cell makes functional contact with another nerve or muscle (neuromuscular junction). In chemical synapses transmission involves the diffusion, across the small gap between the pre- and post-synaptic membranes, of small quantities of specific transmitter substances (eg acetylcholine, noradrenaline) which are released from the pre-synaptic nerve terminals and bind to receptors on the post-synaptic membrane. In electrical synapses transmission is direct, probably involving a gap junction.

synapse Part sectioned to show structures.

synapsid (*BioSci*) (1) A member of the Synapsida, a class of mammal-like reptiles of the Carboniferous, Permian and Triassic periods having a single pair of lateral temporal openings in the skull. (2) In the skull of reptiles, the condition when there is one temporal vacuity, this being low behind the eye, with the post-orbital and squamosal meeting above. Found in Pelycosauria and mammal-like reptiles. Cf DIAPSID

synapsin (*BioSci*) Any of a group of phosphoproteins that coat synaptic vesicles and which are thought to be involved in the regulation of neurotransmitter release.

synapsis (*BioSci*) The pairing of strictly homologous regions of homologous chromosomes during meiosis.

synaptic vesicles (*BioSci*) Structures about 50 nm in diameter found in pre-synaptic nerve terminals and concerned with the storage of the chemical transmitters. See fig. at SYNAPSE.

synaptonemal complex (*BioSci*) A ladder-like structure of DNA and protein observed to lie between the synapsed homologues of a PACHYTENE bivalent in first meiotic prophase. It is essential for crossing-over and chiasma formation to occur.

synarthrosis (*BioSci*) An immovable articulation, esp an immovable junction between bones. Cf AMPHIARTHROSIS, DIARTHROSIS.

sync (*ImageTech*) See SYNCHRONIZATION.

syncarpous (*BioSci*) A GYNOECIUM consisting of two or more fused carpels.

syncaryon (*BioSci*) See SYNKARYON.

synchondrosis (*BioSci*) The connection of two bones by cartilage, usually with little possibility of relative movement.

synchro (*ICT*) A general term used for a family of self-synchronous angle data transmitters and receivers. Also *selsyn*.

synchro Power goes to only two of the armature windings.

synchrocyclotron (*Phys*) A CYCLOTRON in which the frequency of the accelerating voltage is varied to ensure that, despite the relativistic increase of mass of the particle with speed, the particles still arrive in synchronism with the accelerating voltages. Energies up to 700 MeV for protons can be achieved. The output is not continuous but is emitted in bursts of particles lasting about 100 μs.

synchro-edit (*ImageTech*) A remote-control system enabling two videotape recorders or one and a CAMCORDER to both be controlled from one machine for semi-automatic editing. Can be used with an EDIT CONTROLLER.

synchrology (*EnvSci*) The study of the distribution of plants, relationships between areas and species, regional biodiversity and current plant migration patterns.

synchromesh gear (*Autos*) A gear in which the speeds of the driving and of the driven members which it is desired to couple are first automatically synchronized by small CONE CLUTCHES before engagement of the dogs or splines, thus avoiding shock and noise in gear-changing.

synchronism (*ICT*) Said of two signals of the same frequency when the phase angle between them is zero, or of pulsed signals that are in step with each other.

synchronization (*ImageTech*) In general, the matching of signals in precise time relation; in TV, esp establishing the identity of scanning frequency and phase of picture signals between transmitter and receiver. Abbrev *sync*.

synchronization of oscillators (*ICT*) A phenomenon when two oscillators, having nearly equal frequencies, are coupled together. When the degree of coupling reaches a certain point, the two suddenly pull into step.

synchronization supply unit (*ICT*) Equipment used in a SYNCHRONOUS DIGITAL HIERARCHY network to supply timing for all functions and signals at a node, using either external timing sources or, if these fail, an accurate internal clock.

synchronized clock (*Genrl*) A clock driven by any means but having its accuracy corrected electrically at given intervals.

synchronizer (*ElecEng*) See SYNCHROSCOPE.

synchronizer (*ICT*) A unit used to maintain synchronism when transmitting information between two devices. It may merely control the speed of one (eg by clutch) or if the speeds are very different, may include buffer storage.

synchronizing (*ElecEng*) The operation of bringing a machine into synchronism with an ac supply.

synchronizing power (*ElecEng*) The power developed in a synchronous machine that keeps it in synchronism with the ac supply system to which it is connected.

synchronizing torque (*ElecEng*) Also SYNCHRONIZING POWER.

synchronometer (*ElecEng*) A device which counts the number of cycles in a given time. If the time interval is unity the device becomes a digital FREQUENCY METER.

synchronous–asynchronous motor (*ElecEng*) A slip-ring type of induction motor whose rotor is fed from a dc exciter coupled to it. The machine operates asynchronously during start-up, and runs on load as a synchronous motor.

synchronous booster (*ElecEng*) An ac generator coupled to a synchronous converter and having its armature connected in series with that of the converter.

synchronous capacitor (*ElecEng*) A lightly loaded synchronous motor supplying a leading current for power-factor correction.

synchronous capacity (*ElecEng*) The synchronizing power of an interconnector linking two ac power systems. It is defined as the change of kilowatts transmitted over the interconnector per radian change of angular displacement of the voltages of the two systems.

synchronous carrier system (*ICT*) Simultaneous broadcasting by two or more transmitters having the same carrier frequency, the various drive circuits being interlocked so as to avoid heterodyne beats between them.

synchronous cell population (*BioSci*) A culture of cells that all divide in synchrony. Particularly useful for certain studies of the cell cycle, cell populations can be made synchronous for a time by deprivation of essential molecules followed by their restoration.

synchronous computer (*ICT*) One in which all operations are timed by a master CLOCK.

synchronous converter (*ElecEng*) A synchronous machine for converting polyphase ac to dc. It comprises a double-purpose armature, rotating within a salient-pole dc field system.

synchronous detector (*ICT*) One that inserts a missing carrier signal in exact synchronism with the original carrier at the transmitter. Particularly important in colour TV signals for the extraction of the colour information. May also refer to receivers that have a MATCHED FILTER for the selective detection of signals coded in a certain way.

synchronous digital hierarchy (*ICT*) A standard for multiplexed OPTICAL FIBRE networks which, by specifying truly synchronous digital channels, allows efficient addition and extraction of individual channels without demultiplexing an entire level of the hierarchy. It also makes specific provision for network management. Cf PLESIOCHRONOUS DIGITAL HIERARCHY.

synchronous gate (*ICT*) A gate controlled by clock pulses and used to synchronize operations.

synchronous generator (*ElecEng*) See ALTERNATOR.

synchronous homodyne (*ICT*) Reception in which the incoming modulated carrier has added to it a local oscillation of correct phase, with possible locking.

synchronous impedance (*ElecEng*) The ratio of the open-circuit emf to the short-circuit current of a synchronous machine, both values referred to the same field excitation.

synchronous induction motor (*ElecEng*) An induction motor in which dc is passed into the rotor winding after it has run up to speed, so that, after starting as an induction motor (with a high starting torque), it runs as a synchronous motor. Also *autosynchronous motor*.

synchronous machine (*ElecEng*) An ac machine which rotates at a constant speed which is harmonically related to the frequency of the supply to which it is connected. If the machine is two-pole, it will rotate at the supply frequency: if four-pole, at half supply frequency and so on.

synchronous motor (*ElecEng*) An ac electric motor designed to run in synchronism with supply voltage. See SYNCHRONOUS MACHINE.

synchronous orbit (*Space*) Circular orbit of a satellite of a body, moving in the same direction and with the same period as the parent body. If the latter is the Earth, the term *geosynchronous* is used.

synchronous phase modifier (*ElecEng*) A large synchronous machine used solely for varying the power factor at the receiving end of a transmission line to maintain the voltage constant under all conditions of loading.

synchronous reactance (*ElecEng*) The vector difference between the synchronous impedance and the effective armature resistance of a synchronous machine.

synchronous transmission (*ICT*) A method of transmitting data between two devices that are operating continuously and are controlled by the same CLOCK.

synchronous transport module 1 (*ICT*) The basic data rate in the synchronous digital hierarchy, 155·52 Mbps. Higher rates, collectively STM-N, are integer multiples of this: eg STM-4 represents 622·08 Mbps. Abbrev *STM-1*.

synchronous watt (*ElecEng*) A unit of torque used loosely in connection with ac machines. It is defined as the torque which, at the synchronous speed of the machine, would develop a power of 1 watt.

synchroscope (*ElecEng*) An instrument indicating the difference in frequency between two ac supplies. Also *synchronizer*.

synchrotron (*Phys*) A machine for accelerating charged particles to very high energies. The particles move in an orbit of constant radius guided by a magnetic field. The acceleration is provided at one point in their orbit by a high-frequency electric field whose frequency increases to insure that particles of increasing velocity arrive at the

correct instant to be further accelerated. Proton synchrotrons can produce energies greater than 200 GeV. Electron synchrotrons give energies up to 12 GeV. See BETATRON, CYCLOTRON.

synchrotron magnets (*Phys*) The shaped magnets which ensure that faster or slower particles moving at greater or lesser radii tend to move back to the mean radius in a synchrotron track. An essential prerequisite of these accelerators.

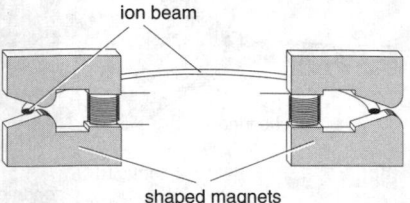

ion beam

shaped magnets

synchrotron magnets

synchrotron radiation (*Phys*) (1) The very intense, highly collimated, polarized beam of electromagnetic radiation produced by electrons accelerated in a SYNCHROTRON; wavelength ranges continuously from 10^{-2} mm to 10^{-2} nm. Used with a monochromator, it is an important source for research purposes. (2) Cosmic radio waves with a similar spectrum; it is suggested that the electrons moving in an orbit in a magnetic field are accelerated as in a synchrotron, but on a vastly larger scale. See panel on QUASAR.

synchysis (*Med*) Abnormal softening and fluidity of the vitreous body of the eye.

syncline (*Geol*) A concave-upwards fold with the youngest rocks in the centre. See fig. at FOLDING.

synclitism (*Med*) The compensatory difference in the rates of descent of the anterior and posterior portions of the presenting fetal part in the pelvis during labour.

sync modulation (*ImageTech*) The range of modulation depth reserved for the synchronizing pulses, as distinct from that for picture (video) signals.

syncope (*Med*) A fainting attack or sudden loss of consciousness due to sudden reduction of blood flow to the brain, as a result of rhythm disturbance of the heart or mechanical obstruction to the pump action of the heart.

sync pulse (*ImageTech*) A pulse transmitted at the beginning of each line and field to ensure correct scanning rate on reception.

sync pulse generator (*ImageTech*) A device for generating the synchronization pulses to maintain accurate SYNCHRONIZATION in video signals.

sync tip (*ImageTech*) The base level of the synchronization pulses in a video signal. See SYNC MODULATION.

syncytium (*BioSci*) A tissue containing many nuclei, which is not divided into separate compartments by cell membranes. Adj *syncytial*.

syndactyl (*BioSci*) Showing fusion of two or more digits, as some birds. N *syndactylism*.

syndesmochorial placenta (*BioSci*) A chorio-allantoic placenta in which the uterine epithelium disappears so that the chorion is in contact with the endometrium or glandular epithelium of the uterus. Usually COTYLEDONARY as in sheep.

syndesmosis (*BioSci*) The connection of two bones by a ligament, usually with little possibility of relative movement.

syndiazo compounds (*Chem*) The stereoisomeric forms of diazo compounds in which the groups attached to the nitrogen atoms are on the same side of the plane of the double bond.

syndiotaxy (*Plastics*) Polymerization exhibiting regular alternation of differences in stereochemical structure. Also *syndyotaxy*. Adj *syndiotactic*. See fig. at STEREOREGULAR POLYMERS. Cf ISOTAXY.

syndrome (*Med*) A concurrence of several symptoms or signs in a disease which are characteristic of it, but do not in themselves constitute a disease and may be associated with several conditions; a set of concurrent symptoms or signs.

syndyotaxy (*Chem*) See SYNDIOTAXY.

synechia (*Med*) A morbid adhesion of the iris of the eye to the cornea or to the lens.

synecology (*BioSci*) The study of relationships between communities and their environment. Cf AUTECOLOGY.

syneresis (*Chem*) Spontaneous expulsion of liquid from a gel.

synergetic (*BioSci*) Working together; said of muscles which co-operate to produce a particular kind of movement. Also *synergic*.

synergid (*BioSci*) Either of the two nuclei that with the egg nucleus constitute the egg apparatus at the micropylar end of the embryo sac of an angiosperm.

synergism (*BioSci*) (1) The condition in which the result of the combined action of two or more agents, eg two growth substances, is greater than the sum of their separate, individual, actions. (2) A type of SOCIAL FACILITATION in which the nearby presence of another organism enhances the efficiency or intensity of a physiological process or behaviour pattern in an individual.

synergist (*Chem*) A substance which increases the effect of another.

syngamiasis (*Vet*) Infection of the trachea, bronchi and lungs of birds by the nematode worm *Syngamus trachea*.

syngamy (*BioSci*) Sexual reproduction; fusion of gametes.

syngeneic (*BioSci*) Genetically identical. Usually applied to grafts made or cells transferred within an inbred strain.

syngenesis (*BioSci*) Lateral fusion of plant members, eg the anthers, which unite laterally to form a hollow tube round the style in the Compositae.

syngenetic (*Min*) A category of ore bodies comprising all those which were formed contemporaneously with the enclosing rock. Cf EPIGENETIC.

syngnathous (*BioSci*) Of certain fish, having the jaws fused to form a tubular structure.

synkaryon (*BioSci*) A pair of nuclei in close association in a fungal hypha, dividing together to give the same close association. In animals a zygote nucleus resulting from the fusion of two pronuclei.

synkinesia (*Med*) The occurrence simultaneously of both voluntary and involuntary movements, eg in the movement of a partly paralysed muscle in conjunction with the voluntary movement of healthy muscle.

synodic month (*Astron*) The interval (amounting to 29·530 59 days) between two successive passages of the Moon through conjunction or opposition respectively; therefore, the period of the phases. Also *lunar month*, *lunation*.

synodic period (*Astron*) An interval of time between two similar positions of the Moon or a planet, relative to the line joining the Earth and Sun; hence the length of time from one conjunction or opposition to another, and the period of the phases of the Moon or a planet.

synoptic chart (*EnvSci*) A chart or map on which are marked synchronous observations of atmospheric pressure, temperature, strength and direction of the wind, the state of the weather, cloud and visibility. Synoptic charts are used as a basis for weather forecasting.

synoptic meteorology (*EnvSci*) That part of the science of meteorology which deals with the preparation of a *synoptic chart* of the observed *meteorological elements* and, from consideration of this chart, the production of a weather forecast.

synosteosis (*BioSci*) See ANKYLOSIS.

synovia (*BioSci*) In vertebrates, a viscous transparent lubricating fluid, occurring typically within tendon

sheaths and the capsular ligaments surrounding movable joints.

synovial membrane (*BioSci*) The delicate connective tissue layer which lines a tendon sheath or a capsular ligament, and is responsible for the secretion of the SYNOVIA.

synovitis (*Med*) Inflammation of the synovial membrane of a joint.

synroc (*NucEng*) Nuclear waste products incorporated into a mixture of crystalline structures known to be stable over geological time.

synsacrum (*BioSci*) In birds, part of the pelvic girdle formed by the fusion of some of the dorsal and caudal vertebrae with the sacral vertebrae. See SACRUM.

syntactic foam (*Eng*) FOAM in which the voids are hollow fillers such as HOLLOW GLASS MICROSPHERES.

syntax (*ICT*) A set of rules for combining the elements of a programming language (eg words) into permitted constructions (eg program statements). See SEMANTICS.

syntax analysis (*ICT*) Second stage during COMPILATION where language statements are checked for compliance with the rules of the language.

syntax error (*ICT*) One that results from an incorrect use of the rules governing the structure of the language.

syntechnic (*BioSci*) Said of unrelated forms showing resemblance due to environmental factors; convergent.

syntenic (*BioSci*) A term used to describe genes that lie on the same chromosome. Some loci are syntenic in both humans and mice, others are not.

syntenosis (*BioSci*) The union of bones by means of tendons, as in the phalanges of the digits.

synthetase (*BioSci*) Enzymes of Class 6 in the E CLASSIFICATION. They catalyse synthesis of molecules, their activity being coupled to the breakdown of a nucleotide triphosphate.

synthetic aperture radar (*Radar*) One in which an aircraft flying a straight path emits pulses continuously and at a precisely controlled frequency such that the transmitted power is coherent. All the echoes are processed in such a way as to simulate an antenna with an aperture as long as the flight path. Extremely fine resolution is attainable, giving fine detail; used for terrain mapping etc.

synthetic fibre (*Textiles*) A fibre that has been manufactured by synthesis of a suitable linear polymer followed by extrusion of the molten material through a spinneret.

synthetic oligonucleotide (*BioSci*) A defined DNA sequence chemically polymerized *in vitro*.

synthetic paints (*Build*) Paints which contain a proportion of synthetic resin.

synthetic paper (*Paper*) Paper made by conventional means but utilizing synthetic fibre as the whole or a substantial part of the furnish. Generally a paper of considerable durability, permanence and chemical resistance. The term is (incorrectly) applied to other sheet materials of similar appearance to paper and used for similar purposes.

synthetic-resin adhesive (*Plastics*) One made from a thermosetting resin, eg urea or phenol formaldehyde, and used with an accelerator to regulate setting conditions, or from a thermoplastic resin, eg polymethyl methacrylate or polyvinyl acetate.

synthetic resins (*Chem*) Resinous compounds made from synthetic materials, as by the condensation or polymerization of phenol and formaldehyde, formaldehyde and urea, glycerol and phthalic anhydride, polyamides, vinyl derivatives, etc. See PHENOLIC RESINS, VINYL POLYMERS and panel on THERMOSETS.

synthetic rubber (*Eng, Plastics*) Any of numerous synthetic elastomeric materials, eg polymers of isoprene or its derivatives. See BUTYL RUBBER, NEOPRENE, THIOKOL.

synthetic ruby (*Min*) In chemical composition and in all physical characteristics, including optical properties, true crystalline ruby; but produced in quantity in the laboratory by fusing pure precipitated alumina with the predetermined amount of pigmentary material. Can be distinguished from natural stones only by the most careful expert examination. Similarly for synthetic sapphire.

synthetic sands (*Eng*) Sands deficient in clay which have been blended with bentonite or other clay-like material to make them suitable for moulding.

synthetic sapphire (*Min*) See SYNTHETIC RUBY.

synthetic spinel (*Min*) Spinel produced, in a wide variety of fine colours, by the VERNEUIL PROCESS; in chemical and optical characters identical with natural magnesian spinel, it is widely used as a gemstone.

syntony (*ElecEng*) See CURRENT RESONANCE.

synusia (*BioSci*) A group of plants with similar life form and of the same or unrelated species, occupying a similar habitat, eg woodland herbs.

syphilid (*Med*) Any skin affection caused by syphilis. Also *syphilide, syphiloderm, syphiloderma*.

syphilis (*Med*) A contagious venereal disease due to infection with the micro-organism *Spirochaeta pallida* (*Treponema pallidum*); contracted in sexual intercourse, by accidental contact or (by the fetus) from an infected mother.

syphiloma (*Med*) A syphilitic tumour. See GUMMA.

syphon (*Genrl*) See SIPHON.

Syrian garnet (*Min*) A name for ALMANDINE of gemstone quality.

syringitis (*Med*) Inflammation of the Eustachian tube.

syringobulbia (*Med*) A disease characterized by increase of neuroglia and the presence of cavities in the medulla oblongata, giving rise to such nervous phenomena as paralysis of the palate, pharynx and larynx. See SYRINGO-MYELIA.

syringomyelia (*Med*) A chronic, progressive disease of the spinal cord in which increase of neuroglia and the formation of irregular cavities cause paralysis and wasting of muscles and loss of skin sensibility to pain and to temperature. See SYRINGOBULBIA.

syringomyelocele (*Med*) A form of spina bifida in which the part protruding through the defective spinal column consists of the greatly distended central canal of the spinal cord.

syrinx (*BioSci*) The vocal organs in birds, situated at the posterior end of the trachea. Pl *syringes*. Adj *syringeal*.

syrinx (*Med*) A fistula or a fistulous opening.

systaltic (*BioSci*) Alternately contracting and dilating; pulsatory, as the movements of the heart. Cf PERISTALTIC. N *systalsis*.

system (*BioSci*) (1) Tissues of the same histological structure, eg the osseous system. (2) Tissues and organs uniting in the performance of the same function, eg the digestive system. (3) A method or scheme of classification, eg the *Linnaean system*. (4) A systematic treatise on the animal or plant kingdom, or any part of either. Adj *systematic*.

system (*Chem*) A portion of matter, or a group or set of things that forms a complex or connected whole.

system (*ElecEng*) General term used to describe: (1) an entire arrangement of equipment, eg the grid system; (2) a collection of standards or definitions, eg SI system; (3) a set or field of technology, eg digital systems.

system (*Genrl*) Generally, anything formed of parts placed together or adjusted into a regular or connected whole.

system (*Geol*) (1) The chronostratigraphical equivalent of a PERIOD of geological time. (2) The name given to the succession of rocks which were formed during a certain period of geological time, eg *Jurassic system*. (3) The sum of the phases which can be formed from one or more components of minerals under different conditions of temperature, pressure and composition.

systematic (*BioSci*) See SYSTEMIC.

systematic desensitization (*Psych*) A therapeutic approach to anxiety disorders in which there is exposure to gradually increasing anxiety-provoking stimuli under relaxing and reassuring conditions.

systematic errors (*CivEng, MathSci*) Errors which are always in the same direction, ie errors which are always positive or always negative. Sometimes known as cumulative errors. In eg calculations such errors can arise by always rounding fives upwards.

systematics (*BioSci*) The branch of biology that deals with classification and nomenclature.

system building (*Build, CivEng*) Methods designed to increase the speed of construction by preparing component parts of the building in a factory before assembly on site.

system crash (*ICT*) Occurs when the operating system is unable to control the computer and human intervention is needed to restart.

Système International d'Unités (*Genrl*) See SI UNITS.

system engineering (*Space*) A logical process of activities which transforms a set of REQUIREMENTS arising from a specific mission objective into a full description of a system which fulfils the objective in an optimum way. It ensures that all aspects of a project have been considered and integrated into a consistent whole.

system flowchart (*ICT*) See DATA FLOWCHART.

systemic (*BioSci*) Something distributed throughout the organism, not limited to a particular place. Thus the systemic circulation carries blood to the body as a whole; systemic insecticides are distributed though all the tissues of a plant.

systemic arch (*BioSci*) In vertebrates, the main vessel or vessels carrying blood from the heart to the body as a whole.

systemic lupus erythematosus (*Med*) A disease of humans characterized by widespread focal degeneration of connective tissue and disseminated lesions in many tissues including skin, joints, kidneys, pleura, peripheral vessels, peripheral nervous system and transient abnormalities of the central nervous system. Numerous auto-antibodies are present in the blood, of which the most constant are anti-nuclear antibodies. The lesions are mainly the result of the deposition of immune complexes. Abbrev *SLE*.

systemic pesticides (*Agri*) Pesticides that are translocated from the point of application to other sites where the activity is expressed.

systems analysis (*Genrl*) Complete analysis of all phases of activity of an organization, and development of a detailed procedure for all collection, manipulation and evaluation of data associated with the operation of all parts of it.

systems analysis and design (*ICT*) Feasibility study of a potential computer involvement and the design of appropriate system to do a job.

systems analyst (*ICT*) Person responsible for the analysis of a project to assess its suitability for computer application and who may also design the necessary computer system.

systems network architecture (*ICT*) An IBM network standard for DISTRIBUTED PROCESSING systems. It provides communication between terminals and a host computer.

systems of crystals (*Crystal*) The seven large divisions into which all crystallizing substances can be placed: cubic, tetragonal, hexagonal, trigonal, orthorhombic, monoclinic, triclinic. This classification is based on the degree of SYMMETRY displayed by the crystals. See panel on CRYSTAL LATTICE.

systems programmer (*ICT*) A programmer who writes SYSTEMS SOFTWARE.

systems software (*ICT*) The collection of programs that make the computer system usable and control its performance.

system testing (*ICT*) The phase of a testing cycle in which an every part of a system is used to carry out the whole set of processes for which it has been designed, in order to evaluate its suitability for purpose.

System X (*ICT*) Name given by British Telecom plc to describe its fully electronic computerized exchange switching system.

systole (*BioSci*) Rhythmical contraction, as of the heart, or of a contractile vacuole. Cf DIASTOLE.

systole (*Med*) The period when a chamber of the heart is contracting.

systyle (*Arch*) A colonnade in which the space between the columns is equal to twice the lower diameter of the columns.

syzygy (*Astron*) A word applied to the Moon when in conjunction or opposition.

Szilard–Chalmers process (*Phys*) A process in which anuclear transformation occurs with no change of atomic number, but with breakdown of chemical bond. This leads to formation of free active radicals from which material of high specific activity can be separated chemically.

T

T (*BioSci*) Symbol for THYMINE.

T (*Genrl*) Symbol for TERA-.

T (*Phys*) Symbol for TESLA.

T (*Chem*) With subscript, a symbol for TRANSPORT NUMBER.

T (*Phys*) Symbol for TEMPERATURE.

T- (*Chem*) A symbol indicating the presence of a triple bond which begins on the corresponding carbon atom.

T_b (*Chem*) Temperature of boiling.

T_g (*Eng, Glass*) Symbol for GLASS TRANSITION TEMPERATURE.

T_m (*BioSci*) Symbol for the temperature at which DNA is half denatured. Also *melting temperature*.

T_m (*Chem, Glass*) Symbol for CRYSTAL MELTING POINT.

t (*Genrl*) Symbol for tonne (metric TON).

t (*Chem*) Abbrev for (1) *trans-*, ie containing the two radicals on opposite sides of the plane of a double bond or alicyclic ring; (2) *tertiary*, ie substituted on a carbon atom which is linked to three other carbon atoms.

τ (*Chem*) Symbol for a time interval, esp half-life or mean life.

T15 (*NucEng*) Large TOKAMAK experiment, Moscow. See TOKAMAK.

Ta (*Chem*) Symbol for TANTALUM.

tab (*Aero*) Hinged rear portion of a flight control surface. See BALANCE TAB, SERVO TAB, SPRING TAB, TRIMMING TAB.

tabescent (*BioSci*) Shrivelling.

tabes dorsalis (*Med*) A disease of the nervous system marked by attacks of pain in the legs, anaesthesia of certain areas of the skin, ataxia, loss of the pupil reflex to light, and other nervous affections; due to degenerative changes in the nerves, esp of the sensory roots of the spinal cord, as a late result of syphilis. Also *locomotor ataxia (ataxy)*.

tabetic (*Med*) Pertaining to, affected by, or caused by TABES DORSALIS; an affected person.

table (*Eng*) The horizontal portion of a drilling, milling, shaping or other machine, which supports the workpiece and is adjustable relative to the tool.

table (*ICT, MathSci*) A collection of data (eg square-root values) laid out in rows and columns for reference, or stored in a computer memory as an ARRAY.

table matter (*Print*) See TABULAR MATTER.

table of contents (*ImageTech*) The digital index on a disk which the player must first read in order to access the contents.

table of strata (*Geol*) A column which depicts a series of rocks arranged in chronological order, the oldest being at the bottom. It is usual to draw this to scale so that the average thicknesses of the beds are also shown.

table rolls (*Paper*) Rolls from 5 to 30 cm (2 to 12 in) in diameter, which carry the upper face of the machine wire, and also assist in the removal of water by capillary action.

tablet (*ICT*) See GRAPHICS TABLET.

tablet (*Plastics*) A piece of moulding composition of the correct weight and density, and of suitable diameter and thickness to fit the mould: not preformed to the approximate shape of the moulding.

tabloid newspaper (*Print*) A newspaper with a smaller page size, usually half the normal broadsheet size, using a cross-fold to form the spine.

taboo (*Psych*) An anthropological term for the prohibition of some class of people, objects or acts, because they violate the fundamental beliefs and values of a culture.

taboparalysis (*Med*) General paresis and TABES DORSALIS affecting the same person. Also *taboparesis*.

tabs (*ImageTech*) Curtains which can be opened or closed in front of a cinema screen.

tabular (*BioSci*) Having the form of a tablet or slab.

tabular matter (*Print*) Text arranged in accurately spaced columns; if there are rules between the columns, it is *table matter*.

tabulator (*ICT*) Early computer printer.

TAB vaccine (*Med*) A vaccine used in the prophylaxis of enteric fevers. It contains heat-killed *Salmonella typhi* and *S. paratyphi* A and B in the smooth specific phase and possessing their normal complement of O antigens. Because the O antigens contain lipopolysaccharides, some fever and local inflammation commonly follow injection.

tach-, tache-, tachy- (*Genrl*) Prefixes from Gk *tachys* denoting speed or speedy.

tacheometer (*Surv*) A theodolite fitted with a telescope which measures distances by sighting on levelling staff, the space between two cross-lines in the eyepiece being an index (after correction) of the distance concerned. Also *tachymeter*. See ADDITIVE CONSTANT, MULTIPLYING CONSTANT.

tacheometry (*Surv*) The process of surveying and levelling by means of angular measurements from a known station, combined with determination of distances from the station.

tachistoscope (*Psych*) A mechanical instrument capable of flashing visual displays on a screen for very short periods of time; used in perceptual research.

tachometer (*Eng*) An instrument for measuring speed of rotation.

tachy- (*Genrl*) See TACH-.

tachycardia (*Med*) A heart rate faster than normal, usually defined as >100 min^{-1} for adults.

tachylite (*Geol*) A black glassy igneous rock of basaltic composition, which occurs as a chilled margin of dykes and sills. In Hawaii it forms the bulk of certain lava flows. Also *basalt glass, tachylyte*.

tachymeter (*Surv*) See TACHEOMETER.

tachyon (*Phys*) A theoretical particle moving faster than the speed of light.

tachyphylaxis (*Pharmacol*) A condition in which repeated administrations of a drug result in progressively smaller responses.

tachypnoea (*Med*) Excessive frequency of respiration. US *tachypnea*.

tack (*Build*) A small clout nail.

tack (*Eng*) The ability of a material to bond with another when contact is brief and pressure is light. Important in rubber industry when bringing parts together to make products such as tyres.

tack (*Print*) A measure of an ink's internal cohesion; the tack value of an ink gives a guide to an ink film's resistance to splitting at impression.

tack marks (*Print*) (1) Small dots printed in the centre of a back margin to indicate to the binder the lay edges of the sheet, two dots indicating the second side of a work-and-turn sheet; only seen occasionally. (2) A dot printed to overlap the side lay edge of the sheet when the first printing is being run off, so that it shows on the edge of the pile of sheets, to indicate the lay required for subsequent printings, and which will be cut off when the job is trimmed. When wood furniture is used the mark may be made literally by a tack.

tack rag (*Build*) A cotton material impregnated with a non-drying oil for removing dust prior to painting.

tack welding (*Eng*) Making short, provisional welds along a joint to hold it in position and prevent distortion during subsequent continuous welding.

tacky (*Build*) Descriptive of paint or varnish which has not quite dried and is in a sticky condition.

Taconic orogeny (*Geol*) A period of intense folding which affected the eastern parts of N America at the end of the Ordovician period. The effects are best seen in the Taconic Mts on the borders of New York State and Massachusetts.

tacrine (*Pharmacol*) A reversible *cholinesterase* inhibitor used in the treatment of Alzheimer's disease.

tacrolimus (*Pharmacol*) A drug that inhibits T-cell activation and is used as an immunosuppressant to prevent transplant rejection.

TACS (*ICT*) Abbrev for TOTAL ACCESS COMMUNICATIONS SYSTEM.

tacticity (*Chem*) The stereochemical arrangement of units in the main chain of a polymer. See ATACTIC POLYMERS, ISOTACTIC POLYMERS, STEREOREGULAR POLYMERS, SYNDIOTACTIC POLYMERS.

tactic movement (*BioSci*) See TAXIS.

tactile (*BioSci*) Pertaining to the sense of touch.

tactile bristle (*BioSci*) A stiff hair that transmits a contact stimulus.

tactile perception (*Med*) The perception of vibration by the sense of touch; developed particularly in deaf persons, who can be trained to detect and interpret vibrations in another person's larynx, or to interpret vibrations applied selectively to their fingers by vibrators operated through filters by a microphone.

tactite (*Geol*) See SKARN.

tactoid (*Phys*) A rod-shaped droplet or flat particle appearing in colloidal solutions which exhibit double refraction.

tactosol (*Chem, Phys*) Sol containing TACTOIDS.

taenia (*BioSci*) (1) A ribbon-shaped structure, such as the *taenia pontis*, a bundle of nerve fibres in the hind-brain of mammals. (2) A tapeworm. US *tenia*. Pl. *taeniae*.

taeniasis (*Med*) The state of infestation of the human body with tapeworms (genus *Taenia*), which as adults may inhabit the intestine and, as larvae, the muscles and other parts of the body. US *teniasis*.

taenite (*Min*) A solid solution of iron and nickel occurring in iron meteorites; it appears as bright white areas on a polished surface. It crystallizes in the cubic system and has 27–65% Ni.

Tafel plot (*Chem*) A graph of OVERPOTENTIAL against logarithm of the current in an electrochemical cell. Useful in evaluating the kinetics of cell reactions.

taffeta (*Textiles*) A lightweight, plain-weave, crisp fabric with a faint weft rib produced from filament yarns and used for blouses etc.

taffrail log (*Ships*) See NAUTICAL LOG.

taft joint (*Build*) See WIPED JOINT.

tag (*ICT*) A term used of *metadata*, a label, keyword or name that can be associated with a website, data set or web page and identified by a search engine, thus easing the indexing and retrieval of information. See also SENTINEL.

tag block (*ElecEng*) Terminal block, holding varying numbers of double-ended solder tags, which is fitted to every panel of apparatus supported on standard apparatus racks. External wiring to a unit can then be connected without interference with the internal wiring, which is completed during manufacture. External connections to

bays of apparatus are also made to tag blocks mounted at the top of the racks by cable forms.

tagged atom (*Phys*) See LABELLED ATOM, RADIOACTIVE TRACER.

tagger (*Eng*) Thin-sheet iron or tinplate.

T-agglutinin (*BioSci*) Antibody present in the blood of normal persons that agglutinates erythrocytes which have been incubated with NEURAMINIDASE or acted on by bacteria which produce the enzyme.

tag image file format (*ICT*) A standard file format in MS-DOS for transferring graphics images between application programs. Abbrev *TIFF*.

tagma (*BioSci*) A distinct region of the body of a metameric animal, formed by the grouping or fusion of somites, as the thorax of an insect. Pl *tagmata*.

tagmosis (*BioSci*) In a metameric animal, the grouping or fusion of somites to form definite regions or tagmata.

tail (*Aero*) See TAIL UNIT.

tail (*BioSci*) See CAUDA.

tail (*Build*) That end of a stone step which is built into a wall.

tail (*ImageTech*) The end of a roll of film or tape.

tail (*Print*) The bottom or foot margin of a page or volume.

tailband (*Print*) See HEADBAND.

tail bay (*Build*) The part of a canal lock immediately below the tail gates.

tail beam (*Build*) A floor joist which at one end is framed into a TRIMMER. Also *tail joist*.

tail boom (*Aero*) One or more horizontal beams which support the tail unit where the fuselage is truncated; commonly used for cargo aircraft to facilitate loading trucks and bulky freight through full-width rear doors.

tail cap (*Print*) See HEAD CAP.

tail chute (*Aero*) A parachute mounted in an aircraft tail. Cf ANTI-SPIN PARACHUTE, BRAKE PARACHUTE.

tail cone (*Aero*) The tapered streamline FAIRING which completes a fuselage or TAIL BOOM.

tailerons (*Aero*) Two-piece tailplane whose two halves can operate either together, performing the function of an elevator or differentially, causing rolling moments as does an aileron.

tail-first aircraft (*Aero*) An aircraft in which the horizontal stabilizer (ie tailplane) is mounted ahead of the main plane; common on pioneer aircraft and reintroduced for supersonic flight; sometimes *canard* because of its similarity to a planform of a duck in flight.

tail gate (*Build*) The gates at the low-level end of a lock.

tail heaviness (*Aero*) That state in which the combination of forces acting upon an aircraft in flight is such that it tends to pitch up.

tailing (*Build*) The operation of building in and fixing the end of a timber which projects from a wall. Also *tailing down, tailing in*.

tailing hangover (*ImageTech*) Blurring of reproduced picture because of slow decay in electronic circuits. Also *submerged resonance*.

tailing iron (*Build*) A steel section built into a wall, across the top of the fixed end of a projecting member.

tailings (*MinExt*) (1) Rejected portion of an ore; waste, *gangue*. (2) Portion washed away in water concentration. May be impounded in a TAILINGS DAM or pond, or stacked dry on a dump. US *tails*.

tailings dam (*MinExt*) One used to hold mill residues after treatment. These arrive as fluent slurries. Dam may include arrangements for runoff or return of water after the slow-settling solids have been deposited.

tail joist (*Build*) See TAIL BEAM.

tailless aircraft (*Aero*) An aircraft, or glider, in which longitudinal stability and control in flight are achieved without a separate balancing horizontal aerofoil. This balance is achieved by SWEEP, and many *delta-wing* aircraft are tailless because their sharp angle of sweepback renders a tailplane unnecessary.

tailpiece (*Print*) An engraving, design, etc occupying the bottom of a page, as at the end of a chapter.

tailplane (*Aero*) A horizontal surface, fixed or adjustable, providing longitudinal stability for an aircraft or glider. See STABILIZER.

tail race (*Build*) A channel conveying water away from a hydraulically operated machine.

tail race (*MinExt*) The launder or trough for the discharge of waterborne tailings.

tail rotor (*Aero*) See AUXILIARY ROTOR.

tails (*MinExt*) US for TAILINGS.

tails (*NucEng*) The depleted uranium produced at an enrichment plant, containing typically 0·25% of ^{235}U.

tail slide (*Aero*) A difficult aerobatic manoeuvre in which an aircraft is pulled up into a zoom and allowed to slide backward along its longitudinal axis after the vertical speed drops to zero. Flying surfaces are specially strengthened to withstand the reverse airflow encountered.

tailstock (*Eng*) A component mounted to slide on a lathe bed. It carries a spindle in true alignment with the centre of the headstock, is longitudinally adjustable, and is coned internally to receive a centre. See LATHE.

tail trimmer (*Build*) A trimmer close to a wall, used in cases where it is not desired to build the joists into the wall.

tail unit (*Aero*) Hindmost parts of an aircraft, the horizontal *tailplane*, fin rudder and any strakes of an aircraft. Would include oblique (V) surfaces as in butterfly tail. Also *empennage*.

tail wheel landing gear (*Aero*) That part of the alighting gear taking the weight of the rear of the plane when on the ground. It consists of a shock-absorber carrying a wheel (*tail wheel*) or a shoe (*tail skid*).

taint (*FoodSci*) An uncharacteristic aroma or taste in a product arising from contact with another aromatic substance. See OFF-FLAVOUR.

Takayasu's disease (*Med*) A disease with progressive obliteration of the major arteries within the chest. Also *pulseless disease*.

take (*ImageTech*) A record of the action of a scene or part of a scene, repeated if necessary to improve the performance as a different take number.

take-off rocket (*Aero*) A rocket, usually jettisonable, used to assist the acceleration of an aircraft. Cordite rockets were introduced for naval aircraft during World War II and replenishable liquid-fuel rockets, some with controllable thrust, thereafter. Sometimes referred to as a *booster rocket*, although strictly this is for the acceleration of missiles. Abbrev *RATOG* (rocket-assisted take-off gear). US abbrev *JATO* (jet-assisted take-off).

take-up (*ImageTech*) That part of a machine where film or tape is wound up after passing through the operation or process concerned.

take-up motion (*Textiles*) The mechanism that arranges for the fabric produced in weaving to be wound on a roller at the correct speed.

taking-off (*Build, CivEng*) The first process involved in drawing up a bill of quantities. It consists of obtaining the dimensions of each item on the drawing, and entering them in a systematic manner on sheets ruled for the purpose.

talbot (*Phys*) Unit of luminous energy, such that 1 lumen is a flux of 1 talbot per second.

Talbot process (*Build*) An anticorrosion process applied to cast-iron pipes which are rotated to coat them internally with a mixture of bitumen and sand.

talc (*Min*) A monoclinic hydrated magnesium silicate, $MgSi_8O_{20}(OH)_4$. It is usually massive and foliated and is a common mineral of secondary origin associated with serpentine and schistose rocks; also found in metamorphosed siliceous dolomites. Purified talc is used medically, in toilet preparations and in many other ways. See fig. at SILICATES. See STEATITE.

talipes (*Med*) A general term for a number of deformities of the foot: *talipes calcaneus*, in which the toes are drawn up from the ground and the patient walks on the heel; *talipes equinus*, in which the heel is drawn up and the toes point

downwards; *talipes equinovarus*, in which the foot is inverted and turned inwards, with the toes pointing down; *talipes valgus*, in which the foot is abducted and everted, so that the patient walks on its inner side. Also *club-foot*.

talk-back circuit (*ImageTech*) One which enables the controller of a programme to give directions to those originating a performance or rehearsal in a studio or location.

talk-down (*Radar*) See GROUND-CONTROLLED APPROACH.

talkspurt (*ICT*) A short period of time during which sound energy is actually being produced by the user of a telephone channel. Typical talkspurts are 170 ms long, with inter-spurt silences averaging 105 ms.

tall-boy (*Build*) A fitting added to the top of a chimney to prevent downdraught.

tallow wood (*For*) Wood of the tree *Eucalyptus microcorys*, a native of Australasia; it is durable, impermeable to preservative fluids and difficult to work.

tally (*Surv*) A brass tag attached to a chain at every tenth link, and so marked or shaped as to enable the position of the tally along the chain to be immediately read. Also a *teller*.

tally light (*ImageTech*) A red light that indicates which of two or more video cameras is providing the live signal, or that a single camera is operating.

talon (*BioSci*) A sharp-hooked claw, as that of a bird of prey.

talose (*Chem*) An *aldohexose*.

talus (*BioSci*) See ASTRAGALUS.

talus (*CivEng*) An earthwork or batter wall slope.

talus (*Geol*) See SCREE.

talus wall (*Build*) A wall the face of which is built on a BATTER.

tamarugite (*Min*) A hydrated sulphate of sodium and aluminium, crystallizing in the monoclinic system.

Tamiflu (*Pharmacol*) TN for the antiviral drug OSELTAMIVIR.

Tamman's temperature (*Phys*) The temperature at which the mobility and reactivity of the molecules in a solid become appreciable. It is approximately half the melting point in kelvin.

tamoxifen (*Pharmacol*) Estrogen antagonist which blocks receptor sites on cells, used in the treatment of breast cancer.

tamp (*CivEng*) (1) To fill a charged shot hole with clay or other stemming material to confine the force of the explosion. (2) To ram or pound down ballast on a railway track, or road metal. Also *punning*.

tampin (*Build*) A conical plug of boxwood used in opening out the end of a lead pipe.

tampon (*Med*) A plug or packing made of gauze, cotton wool and the like, for insertion into orifices or cavities (esp the vagina and uterus) for the control of haemorrhage, the removal of secretions or the dilating of passages.

tamsulosin (*Pharmacol*) An alpha-blocker used to treat benign prostatic hyperplasia.

tan delta (*ElecEng*) Equivalent to LOSS FACTOR, LOSS TANGENT.

tandem (*ICT*) The connection of the output of one four-terminal network to the input of a second.

tandem engine (*Eng*) An engine in which the cylinders are arranged axially, or end to end, with a common piston rod.

tandem exchange (*ICT*) One used primarily as a switching point for traffic between other exchanges.

tandem mill (*Eng*) A rolling mill with two or more stands operating continuously and synchronized so that successive stands can accommodate the increasing velocity of the rolled material as its thickness is reduced.

tandem mirror (*NucEng*) See MAGNETIC MIRROR.

tandem selection (*ICT*) The selection of outlets by two uniselectors in series, so that the maximum possible availability of outlets is obtained.

tandem working (*ICT*) The using of an intermediate exchange during the transition period when a manual system is being converted to automatic working. Working is effected by trains of impulses that are set up on key senders on instructions from A-operators in originating exchanges.

tang (*Build*) The end of a tool which is driven into its handle. See SOCKET CHISEL.

tangent (*MathSci*) (1) See TRIGONOMETRICAL FUNCTION. (2) A line or plane which touches but does not intersect a curve or surface. See MOVING TRIHEDRAL. The problem of finding tangents was first considered by the Greeks and is now solved by differential calculus.

tangent distance (*Surv*) The distance between the intersection point and one of the tangent points of a railway or highway curve.

tangent galvanometer (*ElecEng*) A vertical circular coil with its plane parallel to the meridian. If I is the current flowing in amperes, and r the effective coil radius in cm, then

$$I = \frac{rH}{50n}\tan\theta$$

where θ is the angle of deflection of a magnetometer needle placed at the centre of the coil, n is the number of turns in the coil, and H is the horizontal component of the Earth's magnetic field in amperes per metre.

tangential field (*Phys*) The image surface formed by the tangential foci of a series of object points lying in a plane at right angles to the axis.

tangential focus (*Phys*) The focus of an object point lying off the axis of an optical system, in which the image is drawn out by the astigmatism of the system into a line tangential to a circle centred on the optical axis.

tangential longitudinal section (*BioSci*) A section cut longitudinally along a more or less cylindrical organ parallel to a tangent at its surface. Abbrev *TLS*.

tangential wave path (*EnvSci*) That of a direct wave, tangential to the surface of the Earth and which is curved by atmospheric refraction.

tangent modulus (*Eng*) Elastic (or quasi-elastic) modulus derived from a non-linear STRESS–STRAIN CURVE by taking the slope of curve at a specified level of stress or strain, eg '0·5% strain tangent modulus'. This is usually less than the SECANT MODULUS at the same point.

tangent point (*Surv*) The point of commencement, or of termination, of a railway or highway curve.

tangent scale (*ElecEng*) The scale of an electrical instrument in which the measured quantity varies as the tangent of the angle of deflection.

tangent screw (*Surv*) A screw by which a fine adjustment may be made to the setting of a theodolite about its axis, either in order to bring the line of sight into coincidence with a signal, or to adjust the vernier reading to a given value.

tanh (*MathSci*) See HYPERBOLIC FUNCTIONS.

tank circuit (*ICT*) The section of a resonating coaxial transmission line or a tuned circuit that accepts power from an oscillator and delivers it, harmonic-free, to a load.

tanker (*Ships*) A term covering all types of ships carrying liquid in bulk, from light acids and oils to molasses and latex.

tank furnace (*Glass*) Essentially a large 'box' of refractory material holding from 6 to 200 tons of glass, through the sides of which are cut 'ports' fed with a combustible mixture (producer gas and air, coke-oven and air, or oil spray and air), so that flame sweeps over the glass surface. With the furnace is associated a regenerative or recuperative system for the purpose of recovering part of the heat from the waste gases.

tanking (*Build, CivEng*) Waterproof material usually applied outside an underground structure to prevent infiltration of subsoil water.

tank line (*ICT*) See TANK CIRCUIT.

tank mix (*Agri*) The final composition of a mixture of agrochemicals held in the sprayer tank immediately before application.

tank reactor (*NucEng*) Covered type of SWIMMING-POOL REACTOR.

tank rectifier (*ElecEng*) Mercury-arc rectifier enclosed in a metal tank with vitreous seals for the conductors.

tank vent pipe (*Aero*) The pipe leading from the air space in an aircraft fuel, or oil, tank to atmosphere, for equalizing changes in pressure due to alterations in altitude; in *aerobatic* aircraft a non-return valve is fitted to prevent liquid escaping when inverted.

tannic acid (*Chem*) See TANNIN.

tannin (*Chem*) A mixture of derivatives of polyhydroxy-benzoic acids. When pure it forms a colourless amorphous mass, easily soluble in water, of bitter taste and astringent properties. Occurs in many trees, eg *Quebracho colorado*.

tanning (*BioSci*) In newly formed cuticle of terrestrial arthropods, the process in which the spaces between the chitin micelles are filled with sclerotin, which consists of protein molecules linked together, or tanned by quinones. This makes the cuticle tougher and darker.

tanning (*Genrl*) The process of converting animal hide into LEATHER by treating with eg TANNIN.

tanning developer (*ImageTech*) A developer which produces hardening or insolubility of the gelatine emulsion in proportion to the silver image formed; used to prepare the MATRIX for IMBIBITION printing.

tannin sac (*BioSci*) A cell containing much tannin.

tantalite (*Min*) Tantalate and niobate of iron and manganese, crystallizing in the orthorhombic system. The principal ore of tantalum, occurring in pegmatites and granitic rocks and in alluvial deposits. When the Ta content exceeds that of Nb, the ore is called tantalate. See COLUMBITE.

tantalum (*Chem*) A metallic element, symbol Ta, at no 73, ram 180·948, rel.d. (at 20°C) 16·6, mp 2850°C, electrical resistivity $15\cdot5 \times 10^{-8}$ ohm metres, Brinell hardness 46. It occurs in crystals and grains (usually containing, in addition, small amounts of niobium) in the Ural and Altai Mts. It is used as a substitute for platinum for corrosion-resisting laboratory apparatus, as acid-resisting metal in chemical industry, and in the form of carbide in cemented carbides. Used in surgical insertions because of its lack of reaction to body fluids.

tantalum capacitor (*Electronics*) Miniature electrolytic capacitor employing tantalum foil.

T-antenna (*ICT*) One comprising a top conductor with a vertical download attached at the centre; much used for long waves.

T-antigens (*BioSci*) A group of surface antigens defining subpopulations of human T-lymphocytes. Many of these have been defined using monoclonal antibodies and are classified in the CD SYSTEM. Corresponding antigens have been identified on mouse, rat and bovine T-lymphocytes, and are presumably ubiquitous in mammals.

tap (*Eng*) A screwed plug of accurate thread, form and size, on which cutting edges are formed along longitudinal grooves; screwed into a hole, by hand or power, to cut an internal thread.

tap changing (*ElecEng*) A method of varying the voltage ratio of a transformer by tapping the windings in such a way that the TURNS RATIO is altered.

tap density (*PowderTech*) The apparent powder density of a powder bed formed in a container of stated dimensions when a stated amount of the powder is vibrated or tapped under stated conditions.

tape (*Build*) A long flexible measuring scale of thin strip steel, linen, linen in which wire is interwoven to increase its strength or plastic reinforced with glass fibre, coiled up in a circular case fitted with a handle for winding purposes.

tape (*ICT*) See MAGNETIC TAPE.

tape (*Textiles*) A woven narrow fabric, frequently used for reinforcing garments.

tape cassette (*ICT*) A device for holding MAGNETIC TAPE. Both tape and cassette may be similar to those used in domestic tape recorders.

tape deck (*Acous*) A platform incorporating essentials for magnetic recording (motor(s), spooling, recording and erasing heads) for adding to amplifier, microphone and loudspeaker, to form a complete set of recording and reproducing equipment.

tape deck (*ICT*) See TAPE DRIVE.

tape density (*ICT*) Packing density of a magnetic tape.

tape drive (*ICT*) A mechanism that transports MAGNETIC TAPE between spools across the read/write heads. Also *tape deck*.

tape joint (*Build*) Joints between plaster, wall or centring boards covered by a paper or gauze strip before a decorative surface is applied. Often applied by a machine which runs the tape and plaster onto the wall or ceiling in one operation.

taper (*Eng*) In conical parts, the difference in diameter per unit length.

tape recording (*Acous*) Longitudinal recording on magnetic particles dispersed in a medium carried on plastic tape. There is a residual magnetization of a high-frequency biasing current, modulated by the signal current. The residual MAGNETOMOTIVE FORCE in the particles allows the modulation to be reproduced by induction in a magnetic circuit. See TAPE DECK.

tapering gutter (*Build*) A parapet gutter having an increased width in the direction of flow, so as to secure the necessary fall.

taper key (*Eng*) A rectangular key having parallel sides, but slightly tapered in thickness along its depth. See KEY.

taper line gratings (*Textiles*) Transparent plastic plates engraved with lines that are more widely spaced at one end than the other. By placing the appropriate grating over a fabric the number of threads per cm (for woven fabrics) or courses per cm (in knitted fabrics) may be determined from the resulting diffraction pattern.

taper pin (*Eng*) A pin, used as a fastener, very slightly tapered to act as a wedge.

taper roller bearing (*Eng*) A roller bearing rendered capable of sustaining end thrust by the use of tapered rollers running between internally and externally coned races.

taper tap (*Eng*) The first tap used in threading a hole. The first few threads are ground down to the core diameter to provide a guide, gradually increasing to the full thread size. See PLUG TAP, SECOND TAP, TAP.

taper-turning attachment (*Eng*) An attachment bolted to the back of a lathe, with a guide bar which may be set at the angle or taper it is desired to impart to the turned part, the lathe tool being guided by the bar via a slide.

tapes (*Print*) Strips of leather, plastic or fabric for carrying paper from one part of press to another.

tape slap (*ImageTech*) Videotape instability caused as it makes and breaks contact with the heads on the DRUM, distorting verticals in the picture. See BULGE CYLINDER.

tape splice (*ImageTech*) A join in film or magnetic tape using transparent adhesive tape applied to the butted ends.

tapestry (*Textiles*) Originally a handwoven furnishing fabric in which a design was produced by hand stitches to form the required pattern. Now employed for JACQUARD figured design on machine-woven cloths.

tapestry brick (*Build*) See RUSTICS.

tapetum (*BioSci*) (1) A layer of cells in a sporangium of a vascular plant surrounding the spore mother cells, becoming absorbed as the spores mature. (2) In the eyes of certain night-flying insects, a reflecting structure. (3) In some vertebrates, a reflecting layer of the retinal side of the choroid. (4) In the vertebrate brain, a tract of fibres in the corpus callosum.

tapeworm (*BioSci*) Parasitic worms of the class Cestoda, generally taking the form of a scolex with hooks and/or suckers for attachment to the host and a chain of individual proglottids in successive stages of development. Species infecting humans are *Hymenolepis nana* or dwarf tapeworm, *Taenia solium* from infected pork,

Diphyllobothrium latum from infected fish, and *Echinococcus granulosus* which spends the larval stage only in humans, causing hydatid cysts, the adult worm inhabiting dogs.

tape wrap (*ImageTech*) The extent that the tape wraps around a videotape recorder's DRUM, usually just over 180°.

tap-field control (*ElecEng*) A method of controlling the speed of a series motor; the field excitation is varied by means of tappings on the field windings.

tap-field motor (*ElecEng*) A series motor whose field windings are arranged for tap-field control.

taphephobia (*Psych*) Morbid fear of being buried alive.

taphrogenesis (*Geol*) Vertical movements of the Earth's crust, resulting in the formation of major faults and rift valleys.

tapioca (*FoodSci*) A starch-rich substance obtained from the root of cassava or manioc.

tapiolite (*Min*) A tantalate resembling TANTALITE but crystallizing in the tetragonal system.

tappet (*Eng*) A sliding member working in a guide; interposed between a cam and the push rod or valve system which it operates, to eliminate side thrust.

tapping (*ElecEng*) An intermediate connection on a circuit element such as a resistor, often used to vary the potential applied to another electrical system.

tapping (*Eng*) (1) The operation of running molten metal from a furnace into a ladle. (2) Using a TAP to thread a hole.

tapping (*Med*) See PARACENTESIS.

TAPPI standard methods (*Paper*) Laboratory test methods conforming to the *Technical Association of the Pulp and Paper Industry* of the USA. Widely used internationally.

taproot (*BioSci*) The first (primary) root of a plant developed directly from the radicle. Sometimes develops as a fleshy storage organ, eg carrot.

taproot system (*BioSci*) A root system, characteristic of dicotyledons and conifers, based on a tap root with laterals of various orders. Cf FIBROUS ROOT SYSTEM.

Taq polymerase (*BioSci*) A heat-stable DNA polymerase, isolated from *Thermus aquaticus*, that is important in the polymerase chain reaction.

tar (*Chem*) See COALTAR, GAS TAR.

Tarantula Nebula (*Astron*) A bright EMISSION NEBULA in the Large Magellanic Cloud. Consists of an extensive region of hydrogen gas, lit and ionized by hot young underlying stars. Also *30 Doradus*.

tarbuttite (*Min*) Hydrated zinc phosphate, which crystallizes in the triclinic system. The crystals are often found in sheaf-like aggregates.

Tardigrada (*BioSci*) A subphylum of minute arthropods with suctorial mouthparts and four pairs of stumpy clawed legs. Common forms, of wide distribution, are found among moss and debris in ditches and gutters, and on tree trunks, and can survive desiccation.

tardive dyskinesia (*Psych, Med*) A disorder, often a side effect of long-term use of antipsychotic drugs, in which there are involuntary, stereotyped, repetitive movements of body and face – tongue protrusion, lip-smacking, head-rolling, etc.

tare (*Genrl*) The weight of a vessel, wrapping, or container, which subtracted from the gross weight gives the net weight.

target (*Electronics*) In general any surface on which a beam impinges. In sputtering, the surface which is eroded as distinct from the substrate, on which sputtered material is deposited.

target (*ImageTech*) A plate in a TV camera tube on which external scenes are focused and scanned by an electron beam.

target (*NucEng*) Material irradiated by beam from accelerator.

target (*Radar*) Reflecting object, eg an aircraft, which returns a minute portion of radiated pulse energy to the receiver of a radar system.

target cell (*BioSci*) (1) An antigen-bearing cell that is the target of attack by lymphocytes or by specific antibody. (2) In haematology, used to describe an abnormally shaped and unusually thin red cell with central stained area seen in blood films, esp in certain disorders of haemoglobin formation.

target pest recurrence (*Agri*) The recovery of a pesticide-treated population to a damaging level, with destruction of natural predators by the treatment as a significant causal factor.

target rod (*Surv*) A type of levelling staff provided with a sliding target, which can be moved by the staff-holder, under direction from the leveller, to a position in which it is in line with the line of sight of the level, the staff reading being recorded by the staff-holder.

target strength (*Acous*) Defined in decibels by $T = E - S + 2H$, where E is the echo level, S is the source level and $2H$ is the transmission loss.

target theory (*Radiol*) Proposed explanation of radio-biological effects, in which only a small sensitive region of each cell is susceptible to ionization damage.

tarif (*ImageTech*) Equipment for modifying colour rendering when motion picture film is reproduced on TV. (Said to be acronym for *technical apparatus for rectification of inferior film*.)

tarmacadam (*CivEng*) A road or runway surfacing of broken stone which has been covered with tar, spread in a layer of uniform thickness and well rolled. Two layers are usually applied, the upper one being of stone of smaller size. In the USA *Tarmac* is a widely used proprietary mixture. Cf MACADAMIZED ROAD.

tarnish (*Chem*) The discoloration produced on the surface of an exposed metal or mineral, esp silver, generally as the result of the formation of an oxide or a sulphide film.

tar pit (*Geol*) An outcrop where natural bitumen occurs. The tar frequently contains the skeletons of trapped animals.

tarsalgia (*Med*) Pain in the instep of the foot.

tar-sands (*Geol, MinExt*) Sedimentary deposits of oil-bearing sands. The oil may be separated by steam heating or solvent extraction, and REFORMING PROCESSES are used to recover normal oil products. The Athabasca Tar Sands are extensive deposits in Canada and are estimated to contain recoverable oil equivalent to three-quarters of the present world crude reserves, but extraction is expensive and difficult.

tarsia (*Build*) Wood inlay, of geometric or architectural patterns, in which comparatively large pieces of wood are used.

tarsus (*BioSci*) (1) In vertebrates, an elongate plate of dense connective tissue that supports the eyelid. (2) In insects, myriapods and some Arachnida (as mites), the terminal part of the leg, consisting typically of five joints. (3) In land vertebrates, the basal podial region of the hind-limb; the ankle. Adj *tarsal*.

tarsus (*Build*) A cylindrical projection along the intersection between the two sloping surfaces on one side of the ridge of a mansard roof.

tartar emetic (*Chem*) Potassium antimonyl tartrate, $2[KSbOC_4H_4O_6] \cdot H_2O$.

tartaric acid (*Chem, FoodSci*) Dihydroxy-succinic acid. HOOCCH(OH)CH(OH)COOH. It exists in four modifications: (+)-tartaric acid, mp 170°C; (−)-tartaric acid, mp 170°C; *racemic* or (±)-tartaric acid, mp 206°C; *meso*tartaric acid, mp 143°C. The (+)-tartaric acid is found in nature, free and as salts of potassium, calcium and magnesium. It occurs in a large number of plants and fruits; the acid potassium salt is deposited from wine.

tartrates (*Chem*) The salts of tartaric acid. Extensively used in medicine, the potassium salts and Rochelle salt as saline purges. Also *cream of tartar*.

tarviated (*CivEng*) A term applied to macadam road surfacings in which the stone is bound together with tar.

TAS (*Aero*) Abbrev for TRUE AIRSPEED.

TAS diagram (*Min*) Abbrev for *total alkali silica* diagram. A graph on which chemical analyses of volcanic rocks may be plotted and used as the basis for classification when modes cannot be determined; a recommended method of the *IUGS system*.

TASI (*ICT*) Abbrev for TIME-ASSIGNMENT SPEECH INTERPOLATION.

taskbar (*ICT*) An area on a computer screen that displays details of all programs currently running.

task swapping (*ICT*) An operation carried out by the OPERATING SYSTEM whereby control will be swapped between two or more application programs. This will be done either in response to a command from the user or by the operating system itself in a MULTIPROGRAMMING or MULTITASKING environment. Also *task switching*.

tasmanite (*Geol*) A type of practically pure spore coal; a variety of CANNEL COAL. See BOGHEAD COAL.

taste bud (*BioSci*) In vertebrates, an aggregation of superficial sensory cells subserving the sense of taste; in higher forms, usually on the tongue.

TAT (*Psych*) Abbrev for THEMATIC APPERCEPTION TEST.

TATA box (*BioSci*) A sequence of bases (thymine, adenine) found upstream of the promoter region of eukaryotic genes to which RNA polymerase II binds; it may be important in determining the exact position at which transcription starts. Also *Goldberg–Hogness box*. See panel on STEROID HORMONES for an example.

TATP (*Chem*) Abbrev for *triacetone triperoxide*, a powerful explosive.

T-attenuator (*ElecEng*) One comprising three resistors, one end of each being connected together. The free ends of two are connected respectively to an input and an output terminal while the free end of the third is connected to the common input and output terminals.

tau (*Phys*) A lepton (τ^-) and its antilepton (τ^+) of exceptionally high mass, 1·78 GeV and a short lifetime of 0·3 ps. Discovered at the Stanford Linear Accelerator Center (SLAC) in electron–positron colliding-beam experiments. Its existence necessitated the postulation of the TOP and BOTTOM quarks to preserve the LEPTON–QUARK SYMMETRY.

taungya (*EnvSci*) A form of land use in the humid tropics, in which villagers are given the right to farm on good forest soils in exchange for their services in tending young trees on the same land. A practical form of AGROFORESTRY.

Taurids (*Astron*) A minor meteor shower which shows maximum activity on 8 November with a rate of around twelve per hour.

Taurus (Bull) (*Astron*) A prominent northern constellation, lying between Aries and Gemini. It includes ALDEBARAN, the CRAB NEBULA and the PLEIADES.

taut-, tauto- (*Genrl*) Prefixes from Gk *tauto-* meaning the same.

tautomerism (*Chem*) The existence of a substance as an equilibrium mixture of two interconvertible forms, usually because of the mobility of a hydrogen atom. Thus tautomeric compounds can give rise to two series of derivatives. See ETHYL ACETO-ACETATE.

tawa (*For*) New Zealand native hardwood (*Beilschmiedia tawa*), whose heartwood is white to yellow, straight-grained and fine-textured.

taxi-channel markers (*Aero*) See AIRPORT MARKERS.

taxis (*BioSci*) Orientation with respect to environmental stimuli; often combined with locomotion, so that the animal moves towards, or away, or at a fixed angle to the source. There are various classifications of types of taxis, eg *positive taxis* when the organism moves towards the stimulus, and away in *negative taxis*. Pl *taxes*. See KINESIS.

taxi track (*Aero*) A specially prepared track on an aerodrome used for the ground movement of aircraft. See PERIMETER TRACK.

taxi-track lights (*Aero*) Lights so placed as to define manoeuvring areas and tracks.

Taxol (*Pharmacol*) TN for paclitaxel, a drug isolated from yew (*Taxus brevifolis*) that stabilizes microtubules and by disrupting the mitotic spindle is cytotoxic. Used in the treatment of cancer, esp breast carcinoma.

taxon (*BioSci*) Any group of organisms to which any rank of taxonomic name is applied.

taxonomic series (*BioSci*) The range of extant living organisms, ranging from the simplest to the most complex forms.

taxonomy (*BioSci*) The science of classification as applied to living organisms, including study of means of formation of species etc.

Taylor process (*Eng*) Making very fine wire by inserting it into a closely fitting glass tube and drawing out the whole during heating.

Taylor's series (*MathSci*) Series expansion for a continuous function, giving the value of the function for one value of the independent variable in terms of that for another value. Under specified conditions the series is

$$f(a+h) = f(a) + hf'(a) + \frac{h^2 f''(a)}{2!} + \cdots$$

For a function of two variables,

$$f(a+h, b+h) = \sum_{n=0}^{\infty} \frac{1}{n!} \left(h\frac{\partial}{\partial x} + k\frac{\partial}{\partial y} \right)^n f(a,b)$$

$$= f(a,b) + h\frac{\partial f(a,b)}{\partial x} + k\frac{\partial f(a,b)}{\partial y} + \cdots$$

Tay–Sachs disease (*Med*) A rare inherited neurological disease of infants, prevalent in Ashkenazi Jews and characterized by progressive paralysis of the body, seizures, blindness, deafness and death before the age of 2. A result of the progressive degeneration of the nerve cells of the brain and spinal cord, due to a deficiency in an iso-enzyme which catalyses the conversion of GANGLIOSIDE.

TB (*Med*) Abbrev for tubercle bacillus, TUBERCULOSIS.

Tb (*Chem*) Symbol for TERBIUM.

T-banding (*BioSci*) A method of staining chromosomes. See BANDING TECHNIQUES and panel on CHROMOSOME.

TBC (*ImageTech*) Abbrev for *time-base corrector*, equipment for correcting timing errors in videotape recording and reproduction.

T-beam (*CivEng*) A beam forming part of the construction of a reinforced concrete floor; regarded as being composed of the beam part projecting below the floor slab, and portions of the floor slab on both sides, the whole having the form of a letter T.

TBO (*Aero*) Abbrev for TIME BETWEEN OVERHAULS. Also *tbo*.

T-bolt (*Eng*) A clamping bolt having a head in the form of a short rectangular crosspiece, and sliding in slots in a machine tool table cut in the shape of an inverted T. Access is provided by wells cast or machined near the ends of the table.

TBP (*Chem*) Abbrev for TRI-N-BUTYL PHOSPHATE.

TBT (*Chem*) Abbrev for TRIBUTYLTIN.

Tc (*Chem*) Symbol for TECHNETIUM.

TCA cycle (*BioSci, Chem*) Abbrev for TRICARBOXYLIC ACID CYCLE, the biochemical pathway whereby, in the presence of oxygen, pyruvic acid formed by GLYCOLYSIS is broken down to form carbon dioxide and water, with the release of large amounts of energy in the form of ATP. It takes place in the mitochondria. Also *Krebs' cycle*. Fig. ▷

TCAS (*Genrl*) Abbrev for *traffic alert and collision avoidance system*, used to warn pilots of potential midair collisions.

T-cell (*BioSci*) Common abbrev for T-LYMPHOCYTE. See panel on IMMUNE RESPONSE.

T-cell epitope (*BioSci*) A sequence of amino acids in a protein which can be recognized by a T-cell receptor.

T-cell growth factor (*BioSci*) See INTERLEUKIN-2. Abbrev TCGF.

T-cell leukaemia viruses (*BioSci*) A group of RETROVIRUSES that infect T-lymphocytes and cause leukaemias. Some cause malignant transformation of T-lymphocytes (eg HTLV-1 or feline leukaemia virus) whereas others cause AIDS (HTLV-3, now renamed HIV for human immunodeficiency virus). Similar viruses are present in monkeys.

T-cell receptor (*BioSci*) The antigen-recognizing receptor on the surface of T-cells. Antigen is recognized in association with MHC antigen at the surface of an antigen-presenting cell.

T-cell repertoire (*BioSci*) The number of different antigenic determinants to which the T-lymphocytes of an individual animal are capable of responding, thought to be comparable in size with the B-lymphocyte repertoire. There are, however, certain peptide sequences which are not recognized, either because they are not recognizably different from 'self' or because they are unable to associate with a particular MHC determinant. See panel on IMMUNE RESPONSE.

T-cell replacing factor (*BioSci*) A soluble factor, derived from helper T-lymphocytes, that can replace the presence of T-lymphocytes in stimulating antibody production by B-lymphocytes which have been activated by antigen.

TCGF (*BioSci*) Abbrev for *T-cell growth factor*. See INTERLEUKIN-2.

TCNE (*Chem*) Abbrev for TETRACYANOETHENE.

TCP/IP (*ICT*) Abbreviation for transmission control protocol/internet protocol. This protocol was developed by the Defense Advanced Research Projects Agency (DARPA). TCP is the transport layer and IP corresponds to the network layer within the open systems interconnection (OSI) model. See panel on INTERNET.

TDI (*Chem*) Abbrev for TOLYLENE-2,4-DIISOCYANATE.

t-distribution (*MathSci*) The sampling distribution of the mean of a set of observations from a normal distribution with unknown variance. The (central) *t*-distribution has one parameter, the degrees of freedom, and describes the sampling distribution of the deviation of the sample mean from the population mean.

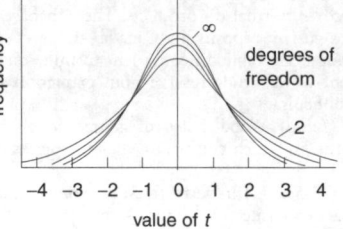

t-distribution

TDM (*ICT*) Abbrev for TIME-DIVISION MULTIPLEXING.

TDMA (*ICT*) Abbrev for TIME-DIVISION MULTIPLE ACCESS.

TDN (*Agri*) See TOTAL DIGESTIBLE NUTRIENT.

TD nickel (*Eng*) A dispersion-strengthened material where high yield and tensile strength are achieved by fine particles of THORIA which act as reinforcement throughout the nickel matrix and, being essentially insoluble, retain their strengthening effect at high temperature.

TdT (*BioSci*) Abbrev for TERMINAL DESOXYNUCLEOTIDYL TRANSFERASE.

Te (*Chem*) Symbol for TELLURIUM.

tea (*FoodSci*) A drink produced from the infusion in hot water of the dried, fermented leaves of the shrub *Camellia sinensis* (China tea) or *Camellia assamica* (India tea). Infusions of dried herbs and other botanicals are also termed teas.

teak (*For*) Valuable and very durable timber from *Tectona grandis*, a HARDWOOD tree of the verbena family, native to India and SE Asia. It is a rich brown with darker-brown markings, straight- to wavy-grained and coarse-textured.

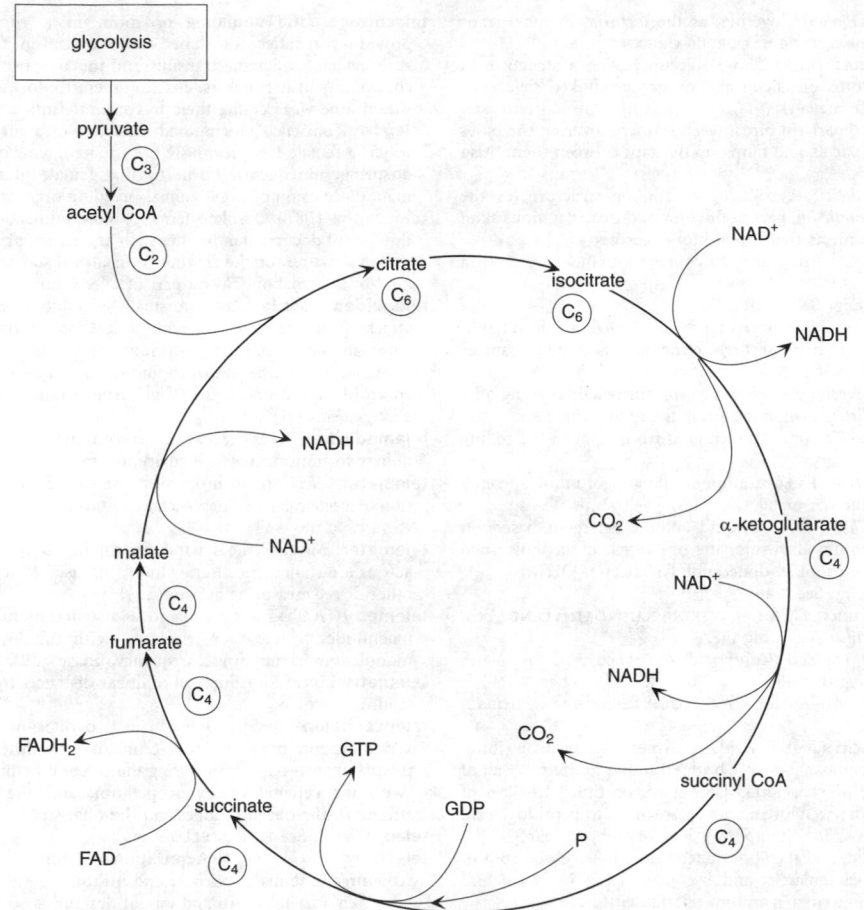

TCA cycle Citric acid cycle. NAD^+ = oxidized-, NADH = reduced-nicotinamide adenine dinucleotide, FAD = flavine adenine dinucleotide, $FADH_2$ = its reduced form, GTP = guanosine triphosphate, GDP = guanosine diphosphate, P = phosphate. The numbers in the circles refer to the number of carbon atoms in the compound.

tear factor (*Paper*) The ratio of the lateral tearing strength of a paper to its GRAMMAGE.

tear fault (*Geol*) A horizontal displacement of a series of rocks along a more or less vertical plane. Cf NORMAL FAULT, THRUST PLANE. See panel on PLATE TECTONICS.

tear gases (*Chem*) Volatile compounds which even in low concentration make vision impossible by their irritant action on the eyes. They are halogenated organic compounds, eg *xylyl bromide*, $CH_3C_6H_4CH_2Br$, and *ethyl iodoacetate*, $CH_2ICOOC_2H_5$.

tear gland (*BioSci*) See LACRIMAL GLAND.

tearing energy (*Eng*) The equivalent of TOUGHNESS when obtained from a TEAR TEST on eg an elastomer. See STRENGTH MEASURES.

tear test (*Eng*) The resistance to tearing in eg a trouser-leg specimen of elastomers, textiles and paper.

teasel (*Textiles*) The dried seed head of the thistle-like plant *Dipsacus fullonum*, fitted on machines to raise the pile of certain fabrics. Also *teazle*.

teat cup (*Agri*) The part of a milking machine that attaches to a cow's teat by vacuum suction. It has a rubber lining that pulsates against the teat to stimulate letdown.

teats (*BioSci*) In female mammals, paired projections from the skin on which the lactiferous tubules of the mammary glands open.

technetium (*Chem*) Radioactive element not found in ores. First produced as a result of deuteron and neutron bombardments of molybdenum. Symbol Tc, at no 43, the most common isotope, ^{99}Tc, has a half-life of $2 \cdot 1 \times 10^6$ years; ^{99m}Tc has a half-life of $6 \cdot 1$ h and is used in nuclear medicine. Found among fission products of uranium, and (unexplained) in the spectra of some stars.

Technical Association of the Pulp and Paper Industry (*Genrl*) US industrial association specifying widely used test procedures known as *TAPPI standard methods*.

technical jack plane (*Build*) Small type of JACK PLANE, with low-cut rear section.

Technicolor (*ImageTech*) TN for a number of systems of colour cinematography, one of which used a special three-strip camera exposing three separation negatives (1932–55). Prints were made by photomechanical imbibition methods (DYE TRANSFER) until 1978.

technology (*Genrl*) The practice, description and terminology of any or all of the applied sciences which have practical value and/or industrial use.

tectonic (*Geol*) Descriptive of rock structures which are directly attributable to earth movements involved in folding and faulting.

tectonics (*Geol*) The study of the major structural features of the Earth's crust.

tectorial (*BioSci*) Covering, as the tectorial membrane or membrana tectoria of CORTI'S ORGAN.

tectosilicates (*Min*) Those silicates having a structure in which atoms of silicon and oxygen are linked in a continuous framework.

tectrices (*BioSci*) In birds, small feathers covering the bases of the REMIGES and filling up the gaps between them. Also *auriculars*.

tectum (*BioSci*) A covering or roofing structure, as the *tectum synoticum*, part of the roof of the cartilaginous skull which connects the two auditory capsules.

tee (*Build*) A fitting used to connect a branch pipe into a pipe run at 90°.

tee bolt (*Eng*) See T-BOLT.

tee hinge (*Build*) A large strap hinge shaped like the letter T, the long arm being secured to the door, and the crosspiece to the hinging post.

tee joint (*ElecEng*) A joint in a cable formed by tapping off a branch circuit, without cutting the main cable.

teem (*Glass*) To pour molten glass from a pot in the rolling process.

teeming (*Eng*) The operation of filling ingot moulds from a ladle of molten metal.

Teepol (*Chem*) TN of liquid anionic detergent based on mixed sodium alkyl sulphates of long-chain alcohols, such as lauryl alcohol [*n*-dodecanol, $CH_3(CH_2)_{11}OH$].

tee rest (*Eng*) See T-REST.

Teflon (*Plastics*) TN for POLYTETRAFLUOROETHYLENE (US).

tegula (*Build*) A roofing tile.

tegulated (*BioSci*) Composed of or covered by plates overlapping like tiles.

tehp (*Aero*) Abbrev for TOTAL EQUIVALENT BRAKE HORSE-POWER.

teichoic acid (*BioSci*) Acidic polymers (glycerol or ribitol linked by phosphodiester bridges) found in the cell wall of Gram-positive bacteria. They may constitute 10–50% of the wall dry weight, and are cross-linked to peptidoglycan.

teichopsia (*Med*) Temporary loss of sight in part of the visual field, and the appearance before the eye of a spot of light which enlarges and becomes zigzag in shape and many-coloured; a symptom of MIGRAINE.

Teklan (*Plastics*) TN for synthetic textile fibre based on a modified acrylic (modacrylic) base. Chiefly noted for its inherent flameproof properties.

tektites (*Min*) A group term which covers moldavites, billitonites and australites. They are natural glasses of non-volcanic origin and may be of extraterrestrial origin.

tektosilicates (*Min*) See TECTOSILICATES.

telangiectasis (*Med*) Morbid dilatation of capillaries and arteries. Adj *telangiectatic*.

telco (*ICT*) Colloq term for *telephone company*.

tele-, teleo- (*Genrl*) Prefix from Gk *telos*, end.

telebanking (*ICT*) Electronic banking service accessed by telephone.

telecentric stop (*Phys*) A stop placed in the second focal plane of a positive lens, forming a viewing system by which a scale can be read without parallax error.

teleceptor (*BioSci*) A sense organ that responds to stimuli of remote origin. Also *telereceptor*.

telecine (*ImageTech*) Apparatus for producing video signals from motion picture film for TV broadcasting or videotape recording.

telecommunication (*Space*) (1) The transmission and reception of data-carrying signals, usually between two widely separated points with the aid of a communications satellite. (2) More generally any communication between two distant points by electrical means.

teleconference (*ICT*) A discussion in which lines remain open between multiple subscribers. Using computers, images can be exchanged using eg a WHITEBOARD.

teleconferencing (*ICT, ImageTech*) See CONFERENCE CALL.

teleconverter (*ImageTech*) Supplementary lens system with extension-tube mounting to convert a camera lens to much greater focal length. Also *range-extender* or *tele-extender*.

telecottage (*ICT*) Building or room in a rural area providing a variety of shared communication facilities such as fax and data transmission for use by a local community of teleworkers, enabling them to do work that would otherwise require them to commute into a city.

telegony (*BioSci*) The supposed influence of a male with which a female has previously been mated, as evinced in offspring subsequently borne by that female to another mate. A delusion of some animal-breeding organizations.

telegraphy (*ICT*) Obsolete term for communication at a distance of documentary matter such as written, printed or pictorial matter, or the reproduction at a distance of any kind of information. Now a part of telecommunication.

teleguided missile (*Aero*) A small subsonic missile for attacking surface targets, eg tanks and ships, controlled by command guidance from an operator, or automatic device, by signals transmitted through fine wires connected to the control box and uncoiled in flight from the missile. Also *wire-guided*.

telemedicine (*ICT*) The use of telecommunications technology to transmit medical advice, diagnoses, etc.

telemeter (*ElecEng*) An instrument for the remote indication of electrical quantities, such as voltage, current, power, etc.

telemeter (*Surv*) General name for an instrument which acts as a distance measurer, without the use of a chain or other direct-measuring apparatus.

telemetry (*ICT*) Transmission to a distance of measured magnitudes by radio or telephony, with suitably coded modulation, eg amplitude, frequency, phase, pulse.

telemetry (*Surv*) Measurement of linear distances by use of tellurometer.

telencephalon (*BioSci*) One of the two divisions of the vertebrate fore-brain or prosencephalon (the other being the DIENCEPHALON), comprising the cerebral hemispheres (with the cerebral cortex or pallium, and the corpus striatum), the olfactory lobes and the olfactory bulbs.

teleo- (*Genrl*) See TELE-, TELEO-.

teleology (*BioSci*) The interpretation of animal or plant structures in terms of purpose and utility. Rightly abjured but often a useful shorthand way of describing something that confers selective advantage. Adj *teleological*.

teleonomy (*BioSci*) The impression of purpose arising from adaptation through natural selection.

Teleostei (*BioSci*) An infraorder of Osteichthyes, including fish with a wide diversity of form and physiological adaptation. The gills are fully filamentous and the tail is externally (and in many cases internally) homocercal. The endoskeleton is completely ossified, and the fins are completely fan-like with no trace of an axis. Bony fishes.

telepathy (*Psych*) Communication between individuals that takes place independently of all known sensory channels.

telephone-answering machine (*ICT*) A device attached to a telephone line, capable of recognizing ringing currents and closing the line circuit as if the SWITCH-HOOK had been operated. An outgoing message then invites the caller to record an incoming message, usually on a tape cassette but more recently by digital means.

telephone influence factor (*ICT*) The weighting factor required for determining the total interference of induced emfs arising from harmonic induction in telephone lines from adjacent power lines. The factor takes into account the average relative sensitivity of the ear for varying frequency, and also the average response curves of telephone receivers. Also *telephone interference factor*. Abbrev TIF.

telephony (*ICT*) The conversion of a sound signal into corresponding variations of electric current (or potential), which is then transmitted by wire or radio to a distant point where it is reconverted into sound.

telephoto lens (*ImageTech*) A long-focus lens having a comparatively short BACK-FOCUS, one-half or one-third of its focal length, giving high magnification without undue extension.

telepoint (*ICT*) Old system of mobile telephony in which base stations, able to accept calls from a pocket handset within a range of about 100 m, were installed in public places. The last UK network closed down in 1993.

telepresence (*ICT*) General term for any system that claims to replace face-to-face meeting, eg VIDEOCONFERENCING or an experimental system in which a head-mounted camera transmits to a head-mounted video display, so that the recipient sees whatever the sender sees.

teleprinter (*ICT*) See TELEX.

teleprocessing (*ICT*) Processing carried out from a remote terminal.

teleradiography (*Radiol*) A technique to minimize distortion in taking X-ray photographs by placing X-ray tube some distance from the body.

telereceptor (*BioSci*) See TELECEPTOR.

telerecording (*ImageTech*) Transferring a TV or videotape programme to motion picture film.

telescience (*Space*) A fully interactive mode of scientific operations where the experiment is performed remotely, using data presented to the experimenter (eg by TV) who can remotely control elements of the experimental equipment.

telescope (*Astron*) An instrument specifically designed to collect, detect and record electromagnetic radiation from any cosmic source. See panel on ASTRONOMICAL TELESCOPE.

telescopic shaft (*Eng*) An assembly of two or more tubes, often keyed together, sliding within each other to provide a hollow shaft of variable length.

telescopic star (*Astron*) A star whose apparent magnitude is numerically greater than the sixth and which is too faint to be seen with the naked eye.

Telescopium (Telescope) (*Astron*) A small southern constellation.

Telesto (*Astron*) The 13th natural satellite of Saturn, discovered in 1980. Distance from the planet 295 000 km; diameter 24 km.

telestrator (*ICT*) A device that enables a TV presenter to superimpose diagrams, writing, etc, on the screen during a broadcast.

Teletex (*ICT*) An international business correspondence defined by CCITT, and offered by telecommunications authorities and operators in many countries; the terminals are generally sophisticated typewriters and word processors.

teletext (*ICT*) A method of transmitting computer-stored information to suitably adapted domestic TV receivers; information is transmitted as static pages of alphanumeric information at a low rate within the conventional TV signal. In the UK, Ceefax (BBC) continues to be used but the ITV version (Oracle) ceased to be broadcast in 1992. See VIEWDATA.

teletherapy (*Radiol*) Treatment by X-rays from a powerful source at a distance, ie by high-voltage X-ray tubes or radioactive sources, such as ^{60}Co, up to 80 000 GBq.

teletypesetting (*Print*) A method of operating LINECASTING MACHINES by using a six-unit punched tape produced on a separate keyboard; can be used within the same office where increased output results through specialization and the optimum use of equipment and skill; or the keyboard output can be transmitted by wire or radio to be converted to tape at the receiving end.

teleutospore (*BioSci*) A thick-walled resting spore of rust and smut fungi.

television (*ICT*) The electronic transmission, reception and reproduction of transient visual images. Abbrev *TV*.

television cable (*ImageTech*) One capable of transmitting frequencies sufficiently high to accommodate TV signals without undue attenuation or relative phase delay; usually coaxial, with as much air insulation as possible.

television camera (*ImageTech*) A camera converting the optical image of an external scene into electronic signals for TV transmission or recording.

television channel (*ImageTech*) Radio-frequency band of sufficient width to accommodate TV transmission signals and allocated for that specific use.

television receiver (*ImageTech*) That part of a TV system in which the picture and associated sounds are reproduced from the input signal. Also *television set*.

television transmitter (*ImageTech*) One which radiates video, audio and synchronizing components of a TV signal as a modulated radio wave.

teleworking (*ICT*) The use of telecommunications to allow employees to stay at home or in a local TELECOTTAGE while they do their work. An example is the home operation of a directory enquiry system.

telex (*ICT*) Abbrev for *automatic teletypewriter exchange service*. An audio-frequency teleprinter system for use over telephone lines (provided in the UK by British Telecommunications plc).

teller (*Surv*) See TALLY.

telluric bismuth (*Min*) An intermetallic compound, Bi_2Te_3, crystallizing in the trigonal system. The name has also been used as a synonym for TETRADYMITE.

telluric current (*MinExt*) Current in, or put into, the Earth, which is used in exploration of strata.

telluric line (*Astron*) Absorption line or band in stellar and planetary spectra, caused by absorption in the Earth's atmosphere, mainly by water vapour and oxygen.

tellurides (*Chem*) Compounds of divalent tellurium [Te (II)], analogous to SULPHIDES.

tellurite (*Min*) Orthorhombic tellurium dioxide.

tellurium (*Chem*) A semi-metallic element, tin-white in colour. Symbol at no 52, ram 127·60, rel.d. (at 20°C) 6·24, mp 452°C, valencies 2, 4, 6, electrical resistivity 2×10^{-8} Ω m. Used in the electrolytic refining of zinc in order to eliminate cobalt; alloyed with lead to increase the strength of pipes and cable sheaths. The chief sources are the slimes from copper and lead refineries and the fine dusts from telluride gold ores. There are numerous but rare silver, gold, bismuth and iron tellurides.

tellurobismuth (*Min*) See TELLURIC BISMUTH.

tellurometer (*Surv*) Electronic instrument used to measure survey lines for distances up to 40 miles (64 km) by measurement of time required for a radar signal to echo back, accuracy being of order of 1:100 000 in good weather. See TRILATERATION.

Telnet (*ICT*) TN for a system of remote access to computers by means of telecommunications.

telo- (*Genrl*) Prefix from Gk *telos*, end.

teloblast (*BioSci*) A large cell from which many smaller cells are produced by budding, as one of the primary mesoderm cells in developing Polychaeta.

telocentric (*BioSci*) Having the CENTROMERE at one end of the chromosome.

telolecithal (*BioSci*) A type of egg that is large in size, with yolk constituting most of the volume of the cell, and with the relatively small amount of cytoplasm concentrated at one pole. Found in sharks, skates, reptiles and birds. Cf MESOLECITHAL, OLIGOLECITHAL.

telome (*BioSci*) Hypothetical morphological unit of primitive vascular plants; an ultimate branch of an axis that repeatedly branches dichotomously.

telomerase (*BioSci*) A DNA polymerase that will only elongate oligonucleotides from the telomere and not other sequences. Maintenance of telomere length may be important for continued proliferation, and upregulation of telomerase may cause or assist neoplasia. Also *telomere terminal transferase*.

telomere (*BioSci*) The structure that terminates the arm of a chromosome.

telome theory (*BioSci*) The proposal that the shoots of modern land plants have evolved from repeatedly dichotomously branched axes.

telomorph (*BioSci*) The sexual or perfect stage of a fungus. Cf ANAMORPH.

telophase (*BioSci*) The final phase of mitosis with cytoplasmic division; the period of reconstruction of nuclei that follows the separation of the daughter chromosomes in mitosis. See diagram at MITOSIS and panel on CELL CYCLE.

telson (*BioSci*) The post-segmental region of the abdomen in some Crustacea and Chelicerata.

Telstar (*ICT*) First transatlantic telecommunications satellite, launched 1962.

TEM (*BioSci, Phys*) Abbrev for TRANSMISSION ELECTRON MICROSCOPE (or *microscopy*).

temazepam (*Pharmacol*) Short-acting *benzodiazepine* drug useful for inducing sleep, particularly in the elderly. Now a controlled drug because of recreational abuse.

témoin (*CivEng*) An undisturbed column of earth left on an excavated site, as an indication of the depth of the excavation.

temozolomide (*Pharmacol*) A drug that interferes with DNA synthesis, used in the treatment of brain tumours.

temper (*Eng*) Vague term describing the relative condition of the hardness and mechanical properties of a metal: eg a solid solution alloy which can be extensively cold-worked may exhibit a range of properties from its softest state when annealed to fully work-hardened; this range is divided into hardness bands, referred to as *tempers*, eg quarter hard, half hard, etc. A given temper can be achieved either by full annealing followed by careful work-hardening or by full work-hardening followed by careful annealing. The latter is referred to as *temper annealing* or *back annealing*. See ANNEALING.

temper (*Glass*) The amount of residual stress in annealed glassware measured by comparison with strain disks.

temperament (*Psych*) An individual's general disposition and typical way of responding to the environment; it may be significantly heritable.

temper annealing (*Eng*) See TEMPER.

temperate phage (*BioSci*) A bacteriophage which, after infecting its host, replicates its genome in step with that of its host without destroying it. UV irradiation or chemical treatment can induce a *lytic phase* in which phage multiplication occurs.

temperature (*Phys*) A measure of whether two systems are relatively hot or cold with respect to one another. Two systems brought into contact will, after sufficient time, be in thermal equilibrium and will have the same temperature. A thermometer using a temperature scale established with respect to an arbitrary zero (eg CELSIUS SCALE) or to absolute zero (KELVIN THERMODYNAMIC SCALE OF TEMPERATURE) is required to establish the relative temperatures of two systems. See THERMODYNAMICS.

temperature coefficient (*BioSci*) The ratio of the rate of progress of any reaction or process, at a given temperature, to the rate at a temperature $10°C$ lower. Also Q_{10}.

temperature coefficient (*Eng, Phys*) The fractional change in any particular physical quantity per degree rise of temperature.

temperature coefficient of resistance (*ElecEng*) In any conductor,

$$R = R_0[1 + \alpha(T - T_0)]$$

where R is the resistance at temperature T, compared with the resistance R_0 at temperature T_0, and the mean coefficient in the range T_0 to T is α. This is useful only for *linear resistances*, ie pure metals.

temperature correction (*Surv*) A correction applied to the observed length of a baseline to correct for any difference between the temperature of the tape during the measurement and that at which it was calibrated.

temperature cycle (*Phys*) A method of processing thick photographic nuclear research emulsions to ensure uniform development. They must be soaked in solutions at refrigerated temperatures and then warmed for the required processing period.

temperature-dependent sex determination (*BioSci*) Sex determination by environmental temperature, seen in

reptiles where sex chromosomes do not occur. Incubation of eggs at low temperatures produces one sex, higher temperatures the other.

temperature inversion (*EnvSci*) Anomalous increase in temperature with height in the troposphere. See panel on TROPOSPHERE.

temperature lapse rate (*EnvSci*) The rate of decrease of temperature with height.

temperature-limited (*Electronics*) Said of a thermionic device operated under saturation conditions, ie with the electrode currents limited by the cathode temperature.

temperature-sensitive mutant (*BioSci*) A mutant organism able to grow at one temperature, the *permissive temperature*, but unable to do so at another, which may be higher or lower than the permissive. This class of mutant has been particularly important in the analysis of mutants affecting vital functions like cell division. See panel on CELL CYCLE.

temper brittleness (*Eng*) A type of brittleness that is shown by notched bar tests, but not by the tensile test, in certain types of steel after tempering; influenced to a marked extent by the composition of the steel, the tempering temperature and the subsequent rate of cooling.

temper colour (*Eng*) In tempering hardened-steel cutting tools etc, the colour of the oxide layer which forms on reheating and which indicates approximately the correct quenching temperature for a particular purpose. See TEMPERING.

tempered scale (*Acous*) The musical scale of keyboard instruments, and, by implication, any other instruments or voices which are concerted with them, in which all semitones have frequencies of the same ratio (1:1·059 463 09), so that twelve semitones amount to one octave. Also *equal-tempered scale, equi-tempered scale*. See NATURAL SCALE.

temper-hardening (*Eng*) A term applied to alloys that increase in hardness when worked or reheated after rapid cooling, and to the operations of producing this effect. Also *artificial ageing*; distinguished from ageing, which occurs at atmospheric temperature. See PRECIPITATION HARDENING.

tempering (*Eng*) The reheating of hardened steel at any temperature below the critical range, in order to decrease the hardness. Also *drawing*. May be applied to reheating after rapid cooling, even when this results in increased hardness, eg in the case of steels that exhibit secondary hardening. See AUSTEMPERING and panel on STEELS.

tempering (*FoodSci*) Gentle agitation of chocolate for a given time at fixed temperature (29–33°C) to induce cocoa butter to crystallize in its most stable form.

template (*Build*) (1) A long flat stone supporting the end of a beam, to spread the load over several joints in the brickwork. (2) A framework of timber or steel used for the final setting out of girders, roof trusses, etc. Also *templet*.

template (*Eng*) A thin plate, cut to the shape or profile required on a finished surface, by which the surface is marked off or gauged during machining or other operation. Also *templet*.

template file (*ICT*) A file commonly used in connection with a WORD PROCESSOR or DESKTOP PUBLISHING system that contains information about the page layout, typefaces and fonts to be used within a document. This is to ensure that each page has a consistent layout. Also *style sheet*. See OUTLINER.

temple (*Textiles*) Devices fitted to weaving machines to keep the cloth at full width and suitably tensioned from side to side as it is woven.

templet (*Build, Eng*) See TEMPLATE.

tempolabile (*Chem*) Tending to change with time.

temporal (*BioSci*) A cartilage bone of the mammalian skull formed by the fusion of the petrosal with the squamosal.

temporal lobe (*Psych, Med*) One of the four lobes of the fore-brain. Contains the auditory cortex and is involved in language reception as well as memory and emotion. See panel on BRAIN STRUCTURE.

temporal openings (*BioSci*) See TEMPORAL VACUITIES.

temporal summation (*Psych*) The phenomenon in which a subthreshold stimulus presented over an extended period of time may produce a response. Can also be thought of as time-averaging, a way to improve signal/noise discrimination.

temporal vacuities (*BioSci*) In reptiles, openings in the skull, varying in number (none, one or two) and position, and used in classification. The various conditions found in different groups are ANAPSID, SYNAPSID, PARAPSID, euryapsid and DIAPSID.

temporary film (*Eng*) A soluble removable protective coating for metals, usually prepared from a lanolin derivative combined with suitable resins and dyed to a distinctive colour. The whole is dispersed in petrol or other solvent.

temporary hardness (*Chem*) Hardness of water caused by the presence of the hydrogen carbonates of calcium and magnesium and therefore removable by boiling to precipitate the carbonate.

temporary memory (*ICT*) See BUFFER.

temporary threshold shift (*Acous*) Temporary hearing loss after exposure to high sound level. Abbrev *TTS*.

temporary way (*CivEng*) The ballast, sleepers and rails laid temporarily by a contractor for transporting material on constructional works.

temporomandibular (*BioSci*) Relating to the joint that connects the lower jawbone to the skull.

tenacity (*Eng*) See ULTIMATE TENSILE STRESS. See STRENGTH MEASURES.

tendency (*Psych*) A general term referring to some measure of the probability that a behaviour will occur, without specifying the nature of the underlying causal factors.

tender ship (*Ships*) A ship with a small transverse metacentric height and long rolling period. Cf STIFF SHIP.

tendinous (*BioSci*) See TENDON.

tendo calcaneus (*BioSci*) See ACHILLES TENDON.

tendon (*BioSci*) A cord, band or sheet of non-elastic fibrous tissue by which a muscle is attached to a skeletal structure, or to another muscle. Adj *tendinous*.

tendril (*BioSci*) A slender, simple or branched, elongated organ used in climbing, at first soft and flexible, later becoming stiff and hard. May be a modified stem, leaf, leaflet or inflorescence.

tenebrescence (*Min*) The reversible bleaching observed in the hackmanite variety of sodalite. This mineral has a pink tinge when freshly fractured; the colour fades on exposure to light, but returns when the mineral is kept in the dark for a few weeks or is bombarded by X-rays.

tenesmus (*Med*) Painful and ineffectual straining at stool.

tenia, teniasis (*BioSci*) See TAENIA, TAENIASIS.

Tenite (*Plastics*) TN for cellulose acetate-butyrate (US).

tennantite (*Min*) Sulphide of copper and arsenic, which crystallizes in the cubic system. It is isomorphous with TETRAHEDRITE. The crystals frequently contain antimony, and grade into tetrahedrite. Also *fahlerz*.

tenon (*Build*) A tongue formed on the end of a timber by cutting away from both sides one-third of the thickness. The projecting part fits into a mortise in a second timber in order to make a joint between them. See TUSK TENON.

tenon-and-slot mortise (*Build*) A joint, such as that made between the posts and heads of solid door frames, in which a tenon cut on the end of the head fits into a SLOT MORTISE on the end of the post.

tenon saw (*Build*) A saw with a very thin parallel blade, having fine teeth (4–6 cm^{-1}) and a stiffened back along its upper edge. Also *mitre saw*.

tenorite (*Min*) Copper oxide, crystallizing in the triclinic system. Occurs in minute black scales as a sublimation product in volcanic regions or associated with copper veins. *Melaconite* is a massive variety.

tenosynovitis (*Med*) Inflammation of the sheath of a tendon. Also *tenovaginitis*.

tenotomy (*Med*) The cutting of a tendon for the correction of deformity.

tenovaginitis (*Med*) See TENOSYNOVITIS.

Tensar (*Plastics*) TN for polymer mesh used for fencing, supporting road foundations, etc. Made by punching holes in extruded polyolefin sheet and hot-stretching along one or two axes.

tensile creep test (*Eng*) See TENSILE TEST, VISCOELASTICITY and panel on CREEP AND DEFORMATION.

tensile modulus (*Eng*) The quantity corresponding to YOUNG'S MODULUS in non-linear and/or viscoelastic materials such as polymers. Cf CREEP MODULUS, SECANT MODULUS.

tensile strength (*Eng*) The maximum tensile force in a tensile test divided by the original cross-sectional area. See STRENGTH MEASURES.

tensile strength (*Textiles*) The maximum force per unit cross-sectional area that a thread or fabric can withstand until it breaks or pulls apart.

tensile stress (*Eng*) See STRESS, ULTIMATE TENSILE STRESS.

tensile test (*Eng*) Main forms are those in which: (1) a static increasing load is applied to a specimen until fracture results; from this a STRESS–STRAIN CURVE may be plotted and the ELONGATION, PROOF STRESS, ULTIMATE TENSILE STRESS, YIELD POINT and YOUNG'S MODULUS determined; (2) a dynamic load is applied giving data on fatigue and IMPACT; (3) a constant load is applied and the increasing deformation with time recorded (tensile creep test). See STRENGTH MEASURES and panels on FATIGUE and IMPACT TESTS.

tensile testing machine (*Eng*) A machine for applying a tensile or compressive load to a test piece, by means of hand- or power-driven screws, or by a hydraulic ram. The load is usually measured by a poise weight and calibrated lever.

tensimeter (*Chem*) Apparatus for the determination of transition points by observation of the temperature at which the vapour pressures of the two modifications become equal.

tensiometer (*Eng*) A versatile, portable testing machine used for a variety of mechanical tests, including tensile tests.

tension (*ElecEng*) Former term used to designate a potential difference, eg low-tension (such as an accumulator) or high-tension supplies (such as a power-distribution cable). Now obsolete.

tension (*Phys*) The state of a wire or string stretched between two points, or the force exerted by it on a support.

tension (*Psych*) A state of barely suppressed emotion, such as excitement, anxiety or hostility; a feeling of strain with resultant symptoms. Can be found in groups as well as in individuals.

tension insulator (*ElecEng*) A suspension insulator for overhead transmission lines, which is designed to withstand the pull of the conductors; it is used, therefore, at terminal, anchor or angle towers.

tension pin (*Eng*) A hollow dowel pin of elastic material, slotted so as to be deformed on assembling, to take up misalignment or shape irregularity, or to exert a force.

tension plate lock-up (*Print*) A method of locking plates to the cylinder by fingers which grip in slots on the underside of the plate. Cf COMPRESSION PLATE LOCK-UP.

tension rod (*Eng*) A structural member subject to tensile stress only. Also *tie rod*.

tension wood (*For*) A form of REACTION WOOD developed in hardwoods, with a higher cellulose and lower lignin content, hence more rubbery, and with lower compressive but higher tensile strength than normal. Formed eg on the upper side of horizontal branches. Cf COMPRESSION WOOD. See panel on WOOD.

tensometer (*Eng*) A versatile testing machine used for a variety of mechanical tests, including tensile tests.

tensor (*BioSci*) A muscle that stretches or tightens a part of the body without changing the relative position or direction of the axis of the part.

tensor (*MathSci*) The generalization of a vector. A mathematical entity specifiable by a set of components with respect to a system of co-ordinates and such that the transformation that has to be applied to the components to obtain components with respect to a new system of co-ordinates is related in a certain way to the transformation that had to be applied to the system of co-ordinates.

tensor force (*Phys*) A non-central nuclear force which depends on the spin orientation of the nucleons.

tent (*Med*) A roll or plug of soft absorbent material, or of expansible material, for keeping open a wound or dilating an orifice.

tentacle (*BioSci*) An elongate, slender, flexible organ, usually anterior, fulfilling a variety of functions in different forms, eg feeling, grasping, holding and sometimes locomotion. Also *tentaculum*. Adjs *tentacular, tentaculiferous, tentaculiform*.

tenter (*Textiles*) See STENTER.

tenth-value thickness (*NucEng*) The thickness of absorbing sheet which attenuates the intensity of a beam of radiation by a factor of ten.

tentorium (*BioSci*) (1) In the mammalian brain, a strong transverse fold of the dura mater, lying between the cerebrum and the cerebellum. (2) In insects, the endoskeleton of the head.

tepal (*BioSci*) One of the members of a perianth which is not clearly differentiated into a calyx and a corolla.

tepee buttes (*Geol*) Conical hills of Cretaceous shale, with steep, smooth slopes of talus and a core of shell-limestone, formed *in situ* by the growth of successive generations of lamellibranchs (*Lucina*). Found in the Great Plains of the US.

tephigram (*EnvSci*) AEROLOGICAL DIAGRAM in which the principal rectangular axes are temperature (T) and entropy (Φ): hence $T\Phi$-gram. Equal area represents equal energy at all points.

tephra (*Geol*) A general term for all fragmental volcanic products, eg *ash, bombs, pumice*.

tephrite (*Geol*) A fine-grained igneous rock resembling basalt and normally occurring in lava flows; characterized by the presence of a feldspathoid mineral in addition to, or in place of, feldspar. See VOLCANIC ROCKS.

tephroite (*Min*) An orthosilicate of manganese, which crystallizes in the orthorhombic system. It forms a member of the olivine isomorphous group, and occurs with zinc and manganese minerals.

ter- (*Genrl*) Prefix meaning thrice, three, threefold.

tera- (*Genrl*) Prefix, denoting 10^{12} times, eg a *terawatt* is 10^{12} watts. Symbol T.

terabyte (*ICT*) Commonly one million million or 10^{12} BYTES.

teratogen (*BioSci*) An agent that raises the incidence of congenital malformations.

teratogenic (*Med*) A substance or drug producing abnormal embryos.

teratology (*BioSci*) The study of monstrosities, as an aid to the understanding of normal development. (Gk *teras*, gen. *teratos*, a wonder.) See TERATOMA.

teratoma (*Med*) A tumour in the body consisting of tissues believed to be derived from the three germ layers (ectoderm, mesoderm and endoderm). May occur in a variety of locations but commonly in the testis and mediastinum. If it contains a predominance of ectodermal elements it is called a *dermoid cyst*.

terbium (*Chem*) A metallic element, a member of the rare earth group. Symbol Tb, at no 65, ram 158·9254. It occurs in the same minerals as dysprosium, europium and gadolinium.

tercom (*Aero*) Abbrev for *terrain comparison, terrain contour matching*. Stored digital data of ground contours are compared with those detected below the aircraft in flight and used to identify the aircraft's present position and track.

terebine (*Build*) Volatile solvents and thinners derived from heavy petroleum and rosin oils, or petroleum and rosin oils mixed with turpentine, used to accelerate the drying process of oil paints.

terebrate (*BioSci*) Possessing a boring organ or a sting.

terephthalic acid (*Chem*) *Benzene-1,4-dicarboxylic acid*. $C_6H_4(COOH)_2$. A powder, hardly soluble in water or ethanol, which sublimes unchanged. It is prepared by the oxidation of 4-toluic acid, or industrially by the catalytic oxidation of 4-xylene. Used in the manufacture of polyethylene terephthalate (PET). See STEP POLYMERIZATION.

terete (*BioSci*) Rounded, more or less cylindrical, and not ridged, grooved or angled.

tergum (*BioSci*) The dorsal part of a somite in Arthropoda; one of the plates of the carapace in Cirripedia. Adj *tergal*.

term diagram (*Phys*) Energy-level diagram for isolated atom in which levels are usually represented by corresponding quantum numbers.

terminal (*ElecEng*) A point in an electrical circuit at which any electrical element may be connected.

terminal (*ICT*) A term used to describe any input/output device that is used to communicate with the computer from a remote site. See KEYBOARD, PRINTER, VDU.

terminal adapter (*ICT*) Circuitry or equipment that provides the interface between a user terminal and a network, ranging from a simple line card in PLAIN OLD TELEPHONE SERVICE to a VIDEO CODEC for digital video services.

terminal bar (*ElecEng*) A bar to which a group of plates of an accumulator is attached. Also *connector bar, terminal yoke*.

terminal deoxynucleotidyl transferase (*BioSci*) An enzyme found in pre-T and pre-B lymphocytes and cortical thymocytes but absent from their progeny. Has a key role in V(D)J recombination during lymphocyte development, contributing to the generation of diversity. The presence of the enzyme is used as a marker for cells which contain it. Abbrev *TdT*.

terminal emulation (*ICT*) Software that allows a PC to act like a DUMB TERMINAL.

terminal equipment (*ICT*) The special apparatus required for connecting the normal telephone exchange pairs to special transmission systems (such as radio-telephone links, carrier systems) or to trunk lines.

terminal impedance (*Phys*) End or load impedance.

terminal lug (*ElecEng*) A projection on a group of accumulator plates for connection to an external circuit.

terminal pillar (*ElecEng*) See POST HEAD.

terminal pole (*ElecEng*) A pole at the end of a power-transmission or telephone line so designed as to withstand the longitudinal load of the conductors as well as the vertical load.

terminal server (*ICT*) A device that connects DUMB TERMINALS to a LAN and hence to a host computer.

terminal sire (*Agri*) A sire brought into a livestock herd from outside to introduce desirable characteristics to the progeny.

terminal tower (*ElecEng*) The transmission-line tower at the end of an overhead transmission line; arrangements must be made for taking the pull of the conductors, and for connecting them to the substation or other apparatus at the end of the line.

terminal velocity (*Aero*) The maximum LIMITING VELOCITY attainable by an aircraft as determined by its total DRAG and thrust.

terminal velocity (*Phys*) The constant velocity acquired by a body falling through a fluid when the frictional resistance is equal to the gravitational pull.

terminal-velocity dive (*Aero*) A nose dive to the greatest obtainable velocity of the machine at that altitude.

terminal voltage (*ElecEng*) The voltage at the terminals of a piece of electrical equipment, eg an electric machine or a power supply.

terminal yoke (*ElecEng*) See TERMINAL BAR.

terminate-and-stay-resident program (*ICT*) See MEMORY-RESIDENT PROGRAM.

terminated level (*ICT*) The reading of a level-measuring set at a point in a system when terminated at that point by a resistance equal to the nominal impedance of the system.

termination (*Chem*) Ending of polymerization reaction by several possible mechanisms. See CHAIN POLYMERIZATION and panel on POLYMER SYNTHESIS.

termination (*ElecEng*) That which is connected at the end of an electrical transmission system or to output of an amplifier. It can be regarded partly as the electrical load, but also includes its physical nature, eg a termination could be a loudspeaker whereas the associated load would be, say, 8 Ω.

termination codon (*BioSci*) See STOP CODON.

terminator (*Astron*) The border between the illuminated and dark hemispheres of the Moon or planets. Its apparent shape is an ellipse and it marks the regions where the Sun is rising or setting. See CHAIN POLYMERIZATION and panel on POLYMER SYNTHESIS.

terminator (*Chem*) Chemical compound which ends a chain-polymerization reaction.

terminator (*ICT*) See ROGUE VALUE.

termitarium (*BioSci*) A mound of earth built and inhabited by termites and containing an elaborate system of passages and chambers.

termite shield (*Build*) A sheet of copper or other non-corroding metal inserted between the foundation and woodwork in buildings to prevent termites entering and destroying the timbers. Also *antproof course*.

termolecular (*Chem*) Pertaining to three molecules.

ternary (*Chem*) Consisting of three components etc.

ternary (*MathSci*) Using POSITIONAL NOTATION with base 3.

ternary diagram (*Eng*) Phase diagram or alloy system formed by three components, represented by the ternary constitutional or equilibrium diagram. Also *ternary system*. See PHASE DIAGRAM.

ternate (*BioSci*) Arranged in threes; esp a compound leaf with three leaflets.

terne metal (*Eng*) Lead alloy with up to 18% tin and 1·5–2% antimony, used to coat steel to improve its corrosion and working properties.

terne plate (*Eng*) Iron or steel sheet coated by hot-dipping with TERNE METAL which acts as a carrier for lubricant used in subsequent drawing operations, as rust protection, as a paint base or to facilitate soldering.

terpadienes (*Chem*) $C_{10}H_{16}$, isocyclic compounds, containing two double bonds. Numerous compounds in the terpene series are terpadienes.

terpenes (*Chem*) Compounds of the formula $(C_5H_8)_n$, the majority of which occur in plants. The number of isoprene units is used as a basis for classification although, slightly confusingly, monoterpenes ($C_{10}H_{16}$) have two isoprene subunits; sesquiterpenes ($C_{15}H_{24}$) have three; diterpenes ($C_{20}H_{32}$) four; triterpenes ($C_{30}H_{48}$) six, and so on. NB Currently sesquiterpenes are pure carbon!

terpenoids (*BioSci*) A group of plant secondary metabolites based on one to four or more isoprene (C_5) units, including many essential oils, the gibberellins, carotenoids, plastoquinone, rubber.

terpineol (*Chem*) $C_{10}H_{17}OH$, colourless crystals; mp 37°C, bp 218°C; it can be obtained from limonene hydrochloride by the action of potassium hydroxide. Terpineol is used extensively as the basis of certain perfumes and in soap perfumery.

terpolymer (*Chem*) Copolymer of three different monomers. See COPOLYMER.

terprom (*Aero*) Abbrev for *terrain profile matching*. See TERCOM.

terrace (*Geol*) See RIVER TERRACE.

terracotta (*Build*) From the Italian for 'baked earth'. A hard, unglazed ceramic ware of characteristic reddish-brown colour and used for decorative tiles and bricks, statues, vases, kitchenware, etc. Cf BONE CHINA, EARTHENWARE, FAIENCE, PORCELAIN, STONEWARE.

terrain-avoidance system (*Aero*) A system providing the pilot with a situation display of the ground or obstacles which project above a plane containing the aircraft so that the pilot can avoid the obstacles. Cf TERRAIN-CLEARANCE SYSTEM.

terrain-clearance system (*Aero*) A fully automatic system for sensing ground obstructions and guiding the aircraft away from them without pilot intervention. Cf TERRAIN-AVOIDANCE SYSTEM.

terrane (*Geol*) A geologically consistent area, discontinuous with that of its neighbours. A *displaced* or *suspect* terrane is one whose distinctive stratigraphical or structural features and its geological history indicate that it is *foreign* to the region.

terrazzo (*Build*) A rendering of cement (white or coloured) and marble or granite chippings, used as a covering for concrete floors, on which it is floated and finally polished by abrasive machine; frequently precast (particularly for sills). Also *Venetian mosaic*.

terrestrial equator (*Geol*) An imaginary circle on the surface of the Earth, the latter being regarded as being cut by the plane through the centre of the Earth perpendicular to the polar axis; it divides the Earth into the northern and southern hemispheres, and is the primary circle from which terrestrial latitudes are measured.

terrestrial flight telephone system (*ICT*) A system allowing aircraft passengers to use mobile telephones in flight, with a technical specification similar to that of the GLOBAL SYSTEM FOR MOBILE COMMUNICATION except that the speech data rate is reduced to 9·6 Kbps.

terrestrial latitude (*Genrl*) See LATITUDE AND LONGITUDE, TERRESTRIAL.

terrestrial magnetism (*Geol, Phys*) The magnetic properties exhibited within, on and outside the Earth's surface. There is a nominal (magnetic) north pole in Canada and a nominal south pole opposite, the positions varying cyclically with time. The direction indicated by a compass needle at any one point is that of the horizontal component of the field at that point. Having the characteristics of flux from a permanent magnet, the Earth's magnetic field probably depends on currents within the Earth and also on those arising from ionization in the upper atmosphere, interaction being exhibited by the aurora borealis.

terrestrial poles (*Geol*) The two diametrically opposite points in which the Earth's axis cuts the Earth's surface are the geographical poles. The magnetic poles, the positions to which the compass needle will point, are unstable and differ from the geographical. N magnetic pole: approx 76°N, 101°W; S magnetic pole: approx 66°S, 139°E. See TERRESTRIAL MAGNETISM.

terrestrial radiation (*EnvSci*) At night the Earth loses heat by radiation to the sky, the maximum cooling occurring when the sky is cloudless and the air dry. Dew and hoar frost are the result of such cooling.

terrestrial telescope (*Phys*) Telescope consisting of an objective and a four-lens eyepiece (*terrestrial eyepiece*), giving an erect image (see ERECTING PRISM) of a distant object.

terrigeneous sediments (*Geol*) Sediments derived from the erosion of the land. They include sediments deposited on land and land-derived material deposited in the sea.

territory (*BioSci*) Areas defended against other individuals, usually of the same species. Territoriality exists among many, but not all, species and may involve aggressive encounters, but it is often regulated by less overt behaviour.

terry fabric (*Textiles*) Woven cloth using an additional warp to produce loops on one or both sides of the ground fabric. Often used for towels. See PLUSH (weft-knitted).

Tertiary (*Geol*) The first period or sub-era of the Cenozoic era, covering an approx time span from 65 to 2 million years ago.

tertiary alcohols (*Chem*) Alcohols containing the group $C≡OH$. When oxidized, the carbon chain is broken, resulting in the formation of two or more oxidation products containing a smaller number of carbon atoms in the molecule than the original compound.

tertiary amines (*Chem*) Amines containing the nitrogen atom attached to three groups. Tertiary aliphatic–aromatic amines yield 4-nitroso compounds with nitrous [nitric (III)] acid.

tertiary amyl methyl ether (*Autos*) A blending component for low-lead or unleaded high-octane motor fuel. It is prepared from *iso*-amylene and methanol. Ethers contain an oxygen atom in the molecule and this is a disadvantage as the oxygen does not contribute to the calorific value of the fuel.

tertiary carbon atom (*Chem*) Carbon atom linked to three other carbon atoms and one hydrogen atom. Generally less stable than secondary carbon atoms, and a weak point in polymer chains, making them more susceptible to degradation and oxidation.

tertiary colours (*Genrl*) A term for the colour resulting from the mixture of two secondary colours, eg olive produced by mixing green and orange.

Tertiary igneous rocks (*Geol*) The various types of igneous rocks which were intruded or extruded during early Tertiary times, esp over a region stretching from the UK to Iceland; eg in the Inner Hebrides and NE Ireland (the Thulean Province).

tertiary nitro compounds (*Chem*) Nitro compounds containing the group $\equiv CNO_2$. They contain no hydrogen atom attached to the carbon atom next to the nitro group, and they have no acidic properties.

tertiary production (*MinExt*) Special methods of increasing oil flow after PRIMARY PRODUCTION (natural flow by gravity or intrinsic pressure) and SECONDARY PRODUCTION (pressurizing the reservoir) are exhausted. Includes chemical treatment and water injection. Also *tertiary recovery*.

tertiary structure (*BioSci*) The three-dimensional configuration of polymers or biomaterials (eg proteins) that is a stable folding of the sequence of units (groups, bases or peptides) along the polymer, ie of their secondary structure.

tertiary wall (*BioSci*) A deposit of wall material on the inner surface, next to the lumen, of a secondary wall, often in the form of helical strips, as in the tracheids of the yew. Also *tertiary thickening*.

tertiary winding (*Phys*) A third winding on a transformer core, linking the same flux as the primary and secondary windings. The functions, and the names, of all three are interchangeable. May be used for additional output or monitoring.

tervalent (*Chem*) See TRIVALENT.

Terylene (*Chem*) TN for straight-chain polyester fibres derived from condensation of TEREPHTHALIC ACID with ethan-1,2-diol, used widely in the manufacture of fabrics, clothing materials and other textiles. See STEP POLYMERIZATION.

Teschen disease (*Vet*) Virus encephalomyelitis of swine. Characterized by mild fever, nervous excitement and paralysis. Also *infectious pig paralysis*.

teschenite (*Geol*) A coarse-grained basic (gabbroic) igneous rock consisting essentially of plagioclase, near labradorite in composition, titanaugite, ilmenite and olivine (or its decomposition products); primary analcite occurs in wedges between the plagioclase crystals, which it also veins. An analcime (foid) gabbro or dolerite. See PLUTONIC ROCKS.

tesla (*Phys*) SI unit of magnetic flux density or magnetic induction equal to 1 Wb m^{-2}. Equivalently, the magnetic induction for which the maximum force it produces on a current of unit strength is 1 N. Symbol T.

Tesla coil (*ElecEng*) Simple source of high-voltage oscillations for rough testing of vacua and gas (by discharge colour) in vacuum systems.

tessella (*Arch*) See TESSERA.

tessellated pavement (*Build*) A pavement formed of small pieces of stone, marble, etc, in the manner of a mosaic. Also *Roman mosaic*.

tessera (*Arch*) One of the small pieces of stone, marble, etc, used in the mosaic of a tessellated pavement. Pl *tesserae*. Also *tessella*, pl *tessellae*.

test (*BioSci*) The external shell of many invertebrates, eg echinoderms. See TESTA.

test (*MathSci*) A procedure for deciding whether to reject or otherwise a HYPOTHESIS in favour of another hypothesis.

testa (*BioSci*) (1) In plants, the seed coat, several layers of cells in thickness, derived from the integuments of the ovule. (2) In animals, a hard external covering, usually calcareous, siliceous, chitinous, fibrous or membranous; an *exoskeleton*; a *shell*; a *lorica*. Adjs *testacean, testaceous*.

test bed (*Eng*) Area with full monitoring facilities for testing new or repaired machinery under full working conditions.

test board (*ElecEng*) A switchboard carrying instruments and switches for connecting up to apparatus to be tested.

test case (*ICT*) A short description of a process, using specific data or conditions, to evaluate the correct functioning of an individual component within a system.

test data (*ICT*) Data, including expected results, used to test a program or flowchart.

test desk (*ICT*) The special position where tests can be applied to faulty lines to discover the cause of faults and to issue instructions for remedying them.

testes (*BioSci*) See TESTIS.

test final selector (*ICT*) The selector, following the test selector, that enables test clerks at the test desk to get onto a subscriber's line.

testicle (*BioSci*) See TESTIS.

testicular feminization (*BioSci*) The situation that arises if genetic males lack receptors for testosterone; they develop as females and are unresponsive to male hormones.

testing machine (*Eng*) A machine for applying accurately measured loads to a TEST PIECE, determining its suitability for a particular purpose.

testing position (*ICT*) A position equipped for testing purposes and forming part of a test desk or test rack.

testing set (*ElecEng*) A self-contained set of apparatus, including switches, instruments, etc, for carrying out certain special tests.

testing transformer (*ElecEng*) A specially designed transformer providing a high-voltage supply for testing purposes.

testis (*BioSci*) A male gonad or reproductive gland responsible for the production of male germ cells or sperms. Also *testicle, testicular*. Pl *testes*.

test jack (*ICT*) One with contacts in series with a circuit, so that a testing device can be immediately introduced for locating faults.

testosterone (*BioSci*) A steroid hormone obtained from the testis. It is a very active androgenic substance and the chief male sex hormone. See panel on STEROID HORMONES.

test pattern (*ImageTech*) A transmitted chart with lines and details to indicate particular characteristics of a transmission system, used in TV for general testing purposes. Also *test card, test chart*.

test piece (*Eng*) A piece of material accurately turned or shaped, often to specified standard dimensions, for subjecting to a tensile test, shock test, etc, in a TESTING MACHINE.

test point (*ElecEng*) Designated junction between components in equipment, where the voltage can be stated (as a minimum or with tolerance) for a quick verification of correct operation.

test record (*Acous*) A recording specially made for the testing of reproducing equipment, having constant-frequency or gliding tones, or selected recordings of speech or music, to emphasize particular faults in the subsequent reproduction.

test–retest reliability (*Psych*) The correlation between scores of the same measuring device administered to the same people on two different occasions.

test selector (*ICT*) A selector operated by a test clerk in an automatic exchange; by means of it, through a TEST

FINAL SELECTOR, it is possible to get onto any line in the exchange.

test terminals (*ElecEng*) Circuit terminals to which a connection is made for purposes of testing.

test vehicles (*Aero*) Aircraft for aerodynamic, control and other tests in guided weapon development. They may simply be for gathering basic information or they may be actual missiles without a warhead. They are known by their initials: CTV, command (control) test vehicle; GPV, general purpose vehicle; MTV, missile test vehicle; RJTV, ramjet test vehicle; RTV, rocket test vehicle.

tetanic contraction (*BioSci*) See TETANUS.

tetanus (*BioSci*) The state of prolonged contraction which can be induced in a muscle by a rapid succession of stimuli.

tetanus (*Med*) A disease due to infection with the tetanus bacillus, *Clostridium tetani*. The toxin it secretes causes the symptoms and signs of the disease, such as painful tonic spasms of the muscles, which usually begin in the jaw and then spread to other parts. Also *lockjaw*.

tetanus antitoxin (*Med*) Antibody to tetanus toxin, usually prepared in horses which have been hyper-immunized against the exotoxin of *Clostridium tetani*. It is used for the prevention of tetanus in humans and animals following possible contamination of wounds, but with the risk of SERUM SICKNESS.

tetanus toxin (*BioSci*) The A–B-type toxin produced by *Clostridium tetani*. A neurotoxin that blocks synaptic transmission in the spinal cord and also blocks neuromuscular transmission. Tetanus toxin binds to a glycolipid, disialosyl ganglioside, which is particularly rich in the membranes of nerve cells. After inactivation by treatment with formaldehyde to form tetanus toxoid this is used for prophylactic immunization against tetanus.

tetany (*Med*) A condition characterized by heightened excitability of the motor nerves and intermittent painful muscular cramps, occurring in many abnormal states, esp those associated with hypocalcaemia.

tethered satellite (*Space*) A satellite which is deployed from a spacecraft and attached to it by a wire or tether which may measure over 100 km.

Tethys (*Astron*) The third natural satellite of Saturn, discovered in 1684. Distance from the planet 295 000 km; diameter 1050 km.

Tethys (*Geol*) An east–west ocean lying between LAURASIA (to the north) and GONDWANALAND (to the south) during Palaeozoic and Mesozoic times, from which the Alpine and Himalayan mountains arose; a Mesozoic geosyncline.

TETRA (*ICT*) Abbrev for *terrestrial trunked radio*, a European standard for mobile communications used by the emergency services.

tetra- (*Genrl*) Prefix from Gk *tetra-*, four.

tetraboric acid (*Chem*) See BORIC ACID.

tetrachloroethane (*Chem*) $C_2H_2Cl_4$. Bp 146°C. Used as a solvent and also as a vermifuge. Prepared by the chlorination of ethyne in the presence of antimony (III) chloride. Intermediate in the manufacture of TRICHLOR-OETHANE.

tetrachloromethane (*Chem*) See CARBON TETRACHLORIDE.

tetrachlorosilane (*Chem*) See SILICON TETRACHLORIDE.

tetrachoric correlation (*Genrl*) A correlational technique used to estimate the Pearson product-moment correlation of two continuous variables that have been dichotomized (eg age is continuous, but when it is split into two groups such as over 40 and under 40 it becomes dichotomous).

tetracycline antibiotics (*Pharmacol*) A group of closely related bacteriostatic antibiotics, with similar antibacterial spectrum and toxicity. They act by binding to the bacterial ribosome and inhibiting protein synthesis. Effective against many streptococci, Gram-negative bacilli, rickettsiae, spirochetes, Mycoplasma and Chlamydia.

tetrad (*BioSci*) The four haploid cells formed at the end of MEIOSIS. The term was formerly used for the four chromatids making up a chromosome pair at the first division of meiosis.

tetradactyl (*BioSci*) Having four digits.

tetrad analysis (*BioSci*) The genetic analysis of TETRADS in studies of mapping, recombination, etc.

tetradymite (*Min*) An ore of tellurium, of composition Bi_2Te_2S; crystallizes in the trigonal system. Bismuth tellurides are commonly found in gold–quartz veins.

tetra-ethyl lead (*Chem*) An organic derivative of lead, formerly used as an antiknock additive to petroleum for use in internal-combustion engines.

tetrafluorosilane (*Chem*) See SILICON TETRAFLUORIDE.

tetragonal system (*Crystal*) The crystallographic system in which all the forms are referred to three axes at right angles; two are equal and are taken as the horizontal axes, whilst the vertical axis is either longer or shorter than these. It includes such minerals as zircon and cassiterite. Also *pyramidal system*. See panel on CRYSTAL LATTICE.

tetragonous (*BioSci*) A stem etc having four angles and four convex faces.

tetrahedrite (*Min*) A sulphide of copper and antimony, crystallizing in the tetrahedral division of the cubic system; frequently contains arsenic and other metals. It is used as an ore of copper and, in some cases, of other metals. Also *fahl ore, fahlerz, grey copper ore*. See TENNANTITE.

tetrahedron (*MathSci*) A polyhedron having four triangular faces. Adj *tetrahedral*.

tetrahydrocannabinol (*Pharmacol*) A cannabinoid; one of the more psychoactive components of cannabis. Abbrev THC.

tetrahydrofuran (*Chem*) Tetramethylene oxide. Bp 64–66°C, rel.d. (at 20°C) 0·888. Organic solvent used in refining lubricating oils, a solvent for a wide range of polymers, and as source of POLYTETRAMETHYLENE ETHERGLYCOL, PTMEG.

tetrahydrogestrinone (*Pharmacol*) A synthetic steroid, illegally used as a performance-enhancing drug by some athletes. Abbrev THG.

tetrahydronaphthalene (*Chem*) See TETRALIN.

tetralin (*Chem*) Organic solvent with a high boiling point (206°C), used in a wide variety of products, such as floor polishes, paints and varnishes. Also *tetrahydronaphthalene*.

tetramerous (*BioSci*) Having four parts; arranged in fours; arranged in multiples of four.

tetramethylene oxide (*Chem*) See TETRAHYDROFURAN.

tetramethyl rhodamine isothiocyanate (*BioSci*) A red fluorescent dye used in immunofluorescence techniques. In conjunction with FITC it allows two colours to be used together. Abbrev TRITC.

tetramethylsilane (*Chem*) $(CH_3)_4Si$. A reference standard for *proton magnetic resonance*.

tetraoses (*Chem*) See TETROSES.

tetraparental chimera (*BioSci*) A chimera (usually mouse) resulting from the artificially induced fusion of two blastocysts at the four- or eight-cell stage. The resulting animal contains cells from both parents in all its tissues, and is an example of the maintenance of mutual immunological tolerance.

tetraploid (*BioSci*) Possessing four sets of chromosomes, each chromosome of a set being represented four times. Cf DIPLOID, HAPLOID.

tetrapod (*BioSci*) Having four feet.

tetrapterous (*BioSci*) Having four wings.

tetrarch (*BioSci*) A stele having four strands of protoxylem, as in the roots of many dicotyledons.

tetrasomic (*BioSci*) A tetraploid nucleus (or organism) having one chromosome four times over, the others in duplicate.

tetrasporophyte (*BioSci*) The typically diploid phase of the red algal life cycle, developing from carpospores and producing haploid tetraspores.

tetratohedral (*Crystal*) Containing a quarter of the number of faces required for the full symmetry of the crystal system.

tetravalent (*Chem*) Capable of combining with four hydrogen atoms or equivalent. Having an oxidation or a co-ordination number of four.

tetrazo dyes (*Chem*) See DISAZO DYES.

tetrazole (*Chem*) A five-membered heterocyclic compound containing four nitrogen atoms and one carbon atom in the ring.

tetrode (*Electronics*) A four-electrode thermionic valve, incorporating a screen grid.

tetrode transistor (*Electronics*) A transistor with additional base contact to improve high-frequency performance.

tetrodotoxin (*BioSci*) A very potent neurotoxin found in Japanese puffer-fish, used to block the SODIUM CHANNELS in nerves. Abbrev *TTX*.

tetroses (*Chem*) *Tetraoses*. Monosaccharides containing four carbon atoms in the molecule, eg $HOCH_2(CHOH)_2CHO$.

tetryl (*Chem*) *N,2,4,6-tetranitromethylaniline*. A yellow crystalline compound used as a detonator.

TE-wave (*ICT*) Abbrev for *transverse electric wave*, having no component of electric force in the direction of transmission of electromagnetic waves along a waveguide. Also *H-wave* (since it must have a magnetic field component in the direction of transmission).

tex (*Textiles*) The basic unit of the tex system, used to express the LINEAR DENSITY (mass per unit length) of fibres, filaments, yarns or other linear material. The tex of a material is the mass in grams of a 1 km length of material (hence *decitex, kilotex*, etc). This system has now replaced the traditional ones based on the DENIER and other units.

Texas fever (*Vet*) A disease of cattle caused by the protozoon *Babesia bigemina*. See REDWATER (1). Also *tick fever*.

text (*Print*) The body of matter in a printed book, exclusive of preliminary matter, end matter, notes and illustrations; words set to music, as distinct from the accompanying music.

text editor (*ICT*) An APPLICATION PROGRAM that is used for editing text files such as SOURCE CODE, eg Pascal, without special formatting instructions. For conventional printed documents that require formatting and other enhancements the editor would be called a word processor.

text file (*ICT*) See ASCII FILE.

textiles (*Textiles*) A term used to describe any fibres, filaments and yarns and any products produced from them.

texting (*ICT*) The practice of communicating in highly abbreviated form using the SMS messaging protocol between, for instance, two mobile-phone users.

text message (*ICT*) A short message, often using abbreviations, typed and sent by means of a mobile phone.

text mining (*ICT*) In computing, the application of the principles of DATA MINING to large amounts of written text.

text mode (*ICT*) A mode of operation of a VIDEO DISPLAY UNIT whereby only text characters may be displayed, ie no graphics images. See GRAPHICS MODE.

text processing (*ICT*) The storing, revising and outputting of text using a computer system. See DESKTOP PUBLISHING, WORD PROCESSING.

text programming (*ImageTech*) Using TELETEXT TV listings pages to enter programme details automatically into a video timer. See STARTEXT.

text-to-speech (*ICT*) Automatic translation of written messages into intelligible speech, using synthesis by rule. The PHONEME sequence represented by the message is deduced using the orthographic rules of the language, after which further rules prescribing the transitions between one phoneme and the next are invoked. Finally, a prosodic component (variation of pitch and stress) is added to improve intelligibility and naturalness.

texture (*Eng*) Physical appearance in terms of roughness and shape of surface features; in microscopical examination it relates to microstructural features such as grain shape, distribution of phases and crystallographic orientation. See BANDING, FRACTOGRAPHY.

texture (*Genrl*) The mode of union or disposition, in regard to each other, of the elementary constituent parts in the structure of any body or material.

texture (*Geol*) The physical quality of a rock which is determined by the relative sizes, disposition and arrangement of the component minerals. The nomenclature and classification of rocks are governed by mineral composition and texture. See eg GRAPHIC TEXTURE, OPHITIC TEXTURE, POIKILITIC TEXTURE.

texture (*ImageTech*) The quality of the surface of a photograph.

texture brick (*Build*) See RUSTICS.

textured vegetable protein (*FoodSci*) Vegetable protein (usually from soya) processed and extruded to solids of various size, shape and density. Widely used meat analogue in vegetarian products and a meat extender in pet foods. Abbrev *TVP*.

textured yarn (*Textiles*) Yarn of synthetic continuous filaments that has been treated to make its surface highly irregular by crimping or by introducing loops having a variety of shapes. Widely used methods include: (1) false-twist texturing in which the yarn is continuously twisted, heat-set and untwisted (no overall twist is produced in the filaments); (2) STUFFER BOX crimping; (3) in air-jet texturing the entangled loops are formed by subjecting the yarn to a turbulent airstream. The yarns have increased bulk and are sometimes described as bulked continuous filaments (BCF). They are also more easily extended and are used in articles (eg tights) that closely conform to the shape of the body. Formerly *bulked yarn*.

texture paints (*Build*) Paints with an added aggregate which can be manipulated with rubber stipplers, serrated combs or brushes to create a pattern or design, raised above the surface.

TFTR (*NucEng*) Abbrev for *tokamak fusion test reactor*. Large TOKAMAK experiment at Princeton, New Jersey.

TFT screen (*ICT*) Abbrev for THIN-FILM TRANSISTOR screen. A flat screen giving a good display, used in laptop computers.

TGA (*Chem*) Abbrev for THERMAL GRAVIMETRIC ANALYSIS.

T-gauge (*Build*) A type of marking gauge with long marker pin and large flat fence; used for marking jobs where mouldings, beading or other surface items have to be cleared.

TGF (*BioSci*) Abbrev for TRANSFORMING GROWTH FACTOR.

T-grain (*ImageTech*) Silver halide grains in tablet form, offering greater area and hence greater sensitivity in a photographic emulsion.

T-group (*Psych*) An encounter group whose purpose is to improve the communication skills of individual members by discussion and analysis of the roles that they each adopt habitually in their dealings with others. Also *sensitivity-training group*.

TGT (*Aero*) See GAS TEMPERATURE.

Th (*Chem*) Symbol for THORIUM.

thalamus (*BioSci*) (1) In the vertebrate brain, the larger, more ventral part of the dorsal zone of the DIENCEPHALON. It is considered the central switching station of the brain because all of the body's senses (except the olfactory senses) pass through this before being relayed to the brain. (2) A synonym for the RECEPTACLE of a flower.

thalassaemia (*Med*) A group of inherited anaemias in which there is a defect in the alpha or beta chains of haemoglobin, *alpha-, beta-thalassaemia*. Thalassaemia *minor* is used to describe heterozygotes and thalassaemia *major* for the homozygotes. US *thalassemia*.

thalasso- (*Genrl*) Prefix from Gk *thalassa*, sea.

thalassophyte (*BioSci*) A seaweed.

thalidomide (*Pharmacol*) A non-barbiturate sedative drug; withdrawn (UK) in 1961 because of teratogenic effects. Now used in treatment of leprosy, AIDS and certain types of cancer.

thallium (*Chem*) Symbol Tl, at no 81, ram 204·37, rel.d. 11·85, mp 303·5°C, bp 1650°C. White malleable metal like lead. Several thallium isotopes are members of the uranium, actinium, neptunium and thorium radioactive series. Thallium isotopes are used in scintillation crystals. Its compounds are very poisonous and it has common valencies of 1 and 3. There are a few rare independent minerals.

thallofide cell (*Electronics*) A photoconducting cell employing thallium oxy-sulphide as the light-sensitive agent; sensitive to far red and infrared.

thallus (*BioSci*) A plant body not differentiated into leaves, stems and roots but consisting of a single cell, a colony, a filament of cells, a mycelium or a large branching multicellular structure. The plant body of the algae, fungi and thalloid liverworts. Adj *thalloid*.

thalweg (*Geol*) (Ger 'valley way'.) The name frequently used for the longitudinal profile of a river, ie from source to mouth. See VALLEYS.

thanatocoenosis (*Geol*) Assemblage of fossil remains of organisms which were not associated during their life but were brought together after death.

thanatoid (*BioSci*) Poisonous, deadly, lethal, as some venomous animals.

thanatosis (*BioSci*) Same as SHAM DEATH.

Thanetian (*Geol*) A stage in the Palaeocene. See TERTIARY.

thatching (*Build*) A form of roof-covering composed of courses of reeds, straw or heather, laced together.

thawing (*Phys*) The beginning of the fusion process of a solid. The corresponding temperature is the *thaw-point* of the solid.

Thebe (*Astron*) A tiny natural satellite of Jupiter, discovered in 1979 by the Voyager 2 mission. Distance from the planet 222 000 km; diameter 100 km.

Thebesian valve (*BioSci*) An auricular valve of the mammalian heart.

theca (*BioSci*) A case or sheath covering or enclosing an organ, as the *theca vertebralis* or dura mater enclosing the spinal cord; a tendon sheath; the wall of a coral cup. Adjs *thecal, thecate*.

thecodont (*BioSci*) Having the teeth implanted in sockets in the bone which bears them.

T-helper cell (*BioSci*) T-lymphocytes of the CD4-positive subset that provide 'help' in the form of INTERLEUKINS to other lymphocytes, allowing them to differentiate to perform their immune effector functions. This subset is attacked by HIV and is severely depleted in AIDS patients.

thelytoky (*BioSci*) Parthenogenesis resulting in the production of females only.

thematic apperception test (*Psych*) A PROJECTIVE TECHNIQUE in which persons are shown a set of pictures and asked to write a story about each. Abbrev *TAT*.

thenardite (*Min*) Sodium sulphate, crystallizing in the orthorhombic system and occurring in saline residues of alkali lakes.

theodolite (*Surv*) An instrument for measuring horizontal and vertical angles by means of a telescope mounted on an axis made vertical by levelling screws, and rotated both horizontally on this axis and in horizontal bearings. Circular graduated plates are used to measure amount of rotatory motion when the telescope is sighted on successive signal stations.

theophylline (*Pharmacol*) Xanthine (phosphodiesterase inhibitor), a strong diuretic and muscle relaxant that is found in tea. Little used, its more soluble derivative, aminophylline, being preferred.

theorem (*MathSci*) A statement of a mathematical truth that can be proved.

theorem of the equipartition of energy (*Chem*) See PRINCIPLE OF THE EQUIPARTITION OF ENERGY.

theoretical plate (*ChemEng*) A concept, used in distillation design, of a plate in which the vapour and liquid leaving the plate are in equilibrium with each other.

theoretical stiffness (*Eng*) Estimate of the stiffness of any material from atomic/molecular data. It can be done for eg completely oriented polymer chains from a knowledge of chain structure, bond lengths and angles, and force constants. For high-density polyethylene, calculation gives a theoretical tensile modulus along the chains of 31 GPa and about 9 GPa perpendicular to the chains. See THEORETICAL STRENGTH and panel on HIGH-PERFORMANCE POLYMERS.

theoretical strength (*Eng*) An estimate of the strength of any material from atomic/molecular data. It can be done for eg oriented polymer chains using molecular data such as the MORSE EQUATION for the intermolecular potential energy. Calculation for high-density polyethylene gives a theoretical strength along the chains of 31 GPa. See THEORETICAL STIFFNESS and panel on HIGH-PERFORMANCE POLYMERS.

theory of indicators (*Chem*) See OSTWALD'S THEORY OF INDICATORS.

theralite (*Geol*) A coarse-grained, holocrystalline igneous rock composed essentially of the minerals labradorite, nepheline, purple titanaugite, and often with soda-amphiboles, biotite, analcite or olivine. A nepheline (foid) gabbro. See PLUTONIC ROCKS.

therapeutic (*Med*) Pertaining to the medical treatment of disease (therapy); remedial; curative; preventive. Hence therapeutics, that part of medical science which deals with the treatment of disease; the art of healing.

therapeutic index (*Pharmacol*) The ratio of the lethal dose (LD_{50}) to the effective dose (ED_{50}) for a drug. Ideally the difference should be large.

therapeutic ratio (*Radiol*) The ratio between tumour-lethal dose and tissue tolerance. In radio-resistant tumours, the tumour-lethal dose equals, or is greater than, the dose required to destroy normal tissues.

Theria (*BioSci*) A subclass of the mammals, containing the extinct Patriotheria, the Metatheria (marsupials) and the Eutheria (placentals).

therm (*Phys*) A unit of energy used mainly for the sale of gas; equal to 10^5 Btu or 105·5 MJ.

therm-, thermo- (*Genrl*) Prefixes from Gk *therme*, heat.

thermal (*Aero, EnvSci*) An ascending current due to local heating of air, eg by reflection of the Sun's rays from a beach.

thermal ammeter (*ElecEng*) One in which the deflection of the pointer depends on the sag of a fine wire, carrying the current to be measured, due to thermal expansion.

thermal analysis (*Eng*) The use of cooling or heating curves in the study of physical changes in materials. The freezing points and the temperatures of any polymorphic changes occurring in pure materials may be determined, as well as ranges and temperatures for changes in solid alloys. The data obtained are used eg in constructing PHASE DIAGRAMS. See DIFFERENTIAL SCANNING CALORIMETRY, DIFFERENTIAL THERMAL ANALYSIS.

thermal capacity (*Phys*) The amount of heat required to raise the temperature of a system through 1 K. SI unit is $J K^{-1}$. See MOLAL SPECIFIC HEAT CAPACITY, SPECIFIC HEAT CAPACITY.

thermal circuit breaker (*ElecEng*) A miniature-type circuit breaker whose overload device operates by thermal expansion.

thermal column (*NucEng*) A column or block of moderator in the reactor which guides large thermal neutron flux to a given experimental region.

thermal comparator (*Phys*) A device which enables comparative measurements of the thermal conductivities of solids to be made rapidly. The instrument is also used to make comparative measurements of eg foil thickness and surface deposits.

thermal conductivity (*Phys*) A measure of the rate of flow of thermal energy through a material in the presence of a temperature gradient. If dQ/dt is the rate at which heat is transmitted in a direction normal to a cross-sectional area A when a temperature gradient dT/dx is applied, then the thermal conductivity is

$$k = -\frac{(dQ/dt)}{A(dT/dx)}$$

SI unit is $W m^{-1} K^{-1}$. Materials with high electrical conductivities tend to have high thermal conductivities.

thermal converter (*ElecEng*) The combination of a thermoelectric device, eg thermocouple, and an electric heater to convert an electrical quantity into heat and then into a voltage. Used in telemetering systems. See THERMO-COUPLE METER. Also *thermoelement*.

thermal cross-section (*Phys*) Effective nuclear cross-section for neutrons of thermal energy.

thermal cueing unit (*Aero*) Visual display presented to the pilot showing likely targets detected by a FLIR system. Targets can be classified by temperature 'signature' and after selection the co-ordinates can be fed to the attack system.

thermal cut-out (*ElecEng*) A thermal circuit breaker designed to screw into a standard barrel-type fuse-holder.

thermal cycle (*Phys*) An operating cycle by which heat is transferred from one part of a system to another. In reactors, separate heat transfer and power circuits are usual to prevent the fluid flowing through the former, which becomes radioactive, from contaminating the power circuit.

thermal cycling (*Phys*) The subjection of a substance to a number of temperature and pressure cycles in succession. Used in petrol refining (CRACKING).

thermal death-point (*BioSci*) The temperature at which an organism is killed or a virus inactivated.

thermal detector (*ElecEng*) Any detector of high-frequency currents which operates by virtue of their heating effect when passed through a resistance.

thermal diffusion (*Phys*) A process in which a temperature gradient in a mixture of fluids tends to establish a concentration gradient; may be used for isotope separation. Also *Soret effect*.

thermal diffusivity (*Phys*) THERMAL CONDUCTIVITY divided by the product of specific heat capacity and density; more generally applicable than thermal conductivity in most heat transfer problems. Unit is $m^2\,s^{-1}$. Symbol κ.

thermal dissociation (*Chem*) The dissociation of certain molecules under the influence of heat.

thermal dye transfer printer (*ICT*) See DYE SUBLIMATION PRINTER.

thermal efficiency (*Eng*) Of a heat engine, the ratio of the work done by the engine to the mechanical equivalent of the heat supplied in the steam or fuel.

thermal effusion (*Phys*) The leaking of a gas through a small orifice, the gas being at a low pressure so that the mean free path of the molecules is large compared with the dimensions of the orifice.

thermal electromotive force (*ElecEng*) That which arises at the junction of different metals because of a temperature different from the rest of the circuit. Widely used for measuring temperatures relative to that of the cold junction, eg ice in a vacuum flask.

thermal equilibrium diagram (*Eng*) See PHASE DIAGRAM.

thermal excitation (*Phys*) Collision processes between particles by which atoms and molecules can acquire extra energy.

thermal expansion coefficient (*Phys*) The fractional expansion of matter (ie the expansion of the unit length, area or volume) per degree rise in temperature. If the coefficients of linear, areal and cubical expansion of a substance are α, β, and γ respectively, β is approximately twice, and γ three times, α.

thermal fatigue (*Eng*) Fatigue failure resulting from strains caused by expansion and contraction during thermal cycling. See panel on FATIGUE.

thermal flasher (*Phys*) A switch for which the operation depends upon the heating effect of the current which it is controlling, eg a wire or bimetallic strip, which when heated by the current interrupts the circuit, which is not remade until it has cooled.

thermal gravimetric analysis (*Chem*) See THERMAL ANALYSIS.

thermal imaging (*Phys*) Imaging based on the detection of weak infrared radiation from objects. Applications include

the mapping of the Earth's surface from the air, weather mapping and medical thermography (thermal contours on the surface of the human body). See OPTICAL–ELECTRONIC DEVICES.

thermal inertia (*NucEng*) See THERMAL RESPONSE.

thermal inkjet printer (*ICT*) See BUBBLEJET PRINTER.

thermal instability (*ElecEng*) The condition in any system when an increase in temperature causes energy losses to increase more rapidly than they can be removed. There is therefore positive feedback leading to thermal runaway, damage and even destruction.

thermal instrument (*ElecEng*) An instrument the operation of which depends upon the heating effect of a current. See HOT-WIRE, THERMOCOUPLE INSTRUMENT.

thermalite (*Build*) TN for aerated lightweight concrete blocks.

thermalization (*Phys*) The process of slowing fast neutrons to thermal energies. Normally the function of a moderator in nuclear reactors.

thermal leakage factor (*NucEng*) The ratio of number of thermal neutrons lost from the reactor by leakage, to the number absorbed in the reactor core. Also used is the *thermal non-leakage probability*, ie the fraction of thermal neutrons which do not leak out of the reactor.

thermal limit (*ElecEng*) Maximum permissible power associated with a piece of electrical equipment, eg the output power of an electric generator, which is set by consideration of safe temperature rise.

thermally activated process (*Chem, Phys*) Any physical or chemical process whose rate increases with the amount of thermal energy supplied to the system. See ARRHENIUS'S (RATE) EQUATION.

thermally bonded non-woven fabric (*Textiles*) Fabric formed from a WEB of fibres that have been partially melted by heat and thereby converted into a coherent sheet.

thermal metamorphism (*Geol*) Metamorphism resulting from the action of heat, involving chemical changes in the rock without the introduction of a material from elsewhere. Pressure is not significant. See REGIONAL METAMORPHISM.

thermal microphone (*Acous*) See HOT-WIRE MICROPHONE.

thermal neutron (*Phys*) See NEUTRON.

thermal noise (*Electronics*) That arising from random (Brownian) movements of electrons in conductors and semiconductors, and which limits the sensitivity of electronic amplifiers and detectors. The noise voltage V is given by

$$V = \sqrt{4RkT\,\delta f}$$

where δf = frequency bandwidth, R = resistance of source, k = BOLTZMANN'S CONSTANT and T = absolute temperature. Also *circuit noise, Johnson noise*. See NYQUIST NOISE THEOREM.

thermal ohm (*Phys*) See THERMAL RESISTANCE.

thermal pollution (*EnvSci*) The release into the environment of substances that are innocuous in themselves but at a temperature higher than the ambient and alter the physical characteristics of the air or water with which they mix.

thermal precipitator (*PowderTech*) A device for sampling aerosol particles. The gas stream is drawn through a chamber containing a hot wire. The particles near the wire are bombarded by gas molecules of high kinetic energy, then driven away from it and collected on microscope slides located on opposite sides of the hot wire.

thermal printer (*ICT*) One that uses heat-sensitive paper, producing visible characters by the action of heated wires.

thermal radiation (*Med*) Analysis of the heat produced by the human body, by the use of infrared radiation to penetrate surface tissue, in order to diagnose certain diseases affecting deeper tissue and to locate superficial tumours. Thermography refers to the images so made.

thermal reactor (*NucEng*) One for which the fission chain reaction is propagated mainly by thermal neutrons and therefore contains a moderator. Formerly sometimes called a *slow reactor*. See REACTOR CLASSIFICATION and panel on NUCLEAR REACTORS.

thermal receiver (*Acous*) See THERMOPHONE.

thermal reforming (*ChemEng*) A heat treatment process that improves the quality of petroleum fractions for use in motor fuel: alkenes are changed to alkanes. Now mainly superseded by CATALYTIC REFORMING.

thermal relay (*ElecEng*) A relay the operation of which depends upon the heating effect of an electric current.

thermal resistance (*Phys*) Resistance to the flow of heat. The unit of resistance is the *thermal ohm*, which requires a temperature difference of 1°C to drive heat at the rate of 1 W. If the temperature difference is θ°C, the resistance S thermal ohms, and the rate of driving heat W watts, then $\theta = SW$.

thermal response (*NucEng*) The rate of temperature rise in a reactor if no heat is withdrawn by cooling. Its reciprocal is the *thermal inertia*.

thermal runaway (*Electronics*) The effect arising when the current through a semiconductor creates sufficient heat for its temperature to rise above a critical value. The semiconductor has a negative temperature coefficient of resistance, so the current increases and the temperature increases again, resulting in ultimate destruction of the device.

thermal shield (*NucEng*) Inner shield of a reactor, used to protect the biological shield from excess heating.

thermal shock (*Phys*) The transient thermal stresses resulting when a body is subjected to sudden changes in temperature. See THERMAL-SHOCK RESISTANCE.

thermal-shock resistance (*Phys*) Merit index of a material's ability to withstand thermal shock without cracking, or otherwise failing; different indices apply to different heat transfer conditions.

thermal signature (*ImageTech*) The characteristic appearance of a substance or object when viewed with thermal-imaging equipment.

thermal siphon (*Phys*) The system causing flow round a vertical loop of fluid. When the bottom of one column is heated, the fluid rises, cools at the top and falls down the other column. Used for applications such as heating buildings and isotope separation.

thermal spike (*Phys*) See SPIKE.

thermal spray (*Eng*) Spraying molten and finely divided metals or non-metallic particles onto a material to form a coating.

thermal station (*ElecEng*) An electric generating station in which the prime movers are steam turbines or internal-combustion engines.

thermal transfer printer (*ICT*) Resembles DYE SUBLIMATION PRINTER but uses coloured waxes which melt on heating. It has lower resolution but good colour saturation.

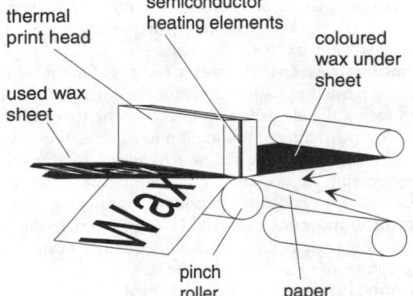

thermal transfer printer The wax sheet has different coloured areas which can be passed in sequence over the same sheet of paper.

thermal trip (*ElecEng*) A tripping relay for a large circuit breaker, which operates by thermal expansion.

thermal tuning (*Electronics*) The change of resonant frequency in an oscillator or amplifier produced by controlled temperature change. Used eg with crystal or resonant cavity in microwave tube.

thermal unit (*Phys*) See BRITISH THERMAL UNIT, CALORIE, JOULE.

thermal utilization factor (*Phys*) The probability of a thermal neutron being absorbed by fissile material (whether causing fission or not) in an infinite reactor core. Symbol f.

thermal vibration (*Phys*) The motion of atoms, vibrating about their equilibrium positions in a crystalline solid. Representing the vibrations by a set of harmonic oscillators which have zero point energy at 0 K, the oscillators increase their energy by discrete amounts, a quantum of energy, as the temperature increases. The specific heat of a solid can be explained in these terms. Detailed information about the thermal vibrations can be obtained from neutron diffraction, X-ray diffraction and other techniques. See DEBYE THEORY OF SPECIFIC HEATS OF SOLIDS, DEBYE–WALLER FACTOR, LATTICE DYNAMICS, PHONON.

thermal wind (*EnvSci*) The vector difference between the geostrophic wind at some level in the upper air and the wind at some lower level.

thermion (*Phys*) A positive or negative ion emitted from incandescent material.

thermionic amplifier (*Electronics*) Any device employing thermionic vacuum tubes for amplification of electric currents and/or voltages.

thermionic cathode (*Electronics*) One from which electrons are liberated as a result of thermal energy, due to high temperature.

thermionic current (*Electronics*) One represented by electrons leaving a heated cathode and flowing to other electrodes.

thermionic emission (*Electronics*) The emission of electrons from a heated material as a consequence of thermal energy. See RICHARDSON–DUSHMAN EQUATION.

thermionic rectifier (*Electronics*) Thermionic valve used for rectification or demodulation.

thermionics (*Genrl*) Strictly, the science dealing with the emission of electrons from hot bodies. Applied to the broader subject of subsequent behaviour and control of such electrons, esp in a vacuum.

thermionic valve (*Electronics*) One containing a heated cathode from which electrons are emitted, an anode for collecting some or all of these electrons, and generally additional electrodes for controlling flow to the anode. Normally the glass or metal envelope is evacuated but a gas at low pressure is introduced for special purposes. Also *thermionic tube*.

thermionic work function (*Electronics*) Thermal energy surrendered by an electron liberated through THERMIONIC EMISSION from a hot surface. Cf PHOTOELECTRIC WORK FUNCTION.

thermistor (*Electronics*) Contraction of *thermal resistor*. A semiconductor, a mixture of cobalt, nickel and manganese oxides with finely divided copper, of which the resistance is very sensitive to temperature. Sometimes incorporated in a waveguide system to absorb and measure all the transmitted power. Also used for temperature compensation and measurement.

thermistor bridge (*Electronics*) One used for measuring microwave power absorbed by a thermistor, in terms of the resulting change of resistance. See BOLOMETER.

thermite (*Eng*) A mixture of aluminium powder and half an equivalent amount of iron oxide (or other metal oxides) which gives out a large amount of heat on igniting with magnesium ribbon. The molten metal forms the medium for welding iron and steel (thermit welding). See ALUMINOTHERMIC PROCESS.

Thermit process (*Eng*) See ALUMINOTHERMIC PROCESS.

thermo- (*Genrl*) See THERM-.

thermoammeter (*ElecEng*) An ac type in which current is measured in terms of its heating effect, usually with a thermocouple.

thermobaric (*Genrl*) Of a weapon, using a combination of heat and pressure to create a blast more powerful than that of conventional weapons.

thermochemistry (*Chem*) The study of the heat changes accompanying chemical reactions and their relation to other physicochemical phenomena.

thermocline (*EnvSci*) In lakes, a region of rapidly changing temperature, found between the EPILIMNION and the HYPOLIMNION.

thermocouple (*Phys*) A device consisting of two wires of different metals joined at their ends to form a loop; when a temperature difference between the two junctions unbalances the contact potentials, a current will flow round the loop. If the temperature of one junction is kept constant, that of the other is indicated by measuring the current, or more usually the potential difference set up when the circuit is open. Conversely, a current driven round the loop will cause a temperature difference between the junctions. See PELTIER EFFECT.

thermocouple instrument (*ElecEng*) An instrument the operation of which depends upon the heating of a thermocouple by an electric current.

thermocouple meter (*ElecEng*) A combination of a thermocouple and an ammeter or voltmeter. The current in the external circuit passes through a coil of suitable gauge wire which is electrically insulated from the thermojunction but in very close thermal contact. If the heating current is alternating and such devices may be used for radio-frequency currents, then the meter indications are rms values.

thermocouple wattmeter (*ElecEng*) One which uses thermoelements in a suitable bridge to measure average ac power.

thermoduric (*Phys*) Resistant to heat.

thermodynamic concentration (*Chem*) See ACTIVITY (2).

thermodynamic potential (*Chem*) See FREE ENERGY.

thermodynamics (*Phys*) The mathematical treatment of the relation of heat to mechanical and other forms of energy. Its chief applications are to heat engines (steam engines and internal-combustion engines; see CARNOT CYCLE) and to chemical reactions (see THERMOCHEMISTRY). The laws of thermodynamics are: (1) (*zeroth law*) if two systems are each in thermal equilibrium with a third system then they are in thermal equilibrium with each other (tacitly assumed in every measurement of temperature); (2) (*first law*) the total energy of a thermodynamic system remains constant although it may be transformed from one form to another (conservation of energy); (3) (*second law*) heat can never pass spontaneously from a body at a lower temperature to one at a higher temperature (Clausius), or equivalently no process is possible whose only result is the abstraction of heat from a single heat reservoir and the performance of an equivalent amount of work (Kelvin–Planck); (4) (*third law*) the entropy of a substance approaches zero as its temperature approaches absolute zero.

thermodynamic scale of temperature (*Phys*) See KELVIN THERMODYNAMIC SCALE OF TEMPERATURE.

thermoelectric cooling (*Phys*) Abstraction of heat from electronic components by the PELTIER EFFECT, greatly improved and made practicable with solid-state materials, eg Bi_2Te_3. Devices utilizing this effect, eg frigistors, are used for automatic temperature control and are energized by dc.

thermoelectric effect (*ElecEng*) See SEEBECK EFFECT.

thermoelectricity (*ElecEng*) Interchange of heat and electric energy. See PELTIER EFFECT, SEEBECK EFFECT, THOMSON EFFECT.

thermoelectric materials (*Eng*) Any set of materials (metals) which constitute a thermoelectric system, eg binary (bismuth and tellurium), ternary (silver, antimony

and tellurium), quaternary (bismuth, tellurium, selenium and antimony, called *Neelium*).

thermoelectric power (*ElecEng*) Defined as dE/dT, ie the rate of change with temperature of the thermo emf of a thermocouple.

thermoelectric pyrometer (*ElecEng*) The combination of apparatus forming a temperature-indicating instrument whose action derives from a thermoelectric current; for high temperatures.

thermoelectroluminescence (*Phys*) See ELECTROTHERMO-LUMINESCENCE.

thermoelement (*ElecEng*) See THERMAL CONVERTER.

thermoforming (*Eng*) Polymer shaping method where extruded sheet is heated to above the glass-transition temperature and pressed into shape by a male or female tool.

thermogalvanometer (*ElecEng*) A very sensitive THERMO-COUPLE METER.

thermogenesis (*BioSci*) Production of heat within the body.

thermograph (*EnvSci*) A continuously recording thermometer. In the commonest forms the record is made by the movement of a bimetallic spiral, or by means of the out-of-balance current in a Wheatstone bridge containing a resistance thermometer in one of its arms.

thermographic (*Print*) A printing process in which an impression from a ribbon is transferred by heated styli to paper. Used now in computer printing.

thermography (*Print*) A term applied to printing effect which simulates the result of printing from steel-die engravings. The printed image is dusted with resinous powder while the ink is still wet on the paper, causing the powder to adhere to the print. Heating causes the powder and ink to fuse, creating a raised image.

thermography (*Radiol*) The use of radiant heat emitted by the body to construct images of increased heat emission which can indicate tumours or inflammation. Reduced heat emission indicates reduced blood supply.

thermogravimetric analysis (*Eng*) A method of thermal analysis of materials, where sample is heated at a controlled rate and the change in mass recorded as a function of time. Used for studying degradation of polymers, oxidation of metals, etc. Abbrev *TGA*.

thermohaline (*EnvSci*) Involving both temperature and salinity.

thermojunction (*ElecEng*) See THERMOCOUPLE.

thermolabile (*Chem*) Tending to decompose on being heated.

thermoluminescence (*Phys*) The release of light by previously irradiated phosphors upon subsequent heating.

thermoluminescent dating (*Geol, Phys*) Radiation emitted by the decay of an unstable isotope in crystalline material like quartz will cause electrons to be trapped in regions of imperfection in the crystal lattice. Subsequent heating will cause these electrons to be released as light (thermoluminescence) which can be measured. Knowledge of the amount of radioactivity present in the crystal will then allow the time to be determined since the crystals were last heated, eg as in the firing of pottery. See panel on RADIOMETRIC DATING.

thermoluminescent dosimeter (*Radiol*) One which registers integrated radiation dose, the readout being obtained by heating the element and observing the thermoluminescent output with a photomultiplier. It has the advantage of showing very little fading if readout is delayed for a considerable period, and forms an alternative to the conventional film badge for personnel monitoring.

thermoluminescent material (*Phys*) Material which, after radiation, releases light when subsequently heated in proportion to the radiation absorbed.

thermolysis (*BioSci*) Loss of body heat.

thermolysis (*Chem*) The dissociation or decomposition of a molecule by heat.

thermomagnetic effect (*Phys*) See MAGNETOCALORIC EFFECT.

thermometer (*Phys*) An instrument for measuring temperature. A thermometer can be based on any property of a substance which varies predictably with change of temperature. For instance, the *constant-volume gas thermometer* is based on the pressure change of a fixed mass of gas with temperature, while the *platinum resistance thermometer* is based on a change of electrical resistance. The commonest form relies on the expansion of mercury or other suitable fluid with increase in temperature.

thermometric scales (*Phys*) See CELSIUS SCALE, CENTIGRADE SCALE, FAHRENHEIT SCALE, FIXED POINT, INTERNATIONAL PRACTICAL TEMPERATURE SCALE, KELVIN THERMODYNAMIC SCALE OF TEMPERATURE, RANKINE SCALE, RÉAUMUR SCALE.

thermometry (*Phys*) The measurement of temperature.

thermonasty (*BioSci*) A nastic movement in response to a change in temperature, eg the opening and closing of crocus flowers.

thermonuclear bomb (*Phys*) See HYDROGEN BOMB.

thermonuclear energy (*Phys*) Energy released by a NUCLEAR FUSION reaction that occurs because of the high thermal energy of the interacting particles. The rate of reaction increases rapidly with temperature. The energy of most stars is believed to be acquired from exothermic thermonuclear reactions. In the hydrogen bomb, a fission bomb is used to obtain the initial high temperature required to produce the fusion reactions. See JET, STELLARATOR, TOKAMAK.

thermonuclear reaction (*Phys*) A reaction involving the release of THERMONUCLEAR ENERGY.

thermoperiodism (*BioSci*) The response of a plant to daily (or other) cycles of temperature.

thermophile (*BioSci*) Requiring, adapted to, or sometimes tolerating high temperatures. Also *thermophilic, thermophilous*. Cf THERMOTOLERANT.

thermophone (*Acous*) Electro-acoustic transducer in which fluctuating temperature changes produce a sound pressure wave. Used for calibration of microphones. Now obsolete. Also *thermal receiver*.

thermophyllous (*BioSci*) Having leaves only in the warmer part of the year; deciduous.

thermopile (*ElecEng*) See PILE.

thermoplastic (*Chem*) Becoming plastic on being heated. Specifically in plastics, any resin which can be melted by heat and then cooled, the process being apparently repeatable in theory any number of times without appreciable change in properties, eg cellulose polymers, vinyl polymers, polystyrenes, polyamides and acrylic polymers. Also *thermosoftening*. Cf THERMOSETS (panel).

thermoplastic binding (*Print*) A synonym for *unsewn binding*, whether the adhesive used is thermoplastic or not.

thermoplastic elastomer (*Plastics*) A class of rubbers not needing vulcanization, so relatively easy to injection-mould, extrude, etc. Includes polyester rubber, some polyurethanes, ethylene vinylacetate and styrene–butadiene–styrene block copolymers. See COPOLYMER and panel on ELASTOMERS.

thermoplastic plates (*Print*) Duplicate printing plates made from thermoplastic material, usually flexible and suitable for rotary printing; normally recoverable for subsequent plates in contrast to THERMOSETTING PLATES.

thermoprinting machine (*Print*) A type of *blocking press*, used in conjunction with automatic food packaging machines using heat-sealing wraps. These are overprinted with details of the contents.

thermoregulator (*Phys*) A type of thermostat which keeps a bath at a constant temperature by regulating its supply of heat.

thermoscopic (*Phys*) Perceptive of change of temperature.

thermosets (*Plastics*) Plastics in which a high-temperature stability has been induced by heating to form a vast network molecule held by covalent bonds between the chains. See panel on THERMOSETS.

thermosetting (*Plastics*) Describes polymer materials in which chemical reactions, including cross-linking, take place while the resins are being moulded; the appearance and chemical and physical properties are entirely changed, and the product is resistant to further applications of heat (up to charring point). See panel on THERMOSETS.

thermosetting compositions (*Plastics*) Compositions in which a chemical reaction takes place while the resins are being moulded under heat and pressure; the appearance and chemical and physical properties are entirely changed, and the product is resistant to further applications of heat (up to charring point), eg phenol formaldehyde, urea formaldehyde, aniline formaldehyde, glycerol-phthalic anhydride.

thermosetting plates (*Print*) Duplicate printing plates made from thermosetting plastic material, not flexible and must be used flat or be curved; not recoverable like THERMOPLASTIC PLATES. Both have certain advantages over metal plates, such as good inking qualities, long life, light weight, easy storage.

thermosiphon (*Eng*) The method of establishing circulation of a cooling liquid by utilizing the slight difference in density of the hot and the cool portions of the liquid.

thermosoftening (*Chem*) Synonym for THERMOPLASTIC.

thermosphere (*Astron*) The region of the Earth's atmosphere, above the MESOSPHERE, in which the temperature rises steadily with height.

thermostable (*Chem*) Not decomposed by heating.

thermostat (*Phys*) Apparatus which maintains a system at a constant temperature which may be preselected. Frequently incorporates a BIMETALLIC STRIP.

thermotaxis (*BioSci*) A directed motile response to temperature. The grex (migrating multicellular phase) of *Dictyostelium discoideum* shows a positive thermotaxis.

thermotolerant (*BioSci*) Able to endure high temperatures, but not growing well under such conditions.

thermotropic (*Phys*) Said of a material for which temperature determines the phase.

therophyte (*BioSci*) Plant that passes the unfavourable season as seeds, and thus has no perennating vegetative buds. See RAUNKIAER SYSTEM.

theta pinch (*NucEng*) Cylindrical plasma constricted by an external current flowing in the θ direction to produce a solenoidal magnetic field.

theta solvent (*Chem*) Organic solvent for polymers producing RANDOM COIL conformation of individual chains.

Thévenin's theorem (*ElecEng*) That the source behind two accessible terminals may be regarded as a constant-voltage generator in series with a source impedance. The value of the voltage is that appearing with terminals open-circuited and the impedance is that measured at the terminal with all voltage sources open-circuited. Often regarded as the dual of NORTON'S THEOREM.

THF (*Chem*) Abbrev for TETRAHYDROFURAN.

THG (*BioSci*) Abbrev for TETRAHYDROGESTRINONE.

THI (*EnvSci*) Abbrev for *temperature–humidity index*.

thi-, thio- (*Chem*) Prefixes denoting a compound in which a sulphur atom occupies a position normally filled by an oxygen atom.

thiamides (*Chem*) A group of compounds derived from amides by the substitution of sulphur for oxygen, eg CH_3CSNH_2.

thiamin (*BioSci*) See vitamin B complex in panel on VITAMINS.

thiazide diuretics (*Pharmacol*) A group of drugs with moderate diuretic effects, used for the long-term treatment of hypertension or oedema associated with congestive heart failure.

thiazines (*Chem*) Six-membered heterocyclic compounds, containing in the ring four carbon, one sulphur and one nitrogen atoms.

thiazole (*Chem*) C_3H_3NS. Bp 117°C. A colourless, very volatile liquid which closely resembles pyridine. It forms salts, but is hardly affected by concentrated sulphuric acid.

Thermosets

THERMOPLASTICS soften or melt when heated and harden on cooling, a process which can apparently be repeated indefinitely without change in properties (but see RECYCLING RATIO). Since the T_g or T_m of most common plastics is relatively low (80–120°C), their use in many products is severely limited. One way of increasing these values is the synthesis of much stiffer chains with higher transitions such as polycarbonates or polysulphones. Some of these polymers (eg aramids, POLYIMIDES) show no transitions, but rather degrade chemically at high enough temperatures. The other approach is to tie the chains together with covalent bonds into a network comprising one single molecule. The cross-linking agents needed may be activated by simple mixing with the monomer, or prepolymer, by photoinitiation or by the action of heat. It is these last cross-linked materials that are called *thermoset*. Apart from vulcanized rubbers, conventional thermosets include phenolics, aminoplastics, epoxies, polyesters and some polyurethanes, although it is important to note that most thermoplastics can be cross-linked using an appropriate reagent. See ION-EXCHANGE RESINS, SCLAIR polyethylene.

Bakelite, a phenolic resin, is one of the oldest thermosets and is formed by reacting phenol and formaldehyde to form NOVOLAK or RESOL prepolymers in the A stage. Further polymerization gives B-stage resins and final heat cure gives a rigid, non-crystalline network with a very high CROSS-LINK DENSITY (see diagram). Since the network is permanent, analysis of thermosets is difficult because most methods involve dissolution in a suitable organic fluid. They are used in products subjected to severe conditions such as heat-resistant surfaces (laminates etc), electrical insulators (eg printed circuit boards) and composites, where they can be combined with inorganic fibres such as glass or higher-performance materials like carbon fibre.

See panels on HIGH-PERFORMANCE POLYMERS and POLYMER SYNTHESIS.

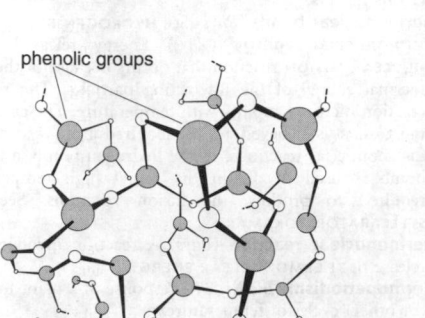

phenolic groups

bridges from formaldehyde

Thermosets Schematic drawing of cross-linked phenol formaldehyde resin in three dimensions.

---

thickener (*FoodSci*) An ADDITIVE for increasing the viscosity of liquids or slurry, eg gums. Added to reduced-calorie drinks in order to improve 'mouthfeel', eg carboxymethylcellulose.

thickener (*MinExt*) Apparatus in which water is removed from ore pulp by allowing solids to settle. To obtain continuous working the solids are worked towards a central hole in the bottom by means of revolving rakes.

thick filament (*BioSci*) Bipolar filament, 12–14 nm in diameter, in striated muscle, made up of myosin. The heads of the myosin undergo cyclical conformational changes associated with binding and release from the thin actin filaments and the hydrolysis of ATP and cause the thick and thin filaments to slide relative to one another.

thick film (*Electronics*) Film deposited by a screen-printing or similar process. For eg thick-film resistors, an 'ink' carries a mixture of glass and semiconductor oxide particles. On firing, the ink is driven off and the particles sinter. See BIROX RESISTOR.

thick-film lubrication (*Eng*) The state of fluid-medium lubrication which exists when the lubricant separates the JOURNAL and bearing surfaces to prevent metal-to-metal contact. The bearing friction is then dependent largely on the force necessary to shear the lubricant film, and perfect lubrication is said to prevail.

thick leg disease (*Vet*) See OSTEOPETROSIS GALLINARUM.

thick lens (*Phys*) Any lens, or system of lenses, in which the distance between the outer faces is not small compared with the focal length.

thickness chart (*EnvSci*) A chart of upper air showing the difference in geopotential between two particular pressure levels. Centres of low thickness are *cold pools*. See CONSTANT-PRESSURE CHART.

thickness–chord ratio (*Aero*) The ratio of the maximum depth of an *aerofoil*, measured perpendicular to the *chord line*, to the *chord length*; usually expressed as a percentage.

thickness dummy (*Print*) A book made up of blank leaves to show size and physical appearance in advance.

thickness moulding (*Build*) A moulding serving to fill up the bare space beneath a projecting cornice.

thick source (*Phys*) Radioactive source with appreciable self-absorption.

thick space (*Print*) The normal standard space between words set in lower case, one-third of the type size; reduced if possible, rather than increased, when justifying text matter.

thick target (*Phys*) A target which is not penetrated by primary or secondary radiation beam.

thick-wall chamber (*NucEng*) Ionization chamber in which build-up of ion current is produced by contribution of knock-on particles arising from wall material.

thick wire Ethernet (*ICT*) See 10 BASE 5.

thief sampler (*PowderTech*) An instrument for sampling powders, comprising two tubes, one fitting closely inside the other, with rectangular slits in corresponding positions. The instrument is thrust into the powder with the holes closed; the inner tube is rotated to open the holes to collect the sample, which is withdrawn with the holes closed.

thigmo- (*Genrl*) Prefix from Gk *thigma*, touch.

thigmocyte (*BioSci*) A type of leucocyte found in some Crustacea.

thigmotropism (*BioSci*) Turning of an organism (or of part of it) towards or away from object providing touch stimulus. See HAPTOTROPISM.

thimble (*Build*) See SLEEVE PIECE.

thimble ionization chamber (*NucEng*) A small cylindrical, spherical or thimble-shaped ionization chamber, volume less than 5 cm^3 with air-wall construction. Used in radiobiology.

thimble-tube boiler (*Eng*) A heat-recovery boiler, consisting of an annular water drum from which thimble-like tubes or pockets project into the central flue, through which the hot exhaust products are passed.

thin filament (*BioSci*) A filament, 7–9 nm in diameter, in striated muscle, made up of F-actin associated with tropomyosin and troponin.

thin film (*Electronics*) Film made by CHEMICAL or PHYSICAL VAPOUR DEPOSITION, or electrolysis, typically no more than a few micrometres thick.

thin-film capacitor (*ElecEng*) One constructed by evaporation of two conducting layers and an intermediary dielectric film (eg silicon monoxide) on an insulating substrate.

thin-film circuit (*Electronics*) A circuit in which all active and passive elements are made from thin films; more commonly referred to as *integrated circuit*.

thin-film lubrication (*Eng*) When metal-to-metal contact exists in a JOURNAL bearing for all or part of the time, thin-film or imperfect lubrication is said to prevail and the materials and surface characteristics of journal and bearing affect the bearing friction significantly.

thin-film memory (*ICT*) The use of an evaporated thin film of magnetic material on glass as an element of a computer memory when a dc magnetic field is applied parallel to the surface. A large-capacity memory will contain thousands of these elements which can be produced in one operation. Also *magnetic-film memory*.

thin-film resistor (*ElecEng*) Modern high-stability type formed by a conducting layer a few tens of nanometres thick on an insulating substrate.

thin-film transistor (*Electronics*) A METAL–OXIDE–SILICON FIELD-EFFECT TRANSISTOR (MOSFET) in which the semiconductor is not a single crystal but polycrystalline or amorphous. Used as switches for matrix displays and in solar cells, having the advantage that they can be deployed over wider-area substrates than feasible in current single-crystal technology.

thin-layer chromatography (*Chem*) A form of chromatography in which compounds are separated by a suitable solvent or solvent mixture on a thin layer of adsorbent material coated on a glass plate. It is rapid, and excellent separations can be obtained. Abbrev *TLC*.

thinners (*Build*) Volatile liquids added to paints to dilute and make the material more fluid and workable. They should evaporate on drying and not alter the original colour of the paint. They aid penetration of priming coats but finishing coats may be damaged by excess.

thin source (*Phys*) Radioactive source with negligible self-absorption.

thin space (*Print*) The narrowest of the justifying spaces, one-fifth of the type size; the HAIR SPACE is not a justifying space.

thin target (*Phys*) A target penetrated by a primary radiation beam so that detecting instrument(s) may be used on the opposite side of the target to the source.

thin-wall bearings (*Autos*) Bearings made of steel sheet coated with a thin layer of soft alloy.

thin-wall chamber (*NucEng*) Ionization chamber in which the number of knock-on particles arising from the wall material and absorption therein is small.

thin wire Ethernet (*ICT*) See 10 BASE 2.

thio- (*Chem*) See THI-.

thio-acids (*Chem*) Acids in which the hydroxyl of the carboxyl group has been replaced by SH, thus forming the group –COSH.

thio-alcohols (*Chem*) See MERCAPTANS.

thiocarbamide (*Chem*) See THIOUREA.

thiocyanates (*Chem*) Compounds formed when alkaline cyanides are fused with sulphur. They contain the group =N=C=S or the ion S=C=N$^-$.

thio-ethers (*Chem*) Compounds in which the ether oxygen has been replaced by sulphur; general formula RSR$'$. They form additive crystalline compounds with metallic salts; they are capable of combining with halogen or oxygen, which becomes attached to the sulphur atom, thereby converting the latter from the bivalent to the tetravalent state; and they form additive crystalline compounds with alkyl halides, eg $(CH_3)_3SI$.

thioglycollic acid (*Chem*) $CH_2(SH)COOH$. A colourless liquid, with a slight odour when pure but extremely unpleasant in impure form. A strong reducing agent, used as a reagent for detecting iron, with which it gives a violet colour in ammoniacal solution. Used in cold-waving treatment of hair, and in certain textile treatments for crease-resistant finishes.

Thiokol (*Plastics*) TN for a synthetic rubber of the polysulphide group, derived from sodium tetrasulphide and organic dichlorides, with sulphur as an accelerator.

thiols (*Chem*) See MERCAPTANS.

thionyl (*Chem*) Containing the group –SO.

thiopental (*Pharmacol*) A barbiturate drug in the form of a yellow powder used intravenously in solution to give general anaesthesia of short duration. Formerly *thiopentone*.

thiophen (*Chem*) C_4H_4S. A five-membered heterocyclic compound with sulphur.

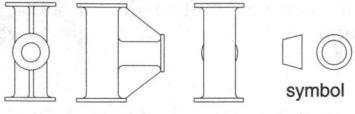

thiophen

thiophil (*Chem*) Having an affinity for sulphur and its compounds, eg bacteria.

thiosulphuric acid (*Chem*) $H_2S_2O_3$. Its salts are thiosulphates.

thiourea (*Chem*) *Thiocarbamide*. NH_2CSNH_2. Colourless prisms, mp 180°C; it is slightly soluble in water, ethanol and ethoxyethane. Used in organic synthesis and as a reagent for bismuth.

thiourea resins (*Plastics*) Formerly resins made from thiourea and an aldehyde, but now superseded by MF thermoset resins. See panel on THERMOSETS.

third-angle projection (*Eng*) A system of projection used in engineering drawing, in which each view shows what would be seen by looking on the near side of an adjacent view.

symbol

third-angle projection Three of possible five views are shown with the standard conventional symbol.

third-generation computer (*ICT*) A machine produced after 1965, probably with some integrated circuits, small-scale and medium-scale integration, replacing transistors and ferrite cores, giving another big jump in computing power and reduction in size. Time sharing and interactive computing resulted from the considerable development in software. Families of machines were produced. ICL 1900 series, IBM 300 series. See COMPUTER GENERATIONS, MAINFRAME, MINICOMPUTER, PARALLEL PROCESSING.

third isomorphism theorem (*MathSci*) If A is a subgroup and N is a normal subgroup of a group G, then the quotient group AN/N is isomorphic to $N/(A \cap N)$.

third-party software (*ICT*) SOFTWARE that is obtained from a software company other than the one which supplied the original software.

third-rail insulator (*ElecEng*) See CONDUCTOR-RAIL INSULATOR.

third-rail system (*ElecEng*) The system of electric traction supply by which current is fed to the electric tractor from an insulated conductor rail running parallel with the track.

thirds (*Paper*) A pre-metrication size of cut card, 38×76 mm ($1\frac{1}{2} \times 3$ in).

third ventricle (*BioSci*) In vertebrates, the cavity of the DIENCEPHALON, joining the two lateral ventricles of the cerebral hemispheres via the foramen of Monro, and the fourth ventricle in the medulla oblongata by the cerebral aqueduct.

thirst (*Psych*) A state of motivation which arises primarily as a result of dehydration of body tissues.

thistle funnel (*Chem*) A glass funnel with a thistle-shaped head and a long narrow tube.

thixotrope (*Chem*) A colloid whose properties are affected by mechanical treatment.

thixotropy (*Chem, Geol*) Rheological property of a fluid where applied stress lowers viscosity, which returns to normal when the stress is released, and of some colloidal materials (particularly in geology, clays and fine sediments) to change from a gel to a sol when under stress and to revert to high viscosity at low stress. The clay thus strengthens when undisturbed. Property of many polymer fluids and mixtures (eg ketchup), and widely used in paints. Fluids which behave in the opposite way are known as *rheopectic*.

tholeiite (*Geol*) A silica-rich basalt abundant in mid-oceanic ridges and continental rifts.

Thomas–Gilchrist process (*Eng*) The use of basic-lined Bessemer converters to remove phosphorus during steel production.

Thomas resistor (*ElecEng*) A standard manganin resistor which has been annealed in an inert atmosphere and sealed into an envelope.

Thomas's splint (*Med*) A skeleton splint consisting of two parallel metal rods and a padded leather ring, used for maintaining the hip and the knee joint in fixed extension.

Thomsen's disease (*Med*) See MYOTONIA CONGENITA.

Thomson compass (*Ships*) See KELVIN COMPASS.

Thomson effect (*ElecEng*) The emf produced by temperature differences in a single conductor, and the heat change associated with current flow between temperature differences.

thomsonite (*Min*) An orthorhombic zeolite; a hydrated silicate of calcium, aluminium and sodium, found in amygdales and crevices in basic igneous rocks.

Thomson scattering (*Phys*) The scattering of electromagnetic waves by free electrons. In a classical interpretation, the electron is set into oscillatory motion by the transverse electric field of the wave and radiates at the same frequency as the wave. The scattering cross-section of an electron is

$$\sigma = \frac{8}{3}\pi(e^2/4\pi\,\varepsilon_0\,mc^2)^2 = 0{\cdot}66 \times 10^{-28} \text{ m}^2$$

where m and e are the mass and charge of the electron, c is the speed of light and ε_0 is the permittivity of free space.

thoracentesis (*Med*) The operation of drawing off a morbid collection of fluid in the pleural cavity through a hollow needle stuck through the wall of the chest. Also *thoracocentesis*.

thoracicolumbar system (*BioSci*) See SYMPATHETIC NERVOUS SYSTEM.

thoracocentesis (*Med*) See THORACENTESIS.

thoracoplasty (*Med*) The operation for collapsing a diseased lung by removal of portions of the ribs.

thoracoscope (*Med*) An instrument for viewing the pleura covering the lung and the chest wall. It is inserted through the chest wall into a pleural cavity previously filled with air.

thoracotomy (*Med*) Incision of the wall of the chest, for draining pus from the pleural cavity or from the lung.

thorax (*BioSci*) (1) In Crustacea and Arachnida, a region of the body lying between the head and the abdomen and usually fused with the former. (2) In insects, one of the three primary regions of the body, lying between the head and the abdomen, and bearing in the adult three pairs of legs and the wings (if present). (3) In some tubicolous Polychaeta, a region of the body behind the head, distinguished by the form of its segments and the nature of its appendages. (4) In land vertebrates, the region of the trunk between the head or neck and the abdomen which contains heart and lungs and bears the fore-limbs, esp in the higher forms, in which it is enclosed by ribs and separated from the abdomen by the diaphragm. Adj *thoracic*.

thorianite (*Min*) Thorium dioxide; crystallizes in the cubic system and is found in pegmatites and gem gravel washings, as in Sri Lanka. An important source of thorium and uranium.

thoriated cathode (*Electronics*) Tungsten cathode containing a small proportion of thorium to reduce the temperature at which copious electronic emission takes place. With heat, the thorium diffuses to the surface, forming a tenuous emitting layer.

thoride (*Phys*) A naturally occurring radioactive isotope in the radioactive series containing thorium.

thorite (*Min*) Tetragonal thorium orthosilicate, found in syenites and syenitic pegmatites.

thorium (*Chem*) Symbol Th, at no 90, ram 232·0381, rel.d. 11·2, mp 1845°C. A metallic radioactive element, dark-grey in colour. Its abundance in the Earth's crust is 8·1 ppm and there are few independent thorium minerals. Its commercial sources are *monazite*, which occurs widely in beach sands where it is derived from acid igneous rocks and pegmatites, and *thorite*. The thorium radioactive series starts with thorium of mass 232. It is FISSILE on capture of fast neutrons and is also FERTILE, ^{233}U (fissile with slow neutrons) being formed from ^{232}Th by neutron capture and subsequent beta decay.

thorium reactor (*NucEng*) Breeder reactor in which fissile ^{233}U is bred in a blanket of fertile ^{232}Th. See panel on NUCLEAR REACTORS.

thorium series (*Phys*) The series of nuclides which result from the decay of ^{232}Th. The mass numbers of the members of the series are given by $4n$, n being an integer. The series ends in the stable isotope ^{208}Pb. See RADIOACTIVE SERIES.

thorn (*BioSci*) A woody sharp-pointed structure; usually restricted to those representing modified branches, as in the hawthorn. Cf PRICKLE, SPINE.

thornproof (*Textiles*) A closely woven suiting made of highly twisted yarns (frequently two-fold) that is strong and firm and resistant to damage (eg by thorns).

thoron (*Phys*) A heavy inert radioactive gas resulting from the decay of thorium; an isotope of radon (radon-220), half-life 54·5 s. Symbol Tn.

thoroughpin (*Vet*) A swelling on the hock of the horse caused by distension of the synovial sheath of the flexor perforans tendon. Also *through-pin*.

THORP (*NucEng*) Abbrev for *Thermal (reactor mixed) Oxide Reprocessing Plant*, at Sellafield, Cumbria. Fig. ▷

thousand (*Build*) Formerly a trade term for 1200 slates; now for 1000, so also *thousand actual*.

Thr (*Chem*) Symbol for THREONINE.

thrashing (*ICT*) Excessive inefficient activity, as of the heads of a hard disk attempting to seek disjointed information.

thread (*Eng*) See SCREW-THREAD.

thread (*ICT*) A series of postings on an Internet message board, each concerning the same subject.

thread (*Textiles*) The general name for a yarn, although commonly used specifically for sewing thread.

thread-dial indicator (*Eng*) A device which forms part of a screw-cutting lathe. It indicates, while the lathe is running, the correct time for engaging the half-nut so that the tool will repeatedly follow the existing thread cut.

thread grinding (*Eng*) The accurate production of screw-threads by a form grinding wheel, profiled to the thread section and automatically traversed along the revolving work.

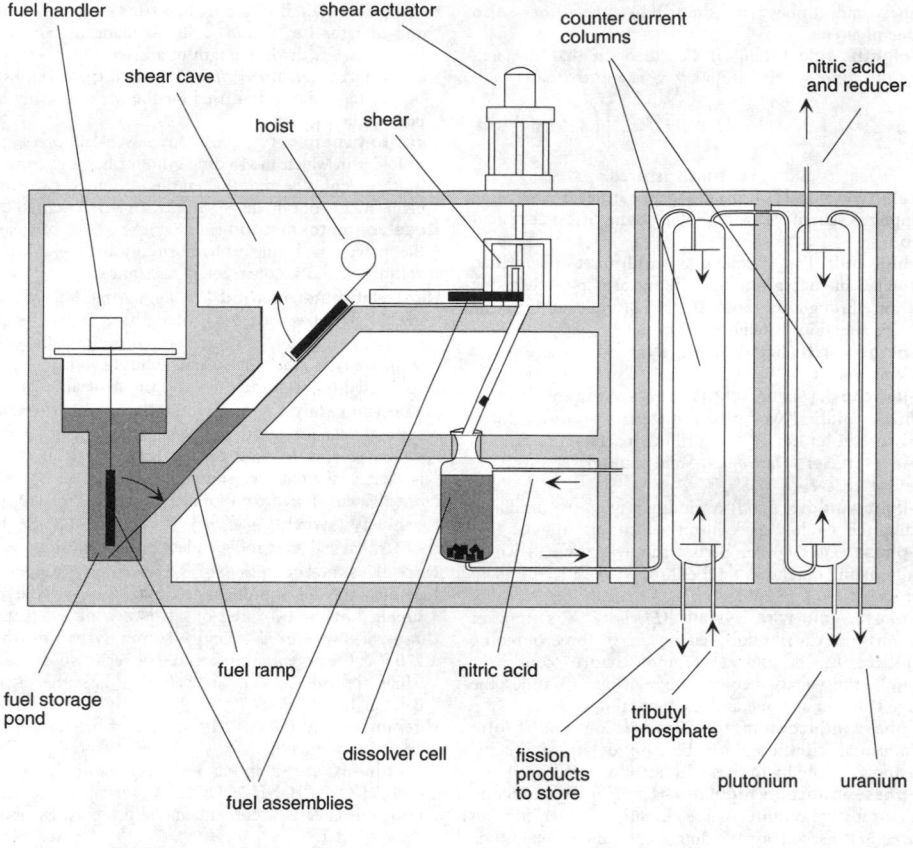

fuel handler

shear cave

hoist

shear

shear actuator

counter current columns

nitric acid and reducer

fuel storage pond

fuel ramp

fuel assemblies

dissolver cell

nitric acid

fission products to store

tributyl phosphate

plutonium   uranium

THORP Fuel elements stored under water are hoisted up to the shear and cut to pieces. These are then dissolved and extracted to separate uranium and plutonium from the fission products.

threading (*ImageTech*) The operation of inserting the start of the film into the mechanism of the camera, or projector, as it leaves the feed reel, and attaching it to the take-up reel. Also *threading up*.

thread rolling (*Eng*) Producing a screw-thread by rolling between flat or cylindrical dies an alloy sufficiently plastic to withstand the cold-working forces without disintegrating.

thread-sewn (*Print*) The traditional method of securing and joining to each other the sections of a book, by hand and on tapes or cords for the best work, by machine for EDITION BINDING.

thread-stitching (*Print*) The traditional method, by hand or machine, of securing the leaves of insetted (or quirewise) books or pamphlets; better and more expensive than WIRE-STITCHING.

threadworm (*Med*) Although can refer to any slender NEMATODE, usually refers to *Enterobius vermicularis*, which is a common parasite in children and causes PRURITUS of the anus.

threat behaviour (*Psych*) A form of communication usually occurring in situations of conflict between fear and aggression; used to repel conspecifics or members of other species without undue risk or injury. Threat displays are very varied, often involving ritualized postures and expressions, as well as specialized morphological features (eg the rattle of the rattlesnake).

three-ammeter method (*ElecEng*) A method of measuring the power carried by a single-phase circuit making use of three ammeters. Cf THREE-VOLTMETER METHOD.

three-axis stabilized (*Space*) Descriptive of a spacecraft which can be held in any position by the application of small torques provided by reaction control thrusters about three orthogonal axes. See GRAVITY STABILIZATION, SPIN STABILIZED.

three-body problem (*Astron*) The problem of the behaviour of three bodies which mutually attract each other; no general solution is possible but certain particular solutions are known. See TROJAN GROUP.

three-centred arch (*Arch*) An arch having the form of a false ellipse struck from three centres.

three-coat work (*Build*) Plastering in three successive coats. See FLOATING, PRICKING-UP, ROUGHING-IN, SETTING.

three-colour process (*ImageTech*) A colour reproduction system, photographic or electronic, in which the subject is analysed into three colour components, red, green and blue, for recording or transmission. Reproduction is similarly in three colours, usually by *subtractive* methods in photography and cinematography and by *additive* means for TV and video.

three-colour process (*Print*) The subtractive process applied to printing. The yellow is, as a rule, printed first, followed by magenta (red), then cyan (blue). A fourth printing, in black, is usually added to enhance the final result.

three-core cable (*ElecEng*) A cable having three conducting cores arranged symmetrically about the axis of the cable and insulated.

three-day sickness (*Vet*) A benign, non-contagious virus disease of cattle in tropical countries characterized by fever, lameness and stiffness; transmitted by mosquitoes. Also *ephemeral fever.*

three-eighths rule (*MathSci*) The theorem that the area under the curve $y = f(x)$ from $x = x_0$ is approximately

$$\frac{3}{8}\left[f(x_0) + 3f(x_1) + 3f(x_2) + f(x_3) \right]\left(x_3 - x_0 \right)$$

where x_0, x_1, x_2 and x_3 are equally spaced.

three-electron bond (*Chem*) Resonance structure involving an unpaired shared electron, similar to the ONE-ELECTRON BOND.

three-high mill (*Eng*) A rolling mill with three rolls, which are rotated in such a way that the metal is passed in one direction through the bottom pair of rolls and in the opposite direction through the top pair.

three-hinged arch (*Build*) Arch ribs hinged at the top and springing points.

three-jaw chuck (*Eng*) A SCROLL CHUCK with three jaws for holding cylindrical workpieces, materials or tools, particularly useful on the lathe or drilling machine.

three-level maser (*Electronics*) Solid-state maser involving three energy levels.

three-light window (*Build*) A window having two mullions dividing the window space into three compartments.

three-phase (*ElecEng*) An electric supply system in which the alternating potentials on the three wires differ in phase from each other by 120°.

three-phase, four-wire system (*ElecEng*) A system of three-phase ac distribution making use of three outgoing conductors (lines) and a common return conductor (neutral), the voltage between lines being $\sqrt{3}$ times the voltage between any line and the neutral.

three-phase induction motor (*ElecEng*) Commonest form of industrial electric motor because of its simplicity, robustness and ability to start without additional windings.

three-phase induction regulator (*ElecEng*) An induction regulator for use on three-phase circuits, in which the emf induced in the secondary winding is constant in magnitude but variable in phase, so that the total emf in the secondary side bears a small phase displacement to the primary voltage.

three-phase, six-wire system (*ElecEng*) A system of three-phase ac distribution in which each phase has separate outgoing and return conductors.

three-point adder (*ICT*) See FULL ADDER.

three-point landing (*Aero*) The landing of an aircraft so equipped on the two wheels and tail skid (or wheel) simultaneously; the normal 'perfect landing'.

three-point problem (*Surv*) A field problem, arising in plane table and hydrographical surveying, in which it is required to locate on the plan the position of the instrument station, given that only three points represented on the plan are in fact visible from the station.

three-point switch (*ElecEng*) See THREE-WAY SWITCH.

three-quarter bat (*Build*) A brick made or cut equal to three-quarters the full length of a brick.

three-quarter-bound (*Print*) Similar to QUARTER-BOUND, but having the material used for the back covering a large part of the sides too.

three-to-two folder (*Print*) On a web-fed press, a type of folder in which the folding cylinder has a circumference of three cut-offs and the cutting cylinder two cut-offs, giving a ratio of cylinder sizes 3:2. Cf TWO-TO-ONE FOLDER.

three-voltmeter method (*ElecEng*) A method of measuring the power in a single-phase circuit by means of three voltmeters and a non-reactive resistance.

three-wattmeter method (*ElecEng*) A method of measuring the power carried by a three-phase, four-wire circuit, making use of three wattmeters whose current coils are connected in the lines and whose voltage coils are connected between the lines and the neutral.

three-way catalyst (*Autos*) A catalyst used in a CATALYTIC CONVERTER for purifying car exhaust emissions. The catalyst is only active at temperatures above 300°C, which is some time after start-up, and is poisoned and deactivated by lead compounds.

three-way switch (*ElecEng*) A rotary-type single-pole switch having three independent contact positions.

three-wire meter (*ElecEng*) An electricity supply meter performing the simultaneous integration of the energy supplied by the two sides of a THREE-WIRE SYSTEM.

three-wire system (*ElecEng*) A supply system in which eg 220 V is the potential difference between two of the wires, while approx 110 V exists between the other two-wire combinations.

thremmatology (*Genrl*) The science of breeding domestic animals and plants.

threonine (*Chem*) *2-amino-3-hydroxybutanoic acid.* $CH_3CH(OH)CH(NH_2)COOH$. A polar amino acid. The L- or S-isomer is a constituent of proteins. Symbol Thr, short form T.

threose (*Chem*) A tetraose.

threshing (*Agri*) A procedure to separate grain from the harvested seed head.

threshold (*EnvSci*) In environmental terms, the level of exposure at which physical damage or harm begins to occur. For some hazards there is doubt as to whether any true threshold exists, merely a level at which it is impossible to detect harm.

threshold (*ICT*) The lowest value of a current, voltage or any other quantity that produces the minimum detectable response.

threshold (*Med*) Generally, the lowest intensity of an effect which is detectable, eg of visibility, below which neither the cones nor the rods in the retina of the eye respond to a light stimulus.

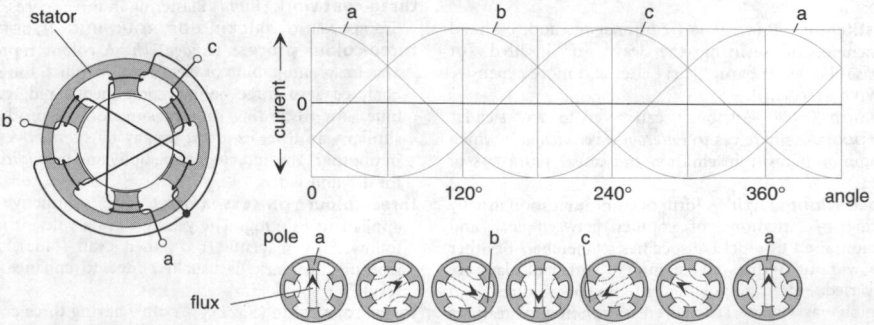

three-phase induction motor Winding of simple 6-pole stator (left) and the rotation of the flux field through one cycle of alternating current.

threshold amplitude (*Electronics*) The lowest amplitude level which a pulse-height selector or window discriminator will accept.

threshold current (*Electronics*) That at which a gas discharge becomes self-sustaining.

threshold dose (*Radiol*) The smallest dose of radiation that will produce a specified result.

threshold effect (*Electronics*) The marked increase in background noise which occurs in a valve circuit when on the verge of oscillation.

threshold energy (*Phys*) The minimum energy that can just initiate a given endoergic nuclear reaction. Exoergic reactions may also have threshold energies.

threshold frequency (*Electronics*) The minimum frequency in a photon which can just release an electron from a surface.

threshold lights (*Aero*) A line of lights across the ends of a runway, strip, or landing area to indicate the usable limits.

threshold limit value (*Genrl*) The maximum concentration of a named pollutant that a worker should be exposed to in a given period of time. Times vary for particular pollutants and are modified as standards change. In the UK standards are supervised by the Factory Inspectorate.

threshold limit value–short-term exposure limit (*BioSci*) The level of continuous exposure to a substance that it is believed will not cause irritation, chronic or irreversible tissue damage, or narcosis such as to increase the likelihood of accidental injury, impair self-rescue or materially reduce work efficiency. Time-weighted averages of acceptable exposure levels are used when exposure is frequent but discontinuous. Abbrev *TLV-STEL*.

threshold of hearing (*Acous*) The minimum rms pressure of a sound wave which an average human listener can just detect at any given frequency. It is $2 \cdot 10^{-5}$ Pa at a frequency of 1000 Hz. This value is often used as a reference pressure so that the threshold of hearing is at 0 dB (see DECIBEL) at 1000 Hz.

threshold of pain (*Acous*) The minimum intensity or pressure of sound wave which causes sensation of discomfort or pain in average human listener. It is between 130 and 140 dB.

threshold of sound audibility (*Acous*) The minimum intensity or pressure of sound wave which the average human listener can just detect at any given frequency. Commonly expressed in decibels relative to 2×10^{-5} pascals (Pa).

threshold treatment (*Chem*) The addition of minute quantities of dehydrated phosphates to water, to inhibit furring and corrosion in pipes, containers, etc.

threshold voltage (*Electronics*) The voltage which must be applied between the gate and substrate of a METAL-OXIDE–SILICON (MOS) structure to cause the formation of an inversion channel. Also *starting voltage*.

thrill (*Med*) A tremor or vibration palpable at the surface of the body, esp in valvular disease of the heart.

throat (*Build, CivEng*) See DRIP.

throat (*Eng*) (1) The C-shaped aperture of a gap press. (2) The root portion of a saw tooth.

throat (*Glass*) The submerged channel through which glass passes from the melting end to the working end of a tank furnace.

throat microphone (*Acous*) A microphone worn against the throat and actuated by contact pressure against the larynx. Used eg by air pilots and deep-sea divers. Also *laryngophone*.

thromb-, thrombo- (*Genrl*) Prefixes from Gk *thrombos*, lump, clot.

thrombectomy (*Med*) The operation of removing a venous THROMBUS.

thrombin (*BioSci*) The proteolytic enzyme that converts fibrinogen to fibrin resulting in blood clotting.

thrombo- (*Genrl*) See THROMB-.

thrombocyte (*BioSci*) A minute greyish circular or oval body found in the blood of higher vertebrates; it plays an important role in coagulation; a blood platelet.

thrombocytopenia (*Med*) Abnormal decrease in the number of platelets (thrombocytes) in the blood.

thrombopenia (*Med*) See THROMBOCYTOPENIA.

thrombophilia (*Med*) Combined inflammation and thrombosis of a vein.

thrombosis (*Med*) (1) Coagulation; clotting. (2) The formation of a clot in a blood vessel during life.

thrombus (*Med*) A clot formed in a blood vessel during life and composed of thrombocytes (platelets), fibrin and blood cells.

throttle valve (*Eng*) (1) In steam engines and turbines, a governor-controlled steam valve, usually a DOUBLE-BEAT VALVE. (2) In petrol engines, the BUTTERFLY VALVE. (3) In refrigerators, the regulating valve controlling the pressure and temperature range of the working agent.

throttling (*Eng*) The process of reducing the pressure of a fluid by causing it to pass through minute or tortuous passages so that no kinetic energy is developed and the total heat remains constant. See REFRIGERATOR, THROTTLING CALORIMETER.

throttling calorimeter (*Eng*) A device for measuring the DRYNESS FRACTION of wet steam by throttling to a measured lower pressure and measuring the resulting superheat.

through bridge (*CivEng*) A bridge in which the track is carried by the lower stringers. Cf DECK BRIDGE.

through level (*ICT*) The reading of a high-impedance level-measuring set at a point in a system, no correction being made for any difference between the actual impedance of the system and the impedance with respect to which the set is calibrated.

through path (*ICT*) The forward path from loop input to loop output in a feedback circuit.

throughput (*ChemEng*) A measure of the rate of production, in terms of mass of product per unit time per unit volume of plant. In vacuum technology, the quantity of gas or vapour passing a given section of a pump or pipeline in unit time; measured as the product of the pressure and the volume per second at the pressure at that section. Units: litre torrs per second or litre pascals per second. Symbol *Q*.

through-stone (*Build*) A bondstone whose length is equal to the full thickness of the wall in which it is laid as a header. Also *jumper*, *perpend*.

throw (*ElecEng*) See SPAN.

throw (*Eng*) (1) The total travel of a crank or similar element, being twice the radius of eccentricity. Sometimes half this distance is called the throw. (2) A hand-driven, dead-centre lathe, used by clock-makers.

throw (*Geol, MinExt*) (1) The amount of vertical displacement (*upthrow* or *downthrow*) of a particular rock, vein or stratum due to faulting. Cf LATERAL SHIFT. See FAULT. (2) The amplitude of shake of a concentrating table. (3) Deviation of a deep borehole from the planned path.

throw (*ImageTech*) The distance between the projector and the screen.

throwback (*Acous*) In a public-address system, when the microphone is near the reproducers, the sound intensity which is applied to the microphone by the reproducers. If this is excessive, the system becomes paralysed with self-sustained oscillations. See FEEDBACK.

throwing (*Textiles*) In silk manufacture, the processes of reeling, doubling and twisting, to bring the silk filaments into the form of a thread. Now extended to include similar processes carried out with continuous filament synthetic fibres.

throwing power (*Chem*) The property of a solution in virtue of which a relatively uniform layer of metal may be electrodeposited on a relatively irregular surface.

throw-off trip (*Print*) An attachment on a printing machine which allows the impression to be suspended without stopping the machine.

throw-out (*Print*) See FOLD-OUT.

throwster (*Textiles*) A company that carries out the THROWING operations, but also now used to describe a manufacturer of false-twist TEXTURED YARNS.

thrum (*BioSci*) The short-styled form of such heterostyled flowers as the primrose, with the anthers visible at the top of the corolla tube. See HETEROSTYLY. Cf PIN.

thrum (*Textiles*) Waste lengths of yarn produced when a loom is being prepared for weaving.

thruput (*ChemEng*) See THROUGHPUT.

thrush (*Med, Vet*) Infection with the fungus *Candida albicans*, characterized by the appearance of white patches on the mucous membranes and the tongue. See also EQUINE THRUSH, HORSE POX.

thrust (*Aero*) Propulsive force developed by a jet or rocket motor.

thrust (*CivEng*) The equal horizontal forces acting upon the abutments of an arch, due to the loading carried by the arch.

thrust (*Eng*) (1) The reaction to a compressive force on a rod. (2) The axial resultant force from a propeller or jet engine.

thrust (*Space*) Reaction force produced on a rocket vehicle as a result of the expulsion of a high-velocity exhaust gas. It is related to the acceleration (*a*) produced according to Newton's second law: $T = ma$, where T is the thrust and m the mass of the rocket vehicle. Thrust is measured in newtons but is often quoted in kilograms.

thrust bearing (*Eng*) A shaft bearing designed to take an axial load. In its simplest form it consists of a loose bronze washer interposed between moving parts on the shaft. Other types range from ball bearings with lateral races, solely for thrust, to taper roller bearings which can sustain both radial and axial loads. Also *thrust block*. See MICHELL BEARING.

thrust chamber (*Aero*) The compartment in a rocket where the propulsive forces are developed before ejection; usually, but not necessarily, the *reaction chamber*. See ROCKET PROPULSION.

thrust deflector (*Aero*) A device, usually a combination of doors closing the jet pipe and a cascade of guide vanes, for deflecting the efflux of a turbojet downwards to provide upward thrust for *STOL* or *VTOL*.

thrust deflexion (*Aero*) The direction of the efflux from a RAMJET, ROCKET or TURBOJET in a direction other than along its axis, for the purpose of obtaining a thrust component normal to this axis. Generally used for the guidance of rockets and for STOL and VTOL aircraft. See JET DEFLECTION.

thrust loading (*Aero*) The sea-level static thrust of the engine(s) of a jet-propelled aircraft divided by its gross weight.

thrust plane (*Geol*) A thrust plane, or thrust, is a *reversed fault* which dips at a low angle.

thrust reverser (*Aero*) A device for deflecting the efflux of a turbojet forwards in order to apply a positive braking thrust after landing. There are two basic types: mechanical ones in which the jet is blocked by hinged doors, which also direct the gases forwards; and aerodynamic ones wherein high-pressure air injected into the centre of the jet causes it to impinge upon peripheral louvres that turn it forwards.

thrust spoiler (*Aero*) A controllable device mounted on, or just behind, the nozzle of a jet-propulsion engine to deflect and thus negate the thrust. See THRUST REVERSER.

thrust-to-weight ratio (*Aero, Space*) (1) The thrust of an aircraft's power plant(s) divided by the gross weight at take-off. (2) The ratio thrust over weight which must be greater than one (both factors expressed in kg) if a rocket system is to leave the ground.

Thudichum speculum (*Med*) An instrument used to inspect the inside of the nose from the face.

thulite (*Min*) A variety of zoisite, pink in colour due to small amounts of manganese.

thulium (*Chem*) Symbol Tm, at no 69, ram 168·9342. A metallic element, a member of the rare earth group. One of the rarest elements, occurring in small quantities in euxenite, gadolinite, xenotime, etc. Radioactive isotope emits 84 keV gamma rays, frequently used in radiography.

thumb latch (*Build*) A latch which is operated by the pressure of the thumb. See NORFOLK LATCH.

thumb plane (*Build*) Rebate plane with curved sole; used for rounded rebates.

thumbscrew (*Build*) A small type of G CRAMP used for light work.

thunder (*EnvSci*) The crackling, booming or rumbling noise which accompanies a flash of lightning. The noise has its origins in the violent thermal changes accompanying the electrical discharge, which cause non-periodic wave disturbances in the air. Its reverberatory characteristic arises mainly from the continuous arrival of the brief noise from sections of the discharge at increasingly remote locations, since the spark may be many kilometres long. Claps of thunder occur when the spark is, roughly, normal to the line of observation. The time interval between lightning and the corresponding thunder (in seconds divided by three) gives the distance of the storm centre (in kilometres) from the observer.

thunderstorm (*EnvSci*) A storm in which LIGHTNING and THUNDER occur, usually associated with CUMULONIMBUS cloud. The mechanism by which the cloud becomes electrically charged is not fully understood.

thuringite (*Min*) A variety of CHLORITE.

Thurnian (*Geol*) A stage in the Pleistocene. See QUATERNARY.

Thury screw-thread (*Eng*) See SWISS SCREW-THREAD.

THX (*ImageTech*) Abbrev for *Tom Holman's experiment*. TN for a cinema SURROUND SOUND installation, subsequently expanded to include parameters for commercial and home cinema audio systems.

thylakoid (*BioSci*) A flattened, membrane-bounded sac in a chloroplast. Thylakoids may be single or associated in pairs or threes or more. See GRANUM, STROMA LAMELLA.

thymectomy (*Med*) Surgical removal of the thymus.

thymic epithelial cells (*BioSci*) Epithelial cells that ramify throughout the cortex and medulla of the thymus. They are believed to control the maturation of lymphocyte precursors by secretion of peptide hormones (*thymopoietin*, *thymosins*).

thymic hypoplasia (*Med*) A congenital cell-mediated immunodeficiency syndrome in human infants. Antibody production and blood immunoglobulin levels are normal.

thymidine (*BioSci*) Casual term for the nucleoside thymine deoxyriboside (not the riboside, despite the naming convention for other nucleosides).

thymine (*Chem*) *5-methyl-2,6-dioxytetrahydropyrimidine*. One of the two pyrimidine bases in DNA in which it pairs with adenine. See panel on DNA AND THE GENETIC CODE.

thymine

thymocyte (*BioSci*) A T-cell precursor found within the thymus.

thymol (*Chem*) *1-methyl-4-methylethyl-3-hydroxybenzene*. $C_{10}H_{14}O$. Large crystals; mp 51°C, bp 230°C; it occurs in thyme oil, and can be synthesized from propan-2-ol and 1-methyl-3-hydroxybenzene. It is used as a disinfectant, and in mouthwashes and in dentistry. Isomeric with CARVACROL.

thymol blue (*Chem*) *Thymol-sulphon-phthalein*, used as an indicator with two pH ranges, 1·2–2·8 (red → yellow) and 8·0–9·6 (yellow → blue).

thymolphthalein (*Chem*) An indicator obtained by reaction between thymol and phthalic anhydride, having a pH range of 9·3–10·5, over which it changes from a colourless to a blue solution.

thymoma (*Med*) A rare thymic tumour. About half are associated with *myasthenia gravis*, and a few others with red-cell aplasia, immunoglobulin deficiency, rheumatoid arthritis or polymyositis. There appears to be some abnormality of immune regulation.

thymopoietin (*BioSci*) A factor that derives from the thymus and that is believed to influence the maturation of T-lymphocytes. It induces the appearance of T-lymphocyte markers on resting lymphocytes both *in vitro* and after administration *in vivo*.

thymosins (*BioSci*) A group of peptides derived from thymus epithelial cells (not the same as THYMOPOIETIN) that are present in blood. They can partially restore T-lymphocyte function in thymectomized animals, and induce differentiation and maturation of immature T-lymphocytes.

thymus (*BioSci*) The lymphoid organ in which T-lymphocytes are educated and mature, composed of stroma (thymic epithelium) and lymphocytes, almost entirely of the T-cell lineage. In mammals the thymus is just anterior to the heart within the rib cage; in other vertebrates in rather undefined regions of the neck or within the gill chamber in teleost fish. The thymus regresses as the animal matures. See panel on IMMUNE RESPONSE.

thymus-dependent antigen (*BioSci*) An antigen that fails to stimulate an antibody response if T-lymphocytes are absent. Co-operation between B-lymphocytes and helper T-lymphocytes is required for the B-lymphocytes responding to such antigens to differentiate into antibody-secreting cells. Most proteins and complex antigens are thymus-dependent. Also *T-dependent antigen*.

thymus-dependent area (*BioSci*) Areas in peripheral lymphoid organs that are predominantly occupied by T-lymphocytes which circulate through them, and are depleted if the thymus fails to liberate T-thymocytes into the circulation. These areas are anatomically segregated, and contain dendritic interdigitating cells with which the T-lymphocytes come into close contact and that are involved in antigen presentation. Often *T-cell area*.

thymus-derived cells (*BioSci*) Lymphocytes derived from the thymus. They are usually identified by the presence of thymus-specific surface antigens.

thymus-independent antigen (*BioSci*) An antigen able to stimulate B-lymphocytes to produce antibody without the co-operation of T-lymphocytes, eg in animals lacking a thymus. Such antigens are usually polymers which are poorly digestible by macrophages and which carry a repeated array of antigenic determinants, enabling them to bind firmly to Ig receptors on B-lymphocytes. They probably stimulate a subset of B-lymphocytes, and do not stimulate memory cells or cell-mediated immunity. Some thymus-dependent antigens (including lipopolysaccharides) can also stimulate non-specific T-cell help, and elicit larger responses in normal than in thymus-deprived animals. These are designated TI-1, whereas those which do not are TI-2. Also *T-independent antigen*.

Thy-1 antigen (*BioSci*) Old name for a differentiation antigen on the surface of mouse T-cells, now assigned a *CD number* (CD90).

thyratron (*Electronics*) A gas-filled triode operating in an atmosphere of mercury vapour, argon, helium, hydrogen or neon. Ionization starts with sufficient positive swing of the negative grid potential, and anode and grid potentials lose control.

thyratron firing angle (*Electronics*) Phase angle of ac anode voltage supply to thyratron (measured relative to zero) at instant when it strikes.

thyristor (*Electronics*) A semiconductor power switch capable of bistable operation. It can have from two to four terminals and can be triggered from its off state to on at any chosen point within a single 90° quadrant of the applied ac voltage. The semiconductor rectifier (SCR) is the most common unidirectional thyristor. See TRIAC.

thyroglossal (*Med*) Of tongue and thyroid, eg *thyroglossal duct*, an embryonic structure from which the thyroid develops.

thyroid antibodies (*BioSci*) Organ-specific auto-antibodies found in a variety of thyroid diseases, esp Hashimoto thyroiditis and thyrotoxicosis. The major antibodies are against thyroglobulin, against another antigen in thyroid colloid, against a microsomal antigen of thyroid acinar cells, and against the receptors for thyroid-stimulating hormone (TSH). The last may mimic TSH and cause thyroid overactivity.

thyroidectomy (*Med*) The surgical removal of part of the thyroid gland.

thyroid gland (*BioSci*) In vertebrates, a ductless gland originating as a median ventral outgrowth from a point well forward on the floor of the pharynx. It may be a single structure, bilobed or paired, and there may be small accessory masses of thyroid tissue in other places. The gland consists of spherical follicles composed of an outer layer of cuboidal secretory cells surrounding and discharging into a central cavity. In this are found the hormones thyroxine and 3,5,3,-tri-iodothyronine, which are concerned with the rate of tissue metabolism, and development of the nervous system and behaviour (deficiency causing CRETINISM), and, in Amphibia, with the control of metamorphosis. Evolutionarily the thyroid originates from the endostyle of amphioxus (of the Cephalochordata) and Tunicata, and the ammocoete larva of lampreys.

thyroiditis (*Med*) Inflammation of the thyroid gland. See RIEDEL'S DISEASE.

thyrotomy (*Med*) See LARYNGOFISSURE.

thyrotoxicosis (*Med*) The condition resulting from an excess of circulating thyroid hormones (T_4 and/or T_3) leading to either a diffuse hyperplasia of the thyroid (GRAVE'S DISEASE) or toxic single or multiple nodules of the thyroid (Plummer's disease).

thyrotrophic (*Med*) Maintaining or nourishing the thyroid; used of the hormone of the anterior lobe of the pituitary gland, which stimulates growth and function of the thyroid gland.

thyroxine (*Med*) One of the active principles of the thyroid gland which controls metabolic rate.

Thysanoptera (*BioSci*) An order of minute insects with asymmetrical piercing mouthparts, a large free prothorax and free tarsi that have a protrusible adhesive terminal vesicle. Some are serious pests causing malformation of plants and sometimes inhibiting the development of fruit. Thrips.

Ti (*Chem*) Symbol for TITANIUM.

tibia (*BioSci*) (1) In land vertebrates, the pre-axial bone of the crus. (2) In insects, Myriapoda and some Arachnida, the fourth joint of the leg.

tic (*Med*) See HABIT SPASM.

tic douloureux (*Med*) An affection of the fifth cranial nerve characterized by paroxysmal attacks of pain in the face and the forehead. Also *trigeminal neuralgia*.

tick-borne fever (*Vet*) A febrile disease of sheep and cattle caused by infection by *Rickettsia phagocytophila*; transmitted by ticks.

tick fever (*Vet*) See REDWATER (1). Also *Texas fever*.

tick pyaemia (*Vet*) A pyaemic disease of lambs caused by bacterial infection by *Staphylococcus aureus*, which causes abscesses in various parts of the body, esp the joints and muscles; the infection is believed to enter via tick bites.

ticks (*BioSci*) Blood-sucking arachnids related to mites which are vectors for a number of diseases in humans and animals, eg RELAPSING FEVER, TYPHUS, VIRUS ENCEPHALITIS. See ACARINA.

tidal flap (*Build*) A sluice in an embankment impounding tidal waters, which can pass through it when the tide is out. Also *go-out*.

tidal friction (*Astron*) The friction caused by the ebb and flow of the tides, esp in narrow channels; it is responsible for a reduction of the Earth's rate of rotation, and for an acceleration of the Moon's motion, so that eventually the day and the month will be equal in length. This effect in the past has caused the Moon to present the same face to the Earth (rotation and revolution being equal).

tidal volume (*BioSci*) The volume of air moving in and out of the lungs of vertebrates (and of the tracheal system of Insects) during normal (unforced) breathing; in humans about 500 ml.

tidal wave (*Phys*) See TSUNAMI.

tide (*Astron*) The distortion of the surface layers (whether liquid or solid) of a planet or natural satellite resulting from differences between the gravitational forces acting on its various parts. The most familiar is the ocean tide on the Earth, produced mainly by the attraction of the Moon. As well as a bulge of water directed towards the Moon, there is a bulge on the opposite side of the Earth because the gravitational force there is less than that acting on the solid sphere of the Earth – so in any one place there are two high tides each lunar day. The Sun contributes a small effect as well. When it reinforces the Moon (at full or new Moon) there are high *spring tides*, and when at 90° to the Moon weak *neap tides*.

tide gauge (*Surv*) An apparatus for determining the variation of sea level with time.

tie (*Eng*) A frame member sustaining only a tensile load. US for SLEEPER.

tie bar (*Eng*) One of four massive horizontal columns on which the tools, bolster and other parts are mounted on an injection-moulding machine. See INJECTION MOULDING.

tie-beam (*Build, Eng*) A structural timber designed to prevent timbers moving apart under load, eg the lower ends of a pair of principal rafters.

tied letters (*Print*) A synonym for LIGATURE.

tie line (*Eng*) See PHASE DIAGRAM.

tie line (*ICT*) A line that may pass through exchanges, but that is used solely for connecting private branch exchanges, and over which incoming calls cannot be extended. Also *interswitchboard line*.

tie line (*Surv*) A survey line forming part of a *skeleton* and serving to fix its shape, eg a diagonal of a four-sided skeleton.

Tiemann–Reimer reaction (*Chem*) See REIMER–TIEMANN REACTION.

tie molecule (*Chem*) Single-polymer chain connecting adjacent single-crystal lamellae, providing bulk material with structural integrity. See CRYSTALLIZATION OF POLYMERS and panel on POLYMER SYNTHESIS.

tie rod (*Eng*) See TENSION ROD.

tie-rod stator frame (*ElecEng*) A form of stator frame for large electric machines in which several frame sections are laterally secured by means of tie rods parallel with the axis of the machine.

tie wall (*Build, CivEng*) A cross-wall built upon the extrados of an arch at right angles to the spandrel wall or walls.

tie wire (*ElecEng*) A wire used to attach a transmission or telephone line conductor to a supporting insulator. Also *binding wire*.

TIF (*ICT*) Abbrev for *telephone interference* (or *influence*) *factor*.

TIFF (*ICT*) Abbrev for TAG IMAGE FILE FORMAT.

TIG (*Eng*) See TUNGSTEN INERT GAS WELDING.

tiger's eye (*Min*) A form of silicified crocidolite stained yellow or brown by iron oxide.

tight binding model (*Phys*) A model from which the electronic band structure of certain transition elements can be calculated. The overlap of wavelengths from atom to atom is regarded as small enough to permit the crystal wavefunction to be calculated by a linear combination of atomic orbital (*LCAO*) approach. See BAND THEORY OF SOLIDS.

tight coupling (*ElecEng*) That between two circuits which causes alteration of the current in either to affect materially the current in the other. In mutual reactance coupling, coupling is said to be *tight* when the ratio of mutual reactance to the geometric mean of the individual reactances (of the same sign) of the two circuits approaches unity. Also *close coupling*.

tight-edged (*Print*) Said of a reel of paper which has dried out at the edges, resulting in a web which is slack in the middle.

tight junction (*BioSci*) A junction between epithelial cells where the membranes are in close contact, with no intervening intercellular space. Tight junctions can bind epithelial cells into sheets which permit no leakage of solutes across the sheets between the cells. Also *zonula occludens*.

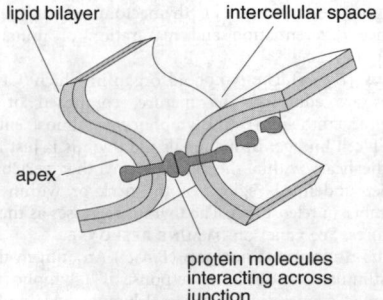

lipid bilayer   intercellular space

apex

protein molecules interacting across junction

tight junction

tile (*Build*) A thin slab, often highly ornamental, of baked clay, terracotta, glass or cement, used for roofing or for covering walls or floors.

tile (*ICT*) A way of arranging open WINDOWS on the screen so that none overlap but all are visible.

tile (*ImageTech*) Video effect in which the picture is broken up into small rectangular areas of uniform tone and colour.

tile-and-a-half tile (*Build*) A purpose-made tile of extra width.

tile creasing (*Build*) A course formed of two or three thicknesses of plain roofing tiles, set in mortar and breaking joint. Laid immediately below a brick-on-edge coping and projecting about 2 in (5 cm) over each side of the wall, with the top surface sloped in cement, in order to prevent the percolation of water into the wall below the coping.

tile hanging (*Build*) See WEATHER TILING.

tile lintel floor (*Build*) A type of fire-resisting floor having a steel framework similar to the *filler joist floor*, but with hollow tile, terracotta or fireclay lintels filling in the panels between filler joists, thus reducing the amount of concrete required for encasement.

tile ore (*Min*) The earthy brick-red variety of cuprite; often mixed with red oxide of iron.

tiling batten (*Build*) See SLATING AND TILING BATTENS.

till (*Geol*) A poorly sorted mixture of unconsolidated sediment produced by glacial action. Also *boulder clay*.

tiller (*BioSci*) A shoot that develops from an axillary or adventitious bud at the base of a stem; characteristic of grasses, including cereals.

tillering (*Agri*) The stage of grain crop development when additional shoots are produced from a single crown.

tillite (*Geol*) Consolidated and lithified till.

tilting (*ImageTech*) Swinging the camera vertically.

tilting fillet (*Build*) A strip of wood laid beneath a DOUBLING COURSE to tilt it up slightly, so that the slates may rest properly in the roof. Also *eave-board, skew fillet*.

tilting level (*Surv*) A type of level whose essential characteristic is that the telescope and attached level tube may be levelled without the necessity for setting the rotation axis truly vertical.

tilt roof (*Build*) A roof having the form of a circular arc in which the rise is small compared with the span.

tilt-rotor (*Aero*) A large retractable propeller on the wing of an aircraft that enables vertical take-off and landing; an aircraft fitted with such propellers.

tilt wing (*Aero*) VERTICAL TAKE-OFF AND LANDING (VTOL) aircraft whose wing, complete with propulsion units, propellers or nozzles, can be rotated through 90° about a transverse axis, so that thrust acts vertically. Such arrangements have flown but have proved impractical.

timber (*For*) Felled trees or logs suitable for conversion by sawing or otherwise.

timber brick (*Build*) See WOOD BRICK.

timber connectors (*Build*) (1) Metal plates with spikes formed on one or both sides used to fix timbers together. (2) Rings of metal which can be housed or grooved into the faces of adjoining timbers held with bolts.

timbering (*Build, CivEng*) Temporary timbers arranged for the support of the earth in excavations, to prevent collapse of the sides.

timber line (*BioSci*) The line or zone on a mountain or at high latitudes beyond which trees do not grow to normal size or form. See KRUMMHOLZ.

timbre (*Acous*) The characteristic tone or quality of a sound, which arises from the presence of various harmonics or overtones of the fundamental frequency.

time (*Astron*) Originally measured by the HOUR ANGLE of a selected point of reference on the celestial sphere with respect to the observer's meridian. The fundamental unit of time measurement now is the second based on an atomic oscillation. See APPARENT SOLAR TIME, EPHEMERIS TIME, GREENWICH MEAN TIME, LOCAL TIME, MEAN SOLAR TIME-, SIDEREAL TIME, STANDARD TIME, UNIVERSAL TIME.

time-assignment speech interpolation (*ICT*) A method for increasing the capacity of submarine cables. The intervals of silence (pauses in speech, intervals between data, etc) are used temporarily for the transmission of information for other channels, with channel switching occurring so fast that users are unaware of it.

time base (*Electronics*) A line (usually horizontal) formed by a particular waveform applied to a CRT deflection system. For a linear time base, the waveform is a sawtooth form though circular time-base patterns may be used in some specialized applications.

time-base generator (*Electronics*) A circuit or equipment which produces the waveform necessary to deflect a CRT beam in the desired time-base pattern. TV sets have two time-base generators, both providing sawtooth waveforms, one operating rapidly to scan the lines, the other operating more slowly at the frame frequency, thus ensuring that each line on the screen is scanned below the previous one.

time between overhauls (*Aero*) The period in hours of running time between complete dismantling of an aero-engine. Abbrev TBO or tbo. Also *overhaul period*.

time code (*ImageTech*) A coding system employing units of hours, minutes, seconds and frames for the identification and automatic location of each frame by an EDIT CONTROLLER. The *8 mm time code* and *rewritable consumer time code* (RCTC) are recorded on extensions of the video tracks; *linear time code* (LTC) is recorded on a time code or spare audio track; and *vertical interval time code* (VITC) is recorded in spare lines at the beginning of video tracks. Also applied to motion picture film for the same purpose.

time constant (*ElecEng*) When any quantity varies exponentially with time, the time required for a fractional change of amplitude equal to:

$$100\left(1 - \frac{1}{e}\right) = 63\%$$

where *e* is the exponential constant (the base of natural logarithms). For a capacitance C to be charged from a constant voltage through a resistance R, the time constant is RC. For current in an inductance L being fed from a constant voltage through a resistance R, the time constant is L/R. See RESPONSE TIME.

time-delay relay (*ElecEng*) One which closes contacts in one circuit a specified time after those in a second circuit have been closed.

time dilation (*Phys, Space*) The increase in time interval between two events when they occur in a reference frame which is moving at very high velocity relative to the observer's reference frame rather than in the observer's rest frame. Time dilation is a consequence of the LORENTZ TRANSFORMATIONS in the SPECIAL RELATIVITY theory and can be expressed as

$$t = t_0\sqrt{1 - \left(\frac{v}{c}\right)^2}$$

where t is the time of a clock moving with velocity v, t_0 is the elapsed time of a stationary clock and c is the speed of light.

time discriminator (*ICT*) A circuit that gives an output proportional to the time difference between two pulses, its polarity reversing if the pulses are interchanged.

time-division duplex (*ICT*) A method of making a HALF-DUPLEX channel appear to users as full DUPLEX by automatically switching the channel direction according to demand. It is used eg in the CT2 cordless telephone system, where digital code is sent in each direction in the intervals between TALKSPURTS of the other direction.

time-division multiple access (*ICT*) A technique used extensively in satellite communications in which a stream of time slots in a time-division multiplex system is allocated to users in accord with the demands they are making at any time. Cf SINGLE-CHANNEL PER CARRIER.

time-division multiplex (*ICT*) A form of multiplex transmission that follows logically from the adoption of PULSE MODULATION and processes involving SAMPLING. The gaps between pulses that constitute a signal allow other pulses to be interleaved; extraction of the desired signal at the receiver requires a system operating in synchrony with the transmitter. Cf FREQUENCY-DIVISION MULTIPLEX.

time-domain reflectometer (*ICT*) An instrument that detects the transmission properties of wide-band systems, components and lines by feeding in a voltage step and displaying the pulses reflected from any discontinuities on a suitable oscilloscope. The display can be calibrated to reveal the nature and location of the defects.

time exposures (*ImageTech*) Exposures in cameras for periods longer than about 1/25 of a second, longer than the so-called instantaneous exposure.

time lapse (*ImageTech*) Cinematography or videotape recording with a controlled delay between the exposure of each frame, so that in normal presentation the action appears very greatly speeded up. An important technique for observing the movement of tissue cells.

time-limit attachment (*ElecEng*) The mechanical device whereby a circuit breaker opens only after a predetermined time delay. Cf TIME-LIMIT RELAY.

time-limit relay (*ElecEng*) An electric relay which comes into action some time after it has received the electrical operating impulse.

time-meter (*ElecEng*) An instrument for measuring the time during which current flows in a circuit. Also *hour-counter*, *hour-meter*.

time-of-day routing (*ICT*) Automatic diversion of telephone calls in a private network according to the time of day, as an alternative to simple CALL FORWARDING.

time-of-flight spectrometer (*NucEng*) A way of measuring a neutron spectrum in which the energy or speed of neutrons is determined by the time taken by the neutrons to travel a known distance. A *chopper* admits neutrons in short bursts and the travel time is determined either by a second chopper which passes only neutrons of the correct velocity or by electronic delay measuring equipment. See NEUTRON VELOCITY SELECTOR.

time of operation (*ICT*) In relays, the time between application of current or voltage and the occurrence of a definite change in circuits controlled by its contacts.

time of operation (*Phys*) The time between the occurrence of a primary ionizing event and the occurrence of the count in the detector system.

timeout (*ICT*) The amount of time a computer will wait for a peripheral device to respond before it detects and reports this as an error.

timer (*ElecEng*) A device which opens or closes at specified times control circuits for electrical or other equipment.

time-scale (*Geol*) A chronological sequence of geological events.

time series (*MathSci*) A set of observations taken sequentially over time.

time-shared amplifier (*ICT*) A form of multiplex system in which one amplifier handles several signals simultaneously using successive short intervals of time for each.

time sharing (*ICT*) Means of providing multi-access to a computer system. Each user is, in turn, allowed a TIME SLICE of the central processor although each appears to have continuous use of the system. See JOB QUEUE, MULTIPROCESSOR, MULTIPROGRAMMING.

time shift (*ImageTech*) Recording a TV broadcast programme on videotape for viewing at a later more convenient occasion.

time signal (*ICT*) Radio transmission indicating standard time by reference to an atomic standard.

time slice (*ICT*) Predetermined maximum length of time during which each program is allowed to run during MULTIPROGRAMMING. See INTERRUPT.

timestamp (*ICT*) (1) A record of the time of an event or transaction, automatically created by and stored on a computer. (2) To add a record of the time of an event or transaction to data on a computer.

time switch (*ElecEng*) An electrical or other switch arranged to open or close a circuit at a predetermined time.

time switching (*ICT*) A method of routing calls through a network without establishing an end-to-end physical connection for the duration of the call as with SPACE SWITCHING. Each call is sent as a stream of short code segments interleaved with many others in a repetitive pattern, and the route that a call takes is determined by the time slot its particular set of segments occupies.

time–temperature transformation diagram (*Eng*) See ISOTHERMAL TRANSFORMATION DIAGRAM.

timing (*Autos*) The process of setting the valve-operating mechanism of an engine so that the valves open and close in correct relation to the crank during the cycle; a similar adjustment of the magneto or distributor drive; the actual valve or magneto or distributor drive; the actual valve or magneto setting, called *ignition timing*, *valve timing*.

timing chain (*Autos*) A chain which drives the camshaft from the crankshaft in some types of timing gear.

timing gear (*Autos*) The drive between the crankshaft and the camshaft, by direct gearing, bevel shaft, chain and sprocket wheels or toothed belt, giving a reduction ratio of 1:2.

tin (*Chem*) Symbol Sn (Lt *stannum*, tin), at no 50, ram 118·69, rel.d. (at 20°C) 7·3, mp 231·85°C. A soft, silvery-white metallic element, ductile and malleable, existing in three allotropic forms. Not affected by air or water at ordinary temperatures. Electrical resistivity is $11·5 \times 10^{-8}$ Ω m at 20°C. Occurs very rarely as native metal and in independent tin minerals but dominantly in *cassiterite*, SnO_2. This occurs in lodes of tin associated with granite as in Cornwall, and in placer deposits. The principal use is as a coating on steel in tinplate; 'tin' cans are made of tin-plated steel. Also used as a constituent in alloys and with lead in low-melting-point solders for electrical connections. Non-toxic, but organic compounds of tin are toxic. See TIN ALLOYS.

tin alloys (*Eng*) Tin is an essential constituent in soft solders, type metals, fusible alloys and certain bearing metals. These

last contain 50–92% of tin alloyed with copper and antimony and sometimes lead. Tin is also a constituent of BRONZE and PEWTER.

Tinamiformes (*BioSci*) An order of birds, containing small, superficially partridge-like, almost tailless birds that are essentially cursorial but can fly clumsily for short distances, and that have a keeled sternum. Tinamus.

tincal (*Min*) The name given since early times to crude borax obtained from salt lakes, eg in Kashmir and Tibet. See BORAX.

tinea (*Med*) See RINGWORM.

tingle (*Build*) A flat strip of lead or copper used as a clip between jointing sheets of lead.

tinman's solder (*Eng*) A tin–lead solder melting below a red heat, used for tinning. The most fusible solder contains 65% tin, the eutectic composition.

tin–nickel (*Eng*) Metal finish which results from the simultaneous electrodeposition of tin and nickel on a polished surface of brass in a carefully controlled bath, resulting in a non-tarnishable and non-corrodible polished surface of low friction.

tinnitus (*Med*) Persistent sensation of ringing noises in the ear. Also *tinnitus aurium*.

tinplate (*Eng*) Thin-sheet steel covered with an adherent layer of tin formed by passing the steel though a bath of molten tin or by electrodeposition. Resists atmospheric oxidation and attack by many organic acids. Used for food containers etc.

tin pyrites (*Min*) See STANNITE.

tinsel yarn (*Textiles*) See METALLIZED YARN.

tinstone (*Min*) See CASSITERITE.

tinting (*Print*) (1) A MECHANICAL STIPPLE esp when used for line-colour purposes. (2) A fault occurring in lithographic printing where a coloured tint prints over the non-image area of the sheet, caused by the pigment being dispersed in the damping solution.

tin–zinc (*Eng*) Metal finish which results from the simultaneous electrodeposition of tin and zinc on a clean steel surface, giving a non-corrodible finish to chassis for electronic apparatus.

tip (*Eng*) The part of a cutting tool containing the cutting edges, if made of a material of superior quality to that of the remainder of the tool, eg tungsten carbide, and securely fastened to the latter by brazing or otherwise.

tip (*MinExt*) See DUMP.

tip-cat folder (*Print*) A type of folder in which the folding blade shaft is actuated by a specially shaped cam and forms the transverse fold by pushing the copy through the folding rollers.

Ti plasmid (*BioSci*) A plasmid carried by virulent strains of the crown gall bacterium, *Agrobacterium tumefaciens*. Part of it (the T-DNA) may, when the bacterium infects a plant, become transferred and incorporated into the nuclear genome of the host cells, inducing them to grow and form the characteristic galls. Ti plasmids are vectors for the 'genetic engineering' of dicotyledonous plants. Also *tumour-inducing principle*. See CROWN GALL.

tip-path plane (*Aero*) The plane of rotation of the tips of a rotorcraft's blades, which is higher than the rotor hub in flight. See CONING ANGLE.

tip penetration (*ImageTech*) The protrusion of the magnetic heads of a VTR into the videotape.

tipping point (*Genrl*) The point in a process at which an irreversible momentum is reached.

tipple (*MinExt*) A frame into which ore trucks are run, gripped and rotated to discharge contents.

tippy wool (*Textiles*) Wool with the fibre tips damaged (eg by PHOTODEGRADATION) while on the sheep. The defect is often revealed at the dyeing stage because the damaged tips take up different amounts of dyes compared with the undamaged parts.

Tiptronic transmission (*Autos*) A proprietary electronic automatic transmission system that allows the driver to select an individual gear by a manually operated switch,

tirodite (*Min*) A rare honey-yellow monoclinic amphibole, the name being used for the manganese-rich, magnesium-bearing end-member. See DANNEMORITE.

T-iron (*CivEng*) A structural member of wrought-iron or rolled mild steel having a T-shaped cross-section.

tissue (*BioSci*) An aggregate of similar cells forming a definite and continuous fabric, and usually having a comparable and definable function; as *epithelial tissue, nervous tissue, vascular tissue.*

tissue culture (*BioSci*) The growth of cells, including tissues and organs, outside the organism in artificial media of salts and nutrients. Depending on the cell type, the cells may be capable of a limited number of divisions or may divide indefinitely. Under appropriate conditions, cultured plant tissues can often be made to regenerate new plants. See TRANSFORMATION (2).

tissue dose (*Radiol*) Absorbed DEPTH DOSE of radiation received by specified tissue. Cf SKIN DOSE.

tissue engineering (*BioSci*) The creation of new body parts for transplantation by *in vitro* culture of cells, on an artificial matrix or support. There are considerable practical difficulties and few tissues have yet been successfully manufactured in this way.

tissue equivalent material (*Radiol*) See PHANTOM MATERIAL.

tissue-specific antigen (*BioSci*) A cell antigen present in a given tissue but not found in other tissues, eg thyroglobulin is specific for thyroid.

tissue tensions (*BioSci*) The mutual compression and stretching, of deeper and more superficial tissues respectively, exerted by the tissues of a living plant.

tissue typing (*BioSci*) The identification of histocompatibility antigens, usually done on blood leucocytes from prospective donor and recipient prior to tissue or organ transplantation.

Titan (*Astron*) Saturn's largest natural satellite, and the second-largest moon in the solar system, discovered in 1655. It is the only satellite with a substantial atmosphere, principally composed of nitrogen and methane. Distance from the planet 1 222 000 km; diameter 5150 km.

titanates (*Chem*) Compounds found in minerals, or formed when titanium (IV) oxide is fused with alkalis; containing the anion TiO_4^{4-} or TiO_3^{3-}.

titanaugite (*Min*) A titaniferous variety of the monoclinic pyroxene AUGITE.

Titania (*Astron*) The largest natural satellite of Uranus, discovered in 1787. Distance from the planet 436 000 km; diameter 1580 km.

titania (*Chem*) See TITANIUM (IV) OXIDE.

titaniferous iron ore (*Min*) See ILMENITE.

titanite (*Min*) See SPHENE.

titanium (*Chem*) Symbol Ti, at no 22, ram 47·90, rel.d. (at 20°C) 4·5, mp 1850°C, bp above 2800°C. A metallic element resembling iron. Abundance in the Earth's crust 0·6% (ninth commonest element). Occurs in accessory oxides and oxide minerals in igneous and metamorphic rocks (and in beach sands). The principal sources are *ilmenite* $FeTiO_3$, *rutile* TiO_2 and from other industrial processes. Manufactured commercially since 1948, it is characterized by strength, lightness and corrosion resistance. Widely used in aircraft manufacture, for corrosion resistance in some wet extraction processes, as a deoxidizer for special types of steel, in stainless steel to diminish susceptibility to intercrystalline corrosion and as a carbide in cemented carbides. Sometimes used for the solid horns of magnetostrictive generators.

titanium alloys (*Eng*) Titanium metal remained a laboratory curiosity until the mid-1950s, but it is now produced in large quantities by electrode melting processes in a vacuum or under argon. In commercially pure form it is widely used for corrosion-resistant applications; the metal itself is highly reactive but a tenacious oxide film on the surface renders it practically inert to atmospheric attack and resistant to the majority of industrial chemicals. Controlled amounts of oxygen (up to 0·2% by weight) significantly raise the yield and tensile strength, giving rise to a number of commercial grades. With half the density of steel and high tensile strength these are very important alloys in the aerospace industry.

titanium dioxide (*Chem*) TiO_2. Pure white pigment of great opacity. Widely used industrially in paints, plastics, etc. See TITANIZING. Forms titanites when fused with alkalis. Also *titania, titanium* (IV) *oxide.* Minerals called *anatase, brookite* and *rutile.*

titanizing (*Glass*) Use of titanium dioxide, TiO_2, in the manufacture of heat-resisting and durable glass by incorporating it to replace certain proportions of soda, the viscosity increasing proportionately.

Titan rocket (*Space*) One of a series of satellite launchers developed from US Air Force missiles; the most powerful is the Titan 4, first launched in June 1989. The final Titan launch was in late 2005.

Titius–Bode law (*Astron*) See BODE'S LAW.

title signature (*Print*) The first SECTION (or signature) of a book, containing the title page and other prelims, and not normally requiring a SIGNATURE MARK.

titling font (*Print*) Consists of capitals, figures and punctuation marks only, no lower case, a narrow beard being sufficient; used for dropped initials and when close spacing between lines is required.

titration (*Chem*) The addition of a solution from a graduated vessel (burette) to a known volume of a second solution, until the chemical reaction between the two is just completed. A knowledge of the volume of liquid added and of the strength of one of the solutions enables that of the other (the *titre*) to be calculated.

titre (*BioSci*) In serological reactions involving the use of serial dilutions of antiserum, describes the highest dilution at which the measured effect is detected.

titre (*Chem*) See TITRATION.

titrimeter (*Chem*) Apparatus for electrometric titrations in which potential changes are followed continuously and automatically.

titubation (*Med*) Staggering and reeling movements of the body, due to disease of the nervous system.

TiVo (*ICT*) TN of a system for making digital recordings of TV programmes on a hard disk.

TL (*Acous*) Abbrev for TRANSMISSION LOSS.

Tl (*Chem*) Symbol for THALLIUM.

TLC (*Chem*) Abbrev for THIN-LAYER CHROMATOGRAPHY.

T-lymphocyte (*BioSci*) Also T-CELL. See panel on IMMUNE RESPONSE.

Tm (*Chem*) Symbol for THULIUM.

T-maze (*Psych*) A maze shaped like a T with an approach path and two arms, left and right, one of which will have the reinforcing stimulus (reward) at the end.

T-memory cell (*BioSci*) A subset of T-cells that have differentiated after exposure to their specific antigens. T-memory cells are relatively long-lived and activate rapidly if they encounter their antigens again allowing immune effector responses to be generated in shorter times. See panel on IMMUNE RESPONSE.

TMS (*Chem*) Abbrev for TETRAMETHYLSILANE.

TMS (*Med*) Abbrev for transcranial magnetic stimulation.

TM-wave (*ICT*) Abbrev for *transverse magnetic wave*, having no component of magnetic force in the direction of transmission of electromagnetic waves along a waveguide. Also *E-wave* (it must have electric field component in direction of transmission).

Tn (*Phys*) Symbol for THORON.

T-network (*ICT*) One formed of two equal series arms with a shunt arm between.

TNF (*BioSci*) Abbrev for TUMOUR NECROSIS FACTOR.

TNT (*Chem*) Abbrev for TRINITROTOLUENE.

T-number (*ImageTech*) A rating of the actual light transmission of a lens at a given stop, in contrast to the calculated F-NUMBER.

toad's-eye tin (*Min*) A variety of CASSITERITE occurring in botryoidal or reniform shapes which show an internal concentric and fibrous structure. It is brownish in colour.

toadstone (*Geol*) An old and local name for the basalts found in the Carboniferous Limestone of Derbyshire. The name may be derived from the rock's resemblance in appearance to a toad's skin, or from the fact that it weathers into shapes like a toad, or from the Ger *todstein* ('dead stone') in reference to the absence of lead.

Toarcian (*Geol*) A stage in the Lower Jurassic. See MESOZOIC.

tobacco amblyopia (*Med*) A visual defect which can cause blindness associated with smoking tobacco and may be due to cyanide in smoke.

Tobin bronze (*Eng*) A type of alpha–beta brass or Muntz metal containing tin. It contains 59–62% copper, 0·5–1·5% tin, the remainder being zinc. Used when resistance to sea water is required. Also *Admiralty brass, naval brass*. See COPPER ALLOYS.

tobramycin (*Pharmacol*) An aminoglycoside antibiotic used for serious infections that are resistant to gentamicin.

TOC (*ImageTech*) Abbrev for TABLE OF CONTENTS.

tocopherol (*BioSci*) See vitamin E in panel on VITAMINS.

Todd-AO (*ImageTech*) TN of an older system of WIDE-SCREEN cinematography, using 65 mm width negative in the camera, with 70 mm prints and six magnetic tracks for presentation.

todorokite (*Min*) Hydrated manganese oxide, enriched in other elements. It is one of the dominant minerals of the deep-sea MANGANESE NODULES.

toe (*ImageTech*) The lower end of the photographic CHARACTERISTIC CURVE, departing from the straight-line portion, where the densities begin to approach the basic FOG LEVEL.

toe-in (*Autos*) A slight forward convergence given to the planes of the front wheels to promote steering stability and equalize tyre wear.

toe-picking (*Vet*) The habit, acquired by individual budgerigars, of biting the feet of birds of other species, particularly finches, within the same aviary.

Togaviridae (*BioSci*) Viruses with a single positive strand RNA genome enclosed in a bullet-shaped capsid, enveloped by a membrane formed from the host cell plasma membrane. Includes alphaviruses such as Semliki Forest virus and Flaviviridae such as yellow fever virus and rubella (German measles) virus.

toggle (*ICT*) (1) Bistable trigger circuit, a multivibrator with coupling capacitors omitted, that switches between two stable states depending on which valve (or transistor) is triggered. Formerly *flip-flop* in the UK. (2) The ability to switch between two possible states of a system.

toggle joint (*Eng*) A mechanism comprising two levers hinged to each other, the end of one being hinged on a fixed point, the opposite end of the other being hinged on a press ram or other load point. If the levers form an obtuse angle and an effort tending to increase this angle is applied at their common hinge, a considerable force is produced at the load point.

toggle press (*Eng*) A power press, often double-acting, incorporating a TOGGLE JOINT, used for deep-drawing operations.

toilet (*Med*) The cleaning and dressing of a wound or injured part.

tokamak (*NucEng*) (1) In nuclear fusion, a toroidal apparatus for containing plasma by means of two magnetic fields: (a) a strong toroidal magnetic field created by coils surrounding the vacuum chamber and (b) a weaker poloidal field created by an intense electric current through the plasma. The resultant magnetic field is in the form of a helix surrounding the intense electric current through the plasma. Tokamak is an acronym of the Russian words meaning toroidal magnetic chamber. Cf POLOIDAL FIELD, TOROIDAL FIELD. See TOKAMAK FIELD. (2) General name for the fusion reactors which use this principle. Fig. ▷

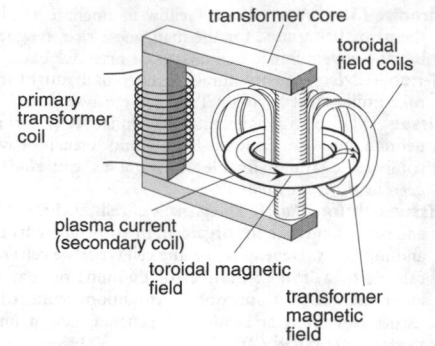

tokamak Arrangement of components.

tokamak field (*NucEng*) The field induced by combining a toroidal field with a poloidal field. See TOKAMAK.

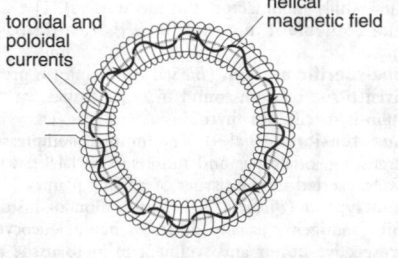

tokamak field The helical magnetic field.

token (*ICT*) A unique combination of BITS which passes from device to device connected to a LAN. Each device may only transmit data when it receives the token.

token bus (*ICT*) A NETWORK based on BUS TOPOLOGY in which a TOKEN is passed between NODES to grant permission for each to transmit.

token economy (*Psych*) A behaviour modification procedure, based on operant conditioning principles, in which patients are given artificial rewards for socially desirable behaviour; the rewards or tokens can be exchanged for desirable items.

Token Ring (*ICT*) A proprietary NETWORK based on RING TOPOLOGY to which access is controlled by a digital code called a token. An idle terminal receiving the token sends it to the next on the ring, but one that wishes to communicate sends its data instead, followed by the token. Since terminals are not allowed to send data until they have received the token, data clashes are eliminated.

tolbutamide (*Pharmacol*) A *sulphonylurea* drug used for oral treatment of diabetes.

tolerable daily intake (*BioSci*) An estimate of the amount of a potentially harmful substance that can be ingested daily on a lifetime basis without an appreciable health risk. Abbrev *TDI*.

tolerance (*Eng*) The range between the permissible maximum and minimum limits of size of a workpiece or of distance between features (eg hole centres) on a workpiece.

tolerance (*Psych, Med*) (1) Acceptance of the beliefs, behaviours and idiosyncrasies of others. (2) The development of insensitivity to a drug or substance so that increasing doses are required to elicit a response. Characteristic of addiction but also of many receptor systems where there is down-regulation of receptor number following prolonged exposure. (3) See IMMUNO-LOGICAL TOLERANCE.

tolerance dose (*Radiol*) The maximum dose which can be permitted to a specific tissue during radiotherapy involving irradiation of any other adjacent tissue.

tolerogen (*BioSci*) Capable of inducing immunological tolerance.

toll TV (*ImageTech*) Programme service which, through technical scrambling devices, is available only by *ad hoc* payment. Also *pay as you view, pay-TV, subscription TV*.

tolterodine (*Pharmacol*) A muscarinic receptor antagonist that acts as an antispasmodic on the muscle of the bladder.

Tolu balsam (*Chem*) See BALSAM OF TOLU.

toluene (*Chem*) $C_6H_5CH_3$, a colourless liquid, mp $-94°C$, bp $110°C$. It occurs in coal- and wood-tar; insoluble in water; miscible with ethanol, ethoxyethene, trichloromethene. Used as a solvent and as an intermediate for its derivatives. Also *methylbenzene, toluol*.

toluene di-isocyanate (*Chem*) $CH_3C_6H_3(NCO)_2$. The most commonly used isocyanate in the production of polyurethane foams. The isocyanate is reacted with a high-molecular-weight glycol to form a liquid prepolymer. The prepolymer reacts with water to give the polyurethane and carbon dioxide, which cause the foam. Abbrev *TDI*.

toluidines (*Chem*) *Methyl*(*aminobenzenes*). $H_3CC_6H_4NH_2$. Homologues of aniline. There are three isomers: 2-toluidine, a liquid, bp $197°C$; 4-toluidine, crystals, mp $43°C$, bp $198°C$; 3-toluidine, a liquid, bp $199°C$.

tolylene-2,4-diisocyanate (*Chem*) A type of isocyanate monomer used for making polyurethanes. They are reacted with polyglycols or polyesters, forming urethane links with terminal hydroxy groups. Abbrev *TDI*. See DIPHENYL-METHANE DIISOCYANATE.

tombolo (*Geol*) A *bar* or *spit* of sand, shingle or gravel which joins an island to the mainland or to another island.

tomentum (*BioSci*) A covering of felted cotton hairs. Adj *tomentose*.

tommy bar (*Eng*) See BOX SPANNER.

tomography (*Med*) An X-ray technique in which the source and detector rotate round a structure at a given axial position. A single 'exposure' thus records the signal from a narrow cylinder through the tissue. A computer reconstructs the two-dimensional image from many such cylinders at each position. See COMPUTER-AIDED TOMOGRAPHY.

-tomy (*Genrl*) Suffix from Gk *tome*, a cut.

ton (*Genrl*) A unit of mass for large quantities. The *long ton*, once commonly used in the UK, is 2240 lb. The *short ton*, commonly used in the US, is 2000 lb. The *metric ton* or *tonne* (1000 kg; symbol t) is 2204·6 lb. In the UK the *short ton* was used in metalliferous mining, the *long ton* in coal mining.

ton (*Ships*) See TONNAGE.

tonalite (*Geol*) A coarse-grained igneous rock of dioritic composition carrying quartz as an essential constituent, usually with biotite and hornblend. See PLUTONIC ROCKS.

tone (*Acous*) Sound signal of a single frequency. In the terminology of music, tone is often used to specify a complex note having a constant fundamental frequency. To avoid misunderstanding, a single-frequency sound is termed a pure tone.

tone (*BioSci*) The resting level of muscle contraction due to background neuromuscular activity.

tone dialling (*ICT*) Telephone dialling system in which digits are transmitted as a combination of tones. Cf PULSE DIALLING.

toner (*ICT*) A fine, black, carbon-based powder that is used in laser printers and photocopiers. The powder is attracted to an electrostatically charged drum, transferred to the paper sheet and fused to the paper by heat.

tongs (*Build*) See STONE TONGS.

tong-test ammeter (*ElecEng*) An ac ammeter and current transformer combination whose iron core can be opened and closed round a cable which thus forms the single-turn primary winding of the transformer. Fig. ▷

secondary winding
iron tongs
meter
insulated cable
handle

tong-test ammeter Essential components only shown.

tongue (*BioSci*) (1) In vertebrates, the movable muscular organ lying on, and attached to, the floor of the buccal cavity; it has important functions in connection with tasting, mastication, swallowing, and (in higher forms) sound production. (2) In invertebrates, esp insects, any conformation of the mouthparts which resembles the tongue in structure, appearance or function, eg proboscis, antlia, haustellum, radula. (3) Generally, any structure which resembles the tongue.

tongue (*Build*) A slip feather.

tongue-and-groove joint (*Build*) A joint formed between the butting edges of two boards. There is a projecting fin on the edge of one board which fits into a groove on the other.

tongue-worm (*Vet*) An aberrant arthropod (*Linguatula serrata*) belonging to the class Pentastomida, which occurs in the nasal passages of the dog, fox and wolf, and more rarely in other animals; the larval and nymphal stages occur in herbivorous animals.

tonicity (*BioSci*) See TONE.

toning (*ImageTech*) See CHEMICAL TONING, DYE TONING.

tonnage (*Ships*) A measurement assigned by the relevant authority for assessing dues etc: 1 ton equals 100 cubic feet. See GROSS TONNAGE, NET REGISTER TONNAGE.

tonnage breadth (*Ships*) A number of tonnage breadths are measured horizontally inside the frames or SPAR CEILING if fitted. Internal cross-sectional areas are calculated from these by SIMPSON'S RULE. These areas are then used in the calculation of TONNAGE.

tonnage depth (*Ships*) A number of tonnage depths are measured at certain points throughout the length of the ship from the top of the inner bottom or CEILING, if fitted, to the top of the deck, deducting one-third of the CAMBER. Used, together with TONNAGE BREADTHS, in the calculation of cross-section areas.

tonnage dimensions (*Ships*) The internal dimensions used in the calculation of TONNAGE. These are TONNAGE LENGTH, TONNAGE BREADTH and TONNAGE DEPTH.

tonnage length (*Ships*) The length measured along the uppermost continuous deck in ships having less than three decks and the second continuous deck from below in others. It is measured from a point where the line of the inside of the frames or spar ceiling cuts the centre line forwards to a similar point aft.

tonne (*Genrl*) Metric ton, 1000 kg. See TON.

tonofilament (*BioSci*) Filaments composed of cytokeratin and found in epithelial cells.

tonometer (*Acous*) A device consisting of a series of tuning forks for determining the frequencies of tones.

tonometer (*Med*) An instrument for measuring hydrostatic pressure within the eye.

tonometer (*Phys*) An instrument for measuring vapour pressure.

tonoplast (*BioSci*) The membrane around a vacuole in a plant cell.

tonsil (*BioSci*) Lymphoid organs found at the back of the mouth and in the throat. While the term is commonly used to refer to the two palatine tonsils located one on each side of the throat, in fact there is also a collection of tonsils on the back of the tongue, on the rear of the pharynx (called adenoids when enlarged) and where the Eustachian tubes open into the nasal cavity. See also WALDEYER'S RING.

tonsillectomy (*Med*) The surgical removal of the tonsils.

tonsillitis (*Med*) Inflammation of the tonsils.

tonus (*BioSci*) A state of prolonged tension in a muscle without change in length.

tooled ashlar (*Build*) A block of stone finished with parallel vertical flutes.

tooling (*Print*) The decoration by hand of a book cover, usually leather.

toolmaker (*Eng*) A highly skilled technician employed to make press tools, cutters and other precision equipment.

tool post (*Eng*) The clamp by which a lathe or shaping-machine tool is held in the slide rest or ram. In its simplest form it consists of a slotted post, the end of which carries a clamping screw.

tool-post grinder (*Eng*) A small grinding machine held on the tool post of a lathe and fed across the work by means of the regular longitudinal or compound rest. It is used to grind the odd small workpiece in the lathe and to true lathe centres but lathe ways need protection from grinding dust.

toolpusher (*MinExt*) Field supervisor of drilling operations.

tool steel (*Eng*) Steel suitable for use in tools, usually for cutting or shaping wood or metals. The main qualities required are hardness, toughness, ability to retain a cutting edge, etc. Contains 0·6–1·6% carbon. Many tool steels contain high percentages of alloying metals: tungsten, chromium, molybdenum, etc. (See HIGH-SPEED STEEL.) Usually quenched and tempered, to obtain the required properties. See panel on STEELS.

tooth (*BioSci*) (1) Any small pointed projection as on the margin of a leaf. (2) In vertebrates, a hard calcareous or horny body attached to the skeletal framework of the mouth or pharynx, used for trituration or fragmentation of food. (3) In invertebrates, any similar projection of chitinous or calcareous material used for mastication or trituration.

toothed wheel (*Eng*) See BEVEL GEAR, HELICAL GEAR, SPIRAL GEAR, SPUR GEAR, WORM GEAR.

toothing plane (*Build*) One with a serrated blade set perpendicularly; used for providing a key for gluing veneers etc.

toothings (*Build*) The recesses left in alternate courses of a wall when later extension is expected, so that the extension can be properly bonded in.

tooth ratio (*ElecEng*) The ratio of slot width to tooth width as measured at the circumference of the armature.

tooth ripple (*ElecEng*) See SLOT RIPPLE.

tooth thickness (*Eng*) Of a gear wheel, the length of arc of the pitch circle between opposite faces of the same tooth. See PITCH DIAMETER.

top (*Phys*) The heaviest of the six FLAVOURS of QUARKS with a mass of 170 GeV and a charge of $+2e/3$. Also the property of the top quark (topness), which is zero for all other particles, 1 for the top, and -1 for the antitop quark. Also *truth*.

top (*Textiles*) Originally used to describe a sliver of combed wool ready for spinning into worsted yarns. Now also applied to the material produced by cutting or breaking continuous filament TOWS of synthetic fibres.

topaz (*Min*) Silicate of aluminium with fluorine, usually containing hydroxyl, which crystallizes in the orthorhombic system. It usually occurs in veins and druses in granites and granite-pegmatites. It is colourless, pale-blue or pale-yellow, and is used as a gemstone. Cf CITRINE, ORIENTAL TOPAZ, SCOTTISH TOPAZ, SPANISH TOPAZ.

topazolite (*Min*) A variety of the calcium–iron garnet ANDRADITE, which has the honey-yellow colour and transparency of topaz.

top beam (*Build*) The horizontal beam connecting the rafters of a COLLAR-BEAM ROOF.

top blanket (*Print*) On web-fed relief presses the outside dressing of an impression cylinder. Also *top sheet*. Cf UNDER BLANKET.

top casing (*MinExt*) The topmost part of the casing of an oil well to which either the drilling gear or the flow-control gear is attached. Also *anchor string*, *conductor*.

top dead centre (*Eng*) See INNER DEAD CENTRE.

top-down programming (*ICT*) See JACKSON STRUCTURED PROGRAMMING, STRUCTURED PROGRAMMING.

top dressing (*Agri*) Application of a fertilizer onto a growing crop.

top girth (*For*) In standing trees, the girth at the top of the merchantable length of the bole.

tophaceous (*Med*) Sandy, gritty, of the nature of TOPHUS, eg *tophaceous gout*.

top hamper (*Ships*) That part of the structure of the ship which is above the SUPERSTRUCTURE. Also *deck houses*, *erections*.

top-hung (*Build*) Said of a window sash arranged to open outwards about hinges on its upper edge.

tophus (*Med*) A hard nodule composed of crystals of sodium biurate which are deposited in bodily tissues in gout.

topical application (*Agri*) Placement of small amounts of pesticide onto target sites on cultivated plants or livestock to reduce infestation, or directly onto pest organisms to achieve a lethal effect.

topiramate (*Pharmacol*) A drug used as an adjunct in the treatment of partial seizures of epilepsy.

topochemistry (*Chem*) The study of reactions which occur only at definite regions in a system.

topographic map (*BioSci*) The ordered pattern of projections of neurons onto their target tissue.

topography (*Surv*) The delineation of the natural and artificial features of an area.

topoisomerase (*BioSci*) Any of a group of enzymes that alter the degree of supercoiling of DNA by cleaving one or both strands of the double helix.

topological space (*MathSci*) A non-empty set of points with some of its subsets defined to be open sets in such a way that (1) the whole space and the empty sets are both open, (2) the intersection of any two open sets is open, and (3) the union of any collection of open sets is open.

topology (*MathSci*) The study of those properties of shapes and figures which remain invariant under homeomorphic mapping.

topotype (*BioSci*) A specimen collected in the same locality as the original type specimen of the same species.

topping (*Agri*) Reducing the foliage of the standing crop chemically or mechanically to stimulate subsequent growth or aid further treatments.

top rake (*Eng*) In cutting tools, the angle which that part of the cutting face which lies immediately behind the tool point makes with the horizontal.

topset beds (*Geol*) Gently inclined strata deposited on the subaerial plain or the just-submerged part of a delta. They are succeeded seawards by the *foreset beds* and, in deep water, by the *bottomset beds*.

top sheet (*Print*) See TOP BLANKET.

top shot (*ImageTech*) One taken with axis of camera nearly vertical.

top yeast (*BioSci*) See YEAST.

torbanite (*Min*) A variety of BOGHEAD COAL or oil shale containing 70–80% of carbonaceous matter, including an abundance of spores; dark-brown in colour.

torch igniter (*Aero*) A combination igniter plug and fuel atomizer for lighting up gas turbines.

torching (*Build*) The operation, sometimes performed on slates which have been laid on battens only and not on boarding, of pointing the horizontal joints from the inside with hair mortar or cement. Similarly a method of repairing loose tiles or slates by injecting adhesive resin. Also, regionally, *pargeting*.

tornado (*EnvSci*) (1) An intensely destructive, advancing whirlwind formed from strongly ascending currents. When over the sea, the apparent drawing up of water is actually water vapour condensing in the vacuous core. (2) In W

Africa, the squall following thunderstorms between the wet and dry seasons.

toroid (*ElecEng*) Magnetic component (a coil or transformer), made in the shape of an anchor ring. Adopted because, with this construction, most of the magnetic field is contained within the core and leakage is minimal; thus, there is little or no interaction with adjacent components and circuits. Also *torus*.

toroid (*MathSci*) A solid generated by rotating a circle about an external point in its plane. Also *anchor ring, torus*.

toroidal field (*NucEng*) Magnetic field generated by a current flowing in a solenoid round a torus. Cf POLOIDAL FIELD, TOKAMAK.

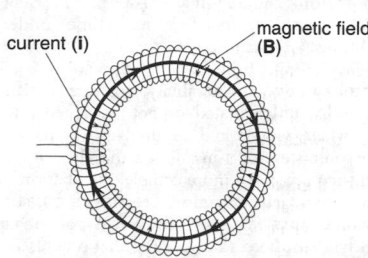

toroidal field

toroidal-intake guide vanes (*Aero*) The flared annular guide vanes which guide the air evenly into the intake of a centrifugal *impeller*.

toroidal surface (*Phys*) A lens surface in which the curvature in one plane differs from that in a plane at right angles.

toroidal winding (*ElecEng*) See RING WINDING (1).

torque (*Aero*) A measure of the total air forces on the propeller blades, expressed as a moment about its axis.

torque (*Phys*) The turning MOMENT exerted by a tangential force acting at a distance from the axis of rotation. It is measured by the product of the magnitude of the force and the distance. Units N m. Also *moment of force*.

torque amplifier (*ElecEng*) (1) Mechanical amplifier in which torque (not angle) is varied by differential friction of belts on drums. (2) Electrical servo system performing same function.

torque converter (*Eng*) A device which acts as an infinitely variable gear, but generally at varying efficiency, eg a centrifugal pump in circuit with an inward-flow turbine.

torque limiter (*Aero*) Any device which prevents a safe torque value from being exceeded, but specifically one which is used on a constant-speed TURBOPROP to prevent it from delivering excess power to its propeller.

torque link (*Aero*) A mechanical linkage, usually of simple scissor form, which prevents relative rotation of the telescopic members of an aircraft LANDING GEAR shock absorber.

torquemeter (*Aero*) A device for measuring the torque of a reciprocating aero-engine or turboprop, the indication of which is used by the pilot, together with rpm and other readings, to establish any required power rating.

torquemeter (*Eng*) A torsion meter attached to a rotating shaft, the angle of twist of a known length of shaft between the gauge points of the meter being indicated by optical or electrical means, thus enabling the power transmitted to be calculated. A form of transmission DYNAMOMETER.

torque motor (*ElecEng*) One exerting high torques at low speeds.

torque spanner (*Eng*) A spanner with a special attachment whereby a prescribed force can be applied to a bolt through turning the nut.

torr (*Phys*) Unit of low pressure equal to head of 1 mm of mercury or $133 \cdot 3$ N m^{-2}.

Torricellian vacuum (*EnvSci, Phys*) See MERCURY BAROMETER.

Torricelli's theorem (*Phys*) A theorem which states that a fluid flows from a hole in an open-topped container with a speed equal to the speed that any object would acquire were it to be dropped through a height equal to the head of water above the opening; stated by Italian physicist E Torricelli.

Torridonian (*Geol*) A succession of conglomerates and red sandstones and arkoses forming part of the Precambrian system in the NW highlands of Scotland. They rest on the Lewisian schists and gneisses.

torsion (*BioSci*) The preliminary twisting of the visceral hump in gastropod larvae which results in the transfer of the mantle cavity from the posterior to the anterior face, as distinct from the secondary or spiral twisting of the hump exemplified by the spiral form of the shell.

torsion (*Eng*) The state of strain set up in a body by twisting or applying a TORQUE. The external twisting effort is opposed by the shear stresses induced in the material.

torsion (*MathSci*) See CURVATURE.

torsional braid analysis (*Chem, Plastics*) A method of following the CURING reaction of THERMOSETS (panel) by coating an absorbent braid with the starting materials in liquid form, and making this the specimen in a TORSION PENDULUM. The oscillation characteristics of the system are affected by the degree of cure.

torsional wave (*Phys*) A wave of flexure in which the elements of the medium carrying the wave perform torsional oscillations about an axis parallel to the direction of propagation. Observed in solid structures such as long thin lamina, eg the roadway of a suspension bridge.

torsion balance (*Phys*) A delicate device for measuring small forces such as those due to gravitation, magnetism or electric charges. The force is caused to act at one end of a small horizontal rod, which is suspended at the end of a fine vertical fibre. The rod turns until the turning moment of the force is balanced by the torsional reaction of the twisted fibre, the deflection being measured by a lamp and scale using a small mirror fixed to the suspended rod.

torsion bar (*Eng*) A steel bar usually with splined ends acting as a spring resisting torque.

torsion bar suspension (*Autos*) A springing system, used in some independent suspension designs, in which straight bars, anchored at one end, are subjected to torsion by the weight of the car, thereby acting as springs.

torsion bush (*Eng*) Rubber spring comprising a cylinder of rubber enclosing a shaft and enclosed by a steel cover, for resisting torsion, tilt, etc. Commonly used for flexible couplings.

torsion galvanometer (*ElecEng*) A galvanometer in which the controlling torque is measured by the angle through which the suspension head must be rotated in order to bring the pointer back to zero.

torsion meter (*Eng*) A device for measuring TORQUE. See TORQUEMETER.

torsion pendulum (*Eng*) A DYNAMIC MECHANICAL TEST for polymeric materials, where the tape, ribbon or fibre is attached to an inertial mass at one end and a fixed support at the other. The mass is rotated and its amplitude of oscillation measured as a function of time. The ratio of successive amplitudes gives the logarithmic decrement and hence the LOSS FACTOR.

torticollis (*Med*) Wry neck, stiff neck. A disease of the cervical vertebrae or to affections (esp rheumatic) of the muscles of the neck. See SPASMODIC TORTICOLLIS.

Tortonian (*Geol*) A stage in the Miocene. See TERTIARY.

torulosis (*Med*) A disease due to infection with the yeast-like micro-organism *Torula histolytica*; affects esp the central nervous system.

torus (*BioSci*) (1) The thickened central part of the pit membrane of the bordered pits of many gymnosperm tracheids. It seems to act as a valve closing a pit to prevent the threatened spread of an embolism. (2) The receptacle, or part of the flower on which the carpels stand. (3) A ridge or fold, as in some Polychaeta, with rows of uncini (hooks).

torus (*ElecEng, MathSci*) See TOROID.

torus (*NucEng*) A surface of revolution shaped like the ring of an anchor or an American doughnut. It is generated by rotating a circle about a non-intersecting coplanar line as axis. The shape of the containment vessel in a TOKAMAK fusion reactor. Adj *toroidal*. Also *anchor ring*.

toss-bombing (*Aero*) A manoeuvre for the release of a bomb (usually with a nuclear warhead) which allows the pilot to evade the blast. A special computing sight enables the pilot to loop before reaching the target, when the bomb is lobbed forwards; or the pilot can overfly the target and loop to toss the bomb 'over-the-shoulder'.

tossing (*MinExt*) The operation of raising the grade or purity of a concentrate by violent stirring, followed by packing in a *kieve*. Also *kieving*.

tosyl (*Chem*) The toluene-4-sulphonyl group.

total absorption coefficient (*Phys*) See ATTENUATION COEFFICIENT.

total access communications system (*ICT*) The original UK analogue CELLULAR RADIO mobile-telephone system which opened in 1985, with two independent networks operated by Cellnet and Vodaphone. The system offers 600 two-way channels in the bands 890–905 MHz (mobile) and 935–950 MHz (base station). Channels are spaced 25 kHz apart and speech is conveyed by FREQUENCY MODULATION. Abbrev TACS.

total air–gas mixture (*Eng*) Air–gas mixture in which the proportion of air is the amount needed for perfect combustion.

total body burden (*Radiol*) (1) The summation of all radioactive materials contained in any person. (2) The maximum total amount of radioactive material any person may be permitted to contain.

total cross-section (*Phys*) The sum of the separate cross-sections for all processes by which an incident particle can be removed from a beam. If all the atoms of an absorber have the same total cross-section, then it is identical to the ATOMIC ABSORPTION COEFFICIENT.

total curvature (*MathSci*) See CURVATURE (3).

total differential (*MathSci*) See COMPLETE DIFFERENTIAL.

total digestible nutrients (*Agri*) Approximate estimate of the total energy value of feed based on the digestible fats, proteins, carbohydrates and fibres it contains.

total dissolved solids (*EnvSci*) The amount of solid material remaining after a sample of water has been evaporated to dryness, usually expressed as mg per l.

total drift (*Agri*) The volume or mass of sprayed, formulated agrochemical that was delivered outside of the defined target area.

total electron binding energy (*Electronics*) That required to remove all the electrons surrounding a nucleus to an infinite distance from it and from one another.

total emission (*Electronics*) See SATURATION CURRENT.

total equivalent brake horsepower (*Aero*) The brake horsepower at the propeller shaft plus the bhp equivalent of the residual jet thrust of a turboprop. Abbrevs *tehp* or *ehp*.

total head (*Aero*) In fluid flow, the algebraic sum of the DYNAMIC PRESSURE and the STATIC PRESSURE.

total impulse (*Aero*) That available from a self-contained rocket expressed as the product of the mean thrust, in newtons (N), and the firing time, in seconds (s), expressed as N s.

total internal reflection (*Phys*) Complete reflection of incident wave at boundary with medium in which it travels faster, under conditions where SNELL'S LAW of refraction cannot be satisfied. The angle of incidence at which this occurs (corresponding to an angle of refraction of 90°) is known as the *critical angle*. See GEM.

total losses (*ElecEng*) The power loss in an electric machine, equal to the difference between the input and the output powers.

totally enclosed motor (*ElecEng*) A motor with no provision for ventilation but not necessarily water- or gastight.

total herbicide (*Agri*) See NON-SELECTIVE HERBICIDE.

total normal curvature (*MathSci*) See CURVATURE (3).

total parenteral nutrition (*Med*) Giving the complete nutritional requirements by an intravenous line, usually centrally placed.

total response (*Acous*) See MEAN-SPHERICAL RESPONSE.

total soluble solids (*FoodSci*) The sum of the solids which are in solution. As it increases, water activity (a_w) is reduced and survival of micro-organisms becomes less likely. Abbrev TSS. See REFRACTOMETRIC SOLIDS, WATER ACTIVITY.

total temperature (*Aero*) See STAGNATION TEMPERATURE.

total viable count (*FoodSci*) The number of viable colonies of micro-organisms from a gram or millilitre sample of food or drink cultured in a non-specific nutrient medium at a specified temperature and time under aerobic conditions. Also *aerobic plate count*.

toti- (*Genrl*) Prefix from Lt *totus*, all, whole.

totipotency (*BioSci*) The ability, possessed by most living plant cells, differentiated or not, to regenerate a whole plant when isolated and cultured in a suitable medium. Embryonic stem cells in animals may be comparable but would require much more complex incubation *in utero* to regenerate a viable organism. See TISSUE CULTURE.

touch and close fastener (*Textiles*) A pair of tapes made from synthetic fibres that stick together on pressing by hand and then separate by peeling when required, eg Velcro. The tapes have pile surfaces, one formed of hooks and the other of loops, which engage on being pressed together.

touch screen (*ICT*) A visual display that accepts COMMANDS when the USER points at or lightly touches parts of the screen.

touchstone (*Min*) See LYDIAN STONE.

touch-tone (*ICT*) US for DUAL-TONE MULTIFREQUENCY signalling.

touchwood (*BioSci*) Wood much decayed as a result of fungal attack; it crumbles readily, and when dry is easily ignited by a spark.

toughened glass (*Glass*) See SAFETY GLASS.

toughness (*Eng*) Qualitatively used to denote a condition intermediate between BRITTLENESS and softness, as indicated in tensile tests by high ULTIMATE TENSILE STRESS and low to moderate ELONGATION and reduction in area, or by high values of energy absorbed by IMPACT TESTS (panel). More precisely it is the value of the critical STRAIN ENERGY RELEASE RATE, G_C, required for propagation in a material. In a fracture test on a cracked specimen of thickness t, it is evaluated from

$$G_C = \frac{1}{2t}\frac{\partial C}{\partial a}$$

where C is the COMPLIANCE of the specimen and a is the crack length. Formally G_C is twice the FRACTURE ENERGY of the material. Also *crack driving force*. Cf FRACTURE TOUGHNESS. See GRIFFITH EQUATION, STRENGTH MEASURES.

toughness (*MinExt*) The ability of mineral to withstand disruption, assessed empirically by comparison with standard minerals under controlled test conditions.

tough pitch (*Eng*) A term applied to copper in which the oxygen content has been correctly adjusted at about 0·03% by poling. Distinguished from 'overpoled' and 'underpoled' copper. See POLING.

tourmaline (*Min*) A complex silicate of sodium, boron and aluminium, with, in addition, magnesium (*dravite*), iron (*schorl*) or lithium (*elbaite*), and fluorine in small amounts, which crystallizes in the trigonal system. It is usually found in granites or gneisses. The variously coloured and transparent varieties are used as gemstones, under the names *achroite* (colourless), *indicolite* (blue), *rubellite* (pink). The common black variety is *schorl*.

tourmalinization (*Geol*) The process whereby minerals or rocks are replaced wholly or in part by tourmaline. See PNEUMATOLYSIS.

Tournaisian (*Geol*) The lowermost series of the Carboniferous of Europe. See PALAEOZOIC.

tourniquet (*Med*) Any instrument or appliance which, by means of a constricting band, exerts pressure on an artery to control bleeding, or, at a lesser pressure, to produce dilatation of a vein for PHLEBOTOMY.

tow (*Textiles*) A collection of a large number of parallel continuous filament fibres, not twisted together. By cutting or breaking (ie CONVERTING) a TOP is produced, the process being known as *tow-to-top conversion*.

tower (*ElecEng*) The lattice-type steel structure used to carry the several conductors of a transmission line at a considerable height above the ground. Also *pylon*.

tower (*Eng*) See PYLON.

tower bolt (*Build*) Large type of BARREL BOLT.

tower crane (*Eng*) A rotatable cantilever pivoted to the top of a steelwork tower, either fixed or carried on rails. The load is balanced by the lifting machinery carried on the opposite side of the pivot. Formerly used in shipbuilding as the *hammerhead crane* but now widely used for building construction where the tower is often extensible and attached periodically to the building with the lifting machinery within the tower and about half the maximum overturning moment counterbalanced by KENTLEDGE on the side remote from the load.

tower systems (*ICT*) A shape of the case for a PERSONAL COMPUTER in which the EXPANSION CARDS are stacked vertically rather than horizontally. The case is taller but occupies a smaller floor area.

Townend ring (*Aero*) A cowling for radial engines consisting of an aerofoil section ring, which ducts the air onto the engine cylinders and directs a streamlined flow onto the fuselage or nacelle, thus reducing drag; now obsolete.

town gas (*MinExt*) Usually a mixture of coal gas and carburetted water gas; the energy density is about 500 Btu ft^{-3} or 20 MJ m^{-3}, which is about half that of *natural gas* which in most areas has superseded town gas.

Townsend avalanche (*Phys*) Multiplication process whereby a single charged particle, accelerated by a strong field, causes, through collision, a considerable increase in ionized particles.

Townsend discharge (*Phys*) A TOWNSEND AVALANCHE initiated by an external ionizing agent.

toxaemia (*Med*) The condition of a patient caused by the absorption into the tissues and into the blood of toxins formed by micro-organisms at the site of infection. US *toxemia*.

toxicodynamics (*BioSci*) By analogy with *pharmacodynamics*, the interaction of potentially toxic substances with target sites, and the biochemical and physiological consequences that cause adverse effects.

toxicogenetics (*BioSci*) The study of the influence of genetic factors on the effects of potentially toxic substances on individual organisms.

toxicogenomics (*Pharmacol*) The study of how differences in the genome affect the response of organisms to stressors or toxins.

toxicokinetics (*BioSci*) The study of the uptake, metabolism (biotransformation), distribution and elimination of potentially toxic substance by the body. See PHARMACO-KINETICS.

toxicology (*Med*) The branch of medical science dealing with the nature and effects of poisons.

toxic shock syndrome (*Med*) Endotoxic shock caused by bacterial contamination of tampons; the toxin responsible is produced by strains of *Staphylococcus aureus*. Abbrev TSS.

toxigenicity (*BioSci*) The ability of a pathogenic organism to produce injurious substances that damage the host.

toxin (*BioSci*) A poisonous substance of biological origin.

toxoid (*BioSci*) Bacterial exotoxin that has been treated (usually with formaldehyde) so that it has lost its toxic properties but retains its ability to stimulate an immune response against the toxin.

toxoplasmosis (*Vet*) Caused by *Toxoplasma gondii* and found mainly in sheep and goats although humans and other species can be infected. Common cause of ovine abortion. The cat is the natural host for this parasite.

TP (*Surv*) Abbrev for *turning point*. See CHANGE POINT.

TPE (*Plastics*) Abbrev for THERMOPLASTIC ELASTOMER.

TPI (*Eng*) Abbrev for THREADS PER INCH, the inverse of the pitch of a screw-thread.

TPI (*Ships*) Abbrev for *tons per inch*: weight required to increase the MEAN DRAUGHT by 1 inch.

TPX (*Plastics*) TN for poly(4-methylpent-1-ene) (Japan).

trabecula (*BioSci*) A rod-like structure, eg of cell-wall material across the lumen of a cell, or of a cell or cells across some larger cavity.

trabecular bone (*BioSci*) See CANCELLOUS BONE.

TRACE (*Aero*) (1) Abbrev for *test-equipment for rapid automatic check-out and evaluation*. A computerized general purpose diagnostic testing rig for aircraft electrical and electronic systems. (2) Trace: visible record of instrument reading.

trace (*Electronics*) The image on a phosphor on electron-beam impact, forming the display.

trace (*ICT*) Means of checking the logic of a program by inserting statements that cause the values of the variables and other information to be printed out as the program is executed.

trace element (*BioSci, Med*) See MICRONUTRIENT.

trace element (*Geol*) A non-essential element (<1%) in a mineral.

trace element analysis (*Chem, Phys*) A type of elemental analysis which seeks to identify origin of materials by finding minor traces of elements accidentally present in samples (eg impurities). The principle can be extended to compound analysis using such hypersensitive methods as NUCLEAR MAGNETIC RESONANCE. See panel on TRACE ELEMENT ANALYSIS.

trace fossil (*Geol*) Any sedimentary structure caused by the activity of a fossil organism during its life, eg trails, burrows.

tracer atom (*Phys*) Labelled atom introduced into a system to study structure or progress of a process. Traced by variation in isotopic mass or as a source of weak radioactivity which can be detected, there being no change in the chemistry. See RADIOACTIVE ISOTOPE.

tracer chemistry (*Chem*) The use of isotopic compounds to study chemical and biological processes. Radioactive isotopes are often employed, but stable isotopes may also be used if their nuclear magnetic resonance spectra are different from those of the most abundant isotope. Studies using ^{13}C are often done in this way.

tracer compound (*Phys*) A compound in which a small proportion of the molecules are labelled with a radioactive isotope.

tracer element (*Phys*) One of the RADIOELEMENTS, eg radio-phosphorus, used for experiments in which its radioactive properties enable its location to be determined and followed. Tracer technique may be applied to physiological, biological, pathological and technological experiments. For some purposes stable isotopes, eg carbon-13 and heavy hydrogen (DEUTERIUM), are more conveniently used than radioisotopes.

tracers (*MinExt*) (1) Elements diffused as a HALO around an ore body, identified by GEOCHEMICAL PROSPECTING. These may be dilute concentrations of metals also present in the ore body, or elements found in association with ore bodies but not of primary interest themselves. (2) Minerals identified in soil and stream gravel which are common accessories to minerals of economic significance.

trace table (*ICT*) A tabular method of recording the expected values of variables as TEST DATA are processed by a program. See DRY RUN, TRACE.

Trace element analysis

The ready availability of neutrons and other atomic particles has led to the development of techniques in which very small samples are bombarded and their elemental composition measured from the energies of the gamma rays and X-rays emitted.

One method is called *neutron activation analysis* (NAA) in which the sample is first irradiated with thermal neutrons to convert stable nuclei into radioactive ones by neutron capture and beta decay. After irradiation the sample is placed in a germanium-crystal gamma detector and the energies and intensities of the gamma rays emitted carefully measured. The results can then be used to calculate how much of the original stable element was present.

The method is both very sensitive and essentially non-destructive. It is therefore very useful for very small samples of rare and valuable material such as paint from an old picture or glaze from porcelain, the results helping to determine the age and authenticity of the original. Sensitivity depends on the number of counts coming from the radioactive isotope and varies from about a millionth of a gram in the irradiated volume of the sample for zirconium, to as little as a million times less for manganese. Only about twelve elements, including the eight lightest, cannot be measured accurately in this way.

A second rather similar method is *particle-induced X-ray emission* (PIXE), but here protons and sometimes alpha particles or heavy ions are made to bombard a similar thin sample. Bombardment causes the innermost electrons in the K- or L-shell to be removed and the atom ionized. This is followed by outer electrons falling into the vacancies with the emission of X-rays, whose energies and amounts are measured. The sensitivity of the method decreases as Z, the mass number, increases; it therefore complements neutron activation analysis. The overall precision of the method is about the same as neutron activation analysis although some combinations of elements in the sample may give confusing results because of the overlap in energies between the X-rays produced.

---

trachea (*BioSci*) (1) An air tube of the respiratory system in certain Arthropoda, as ansects. (2) In air-breathing vertebrates, the windpipe leading from the glottis to the lungs.

tracheal gills (*BioSci*) In some aquatic insect larvae, filiform or lamellate respiratory outgrowths of the abdomen richly supplied with TRACHEAE and TRACHEOLES.

tracheal system (*BioSci*) In certain Arthropoda, as insects and myriapods, a system of respiratory tubules containing air and passing to all parts of the body.

tracheary elements (*BioSci*) Water-conducting cells of the xylem of a vascular plant. After the death and loss of their protoplasmic contents, the cell walls become thickened with lignin so that they can act as tubes for the conduction of water under tension. Cf HYDROID.

tracheid (*BioSci*) An elongated element with pointed ends, occurring in wood. It is derived from a single cell, which lengthens and develops thickened pitted walls, losing its living contents. Tracheids conduct water. Also *tracheide*.

tracheitis (*Med*) Inflammation of the mucous membrane of the trachea.

trachelate (*BioSci*) Neck-like, from Gk *trachelos*, neck.

trachelorrhaphy (*Med*) The surgical repair of lacerations of the cervix uteri.

tracheobronchitis (*Med*) Inflammation of the mucous membrane of both the trachea and the bronchi.

tracheobronchitis (*Vet*) An inflammation of the trachea and bronchi of dogs due to infection by the nematode worm *Oslerus osleri*.

tracheocele (*Med*) An air-containing swelling in the neck due to the bulging of the wall of the trachea between the cartilages of the trachea.

tracheole (*BioSci*) The ultimate branches of the tracheal system.

tracheophyte (*BioSci*) A vascular plant (ie a pteridophyte or seed plant). Sometimes made a division, Tracheophyta.

tracheotomy (*Med*) The operation of cutting into the trachea, usually for the relief of respiratory obstruction.

trachoma (*Med*) A highly contagious infection of the conjunctiva covering the eyelids, characterized by the presence of small elevations on the inner side of the lid and leading often to blindness. The disease is hyperendemic in rural communities of the Middle East, Asia, Africa and C and S America. The infecting organism is *Chlamydia trachomatis*, transmitted by house flies and by poor hygiene.

trachyandesite (*Geol*) Fine-grained igneous rock, commonly occurring as lava flows, intermediate in composition between trachyte and andesite; ie containing both orthoclase and plagioclase in approximately equal amounts.

trachybasalt (*Geol*) A fine-grained igneous rock commonly occurring in lava flows and sharing the mineralogical characters of trachyte and basalt. The rock contains sanidine (characteristic of trachyte) and calcic plagioclase (characteristic of basalt).

trachyte (*Geol*) A fine-grained igneous rock type, of intermediate composition, in most specimens with little or no quartz, consisting largely of alkali feldspars (sanidine or oligoclase) together with a small amount of coloured silicates such as diopside, hornblende or mica. See VOLCANIC ROCKS.

tracing (*Eng*) An engineering drawing transferred to transparent tracing paper or cloth, in Indian ink for permanence and for making good dye-line prints. A tracer's work carries no design responsibility. Cf DRAUGHTSPERSON.

track (*Acous*) (1) The groove which is cut on a blank during vinyl disk recording. (2) Single circumferential area on a magnetic drum, or longitudinal area on a magnetic tape, alongside other tracks, allocated to specific recording and reproduction channel. (3) Space on a disk or sound film allocated to one channel of sound recording and hence reproduction.

track (*Aero*) (1) The distance between the outer points of contact of port and starboard main wheels. (2) The distance between the vertical centre lines of port and starboard undercarriages where the wheels are paired. (3) More generally, the projection of a flight path upon the Earth's surface.

track (*Electronics*) A conducting path in a circuit, forming interconnections.

track (*ICT*) The path on a DISK, DRUM or TAPE along which data are stored. See fig. at HARD DISK.

track (*Phys*) The path followed by particle, esp when rendered visible in photographic emulsion by cloud chamber, bubble chamber or spark chamber.

track ball (*ICT*) A device used to control the position of a CURSOR or POINTER on the screen by moving a captive ball with the palm of the hand. Also *tracker ball*.

track-circuit signalling (*ElecEng*) An electric signalling system making use of the change in resistance of a track circuit when a train passes over a section of railroad track and thus completes a circuit between the rails.

tracker ball (*ICT*) See TRACK BALL.

tracking (*ElecEng*) Surface breakdown on an electrical insulator, eg due to moisture or other impurities, often leading to waxing and carbon formation.

tracking (*Radar*) Automatic holding of radar beam onto target through operation of return signals.

tracking (*Space*) The continuous process of following, from a distance, an object to determine its position in space.

tracking control (*ImageTech*) Used to adjust the tape servo during playback for correct head to video track alignment with tapes recorded on another machine. See AUTOMATIC TRACKING, DIGITAL TRACKING.

tracking resistance (*ElecEng*) The ability of an insulator to withstand surface discharge of electricity, esp over long times. Track formation often occurs at a critical voltage and consists of paths of carbonized material formed by degradation in the case of polymers. Standard tests include measurement of the COMPARATIVE TRACKING RESISTANCE (CTR).

tracking shot (*ImageTech*) A shot in which the camera, and its operators, are moved to follow the chosen part of the action.

track rail bond (*ElecEng*) A rail bond for preserving the electrical continuity of the track rails when these are used for carrying traction or other currents.

track relay (*ElecEng*) A relay used in track-circuit signalling for controlling the electrically operated signals.

track rod (*Autos*) A transverse link which, through ball joints, connects arms carried by the stub axles, in order to convey angular motion from the directly steered axle to the other.

track-sectioning cabin (*ElecEng*) A cabin housing switchgear by means of which the supply to different sections of an electrified railway line may be disconnected.

track switch (*ElecEng*) A switch controlling the supply of current to a section of an electrified railway line.

track-while-scan (*Radar*) Electronic process for detecting a target, computing its velocity and predicting its future position without interfering with the process of continuous scanning.

tract (*BioSci*) (1) The extent of an organ or system, as the *alimentary tract*. (2) An area or expanse, as the *ciliated tracts* of some Ctenophora. (3) A band of nerve fibres, as the *optic tract*.

traction (*Med*) Treatment involving tension on affected parts (eg fractures) by means of suitable applied weights or otherwise.

traction engine (*Eng*) A road locomotive in which large road wheels are gear-driven from a simple or compound steam engine mounted on top of the boiler, a rope drum being provided for additional haulage purposes.

traction generator (*ElecEng*) A dc generator used solely for supplying power to an electric traction system.

traction lamp (*ElecEng*) An electric lamp having a specially robust filament to withstand vibration: used on trains or road vehicles.

traction load (*ElecEng*) That part of the load carried by a dc generating station which is formed by the traction system which it supplies.

traction load (*Geol*) That part of the load of solid material carried by a river which is rolled along the bed.

traction motor (*ElecEng*) An electric motor specially designed for traction service.

traction rope (*CivEng*) The endless rope in an aerial ropeway system which moves the carriers transporting the loads. Also *hauling rope*.

tractive effort (*ElecEng*) The pull necessary to detach the armature from an excited electromagnet. Also *tractive force*.

tractive effort (*Eng*) Of a locomotive, the pull which the engine is capable of exerting at the draw-bar, the limiting value of which is given by the product of the weight on the coupled wheels and the coefficient of friction between wheels and rails. Also *tractive force*.

tractor (*Aero*) A propeller which is in front of the engine and the structure of the aircraft, as contrasted with a *pusher*, which is behind the engine and pushes the aircraft forwards. See PROPELLER.

tractor feed (*ICT*) Mechanism for advancing paper by use of perforations and a toothed wheel or sprocket.

tractrix (*MathSci*) The involute of a catenary.

tractrix horn (*Acous*) A horn which is so shaped that the area at a distance from the throat is dependent on the tractrix curve, as contrasted with the exponential horn.

trade effluent (*Build*) The liquid discharge, other than soil or waste, from a manufacturing process.

trade-off study (*Aero, Space*) A logical evaluation during the preliminary design process of the pros and cons of alternative concepts or approaches and/or parameters which leads to the choice of the preferred ones; typical criteria for the analysis are performance, schedule, risk and cost.

trade winds (*EnvSci*) Persistent winds blowing from the NE in the northern hemisphere and from the SE in the southern hemisphere between the horse latitudes (calm belts at 30°N and S of the equator) towards the DOLDRUMS. See panel on TROPOSPHERE.

traffic engineering (*ICT*) A term used to cover all areas of traffic, capacity and performance analysis, including traffic modelling, grade of service prediction, DIMENSIONING methods and traffic measurements.

traffic flow (*ICT*) The number of calls that an exchange, or a set of switches, is carrying at any instant.

traffic meter (*ICT*) A meter that, inserted at any part of an automatic telephone exchange, totals the number of calls passing through.

traffic mix (*ICT*) On an integrated digital network, the variety of different traffic types that may occur together at different times, eg homogeneous; various combinations of BURSTINESS and PEAK CELL RATE; mixtures of constant and variable bit-rate sources.

traffic unit (*ICT*) The measure of the occupancy of telephonic apparatus during conversation. One traffic unit equals the use of one circuit for 1 minute or for 1 hour. Abbrev *TU*. See CONGESTION TRAFFIC-UNIT METER.

tragacanth (*FoodSci*) A gum derived from the plant genus *Astragalus*. It swells slowly in cold water to produce a viscous sol with maximum viscosity at pH 5. A STABILIZER and THICKENER. Gum tragacanth is used to stabilize sauces, salad dressings and similar products.

tragus (*BioSci*) In the ear of some mammals, including the Microchiroptera (bats), an inner lobe to the pinna.

trail (*Autos*) The distance by which the point of contact of a steered wheel with the ground lies behind the intersection of the swivel-pin axis and the ground. See CASTER ACTION.

trailer (*ImageTech*) (1) A short film or video production advertising a forthcoming presentation. (2) A protective and identification section at the end of a reel of film (tail leader).

trailing action (*Autos*) See CASTER ACTION.

trailing axle (*Eng*) In a locomotive, the rearmost axle.

trailing edge (*Aero*) The rear edge of an aerofoil, or of a strut, wire, etc.

trailing edge (*ElecEng*) See LEAVING EDGE.

trailing edge (*ICT*) The falling portion of a pulse signal. See LEADING EDGE.

trailing flap (*Aero*) A FLAP which is mounted below and behind the wing trailing edge so that it normally trails at neutral incidence and is rotated to various positive angles of incidence to increase lift, there always being a gap between the wing undersurface and the flap leading edge.

trailing points (*CivEng*) See POINTS.

trailing pole horn (*ElecEng*) The edge of a field pole which is passed last by an armature conductor, irrespective of the direction of rotation of the armature. Also *trailing pole tip*.

trailing springs (*Eng*) In a locomotive, the springs belonging to the rearmost axle.

trailing vortex (*Aero*) The vortex passing from the tips of the main surfaces of an aircraft and extending downstream and behind it.

trailing wheels (*Eng*) In a locomotive, the wheels of the rearmost axle.

train (*Eng*) Similar or identical parts in a machine, arranged in series, such as a simple or compound train of gears.

train brake (*Eng*) See AIR BRAKE, VACUUM BRAKE.

train control (*ElecEng*) The way in which the train driver's control operations alter the electrical or fuel supply to the traction motors.

train describer (*ElecEng*) An automatic or semi-automatic device for giving information regarding the destination of trains. Also *destination indicator*.

training (*Psych*) (1) Generally, the application of learning principles to improve social or technical skills through some systematic activity. (2) *Operant conditioning*, the procedure of conditioning an animal through the use of reinforcements in order to establish a desired behaviour.

training check frame (*ICT*) A sequence of bits transmitted by a digital fax machine at the start of a call to assess the quality of the channel. Data are sent initially at 9600 bps, but if the frame is incorrectly received at this rate the communicating machines successively fall back to rates of 7200, 4800 and 2400 bps.

training works (*Genrl*) Works undertaken to remedy instability and eccentricity of flow in channels. See DYKE, GROUND SILLS, GROYNES, LEVEE.

trait (*Psych*) A stable and enduring attribute of a person or animal which varies from one individual to another; traits may be physical (eye colour) or psychological (spatial intelligence) and are often used in the study of individual differences in personality.

trajectory (*EnvSci*) The actual path along which a small quantity, or parcel, of air travels during a definite time interval.

trajectory (*Genrl*) The path of a moving object, esp a projectile.

trajectory (*Space*) The path of a rocket or space vehicle. Implies a path which is limited in length, eg a *trajectory* to the Moon, whereas *orbit* usually refers to a path which is closed and repetitive.

tram (*MinExt*) A small wagon, tub, cocoa-pan, corve, corf or hutch, for carrying mineral.

trammels (*Eng*) See BEAM COMPASSES.

tramp alloys (*Eng*) See TRAMP ELEMENT.

tramp element (*Eng*) (1) Noble metals (eg copper) present in steel scrap which are difficult to remove by refining. (2) General term for any elements present in scrap material which are difficult to remove before the material can be recycled. By extension, *tramp alloys, tramp compounds, tramp metals*. (3) Metal contaminating a structure, eg TRAMP URANIUM on nuclear reactor pipework. (4) Stray metal pieces which are accidentally entrained in food or other processed materials or products, eg a piece of steel broken from a machine in a load of rock from a mine.

tramp iron (*MinExt*) See TRAMP ELEMENT.

tramp metal (*Eng*) See TRAMP ELEMENT.

tramp uranium (*NucEng*) Uranium dissolved from exposed fuel and plated out onto the structure in a liquid-cooled nuclear reactor.

trance (*Psych*) A state of disassociation which occurs under hypnosis, and in various conditions such as sleepwalking, in which the individual's will is suspended and he or she acts on wishes or fantasies that are otherwise kept under control.

trans- (*Chem*) That geometrical isomer in which like groups are on opposite sides of the bond with restricted rotation, or in which like ligands are on opposite sides of the central atom of a co-ordination compound.

transaction (*ICT*) A single event involving the recording and processing of data.

transactional analysis (*Psych*) A form of psychotherapy carried out in a group setting or between therapist and client, aimed at adjusting behaviour towards a mature and realistic attitude with improved interpersonal relations.

transaction file (*ICT*) Records used in batch processing to update a MASTER FILE. Also *change file, update file*. See GRANDFATHER FILE, FATHER FILE, SON FILE.

transaction processing (*ICT*) Use of an on-line computer system to interrogate or update files as requested rather than batching such requests together for subsequent processing.

transactivation (*BioSci*) Stimulation of transcription by a transcription factor binding to DNA and activating adjacent proteins.

transadmittance (*Phys*) Output ac current divided by input ac voltage for electronic device when other electrode potentials are constant.

transaminase (*BioSci*) An enzyme that catalyses the transfer of amino groups from amino acids to keto acids, thus converting a keto acid into an amino acid, eg the conversion of α-ketoglutarate to glutamate, *transamination*.

transatmospheric vehicle (*Aero*) Aircraft capable of normal wingborne flight through the atmosphere and also of travelling into space orbits. See AEROSPACEPLANE, HOTOL.

transaxial tomography (*Radiol*) A process whereby serial radiographs are taken transverse to the vertical axis of the body. See COMPUTER-AIDED TOMOGRAPHY.

transceiver (*ICT*) Equipment in which circuitry is common to transmission or reception. Also *transreceiver*.

transcendental function (*MathSci*) Any non-ALGEBRAIC FUNCTION, eg trigonometric, exponential, logarithmic, Bessel or gamma functions.

transcendental number (*MathSci*) Any real number which cannot be obtained as a solution of some polynomial equation with rational coefficients, eg π or e. Cf ALGEBRAIC NUMBER.

transcode (*ICT*) To convert (data) from one digital format to another.

transconductance (*Electronics*) Applied to valves, or to field-effect transistors, it is the change in anode or drain current divided by the increment of grid or gate voltage which initiated the change. Also *mutual conductance*.

transcribing genes (*BioSci*) Genes which are being actively transcribed into RNA. See CODING SEQUENCES.

transcription (*BioSci*) The process by which an RNA polymerase produces single-stranded RNA complementary to one strand of the DNA or, rarely, RNA.

transcription (*ICT*) The recording of a broadcast performance for subsequent rebroadcast or other use.

transcriptionally active chromatin (*BioSci*) CHROMATIN which is being transcribed into RNA.

transcription complex (*BioSci*) Functional association of DNA, nascent RNA, protein and ribonucleoprotein actively transcribing and processing RNA.

transcription factor (*BioSci*) A protein required for recognition by RNA polymerases of specific stimulatory sequences in eukaryotic genes. A wide range of different transcription factors are known and the cell-specific expression or activation of these determines the selective read-out of genes and thus the activities of the differentiated cell.

transcrystalline failure (*Eng*) The normal type of failure observed in metals. The line of fracture passes through the crystals, and not round the boundaries as in INTERCRYSTALLINE FAILURE.

transcurrent fault (*Geol*) A strike–slip fault.

transdermal (*Med*) Absorbed or injected through the skin. See PATCH.

transducer (*ElecEng*) A device which converts a physical quantity into an electrical signal, either proportionally or according to a specified formula. Examples include accelerometers, microphones and photocells. The term also covers devices which convert electrical signals into some physical phenomenon, eg a loudspeaker.

transducer translating device (*ElecEng*) In a servo system, the device which measures error in the controlled output and converts it to an electrical signal which can be fed back to correct the error.

transduction (*BioSci*) (1) During phage infection and consequent bacterial lysis, the integration of segments of host DNA into that of the phage. It can then be transferred to another host. (2) The conversion of a signal from one form to another by a cell, eg the conversion of light into nerve impulses by sensory cells; SIGNAL TRANSDUCTION.

transductor (*ElecEng*) An arrangement of windings on a laminated core, which, when excited, permits current amplification. Part of a MAGNETIC AMPLIFIER.

transect (*BioSci*) A line or belt of vegetation marked off for study.

transept (*Arch*) Part of a church at right angles to the NAVE, or of another building to the body: either wing of such a part where it runs right across.

trans fatty acid (*FoodSci*) A stereoisomeric form of a monounsaturated fatty acid. Trans fatty acids are thought to cause an increase in blood cholesterol.

transfection (*BioSci*) The alteration of the host genome after infection by phage.

transfer (*ICT*) See PRINT-THROUGH.

transfer (*Print*) An impression taken of a non-lithographic surface (intaglio, letterpress), on one of several specially coated papers, for transferring to a lithographic surface; a transfer from an existing lithographic surface is called a *retransfer*.

transfer admittance (*Phys*) The ratio of the current driving at one node in a network to the resulting voltage at another node, all other sources being set to zero.

transferase (*BioSci*) An enzyme that catalyses the transfer of chemical groups between compounds, eg glycosyl transferase transfers a sugar residue onto the growing oligosaccharide complex of a glycoprotein.

transfer cell (*BioSci*) A parenchymatous cell with elaborate ingrowths of the cell wall, greatly increasing the area of the plasmalemma. Such cells occur where there is substantial movement of solutes between SYMPLAST and APOPLAST in scattered families of bryophytes and vascular plants.

transfer characteristic (*ImageTech*) The relationship between TV camera tube illumination and corresponding signal current.

transfer ellipse (*Space*) The trajectory (part of an ellipse) by which a space vehicle may transfer from an orbit about one body (eg the Earth) into an orbit about another (eg the Sun).

transference (*Psych*) In psychoanalytic theory, the tendency of the client to displace onto the analyst feelings and ideas derived from previous experience with other figures (eg one's parents).

transfer factor (*Med*) A measure of the lungs' ability to take up oxygen which is depressed by a variety of lung diseases.

transfer function (*ICT*) (1) The complex ratio of the output (current or voltage) of a circuit or device to its input. This gives the phase and frequency response. (2) A mathematical expression relating the output of a closed-loop system to its input.

transfer impedance (*Phys*) The ratio of the driving voltage in one mesh of a network to the resulting current in another mesh, all other sources being set to zero.

transfer instrument (*ElecEng*) An instrument which gives dc indication independent of frequency, including zero frequency, so that when calibrated for dc it can be used for calibrating ac instruments, as with electrostatic wattmeters.

transfer lettering systems (*Print*) Alphabets, symbols, rules and tints which can be stripped into position on layouts or film for photoreproduction, eg Artype, Chart-Pak, Craf.Type, Letraset, Letter-on, Prestype, Zip-a-Tone.

transfer line (*Eng*) A long series of machines operating on a succession of similar parts, eg car cylinder blocks, automatically or semi-automatically.

transfer machine (*Eng*) A machine in which a workpiece or an assembly passes automatically through a number of stations, at each of which it undergoes one or more production processes.

transfer moulding (*Plastics*) The development of compression moulding, where powder polymer is placed in a small chamber adjacent to the tool mould and, when heated, is forced into the tool cavity. Not to be confused with INJECTION MOULDING.

transfer of training (*Psych*) The facilitation (positive) or hindrance (negative) of performance on a learning or training task as a result of previous activity. The positive and negative transfer effects appear to be a function of the similarity of the tasks.

transfer orbit (*Space*) The path for a spacecraft to take from one planet to another which requires least fuel and energy; follows elliptical path tangent to orbits of both the departure and the target planet. Also *Hohmann orbit*.

transfer port (*NucEng*) The aperture through which items are inserted into or removed from a dry box, glove box or shielded box (eg by sealing into plastic sac attached to the rim of the port).

transfer printing (*Textiles*) A process for transferring a design from paper to fabric. Usually the paper is coloured with disperse dyes which sublime onto fabrics made from synthetic fibres when the paper and fabric are pressed together in a heated press or calender.

transfer process (*ImageTech*) Any means whereby an image, dyed or pigmented, is transferred to a new emulsion.

transferrin (*BioSci*) A mammalian serum β-globulin that binds and transports ferric ions; it is an important constituent of cell culture media.

transfer RNA (*BioSci*) An RNA molecule about 80 nucleotides long, with complementary sequences which result in several short hairpin-like structures. The loop at the end of one of these carries the anticodon triplet, which binds to the codon of the mRNA. The corresponding amino acid is bound to the $3'$ end of the molecule. Abbrev *tRNA*. See ADAPTOR HYPOTHESIS.

transfinite numbers (*MathSci*) A system of CARDINAL and ORDINAL numbers, invented by G Cantor (1845–1918), to describe infinite sets. In effect Cantor classifies different types of INFINITY: eg the infinite associated with the set of positive integers is different from that associated with the set of all real numbers.

transfixion (*Med*) A cutting through, as in amputation.

transform (*MathSci*) A process or rule for deriving from a given mathematical entity (eg point, line, function, etc) a corresponding entity. The word 'transform' is also sometimes used in respect of the corresponding entity itself (eg Fourier and Laplace transform). Certain problems (eg differential equations) can be transformed so as to obtain a simpler problem whose solution can then be transformed back to give the solution of the original problem. Also *transformation*. See AFFINE TRANSFORMATION, CONJUGATE ELEMENTS OF A GROUP, FOURIER TRANSFORM, ISOGONAL TRANSFORMATION, LAPLACE TRANSFORM, LINEAR TRANSFORMATION.

transformation (*BioSci*) (1) The alteration of the bacterial or eukaryotic cell genotype following the uptake of purified DNA. (2) The alteration of cells in tissue culture by various agencies so that they behave in many ways like cancer cells, eg their lack of growth control and the ability to divide indefinitely. See CONTACT INHIBITION, GROWTH IN SOFT AGAR.

transformation (*Chem, Eng*) A constitutional change in a solid material, eg the change from gamma to alpha iron, the formation of pearlite, bainite or martensite from austenite in steels, or transformation from alpha to beta quartz in a silica refractory.

transformation (*MathSci*) See TRANSFORM.

transformation (*Phys*) See ATOMIC TRANSMUTATION.

transformation constant (*Phys*) See DISINTEGRATION CONSTANT.

transformation points (*Glass*) In the measurement of thermal expansion coefficient, the two temperatures (M_g point and T_g point) at which the slope of the graph of expansion against temperature changes fairly sharply.

transformation ratio (*ElecEng*) See TURNS RATIO.

transformation temperature (*Eng*) The temperature at which phase changes occur during the heating of iron and steels. Denoted by various symbols: A_c, the temperature on heating; A_e, the temperature at equilibrium; A_r, the temperature on cooling. Subscripts refer to the transformation in question.

transformation theory (*Eng*) A theory in which the factor relating to yield is assumed to be the shear–strain energy stored in unit volume.

transformer (*ElecEng*) An electrical device without moving parts, which transfers energy of an ac in the primary winding to that in one or more secondary windings, through electromagnetic induction. Except in the case of the AUTOTRANSFORMER there is no electrical connection between the two windings and, except for the isolating transformer, the voltage is changed.

transformer (*Phys*) A mechanical device involving lever arms of different lengths, used to vary MECHANICAL ADVANTAGE.

transformer booster (*ElecEng*) A transformer connected with its secondary in series with the line, so that its voltage is added to that of the circuit; used to compensate the voltage drop in a feeder or distributor.

transformer core (*ElecEng*) The structure, usually of laminated iron or FERRITE, forming the magnetic circuit between windings of a transformer, of any degree of coupling. Also *mutual coupling*.

transformer coupling (*ElecEng*) Transference, in either direction, of electrical energy from one circuit to another by a transformer, of any degree of coupling. Also *mutual coupling*.

transformer oil (*ElecEng*) A mineral oil of high dielectric strength, forming the cooling and insulating medium of electric power transformers.

transformer plate (*ElecEng*) Sheet iron of low magnetic loss, for transformer core laminations.

transformer ratio (*ElecEng*) See TURNS RATIO.

transformer-ratio bridge (*ElecEng*) An ac bridge similar to a Wheatstone bridge but with two transformer windings used for the two ratio arms.

transformer stampings (*ElecEng*) The laminations, stamped out of transformer plate, which are assembled to form the TRANSFORMER CORE.

transformer tank (*ElecEng*) The steel tank encasing the core and windings of a transformer and holding the transformer oil.

transformer tapping (*ElecEng*) A means of varying the voltage ratio of a transformer by making a connection to a point on one winding intermediate between the ends.

transformer tube (*ElecEng*) One of a number of steel tubes on the outside of a transformer tank to provide a vertical path of circulation for the transformer oil.

transformer winding (*ElecEng*) The electrically active part of a transformer, which surrounds the magnetically active transformer core.

transform fault (*Geol*) A strike–slip fault along which two plates slide past each other, eg the San Andreas fault.

transforming growth factor (*BioSci*) Either of two proteins secreted by transformed cells that stimulate growth, and some other characteristics of the transformed phenotype, in normal cells grown in culture. Abbrev *TGF*.

transforming station (*ElecEng*) A point on an electricity supply system where a change of supply voltage occurs.

transfusion (*Med*) Blood transfusion. The operation of transferring the blood (or any required constituent of it) of one person into the veins of another, either to make good loss or counteract deficiency.

transfusion reaction (*BioSci*) The disturbance following transfusion of blood, due to antibodies in the recipient reactive with donor blood cells or, more rarely, to antibodies in the transfused blood reactive with the recipient's blood cells. See ABO BLOOD GROUP SYSTEM.

transfusion tissue (*BioSci*) A tissue of short tracheids and parenchyma cells surrounding or associated with the vascular bundle(s) in the leaves of many gymnosperms, presumably functioning in the distribution of water and collection of photosynthate.

transgenic (*BioSci*) Used to describe animals, eg *transgenic mice*, that are derived from embryos into which isolated genomic DNA from another species has been introduced at an early stage of development. Such foreign genes may be incorporated into the nucleus and chromosomes so that the animal can express the foreign gene product. Also applied to plants. See PLANT GENETIC MANIPULATION.

transgression (*Geol*) The gradual submergence of land caused by a relative rise in sea level.

transient (*Acous*) A sound of short period and irregular non-repeating waveform, which implies a continuous spectrum of sound-energy contributions.

transient (*ElecEng*) A short surge of voltage or current. The voltage or current before steady-state conditions have become established.

transient (*ICT*) Any non-cyclic change in a part of a communication system. The most general transient is the step, while the steady state is represented by any number of sinusoidal variations. See HEAVISIDE UNIT FUNCTION.

transient analyser (*ICT*) Test instrument that generates repeated transients and displays their waveform (which is usually adjustable) at different points in the system under investigation, on a CRT screen.

transient distortion (*ICT*) Distortion arising only when there is a rapid fluctuation in frequency and/or amplitude of the stimulus.

transient effects (*Phys*) The damped vibrations at the natural frequency of the system which initially occur when a system is set into FORCED VIBRATIONS. The damping causes these to die away, as the system settles down to vibrate at constant amplitude at the driving frequency. Transient effects are important in music for they help to give a musical instrument its characteristic sound.

transient equilibrium (*Phys*) Radioactive equilibrium between daughter product(s) and parent element for which activity is decaying at an appreciable rate. Characterized by ratios of activity, but not magnitudes, being constant.

transient flow permeability (*PowderTech*) A permeametry technique in which sample of the powder under test, contained in a long vertical column, is evacuated. Then gas at a known pressure is introduced at the foot of the column. The time required for the gas to diffuse through the column is used to calculate the average pore diameter of the powder bed and hence the surface area.

transient overpower (*NucEng*) Running a reactor accidentally at a power greater than the design level but without loss of coolant.

transient reactance (*ElecEng*) The reactance of the armature winding of a synchronous machine which is caused by the leakage flux. Cf SYNCHRONOUS REACTANCE.

transient stability (*ElecEng*) That of a power system under transient current conditions.

transient state (*ICT*) Transition period and associated phenomena between steady states in the repetition of a waveform.

transillumination (*Med*) The passing of a strong light through the walls of a cavity so that its outlines become visible and abnormalities detected.

transistor (*Electronics*) Three-electrode semiconductor device with thin layer of n- (or p-)type semiconductor sandwiched between two regions of p- (or n-)type, thus forming two p–n junctions back to back. The emitter junction is given a forward bias and the collector junction a reverse bias. Owing to the low forward resistance of the emitter junction and the high reverse resistance of the collector junction considerable power gain is possible for signals in the emitter or base leads. The latter arrangement also gives current gain. Amplification in the p–n–p transistor is due to hole conduction, that in an n–p–n transistor to electron conduction. See BIPOLAR TRANSISTOR, FIELD-EFFECT TRANSISTOR.

transistor (*ICT*) See SECOND-GENERATION COMPUTER.

transistor amplifier (*Electronics*) One which uses transistors as the source of current amplification. Depending on impedance considerations, there are three types, with base, emitter or collector grounded.

transistor characteristics (*Electronics*) General name for graphs relating electrode currents and/or voltages for transistors connected in various configurations: eg collector current plotted against collector–emitter voltage, base current being kept at a fixed value; collector current against base current, with collector–emitter voltage fixed; or h_{fe} against collector current. See TRANSISTOR PARAMETERS.

transistor construction (*Electronics*) See ALLOY JUNCTION, EPITAXIAL TRANSISTOR, MASK, MESA TRANSISTOR, METAL–OXIDE–SILICON, PLANAR TRANSISTOR.

transistor current gain (*Electronics*) The slope of the output current against input current characteristic for constant output voltage. In a common-base circuit it is inherently a little less than unity but in a common-emitter circuit it may be relatively large. The current gain is the hybrid parameter h_{21}.

transistor equivalent circuit (*Electronics*) For the purpose of circuit analysis, a transistor may be represented by a four-terminal network having two common terminals. In this way the transistor characteristics may be expressed in terms of four independent variables: the input and output voltages and currents of the equivalent circuit. Such possible circuits are called the common base, the common emitter and the common collector.

transistor parameters (*Electronics*) In circuit analysis the performance of transistors is calculated from parameters obtained from the slope of the various characteristic curves. Many such sets of parameters have been used, the most widely adopted probably being the hybrid parameters, h_{11}, h_{12}, h_{21} and h_{22} (also *h-parameters*, *small-signal parameters*). These are given by the slopes of the input, feedback, transfer and output characteristics respectively, at the selected working point. The symbols h_{fe} and h_{fb} may also be used.

transistor power-pack (*Electronics*) One in which high-tension supply of low power is obtained by rectifying transformed high-voltage current from a transistor oscillator fed at low voltage.

transit (*Astron*) (1) The apparent passage of a heavenly body across the meridian of a place, due to the Earth's diurnal rotation. See CULMINATION. (2) The passage of a smaller body across the disk of a larger body as seen by an observer on the Earth, eg of Venus or Mercury across the Sun's disk, or of a satellite across the disk of its parent planet.

transit (*Surv*) Rotation of the telescope of a theodolite about its trunnion axis, so that the positions of the ends of the telescope are reversed. See CHANGE FACE, TRANSIT THEODOLITE.

transit angle (*Electronics*) The product of delay or transit time and angular frequency of operation. In a velocity-modulated valve the transit time corresponds to the time taken for an electron to pass through a drift space.

transit circle (*Astron*) See MERIDIAN CIRCLE.

transition (*Aero*) In VTOL aircraft flight, the action of changing to or from the vertical lift mode (jet, fan or rotor) to forward flight with wing lift.

transition (*Phys*) In atomic and nuclear physics, the change from one quantum state to another. A transition from a higher to a lower energy state may be accompanied by the emission of a PHOTON, while a transition from a lower to a higher state requires the absorption of a photon. Transitions are governed by SELECTION RULES which make certain transitions highly improbable.

transitional epithelium (*BioSci*) A stratified epithelium consisting of only three or four layers of cells. Able to stretch, it lines eg the ureters, the bladder, and the pelvis of the kidney in vertebrates.

transitional object (*Psych*) Winnicott's concept of objects (eg a doll or piece of cloth) which act as comforters during the child's initial development from total dependence to self-reliance.

transition curve (*Surv*) A curve of special form connecting a straight and a circular arc on a railway or road. Designed to eliminate sudden change of curvature between the two, and to allow of superelevation being applied gradually to the outer rail or outer part of the curve. Also *easement curve*.

transition element (*Phys*) See TRANSITION METAL.

transition energy (*Electronics*) That at which phase focusing changes to defocusing in a synchrotron accelerator. This necessitates a sharp arbitrary change of phase in the radio-frequency field.

transition fit (*Eng*) A class of fit intermediate between a clearance fit and an interference fit.

transition frequency (*Electronics*) The term used for the GAIN–BANDWIDTH PRODUCT for a transistor used in the common-emitter configuration.

transition metal (*Phys*) Any one of a large group of elements in which the filling of the outermost electron shell to 8 electrons is interrupted to bring the penultimate shell (which can be used in bonding) from 8 to 18 or 32 electrons. This has profound effects upon the properties, and many metals of geological importance occur as transition elements. Also *transition element*.

transition point (*Aero*) The point where the flow in a BOUNDARY LAYER changes abruptly from *laminar* to *turbulent*.

transition point (*Chem*) The temperature at which one crystalline form of a substance is converted into another solid modification, ie that at which they can both exist in equilibrium.

transition probability (*Phys*) In atomic and nuclear physics, the probability per unit time that a system in quantum state k will undergo a transition to quantum state l. For an atom in an excited state there will be certain probability that the atom will spontaneously undergo a transition to a lower energy state. In the presence of photons of energy equal to the difference in energy between the states, there will be in addition a stimulated transition probability.

transition region (*BioSci*) The region of the axis of a plant in which the change from root to shoot structure occurs.

transition region (*Electronics*) The part of a doped semiconductor over which the impurity concentration varies.

transition resistor (*ElecEng*) A resistor connected across that part of a transformer winding short-circuited during onload TAP CHANGING.

transition state (*Chem*) The atomic arrangement of highest energy in the course of a one-step chemical reaction.

transition stops (*ElecEng*) In a traction-motor controller, intermediate electrical positions inserted between the main circuit positions in order to avoid breaking the circuit.

transition temperature (*Electronics*) The temperature at which some critical change occurs such as in magnetism (CURIE TEMPERATURE) and superconductivity.

transitive (*MathSci*) A relationship is transitive if when it applies from A to B and from B to C it also applies from A to C, ie whenever R(x,y) and R(y,z), then also R(x,z). Examples of transitive relations include 'is less than', 'is similar to', 'is a subset of', 'belongs to the same family as'. 'Is the father of' is not a transitive relation.

transitman (*Surv*) US term for a man operating a *transit theodolite*.

transit tetany (*Vet*) A form of hypocalcaemia and hypophosphataemia occurring after a period of transport. Affects cows in late pregnancy, but also sheep and horses. Symptoms include restlessness, staggering, paresis and recumbency which can start during transit or within 2 days thereafter. Also *railroad disease*.

transit theodolite (*Surv*) A theodolite whose telescope is capable of being completely rotated about its horizontal axis. See EVEREST THEODOLITE, WYE THEODOLITE.

transit time (*Electronics*) The time required for an electron or other charge carrier to travel between the electrodes of a valve, transistor or other active device. In some devices, the transit time may be undesirable, in that it limits speed or frequency of operation; in others (transit-time devices) the delay actually permits manipulation of the flow of electrons (eg in the travelling-wave tube).

translation (*BioSci*) The process by which ribosomes and tRNA decipher the genetic code in a messenger RNA in order to synthesize a specific polypeptide. See panel on DNA AND THE GENETIC CODE.

translational control (*BioSci*) The control of protein synthesis by regulation of the translation step rather than transcription, eg by selective usage of preformed mRNA or instability of the mRNA.

translator (*ICT*) Computer program used to convert a program from one language to another, usually from a low-level language to machine code.

translocated herbicide (*BioSci*) A plant poison which, if absorbed in one region, will be conducted to all parts of the plant and eg kill the roots as well. Cf CONTACT HERBICIDE.

translocated injury (*BioSci*) An injury occurring in an area remote from the original directly affected part of an animal or plant, but associated with it in type and extent.

translocation (*BioSci*) (1) Transport of solutes about the plant, including the upward movement of inorganic salts in the transpiration stream in the xylem and the movement of sugars in the phloem. See MASS–FLOW HYPOTHESIS. (2) An exchange between non-homologous chromosomes whereby a part of one becomes attached to the other, or a rearrangement within one chromosome.

translucent (*Min*) A mineral which is capable of transmitting light but through which no object can be seen.

transmembrane protein (*BioSci*) An integral membrane protein with polypeptide chain exposed on both sides of the membrane.

transmissibility (*Eng*) The ratio of the amplitude of motion of isolated structure to that of its foundation, and a key starting point for analysis of vibration isolation. Main problem is minimizing it at and near the natural frequency of the system, where it rises, often catastrophically. Often solved by using rubber springs.

transmissible spongiform encephalopathy (*Med*) A group of degenerative diseases of the central nervous system in mammals. See panel on TRANSMISSIBLE SPONGIFORM ENCEPHALOPATHY.

transmission (*Autos*) The means by which power is transmitted from the engine of an automobile to the *live axle*. Includes the change gear, propeller shaft, clutch, differential gear, etc. See AUTOMATIC TRANSMISSION.

transmission (*ICT*) (1) The process of transferring information (speech, code or data, still or moving pictures, control instructions, etc) from one location to another, or to several others (as in broadcasting) by electronic or optical means. (2) The actual information being transmitted over a communication or broadcasting system.

transmission (*Phys*) See TRANSMITTANCE.

transmission band (*ICT*) The section of a frequency spectrum over which minimum attenuation is desired, depending on the type and speed of transmission of desired signals.

transmission bridge (*ICT*) A device for separating a connection into incoming and outgoing sections for the purpose of signalling, at the same time permitting the through transmission of voice frequencies.

transmission chain (*Eng*) Roller or inverted-tooth chain designed for transmitting power.

transmission coefficient (*Phys*) The probability of penetration of a nucleus by a particle striking it.

transmission delay (*ICT*) That part of the delay in a digital communications system that is due to physical propagation of the signal as distinct from any digital processing. In eg an OPTICAL FIBRE, this amounts to 6 μs km^{-1}.

transmission dynamometer (*Eng*) A device for measuring the torque in a shaft, and hence the power transmitted, either (1) by inference from the measured twist over a given length of shaft, obtained by a torsion meter, or (2) by direct measurement of the torque acting on the cage carrying the planetary pinions of an interposed differential gear. Cf ABSORPTION DYNAMOMETER.

transmission electron microscope (*Phys*) A form of high-resolution electron microscope in which the specimen is usually either a thin (<70 nm) section of fixed, embedded material, virus particles or macromolecules. Abbrev TEM. See panel on ELECTRON MICROSCOPE.

transmission experiment (*Phys*) An experiment in which radiation transmitted by a THIN TARGET is measured, to investigate the interaction which takes place. Such experiments are used in the measurement of total cross-sections for neutrons.

transmission gain (*ICT*) The increase of power (usually expressed in dB) in a transmission from one point to another.

transmission level (*ICT*) Electric power in a transmission circuit, stated as the decibels or nepers by which it exceeds a reference level. Also *power level*.

transmission line (*ElecEng*) General name for any conductor used to transmit electric or electromagnetic energy, eg power line, telephone line, coaxial feeder, G-string, waveguide, etc. Also for the acoustic equivalent.

transmission-line amplifier (*ICT*) A wide-band amplifier in which the amplifying devices are distributed along a real or artificial transmission line. Also *distributed amplifier*.

transmission loss (*Acous*) The quantity given by ten times the logarithm to base 10 of a power ratio, describing the transmission of sound through eg walls and windows. The numerator is the power of the transmitted sound and the denominator the power of the incident sound. Abbrev TL or TR in continental Europe. Also *sound reduction index*.

transmission loss (*ICT*) The difference between the output power level and the input power level of the whole, or part, of a transmission system in decibels or nepers.

transmission measuring set (*ICT*) Apparatus consisting essentially of a sending circuit and a level measuring set. The sending circuit has a specified impedance (eg 600 Ω) and its output power is known.

transmission media layer (*ICT*) In a generalized telecommunications network, that part which represents the physical transmission plant, eg multiplexers, lines or satellites.

transmission mode (*ICT*) Field configurations by which electromagnetic or acoustic energy may be propagated by transmission lines, esp waveguides. See MODE.

transmission primaries (*ImageTech*) In colour TV, the set of three primaries chosen so that each one corresponds to one of the independent signals which comprise the colour signal.

transmission ratio (*ImageTech, Phys*) The ratio of the transmitted luminous flux to that incident upon a transparent medium. Reciprocal of *opacity*.

Transmissible spongiform encephalopathy

A group of degenerative diseases of the central nervous system of mammals, causing erratic behaviour followed by death. It includes CREUTZFELDT–JAKOB DISEASE and KURU in humans, BOVINE SPONGIFORM ENCEPHALOPATHY (BSE) in cattle and SCRAPIE in sheep.

Creutzfeldt–Jakob disease (CJD) was first recognized in 1920. About 85% of cases occur sporadically in a small number of patients between 50 and 75 years old who have a rapidly progressive dementia resulting in death, and whose brain cortex can subsequently be shown to have a sponge-like appearance. There has been a revealing series of cases where the disease was infectious, and it is occasionally found to be heritable.

The infectious disease was called *kuru* and occurred in Papua New Guinea where the local people once ritually ate the brains of their neighbours. It had the same pathology as CJD and was eliminated when the custom was abolished. CJD has also been transmitted in human growth hormone and gonadotrophin given for medical purposes, and through the use of contaminated needles and other surgical instruments. These unsafe medical procedures have accounted for most of the remaining 15% of recorded cases. A very few patients have the inherited form of the disease, which is linked to dominant mutations on the short arm of chromosome 20. (See panel on CHROMOSOME.)

New variant Creutzfeldt–Jakob disease
In the last few years a new type of CJD, called *new variant* or *nvCJD*, has been found in relatively small numbers (between 1990 and 2006, 158 deaths in the UK due to nvCJD, cf 878 due to the sporadic form). Unlike the original disease, the peak age of incidence is around 25–30 years of age. In all these patients the strain of the causative agent is exactly like that isolated from cattle with BSE and unlike that of the sporadic CJD of the more elderly human patients.

It is this clear linkage to the cattle disease that has resulted in various controls on the consumption of cattle products and has reinforced the necessity of eradicating the disease from cattle. There seems little doubt that the epidemic in cattle resulted from the practice of supplementing feedstock with protein extracted from carcasses by relatively mild methods which did not denature all the proteins.

The control of these conditions in humans or animals is made more difficult by the lack of diagnostic procedures which can determine the exact nature of the infectious material during life, and this in turn stems from the highly unusual nature of the causative agent.

Prions
When the only well-investigated spongiform encephalopathy was the disease *scrapie*, known in older sheep from the late 17th century, it was thought to be a 'slow virus' because of its long latency, although no virus has ever been linked to the disease and it shows no immunological response. Recently evidence has been accumulating that the agents causing all these diseases may be infectious protein particles or *prions* which are derived from a protein in the normal membrane of neural cells. The normal proteins are protease-sensitive and soluble and are called *PrPc particles*, but rare mutants synthesize abnormal particles which are protease-insensitive and insoluble. These *PrPsc particles* have an additional remarkable and unusual property: they force normal PrPc particles in their vicinity to change their conformation to their own aberrant form.

Thus, an initial genetic effect produces an abnormal prion which quickly multiplies exponentially in the cell membrane and can be transferred rather easily to other individuals and across species, helped by the extreme stability of the PrPsc particle. This attractive hypothesis explains much of the behaviour of the prions from humans and from cattle although the particular behaviour of strains of scrapie do not fit in quite so well.

Transmission between different species
Despite the early finding that kuru could be transmitted experimentally to chimpanzees, it was thought at first that the bovine infectious agent could not pass to humans. This was because scrapie, the first known spongiform encephalopathy, does not seem to cross species boundaries easily. The best evidence for this is that scrapie has been endemic in UK sheep for centuries but Australia is without it. Despite this, the incidence of the sporadic form of CJD is no different in the two countries. On the other hand it is now known that the BSE prion is transmissible to a wide range of mammals in addition to humans and including sheep.

Much still needs to be understood about these diseases: the exact method by which the abnormal prion causes those with normal properties to refold themselves to the new insoluble form; the length of the incubation period; the number of abnormal prions needed to establish the disease in another animal. Reliable early diagnostic procedures for this group of diseases must also be established.

See panel on IMMUNE RESPONSE.

transmission reference system (*ICT*) See MASTER TELE-PHONE TRANSMISSION REFERENCE SYSTEM.

transmission speed (*ICT*) The number of bits or elements of information transmitted in unit time. See BAUD.

transmission tower (*ElecEng*) The steel structure that carries a high-voltage transmission line.

transmission voltage (*ElecEng*) The nominal voltage at which electric power is transmitted from one place to another.

transmissivity (*Genrl*) The rate at which something passes through another medium. The term is used eg of water movement through an aquifer (hydraulic conductivity) or of light through the atmosphere (transparency).

transmissivity (*Phys*) See TRANSMISSION COEFFICIENT.

transmit–receive tube (*Electronics, Radar*) A switch which does or does not (anti-transmit–receive tube) permit flow of high-energy radar pulses. It is a vacuum tube containing argon for low striking, and water vapour to assist recovery after the passage of a pulse. Used to protect a radar receiver from direct connection to the output of the transmitter when both are used with the same scanning aerial through a common waveguide system. Abbrev *TR tube*.

transmittance (*Phys*) The ratio of energy transmitted by a body to that incident on it. If scattered emergent energy is included in the ratio, it is termed *diffuse transmittance*, otherwise *specular transmittance*. Also *transmission*.

transmitted carrier system (*ICT*) An AMPLITUDE-MODU-LATED SYSTEM in which the carrier wave is radiated. Cf SUPPRESSED-CARRIER SYSTEM.

transmitter (*ICT*) Strictly, complete assemblage of apparatus necessary for production and modulation of radio-frequency current, together with associated antenna system; but frequently restricted to that part concerned with the conversion of dc or mains ac into modulated radio-frequency current.

transmitter frequency tolerance (*ICT*) The maximum permitted FREQUENCY DEPARTURE.

transmitting valve (*ICT*) One which handles output power of a radio transmitter; may be in parallel or push–pull with others.

transmittivity (*Phys*) The TRANSMITTANCE of unit thickness of a non-scattering medium.

transmutation (*Phys*) See ATOMIC TRANSMUTATION.

transom (*Build*) An intermediate horizontal member of a window frame, separating adjacent panes. Also *transome*.

transonic range (*Aero*) The range of airspeed in which both SUBSONIC and SUPERSONIC airflow conditions exist round a body. Largely dependent upon body shape, curvature and THICKNESS–CHORD RATIO, it can be broadly taken as Mach 0·8–1·4. See panel on AERODYNAMICS.

transparency (*ICT*) The attribute of a system that allows users to disregard its physical implementation, 'seeing through' it to their final goal. For instance, in an INTELLIGENT NETWORK that allows users to update their database entries, the fact that the database may be replicated at several sites should not require users to alter any procedure.

transparency (*ImageTech*) (1) The measure of light transmitted by a transparent medium. See TRANSMISSION RATIO. (2) A picture on a clear support in which the tones and colours of the image are transparent, intended to be viewed by transmitted light or by projection on a screen.

transparency (*Phys*) The proportion of energy or number of incident photons or particles which pass through the window of an ionization chamber or Geiger counter.

transparent software (*ICT*) SOFTWARE which has been modified or through which the user accesses other software such that any changes or an intermediate program are not noticed. The phrase 'transparent to the user' is commonly used.

transpiration (*Aero*) The flow of gas along relatively long passages, the flow being determined by the pressure difference and the viscosity of the gas, surface friction being negligible.

transpiration (*BioSci*) The loss of water by evaporation, mainly through the stomata in vascular plants.

transpiration stream (*BioSci*) The flow of water from the soil through the tissues of the plant to the evaporating surfaces, all driven by transpiration. Cf COHESION THEORY.

transplant (*BioSci*) (1) In surgery and experimental zoology, the process of transferring a part or organ from its original position to another position in the same individual (autologous) or to a position in another individual (heterologous). Also *transplantation*. (2) The part or organ transferred in this way.

transpolarizer (*ElecEng*) Ferroelectric dielectric impedance, controlled electrostatically.

transponder (*ImageTech, ICT*) Equipment forming part of a communications satellite, which receives signals from a ground station at one frequency and retransmits them to another ground station or to domestic satellite receivers at another frequency.

transponder (*Radar*) A form of transmitter–receiver which transmits signals automatically when the correct interrogation is received. An example is a radar beacon mounted on a flight vehicle (or missile), which comprises a receiver tuned to the radar frequency and a transmitter which radiates the received signal at an intensity appreciably higher than that of the reflected signal. The radiated signal may be coded for identification.

transport (*ICT*) See TAPE DRIVE.

transport (*Phys*) The rate at which desired material is carried through any section of processing plant, eg isotopes in isotope separation.

transport cross-section (*NucEng*) The reciprocal of TRANSPORT MEAN FREE PATH.

transport disease (*Med*) Single-gene defect diseases in which there is an inability to transport particular small molecules across membranes. Examples are aminoacidurias such as *cystinuria*, *Fanconi syndrome*, *Hartnup disease*, *iminoglycinuria*.

transporter bridge (*CivEng*) A bridge consisting of two tall towers, one on each side of the river, connected at the top by a supporting girder along which a carriage runs. A small platform at the ordinary road level is suspended from the carriage, and this system can be made to travel along the girder across the river. Such bridges, inadequate for modern traffic, are obsolescent, although a few have been preserved.

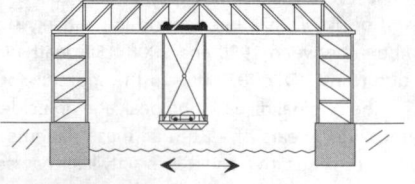

transporter bridge

transport host (*BioSci*) See PARATENIC HOST.

transport layer (*ICT*) Level 4 of the OPEN SYSTEMS INTERACTION (OSI) model that provides an end-to-end service between devices and between HOST COMPUTERS. It deals with addressing and error controls and regulates data transfers.

transport mean free path (*NucEng*) If FICK'S LAW of diffusion is applicable to the conditions in a nuclear reactor then the mean free path is three times the diffusion coefficient of neutron flux. In practice the theory usually has to be modified to take account of anisotropy of scattering and persistence of velocities.

transport number (*Chem*) The fraction of the total current flowing in an electrolyte which is carried by a particular ion.

transport theory (*NucEng*) Rigorous theoretical treatment of neutron migration which must be used under conditions where FICK'S LAW does not apply. See DIFFUSION THEORY.

transposable element (*BioSci*) See TRANSPOSON.

transposase (*BioSci*) An enzyme that brings about the transposition of a sequence of DNA within a chromosome or between chromosomes.

transpose (*MathSci*) The matrix whose rows are the columns of the given matrix. Formerly, *conjugate matrix*.

transposition (*ElecEng*) Ordered interchange of position of the lines on a pole route, and also of phases in an open power line, so that effects of mutual capacitance and inductance, with consequent interference, are minimized or balanced. See BARREL.

transposition insulator (*ElecEng*) A special type of insulator used at transposition points on a transmission line.

transposition tower (*ElecEng*) A transmission tower specially designed to allow of the transposing of the conductors.

transposon (*BioSci*) A sequence of DNA that is capable of inserting itself into many different sites in the host's chromosome. Also *transposable element*.

transputer (*ICT*) A microprocessor chip with some inbuilt memory, but able to function in parallel with a number of similar chips to form a PARALLEL ARRAY.

transreceiver (*ICT*) See TRANSCEIVER.

transrectification factor (*ElecEng*) The ratio of change in average output current to change in alternating voltage applied to a rectifier.

trans-sexualism (*Psych*) Gender identification with the opposite sex.

transudate (*Med*) A passive effusion of fluid from blood vessels, the fluid containing little protein, few cells, and not clotting outside the body.

transuranic elements (*Chem*) The artificial elements 93 and upwards, which possess heavier and more complex nuclei than uranium, and which can be produced by the neutron bombardment of uranium. More than twelve of these have been produced, including neptunium, plutonium, curium, lawrencium, etc.

Transvaal jade (*Min*) Massive light-green hydrogrossular garnet, used as a simulant for jade. Also *South African jade*.

transversal (*BioSci*) See TRANSVERSE.

transversal (*MathSci*) A line which intersects a system of other lines.

transverse (*BioSci*) Broader than long; lying across the long axis of the body or of an organ; lying cross-wise between two structures; connecting two structures in cross-wise fashion. Also *transversal*.

transverse architrave (*Build*) The moulding across the top of a door or window opening.

transverse-beam travelling-wave tube (*Electronics*) One in which the directions of propagation of the electron beam and the electromagnetic wave carrying the signal are mutually perpendicular. The cavity magnetron is an example.

transverse electric wave (*ICT*) See TE-WAVE.

transverse fracture (*Med*) A fracture caused by an impact at right angles to the bone.

transverse frame (*Aero*) The outer-ring members of a rigid airship frame. It may be of a stiff-jointed type, or braced with taut radial members to a central fitting. It connects the main longitudinal girders together.

transverse frame (*Ships*) A stiffening member of a ship's hull, disposed transversely to the longitudinal axis. In double-bottom construction, it is that portion above the tank margin.

transverse heating (*ElecEng*) Dielectric heating in which electrodes impose a high-frequency electric field normal to layers of laminations.

transverse joint (*Build*) Any joint in a brick wall which cuts across the bed from the front to the back surface, such joint in the best practice being always a continuous one in order to avoid the setting up of STRAIGHT JOINT. Also *cross joint*.

transverse magnetic wave (*ICT*) See TM-WAVE.

transverse magnification (*Phys*) See MAGNIFICATION.

transverse metacentre (*Ships*) The METACENTRE obtained by inclining the vessel through a small angle about a longitudinal axis. Cf LONGITUDINAL METACENTRE.

transverse section (*Genrl*) Generally a cross-section taken at right angles to the longest axis of the item being examined or drawn, but since multiple locations are possible further specification is usually required eg 'at the mid-point'.

transverse springs (*Autos*) Laminated springs arranged transversely across the car, parallel to the axles, instead of longitudinally; usually SEMI-ELLIPTIC SPRINGS and anchored centrally to the chassis.

transverse tubule (*BioSci*) An invagination of the plasma membrane of a striated muscle cell that depolarizes after stimulation at the neuromuscular synapse, and thus spreads the activation throughout the muscle cell. Also *T-tubule*.

transverse wave (*Phys*) A wave motion in which the disturbance of the medium occurs at right angles to the direction of wave propagation; eg waves on a stretched string, electromagnetic waves.

transvestism (*Psych*) Sexual gratification through dressing in the clothes of the opposite sex.

tranverse-field travelling-wave tube (*Electronics*) One in which the electric fields associated with the signal wave are normal to the direction of motion of the electron beam.

trap (*Build*) See AIR TRAP.

trap (*Electronics*) Crystal lattice defect at which current carriers may be trapped in a semiconductor. This trap can increase recombination and generation or may reduce the mobility of the charge carriers.

trap (*Geol*) (1) Imprecise and obsolete term for dark fine-grained volcanic and hypabyssal rocks, eg basalt. (2) Plateau basalts, eg the Deccan Traps of India. (3) A structure in which oil or gas may accumulate.

TRAPATT diode (*Electronics*) Abbrev for *trapped plasma avalanche transit-time diode*. A diode which can operate as an oscillator at microwave frequencies, the frequency being determined by the thickness of the active layer in the diode. The avalanche zone moves through the drift region, a trapped space-charge plasma within the p–n junction.

trap crop (*Agri*) An additional crop planted among a main crop to attract invertebrate pests. It is pesticide-treated or removed and destroyed when infested.

trapezium (*BioSci*) In the mammalian brain, a part of the MEDULLA OBLONGATA consisting of transverse fibres running behind the pyramid bundles of the PONS VAROLII.

trapezium (*MathSci*) (1) Quadrilateral with one pair of opposite sides parallel. US *trapezoid*. (2) US for TRAPEZOID. Euclid uses the term trapezium for any quadrilateral which is not a parallelogram. Clearly these terms should never be used without explanation.

trapezium diagram (*Electronics*) The pattern on the screen of a CRO when an amplitude-modulated radio-frequency voltage is applied to one pair of plates and the modulating voltage is applied to the other pair.

trapezium distortion (*Electronics*) That associated with the TRAPEZIUM EFFECT.

trapezium effect (*Electronics*) A phenomenon in which the deflecting voltage applied to the deflector plates of a cathode-ray tube is unbalanced with respect to the anode. If equal alternating voltages, of different frequencies, are applied to the two sets of plates, the resulting pattern on the screen is trapezoidal instead of square.

trapezoid (*MathSci*) (1) Quadrilateral with no parallel sides. US *trapezium*. (2) US for TRAPEZIUM.

trapezoidal rule (*MathSci*) A rule which approximates the definite integral

$$\int_a^b f(x)\,dx$$

The interval $[a,b]$ is divided into n subintervals $[x_i, x_{i+1}]$ ($i=0,1,2,\ldots,n-1$) of length $h=(b-a)/n$, and the integral may then be approximated by

$$\int_a^b f(x)\,dx = \frac{h}{2}[\,f(a) + 2f(x_1) + 2f(x_2) + \cdots$$

$$+ 2f(x_{n-1}) + f(b)]$$

The difference between the approximation and the value of the integral is not greater than $Mh^2(b-a)/12$, where M is the least upper bound of the second derivative of $f(x)$ on the interval a to b.

trapezoidal rule (*Surv*) A rule for the estimation of the area of an irregular figure. For this purpose it is divided into a number of parallel strips of equal width. The lengths of the boundary ordinates of the strips are measured, and the area is calculated from the rule stating that the area is equal to the common width of the strips multiplied by the sum of half the first and half the last ordinates plus all the others.

trapezoidal speed–time curve (*Eng*) A simplified form of speed–time curve used in making preliminary calculations regarding the energy consumption and average speed of moving vehicles. The acceleration and braking portions of the curve are sloping straight lines, while the coasting portion is a horizontal straight line, so that the complete curve becomes a trapezium. Cf QUADRILATERAL SPEED–TIME CURVE.

trapped mode (*Phys*) Radio-wave propagation in which the radiated energy is substantially confined within a tropospheric duct in the Earth's atmosphere.

trapping (*ICT*) A feature of some computers by which they make an unscheduled jump to some specified location if an abnormal arithmetic situation arises.

trapping region (*Phys*) Three-dimensional space in which particles from the Sun are guided into paths towards the magnetic poles, giving rise to AURORA, and otherwise forming ionized shells high above the ionosphere. Also *magnetic tube*.

trappoid breccias (*Geol*) A succession of breccias found near Nuneaton, Charnwood and Malvern, UK, consisting of angular blocks of rhyolite and feldspathic tuffs which are of Permian age. They probably represent fossil scree material.

trash (*Genrl*) Generic term (US) for post-consumer waste.

trash (*Textiles*) The unwanted material present in bales of raw cotton.

trass (*Build, Geol*) A material similar to pozzuolana, found in the Eifel district of Germany; used to give additional strength to lime mortars and plasters.

trass mortar (*Build, CivEng*) A mortar composed of lime, sand and TRASS or brick-dust, or of lime and trass without sand, the trass making the mortar more suited for use in structures exposed to water.

trastuzumab (*Pharmacol*) A therapeutic antibody used in the treatment of advanced breast cancer. Proprietary name *Herceptin*.

trauma (*Psych, Med*) (1) Any totally unexpected experience, physical or psychological, which the person cannot assimilate. See SHOCK. (2) A wound or body injury. Adj *traumatic*.

traumatic neurosis (*Psych*) A psychiatric illness resulting from severe and unexpected experience; characterized by periods of trance when the events are re-experienced, and often by traumatic dreams. Differs from other neuroses in that the symptoms have no unconscious meaning, but are an attempt to assimilate the experience by repeating it.

travel (*Eng*) The distance between the extreme positions reached by a mechanism executing a reciprocating or other reversing motion.

traveller (*Textiles*) See RING SPINNING.

travelling block (*MinExt*) Also *traveling block*. See panel on DRILLING RIG.

travelling matte shot (*ImageTech*) A composite shot in which the components are printed together at the laboratory by the use of film MATTES with varying outlines matching the action.

travelling wave (*Phys*) A wave carrying energy continuously away from the source.

travelling-wave amplifier (*ICT*) One using a travelling-wave tube.

travelling-wave antenna (*ICT*) One in which many radiating elements are excited progressively as the result of a single wave traversing its length in one direction only.

travelling-wave magnetron (*Electronics*) Multiple-cavity magnetron in which cavities are coupled by travelling-wave systems.

travelling-wave maser (*Electronics*) See SOLID-STATE MASER.

travelling-wave tube (*Electronics*) One in which energy is interchanged between a helix delay line and an electron beam, which can be at an angle. Used to amplify ultrahigh and microwave frequencies. See TRANSVERSE-BEAM TUBE, TRANSVERSE-FIELD TUBE.

traverse (*Surv*) A survey consisting of a continuous series of lines whose lengths and bearings are measured.

traverse tables (*Surv*) Tables from which the differences of latitude and departure of a line of any length and bearing may be read off.

traversing (*Eng*) The sliding motion in a self-acting lathe or, more generally, the sideways movement of part of a machine.

traversing bridge (*Eng*) A type of movable bridge which is capable of rolling backwards and forwards across an opening, such as a dock entrance, to allow of the passage of a vessel.

travertine (*Geol*) A variety of calcareous tufa of light colour, often concretionary and varying considerably in structure; some varieties are porous. A deposit characteristic of hot springs in volcanic regions.

tread (*Build*) The horizontal part of a step.

tread (*Eng*) In the wheels of a vehicle, the part of the tyre in contact with the road or rail. See panel on TYRE TECHNOLOGY.

tread (*Vet*) An injury of the coronet of a horse's hoof due to striking with the shoe of the opposite foot.

trebles (*MinExt*) See COAL SIZES.

tree (*BioSci*) A tall, woody perennial plant having a well-marked trunk and few or no branches persisting just above the base. A form known as EXCURRENT.

tree (*Chem*) Crystal growth structure. See DENDRITE, LEAD TREE.

tree (*ICT*) (1) Non-linear hierarchic data structure, where each data item is thought of as a 'node', and links from it to other items as 'branches'. (2) In telecommunications, a number of connected circuit branches that do not include meshes.

tree (*MathSci*) A connected graph which does not contain cycles, loops or multiple edges. This idea has many applications including the classification of isomers in chemistry.

tree ferns (*BioSci*) Ferns (*Cyathea, Dicksonia* and several extinct genera) that form a trunk up to 20 m high, typically unbranched and with a relatively slender stem surrounded and supported by matted adventitious roots and persistent leaf bases.

tree topology (*ICT*) A NETWORK LAYOUT where there is only one route between any two NODES. See fig. at NETWORK TOPOLOGY.

trega- (*Genrl*) A prefix signifying 10^{12} times. It is replaced in the SI system of unit notation by the prefix TERA-.

trehalose (*BioSci*) A disaccharide of glucose found in the haemolymph of insects and used as a storage compound in some bacteria, fungi and plants.

Tremadoc (*Geol*) The oldest epoch of the Ordovician period, sometimes classed as Upper Cambrian. See PALAEOZOIC.

trematic (*BioSci*) Pertaining to the gill clefts.

Trematoda (*BioSci*) A class of Platyhelminthes, all the members of which are either ectoparasites or endoparasites, and have a tough cuticle, a muscular non-protrusible pharynx, and a forked intestine. A ventral sucker for attachment is usually present, and a sucker surrounds the mouth. Such organisms are sometimes divided into three classes: Digenea, Aspidogastrea, Monogenea. Liver flukes.

trembling (*Vet*) See MYOCLONIA CONGENITA.

tremolite (*Min*) A monoclinic amphibole, hydrated calcium magnesium silicate. It is usually white or grey and occurs in bladed crystals or fibrous aggregates in metamorphic rocks. It differs from *actinolite* in having less iron; the name is used for the magnesium end-member. Also *grammatite*.

tremor (*Med*) Involuntary agitation of the muscles of the body, or of a limb, due to emotional disturbance, old age or disease of the nervous system.

trenail (*Build*) A hardwood pin driven transversely through a mortise and tenon to secure the joint. Also *trunnel*. See DRAW-BORE.

trench fever (*Med*) A disease common among troops in World War I; symptoms were relapsing fever, headache, pains in the back and in the limbs, often a rose-red eruption, due to infection with a virus conveyed by lice.

trenching plane (*Build*) See DADO PLANE.

trepan (*Med*) (1) To TREPHINE. (2) A form of trephine no longer in use.

trepanning (*Eng*) Producing a hole by removing a ring of material, as opposed to disintegrating all the material corresponding to the hole.

trephine (*Med*) (1) To operate with the trephine; to remove by surgical means a part of the skull; to remove by surgical means a disk from any part, eg from the globe of the eye in the treatment of glaucoma. (2) A crown saw which is designed to remove a circular area of bone from the skull.

Treponemataceae (*BioSci*) A family of mainly parasitic, small spirochaetes, many of which are pathogenic, eg *Treponema pallidum* (syphilis), *Treponema pertenue* (yaws).

Tresca yield criterion (*Eng*) When a triaxial stress system acts in a material, yield will occur when the maximum SHEAR STRESS reaches a critical value τ_c. Since by convention, σ_1 is the largest PRINCIPAL STRESS and σ_3 the smallest, the criterion can be stated as

$$\tau_c = \frac{\sigma_1 - \sigma_3}{2}$$

In terms of the uniaxial YIELD STRESS, σ_Y, obtained from a TENSILE TEST, the criterion becomes

$$\sigma_1 - \sigma_3 = \sigma_Y$$

T-rest (*Eng*) A T-shaped rest clamped to the bed of a wood-turning lathe for supporting the tool. Also *tee rest*. See L-REST.

TRF receiver (*ICT*) Abbrev for TUNED RADIO-FREQUENCY RECEIVER.

tri- (*Genrl*) Prefix from Lt *tres*, Greek *tria*, three.

triac (*Electronics*) A bi-directional gate-controlled THYRISTOR for full-wave control of ac power. By altering the phase of the gate-switching signal, load current can be adjusted from a few per cent to nearly 100% of the full load current.

triacetate (*Textiles*) Textile fibres of CELLULOSE ACETATE of which nearly all (92%) of the hydroxyl groups are acetylated.

triacetin (*Chem*) *Glycerol triacetate*. Chiefly used as a plasticizer for cellulose acetate and cellulose ether plastics.

triacidic (*Chem*) Of a base, capable of reacting with three hydrogen ions per molecule.

triad (*Chem*) The three groups of similar atoms often given as group VIII in the periodic table: Fe, Co and Ni; Ru, Rh and Pd; and Os, Ir and Pt.

triad (*ImageTech*) The unit image component of the screen of a SHADOW MASK TUBE, comprising one dot of each of the red, green and blue phosphors.

trial and error learning (*Psych*) In learning theory, refers to an essentially passive type of learning in which behaviour changes occur as a result of their association with positive or negative consequences (REINFORCEMENTS). Originally the term was used to compare it with INSIGHT LEARNING and is now more frequently referred to as operant conditioning.

trial pit (*CivEng*) A pit sunk into the ground to obtain information as to nature, thickness and position of strata.

triamcinolone acetonide (*Pharmacol*) A potent *glucocorticoid* similar to prednisolone, used for the treatment of inflammatory disorders.

triamterene (*Pharmacol*) A potassium-sparing diuretic used to treat oedema.

triandrous (*BioSci*) Having three stamens.

triangle (*MathSci*) A three-sided rectilineal plane figure. See EQUILATERAL, ISOSCELES, SCALENE.

triangle of error (*Surv*) The triangle formed in the trial and error solution of the three-point problem when, on drawing back rays through the three known points on plan, they form a small triangle instead of intersecting at a single point, as a result of the positioning of the survey instrument or the measurement of one or more angles being incorrect.

triangle of forces (*Phys*) A particular case of the polygon of forces drawn for three forces in equilibrium at a point. See POLYGON OF FORCES.

triangular (*Genrl*) Having three angles.

triangular matrix (*MathSci*) A square matrix for which either all the entries below or all those above the main diagonal are zero. The former case is called *upper triangular*, and the latter *lower triangular*.

triangular numbers (*MathSci*) The numbers 1, 3, 6, 10, 15,... which can be represented in the form of a triangle:

    1   3   6

triangulation (*Surv*) The process of dividing up a large area for survey purposes into a number of connected triangles with their apexes (triangulation or 'trig' stations) mutually visible, measuring one side of one of the triangles (the 'baseline') and all the angles. See INTERSECTION, TRILATERATION.

Triangulum (Triangle) (*Astron*) A small but prominent northern constellation.

Triangulum Australe (Southern Triangle) (*Astron*) A small southern constellation.

triarch (*BioSci*) A stele having three strands of protoxylem, eg the roots of some dicotyledons.

Triassic (*Geol*) The geological period between Permian and Jurassic. It is the oldest period of the Mesozoic era and has a time span from approx 245 to 210 million years. It was named by von Alberti from the threefold division in Germany. The corresponding system of rocks. See MESOZOIC.

triazine (*Chem*) A class of organic heterocyclic compounds containing three nitrogen atoms in a six-membered ring; often used in herbicides.

triazole (*Chem*) $C_2H_3N_2$. A heterocyclic compound consisting of a five-membered ring.

tribasic (*Chem*) Descriptive of an acid, capable of reacting with three hydroxide ions per molecule.

tribe (*BioSci*) A section of a family consisting of a number of related genera.

tribo- (*Genrl*) Prefix from Gk *tribein*, to rub, denoting rubbing or friction.

triboelectric series (*ElecEng*) The ranking of insulators, esp polymers, on a linear scale of severity with positive charging at one end, negative charging at other. See STATIC ELECTRIFICATION.

triboelectrification (*Phys*) The separation of charges through surface friction. If glass is rubbed with silk, the glass becomes *positive* and the silk *negative*, ie the silk takes electrons. The phenomenon is that of contact potential, made more evident in insulators by rubbing.

tribology (*Phys*) The science and technology of interacting surfaces in relative motion (and the practices related thereto), including the subjects of friction, lubrication and wear.

triboluminescence (*Phys*) Luminescence generated by friction.

tributary unit (*ICT*) In a SYNCHRONOUS DIGITAL HIERARCHY network, the term used for a VIRTUAL CONTAINER when it has had a pointer added to identify its starting point in relation to the frame in which it will be carried. Tributary units are interleaved to form tributary unit groups that are further interleaved to form higher-order virtual containers.

tributyltin (*Chem*) A toxic pesticide/biocide, an organotin compound previously used as an antifouling agent for boat hulls and other marine structures; an endocrine-disrupting compound that can affect reproduction in aquatic and other organisms (see IMPOSEX). Abbrev *TBT*.

tricarboxylic acid cycle (*BioSci*) See TCA CYCLE

tricarpellary (*BioSci*) Consisting of three carpels.

triceps (*BioSci*) A muscle with three insertions.

trich-, tricho- (*Genrl*) Prefixes from Gk *thrix*, gen *trichos*, hair.

trichinosis (*Med*) Infestation of the human intestine, as a result of eating raw or underdone pork, with the nematode worm *Trichinella* (or *Trichina*) *spiralis*, the larvae of which migrate to, and become encysted in, the muscles of the body. Also *trichiniasis*.

trichlorethene (*Chem*) C_2HCl_3. Used as a solvent in drycleaning, in the extraction of fat from wool, and in the manufacture of paints and varnishes. Used in surgery to give general analgesia, and, with nitrogen (I) oxide, light general anaesthesia. TN *Trilene*.

trichloroacetic acid (*Chem*) CCl_3COOH. Organic acid prepared by oxidizing chloral (CCl_3CHO) with nitric (V) acid. Acid with a high DISSOCIATION CONSTANT. Also *trichloroethanoic acid*.

3:4:4′-trichlorocarbanilide (*Chem*) $ClC_6H_4NHCONHC_6H_3Cl_2$. Used as a germicide in toilet soaps. Low toxicity.

1,1,1-trichloroethane (*Chem*) Methyl chloroform. CH_3CCl_3. Ram 133·5; bp 74°C. Chlorinated solvent with low toxicity (much safer in use than tetrachloromethane). Nonflammable. Widely used industrially for cleaning electrical equipment. TNs Chlorothene NU, Genklene.

trichloroethanoic acid (*Chem*) See TRICHLOROACETIC ACID.

trichloromethane (*Chem*) See CHLOROFORM.

tricho- (*Genrl*) See TRICH-.

trichocephaliasis (*Med*) See TRICHURIASIS.

trichocyst (*BioSci*) In some Ciliophora, a minute hair-like body lying in the subcuticular layer of protoplasm; it is capable of being shot out, and is an organ of attachment.

trichogyne (*BioSci*) An outgrowth from the female sex organ of the red algae, some fungi and lichens and a few green algae for the reception of the male gamete.

trichoid (*BioSci*) Hair-like.

trichology (*Genrl*) The scientific study of hair and its disorders.

trichome (*BioSci*) Any outgrowth of the epidermis of a plant, composed of one or more cells but without vascular tissue.

trichomoniasis (*Med*) Infection with mobile flagellated protozoal organisms. *T. vaginalis* causes irritation and discharge from the vagina and the male urethra.

trichosis (*BioSci*) Arrangement or distribution of hair.

trichotomous (*BioSci*) Branching into three. Cf DICHOTOMOUS.

trichromatic coefficients (*Phys*) The relative intensities of three primary colours of a given trichromatic system of colour specification required to match a colour sample. Generally add to unity.

trichromatic filter (*ImageTech*) See TRICOLOUR FILTERS.

trichromatic process (*ImageTech*) See THREE-COLOUR PROCESS.

trichuriasis (*Med*) Infestation of the human intestine with the nematode whip-worm *Trichuris trichiura* (also *Trichocephalus dispar*).

tricipital (*BioSci*) Adj. from TRICEPS.

trick valve (*Eng*) See ALLAN VALVE.

triclinic system (*Crystal*) The lowest system of crystal symmetry containing crystals which possess only a centre of symmetry. Also *anorthic system*. See fig. at BRAVAIS LATTICES.

tricolour filters (*ImageTech*) A set of three filters covering the visible spectrum as the red, green and blue regions respectively.

tricot (*Textiles*) Fabric knitted on a WARP-KNITTING MACHINE.

tricotine (*Textiles*) (1) A fine-ribbed, plain-weave fabric with a silk warp and cotton weft. (2) A weft-face woven fabric with a cotton warp and worsted wool weft containing a fine twill line.

tricusp (*MathSci*) See STEINER'S TRICUSP.

tricuspid (*BioSci*) Having three points, as the tight auriculoventricular valve of the mammalian heart.

tricycle landing gear (*Aero*) A landing gear with a nosewheel unit.

tricyclic antidepressants (*Pharmacol*) Somewhat misleading term (as there are now other ring compounds with broadly similar properties) for a group of drugs useful in the treatment of moderate to severe depressive illness. Some may have additional sedative properties (*amitriptyline*) and others are reputed to have fewer cardiac side effects (*mianserin*).

tridymite (*Min*) A high-temperature form of silica, SiO_2, crystallizing in the orthorhombic system, but possessing pseudohexagonal symmetry. The stable form of silica from 870 to 1470°C. Typically occurs in acid volcanic rocks.

triene (*Chem*) Any chemical compound containing three carbon–carbon double bonds.

triethanolamine (*Chem*) $(HOC_2H_4)_3N$. Strongly alkaline organic solvent used in some paint strippers. Also used as a stabilizer for chlorinated hydrocarbon solvents.

trifacial (*BioSci*) The fifth cranial or trigeminal nerve of vertebrates.

trifid (*BioSci*) Split into three parts but not to the base.

Trifid Nebula (*Astron*) A colourful emission nebula in the constellation Sagittarius. Distance approx 2000 pc.

trifoliate (*BioSci*) Having three leaves or, sometimes, three leaflets.

trifoliolate (*BioSci*) A compound leaf having three leaflets, eg clover.

trifurcate (*BioSci*) Having three branches.

trigatron (*Electronics*) An envelope with an anode, cathode and trigger electrode, containing a mixture of argon and oxygen. The device operates as an electronic switch, in which a low-energy pulse ionizes the gas in the switch, and permits discharge of a much higher-energy pulse across the main electrodes.

trigeminal (*BioSci*) (1) Having three branches. (2) The fifth cranial nerve of vertebrates, dividing into the ophthalmic, maxillary and mandibular nerves. Also *trifacial*.

trigeminal neuralgia (*Med*) See TIC DOULOUREUX.

trigger (*Chem*) The agent which causes the initial decomposition of a chain reaction.

trigger (*ICT*) Manual or automatic signal for an operation to start.

trigger circuit (*Electronics*) A circuit having a number of states of electrical condition which are either stable (or quasi-stable) or unstable with at least one stable state, and so designed that desired transition can be initiated by the application of suitable trigger excitation.

trigger electrode (*Electronics*) See PILOT ELECTRODE.

trigger level (*Electronics*) The minimum input level at which a trigger circuit will respond.

trigger pulse (*Electronics*) One which operates a trigger circuit.

trigger relay (*ICT*) A relay which can be mechanical, thermionic, eg a gas-filled triode, or solid-sate and that, when operated, remains in its operated condition when the operating current or other control is removed, because of a mechanical latch or other property.

trigger valve (*Electronics*) A thermionic or gas-discharge valve used as a trigger relay.

triglyceride (*Chem*) A term applied to a fatty acid ester of glycerol in which all three hydroxyl groups are substituted.

triglyph (*Arch*) A group of three GLYPHS, or of two glyphs and two half-glyphs, used as a decoration for a flat surface.

trigonal system (*Crystal*) A style of crystal architecture characterized essentially by a principal axis of threefold symmetry; otherwise resembling the hexagonal system. Such important minerals as calcite, quartz and tourmaline crystallize in this system.

trigone (*Med*) Triangular area of interior of urinary bladder between the openings of the ureters and of the urethra.

trigonitis (*Med*) Inflammation of the trigone.

trigonometrical function (*MathSci*) Any of a number of functions initially defined in terms of the ratios of sides in a right-angled triangle: if θ is any angle, and ABC is the right-angled triangle formed by dropping a perpendicular BC from a point B in one of the lines enclosing the angle to the other, the trigonometrical functions, or ratios, are as follows:

$$\sin\theta = \frac{BC}{AB}, \quad \operatorname{cosec}\theta = \frac{1}{\sin\theta}$$

$$\cos\theta = \frac{AC}{AB}, \quad \sec\theta = \frac{1}{\cos\theta}$$

$$\tan\theta = \frac{BC}{AC}, \quad \cotan\theta = \frac{1}{\tan\theta}$$

Usually abbreviated to *sin, cosec, cos, sec, tan, cot*, the graphical representations of four of them are shown in the figure. Independent arithmetic definitions of *sin, cos* and *tan* are as follows:

$$\sin x = \sum_{0}^{\infty}(-1)^r\frac{x^{2r+1}}{(2r+1)!}$$

$$\cos x = \sum_{0}^{\infty}(-1)^r\frac{x^{2r}}{(2r)!}$$

$$\tan x = \frac{\sin x}{\cos x}$$

See CONJUGATE ELEMENT, INVERSE TRIGONOMETRICAL FUNCTION. Cf HYPERBOLIC FUNCTIONS.

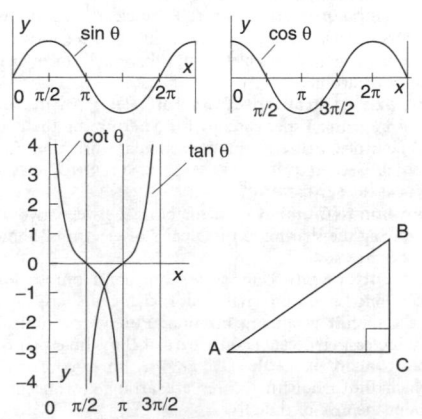

trigonometrical functions The triangle ABC and the graphs of four functions.

trigonometrical station (*Surv*) A survey station used in a triangulation.

trigonometrical survey (*Surv*) A survey based on a triangulation.

trigonometry (*Genrl*) The branch of mathematics that deals with the relations between the sides and angles of triangles.

trigonous (*BioSci*) A triangular stem but obtusely angled and with convex faces. Cf TRIQUETROUS.

trihydric alcohols (*Chem*) Alcohols containing three hydroxyl groups attached to three different carbon atoms, eg GLYCERINE.

trilateration (*Surv*) Land survey, by triangulation, in which distances are measured direct by a TELLUROMETER. In the *shoran* system, up to 800 km can be thus measured by aid of airborne magnetometer flown between two ground stations.

Trilene (*Chem*) See TRICHLORETHENE.

trillion (*MathSci*) The cube of a million, 10^{18}; (US) the cube of 10 000, 10^{12}. Colloq only.

trilobites (*Geol*) Extinct marine arthropods, the most important of the Cambrian faunas that became extinct in the Permian. They had a segmented oval body long-itudinally divided into three lobes, hence the name. Commonly from under 1 cm to well over 20 cm long. Entirely marine and most living in or near the sea bottom.

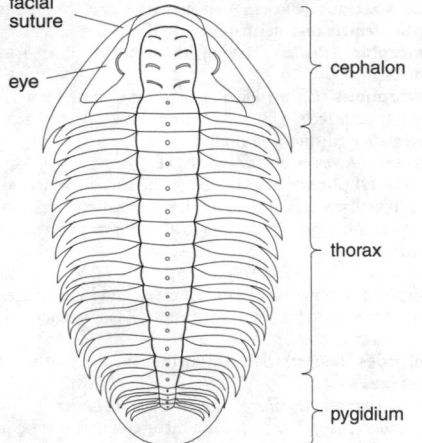

trilobite

trim (*Aero*) Adjustment of an aircraft's controls to achieve stability in a desired condition of flight. See TRIMMING STRIP, TRIMMING TAB.

trim (*Build*) Architraves and other finishings around a door or window opening.

trim (*Ships*) The difference between the draughts measured at the forward and after perpendiculars. May be expressed as an angle.

Trimask (*Print*) TN for an all-purpose film used for COLOUR MASKING.

trimer (*Chem*) A substance in which molecules are formed from three molecules of a monomer.

trimeric (*Chem*) Having the same empirical formula but a relative molecular mass three times as great.

trimerous (*BioSci*) Arranged in threes or in multiples of three.

trimethylene glycol (*Chem*) Propan-1,3-diol. $CH_3OHCH_2CH_2OH$. Organic solvent. Bp 214°C (with decomposition); rel.d. (at 20°C) 1·060.

trimix (*Genrl*) A mixture of nitrogen, helium and oxygen, used by deep-sea divers.

trimmed edges (*Print*) See CUT EDGES.

trimmed size (*Print*) A necessary specification for books and for any subdivision of a sheet, the standard trim being $\frac{1}{8}$ in (3 mm) from each edge that requires it. International PAPER SIZES are always given as trimmed.

trimmer (*Aero*) See TRIMMING TAB.

trimmer (*Build*) The cross-member which is framed between the full-length members to afford intermediate support to the shortened joists in a TRIMMING.

trimmer (*ElecEng*) See TRIMMING CAPACITOR.

trimmer (*ICT*) See PAD.

trimmer joint (*Build*) A joint formed with a TUSK TENON.

trimming (*Build*) The operation by which bridging joists or rafters are shortened and given intermediate support around a fireplace or chimney.

trimming (*Eng*) The removal of flash from the edge of a workpiece, eg a plastic moulding or metal casting.

trimming capacitor (*ElecEng*) Variable capacitor of small capacitance used to take up discrepancies between self- and stray capacitances in a circuit. Also *trimmer*.

trimming joist (*Build*) One of the two full-length members between which the trimmer is framed. As these members have to carry more than the other bridging joists, they are thicker.

trimming strip (*Aero*) A metal strip, or a cord of wire doped in place with fabric, on the trailing edge of a control surface to modify its balance or trim; it is adjustable only on the ground.

trimming tab (*Aero*) A TAB, which can be adjusted in flight by the pilot, for trimming out control forces. Colloq *trim tab, trimmer*.

trimonoecious (*BioSci*) A species in which the plants bear male, female and hermaphrodite flowers.

trimorphic (*BioSci*) Having three forms. See HETEROSTYLY.

trimorphous (*Chem*) Having three crystalline forms.

trims (*ImageTech*) Unused portions of a selected scene left over after editing is complete.

trim tab (*Aero*) See TRIMMING TAB.

tri-*n*-butyl phosphate (*NucEng*) The solvent which, diluted with kerosene, is now used for separating uranium and plutonium from spent fuel in the Purex process. Abbrev *TBP*.

triniscope (*ImageTech*) System of colour-TV display using optical combination of the red, green and blue images from the screens of three separate CRTs, particularly for TELERECORDING.

trinitrides (*Chem*) Salts of hydrazoic acid. Also *azides, hydrazoates*.

trinitroglycerine (*Chem*) See NITROGLYCERINE.

Trinitron (*ImageTech*) TN for a three-colour TV tube using a vertical grating and vertical phosphor stripes. A common electron-gun assembly and deflexion system is used with three cathodes to provide the electron beams. Cf SHADOW MASK TUBE.

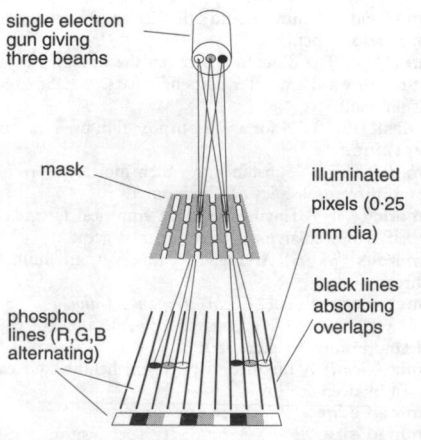

single electron gun giving three beams

mask

illuminated pixels (0·25 mm dia)

black lines absorbing overlaps

phosphor lines (R,G,B alternating)

trinitron The mask is immediately in front of the phosphors in practice.

trinitrophenol (*Chem*) *2,4,6-trinitro-1-hydroxybenzene*. See PICRIC ACID.

trinitrotoluene (*Chem*) The symmetrical isomer, *2,4,6-trinitrotoluene*, $C_6H_2(CH_3)(NO_2)_3$, is a solid, mp 82°C. It is manufactured by slowly adding toluene to a mixture of nitric and sulphuric acids containing oleum. It is used as a high explosive, and is known as *TNT*.

triode valve (*Electronics*) Thermionic vacuum tube containing an emitting cathode, an anode and a control electrode or grid, whose potential controls the flow of electrons from cathode to anode.

trioecious (*BioSci*) A species in which some individuals bear male flowers only, others female only and the rest hermaphrodite.

triolein (*Chem*) Naturally occurring triglyceride in which all three fatty acid chains are oleic acid.

trioses (*Chem*) The simplest monosaccharides. They contain three carbon atoms in the molecule, eg $HOCH_2CO$-CH_2OH, *glyceraldehyde*.

trip (*MinExt*) See ROUND TRIP.

trip (*NucEng*) Automatic shutdown of reactor power initiated by signal from one of the safety circuits when one of the operational characteristics of the reactor deviates beyond a certain limit.

tripack (*ImageTech*) A photographic material whose base carries three emulsion layers appropriately sensitized to specific spectral regions, with intermediate filter layers, for recording images of separate colours.

trip amplifier (*NucEng*) One operating the trip mechanism of a nuclear reactor. Also termed a *shutdown amplifier*.

trip circuit (*ElecEng*) The electric circuit operating the tripping mechanism of a circuit breaker. Cf SHUNT TRIP.

trip coil (*ElecEng*) Any magnet coil with a moving armature which operates some other circuit, eg a coil which operates a circuit breaker.

trip gear (*Eng*) A valve-actuating gear, used for drop valves and rocking (Corliss) valves of large steam engines, in which the valve is opened by a trigger mechanism, which is then tripped out of engagement to allow the valve to close under a heavy spring. See CORLISS VALVE, DROP VALVE.

triphenylmethane dyes (*Chem*) A group of dyestuffs derived from triphenylmethane. They comprise the malachite green group derived from diaminotriphenylmethane, the rosaniline group derived from triaminotriphenylmethane, the aurine group derived from trihydroxytriphenylmethane, the phthalein group derived from triphenylmethane-carboxylic acid.

triphylite (*Min*) An orthorhombic lithium iron phosphate isomorphous with LITHIOPHILITE.

tripinnate (*BioSci*) Three times PINNATE.

triple alpha process (*Astron*) A nuclear reaction which occurs in late stages of stellar evolution after most of the core hydrogen has been used up; three helium nuclei fuse to form carbon.

triple-axis neutron spectrometer (*Phys*) An instrument used in neutron spectroscopy for determining the energies of neutrons scattered in a particular direction from a crystal. See NEUTRON ELASTIC SCATTERING, NEUTRON INELASTIC SCATTERING.

triple bond (*Chem*) A covalent bond between two atoms involving the sharing of three pairs of electrons. Shown as ≡. See ALKYNE.

triple bottom line (*EnvSci*) A shorthand phrase for the economic, environmental and social gains arising from the activities of an organization. Purely economic gains are increasingly seen as only part of the whole picture.

triple-concentric cable (*ElecEng*) A three-core cable in which the conducting cores are arranged concentrically about the axis of the cable.

triple effect evaporator (*FoodSci*) A method for the bulk dehydration of food products at moderate temperature and low pressure. Fig. ▷

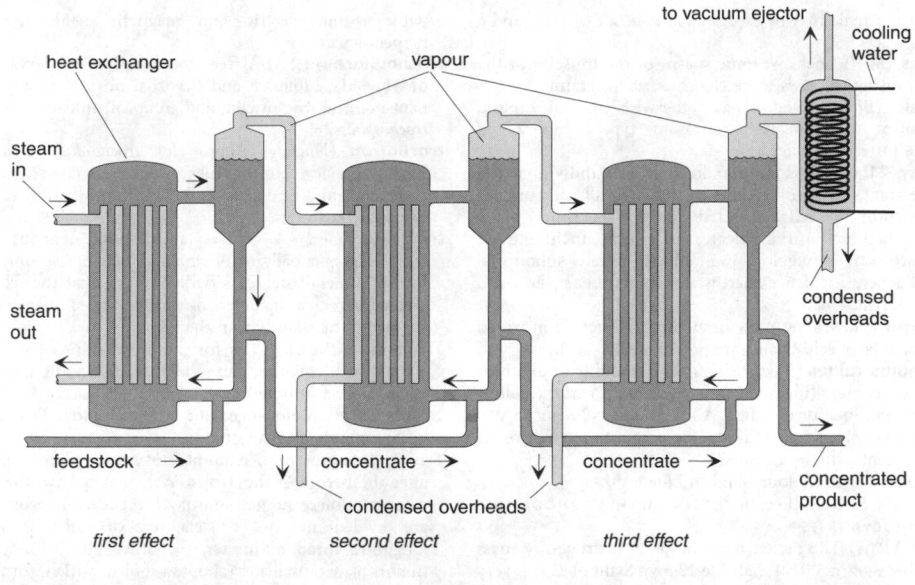

to vacuum ejector

cooling water

heat exchanger

vapour

steam in

steam out

condensed overheads

feedstock →

concentrate →

concentrate →

concentrated product

condensed overheads

first effect   *second effect*   *third effect*

triple effect evaporator Steam heats the first effect; vapour from the feedstock heats those following.

triple-expansion engine (*Eng*) An engine in which the steam expands, successively, in a high-pressure, intermediate-pressure and low-pressure cylinder, working on the same crankshaft.

triple fusion (*BioSci*) The fusion of the two POLAR NUCLEI with the second male gamete in angiosperms.

triple junction (*Geol*) The focal point of three tectonic plates, eg at CONVERGENCE ZONES, DIVERGENCE ZONES or TRANSFORM FAULTS.

triple point (*Phys*) The temperature and pressure at which the three phases of a substance can co-exist (see fig. at PHASE DIAGRAM). The triple point of water is the equilibrium point (273·16 K and 610 N m^{-2}) between pure ice, air-free water and water vapour obtained in a sealed vacuum flask. It is one of the fundamental FIXED POINTS of the international practical scale of temperature. See KELVIN THERMODYNAMIC SCALE OF TEMPERATURE.

triple-pole switch (*ElecEng*) A switch for simultaneously making or breaking a three-wire electric circuit.

triple superphosphate (*ChemEng*) The product obtained by reacting phosphate rock with phosphoric acid giving a higher concentration of soluble calcium phosphate than obtains in ordinary or 'single' *superphosphate*.

triplet (*BioSci*) The sequence of three bases in an mRNA which specify a particular amino acid. See panel on DNA AND THE GENETIC CODE.

triplet (*Chem*) A state in which there are two unpaired electrons.

triplet (*Phys*) The three resonance peaks in spectroscopy, often produced by two unpaired electrons (electron spin resonance) or two hydrogen bonds (nuclear magnetic resonance). Singlet, doublet, etc, are similarly defined.

triple vaccine (*Med*) A vaccine containing a mixture of diphtheria toxoid, tetanus toxoid and pertussis vaccine. It is routinely used to produce active immunity against diphtheria, tetanus and whooping cough in infants.

Triplex glass (*Glass*) A patented form of laminated glass. See SAFETY GLASS.

triplex winding (*ElecEng*) A dc armature winding having three parallel paths per pole between positive and negative terminals.

triploblastic (*BioSci*) Having three types of tissue in the body, there being mesoderm between the ectoderm and endoderm, which gives rise to connective, skeletal and muscular tissues etc. Cf DIPLOBLASTIC.

triploid (*BioSci*) Having three times the haploid number of chromosomes for the species.

tripod (*Surv*) A device by which some surveying and other instruments are supported firmly off the ground. It consists of three legs hinged to a common head on which the instrument is secured. In photography, used for steadying a camera for long exposure, telephoto shots, etc.

tripod bush (*ImageTech*) Part of a camera body with a threaded hole to accept the screw of a tripod head.

tripod drill (*MinExt*) A rock drill on a heavy tripod.

tripod head (*ImageTech*) The mounting which attaches the camera to a tripod, usually including means for its rotation and tilt; for a motion picture camera, continuous smooth movement may be provided by geared handles.

Tripoli powder (*Min*) See TRIPOLITE.

tripolite (*Min*) A variety of opaline silica which is formed from the siliceous frustules of diatoms. It looks like earthy chalk or clay, but is harsh to the feel and scratches glass. When finely divided it is sometimes called *earthy tripolite*. Also *diatomite, infusorial earth, Tripoli powder*.

tripotassium dicitratobismuthate (*Pharmacol*) A bismuth chelate which promotes healing of gastric and duodenal ulcers.

trippage (*Genrl*) The number of times a product is reused as part of its normal product life cycle; thus the UK milk bottle is reused on average eight times before being scrapped for CULLET.

trip relay (*ElecEng*) A relay controlling the electromagnetic tripping mechanism of a circuit breaker.

trip switch (*ElecEng*) A control switch for closing the tripping circuit of a circuit breaker.

triptan (*Pharmacol*) Any of a family of drugs used for the acute treatment of migraine attacks.

trip value (*ElecEng*) The current or voltage required to operate a relay.

triquetrous (*BioSci*) A triangular stem with acute angles and concave faces. Cf TRIGONOUS.

trisaccharides (*Chem*) Carbohydrates resulting from the condensation of three monosaccharides with the elimination of two molecules of water, eg maltotriose in $G\alpha1 \rightarrow 4G\alpha1 \rightarrow 4G$ (G=glucose).

trisecting the angle (*MathSci*) The ancient problem of dividing a given angle into three equal parts using only ruler and compasses. This is impossible except for the special case when the angle may be expressed as $a90°/2^b$, where a and b are integers.

trismus (*Med*) Lockjaw; tonic spasm of the muscles of the jaw, causing the jaws to be clenched, as in tetanus.

trisomic (*BioSci*) Said of an otherwise normal diploid organism in which one chromosome type is represented thrice instead of twice.

trisomy 21 (*Med*) Condition in which an individual has three copies of chromosome 21, either in all cells or in a proportion of cells. Affected individuals show many abnormalities to varying degrees of severity, including the characteristic 'mongoloid' eye fold, and usually subnormal intelligence. Incidence increases with mother's age. Also *Down's syndrome*.

tristearin (*Chem*) Naturally occurring triglyceride in which all three fatty acid chains arise from stearic acid.

tristimulus values (*Phys, Textiles*) The amounts of the three primary colours (blue, red and green) that form the colour being examined or matched. A tristimulus colorimeter can analyse a colour and indicate the amounts present of its constituent primary colours.

tritanopic (*Med*) Colour blind to blue.

TRITC (*BioSci*) Abbrev for TETRAMETHYL RHODAMINE ISOTHIOCYANATE.

tritium (*Phys*) The radioactive isotope of hydrogen of mass number 3, ram 3·0221, half-life 12·3 yr. Symbol T. It is very rare, the abundance in natural hydrogen being 1 atom in 10^{17}, but can be produced artificially by neutron absorption in lithium. Used in thermonuclear weapons and fusion reactors. It also can be used to label any compound containing hydrogen and is consequently of great importance in biology.

tritium unit (*Chem*) A unit corresponding to the presence of 1 atom of tritium in 10^{18} atoms of hydrogen which represents seven disintegrations per minute in 1 litre of water. Pure tritium is 10^{18} tritium units.

tritolylphosphate (*Chem*) Polyvinyl chloride plasticizer with high-temperature resistance, although sensitive to hydrolysis. Abbrev *TTP*.

Triton (*Astron*) The principal natural satellite of Neptune, discovered in 1846. Distance from the planet 355 000 km; diameter 2700 km. Uniquely, for a large moon, its orbit is retrograde about Neptune.

triton (*Phys*) The tritium nucleus, consisting of one proton combined with two neutrons.

triton X-100 (*BioSci*) Iso-octylphenoxypolyethoxyethanol. A non-ionic detergent which is commonly used to solubilize membrane proteins in their biologically active state.

tritor (*BioSci*) The masticatory surface of a tooth.

triturate (*Chem*) To grind to a fine powder, esp beneath the surface of a liquid.

trityl (*Chem*) The *triphenylmethyl* group, $C(C_6H_5)_3$; the triphenylmethyl radical is generally accepted as being the first organic radical to be obtained in a free state.

trivalent (*BioSci*) Said of association of three chromosomes at meiosis.

trivalent (*Chem*) Capable of combining with three atoms of hydrogen, or their equivalent.

trixylylphosphate (*Chem*) Polyvinyl chloride plasticizer with high-temperature resistance, although sensitive to hydrolysis. Abbrev *TXP*.

tRNA (*BioSci*) Abbrev for TRANSFER RNA.

trocar (*Med*) A sharp-pointed perforator which is used to introduce a cannula into the body. Also *trochar*.

trochal (*BioSci*) Wheel-shaped.

trochanter (*BioSci*) (1) The second joint of the leg in insects. (2) A prominence for muscled attachment near the head of the femur in vertebrates.

trochlea (*BioSci*) Any structure shaped like a pulley, esp any foramen through which a tendon passes. Adj *trochlear*.

trochoid (*MathSci*) See ROULETTE.

trochoidal mass analyser (*NucEng*) A form of mass spectrometer in which the ion beams traverse trochoidal

paths within electric and magnetic fields mutually perpendicular.

trochophore (*BioSci*) A free-swimming pelagic larval form of Annelida, Mollusca and Bryozoa, possessing a prominent pre-oral ring of cilia and an apical tuft of cilia. Also *trochosphere*.

trochotron (*NucEng*) Abbrev for *trochoidal magnetron*. High-frequency counting tube, which uses crossed electric and magnetic fields to deflect a beam onto radially disposed electrodes.

troctolite (*Geol*) A coarse-grained basic igneous rock, consisting essentially of olivine and plagioclase only. The former mineral occurs as dark spots against the feldspar, giving the rock a characteristic spotted appearance, whence the name *troutstone* (from the Gk).

Trogamid (*Plastics*) TN for non-crystalline and glassy NYLON with complex, mixed aliphatic/aromatic backbone chain. Used for mouldings.

troilite (*Min*) A non-magnetic iron sulphide, FeS, which occurs mainly in meteorites.

Trojan group (*Astron*) A number of minor planets, named after the heroes of the Trojan War, which have the same mean motion as Jupiter and travel in the same orbit. They are divided into two clusters, one of which is 60° of longitude ahead of Jupiter, the other 60° behind; each minor planet oscillates about a point which forms an equilateral triangle with Jupiter and the Sun. These are two particular solutions of the THREE-BODY PROBLEM.

trojan horse (*ICT*) A destructive piece of PROGRAM CODE introduced to a system maliciously which appears to function correctly. At some predetermined time or when some predetermined conditions are met the software damages the computer system by destroying other programs and data. See HACKER, VIRUS.

trolleybus (*ElecEng*) A rail-less passenger vehicle, powered by current transmitted from two overhead trolley wires through two roof-mounted *trolley poles*.

trolley system (*ElecEng*) The overhead current-collecting system used on tramcars and trolley buses, in which a small grooved wheel or skid runs under the contact wire.

trombone (*ICT*) A U-shaped length of waveguide or transmission line that is of adjustable length for use in a waveguide circuit.

trommel (*MinExt*) A cylindrical revolving sieve for sizing crushed ore or rock.

trondhjemite (*Geol*) A coarse-grained igneous rock consisting essentially of plagioclase (ranging from oligoclase to andesine), quartz and small quantities of biotite.

troph-, tropho- (*Genrl*) Prefixes from Gk, *trophe*, nourishment.

trophallaxis (*BioSci*) Mutual exchange of food between *imagines* (sing *imago*) and their larvae, as in some social insects.

trophectoderm (*BioSci*) The extra-embryonic part of the ectoderm of mammalian embryos at the blastocyst stage.

trophic (*BioSci*) Pertaining to nutrition.

trophic level (*BioSci*) Broad class of organisms within an ecosystem characterized by mode of food supply. The first trophic level comprises the green plants, the second is the herbivores and the third is the carnivores which eat the herbivores.

trophic structure (*BioSci*) A characteristic feature of any ecosystem, measured and described in terms of either the standing crop per unit area, or energy fixed per unit area per unit time, at successive trophic levels. It can be shown graphically by the various ECOLOGICAL PYRAMIDS.

tropho- (*Genrl*) See TROPH-.

trophoblast (*BioSci*) The differentiated outer layer of epiblast in a segmenting mammalian ovum.

trophozoite (*BioSci*) In Protozoa, the trophic phase of the adult, which generally reproduces by schizogony.

tropical month (*Astron*) The period of lunar revolution with respect to the equinox (27·321 58 days).

tropical revolving storm (*EnvSci*) A small intense cyclonic depression originating over tropical oceans. Also *cyclone, hurricane, typhoon*, depending on the locality.

tropical switch (*ElecEng*) A switch mounted on feet or bosses with an air space between its base and mounting surface used in excessively damp climates. Also *feet-switch*.

tropical year (*Astron*) The interval between two successive passages of the Sun in its apparent motion through the First Point of Aries; hence the interval between two similar equinoxes or solstices and the period of the seasons; its length is 365·242 194 mean solar days.

tropism (*BioSci*) A reflex response of a cell or organism to an external stimulus; movement that orients an organism to achieve a certain distribution of stimulation. See GEO-TROPISM, PHOTOTROPISM. Cf TAXIS.

tropomyosin (*BioSci*) A filamentous protein aligned along the actin fibres of muscle. Under the influence of troponin it controls the interaction of actin and myosin.

troponin (*BioSci*) A complex of three polypeptide chains which mediates the effect of calcium on muscle contraction.

tropopause (*EnvSci*) The upper limit of the troposphere, where the lapse rate of temperature becomes small ($\leqslant 2°C$ km^{-1}). Sometimes a single unique tropopause cannot be defined and there is a multiple tropopause structure. See panel on TROPOSPHERE.

troposphere (*EnvSci*) The lower part of the atmosphere extending from the surface up to a height varying from about 9 km at the poles to 17 km at the equator, in which the temperature decreases fairly regularly with height. See panel on TROPOSPHERE.

tropospheric (*ICT*) Said of reflection, absorption or scattering of a radio wave when encountering variations in the troposphere.

tropospheric scatter (*ICT*) Propagation in which radio waves are scattered by the troposphere. It does not depend critically on frequency, but is generally used for communication over several hundred km at ultrahigh frequency and low microwave frequencies. High power is normally used because of the process's inefficiency.

tropospheric wave (*ICT*) A radio wave whose path between two points at or near the Earth's surface lies wholly within the troposphere and will be governed by meteorological conditions.

Trotter photometer (*Phys*) A portable photometer in which the brightness of the comparison screen is varied by tilting.

trouble-shooting (*Electronics*) See FAULT-FINDING.

trough (*EnvSci*) A V-shaped extension of the isobars from a centre of low pressure.

troughed belt conveyor (*Eng*) A CONVEYOR comprising an endless belt which is made hollow or troughed on the carrying run to increase the carrying capacity and to obviate spillage.

trough gutter (*Build*) A gutter used along roof valleys or parapets. Also *box gutter*.

trouser tear test (*Eng*) Test method for measuring tearing energy of various materials of film or sheet form, after shaping the test specimen the legs are grasped and pulled in tension. See STRENGTH MEASURES.

Trousseau's phenomenon (*Med*) Spasm of the muscles of a limb whose blood vessels are compressed, occurring in tetany.

Troutonian fluid (*Phys*) A fluid whose coefficient of tensile viscosity is unaffected by increasing stress. Equivalent to Newtonian fluids tested in shear. Most polymer melts are Troutonian, but low-density polyethylene is tension-stiffening.

Trouton's rule (*Chem*) For most non-associated liquids, the ratio of the latent heat of vaporization per mole, measured in joules, to the boiling point, on the absolute scale of temperature, is approx equal to 88 at atmospheric pressure.

troutstone (*Geol*) See TROCTOLITE.

trowel (*Build*) A flat steel tool used for spreading and smoothing mortar or plaster.

Trp (*Chem*) Symbol for TRYPTOPHAN.

TR tube (*Electronics*) See TRANSMIT–RECEIVE TUBE.

truck (*Eng*) US for BOGIE.

truck-type switchgear (*ElecEng*) Switchgear in which each circuit breaker and associated equipment is mounted on a truck that can be completely removed from the gear for maintenance and repair. Also *carriage-type switchgear*.

true absorption coefficient (*Phys*) The absorption coefficient applicable when scattered energy is not regarded as absorbed. Applicable to broad-beam conditions. Also *real absorption coefficient*.

true airspeed (*Aero*) The actual speed of an aircraft through the air, computed by correcting the indicated airspeed for altitude, temperature, position error and compressibility effect. Abbrev *TAS*

true altitude (*Astron*) The ALTITUDE of a heavenly body as deduced from the apparent altitude by applying corrections for atmospheric refraction, for instrumental errors, and where necessary for geocentric parallax, Sun's semi-diameter, and dip of horizon.

true azimuth (*Surv*) That measured relative to true geographical north.

true bearing (*Surv*) That measured clockwise from true geographical north.

true bias (*Textiles*) The direction in a fabric along which it will show the greatest elongation in response to a given load.

true coincidence (*Phys*) Coincidence in a COINCIDENCE COUNTER produced by a single particle discharging both or all counters. Cf SPURIOUS COINCIDENCE.

true course (*Ships*) The angle between the true meridian and the direction of the ship's head.

true density (*PowderTech*) The mass of the particle divided by its volume, excluding open and closed pores.

true horizon (*Surv*) A great circle of the celestial sphere parallel to the horizon and passing through the Earth's centre. Also *rational horizon*.

true north (*Surv*) The direction of the geographical north pole.

true resistance (*ElecEng*) See DC RESISTANCE.

true section (*Surv*) A section which has been drawn, with the same scales, horizontally and vertically.

true strain (*Eng*) See STRAIN.

true stress (*Eng*) See STRESS.

true-type fonts (*ICT*) An alternative standard to POST-SCRIPT or PCL for the specification of TYPEFACES and fonts. These fonts are SCALABLE and are available both as PRINTER FONTS and SCREEN FONTS.

true watts (*Phys*) Power dissipated in an ac circuit.

trumpet arch (*Build*) See SPLAYING ARCH.

truncate (*BioSci*) Square-ended base or apex of a structure, as if cut off.

truncated spur (*Geol*) See VALLEY.

truncation (*ICT*) (1) Ending of a computational procedure in accordance with some program rule as soon as a specified accuracy has been reached. (2) Rejection of final digits in a number, thus lessening precision (but not necessarily accuracy). See ROUNDING.

truncation error (*ICT*) Error introduced by TRUNCATION.

truncus (*BioSci*) A main blood vessel; as the *truncus transversus* or CUVIERIAN DUCT and the *truncus arteriosus* or great vessel, through which blood passes from the ventricle. Also *trunk*.

trunk (*Arch*) Shaft of a column.

trunk (*BioSci*) (1) The upright, massive main stem of a tree. (2) The body, apart from the limbs. (3) The proboscis of an elephant. (4) See TRUNCUS.

trunk (*ElecEng*) See TRUNK FEEDER.

trunk (*ICT*) US for LINK.

trunk call (*ICT*) In the UK, a telephone call from one telephone area to another, involving links between two trunk centres each dealing with all calls passing in or out of its area.

trunk circuit (*ICT*) In the UK, a two- or four-wire connection between trunk centres, for establishing trunk

Troposphere

The lower part of the atmosphere which includes the ATMOSPHERIC BOUNDARY LAYER (see panel) in which virtually all life on Earth exists. It extends from the surface up to a height varying from about 9 km at the poles to 17 km at the equator, in which the temperature decreases fairly regularly with height.

Almost all the Sun's energy, which drives the general circulation of the troposphere and creates the weather, enters the atmosphere by means of the troposphere. At low latitudes an excess of solar heat energy is received at the surface of the Earth and is then distributed upwards and polewards by the wind until it is lost by radiation out to space. The Sun's radiative energy is contained mainly in the visible region of the spectrum to which the atmosphere, apart from clouds, is largely transparent. This energy is absorbed by soil, vegetation and the upper layers of the seas and oceans, and it then either heats the lowest part of the atmosphere or evaporates large quantities of water. Outward radiation by the atmosphere takes place at infrared wavelengths and is inadequate to compensate

In reality, continental land masses and major mountain ranges cause much divergence from this simple picture, especially in the northern hemisphere; there are also seasonal variations.

Schematic representation of sea level winds and pressure systems on an idealized Earth The mean upper-level flow in a meridional verticle cross-section is also indicated.

calls between telephone areas. In the US, a circuit between exchanges in the same telephone area. Also *trunk line.*

trunk conveyor (*MinExt*) In colliery, belt conveyor in main road.

trunk exchange (*ICT*) An exchange in a telephone area that is connected by links to other trunk exchanges, and to subscribers through local exchanges.

trunk feeder (*ElecEng*) A feeder connecting two generating stations, or a generating station and a large substation. Also *trunk main.*

trunking (*ICT*) The cables that contain the links between one rank of selectors and others in the sequence of operation, the cables taking a common route through the exchange building.

trunk junction circuit (*ICT*) The junction between an exchange and the trunk exchange for routing subscribers to the trunk exchange system.

trunk line (*ICT*) See TRUNK CIRCUIT.

trunk main (*ElecEng*) See TRUNK FEEDER.

trunk piston (*Eng*) A piston, long in relation to its diameter, used where there is no piston rod or crosshead, the piston having to take the connecting-rod thrust; most internal-combustion engine pistons are of this type.

trunnion axis (*Surv*) The horizontal axis about which the telescope of a theodolite or tacheometer may be rotated on its trunnion bearings.

trunnion mounting (*Eng*) A pair of short journals, supported in bearings, projecting coaxially from opposite sides of a vessel or cylinder required to pivot about their axis.

truss (*CivEng*) A framed structure built up entirely from tension and compression members, arranged in panels so as to be stable under load; used for supporting loads over long spans.

truss Whipple–Murphy or N-type.

for received solar radiation roughly between latitudes 40°N and 40°S. The excess of latent and sensible heat in the lowest atmospheric layer between these latitudes induces systematic wind fields, vertical as well as horizontal, which dominate global weather patterns.

Because surface winds are subject to friction and loss of KINETIC ENERGY at the Earth's surface, they exchange angular momentum with the solid rotating Earth. There are also a number of constraints on their general circulation: (1) the need to conserve the angular momentum of the global wind system; (2) all winds are subject to the CORIOLIS FORCE because the Earth rotates; (3) total energy must be conserved, though it is transformed between kinetic and potential energy, the latent heat of water, etc. The net effect of all these processes is roughly as follows.

Over the tropics, air rises by vertical convection causing heavy showers and thunderstorms with much release of latent heat up to heights of 12 km, and then it spreads north and south. In the subtropics, air subsides gently to levels near the surface where some returns equatorwards to form the TRADE WINDS. Polewards of this subtropical high-pressure belt, the potential energy derived from the contrast between warm tropical air and cold polar air causes the formation of typical travelling mid-latitude depressions and anticyclones (H in the diagram), in which vast quantities of warm moist air are carried upwards, polewards and eastwards, with cold air drawn down equatorwards as a replacement.

These mid-latitude weather systems are not minor eddies superimposed on a much greater general circulation pattern. Where they occur they *are* the general circulation, and charts of mean values can be very misleading. In polar regions tropospheric winds are generally much lighter again and air subsides with the strong radiational cooling which dominates the energy balance.

The various constraints and physical laws already mentioned give rise to two important tropospheric jet streams in each hemisphere: one in the subtropics at about 12 km altitude and fairly constant latitude of 30°; and one associated with the polar front and the formation of mid-latitude depressions. The polar frontal jet is much stronger in winter than in summer and varies considerably in position and orientation, often forming a wave-like pattern around the hemisphere (see diagram).

Systematic complications to all these patterns are caused by the geographical features of mountain masses and chains, and the relative distribution of land and sea, esp in the northern hemisphere; the MONSOONS are typical examples.

The oceans, which cover 71% of the Earth's surface, absorb twice as much solar energy as the land. This energy is eventually given up to the atmosphere as sensible and latent heat, much of it in higher latitudes. Probably about 40% of the total heat which is transported out of the tropics and subtropics is provided directly by the oceans and there is good evidence that anomalies of sea-surface temperature have far-reaching effects on weather patterns.

The motions described so far are largely confined to the troposphere, which accounts for about 80% of the total atmospheric mass and is bounded by the TROPOPAUSE, where the temperature decrease is $\leqslant 2°C\ km^{-1}$.

See panels on ATMOSPHERIC BOUNDARY LAYER, EARTH and STRATOSPHERE AND MESOSPHERE.

---

truss *(Med)* A surgical appliance consisting of a pad incorporated in a spring or belt for retaining a reduced hernia in place.

truss-beam *(Build)* A framework acting as a beam.

trussed partition *(Build)* A partition which is framed so as to be self-supporting between its ends, used when the floor cannot carry the weight.

truth *(Phys)* See TOP.

truth condition *(Genrl)* In logic, the circumstances that must be satisfied for a statement to be true.

truth function *(Genrl)* In formal logic, a function that determines whether or not a complex statement is true, depending on the truth values of the component parts of the sentence. 'And' is a truth function; 'but' is not.

truth mark *(Textiles)* Some permanent mark near the end of a piece of fabric which should remain until the fabric is delivered to the customer. Its presence shows that no end of fabric has been improperly retained by anyone.

truth value *(ICT)* The truth values of BOOLEAN ALGEBRA are TRUE and FALSE (often abbrev *T* and *F*) which may be represented by the binary digits 1 and 0.

truth-value *(MathSci)* In logic, of a statement: either of the values *true* or *false*.

trying plane *(Build)* A tool similar to the jack plane but about 22 in (56 cm) long; used after the jack plane to obtain a straight and true surface.

trypan blue *(BioSci)* An azo dye that is used as a test for cell viability; dead cells stain blue but live cells exclude the dye.

trypano- *(Genrl)* Prefix from Gk *trypanon*, borer.

trypanosomes *(BioSci)* A group of flagellate Protozoa. Many cause disease in humans (see TRYPANOSOMIASIS) and animals.

trypanosomiasis *(Med)* A disease, occurring in parts of S America (*Chagas' disease*), caused by *Trypanosoma cruzi*, and Africa, caused by *T. brucei* and related species. The S American form is transmitted by reduviid bugs, the African by the tsetse fly (see also *nagana*). Also *sleeping sickness*.

trypsin *(BioSci)* A peptidase, secreted by the pancreas, that is specific for peptide bonds adjacent to lysine and arginine residues.

tryptophan *(Chem)* *2-amino-3-indolepropanoic acid*. An amino acid. The L- or s-isomer is a constituent of proteins. Symbol Trp, short form W.

try square *(Build)* A tool having the blade fixed at 90° for testing squareness.

tschermakite (*Min*) An end-member subspecies in the hornblende group of amphiboles, rich in aluminium and calcium.

T-section cramp (*Build*) Strong form of SASH CRAMP having a T-section steel bar for the sliding members.

T-section filter (*ICT*) T-network ideally formed of non-dissipative reactances, having frequency pass band over which attenuation is theoretically zero or very low. See BUTTERWORTH FILTER, CHEBYSHEV FILTER.

tsetse fly disease (*Vet*) See NAGANA.

TSR program (*ICT*) See MEMORY-RESIDENT PROGRAM.

TSS (*FoodSci*) Abbrev for TOTAL SOLUBLE SOLIDS.

tsunami (*Geol*) A destructive sea wave caused by an earthquake or submarine eruption. Because of its very long wavelength it behaves as a 'shallow' SURFACE WAVE. Its amplitude in mid-ocean is very small; as it approaches land, the amplitude builds up and all the energy of the original disturbance is concentrated into a few wavelengths with devastating results. Sometimes erroneously called a *tidal wave*.

T-suppressor cell (*BioSci*) A set of T-cells involved in suppressing specific B-cell differentiation into antibody-secreting cells. There is also evidence for T-suppressors of T-cell functions.

tsutsugamushi fever (*Med*) See SHIMAMUSHI FEVER.

T-tail (*Aero*) A *tail unit* characterized by positioning the horizontal stabilizer at or near the top of the vertical stabilizer. The unit is employed on aircraft having rear-mounted engines, and has variable incidence. See ALL-MOVING TAIL.

T Tauri star (*Astron*) A type of star showing irregular variability (prototype T Tauri), with a spectrum dominated by strong emission lines due to a powerful stellar wind. Occurs in groups, usually embedded in clouds of gas and dust, and is thought to represent one of the early evolutionary stages of a star of similar mass to the Sun.

t-test (*MathSci*) A group of statistics used to determine if a significant difference exists between the means of two sets of data.

TTL (*Electronics*) Abbrev for *transistor–transistor logic*. Referring to logic circuits consisting of two or more directly interconnected transistors intended to drive capacitive loads at high rates.

TTL (*ImageTech*) Abbrev for *through the lens*. Referring to cameras in which the viewfinder picture is provided by the same lens which forms the exposed image; such cameras will then incorporate *TTL autofocus*, *TTL exposure metering* and *TTL flash metering*.

TTP (*Chem*) Abbrev for TRITOLYLPHOSPHATE.

TTS (*Acous*) Abbrev for TEMPORARY THRESHOLD SHIFT.

TTT diagram (*Eng*) Abbrev for *time–temperature transformation diagram*. See ISOTHERMAL TRANSFORMATION DIAGRAM.

T-tubule (*BioSci*) See TRANSVERSE TUBULE.

TU (*ICT*) Abbrev for *traffic unit, transmission unit*.

tub (*MinExt*) A tram, wagon, corf or corve.

tubbing (*MinExt*) The lining of a circular shaft, formed of timber or by steel segments.

tube (*BioSci*) (1) The cylindrical proximal part of a calyx or corolla in which the sepals or petals are fused at their edges. (2) See TUBICOLOUS.

tube (*Electronics*) (1) Enclosed device with gas at low pressure, depending for its operation on ionization originated by electrons accelerated from a cathode by a field applied by an anode. (2) US for all vacuum and gas-discharge devices. The term now widely used in the UK where formerly 'valve' was almost universal.

tube drawing (*Eng*) The production of seamless tubes by drawing a large, roughly formed tubular piece of material through dies of progressively decreasing size, usually using a mandrel or plug to control the internal diameter. Cf MANNESMANN PROCESS.

tube extrusion (*Eng*) A method of producing tubes by direct extrusion from a billet, using a mandrel to shape the inside of the tube.

tube feet (*BioSci*) See PODIUM (2).

tube fuse (*ElecEng*) A fuse in which the fuse wire is enclosed in an insulating tube. Also *cartridge fuse*.

tubeless tyre (*Autos*) One in which the air seal is provided by adhesion between the beads and the wheel rim.

tube mill (*MinExt*) Horizontal mill in which diameter/length ratio is usually high compared with that of standard ball mill, and which has high discharge.

tube of force (*Phys*) A space enclosed by all the lines of force of an electric or magnetic field passing through a closed contour. Of unit magnitude when it contains unit flux.

tube plate (*Eng*) End wall of a surface condenser, between which the water tubes are carried; they are bolted between the casing and water-chamber covers. See CONDENSER TUBES.

tuber (*BioSci*) A swollen underground stem acting as a storage and perennating organ, eg the potato.

tubercle (*BioSci*) (1) Small circular swelling or nodule. (2) The dorsal articulator process of a rib. (3) A cusp of a tooth. Also *tuberculum*. Adjs *tubercled, tubercular, tuberculate, tuberculose*.

tubercle (*Med*) (1) Any small rounded projection on a bone or other part of the body. (2) A solid elevation of the skin larger than a papule. (3) A small mass or nodule of cells resulting from infection with the bacillus of tuberculosis. (4) Loosely, tuberculosis; the tubercle bacillus.

tubercular (*Med*) Of, pertaining to, resembling, or affected with, nodules (tubercles); less correctly, affected with tuberculosis (ie tuberculous).

tuberculate (*BioSci*) Covered with small wart-like projections.

tuberculid (*Med*) Any skin lesion due to infection with bacillus of tuberculosis. Also *tuberculide*.

tuberculin (*Med*) A protein or mixture of proteins derived from *Mycobacterium tuberculosis*, which is employed in the tuberculin test as a diagnostic reagent for detecting sensitization by, or infection with, *M. tuberculosis*. Old tuberculin (OT) is a heat-concentrated filtrate from the medium in which the organism has been grown. Purified protein derivative (tuberculin PPD) is a soluble protein fraction, precipitated by trichloroacetic acid from a synthetic medium in which *M. tuberculosis* has been grown. Tuberculins can be derived from human, bovine or avian strains of the bacillus but show extensive antigenic cross-reactivity.

tuberculin test (*BioSci*) A test for delayed hypersensitivity to TUBERCULIN in humans or other animals. Positive reactions are presumptive evidence of cell-mediated immunity to, and therefore of past or present exposure to, *Mycobacterium tuberculosis*, but do not necessarily indicate active disease. See MANTOUX TEST.

tuberculoma (*Med*) A slow-growing, circumscribed tuberculous lesion, sometimes present in the brain.

tuberculosis (*Med*) Infection by *Mycobacterium tuberculosis*, esp of the lungs; characterized by the development of tubercles in the bodily tissues and by fever, anorexia and loss of weight. Spread by air droplets and the bovine form by raw milk. Abbrev *TB*.

tuberculous (*Med*) Pertaining to, affected with, or caused by, tuberculosis.

tuberose sclerosis (*Med*) See TUBEROUS SCLEROSIS.

tuberosity (*BioSci*) A prominence on a bone, generally from muscle attachment, esp prominences near the head of the humerus.

tuberous (*BioSci*) Of or like a tuber; having tubers.

tuberous sclerosis (*Med*) A condition in which hyperplasia of the neuroglia gives rise to hard, tumour-like masses in the brain, associated with epilepsy and mental deficiency; the disease is part of the developmental defect known as EPILOIA. Also *tuberose sclerosis*.

tube sinking (*Eng*) Drawing an existing tube through a die or rolls to reduce its diameter without an interior plug or mandrel.

tubicolous (*BioSci*) Living in a tube.

tubifacient (*BioSci*) Tube-building, as certain Polychaeta.

tub sizing (*Paper*) The action of applying to the surfaces of the sheet or web a solution of gelatine size contained in a bath. The film of gelatine should ideally be gently dried by hot air to confer the maximum benefits of ink resistance, strength and durability.

tubular rivet (*Eng*) A rivet with a shank from which the centre has been removed to leave a thin wall, so that the rivet can be used to punch its own hole in thin, soft materials and to facilitate setting.

tubular scaffold (*Build*) A form of scaffold constructed of steel tubes which can be clamped together in any desired manner by special steel collar pieces with screw fixings.

tubule (*BioSci*) Any small tubular structure. Also *tubulus*. Adjs *tubulate, tubuliferous, tubuliform, tubulose*. See MICRO-TUBULE.

tubulin (*BioSci*) A highly conserved globular protein abundant in the cytoplasm of eukaryotic cells. There are three closely related variants: α and β tubulin form a heterodimer from which microtubules are assembled. γ-tubulin is restricted to the centrosome.

tubulus (*BioSci*) See TUBULE.

Tucana (Toucan) (*Astron*) A faint southern constellation, which includes the prominent Small Magellanic Cloud.

tucker blade (*Print*) Also *tucking blade*. See FOLDING BLADE.

tuck pointing (*Build*) Pointing finished by cutting a groove in the surface at the joints and tucking into the groove a narrow projecting artificial joint of putty.

tufa (*Geol*) A porous, concretionary or compact form of calcium carbonate which is deposited from solution around springs, of which the dense variety is called tufa.

tuff (*Geol*) A rock formed of compacted volcanic fragments, some of which can be distinguished by the naked eye. If the fragments are larger, then the rock grades into an agglomerate.

Tufnol (*Plastics*) A proprietary laminated plastic; light-weight, tensile strength approx 55–110 MN m^{-2}; strong insulation qualities, but increasingly replaced as an engineering plastic for bearings, gear wheels and pulleys by ACETAL RESIN.

tufted (*BioSci*) Grass shoots, clustered or clumped rather than scattered. Also *caespitose*.

tufted carpet (*Textiles*) A carpet formed by inserting with needles U-shaped lengths of yarn, or similar, into a strong backing material (eg a hessian or polyolefin fabric or a plastic foam).

tularaemia (*Vet*) A disease of rodents due to infection with *Pasteurella tularensis*. Spread by fleas and ticks and can infect humans. The human disease is characterized by prolonged fever, enlargement of the lymph glands, depression and emaciation. US *tularemia*.

tulip tree (*For*) See AMERICAN WHITEWOOD.

tulipwood (*For*) See BRAZILIAN TULIPWOOD.

tulle (*Textiles*) Traditionally, a fine plain-woven silk net. Now also applied to a net with hexagonal holes produced on a warp-knitting machine.

tumble-home (*Ships*) A term defining the narrowing of a ship's breadth. It is the measure of the inward fall when the deck breadth is less than the maximum breadth.

tumbler gear (*Eng*) A gear in a train, mounted on a pivot arm so that it can be swung into and out of engagement with an adjacent gear.

tumbler switch (*ElecEng*) A small single-pole switch having a quick-break action, universally used in electric-lighting installations for controlling individual lamp circuits.

tumbling (*Eng*) A method of removing sand, irregularities, etc, from castings or forgings by rotating them in a box with abrasives or special metal slugs.

tumbling-in (*Build*) The brickwork forming the top surface of a pier and sloping in towards the general face of the wall.

tumbu disease (*Med*) A disease, common in C and W Africa, due to invasion of the surface of the body by the larvae of the tumbu fly, *Cordylobia anthropophaga*; characterized by the formation of a boil or a warble in the skin.

tumefaction (*Med*) The process or act of swelling; the state of being swollen. Also *tumescence*. Adj *tumescent*.

tumid (*BioSci*) Swollen; inflated.

tumour (*Med*) Any swelling or morbid enlargement. The term now usually denotes neoplasm, a non-inflammatory mass formed by the growth of new cells in the body and having no physiological function. An *innocent tumour* or *benign tumour* is encapsulated and usually solitary, pressing upon, but not invading, adjacent tissues; a *malignant tumour* (CARCINOMA, SARCOMA) invades tissues, tends to recur and spreads to other parts of the body. US *tumor*.

tumour angiogenesis factor (*BioSci*) A substance(s) released from a tumour that promotes vascularization of the mass of neoplastic cells; once vascularized, the tumour will grow more rapidly, and is more likely to metastasize.

tumour-inducing principle (*BioSci*) See TI PLASMID.

tumour necrosis factor (*BioSci*) A type of pro-inflammatory cytokine that has a wide range of actions, including the inhibition of some tumour cells. TNFα was formerly known as *cachectin*, TNFβ as *lymphotoxin*.

tumour promoter (*BioSci*) An agent that increases the probability of tumour formation by a previously applied primary carcinogen, but does not induce tumours when used alone. An important example is the phorbol ester, phorbol myristate acetate.

tumour-specific antigen (*BioSci*) Antigen present on tumour cells that is not expressed (or only very minimally) by their normal counterparts. No universal set of such antigens exists. Sometimes they are simply characteristic of rapidly proliferating cells, sometimes they are coded for by a tumour virus.

tumour suppressor gene (*BioSci*) A normal gene that codes for a product that regulates the CELL CYCLE (panel) so as to suppress cell division and growth. If the gene is inactivated or mutated, the cell switches to rapid division and tumour formation. Also *anti-oncogene*.

tumour virus (*BioSci*) Any virus that will induce tumours.

tunable dye laser (*Phys*) A laser in which the excited material is a dye in an organic solvent; the dye fluoresces over a wide range of wavelengths, so the laser output can be tuned by altering the parameters of the system.

tunable magnetron (*ICT*) A MAGNETRON in which the frequency can be altered electronically, eg by altering the anode voltage, or by mechanically changing the resonant frequencies of the cavities.

tunance (*ElecEng*) See SHUNT RESONANCE.

tundra (*EnvSci*) The vast, relatively flat treeless zone, with permanently frozen subsoil, found mainly in the Arctic regions of Alaska, N Canada and Siberia, but also on the fringes of the Antarctic region. Its vegetation consists of dwarf trees and shrubs, grasses, sedges, mosses and lichens.

tune (*Acous*) To adjust for resonance or syntony, esp musical instruments or radio receivers. See TUNING.

tuned amplifier (*ICT*) One containing tuned circuits, and therefore sharply responsive to particular frequencies.

tuned anode (*Electronics*) An inductor shunted by a capacitor (either or both of which may be variable) in series with the lead to the anode of a thermionic valve.

tuned-anode coupling (*Electronics*) That between stages of a high-frequency thermionic valve amplifier, in which the coupling impedance is a tuned anode circuit.

tuned antenna (*ICT*) One operating at its natural resonant frequency.

tuned-base oscillator (*Electronics*) One in which the TUNED CIRCUITS are in series with the base of a transistor.

tuned cell (*ICT*) Adjustable cavity in a waveguide structure, particularly in a filter section.

tuned circuit (*Electronics*) One comprising an inductor (*L*, henrys) and a capacitor (*C*, farads) in series (parallel) which offers a low (high) impedance to ac at the resonant frequency given by

$$f = \frac{1}{2\pi}\sqrt{LC} \text{ Hz}$$

See Q.

tuned-emitter oscillator (*ICT*) One in which the TUNED CIRCUITS are in series with the emitter of a transistor.

tuned radio-frequency receiver (*ICT*) A receiver that does not use frequency changing before detection. Abbrev *TRF receiver*.

tuned-rate gyro (*Aero, Space*) An advanced gyro used in high-performance inertial platforms in INERTIAL NAVIGATION SYSTEMS. A rotor within its casing is spun by an electric motor; applied angular rotations cause the rotor to oscillate which is nullified by pivot springs, so reducing random errors. Simpler and more reliable than FLOATED RATE-INTEGRATING GYRO.

tuned relay (*ICT*) One that responds only at a resonant frequency.

tuned transformer (*ElecEng*) Interstage coupling transformer in an amplifier in which one, or more usually both, windings are tuned to resonate with the signal frequency, giving a higher secondary voltage than without resonance.

TUNEL method (*BioSci*) Abbrev for *transferase-mediated dUTP nick-end labelling*, a method to identify cells undergoing apoptosis by labelling the ends of their fragmented DNA.

tuner (*ICT*) (1) Assembly of one or more tuned circuits to form a unit sharply responsive to particular frequencies. (2) Term often used to describe the front end of a receiver; eg an FM tuner precedes the audio amplifier for domestic entertainment, and the section of a TV receiver devoted to receiving and selecting the radio-frequency signal may be so described.

tungsten (*Chem*) Symbol W, at no 74, ram 183·85, rel.d. 19·1, mp 3370°C. A hard grey metal which is resistant to corrosion and is used in high-speed tool steel, cemented carbides for drills and grinding tools, and as wire in incandescent electric lamps. The main tungsten ores are wolfram (wolframite), $(MnFe)WO_4$, and scheelite, $CaWO_4$.

tungsten alloy (*Eng*) A protective material containing tungsten, copper and nickel, and having a density about 50% greater than that of lead, and thus providing better protection from ionizing radiation.

tungsten arc (*ElecEng*) A high-intensity arc of small dimensions, obtained between tungsten electrodes enclosed in a glass bulb.

tungsten bronze (*Eng*) Malleable, machinable alloys with high densities based on tungsten alloyed with 6–7% nickel and 3–4% copper or iron.

tungsten bronzes (*Chem*) Partially reduced WO_3; compounds of varying composition and colour with metallic lustre and conductivity. Similar compounds are formed by other elements capable of displaying two valencies.

tungsten–halogen lamp (*ElecEng*) See QUARTZ–IODINE LAMP.

tungsten inert gas welding (*Eng*) Electric welding in which the tungsten electrode is not consumed and a filler rod supplies the metal to the joint which is protected from reaction by an inert gas, eg argon. Abbrev TIG.

tungsten lamp (*ElecEng*) An electric lamp employing an incandescent tungsten filament.

tungstic acid (*Chem*) WO_3. The starting point for the preparation of tungsten metal. Also *tungsten oxide*, *tungsten (VI) oxide*.

tungstic ochre (*Min*) Hydrated oxide of tungsten. It is usually earthy and yellow or greenish in colour, and is a mineral of secondary origin, usually associated with wolframite. Also *tungstite*.

tunic (*BioSci*) An investing layer. Adj *tunicate*.

tunica (*BioSci*) The outer layer(s) of cells in the shoot apical meristem of many angiosperms, which give rise to the epidermis and which divide anticlinally and thus do not displace the underlying cells from the meristem. See TUNICA–CORPUS CONCEPT.

tunica–corpus concept (*BioSci*) The concept that the shoot apex in many angiosperms is organized into a TUNICA and a CORPUS the distinctness of which is maintained more or less indefinitely. The concept accounts for the existence of PERICLINAL CHIMERAS. See HISTOGEN.

tunicamycin (*BioSci*) A nucleoside antibiotic from *Streptomyces lysosuperificus* that inhibits N-glycosylation of proteins in eukaryotic cells.

Tunicata (*BioSci*) See UROCHORDATA.

tunicate (*BioSci*) Having a coat or covering. Also *tunicated*.

tunicate bulb (*BioSci*) A bulb composed of a number of swollen leaf bases each of which completely encloses the next younger, as in the onion.

tunicated (*BioSci*) Enclosed by a non-living test or mantle. See TUNICATE.

tuning (*Acous*) (1) The adjustment of tension in the strings of stringed instruments (piano, harp, violin) so that the specified notes emitted coincide in frequency with a standard scale, eg concert pitch. (2) The adjustment of length of pipes in organs to obtain the correct emitted pitch.

tuning (*ElecEng*) See CURRENT RESONANCE, VOLTAGE RESONANCE.

tuning (*ICT*) (1) Operation of adjusting circuit settings of a radio receiver so as to produce maximum response to a particular signal, generally by varying one or more capacitors and/or inductors. (2) Carrying out a similar process by electronic or thermal means. Also *tuning-in*.

tuning capacitor (*ICT*) Variable capacitor for tuning purposes, generally consisting of air-spaced vanes; several can be ganged.

tuning coil (*ICT*) See TUNING INDUCTANCE.

tuning control (*ICT*) Mechanical means for tuning a resonant circuit.

tuning curve (*ICT*) That relating the resonant frequency of a tuned circuit to the setting of the variable element, eg a capacitor.

tuning fork (*Acous*) A fork with two prongs and heavy cross-section, generally made of steel. Expressly designed to retain a constant frequency of oscillation when struck. Widely used for tuning musical instruments because its frequency is very insensitive to changes in temperature, atmospheric pressure and humidity. See MAINTAINED TUNING FORK.

tuning-in (*ICT*) See TUNING.

tuning indicator (*ICT*) See MAGIC EYE.

tuning inductance (*ICT*) Fixed or variable inductor used for tuning. Also *tuning coil*.

tuning screw (*Electronics*) A screw used to provide a variable reflection coefficient in a waveguide matching system. An alternative system to STUB TUNING.

tunnel burners (*Eng*) Industrial gas burners using a refractory tunnel at the burner exit for the main purpose of positive flame retention. The tunnel serves as an ignition zone, and accelerates the rate of flame propagation through turbulence and temperature rise, to a point where it is in equilibrium with the relatively high air–gas mixture velocity employed.

tunnel diode (*Electronics*) Junction diode with such a thin depletion layer that electrons bypass the potential barrier. See TUNNEL EFFECT. Negative resistance characteristics can be exhibited and such diodes can be used as low-noise amplifiers or as oscillators, up to microwave frequencies. Also *Esaki diode*.

tunnel effect (*Electronics, Phys*) Piercing of a narrow POTENTIAL BARRIER by a current carrier which is impossible according to classical physics, but has a finite probability according to wave mechanics.

tunnel furnace (*Eng*) Kiln through which material moves slowly on cars, racks or suspending gear.

tunnelling (*Phys*) See POTENTIAL BARRIER, TUNNEL EFFECT.

tunnel slots (*ElecEng*) See CLOSED SLOTS.

tunnel vault (*Build*) See BARREL VAULT.

tunnel windings (*ElecEng*) A term sometimes applied to armature windings in which the conductors are inserted, end-on, into closed slots.

tup (*Agri*) (1) Alternative name for a ram, an entire male sheep. (2) Of sheep, to mate.

Turbellaria (*BioSci*) A class of Platyhelminthes comprising forms of free-living habit, which may be marine, fresh water or terrestrial. They have a ciliated ectoderm and usually have a muscular protrusible pharynx and a pair of eyespots. They rarely have suckers. Planarians.

turbidimeter (*PowderTech*) Equipment for determining the surface area of a powder by measuring the light scattering properties of a fluid suspension.

turbidimetric analysis (*Chem*) See NEPHELOMETRIC ANALYSIS.

turbidite (*Geol*) A sediment deposited from a turbidity current, frequently poorly sorted as in a *greywacke*, it often shows GRADED BEDDING.

turbidity (*Genrl*) Haziness or cloudiness of an otherwise transparent medium (eg air or water) resulting from small suspended particles that scatter the light.

turbidity (*ImageTech*) A property of a photographic emulsion whereby light is scattered by the silver halide grains in the immediate vicinity of the image.

turbidity current (*Geol*) A density flow of mixed water and sediment, capable of rapid movement downslope. See TURBIDITE.

turbinal (*BioSci*) (1) Generally, coiled in a spiral. (2) One of certain bones of the nose in vertebrates which support the folds of the olfactory mucous membrane.

turbinate (*BioSci*) In the form of a whorl or an inverted cone, as certain gastropod shells.

turbinate bone (*BioSci*) See TURBINAL.

turbine (*Aero*) See AXIAL-FLOW TURBINE.

turbine aero-engine (*Aero*) See BYPASS TURBOJET, DUCTED FAN, TURBOJET, TURBOPROP.

turbine blade temperature (*Aero*) The temperature of the metal blades caused by the hot gases in a gas turbine.

turbinectomy (*Med*) Removal of a turbinal.

turbo (*ICT*) A facility once found on most IBM-COMPATIBLE COMPUTERS whereby the CLOCK RATE could be reduced to 8 MHz in order to run software which required the computer to run at this speed. Early IBM-compatible computers ran at this speed. The 'turbo button' was usually mounted on the case of the computer in an accessible position.

turbo- (*Genrl*) Prefix from Lt *turbo*, *-inis* meaning having, connected to, or driven by a turbine.

turbocharger (*Autos*) A form of SUPERCHARGER, used for internal-combustion engines in which the power of the compressor comes from a turbine driven by the exhaust gases.

turbodynamo (*ElecEng*) A specially designed dc generator for direct coupling to a high-speed steam turbine.

turbo-electric propulsion (*ElecEng*) A form of electric drive, once used in marine work, in which turbine-driven generators supply electric power to motors coupled to the propeller.

turbofan (*Aero*) See DUCTED FAN.

turbogenerator (*ElecEng*) The arrangement of a steam turbine coupled to an electric generator for electric power production.

turbojet (*Aero*) An internal-combustion aero-engine comprising compressor(s) and turbine(s), of which the net gas energy is used solely for reaction propulsion through propelling nozzle(s). See BYPASS TURBOJET, DUCTED FAN, SPLIT COMPRESSOR. Fig. ▷

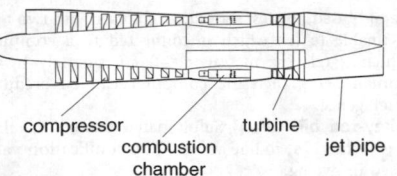

compressor   turbine

combustion   jet pipe
chamber

turbojet

turboprop (*Aero*) A SHAFT TURBINE where the torque output is transmitted to a propeller through a reduction gearbox; it may be of *single-shaft*, *twin-shaft* or *free turbine* form. A constant-power, or supercharged, turboprop has an oversize compressor/turbine assembly which enables it to maintain full power up to a considerable altitude.

turbopump (*Aero*) A combination RAM-AIR TURBINE and hydraulic, or fuel, pump for a guided weapon or aircraft in emergency.

turboramjet (*Aero*) An engine consisting of a TURBOJET mounted within a RAMJET duct, so that the efficiency of the former in subsonic flight is combined with the advantages of the latter at high supersonic speeds.

turborocket (*Aero*) A composite engine in which a rocket propellant (an example would be high-test peroxide catalysed to superheated steam and oxygen) is used to energize a turbine, which in turn drives a compressor, its air delivery joining the products from the turbine for combustion with a fuel to produce a propulsive jet. The object is eg to obtain a high ceiling, say 100 000 ft (30 000 m), without the enormous propellant consumption of a rocket. It could be basis of a hypersonic space-launching aircraft.

turbostarter (*Aero*) An aero-engine starter in which rotation is imparted by a turbine motivated by compressed air, a gas source, or the decomposition by catalysis of an unstable chemical, such as hydrogen peroxide.

turbosupercharger (*Aero*) See EXHAUST-DRIVEN SUPERCHARGER.

turbulence (*Phys*) See TURBULENT FLOW.

turbulent flow (*Phys*) Fluid flow in which the particle motion at any point varies rapidly in magnitude and direction. This irregular eddying motion is characteristic of fluid motion at high REYNOLDS NUMBERS. Gives rise to high drag, particularly in the BOUNDARY LAYER of aircraft. Also *turbulence*. See LAMINAR FLOW, REYNOLDS NUMBER and panel on AERODYNAMICS.

turf (*Agri*) A sod removed from the soil surface containing the shoots and the underground portions of the low-growing, dominant herbaceous plants in the habitat.

turgescence (*Med*) The act or condition of swelling up; the state of being swollen. Adj *turgescent*.

turgid (*BioSci*) (1) A cell that is distended and stiff as a result of the osmotic uptake of water, having a positive turgor pressure. (2) A non-woody tissue which is stiff as a result of the cells being turgid. Cf FLACCID.

turgite (*Min*) See HYDROHAEMATITE.

turgor movement (*BioSci*) Movement of a plant part resulting from changes in the turgor of its cells or the cells of its support. See PULVINUS. Cf GROWTH MOVEMENT.

turgor potential (*BioSci*) That component of the WATER POTENTIAL due to the hydrostatic pressure; equal to the TURGOR PRESSURE. An important component in turgid cells and in the xylem. Also *pressure potential*. Symbol ψ_P.

turgor pressure (*BioSci*) The hydrostatic pressure of the contents of a cell; normally positive in most plant cells; normally negative in the conducting cells of the xylem of transpiring plants.

Turing machine (*ICT*) A finite-state automaton with an unbounded memory. It is an abstract computer and is used to define the concept of COMPUTABILITY.

Turing test (*ICT*) A widely accepted definition of ARTIFICIAL INTELLIGENCE devised by A Turing in 1950.

Paraphrased, the test states that if a user with two terminals is unable to tell which is connected to a computer and which is relaying messages entered remotely by another human being then the computer can be credited with intelligence.

Turkey-red oil (*Chem*) Sulphonated castor oil, rel.d. 0·95, acid value 174, iodine value 82, saponification value 189. Used in dyeing.

turn (*Eng*) A small dead-centre lathe used by watch-makers. Usually held in a vice, and driven by a hand wheel or a BOW. Used for pivoting, polishing and turning small parts.

turn (*Glass*) A work shift in which a definite number of articles, usually two MOVES, is produced.

turn-and-slip indicator (*Aero*) A pilot's instrument for blind flying which indicates the rate of turn and sideslip, or error, in banking. Also *turn-and-bank indicator*.

turnaround document (*ICT*) A document that, after being output by the computer, can be used to record data; these data can then be input to the computer using a document reader.

turnbuckle (*Eng*) See SCREW SHACKLE.

turndown ratio (*EnvSci*) The ratio of maximum and minimum output of an electricity plant that operates continuously.

turner bars (*Print*) See ANGLE BARS.

Turner's syndrome (*Med*) A condition in humans in which a person looks superficially like a female but has only one X chromosome.

turn indicator (*Aero*) Any instrument that indicates the departure of an aircraft from its set course in a horizontal plane. Necessary for flying in clouds or at night.

turning (*Build*) The process of building an arch.

turning (*Eng*) Producing cylindrical, flat or tapered workpieces in a lathe.

turning-bar (*Build*) An iron bar supporting the arch over a fireplace opening.

turning-piece (*Build*) A simple form of centring, consisting of a single solid wooden piece shaped to the form of an intrados, and supported in its temporary position by wooden struts.

turning point (*Surv*) (1) The point at which consecutive straight lines of a traverse meet at an angle. (2) See CHANGE POINT.

turning point on a curve (*MathSci*) A peak (maximum) or a trough (minimum) on a curve. For the curve $y = f(x)$ the point where $x = a$ is a turning point if $f(a+h) - f(a)$ is of constant sign for all values of h sufficiently small.

turnings (*Eng*) Chips or swarf produced as waste in turning.

turning-saw (*Build*) See SWEEP-SAW.

turning tools (*Eng*) See LATHE TOOLS.

turnkey system (*ICT*) A complete computerized system to meet a customer's specification. It may include software, hardware, ancillary equipment and staff.

turnout (*CivEng*) The movable tapered rails or points by which a train or tram is directed from one set of rails to another. Also *crossing*, *point*, *switch*.

turnover (*Phys*) In isotope separation, the total flow of material entering a given stage in a cascade.

turnover (*Print*) The part of a divided word turned over into the next line, or a short line at the end of a paragraph.

turnover (*Radiol*) The rate of renewal of a particular chemical substance in a given tissue.

turnover board (*Eng*) A smooth square board on which an inverted bottom-half box is placed and rammed up round a pattern having a flat joint, thus saving the labour of making the facing joint. After turning over, removing the board, and adding facing sand, the top half may be rammed up at once.

turnover frequency (*Acous*) In vinyl disk recording, the frequency, generally between 200 and 500 Hz, where the change from constant-amplitude to constant-velocity recording takes place. Also *crossover frequency*.

turnsick (*Vet*) See COENURIASIS.

turns ratio (*ElecEng*) The ratio of turns on any pair of windings of a transformer; usually has symbol N. Primary and secondary voltages and currents are related by N and $1/N$ respectively; thus impedances are related by N^2. See TRANSFORMER, TRANSFORMATION RATIO, VOLTAGE RATIO.

turnstile antenna (*ICT*) Two normal dipoles, crossed over at their centre, driven with equal currents in quadrature.

turntable (*Acous*) The rotating table which supports the lacquer-blank during cutting and the processed record while being reproduced. It is of relatively high inertia, to keep down fluctuations of speed.

turntable (*CivEng*) A circular platform capable of rotation about its centre; used to reverse steam locomotives, which are driven on, turned through a half-circle, and driven off pointing the opposite way. In general, any such rotating platform, or system of rings rotating one inside the other.

turn tread (*Build*) A tread, generally triangular in plan, to form a step at a change of direction of the stair.

Turonian (*Geol*) A stage in the Upper Cretaceous. See MESOZOIC.

turpentine (*Chem*) An essential oil, $C_{10}H_{16}$, obtained by the steam distillation of rosin. It is a colourless liquid, of aromatic pine-like odour; bp 155–165°C, rel.d. 0·85–0·91; the chief constituent is pinene. American turpentine is dextrorotatory, others are usually laevorotatory. An important solvent for lacquers, polishes, etc. *Turpentine substitute* is a petroleum fraction of similar boiling point.

turquoise (*Min*) A hydrated phosphate of aluminium and copper which crystallizes in the triclinic system. It is a mineral of secondary origin, found in thin veins or small masses in rocks of various types, and used as a gemstone. The typical sky-blue colour often disappears when the mineral is dried. Much of the gem turquoise of old was fossil bone of organic origin and not true turquoise.

turret (*Eng*) A turntable or wheel for carrying a number of alternative tools, eg in a TURRET LATHE or TURRET PRESS.

turret (*ImageTech*) A rotatable mounting for several lenses on a camera or projector, allowing rapid changes between different focal lengths.

turret lathe (*Eng*) A high-production lathe for long workpieces, using a large number of tools carried on the revolving tool-holder or turret and on the cross-slide. The turret is mounted on a saddle which slides on the lathe bed.

turret press (*Eng*) A power press in which pairs of punches and dies of various sizes are held in upper and lower turrets, the turrets being geared together to bring corresponding punches and dies into positions of exact alignment in which they are then locked.

turtle (*ICT*) Drawing device used by LOGO and related languages. It may be an electromechanical device drawing on the floor (floor turtle), or may be simulated by graphics on a visual display screen (screen turtle).

turtle-shell (*BioSci*) The horny plates of the hawk's-bill turtle. Commonly *tortoise-shell*.

tusks (*Build*) See TUSSES.

tusk tenon (*Build*) A form of tenon used for framing one horizontal piece into another, eg a trimmer into a trimming joist. The tenon is strengthened by a short projection underneath, and by a bevelled shoulder above, both fitting into a suitably cut mortise in the other piece.

tusk tenon Tightened by wedge.

tussah silk (*Textiles*) A coarse silk produced by a wild silkworm, eg *Antheraea mylitta* (from India), *A. pernyi* (from China) and *A. yama-mai* (from Japan). The fibres are pale-brown and are usually rather short. They are spun into a yarn which has many irregular slubs.

tusses (*Build*) Stones left projecting from the face of a wall, when later extension is allowed for. Also *tusks*.

tussive (*Med*) Pertaining to, or caused by, a cough.

tussore (*Textiles*) A fabric woven from tussah silk.

tuyère (*Eng*) A nozzle through which air is blown into a blast furnace. May be kept cool by circulating water. Also TWYERE. See fig. at BLAST FURNACE.

TVC (*FoodSci*) Abbrev for TOTAL VIABLE COUNT.

TVP (*FoodSci*) Abbrev for TEXTURED VEGETABLE PROTEIN.

TVRO (*ImageTech*) Abbrev for *TV receive only*. A satellite dish which can only receive, not send, signals.

TWAIN (*ICT*) Technology that allows the creation of TAG IMAGE FORMAT FILES directly from devices such as SCANNERS and digital cameras. Such a device is said to be 'TWAIN-compliant'.

TW antenna (*ICT*) See TRAVELLING-WAVE ANTENNA.

Twaron (*Plastics*) TN for aramid fibre (Europe).

tweed (*Textiles*) Traditionally a coarse, heavy, rough wool outerwear fabric. Now applies to other wool fabrics having a wide range of weights and weave effects (eg TWILLS).

tweeter (*Acous, ICT*) Colloq term for the higher-frequency (usually >5 kHz) loudspeaker in a sound-reproduction system, normally fed via a CROSSOVER to prevent it being overloaded by low-frequency signals. See WOOFER.

twilight (*Astron*) The period after sunset, or before sunrise, when the sky is not completely dark. Astronomical twilight is defined as beginning (or ending) when the Sun is $18°$ below the horizon; hence twilight will last all night for a period in the summer months in all latitudes greater than about $48°$. See CIVIL TWILIGHT, NAUTICAL TWILIGHT.

twilight sleep (*Med*) A state of semi-consciousness produced by the administration of morphine and scopolamine.

twills (*Textiles*) Woven fabrics with diagonal lines on the face. Regular twills have continuous lines; zigzag twills have the lines reversed at intervals.

twin (*Genrl*) One of a pair of two and related entities similar in structure or function; often synonymous with *double*. See TWINS.

twin cable (*ElecEng*) A cable comprising two individually insulated conductors twisted together. A twin cable for telecommunication may have a large number of such pairs, eg up to 2400 pairs for telephone connections between large exchanges.

twin check (*ICT*) Continuous check achieved by duplication of hardware and comparison of results. See RAID ARCHITECTURE, REDUNDANCY.

twin columns (*Arch*) Two columns springing from one base.

twin-concentric cable (*ElecEng*) A two-core cable in which the conducting cores are concentrically arranged about the axis of the cable.

twin crystal (*Crystal, Min*) A crystal composed of two or more individuals in a systematic crystallographic orientation with each other. See panel on TWINNED CRYSTALS.

twiner (*BioSci*) A plant that climbs by winding around a support.

twin feeder (*ICT*) A transmission line, leading to or from an antenna, consisting of two parallel conductors. In high-power transmitter applications, the two wires may be separated by insulating rods; alternatively they may be moulded into solid polythene. Impedance is determined by conductor diameter, spacing and the dielectric used.

twin lamb disease (*Vet*) See PREGNANCY TOXAEMIA.

twin-lens reflex camera (*ImageTech*) A camera with matched lenses, one for exposing, the other for focusing, generally with a reflex mirror.

twinned crystal (*Crystal*) See panel on TWINNED CRYSTALS.

twinning (*Crystal*) Intergrowth of crystals of near symmetry, such that (in quartz) the piezoelectric effect is not sufficiently determinate. See panel on TWINNED CRYSTALS.

twinning (*ImageTech*) See PAIRING.

twin paradox (*Phys*) See CLOCK PARADOX.

twin-plate process (*Glass*) A process for making polished plate glass in which rolling, annealing and grinding are carried out on a continuously produced ribbon of glass without first cutting it into sections in which top and bottom surface are ground simultaneously. Also *Pilkington twin process*.

twin-quad (*ElecEng*) See QUAD.

twins (*BioSci*) (1) *Identical twins* arise from the same fertilized egg which has subsequently divided into two, each half developing into a separate individual. Also *homozygous twins*. (2) In mammals, *non-identical twins* are produced from separate eggs fertilized at the same time (*heterozygous twins*).

twin-screw extruder (*Eng*) Extruder fitted with intermeshing, counter-rotating screws which provide better mixing of polymers and additives, esp polyvinyl chloride compounds.

twin-shaft turbine (*Aero*) See SPLIT COMPRESSOR.

twin-T network (*ICT*) One consisting of two T-networks that have their input terminal pairs connected in parallel and their output terminal pairs connected in parallel.

twin triode (*Electronics*) A combination of two triode valves within the same envelope.

twist (*Textiles*) Fibres in a yarn are held together by the degree of twist introduced in spinning. This may be quantified as the twist level, ie the number of revolutions per unit length. Lively yarns have a tendency to untwist.

twist and steer (*Aero*) Control of a guided weapon or drone about the pitch and roll axes only, turns being achieved by rolling into a bank so that the elevator can provide the required turning moment. The system simplifies the autopilot and power requirements and is sometimes used with differentially mounted variable-incidence wings.

twist bit (*Build*) A bit with a long spiral cutting section, used for deep holes for dowels etc.

twist direction (*Textiles*) If the direction of twist as viewed on a yarn held vertically goes diagonally from left to right it has S-twist; if from right to left it has Z-twist.

twist drill (*Eng*) A hardened-steel drill in which cutting edges, of specific rake, are formed by the intersection of helical flutes with the conical point which is backed off to give clearance; of universal application.

twisted aestivation (*BioSci*) Same as CONTORTED AESTIVATION.

twisted pair cable (*ICT*) A cable that consists of individual wires wrapped around each other for carrying telephone data and computer data. It is available in two forms: SHIELDED TWISTED PAIR CABLE and UNSHIELDED TWISTED PAIR CABLE.

twister (*Radar*) A plate with slats giving double reflection of a radar wave, one being half-wave retarded, to give a twist in direction of polarization of electric component of wave.

twisting frame (*Textiles*) See DOUBLING FRAME.

twisting paper (*Paper*) A long-fibred paper suitable for waxing and intended for wrapping around sweets, toffee, etc.

twistor (*Phys*) One of a number of complex variables representing the space–time co-ordinates, retaining their relevant symmetries.

twistor theory (*Phys*) A candidate for the so-called 'theory of everything', which attempts to describe the structure of space as eight-dimensional, using complex numbers and particle spin, the building blocks of this structure being TWISTORS.

twitch (*Vet*) A noose for compressing the lip of a horse as a means of restraint.

twitch muscle (*BioSci*) Striated muscle innervated by a single motoneuron and having an electrically excitable membrane that exhibits an all-or-none response: in mammals almost all skeletal muscles are twitch muscles.

two-address program (*ICT*) Used in early computers where each instruction had to include the address of two registers, one for the operand and one for the result of the operation.

Twinned crystals

Many crystals are found with two parts that are reversed on each other, but related in a definite geometrical manner, called TWINNING. Both parts are always of the same mineral species and re-entrant angles between them are common. Although the crystals have developed in this manner throughout their growth from a very early stage, the twinned form may be described as if it had resulted from a purely geometrical operation of symmetry acting on one individual to produce the other. If one part is thus considered to be produced from the other by rotation, it is usually through 180° about a *twin axis*. The plane of reflection when it occurs is the *twin plane* which is often identical with the *composition plane* along which the two parts are joined.

Simple twins have two component parts (calcite, in the diagram) and *multiple twins* involve more than two individuals. When the individuals appear to penetrate each other they form an *interpenetration twin* (fluorite, in the diagram), where the twin axis is a diagonal of the cube. In the interpenetration *Carlsbad twin* of the monoclinic feldspar (orthoclase, in the diagram) the twin axis is the vertical crystallographic axis and the composition plane is the *clinopinacoid*, ie parallel to the vertical and the clino axis.

A *geniculate twin* produces a knee- or elbow-shaped crystal (rutile, in the diagram). Repeated twinning with parallel twin planes is *polysynthetic twinning*, the *twin lamellae* sometimes appearing only as fine lines on the crystal faces.

In quartz, interpenetrant twinning is almost always present but usually difficult to detect. Quartz has two related forms, right- and left-handed, and twins may be right-handed or left-handed, twinned alone about the vertical axis, or interpenetrant right- and left-handed twins twinned about the vertical axis and reflected over the horizontal plane. This is known as the *Brazil twin law* and is revealed in etched basal sections.

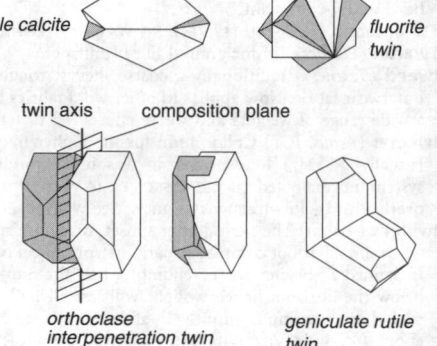

Twinned crystals

See panel on CRYSTAL LATTICE.

---

two-body force (*Phys*) A type of interaction between two particles which is unmodified by the presence of other particles.

two-circuit winding (*ElecEng*) See WAVE WINDING.

two-coat work (*Build*) Plastering in two coats; a first coat of coarse stuff, and a second coat of fine stuff.

two-colour process (*ImageTech*) A colour process recording and reproducing only two broad regions of the spectrum, usually blue-green and orange-red; now effectively obsolete.

two-colour process (*Print*) The application of the subtractive process to printing for the reproduction of a two-colour original.

two-dimensional gas (*Chem*) A unimolecular film whose behaviour in two dimensions is analogous, qualitatively and quantitatively, to that of an ordinary gas in three dimensions.

two-dimensional gel electrophoresis (*BioSci*) A high-resolution separation technique, much used in proteomics, in which protein samples are separated by isoelectric focusing in one dimension and then laid on an SDS gel for size-determined separation in the second dimension. Hundreds of components in a mixture can be resolved on a single gel.

two-electrode valve (*Electronics*) See DIODE.

two-group theory (*NucEng*) Simplified treatment of neutron diffusion in which only two energy groups are considered. Only the high-energy fission and the fully thermalized neutrons are considered, with the partly thermalized neutrons neglected.

two-hybrid assay (*BioSci*) A screening assay system to identify genes encoding proteins that interact specifically with other proteins; the interaction of the two expressed proteins in the yeast hybrid allows a functional promoter of a reporter gene to assemble and operate. Also *yeast two-hybrid system*.

two-light frame (*Build*) A window frame having one mullion dividing the window space into two compartments.

two pack materials (*Build*) Paints or fillers which are supplied in separate containers consisting of a base and a catalyst or activator. When these materials are mixed, gelation takes place rapidly and the pot life of the mixture is very short. This chemical process is known as curing, eg of epoxy ester filling.

two-part coatings (*Build*) Compositions of brushing viscosity which 'dry' by the reaction of two parts, mixed immediately before use.

two-phase (*ElecEng*) A term applied to ac systems employing two phases, whose voltages are displaced from one another by 90 electrical degrees. Also *bi-phase*.

two-phase, four-wire system (*ElecEng*) A system of two-phase ac distribution employing two conductors per phase.

two-phase, three-wire system (*ElecEng*) A system of two-phase ac distribution in which two conductors (lines) each carry a phase, and the third (neutral) is common to both phases.

two-pipe system (*Build*) One in which soil and waste discharges are piped separately with or without ventilating pipes depending on system size.

two-reaction theory (*ElecEng*) A theory used in calculations on salient-pole synchronous machines; the mmfs in the machine are assumed to be divided into two components, one acting along the axis of the main poles, the other at 90° to this.

two-revolution (*Print*) A type of letterpress machine in which the cylinder revolves continuously, making two revolutions while the carriage reciprocates once; the cylinder is pulled down on the bearers for the printing revolution, and rises clear of the forme during a second revolution, while the forme returns to the printing position. Cf SINGLE-REVOLUTION, STOP-CYLINDER.

two-roll mill (*Eng*) Mixing process for polymers, esp rubbers where materials are mixed with additives and fillers on two adjacent rollers. DISPERSIVE MIXING occurs at the nip, and DISTRIBUTIVE MIXING by cutting and overlap of the polymer sheet.

two's complement (*ICT*) Formed from a binary number, it is always one greater than the corresponding ONE'S COMPLEMENT (eg −43 is 11 010 101). It is possible to effect subtraction by addition, using two's complement.

two set (*Print*) Plating a rotary press with two sets of plates to produce two copies for each cylinder revolution.

two-spool compressor (*Aero*) See SPLIT COMPRESSOR.

two-stage pressure-gas burner (*Eng*) Natural-draught type designed for operating with gas under pressure, normally about 35 kN m^{-3}, and having primary and secondary air inspirating stages in the injector.

two-start thread (*Eng*) See DOUBLE-THREADED SCREW.

two-stroke cycle (*Autos*) An engine cycle completed in two piston strokes, ie in one crankshaft revolution, the charge being introduced by a blower or other means, compressed, expanded and exhausted through ports in the cylinder wall, before and during the entry of the fresh charge. See DIESEL CYCLE, OTTO CYCLE.

two-terminal pair network (*ICT*) See QUADRIPOLE.

two-tone keying (*ICT*) Keying of modulated continuous wave through a circuit that changes the modulation frequency only.

two-to-one folder (*Print*) On a web-fed press, a type of folder in which the folding cylinder has a circumference of two cut-offs and the cutting cylinder one cut-off giving a ratio of cylinder sizes 2:1. Cf THREE-TO-TWO FOLDER.

two-to-one interlace (*ImageTech*) The conventional TV scanning arrangement in which two successive FIELDS of odd and even LINES are interlaced to produce each FRAME. Also *2:1 interlace*.

two-up (*Print*) A printing surface made up to print two copies at one impression; can also be arranged for three-up, four-up or any suitable number.

two-way ANOVA (*MathSci*) An ANALYSIS OF VARIANCE (ANOVA) used when there are two independent variables.

two-way circuit (*ICT*) Bi-directional channel that operates stably in either direction.

two-wire circuit (*ICT*) A circuit in which go and return wires take equal currents, with potentials balanced with respect to earth.

two-wire system (*ElecEng*) A two conductor system of dc transmission and distribution.

TWT (*MinExt*) Abbrev for *two-way time*. See SEISMIC PROSPECTING.

TWTT (*MinExt*) Abbrev for *two-way travel time*. See SEISMIC PROSPECTING.

twyere (*Eng*) See TUYÈRE.

TXP (*Chem*) Abbrev for TRIXYLYLPHOSPHATE.

Tycho's star (*Astron*) A supernova explosion observed in the constellation Cassiopeia (1572) by Tycho Brahe; its remnant is a radio and X-ray source.

tye (*MinExt*) STRAKE in which a considerable thickness of low-grade concentrate is collected.

Tyler sieves (*MinExt*) Widely used series of laboratory screens in which mesh sizes are in $\sqrt{2}$ progression with respect to linear distance between wires.

Tylose (*Plastics*) TN for regenerated cellulose with properties resembling flesh.

tylose (*BioSci*) A bladder-like expansion of the wall of a living parenchyma cell through a pit into the lumen of a

xylem tracheid or vessel. Tyloses apparently form in non-functional conduits after spontaneous or wound-induced EMBOLISM, and may restrict the spread of pathogens. Also *tylosis*.

tympan (*Print*) In a hand-press, the frame on which the paper is placed when printing. In printing machines the sheets of paper used to adjust impression, the outermost being tympan paper (a strong sulphite or manilla), oiled or plain.

tympan hooks (*Print*) In a hand-press, thumb-hooks used for locking the outer and inner tympans together.

tympanic bulla (*BioSci*) In some mammals, a bony vesicle surrounding the outer part of the tympanic cavity and external auditory meatus formed by the expansion of the tympanic bone.

tympanites (*Med, Vet*) In humans the distension of the abdomen by accumulation of gas in the intestines or in the peritoneal cavity. In cattle that of the rumen and reticulum.

tympanum (*Arch*) The triangular or segmental space forming the central panel of a pediment.

tympanum (*BioSci*) (1) Generally, a drum-like structure. (2) In some insects, the external vibratory membrane of a chordotonal organ. (3) In some birds, an inflatable air sac of the neck region. (4) In vertebrates, the middle ear, or the resonating membrane of the middle ear. (5) In birds, the resonating sac of the syrinx. Adjs *tympanal*, *tympanic*.

Tyndall effect (*Phys*) Scattering of light by very small particles of matter in the path of the light, the scattered light being mainly blue.

tyndallimetry (*Chem*) The determination of the concentration of suspended material in a liquid by measurement of the amount of light scattered from a Tyndall cone. See NEPHELOMETRIC ANALYSIS.

type A personality (*Psych*) A personality type characterized by urgency, impatience, ambition and excessive competitiveness. Thought by some to predispose to coronary heart disease.

type B personality (*Psych*) A personality type characterized by relaxed approach to life, a lack of urgency, self-reflective and relatively non-competitive.

type I superconductor (*Phys*) See panel on SUPERCONDUCTORS.

type II superconductor (*Phys*) See panel on SUPERCONDUCTORS.

typeface (*Print*) A particular family or font of type in which the characters have distinctive features. Typefaces used in modern English bookwork include Aldine, Bembo, Baskerville, Caslon, Ehrhardt, Fournier, Garamond, Gill Sans, Imprint, Perpetua, Plantin, Times Roman, Univers. Each typeface has its own special characteristics. Characters in a particular POINT size or WEIGHT such as bold or italic are called FONTS.

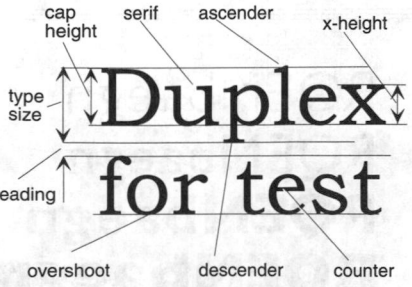

typeface Set in 30 point Bookman with 1·5 point leading.

type family (*Print*) A range of variations on the basic type design: light, medium, bold, extra bold, condensed, etc. A notable example is the Gill Sans family: Fig. ▷

Tyre technology

Over 60% of all rubber produced today, whether synthetic or natural, is used in car or lorry tyres. Although when cut open a tyre reveals an apparently homogeneous matrix material, it is in fact composed of a mixture of different blends (Fig. 1). The obvious reinforcing radial plies are composed of high-tenacity rayon, polyethylene terephthalate or nylon 6,6 fibre. The tread breaker, which keeps the tread firmly gripping the road, is composed of bias-belted steel wire and often aramid fibre, while the steel cord of the bead keeps the tyre firmly attached to the wheel. The ELASTOMERS used in the tread are usually styrene–butadiene and natural rubber blends for high HYSTER-ESIS and hence good road grip. The requirements of the sidewall are quite the opposite: since they must flex to absorb vibrations caused by adverse road conditions, cornering, etc, heat build-up must be reduced. Low-hysteresis blends of natural and polybutylene rubbers are therefore used. The tyre must also be impermeable to the pressurized air contained within it, a function of the butyl (often bromo-butyl) rubber inner lining.

All the rubber components are compounded with CARBON BLACK, sulphur accelerator, activator, anti-oxidants, anti-ozonants, oil-extenders, etc. Apart from the key cross-linking components, carbon black is the most important modifier of mechanical properties; a more recent development is the inclusion of carbon nanotubes together with the carbon black. When used as a pigment in thermoplastics, there is little reinforcing action (rather, the reverse) but in elastomers, extremely fine particles are used, eg *high abrasion fine* (HAF) of mean diameter 30 nm. If well dispersed, the rubber chains are adsorbed on the particle surfaces (Fig. 2).

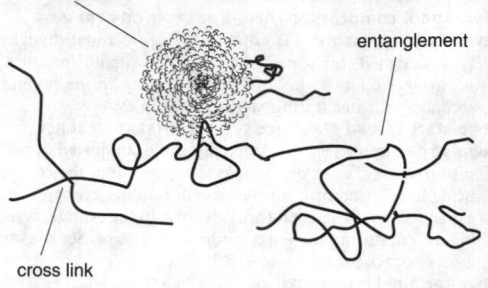

Fig. 2 **Effect of carbon black on the rubber elastomer**.

Some chemical reaction occurs between the free radicals present on the carbon black surface and the highly reactive double bonds of the rubber chains, so that the cross-link density is effectively increased. The maximum elongation is reduced by the rigidity of the graphitic particles together with their high loading. The carbon black does not affect the position of the glass transition (T_g), but smears it out over a wider temperature range rather than increasing it as in EBONITE. This is important as tyres must remain flexible at low temperatures. An additional benefit is that the aromatic rings of the carbon black help absorb ultraviolet radiation from the Sun and protect against degradation without affecting the rubber molecules.

See panels on HIGH-PERFORMANCE POLYMERS and RUBBER TOUGHENING.

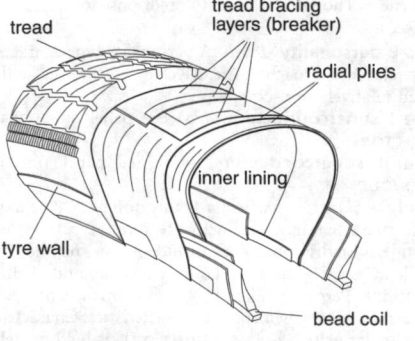

Fig. 1 **Tyre technology**.

RQENbaegn
RQENbaegn
RQENbaegn
RQENbaegn
RQENbaegn
RQENbaegn

type-high (*Print*) When a printing plate or block is mounted on wood or metal and brought to the proper height for printing it is said to be *type-high*, the UK/US standard being 0·918 in (23·32 mm). N *type-height*.

type holder (*Print*) A hand tool for holding the letters to be impressed on the cover of a book by the finisher.

type locality (*Geol*) The locality from which a rock, formation, fossil, etc, has been named and described.

type metal (*Eng*) A series of alloys of lead, antimony and tin, used for type. One composition is antimony 10–20%, tin 2–12% and the remainder lead. In another, tin may reach 26%, and up to 1% of copper may be added.

typesetting machines (*Print*) See COMPOSING MACHINES.

type specimen (*BioSci*) The actual specimen from which a given species was first described.

typewriter composition (*Print*) The use of electric typewriters to produce typesetting for reproduction by printing processes. The product has lower typographic standards than normal typesetting or the major phototypesetting systems but usually has justified lines and a difference of set between wide and narrow letters. Now effectively superseded by WORD PROCESSORS and DESKTOP PUBLISHING systems.

typhlitis (*Med*) Inflammation of the caecum.

typhlosole (*BioSci*) In some invertebrates, a longitudinal dorsal inwardly projecting fold of the wall of the intestine, by which the absorptive surface is increased.

typhoid (*Med*) The most serious form of enteric fever (which includes paratyphoid) due to infection with the bacillus *Salmonella typhi*. Prolonged fever, a rose rash and inflammation of the small intestine with ulceration occur. Infection is by fecal contamination of food or water. Also *typhoid fever*.

typhoon (*EnvSci*) A TROPICAL REVOLVING STORM, in the China Sea and western N Pacific.

typhus (*Med*) Infection with *Rickettsia prowazekii* derived from the bite of a louse, an acute epidemic fever with high mortality. There are varieties of lesser severity (scrub, murine typhus, etc) due to related organisms. Also *typhus fever*.

typical intensity (*Psych*) The high degree of stereotyping observed in many patterns of behaviour that have a communicative function.

typographer (*Print*) A specialist in typographic design who may also be a compositor.

typographic quality (*Print*) A term describing text that is equivalent to that produced by an experienced compositor using normal printer's characters, and which complies with the normal rules of good typography.

typography (*Print*) The art of arranging the printed page, including choice of type, illustration and method of printing.

typology (*Genrl*) The study of types and their succession in eg biology and archaeology.

Tyr (*Chem*) Symbol for TYROSINE.

tyre (*Eng*) (1) A renewable, forged steel, flanged ring shrunk on the rim of a locomotive wheel. (2) A steel or rubber band or air-filled tube of rubber (pneumatic tyre) surrounding a wheel to strengthen it or to absorb shock. See CROSS-PLY, RADIAL PLY, TUBELESS TYRE.

tyre-building machine (*Eng*) The centrepiece of a tyre factory, where all the parts are assembled to make the green tyre. See panel on TYRE TECHNOLOGY.

tyre carcass (*Eng*) See CARCASS.

tyre disposal (*Eng*) Problem created by growth in the use of cars and increasingly stringent regulations on tread depth (min 1·6 mm). As a thermoset, little of the bulk of a tyre is recyclable, but incineration can yield useful energy. Care is needed to limit SO_2 emissions.

tyre lining (*Eng*) The inner impermeable layer. See panel on TYRE TECHNOLOGY.

tyre technology (*Eng*) The making of tyres for transport. See panel on TYRE TECHNOLOGY.

tyre textiles (*Textiles*) Pneumatic tyres are made from composite materials of rubber reinforced with several layers of textile fabrics or steel wire. The fabrics are mostly made of synthetic fibres and include some, the tyre cord, where the warp threads predominate and the few weft threads are present only to assist processing. Radial and cross-ply denote the way the cords are arranged in forming the carcass. See panel on TYRE TECHNOLOGY.

tyrosine (*Chem*) *2-amino-3-(4-hydroxyphenyl)propanoic acid*. $OHC_6H_4CH_2CH(NH_2)COOH$. An amino acid, an oxidation product of PHENYLALANINE. The L- or s-isomer is a constituent of proteins. Symbol Tyr, short form Y.

Tyvak (*Plastics*) TN for bonded, non-woven paper made from synthetic polymer fibre such as polyethylene (US).

Tzaneen disease (*Vet*) A febrile disease, usually mild, of cattle and buffalo in Africa, caused by the protozoon *Theileria mutans*; transmitted by ticks.

U

U (*Chem*) Symbol for URANIUM.
U (*Chem*) Symbol for INTERNAL ENERGY.
U (*Phys*) Symbol for: (1) POTENTIAL DIFFERENCE; (2) tension.
u (*Genrl*) Symbol for ATOMIC MASS UNIT of unified scale.
u (*Chem*) (1) With subscript, a symbol for velocity of ions. (2) Symbol for *specific internal energy*.
UART (*ICT*) Abbrev for UNIVERSAL ASYNCHRONOUS RECEIVER/TRANSMITTER.
UAV (*Aero*) Abbrev for UNMANNED AERIAL VEHICLE.
Ubbelohde viscometer (*Chem*) A type of viscometer used for measuring dilute solution viscosity of polymers. Consists of glass capillary and reservoir on which are enscribed two fixed marks. With the instrument held vertically, the solution is allowed to flow through the capillary from the reservoir, and the time for movement between the fixed marks recorded. This is repeated for several different concentrations, from which data the intrinsic viscosity of the polymer can be calculated.
U-bend (*Build*) See AIR TRAP.
ubiquinone (*BioSci*) A small highly mobile electron carrier mediating the transfer of electrons from flavoprotein to cytochrome in the *electron transfer chain*.
ubiquitin (*BioSci*) A polypeptide of wide distribution in both prokaryotes and eukaryotes. It is attached to proteins prior to their degradation in the course of cellular protein turnover.
U-bolts (*Autos*) Bars bent into U-shape and threaded at each end; used for anchoring a semi-elliptic spring to an axle beam, a plate being threaded over the ends and secured by nuts.
UCR (*Psych*) Abbrev for *unconditioned response*. See CLASSICAL CONDITIONING.
UCS (*Psych*) Abbrev for *unconditioned stimulus*. See CLASSICAL CONDITIONING.
udder (*Vet*) The popular name for the mammary glands of certain animals, eg of the cow, mare, sow and ewe.
UDP (*Agri*) See UNDEGRADABLE PROTEIN.
UF (*Plastics*) Abbrev for *urea–formaldehyde* plastics. See UREA RESINS.
UFO (*Astron*) Abbrev for *unidentified flying object*. Applied to any sighting in the sky which the observer is unable to account for in terms of known phenomena.
Uganda mahogany (*For*) See AFRICAN MAHOGANY.
ugrandite (*Min*) A group name for the *uvarovite*, *grossular* and *andradite* garnets.
UHF (*ICT*) Abbrev for ULTRAHIGH FREQUENCY.
UHMPE (*Plastics*) Abbrev for ULTRAHIGH MOLECULAR MASS POLYETHYLENE.
Uhuru (*Astron*) The first X-ray astronomy satellite. Launched from Kenya in 1970, it made the first detailed map of the X-ray sky.
uintaite (*Min*) A variety of natural asphalt occurring in the Uinta Valley, Utah, as rounded masses of brilliant black solid hydrocarbon. Also *gilsonite*.
UL (*Genrl*) Abbrev for UNDERWRITERS LABORATORIES.
ULA (*ICT*) See UNCOMMITTED LOGIC ARRAY.
Ulbricht sphere photometer (*Phys*) A photometer for directly measuring a lamp's mean-spherical candle power. It comprises a hollow sphere, whitened inside, with the lamp under test at the centre. Owing to the internal

reflection, the illumination on any part of the sphere's inside surface is proportional to the lamp's total light output, measured through a small window.
ulcer (*Med*) A localized destruction of an epithelial surface (eg of the skin or of the gastric mucous membrane), forming an open sore; it is usually a result of infection.
ulceration (*Med*) The process of forming an ulcer; the state of being ulcerated.
ulcerative (*Med*) Of the nature of, or pertaining to, ulcers; causing ulceration; associated with ulceration (eg *ulcerative colitis*).
ulcerative cellulitis (*Vet*) See ULCERATIVE LYMPHANGITIS.
ulcerative colitis (*Med*) Inflammation of the colon and rectum: the cause is unclear, although there are often antibodies to colonic epithelium and *E. coli* strain 0119 B14.
ulcerative dermal necrosis (*Vet*) A fungus disease of salmon.
ulcerative lymphangitis (*Vet*) A chronic contagious lymphangitis of the horse, due to infection by *Corynebacterium pseudotuberculosis* (*C. ovis*). Also *ulcerative cellulitis*.
U-leather (*Eng*) See U-PACKING.
ulexite (*Min*) A hydrated borate of sodium and calcium occurring in borate deposits in arid regions, as in Chile and Nevada, where it forms rounded masses of extremely fine acicular white crystals. Also *cotton ball*.
UL 94 flammability (*Genrl*) Burning test for polymers where a rod, held in a flame under test conditions, is rated on a scale depending on eg the rate of burning. See NBS SMOKE TEST.
uliginose (*BioSci*) Growing in places which are wet. Also *uliginous*.
U-links (*ICT*) The spring links used to join isolated sections of communication channels, the ends of which are brought to a special link-board. Removal of a link opens the circuit, so that test equipment can be rapidly inserted.
ullage (*Ships*) The spare capacity of a partially full container, eg a tank for liquid cargo or fuel.
ullmanite (*Min*) Nickel antimony sulphide. It crystallizes in the cubic system and occurs in hydrothermal veins.
ulna (*BioSci*) The post-axial bone of the antebrachium in land vertebrates. Adj *ulnar*.
ulotrichous (*BioSci*) Having woolly or curly hair.
ULSI (*Electronics*) Abbrev for ULTRA-LARGE-SCALE INTEGRATION.
ultimate analysis (*Chem*) QUANTITATIVE ANALYSIS of the elements in the materials being examined.
ultimate limit switch (*ElecEng*) See LIMIT SWITCH.
ultimate load (*Aero*) The maximum load which a structure is designed to withstand without a failure. See LIMIT LOAD, LOAD FACTOR, PROOF LOAD.
ultimate tensile stress (*Eng*) The highest load applied to a material in the course of a TENSILE TEST, divided by the original cross-sectional area. In brittle or very tough materials it coincides with the point of fracture, but usually extension continues under a decreasing stress, after the ultimate stress has been passed. Also *tenacity* (obsolete). See STRENGTH MEASURES.
ultor (*Electronics*) Anode, esp in a cathode-ray tube, which has highest potential with respect to cathode. Also *second anode*.

ultra- (*Genrl*) Prefix from Lt *ultra*, beyond.

ultrabasic rocks (*Geol*) Igneous rocks containing less silica than the basic rocks (ie less than 45%), and characterized by a high content of mafic constituents, particularly olivine (in the peridotites) and amphiboles and pyroxenes (in the perknites and picrites). See BASIC ROCKS.

ultracentrifuge (*BioSci*) A high-speed centrifuge much used for molecular separation and capable of creating forces up to 500 000 times gravity.

ultrafiltration (*Chem*) The separation of colloidal or molecular particles by filtration, under suction or pressure, through a colloidal filter or semipermeable membrane. Cf REVERSE OSMOSIS.

ultrahigh frequencies (*ICT*) Those frequencies between 3×10^8 Hz and 3×10^9 Hz. Abbrev *UHF*.

ultrahigh molecular mass polyethylene (*Plastics*) Linear polyethylene with molecular mass of several million, so difficult to process by normal means. Sintered to shape in hip joint socket, and raw material for gel spinning into high-performance fibre. Abbrev *UHMPE*. See MOLECULAR MASS DISTRIBUTION.

ultra-large-scale integration (*Electronics*) Grouping of many electronic components on a single chip beyond that of VERY-LARGE-SCALE INTEGRATION. Abbrev *ULSI*.

ultralinear (*ElecEng*) Said of a power amplifier with a very low non-linear distortion.

ultramafic rocks (*Geol*) Igneous rocks containing more than 90% of dark minerals (olivine, pyroxene, amphiboles, micas, opaque minerals, etc). *Plutonic* rocks are classified according to the proportions of olivine, pyroxene (ortho-pyroxene and clinopyroxene) and hornblende, and include dunite, peridotite, pyroxenite, hornblendite, etc. *Volcanic* ultramafic rocks are much rarer.

ultramicroscope (*Phys*) An instrument for viewing particles too small to be seen by an ordinary microscope, eg fog or smoke particles. An intense light projected from the side shows, against a dark background, the light scattered by the particles. See DARK GROUND ILLUMINATION.

ultramicrotome (*BioSci*) A modified microtome developed for cutting ultra-thin sections for examination with the electron microscope. The cutting surface may be of steel, but is more usually glass or diamond, and the movement of the specimen block towards the knife is very delicately controlled, eg by the thermal expansion of a rod.

ultrasonic (*Acous*) Said of frequencies above the upper limit of the normal range of hearing, at or about 20 kHz.

ultrasonic cleaning (*Eng*) A cleaning process used in conjunction with water or solvents and effective for small crevices, blind holes, etc. Ultrasonic frequency vibrations are transferred to the cleaning fluid producing a turbulent penetrating action.

ultrasonic coagulation (*Phys*) The coalescence of particles into large aggregates by ultrasonic irradiation under suitable conditions.

ultrasonic delay line (*ICT*) A device that utilizes the finite time for the propagation of sound in liquids or solids to produce variable time delays. Such systems may also be used for storage in digital computers. Mercury or quartz are used as transmitting media, and nickel wire for magnetostrictive delay lines.

ultrasonic depth finder (*Eng*) Instrument for measuring or displaying the depth of water under a ship by measuring the time of propagation of a pulse of ultrasonic waves to the sea bed and back.

ultrasonic detector (*ICT*) Electro-acoustic transducer for the detection of ultrasonic radiation.

ultrasonic dispersion (*Chem*) High-intensity ultrasonic waves can produce a dispersion of one medium in another, eg mercury in water.

ultrasonic generator (*ElecEng*) One for the generation of ultrasonic waves, eg quartz crystal, ceramic transducer, supersonic air jet, magnetostrictive vibrator.

ultrasonic grating (*Phys*) The region of a medium in which the presence of acoustic waves has produced a periodic spatial variation of density with a corresponding variation of refractive index. Diffraction spectra can be obtained in passing a light beam through such a sound field.

ultrasonic imaging (*Acous*) See ACOUSTIC MICROSCOPE.

ultrasonic machining (*Eng*) The process of removing material by abrasive bombardment and crushing in which a relatively soft tool, matching the shape produced, is made to oscillate at ultrasonic frequencies and drives the abrasive grit, suspended in a liquid, against the work, blasting away fine particles of the material. Operations include drilling round and odd-shaped holes, die sinking and forming wire drawing dies. Primarily for machining hard brittle materials like carbides, ceramics and sintered metals.

ultrasonics (*Acous*) The study and application of ULTRASONIC sound and vibrations.

ultrasonic soldering (*Eng*) A form of soldering with a specially designed soldering bit which emits ultrasonic vibrations. Soldering is particularly difficult with aluminium and the application of ultrasonics is supposed to break up the aluminium oxide layer.

ultrasonic stroboscope (*Eng*) One in which an ultrasonic field is applied to the modulation of a light beam to obtain stroboscopic illumination. See STROBOSCOPE.

ultrasonic testing (*Eng*) A method of testing for flaws in which an ultrasonic source is pressed against the part to be tested, using some form of gel to act as a sonic coupling to the surface, and sound is passed into the material. Reflections or echoes occur from the back face and, additionally, any internal discontinuity will reflect the sound wave and generate a signal in the receiver. The timelags of the echoes are measured to determine the thickness of the workpiece and the distance to the discontinuity. The appropriate probe and type of sonic wave permits the size and shape of a flaw to be determined as well as its position within the material.

ultrasonic welding (*Eng*) A solid-state process for bonding sheets of similar or dissimilar materials, usually with a lap joint. No heat is applied but vibratory energy at ultrasonic frequencies is applied in a plane parallel to the surface of the weldment.

ultrasonography (*Radiol*) Use of reflected high-frequency sound waves to image organs of the body. Widely used in the diagnosis of disease of the abdomen and heart and in the management of pregnancy.

ultrasound (*Acous*) Sound with frequency above the upper limit of the normal range of hearing (greater than 20 kHz). Used by some animals (eg bats, dolphins) for localization and communication, and in a variety of industrial applications.

ultrastructure (*BioSci*) The submicroscopic structure of a cell, particularly as shown by the electron microscope.

ultraviolet (*ImageTech*) Photography using radiations beyond the blue/violet end of the visible spectrum, in practice generally between 400 and 200 nm; of special application in forensic and medical recording and for measuring the distribution of nucleic acids in cells.

ultraviolet astronomy (*Astron*) The detection and analysis of radiation from cosmic sources at wavelengths between 25 and 350 nm. The hottest stars emit the bulk of their radiation in this waveband.

ultraviolet cell (*Phys*) A cell having a maximum response to light in the ultraviolet end of the spectrum.

ultraviolet microscope (*BioSci*) An instrument using ultraviolet light for illuminating the object. Its resolving power is therefore about doubled as the resolution varies inversely with the wavelength of the radiation, but more usefully the nucleic acids absorb strongly in this region and can therefore be localized and measured.

ultraviolet radiation (*Phys*) Electromagnetic radiation in a wavelength range from 400 to 10 nm approximately, ie between the visible and X-ray regions of the spectrum. The *near* ultraviolet is from 400 to 300 nm, the *middle* from 300 to 200 nm and the *extreme* from 200 to 10 nm. Abbrev *UV*. See appendix on Electromagnetic spectrum.

ultraviolet spectrometer (*Phys*) An instrument similar to an optical spectrometer but employing non-visual detection and designed for use with ultraviolet radiation.

ultraviolet spectroscopy (*Chem*) A method of detecting ultraviolet absorption by aromatic groups or conjugated bonds, including polymers in a suitable solvent. Complementary tool to infrared and nuclear magnetic resonance spectroscopy.

ultraviolet therapy (*Med*) The treatment of disease by ultraviolet rays. The therapeutic rays (300–400 nm) are now usually generated by quartz mercury-vapour lamps. The carcinogenic effect of over-dosage is now recognized.

ulvöspinel (*Min*) An end-member species in the magnetite series, with composition ferrous and titanium oxide. First recognized in ore from Södra Ulvön, in N Sweden.

Ulysses (*Space*) A joint ESA–NASA space exploration mission (1990–5) to observe the Sun and solar wind from a high solar latitude perspective.

U-matic (*ImageTech*) TN of an industrial COMPOSITE format using $\frac{3}{4}$ in tape in two sizes of cassette. There is also an upgraded *superior performance* (*SP*) version. *Broadcast video U-matic* (*BVU*) is a HIGH-BAND, sub-BROADCAST STANDARD version, now in SP form.

umbel (*BioSci*) Inflorescence or simple umbel of many flowers borne on stalks arising together from the top of a main stalk. Often this sort of branching is repeated in a compound umbel with several main stalklets arising together from the top of a larger stalk.

umbellate (*BioSci*) Having the characters of an umbel; producing umbels.

umbellifer (*BioSci*) A plant that has its flowers in umbels, esp a member of the family Umbelliferae.

Umbelliferae (*BioSci*) The carrot family, c.3000 spp of dicotyledonous flowering plants (superorder Rosidae). They are mostly herbs, and more or less cosmopolitan, but are found esp in temperate and upland regions. The flowers are in simple or compound umbels; they have five free petals, five stamens and an inferior ovary of two carpels. Includes several vegetables and flavouring plants, eg carrot, parsnip, celery, parsley, coriander, cumin. Several are poisonous including hemlock. Also *Apiaceae*.

umber (*Geol*) Naturally occurring brown iron and manganese oxides or clays strongly coloured by oxides, formed by residual weathering and valued as a pigment. *Raw umber* has a greenish tinge; *burnt umber* that has been calcined is dark-brown. See OCHRE.

umbilectomy (*Med*) Removal of the umbilicus.

umbilic (*MathSci*) The limit point of circular sections of a quadric as the radius tends to zero. Every quadric has four real and eight imaginary umbilics.

umbilical cord (*Space*) A term (frequently simply *umbilical*) applied to any flexible and easily disconnectable cable, eg for conveying information, power or oxygen to a missile or spacecraft before launching, for connecting an operational spacecraft with an external astronaut.

umbilical cord (*BioSci*) In eutherian mammals, the vascular cord connecting the fetus with the placenta.

umbilical point on a surface (*MathSci*) One at which the curvatures of all normal sections are positive and constant. Cf ELLIPTICAL, HYPERBOLIC and PARABOLIC POINT ON A SURFACE.

umbilicate (*Med*) Having a depression which resembles the umbilicus. Also *umbilicated*.

umbilicus (*BioSci*) (1) In gastropod shells, the cavity of a hollow columella. (2) In birds, a groove or slit in the quill of a feather. (3) In mammals, an abdominal depression marking the position of former attachment of the umbilical cord. Pl *umbilici*.

umbo (*BioSci*) A boss or protuberance; the beak-like prominence which represents the oldest part of a bivalve shell. Pl *umbones*. Adj *umbonate*.

umbra (*Astron*) A region of complete shadow of an illuminated object, eg dark central portion of the shadow of the Earth or Moon. Generally applied to eclipses of the Moon or of the Sun, the term is also applied to the dark central portion of a sunspot. The outer, less dark, shadow is known as the *penumbra*.

umbrella (*BioSci*) A flat cone-shaped structure, esp the contractile disk of a medusa.

umbrella antenna (*ICT*) An antenna comprising a vertical uplead from the top of which a number of wires extend radially towards the ground.

Umbriel (*Astron*) A natural satellite of Uranus, discovered in 1851. Distance from the planet 266 000 km; diameter 1170 km.

umen (*MinExt*) The non-mineralized substances of coal etc, and their distillation residues.

umkehr effect (*EnvSci*) An effect used to derive, on certain assumptions, the vertical distribution of ozone from a series of measurements of the relative intensities of two wavelengths in light scattered from the zenith sky; one wavelength is more, and the other less, strongly absorbed by ozone. As the Sun's zenith angle varies, a reversal (Ger *Umkehr*) occurs in the variation of the ratio of the intensities.

Umklapp process (*Phys*) A type of collision between PHONONS, or between phonons and electrons, in which crystal momentum is not conserved. Such processes provide the greater part of the thermal resistance in solid dielectrics.

UMTS (*ICT*) Abbrev for *universal mobile telecommunications system*, a third-generation mobile communications system.

umwelt (*Psych*) The relevant aspects of the environment which constitute the subjectively significant, or meaningful, surroundings for an animal or individual, ie that class of environmental variables capable of influencing behaviour.

unarmed (*BioSci*) Without spines, thorns, prickles, sharp teeth, etc.

unarmoured cable (*ElecEng*) A cable without an outer covering of steel wire (armouring).

unary (*MathSci, Chem*) Applied to or involving a single component, such as a function of a single variable. In chemistry, consisting of one component.

unavailable (*BioSci*) An element in plant mineral nutrition present in the soil but not in a form which the plant can take up.

unavailable energy (*Eng*) The energy which becomes unavailable to do work, in the course of an irreversible process.

unbalanced (*ElecEng*) (1) Of a bridge circuit in which detector signal is not zero. (2) Of a pair of conductors with different voltages or currents with reference to earth.

unbalanced circuit (*ElecEng*) One whose two sides are inherently unlike. See BALANCED CIRCUIT.

unbalanced load (*ElecEng*) A load which is unequal on the two sides of a three-wire dc system, or on the three phases of a symmetrical three-phase ac system.

unbalanced network (*ElecEng*) One arranged for insertion into an unbalanced circuit, the earthy terminal of the input being directly connected to the earthy terminal of the output.

unbalanced system (*ElecEng*) A three-phase ac system carrying an unbalanced load.

unbleached kraft paper (*Paper*) Brown packaging paper made from unbleached sulphate. Wood pulp suitable for use as wrapping material or for conversion into sacks, bags, gummed tape, etc.

UNC (*Eng*) Abbrev for *unified normal coarse*. See UNIFIED SCREW-THREAD.

uncate (*BioSci*) Hooked; hook-like. Also *unciform, uncinate*.

uncertainty factor (*Genrl*) (1) In assay methodology, the confidence interval or fiducial limit used to assess the probable precision of an estimate. (2) In toxicology, a value used in extrapolation from experimental animals to humans (assuming that they may be more sensitive) or from selected individuals to the general population. Abbrev *UF*.

uncertainty principle (*Phys*) The principle that there is a fundamental limit to the precision with which a position co-ordinate of a particle and its momentum in that direction can be simultaneously known. Also, there is a fundamental limit to the knowledge of the energy of a particle when it is measured for a finite time. In both statements, the product of the uncertainties in the measurements of the two quantities involved must be greater than $h/2\pi$, where h is PLANCK'S CONSTANT. The principle follows from consideration of the wave nature of particles. Also *Heisenberg uncertainty principle*.

unciform (*BioSci*) Also *uncinate*. See UNCATE.

uncinariasis (*Med*) Infestation of the small intestine by hookworms.

uncinate fit (*Med*) A hallucination of smell or of taste, due to a cerebral tumour or to epilepsy.

uncinus (*BioSci*) (1) A hook, or hook-like structure, eg a hook-like chaeta of Annelida. (2) In Gastropoda, one of the marginal radula teeth.

uncommitted logic array (*ICT*) An array of standard logic GATES with all possible circuits present. Each element is identical, manufactured on a single LSI CHIP. The circuits not required for a particular application are burnt out. Abbrev *ULA*.

unconditionally stable (*ICT*) Of amplification in a system that continues to satisfy the Nyquist criterion when the gain is reduced.

unconditioned response (*Psych*) Any response that is naturally occurring and elicited by an unconditioned stimulus.

unconditioned stimulus (*Psych*) Any stimulus that reliably elicits a response in a naive unconditioned responder.

unconformity (*Geol*) A substantial break in the succession of stratified sedimentary rocks, following a period when deposition was not taking place. If the rocks below the break were folded or tilted before deposition was resumed, their angle of dip will differ from that of the overlying rocks, in which case the break is an *angular unconformity*.

unconscious (*Psych*) (1) State of being unaware, comatose or deeply asleep. (2) Mental processes of which the subject is unaware, but which influence thought and action. Also *unconscious mind*. (3) In Freudian psychology, that part of the psyche that holds unknown wishes, needs and fears that play a significant role in conscious behaviour.

uncoupling agent (*BioSci*) Agents that uncouple electron transport from oxidative phosphorylation. Ionophores can do this by discharging the ion gradient generated by electron transport across the mitochondrial membrane. More generally, any agent capable of dissociating two linked processes.

uncoursed (*Build*) See RANDOM.

uncut (*Print*) Said of a book whose edges have been left untrimmed, the bolts therefore remaining *uncut*.

undamped oscillations (*ICT*) Same as CONTINUOUS OSCILLATIONS.

undecagon (*MathSci*) A closed plane figure (polygon) with eleven angles and sides.

undegradable protein (*Agri*) The protein in feed that is not digested in the rumen but is enzymically degraded and absorbed in the small intestine. It may be referred to as bypass protein. Abbrev *UDP*.

under blanket (*Print*) (1) On web-fed relief presses, the first dressing on the impression cylinder. Cf TOP BLANKET. (2) In offset printing, a special packing placed beneath the top rubber blanket.

underbunching (*Electronics*) Less than optimum efficiency in a velocity-modulation system.

undercarriage (*Aero*) (1) Each of the units (consisting of wheel(s), shock absorber(s) and supporting struts) of an aircraft's alighting gear, ie two main and either tail or nose undercarriages. (2) Colloq term for the whole *landing gear*.

underclay (*Geol*) See SEAT EARTH.

undercloak (*Build*) The first or lower sheet of lead in a ROLL. Cf OVERCLOAK.

undercoat (*Build*) Paint applied before the finishing coat. It is usually highly pigmented in order to provide good hiding power.

undercompensated meter (*ElecEng*) An induction-type meter provided with insufficient phase compensation which therefore reads high with leading currents and low with lagging currents.

undercure (*Eng*) Rubber technology term for state of rubber compound prior to desired cross-link density has been achieved.

undercut (*Eng*) (1) A mould or pattern is undercut when it has a re-entrant portion and thus cannot be opened or withdrawn in a straight motion. (2) A short portion of a bolt, stepped shaft, or similar part, with a diameter smaller than the peripheries or diameters of the adjacent features. It obviates difficulties of machining sharp corners and facilitates subsequent assembling. (3) A fault appearing as a cutting or washing away of parent plate adjacent to a weld fillet, which reduces the effective section thickness of one of the components joined.

undercutting (*Electronics*) In semiconductor processing, the etching of material beneath the edge of a mask due to lateral etching from an adjacent unmasked region; it is an inevitable consequence of isotropic etching.

undercutting (*Print*) (1) A fault to be guarded against in the etching of line blocks, traditionally by the use of DRAGON'S BLOOD, or by POWDERLESS ETCHING. (2) Cutting away part of the edge of a mount to allow the block to overhang a crossbar.

underdamping (*ICT*) Sometimes synonymous with periodic damping, but often restricted to cases where a critically damped response would be preferable.

underdispersion (*BioSci*) A regular (homogeneous) distribution of organisms; the pattern is best described empirically by the positive binomial probability equation, and the variance is smaller than the mean.

underdispersion (*MathSci*) The decreased variability in a set of data below that which may be expected under a particular model.

underexposure (*ImageTech*) Inadequate exposure to light of a photosensitive surface, photographic or electronic, resulting in an image of unsatisfactory tonal reproduction, particularly in the lack of gradation in the shadow areas of the picture.

underfeed stoker (*Eng*) A MECHANICAL STOKER in which the fuel is fed automatically and progressively from below the fire, and gradually forced up into the active zone, air being injected into the fuel bed just below the combustion level. See MULTIPLE-RETORT UNDERFEED STOKER, SINGLE-RETORT UNDERFEED STOKER.

underflow (*ICT*) Occurs when a number to be stored is less than the smallest number that can be represented in the WORD available for it.

underfold (*Print*) See OVERFOLD.

underground gasification (*MinExt*) A technique for the remote extraction of hydrocarbons from coal by the deliberate combustion of coal seams in a controlled and restricted oxygen environment. Combustion and distillation products include carbon monoxide and volatile hydrocarbons, extracted via boreholes. Also *pyrolytic mining*.

underground volatilization (*MinExt*) A technique for remote extraction of hydrocarbons from coal. Solvents are introduced through boreholes and dissolved volatile hydrocarbons are extracted via a second set of boreholes.

underhand stopes (*MinExt*) Stopes in which excavation is carried downslope from access level.

underlay (*MinExt*) The departure of a vein or thin tabular deposit from the vertical; it may be measured in horizontal feet per fathom of inclined depth. Also *underlie*.

underlay (*Print*) To paste paper or card under the mount of a printing plate in order to bring it to type-height, or to remedy a defect in the mount.

underleaf (*BioSci*) One of a row of leaves on the underside of the stem of a liverwort.

underlining felt (*Build*) See SARKING FELT.

undermodulation (*ICT*) That state of adjustment of a radio-telephone transmitter at which the peaks of speech or music do not produce 100% modulation, so that carrier power is not used to full advantage.

undermodulation (*ImageTech*) Inadequate modulation in a sound recording for satisfactory reproduction.

underpinning (*Build, CivEng*) The operation of propping part of a building to avoid damaging or weakening the superstructure.

underpitch groin (*Build*) See WELSH GROIN.

underpoled copper (*Eng*) See POLING.

undersaturated (*Geol*) Refers to an igneous rock in which there is a deficit of silica. This is normally shown by the presence of a feldspathoid. See OVERSATURATED.

under-sea branching multiplexer (*ICT*) Equipment positioned on the sea bed in a submarine cable network, enabling groups of messages to be switched from one cable to another.

undershoot (*ICT*) See OVERSHOOT.

undershot wheel (*Eng*) A water wheel used for low heads, in which the power is obtained almost entirely from the impulse of the water on the vanes. See PONCELET WHEEL.

undersize (*MinExt*) See RIDDLE.

undersowing (*Agri*) The introduction of rapidly growing and dense plant cover, typically a prostrate legume, below the main crop to suppress weeds and boost fertility by acting as a green manure.

understeer (*Autos*) The tendency of a vehicle to preserve directional stability by reacting against external forces applied to the steering mechanism. Cf OVERSTEER.

under-voltage no-close release (*ElecEng*) A device which acts upon the trip coil of a circuit breaker so that the circuit breaker cannot close if the voltage is below a certain predetermined value.

under-voltage release (*ElecEng*) A device which trips an electric circuit if the voltage falls below a predetermined value. Also *low-volt release*.

underwater cutting (*Eng*) A method of cutting iron or steel under water, using a combusted mixture of oxygen and hydrogen (cf OXYHYDROGEN WELDING), the flame being protected by an airshield formed by a hood.

underwater transducer (*Acous*) A transducer (eg a microphone or loudspeaker) designed for operation under water. See HYDROPHONE.

Underwriters Laboratories (*Genrl*) Insurers' organization which sets standards for testing materials, as well as performing tests (US). Abbrev *UL*.

undinism (*Psych*) A preoccupation with water, specifically with urine and the act of urination.

undistorted output (*ICT*) That of an amplifier free from non-linear distortion. The maximum undistorted output is defined as that obtained with a specified percentage of non-linear distortion.

undistorted transmission (*ICT*) That of any type of line for which the velocity of propagation and coefficient of attenuation are both independent of frequency.

undocumented feature (*ICT*) A form of BUG that produces unexpected behaviour in a system or program not identified in system documentation.

undrawn yarn (*Textiles*) Extruded manufactured filament yarn not yet subjected to the drawing process that orients the linear molecules and gives strength to the yarn.

unducted fan (*Aero*) Recent term for PROPELLER used in high-speed civil turbine engines. *Propellers* when mounted within a circular duct are called *fans* but when advanced gas turbines drive fans without surrounding ducts they are called *unducted fans*. This apparently illogical terminology may be excused as the blade shapes of the new *propellers* are closer to those of fans than to those of the 1960s.

undulant fever (*Med*) (1) A disease characterized by alternating febrile and afebrile periods, splenomegaly, transient painful swelling of joints, neuralgia and anaemia; due to infection with *Brucella melitensis*, conveyed to humans by infected goats or their milk. Also *Malta fever, Mediterranean fever, Gibraltar fever*. (2) A disease with the same SIGNS as (1), due to infection by *Brucella abortus*, a micro-organism which causes abortion in cows and is conveyed to humans in cows' milk or by contact with infected animals. Also *brucellosis*.

undulating membrane (*BioSci*) An extension of the flagella membrane in some Protozoa (eg *Trypanosoma*) by which it is attached to the cell for part of its length.

uneven working (*Print*) See EVEN WORKING.

unexcited (*Phys*) Said of an atom in its ground state. See EXCITATION.

UNF (*Eng*) Abbrev for *unified normal fine*. See UNIFIED SCREW-THREAD.

unfired (*Electronics*) Said of any gas-discharge device when in an un-ionized state.

ungual (*Med*) Pertaining to or affecting the nails. See UNGUIS.

unguiculate (*BioSci*) (1) Generally, provided with claws. (2) Specifically, in plants, applied to a petal with an expanded limb supported on a long, narrow stalk-like base.

unguis (*BioSci*) (1) In insects, one of the tarsal claws. (2) In vertebrates, the dorsal scale contributing to a nail or claw. (3) More generally, a nail or claw. Pl *ungues*. Adjs *ungual, unguinal*.

ungula (*BioSci*) A hoof. Adj *ungulate*.

ungulate (*BioSci*) The term applied to several groups of superficially similar hoofed animals which are not necessarily closely related taxonomically. Horses, cows, deer, tapirs.

unguligrade (*BioSci*) Walking on the tips of enlarged nails of one or more toes, ie hoofs, as in horses etc. Cf DIGITIGRADE, PLANTIGRADE.

uni- (*Genrl*) Prefix from Lt *unus*, one.

uniaxial (*BioSci*) Having a main axis consisting of a single row of large cells with only clearly subordinate branches. Cf MULTIAXIAL.

uniaxial (*Min*) A term describing crystalline minerals in which there is only one direction of single refraction (parallel to the principal crystal axis and known as the optical axis). All minerals which crystallize in the tetragonal, trigonal and hexagonal systems are *uniaxial*. Cf BIAXIAL. See REFRACTIVE INDEX.

unicellular (*BioSci*) Consisting of a single cell.

Unicode (*ICT*) International standard for representing symbols and characters in 16 bit codes; 8 bit codes are limited to 256 characters, but Unicode has space for all foreseeable languages and symbols.

unidirectional antenna (*ICT*) One in which the radiating or receiving properties are largely concentrated in one direction.

unidirectional current (*ElecEng*) One which, although its amplitude may vary, never changes sign.

unified field theory (*Phys*) A theory which unites field theories into a single unified framework, esp the attempt to generalize Einstein's theory of GENERAL RELATIVITY to describe electromagnetism as well as gravitation. The WEINBERG–SALAM THEORY (or electroweak theory) has successfully unified the electromagnetic and weak interactions. See panel on GRAND UNIFIED THEORIES.

unified model (*Phys*) A model of the nucleus incorporating many valuable features of both the COLLECTIVE MODEL and the INDEPENDENT PARTICLE MODEL.

unified scale (*Chem*) The scale of atomic and molecular weights which is based on the mass of the ^{12}C isotope of carbon being taken as twelve exactly; hence the atomic mass unit equals $1 \cdot 660 \times 10^{-27}$ kg. This scale was adopted in 1960 by the International Unions of both Pure and Applied Physics and Pure and Applied Chemistry, hence the name *unified scale*. See ATOMIC MASS UNIT, ATOMIC WEIGHT.

unified screw-thread (*Eng*) A screw-thread form adopted by Canada, the UK and the US. It combines features of the US *Standard Screw-Thread* and the *British Standard*

Whitworth screw-thread. Of 60° angle, the thread has radiused roots and crests while the crests of the nut are flat. There are Unified Normal Fine (UNF) and Unified Normal Coarse (UNC) variants.

uniform convergence (*MathSci*) (1) A sequence $a_1(x),a_2(x),\ldots,a_n(x),\ldots$ converges uniformly to the limit $a(x)$ in the interval (a,b) if, given any $\varepsilon > 0$, there exists for all x in (a,b), N such that $|a_n(x)-a(x)|<\varepsilon$ when $n\geqslant N$. (2) A series converges uniformly if the sequence of partial sums converges uniformly. (3) A product converges uniformly if the sequence of partial products converges uniformly. See CONVERGENCE, M-TEST OF WEIERSTRASS FOR UNIFORM CONVERGENCE.

uniform extension (*Eng*) The plastic extension produced along the gauge length of a tensile test piece before localized necking commences, ie up to the stage when the maximum load is reached. Thereafter further plastic deformation to fracture is confined to the region of the neck. See STRENGTH MEASURES.

uniform field (*MathSci*) A field that is described by the same vector at all points. A constant field.

uniformitarianism (*Geol*) The concept that the processes that operate to modify the Earth today also operated in the geological past. In its more extreme form the concept also infers uniformity of rates as well as of processes.

uniform line (*ElecEng*) One with electric properties identical throughout its length.

Unihi (*ImageTech*) TN of a WIDE-SCREEN, high-definition broadcast-standard component format using $\frac{1}{2}$ in metal tape in a cassette.

unilateral conductivity (*Phys*) The property of unipolarity by which current can flow in one direction only; exhibited by a perfect rectifier.

unilateral impedance (*Phys*) Any electrical or electromechanical device in which power can be transmitted in one direction only, eg a thermionic valve or carbon microphone.

unilateralization (*Phys*) Neutralization of feedback so that a transducer or circuit has a unilateral response, ie there is no response at the input if the signal is applied to the output terminals. While many valve circuits are inherently unilateral, equivalent transistor ones require external neutralization.

unilateral tolerance (*Eng*) A tolerance with dimensional limits either entirely above or entirely below the basic size.

unilateral transducer (*ElecEng*) One for which energy can be transmitted only forwards.

unilocular (*BioSci*) Having a single compartment. Cf BILOCULAR, MULTILOCULAR.

unimolecular layer (*Chem*) See MONOMOLECULAR LAYER.

unimolecular reaction (*Chem*) See MONOMOLECULAR REACTION.

uninemy hypothesis (*BioSci*) The hypothesis that each chromatid contains a single, DNA, double-helical molecule organized linearly with respect to the chromosomal axis.

uninstall (*ICT*) To remove an application with all its elements in different folders from a computer or a disk drive.

uninsulated conductor (*ElecEng*) A conductor at earth potential, such that no care need be taken to insulate it from earth.

uninucleate (*BioSci*) Containing one nucleus.

union (*Build*) A connection for pipes.

union (*MathSci*) Of sets A and B, the set of elements which are in A or B or in both A and B.

union (*Med*) In the process of healing, the growing together of parts separated by injury (eg the two ends of a broken bone, the edges of a wound).

union fabric (*Textiles*) A woven fabric with the warp of one fibre (eg linen) and the weft of another (eg cotton).

union kraft (*Paper*) A packaging material comprising two layers of kraft paper bonded together by means of a laminant that is resistant to the transmission of water in liquid or vapour form, eg bitumen or polythene.

uniparous (*BioSci*) Giving birth to one offspring at a time.

Unipivot instrument (*ElecEng*) An instrument whose moving-coil system is balanced on a single pivot passing through its centre of gravity.

unipolar (*BioSci*) Said of nerve cells having only one process. Cf BIPOLAR, MULTIPOLAR.

unipolar transistor (*Electronics*) A transistor with one polarity of carrier.

unipole antenna (*ICT*) Isotropic antenna conceived as radiating uniformly in phase in all directions. Theoretically useful, but not realizable in practice. See ISOTROPIC RADIATOR.

unipotent (*BioSci*) Of embryonic cells, capable of forming a single cell type only. Cf PLURIPOTENT, TOTIPOTENT.

unique sequence DNA (*BioSci*) DNA sequences that are only represented once in the HAPLOID genome. Most genes are in this category.

uniramous (*BioSci*) Having only one branch, as some crustacean appendages. Cf BIRAMOUS.

uniselector (*ICT*) A selector switch that only rotates its wipers about an axis, in contrast with the TWO-MOTION SELECTOR, in which wipers are raised to a specified level in the rows of contacts by the impulse trains, and then enter the bank of contacts, either by hunting or by a further train of impulses. See SELECTOR.

uniseriate (*BioSci*) Arranged in a single row, series or layer.

unisexual (*BioSci*) Showing the characters of one sex or the other; distinctly male or female. Cf HERMAPHRODITE.

unit (*Genrl*) A dimension or quantity which is taken as a standard of measurement.

unit (*MinExt*) Equal to 1% of a specified element or compound in a parcel of ore, concentrates or metal being sold.

unit arch (*Print*) On a web-fed press, a perfecting unit arranged in an arch or inverted-U design.

unit cell (*Crystal*) The smallest group of atoms, ions or molecules, whose repetition at regular intervals, in three dimensions, produces the lattice of a given crystal. See panel on CRYSTAL LATTICE.

unit character (*BioSci*) A character that can be classified into two distinct types, usually the normal and the mutant, and displaying *Mendelian inheritance*.

unit heater (*Eng*) A combination of air heater and circulator, often in the form of a heated cellular core or finned tube over which air is blown by a fan.

unit interval (*ICT*) In a system using an equal-length code or in a system using an isochronous modulation, the interval of time such that the theoretical duration of the significant intervals of a telegraph modulation (or restitution) are whole multiples of this interval.

unitized (*Autos*) See CHASSIS.

unit leaf rate (*BioSci*) See NET ASSIMILATION RATE.

unit matrix (*MathSci*) A square matrix satisfying the following conditions: (1) the leading diagonal entries, ie the entries on the diagonal from top left to bottom right, are all one; and (2) the entries not on the leading diagonal are all zero. The unit matrix of order n is an identity element for multiplication in the set of all square matrices of order n. A unit matrix is usually denoted by I.

unit membrane (*BioSci*) The ubiquitous, approximately 7 nm wide, membrane structure found in all cells, composed of a fluid lipid bilayer with intercalated proteins. In electron microscopic images it appears to be three-layered because of the distribution of stain.

unit of attenuation (*ICT*) See DECIBEL, NEPER.

unit of bond (*Build*) That part of a brickwork course which, by being constantly repeated throughout the length of the wall, forms a particular bond.

unit plane (*Phys*) See PRINCIPAL PLANES OF A LENS.

unit pole (*Phys*) A magnetic pole which experiences a repulsive force of 1 newton when 1 metre (or 1 dyne when 1 centimetre) apart from a like pole in a vacuum. A mathematical concept formerly used for establishing magnetic and electrical units.

unit testing (*ICT*) The phase of a testing cycle in which individual components of software or hardware are tested in isolation to establish their fitness for purpose before being brought together.

unit type press (*Print*) A web-fed press with one or more printing units in line on the bed plate.

univalent (*BioSci*) One of the single chromosomes which separate in the first meiotic division.

univalent (*Chem*) See MONOVALENT.

univariant (*Chem*) Having one DEGREE OF FREEDOM.

universal asynchronous receiver/transmitter (*ICT*) Computer component that manages serial communication. Abbrev *UART*.

universal beam (*Eng*) H-shaped steel joists with parallel flanges with deep web and narrow flanges to carry a bending load. Maximum size 914×419 mm.

universal bearing piles (*Eng*) H-shaped steel joists with parallel flanges with web and flanges of equal thickness to withstand corrosion. Maximum size 356×368 mm.

universal chuck (*Eng*) See SELF-CENTRING CHUCK.

universal column (*Eng*) H-shaped steel joists with parallel flanges which are thickened to carry an axial load. Maximum size 356×406 mm.

universal combustion burner (*Eng*) Natural-draught gas burner having one injector for the entrainment of primary air prior to combustion, and a secondary injector through which the flow of additional air into the combustion chamber can be regulated.

universal grinder (*Eng*) A machine in which the work rotates against a power-driven grinding wheel with the axes of both mounted parallel with a reciprocating table carrying the workhead and support tailstock. A wide range of movements and attachments allow plain and tapered external and internal cylindrical grinding and also surface grinding.

universal indicator (*Chem*) A mixture of *indicators* which gives a definite colour change for each integral change of pH value over a wide range.

universal joint (*Autos*) A device, usually of the modified *Hooke's* type, which allows rotary drive to be transmitted through an angle. Used on propeller shafts and independently suspended driven wheels to accommodate suspension movement. See CONSTANT VELOCITY JOINT.

universal milling machine (*Eng*) A MILLING MACHINE similar to a plane milling machine but with the additional feature that the table swivels horizontally, and is provided with a dividing head as standard equipment.

universal motor (*ElecEng*) A fractional horsepower commutator motor for use with both dc and single-phase ac.

universal plane (*Build*) Multipurpose plane adaptable for rebating, grooving, trenching, cutting mouldings, etc.

universal planer (*Eng*) A planer that will cut on the forward and on the reverse strokes.

Universal Product Code (*ICT*) Standard BAR CODE now adopted in Europe. Abbrev *UPC*.

universal resource locator (*ICT*) The system of unique addresses that allows an Internet site to communicate with any other. Abbrev *URL*. See panel on INTERNET.

universal serial bus (*ICT*) A versatile bus system for PCs with about ten times the transfer speed of older standards. Abbrev *USB*.

universal set (*MathSci*) The set containing all the elements relevant to a particular mathematical study. It has been shown that there can be no all-inclusive universal set whose numbers include its own subsets.

universal shunt (*Electronics*) A series of high-stability resistors used to shunt galvanometers to provide different ranges of measurement.

universal terminal (*ICT*) The hand-held terminal used with a LOW EARTH ORBIT SATELLITE system for worldwide mobile communication.

universal time (*Astron*) A name for *Greenwich Mean Time*, recommended in 1928 by the International Astronomical Union to avoid confusion with the pre-1925 GMT which began at noon, not midnight. Abbrev *UT*. *Co-ordinated universal time* (*UTC*), the one used by most broadcast time services, is based on the uniform scale of atomic time. See EPHEMERIS TIME.

universal veil (*BioSci*) The membrane that encloses the developing fruiting body of some AGARICS, rupturing, as the stalk grows, to leave the VOLVA.

universal viewfinder (*ImageTech*) One with objectives for lenses of varying focal length.

universal vise (*Eng*) A vice which has two or three swivel settings allowing the workpiece to be set at a compound angle. Also *toolmaker's vice*.

universe (*Astron*) The totality of all that is in the cosmos and which can affect us by means of physical forces. The definition excludes anything which is in principle undetectable physically, such as regions of SPACE-TIME that have been irreversibly cut off from our own space-time.

univibrator (*ICT*) Term for monostable multivibrator circuit. See FLIP-FLOP.

univoltine (*BioSci*) Producing only one set of offspring during the breeding season or year. Cf MULTIVOLTINE.

UNIX (*ICT*) TN for a well-known OPERATING SYSTEM not tied to a particular computer manufacturer. It is a trademark of AT&T Laboratories.

unloaded antenna (*ICT*) One with no inductance coils to increase its natural wavelength.

unmod (*ImageTech*) Unmodulated, describing a photographic sound track without any recorded signal.

unmodulated waves (*ICT*) Waves that do not vary in amplitude with time, such as those radiated from a radiotelephone transmitter when no sound enters the microphone.

unnilenium (*Chem*) See MEITNERIUM.

unnilhexium (*Chem*) See RUTHERFORDIUM.

unniloctium (*Chem*) See HAHNIUM.

unnilpentium (*Chem*) See JOLIOTIUM.

unnilquadium (*Chem*) See DUBNIUM.

unnilseptium (*Chem*) See BOHRIUM.

unpitched sound (*Acous*) Any sound or noise which does not exhibit a definite pitch, but consists of components spread more or less continuously over the frequency spectrum.

unsaturated (*Chem*) (1) Less concentrated than a saturated solution or vapour. (2) Containing a double or a triple bond, esp between two carbon atoms; unsaturated molecules can thus add on other atoms or radicals before saturation is reached.

unsaturated fatty acid (*BioSci*) A fatty acid with one or more double bonds.

unsealed source (*Radiol*) A radioactive source of any kind from which the radiation can escape. It is therefore of very low activity such as a tracer being used during a medical investigation.

unsewn binding (*Print*) The sections of the book are gathered and fed to a machine; part of the back is cut off, the edge roughened and adhesive applied and, when it sets, the sections compacted together. Used mainly for paperbacks. Often PERFECT BINDING, THERMOPLASTIC BINDING.

unshielded twisted pair cable (*ICT*) A data cable consisting of pairs of wires twisted together but not surrounded by a metal screening sheath. See TWISTED PAIR CABLE.

unsoundness (*Eng*) The condition of a solid metal which contains blowholes or pinholes due to gases, or cavities resulting from its shrinkage during contraction from liquid to solid state, ie contraction cavities.

unsqueezed print (*ImageTech*) A print from an anamorphic motion picture negative in which the image has been optically corrected for normal projection.

unstability (*ElecEng*) See STABILITY TEST.

unstable (*Build*) A term applied to a structural framework having fewer members than it would require to be perfect. Also *deficient*.

unstable (*Chem*) Subject to spontaneous change.

unstable equilibrium (*Phys*) The state of equilibrium of a body when any slight displacement decreases its potential energy. The instability is shown by the fact that, having been slightly displaced, the body moves farther away from its position of equilibrium.

unstable oscillation (*Genrl*) Any oscillation, eg in a mechanical body or electric circuit, which increases in amplitude with time.

unstick (*Aero*) See LIFT-OFF.

unstirred layer (*BioSci*) See BOUNDARY LAYER.

unstriated muscle (*BioSci*) A form of contractile tissue composed of spindle-shaped fibrillar uninucleate cells, occurring principally in the walls of the hollow viscera. Also *smooth muscle*. Cf STRIATED MUSCLE. See VOLUNTARY MUSCLE.

unsymmetrical grading (*ICT*) Grading in which subscribers originating higher than average traffic are given access to a greater proportion of individual trunks.

unsymmetrical oscillations (*ICT*) Oscillations in which the positive and negative parts of the waveform are unequal and of different shape.

untrimmed floor (*Build*) A floor consisting of bridging joists only.

untuned antenna (*ICT*) One not separately tuned to the operating frequency, although effectively tuned by coupling to one or more resonant circuits.

untuned circuit (*ICT*) One not sharply resonant to any particular frequency.

unvoiced sound (*Acous*) In speech, any elemental sound which has no discrete harmonic frequencies but consists of a wide frequency band generated by air rushing through the mouth and nasal cavities; these are modified in spectral energy distribution by the posture of the cavities in the mouth and the resultant broad resonances.

up (*ICT*) A term describing a computer when it is functioning and ready for use.

up (*Phys*) One of the six FLAVOURS of QUARKS with a mass of 5 MeV and a charge of +2e/3. Protons and neutrons are made up of UP and DOWN quarks. See appendix on Subatomic particles.

U-packing (*Eng*) A flexible annular ring of U-section used to pack the glands of fluid power pistons, rams. The hollow of the U faces the pressure side so that the pressure expands the legs of the U, forming a seal. Modern designs use neoprene or similar and may have a toroidal spring in the arms of the U. Also *U-leather*.

upcast shaft (*CivEng*) A ventilating shaft through which the vitiated air passes in an upward direction.

update (*ICT*) (1) To bring a file up to date by modifying entries and adding new entries in accordance with a specified procedure. (2) To modify a computer instruction so that the address specified is increased every time the instruction is carried out.

updraught (*Autos*) Said of a carburettor in which the mixture is drawn upwards against the force of gravity.

upgrade (*ICT*) To add additional power, speed, memory capacity or features to a COMPUTER SYSTEM. In the case of SOFTWARE it means to improve the features and performance of the software.

upholsterer's hammer (*Build*) Lightly built claw hammer for use with tacks etc, often magnetized for placing the tacks.

uplink (*ICT*) (1) The radio link from a ground station to a communications satellite. (2) The radio link from a mobile telephone to a mobile-telephone base station.

upload (*ICT*) The process of transferring FILES from a small computer to a larger HOST machine. The reverse process is referred to as download.

up locks (*Aero*) Safety locks which hold the units of a retractable landing gear up in flight; *down locks* hold it down when on the ground.

upmake (*Print*) (1) To arrange lines of print into columns or pages. (2) Print so arranged.

upper atmosphere (*Astron*) A term used somewhat loosely for the region of the Earth's atmosphere above about 30 km

(20 miles), which is not normally explored by sounding balloons, but can be studied by rockets and artificial satellites.

upper bound (*MathSci*) See BOUNDS OF A FUNCTION.

upper case (*Print*) Frequently used as a synonym for capital, the name deriving from the arrangement (now largely superseded) of the capitals in an upper and the small letters in a LOWER CASE.

upper culmination (*Astron*) See CULMINATION.

upper deck (*Ships*) The term correctly denotes the main strength deck of a ship. From this deck all scantlings are determined, freeboard assigned and subdivision arranged, according to type of vessel.

upper mantle (*Geol*) The upper part of the mantle from the MOHOROVIČIĆ DISCONTINUITY (the *Moho*) at the base of the crust to a depth of perhaps 1000 km. It is thought to be peridotitic in composition. See MANTLE.

upper mean hemispherical candle-power (*Phys*) See MEAN HEMISPHERICAL CANDLE-POWER.

upper memory (*ICT*) In the context of an IBM-COMPATIBLE COMPUTER using the MS-DOS OPERATING SYSTEM, the 384 Kbytes of ADDRESS space adjacent to the 640 Kbytes of conventional memory.

upper quartile (*MathSci*) The argument of the cumulative distribution function corresponding to a probability of 0·75; (of a sample) the value below which occur three-quarters of the observations in the ordered set of observations.

upright (*Build*) A vertical member in a structure.

UPS (*ICT*) Abbrev for *un-interruptible power supply*. In the context of a FILE SERVER it provides back-up mains power from rechargeable batteries for a period of time should the mains supply fail. Thus the user may continue to gain access to files but, more importantly, the file server will be able to warn users and close any open files.

upsetting (*Eng*) Metalworking so as to produce an increase in section of part of a component over and above its starting size, as in the forming of the head of a bolt or rivet from round bar.

upstream (*BioSci*) In the direction opposite to that of DNA transcription. The term is also used generally for the earliest events in a chain of sequential reactions.

upstream injection (*Aero*) A gas turbine fuel system in which the fuel is injected towards the compressor in order to achieve maximum vaporization and turbulence.

uptake (*Eng*) The flue or duct which leads the flue gases of a marine boiler to the base of the funnel.

uptake (*Radiol*) In radiobiology, the quantity (or proportion) of administered substance subsequently to be found in a particular organ or tissue.

upturn (*Build*) The part of a lead flashing which is dressed up against a wall face. See FLASHINGS.

UPVC (*Plastics*) Abbrev for UNPLASTICIZED POLYVINYL CHLORIDE. Also *uPVC*. See panel on POLYVINYL CHLORIDE.

uracil (*Chem*) 2,6-dioxypyrimidine. One of the four bases in RNA and the only one which does not occur in DNA. Pairs with adenine. See panel on DNA AND THE GENETIC CODE.

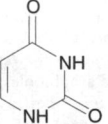

uracil

uraemia (*Med*) The state resulting from failure of a diseased kidney to perform its normal functions; associated with retention of urea in the blood, and characterized by varied symptoms, among which are headache, foul breath, diarrhoea and vomiting, visual disturbances, lethargy, convulsions and coma. US *uremia*.

Uralian emerald (*Min*) Not an emerald, but a green variety of ANDRADITE garnet (*demantoid*), occurring as nodules in

Uranium isotope enrichment

The natural abundance of the fissionable isotope of uranium (uranium-235) is 0·72%, and this must be increased to around 2% to 5% for use in a power station and to over 50% for use in a bomb or a fast-breeder reactor. The first method used hundreds of mass spectrometers in which powerful magnets deflected the lighter uranium-235 to a greater extent than uranium-238 (the Calutron method). Later, gaseous diffusion plants were developed in the USA, and in their heyday in the 1970s consumed 4% of all the electricity generated in the country. More recently a centrifuge and a laser method have been developed which are somewhat cheaper to build and operate.

Gaseous diffusion

Uranium hexafluoride is a highly reactive gas above 56°C and has the advantage that fluorine has a low molecular weight (19) and only one isotope. The whole method depends on the fact that the two forms of UF$_6$ have molecular weights of 352 and 349 and will therefore diffuse through a barrier made of submicroscopic holes at different rates. This needs a plant of some 2000 stages in which the material from one stage is *cascaded* up or down the stages.

Four of these stages are shown in Fig. 1. As drawn, the gas becomes more enriched for uranium-235 as it proceeds upwards and more depleted as it proceeds downwards towards the *tails*. The capacity of such a plant is defined in terms of SEPARATIVE WORK UNITS (SWUs), which in turn depends on the throughput of natural uranium and the concentration of tails which the plant produces. Typically, a diffusion plant producing 0·2% of uranium-235 in the tails will make 200 g of 3·2% enriched uranium-235 per SWU.

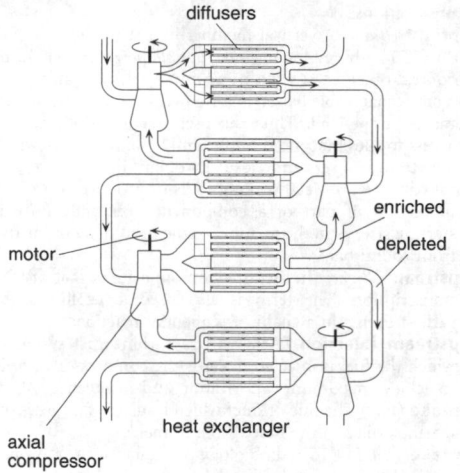

Fig. 1 **Diffusion plant**.

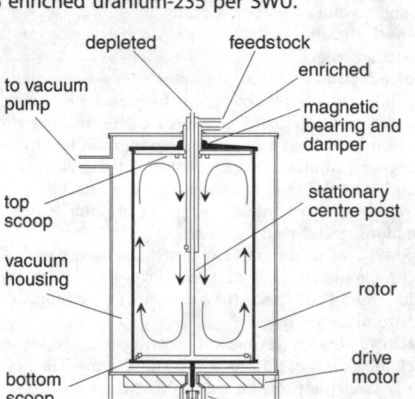

Fig. 2 **Centrifuge enrichment** Cross-section.

The large US diffusion plants produced about 6 million SWUs per year and other plants have been built

---

ultrabasic rocks in the Urals; used as a semiprecious gemstone, though rather soft for this purpose.

uralite (*Min*) A bluish-green monoclinic amphibole, generally actinolitic in composition, resulting from the alteration of pyroxene.

uralitization (*Geol*) A type of alteration of pyroxene-bearing rocks, involving the replacement of the original pyroxenes by fibrous amphiboles, as in some epidiorites.

uranic (*Chem*) Referring to (VI) uranium. See URANOUS.

uranides (*Chem*) The name for elements beyond protactinium in the periodic system. Cf ACTINIDES.

uraninite (*Min*) Uranium oxide, UO$_2$, often more or less hydrated and containing also lead, thorium and the metals of the lanthanum and yttrium groups. It occurs as brownish to black cubic crystals and is an accessory mineral in granite rocks and in metallic veins. When massive, and apparently amorphous, known as *pitchblende*.

uranium (*Chem*) Symbol U, at no 92, ram 238·03, rel.d. 18·68, mp 1150°C. A hard grey metal with seven isotopes; ^{235}U is the only naturally occurring, readily fissile isotope

and comprises 0·72% of natural uranium. The majority isotope ^{238}U is a fertile material (half-life 4·5 × 10^9 yr) which has a small fission cross-section for fast neutrons; ^{233}U is a fissile material which can be produced by the neutron irradiation of ^{232}Th; ^{235}U is used in nuclear reactors and nuclear weapons. Because the half-life of ^{235}U is very much less than that of ^{238}U the relative abundance of these two isotopes has varied over time with ^{235}U being some 3% about 2 × 10^9 years ago. See panel on OKLO NATURAL FISSION REACTOR.

uranium enrichment (*NucEng*) A process to increase the isotopic content of ^{235}U in uranium for reactor use. See panel on URANIUM ISOTOPE ENRICHMENT.

uranium fuel cycle (*NucEng*) The stages from mining to the ultimate disposal of the products of fission and other reactions. Fig. ▷

uranium hexafluoride (*Chem*) *Uranium (VI) fluoride*. A volatile compound of uranium with fluorine, used in the gaseous diffusion process for separating the uranium isotopes. Very corrosive.

elsewhere. The diffusers are ceramic cylinders and contain millions of minute channels with an average diameter of about 10×10^{-9} m. Diffusion is assisted by pumps with valves in the depleted flow circuits to equalize their gaseous resistance with that through the diffusers. Heat exchangers remove the surplus heat introduced from compressing and pumping the gas.

Centrifuge separation
This method uses a cascade of hundreds of centrifuges with specially designed rotors (see Fig. 2). The degree of separation is higher than with diffusion because it depends on the difference in mass of UF_6 rather than on the ratio, but the amount processed at each stage is small. A plant with about half the capacity of a US diffusion plant would need about 2 million centrifuges.

Uranium isotope enrichment *(Cont.)*

The three European plants have a total capacity of about a third of one such diffusion plant.

Laser enrichment
Beams of light from a dye laser can be tuned to provide energy at a wavelength which will ionize uranium-235 but not uranium-238. Ionized uranium is then collected by applying a large negative potential to a plate parallel to the vapour stream (see Fig. 3). Theoretical separation factors are much higher than in the other two methods but yields depend on both the power of available lasers and the need to handle the molten and enriched uranium. The USA has put great effort into the development of this method as a replacement for the old diffusion plants.

See panels on NUCLEAR REACTORS and OKLO NATURAL FISSION REACTOR.

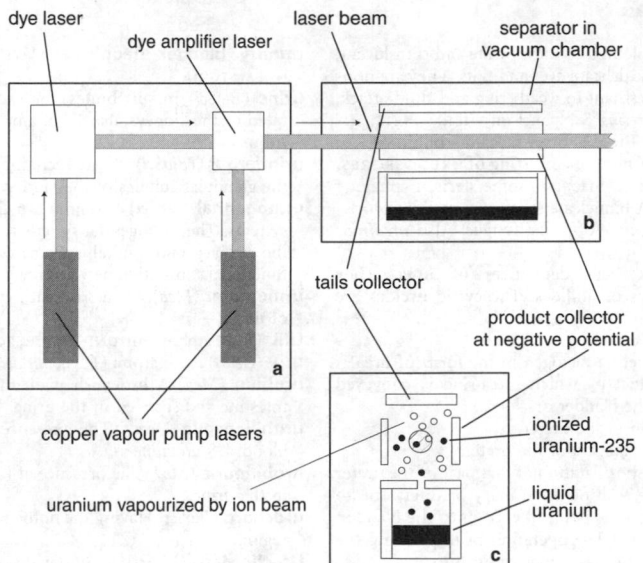

Fig. 3 **Laser enrichment facility** where (c) is a cross-section of the separator chamber (b) and (a) houses the lasers.

uranium–lead dating (*Geol*) A method of determining the age in years of geological material, based on the known decay rate of ^{238}U to ^{206}Pb and ^{235}U to ^{207}Pb. See panel on RADIOMETRIC DATING.

uranium–radium series (*Phys*) The series of radioactive isotopes which result from the decay of ^{238}U. The mass numbers of the members of the series are $4n + 2$, where n is an integer. Series ends in the stable isotope ^{206}Pb. See RADIOACTIVE SERIES.

uranophane (*Min*) An ore of uranium, a yellow secondary hydrated calcium uranium silicate. Also *uranotile*.

uranoplasty (*Med*) Plastic operation for closing a cleft in the hard palate.

uranous (*Chem*) Referring to (IV) uranium. See URANIC.

Uranus (*Astron*) The seventh major planet from the Sun, discovered in 1781 by William Herschel in the course of a systematic survey of the heavens. Orbital period 84·01 years, mass 14·56 times the Earth. There are 17 natural satellites as well as a ring system discovered in 1977. See appendix on Planets.

uranyl (*Chem*) The radical UO_2^{++}, eg uranyl nitrate, $UO_2(NO_3)_2$.

urate (*Chem*) Salt of uric acid.

urea (*Chem*) Carbamide. H_2NCONH_2; mp 132°C; very soluble in water, insoluble in ether. Found in the urine of mammals. Wöhler synthesized (1828) urea from ammonium isocyanate, which undergoes an intramolecular transformation when its aqueous solution is heated, forming urea. Manufactured by heating carbon dioxide and ammonia under high pressure, it is highly nitrogenous, and widely used for fertilizer and animal feed additive.

urea cycle (*BioSci*) The cyclic interconversion of four amino acids that converts carbamyl phosphate into urea. It represents the major pathway for the excretion of nitrogenous waste in terrestrial vertebrates.

urea–formaldehyde plastics (*Plastics*) Abbrev UF. See UREA RESINS.

urea resins (*Plastics*) The thermosetting resins manufactured by heating together urea and an aldehyde, generally

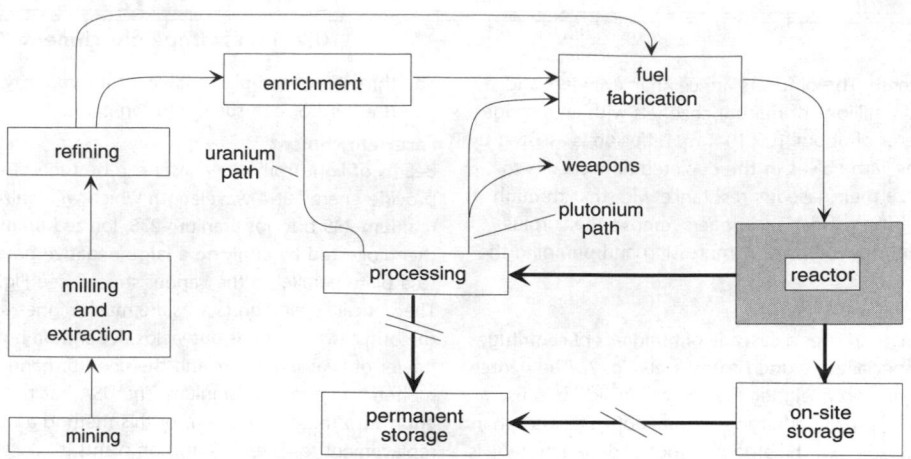

uranium fuel cycle The two breaks in the heavy lines indicate steps not yet achieved.

formaldehyde. Pale-coloured or water-white and translucent, they can therefore take delicate dyes and tints. They are non-flammable, and are resistant to weathering and fluid attack. Also *aminoaldehydic resins*. See panel on THERMOSETS.

ured-, uredo- (*Genrl*) Prefixes from Lt *uredo*, blight.

uredosorus (*BioSci*) A pustule consisting of UREDOSPORES, with their supporting hyphae, and some sterile hyphae.

uredospore (*BioSci*) A binucleate spore that rapidly propagates the dikaryotic phase of a rust fungus. Also *uredinio-spore, urediospore*.

ureides (*Chem*) The acid derivatives of urea. They correspond to amides or anilides. The cyclic ureides are known as the *purine group*.

uremia (*Med*) US for URAEMIA.

ureotelic (*BioSci*) Excreting nitrogen in the form of urea.

ureter (*BioSci*) The duct by which the urine is conveyed from the kidney to the bladder or cloaca.

ureteralgia (*Med*) Pain in the ureter.

ureteritis (*Med*) Inflammation of the ureter.

ureterocele (*Med*) Cystic dilatation of that part of the ureter which lies within the wall of the urinary bladder, due to congenital narrowing at its point of entry into the bladder.

ureterocolostomy (*Med*) The operation of implanting the ureter into the colon so that it may drain into it.

ureterolithotomy (*Med*) The operation of cutting into the ureter to remove a stone from it.

ureteropyelitis (*Med*) Inflammation both of a ureter and of the pelvis of the kidney on the same side.

ureterotomy (*Med*) Surgical incision of a ureter.

urethra (*BioSci*) The duct by which the urine is conveyed from the bladder to the exterior, and which in male vertebrates serves also for the passage of semen. Adj *urethral*.

urethritis (*Med*) Inflammation of the urethra.

urethrocele (*Med*) Prolapse of the floor of the female urethra; usually associated with hernia of the bladder.

urethrocystitis (*Med*) Inflammation of both the urethra and the urinary bladder.

urethrospasm (*Med*) Spasmodic contraction of the muscular tissue of the urethra.

urethrotomy (*Med*) The operation of cutting a stricture of the urethra.

uric acid (*Chem*) 2,6,8-trihydroxypurine. $C_5H_4N_4O_3$. An acid of the purine group. It is a white crystalline powder, insoluble in cold, hardly soluble in hot, water. Uric acid deposits in the organism are the cause of gout and rheumatism. It forms soluble lithium and piperazine salts. It can be recognized by the MUREXIDE test.

uricotelic (*BioSci*) Excreting nitrogen as uric acid.

urinary bladder neoplasia (*Vet*) See BOVINE CYSTIC HAEMATURIA.

urine (*BioSci*) In vertebrates, the excretory product elaborated by the kidneys, usually of a more or less fluid nature. Adj *urinary*.

uriniferous (*BioSci*) Urine-secreting, urine-producing, as the glandular tubules of the kidney. Also *uriniparous*.

urinogenital (*BioSci*) Pertaining to the urinary and genital systems. The urinogenital system consists of the organs of the urinary and genital systems when there is a direct functional connection between them, as in male vertebrates.

urinometer (*Med*) An instrument for measuring the density of urine.

URL (*ICT*) Abbrev for UNIFORM RESOURCE LOCATOR.

uro- (*Genrl*) Prefix from: (1) Gk *ouron*, urine; (2) Gk *oura*, tail.

urobilin (*Med*) A brownish pigment reabsorbed from the intestine and secreted in the urine.

urobilinaemia (*Med*) The presence of UROBILIN in the blood. US *urobilinemia*.

urobilinuria (*Med*) The presence of (an excess of) UROBILIN in the urine.

urochord (*BioSci*) Having the notochord confined to the tail region.

Urochordata (*BioSci*) A subphylum of Chordata, in which only the larvae have a hollow dorsal nerve cord and a notochord, the adults being without coelom, segmentation and bony tissue, and having a dorsal atrium, a reduced nervous system, and a test composed of tunicin, a substance closely related to cellulose. Sea squirts. Also *Tunicata*.

Urodela (*BioSci*) An order of amphibians in which the adults have four similar pentadactyl limbs and a prominent tail. The larvae have external gills which persist in the adults of neotenous forms, and in some others gill slits persist. Newts and salamanders. Also *Caudata*.

urodelous (*BioSci*) Having a persistent tail, as salamanders.

urodynamics (*Med*) The study of urine flow.

urography (*Radiol*) The radiological examination of the urinary tract. See PYELOGRAPHY.

urolithiasis (*Med*) The occurrence of stones or CALCULI in the urinary tract.

urology (*Med*) That part of medical science which deals with diseases and abnormalities of the urinary tract and their treatment. Hence *urologist*.

uropod (*BioSci*) In Malacostraca, an appendage of the abdominal somite preceding the telson.

uropygial gland (*BioSci*) See OIL GLAND.

uropygium (*BioSci*) In birds, the short caudal stump into which the body is prolonged posteriorly.

urosome (*BioSci*) (1) In aquatic vertebrates, the tail region. (2) In Crustacea, the hinder part of the abdomen.

urostyle (*BioSci*) (1) In fish, the hypural bone. (2) In Anura, a rod-like bone formed by the fusion of the caudal vertebrae.

Ursa Major (*Astron*) A large conspicuous constellation, containing the Plough. Also *Great Bear*.

Ursa Minor (*Astron*) The constellation around the northern celestial pole. Its brightest star is Polaris. Also *Little Bear*.

urticant (*BioSci*) Irritating; stinging. Also *urticating*.

urticaria (*Med*) A condition in which smooth, elevated, whitish patches (weals) appear on the skin and itch intensely, as a result of taking drugs or certain foods (eg shellfish), or as a reaction to the injection of serum, insect bites or the stings of plants (nettle-rash). Also *hives*.

urtite (*Geol*) An intrusive igneous rock composed mainly of nepheline.

usage parameter control (*ICT*) Software in an ASYN-CHRONOUS TRANSFER MODE network that protects quality of service by ensuring that users do not violate their contracts in respect of peak network usage and by taking action to deal with any such illegal traffic.

USB (*ICT*) Abbrev for UNIVERSAL SERIAL BUS.

useful life (*ElecEng*) The life which can be expected from a component before the chance of failure begins to rise.

useful load (*Aero*) The gross weight of an aircraft, less the tare weight. Usually includes fuel, oil, crew, equip-ment not necessary for flight (such as parachutes) and payload.

Usenet (*ICT*) Worldwide collection of newsgroups. See panel on INTERNET.

user (*ICT*) A person who uses a computer. An *end user* is someone who uses just APPLICATION PROGRAMS rather than being a PROGRAMMER.

user-defined (*ICT*) Features of a computer system that may be chosen by the user or by the programmer; eg the function of certain keys on the keyboard, the shape of typefaces, the colours displayed on a screen.

user-friendly (*ICT*) The degree to which a system is easy to use. See USER INTERFACE.

user interface (*ICT*) The communication between compu-ter systems and the people who use them. Also *human–computer interface, man–machine interface*. See GRAPHICAL USER INTERFACE, USER-FRIENDLY.

username (*ICT*) The name or code by which a person or group is identified when gaining access to a computer network. Also *user ID*.

user-to-user signalling (*ICT*) The capability of a network to pass control signals, as distinct from speech, from terminal to terminal, eg to display the name of a caller.

U-shaped valley (*Geol*) See VALLEY.

USS thread (*Eng*) See SELLERS SCREW-THREAD.

Ustilaginales (*BioSci*) An order of the Basidiomycotina containing the SMUT fungi.

UT (*Astron*) Abbrev for UNIVERSAL TIME.

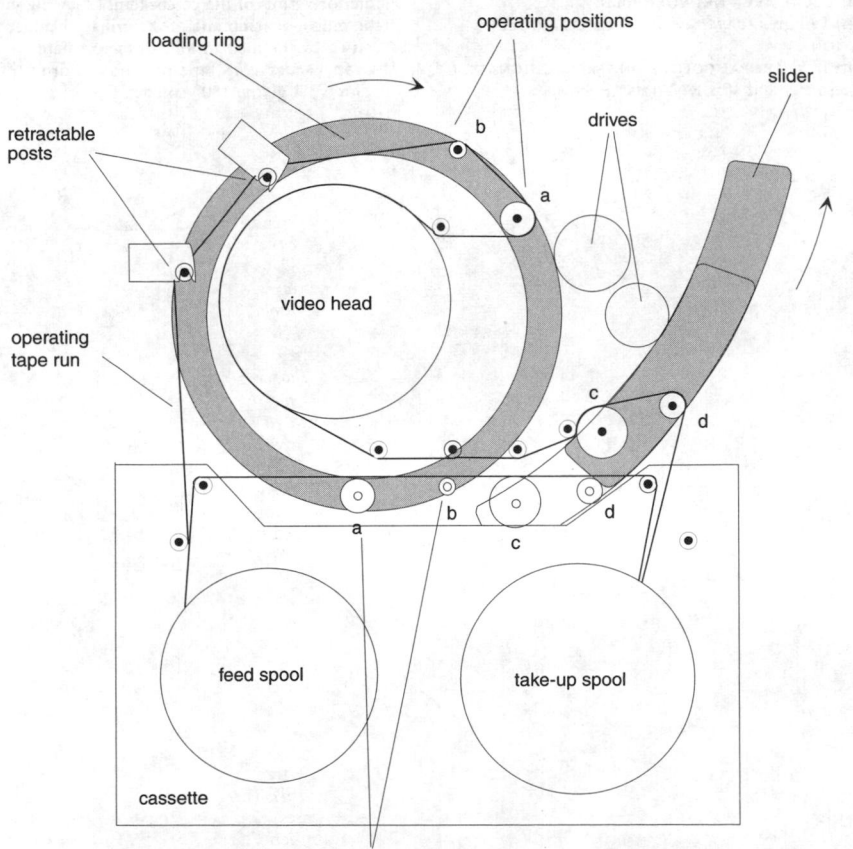

U-wrap Components a, b, c, d are shown in both loading and operating positions. The two retractable posts fold down in the loading position. The shaded elements are moved by the drives.

UTC (*ICT*) Abbrev for *universal time co-ordinates*. See ZULU
TIME.

uterus (*BioSci*) (1) In female mammals, the muscular
posterior part of the oviduct in which the fetus is lodged
during the prenatal period. (2) In lower vertebrates and
invertebrates, a term loosely used to indicate the lower part
of the female genital duct, or in certain cases (as in
Platyhelminthes) a special duct in which eggs are stored or
young developed. Adj *uterine*.

utility program (*ICT*) SYSTEMS PROGRAM designed to
perform a commonplace task such as the transfer of
data from one storage device to another or sorting a set of
data.

utilization factor (*ElecEng*) The ratio of the luminous flux
reaching a specified plane to the total flux emanating from
an electric lamp.

UTP (*ICT*) Abbrev for UNSHIELDED TWISTED PAIR cable.

utricle (*BioSci*) (1) Any one of a variety of small inflated
bladder-like structures. (2) In vertebrates, the upper
chamber of the inner ear from which arise the semicircular
canals. Also *utriculus*.

utricular (*BioSci*) Like a bladder; pertaining to a utricle. Also
utriculiform.

utriculoplasty (*Med*) The operation of excising a portion of
the body of the uterus; done for the treatment of uterine
haemorrhage.

UUCP (*ICT*) Abbrev for *UNIX-to-UNIX copy*. Widely used
mail network connecting machines running the UNIX
operating system, often over ordinary telephone dial-up
lines, although the connection can be made over dedicated
lines or LOCAL AREA NETWORK lines.

UV (*Chem*) Abbrev for *ultraviolet*. See ULTRAVIOLET
RADIATION.

UV (*ImageTech*) The PAL COLOUR DIFFERENCE SIGNALS. *U* is
B−Y (also P_B) and *V* is R−Y (also P_R).

UV absorbers (*Chem, Plastics*) Aromatic chemicals added to
polymers exposed to sunlight so that ultraviolet degradation can be inhibited.

uvarovite (*Min*) A variety of garnet, of an attractive green
colour; essentially silicate of calcium and chromium.

UV-A, UV-B, UV-C (*Phys*) Regions of the ultraviolet
spectrum defined by the Commission Internationale de
l'Eclairage (CIE) as follows: UV-A, 315–400 nm; UV-B,
280–315 nm; UV-C < 280 nm. Used primarily in relation
to environmental exposure and consumer products.

UV degradation (*Chem, Plastics*) Ultraviolet-initiated
breakdown of polymers due to photon absorption at chain
defects, such as carbonyl groups in backbone chain.
Commonly caused by effect of sunlight on products like
garden furniture, resulting in surface cracking. It is closely
related to the oxidation of polymers. Effect is exploited,
however, in PHOTORESISTS and PHOTOLITHOGRAPHY.

uvea (*BioSci*) (1) In vertebrates, the posterior pigment-bearing
layer of the iris of the eye. (2) The iris, the ciliary body and
the choroid considered as one structure. Also *uveal tract*.

uveitis (*Med*) Inflammation affecting the iris, the ciliary
body and the choroid.

uveoparotid fever (*Med*) A condition characterized by
inflammation of the parotid glands and bilateral iridocyclitis, often with paralysis of the seventh cranial nerve.

UV filter (*ImageTech*) A filter absorbing ultraviolet radiation, used to prevent it affecting a blue-sensitive photographic emulsion.

uvula (*Med*) (1) A small conical process hanging from the
middle of the lower border of the soft palate; part of the
inferior vermis of the cerebellum. (2) A slight elevation of
the mucous membrane of the urinary bladder in the male
caused by the median lobe of the prostate.

U-wrap (*ImageTech*) Tape path on the drum of a HELICAL
SCAN VTR giving 180° contact. Fig. ◁

V (*Aero*) Subscripted symbol used in aircraft documentation. V_1: abbrev for *critical speed*. V_{lo}: see LIFT-OFF. V_{ne}: abbrev for the maximum permissible *indicated airspeed*: a safety limitation (the subscript means *never exceed*) because of strength or handling considerations. The symbol is used mainly in operational instructions. V_{no}: *normal operating speed*, usually of an airliner or other civil aircraft; this term is used mainly in flight operation documents and may be quoted in *EAS, IAS* or *TAS*. V_r: abbrev for *rotation speed*.

V (*Chem*) Symbol for VANADIUM.

V (*Phys*) Symbol for VOLT.

V (*Phys*) Symbol for: (1) ELECTROMOTIVE FORCE; (2) POTENTIAL; (3) POTENTIAL DIFFERENCE; (4) volume.

V_{max} (*BioSci*) The maximum initial velocity of an enzyme-catalysed reaction, ie at saturating substrate levels.

v (*Phys*) Symbol for: (1) VELOCITY; (2) specific volume of a gas.

v- (*Chem*) See VICINAL.

vacancy (*Phys*) Site unoccupied by an atom or ion in a crystal lattice.

vaccinal (*Med*) Of, pertaining to or caused by vaccine or vaccination.

vaccination (*Med*) Immunization against an infectious disease by exposure to the appropriate VACCINE.

vaccine (*BioSci, Med*) Therapeutic material, treated to lose its virulence and containing antigens derived from one or more pathogenic organisms. On administration to humans or other animals, the antigens will stimulate active immunity and protect against infection with these or related organisms.

vaccine (*ICT*) SOFTWARE that is used to detect, eradicate and repair the effects of computer VIRUSES.

vaccinia (*BioSci*) Synonym for the virus used in vaccine procedures to produce immunity to smallpox, probably derived originally from cowpox virus. Genes for other viral antigens have been introduced into vaccinia virus by recombinant DNA techniques for use in immunization against those viruses.

vaccinial (*Med*) Of, pertaining to, or caused by, VACCINIA.

vacuolar membrane (*BioSci*) The membrane surrounding an intracellular VESICLE.

vacuole (*BioSci*) A cavity, containing sap and separated from the cytoplasm by a membrane, the tonoplast.

vacuum (*Phys*) Literally, a space totally devoid of any matter. In practice this is impossible, but is approached in interstellar regions. On Earth, the best vacuums produced have a pressure of about 10^{-8} N m^{-2}. Used loosely for any pressure lower than atmospheric, eg in train braking systems and 'vacuum' cleaners.

vacuum activity (*Psych*) Behaviour manifested in the apparent absence of the external stimuli that normally elicit the activity, presumably because of internal factors governing the motivation to perform the behaviour.

vacuum arc furnace (*Chem*) One in which a small specimen is heated by a high-voltage arc in an inert gas, eg argon, at low pressure.

vacuum arc melting (*Eng*) Using a consumable electrode to melt an ingot with the arc *in vacuo*, the reduced pressure causing the metal to outgas.

vacuum augmenter (*Eng*) An AIR EJECTOR placed in a steam condenser to produce a higher degree of vacuum than is obtainable by the use of an air pump alone.

vacuum bag moulding (*Eng*) See BAG MOULDING.

vacuum brake (*Eng*) A brake system used on railway trains and goods vehicles, in which a vacuum, maintained in reservoirs by exhausters, and under the control of the driver, simultaneously operates all brake cylinders. See CONTINUOUS BRAKE.

vacuum concrete (*CivEng*) Concrete enclosed in specially prepared shuttering which incorporates fine filters and air ducts. After pouring, strong suction is applied to the ducts by means of pumps and the excess water is extracted. This enables the shuttering to be removed and reused much sooner than by traditional methods.

vacuum cooking (*FoodSci*) Using the reduction of boiling point of water in a partial vacuum to boil it off without adversely affecting flavour. Commonly used for boiling jams and the production of sauces.

vacuum crystallization (*Chem*) Crystallization of a solution in vacuum at a temperature lower than its bp at ordinary pressure; used in sugar refineries to separate sugar from syrups.

vacuum distillation (*Chem*) Distillation under reduced pressure. As a reduction of pressure effects a lowering of the boiling point, many thermolabile substances can be distilled.

vacuum evaporation (*Phys, Space*) The net loss of molecules from a surface in space or in near-vacuum conditions. Under normal conditions, there is an equilibrium of molecular exchange at the surface of a solid body – molecules leaving the surface and others captured by it.

vacuum filtration (*ChemEng*) A process of filtration where a partial vacuum is applied to increase the rate of filtration by increasing the pressure gradient across the filter.

vacuum forming (*Eng*) A method of thermoforming thermoplastic polymers, where shaping is effected by a vacuum applied through holes in a female tool. The normal process for making polymethyl methacrylate domestic baths.

vacuum furnace (*Eng*) One in which material, either solid or liquid, may be outgassed by heating *in vacuo*. See VACUUM ARC MELTING.

vacuum impregnation (*ElecEng*) The process of treating armature and transformer windings by applying moisture-resisting varnish to the insulation, under vacuum, thereby ensuring that the varnish penetrates the pores of the insulating material when normal atmospheric conditions are restored.

vacuum induction melting (*Eng*) Using induction to melt metal *in vacuo* preparatory to casting.

vacuum melting (*Eng*) See VACUUM INDUCTION MELTING.

vacuum oven (*ElecEng*) An oven for heating armature and transformer windings under vacuum, so as to drive off all moisture from the insulation prior to impregnation.

vacuum packing (*FoodSci*) Using a plastic barrier packaging material to exclude air, particularly oxygen. Such packed products, usually stored and sold under refrigeration, have an extended shelf life because spoilage due to oxidation and bacterial growth is inhibited.

vacuum photocell (*Electronics*) High-vacuum photoemissive cell in which anode current equals total photoemission currents, so that strict proportionality between current and incident illumination is obtained.

vacuum printing frame (*Print*) A frame from which air can be exhausted to ensure close contact between the film image and plate when printing down for any of the printing processes.

vacuum pump (*Eng*) General term for apparatus which displaces gas against a pressure.

vacuum servo (*Autos*) Servo mechanism operated by a vacuum provided by the induction pipe of the engine; used in power-assisted brake systems.

vacuum switch (*ElecEng*) One whose contacts separate in vacuum.

vacuum tube (*Electronics*) See VALVE.

vacuum tube rectifier (*ElecEng*) One which exploits the unidirectional movement of electrons flowing from heated electrode to gathering electrode.

vacuum tube voltmeter (*ElecEng*) US for *valve voltmeter*.

vadose zone (*Geol*) The unsaturated zone between the water table and the surface of the ground.

vagina (*BioSci*) Any sheath-like structure, eg the leaf sheath of grasses; the terminal portion of the female genital duct leading from the uterus to the external genital opening. Adjs *vaginal, vaginant, vaginate, vaginiferous.*

vaginal plug (*BioSci*) In female rodents and insectivores, the coagulated secretion of Cowper's glands which blocks the vagina and prevents premature escape of seminal fluid and further mating.

vaginismus (*Med*) Painful spasmodic contraction of the muscles of the vagina and/or of the muscles forming the pelvic floor. See DYSPAREUNIA.

vaginitis (*Med*) Inflammation of the vagina.

vagotomy (*Med*) See VAGUS.

vagotonia (*Med*) The condition of heightened activity of the vagus nerve. Also *vagotony*.

vagus (*BioSci*) The tenth cranial nerve of vertebrates, supplying the viscera and heart and, in lower forms, the gills and lateral line system. In mammals, also innervates the larynx.

Val (*Chem*) Symbol for VALINE.

Valanginian (*Geol*) A stage in the Lower Cretaceous. See MESOZOIC.

valence band (*Chem*) A range of energy levels of electrons which bind atoms of a crystal together.

valence electrons (*Chem*) Those in the outer shell of an *atom*, which, by gaining, losing, or sharing such electrons, may combine with other atoms to form *molecules.*

valency (*BioSci*) The number of antigen binding sites on an antibody molecule. Those belonging to most Ig classes have two, but IgM has ten combining sites and IgA, which can exist as monomer, dimer and higher polymers, has multiples of two. The valency of an *antigen* can likewise be expressed in terms of the number of antigen combining sites with which it can combine. Most large antigen molecules are multivalent. See LATTICE HYPOTHESIS.

valency (*Chem*) The combining power of an atom or group in terms of hydrogen atoms (or equivalent). The valency of an ion is equal to its charge. For *valency bond, valence*, see CHEMICAL BOND.

valentinite (*Min*) Antimony trioxide, Sb_2O_3, occurring as orthorhombic crystals or radiating aggregates; snow-white when pure. It is formed by the decomposition of other ores of antimony.

valeric acids (*Chem*) C_4H_9COOH, monobasic fatty acids, of which four isomers are known: n-valeric acid (*pentanoic acid*), $CH_3(CH_2)_3COOH$, bp 185°C; isovaleric acid (*3-methylbutanoic acid*), $(CH_3)_2 = CHCH_2COOH$, bp 175°C; methylethylacetic acid (*2-methylbutanoic acid*), $(CH_3)(C_2H_5)CHCOOH$, bp 177°C; pivalic acid (*2,2-dimethylpro;panoic acid*), $(CH_3)_3CCOOH$, bp 164°C.

validation (*ICT*) Input control technique used to detect any data which are inaccurate, incomplete or unreasonable.

valine (*Chem*) 2-amino-3-methylbutanoic acid. $(CH_3)_2 CH(NH_2)COOH$. An amino acid. The L- or s-isomer is a constituent of proteins. Symbol Val, short form V.

valinomycin (*Pharmacol*) A polypeptide antibiotic that is a potassium IONOPHORE.

valley (*Build*) The re-entrant angle formed between two intersecting roof slopes.

valley (*Geol*) Any hollow or low-lying tract of ground between hills or mountains, usually traversed by streams or rivers, which receive the natural drainage from the surrounding high ground. Usually valleys are developed by stream erosion, but in special cases faulting may also have contributed, as in rift valleys.

valley board (*Build*) A board nailed along the top of the valley rafter as a support for a LACED VALLEY.

valley bog (*EnvSci*) A type of *Sphagnum* bog forming where water draining from relatively acid rocks stagnates in a flat-bottomed valley so as to keep the soil constantly wet.

valproate (*Pharmacol*) An anticonvulsant used to treat the manic phase of bipolar disorder as well as epilepsy and migraine.

value-added reseller (*ICT*) A company that packages other companies' products together and hence offers an overall solution to a system requirement; eg selling a computer, printer and software all from different original equipment manufacturers. Abbrev *VAR*.

value-added service (*ICT*) A commercial service delivered by an INTELLIGENT NETWORK, eg weather forecasting or other advice. The operator's costs are recouped from a share of the call charges.

value engineering (*Genrl*) A total approach to engineering design that seeks to achieve required performance, reliability and quality at minimum cost by attention to simplicity, avoidance of unnecessary functions and integration of design and manufacturing techniques.

value function (*NucEng*) See SEPARATION POTENTIAL.

valvate (*BioSci*) (1) Organs having margins touching but not overlapping. See AESTIVATION, VERNATION. Cf IMBRICATE. (2) Opening by valves.

valve (*BioSci*) (1) Any structure that controls the passage of material through a tube, duct or aperture, usually in the form of membranous folds, as the atrio-ventricular valves of the heart. (2) In Mollusca, Cirripedia and Brachiopoda, one of several separate pieces composing the shell. (3) In insects, a covering plate or sheath, esp one of a pair that can be opposed to form a tubular structure, as the valves of the ovipositor. (4) The flattened part of a theca of a diatom frustule. (5) That part of a fruit wall that separates at dehiscence. Also *valva.*

valve (*Electronics*) Older term for a simple vacuum device for amplification by an electron stream, of many types. Now more commonly (and US) *tube.*

valve (*Eng*) Any device which controls the passage of a fluid through a pipe.

valve bounce (*Eng*) The unintended secondary opening of an engine valve due to inadequate rigidity of various parts in the valve gear. The deflected parts allow the valve to seat too early and when their strain energy is released, the valve opens again. This bounce causes valve breakage, seat wear and irregular functioning.

valve box (*Eng*) In a force pump or steam engine, the chamber which contains the valves or valve: the STEAM CHEST of a steam engine. Also *valve chest.*

valve characteristic (*ElecEng*) Graphical relation between voltage and current for specified electrodes, all other potentials being maintained constant.

valve chest (*Eng*) See VALVE BOX.

valve diagram (*Eng*) For a steam-engine slide valve, a graphical method of correlating the throw and angle of advance of the eccentric, the lead and laps of the valve, and the points of admission, cut-off, compression and release.

valve effect (*ElecEng*) The unilateral conductivity of certain electrodes (notably aluminium) in suitable solutions.

As anodes, they may withstand several hundred volts, although current will pass freely in the opposite direction.

valve face (*Eng*) The sealing surface of a valve which slides over, or beds onto, the seating.

valve gear (*Eng*) The linkage by which the valves of an engine derive their motion and timing from the crankshaft rotation.

valve inserts (*Autos*) Valve seatings of special heat- and tetra-ethyl lead-resisting steel which are pressed into the alloy heads of high-duty petrol engines.

valve-opening diagram (*Autos*) A diagram showing the lift or the opening area of a valve to a base of engine crank angle or piston displacement.

valve parameters (*Electronics*) Numerical quantities obtained from the characteristic curves and used in circuit analysis. See, DIFFERENTIAL ANODE RESISTANCE, MUTUAL CONDUCTANCE, TRANSISTOR PARAMETERS.

valve rectifier (*ElecEng*) A rectifier of the vacuum or the gas-discharge type.

valve rockers (*Autos*) See ROCKER ARMS.

valve spring (*Eng*) The helical spring (or springs) used to close a poppet valve after it has been lifted by the cam; generally, any spring which closes a valve after it has been lifted mechanically or by fluid pressure.

valve timing (*Autos*) See TIMING.

valve voltmeter (*ElecEng*) Valve used for measuring voltages, rectified output current being dependent on voltage applied to the input. It takes negligible power from measured circuit; can be calibrated at low frequencies for use at very high frequencies, when other means are impossible. See DIODE VOLTMETER. US term *vacuum tube voltmeter*.

valvular heart disease (*Med*) Disease affecting the valves of the heart, making them either too tight (stenotic) for normal blood flow or too loose (incompetent) to prevent regurgitation of blood.

valvulitis (*Med*) Inflammation of a valve of the heart.

VAM (*BioSci*) Abbrev for VESICULAR–ARBUSCULAR MYCOR-RHIZA.

vanadate (*Chem*) An ion containing vanadium (V). Orthovanadate is VO_4^{3-} and metavanadate is VO_3^{-}.

vanadic (*Chem*) Referring to trivalent vanadium. See VANADOUS.

vanadinite (*Min*) Vanadate and chloride of lead, typically forming brilliant reddish hexagonal crystals or globular masses encrusting other minerals in lead mines.

vanadium (*Chem*) A very hard, whitish metallic element. Symbol V, at no 23, ram 50·941, rel.d. (at 20°C) 5·5, mp 1710°C, electrical resistivity 22×10^{-8} Ω m at 20°C. The principal ore minerals are *carnotite, patronite, roscoelite* and *vanadinite*. Vanadium also occurs in phosphate rock. Its principal use is as a constituent of alloy steel, eg in chromium–vanadium, manganese–vanadium and high-speed steels.

vanadium steel (*Eng*) Those containing small quantities of vanadium as alloying ingredient though seldom as the sole addition. A typical composition might be carbon 0·4–0·5%, chromium 1·1–1·5%, vanadium 0·15–0·2%, balance iron. See panel on STEELS.

vanadous (*Chem*) Referring to divalent vanadium. See VANADIC.

vanadyl (*Chem*) The cations VO^{2+} and VO^{3+} containing vanadium (IV) and vanadium (V) respectively.

Van Allen radiation belts (*Astron*) Two belts encircling the Earth within which electrically charged particles are trapped. The lower Van Allen belt extends from 1000 to 5000 km above the equator with the second at about 20 000 km. Within these zones electrons originally captured from the SOLAR WIND are trapped. The belts are named after the US space scientist who discovered them.

vancomycin (*Pharmacol*) A bactericidal antibiotic from the bacterium *Streptomyces orientalis* that inhibits cell-wall synthesis. It is generally active against all Gram-positive cocci and bacilli, including many staphylococcal strains that are resistant to penicillins and cephalosporins.

van de Graaff generator (*ElecEng*) Very high-voltage electrostatic machine, using a high-speed belt to accumulate charge in a large Faraday cage, which takes the form of a metal globe. Recent models use Freon or nitrogen gas under high pressure. Used as voltage source for accelerator tubes, eg in neutron sources.

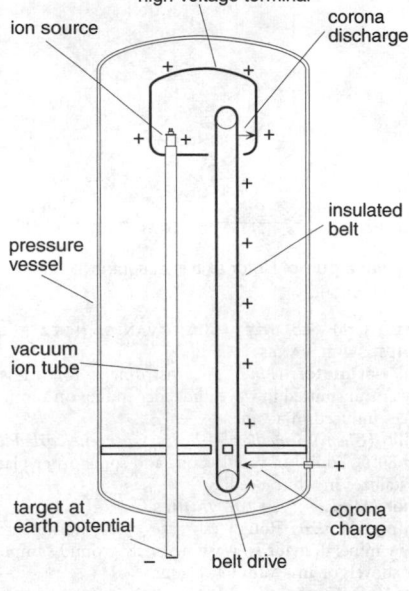

van de Graaff generator

Vandermonde determinant (*MathSci*) The determinant of a particular $n \times n$ matrix which has one as each element of the first column, $x_1, x_2, \ldots, x_n$ as the elements of the second column, and $x'_1, x'_2, \ldots, x'_n$ as the elements of column $r + 1$. The value of this determinant is the product of all the $(x_i - x_j)$ where $i > j$. The Vandermonde determinant of eg order 3 is

$$\begin{vmatrix} 1 & x_1 & x_1^2 \\ 1 & x_2 & x_2^2 \\ 1 & x_3 & x_3^2 \end{vmatrix} = (x_3 - x_2)(x_3 - x_1)(x_2 - x_1)$$

van der Waals' equation (*Chem*) An equation of state which takes into account the effect of intermolecular attraction at high densities and the reduction in effective volume due to the actual volume of the molecules: $(P + av^{-2})(v - b) = RT$, a and b being constant for a particular gas. See GAS LAWS.

van der Waals' forces (*Chem*) Weak attractive forces between atoms or molecules which vary inversely as the sixth power of the interatomic or intermolecular distance, and are due to momentary dipoles caused by fluctuations in the electronic configuration of the atoms or molecules. Also *London forces*. See panel on BONDING.

vane (*BioSci*) The web of a feather, composed of the barbs and barbules. Also *vexillum*.

vane (*Eng*) (1) One of the elements which variably divide the fluid space in a VANE PUMP. (2) An alternative name for blade in turbines, flow meters and similar rotary devices.

vane (*Surv*) A disk attachment to a levelling staff; it provides a sliding target which can be moved into the line of sight of the level. See TARGET ROD.

vane pump (*Eng*) A type of pump used eg as a vacuum or oil pump or as a compressor, in which a slotted rotor is mounted eccentrically in a circular stator (or a similar geometrical arrangement), and vanes sliding in the rotor slots divide the crescent-shaped fluid space into variable volumes.

vanes   rotor   stator

vane pump Rotor axis is eccentric to pump chamber.

vanes (*Aero*) See INLET GUIDE VANES, NOZZLE GUIDE VANES, SWIRL VANES.

vane wattmeter (*Phys*) An instrument for measuring power transmitted in waveguide, depending on mechanical forces induced in a vane.

vanillin (*Chem*) 3-methoxy-4-hydroxy-benzene carbaldehyde. Mp 80°C. Found in vanilla pods and some other plants. It crystallizes in white needles.

vanner (*MinExt*) See FRUE VANNER.

vanning (*MinExt*) Rough estimate of cassiterite or other heavy mineral, made by washing finely ground sample on a flat shovel, or in a vanning plaque.

van't Hoff factor (*Chem*) The ratio of the number of dissolved particles (ions and undissociated molecules) actually present in a solution to the number there would be if no dissociation occurred. Symbol *i*.

van't Hoff's law (*Chem*) A law stating that the osmotic pressure of a dilute solution is equal to the pressure which the dissolved substance would exert if it were in the gaseous state and occupied the same volume as the solution at the same temperature.

van't Hoff's reaction isochore (*Chem*) For a reversible reaction taking place at constant volume,

$$\frac{d \ln K}{dT} = \frac{\Delta U}{RT}$$

where K is the equilibrium constant, T is the absolute temperature, R is the gas constant, ΔU is the heat absorbed in the complete reaction.

van't Hoff's reaction isotherm (*Chem*) For a reversible reaction taking place at constant temperature,

$$-\Delta A = RT \ln K - RT \sum n \ln c$$

where $-\Delta A$ is the decrease in free energy, R the gas constant, T the absolute temperature, K the equilibrium constant, and $\sum n \ln c$ is of the same form as $\ln K$, but with the equilibrium concentrations replaced by the initial values.

vaporization (*Chem*) The conversion of a liquid or a solid into a vapour.

vapour (*Phys*) A gas which is at a temperature below its critical temperature and can therefore be liquefied by a suitable increase in pressure.

vapour barrier (*Chem*) A covering which prevents water condensing within the insulation around a cold surface. Often a pigmented vinyl polymer solution applied by brush.

vapour compression cycle (*Eng*) To operate this refrigeration cycle requires a compressor, throttle valve, evaporator and condenser. The working fluid is one which changes readily between the liquid and vapour phases. The fluid passes through the evaporator and heat is transferred to it, changing it from the saturated to the dry, superheated state. Ideally this compression is reversibly adiabatic. From the compressor it flows through the condenser, where heat is removed and the gas liquefies. It is then throttled (constant enthalpy) to the initial conditions. A suitable working fluid is Freon 12.

vapour concentration (*EnvSci*) The ratio of the mass of water vapour in a sample of moist air to the volume of the sample. Also *absolute humidity*.

vapour deposition (*Electronics*) See CHEMICAL VAPOUR DEPOSITION, PHYSICAL VAPOUR DEPOSITION.

vapour–liquid–solid mechanism (*Chem*) A method, applicable to most crystalline substances, of growing different near-perfect crystalline forms. Abbrev *VLS mechanism*.

vapour lock (*Eng*) The interruption of the flow of a volatile fluid in a pipe, eg the formation of vapour in a petrol feed pipe to a carburettor caused by undue heating.

vapour permeability (*Paper*) The rate of passage of a vapour, eg water vapour, through the paper.

vapour phase epitaxy (*Electronics*) Growing a layer on a crystalline substrate by introducing it as a gas or vapour. See panel on SEMICONDUCTOR FABRICATION.

vapour phase inhibitor (*Chem*) Stable organic chemicals coated on paper or board which slowly evaporate, so preventing air reaching the metallic articles enclosed in the package and preventing corrosion, eg ethanolamine benzoate, cyclohexylamine nitrite (nitrate (III)). Abbrev *VPI*.

vapour pressure (*Phys*) The pressure exerted by a vapour, either by itself or in a mixture of gases. The term is often taken to mean saturated vapour pressure, which is the vapour pressure of a vapour in contact with its liquid form. The saturated vapour pressure increases with rise of temperature. See SATURATION OF THE AIR.

vapourware (*ICT*) Computer software or systems that are announced to the marketplace but that are never actually designed or delivered.

VAR (*ICT*) Abbrev for VALUE-ADDED RESELLER.

VAr (*ElecEng*) Abbrev for *volt-amperes reactive*. Unit of reactive power. See REACTIVE VOLT-AMPERES.

varactor (*Electronics*) Two-electrode semiconductor with a non-linear capacitance instantaneously dependent on voltage. Cf VARISTOR.

variability (*Chem*) The number of DEGREES OF FREEDOM (1) of a system.

variable (*ICT*) A name or label declared as a DATA TYPE that during the execution of a program becomes bound to an actual but changing value in a particular storage location. See GLOBAL VARIABLE, LOCAL VARIABLE.

variable (*MathSci*) (1) In mathematics, a symbol that does not have a fixed value but can be assigned any of a specified range of values. A variable may represent a specific but unspecified value, as in $x^2 + 2x = 4$, or range over all the elements of a domain, as in $x y = y + x$. If x can be replaced by elements of a set S, then x is said to be a variable in the set S. (2) In statistics, a quantity, measurement or attribute which is the subject of statistical analysis. Also *variate*.

variable-aperture shutter (*ImageTech*) See FADE SHUTTER.

variable-area propelling nozzle (*Aero*) A turbojet *propelling nozzle* which can be varied in effective outlet area, either mechanically or aerodynamically, to match it to the optimum engine operating conditions (principally thrust), thereby improving fuel economy: essential for the efficient use of an *afterburner* (see REHEAT) and in supersonic flight.

variable-area track (*ImageTech*) Photographic sound track in which modulation is represented by variations of the image width.

variable bit rate (*ICT*) A method of data compression used in some VIDEO CODECS in which the bit rate is allowed to vary with the information content of the source.

variable-contrast printing paper (*ImageTech*) A photographic paper coated with an emulsion that is partly sensitive to blue and partly to green light, the former giving higher contrast and the latter lower, enabling different grades to be obtained by employing filters in the ENLARGER.

variable coupling (*ElecEng*) An electromagnetic coupling between two ac circuits in which the mutual inductance is continuously variable between wide limits.

variable cycle engine (*Aero*) A gas turbine in which the gas path can be changed by diverters or valves so it can operate in different modes at different flight speeds. Examples are the tandem fan turbojet which operates with four nozzles in hovering flight but with only one as a straight turbojet at high speed. Similarly in hypersonic propulsion systems the airflow has to bypass the turbojet completely at Mach numbers over 4.

variable-density wind tunnel (*Aero*) A closed-circuit wind tunnel wherein the air may be compressed to increase the REYNOLDS NUMBER. Also *compressed-air wind tunnel*.

variable geometry (*Aero*) See VARIABLE SWEEP.

variable inductor (*ElecEng*) An INDUCTOR whose self-inductance is continuously variable.

variable-inlet guide vanes (*Aero*) See INLET GUIDE VANES.

variable-interval schedule (*Psych*) See INTERVAL SCHEDULE OF REINFORCEMENT.

variable-length record (*ICT*) The number of bits (or characters) is not predetermined.

variable-pitch propeller (*Aero*) See PROPELLER.

variable ratio schedule (*Psych*) See RATIO SCHEDULE OF REINFORCEMENT.

variable-ratio transformer (*ElecEng*) A transformer whose voltage ratio can be varied by altering the number of active turns in either the primary or the secondary winding.

variable region (*BioSci*) Regions of the antigen-binding portions of antibody and T-cell receptor that show marked variability, hence the ability to bind a diversity of antigens.

variable resistance (*ElecEng*) See RHEOSTAT.

variable-speed drive (*ElecEng*) An electric drive whose speed is continuously variable between wide limits.

variable-speed motor (*ElecEng*) An electric motor whose speed is continuously variable between wide limits.

variable-speed shutter (*ImageTech*) (1) Any camera shutter by which the exposure can be altered. (2) See FAST SHUTTER.

variable star (*Astron*) Any star with a luminosity not constant with time; the variation can be regular or irregular. In an *eclipsing binary* the pair of stars periodically eclipse, as seen from the Earth, and the apparent MAGNITUDE of the pair falls when one member conceals the other. Variability also appears in pulsating stars, where the change in size and surface temperature leads to a change in luminosity. The principal types are CEPHEID VARIABLES, the long-period MIRA STARS and RR LYRAE VARIABLES.

variable sweep (*Aero*) An aircraft with wings so hinged that they can be moved backwards and forwards in flight to give high SWEEPBACK for low DRAG in supersonic flight and high ASPECT RATIO, with good lifting properties, for take-off and landing. Colloq *swing-wing*.

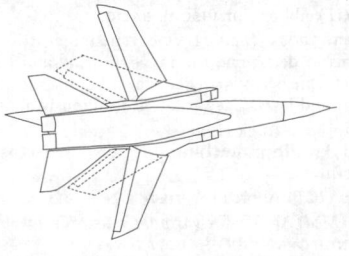

variable sweep Tornado, showing both wing positions.

variable-voltage control (*ElecEng*) A system of controlling speed by varying the voltage applied at the motor terminals.

Variac (*ElecEng*) TN of an autotransformer in the form of a toroidal winding on ring laminations, the output voltage being varied by a rotating brush contact on the turns.

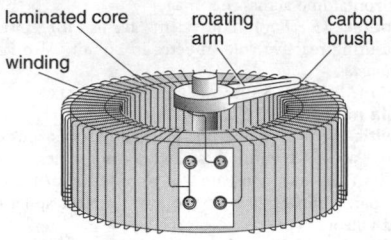

Variac Mounting not shown.

variance (*MathSci*) The ratio of the sum of the squared deviations from the mean of a set of observations to either the sample size (the *population variance*) or the sample size minus one (the *sample variance*).

variant (*BioSci*) A specimen differing in its characteristics from the type and produced either by changed environmental conditions and/or by mutation.

variate (*MathSci*) See VARIABLE.

variation (*Astron*) The fourth principal periodic term in the mathematical expression of the Moon's motion, caused by the variation of the residual attraction of the Sun on the Earth–Moon system during a synodic month; it has a maximum value of 39′ and a period of 14·77 days.

variation (*BioSci*) The differences between the offspring of a single mating; the differences between the individuals of a race, subspecies, or species; the differences between analogous groups of higher rank.

variation (*Surv*) See MAGNETIC DECLINATION.

variation factor (*ElecEng*) The ratio between the maximum and the minimum illumination along a street or roadway illuminated at intervals by overhead lamps.

variation of latitude (*Astron*) A phenomenon which occurs due to the spheroidal form and non-rigid consistency of the Earth; its axis of rotation does not remain constant in direction but varies in a regular manner about a mean position, so that the latitude of a given point on the surface also undergoes periodic variations.

variation order (*CivEng*) A document giving authority for some alteration in work being done under contract.

variations (*Build*) See EXTRAS.

varicella (*Med*) See CHICKENPOX.

varicocele (*Med*) A varicose condition of the plexus of veins which leave the testis to form the spermatic vein, forming at the upper part of the testis a swelling which feels like a mass of spaghetti.

varicose (*Med*) Of the nature of, pertaining to or affected by a varicose vein or varix; of veins, abnormally dilated, lengthened and tortuous.

variegated copper ore (*Min*) A popular name for BORNITE. So named from the characteristic tarnish that soon appears on the freshly fractured surface.

variegation (*BioSci*) The occurrence of differently coloured areas on leaves or petals due to virus infection (streaks, spots, mottles), mineral deficiency (veinal or interveinal chlorosis), genetically determined patterns as in leaves of *Coleus* cultivars, chimerical structure (light-coloured borders on leaves), a TRANSPOSON (transposable element).

variety (*BioSci*) A race; a stock or strain; a sport or mutant; a breed; a subspecies; a category of individuals within a species which differ in constant transmissible characteristics from the type, but which can be traced back to the type by a complete series of gradations; a geographical or biological race.

varifocal lens (*ImageTech*) A camera lens in which changing the focal length causes the focus position to alter. Cf ZOOM LENS.

varifocals (*Genrl*) Spectacles with varifocal lenses in which the refractive index and/or curvature varies vertically, and can thus accommodate a range of defects without a horizontal line across the image.

varimeter (*ElecEng*) Equivalent terms for instrument measuring reactive volt-amperes in circuit. Also *varmeter*, *varometer*.

variola (*Med, Vet*) See SMALLPOX, SWINE POX.

variola minor (*Med*) See ALASTRIM.

variolitic (*Geol*) Said of a fine-grained igneous rock of basic composition containing small, more or less spherical, bodies (*varioles*), consisting of minute radiating fibres of feldspar, comparable with the more perfect spherulites in acid igneous rocks.

variometer (*ElecEng*) Variable inductor with two coils connected in series and arranged one inside the other. Rotating the inner coil varies their mutual inductance and hence the total inductance.

Variscan orogeny (*Geol*) The Late Palaeozoic mountain-building period of Europe extending from the Carboniferous to the Permian. Also *Hercynian orogeny*.

variscite (*Min*) A greenish hydrated phosphate of aluminium ($AlPO_4 \cdot 2H_2O$) occurring as nodular masses.

varistor (*Electronics*) Two-electrode semiconductor with a non-linear resistance dependent on instantaneous voltage. Used to short-circuit transient high voltages in delicate electronic devices.

varix (*Med, BioSci*) (1) An enlarged and convoluted vein (varicose vein), artery or lymphatic vessel. (2) One of the longitudinal ridges on the surface of a gastropod shell. Pl *varices*.

Varley loop test (*ElecEng*) A method of determining the position of a cable fault. Resistance measurements are made with a bridge, first, so that the fault forms one junction of the bridge, and, second, so that the conductor resistance of the cable is measured directly.

varmeter (*ElecEng*) See VARIMETER.

varnish (*Build*) Formerly a transparent solution of a resin or resinous gum in spirits or oil but now more commonly a polymer, eg polyurethane in solvent, applied as a protective, decorative coating to enhance the underlying surface. Available in a range of sheen from flat to high gloss and compounded for interior or exterior use.

varometer (*ElecEng*) See VARIMETER.

varved clays (*Geol*) Distinctly and finely stratified clays of glacial origin, deposited in lakes during the retreat stage of glaciation. The stratification is thought to be a seasonal banding, and its study enabled Baron de Geer to work out the chronology of the Pleistocene Ice Age.

vas (*BioSci*) A vessel, duct or tube carrying fluid. Pl *vasa*. Adj *vasal*. See VAS DEFERENS.

vasa efferentia (*BioSci*) A series of small ducts by which the semen is conveyed from the testis to the vas deferens.

vasa vasorum (*BioSci*) In vertebrates, small blood vessels ramifying in the external coats of the larger arteries and veins.

vascular (*BioSci*) Relating to vessels which convey fluids or provide for the circulation of fluids, eg xylem and phloem; provided with vessels for the circulation of fluids.

vascular area (*BioSci*) See AREA VASCULOSA.

vascular bundle (*BioSci*) A strand of conducting tissue composed of xylem and phloem and, usually in dicotyledons, cambium. See EUSTELE.

vascular cylinder (*BioSci*) Same as STELE.

vascular plant (*BioSci*) A member of those plant groups that have a vascular system of xylem and phloem, the pteridophytes and seed plants. Also *tracheophyte*.

vascular ray (*BioSci*) A RAY in secondary xylem or phloem.

vascular system (*BioSci*) All the conducting tissues (xylem and phloem) in a vascular plant and those responsible for the circulation of blood and lymph (blood vessels and lymphatics) in animals.

vasculitis (*Med*) Inflammation of the blood vessel wall. It may be caused by immune complex deposition in or on the vessel wall.

vasculum (*BioSci*) A receptacle for collecting botanical specimens.

vas deferens (*BioSci*) A duct leading from the testis to the ejaculatory organ, the urino-genital canal, the cloaca or the exterior. Pl *vasa*.

vasectomy (*Med*) Excision of the VAS DEFERENS, or of part of it, either therapeutically or for sterilization.

Vaseline (*Chem*) TN for high-boiling residues obtained from the distillation of petroleum; a *petroleum jelly*.

VASI (*Aero*) Abbrev for VISUAL APPROACH SLOPE INDICATOR.

vasifactive (*BioSci*) See VASOFORMATIVE.

vaso- (*Genrl*) Prefix from Lt *vas*, vessel.

vasochorial placenta (*BioSci*) A chorioallantoic placenta in which the epithelium and the endometrium of the uterus disappear, and the chorion is in intimate contact with the endothelial wall of the maternal capillaries as in some Carnivora. Also *endotheliochorial placenta*.

vasoconstrictor (*BioSci*) Of certain autonomic nerves, or substances, causing constriction of blood vessels.

vasodilator (*Pharmacol*) (1) A drug, eg glyceryl nitrate, which effects expansion of the blood vessels. (2) Of certain autonomic nerves, or substances, causing expansion (relaxation) of blood vessels.

vaso-epididymostomy (*Med*) The operation of anastomosing the vas deferens to the upper part of the epididymis, forming a communication between the two; performed for the treatment of sterility in the male.

vasoformative (*BioSci*) Pertaining to the formation of blood or blood vessels. See ANGIOBLAST

vasohypertonic (*BioSci*) See VASOCONSTRICTOR.

vasohypotonic (*BioSci*) See VASODILATOR.

vasoinhibitory (*BioSci*) See VASODILATOR.

vasomotor (*BioSci*) Causing constriction or expansion of the arteries; as certain nerves of the autonomic nervous system.

vasopressin (*BioSci*) A nonapeptide secreted by the posterior pituitary gland. It elevates blood pressure by the contraction of small blood vessels and aids water resorption by the kidney. It is used in the diagnosis and treatment of DIABETES INSIPIDUS.

vasopressor (*Med*) A substance which causes a rise of blood pressure.

vat dye (*Chem, Textiles*) A water-insoluble dye which is converted into a soluble colourless form by treatment with a reducing agent in alkali. The textile material is immersed in the solution; the dye on the fabric is then converted by oxidation into its insoluble form. Vat dyes usually have good fastness properties.

vaterite (*Min*) A less common polymorph of calcium carbonate, crystallizing in the hexagonal system; forms artificially. Cf ARAGONITE, CALCITE.

vat machine (*Paper*) See CYLINDER MOULD MACHINE.

vault (*Build*) (1) An arched roof or ceiling. (2) A room or passage covered by an arched ceiling. (3) An underground room.

vault light (*Build*) A form of PAVEMENT LIGHT.

VB (*ICT*) Abbrev for VISUAL BASIC.

V-beam radar (*Radar*) One which uses two fan-shaped beams to determine the range, bearing and height of the target. One beam is vertical, the other inclined, intersecting at ground level. They rotate continuously about a vertical axis.

V-bed knitting machine (*Textiles*) See FLAT KNITTING MACHINE.

V-chip (*ICT*) Abbrev for VIEWER or VIOLENCE CHIP.

vCJD (*Med*) Abbrev for *variant Creutzfeldt–Jakob disease*. See SPONGIFORM ENCEPHALOPATHY.

VCO (*Electronics*) Abbrev for VOLTAGE-CONTROLLED OSCILLATOR.

V-connection (*ElecEng*) The open delta connection of two phases of a three-phase ac system.

VCR (*ImageTech*) Abbrev for *video cassette recorder*, equipment for recording and reproducing TV programme material on magnetic tape contained in an enclosed cassette.

V-curve (*ElecEng*) The power-factor/temperature curve of a cable with moisture, which shows a pronounced minimum at about 40°C.

VDR (*ImageTech*) Abbrev for *video disk recorder*. May record video in analogue or digital form, using either write-once disks, such as PHOTO CD, or reusable MAGNETO-OPTICAL DISKS. See INSTANT REPLAY, ORANGE BOOK.

VDRL test (*Med*) A rapid screening test for syphilis which depends on the flocculation of a cardiolipin–cholesterol–lecithin preparation if the serum being tested contains anti-cardiolipin antibody. Used in venereal disease reference laboratories, hence *VDRL*.

VDSL (*ICT*) Abbrev for *very-high-speed digital subscriber line*.

VDU (*ICT*) Abbrev for VIDEO DISPLAY UNIT or *visual display unit* (UK).

vector (*Aero*) The course or track of an aircraft, missile, etc, but generally a quantity possessing both magnitude and direction, eg wind velocity.

vector (*BioSci*) (1) An agent (usually an insect) that transmits a disease caused by a parasite or micro-organism from one host to another. (2) Insect, bird, wind, etc, carrying pollen from stamen to stigma. (3) A DNA molecule derived from a self-replicating phage, virus, plasmid or bacterium that can accept inserted DNA sequences, and is used to transfer DNA from one organism to another.

vector (*ICT*) (1) A technique for passing control in a program through an intermediate ADDRESS or *vector*. As when dealing with an INTERRUPT the computer calls the service ROUTINE not by direct means but through an intermediate address. By changing the contents of this address, different service routines may be called. See INDIRECT ADDRESSING. (2) In the context of computer graphics, a line segment on a surface, esp one that is solid and of minimum width.

vector (*MathSci*) A vector or vector quantity is one which has direction as well as magnitude, eg force or velocity; two such quantities of the same kind obey the parallelogram law of addition. A *localized vector* is one in which the line of action is fixed, as contrasted with a *free vector*, in which only the direction is fixed. Compare SCALAR.

vector addition (*MathSci*) Compounding of two vector quantities according to the parallelogram law.

vector algebra (*MathSci*) The mathematical theory of vector quantities according to defined laws of addition, subtraction and multiplication.

vectorcardiography (*Med*) The recording of the electrical signal from the heart (ELECTROCARDIOGRAM) in a series of mean vectors.

vectored thrust (*Aero*) The deflection of the thrust from turbojet(s) to provide a jet-lift component. Particularly applied to a system using a ducted fan engine with bifurcated nozzles on fan and jet pipe so that there are four sources of thrust, thereby contributing to the balance of the system. The swivelling nozzles which deflect the thrust at any angle from horizontally aft to several degrees forwards of the vertical are under the control of the pilot. Also used in missile rocket-propulsion and control systems.

vector font (*ICT, Print*) A font in which each symbol is stored as the instruction set for drawing its outline and fill at 1-pt size. The printer or imagesetter has an integral computer that converts this information into a suitable raster image at the requested font size. The interpreted PostScript page description language is a well-known example. Cf BIT-MAPPED FONT.

vector graphics (*ICT*) COMPUTER GRAPHICS in which the electron beam in a cathode-ray tube is made to draw onto the screen directly from co-ordinates calculated in the computer.

vector potential (*ElecEng*) The potential postulated in electromagnetic field theory. Space differentiation (curl) of the vector potential yields the field. Magnetic vector potential is due to electric currents, while electric vector potential is assumed to be due to a flow of magnetic charges.

vector product (*MathSci*) Of two vectors, the vector perpendicular (right-hand screw convention) to both the given vectors, of magnitude equal to the product of the magnitudes of the two given vectors multiplied by the sine of the angle between them. Vector products are usually denoted by $a \times b$, or $a \wedge b$, while SCALAR PRODUCTS are denoted by $a \cdot b$. Scalar products alone are used in electrical engineering.

vectorscope (*ImageTech*) An instrument which displays phase and amplitude of an applied signal, eg of chrominance signal in a colour TV system.

vector space (*MathSci*) An algebra consisting of two sets and their operations, such that one is an ABELIAN GROUP (the 'vectors') and the other a FIELD (the 'scalars') together with another operation ('scalar multiplication') that yields a vector from a vector and a scalar. The field is often the real or complex numbers.

vee antenna (*ICT*) A pair of wires fed in a vee formation, fed by a twin feeder at the apex. Maximum radiation is between the two wires in the direction of the apex.

vee belt (*Eng*) A power transmission belt having a cross-section of truncated vee form with a 40° included angle, usually running in corresponding vee grooves in pulleys.

vee gutter (*Build*) A gutter of V-shape, as required eg along the valley between two roofs sloping towards each other.

vee joint (*Build*) A joint between MATCHED BOARDS which have been chamfered along their edges to present a vee depression at their junction.

vee notch (*CivEng*) A notch plate having a triangular notch cut in it, used for the measurement of small discharges.

veering (*EnvSci*) A clockwise change in the direction from which the wind comes. Cf BACKING.

vee roof (*Build*) A roof formed by two lean-to roofs meeting to enclose a valley.

vee-tail (*Aero*) An aircraft tail unit consisting of two surfaces on each side of the centre line, usually at about 45° to the horizontal, which serve both as tailplane and fin. The associated hinged control surfaces are so actuated that they move in unison up/down as elevators and left/right as rudders, following conventional movements of the control column and rudder bar respectively. Also *butterfly tail*.

vee thread (*Eng*) An angular screw-thread in which the thread profile is V-shaped (as for metric or Whitworth threads), as distinct from other forms, eg square thread.

Vega (*Astron*) A prominent white star in the constellation Lyra. Distance approx 8·1 pc. Also *Alpha Lyrae*.

Vega (*Space*) Soviet mission to Venus and Comet Halley (1984–6).

vegan (*Med*) A very strict vegetarian who abstains from all food of animal origin.

vegetable oils (*Chem*) Oils obtained from plants, seeds, etc. Cf MINERAL OILS.

vegetable parchment (*Paper*) A dense, grease-resistant, packaging paper with a high wet strength, produced by treating a base paper with sulphuric acid followed by washing and/or neutralizing.

vegetal pole (*BioSci*) The lower portion or pole of an ovum in which cleavage is slow owing to the presence of yolk. Cf ANIMAL POLE.

vegetation (*Med*) A term used to describe warty aggregations of blood components and bacteria that accumulate on heart valves in ENDOCARDITIS.

vegetation survey (*MinExt*) See GEOBOTANICAL SURVEYING.

vegetative (*BioSci*) Not reproducing sexually; not carrying flowers or other sexually reproducing structures.

vegetative functions (*BioSci*) The autonomic or involuntary functions, as digestion, circulation.

vegetative propagation (*BioSci*) The natural and esp the horticultural production of new plants from bulbs, offsets, stolons, rhizomes, etc, and by layering, taking cuttings, grafting, etc. In the absence of mutation, the offspring will be genetically identical to the parent plant. See ASEXUAL REPRODUCTION, MICROPROPAGATION.

vegetative reproduction (*BioSci*) Propagation by budding, as in yeasts and hydroids, or by sending out stolons or suckers that become independent. Artificial vegetative propagation by taking cuttings and by grafting is important in horticulture.

vehicle (*Build*) The liquid substance which, when mixed with a pigment, forms a paint.

veil (*BioSci*) See PARTIAL VEIL, UNIVERSAL VEIL, VELUM.

veiled cell (*BioSci*) A type of DENDRITIC CELL characterized by large veil-like processes and found in the lymph draining skin, esp after local antigenic stimulation. They express large amounts of surface MHC Class II antigens on their surface, and represent Langerhans cells in transit from the skin to the draining lymph node, where they take the form of interdigitating cells, which are very effective in antigen presentation to T-lymphocytes.

vein (*BioSci*) (1) A vascular bundle and its supporting tissues in a leaf. (2) A vessel conveying blood back to the heart from the various organs of the body. (3) In insects, a wing nervure. Adj *venous*.

vein (*Geol*) A tabular or sheet-like body of rock, penetrating a different type of rock. Sometimes applied to particularly narrow igneous intrusions (DYKES, SILLS), the term is more often applied to material deposited by solutions, such as quartz veins or calcite veins. Many ore deposits consist of veins in which the ore mineral is one of several constituents.

vein islet (*BioSci*) See AREOLE.

vein stuff (*MinExt*) The minerals occurring in veins or fissures.

Vela (Sails) (*Astron*) A southern constellation.

Velcro (*Textiles*) TN for fabric fastening system, relying on mechanical bond between looped nylon filaments and hooked nylon filament ends. Developed by mimicking natural burrs. It is a TOUCH AND CLOSE FASTENER.

veliger (*BioSci*) The secondary larval stage of most Mollusca, developing from the trochophore and characterized by the possession of a velum.

vellum (*Paper*) An early form of writing surface made from the skins of calves, lambs or goats. In modern usage, a thick writing paper.

vellus (*BioSci*) In humans, the widespread short downy hair which replaces the fine lanugo which almost covers the fetus from the fifth or sixth month until shortly before birth.

velocity (*Phys*) (1) The rate of change of displacement of a moving body with time; a vector expressing both magnitude and direction. Cf SPEED, which is scalar. (2) For a wave, the distance travelled by a given phase divided by the time taken. Symbol v.

velocity amplitude (*Acous*) Amplitude of the velocity of the volume elements oscillating with the sound wave.

velocity budget (*Space*) The sum of the CHARACTERISTIC VELOCITIES involved in a complete space mission.

velocity constant (*Chem*) See RATE CONSTANT.

velocity microphone (*Acous*) See PRESSURE-GRADIENT MICROPHONE.

velocity-modulated oscillator (*ICT*) One in which an electron beam is velocity-modulated (bunched) by passing through a toroidal cavity resonator (rhumbatron), the energy exciting a further cavity (collecting) and feeding back into the first.

velocity modulation (*Electronics*) Modulation in a klystron in which the velocities of the electrons, and hence their BUNCHING, is related to radio signals to be amplified.

velocity of light (*Phys*) See SPEED OF LIGHT.

velocity of propagation (*ElecEng*) The velocity of an electromagnetic wave, or the velocity with which a wave

travels along a transmission line. For free space, or an air-spaced cable, it equals the SPEED OF LIGHT.

velocity of sound (*Phys*) See SPEED OF SOUND.

velocity rate constant (*Chem*) See RATE CONSTANT.

velocity ratio (*Phys*) The ratio of the distance moved through by the point of application of the effort to the corresponding distance for the load in a machine. The ratio of the MECHANICAL ADVANTAGE to the velocity ratio is termed the efficiency of the machine.

velocity resonance (*ICT*) See PHASE RESONANCE.

velodyne (*ElecEng*) Tachogenerator in which the rotational speed of an output shaft is made proportional to the applied voltage through feedback.

velour (*Textiles*) A heavy pile or napped woven fabric or felt with the surface fibres all lying in the same direction. Also, a warp-knitted fabric with long raised loops. The velour used for hats is usually a rabbit fur that has been felted and raised.

velour paper (*Paper*) Paper made by depositing short wool fibres on an adhesive coated paper. Also *flock paper*.

velum (*BioSci*) (1) A veil-like structure, as the *velum pendulum* or posterior part of the soft palate in higher mammals. (2) In some Ciliophora, a delicate membrane bordering the oral cavity. (3) In Porifera, a membrane constricting the lumen of an incurrent or excurrent canal. (4) In hydrozoan medusae, an annular shelf projecting inwards from the margin of the umbrella. (5) In Rotifera, the trochal disk. (6) In Mollusca, the ciliated locomotor organ of the veliger larva. (7) In Cephalochordata, the perforated membrane separating the buccal cavity from the pharynx.

velvet (*BioSci*) The tissue layers covering a growing antler, consisting of periosteum, skin and hair.

velvet (*Textiles*) Woven fabric with a dense short pile, formed in loops which are then cut. In one method of manufacture two cloths are woven face to face with the pile warp weaving through both. The double cloth is then sliced in the loom to produce two separate fabrics.

velveteen (*Textiles*) A woven pile fabric, usually of cotton, that is made by cutting extra floating weft threads after weaving.

venae cavae (*BioSci*) The caval veins; in higher vertebrates, three large main veins conveying blood to the right auricle of the heart.

venation (*BioSci*) The arrangement of the veins or nervures; by extension, the veins themselves considered as a whole.

V-end connections (*ElecEng*) V-shaped conductors connecting the ends of corresponding pairs of bars in a bar-wound armature.

Vendian (*Geol*) The uppermost part of the Proterozoic. See PRECAMBRIAN.

veneer (*For*) Thin layer of decorative timber glued to the surface of a less expensive wood.

veneered construction (*Build*) A mode of construction in which a thin external layer of facing material is applied to the steel or reinforced concrete framework.

veneering hammer (*Build*) One with a flat wooden head in which a brass strip is inserted. Used for smoothing veneers and forcing out the glue.

veneer saw (*Build*) Small saw with curved cutting edge, unset teeth, and a wooden grip running along the back. Used for cutting veneers.

Venera (*Space*) Soviet series of space missions to Venus, 1961–83.

venereal disease (*Med*) One of a number of contagious diseases usually contracted in sexual intercourse. Abbrev VD. See CHANCROID, GONORRHOEA, GRANULOMA INGUINALE, LYMPHOGRANULOMA INGUINALE, SYPHILIS and panel on ACQUIRED IMMUNODEFICIENCY SYNDROME (AIDS).

venereology (*Genrl*) The study of venereal diseases.

Venetian (*Print*) A style of type based on the 15th-century original of Nicolas Jenson, and characterized by strong colour, prominent serifs and oblique bar to '*e*'.

Venetian arch (*Arch*) (1) A *Queen Anne arch*. (2) A pointed arch in which the extrados and the intrados are not parallel.

Venetian fabric (*Textiles*) (1) A satin lining fabric made of mercerized cotton. (2) A satin weave, woollen cloth used in overcoats.

Venetian mosaic (*Build*) See TERRAZZO.

Venetian shutters (*Build*) See JALOUSIES.

Venetian window (*Build*) A window having two mullions dividing the window space into three compartments, usually a large centre light and two narrow side lights.

Venn diagram (*MathSci*) A method of representing the relationships between sets in terms of intersecting circles (or other shapes).

venography (*Med*) The intravenous injection of CONTRAST MATERIAL to allow X-ray visualization of the veins.

venomous (*BioSci*) Having poison-secreting glands.

veno-occlusive disease of the liver (*Med*) Syndrome of liver failure in Jamaica caused by occlusion of small blood vessels in the liver, produced by ingestion of plant alkaloids in 'bush teas'.

venosclerosis (*Med*) Hardening of a vein due to thickening of its walls.

venous system (*BioSci*) That part of the circulatory system responsible for the conveyance of blood from the organs of the body to the heart.

vent (*Aero*) (1) The opening (usually at the centre) in a parachute canopy which stabilizes it by allowing the air to escape at a controlled rate. (2) Opening to atmosphere from eg a fuel tank.

vent (*BioSci*) The aperture of the anus or cloaca in vertebrates.

vent (*Eng*) To allow air to enter, or escape from, a confined space to facilitate movement (of liquid, a piston, etc) within the space.

vent (*Geol*) See VOLCANIC VENT.

venter (*BioSci*) (1) A protuberance; a median swelling. (2) The abdomen in vertebrates. (3) The ventral surface of the abdomen.

vent gleet (*Vet*) An infectious disease of fowls, characterized by inflammation of the cloaca.

ventifact (*Geol*) A wind-faceted pebble. See DREIKANTER, ZWEIKANTER.

ventilated wind tunnel (*Aero*) A wind tunnel for TRANSONIC testing in which part of the walls in the working section are perforated, slotted or porous, to prevent choking by the presence of the model, which would otherwise render measurements unreliable in the range from Mach 0·9 to 1·4.

venting (*Eng*) The process of making holes through the rammed sand of a mould or core in order to allow gases to escape during pouring and so avoid blown castings. See VENT WIRES, WAX VENT.

venting (*Eng, Plastics*) Special features which allow gases trapped in barrel or tool to escape without causing burn marks etc.

vent pipe (*Build*) A small escape pipe which carries off foul gases from a sanitary fixture and leads into the vent stack. Abbrev VP.

ventral (*BioSci*) (1) The undersurface of plants with creeping stems next to the substrate. (2) The adaxial side of aerial shoots. Thus, the ventral surface of a leaf is normally the upper surface. The term in this sense is not always used consistently and is better avoided. (3) That aspect of a bilaterally symmetrical animal which is normally turned towards the ground.

ventral fins (*Aero*) Fins mounted under the rear fuselage to increase directional stability, usually under high incidence conditions when the main fin may be blanketed.

ventral suture (*BioSci*) The presumed line of junction of the edges of the infolded carpel.

ventral tank (*Aero*) An auxiliary *fuel tank*, fixed or jettisonable, mounted externally under the fuselage. Also *belly tank*.

ventricle (*BioSci*) A chamber or cavity, esp the cavities of the vertebrate brain and the main contractile chamber or chambers of the heart (in vertebrates or invertebrates). Adj *ventricular*.

ventricose (*BioSci*) (1) Swollen in the middle. (2) Having an inflated bulge to one side.

ventricular fibrillation (*Med*) Uncoordinated rapid electric activity of the ventricle of the heart. There is no effective pulse and death ensues rapidly unless the abnormal heart rhythm is reversed by electrical DEFIBRILLATION.

ventricular septal defect (*Med*) A congenital abnormality where there is an opening in septum between left and right ventricles allowing blood to shunt from left to right ventricles. May also occur as a complication of MYOCARDIAL INFARCTION.

ventriculography (*Radiol*) The radiological visualization of either the cerebral ventricles or the left and right ventricles of the heart following the injection of a CONTRAST MEDIUM. Can also be done after the injection of RADIONUCLIDES.

ventrifixation (*Med*) The operation of stitching the uterus to the anterior wall of the abdomen, for the treatment of retroversion of the uterus.

ventrisuspension (*Med*) An operation for replacing the retroverted uterus by transplanting the round ligaments of the uterus into the anterior abdominal wall in such a way that they exert a strong pull on the uterus.

ventrofixation (*Med*) See VENTRIFIXATION.

ventrosuspension (*Med*) See VENTRISUSPENSION.

vent stack (*Build*) A vertical pipe carried up from the highest point in a system of house drains to a level clear of all windows and opening skylights; it provides a safe escape for foul gases from drains and sanitary fixtures.

ventube (*MinExt*) A flexible ventilating duct some distance away from the source of fresh air.

venturi (*Aero*) A convergent–divergent duct in which the pressure energy of an airstream is converted into kinetic energy by the acceleration through the narrow part of the wasp-waisted passage. It is a common method of accelerating the airflow at the working section of a supersonic wind tunnel. Small venturis are used on some aircraft to provide a suction source for vacuum-operated instruments, which are connected to the low-pressure neck of the duct.

venturi flume (*CivEng*) A flume which is constricted at one section with convergent upstream and divergent downstream walls, the difference in water level at the constriction and at a point in the full channel upstream affording a means of measuring the rate of flow. See FLUME.

venturi meter (*Eng*) One in which flow rate is measured in terms of pressure drop across a venturi (or tapered throat) in a pipe.

vent wires (*Eng*) Wires ranging from $\frac{1}{16}$ to $\frac{3}{8}$ in diameter, used for making vent holes in the rammed sand of a mould or core.

venule (*BioSci*) In Chordata, small blood vessels which receive blood from the capillaries and unite to form veins.

Venus (*Astron*) Second planet from the Sun and that attaining the greatest brilliancy in the sky, outshining all stars, hence its poetic names *morning star* and *evening star*. It approaches nearer to Earth than any other planet. The visible disk is actually a blanket of opaque cloud overlying an atmosphere rich in carbon dioxide, water vapour and sulphur dioxide. Surface pressure is 90 atmospheres and temperature 470°C. The temperature is elevated by the *greenhouse effect*; the atmosphere is transparent to short-wave infrared from the Sun, but opaque to long-wave infrared from the surface. The surface is a rocky desert. Radar mapping has shown the planet has mountain ranges, craters, extinct volcanoes and a deep rift valley. There are no natural satellites. See appendix on Planets.

Venus's hair stone (*Min*) Variety of RUTILE. Also *Veneris crinis*. See FLÈCHES D'AMOUR.

verapamil (*Pharmacol*) A calcium channel blocker used as a coronary vasodilator and anti-arrhythmic.

verbal test (*Psych*) Mental test consisting primarily of items measuring vocabulary, verbal reasoning, comprehension, etc. Cf PERFORMANCE TEST.

Verbenaceae (*BioSci*) A family of c.3000 spp of dicotyledonous flowering plants (superorder Asteridae), comprising trees, shrubs and herbs. Almost all are tropical and subtropical. The flowers are gamopetalous, usually zygomorphic, and have a superior ovary. Species include teak and other important timber trees.

verde antico (*Chem*) A green patina formed on old bronze by oxidation; it is imitated artificially by pickling.

Verdet's constant (*Phys*) For a given transparent material, the angle through which a ray of polarized light rotates due to the FARADAY EFFECT divided by the product of the magnetic field and the thickness traversed. For light of 589 nm in water the rotation is $0·000\ 477$ rad A^{-1}.

verdigris (*Chem*) The green basic copper (II) carbonate (ie $CuCO_3Cu(OH)_2$) formed on copper exposed to moist air.

verdite (*Min*) A green rock, consisting chiefly of green mica (fuchsite) and clayey matter, occurring as large boulders in the North Kaap River, S Africa; used as an ornamental stone.

verge (*Build*) The edge of the roof covering projecting beyond the gable of a roof.

verge board (*Build*) See BARGE BOARD.

verge tile (*Build*) A tile which is purpose-made to a wider size than normal, to assist in forming the bond at the end of a roof. Also *tile-and-a half*.

verification (*ICT*) The act of checking transferred data, usually at the stage of input to a computer, by comparing copies of the data before and after transfer.

vermiculation (*Build*) A variety of rustication, distinguished by worm-shaped sinkings.

vermicule (*BioSci*) A small worm-like structure or organism, as the motile phase of certain Sporozoa.

vermiculites (*Min*) A group of hydrated sheet silicates, closely related chemically to the chlorites, and structurally to talc. They occur as decomposition products of biotite mica. When slowly heated, they exfoliate and open into long worm-like threads, forming a very lightweight water-absorbent aggregate used in seed planting and, in building, as an insulating material.

vermiform (*BioSci*) Worm-like, as the *vermiform appendix.*

vermifuge (*Med*) Having the power to expel worms from the intestines; any drug which has this power. See ANTHELMINTHIC.

vermis (*BioSci*) In lower vertebrates, the main portion of the cerebellum; in mammals, the central lobe of the cerebellum.

vernal (*EnvSci*) Of, or belonging to, spring.

vernal equinox (*Astron*) See EQUINOX.

vernalization (*BioSci*) The natural or artificial promotion of flowering by a period of low temperature, around 4°C.

vernation (*BioSci*) (1) The arrangement of unexpanded leaves in the vegetative bud. See AESTIVATION. (2) Same as PTYXIS.

Verneuil process (*Min*) The technique invented by the French chemist Verneuil for the manufacture of synthetic corundum and spinel by fusing pure precipitated alumina, to which had been added an amount of the appropriate oxide for colouring, in a vertical, inverted blowpipe type of furnace.

vernier (*Eng*) A small, movable auxiliary scale attached to, and sliding in contact with, a scale of graduation. Usually graduated at $\frac{9}{10}$ of the scale on the main scale, it enables readings on the latter to be made to a fraction (usually a tenth) of a division by noting which member of the auxiliary scale is aligned with any line on the main scale.

vernier arm (*Surv*) The part of an instrument carrying the vernier or verniers.

vernier capacitor (*ElecEng*) A variable capacitor of small capacitance, connected in parallel with a larger fixed one, used for fine adjustment of total capacitance.

vernier potentiometer (*ElecEng*) Precision pattern based on the KELVIN–VARLEY SLIDE. Balance can be attained purely by the operation of switches so that the possibility of wear associated with sliding contacts is avoided.

veronal (*Chem*) See BARBITAL.

verruca (*BioSci*) A wart-like process; esp one of a number of wart-like processes situated around the base of certain kinds of alcyonarian polyp.

verrucose (*BioSci*) Warty; covered with wart-like outgrowths.

versatile (*BioSci*) (1) Of an anther, attached to the tip of the filament by a small area on its dorsal side, so that it turns freely in the wind, facilitating the dispersal of the pollen. (2) Capable of free movement, as the toes of birds when they may be turned forwards or backwards.

versed sine (*CivEng*) See RISE (1).

versene (*Chem*) Sodium versenate, the sodium salt of EDTA, used for the COMPLEXOMETRIC TITRATION of calcium ion.

versicolorous (*BioSci*) Not all of the same colour; changing in colour with age.

versine (*MathSci*) A trigonometrical function of an angle, required for the solution of spherical triangles. It is given by $\mathrm{vers(ine)}\ \theta = 1 - \cos(\mathrm{ine})\ \theta$.

version (*Med*) The act of turning manually the fetus *in utero* in order to facilitate delivery.

version control (*ICT*) Arrangements for the monitoring and management of changes to software components such that the state of the system they comprise is known and can be consistently copied and recreated.

verso (*Print*) A left-hand page of a book, bearing an even number. Cf RECTO.

verst (*Genrl*) Russian measure of length, $0·6629$ miles ($1·065$ km).

vertebra (*BioSci*) One of the bony or cartilaginous skeletal elements of mesodermal origin which arise around the notochord and compose the backbone. Pl *vertebrae*. Adjs *vertebral*, *vertebrate*.

Vertebrata (*BioSci*) A subphylum of Chordata in which the notochord stops beneath the fore-brain and a skull is always present. See panel on VERTEBRATE EVOLUTION. Also *Craniata*.

vertebraterial canals (*BioSci*) In vertebrates, small canals found one on each side of all or most of the cervical vertebrae. They are formed by the articulation or fusion of the two heads of the small or vestigial cervical ribs to the centra and transverse processes, and the vertebral arteries run through them.

vertex (*Build, CivEng*) See CROWN.

vertex (*MathSci*) Of a polygon or polyhedron: one of the points in which the sides or faces intersect. Of a conic, the points in which it is cut by its axes. See CONE, PENCIL.

vertex (*BioSci*) (1) In higher vertebrates, the top of the head, the highest point of the skull. (2) In insects, the dorsal area of the head behind the epicranial suture.

vertical aerial photograph (*Surv*) A photograph taken from the air, for purposes of aerial survey work, with the camera pointing directly at the ground so that the optical axis is vertical or nearly so.

vertical boiler (*Eng*) A steam boiler having a vertical cylindrical shell and domed or spheroidal firebox, from which (generally) vertical flue tubes lead to the smoke-box and chimney.

vertical circle (*Surv*) The graduated circular plate used for the measurement of vertical angles by theodolite.

vertical component (*ElecEng*) The vertical component of the force experienced by a unit magnetic pole as the result of the action of the Earth's magnetic field. Cf HORIZONTAL COMPONENT.

vertical curve (*Surv*) The curve, generally parabolic, which is introduced between two railway or highway gradients in order to provide a gradual change from one to the other.

vertical engine (*Eng*) Any engine in which the cylinders are arranged vertically above the crankshaft.

vertical flash tool (*Eng*) Injection-moulding tool where mating surfaces between moving parts are so designed that the tool can open slightly during maximum pressurization. See MOULD BREATHING.

Vertebrate evolution

Vertebrates are a subphylum of Chordata in which the notochord stops beneath the fore-brain, a skull is always present and there are usually paired limbs. The brain is complex and associated with specialized sense organs, and there are at least ten pairs of cranial nerves. The pharynx is small and there are rarely more than seven gill slits. The heart has at least three chambers and the blood has corpuscles containing haemoglobin. The phylum Chordata includes the GRAPTOLITES which existed only in the Palaeozoic era. There are several classes of vertebrates, of which the earliest was a group of fishes, the *agnathans*, a name meaning 'without jaws'. Originally these fishes were armoured but later lost this characteristic. They range from the Ordovician period to the present day although most genera died out by the end of the Devonian period; lampreys are modern examples. Fossil agnathans are sometimes called ostracoderms (not to be confused with OSTRACODS which are crustaceans). '*Pisces*' (fishes) is an old taxonomical class that loosely includes a number of aquatic vertebrates, including the spiny *acanthodians* (Silurian to Permian), *placoderms*, heavily armoured fish which existed mainly in the Devonian, and other classes. There are various classificatory groupings. During the Devonian the fishes became dominant. The cartilaginous fishes are the *chondrichthyans* (now represented by the sharks, rays and skates); the bony fishes, the *osteichthyans*, also first appeared in the Devonian in fresh-water deposits and evolved into the marine environment. They became the most successful of aquatic vertebrates and are the dominant fishes of the present day. The lungfish, the *crossopterygians*

and the *dipnoans* are classes that evolved in the Devonian.

Amphibians, members of the class Amphibia, are cold-blooded tetrapods and include frogs, newts and salamanders as living examples. They are semi-aquatic, breathing by means of gills in the early stages of life and lungs in the later stages, and probably evolved from crossopterygian fish in the Upper Devonian. They were dominant in the Carboniferous and Permian periods, the 'age of amphibians'.

Reptiles, vertebrates of the class Reptilia, are also cold-blooded tetrapods and first appeared in the Late Carboniferous. They were able to live completely on land and came to dominate life in the Mesozoic era (the 'age of reptiles'). Present-day representatives include snakes, lizards, crocodiles and turtles. The dinosaurs included the largest terrestrial animals.

Birds (Aves) evolved from flying reptiles in the Jurassic. One of the first, *Archaeopteryx*, had reptilian features including teeth and a lizard-like tail.

Mammals, members of the class Mammalia, are warm-blooded, generally covered with hair. The young of most genera are born fully developed and initially nourished by milk. They evolved as small, primitive, furry shrew-like animals in the Triassic period during the dominance of the reptiles, but for about 150 million years remained as small animals with the class not expanding greatly until the Cenozoic era (the 'age of mammals'). In the Tertiary period, the modern groups of mammals (including horses, pigs, camels, elephants and primates) developed. The primates are an order of mammals including lemurs, monkeys, apes and humans. All living races of humans belong to the genus *Homo*, of the suborder Anthropoidea. *Homo sapiens*, modern man, appeared during the Pleistocene series.

See appendices for details of the various geological eras.

---

vertical force instrument (*Ships*) An instrument used in adjusting magnetic compasses, particularly in the correction of HEELING ERROR.

vertical frequency (*ImageTech*) The number of TV FIELDS per second. Also *field frequency*.

vertical gust (*Aero*) A vertical air current, which can be of dangerous intensity, particularly when met by aircraft flying at high speed.

vertical-gust recorder (*Aero*) An ACCELEROMETER which records graphically the intensity of accelerations due to vertical gusts and, simultaneously, the airspeed; used in assessment of aircraft fatigue life. Abbrev *v.g. recorder*.

vertical interval (*ImageTech*) The period of FIELD BLANKING between successive TV pictures.

vertical-lift bridge (*CivEng*) A bridge consisting of two towers connected by a span which can be raised and lowered vertically, maintaining its horizontal position between the towers.

vertical polarization (*ICT*) The transmission of radio waves in such a way that the electric lines of force are vertical and the magnetic lines horizontal; transmitting and

receiving dipoles are mounted vertically to handle signals polarized in this way. Cf HORIZONTAL POLARIZATION.

vertical scanning (*ImageTech*) That in which individual lines are vertical, not, as normal, horizontal.

vertical separation (*Aero*) See SEPARATION.

vertical shaft alternator (*ElecEng*) A water-turbine-driven alternator designed to operate with its shaft vertically above, and directly coupled to, the turbine shaft.

vertical smearing (*ImageTech*) Lights being reproduced with vertical streaks, caused by local overloading of the PIXELS in a SOLID-STATE IMAGE SENSOR.

vertical speed indicator (*Aero*) A sensitive form of differential pressure gauge which measures variations in pressure sensed at the STATIC-PRESSURE TUBE and indicates them in terms of rates of climb and descent. Mainly used in high-performance gliding.

vertical take-off and landing (*Aero*) See VTOL.

vertical tiling (*Build*) See WEATHER TILING.

vertical wind tunnel (*Aero*) A wind tunnel wherein the airflow is upwards and which is used principally for testing freely spinning models. Also *spinning tunnel*.

verticil (*BioSci*) A whorl.

verticillaster (*BioSci*) A kind of inflorescence found in dead nettles and related plants. It looks like a dense whorl of flowers, but is really a combination of two crowded dichasial cymes, one on each side of the stem.

verticillate (*BioSci*) Arranged in whorls.

vertigo (*Med*) Dizziness: a condition in which the person has the sensation of turning or falling, or of surrounding objects turning about the person. See MÉNIÈRE'S DISEASE.

very fine screen (*ImageTech, Print*) A term for half-tones used only for esp fine detail, from 175 to 400 lines per in (7–16 lines per mm); first figure is limit for half-tone blocks, second is rarely used.

very high frequencies (*ICT*) Those between 30 and 300 MHz. Abbrev *VHF*.

Very Large Array (*Astron*) Elaborate full synthesis radio telescope at Socorro, New Mexico, consisting of 27 antennas arranged on rail tracks forming a Y shape. Used to investigate the structure of gaseous nebulae as well as remote galaxies and quasars. Abbrev *VLA*.

very-large-scale integration (*ICT, Electronics*) The term refers to large and complex INTEGRATED CIRCUITS which may contain one or more MICROPROCESSORS. Abbrev *VLSI*. See panel on PRINTED, HYBRID AND INTEGRATED CIRCUITS.

very long baseline interferometry (*Astron*) See VLBI.

very low frequencies (*ICT*) Those between 10 and 30 kHz. Abbrev *VLF*.

VESA (*ICT*) Abbrev for VIDEO ELECTRONICS STANDARDS ASSOCIATION.

vesica (*BioSci*) The urinary bladder.

vesicant (*Med*) Causing blisters; any agent which does this. See WAR GAS.

vesicle (*BioSci*) (1) Any small cavity containing fluid or gas. (2) One of the three primary cavities of the vertebrate brain. (3) Any of a variety of small membrane-enclosed units within a cell, eg synaptic vesicles, coated vesicles. Also *vesicula*. Adj *vesicular*.

vesicle (*Geol*) See VESICULAR STRUCTURE.

vesicular–arbuscular mycorrhiza (*BioSci*) Endotrophic MYCORRHIZA in which the fungus invades the cortical cells to form vesicles and arbuscules (finely branched structures). They are very common among herbaceous plants, including many crop plants, and may significantly improve the mineral nutrition of the host. Abbrev *VAM*.

vesicular exanthema (*Vet*) A febrile, virus disease of pigs in which vesicles develop on the snout, lips, tongue and feet.

vesicular stomatitis (*Vet*) A virus disease of horses, and occasionally cattle, characterized by vesicle formation on the tongue and mucosa of the mouth.

vesicular structure (*Geol*) A character exhibited by many extrusive igneous rocks, in which the expansion of gases has given rise to more or less spherical cavities (*vesicles*). The latter may become filled with such minerals as silica (chalcedony, agate, quartz), zeolites, chlorite, calcite, etc.

vesicula seminalis (*BioSci*) In many animals, including humans, a sac in which spermatozoa are stored during the completion of their development.

vesiculate (*BioSci*) Having VESICLES or vacuoles. Also *vacuolate*.

vesiculitis (*Med*) Inflammation of the vesiculae seminales.

vessel (*BioSci*) (1) A channel or duct with definitive walls, as one of the principal vessels through which blood flows. (2) An unbranched, water-conducting tube, from 1 cm to 10 m long, in the xylem, formed from a longitudinal file of cells by the perforation of their common end walls. Water moves through perforations within a vessel but through pits into and out of vessels and from one vessel to the next. Vessels are found in very few pteridophytes, a few gymnosperms and most angiosperms.

vessel element (*BioSci*) A tracheary element of the xylem that with others in a file forms a vessel. Also *vessel member, vessel segment*. Cf TRACHEID.

Vesta (*Astron*) The fourth asteroid to be discovered and the second largest.

vestibule (*BioSci*) (1) Generally, a passage leading from one cavity to another or leading into a cavity from the exterior. (2) In Protozoa, a depression in the ectoplasm at the base of which is the mouth. (3) In a female mammal, the space between the vulva and the junction of the vagina and the urethra (urinogenital sinus). (4) in birds, the posterior chamber of the cloaca. (5) In vertebrates generally, the cavity of the internal ear. Adjs *vestibular, vestibulate*.

vestibule (*Build*) A small antechamber at the entrance to a building, or serving as an entrance room to a larger room.

vestibulectomy (*Med*) Surgical removal of the membranous labyrinth of the inner ear.

vestibulitis (*Med*) A condition characterized by slight fever, vertigo, vomiting and ataxia, resulting in complete deafness; due to an inflammation of the labyrinth and cochlea of the inner ear.

vestigial (*BioSci*) Of small or reduced structure; of a functionless structure representing a useful organ of a lower form. N *vestige*.

vestigial sideband (*ICT*) A type of amplitude-modulated transmission in which the whole of one sideband is transmitted, but only part of the other; used generally in TV transmitters.

vestiture (*BioSci*) A covering, eg of hairs, feathers, fur or scales.

vesuvianite (*Min*) Hydrated silicate of calcium and aluminium, with magnesium and iron, crystallizing in the tetragonal system. It occurs commonly in metamorphosed limestones. Also *idocrase*.

veterinary (*Genrl*) Relating to, concerned with, the diseases of domestic animals.

VFR (*Aero*) Abbrev for VISUAL FLIGHT RULES.

VGA (*ICT*) Abbrev for *video graphics array*. The IBM PC video ADAPTER standard that gives both text and graphics in 256 colours on a screen 640 PIXELS wide by 480 pixels high. SVGA (super VGA) is a development of this standard. Cf CGA, EGA.

V-gene (*BioSci*) The gene coding for the variable region of immunoglobulin light or heavy chain. During maturation of both T- and B-cells it is rearranged by translocation to a position close to the gene for the constant region (C-gene).

v.g. recorder (*Aero*) Abbrev for aircraft speed (*v*) and normal acceleration (*g*) in a vertical-gust recorder.

VHF (*ICT*) Abbrev for VERY HIGH FREQUENCY.

VHS (*ImageTech*) Abbrev for *video home system*. TN of a domestic COMPOSITE format using $\frac{1}{2}$ in tape in a cassette. See S-VHS, VHS-C, W-VHS.

VHS-C (*ImageTech*) Abbrev for VHS-COMPACT.

VHS-Compact (*ImageTech*) TN of a smaller cassette used in camcorders, which otherwise conforms to the VHS format specifications. An adapter allows its use in full-size equipment, although some equipment accepts both sizes directly – full and compact (F/C). Also *S-VHS-C*. See S-VHS.

viability test (*BioSci*) A test to determine the proportion of living cells or organisms in a sample; an exclusion test depends on their ability to exclude a dye such as TRYPAN BLUE while an inclusion test depends on their ability to take up a test substance.

viable (*BioSci*) Capable of living and developing normally.

Viagra (*Pharmacol*) TN. See SILDENAFIL.

vial (*Surv*) The glass tube containing the liquid in a LEVEL TUBE.

viameter (*Surv*) See PERAMBULATOR.

Vi antigen (*BioSci*) A surface somatic antigen present in freshly isolated strains of *Salmonella typhi* and *S. paratyphi*, which masks the O antigen and renders the organisms relatively unable to combine with antibody against the O antigen. Vi antigen is associated with virulence, possibly for this reason.

Vibram (*Plastics*) TN of a type of vulcanized rubber sole with characteristic tread pattern, used for climbing boots etc.

vibrating capacitor (*ElecEng*) One in which the potential on the electrode is varied by mechanical oscillation, so that the steady applied potential is converted to an alternating potential, which can be more easily amplified. Also *oscillating capacitor.*

vibrating conveyor *(Eng)* A tubular or flat trough with sides, to which vibrators are attached. The latter impart an upward- and forward-conveying movement to granular materials in the trough.

vibrating reed *(Eng)* A method of measuring dynamic mechanical properties of materials, esp polymers, using resonance vibration of the sample. Cf forced vibration experiments, where specimen is controlled by apparatus, eg TORSION PENDULUM.

vibration *(Phys)* A repetitive periodic change in displacement with respect to some reference point.

vibrational energy *(Chem)* Energy due to the relative oscillation of two contiguous atoms in the molecule.

vibration dampers *(Eng)* Devices fitted to an engine crankshaft in order to suppress or minimize stresses resulting from torsional vibration at critical speeds. See DYNAMIC DAMPER, FRICTIONAL DAMPER.

vibration galvanometer *(ElecEng)* Moving-coil taut-suspension galvanometer with natural frequency of vibration of coil, tunable usually over range 40–1000 Hz. Small ac at the resonant frequency excites a large response; hence these instruments form sensitive detectors for circuits such as ac bridges or potentiometers.

vibration isolation *(Eng)* Design method for selecting best device for the absorption of unwanted vibrations. See RUBBER SPRING, TRANSMISSIBILITY.

vibration pickup *(ElecEng)* One which uses some form of microphone or transducer, eg crystal, capacitance, electromagnetic, to transform the oscillatory motion of a surface, eg of machinery, into an electrical voltage or current.

vibration–rotation spectrum *(Phys)* The infrared end of the electromagnetic spectrum which arises from vibrational and rotational transitions within a molecule.

vibration white finger *(Med)* See WHITE FINGER DISEASE.

vibrator *(CivEng)* Equipment for: (1) consolidating loose ground, eg granular soils; (2) driving or extracting piles; or (3) consolidating concrete at time of placing.

vibrator *(Eng)* (1) An oscillating mechanism, usually electromagnetically excited, which imparts vibrations to hoppers, conveyors or other parts of machines, usually for the purposes of dislodging, loosening or propelling materials or workpieces. (2) A machine for driving and extracting piles often more efficient than percussion methods.

vibratory bowl feeder *(Eng)* A small parts feeder, comprising a bowl, a spring suspension system and an electromagnetic or hydraulic exciter, used for separating, orientating and feeding workpieces to an assembling machine, machine tool, etc.

vibrionic abortion *(Vet)* A contagious form of abortion in cattle and sheep caused by infection with *Campylobacter fetus,* varieties *fetus* and *intestinalis.*

vibrissa *(BioSci)* (1) In mammals, one of the stiff tactile hairs borne on the sides of the snout and about the eyes. (2) One of the vaneless rictal feathers of certain birds, eg flycatchers. Pl *vibrissae.*

vibrometer *(ElecEng)* An instrument used for the measurement of the displacement, velocity or acceleration of a vibrating body.

vibromill *(ChemEng)* A BALL MILL in which the impacts between the balls and the material to be ground are achieved by vibrating the mill at high frequency.

vibrotron *(Electronics)* A special form of triode valve in which the anode can be vibrated by a force external to the envelope.

Vicat needle *(CivEng)* An apparatus which tests the setting time of a cement specimen by measuring the effect produced by a specially shaped and loaded needle which is pressed against the surface of the specimen.

vice *(Eng)* A clamping device, usually consisting of two jaws which can be brought together by means of a screw, toggle or lever, used for holding work that is to be operated on. Generally named after the trade on which it is used. Also *vise.*

vicinal *(Chem)* Substituted on adjacent carbon atoms, eg on the 1,2,3,4 atoms in a naphthalene nucleus.

vicinal faces *(Min)* Facets modifying normal crystal faces, but themselves abnormal, as their indices cannot be expressed in small whole numbers; they usually lie nearly in the plane of the face they modify.

Vickers hardness number *(Eng)* See VICKERS HARDNESS TEST and panel on HARDNESS MEASUREMENTS.

Vickers hardness test *(Eng)* A common method of determining the hardness of metals by indenting them with a diamond pyramid under a specified load and measuring the size of the impression produced. A 136° diamond pyramid is pushed with a constant force, F, into the surface of a specimen for a specified period during which material flows plastically away from the indenter. After the diamond is withdrawn, the diagonal lengths, d, of the indentation (are measured. The Vickers hardness number, H_V (also VHN, VPN or DPH), is the force divided by the contact surface area of the indentation: $H_V = 2F \sin(136°/2)/d^2$. See panel on HARDNESS MEASUREMENTS.

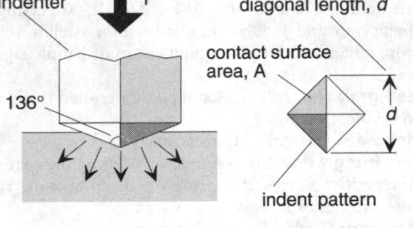

Vickers hardness test

Victrex PEEK *(Plastics)* TN for a polyether ether ketone polymer (UK).

vicunia *(Textiles)* The fine hair obtained from the undercoat of the Vicunia, *Lama vicugna,* a kind of Llama from S America. The fabric woven from it is the softest and finest of any made of wool or hair.

video *(ImageTech)* (1) Originally, the picture component of a TV signal, but now used generally to describe the electronic handling of visual images. (2) Popular abbreviation for both videotape recording and video software.

video adapter *(ICT)* A circuit that generates the output required to drive a display or screen. Several different standards for visual displays are used. See ADAPTER CARD, CGA, EGA, SVGA, VGA.

video amplifier *(ImageTech)* Wide-band amplifier which, in a TV system, passes the picture signal.

video assist *(ImageTech)* Diverting some of the light from a motion picture camera's viewfinder to a small video camera, which feeds pictures to a monitor and videotape recorder for directorial convenience and instant rushes. A single-lens reflex camera can have a video camera attached to its eyepiece for remote usage.

video card *(ICT)* A GRAPHICS CARD.

video carrier *(ImageTech)* The carrier wave which is modulated with a video signal.

video-cassette *(ImageTech)* Enclosed CASSETTE for handling magnetic videotape.

video CD *(ImageTech)* A CD-size disk carrying FULL-MOTION VIDEO (interactive or linear) and stereo audio. See WHITE BOOK.

video codec *(ICT)* Equipment used to extract from a video signal a set of codes sufficient to reconstruct a moving picture (coding), or alternatively to reconstruct a picture from these codes (decoding). By exploiting the similarity between space-adjacent picture blocks and time-adjacent video frames, codecs are capable of considerably reducing the BANDWIDTH required for transmission.

videoconferencing (*ICT*) CONFERENCE CALL in which there is the transmission of video images.

video conforming (*ImageTech*) The assembly, usually by rerecording, of selected sections from a quantity of videotape recordings to match the finally edited programme instructions.

video disk (*ICT*) See CD-ROM, CD-V.

video disk (*ImageTech*) A rotating flat circular plate carrying video and, usually, audio signals, for reproduction on a TV. It may be play-only (eg LASER DISK), write-once (eg PHOTO CD), or reusable (eg MAGNETO-OPTICAL DISK). Rapid access to different parts of the contents makes it particularly suitable for interactive use.

video display unit (*ICT*) Terminal device usually incorporating a CATHODE-RAY TUBE with a screen on which text and GRAPHICS can be displayed. Used as an I/O DEVICE in conjunction with a keyboard or a mouse for INTERACTIVE COMPUTING. Abbrev *VDU*. Also *monitor, visual display unit*.

Video Electronics Standards Association (*ICT*) Responsible *inter alia* for an obsolescent graphics bus standard. Abbrev *VESA*.

video frequency (*ImageTech*) The radio-frequency bands used for European TV broadcasting are generally ultrahigh frequency, 470–582 and 614–854 MHz; super-high frequency around 12 GHz may be used for satellite transmission; circuits within TV equipment must handle signals up to 5·5 MHz at least.

video grab (*ImageTech*) A still picture created from a frame in a video recording.

videogram (*ImageTech*) See SOFTWARE.

video integration (*ICT*) A method of using the redundancy of repetitive signals to improve the signal-to-noise ratio, by summing the successive video signals.

video map (*Radar*) Electronic system for transferring a map of any chosen territory, which may be on a transparency or store in computer memory, onto a radar display. See CHART COMPARISON UNIT.

video on demand (*ICT*) A TV system in which, unlike broadcasting or community antenna TV, selection of the programme source occurs at a central switch rather than at the viewer's receiver, allowing an unlimited number of channels to be made available over a link of bandwidth sufficient for only one channel.

video on sound (*ImageTech*) See BACKGROUND VIDEO.

video overlay board (*ImageTech*) A board (or card) which allows a computer to integrate video images for DESKTOP VIDEO etc.

videophone (*ICT*) A telephone that provides visual communication via the speech channel. A standard 64 Kbps link requires a VIDEO CODEC capable of extreme DATA RATE COMPRESSION, with consequent loss of realism, but videophones using the INTEGRATED SERVICES DIGITAL NETWORK can produce acceptable pictures.

VideoPlus (*ImageTech*) TN of a video timer programming system using coded numbers (PlusCodes) printed beside programmes in TV listings guides and keyed into the timer to set it automatically. US *VCR Plus*.

video printer (*ImageTech*) Any printer capable of producing hard copy from a video input, but usually a DYE SUBLIMATION PRINTER.

video pulse (*ICT*) Colloq term for fast rise-time pulse, ie one occurring in TV.

video RAM (*ICT*) A dedicated area of RAM used to hold the current image for the screen. The larger this MEMORY, the faster the screen may be rewritten since perhaps only half of it is required for the current image; the other can be used to prepare for the next image.

videosender (*ICT*) A small TV transmitter that can be plugged into the back of a video recorder or satellite receiver and used to broadcast the signal received there to other TV sets in the same premises rather than connecting these by cable. Also *video sender*.

video signal (*ImageTech*) That part of a TV signal conveying all the information for the picture image, including colour coding and synchronization.

video stretching (*ICT*) A method of increasing the duration time of a video pulse.

videotape (*ImageTech*) (1) Magnetic tape suitable for video recording and reproduction. (2) A video programme recorded on magnetic tape.

videotape recording (*ImageTech*) Recording signals originated by a TV camera on moving magnetic tape for subsequent reproduction or transmission; a WRITING SPEED much greater than the rate of tape transport is necessary, so heads are mounted on a rotating drum moving across the tape. See HELICAL SCAN, QUADRUPLEX.

videotex (*ICT*) (1) A term proposed for international use in place of VIEWDATA. (2) Generic term covering TELETEXT, a broadcast videotex service as well as viewdata, a wired service.

videotext (*ICT*) Communications system that uses an ordinary TV set to link with databanks through telephone lines, eg PRESTEL (previously *Viewdata*). Cf TELETEXT.

video typewriter (*ImageTech*) A keyboard device which produces text for captions etc. See CHARACTER GENERATOR.

videowall (*ImageTech*) A vertical assembly of a number of TV monitor screens, eg in a block four high and six wide, for the display of images, sometimes integrated over the whole area and sometimes in smaller groupings or individual units.

Vidicon (*ImageTech*) TN of a camera tube having a photoconductive image target layer of antimony trisulphide, scanned by a low-velocity electron beam.

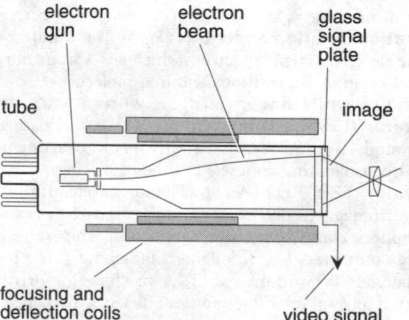

electron gun electron beam glass signal plate

tube image

focusing and deflection coils video signal

Vidicon The signal plate has a photoconductive layer which builds up a charge proportional to the light received. The electron beam neutralizes this charge and transfers it to the signal electrode.

Vierendele girder (*Eng*) Also *Vierendele truss*. See OPEN-FRAME GIRDER.

viewdata (*ICT*) Interactive information service using a telephone line between the user and a central computer, the information being displayed on a suitably adapted domestic TV receiver, developed by the UK Post Office in the 1970s. Now called Prestel in the UK.

viewer (*ImageTech*) A device for the examination of a slide transparency or of motion picture film, frame by frame or in movement.

viewer chip (*ICT*) A chip installed in a TV receiver to control its use, esp to limit use by young viewers. Same as *violence chip*. Abbrev *V-chip*.

viewfinder (*ImageTech*) An optical or video device, forming part of a camera or accessory to it, showing an image of the scene being recorded, with indication of the exact limits of the field of view involved.

vigia (*Ships*) A reported navigational danger of unknown nature, marked on charts although its existence has not been confirmed.

vigilance (*Psych*) A term for the state of readiness to detect changes in the environment.

vignette (*Print*) A half-tone illustration whose edges are undefined and shaded off gradually.

vignetting (*ImageTech*) Darkening at the corners of an image.

Viking project (*Space*) The first successful landing mission to Mars (July 1976).

Vikoma (*MinExt*) TN of equipment used for recovering and cleaning up oil spills. See BOOM.

Villari effect (*Phys*) Temporary change in magnetization of a material, arising from longitudinal stretching.

villose (*BioSci*) Shaggy. Also *villous*.

villus (*BioSci*) (1) A hair-like or finger-shaped process, such as the absorptive processes of the vertebrate intestine. (2) One of the vascular processes of the mammalian placenta which fit into the crypts of the uterine wall. Pl *villi*. Adjs *villiform, villous*.

vimentin (*BioSci*) An INTERMEDIATE FILAMENT protein characteristic of fibroblasts.

vinca alkaloid (*Pharmacol*) Alkaloids used in the treatment of cancer, LEUKAEMIA and LYMPHOMA. They interfere with cell division by causing metaphase arrest. Vincristine and vinblastine are common examples.

vinegar (*Chem, FoodSci*) The product of the alcoholic and acetic fermentation of fruit juices, eg grape juice, cider, etc, or of malt extracts. Vinegar consists of an aqueous solution of acetic (ethanoic) acid (3–6%), mineral salts and traces of esters.

vinhatico (*For*) A hardwood tree (*Plathymenia reticulata*) principally from Brazil. The heartwood is a lustrous, yellow-orange brown with darker and lighter streaks. Also *Brazilian/yellow mahogany*.

vinification (*FoodSci*) Producing wine from grapes which are crushed to release the juice (must). White wines are produced by fermenting only the juice of the grape. Fermenting in the presence of the skins allows the anthocyanins to cause the wine to be red. Fermentation is carried out in closed vessels when an insoluble sediment called lees containing dead yeasts, cell tissues, tartrates, pectin and other solids is formed. The wine is usually removed from the lees before maturation and always before bottling. Colloidal suspensions are removed by the use of FILTRATION and FINING.

vinquish (*Vet*) See PINE.

vinyl coatings (*Build*) Paint coatings based on polyvinyl chloride or acetate or a combination of both. Many of today's emulsion paints are produced with vinyls.

vinyl foils (*Print*) Blocking foils which have no final adhesive layer and transfer exclusively to thermoplastics. Also *cello foils*.

vinyl group (*Chem*) The unsaturated monovalent radical $CH_2=CH-$.

vinyl polymers (*Plastics*) Thermoplastic polymers formed by the copolymerization of vinyl chloride, $CH_2=CHCl$, and vinyl ethanoate, $CH_2COOCH=CH_2$. They are odourless and tasteless. PVC (*polyvinyl chloride*) is used instead of rubber in electric cables; resists oil and some chemicals, but is slightly inferior to rubber in electrical properties. PVA (*polyvinyl acetate*) is similarly resistant and of wide application (sheets, hose, belts, etc). See panel on POLYMERS.

violane (*Min*) Massive violet-blue DIOPSIDE, used as an ornamental stone.

violence chip (*ICT*) Same as VIEWER CHIP.

violetwood (*For*) See PURPLEHEART.

Vioxx (*Pharmacol*) TN for rofecoxib, a selective COX-2 inhibitor, now withdrawn because of cardiovascular complications.

viral transformation (*BioSci*) Virally induced transformation of an animal cell in culture into a malignant phenotype.

virescence (*BioSci*) An abnormal, usually pathogenic, condition in which flowers remain green.

virga (*EnvSci*) Slight rain or snow which evaporates before reaching the ground.

virgin metal (*Eng*) Metal or alloy first produced by smelting, as distinct from secondary metal containing recirculated scrap.

virgin neutrons (*Phys*) Neutrons which have not yet experienced a collision and therefore retain their energy of birth.

virgin state (*Phys*) See NEUTRAL STATE.

vinification In sweet and port wines fermentation is stopped when about 50% sugar remains. Sulphur dioxide is used as a bactericide and most wines are aged in barrels.

Virgo Cluster (*Astron*) The nearest giant cluster of galaxies, in the constellation Vega, and the centre of the local supercluster of which our own Galaxy is a member. Around 2500 members have been observed.

Virgo (Virgin) (*Astron*) A large northern constellation, lying between Leo and Libra. It contains the first QUASAR (panel) to be recognized as such (3C273), and the bright star SPICA.

viridine (*Min*) A green iron- and manganese-bearing variety of ANDALUSITE.

virilism (*Med*) The development of masculine characteristics, physical and mental, in the female, often due to hyperplasia of, or the presence of a tumour in, the cortex of the adrenal gland.

virion (*BioSci*) A single viral particle comprising the nucleic acid core and the surrounding capsid.

viroid (*BioSci*) Extremely small plant viruses with a 240–350 nucleotide circular RNA genome. The term has also been used casually of self-replicative particles such as the kappa particle in *Paramecium*.

virola (*For*) C American hardwood tree with rather featureless pinkish-brown wood. Also *banak*.

virology (*Genrl*) The study of viruses.

viropexis (*BioSci*) The non-specific phagocytosis of viral particles.

virtual channel identifier (*ICT*) A field in the ASYNCHRONOUS TRANSFER MODE CELL header that provides the fine level of routing for cells. Virtual channels are individual channels between users in asynchronous transfer mode. It is possible to divide a virtual path into several virtual channels, each carrying a different type of traffic, within the same call.

virtual circuit (*ICT*) (1) A link in a network that behaves like a dedicated point-to-point link. (2) A system that delivers PACKETS in sequence.

virtual community (*ICT*) A community or group that exists solely through computer communication.

virtual container (*ICT*) A modular unit into which the payload is packaged in a SYNCHRONOUS DIGITAL HIERARCHY network, formed by adding the path overhead (container identification and management information) to a container of size appropriate to the source data rate.

virtual drive (*ICT*) See RAM DRIVE.

virtual earth (*ElecEng*) A point maintained close to ground potential by negative feedback, although not directly connected to ground; eg the input terminal of an operational amplifier to which negative shunt voltage feedback is applied.

virtual height (*ICT*) The apparent height of an ionized layer as deduced from the time interval between the transmitted signal and the resulting ionospheric echo at normal incidence.

virtual image (*Phys*) See IMAGE.

virtually inert anode (*Ships*) An anode, used in CATHODIC PROTECTION, with an impressed dc from mains or battery. Silicon iron or graphite may be used. Has a limited life but lasts much longer than a SACRIFICIAL ANODE.

virtual machine (*ICT*) A consistent cross-platform environment for the execution of a program or application. In the case of the Java Virtual Machine this allows the same code to run independently of the platform on which it was created. In the case of previous 'VM' systems this approach gave users sharing a computer the appearance of being the sole user of the machine.

virtual memory (*ICT*) Memory that appears to be on RAM but is actually transferred from or to hard disk storage.

virtual on-net calling (*ICT*) Facility within a VIRTUAL PRIVATE NETWORK whereby a number on the public switched telephone network can be reached by dialling a (shorter) private number.

virtual particle (*Phys*) A particle appearing as an intermediary in a subatomic particle reaction, which borrows energy according to the Heisenberg UNCERTAINTY PRINCIPLE, and in so doing temporarily violates the mass–energy conservation law; not directly observable.

virtual path (*ICT*) A path established through one or more high-level DIGITAL CROSS-CONNECTS in an ASYNCHRONOUS TRANSFER MODE network. A virtual path connection carries a bundle of one or more virtual channels.

virtual path identifier (*ICT*) A field of 8 or 12 bits in the ASYNCHRONOUS TRANSFER MODE CELL header that provides the coarse level of routing for cells, thus establishing a VIRTUAL PATH for a group of channels.

virtual pet (*Genrl*) See CYBERPET.

virtual private network (*ICT*) A service provided by a public switched network operator to interconnect the various sites of an organization as in a dedicated private network. They use SWITCHES shared by public traffic but have their own numbering schemes and may offer intelligent features, eg call screening and forwarding.

virtual process (*Phys*) A process which temporarily violates the mass–energy conservation law as a consequence of the UNCERTAINTY PRINCIPLE; this may occur for a time t by an amount E such that $Et \approx h$ (DIRAC'S CONSTANT). A transition to a higher quantum state could take place provided this condition was satisfied, but could not be observed. A particle created in such a process is called a *virtual particle*. This is an important mechanism of nuclear forces.

virtual quantum (*Phys*) A photon or quantum which occurs in an intermediate state in which energy is not conserved; this appears in higher-order perturbation theory, between an initial state and a final state connected by a matrix element. The concept enables the coulomb energy between two electrons to be regarded as arising from the emission of virtual quanta by one of the electrons and their absorption by the other.

virtual reality (*ICT*) Computer-simulated environment used eg for training astronauts and for video games, which gives the operator the impression of being in an actual environment, interacting with it by means of goggles, joystick, data gloves or other special equipment. Abbrev *VR*.

virtual reality modelling language (*ICT*) (originally MARKUP LANGUAGE). Abbrev *VRML*.

virtual storage (*ICT*) A way of apparently extending MAIN MEMORY, by allowing the programmer to access backing storage in the same way as immediate access store.

virtual telecommunications access method (*ICT*) A method of using a FRONT-END PROCESSOR or telecommunications controller to deal with the telecommunications processing requirements for access to a HOST COMPUTER by remote TERMINALS often connected to a WIDE AREA NETWORK.

virtual temperature (*EnvSci*) The virtual temperature of a sample of moist air is the temperature at which dry air of the same total pressure would have the same density as the sample. Use of the virtual temperature obviates the need for a variable *gas constant* in applying the usual equation of state to moist air.

virulence (*BioSci*) The capacity of a pathogen to cause disease.

virulent phage (*BioSci*) A bacteriophage that always kills its host, as opposed to *lysogenic phage* (see LYSOGENY).

virus (*BioSci*) A particulate infectious agent smaller than accepted bacterial forms, invisible by light microscopy, incapable of propagation in inanimate media and multiplying only in susceptible living cells, in which specific cytopathogenic changes frequently occur. Viruses are the causative agent of many important diseases of humans, lower animals and plants, eg poliomyelitis, foot and mouth disease, tobacco mosaic. See also BACTERIOPHAGE.

virus neutralization tests (*BioSci*) Tests used to identify antibody response to a virus or, using a known antibody, to identify a virus. They depend on specific antibody neutralizing the infectivity of a virus by preventing it from binding to the target cell. They may be carried out *in vivo* in susceptible animals or chick embryos or, more usually, in tissue culture.

virus pneumonia of pigs (*Vet*) A contagious, pneumonic disease of pigs, usually chronic in form. Abbrev *VPP*.

visceral (*Med*) See VISCUS.

visceral arch (*BioSci*) See GILL ARCH.

visceral clefts (*BioSci*) The gill clefts, esp the abortive gill clefts of higher vertebrates.

visceral gout (*Vet*) See AVIAN GOUT.

viscoelasticity (*Phys*) A property of a solid or liquid which when deformed exhibits both viscous and elastic behaviour through the simultaneous dissipation and storage of mechanical energy. Shown typically by polymers.

viscometer (*Phys*) An instrument for measuring viscosity. Many types of viscometer employ POISEUILLE'S FORMULA for the rate of flow of a viscous fluid through a capillary tube.

viscose (*Chem*) See CELLULOSE XANTHATE.

viscose process (*Textiles*) A method of producing regenerated cellulose fibre (viscose rayon) or film (cellophane) by digestion of natural cellulose. Treatment with carbon disulphide (CS_2) yields cellulose xanthate with approx one xanthate group for every two glucose units. Digestion in caustic soda yields a solution of viscose polymer which is coagulated in acid solution (H_2SO_4) and then spun or extruded.

viscosity (*Phys, Eng*) The resistance of a fluid to shear forces, and hence to flow. In a NEWTONIAN FLUID, such shear resistance is proportional to the relative velocity between the two surfaces on either side of a layer of fluid, the area in shear, the viscosity of the fluid and the reciprocal of the thickness of the layer of fluid:

$$\eta = \tau \left(\frac{d\gamma}{dt}\right)^{-1}$$

The value is given by the COEFFICIENT OF VISCOSITY η. For normal ranges of temperature, η for a liquid decreases with increase in temperature and is independent of the pressure. Important non-Newtonian fluids include polymer melts, whose viscosities decrease with increasing shear rate (pseudo-plastic fluids) and dilatant fluids for which the viscosity increases with shear rate. For comparing the viscosities of liquids, various scales have been devised, eg Redwood No. 1 seconds (UK), Saybolt Universal seconds (US), and Engler degrees (Germany).

viscosity-average molecular mass (*Chem*) The measure of molecular mass derived from dilute solution viscometry, and usually falling between M_n and M_w. Symbol M_v. See MOLECULAR MASS DISTRIBUTION.

viscosity of paint (*Build*) The ability of a paint to flow, affecting its ease of application.

viscosity ratio (*Phys*) See RELATIVE VISCOSITY.

viscountess (*Build*) A slate size, 18×10 in (457×254 mm).

viscous damping (*Phys*) Opposing force, or torque, proportional to velocity, eg resulting from viscosity of oil or from eddy currents.

viscous flow (*Phys*) A type of fluid flow in which fluid particles, considered to be aggregates of molecules, move along streamlines so that at any point in the fluid the velocity is constant or varies in a regular manner with respect to time, random motion being only of a molecular nature. The name is also used to describe *laminar flow* or *streamline flow*.

viscous hysteresis (*ElecEng*) The phenomenon of time-lag between the intensity of magnetization and the magnetizing force producing it.

viscus (*BioSci*) Any one of the organs situated within the chest and the abdomen: heart, lungs, liver, spleen, intestines, etc. Pl *viscera*. Adj *visceral*.

vise (*Eng*) See VICE.

Viséan (*Geol*) A series name in the Lower Carboniferous of Europe. See PALAEOZOIC.

vishnevite (*Min*) The sulphate-bearing equivalent of CANCRINITE, found in nepheline syenite.

visibility (*EnvSci*) The maximum distance at which a black object of sufficient size can be seen and recognized in normal daylight.

visibility (*ImageTech*) The ratio of the luminous flux, in lumens, to the corresponding energy flux, in watts.

visibility curve (*ImageTech*) The relation between visibility and wavelength. Owing to varying sensitivity of the eye, this curve indicates a maximum at 555 nm, which is a bright green.

visibility factor (*ICT*) The ratio of the minimum signal input power to a radar, TV or facsimile receiver for which an ideal instrument can detect the output signal, to the corresponding value when the output signal is detected by an observer watching the cathode-ray tube.

visibility meter (*EnvSci*) A meter which attenuates visibility to a standardized value, and measures such visibility on a scale.

visible horizon (*Surv*) The junction of sea or land with sky as seen from observer's position. Also *apparent horizon, sensible horizon*.

visible radiation (*Phys*) Electromagnetic radiation which falls within the wavelength range of 780–380 nm, over which the normal eye is sensitive.

colour	λ_0(nm)	$\nu(10^{12}$Hz)
Red	780–622	384–482
Orange	622–597	482–503
Yellow	597–577	503–520
Green	577–492	520–610
Blue	492–455	610–659
Violet	455–390	659–769

visible radiation The vacuum wavelength and frequency ranges for various colours.

visible speech (*Acous*) (1) The display of oscillogram patterns corresponding to characteristic speech sounds; used as an aid to speech training of the totally deaf. (2) See SONOGRAM.

vision mixer (*ImageTech*) Equipment, and its operator, for the combination of several picture sources to create the visual effects required by the director, ranging from a simple CUT to complex WIPE transitions and KEYING effects.

vision modulation (*ImageTech*) The modulation of the carrier effected by the picture signal, as distinct from that reserved for the synchronizing impulses.

visitor location register (*ICT*) The database in a MOBILE SERVICES SWITCHING CENTRE of a PUBLIC LAND MOBILE NETWORK that stores a copy of the data held in the HOME LOCATION REGISTER of each mobile currently registered on that switching centre.

VISS (*ImageTech*) Abbrev for *video/VHS index search system*. See INDEX SEARCH.

visual acuity (*Phys*) A term used to express the spatial resolving power of the eye. Measured by determining the minimum angle of separation which has to be subtended at the eye between two points before they can be seen as two separate points.

visual approach slope indicator (*Aero*) A luminous device for day and night use, consisting of red, green and amber light bars on each side of a runway which, by being directed through restricting visors, show pilots if they are below, on or above, and in line with, the approach path for an accurate touchdown. Developed by the Royal Aircraft Establishment from a World War II night-lighting system known as the *Visual Glide Path Indicator*. Abbrev *VASI*.

Visual Basic (*ICT*) A programming language commonly used in client–server or object-oriented systems development.

visual binary (*Astron*) A double star whose two components may be seen as separate in a telescope of sufficient resolving power.

visual cliff (*Psych*) An experimental set-up in which there is a vertical drop, over which an animal is prevented from falling by a sheet of glass. Some animals avoid this area as a result of the visual perception of the drop.

visual display unit (*ImageTech*) UK version of VIDEO DISPLAY UNIT.

visual fatigue (*ImageTech*) Partial loss of visual perception or discrimination as a result of prolonged exposure of the eye to high levels of illumination or light of a dominant colour.

visual flight rules (*Aero*) The regulations set out by the controlling authority stating the conditions under which flights may be carried out without radio control and instructions. The regulations usually specify minimum horizontal visibility, cloud base, and precise instructions for the distance to be maintained below and away from cloud. Abbrev *VFR*.

visual meteorological conditions (*Aero*) Weather conditions in which an aircraft can fly under freedom from air-traffic control except in controlled air space. Abbrev *VMC*.

visual purple (*Phys*) See RHODOPSIN.

visual range (*Phys*) Observable range of ionizing particles in a bubble chamber, cloud chamber or photographic emulsion.

visual violet (*Phys*) A photosensitive retinal cone pigment; iodopsin.

vital actions (*Aero*) Sequences of pilot actions to be performed in preparation for flight and learned as part of good airmanship practice. Learned by mnemonics such as 'BUMPF' standing for 'Check brakes, undercarriage, mixture, pitch, flaps'.

vital capacity (*Med*) The volume of gas that can be expelled from the lungs after a maximal inspiration, usually of the order of 4 litres in humans.

vitallium (*Med*) A metallic alloy used in orthopaedic surgery because of its non-rust, non-reactive properties. These alloys, based on 62·5–65% cobalt, carry such other elements as chromium (27–35%), molybdenum (5–5·6%), manganese (0·5–0·6%), iron (up to 1%) and nickel (up to 2%). They show good resistance to heat and corrosion.

vital stain (*BioSci*) A stain that can be used on living cells without killing them.

vitamin (*BioSci, FoodSci*) An organic substance required in relatively small amounts in the diet for the proper functioning of the organism. See panel on VITAMINS.

vitamin A (*BioSci, FoodSci*) Also *retinol*. See panel on VITAMINS.

vitamin B complex (*BioSci, FoodSci*) B_1 (*thiamin*), B_2 (*riboflavin*), niacin (*nicotinic acid*), B_6 (*pyridoxal*), pantothenic acid, biotin, folic acid (*tetrahydrofolate*), B_{12} (*cobalamin*). See panel on VITAMINS.

vitamin C (*BioSci, FoodSci*) Also *ascorbic acid*. See panel on VITAMINS.

vitamin D (*BioSci, FoodSci*) Also *calciferol*. See panel on VITAMINS.

vitamin E (*BioSci, FoodSci*) Also α-*tocopherol*. See panel on VITAMINS.

vitamin K (*BioSci, FoodSci*) See panel on VITAMINS.

VITC (*ImageTech*) Abbrev for *vertical interval time code*. See TIME CODE.

vitellarium (*BioSci*) A yolk-forming gland.

vitelligenous (*BioSci*) Yolk-secreting or yolk-producing.

vitellin (*Chem*) A phosphoprotein present in the yolk of the egg.

vitelline (*BioSci*) Egg-yellow; pertaining to yolk.

vitelline membrane (*BioSci*) A protective membrane formed around a fertilized ovum to prevent the entry of further sperm.

vitellus (*BioSci*) Yolk of egg.

Viterbi algorithm (*ICT*) A method of decoding convolutional codes, ie those in which the code block generated in a given time slot depends not only on the message block present at that time but also on the message digits within a previous span of $N-1$ time units, where N is the length of the shift register used to generate the code. The algorithm is highly efficient for $N < 10$, but because computational complexity increases as 2^N, it is less suitable for codes offering extremely low error rates, which have high values of N.

vitiligo (*Med*) Patchy de-pigmentation of the skin often with a sharp demarcation line and associated with auto-immune disease. See panel on AUTO-IMMUNITY.

Viton (*Plastics*) TN for synthetic rubber based on a copolymer of vinylidene fluoride and hexafluoropropene. Maintains its rubber-like properties over a very wide temperature range and is chemically very resistant. Used for O-rings etc.

vitreous enamel (*Eng*) Glazed coating fused onto steel surface for protection and/or decoration. Also *enamel*.

vitreous humour (*BioSci*) The jelly-like substance filling the posterior chamber of the vertebrate eye, between the lens and the retina.

vitreous silica (*Glass*) Glassy material consisting almost entirely of silica, made in translucent and transparent forms. The former has minute gas bubbles disseminated in it. Compared with ordinary soda–lime–silica glass, it has a much lower thermal expansion coefficient, a higher thermal shock resistance and a higher temperature limit. Also *fused silica, quartz glass, silica glass*. See SILICA, SILICATES.

vitreous state (*Phys*) A non-crystalline solid or rigid liquid, formed by supercooling the melt. Also *glassy state*.

vitric tuff (*Min*) A tuff in which vitric (glassy) fragments are more abundant than lithic or crystal fragments.

vitrification (*NucEng*) The incorporation of radioactive waste products (particularly from nuclear fuel processing) into glass. Also *glassification*. Other techniques under study include ceramics, glass-ceramics, composite materials (eg glass beads in a metal matrix) and synthetic minerals.

vitrinite (*Geol*) An oxygen-rich maceral that is found in coal.

vitriol (*Chem*) Sulphuric acid. Also *oil of vitriol*.

vitriol (*Min*) See BLUESTONE, MELANTERITE, WHITE VITRIOL.

vitro- (*Genrl*) Prefix from Lt *vitrum* denoting glass.

vitroclastic structure (*Geol*) The characteristic structure of volcanic ashes which have been produced by the disruption of highly vesicular glassy rocks, most of the component fragments thus having concave outlines.

vivianite (*Min*) Hydrated iron phosphate, $Fe_3P_2O_8 \cdot 8H_2O$. Monoclinic.

vivipary (*BioSci*) (1) Giving birth to living young which have already reached an advanced stage of development; n *viviparity*, adj *viviparous*. Cf OVIPAROUS. (2) The production of bulbils or small plants in place of flowers, as in eg *Festuca vivipara*. (3) The premature germination of seeds or spores before they are shed from the parent plant, as in many mangrove trees.

VLA (*Astron*) See VERY LARGE ARRAY.

VLBI (*Astron*) Abbrev for *very long baseline interferometry*. A technique of APERTURE SYNTHESIS used in radio astronomy to link telescopes separated by thousands of kilometres.

VLF (*ICT*) Abbrev for VERY LOW FREQUENCIES.

VL-mount (*ImageTech*) A bayonet video lens mount which incorporates electrical connections for serial communication between the lens and camera microcomputers to control autofocus, iris and zoom mechanisms.

vlog (*ICT*) A video-based BLOG.

VLSI (*ICT, Electronics*) Abbrev for VERY-LARGE-SCALE INTEGRATION.

VLS mechanism (*Chem*) Abbrev for VAPOUR–LIQUID–SOLID MECHANISM.

VMC (*Aero*) Abbrev for VISUAL METEOROLOGICAL CONDITIONS.

VM/CMS (*ICT*) TN of an IBM OPERATING SYSTEM.

VME (*ICT*) The proprietary OPERATING SYSTEM of ICL (International Computers Ltd).

Vitamins

Organic substances required in relatively small amounts in the diet for the proper functioning of the organism. Lack causes DEFICIENCY DISEASES curable by administration of the appropriate vitamin. There are two main groups: the *fat-soluble*, vitamins A, D, E and K; and the *water-soluble*, vitamin C and the vitamins of the B complex.

Vitamin A or *retinol*. A precursor of the prosthetic group of the light-sensitive protein, RHODOPSIN. Deficiency of vitamin A causes night blindness. It is also required by young animals for growth. Fish liver oils and dairy products are rich sources of vitamin A.

vitamin A

Vitamin B complex. (1) B_1 (*thiamin*). As its pyrophosphate it functions as a coenzyme of various enzymes. Deficiency results in the disease BERI-BERI. Present in yeast and cereal germs. (2) B_2 (*riboflavin*). Forms part of the prosthetic group of flavoproteins. Deficiency causes skin and corneal lesions. (3) Niacin (*nicotinic acid*). Component of the coenzyme nicotinamide adenine dinucleotide, NAD. Deficiency results in the disease PELLAGRA. (4) B_6 (*pyridoxal*). As its phosphate it acts as a coenzyme for TRANSAMINASES. (5) Pantothenic acid. A component of coenzyme A. (6) Biotin.

The prosthetic group of the enzyme CARBOXYLASE. (7) Folic acid (*tetrahydrofolate*). Serves as a donor of 1-carbon fragments for several biosyntheses. Deficiency inhibits these reactions which include the synthesis of PURINES. (8) B_{12} (*cobalamin*). Component of the coenzyme cobalamin which takes part in enzymic interconversions of acyl CoAs and methylations. Used in the treatment of PERNICIOUS ANAEMIA. Liver is a rich source.

Vitamin C or *ascorbic acid*. Important in the hydroxylation of collagen which in its absence is inadequately hydroxylated. The defective collagen produces the skin lesions and blood vessel weaknesses which are characteristic of SCURVY, the deficiency disease of this vitamin. Fresh fruit and green vegetables are important sources.

vitamin C

Vitamin D or *calciferol*. The vitamin involved in calcium and phosphorus metabolism. Deficiency impairs bone growth and causes the disease RICKETS. Fish liver oils are a rich source and the vitamin can, in sunlight, be synthesized in the skin from cholesterol.

Vitamin E or α-*tocopherol*. A vitamin involved in reproduction. Its absence leads to sterility in both sexes.

Vitamin K. A necessary requirement for the production of prothrombin, the precursor of thrombin, and consequently essential for normal blood coagulation.

VMOS (*Electronics*) A METAL–OXIDE–SILICON (MOS) technology in which diffused layers are formed in silicon and V-shaped grooves are precisely etched through the layers. INVERSION channels form along the arms of the V under gate control (in contrast with other MOS technologies where the channel is parallel to the wafer surface). Higher densities of current and components per chip are possible than with other MOS techniques.

VMS (*ICT*) TN for the OPERATING SYSTEM used in the VAX range of computers originally produced by DEC (Digital Equipment Corporation).

V–n diagram (*Aero*) See FLIGHT ENVELOPE.

VOC (*EnvSci*) Abbrev for VOLATILE ORGANIC COMPOUNDS.

vocal cords (*BioSci*) In air-breathing vertebrates, folds of the lining membrane of the larynx by the vibration of the edges of which, under the influence of the breath, the voice is produced.

vocal sac (*BioSci*) In many male frogs, loose folds of skin at each angle of the mouth which can be inflated from within the mouth into a globular form, and act as resonators.

vocoder (*Acous*) Abbrev for *voice coder*. System for synthetic speech using recorded speech elements.

VOD (*ImageTech*) Abbrev for VIDEO ON DEMAND, a system in which films can be stored centrally and downloaded anytime.

vodas (*ICT*) Abbrev for *voice-operated device anti-sing*; used for the suppression of echoes in transoceanic radio telephony.

vodcast (*ICT*) The on-line distribution of video material to a client device for subsequent playback. See PODCAST.

voder (*Acous*) Abbrev for *voice operation demonstrator*. System for producing synthetic speech through keyboard control of electronic oscillators.

vogad (*ICT*) Abbrev for *voice-operated gain-adjusting device*. Used in telephone systems to give an approximately constant volume output for a wide range of input signals.

vogesite (*Min*) A hornblende-lamprophyre, the other essential constituent being orthoclase. Cf SPESSARTITE.

voice bank (*ICT*) Digital equipment used to store incoming calls for mobile telephones that are unavailable because they are out of the network area or switched off. When a user logs on again, voice bank messages are relayed to his or her mobile.

voice coder (*Acous*) See VOCODER.

voice coil (*Acous*) The coil attached to the cone of a loudspeaker. The coil currents react with the magnetic field to drive the cone. Also used in microphones to generate the signal.

voiced sound (*Acous*) In speech, an elemental sound in which the component frequencies are exact multiples of a fundamental frequency which is determined by the tension of the oscillating muscles in the larynx.

voice filter (*Acous*) A device which deliberately distorts speech for a specific purpose, eg telephonic imitation.

voice frequency (*ICT*) One in the approximate range 200–3500 Hz (that required for the normal human voice). Also *speech frequency*.

voice guidance (*ICT*) The use of digitized speech in conjunction with DUAL-TONE MULTIFREQUENCY signalling to enable a computer to carry out complex transactions with callers. For instance, to reach the correct department of a large organization, the caller may be asked a series of questions to which he or she can respond by pressing one of a number of keys, each of which leads to a further question, until the correct path on a decision tree has been traversed.

voice mail (*ICT*) Use of a centralized digital facility to record speech messages from telephone callers. Each is stored in an individual voice 'mailbox'. The messages may be accessed remotely by means of a special code and the messages replayed, forwarded to other people or transcribed to a text document.

voice messaging (*ICT*) See VOICE MAIL.

voice operation demonstrator (*Acous*) See VODER.

voice over (*ImageTech*) Speech accompanying a film or video programme in which the speaker is not seen in the picture.

void (*Eng*) (1) Hollow formed within a moulding, esp at thick sections. (2) A defect lowering product strength by concentrating stress.

void (*PowderTech*) In a powder compact or in a powder fluid system, the space between particles.

voidage (*PowderTech*) In a powder COMPACT or powder fluid system, the volume fraction of voids in the system. For a system containing dense particles, the voidage and the POWDER POROSITY are numerically equal.

Voigt effect (*Phys*) DOUBLE REFRACTION of electromagnetic waves passing through a vapour when an external transverse magnetic field is applied. The vapour acts as a uniaxial crystal with its optical axis parallel to the field direction.

Voigt model (*Eng*) Conceptual model for a viscoelastic material consisting of a spring and dashpot in parallel. See MAXWELL MODEL.

voile (*Textiles*) A light, open, plain-weave fabric made from highly twisted yarns.

VOIP (*ICT*) Abbrev for *voice over Internet protocol*, a telephony standard enabling voice communication via the Internet.

Volans (Flying Fish) (*Astron*) A small inconspicuous southern constellation.

volant (*BioSci*) Flying; pertaining to flight.

volatile (*Chem*) Changing readily to a vapour.

volatile acidity (*FoodSci*) The proportion of the total acidity of a wine which is acetic acid. A small amount of acetic acid can be naturally present but a high volatile acidity is an indication of bacterial spoilage.

volatile memory (*ICT*) Stored information that can be destroyed by a power failure. See DYNAMIC MEMORY.

volatile organic compounds (*EnvSci*) Compounds such as ethylene, benzene, styrene and acetone that contribute to air pollution directly or, following photochemically driven reactions, to the formation of secondary pollutants. Methane is often treated as being a special case. Abbrev VOC.

volatilization (*Chem*) See VAPORIZATION.

volcanic ash (*Geol*) The typical product of explosive volcanic eruptions, consisting of comminuted rock and LAVA, the fragments varying widely in size and in composition, and including deposits of the finest dust, lapilli and bombs. See AGGLOMERATE, PYROCLASTIC ROCKS, TUFF.

volcanic bomb (*Geol*) A spherical or ovoid mass of lava, in some cases hollow, formed by the disruption of molten lava

by explosions in an active volcanic vent. See BREAD-CRUST BOMB.

volcanic muds (*Geol*) The products of explosive volcanic eruptions (*volcanic ash*) which have been deposited under water and have consequently been sorted and stratified, thus showing some of the characters of normal sediments, into which they grade. May also be the product of a mud flow down the side of a volcano. Also *volcanic sands*.

volcanic neck (*Geol*) A vertical plug-like body of igneous rock or volcanic ejectamenta, representing the feeding channel of a volcano.

volcanic rocks (*Geol*) Fine-grained crystalline or glassy igneous rocks formed by volcanic action at or near the Earth's surface and named after the classical god Vulcan. Extruded through volcanoes as molten lava or ejected explosively to form pyroclastic deposits. Included with volcanic rocks are the associated minor intrusions of DYKES and SILLS. Basaltic rocks are by far the commonest type, forming eg the great modern shield volcanoes and lava flows of eg Hawaii, Iceland and the rocks of the sea bed. Their compositions vary considerably and there are many rock names relating to their mineralogical composition and textural characteristics.

volcanic sands (*Geol*) See VOLCANIC MUDS.

volcanic vent (*Geol*) The pipe which connects the crater with the source of magma below; it ultimately becomes choked with agglomerate or volcanic ash, or with consolidated lava.

volcano (*Geol*) (1) A centre of volcanic eruption, having the form typically of a conical hill or mountain, built of ashes and/or lava flows, penetrated irregularly by dykes and veins of igneous rocks, with a central crater from which a pipe leads downwards to the source of magma beneath. Volcanoes may be active (periodically), dormant or extinct; the eruptions may involve violent explosions (eg Krakatoa) or the relatively quiet outpouring of lava, particularly in those cases where the lava is basaltic (eg Hawaii). See figs at PLATE TECTONICS (panel) and SUBDUCTION ZONE. See LAVA. (2) A conical hill producing mud or sand. See MUD VOLCANO.

Volkmann's contracture (*Med*) A contracture of the flexor muscles of the forearm and leg due to the pressure of splints or tight bandages used in the treatment of fractures etc causing diminished blood supply and muscle necrosis.

volsella forceps (*Med*) Forceps whose blades have prolonged ends.

volt (*ElecEng*, *Phys*) The SI UNIT of *potential difference*, electrical potential, or emf, such that the potential difference across a conductor is 1 volt when 1 ampere of current in it dissipates 1 watt of power. Equivalent definition: if, in taking a charge of 1 coulomb between two points in an electric field, the work done on or by the charge is 1 joule, the potential difference between the points is 1 volt. Named after Count Alessandro Volta (1745–1827). Symbol V.

Volta effect (*Phys*) Potential difference which results when two dissimilar metals are brought into contact: the basis of voltage cells and corrosion.

voltage (*Phys*) The value of an emf or potential difference expressed in VOLTS.

voltage amplifier (*ICT*) An amplifier whose function is to increase the voltage of the applied signal, without necessarily increasing its power. The output impedance must therefore be high.

voltage between lines (*ElecEng*) The voltage between any two of the line wires in a single- or three-phase system; between the two lines of the same phase in a two-phase system; between any two lines which are consecutive as regards phase sequence in a symmetrical six-phase system. Also *line voltage*, *voltage between phases*, *voltage of the system*.

voltage circuit (*ElecEng*) The circuit of an instrument or relay which is connected across the lines of the circuit under test, and which therefore carries a current proportional to the voltage of this circuit. Also *pressure circuit*. See SHUNT CIRCUIT.

voltage clamp (*BioSci*) An electrophysiological technique in which a MICROELECTRODE is inserted into a cell and a current passed through the electrode to hold the cell's membrane potential at a particular level. The technique is particularly used in the study of ION CHANNELS.

voltage coefficient (*ElecEng*) The constant by which the product of the armature speed in revolutions per minute, the flux in volt-lines, and the number of armature conductors in series must be multiplied in order to obtain the emf of a dc generator.

voltage-controlled oscillator (*Electronics*) An oscillator whose frequency is controlled by a bias signal; a VARACTOR diode may be used as the controlling element. Abbrev VCO.

voltage divider (*ElecEng*) A chain of impedances, most commonly resistors or capacitors, such that the voltage across one or more is an accurately known fraction of that applied to all; used for calibrating voltmeters. Also *potential divider, volt-box*. See KELVIN–VARLEY SLIDE.

voltage doubler (*ElecEng*) Power-supply circuit in which both half-cycles of ac supply are rectified, and the resulting dc voltages are added in series.

voltage drop (*Phys*) (1) The diminution of potential along a conductor, or over apparatus, through which a current is passing. (2) The possible diminution of voltage between two terminals when current is taken from them.

voltage-fed antenna (*ICT*) One that is fed with power from a line at a point of high impedance, where, through resonance, there is a voltage loop in the standing-wave system.

voltage feedback (*ElecEng, Electronics*) In amplifier circuits, a feedback voltage directly proportional to the load voltage. It may be applied in series or shunt with the source of the input signal. See CURRENT FEEDBACK, NEGATIVE FEEDBACK, POSITIVE FEEDBACK.

voltage gain (*ElecEng, Electronics*) The ratio of the change in output voltage (for a network or amplifier) to the change in input voltage which produces it. Often expressed in dB as $A_V = 20\log(V_o/V_i)$, although this definition applies strictly only to the case when the input and output impedances are equal.

voltage-gated ion channel (*BioSci*) An ion channel for which the permeability to ions is very sensitive to the transmembrane potential difference, eg SODIUM CHANNEL.

voltage gradient (*ElecEng*) The difference in potential per unit length of a conductor, or per unit thickness of an insulating medium.

voltage level (*ICT*) Peak-to-peak value at any point in a network, expressed relative to a specified reference level. When this is 1 volt, the symbol dBv is used.

voltage multiplier (*ElecEng*) A circuit for obtaining high dc potential from low-voltage ac supply, effective only when load current is small, eg for anode supply to cathode-ray tube. A ladder of half-wave rectifiers charges successive capacitors connected in series on alternate half-cycles.

voltage ratio (*ElecEng*) Same as *turns ratio*.

voltage reference tube (*Electronics*) A glow-discharge tube designed to operate with anode–cathode voltage as nearly as possible constant regardless of the anode current, and hence suitable for use as a standard of potential difference.

voltage regulation (*ElecEng*) The percentage variation in the output voltage of a power supply for either a specified variation in supply voltage or a specified change of load current.

voltage-regulator tube (*Electronics*) One in which, over a practical range of current, the voltage between electrodes in a glow discharge remains substantially constant.

voltage resonance (*ElecEng*) The condition of a circuit when the magnitude of a voltage passes through the maximum as the frequency is changed; obsolete terms: *syntony, tuning*.

voltage ripple (*ElecEng*) The peak-to-peak ac component of a nominally dc supply voltage.

voltage stabilizer (*Electronics*) Also *voltage-stabilizing tube*. See VOLTAGE-REGULATOR TUBE.

voltage standing-wave ratio (*ICT*) The ratio between a maximum and a minimum in a standing wave, particularly on a transmission line or in a waveguide, arising from inexact impedance terminations. Abbrev VSWR.

voltage to neutral (*ElecEng*) The voltage between any line and neutral of a three- or six-phase system. Also *phase voltage, star voltage, Y-voltage*.

voltage transformer (*ElecEng*) A small transformer of high insulation for connecting a voltmeter to a high-tension ac supply.

voltaic cell (*Chem*) Any device with electrolyte (ionized chemical compound in water), and two differing electrodes which establish a difference of potential.

voltaic current (*Phys*) Current (direct) produced by chemical action.

voltaic pile (*ElecEng*) A source of dc supply. It comprises a battery of primary cells in series, arranged in the form of a pile of disks, successive disks being of dissimilar metals separated by a pad soaked in the chemical agent.

voltameter (*ElecEng*) An instrument for measuring a current by means of the amount of metal deposited, or gas liberated, from an electrolyte in a given time due to the passage of the current.

volt-ampere-hour (*ElecEng*) Unit of apparent power, equal to watt-hour divided by POWER FACTOR.

volt-amperes (*ElecEng, Phys*) Apparent power in an electric circuit, equal to the product of rms current and voltage. See ACTIVE VOLT-AMPERES, REACTIVE VOLT-AMPERES.

volt-box (*ElecEng*) See VOLTAGE DIVIDER. Also *volt ratio-box*.

voltinism (*BioSci*) Breeding rhythm; brood frequency. See BIVOLTINE, MULTIVOLTINE, UNIVOLTINE.

voltmeter (*ElecEng*) An instrument for measuring potential differences.

volt-ohm-milliammeter (*ElecEng*) An electrical dc test instrument measuring voltage, resistance and current. Usually ac volts can also be measured.

volume (*Acous*) A general term comprehending the general loudness of sounds, or the magnitudes of currents which give rise to them. Volume is measured by the occasional peak values of the amplitude, when integrated over a short period, corresponding to the time constant of the ear. See VOLUME INDICATOR, VOLUME UNIT.

volume (*ICT*) An identifiable area of BACKING STORE treated as an entity, eg a FLOPPY DISK or HARD DISK.

volume (*Phys*) The amount of space occupied by a body; measured in cubic units, eg m^3. Symbol V.

volume bottles (*ChemEng*) A *dampener*, in the form of empty pressure vessels (usually steel), depending on the relationship between their volume and the volume of gas passing to produce their dampening effect.

volume compression (*Acous*) Automatic compression of the volume range in any transmission, particularly in speech for radio-telephone transmission, so that the envelope of the waveform is transmitted at a higher average level with respect to interfering noise levels. After expansion at receiving end, resulting transmission is freer from noise.

volume compressor (*Acous*) In communication systems depending on amplitude modulation, intelligence transmitted as a modulation is limited to 100%. So that this is not exceeded with very loud sounds in the modulation, the original transmission has to be compressed into a relatively small dynamic range to maintain a high signal-to-noise ratio.

volume control (*ICT*) A manually operated control used to regulate communication transmission levels.

volume expansion (*Acous*) See VOLUME COMPRESSION.

volume fraction (*Plastics*) An expression of concentration, esp for a component in COMPOSITE MATERIALS; it is the fraction of the total volume occupied by the component, eg the fibres. Cf WEIGHT FRACTION.

volume indicator (*Acous*) An instrument for measuring VOLUME, VOLUME COMPRESSION, VOLUME EXPANSION.

volume ionization (*Phys*) The mean ionization density in any given volume without reference to the specific ionization of the particles.

volume lifetime (*Electronics*) That of current carriers in the bulk of a semiconductor. Cf SURFACE LIFETIME.

volume range (*Acous*) The difference between the maximum amplitude and the minimum useful amplitude of a transmitted signal, expressed in decibels. In speech it is generally taken to be 15–20 dB, and for a full orchestra 60–70 dB.

volume resistivity (*ElecEng*) See RESISTIVITY.

volume shape factor (*PowderTech*) The ratio $d_v = W/Pd_m^3$ where W is the mass of a particle; P is thedensity of the particle; and d is the diameter of the particle as measured by a specified technique. Since d_m depends upon the method of measurement there are many shape factors for a given particle. If the value of volume shape factor is averaged over several particles, this value is called the volume shape factor of the powder.

volumetric analysis (*Chem*) A form of chemical analysis using standard solutions for the estimation of the particular constituent present in solution by titrating the one against a known volume of the other. See BURETTE, END-POINT, INDICATOR (1), PIPETTE, TITRIMETER analysis.

volumetric efficiency (*Eng*) In an internal-combustion engine or air compressor, the ratio of the weight of air actually induced per unit time to the weight which would fill the swept volume at STP.

volumetric heat (*Phys*) See MOLAL SPECIFIC HEAT CAPACITY.

volumetric strain (*Eng*) The algebraic sum of three mutually perpendicular principal strains in a material.

volume unit (*Acous*) A unit used in measuring variations of modulation in a communication circuit, eg telephone or broadcasting. The unit is the decibel expressed relative to a reference level of 1 mW in 600 ohms, and standard *volume indicators* are calibrated in these units. Symbol VU.

voluntary muscle (*BioSci*) Any muscle controlled by the motor centres in the brain, the skeletal muscles. All such muscles are striated. Cf INVOLUNTARY MUSCLES.

volunteer (*Agri*) A self-seeded representative of a previous crop that can cause contamination and competition, and harbour pests and disease.

volva (*BioSci*) A cup-like structure at the base of the fruiting bodies of many Basidiomycetes, eg mushroom, representing the remains of the UNIVERSAL VEIL.

volvulus (*Med*) Torsion of an abdominal viscus, esp of a loop of bowel, causing internal obstruction.

vomer (*BioSci*) A paired membrane bone forming part of the cranial floor in the nasal region of the vertebrate skull; believed not to be homologous in all groups. Adj *vomerine*.

vomerine teeth (*BioSci*) In most fish and amphibia, teeth, sometimes atypical, borne on the VOMERS.

von Mises criterion (*Chem, Crystal*) In a polycrystalline material, SLIP can only occur when five independent slip systems (ie sets of planes in the crystal) can operate. *Independent* means that the change in shape of the crystal due to slip in any one system cannot be achieved by slip in any combination of the other systems.

VOR (*Aero*) Acronym for Very-High-Frequency Omnidirectional Radio Range, a navigational radio aid that is used to determine bearings of an aircraft by comparison with a signal transmitted from the ground.

vortex (*Aero*) An eddy, or intense spiral motion in a limited region; a *vortex sheet* is a thin layer of fluid with intense vorticity; *tip vortices* are a form of *trailing vortex* from aerofoils, caused by shedding of lateral and line-of-flight airflows.

vortex generators (*Aero*) Small aerofoils, mounted normal to the surface of a main aerofoil and at a slight angle of incidence to the main airflow, which re-energize the BOUNDARY LAYER by creating vortices. Used on the wings

and tail surfaces of high-speed aircraft to reduce BUFFETING caused by compressibility effects, so raising the critical MACH NUMBER, and sometimes to improve the airflow over control surfaces near the stall, thereby improving controllability. Cf WING FENCE.

vortex street (*Aero*) A regular procession of vortices forming behind a bluff or rectangular body in two parallel rows. The vortices are staggered and each vortex is in the opposite direction from its predecessor. Also *Kármán street*.

vorticity equation (*EnvSci*) An equation for the rate of change of the vorticity, or curl of the velocity for atmospheric flow, esp the vertical component which is the dominating one. If f is the CORIOLIS PARAMETER,

$$\zeta = \frac{\partial v}{\partial x} - \frac{\partial u}{\partial y}$$

is the vorticity,

$$D = \frac{\partial u}{\partial x} + \frac{\partial v}{\partial y}$$

is the horizontal DIVERGENCE, and small terms are neglected, then

$$\frac{d}{dt}(\zeta + f) = -(\zeta + f)\,D$$

VOS (*ImageTech*) Abbrev for *video on sound*. See BACKGROUND VIDEO.

Voskhod (*Space*) A three-cosmonaut Soviet spacecraft developed from VOSTOK. Alexei Leonov performed the first space walk from Voskhod 2 in March 1965.

Vostok (*Space*) A series of Soviet manned Earth-orbiting spacecraft, first used successfully in April 1961. Vostok 1 carried Yuri Gagarin, the first man to travel in space.

vough (*MinExt*) See VUG.

voussoir (*CivEng*) See ARCH STONE.

vowel articulation (*ICT*) See ARTICULATION.

Voyager (*Space*) Two unmanned spacecraft (Voyager 1 and 2) developed from the spacecraft used in the MARINER PROGRAM, to explore the outer planets of the solar system. Both launched in 1977, they returned spectacular images of the outer planets (Jupiter, Saturn, Uranus and Neptune) and their satellites and ring systems; they continue (in 2006) to transmit data on conditions in space at the edge of the solar system.

voyeurism (*Psych*) Sexual gratification through the clandestine observation of other people's sexual activities or anatomy.

voxel (*ICT*) The smallest computer-addressable volume in a three-dimensional object, from volume element. Cf PIXEL.

VP (*Build*) Abbrev for VENT PIPE.

VPE (*Electronics*) Abbrev for VAPOUR PHASE EPITAXY.

VPI (*Chem*) Abbrev for VAPOUR PHASE INHIBITOR.

VPP (*Vet*) Abbrev for VIRUS PNEUMONIA OF PIGS.

VR (*ICT*) Abbrev for VIRTUAL REALITY.

VRAM (*ICT*) Abbrev for VIDEO RAM.

V-rings (*ElecEng*) V-shaped mica rings insulating the segments of a commutator from the end rings.

VRML (*ICT*) Abbrev for VIRTUAL REALITY MODELLING LANGUAGE.

VS (*Chem*) Abbrev for VOLUMETRIC SOLUTION.

V-series (*ICT*) The CCITT recommendations for data transmission over telephone networks.

VSI (*Aero*) Abbrev for VERTICAL SPEED INDICATOR.

VSWR (*ICT*) Abbrev for VOLTAGE STANDING-WAVE RATIO.

VTAM (*ICT*) Abbrev for VIRTUAL TERMINAL ACCESS METHOD.

VTB curve (*ElecEng*) Abbrev for VOLTAGE/TIME-TO-BREAKDOWN CURVE, ie a curve connecting the time and voltage for breakdown in this time. See SHORT-TIME BREAKDOWN VOLTAGE

VTOL (*Aero*) Abbrev for *vertical take-off and landing*. A general term for aircraft, other than conventional helicopters, capable of vertical take-off and landing: in the UK, the initials *VTO* were originally used, but the US expression *VTOL* has become general.

VTR (*ImageTech*) Abbrev for VIDEOTAPE RECORDER.

V-type commutator (*ElecEng*) A commutator whose segments are provided with projecting spigots, which dovetail into the end rings.

VU (*Acous*) Abbrev for VOLUME UNIT.

vug (*MinExt*) A cavity in a rock or lode, usually lined with crystals. Also *vough*.

Vulcan (*Astron*) A hypothetical planet, believed during the 19th century to lie inside the orbit of Mercury.

Vulcan coupling (*Eng*) A hydraulic shaft coupling, of the Föttinger type, used for connecting marine diesel engines to the propeller shaft in order to avoid torsional vibration troubles. See FÖTTINGER COUPLING.

vulcanite (*Chem*) Hard vulcanized rubber, in the making of which a relatively high proportion of sulphur is used. EBONITE is one form; coloured varieties are obtained by adding various ingredients, such as the sulphides of antimony and mercury. See VULCANIZATION OF RUBBER.

vulcanites (*Geol*) A general name for igneous rocks of fine grain size, normally occurring as lava flows. Cf PLUTONITES.

vulcanization (*Eng, Chem*) The treatment of rubber with sulphur or sulphur compounds to produce cross-links between the polymer chains, resulting in a change in the physical properties of the rubber. Sulphur is absorbed by qthe rubber, and the process can be carried out either by heating raw rubber with sulphur at a temperature between 135 and 160°C, or by treating rubber sheets in the cold with a solution of S_2Cl_2. To increase the rate of vulcanization, ACCELERATORS may be used.

vulcanized fibre (*Chem*) Neither a true fibre nor a product of vulcanization. Zinc chloride solution causes absorbent paper to swell and the cellulose fibres to be coated with amyloid gel. With removal of the inorganic salt, laminate is created by pressing the plies together. Electrical insulator.

vulcanizing agents (*Chem*) Speciality chemicals for effecting cross-linking of rubbers. May be mixture of sulphur, accelerator, process aids, etc, esp for vulcanizing natural rubber, butadiene rubber. Chloroprene rubber's can be vulcanized with metal oxides alone.

vulgar fraction (*MathSci*) See DIVISION.

Vulkollan (*Eng, Plastics*) European TN for cross-linkable polyurethane rubber which can be cast to shape. Extremely tough, but susceptible to hydrolysis.

Vulpecula (Fox) (*Astron*) A small northern constellation in the Milky Way.

vultex (*Chem*) See LATEX.

vulva (*BioSci*) The external genital opening of a female mammal. Adj *vulviform*.

vulvitis (*Med*) Inflammation of the vulva.

vulvovaginitis (*Med*) Inflammation of both the vulva and the vagina.

VU meter (*Acous*) An instrument calibrated to read intensity of electro-acoustic signals directly in VOLUME UNITS. See VOLUME INDICATOR.

VX gas (*Chem*) An organic compound developed as a lethal nerve gas. *Methylphosphonothioic acid; S-2-bis(1-methylethyl)aminoethyl-O-ethyl ester.*

W (*Chem*) Symbol for TUNGSTEN.

W (*Eng, Phys*) Symbol for WATT.

W (*CivEng*) Symbol for total load.

W (*ElecEng*) Symbol for electrical energy.

W (*Phys*) Symbol for: (1) WEIGHT; (2) WORK.

w (*CivEng*) Symbol for load per metre run or for weight per cubic metre.

w (*Phys*) Symbol for WORK.

wacke (*Geol*) A sandstone in which the grains are poorly sorted with respect to size.

wad (*Min*) Bog manganese, hydrated oxide of manganese. See ASBOLANE.

Wadell's sphericity factor (*PowderTech*) A shape factor used in particle size analysis, defined by the equation $S = d_c/D_c$, where S is Wadell's sphericity factor, d_c the diameter of a circle equal in area to the projected image of the particle when the particle rests on its larger face, and D_c the diameter of the smallest circle circumscribing the defined projection diameter.

Wadsworth mounting (*Phys*) A form of stigmatic mounting used for large concave gratings. The grating is illuminated by parallel light and the spectrum is focused at a distance of approximately one-half the radius of curvature of the grating.

wafer (*ICT*) See CHIP.

waggle dance (*Psych*) Semicircular movements of a hive bee, including an *abdominal waggle*, on returning from a foraging trip of more than 150 m from the hive; conveys information about the direction and distance of a food source to worker bees, and stimulates them to visit the site. Cf ROUND DANCE.

Wagner earth (*ElecEng*) A pair of impedances with their common point earthed, connected across an ac bridge network in order to neutralize the effect of stray capacitances. This is done by simultaneously balancing the normal bridge and that formed by the Wagner earth with the ratio arms.

wagon retarder (*CivEng*) See RETARDER.

wagon vault (*Build*) See BARREL VAULT.

wagtail (*Build*) See PARTING SLIP.

wainscot (*Build*) A wooden lining, usually panelled, applied to interior walls. Also *wainscoting*.

wainscoting cap (*Build*) A moulding surmounting a given piece of wainscoting.

wainscot oak (*Build*) Selected oak, cut radially to display the silver grain; much used for panelling.

WAIS (*ICT*) Abbrev for WIDE AREA INFORMATION SERVER.

waist anchor (*Ships*) See SHEET ANCHOR.

wait state (*ICT*) A small period of time when the computer's MICROPROCESSOR is idle owing to the fact that the microprocessor operates faster than the MAIN MEMORY and so has to wait for data to be retrieved.

wait time (*ICT*) The time interval during which a process is suspended.

wake (*Aero*) The region behind an aircraft in which the TOTAL HEAD of the air has been modified by its passage.

Waldegg valve gear (*Eng*) See WALSCHAERT'S VALVE GEAR.

Walden inversion (*Chem*) The transformation of certain optically active substances into their stereoisomeric derivatives by chemical reactions.

Waldenstrom's macroglobulinaemia (*Med*) A disease occurring mainly in elderly males, characterized by the presence of large amounts of monoclonal IgM in the blood, lymphoid tissue enlargement, splenomegaly, a haemorrhagic tendency and depression. The disease is probably a relatively benign and slowly progressing form of myelomatosis.

Waldeyer's ring (*BioSci*) The ring of lymphoid tissue comprising the tonsils, lymph nodes and lymphatic ducts circling the back of the mouth and the throat.

waldsterben (*EnvSci*) Symptoms of tree decline in C Europe from the 1970s, not attributable to known diseases, and widely held to be caused by atmospheric pollution.

wale (*Textiles*) A column of loops running along the length of a knitted fabric. The number of wales per cm helps to characterize the fabric.

walings (*CivEng*) Horizontal beams commonly used to maintain alignment of sheet-pile walls, often a pair of steel channels placed back to back and bolted through the piles. In light work, eg trenches, planks kept apart by struts may be used.

walk-about disease (*Vet*) See KIMBERLEY HORSE DISEASE.

walking beam (*Eng*) A mechanism for conveying solid articles. A fixed set of parallel bars supports the articles, while another set fills the gaps between the fixed set and moves cyclically through a rectangular path. If the second set moves forwards when its tops are above those of the fixed set and backwards when below, then the articles will move forwards.

walking beam (*MinExt*) Rocking beam used for transmitting power, eg for actuating the cable in cable drilling for oil.

walking dragline (*MinExt*) A large power shovel mounted on pads which are mechanically worked to manoeuvre it as required.

walking line (*Build*) An imaginary line, always 18 in from the centre line of the hand rail, used in setting out winders for a stair, the width of the winder measured on the line being made approx the same as the going of the normal treads. Also *going line*.

wall (*BioSci*) See panel on CELL WALL.

Wallace's line (*BioSci*) An imaginary line passing through the Malay Archipelago and dividing the Oriental faunal region from the Australasian region.

wallboard (*Build*) FIBREBOARD, usually of laminated construction.

wall box (*Build*) A support built into a wall to carry the end of a timber.

wall effect (*Phys*) (1) The contribution of electrons liberated in the walls of an ionization chamber to the recorded current. (2) The reduction in the count rate recorded with a Geiger tube due to ionizing particles not having the energy to penetrate the walls of the tube.

wall energy (*Phys*) The energy per unit area stored in the DOMAIN wall bounding two oppositely magnetized regions of a ferromagnetic material.

wall frame (*Build*) A method of system building for high buildings, comprising large precast concrete components.

wall hanger (*Build*) A support partly built into a wall to carry the end of a structural timber, which itself is not to be built into the wall.

wall hook (*Build*) An L-shaped nail used as a means of attachment to a wall.

wall insulator (*ElecEng*) An insulator specifically designed to enable a conductor at high potential to earth to pass through a brick or concrete wall.

wall-less counter (*NucEng*) A low-level proportional counter, the cathode of which consists of a cylindrical cage of thin wires, parallel to the cathode, which considerably reduces the background arising from electrons ejected by gamma rays. The counting volume may also be accurately defined by use of special *field tubes* at each end of the counter.

Wallner lines (*Eng*) Fracture surface markings, usually curved, caused by interaction between a propagating crack front and elastic waves released when pre-existing flaws or cracks are unloaded as the crack passes them. Only observed in brittle materials such as glass, but can be used to determine the speed of the crack. See FRACTOGRAPHY.

wallpaper colour (*Print*) A web, usually of advertising, preprinted in more than one colour on one or both sides in a repeating design, enabling it to be run through rotary presses of any cut-off.

wall plate (*Build*) The top horizontal timber of a wall, supporting parts of the structure.

wall-rock (*MinExt*) Country rock to either side of a vein or lode.

wall-rock alteration (*MinExt*) Mineralogical and/or chemical alteration of the country rock adjacent to a mineralized vein or lode. The result either of diffusion of elements from the mineralizing fluid in the vein, or of leaching of material from the country rock by the mineralizing fluid. See HALO, PRIMARY DISPERSION.

wall-sided (*Ships*) A term signifying absence of TUMBLE-HOME, and indicating that the maximum breadth is maintained to deck level.

wall string (*Build*) A string, generally a HOUSED STRING, positioned against a wall and supporting the inner ends of the steps.

wall tie (*Build*) A galvanized iron or stainless-steel component built into the two parts of a cavity wall or between an outer cladding and internal timber frame to bond them together.

walnut (*For*) European and American varieties of hardwood tree (*Juglans*) with a rich brown heartwood, straight- to wavy-grained and coarse-textured.

Walschaert's valve gear (*Eng*) A valve gear of the radial type used in some steam locomotives. The valve is driven through a 'combination lever' whose oscillation is the resultant of sine and cosine components of the piston motion, derived from connections with the engine crosshead and with an eccentric or return crank at 90° to the main crank. Sometimes *Waldegg valve gear*.

Waltonian (*Geol*) The lowest stage in the Pleistocene, above the Pliocene. Of long duration, it had a variable climate. See QUATERNARY.

WAN (*ICT*) Abbrev for WIDE AREA NETWORK.

wand (*ICT*) See BAR CODE READER.

wander (*ICT*) The long-term variation of the significant instants of a digital signal from their ideal positions in time, arising from changes in the propagation delay of transmission media and equipment.

wandering cells (*BioSci*) Migratory amoeboid cells; these may be leucocytes or phagocytes.

wane (*Astron*) To decrease; said of the Moon as the crescent decreases in size during the lunar cycle of phases.

wane (*For*) A defect in converted timber; some of the original rounded surface of the tree is left along an edge.

Wankel engine (*Autos*) Rotary automobile engine having an approximately triangular central rotor geared epitrochoidally to the central driving shaft and turning in a close-fitting oval-shaped chamber so that the power stroke is applied to each of the three faces of the rotor in turn as they pass a single sparking plug.

WAP (*ICT*) Abbrev for *wireless application protocol*, a technology that enables the Internet to be accessed on a mobile phone.

warble (*Vet*) Parasitic infestation caused by *Hypoderma bovis* and *H. lineatum* (gad flies). Adult flies lay on the hairs of the limb. The larvae then penetrate the skin and migrate through the tissues to arrive (about nine months later) subcutaneously on the back. The third stage larvae then emerge and drop to the ground. Cattle and horses are mainly infected. Compulsory treatment of bovines has limited the problem.

warble tone (*Acous*) Narrow-frequency-band current for testing devices such as microphones. To minimize errors due to standing waves in a room, the frequency is varied cyclically.

Ward–Leonard control (*ElecEng*) A method of speed control for large dc motors, employing a variable-voltage generator to supply the motor armature, driven by a shunt motor.

Ward–Leonard–Ilgner system (*ElecEng*) A modification of the Ward–Leonard system of speed control, in which a flywheel is included on the motor generator shaft to smooth out peak loads otherwise taken from the supply.

warehouse (*Print*) The department of a printing works where cutting, folding and the simpler methods of binding are undertaken.

warfarin (*Chem*) 3-(α-acetonyl-benzyl)-4-hydroxycoumarin. Used (usually in the form of its sodium derivative) as a blood anticoagulant; also widely used as a selective rodenticide.

warfarin sodium (*Pharmacol*) An orally active anticoagulant which antagonizes the effect of vitamin K. Used in the treatment of deep venous thrombosis, pulmonary embolism and myocardial infarction.

war gas (*Chem*) Any gaseous chemical substance used in warfare (or in riot control) to produce poisonous or irritant effects upon the human body. War gases are classified according to the length of time they are effective, *non-persistent* or *persistent* (less or more than 10 min in normal atmospheric conditions), and according to their effect: *lung irritants* or choking gases (phosgene, diphosgene); *lachrymators* or tear gases (chloracetophone, CS gas (orthochlorobenzylidene malononitrile), bromobenzylcyanide); *nerve gases* (derivatives of fluorophosphoric acid); *paralysants* or blood gases (hydrocyanic acid); *sternutators* or irritant smokes (diphenylaminochlorarsine); *vesicants* or blister gases (lewisite, mustard gas). The lachrymators and sternutators, being less toxic, are used in riot and crowd control.

warm-blooded (*BioSci*) Said of animals that have the bodily temperature constantly maintained at a point usually above the environmental temperature, of which it is independent; *homoiothermous, idiothermous*.

warm boot (*ICT*) A method of restarting a computer or a program without turning off the power or without losing the data required by the program. Also *soft boot*.

warm front (*EnvSci*) The leading edge of a mass of advancing warm air as it rises over colder air. There is usually continuous rain in advance of it.

warm start (*ICT*) Restart of a program after stoppage, without losing data from the previous run. Cf COLD START.

warm working (*Eng*) Working metal at a temperature below its recrystallization temperature but above room temperature.

war neurosis (*Psych*) A preferable synonym for *shellshock*. The term was originally used (World War I) for all types of nervous conditions resulting from war experiences, esp those caused by a bursting shell, which might result in (1) a condition of physical shock or concussion to the nervous system, (2) the precipitation of a psychoneurosis in a predisposed individual, (3) a combination of these conditions.

warning coloration (*BioSci*) See APOSEMATIC COLORATION.

warning pipe (*Build*) An overflow pipe fitted to cisterns etc to warn of a defective valve.

warp (*For*) Permanent distortion of a timber from its true form, due to causes such as exposure to heat or moisture.

warp (*Textiles*) The threads of a woven fabric that run continuously along the length of the fabric.

warp-knitted fabric (*Textiles*) Fine, knitted fabric formed wholly from warp yarns wound on a beam the width of the machine. The loops from each warp thread run the whole length of the fabric.

warp streak (*Textiles*) A visible fault in a fabric that arises from a warp yarn being different from the normal yarns, eg the thread may be of the wrong fibre, size, twist, colour. Also *warp stripe*.

Warren girder (*Build, Eng*) A form of girder consisting of horizontal upper and lower members, connected by members inclined alternately in opposite directions.

Warrington hammer (*Build*) A type of cross-pane hammer with slightly convex face.

wart (*Med*) A tumour of the skin formed by overgrowth of the prickle-cell layer, with or without hyperkeratosis; due to infection with a virus. See VERRUCA.

wash (*Arch, Eng*) A thin coat of water-colour paint applied to part of a drawing as an indication of the nature of the material to be used for the part represented, particular colours conventionally indicating particular materials.

wash (*Chem*) The removal of impurities from a gas or vapour by passing it through a reactant solution or solvent that retains or dissolves the impurities. Cf SCRUBBER. Liquids can be washed using liquids or solutions that are immiscible with the product liquid.

washboard (*Build*) See SKIRTING BOARD.

wash box (*MinExt*) A box in which raw coal is jigged in a coal washery.

washed clay (*Build*) See MALM.

washer (*Build, Eng*) Annular piece, usually flat, used under a nut to distribute pressure, or between jointing surfaces for a tight joint etc.

wash gravel (*MinExt*) Alluvial sands worth exploitation for mineral values contained.

wash-in (*Aero*) Increase in the ANGLE OF INCIDENCE from the root towards the tip of an aerofoil. With *wash-out* (the corresponding decrease) it is principally used on wings to ensure that the wing tips stall last so as to maintain aileron control.

washing (*ImageTech*) The removal of chemical residues and soluble components from photographic emulsions by water in the course of processing; the elimination of HYPO and silver compounds is particularly important for image permanence.

wash-out (*Aero*) See WASH-IN.

wash-out valve (*CivEng*) A valve inserted in a pipeline at the bottom of a valley, in order to enable a particular length of the pipe to be emptied as required.

wash plate (*Ships*) Fitted in tanks to prevent large quantities of water or oil rushing from side to side when the vessel rolls.

Wassermann reaction (*Med*) A complement fixation test formerly used in the diagnosis of syphilis.

waste (*CivEng*) See SPOIL.

waste (*Genrl*) Generic term for materials or products discarded during manufacture or following useful life.

waste (*MinExt*) Waste rock, either host (enclosing) rock mined with the true lode, or ore too poor to warrant further treatment.

waste (*NucEng*) (1) Depleted material rejected by an isotope separation plant. (2) Unwanted radioactive material for disposal.

waste (*Textiles*) Textile wastes are frequently recycled usually by degrading them to fibres. *Soft* waste is most easily processed because it is obtained in the earlier processes before twisting etc has been carried out. *Hard* waste includes all materials such as thread and rags which have to be *pulled* before they can be reused.

waste disposal (*NucEng*) See DISPOSAL.

waste heat recovery (*Eng*) The recovery of heat from furnaces, kilns, combustion engines, flue gases, etc, for utilization in eg air preheating, feed-water heating, waste heat boilers.

waste-light factor (*ElecEng*) A factor used in the design of floodlighting installations to allow for the light which, although emitted along the beam from the projector, does not fall on the area to be illuminated.

waster (*Build*) A mason's chisel, sometimes with claw head.

waste storage (*NucEng*) See STORAGE.

waste weir (*CivEng*) The weir provided in reservoir construction to discharge all surplus water in flood-time, preventing the water level from rising above the design limit for the dam.

wasting (*Build*) The operation of removing stone from a block by blows with a pick, prior to squaring and dressing.

WAT (*Psych*) Abbrev for WORD ASSOCIATION TEST.

WAT curves (*Aero*) Abbrev for *weight–altitude–temperature curves*. Complicated graphs relating the take-off and landing behaviour of an aircraft to its weight, airfield altitude and ambient temperature. Their preparation and use is mandatory for UK public transport aircraft. Many other countries use this information in tabular form.

water (*Phys*) Hydrogen oxide, H_2O. A colourless, odourless, tasteless liquid, mp 0°C, bp 100°C, whose molecules associate extensively through hydrogen bonding, which gives it its unique properties. On electrolysis it yields two volumes of hydrogen and one of oxygen. It forms a large proportion of the Earth's surface, occurs in all living organisms, and combines with many salts as water of crystallization. Water has its maximum density of $1000\ kg\ m^{-3}$ at a temperature of 4°C. This fact has an important bearing on the freezing of ponds and lakes in winter, since the water at 4°C sinks to the bottom and ice at 0°C forms on the surface. Besides being essential for life, water has a unique combination of solvent power, thermal capacity, chemical stability, permittivity and abundance. See CHEMICAL BOND, HARD WATER, HEAVY WATER, ION EXCHANGE, SOFT WATER, TRIPLE POINT.

water activity (*FoodSci*) An expression of the amount of water present in a food, raw material or product which is available to support microbial growth. As it is reduced, the rate of growth of micro-organisms declines. The key food preservation principles are based on the reduction of water activity by removing water or by adding solutes such as sugar or salt. Symbol a_w.

water ballast (*Ships*) Water carried for purposes of stability. Also, water taken into ballast tanks to balance or redress change of draught due to consumption of fuel etc.

water bar (*Build*) A galvanized iron (or non-ferrous metal) bar set in the joint between the wood and stone sills of a window, to prevent penetration of water. Also *weather bar*.

water blast (*MinExt*) A sudden escape of confined air due to water pressure, eg in rise workings.

water bomber (*Aero*) An aircraft, usually a flying boat, designed to combat forest fires by collecting water whilst planing on eg a lake and jettisoning its load from low altitude over the fire zone.

waterbound macadam (*CivEng*) A road surfacing formed of broken stone, well-rolled and covered with a thin layer of HOGGING, which is watered in and binds the stones together.

waterbrash (*Med*) A sudden gush into the mouth of a watery secretion containing acid fluids from the stomach; often sign of oesophageal reflux. Also *pyrosis*.

water-carriage system (*Build*) The system of disposing of waste matter from buildings by water closets etc, involving the use of water to carry away the waste matter. Cf CONSERVANCY SYSTEM.

water channel (*Aero*) An open channel in which the behaviour of the surface of water flowing past a stationary body gives a visual simulation of supersonic airflow.

water-checked (*Build*) Said of a casement, the stiles and mullions of which have grooves cut in the meeting edges to prevent entry of rain. See ANTI-CAPILLARY GROOVE.

water-cooled engine (*Autos*) An engine cooled by the circulation of water through jackets, which are usually cast integral with the cylinder block.

water-cooled motor (*ElecEng*) A motor employing water as a cooling medium.

water-cooled resistance (*ElecEng*) A resistance kept cool by immersion in water, which circulates in channels provided for the purpose.

water-cooled transformer (*ElecEng*) A transformer in which the oil is kept cool by means of water circulating in pipes immersed in the oil.

water-cooled valve (*Electronics*) Large thermionic vacuum tube in which the heat generated by the electronic bombardment of the anode is carried away by water circulating around or through it. In the former the anode is made an integral part of the envelope. Cf COOLED-ANODE VALVE.

water culture (*BioSci*) See HYDROPONICS.

water-displacing liquid (*Chem*) A solvent containing surface active materials capable of removing water from moist surfaces and substituting a thin film of rust-inhibitive chemicals. A typical system would consist of organic fatty acids, non-ionic surfactants and barium petroleum sulphonate, dissolved in kerosine.

water equivalent (*Phys*) The mass of water which would require the same amount of heat as a given body to raise its temperature by 1 degree. It is its THERMAL CAPACITY (the product of its mass and its SPECIFIC HEAT CAPACITY) divided by the specific heat capacity of water $(4\cdot186\ \text{kJ kg}^{-1}\ \text{K}^{-1})$.

water finish (*Paper*) The high surface finish produced by the machine CALENDERS when fitted with water doctors which apply a thin film of water at the NIP.

water flooding (*MinExt*) Secondary oil recovery operation in which water is injected into a petroleum reservoir to displace additional oil and enhance production.

water gauge (*Eng*) A vertical or inclined protected glass tube connected, at its upper and lower ends respectively, to the steam and water spaces of a boiler, for showing the height of the water level.

water gauge (*MinExt*) An instrument (eg PITOT-STATIC TUBE) for measuring the difference in pressure produced by a ventilating fan or air current.

water glass (*Chem*) A concentrated and viscous solution of sodium or potassium silicate in water. It is used as an adhesive, as a binder, as a protective coating in water-proofing cement, as a preservative for eggs, and in the bleaching and cleaning of fabrics.

water hammer (*Eng*) A sharp hammer-like blow from a steep-fronted pressure wave in water caused by the sudden stoppage of flow in a long pipe when a valve is closed sufficiently rapidly.

Waterhouse stops (*ImageTech*) Removable metal plates, each with a hole giving the required aperture, which can be inserted in front of the camera lens. Used in process cameras.

water-in-oil emulsion adjuvant (*BioSci*) An adjuvant in which the antigen, dissolved or suspended in water, is enclosed in tiny droplets within a continuous phase of mineral oil. The antigen solution constitutes the dispersed phase, stabilized by an emulsifying agent such as mannitol mono-oleate.

water-jet driving (*CivEng*) A process of pile driving sometimes adopted when the piles have to be sunk into alluvial deposits; a pressure water jet is used to displace the earth around the point of the pile.

water-jet pump (*Eng*) A simple suction pump, capable of producing a moderate degree of vacuum, in which air is drawn through the branch of a T-pipe by the action of a fast jet of water passing through the straight section. The principle is similar to that of an EJECTOR and the pump has no moving mechanical parts.

waterleaf (*Paper*) Paper which has not been sized with rosin prior to the tub-sizing operation.

water lime (*Build*) See HYDRAULIC CEMENT.

water lines (*Ships*) The intersection of the various water planes with the ship's form.

watermark (*Paper*) A device in paper visible by transmitted light as a lighter or darker local area produced by an appropriate relief or intaglio design on a hand mould, cylinder mould or dandy roll.

water/methanol injection (*Aero*) (1) The use of the *latent heat of evaporation* of water (the methanol is an antifreeze agent) injected into a piston engine intake to cool the charge, thereby permitting the use of greater power without detonation for take-off. (2) The injection of water into the airflow of the compressor of a TURBOJET or TURBOPROP to restore take-off power by cooling the intake air at high ambient temperatures.

water monitor (*NucEng*) One for measuring the level of radioactivity in a water supply, similar to, but much more sensitive than, an effluent monitor.

water of capillarity (*Build*) The moisture drawn up by capillary action from the soil into the walls of a building. Also *rising damp*.

water of crystallization (*Chem*) The water present in hydrated compounds. When crystallized from solution in water these compounds retain a definite amount of water, eg copper (II) sulphate, $CuSO_4\cdot5H_2O$. Also *water of hydration*.

water of hydration (*Chem*) See WATER OF CRYSTAL-LIZATION.

water paint (*Build*) Any paint that can be diluted with water, eg emulsion and acrylic paints.

water plane (*Ships*) A horizontal section through a ship's hull. Usually named by measurement from the baseline, but sometimes from the load water plane.

water pore (*BioSci*) In plants, an opening in the epidermis, associated with a hydathode, through which water exudes. It is often a modified stoma.

water potential (*BioSci*) A measure of the free energy of water in a solution, as in a cell or soil sample, and hence of its tendency to move by diffusion, osmosis or as vapour. It is the chemical potential of water in solution minus that of pure water (zero at standard temperature) divided by the partial molar volume of water, and is expressed as units of pressure, MPa or bar. Water diffusing or osmosing always moves down a water potential gradient. The components of water potential are MATRIC POTENTIAL, OSMOTIC POTENTIAL and TURGOR POTENTIAL. Symbol ψ_W.

water proof (*Textiles*) A fabric, usually plastic surfaced, which is impervious to water and air. Cf WATER REPELLENT.

waterproof paper (*Paper*) Packaging paper that has been treated eg by impregnation, coating or lamination to render it resistant to the transmission of liquid water.

water reactor (*NucEng*) A nuclear reactor in which water (including heavy water) is the moderator and/or coolant.

water recovery (*Aero*) The recovery, principally by condensation, of the water in the exhaust gases of an aero-engine. Used in airships for ballast purposes, as a partial set-off against the loss of weight due to the consumption of fuel during flight.

water repellent (*Textiles*) A fabric which resists the penetration of rain by a surface tension effect. The water droplets do not spread but roll off as they strike the fabric. Cf WATER PROOF.

water repellent solutions (*Build*) Often based on silicone resin, these solutions prevent penetration of rainwater on porous or absorbent masonry-type surfaces. Sometimes they are coloured with a fugitive dye to aid visibility of application.

water resistor (*ElecEng*) One made by immersing two electrodes in an aqueous solution. The resistance depends on the strength of the solution, and dimensions of the conducting path.

water rheostat (*ElecEng*) A WATER RESISTOR whose resistance can be varied, usually by moving one electrode relative to the other.

water-rib tile (*Build*) A purpose-made tile having a projecting rib that serves to prevent entry of rain or snow.

water sapphire (*Min*) See SAPHIR D'EAU.

water seal (*Build*) See SEAL.

water softening (*Chem*) The removal of 'hardness' in the form of calcium and magnesium ions, which form precipitates with soap. See CALGON, DOUBLE DECOMPOSITION, HARD WATER, ION EXCHANGE, PERMUTIT.

waterspout (*EnvSci*) See TORNADO.

water stain (*Build*) A stain for wood, consisting of colouring matter dissolved in water.

water stoma (*BioSci*) See WATER PORE.

water-storage tissue (*BioSci*) Tissue of large, highly vacuolate cells with relatively extensible walls, which can buffer the water supply. Water can also be stored in tree trunks, as in tracheids which can be emptied and refilled.

water table (*Build*) See CANTING STRIP.

water table (*Geol*) The surface below which fissures and pores in the strata are saturated with water. It roughly conforms to the configuration of the ground, but is smoother. Where the water table rises above ground level a river, spring or lake is formed.

watertight fitting (*ElecEng*) An electric-light fitting designed to exclude water under certain prescribed conditions. Cf WEATHERPROOF FITTING.

watertight flat (*Ships*) The part of a deck between two watertight bulkheads which are not in the same vertical plane.

water torch (*Build*) A machine which cuts through concrete and steel without generating heat and with relatively little noise. Uses pumping equipment in order to provide water supply to a pressure of 150 MN m^{-2}, and has a specially designed nozzle which produces a fine, penetrating needle of water.

water-tube boiler (*Eng*) A boiler consisting of a large number of closely spaced water tubes connected to one or more drums, which act as water pockets and steam separators, giving rapid water circulation and quick steaming. See BABCOCK AND WILCOX BOILER, FORCED-CIRCULATION BOILERS, YARROW BOILER.

water tunnel (*Aero*) A tunnel in which water is circulated instead of air to obtain a visual representation of flow at high Reynolds numbers with low stream velocities.

water turbine (*Eng*) A prime mover in which a wheel or runner carrying curved vanes is supplied with water directed by a number of stationary guide vanes; usually direct-coupled to large alternators.

water vapour pressure (*EnvSci*) That part of the atmospheric pressure which is due to the water vapour in the atmosphere.

water-vascular system (*BioSci*) (1) In Echinodermata, a system of coelomic canals, associated with the tube feet, in which water circulates. (2) In platyhelminthes, the excretory system.

water wheel (*Eng*) A large wheel carrying peripheral buckets or shrouded vanes on which water is caused to act, either by falling under gravity or by virtue of its kinetic energy. See OVERSHOT WHEEL, UNDERSHOT WHEEL.

watt (*Phys*) SI unit of power equal to 1 J s^{-1}. Thus 1 horsepower (hp) equals 745·70 watts. Symbol W.

wattful loss (*ElecEng*) See OHMIC LOSS.

Watt governor (*Eng*) A simple PENDULUM GOVERNOR in which a pair of links are pivoted to the vertical spindle and terminate in heavy balls. Shorter links are pivoted to the mid-points of the first, and to the sleeve operating the engine throttle.

watt-hour (*Phys*) A unit of energy, being the work done by 1 watt acting for 1 hour, and thus equal to 3600 J.

watt-hour efficiency (*Phys*) The ratio of the amount of energy available during the discharge of an accumulator to the amount of energy put in during charge. Cf AMPERE-HOUR EFFICIENCY.

watt-hour meter (*ElecEng*) Integrating meter for measurement of total electric energy consumed in a circuit. The conventional domestic electricity meter is of this type.

wattle and daub (*Build*) A type of wall construction in which wicker is interlaced about a rough timber framework and the whole is covered with plaster. Also *wattle and dab*.

wattless component (*ElecEng*) See REACTIVE COMPONENT OF CURRENT (VOLTAGE).

wattmeter (*ElecEng*) An instrument for measuring the active power in a circuit.

wattmeter method (*ElecEng*) A method of testing the electrical quality of iron specimens by measuring the power loss with ac magnetization.

wave (*Phys*) A time-varying quantity which is also a function of position. The characteristic of a wave is to transfer energy from one point to another without any particle of the medium being permanently displaced; particles merely oscillate about their equilibrium positions. In electromagnetic waves it is the changes in electric and magnetic fields which represent the wave disturbance. The progress of the wave is described by the passage of a *waveform* through the medium with a certain velocity, the *phase* or *wave velocity*. The energy is transferred at the *group velocity* of the waves making the waveform.

wave analyser (*ElecEng*) One for determining the frequency components in a continuously repeated signal. Also *spectrum analyser*.

wave angle (*ICT*) Either angle of elevation or azimuth of arrival or departure of a radio wave with respect to the axis of an antenna array.

wave antenna (*ICT*) Directional receiving antenna comprising a long wire running horizontally to the direction of arrival of the incoming waves at a small distance above the ground. The receiver is connected to one end, and the other end is connected to earth through a terminating resistance. Also *Beverage antenna*.

waveband (*ICT*) A range of wavelengths occupied by transmissions of a particular type, eg the medium waveband used mostly for broadcasting.

wave cloud (*EnvSci*) A cloud that appears at the crest of a LEE WAVE and thus remains more or less stationary relative to the ground. Wave clouds are usually rather smooth in appearance and often occur in regularly spaced bands demonstrating the lee waves causing them.

wave clutter (*Radar*) See SEA CLUTTER.

wave drag (*Aero*) The drag caused by the generation of SHOCK WAVES, applied to the aircraft as a whole. See COMPRESSIBILITY DRAG.

wave equation (*Phys*) A differential equation which describes the passage of harmonic waves through a medium. The form of the equation depends on the nature of the medium and on the process by which the wave is transmitted. The solutions to the equation depend on the circumstances in which the wave is propagated. See SCHRÖDINGER EQUATION.

wave filter (*ElecEng*) A four-terminal network, consisting of pure reactances, designed to pass particular bands of frequency and to reject others.

waveform (*Phys*) The shape, contour or profile of a wave; described by a phase relationship between successive particles in a medium. A waveform may be periodic, TRANSIENT or random. Also *waveshape*.

wave-formed mouth (*Vet*) A variation in height of the molar teeth of horses.

waveform monitor (*ImageTech*) CRO used for the display and measurement of TV signal waveform.

wavefront (*Phys*) Imaginary surface joining points of constant phase in a wave propagated through a medium. The propagation of waves may conveniently be considered in terms of the advancing wavefront, which is often of simple shape, such as a plane, sphere or cylinder.

wavefunction (*Phys*) Mathematical equation representing the space and time variations in amplitude for a wave system. The term is used particularly in connection with the SCHRÖDINGER EQUATION for particle waves.

waveguide (*Phys*) A hollow metal conductor in which electromagnetic waves in the microwave region can be transmitted efficiently from a source to other parts of a circuit. The transmission can be described by the patterns of electric and magnetic fields produced inside the guide, different modes being characterized by different electric and magnetic field configurations. Dielectric guides operate similarly but generally have higher losses.

waveguide attenuator (*ElecEng*) Conducting film placed transversely to the axis of the waveguide.

waveguide choke flange (*ElecEng*) Coupling flange between waveguide sections which offers zero impedance to signal without requiring metallic continuity.

waveguide coupler (*ElecEng*) An arrangement for transferring part of the signal energy from one waveguide into a second, crossing or branching off from the first; eg in a *directional coupler*, the direction of flow of the energy transferred to the second guide reverses when the direction of propagation in the first guide is reversed.

waveguide filter (*ElecEng*) One having distributed properties, giving frequency discrimination in a waveguide in which it is inserted.

waveguide impedance (*Phys*) The ratio Z derived from $W = V^2/Z$, or I^2Z, where W is power, V a voltage and I a current, the last being defined in relation to type of wave and shape of waveguide.

waveguide iris (*ElecEng*) A diaphragm placed across a waveguide forming a reactance.

waveguide junction (*ElecEng*) A unit joining three or more waveguide branches, eg HYBRID JUNCTION.

waveguide lens (*Phys*) An array of short lengths of waveguide which convert an incident plane wavefront into an approximately spherical one by refraction.

waveguide modes (*Phys*) Modes of propagation in a waveguide. Classified as TE$_{mn}$ (transverse electric) and TM$_{mn}$ (transverse magnetic). In a rectangular guide the subscripts refer to the number of half-cycles of field variation along the axes parallel to the sides. In a circular guide they refer to the field variation in the angular direction and in the radial direction. Waves below the cut-off frequency are said to be *evanescent*.

waveguide stub (*Phys*) A short-circuited length of waveguide used as a reactance for matching.

waveguide switch (*ElecEng*) One which switches power from waveguide A to B or C, with considerable loss between A and C or B, and between B and C.

waveguide tee (*ElecEng*) A T-shaped junction for connecting a branch section of a waveguide in parallel or series with the main waveguide transmission line.

waveguide transformer (*ElecEng*) A unit placed between waveguide sections of different dimensions for impedance matching.

wave impedance (*Phys*) Complex ratio of transverse electric field to transverse magnetic field at a location in a waveguide. For an acoustic wave, the ratio is of pressure to particle velocity.

wave interference (*Phys*) Relatively or completely stationary patterns of amplitude variation over a region in which waves from the same source (or two different coherent sources) arrive by different paths of propagation: *constructive interference* arises when the two waves are in phase and their amplitudes add; *destructive interference* arises when they are out of phase and their amplitudes partly or totally neutralize each other.

wavelength (*Phys*) Symbol λ. (1) Distance, measured radially from the source, between two successive points in free space at which an electromagnetic or acoustic wave has the same phase; for an electromagnetic wave it is equal in metres to

c/v where c is the speed of light (in m s^{-1}) and v is the frequency (in hertz). (2) Distance between two similar and successive points on a harmonic (sinusoidal) wave, eg between successive maxima or minima. (3) For electrons, neutrons and other particles in motion when considered as a wavetrain, $\lambda = h/p$, where p is the momentum of the particle and h is PLANCK'S CONSTANT. See DE BROGLIE WAVELENGTH, WAVE MECHANICS.

wavelength chirp (*ICT*) Unwanted wavelength shift in the output of a laser diode when modulated by a signal.

wavelength constant (*Phys*) The imaginary part of the PROPAGATION CONSTANT.

wavelength division multiplexing (*ICT*) Transmission of several independent signals via a single optical fibre by sending each signal at a slightly different optical frequency.

wavellite (*Min*) Orthorhombic hydrated phosphate of aluminium, occurring rarely in prismatic crystals, but commonly in flattened globular aggregates, showing a strongly developed internal radiating structure.

wave mechanics (*Phys*) The modern form of the quantum theory in which events on an atomic or nuclear scale are explained in terms of the interactions between wave systems as expressed by the SCHRÖDINGER EQUATION. For a bound particle, eg an electron in an atom, standing-wave solutions are found for which only certain wavelengths are permitted, and consequently the energy is quantized. See STATISTICAL MECHANICS and panel on QUANTUM THEORY.

wavemeter (*Electronics*) See FREQUENCY METER.

wavenumber (*Phys*) In an electromagnetic wave, the reciprocal of the wavelength, ie the number of waves in unit distance.

wave packet (*Phys*) A wavetrain; the DE BROGLIE WAVE associated with a particle.

wave parameter (*Phys*) See WAVELENGTH CONSTANT.

wave–particle duality (*Phys*) The observed phenomenon that light and other electromagnetic radiations behave like a wave motion when being propagated, and like particles when interacting with matter. Interference, diffraction and polarization effects can be described in terms of waves. The PHOTOELECTRIC EFFECT and the COMPTON EFFECT can be described in terms of *photons*, quanta of energy $E = hv$ where h is PLANCK'S CONSTANT and v is the frequency.

wavers (*Print*) Ink rollers that reciprocate to distribute the ink.

waveshape (*Phys*) See WAVEFORM.

wave soldering (*Eng*) A method of soldering large pre-assembled circuit boards by passing through a standing wave of molten solder.

wave tail (*Phys*) The portion of a waveform that follows the peak or crest.

wave theory of light (*Phys*) Macroscopic explanation of diffraction, interference and optical phenomena as an electromagnetic wave, predicted by Maxwell and verified by Hertz for radio waves. Cf CORPUSCULAR THEORY OF LIGHT.

wave tilt (*ElecEng*) The angle between the normal to the ground and the electric vector, in a ground wave polarized in the plane of propagation.

wavetrain (*Phys*) A group of waves of limited duration, such as those which result from a single spark discharge occurring in an oscillatory circuit.

wave trap (*ICT*) A circuit tuned to parallel resonance connected in series with the signal source to reject an unwanted signal, eg between a radio receiver and the aerial. See WAVE FILTER.

wave velocity (*Phys*) See PHASE VELOCITY.

wave winding (*ElecEng*) A type of armature winding in which there are only two parallel circuits through the armature, irrespective of the number of poles. Also *two-circuit winding*.

waving groin (*Build*) A groin which is not straight in plan.

wavy paper (*Paper*) A defect of paper showing as undulations, esp near the edges, due to the moisture content of the paper not being in equilibrium with the surrounding atmosphere.

wawa (*For*) See OBECHE.

wax (*Astron*) To increase; said of the Moon as the crescent increases in size during the lunar cycle of phases.

wax (*Chem*) Esters of monohydric alcohols of the higher homologues; eg *beeswax* is the myricyl (melissyl) ester of palmitic acid, $C_{30}H_{61}OCOC_{15}H_{31}$. For properties and uses, see BEESWAX, CABLE WAX.

waxing (*ImageTech*) Application of a thin layer of wax or silicone to the edges of motion picture prints to provide lubrication during projection.

WAXS (*Phys*) Abbrev for WIDE-ANGLE X-RAY SCATTERING.

wax vent (*Eng*) A pliable wax taper with a cotton core, placed in intricate cores during moulding. This wax melts when the core is dried, leaving a clear hole for the escape of gases.

wax wall (*MinExt*) A wall of clay built round the gob or goaf, to prevent the entry of air or egress of gas.

way (*MinExt*) See WIND ROAD.

ways (*Eng*) (1) The machined surfaces of the top of a lathe bed on which the carriage and tailstock slide; sometimes called *shears*. (2) The framework of timbers on which a ship slides when being launched.

way up (*Geol*) The upward direction of a succession of strata in an area of strong folding. The direction of *way up* is most usually determined by bottom structures or by CROSS BEDDING. See SOLE MARK.

W boson (*Phys*) A particle that mediates the WEAK INTERACTION; mass 81 GeV; charge +1 (W^+) or −1 (W^-); spin 1; decays to an electron or muon plus neutrino. Predicted by the WEINBERG–SALAM THEORY (or electroweak theory), it was discovered in 1983 in proton–antiproton collisions at CERN in Geneva.

W-chromosome (*BioSci*) See SEX DETERMINATION and panel on CHROMOSOME.

weak coupling (*ElecEng*) An inductive coupling in which the mutual inductance between two circuits is small; more generally known as *loose coupling*, *tight coupling*.

weak electrolyte (*Chem*) An electrolyte which is only slightly ionized in moderately concentrated solutions.

weak force (*Phys*) See WEAK INTERACTION.

weak inequality (*MathSci*) See STRICT INEQUALITY.

weak interaction (*Phys*) One of the four fundamental forces affecting QUARKS and LEPTONS. It acts through their property of weak charge, causing changes in the quark FLAVOURS, and can transform neutrons into protons and vice versa. The weak interaction is transmitted by massive GAUGE BOSONS, W^+, W^- and Z^0, and completed in 10^{-11} s. See panel on GRAND UNIFIED THEORIES.

weapons system (*Aero*) The overall planned equipment and backing required to deliver a weapon to its target, including production, storage, transport, launchers, aircraft, etc.

wear (*Eng*) The process of losing material from two surfaces that have been rubbed against one another. Mechanisms include ABRASIVE WEAR, ADHESIVE WEAR, CHEMICAL (or corrosive) WEAR and FATIGUE WEAR.

wearing course (*CivEng*) Upper layer of bituminious or asphalt carriageway construction. Usually placed in two layers, a 60 mm base course and a 40 mm wearing course. Also *crust*.

wearing depth (*ElecEng*) The permissible amount of radial wear on a commutator, prior to renewing the segments.

weather bar (*Build*) Patent methods of sealing the bottoms of external doors sometimes with a flap-like plate. See WATER BAR.

weatherboard (*Build*) A board used with others for covering sheds and similar structures. Weatherboards are fixed horizontally and usually overlap each other.

weather check (*Build*) A drip. Mastic or special metal strips are frequently incorporated.

weathercock stability (*Aero*) The tendency for an aircraft to turn into the relative wind, due to the side areas aft of the centre of gravity exceeding the value for directional stability

(as with aircraft designed for flying at low air speeds); excessive weathercock stability causes an oscillating yawing motion when flying in a cross-wind.

weathered pointing (*Build*) The method of pointing in which, in order to throw the rain off the horizontal joints, the mortar is sloped inwards, either from the lower edge of the upper brick, or from the upper edge of the lower brick, the latter method being preferred by bricklayers. Also *struck-joint pointing*.

weather fillet (*Build*) See CEMENT FILLET.

weather forecast (*EnvSci*) See FORECAST, WEATHER MAP and panel on NUMERICAL WEATHER PREDICTION.

weathering (*Build*) (1) The deliberate slope at which an approximately horizontal surface is built or laid so that it may be able to throw off the rain. See COPING (1). (2) The gradual process by which materials on the external faces of a building are affected by natural climatic conditions.

weathering (*Geol*) The processes of disintegration and decomposition effected in minerals and rocks as a consequence of exposure to the atmosphere and to the action of frost, rain and insolation. These effects are partly mechanical, partly chemical, partly organic and for their continuation depend upon the removal, by transportation, of the products of weathering. *Denudation* involves both weathering and transportation.

weathering (*Textiles*) The action of the weather on exposed materials.

weather map (*EnvSci*) See SYNOPTIC CHART.

weather minima (*Aero*) The minimum horizontal visibility and cloud base stipulated by (1) the air traffic authority and (2) the standing orders of each airline, under which take-off and landing are permitted.

weather moulding (*Build*) See DRIPSTONE.

weatherproof fitting (*ElecEng*) An electric-light fitting having an enclosure which excludes rain, snow, etc. Also *splashproof fitting*.

weather radar (*EnvSci*) A radar installation, either PPI or RHI, designed to be useful for the detection of PRECIPITATION and utilizing a wavelength of 3–20 cm. As the strength of the echo varies as the sixth power of the diameter, heavy showers and thunderstorms are much more conspicuous than widespread light rain or drizzle.

weather slating (*Build*) See SLATE HANGING.

weather strip (*Build*) See DOOR STRIP.

weather-struck (*Build*) A term applied to mortar joints finished by the method of *weathered pointing*.

weather tiling (*Build*) Tiles hung vertically to the face of walls to protect them against wet. Also *tile hanging*.

weaving (*Textiles*) The interlacing of warp and weft threads running at right angles to each other to form a fabric. See panel on FIBRE ASSEMBLIES.

weaving machine (*Textiles*) A machine or *loom* that produces woven fabrics. In most, a device transports the WEFT threads and interlaces them with WARP threads. In many looms this is done by a shuttle but in more modern machines other means are used such as projectiles, rapiers, and air or water jets.

web (*BioSci*) (1) The mesh of silk threads produced by spiders, some insects, and other forms. (2) The vexillum of a feather. (3) The membrane connecting the toes in aquatic vertebrates, such as frogs, penguins, otters. Adj *webbed*.

web (*Build, Eng*) The relatively slender vertical part or parts of an I-beam or built-up girder (such as a box girder) separating the two flanges.

web (*Paper*) A continuous sheet of paper on the paper machine, converting machine, etc.

web (*Textiles*) The loosely coherent sheet of fibres produced by a card and used for making non-woven fabrics. Also *batt*. See panel on PAPER AND PAPER-MAKING.

webbing (*Textiles*) A woven narrow fabric that is strong and able to sustain loads. Used in the manufacture of upholstery and of seat belts for cars and aeroplanes.

web-break detector (*Print*) Electronic equipment to stop the web press immediately a web breaks.

webcasting (*ICT*) See PUSH TECHNOLOGY.

webcam (*ICT*) A small digital video camera attached to a computer that can be used to send visual images across the Internet.

webcast (*ICT*) An audio or video programme that is broadcast live over the Internet. Also to broadcast such a programme over the Internet.

webdesign (*ICT*) The design and creation of websites.

weber (*Phys*) The SI unit of magnetic flux. An emf of 1 volt is induced in a circuit through which the flux is changing at a rate of 1 weber per second; $1\,Wb = 1\,V\,s = 1\,J\,A^{-1}$. Equivalent definition: 1 weber is the magnetic flux through a surface over which the integral of the normal component of the magnetic induction is $1\,T\,m^{-2}$.

Weber–Fechner law (*Med*) A law stating that the physiological sensation produced by a stimulus is proportional to the logarithm of the stimulus. See WEBER'S LAW.

Weberian apparatus (*BioSci*) See WEBERIAN OSSICLES.

Weberian ossicles (*BioSci*) In some Teleostei, eg carp, catfish and others, a chain of small bones, derived from processes of the anterior vertebrae, which connect the air bladder to the ear, transmitting vibrations from the former to a perilymphatic sac from which they pass to the endolymph of the inner ear. They correspond functionally to the middle ear ossicles of higher vertebrates.

Weber photometer (*ElecEng*) A transportable photometer in which a direct comparison is made between the brightness of two screens, one illuminated by an unknown light source and the other by a standard lamp.

Weber's law (*Psych, BioSci*) A law stating that the smallest change detectable in a stimulus is proportional to the magnitude of the original stimulus, eg a 1 kg change in the weight of a pencil would be more easily recognized than 1 kg added to a 100 kg barbell. The law works well in the mid-range but less well at the extremes.

web-fed (*Print*) A term indicating that the printing machine uses a reel of paper and not single sheets.

web frame (*Ships*) An extra strong frame usually made up of a plate with double angle bars on outer and inner edges. Commonly used to resist PANTING.

webhead (*ICT*) Casual term for an enthusiastic user of the World Wide Web.

webification (*ICT*) Slang for converting data into a form that can be published on the World Wide Web. Also *webify*.

webliography (*ICT*) A list of websites relating to a particular subject or person; a list of the websites referred to in the process of writing a book, article, etc.

weblog (*ICT*) A BLOG.

webmail (*ICT*) Electronic mail accessed via the World Wide Web.

webmaster (*ICT*) A person who creates, manages or maintains a website.

web offset (*Print*) An offset litho machine using a reel of paper.

web page (*ICT*) An electronic document on the World Wide Web. A component of a website, presented in HTML and usually including additional scripts or graphics.

webring (*ICT*) A set of linked websites, usually with a common theme or purpose, reached in turn via a hyperlink embedded in member sites.

websterite (*Geol*) A coarse-grained ultramafic igneous rock, consisting essentially of hypersthene and diopside.

Wechsler Adult Intelligence Scale, 3rd edition (*Psych*) A proprietary objective measure of intelligence based upon both verbal and performance subtests. The 3rd edition is a 1981 revision. The *Stanford–Binet test* has very similar validity, but is not as popular. See BINET INTELLIGENCE TEST. Abbrev *WAIS III*.

weddelite (*Min*) Hydrated calcium oxalate, $CaC_2O_4 \cdot 2H_2O$, crystallizing in the tetragonal system. It occurs uncommonly in the mineral world but freely in human *calculi*.

Weddle's rule (*MathSci*) The theorem that the area under the curve $y = f(x)$ from $x = x_0$ to $x = x_6$ is approximately

$$\frac{3}{10}[f(x_0) + 5f(x_1) + f(x_2) + 6f(x_3) + f(x_4) + 5f(x_5) + f(x_6)](x_6 - x_0)$$

where $x_0, x_1, x_2, x_3, x_4, x_5$ and x_6 are equally spaced.

wedge (*ElecEng*) (1) Total attenuator, in the form of a wedge of absorbing material, for terminating a waveguide. (2) Insertion of various lossy materials, put into a section of waveguide, to add fixed or variable attenuation in the circuit.

wedge (*EnvSci*) See RIDGE.

wedge (*ImageTech*) A strip of material showing gradation of transmission from clear to opaque along its length; the gradation may be continuous or in recognizable steps (step wedge). The material may be dyed or pigmented gelatine or a processed photographic image in silver or dye and may be neutral or in a single colour. The SENSITOMETRY of photographic materials given known controlled wedge exposures is an important part of processing control.

wedge aerofoil (*Aero*) A supersonic aerofoil section (much used for missiles) comprising plane, instead of curved, surfaces tapering from a very sharp leading edge at an acute included angle to give a THICKNESS–CHORD RATIO of 5% or less; the aerofoil may have a blunt trailing edge, or it may have the section of a very elongated lozenge, or it may have a parallel mid-portion with leading- and trailing-edge wedges, the two latter cases being known as *double-wedge aerofoils*.

wedge contact (*ElecEng*) A contact consisting of two fingers between which a wedge-shaped contact on the moving element is forced; used for circuit breakers etc.

wedge spectrogram (*ImageTech*) A spectrogram made with a neutral wedge whose transmission increases with the slit length of the spectrometer. The resultant photographic image indicates, by the height of the density contours, the differential colour sensitivity of the emulsion.

wedging (*MinExt*) Use of deflecting wedge near bottom of deep diamond boreholes, either to restore direction or to obtain further samples. Also *whipstocking*.

wedging crib (*MinExt*) A segmented steel ring on which shaft tubbing is built up and wedged in place. Also *wedging curb, wedging ring*.

weed (*BioSci*) A plant growing where it is not wanted by humans. Weeds of cultivated land are often natural plants of disturbed habitats and are often apomictic or self-pollinating, or spread vegetatively. See R-STRATEGIST.

weed (*Vet*) See SPORADIC LYMPHANGITIS.

weephole (*CivEng*) A pipe laid through an earth-retaining wall, with a slope from back to front to allow the escape of collected water.

weft (*Textiles*) The threads that run across the width of woven fabrics. In the loom they are interlaced with the WARP threads.

weft detector (*Textiles*) A mechanical or electronic device that indicates when the weft thread in a shuttle is becoming exhausted, or the absence of weft in shuttleless looms.

weft insertion machine (*Textiles*) A special warp-knitting machine which incorporates weft threads right across the fabric.

weft knitting (*Textiles*) A method of making a fabric by normal knitting with the loops being formed right across the fabric in straight lines at right angles to the direction in which the fabric is produced.

weft streak (*Textiles*) A fault in a woven fabric that runs across the fabric and results from a lack of uniformity in the weft threads (eg different fibres, colour, twist or thickness).

Weg rescue apparatus (*MinExt*) Portable breathing apparatus with self-contained oxygen supply controlled automatically by wearer's breathing action.

Weibull's modulus (*Eng*) A measure of the degree of scatter in an extreme-value distribution, eg the results of repeated measurements of the tensile strength of a brittle material. Symbol *m*. Cf STRENGTH MEASURES.

Weierstrass' test for uniform convergence (*MathSci*) See M-TEST OF WEIERSTRASS.

weigh batching (*CivEng*) Batching concrete by weight rather than volume of its constituents; more accurate and producing more consistent concrete.

weighing bottle (*Chem*) A thin-walled cylindrical glass container with a tightly fitting lid, used for weighing accurately hygroscopic etc materials.

weight (*Aero*) *Maximum weight*, or *gross weight* (colloq *max gross*) is the total weight of an aircraft as authorized for flight under the current regulations; *maximum take-off weight* is the highest allowable for the engine power available under the given conditions; *maximum landing weight* is the highest safe weight for landing because of structural strength; *tare weight* is the design weight of an aircraft type in flying condition, without fuel, oil, crew, removable equipment not necessary for flight and payload; *zero fuel weight*, used in airline load calculations, is the weight of the loaded aircraft after all usable fuel has been consumed.

weight (*ICT*) With TYPEFACES, each FONT has a defined weight such as normal, bold, italic, etc.

weight (*Phys*) The gravitational force acting on a body at the Earth's surface. Units of measurement are the newton, dyne or pound-force. Weight is equal to mass multiplied by acceleration due to gravity, and must therefore be distinguished from MASS, which is determined by the quantity of material and measured in kilograms. Symbol *W*.

weight-average molecular mass (*Chem*) See MOLECULAR MASS DISTRIBUTION.

weight coefficient (*ElecEng*) The ratio of the weight of an electric machine to its rated output.

weight fraction (*Chem, Eng*) Less frequently used alternative to VOLUME FRACTION to express concentration of components in COMPOSITE MATERIALS. It is the weight of the component present divided by the total weight of material.

weighting factor (*Phys*) See STATISTICAL WEIGHT.

weighting network (*ICT*) One designed to produce unequal attenuation for different frequency components of a signal, thereby weighting these differently in the final output.

weighting observations (*Surv*) The operation of assigning factors or 'weights' to each of a number of observations to represent their relative liability to error under their individual conditions of measurement.

weightlessness (*Space*) A condition obtained in FREE FALL when reaction is absent; a body has then no 'weight', only inertia.

weightometer (*MinExt*) A device which automatically weighs and records the tonnage of ore in transit on a belt conveyor.

weights (*Chem*) Standardized masses used for comparison with unknown masses, balances of various grades of sensitivity and sensibility being employed.

Weil–Felix reaction (*Med*) An agglutination test used in the diagnosis of rickettsial infections such as typhus. It depends upon a cross-reacting carbohydrate antigen shared by rickettsiae and certain strains of Proteus.

Weil's disease (*Med*) See LEPTOSPIROSIS.

Weinberg–Salam theory (*Phys*) A unified theory of weak and electromagnetic interactions between particles. It predicted the existence and behaviour of the W^+, W^- and Z^0 intermediate vector bosons as the agents for the weak interaction. These particles were discovered later (1983) using the CERN Super Proton Synchrotron modified to produce colliding-beam experiments with 270 GeV protons and antiprotons. Also *electroweak theory*.

weir (*CivEng*) A dam placed across a river to raise its level in dry weather.

Weisbach triangle (*Surv*) A method used in orientating underground workings, in which the theodolite is deliberately set up off the line of the two hanging wires used to transfer direction from above-ground to below-ground, so that the triangle between the instrument and the wires may be solved to enable the setting-out to proceed.

Weismann's ring (*BioSci*) In the larvae of some Diptera (insects), a small ring-like structure behind the brain, containing three types of glandular cell homologous with the corpora allata, corpora cardiaca and prothoracic glands of other insects, and controlling metamorphosis in a similar manner. Also *ring gland*.

Weissenberg effect (*Phys*) The effect shown by elastic liquids during rotary stirring. The melt or solution climbs up the stirring rod, in some cases winding itself entirely around the rod. Caused by normal stresses developed in the sheared fluid.

Weissenberg method (*Crystal*) A technique of X-ray analysis in which the crystal and photographic film are rotated in the beam of X-rays while the film is moved parallel to the axis of rotation.

Weiss theory (*Phys*) Early theory of ferromagnetism based on the concept of independent molecular magnets.

weld decay (*Eng*) A form of pitting corrosion which takes place in heat-affected zones adjacent to welds in non-stabilized stainless steels. Overcome by small additions of titanium or niobium which prevent separation of chromium carbides.

welded joint (*Eng*) A joint between two metals or plastics, made by fusion or diffusion to create interatomic bonding between the parts joined.

welded tuff (*Geol*) A TUFF composed of glass fragments which have partially fused together as a result of being deposited while still at a high temperature. Also *ignimbrite*.

welding (*Eng*) (1) Joining pieces of suitable metals or plastics, usually by raising the temperature at the joint so that the pieces may be united by fusing or by forging or under pressure. The welding temperature may be attained by external heating, by passing an electric current through the joint or by friction. (2) Joining pieces of suitable metals by striking an electric arc between an electrode or filler metal rod and the pieces. (3) Joining of thermoplastic parts using solvent, heat (eg hot-plate welding), friction (eg spin welding), adhesives, radio waves, ultrasound, etc. See ARC WELDING, COLD WELDING, RESISTANCE WELDING, SEAM WELDING.

welding manipulator (*Eng*) Support to which a workpiece can be clamped and which moves manually or automatically relative to a welding head.

welding regulator (*ElecEng*) A reactance by means of which the welding current may be varied in an ac welding set; it is variable by tappings controlled by a handwheel.

welding rod (*Eng*) Filler metal in the form of a wire or rod; in electric welding the electrode supplies the filler metal to the joint. Also *filler rod*. See METAL INERT GAS WELDING, TUNGSTEN INERT GAS WELDING.

welding set (*ElecEng*) The apparatus for electric arc welding, either ac or dc, comprising a supply unit and a regulator, which may be combined or separate.

welding transformer (*ElecEng*) A transformer specially designed to supply one or more welding regulators.

weld line (*Eng*) A ine on the surface of moulding where melt fronts have met, but fused poorly. A kind of mould defect, often inevitable in complex injection mouldings, but may seriously weaken product owing to stress concentration.

weldment (*Eng*) A welded assembly.

well (*Phys*) See POTENTIAL WELL.

well-conditioned (*Surv*) A term used in triangulation to describe triangles of such a shape that the distortion resulting from errors made in measurement and in plotting is, or is nearly, a minimum; achieved in practice by making the triangles equilateral or approximately so.

well counter (*NucEng*) One used for measurements of radioactive fluids placed in a cylindrical container surrounded by the detecting element (hollow scintillation crystal or sensitive volume of special Geiger tube).

well foundation (*CivEng*) A type of foundation formed by sinking MONOLITHS to a firm stratum, plugging the open wells at the bottom with concrete and filling with granular material. Smaller wells may be filled entirely with concrete.

well head (*MinExt*) The top of the casing of a production oil well, with its control valves.

well-hole (*Build*) The vertical opening enclosed between the ends of the flights in a winding or geometrical stair.

well logging (*Geol*) The recording of the composition and physical properties of the rocks encountered in a borehole, particularly one drilled during petroleum exploration. Well logging includes a variety of techniques, eg resistivity log, gamma-ray log, neutron log, spontaneous or self-potential log, temperature log, calliper log, photoelectric log, acoustic velocity log, etc.

well-ordered set (*MathSci*) A set ordered in such a way that any subset has a first element.

Welsh groin (*Build*) A groin formed by two intersecting cylindrical vaults of different rises. Also *underpitch groin*.

welt (*Build*) A joint made between the edges of two lead sheets on the flat. Made by turning up each edge at right angles to the flat surface, bringing the two turned-up parts together, doubling them over and dressing them down flat. Also *seam*.

welt (*Textiles*) A strengthened edge to a knitted fabric. If made at the start or finish of the knitting process the welts run across the fabric, but if made later by seaming they may run along any edge. Probably best known are stocking welts which are a double layer of plain fabric at the top of a stocking made by the machine.

Weltanschauung (*Psych*) Ger for 'world outlook', 'philosophy of life', for neither of which is there a single English word.

Welvic (*Plastics*) TN for polyvinyl chloride (UK).

wen (*Med*) See SEBACEOUS CYST.

Wenlock (*Geol*) The middle series of the Silurian period. See PALAEOZOIC.

Wenner winding (*ElecEng*) A form of winding used in wirewound resistors to construct standard resistances of low residual reactance for use at relatively high frequencies.

Wentworth scale (*Geol*) The size of grains in a sediment or sedimentary rock. It is an extension of the *Udden grade scale*. See PARTICLE SIZE, PHI GRADE SCALE.

Wernerism (*Geol*) See NEPTUNISM.

Werner sedimentation techniques (*PowderTech*) A two-layer method of particle size analysis. Particles sediment through a vertical column of liquid and collect in a narrow capillary at the foot of the column. The height of the column of particles is used as a measure of the particles sedimented out.

Werner's theory (*Chem*) A method of formulation of complex inorganic compounds based on the assumption that saturated groups are held to the central atom by residual valencies. The total number of such groups and ordinary unsaturated radicals which surround the central atom to form an unionizable co-ordination complex is characteristic of the central atom.

Wernicke's aphasia (*Psych, Med*) Aphasia caused by damage to Wernicke's area in the frontal lobe. Affects written and spoken language.

Wernicke's encephalopathy (*Med*) Brain damage due to THIAMINE deficiency, often as a result of alcohol abuse, causing abnormality in eye movements. Usually associated with KORSAKOFF'S PSYCHOSIS.

Wertheim's operation (*Med*) The operation of removing the uterus, the glandular tissue in the pelvis and the upper part of the vagina, in the treatment of cancer of the cervix uteri.

Westcott convention (*NucEng*) An approximation applied to the neutron flux in thermal reactor design. The flux is divided into a thermal component with Maxwellian distribution and a fast component with distribution proportional to dE/E, where E is the neutron energy. Sometimes a third component covering the thermalization region may be included.

Westcott flux (*NucEng*) Theoretical neutron flux defined as equal to the reaction rate of a detector with cross-section which is unity for thermal electrons (velocity 2200 m s^{-1}) and varies inversely with the neutron velocity.

western blotting (*BioSci*) See BLOTTING.

western duck sickness (*Vet*) See ALKALI DISEASE.

Western red cedar (*For*) N American softwood (*Thuja plicata*) which is not a true cedar. Reddish-brown heartwood, straight-grained and coarse-textured.

West Indian ebony (*For*) See COCUSWOOD.

westing (*Surv*) A west departure.

Westmoreland slates (*Build*) Thick slates (6·5–16 mm; $\frac{1}{4}-\frac{5}{8}$ in) of varying size.

West Nile virus (*Med*) An arthropod-borne virus that is a cause of severe meningitis and encephalitis in humans.

Weston standard cadmium cell (*Phys*) A practical portable standard of emf, in which the cathode is $12\frac{1}{2}\%$ cadmium and $87\frac{1}{2}\%$ mercury by weight, with an anode of amalgamated platinum or highly purified mercury. A saturated solution of aqueous Cd_2SO_4 is used as electrolyte. The cell emf at 20°C is 1·018 636 V; temperature coefficient only 0·000 04 V °C^{-1}. Used for calibrating potentiometers and hence all other voltage-measuring devices. Cells with unsaturated solutions have a lower temperature coefficient of emf, but do not give an equally high absolute standard of reproducibility. Also *standard cell*.

Westphal balance (*Phys*) A device in which relative density is determined by suspending a solid specimen from a balance beam in liquid of known density, or vice versa.

Westphalian (*Geol*) A stratigraphical stage in the Carboniferous rocks of Europe, approximately corresponding to the Coal Measures in England and Wales. See PALAEOZOIC.

wet and dry bulb hygrometer (*EnvSci*) A pair of similar thermometers mounted side by side, one having its bulb wrapped in a damp wick dipping into water. The rate of evaporation of water from the wick and the consequent cooling of the 'wet bulb' is dependent on the relative humidity of the air; the latter can be obtained by means of a table from readings of the two thermometers. Also *psychrometer*.

wet assay (*MinExt*) Qualitative or quantitative analysis of ores or their constituents in which dissolution and digestion with suitable solvents plays a part. See CUPELLATION, DRY ASSAY, SCORIFICATION.

wet beaten stuff (*Paper*) Heavily beaten stuff in which hydration is developed. The resultant paper is dense and inclined to be translucent. Wet stock drains less readily on the machine wire and is more difficult to dry.

wet-bulb potential temperature (*EnvSci*) The wet-bulb potential temperature of a sample of moist air at any level may be found on an AEROLOGICAL DIAGRAM by following the SATURATED ADIABATIC curve through the WET-BULB TEMPERATURE of the sample until it intersects the 1000 mb isobar and then reading off the temperature there. It is, for all practical purposes, a conservative quantity for such processes as evaporation, condensation and dry and saturated adiabatic temperature changes, and is thus a useful quantity for AIR-MASS analysis.

wet-bulb temperature (*EnvSci*) The temperature at which pure water must be evaporated adiabatically at constant pressure into a given sample of air in order to saturate the air under steady-state conditions. It is approximated closely by the temperature indicated by a thermometer, freely exposed to the air (but shielded from radiation), whose bulb is covered with muslin wetted with pure water.

wet cell (*Chem*) A primary cell which contains liquid electrolyte, in contrast to the paste of a dry cell.

wet deposition (*EnvSci*) The process that returns pollutants in the atmosphere to the Earth's surface in rain or snow. See ACID RAIN, OCCULT DEPOSITION and panel on ATMOSPHERIC POLLUTION.

wet electrolytic capacitor (*ElecEng*) One in which the negative electrode is a solution of a salt, eg aluminium borate, which is suitable for maintaining the aluminium oxide film without spurious corrosion.

wet end (*Paper*) The wire and press parts of a paper or board machine where the sheet is formed and water removed by drainage, suction and pressure.

wet etching (*Electronics*) Etching with wet chemicals as distinct from those processes based on gas discharges. Cf DRY ETCHING.

wet expansion (*Paper*) The percentage increase in the length of a strip of paper after immersion in water under specified test conditions.

wet felt (*Paper*) A continuous felt or synthetic fabric used in the press section of the paper- or board-making machine to convey the web, assist in water removal and prevent crushing at the press nip.

wet flashover voltage (*ElecEng*) The voltage at which the air surrounding a clean wet insulator completely breaks down. Also *wet sparkover voltage*.

wether (*Agri*) A male sheep castrated as a lamb, before developing secondary sexual characteristics.

wet laying (*Textiles*) Production of non-woven fabric by using the wet laying technique employed for making paper.

wet lease (*Aero*) Hire of commercial aircraft, complete with original crew, serviced by the original owner, but perhaps carrying the new operator's logo.

wet-on-wet (*Print*) Printing two or more colours in quick succession, particularly coloured illustrations, the tack of the ink being graded in order to prevent pickup by the succeeding colours.

wet-plate process (*ImageTech*) See COLLODION PROCESS.

wet rot (*BioSci*) (1) A rot in which the tissue is rapidly broken down with the release of water from the lysed cells, as in the brown rots of stored fruits. (2) The rot of timber that is often wet, caused by the fungus *Coniphora puteana*.

wet sparkover voltage (*ElecEng*) See WET FLASHOVER VOLTAGE.

wet spinning (*Textiles*) (1) Making filaments from a solution of the polymer by EXTRUSION followed by precipitation with a liquid, eg VISCOSE RAYON. (2) A method for making fine *flax* yarns in which the ROVING passes through hot water to soften it and so assist DRAFTING.

wet steam (*Eng*) A steam–water mixture, such as results from partial condensation of dry saturated steam on cooling.

wet strength paper (*Paper*) Any paper so treated that it retains an appreciable proportion of its dry strength when completely wet and has good wet rub resistance. These properties may be obtained by adding suitable resins to the stock and curing (eg urea or melamine formaldehyde) or by PARCHMENTIZING.

wettability (*Chem*) The extent to which a solid is wetted by a liquid, measured by the force of adhesion between the solid and the liquid phases.

wetted area (*Aero*) Total surface area of body immersed in an airflow and over which a boundary develops.

wetting agent (*Chem*) Surface active agent which lowers the surface tension of water by a considerable amount, although present only in very low concentration.

wetwood (*For*) Wood with an abnormally high water content and a translucent or water-soaked glassy appearance. This condition develops only in living trees and not through soaking in water.

wf (*Print*) The former standard mark for *wrong font*. It is written in the proof margin and the letter is underlined or struck through. Replaced by a circled cross.

whalebone (*BioSci*) See BALEEN.

Wharfedale machine (*Print*) The first successful STOP-CYLINDER printing machine, invented in 1858 by William Dawson and David Payne at Otley, in Wharfedale, Yorkshire.

what you see is what you get (*ICT*) See WYSIWYG.

wheatmeal flour (*FoodSci*) Sometimes called brown flour. It is an admixture of white flour and a high proportion of

bran. To comply with UK regulations wheatmeal flour must contain not less than 0·6% crude fibre.

wheatsheaf (*Chem*) Growth form of polymer lamellae. See CRYSTALLIZATION OF POLYMERS.

wheat starch (*FoodSci*) Starch obtained from wheat flour by extraction with water followed by drying. Used as a thickener.

Wheatstone bridge (*ElecEng*) Apparatus for measuring electrical resistance using a NULL INDICATOR, comprising two parallel resistance branches, each branch consisting of two resistances in series. Prototype of most other bridge circuits. See fig. at BRIDGE.

wheel animalcules (*BioSci*) See ROTIFERA.

wheel base (*Eng*) The distance between the leading and trailing axles of a vehicle.

wheeling (*Eng*) A sheet-metal working process, using a machine with one flat and one convex wheel, for producing curved panels or for finishing after PANEL BEATING.

wheeling step (*Build*) See WINDER.

wheel ore (*Min*) See BOURNONITE.

wheel-quartering machine (*Eng*) A horizontal drilling machine having two opposed spindles at opposite ends of the bed; used to drill the crank-pin holes in both wheels on a locomotive coupled axle simultaneously, and in precise angular relationship.

wheel window (*Arch*) See ROSE WINDOW.

wheel wobble (*Autos*) A periodic angular oscillation of the front wheels, resulting generally from poor balance, insufficient castor action or from backlash in the steering gear.

whetstone (*Geol*) See HONE.

whewellite (*Min*) $CaC_2O_4 \cdot H_2O$, hydrated calcium oxalate crystallizing in the monoclinic system. It occurs uncommonly in the mineral world but is abundant in human calculi.

whey (*FoodSci*) The watery liquid or serum which is separated from the curds produced in cheese-making. Whey can be concentrated and dried to produce WHEY POWDER which is a useful source of milk solids in food manufacture.

whin (*Geol*) A popular term applied to doleritic intrusive igneous rock resembling that of the well-known Whin Sill. Also *whinstone*.

whine (*Acous*) The fluctuation in apparent loudness and pitch of a reproduced sound when the speed of the recording or reproducing machine is varying at a slow rate. See WOW.

Whin Sill (*Geol*) A sheet of intrusive quartz-dolerite or quartz-basalt, unique in the UK, as it is exposed almost continuously for over 300 km from the Farne Islands to Middleton-in-Teesdale. See SILLS.

whinstone (*Geol*) See WHIN.

whipcord (*Textiles*) Fabrics woven from cotton or worsted with a bold steep warp twill, used chiefly for dresses, suits and coats.

whiplash (*Med*) A term descriptive of an extension flexion injury to the soft tissues of the neck including the ligaments and apophyseal joints, causing instability and chronic pain; frequent in car accidents involving sudden collision from behind.

whiplash flagellum (*BioSci*) A flagellum without hairs on its surface. In contrast, *tinsel flagellum* is a decorated flagellum. Also *acronematic flagellum*.

whipping (*Print*) See OVERCASTING.

Whipple–Murphy truss (*Eng*) A bridge truss having horizontal upper and lower chords connected by vertical and diagonal members, so that the panels resemble the letter N. Also *Linville truss, N-truss, Pratt truss*.

whipstitching (*Print*) See OVERCASTING.

whipstocking (*MinExt*) See WEDGING.

whirler (*Print*) Mechanical or hand-operated equipment using centrifugal force to spread an even coating on plates.

whirl-gate (*Eng*) See SPINNER-GATE.

whirling arm (*Aero*) Apparatus for making certain experiments in aerodynamics, the model or instrument being carried round the circumference of a circle, at the end of an arm rotating in a horizontal plane.

Whirlpool Galaxy (*Astron*) A bright and well-defined spiral galaxy in the constellation Canes Venatici, oriented face-on to us. Distance approx 6 Mpc.

whirlwind (*EnvSci*) A small rotating windstorm which may extend upwards to a height of several hundred metres; a small tornado.

whisker (*Chem*) A thin strong filament or fibre made by growing a crystal, eg of silicon carbide, sapphire, etc.

whispering gallery (*Acous*) A room in which a whisper can be heard over a surprisingly large distance. The classical example is in the dome of St Paul's Cathedral in London, where sound is reflected on curved walls with high reflectivity and with the property of converging the sound.

whistle (*Acous*) A flow-noise device generating EDGE TONES.

whistlers (*Acous*) Atmospheric electric noises which produce relatively musical notes in a communication system.

whistling (*Vet*) See ROARING.

Whitby method (*PowderTech*) A sedimentation technique in which a bucket centrifuge is used to measure sizes of very fine particles. The particles accumulate in a fine-bore capillary at the bottom of the tube and the weight of particles is estimated from the height of the sediment in the tube.

white arsenic (*Chem*) See ARSENIC.

white balance (*ImageTech*) Balancing a video camera's colour to give a correct rendition of scenes in different types of lighting. See AUTOMATIC WHITE BALANCE.

whiteboard (*ICT*) A board used for teaching or presentation purposes, similar to a blackboard but with a white plastic surface for writing on; also a reserved area on a computer screen on which several users can write.

white bombway (*For*) Timber from the *Terminalia procera*, a native of the Andaman Islands. Planed surfaces of the wood are mildly silky, the texture moderately open and rather uneven, while the grain is straight.

White Book (*ImageTech*) The technical specifications that define the VIDEO CD and *Karaoke CD* standard.

white cell (*BioSci*) See LEUCOCYTE.

white clip (*ImageTech*) Clipping the top of a preemphasized LUMINANCE SIGNAL to prevent OVERMODULATION. See PRE-EMPHASIS.

white coat (*Build*) The last or finishing coat of plaster.

white comb (*Vet*) See AVIAN FAVUS.

white copperas (*Min*) See GOSLARITE.

white corundum (*Min*) See WHITE SAPPHIRE.

white damp (*MinExt*) Carbon monoxide. Produced by the incomplete combustion of coal in a mine fire or by gas or dust explosions. Invisible; very poisonous.

white deal (*For*) See WHITEWOOD.

white dwarf (*Astron*) Small dim star in the final stages of its evolution. The masses of known white dwarfs do not exceed 1·4 solar masses. They are defunct stars, collapsed to about the diameter of the Earth, at which time they stabilize, with their electrons forming a DEGENERATE GAS, which has a pressure sufficient to balance their gravitational force.

white fibres (*BioSci*) Unbranched, inelastic fibres of connective tissue, mostly collagen, occurring in wavy bundles. Cf YELLOW FIBRES.

white fibrocartilage (*BioSci*) A form of FIBROCARTILAGE in which WHITE FIBRES predominate.

white finger disease (*Med*) A loss of colour in the fingers caused by arterial spasms that reduce blood flow, often (*vibration white finger*) caused by prolonged use of vibrating machinery.

white frost (*EnvSci*) See HOAR FROST.

white glass (*Glass*) See OPAL GLASS.

white gold (*Chem, Eng*) Gold alloyed with nickel or platinum in order to render it silvery white in colour.

white-hat hacker (*ICT*) A computer hacker who tries to break into a system in order to test its security (cf BLACK-HAT HACKER).

white-heart process (*Eng*) See MALLEABLE CAST-IRON.

white heat (*Eng*) As judged visually, temperature exceeding 1000°C.

white heifer disease (*Vet*) A condition in cattle in which the vagina, cervix and uterus develop abnormally; associated with the gene for white coat colour and occurring mainly in white shorthorn heifers.

white iron (*Eng*) Pig iron or cast-iron in which all the carbon is present in the form of cementite (Fe_3C). White iron has a white crystalline fracture, and is hard and brittle.

white iron pyrites (*Min*) See MARCASITE (1).

white lead (*Chem*) Basic *lead (II) carbonate* or *hydroxycarbonate*. Made by several processes of which the oldest and best known is the *Dutch* (or stack) *process*. Formerly used extensively as a paint pigment and for pottery glazes.

white lead ore (*Min*) See CERUSSITE.

white-leg (*Med*) See PHLEGMASIA ALBA DOLENS.

white level (*ImageTech*) The level of the TV signal representing normal maximum picture luminance.

white light (*Phys*) Light containing all wavelengths in the visible range at the same intensity. This is seen by the eye as white. The term is used, however, to cover a wide range of intensity distribution in the spectrum and is applied by extension to continuous spectra in other wavelength bands. Also *white radiation*.

white line (*Print*) A line of space.

white matter (*BioSci*) An area of the central nervous system, mainly composed of cell processes, and therefore light in colour.

white metal (*Eng*) Usually denotes tin-base alloy (over 50% tin) containing varying amounts of lead, copper and antimony; used for bearings, domestic articles and small castings; sometimes also applied to alloys in which lead is the principal metal. Also *anti-friction metal*, *bearing metal*.

white muscle disease (*Vet*) A muscular dystrophy. A disease primarily of young calves and lambs but adults can be affected. Symptoms include sudden death, tachycardia, pyrexia, stiffness and recumbency. It is due to lack of vitamin E and/or selenium in diet or failure in their absorption. Also *stiff lamb disease*.

white nickel (*Min*) A popular name for the cubic diarsenide of nickel, $NiAs_2$, CHLOANTHITE.

white noise (*Acous, ICT*) Noise, which may be of the random or impulse variety, having a flat frequency spectrum over the range of interest. May be simulated during development and testing of equipment. Also *flat random noise*.

white oil (*MinExt*) (1) A term for oils that are substantially colourless and without bloom. Usually made from light lubricating oils by acid treatment or hydrogenation, they may be used medicinally (liquid paraffin) or in toilet preparations. (2) Also a term for oils that do not contaminate the tanker or other transport vehicle. Tankers are usually dedicated to one class of oil cargo. Cf BLACK OIL.

white-out (*EnvSci*) A situation where the horizon is indistinguishable, when the sky is overcast and the ground is snow-covered.

white-out (*ImageTech*) Highlight areas which are so overexposed that they reproduce as featureless white. See ZEBRA PATTERN VIDEO LEVEL INDICATOR.

white-out (*Print*) To open out composed type-matter with spacing, in order to fill the allotted area or improve the appearance.

white-out lettering (*Print*) A term indicating that, in the reproduction required, the lettering is to be reversed to white on black.

white radiation (*Phys*) See WHITE LIGHT.

white reference level (*ICT*) Signal modulation level corresponding to maximum brightness (white) in monochrome facsimile or TV transmissions.

whiter-than-white (*ImageTech*) TV luminance signal at a level exceeding the normal white level.

whites (*Vet*) Leucorrhoea of cows.

white sapphire (*Min*) More reasonably called *white corundum*, it is the colourless pure variety of crystallized corundum, Al_2O_3, free from those small amounts of impurities which give colour to the varieties 'ruby' and 'sapphire'; when cut and polished, it makes an attractive gemstone.

white scour (*Vet*) Diarrhoea affecting calves during the first few weeks of life, caused usually by *Escherichia coli* infection, probably in conjunction with nutritional and environmental factors. Vaccination widely used. Also *calf scour*.

white smoker (*Geol*) A plume of hydrothermal fluid, white with mineral precipitates, at the crest of an ocean ridge. Most of the mineral is barytes and silica. See BLACK SMOKER, CHIMNEY.

white spirit (*Build*) A petroleum distillate used as a substitute for turpentine in mixing paints, and in paint and varnish manufacture.

white subject (*ImageTech*) One which reflects all wavelengths of light of the visible spectrum to a substantially equal extent and thus appears to the eye without colour.

white vitriol (*Min*) A popular name for GOSLARITE, $ZnSO_4 \cdot 7H_2O$.

white water (*Paper*) See BACKWATER.

whitewood (*For*) Timber from the softwoods, silver fir (*Abies alba*) and Norway spruce (*Picae abies*). It varies from white to yellowish-brown, with a straight grain and fine texture. Widely used, general purpose timber. Also *European silver pine, white deal*. Not to be confused with AMERICAN WHITEWOOD.

whiting (*Chem*) Powdered chalk or calcium carbonate used as filler for putty, thermoplastic and thermoset resins, etc.

whitlockite (*Min*) Trigonal calcium phosphate, occurring in sedimentary phosphate deposits.

whitlow (*Med*) See PARONYCHIA.

Whitworth screw-thread (*Eng*) See BRITISH STANDARD WHITWORTH THREAD.

whizz-pan (*ImageTech*) Rapid PAN from one point of interest to another. Also *whip pan*.

whole-body monitor (*NucEng*) Assembly of large scintillation detectors, heavily shielded against background radiation, used to identify and measure the gamma radiation emitted by the human body.

whole-bound (*Print*) See FULL-BOUND.

whole-brick wall (*Build*) A wall whose thickness is the length of a whole brick.

whole-circle bearing (*Surv*) The horizontal angle measured from 0° to 360° clockwise, from true north to a given survey line.

whole-coiled winding (*ElecEng*) An armature winding for an alternator having one armature coil per pole, the two sides of the coil being separated by a distance which is equal to the pole pitch.

wholemeal flour (*FoodSci*) Flour containing all the product derived from milling wheat after cleaning.

whole number (*MathSci*) See NATURAL NUMBER, INTEGER.

whole plate (*ImageTech*) A standard format for still photography of dimensions $8\frac{1}{2} \times 6\frac{1}{2}$ in.

whooping cough (*Med*) See PERTUSSIS.

whorl (*BioSci*) (1) A group of three or more plant structures arising at the same level on a stem and forming a ring around it. (2) A ring of floral organs round the receptacle of a flower. (3) A single turn of a spirally coiled shell or other spiral structure.

WI (*Eng*) Abbrev for WROUGHT-IRON.

wick (*Textiles*) A yarn, group of yarns, or a narrow woven fabric or braid with good capillary properties. Used particularly in candles and oil lamps. See CANDLEWICK.

Widal reaction (*BioSci*) A bacterial agglutination test used in the diagnosis of enteric fevers.

wide-angle lens (*ImageTech*) A camera lens of comparatively short focal length having a wide angle of view, of the order of 80°–100° for still cameras and 50°–70° for cinematography.

wide-angle X-ray scattering (*Phys*) A method of structure determination in materials. Abbrev *WAXS*. See X-RAY CRYSTALLOGRAPHY.

wide area information server (*ICT*) Software that enables large databases to be created and searched remotely by users over the Internet. Abbrev *WAIS*. See panel on INTERNET.

wide area network (*ICT*) A NETWORK whose devices may be distributed over large distances and the links are formed using telephone lines, radio and microwave links or satellites. Abbrev *WAN*. Cf LOCAL AREA NETWORK, METROPOLITAN AREA NETWORK, NETWORK.

wide-band amplifier (*ICT*) One that amplifies over a wide range of frequencies, normally with low gain.

wide-cut fuel (*Aero*) Low-octane petrol (gasoline) obtained from wide-cut distillation used in turbojets in order to conserve aviation kerosine. Abbrev *avtag*.

wide-screen (*ImageTech*) General term for systems of motion picture and video presentation having pictures of ASPECT RATIO 1·65:1 or greater.

Widmanstätten structure (*Eng*) A mesh-like distribution of a precipitating phase in a solid-state transformation which occurs along preferred crystal planes. Usually produced by rapid cooling and when the transforming phase has a large grain size. Originally observed in iron meteorites. Also *basket-weave structure*.

widow (*Print*) A single word or part of a word occupying the last line of a paragraph; to be avoided, particularly as the first line of a page. See CLUB LINE.

width (*Phys*) The spread of uncertainty in a specified energy level, arising as a result of the Heisenberg UNCERTAINTY PRINCIPLE, proportional to the instability of the state concerned.

Wiedemann effect (*ElecEng*) The tendency to twist in a rod carrying a current when subject to a magnetizing field.

Wiedemann–Franz law (*Phys*) A law stating that the ratio of thermal to electrical conductivity of any metal equals the absolute temperature multiplied by

$$\frac{\pi^2}{3}\left(\frac{k}{e}\right)^2 = 2 \cdot 5 \times 10^{-8} \, V^2 \, K^{-2}$$

where k is BOLTZMANN'S CONSTANT and e is the electronic charge.

Wien bridge (*ElecEng*) A four-arm ac bridge circuit used for measurement of capacitance, inductance and power factor.

Wien-bridge oscillator (*ElecEng*) One in which positive feedback is obtained from a Wien bridge, the variable frequency being determined by a resistance in an arm of the bridge.

Wien effect (*ElecEng*) The increase in the conductivity of an electrolyte observed with very high voltage gradients.

Wien's laws (*Phys*) Laws relating to radiation from a black body. (1) The *displacement law* states that $\lambda_m T = C$, where λ_m is the wavelength at which the maximum amount of radiation occurs, T is the temperature (in kelvin) and C is a constant equal to $0 \cdot 0029$ mK. (2) The *emissive power* E_λ within the maximum intensity wavelength interval $d\lambda$ is given by $E_\lambda = CT_5 d\lambda$, where $C = 1 \cdot 288 \times 10^{-5} \, W \, m^{-2} \, K^{-5}$ and T is temperature (in kelvin). (3) The *distribution law* gives the emissive power (dE) in the interval $d\lambda$: $dE = A\lambda^{-5} \exp(-B/\lambda T) d\lambda$, where $A = 4 \cdot 992$ mJ and $B = 0 \cdot 01444$ mK.

wi-fi (*ICT*) Abbrev for *wireless fidelity*, a term used casually for any wireless-based networking system, although strictly refers to any type of 802.11 network, whether 802.11b, 802.11a, dual-band, etc.

Wigner effect (*NucEng*) Changes in physical properties of graphite resulting from the displacement of lattice atoms by high-energy neutrons and other energetic particles in a reactor. It results in the building up of stored energy (Wigner energy) in the change of crystal lattice dimensions and hence in the change of overall bulk size.

Wigner energy (*NucEng*) Energy stored within a crystalline substance, due to the WIGNER EFFECT.

Wigner force (*Phys*) Ordinary (non-exchange) short-range force between nucleons.

Wigner nuclides (*Phys*) Isobars of odd mass number in which the atomic number and neutron number differ by one.

wig-wag (*CivEng*) A level-crossing signal which gives its indication, with or without a red light, by swinging about a fixed axis.

wiki (*ICT*) A type of computer software that enables any user of a website to edit and restructure its contents. Term coined by Ward Cunningham (born 1949), creator of the software.

wildcard (*ICT*) A single character that may be used to represent any other character or a group of characters. Often * is used to represent a group of characters whilst ? or # is used to represent a single character. Used eg to help search for particular file names or data where there is doubt as to the exact name or data.

wildcatting (*MinExt*) Prospecting at random, particularly in speculative boring for oil.

wild shooting (*ImageTech*) Recording pictures without synchronized sound.

wild silk (*Textiles*) See TUSSAH.

wild track (*ImageTech*) A sound track recorded independently of any associated picture.

wild type (*BioSci*) The normal phenotype with respect to a specified gene locus; usually symbolized by +.

Wilfley table (*MinExt*) A flat rectangular desk, adjustable about the long axis for tilt, is given rapid but gentle throwing motion along this horizontal axis, while classified sands are washed across- and down-tilt, against restraint imposed by horizontal riffles. Heavy minerals work across and progressively lighter ones gravitate down to separate discharge zones. Also *concentrating table, shaking table.*

willemite (*Min*) Orthosilicate of zinc, Zn_2SiO_4, occurring as massive, granular or in trigonal prismatic crystals; white when pure but commonly red, brown, or green through manganese or iron in small quantities. In New Jersey and elsewhere it occurs in sufficient quantity to be mined as an ore of zinc. Noteworthy as exhibiting an intense bright-yellow fluorescence in ultraviolet light.

Williams–Landel–Ferry equation (*Chem*) An equation which relates the shift factor (a_T) to glass-transition temperature of polymer, T_g. It is

$$\log\left(a_T\right) = -\frac{c_1(T - T_g)}{(c_2 + T - T_g)}$$

where T is the temperature, c_1 and c_2 are constants. Abbrev *WLF equation.* See VISCOELASTICITY.

Williot diagram (*Eng*) A graphical construction for finding the deflection of a given point in a structural framework under load.

willow (*For*) Timber from hardwood trees of widespread northern hemisphere species (*Salix*) with pinkish-white heartwood and white sapwood, straight-grained and fine-textured.

Wilson chamber (*NucEng*) CLOUD CHAMBER of expansion type.

Wilson effect (*ElecEng*) Production of electric polarization when dielectric material is moved through region of magnetic field, due to Faraday-induced emf in the dielectric.

Wilson's disease (*Med*) A rare disease in which excessive amounts of copper are deposited in the brain and liver.

wilt (*BioSci*) A type of plant disease characterized by wilting as an early symptom and usually caused by the infection of the vascular system by a fungus or bacterium, eg Dutch elm disease, caused by the fungus *Ceratocystis ulmi.*

wilting (*BioSci*) The loss of stiffness due to shortage of water and the loss of turgor by the cells. See PERMANENT WILTING POINT.

WIMP (*Astron*) Abbrev for *weakly interacting massive particle.* Such subatomic particles have been proposed as a possible form of the dark matter which is thought to make up more than 90% of the universe, their weakly interacting nature accounting for the fact that they have not yet been detected.

WIMP (*ICT*) An interface for microcomputers that is designed to be 'user-friendly'; it uses WINDOWS, ICONS, menus and pointers for showing and operating the normal system commands. See fig. at WINDOWS.

WIMS (*NucEng*) Abbrev for WINFRITH IMPROVED MULTI-GROUP SCHEME.

Wimshurst machine (*ElecEng*) Early type of electrostatic induction generator.

wince (*Textiles*) See WINCH.

wincey (*Textiles*) A light woven flannel made from mixed yarns containing wool.

winceyette (*Textiles*) A plain or simple twill cotton cloth of light weight, raised slightly on both sides; used for pyjamas, nightgowns or underwear. May be colour-woven or piece-dyed.

winch (*Eng*) Mechanism in which rotation of a drum causes a rope, cable or chain to hoist or lower a load, as in a crane.

winch (*Textiles*) A machine that draws long lengths of fabric through a dyebath and winds the wet fabric on a reel or drum above the liquid. Also *wince.*

winchester (*Glass*) A narrow- or wide-mouthed cylindrical bottle used for the transportation of liquids, capacity 2·5 litres.

winchester drive (*ICT*) Early name for a disk drive suitable for personal computer systems.

winch launch (*Aero*) Launching a glider by towing it into the air by a cable on a motorized winch, or pulling through a pulley by a car.

wind (*ElecEng*) A stream of air arising at any sharply pointed electric conductor charged to a high potential.

wind (*EnvSci*) The horizontal movement of air over the Earth's surface. The direction is that from which the wind blows. The speed may be given in metres per second, miles per hour or knots.

windage (*Eng*) In any machine, the energy dissipated in overcoming air resistance to motion.

wind axes (*Aero*) Co-ordinate axes, having their origin within the aircraft, and directionally orientated by the relative airflow.

wind bag (*Acous*) A bag of thin cloth or silk placed over a microphone when the latter is used out of doors, to eliminate hissing noises due to wind.

wind chill factor (*EnvSci*) Assessment of the power of a cold wind to chill objects and esp living beings which combines the wind speed with the temperature and RELATIVE HUMIDITY of the air.

wind dispersal (*BioSci*) The dispersal of spores, seed and fruits by the wind.

wind-driven generator (*ElecEng*) A generator driven by a prime mover of the windmill type, or directly (in the case of aircraft) by an airscrew carried on the generator shaft.

winder (*Build*) A step, generally triangular in plan, used at a change in direction of the stair. Also *wheeling step.*

wind frost (*EnvSci*) An air frost where the cold air has been propelled by the wind.

windgall (*Vet*) Distension of the joint capsule, articular windgall, or tendon sheath, tendinous windgall, of the fetlock joint of the horse.

winding (*ElecEng*) The system of insulated conductors forming the current-carrying element of an electric machine or static transformer.

winding (*Textiles*) Coiling thread on a spindle or bobbin, after spinning or doubling to form a package convenient to handle.

winding coefficient (*ElecEng*) See WINDING FACTOR.

winding diagram (*ElecEng*) A diagram showing in schematic form the arrangement and sequence of an armature winding and its circuit connections.

winding drum (*Eng*) An engine or motor-driven drum onto which a haulage rope is wound, as the wire rope of a mine cage. The drum may be cylindrical or conical, with a plain or a helically grooved surface for the rope.

winding factor (*ElecEng*) A factor which takes account of the difference between the vector and arithmetic sums of the emfs induced in a series of armature coils occupying successive positions round the periphery of the armature. Also *winding coefficient*.

winding gear (*ElecEng*) The mechanical gear associated with an electric winder.

winding pitch (*ElecEng*) The distance, measured as the number of slots, separating an armature coil from its successor in the winding sequence.

winding plant (*ElecEng*) The equipment needed for an electrically driven winding machine.

winding space (*ElecEng*) The cross-sectional space available in an armature slot for the insertion of the insulated conductors.

winding stair (*Build*) A stair formed in a circular, spiral, elliptical or other mathematical plan.

wind load (*Build, Eng*) The force acting on a structure due to the pressure of the wind upon it.

window (*ElecEng*) (1) The winding space of a transformer, ie the cross-sectional spaces between the limbs and yokes of a multicore transformer. (2) Conducting diaphragms in a waveguide which act inductively or capacitively depending on position.

window (*Geol*) A closed outcrop of strata lying beneath a thrust plane and exposed by denudation. The strata above the thrust plane surround the 'window' on all sides.

window (*ICT*) (1) A part of a video display screen that carries a display different from the rest of the screen, eg a text window in a graphics screen. (2) A portion of a file or image currently visible on the screen.

window (*NucEng*) A thin portion of wall or radiation counter through which low-energy particles can penetrate.

window (*Plastics*) A type of mould defect in partly crystalline polymers where a clear zone is left in opaque or translucent material, due to incomplete crystallization.

window (*Radar*) Strips of metallic foil of dimensions calculated to give radar reflections and hence confuse locations derived therefrom. Also *chaff*, *rope*.

window lock (*Build*) See SASH FASTENER.

Windows (*ICT*) TN of a GRAPHICAL USER INTERFACE produced by Microsoft using a WIMP environment. The system runs on PC-COMPATIBLE computers.

wind pollination (*BioSci*) The conveyance of pollen from anthers to stigmas by means of the wind. Also *anemophily*.

wind road (*MinExt*) An underground passage used for ventilation. Also *wind way*.

wind rose (*EnvSci*) A star-shaped diagram showing, for a given location, the relative frequencies of winds from different directions and of different strengths.

windrow (*Agri*) A row of harvested crop left on the field to dry before picking up and transporting to storage or processing.

wind shear (*EnvSci*) Rate of change of the vector wind in a direction (horizontal or vertical) normal to the wind.

wind sock (*Aero*) A truncated conical fabric sleeve, on a 360° free pivot, which indicates local wind direction. Also *wind stocking*.

windsucking (*Vet*) A habit acquired by certain horses of swallowing air when 'cribbing' or gripping an object with the incisor teeth.

wind T (*Aero*) A T-shaped device displayed at airfields to indicate the direction of the surface wind. The leg of the T corresponds to the wind direction, with wind arrowhead at top of T.

wind tunnel (*Aero*) Apparatus for producing a steady airstream past a model for aerodynamic investigations. See panel on AERODYNAMICS.

wind way (*MinExt*) See WIND ROAD.

wine (*FoodSci*) The alcoholic product of the fermentation of grape juice. More generally, the term can be applied to the product of the alcoholic fermentation of any fermentable fruit or vegetable liquor. See VINIFICATION.

Winfrith Improved Multigroup Scheme (*NucEng*) Widely used set of computer codes for predicting the properties of thermal reactors. Abbrev *WIMS*.

Windows A representation of the main elements of a multiple windowing graphical user interface.

wing (*Aero*) The main supporting surface(s) of an aeroplane or glider.

wing (*BioSci*) (1) Any broad flat expansion. (2) An organ used for flight, as the fore-limb in birds and bats, the membranous expansions of the mesothorax and metathorax in insects. (3) A longitudinal flange on a stem or stalk; the downwardly continued lamina of a decurrent leaf. (4) A flattened outgrowth of a seed or fruit aiding in wind dispersal.

wing area (*Aero*) See GROSS WING AREA, NET WING AREA.

wing car (*Aero*) See CAR.

wing compasses (*Build*) A form of QUADRANT DIVIDERS.

wing coverts (*BioSci*) See TECTRICES.

wing fence (*Aero*) A projection extending chordwise along a wing and projecting from its upper surface. It modifies the pressure distribution by preventing a spanwise flow of air which would otherwise cause a breakaway of the flow near the wing tips and lead to tip stalling. Also *boundary layer fence*. Cf VORTEX GENERATORS.

wing loading (*Aero*) The gross weight of an aeroplane or glider divided by its GROSS WING AREA.

wing nut (*Eng*) A nut having radial lugs or wings to enable it to be turned by thumb and fingers. Also *bow nut, butterfly nut, fly nut*.

wingover (*Aero*) A turning manoeuvre in which an aircraft is rolled onto its side and the nose is allowed to fall.

wing rail (*CivEng*) See CHECK RAIL.

wing shafts (*Ships*) The port and starboard propeller shafts of a triple- or quadruple-screw steamship.

wing-tip float (*Aero*) A watertight float which gives stability and buoyancy on the water; placed at the extremities of the wings of a seaplane, flying boat or amphibian.

wing valve (*Eng*) A mitre-faced or conical-seated valve guided by three or four radial vanes or wings fitting inside the circular port.

wing wall (*CivEng*) A lateral wall built on an abutment and serving to retain earth in embankment.

Winkler reagent for oxygen (*Chem*) Quantitative absorption by a solution of alkaline pyrogallol, with formation of a brown colour.

Winner winding (*ElecEng*) A form of winding used in constructing standard resistances of low residual reactance for use at relatively high frequencies.

Winslow's foramen (*BioSci*) A small opening by which the cavity of the bursa omentalis communicates with the rest of the abdominal cavity in mammals.

winter annual (*BioSci*) A plant that completes its life cycle over a few months in the coldest part of the year, surviving the summer as seeds. See ANNUAL.

winter cereals (*Agri*) Cereals that are autumn planted for harvesting in the subsequent late summer.

winter dysentery (*Vet*) Thought to be an intestinal infection with *Campylobacter fetus* but viruses have also been incriminated. Usually a herd problem in cattle, showing as a sudden outbreak of diarrhoea affecting several animals. Also *vibrionic scour, winter scour*.

winter egg (*BioSci*) In some fresh-water animals, a thick-shelled egg laid at the onset of the cold season which does not develop until the following warm season. Cf SUMMER EGG.

wintergreen (*Chem*) The methyl ester of *salicylic* (2-hydroxybenzoic) acid. HOC_6H_4COOMe. Characteristic smell. Widely used in liniments and other medicinal products.

winterization (*FoodSci*) Fractional crystallization of oils by controlled cooling to remove higher-melting-point glycerides and thus prevent or retard cloudiness during subsequent storage at chill temperatures.

winter solstice (*Astron*) See SOLSTICE.

winze (*MinExt*) Internal shaft, usually between two underground levels in plane of lode, used in exploration and subsequent extraction of valuable ore. See fig. at MINING.

wipe (*ImageTech*) Transition effect in film or video where one picture image is replaced by another at a defined edge moving across the frame area.

wipe (*Print*) A defect in which, instead of an even film, the ink, because of unsuitable make-up, forms a ridge at the edge of the type.

wiped joint (*Build*) A joint formed between two lengths of lead pipe, one of which is opened out with a TAMPIN while the other is tapered to fit into the first. Molten solder in a plastic condition is then wiped around the joint with a pad. Also *taft joint*.

wiper (*ElecEng*) See BRUSH.

wiper (*ICT*) In a uniselector or selector, the conducting arm that is rotated over a row of contacts and comes to rest on an outlet.

wipe test (*NucEng*) See SMEAR TEST.

wiping solder (*Build*) Lead-based soft solder used in plumbing, containing up to 35% tin.

wirebar (*Eng*) High-purity copper cast to a tapered ingot suitable for processing into wire.

wire cloth (*Paper*) See MACHINE WIRE.

wire comb (*Build*) A form of SCRATCHER.

wire-cut bricks (*Build*) Bricks made by forcing the clay through a rectangular orifice, and cutting suitable lengths off the resulting bar of clay by pressing wires through the plastic mass, before burning.

wired glass (*Glass*) A form of FLAT GLASS which incorporates a wire mesh into the sheet to act as a reinforcement by holding the fragments together in the event of fracture. This makes it esp useful in fire-resistant glazing. It is not as strong as non-wired glass of the same thickness. Also *Georgian wired glass* after the wire mesh pattern.

wire-drawing (*Eng*) (1) The process of reducing the diameter of rod or wire by pulling it through successively smaller holes. (2) The fall in pressure when a fluid is throttled by passing it through a small orifice or restricted valve opening.

wire gauge (*Eng*) Any system of designating the diameter of wires by means of numbers, which originally stood for the number of successive passes through the die blocks necessary to produce the given diameter. See BIRMINGHAM WIRE GAUGE, BROWN AND SHARPE WIRE GAUGE, STANDARD WIRE GAUGE.

wire guide (*Paper*) A device located on the return wire run on a FOURDRINIER paper-making machine to control and correct the lateral position or movement of the wire.

wire-guided missile (*Aero*) See TELEGUIDED MISSILE.

wireless (*Genrl*) Obsolete term for RADIO.

wireless LAN (*ICT*) A *local area network* (LAN) that uses wireless connections. See 802.11 and WI-FI. BLUETOOTH is another protocol that supports this sort of networking.

wireless local loop (*ICT*) The use of radio or infrared signals for some part of the connection between a CENTRAL OFFICE and the CUSTOMER PREMISES EQUIPMENT. See RADIO IN THE LOOP.

wireline tool (*MinExt*) Small tools or measuring instruments designed to be lowered into a well on a wire line.

wire recorder (*Acous*) Early type of magnetic recorder with recording medium in the form of iron wire.

wire rope (*Eng*) Steel rope made by twisting or laying a number of strands over a central core, the strands themselves being formed by twisting together steel wires. The advantages are flexibility and high tensile strength. A fibre core impregnated with lubricant is incorporated at the centre of ropes which are subjected to repeated flexing. See LANG LAY.

wire stabbing (*Print*) A simple method of binding in which one or more wire staples are passed through the back margin of all the leaves or sections.

wire-stitching (*Print*) The securing of a booklet by means of wire staples. See SADDLE-STITCHING, STABBING.

wirewound resistor (*ElecEng*) One with metallic wire elements.

wiring point (*ElecEng*) A point in an interior wiring installation where an external connection can be made to the electric circuit.

Wood – structure and properties

There is little in the external appearance of a tree to suggest that wood is not only a cellular composite material, but also highly ANISOTROPIC. Essentially, it is a foam with anisotropic cells, whose walls are themselves an anisotropic composite. Wood is itself used to make a range of composites as well as paper. Being based on CELLULOSE, the cell-wall structure of wood shares common features with cotton, but there are significant differences.

Since it is made up of a very large number of cells, wood has a more complex structure than single-celled cotton, and at different levels of scale. The diagram shows some of the main structural features for a typical softwood, which has a simpler structure than a hardwood. This is apparent at scales between a complete tree (approx 50 m) and the unit cell in crystalline cellulose (approx 1 nm).

The bulk of a tree is made up of dead cells, the live and growing region being confined to the CAMBIUM immediately beneath the bark. The cell-wall structure contains about 40–50% by weight of cellulose and 20–25% of hemicelluloses. The remaining material is largely composed of LIGNIN, a complex, tarry, aromatic compound. The hemicelluloses mainly act as the cement for the cellulose, but the bulk of the lignin is found deposited as a final layer around the outside of the cell wall before the cell dies. The high stiffness of wood relative to other cellulosics such as cotton is due to the higher content of hemicelluloses and the presence of lignin.

A horizontal slice through the trunk reveals the annual growth rings, whilst removing a pie-slice from this and examining it at higher magnification shows that the main structural elements are an array of hollow, vertical cells, known as TRACHEIDS, and these are connected by a much smaller number of horizontal, radial cells known as PARENCHYMA, located in the rays. Tracheids have ASPECT RATIOS of the order of 100, and at higher magnification they have a layered structure.

The outermost primary (P) layer is the first one to form as the cell is growing. In this layer the MICROFIBRILS of cellulose are arranged in an irregular network. When a cell has stopped growing, wall thickening starts and the three secondary layers are formed. In the outer (S_1) layer the microfibrils lie almost perpendicularly to the long axis of the cell. The middle layer (S_2) is very much thicker and has the microfibrils running at about 20° to the long axis. Finally, the inner layer (S_3) has the microfibrils oriented almost at right angles to the long axis again. In each of these layers the cellulose microfibrils are embedded in a mainly hemicellulose matrix.

Finally, the individual cells are joined together with an amorphous, lignin-rich material which contains practically no cellulose. The parenchyma in softwood constitute some 5–10% of the cells and are used for food storage. Hardwoods have a similar proportion of parenchyma but very few tracheids. Instead the tracheids' roles of conduction and support is carried out by VESSEL and FIBRE cells, respectively. This is the main difference between hardwoods and softwoods and although some hardwoods, like oak, are undoubtedly hard, others, like balsa, are soft and light.

See panels on C4 PHOTOSYNTHETIC PATHWAY and CELL WALL.

Wirsung's duct (*BioSci*) The ventral or main pancreatic duct of mammals.

wisdom teeth (*Med*) The third molars, which do not usually erupt until adulthood.

wishbone (*Autos*) V-shaped member used in independent suspension systems.

wishful thinking (*Psych*) Thinking in which the individual substitutes the fantasy of the fulfilment of the wish for the actual achievement; a belief that a particular thing will happen, or is so, engendered by desire that it should happen, or be so.

Wiskott–Aldrich syndrome (*Med*) A sex-linked recessive disease of infants characterized by haemorrhagic diathesis, eczema and recurrent infections. A protein normally present in platelet membranes is missing. There is a combined defect of cell-mediated and humoral immunity.

witches' broom (*BioSci*) A dense tuft of twigs formed on a woody plant as a response to infection.

witch of Agnesi (*MathSci*) The curve whose cartesian equation is $x^2 y = 4a^2 (2a - y)$.

withamite (*Min*) A mineral belonging to the epidote group, containing about 1% of manganese oxide.

withdrawal symptoms (*Psych*) Temporary psychological and physiological disturbances resulting from the body's attempt to readjust to the absence of a drug.

withe (*Build*) The partition wall between adjacent flues in a chimney stack. Also *bridging*, *mid-feather*.

witherite (*Min*) Barium carbonate, $BaCO_3$, crystallizing in the orthorhombic system as yellowish or greyish-white complex crystals of hexagonal appearance due to twinning; also massive. Occurs with galena in lead mines. Exploited as an important source of barium.

withers (*Vet*) The region of the horse's back above the shoulders.

witness mark (*Eng*) (1) A remnant of original surface or scribed line, left during machining or hand working to prove that a minimum quantity of material has been removed or an outline accurately preserved. (2) A mark left on the surface of a moulding by tool parts, eg ejector pin marks. Also *witness*.

WLF equation (*Chem*) Abbrev for WILLIAMS–LANDEL–FERRY EQUATION.

Wobbe index (*EnvSci*) A measure of the energy produced by gases being burned under standard conditions; gases that have the same Wobbe index will produce the same thermal output without a change in the relative air–fuel ratio, at the same fuel metering settings. Usually given in MJ m^{-3}.

wobble crank (*Eng*) A short-throw crank in which the pin, machined from and at an angle to the axis of the

Wood – structure and properties *(Cont.)*

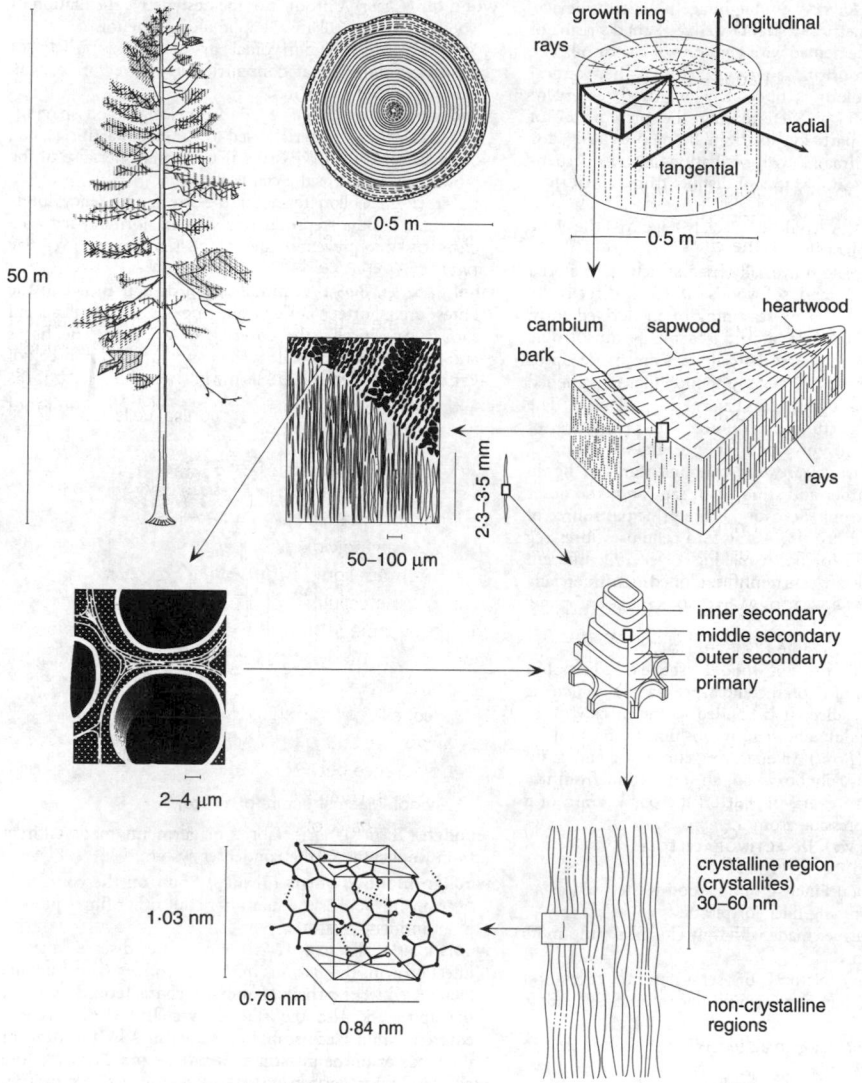

Scale and nature of structural features in a softwood

crankshaft, has been used to give an elliptical motion to a sleeve valve by a short connecting-rod and ball joint.

wobble plate (*Eng*) See SWASH PLATE.

wobble-plate engine (*Eng*) A multicylinder engine in which a wobble-plate or swash-plate mechanism replaces cranks and connecting rods. The cylinders are arranged axially round the shaft, their pistons operating on the wobble plate through sliding blocks. The arrangement is very compact but the mechanical efficiency is usually low.

wobble saw (*Build*) See DRUNKEN SAW.

Wolffian body (*BioSci*) The MESONEPHROS.

Wolffian duct (*BioSci*) The kidney duct of vertebrates. In adult anamniotes (Agnatha, fish and Amphibia) it serves as a kidney duct and a sperm duct in males, while in adult amniotes (reptiles, birds and mammals) whose metanephric kidney has a separate duct, it is present only in males, forming the VAS DEFERENS.

wolf note (*Acous*) Non-harmonic note made by a bow on a violin or cello string at some frequencies.

wolfram (*Min*) See WOLFRAMITE.

wolframite (*Min*) Tungstate of iron and manganese (FeMn)WO$_4$, occurring as brownish-black monoclinic crystals, columnar aggregates, or granular masses. It forms a complete series from ferberite (FeWO$_4$) to hübnerite (MnWO$_4$). An important ore of tungsten. Also *wolfram*.

Wolf–Rayet star (*Astron*) A rare class of stars, with spectra similar to those of novae, strong broad emission lines predominating, indicating violent motion in the stellar atmosphere.

wollastonite (*Min*) A triclinic silicate of calcium, $CaSiO_3$, occurring as a common mineral in metamorphosed limestones and similar assemblages, resulting from the reaction of quartz and calcite. Also *tabular spar*.

Wollaston prism (*Phys*) A double-image polarizing device, made of two geometrically similar wedge-shaped prisms of calcite or quartz, cemented with glycerine or castor oil. The cutting of the two prisms is arranged so that their optical axes cause two coloured, oppositely polarized, emergent beams. It is useful in determining the proportion of polarization in a partially polarized beam providing, for examination, two images with oscillations in two perpendicular directions. Similar to, but distinct from, a ROCHON PRISM.

Wolstonian (*Geol*) A cold to glacial stage of the Late Pleistocene. See QUATERNARY.

wood (*BioSci, For*) The universally used structural material derived from hard- and softwoods, whether directly by machining etc followed by seasoning or in derived form eg blockboard, chipboard, etc. It is a highly anisotropic cellular composite in which the cell walls are made up of 40–50% cellulose and 20–25% hemicelluloses stiffened by the deposit of lignin just before the cells die. The complex macrostructure determines the properties of the individual kinds of wood; the general structure is shown in the figure. Heartwood contains a greater lignin content and is harder and stiffer than sapwood, the outer layers of the tree trunk. Softwood is the principal source of cellulose fibre both for processing into cellulose fibres (eg viscose rayon) and for paper-making. Over 100 different species of wood are currently exploited commercially worldwide. See HEARTWOOD, SAPWOOD, XYLEM and panel on WOOD.

wood alcohol (*Chem*) Same as METHANOL.

wood brick (*Build*) A piece of wood the shape of a brick but larger by the amount of the mortar joints to which eg skirtings may be nailed. It is bonded to the brickwork in the course of building and held in position by friction.

wood engraving (*Print*) An engraving cut on the end-grain of a hardwood, usually boxwood; an impression from it.

woodcut (*Print*) An engraving cut on the plank grain of a softwood; an impression from it.

wooden tongue (*Vet*) See ACTINOBACILLOSIS.

wood fibre (*BioSci*) See XYLEM.

wood flour (*Plastics*) Finely ground wood waste used as an inexpensive reinforcing filler for plastics.

woodfree (*Print*) Paper made with only chemical pulp in its fibre composition.

woodland (*EnvSci*) Natural or semi-natural vegetation containing trees, but not forming a continuous canopy. See FOREST.

wood letters (*Print*) Large type-letters cut in wood, used in some poster work.

wood nog (*Build*) See NOG.

wood opal (*Min*) A form of common opal which has replaced pieces of wood entombed as fossils in sediments, in some cases retaining the original structure.

wood parenchyma (*BioSci*) See XYLEM PARENCHYMA.

wood preservatives (*Build*) Special coatings intended to afford external timbers protection from decay, fungal attack, insects and inclement weather. Produced in eg oil, spirit and water using constituents such as coaltar derivatives, naphthenates and metallic salts.

wood pulp (*Paper*) Pulp produced from wood by any pulping method.

wood ray (*BioSci*) See RAY.

Woodruff key (*Eng*) A key consisting of a segment of a disk, restrained in a shaft key way milled by a cutter of the same radius, and fitting a normal key way in the hub.

Wood's glass (*Glass*) A glass with very low visible and high ultraviolet transmission.

Wood's metal (*Eng*) Fusible alloy based on the quaternary eutectic of lead–bismuth–tin–cadmium system. Contains 50% bismuth, 25% lead, 12·5% each of tin and cadmium and

melts at 69°C. There are other compositions with different melting ranges but all below the boiling point of water.

wood spirit (*Chem*) METHANOL.

wood sugar (*Chem*) XYLOSE.

wood tar (*Chem*) A product of the destructive distillation of wood, containing alkanes, naphthalene, phenols.

wood tin (*Min*) A botryoidal or colloform variety of cassiterite showing a concentric structure of brown, radiating, wood-like fibres.

wood-wool slabs (*Build*) Made from long wood shavings with a cementing material; used for linings, partitions, etc.

woody tissues (*BioSci*) Tissues that are hard because of the presence of lignin in the cell walls.

woofer (*ICT*) Colloq term for the lower-frequency loudspeaker in a sound-reproduction system, normally fed via a CROSSOVER to prevent waste of high-frequency power. See TWEETER.

wool (*Textiles, BioSci*) A modification of hair in which the fibres are shorter, curled, and possess an imbricated surface. Specifically, the covering of a sheep. The fibres are covered with small scales and are composed of KERATIN. Also *fleece wool*. See HAIR.

wool Macrostructure of a fibre.

woollen (*Textiles*) Yarns, fabrics or garments made entirely from wool spun on the condenser system.

woollen blended yarns (*Textiles*) Spun on the condenser system with wool as the main fibre but other fibres present in an intimate mixture.

woolsorter's disease (*Med*) An acute disease due to infection with the *Bacillus anthracis*, conveyed to humans by infected wool or hair of animals; characterized by fever, the appearance on the skin of vesicles which become covered with a black scab, and sometimes by infection of the lungs or of the intestines. See ANTHRAX.

word (*ICT*) A collection of BITS treated as a single unit by the central processor.

word association test (*Psych*) A psychological test in which the subject is presented with a stimulus word and asked to produce the first word that comes to mind; latency to response and the nature of the association word are interpreted as revealing verbal habits, thought processes, personality characteristics and emotional state.

word length (*ICT*) The number of bits in each word of a particular computer. In most computers this is fixed.

word processor (*ICT*) (1) A computer system designed to help in the preparation, editing, printing and sending of textual data. (2) The software for doing the above which is usually implemented on a standalone personal computer with possible network connections. Cf DESKTOP PUBLISHING.

word salad (*Psych*) A schizophrenic speech pattern in which words and phrases are combined in a disorganized fashion, apparently devoid of logic and meaning.

word-wrap (*ICT*) A facility of WORD PROCESSORS to insert a LINE FEED automatically between words. When the text reaches beyond the right-hand end of a line the whole of the last word is moved to the start of the next line. Also *wrap-around*. See SOFT RETURN.

work (*Phys*) One manifestation of energy. The work done by a force is defined as the product of the force and the distance moved by its point of application along the line of action of the force: eg a tensile force does work in increasing the length of a piece of wire; work is done by a gas when it expands against a hydrostatic pressure. As for all forms of energy, the SI unit of work is the JOULE. Symbols w, W. See KILOWATT-HOUR.

work-and-tumble (*Print*) A method of turning printed sheets in which, after printing the first side, the opposite edge must be used as the front lay for the second side.

work-and-turn (*Print*) A process in which one forme or plate is used to print on both sides of the paper which is cut after printing to give two copies of the job or section. Cf SHEET-WORK.

work-and-twist (*Print*) Two impressions are taken on the same side of the paper using the same forme, the paper being turned 180° before taking the second impression. Useful when printing rule formes by letterpress.

work coil (*ElecEng*) Also *heating inductor*.

work electrode (*ElecEng*) See APPLICATOR.

worker (*BioSci*) In social insects, one of a caste of sterile individuals which do all the work of the colony.

work function (*Phys*) The minimum energy that must be supplied to remove an electron so that it can just exist outside a material under vacuum conditions. The energy can be supplied by heating the material (*thermionic* work function) or by illuminating it with radiation of sufficiently high energy (*photoelectric* work function). Also *electron affinity*. See EINSTEIN PHOTOELECTRIC EQUATION.

work hardening (*Eng*) The increase in strength and hardness (ie resistance to plastic deformation) produced by plastic deformation of metals at temperatures below about $0.5T_m$ that results from increasing numbers of dislocations and their entanglement and is accompanied by reduction in ductility. When extensive cold-working is necessary to form a product, it is essential to anneal from time to time to remove the work-hardening effects and permit further deformation. However, controlled work hardening may be used to advantage since it allows substantial increases in yield and tensile strength for a given cross-section, eg in hard drawn steel wire or temper rolled aluminium alloy sheet.

workhead transformer (*ElecEng*) One associated with the workpiece in induction heating when the power supply is at a distance and feeds power though a cable.

working (*ICT*) The technique of routing calls over a telephone system.

working chamber (*CivEng*) The compressed-air chamber at the base of a hollow caisson, being the part in which excavation proceeds. See AIR LOCK.

working edge (*Build*) An edge of a piece of wood trued square with the working face to assist in truing the other surfaces. Also *face edge*.

working face (*Build*) That face of a piece of wood which is first trued and then used as a basis for truing the other surfaces. See FACE MARK.

working flux (*ElecEng*) That part of the total flux produced by the magnetic system of an electric machine which links the armature winding; numerically equal to the difference between the total flux and the leakage flux.

working point (*Glass*) One of the reference temperatures in glass-making. See panel on GLASSES AND GLASS-MAKING.

working standard (*ElecEng*) A standard for everyday use, calibrated against a SECONDARY STANDARD.

working stress (*Eng*) The (safe) working stress is the ultimate strength of a material divided by the applicable factor of safety.

work lead (*Eng*) See BASE BULLION.

work-over (*MinExt*) A term for any operation on an oil well after completion, usually for maintenance.

workpiece (*ElecEng*) The material requiring induction heating by high-frequency electric currents which is placed within a *heating inductor*.

workstation (*ICT*) A desktop computer usually with a large graphical screen, fast microprocessor, running the UNIX operating system and frequently networked with other workstations and MINICOMPUTERS. Once distinct from personal computers but now overlapping in versatility with the more 'high-end' examples of the latter.

work study (*Eng*) A generic term for those techniques, particularly method study and work measurement, which are used in the examination of human work in all its contexts and which lead systematically to the investigation of all the factors which affect the efficiency and economy of the work reviewed, in order to effect improvement.

World Administrative Radio Conference (*ICT*) A conference convened periodically by the International Telecommunications Union and empowered to revise the internationally agreed regulations relating to the use of the radio spectrum and to the positioning of geostationary communications satellites.

World Wide Web (*ICT*) A method of finding text or moving and still images on the Internet using hypertext links. Abbrev *WWW*. See panel on INTERNET.

WORM (*ICT*) Abbrev for *write once, read many*. A type of OPTICAL DISK whose contents can be written once by a computer but that can be read many times afterwards. Cf CD-ROM.

worm (*BioSci*) An imprecise term applied to elongated invertebrates with no appendages, as in flatworm (Platyhelminthes), roundworm, eelworm (Nematoda), earthworm (Lumbricus spp), etc. Also applied to immature forms of some insects, as in mealworm, (Tenebrio, Coleoptera), cutworm (some Lepidoptera), wireworm (Elateridae, Coleoptera), click-beetles and also millipedes (Diplopoda).

worm (*Eng*) A gear of high-reduction-ratio connecting shafts whose axes are at right angles but do not intersect. It consists of a cylindrical core carrying a single- or multi-start helical thread of special form, the worm, meshing in sliding contact with a concave face gear wheel, the worm wheel. Also *worm gear*, *worm wheel*.

worm (*ICT*) A type of computer virus.

worm-and-wheel steering gear (*Autos*) Steering gear in which the steering column carries a worm, in mesh with a worm wheel or sector, attached to the spindle of the drop arm.

worm conveyor (*Eng*) A conveyor in which loose material such as grain, meal, etc, is continuously propelled along a narrow trough by a revolving worm or helix mounted within it. Also *screw conveyor*.

worm gear (*Eng*) Also *worm wheel*. See WORM.

worsted (*Textiles*) Woollen yarns, and fabrics and garments made from them, spun from fibres that have been combed and are fairly parallel.

worsted-spun (*Textiles*) Yarn spun from any staple fibres on the machinery used for making worsted yarns. The fabric made from these yarns is known as worsted-type.

wort (*FoodSci*) A solution of sugars and dextrins which have been enzymically converted from starch. The product of 'mashing' malted barley. The starting material of the production of beer or malt vinegar.

worth (*NucEng*) See REACTIVITY WORTH.

wound rotor (*ElecEng*) An alternative term for SLIP-RING ROTOR.

wound tissue (*BioSci*) In plants, tissue formed in response to wounding, like the vessels that reconnect severed xylem strands. See CALLUS.

wove paper (*Paper*) Paper which does not exhibit a laid design when viewed by transmitted light.

wow (*Acous*) Low-frequency modulation introduced in a sound-reproduction system as a result of speed variation. Similar to, but lower in frequency than, FLUTTER.

W particle (*Phys*) See W BOSON.

wrap-around (*ICT*) See WORD-WRAP.

wrapper (*ICT*) In object-oriented programming, the code within which an object may be ENCAPSULATED.

wrapper plate (*Eng*) (1) In a locomotive boiler, the plate bent round and riveted to the tube plate and back plate, forming the sides and crown of the fire-box. (2) The outer casing of the fire-box.

wrap round (*Print*) See OUTSET.

wrap-round plate (*Print*) A flexible relief printing plate which is clamped round the cylinder.

wreath (*Build*) The part of a continuous hand rail curving in plan around the well-hole of a geometrical stair.

wreathed string (*Build*) The continuous curved outer string around the well-hole of a wooden stair.

wreath filament (*ElecEng*) The usual type of filament in large gas-filled electric lamps; the filament wire is suspended from a horizontal supporting spider.

wrench (*Build*) See PIPE WRENCH.

wrench (*Phys*) A system comprising a force and a couple whose axis is parallel to the force. The force is called the *intensity* of the wrench and the ratio of the moment of the couple to the force of its *pitch*. Any system of forces can be reduced to a wrench. Also *wrench on a screw*.

wringing (*Eng*) A method of temporarily combining several SLIP GAUGES by pressing them together with a slight twisting motion until they adhere, thus ensuring that the combined length equals the sum of the individual lengths.

wrinkle finish (*Build*) A paint with a deliberate pattern of small, fairly uniformly distributed wrinkles. Much used on industrial metal articles to hide surface imperfections of the metal. Also *ripple finish*.

wrinkling (*Eng*) Uneven texture developing on surface during working, eg metallic uranium.

wrist-drop (*Med*) Paralysis of the extensor muscles of the hand and of the fingers.

write (*ICT*) To output data and transfer them to a memory location.

writer's cramp (*Med*) A condition in which writing becomes irregular and difficult or even impossible owing to spasm of the muscles of the hand and forearm; an occupational neurosis: not the result of organic disease.

writing speed (*Electronics*) The speed of deflection of a trace on a phosphor, or rate of registering signals on a charge storage device.

writing speed (*ImageTech*) The speed at which the recording/replay head of a video recorder traverses the surface of the moving magnetic tape.

wrong font (*Print*) The use in error of a character from a type FONT other than that currently being used in composition. Abbrev WF.

wrong lead (*Print*) See LEAD OF THE WEB.

wrong-reading (*Print*) See RIGHT-READING.

Wronskian (*MathSci*) Of *n* functions u_i of an independent variable *x*, the determinant whose *i*,*j*th element is

$$\frac{d^{i-1} u_j}{dx^{i-1}}$$

It is zero when the functions u_i are linearly dependent.

wrought grounds (*Build*) Planed strips of wood used as GROUNDS when they will be incompletely covered by the attached joinery.

wrought-iron (*Eng*) Composite material consisting of stringers of slag in an almost pure ferrite matrix, formerly made by the puddling process. Obsolete.

wryneck (*Med*) See TORTICOLLIS.

wulfenite (*Min*) Molybdate of lead, $PbMoO_4$, occurring as yellow tetragonal crystals in veins with other lead ores.

Würm (*Geol*) The fourth and last stage of the Pleistocene epoch of the Alps. See QUATERNARY.

wurtzite (*Min*) Sulphide of zinc, ZnS, of the same composition as sphalerite, but crystallizing at higher temperatures and in the hexagonal system, in black hemimorphic, pyramidal crystals.

Wurtz synthesis (*Chem*) The reduction of solutions of alkyl halides (in ether) with metallic sodium to yield the corresponding hydrocarbons. If mixtures of different alkyl halides are used, mixtures of hydrocarbons formed by different combinations of the alkyl groups are obtained.

W-VHS (*ImageTech*) Abbrev for *wide-VHS*. TN for an S-VHS variant which records WIDE-SCREEN HIGH DEFINITION signals by dividing the pictures vertically and multiplexing the two halves of each FIELD onto individual tracks. With normal definition signals it is possible to record from two sources simultaneously. (S-VHS/VHS machines with 16:9 SWITCHING can record normal definition wide-screen programmes in ANAMORPHIC form.)

wye (*Build*) A *branch pipe* having only one branch, which is not at right angles to the main run. Also *y-pipe*.

wye level (*Surv*) A type of level whose essential characteristic is the support of the telescope, which is similar to that of the wye theodolite.

wye rectifier (*ElecEng*) Full-wave rectifier system for a three-phase supply

wye theodolite (*Surv*) A form of theodolite differing from the transit in that the telescope is not directly mounted on the trunnion axis but is supported on two Y-shaped forks, in which it may be turned end for end in order to reverse the line of sight.

wyomingite (*Geol*) An alkaline volcanic rock, composed of leucite, phlogopite and diopside.

WYSIWYG (*ICT*) Abbrev for *what you see is what you get*. After editing on the full screen of a VIDEO DISPLAY UNIT, the user can print a replica of what appears on the screen.

WWW (*ICT*) Abbrev for WORLD WIDE WEB. See panel on INTERNET.

X (*Chem*) A general symbol for an electronegative atom or group, esp a halogen.

X (*Phys*) Symbol for REACTANCE.

x, 2x, 3x,... (*BioSci*) Symbols for the number of copies of the haploid chromosome number or basic chromosome set.

x (*Chem*) Symbol for MOLE FRACTION.

xalostocite (*Min*) A pale rose-pink grossular which occurs embedded in white marble at Xalostoc in Mexico. Also *landerite*.

xanth-, xantho- (*Genrl*) Prefixes from Gk *xanthos*, yellow.

xanthan gum (*FoodSci*) A gum formed by the microbial fermentation of the organism *Xanthomonas campestris*. Used as a STABILIZER. It forms a stable viscous sol over a wide range of temperature and pH conditions.

xanthates (*Chem*) The salts of xanthic acid, $CS(OC_2H_5)SH$. Potassium xanthate, obtained by the action of potassium ethoxide on carbon disulphide. CELLULOSE XANTHATE is the basis of the viscose rayon process.

xanthene (*Chem*) *Diphenylene-methane oxide*. Colourless plates, mp 98·5°C.

xanthene dyestuffs (*Chem*) Dyestuffs which may be regarded as derivatives of xanthene containing the pyrone ring. They comprise the pyronines, derivatives of diphenylmethane, and the phthaleins, derivatives of triphenylmethane.

xanthine (*Chem*) *2,6-dihydroxy-purine*. A white amorphous mass which is both basic and acidic, and can be obtained by the action of nitrous acid upon guanine.

xanthine bronchodilators (*Pharmacol*) A group of drugs which can alleviate asthma and chronic obstructive airways disease by bronchodilatation. Aminophylline can be given intravenously and choline theophyllinate as a sustained release oral preparation.

xantho- (*Genrl*) See XANTH-.

xanthochroism (*BioSci*) A condition in which all skin pigments other than golden and yellow ones disappear, as in goldfish.

xanthochromia (*Med*) Any yellowish discoloration, esp of the cerebrospinal fluid.

xanthoma (*Med*) Yellow irregular swellings composed of fibrous tissue and of cells containing cholesterol ester, occurring on the skin (eg in diabetes) or on the sheaths of tendons, or in any tissue of the body (*xanthoma multiplex*).

xanthophore (*BioSci*) A cell occurring in the integument and containing a yellow pigment, as in goldfish. Also *guanophore, ochrophore*.

Xanthophyceae (*BioSci*) The yellow-green algae, a class of eukaryotic algae in the division Heterokontophyta. They lack fucoxanthin and may be naked or walled. There are flagellated and amoeboid unicellular, palmelloid, coccoid, dendroid, simply filamentous and siphoneous types; many are remarkably similar to analogous green algae. They are mostly phototrophs, and live mostly in fresh water or in soil.

xanthophyll (*BioSci*) $C_{40}H_{56}O_2$. One of the two yellow pigments present in the normal chlorophyll mixture of green plants; a yellow pigment occurring in some Phytomastigina.

xanthophyllite (*Min*) A brittle mica, crystallizing in the monoclinic system; hydrated calcium, magnesium, aluminium silicate.

xanthophylls (*BioSci*) Yellowish, oxygenated carotenoids acting as minor or, in a few cases, major ACCESSORY PIGMENTS in photosynthesis. Each major algal and plant group has its characteristic set of xanthophylls.

xanthopicrite (*Chem*) See BERBERINE.

xanthopsia (*Med*) The condition in which objects appear yellow to the observer, as in jaundice. Also *yellow vision*.

x-axis (*Aero*) The longitudinal, or roll, axis of an aircraft. Cf AXIS.

x-axis (*MathSci*) Conventionally, the horizontal axis of a coordinate system.

X-band (*Radar, ICT*) Microwave band lying roughly between 8 and 12 GHz; slight discrepancy between US and UK band limits. Widely used for 3 cm radar which is now correctly designated *Cx-band*.

X chromosome (*BioSci*) See SEX DETERMINATION.

X-disease (*Vet*) See BOVINE HYPERKERATOSIS.

Xe (*Chem*) Symbol for XENON.

Xena (*Astron*) Possibly the tenth planet (depending on how a planet is defined), approximately three times further from the Sun than Pluto and about 30% larger (3000 km wide). Originally numbered UB313.

xenia (*BioSci*) The influence of the pollen on the seed through its effect, by double fertilization, on the nature of the endosperm. Cf METAXENIA.

Xenical (*Pharmacol*) TN for ORLISTAT.

Xenix (*ICT*) A version of UNIX that runs on IBM PERSONAL COMPUTERS.

xeno- (*Genrl*) Prefix from Gk *xenos*, strange, foreign.

xenobiotic (*BioSci*) Any substance that does not occur naturally but that will affect living systems.

xenocryst (*Geol*) A single crystal or mineral grain of extraneous origin which has been incorporated by magma during its uprise and which therefore occurs as an inclusion in igneous rocks, usually surrounded by reaction rims and more or less corroded by the magma. Cf XENOLITH.

xenogamy (*BioSci*) Fertilization involving pollen and ovules from flowers on genetically non-identical plants of the same species (ie different genets). See CROSS-FERTILIZATION, CROSS POLLINATION, OUTBREEDING.

xenogeneic (*BioSci*) Grafted tissue that has been derived from a species different from the recipient. Hence *xenograft*.

xenolith (*Geol*) A fragment of rock of extraneous origin which has been incorporated in magma, either intrusive or extrusive, and occurs as an inclusion, often showing definite signs of modification by the magma.

xenomorphic (*Min*) A textural term implying that the minerals in a rock do not show their own characteristic shapes, but are without regular form by reason of mutual interference. See GRANITOID TEXTURE.

xenon (*Chem*) Symbol Xe, at no 54, ram 131·30, mp −140°C, bp −106·9°C, critical temp +16·6°C, density at stp 5·89 g dm⁻³. A zero-valent element, one of the noble gases, present in the atmosphere in the proportion of 1:170 000 000 by volume. Its isotope ^{135}Xe, which has the highest known capture cross-section for thermal neutrons (2.7×10^6 barns), causes XENON POISONING and is therefore of considerable importance in the design of nuclear reactors. Forms several compounds, eg XeF_2, XeF_6, XeO_3.

xenon lamp (*ImageTech*) A compact high-intensity discharge lamp, with the arc operating in a quartz envelope containing xenon gas at high pressure; widely used for motion picture projection.

xenon override (*NucEng*) The provision of a means of compensating for the effect of xenon poisoning after power reduction in a reactor.

xenon poisoning (*NucEng*) In a fission reactor about 5% of the xenon-135 comes from direct fission and the remainder from the decay of iodine-135 with a half-life of 7·2 h. In a reactor running at a steady state, production and depletion balance out, but when power is *reduced* the long half-life of the iodine-135 results in a continued build-up of xenon-135 which reaches its peak after about 10 h. Depending on the reactor this peak may be as much as 1·4 times the initial steady-state value. This results in increased capture of neutrons, a further loss of power and a potentially unstable state in the reactor, as occurred at Chernobyl. See XENON OVERRIDE.

xenotime (*Min*) Yttrium phosphate, YPO_4, often containing small quantities of cerium, erbium and thorium, closely resembling zircon in tetragonal crystal form and general appearance, and occurring in the same types of igneous rock, ie in granites and pegmatites as an accessory mineral. An important source of the rare elements named.

xenotransplantation (*BioSci*) Transplantation of organs derived from an individual of one species into an individual of a different species.

xenotropic (*BioSci*) A retrovirus that cannot replicate in its carrier host but can infect and replicate in the cells of a host of a different species.

Xenoy (*Plastics*) TN for polycarbonate/polybutylene terephthalate blend or alloy used for car bumpers.

xer-, xero- (*Genrl*) Prefixes from Gk *xeros*, dry.

xeric (*BioSci*) Dry conditions in which plant growth may be limited by water shortage.

xero- (*Genrl*) See XER-.

xeroderma (*Med*) See ICHTHYOSIS.

xeroderma pigmentosum (*Med*) A heritable disease of young children in which prolonged exposure to sunlight on the skin causes erythematous patches which later become pigmented, scaly, wart-like and finally cancerous.

xerodermia (*Med*) See ICHTHYOSIS.

xerographic printer (*ICT*) A high-speed printer using xerographic techniques as in many photocopiers.

xerography (*Print*) Non-chemical photographic process in which light discharges a charged dielectric surface. It is the basis of most office photocopiers. An image is projected onto a drum whose surface is a charged dielectric. This is dusted with a dielectric polymer powder, which adheres to the charged areas, rendering the image visible. Permanent images can be obtained by transferring particles to a suitable backing surface (eg paper or plastic) and fixing, usually by heat. Used for document copying, for making lithographic surfaces, usually on paper, for small-offset printing and as a photoresist.

xeromorphic (*BioSci*) Of a feature typical of XEROPHYTES.

xerophthalmia (*Med*) A dry lustreless condition of the conjunctiva with or without keratomalacia, due to deficiency of vitamin A.

xerophyte (*BioSci*) A plant adapted to a dry habitat, where growth may be limited by water shortage.

xeroradiography (*Radiol*) Radiography in which a xerographic, and not photographic, image is produced.

xerosere (*BioSci*) A succession beginning on dry land, as opposed to under water. Cf HYDROSERE.

xerosis (*Med*) See XEROPHTHALMIA.

xerostomia (*Med*) Excessive dryness of the mouth.

xerothermic (*BioSci*) (1) A region or climate that is both dry and hot. (2) Descriptive of an animal or plant living in a xerothermic environment.

X-guide (*ElecEng*) A transmission line with an X-shaped cross-section dielectric, used for guiding surface waves.

x-height (*Print*) The height of the lower case letters from the baseline excluding extenders; specifically the height of the lower case x. See fig. at TYPEFACE.

X-inactivation (*BioSci*) See LYON HYPOTHESIS.

xiphi-, xipho- (*Genrl*) Prefixes from Gk *xiphos*, sword.

xiphisternum (*BioSci*) A posterior element of the sternum, usually cartilaginous.

xipho- (*Genrl*) See XIPHI-.

xiphoid (*BioSci*) Sword-shaped.

X-linkage (*BioSci*) See SEX LINKED.

Xmas tree (*MinExt*) See CHRISTMAS TREE.

XML (*ICT*) Abbrev for *extensible markup language*, a data definition and manipulation language, widely used for the transfer of data between Internet systems. Also *XTML*.

XMS (*ICT*) Abbrev for EXTENDED MEMORY SPECIFICATION.

XNOR (*ICT*) A logic circuit that has two or more inputs and one output, the output signal being 1 if the inputs total an even number, and 0 if the inputs total an odd number.

xonotlite (*Min*) A hydrous calcium silicate, of composition $Ca_6Si_6O_{17}(OH)_2$.

XOR (*ICT*) A logic circuit that has two or more inputs and one output, the output signal being 1 if the inputs total an odd number, and 0 if the inputs total an even number.

XOR (*MathSci*) A logical operator where (*p* XOR *q*) takes the value FALSE if *p* and *q* are the same. If *p* and *q* differ then *p* XOR *q* is TRUE. Also EXOR. See LOGICAL OPERATION.

X-organ (*BioSci*) A neurosecretory organ in the eye-stalks of certain Crustaceans.

XOR gate (*ICT*) A GATE with two input signals. If both are 0, output is 0. If both are 1, output is 0. When incoming signals differ, output is 1. Also *NEQ gate*, *non-equivalence gate*.

XP (*ICT*) Abbrev for EXTREME PROGRAMMING.

X-plates (*Electronics*) A pair of electrodes in a cathode-ray tube to which a horizontal deflecting voltage is applied in accordance with cartesian co-ordinate system.

X-ray (*Phys*) An electromagnetic wave of short wavelength (approx 10^{-3}–10 nm) produced when high-speed electrons strike a solid target. Electrons passing near a nucleus in the target are accelerated and so emit a continuous spectrum of radiation (*bremsstrahlung*) ranging up from a minimum wavelength. In addition, the electrons may eject an electron from an inner shell of a target atom, and the resulting transition of an electron of a higher energy level to this level produces radiation of specific wavelengths. This is the characteristic X-ray spectrum of the target and is specific to the target element. X-rays may be detected photographically or by a counting device. They penetrate matter which is opaque to light; this makes X-rays a valuable tool for medical investigations. See appendix on Electromagnetic spectrum. See COMPTON EFFECT, K-CAPTURE, L-CAPTURE, MOSELEY'S LAW, SYNCHROTRON RADIATION.

X-ray astronomy (*Astron*) The study of the emission of astronomical objects at X-ray wavelengths. Observations are made at high altitude (above approx 150 km) as the Earth's atmosphere is opaque to X-rays.

X-ray crystallography (*Crystal, Phys*) The study of crystalline structures using diffraction of X-rays. The *Debye and Scherrer method* uses diffraction of monochromatic rays from a powdered crystalline sample (giving a range of angles of incidence with respect to crystallographic axes). The arrangement of atoms in space can be inferred from the distribution of intensity in the diffraction pattern. The unit cell determines the geometry of diffraction maxima while the distribution of atoms within the unit cell controls the intensity of each reflection. The *Laue method* examines polychromatic diffraction from a single crystal. See BRAGG EQUATION.

X-ray diffractometer (*Crystal*) An instrument containing a radiation detector used to record the X-ray diffraction patterns of crystals, powders or molecules. See X-RAY CRYSTALLOGRAPHY.

X-ray fluorescence spectrometry (*Chem*) A method of chemical analysis in which the sample is bombarded by very hard X-rays or gamma rays, and secondary radiations, characteristic of the elements present, are studied spectroscopically. Abbrev *XRF*. Also *X-ray spectrography*. See ENERGY DISPERSIVE ANALYSIS OF X-RAYS (EDAX).

X-ray focal spot (*Electronics*) That small area of the target (anode) of an X-ray tube on which the electron beam is incident, and from which emitted X-rays emerge. High-power tubes frequently have a line focus to minimize localization of the heat dissipated at the anode.

X-ray laser (*Phys*) A laser with an output in the X-ray region of the spectrum. Using highly ionized selenium plasma as the lasing medium, laser output at 20·6 and 20·9 nm has been recently reported.

X-ray microanalysis (*BioSci*) See ELECTRON MICROPROBE.

X-ray microscope (*Phys*) A microscope using soft X-rays and a Fresnel ZONE PLATE as the focusing device; a resolution of 200 nm has been obtained using a scanning technique. The zone plate is made using a scanning transmission electron microscope, and synchrotron radiation is used as the source of X-rays. See panel on ELECTRON MICROSCOPE.

X-ray photon (*Phys*) A quantum of X-ray energy given by hv, where v is the frequency and h is PLANCK'S CONSTANT.

X-ray protective glass (*Glass*) Glass containing a high percentage of lead and sometimes also barium, with a high degree of opacity to X-rays. See LEAD GLASS.

X-ray radiography (*Radiol*) The use of photographic film, often with an IMAGE INTENSIFIER screen, to record the intensities of X-rays which have passed through tissue in medicine or eg some mechanical structure in engineering. The X-rays are emitted by essentially a point source and the image is the shadow formed by the specimen.

X-ray source (*Astron*) One of several sources of cosmic X-rays: (1) the solar CORONA; (2) interacting BINARY STARS, in which one member is a black hole or NEUTRON STAR; (3) SUPERNOVA remnants, such as the CRAB NEBULA; (4) some RADIO GALAXIES, such as CYGNUS A, some SEYFERT GALAXIES and some quasars. See panels on BLACK HOLE and QUASAR.

X-ray spectrography (*Chem*) See X-RAY FLUORESCENCE SPECTROMETRY.

X-ray spectrometer (*Crystal*) The name originally used for the *X-ray diffractometer*, but now abandoned in order to avoid confusion with *X-ray fluorescent spectrometry*.

X-ray telescope (*Astron*) An instrument deployed at high altitude by balloon, rocket or spacecraft to detect and record cosmic X-rays from space. To focus the X-rays by mirrors, grazing incidence techniques are used. X-ray detectors include gas counters and scintillation counters. See panel on ASTRONOMICAL TELESCOPE.

X-ray therapy (*Radiol*) The use of X-rays for medical treatment.

X-ray transformer (*Radiol*) A special type of high-voltage transformer for use with X-ray tubes.

X-ray tube (*Electronics*) The vacuum tube in which X-rays are produced by a cathode-ray beam incident on an anode (or anti-cathode). Such tubes may be sealed high-vacuum or continuously pumped. See COOLIDGE TUBE.

X-ray unit (*Radiol*) A unit in which X-ray wavelengths and cell dimensions were published up to about 1948; 1 kX unit = 1·002 02 Å. Symbol kX.

XRF (*Chem*) Abbrev for X-RAY FLUORESCENCE SPECTROGRAPHY.

X-series (*ICT*) The CCITT recommendations for data transmission over DIGITAL DATA NETWORKS.

X-synchronization (*ImageTech*) See FLASH-SYNCHRONIZED.

X-tgd (*Build*) Abbrev for *cross-tongued*.

xyl-, xylo- (*Genrl*) Prefixes from Gk *xylon*, wood.

xylem (*BioSci*) The VASCULAR TISSUE in plants with the prime function of water transport; it consists of tracheids and vessels and associated parenchyma and fibres. Secondary xylem (wood) may also be important for support (tracheids and fibres) and storage (xylem parenchyma). See APOPLAST, COHESION THEORY, CONDUIT, PRIMARY XYLEM, RAY, SECONDARY XYLEM.

xylem parenchyma (*BioSci*) Parenchyma cells chiefly within the secondary xylem, mostly with lignified walls and with living contents in the sap wood. It has storage and defensive functions.

xylenes (*Chem*) $C_6H_4(CH_3)_2$, dimethylbenzenes. There are three isomers which all occur in coaltar but cannot be separated by fractional distillation; an important starting material for polyester synthetic fibres such as Terylene. Commercial preparation termed *xylol*. Used, in microscopy, as a clearing agent, in the preparation of specimens for embedding, and also in the preparation of tissue sections etc for mounting.

xylenol resin (*Chem*) Thermoset resin of the phenolic type. Produced by the condensation of a xylenol with an aldehyde.

xylenols (*Chem*) $(CH_3)_2C_6H_3OH$. Monohydric phenols derived from xylenes: 1,2,3-xylenol (*adj. ortho*), mp 73°C, bp 213°C; 1,3,4-xylenol (*asym. ortho*), mp 65°C, bp 222°C; 1,2,5-xylenol (*para*), mp 75°C, bp 209°C.

xylitol (*FoodSci*) A naturally occurring polyalcohol produced from hemi-cellulose which is obtained from wood or straw. Used as a low-calorie sweetener in sugar-free confectionery.

xylo- (*Genrl*) See XYL-.

xylogenous (*BioSci*) Growing on wood; living on or in wood. Also *xylophilous*.

xylol (*Chem*) See XYLENES.

Xylonite (*Plastics*) TN for a thermoplastic of the nitrocellulose type.

xylophagous (*BioSci*) Wood-eating.

xylophilous (*BioSci*) See XYLOGENOUS.

xylose (*Chem*) A pentose found in many plants and a stereoisomer of arabinose. Also *wood sugar*.

xylotomous (*BioSci*) Wood-boring; wood-cutting.

x–y recorder (*Eng*) One which traces on a chart the relation between two variables, not including time. Time may be introduced by moving the chart linearly with time and controlling one of the variables.

XYY syndrome (*BioSci*) A condition in which the human male has an extra Y chromosome. They are normal males, except for slight stature and sometimes minor behavioural abnormalities.

Y

Y (*Chem*) Symbol for YTTRIUM.

Y (*Phys*) Symbol for ADMITTANCE.

YAC (*BioSci*) Abbrev for YEAST ARTIFICIAL CHROMOSOME.

YAG (*Min*) Abbrev for YTTRIUM ALUMINIUM GARNET.

Yagi antenna (*ICT*) An END-FIRE ARRAY, characterized by directors in front of the normal dipole radiator and rear reflector.

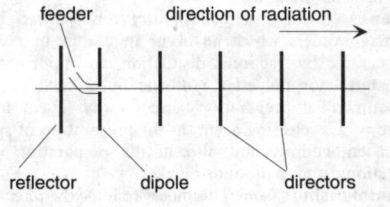

Yagi antenna

Yankee machine (*Paper*) See MACHINE GLAZED.

y-antenna (*ICT*) A delta-matched antenna.

yapp (*Print*) A style of binding with overlapping limp covers, much used for pocket bibles to protect the fore-edges.

yard (*Genrl*) Unit of length in the foot–pound–second system formerly fixed by a line standard (Weights and Measures Act of 1878), redefined in 1963 as 0·9144 m.

yardage (*CivEng*) The volume of excavation in cubic yards.

yard trap (*Build*) See GULLEY TRAP.

yarn (*Textiles*) A thread, ie a long thin material made of fibres or filaments usually twisted together.

yarn count (*Textiles*) See COUNT OF YARN.

Yarrow boiler (*Eng*) A marine water-tube boiler employing an upper steam drum connected by banks of inclined tubes to three lower water drums, between two of which superheating elements are arranged.

yaw (*Aero, Ships*) Angular rotation of an aircraft or other vessel about a vertical axis. The yaw angle is measured between the relative wind and the axis of the vessel.

yaw damper (*Aero*) See DAMPER.

yawing moment (*Aero*) The component about the normal axis of an aircraft due to the relative airflow.

yaw meter (*Aero*) An instrument, usually on experimental aircraft or missiles, which detects changes in the direction of airflow by the pressure changes induced thereby, or by a weather recording vane transmitting to instruments.

yaws (*Med*) A contagious tropical disease occurring in children living in hot humid climates, due to infection with *Treponema pertenue*, and characterized by raspberry-like papules on the skin; as in syphilis, the bones and joints may later become infected. Also *framboesia*, *pian*.

yaw vane (*Aero*) A small aerofoil on a pivoted arm at the end of a long boom or probe, attached to the nose of an aircraft or missile, which measures the angle of the relative airflow and transmits it to recording instruments.

y-axis (*Aero*) The lateral, or pitch, axis of an aircraft. Cf AXIS.

y-axis (*MathSci*) Conventionally, in two-dimensional coordinates, the axis perpendicular to the *x*-AXIS in any type of graph, and in three or more dimensions in the horizontal plane through the *x*-axis.

Yb (*Chem*) Symbol for YTTERBIUM.

Y/B ratio (*Phys*) A term used to classify a type of DICHROMATISM. An observer sees only two colours when examining the solar spectrum, blue and yellow, separated by a white patch. The relative extent of the two colours is the Y/B ratio.

Y/C (*ImageTech*) Separated LUMINANCE SIGNAL (Y) and CHROMINANCE (C) to reduce mutual interference. Also *S-video*. See BETA, HI8, S-VHS, S-TERMINAL.

Y-chromosome (*BioSci*) See SEX DETERMINATION.

Y-class insulation (*ElecEng*) A class of insulating material which is capable of withstanding a temperature of 90°C.

Y-connection (*ElecEng*) See STAR CONNECTION.

year (*Astron*) The civil or calendar year as used in ordinary life, consisting of a whole number of days, 365 in ordinary years, and 366 in leap years, and beginning with 1 January. See ANOMALISTIC YEAR, ECLIPSE YEAR, LEAP YEAR, SIDEREAL YEAR, TROPICAL YEAR.

yearling (*Agri*) A domesticated animal in its second year of life.

yeast (*BioSci, FoodSci*) Various unicellular fungi that reproduce asexually by budding or division, esp the genera *Saccharomyces* and *Schizosaccharomyces*. They are facultative anaerobes, growing in sugar-containing products, that have poor tolerance to heat and can be killed by brief exposure to temperatures of 74°C. With the exception of some osmophilic yeasts, their growth is inhibited by sugar concentrations over 65%. *S. cerevisiae* and *S. ellipsoideus* are the yeasts used for controlled fermentation in baking, brewing and vinification. Other types of yeast cause spoilage in foods. Top yeasts accumulate at the top of the medium during fermentation and are used in brewing traditional ales in the UK, while bottom yeasts do the opposite and are used for brewing lagers.

yeast artificial chromosome (*BioSci*) A cloning VECTOR that allows long segments of DNA to be cloned, and that is useful in chromosome mapping. Abbrev YAC.

yeast extract (*FoodSci*) A by-product of the brewing industry. Spent yeasts are hydrolysed and concentrated to produce a food which is high in protein and B vitamins, and can be used as a flavour enhancer.

yeast genetics (*BioSci*) Yeasts, as simple microbial eukaryotes, are much used in genetic and biochemical studies, esp of fundamental cellular processes. *Saccharomyces cerevisiae* and the fission yeast *Schizosaccharomyces pombe* have been most used. Although they can proliferate by vegetative growth a sexual system is also present allowing classical genetic analysis. A wide range of mutations are known and the genome was one of the first to be fully sequenced.

yeast two-hybrid assay (*BioSci*) See TWO-HYBRID ASSAY.

Yellow Book (*ImageTech*) The technical specifications that define the CD-ROM standard.

yellowcake (*MinExt*) A yellow powder which is the first stage in the production of uranium from ore concentrates, and is often produced at the mine itself. Approx composition U_3O_8.

yellow cells (*BioSci*) In Oligochaeta, yellowish cells forming a layer investing the intestine and playing a role in connection with nitrogenous excretion; chloragen cells.

yellow fever (*Med*) An acute infectious disease caused by a virus, conveyed to humans by the bite of the mosquito *Aëdes aegypti* (*Stegomyia fasciata*); characterized by high fever, acute hepatitis, jaundice, and haemorrhages in the skin and from the stomach and bowels; it occurs in tropical America and W Africa. Also *yellow jack*.

yellow fibres (*BioSci*) Straight, branched, elastic fibres occurring singly in areolar connective tissue. Cf WHITE FIBRES.

yellow fibrocartilage (*BioSci*) A form of FIBROCARTILAGE in which yellow (elastin-rich) fibres predominate.

yellow ground (*MinExt*) Yellowish or buff-coloured, loose clay-rich material formed at the top of a KIMBERLITE pipe by oxidation and alteration of BLUE GROUND.

yellowing (*Textiles*) The yellow discoloration of textile materials, particularly white ones, in use or storage.

yellow pine (*For*) A very soft, even-grained wood from Canada.

yellow quartz (*Min*) See CITRINE.

yellows (*BioSci*) (1) A plant disease in which there is considerable yellowing (chlorosis) of normally green tissue, caused by viruses or mycoplasmata. (2) In veterinary medicine, see CANINE LEPTOSPIROSIS.

yellowses (*Vet*) A form of photosensitization in sheep characterized by dermatitis, subcutaneous oedema affecting mainly the head, and sometimes jaundice (hence yellowses). Associated with liver dysfunction and porphyrins in the blood. Also *head grit*.

yellow spot (*BioSci*) The small area at the centre of the retina in vertebrates at which day vision is most distinct. Also *macula lutea*.

yellow tellurium (*Min*) See SYLVANITE.

yellow vision (*Med*) See XANTHOPSIA.

Yerkes–Dodson law (*Psych*) Generalization that, for performance to be optimal, the amount of arousal required must also be optimized: that too much or too little stimulation will worsen performance.

yew (*For*) A softwood tree (*Taxus baccata*), indigenous to Europe and Asia, whose heartwood is orange gold-brown streaked with mauve and dark brown, straight- or curly grained and medium-textured.

yield (*MinExt*) Tonnage extracted, or ratio of known tonnage, to that recoverable profitably.

yield (*Phys*) (1) Ion pairs produced per quantum absorbed or per ionizing particle. (2) See FISSION YIELD.

yield extension (*Eng*) The amount of plastic strain which occurs at constant or decreasing load in metals which exhibit a pronounced YIELD POINT.

yielding prop (*MinExt*) Support used just behind coal face, which shortens slightly under load, but can be reclaimed and reused.

yield point (*Eng*) The stress at which a substantial amount of plastic deformation takes place under constant or reduced load. This sudden yielding is a characteristic of iron and annealed steels. In other metals, plastic deformation begins gradually and its incidence is indicated by measuring the proof stress, which, however, is frequently called the yield point. See STRENGTH MEASURES.

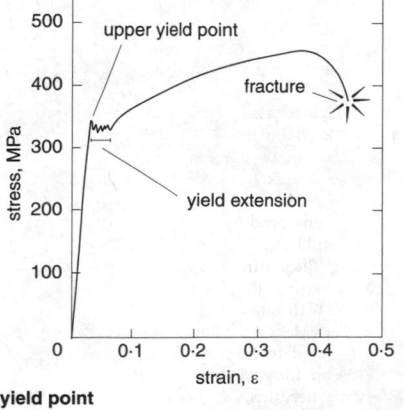

yield point

yield stress (*Eng*) The stress at the onset of plastic deformation determined from the yield point or from a defined amount of plastic strain called the proof stress. See STRENGTH MEASURES.

YIG (*ElecEng*) Abbrev for *yttrium iron garnet*, a material which has a lower acoustic attenuation loss than quartz and which has been considered for use in delay lines.

YIG filter (*ElecEng*) One using a YIG crystal which is tuned by varying the current in a surrounding solenoid, a permanent magnet being used to provide the main field strength.

-yl (*Chem*) A suffix denoting (1) a monovalent organic radical; (2) an electropositive inorganic radical which contains oxygen.

ylem (*Phys*) The basic substance from which it has been suggested that all known elements may have been derived through nucleogenesis (fusion of fundamental particles to form nuclei). It would have a density of 10^{16} kg m^{-3}, and would consist chiefly of neutrons.

Y-level (*Surv*) See WYE LEVEL.

Y-maze (*Psych*) A maze similar to a T-MAZE, but in which the arms are not at right angles to the stem.

Y-network (*ICT*) Same as a T-NETWORK, a three-branch star network.

Yngel trawl (*Genrl*) See YOUNG-FISH NET.

yoderite (*Min*) A silicate of aluminium, iron and magnesium, crystallizing in the monoclinic system.

yoghurt (*FoodSci*) A food produced by inoculating cultures containing *Lactobacillus bulgaricus* and *Streptococcus thermophilus* into warmed homogenized, pasteurized milk. The resulting fermentation causes lactose to be converted to mainly laevorotary lactic acid. Use of the organisms *L. acidophilus* and *L. bifidus* produces mainly dextrorotary lactic acid.

yohimbine (*Chem*) $C_{21}H_{26}N_2O_3$. An alkaloid with a pentacyclic nucleus, obtained from the bark of *Corynanthe johimbe*; colourless needles; mp 234°C; soluble in ethanol and trichloromethane; slightly soluble in ethoxyethane. It is poisonous in excess, allegedly acts as an aphrodisiac, and also exerts a local anaesthetic action.

yoke (*Electronics*) A combination of current coils for deflecting the electron beam in a cathode-ray tube.

yoke (*Phys*) Part or parts of a magnetic circuit not embraced by a current-carrying coil, esp in a generator or motor, or relay.

yoke suspension (*ElecEng*) See BAR SUSPENSION.

yolk (*BioSci*) The nutritive non-living material contained by an ovum.

yolk duct (*BioSci*) The vitelline duct.

yolk epithelium (*BioSci*) The epithelium surrounding the yolk sac.

yolk gland (*BioSci*) See VITELLARIUM.

yolk plug (*BioSci*) A mass of yolk-containing cells which partially occludes the blastopore in some amphibians.

yolk sac (*BioSci*) The yolk-containing sac which is attached to the embryo by the yolk stalk in certain forms.

Yorkshire bond (*Build*) See MONK BOND.

young-fish net (*Genrl*) A large tow-net the mouth of which is kept open by OTTER BOARDS, used for capturing small fishes at the surface or in mid-water. Also *Yngel trawl*.

Young–Helmholtz theory (*Phys*) A theory of colour vision which proposes that the eye contains three systems of colour perception, with maximum response to three primary colours. It is the theory adopted for the realization of colour photography. There is little biochemical support for the theory, but the practice is justified by the physical possibility of matching practically every natural colour by the addition of contributions from three primary colours. See COLOUR.

younging, direction of (*Geol*) The direction in which a series of inclined sedimentary rocks becomes younger. See WAY UP.

Youngman flap (*Aero*) A trailing-edge flap which is extended below the main aerofoil to form a slot before being traversed rearwards to increase the wing area and before being deflected downwards to increase lift and drag coefficient.

Young's equation (*MinExt*) Index of surface wettability, used in flotation research on minerals: $\gamma_S = \gamma_{SL} + \gamma_L \cos\theta$, where θ is contact angle (that between water and air bubble adhering to mineral), γ is free energy per unit area, S and L are solid and liquid phases.

Young's modulus (*Phys*) One of the most important ELASTIC CONSTANTS of materials, it is the ratio (E) of the tensile STRESS (σ) to the tensile STRAIN (ε) in a linear elastic material at loads less than the elastic limit of the material, ie $E = \sigma/\varepsilon$. It is related to the other elastic constants, G (SHEAR MODULUS), v (POISSON'S RATIO) and K (BULK MODULUS) by

$$E = 2G(1+v) + 3K(1-2v) = \frac{9KG}{3K+G}$$

Y-parameter (*Electronics*) The short-circuit admittance parameter of a transistor.

y-pipe (*Build*) See WYE.

Y-plates (*Electronics*) A pair of electrodes to which voltage producing vertical deflection of spot is applied in accordance with cartesian co-ordinate system.

Ypresian (*Geol*) A stage of the Eocene. See TERTIARY.

Y rectifier (*ElecEng*) See WYE RECTIFIER.

yrneh (*Phys*) Unit of reciprocal inductance (HENRY backwards).

Y signal (*ImageTech*) The colour TV signal which conveys the LUMINANCE of the picture.

Y theodolite (*Surv*) See WYE THEODOLITE.

ytterbium (*Chem*) A metallic element, a member of the rare earth group. Oxide Yb_2O_3 white, giving colourless salts. Symbol Yb, at no 70, ram 173·04, mp of metal about 1800°C. Obtained from EUXENITE.

yttrium (*Chem*) A metallic element usually classed with the rare earths because of its chemical resemblance to them. Oxide, Y_2O_3, white, giving colourless salts. Symbol Y, at no 39, ram 88·9059, mp of metal 1250°C.

yttrium aluminium garnet (*Min*) A heavy colourless, isotropic artificial gemstone that is used as a simulant for diamond. Abbrev *YAG*.

yttrocerite (*Min*) A massive, granular or earthy mineral, essentially a cerian fluorite, with the metals of the yttrium and cerium groups, commonly violet-blue in colour, and of rare occurrence.

Yukawa potential (*Phys*) A potential function of the form

$$V = \frac{V_\theta \exp(-kr)}{r}$$

r being distance. Characterizes the meson field surrounding a nucleon. The exponential tail of the Yukawa potential extends with appreciable strength to larger values of r than does that of the coulomb potential.

yu-stone (*Min*) Yu or yu-shih, the Chinese name for the highly prized jade of gemstone quality.

YUV signals (*ImageTech*) In the PAL colour TV system, Y is the LUMINANCE component and U and V are the COLOUR DIFFERENCE SIGNALS, B−Y and R−Y respectively.

Y-voltage (*ElecEng*) See VOLTAGE TO NEUTRAL.

Z

Z (*ICT*) A language for FORMAL SPECIFICATION.

Z (*Min*) The number of formula units per UNIT CELL of a mineral.

Z (*Chem*) Symbol for: (1) ATOMIC NUMBER; (2) number of molecular collisions per second.

Z (*Eng*) Symbol for SECTION MODULUS.

Z (*Phys*) Symbol for IMPEDANCE.

Z- (*Chem*) Prefix denoting 'on the same side' (Ger *zusammen*), and roughly equivalent to *cis-*. See CAHN–INGOLD–PRELOG SYSTEM.

z (*ElecEng*) Symbol for FIGURE OF MERIT.

z (*Chem*) Symbol for the valency of an ion.

ζ (*Chem*) Symbol for electrokinetic potential.

zaffer (*Eng*) Impure cobalt oxide remaining when arsenic and sulphur have been removed by roasting. Also *zaffre*.

Zamboni (*Genrl*) TN of a machine used to clean, resurface and smooth the ice at skating rinks, ice hockey arenas, etc.

zanamivir (*Pharmacol*) An antiviral drug used by inhalation in the treatment of influenza. TN *Relenza*.

Zanclian (*Geol*) A stage in the Pliocene. See TERTIARY.

Zantac TN for RANITIDINE.

z-average molecular mass (*Chem*) Symbol M_z. Molecular mass measure giving greater emphasis to bigger polymer chains, usually obtained by sedimentation analysis in an ultracentrifuge. See MOLECULAR MASS DISTRIBUTION.

zawn (*MinExt*) Cavern, natural or created, in Cornwall.

zax (*Build*) See SAX.

z-axis (*Aero*) The normal, or yaw, axis of an aircraft. Cf AXIS.

z-axis (*MathSci*) Conventionally, the vertical axis in any three-dimensional co-ordinate system, perpendicular to both the *x*- and *y*-axes.

Z-axis modulation (*Electronics*) Variation of intensity of beam, producing varying intensity of brightness of trace.

Z-blade mixer (*Plastics*) A type of mixer for viscous polymer (eg prepolymer) and filler plus additives etc. Often involves two counter-rotating Z-shaped blades, esp for dough moulding compound mixtures for composite products.

Z boson (*Phys*) A particle that mediates the WEAK INTERACTION; symbol Z; mass 94 GeV; charge 0; spin 1; decays to an electron plus positron, or a muon plus antimuon. Predicted by the WEINBERG–SALAM THEORY, it was discovered in 1983 in proton–antiproton collisions at CERN in Geneva.

Z chromosome (*BioSci*) See SEX DETERMINATION.

Z-disc (*BioSci*) A region of the sarcomere of striated muscle into which thin filaments are inserted and where alpha-actinin is located. Also *Z-line*.

Z-DNA (*BioSci*) A form of duplex DNA, in which purines and pyrimidines alternate in a strand and which results in a left-handed helix.

zeatin (*BioSci*) A plant hormone that regulates growth, a type of CYTOKININ.

zebra pattern video level indicator (*ImageTech*) A striped pattern which is superimposed as a warning on overexposed areas of an ELECTRONIC VIEWFINDER picture (it is not recorded). See WHITE-OUT.

Zechstein (*Geol*) The upper series of the Permian. See PALAEOZOIC.

Zeeman effect (*Phys*) The splitting of spectrum lines into a number of components by strong magnetic fields. The field splits the atomic energy levels into several components associated with different quantized orientations of the total magnetic moment with respect to the field. See PASCHEN–BACK EFFECT. Cf STARK EFFECT.

zein (*Plastics*) The maize protein used for experimental production of thermoset plastics by cross-linking with formaldehyde. See CASEIN. Other proteins hitherto unsuccessfully used for plastics include those extracted from ground nuts (Ardil), soyabean, castor bean and blood.

Zeisel's method (*Chem*) A method for the determination of methoxyl and ethoxyl groups in organic compounds, in which the substance is heated with hydriodic acid; the iodoalkane thus formed is passed into an ethanolic solution of silver (I) nitrate, and the resulting silver (I) iodide weighed.

Zeiss–Endter particle size analyser (*PowderTech*) A device for measuring particle size distribution using a photomicrograph, on which the image of the particle is compared directly with a circular spot of light, the area of which can be varied by an iris diaphragm.

zeitgeber (*Psych*) Literally, a *time-giver* that synchronizes various rhythmic behaviours with external events.

Zener breakdown (*Electronics*) Temporary and nondestructive increase of current in diode because of critical field emission of holes and electrons in depletion layer at definite voltage.

Zener diode (*Electronics*) One with a characteristic showing a sharp increase in reverse current at a certain critical voltage; the current can increase indefinitely at this point, unless limited. This makes such diodes suitable for use as a voltage reference. Diodes which show this effect up to about 6 V depend on ZENER BREAKDOWN. At higher voltages, avalanche effects are more prevalent, though the diodes are still called Zener diodes.

Zener effect (*Electronics*) Pronounced and stable curvature in the reverse voltage/current characteristic of a semiconductor point-contact diode; predicted by Zener, and widely used as a reference voltage in stabilizing circuits.

Zener voltage (*Electronics*) That at which ZENER BREAKDOWN occurs in certain types of diode. It denotes the negative voltage at which the reverse current increases very rapidly due to Zener breakdown and avalanche effects, and the voltage at which the diode can provide a reference source. See ZENER DIODE.

zenith (*Astron*) The point on the celestial sphere vertically above the observer's head; one of the two poles of the horizon, the other being the NADIR.

zenith distance (*Astron*) The angular distance of a heavenly body from the zenith, measured as the arc of a vertical great circle; hence the complement of the altitude (*co-altitude*) of the body.

zenith telescope (*Astron*) An instrument similar to the meridian circle, but fitted with an extremely sensitive level and a declination micrometer; used to determine latitude, by observing the difference in zenith distance of two stars whose meridian transit is at a small and equal distance from the zenith, one north and one south.

Zenker's degeneration (*Med*) Hyaline degeneration of STRIATED MUSCLE, occurring eg in the abdominal muscles in typhoid fever.

Zeno's paradoxes (*MathSci*) Four paradoxes intended to show that both the hypothesis that space and time consist of indivisible quanta and that space and time are divisible *ad infinitum* leads to contradiction, ascribed to Zeno (c.450 BC). The paradox of Achilles and the tortoise is best known.

zeolite process (*Chem*) Standard water-softening process, formerly based on using naturally occurring ZEOLITES, but nowadays on synthetic zeolites (insoluble synthetic resins). See ION EXCHANGE, PERMUTIT.

zeolites (*Min*) A group of alumino-silicates of sodium, potassium, calcium and barium, containing very loosely held water, which can be removed by heating and regained by exposure to a moist atmosphere, without destroying the crystal structure. They occur in geodes in igneous rocks, and as authigenic minerals in sediments, and include chabazite, natrolite, mesolite, stilbite, heulandite, harmotome, phillipsite, thomsonite, etc.

zephyr (*Textiles*) Lightweight, plain-weave, fine cotton fabric ornamented with stripes, checks, cords, etc, and used in dresses, blouses and shirts.

Zepp antenna (*ICT*) Horizontal half-wavelength antenna fed from a resonant transmission line. It is connected at one end to one wire of the transmission line and the transmitter or receiver is connected between the two wires, the length of the line being critical.

zero (*MathSci*) (1) The identity element under any addition operation, such that for all elements $x + 0 = x$. (2) A root of an equation. (3) If a function, analytic in a domain D, is expanded in a TAYLOR'S SERIES about any point $z = a$ in D

$$f(z) = \sum_{n=0}^{\infty} b_n (z-a)^n$$

and if all values of b_n up to b_{m-1} vanish but b_m does not vanish, then $f(z)$ is said to have a zero of order m at $z = a$.

zero-address instruction (*ICT*) Operation in computing where the location of the operands is defined by the order code and not specified independently.

zero-beat reception (*ICT*) In SUPPRESSED-CARRIER SYSTEMS, reception in which a locally generated oscillation having the same frequency as the incoming carrier is impressed simultaneously on the detector. See HETERODYNE CONVERSION.

zero-cut crystal (*Electronics*) Quartz crystal cut at such an angle to the axes as to have a zero frequency/temperature coefficient. Used for accurate frequency and time standards.

zero-energy reactor (*NucEng*) See ZERO-POWER REACTOR.

zero error (*Eng*) (1) Residual time delay which has to be compensated in determining readings of range. (2) Error of any instrument when indicating zero, by pointer, angle or display.

zero frequency (*ElecEng*) The component of a complex signal corresponding to the dc level. Abbrev *zf*.

zero fuel weight (*Aero*) See WEIGHT.

zero g (*Space*) The state of weightlessness or FREE FALL.

zero grazing (*Agri*) A farming system in which animals are restricted to prevent grazing and forage is transported to them.

zero level (*ElecEng*) Any voltage, current or power reference level when other levels are expressed in dB relative to this.

Zerol gear (*Eng*) A type of BEVEL GEAR, having curved teeth and a zero helical angle.

zero method (*ElecEng*) A measuring system in which an unknown value can be deduced from other values when a sensitive but not necessarily calibrated instrument indicates zero deflection, as in a Wheatstone bridge or potentiometer. Also *null method*.

zero-order reaction (*Chem*) One in which the rate is independent of the concentration of the reacting species.

zero pause (*ElecEng*) The momentary cessation of ac when passing through a zero value between successive half-cycles on which the action of ac circuit breakers largely depends.

zero phase-sequence component (*ElecEng*) One of three phasors forming a zero phase-sequence system, and one of three components into which any phasor forming part of

an unbalanced three-phase system can be resolved. See PHASE SEQUENCE.

zero-point energy (*Phys*) Total energy at the absolute zero of temperature. The uncertainty principle does not permit a simple harmonic oscillator particle to be at rest exactly at the origin, and by the quantum theory, the ground state still has one half-quantum of energy (ie $hv/2$, where h is PLANCK'S CONSTANT and v is the natural oscillation frequency) and the corresponding kinetic energy.

zero-point entropy (*Phys*) As follows from the third law of THERMODYNAMICS, the entropy of a system in equilibrium at absolute zero must be zero.

zero potential (*ElecEng*) Theoretically, that of a point at infinite distance, used for defining capacitance. Practically, the Earth is taken as being of invariant potential. That of any large mass of metal, eg equipment chassis.

zero power-factor characteristic (*ElecEng*) A curve obtained by plotting the terminal voltage of a synchronous generator delivering full-load current at zero power-factor lagging, against the field excitation.

zero power level (*ICT*) See LEVEL.

zero-power reactor (*NucEng*) An experimental reactor for reactor physics studies with an extremely low neutron flux so that no forced cooling is required and there is insignificant build-up of fission products.

zero stability (*ICT*) Drift in no-signal output level of amplifier or indicator, either with time or with operating conditions, eg mains voltage supply.

zero-sum game (*Psych*) A game in which, over the long term, there is neither net loss nor net gain.

zero suppression (*ICT*) A technique of data processing used to eliminate the storing of non-significant leading zeros.

zero tilling (*Agri*) A crop production system where seeds are drilled directly into the soil without any disturbance such as ploughing or harrowing.

zero-type dynamometer (*ElecEng*) A dynamometer in which the electrical forces are balanced by mechanical forces, in such a manner as to bring the indicating pointer back to zero, before a reading can be taken.

zero-valent (*Chem*) Incapable of combining with other atoms. Also *non-valent*.

zeta function (*MathSci*) See RIEMANN ZETA FUNCTION.

zeta potential (*Chem*) See ELECTROKINETIC POTENTIAL. Abbrev ζ.

Zeta Ursae Majoris (*Astron*) See MIZAR.

zeugopodium (*BioSci*) The second segment of a typical pentadactyl limb, lying between the stylopodium and the autopodium; ante-brachium or crus; forearm or shank.

zf (*ElecEng*) Abbrev for *zero frequency*.

Z-glass (*Glass*) An alkali-resistant zirconia glass, composition (percentage by weight) SiO_2 70, Zr_2O 16, Na_2O 12, Li_2O 1, Al_2O_3 0·2, developed for the reinforcement of PORTLAND CEMENT (where the equilibrium pH is 12·5–13) and concrete.

Z-helix (*BioSci*) A helix winding in the sense of a conventional, right-handed, screw.

zibeline (*Textiles*) Heavy, usually woollen, fabric for overcoats, with a lustrous, satin-type appearance due to a long hairy nap laid in one direction and pressed flat.

zidovudine (*Pharmacol*) See AZT.

Ziegler–Natta catalyst (*Chem*) A catalyst, discovered by Ziegler, which induces specific steric orientation. Used for organic polymerization, notably the low-pressure polyethylene process (for high-density polyethylene). Prepared by the reaction of compounds of strongly electropositive transition metals with organo-metallic compounds, eg titanium trichloride with aluminium alkyls. See panel on POLYMER SYNTHESIS.

ziggurat (*Arch*) A stepped pyramidal structure, the diminishing stages being served by a ramp, or alternatively the stages form a continuous inclined ramp. They were built in Mesopotamia during the Babylonian period, (c.3000–1250 BC), and were used for religious ceremonial purposes, there being a small temple or shrine surmounting the uppermost terrace.

zigzag chain (*Plastics*) Conformation adopted by polyethylene chain in crystalline state; it is the shape of lowest energy in which the steric hindrance is minimal.

zigzag connection (*ElecEng*) A symmetrical three-phase star connection of six windings, situated in pairs on three cores. Each leg of the star consists of two of the windings in series; these windings, being on different cores, have emfs in them differing in phase by 120°. Used in transformers for eliminating harmonics, and in reactors to obtain an artificial neutral.

zigzag leakage (*ElecEng*) Magnetic leakage occurring along the zigzag path between stator and rotor teeth when a stator tooth is opposite to a rotor slot.

zinc (*Chem*) A hard white metallic element with a bluish tinge. Symbol Zn, at no 30, ram 65·37, rel.d. (at 20°C) 7·12, mp 418°C, electrical resistivity $6·0 \times 10^{-8}$ Ω m. Its principal ores are *sphalerite* (zinc blende, ZnS) and *smithsonite* ($ZnCO_3$) but there are many oxy-salts. Because of its good resistance to atmospheric corrosion, zinc is used for protecting steel (see GALVANIZED IRON, SHERADIZING, SPELTER). It is also used in the form of sheet and as a constituent in alloys (see ZINC ALLOYS). Used as an electrode in a Daniell cell and in dry batteries. It is important nutritionally, trace amounts being present in many foods.

zinc alkyls (*Chem*) Organo-metallic compounds such as zinc dimethyl and zinc diethyl prepared by treating the alkyl iodides with zinc–copper couple. Used occasionally as GRIGNARD REAGENTS, and to replace the halogen attached to a tertiary carbon atom by an alkyl group.

zinc alloys (*Eng*) Alloys based on the crystallographic form of zinc (HCP). The most important are those used for die castings, which contain essentially 3–4% aluminium, sometimes with copper 0–3·5% and magnesium 0·02–0·1%. Zinc is also the principal alloying ingredient of brasses and is used in lesser amounts in heat-treatable aluminium alloys, but the amounts do not merit them being classed as zinc alloys.

zincate (*Chem*) The anion ZnO_2^- or $Zn(OH)_4^-$.

zinc blende (*Min*) A much used name for SPHALERITE, the common sulphide of zinc.

zinc bloom (*Min*) See HYDROZINCITE.

zinc chromate primer (*Build*) A yellow coloured primer based on zinc, used on iron and steel because of its rust-inhibiting properties. The recommended primer on aluminium surfaces. Hard drying and lead free.

zinc dust (*MinExt*) Finely divided powder produced either by condensation of zinc vapour or by atomization of molten zinc. Once widely used to precipitate gold and silver from pregnant solution in the cyanidation of gold ore.

zinc ferrite (*Min*) Non-magnetic ferrite, having normal as opposed to inverse SPINEL structure.

zinc finger (*BioSci*) A loop of amino acids in a DNA-binding protein that directly co-ordinates with a zinc atom, and intercalates with the DNA helix.

zincite (*Min*) Oxide of zinc, crystallizing in the hexagonal system and exhibiting polar symmetry; occurring rarely as crystals, usually as deep-red masses; an important ore of zinc. Also *red oxide of zinc*.

zinckenite (*Min*) A steel-gray mineral, essentially sulphide of lead and antimony, $PbSb_2S_4$, occurring as columnar hexagonal crystals, sometimes exceptionally thin, forming fibrous masses.

zinco (*Print*) A line block executed in zinc.

zinc oxide (*Pharmacol*) Oxide of zinc used, for its astringent and soothing qualities, as a constituent of creams, baby ointments, etc.

zinc phosphate (*Build*) Non-toxic primer based on metallic zinc. Excellent properties of adhesion, rust inhibition. Light grey in colour, suitable for application to iron and steel. Often used as a substitute for red lead.

zinc protector (*Ships*) Introduced originally for the CATHODIC PROTECTION of copper sheathing on wooden ships. Now used in steel ships near bronze propellers.

zinc-rich paint (*Build*) A paint containing an extremely high proportion of metallic zinc dust in the dry film (about 95% by weight), applied to iron and steel as an anticorrosive primer. It may be regarded as a less durable form of COLD GALVANIZING.

zinc spinel (*Min*) See GAHNITE.

zinc telluride (*Eng*) A semiconductor capable of high-temperature operation (up to about 750°C) without excessive intrinsic conductivity.

zinnwaldite (*Min*) A mica related in composition to lepidolite (ie containing lithium and potassium) but including iron as an essential constituent; occurring in association with tinstones ores at Zinnwald in the Erzgebirge, in Cornwall, and elsewhere.

zip (*ICT*) To compress a file into a standard format from which it can be unzipped.

Zip disk (*ICT*) TN of a floppy disk with a very high capacity on which data are stored in compressed form using an independent specialized hard drive.

zircaloy (*NucEng*) TN for an alloy of zirconium with small amounts of tin, iron and chromium, used to clad fuel elements in water reactors. There are two kinds commonly used, zircaloy-2 in boiling-water reactors and zircaloy-4 in pressurized-water reactors. The latter has less nickel in the alloy and is significantly less brittle after irradiation. Each contains 98·2% zirconium and has neutron capture cross-sections of 193 millibarns. The fuel cladding is made by compressing and sintering the powdered alloy in the absence of air which would otherwise cause spontaneous combustion.

zircon (*Min*) A tetragonal accessory mineral widely distributed in igneous, sedimentary and metamorphic rocks. It varies in colour from brown to green, blue, red, golden yellow, while colourless zircons make particularly brilliant stones when cut and polished. In composition, it is essentially silicate of zirconium, but often contains yttrium and thorium. A small amount of the rare element hafnium is present.

zirconate (IV) (*Chem*) The anion ZrO_3^-.

zirconia (*Chem*) *Zirconium (IV) oxide*, ZrO_2, used as an opacifier in vitreous enamels, as a pigment and as a refractory.

zirconium (*Chem*) A metallic element, symbol Zr, at no 40, ram 91·22, rel.d. 4·15, mp 2130°C. The principal ores are *zircon* ($ZrSiO_4$), which is a very common accessory mineral of igneous rocks and concentrated in beach sands, and *baddeleyite* (ZrO_2). When purified from hafnium, its low neutron absorption and its retention of mechanical properties at high temperature make it useful for the construction of nuclear reactors. Also used as a refractory, as a lining for jet engines, and as a getter in the manufacture of vacuum tubes. The zirconates are finding application as acoustic transducer materials. Tritium adsorbed in zirconium is a possible target in accelerator neutron sources.

zirconium lamp (*Phys*) A lamp having a zirconium oxide cathode in an argon-filled bulb. It provides a high-intensity point source with only a small emission of the longer visible wavelengths.

Z-line (*BioSci*) See Z-DISC.

Z marker beacon (*Aero*) A form of marker beacon radiating a narrow conical beam along the vertical axis of the cone of silence of a radio range.

ZMC (*Eng, Plastics*) TN of a method of injection moulding long-fibre reinforced polymers. High-locking-force machines are used to produce external car-body parts.

z-modulation (*ImageTech*) The variations in intensity in the electron beam of a cathode-ray tube which form the display or picture on a sweep or raster.

Zn (*Chem*) Symbol for ZINC.

Zodiac (*Astron*) The belt of stars, about 18° wide, through which the ecliptic passes centrally. The Zodiac forms the background of the motions of the Sun, Moon and planets.

zodiacal light (*Astron*) A faint illumination of the sky, lenticular in form and elongated in the direction of the ecliptic on either side of the Sun, fading away at about 90°

from it; best seen after sunset or before sunrise in the tropics, where the ecliptic is steeply inclined to the horizon; it is caused by small particles reflecting sunlight, and appears to be an extension of the solar corona to a distance well beyond the Earth's orbit.

zoidiophilous (*BioSci*) Pollinated by animals.

zoisite (*Min*) Hydrated alumino-silicate of calcium crystallizing in the orthorhombic system and occurring chiefly in metamorphic schists; also a constituent of so-called saussurite. Clinozoisite has the same composition, but crystallizes in the monoclinic system.

zona (*BioSci*) An area, patch, strip or band; a zone. Adjs *zonal, zonary, zonate*.

zona granulosa (*BioSci*) The mass of membrana granulosa cells of the Graafian follicle around the ovum. Also *cumulus oophorus, discus proligerus*.

zonal index (*EnvSci*) Numerical index measuring the strength of the westerly zonal flow in middle latitudes, eg between 35° and 65°. A common type is the mean pressure or GEOPOTENTIAL HEIGHT difference between latitude circles. Indices may be defined for various levels in the atmosphere.

zonal index (*Geol*) See ZONE.

zona pellucida (*BioSci*) A thick transparent membrane surrounding the fully formed ovum in a Graafian follicle.

zona radiata (*BioSci*) The envelope of the mammalian egg outside the vitelline membrane.

zonary placentation (*BioSci*) The condition in which the villi are on a partial or complete girdle around the embryo, as in Carnivora and Proboscidea.

zonation (*BioSci*) The occurrence, in an area, of distinct bands of vegetation each with its own characteristic dominant and other species, as the seaweeds on a shore or the vegetation on a mountain side. Cf SUCCESSION.

zone (*Geol*) A stratigraphical unit with recognizable characteristics. The term has attracted many confusing definitions and is perhaps best used with an appropriate qualifier, eg *Dibunophyllum* zone, marine zone, ash zone. Also *zonal index*.

zone levelling (*Electronics*) An analogous process to ZONE REFINING carried out during the processing of semiconductors in order to distribute impurities evenly through the sample.

zone melting (*Eng*) Localized melting of a narrow band or portion of, usually, a column of metal by an induction coil which is moved along the axis. It is possible with certain alloys to arrange the rate of the passage of the molten zone so that the impurities concentrate in the liquid, allowing considerable purification of the remainder.

zone of audibility (*Acous*) The hearing of explosions at great distances from the source, although, nearer, there is a ZONE OF SILENCE.

zone of avoidance (*Astron*) The region (corresponding roughly to the plane of the Milky Way) in which no galaxies are observed owing to the presence of obscuring clouds of dust in the galactic plane.

zone of cementation (*Geol*) That 'shell' of the Earth's crust lying immediately below the zone of weathering, within which loose sediments are cemented by the addition of such minerals as calcite, introduced by percolating meteoric waters.

zone of silence (*Acous*) Local region where sound or electromagnetic waves from a given source cannot be received at a useful intensity. Also *shadow zone*.

zone of weathering (*Geol*) An 'earth shell' comprising the exposed surface and that part which, through porosity, fracturing and jointing, is subject to the destructive action of the atmosphere, rain and frost. Soil develops in this zone.

zone plate (*Phys*) A transparent plate divided into a series of zones by circles whose radii are in the ratio $\sqrt{1}:\sqrt{2}:\sqrt{3}:\sqrt{4}\ldots$, the alternate zones being blacked. If a plane wave is incident normally on the plate, a maximum of light intensity is formed at a point on the axis as if the

plate were acting as a lens of focal length *f*. Subsidiary focal points at *f*/3, *f*/5, etc, are also formed with progressively much weaker concentrations of light. See FRESNEL ZONE, HALF-PERIOD ZONES.

zone purification (*Eng*) The use of ZONE MELTING for purification. By making several sweeps of the molten zone in the same direction, a bar of metal or semiconductor can be brought to an extremely high degree of purity, by concentrating the impurities at one end which is finally discarded. The very pure silicon required for semiconductors is refined in this way. Also *zone refining*.

zone time (*Astron*) See STANDARD TIME.

zoning (*Aero*) (1) The specification of areas surrounding an airfield in which there is a known clearance above obstruction for the safe landing and take-off of aircraft. (2) The division of an aircraft's fuselage, wings and engine nacelles into specific areas for precise location of equipment, identification and fire protection purposes.

zoning (*Min*) Concentric layering parallel to the periphery of a crystalline mineral, shown by colour banding in such minerals as tourmaline, and by differences of the optical reactions to polarized light in colourless minerals like feldspars; it is due to the successive deposition of layers of materials differing slightly in composition.

zonula adherens (*BioSci*) See ADHERENS JUNCTION.

zonula ciliaris (*BioSci*) In the vertebrate eye, a double fenestrated membrane connecting the ciliary process of the CHOROID with the capsule surrounding the lens.

zonula occludens (*BioSci*) See TIGHT JUNCTION.

zonule (*BioSci*) A small belt or zone, such as the zonula ciliaris of the vertebrate eye.

zoo- (*Genrl*) Prefix from Gk *zoon*, animal.

zoobiotic (*BioSci*) Parasitic on, or living in association with, an animal.

zooblast (*BioSci*) An animal cell.

zoochlorellae (*BioSci*) Symbiotic green algae found in various animals.

zoochorous (*BioSci*) Spores or seeds dispersed by animals.

zoocyst (*BioSci*) See SPOROCYST.

zoogamete (*BioSci*) A motile gamete.

zoogamy (*BioSci*) Sexual reproduction of animals.

zoogeography (*BioSci*) The study of animal distribution.

zooid (*BioSci*) (1) An individual forming part of a colony in Protozoa (Volvocina), Cnidaria, Hemichordata, Urochordata and Bryozoa. (2) In Polychaeta, a posterior sexual region formed by asexual reproduction. (3) A polyp or polypide.

zooidogamy (*BioSci*) Fertilization by motile spermatozoids. Cf SIPHONOGAMY.

zoology (*Genrl*) The science of animal life, included along with botany in the science of biology; the animal life of a region.

zoom (*ImageTech*) The effect produced by the rapid movement of the camera towards or away from the subject, or the equivalent obtained by the use of a ZOOM LENS or by electronic means in video.

zooming (*Aero*) Utilizing the kinetic energy of an aircraft in order to gain height. *Zoom-bombing* involves the release of a nuclear bomb during a zooming manoeuvre to give the aircraft time to escape the blast. See TOSS-BOMBING.

zoom lens (*ImageTech*) A camera lens whose focal length is continuously variable while maintaining a fixed focal plane, thus providing variable magnification of the subject. Cf VARIFOCAL LENS.

zoonomy (*Genrl*) Animal physiology.

zoonosis (*Vet*) A disease of animals communicable to humans.

zooplankton (*BioSci*) Floating and drifting animal life.

zoosperm (*BioSci*) A spermatozoid.

zoosporangium (*BioSci*) A sporangium in which zoospores are formed.

zoospore (*BioSci*) A motile, usually naked, asexual (ie not a gamete) reproductive cell found in some algae and fungi, swimming by means of one to several flagella.

zootaxy (*Genrl*) Zoological classification.

zootechnics (*Genrl*) Animal husbandry.

zootomy (*BioSci*) See ANATOMY.

zoster (*Med*) See HERPES ZOSTER.

Z particle (*Phys*) See Z BOSON.

Zr (*Chem*) Symbol for ZIRCONIUM.

z scheme (*BioSci*) Scheme linking PHOTOSYSTEM II and PHOTOSYSTEM I such that oxygen is produced by the former, NADP is reduced by the latter and ATP is generated by (non-cyclic) photophosphorylation by electron transport from PS II to PS I.

Z-twist (*Textiles*) See TWIST DIRECTION.

zulu time (*ICT*) Used in telecommunications for GMT. See UTC TIME.

Zürich sunspot number (*Astron*) A rather arbitrary index for describing the total numbers of sunspots and sunspot groups; given by $R = k(f + 10g)$, where g is the number of sunspot groups, f is the total number of spots, and k is a constant which depends on the estimated efficiency of the observer and equipment used. Its value has been recorded since the mid-18th century.

zussmanite (*Min*) A pale-green, tabular, trigonal hydrated silicate of iron, magnesium and potassium.

Z-value (*FoodSci*) The empirical value in degrees Celsius or Fahrenheit required to achieve a tenfold reduction of the rate of inactivation of a specific micro-organism. The Z-value is used to calculate equivalence of sterilization processes. Often used in conjunction with D-VALUE.

zweikanter (*Geol*) A wind-faceted pebble with two curved surfaces intersecting at two sharp edges (whence its name, Ger *zwei*, two). See DREIKANTER, VENTIFACT.

zwitterion (*Chem*) An ion carrying both a positive and a negative charge, eg present in solid and liquid amino acids such as glycine (aminoethanoic acid), $N^+ H_3CH_2COO^-$.

Zyban (*Pharmacol*) TN for a drug (*bupropion*) used in the treatment of nicotine addiction.

zyg-, zygo- (*Genrl*) Prefixes from Gk *zygon*, yoke.

zygapophyses (*BioSci*) Articular processes of the vertebrae of higher vertebrates, arising from the anterior and posterior sides of the neurapophyses.

zygo- (*Genrl*) See ZYG-.

zygodactylous (*BioSci*) Said of birds that have the first and fourth toes directed backwards, as parrots.

zygogenetic (*BioSci*) A product of fertilization.

zygoma (*BioSci*) The bony arch of the side of the head in mammals which bounds the lower side of the orbit.

zygomatic (*BioSci*) Pertaining to the zygoma. See JUGAL.

zygomatic arch (*BioSci*) See ZYGOMA.

zygomatic bone (*BioSci*) See JUGAL.

zygomorphic (*BioSci*) A bilaterally symmetrical flower or corolla that can only be divided into two halves by one vertical plane. Also *irregular*. Cf ACTINOMORPHIC.

Zygomycetes (*BioSci*) Formerly one of the two classes of the ZYGOMYCOTA but, given the probably polyphyletic nature of the group, the classification is evolving as new molecular phylogentic evidence emerges.

Zygomycota (*BioSci*) A division of the Eumycota or true fungi defined and distinguished from all other fungi by sexual reproduction via zygospores and asexual reproduction by uni-to-multispored sporangia. It includes fungi that have no motile stages and that are usually mycelial and aseptate. They are ecologically diverse but mostly saprophytic, eg *Mucor*, the pin-mould; some insect parasites. The species, such as *Glomus*, which form VESICULAR ARBUSCULAR MYCORRHIZAS with very many plants, have recently been removed into a separate phylum, the Glomeromycota. Also *Zygomycetes, Zygomycotina*.

zygonema (*BioSci*) The zygotene phase of meiosis.

zygospore (*BioSci*) (1) Any thick-walled resting spore formed directly from a zygote, as in many algae and some fungi. See OÖSPORE. (2) A thick-walled resting spore formed from the zygote resulting from the union of isogametes as in Zygomycota.

zygote (*BioSci*) The cell that results from the fusion of two gametes. Adj *zygotic*.

zygotene (*BioSci*) The second stage of meiotic prophase, intervening between leptotene and pachytene, in which the chromatin threads approximate in pairs and become looped. See fig. at MEIOSIS.

Zyklon B (*Chem*) A commercial form of hydrocyanic acid (HCN, prussic acid); a toxic gas used in World War II extermination camps.

Zürich sunspot number

zym-, zymo- (*BioSci*) Prefixes from Gk *zyme, zymosis*, relating to fermentation.

zymogen (*BioSci*) Any inert precursor of many active proteins and degradative enzymes. The zymogen is converted into the active form at the required site of activity. Thus trypsin is formed in the intestinal lumen from the inactive trypsinogen fibrin generated at the site of blood clotting from the inactive fibrinogen.

zymogen granule (*BioSci*) Secretory vesicle containing an inactive precursor (*zymogen*). The contents are often very condensed.

zymosan (*BioSci*) The cell-wall fraction of yeast that activates the alternative COMPLEMENT pathway, and thus binds C3b. It is frequently used for study of the capacity of cells to phagocytose opsonized materials.

Zytel (*Plastics*) TN for toughened nylon (US).

APPENDICES

SI prefixes

The following prefixes are used to indicate decimal multiples and sub-multiples of the great majority of the SI units.

Symbol	Prefix	Factor
Y	yotta	10^{24}
Z	zetta	10^{21}
E	exa	10^{18}
P	peta	10^{15}
T	tera	10^{12}
G	giga	10^{9}
M	mega	10^{6}
k	kilo	10^{3}
h	hecto	10^{2}
da	deca	10
d	deci	10^{-1}
c	centi	10^{-2}
m	milli	10^{-3}
μ	micro	10^{-6}
n	nano	10^{-9}
p	pico	10^{-12}
f	femto	10^{-15}
a	atto	10^{-18}

SI conversion factors

Exact values are printed in bold type.

Quantity	Unit		Conversion factor
Length	1 in	=	**0·0254 m**
	1 ft	=	**0·3048 m**
	1 yd	=	**0·9144 m**
	1 fathom	=	**1·8288 m**
	1 chain	=	**20·1168 m**
	1 mile	=	1609·34 m
	1 International nautical mile	=	**1852 m**
	1 UK nautical mile	=	1853·18 m
Area	1 in^2	=	**6·4516 × 10^{-4} m^2**
	1 ft^2	=	0·0929 m^2
	1 yd^2	=	0·8361 m^2
	1 acre	=	4046·86 m^2
	1 mile^2	=	2·589 × 10^6 m^2
	1 ha (hectare)	=	**10^4 m^2**

Quantity	Unit		Conversion factor
Volume	1 UK fluid ounce (fl. oz)	=	$2 \cdot 841 \times 10^{-5}$ m^3
	1 US fluid ounce	=	$2 \cdot 957 \times 10^{-5}$ m^3
	1 US liquid pint	=	$4 \cdot 731 \times 10^{-4}$ m^3
	1 US dry pint	=	$5 \cdot 506 \times 10^{-4}$ m^3
	1 UK pint	=	$5 \cdot 682 \times 10^{-4}$ m^3
	1 UK gallon	=	$1 \cdot 201$ US gallon
		=	$4 \cdot 546 \times 10^{-3}$ m^3
	1 US gallon	=	$0 \cdot 833$ UK gallon
		=	$3 \cdot 785 \times 10^{-3}$ m^3
	1 litre (1 dm^3)	=	10^{-3} m^3
	1 in^3	=	$1 \cdot 638 \times 10^{-5}$ m^3
	1 ft^3	=	$0 \cdot 028\,31$ m^3
	1 yd^3	=	$0 \cdot 7645$ m^3
	1 board foot (timber)	=	$2 \cdot 359 \times 10^{-3}$ m^3
	1 cord (timber)	=	$3 \cdot 624$ m^3
Mass	1 grain	=	$0 \cdot 064\,79$ g
	1 dram (avoir.)	=	$1 \cdot 771$ g
	1 ounce (troy or apoth.)	=	$31 \cdot 103$ g
	1 oz (avoir.)	=	$28 \cdot 349$ g
	1 lb	=	**$0 \cdot 453\,592\,37$ kg**
	1 slug	=	$14 \cdot 59$ kg
	1 short cwt (US hundredweight)	=	$43 \cdot 35$ kg
	1 cwt (UK hundredweight)	=	$50 \cdot 8$ kg
	1 UK ton	=	1016 kg
	1 tonne	=	1000 kg
	1 short ton (2000 lb)	=	$907 \cdot 1$ kg
Mass per unit length	1 lb yd^{-1}	=	$0 \cdot 496$ kg m^{-1}
	1 UK ton mile^{-1}	=	$0 \cdot 6313$ kg m^{-1}
	1 UK ton per 1000 yd	=	$1 \cdot 111$ kg m^{-1}
	1 oz in^{-1}	=	$1 \cdot 116$ kg m^{-1}
	1 lb ft^{-1}	=	$1 \cdot 488$ kg m^{-1}
	1 lb in^{-1}	=	$17 \cdot 85$ kg m^{-1}
Mass per unit area	1 lb acre^{-1}	=	$1 \cdot 12 \times 10^{-4}$ kg m^{-2}
	1 UK cwt acre^{-1}	=	$1 \cdot 2553 \times 10^{-2}$ kg m^{-2}
	1 oz yd^{-2}	=	$3 \cdot 3905 \times 10^{-2}$ kg m^{-2}
	1 UK ton acre^{-1}	=	$0 \cdot 251\,07$ kg m^{-2}
	1 oz ft^{-2}	=	$0 \cdot 305\,15$ kg m^{-2}
	1 lb ft^{-2}	=	$4 \cdot 882$ kg m^{-2}
	1 lb in^{-2}	=	703 kg m^{-2}
	1 UK ton mile^{-2}	=	$3 \cdot 922 \times 10^{-4}$ kg m^{-2}
Density	1 lb UK gal^{-1}	=	$99 \cdot 77$ kg m^{-3}
	1 lb US gal^{-1}	=	$119 \cdot 82$ kg m^{-3}
	1 slug ft^{-3}	=	$515 \cdot 37$ kg m^{-3}
	1 ton yd^{-3}	=	$1328 \cdot 94$ kg m^{-3}
	1 lb in^{-3}	=	$2 \cdot 767 \times 10^{4}$ kg m^{-3}
Specific volume	1 in^3 lb^{-1}	=	$3 \cdot 613 \times 10^{-8}$ m^3 g^{-1}
	1 ft^3 lb^{-1}	=	$6 \cdot 243 \times 10^{-5}$ m^3 g^{-1}
Velocity	1 in min^{-1}	=	$4 \cdot 233 \times 10^{-4}$ m s^{-1}
	1 ft min^{-1}	=	**$0 \cdot 005\,08$ m s^{-1}**
	1 ft s^{-1}	=	**$0 \cdot 3048$ m s^{-1}**
	1 mile h^{-1}	=	**$0 \cdot 447\,04$ m s^{-1}**
	1 UK knot	=	$0 \cdot 5147$ m s^{-1}
	1 International knot	=	$0 \cdot 5144$ m s^{-1}
Acceleration	1 ft s^{-2}	=	**$0 \cdot 3048$ m s^{-2}**
Mass flow rate	1 lb h^{-1}	=	$1 \cdot 259 \times 10^{-4}$ kg s^{-1}
	1 UK ton h^{-1}	=	$0 \cdot 282\,23$ kg s^{-1}

Quantity	Unit		Conversion factor
Force or weight	1 ozf (ounce)	=	0·278 N
	1 kgf	=	**9·806 65 N**
	1 tonf	=	$9·964 \times 10^3$ N
Force or weight per unit length	1 lb ft^{-1}	=	14·5939 N m^{-1}
	1 lbf in^{-1}	=	175·127 N m^{-1}
	1 tonf ft^{-1}	=	$3·269 \times 10^4$ N m^{-1}
Force (weight) per unit area or pressure or stress	1 pdl ft^{-2}	=	1·488 N m^{-2}
	1 lbf ft^{-2}	=	47·88 N m^{-2}
	1 mm Hg	=	133·32 N m^{-2}
	1 in H_2O	=	249·08 N m^{-2}
	1 ft H_2O	=	2989·07 N m^{-2}
	1 in Hg	=	3386·39 N m^{-2}
	1 lbf in^{-2}	=	$6·894 \times 10^3$ N m^{-2}
	1 bar	=	$\mathbf{10^5}$ N m^{-2}
	1 std. atmos.	=	$1·013\,25 \times 10^5$ N m^{-2}
	1 ton ft^{-2}	=	$1·072 \times 10^5$ N m^{-2}
	1 tonf in^{-2}	=	$1·544 \times 10^7$ N m^{-2}
Force density	1 lbf ft^{-3}	=	157·08 N m^{-3}
	1 lbf UK gal^{-1}	=	978·472 N m^{-3}
	1 tonf yd^{-3}	=	$1·303 \times 10^4$ N m^{-3}
	1 lbf in^{-3}	=	$2·714 \times 10^5$ N m^{-3}
Moment, torque or couple	1 ozf in (ounce–force inch)	=	$7·061 \times 10^{-3}$ N m
	1 pdl ft	=	0·042 14 N m
	1 lbf in	=	0·1129 N m
	1 lbf ft	=	1·355 N m
	1 tonf ft	=	$3·037 \times 10^3$ N m
Energy	1 erg	=	$\mathbf{10^{-7}}$ **J**
	1 hp h (horsepower hour)	=	$2·684\,52 \times 10^6$ J
	1 thermie $= 10^6$ cal$_{15}$	=	$4·1855 \times 10^6$ J
	1 therm $= 100\,000$ Btu	=	$105·506 \times 10^6$ J
Electric energy	1 kWh	=	$3·6 \times 10^6$ J
Power	1 hp $= $ **550** ft lbf s^{-1}	=	745·700 W
	1 metric horsepower (ch, PS)	=	735·499 W
Heat	1 cal$_{IT}$	=	**4·1868 J**
	1 Btu	=	1055·06 J
Heat power per unit area	1 Btu ft^{-2} h^{-1}	=	3·154 59 W m^{-2}
Cooling power	1 ton$_{ref}$ (refrigeration ton) $= 12\,000$ Btu h^{-1}	=	3516·86 W
Specific heat	1 Btu lb^{-1} °F^{-1}		
	1 Chu lb^{-1} °C^{-1}	=	**418·68** J kg^{-1} K^{-1}
	1 cal g^{-1} °C^{-1}		
Heat flow rate	1 Btu h^{-1}	=	0·293 071 W
	1 kcal h^{-1}	=	**1·163 W**
	1 cal s^{-1}	=	**4·1868 W**
Calorific value or specific enthalpy	1 Btu ft^{-3}	=	$3·725\,89 \times 10^4$ J m^{-3}
	1 Btu lb^{-1}	=	**2326** J kg^{-1}
	1 cal g^{-1}	=	**4·1868** J g^{-1}
	1 kcal m^{-3}	=	**4186·8** J m^{-3}
Thermal conductivity	1 cal cm^{-1} s^{-1} °C^{-1}	=	**4·1868** W cm^{-1} K^{-1}
	1 Btu ft^{-1} h^{-1} °F^{-1}	=	1·730 73 W m^{-1} K^{-1}
Plane angle	1 rad (radian)	=	57·2958°
	1 degree	=	0·017 453 3 rad
	1 minute	=	$2·908\,88 \times 10^{-4}$ rad
	1 second	=	$4·848\,14 \times 10^{-6}$ rad
	1 grade	=	0·9°

Conversion factors between SI, CGS and FPS units

After SI, the two most important systems of units are the *CGS* and *FPS systems*, the latter being still very frequently used in the US. Like SI they are both based on the second as unit of time, but differ in their base units for length (centimetre and foot) and for mass (gram and pound). Some of the more important conversion factors between SI units and the other two systems are set out below.

Physical quantity	SI unit	CGS unit	FPS unit
length	1 m	*centimetre*, $1\,\text{cm} = 10^{-2}\,\text{m}$	*foot*, $1\,\text{ft} = 0\cdot3048\,\text{m}$ *inch*, $1\,\text{in} = 2\cdot54 \times 10^{-2}\,\text{m}$
mass	1 kg	*gram*, $1\,\text{g} = 10^{-3}\,\text{kg}$	*pound*, $1\,\text{lb} = 0\cdot4536\,\text{kg}$ *ounce*, $1\,\text{oz} = 2\cdot835 \times 10^{-2}\,\text{kg}$ *ton*, $1\,\text{ton} = 1\cdot016 \times 10^{3}\,\text{kg}$
area	$1\,\text{m}^2$	$1\,\text{cm}^2 = 10^{-4}\,\text{m}^2$	$1\,\text{ft}^2 = 9\cdot290 \times 10^{-2}\,\text{m}^2$ $1\,\text{in}^2 = 6\cdot452 \times 10^{-4}\,\text{m}^2$
volume	$1\,\text{m}^3$	$1\,\text{cm}^3 = 10^{-6}\,\text{m}^3$ $1\,\text{litre} = 10^{-3}\,\text{m}^3$	$1\,\text{ft}^3 = 2\cdot832 \times 10^{-2}\,\text{m}^3$ $1\,\text{in}^3 = 1\cdot639 \times 10^{-5}\,\text{m}^2$ *gallon*, $1\,\text{gal (UK)} = 4\cdot546 \times 10^{-3}\,\text{m}^3$ $1\,\text{gal (US)} = 3\cdot786 \times 10^{-3}\,\text{m}^3$ *fluid ounce*, $1\,\text{fl oz} = 2\cdot841 \times 10^{-5}\,\text{m}^3$
density	$1\,\text{kg m}^{-3}$	$1\,\text{g cm}^{-3} = 10^{-3}\,\text{kg m}^{-3}$	$1\,\text{lb ft}^{-3} = 16\cdot02\,\text{kg m}^{-3}$ $1\,\text{lb in}^{-3} = 27\cdot6799\,\text{Mg m}^{-3} = 27\cdot6799\,\text{t m}^{-3}$
velocity or speed	$1\,\text{m s}^{-1}$	$1\,\text{cm s}^{-1} = 10^{-2}\,\text{m s}^{-1}$	$1\,\text{ft s}^{-1} = 0\cdot3048\,\text{m s}^{-1}$
momentum	$1\,\text{kg m s}^{-1}$	$1\,\text{g cm s}^{-1} = 10^{-5}\,\text{kg m s}^{-1}$	$1\,\text{lb ft s}^{-1} = 0\cdot1383\,\text{kg m s}^{-1}$
moment of inertia	$1\,\text{kg m}^2$	$1\,\text{g cm}^2 = 10^{-7}\,\text{kg m}^2$	$1\,\text{lb ft}^2 = 4\cdot214 \times 10^{-2}\,\text{kg m}^2$
force	1 N	*dyne*, $1\,\text{dyn} = 10^{-5}\,\text{N}$	*poundal*, $1\,\text{pdl} = 0\cdot1383\,\text{N}$ *pound force*, $1\,\text{lbf} = 4\cdot448\,\text{N}$
pressure or stress	1 Pa	$1\,\text{dyn cm}^{-2} = 10^{-1}\,\text{Pa}$ *bar*, $1\,\text{bar} = 10^{5}\,\text{Pa}$	$1\,\text{lbf in}^{-2}\,\text{(psi)} = 6\cdot895 \times 10^{3}\,\text{Pa}$
energy or work	1 J	*erg*, $1\,\text{erg} = 10^{-7}\,\text{J}$	$1\,\text{ft pdl} = 4\cdot214 \times 10^{-2}\,\text{J}$ $1\,\text{ft lbf} = 1\cdot356\,\text{J}$
power	1 W	$1\,\text{erg s}^{-1} = 10^{-7}\,\text{W}$	$1\,\text{ft pdl s}^{-1} = 4\cdot214 \times 10^{-2}\,\text{W}$ $1\,\text{ft lbf s}^{-1} = 1\cdot356\,\text{W}$ *horsepower*, $1\,\text{hp} = 745\cdot7\,\text{W}$ $1\,\text{PS} = 1\,\text{CV (metric hp)} = 75\,\text{kp m s}^{-1}$
dynamic viscosity	$1\,\text{kg s m}^{-2}$	*poise*, $1\,\text{P} = 10^{-1}\,\text{N s m}^{-2}$	$1\,\text{lbf s ft}^{-2} = 47\cdot88\,\text{N s m}^{-2}$
kinematic viscosity	$1\,\text{m}^2\,\text{s}^{-1}$	*stoke*, $1\,\text{st} = 10^{-4}\,\text{m}^2\,\text{s}^{-1}$	$1\,\text{ft}^2\text{s}^{-1} = 9\cdot29 \times 10^{-2}\,\text{m}^2\,\text{s}^{-1}$
energy (thermal)	1 J	*calorie*, $1\,\text{cal} = 4\cdot187\,\text{J}$	*British thermal unit*, $1\,\text{BTU} = 1\cdot055 \times 10^{3}\,\text{J}$
force density		$1\,\text{dyn cm}^{-3} = 10\,\text{N m}^{-3}$	$1\,\text{lbf UK gal}^{-1} = 978\cdot472\,\text{N m}^{-3}$
thermal conductivity	$1\,\text{J m}^{-1}\,\text{k}^{-1}$	$1\,\text{cal cm cm}^{-2}\,\text{s}^{-1}\,^\circ\text{C}^{-1}$ $= 4\cdot187\,\text{W cm}^{-1}\,^\circ\text{C}^{-1}$	$1\,\text{BTU ft}^{-2}\,\text{h}^{-1}\,^\circ\text{F}^{-1} = 1\cdot73073\,\text{W m}^{-1}\,^\circ\text{C}^{-1}$
heat	1 J	*thermie*, $1\,\text{th} = 10^{6}\,\text{cal}$	$1\,\text{therm} = 100\,000\,\text{BTU} = 1\cdot055 \times 10^{8}\,\text{J}$

Physical constants, standard values and equivalents in SI units

Constant name	Quantity symbol	Algebraic relation	Numerical value (Z)	SI unit*
acceleration of gravity at Greenwich	g		9·818 83	$m\,s^{-2}$
atomic mass unit	u		$1·660\,565\,5 \times 10^{-27}$	kg
Avogadro constant	N_A		$6·022\,045 \times 10^{23}$	mol^{-1}
Bohr magneton	μ_B	$\mu_B = e\hbar/2m_e$	$9·274\,078 \times 10^{-24}$	$A\,m^2$ or $J\,T^{-1}$
Bohr radius	a_0	$a_0 = \alpha/4\,\pi\,R_\infty$	$0·529\,177\,06 \times 10^{-10}$	m
Boltzmann constant	k	$k = R_0/N_A$	$1·380\,662 \times 10^{-23}$	$J\,K^{-1}$
charge to mass ratio of electron		e/m_e	$1·758\,804\,7 \times 10^{11}$	$C\,kg^{-1}$
Compton wavelength of electron	λ_{ce}	$\lambda_{ce} = \alpha^2/2\,R_\infty$	$2·426\,308\,9 \times 10^{-12}$	m
Compton wavelength of proton	λ_{cp}	$\lambda_{cp} = h/c\,m_p$	$1·321\,409\,9 \times 10^{-15}$	m
electron charge	e		$1·602\,189\,2 \times 10^{-19}$	C
electron radius	r_e	$r_e = \mu_0\,e^2/4\,\pi\,m_e$	$2·817\,938\,0 \times 10^{-15}$	m
electron rest mass	m_e		$0·910\,953\,4 \times 10^{-30}$ $5·485\,802\,6 \times 10^{-4}$	kg u
Faraday constant	F	$F\,N_A\,e$	$9·648\,456 \times 10^4$	$C\,mol^{-1}$
fine-structure constant	α		$0·007\,297\,350\,6$	–
	α^{-1}	$1/\alpha = \mu_0\,c\,e^2/2\,h$	$137·036\,04$	–
first radiation constant	c_1	$c_1 = 2\pi\,h\,c^2$	$3·741\,832 \times 10^{-16}$	$W\,m^2$
gas constant	R_0	$R_0 = p_0\,V_m/T_0$	$8·314\,41$	$J\,mol\,K^{-1}$
gravitational constant	G		$6·6720 \times 10^{-11}$	$N\,m^2\,kg^{-2}$ or $m^3\,kg\,s^{-2}$
gyromagnetic ratio of proton	γ_p		$2·675\,198\,7 \times 10^8$	$A\,m^2\,J^{-1}\,s^{-1}$ $rad\,s^{-1}\,T^{-1}$
normal acceleration of gravity	g_n		$9·806\,65$	$m\,s^{-2}$
neutron rest mass	m_n		$1·674\,954\,3 \times 10^{-27}$ $1·008\,665\,012$	kg u
normal pressure	p_0		$101\,325$	Pa
normal temperature	T_0		$273·15$	K
Planck constant	h		$6·626\,176 \times 10^{-34}$	J s
proton rest mass	m_p		$1·672\,648\,5 \times 10^{-27}$ $1·007\,276\,470$	kg u
Rydberg constant	R_∞		$1·097\,373\,177 \times 10^7$	m^{-1}
second radiation constant	c_2	$c_2 = h\,c/k$	$0·014\,387\,86$	m K
standard volume of ideal gas	V_m	$V_m = R_0\,T_0/p_0$	$0·022\,413\,83$	$m^3\,mol^{-1}$
Stefan–Boltzmann constant	σ	$\sigma = (\pi^2/60)\,k^4/\hbar^3\,c^2$	$5·670\,32 \times 10^{-8}$	$W\,m^{-2}\,K^{-4}$
velocity of light *in vacuo*	c		$299\,792\,458$	$m\,s^{-1}$

*u is not an SI unit, but it is accepted by the Comité International des Poids et Mesures.

Units of measurement

The International System (SI) of units has seven *base units*, two *supplementary units* (the radian and the steradian) and a variety of *derived units*. The seven physical quantities on which the system is based, together with their dimensions, the SI base units and their symbols, are listed in the table below.

Basic physical quantity	Dimension	SI base unit	Symbol
length	L	metre	m
mass	M	kilogram	kg
time	T	second	s
electric current	I	ampere	A
temperature	θ	kelvin	K
luminous intensity	J	candela	cd
amount of substance	n	mole	mol

Although the mole is strictly dimensionless, it is usually expressed in the form of g-mole, and represents the amount of substance which contains as many atoms or molecules as there are atoms in 0.012 kg of the carbon isotope carbon-12 ($^{12}_{6}$C).

SI derived units

The SI derived units for other physical quantities are formed from the base units via the equation defining the quantity involved. Thus for example, force = mass × acceleration, and the unit of force, the newton, is equivalent to $kg\,m\,s^{-2}$. The two supplementary units and the principal derived units, together with some of the more important equivalents, are shown below.

Quantity	Dimension	Unit	Symbol	Equivalent
plane angle	–	radian	rad	$(=180°/\pi)$
solid angle	–	steradian	sr	–
density	ML^{-3}	–	$kg\,m^{-3}$	–
velocity or speed	LT^{-1}	–	$m\,s^{-1}$	–
acceleration	LT^{-2}	–	$m\,s^{-2}$	–
momentum	MLT^{-1}	–	$kg\,m\,s^{-1}$	–
moment of inertia	ML^{2}	–	$kg\,m^{2}$	–
force	MLT^{-2}	newton	N	$kg\,m\,s^{-2}$
pressure, stress	$ML^{-1}T^{-2}$	pascal	Pa	$N\,m^{-2}$
energy, work	$ML^{2}T^{-2}$	joule	J	$N\,m$
power	$ML^{2}T^{-3}$	watt	W	$J\,s^{-1}$
dynamic viscosity	$ML^{-1}T^{-1}$	–	$N\,s\,m^{-2}$	$kg\,m^{-1}s^{-1}$
kinematic viscosity	$M^{2}T^{-1}$	–	$m^{2}\,s^{-1}$	–
frequency	T^{-1}	hertz	Hz	s^{-1}
angular frequency	T^{-1}	–	$rad\,s^{-1}$	$(=2\pi\,Hz)$
electric conductance	$L^{-2}M^{-1}T^{3}I^{2}$	siemens	S	Ω^{-1}
electric charge	TI	coulomb	C	$A\,s$
electric potential difference	$L^{2}MT^{-3}I^{-1}$	volt	V	$W\,A^{-1}$
electric capacitance	$T^{4}I^{2}L^{-2}M^{-1}$	farad	F	$C\,V^{-1}$
electric resistance	$L^{2}MT^{-3}I^{-2}$	ohm	Ω	$V\,A^{-1}$
magnetic flux	$L^{2}MT^{-2}I^{-1}$	weber	Wb	$V\,s$
magnetic flux density	$MT^{-2}I^{-1}$	tesla	T	$V\,s\,m^{-2}$
inductance	$L^{2}MT^{-2}I^{-2}$	henry	H	$V\,s\,A^{-1}$
luminous flux	J	lumen	lm	$cd\,sr$
illuminance	JL^{-2}	lux	lx	$lm\,m^{-2}$
radioactivity	T^{-1}	becquerel	Bq	s^{-1}
absorbed dose, ionizing radiation	$L^{2}T^{-2}$	gray	Gy	$J\,kg^{-1}$
dose equivalent	$L^{2}T^{-2}$	sievert	Sv	$J\,kg^{-1}$
catalytic activity	ΔQuantity T^{-1}	katal	kat	$mol\,s^{-1}$

Subatomic particles

	Symbol	Charge (*e*)	Mass (MeV c^{-2})*	Spin	Discovered
Hadrons					
baryons					
proton	p	+1	938·3	$^1/_2$	1919
neutron	n	0	939·6	$^1/_2$	1932
mesons					
pion	π^+, π^0, π^-	+1, 0, −1	140, 135, 140	0	1947
kaon	K^+, K^0, K^-	+1, 0, −1	494, 498, 494	0	1947
J/psi	J/ψ	0	3095	1‡	1974
upsilon	Y	0	9450	1‡	1977
Quarks†					
up quark	u	$+^2/_3$	5	$^1/_2$	1967
down quark	d	$-^1/_3$	10	$^1/_2$	1967
charmed quark	c	$+^2/_3$	1400	$^1/_2$	1974–6
strange quark	s	$-^1/_3$	150	$^1/_2$	1974–6
bottom quark	b	$-^1/_3$	4800	$^1/_2$	1977
top quark	t	$+^2/_3$	170 000	$^1/_2$	1994
Leptons					
electron	e	−1	0·511	$^1/_2$	1897
positron	e^+	+1	0·511	$^1/_2$	1932
muon	μ	−1	106	$^1/_2$	1937
tau lepton	τ	−1	1777	$^1/_2$	1975
neutrino (electron)	ν_e	0	$<10^{-5}$	$^1/_2$	1956
neutrino (muon)	ν_μ	0	$<0·17$	$^1/_2$	1962
neutrino (tau)	ν_τ	0	<25	$^1/_2$	1997§
Gauge bosons					
photon	γ	0	0	1	1905
W	W^+, W^-	+1, −1	80 600	1	1983
Z	Z^0	0	90 161	1	1983
gluon	g	0	<10	1	–
graviton	g	0	0	2	–

* 1 MeV c^{-2} = 1·782 677 × 10^{-30} kg.

† Discovery means identification of a particle containing the quark; evidence suggests that quarks cannot exist as free particles.

‡ At ground state.

§ Discovery to be verified.

Electromagnetic spectrum

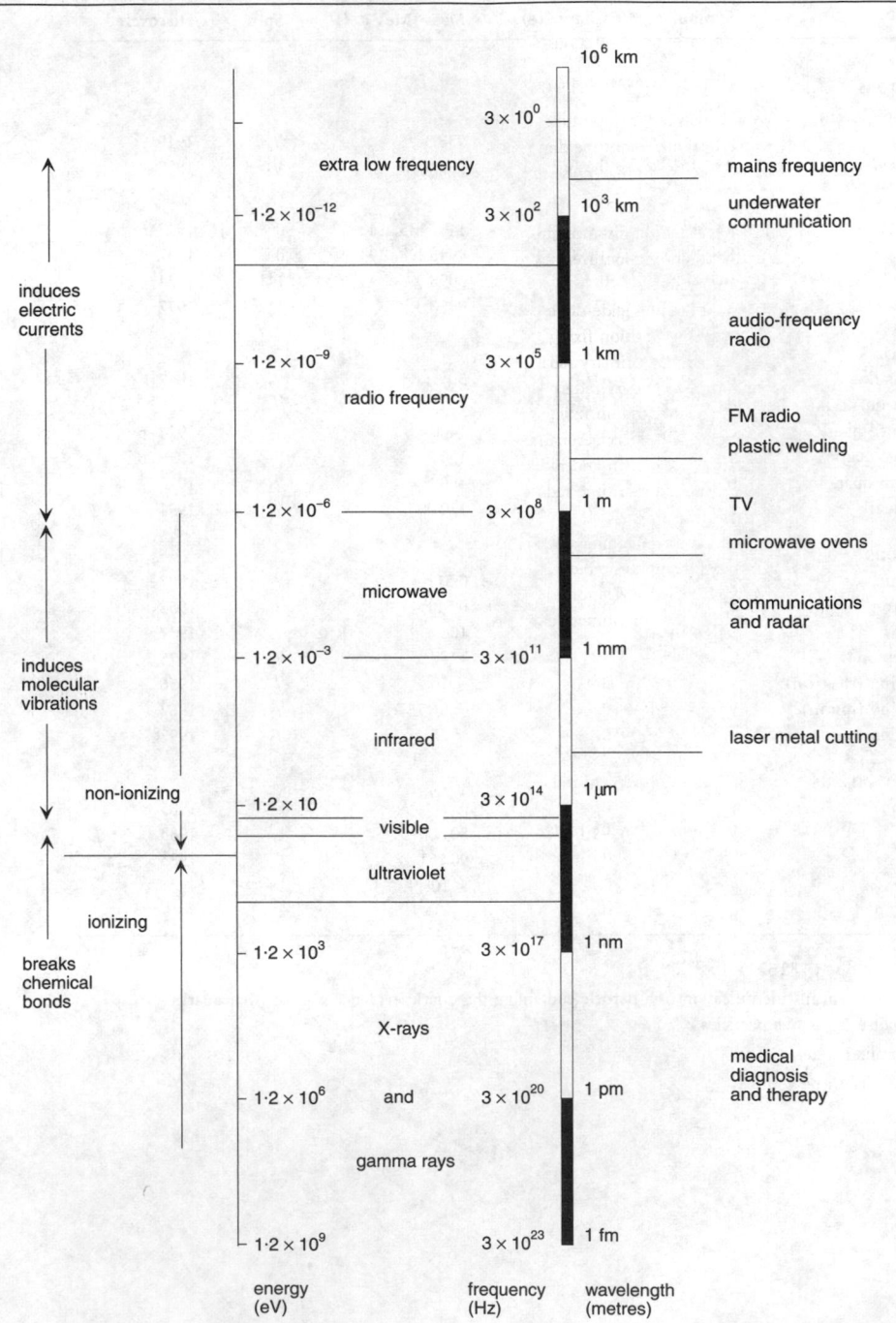

Effects	Name			Uses

Radio and radar frequencies used in aircraft navigation

Band	Application
kHz	
9–14	Omega long-range continuous-wave phase-comparison location
70–90 110–130	Decca continuous-wave hyperbolic system
90–110	Loran-C long-range pulse time-comparison location system
190–526*	Aeronautical and maritime direction finding; fixed beacons
1600–2000*	Loran-A medium-range pulse time-comparison location system
MHz	
75	Instrument landing distance fan-markers
108–112	Instrument landing localizer beams
112–118	VHF omni range
328–336	Instrument landing glide-slope beams
150 & 400	Transit satellite position fixing
582–606*	Air-traffic control primary radar
962–1215*	Aeronautical secondary radar; Aero distance measuring equipment; Tacan
1215–1240	Navstar satellite position fixing system
1300–1350	Primary and secondary aeronautical radar
1559–1626*	Navstar satellite system and radio altimeters
2700–3100*	Aeronautical and maritime radar
4200–4400	Radio altimeters
5000–5250	Future microwave landing system
5350–5650*	Airborne and shipborne radar, beacons and transponders
8750–8850	Airborne dopplet navigation
8850–9800	Aeronautical and maritime radar
GHz	
13·25–13·4	Airborne doppler navigation
14·0 –14·3*	Docking radar for ships
15·4 –15·7	Aeronautical navigation
24·25–25·25	Navigation generally
31·8 –33·4	Airfield surface movement indicator (radar)

* There are restrictions on the use of some parts of these bands for some types of navigation aid.

Letter designations of the frequency bands

Frequency bands are identified by defence organizations by capital letters, and these are often found in radar documents. In 1972 the bands indicated by the letters were changed, leading to confusion. The system is not internationally standardized, but the lists given here are often used.

US system		UK system	
Letter	Frequency band	Letter	Frequency band
HF	3–30 MHz		
VHF	30–300 MHz	A	0–250 MHz
		B	250–500 MHz
UHF	300–1000 MHz	C	500–1000 MHz
L	1000–2000 MHz		
S	2000–4000 MHz		
C	4000–8000 MHz		
X	8000–12 000 MHz	D	1–2 GHz
K_u	12.0–18 GHz		
E	2–3 GHz		
		F	3–4 GHz
		G	4–6 GHz
		H	6–8 GHz
		I	8–10 GHz
		J	10–20 GHz
K	18–27 GHz		
K_a	27–40 GHz		
V	40–75 GHz	L	40–60 GHz
W	75–110 GHz	M	60–100 GHz
mm	110–300 GHz		

MHz = Megahertz   GHz = Gigahertz

Graphical symbols used in electronics

alternating current		negative pulse	
amplifier		non-ionizing radiation	
AND gate		NOR gate	
battery (single cell)		NOT gate	
battery (multicell)		OR gate	
bridge, full-wave rectifier		photodiode	
capacitor		piezoelectric crystal	
diode		positive pulse	
earth/ground		resistor	
fuse		shielded single conductor	
inductor		shielded five conductor	
inductor, metal core		switch	
lamp		transformer, iron core	
light-emitting diode		transistor (npn)	
meter, ammeter, voltmeter, etc		transistor (pnp)	
microphone		twisted conductors	
motor		variable capacitor	
NAND gate		variable resistor	

Greek alphabet

A	α	alpha	I	ι	iota	P	ρ	rho
B	β	beta	K	κ	kappa	Σ	σ	sigma
Γ	γ	gamma	Λ	λ	lambda	T	τ	tau
Δ	δ	delta	M	μ	mu	Y	υ	upsilon
E	ε	epsilon	N	ν	nu	Φ	φ	phi
Z	ζ	zeta	Ξ	ξ	xi	X	χ	chi
H	η	eta	O	o	omicron	Ψ	ψ	psi
Θ	θ	theta	Π	π	pi	Ω	ω	omega

Mathematical symbols

+	plus; positive	⊥	perpendicular
−	minus; negative	‖	parallel
±	plus or minus; error margin	≅	congruent to
∓	minus or plus	∴	therefore
×	multiplied by	∵	because
•	multiplied by; scalar product of two vectors	∀	for all
		∃	there exists
÷, /	divided by	{ }	set
=	equal to	⟨ ⟩	mean
≠	not equal to	∪	union
≡	defined as; identical to	∩	intersection
≈, ≃	approximately equal to	⊂	is a subset of
:	ratio; such that	⊄	is not a subset of
⩾, ≥	greater than or equal to	∂	partial derivative
>	greater than	∈	is an element of
≫	much greater than	∉	is not an element of
≯	not greater than	⇒	implies that
⩽, ≤	less than or equal to	⇐	is implied by
<	less than	⇔	if and only if
≪	much less than	...	etc
≮	not less than	Λ	vector cross product
∝	directly proportional to	*	convolution
∞	infinity	○	composite function
→	approaches the limit	Δ	increment
√	square root	Σ	sum
!	factorial	Π	product
%	per cent	∇	del (differential operator)
'	first derivative; arcminutes; feet	∫	integral
"	second derivative; arcseconds; inches	∮	line integral around closed path
°	degrees	ℑ	imaginary part
∠	angle	ℜ	real part

Planets

Planet	Orbit period	Rotation period (d)	Equatorial diameter (km)	Mass (Earth = 1)	Perihelion (AU)	Aphelion (AU)
Mercury	87·97 d	58·646	4 878	0·06	0·31	0·47
Venus	224·70 d	243 R	12 104	0·82	0·72	0·73
Earth	365·26 d	0·997	12 756	1·00	0.98	1·02
Mars	686·98 d	1·026	6 794	0·11	1·38	1·67
Jupiter	11·86 yr	0·410*	142 800	317·83	4·95	5·45
Saturn	29·46 yr	0·426*	120 536	95·17	9·01	10·07
Uranus	84·01 yr	~0·67 R	51 118	14·50	18·28	20·09
Neptune	164·79 yr	~0·75	49 492	17·20	29·80	30·32
Pluto†	247·7 yr	6·387	2 300	0·002	29·6	49·3

R indicates retrograde motion
* Equatorial value
† Pluto's planetary status has been disputed. See entry.
1 AU=1·496 × 10^{11} m

Planetary satellites

Satellite	Year of discovery	Distance from planet (km)	Diameter (km)	Satellite	Year of discovery	Distance from planet (km)	Diameter (km)
Earth							
Moon	—	384 000	3 476	Rhea	1672	527 000	1 530
				Titan	1655	1 222 000	5 150
Mars				Hyperion	1848	1 481 000	300
Phobos	1877	9 380	27	Iapetus	1671	3 560 000	1 460
Deimos	1877	23 460	15	Phoebe	1898	12 950 000	220
				Pan	1990	134 000	20
Jupiter							
Metis	1979	128 000	40	**Uranus**			
Adrastea	1979	129 000	24	Miranda	1948	130 000	470
Amalthea	1892	181 000	270	Ariel	1851	191 000	1 160
Thebe	1979	222 000	100	Umbriel	1851	266 000	1 170
Io	1610	422 000	3 630	Titania	1787	436 000	1 580
Europa	1610	671 000	3 138	Oberon	1787	583 000	1 520
Ganymede	1610	1 070 000	5 260	Cordelia	1986	49 750	15
Callisto	1610	1 883 000	4 800	Ophelia	1986	53 800	20
Leda	1974	11 100 000	20	Bianca	1986	59 100	50
Himalia	1904	11 480 000	180	Cressida	1986	61 750	70
Lysithea	1938	11 720 000	40	Desdemona	1986	62 700	50
Elara	1905	11 740 000	80	Juliet	1986	64 350	70
Ananke	1951	21 200 000	30	Portia	1986	66 090	90
Carme	1938	22 600 000	40	Rosalind	1986	69 920	50
Pasiphae	1908	23 500 000	50	Belinda	1986	75 300	50
Sinope	1914	23 700 000	40	Puck	1986	85 890	170
				Caliban	1997	7 200 000	60
Saturn				Sycorax	1997	12 200 000	120
Atlas	1980	138 000	40				
Prometheus	1980	139 000	100	**Neptune**			
Pandora	1980	142 000	100	Triton	1846	355 000	2 700
Epimetheus	1980	151 000	140	Nereid	1949	5 515 000	340
Janus	1980	151 000	200	Naiad	1989	48 200	50
Mimas	1789	186 000	390	Thalassa	1989	50 000	80
Enceladus	1789	238 000	500	Despina	1989	52 500	150
Calypso	1980	295 000	30	Galatea	1989	62 000	180
Telesto	1980	295 000	24	Larissa	1989	73 600	190
Tethys	1684	295 000	1 050	Proteus	1989	117 600	400
Dione	1684	377 000	1 120				
Helene	1982	378 000	35	**Pluto**			
				Charon	1978	19 700	1 200

The constellations

Latin name	English name	Abbrev	Latin name	English name	Abbrev
Andromeda	Andromeda	And	Leo	Lion	Leo
Antlia	Air Pump	Ant	Leo Minor	Little Lion	LMi
Apus	Bird of Paradise	Aps	Lepus	Hare	Lep
Aquarius	Water Bearer	Aqr	Libra	Scales	Lib
Aquila	Eagle	Aql	Lupus	Wolf	Lup
Ara	Altar	Ara	Lynx	Lynx	Lyn
Aries	Ram	Ari	Lyra	Harp	Lyr
Auriga	Charioteer	Aur	Mensa	Table	Men
Boötes	Herdsman	Boo	Microscopium	Microscope	Mic
Caelum	Chisel	Cae	Monoceros	Unicorn	Mon
Camelopardalis	Giraffe	Cam	Musca	Fly	Mus
Cancer	Crab	Cnc	Norma	Level	Nor
Canes Venatici	Hunting Dogs	CVn	Octans	Octant	Oct
Canis Major	Great Dog	CMa	Ophiuchus	Serpent Bearer	Oph
Canis Minor	Little Dog	CMi	Orion	Orion	Ori
Capricornus	Sea Goat	Cap	Pavo	Peacock	Pav
Carina	Keel	Car	Pegasus	Winged Horse	Peg
Cassiopeia	Cassiopeia	Cas	Perseus	Perseus	Per
Centaurus	Centaur	Cen	Phoenix	Phoenix	Phe
Cepheus	Cepheus	Cep	Pictor	Easel	Pic
Cetus	Whale	Cet	Pisces	Fishes	Psc
Chamaeleon	Chameleon	Cha	Piscis Austrinus	Southern Fish	PsA
Circinus	Compasses	Cir	Puppis	Ship's Stern	Pup
Columba	Dove	Col	Pyxis	Mariner's Compass	Pyx
Coma Berenices	Berenice's Hair	Com	Reticulum	Net	Ret
Corona Australis	Southern Crown	CrA	Sagitta	Arrow	Sge
Corona Borealis	Northern Crown	CrB	Sagittarius	Archer	Sgr
Corvus	Crow	Crv	Scorpius	Scorpion	Sco
Crater	Cup	Crt	Sculptor	Sculptor	Scl
Crux	Southern Cross	Cru	Scutum	Shield	Sct
Cygnus	Swan	Cyg	Serpens	Serpent	Ser
Delphinus	Dolphin	Del	Sextans	Sextant	Sex
Dorado	Swordfish	Dor	Taurus	Bull	Tau
Draco	Dragon	Dra	Telescopium	Telescope	Tel
Equuleus	Little Horse	Equ	Triangulum	Triangle	Tri
Eridanus	River Eridanus	Eri	Triangulum Australe	Southern Triangle	TrA
Fornax	Furnace	For			
Gemini	Twins	Gem	Tucana	Toucan	Tuc
Grus	Crane	Gru	Ursa Major	Great Bear	UMa
Hercules	Hercules	Her	Ursa Minor	Little Bear	UMi
Horologium	Clock	Hor	Vela	Sails	Vel
Hydra	Sea Serpent	Hya	Virgo	Virgin	Vir
Hydrus	Water Snake	Hyi	Volans	Flying Fish	Vol
Indus	Indian	Ind	Vulpecula	Fox	Vul
Lacerta	Lizard	Lac			

Time

Originally measured by the hour angle of a selected point of reference on the celestial sphere with respect to the observer's meridian, four distinct time-scales are now used in astronomy. Each has different applications.

Solar time is defined by reference to the passage of the Sun across the local meridian. *Apparent solar time* is that shown by the true Sun. However, the Earth's elliptical orbit leads to gross changes in the Suns progress through our sky, and so a fictitious *mean Sun* is used to define a uniformly progressing *mean solar time*. *Universal time* corresponds closely with solar time and is also defined in terms of the Sun. It serves as the basis of all civil time-keeping.

Sidereal time is the hour angle of the equinox. It measures the rotation of the Earth relative to the distant stars, rather than the Sun, and is the time base used in planning and making astronomical observations. It has irregularities due to the irregular rotation of the Earth so there is an unevenness in the passage of sidereal time. The sidereal second is about 0·9973 solar seconds.

International atomic time, introduced in 1972, is a fundamental time-scale based on the second. In the SI system, a second is defined in terms of the hyperfine transition in ^{133}Cs (caesium) atoms.

Leap seconds are used to ensure that universal time and atomic time do not differ by more than 0·9 seconds. This is accomplished by adding a leap second, usually at the end of June or December, whenever wobbles in the Earth's motion make it necessary to bring both systems more closely in line.

Dynamical time is the time factor that occurs in gravitational equations of motion. It is the time that is the fourth dimension in the general theory of relativity. Since 1984 it has replaced ephemeris time which was defined in terms of an ephemeris mean Sun moving uniformly around the mean equator. Two dynamical time-scales are important, terrestrial time and barycentric dynamical time. Terrestrial time is defined by an ideal Earthbound clock at sea level, and for practical purposes it is tied to atomic time with equal seconds and an offset due to the introduction of leap seconds in atomic time. Dynamical time currently runs 32·184 seconds ahead of atomic time. Barycentric dynamical time takes into account relativistic effects and cycles back and forth by a few milliseconds compared with terrestrial time. This difference becomes important in the observation of distance objects such as pulsars or quasars. The differences between dynamical, or ephemeris, time and universal time depend on the Earth's rotation.

Time units:

Anomalistic year: 1 revolution of Earth from perihelion to perihelion, 365·259 641 34 ephemeris days.

Sidereal year: 1 revolution of Earth relative to stars, 365·256 365 56 ephemeris days or 365·256 360 42 mean solar days.

Tropical year: 1 revolution of Earth around Sun, 365·242 198 78 ephemeris days.

Ephemeris second: 1/31 556 925·9747 of the tropical year 1900·0.

Solar year: in universal time, 365 days, 48 min and 45·5 s.

Solar second: 1/86 400 of a solar day.

Geological time

The subdivisions of geological time are arranged in a vertical sequence from the oldest at the bottom to the youngest at the top or as the corresponding stratigraphic units of these subdivisions. The Proterozoic and Archean are considered as eons. Time is shown as millions of years ago (mya).

Eon	Era		Period		Age, mya
Phanerozoic	Cenozoic	Tertiary	Quaternary		
					1·64
			Neogene	Pliocene	
					5·2
				Miocene	
					23
			Palaeogene	Oligocene	
					35
				Eocene	
					56
				Palaeocene	
					65
	Mesozoic		Cretaceous		
					146
			Jurassic		
					208
			Triassic		
					245
	Palaeozoic		Permian		
					290
			Carboniferous	Pennsylvanian	
					323
				Mississippian	
					362
			Devonian		
					408
			Silurian		
					439
			Ordovician		
					510
			Cambrian		
					570
Precambrian	Proterozoic				
					2500
	Archaean				
					4600

Cenozoic

This era is divided into two sub-eras or periods, Tertiary and Quaternary. In the early part of the former, the flora and fauna of the Cretaceous gave way to more modern forms, with the mammals replacing the reptiles as the dominant fauna. Later, in the Miocene, large-scale earth movements built many of the mountain ranges of the world, followed by wide-scale volcanic activity.

In the much shorter Quaternary, modern landscape and geography were laid down. The many stages were much influenced by eg the ice ages, and differ in name and typical climate around the world.

Quaternary sub-era

The Quaternary is completely different from any previous period. It has had a much shorter time span than any earlier period, less than 2 million years, but the period has exerted a profound influence on humankind: its processes and deposits mould the modern landscape and geography. The British Isles, north of a line from the Bristol Channel to the mouth of the Thames, and much of Europe were affected by glaciation which also markedly influenced sea levels. The record of past climatic fluctuations and the history of modern faunas, floras and the human race during and since the last glaciation lie in the Quaternary deposits.

In all previous periods it is possible to establish correlations based on the evolution and disappearance of species, but this method is of limited use in the Quaternary where climatic fluctuations are predominantly used in its chronology. The enormous variations between different regions at any one time, involving latitude and altitude, frequently render precise correlations of deposits and events difficult, and impossible over longer distances. The table shows the stage names that are widely used for the British Quaternary although there is no complete agreement on their use or validity.

Throughout the world, differently named, locally based, stages have been established, with which the British stages cannot be firmly correlated. The most recent stage, the Flandrian, is correlated by general agreement with the Holocene of the Continent, and the Devensian cold stage with the Weichselian of north-west continental Europe. In the Alps, successive glaciations are named as Günst (oldest), Mindel, Riss and Würm (youngest). The Würm glaciation can be correlated with the Devensian.

Era	Period	Series	British stages	Climate	Age
Cenozoic	Quaternary	Holocene or Recent	Flandrian	temperate	10 000 ya
		Pleistocene	Devensian	last glacial	
			Ipswichian	temperate	
			Wolstonian	cold, glacial	
			Hoxnian	temperate	
			Anglian	glacial	
			Cromerian	temperate	
			Beestonian	glacial	
			Pastonian	temperate	
			Baventian	cold	
			Antian	temperate	
			Thurnian	cold	
			Ludhamian	temperate	
			Waltonian	variable	1·64 mya

Tertiary sub-era

The Tertiary derives its name from an old and disused division of all geological time into three parts. It forms the lower part of the Cenozoic era (also *Cainozoic*, *Kainozoic*, meaning 'recent life') and consists of five epochs or series: Palaeocene, Eocene, Oligocene, Miocene and Pliocene.

There was a very marked change from the flora and fauna of the Cretaceous to plants and animals of more modern aspect. On land there was a reversal of roles of reptiles and mammals, the latter becoming predominant.

During the Eocene, lavas, mainly basaltic, erupted from a series of fissures in many parts of the world including north-east Ireland and western Scotland, and plutonic centres developed at a number of localities including Skye, Rhum, Ardnamurchan, Mull, Arran, the Mourne Mountains and Rockall in the North Atlantic. There were also extensive dyke swarms.

Era	Sub-era, Period	Series	Stage	Age, mya
Cenozoic	Quaternary	Holocene		
		pleistocene		
				1·64
	Neogene	Pliocene	Piacenzian	3·4
			Zanclian	
				5·2
		Miocene	Messinian	6·7
			Tortonian	10·4
			Serravallian	14·2
			Langhian	16·3
			Burdigalian	21·5
			Aquitanian	23·3
	Palaeogene	Oligocene	Chattian	29·3
			Rupelian	35·4
		Eocene	Priabonian	38·6
			Bartonian	42·1
			Lutetian	50·0
			Ypresian	56·5
		Palaeocene	Thanetian	60·5
			Danian	65·0

The Palaeogene had a temperate to warm climate and forests became widespread in the Tertiary. Lamellibranchs, gastropods and echinoderms were abundant. In the Neogene (Miocene and Pliocene), the climate was temperate and warm, cooling in the Pliocene of more northern latitudes.

Following earlier Palaeogene tremors, earth movements during the Miocene built many of the mountain ranges of the world (Himalayas, Rockies, Alps, etc). The British Isles were on the edge of the area affected and folding occurred in southern England, including the anticline of the Wealden axis and the syncline of the London basin.

Mesozoic

The Mesozoic ('middle life') was characterized by ammonites and reptiles, together with brachiopods, lamellibranchs, gastropods and corals. The first period, the Triassic, had an impoverished fauna and flora that followed the extinctions at the end of the Palaeozoic era. During the second period, the Jurassic, there was a rich flora in the warm climate, when reptiles, notably dinosaurs, were dominant on land. In the Cretaceous, flowering plants spread and many large reptiles, ammonites, most belemnites and many brachiopod species became extinct, and chalk was the most important formation.

Era	Period	Series		Stage	Age, mya
Mesozoic	Cretaceous	Upper	Senonian	Maastrichtian	74
				Campanian	83
				Santonian	87
				Coniacian	89
				Turonian	90
				Cenomanian	97
		Lower		Albian	112
				Aptian	125
				Barremian	132
				Hauterivian	135
				Valanginian	141
				Ryazanian	146
	Jurassic	Upper (Malm)		Portlandian	152
				Kimmeridgian	155
				Oxfordian	157
				Callovian	161
		Middle (Dogger)		Bathonian	166
				Bajocian	174
				Aalenian	178
		Lower (Lias)		Toarcian	187
				Pliensbachian	194
				Sinemurian	204
				Hettangian	208
	Triassic	Upper		Rhaetian	210
		Middle			235
					241
		Lower			245

Palaeozoic

The oldest of the Phanerozoic eras. The Palaeozoic ('ancient life') began with the Cambrian period, when there was a great expansion of animal life, now recorded by the fossils, especially trilobites, brachiopods, graptolites and molluscs, as well as early plant life. Graptolites and trilobites reached their acme in the Ordovician, and in the Silurian lamellibranchs became abundant. Amphibians evolved by the end of the Devonian when the graptolites had become extinct. During the Carboniferous there was a rich flora in Coal Measure forests, but a glacial climate existed in Gondwana continents. The trilobites died out in the Permian, a period of desert conditions in the British Isles.

Era	Period		Series	Stage	Age, mya
Palaeozoic	Permian		Zechstein		256
			Rotliegendes		290
	Carboniferous	Pennsylvanian, Silesian	Stephanian		
			Westphalian		318
			Namurian		333
		Mississippian, Dinantian	Viséan		350
			Tournaisian		362
	Devonian		Upper	Famerinian	367
				Frasnian	377
			Middle	Givetian	381
				Eifelian	386
			Lower	Emsian	390
				Siegenian	398
				Gedinnian	408
	Silurian		Ludlow		424
			Wenlock		430
			Llandovery		439
	Ordovician		Ashgill		443
			Caradoc		464
			Llandeilo		469
			Llanvirn		476
			Arenig		493
			Tremadoc		510
	Cambrian		Uppper		517
			Middle		536
			Lower		570

Precambrian

All rocks which were formed before the Cambrian. They consist of two divisions, the older a series of highly metamorphosed rocks, crystalline schists and gneisses with intrusive rocks, largely of Archaean age, eg the Lewisian Complex. Unconformably overlying this basement complex are Proterozoic sediments, eg the Torridonian (of Riphean age). There is little agreement on how Precambrian rocks should be divided, and almost no formal divisions. The table below shows a number of the more widely used terms.

In the highest Proterozoic there are impressions of soft-bodied animals and trace fossils (burrows and tracks) indicating a long period of earlier evolution. Primitive plant life existed well back into the Archaean, and bacteria may have existed 3800 million years ago. The algae are the only fossil group to have had a widespread development in the Precambrian.

Eon		Era	Age, mya
Phanerozoic			570
Precambrian	Proterozoic	Vendian	610
		Riphean	
		Aphebian	1650
			2500
	Archaean		4600

Earthquake severity measurement scales

(Mercalli and Richter)

Mercalli	Description	Richter
1	detected only by seismographs	<3
2	**feeble** just noticeable by some people	3–3·4
3	**slight** similiar to passing of heavy lorries	3·5–4
4	**moderate** rocking of loose objects	4·1–4·4
5	**quite strong** felt by most people even when sleeping	4·5–4·8
6	**strong** trees rock and some structural damage is caused	4·9–5·4
7	**very strong** walls crack	5·5–6
8	**destructive** weak buildings collapse	6·1–6·5
9	**ruinous** houses collapse and ground pipes crack	6·6–7
10	**disastrous** landslides occur, ground cracks and buildings collapse	7·1–7·3
11	**very disastrous** few buildings remain standing	7·4–8·1
12	**catastrophic** ground rises and falls in waves	>8·1

The periodic table

All the naturally occurring elements can be classified in terms of their **atomic structure** and physical or chemical properties. Physical properties include melting and boiling points, atomic volume (relative atomic mass divided by density) and mechanical properties like compressibility (reciprocal of **bulk modulus**). Chemical properties cover such characteristics as oxidation number or **valency**, ionic and **atomic radii** as well as **electronegativity**. These properties (see **Properties of the elements**) all show periodic associations (or *periodicity*) as a function of atomic number. If the elements are arranged in sequence of atomic number starting with hydrogen ($Z = 1$) from left to right as in the table below, elements in vertical columns (or groups) tend to show similar properties. The first element hydrogen is ambiguous as it shows both metallic and non-metallic behaviour, forming a cation or an anion depending on its environment.

In the periodic table metals are found on the left of the bold line and non-metals on the right of the dotted line (tinted area). Between metals and non-metals are a cluster of six semi-metals, starting with boron, which have intermediate properties such as semiconductivity. There are also a group of 15 lanthanides or rare earth elements with very similar properties and four radioactive heavy metals, the actinides.

IA	IIA	IIIB	IVB	VB	VIB	VIIB	VIII			IB	IIB	IIIA	IVA	VA	VIA	VIIA	0
1 H																1 H	2 He
3 Li	4 Be											5 B	6 C	7 N	8 O	9 F	10 Ne
11 Na	12 Mg											13 Al	14 Si	15 P	16 S	17 Cl	18 Ar
19 K	20 Ca	21 Sc	22 Ti	23 V	24 Cr	25 Mn	26 Fe	27 Co	28 Ni	29 Cu	30 Zn	31 Ga	32 Ge	33 As	34 Se	35 Br	36 Kr
37 Rb	38 Sr	39 Y	40 Zr	41 Nb	42 Mo	43 Tc	44 Ru	45 Rh	46 Pd	47 Ag	48 Cd	49 In	50 Sn	51 Sb	52 Te	53 I	54 Xe
55 Cs	56 Ba	57* to 71	72 Hf	73 Ta	74 W	75 Re	76 Os	77 Ir	78 Pt	79 Au	80 Hg	81 Tl	82 Pb	83 Bi	84 Po	85 At	86 Rn
87 Fr	88 Ra	89+ to 92															

*Lanthanides	57 La	58 Ce	59 Pr	60 Nd	61 Pm	62 Sm	63 Eu	64 Gd	65 Tb	66 Dy	67 Ho	68 Er	69 Tm	70 Yb	71 Lu	
+Actinides	89 Ac	90 Th	91 Pa	92 U												

The periodic table with the elements arranged in 18 groups, IA–0

(See the appropriate entries for the following transuranic elements: neptunium; plutonium; americium; curium; berkelium; californium; einsteinium; fermium; mendelevium; nobelium; lawrencium; dubnium; joliotium; rutherfordium; bohrium; hahnium; meitnerium.)

The principal properties of the elements

The tables on the following three pages list all the naturally occurring elements alphabetically together with some of their properties. The following abbreviations are used.

bc = body centred; bcc = body-centred cubic
cubic (diam) = diamond structure
Electroneg. = Electronegativities as assigned by
 Pauling on a scale of 0 to 4
fc = face centred; fcc = face-centred cubic
graph = graphite
hcp = hexagonal close-packed
hex = hexagonal
mon = monoclinic
ortho = orthorhombic
Ox. Nos = principal oxidation numbers
r = red and y = yellow

r_a, r_i = atomic, ionic radii calculated from
 closest distances between atoms in crystal
 structure
ram = relative atomic mass
rhombic = rhombohedral
tetra = tetragonal
T_m, T_b = temperatures of melting and boiling,
 all exc. arsenic, helium at atmospheric
 pressure
T_{TRANS} = temperature for transformations in
 crystal structure column
Z = atomic number

Properties of the elements

Element, symbol	Z	r$_{am}$	Crystal structure	T$_{TRANS}$, K	r$_a$, pm	Ox. Nos	r$_i$, pm	Electroneg.	Density, ρ, kg m^{-3}	T$_m$, K	T$_b$, K
Actinium, Ac	89	227	fcc		188	3+	118	1·1	10100	1320	3470
Aluminium, Al	13	26·98	fcc		142	3+	51	1·5	2700	933·2	2740
Antimony, Sb	51	121·75	rhombic		145	3+, 5+	76/62	1·9	6700	903·7	1650
Argon, Ar	18	39·95	fcc		174	0	154	—	1·66	83·7	87·4
Arsenic, As	33	74·92	rhombic		125	3+, 5+	58/46	2·0	5730	1090	886
Astatine, At	85	210	—		—	7+	62	2·2	—	520	623
Barium, Ba	56	137·34	bcc		217	2+	134	0·9	3600	1000	1910
Beryllium, Be	4	9·01	hcp/cubic	1527	112	2+	35	1·5	1800	1550	3243
Bismuth, Bi	83	208·98	rhombic		155	3+, 5+	96/74	1·9	9800	544·4	1830
Boron, B	5	10·81	tetra/rhombic		88	3+	23	2·0	2500	2600	2820
Bromine, Br	35	79·90	ortho		114	1−, 5+	196/47	2·8	3100	265·9	331·9
Cadmium, Cd	48	112·40	hcp		148	2+	97	1·7	8650	594·2	1038
Caesium, Cs	55	132·90	bcc		262	1+	167	0·7	1870	302·6	960
Calcium, Ca	20	40·08	fcc/bcc	737	196	2+	99	1·0	1540	1120	1760
Carbon, C	6	12·01	hex/cubic graph/diam	95/263/998	71/77	4+, 4−	16/260	2·5	2300	>3800	5100
Cerium, Ce	58	140·12	fcc/hex/fcc/bcc		183	3+, 4+	103/92	1·1	6800	1070	3740
Chlorine, Cl	17	35·45	tetra		91	1−, 5+	181/34	3·0	3·21 (273 K)	172·1	283·5
Chromium, Cr	24	52·00	bcc		125	3+, 6+	63/52	1·6	7200	2160	2755
Cobalt, Co	27	58·93	hcp/fcc	690	125	2+, 3+	72/63	1·8	8900	1765	3170
Copper, Cu	29	63·55	fcc		128	1+, 2+	96/72	1·9	8930	1356	2868
Dysprosium, Dy	66	162·50	rhombic/hcp	86	175	3+	91	1·2	8500	1680	2900
Erbium, Er	68	167·26	hcp		173	3+	88	1·2	9000	1770	3200
Europium, Eu	63	151·96	bcc		198	3+, 2+	95/109	1·1	5200	1100	1712
Fluorine, F	9	19·00	—		60	1−, 7+	133/8	4·0	1·7 (273 K)	35·5	85·01
Francium, Fr	87	223	—		—	1+	187	0·7	—	303	920
Gadolinium, Gd	64	157·25	hcp/bcc	1537	178	3+	94	1·2	7900	1585	3000
Gallium, Ga	31	69·72	fcc or ortho		121	1+, 3+	81/62	1·6	5950	302·9	2676
Germanium, Ge	32	72·59	cubic (diamond)		122	4+	53	1·8	5400	1210·5	3100
Gold, Au	79	196·97	fcc		144	1+, 3+	137/85	2·4	19300	1336·1	3239
Hafnium, Hf	72	178·49	hcp/bcc	2050	158	4+	78	1·3	133000	2423	5700
Helium, He	2	4·003	hcp/cubic		176	0	—	—	0·166	0·95	4·21
Holmium, Ho	67	164·93	hcp		176	3+	89	1·2	8800	1734	2900

The abbreviations used are listed on p.25.

Properties of the elements contd

Element, symbol	Z	ram	Crystal structure	T_{TRANS}, K	r_a, pm	Ox. Nos	r_i, pm	Electroneg.	Density, ρ, kg m^{-3}	T_m, K	T_b, K
Hydrogen, H	1	1·008	hcp/cubic		46	1+	154	2·1	0·08987 (273 K)	14·01	20·4
Indium, In	49	114·82	bc tetragonal		162	3+	81	1·7	7310	429·8	2300
Iodine, I	53	126·90	ortho		135	1-	216	2·5	4940	386·6	457·4
Iridium, Ir	77	192·2	fcc		135	4+	68	2·2	22420	2716	4800
Iron, Fe	26	55·85	bcc/fcc/bcc	1180/1670	123	2+, 3+	74	1·8	7870	1808	3300
Krypton, Kr	36	83·80	fcc		201	0	-	-	3·49	116·5	120·8
Lanthanum, La	57	138·91	hcp/fcc/bcc	583/1 137	187	1+, 3+	139/102	1·1	6150	1190	3 742
Lead, Pb	82	207·19	fcc		174	2+, 4+	120/84	1·8	11340	600·4	2017
Lithium, Li	3	6·94	hcp/fcc/bcc	74/140	152	1+	68	1·0	534	462	1500
Lutetium, Lu	71	174·97	hcp		173	3+	85	1·2	9800	1925	3600
Magnesium, Mg	12	24·31	hcp		160	1+, 2+	82/66	1·2	1741	924	1380
Manganese, Mn	25	54·94	cubic		112	2+, 3+	80/66	1·5	7440	1517	2370
Mercury, Hg	80	200·59	rhombic		156	1+, 2+	127/110	1·9	13590 (273 K)	234·3	629·7
Molybdenum, Mo	42	95·94	bcc		136	4+, 6+	70/62	1·8	10200	2880	5830
Neodymium, Nd	60	144·24	hcp/bcc	1 135	181	3+	100	1·1	6960	1296	3300
Neon, Ne	10	20·18	fcc		160	0	-	-	0·839	24·5	27·2
Nickel, Ni	28	58·71	fcc		124	2+, 3+	69	1·8	8900	1726	3005
Niobium, Nb	41	92·91	bcc		143	5+	69	1·6	8570	2741	5200
Nitrogen, N	7	14·01	cubic/hcp	35·4	71	3+, 5+	16/13	3·0	1·165	63·3	77·3
Osmium, Os	76	190·2	hcp		135	4+	69	2·2	22480	3300	4900
Oxygen, O	8	16·00	rhombic		60	2-	132	2·1	1·33	54·7	90·2
Palladium, Pd	46	106·4	fcc		137	2+, 4+	80/65	2·2	12000	1825	3200
Phosphorus, P	15	30·97	cubic		120	3+, 4+	44/35	2·1	2 200 (r) 1 800 (y)	317·2	552
Platinum, Pt	78	195·09	fcc		138	2+, 4+	80/65	2·2	21450	2042	4100
Polonium, Po	84	209	monoclinic		168	2+, 6+	67	2·0	9400	527	1235
Potassium, K	19	39·10	bcc		231	1+	133	0·8	860	336·8	1047
Praseodymium, Pr	59	140·91	hcp/bcc	1065	182	3+	101	1·1	6800	1208	3400
Promethium, Pm	61	145	-		-	3+	98	1·1	-	1308	3000
Protactinium, Pa	91	231·04	tetra		160	3+, 4+	113/98	1·5	15400	1500	4300
Radium, Ra	88	226·03	-		-	2+	143	0·9	5000	970	1410
Radon, Rn	86	222	-		-	0	-	-	9·73 (273 K)	202	211·3

The abbreviations used are listed on p 25.

Properties of the elements *contd*

Element, symbol	Z	ram	Crystal structure	T_{TRANS}, K	r_a, pm	Ox. Nos	r_i, pm	Electroneg.	Density, ρ, kg m^{-3}	T_m, K	T_b, K
Rhenium, Re	75	186·2	hcp		137	4+	72	1·9	20500	3450	5900
Rhodium, Rh	45	102·91	fcc		134	3+	68	2·2	12440	2230	4000
Rubidium, Rb	37	85·47	bcc		246	1+	147	0·8	1530	315·0	961
Ruthenium, Ru	44	101·07	hcp		133	4+	67	2·2	12400	2520	4200
Samarium, Sm	62	150·35	rhombic/fcc	1190	179	3+	96	1·1	7500	1345	2200
Scandium, Sc	21	44·96	hcp/fcc	1223	160	3+	73	1·3	3000	1812	3000
Selenium, Se	34	78·96	hcp		116	2-	191	2·4	4810	490	958
Silicon, Si	14	28·09	cubic		118	4+, 4-	42/38	1·8	2300	1680	2628
Silver, Ag	47	107·87	fcc/hcp	5	144	1+	126	1·9	10500	1234	2485
Sodium, Na	11	22·99	bcc		185	1+	97	0·9	970	371	1165
Strontium, Sr	38	87·62	fcc/hcp/bcc	506/813	215	2+	112	1·0	2600	1042	1657
Sulphur, S	16	32·06	fc ortho		106	2-, 4+	184/37	2·5	2070	386	717·7
Tantalum, Ta	73	180·95	bcc		143	5+	68	1·5	16600	3269	5698
Technetium, Tc	43	98·91	hcp		135	7+	98	1·9	11400	2500	4900
Tellurium, Te	52	127·60	hcp		143	2-	211	2·1	6240	722·6	1260
Terbium, Tb	65	158·92	hcp/rhombic	1590	177	3+	92	1·2	8300	1629	3100
Thallium, Tl	81	204·37	hcp/fcc	503	171	1+	147	1·8	11860	576·6	1730
Thorium, Th	90	232·04	fcc/bcc	1673	180	4+	102	1·3	11500	2000	4500
Thulium, Tm	69	168·93	hcp/bcc	1158	174	3+	87	1·2	9300	1818	2000
Tin, Sn	50	118·69	cubic (diam)/bcc		140	2+, 4+	93/71	1·8	7300	5051	2540
Titanium, Ti	22	47·90	hcp/bcc	1158	146	4+	68	1·5	4540	1948	3530
Tungsten, W	74	183·85	bcc		137	6+	62	1·7	19320	3650	6200
Uranium, U	92	238·03	rhombic/tetra	941	138	4+, 6+	97	1·7	19050	1405·4	4091
Vanadium, V	23	50·94	bcc		131	3+, 5+	74/59	1·6	6100	2160	3300
Xenon, Xe	54	131·30	fcc		221	0, (3+, 5+)	–	–	5·50	161·2	166·0
Ytterbium, Yb	70	173·04	fcc/bcc	1071	193	3+	86	1·2	7000	1097	1700
Yttrium, Y	39	88·91	hcp/bcc	1763	181	3+	89	1·2	4600	1768	3200
Zinc, Zn	30	65·37	hcp		133	2+	74	1·6	7140	692·6	1180
Zirconium, Zr	40	91·22	hcp/bcc	1100	160	4+	79	1·4	6500	2125	3851

The abbreviations used are listed on p 25.

Taxonomy of living organisms

A fairly generally accepted view is that there are six 'kingdoms': Archaea, Eubacteria, Protista, Animalia, Plantae and Fungi. Many alternative schemata have been proposed and quite commonly Monera are considered as including Eubacteria and Archaea, making only five kingdoms. Given, however, the major differences between Archaea and Eubacteria, this is probably not appropriate. The most fundamental subdivision is between **prokaryotic** and **eukaryotic** organisms, the Archaea and Eubacteria being prokaryotic. On the basis of molecular taxonomy (based particularly on ribosomal RNA) some authorities propose three super-kingdoms or domains: Archaea, Bacteria and Eukaryota. The eukaryotic kingdoms are hierarchically subdivided into phylum (division in plant taxonomy), class, order, family, genus and species with subgroups (eg subphylum) of some of these according to the scheme adopted.

The six kingdoms

Kingdom/domain	Subdomains/classes
Archaea	Methanogens Extreme thermophiles Extreme halophiles
Eubacteria	Actinobacteria Bacteroidetes Chlamydiae Cyanobacteria (formerly blue-green algae) Firmicutes Proteobacteria – subdivided into alpha, beta, gamma, delta/epsilon sub-classes Spirochaetes
Protista	Flagellates (Mastigophora) Amoebae (Sarcodina) – subdivided into Actinopoda and Rhizopoda Algae Parasitic protists (Sporozoa)
Animalia	See 'Classification of the animal kingdom'
Plantae	See 'Classification of the plant kingdom'
Fungi	See 'Classification of fungi'

Classification of the animal kingdom

Phylum	Sub-phylum	Class	Sub-class
Porifera		Calcarea	
		Hexactinellida	
		Demospongiae	
Cnidaria		Hydrozoa	
		Scyphozoa	
		Anthozoa	
Ctenophora		Tentaculata	
		Nuda	
Platyhelminthes		Turbellaria	
		Digenea	
		Aspidogastrea	
		Monogenea	
		Cestoda	
Nemertini			
Aschelminthes		Nematoda	
		Nematomorpha	
		Rotifera	
		Gasterotricha	
		Kinorhyncha	
Acanthocephala			
Entoprocta			
Annelida		Polychaeta	
		Archiannelida	
		Oligochaeta	
		Hirudinea	
Echiurida			
Sipunculida			
Arthropoda	Onychophora		
	Tardigrada		
	Pentastomida		
	Trilobitomorpha		
	Chelicerata	Merostomata	
		Arachnida	
	Pycnogonida		
	Mandibulata	Crustacea	
		Pauropoda	
		Diplopoda	
		Chilopoda	
		Symphyla	
		Insecta	Apterygota
			Exopterygota
			Endopterygota

Phylum	Sub-phylum	Class	Sub-class
Mollusca		Monoplacophora	
		Amphineura	
		Scaphypoda	
		Gastropoda	
		Bivalvia	
		Cephalopoda	
Priapuloidea			
Bryozoa		Stenolaemata	
		Gymnolaemata	
		Phylactolaemata	
Phoronida			
Brachiopoda		Inarticulata	
		Articulata	
Chaetognatha			
Pogonophora			
Echinodermata		Asteroidea	
		Ophiuroidea	
		Echinoidea	
		Holothuroidea	
		Crinoidea	
Chordata	Hemichordata	Enteropneusta	
		Pterobranchia	
	Urochordata		
	Cephalochordata		
	Vertebrata	Agnatha	
		Gnathostomata	
		Chondrichthyes	
		Osteichthyes	
		Amphibia	
		Reptilia	
		Aves	
		Mammalia	

Classification of the plant kingdom

(Some of the smaller classes of eukaryotic algae are omitted.)

	Division	Class	Sub-class/super-order
Algae	Rhodophyta	Rhodophyceae	
	Cryptophyta	Cryptophyceae	
	Dynophyta	Dinophyceae Desmophyceae	
	Heterokontophyta	Xanthophyceae Chrysophyceae Bacillariophyceae Phaeophyceae Oomycetes[*] Hypochytridiomycetes[*]	
	Haptophyta	Haptophyceae	
	Euglenophyta	Euglenophyceae	
	Chlorophyta	Chlorophyceae Ulvophyceae Charophyceae	
Bryophytes (mosses)	Bryophyta	Hepaticopsida Anthocerotopsida Bryopsida	
Tracheophytes (vascular plants)	Pteridophyta (ferns)	Rhyniopsida[†] Psilotopsida Zosterophyllopsida[†] Lycopsida Trimerophytopsida[†] Sphenopsida Filicopsida Progymnospermopsida[†]	
	Spermatophyta (gymnosperms; non-flowering plants with unencased seeds)	Pteriodspermopsida[†] Cycadopsida Coniferopsida Gnetopsida	
	Magnoliophyta (angiosperms; flowering plants with encased seeds)	Dicotyledones (Magnoliopsida)	Magnoliidae Hamamelidae Caryophyllidae Dilleniidae Rosidae Asteridae
		Monocotyledones (Liliopsida)	Alismatidae Arecidae Liliidae Commelinidae

[*] Oomycetes and Hypochytridomycetes are often considered Mastigomycotina (fungi).
[†] All members of these classes are extinct.

Classification of fungi

Division	Subdivision	Class
Myxomycota		Acrasiomycetes
		Hydroxymycetes
		Myxomycetes
		Plaxmodiophorales
Eumycota	Mastigomycotina	Chytridiomycetes
	Zygomycotina	Zygomycetes
	Ascomycotina	Hemiascomycetes
		Plectomycetes
		Pyrenomycetes
		Discomycetes
	Basidiomycotina	Teliomycetes
		Hymenomycetes
		Gasteromycetes
	Deuteromycotina	

Amino acids

The full names are on the left with the short names and single-letter abbreviations in brackets. These are followed by the conventional chemical formulae and finally by drawings which give a better impression of their shape.

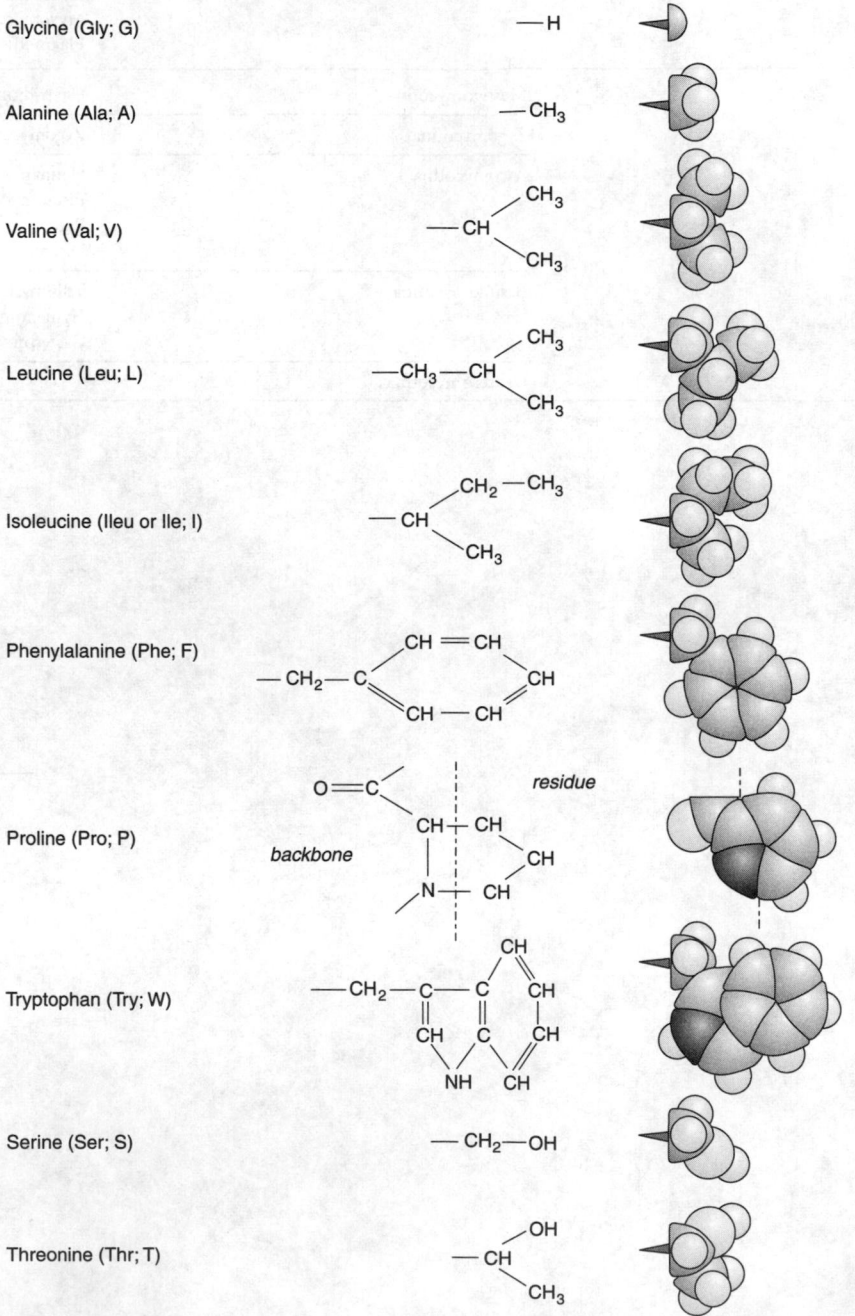

Glycine (Gly; G)   —H

Alanine (Ala; A)   —CH₃

Valine (Val; V)   —CH(CH₃)CH₃

Leucine (Leu; L)   —CH₃—CH(CH₃)CH₃

Isoleucine (Ileu or Ile; I)   —CH(CH₂—CH₃)CH₃

Phenylalanine (Phe; F)   —CH₂—C(CH=CH)(CH—CH)CH

Proline (Pro; P)   backbone / residue

Tryptophan (Try; W)   —CH₂—C ...

Serine (Ser; S)   —CH₂—OH

Threonine (Thr; T)   —CH(OH)CH₃

Amino acids *contd*

For each amino acid residue the small triangular symbol indicates the point of attachment to the backbone, except for proline (strictly, an imino acid) where the segment of the backbone with its double attachment is shown.

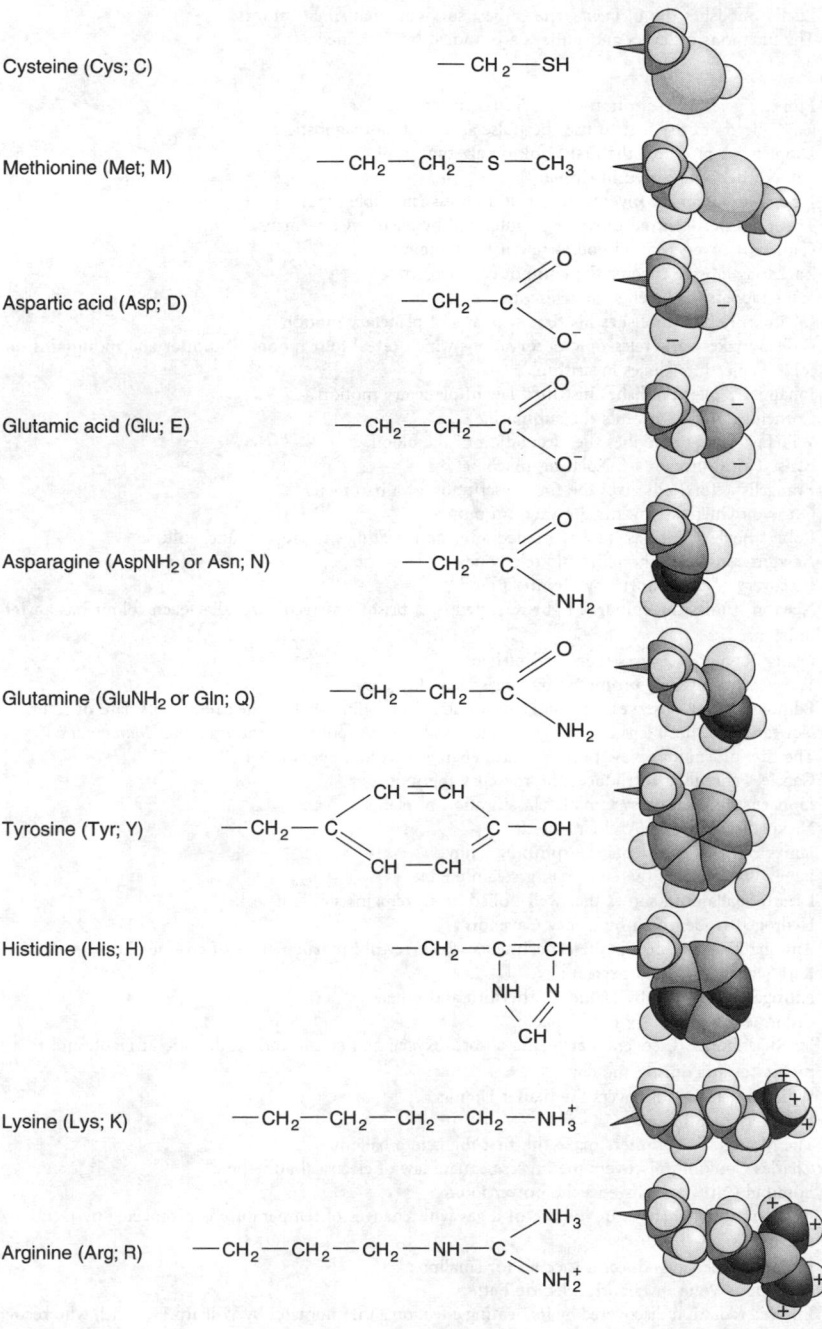

Cysteine (Cys; C) $-CH_2-SH$

Methionine (Met; M) $-CH_2-CH_2-S-CH_3$

Aspartic acid (Asp; D)
$$-CH_2-C\begin{smallmatrix}O\\O^-\end{smallmatrix}$$

Glutamic acid (Glu; E)
$$-CH_2-CH_2-C\begin{smallmatrix}O\\O^-\end{smallmatrix}$$

Asparagine (AspNH$_2$ or Asn; N)
$$-CH_2-C\begin{smallmatrix}O\\NH_2\end{smallmatrix}$$

Glutamine (GluNH$_2$ or Gln; Q)
$$-CH_2-CH_2-C\begin{smallmatrix}O\\NH_2\end{smallmatrix}$$

Tyrosine (Tyr; Y)
$$-CH_2-C\begin{smallmatrix}CH=CH\\CH-CH\end{smallmatrix}C-OH$$

Histidine (His; H)
$$-CH_2-C=CH$$
$$\quad\quad | \quad\quad |$$
$$\quad NH \quad N$$
$$\quad\quad \backslash \quad \nearrow$$
$$\quad\quad CH$$

Lysine (Lys; K) $-CH_2-CH_2-CH_2-CH_2-NH_3^+$

Arginine (Arg; R)
$$-CH_2-CH_2-CH_2-NH-C\begin{smallmatrix}NH_3\\NH_2^+\end{smallmatrix}$$

Chronology of discoveries and inventions

* Approximate date

BC

300*	Euclid publishes the *Elements*, the earliest surviving mathematical text
250*	The functions of levers and pulleys are studied by Archimedes

AD

150*	Ptolemy gives a description of an Earth-centred universe
175*	Galen becomes the first to use the pulse as a medical diagnostic aid
250*	Diophantus produces the first book on algebra
1000*	Gunpowder is invented in China
1440	Johannes Gutenberg invents the printing press (movable type)
1543	Theory of Sun-centred universe is published by Nicolas Copernicus
1590	Galileo discovers that all bodies fall at the same rate
1590	Zacharias Janssen invents the compound microscope
1608	Hans Lippershey invents the telescope
1609	Johannes Kepler publishes his first two laws of planetary motion
1610	Galileo makes early telescopic observations which reveal four moons of Jupiter and mountains on the Moon
1614	John Napier introduces logarithms
1619	Johannes Kepler publishes his third law of planetary motion
1620	Cornelius Drebbel invents the submarine
1628	William Harvey describes the circulation of the blood
1642	Blaise Pascal invents a calculating machine
1644	Evangelista Torricelli gives the first description of a barometer
1650	Otto von Guericke invents the vacuum pump
1662	Robert Boyle publishes the law named after him relating gas pressure and volume
1668	Newton constructs the reflecting telescope
1669	Discovery of phosphorus by Hennig Brand
1671	Newton studies the coloured light refracted by a prism and recognizes that each colour has an inherent difference
1675	Olaus Roemer measures the speed of light
1678	Wave theory of light promoted by Christiaan Huygens
1682	Edmond Halley observes the comet now named after him; in 1705 he predicts its date of return
1687	Newton publishes his theory of gravitation and laws of motion in the *Principia Mathematica*
1712	The first of Thomas Newcomen's steam engines becomes operational
1714	Gabriel Fahrenheit introduces the mercury thermometer
1735	Linnaeus publishes a systematic classification of plants and animals
1751	Nickel discovered by Axel Cronstedt
1764	James Hargreaves invents the spinning jenny
1765	James Watt invents the condensing steam engine
1765	Lazaro Spallanzani shows that well-boiled broth remains sterile if sealed
1766	Hydrogen is identified by Henry Cavendish
1772	Antoine Lavoisier begins the experiments which reveal the true nature of combustion
1772	Karl Scheele discovers oxygen
1772	Nitrogen discovered by Daniel Rutherford and others
1774	Chlorine discovered by Karl Scheele
1779	Jan Ingenhousz discovers that plants absorb oxygen and exhale carbon dioxide at night, and that the reverse process occurs during the day
1781	William Herschel discovers the planet Uranus
1783	Discovery of tungsten
1783	The Montgolfier brothers make the first flight in a balloon
1785	Charles Coulomb discovers the inverse-square law of electrical attraction
1785	Edmund Cartwright invents the power loom
1787	Law describing changes in volume of a gas with changes of temperature is discovered by Jacques-Alexandre Charles
1796	Edward Jenner introduces a vaccine for smallpox
1800	Alessandro Volta invents the electric battery
1800	Infrared radiation discovered by its heating effect on a thermometer by William Herschel, who recognizes that it is similar to visible light
1800	William Nicholson discovers electrolysis
1803	John Dalton proposes his atomic theory of matter
1804	Richard Trevithick develops the steam locomotive

1811	Amedeo Avogadro proposes that equal volumes of all gases at the same temperature and pressure contain equal numbers of molecules
1820	Hans Christian Oersted discovers the magnetic effect of an electric current
1825	Aluminium discovered by Hans Christian Oersted
1825	William Sturgeon invents the electromagnet
1826	Antoine Balard discovers bromine
1826	Joseph Niepce introduces the first permanent photographs
1827	André-Marie Ampère publishes his work on electromagnetism
1827	Georg Ohm announces the relation between voltage, current and resistance in an electric circuit
1831	Michael Faraday discovers electromagnetic induction (previously discovered by Joseph Henry but not published)
1837	Samuel Morse patents the telegraph
1839	Charles Goodyear invents vulcanized rubber
1839	Ozone discovered by Christian Schönbein
1846	Following predictions of the existence of a planet beyond Uranus, Johann Galle discovers Neptune
1851	Lord Kelvin formulates the second law of thermodynamics
1852	Léon Foucault constructs the first gyroscope
1855	Alexander Parkes invents celluloid
1855	Henry Bessemer introduces the Bessemer converter for the manufacture of steel
1859	Charles Darwin publishes *The Origin of Species*
1859	Etienne Lenoir develops the first internal-combustion engine
1859	Gustav Kirchhoff states that the ratio of absorption to emission for energy at any wavelength is the same for all bodies at the same temperature
1862	Louis Pasteur finally disproves the theory of spontaneous generation of bacteria from inorganic matter
1864	James Clerk Maxwell publishes his theory of electromagnetism (extended in 1873) which identifies light as electromagnetic waves
1865	Gregor Mendel publishes the laws of heredity
1868	Jules Janssen and Normal Lockyer discover helium in the Sun's spectrum
1869	Dmitri Mendeleyev develops the periodic table of the elements
1876	Alexander Graham Bell patents the telephone
1876	Josiah Willard Gibbs introduces the concept of chemical potential and extends the science of thermodynamics to chemistry
1877	Nikolaus Otto develops the four-cycle internal-combustion engine
1877	Thomas Edison invents the phonograph
1879	The electric light bulb is invented independently by Thomas Edison and Joseph Swan
1880	Charles Hollerith invents the punched-card machine for storing and analysing data
1884	Charles Parsons invents the steam turbine
1885	James Dewar invents the vacuum flask
1885	William Stanley invents the electric transformer
1886	Fluorine isolated by Henri Moissan
1887	Albert Michelson and Edward Morley's experiment disproves the existence of the ether
1887	Gottlieb Daimler introduces the first petrol-engined car
1888	Heinrich Hertz discovers radio waves, confirming the existence of electromagnetic waves possessing the same properties as light
1895	Guglielmo Marconi invents a system of radiotelegraphy
1895	Helium discovered on Earth by William Ramsay
1895	Wilhelm Röntgen discovers X-rays
1896	Radioactivity discovered by Antoine Becquerel
1897	J(oseph) J(ohn) Thomson discovers and characterizes the electron
1897	Rudolf Diesel invents the diesel engine
1898	Pierre and Marie Curie discover radium
1900	Max Planck publishes his theory of black-body radiation and gives the first indications that radiation may be quantized
1900	Paul Villard discovers gamma rays
1903	Konstantin Tsiolkovsky publishes the first description of how rockets work
1903	Wilbur and Orville Wright make the first powered flight
1904	John Fleming invents the diode valve
1905	Einstein demonstrates the quantum nature of the photon in studies of the photoelectric effect, and publishes the special theory of relativity
1906	Reginald Fessenden makes the first sound radio broadcast
1908	Leo Baekeland invents Bakelite, the first plastic
1910	Thomas Morgan's work on the fruit fly demonstrates that the genes lie linearly along the chromosomes
1911	Ernest Rutherford introduces the concept of the nuclear atom following studies of alpha-particle recoil
1911	Heike Kamerlingh Onnes discovers superconductivity
1913	Frederick Soddy proposes the existence of isotopes

1913	Henry Moseley establishes that the chemical properties of elements are determined by atomic number rather than atomic weight
1913	Niels Bohr introduces the quantized energy level theory of the atom
1913	William Coolidge invents the hot-cathode X-ray tube
1914	James Franck and Gustav Hertz experimentally confirm the quantized nature of atomic energy levels
1915	Alfred Wegener proposes the theory of continental drift
1916	Einstein publishes the general theory of relativity
1919	Ernest Rutherford proves that protons are constituents of atomic nuclei
1920	Ernest Rutherford proposes the existence of the neutron
1923	Louis de Broglie suggests that particles such as the electron can display wave-like behaviour
1924	Edwin Hubble discovers that galaxies exist beyond our own
1926	Erwin Schrödinger introduces wave mechanics
1926	Hermann Müller demonstrates that X-rays produce mutations in the fruit fly
1926	John Logie Baird and others invent the first working television systems
1926	Robert Goddard launches the first liquid-fuelled rocket
1927	Werner Heisenberg introduces the uncertainty principle
1928	Alexander Fleming discovers penicillin
1929	Edwin Hubble discovers that the universe is expanding
1930	Clyde Tombaugh discovers the planet Pluto following predictions by Percival Lowell
1930	Frank Whittle patents the jet engine
1931	Ernest Lawrence constructs the first cyclotron particle accelerator
1931	Wolfgang Pauli predicts the existence of the neutrino
1932	Carl Anderson discovers the positron
1932	Harold Urey discovers deuterium (heavy hydrogen)
1932	James Chadwick discovers the neutron
1932	John Cockcroft and Ernest Walton bombard lithium with protons causing the first artificial nuclear disintegration
1933	Ernst Ruska invents the electron microscope
1934	Irène and Frédéric Joliot-Curie produce radioisotopes by alpha-particle bombardment
1935	Robert Watson-Watt invents radar
1935	Wallace Carothers patents nylon
1937	Chester Carlson invents the process of xerography
1938	Lise Meitner and Otto Frisch suggest that fission has occurred in neutron bombardment experiments and postulate that mass has been converted into energy
1939	Igor Sikorsky designs the helicopter for mass production
1941	Plutonium is discovered by Glenn Seaborg and others
1942	George Beadle and Edward L(awrie) Tatum show that one enzyme is made by one gene
1942	John Atanasoff and Clifford Berry construct one of the earliest computers
1942	The first controlled nuclear chain reaction takes place at the University of Chicago
1945	Detonation of the first atomic bomb, codenamed Trinity, on 16 July
1945	Nuclear weapons first used on Japan, 6 and 10 August
1948	George Gamow and others develop the Big Bang theory of the creation of the universe
1948	Transistor invented by John Bardeen, Walter Brattain and William Shockley
1953	Charles Townes develops the maser
1953	Francis Crick and James Watson determine the double-helix structure of DNA
1955	Emilio Sègre and Owen Chamberlain discover the antiproton
1957	The first artificial satellite (Sputnik 1) is launched by the Soviet Union
1960	Allan Sandage and Thomas Matthews discover the first quasar
1960	First full-scale nuclear power station commissioned
1960	Theodore Maiman constructs the first laser
1961	Soviet cosmonaut Yuri Gagarin becomes the first man in space
1965	Arno Penzias and Robert Wilson discover the microwave background radiation, a remnant of the Big Bang
1966	The genetic code is elucidated by Marshall Nirenberg and others
1968	Experimental evidence increasingly suggests the existence of quarks
1969	Neil Armstrong and Edwin Aldrin land on the Moon (21 July)
1971	First microprocessor (Intel 4004) introduced
1975	Digital camera developed by Steven J Sasson
1975	First personal computer (Altair 8800) commercially available
1975	Frederick Sanger demonstrates dideoxy-method for DNA sequencing
1975	Herbert Boyer and Stanley Cohen selectively clone genes in bacteria
1977	Bell Labs produce cellular mobile phone
1978	First 'test tube baby' born
1979	Personal stereo (Sony Walkman) appears on the market
1979	Three Mile Island accident, Pennsylvania, USA
1981	First flight of the NASA Space Shuttle
1982	AIDS first properly defined by the US Centers for Disease Control
1982	Philips and Sony introduce the compact disc

1983	CERN team discover W and Z bosons, providing experimental evidence for electroweak theory
1983	Idea of polymerase chain reaction thought to have been conceived by Kary Mullis
1985	Harry Kroto and others isolate buckminsterfullerene and other fullerenes
1985	Gerl Binnig, Calvin Quate and Christoph Gerber develop the atomic force microscope
1986	Chernobyl reactor disaster, Ukraine, USSR
1986	The Soviet space station Mir is launched as the first permanently manned space station
1990	Casolyn Napoli, Claude Lemieux and Richard Jorgensen discover RNA interference
1991	The Joint European Torus fusion reactor achieves a peak output of 14.3 MW for 2 seconds in its first experiments using tritium
1991	Tim Berners-Lee devises protocol for multimedia data transmission, the start of the World Wide Web
1993	Global positioning system developed by US Department of Defense
1995	Eric Cornell and Carl Wieman synthesize Bose–Einstein condensate
1995	Publication by The Institute for Genomic Research of the first full DNA sequence of a free-living organism, the bacterium *Haemophilus influenzae*
1996	First DVDs and players available
1997	First adult mammal cloned at Roslin Institute, Scotland (Dolly the sheep)
1998	Full genome of *Caenorhabditis elegans* published by Sanger Institute and Genome Sequencing Unit at Washington University
1998	Researchers at Lawrence Berkeley National Laboratory provide observational evidence that expansion of the universe is accelerating
2000	Almost complete genome of *Drosophila* published
2001	Digital satellite radio becomes available
2002	Raymond Davis and Masatoshi Koshiba detect cosmic neutrinos and show they have mass
2002	Scramjet developed at University of Queensland
2003	Completion of sequencing of human genome

Nobel Prize winners 1950–2006

Year	Chemistry	Physics	Physiology or Medicine
1950	Kurt Alder Otto P H Diels	Cecil F Powell	Philip S Hench Edward C Kendall Tadeus Reichstein
1951	Edwin M McMillan Glenn T Seaborg	John D Cockcroft Ernest T S Walton	Max Theiler
1952	Archer J P Martin Richard L M Synge	Felix Bloch Edward M Purcell	Selman A Waksman
1953	Hermann Staudinger	Frits Zernike	Fritz A Lipmann Hans A Krebs
1954	Linus C Pauling	Max Born Walther Bothe	John F Enders Thomas H Weller Frederick C Robbins
1955	Vincent du Vigneaud	Willis E Lamb Polykarp Kusch	A Hugo Thorell
1956	Cyril N Hinshelwood Nikolai N Semenov	William B Shockley John Bardeen Walter H Brattain	André F Cournand Werner Forssmann Dickinson W Richards
1957	Alexander R Todd	Chen Ning Yang Tsung-Dao Lee	Daniel Bovet
1958	Frederick Sanger	Pavel A Cherenkov Ilya M Frank Igor Y Tamm	George W Beadle Edward L Tatum Joshua Lederberg
1959	Jaroslav Heyrovsky	Emilio Segrè Owen Chamberlain	Arthur Kornberg Severo Ochoa
1960	Willard F Libby	Donald A Glaser	F Macfarlane Burnet Peter B Medawar
1961	Melvin Calvin	Robert Hofstadter Rudolf L Mössbauer	Georg von Békésy
1962	Max F Perutz John C Kendrew	Lev D Landau	Francis H C Crick James D Watson Maurice H F Wilkins
1963	Karl Ziegler Giulio Natta	Eugene P Wigner Maria Goeppert-Mayer J Hans D Jensen	John C Eccles Alan L Hodgkin Andrew F Huxley
1964	Dorothy C Hodgkin	Charles H Townes Nikolai G Basov Aleksandr M Prokhorov	Konrad Bloch Feodor Lynen
1965	Robert B Woodward	Sin-Itiro Tomonaga Julian Schwinger Richard P Feynman	François Jacob André Lwoff Jacques Monod
1966	Robert S Mulliken	Alfred Kastler	Peyton Rous Charles B Huggins
1967	Manfred Eigen Ronald G W Norrish George Porter	Hans A Bethe	Ragnar Granit Haldane K Hartline George Wald
1968	Lars Onsager	Luis W Alvarez	Robert W Holley Har Gobind Khorana Marshall W Nirenberg

Year	Chemistry	Physics	Physiology or Medicine
1969	Derek H R Barton Odd Hassel	Murray Gell-Mann	Max Delbrück Alfred D Hershey Salvador E Luria
1970	Luis F Leloir	Hannes Alfvén Louis Néel	Bernard Katz Ulf von Euler Julius Axelrod
1971	Gerhard Herzberg	Dennis Gabor	Earl W Sutherland
1972	Christian B Anfinsen Stanford Moore William H Stein	John Bardeen Leon N Cooper John R Schrieffer	Gerald M Edelman Rodney R Porter
1973	Ernst O Fischer Geoffrey Wilkinson	Leo Esaki Ivar Giaever Brian D Josephson	Karl von Frisch Konrad Lorenz Nikolaas Tinbergen
1974	Paul J Flory	Martin Ryle Antony Hewish	Albert Claude Christian de Duve George E Palade
1975	John W Cornforth Vladimir Prelog	Aage Bohr Ben Mottelson James Rainwater	David Baltimore Renato Dulbecco Howard M Temin
1976	William N Lipscomb	Burton Richter Samuel C C Ting	Baruch S Blumberg D Carleton Gajdusek
1977	Ilya Prigogine	Philip W Anderson Nevill F Mott John H Van Vleck	Roger Guillemin Andrew V Schally Rosalyn Yalow
1978	Peter D Mitchell	Pyotr L Kapitsa Arno A Penzias Robert W Wilson	Werner Arber Daniel Nathans Hamilton O Smith
1979	Herbert C Brown Georg Wittig	Sheldon L Glashow Abdus Salam Steven Weinberg	Allan M Cormack Godfrey N Housefield
1980	Paul Berg Walter Gilbert Frederick Sanger	James W Cronin Val L Fitch	Baruj Benacerraf Jean Dausset George D Snell
1981	Kenichi Fukui Roald Hoffmann	Nicolaas Bloembergen Arthur L Schawlow Kai M Siegbahn	Roger W Sperry David H Hubel Torsten Wiesel
1982	Aaron Klug	Kenneth G Wilson	Sune K Bergström Bengt I Samuelsson John R Vane
1983	Henry Taube	Subrahmanyan Chandrasekhar William A Fowler	Barbara McClintock
1984	Robert B Merrifield	Carlo Rubbia Simon van der Meer	Niels K Jerne Georges J F Köhler César Milstein
1985	Herbert Hauptman Jerome Karle	Klaus von Klitzing	Michael S Brown Joseph L Goldstein
1986	Dudley R Herschbach Yuan Tseh Lee John C Polanyi	Gerd Binnig Heinrich Rohrer Ernst Ruska	Stanley Cohen Rita Lévi-Montalcini
1987	Donald J Cram Jean-Marie Lehn Charles Pedersen	Georg Bednorz Alex Müller	Susumu Tonegawa

Year	Chemistry	Physics	Physiology or Medicine
1988	Johann Deisenhofer Robert Huber Hartmut Michel	Leon Lederman Melvin Schwartz Jack Steinberger	James Black Gertrude Elion George Hitchings
1989	Sydney Altman Thomas R Cech	Hans Dehmelt Wolfgang Paul Norman Ramsey	J Michael Bishop Harold E Varmus
1990	Elias James Corey	Jerome Friedman Henry Kendall Richard Taylor	Joseph E Murray E Donnall Thomas
1991	Richard R Ernst	Pierre-Gilles de Gennes	Erwin Neher Bert Sakmann
1992	Rudolph A Marcus	Georges Charpak	Edmond H Fischer Edwin G Krebs
1993	Kary Banks Mullis Michael Smith	Russell Hulse Joseph Hooton Taylor Jr	Richard Roberts Phillip Allen Sharp
1994	George Olah	Clifford Shull Bertram Brockhouse	Alfred G Gilman Martin Rodbell
1995	Paul Crutzen Mario Molina F Sherwood Rowland	Martin L Perl Frederick Reines	Edward B Lewis Eric F Wieschaus Christiane Nüsslein- Volhard
1996	Robert Curl Jr Harold Kroto Richard Smalley	David Lee Douglas Osheroff Robert Richardson	Peter Doherty Rolf M Zinkernagel
1997	Paul Boyer Jens Skou John E Walker	Steven Chu William D Phillips Claude Cohen-Tannoudji	Stanley Prusiner
1998	Walter Kohn John A Pople	Robert B Laughlin Horst L Störmer Daniel C Tsui	Robert F Furchgott Louis J Ignarro Ferid Murad
1999	Ahmed Zewail	Gerardus 't Hooft Martinus J G Veltman	Günter Blobel
2000	Alan Heeger Alan G MacDiarmid Hideki Shirakawa	Zhores I Alferov Jack S Kilby Herbert Kroemer	Arvid Carlsson Paul Greengard Eric R Kandel
2001	William S Knowles Ryoji Noyori K Barry Sharpless	Eric A Cornell Wolfgang Ketterle Carl E Wieman	Leland H Hartwell Tim Hunt Paul Nurse
2002	John B Fenn Koichi Tanaka Kurt Wüthrich	Raymond Davis Jr Ricardo Giacconi Masatoshi Koshiba	Sydney Brenner H Robert Horvitz John E Sulston
2003	Peter Agre Roderick MacKinnon	Alexei A Abrikosov Vitaly L Ginzburg Anthony J Leggett	Paul C Lauterbur Peter Mansfield
2004	Aaron Ciechanover Avram Hersko Irwin Rose	David J Gross H David Politzer	Richard Axel Linda B Buck
2005	Yves Chauvin Robert H Grubbs Richard R Schrock	Roy J Glauber John L Hall Theodor W Hänsch	Barry J Marshall J Robin Warren
2006	Roger D Kornberg	John C Mather George F Smoot	Andrew Z Fire Craig C Mello